RSGB YEAR-BOOK

1998 EDITION

Covers callsigns up to:

M0BGP, M1CEJ
and
2E0AQG, 2E1GBS

Editor:
Brett Rider

Design:
Jennifer Crocker

Advertising:
Malcolm Taylor Associates

Published by the Radio Society of Great Britain, Lambda Ho, Cranborne Road, Potters Bar, Herts. EN6 3JE.

Internet: http://www.rsgb.org

Tel: (01707) 659015.
Fax: (01707) 645105.

ISBN 1 872309 44 5
ISSN 1460-454X

Printed in Great Britain at The Nuffield Press Ltd, Nuffield Way, Abingdon, Oxon. OX14 1RL.

INFORMATION SECTION

CALLSIGN SECTION

FOREWORD

Welcome to the RSGB 1998 Year Book.

This is the first edition of what will be an annual publication bringing you all the news on the work of the RSGB and the information you require to operate Amateur Radio world-wide.

Just like the old RSGB Call Book, it contains the full United Kingdom and Eire Callsign listings. The UK Amateur Radio band plans plus much much more really useful information that is to hand in one publication.

The Yearbook is not just an Information Directory, it is a good read and every shack should have a copy. I recommend it to you.

Peter Kirby, G0TWW
General Manager

Introduction

Welcome to the 1998 edition of the *RSGB Yearbook*. We have continued with the traditions of the *Call Book* and made yet more improvements. Most notable this year is the 25% increase in the number of information pages. These now include the Society's annual report previously published in *RadCom*, a pocket history of amateur radio events, dates of forthcoming rallies. The are many other changes such as the inclusion of the 6m repeaters, a much enlarged list of beacons, more explanation of the propagation terms used in GB2RS, and by popular demand, some notes pages at the end.

As in previous years we have we included the 10 km NGR squares on which the Worked All Britain Group base their system and the IARU locators. All the calculations are derived from post codes and there is obviously a limit to their resolution. A few addresses will unfortunately fall just the other side of the boundary of a square, often by just a few meters. For those affected, we would be delighted to override the figures we calculate with the locator and square they inform us of. No charge will be made.

If you are checking your own locator, do take great care in determining your correct latitude and longitude. Due to the projection used on Ordnance Survey maps, lines of latitude and longitude are bowed and do not follow the grid lines. Instead you should work from the 5' intersection crosses. If you are using a program to calculate from NGRs, do check that it takes the oblateness of the earth into account and that you have looked up your NGR as carefully as you can – small distances are important. You might wish to check the sensitivity of the calculation, by entering NGRs 100m away to see if this changes the result. If they do, look very carefully at the map and double check your figures!

The structure of local government in the UK has altered in recent years, and yet more changes are in the pipeline for April 1998. These changes have made it impossible to retain the county listing, as counties are no longer defined in a way which relates to popular understanding. Instead, this year, amateurs are listed by district authority. A fuller explanation and look up tables are given on page 458 to help the translate between old and new systems. The new listing may help some award chasers, however in future years, a listing based on post town may be more convenient for most readers.

Information Section

The Information Section was, as far as possible, accurate at 15 August 1997. Obviously with a publication lasting until December 1998, changes will happen during the lifetime of the edition – so please check the pages of *Radio Communication* for details which came in since we closed for press. This particularly applies to items such as the QSL sub-managers, GB2RS newsreaders, rally list and the composition of Council and volunteer posts.

UK Call Sign Entries

The callsigns listed in this *Call Book* reflect the official records held at SSL on behalf of the RA. They now show the correspondence address, not the licence address as was the case several years ago. Please do not be misled into thinking that because a foreign address is quoted, that the station is operating from there. The amateur concerned will have different station address which is in the UK.

At 13 August, some 61,018 callsigns were current and are listed in this *Yearbook*. This is a decrease of 825 over last year's *Call Book* and represents a fall of 1.3% in the number of callsigns listed. In order to make the book readable and compact, the Society case converts the data and makes a few standard abbreviations. It does not make changes to the substance of any entries.

Special Entries

Some individuals have contacted the Society and made arrangements for some special text to be added after *whatever license details are published*. This allows mention of their interests, telephone number, ratable district, e-mail address, etc. In other words whatever the licensing record shows (either withheld or released), will be published – it will not be deleted. However you have the choice of what *extra* words appear *after* your standard entry. To assist readers in determining the source of information, items special to the *RSGB Yearbook* such as these special entries are given in square brackets [].

The appended text must not be longer than 252 characters (including spaces and punctuation) and may be edited by the Society so as to be consistent with the style and content used in the *Yearbook*. The additional entry will apply for all future editions of the *Yearbook*, until such time as the call sign lapses or changes, or the licensee requests a change, or the Society publishes a change to the procedure in a future edition of the *Yearbook*. This will save you from having to re-apply each year. A one-off charge of £5 is made for each 'appended entry'.

If you want a special entry, please write to: Yearbook Editor, RSGB HQ, Lambda Ho, Cranborne Road, Potters Bar, Herts, EN6 3JE.

'Withheld' Entries

Should you desire your details to be withheld from publication, or released if they are not, or there is an error to the substance of your entry, ***you must write*** to Subscription Services Ltd, and request them to take the necessary action. Their address is:

Radio Licensing Centre,
Subscription Services Ltd,
PO Box 885, Bristol, BS99 5LG.
Tel: (0117) 925 8333.

There are now several other publishers of the data and so it is vital that you contact the source of data such that it is either blocked, or released in all call books. For these reasons, the Society will not accept direct input regarding an individual's entry except that for his/her special appended text.

Common Questions

Throughout the year we receive a number of standard queries about an individual's entry in the *Yearbook*. Here are the answers we give:

Q: I have just gained my new licence, please include my new details in the next Yearbook.

A: Your details will automatically be passed on to us by the RA from the data you supply on your licence application form. There is no need to contact the Society directly!

Q: I am an RSGB member. Thank you for sending RadCom to my new address, but why didn't you change my Yearbook Entry?

A: The *Yearbook* lists all UK amateurs, not just RSGB members and so we keep the records entirely separate. Your call book address details come from the RA – did you let them know of your new address?

Q: My entry shows me as being particular's withheld, but I want my full address to be listed.

A: Please write direct to the Subscription Services Ltd and ask them to release your details. The Society will not accept direct input from licensees.

Q: You have published my address, but I would like it withheld.

A: As above, but ask them to withhold it from all call book publishers.

Q: My call sign is not shown at all, please include it!

A: If a call sign is not shown in the *Yearbook* it was not licensed at the time the data was supplied to the RSGB. Sometimes, through an administrative error or misunderstanding, an amateur believes that he is licensed but the licensing records show that the licence has lapsed. If this is the case, speak to the RA and discuss the matter with them as it is important that you keep your licence current.

Q: The text in [square brackets] at the end of my entry is wrong! Why?

A: The information given in square brackets is particular to just RSGB, so please contact the Society and not SSL about it. If it is your IARU locator or 10km NGR square, please read the opposite column and we will gladly incorporate the amendment in the next edition. If the error relates to text, then the onus is on you to specifically notify the Yearbook Editor – after all is *your* customised entry!

International listings

Amateur sometimes ask us about their entry in the *International Listings* of the Radio Amateur Call Book, sometimes known as the *DX Listings*. This is an entirely separate work published annually by an independent company.

If you want to be listed in their publication, please write direct to them, and not to the RSGB. Corrections are free, but they do make a small charge for special entries. Their address is : Radio Amateur Callbook Inc, 1685 Oak Street, Lakewood, NJ 08701, USA. Tel: 001-908 905-2961. Fax: 001-908 363-0338.

Acknowledgments

Last but definitely not least, thank you to all those who make this book possible, be they secretaries, chairmen, volunteers, staff, or just readers of the *RSGB Yearbook*.

We are particularly fortunate in receiving the co-operation of the both Radiocommunications Agency and Subscription Services Ltd in supplying the basic data on and we extend our grateful thanks to them. Also to the Irish Radio Transmitting Society for supplying the data for the EI listings.

Brett Rider, G4FLQ
Editor

The RSGB – Your Questions Answered

Why Should I Join?

To promote and protect amateur radio when there are so many demands from other band users to have their own slice of the bandwidth, the RSGB needs an extremely strong voice and this can only come from having a strong membership base. The more members we have the more likely we are to be able to argue and support our case to keep the amateur radio bands which we currently enjoy. We are continually meeting with the Radiocommunications Agency of the DTI to 'fight our corner' and negotiate on your behalf.

By joining the RSGB you will add your voice to those already working to ensure that amateur radio continues into the next century. A strong Society means a successful hobby – we are here to help you enjoy your hobby to the full – help us to help you!

What Do I Get?

By becoming a member of the RSGB you will immediately receive many members' only benefits at no extra cost including:

* *Radio Communication* (*RadCom*)

RSGB's 100 page colour magazine sent direct to your home address every month. *RadCom* is undeniably the market leader in amateur radio publications, featuring every aspect of this unique and broad subject, whether its news or technical features *RadCom* has got it all.

* Membership Discount

Save money off a wide range of amateur radio publications (over 100 items on our current list), books, software and sundry items such as ties, badges and stickers.

* QSL Bureau

The RSGB QSL Bureau sorts and despatches these cards free of charge for members saving you a small fortune in postage stamps. The Bureau handles approximately three million cards a year.

* EMC advice

Specialist leaflets, prepared by a team of EMC experts, are available to members advising how to rectify any interference problems you may have.

* Antenna Planning Permission Advice

A special booklet is available free to members advising how to go about obtaining planning permission and a panel of experts to provide extra help for that all important antenna system.

* Novice Training

The RSGB co-ordinates the training of all Novice radio amateurs via a network of specially appointed Novice Instructors. Over 1,000 instructors do marvellous work on behalf of the Society by training new radio amateurs all over the UK.

* Contests

Only members are eligible to enter the 80+ different RSGB contests held each year.

* Specialist advice

The Society is blessed with a large number of experts in different aspects of the hobby who are willing to help members with their specific queries. This is mainly through the committee structure, and details of the various Chairmen are regularly published in *RadCom*. Over 2,000 volunteers work behind the scenes helping radio amateurs pursue their hobby to the full.

* Government Liaison

The RSGB is engaged in on-going discussions with the Radiocommunications Agency to improve and protect the amateur radio bands. These bands are under continual threat from other band users, cable, cellular phone,

* Specialised Equipment Insurance

Amateur radio equipment costs you, the amateur, a great deal of money, and it is only wise to protect your investment. We have negotiated specialist discounted insurance cover to meet this need.

* RSGB Credit Card

A special affinity credit card organised in conjunction with the Bank of Scotland which you can use for all your normal purchases.

1998 Rally Dates

The following dates were notified to us at the time of going to press, but see RadCom for the very latest information of forthcoming events.

18 Jan	Oldham ARC Mobile Rally. Details: (01706) 846143 or (0161) 652 4164.
1 Feb	Harwell Radio & Computing Rally. Details: (01235) 815399.
	South Essex ARS Radio Rally: Details: (01268) 697978.
15 Feb	Northern Cross Rally. Details (01924) 379680.
22 Feb	Barry Amateur Radio and Computer Rally. Details (01222) 832253.
	RSGB National VHF Convention. Details: Norman, G3MVV, on (01277) 225563 or Marcia, 2E1DAY, at RSGB HQ on (01707) 659015.
1 Mar	Red Rose Rally. Details: (01204) 62980.
7/8 Mar	London Amateur Radio and Computer Show. Details: (01923) 893929.
15 Mar	Norbreck Amateur Radio, Electronics and Computing Exhibition. Details: (0151) 630 5790.
29 Mar	Pontefract & DARS Component Fai. Details: (01977) 606345 *(office hours)* or (01977) 616935 *(evenings)*.
19 Apr	Yeovil ARC QRP Convention: Details: (01963) 250594.
25 Apr	International Marconi Day run by Cornish Radio Amateurs. Details: (01209) 212314.
4 May	Dartmoor Radio Rally. Details: (01822) 852586.
10 May	Drayton Manor Radio & Computer Rally. Details (0121) 422 9787 or (0121) 443 1189.
14 Jun	Aldershot Amateur Radio Rally. Details: (01252) 837860.
11 Jul	Cornish Radio Rally and Computer Fair. Details (01209) 820118.
12 Jul	Sussex Amateur Radio and Computer Fair. Details: (01903) 763978 or (01273) 417756 *(office hours)*.
	Three Counties Radio & Computer Rally. Details: (01905) 773181.
19 Jul	Humber Bridge Radio & Computer Rally. Details: (01482) 837042.

* Discounted personal/medical insurance/car breakdown cover

This has been arranged through HMCA (Hospital and Medical Care Association) an organisation who specialises in providing a wide range of personal insurance for members of societies and institutes. They number among their clients The Royal British Legion, Federation of Small Businesses, Institute of Advanced Motorists, Yorkshire County Cricket Club, the Royal Yachting Association, etc.

* Preferential personal loans

Arranged through NWS Bank (a subsidiary of the Bank of Scotland) – whether you want to buy radio equipment, have a holiday or just do some house improvements, then this loan is well worth considering.

You will see that there are many benefits to be gained by becoming a member of the RSGB. However, probably the most important the work that we do is in looking after your interests by protecting the amateur radio frequency bands and privilages so enabling amateur radio to prosper and grow into the 21st Century.

How Do I Join?

Just complete the application form, and send it to the RSGB at the address given. If you do not want to tear out the page, then we will accept a photocopy instead. Send us the form and we will get your membership set up – that's all you have to do!

Many radio amateurs already enjoy the facilities offered by the RSGB, and by completing the application form you can join them – just fill in the form and pop it in the post today – it couldn't be easier.

We have even made provision for you to spread the cost of your subscription by quarterly Direct Debit payments – at no extra cost to you – or you can send in a cheque or include your credit card number – the choice is yours.

JOIN US TODAY AND HELP MAKE AMATEUR RADIO EVEN STRONGER!

RSGB Membership Application Form

Lambda House, Cranborne Road, Potters Bar, Herts EN6 3JE. Tel: (01707) 659015

Please complete this form and return it to RSGB HQ:

Callsign (one only, please) ______ (Leave blank if not a licence holder.)

Title (Mr, Mrs, Dr etc) ______

Initials (first names only) ______

Surname ______

Address ______

Postcode ______

Date of birth (dd/mm/yy) ______

Membership type (please tick)	***Notes***	***UK membership rate***
☐ Corporate member	You must be over 18 years of age *OR* hold a UK amateur radio transmitting licence (UK and overseas).	£36.00
☐ Corporate (concessionary)	You must be over 65 years of age. Proof of age must be submitted with the application.	£27.00
☐ Corporate (family)	You must reside with a relative who is a member of the Society. Family members do not receive *RadCom.* Please insert the callsign/RS number of your relative below: ______	£12.00
☐ Corporate (student)	You must supply evidence of full-time student status.	£22.00
☐ Associate (Junior HamClub)	You must be under 18 years of age. Junior HamClub members receive *RadCom*, but do not have voting rights at the Society's meetings.	£12.00

I enclose cheque/PO ☐ **credit card payment** ☐ (please tick) **for** £ ______

Credit Card No. ______ **Expiry date** ______

Amex start date ______ **Switch issue No.** ______

Special arrangements exist for blind and disabled persons. Details are available from RSGB HQ.

Please sign this form below

I, the undersigned, agree that in the event of my election to Membership of the Radio Society of Great Britain, I will be governed by the Memorandum and Articles of Association of the Society and the rules and regulations thereof as they now are or as they may hereafter be altered; and that I will advance the objects of the Society as far as may be in my power; provided that whenever I signify in writing to the Society addressed to the Secretary that I desire to withdraw from the Society, I shall at the end of one year thereafter, after the payment of any arrears which may be due by me at that period, be free from my undertaking to contribute to the assets of the Society in accordance with Clause 6 of the Memorandum of Association of the Society.

Signed ______ **Date** ______

Amateur Radio – Fact Sheet

We are often asked to help radio amateurs when approached by the local press or local radio stations. This is always available from RSGB HQ, but you might appreciate some basic information in case you are faced with a situation that requires immediate action.

AMATEUR RADIO

* Amateur radio as a hobby is as old as radio itself, and since there were originally no radio professionals it is arguable that Marconi was the first radio amateur. There are over 60,000 licensed radio amateurs in the UK and over two million world-wide.

* Amateur Radio celebrates its 100th Anniversary in 1998.

* Radio amateurs are qualified radio operators. They are enthusiasts who have passed a City & Guilds examination in radio theory and practice which allows them to hold a transmitting licence issued by the Radiocommunications Agency of the DTI.

* Amateur radio should not be confused with CB radio which allows any unqualified member of the public to chat to his or her friends over short distances using low power equipment on very limited frequencies. It should also not be confused with eavesdropping into the commercial uses of radio.

* The licence, which allocates a callsign, lists the rules under which radio amateurs are allowed to transmit. This includes the permitted frequencies - there are more than 25 bands of frequencies available (depending on the class of licence held) covering the short waves, VHF bands and Microwaves.

* There are two types of licence; one for the beginner - the Novice Licence, and one for the more qualified operator. A Morse test must be passed to qualify for a Class A licence which allows the use of frequencies where world-wide contacts are available, but lots of fun can still be had without the need for Morse.

* Although a wide range of types of transmission can be made, from Morse code (still widely used) to computer data and even television pictures, radio amateurs may not transmit such things as music, commercial or political messages. There's still plenty to talk about, though, including of course radio itself.

* Radio amateurs are the only users of the radio spectrum who are permitted to build their own transmitters. This is because they are exam-qualified. Most amateur radio stations have a mix of home-built and commercial equipment.

* Radio amateurs are frequently called upon to assist in times of disaster. Their compact and simple equipment is frequently more flexible in an emergency than today's complex commercial gear. Help has been provided at earthquake sites, train disasters, plane crashes, and so on. Most recently amateurs have provided the only means of communication to and from some of the besieged towns in Bosnia.

* Radio amateurs have designed and built over 25 communications satellites. Astronauts and cosmonauts, including the UK's own Helen Sharman, frequently operate amateur radio stations from space. Many schools have been able to talk directly to space in this

Press Enquiries

RSGB HQ constantly receives requests from the national press and broadcast stations on a variety of radio-related subjects. If you receive any such requests direct, it would be wise to refer this to HQ so that the correct, and of course, positive reply can be given. In many cases we recommend the relevant expert on any given subject to the media so that information provided is up to date and accurate. This advice is given in an effort to ensure that damaging publicity is minimised and so that if an official press release is available this is sent to the enquirer.

The RSGB

The Radio Society of Great Britain represents UK radio amateurs at national and international level. Its 30,000 members receive a monthly 100-page colour magazine *Radio Communication* (*RadCom*) and benefit from technical advice. The RSGB publishes technical books, and co-ordinates most amateur radio activities, including carrying out Morse testing on behalf of the DTI.

Amateur Radio and the Young

The RSGB is keen to encourage more young people to enjoy the hobby. In 1988 HRH Prince Philip launched the RSGB's Project YEAR *(Youth into Electronics via Amateur Radio)*. This ongoing project, which is supported by the DTI and industry, is aimed at encouraging newcomers into amateur radio. The enjoyment which many young people get from the hobby can lead to a professional interest in electronics and communications.

The project includes the Novice Licence, which can be obtained relatively easily, various beginner's books and a special column in *RadCom* entitled 'Down to Earth'.

More information on amateur radio, the Novice Licence and the RSGB can be obtained from our web pages on www.rsgb.org or, in writing, from: Radio Society of Great Britain, Lambda House, Cranborne Road, Potters Bar, Herts EN6 3JE. Tel: (01707) 659015.

Famous Radio Amateurs

Notable radio amateurs who have made their name in other fields include:

9K2CS	Prince Yousef Al-Sabah
A41AA	Qaboos Bin Said Al-Said, Sultan of Oman
EA0JC	Juan Carlos, King of Spain
G2DQU	Lord Rix (formerly Sir Brian), actor and charity head
G3TZH	Tony Dolby, brother of 'the' Dolby
G3YLA	Jim Bacon OBE, Weather broadcaster
GB1MIR	Helen Sharman, astronaut
I0FCG	Francesco Cossiga, former President of Italy
JY1	King Hussein of Jordan
JY2	Queen of above
K1JT	Dr Joseph Taylor Jr, 1993 Nobel Prize in Physics
KA6HV	Burl Ives, singer *(silent key)*
KB2GSD	Walter Cronkite, news reader
KB6LQR	Jeana Yeager, Voyager '86 pilot
KB6LQS	Dick Rutan, Voyager '86 pilot
KN4UB	Larry Junstrom, rock musician
LU1SM	Carlos Saul Menem, President of Argentina
N5YYV	Kathy Sullivan, Chief Scientist NOAA (astronaut)
N6KGB	Stewart Granger (born James Stewart) actor *(silent key)*
S21A	Saif D Shahid, Head of Bangladeshi PTT
U2MIR/UV3AM	Musa Manarov, cosmonaut
VK2BL	Graham Connelly, radio announcer
VK2KB	Sir Allan Fairhall, politician
VR6TC	Tom Christian, great grandson of Fletcher Christian
VU2RG	Rajiv Gandhi, Prime Minister of India *(silent key)*
VU2SON	Sonia Gandhi, XYL of VU2RG
W0ORE	Tony England, astronaut
W3ACE	Armin Meyer, US Ambassador to Japan
W6FZZ	Samuel FB Morse III
W6ZH	Herbert Hoover Jr (Grandson of US Pres)
WA4CZD	Chet Atkins, guitar player
WB6RER	Andy Devine, actor *(silent key)*
WD4SKT	Donny Osmond, singer
YN1AS	General Anastasia Somoza Debayle, President Republic of Nicaragua *(silent key)*

This list has been compiled in good faith from a variety of sources. If anyone knows of any additions or corrections please send these to Marcia Brimson, 2E1DAY at RSGB HQ or e-mail: marcia.brimson@rsgb.org.uk, so that a master list can be kept for inclusion in subsequent issues.

Notable Amateur Radio Dates

Tempus Fugit! **It is all too easy to think that amateur radio was always the same as when we were first licensed (whenever that was) except for the very first experimenters. The fact is that amateur radio has always been constantly developing – some things happened much earlier than one would have thought – others much later! This 'pocket history' puts some of the key dates into perspective.**

Of necessity, it cannot be complete and actual dates are often not reported in the cronicles as their significance is rarely known at the time. If you have had first hand experience of other relevant dates or corrections to the list –please let us know!

1898 First amateur radio station in the world *(The station was in London and was that of M J C Dennis who later had the callsign DNX, then EI2B).*

1904 WT Act passed requiring radio stations to be licensed.

Patent for thermionic valve granted to Fleming.

1905 2nd reading of Bill to encourage wireless experiments.

1910 Wireless Institute of Australia founded – the first national society.

Postmaster General issues distinctive callsigns.

1913 Clubs in Derby, Liverpool, Birmingham and Nottingham formed.

London Wireless Club founded *(July).*

London Wireless Club renamed Wireless Society of London *(Sept).*

Club licence applied for *(Nov).*

1914 Experimental Licence for 850m granted to Society *(June).*

ARRL founded *(April).*

All licences suspended for First World War *(Sept).*

First AGM of Society *(Sept).*

Society suspended until end of war.

1919 Society revived.

1920 New transmitting licences issued as amateur radio commenced after WW1.

1921 First amateur signals to span the Atlantic *(Signals from amateur radio stations in the US were heard in Britain and other European countries).*

1922 Spark transmissions forbidden.

440 m band introduced.

Society name changed to Radio Society of Great Britain.

Edward Prince of Wales accepts position of Patron of the Society.

1923 T & R Section of society formed.

First 2-way contact USA to UK on 100m.

First G2+3 letter callsign issued

1924 International Amateur Radio Union (IARU) founded in Paris.

First amateur radio contact with antipodes (Britain and New Zealand).

1925 First publication of RSGB's journal - *T & R Bulletin (Transmitting & Research).*

Trans oceanic tests. *(Contacts made with ZL and VK using new 23 and 45 m bands)*

G prefix used for first time.

1926 Bands used 23, 45, 150/200 and old 440m. *(All with 10 W power limit)*

QSL bureau set up.

Diplomas for members intrduced for good work to the radio art and services to the Bulletin.

First British Amateur Radio convention held at Inst of Electrical Engineers.

RSGB becomes an incorporated company.

1927 First woman licensed, Barbara Dunn G6YL.

International Radio Telegraphic Conference proposed harmonically related bands 1.7 - 56MHz.

British Empire Radio Union (BERU) formed.

1928 Washington conference, granted 1.7 - 56MHz for international use. Each nation's telecommunications authority allowed to fix input power to PA.

System of callsigns allocated by ITU.

RSGB subscription: 10 shillings.

1929 BRS numbers issued to non-transmitting members.

Miss May Gadsden first paid RSGB member of staff.

1931 Empire Radio week. First BERU Trophy contest.

1932 *A Guide to Amateur Radio* first published – 2,000 copies sold.

John Clarricoates appointed secretary.

1933 First National Field Day (NFD).

RSGB HQ at 53Victoria Street opened.

First contact from aircraft on 56MHz by G6JP and G5CV.

RSGB membership up to 2,000.

1934 First amateur TV licences issued.

RSGB slow morse transmissions started.

RST reports introduced.

1935 Use of /P for portable stations.

FM demonstrated by Maj H Armstrong in Columbia, USA and from Empire State Building.

1936 First international contest on 28MHz.

King Edward VIII ceased to be Patron on accession to the throne.

First F to G contact on 56MHz.

First G8+2 calls issued.

Longer aerials up to 150 feet allowed.

1937 First RSGB Secretary/Editor J Clarricoats appointed.

TVI cases appear as TV transmissions on 45MHz start.

1938 International Radio Conference Cairo.

First edition of Amateur Radio Handbook published.

1939 First G4+2 calls issued.

All amateur activity closed down *(31 Aug).*

WW 2 broke out, all equipment confiscated *(3 Sept).*

1940 Month On The Air column changed to Month Off The Air.

2nd edition of Handbook becomes standard radio text book for the services.

RSGB HQ building bombed.

1941 Society business transferred to home of Secretary G6CL in Palmers Green, London.

Further 10,000 copies of Handbook printed.

1942 Handbook supplement printed.

T & R Bulletin becomes *RSGB Bulletin.*

1943 HQ moved to New Ruskin House, 28 Little Russell St *(July).*

1944 Agreement with Post Office that all new applicants for amateur radio licences should take an exam, under the supervision of the City & Guilds, and a morse test.

1945 Statement of post war licensing policy drawn up including exemption from exam for certain service personnel.

WW2 ends. 28-29MHz & 56-60MHz released.

DXCC introduced

1946 1.8-2.0MHz & 29-30MHz bands released *(March).*

7.13-7.3MHz & 14.1-14.3MHz bands released *(May).*

First Radio Amateur exam held *(June).*

Also Television broadcasts resumed on 45MHz.

3.5-3.8MHz band released *(Sept).*

1947 World Telecommunications Conference in Atlantic City allocated amateur bands world wide, added 2m, 70cms, 23cms and microwave bands (all harmonically related).

First post war NFD held, won by Southgate ARC *(June).*

First transistor made.

1948 145-146MHz, 420-450MHz, 2,350MHz released.
Use of FM on amateur frequencies suggested by G5CD.
Start of amateur radio development of single sideband.
Amateur Radio Exhibition, Royal Hotel, Woburn Place, London WC1.
RSGB TVI/BCI Committee setup.

1949 144-145MHz, 1,215MHz, 5,650MHz and 10,000MHz bands released.
Old 5m band lost to television.
RSGB membership: 14,038.
GPO grants permission for power input increased to 150W on bands above 28 MHz except 420-460 MHz and for the use of SSB.

1950 GB1RS beacon on 3,500.25 kHz setup at HQ.
ATV on 70cms permitted /T.
Portable operation /P permitted on payment of 10 shillings fee.
No /MM operation yet.
IARU conference in Paris, 21MHz band proposed.

1951 GB3FB station at Festival of Britain.
ATV permitted on 1225-1290MHz.
FM permitted on 144.5-145.5MHz.
First RSGB Call Book published.
SSB equipment displayed.

1952 GPO lifts restrictions on operation away from home address.
21-21.45MHz CWoperation allowed.
150W i/p allowed on all bands 3.5-28 and 420 MHz.
HRH Prince Philip, accepts to be Patron of Society.

1953 East coast floods leads to formation of Radio Amateurs' Emergency Network.
First meeting of VHF Convention held at Bedford Corner Hotel organised by LondonUHF Group.
New constitution for Society adopted, council members elected for 3yrs and 6 Zones, each to elect a council member.

1954 First mobile licenses issued.
Telephony permitted on 21MHz.
First amateur transistor transmitter *(used by Yeovil ARC)*.

1955 New style council in being.
GB2SM Science Museum station set up.
First GB2RS news broadcast on 3.6MHz from G6MB.
First International VHF/UHF Convention.

1956 UK Amateurs allowed to use 70.2-70.4MHz.
Amateur (sound) licence amended to include self training and use during disaster relief.
First RSGB 21/28MHz contest.

1957 Certain amateurs allowed to operate on 52.5MHz as part of the International Geophysical Year (IGY) project.
First gathering of mobile enthusiasts at Woburn Rally.
First Sputnik launched *(heard on 21 MHz)*.

1958 Exemptions from RAE and morse test ended.
RAOTA formed at 3rd OT dinner at Horse Shoe Hotel, WC1.
First Scout Jamboree on the air (JOTA) from Gillwell GB3BP.
Pat Hawkers first writes *Technical Topics* column *(Apr)*.

1959 World Radio Conference Geneva, loss of 7.1-7.15 MHz.
RSGB membership: 9,540.
Dr R L Rose-Smith (former Director of Scientific & Industrial Research) elected President for 1959
First 144MHz Field Day.
First CQWW contest
First G2DAF design for amateur SSB equipment published – lead to widespread uptake of amateur SSB.
RTTY to be permitted.

1960 IARU Region 1 conference held in UK.
Membership up to 10,036.
/A and /M suffixes to callsigns agreed.
GB3VHF commences on 144.5MHz for propagation and frequency studies.
First amateur radio moonbounce contact.

1961 Official recognition of use of RTTY in all bands except 160m.
US amateurs given permission to operate /MM in 20m band.
No G3Q calls to be issued.

1962 First Oscar satellite launched, beacon on 2m.
Meeting of ITU to allocate bands for space radio.

1963 Golden Jubilee Celebrations held.
KW Vesper and 2000 SSB transceivers shown at The RSGB International Radio Communications Exhibition, Seymour Hall *(Oct)*.
Discussions with GPO regarding VHF only license.

1964 G6CL John Clarricoats, General Secretary retires and is elected Honorary member.
Class B licenses issued permits operation on 420MHz and above, phone only, no morse, G8+3 calls.
Use of NATO phonetics recommended.
4 m band reduced to 70.1-70.7 MHz.

1965 Withdrawal of 420-427MHz.
RSGB subscriptions increased to £2/10 (£2.50).
GPO agrees to principle of reciprocal licensing – previously no foreign operation in UK.
2m bandplan by counties introduced. *(Different groups of counties had seperate portions of the band to call in.)*

1966 Maritime mobile licenses /MM and reciprocal licenses G5+3 first issued.
Society IGY aurora program set up.

1967 3rd Edition *Amateur Radio Handbook* published.
GPO say no to Citizens Band due to experience in USA.
Wireless Telegraphy Act to ban importation of equipment to transmit on 27MHz.
Publication of Society History *"World at their fingertips"* by John Clarricoates.
Lambda Investment Co formed for purchase of Doughty Street.

1968 RSGB Bulletin renamed "*Radiocommunication*".
Agreement signed with Denmark and Italy for reciprocal license.
GB2LO City of London Festival station.
Move of RSGB HQ to 35 Doughty Street.

1969 Amateur Radio licensing passed from GPO to Ministry of Posts and Telecomms.
AMSAT USA formed.
Oscar 5 Australis, First amateur satellite built outside of US, launched.
QRA locator system retained.

1970 New bandplans for 4, 2, 70 and 23cms.
WARC Space Radio Comms held.
Reciprocal licence with USA extended.
(circa) First Silicon IC made

1971 First G4+3 calls issued.
First SSTV on 20m sent.
Repeaters proposed on 2m.
First 23cms beacon GB3LDN operational.
Wireless receiving licences abolished.
Amateur Satellite Service defined on HF & VHF bands following WARC Space Conference.
First UK adverts for Yaesu FT2F *(First popular Japanese black box)*.

1972 First 2m repeater licensed - GB3PI Cambridge.
First UK amateur RTTY Convention held.

1973 VAT charged on all equipment and components.
15,000 amateurs in UK.
2m bandplan using R1-R7 & S8-S23 using 25kHz spacing introduced.
VAT on RSGB subscriptions charged.

1974 First repeater on 70cms proposed.
Licence renewals taken over by Home Office.
G-QRP club formed.
RNARS station established on HMS Belfast.

1975 First 10GHz beacon operational – GB3IOW.
Simplified logs for mobile operation.

VAT on amateur radio equipment raised to 25%.

License fee up to £4.80. RSGB subscriptions: £8.

(circa) First microprocessor made

1976 First batch of 20 repeaters on 70cms licensed.

Home Office issue new licences covering all forms of Amateur Transmissions for fee of £5.50.

RSGB places membership records on its own IBM32 mini-computer.

VAT reduced to 12.5%.

Digital communications permitted above 144 MHz.

First fully transistorised HF rig introduced.

1977 GE prefix allowed for Queen's Jubilee.

Lord Wallace of Coslany (former Postmaster General) appointed President for 1977

Arthur Milne G2MI retires as QSL Bureau manager after 38yrs.

RSGB International Exhibition at Alexandra Palace.

1978 First SSB repeater proposed - Sheffield *(GB3SF)*.

GB calls issued for special event stations by RSGB.

RAE changes from written to multiple-choice answers.

Peter Martinez proposes AMTOR system.

Launch of first russian RS satellite.

1979 WARC held in Geneva. 2m bandplan allocates 145.8 -146MHz to satellites.

1980 Result of WARC. Morse test requirement reduced from 144 to 30 MHz. New bands at 10, 18 and 24MHz allocated to amateurs.

First 50MHz beacon operational - GB3SIX.

First Phase III satellite launched but failed to orbit.

1981 CB radio legalised on 27 & 934 MHz FM.

FM can be used on 29.6-29.690MHz.

Class B licensees permitted to use morse under supervision.

First 70MHz contact across Atlantic.

Launch of UoSat 1.

First G6 + 3 callsigns issued.

1982 10.1-10.15MHz band released.

First National HF Convention at Belfrey Hotel, Oxford.

New license schedule requires dBW's for power.

Loss of 431-432MHz for 100kms around Charing Cross.

RSGB HQ moves from Doughty St. to Alma Ho, Potters Bar.

1983 40 permits issued for operation on 50/52MHz outside TV hours.

Phase IIIB Oscar 10 launched.

RSGB issues SES GB calls on behalf of Home Office.

Don Baptiste CBE (former head of DTI Radio Regulatory Div) elected President for 1983.

G5+3 calls abolished.

First G1 + 3 callsigns issued.

RSGB Membership: 33,000 members.

First Guides Thinking Day on the Air

First Space Shuttle QSO – op: Owen Garriot *(Dec)*.

1984 60 more permits issued for 50/52MHz.

New (Maidenhead) locator system introduced.

BARTG Silver Jubilee.

RSGB backs AX25 packet standard.

DTI takes over from Home Office radio regulatory work.

1985 Class B holders permitted to use CW under NoV *(1 year experiment)*.

RSGB HQ renamed Lambda House.

Joan Heathershaw, G4CHH elected first woman president of RSGB.

First ATV repeater GB3GV operational.

Limited CW operation permitted on 24MHz band.

3rd party greetings messages permitted from GB stations.

First G0 + 3 callsigns issued.

First colour SSTV with shuttle - Tony England *(Aug)*.

1986 50-50.5MHz generally released to all class A licensees.

First applications for digital repeaters on 2m.

RSGB wins amateur radio morse testing contracts from British Telecom International.

1987 DTI to sponsor "Young amateur of the Year" award.

50 and 70MHz bands made available to Class B holders.

9 packet radio repeaters licensed on 2m UoSat II launched with packet facilities.

1988 75th Anniversary celebrations opened by HRH Prince Philip at NEC.

First YAOTY awarded to Andrew Keeble, G1XYE.

Radcom Editor Alf Hutchinson retires 19 years service.

DIY radio magazine introduced.

First NoVs issued for GB7+3 unattended packet radio mailboxes.

First G7 + 3 callsigns issued.

IOTA award scheme introduced.

1989 18 and 24MHz bands fully released.

Introduction of CEPT T/R 61-1international licensing agreed.

First UK amateur Aeronautical Mobile in recent times from an RAF Canberra.

Novice license discussed with DTI.

1990 QSL bureau moved to HQ.

Novice license instructors appointed.

Special club callsigns GX etc introduced.

Unattended packet operation permitted.

Formation of Radiocommunications Agency as an executive agency of the DTI.

First 2m/70cm hand held rig introduced.

1991 Novice A & B licenses issued.

RSGB Subscriptions up to £30.

Packet DX Clusters set up.

M0RSE call issued to celebrate 200yrs of morse.

Helen Sharman operates GB1MIR from Mir space station.

CTCSS tones proposed on repeaters.

1992 First 4 novices visit HRH Prince Philip at Buckingham Palace.

26dBW allowed on 1.83-1.85 MHz.

Warwick Business Strategy Conference.

Queen visits UOSAT command station at Surrey University.

Subscription Services Ltd win amateur radio and CB licensing contract.

Satellite packet gateways introduced.

1993 RSGB respond to Spectrum review of 28-470MHz.

EEC directive on EMC proposed.

Helen Sharman, first british astronaut visits RSGB.

Morse test format changed to QSO type.

Unattended operation for packet allowed.

No of amateur licences fall by 2200.

1994 Amateur Radio in schools, STELLA started.

Celebrations of D-Day+50yrs.

Removal of power and aerial restrictions on 50MHz.

Higher power permitted on 1.8MHz.

GB2SM to close down.

Age for obtaining full license reduced to 10 years.

1995 RSGB donates £3,000 to AMSAT UK for Phase IIID fund.

GR calls to celebrate 50th anniversary of VE Day.

RA move to Docklands.

Loss of part of 10GHz band.

RA move to Docklands.

First amateur radio DSP equipment introduced (TS950SDX).

1996 Further RSGB donation of £25,000 to AMSAT.

RSGB on Internet.

First M calls issued.

73kHz band released. RSGB assists in issueing of NoVs.

Contest callsigns permitted. RSGB assists in issueing of NoVs.

IARU conference agrees 12.5kHz spacing on 2m.

RA move to New Kingsbeam House following terrorist bomb damage to Docklands building.

RSGB HQ Museum, Library & Shack

The Society's Museum, Shack and Library are run by John Crabbe, G3WFM. They are open between 10.00 am to 4.00pm on Mondays and Thursdays, and by prior appointment. To avoid disappointment, please confirm your intention to visit by first ringing him at HQ during these times.

National Amateur Radio Library

The books cover many technical subjects associated with radio and television both British and foreign and unfortunately, are not available for loan. There is a historical section with articles about early radio and some of the pioneers of amateur radio. There are call books dating back to 1951, and official lists of callsigns dating back to 1914. Also on the shelves are bound volumes of *Bulletins*, *Radio Communication* and *QST* dating back to 1925. There are some early volumes of *Wireless World* and *Short Wave* Magazine, but these are incomplete. We would like to obtain volumes of *Wireless World* before 1925 – please contact John Crabbe.

There are copies of many European, African, Asian and Australasian amateur radio societies' magazines. Also included are data reference books on many valves and cathode ray tubes and out of print *RSGB Handbooks* and publications. Some *ARRL Handbooks* and publications are on the shelves together with special interest group magazines.

RSGB Shack

The headquarters station GB3RS is available for use by members and overseas visitors. Operators must have a current licence and are required to sign the logbook against any QSOs they may make. It helps if they also write out QSL cards for all contacts they make.

The shack is equipped with three modern HF Transceivers and one VHF/UHF transceiver. Antennas for HF are a 3 element triband beam, dipoles for 40 and 80 m and crossed yagis with azimuth and elevation rotators for satellite use. There are also vertical antennas for 2 m and 70 cm.

Operators must be prepared to suffer some QRM as the site is adjacent to an industrial area and many computers.

RSGB Museum

The earliest items on show are from the Maurice Child, G2DC, collection including a coherer receiver 1898, a spark transmitter 1915, a Marconi aerial tuner 1903, a crystal short wave receiver 1915, a morse linker 1900, and a large brass key made in 1917. There are several examples of valve TRF receivers made in the 1920's and 2 self-excited transmitters, one for 45m and one for 90m. Also on show is a fine example of a crystal set made from a kit supplied by Gamages in 1924. (The first meetings of the London Wireless Club were held in this London store in the twenties).

For the 30's there is a compact Tx/Rx for 160m made by G8TL, all built in a wooden cabinet using a 4 valve superhet with a 465 kHz IF, for CW and AM, and a single valve 6F6 crystal controlled transmitter modulated by a 6V6 and an inbuilt ATU. George Jessop G6JP has donated a replica 4 valve transmitter for 5m used in 1934 for air to ground tests. These transmissions where received by amateurs all over London and proved the effectiveness of VHF transmissions from aircraft. A National HRO with a set of 6 coils, is on display together withHallicrafters Skyrider 23, SX16, SX24 and SX28.

The 40's of course was the heyday of Government surplus and here we have on display an AR88 from RCA made for the US Navy and purchased in their hundreds by hams for their main station receivers. It covers 75 kHz to 30 MHz in 6 bands and, in spite of the heat generated by the inumerable valves, can resolve single sideband. There is a typical Transmitter to go with the AR88 as used just after the war, and this is a 40 watt xtal controlled CW and AM Tx made by Webbs radio who used to have a shop in Soho West End of London. This is rack-mounted, in working condition and has put out a signal on 40 mtrs giving 30 watts from its single 807 valve. An 1154 Tx and an 1155 Rx are on display, together with a BC348 Rx, all war surplus equipment, bought in their hundreds by post war amateurs.

In the 50's a British company from Cambridge made a 100 watt Tx called the Labgear LG 300. This is built into 2 large cabinets, one the Tx and the other the modulator and power supply. The Tx has an 813 valve in the PA and gives out a good 100 watts on AM and CW. The vfo operates on 3.5 to 4 MHz and multiplies up to the other bands, with a series of wideband couplers. The modulator has a pair of KT66s and 4 enormous transformers each weighing about 10kgms.

A Minimitter 150W AM/CW transmitter is on display and a typical working shack from this period using a Panda Cub AM/CW transmitter an HRO receiver, with 6 plug in coils, and a BC221 wavemeter.

To complete the show for the fifties, there are Eddystone general coverage receivers S640, 740, 670 and 888A, all in working condition.

For the 60's the Society has been loaned an Eddystone EA12 built in Birmingham and is one of the first receivers designed to resolve SSB. Also from the same decade are a Tx and Rx pair designed by the late G2DAF. These are specifically for amateur SSB with low drift VFO's and narrow band Xtal filters and high image rejection in the receivers. These designs were published in the *'Bull'*, the RSGB journal at the time, as a construction project and many were built putting many hams on the air with SSB.

A small section is devoted to VHF and UHF gear which comprises of converters and transmitters for 4 mtrs, 2 mtrs, 70 cms and 23 cms. These were made by W.Scarr G2WS and the designs were published in the *'Bull'*.

The last decade of the 70's is represented by a Transceiver for 5 HF bands SSB and CW giving about 100 watts PEP. This is the Heathkit HW101 which could be built from a kit. It was manufactured in the USA and sold in this country for about £100. Also designed and built about this time, and kindly donated by Rowley Shears G8KW, is a very rare KW2000D transceiver with digital frequency readout, one of the last British made HF transceivers.

That completes the main show as the 80's decade is represented by the expensive and complex products, which are in the main shack. There is a large collection of radio valves (tubes) ranging from a 1908 DeForest 'Audion' to miniatures from the last days of thermionics. The curator is always looking for typical amateur radio equipment from the past, particularly from the first three decades of the century, and would be pleased to hear from anyone who wishes to donate them to the Society for the display.

Other Museums

Amberley Museum, GB2CPM

3 miles north of Arundel, Sussex on B2139 (adjacent to Amberley BR Station). Industrial open air museum including an amateur radio station GB2CPM and a substantial vintage wireless exhibition. Open 10:00 to 18:00, Easter to October, Wednesday to Sunday and bank holiday Mondays. Daily during local school holidays. Admission charge. Tel: (01798) 831370. Further details from Margaret Brownlow, G4LCU.

The Godfrey Manning Aircraft Museum

This collection contains no amateur radio equipment, however as the main emphasis of the museum is on contemporary airline instruments, avionics and radio, it may be of interest to many amateurs. Viewing is for individuals or small groups and is by appointment only. Details from Godfrey Manning, G4GLM, 63 The Drive, Edgware, Middx HA8 8PS. Tel: (0181) 958 5113.

Isle of Wight Wireless Museums

There are now two wireless museums on the Isle of Wight, both under the aegis of the Communications & Electronics Museum Trust.

One has been at the Arreton Manor near Newport, for the past decade, and the other is at Puckpool Park, Seaview, near Ryde. The special callsign GB3WM is used on HF and VHF. They are open daily during the summer months, and by arrangement during the winter. More details from the curator, Douglas Byrne, G3KPO, on Ryde (01983) 567665.

Mosquito Air Museum – home of the deHavilland Aircraft Heritage

Located at Salisbury Hall, nr London Colney, Herts (J22, M25; 400m down B556 towards South Mimms). The DH98 Mosquito was designed and prototyped here in secret. Many old deHavilland aircraft in preservation. Extensive collection of WW2 military radio equipment.

Open from March to the end of October on Tuesdays, Thursdays and Saturdays 1400-1730 hrs. Sundays and bank holidays from 1030-1730 hrs.

Royal Signals Museum

This museum is situated within Blandford Camp near Blandford Form in Dorset (signed from the A354). It displays a unique collection of military radio and other signalling equipment as well as other items. Open Monday to Friday all year 10.00 to 17.00 and weekends June to September 10.00 to 16.00. Admission charge. Further details on (01258) 482248.

World of Wireless & Communication

The Mill, Atwick Rd, Hornsea, East Yorkshire, HU18 1DZ. Collection of vintage wireless but including amateur, military, and commercial equipment. Viewing by appointment. Details from Jeff Southwell, G4IGY, tel: (01964) 533331 (24 hours)

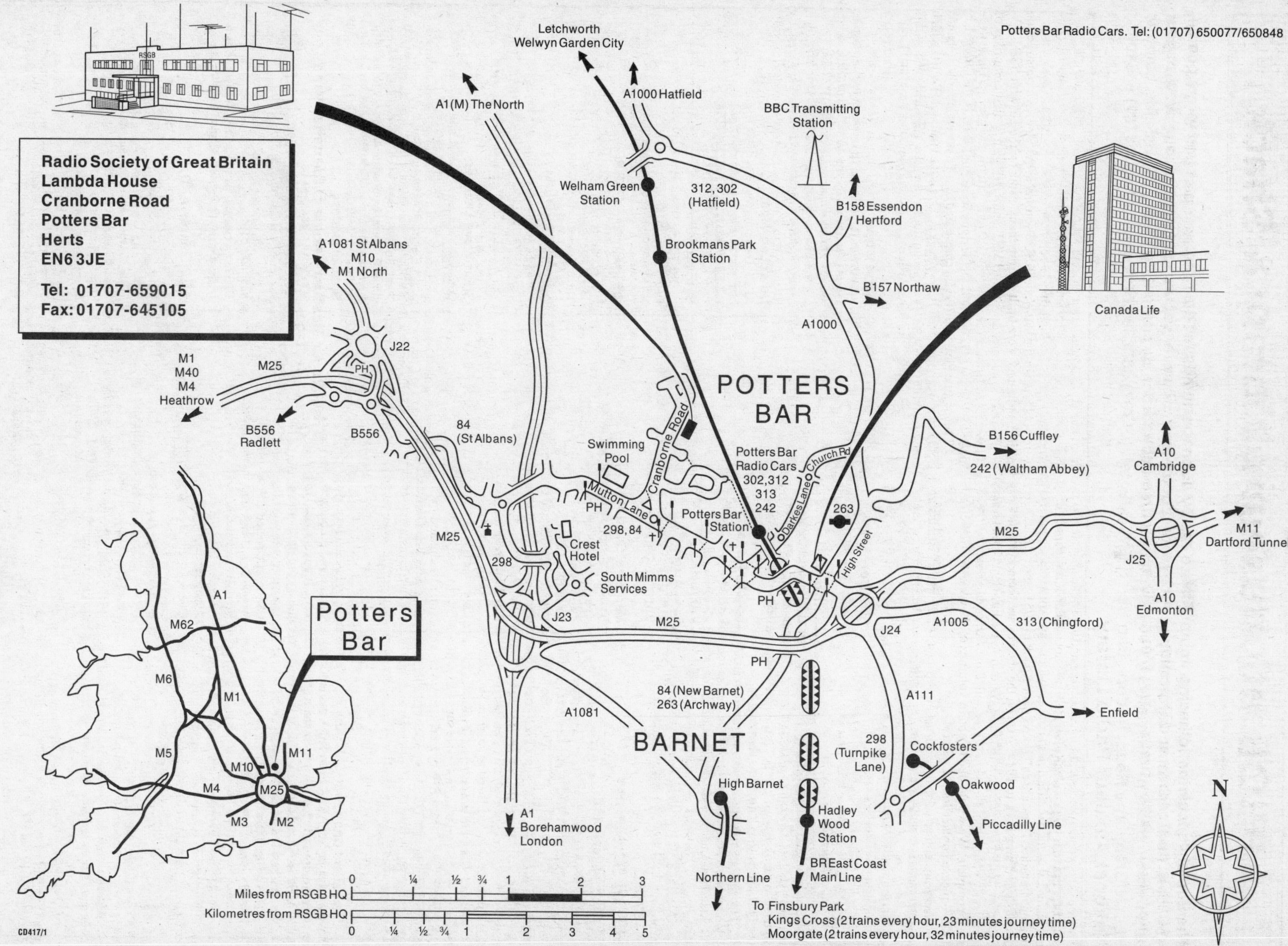
Radio Society of Great Britain
Lambda House
Cranborne Road
Potters Bar
Herts
EN6 3JE
Tel: 01707-659015
Fax: 01707-645105
Potters Bar Radio Cars. Tel: (01707) 650077/650848
Letchworth
Welwyn Garden City
A1(M) The North
A1000 Hatfield
BBC Transmitting Station
Welham Green Station
312, 302 (Hatfield)
B158 Essendon Hertford
Brookmans Park Station
A1081 St Albans
M10
M1 North
B157 Northaw
Canada Life
A1000
J22
M1
M40
M4
Heathrow
M25
PH
B556 Radlett
B556
84 (St Albans)
POTTERS BAR
Swimming Pool
Cranborne Road
Potters Bar Radio Cars
302,312
313
242
Church Rd
Darkes Lane
B156 Cuffley
242 (Waltham Abbey)
A10 Cambridge
Mutton Lane
PH
298, 84
Potters Bar Station
263
M25
M11
Dartford Tunnel
Crest Hotel
High Street
J25
298
South Mimms Services
PH
A10 Edmonton
J23
M25
J24
A1005
313 (Chingford)
PH
84 (New Barnet)
263 (Archway)
A1081
A111
Enfield
BARNET
298 (Turnpike Lane)
Cockfosters
High Barnet
Oakwood
Hadley Wood Station
Piccadilly Line
A1
Borehamwood
London
BR East Coast Main Line
Northern Line
To Finsbury Park
Kings Cross (2 trains every hour, 23 minutes journey time)
Moorgate (2 trains every hour, 32 minutes journey time)
Potters Bar
A1
M62
M6
M1
M5
M10
M11
M4
M25
M3
M2
Miles from RSGB HQ
0 ¼ ½ ¾ 1 2 3
Kilometres from RSGB HQ
0 ¼ ½ ¾ 1 2 3 4 5
N
CD417/1

Publications Available From RSGB

We normally stock the following items. Please see *RadCom* or ring us for the current prices and terms of sale. You can find out more information about our books and place a secure order from our web pages: http://www.rsgb.org. Alternatively you can place a credit card order by contacting us at: RSGB Sales, Lambda House, Cranborne Road, Potters, Bar, Herts, EN6 3JE, tel: (01707) 660888 (*24 hour*), or fax: (01707) 645105.

RSGB Publications

Antennas:
The Antenna Experimenter's Guide
HF Antenna Collection
HF Antennas For All Locations
Practical Wire Antennas

Awards:
IOTA Directory & Yearbook

Beginners & Novices:
Amateur Radio for Beginners
How to Pass the Radio Amateurs Examination
The Novice Licence Student's Notebook
Practical Antennas for Novices
Practical Receivers for Beginners
Practical Transmitters for Novices
Radio Amateurs Examination Manual
RAE Revision Notes
Revision Questions for the Novice RAE
Training for the Novice Licence - Instructor's Manual
Your First Packet Station
Your First Amateur Radio Station

Callbooks:
1998 RSGB Yearbook

EMC:
The Radio Amateur's Guide to EMC

General Technical:
PMR Conversion Handbook
Radio Communication Handbook
Technical Topics Scrapbook
Test Equipment for the Radio Amateur
Radio Data Reference Book
LF Experimenter's Source Book

History:
World at Their Fingertips

Log Books & Log Sheets
Transmitting Log Book
Receiving Log Book
Log Book Cover
Log Sheets - HF Contest
Log Sheets - VHF Contest

Maps and Charts:
Region 1 Beacons, UK Packet & Repeaters
Countries/Awards List
Great Circle DX A4 Map
Great Circle DX Wall Map
Locator Map of Europe - A4
Locator Map of Europe - Wall
Meteor Scatter Data Sheets
International QSL Bureau List
Prefix Guide

Microwaves:
Microwave Handbook Vol.1
Microwave Handbook Vol.2
Microwave Handbook Vol.3

Morse:
Morse Code For Radio Amateurs

Operating Aids:
Amateur Radio Operating Manual
The RSGB Rig Guide
ARDF Handbook

QRP (Low Power):
G-QRP Club Circuit Handbook

RadCom Back Issues & Binders:
Radio Communication Easibinder
Bound Volumes, 1996; '87; '94; '95; '96
Back issues (phone for availability)

RSGB Newsletters:
DX News Sheet
Microwave Newsletter
RLO Newsletter

Satellite:
The Space Radio Handbook

Software:
Instant Morse CD ROM
RadCom on CD ROM
CallSeeker '98 CD ROM

Special Modes:
Packet Radio Primer

VHF/UHF:
Radio Auroras
VHF/UHF Handbook
VHF/UHF DX Book

Members' Sundries:
Badges: Callsign standard inc engraving
Callsign deluxe inc engraving
Lapel Mini
Lapel Standard
Ties: New Design (red or blue)
Car Stickers: RSGB Diamond

Car Stickers:
I Love Amateur Radio
I'm on the Air....

Other Publications

Antennas:
Antenna Compendium Vol.1 *(ARRL)*
Antenna Compendium Vol.2 *(ARRL)*
Antenna Compendium Vol.3 *(ARRL)*
Antenna Compendium Vol.4 *(ARRL)*
Antenna Compendium Vol.5 *(ARRL)*
Antenna Impedance Matching *(ARRL)*
The ARRL Antenna Book (18th Edn) *(ARRL)*
Low Profile Amateur Radio *(ARRL)*
Simple Low Cost Wire Antennas *(ARRL)*
Vertical Antenna Classics *(ARRL)*

Awards:
K1BV Awards Directory *(K1BV)*

Beginners & Novices:
Basic Radio and Electronic Calculations *(RP)*
First Steps in Radio *(ARRL)*
Novice RAE - Additional Worksheets *(RS)*
Now You're Talking (2nd Edn) *(ARRL)*
Radio Amateurs Q & A Ref Manual *(RP)*
Operating an Amateur Radio Station *(ARRL)*

EMC:
Radio Frequency Interference *(ARRL)*

General Technical:
ARRL Handbook *(ARRL)*
Introduction to RF Design *(ARRL)*
W1FB's Design Notebook *(ARRL)*
Radio Buyers Source Book *(ARRL)*
Solid State Design *(ARRL)*
The New Shortwave Propagation Handbook *(CQ)*

Maps and Atlases:
Grid Locator Atlas *(ARRL)*
World Prefix Map (A3 Desk) *(GR Pub)*
DXCC Countries List *(ARRL)*

Morse Code:
Your Intro. to Morse Code Tapes (0-5 WPM)*(ARRL)*
Morse Instruction Tapes (5-10 WPM) *(ARRL)*
Morse Instruction Tapes (10-15 WPM) *(ARRL)*
Morse Instruction Tapes (15-22 WPM) *(ARRL)*
Morse Code the Essential Language *(ARRL)*

Operating Aids:
ARRL Operating Manual *(ARRL)*
Low Band DXing (2nd Edn) *(ARRL)*
DX Edge Propagation Aid *(XANTEK)*

QRP (Low Power):
G-QRP Club Antenna Handbook *(G-QRP)*
QRP Power *(ARRL)*
W1FB's QRP Notebook *(ARRL)*

QST Magazine: (ARRL)
One Year (Air or Surface Mail)
Two Years (Surface Mail)
Three Years (Surface Mail)

Satellite:
Satellite Anthology (3rd Edn) *(ARRL)*
Satellite Experimenter's Handbook *(ARRL)*
The Weather Satellite Handbook *(ARRL)*

Short Wave Listener:
Complete Shortwave Listener's Handbook *(TAB)*
Passport to World Band Radio *(IBS)*

Software:
Super Frequency List CD ROM *(Klingenfuss)*
Radio Amateur World Callbook CD ROM *(DARC)*
Conversation Guide CD ROM *(DARC)*

Special Modes:
Am Packet Radio Link Layer Protocol *(ARRL)*
RTTY Awards *(BARTG)*
Slow Scan TV Explained *(BATC)*
Packet, Speed and More Speed *(ARRL)*

VHF/UHF:
UHF/Microwave Experimenter's Manual *(ARRL)*
UHF/Microwave Projects Manual *(ARRL)*

EMC Filters: (AKD)
2 x Ferrite Rings (Fair-rite 43 Material)
Filter 1 Braid Breaker
Filter 2 HPF for FM Band 2
Filter 3 HPF & Braid Breaker
Filter 4 Notch at 145MHz
Filter 5 Notch at 435MHz
Filter 6 Notch at 50MHz
Filter 7 Notch at 70MHz
Filter 8 High Pass 6-section
Filter 10 28MHz Notch
Filter 15 21MHz Notch

At Your Service!

RSGB HQ, Lambda House, Cranborne Road, Potters Bar, Herts, EN6 3JE.
Tel: (01707) 659015. Fax: (01707) 645105. Web: http://www.rsgb.org

The Society provides an extremely diverse range of services for its members by its blend of professional staff and management team at Headquarters and a country-wide force of skilled, dedictaed and knowledge volunteers. In order to get the best out of the RSGB it is important that you approach the correct part of the Society. This list is a practical guide to find the right person for your enquiry.

Society Policy – *RSGB Council*

The Society's affairs are directed by its Council who are chosen by a postal ballot each autumn. If you have a query about Society policy, you should first contact your Zonal Council Member.

President: Ian Kyle, GI8AYZ
Immediate Past President: Peter Sheppard, G4EJP
Executive Vice-President: John Greenwell, G3AEZ
Company Sec & General Manager: Peter Kirby, G0TWW, RSGB HQ

Zonal Members:

Zone A (North of England): Gordon Adams, G3LEQ, Home: (01565) 652652.

Zone B (Midlands): David Whalley, G4EIX, Home: (01952) 588878.

Zone C (SE England and East Anglia): Fred Stewart, G0CSF, Home: (01732) 780721.

Zone D (SW England): Julian Gannaway, G3YGF, Home: (01794) 340895.

Zone E (Wales): Paul Essery[2], GW3KFE, Home: (01686) 628958

Zone F (Northern Ireland): Terry Barnes, GI3USS, Home: (01247) 473948.

Zone G (Scotland): Tommy Menzies, GM1GEQ, Home: (0131) 445 3928
E-mail: tmenzies@netcomuk.co.uk

Ordinary Members:

John Allaway[1], G3FKM
Don Beattie, G3OZF
Dick Biddulph, G8DPS
Peter Chadwick, G3RZP
Hilary Claytonsmith, G4JKS
Geoff Dover, G4AFJ
John Greenwell, G3AEZ
Richard Horton[2], G3XWH

[1] *Retires from Council on 31 Dec 1997, but is not eligible for re-election.*
[2] *Retires from Council on 31 Dec 1997 and is eligible for re-election.*

Full details of Council's and Committees' terms of reference are published in the Society's Green Book available from RSGB HQ.

Society Activities – *Committees, Honorary Officers, HQ*

The many different actvities of the Society are run by its committees, honorary officers and HQ staff. If you wish to take advantage of one of these services or have an administrative enquiry about any one of them, contact the person(s) listed below.

Antenna Planning

Booklet free to members from RSGB HQ. Planning application refused - specific advice to RSGB members: RSGB Planning Panel, via AR Dept, RSGB HQ. Planning Advisory Committee Chairman: Geoff Bond, G4GJB, QTHR.

Audio-visual library

Videos etc, for short term loan to RSGB affiliated clubs. List published in *RSGB Yearbook.* Co-ordinator: John Davies, G3KZE at RSGB HQ.

Awards

For contest awards, refer to the appropriate contest committee. Enquiries and applications for operating awards - HF: Fred Handscombe, G4BWP, QTHR. VHF: Ian Cornes, G4OUT, QTHR. IOTA: see under IOTA. Trophies Manager: David Simmonds, G3JKB, QTHR.

Band Plans

See *RSGB Yearbook* or April *RadCom* for the latest bandplans. For policy, contact the appropriate committee chairman or spectrum manager. HF Committee Chairman: Colin Thomas, G3PSM. VHF Committee Chairman: Mike Adcock, GW8CMU. VHF Manager: Dave Butler, G4ASR. Microwave Committee Chairman: Andy Talbot, G4JNT. Microwave Manager: Mike Dixon, G3PFR. (All QTHR)

Beacons [3]

List of current beacons: *RSGB Yearbook*, or from RSGB Sales. Application leaflet *'Guide to Beacon Licensing'* free from AR Dept, RSGB HQ. Policy and proposals to appropriate Beacon Co-ordinator: HF: Prof Martin Harrison, G3USF. VHF: John Wilson, G3UUT. Microwave: Graham Murchie, G4FSG. (All QTHR)

Book Sales

Current price list: see *RadCom.* Orders and enquiries: Sales Dept, RSGB HQ. Tel: (01707) 660888 *(24 hour),* e-mail: sales@rsgb.org.uk. Secure server order form available on web site.

Call Book

Change of address, details to be withheld or released, callsign not listed: Subscription Services Ltd, PO Box 885 Britsol, BS99 5LG. Appended text [in square brackets]: Yearbook Editor, RSGB HQ. Club listing: Sales Dept, RSGB HQ, e-mail: subscriptions@rsgb.org.uk. International Listings: all enquiries Radio Amateur Callbook Inc, 1685 Oak St, Lakewood, NJ 08701.

Credit Card

(RSGB Affinity Card) Application forms and details from Sales Dept, RSGB HQ. Other enquiries: Bank of Scotland, Card Services, Pitreavie Business Park, Dunfermline, Fife, KY99 4BS.

Contests (RSGB)

See *RSGB Yearbook* or *'Contesting Guide'* (September *RadCom)* for general rules, other issues of *RadCom* for results DF Contest rules/dates and late changes. For queries / interpretations, contact the adjudicator mentioned in the rules for the specific contest. For policy, contact the appropriate committee chairman. HF: Chris Burbanks, G3SJJ. VHF: Steve Thompson, G8GSQ. ARDF (direction finding) Geoff Foster, G8UKT. (All QTHR)

DX-News Sheet

Editorial content: Chris Page, G4BUE. Subscriptions: Sales Dept, RSGB HQ. Tel: (01707) 66088 *(24 hour).* E-mail: sales@rsgb.org.uk.

DX News Sheet Voice Bank

Five most recent tips recorded by DXers. To listen:(01426) 925240, to record: (01426) 910240.

EMC

(Advice on solving breakthrough and other electromagnetic compatibility matters.) Information sheets: free to members from AR Dept, RSGB HQ. Specific advice from your local EMC Co-ordinator - see *RSGB Yearbook.* For policy: contact the committee chairman, Robin Page-Jones, G3JWI, QTHR.

Exhibition & Rally Committee

(Organises Woburn Rally and trade show at VHF Convention) Committee Chairman: Norman Miller, G3MVV, QTHR.

GB2RS

Schedule of newsbroadcasts: *RSGB Yearbook, RadCom.* Telephone *(main news items only)*: (0336) 407394 *(premium rate).* News for insertion *(to arrive by 10.00 am Wed prior to broadcast at latest):* RadCom Dept, RSGB HQ. E-mail:GB2RS@rsgb.org.uk. Fax: (01707) 649503. News reader appointments and brodcast policy: Gordon Adams, G3LEQ, QTHR.

General Advice

How to get started, licensing queries, RAE courses: AR Dept, RSGB HQ. E-mail: AR.Dept@rsgb.org.uk

Health Insurance
HMCA, Scriven Park, Ripley Road, Knaresborough, Yorks, HG5 9YX. Tel: (01423) 866985

History
Society Historian: George Jessop, G6JP, QTHR.

IARU Committee
(RSGB liaison with sister national societies overseas) Chairman: Tim Hughes, G3GVV, QTHR.

IEE Liaison Officer
Peter Saul, G8EUX, QTHR.

Insurance
(*Policies tailored to amateur radio - discounted to members.)* Amateur Radio Insurnace Services Ltd: Freepost, 10 Philpot Lane, London, EC3B 3PA. Tel: (0171) 338 0111. Fax: (0171) 338 0112.

IOTA
See *IOTA Directory & Yearbook* for explanation of scheme, contact points, honbour roll and island list. Further details IOTA Co-ordinator, RSGB HQ, e-mail: iota.hq@rsgb.org.uk. Committee chairman: Martin Atherton, G3ZAY, IOTA Manager: Roger Balister, G3KMA.

Licensing
Policy: Ian Suart, GM4AUP, QTHR.

Management Committee
Chairman: Hilary Claytonsmith, G4JKS, QTHR.

Membership
Application forms, new members, renewals, changes of address, club details: Sales Dept, RSGB HQ. Tel: (01707) 66088 *(24 hour)*. E-mail: sales@rsgb.org.uk.

Membership Liaison Committee
Chairman: Paul Essery, GW3KFE, QTHR.

Microwave Newsletter
Editorial content: Peter Day, G3PHO and Barry Chambers, G8AGNSubscriptions: Sales Dept, RSGB HQ. Tel: (01707) 66088*(24 hour)*. E-mail: sales@rsgb.org.uk.

Morse Practice
Morse Practice Transmissions (GB2CW): George Allan, GM4HYF, QTHR, Home: (01416) 344597.

Morse Tests [3]
Application forms and dates from AR Dept, RSGB HQ. Policy, new examiner appoitments: Roy Clayton, G4SSH, QTHR.

Novice Licence [3]
Local Instructors: *RSGB Yearbook* or contact AR Dept, RSGB HQ. Instructor appointments: *(Project YEAR Co-ordinator)* Phil Mayer, G0KKL, QTHR.

Packet Radio [3]
A Guide to Data Communications Licensing in the UK from AR Dept, RSGB HQ. Datacomms Committee: Chairman, Paul Overton. Mailboxes: Martin Green, G1DVU, site clearance applications, Steve Morton, G8SFR. (All QTHR)

Publications Management Group
Chairman: Peter Kirby, G0TWW, RSGB HQ.

Propagation
Week's forecast: GB2RS, explanation of terms used: *RSGB Yearbook*. Monthly predictions: *RadCom*. Propagation Studies Committee chairman: Prof Martin Harrison, G3USF, QTHR.

QSL Bureau
Outgoing cards *(RSGB Members only)* PO Box 1773, Potters Bar, Herts EN6 3EP. Incoming cards: SAEs to your sub-manager - see *RSGB Yearbook / RadCom*. QSL Bureau Liaison Officer: John Hall, G3KVA, QTHR.

Raynet Talk-Through Permits
(For RSGB affiliated and independant groups.) Applications: AR Dept, RSGB HQ.

RadCom
Delivery: Sales Dept, RSGB HQ. E-mail: subscriptions@rsgb.org.uk. Articles, editorial policy, members ads: RadCom Dept, RSGB HQ, e-mail: RadCom@rsgb.org.uk. Display and classified advertising: Malcolm Taylor Associates, Ashley Business Centre, Briggs Ho, Ashley Cross, Poole Dorset, BH14 0JR. Tel: (01202) 735999. Fax: (01202) 735585. Advisory Panel Chairman: Peter Kirby, G0TWW

Repeaters [3]
List of current repeaters *RSGB Yearbook*, or from Sales Dept, RSGB HQ. Leaflet: *Guide to Repeater Licensing* free from AR Dept, RSGB HQ. Policy and all other matters: Repeater Management Committee Chairman: Chris Goadby, G8HVV. (QTHR)

RSGB Liaison Officers (RLO)
Local help, knowledge and advice: See *RSGB Yearbook* or *RadCom*.

Shack , Museum, Library
(Times????) Librarian/ Curator: John Crabb, G3WFM, RSGB HQ.

Special Contest Callsigns NoV [3]
(For club callsigns in certain contests.) NoV applications forms: AR Dept, RSGB HQ. Policy: RSGB HF Contest Committee Chairman: Chris Burbanks, G3SJJ. QTHR

Special Event Station NoV [3]
Applications: AR Dept, RSGB HQ.

Spectrum Abuse
Caused by amateurs: Amateur Radio Observation Service: David Peters, AROS Co-ordinator, c/o PO Box ???, Potters Bar, Herts. E-mail: AROS@rsgb.org.uk. Caused by non-amateurs: Intruder Watch Co-ordinator: Chris Cummings, G4BOH, QTHR.

Technical & Publications Advisory Committee
(Reviews technical articles prior to publication) Chairman: Dick Biddulph, G8DPS, QTHR.

Training & Education
Chairman: details not available at time if going to press.

73 kHz NoV [3]
NoV application forms: AR Dept, RSGB HQ.

[3] *These services are conducted on behalf of the RA and are provided to both members and non-members alike.*

RSGB Liaison Officers (RLOs)

AVON (D)
Dave Collins, G4ZYF, 63 Church Road, Hanham, Bristol, BS15 3AF. Tel: 0117 967 6381.

BEDFORDSHIRE (B)
Geoff Linssen, G0PIZ, 401 Dallow Road, Luton, Bedfordshire, LU1 1UL. Tel: 01582 619981.

BERKSHIRE (D)
Dave Chislett, G4XDU, Hilltops, 2A St Mark's Road, Maidenhead, Berks, SL6 6DA. Tel: 01628 25720.

BORDERS (G)
Ian Wilson, GM4UPX, 30 Howdenburn Court, Jedburgh, Roxburghshire, TD8 6NP. Tel: 01835 862656.

BUCKINGHAMSHIRE (D)
Ron Ray, G3NCL, 54 Gladstone Road, Chesham, Bucks, HP5 3AD. Tel: 01494 776420.

CAMBRIDGESHIRE (B)
Mr W Felton, G3XZF, 31 Mountbatten Drive, Leverington, Wisbech, Cambs, PE13 5AF. Tel: 01945 588102.

CENTRAL (G)
Mr Stockton, GM4ZNX, 13,Dunvegan Court, Crossford, Dunfermline, Fife, KY12 8YL. Tel: 01383 731279.

CHESHIRE (A)
Post Vacant, refer to Zonal Council Member.

CLEVELAND (A)
Chris Flanagan, G7NRO, 21 Pentland Avenue, Billingham, Cleveland, TS23 2PG. Tel: 01642 553345.

CLWYD (E)
Post Vacant,refer to Zonal Council Member..

CO ANTRIM (F)
Albert Henry, GI4CRL, 23 Long Common, Dunluce Park, Ballymena, Co Antrim, Northern Ireland, BT42 2NU. Tel: 01266 41068.

CO ARMAGH (F)
Raymond Ashe, GI8RLE, 49 Deans Walk, Sleepy Valley, Richhill, Co. Armagh, N Ireland, BT61 9LD. Tel: 01762 870423.

CO DOWN (F)
Gordon Curry, GI6ATZ, 91 Burren Road, Ballynahinch, Co Down, Northern Ireland, BT24 8LF. Tel: 01238 533092.

CO DURHAM (A)
John Deamer, G4SJY, 28 Brackendale Road, Belmont, Durham, DH1 2AB. Tel: 0191 384 9281.

CO LONDONDERRY (F)
John Crichton, GI4YWT, 10 Bann Drive, Lisagelvin, Londonderry, BT47 2HW. Tel: 01504 43970.

CO TYRONE (F)
John Crichton, GI4YWT, 10 Bann Drive, Lisagelvin, Londonderry, BT47 2HW. Tel: 01504 43970.

CORNWALL (D)
Mr E J White, G0RJH, Hillbrook, 1 Brook Close, Helston, Cornwall, TR13 8NY. Tel: 01326 564690.

CUMBRIA (A)
Mike Gibbings, G3FDW, 5 Meadowbank Lane, Grange Over Sands, Cumbria, LA11 6AT. Tel: 01539 532433.

DERBYSHIRE (B)
Ken Frankcom, G3OCA, 1 Chesterton Road, Spondon, Derby, DE21 7EN. Tel: 01332 720976.

DEVON (D)
Post Vacant, refer to Zonal Council Member..

DORSET (D)
Phil Mayer, G0KKL, 16 Haig Avenue, Canford Cliffs, Poole, Dorset, BH13 7AJ. Tel: 01202 700903.

DUMFRIES & GALLOWAY (G)
Post Vacant,refer to Zonal Council Member..

DYFED (E)
Martin Goodall, GW8ZMU, 91 Uzmaston Road, Haverfordwest, Dyfed, SA61 1UA. Tel: 01437 764009.

EAST SUSSEX (C)
Jim Harris, G4DRV, 11 Boscawen Close, Eastbourne, East Sussex, BN23 6HF. Tel: 01323 728479.

ESSEX (C)
Malcolm Salmon, G3XVV, 54 Church Road, Rivenhall, Witham, Essex, CM8 3PH. Tel: 01376 514377.

FIFE (G)
Post Vacant,refer to Zonal Council Member..

GLOUCESTERSHIRE (D)
Derek Thom, G3NKS, 9 Southern Road, Cheltenham, Glos, GL53 9AW. Tel: 01242-241099.

GRAMPIAN (G)
John Rooney, GM1TDU, 2 Slains Road, Bridge of Don, Aberdeen, AB22 8TT.

GTR LONDON NORTH (C)
Roy Charlesworth, G4UNL, 6 Curzon Ave, Enfield, Middlesex, EN3 4UD. Tel: 0181 245 8119.

GTR LONDON SOUTH (C)
Phillip Stanley, G3BSN, 1 Thames View, Cliffe Woods, Rochester, Kent, ME3 8LR. Tel: 0181 318 7437.

GTR MANCHESTER (A)
Lynda Jopson, G6QA, 68 Greenmount Park, Kearsley, Bolton, BL4 8NT. Tel: 01204 796 564.

GUERNSEY (D)
Post Vacant, refer to Zonal Council Member..

GWENT (E)
Post Vacant, refer to Zonal Council Member..

GWYNEDD NORTH (E)
Dewi Roberts, GW0ABL, 23 Lon Hedydd, Siglan Farm Est, Llanfaipwll, Anglesey, Gwynedd, LL61 5JY. Tel: 01248 713647.

GWYNEDD SOUTH (E)
Mr PEW Allely, GW3KJW, 'Dwyfor', Rhiw, nr Pwllheli, Gwynedd, LL53 8AE.

HAMPSHIRE (D)
Post Vacant,refer to Zonal Council Member..

HEREFORDSHIRE (B)
Mr T R Bridgland-Taylor, G0JWJ, 1 Overbury Court, Venns Lane, Hereford, HR1 1DG. Tel: 01432 279435.

HERTFORDSHIRE (C)
John Rudd, G7OCI, 23 Grange Gardens, Ware, Hertfordshire, SG12 9NE. Tel: 01920 466639.

HIGHLAND (G)
Post Vacant, refer to Zonal Council Member..

NORTH HUMBERSIDE (A)
Clive Reynolds, G8EQZ, 49 Westborough Way, Anlaby Common, Hull, N.Humberside, HU4 7SW. Tel: 01482 563691.

S H'SIDE & LINCS (B)
Ray Degg, G0JOD, 28 The Spinneys, Welton, Lincoln, LN2 3TU. Tel: 01673 863103.

ISLE OF MAN (A)
Mr A C Kissack, GD0TEP, 30 High View Road, Douglas, Isle of Man, IM2 5BH. Tel: 01624-626080.

ISLE OF WIGHT (D)
Doug Byrne, G3KPO, Lynwood, 52 West Hill Road, Ryde, Isle of Wight, PO33 1LN. Tel: 01983 67665.

JERSEY (D)
Syd Smith-Gauvin, GJ0JSY, 31 Jardin-A-Pommiers, Patier Road, St Saviour, Jersey, C.I.., JE2 7LT. Tel: 01534 38996.

KENT (C)
Ray Petri, G0OAT, Tarnwood, Denesway, Meopham, Kent, DA13 0EA. Tel: 01474 812682.

LANCASHIRE (A)
Laurie Bradshaw, G0MRL, 342 Manchester Road, Blackrod, Bolton, Lancs, BL6 5BG. Tel: 01204 697023.

LEICESTERSHIRE (B)
Post Vacant, refer to Zonal Council Member..

LINCOLNSHIRE (B)
Ray Degg, G0JOD, 28 The Spinneys, Welton, Lincoln, LN2 3TU. Tel: 01673 863103.

LOTHIAN (G)
Kenneth Traill, GM0TQK, 2 Third Street, Newtongrange, Midlothian, EH22 4PU. Tel: 0131 663 8240.

MERSEYSIDE (A)
David Johnson, G1GNS, 31 Coniston Avenue, Penketh, Warrington, Cheshire, WA5 2QY. Tel: 01925 726821.

MID GLAMORGAN (E)
Mr G B Brown, GW0PUP, 17 High Street, Senghenydd, nr Caerphilly, Mid Glamorgan, CF8 2GG. Tel: 01222 832253.

NORFOLK (C)
Bill Higgins, G3PNR, 91 Hayden Court, Eleanor Road, Norwich, NR1 2RG. Tel: 01603 629150.

NORTH YORKSHIRE (A)
Post Vacant, refer to Zonal Council Member..

NORTHAMPTONSHIRE (B)
Post Vacant, refer to Zonal Council Member..

N IRE'D- BELFAST (F)
Gordon Curry, GI6ATZ, 91 Burren Road, Ballynahinch, Co Down, Northern Ireland, BT24 8LF. Tel: 01238 533092.

N IRE'D- SOUTH (F)
Raymond Ashe, GI8RLE, 49 Deans Walk, Sleepy Valley, Richhill, Co. Armagh, N Ireland, BT61 9LD. Tel: 01762 870423.

NORTHUMBERLAND (A)
Post Vacant, refer to Zonal Council Member..

NOTTINGHAMSHIRE (B)
John Coates, G4GYU, 30 Abbott Road, Mansfield, Notts, NG19 6DD. Tel: 01623 27257.

ORKNEY (G)
George Christie, GM7GMC, Burnbank, Hillside Road, Stromness, Orkney, KW16 3HR. Tel: 01856 850270.

OXFORDSHIRE (D)
Post Vacant, , refer to Zonal Council Member..

POWYS (E)
Gordon Rogers, GW0RJV, Maes Gwersyl, Garthmyl, Montgomery, Powys, SY15 6RS. Tel: 01686 630327.

SHETLAND (G)
Robert Miles, GM4CAQ, 58 Fogralea, Lerwick, Shetland, ZE1 0SE. Tel: 01595 696411.

SHROPSHIRE (B)
Tony Colton, G0UYE, 9 Pineway, Bridgnorth, Shropshire, WV15 5DS. Tel: 01746 761203.

SOMERSET (D)
Dick Atterbury, G4NQI, 14 Holway Road, Taunton, Somerset, TA1 2EY. Tel: 01823 333009.

SOUTH GLAMORGAN (E)
Mr G O Jones, GW0ANA, Nirvana, 2 Castle Precinct, Llandough, Cowbridge, South Glamorgan, CF7 7LX.

SOUTH YORKSHIRE (A)
Post Vacant, refer to Zonal Council Member..

STAFFORDSHIRE (B)
Ken Parkes, G3EHM, 41 Golborn Avenue, Meir Heath, Stoke on Trent, Staffs, ST3 7JQ. Tel: 01782 397240.

STRATHCLYDE - NW (G)
Post Vacant, refer to Zonal Council Member..

STRATHCLYDE - SE (G)
Gordon Hunter, GM3ULP, 12 Airbles Drive, Motherwell, Strathclyde, ML1 3AS. Tel: 01698 253394.

SUFFOLK (C)
Dave Ferguson, G6FS, 3 Aldeburgh Road, Leiston, Suffolk, IP16 4JY. Tel: 01728 832924.

SURREY (C)
Robin Sykes, G3NFV, 16 The Ridgeway, Fetcham, Leatherhead, Surrey, KT22 9AZ. Tel: 01372 372587.

TAYSIDE (G)
Alfred Low, GM4UZP, 21 Earn Crescent, Menzieshill, Dundee, DD2 4BS. Tel: 01382 644597.

TYNE & WEAR (A)
Keith Ritson, G0PKR, 14 Dunsdale Road, Holywell, Whitley Bay, Tyne and Wear, NE25 0NG. Tel: 0191 237 1963.

WARWICKSHIRE (B)
Post Vacant, refer to Zonal Council Member..

WEST GLAMORGAN (E)
Post Vacant, refer to Zonal Council Member..

WEST MIDLANDS (B)
Tony Faulkner, G0SKG, 105 Corbyn Road, Russells Hall Est, Dudley, West Midlands, DY1 2JZ. Tel: 01384 820616.

WEST SUSSEX (C)
Peter Howard, G0AFN, 12 Meadow Way, Westergate, Chichester, West Sussex, PO20 6QT. Tel: 01243 543399.

WEST YORKSHIRE (A)
Derek Allan, G0RZP, 283 Cliffe Lane, Gomersal, Cleckheaton, West Yorkshire, BD19 4SB. Tel: 01274 872244.

WESTERN ISLES (G)
Post Vacant, refer to Zonal Council Member..

WILTSHIRE (D)
Ivan Rosevear, G3GKC, 20 Christchurch Road, Bradford-on-Avon, Wilts, BA15 1TB. Tel: 01225 863622.

WORCESTERSHIRE (B)
Mr K J Parsons, G4VZA, Appartment 7, Crellin House, Priory Road, Malvern, Worcs, WR14 3DN. Tel: 01684 891596.

Society Report

The work of the Society during its financial year 1 July 1996 to 30 June 1997.

Each year, the Society details the work of its volunteers and staff in an Annual Report. This has in the past been published exclusively to members in *RadCom*, but as part of the new *Yearbook*, we are pleased to make this information available to all.

Licensing and Co-ordination

The principal work of the Society is maintaining and improving the regulations under which radio amateurs operate. This is carried out both nationally, mainly by negotiations with the Radiocommunications Agency (RA), and internationally through membership of the International Amateur Radio Union (IARU). Monitoring of our exclusive frequency allocations and the reporting of intruders is another important function carried out by the Society.

A major task of the **Licensing Advisory Committee** was the creation of new procedures for liaison with the Radiocommunications Agency (RA) and setting the agenda for forthcoming negotiations.

The committee distributed a 'Future of Amateur Radio Opinion Survey' with the July 1996 *RadCom* and reported the results in December. The exercise was an important stage in being able to respond to proposals to change the ITU regulations defining the Amateur Services at the 1999 World Radio Conference. The full 44-page analysis was made available to members at cost.

A letter from the RA was published clarifying which messages were permissible on the packet radio network. There were discussions as to how to monitor packet radio traffic.

In July '96, the RA announced the launch of the Harmonized Amateur Radio Examination Certificate, which made it much easier to obtain a full overseas licence in countries signed up to the agreement. This was the result of much national and international work by the RSGB and other IARU member societies.

Following input by the RSGB to discussion documents, the RA's White Paper *Spectrum Management: into the 21st Century*, which proposed pricing spectrum use, stated that it was unlikely that this would lead to higher charges for amateur radio. The progress and implications of this important piece of legislation are constantly being monitored by the Society.

The work of the IARU and therefore RSGB's **IARU Committee**, is ongoing and continuous. Indeed, the former's stated aims of "the protection, promotion and advancement of the Amateur and Amateur Satellite Services" are equally applicable to the latter. By reason of the Committee's composition, many important facets of amateur radio are represented.

Tangible and public examples of the work which the Committee carries out, are its presence at Regional Conferences of the IARU. In September 1996, at the Region 1 Conference in Tel Aviv, the Society was represented by G3GVV (Delegation Leader and Chairman of the IARU Committee), G3ZNU (Vice Chairman of the IARU Committee), G3HTA (HF Manager), G3PFR (Microwave Manager), GM4ANB (VHF Manager), GM4AUP (Chairman LAC), and G6LX. G3VZV, representing the British Amateur Television club (BATC), attended at his own expense. Once again, the Radiocommunications Agency indicated its interest in the amateur service by sending an observer, Mr Stuart Cooke. Several papers submitted by the RSGB were considered and discussed, one of which made proposals for the rationalisation of ITU Zones. In order to obtain universal agreement the latter will be brought to the notice of Regions 2 and 3, an example of world-wide co-operation on a matter which affects so many amateurs. Several RSGB representatives were re-elected for the following three years: G3FKM will continue as IARU Region 1 Secretary, G3USF as International Beacon Co-ordinator, and G3HCT as Chairman of the Common Licence Group (CLG). The last named is concerned with the ultimate establishment of an amateur radio licence recognised in all countries. G3HCT continues to serve as a member of an international working party, the Future of Amateur Services Committee (FASC), which looks and plans ahead for several years. G6LX was presented with the Region 1 medal, together with gifts from other Societies, for his long service as Chairman of the IARU HF Contests Sub-group. A comprehensive report of the Conference, and the RSGB's contribution, was published in the December '96 and January '97 editions of *RadCom*. In September 1997, Region 3 (Asia and Australasia) holds its Conference in Beijing, organized by the Chinese Radio Sports Association; RSGB has prepared discussion documents for its representatives to present.

The IARU Committee's work is not confined to Conferences. From the European Union come directives which could have a radical effect on amateur radio. One of these would demand that our equipment was 'type approved': this would have the immediate effect of stifling modification, experimentation, and development. Needless to say, this is being actively opposed by experts in this field. The probability of similar restrictions emerging from the EU is very real. We are mobilising reasoned opposition to these dangers.

The work of both the IARU Committee and IARU throughout the world is publicized by the bi-monthly articles in *RadCom*.

Once again there has been a decrease in the number of intruders into the exclusive amateur bands, but this is likely to change as propagation on the higher HF bands begins to pick up once again. The **RSGB Intruder Watch** monitoring team continues to compile a steady stream of reports, albeit at a considerably reduced level compared with several years ago.

The worst problems suffered by the Amateur Service tend to be caused by broadcast stations, either from defective transmitters or illegal use of amateur frequencies; both have been experienced during the last year. The most blatant example of a defective transmitter was Radio Exterior de Espana on 7105kHz which was causing splatter down to 7085kHz, rendering the top part of the amateur band unusable in countries as far away as New Zealand. From the first reports of this situation in late December 1996 to the eventual clearing of the problem early the following March, the RA Monitoring Station at Baldock and their equivalents in other countries sent numerous complaints and reports in an effort to get the Spanish authorities to attend to their problem.

Another defective transmitter, this time Radio Tirana (Albania) on 7150kHz, was putting out an unstable harmonic on 14300kHz for many months. This has apparently been sorted out; Baldock reported the situation to the Albanian authorities on a regular basis to ensure that the problem eventually received effective action. A strange problem occurred briefly on 7050kHz when the internal service of Radio Jordan was broadcast on a daily basis. Immediate action from this country and a number of others led to a fairly abrupt QSY.

A brief reappearance of a CW idler (FDY) from the French Air Force on 7028kHz was immediately terminated when the IW Co-ordinator phoned Baldock with a request that they inform their French counterparts of the problem. The occasional nudge from Baldock led to a number of 12-tone Piccolo transmissions being removed from the 18MHz band during the period of this review.

As in previous years, the Co-ordinator continued to receive numerous enquiries and complaints concerning the presence of signals which appear on our bands. All correspondence is welcomed and acknowledged.

Membership Representation

The Society is governed by a Council comprising mainly those elected by the membership. Also elected are RSGB Liaison Officers (RLOs) who assist members in an area, usually a county. Representation and member services are monitored by the **Membership Liaison Committee** (MLC).

In October 1996, the triennial RLO elections were held for contested posts in Cornwall, Essex and South Gwynedd. Other RLO posts were filled by unopposed nominees. The prime business of the committee was the ongoing search for an Emergency Radio Liaison Officer (ERLO); an open meeting was held in London to which emergency communication groups were invited. Other matters discussed have included RLO Election procedures, updating the *RLO Handbook*, the *'Green Book'*, posters for use at public events, communication with grassroots membership, GB2RS, certificates of merit, plus any points raised with Zonal Council members either by individual members or by RLOs. Meetings were reduced to the minimum, with much business being conducted by telephone. A local open meeting was held in Belfast.

Network Management

Co-ordination and licensing work is carried out for the very large formal packet radio and repeater networks, comprising many hundreds of individual stations. These are detailed elsewhere in this book.

The repeater network was overseen by the **Repeater Management Committee** (re-named during the year from the Repeater Management Group). The management of a complex network of some 300 units spread throughout the UK and consisting of all the UK's voice and TV amateur repeaters which use three modes and operate on five very different bands is a very considerable administrative task. In conjunction with HQ staff a major effort was made to improve the service to the members and the benefits are now beginning to become apparent. At the same time cost-effectiveness was improved.

To try to improve communications with members, a telephone 'hot-line' system was set up for enquiries but low usage led to its termination. Currently a programme of ensuring that repeater keepers can easily contact their RMC Zonal Manager is in place. Ordinary members are being encouraged to first contact either their RMC Zonal Manager or HQ, who will then ask the relevant member of the RMC to return the call. This is having greater success. Another means of communication was the occasional newsletter *Repeater Report* sent out to all repeater groups

Pro-active network management meant that occasionally repeater groups whose repeater's performance was considerably different from the agreed conditions found that they were very soon being asked to improve matters with some urgency. Similarly groups that had, in some way, caused their site's owners to become unhappy were speedily approached to mend their ways, particularly where the problems on one site might have led the RSGB to lose facilities on other sites. In consequence, despite numerous snap inspections by the RA / RIS throughout the year, only one unit had any sort of Conformity Notice served upon it.

Dialogue with the RA was considerably improved during the year and regular meetings to discuss problems, iron out difficulties and explore future possibilities are now a regular feature of the committee officers' work.

A specification for proposed repeaters on the 50MHz band was finalised, and application to the RA was made for 11 units.

To improve its response to queries, the **Data Communications Committee** set up new rules for dealing with correspondence and consultation with members.

Following discussion between the Society and the RA, and an RA survey of SysOps, the definition of a Mailbox was clarified in a *Gazette* Notice which was reproduced in the May *RadCom*.

Replanning of the data sub-bands became necessary following decisions taken at the IARU Region 1 Conference.

During the year the committee processed 60 applications for formal nodes. 78 were approved by the RA; this included some held over from the previous year.

Spectrum Management

Three committees, and three Managers (HF, VHF and Microwave) deal with matters related to our use of the radio spectrum, most notably bandplanning and international liaison.

The **HF Committee** was instrumental in making the 1996 RSGB HF and IOTA Convention a major attraction, despite an unavoidable clash with both the IARU Region 1 Conference and the annual FOC Dinner. Almost £1300 was raised for the RSGB DXpedition Fund as a result of the Convention raffle. The major prize, again donated by Yaesu UK, was won by the late Etienne Heritier, HB9DX, who immediately donated the prize to the IOTA programme. Both Martin Lynch & Son and Yaesu (UK) have again agreed to continue their sponsorship at the same level as last year.

Donations of £200 each were made from the DXpedition Fund for the expeditions to Heard Island, St Peter and St Paul Rocks, Annobon Island and the North Cook Islands. This figure is matched by donations from CDXC (Chiltern DX Club) - The UK DX Foundation. It should be noted that the DXpedition Fund is sustained through the raffle at the HF and IOTA Convention and does not involve RSGB central funds.

Interest in 73kHz continued with the number of NOVs issued at the time of writing being 270. The distance record at the end of the review period was 99.9 kilometres between G3YGF and G4JNT, known primarily for their work with microwaves. A 73kHz Award was introduced to encourage more on the air activity and Datong Electronics kindly agreed to sponsor an LF Experimenters Award over the next five years. In addition to 73kHz the Committee pursued the possibility of a 136kHz pan-European allocation through the RA.

Following a meeting between representatives of the Society, the RA, the BBC and MoD it was agreed to pursue the expansion of the 7MHz band to a width of 300kHz and to pass on the feelings of the meeting to the CEPT Conference Preparatory Group who are formulating a common proposal for WRC99.

To keep costs down, it was agreed to reduce the number of meetings held by the committee and to carry out much of the work via E-mail where possible.

The HF Committee also publicized their work through the Society's Web Page with links to pages about the HF Convention page and 73kHz. All of these have proven to be of great interest.

The use of data modes on HF was clarified in the RSGB Bandplan, bringing the Society in line with agreed IARU Region 1 recommendations.

The subject of Progressive /Incentive licensing and the licensing structure in general remains high on the priority list. A paper on this subject is to be prepared.

The HF Committee recommended the award of the following: ROTAB Trophy - Martin Atherton, G3ZAY; G5RP Trophy - Emma Wills, 2E0AAX.

The principle activity of the **HF Manager** was preparation for, and attendance at, the IARU Region 1 Conference in Israel last October. There were 17 recommendations which directly related to HF matters. Most have been implemented but work is still progressing on some matters, including the RSGB paper on normalisation of ITU Zones. To obtain world-wide agreement, a paper will be submitted to conferences in each of the IARU Regions. It is hoped that by 1998 an IARU-approved document on ITU Zones will be available.

Discussion on proposals for seeking expansion of the 7MHz band was one of the most important HF issues. A strategy Committee was established by the IARU under the Chairmanship of Fred Johnson, ZL1AMJ, with each Region being represented on it. The objective is to obtain an allocation of at least 300kHz in the vicinity of 7MHz. Organization and planned strategy look good for discussion at the 1998 World Radio Conference. Administrations are already considering how an additional 200kHz many be accommodated and in January, the HF Manager attended a Radiocommunications Agency exploratory meeting along with representatives from HF broadcasting and the Ministry of Defence.

Failure to observe the IARU band plan recommendations continued to cause problems for many. Almost all the problems resulting in complaints arose from unattended digital mode systems such as BBSs using packet, AmTOR, Clover and PacTOR. Some users of these modes are not keeping to the band segments assigned to digital modes and may be found anywhere, especially in the CW allocations. Most of the offenders are not members of their national societies and are therefore not always aware of the band plans which are published in journals of IARU member societies.

The **VHF Committee's** year was dominated by the IARU Region 1 Conference - preparations, attendance, and dealing with the aftermath. It was a successful conference, achieving three main goals. The first was a re-organization of 144 to 145MHz, with a properly defined digital section and the beacons moved to the narrow-band segment. At last the increasingly popular digital modes have an agreed home in our most popular band. A knock-on effect of these changes was on the Emergency Communications Priority frequencies. Many people have been asked to change frequency and this has required some careful shuffling and planning. The moves are now taking place according to plan. The VHF Committee would like to thank all those affected - beacon keepers, BBS and Node SysOps, and Raynet® groups - for their patience and co-operation during the changeover.

Committee member John Wilson, G3UUT, co-ordinated the beacon moves not only in the UK but throughout the whole of Region 1.

The second Conference goal was a footnote allowing UK amateurs to put 50MHz repeaters on the band without following the IARU standard plan, which could not be licensed here. This enabled proposals for 11 units to be submitted to the RA.

The third Conference goal was to initiate a change-over to 12.5kHz channel spacing on 145MHz phone. This is to be a slow, gradual changeover, extending at least to the year 2000 and perhaps beyond. It will let us catch up with current good technical practice and remove what has been something of an embarrassment: being the only VHF radio service to use what most others consider to be obsolete technology.

On 432MHz the band plan was opened up to allow digital links in the top and bottom 2MHz segments. There is now room for high-speed trunk links, if any group would like to take up a challenge.

Moving a lot of the committee's work to E-mail produced savings in time and expense, and improved responsiveness.

The **Microwave Committee's** main task over the year was the implementation and final ratification of the new 10GHz bandplan. The region 10,150 to 10,300MHz became unavailable to amateurs on 1 April 1997 and existing users of this part of the spectrum – primarily TV repeaters – are being relocated.

A provisional bandplan for 76GHz was produced so that some compatibility can be made with the alternative frequencies for this band allocated in some European countries.

The committee reviewed the rules of the microwave contests, in particular the 10 and 24GHz cumulatives, in an attempt to foster more portable activity. A bonus multiplier for locator squares was introduced and was liked by the participants, so it is to be continued.

There was considerable discussion on ways to attract newcomers to the bands. One of the Microwave Roundtables suggested the re-introduction of a wideband section to the 10GHz cumulatives to encourage newcomers with simple equipment. It was decided to appoint new committee members specifically for investigating and promoting alternative means of using the microwave bands, such as packet linking, spread spectrum, wideband modes etc. All the current members are interested primarily only in narrowband DX aspects and cannot do full justice to the large bandwidth potential of the bands.

Lectures on microwave matters were arranged for Sandown and two Microwave Roundtables.

Advice and Assistance

A major benefit of RSGB membership is specialist advice from experts in a number of subjects. In particular this covers electromagnetic compatibility (EMC) and planning permission for antennas. Additionally, all of the specialist committees whose work is detailed on these pages can provide advice to members, as can a number of HQ staff members.

A network of EMC Co-ordinators fielded the majority of routine EMC enquiries, leaving the **EMC Committee** with more time to deal with the long term problems of the general EMC environment, and how amateurs are going to survive in it. The Co-ordinators also had an important role in 'spreading the word' about good EMC practice, and how to avoid interference problems by sensible operating practices.

Five new advice leaflets were produced during the year. These were detailed in the April 97 *RadCom* and are available to members on request from HQ.

The EMC Committee was a key member of the working group which produced the Radiocommunications Agency booklet RA-323, *Guidelines for Improving Television and Radio Reception*. The group was chaired by the RA, and included representatives of BBC, ITC, BREEMA and CAI (Confederation of Aerial Industries) as well as the RSGB (EMC Committee). The booklet was issued in May 1997, and is intended for the radio and TV trade. Amateur radio aspects are covered in a fair and reasonable way.

Newspaper articles incorrectly implying that radio amateurs were jamming radio operated car security systems led to the committee having informal discussions with the RAC and the AA. Out of this arose the RAKE (Radio Activated Key Entry) Committee which expanded to formal meetings involving device manufacturers, MIRA (Motor Industries Research Association), insurance interests, and the RA. RAKE is in the process of publishing its 'Guidelines' giving advice on the design and operation of radio key systems for vehicles.

The CTE Directive was issued by the European Union (EU), in 'preliminary draft' form in May 1997. It is intended to simplify procedures for approving commercial telecommunications equipment. There is, however, a risk that it could be interpreted as including amateur equipment. The EMC Committee, is working in conjunction with the LAC and the IARU Committee, to press for amateur radio equipment to be exempt from the Directive.

The Committee continued to keep an eye on various developments which could have a very significant impact on amateur operation particularly in the HF bands. Of these, the most important are the proposals for high speed data transmission over unscreened lines.

At its November 1996 meeting, the **Planning Advisory Committee** defined seven objectives for the year, all of which have since been met. The booklet *Planning Permission - Advice for Members* continued to be popular and about 10 to 15 were sent out to members by HQ each week. It was also sent to each Local Planning Authority in the country. Individual and more detailed help was given by the Planning Panel to no less than 60 members throughout the year. In all such cases, members have been encouraged to do the actual work themselves, as the Panel is no longer able to do advocacy work. Committee members were available to give out advice at the RSGB VHF Convention and at the London Amateur Radio and Computer Show. An article looking at the different points to consider when a house move is contemplated was submitted to *RadCom* for publication, as was a position statement on an automatic right for planning permission for the holder of an amateur radio licence. No further clarification was made on the need for a specific planning permission for a mobile mast.

Close links were maintained with all of the Society's committees but with the EMC Committee in particular, as some of their problems are closely related to planning ones. Inevitably, a small number of members did not achieve success in their quest for planning permission, and this is to be regretted. However, whilst the Committee and Panel exist to help all members, the final decision on every application lies with either the Local Council or with the Planning Inspectorate.

Education and Training

The Society is responsible to the Radiocommunications Agency for running the Novice Licence Training Scheme which is a pre-requisite for taking the City & Guilds examination. The RSGB also sits on advisory committees for the Radio Amateurs Examination (RAE) and the Novice RAE.

The **Training and Education Committee** dealt with the following:

Air Training Corps: Liaison with the organizers of the ATC (Radio Section) continued. It is expected that the incorporation of the Novice licence into the ATC training scheme could be in place by the latter half of 1997. This should result in a large increase in the number of Novices, producing some extra income from books etc with no cost to the Society and with the possibility of new members.

Amateur radio in the school curriculum: This is being progressed slowly and is an on-going item of business.

Books: *Practical Receivers for Beginners* was published in autumn 1996. Two simple guides to the Licence conditions, for the full and Novice licences, were written and are available on application [A5 SASE to HQ please - *Ed*].

Support Packages for RAE Instructors: Considerable work was carried out in preparing a set of ten schemes of work for RAE instructors. Unfortunately these already need modifying in the light of the RAE syllabus changes.

STELAR (Science and Technology through Educational Links with Amateur Radio): Close liaison continued with this group, with its chairman and other members of the group being active members of the T&EC.

Throughout the year considerable work has been carried out on an alternative and interactive method of RAE training. This is ongoing.

Maintaining and improving the efficiency of the Novice training scheme is the core task of the **Project YEAR Co-ordinator**. This was achieved by filling gaps in the instructor coverage across the country, and by weeding out inactive instructors and replacing them with enthusiastic 'new blood'. During the year some 130 new instructors have been appointed and about half that number deleted from the list, which now includes 1200 names. The link between them and the Co-ordinator is the network of county Senior Instructors. Ill health, change of location or pressure of other work have caused some resignations. Five new appointments have been made during the year and the search for suitable candidates to fill seven other vacancies is continuing.

The Air Training Corps (ATC) completed the task of incorporating the essence of the RSGB's *Novice Instructor's Manual* into their *Air Cadet Publication 35* and this became available to Squadrons towards the end of 1996. To meet the requirements of their command structure it was agreed that the Society should, in consultation with the Wing CO, appoint a Senior Instructor (SI) for each Wing, to whom the Instructors on the Squadrons would report. SIs were appointed for six of the 42 Wings and recommendations are awaited from the others.

In order to develop further the arrangement made with the ATC, the Project YEAR Co-ordinator attended a Cadet Signals Symposium at Royal Signals School, Blandford, in November and gave a talk about the Novice licence training scheme. This was followed up in correspondence to the officers who had expressed interest. RSGB General Manager Peter Kirby, G0TWW, who is the CO of a Sea Cadet unit, approached the Eastern Area CO's conference in November and secured approval to run a pilot Novice course for a group of his Sea Cadets based on the ATC manual. If this goes well other units may follow.

The new format for the City & Guilds RAE comprising a single paper of 80 questions was adopted in 1997 and will be in use for exams from May 1998 onwards.

Morse Code

The Society is licensed to broadcast Morse code practice transmissions under the callsign GB2CW (see schedule elsewhere in this book). It is also contracted to the licensing authority to carry out 5wpm and 12wpm Morse code testing in the UK.

The **Morse Practice Service** has approximately 45 volunteer operators who broadcast practice Morse on the HF bands for national listeners and the VHF and UHF bands for local users. In addition many of the operators offer help with sending practice and other Morse related subjects using their own callsigns following the end of their scheduled broadcasts.

During the year the volunteers transmitted some 4000 hours of practice Morse from locations ranging from Stornoway on the Isle of Lewis down to the far southwest of England, as well as from Northern Ireland.

We are most anxious to recruit operators to the service and anyone interested should contact the co-ordinator, George Allan, GM4HYF (QTHR).

The Society's network of Morse testers is headed by the **Chief Morse Examiner**, who chairs the **Morse Test Steering Committee**. During the year the Morse Test Service examined 640 candidates, with the pass rate again remaining pretty constant at 75%. Feedback from examiners continued to indicate that candidates who had received personal tuition were much better prepared for the QSO format test, whereas self-taught candidates had a much higher failure rate.

The number of members motivated and competent enough to teach Morse code is rapidly dwindling with the reduction in the number of ex-professional Morse operators left in the hobby. This often means that anyone who has recently passed the 12wpm Morse test and received an A class licence is immediately appointed Morse tutor at the local club and elevated to instant 'expert' status, with the blind leading the blind. The ability to teach Morse to QSO format standards, especially by anyone with little HF amateur radio experience, cannot be picked up overnight, and this is beginning to be reflected in the returns, which show instances of poor instruction. One common fault is that some candidates are being taught to write the break sign BT as two vertical lines.

There was also evidence that some candidates who had been taught using the 'Farnsworth' method, where individual characters are sent at a faster rate (such as 15wpm) with longer than normal gaps in order to reduce the overall speed down to 12wpm, were incapable of receiving correctly-proportioned Morse. These candidates often complained that the Morse sent during the test was too slow, with insufficient spacing between letters.

Morse tests on demand continued to be popular with the general public, who appear to appreciate this more informal arrangement. County Morse testing teams attended more than 40 venues during the 12 months. There are currently 298 examiners in 70 regional teams throughout the UK and four overseas centres operated by UK examiners on duty with the Forces or Government Service.

DTI inspectors have again continued with spot checks, arriving unannounced both at scheduled test sessions and at on-demand tests. Again, there were no complaints from these inspections.

Recognising Achievement

The Society offers many awards to mark personal operating achievements (see elsewhere in this book). Three awards managers presided over the bands below 30MHz, those above 30MHz, and for the RSGB Islands on the Air (IOTA) awards programme. A new committee was formed to deal with IOTA work.

During the year ending 30 June 1997, 84 award applications were processed by the **VHF / UHF Awards Manager**. This is a slight reduction compared with last year. Applications were received from amateurs and listeners representing nine different countries including Eire, France, Germany, Gibraltar and Holland. One Supreme Transmitting Award was issued to Roger Piper, G3MEH, in recognition of his Senior Awards on 144MHz, 432MHz and 1296MHz. After the initial interest in the first two years of the 50MHz Standard and Senior Transmitting awards, where applicants were competing to get certificates endorsed with low numbers, the interest in collecting counties on this band appears to have all but ceased. This may indicate a perception of 50MHz being a band for working DX rather than for intra-UK contacts. Since the inception of the RSGB VHF / UHF Countries and Counties Awards in 1961 the award scheme has successfully taken account of several changes in County structure. However, the complex local government changes, due to be completed in April 1998 and which involve the restructuring of Scottish and Welsh counties together with the creation of new unitary authorities and counties, is proving very difficult to assimilate, and the future of these awards is under review. The total number of 50MHz awards remained at a similar level to the previous year, as did the number of 144MHz square awards, though the numbers applying for awards on the higher frequencies has reduced. Award details to 30 June 1997 are as follows (last year's issues in parentheses):

Four Metres and Down Certificates:

50MHz Standard	0	(0)	144MHz Standard	4	(5)
50MHz Senior	0	(0)	144MHz Senior	1	(2)
70MHz Standard	0	(1)	432MHz Standard	2	(1)
70MHz Senior	1	(0)	432MHz Senior	2	(2)
1.3GHz Standard	0	(0)	Supreme Award	1	(1)
1.3GHz Senior	0	(0)	M/wave Distance Awards	2	(4)

Total: 13 (16)

Squares/Countries AwardsRSGB 50MHz Award:

70MHz	1	(1)	Two-way Countries	18	(24)
144MHz	16	(18)	Squares Award	25	(26)
432MHz	2	(11)	DX Award	5	(2)
Microwave	4	(9)	Total: 48 (52)		

Total: 23(39)

The year in review has again been a busy one for the **HF Awards Manager**, although the overall numbers of awards issued were once again down when compared with the previous year, reflecting the state of HF propagation conditions. We all look forward to improving conditions during the next 12 months. A total of 41 certificates, 18 endorsement stickers and 3 plaques were issued during the year. Claims came in from all continents including a large number from Taiwan. The most popular award is the IARU Region 1 Award; 27 were issued this year. This was closely followed by the SWL DX Listeners' Century Award with 3 certificates and 17 endorsement stickers. In addition 21 WAC award applications were verified and forwarded to IARU HQ, and 27 applications by RSGB members for other overseas were checked. The fourth Supreme Commonwealth Century Club Award was claimed this year by G3ZBA. This was for confirming contacts with contacts with all of the Commonwealth Call Areas. G3IFB received a 5 Band CCC Supreme Award. The top SWL for this year was from Belgium. ONL-7681 claimed a 300 country sticker for the DX Listeners' Century Award. The highest SWL claim from an RSGB member was again from Ken Creamer, BRS10167, who increased his country score to 280 during the year. Award checking was carried out during the HF Convention in October. 133 items of correspondence were dealt with in order to assist members with their award hunting problems. The rules for all RSGB Awards were reviewed during the year and some minor revisions were made. In addition a new award for 73kHz contacts was introduced. As yet no-one has claimed the first award in this series. Information about RSGB awards was sent to awards managers of other societies, as well as award publications and overseas magazines. This resulted in the IARU Region 1 Award rules being published in Japanese by the JARL. Members can assist the smooth operation of the awards programme by ensuring their applications conform to the current set of rules and are sent to the correct address.

The IOTA programme continued to be a catalyst for amateur radio activity from islands worldwide. There were countless operations ranging from holidays on well-trodden locations to DXpedition to remote and uninhabited islands. **The IOTA Manager** exercises responsibility for day to day management of the programme. This included safeguarding the integrity of the programme, maintaining contact with the international IOTA community, maintaining the 'island list' - this has involved determining island qualification in several hundred cases as well as issuing new reference numbers, and validating operations to new or rare island groups post-event. Where possible, the IOTA Manager maintained contact with expeditions from the planning stage through to their safe return home. This gives support and encouragement and ensures the validity of the island and the operation for IOTA credit. It also establishes a close relationship with the growing body of enthusiasts who love to activate islands and gives the programme a high international profile recognised by active amateurs. *The RSGB IOTA Directory and Yearbook*, listing the rules as well as all the island groups, was updated and enlarged. Almost 2000 copies have been sold since publication in September 1996. The IOTA *Yearbook* has also been translated into Japanese; the first RSGB publication in an overseas language. The **IOTA Committee** was formed in September 1996. Achievements during the year included: IOTA stream and activities at the RSGB HF and IOTA Convention; IOTA stand at the London Amateur Radio Show; IOTA presentations / Q&A at Fresno DX Convention, Dayton Hamvention, and Friedrichshafen (the latter two being the first ever); new record-keeping software developed for members / checkpoints / HQ and scheduled for implementation in November 1997; 1997 Honour Roll, Annual Listing, and commentary prepared for *RadCom*; IOTA World Wide Web pages prepared and maintained on regular basis; 13 new IOTA Reference Numbers issued; 26 IOTA expeditions validated; administration related to unlicensed operations by HH2HM/F, QSLing practices of NL7TB; opposition to Scottish Natural Heritage policy of forbidding amateur operations from Hebridean islands, and persuading owners of Treshnish Islands and St Tudwal's Islands to allow amateur visits; preliminary design work on new certificate range; addition of Brazilian checkpoint / regional team and preparation for admission of Hungarian and Japanese checkpoints.

The **Trophies Manager** ensured that the various cups, keys, rose bowls and other award hardware were in good order, with repairs effected where required and new container boxes made up. The records were updated and stored on computer. These records contain details of each trophy, with reason for, and to whom, it was awarded, from its inception. Additionally there is a photograph, accompanied by brief construction details, weight, material used etc. A master 'what, where, who' spreadsheet is maintained, which details the current whereabouts of every individual trophy.

In addition to awards, *RadCom* recorded a number of operating achievements by RSGB members. These included: the first G to VK on 10GHz moonbounce (EME), a 10GHz tropo distance record of 1275km, 391km two-way on 24GHz, a world record 425km one-way on 24GHz, 41km on 47GHz, and the first ever two-way contact and a number of distance records on the 73kHz band.

Christopher Davies, M0AAU, was chosen as the 1996 Young Amateur of the Year, a prestigious award jointly sponsored by the RSGB, the RA and the electronics and radio industry.

Radio as a Sport

The RSGB sponsors the vast majority of contests in the UK, including unique international events such as 7MHz DX, Commonwealth ('BERU'), and IOTA. Contests were arranged for all bands from 1.8MHz to 248GHz, and there were several Direction Finding (ARDF) competitions.

The **HF Contests Committee** organized a total of 18 contests during the year. Encouragingly, the number of entries in many contests continued to rise. This was most spectacular in the IOTA Contest in July, which in 1996 attracted over 800 entries, making this one of the world's most popular HF contests.

Another highlight was the 60th anniversary of the Commonwealth Contest ('BERU') in March 1997, which - thanks to some 'high-profile' publicity in *RadCom* and elsewhere - attracted the highest number of entrants since 1947! Special anniversary Gold and Silver certificates were awarded for working 60 and 50 Commonwealth band / call areas.

It is interesting to note that both the IOTA and Commonwealth Contests provided the impetus for a number of DXpeditions by UK operators to exotic locations as varied as Antigua, Anguilla, Malta and even Ghana.

The second annual 'RSGB Contesting Guide' was published with *RadCom* and proved useful to members and overseas national societies. It was also the second year of the special short contest callsigns of the format G7X and M7X which have proved popular with contesters throughout the world.

A new trophy, the David Hill, G4IQM, Memorial Trophy, was purchased from donations following David's sad and premature death and will be awarded in this year's Club Calls Contest.

Entries for VHF contests, organized and adjudicated by **the VHF Contest Committee**, also continue on an upward trend overall, even allowing for dreadful weather and propagation conditions during several of the major events.

New contests were introduced during the year, and these proved to be very popular. Other events, such as the 144MHz Backpackers series and the Christmas Fun Contests continued to grow in popularity. In many cases, the modest number of entries for contests do not give a true picture of the activity which they promote on bands which would otherwise be quiet.

The relationship between the VHFCC and contest participants was excellent, typified by the highly constructive forum held at the VHF Convention. As a service to participants an Internet web page carried information and results, avoiding the publication delays arising from restricted space in *RadCom*.

There was a welcome growth in the number of logs submitted on disk and by E-mail, supported by the availability of free logging software from G0GJV, amongst others. G0GJV is assisting with development of software to introduce greater automation into contest adjudication.

Work continued to simplify and streamline the rules, and steps were introduced to maintain the character of some events, such as Backpackers, and promote participation by those with modest stations.

Several new trophies were offered to the committee during the year: The Low Power Championship Trophy (G4DHF), The Cockenzie Quaiche (Cockenzie & Port Seton ARC for the leading resident Scottish station in VHF NFD restricted section). Additionally, there will be trophies from Guildford RS to promote Novice activity, and in memory of GU2HML and G4CJG; the award conditions for these are not yet fully defined.

VHF contesting remains an important area where participants improve their technical and operating skills through 'self learning'.

ARDF, or 'Foxhunting' as it is sometimes called both here and overseas, continued to enjoy increased popularity. In the UK, the events are co-ordinated by the **ARDF Committee**. The eight qualifying events for participation in the RSGB 160m National Final were well organized by the Torbay, South Manchester, Coventry, Mid-Thames, Ripon, Salisbury, Echelford and Banbury clubs. The 16 qualifiers plus last year's winner assembled in Essex to take part in the National Final, ably run by the Chelmsford / Colchester radio club.

The event was won by Brian Bristow, who was presented with the RSGB 1950 Council Trophy. Chris Wells came second and Andy Collett third.

On 2m, two weekend events were held by the Basingstoke Amateur Radio Club in the New Forest; the second of these was designated the RSGB National 2m Championship. This two-station event was won by Dave Dean, G3ZOI.

The Swansea DF Group organized their annual July weekend in the Forest of Dean with three two-station contests.

Many 2m events were organized by clubs around the country for their membership but very few amateurs seemed to want to travel to take part in less local events. Perhaps an inter-club competition would encourage them to show their skills to a wider audience?

On an international level, teams were entered in the German National Championships, the Brussels International Championships and the Belgian International Championships. All three featured contests on 80m and 2m with five hidden stations on each band. Attendance was high with about 150 participants in Germany. Notable results were 2nd and 3rd places in Brussels, while in the Belgian event there were two third places, and a fourth.

The Committee ran a stand at the RSGB National Mobile Rally at Woburn Abbey and organized a DF hunt in the grounds - many thanks to George Whenham, G3TFA, for his help with this.

The Science of Radio

Throughout the year the **Propagation Studies Committee** (PSC) provided members - and radio amateurs generally - with timely information about propagation and the various solar and geophysical factors affecting it. G0CAS produced daily packet bulletins and the solar factual data and forecasts for the *GB2RS* news service. The growing amount of relevant satellite data led to the introduction of new and still not entirely familiar elements into these reports and G0CAS prepared a guide to the propagation terms used in *GB2RS* broadcasts (see Propagation section). For the many amateurs and listeners without packet or the Internet, the PSC introduced updated *GB2RS* propagation bulletins, transmitted from G4FKH at 0900, 1200, 1500 and 1800 local time each Sunday on 3518kHz CW.

In addition, the PSC Web page was introduced, making generally available an unrivalled range of information on propagation-related matters, including beacon lists, near-real time propagation maps, auroral plots and magnetic data, prediction programs and the like at http://www.keele.ac.uk/depts/por/psc.htm. The HF and 50MHz beacon lists maintained by G3USF were also mirrored at Websites on every continent as well as Britain.

The Chairman served on the IARU President's Ad-Hoc Committee on HF Beacons, whose recommendations on future HF beacon policy will be discussed at the next Region 1 conference. He is assisting in the preparation of the IARU President's 'Challenge' for the best automated monitoring system for the IBP HF beacon network.

The PSC supported *The Six & Ten Report* - observations, analysis and discussion of 6m and 10m propagation, produced by two of its members. With the arrival of a full licence for GB3IFX, G2AHU and G4IFX completed a full year's observations of the 6m tropospheric path between Darlington and Leominster, with complementary observations by G2BDV and others from the Bournemouth area. It is intended to write this project up for publication. G0DJA continued to build up the database of VHF Sporadic-E reports in addition to experimentation at 73kHz. RS87676 completed an audio tape lecture and diagrams on propagation for the blind and partially-sighted, which is available in the Society's tape library. The committee was represented at major rallies, and members gave numerous lectures to clubs and societies and answered many queries from both Society members and official and other outside organizations about unusual propagation phenomena.

Promotion, Publicity, Recruitment

Throughout the year, HQ staff and volunteers put in great efforts to promote amateur radio in general and membership of the Society in particular. This involved advertisements and articles in amateur radio, electronics and general publications, the Internet pages, and rally attendance.

The organization of the RSGB National Mobile Rally at Woburn Abbey in August and the VHF Convention at Sandown Park in February falls to the **Exhibition and Rallies Committee**.

Woburn '96 was again well attended and the good weather attracted some 3000 visitors. The 100 exhibitors reported good results from their day's trading. This Society event of 43 years also has a good social gathering when old friends meet, maybe just once a year.

Nearly 3000 visitors attended the VHF Convention, despite it being held on a cold February day. An excellent programme of lectures and discussions arranged by the two VHF Committees had good crowds. The VHF and Microwave Awards were presented by the President, Ian Kyle, MI0AYZ. As this is usually the first radio event of the year, the dealers enjoyed a profitable day's trading.

The committee members were able to attend events around the country and discuss with other organizers their thoughts about the rally scene. "We see too much computer equipment and not enough radio!" is often the cry.

The HF Convention was the responsibility of the HF Committee (see elsewhere).

A headquarters' bookshop and information stand was on show at many events around the country, including: RSGB Woburn Rally, Red Rose Rally, Great Eastern Rally, Telford Rally, Scottish Convention, RSGB HF Convention, Leicester Amateur Radio Show, Martin Lynch Open Day, North Wales Radio & Electronics Exhibition, RSGB Annual Meeting, RSGB VHF Convention, London Amateur Radio & Computer Show, Norbreck Amateur Radio Electronics & Computing Exhibition, Dayton (Ohio) Hamvention, Friedrichshafen Ham Radio 97, Waters & Stanton Open Day and the Longleat Rally.

Publishing Information

One of the main functions of the RSGB is to publish information for the benefit of members and others. This has been accomplished in the main by the 100-page monthly magazine*RadCom*, a bi-monthly beginner's magazine *D-i-Y Radio*, and a wide variety of books.

In July 1996, following an experimental period, the Society launched its **Internet** pages (http://www.rsgb.org). These grew to the equivalent of over 150 A4 pages of information, and featured rapidly updated news material, an overview of the work of the RSGB, a full-colour book catalogue and an introduction to amateur radio.

The most obvious tangible benefit of RSGB membership is the monthly magazine, ***RadCom***, sent free of charge to each member. Each comprised at least 100 pages and three special editions were produced: one to encourage activity on the new 73kHz band, an other dedicated to short-wave listeners and the third explaining how amateur radio is ideal for disabled people and how the Society helps its disabled members. The most valuable amateur radio prize ever offered - a TS-570D worth £1500 - was the subject of a competition in *RadCom*, sponsored by Kenwood UK. Notable technical projects were a Third Method SSB Transceiver and a 600W Solid State 50MHz Linear Amplifier. The number of equipment reviews increased and an additional writer, Chris Lorek, G4HCL, was recruited to provide a greater number of in-depth reviews.

Three specialist publications were produced for subscribers: *D-i-Y Radio* for beginners, *DX News Sheet* for HF and 6m DXers and the *Microwave Newsletter* for those operating above 1GHz.

New books launched during the financial year included: *The LF Experimenter's Source Book*; *Practical Receivers for Beginners*; *The RSGB*

Rig Guide; *Your First Packet Station* (the first of a series of 'pocket guides'); and *The PMR Conversion Handbook*. New editions were published of the *RSGB Amateur Radio Call Book and Information Directory*; the *RSGB Amateur Radio & SWL Diary*; *RSGB Prefix Guide*; *Antenna Experimenter's Guide*; and the *RSGB IOTA Directory & Yearbook*. In addition, the pages of *RadCom* and *D-i-Y Radio* for 1996 were published on CD-ROM, the first product of its kind from the RSGB.

The RSGB as a Business

Although seen as a large radio club by many of its members, the RSGB is in fact a business with a £1.6M annual turn-over. In addition to co-ordinating the amateur radio liaison work, the Society is a magazine and book publisher, book seller, event organizer, QSL card sorting office and licensing agent. This work is carried out by 28 HQ staff members led by the General Manager, and is overseen by the **Management Committee** (MC).

At the start of a new financial year the composition and function of the then Executive Committee was examined. In the past, committee members were drawn almost entirely from Council but this was now beginning to be rather restrictive. The needs of the Society were examined and appropriate people were invited on to the committee to fulfil certain business criteria. By the end of the year just under the half the current members were Council members.

The financial year began with a deficit of £180k which concentrated the mind. After a degree of 'fire fighting' in the first few weeks the longer term function and strategy of the committee were addressed urgently. At this point it was felt that the name of the committee should be changed to the Management Committee to more accurately reflect its role of ensuring the business affairs of the Society are maintained on a sound financial footing. The committee would monitor the day to day operation of financial affairs in accordance with a written strategic plan approved by Council. There were three areas of concern as follows:

Accounting Procedures and Financial Control: this year the committee benefited from the advice of Ken Ashcroft, G3MSW, acting as the Society's financial advisor is the following areas: presentation of accounts, membership module, income accrual calculations, cash reconciliation, audit preparation and the drawing up of a forecast module. At the same time the Society changed its computerised accounting procedures. Every area of the Society from the running of the various departments at RSGB HQ, through Council to the committees was studied with a view to cutting back on expenditure. Emerging from this exercise came an Expenses Policy with guidelines to assist RSGB staff and volunteers in line with good commercial practice.

Falling Membership: As with many other membership-based organizations there has been a trend towards declining numbers. This has been addressed at several levels during the year. However, the bigger issue of broadening the membership base was considered along with the marketing initiatives which would be required. To widen discussion on the issue, HQ department heads and staff members were involved with the MC in a SWOT (strength, weakness, opportunity, threat) analysis of the RSGB, through which a number of projects emerged to be included in the dynamic Business Plan.

Revenue streams: Books and publications, being a major revenue stream, have been analysed in detail especially in respect of marketing and sales, performance targets, advertising, stock levels, trade discount, commercial outlets, new authors and contracts, as well as the content and design of *RadCom*. New products launched during the year such as *RadCom on CD-ROM*, the *PMR Conversion Handbook*, the *Rig Guide* and the books in the 'pocket guide' series were designed to boost sales figures. More new titles are in the pipeline. The SWOT analysis also identified other areas which will be exploited in the next financial year.

The year end looks like showing a modest profit which, considering the situation in July 1996, is remarkable and a testimony to the hard work and determination of all concerned with ensuring the well-being of the RSGB. However, we are not complacent and strict controls will remain next year.

Headquarters

The HQ building in Potters Bar houses offices for the 28 staff members, the publishing operation, the QSL Bureau, the Audio-Visual Library and meeting rooms. In addition there are several public facilities: the bookshop, the National Amateur Radio Museum and Library, and the GB3RS shack. These were open on weekdays and on several Saturdays and an amateur radio car boot sale was introduced on an experimental basis. In addition to members, many overseas visitors came to HQ. To keep costs down, there was no formal Open Day during the year. Headquarters is divided into six departments:

Accounts ensure that income and expenditure are properly recorded and the books balance. For the year under review, this involved a turn-over of about £1.6M.

The **Sales** Department dealt with 30,000 subscription renewals and applications, plus the sale of over 48,000 books by cash-with-order or to dealers. This provides a major contribution to the Society's income. The department is also responsible for the Society's marketing activities.

HQ Services is responsible for the building itself, as well as the QSL Bureau. It also sends out many thousands of letters and parcels from HQ, runs the print facility and looks after the warehouse.

The **Amateur Radio** section receives many members' enquiries and processes licence applications for repeaters, beacons, packet radio and special event stations. It also deals with the RSGB Morse Testing and Novice Licence Training Services as well as the aspects of the IOTA awards programme. During 1996/97 the staff processed: 4227 information requests, 938 Morse test applications, 910 Novice course completion slips and 1290 special event call applications.

During the year **RadCom** produced 12 editions of the monthly magazine, 51 weekly GB2RS news scripts, 11 press releases and more than 100 regularly updated Internet pages. The team also provided a central design resource for HQ.

The smooth operation of all of the functions of the Society and its responsiveness to members' needs is the responsibility of the **General Manager**.

The Committees of Council

The following is a list of those volunteers who comprised the committees during the financial year 1996/97. Note that the name of each Chairman is shown in **bold** and corresponding members are shown in *italics*. The President is an ex-officio member of all committees.

ARDF

P J Smith, GW1XBG, G A Whenham, G3TFA; M P Hawkins, G3WMM; G W Dover, G4AFJ; D A Burleigh, G4WIZ; C D Plummer, G8APB; Mrs D Pechy, G8NMO; **G C Foster, G8UKT**; *J D Forward, G3HTA; D C Holland, G3WFT; C D Merry, G4CDM; G Nicholls, G4DLB; C Mott-Gotobed, G4ODM.*

DATA COMMUNICATIONS

F C Stewart, G0CSF; **P C Overton, G0MHD**; I Phillips, G0RDI; J M Green, G1DVU; P R Maile, GI1ONL; M J Salmon, G3XVV; I R Brothwell, G4EAN; S A Morton, G8SFR; *R J Cooke, G3LDI; R G Harris, G3ZFR; L W Gurney, G4LBJ; J M Short, G8DQN; R B Woods, G8XAN; J M Dundas, GM0OPS; J D Forward, G3HTA.*

EXECUTIVE / MANAGEMENT

P A Kirby, G0TWW; T M Taylor, G0UCX; J C Hall, G3KVA; K Ashcroft, G3MSW; E N Cheadle, G3NUG; D F Beattie, G3OZF; D I Field, G3XTT; R Horton, G3XWH; N Gannaway, G3YGF; R P Horton, G4AOJ; **M H Claytonsmith, G4JKS.**.

EMC

S N Lloyd-Hughes, GW0NVN; D M Lauder, G0SNO; F Robins, G3GVM; **R M Page-Jones, G3JWI**; M H Claytonsmith, G4JKS; M J Culling, G8UCP; *A K Chamberlain, G0AKC; J D S Malits, G0KCT; R E G Petri, G0OAT; J Greenwell, G3AEZ; J D Forward, G3HTA; G F Firth, G3MFJ; A H Hammett, G3VWK; D Cossar, GM3WIL; A S Kessler, G4DXA; C A Webb, G4FWM; G M Allan, GM4HYF; C R Caine, G4MJZ; L Hawkyard, G5HD; P A Burfoot, G8GGM; M R Hobson, GM8KPH; N R Hooper, G8NLY.*

EXHIBITIONS & RALLIES

J Greenwell, G3AEZ; D E Simmonds, G3JKB; **N Miller, G3MVV**; M Shardlow, G3SZJ; R S Hewes, G3TDR; D Lund, RS94770; *D Whalley, G4EIX; R Kingstone, G4HHB; L Hawkyard, G5HD.*

HF

E J Allaway, G3FKM; J D Forward, G3HTA; E N Cheadle, G3NUG; **C J Thomas, G3PSM**; K Kahn, G3RTU; J W Gould, G3WKL; F C Handscombe, G4BWP; R J Nash, G4GEE; D Hill, G4IQM; *R Balister, G3KMA; G H Grayer, G3NAQ; C Cummings, G4BOH; C Page, G4BUE; J M S Snow, G4TSH; B J Waddell, GM4XQJ; R L Glasher, G6LX.*

HF CONTESTS

K J Chandler, G0ORH; D F Beattie, G3OZF; **J C Burbanks, G3SJJ**; D J Lawley, G4BUO; T G Wylie, GM4FDM; L E Mason, G4HTD; D Hill, G4IQM; J B Coyne, G4ODV; J M S Snow, G4TSH; *H Owen, G2HLU; C J Thomas, G3PSM; D J Mason, G3RXP; S V Knowles, G3UFY; B H J Pickford, G4DUS; A R J Cook, G4PIQ; R A Treacher, BRS32525; J D Forward, G3HTA; R L Glaisher, G6LX.*

IARU

R J C Broadbent, G3AAJ; E J Allaway, G3FKM; **R J Hughes, G3GVV**; J Bazley, G3HCT; J D Forward, G3HTA; L W Barclay, G3HTF; M W Dixon, G3PFR; G Shirville, G3VZV; M S Appleby, G3ZNU; I D Suart, GM4AUP; R L Glaisher, G6LX; *W M Dunell, G3BYW; C Cummings, G4BDH; N Roberts, G4IJF; I L Cornes, G4OUT.*

IOTA

A Williamson, G0NWG, (From June 1997); J D Kay, G3AAE; R Balister, G3KMA; E N Cheadle, G3NUG; D F Beattie, G3OZF; I Buffham, G3TMA, **M J Atherton, G3ZAY**; M Pregliasco, I1JQJ; *J M J Krzymuski, A4DQW; S Kahn, G0STU; A R Williamson, GI0NWG; R Small, G3ALI; J D Forward, G3HTA; J L Hall, G3TOK; P Marsh, G4WFZ;* D L Jones, W4BAA.

LICENSING ADVISORY

D A Peters, G0NSX; E J Allaway, G3FKM; T I Lundegard, G3GJW; J Bazley, G3HCT; J D Forward, G3HTA; P Chadwick, G3RZP; J N Gannaway, G3YGF; I D Suart, GM4AUP; B Rider, G4FLQ; *P C Overton, G0MHD; P A Kirby, G0TWW; R J Hughes, G3GVV; M W Dixon, G3PFR; A C Talbot, G4JNT; I L Cornes, G4OUT; C M Goadby, G8HVV; M J Adcock, GW8CMU.*

MEMBERSHIP LIAISON

F C Stewart, G0CSF; T W G Menzies, GM1GEQ; **E P Essery, GW3KFE**; G L Adams, G3LEQ; J N Gannaway, G3YGF; D Whalley, G4EIX; J T Barnes, GI3USS; *J C B Rider, G4FLQ.*

MICROWAVES

M H Walters, G3JVL; P E H Day, G3PHO; Dr M W Dixon, G3PFR; Dr C W Suckling, G3WDG; Dr J N Gannaway, G3YGF; S T Jewell, G4DDK; **A C Talbot, G4JNT**; Mrs P E F Suckling, G4KGC; S J Davies, G4KNZ; D R Edwards, G8BFV; L P D Kellett, G8KMH; *M G Kinder, G0CZD; R W L Limebear, G3RWL; G Shirville, G3VZV E R Jewell, G4ELM, D J Robinson, G4FRE; P G Murchie, G4FSG; R A Stewart, G4PBP; A R J Cook, G4PIQ; Dr B Chambers, G8AGN.*

MORSE TEST STEERING

F C Stewart, G0CSF; P A Kirby, G0TWW; **R Clayton, G3SSH**; P R Sheppard, G4EJP; M H Claytonsmith, G4JKS; G P Pritchard, G4ZGP.

PLANNING ADVISORY

J W E Jackson, G3TZZ; M O Kennett, G4DVX; D Whalley, G4EIX; **G J Bond, G4GJB**; S J Purser, G4SHF; L F G Thomas, GW4ZXG; B K Sankey, G7WRY; *M H Claytonsmith, G4JKS, J I Batley, G0IID; P Whitworth, G4CTO; G W Peck, G4OIG; G A Vallely, G4YRS; L Hawkyard, G5HD; H R Wignall, GM0TFQ; C J Lewis, GW3YTL.*

PROPAGATION STUDIES

S Reed, G0AEV; N Clarke, G0CAS; D Ackrill, G0DJA; R G Cracknell, G2AHU; C E Newton, G2FKZ; G H Grayer, G3NAQ; **M Harrison, G3USF**; G Williams, G4FKH; K Feldmesser, BRS87676; *I D Brotherton, G2BDV; W M Dunell, G3BYW; Prof L W Barclay, G3HTF; S J M Whitfield, G3IMW; M H Walters, G3JVL; G L Adams, G3LEQ; R G Flavell, G3LTP; C J Deacon, G4IFX.*

REPEATER MANAGEMENT

F C Stewart, G0CSF; A R Horsman, G0MBA; M E C Eavis, G0AKI; K G Baker, G3SPX; W L Mahoney, G3TZM; **G W Dover, G4AFJ (to Jan 97)**; E Bailey, G4LUE; D W McQue, G4NJU; M S Voss, GW8ERA; **C M Goadby, G8HVV (from Jan 97);** C Dalziel, GM8LBC; A Marwood, G8SSL; F McGilp, G8URB; *J T Barnes, GI3USS; G Shirville, G3VZV; S A Morton, G8SFR.*

TECHNICAL AND PUBLICATIONS ADVISORY

R H Biddulph, G8DPS; *M J Willis, G0MJW; M C Headey, G0VVV; R J Newstead, G3CWI; P B Buchan, G3INR; R M Page-Jones, G3JWI; J D Harris, G3LWM; A B Plant, G3NXC; D J Walker, G3OLM; P Chadwick, G3RZP; C V Smith, G4FZH; J Wilkinson, G4HGT; E David, G4LQI; W F Floyd, GW5AF; P H Saul, G8EUX; P J Swallow, G8EZE; J D Forward, G3HTA.*

TRAINING AND EDUCATION

P W Mayer, G0KKL; M L Cayless, G0KRV; R Horton, G3XWH; **E J Case, GW4HWR**; Mrs M H Claytonsmith, G4JKS; G J Garrity, G4TPA; M J R Wade, G8OGO; *G L Benbow, G3HB; J D Forward, G3HTA; Dr M W Dixon, G3PFR, R Clayton, G4SSH; C N Trotman, GW4YKL; J F Badger, G4YZO.*

VHF

J P H Burden, G3UBX; J F Wilson, G3UUT; M S Appleby, G3ZNU; D W McQue, G4NJU; I L Cornes, G4OUT; **M J Adcock, GW8CMU**; Dr R H Biddulph, G8DPS; C M Goadby, G8HVV; Dr J R Morris, GM4ANB; J H Nelson, GW4FRX; *R G Cracknell, G2AHU; N A S Fitch, G3FPK; B Hummerstone, G3HBR; A A McKenzie, G3OSS; A J T Whitaker, G3RKL; I F White, G3SEK; D Lemin, G4TDL; R E S Evans, G8KHV; R J Hughes, G3GVV; M W Dixon, G3PFR; R W L Limebear, G3RWL; G Shirville, G3VZV; P A Howarth, G3YAC; G W Dover, G4AFJ; S J Davies, G4KNZ; A G Hobbs, G8GOJ; S R Thompson, G8GSQ.*

VHF CONTESTS

I W N Pawson, G0FCT; J Greenwell, G3AEZ; D Johnson, G4DHF; P C C Bowyer, G4MJS; I L Cornes, G4OUT; A R J Cook, G4PIQ; L E Adams, G4RKV; M J Platt, G4XUM; **S R Thompson, G8GSQ**; *M Gibbings, G3FDW; D Mawhinney, GI4KSO; C W Tran, GM3WOJ; S G Cooper, GM4AFF; R L Glaisher, G6LX; L P D Kellett, G8KMH.*

Honorary Officers (at 1 July 1997)

Chief Morse Examiner:	R Clayton, G4SSH
EMC Liaison Officer:	F Robins, G3GVM
HF Awards Manager:	F C Handscombe, G4BWP
HF Manager:	J D Forward, G3HTA
IEE Liaison Officer:	P H Saul, G8EUX
Intruder Watch Coordinator:	C Cummings, G4BOH
Microwave Manager:	M W Dixon, G3PFR
Morse Practice Transmissions Coordinator:	G Allan, GM4HYF
Project YEAR Coordinator:	P W Mayer, G0KKL
QSL Bureau Liaison Officer:	J Hall, G3KVA
Society Historian:	G R Jessop, G6JP
Trophies Manager:	D E Simmonds, G3JKB
VHF (and Microwaves) Awards Manager:	I L Cornes, G4OUT
VHF Manager:	D J Butler, G4ASR

Trophy Winners – 1996

The Society is fortunate to have a large number of trophies. Many of these are awarded for winners of various contests, however others give public recognition to some particular aspect of society work. They are awarded at a number of RSGB events – typically at the AGM, HF and VHF Conventions. The winners of trophies awarded for activities in 1996 are listed below.

Council

Bennett Prize
(For any significant contribution or innovation which furthers the art of radio communication)
Pat Hawker, G3VA

Calcutta Key
(For work associated with international friendship through amateur radio)
Ahron Kirschner, 4X1AT

Founder's Trophy
(For outstanding service to the Society)
Roger Balister, G3KMA

HF Contest

Beru Junior Rose Bowl
(Runner-up in the Commonwealth Contest)
Bob Whelan, G3PJT

Beru Senior Rose Bowl
(Winner of the Commonwealth Contest)
A Sluymer, VE3EJ

Braaten Trophy
(Leading G station in the ARRL DX CW Contest)
Dave Lawley, G4BUO

Bristol Trophy
(Highest score in the other section of NFD)
Gravesend RS

Col Thomas Rose Bowl
(Highest placed G station in BERU)
Dave Lawley, G4BUO

Edgware Trophy
(Winning team in AFS CW Contest)
Lichfield ARS

Flight Refuelling ARS Trophy
(Winning team in AFS SSB Contest)
Martlesham DX & CG

Frank Hoosen Trophy
(Leading 14MHz score in NFD)
Orkney RG

G3XTJ Memorial Trophy
(Most accurate log in ROPOCO II Contest)
Dave Lawley, G4BUO

G5MY Trophy
(Highest aggregate score in the ROPOCO Contests)
Jan Fisher, G0IVZ

G6ZR Memorial Trophy
(Runner-up in the open section of NFD)
Newbury ARC

Gravesend Trophy
(Runner-up in the restricted section of NFD)
Orkney RG

HFCC Trophy
(Winner of the LF (3.5/7MHz) SSB Contest.)
Orkney RG

Houston Fergus Trophy
(Winner of the 10W section of Low Power Field Day)
Peter Crooks, G4KGG/P

L H Thomas G6QB Memorial Trophy
(Winner of the 7MHz CW DX Contest)
Jan Fisher, G0IVZ

Maitland Trophy
(Scottish station with highest aggregate points in both Top Band Contests)
W D Stirling, GM4DGT

Marconi Trophy
(Highest individual score in AFS CW Contest)
Fraser Robertson, G4BJM

NFD Shield
(Winner/overall highest score in NFD.)
Lichfield RS

R Whelan G3PJT Medal
(Most improved score in the Commonwealth Contest)
J Bautista, ZB2EO

Reading QRP Trophy
(Leading station in the low power section of NFD)
Canberra CG

RSGB Lichfield Trophy
(Highest individual score in AFS SSB Contest)
Andy Cook, G4PIQ

Scottish NFD Trophy
(Leading GM station in NFD)
Orkney RG

Somerset Trophy
(Winner of first 1.8MHz CW Contest)
Ron Stone, GW3YDX

Southgate Trophy
(Winner of the 3W section of Low Power Field Day)
Frank Claytonsmith, G3JKS

Verulam Silver Jubilee Trophy
(Most accurate log in ROPOCO 1 Contest)
Derek Webber, G3LHJ

1930 Committee Cup
(Winner of Low Power 80m Contest single operator)
Derek Stanners, G3HEJ

Microwave

G3VVB Memorial Trophy
(For the best home constructed microwave equipment exhibited at a microwave round table or convention)
Chris Whitmarsh, G0FDZ

Les Sharrock G3BNL Memorial Award
(For innovation or technical development of microwave equipment or techniques)
Jointly to:
Tony Horsfall, G4CBW;
Dave Hall, G8VZT;
Martin Farmer, G7MRF

10GHz Trophy
(Winner of the May 10GHz Trophy event organised by the VHF Contest Committee)
D. J. Bartlett G4VIX

Tech & Pubs Advisory

Courtenay-Price Trophy
(For the most outstanding published technical contribution to amateur radio)
John Regnault, G4SWX

Ostermeyer Trophy
(For the most meritorious description of a piece of home-constructed radio or electronic equipment published in RadCom)
Roger Blackwell, G4PMK

Wortley-Talbot Trophy
(For outstanding experimental work in Amateur Radio)
Dave Lauder, G0SNO

Training & Ed

Kenwood Trophy
(For making a significant contribution to training and development in amateur radio within the United Kingdom)
Esde Tyler, G0AEC

VHF

Harold Rose Memorial Plate
(To the person making an outstanding contribution to 50MHz)
R Cracknell, G2AHU

Louis Varney Trophy
(For advances in space communication)
Mike Dorsett, G6GEJ

1962 VHF Committee
(Awarded at VHF convention for the best home constructed equipment)
J Matthews, G3WZT

VHF Contests

Arthur Watts Trophy
(Awarded to winner of restricted section of VHF NFD)
Warrington CG

Backpackers Trophy
(Leading station in the Backpackers Trophy Contest)
Windbreakers & Hadrabs CG

G6ZR Memorial Trophy
(Winner of 2.3GHz Contest)
Three Spires CG

John Pilags Memorial Trophy
(Leading single operator fixed station in the RSGB VHF Contests Championship)
A R J Cook, G4PIQ

Martlesham Trophy
(Winner of VHF NFD restricted section)
Parallel Lines CG

Mitchell-Milling Trophy
(Winner of 144MHz Trophy Contest - multi-operator section)
Northern Lights CG

Racal-Decca Radio Cup
(Winner of RSGB VHF Contests Championship - open section)
Northern Lights CG

Scottish VHF NFD Trophy
(Leading GM station in VHF NFD restricted section)
Llion Ogden, GW0RQM and Stephen Marsh, GW7VCH

Six Meter Cup
(Highest scoring single UK operator entry in the 50MHz Trophy Contest)
I R Dixon, G4BVY

Surrey Trophy
(Winner of open section of VHF NFD)
Home Counties & Three Spires CG's

Tartan Trophy
(Leading Scottish station in VHF NFD)
Highland CG

Telford Trophy
(Winner fixed station 50MHz Contest)
Northern Lights CG

Thorogood Trophy
(Winner of 144MHz Trophy Contest - single operator section)
A R J Cook, G4PIQ

UHF Contests Committee Cup
(Overall winners of the 1.3GHz Trophy Contest)
Parallel Lines CG

VHF Managers Cup
(Winner of 70MHz Trophy Contest)
Spalding & District ARS

1951 Council
(Winner of 430MHz Trophy Contest)
Parallel Lines CG

6 Metre Backpackers Trophy
(Leading station in the Backpackers Trophy Contest)
West Kent ARS

Audio Visual Library

The RSGB Audio-Visual Library is a service for RSGB affiliated societies in preparing programmes for their club meetings. Unfortunately the service is not available to individual members. The library consists of video cassettes in VHF format and audio cassettes often supported by 35mm slides. A small charge is made for each item borrowed.

Full details of the Library and how to book can be obtained from the librarian: John Davies, G3KZE, RSGB AV Library at RSGB HQ.

Catalogue

General Interest

A BRIEF HISTORY OF TIME (1995, 80 mins) Based on the book by Stephen Hawking.

ARMADA GB400A (40 mins) Account of Plymouth ARC establishing a station to commemorate defeat of Armada.

CLASSIC MANOEUVRES (1992, 40 mins) The Red Arrows on their North American tour 1983 – the world's greatest aerial display team.

COASTAL COMMAND (1994, 70 mins) A 1944 Crown Film Unit production of the role of RAF Coastal Command in protecting the nation's sea-lanes.

CQ FIELD DAY (1993, 26 mins) Setting up and operating a Field Day station to win – California style. CQ 1992.

EMPIRE OF RADIO (1992, 1 hr 50 mins) A first class professional American video of the pioneering of radio.

GETTING STARTED IN DXing (1993, 52 mins) Shows DX'ers station and how to 'winkle out' the rare DX. CQ 1992.

LANCASTER (1992, 58 mins) The story of the RAF's most famous bomber including rare WW2 film shots.

MELBOURNE RADIO CLUB (1982, 65 mins) Video made by this famous VK club showing the city and amateurs stations.

MY AMATEUR RADIO (1996, 60 mins) Biography G2DPQ – lifetime of amateur radio – recorded just before he died.

NORTH TEXAS CONTEST CLUB – 1983 (1992 40 mins) Big beams and big stations, to win contests in Texas style.

PASSPORT TO FRIENDSHIP (1991 25 mins) World Goodwill Games 1990 – Top Ham Operators 'Go for Gold' contest.

PJ9W - 1990 CQ WW SSB CONTEST. (1992, 45 mins) Fascinating video on preparing and running a winning contest station by a keen team of Finnish operators.

SECRET LISTENERS - BBC (30 mins) Account of Amateurs work during World War 2.

SECRET WAR - WORLD WAR 2 SERIES (1991)

- i Battle of the Atlantic (50 mins).
- ii Battle of the Beams (50 mins), To See for a Hundred Miles - Radar (50 mins).
- iii Deadly Waves – Mag Mine detection (50 mins), Still Secret – Enigma code breaking (50 mins).
- iv Terror Weapons - V1 and V2 (100 mins).

TWO PIONEERS OF RADIO - G2DX & G6CJ (22 mins) Bristol TV Group video.

VHF - All You Need to Know to pass (1992, 45 mins) the VHF Marine Operator's Examination.

VHF THEN AND NOW by G5UM JACK HUM (1987, 85 mins).

WESTERN APPROACHES (1994, 83 mins) A 1944 Crown Film Unit production of the bravery of the Merchant Service. The best British documentary of WW2.

WINNING ON THE HILL (1995, 58 mins) Superb video of VHF NFD winning station in Rochester NY.

WORLD AT THEIR FINGERTIPS – RSGB (45 mins) Growth of the amateur movement in the UK by John Clarricoats.

Technical

AERIAL CIRCUS G6CJ (70 mins) Doug Charmans famous lecture (in b&w only). A little scratchy but still good viewing.

AMATEUR TELEVISION (60 mins) A series of short programmes on amateur TV here and in Australia.

BASIC RADIO MEASUREMENTS (1993, 80 mins) Practical demonstration of key station measurements by G3NYK.

CONSTRUCTION OF THE 'ONER' (60 mins) A G-QRP construction project.

ELECTROMAGNETIC WAVES – THE ELECTRON'S TALE –THIN FILM MICROCIRCUITS (3 titles on one 60 mins) video.

GETTING STARTED IN PACKET RADIO (1993, 45 mins) Very good on how to start, set up equipment and go on the air. CQ 1992.

MANUFACTURE OF JUNCTION TRANSISTORS – SOMETHING BIG IN MICROCIRCUITS. (2 titles on one video 65 mins.)

SECRET LIFE OF RADIO (1992, 35 mins) An easy to watch video on radio techniques from crystal set to complex rig. Very 'watchable' for all ranges of skill and knowledge.

SILICONE GLEN (30 mins) Electronics industry video.

SKYWATCHING (1993, 40 mins) Good guide to daytime sky explaining sun, clouds, wind, precipitation, weather hazards etc. Made in USA.

DX-peditions etc.

DX-pedition TO THE PACIFIC - 1991 (1992, 180 mins) by VK2EKY - KH8/5W1/A35/ZK3/KH2/JA This may seem a long video but it breaks down into interesting and self-contained sections.

DX-pedition 4J1FS MALIYSOTSKII IS (25 mins).

DX-pedition 7J1RL OKINO TORISHIMA (35 mins).

DX-pedition to HEARD ISLAND (1997) VK0IR by ON6TT (53 mins).

DX-pedition to HEARD ISLAND (1992, 50 mins) Donated by Northern California DX Foundation..

DX-pedition to HOWLAND Is. – NO1Z/KH1 (22 mins) DX-pedition to BOUVET Is. – 3Y5X (30 mins) Donated by the Northern California DX Foundation 1990/91.

DX-pedition to HOWLAND ISLAND (1993, 45 mins) The latest expedition to KH1.DX-pedition S0ARSD (40 mins).

DX-pedition to JARVIS Island (1992, 35 mins) Superb account of a successful DX-pedition mounted in 1990. Donated by Northern California DX Foundation.DX-pedition VP8ANT.

DX-pedition VU7 LACCADIVE IS. (58 mins).

Expedition BORNEO (1995) Presence Radioamateur (35 mins) *French spoken.*

JARL VISIT TO CHINA.

Specially Suitable for the Beginner

AMATEUR RADIO FOR BEGINNERS (1991, 43 mins) RSGB video to intro Novice Licence.

AN INTRODUCTION TO THE HOBBY OF AMATEUR (1991, 15 mins) RADIO A personal account G4ZDA.

HAM RADIO HORIZONS (1993, 59 mins) Introduction to the exciting and diverse world of ham radio CQ 1992.

NEW WORLD OF AMATEUR RADIO – ARRL (30 mins).

Space, Satellites etc.

AMATEUR RADIOS NEWEST FRONTIER (30 mins) The W5LFL Space Shuttle mission.

GETTING STARTED IN AMATEUR SATELLITES (1993, 50 mins) Newcomers guide to equipment, techniques and jargon of sat. comms. CQ 1992.

METEORITES (2 videos, 1995) Vol 1: Menace From the Sky (42 mins). Vol 2: Witnesses From Beyond the Times (42 mins).

SATELLITE COMMUNICATION TELECOM (1979, 60 mins).

SATELLITE COMMUNICATIONS (1986, 22mins) – SPACETALK (22mins) – PRECISE GIANTS (18mins).

SATELLITE OSCAR X by DJ5KQ (30 mins).

SATELLITE 'FUJI' by JARL (1986 30 mins) A professional video showing the construction, launch and use of FUJI.

SPACE SHUTTLE W0ORE TONY ENGLAND (53 mins) An account of the first amateur in space by the man himself.

STAR JOURNEY - LIFTOFF! (1996, 52 mins) Preparation and mission with the astronauts aboard the shuttle Columbia. Narrated by Patrick Stewart.

THE DREAM IS ALIVE (1993, 37 mins) A window seat on shuttle: share life with the astronauts: see the earth from space: walk in space etc. 1985, made by The Smithsonian, Lockheed & NASA.

THE UNIVERSE by Smithsonian 1992 (1993, 50 mins) Excellent story of how the Universe started. Quality Cosmology.

Cassette and Slide Section

AERIALS FOR DX – G6CJ (18 slides, 64 mins) Dud Charmans famous lecture.

AURORA - WHAT CAUSES IT by G2FKZ (1982) Part 1 (20 slides, 51 mins). Part 2 (30 slides, 63 mins).

BBC MONITORING SERVICE IN WW2 (60 mins, audio only) Read by Alvar Lidell – 1981.

DX-pedition to CLIPPERTON ISLAND (1978, 30 slides, 60 mins).

DX-pedition to ST. PIERRRE & MIQUELON (1959, 63 slides, 42 mins).

DX-pedition to XF4L by N7NG (1989, 80 slides, 35 mins).

ELECTROMANIA (1983, 30 mins, audio only).

HISTORY OF AMATEUR RADIO G2BTO (75 mins, audio only) First class intro to, and history of, radio.

OSCAR VI G3IOR (43 slides, 90 mins) launched 15 Oct 1972.

OSCAR VII G3IOR launched 15 Nov 1974. Part 1 (48 slides, 60 mins) Part 2 (57 mins, audio only).

SOLAR CYCLE 21 (33 slides, 48 mins).

500 kHz – 'THE END IS NIGH' (1990, 65 mins, audio only) Commentary and recordings of the final closing down CW signals of famous coastal stations. A good, somewhat nostalgic, cassette.

HF Awards

HF Awards: General Rules

The following general rules and conditions apply to HF certificates and awards issued by the Radio Society of Great Britain and should be read in conjunction with the conditions which govern the particular award programme:

Applicant eligibility

1. Claimants from the UK, Channel Is, and Isle of Man must be members of the RSGB and, as proof of membership, should provide a recent address label from *Radio Communication*. Applicants from elsewhere need not be members of the RSGB.
2. Claimants may be either licensed radio amateurs or short wave listeners. All certificates, but not special plaques, are available on a 'heard' basis to listeners.

Claim eligibility

3. Each claim must be submitted in a form acceptable to the HF Awards Manager. Where application forms are provided for particular award programmes, these should be used although a computer generated form including the same headings will generally be accepted. Each claim must include the following signed declaration: *'I declare that all the contacts were made by me personally from the same DXCC country and in accordance with the terms of my radio transmitting licence, and that none of the QSLs have been amended in any way since receipt. I accept that a breach of these rules may result in disqualification from the awards programme. I further accept that the decision of the HF Committee shall be final in all cases of dispute.'*
4. All claims from within the UK, Channel Is and Isle of Man must be accompanied by QSL cards. Claims from elsewhere must also be accompanied by QSL cards but only in the case of those categories of award attracting a plaque. In all other cases a statement from the applicant's national society that the necessary cards have been checked will be accepted except that the HF Awards Manager reserves the right to ask to see some or all of the cards. For IOTA claims special rules apply (see the *IOTA Directory*).
5. Each claim must be accompanied by a fee of £3.00 or $US 6.00 or 9 IRCs per certificate or class of certificate. Applicants submitting cards for checking must include sufficient payment to cover their return. Cards will only be returned by air, recorded delivery (UK only), or registered mail (overseas) if adequate postage is enclosed. (For registered mail add US$ 4.50 or 7 IRCs).

Contact eligibility

6. All contacts must be made by the holder of the call sign.
7. Contacts may be made from any location in the same DXCC country.
8. Except where otherwise indicated, credit will be given for contacts made on or after 15 November 1945 on any of the amateur bands below 30 MHz.
9. Contacts with land mobile stations will be accepted, provided the location at the time of contact is clearly stated on the QSL card.
10. Credit will be given for two way contacts on the same mode and band, ie not cross-mode or cross-band. Certificate endorsements for single mode transmission and/or single band may be made on the submission of cards clearly confirming the mode or frequency of transmission, but the request must be made at the time of application. Special rules apply for IOTA.

Disqualification

11. The submission for credit of any altered or forged confirmations or, equally, bad behaviour on or off the air which is judged by the HF Committee to bring a particular programme into disrepute may result in disqualification of the applicant from all RSGB's award programmes. The decision of the HF Committee on this and other matters of dispute will be final.

Applications For Awards (other than IOTA)

All claims, except IOTA, should be sent to:

RSGB HF Awards Manager
F C Handscombe, G4BWP
'Sandholm'
Bridge End Road
Red Lodge
Bury St. Edmunds
Suffolk IP28 8LQ

Prepare your application in accordance with the requirements of the award being claimed. Send QSL cards when required, and do not forget to enclose details of your name, callsign and full address as well as the certificate fee, adequate postage for the return of QSL cards, and for UK applicants, proof of RSGB membership. Payment may be made by cheque drawn on a UK bank or Eurocheque written in Pounds Sterling and payable to 'Radio Society of Great Britain'.

DX Listeners' Century Award (DXLCA)

This award may be claimed by any short wave listener eligible under the General Rules who can produce evidence of having heard amateur radio stations located in at least 100 DXCC countries. Stickers are available for every 25 additional countries confirmed. Submit a list in radio prefix order with the callsign and country name.

Endorsements are available for hearing 100 countries on 5, 6, 7, 8 and 9 bands (they need not be the same countries on each band).

Commonwealth Century Club (CCC)

This award may be claimed by any licensed radio amateur eligible under the General Rules who can produce evidence of having contacted, since 15 November 1945, amateur radio stations in at least 100 Commonwealth call areas on the list current at the time of application.

The certificate holder may claim, on payment of a contributory charge, a handsome plaque with a plate detailing name, callsign, date and number of the award. Additionally, an amateur providing evidence of having contacted all the Commonwealth call areas on the list current at the time of application may claim the Supreme Plaque in recognition of the magnitude of the achievement, again on payment of a contributory charge.

Notes

a) Credit for South Georgia and the South Sandwich Is will only be given for contacts with stations using a VP8 callsign. Credit for Antarctica and the South Orkney and South Shetland Is will only be given for contacts with stations using a callsign issued by a Commonwealth government.

b) Where, very occasionally, a contact is made with a station using a callsign legitimately outside the geographical area to which the prefix normally applies, it will count for the actual area from which the operation took place. The evidence submitted will need to be clear.

5 Band Commonwealth Century Club (5BCCC)

This award, available in 5 classes, may be claimed by any licensed radio amateur under the General Rules who can produce evidence of having effected two-way communication, since 15 November 1945, with the requisite number of amateur radio stations located in the call areas listed, using all 5 bands, 3.5, 7, 14, 21, and 28 MHz. Each station should be located in a different call area per band. The 5 classes are for contacts as follows:

5BCCC Supreme	500 stations
5BCCC Class 1	450 stations
5BCCC Class 2	400 stations, with a minimum of 50 on each band.
5BCCC Class 3	300 stations, with a minimum of 40 on each band.
5BCCC Class 4	200 stations, with a minimum of 30 on each band.

Certificates will be issued to winners of all classes. Additionally, as in the case of the CCC, winners of the Class 1 award will be eligible to claim a handsome plaque suitably inscribed on payment of a contributory charge, while winners of the Supreme Award will be able to claim an engraved plaque on payment of a contributory charge.

WARC Bands Endorsement

A holder of the basic award who can provide evidence of contact with the required number of call areas on the 10, 18 and 24 MHz bands may claim a WARC Band 'sticker' endorsement. This is available in 5 classes as follows:

5BCCC (WARC)	Supreme	300 call areas.
5BCCC (WARC)	Class 1	275 call areas.
5BCCC (WARC)	Class 2	250 call areas, with a minimum of 50 on each band.
5BCCC (WARC)	Class 3	200 call areas, with a minimum of 40 on each band.
5BCCC (WARC)	Class 4	150 call areas, with a minimum of 30 on each band.

Note: on 10 MHz credit will only be given for contacts on CW and datamodes.

Top Band Endorsement
A holder of the basic award who can produce evidence of contact with the required number of call areas on the 1.8 MHz band may claim a Top Band 'sticker' endorsement. This is available in 5 classes for 30, 40, 50, 60 and 70 call areas.

Credit For Deleted Commonwealth Call Areas
Credit will be given for contacts with stations in Commonwealth countries at the time of contact, and additionally, with stations using a Commonwealth callsign in Antarctic and the S Orkney and S Shetland Is. Since 1945 some countries have left the Commonwealth, others have joined. The list of call areas specifies the relevant dates. For 5BCCC (including endorsements) contacts with 'deleted' call areas made at a time when the countries concerned were in the Commonwealth may count in place of 'missing' credits on any band up to the maximum possible for that band (currently, at July 1996, 132). **Deleted call areas do not count for CCC.**

List of Deleted Commonwealth Call Areas
Contacts with the following may count for a 'missing' credit if made before the date specified. A contact after that date may still count for one of the Commonwealth call areas (see brackets).

AC3	Sikkim	1 May 1975 (India)
P2, VK9	Papua Territory	16 September 1975 (PNG)
P2, VK9	Territory of New Guinea	16 September 1975 (PNG)
VO	Newfoundland, Labrador	1 April 1949 (Newfoundland, Canada)
VQ1	Zanzibar	25 April 1964 (Tanzania)
VQ6	British Somaliland	1 July 1960
VQ9	Aldabra Is	29 June 1976 (Seychelles)
VQ9	Desroches Island	29 June 1976 (Seychelles)
VQ9	Farquhar Group	29 June 1976 (Seychelles)
VS2, 9M2	Malaya	16 September 1963 (W Malaysia)
VS4	Sarawak	16 September 1963 (E Malaysia)
VS9	Aden	1 December 1967
VS9	Kamaran Island	1 December 1967
VS9	Kuria Muria Is	1 December 1967
ZC5	British North Borneo	16 September 1963 (E Malaysia)
ZD4	Gold Coast, Togoland	6 March 1957 (Ghana)
3D2	Fiji	15 October 1987
3D2	Rotuma Is	15 October 1987

Applications for CCC and 5BCCC
Applications for 5BCCC should use the special application form available from the HF Awards Manager. The form allows space for callsigns to be recorded for contacts on each band. A check list of current Commonwealth call areas is also available and can be used for applications for the CCC Award. Please send an A5 size SAE for the check list and application forms.

28 MHz Counties Award

This award may be claimed by any licensed radio amateur eligible under the General Rules who can produce evidence of having effected two-way communication, since 1 April 1983, with amateur radio stations located in 40 counties/regions in the UK, Channel Islands and Isle of Man on the 28 MHz band. Stickers are available for 60 and all 77 counties/regions confirmed.

Worked ITU Zones (WITUZ)

This award may be claimed by any licensed radio amateur eligible under the General Rules who can produce evidence of having contacted, since 15 November 1945, land based amateur radio stations in at least 70 of the 75 broadcasting zones as defined by the International Telecommunications Union (ITU).

The certificate holder may claim, on payment of a contributory charge, a handsome plaque with a plate detailing name, callsign, date and number of the award. Additionally, an amateur providing evidence of having contacted all 75 ITU zones may claim the Supreme Plaque in recognition of the magnitude of the achievement, again on payment of a contributory charge.

5 Band Worked ITU Zones (5BWITUZ)

This card, available in 5 classes, may be claimed by any licensed radio amateur eligible under the General Rules who can produce evidence of having contacted, since 15 November 1945, the required number of land based amateur radio stations located in the 75 ITU broadcasting zones, using all 5 bands, 3.5, 7, 14, 21, and 28 MHz. Each station should be located in a different ITU zone per band. The 5 classes are for contacts as follows:

5BWITUZ Supreme	350 zones
5BWITUZ Class 1	325 zones
5BWITUZ Class 2	300 zones, with a minimum of 50 on each band.
5BWITUZ Class 3	250 zones, with a minimum of 40 on each band.
5BWITUZ Class 4	200 zones, with a minimum of 30 on each band.

Certificates will be issued to winners of all classes. Also, as in the case of the WITUZ, winners of the Class 1 award may claim a handsome plaque suitably inscribed, while winners of the Supreme Award will be eligible for the Supreme Plaque, both on payment of a contributory charge.

WARC Band Endorsement

A holder of the basic award who can produce evidence of contact with the required number of ITU zones on the 10, 18 and 24 MHz bands may claim a WARC Band 'sticker' endorsement. This is available in 5 classes as follows:

5BWITUZ (WARC)	Supreme	210 zones
5BWITUZ (WARC)	Class 1	195 zones
5BWITUZ (WARC)	Class 2	180 zones, with a minimum of 50 on each band.
5BWITUZ (WARC)	Class 3	150 zones, with a minimum of 40 on each band.
5BWITUZ (WARC)	Class 4	120 zones, with a minimum of 30 on each band.

Note: on 10 MHz credit will only be given for contacts on CW and datamodes.

Top Band Endorsement

A holder of the basic award who can produce evidence of contact with the required number of call areas on the 1.8 MHz band may claim a Top Band 'sticker' endorsement. This is available in 5 classes for 20, 30, 40, 50 and 60 zones.

Notes

(a) The number of ITU broadcasting (ie land) zones recognised by the ITU is 75 zones (Zones 1 to 75) and this therefore is the maximum score which can be claimed per band. However, the islands of Minami Torishima (JD1) and Salas-y-Gomez (CE0) lie outside the 75 zones in sea zones 90 and 85 respectively. For 5BWITUZ (including endorsements) contacts with these islands may count in place of one 'missing' credit on any band up to the maximum 75 for that band. **Contacts with Minami Torishima and Salas-y-Gomez do not count for WITUZ.**

(b) In the case of the WITUZ and 5BWITUZ confirmations need not bear the appropriate ITU zone number but in order to count for credit they should give the location of the station in sufficient detail to place it clearly within one particular zone. Doubtful cases indicating possible overlap across two zones will not be given credit.

(c) The HF Awards Manager will use as his reference a list which is based on the *Radio Amateurs' Prefix Map of the World* published by Radio Amateur Call Book Inc., PO Box 2013, Lakewood, New Jersey 08701, USA. In the case of countries which encompass two or more ITU zones, e.g. USA, Russia and Brazil, zonal boundaries will generally follow the longitude/latitude grid lines as shown in the map. In the few instances of discrepancy between the map and the accompanying prefix/country list the decision of the HF Awards Manager will be final.

Applications for WITUZ and 5BWITUZ

Applications for WITUZ and 5BWITUZ should use the special application form available from the HF Awards Manager. A check list of ITU Zones and a map are included with the application form. Please send an A5 size SAE for the check list and application form.

Worked all Continents (WAC)

This award, issued by IARU headquarters, may be obtained by any licensed radio amateur in the UK, Channel Is or Isle of Man who is a member of the RSGB and can produce evidence of having effected two-way communication with amateur radio stations located in each of the six continents - North America, South America, Europe, Africa, Asia and Oceania.

Applicants should send QSL cards to the RSGB HF Awards Manager who will certify the claim to the IARU headquarters society (ARRL) for issuance of the award. They should also enclose a self addressed stamped envelope for return of the cards, and proof of RSGB membership.

All contacts must be made from the same country or separate territory within the same continent. Various endorsements including 'all 1.8 MHz' are available. In addition both a 5 and 6 Band WAC may be claimed.

IARU Region 1 Award

This award, available in 3 classes, may be claimed by any licensed radio amateur eligible under the General Rules who can produce evidence of having contacted amateur radio stations located in the required number of countries whose national societies are members of the Region 1 Division of the International Amateur Radio Union (IARU).

The 3 classes are for contacts as follows:

Class 1	All member countries on the current list
Class 2	60 member countries
Class 3	40 member countries

Members of IARU Region 1 (as of July 1996) are:

Albania
Algeria
Andorra
Austria
Bahrain
Belarus
Belgium
Bosnia
Botswana
Bulgaria
Burkina Faso
Croatia
Cyprus
Czech Republic
Denmark
Djibouti
Egypt
Estonia
Faroe Is
Finland
France
Gabon
Gambia
Germany
Ghana
Gibraltar
Greece
Hungary
Iceland
Iraq
Ireland
Israel
Italy
Ivory Coast
Jordan
Kenya
Kuwait
Latvia
Lebanon
Lesotho
Liberia
Liechtenstein
Lithuania
Luxembourg
Macedonia
Malta
Mauritius
Monaco
Mongolia
Morocco
Namibia
Netherlands
Nigeria
Norway
Oman
Poland
Portugal
Qatar
Romania
Russian Federation
San Marino
Senegal
Sierra Leone
Slovakia
Slovenia
South Africa
Spain
Swaziland
Sweden
Switzerland
Syria
Tadjikstan
Tanzania
Turkey
Turkmenistan
Uganda
Ukraine
United Kingdom
Yugoslavia
Zambia
Zimbabwe

73 kHz Award

This award is to recognise achievements in both transmission and reception on the 73 kHz band, for both one and two-way QSOs.

The award is available in three categories, with endorsements for various distances.

Basic: for a confirmed one-way transmission on 73 kHz from the UK, any mode, with talk-back on another amateur band, by phone or confirmed by a listeners report (UK or otherwise), over a distance in excess of 8 km. This award can be endorsed for distances in excess of 32 and 128 km.

Full: for a confirmed two-way QSO on 73 kHz within the UK by holders of the 73 kHz NoV, any mode, over a distance in excess of 8km. This award can be endorsed for distances in excess of 32 and 128 km.

Listener: for a confirmed report of any UK amateur 73 kHz transmission, in any mode, over a distance in excess of 8km. This award can be endorsed for distances in excess of 32 and 128 km.

In order to confirm the distances for the above awards, the locations of both stations must be given as either a 6 digit Maidenhead QTH locator or a 6 digit OS reference (eg TL123456).

VHF Awards

General Rules for VHF/UHF Awards

1. Awards are available to licensed amateurs and listeners (on a heard basis). All claims must be fully supported by QSL cards. For the various squares awards, these cards must also bear the IARU (Maidenhead) locator details. A card without an IARU locator originally printed on it is acceptable provided that it bears some adequate form of positional information (for example old QTH locator or latitude and longitude), in which case the IARU locator square designation should be clearly added to the card by the award claimant.
2. For all awards with a fixed station category, the applicant must state that all the contacts were made from the same location. In the case of an amateur moving home location he/she may apply for the award using QSL cards gained from more than one location, but the award will be endorsed as 'gained from more than one location'.
3. Endorsements such as all cw, all ssb, all auroral contacts, or all contacts made during the first year of being licensed may be made on application. The appropriate information must be contained on the QSL cards and a declaration signed when applying for endorsements.
4. The charges for awards are:- RSGB members £3, US$6 or 12 ircs. UK residents who are not RSGB members £6, US$12 or 24 ircs. Overseas applicants who are members of their national society £6, US$12 or 24 ircs. Other overseas applicants £9, US$18 or 36 ircs. Where applicable proof of membership of their national (IARU approved) society is required, eg recent *RadCom* label or photocopy of membership card. There is no charge for stickers to update levels of achievement, however, if a new certificate is requested, the charges above will apply.
5. All claims must be submitted to the RSGB VHF/UHF Awards Manager, Ian L Cornes, G4OUT, 6 Haywood Heights, Little Haywood, Stafford, ST18 0UR. Application forms may be obtained from this address by sending an SAE.
6. For the safe return of the QSL cards, adequate postage and a self addressed envelope must be sent with the application.

RSGB 50 MHz Countries Award

The initial qualification for this certificate is proof of completed two-way QSOs on 50 MHz with ten countries. Stickers will be provided for increments of every ten countries worked. Only contacts with countries permitting 50 MHz operation can be considered.

Rules

1. All contacts must have been on or after 1 June 1987.
2. QSL cards submitted must be arranged in alphabetical order of the countries claimed, and a checklist enclosed.
3. Stations are eligible for awards in the following categories:
   (a) Fixed stations.
   (b) Temporary location or portable operation (/P).
   Categories cannot be mixed.

RSGB 50 MHz DX Certificate

This certificate takes into account the considerable potential for cross-band working when transmitting in the 50 MHz band. There is therefore no stipulation on the band used for the incoming signal. The initial qualification is confirmation from 25 different countries of a successful QSO with transmission from the applicant's country taking place within the 50 MHz band. Stickers will be provided for increments of 25 countries confirmed.

Rules

1. All contacts must have been on or after 1 June 1987.
2. QSL cards submitted must be arranged in alphabetical order of the countries claimed, and a checklist enclosed.
3. Stations are eligible for awards in the following categories:
   (a) Fixed stations.
   (b) Temporary location or portable operation (/P).
   Categories cannot be mixed.

RSGB 50 MHz Squares Award

The 50 MHz Squares Award is intended to mark successful vhf achievement. The initial qualification needed for this certificate is proof that 25 different locator squares have been worked with complete two-way QSOs within the 50 MHz band. Squares in any country will qualify provided that operation from that country is formally authorised. Additional stickers will be provided when proof is submitted for increments of 25 squares.

Rules

1. All contacts must have been on or after 1 June 1987.
2. QSL cards submitted must be arranged in alphabetical order of the QTH squares claimed, and a checklist enclosed.
3. Stations are eligible for awards in the following categories:
   (a) Fixed stations.
   (b) Temporary location or portable operation (/P).
   Categories cannot be mixed.

4-2-70 Squares Award

The 4-2-70 Squares Awards are intended to mark successful vhf/uhf achievement. Initially, a certificate and one sticker will be issued. Further stickers will be issued as additional locator squares are claimed. The title of each award gives the number of locator squares and countries needed to qualify for the award. For example, to obtain the 144 MHz 40/10 award you must have QSL cards confirming contact with 40 locator squares including 10 countries on 144 MHz. The following awards are available:

70 MHz	20/4	144 MHz	40/10	432MHz	30/6
70 MHz	25/6	144 MHz	60/15	432 MHz	40/10
70 MHz	30/8	144 MHz	80/18	432 MHz	50/13
70 MHz	35/8	144 MHz	100/20	432 MHz	60/15
70 MHz	40/8	144 MHz	125/20	432 MHz	70/15
70 MHz	45/8	144 MHz	150/20	432 MHz	80/15
70 MHz	50/8	144 MHz	175/20	432 MHz	90/15
		144 MHz	200/30	432 MHz	100/15
		144 MHz	225/30	432 MHz	110/15
		144 MHz	250/35	432 MHz	120/18
		144 MHz	275/35	432 MHz	130/18
		144 MHz	300/40	432 MHz	140/20
		144 MHz	325/40	432 MHz	150/20
		144 MHz	350/45	432 MHz	160/20
		144 MHz	375/45	432 MHz	170/23
		144 MHz	400/50	432 MHz	180/25
		144 MHz	425/50		
		144 MHz	450/50		

Rules

1. All contacts must have been after 31 December 1978.
2. Eligible countries are those shown in the countries list printed elsewhere in this *Information Directory*.
3. Stations are eligible for awards in the following categories:
   (a) Fixed stations.
   (b) Portable stations, any location.
   (c) Mobile stations, any location.
   Categories cannot be mixed.
4. QSL cards submitted must be arranged in alphabetical order of the QTH squares claimed, and a checklist enclosed.

VHF Countries and Counties Awards

The following awards, intended to mark successful vhf/uhf achievements, are available:

Title of Award	Requirement Countries	Counties
50 MHz Standard Transmitting	12	40
50 MHz Senior Transmitting	20	60
50 MHz Standard Receiving	12	40
50 MHz Senior Receiving	20	60
70 MHz Standard Transmitting	3	30
70 MHz Senior Transmitting	6	60
70 MHz Standard Receiving	3	30
70 MHz Senior Receiving	6	60
144 MHz Standard Transmitting	9	40
144 MHz Senior Transmitting	15	60
144 MHz Standard Receiving	9	40
144 MHz Senior Receiving	15	60
432 MHz Standard Transmitting	3	20
432 MHz Senior Transmitting	9	40
432 MHz Standard Receiving	3	20
432 MHz Senior Receiving	9	40

1296 MHz Standard Transmitting	3	20
1296 MHz Senior Transmitting	6	40
1296 MHz Standard Receiving	3	20
1296 MHz Senior Receiving	6	40

Supreme Award (for fixed stations only) For holding: 3 Senior awards or 2 Senior + one 1296 MHz Awards.

Rules

1. All contacts must have been made after 1 January 1961, in respect of old UK counties, or after 1 January 1975, in respect of new counties. Scotland revisions with effect from 1 January 1976. Start date for 50 MHz contacts 1 June 1987.
2. Eligible counties are those listed overleaf.
3. Stations are eligible for awards in the following categories:
   a) Fixed stations;
   b) Portable stations (/P any location);
   c) Mobile stations (/M any location).
   Categories cannot be mixed.
4. Each different confirmed contact with a station in a Scottish region counts up to a maximum of three per region.

Microwave Award

The following awards, intended to mark achievement on the microwave bands, are available. Successful applicants will initially receive a certificate and one sticker; further stickers will be issued as later claims are received.

Locators:

Award	Two-way contact with QTH Locator squares
1.3 GHz / 5	5
1.3 GHz / 10	10
1.3 GHz / 15	15
1.3 GHz / 20	20
1.3 GHz / 25 etc. (up to 80)	80
2.3 GHz / 5	5
2.3 GHz / 10, 15, 20, 25 etc.	as 1.3 GHz
3.4 GHz / 5	5
3.4 GHz / 10, 15, 20, 25 etc.	as 1.3 GHz
5.7 GHz / 5	5
5.7 GHz / 10, 15, 20, 25 etc.	as 1.3 GHz
10 GHz / 5	5
10 GHz / 10, 15, 20, 25 etc.	as 1.3 GHz
24 GHz / 5	5
24 GHz / 10, 15, 20, 25 etc.	as 1.3 GHz

Countries + Counties:

Two-way contact with 3 countries and 20 UK counties on 1.3 GHz, 2.3 GHz, 3.4 GHz, 5.7 GHz, 10 GHz and 24 GHz. For the purposes of the Award a county is defined as that current at the time of introduction of the award.

Rules

1. All claims must be fully supported by QSL cards carrying the relevant IARU Locator information or Country-and-County information.
2. All contacts must be made after 31 December 1978.
3. Eligible countries are those listed overleaf.
4. Stations are eligible for awards in the following categories:
   a) Fixed stations;
   b) Portable and mobile stations. (The applicant must state that the operation was from one site, defined as being anywhere within a 5 km radius of the point.)
   Categories cannot be mixed.
5. QSL cards submitted should be listed and arranged in IARU QTH Locator alphabetical numeric order.

Microwave Distance Award

The following distance awards, intended to mark achievement on the microwave bands, are available.

1.3 GHz	for the first contact made beyond a distance of 600km
2.3 GHz	for the first contact made beyond a distance of 500km
3.4 GHz	for the first contact made beyond a distance of 400km
5.6 GHz	for the first contact made beyond a distance of 300km
10 GHz	for the first contact made beyond a distance of 150km (Basic Class)
10 GHz	for the first contact made beyond a distance of 300km (Intermediate Class)
10 GHz	for the first contact made beyond a distance of 600km (Advanced Class)
24 GHz	for the first contact made beyond a distance of 25km (Basic class)
24 GHz	for the first contact made beyond a distance of 75km (Intermediate class)
24 GHz	for the first contact made beyond a distance

Rules

1. Stations are eligible for awards in the following categories:
   a) Fixed stations;
   b) Portable stations (/P any location);
   c) Mobile stations (/M any location).

4-2-70 Counties List

COUNTY	CALLSIGN	DATE
Avon		
Bedfordshire		
Berkshire		
Buckinghamshire		
Cambridgeshire		
Cheshire		
Cleveland		
Cornwall & Isles of Scilly		
Cumbria		
Derbyshire		
Devon		
Dorset		
Durham		
East Sussex		
Essex		
Gloucestershire		
Greater London		
Greater Manchester		
Hampshire		
Hereford & Worcester		
Hertfordshire		
Humberside		
Isle of Wight		
Kent		
Lancashire		
Leicestershire		
Lincolnshire		
Merseyside		
Norfolk		
Northamptonshire		
Northumberland		
North Yorkshire		
Nottinghamshire		
Oxfordshire		
Shropshire		
Somerset		
South Yorkshire		
Staffordshire		
Suffolk		
Surrey		
Tyne and Wear		
Warwickshire		
West Midlands		
West Sussex		
West Yorkshire		
Wiltshire		
Clwyd		
Dyfed		
Gwent		
Gwynedd		
Mid Glamorgan		
Powys		
South Glamorgan		
West Glamorgan		
Isle of Man		
Alderney		
Guernsey		
Jersey		
Sark		

COUNTY	CALLSIGN	DATE
County Antrim		
County Armagh		
County Down		
County Fermanagh		
County Londonderry		
County Tyrone		
SCOTTISH REGIONS		
Borders		
Borders		
Borders		
Central		
Central		
Central		
Dumfries & Galloway		
Dumfries & Galloway		
Dumfries & Galloway		
Fife		
Fife		
Fife		
Grampian		
Grampian		
Grampian		
Highland		
Highland		
Highland		
Lothian		
Lothian		
Lothian		
Orkney		
Orkney		
Orkney		
Shetland		
Shetland		
Shetland		
Strathclyde		
Strathclyde		
Strathclyde		
Tayside		
Tayside		
Tayside		
Western Isles		
Western Isles		
Western Isles		

Please note: Each different confirmed contact with a station in a Scottish region counts up to a maximum of three per region.
VHF Award chasers should use the Country Checklist for international contacts, which is listed elsewhere in this information directory.

UK Amateur Radio Band Plans

1.8 MHz (160 m)

LICENCE NOTES:

Amateur Service: 1.810 - 1.850 MHz, Primary. Remainder secondary. *Available on the basis of non-interference to other services (inside or outside the UK)*
Satellite Service: No allocation
Power limit: 1.810 - 1.850 MHz: 26 dBW PEP. Remainder 15 dBW
Permitted modes: Morse, telephony, RTTY, data, fax, SSTV

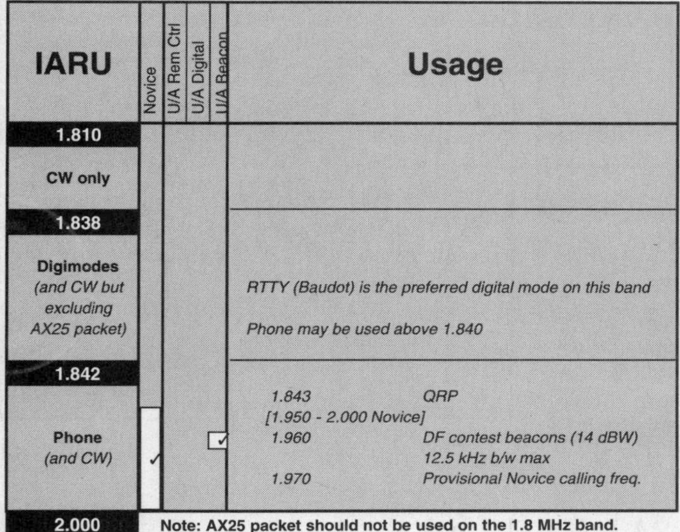

IARU	Novice	U/A Rem Ctrl	U/A Digital	U/A Beacon	Usage
1.810					
CW only					
1.838					
Digimodes *(and CW but excluding AX25 packet)*					*RTTY (Baudot) is the preferred digital mode on this band* *Phone may be used above 1.840*
1.842					
Phone *(and CW)*	✓			✓	*1.843 QRP* *[1.950 - 2.000 Novice]* *1.960 DF contest beacons (14 dBW) 12.5 kHz b/w max* *1.970 Provisional Novice calling freq.*
2.000					

Note: AX25 packet should not be used on the 1.8 MHz band.

3.5 MHz (80 m)

LICENCE NOTES:

Amateur Service: Primary, *Shared with other services*
Satellite Service: No allocation
Power limit: 26 dBW PEP
Permitted modes: Morse, telephony, RTTY, data, fax, SSTV
Unattended beacons: Only for DF contests Sat & Sun only, 14 dBW ERP PEP max

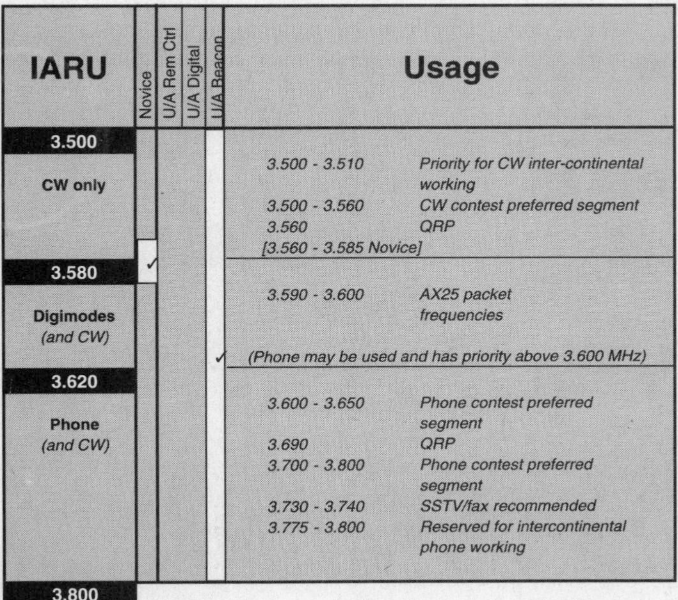

IARU	Novice	U/A Rem Ctrl	U/A Digital	U/A Beacon	Usage
3.500					
CW only	✓				*3.500 - 3.510 Priority for CW inter-continental working* *3.500 - 3.560 CW contest preferred segment* *3.560 QRP* *[3.560 - 3.585 Novice]*
3.580					
Digimodes *(and CW)*				✓	*3.590 - 3.600 AX25 packet frequencies* *(Phone may be used and has priority above 3.600 MHz)*
3.620					
Phone *(and CW)*					*3.600 - 3.650 Phone contest preferred segment* *3.690 QRP* *3.700 - 3.800 Phone contest preferred segment* *3.730 - 3.740 SSTV/fax recommended* *3.775 - 3.800 Reserved for intercontinental phone working*
3.800					

Novice Licence: powers and modes

The power levels shown in these band plans are for the full UK licences. Novice licensees are limited to 5 W DC input or 3 W RF output. Furthermore, the Novice licence schedule makes some restrictions on the modes which are permitted *within* the bands shown in these pages as being available to Novices. Please refer to the Amateur Radio Novice Licence and its schedule for full details.

7 MHz (40 m)

LICENCE NOTES:

Amateur Service: Primary
Satellite Service: Primary
Power limit: 26 dBW PEP
Permitted modes: Morse, telephony, RTTY, data, fax, SSTV

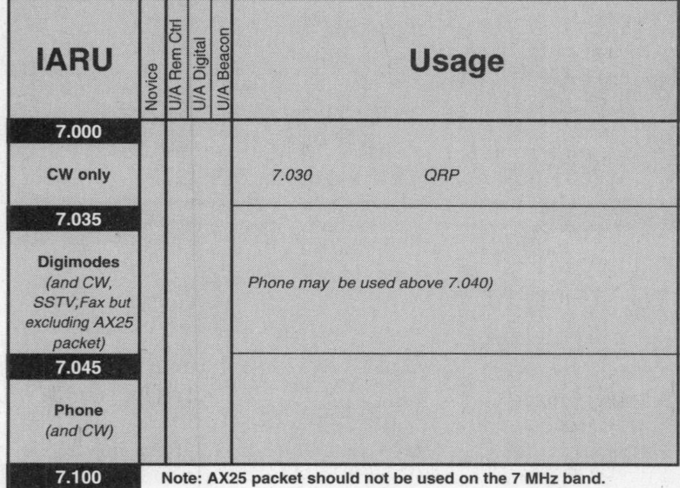

IARU	Novice	U/A Rem Ctrl	U/A Digital	U/A Beacon	Usage
7.000					
CW only					*7.030 QRP*
7.035					
Digimodes *(and CW, SSTV,Fax but excluding AX25 packet)*					*Phone may be used above 7.040)*
7.045					
Phone *(and CW)*					
7.100					

Note: AX25 packet should not be used on the 7 MHz band.

10 MHz (30 m)

LICENCE NOTES:

Amateur Service: Secondary
Satellite Service: No allocation
Power limit: 26 dBW PEP
Permitted modes: Morse, telephony, RTTY, data, fax, SSTV

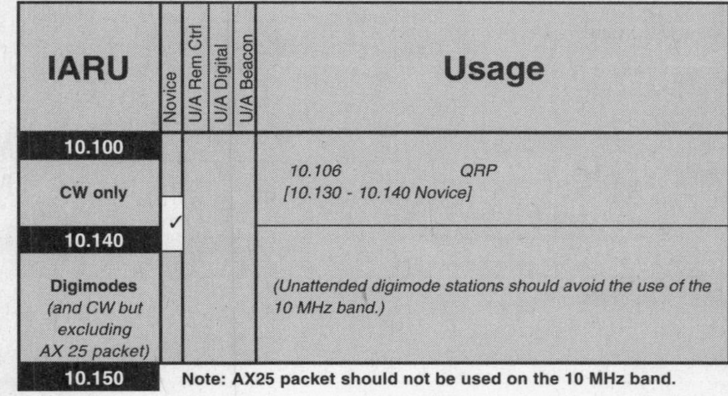

IARU	Novice	U/A Rem Ctrl	U/A Digital	U/A Beacon	Usage
10.100					
CW only	✓				*10.106 QRP* *[10.130 - 10.140 Novice]*
10.140					
Digimodes *(and CW but excluding AX 25 packet)*					*(Unattended digimode stations should avoid the use of the 10 MHz band.)*
10.150					

Note: AX25 packet should not be used on the 10 MHz band.

10 MHz Band Plan notes:

Note: The 10 MHz band is allocated to the amateur service only on a secondary basis. Therefore IARU have agreed on a worldwide basis that only CW and digimodes being narrow bandwidth modes, are to be used on this band. Likewise this band is not to be used for contests or news bulletins.

14 MHz (20 m)

LICENCE NOTES:

Amateur Service : Primary
Satellite Service : 14.000 - 14.250 MHz: Primary
Power limit: 26 dBW PEP
Permitted modes: Morse, telephony, RTTY, data, fax, SSTV

IARU	Novice	U/A Rem Ctrl	U/A Digital	U/A Beacon	Usage
14.000					
CW only					*14.060 QRP* *14.000 - 14.060 CW only contest preferred segment*
14.070					
Digimodes *(and CW)*					*14.089 - 14.099 No digimode mailbox or forwarding AX25 packet preferred frequencies*
14.099					
Beacons only					*14.099 - 14.101 Reserved exclusively for beacons*
14.101					
Digimodes *(+ phone & CW)*					*14.101 - 14.112 Digimode mailbox and forwarding AX25 packet preferred frequencies*
14.112					
Phone *(and CW)*					*14.125 - 14.300 SSB only contest preferred segment* *14.225 - 14.235 Used for SSTV/fax* *14.285 QRP*
14.350					

21 MHz (15 m)

LICENCE NOTES:

Amateur Service: Primary
Satellite Service : Primary
Power limit: 26 dBW PEP
Permitted modes: Morse, telephony, RTTY, data, fax, SSTV

IARU	Novice	U/A Rem Ctrl	U/A Digital	U/A Beacon	Usage
21.000					
CW only					*21.060 QRP*
21.080					
Digimodes *(and CW)*					*21.100 - 21.120 AX25 packet preferred* *[21.100 - 21.149 Novice]*
21.120					
CW only	✓				
21.149					
Beacons only					*21.149 - 21.151 Beacons exclusive*
21.151					
Phone *(and CW)*					*21.285 QRP* *21.335 - 21.345 Used for SSTV / fax*
21.450					

18 MHz (17 m)

LICENCE NOTES:

Amateur Service: Primary
Satellite Service: Primary
Power limit: 26 dBW PEP
Permitted modes: Morse, telephony, RTTY, data, fax, SSTV

IARU	Novice	U/A Rem Ctrl	U/A Digital	U/A Beacon	Usage
18.068					
CW only					
18.100					
Digimodes *(and CW)*					
18.109					
Beacons only					*18.109 -18.111 Exclusively beacons*
18.111					
Phone *(and CW)*					
18.168					

24 MHz (12 m)

LICENCE NOTES:

Amateur Service: Primary
Satellite Service: Primary
Power limit: 26 dBW PEP
Permitted modes: Morse, telephony, RTTY, data, fax, SSTV

IARU	Novice	U/A Rem Ctrl	U/A Digital	U/A Beacon	Usage
24.890					
CW only					
24.920					
Digimodes *(and CW)*					
24.929					
Beacons only					*24.929 - 24.931 Beacons exclusive*
24.931					
Phone *(and CW)*					
24.990					

Band Plans - *Simply being a good neighbour to your fellow amateur!*

28 MHz (10 m)

LICENCE NOTES:

Amateur Service:	Primary
Satellite Service:	Primary
Power limit:	26 dBW PEP
Permitted modes:	Morse, telephony, RTTY, data, fax, SSTV
Unattended beacons:	Only for DF contests (14 dBW PEP max)

IARU	Novice	U/A Rem Ctrl	U/A Digital	U/A Beacon	Usage
28.000					
CW only					
28.050					
Digimodes (and CW)	✓				*[28.060 - 28.190 Novice]* *28.060 QRP* *28.120 - 28.150 AX25 packet preferred*
28.150					
CW only					*28.190 - 28.199 Regional time shared International Beacon Project - Exclusive*
28.199					
Beacons only					*28.199 - 28.201 Worldwide time shared International Beacon Project - Exclusive*
28.201					
Phone (and CW)	✓			✓	*28.201 - 28.255 Continuous duty International Beacon Project - Exclusive* *[28.225 - 28.500 Novice]* *28.360 QRP* *28.675 - 28.685 Used for SSTV / fax*
29.200					
AX25 packet (+ phone and CW)					
29.300					
Satellite downlinks					*29.300 - 29.500 Reserved exclusively for satellite downlinks*
29.550					
Phone (and CW)					*Some experimental FM repeaters may be established in IARU Region 1*
29.700					

50 MHz (6 m)

LICENCE NOTES:

Amateur Service:	50.0 - 51.0 MHz, Primary; 51.0 - 52.0 MHz, Secondary. *Available on the basis of non-interference to other services (inside or outside the UK).*
Satellite Service :	No allocation
Power limit:	50.0-51.0MHz, 26dBW PEP; 51.0-52.0MHz, 20 dBW PEP
Permitted modes:	Morse, telephony, RTTY, data, fax, SSTV

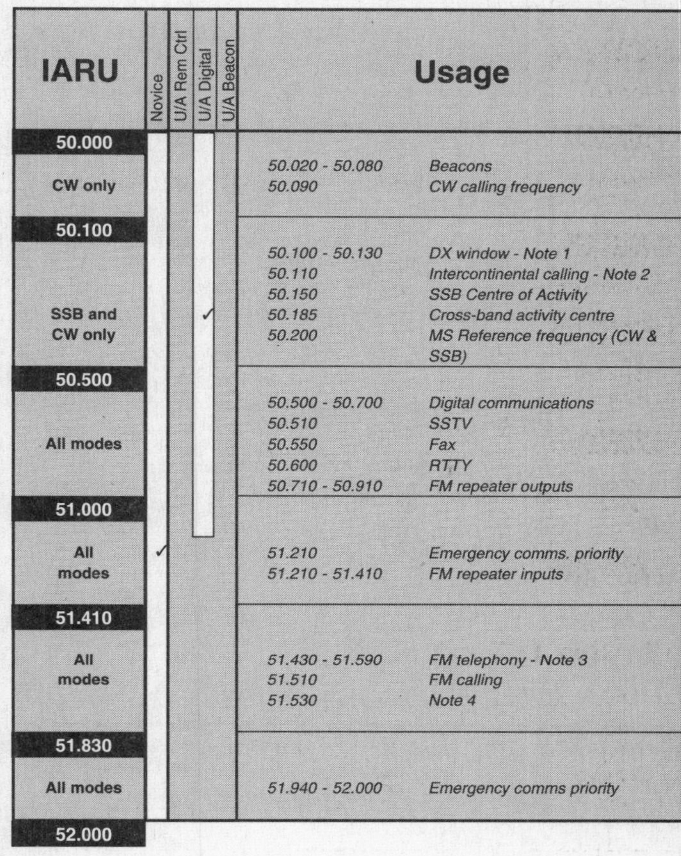

IARU	Novice	U/A Rem Ctrl	U/A Digital	U/A Beacon	Usage
50.000					
CW only					*50.020 - 50.080 Beacons* *50.090 CW calling frequency*
50.100					
SSB and CW only			✓		*50.100 - 50.130 DX window - Note 1* *50.110 Intercontinental calling - Note 2* *50.150 SSB Centre of Activity* *50.185 Cross-band activity centre* *50.200 MS Reference frequency (CW & SSB)*
50.500					
All modes					*50.500 - 50.700 Digital communications* *50.510 SSTV* *50.550 Fax* *50.600 RTTY* *50.710 - 50.910 FM repeater outputs*
51.000					
All modes	✓				*51.210 Emergency comms. priority* *51.210 - 51.410 FM repeater inputs*
51.410					
All modes					*51.430 - 51.590 FM telephony - Note 3* *51.510 FM calling* *51.530 Note 4*
51.830					
All modes					*51.940 - 52.000 Emergency comms priority*
52.000					

50 MHz Band Plan notes:

1. Only to be used for QSOs between stations in different continents.
2. No QSOs on this frequency. Always QSY when working intercontinental DX.
3. 20 kHz channel spacing. Channel centre frequencies start at 51.430 MHz.
4. Used by GB2RS news and for slow Morse transmissions.

Notes to the HF Band Plans

1. The expression "phone" includes all permitted forms of telephony.
2. If transmitting very close to a band edge, take care not to radiate outside of the band.
3. Before transmitting, all operators should check that the frequency is not already occupied. The normal advice is to use the phrase "Is this frequency in use?" on SSB or "QRL?" on CW.
4. Digimodes are defined as including: AmTOR, PacTOR, Clover, ASCII, RTTY (Baudot) and AX25 packet.
5. LSB is recommended on bands below 10 MHz, and USB recommended on bands above 10 MHz.
6. The Region 1 IARU HF band plans are designed to enable the best utilisation of the HF spectrum space available. They achieve this objective because the vast majority of licensed amateurs observe the voluntary recommendations. In some countries (e.g. the USA) licence regulations require that specific modes be confined to specific sections of each band.

Notes on the VHF Band Plans

1. The beacon and satellite services must be kept free of normal communication transmissions to prevent interference with these services.
2. The use of the FM mode within the SSB / CW section and CW and SSB in the FM-only sector is not recommended.
3. Repeater stations are primarily intended as an aid for mobile working and they are not intended to be used for DX communication. FM stations wishing to work DX should use the all-modes section, taking care to avoid frequencies allocated for specific purposes.

70 MHz (4 m)

LICENCE NOTES:

Amateur Service:	Secondary. Available on the basis of non-interference to other services (inside or outside the UK).
Satellite Service:	No allocation
Power limit:	22 dBW PEP
Permitted modes:	Morse, telephony, RTTY, data, fax, SSTV

IARU	Novice	U/A Rem Ctrl	U/A Digital	U/A Beacon	Usage	
70.000						
Beacons					*70.030*	*Personal beacons*
70.030						
SSB and CW only					*70.150*	*Meteor scatter calling*
					70.185	*Cross-band activity centre*
					70.200	*SSB / CW calling*
70.250						
All modes					*70.260*	*AM / FM calling*
70.300						
Channelised operation using 12.5 kHz channels		✓	✓	✓	*70.3000*	*RTTY / fax*
					70.3125	*Packet radio*
					70.3250	*Packet radio*
					70.3375	*Packet radio*
					70.3500	*Emergency comms priority*
					70.3625	
					70.3750	*Emergency comms priority*
					70.3875	
					70.4000	*Emergency comms priority*
					70.4125	
					70.4250	
					70.4375	
					70.4500	*FM calling*
					70.4625	
					70.4750	
			✓		*70.4875*	*Packet radio*
70.500						

144 MHz (2 m)

LICENCE NOTES:

Amateur Service:	Primary
Satellite Service:	Primary
Power limit:	26 dBW PEP
Permitted modes:	Morse, telephony, RTTY, data, fax, SSTV
Unattended beacons:	Only for DF Contests

IARU	Novice	U/A Rem Ctrl	U/A Digital	U/A Beacon	Usage		
144.000							
EME (SSB/CW)					*144.000-144.035*	*Moonbounce (only)*	
144.035							
CW only					*144.050*	*CW calling frequency*	
					144.100	*MS CW ref frequency (Note 1)*	
					144.140 - 144.150	*CW FAI working*	
144.150			✓	✓	*144.150 - 144.160*	*SSB FAI working*	
SSB and CW only					*144.175*	*Microwave talk-back (UK)*	
					144.195 - 144.205	*SSB random MS*	
					144.250	*GB2RS and slow Morse*	
					144.260	*Emerg. comms priority*	
					144.300	*SSB calling frequency*	
					144.390 - 144.400	*SSB random MS*	
144.400							
Beacons					*144.490*	*SAREX uplink*	
144.490							
Guard band							
144.500							
All modes non-channel-ised					*144.500*	*SSTV calling frequency*	
					144.525	*ATV talkback (SSB)*	
					144.600	*RTTY calling frequency*	
					144.600 ±	*RTTY working (fsk)*	
					144.625 - 144.650	*Emergency comms priority*	
					144.700	*Fax calling frequency*	
					144.750	*ATV calling+talk-back*	
					144.775 - 144.800	*Emergency comms priority*	
144.800							
Digital Modes					*144.800 - 144.990*	*Digital Modes (inc unattended)*	
144.990							
Guard band							
145.000			✓	✓	*145.000*	*RV48 (R0)*	
FM Repeater Inputs					*145.025*	*RV50 (R1)*	
					145.050	*RV52 (R2)*	
					145.075	*RV54 (R3)*	*(Note 2)*
					145.100	*RV56 (R4)*	
					145.125	*RV58 (R5)*	
					145.150	*RV60 (R6)*	
					145.175	*RV62 (R7)*	
145.200					*145.200*	*V16 (S8)*	*Emergency comms priority*
FM Simplex Channels					*145.225*	*V18 (S9)*	*Emergency comms priority*
					145.250	*V20 (S10)*	*Used for slow Morse transmissions*
					145.275	*V22 (S11)*	
					145.300	*V24 (S12)*	*RTTY afsk*
					145.325	*V26 (S13)*	
					145.350	*V28 (S14)*	
					145.375	*V30 (S15)*	*(Note 2)*
					145.400	*V32 (S16)*	
					145.425	*V34 (S17)*	
					145.450	*V36 (S18)*	
					145.475	*V38 (S19)*	
					145.500	*V40 (S20)*	*FM calling channel*
					145.525	*V42 (S21)*	*Used for GB2RS*
					145.550	*V44 (S22)*	*Recommended channel for rally & exhibition talk-in*
					145.575	*V46 (S23)*	
145.600					*145.600*	*RV48 (R0)*	
FM Repeater Outputs					*145.625*	*RV50 (R1)*	
					145.650	*RV52 (R2)*	
					145.675	*RV54 (R3)*	*(Note 2)*
					145.700	*RV56 (R4)*	
					145.725	*RV58 (R5)*	
					145.750	*RV60 (R6)*	
					145.775	*RV62 (R7)*	
145.800							
Satellites							
146.000							

Continued in next column

144 MHz Band Plan notes:

1. Meteor scatter operation can take place up to 26 kHz higher than the reference frequency.
2. Additional 12.5 kHz channels will be phased in by the year 2000 (see *RadCom* March 1997 page 16).

UK Repeater CTCSS Tones

A number of UK 2m and 70 cms repeaters now use CTCSS tones on a regional basis to help minimise unwanted access to other co-channel repeaters.

Tone Area	CTCSS Tone (Hz)	Tone Area	CTCSS Tone (Hz)
A	67.1	F	94.8
B	71.9	G	103.5
C	77.0	H	110.9
D	82.5	J	118.8
E	88.5		

430 MHz (70 cm)

LICENCE NOTES:

Amateur Service:	Secondary
Satellite Service:	435-438 MHz, Secondary
Exclusion:	431 - 432 not available for use within 100 km radius of Charing Cross, London. (51° 30' 30"N, 00° 7' 24"W)
Power limit:	430 - 432 MHz: 16 dBW ERP PEP, 432 - 440 MHz: 26 dBW
Permitted modes:	Morse, telephony, RTTY, data, fax, SSTV, FSTV

IARU	Novice	U/A Rem Ctrl	U/A Digital	U/A Beacon	Usage
430.000 All modes					*430.000 - 430.810 Digital communications (Notes 6, 7)* *430.600 - 430.800 Note 5*
430.810 Low power repeater i/p Note 1					*430.810 - 430.990 Low power repeaters*
431.000 All modes Note 1					*430.990 - 431.900 Digital communications (Note 6)*
432.000 CW only					*432.000 - 432.025 Moonbounce* *432.050 CW centre of activity*
432.150 SSB and CW only					*432.200 SSB centre of activity* *432.350 Microwave talk-back calling frequency (Europe)*
432.500 All modes non-chanelised		✓	✓		*432.500 - 432.600 IARU Region 1 linear transponder outputs* *432.600 - 432.800 IARU Region 1 linear transponder inputs* *432.500 SSTV activity centre* *432.600 RTTY (fsk) activity centre* *432.625 Packet radio* *432.650 Packet radio* *432.675 Packet radio* *432.700 Fax activity centre*
432.800 Beacons	✓				*432.800 - 432.990 Beacons*
433.000 FM repeater outputs in UK only Note 1					*433.000 RB0* *433.025 RB1* *433.050 RB2* *433.075 RB3* *433.100 RB4* *433.125 RB5* *433.150 RB6* *433.175 RB7* *433.200 RB8* *433.225 RB9* *433.250 RB10* *433.275 RB11* *433.300 RB12* *433.325 RB13* *433.350 RB14* *433.375 RB15*
433.400					

Continued in next column

IARU	Novice	U/A Rem Ctrl	U/A Digital	U/A Beacon	Usage
433.400 FM simplex channels			✓		*433.400 SU16* *433.425 SU17* *433.450 SU18* *433.475 SU19* *433.500 SU20 FM calling channel* *433.525 SU21* *433.550 SU22 Recommended channel for rally and exhibition talk-in* *433.575 SU23* *433.600 SU24 RTTY afsk* *433.625 Packet radio* *433.650 Packet radio* *433.675 Packet radio* *433.700 Notes 2, 3 and 5* *433.725 Notes 2 and 5* *433.750 Notes 2 and 5* *433.775 Notes 2 and 5*
434.600 FM repeater inputs (in UK only) - note 1; and fast scan television - note 4					*434.600 RB0* *434.625 RB1* *434.650 RB2* *434.675 RB3* *434.700 RB4* *434.725 RB5* *434.750 RB6* *434.775 RB7* *434.800 RB8* *434.825 RB9* *434.850 RB10* *434.875 RB11* *434.900 RB12* *434.925 RB13* *434.950 RB14* *434.975 RB15*
435.000 Satellites and fast scan TV - note 4	✓				
438.000 Fast scan TV					*438.025 - 438.175 Note 5* *438.200 - 439.425 Note 1*
438.425 Low power repeater o/p + fast scan TV					*438.425 R61* *438.450 R62* *438.475 R63* *438.500 R64* *438.525 R65* *438.550 R66* *438.575 R67*
438.575 Fast Scan TV					*438.200 - 439.425 Note 1* *439.440 - 439.575 Peripheral repeater inputs* *439.600 - 439.750 Digital communications (Note 6)*
439.750 Packet radio					*439.750 - 440.000 Digital communications (Note 6)*
440.000					

430 MHz Band Plan notes:

1. In Switzerland, Germany and Austria, repeater inputs are 430.600 - 431.825 MHz with 25 kHz spacing, and outputs are 438.200 - 439.425 MHz. In France and the Netherlands repeater outputs are 430.025 - 430.375 MHz with 25 kHz spacing and inputs at 431.625 - 431.975 MHz. In other European countries repeater inputs are 433.000 - 433.375 MHz with 25 kHz spacing and outputs at 434.600 - 434.975 MHz ie the reverse of the UK allocation.
2. Emergency communications priority.
3. IARU Region 1 fax / afsk.
4. Fast Scan Television carrier frequencies shall be chosen so as to avoid interference to other users, in particular the satellite service and repeater inputs. IARU Region 1 recommends that video carriers should be in the range 434.000 - 434.500 MHz or 438.500 - 440.000 MHz.
5. IARU Region 1 packet radio.
6. The DCC will recommend usage of this sub-band at a later date.
7. Users must accept interference from F/PA repeater output channels in 430.025 to 430.375MHz. Users with sites which allow propagation to other countries (notably F and PA) must survey the proposed frequency before use to ensure that they will not cause interference to users of repeaters in those countries.

1.3 GHz (23 cm)

LICENCE NOTES:

Amateur Service:	Secondary
Satellite Service:	1260 - 1270, Secondary *Earth to space only*
	1296 - 1297, Secondary *Earth to space only*
Power limit:	26 dBW PEP
Permitted modes:	Morse, telephony, RTTY, data, fax, SSTV, FSTV
Unattended operation:	Not permitted in Northern Ireland

IARU	Novice	U/A Rem Ctrl	U/A Digital	U/A Beacon	Usage	
1,240.000						
All modes					*1240.150*	*Packet radio (150 kHz b/w)*
					1240.300	*Packet radio (150 kHz b/w)*
					1240.450	*Packet radio (150 kHz b/w)*
					1240.600	*Packet radio (150 kHz b/w)*
					1240.750	*Packet radio (150 kHz b/w)*
1,243.250						
ATV					*1248.000*	*RT1-3 FM TV input*
					1249.000	*RT1-2 FM TV input*
1,260.000						
Satellites						
1,270.000						
All modes						
1,272.000						
ATV					*1276.500*	*RT1-1 AM TV input*
1,291.000	✓					
Repeater inputs					*1291.000*	*RM0* *(UK) 25 kHz spacing*
					1291.375	*RM15*
1,291.500						
All modes						
1,296.000						
CW only					*1296.000-1296.025*	*Moonbounce*
1,296.150						
SSB and CW					*1296.200*	*Narrow band centre of activity*
					1296.400- 1296.600	*Linear transponder input*
					1296.500	*SSTV*
					1296.600	*RTTY*
					1296.700	*Fax*
					1296.600- 1296.800	*Linear transponder output*
1,296.800						
Beacons exclusive					*1296.800- 1296.990*	*Beacons*
1,297.000						
Repeater outputs - note 1					*1297.000*	*RM0* *(UK) 25 kHz spacing*
					1297.375	*RM15*
1,297.500						
FM simplex - note 1					*1297.500*	*SM20*
					1297.750	*SM30*
1,298.000						
All modes		✓		✓		*Digital communications*
1,298.500						

Continued in next column

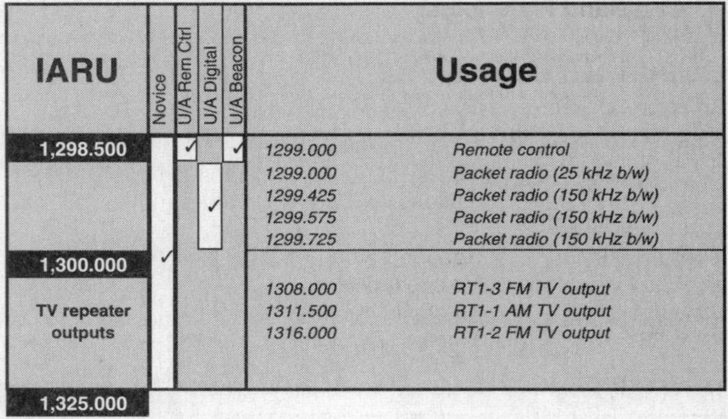

IARU	Novice	U/A Rem Ctrl	U/A Digital	U/A Beacon	Usage	
1,298.500		✓		✓	*1299.000*	*Remote control*
					1299.000	*Packet radio (25 kHz b/w)*
			✓		*1299.425*	*Packet radio (150 kHz b/w)*
					1299.575	*Packet radio (150 kHz b/w)*
					1299.725	*Packet radio (150 kHz b/w)*
1,300.000	✓					
TV repeater outputs					*1308.000*	*RT1-3 FM TV output*
					1311.500	*RT1-1 AM TV output*
					1316.000	*RT1-2 FM TV output*
1,325.000						

1.3 GHz Band Plan notes:

1. Local traffic using narrow-band modes should operate between 1296.500 - 1296.800 MHz during contests and band openings.
2. Stations in countries which do not have access to 1298 - 1300 MHz (eg Italy) may also use the FM simplex segment for digital communications.

2.3 GHz (13 cm)

LICENCE NOTES:

Amateur Service:	Secondary. *Users must accept interference from ISM users*
Satellite Service:	2400 - 2450, Secondary. *Users must accept interference from ISM users.*
Power limit:	26 dBW PEP
Permitted modes:	Morse, telephony, RTTY, data, fax, SSTV, FSTV

ISM = Industrial Scientific and Medical

IARU	Novice	U/A Rem Ctrl	U/A Digital	U/A Beacon	Usage	
2,310.000						
Sub-regional (national band plans)					*2310.000 - 2310.500*	*Repeater links*
					2310.100	*Packet radio (200 kHz b/w)*
					2310.300	*Packet radio (200 kHz b/w)*
					2310.000 - 2310.500	*Remote control*
2,320.000						
CW exclusive					*2320.000 - 2320.025*	*Moonbounce*
2,320.150						
CW and SSB					*2320.200*	*SSB centre of activity*
2,320.800						
Beacons exclusive		✓	✓	✓	*2320.800 - 2320.990*	*Beacons*
2,321.000						
Simplex & repeaters (FM) - note 1						
2,322.000						
All modes					*2322.000 - 2355.000*	*ATV*
					2355.100 - 2364.000	*Repeater links*
					2355.100	*Packet radio (200 kHz b/w)*
					2355.300	*Packet radio (200 kHz b/w)*
					2364.000	*Packet radio (1 MHz b/w)*
					2365.000 - 2370.000	*Repeaters*
					2370.000 - 2390.000	*ATV*
					2390.000 - 2392.000	*Moonbounce*
2,400.000						
Satellites						
2,450.000						

Notes continued in next column

2.3 GHz Band Plan notes:

1. Stations in countries which do not have access to the All Modes section (2,322 - 2,390 MHz), use the simplex and repeater segment 2,321 - 2,322 MHz for data transmission
2. Stations in countries which do not have access to the narrow band segment 2,320 - 2,322 MHz, use alternative narrow band segments: 2,304 - 2,306 MHz and 2,308 - 2,310 MHz.

3.4 GHz (9 cm)

LICENCE NOTES:

Amateur Service:	Secondary
Satellite Service:	No allocation
Power limit:	26 dBW PEP
Permitted modes:	Morse, telephony, RTTY, data, fax, SSTV, FSTV

IARU	Novice	U/A Rem Ctrl	U/A Digital	U/A Beacon	Usage	
3,400.000						
Narrow band CW/EME/SSB					*3400.100* *3400.800 - 3401.000* *3401.000 - 3402.000*	*Centre of activity* *Beacons* *Remote control*
3,402.000						
All modes		✓	✓	✓		
3,456.000						
Narrow band CW/EME/SSB					*3456.000*	*EME to USA*
3,458.000						
All modes						
3,475.000						

5.7 GHz (6 cm)

LICENCE NOTES:

Amateur Service :	5,650 - 5,680, Secondary; 5,755 - 5,765 + 5820 - 5850: Secondary. Users must accept *interference from ISM users*
Satellite Service:	5,650 - 5,670 Secondary *Earth to Space only;* 5,830 - 5,850 Secondary *Users must accept interference from ISM users Space to Earth only*
Power limit:	26 dBW PEP
Permitted modes:	Morse, telephony, RTTY, data, fax, SSTV, FSTV

ISM = Industrial, Scientific & Medical

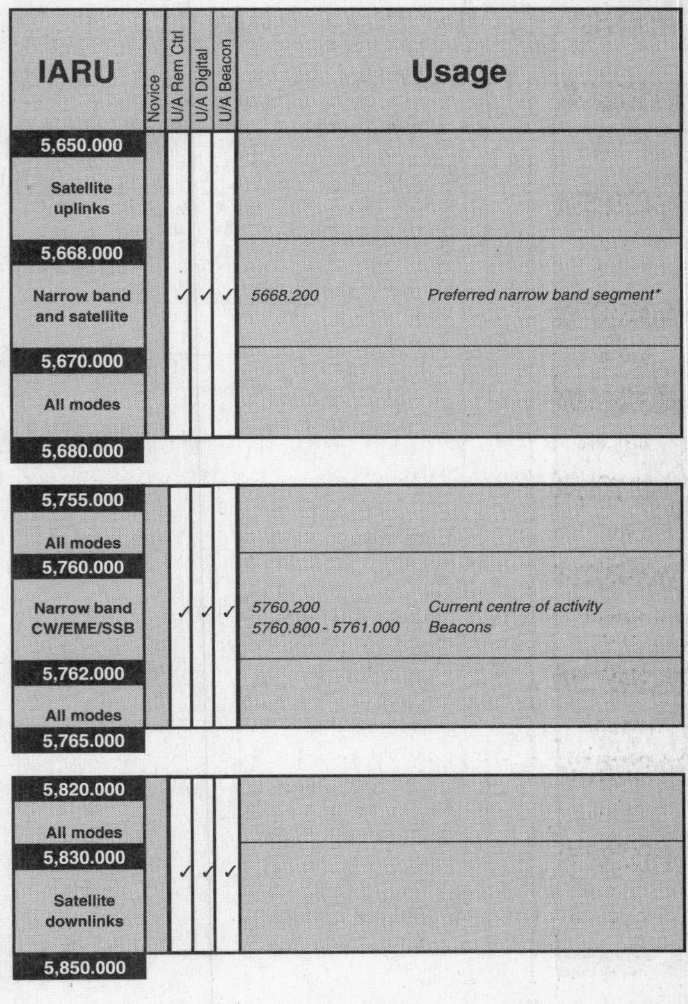

IARU	Novice	U/A Rem Ctrl	U/A Digital	U/A Beacon	Usage	
5,650.000						
Satellite uplinks						
5,668.000						
Narrow band and satellite		✓	✓	✓	*5668.200*	*Preferred narrow band segment**
5,670.000						
All modes						
5,680.000						
5,755.000						
All modes						
5,760.000						
Narrow band CW/EME/SSB		✓	✓	✓	*5760.200* *5760.800 - 5761.000*	*Current centre of activity* *Beacons*
5,762.000						
All modes						
5,765.000						
5,820.000						
All modes						
5,830.000						
Satellite downlinks		✓	✓	✓		
5,850.000						

* IARU aim to move narrow band operation to this segment, but for the time being operation will continue in the 5760 - 5762 band.

Unattended (U/A) Operation

Frequencies on which unattended (U/A) operation is permitted by full licensees are shown in these band plans. Novice licensees can also operate their stations unattended but the frequencies and powers are different – please see the Novice licence for the details. Remember that unattended operation requires the prior consent of the local Radio Investigation Service before operation can begin, to enable close down arrangements to be made.

Unattended beacons are limited to 14dBW ERP max. Do not confuse this type of unattended beacon operation with the normal beacon sections of the bands (these are fully site cleared, have special licences and are co-ordinated on an international basis.

Unattended low power remote control is limited to -20 dBW ERP and should not radiate outside the boundary of the premises from which you are operating.

Unattended digital operation is limited to 10 dBW on the 50 MHz band and 14 dBW on the other bands where it is permitted.

IARU – International Amateur Radio Union

As the RSGB represents the interests of radio amateurs within the UK, so the International Amateur Radio Union (IARU) represents amateur radio on an international scale. Its membership is made up of national societies rather than individuals and it has more than 140 member societies. The RSGB is the UK's IARU member society. The IARU was founded in 1925 and has its headquarters in the USA. It is divided into three regions as is the International Telecommunications Union (ITU). Region 1 comprises the UK, Europe, Africa, the CIS and the Middle East.

The aim of the IARU is to promote, preserve and protect world-wide growth in amateur radio and where necessary represent the movement's interests at the ITU. It also regulates and co-ordinates band plans, and makes recommendations for the operation of specialised activities such as meteor scatter.

Another service provided is the Monitoring System (IARUMS) which monitors unauthorised transmissions by other services within the amateur bands. Reports from the IARUMS are sent to both the ITU and national telecommunication administrations.

10 GHz (3 cm)

LICENCE NOTES:

Amateur Service: Secondary
Satellite Service: 10,450 - 10,500: Secondary
Power limit: 26 dBW PEP
Permitted modes: Morse, telephony, RTTY, data, fax, SSTV, FSTV

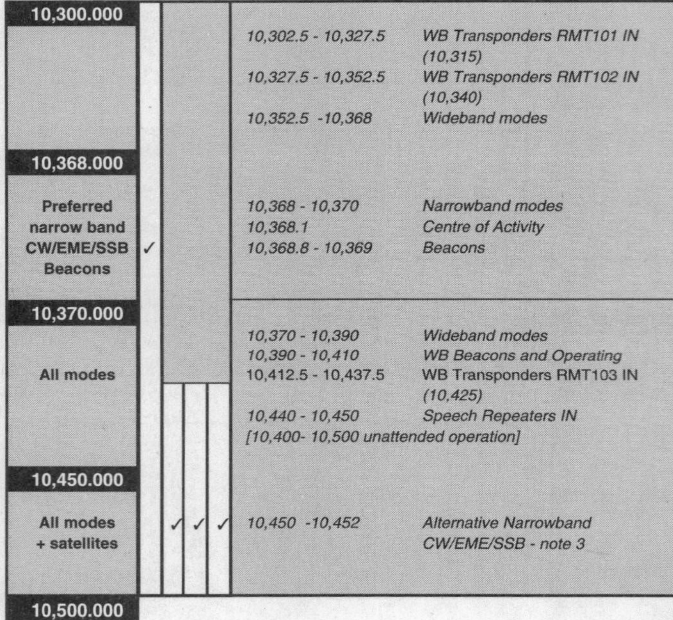

IARU	Novice	U/A Rem Ctrl	U/A Digital	U/A Beacon	Frequency	Usage
10,000.000						
All modes (ATV, data FM simplex, duplex and repeaters)	✓	✓	✓	✓	*10,002.5 - 10,027.5*	*WB Transponders or WB Simplex (10,015)*
					10,027.5 - 10,052.5	*WB Transponders RMT101 OUT (10,040)*
					10,052.5 - 10,077.5	*WB Transponders RMT102 OUT (10,065)*
					10,077.5 - 10,090	*Packet Links*
					10,090 - 10,110	*Wideband Beacons and Operating*
					10,110 - 10,127.5	*Speech Repeaters OUT*
					10,127.5 - 10,147.5	*WB Transponders RMT103 OUT (10,135)*
10,150.000						
10,300.000						
					10,302.5 - 10,327.5	*WB Transponders RMT101 IN (10,315)*
					10,327.5 - 10,352.5	*WB Transponders RMT102 IN (10,340)*
					10,352.5 -10,368	*Wideband modes*
10,368.000						
Preferred narrow band CW/EME/SSB Beacons	✓				*10,368 - 10,370*	*Narrowband modes*
					10,368.1	*Centre of Activity*
					10,368.8 - 10,369	*Beacons*
10,370.000						
All modes					*10,370 - 10,390*	*Wideband modes*
					10,390 - 10,410	*WB Beacons and Operating*
					10,412.5 - 10,437.5	WB Transponders RMT103 IN (10,425)
					10,440 - 10,450	*Speech Repeaters IN*
					[10,400- 10,500 unattended operation]	
10,450.000						
All modes + satellites		✓	✓	✓	*10,450 -10,452*	*Alternative Narrowband CW/EME/SSB - note 3*
10,500.000						

10 GHz Band Plan notes:

1. 10,400 is the preferred frequency for wideband beacons, but 10,100 is still used.
2. Wideband FM is preferred around 10,350-10400 to encourage compatibility with narrowband systems; however, there is still activity around 10,050-10,150.
3. The current NB sub-band is at 10,368, however, a sub-band at 10,450 is being considered as a possible future alternative.
4. Simplex TV operation should take place on RMT inputs which are not used by local transponders.

24 GHz (12 mm)

LICENCE NOTES:

Amateur Service: 24,000 - 24,050 Primary. *Users must accept interference from ISM users*; 24,050 - 24,150 Secondary, *May only be used with the written consent of the Secretary of State. Users must accept intereference from ISM users*; 24,150 - 24,250 Secondary. *Users must accept interference from ISM users.*
Satellite Service: 24,000 - 24,050 Primary. *Users must accept interference from ISM users*
Power limit: 26 dBW PEP
Permitted modes: Morse, telephony, RTTY, data, fax, SSTV, FSTV

ISM = Industrial, Scientific & Medical

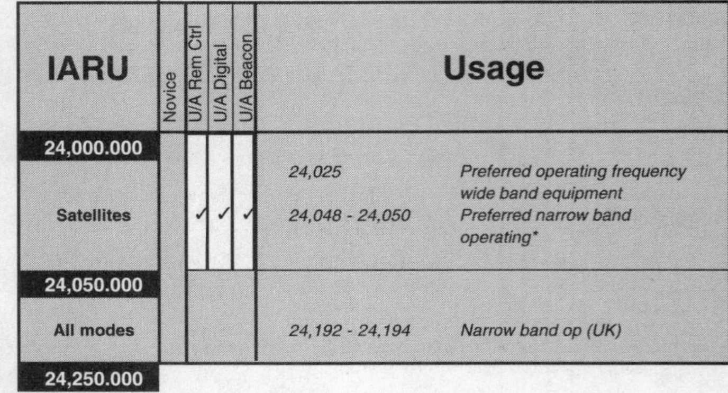

IARU	Novice	U/A Rem Ctrl	U/A Digital	U/A Beacon	Frequency	Usage
24,000.000						
Satellites		✓	✓	✓	*24,025*	*Preferred operating frequency wide band equipment*
					24,048 - 24,050	*Preferred narrow band operating**
24,050.000						
All modes					*24,192 - 24,194*	*Narrow band op (UK)*
24,250.000						

* Will eventually be used if and when allocation changes force this.

47 GHz (6 mm)

LICENCE NOTES:

Amateur Service: Primary
Satellite Service: Primary
Power limit: 26 dBW PEP
Permitted modes: Morse, telephony, RTTY, data, fax, SSTV, FSTV

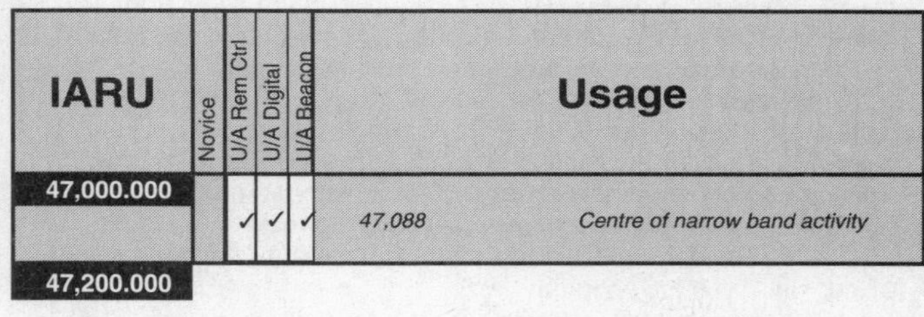

IARU	Novice	U/A Rem Ctrl	U/A Digital	U/A Beacon	Frequency	Usage
47,000.000						
		✓	✓	✓	*47,088*	*Centre of narrow band activity*
47,200.000						

Other amateur bands allocated in the UK are: 71.6 - 74.4kHz (Notice of Variation only) and 75.5 - 76.0, 142.0 - 144.0, 248.0 - 250.0 GHz.

Beacons

Beacons are intended mainly as propagation indicators although, especially on the microwave bands, they may also serve as signal sources for alignment purposes. The table lists a selection, some of which may be heard regularly in the UK, so that variation in strength gives an indication of conditions; others may be heard occasionally, and the appearance of one can indicate exceptional propagation. For example, the 144 MHz beacon GB3VHF can always be heard over much of the UK so, if its strength is above average, then there is a 'lift' on. If the 50 MHz beacon in Newfoundland, not usually audible in the UK, appears then there is a path to North America. Conversely, the 28 MHz beacon GB3RAL is of little interest to UK stations but can indicate to overseas operators the presence of propagation to the UK.

Some 20 years ago, in order to avoid interference between 144 MHz beacons in Europe, the RSGB was asked by the International Amateur Radio Union (IARU) to co-ordinate their frequencies. The Society has done so since and has extended the service to other bands above 30 MHz

On HF, beacons are co-ordinated by IARU. At 28 MHz, 28.190 - 28.199 is reserved for regional networks, 28.200 is shared by the International Beacon Project beacons and 28.201 - 28.225 is allocated to approved continuous cycle beacons. The RSGB beacon co-ordinators are: Martin Harrison, G3USF, for HF RSGB/IARU beacons, John Wilson, G3UUT, for 50, 70, 144, 432MHz RSGB/IARU and Graham Murchie, G4FSG, for RSGB microwave ones. Thanks are due this year to GJ4ICD of the UK 6m group for details on the many 50 MHz beacons.

For information on a proposed UK beacon sending solar and geomagnetic information, please see the propagation section in this *Yearbook*.

Interested in setting up a beacon?

The UK licence does permit a private station to operate as a low-power unattended beacon, but only in some bands above 2.3 GHz, together with 70 MHz and part of 432MHz. Establishing a permanent reliable beacon at a remote site can be a complex undertaking, maybe more suited to a group than an individual. Site clearance by the RA is required before a licence (GB3 + three letters) can be issued. Full details are given in *Guide to Beacon Licensing* available from RSGB HQ on receipt of a large SASE.

Freq MHz	Call	Nearest town	IARU loc	ASL m	Aerial	Beam Direction	ERP W	Info from	Notes
1.801	PY2AMI	Americana SP	GG67IF		Inv. Vertical		5		Operational ?
1.814	NQ3G	Berwick PA	FN11UB						?
1.817	ZS1J	Plettenberg Bay	KF15PF	1	½ Wave Dipole	90°/270°	1	ZS1J	24
1.840	OK0EM	Kromeriz	JN89		LW		5	OK1ADM	Non-op
3.525	PY2AMI	Americana SP	GG67IF		Inv. Vertical		5		Operational ?
3.557	DK0WCY ‡	Scheggerott	JO44VQ		Dipole		30	DL1FL	0700-0800 1530-1700zz
3.600	OK0EN	Kam. Zehrovice	JO70AC		Dipole	90°/270°	150	OK1ADM	24
3.699	VK2RCW	Turramurra	QF56					WIA	24
5.471	LN2A †	Sveio	JO29PO		Vertical Monop	Omni	550	LA9AHA	ITU beacon
5.471	VL8IPS	Nr Darwin NT	PH57PJ		Vertical Monop	Omni	2		ITU 24+
7.025	AS1RMS	George Airport	KF16EA		Dipole	E-W	0.5		24
7.048	PY2AMI	Americana SP	GG67IF		Inv. Vertical		5		Opertational ?
7.049	PY7BCN	Fortaleza CE	HI06RF						?
7.871	LN2A †	Sveio	JO29PO		Vertical Monop	Omni	700	LA9AHA	ITU beacon
7.871	VL8IPS	Nr Darwin NT	PH57PJ		Vertical Monop	Omni	2		ITU 24+
10.144	DK0WCY	Scheggerott	JO44VQ		Horizontal Loop	Omni	30	DL1FL	24
10.408	LN2A †	Sveio	JO29PO		Vertical Monop	Omni	750	LA9AHA	ITU beacon
10.408	VL8IPS	Nr Darwin NT	PH57PJ		Vertical Monop	Omni	2		ITU 24+
14.070	PY2AMI	Americana SP	GG67IF		Ground Plane	Omni	5		Operational ?
14.100	JA2IGY	Mt Asama	PM84JK	540	Vertical	Omni	0-1-100	W6ISQ	IBP cycle
14.100	KH6WO	Honolulu	BL11BK	8	Vertical	Omni	0-1-100	W6ISQ	IBP cycle
14.100	LU4AA	Buenos Aires	GF05		Vertical	Omni	0-1-100	W6ISQ	IBP cycle
14.100	OH2B	Espoo	KP20KE		Vertical	Omni	0-1-100	W6ISQ	IBP cycle
14.100	W6WX	San Jose, CA	CM87		Vertical	Omni	0-1-100	W6ISQ	IBP cycle
14.100	4U1UN	United Nations NY	FN20AS		Vertical	Omni	0-1-100	W6ISQ	IBP cycle
14.100	4X6TU	Tel Aviv	KM72JC	20	Vertical	Omni	0-1-100	W6ISQ	Temp. non op.
14.100	ZS6DN	Pretoria	KF44DC		Vertical	Omni	0-1-100	W6ISQ	IBP cycle
14.100	YV5B	Caracas	FK60NL	1,300	Vertical	Omni	0-1-100	W6ISQ	IBP cycle
14.100	CS3B	Madeira	IM12		Vertical	Omni	0-1-100	W6ISQ	IBP cyle
14.100	5Z4B	Nr Mombasa	KI95		Vertical	Omni	0-1-100	W6ISQ	IBP cycle
14.100	OA4B	Lima	FH17KV		Vertical	Omni	0-1-100	W6ISQ	IBP cycle
14.100	VE8AT	Edmonton AB	DM33		Vertical	Omni	0-1-100	W6ISQ	IBP cycle
14.100	ZL6B	Nr Masterton						W6ISQ	IBP cycle
14.100	4S7B	Wadduwa	MJ96		Vertical	Omni	0-1-100	W6ISQ	IBP cycle
14.100	VK6RBP	SE Perth	OF87BW		Vertical	Omni	0-1-100	W6ISQ	IBP cycle
14.100	VE9BEA	Crabbe Mtn. NB	FN66	3	Vertical	Omni			24
14.406	LN2A †	Sveio	JO29PO		Vertical Monop	Omni	750	LA9AHA	ITU beacon
14.406	VL8IPS	Nr Darwin NT	PH57PJ	2	Vertical Monop	Omni			ITU 24+
18.068	IK6BAK	Montefelcino	JN63KR		2 x Dipole	Omni	12	IK1PCB	24
18.100	PY2AMI	Americana SP	GG67IF		Ground Plane	Omni	5		Operational ?
18.101	VE3RAT	Thornhill, ONT	FN03HO		Vertical	Omni	1		24
18.102	I1M	Bordighera	JN33UT		5/8 Vertical	Omni	10	IK1PCB	24
18.110	DL0AGS	Kassel	JO41NL		Ground Plane	Omni	5	DL1VDL	24
18.110	YV5B	Caracas	FK60NL	1,300	Vertical	Omni	0-1-100	W6ISQ	IBP cycle
18.110	LU4AA	Buenos Aires	GF05		Vertical	Omni	0-1-100	W6ISQ	IBP cycle
18.110	JA2IGY	Mt Asama	PM84JK		Vertical	Omni	0-1-100	W6ISQ	IBP cycle
18.110	ZS6DN	Pretoria	KF44DC	1,573	Vertical	Omni	0-1-100	W6ISQ	IBP cycle
18.110	4X6TU	Tel Aviv	KM72JC		Vertical	Omni	0-1-100	W6ISQ	Temp. non op.
18.110	4U1UN	United Nations NY	FN20AS		Vertical	Omni	0-1-100	W6ISQ	IBP cycle
18.110	OH2B	Espoo	KP20KE		Vertical	Omni	0-1-100	W6ISQ	IBP cycle
18.110	CS3B	Madeira	IM12		Vertical	Omni	0-1-100	W6ISQ	IBP cycle
18.110	5Z4B	Nr Mombasa	KI95		Vertical	Omni	0-1-100	W6ISQ	IBP cycle
18.110	OA4B	Lima	FH17KV		Vertical	Omni	0-1-100	W6ISQ	IBP cycle
18.110	VE8AT	Edmonton AB	DM33		Vertical	Omni	0-1-100	W6ISQ	IBP cycle
18.110	ZL6B	Nr Masterton						W6ISQ	IBP cycle
18.110	4S7B	Wadduwa	MJ96		Vertical	Omni	0-1-100	W6ISQ	IBP cycle
18.110	VK6RBP	SE Perth	OF87BW		Vertical	Omni	0-1-100	W6ISQ	IBP cycle
18.110	VE9BEA	Crabbe Mtn. NB	FN66	3	Vertical	Omni			24
20.948	LN2A †	Sveio	JO29PO		Vertical Monop	Omni	550	LA9AHA	ITU beacon
20.948	VL8IPS	Nr Darwin NT	PH57PJ	2	Vertical Monop	Omni			ITU 24+
21.105	PY2AMI	Americana SP	GG67IF	5	Inv. Vertical				Operational ?

Freq MHz	Call	Nearest town	IARU loc	ASL m	Aerial	Beam Direction	ERP W	Info from	Notes
21.150	4U1UN	United Nations NY	FM20AS		Vertical	Omni	0-1-000	W6ISQ	IBP sycle
21.150	W6WX	San Jose CA	CM87	98	Turnstile	Omni	0-1-100	W6ISQ	IBP cycle
21.150	KH6WO	Honolulu	BL11BK		Vertical	Omni	0-1-100	W6ISQ	IBP cycle
21.150	LU4AA	Buenos Aires	GF05		Vertical	Omni	0-1-100	W6ISQ	IBP cycle
21.150	YV5B	Caracas	FK60NL	1,300	Vertical	Omni	0-1-100	W6ISQ	IBP cycle
21.150	JA2IGY	Mt Asama	PM84JK		Vertical	Omni	0-1-100	W6ISQ	IBP cycle
21.150	4X6TU	Tel Aviv	KM72JC		Vertical	Omni	0-1-100	W6ISQ	Temp. non op.
21.150	ZS6DN	Pretoria	KF44DC	1,573	Vertical	Omni	0-1-100	W6ISQ	IBP cycle
21.150	OH2B	Espoo	KP20KE		Vertical	Omni	0-1-100	W6ISQ	IBP cycle
21.150	CS3B	Madeira	IM12		Vertical	Omni	0-1-100	W6ISQ	IBP cycle
21.150	5Z4B	Nr Mombasa	KI95		Vertical	Omni	0-1-100	W6ISQ	IBP cycle
21.150	OA4B	Lima	FH17KV		Vertical	Omni	0-1-100	W6ISQ	IBP cycle
21.150	VE8AT	Edmonton AB	DM33		Vertical	Omni	0-1-100	W6ISQ	IBP cycle
21.150	ZL6B	Nr Masterton						W6ISQ	IBP cycle
21.150	4S7B	Wadduwa	MJ96		Vertical	Omni	0-1-100	W6ISQ	IBP cycle
21.150	VK6RBP	SE Perth	OF87BW		Vertical	Omni	0-1-100	W6ISQ	IBP cycle
21.151	I1M	Bordighera	JN33UT		2 x 5/8 Vertical	Omni	10	IK1PCB	24
24.915	IK6BAK	Montefelcino	JN63KR		x 2 Dipole	Omni	12	IK1PCB	24
24.930	DK0HHH	Hamburg	JO53AM		Dipole	Omni	10	DL1VDL	24
24.930	YV5B	Caracas	FK60NL	1,300	Vertical	Omni	0-1-100	W6ISQ	IBP cycle
24.930	LU4AA	Buenos Aires	GF05		Vertical	Omni	0-1-100	W6ISQ	IBP cycle
24.930	ZS6DN	Pretoria	KF44DC	1,573	Vertical	Omni	0-1-100	W6ISQ	IBP cycle
24.930	JA2IGY	Mt Asama	PM84JK		Vertical	Omni	0-1-100	W6ISQ	IBP cycle
24.930	4X6TU	Tel Aviv	KM72JC		Vertical	Omni	0-1-100	W6ISQ	Temp. non op.
24.930	4U1UN	United Nations NY	FN20AS		Vertical	Omni	0-1-100	W6ISQ	IBP cycle
24.930	OH2B	Espoo	KP20KE		Vertical	Omni	0-1-100	W6ISQ	IBP cycle
24.930	CS3B	Madeira	IM12		Vertical	Omni	0-1-100	W6ISQ	IBP cycle
24.930	5Z4B	Nr Mombasa	KI95		Vertical	Omni	0-1-100	W6ISQ	IBP cycle
24.930	OA4B	Lima	FH17KV		Vertical	Omni	0-1-100	W6ISQ	IBP cycle
24.930	VE8AT	Edmonton AB	DM33		Vertical	Omni	0-1-100	W6ISQ	IBP cycle
24.930	ZL6B	Nr Masterton						W6ISQ	IBP cycle
24.930	4S7B	Wadduwa	MJ96		Vertical	Omni	0-1-100	W6ISQ	IBP cycle
24.930	VK6RBP	SE Perth	OF87BW		Vertical	Omni	0-1-100	W6ISQ	IBP cycle
24.932	PY2AMI	Americana SP	GG67IF		Ground Plane	Omni	5		Operational ?
28.125	KA5FYI	Austin TX	EM10DI	1	Slope Dip		1	KD4DPC	24
28.175	VE3TEN	Ottawa, ONT.	FN25		Ground Plane	Omni	10		24
28.180	OD5TEN	Tripoli	KM74WK						Non-op
28.180	I1M	Bordighera	JN33UT		5/8 Vertical x 2	Omni		IK1PCB	24
28.182	SV3AQR	Amalias	KM07QS		Ground Plane	Omni	4		24?
28.188	JA7ZMA	Fukushima	QM07	50	Stack Dipole				24
28.193	LU8EHH	Berazategui		5					Irregular
28.186	ZS6PW	Pretoria	KG44DE	1,573	3 ele yagi	337°	15	ZS6PW	Irregulat
28.195	IY4M	Bologna	JN54QK		5/8 Ground Plane	Omni	20	IK1PCB	ROBOT
28.195	LU6DTS	La Plata	GF15AC	5	Ground Plane	Omni	5		24
28.195	VE6YF	Edmonton AB	DO25	10	Ground Plane	Omni	10		24
28.197	LU9AUY	Buenos Aires		10	3 el Yagi				Part time
28.197	VE7MTY	Pitt Meadows BC		5	Vertical	Omni	5		24
28.198	LU5FSY	Rafael	FF98GS				5		24?
28.200	W6WX	San Jose, CA	CM87	98	Turnstile	Omni	0-1-100	W6ISQ	IBP cycle
28.200	KH6WO	Honolulu	BL11BK		Vertical	Omni	0-1-100	W6ISQ	IBP cycle
28.200	JA2IGY	Mt Asama	PM84JK		Vertical	Omni	0-1-100	W6ISQ	IBP cycle
28.200	LU4AA	Buenos Aires	GF05		Vertical	Omni	0-1-100	W6ISQ	IBP cycle
28.200	YB5B	Caracas	FK60NL	1,300	Vertical	Omni	0-1-100	W6ISQ	IBP cycle
28.200	ZS6DN	Pretoria	KF44DC	1,573	Vertical	Omni	0-1-100	W6ISQ	IBP cycle
28.200	4X6TU	Tel Aviv	KM72JC		Vertical	Omni	0-1-100	W6ISQ	Temp. non op.
28.200	OH2B	Espoo	KP20KE		Vertical	Omni	0-1-100	W6ISQ	IBP cycle
28.200	CS3B	Madeira	IM12		Vertical	Omni	0-1-100	W6ISQ	IBP cycle
28.200	5Z4B	Nr Mombasa	KI95		Vertical	Omni	0-1-100	W6ISQ	IBP cycle
28.200	OA4B	Lima	FH17KV		Vertical	Omni	0-1-100	W6ISQ	IBP cycle
28.200	VE8AT	Edmonton AB	DM33		Vertical	Omni	0-1-100	W6ISQ	IBP cycle
28.200	ZL6B	Nr Masterton			Vertical	Omni	0-1-100	W6ISQ	IBP cycle
28.200	4S7B	Wadduwa	MJ96		Vertical	Omni	0-1-100	W6ISQ	IBP cycle
28.200	VK6RBP	SE Perth	OF87BW		Vertical	Omni	0-1-100	W6ISQ	IBP cycle
28.200	KB9FOF	Quincy IL					1		24
28.200	K8UNP	Hollywood FL	EL96UA						24
28.202	ZS1J	Plettenberg Bay	KF15PF	5	½ Vertical	Omni	5	ZS1J	Irregular
28.203	KD6UVN	Laguna Beach CA	DM13CN						?
28.203	WN2A	Budd Lake NJ	FN20	3					24
28.204	W3AW	N Canton OH		1	Loop				?
28.204	AK2F	Randolph NJ	FN20QT						24 ?
28.205	WA4SZE/4	McCaysville GA			Vertical	Omni		KC4DPC	24?
28.205	DL0IGI	Mt Predigstuhl	JN67KQ	100	Vertical Dipole	Omni			Temp. non-op
28.205	ND9X	Forest View IL	EN61CT						?
28.206	KB3BOE	Ridgway PA	FN01PK						?
28.207	KJ4X	Pickens SC	EM84PW	2	Vertical	Omni	2	KC4DPC	24
28.207	KE4NL	Ninety Six SC	EM94AE		Vertical	Omni	10	KC4DPC	Irregular
28.208	W8KFL/4	Venice FL	EL87TB		Vertical	Omni	10		24
28.209	NX2O	Annapolis MD	FN30		Ground Plane	Omni	10		24
28.210	N9YDZ	Brighton IL		3	Vertical	Omni			24 ?
28.210	KC4DPC	Wilmington NC	FM14BF		Indoor Dipole		5	KC4DPC	24
28.210	N7SCQ	Portola CA	CM99ST				5		24
28.211	LA4TEN	Sotra Is	JP20MG		Vertical	Omni	250		24
28.212	LU1UG	Gen. Pico			Ground Plane	Omni	5		Irregular
28.213	PT7BCN	Fortaleza CE	HI06RF		Ground Plane	Omni	5	PY2BBL	Irregular
28.214	KB4SB	Sugarloaf Key FL	EL94FP		Dipole		0.5		?
28.215	GB3RAL	Didcot	IO91IN		¼ Ground Plane	Omni	25	G3NAQ	24
28.215	N6EU	Catoosa OK	EM26CL	1	Fold Dipole				?
28.215	KA9SZX	Champagne IL	EN50VD		CX-1000	Omni	1		24
28.217	N2BJB	Suffern NY	FN21WC						24
28.218	VE2TWO	Radisson PQ	FO13					VE2UG	24
28.218	W8MI	Mackinaw C MI	EN75		Vertical	Omni	0.5	W8MI	24
28.219	PT8AA	Rio Branco AC	FI60CA		Ground Plane	Omni	5		Irregular

Freq MHz	Call	Nearest town	IARU loc	ASL m	Aerial	Beam Direction	ERP W	Info from	Notes
28.219	WB9VMY	Calumet OK	EM05		Vertical	Omni	2		Irregular
28.219	LU4XS	Tierra/d/Fuego	FD65PA	2	Ground Plane	Omni			24 ?
28.220	5B4CY	Zyyi	KM64PR		Ground Plane	Omni	26		24
28.220	KB9DJA	Mooresville IN	EM69RO		Ground Plane	Omni	35		24
28.220	LU4XS	Tierra del Fuego	FD65PA		Ground Plane	Omni	2		24 ?
28.221	K5PF	Apex NC	FM05		Vertical	Omni	8		12 - 0300
28.222	W9BZW	Lake Bluff, IL	EN62BG		Ground Plane	Omni	10	KD4DPC	24
28.222	HG5GEW	Tapolca	JN86NQ	110	Ground Plane	Omni	10		24
28.225	KW7Y	Camano I. WA	CN88SD		Vertical	Omni	4		24
28.225	LW5EJU								?
28.226	PY2AMI	Americana SP	GG67IF		Ground Plane	Omni	5		Op ?
28.230	ZL2MHF	Manuwatu	RE78BU		Vertical dipole	Omni	1	ZL1BAD	24
28.231	KQ4TG	Leland, NC	FM14		Ground Plane	Omni	7	KC4DPC	24
28.232	W7JPI	Sonoita AZ	DM41QP		3 el Yagi	45°	5		24
28.233	N6TWX	Lawrenceville GA	EM73		Double Zep				Irregular
28.233	KD4EC	Jupiter FL	EL96WV		Vertical	Omni			24
28.233	N2VMF	Freehold NJ						N2OPO	24
28.233	N6TWX	Lawrenceville GA	EM73	20	Double Zep				Irregular
28.235	VE1CBZ	Fredericton NB	FN65		Vertical	Omni	5		24
28.237	LA5TEN	Nr Oslo	JO59JV	166	5/8 Ground Plane	Omni	10	OH6DD	24
28.237	NV6A	San Diego CA	DM12		Yagi		0.5		24
28.239	YO2X	Timisoara	KN05OS		Dipole		5	YO2IS	Irregular
28.239	K8UZW	Parma, OH	EN91		Dipole		16		24
28.240	N3SME	Freeland PA	FN21BA						?
28.240	AB8Z	Parma OH	EN91DJ						24 ?
28.240	VA3SBB	Thunder Bay ON	EN58						24
28.244	VE9BEA	Crabbe Mt NB	FN66		Vertical	Omni	3	WB2VUO	24
28.244	WA6APQ	Long Beach CA	DM13		Vertical	Omni	30		24
28.244	KF9N	Gray TN	EM86RJ		Vertical	Omni	5		24
28.245	K0VXU	Stilwell KS	EM28QT		4 el Yagi				Irregular
28.248	K8NHE	Mackinaw C. MI	EN75SN		Vertical	Omni	0.05		24
28.249	N7LT	Bozeman MT	DN45LQ	3	Ground Plane				?
28.249	PI7BQC	Haarlem	JO22HK				2	PI7BQC	24
28.250	EA3JA	Barcelona	JN11BI						24
28.250	K0HTF	Des Moines IA	EN31EN		Ground Plane	Omni	10	K0HTF	24
28.250	Z21ANB	Bulawayo	KG47		Ground Plane	Omni	25		24
28.250	S55ZRS	Mt Kum	JN76MC		Vertical	Omni	1		24
28.251	WJ9Z	St Francis WI	EN62BX		Vertical	Omni	15		24
28.252	WJ7X	Prior Lake, MN	EN34HQ		Vertical	Omni	5		24
28.253	VK3SIX	Wannon Falls	QF02WH		Ground Plane/ 5 el	Varies	25	VK3OT	24
28.254	WA4SLT	Hastings FL	EL99GQ		Vertical	Omni	20		12 - 24h
28.256	KD4BFF	Morrisville NC	FM05OU					WB2VUO	?
28.257	DK0TEN	Konstanz	JN57NP	440	Ground Plane	Omni	40	DL1VDL	24
28.257	KM4Y	Hollywood FL	EL96UA						Irregular
28.259	KA1NSV	Hyannis MA	FN41UP		Vertical Dipole	Omni	10		24
28.260	PY3PAG	Porto Alegre RS							?
28.260	VK5WI	Adelaide	PF95GD		Ground Plane	Omni	10	WIA	24
28.262	VK2RSY	Dural NSW	QF56MH	240	½ Wave Dipole	Omni	25	WIA	24
28.264	VK6RWA	Perth WA	OF78WB	300	Vertical	Omni	20	WIA	24
28.264	JA5ALE	Tokushima	PM74GA	10	½ Ground Plane	Omni			24
28.265	VK4RIK	Cairns QLD	QH23						24
28.265	LU1FHH	Sante Fe	FF98						24
28.266	VK6RTW	Albany WA	OF84		Vertical	Omni	4		24
28.266	LZ1TEN	Nr Sofia	KN12PO		Vertical	Omni	0-1-100		24
28.268	OH9TEN	Pirttikoski	KP36OI		½ Ground Plane	Omni	20	OH6DD	24
28.269	VK8VF	Darwin NT	PH57KP		Vertical	Omni	40	WIA	24
28.270	KF4MS	St Petersburg FL	EL87PS		Ground Plane	Omni	5		24
28.270	VK4RTL	Townsville QLD	QH30JS					WIA	Irregular
28.271	KD4UAI	Smithfield NC	FM05TL		Vertical	Omni	5		QRT ?
28.272	WA9TPZ	Greenfield IN	EM79CS		Dipole		100		?
28.272	KN5H	Las Cruces NM	DM62		Vertical	Omni	5		24
28.275	K4VXP	Campbellsville KY	EM77IH						?
28.275	ZS1LA	Still Bay	KF05QK	15	3 el Yagi	315°	20	ZS6PW	24
28.276	N0JAR	Newton IA	EN31LP				5		24
28.276	NS8V	Grand Rapids MI	EN73EW		Ringo	Omni	5	WS8V	24
28.278	DF0AAB	Kiel	JO54GH	163	Ground Plane	Omni	10	DL1VDL	24
28.279	KG5YB	Tyler TX	EM22						24
28.279	K2SPO	Farmington NY	FN12HW	7	Vertical				24 ?
28.280	NO6J	Thousand Oaks, CA	DM04NF			Omni	5		24
28.280	K5AB	Austin TX	EM10DH		Ground Plane	Omni	20		24
28.280	KD4NOQ	Memphis TN	EM55BE		Ground Plane				24
28.282	LU2HDX	Villa Carlos	FF78RO		Vertical	Omni	10		Irregular
28.282	OK0EG	Hradec Kralove	JO70WE		Ground Plane	Omni	10	OK1ADM	24
28.282	VE2HOT	Montreal PQ	FN35BJ		Vertical dipole	Omni	5		24
28.282	W0ERE	Highlandville MO	EM36IW						24
28.283	K8LKC	3 Rivers MI	EN71EX		¼ Ground Plane	Omni	1		24
28.284	N2JNT	Troy NY	FN32DR		Ground Plane	Omni	1		24
28.284	KD7K	West Jordon UT	DN40AP	5					24
28.285	KJ7AZ	Rawlins WY	DN61JS				5		24
28.285	KB7DQJ	Port Orchard, WA	CN87						24?
28.285	KB7EFZ	Portland, OR	CN85		5/8 Ground Plane	Omni	1	KB7EFZ	0, 15, 30, 45m
28.285	VP8ADE	Adelaide I	FC52WK						Irregular
28.286	N5AQM	Chandler, AZ	DM43AH		Vertical	Omni	2		24?
28.287	KB2YTW	Bergen NY	FN13AC		Vertical	Omni	4	WB2VUC	24
28.287	KE2DI	Rochester NY	FN13AC		Ground Plane	Omni	5?		24
28.289	WJ5O	Corpus Christi, TX	EL17				2		
28.290	SK5TEN	Straengnaes	JO89KK		Ground Plane	Omni	75	OH6DD	24
28.290	KE4YVL	Sophia NC	FN65RR		Vertical	Omni	3		24
28.291	KB9NV	Collinsville IL	EM58AQ		Vertical	Omni	5		24
28.292	LU2FFV	San Jorge			Ground Plane	Omni	5		Irregular
28.292	W3RGQ	Berwick PA	FN11UB		Vertical	Omni	5		Irregular
28.292	KS7K	Des Moines WA							?

Freq MHz	Call	Nearest town	IARU loc	ASL m	Aerial	Beam Direction	ERP W	Info from	Notes
28.293	PY2KC	San Paulo SP		10	Ground Plane				24 ?
28.293	WC8E	Deer Park, OH	EM79SD		Ringo	Omni	10		24
28.294	KE0UL	Greeley, CO	DN70		Vertical	Omni	5		24
28.295	SK2TEN	Kristineberg	JP95HB		Vertical	Omni	5	SSA	24
28.295	W3VD	Laurel MD	FM19NE	130	Vertical dipole	Omni	10		24
28.299	VE9MS	Fredericton NB	FN65		Loop		5	VE9AA	24
28.301	PI7ETE	Amersfoort	JO22QD		Vertical	Omni	0.5	PA0VDV	24?
50.000	GB3BUX	Buxton, Derbys	IO93BF	457	Turnstile	Omni	20	G4IHO	
50.001	VE1SMU	Halifax NS	FN84		3 el Yagi	90°			
50.001	BV2FG	Taihoku	PL05		5/8 Vertical	Omni			QRT Sunday
50.003	7Q7SIX	Malawi	KH74						
50.004	PJ2SIX	Willemstad	FK52		4 x Horz dipole	Omni	22		
50.005	4N0SIX	Belgrade	KN04FU		Dipole	Omni	1		
50.008	VE8SIX	Inuvik NWT	CP38		Double Bay	0°/180°	80		
50.008	HI0VHF		FK58						
50.008	XE2HWB/B	La Paz Baja	DL44		6 el Yagi	0°	5		
50.0095	PY2SFY/B		GG77GA		5/8 Vert	Omni	5		
50.010	SV9SIX	Iraklio	KM25NH		Vertical dipole	Omni	30		
50.010	JA2IGY	Mie	PM84JK		5/8 G/Plane	Omni	10		
50.013	CU3URA	Terceira	HM68		5/8 Vertical		5		
50.014	S55ZRS	Mt. Kum	JN76MC	1219	Ground Plane	Omni	8	S57C	
50.0155	LU9EHF	Lincoln City	FF95		Dipole		15		
50.017	JA6YBR	Miyazaki	PM51		Turnstile		50		
50.018	V51VHF	Namibia	JG87		5/8 Vert	Omni	50		
50.019	CX1CCC	Montevideo	GF15		Ground Plane	Omni	5		
50.020	GB3SIX	Anglesey	IO73TJ	58	3 el Yagi	270°	100		Temp QRT
50.021	OZ7IGY	Tollose	JO55VO	92	Turnstile	Omni	20	OZ7IS	
50.0215	FR5SIX	Reunion Is	LG78	2896	Halo	Omni	2	F5QT	
50.0225	XE1KK/B		EK09			Omni	20		
50.023	LX0SIX	Bourscheid	JN39AV	500	Horizontal Dipole	0°/180°	5	LX1JX	
50.023	SR5SIX	Wesola	KO02OF	130	Ground Plane	Omni	3	SP5TAT	
50.0235	ZP5AA	Asuncion	GG14		Vertical	Omni	5		
50.025	9H1SIX	Attard, Malta	JM75FV	75	Ground Plane	Omni	7	9H1ES	
50.025	OH1SIX	Ikaalinen	KP11QU	157	4 x Turnstile	Omni	40		
50.025	YV4AB	Valensia	FK50		Ringo		15		
50.027	JA7ZMA		QM07		2 x Turnstile	Omni	50		
50.028	SR6SIX	Sztobno / Wolow	JO81HH		Ground Plane	Omni	10	SP6GZZ	
50.028	XE2UZL/B		DM10		2 Sq Loops		25		
50.029	SR8SIX	Sanok	KN19CN						
50.030	CT0WW	Portugal	IN61GE	400	H. Dipole	45°/225°	40		
50.032	JR0YEE	Niigata	PM97		Loop		2		
50.0325	ZD8VHF	Ascension Is	II22TB	723	5/8 Vertical	Omni	50		
50.036	VE4VHF		EN19		Vertical	Omni	35		
50.037	ES0SIX	Muhu Island	KO18PO	30	Hor. dipole	90°/270°	15	ES0CB	
50.037	JR6YAG	Okinawa	PL36		2 x 5/8 G Planes		10		
50.037	FY7THF	French Guiana	GJ35		Ground Plane	Omni	100		
50.038	FP5XAB	St Pierre Miquelon Is	GN16			Omni		FP5EK	
50.039	VO1ZA	St Johns	GN37		1/4 Wave Vert	Omni	10		
50.040	SV1SIX	Athens	KM17UX	130	Vertical Dipole	Omni	30		
50.040	ZL3SIX	Christchurch	RE66			45°/315°	20		
50.041	VE6EMU	Camrose	DO33		4el Yagi	22°	35		
50.042	GB3MCB	St Austell	IO70OJ	320	Dipole	90°/270°	40	G3YJX	
50.042	YB0ZZ	Jakarta	OI33		Ground Plane	Omni	15		
50.043	YO2S	Timisoara	KN05PS		Dipole		2	YO2IS	
50.044	VE6ARC		DO05		Ground Plane	Omni	25		
50.044	ZS6TWB/B		KG46		6 el Yagi	0°	30		
50.046	VK8RAS		PG66		X Dipole		15		
50.047	TR0A		JJ40		5 el Yagi	0°	15		
50.047	4N1SIX	Belgrade	KN04OO		Vee	Omni	10		
50.047	JW7SIX	Svalbard	JQ78TF		4 el Yagi	210°	40	LA0BY	
50.048	VE8BY	Iqaluit NT	FP53				30		
50.050	GB3NHQ	Potters Bar	IO91VQ	35	Turnstile	Omni	15	G3UUT	
50.050	ZS6DN/B	Pretoria	KG44DE		5 el Yagi	135°	100	ZS6DN	
50.051	LA7SIX	Tromso	JP99LO	30	4 el Yagi	190°	25	LA0BY	
50.052	Z21SIX	Zimbabwe	KH52NK		Ground Plane	Omni	8		
50.054	OZ6VHF	Oestervraa	JO57EI		Turnstile	Omni	50	OZ1IPU	
50.053	VK3SIX	Hamilton	QF12		Colinear		12		
50.0555	V44K	St Kitts/Nevis	FK87		5/8 Vertical		3		
50.057	VK7RAE	Lonah	QE38		X Dipoles		20		
50.057	VK8VF	Darwin	PH57		1/4 Vertical		100		
50.058	VK4RGG	Nerang	QG62				6		
50.058	VE3UBL	Brougham	FN03		Turnstile		10		
50.059	PY2AA	Sao Paulo	GG66		Ground Plane		5		
50.059	JH0ZPI		PM96				10		
50.060	KA5FYI		EM10						
50.060	W5VAS		EM40		Squalo		50		
50.060	K4TQR/B		EM63		Dipole		3		
50.060	GB3RMK	Inverness	IO77UO	270	Dipole	0°/180°	10	GM3WOJ	
50.061	KH6HME/B		BK29		Dipole		20		
50.061	KE7NS/B		DN31		Squalo		2		
50.061	WB0RMO		EN10		Squalo		50		
50.062	GB3NGI	Ballymena	IO65PA	240	Dipole	140°/320°	10	GI6ATZ	Back 09/97
50.062	W7HAH		DN28		Halo	Omni	25		
50.062	K8UK/b		EN82			Omni	2		
50.062	KA0NNO		EM24		Halo	Omni	8		
50.064	AA5ZD		EM12						
50.064	GB3LER	Lerwick	IP90JD	104	Dipole	0°/180°	30	GM4IPK	
50.065	AB5L		EM13		Dipole		0.2		
50.065	W0IJR		DM79		2 x Ring Halo	Omni	20		
50.065	KG9AE		EM69		AR6	Omni	10		
50.065	KH6HI/b		BL01		Turnstile	Omni	15		
50.065	W3VD		FM19		Squalo		7		

Freq MHz	Call	Nearest town	IARU loc	ASL m	Aerial	Beam Direction	ERP W	Info from	Notes
50.065	W0MTK		DM59		4 V Dipoles		2		
50.0655	GB3IOJ	St Helier	IN89WE	115	Vertical	Omni	10	GJ4ICD	
50.066	W5OZI		DM90		Dipole		20		
50.066	VK6RPH		OF88		U Dipole		10		
50.066	WA1OJB		FN54		J Pole		30		
50.067	W3HH		EN90		Loop		10		
50.067	KQ4E		EM86		Vertical		10		
50.067	W4RFR		EM66				2		
50.067	OH9SIX	Pirttikoski	KP36OI	192	2 x Turnstile	Omni	35	OH6DD	
50.068	W7US		DM42		4 el		50		
50.069	K6FV		CM87				100		
50.070	EA3VHF	Barcelona	JN01	235	3 el Yagi	315°	0.5	EA3DBQ	
50.070	SK3SIX	Edsbyn	JP71XF	505	Hor X dipole	180°	10	SM3EQY	
50.070	W2CAP/B		FN41		V Dipole		15		
50.070	W7WKR/B		CN87		Beam		10		
50.070	ZS1SES								
50.071	WB5LUA		EM12		Halo		1.5		
50.071	KA5BTP		EM40						
50.072	KS2T		FM29		Ground Plane		10		
50.072	WA4NTF/B		EM81						
50.072	W4IO		EM81		M2 Halo	Omni	1		
50.072	KW2T		FN13		Squalo		0.5		
50.073	WB4WTC/B		FM06		2 Loops		10		
50.073	WR7V/B		CN87		Halo		10		
50.073	ES6SIX	Voeru	KO37MT	85	Ground Plane	Omni	1	ES5MC	
50.073	NN7K		DM09		Ringo Ranger		1		
50.075	W6SKC/7		DM41		Halo		5		
50.075	VR2SIX		OL72		Ground Plane		7		
50.075	NL7XM/2		FN20				1		
50.076	KL7GLK/3		FM18			Omni	4		
50.077	VE3DRL								
50.077	N0LL		EM09		2 x Halo		21		
50.077	WB2CUS		EL98		Loop		1		
50.0775	VK4BRG		QG48		Turnstile		5		
50.078	OD5SIX	Lebanon	KM74WK		1/4 Vertical	Omni	7	OD5SB	
50.078	KE4SIX		EM83		Ringo Ranger	Omni	5		
50.079	JX7DFA	Jan Mayen Island	IQ50		5 el yagi	160°	40	LA7DFA	
50.079	TI2NA		EJ79		Dipole		20		
50.080	ZS1SIX		JF96		Halo		10		
50.082	CO2FRC		EL83		Dipole		2		
50.083	LZ1SIX		KN12						
50.086	VP2MO		FK86		6 el Yagi		10		
50.087	PB0ALN		JO22						
50.0873	YU1SIX		KN03KN		Dipole		15		
50.0875	VE9MS/B		FN65		2 H/Loops		40		
50.089	VE2TWO	Radisson	FO13		Dipole		15		
50.095	PY5XX		GG54		Dipole		50		
50.162	IS0SIX	Sardinia	JM49NG		Dipole		1	IS0AGY	
50.275	GB3IFX	Darlington	IO94FM	70	2 x 6 el Yagi	180°	400	G4IFX	30s at hour
50.283	VK3RMV		QF02		Colinear	Omni	12		
50.306	VK6RBU		OF76		3 el Yagi	260°/80°	100/10		
50.315	FX4SIX	Neuville	JN06CQ	153	Turnstile	Omni	25	F5GTW	
50.480	JH8ZND/B		QN02		Ground Plane		10		
50.485	JH9YHP		PM86		X Dipole		2-Oct		
50.490	JG1ZGW	Tokyo	PM95		Dipole		10		
50.499	5B4CY	Zyghi, Cyprus	KM64PR	30	Ground Plane	Omni	20	5B4BBC	
50.521	SZ2DF		KM25		4 x 16 el	30°/330°	1000		
51.029	ZL2MHB	Hastings	RF80		1/2 Vertical	Omni	1-Oct		
52.345	VK4ABP	Longreach	QG26		1/4 Vertical	Omni	4		
52.420	VK2RSY	Dural	QF56		Turnstile	Omni	25		
52.510	ZL2MHF	Mt Climie	RE78		Dipole		4		
70.000	GB3BUX	Buxton, Derbys	IO93BF	456	2 x Turnstile	Omni	20	G4IHO	
70.010	GB3REB	Camberley	IO91OH	117	2 el Yagi	330°	28	G3ZYV	
70.020	GB3ANG	Dundee	IO86MN	370	3 el Yagi	160°	100	GM4ZUK	
70.025	GB3MCB	St Austell	IO70OJ	320		45°	40	G3YJX	
70.030	G Personal Beacons					2 el Yagi			
70.114	5B4CY	Zyghi, Cyprus	KM64PR	30	4 el Yagi	315°	15	5B4BBC	
70.130	EI4RF	Dublin	IO63WD	120	2 x 5 el Yagi	45°/135° seq	25	EI9GK	

Region 1 2m beacons moved to these new frequencies on 1st July 1997 following a decision of the IARU Region 1 Conference in Tel Aviv

Freq MHz	Call	Nearest town	IARU loc	ASL m	Aerial	Beam Direction	ERP W	Info from	Notes
144.400	Transatlantic beacon								
144.402	EA8VHF	Canary Is	IL28GC			Omni	10		?
144.402	OY6VHF	Faroe Islands	IP62OA	300	2 x 4 el	45°/135°	50/50	OY1A	Not op
144.403	EI2WRB	Portlaw	IO62IG	248	5 el Yagi	95°	200	EI6GY	
144.404	EA1VHF	Curtis	IN53UG	100	5 el Yagi	45°	150	EA1DKV	?
144.405	F5XAR	Lorient	IN87KW	165	9 ele Yagi	270°	500	F6ETI	Trans Atl Plan
144.407	GB3?	Planned UK Transatlantic Beacon							
144.409	F5XSF	Lannion	IN88GS	145	9 el Yagi	90°	50	F6DBI	
144.410	DB0SI	Schwerin DOK V 14	JO53QP	90	Big wheel	Omni	10	DL1SUZ	
144.411	I1G	La Spezia	JN44VC	745	4 el Yagi	135°	4	IK1LBW	Temp QRT
144.412	SK4MPI	Borlaenge	JP70NJ	520	4 x 6 el Yagi	45°/315°	1500	SM4HFI	
144.413	3A2B	Monaco	JN33RR	50	Yagi	90°	50	3A2LF	?
144.414	DB0JW	Wurselen DOK G 05	JO30DU	238	7 el Yagi	22°	50	DL9KAS	
144.415	I1M	Bordighera IM	JN33UT	300	Big wheel	Omni	20	IK1PCB	
144.416	PI7CIS	Delft	JO22DC	40	Omni	Omni	50	PA0CIS	
144.417	OH9VHF	Pirttikoski	KP36OI	310	10 dBd gain	200°	200	OH6DD	
144.418	ON4VHF	Louvain La Neuve	JO20FP	180	Clover leaf	Omni	15	ON7PC	
144.419	I2M	Cremona	JN55AD	46	Big wheel	Omni	10	IK2AWT	
144.420	DB0RTL	DOK P 60	JN48OM	480	Big wheel	Omni	15	DL8SDL	
144.421	OZ7IGY	Tollose	JO55VO	96	Big wheel	Omni	25	OZ7IS	
144.422	DB0TAU	DOK F 11	JO40HG	326	4 x 4 el Yagi	Omni	15	DL3DC	

Freq MHz	Call	Nearest town	IARU loc	ASL m	Aerial	Beam Direction	ERP W	Info from	Notes
144.423	PI7FHY	Heerenveen	JO33WW	52	Halo	Omni	10		
144.424	IN3A	Trento	JN56NB	225	Ground plane	Omni	0.1	IN3IYD	
144.425	F5XAM	Blaringhem	JO10EQ	99	Big wheel	Omni	14	F6BPB	
144.426	EA6VHF	San Jose, Ibiza	JM08PV	150		Omni	20	EA6FB	
144.427	OK0EJ	Frydek-Mistek	JN99FN	1323	4 el Yagi	270°	0.3	OK2UWF	
144.427	PI7PRO	Nieuwegein	JO22NC	20	Halo	Omni	10		
144.428	DB0JT	Oberndorf DOK C 16	JN67JT	785	4 x Dipole	0°	30	DJ8QP	
144.429	IV3A	Manzano UD	JN65QX		Ground plane	Omni	1		
144.430	GB3VHF	Wrotham, Kent	JO01DH	268	2 x 3 el Yagi	315°	40	G8JNZ	
144.431	9A0BVH		JN85JO	489	V Dipole	Omni	1		?
144.432	9H1A	Malta	JM75FV	160	Turnstile	Omni	1.5	9H1BT	QRT
144.434	DB0LBV	DOK S 30	JO61EH	232	2 x Dipole	Omni	0.4	DL1LWM	
144.435	HB9H	Locarno	JN46KE						
144.435	SK2VHG	Svappavara	KP07MV	380	16 el Yagi	180°	800	SK2CP	
144.436	I3Z	Verona	JN55OL		Yagi	180°	50	I3LDP	Planned
144.436	PI7NYV	Holtenburg	JO32EH	80	X dipole	Omni	1	PA3FJY	
144.437	LA1VHF	Oslo	JO49GT	1882	Turnstile	Omni	12	LA4PE	Temp QRT
144.438	LX0VHF	Walferdange	JN39BP	420	Big wheel	Omni	10	LX1JX	QRT
144.438	OK0EO	Olomouc	JN89QQ	602	Ring dipole	Omni	0.05	OK2VLX	Plan 10/97
144.439	SK3VHF	Oestersund	JP73HF	325	Horizontal Yagi	180°	500	SM3PXO	MS beacon
144.440	DL0UH	Melsungen DOK Z 25	JO41RD	385	V Dipole	Omni	1	DJ3KO	
144.441	LA4VHF	Bergen	JP20LG	30	2 x 8 el Yagi	0°	380	LA6LU	
144.442	I4A								
144.443	OH2VHF	Nummi	KP10VJ	76	9 el yagi	0°	150		
144.444	DB0KI	Bayreuth DOK Z42	JO50WC	1025	Dipole	Omni	2.5	DC9NL	
144.444	I5A								
144.445	GB3LER	Lerwick	IP90JD	108	2 x 6 el Yagi	45°/135°	500/500	GM4IPK	
144.446	OK0EB	Ceske Budejovice	JN78DU	1084	3 x Dipole	Omni	0.07/0.007	OK1APG	
144.447	SK1VHF	Klintehamn	JO97CJ	55	2 x Cloverleaf	Omni	20		
144.448	HB9HB	Biel	JN37OE	1300	3 el Yagi	345°	120	HB9AMH	
144.449	I0A	P.Mirteto RI	JN62IG	300	2 x Big wheel	Omni	10	IW0BCF	Temp QRT
144.450	DL0UB	Trebbin	JO62KK	120	4 x Dipole	Omni	10	DL7ACG	
144.451	LA7VHF	Tromso	JP99LO	30	10 el Yagi	190°	500	LA0BY	
144.452	OK0EC	As	JO60CF	778	3 el Yagi	90°	0.7	OK1VOW	
144.453	GB3ANG	Dundee	IO86MN	370	4 el Yagi	160°	20	GM4ZUK	
144.454	IS0A	Olbia SS	JN40QW	350	Turnstile	Omni	1	IW0UGR	
144.455	OH5ADB	Hamina	KP30NN	65	Dipole	135°/315°	0.1		
144.456	DB0GD	Rhoen DOK Z 62	JO50AL	930	Dipole	0°/180°	1	DG6ZX	
144.457	SK2VHF	Vindeln	JP94TF	300	2 x 10 el Yagi	0°/225°	100		
144.457	EA2VHF		IN91DJ			Omni	18		QRT
144.458	FX4VHF	Brive	JN05VE	600	Big wheel 6dB	Omni	25	F6IAL	
144.458	I0G	Foligno PG	JN63IB	1200	4 x dipole	Omni	10	IW0QIT	
144.459	LA5VHF	Bodo	JP77KI	260	2 x 6 el Quad	15°/180°	100	LA1UG	
144.460	HG1BVA	Szentgotthord	JN86CW	370	Hybrid Quad	80°	40	HA1YA	?
144.461	SK7VHF	Falsterbo	JO65KJ	25	2 x Cloverleaf	Omni	10		
144.462	I6A	Ortona CH	JN72FH	150	2 x 5 el Yagi	340°/180°	24	IW6MME	Temp QRT
144.463	LA2VHF	Melhus	JP53EG	710	10 el Yagi	15°	500	LA1K	
144.464	I7A	Bari	JN81EC	685	Big wheel	Omni	8	I7FNW	
144.465	DF0ANN	DOK B 25	JN59PL	630	V Dipole	Omni	0.3	DL8ZX	
144.466	OZ4UHF	Bornholm Island	JO75KC	115					
144.467	HB9RR	Zurich	JN47FI	871					
144.467	I8A	Reggio C.	JM78WD	1778	SqLo	Omni	8	I8GMP	
144.467	OK0ED	Frydek-Mistek	JN99DQ	290	2 x Dipole	Omni	0.1	OK2UWF	
144.468	FX7VHF	Beaune	JN26IX	561	Big wheel	Omni	20	F1RXC	
144.468	LA6VHF	Kirkenes	KP59AL	70	14 el Yagi	210°	250	LA4OO	
144.469	GB3MCB	St Austell	IO70OJ	320	3 el Yagi	45°	40	G3YJX	
144.469	IT9A	Alcamo TP	JM67LX	825	2 x Big wheel	Omni	10	IT9QPF	
144.470	OH2VAN	Vantaa	KP20			Omni			Planned
144.471	OZ?								Planned
144.472	IT9G	Mondello PA	JM68QE	50	5 el Yagi	0°	35	IT9BLB	Planned
144.473	SK2VHH	Lycksele	JP94	300	Horizontal	22°	15000	Scientific	QRV summer
144.474	EA3VHF	Soria	JN11MV	155	Halo	Omni	1		QRT
144.474	OK0EL	Benecko	JO70SQ	900	Dipole		0.004	OK1AIY	
144.475	DL0SG	DOK U 14	JN69KA	1024	4 x 4 el Yagi	Omni	5	DJ4YJ	
144.475	LY2WN	Jonava	KO25GC	120	2 x Dipole	Omni	15	LY2WN	
144.476	F5XAL	Pic Neulos	JN12LL	1100	Big wheel	Omni	0.5/10	F6HTJ	
144.477	DB0ABG	DOK U 01	JN59WI	522	Big Wheel	Omni	4	DJ3TF	
144.478	LA3VHF	Mandal	JO38RA	30	16 el Yagi	180°	100	LA8AK	
144.478	S55ZRS	Kum	JN76MC	1219	Dipole	Omni	1	S57C	
144.479	F6KJD	Bourg/Bresse	JN26QE	250	Big wheel	Omni	50	F6GGX	
144.479	SR5VHF	Wesola	KO02OF	130	Turnstile	Omni	0.75	SP5TAT	
144.479	IT9S	Zafferana CT	JM77NO	800	2 x Big wheel	Omni	3	IW9AFI	Planned
144.482	GB3NGI	Ballymena	IO65VB	528	2 x 4 el Yagi	45°/135°	120/120	GI6ATZ	
144.486	DL0PR	Garding DOK Z 69	JO44JH	75	4 x 6 el Yagi	0°/180°	200	DL8LD	
144.490	DB0FAI	Langerringn DOK T01	JN58IC	590	16 el Yagi	305°	1000	DL5MCG	
432.128	S55ZNG	Trstelj	JN65UU	643	Horizontal Loop	Omni	0.1	S50M	
432.800	DB0GD	Rhoen	JO50AL	930	Dipole	0°/180°	1	DG6ZX	
432.800	OE3XMB	Muckenkogel	JN77TX				2		
432.810	DB0OB	DOK U 17	JN69EQ	825	Schlitz	Omni	1	DC9RK	
432.820	LA8UHF	Tonsberg	JO59DD	30	8 el Yagi	180°	50	LA6LCA	
432.830	FX1UHF	Preaux	JN18KF	166	4 x HB9CV	Omni	10	F6HZA	
432.830	LA7UHF	Bergen	JP20LG	30	4 el Yagi	0°	200	LA6LU	
432.835	ES0UHF	Hiiumaa Island	KO18CW	105	Horizontal	Omni	50	ES0NW	
432.840	DB0KI	Bayreuth	JO50WC	925	Dipole	Omni	10	DC9NL	
432.840	OH6UHF	Uusikaarlepyy	KP13GM	55	3 x Big wheel	Omni	7	OH6UH	
432.845	DB0LBV	DOK S 30	JO61EH	234	Schlitz	Omni	2	DL1LWM	
432.845	LA9UHF	Geilo	JP40CM	1000	2 x 13 el Yagi	33°	250	LA3SP	
432.847	9A0BUH		JN85JO	489	V dipole	Omni	1		
432.850	DL0UB	DOK Z 20	JO62KK	120	Malteser	Omni	10	DL7ACG	
432.852	OH2UHF	Nummi	KP10VJ	76	2 x dipole	90°/270°	50		
432.855	LA5UHF	Bodo	JP66WX	1110	10 el Yagi	15°	100	LA1UG	QRT 12/96
432.855	SK3UHF	Nordingra	JP92FW	200	4 x Double quad	Omni	10	SM3AFT	

Freq MHz	Call	Nearest town	IARU loc	ASL m	Aerial	Beam Direction	ERP W	Info from	Notes
432.860	LA1UHF	Oslo	JO59IX	522	Mini wheel	Omni	10	LA4PE	QRT 12/96
432.863	F5XAG	Lourdes	IN93WC	550	2 x 10 el	22°	40	F5HPQ	
432.870	EI2WRB	Portlaw	IO62IJ	248	5 el Yagi	95°	250	EI9GO	
432.873	PI7HVN	Heerenveen	JO22WW	50	Horizontal	Omni	0.5	PE1HUE	
432.875	DB0FAI	DOK T 01	JN58IC	610		Omni	10	DL5MCG	
432.875	OH7UHF	Kuopio	KP32TW	215	6 dBd	225°	15/1.5/.15		
432.875	SK2UHF	Vindeln	JP94WG	445	2 x 20 el coll	0°/225°	300	SK2AT	
432.880	LA3UHF	Mandal	JO38RA	12	15 el Yagi	180°	29	LA8AK	
432.885	OK0EP	Sumperk	JO80OC	1505	2 x 3 el Yagi	90°	6	OK1VPZ	
432.885	OY6UHF	Faroe Is	IP62OA	300	7 dB Group	135°	50	OY1A	
432.886	FX4UHB	St Savin	JN06KN	144	Big wheel	Omni	50	F5EAN	
432.890	GB3SUT	Sutton Coldfield	IO92CO	270	2 x 8 el Yagi	0°/135°	10		
432.890	LA4UHF	Haugesund	JO29PJ	75	10 el Yagi	200°	50	LA7GN	
432.895	OZ4UHF	Bornholm Island	JO75KC	115	Clover leaf	Omni	30	OZ1HTB	
432.900	DB0YI	Hildesheim Z 35	JO42XC	480	Big wheel	Omni	3	DL4AS	
432.905	PI7QHN	Zandvoort	JO22FH	20	3 dB Gain	Omni	2	PA0QHN	
432.905	SK4UHF	Garphyttan	JO79LK	270	Horizontal	Omni	50	SM4RWI	
432.908	EA8UHF	Canary Is	IL28GC			Omni	10		
432.910	GB3MLY	Emley Moor	IO93EO	600	6 el Yagi	150°	40	G3PYB	
432.918	EA6UHF		JM08PV			Omni	10		
432.918	FX3UHB	Locronan	IN78VC	285	Big wheel	Omni	15	F5MZN	
432.920	DB0UBI	DOK N 59	JO42GE	125	8el Coll	45°	12	DD8QA	
432.920	SK7UHF	Taberg	JO77BQ	350	Big wheel	Omni	15	SM6DHW	
432.925	DB0JG	Bocholt DOK N17	JO31GT	45	Clover Leaf	Omni	1	DL3QP	
432.925	I1M		JN33UT						
432.925	SK6UHF	Varberg	JO67EH	175	Clover Leaf	Omni	10	SM6ESG	
432.930	HG7BUA		JN97KR	690	Slot	Omni	2	HG5ED	
432.930	OK0EA	Trutnov	JO70UP	1355	2 x 15 el Yagi	180°/270°	3	OK1AIY	*432.934
432.930	OZ7IGY	Tollose	JO55VO	93	Omni / 9el yagi	45°/Omni	300/30	OZ7IS	
432.934	GB3BSL	Bristol	IO81QJ	252	4 x 3 el Yagi	90°	250	GW8AWM	
432.940	DL0UH	Melsungen DOKZ25	JO41RD	385	V Dipole	Omni	1	DJ3KO	
432.940	SK7MHH	Faerjestaden	JO86GP	45	Horizontal	0°/Omni	300/30		
432.942	GB3NGI	Ballymena	IO65VB		12 el Yagi	125°	250	GI6ATZ	Planned
432.945	DB0OS	Erndtebruck DOKN32	JO40CW	730	2 el Yagi	270°	0.3	DG6YW	
432.945	OH9UHF	Pirttikoski	KP36OI	307	9 dBd gain	200°	70	OH6DD	
432.947	HG6BUA		KN07AU	1050			2	HG5ED	
432.950	DB0IH	Oberthal DOK Q 18	JN39ML	630	Big wheel	Omni	1	DC8DV	
432.950	S55ZRS	Kum	JN76MC	1219	Slot Dipole	Omni	1	S50M	
432.950	SK1UHF	Klintehamn	JO97CJ	55	2 x Big wheel	Omni	20	SM1IUX	
432.955	OZ1UHF	Frederikshavn	JO57FJ	150	Big wheel	Omni	10	OZ9NT	
432.965	DF0ANN	Altdorf	JN59PL	630	Big wheel	Omni	1	DL8ZX	
432.965	GB3LER	Lerwick	IP90JD	104	12 el Yagi	165°	675	GM4IPK	
432.966	OK0EO	Olomouc	JN89QQ	602	Ring Dipole	Omni	0.05	OK2VLX	
432.970	GB3MCB	St Austell	IO70OJ	320	4 el Yagi	45°	12	G3YJX	
432.970	OK0EB	Ceske Budejovice	JN78DU	1084	Mini Wheel	Omni	0.03/0.16	OK1APG	
432.975	DB0JW	Aachen DOK G 05	JO30DU	238	2 x 11 el Yagi	45°	50	DL9KAS	
432.975	DL0SG	DOK U 14	JN69KA	1024	4 x 11 Yagi	Omni	5	DJ4YJ	
432.975	HG1BUA	Szombathely	JN87GG	370	Hybrid Quad	90°	20	HA1YA	
432.980	GB3ANG	Dundee	IO86MN	370	9 el Yagi	170°	100	GM4ZUK	
432.980	S55ZCE	Sv. Jungert	JN76OH	574	Ground plane (V)	Omni	0.07	S51KQ	
432.982	OZ2ALS	Sonderborg	JO44WX	28	4 x dipole	Omni	40	OZ9DT	
432.982	SR5UHF	Wesola	KO02OF	130	Turnstile	Omni	0.25	SP5TAT	
432.984	HB9F	Interlaken	JN36XN	3573	Corner reflector	0°	15	HB9MHS	
432.990	DB0VC	DOK Z 10	JO54IF	300	4 x DQ	Omni	10	DL8LAO	
432.990	ON4UHF	Brussels	JO20ET	180	Clover leaf	Omni	0.5	ON4LC	
432.995	DL0IGI	Mt Predigstuhl DOKZ57	JN67KQ	1618	2 x DQ	315°	50	DJ1EI	
1296.063	S55ZNG	Trstelj	JN65UU	643	V-J Slot	Omni	0.1	S50M	
1296.380	S55ZRS	Kum	JN76MC	1219	Turnstile	Omni	1	S57C	
1296.642	PI6ASD	Amsterdam	JO22KH	30		Omni	1	PA0AWP	Trnspnder
1296.739	FX6UHY	Strasbourg	JN38UO	144	Big wheel	Omni	4	F6BUF	Planned
1296.800	DB0GD	DOK Z 62	JO50AL	930	Dipole	Omni	1	DG6ZX	
1296.800	DB0JS	DOK T 09	JN59GB	700	Slot	Omni	5	DL2QQ	Non op
1296.800	OE3XMB	Muckenkogel	JN77TX				0.1		
1296.800	SK6UHI	Hallandsaas	JO66LJ	230	Big wheel	Omni	10	SM6IKY	Planned
1296.805	DB0GP	Schwabischgmund	JN48WQ	760	4 x 5 el Yagi	Omni	50	DC1SO	
1296.810	DB0OB	DOK U 17	JN69EQ	825	Slot	Omni	1	DC9RK	
1296.810	GB3NWK	Orpington	JO01BI	180	15/15 Slot Yagi	293°	50	G8BJG	Part time
1296.812	FX6UHX	P. Ballon	JN37NX	1278	4 el Yagi	135°	1	F1AHO	On test
1296.815	DB0VI	Saarbrucken	JN39MF	400	13 el Yagi		1	DK1ME	
1296.818	PI7DIJ	Drachten	JO33BC	15	5.5 dB Gain	Omni	4	PA3DIJ	
1296.820	DB0OT	Lathen	JO32QR	80	Big wheel	Omni	1	DL1BFZ	
1296.820	LA8UHG	Oslo	JO59JW	364	14 el Yagi	160°	10	LA4PE	
1296.825	DB0ABG	DOK U 01	JN59WI	522	Slot	Omni	0.5	DJ3TF	
1296.825	DB0HF	Wandsbek	JO53BO	45	Big wheel	Omni		DK2NH	
1296.825	OE1XTB	Vienna	JN88EE	170		Omni	10	OE1MCU	
1296.830	GB3MHL	Martlesham	JO02PB	80	4 x 16 Slot wg	90°/270°	700	G4DDK	
1296.835	SK0UHG	Vaellingby	JO89WI	60	Horizontal	Omni	10		
1296.840	DB0KI	Bayreuth	JO50WC	925	4 x DQ 23	Omni	5	DC9NL	
1296.840	OH6SHF	Uusikaarlepyy	KP13GM	55	Dipole	Omni	8		
1296.845	DB0LBV	DOK S 30	JO61EH	234	4 x Slot	Omni	2	DL1LWM	
1296.845	SR3SHF	Kalisz	JO91CQ					SP3JBI	
1296.847	FX1UHY	Faviers	JN18IR	160	Alford Slot	Omni	30	F6ACA	
1296.850	DL0UB	Templehof	JO62QL	80	2 x helical	Omni	3	DL7ACG	
1296.850	GB3FRS	Farnborough	IO91PH	120	Disc	Omni	3	G8ATK	Planned
1296.854	DB0JO	Witten	JO31SL	312	4 x 15 el Yagi	270°	350	DC6MR	
1296.855	OZ3UHF		JO56CE	150	5 el Yagi	180°	6	OZ1GMP	
1296.860	GB3MCB	St Austell	IO70OJ	300	15/15	45°	50	G3YJX	
1296.860	LA1UHG	Tonsburg	JO59DD	30	13 dB Horn	180°	60	LA6LCA	
1296.860	OE1XVB	Vienna	JN88EF	191	4 x 6 el	Omni	100		
1296.862	FX9UHZ	Istres	JN23MM	114	Slotted WG	Omni	150	F1AAM	
1296.865	DB0JK	Koln DOK Z 12	JO30LX	260	4 x 8 el Yagi	Omni	40	DK2KA	
1296.865	HB9WW	Neuchatel	JN37LA	1145	15 el loop	125°	30	HB9HLM	

Freq MHz	Call	Nearest town	IARU loc	ASL m	Aerial	Beam Direction	ERP W	Info from	Notes
1296.865	SK7MHG	Veberod	JO65SO	200		Omni	50		
1296.870	DB0IBB	DOK N 49	JO32VG	200	4 x Slot	Omni	170	DB7QW	
1296.875	DB0FAI	DOK T 01	JN58IC	610		Omni	10	DL5MCG	
1296.875	FX3UHX	Landerneau	IN78UK	121	Quad	90°	1	F6CGJ	
1296.875	GB3USK	Bristol	IO81QJ	235	Slotted waveguide	90°	250	GW8AWM	Non-op
1296.880	LA3UHG	Fleckkeroy	JO38XB	5	2 x 15 el Yagi	180°	10	LA8AK	
1296.880	ON4SHF	Ath	JO10UN	130	Slotted	90°	10	ON5PX	Planned
1296.883	DB0INN	DOK C 15	JN68GI	504	Schlitz	Omni	1	DL3MBG	
1296.886	FX4UHY	Loudun	JN06BX	140	Alford Slot	Omni	25	F1AFJ	
1296.890	GB3DUN	Dunstable, Beds	IO91RV	263	Alford slot	Omni	2	G3ZFP	
1296.897	HG6BUB		KN07AU	1050			2	HG5ED	
1296.900	DB0AN	St Mauritz	JO31SX	100	Big wheel	Omni	1	DF1QE	Proposal
1296.900	GB3IOW	Newport, IOW	IO90IO	250	Alford Slot	Omni	100	G3WXC	
1296.900	OK0EA	Trutnov	JO70UP	1355	4 x 15 el Yagi	S/SW/W/NW	1.6	OK1AIY	
1296.902	LX0SHF	Walferdange	JN39BP	420	2 x Big wheel	Omni	3	LX1JX	
1296.905	OH4SHF	Haukivuori	KP31OX	200	Alford Slot	Omni	15		
1296.907	FX9UHX	Pic Neulos	JN12LL	1100	5 el Yagi	0°	90	F6HTJ	
1296.910	DB0UX	Karlsruhe	JN48FX	275	Big wheel	Omni	1	DK2DB	
1296.910	GB3CLE	Clee Hill, Salop	IO82RL	540	2 x 15/15 Yagi	0°/135°	20	G3UQH	
1296.920	9A0BLB		JN83HG	778	Dipole		1		
1296.920	DB0VC	Lutjenberg	JO54IF	244	2 x Big wheel	Omni	3	DL8LAO	
1296.920	PI7QHN	Zandvoort	JO22FH	20	6 dB Gain	Omni	4	PA0QHN	
1296.920	SK7UHG	Taberg	JO77BQ	350	Big wheel	Omni	3	SM6DHW	
1296.925	DB0KME	DOK C 35	JN67HT	800	Vertical	Omni	1	DL8MCG	
1296.925	SK6UHG	Hoenoe	JO57TQ	35	4 x Big wheel	Omni	10	SM6EAN	
1296.930	GB3MLE	Emley Moor	IO93EO	600	Corner Reflector	160°	50	G8AGN	Proposal
1296.930	OK0EL	Benecko	JO70SQ	900	5 dB Horn	135°/270°	0.8	OK1AIY	Proposal
1296.930	OZ7IGY	Tollose	JO55VO	95	Big wheel	Omni	15	OZ7IS	
1296.935	DB0YI	Hucksheim	JO42XC	480	Big wheel	Omni	3	DL4AS	
1296.935	OH5SHF	Kuusankoski	KP30HV	145	Alford Slot	Omni	25		
1296.940	DL0UH	Melsungen	JO41RD	385	V-Dipole	Omni		DJ3KO	
1296.945	DB0OS	Hitchembach	JO40CW	745	6 el array	270°	1	DG6YW	
1296.945	HB9F	Bern	JN46SW	1015	Corner reflector	0°	15	HB9MHS	
1296.945	OH9SHF	Pirttikoski	KP36OI	236	10 dBd	200°	30	OH6DD	
1296.948	FX4UHX	St Aignan	IN94UW	88	2 x Big wheel	Omni	50	F6CIS	
1296.950	DB0HG	Frankfurt am Main	JO40HG	300	Big wheel	Omni	1	DH2FAH	Proposal
1296.950	OZ5UHF	Kobenhavn	JO65GQ	35	Collinear	Omni	1	OZ3TZ	Proposal
1296.955	OZ1UHF		JO57FJ	150	Big wheel	Omni	10	OZ9NT	
1296.960	HG7BUB		JN97KR	690	Slot	Omni	0.5	HA1YA	
1296.960	SK4UHG	Hagfors	JP60VA	440	2 x Helix	Omni	50	SM4DHN	
1296.965	GB3ANG	Angus	IO86MN	319	Slot Yagi	170°	15	GM4ZUK	
1296.965	DF0ANN	Lauf	JN59PJ		4 x DQ 23	Omni	2	DL8ZX	
1296.975	DL0SG	DOK U 14	JN69KA	1024	4 x DQ	Omni	5	DJ4YJ	
1296.975	HG1BSA	Szombathely	JN87GG	370	2 x Hybrid Quad	90°	2.5	HA1YA	
1296.975	OH3RSE	Tampere	KP11UM	247	Big wheel	Omni	7		
1296.975	ON4AZA	Antwerp	JO21EE					ON1BPS	On test
1296.980	DB0JU	DOK L 04	JO31CV	150	Helical	Omni		DF5EO	
1296.980	SK2UHG	Kristineberg	JP95HB	500	Horizontal	Omni/180°	80/500		
1296.984	OZ2ALS	Als	JO44WX	28	2 x slot	Omni	8	OZ9DT	
1296.985	DB0AS	DOK C 29	JN67CR	1565	Dipolfeld	10°	0.5	DL2AS	
1296.990	DB0FB	DOK Z 06	JN47AU	1495	8 el Group	45°	5	DJ3EN	
1296.990	GB3EDN	Edinburgh	IO85HW	117	2 x Corner Refl	45°/315°	25	GM8BJF	
1296.995	DB0WOS	DOK U 16	JN68ST	850	4 x DQ	Omni	5	DF8RU	
1296.965	GB3ANG	Dundee	IO86MN	310	15/15 Slot Yagi	Horizonal	16	GM4ZUK	Lic - non op 5/9/97
1297.010	DB0JW	Ubach-Palenburg	JO30DU	200	4 x 11 el Yagi	45°	50	DL9KAS	
2304.040	S55ZNG	Trstelj	JN65UU	643	V-J Slot	Omni	0.1	S57C	
2320.800	SK6UHJ	Goeteborg	JO57	120	Horizontal	Omni	8	SM6PGP	
2320.805	SK0UHH	Taeby	JO99BM	90	Horizontal	Omni	25		
2320.810	DB0OB	DOK U 17	JN69EQ	825	6 x Slot	Omni	1	DC9RK	
2320.820	DB0OT	Esterwegen	JO32QR	80	Big wheel	Omni	1	DL1BFZ	
2320.825	DB0HF	Harksheide	JO53BO	45	Big wheel	Omni		DK2NH	
2320.825	OE1XTB	Vienna	JN88EE	170	4 x Dipole	Omni	1	OE1MCU	Temp QRT
2320.830	GB3MHS	Martlesham	JO02PB	85	Slotted Waveguide	Omni	15	G4DDK	
2320.830	DB0JX	Willich	JO31FF	115	Double heliax	Omni		DK4TJ	
2320.838	F5XAC	Pic Neulos	JN12LL	1100	Slotted Waveguide	Omni	20	F6HTJ	
2320.840	DB0KI	Bayreuth	JO50WC	925	4 x 6 el	Omni	1	DC9NL	Planned
2320.845	DB0LBV	DOK S 30	JO61EH	234	DQ	135°/225°	1.5	DL1LWM	
2320.845	SR3SHF	Kalisz	Jo91CQ					SP3JBI	
2320.850	DB0GW	DOK L 01	JO31JK	80	2 x Helix	Omni	8	DL4JK	
2320.850	DL0UB	DOK Z 20	JO62KK	120	5 x Dipole	Omni	10	DL7ACG	
2320.850	GB3NWK	Orpington	JO01BI	180	Alford Slot	Omni	5	G8BJG	
2320.855	DB0SHF	DOK Z 46	JN48XS	800	6 x Dipole	260°	0.2	DL1SBE	
2320.857	PI7GHG	Capelle	JO21GW	49	20dB Gain	135°	400	PE1GHG	
2320.860	LA1UHH	Tonsberg	JO59DD	30	13 dB Horn	180°	50	LA6LCA	
2320.862	F1XAH	Istres	JN23MM	114	Slotted Waveguide	Omni	15	F1AAM	
2320.865	OE1XVB	Vienna	JN88EF	191	4 x 6 el	Omni	100		
2320.865	SK7MHG	Veberöd	JO65SO	200		Omni	50		
2320.870	DB01BB	DOK N 49	JO32VG	200	10 x Slot	Omni	4	DB7QW	
2320.879	PI7TGA	Nijmegan	JO21WU	75	10dB Gain	270°/ 315°	10/10t	PA0TGA	
2320.880	DB0GO	DOK N 32	JO41ED	738	10 x Slot	Omni	50	DB1DI	Proposal
2320.880	DB0YI	Hildesheim	JO42XC	480	Big Wheel	Omni	3	DL4AS	
2320.880	LA3UHH	Flekkeroy	JO38XB	5	2 x 6 dB Horn	90°/180°	1	LA8AK	Planned
2320.890	GB3ANT	Norwich	JO02PP	75	Alford slot	Omni	5	G8VLL	
2320.883	DB0INN	DOK C 15	JN68GI	504	Slot	Omni	1	DL3MBG	
2320.900	DB0UX	Grotzingen	JN48FX	275	Big wheel	Omni	1	DK2DB	
2320.900	DB0JW	DOK G 05	JO30DU	238	6 el Array	45°	50	DL9KAS	
2320.902	LX0THF	Walferdange	JN39BP	420	Double quad	Omni	0.5	LX1JX	
2320.905	GB3SCS	Swanage	IO90AP	210	Alford slot	Omni	60	G4GHP	
2320.912	DL0UH	DOK Z 25	JO41RD	385	6 x Dipole	0°	2	DJ3KO	
2320.915	DB0UBI	DOK N 59	JO42GE	165	Collinear	45°	0.5	DD8QA	
2320.920	DB0VC	Albersdorf	JO54IF	244	Big wheel	Omni	5	DL8LAO	Non op
2320.920	PI7QHN	Zandvoort	JO22KH	45	6dB Gain	Omni	4	PA0QHN	Non op

Freq MHz	Call	Nearest town	IARU loc	ASL m	Aerial	Beam Direction	ERP W	Info from	Notes
2320.925	GB3PYS	Newtown	IO82HL	436	Alford Slot	Omni	10	GW4NQJ	Planned
2320.930	OK0EL	Vrchlabi	JO70SQ	900	5dB Horn	135°/270°	0.8	OK1AIY	Op.as of 10/11/95
2320.930	OZ7IGY	Tollose	JO55VO	91	Alford slot	Omni	20	OZ7IS	
2320.935	PI7PLA	Zuidlaren	JO33IC	50	6dB Gain	Omni	1	PA0PLA	Proposal
2320.937	DB0JO	Kamp-Lintfort	JO31SL	312	Horn	270°	0.2	DG8DCI	
2320.945	DB0OS	Hitchinbach	JO40CW	745	8 el array	270°	1	DC1DB	Planned
2320.950	DB0KP	DOK P 09	JN47TS	435	Slot	Omni	0.1	DL1GBQ	
2320.950	OZ9UHF		JO65HP	30	Slot	Omni	5	OZ2TG	
2320.955	GB3LES	Leicester	IO92IQ	220	Slot	160°	30	G3TQF	
2320.955	OZ1UHF		JO57FJ	150	Slot	Omni	8	OZ9NT	
2320.963	HG6BUC		KN07AU	1050		Omni	2	HG5ED	
2320.965	DF0ANN	Lauf	JN59PJ	630	4 x Double quad	Omni		DL8ZX	Proposal
2320.967	DB0AS	Rosenheim	JN67CR	1560	28 el Yagi	337°		DL2AS	On test
2320.975	DB0JL	DOK R 25	JO31MC	195	Slot	Omni	2	DF1EQ	
2320.980	DB0JU	Doesburg	JO31CV	150	Helical	Omni	2	DF5EO	Non op
3400.009	DL0UB	DOK Z 20	JO62KK	120	12 x Slot	Omni	10	DL7ACG	
3400.018	PI7SHF	Schipol Airport	JO22JH	80	10dB Slot	Omni	2	PA0EZ	
3400.020	DB0AS	DOK C 29	JN67CR	1565	Double 8	10°	0.5	DL2AS	
3400.040	DB0KI	Bayreuth DOK z 42	JO50WC	925	SLot	Omni	50	DC9NL	
3400.050	DB0JL	DOK R 25	JO31MC	195	Helical	Omni	1	DF1EQ	
3400.170	PI7CKK	Groningen	JO33GE	50	10dB Slot	Omni	5	PE1CKK	
3400.850	DB0GW	Duisburg DOK L 01	JO31JK	80	Double Helical	Omni	8	DL4JK	
3400.905	GB3SCF	Swanage	IO90AP	210	Alford Slot	Omni	17	G4GHP	
3400.955	GB3LEF	Leicester	IO92IQ	222	Alford slot	135°	8	G3TQF	
3456.005	DB0EZ	Zevengar	JO31BS	110	Slot	115°	0.1	DB9JC	
3456.800	DB0KHT	DOK F 16	JO40FE	247	Horn	Omni	10	DJ1RV	
3456.830	DB0JX	DOK R 21	JO31FF	115	Helical	Omni	0.1	DK4TJ	
3456.855	DB0SHF	DOK Z 46	JN48XS	800	Horn	260°	0.5	DL1SBE	
3456.883	DB0INN	DOK C 15	JN68GI	504	Slot	Omni	1	DL3MBG	
3456.900	GB3OHM	S Birmingham	IO92AJ	171	16 Slot waveguide	Omni	8	G6KOA	
5760.030	OK0EL	Vrchlabi	JO70SQ	1035	5dB Horn	135°/270°	0.08	OK1AIY	
5760.040	PI7EHG	Schipol Airport	JO22JH	80	Horizontal	Omni	6	PA0EHG	
5760.040	OK0EA	Trutnov	JO70UP	1355	12 el Slot	180°/270°	0.5	OK1AIY	
5760.050	DB0EZ	Didam	JO31BS	110	Slot	Omni	1	DB9JC	
5760.060	F1XAO		IN88HL	326	Slotted Waveguide	Omni	10	F1LHC	
5760.070	DB0JL	DOK R 25	JO31MC	195	Slot	Omni	0.8	DF1EQ	
5760.100	DB0AS	DOK C 29	JN67CR	1565	Double 8	10°	0.5	DL2AS	
5760.800	DB0KHT	DOK F 16	JO40FE	247	Horn	Omni	0.5	DJ1RV	
5760.830	DB0JX	DOK R 21	JO31FF	115	Slot	Omni	0.08	DK4TJ	
5760.840	DB0KI	Bayreuth	JO50WC	925	Slot	Omni		DC9NL	
5760.850	DL0UB	DOK Z 20	JO62KK	120	12 x Slot	Omni	0.2	DL7ACG	
5760.855	DB0SHF	DOK Z 46	JN48XS	800	Array	260°	0.4	DL1SBE	
5760.860	DB0ARB	DOK U 02	JN69NC	1456	Slot	Omni	3	DJ4YJ	
5760.860	LA1SHF	Tonsberg	JO59DD	30	13 dB Horn	180°	25	LA6LCA	
5760.865	OE1XVB	Vienna	JN88EF	191	Horn		4		
5760.883	DB0INN	DOK C 15	JN68GI	504	Slot	Omni	1	DL3MBG	
5760.900	DB0CU	DOK A 28	JN48BI	970	Slot	Omni	5	DJ7FJ	
5760.900	HG6BSA		KN07AU	1050			2	HG5ED	
5760.905	GB3SCC	Swanage	IO90AP	210	Alford Slot	Omni	6	G4GHP	
5760.930	OZ7IGY	Tollose	JO55VO	91	Slotted Waveguide	90° /270°	15	OZ7IS	
5760.950	OZ9UHF		JO65HP	30	Slotted Waveguide	Omni	2	OZ2TG	
5760.955	OZ8SHF		JO57FJ	150	Slotted Waveguide	Omni	8	OZ9NT	
10100.000	GB3IOW	Newport, IOW	IO90IO	250	Slotted waveguide	Omni	1	G8MBU	
10120.000	GB3ALD	Alderney	IN89VR	90	Sectoral horn	30°	1		
10368.015	DB0EZ	Kleve DOK L IG	JO31BS	110	Slot	115°	1	DB9JC	
10368.040	PI7SHY	Eindhoven	JO21RK	80	21 dBi	315°	1	PA0SHY	
10368.050	OK0EL	Benecko	JO70SQ	900	Waveguide	135°/270°	0.05	OK1AIY	
10368.050	OZ9UHF		JO65HP	30	Slotted WG	Omni	3	OZ2TG	
10368.060	F1XAI	Orleans	JN07WT	160	Slotted WG	Omni	10	F1JGP	
10368.075	OK0EA	Trutnov	JO70UP	1355	12 el Slot	180°/270°	0.5	OK1AIY	
10368.100	PI7TGA	Nijmegan	JO21WU	75		Omni		PA0TGA	
10368.120	DB0JL	DOK R 25	JO31MC	195	Slot	Omni	0.15	DF1EQ	
10368.140	ON4TNR	Wepion	JO20KJ	250	18 dB Horn	292°	7	ON5VK	
10368.150	OE8XXQ	Dobratsch	JN76UO		Horn		0.1		
10368.175	DB0AS	DOK C 29	JN67CR	1565	Horn	10°	0.5	DL2AS	
10368.180	F1XAP		IN88HL	326	Slotted Waveguide	Omni	5	F1LHC	
10368.205	PI7EHG	Schipol Airport	JO22JH	90	13 dBi Slot	Omni	30	PA0EHG	
10368.240	GB3SWH	Watford	IO91TP	187	Slotted waveguide	45°/ 225°	1	G4KUJ	
10368.270	DL0WY	Rosenheim	JN67CR	1560	10 dB Slot horn	315°		DJ8VY	
10368.270	PI7GHG	Capelle Ijssel	JO21GW	50		Omni	10	PE1GHG	
10368.805	DB0XL	DOK E IG	JO53HU	45	Slot	Omni	1	DK1KR	
10368.820	DB0KHT	DOK F 16	JO40FE	247	Horn	Omni	3	DJ1RV	
10368.825	DB0HRO	DH0HRO	JO64AD	185	Slot	Omni	0.2	DL5CC	
10368.830	DB0JX	Wickrath	JO31FF	115	10 dB	Omni		DK4TJ	
10368.830	GB3MHX	Martlesham	JO02PB	80	12 Slot waveguide	Omni	1	G4DDK	
10368.835	SK0SHG	Kista	JO89XJ	60	Horizontal	Omni	0.5	SM0KAK	
10368.840	DB0JO	Kamp-Lintfort	JO31SL	312	6 x Slot	Omni	1	DG8DC	
10368.840	DB0KI	Bayreuth	JO50WC	925	Slot	Omni		DC9NL	
10368.850	DB0GG	DOK P 24	JN48NS	400	Slot	Omni	0.05	DL5AAP	
10368.850	DL0UB	DOK Z 20	JO62KK	120	12 x Slot	Omni	0.1	DL7ACG	
10368.850	GB3SEE	Reigate	IO91VG	250	Slotted waveguide	Omni	3	G0OLX	
10368.855	DB0SHF	DOK Z 46	JN48XS	800	Horn	260°	0.1	DL1SBE	
10368.860	DB0ARB	DOK U 02	JN69NC	1456	Slot	Omni	3	DJ4YJ	
10368.860	F5XAD	Pic Neulos	JN12LL	1100	Slotted Waveguide	0°	3	F6HTJ	On Test
10368.860	LA6SHG	Tonsberg	JO59DD	30	13 dB Horn	180°	10	LA6LCA	
10368.864	LA1UHG								
10368.865	DB0JK	Koln DOK Z 12	JO30LX	260	Slot	Omni	200	DK2KA	
10368.865	OE1XVB	Vienna	JN88EF	191	Horn		4		
10368.870	GB3KBQ	Taunton	IO80LW	167	Slotted waveguide	Omni	1	G4UVZ	
10368.875	OE5XBM	Hellmonsoedt	JN78DK		Slotted Waveguide		0.9		
10368.875	ON4AZB	Antwerp	JO21FF	60	Slotted	Omni	2	ON1BPS	On Test
10368.880	DB0IS	Bodenwerder	JO51GR	1020	Slot			DK6AB	

Freq MHz	Call	Nearest town	IARU loc	ASL m	Aerial	Beam Direction	ERP W	Info from	Notes
10368.880	GB3CEM	Wolverhampton	IO82WO	165	Slotted waveguide	Omni	30	G4PBP	
10368.883	DB0INN	DOK C 15	JN68GI	504	Slot	Omni	1	DL3MBG	
10368.884	HB9G	Geneva	JN36BK	1600	Slotted waveguide	Omni	2	HB9PBD	
10368.885	DB0TUD	DOK S 07	JO61UA	285	Slot	Omni	5	DL4DTU	
10368.890	DB0KLX	DOK K 16	JN39VK	350	Slot	Omni	1	DC2UG	
10368.890	ON4RUG	Ghent	JO11UB	95	Slotted	Omni	7	ON6UG	
10368.900	DB0UX	DOK A 35	JN48FX	275	Slot	Omni	1	DK2DB	
10368.900	DB0CU	DOK A 28	JN48BI	970	Slot	Omni	5	DJ7FJ	
10368.900	GB3SCX	Swanage	IO90AP	200	Slotted waveguide	Omni	1	G4JNT	
10368.900	HG6BSB		KN07AU	1050			0.2	HG5ED	
10368.900	OZ5SHF		JO45WX	170	Slotted WG	Omni	4	OZ2OE	
10368.910	GB3RPE	Swansea	IO81AO	60	Slotted Waveguide	Omni	4	GW4ADL	
10368.915	OZ4SHF		JO65BV	22	Slotted WG	Omni	10	OZ1UM	
10368.920	DB0VC	DOK Z 10	JO54IF	291	Slot	Omni	1	DL8LAO	
10368.920	OE2X	Salzburg	JN		Slotted Waveguide		0.15		
10368.925	OE3XMB	Muckenkogel	JN77TX		Slotted Waveguide		0.15		
10368.930	GB3MLE	Emley Moor	IO93EO	600	Sectoral horns	0°/180°	1	G8AGN	
10368.940	GB3CCX	Cheltenham	IO81XW	342	Slotted Waveguide	Omni	3	G6AWT	
10368.955	OZ9SHF		JO57FJ	150	Slotted WG	Omni	0.8	OZ9NT	
10368.955	GB3LEX	Leicester	IO92JP	220			1	G3TQF	
10368.960	GB3CMS	Chelmsford	JO01HR	120	Slotted WG	Omni	-3	G1EUC	
10368.965	DF0ANN	DOK B 25	JN59PL	630	12 x Slot	Omni	0.2	DL8ZX	
10368.975	ON4KUL	Leuven	JO20IV	125	Slotted waveguide	Omni	1	ON4ADD	
10368.975	OZ3SHF		JO45NL	58	Slotted WG	Omni	2	OZ1IN	
10368.977	HG7BSA		JN97KR	700		150°	0.2	HG5ED	
24025.000	GB3IOW	Newport, IOW	IO90IO	250	Sectoral horn		8	G8IDZ	
24048.905	GB3SCK	Swanage	IO90AP	210	Alford Slot	Omni	3	G4GHP	
24192.000	ON4RUG	Gent	JO11UB	95	Slotted	Omni	0.1	ON6UG	
24192.050	OE5XBM	Hellmonsoedt	JN78DK		Slotted Waveguide		0.05		
24192.055	DB0JO	DOK Z 03	JO31SL	312	6 x Slot	Omni	0.6	DG8DCI	
24192.075	PI7EHG	Schipol Airport	JO22JH	90	Slot	Omni	0.1	PA0EHG	
24192.114	OK0EL	Vrchlabi	JO70SQ	1035	Waveguide	135°/270°	20μ	OK1AIY	
24192.120	DB0JL	DOK R 25	JO31MC	195	Slot	Omni	0.01	DF1EQ	
24192.150	DL0WY	DOK C 29	JN67AQ	1838	Sectored Horn	45°/270°	0.01	DJ8VY	
24192.250	F1XAQ		IN88HL	326	Slotted Waveguide	Omni	0.05	F1LHC	
24192.405	DB0AS	DOK C 29	JN67CR	1565	Horn	10°	0.5	DL2AS	
24192.830	GB3MHK	Martlesham	JO02PB	85	Slotted Waveguide	Omni	15	G4DDK	
24192.830	F5XAF	Paris	JN18DU		Cornet			F5ORF	On Test
24192.840	DB0KI	Bayreuth	JO50WC	925	Slot	0°	0.5	DC9NL	
24192.860	DB0ARB	DOK U 02	JN69NC	1456	Parabola	225°	0.03	DJ4YJ	
24192.865	DB0JK	DOK Z 12	JO30LX	260	2 x H-Horn	Omni	1	DK2KA	
24192.890	GB3DUN	Dunstable	IO91RV	260	Slotted WG	Omni	1	G3ZFP	
24192.900	DB0CU	DOK A 28	JN48BI	970	Horn	180°	5	DJ7FZ	
24192.910	GB3NUL	Mow Cop	IO83VC	337	Slotted Waveguide	Omni	1	G7MRF	
24192.915	OZ4SHF		JO65BV	22	Slotted WG	Omni	10	OZ1UM	
24192.940	GB3AMU	Cardiff	IO81JN	266	Sectorial Horn	135°	1	GW3PPF	
24192.975	ON4LVN	Leuven	JO20IV	120	Slotted Waveguide	Omni	0.5	ON4AOD	

Notes:

IBP = International Beacon Project, stations transmit 10 seconds on each frequency in turn as follows (only currently operational stations shown):

Country	Call	14.100	18.110	21.150	24.930	28.100
United Nations NY	4U1UN	00.00	00.10	00.20	00.30	00.40
Northern Canada	VE8AT	00.10	00.20	00.30	00.40	00.50
USA (CA)	W6WX	00.20	-	00.40	-	01.00
Hawaii	KH6WO	00.30	00.40	00.50	01.00	01.10
New Zealand	ZL6B	00.40	00.50	01.00	01.10	01.20
West Australia	VK6RBP	00.50	01.00	01.10	01.20	01.30
Japan	JA2IGY	01.00	01.10	01.20	01.30	01.40
China	BY	01.10	01.20	01.30	01.40	01.50
Siberia	UA	01.20	01.30	01.40	01.50	02.00
Sri Lanka	4S7B	01.30	01.40	01.50	02.00	02.10
South Africa	ZS6DN	01.40	01.50	02.00	02.10	02.20
Kenya	5Z4B	01.50	02.00	02.10	02.20	02.30
Istael	4X6TU	02.00	02.10	02.20	02.30	02.40
Finland	OH2B	02.10	02.20	02.30	02.40	02.50
Madeira	CS3B	02.20	02.30	02.40	02.50	00.00
Argentina	LU4AA	02.30	02.40	02.50	00.00	00.10
Peru	OA4B	02.40	02.50	00.00	00.10	00.20
Venezuela	YV5B	02.50	00.00	00.10	00.20	00.30

W6WX and KH6WO are not currently licensed for 18 or 24 MHz operation.

? Activity pattern uncertain

V Variable

† LN2A and VL8IPS are part of an ITU field-strength measuring programme. Their schedule is:

		LN2A			VL8IPS		
1	14409	00-04	20-24	40-44 mins	12-16	32-36	52-56 mins
2	20947	04-08	24-28	44-48	16-20	36-40	56-00
3	5470	08-12	28-32	48-52	00-04	20-24	40-44
4	7870	12-16	32-36	52-56	04-08	24-28	44-48
5	10407	16-20	36-40	56-00	08-12	28-32	48-52

NB. These are the nominal frequencies; morse IDs will be heard at 1650Hz above nominal, and consequently the cw space' frequency is given in the listings above.

zz Normal transmission: DK0WCY beacon (x3) + 4 seconds dash.
During auroras: DK0WCY beacon (x3) aurora + short dashes or DK0WCY beacon (x3) strong aurora + 9 seconds dash. At every full 5 minutes basic solar/geophysical information and forecast on CW updated around 0630utc. May not operate on contest weekends. 1 hour earlier in Summer.

Please notify HF errors/changes to Martin Harrison G3USF, HF Beacon Coordinator, Region 1 of the International Amateur Radio Union, by e-mail, M.Harrison@keele.ac.uk) or to 1 Church Fields, Keele, Staffs ST5 5HP, England. Tel: (home) +44 (0)1782 627396 Fax (work) +44 (0)1782 583452.

The IARU Region 1 VHF/UHF Beacons is copiled by G3UUT of the RSGB VHF Committee. It includes inputs from the VHF/UHF/Microwave managers of radio societies across Europe, beacon keepers, beacon coordinators + VHF/UHF DXers too numerous to mention.

This year, thanks to the excellent beacon list of GJ4ICD of the UK 6m group, we have included world wide 6m beacons in the list although it should be remembered that at this part of the sunspot cycle the chances of hearing many of them are very low.

All inputs are welcome and should be sent to; John F.Wilson G3UUT RSGB/ IARU Region 1 VHF Beacon Coordinator (Email: jfwilson@iee.org. Packet: G3UUT@GB7HXA)

Club and Society Affiliation

Many local amateur radio societies and clubs choose to affiliate with the Radio Society of Great Britain because they see it as an effective way of demonstrating their support for the aims and aspirations of the National Society. The Society welcomes this support because it can only strengthen its claim to speak on behalf of amateurs and amateur radio.

The Society also recognises that much of the vitality of amateur radio at the local level lies in clubs. It sees affiliation as a direct link reflecting the complementary nature of their relationship with the national society – and which it has every intention of trying to strengthen.

There are more tangible benefits for the Affiliated Society. These include:

- ✦ Publicity for club activities through the 'Club News' in RadCom and via broadcasts on the Society's news service GB2RS.
- ✦ Full facilities of the RSGB QSL Bureau for cards bearing the club station call.
- ✦ Purchase of publications at a discount with the RSGB.
- ✦ Receipt of *Radio Communication.*
- ✦ Freedom to participate in RSGB Affiliated Societies contests.
- ✦ Insurance for club-owned equipment under the terms of the Amateur Radio Insurance Scheme.
- ✦ Freedom to borrow RSGB films, tapes and display materials. (This facility is also available to certain non-affiliated groups such as schools).

Tax Position of Members' Clubs

by Richard Horton, FCA

Introduction

The RSGB has received a number of enquiries from radio clubs regarding the tax position in respect of their income where this is derived in part from events such as rallies which are open to non-members.

Summary of the legal position

There are no specific statutory provisions in relation to members' clubs, radio or otherwise. Their liability to tax is determined by the usual rules that apply to trading activities. Liability will invariably be to corporation tax (not income tax) because the club will either be a limited company (unlikely) or an unincorporated association (most likely).

Liability to corporation tax

No liability should arise on surpluses that accrue from members' subscriptions, entrance fees, donations or the sale of equipment to members.

However, there is a liability to tax on interest received and other investment income. In addition, where the club derives income from non-members – for example, entrance fees for rallies or other events, equipment rental or sales, table fees from traders at rallies etc, then any profit arising from these activities will be chargeable to corporation tax. In arriving at the profit chargeable to tax, a deduction for expenses (including any interest payable) will be permitted and this may involve an apportionment of the club's general expenses.

Disclaimer

This note is intended to provide general guidance. It should not be taken as definitive advice and the precise tax position of each case will depend on the circumstances. Clubs that believe they may have a liability to tax should consider taking professional advice from a firm of accountants or lawyers. No responsibility, legal or otherwise, is accepted by the RSGB, its officers or employees for the tax or financial affairs of independently run clubs.

How to Affiliate Your Club, Society or Group to the RSGB

Membership categories

Any UK club, group, society or emergency communication group may affiliate to the RSGB provided it fulfills just a few requirements. The affiliation fee is presently £20.00 and this includes the receipt of *Radio Communication.*

Procedure

There is a special procedure required to affiliate a club or society to the RSGB, or register an amateur radio group. The procedure is as follows:

i) Please complete a society affiliation form obtainable from RSGB Sales Dept. If your organisation has a callsign, please let us know on the application form. If it does not, we will issue a receiving station number for reference purposes.

Note that for UK clubs, groups and societies the RSGB zone and RSGB region will be determined by the administrative county for the address given on the application form. Clubs, groups and societies near to, or spanning county boundaries should decide carefully with which county they wish to be associated and insert the appropriate choice in the address of the club they are entering on the application form.

Please also note that once your club address is on our files, we will regard it as information that can be freely given out to those seeking to contact local clubs.

ii) Please send to the appropriate Zonal Council Member the following:

- Your completed application form signed by the chairman or secretary
- A copy of the club's constitution or rules
- A list of current officers of the club
- A statement of the number of members, and the proportion who are members of RSGB.

A list of Zonal Council Members is published in *Radio Communication* and in the *Yearbook.* The Zonal Council Member will vet your constitution/rules and if suitable will countersign your application form. He or she will then return the form and your constitution or rules to you.

Note that only the RSGB Zonal Council member may countersign an application for a club, group or society. Overseas organisations should send their application form and constitution/rules direct to RSGB HQ addressed to 'The RSGB Secretary'.

iii) Finally, please send your countersigned application form, constitution and remittance to:

New Members
RSGB HQ
Lambda House
Cranborne Road
POTTERS BAR
Herts, EN6 3JE

Objectives of the RSGB

1. **To promote the general advancement of the science and practice of radio communication**
2. **To facilitate the exchange of information and ideas on these subjects amongst its members**
3. **To obtain the maximum liberty of action consistent with safeguarding the interests of all concerned**

The RSGB is a UK registered company - 216431

Model Constitution for Affiliated Societies

The following guidance is intended for those writing a constitution for their local society which will be acceptable for affiliation to the RSGB.

CONSTITUTION

1. Name

The Society (1) shall be known as the.............Amateur Radio (2) Society.

2. Aims

The aims of the Society shall be to further the interests of its members in all aspects of amateur radio and directly associated activities.

3. Membership

Membership shall be open, subject to the discretion of the Committee, to all persons interested in the aims of the Society (3).

(a) **Full members.** Full members must be 18 years of age or over or must hold the permission of any competent authority to install and operate an amateur radio station.

(b) **Student members.** Student members must be under 25 years of age and in full-time formal education.

(c) **Honorary members.** Honorary Life Membership may be granted to any person, who, in the opinion of the Committee, has rendered outstanding service to the Society, either directly or indirectly. Such membership shall carry the rights of full membership but shall be free from subscriptions.

(d) **Guests.** Members may invite guests to meetings. No visitor may attend more than three meetings in each year.

All members shall abide by the constitution of the Society. The Committee shall have power to expel any member whose conduct, in the opinion of at least three-quarters of the full Committee, renders that person unfit to be a member of the Society. No Member shall be expelled without first having been given an opportunity to appear before the Committee.

4. Subscriptions

(a) The annual subscriptions for membership shall be set by the Committee (4).

(b) All subscriptions shall be due and payable at the beginning of the financial year. Members in arrears have no voting rights.

(c) The financial year shall run from 1 September to 31 August (5).

(d) A member shall have deemed to have resigned from the Society, if, by the following 31 August (6), the subscription has not been paid.

(e) The Committee shall have the power to waive or reduce subscriptions in special circumstances for a period not exceeding...years at a time (7).

5. Finance

All money received by the Society shall be promptly deposited in the Society's bank account. Withdrawals require the signature of the Society's Treasurer and one other nominated officer of the Society (8).

6. Membership of the Society's Committee

The Society's affairs shall be administered by a Committee elected at the Annual General Meeting (9). The Committee, in whom the Society's property shall be vested, shall consist of:

(a) A Chairman who will preside at all meetings at which he is present. No member may hold this position for more than two consecutive years. He may be re-elected after a break of one year.

(b) A Vice-Chairman who will act as chairman in the absence of the Chairman.

(c) A Secretary who will be responsible for:
- (i) keeping the minutes of all meetings of the Society.
- (ii) ensuring that all correspondence is correctly handled.
- (iii) maintaining a master roll of members and honorary members.
- (iv) maintaining a register of Society equipment.

(d) A Treasurer, who will be responsible for:
- (i) keeping the Society's accounts.
- (ii) advising the Committee on all financial matters.
- (iii) preparing the accounts for audit and presenting them at the AGM.

(e) ......Ordinary Committee Members (10).

(f) Not more than......co-opted members who have full voting powers (11), and not more than......who are not permitted to vote (12).

7. Committee standing orders

(a) The quorum for the Committee shall be......(13). In the absence of a quorum, business may be dealt with but any decisions taken only become valid after ratification at the next meeting at which a quorum exists.

(b) Committee meetings may be called by the Chairman, the Secretary or any vote.

8. Annual General Meeting

(a) The Annual General Meeting shall normally be held on the first......day of October each year (14). At least 21 days notice shall be given to each member in writing.

(b) The quorum for the meeting shall be......(15).

(c) The agenda for the meeting shall be:
- (i) Apologies for absence
- (ii) Minutes of the previous AGM
- (iii) Chairman's report
- (iv) Secretary's report
- (v) Treasurer's report
- (vi) Election of the new Committee
- (vii) Election of auditors
- (viii) Other business

(d) Items (i) to (v) shall be chaired by the out-going Chairman, item (vi) by an acting Chairman who is not standing for election to office, and the remaining business by the newly elected Chairman.

(e) Nominations for Committee members will only be valid if confirmed by the nominee at the meeting or previously in writing.

(f) Items to be raised by members under other business must be notified to the Secretary not less than 21 days before the AGM.

9. Extraordinary General Meeting

(a) Extraordinary General Meetings may be called by the Committee or not less than......members of the Society, the date of the meeting being the earliest convenient as decided by the Committee (16). At least 28 days notice in writing must be given to the Secretary, who in turn shall give members at least 14 days notice in writing of the agenda. No other business may be transacted at the EGM.

(b) The quorum for the EGM shall be......(15).

10. Amendments to the constitution

The constitution may be amended only at an EGM called for that purpose.

11. Winding up of the Society

(a) The decision to wind up the Society may be taken only at an EGM.

(b) The funds of the Society shall, after the sale of all assets and the payment of all outstanding debts, be disposed of as directed by members at the final EGM (17).

Notes

(1) *Society/Club/Group, etc*

(2) *For a Society to be affiliated to the RSGB, it must include the words "amateur radio" in its heading unless a special dispensation is obtained from the RSGB's Council. This requirement does not apply to Societies already affiliated.*

(3) *This group may include, for example, youngsters, the family and friends of full members whose interest is in the social side rather than amateur radio, or local persons of influence whom it is wished to link with the Society. They may pay a reduced subscription. It is important to specify what voting rights they may or may not have.*

(4) *Alternatively, the subscription may be recommended by the Committee for ratification at the AGM.*

(5) *These dates have the advantage of corresponding to the start of the "winter season", but others can be chosen. Bear in mind that they relate also to the date of the AGM.*

(6) *More specifically, the end of the financial year.*

(7) *This period perhaps should not exceed one to three years to avoid placing an undue burden on future Committees.*

(8) *There are great advantages in running the Society's finances on a strict basis, although a less formal arrangement may still be effective.*

(9) *There are two methods for electing the Committee: the more common is for the meeting to elect the Committee members and for the latter in turn to elect the officers from within the Committee; alternatively, the members may elect individuals to specific offices. The method adopted will need to be specified.*

(10) *The number of Ordinary Committee members should be related to the size of the Society. Remember that being a committee member is an essential part of the training of the future officers of the Society.*

(11) *These can replace elected Committee members who have left the Committee.*

(12) *These can be people who need to be familiar within the work of the Committee such as the editor of the Society magazine or the press officer.*

(13) *This can be expressed either as a fixed number or, for example, as at least half or two-thirds of the full membership of the committee.*

(14) *This date obviously must be related to the Society's financial year.*

(15) *This can be set either as a fixed number or a fixed percentage of the membership (state which members are to be included), or both "whichever is the smaller/greater". It is probably safer to make the numbers on the small side so as to ensure that the meeting can take place.*

(16) *A typical number would be 10.*

(17) *Such as, among its members, to a charity, or to a society of similar interest.*

National Affiliated Societies and Clubs

Amateur Radio Caravan & Camping Club *(G4RCC)*

This club was formed in 1979 by G4EPN, G4MTP and G8RRB with the aim of promoting caravan and camping rallies for radio amateurs, short wave listeners and their families. The club is esssentially a midlands based organisation, where the majority of rallies are held.

Details from the Membership Secretary (SAE please): Mr J. Bennett, G3PVG, 11 Enderby Road, Thurlaston, Leicester, LE9 7PF. Tel: (01455) 888201.

AMSAT-UK *(G0AUK)*

Formed 18 years ago, this is the UK national society specialising in amateur satellite matters. It has approximately 2,000 paid-up members, produces a regular publication, '*Oscar News*', for its members six times per year and holds three weekly nets on 3,780kHz ± QRM. (Mondays and Wednesdays at 1900 and Sunday at 10.15 local time). Membership is by donation, for which there is a suggested minimum. Extra donations are always welcome and can be sent anonymously. AMSAT-UK now has full representation at IARU. Funds raised are used to build satellites for all to use.

Enquiries, and application forms for membership, with an sase to the Honorary Secretary: Ron Broadbent, MBE, G3AAJ, 94 Herongate Road, London E12 5EQ. Tel: (0181) 989 6741. Fax: (0181) 989 3430.

If you use amateur radio satellites, be prepared to pay something to the organisation which designs, builds and launches them – AMSAT!

BARTG *(British Amateur Radio Teledata Gp - G4ATG, GB2ATG, GB4ATG)*

Interested in packet, AMTOR, RTTY, fax or other forms of data comms? If you are then you should find out more about BARTG, the national (and international) specialist group for the data enthusiast. BARTG offers you its quarterly journal '*DATACOM*', contests on HF, runs an annual rally, runs an awards scheme, supplies a range of components, publishes a range of useful books and does lots more.

So you want to find out more? Please contact BARTG's Membership Secretary: Bill McGill, G0DXB, 14 Farquahar Road, Maltby, Rotherham, South Yorkshire, S66 7PD. Tel: (01709) 814010.

British Amateur Television Club *(BATC - RS38114)*

Transmitting and receiving television on the amateur bands is tremendous fun and real satisfaction can be achieved from constructing and operating a television station.

The BATC produces a quarterly 90 page magazine called CQTV which is full of circuits, projects and information for television enthusiasts. Fast Scan, Slow Scan and all forms of television are covered in CQTV which is sent free to all members. A range of PCB's and certain components are available for some of the projects in CQTV and in the television handbooks published by the BATC.

The BATC was founded in 1949 to inform, instruct, co-ordinate and represent the activities of television enthusiasts both in the UK and world-wide. The BATC is run by an elected committee of volunteers.

Membership enquiries and applications should be made to: Dave Lawton, G0ANO 'Grenehurst', Pinewood Road, High Wycombe, Bucks, HP12 4DD. Internet http://www.batc.org.uk The subscription fee for 1997 is £12.00.

British Rail ARS *(G4LMR)*

The Society was formed in 1966 by a few railway radio amateurs and has now over 100 rail and non-rail members. It is part of the international society FIRAC. A quarterly newsletter, spring get-together and yearly congress are some of the activities. XYL's and YL's are made welcome. Details from: Mr G Sims, 85 Surrey Street, Glossop, Derbyshire, SK13 9AJ.

British Young Ladies Amateur Radio Assn *(BYLARA)*

BYLARA was formed in April 1979 to further YL operation in Britain and so promote friendship, stimulate interest, and in particular encourage good operating techniques and courtesy to all operators at all times. A weekly net on Mondays at 7.15pm local time 3.688 or 3.703 MHz ± QRM. OM welcome to join 7.45pm.

Details about BYLARA can be obtained from the Secretary: Mrs M Ritson, G7FYV, 14 Dunsdale Rd., Holywell, Whitley Bay, Tyne & Wear, NE25 0NG.

Chiltern DX Club *(RS92589)*

The CDXC caters for all UK amateurs who have an interest in competitive activity on the HF bands (DXing, contesting, award chasing etc). The club produces a bi-monthly newsletter, supplies a high quality QSL card at an attractive price, supports worldwide DXpeditions and special event activities by members. Membership is open to any amateur holding the equivalent of a British Class A licence and who has worked a minimum of 100 DXCC countries. The club is also open to active SWLs.

Details from the Secretary: Alan Jubb, G3PMR, 30 West Street, Gt. Gransden, Sandy, Beds, SG19 3AU. E-mail: shacklog@aol.com or visit our web site: http://www.g4uol.demon.co.uk/cdxc.htm

Civil Service ARS *(G3CSR & G1CSR)*

Open to any serving or retired civil or public servant interested in amateur radio. HF station available to members. Society nets, usually controlled by G3ENV and to which all are welcome, are held every Tuesday evening at 7.30pm on 144.370 MHz and 8.00pm on 3.720 MHz ±QRM.

Details from the Secretary, CSARS, Civil Service Recreation Centre, 1 Chadwick St, London SW1P 2EP; or the Chairman: John Pinnell, G3XWK, 31 Nunhead Ln, London SE15 3TR. Tel: (0171) 732 8605.

Club of Friendship Between UK and Russian Radio Amateurs *(G4BAS)*

Formed in 1987 by Ken Norvall, G3IFN, the aims of the club are to promote and develop links between radio amateurs in both countries. Newsletters are published and sent to all members.

Further details can be found on the Internet at http://www.proweb.co.uk/~howardk or from the President, Howard Ketley, 1 Tewkesbury Avenue, Mansfield Woodhouse, Notts, NG19 8LA

G-QRP Club *(RS38364)*

This club specialises in low power operation (hence the QRP), primarily on the HF bands. It produces a quarterly magazine for its members called *'Sprat'*. Membership is in excess of 5,000. Further details can be obtained from the Secretary: Rev. George Dobbs, G3RJV, St. Aidans Vicarage, 498 Manchester Road, Rochdale, Lancs, OL11 3HE.

International Listeners' Association *(RS88763)*

The Association was formed in 1985 to encourage radio listening in all its aspects. The quarterly newsletter *'Just Listening'* contains articles, features, news and projects covering the whole field of the hobby. Awards and competitions encourage dedicated listening. Further details are available from Trevor Morgan, GW4OXB, 1 Jersey St, Hafod, Swansea, SA1 2HF.

International Police Assn Radio Club *(IPARC - G4IPA)*

This club is open to any serving or retired member of the police service. Aims of the club are expressed in the motto: 'service through friendship'. Regular weekly nets, both national and international. Further details can be obtained from Mike Hammond, G3PGA, 47 Yelland Road, Fremington, Devon, EX31 3DS. Tel: (01271) 860930.

International Short Wave League *(G4BJC)*

Known as the ISWL, the League was formed in October 1946 and caters for members with interests in both the amateur and broadcast bands, membership being open to SWLs and licensed amateurs. A monthly journal *'Monitor'* covers HF and VHF reception conditions, the SW BC bands, transmitting topics, technical and various general articles. The League holds monthly contests and has a comprehensive awards programme.

Details from Honorary Secretary: Mrs M. H. Carrington, G0WDM, 3 Bromyard Drive, Chellaston, Derby, DE73 1PF.

Military Wireless Amateur Radio Society *(G0PTZ)*

The Society was founded in 1993. The members collect, restore, display at Shows and use ex-military radio and other electronic equipment. The club net is on Saturday mornings at 0930 on 3.620± MHz. An A4 sized club newsletter is produced every other month. Membership now exceeds 300 with members in 18 different countries. Further details are available from John Taylor-Cram, 7 Hart Plain Avenue, Cowplain, Waterlooville, Hampshire, PO8 8RP. Tel: (01705) 250463. WT net: 0900, Sundays on 3.577 MHz.

Moroni ARA (UK) *(G0LDS, G1LDS)*

Details can be obtained from the President: John Wiles, G4TVA, 38 Northwood Lane, Clayton, Newcastle-Under-Lyme, Staffs, ST5 4BN.

Open University ARC *(OUARC - G0OUR)*

Membership is open to both students and staff of the university. Weekly meetings are held in the shack on the Walton Hall campus every Thursday lunch time. The club station is currently active on both HF and VHF. Packet mail can be sent to G0OUR @ GB7BEN. The club is also on the Internet, and can be reached via e-mail at adrian@euroneta.com. Information on the club and amateur radio in general can be found on our Web pages. The home page is: http://www-tec.open.ac.uk/staff/RadioClub/ouarc.html. Further details can be obtained from the Secretary, Adrain Rawlings, M0ANS, RMRG, Venables Building, OpenUniversity, Walton Hall, Milton Keynes, MK7 6AA.

Prudential Amateur Radio Society *(G0PRU, G0PPS, G8PRU)*

This is open to all current, retired and pensioned employees of the Prudential Group of companies together with any swls world wide. We send out a special QSL card for our contacts and for those who QSL us. If you don't want this special card, then please let us know at the time of the contact.

Chairman: Gerald Haines, G4SXY, tel: (0181) 653 7812; Secretary: Dennis Egan, GW4XKE, tel: (01222) 512959; Publicity officer is John Wimble, G4TGK, tel: (01797) 362295; Call-sign Manager: Mike Butler, G0NRK, tel: (01833) 690515; Overseas Liaison: Alan McCullach, ZS6KU. Details about this society can be obtained from the Membership Secretary: David Dyer, G4DNX, 'Highbank Cottage', Underhill, Moulsford, Wallingford, Oxon, OX10 9JH. Tel: (01491) 651310.

Radio Amateur Relief Expeditions *(RARE - RS95155, Reg. charity no: 1037428)*

Following the experiences of radio amateurs in Romania in 1991/2, it was realised that we have a great deal of communication and other expertise to assist relief projects, aid convoys and disaster areas, so the group was formed to achieve this aim.

RARE has been involved not only in Romania, where it gives continued health and education support, but also to other East European countries. Programmes of english summer camps are developing which include, amateur radio, drama, music and english language workshops.

Enthusiastic members are needed to go on expeditions, provide UK home support, help obtain relief aid and raise funds. If you feel you would like to join us, please send brief details of your experiences and skills, and how you think you can help to: Don Sunderland, G6FHM, 1 Allfield Cottages, Condover, Shrewsbury, SY5 7AP. Tel: (01743) 873815 or fax: (01743) 874729. G6FHM@GB7PMB, e-mail: rare@donsun.demon.co.uk.

Radio Amateur Old Timers' Association *(RAOTA - G2OT)*

Have you a quarter century in amateur radio? If so, why not join RAOTA?

The Association seeks to keep alive the pioneer spirit and traditions of the past in today's amateur radio by personal and radio contact, whilst being mindful of any in special need. Membership is open to anyone who has had an active interest in amateur radio for at least 25 years. It is not necessary to have held a call-sign for this period, or indeed at all. A QSL card or a recommendation from another member is all that is required as proof. Associate membership (without voting rights) is available to anyone with an active interest in amateur radio.

RAOTA publishes a quarterly journal *'Old Timer News'* which is also available on cassette. Regular nets are held on 80m ssb and cw under the call-sign G2OT and awards are available for contacts with other RAOTA members.

Applications for membership and requests for a sample copy of *'Old Timer News'* should be addressed to the Hon. Secretary/Treasurer: Mrs Sheila Gabriel, G3HCQ, Millbrook House, 3 Mill Drove, Bourne, Lincs, PE10 9BX.

Radio Amateurs Invalid & Blind Club *(RAIBC - G4IBC, GB0IBC, GB1IBC)*

Founded in 1954, the RAIBC caters for the special needs of handicapped amateurs and short wave listeners. They offer many services and a local representation scheme. For more details, please see 'Services and Facilities Available for Disabled or Blind Amateurs' elsewhere in this book.

The RAIBC chairman is Johnny Clinch, G3MJK while subscriptions and donations should go to the Honorary Treasurer/Membership Secretary: Mrs Shelagh Chambers, 78 Durley Avenue, Pinner, Middx, HA5 1JH. An advice and helpline is provided by Margery Hey on (01953) 454920 (1000-2200 daily).

Remote Imaging Group *(RS88803)*

This is an international group for APT, HRPT and all weather satellite enthusiasts. It was formed in 1985 to promote and further interest in weather satellite watching including the construction, hardware and software of suitable receiving equipment and techniques. The group publishes a quarterly magazine to its 2,000 members which gives satellite predictions, constructional projects, news, views and information on available equipment.

Further details can be obtained from the Secretary: John Tellick, 34 Ellerton Road, Surbiton, Surrey, KT6 7TX. Tel: (0181) 390 3315; or from the Membership Secretary, Ray Godden, PO Box 142, Rickmansworth, Herts, WD3 4RQ.

Rotarians of Amateur Radio, RS170647

The Society was formed in the UK in 1996 by Harold Chadwick, G8ON of the Rotary Club of Worksop. Regular nets operate on Sundays at 0900 (local) on 3.693± MHz and at 1900 (local) by G4YZE. International nets operate at 1200Z on 14.293 MHz on Thursdays and Sundays. A VK Rotary net is at 1000Z on 14.293 MHz on Sunday operated by VK4DP. Regular bulletins are distributed on packet by VK4BB@VK4PKT. The AGM is usually held in May or June, but various social gatherings are arranged during the year at members' QTHs. Details from the Secretary, Don Cliffe, 72 Pares Way, Ockbrook, Derby, DE72 3TL. Tel/Fax: (01332) 675657 or packet @GB7NOT.

Royal Air Force Amateur Radio Society *(RAFARS - G8FC, G3RAF, G8RAF)*

The Royal Air Force Amateur Radio Society (RAFARS) is the national society for amateur radio enthusiasts who are, or have been, serving members of the Royal Air Force, Commonwealth Air Forces, Allied Air Forces, RAF Reserves or civilians directly connected with the RAF. The contact address is: Admin. Secretary, Royal Air Force ARS, RAF Locking, Weston-Super-Mare, Avon, BS24 7AA.

Royal Naval Amateur Radio Society *(RNARS - GB3RN, G3CRS, G1BZU)*

Membership of RNARS is open to members worldwide who have, or have had, connections with the Royal Navy, Commonwealth navies, naval reserves, the Merchant Navy or foreign navies. Further details from: The Secretary, A. G. Walker, G4DIU, Royal Naval Amateur Radio Society, 103 Torrington Road, North End, Portsmouth, PO2 0TN.

Royal Signals Amateur Radio Society *(RSARS - G4RS)*

RSARS membership is open to serving and past members of Royal Corps of Signals, both Regular Army and TA and all other military personnel and civilian staff who have worked closely with army telecommunications. RSARS has members world-wide. The Society was formed in 1961 under the chairmanship of the late Major-General Eric Cole CB CBE, G2EC.

The Society's produces a high quality magazine '*Mercury*', three times a year and runs its own contests and awards scheme. Annual subscription is £6.00 (UK), £8.00 (overseas).

More information is available from the General Secretary, HQ RSARS, Blandford Camp, Dorset, DT11 8RH. Tel: (01258) 482172.

St Dunstan's Amateur Radio Society *(G3STD, G8STD)*

This is a national organisation for men and women blinded in the Services. Details c/o 52 Broadway Avenue, Wallasey, Wirral, L45 6TD.

Science & Technology through Educational Links with Amateur Radio *(STELAR - RS95685)*

STELAR was formed in 1993 to coordinate and promote amateur radio activity in schools and colleges. It publishes a termly magazine *AMRED* (Amateur Radio in Education) which is sent three times a year to affiliated individuals and institutions. During school term time, a weekly Wednesday Stelar Schools & Colleges net is held at 1.00 pm on 3.770 MHz – net control is GB2SR.

One of its main activities is to increase the number of schools which have clubs training students for the RAE/NRAE by running a free residential 'Crash RAE course' for 20 teachers at Easter each year (open to any school who has no licensed teacher).

Further details and an affiliation form may be obtained from the Chairman, Richard Horton, G3XWH at: STELAR, 7 Carlton Road, Harrogate, North Yorkshire, HG2 8DD. Tel/fax: (01423) 871027. Packet: G3XWH@GB7CYM. E-mail: G3XWH@amsat.org.

The Radio Amateurs' Emergency Network *(G4NRC)*

The Network was formed in 1992 and became a registered charity in 1995. It is a national organisation representing the interests of its member RAYNET Groups.

Further details can be found on the Internet at http://reality.sgi.com/csp/raynet or on telephone BBS (01296) 393737. Alternatively, contact the Chairman at 'Hunters Moon', Newton-le-Willows, Bedale, North Yorks, DL8 1SX. Tel: (0141) 620 1000.

The Scout Association *(RS85972)*

Details from: Programme & Training Department, The Scout Association, Gilwell Park, Chingford, London, E4 7QW. Tel: (0181) 524 5246.

UK Six Metre Group *(GB0SIX)*

The group was formed in 1982 with the primary aim of encouraging an interest in the 50 MHz band by all amateurs. It supports beacons in various parts of the world and has supplied equipment to encourage and help 6m enthusiasts activate new countries. Dxpeditions supported include to: JY7, D44, CY0 and 4L6. Through its quarterly journal *'Six News'* it seeks to provide the best information on all aspects of the band including DX and beacon news, propagation and technical articles, equipment reviews, QSL addresses and DXpedition news.

Details from Iain Phillips, G0RDI, 24 Acres End, Amersham, Bucks, HP7 9DZ.

Worked All Britain Awards (WAB) Group *(G4WAB)*

This group was founded in 1969 by the late John Morris, G3ABG, to further greater amateur radio interest in Britain. The group promotes an award programme, contests and activity weekends and makes regular donations to organisations such as the Radio Amateur Invalid and Blind Club who help the less fortunate members of the amateur radio fraternity.

The award scheme, which is open to licensed amateurs and short wave listeners, is based on the geographical and administrative division of the UK. QSL cards are not required, only log entries, and special record books are available to assist in the claiming of awards. Full details and checklists of all the areas and counties for all the WAB awards are contained in the 'WAB Book'.

For further details of the book, awards and newsletter please write to the Membership Secretary: Brian Morris, G4KSQ, 22 Burdell Avenue, Sandhills Estate, Headington, Oxford, OX3 8ED.

World Association of Christian Radio Amateurs & Listeners *(WACRAL - G3NJB)*

WACRAL welcomes all Christian radio amateurs regardless of their denomination worldwide. There are regular nets, a quarterly newsletter and a conference each autumn. Current annual subscription is £5.00.

General Secretary: Rev Phyl Fanning, G6UFI, 333 Lyndon Rd, Solihull, West Mids, B92 7QS. Tel: (0121) 742 5583 / (0121) 688 2767, web: http://ourworld.compuserve.com/homepages/Phyl_Fanning. Membership Secretary: Derek Chivers, G3XNX, 51 Alma Road, Brixham, Devon, TQ5 8QR.

Local Affiliated Societies and Clubs

The following list shows all local societies and clubs which were affiliated to the RSGB at the time of closing for press. Each club is listed under the administrative county or Scottish region in which it regularly meets.

For example the clubs which meet in Bromley whose postal county is Kent can be found in the Greater London section, likewise the Todmorden club is listed under West Yorkshire despite having a Lancashire address.

Some affiliated societies, for example contest groups, do not hold formal meetings, and so these societies have been listed under the county of their RSGB registered address.

Each county is assigned to an RSGB Zone (A to G) and the individual members of which elect a Council Member. Details of Zonal Council Members, and of RLOs (RSGB Liaison Officers) can be found on the *'At Your Service'* pages of the *RSGB Yearbook*.

Please check the pages of *Radio Communication* and our web pages at http://www.rsgb.orgfor the latest information.

Please send any amendments to your club's details to our Sales Dept at RSGB HQ or by e-mail to: subscriptions@rsgb.org.uk.

RSGB Groups

BELFAST RSGB GROUP, RSGBG-BT. Meets 8.00pm on 3rd Wednesday in the month at Belmont Road. Details from: Mr R T Ferris, GI0OUM, 3 Kingsland Drive, Belfast, BT5 7EY.

BRISTOL RSGB GP, G6YB. Meets on the last Tuesday in the month The New Friends Meeting Hall, Purdown Hospital, Bell Hill, Stapleton. Details from: Mr Robin Thompson, Sec, 179 Newbridge Hill, Bath, BA1 3PY. Tel: 01225 420442.

CARDIFF RSGB GP, RSGBG-CF. Details from: E J Case, 2 Abbey Cl. Tyrhiw, Taffs Well, South Glamorgan, CF7 7BW.

ILFORD RSGB GP, G3XRT. Meets 7.15pm on Thursdays at 50 Mortlake Road, Ilford, Essex, IG1 2SX. Details from: Mr J R Hooper, G3PCA, 50 Mortlake Road, Ilford, Essex, IG1 2SX. Tel: 0181 4783741.

Avon

Council Zone: D

BATH UNIVERSITY RC, G0RUB. Details from: Mr Richard Weston, G0LIB, c/o 38 Church Road, Peasedown St John, Bath, Avon, BA2 8AF. Tel: (01225) 826929.

BRISTOL 70 CM REPEATER GRP, GB3BS. Web: www.iridium.demon.co.uk/gb3bs/bstec.htm. Details from: Mr S J Bailey, G4MCQ, 50 Quantock Close, Warmley, Bristol, BS15 5UT.

BRISTOL ARC, G3TAD. Meets at 7.30pm on Thursdays at the Scout HQ, Firtree Lane, St George, Bristol. Web: www.gifford.co.uk/~passim/barc.html. Details from: H.Sec. Mr WJF Humphries, G7IHL, 47 Fouracre Crescent, Downend, Bristol, BS16 6PT. Tel: 0117 956 3023.

GORDANO ARG, G6GRG. Meets 8.00pm on 4th Wednesday at The Ship, Redcliffe Bay, Portishead, Avon. Details from: Mr R T White, G8SPC, c/o 3 Robin Lane, Clevedon, Avon, BS21 7EX. Tel: 01275 874001.

NORTH BRISTOL ARC, G4GCT. Meets 7.00pm on Fridays at the Self Help Enterprise, 7 Braemar Close, Northville, Bristol. Details from: Mr David Coxon, G0GHM, 7 Kingston Way, Nailsea, Bristol, BS19 2RA. Tel: 01275 790448.

SEVERNSIDE TV GROUP, GB3ZZ. Meets informally most Fridays at NBARC, Filton, Bristol. Details from: Paul Stevenson, G8YMM, c/o 14 Camelford Road, Greenbank, Bristol, BS5 6HW. Tel: 01225 873098.

SHIREHAMPTON ARC, G4AHG. Meets 7.30pm on Fridays at the Twyford House, Lower High Street, Shirehampton, Bristol. Details from: Mr R G Ford, G4GTD, Twyford House, High Street, Shirehampton, Bristol, BS11 0DE. Tel: 01179 856253.

SOUTH BRISTOL ARC, G4WAW. Meets 7.30pm on Wednesdays at the Whitchurch Folk House, East Dundry Road, Bristol. Web: ourworld.compuserve.com/homepages/Steven_Nash/southbri.htm. Details from: Mr L F Baker, c/o 62 Court Farm Road, Whitchurch, Bristol, BS14 0EG. Tel: 01275 834282.

THORNBURY & DARC, G4ABC. Meets 7.30pm every Wednesday at the United Reformed Church Hall, Rock Street, Thornbury, Bristol. Details from: Peter Cabban, G4OST, Ivydean, Upper Tockington Road, Tockington, Bristol, BS12 4LQ. Tel: 01454 612689.

WESTON-SUPER-MARE RS, G4WSM. Meets 7.30 for 8.00 1st and 3rd Mondays in the month at the Woodspring Hotel, High Street, Worle, Weston-Super_Mare. Details from: Mr Graham Pinder, G8WAR, c/o 77 Bridge Road, Weston-Super-Mare, Avon, BS23 3PW. Tel: 01934 415700.

Bedfordshire

Council Zone: B

BEDFORD MODERN SCHOOL ARC, G1BYT. Open to pupils only. Meets at Bedford Modern School, Manton Lane, Bedford, MK41 7NT. Details from: Mr N E Kinselley, Manton Lane, Bedford, MK41 7NT.

DUNSTABLE DOWNS RC, G4DDC. Meets 8.00pm on Fridays at the Chews House, 77 High Street South, Dunstable, Beds, LU6 3SF. Details from: Mr Ken Brewer, G4WYO, 14 Poplar Road, Kensworth, Beds, LU6 3RS. Tel: 01582 873 285.

MID BEDS CONTEST ASSN, G4MBC. Details from: Mr M J Down, c/o 95 High Street, Henlow, Beds, SG16 6AB. Tel: 01462 812253.

SHEFFORD & DARS, G3FJE. Meets 8.00pm on Thursdays at the Church Hall, Ampthill Road, Shefford, Beds. Details from: Mr John West, 6 Iveldale Drive, Shefford, Beds, SG17 5AD. Tel: 01462 812739.

ST SWITHUN'S ARC, M0AJV. Meets 7.30 - 9pm on Thursdays at St Swithun's Church, Rectory Rooms, Sandy, Beds. Details from: Mr Kelvyn Darton, G0WOD, 8 Foster Grove, Sandy, Beds, SG19 1HP. Tel: 01767 683179.

Berkshire

Council Zone: D

ARBORFIELD ARC, G3IHH. Meets 7.30pm on Wednesdays. Details from: Mrs E W Harding, 2E1AUQ, School of Elect.Eng., Radio C, Arborfield, Reading, Berkshire, RG2 9NH.

BRACKNELL ARC, G4BRA. Meets 8.00pm on 2nd Wednesday in the month at the Coopers Hill Community Centre, Bagshot Road, Bracknell, Berks. Details from: Mr Steve Baugh, G4AUC, c/o 70 Madingley, Bracknell, Berkshire, RG12 7TF. Tel: 01344 420577.

BROADMOOR HOSPITAL, RS95231. Deatils from: B L Graham, The Radio Shop, Crowthorne, Reading, Berks, RG11 7EG.

BURNHAM BEECHES RC, G3WIR. Meets 8.00pm on 1st and 3rd Mondays in the month at the Farnham Common Village Hall, Victoria Road, Farnham Common, Bucks. Details from: Mrs Eileen Chislett, G6EIL, 2a St Marks Road, Maidenhead, Berks, SL6 6DA. Tel: 01628 25720.

MAIDENHEAD & DARC, G3WKX. Meets 7.45 on 1st Thursday and 3rd Tuesday in the month at the Red Cross Hall, The Crescent, Maidenhead, Berks. Details from: Mr Roy Savin, G4XYN, 7 Bannard Road, Maidenhead, Berks, SL6 4NG. Tel: 01628 25952.

MARY HARE GRAMMAR SCH ARS, G7MHS. Meets 7.30 on Wednesdays at the school. Open only to staff and pupils. Details from: Mr Merton Vaslet, G4JAL, Arlington Manor, Snelsmore Common, Newbury, Berkshire, RG16 9BQ. Tel: 01635 46078.

NEWBURY & DARS, G3WOI. Meets 7.30pm on 4th Wednesday in the month at the Memorial Hall, Upper Bucklebury, nr Newbury. Details from: Mr Ian Trusson, G3RVM, c/o 27A Roman Way, Thatcham, Berks, RG18 3BP. Tel: 01635 826019.

RACAL ARC, G3RAC. Details from: Mr Peter Roberts, G4DJB, c/o Racal Comms, Western Road, Bracknell, Berkshire, RG12 1RG.

READING ARC, G3ULT. Meets 8.00pm on the 2nd and 4th Thursday in the month at the Woodley Pavillion, Woodford Park, Haddon Drive, Woodley, Reading. Web: www.radarc.org/. Details from: Mr Peter Swynford, G0PUB, 219 Wykeham Road, Earley, Reading, RG6 1PL. Tel: 0370 354 054.

Borders Region

Council Zone: G

BORDERS ARS, GM0BRS. Meets 1st Friday in month at the St. John Ambulance Hall, Berwick-upon-Tweed. Details from: Mr A M McCreadie, GM0BPY, c/o 16 Fancove Place, Eyemouth, Borders, TD14 5JQ. Tel: 018907 50492.

GALASHIELS & DARS, GM4YEQ. Meets 7.30pm on Wednesdays at the Focus Centre, Galashiels. Details from: Mr John Campbell, GM0AMB, Focus Centre, Livingston Place, Galashiels, Selkirkshire, TD1 1DQ.

KELSO ARS, GM4KHS. Meets 7.30pm on Mondays at the Abbey Row Community Centre, Kelso. Details from: Margaret Chalmers, GM0ALX, c/o Abbey Row Community Centre, Kelso, Roxburghshire, TD5 7BJ. Tel: 01573 226372.

SCOTTISH BDRS REPEATER GP, RS43855. c/o 85 Forest Road, Selkirk, Selkirkshire, TD7 5DD.

Buckinghamshire

Council Zone: D

AYLESBURY VALE REPEATER GP, GB3VA. Meets in March, June, July and December at the Robin Hood, on the A422 Buckingham to, Brackley Rd. (AGM at Stone, Village Hall, nr Aylesbury). Details from: Mr Mike Marsden, G8BQH, Hunters Moon, Buckingham Road, Hardwick, Aylesbury, Bucks, HP22 4EF. Tel: 01296 641783.

AYLESBURY VALE RS, G4VRS. Meets 8.00pm on 1st and 3rd Wednesdays in the month at the Harwick Village Hall, Aylesbury, Bucks. Details from: Mr Gerry Somers, G7VFV, 76 Fowler Road, Aylesbury, Bucks, HP21 8QG. Tel: 01296 432234.

CHESHAM & DARS, G3MDG, G1MDG. Meets 8.15pm on Wednesdays at the Exhibition Room, White Hill Centre, Chesham, Bucks. Details from: Mr Iain Phillips, G0RDI, c/o 24 Acres End, Amersham, Bucks, HP7 9DZ.

CHILTERN ARC, G3CAR. Meets every 2nd Wednesday in the month, plus 2m net every Tuesday and Thursday 8.00pm. Details from: Mr Roy Page, G4YAN, 26 Colne Road, High Wycombe, Bucks, HP13 7XN. Tel: 01494 534216.

MILTON KEYNES & DRS, G3HIU. Meets 8pm, 2nd and 4th Monday in the month at Faulkner House, Bletchley Park, Milton Keynes. Details from: Linda Taylor, 2E0AON, 18 Folly Lane, North Crawley, Bucks, MK16 9LW. Tel: 01234 391374.

MILTON KEYNES SCOUT ARS, G0SMK. Meets every 1st Saturday of each month 10.30am - 4.00pm at 'The Quarries', M.K. Scout Campsite, Cosgrove. Details from: Mr P A Orchard, G0RYZ, 68 Simpson Road, Bletchley, Milton Keynes, MK1 1BA. Tel: 01908 648186.

Cambridgeshire

Council Zone: B

CAMBRIDGE & DRC, G2XV. Meets 7.30pm on Fridays at the Coleridge Community College, Radegund Road, Cambridge. Details from: Mr Colin Havercroft, G8CTX, 28 Anglers Way, Cambridge, CB4 1TZ. Tel: 01223 420909.

CAMBRIDGESHIRE REPEATER GP, GB3PI. Details from: 10 Quince Road, Hardwick, Cambridge, CB3 7XJ.

DUXFORD ARS, GB2IWM. Meets every Sunday at Building 177, Imperial War Museum, Duxford Airfield, Cambs. Details from: Mrs B I Pope, Imperial War Museum, Duxford Airfield, Duxford, Cambridgeshire, CB2 4QR. Tel: 01279 656149.

GTR PETERBOROUGH ARC, G4EHW. Meets 7.00pm on the 4th Wednesday in the month at the 6th Form Building, Stanground College, Farcet Road, Fletton, Peterborough. Details from: Hon. Sec. Mr D W Mason, G0HPJ, 40 Washingley Road, Folksworth, Peterborough, PE7 3SY. Tel: 01733 245031. E-mail: dmason9932@aol.com.

HUNTINGDONSHIRE ARS, G0HSR. Meets on 1st, 3rd and 5th Thursdays in the month at the Medway Centre, Medway Road, Huntingdon. Details from: Mr David Leech, G7DIU, 4 Rydal Close, Huntingdon, Cambs, PE18 6UF. Tel: 01480 431333.

MARCH & DRAS, G3PMH. Meets 7.30pm on Tuesdays at British Legion Club, Rookswood Road, March, Cambs, PE15 8DP. Details from: Mr J Braithwaite, G3PWK, Old 10 Lawn Lane, Little Downham, Ely, Cambs, CB6 2TS. Tel: 01353 698885.

PETERBOROUGH R & ES, G3DQW. Details from: Mr V Edwards, G8NGZ, 33 Eyrescroft, Bretton, Peterborough, Cambs, PE3 8ES.

WISBECH AR & ELEC. CLUB, G4PQL. Details from: Mr Alan Bridgeland, 17 Oldfield Lane, Wisbech, Cambs, PE13 2RJ.

Central Region

Council Zone: G

DOLLAR ACADEMY ARC, GM0SNG. Meets after 3.30pm most days at the academy. Details from: Mr Geoff Collier, GM0LOD, Dollar, Clackmannanshire, FK14 7DU. Tel: 01259 742126.

STIRLING & DARS, GM4TMS. Meets 7.30pm every Thursday at Bandeath Industrial Estate, Throsk, nr Stirling. Details from: Mr John Sherry, GM0AZC, 26 Grahamshill Terrace, Fankerton, Nr Denny, Stirlingshire, FK6 5HX. Tel: 01324 824709.

Cheshire

Council Zone: A

CHESTER & DRS, G3GIZ. Meets at 8.00pm on 2nd, 3rd and 4th Tuesday in the month at the Upton Recreation Centre, Cheshire County Sports &, Social Club, Plas Newton Lane, Chester, CH2 1PR. Details from: Mr Bob Campbell, G4CMI, 6 Dulverton Avenue, Vicars Cross, Chester, CH3 5LX. Tel: 01244 314083.

MACCLESFIELD & DRS, G4MWS. Meets 8.30pm on Tuesdays at the Pack Horse Bowling Club, Abbey Road, Macclesfield. Details from: Mr Ray Walton, Sec., 4 Cornwall Close, Macclesfield, Cheshire, SK10 3HE. Tel: 01625 426056.

MID CHESHIRE ARS, G3ZTT. Meets 8.00pm on Wednesdays at the Cotebrook Village Hall, Cotebrook, nr Tarporley, Cheshire (NGR: SJ 571 655). Details from: Mr Mike Tyrrell, G6GAK, 189 Runcorn Road, Barnton, Northwich, Cheshire, CW8 4HR. Tel: 01606 784795.

NORTH CHESHIRE RC, G0BAA. Meets 8.00pm on Sundays at the Morley Green Club, Mobberley Road, Wilmslow, Cheshire. Details from: Mrs Jill Gourley, G0OZJ, Morley Green Club, Morley Green, Wilmslow, Cheshire, SK9 5NT. Tel: 0161 485 5036.

UKFM GROUP WESTERN, GB3MP. Meets 8.00pm on 2nd Sunday in the month at the Morley Green Club, Mobberley Road, Wilmslow, Cheshire. Details from: Mr Gordon Adams, G3LEQ, 2 Ash Grove, Knutsford, Cheshire, WA16 8BB. Tel: 01565 652652. Fax: 01565 634560.

WARRINGTON & DARS, G4CDA. Meets 7.45pm for 8.00pm on Tuesdays at the Grappenhall Community Centre, Bellhouse Lane, Grappenhall, Warrington, Cheshire. Details from: Mr John Riley, G0RPG, 1 Chatsworth Avenue, Culcheth, Warrington, Cheshire, WA3 4LD. Tel: 01925 762722.

WIDNES & RUNCORN ARC, G0FWR. Meets 7.30pm every other Tuesday at the Scout Hut, Castle Road, Halton Castle, Runcorn, Cheshire. Details from: Dave Wilson, G7OBW, 12 New Street, Elworth, Sandbach, Cheshire, CW11 3JF. Tel: 01270 761608.

Cleveland

Council Zone: A

EASINGTON ARS, G4APN. Meets Thursday evenings at the Southside Social Club, Southside, Easington Village, Peterlee, Co Durham. Details from: Mr George Ford, G0MHC, 11 Sandbanks Drive, Hartlepool, Cleveland, TS24 9RP. Tel: 01429 264735.

EAST CLEVELAND ARC, G4CRS. Meets at the Jubilee Hall, Gurney Street, New Marske, Cleveland. Details from: Mr Malcolm Brass, G4YMB, 11 Leacholm Way, Guisborough, Cleveland, TS14 8LN. Tel: 01287 638119.

STOCKTON & DARG, G4XXG. Meets Wednesday evenings at the Billingham Community Centre, Billingham, Cleveland. Details from: Mr Malcolm Hotson, G0NRP, 13 Repton Avenue, Stockton-on-Tees, Cleveland, TS19 9BQ.

Clwyd

Council Zone: E

CONWAY VALLEY ARC, GW6TM. Meets 7.30pm on 1st Wednesday in the month at the Studio, Penrhos Road, Colwyn Bay, Clwyd. Details from: Mr R W Evans, GW6PMC, 16 Monmouth Grove, Prestatyn, Clwyd, LL19 8TS. Tel: 01745 855068.

MARFORD & DARS, GW0WXW. Details from: 7 Oak Drive, Marford, Wrexham, Clwyd, LL12 8XT.

NORTH WALES RS, GW0NWR. Meets 7.30pm every Thursday at the Old YMCA, Queen's Drive, Colwyn Bay, Clwyd. Details from: Mr N B Mee, GW7EXH, Anncott, Hylas Lane, Rhuddlan, Clwyd, LL18 5AG. Tel: 01745 591704.

RAF SEALAND ARC, GW4RAF. Meets most lunchtimes. For RAF and MoD personnel. Details from: Mr Vince Priamo, c/o K Slee (OIC), OC SSF, Blg 47A, RAF Sealand, Deeside, Clwyd, CH5 2LS. Tel: 01244 288331.

WREXHAM ARS, GW4WXM. Meets 7.30pm on 1st and 3rd Tuesdays in the month at the Community Centre, Maesgwyn Road, Wrexham. Details from: Mr Mike Bryant, GW6NLP, The Nook, Llanarmon Road, Bwlchgwyn, Wrexham, Clwyd, LL11 5YP. Tel: 01978 755842.

Co Antrim

Council Zone: F

BALLYMENA RC, GI3FFF. Meets 8.00pm on Thursdays at 70 Nursery Road, Gracehill, Ballymena, Co Antrim. Details from: Mr Jeffery Clarke, GI4HCN, 70 Nursery Road, Gracehill, Ballymena, Co Antrim, Northern Ireland, BT42 2QA. Tel: 01266 659769.

CARRICKFERGUS ARG, GI0LIX. Meets every Tuesday at the Downshire Community School, Downshire Road, Carrickfergus. Details from: Mr John Branagh, GI3YRL, c/o 17 Rathmoyle Park West, Carrickfergus, Co Antrim, BT38 7NG. Tel: 01960 367208.

EAST ANTRIM ARC, GI4KKK. Meets 8.00pm on 2nd Wednesday in the month at Torrens Hall, Doagh, Co Antrim. Details from: Mr N Jenkins, GI4RVT, c/o 10 Highgate Drive, Mallusk, Newtownabbey, Northern Ireland, BT38 8WQ.

LAGAN VALLEY ARS, GI4GTY. Meets 8pm on 2nd Wednesday in the month at the Harmony Hall Arts Centre, Harmony Hill, Lisburn, Co Antrim. Details from: Mr Ed Campbell, Hon Sec, c/o 265 Ballynahinch Road, Lisburn, Co Antrim, BT27 5LS. Tel: 01846 607702.

ROYAL NAVAL (ULSTER) ARC, GI0URN. Club affiliated to the Royal Naval Amateur Radio Society. Details from: Mr Alex Miller, GI4SFV, 21 Marmont Drive, Belfast, N Ireland, BT4 2GT.

Co Armagh

Council Zone: F

ARMAGH & DARC, GI0ADD. Meets 8.00pm on 2nd and 4th Wednesday in the month at Co Armagh Golf Club, Newry Road, Armagh. Details from: Mr J A Murphy, GI7DWF, Armagh County Golf Club, The Demesne, Newry Road, Armagh, BT61 7EN. Tel: 01861 522153.

Co Down

Council Zone: F

BANGOR & DARS, GI3XRQ. Meets 8.00pm on 1st Wednesday in the month at the Clandeboye Lodge Hotel, 10 Estate Road, Bangor. Details from: Mr Terry Barnes, GI3USS, 95 Crawfordsburn Road, Bangor, Co Down, BT19 1BJ. Tel: 01247 473948.

BELFAST ROYAL ACADEMY, MI1BRA. Meets at lunchtime at the school. Membership limited to school members and staff. Details from: Mr N Moore, GI7CMC, Radio Club, 164 Ardenlee Avenue, Belfast, BT6 0AE. Tel: 01232 452202.

NEWRY HIGH SCHOOL ARC, MI0AVI, RS174754. Details from: 23 Ashgrove Road, Newry, Co Down, BT34 1QN.

THE MID ULSTER ARC, GI3VFW. Details from: Hon. Tres. Mr I V Gracey, 23 Cascum Road, Banbridge, Co Down, N.I., BT32 4LF.

Co Durham

Council Zone: A

BISHOP AUCKLAND RC, G4TTF. Meets Thursday evenings at the Stanley Village Hall, Rear High Road, Stanley, Crook, Co Durham. Details from: Mr Derek Perrey, G3WUE, 39 Pearson Street, Spennymoor, Co Durham, DL16 6HP. Tel: 01388 537336.

DARLINGTON & DARS, G4ZVH. Meets 7.00pm on Mondays and Wednesdays (construction night) at the Hurworth Community Centre, Hurworth, Darlington, Co Durham. Details from: Mr Mike Wood, G8MTV, Spa Cottage, Croft-on-Tees, Darlington, DL2 2SY. Tel: 01325 721545.

DERWENTSIDE ARC, G4PFQ. Meets Wednesday evenings at the Steel Club, 36 Medomsley Road, Consett, Co Durham. Details from: Mr G Darby, G7GJU, c/o 60 Pine Street, Grange Villa, Chester-le-Street, Co Durham, DH2 3LX. Tel: 0191 3702032.

GREAT LUMLEY AR & ES, G4EUZ. Meets 8.00pm on Wednesdays at the Community Centre, Great Lumley, Chester-le-Street, Co Durham. Details from: Mr Barry Overton, G1JDP, Community Centre, Front Street, Gt Lumley, Chester-le-Street, Co Durham, DH3 4JD. Tel: 0191 3885936.

PETERLEE RADIO CLUB, G0KVJ. Details from: Mr R W Raine, G4RXR, 47 Buckingham Road, Peterlee, Co Durham, SR8 2DT.

Co Londonderry

Council Zone: F

COLERAINE & DARS, GI4NRQ. Details from: 1 Loguestown Court, Coleraine, Co Derry, BT52 2HS.

Co Tyrone

Council Zone: F

THE FOYLE & DARS, MI0AKU. 8.00pm on 3rd Monday in month Upstairs Lounge, Railway Bar, Railway Street, Strabane, Co Tyrone. Details from: Mr Terry White, GI7THH, Sec., c/o Shallamar, 3A Park Road, Strabane, Co Tyrone, BT82 8EL.

Cornwall & Scilly Is

Council Zone: D

CORNISH RAC, G4CRC. Meets 1st Thursday and 2nd Monday in the month at the Perran-ar-Worthal Village Hall, Perranwell, nr Truro, Cornwall. Details from: Mrs Cheryll Hammett, 2E1ADQ, Rosehill, Ladock, Truro, Cornwall, TR2 4PQ. Tel: 01726 882758.

NEWQUAY & DARS, G4ADV. Meets on alternate Fridays at the Treviglas School, Newquay. Details from: Mrs Maggie Reed, G0KEM, Larks Rise, Great Hewas, Grampound Road, Truro, Cornwall, TR2 4EP. Tel: 01726 882752.

POLDHU ARC, GB2GM. Meets every Tuesday and Friday at the Club House, Poldhu Cove, Mullion, Cornwall, TR12 7JB. Details from: Mrs Carolyn Rule, G1ZPC, The Kiteshop, Mullion, Cornwall, TR12 7DN. Tel: 01326 240144.

POLTAIR SCHOOL, G4IAP. Details from: C H Hender, Trevarthian Road, St Austells, Cornwall, PL25 4BZ.

SALTASH & DARC, G4GXK, G8SAL. Meets 1st and 3rd Fridays in the month at the Toc H Hall, Warraton Road, Saltash, Cornwall. Details from: Mr Brian Giles, 171 St Stephens Rd, Saltash, Cornwall, PL12 4NJ. Tel: 01752 844321.

ST AUSTELL ARC, G0ECC. Meets on 1st and 3rd Mondays (except Bank Holidays) during term-time at Poltair School. Details from: Mr Reg Pears, G4TRV, 24 Westbourne Drive, St Austell, Cornwall, PL25 5EA. Tel: 01726 72951.

Cumbria

Council Zone: A

BBC SKELTON ARC, G3ZSK. Open only to BBC employees. Deatils from: BBC Transmitting Station, Skelton Pastures, Skelton, nr. Penrith, Cumbria, CA11 9SY.

CARLISLE & DARS, G4ARS. Meets 7.30pm on Mondays at the Morton Community Centre, Wigton Road, Carlisle. Details from: Mr J A Ennis, G3XWA, c/o 30 Hillcrest Avenue, Carlisle, CA1 2QJ. Tel: 01228 27463.

EDEN VALLEY RS, G0ANT. Meets on last Thursday of the month at the BBC Club, Penrith. Details from: Mr P D Godolphin, G4XTA, c/o 3 Knipe View, Bampton, Penrith, Cumbria, CA10 2RF. Tel: 01931 713359.

FURNESS ARS, G4ARF. Meets 8.30pm on Mondays at the Cavendish Arms Hotel, Dalton-in-Furness. Details from: Mr B Bull, G4AGB, c/o 16 Raleigh Street, Barrow in Furness, Cumbria, LA14 5RH.

Derbyshire

Council Zone: B

BOLSOVER ARS, G4RSB. Meets 8.00pm on Wednesdays at the Horse & Groom, Scarcliffe, Chesterfield, Derbyshire. Details from: Mr Colin Morris, G0RXT, c/o C J Morris, 133 Shuttlewood Rd, Bolsover, Derbys, S44 6NX. Tel: 01246 822856.

BUXTON RA, G4SPA. Meets 8.00pm on 2nd and 4th Tuesday in the month at the Leewood Hotel, Buxton. Details from: Derek Carson, G4IHO, c/o 21 Harris Road, Harpur Hill, Buxton, Derbys, SK17 9JS. Tel: 01298 25506.

DERBY & DARS, G2DJ. Meets 7.30pm on Wednesdays at 119 Green Lane, Derby. Details from: Mr Richard Buckby, G3VGW, 20 Eden Bank, Ambergate, Belper, Derbyshire, DE56 2GG. Tel: 01773 852475.

DERBYSHIRE FS, ARS, G0DFS. Open only to fire service members. Details from: R. Doran, Fire Service HQ, Burton Road, Littleover, Derby, DE23 6EH. Tel: 01332 771221x206.

DRONFIELD & DARC, G0OUT. Meets 1st & 3rd Mondays in month at the Dronfield Sports & Social Club, 114 Carr Lane, Dronfield Woodhouse. Details from: Mr P Cardwell, G7UEX, 2 Hayfield Place, Frecheville, Sheffield, S Yorks, S12 4XH. Tel: 01142 654978.

EREWASH VALLEY ARG, G0PCX. Meets 8.30pm on Tuesdays at the Ilkeston Rugby Club. Details from: 9 Bagot Street, West Hallam, Ilkeston, Derbys, DE7 6HA.

EVETS COM. LTD., ARC, G6ECL, G0JBX. Open to staff of Evets Communications Ltd and Enfield Computers plus associates. Details from: The Secretary, Evets Communications Ltd, Enfield House, 303 Burton Road, Derby, DE23 6AG. Tel: 01332 363981.

LANDAU FORTE COLLEGE ARC, GX0TLF. Details from: John Hackett, Fox Street, Derby, DE1 2LF. Tel: 01332 204040.

MOUNT ST MARY'S ARC, G4MSM. Meets as announced monthly at the College, Spinkhill, Eckington, Sheffield. Details from: Rev. P McArdle, G0DAG, Mount St Mary's College, c/o The Headmaster, Spinkhill, Sheffield, S31 9YL. Tel: 01246 812230.

NOTTS & DERBY BORDER ARC, G4NID. Meets 7.30pm on Tuesdays at Marlpool United Reform Church, Chapel Street, Marlpool, Ilkeston. Details from: Mr Graham Bromley, G4UTN, c/o Plumtree House, Walk Cl, Draycott, Derbys, DE72 3PN. Tel: 01773 834308.

NUNSFIELD HOUSE ARG, G3EEO. Meets 7.45pm on Fridays at the Nunsfield House, Boulton Lane, Alvaston, Derby. Details from: Mr Peter Walker, G6KUI, 23 Denstone Drive, Alvaston, Derby, DE24 0HZ.

STH DERBYS & ASHBY W ARG, G0SRC. Meets 7.00pm on Wednesdays and on Tuesdays (RAE classes) at the Moira Replan Centre, 17 Ashby Road, Moira, Swadlincote, Derbyshire, DE12 6DJ. Details from: Mrs B Walley, 52 Main Street, Rosliston Swadlincote, Derbyshire, DE12 8JW. Tel: 01283 760822.

STH NORMANTON, ALFRETON & DARC, G0CPO. Meets 7.30pm on Mondays (except bank holidays) at the New St Community Centre, New St, South Normanton, Derbyshire. Details from: Mr Russell Bradley, G0OKD, 42 The Croft, South Normanton, Alfreton, Derby, DE5 7BP. Tel: 01773 863892.

THE A1 CONTEST GROUP, G4ZAP. Details from: Thorntree House, Wensley, Matlock, Derbys, DE4 2LL.

Devon

Council Zone: D

APPLEDORE & DARC, G2FKO. Meets 7.30pm on 3rd Monday in the month at the Appledore Football Club. Details from: Mr Den Williams, G0UMT, 5 Little Meadow Way, Bideford, Devon, EX39 3QZ.

AXE VALE ARC, G8CA. Meets 7.30pm on the 1st Friday in the month at the New Commercial Hotel, Axminster, Devon. Details from: Jon Frings, G3FFH, Hill Farm, Gore Lane, Uplyme, Lyme Regis, Dorset, DT7 3RJ. Tel: 01297 445518.

DARTMOOR RADIO CLUB, G1RCD, G0DRC. Meets 7.30pm on 1st Thursday in the month at the Yelverton War Memorial Village, Hall, Meavy Lane, Yelverton, Devon. Details from: Mr Ron Middleton, G7LLG, "Fair Winds", Southella Road, Yelverton, Devon, PL20 6AT. Tel: 01822 852586.

EXETER ARS, G4ARE. Meets on 2nd Monday in the the month at the Moose Centre, Spinning Path Lane, Blackboy Road, Exeter. Details from: Mr Ray Donno, G3YBK, c/o Brixington, College Lane, Longdown, Exeter, Devon, EX6 7SS. Tel: 01392 78710.

EXMOUTH ARC, G0XRC. Meets alternate Wednesdays at The Scout Hut, Marlpool Hill, Exmouth. Details from: Mr Michael Newport, G1GZG, c/o 9 Highbury Park, Exmouth, Devon, EX8 3EJ. Tel: 01395 274172.

NORMAN LOCKYER OBSERVATORY ARG, G0AXC. Meets 7.30pm on 2nd and 4th Tuesdays in the month at the Norman Lockyer Observatory, Salcombe Hill, Sidmouth. Details from: Mr Ron Hamson, G0NOC, c/o 43 Arcot Park, Sidmouth, Devon, EX10 9HU. Tel: 01395 515349.

NORTH DEVON RC, RS37569. Meets 1st Wednesday in the month at the SWEB Social Centre, Victoria Road, Barnstaple, Devon. Details from: Mr Jack Kelly, G4JAK, c/o 20 Philip Avenue, Sticklepath, Barnstaple, North Devon, EX31 3AQ. Tel: 01271 23525.

NTE (PAIGNTON) ARS, G0OSH. Meets every Friday at the Paignton Community College, Upper School, Waterleat Road, Paignton. Details from: Mr Rod Maude, G0SWM, c/o 33 Sandown Road, Paignton, Devon, TQ4 7RL. Tel: 01803 521066.

PLYMOUTH RC, G3PRC. Meets at 7.30pm on Tuesdays at the Royal Fleet Club, Devonport. Details from: Frank Parker, G7LUL, The Royal Fleet Club, Devonport, Plymouth, PL5 4PS. Tel: 01752 563222.

SOUTH DEVON ARC, G4SSD. Details from: 68 Penwill Way, Paignton, Devon, TQ4 5JQ.

TIVERTON RC, G4TSW. Details from: P O Box 3, Tiverton, Devon, EX16 6RS.

TORBAY ARS, G3NJA. Meets on Fridays at the ECC Social Club, Highweek, Newton Abbot, Devon. Details from: Pam Helliwell, G7SME, c/o 31 Haytor View, Heathfield, Newton Abbot, Devon, TQ3 3XG. Tel: 01626 833400.

UNIVERSITY OF PLYMOUTH ARS, G0UOP. Details from: Alan Santillo, G0XAW, c/o Alan Santillo, SEC & EE, Rm 317, Smeaton Building, Drake Circus, Plymouth, Devon, PL4 8AA.

Dorset

Council Zone: D

BLACKMORE VALE ARS, G4RBV. Meets 7.45pm 2nd & 4th Tuesdays in month at the Butt of Shery, Castle Street, Mere, Wilts.

BOURNEMOUTH RS, G2BRS. Meets 7.30pm on 1st and 3rd Fridays in the month at the Kinson Community Centre, Kinson, Bournemouth, Dorset. Details from: Mr Ian Brotherton, G2BDV, 6 Cranfield Avenue, Wimborne, Dorset, BH21 1DE. Tel: 01202 886887.

PLESSEY (CHRISTCHURCH) ARS, G0MUD. All amateurs and SWLs welcome. Meets 8.00pm on Thursdays at the Siemens Plessey Sports &, Social Club, Grange Road, Somerford, Dorset. Deatils from: Mr K P Harris, W4U Block, Grange Road, Christchurch, Dorset, BH23 4JE. Tel: 01202 404147.

POOLE RS, G4PRS. Meets 7.30pm on 2nd Friday in the month at the Lady Russell Cotes House, Constitution Hill Road, Poole, Dorset. Details from: Mr Bob Kendrick, G0SJT, 68 Nansen Avenue, Oakdale, Poole, Dorset, BH15 3DD.

PORTLAND ARC, G0VOP. Meets Thursdays & Sundays each week at the Clubs' own premises. Details from: Mrs C Houlden, 2E1DQZ, 29 Court Barton, Weston, Portland, Dorset, DT5 2HJ. Tel: 01305 823373.

SOUTH DORSET RS, G3SDS. Meets 7.30pm on 1st Tuesday in the month at the Chickerell Church Hall, Weymouth, Dorset. Details from: Mr John Rose, M1BDW, 45 Ringstead Crescent, Weymouth, Dorset, DT3 6PT. Tel: 01305 832057.

WESSEX AMATEUR WIRELESS CLUB, G1WAW. No regular meetings are held. Details from: Mr Ken Powell, G1NCG, 99 Pine Road, Winton, Bournemouth, Dorset, BH9 1LU. Tel: 01202 549376.

Dumfries & Galloway

Council Zone: G

DUMFRIES & GALLOWAY RC, GM4HAA. Meets 7.30pm on 1st Monday in the month at the Edenbank Hotel, Laurie Knowe, Dumfries. Details from: Mr Jim Weatherer, GM4RJF, c/o 20 Gilloch Crescent, Georgetown, Dumfries, DG1 4DW.

WIGTOWNSHIRE ARC, GM4RIV. Meets on Thursdays at the Stranraer Academy, B Block, Stranraer. Details from: Mr Ellis Gaston, GM0HPK, Lochans, Stranraer, Dumfries & Galloway, DG9 9BA. Tel: 01776 820413.

Dyfed

Council Zone: E

ABERPORTH YMCA ARC, GW4SZV. c/o S J Evans, Hazelbrook, Melin Y Coed, Cardigan, Dyfed, SA43 1PG.

ABERYSTWYTH & DARS, GW0ARA. Meets 8.00pm on the 2nd Thursday in the month (except July and August) at the Scout Hut, Plascrug Avenue, Aberystwyth. Details from: Mr John Woodward, GW6IDK, c/o 29 Maes Yr Awel, Ponterwyd, Aberystwyth, Ceredigion, SY23 3JT. Tel: 01970 890657.

CARMARTHEN ARS, GW4YCT. Meets 7.15pm on 1st and 3rd Tuesday in the month at the Club Room, County Civil, Protection Planning Unit, Hill, House, Picton Terrace, Carmarthen, SA31 3BS. Details from: Mr W D Hughes, GW4ZXL, 31 Ystrad Drive, Johnstown, Carmarthen, Dyfed, SA31 3PQ. Tel: 01267 231359.

CLEDDAU ARS, GW0SYG. Details from: Trevor Perry, GW4XQK, The Community Education Centre, St Clememts Road, Neyland, Pembs, SA73 1EH. Tel: 01646 600725.

COLESHILL RAIBC, GW0SZW. Meets on Tuesday and Wednesday mornings at the Coleshill Social Centre, Coleshill Terrace, Llanelli. Details from: Mr E W S Meredith, GW4XLK, Coleshill Social Centre, Coleshill Terrace, Llanelli Dyfed, SA15 3BP.

FISHGUARD & DARS, GW0AQC. Meets 7.30pm on Thursdays at the Radio Shack, Community Centre, Ropewalk, Fishguard, Dyfed. Details from: Mr Paul Stevens, Noddfa, Dwrbach, Fishguard, Dyfed, SA65 9RL. Tel: 01348 840503.

LLANELLI ARS, GW0EZQ. Meets 7.00pm on Mondays at TS. Echo Sea Cadets Unit. Details from: Mr Roy Jones, GW0KJZK, 64 Cleviston Park, Llangennech, Llanelli, Dyfed, SA14 9UP. Tel: 01554 820207.

PEMBROKESHIRE RS, GW0EJE. Meets 7.30pm on Mondays at the Adult Community Centre, off Dew Street behind main, library, Harverfordwest. Details from: Mr Paul Delaney, GW0HPQ, Mount Pleasant, Ambleston, Haverfordwest, Dyfed, SA62 5DP. Tel: 01437 764009.

ST. TYBIE ARS, GW0VPR. Details from: 5 Woodfield Road, Llandybie, Ammanford, Dyfed, SA18 3UR.

TENBY ARC, GW0VJO. Details from: 3 Giltar Terrace, Penally, Tenby, Dyfed, SA70 7QD.

East Sussex

Council Zone: C

BRIGHTON & DRS, G4GQR. Meets 8.00pm on 1st and 3rd Wednesday in the month at the Roast Beef Bar, Brighton Racecourse, Elm Grove, Brighton. Details from: Mr P J Fellingham, c/o 26 Fitch Drive, Brighton, E. Sussex, BN2 4HX.

BRIGHTON COLLEGE ARC, G0BRI. Brighton College, Eastern Road, Brighton, East Sussex, BN2 2AL. Tel: 01273 697131.

CROWBOROUGH DARS, G0CRW. Meets on 4th Thursday in the month at the Plough and Horses, Walshes Road, Jarvis Brook. Details from: Mrs Pauline Moldon, G7SPT, c/o Heston, Western Road, Jarvis Brook, Crowborough, East Sussex, TN6 3EH. Tel: 01892 653782.

HASTINGS ELEC & RC, G6HH, G1HHH, G6LL. Meets 7.30pm on 3rd Wednesday in the month at West Hill Community Centre, Croft Road, Hastings, E Sussex. Details from: Mr Doug Mepham, G4ERA, 8 The Close, Fairlight, Hastings, E Sussex, TN35 4AQ. Tel: 01424 812350.

HASTINGS RG, GB3HE. Details from: 9 Lyndhurst Avenue, Hastings, East Sussex, TN34 2BD.

SOUTHDOWN ARS, G3WQK. Meets 7.30pm on Wednesdays and Fridays (Leisure Centre) and 1st Monday in the month at the Chaseley Home for Disabled, Ex-Servicemen, Bolsover Rd,, Southcliffe Eastbourne or, Lagoon Leisure Centre. Details from: Mr Jim Harris, G4DRV, c/o Lagoon Leisure Centre, Vicarage Lane, Hailsham, E Sussex, BN27 2AX. Tel: 01323 728479.

SUSSEX CONTEST GROUP, G0MSA. c/o Oakwell, Newtons Hill, Hartfield, East Sussex, TN7 4DH.

Essex

Council Zone: C

BRAINTREE & DARS, G4JXG. Meets 8.00pm on 1st and 3rd Mondays in the month at the Braintree Hockey Club, Church Street, Bocking, Braintree. Details from: Mr J Button, G1WQQ, c/o 1 Ross Cottage, Southey Green, Sible Hedingham, Halstead, Essex, CO9 3RN. Tel: 01787 460947.

CHELMSFORD ARS, G0MWT. Meets 7.30pm on 1st Tuesday in the month at the Marconi College, Arbour Lane, Chelmsford, Essex. Details from: Mr Roy Martyr, G3PMX, c/o 1 High Houses, Mashbury Rd, Great Waltham, Essex, CM3 1EL. Tel: 01245 360545.

CLACTON RADIO CLUB, G3CRC. Meets on alternate Wednesdays in the month. Details from: Mr K J Prior, 74 Walden Way, Frinton on Sea, Essex, CO13 0BQ. Tel: 01255 679891.

COLCHESTER ARS, G3CO. Meets 7.30pm on alternate Thursdays at the Colchester Institute, Sheepen Road, Colchester. Details from: Mr Frank R Howe, G3FIJ, 29 Kingswood Road, Colchester, Essex, CO4 5JX. Tel: 01206 851189.

DENGIE HUNDRED ARS, G0UTT, G7SDH. Meets 8.00pm on 1st and 3rd Mondays in the month at the Henry Samuel Hall, Mayland, Essex. Details from: Mr Dave Adams, G7GVK, c/o 85 Wood Road, Heybridge, Maldon, Essex, CM9 4AS. Tel: 01621 852296.

ESSEX REPEATER GROUP, GB3DA. Meets on 1st Wednesday in the month at the Bell Public House, Danbury. Web: www.swsystem.demon.co.uk/erg/. Details from: Mr Richard Merrell, G4GUJ, 40 Fanton Walk, Shotgate, Essex, SS11 8QT. Tel: 01268 769754.

GEC SENSORS (RADIO), G0GEC. Sports & Social Club, Gardiners Lane, Basildon, Essex, SS14 3EL.

HARLOW & DARS, G6UT. Meets 8.00pm on Tuesdays at the Mark Hall Barn, First Avenue, Harlow, Essex. Details from: Mr Ken Mott, G0HRR, Mark Hall Barn, First Avenue, Harlow, Essex, CM20 2LE. Tel: 01279 426647.

HARWICH ARIG, G0RGH. Meets 2nd Wednesday in the month at the Park Pavillion, Barrack Lane, Harwich. Details from: Mr D Cutts, Homeric, Main Road, Lt Oakley, Harwich, Essex, CO12 5JF. Tel: 01255 553510.

LOUGHTON & DARS, G4ONP. Meets 7.45pm on alternate Fridays. Details from: Mr Raymond E Pedley, G0LWF, 65 Mount Pleasant Road, Chigwell, Essex, IG7 5EP. Tel: 0181 500 2811.

MARTLESHAM DX CONTEST GRP, G0KPW. Details from: Fishers Farm, Colchester Road, Tendring, Clacton-on-Sea, Essex, CO16 9AA.

SOUTH ESSEX ARS, G4RSE. Meets 8.00pm on 1st and 3rd Wednesdays in the month at the Paddocks, Long Road, Canvey Island, Essex. Details from: Mrs Betty Maynard, G6LUO, 11 Denham Road, Canvey Island, Essex, SS8 9HB. Tel: 01268 695474.

SOUTHEND & DRC, G5QK. Meets 8.00pm on Thursdays at the Scout Centre, Eastern Avenue, Southend-on-Sea, Essex. Details from: Mr Alan Radley, G0TTM, 16 Kingsley Lane, Thundersley., Essex, SS7 3TU. Tel: 01268 741229.

STANFORD-LE-HOPE & DARC, G4SLH. Meets 8.30pm on alternate Tuesdays at the St Joseph Parish Rooms, Scratton Road, Stanford-le-Hope, Essex. Details from: Mr Ken Thompson, G4PAD, c/o 113 Gordon Road, Stanford-le-Hope, Essex, SS17 7QZ. Tel: 01375 671238.

VANGE ARS, G3YCW. Meets 8.00pm on Thursdays at the Barstable Community Centre, Basildon, Essex. Details from: Mrs D Thompson, 10 Feering Row, Basildon, Essex, SS14 1TE. Tel: 01268 552606.

WEST ESSEX RADIO COM TEAM, RS96543. Meets once a month. Details from: The Secretary, 282A Fullwell Avenue, Barkingside, Ilford, Essex, IG5 0SA.

Fife

Council Zone: G

GLENROTHES & DARC, GM4GRC. Meets 7.30pm on Wednesdays at the Old Nursery Building, Provost Land, Leslie, Fife.

Gloucestershire

Council Zone: D

CHELTENHAM AR ASSN, G5BK. Meets 7.45 for 8pm on the first Friday in the month at the Prestbury Library, Prestbury, Cheltenham. Details from: Mrs Patricia Thom, G1NKS, 9 Southern Road, Cheltenham, GL53 9AW. Tel: 01242 241099.

CHELTENHAM CLUSTER SUPP GP, GB7DXC. Details from: Mr A M Davies, G0HDB, c/o 11 Gravel Pits Close, Bredon, Tewkesbury, Glos, GL20 7QL. Tel: 01684 72178.

GLOUCESTER ARS, G4AYM. Meets 7.30pm on Wednesdays at the St John Ambulance HQ, Heathville Road, Gloucester. Details from: Mrs Jenny Beckingham, G7JUP, c/o 20 Baptist Close, Abbeymead, Gloucester, GL4 5GD. E-mail: GARS@Furcot.demon.co.uk.

QUEEN MARY ARCG, G6QM. Meets at Blazefield, Pateley Bridge, Harrogate, North Yorks, HG9 4DB. Details from: Mr Frank Harris, G4IEY, 4 Merestones Drive, The Park, Cheltenham, Glos, GL50 2SS. Tel: 01242 236715.

SMITHS INDUSTRIES RS, G4MEN. Meets 8.00pm on alternate Thursdays in the month at the Sports & Social Club, Evesham Road, Bishops Cleeve, Cheltenham, GL52 4SF. Details from: Mr A J Hooper, G1JMF, 5 Nine Elms Road, Longlevens, Gloucester, GL2 0HA.

STROUD RS, G4SRS. Meets every other Wednesday at the Minchampton Youth Centre, nr Stroud. Details from: Mr Stuart Goodfield, c/o 47 The Martins, Westrip, Stroud, Glos, GL5 4PG. Tel: 01453 752411.

STROUD VALLEYS ARS, RS174101. Details from: Lyndale, Brimscombe Lane, Brimscombe, Stroud, Glos, GL5 2RF.

WHITE NOISE LISTENING, G0WNL. Details from: Mr Glyn Davies, G7HTS, Amateur Radio Contest Group, c/o Walmore House, Walmore Hill, Minsterworth, Glos, GL2 8LA. Tel: 01452 750307.

Grampian Region

Council Zone: G

ABERDEEN ARS, GM3BSQ. Meets 7.30pm every Friday at the Lady of Aberdeen RC Church, Kincorth, Aberdeen.

BANFF & DARC, GM0PYC. Meets 7.00pm for 7.30pm on the 1st and 3rd Friday in the month at Banff Castle, Castle Street (opposite Job, Centre) Banff, AB45 1DL. Details from: Alex I Duncan, GM3DZB, Drumduan, 19 Bellevue Road, Banff, AB45 1BJ. Tel: 01261 812274.

GRAMPIAN REPEATER GRP, GB3GN. Details from: Mr Pete Weller, GM3XOQ, c/o Mither Tap, Bridge Road, Kemnay, Inverurie, Aberdeenshire, AB51 5QT. Tel: 01467 642148.

SHELL EXPRO ARC, GM0UEP. Meets 4.40pm to 6.00pm on Wednesdays at 1 Altens Farm Road, Nigg, Aberdeen, AB9 2HY. Details from: Mr Brian Newton, FAO The Librarian, 1 Altens Farm Road, Nigg, Aberdeen, AB9 2HY.

Greater London

Council Zone: C

ADDISCOMBE ARC, G4ALE. Meets 9.00pm on 1st Tuesday in the month at the Lion Inn, Pawsons Road, Croydon. Details from: Mr Q G Collier, G3WRR, 19 Grangecliffe Gardens, London, SE25 6SY. Tel: 0181 653 6948.

BARKING R & ES, G3XBF. Meets 7.30pm Thursdays at the Westbury Centre, Ripple Road, Barking, Essex. Details from: Mr Peter Allen, G0IAP, Westbury Recreation Centre, Westbury School Ripple Road, Barking Essex, IG11 7PT. Tel: 01708 474443.

BBC CRYSTAL PALACE CLUB, G0PAL. BBC Transmitting Station, Crystal Palace Parade, London, SE19 1UE.

CHACE COMMUNITY SCHOOL RC, G0VPK. Details from: Churchbury Lane, Enfield, EN1 3HQ.

CLIFTON ARS, G3GHN. Meets 8.00pm on Fridays at the Kidbrooke House, Community Centre, 90 Mycenae Road, London, SE3 7SE. Details from: Mr Keith Lewis, G4TJE, c/o 64 Rider Close, Blackfen, Sidcup, Kent, DA15 8TL. Tel: 0181 859 7630.

CRYSTAL PALACE & DRC, G3VCP. Meets 8.00pm on 3rd Saturday in the month at the All Saints Church, Parish Rooms, Beulah Hill, London. Details from: Mr V H Johnston, G1PKS, c/o 14 Auckland Close, London, SE19 2DA. Tel: 0181 653 2946.

DARENTH VALLEY RADIO, G0KDV. Meets 8.00pm on Wednesdays, twice monthly at the Crockenhill Village Hall, Swanley, Kent. Details from: Mr Ray Rodgers, G1UKH, Carmel, Chelsfield Lane, Orpington, Kent, Orpington, Kent, BR6 7RS. Tel: 01689 826846.

ECHELFORD ARS, G3UES. Meets 8.00pm 2nd & 4th Thursdays in the month at the Community Hall, St Martins Court, Kingston Crescent, Ashford, Middx. Details from: Mr Robin Hewes, G3TDR, c/o 24 Brightside Avenue, Laleham, Staines, Middx, TW18 1NG. Tel: 01784 456513.

EDGWARE & DRS, G3ASR. Meets 8.00pm on 2nd and 4th Thursdays in the month at the Watling Community Centre, 145 Orange Hill Road, Burnt Oak, Edgware, Middlesex. Details from: Mr Stephen Slater, G0PQB, 24 Lullington Garth, Borehamwood, Herts, WD6 2HE. Tel: 0181 953 2164.

HAVERING & DARS, G4HRC. Meets 8.00pm on Wednesdays at the Fairkytes Arts Centre, 51 Billet Lane, Hornchurch, Essex. Details from: Hon. Secretary, Fairkytes Art Centre, 51 Billet Lane, Hornchurch, Essex, RM11 1XA.

IMPERIAL COLLEGE RS, G5YC. Imperial College Union, Prince Consort Road, London, SW7 2BB.

KINGSBURY HIGH SCHOOL ARC, G0WVJ. Details from: Mr N D Purchon, G0WKM, Princes Avenue, Kingsbury, London, NW9 9JR.

KINGSTON COLLEGE, G0VDX.

KINGSTON COLLEGE RC, G7KCR, G0VDX. Meets 7.00pm on the 1st and 3rd Thursday during term time at the College.T Fell, College Radio Club, Kingston Hall Road, Kingston-upon-Thames, KT1 2AQ. Tel: 0181 546 2151.

KODAK ARS, G4FVJ. Open to Kodak employees only. Details from: Robert Parkin W153 3C, Kodak Ltd, Headstone Drive, Harrow, Middx, HA1 4TY.

LONDON AIRWAYS ARC, G4GTT. Details from: Distress & Diversion Flt, LATCC, Porters Way, West Drayton, Middx, UB7 9AU.

NSY AMATEUR RADIO SOCIETY, GX4NSY, GX6NSY. Open only to members of the Metropolitan Police Service. Details from: the Secretary, Room 99, New Scotland Yard, Broadway, London, SW1H 0BG.

RS OF HARROW, G3EFX. Meets at 8.00pm every Friday at the Harrow Arts Centre, Uxbridge Road, Hatch End, Middx. Details from: Mr C Friel, G4AUF, 5 Windmill Hill, Ruislip, Middx, HA4 8QF. Tel: 01895 621310.

SILVERTHORN RC, G3SRA, G2HR, G8CSA. Meets 7.30om on Fridays at the Chingford Adult Education and, Community Centre, Friday Hill, House, Simmons Lane, Chingford, London, E4 6JH.

SOUTH LONDON COLLEGE ARS, G3HFY. Meets at the Lambeth College, Norwood Centre, Knights Hill, West Norwood, London. Details from: Mr M Knott, G0WCR, c/o 76 New Barns Avenue, Mitcham, Surrey, CR4 1LF.

SOUTHGATE RC, G3SFG. Meets 7.30pm on 2nd and 4th Thursdays in the month at the Winchmore Hill Cricket, Pavillion, Firs Lane, London, N21 3ER. Details from: Mr Dave Michael, G0ASA, Edmonton, London, N9 0LL. Tel: 0181 482 6795.

ST DUNSTANS COLLEGE ARS, G4SDC. Meets irregularly. Details from: Mr Sam Kennard, G4OHX, Stanstead Road, Catford, London, SE6 4TY. Tel: 0181 6901274.

SURREY RADIO CONTACT CLUB, G3SRC. Meets 8.00pm on 1st and 3rd Mondays in the month at the T.S. Terra Nova, 34 The Waldrons, Croydon, Surrey. Details from: Mr Maurice Fagg, G4DDY, 113 Bute Road, Wallington, Surrey, SM6 8AE. Tel: 0181 669 1480.

THREE A'S CONTEST GRP, G0AAA. Details from: Mr Roger Western, G3SXW, c/o 7 Field Close, Chessington, Surrey, KT9 2QD. Tel: 0181 3973319.

WHITTON ARG, G0MIN. Meets 7.30pm on Fridays at the Whitton Community Centre, Percy Road, Whitton. Details from: Mr Ian Clabon, G0OFN, Whitton Community Centre, Percy Road, Whitton, Middx, TW2 6JL. Tel: 0181 894 9131.

Greater Manchester

Council Zone: A

BOLTON SCHOOL ARC, G0VUX. Meets on Fridays 12.00pm till 12.40pm, and Mondays 4.00pm till 5.30pm during term time at the school. Details from: Chris Walker, G0BGQ, Bolton School, Chorley New Road, Bolton, Lancs, BL1 4PA. Tel: 01204 840201.

BURY RS, G3BRS. Meets 8.00pm on Tuesdays at the Mosses Centre, Cecil Street, Bury, Lancs, BL9 0SB. Details from: Mr Steve Gilbert, G3OAG, Mosses Centre, Cecil Street, Bury, Lancs, BL9 0SB. Tel: 0161 8811850.

DOUGLAS VALLEY ARS, G3BPK. Meets 8.00pm on 1st and 3rd Thursday in the month at the Wigan Sea Cadet HQ, Training Ship Sceptre, Brookhouse Terrace, off Warrington Lane, Wigan. Details from: Mr D Snape, G4GWG, c/o 30 Culcross Avenue, Highfield, Wigan, WN3 6AA. Tel: 01942 211397.

ECCLES & DARS, G3GXI. Meets 9.30pm on Tuesdays at the Eccles Liberal Club, Wellington Road, Eccles, Manchester. Details from: Dr Chris Harrison, G8KRG, c/o 53 Peveril Close, Whitefield, Manchester, M45 6NS. Tel: 0161 773 7899.

MANCHESTER & DARS, G5MS. Meets 7.00pm on Tuesdays at the Simpson Memorial Community, Hall, Moston Lane, Moston, Manchester. Details from: Mr Harold Jeffrey, G0VJZ, 11 Linden Road, Stalybridge, Cheshire, SK15 2LS. Tel: 0161 338 4412.

MANCHESTER ARIEL RADIO GRP, G3ETK. Open only to BBC employees.BBC Club,New Broadcasting Hse., Oxford Road, Manchester, M60 1SJ.

NORTH TRAFFORD COLLEGE, G4FXP. Details from: Mr J.C. Goddard, A.L.A., Talbot Road, Stretford, Manchester, M32 0XH.

OLDHAM ARC, G4ORC, G1ORC. Meets 8.00pm on Thursdays at the Moorside Conservative Club, Ripponden Road, Moorside, Oldham. Details from: Nick Foster, G0ULA, 5 Lees Grove, Leesbrook, Oldham, OL4 5LG. Tel: 0161 627 1639.

OULDER HILL ARS, G0UQA. Details from: Mr Alan Gale, G4TMV, Oulder Hill Community School, Hudsons Walk, Rochdale, Lancs, OL11 5EF.

ROCHDALE & DARS (R.A.D.A.R.S), G0ROC. Meets 8.00pm on Mondays at the Cementary Hotel, Bury Road, Rochdale. Details from: Mr John Cannell, G7OAI, 53 Thimble Close, Rochdale, Lancashire, OL12 9QP. Tel: 01706 376204.

SOUTH MANCHESTER RC, G3FVA. Meets 8.00pm on Fridays at the Sale Moor Community Centre, Norris Road, Sale. Details c/o 2 Arlington Drive, Woodsmoor, Stockport, SK2 7EB.

STOCKPORT RS, G6UQ. Meets 7.30pm on 2nd and 4th Wednesdays in the month at the T.S. Hawkins, Stockport Sea, Cadets HQ, Pear Mill Ind., Estate, Stockport Rd West,, Lower Bredbury, Stockport. Details from: Hon. Secretary, c/o 2 Linney Road, Bramhall, Stockport, Cheshire, SK7 3JW. Tel: 0161 439 4285.

TAMESIDE ARS, G0SQU. Meets 7.30pm every Wednesday at the ATC Hut, Moorcroft Street, Droylsden, Tameside, M43 7BP. Details from: Mr Arthur N Laughlan, G1YCM, 8 Kempton Close, Droylsden, Tameside, Manchester, M43 7JL.

TRAFFORD ARC, G1TRC. Meets 7.30pm on Thursdays at the Watch House Cruising Club, Canal Bank, Stretford, Manchester, M32 8WE. Details from: Graham, c/o 2 Westwood Avenue, Urmston, Manchester, M41 9NG. Tel: 0161 748 9804.

TRAFFORD RADIO GROUP, G0TRG. Meets 8.00pm on Thursdays at 17th Stretford Scouts HQ, Barton Road, Stretford, Manchester. Details from: Mr Jon Mossman, G7JKK, c/o 31 Marlborough Road, Stretford, Manchester, M32 0AW. Tel: 0161 865 5609.

WEST MANCHESTER RC, G4MWC. Meets 8.00pm on Wednesdays at the Astley & Tyldesley Miners, Welfare Club, Meanly Road, Astley, Tyldesley, Manchester. Details from: Secretary, Mr R Lowe, G0FRL, c/o 12 Cavenham Grove, Bolton, Gtr Manchester, BL1 4UA. Tel: 01204 494308.

WIGAN & DARC, G0HRW. Meets 7.30pm on 1st and 3rd Thursday in the month. Details from: Mr D H Barkley, G0DPI, c/o 39 Fulbeck Avenue, Goose Green, Wigan, Lancs, WN3 5QN. Tel: 01942 237162.

WIGAN DEANERY HIGH SCHOOL, G0TWD. Meets at the school most lunch times and after school on Tuesdays and Thursdays. Details from: Revd. Malcolm Drummond, Wigan Deanery High School, Frog Lane, Wigan, Lancs, WN1 1HR. Tel: 01942 244355.

Guernsey & Dep'ncies

Council Zone: D

GUERNSEY ARS, GU3HFN. Details from: The Secretary, GARS, P O Box 100, The Lodge, la Corbinerie, Oberlands, St Martins, Guernsey, GY1 3EL. Tel: 01481 725450.

Gwent

Council Zone: E

ABERGAVENNY RS, GW4GFL. Meets 7.30pm on alternate Tuesdays at the Hill Residentail College, Pen-y-Pound, Abergavenny, Gwent. Details from: Mr Reg Lloyd, GW4IQA, Llwyn Celyn, Pandy, Abergavenny, Gwent, NP7 8DN. Tel: 01873 890681.

BLACKWOOD & DARS, GW6GW. Meets 7.00pm on Fridays at the Oakdale Comprehensive School, Oakdale, Blackwood, Gwent. Details from: Mr John Evans, GW8ITI, Rosegarth, Tuckers Villas, Woodbine Road, Blackwood, Gwent, NP2 1QH. Tel: 01495 225178.

EBBW VALE COLLEGE RS, GW0IIW. Meets 6.30pm on Wednesdays at the Gwent Tertiary College, Ebbw Vale Campus, College Road, Ebbw Vale, Gwent. Details from: Mr T Hayden, GW0HCN, c/o 7 Attlee Close, Garnlydan, Ebbw Vale, Gwent, NP3 5ES. Tel: 01495 305192.

NEWPORT ARS, GW4EZW. Meets 7.30pm on Mondays at the Brynglas Community Centre, Brynglas Road, Newport, Gwent. Details from: Mr Paul Nicholls, Brynglas Community Educ. Cntr, Brynglas Road, Newport, Gwent, NP1 9FB.

PONTYPOOL ARS, GW3RNH. Meets 7.00pm on Tuesdays at the Settlement, Rockhill Road, Pontypool,/Gwent. Details from: Mr L J La-Traille, c/o 33 Festival Crescent, New Inn, Pontypool, Gwent, NP4 0NB. Tel: 01495 764393.

RED DRAGON CON.GROUP, GW8GT. Details from: Mr Brian Davies, GW3KYA, 16 Vancouver Drive, Penmain, Blackwood, Gwent, NP2 0UQ. Tel: 01495 225825.

TORFAEN SCOUTS ARC, GW0UKT. Meets on 2nd and 4th Wednesday each month (excluding December). Details from: Mr Richard Chatwin, GW0VAW, New Street Post Office, Pontnewydd, Cwmbran, Gwent, NP44 1EE. Tel: 01633 483277.

Gwynedd

Council Zone: E

ARFON REPEATER GROUP, GB3AR. No meetings except AGM in June of each year. Details from: Mr B V Davies, GW4KAZ, Garth, 2 Glanllyn, Bethel, Caernarfon, Gwynedd, LL55 1YL. Tel: 01248 670009.

DX CLUSTER SUPPORT GROUP, GW0VYG. Details from: Mr Tony Jones, GW4VEQ, Penrhiw Bach, Brungwran, Isle of Anglesey, LL65 3RD.

MEIRION ARS, GW4LZP. Meets 8.00pm on 1st Thursday in the month Royal Ship Hotel, Dolgellau, Gwynedd. Details from: Mr M D Fowler, GW3GKZ, c/o Ty Gwyn, Abergwynant, Dolgellau, Gwynedd, LL40 1YF. Tel: 01341 422447.

PORTHMADOG & DARS, GW0MVI. Meets 8.00pm on 3rd Thursday in the month Harbour Cafe, Ffestiniog Railway, Porthmadog, Gwynedd. Details from: Mr Robert Anglesea, GW0WNW, Argraig, Llanfrothen, Gwynedd, LL48 6LJ.

THE DRAGON ARC, GW4TTA. Meets 7.30pm for 8.00pm on 1st and 3rd Monday in the month at the Four Crosses Hotel, Pentraeth Road, Menai Bridge, Gwynedd. Details from: Mr Tony Rees, GW0FMQ, c/o No 2 Bryn Poeth, Tregarth, Bangor, Gwynedd, LL57 4PG. Tel: 01248 660963.

YNYS MON AR USERS GP, RS172926. Last Wednesday of the month 7.30pm - 10pm at Llangnefi Scout Hall, (opp. Kwiksave), Llangnefi. Details from: Mr Tony Anziani, GW4ZWN, Sec, c/o Ty-Coch, Penrhyd, Amlwch, Anglesey, Gwynedd, LL68 1XX. Tel: 01407 832197.

Hampshire

Council Zone: D

ANDOVER RAC, G0ARC. Meets 7.30pm 1st and 3rd Tuesdays in the month at the Village Hall, Wildhern, Andover, Hants. Details from: Terry Cull, G8ALR, Drybrook Cottage, Amesbury Road, Cholderton, Salisbury, Wilts, SP4 0ER. Tel: 01980 629346.

BASINGSTOKE ARC, G3TCR, G8JYN. Meets 7.30pm on 1st Monday in the month at the The Sony Broadcast Social Club, Priestley Road, Basingstoke, Hants. Details from: David Burleigh, G4WIZ, 2 Adam Close, Baughurst, Basingstoke, Hants, RG26 5HG.

FAREHAM & DARC, G3VEF. Meets 7.30pm on Wednesdays at the Portchester Community Centre, Westlands Grove, Portchester, Hants. Details from: Mr Andrew Sinclair, G0AMS, c/o 9 Solent View, Fareham, Hants, PO16 8HE. Tel: 01329 235397.

FLIGHT REFUELLING ARS, G4RFR. Open to all with an interest in amateur radio. Meets 7.30pm Wednesdays and Sundays at the Flight Refuelling Social Club, Merley, Wimborne, Dorset. Details from: Mr John Hart, G4POF, c/o 1 Meadow Court, Fordingbridge, Hants, SP6 1LW. Tel: 01425 653404.

HANTS POLICE ARC, G0LAW. Details from: 49 Estridge Close, Lowford, Southampton, Hants, SO31 8FN.

HIGHFIELD PARK RC, G4BWD. Meets on Thursday evenings Highfield Park RC, National Air Traffic Service, Highfield Park, Heckfield, Hants, RG27 0LD.Mr A K Whillock, G4ZLX, National Air Traffic Services, Room 14, Highfield Park, Heckfield, Hook, Hants, RG27 0LG. Tel: 01734 225019.

HORNDEAN & DARC, G4FBS. Meets 7.30pm on 1st and 4th Tuesday in the month at Lovedean Village Hall, Lovedean Lane, Lovedean, Hants. Details from: Mr Stuart Swain, G0FYX, 35 Mavis Crescent, Havant, Hants, PO9 2AE. Tel: 01705 472846.

IBM UK LABORATORIES ARC, G3YXR. Open only to IBM employees. Meets 7.00pm on Thursdays at the IBM (UK) Laboratories, Hursley House, Hursley Park, Winchester.

ITCHEN VALLEY ARC, G0IVR. Meets 7.30pm on 2nd and 4th Fridays in the month (except August) at the Scout Hut, Brickfield Lane, Chandlers Ford, Eastleigh, Hants. Details from: Sheila Williams, G0VNI, The Croft, Ringwood Road, Bartley, Hants, SO40 7LA. Tel: 01703 813827.

MARS (PORTSMOUTH), G4JMR. Open to employees only. Meets at the Anchorage Road, Portsmouth, PO3 5PU. Details from: Mr V G Scambell, G3FWE, Mr P Rudwick, Marconi Systems, Anchorage Rd, Portsmouth, PO3 5PU. Tel: 01983 616966.

SOLENT REPEATER GRP, GB3PC. c/o 3 Fernie Close, Hill Head, Fareham, Hants, PO14 3SQ. Tel: 01329 661078.

SONY BROADCAST ARC, G4SZC. Accredited C&G RAE centre. Meets 7.30pm at Sony Sports & Social Club, Priestley Road, Basingstoke. Details from: Mr Stephen Harding, G4JGS, Jays Close, Viables, Basingstoke, Hants, RG22 4SB. Tel: 01256 55011.

SOUTH HAMPSHIRE INT.TELE.SOC., G3DIT. Meets 7.30pm on the 1st Wednesday in the month at G3JZV's QTH. Space is limited. Details from: Rev T R Mortimer, G3JZV, 59 First Avenue, Farlington, Portsmouth, PO6 1JL.

SOUTHAMPTON UNIVERSITY RC, G3KMI. Meets 8.00pm on Mondays during term time at the Student Union, West Building (top floor), University Road, Southampton.Students Union, The University, Highfield, Southampton, SO17 1BJ.

THREE COUNTIES ARC, G4WWR. Meets 8.00pm on the 2nd & 4th Thursdays in the month at the Bramshott Parish Inst. & Club, Headley Road, Liphook, Hants. Details from: Mr Damian Kamm, G7RFV, 134A Haslemere Road, Liphook, Hants, GU30 7BX. Tel: 01428 724456.

TORBAY BLOCK SUBMARINE ARC, G0NKL. Meets 7.30pm every Thursday at the Royal Navy Submarine Museum, Haslar Jetty Road, Gosport, Hants.

UKFM SRH GROUP, GB3SN. PO Box 6, Alton, Hampshire, GU34 1XT.

WATERSIDE ARS, G4JYN. Meets 7.30pm on 1st Tuesday in the month at the Applemore Scout HQ, Applemore, Hythe, Southampton. Details from: Mr Bill Warburton, G0WSI, 4 Haynes Way, Dibden Purlieu, Southampton, SO45 5QQ. Tel: 01703 207180.

WINCHESTER ARC, RS95704. Meets 7.30pm on 3rd Friday in the month at Durngate House, North Walls, Winchester.4 Buriton Road, Harestock, Winchester, Hants, SO22 6HX.

Hereford & Worcester

Council Zone: B

ARIEL RADIO GROUP, G3PPG. Open only to BBC personnel. Meets at the Centre for Broadcast Skills, Training, Wood Norton, Evesham, Worcs, WR11 4TF. Details from: R W Penman, BBC Centre for, Broadcast Skills (CBST), Wood Norton, Evesham, Worcs, WR11 4YB.

BROMSGROVE ARS, G4TUI. Meets 8.00pm on 2nd and 4th Tuesday in the month at the Likey End WMC, Bromsgrove, Worcs. Details from: Mr G E Duffin, G2FTY, c/o 20 Byron Road, Redditch, Worcs, B97 5EB.

DROITWICH ARC, G4PVO. Meets 8.00pm on 1st Tuesday in the month in the John Corbett Room, Droitwich Community Hall, Droitwich, Worcs. Details from: Mr J Jackson, G4OPV, c/o 15 Jackson Crescent, Stourport-on-Severn, Worcs, DY13 0EW. Tel: 01922 826188.

HEREFORD ARS, G3YDD. Meets 8.00pm on 1st and 3rd Friday in the month at the Civil Defence HQ, Magistrates Court, Gaol Street, Hereford. Details from: Mr E Wyman, G0UDF, 1 Bridle Road, Hereford, HR4 4PP.

KIDDERMINSTER & DARS, G0KRC. Meets 8.00pm on alternate Tuesdays at the Music Block, Stourport-on-Severn High, School, Worcs. Details from: Mr Jeff Philpotts, G0RJP, c/o 62 Erneley Close, Stourport-on-Severn, Worcs, DY13 0AH. Tel: 01299 822206.

MADLEY COMMUNITY CENTRE RG, G7BTI. Meets at the centre usually every three months.

MALVERN HILLS ARC, G4MHC. Meets 8.00pm on 2nd Tuesday in the month at the Red Lion Inn, St Anne's Road, Malvern, Worcs. Details from: Mr Dave Hobro, G4IDF, Sec., 60 Linksview Crescent, Newtown, Worcester, WR5 1JJ. Tel: 01905 351 568.

REDDITCH RC, G4ACZ. Meets 8.00pm on the 2nd Thursday in the month at the WRVS Centre, Ludlow Road, Redditch, Worcs. Details from: Mr R J Mutton, G3EVT, 'Summerhayes', Mill Lane, Oversley Green, Alcester, Warwickshire, B49 6LF. Tel: 01789 762041.

VALE OF EVESHAM RAC, G0ERA. Meets 8.00pm on 1st Thursday in the month at the BBC Club, High Street, Evesham, Worcs. Details from: Mr A C Lindsay, G4NRD, c/o 21 Willow Road, Four Pools, Evesham, Worcester, WR11 6YW. Tel: 01386 41508.

WOLVERLEY & DARS, G0WWS. Details from: 12 Ashley Road, Broadwaters Heath, Kidderminster, Worcs, DY10 2XD.

WYRE FOREST RG, GB3KR. Details from: 16 Great Western Way, Stourport-on-Severn, Worcs, DY13 8AG.

WYTHALL RADIO CLUB, G4WAC. Meets 8.00pm on Tuesdays at the Wythall House, Silver Street, Wythall, Birmingham. Details from: Mr David Dawkes, G0ICJ, c/o 83 Alcester Road, Hollywood, Birmingham, B47 5NR. Tel: 0121 430 2929.

Hertfordshire

Council Zone: C

BISHOPS STORTFORD ARS, G5ZG. Meets 8.00pm on 3rd Monday in the month at the Royal British Legion Club, Windhill, Bishop's Stortford, Herts. Details from: Mr Tony Judge, G0PQF, 44 Thorley Lane, Bishops Stortford, Herts, CM23 4AD. Tel: 01279 506933.

CHESHUNT & DARC, G4ECT. Meets 8.00pm on Wednesdays at the Church Rooms, Church Lane, Wormley, Herts. Details from: Mr D R French, G3TIK, c/o 37 Warner Road, Ware, Herts, SG12 9JN. Tel: 01920 461711.

DACORUM ARTS, G7RIH, G0WIH. Meets on 1st and 3rd Tuesdays in the month at the Guide Meeting Rooms (next to, the Royal British Legion), Queensway, Hemel Hempstead. Details from: Mr Ian Hamilton, G0TCD, c/o 139 Elstree Road, Hemel Hempstead, Herts, HP2 7QW. Tel: 01442 211925.

HARPENDEN ARC, G0OMY. Meets 8.00pm on 1st Thursday in the month (from September to March) at Science Block, Aldwickbury School, Wheathampstead Road, Harpenden. Details from: Mr John Wedderburn, G4JOV, c/o Aldwickbury School, Wheathampsted Road, Harpenden, Herts, AL5 1AE. Tel: 01582 765821.

HODDESDON RADIO CLUB, G0TSN. Meets 8.00pm on alternate Thursdays at the Rye Park Conservative Club, Rye Road, Hoddesdon, Herts. Details from: Don Platt, G3JNJ, 22 Charcroft Gardens, Ponders End, Enfield, Middx, EN3 7HA. Tel: 0181-292 3678.

MIMRAM CONTEST GP, M0ABC. Details from: Alan Holdsworth, G0SAH, c/o 26 Chelveston, Panshanger, Welwyn Garden City, Herts, AL7 2PW. Tel: 01707 392950.

RADIO SCOUTING TEAM, GB2RST. Meets most weekends at Tolmers Scout Camp, Tolmers Road, Cuffley, Herts, EN6 4JS. Details from: Mr Bill Livens, G2CKB, 10 Cotton Drive, Pinehurst Est, Hertford, SG13 7SU. Tel: 01992 558493.

STEVENAGE & DARS, G3SAD. Meets 7.30pm on Tuesdays at the Stevenage Day Centre, Chells Way, Stevenage, Herts, SG2 0LT. Details from: Mr Peter Bell, 2E1CRK, 122 Howard Drive, Letchworth, Herts, SG6 2DE. Tel: 01462 674505.

TRIO-KENWOOD UK, G0TKU. Details from: Mr David Wilkins, G5HY, Kenwood House, Dwight Road, Watford, Herts, WD1 8EB. Tel: 01923 212044.

UNIVERSITY OF HERTFORDSHIRE RS, G4WTB, G6BOB. Open only to staff & students. Meets 2.00pm on Wednesdays during term time in the Radio Shack (Room E423), University of Hertfordshire. Details from: Mr David Lauder, G0SNO, Elect Engineering Dpt, College Lane, Hatfield, Herts, AL10 9AB. Tel: 01707 284187.

VERULAM ARC, G3VER. Meets 7.45pm on 2nd and 4th Tuesdays in the month at the RAF Association HQ, New Kent Road, St Albans, Herts. Details from: Mr Bob Heath, G3UJV, 26 Lancaster Avenue, Hadley Wood, Barnet, Herts, EN4 0EX.

WELWYN & HATFIELD ARC, G3WGC. Meets 8.00pm on 1st and 3rd Mondays in the month at the Hyde Association, Holly Bush Road, Welwyn Garden City, Hertfordshire. Details from: Mr Kevin Howard, G7VLD, c/o 43 Hazeldell, Watton at Stone, Herts, SG14 3SN. Tel: 01920 830617.

Highland Region

Council Zone: G

BLACK ISLE REPEATER GROUP, GB3BI. Meets at various times.

EASTER ROSS RC, GM4MFL. Meets 7.30pm on Fridays at QTH GM4FDT. Details from: Mr Robert Kerr, GM4FDT, c/o Rosskeen Bridge, Invergordon, Ross-shire, IV18 0PR. Tel: 01349 852332.

FORT WILLIAM ARG, GM0FRG. Details from: 10 Lundy Road, Inverlochy, Fort William, Inverness-shire, Scotland, PH33 6NX. Tel: 01397 703046.

INVERNESS ARC, GM4TPF. Meets 7.30pm on 1st and 3rd Thursdays in the month at the 18 Inverness (Muirtown) Scouts, Old Navy Building, Muirtown, Basin, Inverness, IV3 6LS.

Humberside (North)

Council Zone: A

E YORKS RPTR GP, GB3HA. Details from: Alan Santos, G4PMJ, 17 Elm Garth, Roos, Hull, HU12 0HH.

HORNSEA ARS, G4EKT. Meets 8.00pm on Wednesdays at The Mill, Atwick Road, Hornsea, North Humberside. Details from: Mr Jeff Southwell, G4IGY, Mill House, Atwick Road, Hornsea, North Humberside, HU18 1DZ. Tel: 01964 533331.

HULL & DARS, G3AMW. Meets 8.00pm every Friday at the S.W.L. Centre, Club Room, Goathland Close, Walton St, Hull. Details from: Mr Roly Towler, G0UKS, c/o 77 Glebe Road, Stoneferry, Hull, East Yorkshire, HU7 0DU. Tel: 01482 837042.

NORTH FERRIBY UNITED ARS, G0ECR. Meets 8.00pm Fridays at the Ferriby United Football Club, North Ferriby, North Humberside. Details from: Mr David Boughton, G7PER, 59 Redland Drive, Kirk Ella, Hull, HU10 7UX.

RAYWELL PARK SCOUTS ARS, G0VRM. Details from: Mr Roy Andreang, G4CMT, 6 Beech Avenue, Bilton, Hull, HU11 4EN. Tel: 01482 812115.

Humberside (South)

Council Zone: B

CIBA-GEIGY AR & CG, G0CGG. Open to employees only. Meets on Mondays at Moody Lane, Pyewipe, Grimsby, S Humberside. Details from: 45 Meadowbank, Great Coates, Grimsby, S Humberside, DN37 9PL. Tel: 01472 882165.

GOOLE R & ES, G0OLE. Meets 7.30pm on Fridays at the West Park Pavillion, Goole, South Humberside. Details from: Mr Richard Sugden, G0GLZ, c/o 162 Centenary Road, Goole, North Humberside, DN14 6PE. Tel: 01405 769968.

GRIMSBY ARS, G3CNX. Meets 8.00pm on Thursdays at Cromwell Social Club, Cromwell Road, Grimsby, South Humberside. Details from: Mr G J Smith, G4EBK, c/o 6 Fenby Close, Wybers Wood, Grimsby, South Humberside, DN37 9QL. Tel: 01472 887720.

SCUNTHORPE & DARC, G4FUH. Details from: Mr Ken Coxon, G0HDV, 29 Chapel Road, Broughton, Brigg, DN20 0HW. Tel: 01652 653847.

Isle of Man

Council Zone: A

ISLE OF MAN ARS, GD3FLH. Details from: Mr Chris Wood, GD6TWF, c/o 2 2 Lyndale Avenue, Peel, Isle of Man, IM5 1JY. Tel: 01624 842786.

Isle of Wight

Council Zone: D

ISLE OF WIGHT RS, G3SKY. Meets 8.00pm Fridays at Unity Hall, Wootton Bridge, Isle of Wight. Details from: Mr Alan Reeves, G4ZFQ, Wayside, 41 Nodes Road, Northwood, Cowes, Isle of Wight, PO31 8AD. Tel: 01983 294309.

WEST WIGHT RADIO SOCIETY, G0UZT, GB2GMM. Details from: Mr Robert B. Clegg, G7RER, Cleggs Castle, Monks Lane, Freshwater, Isle of Wight, PO40 9ST.

Jersey

Council Zone: D

CHANNEL ISLAND ARC GROUP, GJ7DGJ. Details from: Mr John Poole, GJ1TJP, Cheriton, Manor Park Road, la Pouquelaye, St Helier, Jersey, C.I, JE2 3GJ. Tel: 01534 53694.

JERSEY AM ELECTRONICS CLUB, GJ4HXJ. No meetings held, club runs the local 6m beacon. Details from: Mr Geoff Brown, GJ4ICD, c/o 1 Belmont Gardens, St Helier, Jersey, C I, JE2 4SD. Tel: 01534 77067.

JERSEY ARS, GJ3DVC. Meets 8.00pm Fridays and on 7.30pm on Wednesdays for Novice classes at the German Signal Station, Rue Baal, La Moye, St. Brelade. Details from: Mrs Anne Mourant, GJ7HTV, P O Box 338, St Helier, Jersey, JE4 9YG. Tel: 01534 34948.

Kent

Council Zone: C

BORDEN ARC, G4LBS. Borden School, Ave of Remembrance, Sittingbourne, Kent, ME10 4DB.

BREDHURST RX & TX SOC, G0BRC. Meets 8.15pm Thursdays at Rock Avenue Working Mans Club, Rock Avenue, Gillingham, Kent. Details from: Mr David Lowe, Sec, c/o 2 Speedwell Close, The Vineries, Gillingham, Kent, ME7 2PR.

CANTERBURY COLLEGE ARC, G4JLN. Canterbury Coll. of Technology, New Dover Road, Canterbury, CT1 3AJ.

CRAY VALLEY RS, G3RCV, G1RCV. Meets 8.00pm on 1st and 3rd Thursday in the month at the Progress Hall, Admiral Seymour Road, Eltham, London. Details from: Mr A R Burchmore, G4BWV, c/o 49 School Lane, Horton Kirby, Dartford, Kent, DA4 9DQ. Tel: 01322 862470.

DOVER RADIO CLUB, G3YMD. Meets 8.00pm Wednesdays during school terms (other meetings as announced) at the Duke of York's School, Guston, Dover. Details from: Mr Brian Hancock, G4NPM, PO Box 73, Dover, Kent, CT16 2FD. Tel: 01304 821007.

EAST KENT RADIO SOCIETY, G0EKR. Meets 8.00pm on the 1st and 3rd Fridays in the month at St Bartholomew's Church Hall, Herne Bay. Details from: Paul Nicholson, G3VJF, 34 Cliff Avenue, Herne Bay, Kent, CT6 6LZ. Tel: 01227 743070. Fax: 01227 742288.

GRAVESEND RS, G3GRS. Meets 8.00pm Mondays at The Coach & Horses, 139 Paddock Street, Shrubbery Road, Gravesend, Kent. Details from: Mr D Blakeley, G3KZN, c/o 34 Mead Road, Gravesend, Kent, DA11 7PP. Tel: 01474 355736.

HILDERSTONE ARS, G0HRS. Meets 7.00pm on Fridays at Hilderstone A.E.C., Broadstairs, Kent. Details from: Mr V B de Rose, G0CLO, 4 Briars Walk, Broadstairs, Kent, CT10 2XR. Tel: 01843 869812.

KENT IP GROUP, RS174339. Details from: Mr Nick Parnell, G4ZXI, 4 Forge Lane, Headcorn, Kent, TN27 9QQ.

KENT REPEATER GROUP, GB3KS. Details from: Mr John Wellard, G6ZAA, 19 South Motto, Park Farm, Kingsnorth, Ashford, Kent, TN23 3NJ. Tel: 01233 503050.

MAIDSTONE YMCA ARS, G3TRF. YMCA Sports Centre, Melrose Close, Maidstone, Kent Details from: Mr J Belling, G0RHO, 'Y' Sports Centre, Melrose Close, Maidstone, Kent, ME15 6BD. Tel: 01622 832259.

MEDWAY ARTS, G5MW, G8MWA. Meets 7.30pm Fridays at Tunbury Hall, Catkin Close, Tunbury Avenue, Walderslade, Chatham. Details from: Mrs Gloria Ackerley, G3VUN, Forty Towers, 40 Linwood Avenue, Strood, Rochester, Kent, ME2 3TR. Tel: 01634 710023.

MEOPHAM PARISH RC, G0FBB. Details from: 30 Highview, Vigo, Meopham, Gravesend, Kent, DA13 0RR.

NORTH KENT RS, G4CW. Meets 8.00pm on 1st and 3rd Tuesday in the month at The Pop-in-Parlour,

Graham Road, Bexleyheath, Kent. Details from: Mr A V Fribbens, G8MLQ, c/o 12 Appleshaw Close, Gravesend, Kent, DA11 7PB. Tel: 01474 365694.

RAINHAM RADIO CLUB, G0WDY. Details from: Mr Ian Buckle, G0MIF, c/o 25 Portsmouth Close, Strood, Rochester, Kent, ME2 2QY. Tel: 01634 723551.

RC OF THANET, G2IC. Meets 7.30pm 2nd and 4th Thursday in the month at Hoverspeed Social Club, High Street, Manston Village, Kent. Details from: Mr Chris Turner, G0VUT, c/o 28 Reading Street, Broadstairs, Kent, CT10 3AZ. Tel: 01843 603065.

SEVENOAKS & DARS, G0SEV. Meets on 3rd Monday in the month at Sevenoaks District Council, Argyll Road, Sevenoaks, Kent, TN13 1HG. Details from: Mr Ted Denman, Council Offices, Argyle Road, Sevenoaks, Kent, TN13 1HG. Tel: 0181 304 3950.

SOUTH FORELAND AR OP GP, G0UJR. Details from: Mrs E M Berridge, Bracklyn, St Clare Road, Walmer, Deal, Kent, CT14 7QB.

ST. OLAVE'S ARC, RS175768. St. Olave's & St. Saviours Sch, Goddington Lane, Orpington, Kent, BR6 9SH.

SWADELANDS SCHOOL RC, G0WFN. Maidstone Road, Lenham, Maidstone, Kent, ME17 2QJ.

SWALE ARC, G4SRC, G6SRC. Meets 8.00pm Mondays at the Ivy Leaf Club, Dover Street, Sittingbourne, Kent. Details from: Mr Dennis Spalding, G1JQH, c/o 171 Minster Road, Minster-on-Sea, Sheerness, Kent, ME12 3LH. Tel: 01795 876091.

THE MORSE CLUB, GX0OXE. Details from: Mr R Francis, 163 Sherwood Park Avenue, Blackfen, Sidcup, Kent, DA15 9JD.

TONBRIDGE SCHOOL RS, G4AJS. Open only to staff and pupils. Meets at Tonbridge School, Tonbridge, Kent, TN9 1JP. Details from: R J Hughes, Communications Div, Tonbridge School, Tonbridge, Kent, TN9 1JP.

WEST KENT ARS, G3WKS. Meets 8.00pm on 1st & 3rd Friday in the month at the St. Peter's School Hall, North Street, Tunbridge Wells, Kent. Details from: Mr A Korda, G4FDC, 5 Windmill Court, North Street, Tunbridge Wells, Kent, TN2 4SU. Tel: 01892 541733.

Lancashire

Council Zone: A

BAY ARG, G0WJW. Details from: Mr T J Newstead, 5 Farnlea Drive, Bare, Morecambe, Lancs, LA4 6JU.

BURNLEY & DARS, RS87674. Meets on Tuesdays at Barden High School, Barden Lane, Burnley, Lancashire. Details from: Mr Bill Scrivener, G0BQC, c/o Mr Scrivener, Sec., 67 Lower Manor Lane, Burnley, Lancs, BB12 0EB. Tel: 01282 39765.

CENTRAL LANCS ARC, G0FDX. Meets 8.00pm on 1st and 3rd Monday in the month at the Priory Club, Broadfield Drive, Leyland, Lancs. Details from: Mr Brian Birkby, G0NEI, 10 Rankin Avenue, Hesketh Bank, Preston, PR4 6PA. Tel: 01772 813340.

DARWEN ARC, G4JS. Meets 7.30pm on the 3rd Wendesday in the month at the Darwen Catholic Club, Wellington Fold, Darwen, Lancashire. Details from: Mr W Lishman, G2AKK, 28 Lightbown Street, Darwen, Lancs, BB3 0DY. Tel: 01254 703767.

EAST LANCS PACKET GP, RS170937. Meets quarterly, as advertised on the Group's BBS (GB7HVU), at the Bowling Club, Willows Lane, Accrington, Lancs. Details from: Andrew Chisholm, G3INL@GB7HVU, 95a Salthill Road, Clitheroe, Lancashire, BB7 1PE. Tel: 01200 24482.

FISTS CW CLUB, G0IPX. Details from: Mr E Longden, G3ZQS, 119 Cemetery Road, Darwen, Lancs, BB3 2LZ. Tel: 01254 703948.

FYLDE ARS, RS53939. Meets 7.45pm on 2nd and 4th Tuesday in the month at the South Shore Lawn Tennis Club, Midgeland Road, Blackpool. Details from: Mr Harold Fenton, G8GG, c/o 5 Cromer Road, St Annes, Lytham St Annes, Lancs, FY8 3HD.

MORECAMBE BAY ARS, G4YBS. Meets 7.30pm on alternate Tuesdays at the Trimpell Sports & Social Club, Outmoss Lane, Morecambe, Lancs. Details from: Mr Brian Watson, G0RDH, c/o 7 Branksome Drive, Morecambe, Lancs, LA4 5UJ. Tel: 01524 424522.

PRESTON ARS, G3KUE. Meets 8.00pm on alternate Thursdays at the Lonsdale Club, Fulwood Hall Lane, Fullwood, Preston. Details from: Mr Eric Eastwood, G1WCQ, 56 The Mede, Freckleton, Preston, PR4 1JB. Tel: 01772 686708.

RED ROSE ARG, G0WMR. Details from: Knowleswood, Wrennalls Lane, Heskin, Chorley, Lancs, PR7 5PW.

ROLLS-ROYCE ARC, G3RR. Meets 8.00pm on Mondays, Wednesdays & Fridays, and at 11.30am on Sundays at the Club Room, Rolls-Royce Sports Ground, Barnoldswick. Details from: Mr J A York, G3KJY, c/o 13 Melville Avenue, Barnoldswick, Lancs, BB8 5JS. Tel: 01282 816033.

ROSSENDALE ARS, G1RRS. Meets 8.00pm Mondays at the Old Fire Station, Burnley Road, Rawtenstall, Rossendale, Lancs, BB4 8EW. Details from: Mr Ken Slaughter, c/o 652 Newchurch Road, Newchurch, Rossendale, Lancashire, BB4 9HG. Tel: 01706 830306.

THORNTON CLEVELEYS ARS, G4ATH. Meets 7.45pm Mondays at the 1st Norbreck Scout HQ, Carr Road, Bispham, Blackpool, Lancs. Details from: Mr J E Duddington, G4BFH, 8 The Grove, Thorton-Cleveleys, Blackpool, FY5 2JD. Tel: 01253 853554.

Leicestershire

Council Zone: B

91ST LEICESTER SCOUT ARS, G4NLS.

DE MONTFORT UNIVERSITY, G3SDC. Open to past and present students. Details from: Mr R G Titterington, c/o Dept of Electronic Eng, Leicester, LE1 9BH. Tel: 0116 257 7059.

HINCKLEY AR & ES, G3VLG. Meets 7.30pm on alternate Wednesdays in the month at the RAF Association HQ, St Marys Road, Hinckley. Details from: Mr R A Bennett, G8BFF, c/o 108 Hereford Close, Barwell, Leicester, LE9 8HP. Tel: 01455 846493.

LEICESTER RS, G3LRS. Meets 8.00pm Mondays at Gilroes Cottage, Groby Road, Leicester. Details from: The Secretary, 190 Braunstone Avenue, Leicester, LE3 1EF. Tel: 0116 291 7250.

LOUGHBOROUGH & DARC, G3RAL. Meets 8.00pm Mondays and Tuesdays at Hind Leys College, Shepshed, Loughborough, Leics. Details from: Mr Alan Hemmings, G0PHT, c/o 35 Ravensthorpe Drive, Loughborough, Leics, LE11 0WA. Tel: 01509 231289.

MELTON MOWBRAY ARS, G4FOX. Meets 7.30pm on 3rd Friday in the month at the St John Ambulance Hall, Asfordby Hill, Melton Mowbray, Leics. Details from: Mr R Winters, G3NVK, c/o 8 Epping Drive, Melton Mowbray, Leics, LE13 1UH. Tel: 01664 63369.

RAF NORTH LUFFENHAM ARC, G6RAF, G3TCQ. Meets 7.30pm Wednesdays. Details from: Mr Roger Hyde, 25 The Pastures, Cottesmore, Oakham, Leics, LE15 7DZ. Tel: 01572 813547.

TAMWORTH ARS, G8TRS. Meets 8.00pm Mondays. Details from: Mr A I Dyson, G0HUW, 24 Newborough Close, Austrey, Atherstone, Warks, CV9 3EX. Tel: 01827 830437.

WELLAND VALLEY ARS, G4WVR. Meets 7.30pm on 1st & 3rd Tuesday in the month at the Village Hall, The Green, Great Bowden, Leics. Details from: The Secretary, c/o 21 Farndon Road, Market Harborough, Leicestershire, LE16 9NW.

Licolnshire

Council Zone: B

FIVE BELLS GROUP, G4SIV. Active in VHF/UHF contest, meteor scatter, eme and DX. Details from: Mr B K Tatnall, G4ODA, c/o 73 Acacia Avenue, Spalding, Lincs, PE11 2LW.

GRANTHAM RC, G0GRC. Meets 8.00pm on 1st & 3rd Tuesday in month at the Kontak Social Club, Barrowby Road, Grantham, Lincs. Details from: the secretary, c/o Treetops, 13 Saltersford Road, Grantham, Lincs, NG31 7HH. Tel: 01476 657436.

LINCOLN SHORT WAVE CLUB, G5FZ. Meets 8.00pm Wednesdays at the City Engineers Club, Waterside South, Lincoln. Details from: Mrs Pam Rose, G4STO, c/o Pinchbeck Farmhouse, Mill Lane, Sturton-by-Stow, Lincoln, LN1 2AS. Tel: 01427 788356.

LOUTH & DARC, G4LRC. Meets 7.30pm on 3rd Tuesday in the month at the Greyhound Public House, Louth. Details from: Mrs Julie Wilson, 112 Upgate, Louth, Lincs, LN11 9HG.

NORTH LINCS COLLEGE ARS, G4LCT. Details from: Mr Richard Merriman, c/o College Library, Monks Road, Lincoln, LN2 5HQ. Tel: 01522 510530.

RAF CONNINGSBY ARC, G3LQS. Meets 6.00pm on 1st & 3rd Sunday in month at 28 Baxter Cl., RAF Conningsby. Details from: Mr D J Bloomfield, G0KUC, c/o 8 Sunningdale Drive, Boston, Lincs, PE21 8HZ.

RAF WADDINGTON ARC, G0RAF. Meets 7.00pm Thursdays at Newell House, High Dyke, Waddington. Details from: Sgt. Andy Lavey, G0MTG, Royal Air Force Waddington, Lincoln, LN5 9NB.

SPALDING & DARS, G4DSP. Meets 7.30pm Fridays at The Old Fire Station, Spalding, Lincs. Details from: Mr Dennis Hoult, G4OO, c/o Chespool Hse, Gosberton Risegate, Spalding, Lincs, PE11 4EU. Tel: 01775 750382.

SPILSBY ARS, RS91468. Meets 7.45pm on 1st Thursday in month. Details from: Mr Clive Ironmonger, G6HYF, 77 Boston Road, Spilsby, Lincs, PE23 5HH. Tel: 01790 752712.

STAMFORD & DARS, G4FPQ. Details from: Mr Bill Ballard, G7IGV, c/o 9 Fife Close, Stamford, Lincs, PE9 1DS. Tel: 01780 56686.

TETNEY COUNTY PRIMARY SCH. ARC, GX0PHA. Details from: Mr Paul Hewitt, Tetney County Primary School, Tetney, Grimsby, South Humberside, DN36 5NG. Tel: 01472 812074.

WILLIAM ROBERTSON SCHOOL ARC, G4WRS. Meets Tuesdays & Thursdays at lunchtimes during term time. Details from: Mr Andrew Kiddle, G4HVC, Welbourn, Lincoln, Lincs, LN5 0PA.

Lothian Region

Council Zone: G

DUNFERMLINE RS, GM3IDS. Meets 7.30pm on Thursdays at Outh Radio Station, nr Knock Hill Racing Circuit. Details from: Mr J Gentles, c/o Culra, 19 Clufflat Brae, South Queensferry, West Lothian, EH30 9YQ. Tel: 0131 3314340.

LOTHIAN RS, GM3HAM. Meets 7.30pm on 2nd & 4th Wednesday in month at the Orwell Lodge Hotel, Polwarth Terrace, Edinburgh, EH11 1NH. Details from: Mr Brian Howie, GM4DIJ, Sec., 36 Clermiston Road, Edinburgh, EH12 6XB. Tel: 0131 334 2247.

Merseyside

Council Zone: A

KIRKBY ARC, G0LKR. 5 Derby Grove, Maghull, Merseyside, L31 5JJ.

LIVERPOOL & DARS, G3AHD. Meets 7.30pm Tuesdays at the Churchill Conservative Club, Church Road, Wavertree, Liverpool, L15. Details from: Mr G Wardale, M0ANN, 25 The Crescent, Huyton Quarry, Merseyside, L36 6ER. Tel: 0151 289 0558.

NORTH SEFTON ARC, G0TSI. Meets 1st and 3rd Wednesday in the month. Details from: Mr Ian Hampson, G1DFT, 57 Cornwall Way, Ainsdale, Southport, PR8 3SG. Tel: 01704 579017.

SOUTH WIRRAL CONTEST GROUP, G3CSA. Details from: Mr T B Saggerson, G4WSE, 18 Ploughmans Way, Great Sutton, South Wirral, L66 2YJ. Tel: 0151 339 0842.

SOUTHPORT & DARC, G2OA. Meets 8.00pm on 3rd Monday in the month at St Marks Church Hall, Scarisbrick, Lancs. Details from: Mr Walter Cliffe, G4YYV, c/o 53 Easdale Drive, Ainsdale, Southport, Merseyside, PR8 2BW. Tel: 01704 79825.

WIRRAL & DARC, G4MGR. Meets 8.00pm 2nd & 4th Wednesdays in month at the Irby Cricket Club, Mill Hill Road, Wirral. Details from: Gerry Scott, 19 Penkett Road, Wallasey, Merseyside, L45 7QF. Tel: 0151 630 1393.

WIRRAL ARS, G3NWR. Meets 8.00pm 1st & 3rd Wednesday in month at the Club Room, Ivy Farm, Arrowe Park Road, Wirral, L49 5LW. Details from: Mr John Phillips, G3PXX, 18 Rockfarm Drive, Little Neston, South Wirral, L64 4DZ.

Mid-Glamorgan

Council Zone: E

BRIDGEND & DARC, GW4LNP. Meets 1st & 3rd Wednesday in month at the Club Brynmenyn, Brynmenyn, Bridgend. Details from: Mr Alun Hulmes, c/o Ashley House, Blackmill Rd, Bryncethin, nr Bridgend, Mid Glamorgan, CF32 9YN. Tel: 01656 721574.

BRISTOL CHANNEL REPEATER GROUP, GB3BC. Details from: T.H. Watkins, Ty Unig, Forest Road, Treharris,Mid Glamorgan, CF46 5HG. Tel: 01222 709613.

HOOVER (MERTHYR) ARC, GW3RDB. Meets 7.30pm Mondays at the Hoover Sports Pavillion, Hoover Ltd, Pentrebach, Merthyr Tydfil, Mid Glam. Details from: Mr Andrew Lipian, GW0TOI, 10 Field Street, Trelewis, Treharris, Mid Glamorgan, Wales, CF46 6AW. Tel: 01443 410964.

MID GLAMORGAN ARG, GW0VJS. Meets every Thursday (September to June) Aberkenfig Sports and Social, Club. Details from: Mr Mervyn Carey, GW4VSE, c/o 47 Heol-Ty-Gwyn, Maesteg, Mid Glamorgan, CF34 0BD. Tel: 01656 734668.

RHONDDA ARS, GW2FOF. Meets 7.30pm Thursdays at the NUM Club, Tonypandy, Mid Glamorgan. Details from: Mr Dennis Tippett, c/o 'Bryn Eglur', Vicarage Rd, Penygraig, Rhondda, Mid Glamorgan, CF40 1HP. Tel: 01443 440680.

ST. CENYDD SCHOOL ARS, GC4CCS. Meets after school between 3.00pm & 4.30pm. Details from: Mr Clive Davies, GW3YBN, 31 Park Prospect, Pontypridd, Mid. Glamorgan, CF37 2HF. Tel: 01222 852504.

Norfolk

Council Zone: C

ANGLIA TELEVISION ARS, G0TXV. Meets between 12.30pm - 2.00pm on Fridays at Anglia TV, Norwich, NR1 3JG. Details from: Mr Jim Bacon, G3YLA, ARS, c/o Weather Dept, Anglia Television, Norwich, NR1 3JG. Tel: 01603 615151.

FAKENHAM ARC, G4LSF. Meets 7.30pm 1st Tuesday in the month at the Trinity Church Room, Hempton. Details from: Mr P J Smyth, G0MQU, c/o 15 Groveside Estate, East Rudham, King's Lynn, Norfolk, PE31 8RL.

GREAT YARMOUTH RS, G3YRC. Meets 8.00pm Thursdays at the Drill Hall (West entrance), York Road, Great Yarmouth, Norfolk. Details from: Mr A D Besford, G3NHU, 2A Halt Road, Caister-on-Sea, Great Yarmouth, Norfolk, NR30 5NZ.

KINGS LYNN ARC, G3XYZ. Meets 7.30pm Thursdays. Details from: Mr Derek Franklin, G0MQL, c/o Laurel Farm, 7 Holly Cl, West Winch, Kings Lynn, Norfolk, PE33 0PW.

NORFOLK ARS, G4ARN. Meets 7.30pm Wednesdays at the Norman Centre, Bignold Road, Norwich. Details from: Mr John Wadman, G0VZD, 11 Ash Close, Wymondham, Norfolk, NR18 0HR. Tel: 01953 604769.

NORTH NORFOLK AR REPEATER GRP, GB3NN. Details from: Mr Frederick Tuck, G8KZP, c/o Whalebone Cottage, Wells Next The Sea, Norfolk, NR23 1EH. Tel: 01328 710057.

NORTH NORFOLK ARG, GB2MC. Details from: Mr Andrew Thomson, G4OLF, 7 Bloomstyles, Salthouse, Holt, Norfolk, NR25 7XJ.

North Yorkshire

Council Zone: A

HAMBLETON ARS, G0JQA. Meets 7.30pm Thursdays at the Allertonshire School, Brompton Road, Northallerton, North Yorkshire. Details from: Mr John Hampson, G0VXH, Marloes, Borrowby, Thirsk, N. Yorks, YO7 4QP. Tel: 01845 537547.

HARROGATE LADIES COLLEGE RS, G0HCA. Open to pupils and staff only. Details from: Mr Richard Horton, G3XWH, Clarence Drive, Harrogate, N Yorks, HG1 2QG. Tel: 01423 504543.

HARROGATE REPEATER GROUP, GB3HG. Details from: Mr Brian Dooks, G0RHI, 7 Manor Drive, Kirby Hill, Boroughbridge, York, YO5 9DY. Tel: 01423 322988.

RICHMOND SCHOOL ARS, GX0RYS. Active 1.00pm - 2.00pm most weekdays during term time. Details from: Mr M S Vann, Richmond Sch, Darlington Road, Richmond, North Yorkshire, DL10 7BQ. Tel: 01748 850111.

RIPON & DARS, G4SJM. Meets 7.30pm Thursdays at The Bunker, rear of Ripon Town Hall, Ripon, North Yorkshire. Details from: Mr Phillip Hughes, G0KHQ, c/o 2 Butterbur Way, Harrogate, N Yorks, HG1 2XH. Tel: 01423 526374.

ROYAL SIGNALS SCARBOROUGH ARC, G0RCS. Meets as and when required. Contact Mr A W W Timme, G3CWW, on Tel: 01484 842330.

SCARBOROUGH ARS, G4BP. Meets 7.30pm on Mondays at the Scarborough Cricket Club, Pavillion, North Marine Road, Scarborough, North Yorks, YO12 7TJ. Details from: Mr D P Tipper, G3JBR, c/o 10 Lowdale Avenue, Scarborough, North Yorks, YO12 6JW. Tel: 01723 377296.

SCARBOROUGH SE GRP, GX0OOO. Meets as and when required. Details from: Roy Clayton, G4SSH, 9 Green Island, Irton, Scarborough, N Yorks, YO12 4RN. Tel: 01723 862924.

YAXPAK, RS92311. We are a group helping with packet links/equipment in Yorks for you.

YORK ARS, G3HWW. Meets 7.30pm Fridays at the Guppy's Enterprise Club, 17 Nunnery Lane, York. Details from: Mr Keith Cass, G3WVO, 4 Heworth Village, York, YO3 0AF. Tel: 01904 422084.

YORK RADIO CLUB (AMATEUR), G4YRC. Meets 7.30pm Thursdays at the Bishopthorpe Social Club, Bishopthorpe Main Street, York. Details from: Mr Andy White, c/o 27 Old Church Green, Kirk Hammerton, York, YO5 8DL. Tel: 01423 330393.

Northamptonshire

Council Zone: B

ARIEL RG, G5XX. Open only to BBC employees.Borough Hill, Daventry, Northamptonshire, NN11 4NB.

KETTERING & DARS, G5KN. Meets 8.00pm Tuesdays at the The Lilacs Public House, 39 Church Street, Isham, Kettering, Northants, NN14 1HD. Details from: Mr Dave Denny, G0XAJ, c/o 6 Hill Street, Kettering, Northants, NN16 8EE. Tel: 01536 520837.

MID NORTHANTS AR EXP, G0ING. Details from: Mr Lionel Parker, G5LP, 128 Northampton Rd, Wellingborough, Northants, NN8 3PJ.

NENE VALLEY RC, G4NWZ. Meets 8.00pm Wednesdays at the Prince of Wales Public House, Well Street, Finedon, Wellingborough, Northants. Details from: Mr P Byles, G6UWS, c/o 108 Kingsway, Wellingborough, Northants, NN8 2EN. Tel: 01933 71189.

NORTHAMPTON RC, G3GWB. Meets 8.00pm Thursdays at the RAFA Club, Northampton. Details from: Mr A H Rowley, G0TML, c/o 32 Spring Lane, Flore, Northampton, NN7 4LS.

NORTHAMPTON SCOUTS ARG, G6NDS. Meets Saturdays Overstone Scout Activity Cntr., Northampton. Details from: Mr Ian Rivett, G8WPU, 30 Millside Close, Kingsthorpe, Northampton, NN2 7TR.

PARALLEL LINES C G, G4LIP. Details from: Mr P S Lidsay, G4CLA, Greystones, Stanford Close, Cold Ashby, Northampton, NN6 7EW.

Northumberland

Council Zone: A

BLYTH ARC, G4VKY. Meets 7.00pm Wednesday at the Newsham Community Centre, Elliott Street, Blyth, Northumberland. Details from: Mr K Stewart, G0TWV, 25 Millfield, Seaton Sluice, Whitley Bay, Tyne & Wear, NE26 4DD. Tel: 0191 237 7011.

NORTHUMBRIA ARC, G4AAX. Meets Thursday evenings at the Old Telephone Exchange, Cresswell Road, Ellington, Morpeth, Northumberland. Details from: Mr D Stansfield, G0EVV, Old Telephone Exchange, Cresswell Rd, Ellington, Morpeth, Northd, NE61 5HS. Tel: 01670 513026.

Nottinghamshire

Council Zone: B

ARC OF NOTTINGHAM, G3EKW. Meets 7.30pm Thursdays at the Sherwood Community Centre, Woodthorpe House, Mansfield Road, Nottingham. Details from: Mr Simon Williams, G0IEG, c/o 55 Arundel Drive, Bramcote Hills, Beeston, Nottingham, NG9 3FN. Tel: 0115 9501733.

ARNOLD & CARLTON FE COLL. ARS, G0ACC. Details from: Mr Frank Skillington, G4DFU, c/o 53 Temple Drive, Nuthall, Nottingham, NG16 1BE. Tel: 01159 278173.

DUKERIES ARS, G4XTL. Meets 7.30pm on 2nd and 4th Tuesday in the month at Ambleside Community Centre, Ambleside, New Ollerton, Notts. Details from: Mr Colin Foster, G7DEX, c/o 13 Breck Bank, New Ollerton, Newark, Notts, NG22 9XQ.

HUCKNALL ROLLS ROYCE ARC, G5RR. Meets 8.00pm on Fridays at the Hucknall Rolls Royce Sports &, Social Club, Watnall Road, Hucknall, Nottingham. Details from: Mr Jon Lee, G4TSN, c/o 46 Little Lane, Huthwaite, Sutton-in-Ashfield, Notts, NG17 2RA. Tel: 01623 558214.

MANSFIELD ARS, G3GQC. Meets 7.30 for 8.00pm on the second Monday in the month at the Polish Catholic Club, Windmill Lane, off Woodhouse Road, Mansfield, Nottingham. Details from: David Peat, G0RDP, c/o Jeswyn, Brookland Avenue, Mansfield, Notts, NG18 5NB. Tel: 01623 631 931.

MANSFIELD DISTRICT SCOUT ARC, G0MDS. Details from: J M Coates, G4GYU, 30 Abbott Road, Mansfield, Notts, NG19 6DD.

NORTH NOTTS DATA GROUP, G0WNN, G7WNN. Details from: Mr Tony Jenkins, G8TBF, 13 Baulk Lane, Worksop, Notts, S81 7DF.

NOTTINGHAM AR REPEATER GROUP, GB3NM. Nottingham AR Repeater Group, c/o 26 Dovecote Lane, Beeston, Nottingham, NG9 1HU.

NOTTINGHAM UNIVERSITY ARS, G3UNU. Chemical Engineering Dept, University of Nottingham, University Park, Nottingham, NG7 2RD.

SIEMENS ARC, G8ZK, G8IGQ. Meets 7.30pm Thursdays and 10am Sundays at the GPT Sports Ground, Beeston, Nottinghamshire. Details from: Mr Chris Archer, G4VFK, Sports Office, Technology Drive, Beeston, Nottingham, NG9 1LA. Tel: 0115 943 3387.

SOUTH NOTTS ARC, G0OAU. Meets 7.00pm Fridays at the Fairham Community College, Farnborough Road, Clifton, Nottingham, NG11 9AE. Details from: Mr Gary Bishop, G0WUG, POB 4, Clifton Estate, Nottingham, NG11 9DE. Tel: 01509 672846.

WELBECK COLLEGE ARS, G7TTW. Welbeck College, Welbeck, Worksop, Notts, S80 3LN.

WORKSOP ARS, G3RCW. Meets 8.00pm Tuesdays at the Club House, 59 - 61 West Street, Worksop, Nottingham, S80 1JP. Details from: Mr Terry Calvert, G4GBS, 59-61 West Street, Worksop, Notts, S80 1JP. Tel: 01302 743130.

Oxfordshire

Council Zone: D

BANBURY ARS, G0BRA. Meets 7.30pm on 2nd and 4th Wednesday in the month at St Johns Church Social Club, South Bar, Banbury, Oxon. Details from: Mr R S Marsden, G1YSY, c/o 25 High Acres, Banbury, Oxon, OX16 9SL. Tel: 01295 253509. Fax: 01295 253509.

BBC CLUB ARIEL RADIO GRP, G8BBC. Open only to BBC employees. Details from: Mr J M Eason, G0INR, Lynwood, Holton, Oxford, OX33 1PU.

HARWELL ARS, G3PIA. Meets 8.00pm 3rd Tuesday in month at the Social Club, Harwell Laboratory, Didcot, Oxon. Details c/o Mr J Durban, Building 346, Harwell Lab, Didcot, Oxon, OX11 0RA. Tel: 012357 68453.

MID THAMES DFC, G4MDF. Meets monthly in the winter at the Clayton Arms, Lane End, Bucks. Details from: Mrs Doreen Pechet, G8NMD, Jays Lodge, Crays Pond, Reading, RG8 7QG. Tel: 01491 681236.

OXFORD & DARS, G5LO. Meets 7.45pm 2nd & 4th Thursdays in month at the Grove House Club, George Street, Summertown, Oxford. Details from: Mr D Walker, G3BLS, c/o 32 South Street, Osney, Oxford, OX2 0BE. Tel: 01865 247311.

RAF BRIZE NORTON ARC, G0RBN. Details from: Mr Dave Shorten, G7SRB, c/o 32 Stoneleigh Drive, Carterton, Oxon, OX18 1ED. Tel: 01993 846975.

RUTHERFORD LABRATORIES ARC, G3RRS. Open to staff members only. Details from: The Secretary, Mr J S Wright, Building R25 1.105, Chilton, Didcot, Oxon, OX11 0QX. Tel: 01235 445809.

VALE OF WHITE HORSE ARS, G5RP, G4VWH, G6VWH. Meets 8.00pm 1st Tuesday in the month at The Fox, Steventon. Details from: Mr Ian White, G3SEK, c/o 52 Abingdon Road, Drayton, Abingdon, Oxon, OX14 4HP. Tel: 01235 531559.

Powys

Council Zone: E

POWYS ARC, GW4HVN. Meets 8.00pm Thursdays at the British Legion Club, Broad Street, Newtown, Powys.

Shetland Is

Council Zone: G

LERWICK RC, GM3ZET. Meets 7.00pm Tuesdays at the Islesburgh Community Centre, King Harald Street, Lerwick, Shetland.

UNST RC, GM3STU. Meets irregularly. Details from: Mr Cedric Auty, GM4GPP, Valsgarth, Haraldswick, Unst, Shetland, ZE2 9EF. Tel: 0195781 349.

Shropshire

Council Zone: B

OSWESTRY & DARC, G4TTO, G1ORA. Meets 7.30pm on the 1st and 3rd Wednesdays in the month at the Sweeney Hall Hotel, Oswestry. Details from: Stan Hutton, G1MAB, c/o Awelon, Turners Lane, Llynclys, Oswestry, SY10 8LL. Tel: 01691 830328.

SALOP ARS, G3SRT, M1AXW. Meets 8.00pm Thursdays at The Oak Hotel, Shelton, Shrewsbury. Web: www.clemalv.demon.co.uk. Details from: Mrs Diane Parslow, G4XBI, c/o Mr J Bumford, 19 Bewdley Avenue, Telford Estate, Shrewsbury, Shropshire, SY2 5UQ.

SEVERN VALLEY RS, G3SVR. Details from: Mr E G Churchyard, G3TVR, c/o 11 Greenfields Drive, Bridgnorth, Shropshire, WV16 4JW.

TELFORD & DARS, G3ZME. Meets 7.30pm Wednesdays at the Dawley Bank Community Centre, Dawley, Telford, Shropshire. Details from: Mr M Vincent, G3UKV, 9 Sleapford, Long Lane, Telford, Shropshire, TF6 6HQ. Tel: 01952 255416.

Somerset
Council Zone: D

PRESTON COMMUNITY SCHOOL ARC, G0PCS. Details from: Craig Douglas, G0HDJ, Monks Dale, Yeovil, Somerset, BA21 3JD. Tel: 01935 71131.

TAUNTON & DARS, G3XZW. Meets 7.30pm 1st & 3rd Fridays in month at The Basement, County Hall, The Crescent, Taunton. Details from: Mr Bill Lindsay-Smith, G3WNI, Way Close, Madford, Hemycock, Cullompton, Devon, EX15 3QY. Tel: 01823 680778.

WEST SOMERSET ARC, G0OWX. Meets 1st Tuesday in the month at the West Somerset Community, College, Minehead, Somerset. Details from: Mr Alan Elliott, G7RSU, c/o Alrikitira, 26 Watery Lane, Minehead, Somerset, TA24 5NZ. Tel: 01643 707207.

WINCANTON ARC, G0WRA. Meets 7.30pm on 1st and 3rd Monday (except Bank Holidays) at King Arthur's Community School, West Hill, Wincanton. Details from: Mr G A Fingerhut, G0ENW, Wild Rose, Behind Hayes, South Cheriton, Templecombe, Somerset, BA8 0BP. Tel: 01963 370506.

YEOVIL & DARC, G3CMH, G8YEO. Meets 7.30pm Thursdays at the British Red Cross HQ, 72 Grove Avenue, Yeovil, Somerset. Details from: Mr Malcolm Sadler, G7WAL, Hillview, Forest Mill Lane, Horton, Ilminster, Somerset, TA19 9QU. Tel: 01460 54657.

South Glamorgan
Council Zone: E

BARRY ARS, GW3VKL. Meets 7.30pm on Thursdays at Sully Sports & Leisure (BP Club), South Road, Sully, South Glamorgan. Details from: Mrs Margaret Beynon, GW4GSH, 16 Hardy Close, Woodfield Heights, Barry, South Glamorgan, CF62 9HJ. Tel: 01446 738756.

HIGHFIELDS ARC, GW4LFO. Meets 7.30pm Thursdays at the Highfields Physically, Handicapped Centre, Allensbank Road, Cardiff. Details from: 26 Allensbank Road, Cardiff, CF4 3RB. Tel: 01222 750856.

WENVOE TRANSMITTER CLUB ARS, GW4WVO. Open only to BBC staff. Details from: Mr Keith Winnard, GW3TKH, Wenvoe Transmitting Station, St Lythans Down, Wenvoe, Cardiff, CF5 6BQ.

South Yorkshire
Council Zone: A

BARNSLEY & DARC, G6AJ. Meets 8.00pm every Monday at the Three Horse Shoes, Barnsely Road, Brierley, Barnsley, South Yorks. Details from: Mr Ernie Bailey, G4LUE, 8 Hild Avenue, Cudworth, Barnsley, South Yorks, S72 8RN. Tel: 0836 748958.

CHAPEL GREEN ARS, GX0VYJ. Details from: 230 Lane End, Chapleton, Sheffield, S30 4UJ.

MALTBY & DARS, G4SKM. Meets 7.30pm Wednesday at the Prospect House, Muglet Lane, Maltby, Rotherham. Details from: Keith Johnson, G1PQW, 20 Rolling Dales Close, Maltby, Rotherham, South Yorks, S66 8EJ. Tel: 01709 814135.

MEXBOROUGH & DARS, G4BTS. Meets 7.30pm Fridays at the Harrop Hall, Mexborough, South Yorks. Details from: Mr R T Sheppard, G0KSK, c/o 4 Lindrick Avenue, Swinton, Mexborough, South Yorks, S64 8TE. Tel: 01709 586329.

RAF FINNINGLEY ARS, G7HAH. Meets Tuesdays and Thursdays, prospective visitors should call in advance. Details from: Mr Vic Lowe, Secretary, 7 Castell Crescent, Cantley 2, Doncaster, DN4 6LG. Tel: 01302 531927.

SHEFFIELD ARC, G0INF. NRAE/RAE tuition provided. Meets 7.00pm on Mondays and occasional Tuesday at the Club, (Sheffield University Staff, Club), 197 Brook Hill, Sheffield. Details from: Mr David Briggs, G0JJR, c/o PO Box 365, Sheffield, S1 1BY. Tel: 0114 244 6282.

SOUTH YORKSHIRE PUG, G0PUG. Details from: Mrs Sables, G4ZJN, 54 Harvey Street, Deepcar, Sheffield, S30 5QB.

Staffordshire
Council Zone: B

BURTON-ON-TRENT & DARS, G3NFC. Meets 8.00pm Wednesdays at the Stapenhill Institute, Main Street, Stapenhill, Burton-on-Trent, Staffs. Details from: Mr M W Cotton, G4HBY, 113 Belvedere Road, Burton-on-Trent, Staffs, DE13 0RF.

CANNOCK CHASE ARS, G6SW. Meets 8.00pm Thursdays at the Four Crosses Inn, Watling Street, Hatherton, Cannock. Details from: Mr Arnold Matthews, G3FZW, c/o 2 The Parchments, Lichfield, Staffs, WS13 7NA. Tel: 01543 262495.

CHAD RC, G4CAR. Meets 8.30pm Mondays at the Swinfen Officer's Club, Swinfen, Lichfield, Staffs. Details from: Mr Bernard Jayne, G8BFL, c/o 38 Townfields, Lichfield, Staffs, WS13 8AA. Tel: 01543 268569.

HILLCREST SCHOOL ARS & CC, G0SPM. Meets 7.30pm 1st & 3rd Thursdays in month at The College, Simms Lane, Netherton, Dudley, West Midlands. Details from: Mrs Megan Fleetwood, G0TMF, 9 Reynolds Close, Swindon, Dudley, West Midlands, DY3 4NQ. Tel: 01384 294804.

ICL KIDSGROVE AR & PCC, G6ICL. Open only to employees of ICL. Details from: B A Morris, ICL Kidsgrove, West Avenue, Kidsgrove, Stoke-on-Trent, ST7 1TL.

LICHFIELD ARS, G3WAS. Meets 8.15pm 1st Monday & 3rd Tuesday in the month at the Queen's Head, Sandford Street, Lichfield. Details from: Mr Roger Smethers, G3NLY, 46 Church Rd, Burntwood, Staffs, WS7 9EA. Tel: 01543 672762.

MOORLANDS & DARS, G4NHT, G1MAD. Meets 8.30pm on Thursdays at the Creda Works, Blythe Bridge, Stoke-on-Trent, Staffordshire, ST11 9LJ. Details from: Mr C F Beesley, G4OUG, 15 Byron Close, Cheadle, Stoke-on-Trent, Staffordshire, ST10 1XB. Tel: 01538 756323.

NEWCASTLE U LYME SCOUT ARCOMGR, G7UQG. c/o Dr R N Bloor, Pinewood Hse, Pinewood Drive, Ashley Heath, Market Drayton, Shropshire, TF9 4PA.

NORTH STAFFS ARS, G4BEM. Meets 7.30pm Mondays at the Blacklake Lodge, Hilderstone Road, Meir Heath, Stoke-on-Trent, ST3 7NS. Details from: Mr Mervyn Bennett, G4HUO, c/o 9 Lavender Avenue, Blythe Bridge, Stoke-on-Trent, ST11 9RN. Tel: 01782 394887.

ST LEONARDS ARS, G3SBL. Meets 8.00pm Thursdays at the GEC Alsthom Protection &, Control, St Leonards Works, Stafford, ST17 4LX.

STOKE-ON-TRENT ARS, G3GBU. Meets 7.30pm Thursdays at The '45' Club, 92 Lancaster Road, Newcastle-under-Lyme, Staffs. Details from: Mr Albert Allen, G4DHO, c/o 3 Wayfield Grove, Harfields, Stoke-on-Trent, Staffordshire, ST4 6DB. Tel: 01782 638801.

STOURBRIDGE & DRS, G6OI. Meets 8.00pm 1st & 3rd Mondays in month at the Robin Woods Centre, School Road (off Enville St), Stourbridge, West Midlands. Details from: Mr J French, G7HEZ, c/o Hyde Bungalow, The Hyde, Kinver, Stourbridge, Staffs, DY7 6LS. Tel: 01384 374354.

SUTTON COLDFIELD RS, G3RSC. Meets 8.00pm 2nd & 4th Mondays in the month at the Rugby Club, Walmley Road, Sutton Coldfield, West Midlands. Details from: The Secretary, c/o 2 Hopleys Close, Tamworth, Staffs, B77 3JU.

Strathclyde Region
Council Zone: G

AYR ARG, GM0AYR. Meets 7.30pm Friday from (Sept to June) at the Connunity Leisure Centre, 24 Wellington Square, Ayr. Details from: Mr Gary Olesen, GM3MQO, 11 Shawfield Avenue, Ayr, KA7 4RE. Tel: 01292 79245.

CENTRAL SCOTLAND FM GROUP, RS38728. Details from: Mr A Hood, GM7GDE, c/o 39 Broomhill Crescent, Freelands, Erskine, Renfrews, PA8 7AN.

CUNNINGHAME & DARC, GM3USL. Meets 7.30pm Thursdays at the Woodlands Centre, Kilwinning Road, Irvine, Ayrshire.

DALRY ARG, MM0ARG. c/o Dalry Community Centre, St Margarets Avenue, Dalry, Ayrshire, KA24 4BA.

DUNOON & DARS, GM0COD. Visitors are welcome. Meets 7.30pm Fridays at the Edward Street Community Centre, Edward Street, Dunoon. Details from: Mr A B Horton, GM0BUL, c/o Thornbank, Blairmore, Dunoon, Argyll, PA23 8TJ. Tel: 01369 840217.

GREENOCK & DARC, GM3ZRC. Meets 7.00pm Tuesdays & Fridays at the Port Glasgow Town Hall, Port Glasgow.

HELENSBURGH ARC, GM4HEL. Details from: Mr G Capstick, GM7OAF, 24 Dalmore Crescent, Helensburgh, Dunbartonshire, G84 8JP. Tel: 01436 675922.

HIGH SCHOOL OF GLASGOW, GM0HSG. 637 Crow Road, Glasgow, G13 1PL.

INVERCLYDE ARG, GM0GNK. Details from: Mr J Bertram, GM0GMN, c/o 47 Bournemouth Road, Gourock, Strathclyde, PA19 1HN. Tel: 01475 631320.

INVERCLYDE RADIO CONTEST GROUP, MM0APF. Details from: Mr John Dunlop, GM0WDF, c/o West Dougliehill Farm, Port Glasgow, Renfrewshire, PA14 5XF.

KILMARNOCK & LOUDOUN ARC, GM0ADX. Meets 7.30pm on alternate Tuesdays at The Community Centre, Cessnock Road, Hurlford, Kilmarnock, Ayrshire. Details from: Mr Bill Strachan, GM3ZRT, Sparnel, Cemetery Road, Galson, Ayrshire, KA4 8LL. Tel: 01563 820052.

LARGS & DARS, GM0VKG. Details from: 3 Arran View, Largs, Ayrshire, KA30 9ER.

LORN ARS, GM0LRA. Meets 1st & 3rd Thursdays in month. Details from: Mr T Olsen, GM0EQW, c/o Tigh An Drochaid, Kilchrevan, By Taynuilt, Argyll, PA35 1HD. Tel: 018662 580.

MID LANARK ARS, GM3PXK. Meets 7.30pm Fridays at the Newarthill Community Ed. Cent., High Street, Newarthill, Motherwell, Lanarkshire, ML1 5GU. Details from: Mr John Neary, GM0XFK, 17 Harkins Avenue, Blantyre, Glasgow, G72 0RQ. Tel: 01698 822860.

MILTON OF CAMPSIE ARS, GM0MOC. Meets 7.30pm every other Wednesday at the Red Cross Hall, Kirkintilloch. Details from: Mr John MacKenzie, GM0HJU, c/o 13 Boghead Road, Kirkintilloch, Glasgow, G66 4EG. Tel: 01360 312954.

PAISLEY ARC, GM0PYM. Meets 7.30pm on 2nd Wednesday in the month at the Paisley YMCA Hall, 5 New Street, Paisley, PA1 1XU. Details from: Mr John Quigley, GM0TQA, c/o John Quigley, 90 George Street, Paisley, PA1 2JR. Tel: 0141 889 6860.

WEST OF SCOTLAND ARS, GS4AGG. Meets 7.30pm for 8.00pm on Fridays at the Multi Cultural Centre, 21 Rose Street, Glasgow. Details from: Hon Sec, P O Box 599, Glasgow, G3 6QH.

Suffolk
Council Zone: C

BURY ST EDMUNDS ARS, G2TO. Meets 7.30pm 3rd Tuesday in month at the Culford School, Culford, Bury St Edmunds, Suffolk. Details from: Peter Brindley, G0HEV, c/o 2 Beech Park, Great Barton, Bury St Edmunds, Suffolk, IP31 2JL.

FARLINGAYE ARC, RS171377. Farlingaye High School, Ransom Road, Woodbridge, Suffolk, IP12 4JX.

FELIXSTOWE & DARS, G4ZFR. Meets 8.00pm on alternate Mondays in month at the Orwell Park School, Nacton, near Ipswich. Details from: Mr Paul Whiting, G4YQC, c/o 77 Melford Way, Felixstowe, Suffolk, IP11 8UH. Tel: 01473 642595.

IPSWICH RC, G4IRC. Meets 8.00pm 1st, 3rd & last Wednesdays in the month at the Rose and Crown, Bramford Road, Ipswich. Details from: Mr Ian Moffatt, G0OZS, c/o 30 Daimler Road, Ipswich, Suffolk, IP1 5PQ.

KESGRAVE HIGH SCHOOL ARC, RS172925. FAO Mr C Bennett, Kesgrave, Ipswich, Suffolk, IP5 7PB.

LEISTON ARC, G0TUQ. Meets 7.30pm 1st Tuesday in the month at Sizewell Sports & Social Club, King George's Avenue, Leiston, Suffolk. Details from: Mr Charles Ormerod, G4JGJ, 5 Leiston Court, High Street, Leiston, Suffolk, IP16 4BZ. Tel: 01728 831618.

LOWESTOFT DARS, G3JRM. Meets 8.00pm on alternate Thursdays in the month at the George Borrow Hotel, Oulton Road, Lowestoft. Details from: Mr Ian Burden, G0RRI, c/o 79 Tonning Street, Lowestoft, Suffolk, NR32 2AL. Tel: 01502 500712.

MARTLESHAM RS, G4MRS. Meets 7.30pm on occassional 1st Wednesday in the month at the BT Laboratories, Martlesham Heath, Ipswich, Suffolk. Details from: Mr Darren Hatcher, 44 Dewar Lane, Kesgrave, Ipswich, IP5 7GJ. Tel: 01473 644475.

SUDBURY & DRA, G0SWI, G7SRA. Meets 1st and 3rd Tuesday in the month at Wells Old School, Great Cornard, or Five Bells Public House, Great Cornard. Details from: Mr Mike Marsh, G4GGC, 21 Stour Gardens, Great Cornard, Sudbury, Suffolk, CO10 0JN. Tel: 01787 371842.

Surrey
Council Zone: C

BENTLEY ARC, G0VZS. Details from: Mr Derek Gilbert, G0NFA, 2 Greenfields Cottages, Farnham, Surrey, GU10 5HZ.

CATERHAM RG, G0SCR. Meets on alternate Fridays evenings at fellow member's houses. Details from: Mr P N Lewis, G4APL, 'Sky-Waves', 20 Annes Walk, Caterham-on-the-Hill, Surrey, CR3 5EL.

COULSDON A TRANS S, G4FUR. Meets 8.00pm on 2nd Monday in the month St Swithuns Church Hall, Grovelands Road, Purley, Surrey. Details from: Mr A R Bartle, 105 Mayfield Road, Thornton Heath, Surrey, CR7 6DP. Tel: 0181 684 0610.

DORKING & DRS, G3CZU, G7DOR. Meets 2nd & 4th Tuesday in month. Details from: Mr John Greenwell, G3AEZ, Eastfield, Henfold Lane, Beare Green, Dorking, Surrey, RH5 4RW. Tel: 01306 631236.

FARNBOROUGH & DRS, G4FRS. Meets 8.00pm on 2nd and 4th Wednesdays in the month at The Community Centre, Meudon Avenue, Farnborough, Hants. Details from: Mr M Hearsey, G8ATK, c/o Halcyon, Lawday Link, Upper Hale, Farnham, Surrey, GU9 0BS. Tel: 01252 715765.

GUILDFORD & DRS, G6GS. Meets 7.30pm for 8.00pm, 2nd & 4th Fridays in the month at the Guildford Model Engineers HQ, Stoke Park, Guildford, Surrey. Details from: Mr Phil Manning, G1LKJ, 39 Ashbury Crescent, Merrow Park, Guildford, Surrey, GU4 7HG.

GUILDFORD COLL.OF TECH AR & EC, RS8600. AR & Elec Club. FAO Mr B Purse, Stoke Park, Guildford, Surrey, GU1 1EZ.

GUILDFORD UHF REPEATER GROUP, GB3GF. Meets unofficially at 8.30pm 1st Thursday in the month at Sanford Arms, Epsom Road, Guildford, Surrey. Details from: Mr Alex Morris, G6ZPR, Follywood, New Road, Wonersh, Surrey, GU5 0SE. Tel: 01483 892348.

REIGATE ATS, G5LK, G7RAT. Meets 8.00pm 3rd Tuesday in month in the Conference Room of the RNIB, College, Philanthropic Road, Redhill, Surrey. Details from: Mr A C Embling, G1LNT, 224 Croydon Road, Caterham, Surrey, CR3 6QG. Tel: 01883 344723.

ROYAL GRAMMAR SCHOOL ARC, G7BAI. Meets usually during lunch times during term time. Details from: Mr F J Bell, High Street, Guildford, Guildford, Surrey, GU1 3BB. Tel: 01483 502427.

SUTTON & CHEAM RS, G2XP. Meets 8.00pm on 3rd Thursday in the month at the Sutton United Football Club, Borough Sports Ground, Gander Green Lane, Sutton, Surrey. Details from: John Puttock, G0BWV, 53 Alexandra Avenue, Sutton, Surrey, SM1 2PA. Tel: 0181 644 9945.

THAMES VALLEY ARTS, G3TVS. Meets 8.00pm 1st Tuesday in the month at the Thames Ditton Library, Watts Road, Giggs Hill, Thames Ditton, Surrey. Details from: Cdr. J Pegler, G3ENI, c/o Brook House, Forest Close, East Horsley, Leatherhead, Surrey, KT24 5BU. Tel: 01483 284279.

UNIVERSIRY OF SURREY E & ARS, G3IGQ. Open only to University students and staff. Meets 1.00pm Wednesdays. For location see notice board, outside shack in AB22. Details from: Mr Mike Blewett, G4VRN, E and Amateur Radio Society, Electronic & Electrical Eng., University of Surrey, Guildford, Surrey, GU2 5XH. Tel: 01483 300800.

WIMBLEDON & DARS, G3WIM. St. Andrews Church Hall, Herbert Road, Wimbledon, London. Details from: Mr P S Horbaczewskyj, G4ZXO, c/o 11 Tadworth Avenue, New Malden, Surrey, KT3 6DJ. Tel: 0181 3970427.

Tayside Region

Council Zone: G

DUNDEE ARC, GM4AAF. Meets 7.30pm Tuesdays at the Dundee College, Graham Street Annex, Dundee. Details from: Allan Martin, GM7ONJ, 11 Langlee Place, Broughty Ferry, Dundee, DD5 3RP. Tel: 01382 739179.

PERTH & DARG, GM4EAF. Meets 8.00pm Wednesdays at the Perth Sports & Social Club, 18 Leonard Street, Perth. Details from: Dr Ron Harkess, GM3THI, Friarton Bank, Rhynd Road, Perth, PH2 8PT. Tel: 01738 643435.

STRATHMORE & DARC, GM3GBZ. Meets Tuesdays 7.30pm at 2231 Sqdn ATC, 1 Lochside Road, Forfar. Details from: Mr Bill Henderson, GM0VIT, Plot 1, Drumglen, Bridge of Cally, Blairgowrie, PH10 7JL. Tel: 01250 886 324.

Tyne & Wear

Council Zone: A

HOUGHTON-LE-SPRING ARC, G3NMD. Meets Wednesday evenings at the Dubmire Royal British Legion, Dubmire, Fencehouses, Tyne & Wear, DH4 6LJ. Details from: Mr Foster Aungles, G0ABF, c/o 158 Burn Park Road, Houghton-le-Spring, Tyne & Wear, DH4 5DH. Tel: 0191 584 4673.

SOUTH TYNESIDE ARS, GX0WKQ. Meets 7.30pm - 9.00pm every other Monday at the Boldon Scout Hut, Grey Horse Car Park, Front Street, Boldon.

TYNEMOUTH ARC, G0NWM. Meets 7.00pm Fridays at the Linskill Centre, Linskill Terrace, North Shields, Tyne & Wear. Details from: Mr Terry Lambert, G8EZL, c/o 40 Deepdale Road, North Shields, Tyne & Wear, NE30 3AN. Tel: 0191 2570799.

TYNESIDE ARS, G3ZQM. Meets Wednesday evenings at the St Teresa's Club, 200b Heaton Road, Newcastle-upon-Tyne, NE6 5HP. Details from: Mr J Pickersgill, G0DZG, c/o 38 Sefton Avenue, Heaton, Newcastle-upon-Tyne, NE6 5QR. Tel: 0191 265 1718.

Warwickshire

Council Zone: B

AVON VALLEY ARA, RS90309. Details from: Mr Robin Harper, G1ZUU, 25 Duttons Close, Snitterfield, Stratford-upon-Avon, Warks, CV37 0JR. Tel: 01789 731637.

CASTLE TRANSMISSION INT. CLUB, G2LO. Open only to CTI employees. Details from: I P Jefferson, P O Box 98, Warwick, CV34 6TN.

MID WARWICKSHIRE ARS, G3UDN. Meets 8.00pm 2nd & 4th Tuesdays in month at the St John Ambulance HQ, 61 Emscote Road, Warwick. Details from: Mr B D Clulee, c/o 11 Ascot Ride, Lillington, Leamington Spa, Warks, CV32 7TT.

RUGBY ATS, G4APD. Meets 7.30pm Tuesdays at the Cricket Pavillion, B Entrance, Rugby Transmitting Station,, off A5 trunk Road, Hillmorton, Warwickshire. Details from: Mr Arthur Gallichan, M0ASD, c/o 4 Wigston Road, Hillmorton, Rugby, Warks, CV21 4LT. Tel: 01788 550778.

STRATFORD-UPON-AVON & DRS, G0SOA. Meets 7.30pm 2nd & 4th Mondays in the month at the Home Guard Club, Tiddington, Stratford-upon-Avon, Warwickshire. Details from: J Porter, G4OHJ, 77 Westholme Road, Bidford on Avon, Alcester, Warks, B50 4AN. Tel: 01789 773286.

WARWICK SCHOOL ARS, G4WKS. (Open only to staff and pupils). Meets most lunchtimes. Details from: Mr G N Frykman, G0GNF, Myton Road, Warwick, CV34 6PP. Tel: 01926 492484.

West Glamorgan

Council Zone: E

PORT TALBOT (BS PLC) ARS, GW3EOP. Meets 7.30pm Thursdays at the British Steel PLC Sports &, Social Club, Margam, Port Talbot, West Glamorgan. Details from: Mr Steven Hill, MW0AES, 1a Rugby Avenue, Neath, West Glamorgan, SA11 1YT. Tel: 01639 645693.

SWANSEA (UNIV COLL) ARS, GW3UWS. (Primarily intended for students.) Meets 7.00pm every Tuesday at the Radio Shack in the, Faraday Building.Malcolm Bowen, GW3KGI, Rm 700, Electrical Engineering Dept, University of Wales Swansea, Singleton Park, Swansea, SA2 8PP. Tel: 01792 295412.

SWANSEA ARS, GW4CC. Meets 7.30pm 1st & 3rd Thursdays in the month at the Applied Sciences Building, Swansea University. Details from: Mr Richard Hope, GW8TVX, 75 Priors Way, Dunvant, Swansea, W Glamorgan, SA2 7UH. Tel: 01792 201111.

SWANSEA RACC, GW4UNV. Meets 7.30pm Fridays at the Swansea Yachting & Subaqua, Club, Swansea Marina. Details from: Mr Mark Pilot, GW1DTA, c/o 92 Llanlienwen Road, Morriston, Swansea, SA6 6LU. Tel: 01792 794505.

WEST WALES REPEATER GROUP, GB3WW. Details from: Mr John Gray, GW6ZUS, 36 Heol Pentre Felen, Llangyfelach , Swansea, SA6 6BY. Tel: 01792 415021.

West Midlands

Council Zone: B

ALDRIDGE & BARR BEACON ARC, G0NEQ. Meets 7.30pm 1st & 3rd Mondays in the month at the The Manor House, Little Aston Road, Aldridge, Walsall. Details from: Mr Charles Baker, G0NOL, 19 Elizabeth Road, Walsall, West Midlands, WS5 3PF. Tel: 01922 36162.

BBC CLUB ARG, G2BBC. Open only to BBC employees.Broadcasting Centre, Pebble Mill Road, Birmingham, B5 7QQ.

BLACK COUNTRY PUG, GB7BHM. Details from: Mr D Wood, G7BNK, 16 Church Road, Pelsall, Walsall, W Midlands, WS3 4QN.

BROMSGROVE & DARC, G3VGG. Meets 8.00pm on 2nd Friday in the month at the Avoncroft Arts Centre, Bromsgrove, Worcs.

COVENTRY ARS, G2ASF. Meets 8.00pm Fridays at 121 St Nicholas Street, Radford, Coventry. Details from: Mr David Stanley, G1ORG, 60 Ainsbury Road, Canley Gardens, Canley, Coventry, CV5 6BB. Tel: 01203 311468.

DUDLEY ARC, G4DAR. Meets 7.45pm 4th Monday in month at the Central Library, St James Road, Dudley. Details from: Mr Tony Lucas, G4LVA, 12 Digby Road, Kingswinford, West Midlands, DY6 7RP. Tel: 01384 277925.

GPT (COVENTRY) ARS, RS172991. c/o 14 George Birch Close, Brinklow, Rugby, Warks, CV23 0NN.

KING EDWARDS SCHOOL ARS, G8ZKE. (Open only to staff and pupils). King Edward's School, Edgbaston Park Road, Birmingham.Edgbaston Park Road, Birmingham, B15 2UA.

KYNOCH R & TVS, G3HPP. Meets Thursday evenings at the Club Workshop, IMI Ltd, Sportsfield, Perry Bar, Birmingham. Details from: Mr G E Nicholls, 27 Canberra Road, Walsall, West Midlands, WS5 3NH. Tel: 01922 35376.

MIDLAND ARS, G3MAR. Meets several times a week at Unit 22, 60 Regent Place, Hockley, Birmingham (jewelry quarter). Details from: Mr John A Crane, G0LAI, 68 Max Rd, Quinton, Birmingham, B32 1LB. Tel: 0121 628 7632.

MIDLANDS AX25 PACKET R.U.G., G7CRS. c/o Mr E R Loach, G4ZXS, 16 Cheriton Grove, Perton, Wolverhampton, WV6 7SP.

MIDLANDS ELECTRICITY RS, G4MEB. Meets 8.00pm twice a month at the MEB Social Club, Mucklow Hill, Halesowen, West Midlands. Details from: Mr Dave Hillyer, G0RQO, Radio Section, Mucklow Hill, Halesowen, West Midlands, B62 8BP. Tel: 0121 421 3207.

OLD SWINFORD HOSPITAL (SCHOOL), G4CVK. Intended for pupils & staff but guests very welcome. Details from: Mr C S Williamson, Old Swinford Hospital, Hagley Road, Stourbridge, West Mids, DY8 1QX.

SANDWELL AMATEUR RADIO CLUB, G0CWC. Monday, Wednesday and Thursday eve 7-9.30pm at Sandwell ARC, Broadway, Oldbury, Warley, W Midlands, B68 9DP. Details from: Mr Clive Binnell, G0TVR, c/o 146 Hales Crescent, Smethwick, Warley, West Midlands, B67 6QX. Tel: 0121 429 6061.

SOLIHULL ARS, G3GEI. Meets 3rd Thursday in the month at The Shirley Centre, Stratford Road, Shirley, Solihull, West Midlands. Details from: Mr Paul Gaskin, G8AYY, 58 Elmcroft Road, South Yardley, Birmingham, B26 1PL. Tel: 0121 783 2996.

SOUTH BIRMINGHAM RS, G3OHM. Meets 8.00pm every Mon, Thurs and Fri nights as well as the 1st Wednesday in the month at Hampstead House, Fairfax Rd, West Heath, Birmingham. Details from: The SBRS Secretary, c/o West Heath Community Assn, Hampstead Ho, Fairfax Road, Birmingham, B31 3QY.

WEST BROMWICH CENTRAL RC, G4WBC. Meets 7.30pm Sundays at The Sandwell Public House, High Street, West Bromwich, West Midlands. Details from: Mr Ian Leitch, G0PAI, 72 Dawes Avenue, West Bromwich, West Midlands, B70 7LS. Tel: 0121 561 2884.

WEST MIDLANDS POLICE ARC, G0COP, G1WMP. Meets Monday evenings. Details from: The Secretary, PC 5795, Steve Jones, Digbeth Police Station, Digbeth, Birmingham, West Midlands, WS9 8XR. Tel: 0121 626 6020.

WILLENHALL & DARS, G4ETW. Meets 8.00pm Wednesdays at The Brewers Droop Inn, Wolverhampton Street, Willenhall, West Midlands. Details from: Mr Maurice Ravenscroft, G0TMR, P.O.Box 252, Willenhall, West Midlands, WV13 3DW. Tel: 01902 632242.

WOLVERHAMPTON ARS, G8TA. Meets 8.00pm Tuesdays at the Electricity Board Sports Club, St Marks Road, Chapel Ash, Wolverhampton. Details from: Mrs M Bentley, G6AKN, c/o Mrs M Bentley, G6AKN, 9 Tinkers Castle Road, Seisdon, Wolverhampton, WV5 7HF.

WORDSLEY RC, G4WRA. Meets at the Brick Maker's Arms, Mount Pleasant, Brierley Hill, West Midlands. Details from: Mr Andy Evans, G1PKZ, c/o 6 West Road South, Halesowen, West Midlands, B63 2UT.

West Sussex

Council Zone: C

ARUN ARC, G0UKR. Details from: Mr P Howard, 12 Meadow Way, Westergate, Chichester, West Sussex, PO20 6QT. Tel: 01243 543399.

CHICHESTER ARC, G3ISO. Meets 7.30pm on 1st & 3rd Tuesdays in the month at the St Pancras Hall, Chichester. Details from: Mr John Stratfull, G3IJS, 55 Craigweil Lane, Bognor Regis, West Sussex, PO21 4XN. Tel: 01243 861578.

CIBA-GEIGY SPORTS & SOC CLB RS, G4CGY. (Only open to employees.) Meets at Radio Section, Ciba Geigy Sports & Social, Club, Wimblehurst Road, Horsham, West Sussex.Wimblehurst Road, Horsham, West Sussex, RH12 4AB.

CRAWLEY ARC, G3WSC. Meets 8.00pm Wednesdays and Sundays 10.30am at the Tilgate Forest Rec. Centre, Hut 18, Tilgate Forest, Crawley, West Sussex. Details from: Mr J S Spence, G0FPI, 60 Railey Road, Northgate, Crawley, West Sussex, RH10 2BZ.

HORSHAM ARC, G4HRS. Meets 8.00pm 1st Thursday in month at the Guide Hall, Denne Road, Horsham, West Sussex. Details from: Mr Alister Watt, G3ZBU, c/o 5 Brambling Road, Horsham, West Sussex, RH13 6AX. Tel: 01403 253432.

LANCING COLLEGE ARC, G0VSI. Lancing College, Lancing, W Sussex, BN15 0RW.

MID SUSSEX ARS, G3ZMS. Meets 7.45pm Fridays at Marle Place, Leylands Road, Burgess Hill, West Sussex. Details from: Mr C Childs, 2E1DCP, c/o 17 Gladstone Road, Burgess Hill, West Sussex, RH15 0QQ. Tel: 01444 244689.

MORSE GROUP BOGNOR, G0MGB. Meets 8.00pm Wednesdays at 2 Silverston Avenue, Bognor Regis, West Sussex. Details from: Mr Fernandez, G0IAF, 1 Milestone Cottages, Fishbourne, Chichester, West Sussex, PO18 8AU.

T.S. VINDICATRIX ASN, G0WVB. Details from: Don Still, G0OOC, Details from: D G Still, G0OOC, 25 Brooks Green Park, Emms Lane, Brooks Green, Horsham, W Sussex, RH13 8QR.

WORTHING & DARC, G3WOR. Meets 8.00pm on Wednesdays at the Lancing Parish Hall, South Street, Lancing, West Susesx.

WORTHING & DISTRICT VIDEO RG, GB3VR. Meets 7.00pm on 1st Tuesday in the month. Details from: The Treasurer, c/o 21 St James Avenue, Lancing, West Sussex, BN15 0NN. Tel: 01903 211919 (w).

West Yorkshire

Council Zone: A

CEN YORKS SCOUT RAD FELLOWSHIP, G0OPZ. c/o 25B Glenhome Road, Farsley, Pudsey, West Yorkshire, LS28 5BY.

CRAVEN RADIO AMATEUR'S GROUP, M0BCQ. Meets 8.00pm on Tuesdays at the White Lion, Kildwick, Keighley, W Yorks. Details from: Mr Andy Jackson, G0VJL, 14 Earls View, Sutton in Craven, Keighley, W Yorks, BD20 7PR. Tel: 01535 637 601.

DENBY DALE & DARS, G4CDD, G8KMK. Meets Wednesdays at the Pie Hall, Denby Dale, West Yorkshire. Details from: Malcolm McKenzie, G8RWN, c/o 9 Broomhouse Close, Denby Dale, Huddersfield, HD8 8UX. Tel: 01484 861782.

HALIFAX & DARS, G2UG. Meets Tuesdays. Details from: Mr S P Ortmayer, G4RAW, c/o 14 The Crescent, Hipperholme, Halifax, W Yorks, HX3 8NQ. Tel: 01422 203062.

KEIGHLEY ARS, G0KRS. Meets 8.00pm Thursdays at the Cricket Club, Ingrow, Keighley, West Yorkshire. Details from: Mr Jack Birse, G4ZVD, 178 Long Lee Lane, Keighley, W Yorkshire, BD21 4TT. Tel: 01535 212985.

LEEDS & DARS, G4LAD. Meets Monday evenings at The Radio Shack, Yarnbury (Horsforth), RUFC Grounds, Brownberrie Lane, Horsforth, Leeds, LS18 5HB. Details from: Mr E Howden, G0IBU, 36 Moseley Wood Green, Cookridge, Leeds, LS16 7HB.

NORTH WAKEFIELD RC, G4NOK. Meets 8.00pm Thursday at the East Ardsley Cricket Club, Nr Wakefield. Details from: Mrs Olga Parker, 2E1ASV, c/o 37 Springbank Crescent, Gildersome, Morely, Leeds, LS27 7DN. Tel: 0113 253 9087.

NORTHERN HEIGHTS ARS, G2SU. Meets 1st & 3rd Wednesday in the month at the Bradshaw Tavern, Halifax, West Yorkshire. Details from: Mr L L N Cobb, G3UI, c/o 144 Harecroft, nr Wilsden, Bradford, BD15 0BP. Tel: 01422 360574.

NORTHERN VHF ACTIVITY GROUP, G7UEG. Details from: 24 Farifield Terrace, Bramley, Leeds, LS13 3DH.

OTLEY ARS, G3XNO. Meets 8.00pm Tuesdays at The RAOB Club, Westgate, Otley, West Yorkshire. Details from: Mr Jack Worsnop, G0SNV, 35 Westwood Avenue, Eccleshill, Bradford, BD2 2NJ. Tel: 01274 636197.

PENNINE ARS, G0DRP. Details from: 31 Dartmouth Avenue, Almondbury, Huddersfield, West Yorks, HD5 8UP.

PONTEFRACT & DARC, G3FYQ. Meets Thursdays at the Carleton Community Centre, Pontefract, West Yorkshire. Details from: Mr Colin Wilkinson, G0NQE, 8 Westfield Avenue, Knottingley, West Yorks, WF11 0JH. Tel: 01977 677006.

RISHWORTH SCHOOL ARC, G0SQA. Meets between 9.30 - 1200am Saturdays during term time in the Physics Laboratory, Rishworth School. Deatils from: Mr Vinters, Rishworth School, Rishworth, Sowerby Bridge, West Yorkshire, HX6 4QA.

SPEN VALLEY ARS, G3SVC. Meets 8.00pm Thursdays at the Old Bank WMC, Mirfield, West Yorkshire. Details from: Mr Derek Allan, G0RZP, c/o 283 Cliffe Lane, Gomersal, Cleckheaton, W Yorks, BD19 4SB. Tel: 01274 872244.

WAKEFIELD & DRS, G3WRS, G1WRS. Meets 8.00pm Tuesdays at the Ossett Community Centre, Ossett, West Yorkshire. Details from: Mr John Carter, G7JTH, 30 Swift Way, Sandal, Wakefield, W Yorks, WF2 6SR. Tel: 01924 251 822.

WAKEFIELD RPTR GP, G0KNR. Details from: 1 Wavell Garth, Sandal Magna, Wakefield, W Yorks, WF2 6JP.

WHITE ROSE ARS, G3XEP. Meets Wednesday evenings at the Moortown RUFC, Moss Valley, Kings Lane, Leeds, LS17 7NT. Details from: Mr M Wilson, G7SDW, PO Box 73, Leeds, LS1 5AR. Tel: 0113 273 6039.

Wiltshire

Council Zone: D

CHIPPENHAM & DARS, G3VRE. Meets 8.00pm Tuesdays at the Sea Cadet HQ, Chippenham. Details from: Mr Jon Ainge, G4LGZ, 15 Wolverton Close, Cepen Park South, Chippenham, Wilts, SN14 0FG. Tel: 01249 462610.

DEVIZES & DARC, G4WIK. Meets 8.00pm Fridays at the Hare & Hounds Inn, Hare & Hounds Street, Devizes, Wilts. Details from: Mr Noel Woolrych, G4TIX, 20 Meadow Drive, Devizes, Wilts, SN10 3BJ. Tel: 01380 724533.

POST OFFICE RES. ARC, G4BPO. Room FW01, Post Office Res., Wheatstone Road, Dorcan, Swindon, Wiltshire, SN3 4RD.

RIDGEWAY REPEATER GROUP, GB3WH, GB3TD, G0RRG. Details from: Mr R Loss, G4XUT (Secretary), c/o 84 Thorne Road, Eldene, Swindon, Wiltshire, SN3 6DU.

SALISBURY ARC, G3FKF. Meets 8.00pm 2nd and 4th Tuesdays in the month The Scout Hut, St Marks Avenue, Salisbury, Wilts. Details from: Mr Jamie Donaghy, G7WAA, Mendips, Attwood Road, Salisbury, Wilts, SP1 3PR. Tel: 01722 334 935.

SWINDON & DARC, G3FEC. Meets 7.00 pm every Thursday at the Eastcott Community Centre, Savenake St, Swindon. Details from: Mr Den Forrest, G7PDV, 19 Burns Way, Swindon, Wiltshire, SN2 6LP.

TROWBRIDGE & DARC, G2BQY. Meets 8.00pm 1st & 3rd Wednesdays in the month at the Southwick Village Hall, Southwick, Trowbridge, Wilts. Details from: Mr Ian Carter, G0GRI, 12 Bobbin Lane, Westwood, Bradford-on-Avon, Wilts, BA15 2DL. Tel: 01225 864698.

Emergency Com

ASSOC NORFOLK RAYNET GROUPS, RS96544. Top-o-Hill, Hospital Road, Wicklewood, Wymondham, Norfolk, NR18 9PR. Tel: 01953 607594.

BRECKLAND RAYNET, G7BRN. 13 Swathing, Cranworth, Thetford, Norfolk, IP25 7SJ.

BRIDGWATER RAYNET GROUP, RS95573. Meets on the air each Wednesday evening at 2000 hrs 144.800. Details from: Mr F Rhodes, G3TWO, 23 Quantock Avenue, Bridgwater, Somerset, TA6 7EB.

BRIDLINGTON RAYNET GRP, G0VGH. Meetings as required. AGM in November. Details from: Mr Leonard C Bacon, G6JFJ, c/o 100 Etherington Drive, Hull, HU6 7JT. Tel: 01482 853276.

BROADLAND RAYNET, G0RBG. Details from: Mr H W Holmes, 7 Parkland Crescent, Old Catton, Norwich, Norfolk, NR6 7RQ.

CALDERDALE RAYNET ARC, G7RRC. Details from: Alan Harvey, G4MUR, 3 High Lane, Norton Tower, Halifax, West Yorkshire, HX2 0NW.

CARRICK RAYNET, RS95707. 39 Gwell An Nans, Probus, Cornwall, TR2 4ND.

CENTRAL GWYNEDD RAYNET, RS95526. Details from: Mr H F Postel, GW6UWV, c/o Ty Isaf, Clynnogfawr, Caernarfon, Gwynedd, LL54 5NH.

CENTRAL LANCS RAYNET, G7CLR. Details from: Mr Eddy Wane, G0OFY, 26 Coniston Avenue, Euxton, Chorley, Lancs, PR7 6NY. Tel: 01257 274368.

CHESHIRE RAYNET GP, G7CRG. Details from: the controller, 5 Llandovery Close, Winsford, Cheshire, CW7 1NA.

CHICHESTER RAYNET, G0CRG. c/o Nebraska, Southover Way, Hunston, Chichester, W Sussex, PO20 6NY.

CONWY COUNTY RAYNET ARG, RS96399. 33 Fford Morfa, Llandudno, Gwynedd, LL30 1ES.

CORNWALL COUNTY RAYNET, G4EQU. Details from: the controller, Northmead, 111 Mount Ambrose, Redruth, Cornwall, TR15 1NW.

DAVENTRY RAYNET GRP, G4RFG. c/o 8 Spring Close, Daventry, Northants, NN11 4HG.

DUMFRIES & GALLOWAY RAYNET GP, RS94807. Meets 8.00pm 2nd Monday in the month at the Emergency Planning Centre, Council Offices, English Street, Dumfries, DG1 2DD. Details from: Mr Tony Regnart, GM8YFA, c/o Lyndhurst, Corstorphine Road, Thornhill, Dumfries, DG3 5NB. Tel: 01848 330340.

EAST BERKSHIRE RAYNET GROUP, G0SRS. Details from: Mr J F Hicks, G8RYW, c/o Wayside, 22 Courthouse Rd, Maidenhead, Berks, SL6 6JB.

EAST DUNBARTONSHIRE RAYNET GP, GM0RED. Details from: Mr Jim Murdoch, GM3JMM, 4 Cedar Drive, Milton of Campsie, Glasgow, G65 8AY. Tel: 01360 310908.

EAST LANCS RAYNET, G0ELR. Details from: 31 Belgrave Street, Nelson, Lancs, BB9 9HR.

ESSEX RAYNET, G4BCV. 95 Washington Road, Maldon, Essex, CM9 6JF. Tel: 01621 859 833.

FYLDE COAST RAYNET GROUP, G7FCR. Meets 7.30pm on 2nd Wednesday in the month at the Wyre Civic Centre, Breck Road, Poulton-le-Fylde, FY6 7PU. Details from: Ray Knighton, G0GER, 262 Victoria Road West, Thornton, Cleveleys, Lancs, FY5 3QB. Tel: 01253 824319.

GLOUCESTERSHIRE COUNTY RAYNET, G6GLO. Meets 2nd Tuesday in the month at Ullenwood. Details from: Mr Jerry Pallister, G1YXF, 1 Third Avenue, Highfields, Dursley, Glos, GL11 4NT. Tel: 01453 545947.

GRAMPIAN RAYNET GRP, RS94808. Details from: D.Wemyss, 24 Brucklay Court, Peterhead, AB42 6UF.

GWENT COUNTY RAYNET, RS95608. Meets 7.30pm on 1st or 2nd Wednesday in the month at County Hall, Cwmbran, Gwent. Details from: Mr Kevin Snelling, GW7BSC, 91 Oakfield Road, Newport, Gwent, NP9 4LP. Tel: 01633 262488.

HAMPSHIRE RAYNET GRP, G1SEH.

HASTINGS & ROTHER RAYNET GROUP, G7ERC. Details from: 15 Kingsley Close, St. Leonards-on-Sea, East Sussex, TN37 7BX.

HORNSEA & DIST RAYNET GROUP, RS175131. Details from: Mr Duncan Heathershaw, G3TLI, c/o The Old School, Mappleton, Hornsea, E Yorks, HU18 1XX. Tel: 01964 532588.

HYNDBURN & ROSSENDALE RAYNET, G0LRR. Details from: Mr B Murray, G4VVK, c/o 30 Middlegate Green, Crawshawbooth, Rossendale, Lancs, BB4 8PY. Tel: 01706 229026.

IOW RAYNET GROUP, G1IWR. Details from: Dennis Chubb, Walls End Cottage, Heathfield, Close, Bembridge, Isle of Wight, PO35 5UG. Tel: 01983 873087. Fax: 01983 874757.

KENT COUNTY RAYNET, G4BRC. Details from: Kent County Council, Emergency Plannning Department, Waterton Lee House, 99-102 Sandling Road, Maidstone, Kent, ME14 1AE.

LEEDS RAYNET GP, G4RDL. Details from: Mr A J Scott, G6NIZ, The Conifers, Back Lane, Newton-on-Ouse, York, YO6 2DF.

LEICESTER RAYNET GROUP, G1LCR. Meets 7.30pm 2nd & 4th Thursday in month at The Emergency Planning Rooms, County Hall, Leicester Road, Glenfield, Leicester. Details from: Mr Ivon Johnson, G1IMJ, c/o PO Box 4, Cosby, Leicestershire, LE9 1ZX.

LIZARD & HELSTON RAYNET GP, RS108286. Details from: Mr D Cusick, G0BQH, Chy Carne, Caravan Park, Kuggar, Ruan Minor, Nr Helston, Cornwall, TR12 7LX. Tel: 01326 290200.

MENDIP RAYNET, RS95597. Details from: 6 Middlemead, Stratton on the Fosse, Bath, BA3 4QH.

MID BEDFORDSHIRE RAYNET GROUP, G0MBR, G6MBR. Meets 8.30pm on Thursdays in a pub in the Bedford area. Details from: Mr Ian McIver, G0BKN, 31 Hartshill, Bedford, MK41 9AL. Tel: 01234 328816.

MID GLAMORGAN RAYNET GROUP, GW1MGR.

MID-HERTS RAYNET GRP, G0MHR. Meets 8.00pm on Thursdays on 144.825 MHz. Details from: Mr Mike Faithfull, G6UBH, 10 Beacon Road, Ware, Herts, SG12 7HY. Tel: 01920 462241.

MID-THAMES RAYNET, G0MTR. Meets on 1st Monday in the month (except bank holidays). Details from: Mr Dunn, G6DOV, 24 Mynchem Road, Beaconsfield, Bucks, HP9 2BA. Tel: 01494 673372.

NE DERBYSHIRE RAYNET GROUP, RS171905. Details from: Mr R Neale, G4OIE, Field House, Recreation Road, New Houghton, Mansfield, Notts, NG19 8TL.

NE ESSEX RAYNET GRP, RS95987. Meets on the air every Monday (except Bank Holidays) at 2100 hrs on 145.200 MHz. Details from: The Controller, Capelside, Chequers Road, Little Bromley, Manningtree, Essex, CO11 2QE.

NE LONDON RAYNET GRP, G4NEL. Details from: 154 Cherrydown Ave, Chingford, London, E4 8DZ.

NORFOLK RAYNET, RS94647. Details from: 7 Parkland Crescent, Old Catton, Norwich, NR6 7RQ.

NORTH CORNWALL RAYNET GP, G0NCD. Details from: Mr A E Warne, G3YJX, Treryn, Trevanson Road, Wadebridge, Cornwall, PL27 7HB. Tel: 01208 812772.

NORTH DEVON RAYNET, RS95524. Meets once a month. Details from: Mr I T S Binding, G4RVG, 40 Parklands, South Molton, N. Devon, EX36 4EW.

NORTH DYFED RAYNET GRP, RS95677. Details from: Mr Bryan Blake, GW1XOT, Brynamlwe, Devils Bridge, Aberystwyth, Dyfed, SY23 4RD. Tel: 01970 890393.

NORTH GWYNEDD RAYNET GRP, GW7GWS. 14 Maeshyfryd, Llangefni, Anglesey, Gwynedd, LL77 7PY.

NORTH HERTFORDSHIRE RAYNET ASS, RS94942. Details from: Mr Stuart Donald, G6EDD, 5 Windsor Road, Royston, Herts, SG8 9JF. Tel: 01763 242876.

NORTH LANCASHIRE RAYNET, G0NLR. Details from: Mr David Andrew, G6OUT, 30 Woodhill Avenue, Morecambe, Lancashire, LA4 4PF. Tel: 01524 416564.

NORTH NORFOLK RAYNET GROUP, G7RNN. Details from: Mr Colin Harrold, G4RRN, Boundary Farm, Felbrigg, Norfolk, NR11 8PD. Tel: 01263 512736.

NORTH POWYS RAYNET GROUP, RS95365. Details from: Mr Dave Brown, GW4NQJ, Kingsdown Cottage, Fron, Montgomery, Powys, SY15 6SB. Tel: 01686 640814.

NOTTINGHAM RAYNET ARS, G7KXR. Nottingham County Council Off, Loughborough Road, West Bridgford, Nottingham Details from: Mr M C Shaw, G4EKW, 50 White Road, Nottingham, NG5 1JR.

ORKNEY RAYNET GROUP, RS95262. c/o Mrs A O Wright, Crosslea, Berstane Road, Kirkwall, Orkney, KW15 1SZ.

PEMBROKESHIRE RAYNET GROUP, RS96009. 40 Lower Quay Road, Hook, Haverfordwest, Dyfed, SA62 4LR.

PENWITH RAYNET GP, G0PEN. Details from: the controller, 25 Gwavas Street, Penzance, Cornwall, TR18 2DF.

RAYNET (N IRELAND), RS95281. Details from: the Sec, c/o 4 Ilford Avenue, Crossnacreevy, Belfast, BT6 9SF.

READING & W BERKS RAYNET GP, G0VPE. Details from: Denis Pibworth, G4KWT, c/o 20 Marathon Close, Woodley, Reading, Berks, RG5 4UN. Tel: 01734 698526.

RESTORMAL RAYNET GP, G0BRR. Meetings held monthly at The Emergency Centre, Restormel Borough HQ, Penwick Road, St Austell. Details from: the controller, 19 Trewanney Road, St Austell, Cornwall, PL25 4JA. Tel: 01726 72951.

S E CORNWALL RAYNET GRP, RS95706. Details from: Mr J A Husk, G8GLI, Bran-Dhu, Commonmoor, Liskeard, Cornwall, PL14 6EP.

SHEFFIELD & ROTHERHAM RAYNET, G6AEN. c/o Mr T Haddon, G4KMA, 11 Lovetot Avenue, Aston, Sheffield, S Yorks, S31 0BQ.

SHROPSHIRE RAYNET, G1SCR. 16 Kynaston Drive, Wem, Shrewsbury, Shropshire, SY4 5DE.

SOUTH DEVON RAYNET, G0SBM. 46 Clarendon Road, Ipplepen, Newton Abbott, Devon, TQ12 5QS.

SOUTH GLOS RAYNET GRP, G7NAR. Details from: Mr J M Davis, 62 Kingscote, Yate, Bristol, BS17 4YE.

SOUTH KENT RAYNET GROUP, G1URG. Details from: John Wellard, G6ZAA, 19 South Motto, Park Farm, Kingsnorth, Ashford, Kent, TN23 3NJ. Tel: 01233 503 050.

SOUTH NORFOLK RAYNET, G0SNR. Weekly net Friday 8.30pm on 145.225 MHz and 433.775 MHz for Novices. Details from: Ms Sue Brodie, G0PSY, Top of The Hill, Hospital Road, Wicklewood, Wymondham, Norfolk, NR18 9PR. Tel: 01953 607594.

SOUTH WEST DURHAM RAYNET ARG, RS95059. Meets 8.00pm 2nd Monday in the month at the Stanley Crook Village Hall, rear of High Road, Stanley Crook, Co. Durham. Details from: Mr Ian Bowman, 22 Bryan Street, Spennymoor, Co. Durham, DL16 6DW. Tel: 01388 812104.

SOUTH WEST LINCOLNSHIRE RAYNET, G0GRN. Details from: Mr Alan Burton, c/o 26 Woffindin Close, Great Gonerby, Grantham, Lincs, NG31 8LP.

SOUTHPORT & DISTRICT RAYNET GP, G0SDR. Meets 1st Wednesday in month at the Royal British Legion, West Street, Southport. Details from: Mr Carl Riley, G4RQX, 80 Fylde Road, Southport, Southport, PR9 9XL. Tel: 01704 25172.

STRATHCLYDE RAYNET GROUP, GM1NET. Details from: Mr C D Ross, GM8HBY, 16 Glebe Crescent, Airdrie, Lanarkshire, Lanarkshire, ML6 7DH. Tel: 01236 755177.

SUFFOLK RAYNET, G0SAR. Details from: Mr Mike Watson, G8CPH, The Tubbery, Henley, Ipswich, Suffolk, IP6 0BR. Tel: 01473 831448.

TAUNTON RAYNET GROUP, RS95600. 1 The Rowans, Victoria Street, Wellington, Somerset, TA21 8HR.

THANET RAYNET GROUP, G1RCT. Meets on the air 6.30pm on Sundays on 144.850 MHz. Net Controller is G0DFI. Details from: Derek Oakley, G0DFI, 6 Staplehurst Gardens, Cliftonville, Margate, Kent, CT9 3JB. Tel: 01843 228718.

WEST CLWYD RAYNET, GW0CCR. Details from: 23 Russell Avenue, Colwyn Bay, Clwyd, LL29 7TR.

WEST GLAMORGAN RAYNET, GW1WGR. Details from: 25 Stepney Road, Cockett, Swansea, West Glamorgan, SA2 0FZ.

WEST KENT RAYNET, G7TYR. Details from: Denis Collins, G0DJC, 71 Trench Road, Tonbridge, Kent, TN10 3HG. Tel: 01732 357125.

WESTON-SUPER-MARE RAYNET, RS95806. Deatails from: 36 Tormynton Road, Worle, Weston-Super-Mare, Avon, BS22 9HT.

WREXHAM & D RAYNET GROUP, RS95364. Details from: Mr Peter Higgs, Hafan, Church Street, Penycae, Wrexham, Clywd, LL14 2RL. Tel: 01244 570212.

Overseas

AGRA RADIO CLUB, VU2AGR. c/o Dr Mukesh Chandra, VU2MCC, Mulberry House, 14/193 Ghatia Azam Khan, Agra 282 003, U. P., India.

BORAS RADIOAMATORER, SK6LK. Box 22137, 504 12 Boras, Sweden.

BRISBANE ARC INC, VK4BA. Details from: the secretary, PO Box 3007, Darra 4076, Queensland, Australia.

CERN AMATEUR RADIO CLUB, F6KAR. Case Postale 16, 1211 Geneva 23, Switzerland.

DONEGAL TIR CONAILL ARS, EI5TCR. Details from: Mr D K McDermott, c/o D K McDermott, Curraghamone, Ballybofey, Co Donegal, Eire.

ELFA RADIO GANG, SK0AY. Details from: the secretary, U.Soder, 17117 Solna, Sweden.

ESBJERG AFDELING, OZ5ESB. Details from: the secretary, Postbox 94, 6701 Esbjerg, Denmark.

GALWAY EXPERIMENTERS CLUB, EI4GRC. c/o Mr T Frawley, Killoughter, Castlegar, Co Galway, Eire.

GIBRALTAR ARS, GARS. Details from: the secretary, PO Box 292, Gibraltar.

HISPANIA CW CLUB, EA3HCC. Details from: the secretary, Av. Roma 10 - 17/2, 08015 Barcelona, Spain.

HISPANIA CW CLUB, EA7HCC. Attn. Carlos Cabeza, Cevita 1, 41530 Moron Ftra. (Sevilla), Spain.

HRVATSKI RADIOAMATERSKI SAVEZ, RS174902. Hrvatski Radioamaterski Savez, Dalmatinska 12, 10000 Zagreb, Croatia.

MARIESTAD AMATORRADIOKLUBB, SK6QW. Box 131, 542 22 Mariestad, Sweden.

OV ISARWINKEL CLUB, DK0IW. Details from: the secretary, c/o Butz, St Johannisrain 9, 82377 Penzberg, Germany.

RADIOKLUBBEN SK0CT, SK0CT. Details from: Ronny Borgstrand, Theodor Berg, Ericsson Radio Systems AB, S-164 80 Stockholm, Sweden.

RAUMAN RADIOKERHO RY, OH1AK. Details from: The Secretary, PO Box 143, Rauma, Finland.

SA OF HONG KONG, VS6EA. Details from: the secretary, Room 915, Hong Kong Scout Ctr, 8 Austin Road, Kowloon, Hong Kong.

SYDKUSTENS ARC, SK7OA. Details from: the secretary, Svaneholmsv 22 B, c/o Nilsson, S-27431 Skurup, Sweden.

TAEBY SANDARAMATORER, SK0MT. Details from: the secretary, Sjoflygvagen 4, S-183 62 Taby, Sweden.

THE RADIO CLUB, 5B4ES. Details from: the secretary, The English School, Nicosia, Cyprus.

RSGB Contests

160 m National ARDF Rules

1. Events

(a) Qualifying Events will be open to members of the RSGB or affiliated societies on the strict understanding that they and the members of their teams take part entirely at their own risk and that neither the RSGB nor the organising club shall be held responsible for any loss or damage resulting from taking part in the contest. Every named person taking part in the contest will be required to sign their name on behalf of themselves and the members of their team on a form provided by the organiser at the start acknowledging this and the rest of the rules. Only the signed-in person may operate the DF receiver outside the vehicle and hand in his form to the transmitter operator. Contests will be held on Sunday afternoons, commencing at 13.20 and concluding at 16.30.

(b) The National Final will be held after the Qualifying Events have been decided and only the following will be allowed to compete:

(i) The winner of the National Final in the previous year.

(ii) Competitors qualifying in the Qualifying Events.

iii) One or more competitors specifically invited by the contest committee.

Only entries under (i) and (ii) will be entitled to win the trophy.

The National Final will be held on a Sunday afternoon commencing at 12.50 and concluding at 16.30. Note: all times are clock times.

2. Transmitters

In the Qualifying Events, competitors will be required to locate two hidden transmitters, and, on the National Final only, will be required to locate three. All transmitters will operate, using CW and amplitude modulation, in the 1.8 MHz band, each with a maximum carrier level of 9 dBW and the power output will remain constant throughout the event.

3. Identification

For identification purposes, the callsigns and frequencies (which will be separated by at least 10 kHz) will be announced at the start of each event. Identification signals will be given in CW for the first four minutes of the first transmission, immediately followed by two minutes of telephony.

4. Signals

(a) After 13.26 (12.56 on the National Final) competitors who are satisfied with their bearings may leave the start at their own discretion.

(b) If any of the competitors fail to detect signals from any of the transmitters they will, at 1335 (1305 on the National Final), be given a bearing(s) which, when drawn on the appropriate 1:50,000 OS map, will pass within 4 km (8 cm) of the transmitter(s) they have not detected.

(c) If none of the competitors detect a signal from any one or more of the transmitters, the starter, in addition to providing the information above, will state whether the transmitter is either

less than 10 km,
more than 10 km but less than 20 km
or more than 20 km

from the start location.

5. Transmission Times

After 14.04 (13.34 on the National Final) transmissions will continue on telephony for not less than two minute periods at irregular intervals, such periods to be not more than 15 minutes apart. Each transmission will be preceded by a short identification signal in CW of the form '*TEST, TEST, TEST DF DE G...../P*. After 16.00,

	Qualifying Events	National Final
CW	13.20 - 13.24	12.50 - 12.54
Telephony	13.24 - 13.26	12.54 - 12.56
CW	14.00 - 14.02	13.30 - 13.32
Telephony	14.02 - 14.04	13.32 - 13.34
Random	14.04 - 16.00	13.34 - 16.00
Telephony	16.00 - 16.02	16.00 - 16.02
Random	16.02 - 16.15	16.02 - 16.15
Continuous	16.15 - 16.30	16.15 - 16.30

transmissions will continue on telephony only. Transmitters will operate on the same fixed schedule of transmissions (ie operating simultaneously) until 14.04 (13.34 on the National Final) subsequently operating independently, except for the 16.00 transmission. Where severe interference is known to exist, slow CW may be used in place of telephony.

Contests will terminate at 16.30 and, in the event of no one finding all the transmitters in the time allowed, the contest will be declared a one or two station contest, as the case may be, and the winner declared on that basis.

6. Locating Hidden Stations

Competitors may locate stations in any order and upon arrival at each station the competitor must hand his numbered entry form directly to a member of the transmitter crew who will initial the form, mark on it the time of arrival and hand it back to the competitor. The transmitter operator, or his assistant, must, if challenged by a competitor holding a DF receiver, admit that his is one of the hidden DF stations.

7. Qualifiers and Winners

(a) If seven or more Qualifying Events are held in any one year the first two competitors, not having previously qualified to locate their second transmitter, will go forward to the National Final. If there are six or less Qualifying Events, the first three competitors, as defined above, go forward to the National Final.

(b) In the National Final the first competitor to locate his third transmitter (or second transmitter if no one has located three) will be declared the winner.

8. General

(a) Competitors searching likely transmitter sites prior to the commencement of a competition will, at the discretion of the ARDF Committee, be disqualified from the competition.

(b) The hidden stations will be located at least 50 yards from any inhabited building and will be directly accessible to competitors without them entering, crossing or trespassing upon property in private occupation.

(c) The hidden stations will be located at least 50 feet from any public highway.

(d) Transmitter locations and starting point shall be covered by one sheet of the Ordnance Survey map (1:50,000 series) and the sheet number must be published prior to the event.

(e) Each competitor must sign on at the starting point and must receive an entry form numbered to confirm with the entry on the starters sheet.

(f) A team shall consist of a competitor plus not more than three others.

(g) Tampering with the transmitter aerial by the competitor or his team is strictly forbidden and may entail disqualification of the competitor. Competitors and their teams must leave the vicinity of each transmitter immediately the signed form has been handed back to the competitor.

(h) Only one portable receiver, capable of being tuned to the 1.8 MHz band, shall be carried by any team during the event, and the competitor at the time of his arrival at each hidden transmitter must have his receiver with him and, if required, must demonstrate that it is in working order. However, there is no objection to having a second portable receiver as a reserve for use in the event of failure of the first receiver provided that the reserve receiver remains in the competitor's vehicle until a failure of the first receiver has occurred. Simultaneous use of two receivers may result in disqualification from the contest. The use of any transmitting equipment by the competitor or his team is expressly forbidden. Any aerial connected to a fixed monitoring receiver in a competitor's car must be of a non-directional type.

(i) The aerial in each case will be directly connected to the transmitter without the use of non-radiating feeders. The transmitter will not be operated by remote control.

(j) Divulging information or assisting other competitors is forbidden and may lead to diqualification.

(k)Competitors who qualify for the National Final on a Qualifying Event who do not wish to compete in the final must notify the ARDF Committee before the start of the next event. The next person in line on that event will qualify for the National Final.

9. National Organiser

The National Organiser, before the start of any event, shall appoint a person whose main function shall be to ensure that the rules are compiled with at the start of the contest.

In the event of any dispute(s) occurring during the contest the competitor may, at his discretion, refer the matter in writing to the Chairman of the RSGB ARDF Committee no later than 14 days after the event. After consideration by the committee, its decision will be final.

2 m National ARDF Rules

1. The recommended frequency will be 144.725 MHz.
2. All teams must start the hunt from a pre-arranged start location.
3. The FM carriers, either voice modulated or unmodulated, will be for 30 seconds every 5 minutes. A warning will *only* be given 2 minutes before the *first* carrier is due. The second fox will start transmission as soon as the first carrier ends.
4. Clues will *not* be given as this could give unfair advantage to some teams. For the same reason, the fox should *not* transmit outside the prescribed times on *any* frequency.
5. There are no restrictions on polarisation or antenna type. The antenna should not be moved or adjusted after the commencement of the foxhunt. If a beam is used, it should be directed to the central point of the foxhunt area.
6. Transmission power should be a minimum of 2.5 W output and remain constant throughout the event. If a change of power is necessary due to technical problems, the fox will make announcements of such on all subsequent carriers.
7. The transmitters should not be in any location that is private, or requires permission or payment.
8. The team is deemed to be the driver and the passengers in one car. When searching for a portable fox, the team should search together and not spread out, increasing the area of search.
9. Only one set of DF equipment is to be used per team at any one time.
10. Apart from the fox, transmission is forbidden on the fox frequency. When a team has found the fox, they should leave the immediate area, and should not transmit on *any* frequency in the immediate vicinity of the fox.
11. The fox is deemed to be the *transmitter*, not the antenna or operator.
12. The winning team will be the one to find *both* transmitters in the shortest time. It is the prerogative of each team to decide in which order they search for the transmitters.

HF Contest Rules 1998

General Rule

1. These rules apply to all RSGB HF Contests, except where superseded by the specific Contest Rules.

2. UK means England, Scotland, Wales, Northern Ireland, Channel Islands and Isle of Man.

3. Entrants must abide by their licence conditions.

4. Contacts:

a. A contact consists of an exchange with incrementing serial number commencing from 001 and acknowledgement of receipt of callsigns and contest data. Incomplete contacts must be logged with zero points claimed. Points are not lost if a non-competing station does not send appropriate information, but a report *must* be logged and any other exchange sent by that station must be recorded. The full contest exchange must be sent to all stations worked.

b. One contact only with the same station per band counts for points, regardless of that station's operator or callsign. More than one contact with the same operator using different callsigns may not be claimed. Contacts with stations who have no other contest contacts may be disallowed. b. Duplicate contacts must be logged, with zero points claimed.

c. Cross-band contacts do not score.

d. Contacts scheduled before the contest do not count for points. Schedules may only be made during the contest.

e. Simultaneous transmissions on more than one frequency are not permitted.

f. Proof of contact may be required.

g. For contest purposes, /AM and /MM stations are treated as /M stations in their own country. Other stations are regarded as being in the call area / country indicated by their callsign as sent.

5. Multipliers, where applicable, are scored per band, and consist of (a) for UK stations: Countries as per the DXCC countries list, except that JA, W, VE, VO, VK, ZL and ZS call areas count as separate countries. (b) for non-UK stations: one for each UK county (c) IOTA and SSB Field Day contests, see specific rules.

6. Scoring. Where multipliers are applicable the Final Score is the total QSO points for all bands added together, multiplied by the number of multipliers from all bands added together. Where multipliers are not applicable, the Final Score is the total QSO points for all bands plus the total Bonus points (if any) for all bands added together.

7. Portable stations:

(a) entrants must operate from the same site for the whole contest;

(b) stations must not be located in a permanent building or shelter;

(c) no permanent building or structure may be used as an aerial support (trees are acceptable);

(d) power must be obtained solely from on-site batteries, portable generators or solar cells, without use of mains;

(e) All equipment, aerials and supports must be transported and set up on site no more than 24 hours before the start of the contest. This does not apply to short term storage of equipment on site.

8. All operators of UK stations must be RSGB members except visiting amateurs, not resident in the UK. UK stations may not use special (eg GB, GX etc) callsigns nor be /MM or /AM.

a. A single-operator station is operated by one person, who receives no assistance whatsoever from any other person in operating, log-keeping, checking and so on, and who does not receive notification from others by radio (including packet), telephone or any other method, of band or contest information during the contest.

b. Multi-operator entries are those not covered by 8a. One operator must act as Entrant and sign the Summary Sheet.

9. Adjudication.

a. Errors in sending / receiving callsigns are penalised by loss of all points for the QSO. Errors in sending / receiving other data result in loss of one third QSO points per error.

b. Duplicate contacts with non-zero points claimed are penalised by deduction of ten times the QSO points. Excessive numbers of such contacts may attract other penalties, including disqualification.

c. Points may be deducted or entries disqualified or excluded for any breach of the rules or spirit of the contest. The decision of the RSGB is final.

10. Entries must be sent to **RSGB - G3UFY, 77 Bensham Manor Road, Thornton Heath, Surrey CR7 7AF, England** and postmarked no more than 16 days after the end of the contest, unless superseded by specific contest rules. Checklogs are welcome where an entrant does not wish to make a formal entry. Acknowledgement will be sent if a stamped, addressed postcard or IRC is enclosed. Logs become the property of the RSGB. **Entries consist of:**

A Summary Sheet (RSGB form HFC2 or equivalent) showing: Contest; Date; Final Score; Station Callsign and address; Name of Club or Group (if applicable); Exchange (County Code) sent; Entrant's Name, Address and Callsign; Equipment and Antennas (and height) used for each band; Output Power; Callsigns of all operators and a Signed Declaration, **plus either:**

Logs on Computer Disk:

a. All files must be on an MS-DOS formatted 3.5in disk.

b. The disk label must indicate the contest name and the name of the log files(s) in the form of (callsign).LOG, eg G9XXX.LOG or G9XXX-P.LOG, (for portable stations).

c. Acceptable formats are CT .Bin, NA .LOG, Super Duper .LOG, G3WGV .LOG, TR .DAT and RSGB standard format for disk logs.

or

Logs on paper:

a. UK stations must use log sheets in RSGB format. Sample forms are printed in the RSGB *Call Book*. Others may use their own National Society's format.

b. Separate logs, with separate page numbers, for each band.

c. Log sheets must be headed with Name of Contest, Date, Band, Callsign and Page x of n.

d. Log pages should contain 40 QSOs, with columns as follows: Time, Callsign worked, RS(T) / serial sent, RS(T) / serial received, Other Data (specific to the contest), New bonus / multiplier, QSO points. Any RS(T) column left blank will be taken as 59(9).

e. A list of multipliers / bonuses for each band.

f. A Duplicate Sheet for each band. This comprises a list of all callsigns worked, sorted into alphabetical order (or alphabetical order of suffix) together with the serial number sent to that station, or the time of the QSO.

E-mail logs:

Logs for the AFS CW, AFS SSB, 7MHz DX, National Field Day, and IARU Region 1 SSB Field Day contests *only* may be E-mailed to **hf.contest.logs@rsgb.org.uk** The .LOG and .SUM (a text file of the summary sheet including declaration) files *only* should be E-mailed.

11. Receiving Contests. The above rules apply, but also:

(a) Only SWLs or holders of licences to transmit *only above* 30MHz may enter.

(b) Entrants should use RSGB SWL Contest forms if possible. The Callsigns of both the 'station heard' (for which points are to be claimed) and the 'station being worked' must be logged.

(c) The same callsign may appear only once in any group of three consecutive entries in the 'Station being worked' column.

(d) The Summary Sheet declaration to include: "I do not hold a licence to transmit on frequencies below 30MHz."

Affiliated Societies Team Contest

This popular club event has something for everyone. What better way to start contesting than in AFS. *You* can contribute by just participating and gathering points for your team and Club. Enjoy the local rivalry, it's great fun! You don't need to have a high or complex antenna either, a simple dipole or doublet works well.

Date	Time UTC	Freq	Mode	Exchange
Sun 11 Jan	1400-1800	3510-3590	CW	RS(T) + serial number
Sat 17 Jan	1400-1800	3600-3750	SSB	

1. Eligible Entrants: (a) Each entering club must be affiliated to the RSGB. (b) Each operator of a team station must be a member of the club they represent. The operator is not required to be a member of RSGB. (c) All stations representing a club must be located within a radius of 50 miles of the normal meeting-place of the club. Where a club has 'branches', eg RNARS, it may define separate 'branch' meeting-places, and the team(s) entered by each branch will be considered to be entirely separate from those entered by other branches, except in respect of affiliation. (d) Each station may be single or multi-operator, but no station or operator may represent more than one affiliated club or branch.

2. Teams: Teams comprise of up to *five* stations for the CW section and *three* for the SSB section. A club may enter as many teams as they wish. Which stations make up each team is determined by the club entering the event, as defined on the summary sheet.

3. Contacts: In the CW section, 3570 to 3590kHz is reserved for slower-speed contacts. It is intended that operators less experienced in CW and contest techniques should be able to make contacts here in a more relaxed environment. Experienced contesters using the segment are required to keep their speed down.

4. Scoring: 10 points per contact including overseas.

5. Entries: (a) Entries must be accompanied by a Summary Sheet signed by an officer of the affiliated society, showing: name of team, callsign of each station in each team, individual scores, team score, the normal meeting place of the club / branch and a declaration that each operator is a member of the affiliated club. Each log within the entry should include a completed Summary Sheet.

6. Awards: Certificates of Merit will to the three leading teams, individual stations and the highest placed Scottish team and individual. **CW:** The Edgware Trophy to the leading team. The Marconi Trophy to the leading individual station. A particular operator will be eligible for the trophy only once in any period of five years; if the leader is not eligible they will receive a certificate of merit, the trophy passing on to the next highest scoring entrant who is eligible. **SSB:** The Flight Refuelling ARS Trophy to the leading team. The RSGB Lichfield Trophy to the leading individual station.

'LF' Cumulative Contests

This series of short contests, each of just two hours duration, will enable you to sharpen your operating skills and develop antenna systems for the three lowest frequency bands. As the events count towards the HF Contests Championship, you will find a good mix of experienced and less-experienced operators taking part.

Date	Time UTC	Freq	Mode	Exchange
Tue 13 Jan Wed 21 Jan Thu 29 Jan	2000-2200	1830-1870 & 1950-1960	CW	RST + serial number commencing with 001 for each session and band
Sun 4 Jan Sat 10 Jan Sun 1 Feb	1600-1800	3530-3580	CW	
Sat 3 Jan Sun 18 Jan Sat 31 Jan	1000-1200	7015-7040	CW	

1: The contest is single or multi operator. Entrants should endeavour to minimise interference caused to SSB users operating above 1840kHz. There is a speed limit of 12WPM maximum in the sub-bands 3560 - 3580 and 1950 - 1960kHz.

2. Scoring: 3 points per contact with any station in each session, except contacts with Novices score 20 points. The final score for each contest is the sum of the best two sessions on that band, as chosen by the entrant.

3. Logs: One cover sheet is required for each band. Entrants should submit logs for every session that they are active to assist in cross checking against other entries.

4. Awards: The 1989 HF Contests Committee Trophy and a Certificate of Merit to the entrant with the highest aggregate score from all three contests combined. Certificates of Merit will be awarded to the leading station in each contest, band leaders and the highest placed Novice station entrant and station licensed during 1997 or 1998. The contest counts towards the HF Contests Championship.

1.8MHz CW Contests

Competitive and closely-fought events. Long haul DX is available and there will be other European 1.8MHz contests running at the same time, increasing activity and interest. A challenging band for antennas and receiving skills.

Date	Time UTC	Freq	Mode	Exchange
1st 7 / 8 Feb	2100-0100	1820-1870	CW	RST + serial number &
2nd 14 / 15 Nov	2100-0100	1820-1870	CW	district code

1. Sections: Single-operator entries. (a) UK. (b) Overseas including EI.

2. Scoring: Section (a) Three points per contact plus a bonus of five points for the first contact with each UK County worked and the first contact with each Country outside the UK worked. Section (b) Three points per contact plus a bonus of five points for the first contact with each UK County worked. Overseas stations may only work UK stations.

3. Awards: 1st The Somerset Trophy to the leading UK station. **2nd** The Victor Desmond Trophy to the leading UK station (c) Certificates of Merit to the first three entrants in section of each event. (d) The Maitland Trophy and a certificate to the Scottish entrant with the highest aggregate number of points in the 1st and 2nd events.

7MHz DX contest

A chance for you to experiment with simple or complex antenna systems and to gain knowledge of propagation during daylight and darkness conditions.

Date	Time UTC	Frequency	Mode	Exchange
21 / 22 Feb	1500-0900	7000-7030	CW	RST + serial number & district code (UK)

1. Eligible entrants: UK and Overseas (including EI). Single and Multi operator entries will be accepted.

2. Sections: (a) UK Open (b) UK Restricted (c) Europe including EI (d) North America, South America, Africa, Asia (e) Oceania. Open section has no antenna limitations. In the Restricted section, only one antenna is allowed which must be a single element with a maximum height of 15m, and a maximum of 100W output.

3. Scoring: UK stations contact only overseas stations. Contacts with stations in section (c) score 5 points, in section (d) 15 points and in section (e) 30 points. Multipliers as per General Rules. **Overseas stations** contact only UK stations. Stations in section (c) score 5 points, section (d) 15 points and section (e) 30 points. **Multipliers:** 1 for each UK County worked. The final score is the total of contact points times the number of Multipliers worked.

4. Closing Date for logs: 30 March 1998.

5. Awards: Single-operator: The Thomas (G6QB) Memorial Trophy to the leading UK station. Certificates of Merit to the leading Open and Restricted section stations, and to the leading entrants in each overseas section. Multi-operator: Certificates of Merit to the leading groups in each section.

61st Commonwealth Contest

The Commonwealth Contest promotes contacts between stations in the Commonwealth and Mandated Territories. A more relaxed contest environment which gives you the opportunity to work some choice DX.

Date	Time UTC	Bands	Mode	Exchange
14 / 15 Mar	1200-1200	3.5, 7, 14, 21, 28MHz	CW	RST + serial number

1. Eligible entrants: UK entrants must be members of the RSGB. Overseas - Licensed radio amateurs within the Commonwealth or British Mandated Territories. Single operator. Entrants may not receive any assistance whatsoever during the contest, including the use of spotting nets, packet clusters or other assistance in finding new bonuses. Headquarters stations, GB or other UK special event callsigns and maritime / aeronautical mobile are not eligible.

2. Sections: (a) Open, no limit on operating time (b) Restricted, operation is limited to 12 operating hours. Off periods must be clearly marked and a minimum of 60 minutes in length. In addition, at least four operating hours must take place after 0000UTC on 15 March.

3. Frequencies: Entrants should operate in the lower 30kHz of each band, except when contacting Novice stations operating above 21030 and 28030kHz.

4. Scoring: Contacts may be made with any station using a British Commonwealth prefix except those within the entrant's own call area. Note that for this contest, the entire UK counts as *one* call area, and therefore UK stations may not work each other. Each contact scores 5 points with a bonus of 20 points for each of the first three contacts with each Commonwealth Call Area on each band.

5. 'Headquarters' stations: A number of Commonwealth Society HQ stations are expected to be active during the contest and will send 'HQ' after their serial number, to identify themselves. Every HQ station counts as an additional call area and entrants may contact any HQ station for points and bonuses.

6. Logs: Separate logs and lists of bonuses claimed are required for each band. Each entry must be accompanied by a summary sheet indicating the section entered and the scores claimed on each band

7. Closing date for logs: Logs must be postmarked no later than 7 April 1998.

8. Awards: (a) **Open:** The Senior Rose Bowl to the overall leader, and the Junior Rose Bowl to the leader of the Restricted Section. The Col Thomas Rose Bowl to the highest-placed UK station. Certificates of Merit to the first three entrants overall, and to the leading station in each Call Area. (b) **Restricted:** Certificates of Merit to the leading three and to the leading stations in each Call Area. (c) A **Commonwealth Medal** will be awarded to the entrant in either section who in the opinion of the HF Contests Committee has most improved their score or contributed to the contest over the years. (d) Special certificates will be awarded to every entrant in each section who makes contact with more than 61 Band-Call Areas in the 1998 contest. One certificate per entrant. For example, VP9 worked on three different bands counts as three Band-Call Areas. Entrants are asked to note their claimed Band-Call Area total on the summary sheet.

COMMONWEALTH CONTEST CALL AREAS

3B6 / 7	Agalega and St Brandon
3B8	Mauritius
3B9	Rodriguez Island
3DA	Swaziland
4S	Sri Lanka
5B	Cyprus
5H	Tanzania
5N	Nigeria
5W	Western Samoa
5X	Uganda
5Z	Kenya
6Y	Jamaica
7P	Lesotho
7Q	Malawi
8P	Barbados
8Q	Maldives
8R	Guyana
9G	Ghana
9H	Malta
9J	Zambia
9L	Sierra Leone
9M0	Spratly Islands
9M2	West Malaysia
9M6 / 9M8	East Malaysia
9V	Singapore
9Y	Trinidad and Tobago
A2	Botswana
A3	Kingdom of Tonga
AP	Pakistan
C2	Nauru
C5	Gambia
C6	Bahamas
C9	Mozambique
CY0	Sable Island
CY9	St Paul Island
G, M, 2 (all prefixes)	United Kingdom (all one area)
H4	Solomon Is
J3	Grenada
J6	St Lucia
J7	Dominica
J8	St Vincent
P2	Papua New Guinea
S2	Bangladesh
S7	Seychelles
T2	Tuvalu
T30	West Kiribati
T31	Central Kiribati
T32	East Kiribati
T33	Banaba Island
TJ	Cameroon
V2	Antigua and Barbuda
V3	Belize
V4	St Kitts and Nevis
V5	Namibia
V8	Brunei
VE1	Nova Scotia
VE2	Quebec
VE3	Ontario
VE4	Manitoba
VE5	Saskatchewan
VE6	Alberta
VE7	British Columbia
VE8	North West Territories
VE9	New Brunswick
VK0	Heard Island
VK0	Macquarie Island
VK1	Australian Capital Territory
VK2	New South Wales
VK3	Victoria
VK4	Queensland
VK5	South Australia
VK6	Western Australia
VK7	Tasmania
VK8	Northern Territory
VK9C	Cocos (Keeling) Islands
VK9L	Lord Howe Island
VK9M	Mellish Reef
VK9N	Norfolk Island
VK9W	Willis Island
VK9X	Christmas Island
VO1	Newfoundland
VO2	Labrador
VP2E	Anguilla
VP2M	Montserrat
VP2V	British Virgin Islands
VP5	Turks and Caicos
VP8	Antarctica (together with VK0, ZL5)
VP8	Falkland Islands
VP8	South Georgia
VP8	South Orkney Islands

Prefix	Country
VP8	South Sandwich Islands
VP8	South Shetland Islands
VP9	Bermuda
VQ9	Chagos
VR6	Pitcairn Island
VU	India
VU4	Andaman and Nicobar Islands
VU7	Laccadives
VY1	Yukon
VY2	Prince Edward Island
YJ	Vanuatu
Z2	Zimbabwe
ZB2	Gibraltar
ZC4	Cyprus (Sovereign Bases)
ZD7	St Helena
ZD8	Ascension Island
ZD9	Tristan da Cunha, Gough Island
ZF	Cayman Islands
ZK1	North Cook Islands
ZK1	South Cook Islands
ZK2	Niue
ZK3	Tokelau
ZL0 or /ZL	New Zealand reciprocal calls
ZL1	New Zealand
ZL2	New Zealand
ZL3	New Zealand
ZL4	New Zealand
ZL7	Chatham Island
ZL8	Kermadec Islands
ZL9	Auckland and Campbell Islands
ZS1	Western Cape Province
ZS2	Eastern Cape Province
ZS3	Northern Cape Province
ZS4	Orange Free State
ZS5	Kwa Zulu / Natal
ZS6	Transvaal
ZS8	Marion and Prince Edward Islands
GB5CC	RSGB HQ Station + various other Commonwealth HQ Stations.

Ropoco Contests

A real test of your CW operating skill. Chinese whispers using Rotating Postcodes!

Date	Time UTC	Frequency	Mode	Exchange
1 Sun 5 Apr	0700-0900	3520-3570	CW	RST + Postcode
2 Sun 2 Aug	0700-0900	3520-3570	CW	Rcvd

1. Exchange: RST plus for the first contact, the entrant's own postcode. For each subsequent contact, the postcode received from the previous contact.

2. Scoring: Ten points for each contact with another UK station.

3. Awards: Trophy and certificate to the highest scoring entrant with the most accurate log; in RoPoCo 1 the Verulam Silver Jubilee Trophy and in RoPoCo 2 the G3XTJ Memorial Trophy. The G5MY Trophy to the entrant with the highest aggregate score from both events. Certificates of Merit to the leading entrants in both contests.

Slow Speed Cumulative Contests

The aim of these events is to provide training and encouragement for those less experienced in CW and contesting. It is intended primarily for Novices and newly-licensed operators; more experienced contesters are asked to support the event by inviting an entrant to guest-operate their station.

Date	Time UTC	Freq	Mode	Exchange
Mon 30 Mar Tue 7 Apr Wed 15 Apr Thu 23 Apr Fri 1 May	1900-2030	3540-3580	CW	RST + First Name
Mon 31 Aug Tue 8 Sep Wed 16 Sep Thu 24 Sep Fri 2 Oct	1900-2030	3540-3580	CW	RST + First Name

1. Sections: (a) Transmitting, single or multi operator. No limit on the number of operators in a team, nor need they be the same for each session. (b) Receiving, single operator only.

2. Speed Limit: No faster than 12WPM.

3. Exchange: RST and First Name. Multi-operator stations must send only one name during any particular session, regardless of who is operating, although different names may be used during different sessions.

4. Maximum Power: 3W RF output for Novices, 10W RF output for Full licensees.

5. Scoring: Section (a) Any station may be worked once during each session. Any contact with a Novice callsign at either or both ends scores 20 points. Contacts between two Full licence-holders score 5 points. The overall score is the total of the best three sessions, as chosen by the entrant. Section (b) Listeners may log only stations actively participating in the contest. Each Novice logged scores 20 points, each Full callsign counts 5 points.

6. Logs: Entrants are requested to submit logs for all sessions during which they are active to assist with checking other entries. The name of the operator worked / heard should be recorded in column 5.

7. Awards: Section a: Certificates of Merit to the leading Novice and Full licence-holder, and also to the highest placed station entering any RSGB HF CW Contest for the first time (please note on your Cover Sheet if you qualify for this last award). Section b: Certificate of Merit to the leading listener.

National Field Day

An excellent club activity with varied areas of expertise required, such as antenna design, construction and erection, generator maintenance and, increasingly, computer expertise. Give your CW operators some support!

Date	Time UTC	Bands	Mode	Exchange
6 / 7 Jun	1500-1500	1.8, 3.5, 7, 14, 21, 28MHz	CW	RST + Serial Number

1. Notification: Each group intending to compete must send details of the site to be used to: D J Lawley, G4BUO, Carramore, Coldharbour Road, Penshurst, Tonbridge, Kent TN11 8EX, to arrive no later than 11 May 1998. Details must include the name and address of the person responsible for the entry; section to be entered; name of group; callsign(s) to be used; national grid reference and sufficient access information for an inspector to locate the site. Contest stationery will be sent on request. Entries to be postmarked no later than Monday 22 June 1998.

2. Sections: All sections are multi-operator. Maximum of 100W output power. This is a Portable Contest as defined in General Rule 7. (a) Open. One transmitter and one receiver (or one transceiver) plus a second receiver. There is no restriction on the number or type of antennas, but the maximum height must not exceed 20m. Power is limited to 100W output. (b) Restricted. One transmitter and one receiver (or one transceiver) plus a second receiver. One antenna only which must be a single element having not more than two elevated supports and not exceeding 11m above ground at its highest point. (c) Low Power. Same equipment and aerial limitations as the restricted section. Power is further restricted to 5W output.

Notes: (i) A transceiver with a second receiver, eg FT-1000, counts as two receivers. (ii) Stand-by equipment is allowed on site, but may not be connected to a power source when the main equipment is in use. (iii) All stations are subject to inspection by representatives of the HF Contests Committee, whose brief will be to ensure that the rules and spirit of the contest are being observed. Should the inspector be unable to locate the site to due inadequate or incorrect information, the entry may be disallowed. In the event of a late change of site, it is the responsibility of the members of the group to make suitable arrangements for the inspector to find the new site. The inspector must be given immediate access to all parts of the site with the right to stay as long as desired, and the ability to return at any time during the contest. The inspector may also visit in the 24 hours before the start of the contest. The presence on site of any amplifier or modified commercial equipment capable of excess power may result in the entry being disallowed, and in the event of such an infringement being proven, all operators listed as being associated with the group in operating the station may be disbarred by the HF Contests Committee from entering any RSGB HF contest for five years.

3. Frequencies: Contest preferred segments, as recommended by the IARU, should be used, ie 3510 - 3560 and 14010 - 14070kHz.

4. Scoring: For contacts with:

Fixed stations in Europe (including UK) .... 2 points.

Fixed stations outside Europe .... 3 points.

Portable and Mobile stations in Europe (including UK) .... 4 points.

Portable and Mobile stations outside Europe . 6 points.

Contacts on 1.8MHz and 28MHz should be scored as above and then multiplied by two to obtain the band score. Points must not be claimed for contacts made by a competing station with members of its own group.

5. Awards: The National Field Day Trophy to the overall leading station. The Bristol Trophy to the station having the leading score in the other section. The Scottish Trophy to the leading Scottish station. The Gravesend Trophy to the runner-up in the Restricted section. The G6ZR Memorial Trophy to the runner-up in the Open section. The Frank Hoosen G3YF Trophy to the leading station on the 14MHz band. Certificates of Merit will be awarded to the first three stations and to the band leaders in each section. A Certificate of Merit to the overseas station in each continent whose checklog shows the most contacts contributed to competitors.

Low Power Contest

This a serious event for the QRPer, providing a choice of fixed station operation or outdoor fun. We even give you a lunch break!

Date	Time UTC	Frequencies	Mode	Exchange
Sun 12 Jul	0900-1200 & 1300-1600	3510-3580 & 7000-7040	CW	RST + Serial No + Power + District Code

1. Sections: Single or multi operator (a) Fixed (b) Portable, both sections 10W RF output maximum. (c) Fixed (d) Portable, both sections 3W RF output maximum.

2. Frequencies: Both bands may be used during each session. Any station may be contacted once on each band.

3. Special conditions for Portable sections: (i) A Portable station is defined in General Rule 7. (ii) Antennas must not exceed 11m above ground and may have no more than two elevated supports.

4. Exchange: RST, serial number, district code and RF output power in Watts. Serial numbers commence at 001 and continue through both sessions. Output power should be expressed as one or two digits plus 'W' in place of the decimal point, eg 1W, 1W5. Participants using more than 10W should send 'QRO'.

5. Scoring: 15 points for each contact with a QRP Portable or Mobile station; 10 points for a QRP Fixed station; 5 points for all other contacts. For the purposes of scoring, 'QRP stations' are those using 10W RF output or less.

6. Awards: The 1930 Committee Cup to the winner of section (a). The Houston-Fergus and Southgate Trophies to the winners of sections (b) and (d) respectively. Certificates of Merit to the first three entrants in each section.

Islands On The Air Contest

The object of the contest is to promote contacts between stations in qualifying Island groups and the rest of the world and to encourage expeditions to IOTA Islands. Many IOTA Islands are very accessible and it is relatively easy for small groups of amateurs or individuals to mount Island expeditions for the contest. A special interest for UK stations is that mainland Britain counts as an Island (EU-005) and GI / EI count as EU-115.

Date	Time UTC	Bands	Mode	Exchange
25 / 26 Jul	1200-1200	3.5, 7, 14, 21, and 28MHz	SSB & CW	RS(T) + serial number + IOTA reference

1. General. The aim of the contest is to promote contacts between stations in qualifying IOTA island groups and the rest of the world and to encourage expeditions to IOTA islands.

2. When. 1200 UTC Saturday 25 July to 1200 UTC Sunday 26 July 1998.

3. Bands and Modes. 3.5, 7, 14, 21 and 28MHz, CW and SSB. IARU bandplans should be observed, with CW contacts being made only in the recognised CW ends of the bands. Contest preferred segments should also be observed, ie no operation taking place on 3.56 - 3.6MHz, 3.65 - 3.7MHz, 14.06 - 14.125 and 14.3 - 14.35MHz.

4. Categories.

(a) Single operator. Only one transmitted signal. Use of *PacketCluster* or other assistance during the contest places the entrant in the multi-operator category. CW only, SSB only or mixed-mode. (b) Single operator limited. CW only, SSB only or mixed-mode. Operation is limited to 12 hours. Off periods must be clearly marked and must be a minimum of 60 minutes in length. (c) Multi-operator, mixed mode. A maximum of two transceivers may be used. The second transceiver may be used to call other stations only if the station is a new multiplier. It may not be used to solicit other contacts, eg by calling "CQ" or "QRZ?" NB: this category is open to **Island Stations only**.

5. Sections.

(a) IOTA Island Stations

Stations on an island with an IOTA reference, for example AS-007, EU-005. This section includes the British Isles. Note: mainland G / GM / GW = EU-005, mainland GI / EI = EU-115. Entrants intending to operate from a location whose IOTA status is not clear are advised to confirm validity by reference to the *RSGB IOTA Directory and Yearbook*, available from RSGB headquarters. Please indicate on the entry whether the station is permanent or a contest DXpedition, ie antennas and equipment installed specifically for the contest.

(b) World (listed by continent) Any station in a location which does not have an IOTA reference.

(c) Short Wave Listener See rule 10. The format of the listings will depend on the number of entries received.

6. Exchange. Send RS(T) and serial number starting from 001, plus IOTA reference number if applicable. Do not use separate numbering systems for CW and SSB. Stations may be contacted on both CW and SSB on each band. Entrants in section (a) *must* send their IOTA reference as part of each contact.

7. Scoring.

(a) QSO Points

Each contact with an IOTA island counts 15 points. Other contacts count 5 points, except contacts with the entrant's own country or own IOTA reference, which count 2 points.

(b) Multiplier

The multiplier is the total of different IOTA references contacted on each band on CW, plus the total of different IOTA references contacted on each band on SSB.

(c) Total Score

The score is the total of QSO points on all bands added together, multiplied by the total of multipliers.

8. Logs. Entries are preferred on disk using recognised contest software. Paper logs are also acceptable with separate sheets per band (but not each mode). Single mode entrants who make contacts on the other mode should submit these separately as checklogs. All entries should include a summary of contacts and points per band and mode. A list of IOTA multipliers contacted would help adjudication.

Logs must show: Time, Callsign, RST / serial number / IOTA reference sent, RST / serial number / IOTA reference received, multiplier claimed, and QSO points. Entrants are encouraged to submit cross-check ('dupe') sheets and a multiplier list. Entries must be postmarked 1 September 1998 at the latest, and mailed to ***RSGB IOTA Contest, PO Box 9, Potters Bar, Herts EN6 3RH, England*** or E-Mailed to iota.contest.logs@rsgb.org.uk IOTA stations must state their location, ie island from where they operated, *as well as* their IOTA reference number. Checklogs from non-entrants are welcome.

9. Penalties. Points may be deducted, or entrants disqualified, for violation of the rules or the spirit of the contest. This includes refusal by IOTA island stations to make contacts with their own country when requested. Use of a third party to make contacts on a list or net is also against the spirit and may lead to disqualification. Duplicate contacts must be marked as such with no points claimed. Unmarked duplicates will be penalised at ten times the claimed points, and excessive duplicates may cause disqualification.

10. SWL Contest. Scoring is as for the transmitting contest. Logs must be separate for each band, and show: Time, Callsign of station heard, RST / serial number / IOTA reference sent, callsign of station being worked, multiplier claimed, and QSO points. Under 'callsign of station being worked', there must be at least two other QSOs before a callsign is repeated, or else ten minutes must have elapsed. If both sides of a QSO can be heard, they can be logged separately for points if appropriate.

11. Awards

(a) Certificates will be awarded to leading stations in each category and section, and in each continent, according to entry.

(b) The CDXC **Geoff Watts Memorial Trophy** (non-returnable) will be presented to the entrant, whether single operator or a multi-operator group, in the IOTA Islands stations section (non-DXpedition subsection) with the overall highest score, regardless of mode.

(c) The **IOTA Trophy** (non-returnable) will be presented by the IOTA Committee to the entrant, whether single operator or a multi-operator group in the IOTA Island Stations Section (DXpedition subsection), with the overall highest score, regardless of mode.

(d) The **DX News Sheet Trophy** (retained for one year) will be presented to the British entrant operating from a location in the UK (including GD, GJ and GU) with the highest checked score in the single operator SSB category, (Category A). The winner of the IOTA Trophy will not be eligible for this award.

(e) The **David King, G3PFS, Trophy**, in memory of Geoff Watts (retained for one year), will be presented to the British entrant operating from a location in the UK (including GD, GJ and GU) with the highest checked score in the single operator SSB category, (Category B). The winner of the IOTA Trophy will not be eligible for this award.

NB: A DXpedition is defined as an operation involving foreign or overseas travel and the transportation of a significant amount of equipment. The final responsibility for determining if an operation is a DXpedition or not rests with the HF Contests Committee.

12. Note from the RSGB IOTA Manager.

Amateurs planning to activate an all-time new one for IOTA over the IOTA Contest weekend should, if possible, arrange to commence their operation in the preceding 24 hours to enable the new reference number to be issued before the start of the contest. Once the contest is under way, it will not be possible to issue a new number and, without this, contacts made will not count as island contacts.

IARU Region 1 SSB Field Day

An increasingly popular club activity. Finding the best ratio of contact rate to country multipliers provides an intriguing back drop to weekend outdoors. The more competitive you are, the greater the fun!

Date	Time UTC	Bands	Mode	Exchange
5 / 6 Sep	1300-1300	3.5, 7, 14, 21, 28MHz	SSB	RS + Serial Number

1. Sections: All sections are multi-operator. This is a Portable contest as defined in General Rule 7. Entrants in both sections may keep standby equipment on site, but it may not be connected to a power source or antenna at the same time as the main equipment. (a) Open: Maximum licensed power. Equipment: one transmitter and one receiver (or one transceiver), plus a second receiver. No antenna restrictions. (b) Restricted: One transmitter and one receiver (or one transceiver) plus a second receiver. One antenna only per band, which must be a single element having not more than two elevated supports and not exceeding 15m above ground at its highest point. Maximum of 100W output power.

2. Scoring: For contacts with:

Fixed stations in IARU Region 1 .... 2 points.
Stations outside IARU Region 1 .... 3 points.
/P or /M stations in IARU Region 1 .... 5 points.

IARU Region 1 countries include those in Europe, Africa, USSR, ITU Zone 39 and Mongolia. For a more precise definition refer to the RSGB *Amateur Radio Operating Manual*. Points must not be claimed for contacts made by a competing station with members of its own group.

3. Multiplier: *One* for each DXCC Country worked on each band.

4. Awards: The leading station in the Open section will receive the Northumbria Trophy and in the Restricted section, the G3PSH Memorial Trophy. Certificates of Merit will be awarded to the first three stations and to the band leaders in each section. A Certificate of Merit to the overseas station in each continent whose checklog shows the most contacts contributed to competitors.

21 / 28MHz Contests

At the minimum of the sunspot cycle, this contest is a challenge to exploit the often short propagation openings on these two bands. We have now added a Restricted section to enable less complex antenna systems to be used.

Date	Time UTC	Frequencies	Mode	Exchange
Sun 4 Oct	0700-1900	21150-21350 & 28450-29000	SSB	RS + Serial Number + District Code (UK)
Sun 18 Oct	0700-1900	21000-21150 & 28000-28100	CW	RST+ Serial Number + District Code (UK)

1. Sections: (a) UK Open (b) UK Restricted (c) UK QRP (d) Overseas Open (e) Overseas Restricted (f) Overseas QRP (g) UK Receiving (h) Overseas Receiving. QRP stations must use 10W RF output maximum. Open section has no antenna limitations. In the Restricted section, only one antenna is allowed, which must be a single element with a maximum height of 15m, and a maximum of 100W output. Single or Multi operator entries accepted in the transmitting sections. Entrants are reminded that stations using packet or other spotting facilities must enter as multi-operator stations.

2. Frequencies: CW: Entrants are requested not to operate in the sub-band 21.075 - 21.125MHz.

3. Scoring: The same station may be contacted on both bands for points and multipliers (a) UK. 3 points per contact with Overseas stations. Multipliers as per General Rules. (b) Overseas. 3 points per contact with UK only stations. 1 Multipliers for each UK County worked on each band.

4. Closing date for logs: to be postmarked by 16 November 1998.

5. Awards: Certificates of Merit to the first three in each section, overall and on each band, also to the highest placed multi operator entries from UK and Overseas. CW: T E Wilson, G6VQ, Trophy to UK single operator overall winner. SSB: the Whitworth Trophy to the UK single operator overall winner. The Powditch Transmitting Trophy to the leading single operator entry on 28MHz.

Receiving Section

Single-operator entries only will be accepted. General Rule 11 and transmitting section rules apply except where specified below.

1. Scoring: UK SWLs log only Overseas stations in contact with UK stations participating in the contest. Overseas SWLs log only UK stations in contact with Overseas stations participating in the contest. Scoring and multipliers as for the transmitting section.

2. Logs: Columns to be headed: time UTC; callsign of station heard; report / serial number sent by that station; County Code sent by that station (if applicable); callsign of station being worked; multiplier (if new); points claimed. **Note**: in the column headed 'station being worked' the same callsign may only appear once in every three contacts except when the logged station counts as a new multiplier.

3. Awards. SSB: the Metcalf Trophy to the overall leading UK entrant. The Powditch Receiving Trophy to the leading 28MHz entrant. Certificates of Merit to the leading entrants from each section and to the leading entrant from each Overseas country.

Club Calls Contest

The aim of this event is to encourage contacts between Affiliated Societies, to give their callsigns an airing and to encourage Class B licensees to operate under appropriate supervision.

Date	Time UTC	Frequencies	Mode	Exchange
Sat 7 Nov	2000-2300	1870-1990	SSB & CW	RS(T) + serial number + other data

1. Eligible Entrants: All licensed amateurs and SWLs in UK. Multi-operator entries accepted in the Transmitting Contest.

2. Frequencies / Mode: CW operation to centre about 1955kHz to encourage QSOs with Novices. Entrants should avoid the JA DX window and should take care to avoid causing unnecessary QRM to non-contest users of the band.

3. Exchange: RS(T) + serial number + name of Club + 'Club Station', 'Club Member' or 'No Club', as appropriate. NB: the name of the club may only be reduced to initials for CW contacts, otherwise it must be given in full. A Club Station *must* use a callsign which is specifically issued to a Club or Society which is affiliated to the RSGB.

4. Scoring: 3 points per contact, plus bonuses of 5 points contacts with any club members, and 25 points for contacts with any Club Station. Each station may only be worked once regardless of mode.

5. Awards: The Ariel Trophy to the leading Club station. Certificates to the leading club, individual club and non-club member. The David Hill, G4IQM, Memorial Trophy will be awarded to the club having the highest aggregate score of five club members (within a 30 mile radius of club HQ).

RECEIVING CONTEST

General Rule 11 and transmitting section rules apply except where specified below.

1. Log column 'Other Data' to show name of Club + 'Member', or 'No Club', or name of Club + 'Club Station' as appropriate. Any station may appear only once in the 'station heard' column, regardless of mode.

2. A certificate will be awarded to the leading entrant.

The RSGB VHF Contests Committee Internet site is at: http://www.blacksheep.org/vhfcc

RSGB HF Contests Calendar 1998

Date	Contest	Mode
3 Jan	LF Cumulative	CW
4 Jan	LF Cumulative	CW
10 Jan	LF Cumulative	CW
11 Jan	AFS	CW
13 Jan	LF Cumulative	CW
17 Jan	AFS	SSB
18 Jan	LF Cumulative	CW
21 Jan	LF Cumulative	CW
29 Jan	LF Cumulative	CW
31 Jan	LF Cumulative	CW
1 Feb	LF Cumulative	CW
7 / 8 Feb	1.8MHz	CW
21 / 22 Feb	7MHz DX	CW
15 / 15 Mar	Commonwealth	CW
30 Mar	Slow Speed Cumulative	CW
5 Apr	RoPoCo 1	CW
7 Apr	Slow Speed Cumulative	CW
15 Apr	Slow Speed Cumulative	CW
23 Apr	Slow Speed Cumulative	CW
1 May	Slow Speed Cumulative	CW
6 / 7 Jun	National Field Day	CW
12 Jul	Low Power Contest	CW
25 / 26 Jul	Islands on the Air (IOTA)	SSB / CW
2 Aug	RoPoCo 2	CW
31 Aug	Slow Speed Cumulative	CW
5 / 6 Sep	SSB Field Day (IARU Reg 1)	SSB
8 Sep	Slow Speed Cumulative	CW
16 Sep	Slow Speed Cumulative	CW
24 Sep	Slow Speed Cumulative	CW
2 Oct	Slow Speed Cumulative	CW
4 Oct	21 / 28MHz	SSB
18 Oct	21 / 28MHz	CW
7 Nov	Club Calls	SSB / CW
14 / 15 Nov	1.8MHz	CW

HF Contests Championship

Every UK single operator station entering two or more of the events listed will automatically be entered for the Championship. For each event, the entrant will be awarded points according their score, expressed as a percentage of the score achieved by the leading UK station in that event. These points will then be multiplied by the appropriate factor for the contest. The winner will be the station with the highest number of points at the end of the year and will be awarded the G2QT Trophy. Certificates of Merit will be awarded to the winner and second placed station.

Event	Factor
LF Cumulatives	20
1st 1.8MHz	10
7 MHz DX	20
Commonwealth	30
RoPoCo 1	10
IOTA	30
RoPoCo 2	10
21 / 28MHz SSB	20
21 / 28MHz CW	20
2nd 1.8MHz	10

Equipment Coding System

This has been designed to give an easily identifiable indication of a station's power and antenna system and will be used in Contest reports.

First character	**power**
0	0 - 1W
1	1.1 - 5W
2	6 - 20W
3	21 - 100W
4	101 - 400W
Second character	**antenna**
C	Centre-fed (dipole, doublet, G5RV etc)
G	Ground Plane or Vertical
Y	Yagi
Q	Quad or Loop
W	Wire (any other type)
Third character	**number of antenna elements**
Fourth character	**max height of antenna above ground**
0	0 - 9ft
1	10 - 19ft
2	20 - 29ft
8	80 - 89ft
9	90 plus ft

General

RSGB Standard for Contest Log Data on Computer Disk

1. All files must be in standard ASCII format (ie no tabs or other control characters).

2. All files must be on an MS-DOS formatted 3.5in disk.

3. The diskette label must clearly indicated contest name and the name of the log file(s).

4. The log file must consist of one logical line of data per QSO. Each contact line *must* be terminated with a carriage return character.

5. The QSO data defined below must appear in each line, except that a hyphen in any field, or a field which is all blanks (spaces) will be taken as indicating a data item which is the same as in the previous contact. Each field except the last *must* be padded out to the correct length with blank characters and neighbouring fields *must* be separated by a blank character. Exact adherence to the start and finish columns given below is not mandatory but all data must be column-aligned within the specified field limits, eg: every callsign must start in the same character column and all must fit between character columns 22 and 36. Characters:

1 - 6 Date in YYMMDD format

8 - 11 Time in HHMM format

13 - 16 Band in MHz (embedded periods are allowed, eg 1.8)

18 - 20 Mode (A1A, J3E, F3E etc)

22 - 36 Callsign (left aligned)

38 - 40 RS(T) sent

42 - 45 Serial number / Power / Zone / State sent (contest-dependent)

47 - 49 RS(T) received

51 - 54 Serial number / Power / Zone / State received (contest-dependent)

56 - 59 New Bonus / Multiplier (country prefix / county code etc)

61 - 64 Points (Note: for VHF contests only, character 60 may be used if the points exceed 9999)

66 - 71 Station callsign / Operator callsign for multi-op events

73 - 128 HF CONTESTS: Further contest specific data, eg postcode, District Code or QTH locator received. THIS FIELD MUST BE TERMINATED BY <CR>. Zero point QSOs such as Duplicates, unfinished contacts etc may be explained here. This field should be left justified and may contain spaces but not tabs.

73 - 78 VHF CONTESTS: IARU Locator.

80 - 80 VHF CONTESTS: 'M' if this locator is a new multiplier (Locator multiplier contests).

82 - 84 VHF CONTESTS: District Code as given in the 'RSGB District Codes for HF and VHF/UHF Contests' table.

86 - 86 VHF CONTESTS: 'M' if this District is a new multiplier (District multiplier contests).

88 - 90 VHF CONTESTS: Country code, eg G, GI, GM, GW, F, DL.

92 - 92 VHF CONTESTS: 'M' if this country is a new multiplier (country multiplier contests).

94 - 256 VHF CONTESTS: Free text for QTH information and / or comments. THIS FIELD MUST BE TERMINATED BY <CR>. Zero point QSOs such as Duplicates, unfinished contacts or poor signal reports etc may be explained here. This field should be left justified and may contain spaces but not tabs.

6. Logs must be submitted as a single contiguous file (in chronological order of contact) for each station. Separate files for each band are NOT required. The log data filename must consist of the call sign and the extension '.LOG', eg: G9XXX.LOG. Entries in contests where more than one callsign is used, eg VHF Field Day or AFS, should contain all the entry's logs on one disk, as separate files. In contests where the same station submits multiple logs on the same band, such as LF Cumulatives, the filename should be suffixed with a distinguishing numeral eg G9XXX1.LOG, G9XXX2.LOG. Only .LOG files should be put on the disk.

7. Standard abbreviations must be used, eg RSGB two-letter District Codes, ITU Country prefixes, US Postal Service two-letter State abbreviations, ARRL sections, etc.

RSGB District Codes for HF and VHF / UHF Contests

Code	District	Code	District	Code	District
AB[1]	Aberdeen	HA	Harrow	PO	Portsmouth
AL	St Albans	HD	Huddersfield	PR	Preston
BM	Birmingham	HG	Harrogate	RG	Reading
BA	Bath	HP	Hemel Hempstead	RH	Redhill
BB	Blackburn	HR	Hereford	RM	Romford
BD	Bradford	HS[1]	Scottish Islands	SD	Sheffield
BH	Bournemouth	HU	Hull	SA	Swansea
BL	Bolton	HX	Halifax	SE	London SE1 - 28
BN	Brighton	IG	Ilford	SG	Stevenage
BR	Bromley	IM	Isle of Man	SK	Stockport
BS	Bristol	IP	Ipswich	SL	Slough
BT[2]	Belfast	IV[1]	Inverness	SM	Sutton
CA	Carlisle	JE	Jersey	SN	Swindon
CB	Cambridge	KA[1]	Kilmarnock	SO	Southampton
CF	Cardiff	KT	Kingston upon Thames	SP	Salisbury
CH	Chester	KW[1]	Orkney	SR	Sunderland
CM	Chelmsford	KY[1]	Kirkcaldy	SS	Southend on Sea
CO	Colchester	LP	Liverpool	ST	Stoke on Trent
CR	Croydon	LA	Lancaster	SW	London SW1 - 20
CT	Canterbury	LD	Llandrindod Wells	SY	Shrewsbury
CV	Coventry	LE	Leicester	TA	Taunton
CW	Crewe	LL	Llandudno	TD[1]	Tweed
DA	Dartford	LN	Lincoln	TF	Telford
DD[1]	Dundee	LS	Leeds	TN	Tonbridge
DE	Derby	LU	Luton	TQ	Torquay
DG[1]	Dumfries	MR	Manchester	TR	Truro
DH	Durham	ME	Medway	TS	Teeside
DL	Darlington	MK	Milton Keynes	TW	Twickenham
DN	Doncaster	ML[1]	Motherwell	UB	Uxbridge
DT	Dorchester	NL	London N1 - 22	WL	London W1 - 14
DY	Dudley	NE	Newcastle upon Tyne	WA	Warrington
EL	London E1 - 18	NG	Nottingham	WC	London WC1 - 2
EC	London EC1 - 4	NN	Northampton	WD	Watford
EH[1]	Edinburgh	NP	Newport	WF	Wakefield
EN	Enfield	NR	Norwich	WN	Wigan
EX	Exeter	NW	London NW1 - 11	WR	Worcester
FK[1]	Falkirk	OL	Oldham	WS	Walsall
FY	Blackpool	OX	Oxford	WV	Wolverhampton
GS[1]	Glasgow	PA[1]	Paisley	YO	York
GL	Gloucester	PE	Peterborough	ZE[1]	Shetland Isles
GU	Guildford	PH[1]	Perth		
GY	Guernsey	PL	Plymouth		

These district codes are based on the first two letters of the postcode of your operating location. (Single letter postcodes eg B have been padded out to 2 letters eg BM.) The description column is for guidance only.

[1] Can be counted three times as multipliers. In the case of DG and TD, regardless of whether the contact is with a station in England or Scotland multipliers in VHF contests. (See VHF General Rules 7d and 7e.)

[2] BT can be counted six times in VHF Contests. (See VHF General Rules 7d and 7e.)

VHF / UHF / SHF Conest Rules 1998

General Rules

1. Entries

a. In submitting an entry to a contest, you agree to be bound by the rules and spirit of the contest, and you agree that the decision of the RSGB shall be final in cases of dispute.

b. All paper and / or disk entries should be addressed to PO Box 2399, Reading RG7 4FB.

c. Alternatively, entries may be submitted by E-mail to vhf-entry@blacksheep.org

d. Entries should be postmarked or e-mailed not later than 16 days after the end of the contest, or, for cumulative contests, the last activity period.

e. Entries become the property of RSGB and cannot be returned.

f. Proof of contact may be required. Any station may be approached, without notice to the entrant, for confirmation of contact details.

g. In case of dispute, in the first instance, the Chairman of the VHF Contests Committee (VHFCC) should be contacted in writing. The VHFCC may refer cases of appeal to RSGB Council. Council's decision shall be final.

h. In multi-band contests, single band entries are always acceptable.

i. Queries about the contests may be addressed to the VHFCC Chairman, Steve Thompson, G8GSQ, PO Box 2399, Reading RG7 4FB; tel: 0118 982 0848, evenings or weekends (fax at other times) or by E-mail: g8gsq@blacksheep.org

2. Paperwork

a. All paper or disk entries should be accompanied by a VHF / UHF contest cover sheet (form 427) for each band used, or a similar form which supplies the same information. Please include a contact telephone number or E-mail address in case of query.

b. The logs for paper contest entries should be made out on current RSGB VHF / UHF log sheets or a close replica. These forms may be photocopied from the RSGB *Call Book,* or small quantities are available from members of the VHFCC upon receipt of an SAE. Larger quantities may be purchased from RSGB HQ. If computer listings are to be submitted, these should be cut to A4 size, and be in RSGB log format, line spaced to contain 25 contacts per sheet, and be correctly collated (not Z fold). Each sheet should be headed with the entrants callsign, IARU locator, contest title and sheet number. Logs should be tabulated as follows:

i. Date / time (UTC)
ii. Callsign of station worked
iii. My report on his / her signal and serial number
iv. His / her report on my signal and serial number
v. IARU Locator received
vi. QTH or District Code received (when required) or comments
vii. Points claimed.

c. Alternatively, entries are encouraged on floppy disk. The VHFCC guarantees that such entries will receive the same level of scrutiny as similar logs submitted on paper. A paper copy of the log is not required, but a VHF / UHF contest cover sheet (form 427) is required.

i. All files must be on an MS-DOS formatted disk, 3.5in (720kb or 1.44Mb) or 5.25in (360kb or 1.2Mb).

ii. The disk label must clearly indicate the contest name and the name of the log file(s), which must consist of the callsign and the extension .LOG, eg G9XXX.LOG or G9XXX-P.LOG.

iii. The log file must consist of one logical line of data per contact. Acceptable formats are Super Duper .LOG, G0GJV .LOG/.GJV, G3WGV .LOG, N6TR .DAT, REG1TEST and RSGB standard log format for disk logs. Details of the RSGB format may be found on page ix of this supplement. We will endeavour to work with any other reasonable format - please contact the VHFCC to discuss this.

iv. All diskettes become the property of the RSGB, unless you enclose an SAE, whereupon the disk will be returned to you.

d. Entries may also be submitted by E-mail to vhf-entry@blacksheep.org

i. The blacksheep system will send you a receipt when your entry is received.

ii. The same formats and naming conventions for log files listed above in rules 2c (ii and iii) are acceptable.

iii. The title of your mail message should contain the contest for which you are submitting the entry and your callsign.

iv. You must also submit a text file which contains the same information as on a paper VHF / UHF contest cover sheet (form 427). Suitable templates can be found at http://www.blacksheep.org/vhfcc/forms

v. The WWW page http://www.blacksheep.org/vhfcc may also contain details of other means of electronic log submission as they become available.

e. In contests with a multiplier scoring system, when submitting paper logs, please also submit a list of multipliers worked, showing at least the callsign, and either serial number sent or time of QSO, for each contact claimed as a new multiplier.

f. Any adverse comments received or made about signal quality must be recorded in the comments column of the paper log or electronic log.

3. Station / Operators

a. All operators must be RSGB members except in VHF NFD and the Affiliated Society contests - see individual rules.

b. Stations entering a fixed station section or contest must operate from permanent and substantial buildings located at the main station address as shown on the licence validation document. The spirit of the contest will be paramount.

c. Entrants must not change their location or callsign during the contest. In multi-band events, all stations forming one entry must be located within a circle of 1km radius.

d. Stations located outside of the UK (G, GW, GM, GI, GD, GU, GJ) may enter a contest, and will be tabulated within the overall results tables, but will only be eligible for their own awards.

e. Entries will not be accepted from stations using special event callsigns (eg GB), or special club callsigns (eg GX, GS etc.) Normal club callsigns can be used - ie G4DSP is OK, but GX4DSP is not.

f. There must be only one frequency used for transmit on any band at any one time.

g. The lower of the contest power limit or the standard

licence power limit must not be exceeded during the contest. Contacts made under a high-power permit will not count for points. Severe action may be taken against infringements of this rule.

h. Stations which persistently radiate poor quality signals, cause deliberate interference to other stations, or otherwise contravene the code of practice for VHF / UHF / SHF contest operation may be penalised.

i. Entrants must permit inspection of their stations by members of VHFCC or its representatives, and give site access information if requested to do so. The inspector must be permitted to remain for as long as desired, and to return to the site for subsequent inspections at any time during the contest. Contestants must demonstrate to the inspector's satisfaction that they are obeying the rules of the contest.

4. Contacts

a. The contest exchange consists of at least both callsigns, RS(T) signal reports followed by a serial number, and the IARU locator. Particular contests may require additional information to be exchanged as described in the individual contest rules.

b. Serial numbers start from 001 on each band and advance by one for each contact. In cumulative contests serial numbers start from 001 for each activity period.

c. Crossband contacts do not count for points below 2.3 GHz. On 2.3 GHz and above, crossband contacts are scored at 50% of the two way score.

d. No points will be lost if a non-competing station cannot provide an IARU locator, serial number, or any other information that may be required. However, the receiving operator must receive and record sufficient information to be able to calculate the score.

e. Contacts with callsigns appearing as operators on any of the cover sheets forming an entry will not count for points or multipliers. In AFS contests, stations within the same AFS team may work each other for points / multipliers.

f. Only one scoring contact may be made with a given station on each band, regardless of suffix (/P, /M etc) during an individual contact or cumulative activity period. All non-scoring contacts must be clearly marked in the log, and unmarked duplicates will be penalised at ten times the claimed score for that contact.

g. Contacts made using repeaters, satellites or moonbounce will not count for points.

h. The IARU / RSGB band-plans must be observed.

i. All information must be copied off air at the time of the QSO. Databases must not be used to fill in missing information.

j. The DX Cluster may be used in all sections of the contest, but deliberate self-spotting by the entrant or close associates is not permitted.

k. Any band may be used for setting up contacts or talkback. No confirmation of QSO details must take place on the talkback frequency. All exchanges for the contest band in use must be made on that band. The talkback channel can be used for antenna alignment signals and confirmation that signals are audible, but not for giving reports and serial numbers.

l. In contests with a section SS, single operator fixed stations may enter and choose any continuous six-hour period in which to operate (eg 1500 - 2300, or 1917 - 0317). Serial numbers must start at 001 for this period, and you cannot enter both this section and the full contest.

5. Scoring

a. Scoring will normally be at 1 point per km. Contacts with stations in the same locator square as your station will score 3 points.

b. For computer purposes a conversion factor of 111.2 km / degree must be used.

c. Multi-band contests will contain an overall results table in addition to the individual band results. The scores in this final tabulation will be formed by taking the sum of the normalised scores on each band. The normalised scores will be calculated by:

Normalised score for each band / session = Score achieved x 1000, divided by Band / session leader score.

6. Awards and Results

a. Certificates will be awarded to the leading and second-placed station in each section of the contest. Additional Certificates of Merit may be awarded at the adjudicator's discretion.

b. In all contests / sections where the power limit is above 25W, a certificate will be awarded to the leading fixed station using 25W or less into a single antenna.

c. Placement certificates showing the result achieved in the contest can be obtained by including an A4 SAE with the entry marked with callsign, contest and (if applicable) group name.

7. Multipliers

a. Where a contest uses multipliers, the score for each band will be the number of points made on that band multiplied by the number of multipliers contacted on that band.

b. The type of multiplier scheme for a particular contest will be referred to in the individual rules for the contest. Not all contests will use multipliers.

c. Each new multiplier must be clearly marked in the log and a summary sheet provided (see rule 2c.)

d. In contests where District Codes are the multipliers (see rule 7.e. below), each Scottish District Code may be worked up to three times for multiplier credit and BT for Northern Ireland may be worked up to six times for multiplier credit.

e. In co-operation with the HFCC, we have agreed that because of the break-down of the existing counties for 1998, we will replace county multipliers with 'District Code' multipliers. The District Codes are based on the entrant's postcode district. In contests using District Code multipliers, the exchange will include the first two letters of the postcode (eg 'EN' for EN6 3JE). Where a postcode consists of only a single initial letter (eg B6 9AA), the exchange will be 'padded out' to two letters (ie, in this case, 'BM'). A complete list of valid exchanges is included on the previous page. This extended exchange is used to keep a common format, but entrants need to be aware that some non-contestants may not be aware of their extended code.

Special Rules

Certain of these rules are invoked for individual contests as listed in the individual contest rules.

S1. Instead of 1 point / km scoring (rule 5a), scores will be calculated at 1 point / QSO.

S2. In addition to the IARU locator, QTH information must be exchanged. This should be given as a point identifiable on an Ordnance Survey route planning map or equivalent (scale 1:625,000) or as a direction and distance not greater than 25km from such a point, eg 10km West of Skegness.

S3. This is an Affiliated Societies contest and is open to both individual entrants (who must be RSGB members), and to teams made up of a number of operators who must all be members of the same affiliated society, but not necessarily RSGB members themselves.

a. All members of the team must operate from within 50km of the normal meeting place of the society. In the case of national societies, each team must define a separate meeting place, and each team member must operate within 50km of that designated meeting place.

b. No station may represent more than one society.

c. No operator is allowed to use more than one callsign during the contest. QSOs with other members of your team will count for points.

d. Multiple teams are encouraged from both local and national societies. The best three or five scores (determined in individual contest rules) of each team will be used to form an entry, but please submit all logs so that the adjudicator can form teams appropriately after checking of the logs is complete.

e. Please mark your RSGB Zone (which can be found in the *Call Book*) on the cover sheet.

f. Logs should be sent as a single package for each club and should include a declaration signed by a club official that all operators are members of the Affiliated Society, and listing the QTH locator of the normal meeting place of the club.

S4. This contest runs concurrently with a Backpackers contest. Stations entering the Backpackers contest only may be worked once from a fixed location and once from their portable location for points.

S5. This is a cumulative contest. The following special rules apply:

a. For cumulative contests the overall score will normally be calculated from the best 3 normalised session scores - the normalised score being calculated as above in rule 5c. It is impossible for you to determine your best sessions without knowing everyone else's scores, so please submit your logs and scores from all sessions in which you were active and allow the adjudicator to calculate your best sessions.

b. Stations may move location between individual cumulative activity periods.

c. For cumulative contests, please summarise your scores from each session on the reverse of the cover sheet.

S6. This contest runs concurrently with the first few hours of an RSGB 24-hour event. You may submit entries to both contests with a single set of logs, but please include two cover sheets - one for the shorter contest and one for the 24-hour event. Entries may be automatically submitted into the 24-hour event unless you specifically request otherwise.

S7. This contest runs concurrently with all or part of an IARU co-ordinated contest. You may submit a single set of logs for entry to both the RSGB and IARU events. Entries may be submitted to the IARU event unless you specifically request otherwise.

Multiplier Types

One of the following rules as defined in the individual contest rules will apply to any contest using multipliers. In each case, a QSO with your own District Code, country or large square, as appropriate to the contest, counts for multiplier credit, and any appropriate QSO can count as more than one multiplier (eg your first G QSO in an M3 contest will count for a new locator, District Code and country).

M1. District Code and Country Multipliers. The multiplier for a band is the sum of the number of different DXCC countries and UK District Codes worked on that band.

M2. QTH Locator Multiplier. The multiplier for a band is the sum of the number of different large locator squares (eg JO01, IO91 etc) worked on that band.

M3. District Code, Country and QTH Locator Multiplier. The multiplier for a band is the sum of the number of different DXCC countries, UK District Codes, and large QTH locator squares (eg JO01, IO91 etc) worked on that band.

M4. Country and QTH Locator Multiplier. The multiplier for a band is the sum of the number of different DXCC countries, and large QTH locator squares (eg JO01, IO91 etc) worked on that band.

The VHF Contests' Championship

1. The VHF Contests' Championship aims to provide an overall result for the year based on a representative selection of contests. The contests which count towards the championship are:

i. March 2m / 70cm (the overall two band normalised score)
ii. 432MHz Trophy
iii. May 144MHz
iv. 50MHz Trophy
v. 144MHz Low Power
vi. 432MHz Low Power
vii. 144MHz Trophy
viii. 70MHz Trophy
ix. 1.3GHz Trophy and 2.3GHz Trophy (combined score)

2. There is a Single Operator Fixed Station section (SF), a low power section (LP) and a section for All Others (O).

3. The low power section is open to single operator fixed stations who enter any of the above contests using 25W or less output and a single antenna. If a station enters some contests with high power or extra antennas, and some with low power and a single antenna, only the low power, single antenna scores will count towards this award.

4. The overall score is calculated from the sum of the normalised scores for each event listed above. The normalised scores are calculated as in general rule 5c. Low power entrants will be normalised against the best low power score.

5. Stations entering the single operator fixed section of a contest may elect to submit their score towards an All Others score if they wish. In this case their score will be normalised against the leader of the All Others section. Please mark your cover sheet clearly with the name of your contest group / club if you wish to do this.

6. The John Pilags Memorial Trophy is awarded to the winner of section SF, the Racal Radio Cup to the winner of section O, and the Low Power Championship Trophy to the winner of section LP.

Code of Practice for VHF / UHF / SHF Contests

1. Obtain permission from the landowner or agent before using the site and check that this permission includes right of access. Portable stations should observe the Country Code.

2. Take all possible steps to ensure that the site is not going to be used by some other group or club. Check with the club and last year's results table to see if any group used the site last year. If it is going to be used by another group, come to an amicable agreement before the event. Groups are advised to select possible alternative sites.

3. All transmitters generate unwanted signals; it is the level of these signals that matters. In operation from a good site, levels of spurious radiation which may be acceptable from a home station may well be found to be excessive to nearby stations (25 miles away or more).

4. Similarly, all receivers are prone to have spurious responses or to generate spurious signals in the presence of one or more strong signals, even if the incoming signals are of good quality. Such spurious responses may mislead an operator into believing that the incoming signal is at fault, when in fact the fault lies in the receiver.

5. If at all possible, critically test both receiver and transmitter for these undesirable characteristics, preferably by air test with a near neighbour before the contest. In the case of transmitters, aim to keep all in-amateur band spurious radiation, including noise modulation, to a level of -100dB relative to the wanted signal. Similarly, every effort should be made to ensure that the receiver has adequate dynamic range.

6. Above all, be friendly and polite at all times. Be helpful and inform stations apparently radiating unwanted signals at troublesome levels, having first checked your own receiver. Try the effect of turning the antenna or inserting attenuators in the feedline; if the level of spurious signal changes relative to the wanted signal, then non-linear effects are occurring in the receiver. Some synthesised equipment has excessive local oscillator phase-noise, which will manifest itself as an apparent splatter on strong signals, even if there is no overloading of the receiver front-end. Pre-amplifiers should always be switched out to avoid overload problems when checking transmissions. If you receive a complaint, perform tests to check for receiver overload and try reducing drive levels and switching out linear amplifiers to determine a cure. Monitor your own signal off-air if possible. Remember that many linear amplifiers may not be linear at high power levels under field conditions with poorly regulated power supplies. The effects of over-driving will be more severe if speech processing is used, so pay particular attention to drive level adjustment. If asked to close down by a Government Official or the site owner, do so at once and without objectionable behaviour.

VHF / UHF Listeners' Contests

1. Listeners contests are open to all non-licensed members of the RSGB and foreign SWLs. Only one entrant may operate the receiving station. Every VHF contest is open to listeners' entries.

2. Logs must show in columns:

i. Date / time (UTC).

ii. Callsign of station heard.

iii. My report on his / her signals.

iv. Report and serial number sent by station heard.

v. Callsign of station being worked.

vi. IARU Locator given by station heard.

vii. QTH given by station heard (if appropriate).

viii. Points claimed.

On 144 MHz, the callsign in column (v) may occur once in every ten contacts logged. CQ and test calls do not count for points and should not be logged. If both sides of the QSO can be heard, both can be claimed for points.

3. The Hanson Trophy will be awarded to the entrant with the highest aggregate score in all SWL contests between March and September inclusive of each year. The aggregate score will be calculated in accordance with General Rule 5c.

The Backpackers Series of Contests

Aims:

a. To promote the fun of contesting and to develop skills in contesting and operating.

b. To increase access to major contesting events.

c. To encourage low power portable operation with operators working fellow low-power enthusiasts from a variety of hill-top sites within the UK.

d. To introduce the art of contesting to those who, for various reasons are unable / unwilling to form / join contest groups or those who simply do not have the time for 'full-blown' contests.

e. To promote innovation, home construction and an awareness of how equipment actually works, particularly in the development of receivers, transmitters, antennas, pre-amplifiers and feeder systems. It is in the spirit of the contests that the equipment should be capable of being carried to the operating site by the operator(s) or being transported / erected outside a car.

Times:

'Socially-acceptable' *four* hour periods. Timing of the contests should allow participants time to (walk) reach their destination, set-up, operate, clear away and return home with a good margin of daylight. Times will be staggered to co-ordinate with existing contests. For dates and times, see the individual contest rules table.

Modes:

All mode.

Sections:

a) 10W Single Operator Portable. b) 10W Multi-Operator Portable. c) 3W Single Operator Portable. d) 3W Multi-Operator Portable.

The listed power is output from the transmitter. Participants will be expected to demonstrate how their power level was determined, particularly where the basic commercial equipment is rated at higher output power.

Restrictions:

1. All operators must be RSGB members.

2. The contest is open to all stations, but only portable stations may submit a contest entry.

3. No fixed or mobile towers, cranes or any other 'significant structure' (in excess of 2in outside diameter) is to be used as an antenna support. The highest part of any antenna system will be limited to 10m above ground level on a single mast. On 2m the antenna system must not exceed 20 halfwave elements or equivalent.

4. All equipment must be battery powered. If a mains rotator is envisaged, they must also be powered from a single source battery (with suitable converter circuitry) supply not exceeding 28V.

5. Petrol / Gas / Diesel generators for charging are not permitted. This includes a motor vehicle engine. If operating from a vehicle supply, the engine must be switched off for the duration of the contest. Wind and solar power generation and charging is permitted.

6. In addition, the General Rules apply.

Scoring:

This is at 1 point per km (general rule 5a) with a multiplier applied. The multiplier type differs between individual contests in order to match the exchange in the main contest running at the same time - check the individual rules table carefully.

Award:

Each session should be treated as a separate contest. Please submit an entry after each session. Session winners and runner's-up certificates will be awarded. In addition, a certificate will be awarded to the leading station running one watt or less into a single antenna for each session.

On 144MHz, The Backpacker's Trophy will be awarded to the leading stations in either category, the best three placings out of a maximum of five sessions. Scores will be normalised as in general rule 5c. In the event of a tie, if appropriate, the remaining session will be taken into consideration. The 50MHz Backpackers Trophy will be determined from the two sessions.

Recommendation:

If stations intend to enter any of these Backpackers contests, they are requested not to call stations in the major events which run alongside from home before the contest, as they may, in effect, appear to be working the same station twice. This in fact is not the case, as the Backpackers series should be seen as separate, independent, events. However, the reality of the situation is such that stations operating in the major events will effectively register the second, portable, contact as a 'dupe', thereby causing some confusion and delay. Should this happen, the second contact should be corrected and scored at a later time. This anomaly has arisen as a result of attempting to create more activity by co-ordinating two quite different contests simultaneously. Backpacker's participants, in particular, are requested to bear this in mind in order to help both contests to run as smoothly as possible.

VHF National Field Day 1998 Rules

General Rules Apply

1. Site Notification:

Each Group intending to compete must send two copies of a completed site registration form (available in the *Call Book* or from G8GSQ) to: VHF Contests Committee, PO Box 2399, Reading RG7 4FB, to arrive no later than 10 June 1998. Each group may only register one site, although changes can be made provided G8GSQ is informed before the contest; tel: 0118 982 0848 evenings / weekends, or E-mail g8gsq@blacksheep.org

2. Bands:

Up to four separate stations may operate simultaneously on the 70, 144, 432 and 1296MHz bands. 70MHz will be CW only from 1400 - 2200UTC, and phone only from 0600 - 1400UTC, with close down between 2200 and 0600UTC. Each station may be worked once on phone and once on CW on 70MHz. Single band entries for any band are acceptable.

3. Operators:

Any RSGB member or group of members operating from the British Isles (excluding the Irish Republic) may enter. Also, affiliated RSGB societies may enter (operators *must* be members of the Affiliated Society (AFS), but not necessarily members of RSGB themselves). In this case, a declaration signed by an officer of the AFS that the operators are members of the society is required with the entry. RSGB members are allowed to operate in AFS groups whether or not they are actually members of that AFS group.

4. Stations:

All equipment including antennas, must be installed on site not more than 24hrs before the contest. Only portable accommodation can be used to house the stations. Power for all equipment must be derived from an on site generator, battery, wind or solar power.

5. Contest exchanges:

a. On each band report, serial number and locator must be exchanged.

b. Additionally, on 70MHz only, QTH information must be exchanged (special rule S2). It must be given in a different form in the CW and phone sections.

6. Sections:

Restricted section (R)

(i) The height of any part of the antenna's driven element must not exceed 10 metres above ground level.

(ii) Only one antenna per band may be used (ie no stacked, bayed or collinear arrays or switching between two or more antennas). A slot-fed Yagi or quad antenna is permitted. Dish or backfire antennas must not exceed 2m diameter.

Low Power section (L):

(i) The power output of any band must not exceed 25W PEP at the transmitter.

(ii) The height of any antenna must not exceed 10 metres above ground level.

(iii) Only one antenna per band may be used (ie no stacked, bayed or collinear arrays or switching between two or more antennas). A slot fed Yagi or quad antenna is permitted. Dish or backfire antennas must not exceed 2m diameter.

Open section (O): as per general rules.

Mix and Match (MM)

(i) A group can elect to place different bands into either Restricted, Low Power or Open sections, eg 4m in Restricted, 2m and 70cm in Open, and 23cm in Low Power. This decision must be made at registration time and the details shown on the registration form.

(ii) Individual band entry will be tabulated in the appropriate main section, and a normalised score for the band produced on this basis.

(iii) The sum of the normalised scores will appear in a separate Mix and Match section table.

SWL section (S): as per general rules.

7. Inspections:

All stations are subject to inspection by members of the VHF Contests Committee or nominated representatives. Should the inspector be unable to locate the site due to inadequate or incorrect information, the entry may be disallowed. In the event of a last minute site change it is the responsibility of the group to make suitable arrangements for the inspector to find the site. The inspector must be given immediate access to all parts of the site with the right to stay as long as desired, and the ability to return at any time during the contest.

8. Entries:

(a) All entries must be postmarked no later than 31 July 1998.

(b) Entries must be addressed to: VHF Contests Committee, PO Box 2399, Reading RG7 4FB.

(c) Please enclose a 427 cover sheet for each band, including separate ones for the 70MHz SSB and CW sections.

9. Awards:

The Surrey, Martlesham, and Arthur Watts Trophies will be awarded to the overall winners of the Open, Restricted and Low Power sections respectively. The Tartan Trophy will be awarded to the leading resident Scottish entry in the Open section, the Scottish Trophy to the leading Scottish entry in the Low power section, and the Cockenzie Quaiche to the leading resident Scottish station in the Restricted section. Certificates will be awarded to the winners and runners-up on all bands in each section, and to the leading stations in each country.

Summary of RSGB VHF / UHF / SHF Contests

Date	Time UTC	Contest Name	Sections	Notes / Special Rules
4 January	1000 - 1600	144MHz CW	SF, O	District Code / Country Multipliers (M1)
18 / 25 January, 8 / 15 February, 1 March	1000 - 1230	70MHz Cumulatives	SF, O	Full QTH Information to be sent (S2). Cumulative contest rules apply (S5).
1 February	0900 - 1500	432MHz Fixed / AFS	SF, MF	AFS rules apply (S3), three stations per team.
7 / 8 March	1400 - 1400	March 144 / 432MHz	S, M, SS	Low power stations running 25W or less at the transmitter will be specially identified in the results and the leading and second placed low power stations in each section will receive certificates.
29 March	0900 - 1300	First 70MHz Fixed	SF, MF	Full QTH Information to be sent (S2).
31 March / 8 / 16 April	1900 - 2100	144MHz SSB Fixed	SF, MF	This contest is scored at 1 pt / QSO (S1), with a QTH locator multiplier (M2). Section 1 for stations with 25W maximum output at the transmitter and section 2 for full legal Station Cumulatives power. Cumulative rules (S5) apply, but the best two sessions count to the final score.
5 April	1700 - 2100	1.3 / 2.3GHz Fixed	SF, MF	These run as separate contests - there will be no overall tabulation.
19 April	0900 - 1300	50MHz Fixed	SF, MF	District Code, Country and QTH Locator Multiplier (M3).
2 May	1300 - 1700	1st 432MHz Backpackers	S, M	District Code & QTH Locator Multiplier (M4). See separate Backpackers rules.
2 May	1400 - 2200	432MHz Trophy	S, M	This contest runs concurrently with the first eight hours of the 432MHz - 248GHz event (S6). The 1951 Council Cup is awarded to the overall winner of this contest.
2 May	1400 - 2200	10GHz Trophy	O	District Code & Country Multiplier (M1) Multiplier to be confirmed by RSGB Microwave Committee (announcement will be made in *Contest Classified*). Crossband contacts will count for multiplier credit. The 10GHz Trophyis awarded to the overall winner of this contest.
2 /3 May	1400 - 1400	432MHz - 248GHz	S, M	
16 /17 May	1400 - 1400	144MHz	SF, SP, M, SS	District Code / Country Multiplier (M1).
17 May	1100 - 1500	First 144MHz Backpackers	S, M	District Code, Country & QTH Locator Multiplier (M3)0. See separate Backpackers rules.
31 May	0900 - 1200	70MHz CW	SF, O	District Code & Country multiplier (M1). Full QTH Information to be sent (S2).
6 / 7 June	1400 - 1400	50MHz Trophy	S, M, SS	Co-ordinated with IARU contest (S7). Country & QTH Locator Multiplier (M4). The Telford Trophy is awarded to the overall winner of this contest, and the Six Metre Cup to the highest-scoring UK single operator entrant.
7 June	1100 - 1500	First 50MHz Backpackers	S, M	Country & QTH Locator Multiplier (M4). See separate Backpackers rules.
14 June	0900 - 1300	Second 144MHz Backpackers	S, M	Country & QTH Locator Multiplier (M4). See separate Backpackers rules. This event is co-ordinated with the first four hours of the *Practical Wireless* QRP contest.
21 June	1800 - 2200	432MHz FM	SF, O	District Code & country multiplier (M1).
4 July	1300 - 1700	2nd 432MHz Backpackers	S, M	Country & QTH Locator Multiplier (M4). See separate Backpackers rules.
4 / 5 July	1400 - 1400	VHF NFD	-	See separate rules.
5 July	1100 - 1500	Third 144MHz Backpackers	S, M	Country & QTH Locator Multiplier (M4). See separate Backpackers rules.
12 July	1100 - 1500	Second 50MHz Backpackers	S, M	District Code, Country & QTH Locator Multiplier (M3). See separate Backpackers rules.
18 July	1400 - 2200	144MHz Low Power	S, M	25W maximum output from the transmitter. District Code, country & QTH locator multiplier (M3).
19 July	0800 - 1400	432MHz Low Power	S, M	25W maximum output from the transmitter. District Code, country & QTH locator multiplier (M3).
2 August	1100 - 1500	Fourth 144MHz Backpackers	S, M	District Code, Country & QTH Locator Multiplier (M3). See separate Backpackers rules.
9 August	0900 - 1500	70MHz Trophy	SF, O	District Code & country multipliers (M1).
17 August, 1 / 16 September, 1 / 16 October	2000 - 2230 LOCAL TIME	144MHz CW Cumulatives	S, M	Cumulative contest rules apply (S5). **Note:** These contests may be replaced with another format of event if support is not high during the 1997 events. A decision will be published in the June 1998 *RadCom*.
23 August	1700 - 2100	432MHz Fixed	SF, MF	District Code, country & QTH locator multiplier (M3).
5 /6 September	1400 - 1400	144MHz Trophy	S, M, SS	Co-ordinated with IARU contest (S7). The Thorogood Trophy is awarded to the winner of section S, and the Mitchell-Milling Trophy to the winner of section M of the contest.
6 September	1100 - 1500	Fifth 144MHz Backpackers	S, M	Country & QTH Locator Multiplier (M4). See separate Backpackers rules.
13 September	1800 - 2200	1.3 / 2.3GHz Fixed	SF, MF	The 1.3GHz and 2.3GHz contests are separate contests - there will be no overall two-band tabulation.
27 September	0900 - 1300	Second 70MHz Fixed	SF, MF	Full QTH Information to be sent (S2).
29 September, 14/29 October, 13 / 30 November	2000 - 2230 LOCAL TIME	1.3 / 2.3GHz Cumulatives	SF, O	The 1.3GHz and 2.3GHz events are separate contests - there will be no overall two-band tabulation. Cumulative contest rules apply (S5).
3 October	1400 - 2200	1.3 / 2.3GHz Trophies	S, M	These contests run concurrently with the first eight hours of the IARU contest (S6, S7). The VHF Contests Committee cup is awarded to the winner of the 1.3GHz contest, and the G6ZR Memorial Trophy to the winner of the2.3GHz event.
3 / 4 October	1400 - 1400	432MHz - 248GHz IARU	S, M	Co-ordinated with IARU contest (S7).
9 / 26 October, 10 / 25 November, 10 December	2000 - 2230 LOCAL TIME	432MHz Cumulatives	SF, O	Cumulative contest rules apply (S5).
18 October	0900 - 1300	50MHz Fixed Station	SF, MF	District Code, Country and QTH Locator Multiplier (M3).
7 / 8 November	1400 - 1400	144MHz CW Marconi	S, M	The RSGB and European Marconi Memorial events run concurrently (S7).
8 November	0800 - 1400	Six-hour 144MHz CW	S, M	This event runs in the last six hours of the European contest (S7).
6 December	0900 - 1700	144MHz Fixed / AFS	SF, MF	AFS rules apply (S3), five stations per team.
26 / 27 / 28 / 29 December	1400 - 1600	50 / 70 / 144 / 432MHz Christmas Cumulatives	SF, O	Cumulative contest rules apply (S5). Score at 1 pt / QSO (S1). QTH locator multipliers (M2) applies, and the same multipliers maybe claimed for credit on each band on each Christmas Cumulatives day.

Key to sections: S - Single Operator; M - Multi Operator; O - All Others; SF - Single Operator Fixed; MF - Multi Operator Fixed; SP - Single Operator Portable; SS - Six Hour Single Operator Fixed.

CALL SIGN

TOTAL CLAIMED SCORE

RSGB

Radio Society of Great Britain

HF CONTEST SUMMARY SHEET

Contest .... Date ....

Phone ☐ Single operator ☐ Single band ☐

CW ☐ Multi operator ☐ Multi band ☐

Section .... Name of group/club ....

Location of station ....

Contest exchange (e., County Code) ....

BAND	1.8MHz	3.5MHz	7MHz	14MHz	21MHz	28MHz	TOTAL
VALID QSOs							
QSO POINTS							
BONUS/ MULT							
HFCC USE							

Transmitter/Receiver .... Output Power.................. dBW

Antennas ....

Operators*..

I declare that this station was operated strictly in accordance with the rules and spirit of the contest and within the conditions of my license. I agree that the decision of the Council of the RSGB shall be final in all cases of dispute.

DATA PROTECTION ACT: I agree to the data from this entry being entered into a computer for the sole purpose of the contest adjudication.¶

Signature .... Date ....

Name (Block letters) .... Callsign ....

Address ....

....

.... Post Code ....

Notes: 1. Use separate log sheets for each band.
2. Include list of bonus/multipliers where applicable.
*3. Multi-operator entries must indicate callsign of operator against each contact.
¶4. Entrants information is erased after checking is completed.

Form HFC2 (Rev 90)

THIS FORM IS SHOWN AS AN EXAMPLE. IT SHOULD BE ENLARGED TO110% BEFORE BEING USED FOR ANY RSGB CONTEST ENTRY.

RADIO SOCIETY OF GREAT BRITAIN

HF CONTEST LOG SHEET (HFC1 Rev90)

CALLSIGN		DATE		BAND MHz	Page of	
TIME (GMT)	CALLSIGN	RST/SERIAL NUMBER		OTHER DATA	BONUS /MULT	POINTS
		SENT	RECEIVED			

THIS FORM IS SHOWN AS AN EXAMPLE. IT SHOULD BE ENLARGED TO 110% BEFORE BEING USED FOR ANY RSGB CONTEST ENTRY.

RADIO SOCIETY OF GREAT BRITAIN

CONTEST SITE REGISTRATION REG2-97

CONTEST GROUP NAME	
ADDRESS FOR CORRESPONDENCE TELEPHONE NUMBER IN CASE OF QUERY:	SITE ADDRESS
MAIDENHEAD LOCATOR OF SITE	COUNTY
SECTION OF CONTEST ENTERED	RLO NAME AND AREA (for contest site)
CALLSIGNS TO BE USED: 50MHz 70MHz 144MHz 432MHz 1.3GHz 2.3GHz	QTH DETAILS TO BE SENT ON: 70MHz (CW): 70MHz (SSB):
FULL DETAILS OF ACCESS TO SITE	Please ensure that this section is totally accurate
Nearest large town:	NGR of site:
Use additional sheets if necessary Please attach map or drawing on reverse of form	
NOTE: Failure to allow free access to the registered site for any reason for the whole period of the contest will invalidate the site registration.	
Please supply TWO (2) copies of this form, all additional sheets and all maps and drawings.	
NO DUAL REGISTRATIONS. Groups may not register more than one site at a time. Changes of site can be registered on a new form up to one (1) week before the contest and this will cancel any previous registrations. Entries not adhering to this rule cannot be accepted.	

RADIO SOCIETY OF GREAT BRITAIN

VHF-UHF-SHF COVER SHEET

Form 427-97

RSGB

ENTRANTS - PLEASE COMPLETE ALL SECTIONS IN CLEAR PRINT OR TYPE

Contest Title			Number of QSOs made
Band MHz	Section		Claimed Points
Callsign Used	Location as sent (QTH / Zone / County)		Multiplier
Locator as sent			Overall Score
Best DX QSO Station Callsign	Distance (km)	in Locator	Adjudicated Score
Name or Group			

Name and Callsign of all operators

NAME	CALLSIGN	NAME	CALLSIGN

TX (brief description)		Power output (Watts)
RX (brief description)		QTH ASL (m)
Antenna Height AGL (m)	Antenna type	

CONDITIONS DURING CONTEST

PLEASE LIST CUMULATIVE SESSION SCORES AND MAKE ANY FURTHER COMMENTS ON THE REVERSE SIDE OF THIS SHEET

IMPORTANT NOTE

BY SUBMITTING AN ENTRY TO THIS CONTEST YOU AGREE TO BE BOUND BY THE RULES OF THIS CONTEST AND YOU AGREE THAT THE DECISION OF THE RSGB SHALL BE FINAL IN CASES OF DISPUTE.

Name .... Callsign ....

Address for correspondence

Telephone No ....

Please tick appropriate box for further stationery supplies and enclose a large stamped addressed envelope

FORM LSVHF LOG SHEETS	
FORM 427 COVER SHEETS	
FORM REG2 SITE REGISTRATION	
FORM MUL1 MULTIPLIER LIST	

Please enclose a stamped addressed post card if you require confirmation that your entry in the contest has been received.

RADIO SOCIETY OF GREAT BRITAIN VHF-UHF-SHF LOG SHEET

RSGB

Form LSVHF/93

Sheet Number ____ of ____ Sheets

CONTEST:	BAND MHz	CALLSIGN:
LOCATOR:	LOCATION:	

DATE GMT	CALLSIGN	RS(T) SERIAL NUMBER SENT	RS(T) SERIAL NUMBER RECEIVED	QTH LOCATOR	LOCATION	POINTS RADIAL	POINTS KMS
					PAGE TOTALS		

USE SEPARATE LOG SHEETS FOR EACH BAND - TOTAL EACH SHEET SEPARATELY

CALL SIGN

TOTAL CLAIMED SCORE

RSGB

Radio Society of Great Britain

HF CONTEST SUMMARY SHEET

Contest .... Date ....

Phone ☐ Single operator ☐ Single band ☐

CW ☐ Multi operator ☐ Multi band ☐

Section .... Name of group/club ....

Location of station ....

Contest exchange (e., County Code) ....

BAND	1.8MHz	3.5MHz	7MHz	14MHz	21MHz	28MHz	TOTAL
VALID QSOs							
QSO POINTS							
BONUS/ MULT							
HFCC USE							

Transmitter/Receiver .... Output Power...................... dBW

Antennas ....

Operators*..............................

I declare that this station was operated strictly in accordance with the rules and spirit of the contest and within the conditions of my license. I agree that the decision of the Council of the RSGB shall be final in all cases of dispute.

DATA PROTECTION ACT: I agree to the data from this entry being entered into a computer for the sole purpose of the contest adjudication.¶

Signature .... Date ....

Name (Block letters) .... Callsign ....

Address ....

....

.... Post Code ....

Notes:
1. Use separate log sheets for each band.
2. Include list of bonus/multipliers where applicable.
*3. Multi-operator entries must indicate callsign of operator against each contact.
¶4. Entrants information is erased after checking is completed.

Form HFC2 (Rev 90)

RADIO SOCIETY OF GREAT BRITAIN **HF CONTEST LOG SHEET (HFC1 Rev90)**

CALLSIGN	DATE	BAND MHz	Page of

TIME (GMT)	CALLSIGN	RST/SERIAL NUMBER		OTHER DATA	BONUS /MULT	POINTS
		SENT	RECEIVED			

CALL SIGN [ ] TOTAL CLAIMED SCORE [ ]

Radio Society of Great Britain

SWL HF CONTEST SUMMARY SHEET

Contest .... Date ....

Phone [ ] Single operator [ ] Single band [ ]

CW [ ] Multi operator [ ] Multi band [ ]

Section .... Name of group/club ....

Location of station ....

Contest exchange (e.g., County Code)..

BAND	1.8MHz	3.5MHz	7MHz	14MHz	21MHz	28MHz	TOTAL
VALID QSOs							
QSO POINTS							
BONUS/ MULT							
HFCC USE							

Receiver ....

Antennas ....

I declare that this station was operated strictly in accordance with the rules and spirit of the contest and that I do not hold a Class A transmitting license. I agree that the decision of the HF Contest Committee and the Council of the RSGB shall be final in all cases of dispute.

DATA PROTECTION ACT: I agree to the data from this entry being entered into a computer for the sole purpose of the contest adjudication. ¶

Signature .... Date ....

Name (Block letters) .... Callsign ....

Address ....

....

.... Post Code ....

Notes: 1. Use separate log sheets for each band.
2. Include list of bonus/multipliers where applicable.
¶3. Entrants information is erased after checking is completed.

Form HFC2 RX

Page of	**RSGB HF RX CONTEST LOG SHEET (HFC1 Rx)**				CALLSIGN	
CONTEST	DATE		BAND MHz		HFCC	
TIME (GMT)	STATION HEARD		STATION WORKED	OTHER DATA	BONUS /MULT	POINTS
	CALLSIGN	RST/No SENT				
(USE SEPARATE LOGS FOR EACH BAND)				PAGE TOTALS		

Services for disabled or blind amateurs

Radio Amateur Invalid & Blind Club

The Radio Amateur Invalid and Blind Club is the national society helping disabled and blind amateurs and short wave listeners.

Membership services

The club appoints local representatives who have kindly agreed to look after individual member's problems concerning the installation, operation and maintenance of their amateur radio or short wave listening station. Zonal Co-ordinators have been appointed to manage activities in each of the 7 RSGB zones and are ex-officio committee members.

The club is most grateful to the many amateurs who donate equipment for use by members, and frequently it receives equipment and donations from legacies in wills, etc. This loan equipment always remains RAIBC's property. The club has arrangements with many manufacturers and dealers for equipment and accessory discounts, but only for full members.

A good selection of useful cassette tapes are available. These are available from Nick Chambers, G0IRM, QTHR. Send sae for list, or phone (0181) 868 2516.

The club can provide information on various types of audio aid, mostly for blind operators. A limited supply of audio gimmicks for RF tuning up, etc., is available as stocks and funds permit. EMC advice is available for club members from G5HD QTHR.

Club nets

There are regular club nets operated by members and supporters. The club callsign is G4IBC (from N. Ireland – GI0IBC). The net controller is G3SKF and the nets are:

Mondays	CW	3.553 MHz	1415 hrs	G5LW
Tuesdays	SSB	3.740 MHz	1000 hrs	G3SKF
(Note: G3MJK operates an early birds net from approx 0900 hrs principally for RAIBC affairs)				
Wednesdays	SSB	3.740 MHz	1400 hrs	G0MZI / G0KME
Fridays	SSB	3.740 MHz	1900 hrs	GI4GVS
Sundays	SSB	7.045 MHz	1500 hrs	

Times are local time. As and when propagation conditions permit, SSB nets will go to 7.050 MHz and the CW net to 7.018 MHz. It should also be noted that towards the end of October the Friday evening net is suspended until March due to propagation and GI4GVS takes the Sunday afternoon net.

Publications

RAIBC publish their newsletter, *Radial*, approximately four times per year. It is also available on cassette for blind members. The editor is Peter Hunter, G0GSZ.

Committee members (include)

Chairman:	Johnny Clinch, G3MJK
Vice-chairman:	Margery Hey, 29 Besthorpe Road, Attleborough, Norfolk, NR17 2AN. Tel: (01953) 454920 (10 am – 10 pm please).
Secretary:	Phillip Stanley, G3BSN, 1 Thames View, Hilton Woods, Cliffe Woods, Rochester, Kent, ME3 8LR.
Treasurer/Membership Sec:	Shelagh Chambers, 78 Durley Avenue, Pinner, Middx, HA5 1JH. Tel: (0181) 868 2516
Loan Equipment Manager:	Terry Stanley, G0GTO, 1 Thomas View, Hilton Woods, Cliffe Woods, Rochester, Kent, ME3 8LR.

Royal National Institute for the Blind

The RNIB has a well established library of items in braille and on tape relating to amateur radio which may be borrowed, as well as a number of items which may be purchased. Items available include Benbow's *Radio Amateurs Examination Manual*, Hawker's *A Guide to Amateur Radio, RSGB Yearbook*, Orrs's *Simple Low Cost Wire Antennas*, as well as various equipment manuals, RSGB operating lists and RA information sheets.

Enquiries should be made to:

RNIB Customer Services,
PO Box 173,
Peterborough, PE2 6WS.
Tel: 0345-023153 – for the price of a local call.

Enterprises by the Blind

This organisation operates a talking newspaper service for the visually impaired. Among the periodicals that they read are, RSGB's *Radio Communication* (monthly, three C90s), which is free to registered blind or partially sighted members of RSGB who are supplied with free cassettes and posting wallets. Further details can be obtained from: Roy Gerrard, G3LAZ, 46 Hadrian Avenue, Dunstable, Beds. LU5 4SP. Tel: (01582) 727588 (daytime).

St Dunstan's

St Dunstan's, the organisation for men and women blinded in the Services, are happy to offer advice and answer queries. Please contact Ray Hazan, Public Relations Officer, St Dunstan's, PO Box 4XB, 12-14 Harcourt Street, London, W1A 4XB. Tel: (0171) 723 5021.

City & Guilds Exams (RAE + NRAE)

City & Guilds of London Institute has for many years made arrangements for candidates with special needs, who would have difficulties taking the examination under normal circumstances. Essentially, those who are housebound, blind, or handicapped in some other way, may be examined at home. Moreover, if the candidate is unable to complete a written paper, it is possible for an examination to be conducted orally. Remember that candidates must first have passed the practical training course before taking the NRAE.

Candidates who wish to take advantage of these special arrangements will be required to provide some official notification which confirms the reason they have given for requesting a home examination. Examples of acceptable documentation include a doctor's certificate or a DSS letter. In all cases, however, the final decision rests with City & Guilds.

Application forms are available from: Division 24, City & Guilds of London Institute, 46 Britannia St, London. WC1X 9RG. Tel: (0171) 278 2468. Further information is also available from the same source. Please note that completed application forms must be received at City & Guilds by 25 October (for examination in December) or 14 March (for the May examination). Candidates must also be prepared to make themselves available for examination in the 10 days following the date of the regular examination.

Morse tests

Morse Tests for the disabled fall into three general categories, which cover most situations:

a) Those who have difficulty with hearing, sight, or normal manipulation of the Morse key. In other words, those who are physically able to attend a centre, but on arrival require individual attention because of their disability.
b) Those who are mobile, but housebound because of their disability.
c) Those who are bedridden, and a home visit is required.

Each case is considered on its merits. There are no exemptions. Although the rules may be relaxed to take into account individual problems, candidates must demonstrate that they have made a sustained effort to learn the Morse code.

Candidates applying for the Morse test as a disabled person are required to supply details of their disability on a supplementary application form. Supporting medical evidence must be supplied and in the case of a request for a home visit, the medical note must additionally confirm the inability of the candidate to travel to a scheduled Morse test centre. Application forms can be obtained from RSGB HQ or from the Deputy Chief Morse Examiner.

RSGB

The Society offers concessionary membership for those who are blind or disabled.

The concessionary rate of £18 pa (ie 50% of the full corporate rate) is available to those who are disabled such that they are unable to permanently follow or obtain full time employment. They must also be UK resident or be an ex-patriot and be under state pensionable age.

An application form is available from RSGB Sales Dept, which needs to be completed and returned to the Society with appropriate documentation.

For those who are blind, similar criteria and application procedures apply except that the entire membership fee is waived. A choice is available of having either a printed copy of *RadCom* , or receiving a cassette tape version produced by Enterprises by The Blind. (An £18 pa subscription fee is charged if both printed and cassette tape versions are desired.)

EMC – Dealing With Interference

Some hints by Robin Page-Jones, CEng MIEE, G3JWI, EMC Committee Chairman, taken mainly from 'The Radio Amateur's Guide to EMC'.

The most common problem in amateur radio is interference caused by the fundamental transmission getting into all types of electronic equipment. The term 'breakthrough' is normally used to describe this phenomenon; emphasising the fact that it is really a shortcoming on the part of the equipment being interfered with, and not a transmitter fault.

Good Radio Housekeeping

The main object of good radio housekeeping is to minimise breakthrough, by making sure that as little as possible of the precious RF energy finds its way into neighbouring TV, videos, telephones, and the multitude of electronic gadgets which are part and parcel of the modern home.

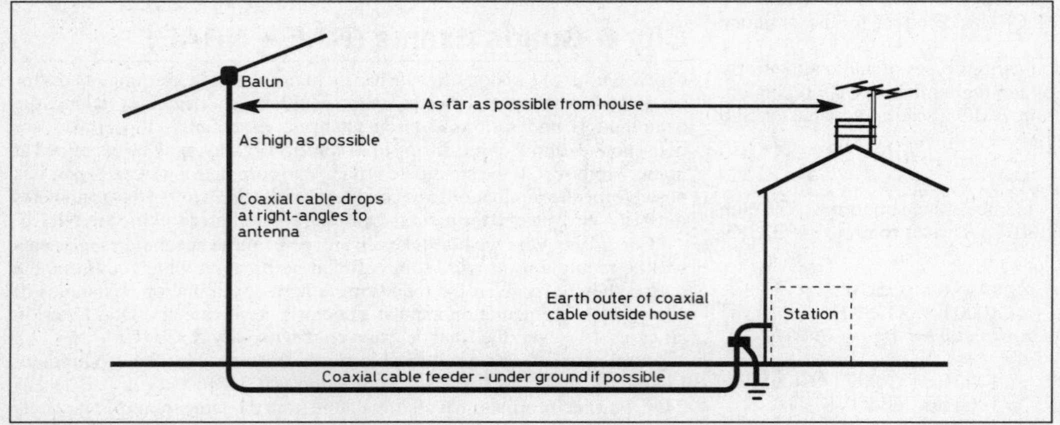

Good radio housekeeping – site your antenna and feeder system well away from the house.

In many cases it can be rightly argued that the immunity of the domestic equipment is inadequate, but this does not absolve the amateur from the responsibility of keeping his RF under reasonable control. Many of the features which contribute to minimising breakthrough also help in reducing received interference, so that the virtue of good neighbourliness has the bonus of better all round station performance.

Antennas

By far the most important factor in preventing both breakthrough and received interference problems is the antenna and its siting. The aim is to site the antenna as high as you can, and as far as possible from your own house, and from neighbouring houses. If there is any choice to be made in this regard give your neighbours the benefit of the increased distance - it is usually much easier to deal with any problems in your own home. It is a sad fact that many amateurs are persuaded by social pressures into using low, poorly sited, antennas only to find that breakthrough problems sour the local relations far more than fears of obtrusive antennas would have done.

HF Antennas

The question of which antenna to use is a perennial topic and the last thing that anyone would want to do is to discourage experimentation, but there is no doubt that certain types of antenna are more likely to cause breakthrough than others. It is simply a question of horses for courses; what you can get away with in a large garden, or on HF field day, may well be unsuitable for a confined city location. Where EMC is of prime importance, the antenna system should be:

(a) Horizontally polarised. TV down leads and other household wiring tend to look like an earthed vertical antenna so far as HF is concerned, and are more susceptible to vertically polarised radiation.

(b) Balanced. This avoids out of balance currents in feeders giving rise to radiation which has a large vertically polarised component. Generally, end fed antennas are unsatisfactory from the EMC point of view and are best kept for portable and low power operation. Where a balanced antenna is fed with coaxial feeder, then a balun must be used.

(c) Compact. So that neither end comes close to the house and consequently to TV down leads and mains wiring. Antennas to be careful with are the extended types such as the W3DZZ trap dipole or the G5RV, because almost inevitably, in restricted situations, one end is close to the house.

On frequencies of 14 MHz upwards it is not too difficult to arrange an antenna fulfilling these requirements, even in quite a small garden. A half wave dipole or small beam up as high as possible, and 15 metres or more from the house is the sort of thing to aim for.

At lower frequencies compromise becomes inevitable, and at 80 metres most of us have no choice but to have one end of the antenna near the house, or to go for a loaded vertical antenna which can be mounted further away. A small loop antenna is another possibility, but in general any antenna which is very small compared to a wavelength will have a narrow bandwidth, and a relatively low efficiency. Many stations use a G5RV or W3DZZ trap dipole for the lower frequencies, but have separate dipoles (or a beam) for the higher frequencies, sited as far down the garden as possible.

VHF Antennas

The main problem with VHF is that large beams can cause very high field strengths. For instance 100 watts fed to an isotropic transmitting antenna in free space would give a field strength of about 3.6 volts/metre at a distance of 15 metres.

The same transmitter into a beam with a gain of 20 dB would give a field strength, in the direction of the beam, of 36 volts/metre the same distance away. Again, it comes down to the fact that if you want to run high power to a high gain beam, the antenna must be kept as far from neighbouring houses as possible, and of course, as high as practical.

Operation in Adverse Situations

The obvious question arises as to what to do if your garden is small or non-existent, or domestic conditions make a simple wire tuned against ground the only possibility.

First of all, and most important, don't get discouraged - many amateurs operate very well from amazingly unpromising locations. It is really a question of cutting your coat according to your cloth. If there is no choice but to have antennas very close to the house, or even in the loft, then it will almost certainly be necessary to restrict the transmitted power. It is worth remembering that it is good radio operating practice not to use more power than is required for satisfactory communication. In many cases relations with neighbours could be significantly improved by observance of this simple rule.

Not all modes are equally 'EMC friendly', and it is worth looking at some of the more frequently used modes from this point of view.

SSB This is one of the least EMC friendly modes, particularly where audio breakthrough is concerned.

FM This is a very EMC friendly mode, mainly because in most cases the susceptible equipment sees only a constant carrier turned on and off every minute of so.

CW This is the old faithful for those with EMC problems, because it has two very big advantages. First, providing the keying waveform is well shaped with rise and fall times of about 10 ms or so, the rectified carrier is not such a problem to audio equipment as SSB. The second is that it is possible to use lower power for a given contact. Of course, low power CW is not everybody's favourite mode, but it does provide a way of staying on the air, even in the most difficult circumstances.

DATA Generally the data modes used by amateurs are based on frequency shift keying (FSK), and should be EMC friendly. All data systems involve the carrier being keyed on and off – when going from receive to transmit, and vice versa – and consideration should be given the carrier rise and fall times; just as in CW. On VHF data systems, interference can occasionally be caused by residual amplitude modulation on an FM data transmission, or by simple over-deviation by the tone modulation.

Earths

From the EMC point of view, the purpose of an earth is to provide a low impedance path for RF currents which would otherwise find their way into household wiring, and hence into susceptible electronic equipment in the vicinity. In effect the RF earth is in parallel with the mains earth path. Good EMC practice dictates that any earth currents should be reduced to a minimum by making sure that antennas are balanced as well as possible. An inductively coupled ATU can be used to improve the isolation between the antenna/RF earth system and the mains earth. The impedance of the mains earth path can be increased by winding the mains lead supplying the transceiver and its ancillaries onto a stack of ferrite cores as described below for breakthrough reduction.

End fed wires tuned against earth should be avoided since these inevitably involve large RF currents flowing in the earth system.

The minimum requirement for an RF earth are several copper pipes 1.5 metres long or more, driven into the ground at least 1 metre apart and connected together by thick cable. The connection to the station should be as short as possible using thick cable or flat copper strip/braid.

Where the shack is installed in an upstairs room, the provision of a satisfactory RF earth is a difficult problem, and sometimes it may found that connecting an RF earth makes interference problems worse. In such cases it is probably best to avoid the need for an RF earth, by using a well balanced antenna system - but don't forget to provide lightning protection.

PME

An increasing number of houses are now being wired on the Protective Multiple Earthing (PME) system, which has a common neutral/earth conductor from the sub-station to the consumer's premises. For safety reasons special regulations apply to earthing in a PME installation. If in doubt consult a qualified electrician or contact your local regional electricity company (REC) for advice.

Harmonics

Harmonics and other spurious emissions in general are much less of a problem than formerly, partly due to the closing down of VHF TV in the UK and partly to the much greater awareness by home-brewers and commercial manufacturers alike of the importance of good design and construction. Not withstanding this, if there is any doubt about the harmonic performance of a transceiver, then a low pass filter should be used. Particular care should be taken where harmonics can fall into broadcast radio or TV bands, in particular:-

1. The harmonics of some HF bands fall into the VHF broadcast band, 88 to 108 MHz, as does the second harmonic of 50 MHz.
2. The fourth harmonic from the 144 MHz band could cause problems on TV channels 34 and 35 and the fifth on channels 52 and 53.
3. The second harmonic of 18 MHz falls into the IF band of TV receivers.

Tackling EMC Problems

By David Lauder, CEng MIEE, G0SNO, EMC Committee

This section covers the following topics:

* Sources of further information on amateur radio EMC problems.
* Characteristics and use of filters and ferrite rings.

RA Information

RA323, 'Guidelines for Improving Television and Radio Reception'

This new 16 page booklet was first published in 1997 and is primarily intended for radio and television dealers, service engineers and aerial installers. It contains some topics of particular interest to radio amateurs including TV aerial amplifiers, CE marking and effects due to lack of immunity. It was produced by the Radiocommunications Agency in consultation with the RSGB EMC Committee, the BBC, ITC, CAI (Confederation of Aerial Industries) and BREMA (British Radio and Electronic Equipment Manufacturers Association).

RA 179, 'Advice on Television and Radio Reception'

This form is used by a householder to request the RA to investigate a reception problem with a domestic television set, video recorder or FM radio. If the source of interference is known or suspected, Part A of RA 179 can be used to report the source for possible investigation by the RA. There is no charge for reporting a source but the RA does not investigate the affected equipment. Alternatively, Part B of RA 179 can be used to request the RA to visit the householder and investigate a reception problem if a fee is paid (£45 in 1997). The RA does not currently investigate satellite television reception problems nor RF breakthrough on equipment such as telephones, tape recorders or alarm systems which are not intended to receive radio signals.

Copies of RA179, RA323 and many other RA publications are available from:

Radiocommunications Agency,
Information and Library Service,
11th Floor, New Kingsbeam House,
22, Upper Ground,
London, SE1 9SA

Tel: (0171) 211 0502/0505. Fax: (0171) 211 0507

Some other RA publications are also available on the World Wide Web. The URL for the index page is http://www.open.gov.uk/radiocom/libind.htm

RSGB Information

The RSGB EMC Committee has produced a series of information sheets which are listed below. RSGB members can obtain copies by writing to the Amateur Radio Department at RSGB HQ. Please state which sheets you require, quote your RSGB membership number and enclose a Self Addressed Stamped Envelope.

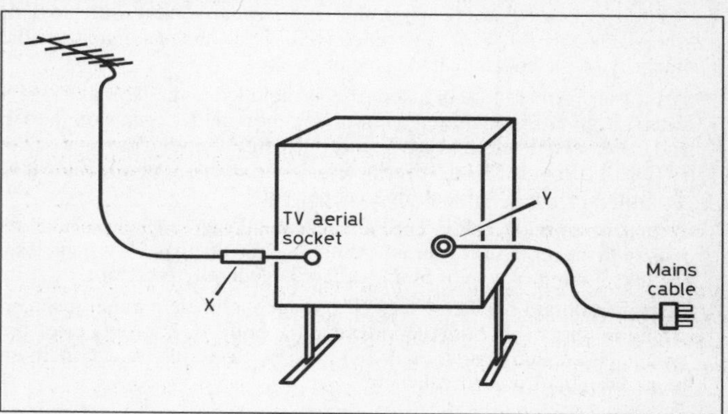

Fig 1: Fitting filters to TV set alone.

EMC 01	Radio Transmitters and Domestic Electronic Equipment
EMC 02	Radio Transmitters and Home Security Systems
EMC 03	Dealing with alarm EMC problems – Advice to RSGB Members
EMC 04	Interference to Amateur Radio Reception
EMC 05	Radio Transmitters and Telephones – Information for Telephone Users
EMC 06	Automotive EMC for Radio Amateurs

Use of filters and ferrite rings

TV and video

With a TV set alone, a suitable filter in the coaxial aerial cable at 'X' in figure 1 may be all that is required to cure breakthrough. If this does not cure the problem, then a ferrite ring (see below) may also be required on the mains cable at 'Y'.

Figure 2 shows a video recorder with its RF output connected to the aerial socket of a TV. A filter and/or 'braid breaker' (see below) should be fitted in the coaxial cable at 'X1' but if this is not sufficient, another should be fitted at either 'X2' or at 'X3'. In some cases, ferrite rings are required on the mains cables at 'Y1' and/or at 'Y2'. The use of SCART cables instead of RF links can offer improved immunity. Further information is available in [1].

Audio systems

RF breakthrough in an audio system can often be cured by fitting ferrite rings to the loudspeaker cables as shown by 'Y1' and 'Y2' in Figure 3. In some cases, additional ferrite rings may be required such as at 'Y3' on the mains cable. Rings may also be required on audio cables or mains cables of other units such as a cassette deck or CD player.

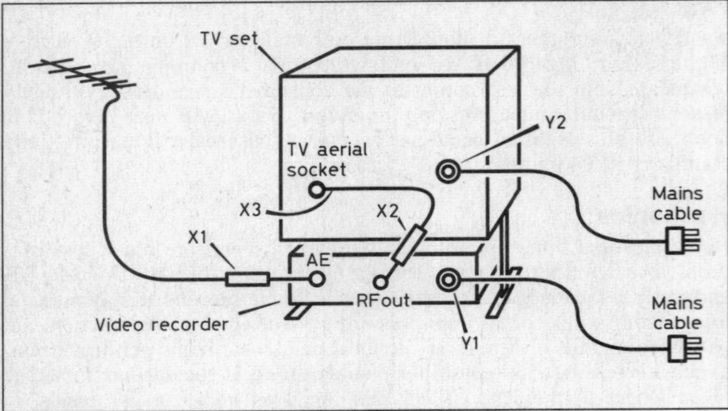

Fig 2: Fitting filters to a TV set with a video recorder.

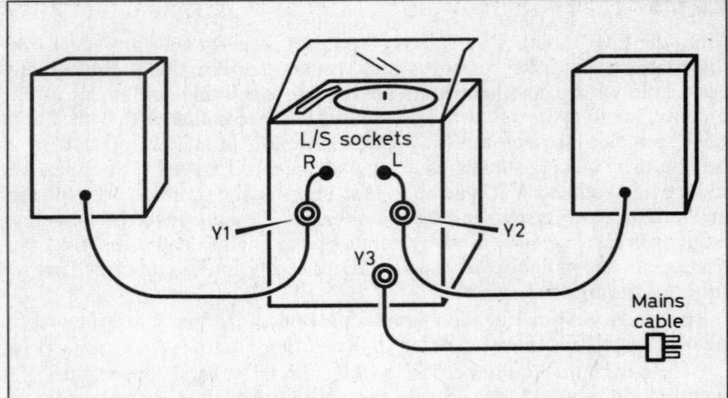

Fig 3: Fitting common mode chokes to an audio system.

Filter characteristics available from RSGB

All the filters listed below allow UHF TV signals to pass through but as with any filter, there is a small loss in the pass band. The HPF2 allows both VHF/FM broadcast radio (Band 2, 88-108 MHz) and UHF TV to pass through.

In many cases, particularly on the HF bands, amateur signals are picked up by the TV aerial downlead rather than by the TV aerial itself. If this causes breakthrough, a so-called 'braid breaker' is required. This could be part of a filter or could be a separate unit. There are several types of 'braid breaker' although some of these do not actually break the braid at DC:

* A ferrite ring choke. Winding a length of coaxial cable onto one or more ferrite rings has the advantage that it introduces very little loss or impedance mismatch for the wanted UHF signal and also maintains the integrity of the coaxial cable screening.
* A 1:1 transformer braid breaker. This is included in the BB1 and HPFS filters. It gives excellent rejection of common mode signals on the HF and lower VHF bands and can be more effective than ferrite rings on the HF bands below 10MHz. Nevertheless, it introduces a significant loss of the wanted UHF TV signal (up to about 4dB).
* A capacitive braid breaker. The HPF1 has small value capacitors in series with both the inner and the braid. Although this filter has a low band loss, the use of another type of braid breaker is generally preferable.
* A resonant braid breaker. The TNF2 range include a parallel resonant circuit in series with both the inner and the braid. This is only effective on one particular amateur band and in some cases, its use can increase breakthrough of other frequencies.

HPF 1 High pass filter and braid breaker (No RSGB order code)

Pass band: UHF TV
Stop band: All bands 1.8 - 70MHz with limited effect on 145 MHz.
'Braid breaking': Capacitive braid breaking on all HF bands.
Remarks: The HPF1 is not stocked separately by RSGB but is included in the RFK1 filter kit (see below).

HPF 2 High pass filter (RSGB order code Filter 2)

Pass band: FM radio broadcast (88 - 108 MHz), up to UHF TV.
Stop band: All HF bands plus limited effect at 50MHz.
'Braid breaking': None.

HPF 6 High pass filter (RSGB order code Filter 8)

Pass band: UHF TV
Stop band: All bands 1.8 - 440 MHz
'Braid breaking': None.
Remarks: This is a high performance six section filter with a very sharp cut-off below 470 MHz. It provides excellent rejection of all amateur bands 1.8 - 440MHz.

BB1 Braid breaker (RSGB order code Filter 1)

Pass band: Most amateur bands plus FM radio broadcast (88 - 108 MHz), and UHF TV.
Stop band: Some rejection below 10 MHz but the BB1 is primarily intended as a 'braid breaker' rather than a high pass filter.
Braid breaking: A 1:1 transformer type braid breaker which is more effective on all HF bands than the HPF1 type. Limited braid breaking up to 144 MHz.
Remarks: A BB1 on its own may be effective on the HF bands. Alternatively, it can be cascaded with other filters such as HPF2 or HPF6 which do not have any braid breaking action, although this increases the total passband loss.

HPFS High pass filter (special) (RSGB order code Filter 3)

Pass band: UHF TV.
Stop band: All bands up to and including 144 MHz.
'Braid breaking': Includes 1:1 transformer type braid breaker (see BB1 above).
Remarks: The HPFS is a BB1 combined with a high pass filter. It is a good all-round filter for all HF and VHF amateur bands but is not effective on 430-440MHz. Due to the relatively high pass-band loss, it is not suitable for areas where the TV signal strength is low. Its electrical characteristics are the same as the Global HP-4A which is sold by other suppliers.

RBF1/70 cms notch filter (RSGB order code Filter 5)

Pass band: UHF TV
Stop band: 430-440 MHz and all HF bands.
Braid breaking: None.
Remarks: The RBF1/70cms is pretuned to reject 435 MHz although an HPF6 gives greater rejection of the 430 - 440 MHz band. The RBF1/70 cms also has a high-pass action with rejection of HF signals.

TNF2 tuned notch filter range.

Type:	RSGB order code:	Notch tuned to:
TNF2/145	Filter 4	145 MHz (2m)
TNF2/70	Filter 7	70 MHz (4m)
TNF2/50	Filter 6	50 MHz (6m)
TNF2/28	Filter 10	28 MHz (10m)
TNF2/21	Filter 15	21 MHz (15m)
TNF2/14	Filter 20	14 MHz (20m)

Pass band: UHF TV.
Stop band: Only the specified band.
'Braid breaking': Resonant, only on the specified band.
Remarks: These filters are designed to provide a high degree of rejection of one particular amateur band. They are not recommended for use at the input of some types of TV distribution amplifier.

Filter kit (RSGB order code FKIT)

A pack which contains the following items: 4 off ferrite rings, 2 off BB1, 2 off HPF1, 2 off HPFS, 1 off TNF2/145, 1 off HPF2, 1 off HPF6.

Ferrite Rings (RSGB Order code: FERR)

A pack of two rings made in the USA by Fair-Rite Corporation in type 43 material, part number 2643802702. The inside diameter is 22.85 mm (0.9 inch) and the width is 12.7 mm (0.5 inch). They can accommodate 12 turns of 5mm diameter cable and are equivalent to FT140-43. Details of their characteristics are given below.

Use of ferrite chokes

RF breakthrough into electronic equipment can be caused by signals being picked up on external cables such as loudspeaker cables, mains cables, etc. This effect can often be reduced or eliminated by winding the affected cable onto a suitable ferrite core which presents a high impedance to unwanted RF signals without affecting the wanted signals. For the best chance of success, it is important to use a suitable grade of ferrite with enough turns in order to obtain the highest possible impedance at the frequencies of interest (3 - 5kΩ or more is possible).

Winding a choke

Above about 10 MHz, it becomes increasingly important to minimise the stray capacitance between the ends of the winding. This stray capacitance can be reduced by using the winding method shown in figure 4. It is vital that

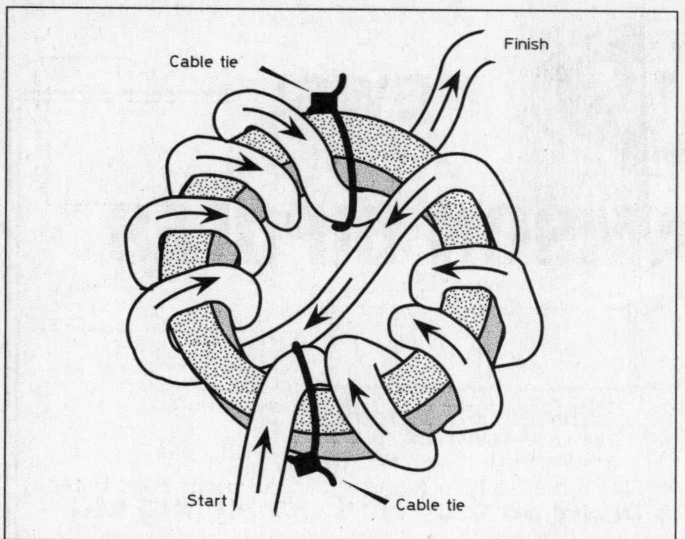

Fig 4: Recommended winding method for ring cores.

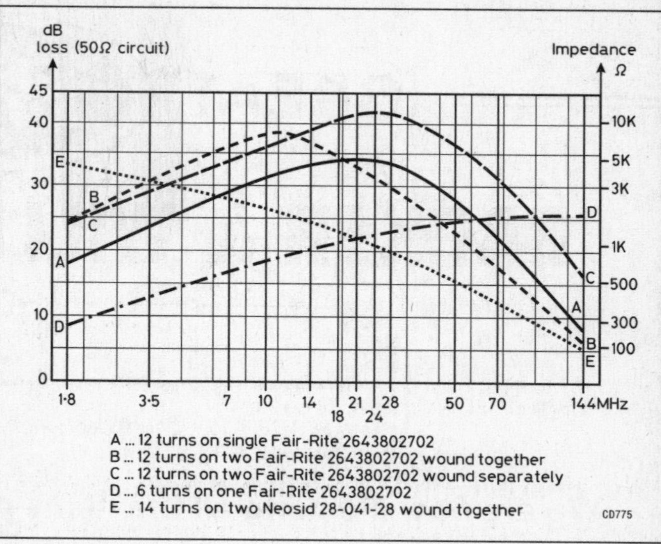

Fig 5: Impedance of various ferrite rings at different frequencies.

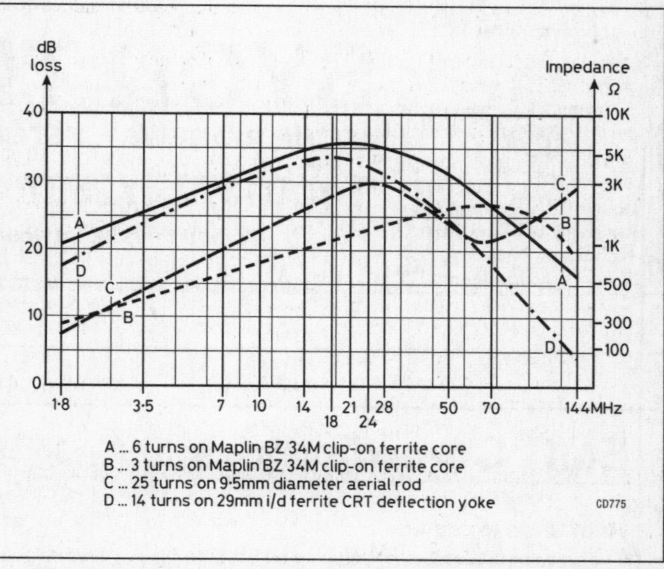

Fig 6: Impedance of some other types of ferrite cores at various frequencies.

the end of the cable is always threaded through the core in the same direction as shown by the arrows. Finally, the ends should be secured with self-locking cable ties as shown.

If the cable is very thick, if it has connectors which cannot easily be removed or if it is not long enough, the best solution is usually to make up a short plug-in extension lead using the thinnest suitable cable. This is wound through a ferrite core and then fitted with suitable connectors. Normal TV coaxial cable should not be wound tightly through a ferrite ring or onto a ferrite rod otherwise it may collapse internally and short-circuit. Instead, a one metre length of miniature 75Ω coaxial cable such as Maplin Electronics XR88V should be used and fitted with coaxial connectors.

Ferrite ring core characteristics

There is no single grade of ferrite which is ideal for EMC use on all HF and VHF amateur bands. The Fair-Rite type 43 material gives good results from 7MHz upwards but for the 1.8 MHz and 3.5MHz bands, better results are obtained with a higher permeability ferrite such as Fair-Rite type 73 material or a pair of Neosid 28-041-28 type rings, if available.

At VHF, minimising stray capacitance becomes particularly important so for 144 MHz, the recommended number of turns should be halved to 6 or 7 on a ring core or 3 for a clip-on core. Further details of various cores and their characteristics are given below.

Figure 5 shows the approximate impedance of various chokes wound on ferrite ring cores. The loss in dB is also shown but this is measured in a 50Ω test circuit (See Ref 2 for details of test methods).

Curve 'A' shows the characteristics of a 12 turn winding on a single Fair-Rite 2643802702 ring core which gives good results from 7-28 MHz. For cables thicker than 5mm, similar results can be achieved using 8-9 turns on 2 rings stacked together or 6 turns on 4 rings stacked together. (Note that halving the number of turns gives **one quarter** of the inductance.)

Curve 'B' shows the characteristics of a 12 turn winding on two Fair-Rite 2643802702 ring cores wound together. This gives excellent performance from 3.5 - 14MHz.

Curve 'C' shows the characteristics of 12 turn windings on two Fair-Rite 2643802702 ring cores wound separately which reduces the overall stray capacitance but requires more cable. This gives excellent performance from 3.5 - 70MHz

Curve 'D' shows the characteristics of a 6 turn winding on a single Fair-Rite 2643802702 ring core. This gives good results on 144 MHz.

Curve 'E' shows the characteristics of a 14 turn winding on a pair of Neosid 28-041-28 rings wound together. These give good results on 1.8MHz and 3.5 MHz but their performance falls off on the higher bands

Another type of ferrite ring is sold by Waters & Stanton Electronics and by Maplin Electronics as a 'ferrite filter' ring AM35Q. Winding 14 turns on two of these rings stacked together gives a result which is in between curves 'A' and 'B' in figure 5.

Other types of ferrite core

Fig 6 shows the characteristics of a clip-on core, a ferrite rod and a TV deflection yoke, all of which can be fitted to a cable without access to the ends. A clip-on ferrite core is sold by Maplin Electronics as a 'Computer Data Line Noise Filter' (Stock No BZ34M). Similar clip-on cores are available from trade suppliers such as Farnell Components and RS Components. Such cores require fewer turns than a ferrite ring but have an inside diameter of only 13mm.

Curve 'A' shows the characteristics of a 6 turn winding on a Maplin BZ34M. This gives good results from 7-28MHz, although a 6 turn winding is only possible with fairly thin cables. For 3.5 MHz, 9 turns on one core or 6 turns on two separate cores are recommended. The core halves **must** be able to close together properly without the slightest air gap.

Curve 'B' shows 3 turns on a BZ34M which gives better results at 144 MHz.

Curve 'C' shows a single layer 25 turn winding on a 9.5 mm diameter ferrite aerial rod at least 150 mm long. This is effective at 21 MHz and above.

Curve 'D' shows the typical characteristics of a 14 turn winding on a ferrite yoke ring core which can be salvaged with care from the deflection coils of a scrap TV or VDU tube. Such cores have an inside diameter of 29 - 48 mm and consist of two halves clipped together. The grade of ferrite used is likely to be effective for EMC use from 7 to 28 MHz. For the HF bands, 18 - 20 turns should be used if possible. Other types of surplus cores such as ferrite transformer cores are generally NOT suitable for amateur radio EMC use except possibly at 1.8MHz.

Reference:

[1] Radio Communication EMC Column, April 1997 and June 1997, item on TVI, Parts 1 and 2.

[2] The Radio Amateurs' Guide to EMC, Appendix 3, R. Page-Jones G3JWI, RSGB

EMC Co-ordinators

What to do for advice

If you are an RSGB member and have an EMC problem, your first point of contact should be your nearest EMC Co-ordinator. In many cases this person will be able to give you all the necessary advice, but where this is not possible, the problem will be passed to a committee member who specialises in that particular type of problem.

Before you contact your EMC Co-ordinator:

1. Make sure that you have done everything possible to solve the problem yourself.
2. Arm yourself with as much information as possible which will be useful to the co-ordinator.
3. Remember that the co-ordinator is a volunteer, so please ring at sociable times.
4. Remember also that the scheme only offers telephone advice at present – no visits will be made.

RSGB ZONE A – *North of England*

Arthur Armstrong, G0FBW, Co. Durham. Tel: (0191) 586 4500.
Neil Carr, G0JHC, Preston, Lancs. Tel: (01772) 423099. Calls after 6 pm.
Sidney Dimmock, GD8COH, Laxey, Isle of Man, Tel: (01624) 862802.
Stan Ellis, GD3LSF, Douglas, Isle of Man, Tel: (01624) 673303.
Raymond Gilchrist, G0TUE, Hazelrigg, Cumbria. Tel: (01229) 770246.
Fred Sawyer, G3SLN, Manchester. Tel: (0161) 643 9014.
Ron Smith, G3SVW, Sale, Cheshire. Tel: (0161) 969 3999.

RSGB ZONE B – *Midlands*

Bob Harrison, G4UJS, Shropshire. Tel: (01948) 880392.
Mrs Sandra Morley, G0MCV, Loughborough. Tel: (0116) 237 4999.
Adrian Jones, G1KEA, Solihull, W Midlands. Tel: (0121) 743 4039.
Gerry Valleley, G4YRS, Bedfordshire, Tel: (01767) 601008.

RSGB ZONE C – *SE England*

Peter Daly, G0GTE, Stevenage, Herts. Tel: (01438) 724991.
George Halse, G3GRV, Hemel Hempstead, Herts. Tel: (01442) 214972.
Ken Hendry, G0BBN, S. Benfleet, Essex. Tel: (01268) 755350.
Andrew Maish, G4ADM, Worcester Park, Surrey. Tel: (0181) 335 3434.

RSGB ZONE D – *SW England*

Peter Bertram, GJ8PVL, Grouville, Jersey, Tel: (01534) 811133.
Geoff Brown, GJ4ICD, St. Helier, Jersey, Tel: (01534) 877067.
Phil Goodfellow, G4KUQ, Bristol. Tel: (0117) 971 6093.
Dave McQue, G4NQU, Milton Keynes. Tel: (01908) 378277
Shaun O'Sullivan, G8VPG, Bristol. Tel: (01225) 873098.
Les Parry, G8AMK, Bracknell, Berks. Tel: (01344) 423704.
Hugh Pearson, G7KET, Bristol. Tel: (01275) 462934.
Ken Watkins, G3AIK, Martock, Somerset. Tel: (01935) 825266.

RSGB ZONE E – *Wales*

Dr Chris Barnes, GW4BZD, Bangor, Gwynedd. Tel: (01248) 602027.
Bill Holt, GW0SGG, Swansea, West Glamorgan. Tel: (01792) 299510.
John Lawrence, GW3JGA, Prestatyn, Clwyd. Tel: (01745) 853255.

RSGB ZONE F – *Northern Ireland*

Des Kernaghan, GI3USK, Holywood, Co Down. Tel: (01232) 426743.

RSGB ZONE G – *Scotland*

R Adam, GM4ILS, Elgin, Morayshire. Tel: (01343) 545842.
Rev S Bennie, GM4PTQ, Stornoway, Isle of Lewis. Tel: (01851) 703609.
G Brooks, GM4NHX, Halkirk, Caithness. Tel: (01847) 831570.
Dave Morris, GM3YEW, Abernethy, Perth. Tel: (01738) 850533.
Tommy Menzies, GM1GEQ, Glasgow, Tel: (0131) 445 3928.

GB2RS News Broadcasts

News Items

Items intended for inclusion should reach RSGB HQ by 9.15 am on the Tuesday prior to broadcast. Send in your copy either by letter (marked GB2RS NEWS), special postcards (available from RSGB HQ) or, in the case of last minute changes *only*, by telephone on Potters Bar (01707) 659015 or by fax on (01707) 649503.

We always welcome items of national importance, but club news items need to be more comprehensive than 'natter-nites'. Clubs with a prepared programme of events are invited to forward a copy to RSGB HQ so as to ensure regular inclusion of their activities.

Routine enquiries about scripts and their content should go to the RadCom Editorial Department at RSGB HQ.

Newsreaders

Provision of the GB2RS news service is part of the remit of the Society's Membership Liaison Committee. All queries about the service itself, or its schedule, or suggestions as to how it might be improved should be sent to the GB2RS Manager c/o RSGB HQ.

Requests for new or additional services should be made through RLOs and Zonal Council Members.

Voice

GB2RS is the news broadcast service of the RSGB. It is provided for the benefit of all radio amateurs in the UK. The script is prepared at RSGB HQ on a weekly basis and transmitted each Sunday by a network of volunteer news readers. These Sunday broadcasts provide almost complete coverage of the UK and are transmitted using the callsign GB2RS.

Packet

As part of the service, the weekly news script can also be down-loaded from the packet network. The script is divided into 13 or 14 small sections to reduce the down-loading time. The usual headings which appear in the subject field of the BBS are:

- Main News
- DX News
- Rally News
- Contest News
- Special Events
- Solar News
- South East News
- South West News
- East Anglia News
- Midland News
- North of England News
- Scottish News
- Welsh News
- Northern Ireland News

The order and the inclusion of local news for specific areas may vary according to the contents of a particular script. The BBS header also shows the date of the script and number of parts it is split into.

To list the news headers, just type `L< GB2RS` at the prompt.

The news is usually put on during Thursday evening prior to the Sunday broadcast and normally reaches most BBSs within 24 hours.

Internet

A copy of the script is also made available on the RSGB's World Wide Web site, http://www.rsgb.org. Late breaking stories can also be found here as well as news in greater depth.

The script also appears in 3 newsgroups: uk.radio.amateur, rec.radio info and rec.radio.amateur.misc.

Telephone

The Society provides a recording of the main GB2RS news items on a premium rate telephone number – (0336) 407394. *(Calls will cost you 39p per minute cheap rate, and 49p per minute at other times. Recordings normally last about 10 minutes.)*

The message line is updated each week on a Wednesday afternoon, so you are able to learn the news some 4 days earlier than normal Sunday broadcasts. Likewise, as the recording is not erased until the following Wednesday, you can still find out what is happening even if you cannot get to a radio on Sunday!

GB2RS Do's and Don'ts

DO:-

* Submit items in writing by letter or fax. Please use the phone only for urgent ***alterations or corrections*** to items already sent in.
* ***Always*** give a contact name, callsign (if any) and phone number.
* Say if a phone number is daytime only or evening only.
* Send GB2RS any last-minute details of your rally.
* Give the proper name of your club - there is a "Wirral and District Amateur Radio Club" and a "Wirral Amateur Radio Society", so saying "the Wirral club" could lead to confusion.
* Always give the callsign of a speaker if he / she has one, ***not*** just "Talk by John on aerials".
* Write full dates, ***not*** "Last Friday in Month".
* Listen to GB2RS to hear for yourself the format used.

DON'T:-

* Forget to include the venue and opening time of a rally, details of talk-in (if any) and a contact name, callsign, and phone number.
* Send in "to be confirmed" items. If they are not confirmed we must assume that they are not taking place and they will therefore not be broadcast. Only send the item in when it ***has been*** confirmed.
* Give more than one contact person or telephone number : there is only time to broadcast one, so ***you*** decide which one to use.
* Expect "natter nites", "get-togethers" or "open nights" to appear. We only publicize specific events, so a "skittles night" is OK, but a "social night" isn't.
* Mix regular club meetings with main news items, such as rallies and special event stations.
* Use GB2RS and *RadCom* as the ***only*** means of publicizing club events to your ***own*** members - the main purpose of GB2RS should be to inform casual listeners and members of ***other*** clubs of the exciting things ***your*** club is doing.
* Use 'cryptic' titles for talks, or 'in-jokes': if we don't know what you mean, no-one else is likely to.

GB2ATG

GB2ATG is the news service run by BARTG. Its broadcasts are normally on 80m and copies are carried on the amateur packet network and the Internet.

Due to a change of editor, the current schedules were not available for publication at the time of going to press.

Up to date information about GB2ATG can be obtained either from BARTG's web site (www.bartg.demon.co.uk) or by contacting the BARTG Secretary, Ian Brothwell, G4EAN.

GB2RS Broadcast Schedule

(Sundays only – all times local)

National *(see notes below)*

TIME	FREQ	MODE	READER	LOCATION
08:00	14.308	USB	G4RKK, G4NZQ	Norwich
09:00[1]	3.518	CW	G4FKH, G2FKZ	Chelmsford, Wakefield
09:00	3.650	LSB	G3RFX, GW3JSV	Bristol, Welshpool
09:30[2]	3.650	LSB	G2CVV, G8QZ, G3SZJ	Derby, Notts
10:00	7.048	LSB	GI3GGY, GI4MJD	Londonderry
12:00[1]	3.518	CW	G4FKH, G2FKZ	Chelmsford, Wakefield
14:15[3]	3.518	CW	GM4HYF, G3LEQ	Glasgow, Knutsford
15:00[1]	3.518	CW	G4FKH, G2FKZ	Chelmsford, Wakefield
18:00[1]	3.518	CW	G4FKH, G2FKZ	Chelmsford, Wakefield
18:00[2]	3.650	LSB	G8QZ, G2CVV, G3SZJ	Derby, Notts
20:00	14.308	USB	G4WCE, G0BAA, G0STA	North Cheshire
21:30[2]	1.990	LSB	G4WCE, G0BAA, G0STA	North Cheshire

South & South East *(RSGB Zone C)*

TIME	FREQ	MODE	READER	LOCATION
09:00	3.640	LSB	G4ARZ, G3WDY	Kent, London
09:30	145.525	FM	G8CKN, G4ODM	Hampshire
09:30	145.525	FM	G4BWJ, G3NDJ	Sussex Coast
09:30	433.200 *(via GB3EA)*	FM	G8CKN	Southampton
09:30	1308.000 *(via GB3HV)*	ATV	G8CKN	High Wycombe
09:30	1316.000 *(via GB3AT)*	ATV	G8CKN	Southampton
10:00	145.525	FM	G0AVY, G3ZVW, G4BWV	Kent, London
10:30	51.530	FM	G8SC, G3EKJ	East Sussex
10:30	145.525	FM	G8LVC, G4COM, G4IDW	Southampton
10:30	145.525	FM	G8SC, G3EKJ	East Sussex
11:00	145.525	FM	G3XVN, G4MEO	Beds, Herts
12:00	51.530	FM	G3MEH, G6NB, G8BQH	Herts, Oxon, Bucks
12:00	145.525	FM	G3MEH, G6NB, G8BQH	Herts, Oxon, Bucks
12:00	433.525	FM	G3MEH	West Herts
18:00	433.150 *(via GB3SK)*	FM	G6DIK	Canterbury
20:00	433.525	FM	G4OBE, G7TAI	Enfield, North London

South West & Channel Is *(RSGB Zone D)*

TIME	FREQ	MODE	READER	LOCATION
09:00	145.525	FM	GJ0PDJ, GJ0JSY, GJ7HTV	Jersey
09:30	51.530	FM	GU0ELF, GU1HTY, GU1WJA	Guernsey
09:30	145.525	FM	GU0ELF, GU1HTY, GU4WMG, GU4XGG, GU1WJA	Guernsey
09:30	145.725 *(via GB3NC)*	FM	G3NPB, G4USB, G4BHD	St. Austell
09:30	433.525	FM	G0NZU, G3RFX	Bristol
10:00	3.640	LSB	G0LRJ, G4ZYF, G0NZU G3PLE	Plymouth
10:00	145.525	FM	G3ZYY, G0LRJ	Plymouth
10:30	145.525	FM	G2HDR, G0NZU	Bristol
11:00	145.525	FM	G1OCN, G1WIK	Portland, Weymouth
19:30	433.525	FM	G1OCN, G1WIK	Portland, Weymouth
20:30	145.525	FM	G0GRI, G4YXS, G0HFX	Bradford-on-Avon

Midlands *(RSGB Zone B)*

TIME	FREQ	MODE	READER	LOCATION
09:30	3.650	LSB	G2CVV, G8QZ, G3SZJ	Derby, Notts
09:30	144.250	USB	G3KQF, G4AFJ	Derby, Hinckley
10:00	145.525	FM	G7RUF, G7RUE	Sutton Coldfield
10:00	433.525	FM	G1ZCW	Louth
10:30	145.525	FM	G1ZCW	Louth
10:30	433.525	FM	G7UBA, G7UGX	Birmingham South
12:00	433.200 *(via GB3TF)*	FM	G3JKX, 2E1DJM	Telford
18:00	145.525	FM	G3USF, G0VVT	Keele, Stoke-on-Trent

Wales *(RSGB Zone E)*

TIME	FREQ	MODE	READER	LOCATION
09:00	3.650	LSB	GW3JSV, G3RFX	Welshpool, Bristol
11:00	145.525	FM	GW8TVX	Swansea
11:00	145.525	FM	GW3KJW, GW0PZT	Pwllheli
12:00	145.775 *(via GB3PW)*	FM	GW3KFE, GW4NQJ, GW3JSV	Newtown
18:30	145.525	FM	GW0AQR, GW4VEQ, GW0ABL	North Wales coast

East Anglia *(RSGB Zone C)*

TIME	FREQ	MODE	READER	LOCATION
09:30	3.640	LSB	G4NZQ, G4RKK	Norwich
09:30	145.525	FM	G0OZS, M0AKK, G4YQC	Ipswich, Felixstowe
19:00	145.525	FM	G4NZQ, G4RKK	Norwich

North of England *(RSGB Zone A)*

TIME	FREQ	MODE	READER	LOCATION
09:00	145.525	FM	G3AVJ, G8NNS	Liverpool, Wallasey
09:00	145.525	FM	G4OLK, G7GJU	Tyne, Tees
09:30	51.530	FM	G4LAA	Carlisle *(beaming N)*
09:30	144.250	USB	G4LAA	Carlisle *(beaming N)*
09:30	145.525	FM	G2FKZ, G4IOD	Wakefield, Huddersfield
10:00	144.250	USB	G3SMT, G3SMM	Stockport, Sale *(beaming NW)*
10:00	144.250	USB	G4KUX	Barnard Castle *(beaming N)*
10:00	145.525	FM	G4NVD, G8EQZ	Grimsby, Hull
10:30	3.640	LSB	G3LEQ, G0MRL	Knutsford, Bolton
10:30	51.530	FM	G3LEQ, G3IJE, G0MRL	Knutsford, Bolton
10:30	51.530	FM	G4LAA	Carlisle *(beaming W)*
10:30	144.250	USB	G4LAA	Carlisle *(beaming W)*
10:30	145.525	FM	G3LEQ, G4GSY, G0MRL	Knutsford, Bury, Bolton
10:30	433.350 *(via GB3MR)*	FM	G3LEQ, G4GSY, G0MRL	Stockport
11:00	3.640	LSB	G5VO	Bridlington
11:30	145.525	FM	G4FCH, G4ZGP	Scarborough
21:00	51.530	FM	G3LEQ, G3IJE, G0MRL G0BAA, G0OZJ	Knutsford, Bolton Cheadle
21:00	145.525	FM	G3LEQ, G4GSY, G0MRL, G0BAA, G0OZJ	Knutsford, Bury, Bolton Cheadle
21:00	433.350 *(via GB3MR)*	FM	G3LEQ, G4GSY,G0MRL, G0BAA, G0OZJ	Stockport

Scotland *(RSGB Zone G)*

TIME	FREQ	MODE	READER	LOCATION
09:00	145.525	FM	GM4DQJ, GM6OFO	Perth
09:30	145.650 *(via GB3OC)*	FM	GM7GMC, GM0IJV, GM0EXN	Kirkwall
09:30	145.525	FM	GM4DTH, GM4AQO	Firth of Forth
10:00	51.530	FM	GM4ILS	Elgin
10:00	145.525	FM	GM4ILS	Elgin
10:00	145.525	FM	GM4OPU	Fort William
10:00	145.525	FM	GM3VTB, GM4COX	Glasgow
10:30	51.530	FM	GM0PKW, GM3JIJ	Isle of Lewis
10.30	144.250	USB	GM1TDU	Aberdeen
10:30	145.525	FM	GM0BPO, GM0HNX	Borders
10:30	145.525	FM	GM0PKW, GM3JIJ	Isle of Lewis
10:30	433.525	FM	GM1GEQ, GM1CNH	Edinburgh
11:00	3.650	LSB	GM4OPU	Fort William
11:30	3.640	LSB	GM3HGA, GM3VEY	Aberdeen
11:30	3.650	LSB	GM3TCW, GM3CIX	Wishaw, Barrhead
11:30	51.530	FM	GM0JKF, GM1TDU	Aberdeen
11:30	145.675 *(via GB3LU)*	FM	GM0ILB	Lerwick
12:30	3.640	LSB	GM0ILB	Shetland

Northern Ireland *(RSGB Zone F)*

TIME	FREQ	MODE	READER	LOCATION
09:00	145.525	FM	GI4MJD, GI3GGY	Londonderry
09:30	145.525	FM	GI3WEM, GI4FUM, GI8AYZ, GI8WBZ	Banbridge, Antrim, Lisburn, Carrickfergus
10:30	3.650	LSB	GI0RYK, GI0DVU, GI3TLT	Lisburn, Newtownards
11:30	51.530	FM	GI8AYZ, GI0RYK, GI8WBZ	Lisburn, Newtownards, Carrickfergus
11:30	144.250	USB	GI8AYZ, GI0DVU, GI3WEM	Lisburn, Banbridge
11:30	433.050 *(via GB3UL)*	FM	GI8AYZ, GI0VTS, GI8WBZ	Belfast
21:00	145.525	FM	GI8LDM	Enniskillen

Notes

[1] Propagation news only, sent at approximately 15 wpm.
[2] National news followed by all regional news.
[3] Selected news items sent at 30, 26, 22, and 18 wpm.

Tel no:
(01438) 351710

AKD

PARSONS GREEN ESTATE
BOULTON ROAD,
STEVENAGE, HERTS SG1 4QG.

Fax no:
(01438) 357591

NEW ADDITION TO THE TARGET HF3 RANGE OF RECEIVERS

BRITISH MADE **MODEL HF3E** CE APPROVED

With parallel computer Interface and Receiver Control Software including Spectral Display plus serial interface for WEFAX, RTTY, Fax, etc.

- ★ Frequency range 30kHz - 30MHz
- ★ USB, LSB, AM
- ★ 1kHz steps with clarify
- ★ Filter bandwidth SSB=2.6kHz wide AM=6kHz wide
- ★ Fully synthesised employing a phase lock loop VCO to ensure stable and accurate signal reception
- ★ Quasi synchronous demodulator
- ★ SO239 antenna connection
- ★ Discreet backlit display
- ★ 10 user programmable memories
- ★ Tilt foot for ease of desktop view of display
- ★ UK 12V PSU and wire aerial
- ★ Audio ouput 2 watts
- ★ Extension speaker socket

TARGET HF3 AND HF3M RECEIVERS

HF3 – £159.95 + £6 P&P

- ★ Fully synthesised employing a phase lock loop VCO to ensure stable and accurate signal reception
- ★ Frequency range 30kHz - 30MHz
- ★ CE approved
- ★ 1kHz steps with clarify
- ★ Audio output 2 watts
- ★ Headphone socket
- ★ UK PSU and wire aerial

HF3M – £209.95 + £6 P&P

Also includes:

- ★ Built-in weatherfax interface
- ★ Wefax disc and software
- ★ Interconnection cable to PC (9 PIN serial)

NEW

BRITISH MADE

TRANSCEIVERS CE APPROVED

£199.95 inc VAT (Add £6 p&P)

NEW

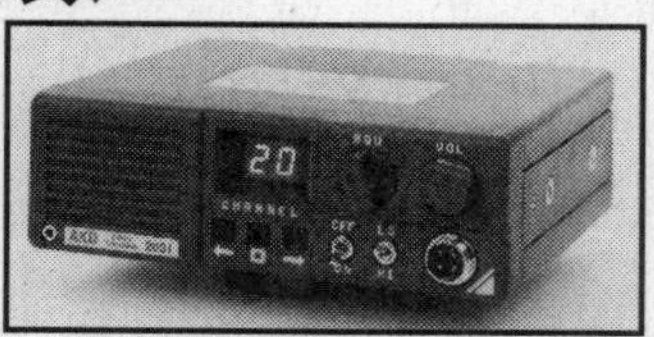

2 MTR MODEL 2001
144·500-145·987S
PTT Tone Burst, Listen on Input Facility, 12.5kHz Spacing 25/5 Watts.

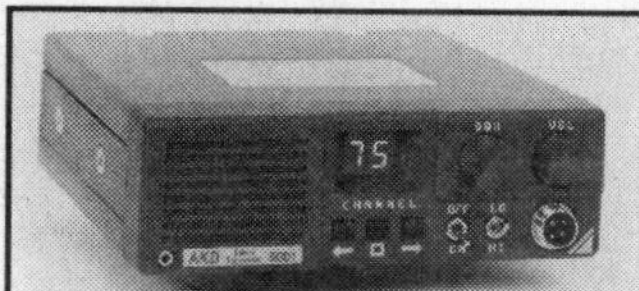

6 MTR MODEL 6001
50·500-51·990
10kHz Spacing Where Applicable. 25/5 Watts CTCSS Tone held in Non-Volatile Memory.

*UPGRADES AVAILABLE FOR OLDER 2001 & 6001 MODELS PHONE FOR FURTHER DETAILS

70CMS MODEL 7003
432·500 to 434·975MHz
25kHz Spacing. Power, 3 Watts, PTT Tone Burst. Listen on Input.

4 MTR MODEL 4001
70·250 to 70·487S
12kHz Spacing.
Power 25/5 Watts.

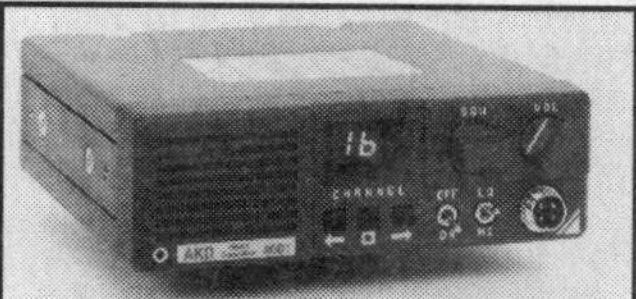

MANUFACTURERS OF TELEVISION INTERFERENCE FILTERS AND AMATEUR RADIO ACCESSORIES. PHONE FOR LATEST CATALOGUE OF FULL PRODUCT RANGE.

Web site:
http://www.kbnet.co.uk/akd
E-mail: akd@kbnet.co.uk

The RSGB's Internet Web Site

http://www.rsgb.org

Our Site

For those with access to the Internet, there is a wealth of information on every subject under the sun. For amateur radio, your first port of call should be the RSGB's site which contains the equivalent of more than 200 A4 printed pages, including the very latest news. The pages are by no means the most flashy that you will find on the worldwide web - we believe you want the information rapidly, rather than have to wait for too many graphics to download - but you will need a fairly recent browser and certainly one which can display tables.

The site is constantly being updated and expanded. It is very rare for two days to pass without some amendment being made. The pages are organised in eight sections, accessed from buttons which appear on every page.

Contents

The Contents page is the one which should be bookmarked on your browser. When you click on the bookmark (or type in www.rsgb.org/contents.htm) this page will download and display the current news headlines. These are changed at least once a week, when the new GB2RS script is posted, but hot news stories are often flagged up here first, so it is well worth checking this page every day or so if you want to be up with the latest news. The rest of the contents page shows an overview of the entire site with links to each part of each section.

News

By far the most popular pages on the site are the GB2RS news bulletin scripts which are usually published at the end of the Wednesday prior to the on-air reading. These are divided into Main and Local pages and include the full text of the Solar and Propagation news. A number of internal and external hypertext links are added to this version of the script. There is no archiving so it is important to check the news once a week.

Late breaking news stories and supplementary information can be found on the News Extra page. Any new stories are headlined on the Contents page (see above) but this page may contain other information which we feel will be of use after it ceases to be part of the GB2RS script. This is another rapidly changing part of the site.

An RSGB Events Calendar lists those rallies, conventions and open days where the Society has a book and information stand. Where appropriate, links are provided to sites dedicated to those events.

Non-members can see the front page and a list of the contents of the latest *RadCom* each month, linked of course to a page showing the benefits of membership.

Books

One of the largest sections is the on-line book catalogue. On display are the covers of all RSGB publications, together with synopses and chapter headings. This gives you a much more comprehensive view of each book and item of software than appears in *RadCom*. Non-RSGB books are also listed.

It couldn't be easier to order books via the Internet using your credit card and our secure server.

Purchasers of *The LF Experimenters Guide* can copy from the web site the BASIC computer listings which appear in the book. This will save a lot of typing.

Every week, the catalogue features a special offer where a selected book or books can be bought at a bargain price. This is another page worth checking regularly.

Membership

One of the aims of the Web site is to encourage radio amateurs to join the Society. This section explains what members get for their money, and has a simple joining form.

Also listed here is information on the hundreds of national and local clubs that are affiliated to the RSGB, listed by county.

Beginners

Another aim of the site is to recruit new people into our hobby. These pages explain why amateur radio is such an exciting and satisfying activity. A list of RAE courses is added for those who have become interested.

Society

Each RSGB committee has a page here, describing what they do and how to get in touch with them. Many committee pages are linked to others, either on our site or external sites, providing comprehensive news and information services.

Amongst the activities covered in this way are: Data Communications, HF, IOTA, Microwaves, Planning advice, Propagation, Repeaters, VHF and VHF Contesting.

The facilities of the Society's Headquarters are described, together with opening times etc. Each page includes the E-mail addresses of Society officers where appropriate.

Licensing

The Radiocommunications Agency has its own Web site which is linked from here. Also available is the full text of the most recent changes to the Amateur Radio Licences.

Operating

The full rules of the IOTA Programme can be found here, together with the very latest news and updates to the *IOTA Directory and Yearbook*. Anyone contemplating operating on the microwave bands will find many pages of advice and information, together with photographs of typical stations.

There is also an article describing why the UK is changing to 12.5kHz channel spacing on the 2m band, and how to measure your FM deviation level.

On-Line

This curious heading covers a page about the site, and - most usefully - a What's New page which is worth looking at every time you visit the site. There are also links to other useful Internet sites.

Search

A simple search facility so that you can find exactly which page contains the information you want.

Having Trouble Downloading the Latest Pages?

Most Internet browsers keep copies of your most recently looked at pages in a 'cache' file. This allows the browser to offer you a locally stored page, rather than taking an age to download the page on-line. The disadvantage of this is that you are likely to see an out-of-date version of a page which is constantly updated. If you are getting old versions of the GB2RS script (for instance) every time you check our pages, try pressing the 'Refresh' or 'Reload' button. Better still set up your browser to check for a revised page each time you go on-line, as follows:

Netscape Navigator: From the Options menu, select Network Preferences and click on the Cache tab. Under Verify Documents choose either 'Once per session' or 'Every time'. Click on OK to finish, and also on Save Options in the Options Menu.

Internet Explorer: From the View menu, select Options, take the Advanced tab and click on the Settings button in the Temporary Internet Files box. Choose either 'Every visit to the page' or 'Every time you start Internet Explorer'. Click on OK to finish.

Be Informed

Thousands of people use our Internet site every week. In fact an RSGB page is requested every minute of every day, 24 hours a day, seven days a week. For the very latest news, buying books or software, finding links to other amateur radio sites, or simply having a good read, the RSGB's site is the best available.

Contents	News	Books	Membership	Beginners
RSGB	Licensing	Operating	On-Line	Search

Operating Abroad

The following list of countries all permit a UK amateur to operate in their country. Some countries do this unilaterally (usually to any foreign radio amateur). Others have a reciprocal agreement whereby they recognise the UK licence, and in return, our RA recognises their licence. In addition to reciprocal licensing, many European and a few other countries also permit temporary operation with the minimum of formality, provided the licence conforms to certain common standards (CEPT recommendation T/R 61-01). Unfortunately no reciprocal agreements nor CEPT operation apply to novice licensees.

CEPT TR/61-01

The European Conference of Postal and Telecommunication Administrations (CEPT) is a group of some 40 European countries. Amongst their many actions, they have agreed a common standard of amateur radio licence (T/R 61-01) so as to facilitate the temporary operation in a fellow member's country.

Once each member's country has confirmed that their amateur radio licence conforms to the CEPT minimum standard, then its amateurs may operate in a fellow member's country which has also confirmed the recommendation. At the time of going to press, some 33 countries have implemented T/R 61-01 including the UK.

CEPT operation does not replace reciprocal licensing – rather it supplements it. Only temporary operation is permitted under CEPT rules e.g. from hotel accommodation or mobile. Therefore, if you seek either long term (over 3 months) residence or additional facilities, then you will still need to apply for a reciprocal licence.

Operating under CEPT regulations means that you are restricted by both the UK regulation's as well as the foreign country's i.e. it is the lowest common denominator of the two licences. For example, your UK licence permits you to operate on 70 MHz, but the Dutch licence does not – therefore you cannot operate on 4m in the Netherlands. However, the Dutch authorities permit full power operation on 430-432 MHz, but the UK licence does not – but you still cannot operate full power on 430 MHz! Incidentally, no CEPT class II licensees (UK 'B' licensees) are allowed to operate below 144MHz.

In order to operate under CEPT regulations, you need to have with you your UK licence validation document, a copy of the UK licensing regulations (BR68) *and a copy of the foreign country's licensing regulations*. (You will need to write to the foreign country's licensing administration to obtain a copy of the latter.) For more details, please see either your UK licence or the RA information sheet RA247 '*Operation Under CEPT*'.

The lists below shows which countries have implemented CEPT T/R 61-01, which class of UK licensee benefits and the prefix to be used.

Reciprocal Licensing

A reciprocal licence is a licence issued by a foreign country to you because that country recognises the standards of the UK licence. Some countries do this unilaterally, others require a two way recognition – a reciprocal licence. The call-sign issued is sometimes your own call-sign with the foreign country's suffix or prefix. In other countries, it is a call-sign allocated in their normal series of call-signs.

In general, due to overseas post and administration delays, it is best to allow at least 2 to 3 months for your application to be processed – longer if it is a third world country where amateur radio is not so sympathically regarded. Airmail does help! Be warned, not all countries recognise the UK class B licence – but these are shown in the table below. The details given are the best available, but some of this data can be elusive to find! Please let RSGB HQ know if your researches discover something new – thank you.

National Societies

Most countries have a national society which looks after the well being of that country's amateurs. Few of the are of the size of RSGB – indeed many are staffed entirely by volunteers. Nevertheless, they will all give you as much assistance as they can. Therefore, if you desire information about their band plans, repeaters, local clubs, or have difficulty in obtaining a reciprocal licence, please contact that country's national society.

Customs

There is usually little problem with customs. It certainly helps to be able to show that the equipment was purchased abroad and is not being exported. Unfortunately neither a reciprocal licence nor operation under CEPT regulations is deemed an exemption from customs formalities. If in-doubt, you should seek additional advice about importing/exporting equipment.

Albania

CEPT T/R 61-01: Not yet implemented.
National Society: Albanian Amateur Radio Assn, PO Box 1501, Tirana, Albania.

Andorra

Reciprocal licence: Unilateral, UK A + B, C30/C31 series.
Licensing address: (1) Delegation Permanent pour l'Andorre, Prefecture des Pyrenees-Orientales, Andorrans, 66 000 Perpignan, France. (2) Vegueria Episcopal, c/prat de la Creu, La Vella, Andorra.
National society: Unio de Radioaficionats, PO Box 150, Principality of Andorra Tel: 00 33 376 825380.

Australia

Reciprocal licence: UK A + B, VK series.
Licensing address: Spectrum Management Agency, Purple Building, Benjamin Offices, Chan Street, Belconnen ACT 2617, Australia Tel: 06 256 5555 Fax: 06 256 5200 Web site: www.sma.gov.au
National society: Wireless Institute of Australia, PO Box 2175, South Caulfield, Victoria 3161, Australia Tel: 03 9528 5962 Fax: 03 9523 8191.

Austria

CEPT T/R 61-01: UK A+B, OE/own call.
Reciprocal licence: UK A + B, OE series.
Licensing address: BMöWV, Sektion IV, Rechtsabteilung, Kelsentr. 7, A1030 - Vienna, Austria. Tel: 00 43 1 79731 4100.
National society: Osterreichischer, Versuchssenderverband, Theresiengasse 11, A-1180 Vienna, Austria. Tel: 43 1 408 55 35.

Bahamas

Reciprocal licence: Unilateral, UK A only, C6A series.
Licensing address: Bahamas Telecomms Corp. PO Box N-3048, Nassau, Bahamas. Tel: 00 1 809 323-4911.
National society: Bahamas ARS, PO Box GT-2318, Nassau, NP, Bahamas.

Barbados

Reciprocal licence: Unilateral, UK A only, 8P9 series.
Licensing address: Senior Telecoms Officer, Ministry of Public Works, Communications & Transportation, Herbert Ho, Fontabelle, St. Michael, Barbados, West Indies. Tel: 00 1 809 426 2669.
National society: Amateur Radio Soc of Barbados, PO Box 814E, Bridgetown, Barbados. E-mail: decarlo@sunbeach.net

Belgium

CEPT T/R 61-01: A + B, ON/own call.
Reciprocal licence: UK A + B, ON9 series.
Licensing address: BIPT, Sterrekundelaan 14, B-1030 Brussel, Belgium, Tel: 00 32 2 20 77777.
National society: UBA, Rue da la Presse 4, B-1000 Brussels, Belgium.

Bermuda

Reciprocal licence: Unilateral, UK A only, own call/VP9.
Licensing address: Dept. of Telecomms, PO Box 101, Hamilton 5, Bermuda. Tel: 00 1 809 29 5-5151 Ex 1120.
National society: Radio Society of Bermuda, PO Box HM 275, Hamilton HM AX, Bermuda. E-mail: gcuoco@ibe.bm

Bosnia-Herzegovina

CEPT T/R 61-01: Not yet implemented.
Licensing address: Savez Radioamatera Bosne i Hercegovine, PO Box 61, Dariela Ozme 7, 71 000 Sarajevo.

Brunei

Reciprocal licence: Unilateral, UK A + B.
Licensing address: Director of Telecommunications, Telecom Dept, Ministry of Communications, Negara, Brunei Darussalam. Tel: 00 673 42324.
National society: Negara Brunei Darussalam ARA, PO Box 73, Gadong, Bandar Seri Begawan 3100, Brunei.

Botswana

Reciprocal licence: UK A only, A25/own call.
Licensing address: Botswana Telecoms Corporation, Radio Spectrum Management Section, PO Box 700, Gaberone, Botswana. Tel: 00 267 358000.
National society: Botswana ARS, PO Box 1873 Gaborone, Botswana.

Brazil

Reciprocal licence: UK A only.
Licensing address: Departamento Nacional, de Telecomicacoes, Dentel 4 Andar, Ministerio das Comunicacoes, 7000 Brasillia DF, Brazil.
National society: Liga de Amadores Brasileiros de Radio Emissao, Setor de Clubes Esportivos Sul, Trecho 04, Lote 1/A, 70359-970 Brasila DF, Brazil. E-mail: py5eg@sul.com.br

Bulgaria

CEPT T/R 61-01: UK A + B, LZ/own call.
Reciprocal licence: Unilateral, A only.

Licensing address: Committe of Posts and Telecomms, Radioregulatory Dept, Gurko Street,No 6, 1000 Sofia, Bulgaria. Tel: 00 3592 88 9511.
National society: Bulgarian Fed of Radio Amateurs, PO Box 830, 1000 Sofia-C, Bulgaria.

Canada
CEPT T/R 61-01: UK A + B, VE/own call.
Reciprocal licence: UK A + B, VEx/own call.
Licensing address: The Industry Canada, Amateur Radio Dept, 300 Slater St, Ottawa, Ontario, K1A 0C8. Tel: 001 613 998 3693 1700.
National society: Radio Amateurs of Canada, 720 Belfast Road, Suite 217, Ottawa, ON K16 0Z5. E-mail: rachq@king.igs.net

Cayman Is
Reciprocal licence: Unilateral, UK A only?, ZF1.
Licensing address: The Postmaster, General Post Office, Grand Cayman, British West Indies. Tel: 00 1 809 94 92474.
National society: Cayman ARS, PO Box 1029, Grand Cayman, Cayman Is. Web site: zfla@candw.ky

Chile
Reciprocal licence: Unilateral, UK A only.
Licensing address: Radio Club de Chile, Casilla 13630, Correo, Santiago, Chile.
National society: Radio Club de Chile, PO Box 13630, Santiago 21, Chile.

China
Reciprocal licence: Unilateral, UK A only.
Licensing address: Liaison Dept, Chinese Radio Sports Assn., PO Box 6106, Beijing, China 100061, Tel: 00 86 10 67025488.
National society: Chinese Radio Sports Assn., 9 Tiyugan Road, PO Box 6106, Beijing, China 100061. Tel: 00 86 10 67025488

Columbia
Reciprocal licence: Unilateral, UK A only, HK7/own call
Licensing address: Ministerio do Communicaciones, Bogata DE1, Columbia.
National society: Liga Colombiana de Radioaficionados, A.A.584, Santafe de Bogata, Columbia.

Costa Rica
Reciprocal licence: Unilateral, UK A only.
Licensing address: Apply via national society.
National society: Radio Club de Costa Rica, PO Box 2412, San Jose 1000, Costa Rica.

Croatia
CEPT T/R 61-01: UK A + B, 9A/own call.
National society: Hrvatski Radio-Amaterski Savez, Dalmatiska 12, HR-10000 Zagreb, Croatia. Tel: 00 385 1 433 025.

Cyprus
CEPT T/R 61-01: UK A = 5B4/own call, UK B = 5B8/own call.
Reciprocal licence: UK A + B.
Licensing address: Chief Comms. Officer, Ministry of Comms & Works, 1424 Nicosia, Cyprus. Tel: 00 357 2 302268 Fax: 00 357 2 360578.
National society: Cyprus ARS, PO Box 1267, Limassol, Cyprus. Tel: 00 357 5 362792.

Czech Republic
CEPT T/R 61-01: UK A + B, OK/own call
Reciprocal licence: UK A + B
Licensing address: Czech Communication Office, Att: Mrs Bockov, Klimentsk 27, 125 02 Praha 1, Czech Rep. Tel: 00 42 2 2491 1605. Fax: 00 420 2 2491 1658.
National society: Czech Radio Club, PO Box 69, 113 27 Praha 1, Czech Republic. Tel:00 420 2 8722 240. Fax: 00 420 2 8722 242.

Denmark
CEPT T/R 61-01: UK A + B, OZ/own call
Reciprocal licence: UK A + B, OZ/own call.
Licensing address: Telestyrelsen, Holsteinsgade 63, DK-2100 Kobenhavn 0, Denmark. Tel: 00 45 35 43 03 33.
National society: Eksperimenterende Danske Radioamatorer, PO Box 172, DK-5100 Odense, Denmark. Tel: 00 45 66 15 6511.

Estonia
CEPT T/R 61-01: ES#/own call.
Reciprocal licence: UK A + B
Licensing address: Inspection of Telecommunications Republic of Estonia, Adala 4D, E0006 Tallinn, Estonia. Tel: 00 372 6399075.
National society: Eesti Raadioamatooride Uhing, PO Box 125, EE-0090 Tallinn, Estonia. Tel: 00 372 2 449312.

Falkland Is
Reciprocal licence: Unilateral, UK A + B, VP8 series.
Licensing address: Superintendent, Post & Telecommunications, The Post Office, Port Stanley, Falkland Is. Tel: 00 500 27135.

Faroe Is
CEPT T/R 61-01: UK A + B, OY/own call.
Reciprocal licence: UK A + B, OY/own call.
Licensing address: National Telecom Agency, Holsteinsgarde 63, DK-2100 Kobenhavn), Denmark. Tel: 00 45 35 43 03 33.
National society: Foroyskir Radioamatorar, PO Box 343, FR-110 Torshavn, Faroe Islands. Tel: 00 45 9 298 10644.

Fiji Islands
Reciprocal licence: Unilateral, UK A+B.
Licensing address: The Director, Reglatory Unit, Ministry of Information, Broadcasting Television & Telecommunications, Goverment Bulidings, Suava, Fiji Islands, Tel: 00 679 211 257.
National society: Fiji ARA, PO Box 184, Suva, Fiji Islands.

Finland
CEPT T/R 61-01: UK A + B, OH/own call.
Reciprocal licence: Reciprocal, A + B, OH/own call.
Licensing address: Telecommunications Administration Centre, PO Box 53, SF-00211 Helsinki, Finland. Tel: 00 358 0 696 61.
National society: Suomen Radioamatooriliitto, PO Box 44, SF-00441 Helsinki, Finland. E-mail: sral@compart.fi Web site: www.compart.fi/sral/

France
CEPT T/R 61-01: UK A + B, F/own call. (Overseas territories use different prefix. Visting and/or operating permission may also be needed.)
Reciprocal licence: UK A + B, F series.
Licensing address: Direction Generale des Telecommunications, Direction des Affairs Industrielles et Internationales, (DA11) - Service des Affairs Internationales (SA1), 7 Boulevard Romain Rolland, F-92128 Montrouge, France. Tel: 00 33 1 45 64 22 22.
National society: REF - Union, BP7429, 37074 Tours Cedex 2, France. Tel: 00 33 47 418873.

Gambia
Reciprocal licence: Unilateral, UK A + B, C53/own call.
Licensing address: The Managing Director, Gambia Telecommunications Co Ltd, PO Box 387, Banjul, The Gambia.Tel: 00 220 29999.
National society: Radio Soc of the Gambia, c/o Jean-Michel Voinot, C53GB, PMB 120, Banjul, Gambia.

Germany
CEPT T/R 61-01: UK A = DL/own call, UK B = DC/own call.
Reciprocal licence: Reciprocal, DL/DC series.
Licensing address: BAPT, Postfach 100353, D-64203, Darmstadt, Germany. Tel: 00 49 6151 301161. Fax: 00 49 6151 301181.
National society: Deutscher Amateur Radio Club, PO Box 1155, D-34216 Baunatal 1, Germany. Tel: 00 49 561 94 98 80. Fax:00 49 561 94988 50.

Gibraltar
Reciprocal licence: UK A + B, ZB2/own call.
Licensing address: Wireless Officer, Gibraltar Post Office, 104 Main St, Gibraltar. Tel/fax: 00 350 75714.
National society: Gibraltar ARS, PO Box 292, Gibraltar. Tel: 00 350 75452. Web site: www.gibnet.com

Greece
CEPT T/R 61-01: UK A + B, SV#/own call.
Reciprocal licence: UK A +B, SV#/own call.
Licensing address: Ministry of Transport & Telecommunications, Administration of Posts & Telecommunication, 49 Avenue Syngrou, GR 117 80 Athens, Greece. Tel: 00 30 1 923 2906.
National society: Radio Amateur Association of Greece, PO Box 3564, GR-10210, Athens, Greece. Tel: 30 1 522 6516.

Greenland
Reciprocal licence: UK A + B, OX/own call.
Licensing address: National Telecom Agency, Holsteinsgarde 63, DK-2100 Kobenhavn 0, Denmark. Tel: 00 45 35 43 03 33.

Hong Kong
Reciprocal licence: UK A + B, VR2 series.
Licensing address: Officer in Charge, Telecommunications Branch, 19th floor, Sincere Building, 173 Des Voeux Road, Central, Hong Kong.
National society: Hong Kong ARTS, PO Box 541, Hong Kong.

Hungary
CEPT T/R 61-01: UK A = HA/own call, UK B = HG/own call.
Reciprocal licence: Unilateral, UK A only.
Licensing address: Frekvencia Gazdalkodasi Intezet, Ostrom U.23/25, H/1012 Budapest, Hungary.
National society: Magyar Radioamator Szovetseg, PO Box 11, H-1400, Budapest, Hungary. Tel: 00 36 1 312 1616.

Iceland
CEPT T/R 61-01: UK A + B, TF/own call.
Reciprocal licence: UK A + B.
Licensing address: National Telecom Inspectorate, Malarhofda 2, IS - 150 Reykjavik, Iceland. Tel: 00 354 587 2424.
National society: Islenkir Radioamatorar, PO Box 1058, IS-121, Reykjavik, Iceland. Tel: 00 354 552 0157. Fax: 00 354 568 0161. Web site: www.itn.is

India
Reciprocal licence: UK A + B, VU series.
Licensing address: Ministry of Comms, Sanchar Bhavan, 20 Ashoka Road, New Dehli 110001, India. Tel: 00 91 11 3355441. Fax: 00 91 11 371 6111.
National society: Amateur Radio Society of India,4 Kurla Industrial Estate, Bombay 400 086, India.

Indonesia
Reciprocal licence: Unilateral, A only.
Licensing address: Apply via national society.
National society: Organisasi Amatir Radio Indonesia, PO Box 6797 JKSRB, Jakarta 12067, Indonesia. Tel: 62 21 582226.

Irish Republic
CEPT T/R 61-01: UK A + B, EI/own call.
Reciprocal licence: UK A + B, EI: maninland, EJ: offshore islands.
Licensing address: Office of Director of Telecommunications Regulation, Radio Section, Abbey Court Irish Life Centre, Lower Abbey Street, Dublin 1. Tel: 00 353 1 804 9600. Fax: 00 353 1 804 9680.
National society: Irish Radio Transmitters Society, PO Box 462, Dublin 9, Ireland. E-mail: jryan@iol.ie

Israel
CEPT T/R 61-01: UK A = 4X/own call, UK B = 4Z/own call
Reciprocal licence: UK A + B.
Licensing address: Ministry of Communications, PO Box 29107, Tel Aviv 61290, Israel. Tel: 03 5198277.
National society: Israel Amateur Radio Club, PO Box 17600, Tel Aviv 61176, Israel.

Italy
CEPT T/R 61-01: UK A = IK#/own call, UK B = IW#/own call.
Reciprocal licence: UK A + B.
Licensing address: Ministero delle Poste e Telecom, Direzione Centrale dei Servizi Radioelettrici, Divisione 6 - Sezione 4, Viale Europa 175, I-00144 Roma, Italy. Tel: 00 39 6 5954894.
National society: Associazione Radioamatori Italiani, Via Scarlatti 31,20124 Milan, Italy. Tel: 00 39 2 669 21 92. Fax: 00 39 2 667 14809.

Jamaica
Reciprocal licence: Unilateral, UK A only, own call/6Y5.
Licensing address: Headquarters, Post & Telegraphs Dept, South Camp Rd, Kingston, Jamaica.
National society: Jamaica Amateur Radio Assn., 76 Arnold Road, Kingston 5, Jamaica. E-mail: 6y5mm@toj.com

Jordan
Reciprocal licence: Unilateral, UK A only? JY9 series.
Licensing address: Apply via national society.
National society: Royal Jordanian Radio Amateur Society, PO Box 2353, Amman, Jordan. Tel: 962 6 666 235.

Kenya
Reciprocal licence: Unilateral, UK A only, 5Z4 series.
Licensing address: Managing Director, Kenya Posts & Telecommunications, PO Box 30301, Nairobi, Kenya. Tel: 00 254 227401 Ex 2385.
National society: Radio Society of Kenya, PO Box 45681, Nairobi, Kenya. Tel/fax: 00 254 35 21400.

Latvia
CEPT T/R 61-01: UK A + B, YL/own call.
Licensing address: Latvis Communication State Inspection, 41/43 Elizabetes Street, Riga LV1010, Latvia. Tel: 371 2 333034.
National society: Latvijas Radioamatieru Liga, PO Box 164, Riga, LV-1098, Latvia. Tel: 00 371 2 333187.

Liechtenstein
CEPT T/R 61-01: UK A + B, HB0/own call.
Reciprocal licence: Unilateral, UK A + B?.
Licensing address: Direction Generale, Radio und Fernsehen, Speichergasse 6, 3030 Bern, Switzerland. Tel: 00 41 75 6 11 11.
National society: Amateurfunk Verein Liechtenstein, PO Box 629, FL-9495 Triesen, Liechtenstein.

Lithuania
CEPT T/R 61-01: UK A & B, LY/own call.
Licensing Administration: Valstybine Radijo Dazniu Tarnbya, Algirdo 27, Vilnius, Lithuania 2006. Tel: 00 370 2 26 31 77.
National society: Lietuvos Radijo Megeju Draugija, PO Box 1000, Vilnius, Lithuania 2001. Tel: 00 370 2 221 836. Fax: 00 37 02 700447. E-mail: hq@lrmd.ktl.mii.it

Luxembourg
CEPT T/R 61-01: UK A + B, LX / own call.
Reciprocal licence: Unilateral, UK A only, LX series.
Licensing address: Ministère des Communications, 18 Montée de la Pétrusse, L-2945 Luxembourg. Tel: 00352 478-1.
National society: Reseau Luxembourgeois des Amateurs d'Ondes Courtes, 23 Route de Noertzange, L-3530 Dudlange, Luxembourg.

Madeira
CEPT T/R 61-01: Not yet implemented.
Reciprocal licence: UK A + B
Licensing address: ICP Instituto Das Comunicaçoes Portugal, Delegaçao do ICP Madeira, Centro Fiscalizacao, Pico da Cruz, 9000 Funchal. Tel: 00 351 91 762868.
National society: Rede dos Emissores Portugueses, Rua D. Pedro V 7-4, P-1200 Lisboa, Portugal. Tel:00 351 1 346 1186. Fax: 00 351 1 342 0448.

Malaysia
Reciprocal licence: UK A only.
Licensing address: IBU Pejabat, Jabatan Telekom Malaysia, (Kementerian Tenega Telekom Dan Pos Malaysia), Wisma Damansra, Jalan Sematan, 50668 Kuala Lumpur, Malaysia. Tel: 00 60 32556687.
National society: Malaysian ARTS, PO Box 10777, 50724 Kuala Lumpur, West Malaysia

Malta
CEPT T/R 61-01: Not yet implemented.
Reciprocal licence: UK A + B, 9H3 series.
Licensing address: Wireless Telegraphy Dept, Evans Laboratory Buildings, Merchants St, Valletta, Malta. Tel: 00 356 24 39 25 / 23 27 28. Fax: 00 356 23 44 94 / 23 36 95.
National society: Malta ARL, PO Box 575, Valletta, Malta.

Mauritius
Reciprocal licence: UK A only.
Licensing address: Mauritius Telecommunication Authority, 6th Floor, Blendax House, Dumas Street, Port Louis, Mauritius. Tel: 00 230 208 5623. Fax: 00 230 211 2871.
National society: Mauritius Amateur Radio Society, PO Box 104, Quatre-Bornes, Mauritius.

Monaco
CEPT T/R 61-01: UK A + B, 3A/own call.
Reciprocal licence: UK A + B, 3A/own call.
Licensing address: Direction Générale de Telecomms, Service Radio-Amateurs, 25 Boulevard de Suisse, MC 98030 Monaco Cedex, Monaco. Tel: 00 33 93 25 05 05.
National society: Association des Radio Amateurs de Monaco, PO Box 2, MC-98001 Monaco Cedex, Monaco.

Montserrat
Reciprocal licence: Unilateral, UK A only? VP2M series.
Licensing address: Ministry of Communications & Works, General Turning Road, Plymouth, Montserrat, West Indies.
National society: Montserrat ARS, PO Box 448, Plymouth, Monserrat.

Morocco
Reciprocal licence: Unilateral, UK A only, own call/CN.
Licensing address: Ministere des PTT, Division des Telecommunications, Rabat, Moroco.
National society: Association Royale des Radio Amateurs du Maroc, PO Box 299, Rabat, Moroco.

Mongolia
Reciprocal licence: Unilateral, UK A , JT/own call.
Licensing address: MRSF, PO Box 639, Ulaanbaatar - 13, Mongolia. E-mail: jt1kaa@magicnet.mn
National society: Mongolian Radio Sports Federation, PO Box 639, Ulaanbaatar 13, Mongolia. Tel: 00 976 1 320058. E-mail: mrsf@magicnet.mn

Netherlands
CEPT T/R 61-01: UK A + B, PA/ own call.
Reciprocal licence: UK A + B, PA /own call
Licensing address: Minister of of Transport & Public Works, Dept of Telecommunication & Post, PO Box 450, 9700 AL Groningen, Netherlands. Tel: 00 31 50 522 2214.
National society: VERON, PO Box 1166, NL-6801 BD Arnhem, Netherlands. Tel: 31 264 426760.

New Zealand
CEPT T/R 61-01: UK A + B, own call/ZL.
Reciprocal licence: UK A + B, own call/ZL
Licensing address: Radio operations, Comms. Div, Ministry of Commerce, PO Box 1473, Wellington, New Zealand. Tel: 00 64 4 472 0030.
National society: New Zealand ART, PO Box 40-525, Upper Hutt, New Zealand. Tel/Fax: 00 64 4 528 2170.

Norway
CEPT T/R 61-01: UK A + B, LA/own call,
Reciprocal licence: UK A only, LA0 series.
Licensing address: NTRA, PO Box 447 Sentrum, N-0104 Oslo, Norway. Tel: 00 47 22 82 48 38. Fax: 00 47 22 48 40.
National society: Norsk Radio Relae Liga, PO Box 20, Haugenstua, N-0915 Oslo, Norway. Tel: 00 47 22 21 37 90. Fax: 00 47 22 21 37 91

Panama
Reciprocal licence: Unilateral, UK A only? own call/HP4X.
Licensing address: Ministerio de Gobierno y Justicia, Direccion Nacional de Medio de Comunicacion Social, Apartado Postal 1628, Zona 1, Panama, Republic de Panama.
National society: Liga Panamena de Radioaficionados, PO Box 9A-175, Panama.

Papua New Guinea
Reciprocal licence: UK A + B, P29V series.
Licensing address: Manager Radio Branch, Post & Telecomms Corp, Spectrum Management Dept, PO Box 1783, Port Moresby, Papua New Guinea. Tel: 00 675 27 4236.
National society: Papua New Guinea ARS, PO Box 204, Port Moresby, NCD, Papua New Guinea.

Phillipines
Reciprocal licence: UK A only?own call/DU1.
Licensing address: Planning Division, Telecommunication Control Bureau, 5th floor, Delos Santos Building, Quezon Avenue, Quezon City, Philippines.
National society: Phillippine Amateur Radio Assn, PO Box 4083, Manila, Philippines.

Peru
CEPT T/R 61-01: UK A + B, OA/own call.
Reciprocal licence: Unilateral, UK A only, OA/ own call.
Licensing address: Ministerio de Transportes Comunicaciones, Vivienda Y Construccion,

Av. Wilson y 28 de Julio, Cercado de Lima, Lima, Peru. Tel: 00 51 1 433 7800.
National society: Radio Club Peruano, PO Box 538, Lima 100, Peru. E-mail: oficina@oabbs.org.pe

Pitcairn + Henderson Is
Reciprocal licence: Unilateral, UK A only?
Licensing address: Office of the Governor of Pitcairn Henderson, Ducie and Oeno Islands, c/o British Consulate General, Auckland, New Zealand.

Poland
CEPT T/R 61-01: Not yet implemented .
Reciprocal licence: UK A + B, SO series
Licensing address: Panstwowa Agencja Radiokomunikacyjna, Zarzad Krajowy ul. Kasprzata 18/20, 01-211 Warsaw, Poland.
National society: Ploski Zwiazek Krotkofalowcow, PO Box 61, 64-100 Leszno 1, Poland. Tel/Fax: 00 48 65 209529.

Portugal
CEPT T/R 61-01: A + B, CT/own call
Reciprocal licence: UK A + B.
Licensing address: ICP Instituto Das Comunicaçoes Portugal, Delegaçao do ICP Portugal, Centro Fiscalizacao, Pico da Cruz, 9000 Funchal. Tel: 00 351 91 762868.
National society: Rede dos Emissores Portugueses, Rua D. Pedro V 7-4, P-1200 Lisboa, Portugal. Tel: 00 351 1 36 11 86.

Romania
CEPT T/R 61-01: UK A + B, YO/own call
Reciprocal licence: Unilateral, UK A + B?
Licensing address: General Inspectorate of Radiocommunications, 202A Splaiul Independentei sector 6, Bucharest, 77208 Romania. Tel: 00 40 386891. Fax: 00 40 1 3124797.
National society: Federatia Romana de Radioamatorism, PO Box 22-50, R-71100 Bucharest, Romania. Tel: 00 40 01 211 97 87.

Russian Federation
CEPT T/R 61-01: Not yet implemented.
Reciprocal licence: Unilateral, UK A only.
Licensing address: Apply via national society.
National society: Union of Radioamateurs of Russia, PO Box 59, Moscow 105122, Russia. Tel: 00 7 095 9393741.

San Marino
CEPT T/R 61-01: Not yet implemented.
Licensing address: Direzione Generale Post e Telecomms., San Marino. Tel: 378 991349.
National society: Associazione Radioamatori Della Repubblica Di San Marino, PO Box 77, RSM 47031, San Marino. E-mail: arrsm@internt.sm/arrsm Web site: www.internet.sm/arrsm

Seycheles
Reciprocal licence: Unilateral, UK A only, SK79 series
Licensing address: Seychelles Licensing Authority, PO Box 3, Victoria Mahe, Republic of Seychelles.

Sierra Leone
Reciprocal licence: UK A + B?
Licensing address: Licensing Office, Posts & Telegraphs Dept, GPO, Freetown, Sierra Leone.
National society: Sierra Leone ARS, PO Box 10, Freetown, Sierra Leone.

Singapore
Reciprocal licence: Unilateral, UK A only, 9V1 series.
Licensing address: Telecoms Authority of Singapore, Radio Licensing Dept, 4-8 George Street #04-00, Singapore 0104. Tel: 00 65 5380402.
National society: Singapore ARTS, GPO Box 2728, Singapore 9047, Singapore.

Slovakia
CEPT T/R 61-01: UK A + B, OM/own call.
Licensing address: Telekomunikacny Vrad SR, Povolovame Radiostamic, Jarosova 1, SK-830 08 Bratislava, Slovakia. Tel: 00 42 7 2792704.
National society: Slovak Amateur Radio Assn, Wolkrova 4, 851 01 Bratislava, Slovakia. Tel: 00 42 7 847 501. Fax: 00 42 7 845 138.

Slovenia
CEPT T/R 61-01: Not yet implemented.
National society: Zveza Radioamaterjev Slovenije, Lepi Pot 6, SI 1000 Ljubljana, Slovenia. Tel: 00 386 61 22 24 59. Fax: 00 386 61 22 24 59. E-mail: zrs-hq@s55tcp.ampr.org

Solomon Is
Reciprocal licence: Unilateral, UK A + B, H44 series.
Licensing address: Director, Ministry of Posts & Communications, PO Box G25, Honiara, Solomon Is.
National society: Solomon Island Radio Soc, PO Box 418, Honiara, Solomon Islands.

South Africa
Reciprocal licence: UK A = ZS#/own call, UK B = ZR#/own call.
Licensing address: Apply via national society.
National society: South African Radio League, PO Box 807, Houghton 2041, South Africa. Tel: 00 11 484 2830.

Spain
CEPT T/R 61-01: UK A = EA/own call, UK B = EB/own call.
Reciprocal licence: UK A = EA/own call, UK B = EB/own call
Licensing address: Direccion General de Telecomunicaciones, Pza. Cibeles, Palacio de Communicacions, 28014 Madrid, Spain. Tel: 00 34 1 3461500. Fax: 00 34 1 3962229.
National society: Union de Radioaficionades Espanoles, PO Box 220, 28080 Madrid, Spain. Tel: 00 34 1 4771413. Fax: 00 34 1 4772071. E-mail: ure@ure.es Web Site: www.ure.es

Sri Lanka
Reciprocal licence: UK A + B.
Licensing address: The Director General of Telecomms. Regulatory Authority, EH Cooray Bldg, 4th floor, 411 Galle Rd, Columbo 3, Sri Lanka.
National society: Radio Society of Sri Lanka, PO Box 907, Colombo, Sri Lanka.

Swaziland
Reciprocal licence: UK A + B.
Licensing address: Attn: Engineer Frequency Management, Managing Director of Posts & Telecoms, PO Box 125, Mbabane, Swaziland.
National society: Radio Society of Swaziland, PO Box 3744, Manzini, Swaziland.

Sweden
CEPT T/R 61-01: UK A + B, SM/own call
Reciprocal licence: UK A + B, SM/own call
Licensing address: Post & Telestyrelsen, Frequency Division, Box 5398, S-102 49 Stockholm, Sweden. Tel: 00 46 8 678 5500. Fax: 00 46 8 678 5505.
National society: Foreningen Sveriges Sandareamatorer, Box 2021, S-123 26 Farsta, Sweden. Tel: 00 46 8 6044006. Fax: 00 46 8 604 4007. E-mail: hq@svessa.se Web site: www.svessa.se

Switzer.land
CEPT T/R 61-01: UK A + B, HB9/own call.
Reciprocal licence: UK A + B
Licensing address: Direction Generale, Radio und Fernsehen, Speichergasse 6, 3030 Bern, Switzerland. Tel: 00 41 31 338 51 91.
National society: Union Schweizerischer Kurzwellen-Amateure, PO Box 9, CH-4539 Rumisberg, Switzerland. Tel: 00 41 65 763676. E-mail: hq@uska.ch Web site: www.uska.ch

Tonga
Reciprocal licence: UK A + B, A35 + 2 letters.
Licensing address: Tonga Telecom. PO Box 46, Nukualofa. Tonga. Tel: 00 676 24255. Fax: 00 676 22200.
National society: Amateur Radio Club of Tonga, c/o Manfred Schuster, PO Box 1078, Nuku'alofa, Tonga.

Turkey
CEPT T/R 61-01: UK A + B, TA/own call,
Reciprocal licence: Unilateral, UK A only?
Licensing address: Directorate General of PTT, Ankara, Turkey. Tel: 00 90 43 12 52 52.
National society: Telsiz Radyo Amatorleri Cemiyeti, PO Box 699, Karakoy 80005, Istanbul, Turkey. Tel/Fax: 00 90 212 245 3942.

Virgin Is
Reciprocal licence: Unilateral, UK A + B, VP2V/own call,
Licensing address: Telecommunications (Radio) Officer, Ministry of Communications & Works, Government of BVI, Road Town, Tortola, British Virgin Islands.
National society: British Virgin Is Radio League, PO Box 4, West End, Tortola, British Virgin Islands.

USA
Reciprocal licence: UK A + B, Wx/own call
Licensing address: Federal Communications Commission, Gettysburg, PA 17326 USA.
National society: American Radio Relay League, 225 Main St, Newington, CT 06111, USA. E-mail: hq@arrl.org Web site: www.arrl.org

Uruguay
Reciprocal licence: Unilateral, UK A only.
Licensing address: Apply via national society.
National society: Radio Club Uruguayo, PO Box 37, Montevideo, Uruguay.

Vanuatu
Reciprocal licence: UK A only.
Licensing address: Director of Posts & Telecomms, Dept of Post & Telecomms, Vila, Vanuatu.
National society: Vanuatu Amateur Radio Society, PO Box 665, Port Vila 3092, Vanuatu.

Western Samoa
Reciprocal licence: Unilateral, UK A only?
Licensing address: Director, Post Office & Radio, Chief Post Office, Apia, Western Samoa.
National society: Western Samoa ARC, PO Box 2015, Apia, Western Samoa.

Zimbabwe
Reciprocal licence: UK A + B, own call/Z2
Licensing address: Manager, National Telecommunications Services, PO Box 8061, Causeway, Harare, Zimbabwe. Tel: 00 263 4 731989.
National society: Zimbabwe ARS, PO Box 2377, Harare, Zimbabwe.

† These countries, although not part of CEPT, have adopted T/R 61-01.
# The appropriate call district must be included as part of the foreign country prefix. Please refer to the appropriate licensing address shown for details.

The Radiocommunications Agency

General

The Radiocommunications Agency is an executive agency of the Department of Trade & Industry and is responsible for the management of that part of the radio spectrum used for civil purposes within the UK. One of their responsibilities is the use of amateur radio and this includes the following:

- Advising government on policy matters.
- Considering the need for changes to licence conditions.
- Interpretation of licence conditions.
- Taking responsibility for the Novice Licence Training Course, the RAE, the NRAE, the 5 and 12 wpm Morse code tests.
- Taking responsibility for the distribution of amateur licences.
- Taking part in international discussions on behalf of UK amateurs.
- Co-ordination of clearance procedures for beacons and repeaters.
- Provision of information.
- Promotion of amateur radio.
- Enforcement of the Wireless Telegraphy Act.

The Amateur Radio Licence

The Agency works closely with the RSGB to ensure that the amateur licence fully meets the needs of today's amateur. Changes normally occur following representations made to us by the RSGB and the Agency meets regularly with them to discuss possible changes.

There are only limited resources allocated to the amateur radio service and channelling of views through a representative organisation allows the RA to make maximum use of those resources. It is of course important to balance these views and hence the Agency will also receive comments from individuals. If you would like clarification of an existing licence condition, you can contact to the RA at the following address:

Radiocommunications Agency
Amateur and Citizens' Band Radio Services,
11th Floor, New King's Beam House,
22 Upper Ground,
London,
SE1 9SA
Tel: (0171) 211 0160

Any changes affecting all amateurs are announced in the London, Edinburgh and Belfast Gazettes as well as in the amateur press. A copy of the licence conditions is provided when a new licence is issued and any subsequent amendments at the time of renewal.

Examinations

The RAE and NRAE are run on the RA's behalf by the City & Guilds of London Institute. The Agency works closely with City & Guilds and the RSGB through advisory committees to ensure that the present high standards are maintained and that the syllabi are kept up to date.

The Novice Licence Training Course and both Morse tests are run on the Agency's behalf by the RSGB and again we are closely involved in both the standard of examination and the format of the actual course and tests.

International Discussions

There are three main areas of international interest:

i) **Reciprocal Agreements.**
The Agency has already negotiated a number of reciprocal agreements which allow UK amateurs to operate abroad. There are still a number of countries with which no agreement currently exists; however the Agency will continue to make contact in these cases.

ii) **Commonality of Licensing Conditions.**
The Agency represents the UK within CEPT and is working towards harmonisation of licence conditions and mutual recognition.

iii) **Control of Interference.**
The RA is responsible for administering the international Radio Regulations in the UK, and as such may not give any station a frequency which would cause interference to stations operating within the International Telecommunications Convention.

The Repeater Network

Although initial vetting of applications is carried out by the Society on the RA's behalf, nevertheless the Agency is responsible for the final clearance of applications.

This can be a relatively long procedure as clearance usually involves consultation both with other Government departments and with the Agency's local offices. This also applies to packet clearances outside of those frequencies listed in BR68.

Provision of Information

A number of free information sheets are available. The following are some of them which relate specifically to the amateur service:

RA0	Current List of Agency Publications
RA67	The Radio Users Guide to the Law
RA165	Application for Novice Licence
RA166	The Novice Licence
RA169	Receive only - Scanners etc.
RA179	Advice on Television and Radio Reception
RA188	Application for an Amateur Radio Licence
RA189	Applications for a Temporary Licence
RA190	How to Become a Radio Amateur
RA206	Addresses of Radiocommunications Agency Local District Offices
RA234	EMC and the Radio Amateur
RA245	Licensing / Morse / Callsigns / Clubs and Examinations
RA247	Operation under CEPT
–	RA Annual Report

Agency General Enquiry Point: (0171) 211 0211

Licensing

The distribution of amateur radio licences is currently handled by Subscription Services Ltd, a wholly owned subsidiary of the Post Office. If you have any queries about your licence, please feel free to contact:

The Radio Licensing Centre,
Subscription Services Ltd,
PO Box 885,
Bristol,
BS99 5LG.
Tel: (0117) 925 8333

Promotion of Amateur Radio

The Agency is well aware of the importance of amateur radio as a valuable training ground for careers in radio and electronics and the RA try to promote amateur radio as much as possible.

The novice licence, for example, was designed specifically to encourage young people into the hobby and it is hoped that novices will progress to becoming full licensees.

Enforcement

The number of people who misuse amateur radio is fortunately small. Nevertheless, there are some individuals who cause considerable problems to other amateurs or to other licensed radio users by transmitting on unauthorised frequency bands, use obscene language or generally use radio in an anti-social way.

Much abuse is directed at the repeater network. Repeater keeper responsibilities include taking steps to prevent and stop abuse. Details of repeater abuse should therefore be sent to RSGB HQ.

Other cases of abuse should also be taken up with the Agency through the Society. Subject to priorities, the Agency does take action in cases of abuse or deliberate interference involving the amateur service.

Having obtained evidence against an offender the Agency will either issue a formal warning, revoke a licence and/or initiate prosecution proceedings, depending on the seriousness of the offence.

Spectrum Abuse

Spectrum abuse is handled in two ways. Unauthorised non-amateur transmissions within our amateur bands are covered by a RSGB's Intruder Watch organisation. While UK amateurs operating outside the terms of their licences or in breach of established operating practices are covered the Society's Amateur Radio Observation Service.

RSGB Intruder Watch

The RSGB Monitoring System, more popularly known as the Intruder Watch, forms part of the IARU Monitoring System. As such it submits reports of non-amateur transmissions heard on the exclusive HF amateur bands to both the RA Monitoring Station at Baldock and the IARU Region 1.

Intruders removed from the 14 MHz as a direct result of Intruder Watch reports being acted on by Baldock include two Russian broadcast station harmonics and a French military data transmission. On 18 MHz the list includes an Argentine weather fax, military stations in Somalia and India, and diplomatic stations located at foreign embassy stations in Paris, Ankara and New Dehli.

Although the Intruder Watch has a sufficient number of monitors for non-data type intruders, the Co-ordinator would like to hear from potential monitors who possess data decoders as manufactured by companies such as 'Hoka', 'Wavecom' and 'Universal' for analysis of data signals. Data intruders are by far the most common and it is an area where we could do with more support.

Other non-data categories of intruding signals include CW, broadcast stations, speech, and over the horizon radar (OTHR). Any report should include as much information as possible, but preferably frequency, date, time (UTC), mode of transmission, any identification signal or callsign, language used, text (where appropriate), and beam heading where possible. For data transmissions a 'zero beat' frequency will be accurate enough for monitors without decoders. This information can then be passed on to a suitably equipped monitor for further investigation.

Most information received by the Co-ordinator arrives from regular monitors, but occasional reports are also welcome from anyone who finds what may be an intruding signal on one of our exclusive amateur bands. All reports are welcome and will be acknowledged.

Anyone interested in assisting in this work should write to Chris Cummings, G4BOH, RSGB Intruder Watch Co-ordinator, Castle View, Childs Lane, Brownlow, Congleton, Cheshire, CW12 4TQ; via e-mail: chris@g4boh.demon.co.uk.

Amateur Radio Observation Service (AROS)

The Amateur Radio Observation Service is an advisory and reporting service of the RSGB which is intended to assist radio amateurs and others who may be affected by problems which occur within the amateur bands or which develop on other frequencies as a result of amateur transmissions. It operates within the remit of the Society's Licensing Advisory Committee (LAC). The Service investigates reports of licence infringements, or instances of poor operating practice which might bring the Amateur Service into disrepute. Reports, complaints and associated supplementary information are accepted from any source and the contents of each communication is regarded as confidential material. The source of any report or supplementary information is not disclosed without the permission of the originator. The originator of any report or complaint should be prepared to respond to further enquiries. Requests for further details may be made to the originator and in addition, independent verification on an individual case basis may be supplied by AROS Observers. A report to AROS should contain details of the alleged infringement and should include: dates, times, modes, details of what was heard – supported, if possible, by a tape recording. The originator should also state where he/she considers the 'offender' to have infringed the terms of the licence or where the 'offender' has acted in a manner contrary to codes of operational practice which have been agreed nationally or internationally. The identity of the 'offender' or his/her location, where this is known positively, should be included. However, reports where the identity or the location of the offender is not known, or not known with certainty are still of value and are required.

After investigation and where there is evidence of deliberate malpractice or malicious abuse of Amateur Radio facilities, a formal report may be made to the appropriate authorities. This report will contain sufficient detail and evidence to enable further investigations to be made and the authorities may take such action as is appropriate. However, AROS prefers to settle problems - great or small – within the Amateur Service. Problems arising are referred to the authorities as a last resort.

Reports may be made to AROS at the following addresses:

AROS, PO Box 113, Potters Bar, Herts. EN6 3ZY
E-mail: AROS@rsgb.org.uk

UK Amateur Callsigns

Prefixes:

1st Character: Novice: 2 Full: G or M

2nd Character:

	Novice	Full	Club (Full)
England	E	(none)	X
Isle of Man	D	D	T
Northern Ireland	I	I	N
Jersey	J	J	H
Scotland	M	M	S
Guernsey	U	U	P
Wales	W	W	C

GB3 + 2 letters: Repeaters — GB7 + 2 letters: Data repeaters
GB3 + 3 letters: Beacons — GB7 + 3 letters: Data mailboxes
GB + other digits: Special event stations (the class of callsign normally corresponds to the appropriate M format)

G0/foreign call: Class A reciprocal licence
G7/foreign call: Class B reciprocal licence

Suffixes:

-/M mobile (includes inland waterways or pedestrian)
-/P operation from a temporary location (ie portable) or address
-/MM maritime mobile

Approximate issue dates:

Class A

Two letters:

G2AA	1920-39
G3AA	1937-8
G4AA	1938-9
G5AA	1921-39
G6AA	1921-39
G8AA	1936-7

Three letters:

G2AAA	Pre-war
G3AAA	1946
G3CAA	1947
G3EAA	1948
G3GAA	1949-50
G3HAA	1950-1
G3IAA	1951-2
G3JAA	1952-4
G3KAA	1954-6
G3LAA	1956-7
G3MAA	1957-8
G3NAA	1958-60
G3OAA	1960-61
G3PAA	1961-2
G3QAA	not issued
G3RAA	1962-3
G3SAA	1963-4
G3TAA	1964-5
G3UAA	1965-6
G3VAA	1966-7
G3WAA	1967
G3XAA	1967-8
G3YAA	1968-9
G3ZAA	1969-71
G4AAA	1971-72
G4BAA	1972-3
G4DAA	1974-75
G4EAA	1975-6
G4GAA	1977
G4IAA	1979
G4MAA	1981
G4QAA	not issued
G4RAA	1982
G4SAA	1983
G4WAA	1984
G0AAA	1985
G0EAA	1986
G0HAA	1987
G0JAA	1988
G0LAA	1989
G0MAA	1990
G0NAA	1991
G0SAA	1992
G0TAA	1993
G0VAA	1994

Class A continued

Three letters:

G0WAA	1995
M0AAA	1996
M1BAA	1997

Class B:

G8AAA	1964-7
G8BAA	1967-8
G8CAA	1968-9
G8DAA	1969-70
G8EAA	1970-1
G8FAA	1971-2
G8HAA	1973
G8IAA	1973-4
G8JAA	1974-5
G8KAA	1975
G8MAA	1976-7
G8NAA	1977
G8OAA	1977-8
G8PAA	1978
G8QAA	not issued
G8TAA	1979
G8ZAA	1981
G6CAA	1981
G6QAA	not issued
G6RAA	1982
G1AAA	1983
G1DAA	1984
G1LAA	1985
G1QAA	not issued
G1SAA	1986
G1XAA	1987
G7AAA	1988
G7EAA	1989
G7FAA	1990
G7HAA	1991
G7MAA	1992
G7OAA	1993
G7SAA	1994
G7TAA	1995
G7WAA	1996
M1AAA	1996
M1CAA	1997

Novice Class A:

2E0AAA	1991+

Novice Class B:

2E1AAA	1991
2E1BAA	1992
2E1CAA	1993
2E1DAA	1994
2E1EAA	1995
2E1GAA	1997

RSGB Morse Practice – GB2CW

Practice morse is broadcast on behalf of the RSGB by morse practice volunteers in many parts of the UK using the callsign GB2CW. The aims are twofold. First to assist those preparing for the amateur radio Morse tests and second to provide practice Morse at speeds which will help qualified operators to improve their standards.

Receiving practice alone is not enough and many volunteers use their own callsigns after the transmissions to provide sending practice. New volunteers are urgently required on hf and vhf to extend cover over the whole of the UK.

For more information on the RSGB Morse Practice Service, please contact the Co-ordinator: George M. Allan, GM4HYF, 22 Tynwald Avenue, High Burnside, Ruthergeln, Glasgow, G73 4RN.

HF Transmissions

DAY	TIME	FREQ	OP	LOCATION
Mon	20:30 [1]	1.9787	G3SJE, G4GYS	Harrow, Bushey
Wed	19:30	3.550	G4XQI	Stockport *(temp suspended)*
	20:00	3.602	GM4HYF	Rutherglen
Thu	19:45 [1,2]	1.9787	G3SJE, G4GYS	Harrow, Bushey
	20:45 [3]	3.527	GM4HYF	Rutherglen
Fri	18:30	3.550	GW0TAF	Neath
Sat	19:30	3.550	G4XQI	Stockport *(temp suspended)*
Sun	18:00	3.562	GW0CVY	Carmarthen
	19:30	3.550	G4XQI	Stockport *(temp suspended)*

VHF Transmissions

DAY	TIME	FREQ	OP	LOCATION
SCOTLAND				
Mon	20:00	145.250	GM0VIY, GM3YCG	Nr Eaglesham
Tue	20:00	145.250	GM0WRR, GM3YCG	Glasgow
	20:30	145.250	GM0LZE	Stornoway
Wed	20:00	145.250	GM0MDX, GM3YCG	Hamilton
Thu	20:00	145.250	GM0UOU, GM3YCG	Paisley
Fri	20:00	145.250	GM0NPS, GM3YCG	Coatbridge
WALES				
Mon	18:30	145.250	GW0KPD	Port Talbot
Tue	18:30	145.250	GW0KPD	Port Talbot
Fri	18:30	145.250	GW0KPD	Port Talbot
NORTH WEST ENGLAND				
Mon	19:00	145.250	G4OTN	Preston
	19:30	145.275	G0IIM	Sale
	21:00	145.250	G3AVJ	Huyton
Tue	19:30	145.275	G4GBK	Atherton
Wed	19:00	145.250	G4OTN	Preston
	19:30	145.275	G4XQI	Stockport *(temp suspended)*
	21:00	145.250	G3AVJ	Huyton
Thu	19:30	145.275	G4GBK	Atherton
	21:00	145.250	G3AVJ	Huyton
Fri	19:30	145.275	G4IAV	Atherton
	20:00	145.250	G3KJY [1]	Barnoldswick
	21:00	145.250	G3AVJ	Huyton
Sat	19:30	145.275	G0IIM	Sale
Sun	12:00	145.575	G0RDH	Morecambe
	19:30	145.275	G4XQI	Stockport *(temp suspended)*
NORTH EAST ENGLAND				
Mon	20:00	145.250	G4RXR	Peterlee
Tue	20:00	145.250	G4RXR	Peterlee
Thu	20:00	145.250	G4RXR	Peterlee
Sat	20:00	145.250	G4RXR	Peterlee

VHF Transmissions (continued)

DAY	TIME	FREQ	OP	LOCATION
WEST MIDLANDS				
Tue	19:30	144.160	G4TDO	Wolverhampton
	21:00	144.250	G3HZL	Upper Tean
Thu	19:30	145.250	G0KCM	Penkridge
Sat	14:00	145.250	G3HVI	Meir Heath
	19:30	144.160	G4TDO	Wolverhampton
Sun	14:00	145.250	G3HVI	Meir Heath
EAST MIDLANDS				
Mon	20:00	145.250	G4NZU	Nottingham
Tue	19:00	145.250	G0FOG	Nottingham
Fri	19:00	145.250	G4NZU	Nottingham
SOUTH MIDLANDS				
Tue	19:00	145.225	G4LHI	Huntingdon *(temp suspended)*
	20:30	144.250	G4PDP	Chawston
Wed	19:00	145.250	G3BLS	Oxford
Thu	20:00	145.250	G4DLB	Banbury
Sun	11:00	145.250	G3BLS	Oxford
SOUTH WEST ENGLAND				
Mon	19:45	145.225	G4ULH	Fishponds
	20:00	145.250	G0JVA	Taunton
Tue	19:30	145.250	G3ZYY, G0NIE	Saltash, Tavistock
	19:45	145.225	G4ULH	Fishponds
Wed	19:45	145.225	G4ULH	Fishponds
	20:00	145.250	G0JVA	Taunton
Thu	19:30	145.250	G3ZYY, G0NIE	Saltash, Tavistock
	19:45	145.225	G4ULH	Fishponds
	20:00	145.250	G0JVA	Taunton
SOUTH EAST ENGLAND				
Mon	20:00	145.250	G0IZU	West Ewell
Tue	19:00	145.250	G0NFJ	Abridge
Wed	20:00	145.250	G0JUD	Aldershot
Thu	20:00	145.250	G0JUD	Aldershot
Fri	19:00	145.250	G0NFJ	Abridge
Sat	20:05	145.250	G0JUD	Aldershot
Sun	10:00	145.250	M0AGQ	Brighton
	19:00	145.250	G0NFJ	Abridge
	20:30	144.250	G3ORP	Maidstone

NOTES

Modes of emmission:

A1A/J3E: 144.160, 144.250 MHz and all HF transmissions
F2A/F3E: 145.225, 145.250, 145.275, 145.575 MHz

[1] Club Transmissions:
The broadcasts by G3SJE and G4GYS are on behalf of Edgware &DARC. Those by G3KJY are on behalf of Rolls Royce ARC.

[2] Only on the first and third Thursdays of the month

[3] The transmission on Thursdays at 2045 on 3.527 MHz provides practice at speeds from 15 to 30 wpm.

Morse Test Service

The service was set up after the Society's successful bid for the DTI contract in 1985. There are now some 290 examiners covering all counties of the UK including Northern Ireland, Shetland, Orkney, Isle of Man and the Channel Isles. Recently, teams have been organised so that tests can now be taken in some overseas places. The examiners work in county teams and each has a senior examiner who organises within the area.

The service organises, conducts the tests and issues the certificate required by the Radiocommunications Agency to satisfy the Morse requirement for a novice 5 wpm and full 12 wpm Class A licence. It carries out this commitment in several ways:

Normal Session: Regular bimonthly tests each area.

Closed Session: Tests by arrangement with a group of candidates who provide their own approved accommodation, etc. Tutors or organisers requiring a closed session should contact the Chief Examiner to make arrangements.

Rally Session: Rally organisers who can provide suitable accommodation without cost to the Society and are able to arrange for tests to be carried out at their event. Likewise, rally organisers should contact the Chief Examiner to make arrangements.

Disabled: Disabled persons are dealt with according to their disability requirements. In extreme cases a home visit can be arranged, in others a relaxation of the requirements may be permitted. Each case is considered on its merits. There are no exemptions. Candidates are required to supply details of their disability on a supplementary application form and provide evidence where applicable.

Special cases not covered by any of the above, such as serving members of the forces on leave, seamen, overseas visitors, etc., can contact the Chief Examiner when every effort will be made to assist. The additional cost (if any) of any special arrangements would be the candidate's responsibility.

The Form of the Morse Test

Radiocommunications Agency Requirements

In the receiving test, the candidate will be required to receive a minimum of 120 letters and 7 figures in the form of a typical exchange between two radio amateurs. A maximum of 6 uncorrected errors will be allowed.

The 5 wpm test will use computer generated Morse from a tape recorder which will also contain voice announcements. The characters will be sent at a speed of 12 wpm but with a longer than normal gap between the letters, so as to reduce the overall reception speed to 5 wpm. (Candidates will not be allowed to write down the Morse symbols for later translation.) The 12 wpm test will use hand sent Morse. The receiving test lasts about 6 minutes for the 5 wpm and 2½ minutes for the 12 wpm test.

In the sending test, the candidate will be given a text to send by hand on a straight Morse key consisting of not less than 75 letters and 5 figures – also in the form if a typical exchange between radio amateurs. There must be no uncorrected errors in the sending and not more than 4 corrected errors will be permitted. This test will last about 3 minutes for the 5 wpm, and 1½ minutes for the 12 wpm test.

Both tests can include any of the commonly use abbreviations, Q-codes or procedural characters shown below. Figures and procedural characters will count as two letter for timing purposes.

Form of the Test

Tests will be as realistic as possible to on-the-air working, with the exception of repetition of key words, such as signal reports, name and QTH, which would make the test too easy. However, callsigns will be repeated at the end of the 'over'. All callsigns, QTH and names will be used in the correct context of the country of origin.

Each QSO will normally contain a basic framework of calls, signal report, name and QTH, though not necessarily in that order. In addition, there will be a selection of typical operating remarks, such as equipment in use, weather, temperature, age, power output etc.

The receiving test will commence with $\overline{CT}$ as a 'stand-by' signal to the candidate that the test will follow immediately afterwards. It is not part of the test and will not be marked if written down by the candidate.

For Novice candidates the first callsign sent will be in the UK Class 'A' Novice callsign allocation (regional prefixes may be used, such as 2E, 2M, 2W). For 12 wpm candidates, the first callsign sent will be in the UK full class 'A' G-series callsign allocation, (such as G, GM, GW etc). The callsign following the DE (as in real life) can be any amateur radio callsign from any country, including the UK M-series callsigns.

Example of 5 wpm Morse Test:

Receiving: $\overline{CT}$
2E4DKZ DE F6JVX
GE OM TNX FER CALL UR RST 579 = QTH IS 15 KM
SOUTH OF PARIS ES NAME ANDRE = RIG IS TS830
ANT IS 4 EL BEAM SO HW CPY? $\overline{AR}$
2E4DKZ DE F6JVX $\overline{KN}$

Sending: 2M0AIZ DE 2E3DNO
GD JACK UR RST 569 QSB = NAME VAL QTH
HALIFAX HW CPY? $\overline{AR}$
2M0AIZ DE 2E3DNO $\overline{KN}$

Example of 12 wpm Morse Test:

Receiving: $\overline{CT}$
G3LKQ DE HB9JV/P
GM OM UR RST 549 = OP FRITZ QTH ZURICH WX
SUNNY WID TEMP 76F = CAMPING HR FER 12 DAYS
WID 80 SCOUTS SO HW CPY? $\overline{AR}$
G3LKQ DE HB9JV/P $\overline{KN}$

Sending: HB9JV/P DE G3LKQ
GM FRITZ TKS CALL = UR RST 579 NAME ROY QTH
EXMOUTH OK? $\overline{AR}$
HB9JV/P DE G3LKQ $\overline{KN}$

No abbreviations, Q-code or procedural signals other than on the approved list will be used, with the exception of commonly written remarks such as 'Temp 15 C' or 'QRP 2 W'.

Procedural signals (with the exception of $\overline{CT}$) will form part of the test and must be written down by the candidate. These too are specified below.

It is not a requirement for the Morse test that the candidate knows the meaning of the listed Q-codes and abbreviations. However, a working knowledge of these will assist in the understanding of the QSO. The meaning of all abbreviations is given below.

Experienced examiners have the ability to very quickly ascertain whether the candidate can send readable Morse, therefore the sending test is much shorter than the receiving test. As the candidate and the examiner have a copy of the sending test, there is no requirement for the candidate to commence sending with $\overline{CT}$. If this is sent by choice, it will not be counted as part of the test.

If the candidate makes a mistake in sending, the error signal consisting of *8 dots* must be sent and the *word or group must be repeated.* Note that the error signal itself is part of the sending test.

General

In the receiving test, the requirement for 120 letters and 7 figures give an overall minimum total of 134 letters (counting figures and procedural symbols as two letters). Therefore in order to achieve standardisation, each receiving test will contain 134 - 137 letters and will normally use every letter of the alphabet.

In the sending test, 75 letters and 5 figures gives a minimum of 85 letters. This test will contain 85-88 letters. The sending test will not normally contain every letter of the alphabet as this could require unusual words to be used.

How to apply

Application forms are available from RSGB HQ. Forms supplied will also include a list of all the available test centres at the time of application and an explanation of the QSO format, with examples of the new test. Candidates should bear in mind that there is a closing date for entries, so please avoid disappointment by applying early.

Disabled persons can also obtain their form plus the additional form from the Deputy Chief Examiner.

Normal applications should be returned to RSGB HQ while those from disabled persons should go to the Deputy Chief Examiner direct.

What happens next

Your application is processed and you will be booked into the system and your booking will be confirmed. You will be advised that you may bring with you your own phones and straight key. It would be helpful if these had ¼in jacks, however, while we cannot promise the impossible, the examiners will make every effort to get you connected!

On the day!

It is easy to say do not worry – *but don't*! Our aim is to put you at ease and provide the right atmosphere for you to present your best. Remember there is no pleasure for the examiners in failing candidates - they feel much better when able to sign pass slips! If anything is worrying you, do tell the examiner – he may be able to help.

Get to the centre in good time so that you are not all tense from trying to park, etc.

Do not forget to bring your two passport photographs, appointment confirmation, and if you wish to use them, your phones and key, and above all, something to write with (perhaps a spare as well)!

The test procedure

This may vary slightly between centres but the general pattern is as follows.

Candidates are taken in threes and after being seated are given their forms to sign and their identity is checked. After this any special connections for phones will be made and you will be given every chance to make yourself comfortable. Although there is no requirement for them to do so the examiner will normally send you a few words practice to really settle you down and give you an opportunity to see that all is well. It is surprising how many pencils break and phones go faulty at this stage! You will then be asked if all is well and the test will be sent. If you are in difficulty during the test piece, try to forget the letter you missed and concentrate on the one just sent, and above all, should you have real trouble please remember the other two candidates; don't spoil their chances. The examiner will note if there is any problem but will continue to the end of the test piece.

At the end of the test you will be given a reasonable period to read through your copy and are permitted to amend doubtful characters. You should not worry unduly about splitting words, the examiners will make allowance for this. Your papers are collected and two of you will be asked to return to the waiting room for a short period.

The remaining candidate will then be invited to use our key or his own if this can be connected. Take your time, there are no extra marks for exceeding the speeds, just remember to correct all errors. For those in doubt the correct method of correction is to send eight dots .... *and* go back to the beginning of the word or group where the error occurred. Any uncorrected errors mean you fail! You will then be thanked for coming and informed that the result will be sent to you as soon as possible. Please do not expect the examiner to give you an instant result: at this stage he does not know.

When you have gone and the session ends the examiners will check and recheck the papers which will then be sent to RSGB HQ and the Deputy Chief Examiner for further checking and processing. Any failures are especially rechecked by the Deputy Chief Examiner. A special report is sent to the Chief Examiner after each session advising of any difficulties and it is thus a good idea to let the examiners know if you feel in any way dissatisfied with your test; they may be able to rectify matters there and then. If, however, you are still unhappy write to the Chief Examiner who will look into your complaint.

What next

A successful candidate will receive a pass slip together with an application form to send off for a Class A licence. The unfortunate will be sent a fail slip and an application form to apply for a further test.

Here are some tips based on results

- ✻ Remember to correct all errors. More candidates fail their sending test because of failure to correct than anything else.
- ✻ Prepare for the test by receiving hand-sent QSO-format Morse from more than one person, using a variety of oscillator notes. Many candidates become used to one particular note during practice and are unsettled if a different note is used at the test.
- ✻ Some candidates learn to receive Morse using the Farnsworth method, with individual characters sent much faster than test speed, using longer than normal gaps in order to reduce the overall speed down to 12 wpm. Many of these candidates have great difficulty receiving correctly proportioned Morse sent by the examiner, because the two methods sound very different.

Queries

If you have a query about any of the above, please contact either:

Chief Examiner:

Mr Roy Clayton, G4SSH
9 Green Island
Irton, Scarborough
North Yorks, YO12 4RN

Tel: (01723) 862924

Deputy Chief Examiner:

Mr Geoff Pritchard, G4ZGP
45 Fairfield Crescent
Newby, Scarborough
North Yorks, YO12 6TL

Tel: (01723) 372275

Abbreviations and Q Codes used for the Morse Test

Abbreviations

ABT	About
AGN	Again
ANT	Antenna
BK	Signal used to interrupt a transmission in progress
CPI	Copy
CPY	Copy
CQ	General call to all stations
CUL	See you later
CW	Continuous wave
DE	From, used to precede the call sign of the calling station
DR	Dear
EL	Element
ES	And
FB	Fine business
FER	For
FM	From
GA	Good afternoon
GD	Good day
GE	Good evening
GM	Good morning
HPE	Hope
HR	Here
HVE	Have
HW	How
K	Invitation to transmit
MNI	Many
MSG	Message
NW	Now
OC	Old chap
OM	Old man
OP	Operator
PSE	Please
PWR	Power
R	Receive
RPRT	Report
RST	Readability, signal-strength, tone-report
RX	Receiver
SIG	Signal
SRI	Sorry
TEMP	Temperature
TKS	Thanks
TNX	Thanks
TU	Thank-you
TX	Transmitter
TXR	Transceiver
UR	Your
VERT	Vertical
VY	Very
WID	With
WX	Weather
XYL	Wife
YL	Young lady
73	Best wishes
88	Love and kisses

Q-Codes

The full international meanings of Q codes can be found in the RSGB's *Amateur Radio Operating Manual*, but the informal amateur-use meanings (which can be used in the Morse test) are as follows:

✻ QRA	Name of Station
QRG	Frequency
✻ QRK	Intelligibility of signals
QRL	Busy (the frequency is in use)
QRM	Interference from other stations
QRN	Interference from atmospherics
QRO	High power
QRP	Low power
QRQ	Send faster (high speed Morse)
QRS	Send slower (slow speed Morse)
QRT	Close down (stop sending)
QRV	Ready (go ahead)
QRX	Stand-by
QRZ	You are being called by
✻ QSA	Signal stength
QSB	Fading
QSL	Confirm contact
QSO	Radio contact
QSY	Change frequency
QTH	Location

✻ Not normally used in the Amateur Service, but may be used by commercial stations to inform the amateur that they are causing interference

Procedural characters and punctuation

$\overline{\text{AR}}$ (+)	[di-dah-di-dah-dit]	end of message (will be used in the test before the final calls and must be written as 'AR' or '+')
$\overline{\text{CT}}$	[dah-di-dah-di-dah]	preliminary call
BT (=)	[dah-di-di-di-dah]	separation signal (will be used in the text and must be written as 'BT' or '=')
$\overline{\text{KN}}$	[dah-di-dah-dah-dit]	transmit only the station called (will be used in the test after the final calls and must be written as KN)
$\overline{\text{VA}}$	[di-di-di-dah-di-dah]	transmission ends (in the test, must be written as 'VA' or 'SK')
?	[di-di-dah-dah-di-dit]	question (in the test, must be written as IMI or ?)
/	[dah-di-di-dah-dit]	oblique stroke (can be used in the test as part of a callsign and must be written as '/')

Packet Radio

What is Packet Radio?

Packet radio is digital communications via radio. It first started back in 1978 in Canada, and it was introduced into UK in the 1980's. Packet radio mailboxes (BBS's) were first licensed in UK in about 1988, and the numbers have continued to grow ever since.

As well as the large number of mailboxes in UK there are a large number of packet nodes, these help to link the mailboxes, and together they form the packet radio network that covers most of the UK.

What Can I do on Packet Radio?

Live Contacts

Like RTTY, packet radio can be used to talk to other amateurs, chatting keyboard to keyboard. Some mailboxes also have a conference mode, so people can log on and chat with many people at once just like an HF net.

Mailboxes

Mailboxes allow amateurs to connect their local mailbox and send and receive text messages. These messages can be sent as personal messages to another amateur anywhere in the World (or Space!) Alternatively, messages can be sent as a bulletin for any amateur to read.

Within the UK your messages will normally be relayed via other mailboxes and nodes on VHF and UHF frequencies. There also exist a limited number of HF and Satellite mail forwarding gateways, these are used to forward mail to more distant continental mailboxes.

So it is possible to exchange mail with amateurs on the other side of the world by simply logging into your local mailbox using a low power VHF transceiver and a simple aerial system, as well as a TNC and computer.

File Transfer

Packet radio also allows you to be able to transfer files between amateur packet stations in both text and binary format.

DXCluster

There are several DXCluster stations around the UK. These allow DXers to exchange valuable DX information on a close to real time network. This information on band conditions and stations heard, is valuable to HF and VHF and above DX operators.

What will I need?

As well as a VHF or UHF FM radio transceiver, you will also need a Terminal Node Controller (know as a TNC), a terminal or computer with some form of terminal software program or a specialist packet radio software program, and finally a great deal of patience and willingness to learn!

Packet Radio User Groups

There are a large number of User Groups throughout the UK. Some of these groups are specific to a particular mailbox whilst others cover all the mailboxes and nodes in an entire county or region. These user groups have been generally setup to help new users and to help fund mailboxes and nodes. Some nodes are located on remote sites, so the cost of maintaining them can be considerable. All packet radio users are encouraged to join and support their local group.

Where will I find my nearest Mailbox or Node?

The packet radio network is a very dynamic network – it changes quite frequently to meet changing needs, new mailboxes and nodes open up, or existing ones expand their number of access ports. An up to date list of packet mailboxes and nodes is available for a small cost from RSGB HQ. This list is updated on a monthly basis. Lists of Mailboxes and Nodes are also published on packet radio, and may be available in the files area of your local mailbox.

Data Communications Committee (DCC)

The Datacommunications Committee was set up to deal with all matters concerning data communications.

The Committee is responsible for the processing of applications for Notices of Variation (NoV's) for mailboxes. The committee is also responsible for the vetting of all requests for mailbox and node site clearances prior to submission to the Radiocommunications Agency.

The DCC also recommend general operating on data communications matters by issuing guidelines.

The DCC has published *'The Guide to Data Communications Licensing in UK'* which is available from RSGB HQ free of charge on request. This includes the necessary forms to apply for a Mailbox NoV and for Site Clearance for a Mailbox or Node.

DCC Committee Members

Enquiries from datacommunincations users and potential mailbox sysops and node sysops should be sent to the appropriate committee member as follows:

Paul Overton	G0MHD	Chairman
Steve Morton	G8SFR	Site Clearance Manager
Iain Philipps	G0RDI	Minute Secretary & PRO
Martin Green	G1DVU	Mailbox Manager
Ian Brothwell	G4EAN	BARTG Liasion
Malcolm Salmon	G3XVV	Technical Manager & Vice Chairman
Peter Maile	GI1ONL	
Fred Stewart	G0CSF	Council Liasion Member

DCC on the Internet

The DCC Web Pages contain a lot of up to date information about packet radio, notes of meetings, bandplans, lists of nodes and mailboxes. The address is as follows: http://www.g0mhd.demon.co.uk/dcc

Guidelines for the Use of the Packet Radio Network

The packet radio network in the UK. and throughout the world is an immensely useful tool for the dissemination of information, the seeking of help and advice and the publication of amateur radio related news. It is not uncommon to find messages giving information on AMSAT, Raynet or other similar amateur radio related activities.

The GB2RS news is also available on the network, as is local club news in the area of a particular mailbox. This use of the network is what was in many operators minds when they spent large amounts of time and money in developing it. With the advent of high speed modems and use of dedicated links, some in the microwave bands, the Packet Radio Network is developing and hopefully will continue to do so for many years to come.

The RSGB Data Communications Committee, in consultation with the Radiocommunications Agency, has devised the following guidelines with which all operators are urged to comply.

SECTION 1: Types of Message

a) All messages should reflect the purposes of the amateur licence, in particular 'self training in the use of communications by wireless telegraphy'.

b) Any messages which clearly infringe licence conditions could result in prosecution, or revocation, or variation of a licence. The Secretary of State has the power to vary or revoke licences if an amateur's actions call into question whether he is a fit and proper person to hold an amateur licence.

c) The Radiocommunications Agency has advised that the Amateur Radio Licence prohibits any form of advertising, whether money is involved or not.

d) Messages broadcast to *all* are considered acceptable but should only be used when of real value to other radio amateurs, in order to avoid overloading the network.

e) *Do not* send anything which could be interpreted as being for the purpose of business or propaganda. This includes messages of, or on behalf of, any social, political, religious or commercial organisation. However, our licence specifically allows news of activities of non-profit making organisations formed for the furtherance of amateur radio.

f) *Do not* send messages that are *deliberately designed to provoke* an adverse response. Debate is healthy but can sometimes lead to personal attacks and animosity which have no place on the Packet Network.

g) Unfortunately the very success of the network has resulted in messages appearing which are of doubtful legality under the terms of the UK. Licence. The use of 7+ and other like programs to pass text and binary based material in compressed form via the Network has become common practice.

Users must always be aware of the licence conditions in BR68 (also copyright, and illegal use of software, or software which when decoded and used may contravene the Licence) when entering such messages into the Network via their local BBS. If in doubt consult your local sysop or your local RSGB Data Comms Committee representative.

SECTION 2: Legal Consequences

a) Do not send any message which is libellous, defamatory, racist or abusive.
b) Do not infringe any copyright or contravene the Data Protection Act.
c) Do not publish any information which infringes personal or corporate privacy e.g. ex-directory telephone numbers or addresses withheld from the *RSGB Yearbook*.

SECTION 3: Action in Cases of Abuse

a) Any cases of abuse noted should be referred in the first instance to the DCC Chairman c/o RSGB HQ.
b) It is worth noting that any transmissions which are considered grossly offensive, indecent or obscene, or contain threatening language, may contravene the Wireless Telegraphy (Content of Transmission) Regulations 1988 and should be dealt with by the police. This action should also be co-ordinated by the RSGB DCC initially.
c) Mailbox Sysops have been reminded by the Radiocommunications Agency that they should review messages, and that they should not hesitate to delete those that they believe to contravene the terms of the license or these guidelines. It is worth remembering that their licence is also at risk as well as your own.

SECTION 4: Unattended Operation

a) As of July 1994 unattended operation of digital communications cannot be carried out without giving 7 days notice in writing of operation to the Manager of the local Radio Investigation Service (RIS). (BR68 Para 2(5)) The Manager may, before the commencement of operation, prohibit the Unattended Operation or allow the operation on compliance with the conditions which he may specify.

The RSGB Data Communications Committee recommend supplying the following information when applying to your local RIS office for permission to operate unattended digital operations.

1) An external close down switch or other means of closing down the station, which is separate from the rest of the premises.
2) A list of 4 persons, including telephone numbers who can close down the Station. (Not all need to be amateurs.)
3) Travelling times and availability times (i.e. 24hrs) of close down operators of the Station.
4) Frequencies of operation (within BR68 clause 2 4(c)) antennas and powers used.
5) Use of Station for digital operation (i.e. PMS, node etc.).
6) Only the licensee can re-activate the Station after permission from the RIS.

SECTION 5: General Advice.

a) With the advance in software writing it is now possible for packet users to set up intelligent software nodes (G8BPQ and similar applications) via a PC and radio. The RSGB DCC strongly recommend that users contact their local packet group and local BBS Sysop before starting operation of such nodes.

Appearance of such nodes without co-ordination causes problems within the local and inter-BBS, DXcluster, TCP/IP Network routing tables.

Network Sysops work closely with each other to determine route qualities and node tables, to aid the fast movement of traffic via the National Trunk System.

The appearance of uncoordinated (rogue nodes) causes in some cases severe problems in traffic routing.

Packet users can experiment with software nodes without affecting the node tables of local network nodes, by setting the software parameters of the node to stop propagation of the node into the Network. Advice can be sought on the setting of software parameters from local node sysops or BBS Sysops.

b) Do not send *'Open Bulletins"* to individuals.
c) Do not write in the heat of the moment. Word process your bulletin first, then re-read it. You may feel differently after a few minutes.
d) Stop to think before sending GIF images and the like which sometimes are in large multi-part files, do you really need to send them, are they amateur radio related, would they be better sent on disc in the post?
e) Please try to show some consideration for your local sysop. Remember that you are using, in most cases his own equipment, which is in his home. Try to comply with any requests he makes of you.
f) When accessing your local Mailbox at busy time and are having problems holding the link, try not to turn your power up just to maintain the link, try later when it is not as busy.
g) Obey the Golden Rule - *If you would not say it on voice do not send it on packet.*

Packet Radio Band Plans

Frequencies currentluy approved for Data Communications.

HF –Please contact DCC for digital mode frequency allocations details.

50 MHz band

50.510	Slow Scan Television	50.610	10 kHz any digital mode
50.530	20 kHz Packet	50.630	20 kHz Packet
50.550	Facsimile (FAX)	50.650	20 kHz Packet
50.570	20 kHz Packet	50.670	20 kHz Packet
50.590	10 kHz Any digital mode	50.690	20 kHz Packet
50.600	RTTY calling channel		

70 MHz band - All channels are 12.5 kHz wide.

70.3000	RTTY calling/FAX	70.3375	
70.3125		70.4875	BBS/Node Linking
70.3250	Packet cluster input		

144 MHz band (User Access Only)

144.825*	High speed data only
144.850	AX25 user access
144.8625	Unallocated, available on NoV subject to local agreement
144.875	TCP/IP user access
144.8875	AX25, priority to DX Cluster access subject to local agreement
144.900	DX custer user access
144.9125	General Packet, one to one users, not available on NoV
144.925	TCP/IP user access
144.9375	Unallocated, available on NoV subject to local agreement
144.950	AX25 user access
144.975*	High speed data only

Channels marked with '' are 25 kHz wide, others are 12.5 kHz wide. In accordance with IARU policies there should be no inter-node or inter-BBS linking in the 144 MHz band.*

432 MHz band

Sub-Band A (Forwarding and Linking only)

430.625	BBS/Node linking
430.650	BBS/Node linking
430.675	BBS/Node linking
430.725	High speed linking 50 kHz channel
430.775	High speed linking 50 kHz channel

This sub-band also used for duplex linking.

Sub-Bands B and C (User Access Only)

432.625	TCP/IP & High speed	433.625	TCP/IP
432.650	AX25	433.650	AX25
432.675	AX25	433.675	AX25

Sub-Band D (Forwarding and Linking only)

439.825	BBS/Node linking	
439.850	BBS/Node linking	
439.875	BBS/Node linking	
439.925	High speed linking	50 kHz channel
439.975	High speed linking	50 kHz channel

This sub-band also used for duplex linking. Except were shown, channels are 25 kHz wide. For details of regenerative node frequencies and other linking frequencies soon to become available, please contact the DCC.

1300 MHz band (Forwarding and Linking only)

1240.150	150 kHz channel	1299.000	25 kHz channel
1240.300	150 kHz channel	1299.925	25 kHz channel
1240.450	150 kHz channel	1299.050	25 kHz channel
1240.600	150 kHz channel	1299.075	25 kHz channel
1240.700	50 kHz channel	1299.100	25 kHz channel
1240.750	50 kHz channel	1299.125	25 kHz channel
1240.800	50 kHz channel	1299.175	50 kHz channel
1240.850	25 kHz channel	1299.225	50 kHz channel
1240.875	25 kHz channel	1299.275	50 kHz channel
1240.900	25 kHz channel	1299.375	50 kHz channel
1240.925	25 kHz channel	1299.425	150 kHz channel
1240.950	25 kHz channel	1299.575	150 kHz channel
1240.975	25 kHz channel	1299.725	150 kHz channel

2320 MHz band

2310.100	200 kHz channel	2355.300	200 kHz channel
2310.300	200 kHz channel	2364.000	1 MHz channel
2355.100	200 kHz channel		

10000 MHz band

10077.5 to 10090.0	10370 to 10390

Prefix List

Callsigns for the world's nations are determined by the International Telecommunications Union (ITU). This is the United Nations agency that co-ordinates radio activity for all spectrum users. The prefixes used by a country for both commercial and amateur radio purposes are determined from one or more ITU allocation blocks issued to that country. The amateur radio callsigns in use for a particular country might use one or a number of combinations derived from the authorised ITU allocation(s) for that country.

The following list shows callsign prefixes currently in use. Most are derived from the callsign blocks allocated to administrations by the ITU for use within the countries, territories and dependencies for which a country is responsible. Also shown are some unauthorised prefixes which may be heard and which may or may not be recognised as a DXCC country. Examples are: 1S Spratly Archipelago, and 1A0 SMOM. Both of these are DXCC countries although the prefixes used are unofficial. 1B the Turkish area of North Cyprus and X5 the Serbian occupied area of Bosnia are unofficial and are not recognised for DXCC purposes. Where this applies, the country details are shown in italics. Information which is not a country name is also printed in italics. Where a prefix is shown that is perhaps unusual, then the common prefix is shown after the country name or details. Full information on prefixes is contained in the *RSGB Prefix Guide*.

Prefix	Country
A2	Botswana
A3	Tonga
A4	Sultanate of Oman
A5	Bhutan
A6	United Arab Emirates
A7	Qatar
A8	Liberia EL
A9	Bahrain
AA - AG	USA W
AH1 - AH0	*USA Pacific Islands* KH1 - KH0
AI - AK	USA W
AL	Alaska KL
AM - AO	*Spain including overseas Territories and Islands* EA 6 8 9
AP AR	Pakistan
AT	India VU
AX	*Australia and Islands*
AY - AZ	Argentina LU
BO	Quemoy Matsu BV
BS	Scarborough Reef
BV	Taiwan
BV9P	Pratas I.
BV9S	Spratly Archipelago 9M0
BY	China BA BD BG BT BZ
C3	Andorra
C4	Cyprus 5B
C5	Gambia
C6	Bahamas
C9	Mozambique
CE	Chile
CE0	Easter I.
CE0	San Felix and San Ambrosio Is
CE0	Juan Fernandez Is
CF - CK	Canada VE
CL CM	Cuba CO
CN	Morocco
CO	Cuba
CP	Bolivia
CT1CQ-CT2 4-8 0	Portugal
CT3 CQ-CS3 CT9	Madeira Is
CU	Azores
CX CV CW	Uruguay
CY CZ	Canada VE
CY9	St Paul Is
CY0	Sable I.
D2 D3	Angola
D4	Cape Verde
D6	Comoros
D7	Korea (Republic of) HL
DL DA-DD DF-DH DJ DK DP	Federal Republic of Germany
DS	Korea (Republic of) HL
DU DV-DZ	Philippines
DU	Spratly Archipelago 9M0
E2	Thailand HS
E3	Eritrea
EA EB-EH1-5 7 0	Spain
EA6 EB6-EH8	Balearic Is
EA8 EB8-EH8	Canary Is
EA9 EB9-EH9	Ceuta and Melilla
EI EJ	Republic of Ireland
EK	Armenia
EL	Liberia
EM EN EO	Ukraine UR
EP	Iran
ER	Moldova
ES	Estonia
ET	Ethiopia
EU EV EW	Belarus
EX	Kyrghyzstan
EY	Tajikistan
EZ	Turkmenistan
F	France
FG	Guadeloupe
FH	Mayotte
FJ	St Barthelemy (French St Martin) FS
FK	New Caledonia
FM	Martinique
FO	French Polynesia
FO8X	Clipperton I.
FP	St Pierre and Miquelon
FR	Reunion I.
FR——/E	Europa I. FR——/J
FR——/G	Glorioso Is
FR——/J	Juan de Nova
FR——/T	Tromelin I.
FS	French St Martin
FTnW	Crozet Is
FTnX	Kerguelen Is
FTnZ	Amsterdam I. and St Paul I.
FW	Wallis and Futuna Is
FY	French Guiana
G GX	England
GB	United Kingdom G GD GI GJ GM GU GW
GD GT	Isle of Man
GI GN	Northern Ireland
GJ GH	Jersey
GM GS	Scotland
GW GC	Wales
H2	Cyprus5B
H3	PanamaHP
H4	Solomon Is
H5	Bophuthatswana ZS
H6 H7	Nicaragua,YN
H8 H9	Panama,HP
HA	Hungary
HB	Switzerland
HB0	Liechtenstein
HC,HD	Ecuador
HC8,HD8	Galapagos Is
HE	Switzerland,HB
HF	Poland,SP
HG	Hungary HA
HH	Haiti
HI	Dominican Republic
HK HJ	Colombia
HK0	Malpelo I.
HK0 HJ0	San Andres and Providencia
HL	Korea (Republic of)
HP HO	Panama
HR HQ	Honduras
HS	Thailand
HT	Nicaragua YN
HU	El Salvador YS
HV	Vatican City
HZ	Saudi Arabia
I IA-IH IK IL IN IP IR IT IV-IX	Italy
IS0 IM0	Sardinia
J2	Djibouti
J3	Grenada
J4	Greece SV
J5	Guinea-Bissau
J6	St Lucia
J7	Dominica
J8	St Vincent and the Grenadines
JA JE-JS	Japan
JD 7J	Minami Torishima
JD 7J	Ogasawara Is
JT JU JV	Mongolia
JW	Svalbard
JX	Jan Mayen
K KA-KZ	*USA and US Islands* W KC6xx KG4xx KH1-0 KP1-5
KC6 x x	Republic of Palau
KG4 x x	Guantanamo Bay
KG6 x x	Guam
KH1	Baker I. and Howland I.
KH2 (KG6)	Guam
KH3	Johnston I.
KH4	Midway Is.
KH5	Palmyra I.
KH5J	Jarvis I. KH5
KH5K	Kingman Reef
KH6 7	Hawaiian Is
KH7K	Kure I.
KH8	American Samoa
KH9	Wake I.
KH0	North Mariana
KL	Alaska
KP1	Navassa I.
KP2	US Virgin Is
KP3 4	Puerto Rico
KP5	Desecheo I.
L2-L9	Argentina LU
LA LB LC LG LI LJ LN	Norway
LU LO-LT LV LW	Argentina
LX	Luxembourg
LY	Lithuania
LZ	Bulgaria
M MD MI MJ MM MU MW	United Kingdom G GD GI GJ GM GU GW GM GU GW
N NA-NG NI-NK NM-NO NQ-NZ	USA W
NH1-NH0	*US Pacific islands* KH1-KH0
NL	Alaska KL
NP1-NP5	*US Caribbean Islands* KP1-KP5
OA OB OC	Peru
OD	Lebanon
OE	Austria
OH OF OG OI	Finland
OH0 OF0 OG0	Aland Is
OJ0 OF0M OH0M	Market Reef
OK OL	Czech Republic
OM	Slovak Republic
ON OO-OT	Belgium
OX	Greenland
OY	Faroe Is
OZ	Denmark
P2	Papua New Guinea
P3	Cyprus 5B
P4	Aruba
P5	Korea (Dem Peoples Rep of)
PA PB PD PE PI	Netherlands
PJ1 PJ2 4 9	Netherlands Antilles
PJ5 PJ6 7 8	Sint Maarten, Saba and St Eustatius
PY PP-PX	Brazil
PY0F	Fernando de Noronha Archipelago
PY0M	Martim Vaz I. PU0T
PY0R	Atol das Rocas PY0F
PY0S	St Peter and St Paul Rocks
PY0T	Trindade I.
PZ	Suriname
R1F	Franz Josef Land
R1M	Malyj Vysotskij I.
R RA RK RN RU-RZ	European Russia UA
R RA RK RN RU-RZ	Asiatic Russia UA9
R2 RA2 RK2 RN2 RY2	Kaliningradsk UA2
S2	Bangladesh
S4	CiskeiZS

Prefix	Country
S5	Slovenia
S6	Singapore 9V
S7	Republic of Seychelles
S8	Transkei ZS
S9	Sao Tome and Principe
S0	Western Sahara
SM SH-SL	Sweden
SP SN-SR	Poland
ST	Republic of the Sudan
ST0	Southern Sudan
SU	Egypt
SV SX-SZ	Greece
SV—/A	Mount Athos
SV5	Dodecanese Is
SV9	Crete
SV0	Non-nationals in Greece or on Greek Is SV SV5 SV9
T2	Tuvalu
T30	West Kiribati
T31	Central Kiribati
T32	East Kiribati
T33	Banaba
T4	Cuba CO
T5	Somalia
T6	Afghanistan YA
T7	San Marino
T9	Bosnia - Hercegovina
TA	Turkey
TD	Guatemala TG
TE	Costa Rica TI
TF	Iceland
TG	Guatemala
TI	Costa Rica
TI9	Cocos I.
TJ	Cameroon
TK	Corsica
TL	Central African Republic
TM	*France including overseas Territories and Departments* F
TN	Congo
TO	*France including overseas Territories and Departments* FG FJ FM FP FR FS FY
TP	*Council of Europe - Strasbourg* F
TR	Gabon
TT	Chad
TU	Cote d'Ivoire
TX	*France including overseas Territories and Departments* FK FO FW
TY	Benin
TZ	Mali
UA U UA UE 1 3 4 6	European Russia
UA2 U UA UE 2	Kaliningrad
UA9 U UA UE 8-0	Asiatic Russia
UK U8 UJ UK7-9 UM	Uzbekistan
UN UN1-0 UP UQ	Kazakhstan
UR US-UZ	Ukraine
V2	Antigua and Barbuda
V3	Belize
V4	Federation of St Kitts and Nevis
V5	Namibia
V6	Micronesia
V7	Marshall Is
V8	Brunei Darussalam
V9	Vendaland ZS
VE VA-VG	Canada
VE0	*Canadian /MM Stations*
VK VI	Australia
VK9C	Cocos Keeling Is
VK9L	Lord Howe I.
VK9M	Mellish Reef
VK9N	Norfolk I.
VK9W	Willis Is
VK9X	Christmas I.
VK0	Heard I.
VK0	Macquarie I.
VO1 VO3 5 7 9	Newfoundland VE
VO2 VO4 6 8 0	Labrador VE
VP2E	Anguilla
VP2M	Montserrat
VP2V	British Virgin Is
VP5	Turks and Caicos Is
VP8	Falkland Is
VP8	South Georgia
VP8	Antarctica
VP8 AZ1 5 ED0 L UnZx	South Orkney Is
VP8	South Sandwich Is
VP8 CE9 CX0 ED0	South Shetland Is
HF0 HL5 LUnZx ZX0 4K1	South Shetland Is cont.
VP9	Bermuda
VQ9	Chagos Is
VR2	Hong Kong
VR6	Pitcairn I.
VU	India
VU	Lakshadweep
VU	Andaman Is and Nicobar Is
VX VY	Canada VE
VY1	Yukon Territory VE
VY2	Prince Edward I. VE
W WA-WG WI-WK WM-WO WQ-WZ	USA
WH1-WH0	*US Pacific Islands* KH1 - KH0
WL	Alaska KL
WP1-WP5	*US Caribbean Islands* KP1 - KP5
X5	*Serbian occupied Bosnia*
XE XB-XH	Mexico
XF4	Revilla Gigedo Is
XJ-XO	Canada VE
XQ XR	*Chile and Islands* CE CE9 CE0
XT	Burkina Faso
XU	Cambodia
XV	Vietnam 3W
XW	Lao Peoples Democratic Republic
XX3	Madeira Is CT3
XX9	Macao
XZ XY	Myanmar
XZ5 XZ9	*Karen State* XZ
YA	Republic of Afghanistan
YBYC YE-YH	Indonesia
YI	Iraq
YJ	Vanuatu
YK	Syria
YL	Latvia
YM	TurkeyTA
YN	Nicaragua
YO YP-YR	Romania
YS	El Salvador
YU YT	Yugoslavia
YV YW-YY	Venezuela
YV0	Aves I.
YZ	Yugoslavia YU
Z2	Zimbabwe
Z3	Macedonia
ZA	Albania
ZB ZG	Gibraltar
ZC	UK Sovereign Bases on Cyprus - Akrotiri and Dhekelia
ZD7	St Helena
ZD8	Ascension I.
ZD9	Tristan da Cunha and Gough I.
ZF	Cayman Islands
ZK1	South Cook Is
ZK1	Northern Cook Is
ZK2 ZK9	Niue
ZK3	Tokelau Is
ZL	New Zealand
ZL7	Chatham Is
ZL8	Kermadec Is
ZL9	Auckland I. and Campbell I.
ZM	*New Zealand and Islands* ZL ZL7 ZL8 ZL9
ZP	Paraguay
ZS ZR ZU	Republic of South Africa
ZS8	Prince Edward I. and Marion I.
ZV-ZZ	*Brazil and Islands* PY PY0
1A0	Sovereign Military Order of Malta (Rome, Italy)
1B	*Turkish area of North Cyprus*
1C	*Chechnya Rep. (Russian Federation)*
1P	*Seborga Principato (Italy)*
1S	Spratly Archipelago 9M0
1Z	*Karen State (Myanmar)*
2D	Isle of Man GD
2E	England G
2I	Northern Ireland GI
2J	Jersey GJ
2M	Scotland GM
2U	Guernsey and Dependencies GU
2W	Wales GW
3A	Monaco
3B6	Agalega Is
3B7	Cargados Carajos (St Brandon) 3B6
3B8	Mauritius
3B9	Rodriguez I.
3C	Equatorial Guinea
3C0	Annobon I.
3D2	Republic of Fiji
3D2	Conway Reef
3D2	Rotuma I.
3DA0	Swaziland
3E-3F	Panama HP
3G	*Chile and Islands* CE CE9 CE0
3V	Tunisia
3W XV	Vietnam
3X	Republic of Guinea
3Y	Bouvet I.
3Y	Peter I Island
3Z	Poland SP
4A-4C	*Mexico and Islands* XE XF4
4D-4I	Philippines DU
4J 4K	Azerbaijan
4L	Georgia
4M	*Venezuela and Islands* YV YV0
4N1 6-0	Yugoslavia YU
4T	Peru OA
4U	United Nations Organization
4U1ITU 4UnITU	United Nations Geneva
4U1SCO	UNESCO, Paris F
4U1UN 4UnUN	United Nations New York
4U1VIC	United Nations Vienna OE
4U1WB	World Bank Washington D.C. W
4V	Haiti HH
4X 4Z	Israel
5A	Libya
5B	Cyprus
5C	Morocco CN
5H	Tanzania
5J 5K	*Colombia and Islands* HK HK0
5L	Liberia EL
5N	Nigeria
5P	Denmark OZ
5R	Madagascar
5T	Mauritania
5U	Niger
5V	Togo
5W	Western Samoa
5X	Uganda
5Z 5Y	Kenya
6C	Syria YK
6D-6J	*Mexico and Islands* XE-XF4
6K	Republic of Korea HL
6O	Somalia T5
6P	Pakistan AP
6T 6U	*Sudan and Southern Sudan* ST ST0
6W 6V	Senegal
6Y	Jamaica
7J-7N	Japan JA
7O	Republic of Yemen
7P	Lesotho
7Q	Malawi
7S	Sweden SM
7X 7W	Algeria
7Z	Saudi Arabia HZ
8A 8B 8E 8I	Indonesia YB
8J	Japan JA
8O	Botswana A2
8P	Barbados
8Q	Maldives
8R	Guyana
8	Sweden SM
9A	Croatia
9G	Ghana
9H	Malta
9J 9I	Zambia
9K	Kuwait
9L	Sierra Leone
9M2	Malaya (Malaysia)
9M6	Sabah (Malaysia) 9M8
9M8	Sarawak (Malaysia)
9M0 BV9S 1S DU	Spratly Archipelago
9N	Nepal
9Q 9R	Zaire
9U	Burundi
9V	Singapore
9W	Malaysia *(including Sabah & Sarawak)* 9M2 8
9X	Rwanda
9Y 9Z	Trinidad and Tobago

GB2RS – Propagation

An explanation of the solar and geophysical information contained in the GB2RS broadcasts.

Time Covered by the Bulletins

The factual data covers the 7 days prior to the preparation of the broadcast – usually Monday to Sunday, while the forecast normally runs from the Sunday of transmission to the following Sunday.

Solar Activity

Sun Spots

At solar minimum, sunspots have little relevance, so these details are not included in the broadcasts, except for the monthly means, and maximum and minimum days.

Solar Activity

The general level of solar activity is referred to as – *Quiet:* no active regions. *Moderate:* active regions erupting but with low intensity. *Active:* one or more active regions erupting with high intensity bursts, particularly on the visible disc.

Eruptive events on the limbs are reported if information is received in time, as these can be aurora producers. Solar flares are divided into five types: *'A', 'B', 'C', 'M'* and *'X'* according to their X-ray energy level. This is measured by satellites in megaelectron volts (MeV), or in proton flux. There are four energy thresholds, 2, 10, 50, 100 MeV, which are classified by numbers, i.e., 'M3' or 'X4', etc. The major 'M' or 'X' events are always referred to in the broadcasts, but the low energy ones ('A', 'B' and 'C') only rarely.

As well as X-ray intensity of flares, there is the *'Optical Importance'* which gives a measure of its size and brilliance. The size is given by a number up to 4 while the brilliance is defined as – *'F':* faint, *'N':* normal, *'B':* bright and *'S'* means sub for example *'SF'* means sub-faint. These are combined to give the complete flare data eg 'M3/2B'.

An *Events Summary* is always given in the GB2RS news. This summary relates to the X-Ray, optical, 10 cm flux and proton particle data. The *'states'* referred to are *'Quiet', 'Alert'* or *'Active'*. These can be coupled with any combination of solar, magnetic or proton activity. A summary of the week's state is always given in the news broadcast.

Major Flares

Those major flares above M9/3B and all X types can be very disruptive to the ionosphere, severely degrading HF bands. They can cause auroral events, usually about 30 to 50 hours later, particularly if they are on the sun's limbs, or have just passed the central meridian. Flare effects are reported as – *SSC: sudden storm commencement* (a rapid increases of magnetic activity); *SID: sudden ionospheric disturbance* (ionospheric blackouts of the higher HF frequencies, lasting minutes to hours); *SWF: short wave fade-outs* (reductions in long range signal strengths, taking or lasting, minutes to hours, compared with expected signal level). A 90 day average of flux levels is given, as this has been found to be best for home computer prediction programs.

Solar Flux

This is the 2,800 MHz radio noise output from the sun at midday. This frequency is chosen because the radio sun looks the same size as the visible sun. The figure given is that obtained at Penticton (BC, Canada) which is the world standard. The level varies from, at the cycle's minimum, about 67 units, to, at maximum of around 300 units. The higher the level, the more intense is the sun's ionising radiation, and the higher the frequency that can be reflected from the ionosphere. Good HF band conditions require a high solar flux, but the magnetic conditions must also be taken into consideration.

Coronal Holes

Coronal holes are holes in the sun's outer corona through which material is ejected by various means. There are always holes at the sun's polar regions, but tongues sometimes extend to the equatorial regions, or small holes can form. The passage of these can cause a magnetic disturbance. This is particularly so if the interplanetary magnetic field is southerly as this couples to the earth's northward field. What have become known as 'Scottish' type auroras can generally be attributed to the passage of a coronal hole. If known about, coronal holes are always referred to in the text due to their importance and are more prevalent around sunspot minimum.

Proton Flares

These are detected by satellite and are reported in GB2RS if the energy level exceeds 10 MeV. Plotting these events on the solar base map will tell you where the active regions are. Proton events cause high absorption in the ionospheric 'D' region, and severely affect the ionosphere particularly over the polar regions, by *'polar cap absorption'*.

Other Solar Events

Now and again reference is made to *'Solar Filaments'*. These are tenuous streams of flare material, protruding into space from the sun's surface which sometimes break off. They are usually referred to in fractions of the solar diameter, eg 0.8 of the Sun's diameter. These events are sudden and unpredictable and can cause widespread auroras, ionospheric blackouts, and worldwide disruption to radio communication. Sometimes *'Eruptive Prominences'* are reported, but these are usually not so disruptive as the Solar Filaments.

Monthly Means

The Belgium observatory issue the previous month's mean provisional sunspot number RI together with the month's maximum and minimum days. As soon as the data is received, a provisional smoothed number for the previous 6 months and forecast for the next 6 months is given. The mean and 6 month smoothed values of the previous month's solar flux and geomagnetic A_p index are usually received from Boulder.

Geomagnetic Activity

With the change in the URSIgram system, the old A index is no longer produced. Instead the more meaningful planetary A_p index is mainly used, however the K_p index may be referred to for highlighting an event. Although these are estimated levels, they differ little from the final figures in practice. If a particular situation occurs, the Chambon la Foret and Wingst A index may be referred to, or the planetary K_p indices. In cases where major magnetic storms occur, all the data may be given together with data from Kerguelen Island (49°S 70°E).

An 'A_p' index of 0 to 10 is *'Quiet'*, 11 to 20 *'Unsettled'*, 21 to 50 *'Sub-storm'*, 51 to 80 *'Storm'*, 81 to 150 *'Severe or major storm'* and the maximum can be 300 or more. The 'A' scale is linear (1 to 400) and is based on the 8 K figures for a day. High levels of geomagnetic activity, say over 30, are associated with poor HF band conditions, and higher than this with auroras as well.

Satellite Data

This is replacing ground based data. The X-ray background flux (more sensitive than solar flux) is averaged for the week. Special mention is made of unusual levels which start from A 1 to 9, then B1 and C1. At present (near solar minimum) they are mainly A. The >2MeV electron fluence is referred to as high, normal or low. High levels affect the state of the HF bands due to ionospheric disruption. The Wind Satellite is now operating experimentally measuring Solar wind speeds – 300 to 400 kms^{-1} being normal. Particle densities can be as low: < 10 pcm^{-3}, moderate: 25 pcm^{-3} and high above this. B_z is the magnetic orientation measured in nanoTeslas. If it is negative, or south, it couples to the earth's northern orientation, resulting in HF disturbances and auroras. If it is positive, then little or nothing happens.

Ionospheric Data

The daily F_2 layer critical frequencies for a UK observatory are no longer available to the URSIgram service. A limited service is now given of the only data available to GB2RS: from Poitiers in central France (46°N, 00°E) and Juliusruh in northern Germany (54°N, 13°E). The 'daily highs' and 'darkness hour lows' of the hourly observations are given and average times when these occur.

The *'critical frequency'* is a measure of the highest frequency that can be reflected from the ionosphere from a signal sent vertically upwards.

In practice, the maximum frequency that can be used from normal communication is about 3 times the critical at 'equal latitudes' but if you work South, then higher levels are possible, and to the North, lower. The actual level depends on prevailing conditions and the time of the day.

During darkness hours, the normally smooth layers can break up during magnetically disturbed times and this is referred to as *'spread F'*. The break-up can be vertically, or horizontally, or both at once. These holes give rise to deep fading (QSB). In practice the more northern circuits are more prone to this effect, and it is also more likely to occur during the early morning periods. The bulletin usually refers to the numbers of hours the spread F has been present, or if very bad, on any particular day.

'Blanketing E' means that the E layer is so heavily ionised that the ionosonde station cannot see through it. The effect is often associated with summertime sporadic-E, or for northern stations, it may be due to auroral conditions.

Absorption

Sometimes, for northern stations, complete absorption of the ionosonde signal occurs. This suggests that the D region is so heavily ionised that it is absorbing all but the strongest radio signals. These events can be associated with proton flares, or high energy 'M' or 'X' flare events, or by electron precipitation from the earth's radiation belts.

Seasonal Changes

The *'daily highs'* tend to be higher in winter, and lower in summer. The darkness hour *'lows'* vary in the opposite way – high in summer and low in winter. The weekly average variations are balanced against these seasonal changes, and reference is made to any discrepancy when this applies. For the HF bands, the higher the daily 'highs', the better is the chance of DX on the higher HF bands. The average times of the highest and lowest frequency recorded is given. The high times vary with the season, being around midday during the winter and early evening (about 2000 UTC) in the summer. The low times do not vary much, being usually about 0400 UTC.

Forecasts

Each week, a solar forecast of expected events for the 7 days from the Sunday of broadcast. The forecast includes expected levels of solar flux, geomagnetic activity and the passage of any expected coronal holes. Due to the absence of UK ionosonde data, a forecast of the expected MUFs for the South of England is estimated based on Poitiers data, and this is given as the most likely daylight level. The darkness hour levels are also given in terms of the best band to use. Scotland and the North will be generally down on these levels the amount depends on other factors and warnings as to how the expected magnetic activity will affect the North more are referred to. In general, working North / South paths will give better results than East / West. During the summer months around sunspot minimum, such as now, the sporadic E season is very prevalent and is often extended: it may well cover from early May to late August with a secondary smaller peak around December. A reminder is given, as well as reports on openings during the previous week, particularly on the 2, 6 and 10 metre bands.

RadCom HF Predictions

The table printed each month in *RadCom* shows path predictions from the UK to 44 locations around the world, three of them by both short path and long path. They are shown in the form of probabilities, where the figures 0-9 represent 0-90 per cent. While there is no guarantee on any particular day that the path will be open, it does provide a useful guide eg a figure 1 = 10% probability i.e. the path should be open about 3 days in the month.

It is important to appreciate that the numbers do not represent a snapshot of signal strengths on a typical day during the month. For example, a weak signal heard nearly every day on 7 MHz might be S1 against a probability rating of 9, whereas on 21 MHz, the band might be closed for all but three days. In the predictions, the highest probability appears on the band which is nearest to the calculated path optimum working frequency. Probabilities are lower on bands above and below.

It should be noted that the HF predictions take no account of sporadic-E propagation, a mode which often, but rather unpredictably, enlivens 28 MHz during the summer, nor do they cater for the 'grey-line' long-distance openings around dusk or dawn, which have too short a duration to appear in the two-hourly steps of our tables.

Please see *RadCom*: March 1994, page 72 for a fuller explanation of the F_2 propagation predictions and in particular how corrections can be made for northern stations.

THE CARRINGTON SERIES SOLAR ROTATION PLOTTING CHART

Regular users of the solar rotation data section of the *Yearbook* will find a change in the format this year. The Solar Rotation Base Maps, which have appeared here in previous years, were born to satisfy an astronomical need, a purpose that they have served very well since the late nineteen-fifties. But most radio amateurs have little interest in knowing where the Earth is in its orbit around the Sun on a given day so it has seemed reasonable to drop the Sun's True Longitude scale of the Base Map in order to provide a simplified diagram – one which presents dates in the form of a neat series of columns and rows instead of a sloping raster, but without losing the benefit of synchronism with the Carrington solar rotation criteria.

The purpose of the chart is still to provide a framework on which may be entered either daily values of some function likely to have links with the rotation period – sunspot numbers, solar radio flux values, geomagnetic indices, etc. – or for flagging days on which have occurred significant isolated events, such as radio aurora, ionospheric disturbances, etc. Then, if the hoped-for links are present they should reveal themselves by horizontal (or nearly so) trends across the chart.

For the ordinate scale the 360° of solar longitudes have been divided into 28 equal segments, similar to the inside of an orange. It is convenient to group them in four quadrants, shown as A1 to A7, B1 to B7, C1 to C7, and D1 to D7 down the left-hand side of the plotting chart. The boxes which represent days have been aligned with these segments according to which solar longitude appeared at the centre of the solar disc, as seen from the Earth, at 12 UT on the day in question. But, because the width of a segment is just under 13° and the daily change in solar longitude is more than that (it varies between 13.24° in June and 13.17° in December), occasionally a segment has to be skipped in order to maintain synchronism. When that happens it has been indicated by an asterisk in the box to show that the box should be ignored when entering data.

Besides providing a simple and convenient grouping for any subsequent statistical analyses of collected data, the system provides an easy way of estimating past and future movements of significant features as they cross the face of the Sun. As an example, should such a feature appear over the east limb of the Sun on 27 September, the chart assigns it to segment B3 on Rotation 1914. It may be expected to cross the central meridian on the date shown against C3 (4 October), and it should leave the disc over the western limb at D3 (11 October). If the feature persists beyond the period when it is out of sight behind the Sun it might be expected to reappear at the east limb again at B3 on the following rotation, i.e., on 24 October. Similarly, of course, for other segments.

The scale of L_o, representing the longitude at the centre of the solar disc on successive days, provides a means of comparing observed data logged on the chart with solar maps of each individual rotation of the Sun, drawn up by professional observers. Such an exercise would be found useful when seeking instances of cause and effect, but patience is required as the solar maps are published, of necessity, a long time in arrears. It is suggested that observers intending to pursue their research that far may find it useful to contact the Society's Propagation Studies Committee for further information.

Packet BBS Service

The Society's Propagation Studies Committee are now running a daily service on the packet system. The message is the joint USAF/ NOAA solar and geophysical activity summary compiled by SESC Boulder. It consists of the previous day's solar data and gives details of:

a) List of energetic events
b) Proton events
c) Geomagnetic activity summary
d) Stratwarm
e) Daily indices
f) Comments
g) The aa indicies

This is put into the packet bulletin boards as soon as possible each morning by G0CAS. To locate this service list under the 'TO' field under 'SUN' or under the 'SUBJECT FIELD' solar indices, or under the callsign G0CAS. This is repeated to all other BBSs.

The aa indicies are supplied by the British Geological Survey and are put on the packet bulletin board every week as soon as they are received. They are listed in full so that the complete daily data is available in nanoTeslas. As a general guide: Quiet levels are up to 20 nT, unsettled: 70 nT, Sub-storm: 120 nT, and severe storm: 500 nT. The series is plotted in 27 day Bartell's rotations with the numbers given.

GAM1/80 Broadcasts

The RA have ruled that the proposed broadcast on 3.812 MHz cannot be licensed. However, a solar data service is now operating under the GB2RS callsign. These broadcasts are on Sundays at 09:00, 12:00, 15:00 and 18:00 hours local time on 3.518 MHz, using CW (12 wpm) sent by G4FKH with G2FKZ acting as a reserve broadcaster. Please send an SASE to G2FKZ for an explanation of the terms used in these broadcasts.

Carrington Series Solar Rotation Plotting Chart

For use in connection with dates or data values known, or thought likely, to be linked to some form of solar activity, such as auroral propagation, geomagnetism, sunspot numbers, etc. The segment numbers A1 - D7 provide a guide to the expected progress of relevant features as they cross the solar disc. For example, an active region appearing over the Sun's east limb on a date corresponding to segment A5 should cross the central meridian at B5 and (if it persists long enough) disappear over the west limb on the date opposite C5, maybe appearing again around A5 on the next rotation.

From time to time a segment has to be skipped (shown on the chart by the symbol ✱), because the solar rotation period is not an exact number of days. The orbital speed of the Earth around the Sun varies throughout the year so those skipped days do not present a regular pattern over the relatively short time span covered by this chart. The resulting format allows of direct comparision against source maps of solar activity scaled against L_0 published by astronomical observatories.

Rotation no:		1927	1928	1929	1930	1931	1932	1933	1934	1935	1936	1937	1938	1939	1940	1941	1942	1943	1944	1945	1946	1947
Starting month:		97 Sep	97 Oct	97 Nov	97 Nov	97 Dec	98 Jan	98 Feb	98 Mar	98 Apr	98 May	98 Jun	98 Jul	98 Aug	98 Aug	98 Sep	98 Oct	98 Nov	98 Dec	99 Jan	99 Feb	99 Mar
A1	L_0 = 360°	8	5	2	29	26	✱	19	18	14	12	8	5	1	29	25	22	18	16	12	8	8
A2		9	6	✱	30	27	23	20	19	15	13	9	6	2	30	26	23	19	17	13	9	9
A3		10	7	3	1	28	24	21	20	16	14	10	7	3	✱	27	24	20	18	14	10	10
A4		11	8	4	2	29	25	22	21	17	15	11	8	4	31	28	25	21	19	15	11	11
A5		12	9	5	3	30	26	23	22	18	16	12	9	5	1	29	26	22	20	16	12	12
A6		13	10	6	4	31	27	24	23	19	17	13	10	6	2	30	27	23	21	17	13	13
A7		14	11	7	5	1	28	25	24	20	18	14	11	7	3	1	28	24	22	18	14	14
B1	L_0 = 270°	15	12	8	6	2	29	26	25	21	✱	15	12	8	4	2	29	25	23	19	15	15
B2		16	13	9	7	3	30	27	26	22	19	16	13	9	5	3	30	26	24	20	16	16
B3		17	14	10	8	4	31	28	27	23	20	17	14	10	6	4	31	27	✱	21	17	✱
B4		18	15	11	9	5	1	1	28	24	21	18	15	11	7	5	1	28	25	22	18	17
B5		19	16	12	10	6	2	2	29	25	22	19	16	12	8	6	2	29	26	23	19	18
B6		20	17	13	11	7	3	3	30	26	23	20	17	13	9	✱	3	30	27	24	20	19
B7		21	18	14	12	8	4	4	31	27	24	21	18	14	10	7	4	1	28	25	21	20
C1	L_0 = 180°	22	19	15	✱	9	5	✱	1	28	25	✱	19	15	11	8	5	2	29	26	22	21
C2		23	20	16	13	10	6	5	2	29	26	22	20	16	12	9	6	3	30	27	23	22
C3		24	21	17	14	11	7	6	3	30	27	23	21	17	13	10	7	4	31	28	24	23
C4		25	22	18	15	12	8	7	4	1	28	24	22	18	14	11	8	5	1	29	25	24
C5		✱	23	19	16	13	9	8	5	2	29	25	23	19	15	12	9	6	2	30	26	25
C6		26	24	20	17	14	10	9	6	3	30	26	24	20	16	13	10	7	3	31	27	26
C7		27	25	21	18	15	11	10	7	4	31	27	25	21	17	14	11	8	4	1	28	27
D1	L_0 = 090°	28	26	22	19	16	12	11	8	5	1	28	✱	22	18	15	12	9	5	2	1	28
D2		29	27	23	20	17	13	12	9	6	2	29	26	23	19	16	13	10	6	3	2	29
D3		30	28	24	21	18	14	13	10	7	3	30	27	24	20	17	14	11	7	✱	3	30
D4		1	29	25	22	19	15	14	11	8	4	1	28	25	21	18	✱	12	8	4	4	31
D5		2	30	26	23	20	16	15	12	9	5	2	29	26	22	19	15	13	9	5	5	1
D6		3	31	27	24	21	17	16	✱	10	6	3	30	27	23	20	16	14	10	6	6	2
D7		4	1	28	25	22	18	17	13	11	7	4	31	28	24	21	17	15	11	7	7	3
	L_0 = 000°																					
Ending month:		97 Oct	97 Nov	97 Nov	97 Dec	98 Jan	98 Feb	98 Mar	98 Apr	98 May	98 Jun	98 Jul	98 Jul	98 Aug	98 Sep	98 Oct	98 Nov	98 Dec	99 Jan	99 Feb	99 Mar	99 Apr

Summary of Terms used in the GB2RS Broadcasts

By Neil Clarke G0CAS

a INDEX. A 3-hourly 'equivalent amplitude' index of local geomagnetic activity; 'a' is related to the 3-hourly K INDEX according to the following scale:

K	0	1	2	3	4	5	6	7	8	9
a	0	3	7	15	27	48	80	140	240	400

A INDEX. A daily index of geomagnetic activity derived as the average of the eight 3-hourly a indices.

ACTIVE REGION (AR). A localized, transient volume of the solar atmosphere in which PLAGES, SUNSPOTS, FACULAE, FLARES, etc. may be observed.

Ap INDEX. An averaged planetary A INDEX based on data from a set of specific stations.

AURORA. A faint visual phenomenon associated with geomagnetic activity, which occurs mainly in the high-latitude night sky; typical auroras are 100 to 250 km above the ground.

AURORAL OVAL. An oval band around each geomagnetic pole which is the locus of structured AURORAE.

AUTUMNAL EQUINOX. The equinox that occurs in September.

BARTEL'S ROTATION NUMBER. The serial number assigned to 27-day rotation periods of solar and geophysical parameters. Rotation 1 in this sequence was assigned arbitrarily by Bartel to begin in January 1833.

CARRINGTON LONGITUDE. A system of fixed longitudes rotating with the sun.

CENTIMETER BURST. A solar radio burst in the centimeter wavelength range.

CENTRAL MERIDIAN PASSAGE (CMP). The passage of an ACTIVE REGION or other feature across the longitude meridian that passes through the apparent centre of the solar disk.

CHROMOSPHERE. The layer of the solar atmosphere above the PHOTOSPHERE and beneath the transition region and the CORONA.

COMPREHENSIVE FLARE INDEX (CFI). The indicative of solar flare importance given by the sum of the following five components

a) Importance of ionizing radiation as indicated by time-associated Short Wave Fade or Sudden Ionospheric Disturbance; (Scale 0-3)

b) Importance of H-Alpha flare; (Scale 0-3)

c) Magnitude of 10 cm flux; (Characteristic of log of flux in units of 10^{-22} Watt/m^2/Hz)

d) Dynamic spectrum; (Type II = 1, Continuum = 2, Type IV with duration > 10 minutes = 3)

e) Magnitude of 200 MHz flux; (Characteristic of log of flux in units of 10^{-22} Watt/m^2/Hz)

CONJUGATE POINTS. Two points on the earth's surface, at opposite ends of a geomagnetic field line.

CORONA. The outermost layer of the solar atmosphere, characterized by low densities ($<1.0 \times 10^9$ cm^{-3}) and high temperatures ($>1.0 \times 10^6$K).

CORONAL HOLE. An extended region of the CORONA, exceptionally low in density and associated with unipolar photospheric regions.

D REGION. A daytime layer of the earth's IONOSPHERE approximately 50 to 90 km in altitude.

DIFFERENTIAL ROTATION. The change in SOLAR ROTATION RATE with latitude. Low latitudes rotate at a faster angular rate (approx. 14° per day) than do high latitudes (approx. 12° per day).

DISAPPEARING SOLAR FILAMENT (DSF). The sudden (timescale of minutes to hours) disappearance of a solar FILAMENT (PROMINENCE).

E REGION. A daytime layer of the earth's ionosphere roughly between the altitudes of 85 and 140 km.

F REGION. The upper layer of the IONOSPHERE, approximately 120 to 1500 km in altitude. The F region is subdivided into the F_1 and F_2 regions. The F_2 region is the most dense and peaks at altitudes between 200 and 600 km. The F_1 region is a smaller peak in electron density, which forms at lower altitudes in the daytime.

FACULA. A bright region of the PHOTOSPHERE seen in white light, seldom visible except near the solar LIMB.

FILAMENT. A mass of gas suspended over the PHOTOSPHERE by magnetic fields and seen as dark lines threaded over the solar DISK. A filament on the LIMB of the sun seen in emission against the dark sky is called a PROMINENCE .

FLARE. A sudden eruption of energy on the solar disk lasting minutes to hours, from which radiation and particles are emitted.

GEOMAGNETIC FIELD. The magnetic field observed in and around the earth. The intensity of the magnetic field at the earth's surface is approximately 0.32 gauss at the equator and 0.62 gauss at the north pole.

GEOMAGNETIC STORM. A worldwide disturbance of the earth's magnetic field, distinct from regular diurnal variations.

- Minor Geomagnetic Storm: A storm for which the A_p index was greater than 29 and less than 50.
- Major Geomagnetic Storm: A storm for which the A_p index was greater than 49 and less than 100.
- Severe Geomagnetic Storm: A storm for which the A_p index was 100 or more.
- Initial Phase: Of a geomagnetic storm, that period when there may be an increase of the MIDDLE-LATITUDE horizontal intensity (H).
- Main Phase: Of a geomagnetic storm, that period when the horizontal magnetic field at middle latitudes is generally decreasing.
- Recovery Phase: Of a geomagnetic storm, that period when the depressed northward field component returns to normal levels.

GRADUAL COMMENCEMENT. The commencement of a geomagnetic storm that has no well-defined onset.

HIGH-SPEED STREAM. A feature of the SOLAR WIND having velocities that are about double average solar wind values.

INTERPLANETARY MAGNETIC FIELD (IMF). The magnetic field carried with the SOLAR WIND .

IONOSPHERE. The region of the earth's upper atmosphere containing a small percentage of free electrons and ions produced by photoionization of the constituents of the atmosphere by solar ultraviolet radiation at very short wavelengths (< 1000 Å). The ionosphere significantly influences radio wave propagation of frequencies less than about 30 MHz.

IONOSPHERIC STORM. A disturbance in the F REGION of the IONOSPHERE , which occurs in connection with geomagnetic activity.

K INDEX. A 3-hourly quasi-logarithmic local index of geomagnetic activity relative to an assumed quiet-day curve for the recording site. Range is from 0 to 9. The K index measures the deviation of the most disturbed horizontal component.

Kp INDEX. A 3-hourly planetary geomagnetic index of activity generated in Gottingen, Germany, based on the K INDEX from 12 or 13 stations distributed around the world.

LIMB. The edge of the solar disk.

MAJOR FLARE. This flare is the basis for the forecast of geomagstorm, cosmic storm and/or protons in the earth's vicinity.

MeV. Mega (million) electronvolt. A unit of energy used to describe the total energy carried by a particle or photon.

MIDDLE LATITUDES. With specific reference to zones of geomagnetic activity, 'middle latitudes' refers to 20° to 50° geomagnetic.

MOUNT WILSON MAGNETIC CLASSIFICATIONS.

- Alpha. Denotes a unipolar SUNSPOT group.
- Beta. A sunspot group having both positive and negative magnetic polarities, with a simple and distinct division between the polarities.
- Beta-Gamma. A sunspot group that is bipolar but in which no continuous line can be drawn separating spots of opposite polarities.
- Delta. A complex magnetic configuration of a solar sunspot group consisting of opposite polarity UMBRAe within the same PENUMBRA .
- Gamma. A complex ACTIVE REGION in which the positive and negative polarities are so irregularly distributed as to prevent classification as a bipolar group.

NANOTESLA (nT). A unit of magnetism 1.0 x 10^{-9} tesla, equivalent to a gamma (1.0 x 10^{-5} gauss).

PENUMBRA. The SUNSPOT area that may surround the darker UMBRA or umbrae. It consists of linear bright and dark elements radial from the sunspot umbra.

PHOTOSPHERE. The lowest layer of the solar atmosphere; corresponds to the solar surface viewed in WHITE LIGHT, SUNSPOT and FACULAe are observed in the photosphere.

PLAGE. An extended emission feature of an ACTIVE REGION that exists from the emergence of the first magnetic flux until the widely scattered remnant magnetic fields merge with the background.

POLAR CAP ABSORPTION (PCA). An anomalous condition of the polar IONOSPHERE whereby HF and VHF (3 - 300 MHz) radiowaves are absorbed, and LF and VLF (3 - 300 kHz) radiowaves are reflected at lower altitudes than normal. In practice, the absorption is inferred from the proton flux at energies greater than 10 MeV, so that PCAs and PROTON EVENTs are simultaneous. Transpolar radio paths may still be disturbed for days, up to weeks, following the end of a proton event.

PROMINENCE. A term identifying cloud-like features in the solar atmosphere. The features appear as bright structures in the CORONA above the solar LIMB and as dark FILAMENTs when seen projected against the solar disk.

PROTON EVENT. By definition, the measurement of at least 10 protons/sq.cm/sec/steradian at energies greater than 10 MeV.

PROTON FLARE. Any FLARE producing significant FLUXes of >10 MeV protons in the vicinity of the earth.

RADIO EMISSION. Emissions of the sun in radio wavelengths from centimetres to decametres, under both quiet and disturbed conditions.

Type I. A noise storm composed of many short, narrow-band bursts in the metric range (300 - 50 MHz).

Type II. Narrow-band emission that begins in the meter range (300 MHz) and sweeps slowly (tens of minutes) toward decametre wavelengths (10 MHz). Type II emissions occur in loose association with major FLAREs and are indicative of a shock wave moving through the solar atmosphere.

Type III. Narrow-band bursts that sweep rapidly (seconds) from decimetre to decametre wavelengths (500 - 0.5 MHz). They often occur in groups and are an occasional feature of complex solar ACTIVE REGIONs.

Type IV. A smooth continuum of broad-band bursts primarily in the meter range (300 - 30 MHz). These bursts are associated with some major flare events beginning 10 to 20 minutes after the flare maximum, and can last for hours.

RECURRENCE. Used especially in reference to the recurrence of physical parameters every 27 days (the rotation period of the sun).

SECTOR BOUNDARY. In the SOLAR WIND , the area of demarcation between sectors, which are large-scale features distinguished by the predominant direction of the interplanetary magnetic field, toward or away from the sun.

SHORT WAVE FADE (SWF). A particular ionospheric solar flare effect under the broad category of sudden ionospheric disturbances (SIDs) whereby short-wavelength radio transmissions, VLF, through HF, are absorbed for a period of minutes to hours.

SMOOTHED SUNSPOT NUMBER. An average of 13 monthly R_I numbers, centred on the month of concern.

SOLAR CYCLE. The approximately 11-year quasi-periodic variation in frequency or number of solar active events.

SOLAR MAXIMUM. The month(s) during the SOLAR CYCLE when the 12-month mean of monthly average SUNSPOT NUMBERS reaches a maximum.

SOLAR MINIMUM. The month(s) during the SOLAR CYCLE when the 12-month mean of monthly average SUNSPOT NUMBERS reaches a minimum.

SOLAR WIND. The outward flux of solar particles and magnetic fields from the sun. Typically, solar wind velocities are near 350 kms^{-1}.

SPORADIC E. A phenomenon occurring in the E REGION of the IONOSPHERE , which significantly affects HF radiowave propagation. Sporadic E can occur during daytime or night-time and it varies markedly with latitude.

SUDDEN COMMENCEMENT (SC, or SSC for Storm Sudden Commencement). An abrupt increase or decrease in the northward component of the geomagnetic field, which marks the beginning of a GEOMAGNETIC STORM .

SUDDEN IMPULSE (SI+ or SI-). A sudden perturbation of several gammas in the northward component of the low-latitude geomagnetic field, not associated with a following GEOMAGNETIC STORM . (An SI becomes an SC if a storm follows.)

SUDDEN IONOSPHERIC DISTURBANCE (SID). HF propagation anomalies due to ionospheric changes resulting from solar FLAREs , PROTON EVENTs and GEOMAGNETIC STORMs.

SUNSPOT. An area seen as a dark spot on the PHOTOSPHERE of the sun. Sunspots are concentrations of magnetic flux, typically occurring in bipolar clusters or groups. They appear dark because they are cooler than the surrounding photosphere.

SUNSPOT GROUP CLASSIFICATION (Modified Zurich Sunspot Classification).

A A small single unipolar SUNSPOT or very small group of spots without PENUMBRA.

B Bipolar sunspot group with no penumbra.

C An elongated bipolar sunspot group. One sunspot must have penumbra.

D An elongated bipolar sunspot group with penumbra on both ends of the group.

E An elongated bipolar sunspot group with penumbra on both ends. Longitudinal extent of penumbra exceeds 10° but not 15°.

F An elongated bipolar sunspot group with penumbra on both ends. Longitudinal extent of penumbra exceeds 15°.

H A unipolar sunspot group with penumbra.

SUNSPOT NUMBER. A daily index of SUNSPOT activity (R), defined as R = k (10g + s) where S = number of individual spots, g = number of sunspot groups, and k is an observatory factor.

UMBRA. The dark core or cores (umbrae) in a SUNSPOT with PENUMBRA , or a sunspot lacking penumbra.

X-RAY FLARE CLASS. Rank of a FLARE based on its X-ray energy output. Flares are classified by the order of magnitude of the peak burst intensity (I) measured at the earth in the 1 to 8 Å band as follows:

Class	I (Wm^{-2})
B	$< 1.0 \times 10^{-6}$
C	$>= 1.0 \times 10^{-6}$ and $< 1.0 \times 10^{-5}$
M	$>= 1.0 \times 10^{-5}$ and $< 1.0 \times 10^{-4}$
X	$>= 1.0 \times 10^{-4}$

RSGB QSL Bureau

The purpose of the Bureau is to exchange QSL cards between RSGB members and other radio amateurs.

Most national radio societies operate bureaus, some making an extra charge for this service. The RSGB provides it as a free service, though only members may make use of the outgoing service.

The bureau is the cheapest way of sending cards.

How the QSL Bureau Operates

Your Outgoing Cards

All cards for distribution should be sent to the RSGB QSL Bureau at Headquarters. There is no limit to the number of cards which may be sent at any one time.

When the cards arrive at the bureau, those destined for abroad are sorted into countries and despatched in bulk to the appropriate overseas QSL bureaux, most of which are operated by members societies of the International Amateur Radio Union.

Cards for stations within the UK are sorted into call-sign groups, each of which is in the charge of a volunteer sub-manager. It is this person's task to associate the cards sent to the sub-bureau from the main QSL Bureau, with the envelopes which are on file.

Collecting Cards from the Bureau

Your Incoming Cards

Supply your sub-manager with stamped self-addressed envelopes of a suitable size and strong material – 8" × 5" is ideal.

Print your callsign or RS number in the top left hand corner of each envelope. Always use 1st or 2nd class stamps and not those bearing monetary amounts.

Envelopes should be numbered and *'Last envelope'* marked on one so that it is known when a fresh batch is needed.

Envelopes are not normally returned until full weight has been reached for the postage paid; those wishing to receive cards at more frequent intervals should mark their envelopes *'Wait 6'* etc.

An up-to-date list of names and addresses of sub-managers is available from HQ on request. Changes to the list are broadcast on GB2RS and in the QSL column in *Radio Communication*.

General notes

1) Licensed UK amateurs who are non-members of the RSGB may send stamped addressed envelopes to their sub-manager for collection of their cards, but they *may not* send cards for distribution.
2) Cards for amateurs who have neglected to send envelopes are retained for three months, after which the cards are destroyed. Amateurs who do not wish to collect cards should notify the QSL Bureau accordingly.
3) Amateurs who operate from a part of the United Kingdom which has a different prefix should deposit envelopes with the appropriate sub-manager for the different prefix. For example a G7 station who operates temporarily from Wales and who wishes to receive cards should leave envelopes with the GW7 submanager.
4) Overseas members of RSGB in countries where there is no QSL service operated by the IARU member society for that country, may send their cards to the RSGB QSL Bureau for distribution.
5) Overseas members who are not members of the RSGB may send cards addressed to UK stations only, direct to the RSGB QSL Bureau.
6) The facilities of the RSGB QSL Bureau are available both to transmitting and receiving members of the Society. Listeners are reminded, however, that their reports should contain sufficient information to be of genuine value to the transmitting amateurs concerned. Reception reports relating to short-wave broadcasting stations unfortunately cannot be accepted.

All QSL cards and correspondence relating to the RSGB QSL Bureau should be sent to the QSL Bureau at RSGB Headquarters.

Adhesive address labels are available free of charge on receipt of an SAE.

Envelopes for the collection of cards and correspondence concerning incoming cards should be sent to the appropriate sub-manager.

Sending Cards Through the Bureau

Choose QSL cards which do not exceed normal post-card size, viz 5.5" × 3.5" (140 x 80 mm). Large cards invariably have to be folded, whilst small ones and those of a thin nature are difficult to handle.

Print the addressee's call-sign on both sides of the cards.

Separate cards destined within the UK from foreign-going ones. Sort all cards alphabetically by prefix. Sort USA cards into call areas regardless of prefix. When a QSL Manager is involved, sort under *his* callsign. Please pack the cards so that they are all the same way up, and do not space cards with markers or similar. Pack cards adequately so that they do not get damaged in transit and with the correct postage to reach RSGB and send them to:

RSGB QSL Bureau
PO Box 1773
POTTERS BAR
Herts.
EN6 3EP

RSGB QSL Bureau Sub-Managers

G0AAA-AZZ
Ken Plumridge, GW4BYY, Swn-y-Gwynt, High Street, Llanberis, Caernarfon, Gwynedd, LL55 4EH

G0BAA-BZZ
Tom Bruin, G0PRN, Seaford, 38 Kirkley Cliff Road, Lowestoft, Suffolk, NR33 0DB

G0CAA-CZZ
Mr R Denton, G4YRZ, 399 Gateford Road, Worksop, Notts, S81 7BN

G0DAA-DZZ
John Purvess, G0FWP, 389 Otley Old Road, Cookridge, Leeds, LS16 6BX

G0EAA-EZZ
Geoffrey Jenner, G3KIW, Pogles Wood Cottage, Paradise Lane, Chapel Row, Bucklebury, Reading, RG7 6NU

G0FAA-FZ
Margaret Burchmore, G0ARQ, 49 School Lane, Horton Kirby, Dartford, Kent, DA4 9DQ

G0GAA-GZZ
Nigel Roberts, G4KZZ, 13 Rosemoor Close, Hunmanby, Filey, North Yorks, YO14 0NB

G0HAA-HZZ
Mr JT Macrae, G4DXI, Park House, 1 Highsted Road, Sittingbourne, Kent, ME10 4PS

G0IAA-IZZ
Mr A Lord, G4KHT, 5 Wasdale Green, Cottingham, Hull, HU16 4HN

G0JAA-JZZ
Mr J A Towle, G4PJZ, 63 Digby Avenue, Mapperley, Nottingham, NG3 6DS

G0KAA-KZZ
Keith Draycott, G3UQT, 28 Ladywood Road, Kirk Hallam, Ilkeston, Derbys, DE7 4NE

G0LAA-LZZ
Mr C H Lennox, G4LXU, Blazefield House Farm, Blazefield, Harrogate, North Yorks, HG3 5DR

G0MAA-MZZ
Harry Foster, G4EZS, 23 Ghyllroyd Drive, Birkenshaw, Bradford, West Yorkshire, BD11 2ET

G0NAA-NZZ
Edward Allen, G3DRN, 30 Bodnant Gardens, Wimbledon, London, SW20 0UD

G0OAA-OZZ
David Bloomfield, G0KUC, 8 Sunningdale Drive, Boston, Lincs, PE21 8HZ

G0PAA-PZZ
Allen Spence, 2M1AGP, 6 Woodend Terrace, Aberdeen, AB2 6YG

G0RAA-RZZ
Mr G P Greatrix, G7HNM, 80 Liquorpond Street, Boston, Lincs, PE21 8UJ

G0SAA-SZZ
Mr E J Otty, G4XRL, Sunny Bank, 103 Fifers Lane, Hellesdon, Norwich, NR6 6EF

G0TAA-TZZ
Jim Taylor, G0RFN, 121 Garesfield Gardens, Burnopfield, Newcastle upon Tyne, NE16 6LQ

G0UAA-UZZ
Mr D L Hughes, G0RVW, 7 Mellor Close, Norton, Runcorn, WA7 6BQ

G0VAA-VZZ
MR P F J Eames, M0AXD, 6 Sirius Close, Seaview, Isle of Wight, PO34 5LH

G0WAA-WZZ
Kathy Catlow, G4ZEP, Ballingroyd Farm, Crossley New Road, Crosstone, Todmorden, Lancs, OL14 8RP

G0XAA-ZZZ
Mr F B Stanbridge, G3PZS, 119 High Street, Earith, Huntingdon, Cambs, PE17 3PN

G1 series
Mr M Marriott, G0OPC, Greenfields View, March Road, Friday Bridge, nr Wisbech, Cambs, PE14 0HA

G2 series
Mr C H Adams, RS10906, 4 Park Gate Gardens, East Sheen, London, SW14 8BQ

G3AA-ZZ
Mr P J Pasquet, G4RRA, 64 Bricksbury Hill, Farnham, Surrey, GU9 0LY

G3AAA-DZZ
Edward Allen, G3DRN, 30 Bodnant Gardens, Wimbledon, London, SW20 0UD

G3EAA-HZZ
Mr E L Simpson, G3GRX, Everdene, Fell Lane, Penrith, Cumbria, CA11 8AW

G3IAA-KZZ
Nigel Entwistle, G0BRM, Park Garden House, Park Garden, West Row, Bury St Edmunds, Suffolk, IP28 8QG

G3LAA-NZZ
Thomas Bartlett, G3ITB, Yew Tree, 19 The Street, Hardley, Norwich, Norfolk, NR14 6BY

G3OAA-PZZ
Jack Brazzill, G3WP, 43 Forest Drive, Chelmsford, Essex, CM1 2TT

G3RAA-TZZ
Mrs P H McVey, G0PXJ, 18 Worlebury Hill Road, Weston-Super-Mare, Avon, BS22 9SP

G3UAA-VZZ
Mr M J Newton, G3UKW, 11 Chestnut Close, Rushmere St Andrew, Ipswich, IP5 7ED

G3WAA-ZZZ
Mr D Talbot, G0JHT, Southways, Tichborne Down, Alresford, Hants, SO24 9PL

G4AA-ZZ
Mr P J Pasquet, G4RRA, 64 Bricksbury Hill, Farnham, Surrey, GU9 0LY

G4AAA-AZZ
Dave Roebuck, G0LJM, 85 Crook Farm, Caravan Park, Glen Road, Baildon, Shipley, W Yorks, BD17 5ED

G4BAA-BZZ
L Harper, G4FNC, Three Oaks, Braydon, Swindon, Wilts, SN5 0AD

G4CAA-CZZ
Mr R Denton, G0JHT, 399 Gateford Road, Worksop, Notts, S81 7BN

G4DAA-DZZ
Deryck Buckley, G3VLX, Little Oaks, Park Road, Marden, Tonbridge, Kent, TN12 9LG

G4EAA-EZZ
Geoffrey Jenner, G3KIW, Pogles Wood Cottage, Paradise Lane, Chapel Row, Bucklebury, Reading, RG7 6NU

G4FAA-FZZ
Margaret Burchmore, G0ARQ, 49 School Lane, Horton Kirby, Dartford, Kent, DA4 9DQ

G4GAA-GZZ
Mr M E Slater, G3NML, 46 Ladywood, Boyatt Wood, Eastleigh, Hants, SO50 4RW

G4HAA-HZZ
Dave Roebuck, G0LJM, 85 Crook Farm, Caravan Park, Glen Road, Baildon, Shipley, W Yorks, BD17 5ED

G4IAA-IZZ
Ian Fugler, G4IIY, 9 Westover Road, Fleet, Hants, GU13 9DG

G4JAA-JZZ
Mr J A Towle, G4PJZ, 63 Digby Avenue, Mapperley, Nottingham, Nottinghamshire, NG3 6DS

G4KAA-KZZ
Keith Draycott, G3UQT, 28 Ladywood Road, Kirk Hallam, Ilkeston, Derbys, DE7 4NE

G4LAA-LZZ
Mr C H Lennox, G4LXU, Blazefield House Farm, Blazefield, Harrogate, North Yorks, HG3 5DR

G4MAA-MZZ
Mr C G Rowe, G4MAR, 29 Lucknow Road, Willenhall, West Mids, WV12 4QF

G4NAA-NZZ
Mr M J Musgrave, G4NVT, 49 Vowler Road, Langdon Hills, Basildon, Essex, SS16 6AQ

G4OAA-OZZ
Mr R Satterthwaite, G6BMY, 47 Aberford Road, Baguley, Manchester, M23 1JY

G4PAA-PZZ
Mr M I Humphrey, G0SWY, 4 Bluebell Road, Bassett, Southampton, Hampshire, SO16 3LQ

G4RAA-RZZ
Deryck Buckley, G3VLX, Little Oaks, Park Road, Marden, Tonbridge, Kent, TN12 9LG

G4SAA-SZZ
Mr J D Harris, G3LWM, 44 Fourth Avenue, Frinton-on-Sea, Essex, CO13 9DX

G4TAA-TZZ
John Porter, G3YZR, 94 Oaken Grove, Haxby, York, YO3 3QZ

G4UAA-UZZ
Mr J D Harris, G3LWM, 44 Fourth Avenue, Frinton-on-Sea, Essex, CO13 9DX

G4VAA-VZZ
Mr G R Marley, G0VFV, Whitestone Farm Cottage, Staintondale, Scarborough, N Yorks, YO13 0EZ

G4WAA-WZZ
Mr L Gaunt, G4MLV, 31 Moat Hill, Birstall, Batley, West Yorks, WF17 0DX

G4XAA-XZZ
Mr S R Tyler, G4UDZ, 2 John Court, Hoddesdon, Herts, EN11 9LZ

G4YAA-YZZ
Mr D J Newbury, G0ENR, 8 Mayfield Road, Pershore, Worcs, WR10 1NW

G4ZAA-ZZZ
John Densem, G4KJV, 'Cotswold', Startley, Chippenham, Wilts, SN15 5HG

G5 & reciprocals
Mr P J Pasquet, G4RRA, 64 Bricksbury Hill, Farnham, Surrey, GU9 0LY

G6AA-ZZ
Frank Harris, G4IEY, 4 Merestones Drive, The Park, Cheltenham, Glos, GL50 2SS

G6AAA-ZZZ
Mr G M Foote, G7NCR, 64 Stable Yard, Tyntesfield Estate, Wraxall, Bristol, BS19 1NS

G7AAA-ZZZ
Mr D J Hudson, G6OVO, 62 Derron Avenue, South Yardley, Birmingham, B26 1LA

G8AA-ZZ
Frank Harris, G4IEY, 4 Merestones Drive, The Park, Cheltenham, Glos, GL50 2SS

G8AAA-ZZZ
John Purvess, G0FWP, 389 Otley Old Road, Cookridge, Leeds, LS16 6BX

GBxAAA-MZZ
Mr G Whaling, G0PPR, 32 The Croft, Little Snoring, Fakenham, Norfolk, NR21 0JS

GBxNAA-ZZZ
Alex Devereaux, G0TTZ, 39 Lower Green Road, Rusthall, Tunbridge Wells, Kent, TN4 8TW

GD series
Mr G W Ripley, GD3AHV, Corlea Bungalow, Ronague Road, Ballasalla, Isle of Man, IM9 3BA

GI series
Edward Barr, GI7FFF, Ed-Mar, 1 Onslow Drive, Bangor, Co Down, Northern Ireland, BT19 7HQ

GJ series
Reginald Allenet, GJ3XZE, Les Sablons, le Bourg, St Clements, Jersey, C.I., JE2 6SE

GM0AAA-LZZ
Mr F A Roe, GM0ALS, 8 South Gyle Gardens, Edinburgh, EH12 7RZ

GM0MAA-ZZZ
Mr J E Clough, GM0MDD, 84A Main Road, Fairlie, Largs, Ayrshire, KA29 0AD

GM1, 6, 7 & 8
Mr G E Bell, GM4LKJ, 21 St Andrew's Crescent, Dumbarton, Strathclyde, G82 3ER

GM2AA-GM3ZZ
James Johnston, GM3LYY, The Dolphins, Montgomerie Dr, Fairlie, Largs, Ayrshire, KA29 0DZ

GM2AAA-GM3ZZZ
James Johnston, GM3LYY, The Dolphins, Montgomerie Dr, Fairlie, Largs, Ayrshire, KA29 0DZ

GM4AAA-ZZZ
Mr G E Bell, GM4LKJ, 21 St Andrew's Crescent, Dumbarton, Strathclyde, G82 3ER

GU series
Mr P F H Cooper, GU0SUP, 1 Clos au Pre, Hougue du Pommier, Castel, Guernsey, GY5 7FQ

GW series
Mr K Hudspeth, GW0ARK, 67 Bloomfield Road, Blackwood, Gwent, NP2 1LX

M0AAA-AZZ
Mr R Hamer, G0MOK, 125 New Street, Blackrod, Bolton, Lancs, BL6 5AG

M0BAA-BZZ
Mr B Mulleady, GM0KWL, 9 Elizabeth Crescent, Camelon, Falkirk, FK1 4JF

M1AAA-ZZZ
Mr M Marriott, G0OPC, Greenfields View, March Road, Friday Bridge, nr Wisbech, Cambs, PE14 0HA

M1BAA-BZZ
Mr M C Clark, 'Lymemore', Madderty, by Crief, Perth, Tayside, PH7 3NY

M1CAA-CZZ
Edward Allen, G3DRN, 30 Bodnant Gardens, Wimbledon, London, SW20 0UD

MI series
Edward Barr, GI7FFF, Ed-Mar, 1 Onslow Drive, Bangor, Co Down, Northern Ireland, BT19 7HQ

MM0AAA-LZZ
Mr FA Roe, GM0ALS, 8 South Gyle Gardens, Edinburgh, EH12 7RZ

MM0MAA-ZZZ
Mr J E Clough, GM0MDD, 84A Main Road, Fairlie, Largs, Ayrshire, KA29 0AD

M1AAA-ZZZ
Mr G E Bell, GM4LKJ, 21 St Andrew's Crescent, Dumbarton, Strathclyde, G82 3ER

Abbreviated Contest
Ian N Fugler, G4IIY, 9 Westover Road, Fleet, Hants, GU13 9DG

Novices
Mike Shread, GM6TAN, 15 Hardie Court, Aberchirder, Huntly, Aberdeenshire, AB54 7TG

RS
David Borne, G4CYW, Roughways, Chub Tor, Yelverton, Devon, PL20 6HY

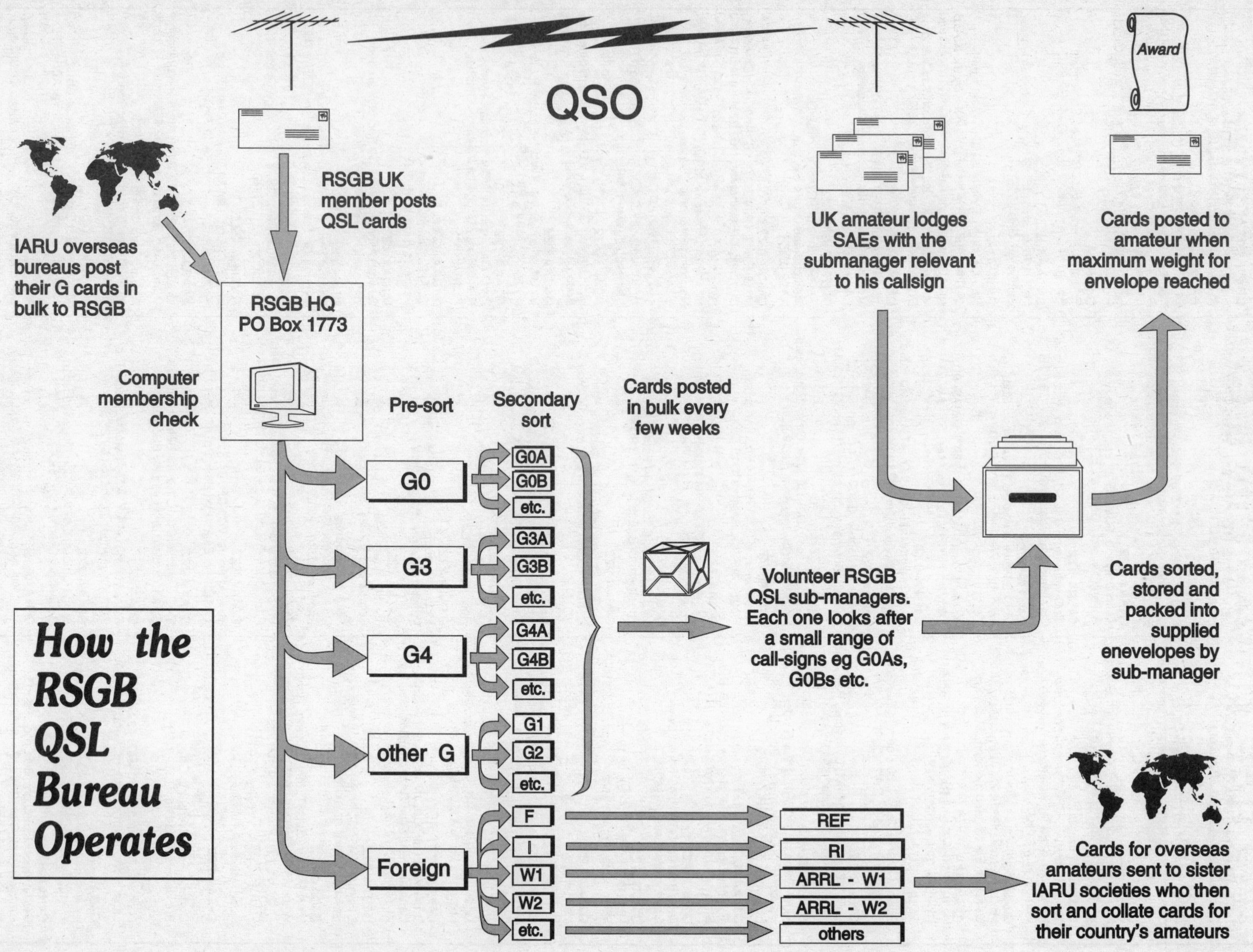
QSO
IARU overseas bureaus post their G cards in bulk to RSGB
RSGB UK member posts QSL cards
RSGB HQ PO Box 1773
Computer membership check
Pre-sort
Secondary sort
Cards posted in bulk every few weeks
G0
G3
G4
other G
Foreign
G0A
G0B
etc.
G3A
G3B
etc.
G4A
G4B
etc.
G1
G2
etc.
F
I
W1
W2
etc.
REF
RI
ARRL - W1
ARRL - W2
others
Volunteer RSGB QSL sub-managers. Each one looks after a small range of call-signs eg G0As, G0Bs etc.
UK amateur lodges SAEs with the submanager relevant to his callsign
Award
Cards posted to amateur when maximum weight for envelope reached
Cards sorted, stored and packed into supplied enevelopes by sub-manager
Cards for overseas amateurs sent to sister IARU societies who then sort and collate cards for their country's amateurs
How the RSGB QSL Bureau Operates

RAE & NRAE City & Guilds Exams

City & Guilds Exams

Both the Radio Amateurs' Examination (RAE No. 7650) and the Novice Radio Amateurs' Examination (NRAE No. 7730) are two of the many examinations conducted by City & Guilds of London Institute. Consequently City & Guilds' standard rules and procedures apply to both exams. The scope of these two exams is kept under periodic review by City & Guilds, RA and RSGB through the appropriate C&G Subject Committees.

Except in a few special cases, all candidates must take their exam at a City & Guilds recognised examinations centre. These are usually, but not always, colleges or other educational establishments. Recently some clubs have gained accreditation to help their local amateurs. A list of UK centres which have in the recent past held either exam is shown below.

It is possible to take the RAE abroad, at sea, or through the Forces. In these cases it is best to contact the Examinations Department of City & Guilds for guidance, but please allow extra time.

There are no exemptions from either the RAE or NRAE on the grounds having passed a higher level professional, City & Guilds or service examination. Consequently even if you have one of these, you will still have to sit the NRAE or RAE if you desire an amateur radio licence.

Copies of sample papers, the syllabus and the general regulations may be purchased from: Sales Section, City & Guilds of London Institute, 1 Giltspur Street, London EC1A 9DD. Tel: (0171) 294 2468.

Candidates with special needs

City & Guilds of London Institute will make special arrangements for candidates with special needs who would otherwise have difficulties taking either the NRAE or RAE. Essentially, those who are housebound, blind, or handicapped in some other way may be examined at home. Moreover, if the candidate is unable to complete a written paper, it is possible for an examination to be conducted orally.

Candidates who wish to take advantage of these special arrangements will need to provide some official documentation which confirms the reason they are requesting a home examination. Examples of acceptable documentation include a doctor's certificate or a DSS letter. In all cases, however, the final decision concerning acceptance (or otherwise) rests with City & Guilds. Candidates must be prepared to make themselves available for testing in the 10 days following the date of the regular examination. Application and further information should be made to: Assesment Services, City & Guilds of London Institute, 1 Giltspur Street, London EC1A 9DD. Tel: (0171) 294 2468.

Please note that for the NRAE, even a disabled candidate *must have passed the training course before* sitting the City & Guilds exam.

Costs

City & Guilds charge an examination centre for every candidate that they enter and these costs are then normally passed on by the centre to the candidate. Presently their fees are £26.05 for the RAE and £12.10 for the NRAE. The centre may make an extra charge to cover their own costs.

It is your responsibility to ensure that you arrive at the examination centre in good time. If you fail to attend, centres can make either a partial or full refund. However, City & Guilds will only refund a centre under certain circumstances. If a centre does give a refund, it is likely that they will apply the same criteria to the candidate as City & Guilds does to them. These City & Guilds' criteria are if the candidate:

i) is prevented by accident or illness from taking the examination (medical certificate required).
ii) dies before the examination.
iii) is in HM Forces and is prevented by the exigencies of the service from taking an examination.

Applications have to be made by the centre to City & Guilds within one month of the exam, so do not delay! Incidentally, City & Guilds will not carry fees forward from one exam to another.

Results

The results are despatched by City & Guilds to all exam centres at the same time, normally one month after the date of the exam. Centres then have the responsibility of forwarding them on to the candidates. If you change address, remember to notify your college, so that they will be able to forward your results without delay.

Candidates are responsible for the safe keeping of their results. If a certificate has been lost or damaged, City & Guilds will issue a replacement certificate. Applications should be made to the Examinations Department at their Britannia Street address given above, enclosing the appropriate fee (August 1997: £21.20) as well as details of the date and location of the examination which was passed.

Where the results of candidates are seriously at variance with the reasonable expectations of their teachers, City & Guilds may give further details, but they will only deal with the college principal. So if you have a query about your results or the conduct of your examination, you should contact the college's principal. He or she will then liaise with City & Guilds on your behalf.

Novice RAE – No. 7730

Training Courses

Before you sit the City & Guilds NRAE, you must have passed the Novice Licence Training Course. This is organised on behalf of the RA by the RSGB and consists of some 30 hours of tuition. It normally lasts about 12 weeks and covers all the material tested in the NRAE. Furthermore it includes some practical experience – candidates learn how to assemble their own receiver and audio amplifier. The intention is to provide a good grounding on how to operate on the bands in a safe and disciplined manner – 'learning by doing'.

The course is continually assessed by the instructor so regular attendance of the course is essential. The assessment is of a general nature, so a weakness in one or two areas will not jeopardise the overall result. On successful completion of the course, the tutor applies to RSGB HQ for a Course Completion Slip for the candidate. Should the candidate desire, an attractive certificate may be purchased for £2.50.

The RSGB Training Courses are run by RSGB registered instructors and are normally held in schools or radio clubs. If you wish to apply for a course, please contact the Senior Instructor for your area. He will know where and when the most appropriate courses are being held. The charges for these courses are quite modest, just sufficient to pay for the instructor's incidental costs.

The Society is always welcome to hear from amateurs who are willing to volunteer their time and expertise as an instructor. If you are interested, please contact RSGB HQ for the details.

The NRAE Exam

Having successfully completed the Practical Course, you will then be allowed to sit the NRAE. It is held four times a year – the normal timetable being:

Exam Date:	Closing Date for Entries:
1st/2nd Monday in March	during January
1st/2nd Monday in June	during March
2nd Monday in September	during July
2nd Monday in December	during October

The NRAE consists of a single paper and lasts 1¼ hours. The 45 multiple choice questions cover:

Receivers and receiving techniques
Components, applications and units
Measurements
Propagation and antennas
Transmitters and transmitting techniques
Operating techniques
Station layout
Construction
Safety
Licensing Conditions

Recommended Reading

The Society publishes a number of books for the Novice. Essential study material are the *RSGB Novice Licence Student's Notebook* and *Revision Questions for the Novice RAE*. The beginner may also appreciate reading *Amateur Radio for Beginners* and *Practical Receivers for Novices, Practical Transmitters for Novices* and *Practical Antennas for Novices* which will assist him or her in the background knowledge necessary to become a good amateur.

All these titles are available from RSGB Publications, Lambda House, Cranborne Road, Potters Bar, Herts, EN6 3JE. Tel: (01707) 659015.

The RA Information Sheets -*Novice Licence Information Sheet* (RA166) and *How to Become a Radio Amateur* (RA190) are also important reading. They are available from the RA Library and Information Service, Radiocommunications Agency, New King's Beam House, 22 Upper Ground, London, SE1 9SA. Tel: (0171) 211 0211.

Passing the NRAE and successfully completing the Practical course will enable you to apply for a Novice Licence 'B'. If you pass the 5wpm Morse test, you can then apply for a Novice Licence 'A'. Details of this can be found in the Morse Code section of this book.

Presently the fee for a Novice Licence (A or B) is £15, but free to those under 21.

RAE – No. 7650

The City & Guilds' Radio Amateurs' Examination (No 7650) is held twice each year, in December and May. The exam dates are usually the first Monday in December, and the second or third Monday in May and they normally start at 6.30pm.

RSGB SENIOR INSTRUCTORS

Prospective NRAE candidates should contact their county's Senior Instructor for details of their nearest training course.

County	Senior Instructor
AVON	Mr S Hartley, G0FUW
BEDS	Brian Elliott, G4MEO
BERKS	Mr P R Swynford
BUCKS	Mr V C Webley, G0RKV
CAMBS	Mr J T Hammond, G0FLP
CARMARTHEN	Mr E W S Meredith, GW4XLK
CENTRAL	Mr G L Collier, GM0LOD
CHESHIRE	Gordon Adams, G3LEQ
CLWYD	Mr R Millward, GW1VCN
CO. ANTRIM	John Branagh, GI3YRL
CO. ARMAGH	Mr C R Blezard, GI4RNC
CO. DOWN	Mr J M Skillen, GI4TSK
CO. DURHAM	Mr J Marr, G4WUI
CORNWALL	Bert Hammett, G3VWK
CUMBRIA	Raymond Gilchrist, G0TUE
DERBYSHIRE	Mr F L Whitehead, G4MLL
DEVON	Roger Quaintance, G0DIZ
DORSET	Phil Mayer, G0KKL
EAST SUSSEX	Mr R C Gornall, G7DME
ESSEX	Mr R Easting, G7NZV
FIFE	Mr K D Horne, GM3YBQ
GRAMPIAN	Mr S Sutherland, GM4BKV
GTR LONDON	Robert Snary, G4OBE
GTR MANCHESTER	Mr P E Maggs, G0OVY
GWENT	Mr F R Clare, GW3NWS
GWYNEDD	Mr RAS Rees, GW0FMQ
HAMPSHIRE	Mr P A Steed, G0VEP
HEREFORD & WORCS	Mr MJ Butler, G4UXC
HERTS	Mr J H Maclagan-Wedderburn, G4JOV
HUMBERSIDE	Mr W A Jackson, G0DLL
ISLE OF WIGHT	Mr A Ash, G3PZB298731.
KENT	Dr K L Smith, G3JIX
LEICS	Gwynne Harries, G4WYN
LINCS (NORTH)	Mr D G Scothern, G1YFQ
LINCS (SOUTH)	Mr R M Coaker, G0LME
LOTHIAN	Mr G R Winchester, GM4CUX
MERSEYSIDE	Mr D G Clifford, G0NVF
MID GLAM	Mr HEJ Clarke, GW0PYU
NORFOLK	David Buddery, G3OEP
NORTH YORKS	Mr A Easom, G4OPI
NORTHUMBERLAND	Mr M Stott, G0NEE
NOTTS	Mr J P Mayfield, G0LXX
SHROPSHIRE	Mr W S Cowell, G0OPL
SOMERSET	Mr G W Davis, G3ICO
SOUTH GLAM	Mr G V Bibby, GW1UOU
SOUTH YORKS	Mr J W Denniss, G0NMJ
STAFFS	Mr A J Matthews, G3UNM
STRATHCLYDE (N)	Mr SM Lewis, GM4PLM
STRATHCLYDE (S)	Mr J W Hambrook, GM1RJS
SUFFOLK	Mr M C Baldry, G6MCB
SURREY	Mr T Fell, G7DGW
TAYSIDE	Mr R Bennett, GM0PTP
TYNE & WEAR	Mr M Stott, G0NEE
WARWICKSHIRE	Mr G N Frykman, G0GNF
WEST GLAM	Mr R Holt, GW6OLS
WEST MIDS	Mr P F Morrall, G4TMK
WEST SUSSEX	Peter Howard, G0AFN
WEST YORKS	Gerald Edinburgh, G3SDY
WILTS	Mr H N Woolrych, G4TIX

As of May 1998, the exam will be held in one part and covers licensing conditions, transmitter interference , EMC, operating procedures, practices and theory The exam lasts 2¼ hours and takes place in the early evening.

How to apply

Many RAE centres also run courses leading to the RAE, and so their students are submitted as internal candidates. Some centres will also accept 'external' candidates but in these cases centres often ask candidates for some means of identification when registering.

There is some variation between centres in their closing date for applications. Typically the closing dates for the December exam are in late September to early October; and for May, sometime in February.

Courses

It is quite common for people to attend an evening course in order to study for the RAE, though many study on their own. Courses usually start in September and prepare candidates for the May examination.

Most of the RAE centres listed below run courses leading to the exam. Alternatively your local club may run a course itself, or know of one in the locality. Your RSGB Liaison Officer may be able to assist you.

A number of correspondence courses are available for RAE candidates, some are run by local clubs. These include:

Friendly Correspondence Course for the RAE
c/o Pete Pennington, G4EGQ
6 Highland Close
Golden Valley
Folkestone
Kent, CT20 3SA
Tel: (01303) 220010

Radio & Telecomms Correspondence School
12 Moor View Drive
Teignmouth
Devon, TQ14 9UN
Tel: (01626) 779398

Rapid Results College
Tuition House
27-37 St. George's Road
London, SW19 4DS
Tel: (0181) 947 7272

In view of the individual attention necessary, correspondence courses can be somewhat expensive. Before committing oneself, it is worthwhile ensuring that the course covers the full syllabus, is suitable to your needs and abilities, and that you are prepared to make the commitment necessary.

Study material & question papers

The questions in the RAE are all multiple choice: the answers are written in pencil (HB is strongly recommended) on the answer grid. In common with all other City & Guilds multiple choice examinations, both the answer sheets and question papers are collected back in at the end. Neither candidates nor centres are allowed to retain any copies of the question booklet.

In order for candidates to have access to at least some questions on which to practise, City & Guilds publish two complete sample papers. These are both reprinted, with permission, in the *Radio Amateurs' Examination Manual* published by the RSGB. In addition the Society also publishes a companion book, *How to pass the Radio Amateurs' Examination,* which contains practical advice on how to approach the exam as well as 9 typical examination papers on which to practise. To assist candidates to revise for the exam, the *RAE Revision Notes* has been republished.

Almost all RAE classes use the *RAE Manual* as their course book. Where candidates need a more detailed reference works, then both the *Amateur Radio Operating Manual* and the *Radio Communication Handbook* can be used. All these books may be obtained from: RSGB Sales, Lambda House, Cranborne Road, Potters Bar, Herts EN6 3JE. Tel: (01707) 659015.

US Licence Exams

The US Federal Communications Commission exams may be taken in this country. This permits interested individuals who plan to visit or reside in the USA to obtain their licence.

Prospective applicants should first contact: Dr. Harvey M. Good, G0NAT/KK6JV, 30 Lower Station Road, Ilkeston, Derbyshire, DE7 4LN. Tel: (01159) 440646 – or for those in the South East: Yves Remedios, AC4WT/G4UDT, 44 Kingsway, Wembley, Middx, HA9 7QR, tel: (0181) 902 5995.

Radio Amateurs' Examination Centres

The following list of City & Guiulds Examination Centres have all held the Radio Amateurs Exam in the recent past. It is therefore likely that they may be accepting candidates for forthcoming RAEs.

*** Centres marked with an asterisk are known to have held the NRAE in the recent past. Please remember that you must have successfully completed the Novice Licence Training Course *before* attempting the NRAE.**

C&G Exam Centres

Avon

BRISTOL CATHEDRAL SCHOOL, College Square, Bristol. Tel: (01179) 291872.
BRUNEL COLL OF A&T*, Ashley Down Rd, Bristol, BS7 9BU. Tel: (0117) 904 5121.
CITY OF BATH COLLEGE, Avon Street, Bath. Tel: (01225) 3112191.
NORTH & WEST BRISTOL CONTINUING ED. AREA, Stoke Lodge, Bristol, BS9 1BN. Tel: (01179) 683112.
WESTON-SUPER-MARE COLL OF FE*, Knightstone Rd, Weston-Super-Mare, BS23 2AL. Tel: (01934) 621301.
SOUTH BRISTOL COLLEGE, Bisport Avenue, Hartcliffe, Bristol, BS13 0RJ. Tel: (0117) 935 8071.

Bedfordshire

BEDFORD COLLEGE, Caulwell Street, Bedford, MK42 7EB. Tel: (01234) 345151.
DUNSTABLE COLLEGE, Kingsway, Dunstable, LU5 4HG. Tel: (01582) 477776.
HASTINGSBURY SCHOOL & COMM COLL, Hill Rise, Kempston, Bedford, MK42 7EB. Tel: (01234) 853636.
LEIGHTON LINSLADE COMM. COLLEGES, Leighton Buzzard, LU7 8HS. Tel: (01525) 735769.
REDBORNE COMM COLL, Redborne Upper School, Flitwick Rd, Ampthill, Bedford, MK45 5NJ. Tel: (01525) 404462.
SANDY UPPER SCHOOL & COMM CEN*, Engayne Ave, Sandy, SG19 1BL. Mrs Terri Creed, Tel: (01767) 680574 / Brian Elliott, G4MEO, Tel: (01767) 680043.

Berkshire

CROWTHORNE CENTRE, Crowthorne, Berks. Tel: (01344) 773111.
NEWBURY CFE, Newbury, Berks, RG13 1PQ. Tel: (01635) 42824.
PRINCESS MARINA COLLEGE*, Aborfield Camp, Reading, RG2 9NJ. Tel: (01734) 760421.
READING & DARC*. Details from Peter Swynford, G0PUB. Tel: (01734) 665981.

Borders

BORDERS COLLEGE OF F.E., Melrose Road, Galashiels, Selkirk, TD1 2AF. Tel: (01896) 57755.

Buckinghamshire

AYLESBURY COLLEGE, Aylesbury, Berks. Tel: (01296) 34111.
BURNHAM ADULT ED CENTRE*, Opendale Rd, Burnham, SL1 7LZ. Tel: (01628) 665513.
MILTON KEYNES COLLEGE*, Chaffron Way, Wolverton Campus, Leadenhall West, MK6 5LP. Tel: (01908) 684444.
WHITE HILL CEN, Chesham, HP5 3AD. Tel: (01494) 776420 (answering machine - please leave full details)

Cambridgeshire

CAMBRIDGE REG. COLLEGE, Kings Hedges Rd, Cambridge, CB4 2QT. Tel: (01223) 418200.
PETERBOROUGH REG COLLEGE.*, Park Crescent, Peterborough, PE1 4DZ. Tel: (01733) 767366.
115 PETERBOROUGH SQN ATC, Details from Robert Maskill, G4JDL, 21 Clayton, Orton Goldhay, Peterborough, PE2 4SB. Tel: (01733) 330830.
SAWSTON VILLAGE COLLEGE, New Rd, Sawston, CB2 4BP. Tel: (01223) 832217.
SOHAM VILLAGE SCHOOL, Sand Street, Soham, Ely, CB7 5AA. Tel: (01353) 720569.

Central

FALKIRK COLLEGE OF TECH*, Grangemouth Road, Falkirk, FK2 9AD. Tel: (01324) 403000.
DOLLAR ACADEMY, 23 West Burnside, Dollar, Clacks, FK14 7DU. Tel: (01259) 742511.

Cheshire

AVONDALE EVENING CENTRE, Heathbank Road, Sheadle Heath, Stockport, SK3 0UP. Tel: (0161) 477 2382.
MID CHESHIRE AMATEUR RADIO SOCIETY, 5 Llandovery Close, Winsford, CW7 1NA. Tel: (0160) 655 3401.
NORTH CHESHIRE COLLEGE, North Campus, Winwick Rd, Warrington, WA2 8QA. Tel: (01925) 814343.
NORTH CHESHIRE RC, Wilmslow. Details from Gordon Adams, G3LEQ. Tel: (01565) 652652. Fax: (01565) 634560.
REDDISH VALE AEC, Stockport, SK5 7HD. Tel: (0161) 477 3544.
WEST CHESHIRE COLLEGE-GRANGE CENTRE, Regent St, Ellesmere Port, Wirral, L65 8EJ. Tel: (0151) 356 2300.
WARRINGTON COLLEGIATE INST., Padgate Campus, Fearnhead, Warrington, WA2 0DB. Tel: (01925) 814343.

Cleveland

CLEVELAND TERTIARY COLLEGE, Redcar, Cleveland, TS10 1EZ. Tel: (01642) 473132.
HARTLEPOOL COLLEGE OF F.E., Hartlepool, Cleveland. Tel: (01429) 275453.
LONGLANDS COLL OF FE*, Douglas Street, Middlesbrough, TS4 2JW. Tel: (01642) 300100.
STOCKTON & BILLINGHAM COLLEGE, The Causeway, Billingham, TS23 2DB. Tel: (01642) 552101.

Clwyd

COLLEGE OF HORTICULTURE, Northop, Nr Mold, CH7 6AA. Tel: (01352) 840861.
CONNAH'S QUAY HIGH SCHOOL, Golftyn Lane, Deeside, CH5 4BH. Tel: (01244) 813491.
GROVES ADULT CENTRE, Wrexham, Clwyd, LL12 7AP. Tel: (01978) 266375.
LLANDRILLO TECH COLLEGE*, Llandudno Road, Colwyn Bay. Tel: (01492) 546666.
NEWI PLAS COCH, Wrexham. Tel: (01978) 290666.
NORTH WALES RADIO RALLY CLUB*, Rhuddlan. Tel: (01745) 591704.
YALE COLLEGE, Grove Park Rd, Wrexham. Tel: (01978) 311794.

Cornwall

CORNWALL COLLEGE, Trevenson Pool, Redruth, TR15 3RD. Tel: (01209) 712911,
FALMOUTH-PENRYN ADULT ED CEN, 2 Trelawney Rd, Falmouth, TR11 3QS. Tel: (01326) 319275.
HELSTON SCHOOL, Church Hill, Heston, TR13 8NR. Tel: (01326) 572133.
PENWITH ADULT ED CEN, St Clare Street, Penzance, TR18 2SA. Tel: (01736) 62604.
St. AUSTELL COLLEGE, Trevarthian Road, St Austell, PL25 4BU. Tel: (01726) 67911.
THE ADULT EDUCATION CENTRE, The Bungalow, Darke Lane, Camelford, PL32 9UJ. Tel: (01840) 213511.
TRURO COLLEGE, College Road, Truro, TR1 1XX. Tel: (01872) 79867.

Co Antrim

NORTH EAST INST. OF FHE*, Farm Lodge Ave, Ballymena, BT43 7DJ. Tel: (01266) 652871.
BELFAST INST OF F&HE*,College Sq., Belfast, BT1 6DJ. Tel: (01232) 265000.
CASTLEREAGH COLLEGE, Montgomery Rd, Belfast, BT6 9JD. Tel: (01232) 797144.
DOWNSHIRE SCHOOL*, Downshire Rd, Carrickfergus, BT38 7DA. Tel: (01232) 46334.

Co. Armagh

PORTADOWN COLLEGE OF FE*, 26-44 Lurgan Rd, Portadown, Craigavon, Co. Armagh, BT63 5BL. Tel: (01762) 337111.

Co. Down

EAST DOWN INST. of F.E., Market St, Downpatrick, BT30 6ND. Tel: (01396) 615815.
NEWRY CFE, Patrick Street, Newry, BT35 8DN. Tel: (01693) 61071.
NORTH DOWN & ARDS COLLEGE OF FE., Castle Pk Rd, Bangor, BT20 3ED. Tel: (01247) 271254.

Co. Tyrone

EAST TYRONE FE COLLEGE, Circular Road, Dungannon, BT71 6BQ. Tel: (01868) 722323.
ORMAGH COLLEGE of F.E., Mountjoy Rd, Ormagh, BT79 7AH. Tel: (01662) 45433.

Co. Durham

BISHOP AUKLAND COLLEGE, Woodhouse Lane, Bishop Auckland. Tel: (01388) 603052.
BISHOP AUCKLAND ARC, 6 Buttermere Grove, West Auckland, Bishop Auckland, DL14 9LG. Tel: (01388) 832948.
DARLINGTON COLLEGE*, Cleveland Ave, Darlington, DL3 7BB. Tel: (01325) 503050.
EASTBOURNE SCHOOL*, The Fairway, Darlington, DL1 8RW. Tel: (01325) 464657/ 464357.
NEW COLLEGE DURHAM, Framwell Gate Moor, Durham, DH1 5ES. Tel: (0191) 375 4000.
NORTHUMBRIA TOURIST BOARD, Akley Heads, Durham, DH1 5XU. Tel: (0191) 384 4905.

Co. Londonderry

CAUSWAY INST. OF FURTHER & HIGHER EDUCATION, Coleraine, BT52 1QA. Tel: (01265) 54717.
NORTH WEST COLLEGE OF TECH, Strand Rd, Londonderry, BT48 7BY. Tel: (01504) 266711.

Cornwall
SALTASH COLLEGE, Church Road, Saltash, Cornwall, PL12 4AE. Tel: (01752) 848147.

Cumbria
CARLISLE COLLEGE, Victoria Place, Carlisle, CA1 1HS. Tel: (01228) 24464.
DOWDALES SCHOOL, Dalton-in-Furness, LA15 8AH. Tel: (01229) 62535.
DTS BUSINESS SERVICES, Whitehaven, CA28 7EB. Tel: (01946) 66636.
FURNESS COLLEGE, Howard St, Barrow in Furness, LA14 1NB. Tel: (01229)825017.
NELSON THOMLINSON SCHOOL, Wigton, Cumbria, CA7 9PX. Tel: (01965) 42160.
KENDAL CFE, Kendal, LA9 5AY. Tel: (01539) 724313.
WEST CUMBRIA COLLEGE*, Park Lane, Workington, CA14 2RW. Tel: (01900) 64331.

Derbyshire
CHESTERFIELD COLLEGE OF A&T*, Infirmary Road, Chesterfield, S41 7NG. Tel: (01246) 500500.
DERBY TERTIARY COLLEGE*, Wilmorton, Derby, DE2 8UG. Tel: (01332) 757570.
HIGH PEAK COLLEGE, Harpur Hill, Buxton. Tel: (01298) 71100.
LANDAU FORTE COLLEGE*, Fox Street, Derby, DE1 2LF. Tel: (01332) 204040.
MACKWORTH COLLEGE, Prince Charles Dr, Mackworth, Derby, DE3 4LR. Tel: (01332) 519951.
MURRAY PARK SCHOOL, Mickleover, DE3 5LD. Tel: (01332) 515922.
NUNSFIELD HSE COMM. ASSN. AMATEUR RADIO GRP., Alvaston, DE24 0FD. Tel: (01332) 755900.
SIMPSON MEMORIAL COMM. CENTRE, Glossop, SK13 8RJ. Tel: (0161) 681 5406.

Devon
BIDEFORD COMMUNITY COLLEGE AEC, The Arts Centre, The Quay, Bideford, EX39 2EY. Tel: (01237) 272462.
BRAUNTON COMMUNITY COLLEGE, Barton Lane, Braunton, EX33 2BP. Tel: (01271) 812221.
BRIXHAM COMM COLLEGE*, Higher Ranscombe Rd, Brixham, TQ5 9HF. Tel: (01803) 858271.
BUDHAVEN SCHOOL, Valley Rd, Bude, EX23 8DQ. Tel: (01288) 353271.
EAST DEVON CFE, Bolham Road, Tiverton, EX16 6SH. Tel: (01884) 254247.
EXETER COLLEGE, Hele Road, Exeter, EX4 4JS. Tel: (01392) 205477.
ILLFRACOMBE SCHOOL AND COMM. COLLEGE, Ilfracombe, EX34 8JB. Tel: (01271) 63427.
KELLY COLLEGE, Tavistock, PL19 0HJ. Tel: (01822) 3005.
NORTH DEVON COLLEGE, Barnstaple, EX31 2BQ. Tel: (01271) 45291.
PLYMOUTH CFE, Kings Road, Devonport, Plymouth, PL1 5QG. Tel: (01752) 385868.
PLYMOUTH RADIO CLUB, c/o The Royal Fleet Club, 9-12 Morice Square, Devonport, Plymouth, PL1 4PQ. Tel: (01752) 562723.
SOUTH DARTMOOR SCHOOL*, Balland Lane, Ashburton, Newton Abbot, TQ13 7EW. Tel: (01364) 52230.
SOUTH DEVON COLLEGE, Newton Road, Torquay, TQ2 5BY. Tel: (01803) 291212.
ST. CLARE'S AEC, Fore Street, Seaton, EX12 2AN. Tel: (01297) 21904.
TORBAY AMATEUR RADIO SOCIETY*, 3 Manor Road, Paignton, TQ3 2HT.

Dorset
BURTON CLIFF HOTEL, Photographic Unit, Burton Bradstock, Nr. Bridport, DT6 4RB. Tel: (01308) 897205.
CANFORD SCHOOL, Canford Magna, Wimborne, BH21 3AD. Tel: (01202) 841254.
CHRISTCHURCH AEC, Soper's Lane, Christchurch, DH23 1JF. Tel: (01202) 482789.
DORCHESTER THOMAS HARDYE SCHOOL, Dorchester, DT1 2ET. Tel: (01305) 208904.
MARTIN KEMP WELCH ADULT ED CEN*, Herbert Ave, Parkstone, Poole, BH12 4HS. Tel: (01202) 721600.
NO.16 ARMY EDUCATION, Bovington Camp, Wareham. BH20 6JA. Tel: (01929) 403420.
PURBECK AEC, The Purbeck School, Worgret Road, Wareham, BH20 4PE. Tel: (01929) 556809.
WEYMOUTH COLLEGE*, Cranford Ave, Weymouth, DT4 7LQ. Tel: (01305) 208904.
SPECTRUM COMMUNICATIONS, Unit 6B, Poundbury West Estate, DT1 2PG. Tel: (01305) 262250.
SHAFTESBURY SCHOOL, Shaftesbury Road, Shaftesbury. Tel: (01747) 54498.

Dumfries & Galloway
DUMFRIES & GALLOWAY COLLEGE OF TECH, Heathhall, Dumfries, DG1 3QZ. Tel: (01387) 61261/5.
LANGHOLM ACADAMY, Langholm. Tel: (013873) 80914.
STRANREAR ACADEMY F.E. CENTRE, McMaster Rd, Stranaer, DG9 9BY. Tel: (01776) 706484.

Dyfed
ABERAERON COMP. SCHOOL, Aberaeron. Tel: (01545) 570217.
BIGYN COUNTY PRIMARY SCH., Drefach, Llanelli, SA15 7AH. Tel: (01269) 831235.
CARMARTHENSHIRE COLLEGE OF T&A, Alban Rd, Llanelli. Tel: (01554) 759165.
COLEG CEREDIGION, Aberystwyth. Tel: (01970) 624511.
CEREDIGION COLLEGE OF FE. Cardigan. Tel: (01239) 612032.
COLES HILL DAY CENTRE*, Llanelli. Tel: (01544) 773610.
LLANDOVERY COLLEGE*, Llandovery, SA20 0EE. Tel: (01550) 20315.
PEMBROKE SCHOOL, Bush, Pembroke, SA71 4RL. Tel: (01646) 682461.

East Sussex
BRIGHTON COLLEGE OF TECH., Pelham St, Brighton, BN1 4FA. Tel: (01273) 667788.
EASTBOURNE COLLEGE OF A&T*, St. Anne's Rd, Eastbourne, East Sussex, BN21 2HS. Tel: (01323) 644711.
HASTINGS COLLEGE OF A&T*, Archery Road, St. Leonards-on-Sea, Sussex. TN38 01HX. Tel: (01424) 423847.
HASTINGS E&RC. Tel: (01424) 751400.
NORTHEASE MANOR SCHOOL*, Lewes, BN7 3EY. Tel: (01273) 477123.
ST LEONARDS ON SEA COLLEGE OF A&T, Archery Road, St Leonards on Sea, TN38 0HX. Tel: (01424) 423847.

Essex
BARKING COLLEGE, Dagenham Rd, Romford, Essex, RM7 0XU. Tel: (01708) 766841.
BARKING RADIO & ELECTRONIC SOCIETY*, Ilford, Essex. Tel: (0181) 478 4758.
BASILDON COLLEGE OF FE. Basildon. Tel: (01268) 532015.
BENFLEET/HADLEIGH CNTR FOR COMM. ED., 114 Benfleet Road, Hadleigh, Benfleet, SS7 1QH. Tel: (01702) 556622.
CASTLE POINT & ROCHFORD ADULT COMM. CENTRE, Rochford, SS4 1DQ. Tel: (01702) 544900.
CHELMSFORD COLLEGE, Upper Moulsham St, Chelmsford, CM2 0JQ. Tel: (01245) 265611.
COLCHESTER INST*, Sheepen Road, Colchester, CO3 3LL. Tel: (01206) 718000.
HARLOW COLLEGE*, West Site, College Sq, The High, Harlow, CM20 3LT. Tel: (01279) 868000.
THE PHILIP MORANT SCHOOL, Rembrandt Way, Gainsborough Rd, Colchester, CO3 4QS. Tel: (01206) 45222.
RAYLEIGH CENTRE FOR COMM ED, 72 Hockley Rd, Rayleigh, SS6 8EB. Tel: (01268) 742500.
ST. HELENA SCHOOL, Sheepen Rd, Colchester, CO3 3LE. Tel: (01206) 572253.
WILLIAM DE FERRES SCHOOL*, Trinity Sq, South Woodham, Ferrers, Chelmsford, CM3 5JU. Details from Mr S. Hook, G4OCP. Tel: (01245) 329555.

Fife
FIFE COLL OF TECH*, St Brycedale Ave, Kirkcaldy, KY1 1EX. Tel: (01592) 268591/262414.
BALWEARIE HIGH SCHOOL*, Balwarie Gardens, Kirkcaldy, KY2 5LY. Tel: (01592) 266262.

Gloucestershire
CHURCHDOWN SCHOOL, Winston Rd, Churchdown, GL3 2RB. Tel: (01452) 713340.
EASTWOOD PARK NHS TRAINING & CONF. CENTRE, Eastwood Park, Falfield, Wotton Under Edge, GL12 8DA. Tel: (01454) 260207.
GLOS. COLLEGE OF A&T*, Park Campus, 73 The Park, Cheltenham, GL50 2RR. Tel: (01242) 532000.
HEYWOOD SCHOOL*, Cinderford, Glos, GL14 2AZ. Tel: (01594) 822257.
ROYAL FOREST OF DEAN SCHOOL, Five Acres Campus, Berry Hill, Coleford, GL16 7JT. Tel: (01594) 33416.
STROUD COLLEGE OF FE*, Stratford Rd, Stroud, GL5 4AH. Tel: (01453) 763424.
SIR WILLIAM ROMNEY'S SCHOOL, Lowfield Road, Tetbury, GL8 8AE. Tel: (01666) 502378.

Grampian
ABERDEEN COLLEGE, Gallowgate, Aberdeen, AB9 1DN. Tel: (01244) 612000.
BANFF & BUCHAN COLLEGE OF F.E., Fraserburgh, AB4 5RF. Tel: (01346) 25777.
MORAY COLLEGE OF FE*, Hay St, Elgin. Tel: (01343) 554321.
THE GORDON SCHOOLS ARC*, Att: Elaine Shread, GM7TZT, Huntly, AB54 4SE. Tel: (01466) 792181.

Greater London
BARKING COLLEGE, Dagenham Road, Romford, Essex. RM7 0XU. Tel: (01708) 766841. Fax: (01708) 731067.
BARKING R&ES. Rainham, Essex, RH13 8BE. Tel: (01708) 557606.
BRENTFORD SCHOOL FOR GIRLS, Clifton Rd, Brentford, TW8 0PG. Tel: (0181) 560 6292.
BROMLEY ADULT ED CEN, Church Lane, Bromley, BR2 8LD. Tel: (0181) 462 9184.
BROMLEY COLLEGE OF F&HE*, Rookery Lane, Bromley Common, Bromley, Kent, BR2 8HE. Tel: (0181) 295 7000.
CITY OF WESTMINSTER COLLEGE*, 25 Paddington Green, London, W2 1NB. Tel: (0171) 723 8826.
CROYDON CONT ED & TRAINING SERVICE, South Croydon Centre, Haling Manor School, Kendra Hall, Croydon. Tel: (0181) 655 0905.
KINGSBURY HIGH SCHOOL, Princess Ave, Kingsbury, NW9 9JR. Tel: (0181) 204 9814.
KINGSTON COLLEGE OF FE*, Kingston Hall Rd, Kingston-on-Thames, Surrey, KT1 2AQ. Tel: (0181) 546 2151.
LAMBETH COLLEGE, 56 Brixton Hill, London, SW2 1QS. Tel: (0171) 501 5204.
MERTON COLLEGE, Morden Park, London Road, Morden, Surrey. Tel: (0181) 640 3001/0835
SOUTHGATE COLLEGE, High Street, Southgate, London, N14 6BS. Tel: (0181) 886 6521, Fax: (0181) 882 9522.
WALTHAM FOREST COLLEGE, Walthamstow, Tel: (0181) 527 2311.
WEALD COLLEGE*, Brookshill, Harrow, Middx, HA3 6RR. Tel: (0181) 420 8886.
YSBYTY & YSTWYTH CONTEST GRP., St Augustines Church, Southborough Lane, Bromley, Kent, BR2 8AT. Tel: (0181) 761 5675.

Greater Manchester

BOLTON MET. COLLEGE*, Manchester Rd, Bolton, BL2 1ER. Tel: (01204) 531411.
BOLTON SCHOOL, Chorley New Road, Bolton, BL1 4PA. Tel: (01204) 840201.
FRED LONGWORTH HIGH SCHOOL, Printshop Lane, Tyldesley, M29 8JN.
MANCHESTER COLL OF A&T*, Openshaw Campus, Whitworth St, Openshaw, Manchester, M11 2EH. Tel: (0161) 953 5995.
NORTH TRAFFORD COLLEGE OF FE, Talbot Road, Stretford, Manchester M32 0XH. Tel: (0161) 872 3731.
OAKWOOD HIGH SCHOOL, Nell Lane, Manchester, M21 2SL. Tel: (0161) 881 4778.
OLDHAM ARC, c/o Mrs K. Catlow. Tel: (0161) 652 8617.
OLDHAM COLLEGE, Rochdale Rd, Oldham, OL9 6AA. Tel: (0161) 624 5214.
REDDISH VALE COMPREHENSIVE SCHOOL, Reddish Vale Rd, Reddish, Stockport, SK5 7HD. Tel: (0161) 477 3544.
SALFORD COLL OF FE, Worsley Campus, Mardale Avenue, Swinton, M27 3QP. Tel: (0161) 886 5070.
MANCHESTER & DARS*, Simpson Memorial Comm. Centre, Manchester. Tel: (0161) 6815406.
123rd MANCHESTER SCOUT HQ*, Details from Paul Maggs, G0OVY. Tel: (0161) 226 4053.

Guernsey

GUERNSEY COLLEGE OF FE, Route des Coutanchez, St. Peter Port, Guernsey, GY1 2TT. Tel: (01481) 727121.

Gwent

BRYN GLAS COMM CENTRE*, Newport, NP9 5QU. Tel: (01633) 858657.
EBBW VALE COLLEGE OF FE, Ebbw Vale. Tel: (01495) 302083.
GWENT TERTIARY COLLEGE, Nash Rd, Newport, NP4 8AU. Tel: (01633) 274861.
OAKDALE COMMUNITY COLL, Oakdale, Blackwood. Tel: (01495) 228289.
PONTYPOOL COMMUNITY EDUCATION, Pontypool. Tel: (01495) 762266.
TORFAEN SCOUTS ARC, New Street Post Office, Pontnewydd, NP44 1EE. Tel: (01633) 483277.

Gwynedd

COLEG MENAI, Bangor, LL57 2TP. Tel: (01248) 370125.
COLEG MEIRION-DWYFOR*, Barmouth Rd, Dolgellau, LL40 2SW. Tel: (01758) 701385.
COLEG MERIONNYDD DWYFOR, Dolgellau. Tel: (01341) 422827.
COLEG PENCRAIG*, Llangefni, Isle of Anglesey, LL77 7HY. Tel: (01248) 750101. Fax: (01248) 772097.
GWYNEDD TECH COLL*, Ffriddoedd Road, Bangor. Tel: (01248) 370125.

Hampshire

BROCKENHURST COLLEGE, Lyndhurst Rd, Brockenhurst, SO42 7ZE. Tel: (01590) 23565.
CRICKLADE COLLEGE, Andover, SP10 1EJ. Tel: (01264) 363311.
FAREHAM COLLEGE*, Bishopsfield Rd, Fareham, PO14 1NH. Tel: (01329) 815290.
FARNBOROUGH COLLEGE OF TECH., Farnborough. Tel: (01252) 515511.
HIGHBURY COLL OF TECH*, Dovercourt Rd, Cosham, Portsmouth, PO6 2SA. Tel: (01705) 383131. Fax: (01705) 383131.
LYMINGTON COMMUNITY CENTRE, New St, Leamington, SO41 9BQ. Tel: (01590) 672337.
NEVILLE LOVETT AEC*, Neville Lovett Comm. School, Fareham. Tel: (01329) 823471.
SONY BROADCAST & COMMS, Basingstoke, RG22 4SB. Tel: (01256) 483454.
SOUTH DOWNS CFE, Purbrook Way, Havant, PO7 8AA. Tel: (01705) 797979.
ST JOHN'S AMBULANCE HQ, Farlington, Portsmouth. Tel: (01705) 371677.

Hereford & Worcester

AVONSCROFT ART CENTRE, 26 Shubbery Road, Bromsgrove, Worcs, B61 7BH. Tel: (01527) 570020.
EVESHAM COLL OF FE*, Cheltenham Rd, Evesham, WR11 6LP. Tel: (01386) 41091.
EVESHAM SEA CADETS*. Details from Mr M. Butler, G4UXC. Tel: (01386) 442899.
HEREFORDSHIRE COLL, Folly Lane, Hereford, HR1 1LS. Tel: (01432) 352235.
KIDDERMINSTER COLL OF FE*, Hoo Rd, Kidderminster, DY10 1LX. Tel: (01562) 820811.
NORTH EAST WORCS COLLEGE, School Drive, Stratford Rd, Bromsgrove, B60 1PQ. Tel: (01527) 570020.
NORTH EAST WORCS COLLEGE, Peakman St, Redditch, B98 8DW. Tel: (01527) 572524.

Hertfordshire

ALDWICKBURY SCHOOL*, Harpenden. Details from John Wedderburn, G4JOV. Tel: (01582) 765821.
BISHOPS STORFORD COLLEGE, Hertford. Tel: (01279) 758575.
HERTFORD REGIONAL COLLEGE, Broxbourne Centre, Turnford, Broxbourne, EN10 6AF. Tel: (01992) 466451.
OAKLANDS COLLEGE, Borehamwood. Tel: (0181) 953 6024.
RADIO SOCIETY OF GREAT BRITAIN*, Lambda House, Cranborne Road, Potters Bar, EN6 3JE. Tel: (01707) 659015.
ST. CHRISTOPHER SCHOOL*, Letchworth, Hertford. Tel: (01426) 679301.
UNIVERSITY OF HERTFORDSHIRE, College Lane, Hatfield, AL10 9AB. Tel: (01707) 284020.
WEST HERTS COLLEGE (DACORUM CAMPUS), Marlowes, Hemel Hempstead, HP1 1HD. Tel: (01442) 221525.

Highland

INVERNESS COLLEGE*, 3 Longman Road, Longman South, Inverness, IV1 1SA. Tel: (01463) 236681.

Humberside

BARTHOLEMEW CENTRE*, Boothferry Rd, Goole, DN14 6AJ. Tel: (01405) 762420.
BOOTHFERRY ADULT EDUCATION, Goole, DN14 5DZ. Tel: (01405) 762714.
BRIDLINGTON COLL OF FE*, St Mary's Walk, Bridlington, YO16 5JW. Tel: (01262) 672676.
GRIMSBY COLLEGE OF T&A*, Nuns' Corner, Grimsby, DN34 5BQ. Tel: (01472) 311222.
HORNSEA ADULT EDUCATION CENTRE, Eastgate, Hornsea, HU18 1DW.
HULL COLLEGE OF FE*, Queen's Gardens, Hull, HU1 3DG. Tel: (01482) 29943.
HULL ADULT EDUCATION SERVICE, Myton Centre, Hull, HU1 2RE. Tel: (01482) 320 539.

Isle of Man

ISLE OF MAN COLLEGE, Homefield Road, Douglas. Tel: (01624) 623113.

Isle of Wight

ATHENA HOUSE CENTRE FOR ED. & TECH.*, 5 John St, Ryde, PO33 2PZ. Tel: (01983) 566734.
ISLE OF WIGHT COLLEGE, Newport, PO30 5TA. Tel: (01983) 526631.

Jersey

HIGHLANDS COLLEGE, PO Box 1000, St Saviour, JE4 9QA. Tel: (01534) 608608.
JERSEY AMATEUR RADIO SOCIETY*, Little Mead, Claremnt Road, St Saviour, JE2 7RT. Tel: (01534 34948).

Kent

BEXLEY COLLEGE, Tower Road, Belvedere, Kent, DA17 6JA. Tel: (01322) 442331.
CANTERBURY COLLEGE*, New Dover Road, Canterbury, CT1 3AJ. Tel: (01227) 811111.
CRANBROOK SCHOOL*, Cranbrook, TN17 3JD. Details from Alistair Hamilton. Tel: (01580) 712163.
DOVER RADIO CLUB*: Deal, Kent, CT14 9RG. Tel: (01304) 361939.
HUNDRED OF HOO SCHOOL*, Main Rd, Hoo St, Werburgh, Rochester, ME2 3HH. Tel: (01634) 251443.
MEDWAY AEC, Rochester. Tel: (01634) 845359.
NORTH-WEST KENT COLLEGE, Dartford. Tel: (01322) 225471.
NORTH-WEST KENT COLLEGE OF TECH., Gravesend. Tel: (01322) 225471.
PENT VALLEY SECONDARY SCHOOL, Folkestone. Tel: (01303) 77161.
SEVENOAKS AEC, Sevenoaks. Tel: (01732) 451618.
SOUTH KENT COLLEGE, Dover. Tel: (01304) 204573.
SUTTON VALENCE SCHOOL*, Valence, Maidstone, ME17 3HL. Tel: (01622) 842281.
TONBRIDGE SCHOOL, Tonbridge. Tel: (01732) 357751.
UNIVERSITY OF KENT RADIO SOC*, Canterbury. Tel: (01227) 66822.
URSULINE CONVENT SCHOOL, Westgate-on-Sea. Tel: (01843) 834431.

Lancashire

ACCRINGTON & ROSSENDALE COLLEGE, Sandy Lane, Accrington, BB5 2AW. Tel: (01254) 354126.
BLACKBURN COLLEGE OF TECH & DESIGN, Fielden Street, Blackburn, BB2 1LH. Tel: (01254) 55144.
BLACKPOOL & THE FYLDE COLLEGE, Ashfield Rd, Bispham, Blackpool, FY2 0HB. Tel: (01253) 352352.
BURNLEY COLLEGE*, Ormerod Rd, Burnley, BD11 2RX. Tel: (01282) 436111.
CALDER COLLEGE OF ADULT ED., Burnley Rd, Todmorden. Tel: (01706) 812743.
DEANERY HIGH, Wigan, Lancs, WN1 1HQ. Tel: (01942) 44355.
HUTTON GRAMMAR SCHOOL*, Liverpool Rd, Preston, PR4 5SN. Tel: (01772) 613112.
LANCASTER & MORECAMBE COLLEGE, Morecambe Rd, Lancaster, LA1 2TY. Tel: (01524) 66215.
LANCASTER UNIVERSITY, Bailrigg, Lancaster, LA1 4YR. Tel: (01524) 65201.
NELSON & COLNE COLLEGE, Scotland Rd, Nelson, BB9 7YT. Tel: (01282) 440200.
ORMSKIRK ADULT ED CENTRE, Ormskirk, Lancs. Tel: (01695) 728744.
OULDER HILL COMM. SCHOOL, Hudsons Walk, Rochdale, OL11 5EF. Tel: (01706) 55222.
PRESTON COLLEGE, St. Vincent's Rd, Fulwood, Preston, PR2 9UR. Tel: (01772) 772200.
ROSSENDALE COMMUNITY CENTRE*, Kearsley, Bolton, BL4 8NT. Tel: (0161) 796564.
1196 SQN HQ, Ashton under Lyne, OL7 0DU. Tel: (0161) 285 2281.

Leicestershire

CHARLES KEENE COLLEGE OF F.E.*, Painter St, Leicester, LE1 3WA. Tel: (0116) 251 6037.
HIND LEYS COLLEGE*, Forest St, Shepshed, Loughborough, LE12 9DB. Tel: (01509) 504511/2/3.
HINCKLEY COLLEGE OF FE*, London Rd, Hinckley, LE12 9DB. Tel: (01509) 504511.
LUTTERWORTH COMM. COLLEGE, Biteswell Rd, Lutterworth. Tel: (01455) 554101.

Lincolnshire

BOSTON COLLEGE OF FE*, Rowley Rd, Boston, PE21 6JF. Tel: (01205) 365701.
GRANTHAM COLLEGE OF FE*, Stonebridge Rd, Grantham, NG31 9AP. Tel: (01476) 63141. Fax: (01476) 64882.
LINCOLN COLLEGE OF TECH*
NORTH LINCS COLLEGE, Monks Road, Lincoln LN2 5HQ. Tel: (01522) 510530.
SKEGNESS GRAMMAR SCHOOL*, Vernon Rd, Skehgness, PE25 2QS. Details from Andrew Rigby, G4BJY. Tel: (01754) 610000.

Lothian

FETTES COLLEGE, Carrington Rd, Edinburgh, EH4 1QX. Tel: (0131) 332 2281.
JEWEL & ESK VALLEY COLLEGE, Eskbank Centre, New Battle Rd, Dalkeith, EH22 3AE. Tel: (0131) 663 1951.
LOTHIAN REGIONAL ED DEPT, Edinburgh. Tel: (0131) 229 9166.
NAPIER UNIVERSITY*, Edinburgh, EH14 1DJ. Tel: (0131) 455 4370 .
WEST LOTHIAN COLLEGE*, Majoribanks St, Bathgate, EH48 1QJ. Tel: (01506) 634300.

Merseyside

CITY OF LIVERPOOL COLLEGE, Old Swan Centre, Broadgreen Rd, Liverpool, L13 5SQ. Tel: (0151) 252 1515.
FAZAKERLEY COMPREHENSIVE SCHOOL, Fazakerley, Liverpool,L10 1LB. Tel: (0151) 5252870.
LIVERPOOL F.E. ED. DEPT, 20 Sir Thomas Street, Liverpool, L1 6BJ. Tel: (0151) 225 2855.
SOUTHPORT COLLEGE*, Mornington Rd, Southport, PR9 0TT. Tel: (01704) 500606.
WIRRAL METROPOLITAN COLLEGE*, Borough Road, Birkenhead, L42 9QD. Tel: (0151) 653 5555.

Mid-Glamorgan

BLAENGAWR COMPREHENSIVE, Aberdare. Tel: (01685) 874341.
BRIDGEND ADULT ED. CENTRE*, c/o Penybont Primary School, Minerva St, Bridgend, CF31 1TD. Tel: (01656) 668279.
ST. CENYDD SCHOOL, Caerphilly. Tel: (01222) 852504.

Norfolk

ACLE ADULT EDUCATION CENTRE, Acle High School, South Walsham Rd, Acle, NR13 3ER. Tel: (01692) 670432.
GREAT YARMOUTH COLLEGE OF FE*. Southtown, Gt. Yarmouth, NR31 0ED. Tel: (01493) 655261.
NORFOLK COLLEGE OF ARTS & TECH*, Tennyson Avenue, King's Lynn, PE30 2QW. Tel: (01553) 761144. Fax: (01553) 764902.
NORFOLK ARC, Norfolk. Contact John Wadman, G0VZD, Tel: (01953) 604769.
NORWICH CITY CFE, Ipswich Rd, Norwich, NR2 2LJ. Tel: (01603) 660011.
THE NORMAN CENTRE, 11 Ash Close, Wymondham, NR18 0HR. Tel: (01953) 604769.
THORPE ADULT ED CENTRE, Longfields Road, Thorpe, St. Andrew, Norwich, NR7 0NB. Tel: (01603) 35857.

Northamptonshire

EDUCATION DEPT. - ST.ANDREWS HOSPITAL* Billing Rd, Northampton. Tel: (01604) 29696.
NENE COLLEGE*, St. Georges Avenue, Northampton, NN2 6JD. Tel: (01604) 714101.
NORTHAMPTON COLLEGE, St Gregory's Rd, Northampton, NN3 3RF. Tel: (01604) 734567.
TRESHAM INST. OF F.E., St Mary's Rd, Kettering, NN15 7BS. Tel: (01536) 410252.

Northumberland

FURTHER EDUCATION & YOUTH CENTRE, Seaton Delaval, NE25 0BP.
THE KING EDWARD VI SCHOOL, Morpeth, NE61 1DN. Tel: (01670) 515415.

North Yorkshire

ASKHAM BRYAN COLLEGE, YORK, YO2 3PR. Tel: (01904) 702121.
HARROGATE COLLEGE OF ARTS & TECH*, Hornbeam Pk, Hookstone Rd, Harrogate, HG2 8QT. Tel: (01423) 879466/829.
HARROGATE LADIES COLLEGE*, Clarence Drive, Harrogate, HG1 2QG. Tel: (01423) 504543.
NORTHALLERTON GRAMMAR SCHOOL, Grammar School Lane, Northallerton, DL6 1DD. Tel: (01609) 773340.
RICHMOND SCHOOL*, Darlington Rd, Richmond, DL10 7BQ. Tel: (01748) 850111.
SCARBOROUGH SIXTH FORM CENTRE, Sandybed Lane, Scarborough, YO12 5LF. Tel: (01723) 365032.
WHITBY SCHOOL, Prospect Hill, Whitby, YO21 1LA. Tel: (01947) 602406.
YORK*. Details from Barry Firth, G4KCT, at the York Radio Club. Tel: (01904) 411864 evenings, (01904) 432510 daytime.
YORKSHIRE COAST COLLEGE,* Lady Edith's Drive, Scalby Rd, Scarborough, YO12 5RN. Tel: (01723) 372105.
UNIVERSITY OF YORK, Heslington, York, YO1 5DD. Tel: (01904) 432510.

Nottinghamshire

ARNOLD & CARLTON COLLEGE, Digby Ave, Mapperley, Nottingham, NG3 6DR. Tel: (0115) 987 6503. Fax: (0115) 987 1489.
BASFORD HALL COLLEGE OF FE, Stockhill Lane, Nottingham, NG6 0NB. Tel: (0115) 970 4541. Fax: (0115) 942 2334.
BROXTOWE COLLEGE, High Road, Chilwell, Beeston, Nottingham, NG9 4AH. Tel: (0115) 922 8161.
FAIRHAM COMMUNITY COLLEGE*, Farnborough Rd, Clifton Est, Nottingham, NG11 9AE. Tel: (0115) 974 4400.
NORTH NOTTS COLLEGE OF FE, Carlton Rd, Worksop, S81 7HP. Tel: (01909) 473561. Fax: (01909) 485564.
RUGELEY ADULT EDUCATION, Broxtowe College, Chilwell. Tel: (01889) 578738.
WEST NOTTS COLLEGE, Derby Rd, Mansfield, NG18 5BH. Tel: (01623) 27191. Fax: (01623) 23063.
WELBECK COLLEGE, Worksop, S80 3LN. Tel: (01909) 476326.

Orkney

KIRKWALL FE CENTRE. Tel: (01856) 872839. Fax: (01856) 872911.

Oxon

ABINGDON COLLEGE, Northcourt Rd, Abingdon, OX11 0RA. Tel: (01235) 555585.
AEA TECHNOLOGY, Building 346, Harwell, Didcot, OX11 0RA. Tel: (01235) 432797.
NORTH OXFORDSHIRE COLLEGE OF A&T, Broughton Rd, Banbury, OX16 9QA. Tel: (01295) 52221.
VC10 TRISTAR MAINTENANCE SCHOOL, RAF Brize Norton, Carteron, OX18 3LX. Tel: (01993) 897669.

Powys

COLEG POWYS*, Montgommery College, Llanidloes Rd, Newtown. Tel: (01686) 622722.

Shetland

SHETLAND COLLEGE OF FE*, Gressy Loan, Lerwick, ZE1 0BB. Tel: (01595) 695514.

Shropshire

PHOENIX SCHOOL*, Manor Rd, Dawley, Telford, TF4 3DZ. Tel: (01952) 591531.
SHREWSBURY COLLEGE OF A&T, London Rd, Shrewsbury, SY2 6PR. Tel: (01743) 231544.

Somerset

BRIDGWATER COLLEGE, Bath Road, Bridgwater, TA6 4PZ. Tel: (01278) 455464.
SOMERSET COLLEGE OF ART & TECH, Wellington Rd, Taunton, TA1 5AX. Tel: (01823) 366366.
STRODE COLLEGE*, Church Rd, Street, BA16 0AB. Tel: (01458) 42277.
KING ARTHUR'S SCHOOL, West Hill, Wincanton, BA9 9BX. Tel: (01963) 32368.
YEOVIL COLLEGE*, Hollands Campus, Yeovil, BA21 3BA. Tel: (01935) 23921.

South Glamorgan

BARRY COLLEGE, Colcot Road, Barry, CF6 8YJ. Tel: (01446) 743519.
HIGHFIELD CENTRE*, Heath, Cardiff. Tel: (01222) 750315.
CARDIFF INST. OF HIGHER ED., Western Avenue, Llandaff, Cardiff, CF5 2SG. Tel: (01222) 551111.

South Yorkshire

BARNSLEY COLLEGE*, Church St, Barnsley, S70 2AX. Tel: (01226) 730191. Fax: (01226) 298514.
CHAPEL GREEN COMM. COLLEGE, Chapeltown, Sheffield, S30 4UT. Tel: (01142) 846720.
DEARNE VALLEY COLLEGE, Wath upon Dearne, Rotherham, S63 7EW. Tel: (01709) 513333.
DONCASTER COLLEGE, Waterdale, Doncaster, DN1 3EX. Tel: (01302) 553553.
INTAKE FIRST & MIDDLE SCHOOL*. Mansfield Rd, Shefield, S12 2AR. Tel: (01142) 399824.
LOXLEY TERTIARY COLLEGE, Myers Grove Lane, Sheffield, S6 5TL. Tel: (0114) 260 2200.
MEXBOROUGH SCHOOL*, Maple Rd, Mexborough, S64 9SD. Tel: (01709) 585858.
MOUNT ST MARY'S COLLEGE, Spinkhill, Sheffield, S31 9YL. Tel: (01142) 433388.
SHEFFIELD COLLEGE. Tel: (0114) 260700.
TRITEC*, Thomas St, Sheffield, S1 4LE. Tel: (01142) 756297.

Staffordshire

CARDINAL GRIFFIN SCHOOL*, Cardinal Way, Cannock, WS11 2AW. Tel: (01543) 52215.
LICHFIELD COLLEGE, The Friary, Lichfield, WS13 6QG. Tel: (01543) 262150.
MOORLANDS &DARS. Ranville, Billington Lane, Derrington, Stafford, ST18 9LR. Tel: (01785) 55925.
RUGELEY ADULT ED. CENTRE, Taylors Lane, Rugeley, WS15 2AA. Tel: (01889) 578738.
STAFFORD COLLEGE, Earl St, Stafford, ST16 2QR. Tel: (01785) 223800. Fax: (01785) 59953.
STOKE-ON-TRENT TECH COLLEGE, Moorland Road, Burslem, Stoke-on-Trent, ST6 1JJ. Tel: (01782) 208208.
STOKE RS*, Stoke-on-Trent. Tel: (01538) 722581.

Strathclyde

ARGYLL F.E. CENTRE, Oban High School, Argyll, PA34 4JB. Tel: (01631) 64231.
CLYDEBANK COLLEGE, Kilbowie Road, Clydebank, G81 2AA. Tel: (0141) 952 7771.
GLASGOW CALEDONIAN UNIVERSITY, Cowcaddens Rd, Glasgow. Tel: (0141) 331 3512.
GLASGOW COLL OF NAUTICAL STUDIES*, 21 Thistle Street, Glasgow, G5 9XB. Tel: (0141) 429 3201.
JAMES WATT COLLEGE, Finnart St, Greenock. Tel: (01475) 24433.

Suffolk

FARLINGAYE HIGH SCHOOL, Ransom Rd, Woodbridge, IP12 4JX. Tel: (01394) 32869.
FELIXSTOWE & DARS*, Ipswich. Tel: (01473) 642595.
KING EDWARD VI SCHOOL, Grove Road, Bury St Edmunds, IP33 3BH. Tel: (01284) 761393.
LOWESTOFT COLLEGE*, St. Peter's St, Lowestoft, NR32 2NB. Tel: (01502) 583521.
SUFFOLK COLLEGE OF H&FE, Rope Walk, Ipswich, IP4 1LT. Tel: (01473) 255885.
THE HIGH SCHOOL, Seaward Ave, Leiston, IP16 4BG. Tel: (01728) 830570.
WEST SUFFOLK COLLEGE, Out Risbygate, Bury St Edmunds, IP33 3RL. Tel: (01284) 701301.

Surrey

BROOKLANDS TECHNICAL COLLEGE, Weybridge. Tel: (01932) 853300.
GUILDFORD COLLEGE OF TECH*, Stoke Park, Guildford, GU1 1EZ. Tel: (01483) 31251.

ROYAL GRAMMAR SCHOOL*, High St, Guildford, GU1 3BB. Tel: (01483) 502424.
RSME GIBRALTAR BARRACKS, Camberley, Surrey, GU17 9LP. Tel: (01252) 867400.
UNIVERSITY OF SURREY*, Guildford, Surrey, GU2 5XH. Tel: (01483) 300800.

Tayside

DUNDEE COLLEGE OF FE*, Graham Street Centre, Dundee, DD4 9RF. Tel: (01382) 834834.
PERTH COLLEGE OF FE*, Brahan Estate, Crieff Road, Perth, PH1 2NX. Tel: (01738) 21171.

Tyne & Wear

CITY OF SUNDERLAND COLLEGE, Durham Road, Sunderland, SR3 4AH. Tel: (0191) 511 6000.
GOSFORTH FE COLLEGE*, Knightsbridge, Gosforth, Newcastle-on-Tyne, NE3 2JH. Tel: (0191) 284 5822.
MUSEUM OF SCIENCE & ENGINEERING, Blandford St, Newcastle-upon-Tyne, NE1 4JA. Tel: (01661) 832020.
NEWCASTLE COLLEGE, Rye Hill Campus, Newcastle-upon-Tyne, NE4 7SA. Tel: (0191) 200 4000.
SOUTH TYNESIDE COLLEGE, St. George's Ave, South Shields, NE34 6ET. Tel: (0191) 427 3500.
PENSHAW HOUSE COMMUNITY HOME, Shiney Row. Tel: (01385) 3852218.
WEARSIDE COLLEGE OF F.E., Sea View Rd West, Sunderland, SR2 9LH. Tel: (0191) 567 0794. Fax: (0191) 564 0620.

Warwickshire

EAST WARWICKSHIRE COLLEGE*, Lower Hillmorton Rd, Rugby, CV21 3QS. Tel: (01788) 541666.
KINETON HIGH SCHOOL, Banbury Rd, Kineton, CV35 0JX. Tel: (01926) 640465.
MID WARWICKSHIRE COLLEGE, Warwick New Rd, Leamington Spa, CV32 5JE. Tel: (01926) 318000.
NORTH WARWICKSHIRE COLLEGE OF TECH & ART, Hinckley Road, Nuneaton, CV11 6BH. Tel: (01203) 349321.
WARWICK SCHOOL*, Myton Rd, Warwick, CV34 6PP. Tel: (01926) 492484.

Western Isles

LEWS CASTLE COLLEGE, Stornoway, Isle of Lewis, PA86 0XR. Tel: (01851) 703311.

West Midlands

BILSTON COMMUNITY COLLEGE, Westfield Rd, Bilston, Wolverhampton, WV14 6ER. Tel: (01902) 821026.
DUDLEY COLLEGE OF TECH, The Broadway, Dudley, DY1 4AS. Tel: (01384) 455433.
ERDINGTON & SUTTON AEC, Westmead Crescent, Erdington, Birmingham, B24 0JS. Tel: (0121) 382 6238.
ELMORE ROW CENTRE, Elmore Row, Bloxwich, Walsall, WS3 2HR. Tel: (01922) 710076.
HALESOWEN COLLEGE, Whittingham Rd, Halesowen, B63 3NA. Tel: (0121) 550 1451.
HENLEY COLLEGE OF FE, Henley Rd, Bell Green, Coventry, CV2 1ED. Tel: (01203) 611021.
HODGE HILL & YARDLEY COMM. CENTRE, Ward End Park House, Heath Road, Birmingham, B8 2HB. Tel: (0121) 327 1335.
KING EDWARD'S SCHOOL, Edgbaston Park Road, Edgbaston, Birmingham, B15 2UA. Tel: (0121) 472 1147.
KINVER WOMBOURNE EVENING INST, Church Rd, Womborne, Wolverhampton, WV5 8BJ. Tel: (01902) 895198.
MIRFIELD CENTRE. Scolars Gate, Lea Village, Birmingham, B33 0DL. Tel: (0121) 783 5898.
MATHEW BOLTON COLLEGE OF F&HE, Sherlock St, Birmingham, B5 7DB. Tel: (0121) 446 4545. Fax: (0121) 446 4324.
OLDBURY EVENING STUDY ASSN. Details from Gordon Adams, G3LEQ. Tel: (0121) 544 0771. Fax: (0121) 552 0051.
OLD SWINFORD HOSPITAL SCHOOL*, Old Swinford, Stourbridge. Tel: (01384) 370025.
RAF COSFORD, Albrighton. Tel: (01902) 372393.
SANDWELL COLLEGE OF F&HE*, Woden Rd South, Wednesbury, Sandwell, WS10 0PE. Tel: (0121) 556 6000.
SPARKBROOK AREA EXAMINATION CENTRE, Stratford Rd, Sparkhill, Birmingham. Tel: (0121) 772 1893.
ST. THOMAS MOORE ROMAN CATHOLIC. COMP. SCH, Bilston Rd, Willenhall, WV13 2JY. Tel: (01902) 368798.
THE CABLE & WIRELESS COLLEGE, Coventry. Tel: (01203) 868600.
TILE HILL CFE, Coventry. Tel: Michael Dixon, (01203) 694200.
UNIVERSITY OF CENTRAL ENGLAND, Inst. of Art & Design, Corporation Street, Birmingham, B4 7DX. Tel: (0121) 331 5887.

West Sussex

ARUN ARC*, Chichester. Tel: (01243) 5433399.
BOGNOR REGIS COMM COLLEGE, Westloats Lane, Bognor Regis, West Sussex, PO21 5LH. Tel: (01243) 827422.
BRIGHTON COLLEGE, Brighton. Tel: (01273) 697131.
CRAWLEY ARC*. , Tilgate, Crawley, RH10 5HR. Tel: (01293) 407469.
CRAWLEY COLLEGE, College Rd, Crawley, RH10 1NR. Tel: (01293) 442200.
CHICHESTER COLLEGE OF TECH, Chichester, PO19 1SB. Tel: (01243) 786321.
CHRIST'S HOSPITAL, SCIENCE SCHOOL, Horsham, RH13 7LS. Tel: (01403) 52547.
LANCING COLLEGE, Lancing, BN15 0RW. Tel: (01273) 452213.
OAKMEEDS COMM SCHOOL*, Burgess Hill. Tel: (01444) 66355.
NORTHBROOK COLLEGE, Worthing, BN14 8HJ. Tel: (01903) 231445.

West Yorkshire

AIREDALE & WHARFEDALE COLLEGE, Calverley Lane, Horsforth, Leeds, LS18 4RQ. Tel: (0113) 239 5800.
ALMONDBURY ADULT ED CENTRE*, Fernside Ave, Huddersfield, HD5 8PQ. Tel: (01484) 536333.
BRADFORD & ILKLEY COMM COLLEGE, Wells Rd, Ilkley, LS29 9RD. Tel: (01943) 609010.
BRIGHOUSE ADULT ED CENTRE, Church Lane, Brighouse, HD6 1AT. Tel: (01484) 714019.
CALDERDALE COLLEGE, Percival Whiteley Centre, Halifax, HX1 3UZ. Tel: (01422) 358221.
DENBY DALE RADIO SOCIETY EXAM CENTRE*, Shelley, HD8 8NL. Tel: (01484) 424776
DEWSBURY COLLEGE, Halifax Rd, Dewsbury, WF13 2AS. Tel: (01924) 465916.
HUDDERSFIELD TECH. COLLEGE, New North Rd, Huddersfield, HD1 5NN. Tel: (01484) 536521.
JOSEPH PRIESTLEY COLLEGE, 71 Queen St, Morley, Leeds, LS27 8DZ. (0113) 253 3749.
KEIGHLEY COLLEGE*, Cavendish St, Keighley, BD21 3DF. Tel: (01535) 618555.
LEEDS COLLEGE OF TECHNOLOGY, Coookridge St, Leeds, LS2 8BL. Tel: (01132) 430381.
LEEDS FE COLLEGE*, Merrion Hse, Leeds, LS2 8DT. Tel: (0113) 246 2706.
OAKBANK SCHOOL, Keighley, BD22 7DU. Tel: (01535) 662787.
RISHWORTH SCHOOL, Sowerby Bridge, HX6 4QA. Tel: (01422) 822217.
WAKEFIELD COLLEGE, Margaret St, Wakefield, WF1 2DH. Tel: (01924) 370501.
WAKEFIELD DISTRICT COLLEGE, Whitwood Cntr., Castleford, WF10 5NF. Tel: (01924) 810405.

Wiltshire

CHIPPENHAM COLLEGE, Cocklebury Rd, Chippenham, SN15 3QD. Tel: (01249) 444501. Fax: (01249) 653772.
SALISBURY COLLEGE, Southampton Rd, Salisbury, SP1 2LW. Tel: (01722) 323711. Fax: (01722) 326006.
SHELDON SCHOOL, Hardenhuish Lane, Chippenham, SN14 6HJ. Tel: (01249) 651216.
SWINDON COLLEGE, Regent Circus, Swindon, SN1 1PT. Tel: (01793) 491591.
THE ROYAL BRITISH LEGION TRAINING CO., Tidworth College, Ordnance Rd, Tidworth, SP9 7QF. Tel: (01980) 844340.
TROWBRIDGE COLLEGE, College Rd, Trowbridge, BA14 0ES. Tel: (01225) 766241.

RAE COURSES 1997/98

GREATER MANCHESTER/ OLDHAM ARC. Courses available for RAE and NRAE. Further details from Mr G Oliver, G0BJR, Tel: (0161) 652 4164.The Hillcrest School & Community College, Simms Lane, Netherton, Dudley, West Midlands. DY2 0PB. Courses starting: 18 Sept 1997, 15 January 1998 and 23 April 1998. Thusday evenings 7-9pm. Further details from Kerry Rudge Tel: (01384) 816503.
MEXBOROUGH & DARS, Harrop Hall, Mexborough. Course starting: September 1997. Every Friday from 7.30pm. Further details from J Denniss (01320) 531011.
NORTH CHESHIRE RC, MORLEY CLUB, Morley Green, Wilmslow, Cheshire. Course starting: September 1997. Every Sunday from 7pm, for both RAE and NRAE. Further details from Gordon Adams, G3LEQ, Tel: (01565) 652652.
TILE HILL COLLEGE, Tile Hill Lane, Coventry, CV4 9SU. Courses available throughout 1997/98. RAE, Post RAE, Morse code & Shortwave Listeners courses. Further details from Michael Dixon, G4GHJ, Tel: (01203) 694200 ext 285.
TROWBRIDGE & DARC, Southwick Village Hall, Trowbridge. Course starting: September 1997 (Morse tuition also available). Further details from Chris Parnell, G0HFX, Tel: (01225) 764874 (evenings only).
WARRINGTON COLLEGAITE INSTITUTE, North Campus, Winwick Road, Warrington, Cheshire, WA2 8QA. Course starting: September 1997. Every Thursday from 7pm. Further details from Gordon Adams, G3LEQ, Tel: (01565) 652652.
115 PETERBOROUGH ATC SQUADRON, Courses available for RAE and NRAE. Further details from Robert Maskill, G4JDL, Tel: (01733) 760005.

Repeaters

What is a repeater

A repeater is an un-manned slave station using a self-contained control system usually called its 'logic'. This controls the operation of the repeater, only allowing signals to be relayed that are of intelligible quality, transmitting the callsign in Morse at appropriate intervals, and some cases, limiting the talk-through time to restrict excessively long transmissions from any source.

UK voice repeaters are generally accessed initially with a 'toneburst' on a frequency of 1750 Hz. This tone must be at least 300 milliseconds duration and should not be more than 500 milliseconds long. Once accessed in this way further tonebursts are not normally required unless the unit has heard nothing intelligible for some time.

Once the repeater logic cannot sense a satisfactory signal to the receiver, it will stay on for a few seconds before identifying itself in Morse code and then closing down. Another burst of tone is required to make use of the unit again once it has closed itself down. An additional means of access may be by use of a sub-audible tone (CTCSS) and those units fitted with this option indicate this by appending a code letter to their ident callsign indicating the tone frequency to be used, as outlined in the table shown on the CTCSS code map .

How do you use a voice Repeater

The repeater receiver continuously monitors its input frequency, whether the repeater transmitter is on or not. In general, upon detecting the 1750 Hz initial access toneburst or the CTCSS tone frequency applicable the logic will open the through audio path and key up the transmitter. A few units additionally require some seconds of speech, and naturally all require an input signal of sufficient strength to operate successfully. Given that you transmit such a signal to the repeater you are being retransmitted and can thus use the repeater.

To avoid re-radiating poor signals or just noise the repeater receiver's squelch signal is used by the logic to tell the repeater whether it should relay any signal heard on the input. If this squelch signal drops below the limit of usability for an appreciable time, usually about two seconds, the repeater will cut off the through audio, send a Morse 'T' or 'K' and unless another signal is detected on the input will, after a short pause, then close down the unit. The 'T' or the 'K' is a recognised signal that another person can proceed. The two or three second delay is essential to prevent rapid sending of 'T' or 'K' under flutter conditions, and it also allows a useful gap ('K-break') for additional stations to announce their presence if they wish to join in.

Almost all repeaters will automatically time the length of an individual over. If this exceeds a pre-set time (usually 2 minutes for VHF and 5 minutes on UHF), the repeater will 'time-out', which will mean it stops relaying its input. and either go completely silent or insert a busy signal to tell other users what has happened. This is a measure to encourage better use of the repeater, and as a jamming-avoidance measure.

All repeaters must identify themselves at least every fifteen minutes when in use; most do this much more frequently and many do it whether in use or not. The callsign, in the form GB3XY, is transmitted in Morse at a speed of about 12 wpm.

Technical Aspects

Repeaters have relatively close input/output frequencies (about 0.5% difference e.g. 600 kHz on VHF, and 1.6 MHz on UHF) so it is essential to have very selective filtering, especially if single antenna working is to be employed. The actual amount of rejection in the transmit and receive paths to achieve this isolation depends on a number of factors, including the physical separation of the antennas, the selectivity of the receiver, the quality of cable screening and the special filters employed.

The level of transmitter noise that appears at the receiver will be about 70 to 80 dB below carrier level. In order for the receiver to be as sensitive as necessary for successful repeater operation, a minimum of 50 dB of rejection is required in the receive path, and at least 60 dB rejection in the transmit path. These severe filter requirements are usually met by employing a number of high-Q cavity filters, connected in series in each path.

The RSGB's Repeater Management Committee can supply further information on these and all the other technical aspect of repeaters.

Repeater Protocol

Unlike amateur contacts on simplex frequencies, the recognised way of making a general call on a repeater is something of the form: '*GM8LBC listening through GB3DG*'. One announcement is usually sufficient. If you are calling another station it is usual to give their callsign followed by your own, for example '*G4AFJ from GM8LBC*'. At the beginning and end of each over, you need only give your own callsign, for example '*from GM8LBC*'.

Keep your overs short or else you may time-out or be unaware that you are no longer in the repeaters service area and if you are answering a call or having a conversation don't forget to wait for the 'K' or 'T' as the logic will not reset the timer unless there has been a 'K' or a 'T'. If your signal is very noisy, it is best not to continue, but to try again later when you are in a better location to gain access. Having a knowledge of where the repeater is located should help you gain an appreciation of where you can access it that is its coverage area.

Repeater Costs

All repeater users are encouraged to either join their local repeater group, and/or contribute to those repeaters they regularly use. This helps the local amateurs maintain the service and pay for the costs of upkeep and administration of the repeater, which will usually include site rental, electricity, and replacement parts

Many repeaters are covered by site-sharing agreements with large organisations for use of their prime radio masts, and costs of £100-300 for site sharing and £100 for electricity are typical.

Furthermore the repeater hardware may need replacing about every five years. Typical costs are: aerials and feeders – £1050-2100; receiver, transmitter and logic – £500; mast fittings – £100; duplexing equipment – £800. This gives an annual cost of between £500-700. In addition many groups also publish a newsletter or magazine to members.

As well as managing the network (see below) the RSGB encourages groups to affiliate to the Society – presently this costs £20. To further assist groups, the RSGB has negotiated a special third party indemnity insurance (£20) especially suitable for repeaters, which protects them against extremely large claims in the event of an accident for which the amateur group are adjudged to be liable.

So all repeater groups do need the *financial* support of their users just to keep the service going!

Problems With Repeaters

From time to time, and mainly on 2m units serving urban areas, there is an outburst of abuse or jamming. The best advice following many years of experience is

- Do not respond in anyway at all on the air to unlicensed transmissions or abusers.
- Do not approach suspected offenders as this can encourage further abuse and may prejudice investigations already underway.
- Help to gather as much information about the problem as possible.

  For example write down dates, times, and frequencies when the interference took place. Note any pattern of operation, suspected location of offenders, details of any bearings obtained with DF equipment etc. Tape recordings of the interference can be useful. Also details of other callsigns, names and addresses of those who have heard the interference as well.

  A copy of this information should be sent to the repeater keeper (see the list below for the callsign of the keeper) or if problems persist then all correspondence and information should be sent to the RMC Repeater Abuse Co-ordinator, c/o RSGB HQ at Potters Bar.

Further information on procedures to be followed can be obtained by contacting the RMC Repeater Zonal Manager.

Repeater Network Management

A committee of the RSGB, the *Repeater Management Committee,* administers the repeater network on behalf of the Society and works closely with the RA on obtaining new repeater licenses, and with other committees on agreeing technical standards. The RMC, which is a full committee of Council, has zonal committee members whose area of responsibility match RSGB Zones and whose details of which are shown in the RSGB section of this book. The RMC also publish various maps, lists, and information on the repeater network.

Specific information about particular repeaters can be obtained from the keeper as listed below or from the RMC Internet Web site http://members.aol.com/rmcweb/rmc.htm

CTCSS Repeater Areas

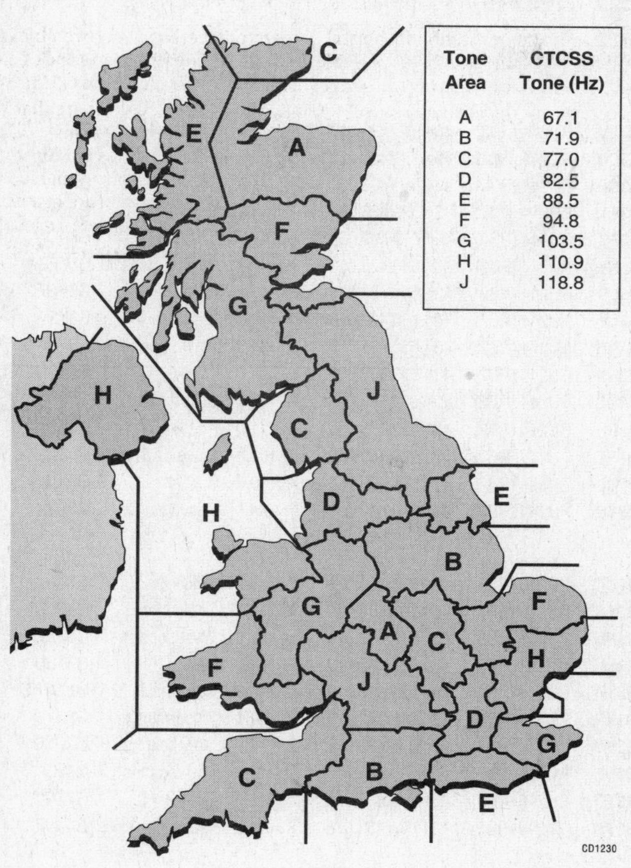

CTCSS REPEATER TONES

CTCSS tones are often used as an optional feature on 2 m and 70 cm repeaters to improve the effectiveness of the UK network.

The principle of CTCSS (Continuous Tone-Coded Squelch System) is that a sub-audible tone is continuously transmitted in addition to the usual signal. Being below the normal speech frequencies, it does not interfere with the received signal.

A repeater user who is on the border of more than one repeater's coverage area can now be selective. By transmitting the appropriate CTCSS tone, only one repeater will be activated rather than another on the same channel. The system operates in parallel with the usual 1750 Hz access tone - so you will be able to use either the appropriate CTCSS tone or 1750 Hz tone-burst to access a repeater.

In addition, a repeater only transmit its CTCSS tone when relaying speech, but not with its periodic idents. A suitably equipped amateur station will thus be able to screen out the annoying idents, so making it more convenient to monitor the repeater.

The UK has been divided into 23 different CTCSS regions, so that generally repeaters in the same area share the same CTCSS tone as shown in the map.

The scheme is optional and will not be used to create 'closed repeaters'. Users will know when a particular repeater has the CTCSS facility available because it will transmit the appropriate letter in morse code after the repeater callsign.

CALL	CHANNEL	INPUT/OUTPUT	CTCSS	LOCATION	KEEPER
GB3AE	RF72 (R50-1)	51.220/50.720	F 94.8	Tenby	GW0WEQ
GB3EF	RF72 (R50-1)	51.220/50.720	H 110.9	Ipswich	G0VDE
GB3UM	RF74 (R50-3)	51.240/50.740	C 77.0	Leicester	G0ORY
GB3UK	RF77 (R50-6)	51.270/50.770	D 82.5	Lancashire	G8NSS
GB3PX	RF78 (R50-7)	51.280/50.780	C 77.0	Royston	G4NBS
GB3SX	RF79 (R50-8)	51.290/50.790	G 103.5	Stoke-on-Trent	G8DZJ
GB3HX	RF80 (R50-9)	51.300/50.800	D 82.5	Huddersfield	G0PRF
GB3FX	RF81 (R50-10)	51.310/50.810	D 82.5	Farnham	G4EPX
GB3RR	RF82 (R50-11)	51.320/50.820	B 71.9	Nottingham	G4TSN
GB3WX	RF83 (R50-12)	51.320/50.830	C 77.0	Shaftesbury	G3ZXX
GB3AM	RF84 (R50-13)	51.340/50.840	C 77.0	Amersham	G0RDI
GB3PD	RF85 (R50-14)	51.350/50.850	B 71.9	Portsmouth	G4JXL
GB3AS	RV48 (R0)	145.000/145.600	C 77.0	Carlisle	G0JGS
GB3CF	RV48 (R0)	145.000/145.600	C 77.0	Leicester	G0ORY
GB3EL	RV48 (R0)	145.000/145.600	D 82.5	East London	G4RZZ
GB3FF	RV48 (R0)	145.000/145.600	F 94.8	Fife	GM0GNT
GB3LY	RV48 (R0)	145.000/145.600	H 110.9	Limavaldy	GI3USS
GB3MB	RV48 (R0)	145.000/145.600	D 82.5	Manchester	G8NSS
GB3SR	RV48 (R0)	145.000/145.600	E 88.5	Brighton Centre	G8VEH
GB3SS	RV48 (R0)	145.000/145.600	A 67.1	Elgin	GM7LSI
GB3WR	RV48 (R0)	145.000/145.600	F 94.8	Wells	G0MBX
GB3YC	RV48 (R0)	145.000/145.600	E 88.5	Scarborough	G0OII
GB3GD	RV50 (R1)	145.025/145.625	H 110.9	Douglas	GD3LSF
GB3HG	RV50 (R1)	145.025/145.625	J 118.8	Northallerton	G0RHI
GB3KS	RV50 (R1)	145.025/145.625	G 103.5	Dover	G4OJG
GB3NB	RV50 (R1)	145.025/145.625	F 94.8	Norwich	G8VLL
GB3NG	RV50 (R1)	145.025/145.625	A 67.1	Fraserburgh	GM8LYS
GB3PA	RV50 (R1)	145.025/145.625	G 103.5	Renfrew	GM0BFW
GB3SC	RV50 (R1)	145.025/145.625	B 71.9	Bournemouth	G0API
GB3SI	RV50 (R1)	145.025/145.625	C 77.0	St Ives, Cornwall	G3NPB
GB3WL	RV50 (R1)	145.025/145.625	D 82.5	West London	G8SUG
GB3AY	RV52 (R2)	145.050/145.650	G 103.5	Ayrshire	GM3YKE
GB3BF	RV52 (R2)	145.050/145.650	C 77.0	Bedford	G1BWW
GB3EC	RV52 (R2)	145.050/145.650	A 67.0	Birmingham	G4YQE
GB3GJ	RV52 (R2)	145.050/145.650	C 77.0	St Helier, Jersey	GJ0NSG
GB3HS	RV52 (R2)	145.050/145.650	E 88.5	Hull	G7JZD
GB3KM	RV52 (R2)	145.050/145.650	F 94.8	King's Lynn	G1HYU
GB3MN	RV52 (R2)	145.050/145.650	D 82.5	Stockport	G8LZO
GB3OC	RV52 (R2)	145.050/145.650	C 77.0	Kirkwall, Orkney	GM0HQG
GB3PO	RV52 (R2)	145.050/145.650	H 110.9	Ipswich	G8CPH
GB3SB	RV52 (R2)	145.050/145.650	J 118.8	Selkirk	GM0FTJ
GB3SL	RV52 (R2)	145.050/145.650	D 82.5	S London	G3PAQ
GB3TR	RV52 (R2)	145.050/145.650	F 94.8	Torquay	G8XST
GB3WH	RV52 (R2)	145.050/145.650	J 118.8	Swindon	G8HBE
GB3BX	RV54 (R3)	145.075/145.675	A 67.1	Wolverhampton	G4JLI
GB3ES	RV54 (R3)	145.075/145.675	G 103.5	Hastings	G7LEL
GB3LD	RV54 (R3)	145.075/145.675	H 110.9	Barrow in Furness	G6LMW
GB3LG	RV54 (R3)	145.075/145.675	G 103.5	Lochgilphead	GM4WMM
GB3LU	RV54 (R3)	145.075/145.675	C 77.0	Lerwick	GM4SWU
GB3NA	RV54 (R3)	145.075/145.675	D 82.5	Barnsley	G4TCG
GB3PE	RV54 (R3)	145.075/145.675	J 118.8	Peterborough	G1ARV
GB3PR	RV54 (R3)	145.075/145.675	F 94.8	Perth	GM8KPH
GB3RD	RV54 (R3)	145.075/145.675	D 82.5	Aldermaston	G8DOR
GB3SA	RV54 (R3)	145.075/145.675	F 94.8	Swansea	GW6KQC
GB3AR	RV56 (R4)	145.100/145.700	H 110.9	Caernarfon	GW4KAZ
GB3BB	RV56 (R4)	145.100/145.700	G 103.5	Brecon	GW0GHQ
GB3BT	RV56 (R4)	145.100/145.700	J 118.8	Berwick	GM1JFF
GB3EV	RV56 (R4)	145.100/145.700	C 77.0	Cumbria	G0IYQ

CALL	CHANNEL	INPUT/OUTPUT	CTCSS	LOCATION	KEEPER
GB3HH	RV56 (R4)	145.100/145.700	B 71.9	Buxton	G4IHQ
GB3HI	RV56 (R4)	145.100/145.700	E 88.5	Mull	GM3RFA
GB3KN	RV56 (R4)	145.100/145.700	G 103.5	Maidstone	G3YCN
GB3VA	RV56 (R4)	145.100/145.700	J 118.8	Aylesbury	G8BQH
GB3WD	RV56 (R4)	145.100/145.700	C 77.0	Plymouth	G6URM
GB3AG	RV58 (R5)	145.125/145.725	F 94.8	Forfar, Angus	GM1CMF
GB3BI	RV58 (R5)	145.125/145.725	A 67.1	Inverness	GM0JFL
GB3DA	RV58 (R5)	145.125/145.725	H 110.9	Chelmsford	G4GUJ
GB3LM	RV58 (R5)	145.125/145.725	B 71.9	Lincoln	G8VGF
GB3NC	RV58 (R5)	145.125/145.725	C 77.0	St Austell	G3IGV
GB3NI	RV58 (R5)	145.125/145.725	H 110.9	Belfast	GI3USS
GB3RA	RV58 (R5)	145.125/145.725	G 103.5	Llan'dod Wells	GW0KQX
GB3SN	RV58 (R5)	145.125/145.725	B 71.9	Alton	G0KVT
GB3TP	RV58 (R5)	145.125/145.725	D 82.5	Keighley	G3RXH
GB3TW	RV58 (R5)	145.125/145.725	J 118.8	Durham	G4GBF
GB3VT	RV58 (R5)	145.125/145.725	G 103.5	Stoke on Trent	G8DZJ
GB3BC	RV60 (R6)	145.150/145.750	F 94.8	Newport, Gwent	GW8ERA
GB3CS	RV60 (R6)	145.150/145.750	G 103.5	Salsburgh	GM4COX
GB3MP	RV60 (R6)	145.150/145.750	H 110.9	Moel-y-Parc	G7OBW
GB3MX	RV60 (R6)	145.150/145.750	B 71.9	Mansfield	G0UYQ
GB3PI	RV60 (R6)	145.150/145.750	C 77.0	Royston	G4NBS
GB3TY	RV60 (R6)	145.150/145.750	J 118.8	Northumberland	G0GXO
GB3WS	RV60 (R6)	145.150/145.750	E 88.5	Crawley	G4EFO
GB3DG	RV62 (R7)	145.175/145.775	G 103.5	Gateh'se of Fleet	GM4VIR
GB3FR	RV62 (R7)	145.175/145.775	B 71.9	Spilsby	G8LXI
GB3GN	RV62 (R7)	145.175/145.775	A 67.1	Aberdeen	GM4NHI
GB3IG	RV62 (R7)	145.175/145.775	E 88.5	Stornoway	GM4PTQ
GB3NL	RV62 (R7)	145.175/145.775	D 82.5	Enfield, London	G3TZZ
GB3PC	RV62 (R7)	145.175/145.775	B 71.9	Portsmouth	G4NAO
GB3PW	RV62 (R7)	145.175/145.775	G 103.5	Newtown, Powys	GW4NQJ
GB3RF	RV62 (R7)	145.175/145.775	D 82.5	Burnley	G4FSD
GB3TE	RV62 (R7)	145.175/145.775	H 110.9	Clacton-on-Sea	G7HJK
GB3WK	RV62 (R7)	145.175/145.775	A 67.1	Leamington	G6FEO
GB3WT	RV62 (R7)	145.175/145.775	H 110.9	Omagh	GI3NVW
GB3WW	RV62 (R7)	145.175/145.775	F 94.8	Cross Hands, Dyfed	GW6ZUS
GB3SF	(R7X)	145.185/145.785	PSSB	Buxton	G4IHO
GB3BN	RU240 (RB0)	434.600/433.000	D 82.5	Bracknell	G4DDN
GB3CK	RU240 (RB0)	434.600/433.000	G 103.5	Ashford	G0GCQ
GB3DT	RU240 (RB0)	434.600/433.000	B 71.9	Blandford Forum	G8BXQ
GB3EX	RU240 (RB0)	434.600/433.000	C 77.0	Exeter	G8UWE
GB3LL	RU240 (RB0)	434.600/433.000	H 110.9	Llandudno	GW8WFS
GB3MK	RU240 (RB0)	434.600/433.000	C 77.0	Milton Keynes	G4NJU
GB3NR	RU240 (RB0)	434.600/433.000	F 94.8	Norwich	G8VLL
GB3NT	RU240 (RB0)	434.600/433.000	J 118.8	Newcastle	G4GBF
GB3NY	RU240 (RB0)	434.600/433.000	E 88.5	Scarborough	G4EEV
GB3PF	RU240 (RB0)	434.600/433.000	D 82.5	Blackburn	G4FSD
GB3PU	RU240 (RB0)	434.600/433.000	F 94.8	Perth	GM8KPH
GB3SO	RU240 (RB0)	434.600/433.000	B 71.9	Boston	G8LXI
GB3SV	RU240 (RB0)	434.600/433.000	H 110.9	Bishops Stortford	G1NOL
GB3US	RU240 (RB0)	434.600/433.000	G 103.5	Sheffield	G3RKL
GB3WN	RU240 (RB0)	434.600/433.000	A 67.1	Wolverhampton	G4OKE
GB3BA	RU242 (RB1)	434.625/433.025	A 67.1	Stonehaven	GM4NHI
GB3BV	RU242 (RB1)	434.625/433.025	D 82.5	Hemel Hempstead	G8BQH
GB3DV	RU242 (RB1)	434.625/433.025	B 71.9	Doncaster	G4LUE
GB3EM	RU242 (RB1)	434.625/433.025	C 77.0	Melton Mowbray	G8WWJ
GB3HJ	RU242 (RB1)	434.625/433.025	J 118.8	Harrogate	G3XWH
GB3HO	RU242 (RB1)	434.625/433.025	E 88.5	Horsham	G7JRV
GB3MA	RU242 (RB1)	434.625/433.025	D 82.5	Bury	G8NSS
GB3TC	RU242 (RB1)	434.625/433.025	C 77.0	Wincanton	G3OOL
GB3AV	RU244 (RB2)	434.650/433.050	D 82.5	Aylesbury	G6NB
GB3CH	RU244 (RB2)	434.650/433.050	C 77.0	Liskeard	G1NSV
GB3CI	RU244 (RB2)	434.650/433.050	B 71.9	Corby	G8MLA
GB3EK	RU244 (RB2)	434.650/433.050	G 103.5	Margate	G4TKR
GB3FC	RU244 (RB2)	434.650/433.050	A 67.1	Blackpool	G6AOS
GB3HK	RU244 (RB2)	434.650/433.050	J 118.8	Hawick, Borders	GM0FTJ
GB3LS	RU244 (RB2)	434.650/433.050	B 71.9	Lincoln	G8VGF
GB3LV	RU244 (RB2)	434.650/433.050	D 82.5	Enfield	G3KSW
GB3NN	RU244 (RB2)	434.650/433.050	F 94.8	Wells, Norfolk	G0FVF
GB3NX	RU244 (RB2)	434.650/433.050	E 88.5	Crawley	G0DSU
GB3OS	RU244 (RB2)	434.650/433.050	A 67.1	Stourbridge	G1PKZ
GB3PH	RU244 (RB2)	434.650/433.050	B 71.9	Portsmouth	G8PGF
GB3ST	RU244 (RB2)	434.650/433.050	G 103.5	Stoke on Trent	G8DZJ
GB3UL	RU244 (RB2)	434.650/433.050	H 110.9	Belfast	GI3USS
GB3YS	RU244 (RB2)	434.650/433.050	F 94.8	Yeovil	G0LHX
GB3CC	RU246 (RB3)	434.675/433.075	E 88.5	Chichester	G3UEQ
GB3ER	RU246 (RB3)	434.675/433.075	H 110.9	Chelmsford	G4GUJ
GB3HL	RU246 (RB3)	434.675/433.075	D 82.5	West London	G8SUG
GB3HU	RU246 (RB3)	434.675/433.075	E 88.5	Hull	G3TEU
GB3KA	RU246 (RB3)	434.675/433.075	G 103.3	Kilmarnock	GM3YKE
GB3KR	RU246 (RB3)	434.675/433.075	A 67.1	Kidderminster	G8NTU
GB3MD	RU246 (RB3)	434.675/433.075	B 71.9	Mansfield	G0UYQ
GB3NH	RU246 (RB3)	434.675/433.075	C 77.0	Northampton	G4IIO
GB3TD	RU246 (RB3)	434.675/433.075	J 118.8	Swindon	G4XUT
GB3VS	RU246 (RB3)	434.675/433.075	F 94.8	Taunton	G4UVZ
GB3GC	RU248 (RB4)	434.700/433.100	E 88.5	Goole	G0GLZ
GB3IH	RU248 (RB4)	434.700/433.100	H 110.9	Ipswich	G8CPH
GB3IW	RU248 (RB4)	434.700/433.100	B 71.9	Newport	G0ISB
GB3KL	RU248 (RB4)	434.700/433.100	F 94.8	Kings Lynn	G3ZCA
GB3LE	RU248 (RB4)	434.700/433.100	C 77.0	Leicester	G0ORY
GB3NK	RU248 (RB4)	434.700/433.100	G 103.5	Wrotham	G8JNZ
GB3OH	RU248 (RB4)	434.700/433.100	F 94.8	Bo'ness	GM6WQH
GB3SP	RU248 (RB4)	434.700/433.100	F 94.8	Pembroke	GW4VRO
GB3UB	RU248 (RB4)	434.700/433.100	J 118.8	Bath	G0LIB
GB3VE	RU248 (RB4)	434.700/433.100	C 77.0	Appleby, Cumbria	G0IYQ
GB3EB	RU250 (RB5)	434.725/433.125	H 110.9	Brentwood	G6IFH
GB3GH	RU250 (RB5)	434.725/433.125	J 118.8	Cheltenham	G6AWT
GB3HY	RU250 (RB5)	434.725/433.125	E 88.5	Haywards Heath	G3XTH
GB3IM	RU250 (RB5)	434.725/433.125	H 110.9	Douglas, IoM	GD3LSF
GB3OV	RU250 (RB5)	434.725/433.125	F 94.8	Huntingdon	G8LRS
GB3WB	RU250 (RB5)	434.725/433.125	F 94.8	Weston-S-Mare	G7KUD
GB3WJ	RU250 (RB5)	434.725/433.125	B 71.9	Scunthorpe	G3TMD
GB3BD	RU252 (RB6)	434.750/433.150	C 77.0	Ampthill, Beds	G1BWW
GB3BR	RU252 (RB6)	434.750/433.150	E 88.5	Brighton	G8VEH
GB3CR	RU252 (RB6)	434.750/433.150	H 110.9	Clywd	GW4GTE
GB3CW	RU252 (RB6)	434.750/433.150	G 103.5	Powys	GW4NQJ
GB3DI	RU252 (RB6)	434.750/433.150	J 118.8	Didcot	G8CUL
GB3HA	RU252 (RB6)	434.750/433.150	E 88.5	Hornsea, Yorks	G4YTV
GB3HC	RU252 (RB6)	434.750/433.150	J 118.8	Hereford	G4JSN
GB3LW	RU252 (RB6)	434.750/433.150	D 82.5	Central London	G8AUU
GB3ME	RU252 (RB6)	434.750/433.150	A 67.1	Rugby	G8DLX
GB3SK	RU252 (RB6)	434.750/433.150	G 103.5	Canterbury	G6DIK
GB3SY	RU252 (RB6)	434.750/433.150	B 71.9	Barnsley	G4LUE
GB3WG	RU252 (RB6)	434.750/433.150	F 94.8	Port Talbot	GW3VPL

CALL	CHANNEL	INPUT/OUTPUT	CTCSS	LOCATION	KEEPER
GB3BL	RU254 (RB7)	434.775/433.175	C 77.0	Bedford	G1BWW
GB3MF	RU254 (RB7)	434.775/433.175	G 103.5	Macclesfield	G0AMU
GB3MG	RU254 (RB7)	434.775/433.175	F 94.8	Bridgend	GW3RVG
GB3NM	RU254 (RB7)	434.775/433.175	B 71.9	Nottingham	G2SP
GB3TS	RU254 (RB7)	434.775/433.175	J 118.8	Middlesbrough	G8MBK
GB3WY	RU254 (RB7)	434.775/433.175	D 82.5	Halifax	G8NWK
GB3AN	RU256 (RB8)	434.800/433.200	H 110.9	Amlwch	GW6DOK
GB3CM	RU256 (RB8)	434.800/433.200	F 94.8	Carmarthen	GW0IVG
GB3EA	RU256 (RB8)	434.800/433.200	B 71.9	Southampton	G4MYS
GB3EH	RU256 (RB8)	434.800/433.200	A 67.1	Banbury	G4OHB
GB3LA	RU256 (RB8)	434.800/433.200	D 82.5	Leeds	G8ZXA
GB3PY	RU256 (RB8)	434.800/433.200	C 77.0	Cambridge	G4NBS
GB3TF	RU256 (RB8)	434.800/433.200	A 67.1	Telford	G3UKV
GB3BE	RU258 (RB9)	434.825/433.225	H 110.9	Bury St Edmunds	G8KMM
GB3CL	RU258 (RB9)	434.825/433.225	B 71.9	Salisbury	G3YWT
GB3AW	RU260 (RB10)	434.850/433.250	B 71.9	Newbury	G8DOR
GB3BS	RU260 (RB10)	434.850/433.250	J 118.8	Bristol	G4SDR
GB3DD	RU260 (RB10)	434.850/433.250	F 94.8	Dundee	GM4UGF
GB3DY	RU260 (RB10)	434.850/433.250	B 71.9	Derby	G3ZYC
GB3LI	RU260 (RB10)	434.850/433.250	D 82.5	Liverpool	G3WIC
GB3LT	RU260 (RB10)	434.850/433.250	C 77.0	Luton	G6OUA
GB3ML	RU260 (RB10)	434.850/433.250	G 103.5	Airdrie	GM3SAN
GB3MW	RU260 (RB10)	434.850/433.250	A 67.1	Leamington Spa	G6FEO
GB3NS	RU260 (RB10)	434.850/433.250	D 82.5	Banstead	G0OLX
GB3PB	RU260 (RB10)	434.850/433.250	C 77.0	Peterborough	G1ARV
GB3AH	RU262 (RB11)	434.875/433.275	F 94.8	Swaffham	G8PON
GB3BK	RU262 (RB11)	434.875/433.275	D 82.5	Aldermaston, Berks	G8DOR
GB3DC	RU262 (RB11)	434.875/433.275	J 118.8	Sunderland	G6LMR
GB3GR	RU262 (RB11)	434.875/433.275	B 71.9	Grantham	G4WFK
GB3GY	RU262 (RB11)	434.875/433.275	E 88.5	Grimsby	G1BRB
GB3HN	RU262 (RB11)	434.875/433.275	D 82.5	Hitchin	G3ZQI
GB3HT	RU262 (RB11)	434.875/433.275	C 77.0	Hinckley	G8SHH
GB3LR	RU262 (RB11)	434.875/433.275	E 88.5	Newhaven	G0ENJ
GB3RE	RU262 (RB11)	434.875/433.275	G 103.5	Maidstone	G4AKQ
GB3SH	RU262 (RB11)	434.875/433.275	F 94.8	Honiton	G6WWY
GB3WP	RU262 (RB11)	434.875/433.275	D 82.5	Hyde, Cheshire	G6YRK
GB3ZI	RU262 (RB11)	434.875/433.275	A 67.1	Stafford	G1UDS
GB3EE	RU264 (RB12)	434.900/433.300	B 71.9	Chesterfield	G6SVZ
GB3GB	RU264 (RB12)	434.900/433.300	A 67.1	Great Barr	G8NDT
GB3GF	RU264 (RB12)	434.900/433.300	E 88.5	Guildford	G4EML
GB3HM	RU264 (RB12)	434.900/433.300	J 118.8	Boroughbridge	G0RHI
GB3MT	RU264 (RB12)	434.900/433.300	D 82.5	Bolton	G8NSS
GB3OX	RU264 (RB12)	434.900/433.300	J 118.8	Oxford	G4WXC
GB3PT	RU264 (RB12)	434.900/433.300	Data	Royston	G4NBS
GB3CA	RU266 (RB13)	434.925/433.325	C 77.0	Carlisle	G0JGS
GB3CY	RU266 (RB13)	434.925/433.325	J 118.8	York	G4FUO
GB3DS	RU266 (RB13)	434.925/433.325	B 71.9	Worksop	G3XXN
GB3GU	RU266 (RB13)	434.925/433.325	C 77.0	St Peter Port, CI	GU4EON
GB3HW	RU266 (RB13)	434.925/433.325	H 110.9	Romford	G4GBW
GB3LC	RU266 (RB13)	434.925/433.325	B 71.9	Louth, Lincs	G6GZS
GB3SM	RU266 (RB13)	434.925/433.325	G 103.5	Leek, Staffs	G8DZJ
GB3VH	RU266 (RB13)	434.925/433.325	D 82.5	Welwyn Garden City	G4THF
GB3XX	RU266 (RB13)	434.925/433.325	C 77.0	Daventry	G1ZJK
GB3AB	RU268 (RB14)	434.950/433.350	A 67.1	Aberdeen	GM1LKD
GB3CB	RU268 (RB14)	434.950/433.350	A 67.1	Birmingham	G8AMD

CALL	CHANNEL	INPUT/OUTPUT	CTCSS	LOCATION	KEEPER
GB3CE	RU268 (RB14)	434.950/433.350	H 110.9	Colchester	G7BKU
GB3ED	RU268 (RB14)	434.950/433.350	F 94.8	Edinburgh	GM3GBX
GB3GL	RU268 (RB14)	434.950/433.350	G 103.5	Glasgow	GM3SAN
GB3HE	RU268 (RB14)	434.950/433.350	G 103.5	Hastings	G4FET
GB3HR	RU268 (RB14)	434.950/433.350	D 82.5	Harrow	G4KUJ
GB3LF	RU268 (RB14)	434.950/433.350	H 110.9	Lancaster	G8UHO
GB3MR	RU268 (RB14)	434.950/433.350	D 82.5	Stockport	G8LZO
GB3ND	RU268 (RB14)	434.950/433.350	F 94.8	Bideford	G4JKN
GB3SD	RU268 (RB14)	434.950/433.350	B 71.9	Weymouth	G0EVW
GB3TL	RU268 (RB14)	434.950/433.350	B 71.9	Spalding	G7JBA
GB3WF	RU268 (RB14)	434.950/433.350	D 82.5	Leeds	G8ZXA
GB3YL	RU268 (RB14)	434.950/433.350	F 94.8	Lowestoft	G4TAD
GB3FN	RU270 (RB15)	434.975/433.375	D 82.5	Farnham	G4VDF
GB3HB	RU270 (RB15)	434.975/433.375	C 77.0	St Austell	G3IGV
GB3LH	RU270 (RB15)	434.975/433.375	A 67.1	Shrewsbury	G3UQH
GB3OM	RU270 (RB15)	434.975/433.375	A 67.1	Omagh	GI4SXV
GB3PP	RU270 (RB15)	434.975/433.375	D 82.5	Preston	G3SYA
GB3SG	RU270 (RB15)	434.975/433.375	F 94.8	Cardiff	GW7KWG
GB3SU	RU270 (RB15)	434.975/433.375	H 110.9	Sudbury	G8AAR
GB3SZ	RU270 (RB15)	434.975/433.375	B 71.9	Bournemouth	G0API
GB3TH	RU270 (RB15)	434.975/433.375	A 67.1	Tamworth	G4JBX
GB3WI	RU270 (RB15)	434.975/433.375	F 94.8	Wisbech	G4NPH
GB3WU	RU270 (RB15)	434.975/433.375	D 82.5	Wakefield	G0COA
GB3MC	RM0	1291.000/1297.000	D 82.5	Bolton	G8NSS
GB3NO	RM0	1291.000/1297.000	F 94.8	Norwich	G8VLL
GB3FM	RM2	1291.050/1297.050	J 118.8	Farnham	G4VDF
GB3PS	RM3	1291.075/1297.075	C 77.0	Royston	G4NBS
GB3SE	RM3	1291.075/1297.075	G 103.5	Stoke on Trent	G8DZJ
GB3CN	RM5	1291.125/1297.125	C 77.0	Northampton	G6NYH
GB3BW	RM6	1291.150/1297.150	C 77.0	Bedford	G1BWW
GB3MM	RM6	1291.150/1297.105	A 67.1	Wolverhampton	G4OKE
GB3WC	RM15	1291.375/1297.375	D 82.5	Wakefield	G0COA
GB3EY		1248/1308	FM TV	Hull	G8EQZ
GB3HV		1248/1308	FM TV	High Wycombe	G8LES
GB3KT		1249/1310	FM TV	Sheerness, Kent	G8SUY
GB3AT		1249/1316	FM TV	Southamtpon	G6HNJ
GB3ET		1249/1316	FM TV	Emley Moor	G8HUA
GB3GV		1249/1316	FM TV	Leicestershire	G8OBP
GB3LO		1249/1316	FM TV	Lowestoft	G4TAD
GB3MV		1249/1316	FM TV	Northampton	G1IRG
GB3PV		1249/1316	FM TV	Cambridge	G4NBS
GB3RT		1249/1316	FM TV	Coventry	G1GPE
GB3TM		1249/1316	FM TV	Amlwch	GW8PBX
GB3TN		1249/1316	FM TV	Fakenham	G4WVU
GB3TT		1249/1316	FM TV	Chesterfield	G4AGE
GB3WV		1249/1316	FM TV	Dartmoor	G6URM
GB3TV		1249/1318.5	FM TV	Dunstable	G4ENB
GB3UD		1249/1318.5	FM TV	Stoke on Trent	G0KBI
GB3VR		1249/1316	FM TV	Brighton	G8KOE
GB3ZZ		1249/1316	FM TV	Bristol	G6TVJ
GB3UT		1276.5/1311.5	AM TV	Bath	G7AUP
GB3TG		10315/10135	FM TV	Bletchley	G4NJU
GB3XG		10315/10135	FM TV	Bristol	G6TVJ
GB3XT		10340/10065	FM TV	Burton on Trent	G8OZP

6m Repeaters

Channel	Input	Output
RF72 (R50-1)	51.220	50.720
RF72 (R50-1)	51.220	50.720
RF74 (R50-3)	51.240	50.740
RF77 (R50-6)	51.270	50.770
RF78 (R50-7)	51.280	50.780
RF79 (R50-8)	51.290	50.790
RF80 (R50-9)	51.300	50.800
RF81 (R50-10)	51.310	50.810
RF82 (R50-11)	51.320	50.820
RF83 (R50-12)	51.320	50.830
RF84 (R50-13)	51.340	50.840
RF85 (R50-14)	51.350	50.850

Frequencies shown on map are repeater outputs.

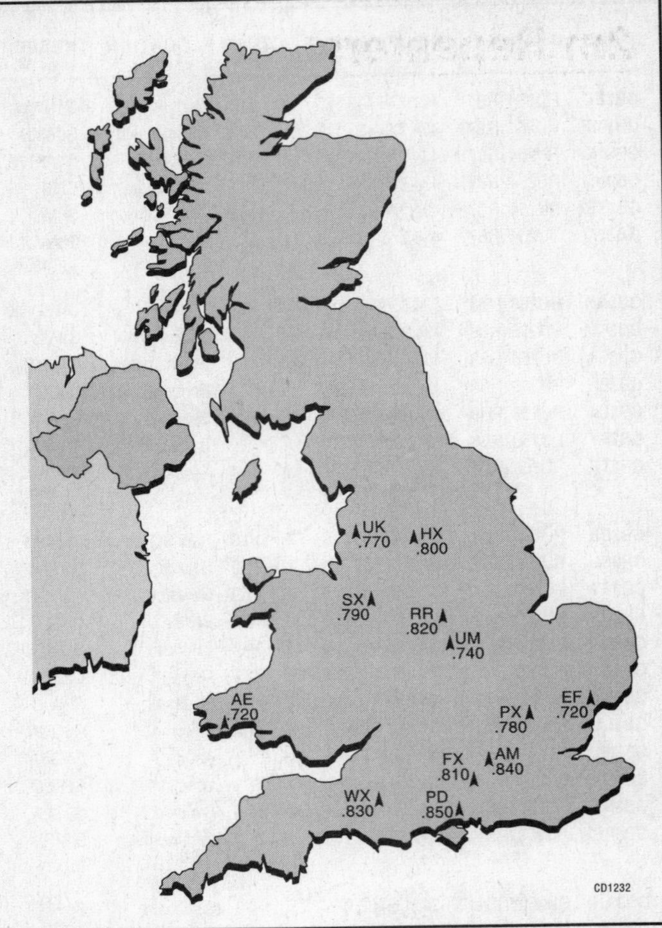

Microwave Repeaters

23 cms Speech

Channel	Input (MHz)	Output (MHz)
RM 0	1291.000	1297.000
RM 2	1291.050	1297.050
RM 3	1291.075	1297.075
RM 5	1291.125	1297.125
RM 6	1291.150	1297.150
RM 9	1291.225	1297.225
RM 12	1291.300	1297.300
RM 15	1291.375	1297.375

When not being used as a repeater, microwave units revert to a beacon mode

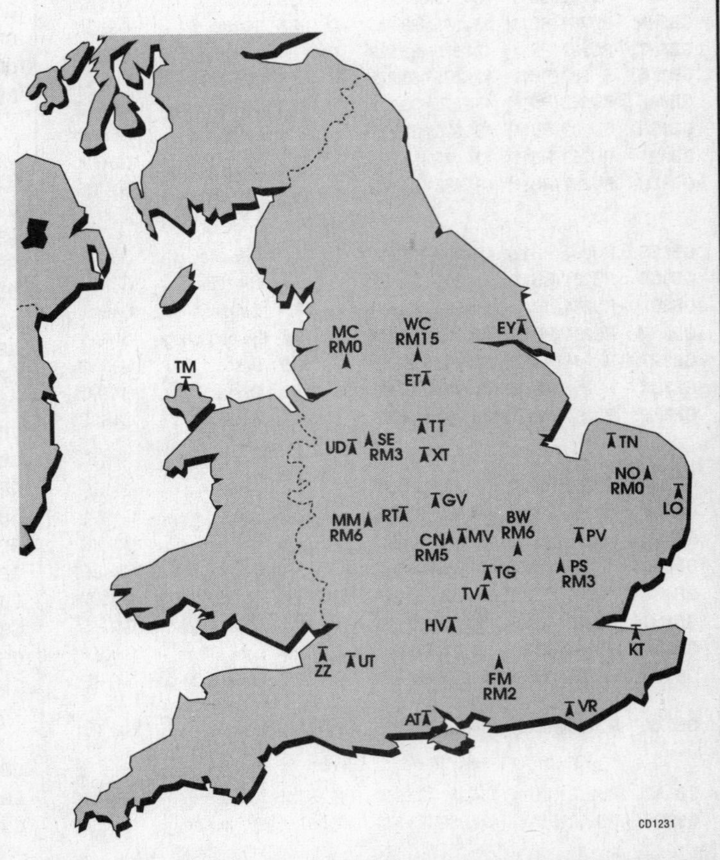

2m Repeaters

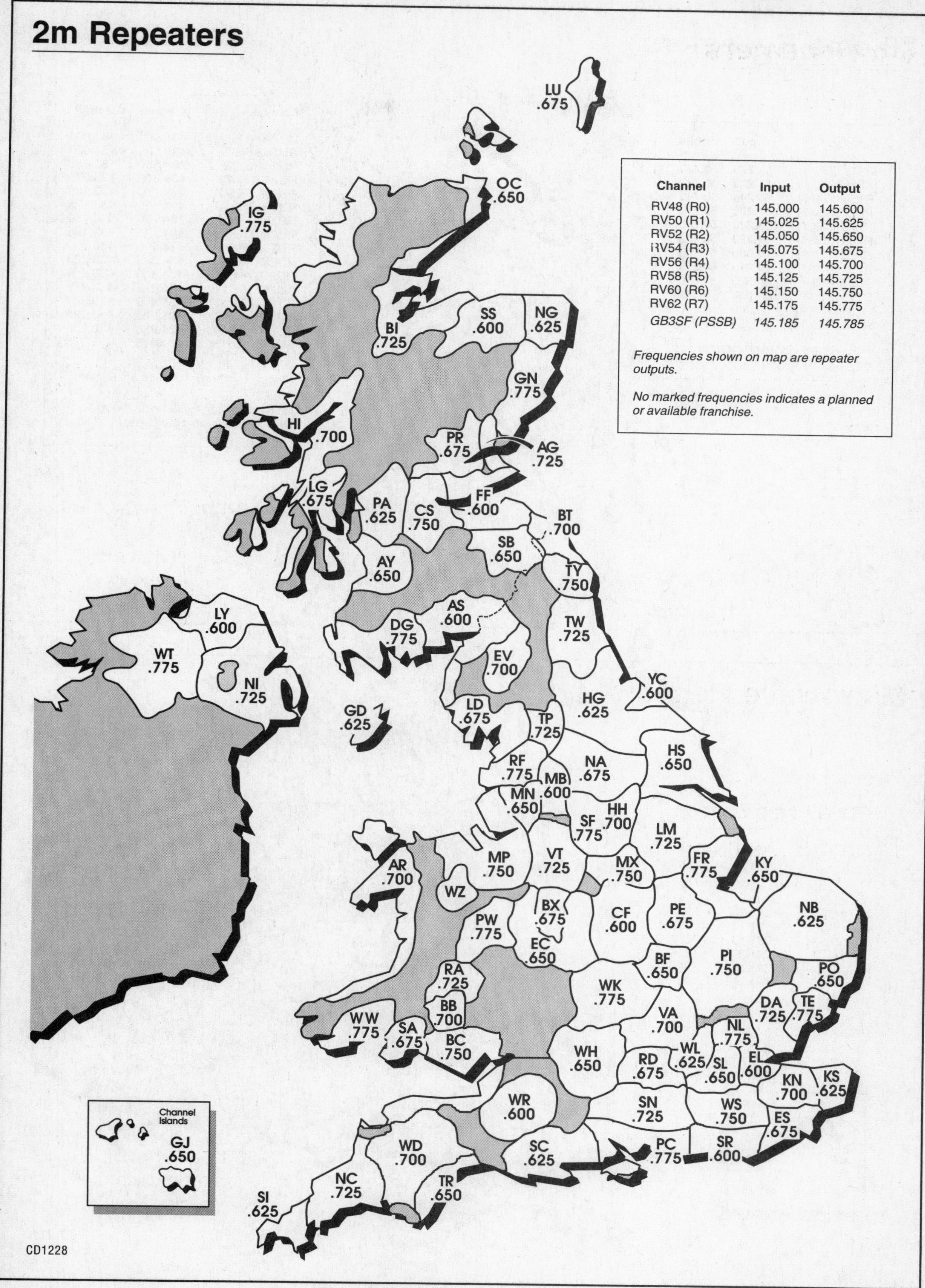

Channel	Input	Output
RV48 (R0)	145.000	145.600
RV50 (R1)	145.025	145.625
RV52 (R2)	145.050	145.650
RV54 (R3)	145.075	145.675
RV56 (R4)	145.100	145.700
RV58 (R5)	145.125	145.725
RV60 (R6)	145.150	145.750
RV62 (R7)	145.175	145.775
GB3SF (PSSB)	*145.185*	*145.785*

Frequencies shown on map are repeater outputs.

No marked frequencies indicates a planned or available franchise.

70 cms Repeaters

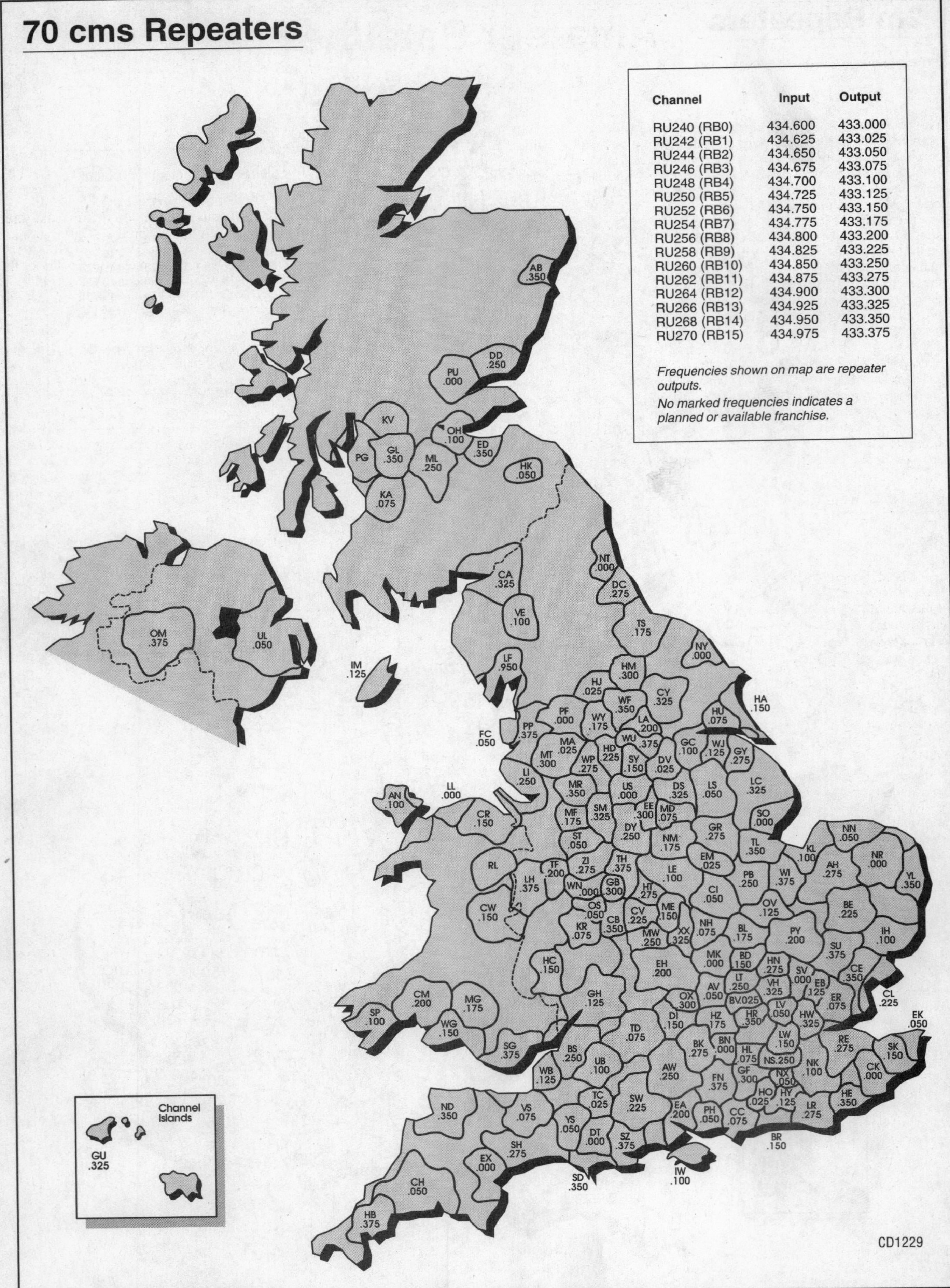

Channel	Input	Output
RU240 (RB0)	434.600	433.000
RU242 (RB1)	434.625	433.025
RU244 (RB2)	434.650	433.050
RU246 (RB3)	434.675	433.075
RU248 (RB4)	434.700	433.100
RU250 (RB5)	434.725	433.125
RU252 (RB6)	434.750	433.150
RU254 (RB7)	434.775	433.175
RU256 (RB8)	434.800	433.200
RU258 (RB9)	434.825	433.225
RU260 (RB10)	434.850	433.250
RU262 (RB11)	434.875	433.275
RU264 (RB12)	434.900	433.300
RU266 (RB13)	434.925	433.325
RU268 (RB14)	434.950	433.350
RU270 (RB15)	434.975	433.375

Frequencies shown on map are repeater outputs.

No marked frequencies indicates a planned or available franchise.

Amateur Satellites

By Ron Broadbent, MBE, G3AAJ, Honorary Secretary AMSAT-UK

Amateur satellites have been likened by many to amateur repeaters. Satellites do receive a signal on one frequency and return it on another, but in truth, that is where the similarity ends. Terrestrial repeaters are at a fixed point, and do not need steer-able arrays to track across the sky. Most repeaters use FM, most satellites use SSB and CW, although FM is used for digital processing on the Microsats now in orbit.

Terrestrial FM repeaters cost a few thousand pounds to build and support, and can be maintained or replaced without much effort. Satellites, on the other hand, cost upwards of a third of a million pounds to design, build and launch into orbit, plus a few thousand a year to command and maintain the internal computer software. Orbiting satellites cannot be serviced with new hardware once launched, and to date, battery supplies have to be provided with solar energy to maintain control and transponder output.

Satellites have a short life span in the region of five to ten years before decay or mishap and once launched, access is not practicable. At several hundreds of thousands of pounds satellites are vastly more expensive than terrestrial repeaters. Repeaters are 'channelled' and local in range; satellites are designed in most cases, to use a bandpass of a few hundred kilohertz, in internationally allocated Amateur-Satellite Service bands, use SSB and CW (analogue) and FM (digital) and provide a global range inside a 'footprint' approximately every 95-120 minutes during the 24 hours, excepting the Phase 3 Series which are designed for a 5-12 hour period.

Current satellites *(As at August 1997)*

RS10

Part of the Russian satellite Cosmos 1861 framework. Uplinks in the 21 MHz and 145 MHz bands. Downlinks into 29 and 145 MHz [2]. Look for the beacons on 29.357, 29.403, 29.407, and 29.453, 145.857, and 145.953 to see if the satellite is in range. Telemetry and schedule is by CW – so should be easy to copy. Carries a SSB and CW transponder. This transponder is not currently available.

RS12

As above, part of another Cosmos satellite. The transponder system is nearly the same as RS10. Launched by the USSR (now CIS) military for the Russian Sport Organisations. Uplinks and downlinks similar to above.

RS15

In the range of 'Radio Sport' satellites, launched on 26 December 1995 from Tyuratam on the 1st orbital flight of 'ROKOT', a launcher derived from a SS-19 missile. Carries a similar transponder system to its predecessors, providing Modes A, K and T. It is in a higher orbit than its siblings and has a potential for greater DX contacts. RS-15 appears to have a weak (low) output TX on 29 MHz. Uplink: 145.858-145.898 MHz. SSB/CW.

RS16.

Another in the ' Russian Sport' series. Launched early 1997. Beacons on 435.504 and 548 MHz. Transponder CW and SSB only. 145.915-145.948 MHz uplink for a 29.415 - 29.448 MHz downlink. Beacon on 29.408 MHz.

FO-20 JAS-1a

Japanese satellite. Transponder for CW and SSB only. On mode JA continuously. Uplink: 145.900 - 146.000 MHz. Downlink: 435.800 - 435.900 MHz.

OSCAR 10

The second spacecraft of the operational AMSAT Phase 3 satellite programme.

OSCAR 10 has received a large dose of radiation from the Van Allen Belt which it enters twice each day. This has caused deterioration of the on-board computers, thus AMSAT command stations cannot now control this spacecraft.

Requests are made for amateurs not to use the satellite if the beacon on 145.810 MHz is heard to FM (warble tone). It is known that at least four cells of the power supply are dead, and the only source of bus power is the solar cell system. Uplink (Mode B) 435.1 MHz, downlink 145.9 MHz. Carries an SSB and CW transponder.

As of this date, Oscar 10 is still of some use to keen radio amateurs who do not send up huge amounts of RF power. ie Your return signal should*never* exceed the signal strength of the on-board beacon on 145.809 MHz. This satellite is *only* operational from the solar cells output. Use with care, we hope it will then last a few more years.

OSCAR 11 (UoSAT 2)

This is the second of the University of Surrey/AMSAT-UK scientific satellites and is for receive only on 145.825, and 435.025 MHz FM. It transmits bulletins, data and 'Digitalker' information to amateurs and schools. News items in ASCII are sent out regularly by AMSAT-UK.

This is not a general communications satellite. There are about three passes in the morning and three in the late afternoon. If you hear a rasping sound on one of the above frequencies you have found UoSAT 2 (OSCAR 11). The next pass will be some 98 minutes later, if in range. You can decode the data with your home computer and a simple decoder [3]. This series of satellites is perhaps the easiest to hear with very simple antennas and FM receiving equipment.

OSCAR 22 (UoSAT 5)

Fully operational on packet and CCD modes in the amateur bands. You will need to obtain a 9.6k baud modem for your normal TNC to use the 'store and forward' transponder on this 'bird'. This transponder was paid for, and maintained from money donated entirely by AMSAT-UK members, and built by University of Surrey.

Launched early 1990, it has been in continuous use with very few breaks in service. It is mainly used for international packet forwarding using the Satgate system. Commanded by SST Ltd at University of Surrey Command Station.

Microsats OSCARS 16, 17, 18, 19

These series of small satellites, just 9 inches cube, were launched in January 1990, with two UoSATs on one platform by an Ariane rocket. Comprise various modes of operation of interest to a wide range of amateur users. Including packet and digital modes.

OSCAR 16

Now operational in PACSAT mode on frequencies of 145.900, 145.920, 145.940, and 145.960 MHz – see the AMSAT-UK Operating Guide from [1] below for full details.

OSCAR 17 (DOVE)

This is a command station only uplink, but records messages from schools across the world via a voice encoder, and by AX25 packet format. Currently intermittent service. Has a bpsk/ssb beacon on 2401.2205 MHz which is a useful tools for setting up equipment for the new P3D satellite be launched at year end 1997 or early 1998.

OSCAR 18 (WEBERSAT)

This was built by the Weber State University in USA and carries a full CCD device which has transmitted pictures in AX25 format from space. Frequencies are downlinks: 437.102, 437.075. MHz. Uplinks: 145.900 MHz, ATV, uplink NTSC: 1265.000 MHz.

OSCAR 19

Built by the LU-AMSAT organisation, this transmits similar information as OSCAR 16 but on frequencies of 145.840, 145.860, 145.880, and 145.900 MHz. Software is normally a requirement for access to BBS, WOD, packet and telemetry on Microsats and UoSATs. This software is obtainable from all AMSAT Groups [1] below.

OSCAR 23 (Kitsat-A)

Launched 11 August 1992. Built for the Korean Amateur Radio AMSAT Group and complies with the IARU specifications for use on amateur radio frequencies.

Similar in all respects to UoSAT 5 (OSCAR 22). Downlink: 435.175 MHz (9600 FSK/FM); uplinks on 145.850 and 145.900 (FSK/FM).

OSCAR 25 (Kitsat-B)

Launched 26 September 1993, this is a clone of KO-23 but in a higher orbit. Transponder details are similar with a downlink on 435.600 MHz (9600 FSK/FM) and a single uplink on 145.870 MHz (FSK/FM).

OSCAR 26. (IT-AMSAT)

This satellite is currently operational on no definite schedule, i.e. catch it when you can.

Downlink: 435.867 and 435.822 MHz. Uplink: 145.875, .900, .925, .950 MHz.

OSCAR 27

Name: EYESAT. Launched September 1993. Near polar orbit. Period 100 minutes. 145.850 MHz. FM. AX25 Experimental. FSK UP to AFSK Down. at 436.800 MHz. This satellite is carried on a commercial bird, developed by AMRAD for experiment with different digital modulation systems. Rarely heard since launch.

OSCAR 29.

Launched in May 1996. named JAS-2 by JARL. Analogue mode: 145.900 - 146.000 MHz (uplink), 435.900 - 435.800 MHz (downlink, inverted). Output power 1 Watt. Same band plan as FO-20 above. Digital Mode: 145.850, .870, .890, and .910 MHz (uplink). 435.910 MHz (downlink). Additionally, 9600 bps FSK is available. Digitalker: FM Voice 25 seconds recording: 435.910 MHz (downlink). CW Telemetry: 435.795 MHz, 12 wpm (downlink).

New Satellites

At least four further satellites (LEO's) are expected to be launched towards the end of 1997. These include another U-AMSAT. A further TECHSAT, and two UoSAT style satellites from University of Surrey (SSTL). Details about these satellites will be available only after a successful launch. The latest information is made available to AMSAT-UK members via *OSCAR NEWS*, the magazine of AMSAT-UK.

PHASE III D

This satellite is a replacement for OSCAR-13 and is now in the final stage. Launch is expected to be in late 1997, or early 1998. It is the largest and most sophisticated satellite planned so far, and is intended to make operating via satellite available to many more enthusiasts using simpler equipment. It is 99% completed as of this date and was a joint effort of all AMSAT groups around the world. It will also be the most expensive satellite ever built and funded by amateurs. AMSAT-UK have raised nearly a quarter of a million pounds for the building funds. Your further donations will also help to maintain this satellite in orbit via the six Command Station across the world after Launch. Donations, however small, will be very welcome to the AMSAT-UK office. [1]. Don't be a freeloader. ie. If you use the satellite, please help pay for it's upkeep.

Band plans

Most satellites are built and are recommended for use for low duty cycle modes of transmission. This means SSB or CW being the norm with recommended use as follows:

CW: The low one third of the received signal section

Mixed CW/SSB: The middle one third section of the band.

SSB: The high one third end of the receive section. There are guard frequencies at each end of every satellite band for beacons and command traffic.

Digital FM: On channelled frequencies only. By international agreement, between each one third section there is a few kilohertz for use of RTTY, packet and SSTV, but these modes are not encouraged due to their wide bandwidth and long transmit 'on time' (high duty cycle).

Antennas

As will be realised, the best communication is obtainable from a fully steer-able elevation and azimuth rotatable beam, or beams for each band. Would it that we could all afford them, but this is not the norm! Many satellite users enjoy this aspect of the hobby by using part fixed, and part rotatable, arrays.

For instance, to use Mode A satellites (RS), a horizontal dipole for two metres, and another for 10 metres, running east and west, should enable you to work 80% of each pass. You will certainly be able to hear the RS series on these two antennas. At a later date you can decide what type of gain antenna you wish to use. Advice is always available on this subject from standard textbooks and, of course, from satellite users.

For the low orbiters (OSCARS 16 to 25) a very simple set-up as regards antennas can be used; it is, however, preferable to use a high gain steer-able system for best results.

Phase 3 Satellites

It is advisable to have a gain type of antenna for each band, steer-able elevation as well as azimuth. It is not always necessary to have an elevation rotator, though. Some stations use a fixed elevation 70 cm and 2m antennas at 20-40 degrees above the horizontal. This combination, on a normal steer-able rotatable mast, enables the stations to use their antennas for terrestrial contacts as well as satellites, at minimal cost.

Code of practice

* **Ensure that your downlink is the very best that you can obtain with the resources you have.** No other factor will assist more than a really low noise, high sensitivity receive system from antenna to audio detector.

  It really is no good trying to make a satellite contact by running high EIRP with a poor receiver system. Use the best coax and the shortest possible run you can obtain on both uplink and downlink systems. If you are thinking in terms of 100 feet of TV coax purchased at the last rally, forget it!

  The very least type of coax for frequencies above 144 MHz is Pope's hardline. Leave your new coil of UR67 for the DC bands (below 30 MHz). You should also invest in N-type connectors, (not the chrome plated type seen at rallies) on all circuits above 144 MHz, and ensure that the connectors are soldered and fitted correctly. If you cannot hear your own uplink, without pumping hundreds of watts into the transponder receiver, how do you expect someone else to do so! Hearing your own signal also assists you in getting onto the correct frequency.

  If you must whistle, send a tone, or hold your Morse key down with your foot, for minutes on end, please do it off someone else's QSO. Better still, do not do it at all! Use a frequency conversion chart [2]. Get near to the frequency, and then slide on to the QSO when invited to do so. This should take you three seconds at most, *if* you are a good operator. Most satellite users will know you are near the frequency whilst they are in QSO, and will then call you when they have finished. The satellite bands are nothing like 20 metres during a contest – yet!

* **Do not use any more power than required to get a clean signal return from the transponder.** You should be able, at all times, to monitor the beacons at the end of every band, and *never* be above the beacon signal strength. You will be branded a 'lid' if you do so! Signals greater than the beacon de-sense the satellite for *all* users.

* **If you can use duplex, please do so.** You will then hear other stations on your frequency and can respond immediately without long CQ calls.

* **Please make meaningful contacts, and act in the spirit of amateur radio, ie, in a gentlemanly manner.** If in doubt about any aspect of satellite communications, just ask the many willing users of the satellites to assist. Most users are only too willing to assist a newcomer. How do you find such a person? Ask on air, or call into the many VHF AMSAT nets on 144.280 MHz. Ask any member of AMSAT-UK to give you the callsign of your nearest 'Elmer' from *Oscar News* list. You can also call into the AMSAT-UK HF Net on 3.780 MHz (G0AUK) at 1015 local time on Sundays. 1900 local on Monday or Wednesday. Give a CQ AMSAT call on 144.280 MHz (AMSAT listening frequency in UK) or [1].

Where to find orbital information?

The Oscalator and Orbital Calendar are all that are required with this information to find your satellite. You do *not* need a computer to use most satellites. The 3.780 MHz net on HF always give pass times for the MIR and Russian satellites.

	Periods of time between passes	Longitude increment
OSCAR 11	98.5 minutes	24.50° longitude
RS12	104.7 minutes	26.20° longitude
RS15	127.0 minutes	31.60° longitude
RS16	100.0 minutes	29.00° longitude

Up-to-date information is provided on the many AMSAT-UK nets (VHF and 3.780 MHz). An up to date Orbital Calendar for all current amateur radio satellites is available to members of AMSAT-UK for a small charge which just covers printing costs. This is not available separately to non-members.

Two-line Kepler elements for satellites from NASA can be obtained via the packet networks. Kepler elements in AMSAT form (as used by most computer programmes), are always printed in *OSCAR News* the bi-monthly journal of AMSAT-UK. *OSCAR News* is printed within one week of receipt of the very latest element sets for all satellites currently in orbit.

A full range of PC tracking software is available from [1]. A full range of books on the subject are also available from RSGB and AMSAT-UK. A stamped SAE, plus *two* loose 31p postage stamps or IRCs will bring information, plus an application form and details of membership. A current copy of the latest *Oscar News* is available for £2:00. plus P&P.

References

[1] Contact AMSAT-UK. Ron Broadbent, MBE. G3AAJ, Honorary Secretary, AMSAT-UK, London, E12 5EQ. Tel: (+44 (0)181) 989 6741. Fax: (+44 (0)181) 989 3430. Compuserve ID: 100024,614 or e-mail: g3aaj@amsat.org.

[2] (a) Frequency chart of all amateur satellites. (b) Satellite Information Sheets, double-sided, A4, for each satellite. Complete set, 17 in number includes MIR. PVC Laminated, £20.00 plus postage and packing the set.

[3] Orbital Calendar, bi-monthly service to members only.

Special Event Stations

Do You Really Need a GB Callsign?

From 1 June 1990, all club stations have been able to pass greetings messages sent by a non-licensed third party. This means that applying for a GB callsign no longer holds any advantges!

Provided the club uses the prefix letters (see below), the club station is allowed to pass greetings messages and to operate simultaneously on more than one band. The club prefixes are very distinctive and create interest through their rarity. If a club regularly operates a special event station using the club callsign, this will help increase the club's identity. It will also benefit by being able to print its QSL cards in larger, more economic quantities.

Another advantage is that any suitably licensed and authorised club member may operate the club station. This gives greater flexibility over a GB callsign which is just a variation to an individual's licence.

Best of all, you don't have to fill in any forms or wait 28 days!

Remember when operating away from the Main Address, if you give prior written notice to the RIS office covering your area, you do not have to sign /P or give your location to within 5km.

Old Club Prefix	New Club Prefix
G/M	GX/MX
GD/MD	GT/MT
GI/MI	GN/MN
GJ/MJ	GH/MH

Old Club Prefix	New Club Prefix
GM/MM	GS/MS
GU/MU	GP/MP
GW/MW	GC/MC

GB Callsigns

Since 1983 the RSGB has despatched on behalf of the Radiocommunications Agency the Notices of Variation issued by the Secretary of State for Trade & Industry authorising special event stations. Consequently, all enquiries and correspondence should be addressed to the Society and not to the RA.

The RA has stated that it "... requires the Society when distributing Notices of Variation allowing radio amateurs to set up special event stations, to ensure that the request is for an event which is of special significance and therefore is generally accepted as one requiring celebration, and that the event is open to viewing by members of the public".

Applying for a GB Callsign

No charge is made by the RSGB for a special event callsign, but application forms must be returned at least 28 days prior to the start of the event!

Applications are normally processed shortly after receipt. If nothing has been received 14 days prior to the event, please contact the Amateur Radio Dept at RSGB HQ immediately. *Please note that no authority exists until this notice has been received.*

A letter of variation will only be issued to individuals who hold a current full UK A or B licence (ie not novices). It will be valid for a maximum of 28 consecutive days. Please note that a notice of variation will only be issued for an individual's licence; variations will not be issued on a club licence.

The station may only be established and operated at one specified location. This must be the address stated on the application form which must be detailed enough for anyone to find easily. Operation of a special event station from a licensee's home address is not normally permitted.

Only the person responsible for the station need sign the form, as the authorisation is by notice of variation to that individual's licence. This person is required to be present to supervise the correct operation of the station. Additional operators need only sign and write their callsigns in the log-book.

If you have not used the callsign before, you can avoid last minute disappointment by first contacting RSGB. We can then check that it is available and reserve it for you. A GB callsign may be reserved for up to six months in advance. When a GB callsign has been used it will not normally be re-issued to another amateur for use at a different event for a period of 24 months.

If operation of the special event station on the HF bands is desired, a Class 'A' licence holder must apply for a Class 'A' GB callsign.

Subject to availability special event callsigns are available in the following formats:-

Class A

GB0 + 2 or 3 letters
GB2 + 2 or 3 letters
GB4 + 2 or 3 letters
GB5 + 2 letters
GB6 + 2 letters, GB8 + 2 letters

Class B

GB1 + 2 or 3 letters
GB5 + 3 letters
GB6 + 3 letters
GB8 + 3 letters

Other Callsign Formats

The RSGB is not authorised to issue callsigns in formats different from the above. Applications for special anniversaries which require the numbers such as 25, 40, 50, 60 or 75 must be specially supported by the RSGB to the RA. The Radiocommunications Agency will only consider applications via the RSGB from the headquarters of a nationally based organisation. Applications cannot be entertained from individuals, local clubs or branches of nationally based organisations.

Greetings Messages

The guidelines agreed with the RA are:

1. Each greetings message should not exceed 2 minutes.
2. Each person may pass only one message to each station with which the originating station is in contact.
3. A non-licensed person may speak into the microphone, but the licensed radio amateur must identify the station and operate the transmitter controls at all times.
4. Greetings messages by third parties may only be sent from and received by stations within the UK or the USA, Canada, Falkland Is and Pitcairn Is. The licensee may exchange greetings as in any QSO, with any station.

Charitable Events

It is recognised that some special event stations will be established at certain charitable events where a major concern will be the raising of funds.

The RA has agreed that the charity (if one is involved) or the reason for establishing the special event station may be mentioned " on-air " provided that under *no* circumstances may a donation be requested during the contact, and sending of QSL cards must *not* be conditional upon the pledge of a donation. It is in the interests of everyone who holds a special event station licence that operators keep within the spirit of this by not asking for any money over the air.

The station may be sponsored per contact, i.e., the licensee may in advance of the event seek from his friends and relatives sponsorship assurances under the usual arrangements for sponsorship. You must *not* seek sponsors "on-air" at any time.

QSL Information

Special event stations generate many QSL cards; it is therefore important that you use the QSL Bureau correctly.

Please choose QSL cards which do not exceed normal post-card size, viz 5½" × 3½". As packets going abroad are sent open-ended at printed paper rate, large cards invariably have to be folded, whilst small ones or thin ones are difficult to handle.

Print the addressee's callsign on both sides of the cards, together with the details of his/her QSL manager if applicable. Sort USA cards into call areas and all others alphabetically by prefix. Do not space the cards with paper markers, etc. Please pack all cards the same way up and ensure they are adequately packed with the correct postage prepaid and send them to:-

RSGB QSL Bureau
PO Box 1773
POTTERS BAR
Herts.
EN6 3EP

Please supply the GB QSL Submanager with stamped addressed envelopes no larger than 7½" x 5" and of strong material, as soon as you receive your letter of variation. Print the Special Event Station Callsign and date in the top left of each envelope and send them to the appropriate sub-manager:-

For SES c/s:
GB?AAA - GB?MZZ
GB?AA - GB?MZ

Grenville Whaling, G0PPR
GB QSL Submanager
32 The Croft
Little Snoring
FAKENHAM
Norfolk
NR21 0JS

For SES c/s :
GB?NAA - GB?ZZZ
GB?NA - GB?ZZ

Alex Devereaux, G0TTZ
GB QSL Submanager
39 Lower Green Road
Rusthall
TUNBRIDGE WELLS
Kent
TN4 8TW

Cards for amateurs who have neglected to send envelopes are retained for three months, after which the cards have to be destroyed for reasons of space. Amateurs who do not wish to collect cards should notify the QSL bureau.

Application for a Special Event Station

Please Return at least 28 Days Prior to the Event

PLEASE READ THE NOTES ACCOMPANYING AND THEN COMPLETE THIS FORM CLEARLY IN BLOCK CAPITALS

Event name ....

Brief details of event and its nature ....

....

....

Start date for letter of variation: .... Duration of licence: .... days (Maximum of 28 days)

Callsign preferred	**Callsign stands for**
1st Choice	....
2nd Choice	....
3rd Choice	....

OFFICE USE ONLY

Call-sign: ....

SES No: ....

Date: ....

From where will the special event station be operated?

....

....

....

.... Post code ....

If no postal address, please give full national grid reference ....

For inclusion in *RadCom* if received in time:

Actual start date of event: ....

What frequency bands/modes do you propose using? (tick boxes only)

160m	❑	30 - 10m	❑	2m	❑	Packet	❑
80/40m	❑	6/4m	❑	70cm	❑	Satellite	❑

Do you want the QSL Submanager to keep cards for this station? (If yes, please supply him with envelopes.) **YES / NO** *Outgoing cards must still be sent to the RSGB QSL Bureau (see notes).*

To whom and where should the letter of variation be sent?

Name: Mr/Mrs/Miss .... Callsign ....

....

....

.... Post code ....

Telephone: Home .... Work .... Ext ....

N.B. In signing this form, you are confirming that your Special Event Station will operate in accordance with the following conditions laid down by the RA:–

"No mobile or maritime mobile operation is permitted."

"The RA requires the Society, when distributing Notices of Variation allowing radio amateurs to set up special event stations, to ensure that the request is for an event which is of special significance and therefore is generally accepted as one requiring celebration, and that the event is open to viewing by members of the public."

Declaration: (To be signed by the person responsible for the callsign and station).

I, the undersigned, understand that failure to complete this form accurately and properly may invalidate my application for a special event station. I also understand that the information on this form may be freely published.

Name: Mr/Mrs/Miss .... Callsign ....

Signature .... Date ....

Please post (fax not acceptable) the completed form at least 4 weeks before the start of the event to:–
Amateur Radio Department, RSGB, Lambda Ho, Cranborne Rd, Potters Bar, Herts. EN6 3JE. Tel: (01707) 659015.

JOTA and Radio Scouting

Jamboree on the Air (JOTA) is an annual event designed to allow Scouts to send greetings messages to each other. Started in 1957, it now involves approximately 600,000 Scouts and Guides, with the help of over 23,000 radio amateurs in over 100 countries.

JOTA takes place on the third full weekend of October each year, officially between 00.00 Saturday and 24.00 Sunday, although most stations run for a period within these hours, to suit their own requirements. The event is organised by the Scout movement, supported by radio amateurs or clubs. Their aim is to bring Scouts around the world closer together, and to introduce them to the capabilities of amateur radio.

All amateur bands are used. Most stations use a special event or a club call, allowing the Scouts to pass greetings messages over the air. A bulletin of information about the current Jamboree on the Air is sent out with the special event Notice of Variation, with last minute information being sent to the same address in the beginning of October. Club stations should request a copy of these bulletins fromh Gilwell Park.

The interest fostered by JOTA and World Jamboree has spread and many Scout camps and campsites boast amateur radio facilities. A number of proficiency badges in the radio, electronics and computer fields are available for Scouts. Several countries have permanent Scout Headquarters stations – for example the World Scout Bureau in Geneva has the call sign HB9S and Gilwell Park in the UK operates under the call sign GB2GP.

Many countries run periodic Scout nets. There are regular weekly UK and European nets aimed at Scouters who are also radio amateurs.

Factsheets and further information about JOTA and other Scout Radio and electronics activities are available from: The Scout Association, Gilwell Park, Chingford, London, E4 7QW. Tel: (0181) 524 5246.

USUAL SCOUT NET FREQUENCIES:

Band	SSB (Phone)	CW
80m	3.740 and 3.940	3.590
40m	7.090	7.030
20m	14.290	14.070
17m	18.140	18.080
15m	21.360	21.140
12m	24.960	24.910
10m	28.990	28.190

UK Scout net is on Saturdays, 3.740 ± MHz at 09:00 local time.
European Scout net is on Saturdays, 14.290 MHz at 0930 GMT.

Thinking Day On The Air

Thinking Day On The Air (TDOTA) is organised by The Guide Association on the weekend nearest to Thinking Day - February 22nd. In 1998 this will be February 21-22th. February 22nd was the birthday of the founder of the Guide Movement, Lord Baden-Powell, and of his wife, Lady Baden-Powell, the World Chief Guide. On Thinking Day members all over the world think about the international nature of Guiding.

The aim of TDOTA is to encourage the girls to make Guiding friendships with members of other units, and to introduce them to amateur radio. Station organisers are asked to keep these objectives in mind. We rely heavily on the goodwill of radio amateurs in setting up stations, though the number of Guiders, Rangers and even Guides with callsigns is increasing. In 1997, at least 36 UK stations operated with 2482 members (Rainbows, Brownies, Guides, Rangers, Young Leaders and Guiders). Some Scouts also visit stations or have their own. Countries outside the UK are beginning to participate, especially Canada with about 100 stations.

Stations complete a report form which is sent to the Guide Headquarters. A report is compiled from information received which is produced by The Guide Association. Copies of the current report are available from The Guide Association on receipt of an A4 SAE.

Guiders interested in setting up a station can obtain a comprehensive information pack which includes the last report, a blank report form, a brief description of TDOTA, a detailed description of how to organise a station, a sheet of 'Guides On The Air' logos and a certificate for organisers to copy and issue to participants. Further information is published from time to time in The Guide Association's magazines.

Guides On The Air activities are increasingly happening at other times, with radio stations at camps, activity days and leader training days.

There is a radio communications badge in both the Brownie and Guide sections. It is hoped that Brownies and Guides will be able to use TDOTA to work towards this badge, though some may prefer to complete the badge at other times.

The Guide Association fully supports Project YEAR and the Novice Licence. A holder of the Guide badge will have a good basis on which she can build during a Novice Licence course.

Further information can be obtained on receipt of an A4 SAE from:

The Youth Programme Team
The Guide Association
17-19 Buckingham Palace Road
London, SW1W 0PT
Tel: (0171) 834 6242

Permanent (Special Event) Callsigns

The following GB callsigns are all permanent stations, as opposed to the more usual 28 day ones. Application for a new permanent SES station should be made to the Radiocommunications Agency enclosing proof that the station is on show to the general public throughout the year and details of the call-sign of the club station. The RA liaise with the RSGB to issue successful applicants with the appropriate Notice of Variation to their club station.

GB0IBC	RAIBC (G4IBC) - c/o Brig. Johnny Clinch, G3MJK, The Pippins, Dummer Rd, Axford, Basingstoke, Hants. RG25 2ED.
GB0SUB	Submarine ARC (G7DOL), Submarine Museum, Haslar Jetty Road, Gosport, Hants PO12 2AS.
GB1IBC	RAIBC (G4IBC) VHF net controls - c/o Brig. Johnny Clinch, G3MJK, The Pippins, Dummer Road, Axford, Basingstoke, Hants. RG25 2ED.
GB2AIR	South Yorkshire Aircraft Museum, Home Farm, Firbeck, nr Worksop, S Yorks, S81 8JR.
GB2ATG	BARTG (G4ATG) RTTY News Service.
GB2BP	Bletchley Park Museum, Bletchley Park, Milton Keynes, MK6 3LF.
GB2CPM	Chalk Pits Museum, Houghton Bridge, Arundel, W Sussex.
GB2CW	RSGB Morse Practice Service.
GB2GM	Guglielmo Marconi, Poldhu ARC (G0PZE), c/o Poldhu Cove, Muillion, Cornwall.
GB2GMM	Guglielmo Marconi Memorial, The Needles Pleasure Park, Alum Bay, Isle of Wight, PO39 0JD.
GB2GP	Gilwell Park Scout Campsite, Gilwell Park, Chingford, London, E4 7QW.
GB2IW	Monitoring Service, RSGB HQ.
GB2IWM	Imperial War Museum, Duxford Airfield, Cambs.
GB2MC	The Muckleburgh Collection, Weybourne Military Camp, Weybourne, Norfolk, NR25 7EG
GB2NLO	Norman Lockyer Observatory, Sidmouth. Details c/o G0AXC
GB2OWM	Orkney Wireless Museum. Kiln Corner, Junction Road, Kirkwall, Orkney, KW15 1LB.
GB2PLY	HMS Plymouth, HMS Plymouth RC (G0SJW), East Float, Birkenhead, Mersdeyside, L41 9DJ
GB2RN	Royal Navy (Crown Use), HMS Belfast, London.
GB2RNR	Royal Naval Reserve, TS Grenville, Sea Cadet Corps, 27 Ferguslie Main Road, Paisley, PA1 2QE.
GB2RS	RSGB News Service, RSGB HQ, Lambda House, Cranborne Road, Potters Bar, Herts, EN6 3JE.
GB2SFL	South Foreland Lighthouse, St Margaret's Bay, Dover, Kent. Details c/o G0UJR.
GB2SM	Science Museum, South Kensington, London SW7 *(Dormant).*
GB2VHF	Special Activities Station, RSGB HQ, Lambda House, Cranborne Road, Potters Bar, Herts, EN6 3JE.
GB3RN	Royal Naval ARS, HMS Collingwood, Newgate Lane, Fareham, Hants, PO14 1AS.
GB3RS	RSGB Headquarters Station. Lambda House, Cranborne Road, Potters Bar, Herts, EN6 3JE.
GB3WM	National Wireless Museum, Arreton Manor, Newport, Isle of Wight. Details from G3KPO.
GB4HMS	HMS Warrior, Portsmouth.
GB4RS	RSGB (reserved for Presidential Installations), RSGB HQ.

STELAR

STELAR, RS95685 (Science and Technology Through Educational Links with Amateur Radio) was formed in 1993 to coordinate and promote amateur radio activity in schools and colleges. It publishes a termly magazine *AMRED* (Amateur Radio in Education) which is sent three times a year to affiliated individuals and institutions. During school term time, a weekly Wednesday Stelar Schools & Colleges net is held at 1.00 pm on 3.770 MHz – net control is GB2SR.

One of its main activities is to increase the number of schools which have clubs training students for the RAE/NRAE by running a free residential 'Crash RAE course' for 20 teachers at Easter each year (open to any school who has no licensed teacher).

Further details and an affiliation form may be obtained from the Chairman, Richard Horton, G3XWH at: STELAR, 7 Carlton Road, Harrogate, North Yorkshire, HG2 8DD. Tel/fax: (01423) 871027. Packet: G3XWH@GB7CYM. E-mail: G3XWH@amsat.org.

The following is a list of affiliated schools/colleges as of July 1997.

Aberdeen Grammar School, Skene Street, Aberdeen, AB9 1HT. Mr A W Craig.

Ailwyn Community School, G7TXA, Hollow Lane, Ramsey, Cambs., PE17 1DG. Mr G Siddle, G7TBX

Alton College, Old Odiham Road, Alton, Hants, GU34 2LX, Mr K Baddeley.

Ansty Junior School, Eastbrook Road, Alton, Hants, GU34 2DR. Mr M R Holland.

Antrim Grammar School, Steeple Road, Antrim, BT41 1AF. Mr V W Hughes.

Barlborough Hall School, G0OBR, Balborough, Chesterfield, S43 4TJ. Mr D Wilson, G0DAG

Barnfields Primary School, G4LQT/G6BQW, Lansdowne Way, Wildwood, Stafford, ST17 4RD. Mr D F Morton, G4LQT

Bedford Modern School, G7BMS, Manton Lane, Bedford, MK41 7NT. Mr N E Kinselley, G1BYT

Belfast Royal Academy, GI0WOO, Cliftonville Road, Belfast, BT14 6JL. Mr N E Moore, GI7CMC

Benenden School, Cranbrook, Kent, TN17 4AA. Mr A K Keir, G4KZO

Blairgowrie High School, Beeches Road, Blairgowrie, Perthshire, PH10 6PW. Mr W Henderson, GM0VIT

Bolton School, GX0VUX, Corley New Road, Bolton, BL1 4PA. Mr C J Walker, G0BGQ

Borden School, G4LBS, Sittingbourne, Kent, ME10 4DB. Mr J T MacRae, G4DXI

Britannia High School, Britannia Road, Rowley Regis, Warley, Sandwell, B65 9NF. Mr C R Margetts.

Broughton Hall High School, Yew Tree Lane, Liverpool, L12 9HJ. Mr J R Mellor.

Brunel College of Technology, G5FS, Ashley Down, Bristol, BS7 9BU. Mr I Anderson.

Caistor Grammar School, GX7TWV, 2 Church Street, Caistor, Lincoln, LN7 6QJ. Mr K Liddle, G7TRG

Callington Community School, G1XIC, Launceston Road, Callington, Cornwall, PL17 7BT. Mr K C Harris, G1FMU

Cardinal Wiseman RC School, G1FEO, Potters Green Road, Coventry, CV33 9TH, Mrs G Jones, G1FEO

Castle House School, GX0OEN, Chetwynd End, Newport, Shropshire, TF10 7JE. Mr L R Davies, G3MVK

Christ's College, 4 St Germans Place, Blackheath, London, SE3 0NJ. Mr R Bellerby, G3ZYE

Churchdown School, Winston Road, Churchdown, Glos., GL3 2RB. Mr J R Kirkman.

City of Bristol College, Bristol, BS7 9BU. Mr I Anderson, G7VVO

Clayton High School, Clayton Lane, Clayton, Newcastle-under-Lyme, ST5 3DN. Mr P Jones.

Colne Park High School, G4CPS, Venables Avenue, Colne, Lancs, BB8 7DP. Mr M Crawshaw, G4BLH

Costessey High School, Richmond Road, Norwich, NR14 7ST. Mr M Hawkes.

Cranleigh School, Horseshoe Lane, Cranleigh, Surrey, GU6 8QQ, Dr D Morrison-Smith.

Cromer High School, G0WPS/M1ABO, Norwich Road, Cromer, Norfolk, NR27 0EX. Mr S M Rutter, G7VAH

De Montfort University, Polhill Avenue, Bedford, Beds, MK41 9AE. Mr N Ash, G7ASH

Dean Close School, Shelburne Road, Cheltenham, Glos., GL51 6HE. Mr K F Downes.

Deben High School, Garrison Lane, Felixtowe, Suffolk, IP11 7RF, Miss Naomi Palmer.

Debenham High School, Gracechurch Street, Debenham, Stowmarket, IP14 6BL. Mr P C Lincoln.

Denton Park Middle School, G0IOQ, Linhope Road, West Denton, Newcastle uopn Tyne, NE5 2NN. Mr T Cull, G1SUM

Devonshire House School, Hampstead, G4KMM, Flat 4, 12 Lymington Road, West Hampstead, London, NW6 1HY. Mr P R Northmore, G4KMM

Dollar Academy, GS0SNG, Dollar, Scotland, FK14 7DU. Mr G L Collier, GM0LOD

Dragon School, Bardwell Road, Oxford, OX2 6SS. Mr J Reece.

Dunfermline High School, St Leonards Place, Dunfermline, Fife, KY11 3B6. Mr G Winchester.

Eton College, MX1BEC, 13A Brocas Street, Eton, Windsor, Berkshire, SL4 6BW. Mr T Woods, M1BDH

Farlingaye High School, MX0ADM, Ransom Road, Woodbridge, Suffolk, IP12 4JX. Mr M B Brown, G7SQE@GB7CFB

Filton High School, New Road, Stoke Gifford, Bristol, BS12 6QT. Mr N J Pratt.

Fitzalan High School , FAO CDT Faculty, Lawrenny Avenue, Leckwith, Cardiff, CF1 8XB. Mr J Woodland.

Gresham's School, G3PXO, Cromer Road, Holt, Norfolk, NR25 6EA, Rev R N Myerscough, G0SIQ

Harrogate Ladies' College, G0HCA, Clarence Drive, Harrogate, N Yorks, HG1 2QG. Mr R Horton, G3XWH

Heather Grove First School, Haworth Road, Heaton, Bradford, BD9 6LP. Mr C Shackleton, G7UST

Hewett School, G0REG, Cecil Road, Norwich, NR1 2PL. Mr A Wright, G0KRU

Hewitt School, Cecil Road, Norwich. Mr T Benson.

Hillcrest School & Community College, GX0SPM, Simms Lane, Netherton, Dudley, West Midlands, DY2 0PB. Mr K Sherman, G0IZF

Holsworthy Community College, Victoria Hill, Holsworthy, EX22 6JD. Mr G B Forster.

Itchen College, Middle Road, Bitterne, Southampton, SO19 7TB. Mr D N Batten.

John Hanson Community School, GX0RBL, Croye Close, Andover, Hants, SP10 3AB. Mr M D Adams, G0AMO

John Spendluffe School, Hanby Lane, Alford, Lincs, LN13 9BL. Mr K Boyd.

Joseph Eastham High School, Hilton Lane, Worsley, Manchester, M38 0SY.

Kelly College, GX4COF, Amateur Radio Society, Tavistock, Devon, PL19 0HX. Mr P S Marshall.

Kesgrave High School, Ipswich, IP5 7PB. Mr C Bennett.

Kilve Court Centre, GB2KSS/GB2KRC, 80A, Preston Grove, Yeovil, Somerset, BA20 2DA. Mr A C Dening, G4JBH

King Aldred's School, West Challow Road, Wantage, OX12 9BU. Mr G Offler.

King Edward VI School, Biry St Edmunds, IP33 3BH. Mr M Hampshire.

King Edward VII High School, Gaywood Road, King's Lynn, Norfolk, PE30 2QB, Ms E Maun.

King Edward's School, GX4SKE, GX8ZKE, Edgbaston Park Road, Birmingham, B15 2UA. Mr D C R Rigby, G4KXV

Kingsbury High School, Princes Avenue, London, NW9 9JR. Mr N D Purchon, G0WKM

Lancing College, G0VSI, Sussex, BN15 0RW, Dr A J Betts, G0VLC

Landau Forte College, GX0TLF, Fox Street, Derby, DE1 2LF. Mr J Hackett, G0OAF

Langholm Academy, Langholm, Dumfrieshire, DG13 0BG. Mr R B McCartney, GM4BDJ

Le Retraite School, Campbell Road, Salisbury, Wilts, SP1 3BQ. Mr N Evans, G7RYG

Leiston High School, Seaward Avenue, Leiston, Suffolk, IP16 4BG. Mr J Heald, G0UEA

Lochgelly High School, Station Raod, Lochgelly, Fife, KY5 8LZ. Mr I G Wilkinson.

Looe School, G6RLS, Sunrising, East Looe, Cornwall, PL13 1NQ. Mr J A Bond, G1ZHH

Ludford CE Primary School, Magna Mile, Ludford, Lincolnshire, LN3 6AD. Mr G A Tuckett.

Maiden Erlegh School, 22 St Georges Road, Aldershot, Hants, Berks, RG45 7HR. Mr D E Kearns, G0HVS

Makini Academy, P O Box 21784, Nairobi, Kenya. Mr D A M Mado.

Manchester Grammar School, Old Hall Lane, Manchester, M13 0XT. Mr M Burbidge.

Mariengymnasium Warendorf, DF0STT, von Ketteler Str 15, Warendorf, Germany, 48231. Mr B Meisel, DF8YM

Marlborough College, Hawkwind, Elcot Lane, Marlborough, Wilts, SN8 2AZ. Mr P A Knight, G6EPN@GB7SDN

Marshland High School, G0WCX, School Road, West Walton, Wisbech, Cambs, PE14 7HA. Mr P J Wells, G4RKN

Mary Hare Grammar School, G0UMH, G7MHS, Arlington Manor, Snelsmore Common, Newbury, Berks, RG16 9BQ. Mr M R Vaslet, G4JAL @ GB7APC

North Bolton Sixth Form College, Smithills Dean Road, Bolton, BL1 6JT. Mr M T Callaghan.

Oakbank School, G0WNE, Oakworth Road, Keighley, West Yorkshire, BD22 7DU. Mr K S G Allack, G0WGC

Otley Amateur Radio Society, G3XNO, 35 Westwood Avenue, Bradford, BD2 2NJ. Mr J Worsnop, G0SNV

Oulder Hill Community School, GX0UQA, G7RQL, Hudsons Walk, Rochdale, OL11 5EF. Mr C Jenkins, G0VRX

Oundle School, Peterborough, PE8 4BN. Mr R Field.

Preston School, G0PCS, Monks Dale, Yeovil, Somerset, BA21 3JD. Mr A C Douglas, G0HDJ

Queen Elizabeth Girls' School, Barnet High Street, Barnet, N6 5HF. Mr N Parmar.

Queensbury School, Langdale Road, Dunstable, Beds, LU6 3BU. Mr A A Brown.

Raddery School, Fortrose, Ross-shire, Scotland, IV10 8SN. Mr A P Thornton.

Richmond School, GX0RYS, Darlington Road, Richmond, N Yorks, DL10 7BQ. Mr M S Vann, G3RLV

Rickstones School & College, Conrad Road, Witham, Essex, CM8 2SD. Mr J H Giles.

Ridge Danyers College, G0VRC, Hibbert Lane, Marple, Stockport, Cheshire, SK6 7PA. Mr I R Smith.

Rishworth School, GX0SQA, Rishworth, Sowerby Bridge, W Yorks, HX6 4QA. Mr A E Vinters, G0WFG

Royal Grammar School Newcastle, Eskdale Terrace, Newcastle-upon-Tyne, NE2 4DX. Mr A N Baker, G7UHV

Rushey Mead School, G4RMS, Melton Road, Leicester, LE4 7PA. Mr G W Dover, G4AFJ@GB7AYI

Ryelands Middle School, Prestbury Road, Duston, Northampton, NN5 6XP. Mr G K Fletcher, G7TEG

Scarborough College, G3LCG, Filey Road, Scarborough, N Yorks, YO11 3BA. Mr G P Bateman, G3LCG

Schlossberg Schule, D77876 Kappelrudeck, Germany, 77876. Mr P Gutmann, DK1SK

Sir William Robertson School, G4WRS, Welbourne, Lincoln, LN5 0PA. Mr A R Kiddle, G4HVC

Sir William Romney's School, Low Field Road, Tetbury, GL8 8AE. Mr A Wilkes.

St Columba's RC Boys' School, Halcot Avenue, Bexleyheath, Kent, DA6 7QB. Mr D Legg.

St Lawrence College, Ramsgate, Kent, CT11 7AE. Mr P Couch, G8UHJ

Stanchester Community School, GX7LBV, Stoke Sub Hamdon, Somerset, TA14 6UG. Mr G Middleton, G6EER

Strathallan School, GM0PSS/GM7NSS, Forgandenny, Perth, PH2 9EG. Mr T S Goody, GM0MXZ

Stromness Academy, Garson, Stromness, Orkney, KW16 3AW. Mr J S Grieve, GM0HTH

Swakeleys School, Clifton Gardens, Hillingdon, Uxbridge, TW17 0AJ. Mr I R Hogg.

Swindon College, 2E1EQI, Northstar Avenue, Swindon, Wiltshire, SN2 1DY. Mr R D Collins, G8BXG

The Bishop of Herford's Bluecoat School, Hampton Dean Road, Tupsley, Hereford, HR1 1UU. Mr N S J Hancocks, G4XTF

The Bishop Stopford's School, 18 Mannock Road, Wood Green, London, N22 6AA, Mrs M L Craine, G0UMF

The Hollins County High School, GW4CCS, Hollins Lane, Accrington, Lancs, BB5 2QY. Mr R Hewson, G4KFF

The James Young High School, Quentin Rise, Dedridge, Livingston, West Lothian, EH54 6NS. Mr N C Bethune, GM4IUS

The Royal Grammar School Guildford, G7BAI, High Street, Guildford, Surrey, GU1 3BB. Mr F Bell, G7CND

The Royal School, Farnham Lane, Haslemere, Surrey, GU27 1HQ, Mrs R Blackburn.

Trinity School, Springfields, 12 Ranmore Avenue, Croydon, CR0 5QA. Mr M J Buckley.

University of Bradford, Dept. of Electronic & Electrical Engineering, Bradford, BD7 1DP, Dr P S Excell.

University of Hertfordshire, G4WTB G6BOB, College Lane, Hatfield, Herts, AL10 9AB. Mr D M Lauder, G0SNO

University of Plymouth, G0UOP, Sec & EE, The Smeaton Building, Drake Circus, Plymouth, PL4 8AA. Mr A Santillo.

University of Wales Swansea, GW3UWS, Dept. of Electrical Engineering, Singleton Park, Swansea, SA2 8PP. Mr M T Bowen.

Ursuline College, GX7UPG, Westgate on Sea, Kent, CT8 8LX. Mr J D Hislop, G7OHO

Wigan Deanery High School, G0TWD, Frog Lane, Wigan, Lancs, WN1 1HQ. Mr J M Drummond, G0RGO

Wilnecote High School, G0EVJ, B77 5LF. Mr S Evans, G0EVJ

Traders

This list details some of the many retailers of amateur radio related equipment in the UK and overseas. We are indebted to Malcolm Taylor Associates for supplying this list and to whom any additions, corrections or amendments should be notified. No significance should be attached to the inclusion or omission of any trader in this list.

Abacus Polar Electronics
Amherst House, 22 London Road, Riverhead, Sevenoaks, Kent, TN13 2BT.

Absolute Value Systems
115 Stedman Street, Chelmsford, MA 01824-1823, USA.
Tel: 001 5082566907.

Adam Bede High Tech Centre
Derby Road, Wirksworth, Derbyshire, DE4 4BG.
Tel: 01629 825926.

Adur Village Press
Alamosa, The Paddocks, Upper Beeding, Steyning, West Sussex, BN44 3JW.
Tel: 01903 879750.
Fax: 01903 814594.

Aerial Techniques
11 Kent Road, Parkstone, Dorset, BH12 2EH.
Tel: 01202 738232.
Fax: 01202 716951.

AKD
Unit 5, Parsons Green Estate, Boulton Road, Stevenage, Herts, SG1 4QG.
Tel: 01438 351710.
Fax: 01438 357591.

Alan Hooker Radio Communications
42 Nether Hall Road, Doncaster, DN1 2PZ.
Tel: 01302 325690.
Fax: 01302 325690.

Allsopp Helikites
Chestnut Lodge, Chalford, Stroud, Glos, GL6 8NW
Tel: 01453 886515.

Alpha Delta Communications Inc.
PO Box 620, Manchester, KY 40962, USA
Tel: 606-598-2029.
Fax: 606-598-4413.

Alpha operations of ETO, Inc.
4975 North 30th Street, Colorado Springs, CO 80919, USA
Tel: 719 260-1191.
Fax: 719 260-0395.

Altron Communications
Unit 1, Plot 20, Cross Hands Business Park, Cross Hands, Dyfed, South Wales, SA14 6RE
Tel: 01269 831431.
Fax: 01269 845348.

Amateur Radio Communications Ltd
38 Bridge Street, Earleston, Newton-le-Willows, Merseyside, WA12 9BA
Tel: 01925 229881.
Fax: 01925 229882.

Ameritron
116 Willow Road, Starkville, MS 39759, USA
Tel: 601 323-8211.
Fax: 601 323-6551.

Amstrutt
Unit 10, Pixon Trading Estate, Tavistock, Devon, PL19 8DJ.

Ant-Ventures
P.O. Box 776, McHenry, IL 60051-0776, USA
Tel: 001 8153441702.

Antex (Electronics) Ltd.
2 Westbridge Industrial Estate, Tavistock, Devon, PL19 8DE
Tel: 01822 613565.
Fax: 01822 617598.

AOR (UK) Ltd.
Adam Bede High Tech. Centre, Derby Road, Wirksworth, Derbyshire, DE4 4BG
Tel: 01629 825926.
Fax: 01629 825927.

Applied Tecnologies & Mf
The Old Mill, Tyne & Wear.

ARE Electronic Engineering
Zwette 7, NL8032 XL Zwolle, The Netherlands

ART Creative Partnership
2nd floor,28-40 Stonehills, Welwyn Garden City, Herts, AL8 6PD
Tel: 01701 396 300.
Fax: 01707 396 888.

ASK Electronics Ltd
248/250 Tottenham Court Road, London, W1P 9AD
Tel: 0171 637 0353.
Fax: 0171 637 2690.

Atlanta Communications
PO Box 5, Chatteris, Cambridgeshire, PE16 6JT
Tel: 01802 422891.
Fax: 01394 460995.

Audio Engineering
3rd Floor Fitzroy House, Abbot Street, London, E8 3LP
Tel: 0171 254 5475.
Fax: 0171 249 0347.

Audon Electronics
36 Attenborough Lane, Chilwell, Nottingham, NG9 5JW
Tel: 01602 259737.

Austin Knight Ltd
20 Soho Square, London, W1A 1DS
Tel: 0161 834 8723

Autogard (UK) Ltd
Unit 3, Acorn Firms Centre, Ablewell Street, Walsall, West Midlands, WS1 2EG
Tel: 01922 613654

Azden
147 New Hyde Park Road, Franklin Square, New York 11010, USA
Tel: 516 328-7501.
Fax: 516 328-7506.

Bangle, M
Badgers End, Addlestone, Surrey, KT25 1DH
Tel: 01932 846139

Barnet Metal & Engin'g Co. Ltd
Stirling Works, Tewin Road, Welwyn Garden City, Herts, AL7 1AG
Tel: 01707 324327.
Fax: 01707 371375.

B. Bamber Electronics
5 Station Road, Littleport, Cambs, CB6 1QE
Tel: 01353 860185

Barton Communications
Barton Park, Richmond, North Yorks, DL10 1LH.
Tel: 01325 377086.
Fax: 01325 377086.

Beezley, Brian, K6STI
3532 Linda Vista, San Marcos, CA 92069, USA
Tel: 619 599-4962.

Bencher, Inc.
831 N. Central Avenue, Wood Dale, IL 60191, USA
Tel: 708 238-1183.
Fax: 708 238-1186.

Benham-Holman, R (G2DYM)
Cobhamden, Uplowman, nr Tiverton, Devonshire, EX16 7PH
Tel: 01398 361215.

B. K. Electronics
Units 1 & 5, Comet Way, Southend-on-Sea, Essex, SS2 6TR
Tel: 01702 527572.
Fax: 01702 420243.

Billington Export Ltd.
Unit 1E Gillmans Ind. Estate, Billingshurst, Sussex, RH14 9EZ
Tel: 01403 784961.
Fax: 01403 783519.

Birkett, J
25 The Strait, Lincoln, LN2 1JF
Tel: 01522 520767

BMK Communications
2 Beacon Close, Seaford, East Sussex, BN25 2JZ

Brandon, P
1 Woodlands Rd, Chester, CH4 8LB
Tel: 01244-683563.

Bring & Buy Database, The
23 North End, Meldreth, Royston, Herts, SG8 6NR
Tel: 01763 262443.
Fax: 01763 262443.

Bull Electrical
250 Portland Road, Hove, Sussex, BN3 5QT
Tel: 01273 203500.
Fax: 01273 323077.

Cambridge Kits
45 Old School Lane, Milton, Cambridge

Castle Electronics
Unit 20, Bobbington nr Stourbridge, West Midlands, DY7 5DY
Tel: 01384 221036.
Fax: 01384 221037.

Caswell Press
11 Baron's Way, Woodhatch, Surrey
Tel: 01737 244916

Cedar Electronics
12 Isbourne Way, Winchcombe, Glos
Tel: 01242 602402

Chelcom Aerials
Riverside House, Homecroft Drive, Cheltenham, Gloucestershire, GL51 9SN
Tel: 01242 680653.
Fax: 01242 680653.

Chevet Supplies Ltd
(Dept. RC), Blackpool, Lancs, FY1 2EU
Tel: 01253 751858
Fax: 01253 302979.

Choice PCB
31 Campbell Road, Caterham, Surrey, CR3 5JP
Tel: 01883 343838.

Cirkit Distribution Ltd.
Park Lane, Broxbourne, Herts., EN10 7NQ
Tel: 01992 448899
Fax: 01992 471314.

Cliveden Recruitment plc
92 Broadway, Bracknell, Berks, RG12 1AR
Tel: 01344 489 489
Fax: 01344 489505.

Coastal Communications
19 Cambridge Road, Clacton-on-Sea, Essex, CO15 3QJ
Tel: 01255 474292

Cole, D. (G3RCQ)
9 Troopers Drive, Romford, Essex, RM3 9DE

Collings, Simon
48 St Michaels Road, Cheltenham, Gloucs, GL51 5RR
Tel: 01242 514429

Colomor (Electronics) Ltd
170 Goldhawk Road, London, W12 8HJ
Tel: 0181 743 0899
Fax: 0181 7493934.

Comet
1275 N. Grove Street, Anaheim, California 92806, USA
Tel: 714 630-4541
Fax: 714 630-7024.

Cooper (Mrs)
Mill Lodge Guest House, Gatwick, London
Tel: 01293 771170

Cricklewood Electronics
40 Cricklewood Broadway, London, NW2 3ET
Tel: 0181 4520161
Fax: 0181 2081441.

Cushcraft Corporation
48 Perimeter Road, Manchester N.H 03108, USA
Tel: 603 627 7877
Fax: 603 627 1764.

Datong Electronics Ltd
Clayton Wood Close, West Park, Leeds, Yorks, LS16 6QE
Tel: 0113 2744822.

DBA Computer Aided Technology
P.O. Box 18285, Shreveport, LA 71138, USA

Dee Comm
Unit 1a, Canal View Industrial Estate, Brierley Hill, West Midlands, DY5 3LQ
Tel: 01384 480565.

Delta
19 Ffordd-y-Morfa, Cross Hands, Carms, SA14 6SL
Tel: 01269 843639
Fax: 01269 843639.

Display Electronics
32 Biggin Way, Upper Norwood, London, SE19 3XF
Tel: 0181 6794414
Fax: 0181 6791927.

Eastern Communications
Cavendish House, Happisburgh, Norfolk, NR12 0RU
Tel: 01692 650077
Fax: 01692 650925.

Economic Devices
32 Temple Street, Wolverhampton, WV2 4AN
Tel: 01902 773122
Fax: 01902 29052.

Eden, John
Dunnet Head Food Inn B & B, Caithness, Scotland, KW14 8YE
Tel: 01847 851774

Electromail
P.O. Box 33, Corby, Northants, NN17 9EL
Tel: 01536 204555.
Fax: 01536 405555.

Electronic Shop, The
29 Hanging Ditch, Manchester, M4 3ES
Tel: 0161 8341185.

Eltac Ltd
77-79 Bath Road, Thatcham, Berks, RG18 3BG
Tel: 01635 868584, Fax: 01635 868584.

EMA Electronics
69 Gorse Cover Road, Severnbeach, Bristol, BS12 3NR
Tel: 01454 633524

EPS Ltd
EPS Technology Centre, Chipping Warden, Oxon Banbury, OX17 1LL
Tel: 01295 660066
Fax: 01295 660073.

ERA
26 Clarendon Court, Winwick Quay, Warrington, WA2 8QP
Tel: 01925 573118.

ESR Electronic Components
Station Road, Cullercoats, Tyne & Wear, NE30 4PQ
Tel: 0191 2514363
Fax: 0191 2522296.

Essex Amateur Radio Services
4 Northern Avenue, Benfleet, Essex, SS7 5SN
Tel: 01268 752522

FBS Ltd
21 Halford Road, Ettington, CV37 7TH
Tel: 01789 740073. Fax: 01892 667473.

Ferromagnetics
PO Box 577, Flintshire, North Wales, CH7 1AH
Tel: 0352 771520.

Force Components Ltd
3 Twyford Business Units, Station Road, Twyford, Berks, RG10 9TU

Frequency Precision Ltd
Shorts, Nothlew, Okehampton, Devon, EX20 3NR

GAP Anteena Products Inc.
6010 N. Old Dixie Highway, Vero Beach, FL 32967, USA
Tel: 561 778-3728.
Fax: 561 778-0717.

Garex Electronics
Unit 8, Sandpiper Court, Harrington Lane, Exeter, Devon, EX4 8NS
Tel: 01392 466899.
Fax: 01392 466887.

Grandata Ltd
K.P House, Unit 15, Pop In Commercial Centre, Southway, Wembley, Middx., HA9 OHB
Tel: 0181 9002329.
Fax: 0181 9036126.

Greenweld
27 Park Road, Southampton, SO1 3TB
Tel: 01703 236363.
Fax: 01703 236307.

Grosvenor Software
2 Beacon Close, Seaford, East Sussex, BN25 2JZ
Tel: 01323 893378

G.W.M. Radio Ltd.
40/42 Portland Road, Worthing, West Sussex, BN11 1QN
Tel: 01903 234897.
Fax: 01903 239050.

G4TNY Amateur Radio
41 Onslow Crescent, Colchester, Essex, CO2 8UN

G4ZPY Paddle Keys
41 Mill Dam Lane, Burscough, Ormskirk, Lancs, L40 7TG
Tel: 01704 894299

Halcyon Electronics
423 Kingston Road, Wimbledon Chase, London, SW20 8JR
Tel: 0181 390 4817

Ham Radio Today
Nexus Special Interests Ltd, Nexus House, Boundary Way, Hemel Hempstead, Herts, HP2 7ST.
Tel: 01442 66551.
Fax: 01442 66998

Hands Electronics
Tegryn, Llanfyrnach, Dyfed, SA35 0BL
Tel: 01239 77427.

Hart Electronic Kits
Penylan Mill, Oswestry, Shropshire, SY10 9AF
Tel: 01691 652894

Haslimann Taylor
Knights House, 2 Parade, Sutton Coldfield, Birmingham, B72 1PD

Hately Antenna Technology
1 Kenfield Place, Aberdeen, AB1 7UW
Tel: 01224 316004

Haydon Communications
132 High Street, Edgware, Middx, HA8 7EL
Tel: 0181 9515781.

HD Promotions
Events House, Wycombe Air Park, Booker, Marlow, Bucks, SL7 3DP

Hedgeways B & B
Highmoor Lane, W. Yorkshire

Hesing Technology
41 Bushmead Road, Eaton Socon, Huntingdon, Cambs, PE19 3BT
Tel: 01480 386156.
Fax: 01480 386157.

Hobbs, D. P. (Norwich) Ltd
13 St. Benedict Street, Norwich, NR2 4PE
Tel: 01603 615786.

Holdings Amateur Electronics
45 Johnston Street, Blackburn, Lancs, BB2 1EF
Tel: 01254 59595

Howes, C. M., Communications
Eydon, Daventry, Northants, NN11 3PT
Tel: 01327 60178

H.S. Wholesale
125-127 Queen Street, Withernsea, HU19 2DJ
Tel: 01964 614439.
Fax: 01964 614439.

Huddersfield Electronics
4a Cross Church Street, Huddersfield, West Yorks, HD1 3PT
Tel: 01484 420774

Icom (UK) Ltd
Sea Street, Herne Bay, Kent, CT6 8LD
Tel: 01227 743000.
Fax: 01227 741742.

ICS Electronics Ltd.
Unit V, Rudford Industrial Estate, Ford, Arundel, West Sussex, BN18 0BD
Tel: 01903 731101.
Fax: 01903 731105.

IFW Technical Services
52 Abingdon Road, Drayton, Abingdon, Oxon, OX14 4HP
Tel: 01253 535981

Index Laboratories
9318 Randall Drive NW, Gig Harbor, WA 98332, USA
Tel: 206 851-5725.
Fax: 206 851-8385.

Infracom
69 Boulevard Albert 1er, 44600 Saint Nazaire, France
Tel: 02 40 70 97 68.
Fax: 02 40 70 98 30.

IQ
16 Park Square, Ossett, West Yorks, WF5 0JF

ISWL
10 Clyde Crescent, Wharton, Winsford, Cheshire, CW17 3LA

JAB Electronic Components
The Industrial Estate, 1180 Aldridge Road, Great Barr, Birmingham, B44 8PE
Tel: 0121 3666928

Jackson Brothers Ltd.
58-72 Dalmain Road, London, SE23 1AX

Javiation
Carlton Works, Carlton Street, Bradford, West Yorkshire, BD7 1DA
Tel: 01274 732146

Jaycee Electronics Ltd
20 Woodside Way, Glenrothes, Fife, Scotland, KY7 5DF
Tel: 01592 756962.
Fax: 01592 610451.

J & P Electronics
Unit 45, Meadowmill Estate, Dixon Street, Kidderminster, DY10 1HH
Tel: 01562 753893

JPS Communications Inc.
PO Box 97757, Raleigh, NC 27624-7757, USA
Tel: 919 790-1048.
Fax: 919 790-1456.

Kalestead
300-302 Cressing Road, Braintree, Essex, CM7 6PF
Tel: 01376 349036. Fax: 01376 348976.

Kanga Products
Seaview House, Crete Road East, Folkestone, Kent, CT18 7EG
Tel: 01303 891106.
Fax: 01303 891106.

Kenwood UK Ltd
Dwight Road, Watford, Herts, WD1 8EB.
Tel: 01923 816444.
Fax: 01923 819131.

Kestrel Network Sciences
Talbot House, High St , Crowthorne, Berks, RG45 7AQ
Tel: 01344 762573

Kiwi Campervans Ltd
RD1 Upper Moutere, Nelson, New Zealand
Tel: 0064-3-543 2022.
Fax: 0064-3-543 2022.

Klingenfuss Publications
Hagenloher Strasse 14, Tuebingen, Germany
Tel: 0049707162830

Labcenter Electronics
53-55 Main Street, Grassington, BD23 5AA
Tel: 01756 753440.
Fax: 01756 752857.

Lake Electronics
7 Middleton Close, Nuthall, Nottingham, NG16 1BX
Tel: 0115 9382509

Lesniak Jones Liddell Ltd
First Avenue, Porthill, Newcastle-under-Lyme, Staffordshire, ST5 8QX
Tel: 01782 715715.
Fax: 01782 715000.

Linear Amp UK
Field Head, Leconfield Road, Leconfield, Beverley, East Yorks, HU17 7LU
Tel: 01964 550921.
Fax: 01964 550921.

Link Electronics
Tandy Millfield, 216 Lincoln Road, Peterborough, PE1 2NE
Tel: 01733 345731.
Fax: 01733 346770.

Lockwood, Peter G8SLB
36 Davington Rd, Dagenham, RM8 2LR
Tel: 0181-595 0823.
Fax: 0181-595 0823.

Lowe Electronics
Chesterfield Road, Matlock, Derbyshire, DE4 5LE
Tel: 01629 580800.
Fax: 01629 580020.

Martin Lynch & Son
140-142 Northfield Avenue, Ealing, London, W13 9SB.
Tel: 0181 566 1120.
Fax: 0181 566 1207.

Mailtech
P. O. Box 16, Ludlow, Shropshire, SY8 4NA
Tel: 0158 474475

Mainline Electronics
P.O. Box 235, Leicester, LE2 9SH
Tel: 0116 278 0891/277 7648.
Fax: 0116 247 7551.

Mancomp
240 Platt Lane, Manchester, M14 7BS
Tel: 0161 2241888.

Maplin Electronics
Maplin Complex, Oak Road South, Benfleet, Essex, SS7 2BB
Tel: 01702 554155 (ext 287)

Masthead Publicity
Tel: 01983 761004.
Fax: 01983 761004.

Mauritron Technical Services
8 Cherry Tree Road, Chinnor, Oxfordshire, OX9 4QY
Tel: 01844 351694.
Fax: 01844 352554.

M D H
Deerbolt, Tenby, Pembrokeshire, SA70 8SG
Tel: 01646 661699.
Fax: 01646 661699.

Media Applications Ltd
Unicorn Business Centre, Chiseldon, Wiltshire, SN4 0HT

Media Division
53 Church Road, Edgbaston, Birmingham, B15 3SJ
Tel: 0121 4549929

MEGS Publicity
37 Clepington Road, Maryfield, Dundee, DD4 7EL

Methodical Engineers Ltd.
Manor Trading Estate, 4/6 Armstrong Road, Benfleet, Essex, SS7 4PW
Tel: 01268 792681.
Fax: 01268 795375.

Miller, Brenda
Lambs Green, Horsham, West Sussex, RH12 4RQ
Tel: 01293 871621.
Fax: 01293 871578.

Milford Instruments
Milford House, 120 High Street, South Milford, Leeds, LS25 5AQ
Tel: 01977 683665.
Fax: 01977 681465.

Mirage
300 Industrial Park Road, Starkville, MS 39759, USA
Tel: 601-323-8287.
Fax: 601-323-6551.

Multicomm 2000 Ltd
Unit 1, Cambridge Street, St. Neots, Cambridgeshire, PE19 1PJ
Tel: 01480 406770.
Fax: 01480 470771.

muTek Limited
278 Bennett Street, Long Eaton, Nottingham, NG10 4JA

National Vintage Communications
2-4 Brook Street, Bampton, Devon, EX16 9LY
Tel: 01398 331532

Nevada Communications
189 London Road, North End, Portsmouth, Hants, PO2 9AE
Tel: 01705 662145.

Niche Software
22 Tavistock Drive, Belmont, Hereford, HR2 7XN
Tel: 01432 355414

Nikko Electronics
Dalbani House, 257 Burlington Rd, New Malden, Surrey, KT3 4NE
Tel: 0181 3360566.
Fax: 0181 3956003.

No Compromise Communications
The SGC Building, P.O. Box 3526, Bellevue, WA 98009, USA
Tel: 0101 206 746 6310.
Fax: 0101 206 746 6384.

NTech Communications
8 The Crescent,, Willingdon, Eastbourne, BN20 9RN
Tel: 01323 483966.

Number 1 Systems
Harding Way, St.Ives, Huntingdon, Cambs, PE17 4WR

O'Kane, Paul (EI5DI)
36 Coolkill, Sandyford, Dublin 18, Ireland
Tel: 00353 1295 3668.

Omni Electronics
174 Dalkeith Road, Edinburgh, EH16 5DX
Tel: 0131 6672611

Optoelectronics
5821 NE 14th Avenue, Ft Lauderdale, FL 33334, USA
Tel: 954-771-2050.
Fax: 954-771-2052.

Palomar Engineers
Box 462222, Escondido, CA 92046, USA
Tel: 619 747-3343.
Fax: 619 747-3346.

PDSL
Winscombe House, Beacon Road, Crowborough, East Sussex, TN6 1UL
Tel: 01892 663298.
Fax: 01892 667473.

Pervisell Ltd
8 Temple End, High Wycombe, Bucks, HP13 5DR
Tel: 01494 443033.
Fax: 01494 448236.

Photo Acoustics Ltd.
58 High Street, Newport Pagnell, Bucks, MK16 9AQ
Tel: 01908 610625.
Fax: 01908 216373.

Pico Technology
Broadway House, 149-151 St. Neots Road, Hardwick, Cambridge, CB3 7QJ
Tel: 01954 211716.
Fax: 01954 211880.

Poole Logic
49 Kingston Road, Poole, Dorset
Tel: 01202 683093

Powerware
14 Ley Lane, Marple Bridge, Stockport, SK6 5DD
Tel: 0161 4497101

PW Publishing Ltd
Arrowsmith Court, Station Approach, Broadstone, Dorset, BH18 8PW
Tel: 01202 659930.
Fax: 01202 659950.

QRP Component Company, The
2 Courts Hill Road, Haslemere, Surrey, GU27 2NG
Tel: 01428 641771.
Fax: 01428 661794.

QSL Communications
Unit 6, Worle Industrial Centre, Coker Road, Worle, Weston-Super-Mare, BS22 0BX
Tel: 01934 512757.
Fax: 01934 512757.

Quantek Electronics
1678 Bristol Road South, Birmingham, B45 9TZ
Tel: 0121 457 7994.
Fax: 0121 4579745.

Quartslab Marketing Ltd.
P. O. Box 19, Erith, Kent, DA8 1LH
Tel: 01322 330830.
Fax: 01322 334904.

R & D Electronics
Beaufort House, 12 Percy Avenue, Kingsgate, Broadstairs, Kent, CT10 3LB
Tel: 01843 866662.
Fax: 01843 866663.

R. A. Kent (Engineers)
243 Carr Lane, Tarletton, Preston, Lancs, PR4 6YB
Tel: 01772 814998.
Fax: 01772 815437.

R. N. Electronics
1 Arnolds Court, Arnold Farm Lane, Mountnessing, Essex, CM13 1UT
Tel: 01277 352219.
Fax: 01277 352968.

Radio Amateur Supplies
3 Farndon Green, Wollaton Park, Nottingham, NG8 1DU
Tel: 0115 9280267

Radio Bygones
9 Wetherby Close, Broadstone, Dorset, BH18 8JB
Tel: 01202 658474

Radio Data & Signalling Ltd.
5 Church Street, Crewkerne, Somerset, TA18 7HR
Tel: 01460 73886.
Fax: 01460 73896.

RadioSport Ltd
126 Mount Pleasant Lane, Bricket Wood, Herts, AL2 3XD
Tel: 01923 893929.
Fax: 01923 678770.

Rapid Results College, The
Tuition House, London, SW19 4DS
Tel: 0181 9477272

Remote Imaging Group (RIG)
P.O.Box 142, Rickmansworth, Herts, WD3 4RQ

Richards, Maurice Engineering
Wayside, Penwithic Road, St. Austell, Cornwall, PL26 8YH

Robinson Marshall (Europe) PLC
Nadella Building, Progress Close, Leofric Business Park, Coventry, Warwickshire, CV3 2TF
Tel: 01203 233216.
Fax: 01203 233210.

S.E.M.
8 Fort William Road, Head Road, Douglas, Isle of Man
Tel: 01624 662131.

S.R.P. Trading
Unit 20, Nash Lane, Forge Lane Belbroughton, Nr. Stourbridge, Worcs
Tel: 01562 730672.
Fax: 01562 731002.

Satellite Surplus (Telford)
PO Box 579, Telford, Shropshire, TF3 1WH

Seaward Mail Order
7 St Olafs Road, Stratton nr. Bude, Cornwall, EX23 9AF
Tel: 01288 355796.
Fax: 01288 355796.

Sendz Components
63 Bishopsteignton, Shoeburyness, Essex, SS3 8AF
Tel: 01702 332992.
Fax: 01702 338805.

SGC Inc.
SGC Building, 13737 SE 26th Street, Bellevue, WA 98005, USA
Tel: 206 746-6310.
Fax: 206 746-6384.

Sharward Promotions
Knightsdale Business Centre, 30 Knightsdale Road, Ipswich, IP1 4JJ
Tel: 01473 741533.
Fax: 01473 741361.

Short Wave Listening Guide
PO Box 151, Abingdon, Oxon, OK1 5DP

SIS Communication
19 Station Street, Mansfield Woodham, Notts, NG19 8AD
Tel: 01623 451919.

Siskin Electronics
Unit 1a, Hampton Lane, Blackfield, Hants, SO45 1WE
Tel: 01703 243400.
Fax: 01703 847754.

Skyview Communications
Skyview House, Arlesford, Essex, CO7 8BZ
Tel: 01206 823185.
Fax: 01206 825328.

SMC
SM House, School Close, Chandlers Ford Ind. Estate, Eastleigh, Hants, SO5 3BY
Tel: 01703 251549/255111.
Fax: 01703 263507.

Software Design Ltd
Elgin House, Sleaford, Lincs., NG34 7PN
Tel: 01526 833042

Southern Scanning & Shortwave
P. O. Box 2126, Bournemouth, Dorset, BH11 9YH

Speaker Builder
PO Box 494, Peterborough, New Hampshire, 03458 0494, USA
Tel: 0016039249464.
Fax: 0016039249467.

Spectronica
8 Bouverie Park, Stanton St Quintin, Chippenham, SN14 6EE
Tel: 01666 837080.
Fax: 01666 838130.

Spectrum Communications
Unit 6B, Poundbury West Estate, Dorchester, Dorset, DT1 2PG
Tel: 01305 262250.
Fax: 01305 262250.

Spencer, Ben Consultants
Enterprise House, 33 New King Street, Bath, BA1 2BL
Tel: 01225 482604.
Fax: 01225 482604.

Spimin Technology Ltd.
Spimin House, Cocker Avenue, Poulton-Le-Fylde, Lancs
Tel: 01253 894111

Standfast Press
5 South Drive, Inskip, Preston, PR4 0UT

Stevens, Mike (G8CUL)
67 New Road, East Hagbourne, Didcot, Oxon, OX11 9JX
Tel: 01235 816379.

Stewart of Reading
110 Wykeham Road, Reading, Berks, RG6 1PL
Tel: 01734 268041.
Fax: 01734 351696.

Strumech Versatower Ltd.
Portland House, Coppice Side, Brownhills, Walsall, West Midlands, WS8 7EX
Tel: 01543 452321.
Fax: 010543 361050.

SureData
45 Wychwood Avenue, Canons Park, Edgware, Middx, HA8 6ED
Tel: 0181 9057488

Sussex Surplus
13 Station Road, Horsham, West Sussex, RH13 5EZ
Tel: 01403 251302.
Fax: 01403 270339.

Svetlana Electron Devices
3000 Alpine Road, Portola Valley, CA 94028-7582, USA
Tel: 0014152330429.
Fax: 0014152330439.

Swift Television Publications
17 Pittsfield, Cricklade, Wilts, SN6 6AN
Tel: 01793 750620.
Fax: 01793 753399.

Syon Trading
16 Ridgeway, Fetcham, Leatherhead, Surrey, KT22 9AZ
Tel: 01372 372587.
Fax: 01372 361421.

Tangent International
Shelduck House, 10 Woodbrook Crescent, Billericay, Essex, CM12 0EQ
Tel: 01277 630055.
Fax: 01277 633133.

Telex Communications, Inc.
8601 East Cornhusker Highway, Lincoln, NE 68505, USA
Tel: 402 467-5321.
Fax: 402 467-3279.

Ten-Tec
1185 Dolly Parton Parkway, Sevierville, TN 37862, USA
Tel: 1-800-833-7373.
Fax: 615-428-4483.

Tennamast Scotland
81 Mains Road, Beith, Ayrshire, KA15 2HT

Thorne, Don
11 Bower Hall Drive, Steeple, Suffolk, CB9 7ED
Tel: 01440 730247.

Those Engineers Ltd
31 Birbeck Road, London, NW7 4BP

Timestep
P. O. Box 2001, Newmarket, CB8 8QA
Tel: 01440 820040.
Fax: 01440 820281.

Timewave Technology Inc.
2401 Pilot Knob Road, St. Paul, MN 55120, USA
Tel: 612-452-5939.
Fax: 612-452-4571.

Track & Parzer
Kapuzinerstraße 7, D-8812 Ravensburg, Germany
Tel: 00497513636920.
Fax: 00497513636929.

Trade Centre P.M.R.
286 Northfield Avenue, Ealing, London, W5 4UB

Turbolog - 2
Tim Kirby G4VXE, 19 Sidney Street, Cheltenham, Glos, GL52 6DJ
Tel: 01242 236723

Twrog Press
Penybont, Gellilydan, Blaenau Ffestiniog, Gwnedd, LL41 4EP
Tel: 01766 590341.

Tynrhos
Mynytho, Pwllheli, North Wales, LL53 7PS
Tel: 01758 740712

U-View
4 South Parade, Bawtry, Doncaster, Yorks, DN10 6JH

Venus Electronics Ltd
26 Pevensey Way, Frimley, Camberley, Surrey, GU16 5YJ
Tel: 01252 837860

Verlag, Theuberger DG0ZB
P.O. Box 73, 10122 Berlin, Germany
Tel: 493044669460.
Fax: 493044669411.

Viewcom Electronics
77 Upperton Road West, Plaistow, London, E13 9LT
Tel: 0181 4719338

Vine Antenna Products
The Vine, Llandrinio, Powys, SY22 6SH
Tel: 01691 831111.
Fax: 01691 831386.

Walford Electronics
Upton Bridge Farm, Long Sutton, Langport, Somerset, TA10 9NJ
Tel: 01458 241224

Waters & Stanton Electronics
Warren House, 22 Main Road, Hockley, Essex, SS5 4QS
Tel: 01702 206835.
Fax: 01702 205843.

Watson, Mike, (G8CPH)
The Tubbery, Ipswich, Suffolk, IP6 0BR

Weathervane G7IMD
Pentlands, Wroot, Doncaster, DN9 2BT
Tel: 01302 770769.

Western Electronics
9 Dorothy Crescent, Skegness, Lincs, PE25 2BU
Tel: 01754 610331

Westlake, W. H.
West Park, Clawton, Holsworthy, Devon, EX22 6QN
Tel: 01409 253758.
Fax: 01409 253458.

Wilson Valves
28 Bank Avenue, Golcar, Huddersfield, Yorks, HD7 4LZ
Tel: 01484 654650/420774.
Fax: 01484 655699.

Wimborne Publishing
Allen House, East Borough, Wimborne, Dorset, BH21 1PF

Wood & Douglas Ltd
Lattice House, Baughurst, Tadley, RG26 5LI
Tel: 0113 2425995

Woodhouse Communication
PO Box 73, Plainwell, MI 49080-0073, USA
Tel: 616 226-9073.
Fax: 616 226-9073.

Yaesu (UK) Ltd.
Unit 2, Maple Grove Business Centre, Lawrence Road, Hounslow, Middx., TW4 6DR
Tel: 0181 814 2001.
Fax: 0181 814 2002.

Zentek
132 Gladstone Street, Darlington, Co. Durham, DL3 6LE
Tel: 01325 482344.
Fax: 01325 255009.

3TH Ltd.
48 Hutchcomb Road, Oxford, OX2 9HL
Tel: 01865 791452.
Fax: 01865 794267.

United Kingdom

G0

G0	AAA	R K Western obo The Three A's Contest G, 7 Field Cl, Chessington, KT9 2QD [IO91UI, TQ16]
G0	AAD	H B Mutter, 13805 Town Line Rd, Silver Spring, Maryland, Usa, 20906
GM0	AAJ	J McLaverty, 75 Lugar Dr, Glasgow, G52 1EY [IO75UU, NS56]
G0	AAK	G P Johnson, 42 Woolley Rd, Southborough, Tunbridge Wells, TN4 0LF [JO01DD, TQ54]
G0	AAM	G H H Willetts, 2 Underlane, Boyton, Launceston, PL15 9RR [IO70TQ, SX39]
G0	AAN	Prof L Schnurr, 6 Blackwater Cl, Heybridge Basin, Maldon, CM9 4SB [JO01IR, TL80]
G0	AAP	C L Gosney, Hywood, Seymour Plain, Marlow, Bucks, SL7 3DA [IO91OO, SU88]
G0	AAR	Details withheld at licensee's request by SSL.
G0	AAT	P M Wheatley, 44 Primrose Cres, Worcester, WR5 3HT [IO82VD, SO85]
G0	AAU	J N Blades, 42 Ellesmere, Burnmoor, Houghton-le-Spring, DH4 6EA [IO94FU, NZ35]
G0	AAV	S D Massey, 35 School Ln, Ashton in Makerfield, Standish, Wigan, WN6 0TG [IO83QO, SD51]
GM0	AAX	G Anthoney, 10 Cedar Rd, Kilmarnock, KA1 2HP [IO75RO, NS43]
G0	AAY	I B Brooks, The Breck, 210 Breck Rd, Poulton-le-Fylde, FY6 7JZ [IO83MU, SD34]
G0	ABB	M J Honeywell, High Winds, 23 Deverell Pl, Widley, Waterlooville, PO7 5ED [IO90LU, SU60]
GW0	ABE	P G Hughes, 59 Jeffreys Rd, Rhosnessney, Wrexham, LL12 7PD [IO83MB, SJ35]
G0	ABF	J F Aungles, 158 Burn Park Rd, Houghton-le-Spring, DH4 5DH [IO94GU, NZ34]
GI0	ABH	H Curry, 1 Ballycraigy Rd, Newtownabbey, BT36 5ZZ [IO74AQ, J38]
G0	ABI	P A Green, Hollowtree, The Challices, Eggesford, Chulmeleigh, Devon, EX18 7QX [IO80BU, SS60]
G0	ABK	R Goldblatt, 17 Ferncroft Ave, London, NW3 7PG [IO91VN, TQ28]
GW0	ABL	D E Roberts, 23 Lon Hedydd, Siglan Farm Est, Llanfairpwllgwyngyll, LL61 5JY [IO73VF, SH57]
G0	ABM	T J Johnson, 29 Meadow Rd, Rusthall, Tunbridge Wells, TN4 8UN [JO01CD, TQ53]
G0	ABN	A M Samuels, 45 Mermaid Cl, Walderslade, Chatham, ME5 7PT [JO01GI, TQ76]
G0	ABP	D Y Orgill, 32 Upland Ave, Chesham, HP5 2EB [IO91QR, SP90]
G0	ABQ	E I Waller, 26 Hermitage Way, Sleights, Whitby, YO22 5HG [IO94QK, NZ80]
G0	ABR	H Cooper-Bland, 334 Mountnessing Rd, Billericay, CM12 0EU [JO01EP, TQ69]
G0	ABS	P Greener The Alternative, Barking Society, 158 Grosvenor Dr, Loughton, IG10 2LE [JO01AP, TQ49]
GW0	ABT	T R Thomas, Tymawr Farm, Llanwern, Brecon, Powys, LD3 7UW [IO81IW, SO12]
G0	ABU	P D Crowther, 115 Snape Hill Cres, Dronfield, Sheffield, S18 6GR [IO93GH, SK37]
G0	ABV	D E Wood, 18 Bankhouse Rd, Nelson, BB9 7RA [IO83VU, SD83]
G0	ABW	J N Harding, Fen End Farm, High St, Abbotsley, Huntingdon, Cambs, PE19 4UE [IO92VE, TL25]
G0	ABZ	Details withheld at licensee's request by SSL.[Station located near Chesham.]
G0	ACA	J A Kliffen, 8 West Park, Minehead, TA24 8AW [IO81GE, SS94]
G0	ACD	M J Amos, 41 Jocelyn Rd, Richmond, TW9 2TJ [IO91UL, TQ17]
GI0	ACE	P Smith, 26 Green Acres, Newtownabbey, BT36 6NL [IO74AQ, J38]
G0	ACK	D T Lamb, 339 Victoria Rd, Ruislip, HA4 0DS [IO91TN, TQ18]
G0	ACL	D M Kipping, 46 Old Hardenwaye, Totteridge, High Wycombe, HP13 6TJ [IO91PP, SU89]
G0	ACM	Details withheld at licensee's request by SSL.
G0	ACN	Details withheld at licensee's request by SSL.
G0	ACO	R Spence, 36 Argyll Ave, Larne, BT40 2JX [IO74BU, D30]
G0	ACP	N Sumner, 56 Severn Rd, Culcheth, Warrington, WA3 5EB [IO83RK, SJ69]
G0	ACQ	M Godden, 20 Channel View Rd, Easton, Portland, DT5 2AY [IO80SN, SY67]
G0	ACR	R Little, 16 Greenwood Ave, Bedlington, NE22 7EE [IO95FD, NZ28]
G0	ACX	C Bolt, 33 Manscombe Rd, Allerton, Bradford, BD15 7AQ [IO93CT, SE13]
G0	ACZ	A Hope, 71 Wolds Rise, Matlock, DE4 3HJ [IO93FD, SK36]
G0	ADA	N R Law, 17 Sycamore Rd, Guildford, GU1 1HJ [IO91RF, SU95]
G0	ADB	C Willis, 5 Gower Dr, Biddenham, Bedford, MK40 4PZ [IO92SD, TL04]
GW0	ADC	K M Shepherd, Hafanedd, Tan Lan, Ffynnongroew, Holywell, CH8 9UU [IO83II, SJ18]
GI0	ADD	J R Ashe obo Armagh and Dungannon Distr, 49 Deans Walk, Sleepy Valley, Richhill, Armagh, BT61 9LD [IO64RJ, H94]
G0	ADE	R E Lepage, 30 Westside, London, NW4 4XB [IO91VO, TQ29]
GM0	ADF	D I Mackinnon, 5 Octavia Terr, Greenock, PA16 7SP [IO75OX, NS27]
G0	ADG	R G Bristeir, 94 Burnthwaite Rd, Fulham, London, SW6 5BG [IO91VL, TQ27]
G0	ADH	R C Razey, 2 Park Farm Cottage, 26 St. Georges Rd, Wallingford, Oxon, OX10 8HP [IO91KO, SU68]
G0	ADI	W J Elsworth, 34 Seal Rd, Bramhall, Stockport, SK7 2JR [IO83WI, SJ88]
G0	ADJ	D C Elsworth, 34 Seal Rd, Bramhall, Stockport, SK7 2JR [IO83WI, SJ88]
G0	ADK	J F C Saueressig, 8 The Ridgeway, River, Dover, CT17 0NX [JO01PD, TR24]
G0	ADL	B W Barlow, 134 Bury Rd, Radcliffe, Manchester, M26 2UX [IO83UN, SD70]
G0	ADM	A W Russell, 49 Well Ln, Horsell, Woking, GU21 4PS [IO91RH, SU95]
G0	ADO	A Hodgson, Arla Burn Farm, Middleton in Teesdale, Co Durham, DL12 0QU [IO84WP, NY92]
G0	ADP	A W Wither, 30 Mersey Rd, Aigburth, Liverpool, L17 6AD [IO83MI, SJ38]
G0	ADQ	J H Walton, 3 Maltkiln Cttgs, Raby Rd, Staindrop, Darlington, DL2 3AG [IO94CO, NZ12]
G0	ADR	Details withheld at licensee's request by SSL.[Station located in London SW16.]
GW0	ADS	J P Jenkins, Maglona, Margaret St., New Quay, Dyfed, SA45 9QJ [IO72TF, SN35]
G0	ADT	I A Houghton, 7 Quebec Gdns, Blackwater, Camberley, GU17 9DE [IO91OH, SU85]
G0	ADU	B Gibson, 55 Ledward St., Wharton, Winsford, CW7 3EN [IO83RE, SJ66]
GD0	ADV	L M Kinvig, 79 Clagh Vane, Ballasalla, IM9 2HF
G0	ADW	P J Radford, 42 Ashbury Dr, Weston Super Mare, BS22 9QS [IO81MI, ST36]
GM0	ADX	W Strachan Kilmarnock and Loudoun ARC, c/o Sparnel, Cemetery Rd, Galston, Ayrshire, KA4 8LL [IO75TO, NS53]
GW0	ADY	M P Jones, Stretford End, Dryslwyn, Carmarthen, Dyfed, SA32 8SA [IO71WV, SN52]
G0	ADZ	D J G Price, Glenmore, Moulton Rd, Cheveley, Newmarket, CB8 9DW [JO02FF, TL66]
G0	AEA	C S Oakley, 2 Broomfield Lodge, Leeds Castle, Broomfield, Maidstone, ME17 1PL [JO01HF, TQ85]
G0	AEB	J A Smith, 31 Ribble St., Bacup, OL13 9RH [IO83VQ, SD82]
G0	AEC	E P Tyler, 43 Nest Est, Mytholmroyd, Hebden Bridge, HX7 5BH [IO93AR, SE02]
GM0	AEG	D J Greatorex, 31 Seatown, Lossiemouth, IV31 6JJ [IO87IR, NJ27]
G0	AEJ	D J F Walker, 23 Trent View, Keadby, Scunthorpe, DN17 3DR [IO93PO, SE81]
G0	AEL	K D Horton, Lorien, 81 Slepe Cres, Brooklands Park, Poole, BH12 4DJ [IO90AR, SZ09]
G0	AEN	S L Webb, 34 Walsham Rd, Walderslade, Chatham, ME5 9HX [JO01GI, TQ76]
G0	AEP	G F W Roper, 192 Howeth Rd, Ensbury Park, Bournemouth, BH10 5NX [IO90BS, SZ09]
G0	AES	N H M Richardson, 2 Edna Rd, Maidstone, ME14 2QJ [JO01GG, TQ75]
G0	AEU	P Tietz, 5 Chevin Rd, Belper, DE56 2UW [IO93GA, SK34]
G0	AEV	Dr S J Reed, Bridlands, Middle Common, Kington Langley, Chippenham, Wilts, SN15 5NN [IO81WL, ST97]
G0	AEW	D T Arlette, 12 Polmear, Par, PL24 2AT [IO70PI, SX05]
G0	AEX	E Hannaby, 32 Lingfield Gate, Moortown, Leeds, LS17 6DD [IO93FU, SE23]
GM0	AEY	D E Palmer, 36 Kilsyth Rd, Haggs, Bonnybridge, FK4 1HE [IO85AX, NS77]
GW0	AEZ	J R Howarth, 17 Howard Pl, Llandudno, LL30 1EB [IO83CH, SH78]
G0	AFB	A C Muscat, 32 Windmill Way, Reigate, RH2 0JA [IO91VF, TQ25]
G0	AFC	J Trout, 2 Balfour Manor, Station Rd, Sidmouth, EX10 8XW [IO80JQ, SY18]
G0	AFD	Details withheld at licensee's request by SSL.
G0	AFF	Details withheld at licensee's request by SSL.
G0	AFG	H J Daly, Firside, 26 Hayters Way, Alderholt, Fordingbridge, SP6 3AX [IO90CV, SU11]
G0	AFH	I T Burns, 30 Highview, Meopham, Gravesend, DA13 0RR [JO01EH, TQ66]
G0	AFJ	A R Brown, 33 Marion Rd, Haydock, St. Helens, WA11 0PY [IO83QL, SJ59]
G0	AFN	P G Howard, 12 Meadow Way, Westergate, Chichester, PO20 6QT [IO90PU, SU90]
G0	AFP	D G Rayner, 13 St. Bartholomews Cl, Cam, Dursley, GL11 5US [IO81TQ, SO70]
G0	AFQ	J Cook, 71 Richmond Ave, Burscough, Ormskirk, L40 7RB [IO83NO, SD41]
G0	AFR	C B Mears, 11 Aberford Cl, Watermead, Reading, RG3 2NX [IO91LJ, SU66]
G0	AFS	P J C McDonnell, 7 Sussex Gdns, Westgate on Sea, Kent, CT8 8AG [JO01QJ, TR37]
G0	AFT	C J Bovey, 12 The Mead, Candlemas Mead, Beaconsfield, HP9 1AW [IO91QO, SU99]
G0	AFU	M C Bousfield, 10 Thorneyford Pl, Ponteland, Newcastle upon Tyne, NE20 9QN [IO95DB, NZ17]
G0	AFV	B J Betambeau, 60 Deerpark Rd, Langtoft, Peterborough, PE6 9RB [IO92UQ, TF11]
G0	AFY	R G Norman, 98 Foxwell Dr, Headington, Oxford, OX3 9QF [IO91JS, SP50]
G0	AFZ	C H Wilson, 38 Wiclif Way, Stockingford, Nuneaton, CV10 8NH [IO92FM, SP39]
GW0	AGA	G D Bonter, 11 Devon Ave, Barry, CF63 1BJ [IO81IJ, ST16]
G0	AGB	R A Ellis, 10 Post View, Storrington, Pulborough, RH20 4EZ [IO90SV, TQ01]
G0	AGC	A G Core, Orchard House, 22 Golf Links Ln, Wellington, Telford, TF1 2DT [IO82RQ, SJ61]
G0	AGD	P M Shortland, 1 Kenilworth Dr, Carlton in Lindrick, Worksop, S81 9BY [IO93KI, SK58]
G0	AGF	D Linskill, Mistletoe Cottage, Rivermead, Herodsfoot, Liskeard, Cornwall, PL14 4QX [IO70RK, SX26]
G0	AGH	Details withheld at licensee's request by SSL.[Station located near Stourbridge, Worcs.]
G0	AGI	M J Bingham, 9 Freda Cl, Gedling, Nottingham, NG4 4GP [IO92KX, SK64]
G0	AGJ	R G Hughes, 90 Oxford Rd, Old Marston, Oxford, OX3 0RD [IO91JS, SP50]
GW0	AGL	E M Williams, Caeabergam Cottage, Llanbeor, Gwynedd, LL45 2HT [IO72WU, SH52]
G0	AGM	B D Afford, 16 Broomhill Dr, Knottingley, WF11 0EG [IO93JQ, SE52]
GM0	AGN	G W A Speirs, 43 Sheuchan View, Stranraer, DG9 7TA [IO74LV, NX05]
G0	AGR	B Ashdown Aylesbury Raynet Grp, 44 Summerleys Rd, Princes Risborough, HP27 9DT [IO91NR, SP80]
G0	AGS	R J Clutterham, 12 Roman Bank, Leverington, Wisbech, PE13 5AN [JO02BQ, TF41]
G0	AGT	M J Felton, 233 Calshot Rd, Great Barr, Birmingham, B42 2BY [IO92AN, SP09]
G0	AGU	B L H Humphries, 22 Leander Cl, Burntwood, WS7 8PW [IO92AQ, SK00]
G0	AGW	Details withheld at licensee's request by SSL.
G0	AGY	S R Talbot, 22 Courtwick Rd, Wick, Littlehampton, BN17 7NE [IO90RT, TQ00]
GW0	AGZ	Cpl J E Squire, Bevlands, Sandy Ln, Bagillt, CH6 6EY [IO83JG, SJ27]
G0	AHA	A J Pannell, 104 Southwood Rd, Ramsgate, CT11 0AW [JO01QI, TR36]
G0	AHB	A W Bennett, 2 Portland Pl, Hertford Heath, Hertford, SG13 7RR [IO91XS, TL31]
G0	AHC	B R Hewitt, 32 Pinehurst Dr, Kings Norton, Birmingham, B38 8TH [IO92AJ, SP07]
G0	AHD	C A Richardson, 122 Elmton Rd, Cresswell, Creswell, Worksop, S80 4DE [IO93JG, SK57]
G0	AHE	P A Moss, Wick Rd, St Margarets Cross, Langham, Essex, CO4 5PE [JO01LW, TM03]
G0	AHI	T R Deacon, 155 Cloonmore Ave, Orpington, BR6 9LN [JO01BI, TQ46]
G0	AHJ	J D Johnson, 8 Park House Dr, Prestbury, Macclesfield, SK10 4HY [IO83WH, SJ97]
G0	AHK	D I Layne, 12 Balmoral Walk, New Milton, BH25 5UF [IO90ES, SZ29]
G0	AHL	R F Sears, 29 Selwyn Cl, Newmarket, CB8 8DD [JO02EF, TL66]
G0	AHM	P N Gregg, 6 Westbrook Ave, Hampton, TW12 2RE [IO91TK, TQ17]
G0	AHN	Details withheld at licensee's request by SSL.
G0	AHQ	J E Harper, 77 Victoria St., Chesterton, Newcastle, ST5 7EP [IO83VA, SJ84]
G0	AHR	P E Cuthbert, 115 Tintern Ave, Whitefield, Manchester, M45 8WY [IO83UN, SD80]
G0	AHU	J E Tolson, 37 Raphael Dr, Elburton, Plymouth, Devon, PL9 8EU [IO70XI, SX55]
G0	AHV	R D Gilling, 24 Bellerby Rd, Skellow, Doncaster, DN6 8PD [IO93JO, SE51]
G0	AHX	F B Limond, 14 The Close, Fairlight, Hastings, TN35 4AQ [JO00HU, TQ81]
GI0	AHZ	E F Brown, 16 Benbraddagh Ave, Rathbeg Est, Limavady, BT49 0AR [IO65MB, C62]
G0	AID	E Dawson, 50 Grendon Rd, Polesworth, Tamworth, B78 1NU [IO92EO, SK20]
G0	AIE	Details withheld at licensee's request by SSL.
G0	AIF	R J Turnbull, 7 Jackson Cl, Cayton, Scarborough, YO11 3RW [IO94TF, TA08]
G0	AIG	M T S Sutherland, 17 Mill Ln, Enderby, Leicester, LE9 5NW [IO92JO, SP59]
G0	AIH	R Baker, 60 Annan Rd, Billingham, TS23 3HQ [IO94IO, NZ42]
G0	AII	D J Fairbank, 2 Sandpit Rd, Welwyn Garden City, AL7 3TN [IO91VS, TL21]
GI0	AIJ	I Greenwood, Deers Leap, 24 Tullyrusk Rd, Dundrod, Crumlin, BT29 4JQ [IO64WO, J27]
G0	AIL	D T Penny, 14 St. Marys Cres, Yeovil, BA21 5RP [IO80QW, ST51]
G0	AIM	S J R Robinson, 164 Leigh Rd, Westhoughton, Bolton, BL5 2LE [IO83RN, SD60]
G0	AIN	S C H Withnell, 85 Headroomgate Rd, Lytham St. Annes, FY8 3BG [IO83LS, SD33]
G0	AIO	A I Owens, 69 Locomotion Way, North Shields, NE29 6XE [IO95GA, NZ36]
GI0	AIQ	F J Holland, 413 Ballyoran Park, Portadown, Craigavon, BT62 1JX [IO64SK, J05]
GM0	AIR	D A Parker, Devon Cottage, Cupar, Newton of Falkland, Fife, KY15 7RZ [IO86JG, NO20]
G0	AIS	J C Borg, 94 Coldershaw Rd, London, W13 9DT [IO91UM, TQ17]
G0	AIT	I D Thompson, Leitholm, Tunstead Rd, Hoveton, Norwich, NR12 8QN [JO02QR, TG31]
G0	AIU	P J Rooney, 10 Baslow Ave, Carlton, Nottingham, NG4 3NT [IO92KX, SK64]
G0	AIX	D S H Westlake, Chyvellin, Newmill, Penzance, Cornwall, TR20 8XW [IO70FD, SW43]
GW0	AIY	R A G Gibbons, Cefnysgwyn Capel, Issac Llandeilo, Dyfed, SA19 7UA [IO71XV, SN52]
G0	AIZ	W D Chesterton, 7 Norfolk Rd, Desford, Leicester, LE9 9HR [IO92IO, SK40]
G0	AJA	A J Astley, Erw Wen, Ffordd Goed, Pen-y-Bont Fawr, Oswestry, Shropshire, SY10 0PF [IO82HT, SJ02]
G0	AJB	A Botherway, 14 Amherst Rd, Kenilworth, CV8 1AH [IO92EI, SP27]
G0	AJC	G A de Bertodand, Raceground Cottage, Dunwich Rd, Westleton, Saxmundham, IP17 3DD [JO02TG, TM46]
G0	AJE	J D Evans, 38 Sidney St., Salisbury, SP2 7AJ [IO91CB, SU13]
G0	AJF	S M Harrison, 25 High Spring Rd, Keighley, BD21 4TF [IO93BU, SE04]
G0	AJG	D B Cattran, 13 Kenstella Rd, Newlyn, Penzance, TR18 5AY [IO70FC, SW42]
G0	AJH	J S Hornsby, 15 Coronation Dr, Hornchurch, RM12 5BL [JO01CN, TQ58]
GW0	AJI	S J Davies, 9 Coed y Llyn, Radyr, Cardiff, CF4 8RF [IO81IM, ST18]
G0	AJJ	L I Leavold, 8 Wilkinson Way, North Walsham, NR28 9BB [JO02QT, TG22]
GM0	AJK	D G M Cameron, 125 Rose St., Dunfermline, KY12 0QT [IO86GB, NT08]
G0	AJL	B Bromsgrove, 34 Boundary Dr, Hunts Cross, Liverpool, L25 0QD [IO83NI, SJ48]
G0	AJM	Details withheld at licensee's request by SSL.
G0	AJO	J Wright, 116 Wynchgate, London, N21 1QU [IO91WP, TQ39]
G0	AJQ	R J Wood, School Farm, Brock Rd, Great Eccleston, Preston, PR3 0XD [IO83NU, SD43]
G0	AJS	J R Ballantyne, 71 St. Andrews Cres, Windsor, SL4 4EP [IO91QL, SU97]
GW0	AJU	A J Underwood, Rock Hill, Llanarthney, Carmarthen, Dyfed, SA32 8LJ [IO71VU, SN51]
G0	AJW	J Welsby, 22 Parkstone Ave, Carleton, Poulton-le-Fylde, FY6 7PF [IO83LU, SD34]
G0	AJX	M G Coles, 13 Dawnay Rd, Bilton, Hull, HU11 4HB [IO93VS, TA13]
GW0	AJY	R Davies, Bryn Canol, Lampeter Velfrey, Narbeth, SA67 8UG [IO71PT, SN11]
G0	AJZ	B J Park, 147 Castle Rd, Ings Farm, Redcar, TS10 2LT [IO94LO, NZ62]
G0	AKA	D Seville, 35 Barnard Rd, Manchester, M18 7RE [IO83VK, SJ89]
G0	AKC	Details withheld at licensee's request by SSL.
G0	AKE	M E K Dixon, 323 Margate Rd, Ramsgate, CT12 6TG [JO01QI, TR36]
G0	AKF	K Farrance, Clarewood, Tabley Rd, Knutsford, WA16 0NE [IO83TH, SJ77]
G0	AKH	K Hale, Apartment, 5-91 St. Davids Hill, Exeter, EX4 3RG [IO80FR, SX99]
G0	AKI	M E C Eavis, 33 Welldon Cres, Harrow, HA1 1QP [IO91TO, TQ18]
GM0	AKJ	P J Seaton, 51 Leachkin Ave, Inverness, IV3 6LH [IO77UL, NH64]
G0	AKK	J B Chivers, 27 Clyde Rd, Worthing, BN13 3LG [IO90SU, TQ10]
G0	AKL	R King, 4 Pinewood Dr, Horning, Norwich, NR12 8LZ [JO02RQ, TG31]
G0	AKM	K D Muchamore, 163 Junction Rd, London, N19 5PZ [IO91WN, TQ28]
G0	AKN	Details withheld at licensee's request by SSL.
G0	AKO	R W Cleveland, 1 Yarrow Rd, Dereham, NR20 3BH [JO02LQ, TG01]
G0	AKR	A F G Poynter, Hill Top Farm, Warren Rd, Blue Bell Hill, Chatham, ME5 9RD [JO01GH, TQ76]
G0	AKS	D A Newell, Westcot, 7 Edward Rd West, Walton St. Mary, Clevedon, BS21 7DY [IO81NK, ST47]
G0	AKU	R A Reanney, 9 Stapleford Ct, Ellesmere Port, South Wirral, L66 1RW [IO83MG, SJ37]
GW0	AKV	J M S Anderson, 173 Gate Rd, Penygroes, Llanelli, SA14 7RW [IO71XT, SN51]
G0	AKW	R Sim, 22 Dene View, Ashington, NE63 8JT [IO95EE, NZ28]
G0	AKY	W G Staples, 8 Lesley Cl, Bexhill on Sea, TN40 2RF [JO00FU, TQ70]
G0	ALA	D Whitehouse, 17 The Terrace, St. Ives, TR26 2BP [IO70GF, SW54]
G0	ALB	C Matcham, 31 Buchanan Dr, Luton, Beds, LU2 0RX [IO91TV, TL12]
G0	ALC	J Richards, 77 Poxon Rd, Walsall Wood, Walsall, WS9 9JR [IO92AP, SK00]
GU0	ALD	Details withheld at licensee's request by SSL.
G0	ALE	P M Madagan Tatsfield ARTS, 40 Lagham Park, South Godstone, Godstone, RH9 8ER [IO91XF, TQ34]
GW0	ALF	O H Edwards, 51 Bro Infryn, Glasinfryn, Bangor, LL57 4UR [IO73WE, SH56]
G0	ALH	S A Allwood, 69 Farhalls Cres, Horsham, RH12 4BT [IO91UB, TQ13]
G0	ALI	A Soars, 84 Ridge Rd, Kingswinford, DY6 9RG [IO82VL, SO88]
G0	ALJ	R E Prior, 78 Chalk Rd, Chalk, Gravesend, DA12 4UZ [JO01EK, TQ67]
GW0	ALM	D Cornick, Cartref, Hospital Rd, Nantyglo, Gwent, NP3 4UT [IO81KS, SO10]
G0	ALP	Details withheld at licensee's request by SSL.
G0	ALQ	J Amer, 45 Horbury Dr, Bury, BL8 2PS [IO83UO, SD71]
GW0	ALR	A J Wagenaar, Gilfach Rhew, Abergwili, Carmarthen, Dyfed, SA32 7ER [IO71UU, SN42]
GM0	ALS	F A Roe, 8 South Gyle Gdns, Edinburgh, EH12 7RZ [IO85IW, NT17]
GM0	ALU	E A Brown, 11 Wadeslea, Elie, Leven, KY9 1EB [IO86OE, NT49]
G0	ALV	C E Blumfield, 43 Stapleton Rd, Meole Brace, Shrewsbury, SY3 9LU [IO82OQ, SJ41]
GM0	ALW	G Chalmers, 38 Gr Hill, Kelso, TD5 7AS [IO85SO, NT73]
GM0	ALX	M A Chalmers, 38 Gr Hill, Kelso, TD5 7AS [IO85SO, NT73]
G0	ALY	D F Mathias, 43 Chalk Farm Rd, Stokenchurch, High Wycombe, HP14 3TB [IO91NP, SU79]
GM0	AMB	J G Campbell, 9 Brunton Park, Bowden, Melrose, TD6 0SZ [IO85PN, NT53]
G0	AMD	A M Dorman, Glenfield, Main St., Cotesbach, Lutterworth, LE17 4HZ [IO92JK, SP58]
G0	AME	J R Spence, 12 Hillyfields, Dunstable, LU6 3NS [IO91RV, TL02]
G0	AMG	R C Rutt, 19 Cloverland, Hatfield, AL10 9ED [IO91VR, TL20]
GM0	AMJ	Details withheld at licensee's request by SSL.
G0	AMM	R J Cooper, Shawcroft School, Crockenhill Rd, St. Mary Cray, Orpington, BR5 4ES [JO01BJ, TQ46]
G0	AMO	M D Adams, 9 Brancaster Ave, Charlton, Andover, SP10 4EN [IO91FF, SU34]
G0	AMP	R E Senft, Mill Hay, Standard Rd, Downe, Orpington, BR6 7HL [JO01AI, TQ46]
G0	AMS	A M Sinclair, 77 Red Barn Ln, Fareham, PO15 6HD [IO90JU, SU50]
G0	AMT	S Weglarz, 4 Bosworth, Killingworth, Newcastle upon Tyne, NE12 0QE [IO95FA, NZ27]
G0	AMU	L Parrott, 55 Brown St., Macclesfield, SK11 6RY [IO83WG, SJ97]
G0	AMV	G Aspden, 164 Stanley St., Accrington, BB5 6QG [IO83TS, SD72]
G0	AMW	C Dunn, 32 School Rd, Quarry Bank, Brierley Hill, DY5 1BE [IO82WL, SO98]
G0	AMX	P A E Appleyard, 28 Romany Cl, Letchworth, SG6 4JZ [IO91VX, TL23]
G0	AMY	J Sheppeck, 25 Kingsleigh Rd, Heaton Mersey, Stockport, SK4 3QF [IO83VK, SJ89]
G0	AMZ	K E Fay, 37 Sandringham Rd, Rainham, Gillingham, ME8 8RP [JO01HI, TQ86]
GW0	ANA	G O Jones, Nirvana, Castle Precinct, Llandough, Cowbridge, CF7 7LX [IO81HN, ST08]
G0	ANE	W Burrows, 35 Bamford Ave, Barnsley, S71 3SJ [IO93GN, SE30]

G0 ANF H E Hogg, 24 Lannoweth Rd, Penzance, TR18 3AB [IO70FC, SW43]
GM0 ANG J M Fish, Senekal, Alma Rd, Fort William, PH33 6HB [IO76KT, NN17]
G0 ANH J D Wright, 79 Coach Rd, Sleights, Whitby, YO22 5EH [IO94QK, NZ80]
G0 ANK I Smallwood, 3 Barnaby House, Sir Evelyn Rd, Rochester, ME1 3PD [JO01FI, TQ76]
G0 ANL P Ellis, 8 Shropshire Cl, Woolston, Warrington, WA1 4DY [IO83RJ, SJ68]
G0 ANN M E Viney, 20 Auckland Rd, Potters Bar, EN6 3ES [IO91VQ, TL20]
G0 ANO D B Lawton, Grenehurst, Pinewood Rd, High Wycombe, HP12 4DD [IO91OP, SU89]
G0 ANP D P Guy, 7 Park Ave, Castle Cary, BA7 7HE [IO81RC, ST63]
G0 ANR Details withheld at licensee's request by SSL.
G0 ANS J M Vye, 16 Breach Cl, Steyning, BN44 3RZ [IO90UV, TQ11]
G0 ANT J G Sutton Eden Valley RS, 15 Lowther St., Penrith, CA11 7UW [IO84OQ, NY53]
G0 ANV D W Burchell, 40 Daisy Cl, Cambridge, CB4 3XH [JO02BF, TL46]
G0 ANW J D Hockley, Two Chimneys, Shottendane Rd, Birchington, Kent, CT7 0HD [JO01QI, TR36]
G0 ANX W F Tully, 8 The Croft, West Hanney, Wantage, OX12 0LD [IO91HP, SU49]
G0 ANY R A Corey, 11 Veor Rd, Newquay, TR7 3BX [IO70LK, SW86]
G0 ANZ N D Paxton, 124 Greenfield, Witham, CM8 2FA [JO01HT, TL81]
G0 AOA T Marten, 10 Chieveley Dr, Tunbridge Wells, TN2 5HG [JO01DC, TQ53]
G0 AOB M R J Beal, 12 Alver Green, Alverdiscott Rd, Bideford, EX39 4DL [IO71VA, SS42]
G0 AOD D P Heathcote, 8 Ferrers Ave, Tutbury, Burton on Trent, DE13 9JR [IO92DU, SK22]
G0 AOE N Evans, 16 Humbledon Park, Sunderland, SR3 4AA [IO94HV, NZ35]
GM0 AOF R B Wallace, 1 Holding West Kincardine, Crieff, PH7 3RP [IO86CI, NN82]
G0 AOG R Tribbeck, 27 Lea Combe, Axminster, EX13 5LJ [IO80MS, SY29]
G0 AOH J V Abbruscato, 8234 Hollyleaf Dr, Spring, Texas, USA 77379
G0 AOI R B Bennett, 44 Flyford Cl, Lodge Park, Redditch, B98 7LU [IO92AH, SP06]
G0 AOJ F Fenwick, 6 School Ln, Bonby, Brigg, DN20 0PP [IO93SO, TA01]
G0 AOK D A Gall, 2 Norham Cl, Wideopen, Newcastle upon Tyne, NE13 7HS [IO95EB, NZ27]
G0 AOL G Parsons, Shangri Ln, Maldon Rd, Margaretting, Ingatestone, CM4 9JW [JO01FQ, TL60]
G0 AOM R Day, 3 Railway Cttgs, Newby Bridge, Ulverston, LA12 8AW [IO84MG, SD38]
G0 AOO J K Butterwick, 45 Fox Howe, Coulby Newham, Middlesbrough, TS8 0RU [IO94JM, NZ51]
G0 AOP P Warriner, 36 Eskdaleside, Sleights, Whitby, YO22 5EP [IO94QK, NZ80]
G0 AOR G F Tombs, Balaton, Estuary Rd, Shotley Gate, Ipswich, IP9 1PZ [JO01PW, TM23]
G0 AOS M D Reynolds, 9 Winscombe Cres, Ealing, London, W5 1AZ [IO91UM, TQ18]
G0 AOT J E Ballard, 48 Elmbridge Dr, Ruislip, HA4 7UT [IO91TO, TQ18]
G0 AOU D J Armitage, 12 Loughborough Cl, Sale, M33 5UF [IO83TK, SJ79]
G0 AOV J H Eaton, Cable House, Cable Gap, Bacton, Norfolk, NR12 0EW [JO02RU, TG33]
G0 AOW S E Sands, Gabledown, Bridgnorth Rd, Stourton, Stourbridge, DY7 6RW [IO82VL, SO88]
G0 AOX A J Sands, Gabledown, Bridgnorth Rd, Stourton, Stourbridge, DY7 6RW [IO82VL, SO88]
G0 AOY T Rudd, Grasmere, Burgh, Woodbridge, Suffolk, IP13 6SU [JO02PC, TM25]
G0 AOZ R D Powell, Town Pond Cottage, Town Pond Ln, Southmoor, Abingdon, OX13 5HS [IO91GQ, SU49]
G0 APA C P Slenk, 653 Main St, Niagara Falls, Ny 14301, USA
G0 APB P Buckley, 7 Callams Cl, Rainham, Gillingham, ME8 9ES [JO01HI, TQ86]
G0 APE Details withheld at licensee's request by SSL.
G0 APF C K Stewart, 6 Chapel Cl, Kibblesworth, Gateshead, NE11 0YE [IO94EV, NZ25]
GI0 APH R J V Patterson, 12B Falcon Rd, Belfast, BT12 6RD [IO74AN, J37]
G0 API J M Fell, 14 Rectory Ave, Corfe Mullen, Wimborne, BH21 3EZ [IO80XS, SY99]
GW0 APJ J M Jones, 5 Clos Llyswen, Bryn Siriol, Penpedairheol, Hengoed, CF8 7TQ [IO81HN, ST08]
GW0 APL R W G Davies, 72 St. Davids Cres, Penarth, CF64 3NB [IO81JK, ST17]
G0 APM Details withheld at licensee's request by SSL.
G0 APO Details withheld at licensee's request by SSL.
G0 APP A P Pamment, 5 New Captains Rd, West Mersea, Colchester, CO5 8QP [JO01KS, TM01]
G0 APQ K Taylor, 25 Glentworth Ave, Middlesbrough, TS3 0QH [IO94JN, NZ51]
G0 APS H W L Cage, 38 Highfield Ave, Birds Edge, Birdsedge, Huddersfield, HD8 8XT [IO93DN, SE20]
G0 APV V M Covell-London, 15 St. Nicholas Way, Potter Heigham, Great Yarmouth, NR29 5LG [JO02SR, TG41]
G0 APW P H Morris, 56 Gloucester Pl, London, W1H 3HL [IO91WM, TQ28]
G0 APY G H Flood, 4 Campbell Cres, Great Sankey, Warrington, WA5 3DA [IO83QJ, SJ58]
G0 APZ E L Killip, Bramley, 10 Orchard Ln, Ditchling, Hassocks, BN6 8TH [IO90WW, TQ31]
G0 AQB J Ireland, The Grange, Grange Ln, Lower Broadheath, Worcester, WR2 6RW [IO82UF, SO85]
GW0 AQC P Stevens Fshgrd & Dst AR, Noddfa, Dwrbach, Fishguard, Dyfed, SA65 9RL [IO71MW, SM93]
GI0 AQD D R J Burn, 110 Ballywalter Rd, Millisle, Newtownards, BT22 2HS [IO74FO, J57]
G0 AQE B D King, 29 Pembroke Rd, Clifton, Bristol, BS8 3BE [IO81QL, ST57]
G0 AQF D Dolphin, Brook House, 16 Golden Cross Ln, Catshill, Bromsgrove, B61 0LQ [IO82XI, SO97]
GM0 AQG A M L Spencer, 34 Oxgangs Rd North, Edinburgh, EH13 9DR [IO85JV, NT26]
G0 AQH G R Griffin, 23 St. Giles Cl, Shoreham By Sea, BN43 6GR [IO90UU, TQ20]
G0 AQI E L P Greenhalgh, 31 The Cres, Friern Barnet, London, N11 3HH [IO91WO, TQ29]
G0 AQJ B James, 81 Birches Rd, Codsall, Wolverhampton, WV8 2JJ [IO82VO, SJ80]
G0 AQK P McMillan, 24 Roman Way, Warfield, Bracknell, RG42 7UT [IO91PK, SU86]
G0 AQL Details withheld at licensee's request by SSL.
G0 AQN J A Thorns, 91 Smeaton Rd, London, SW18 5JJ [IO91VK, TQ27]
G0 AQP Details withheld at licensee's request by SSL.[Op: S R Langley. Station located near Hessle.]
GW0 AQR J C Richards, Gwyndre (No.3), Glan Beuno, Bontnewydd, Gwynedd, LL55 2UB [IO73UC, SH46]
G0 AQS P A Goldthorpe, 164 Henrietta St., Ashton under Lyne, OL6 8QR [IO83WL, SD90]
G0 AQT V Taylor, 5 St. Matthews Dr, St. Leonards on Sea, TN38 0TR [JO00GU, TQ81]
G0 AQU R F Mason, 28 Vandyke, Great Hollands, Bracknell, RG12 8UP [IO91OJ, SU86]
G0 AQW W F Harder, 12 Tindale Rd, Maybush, Southampton, SO16 9EW [IO90GW, SU31]
G0 AQZ D T McDonald, 9 Lime St., Nelson, BB9 7BP [IO83VU, SD83]
GW0 ARA L A France Aberystwyth Dar, 5 Heol y Garth, Penparcau, Aberystwyth, SY23 1TE [IO72XJ, SN58]
G0 ARC M D Adams Andover Rad Club, 9 Brancaster Ave, Charlton, Andover, SP10 4EN [IO91FF, SU34]
GM0 ARD J Hoey, 152 Muirhouse Ave, Motherwell, ML1 2LB [IO85AS, NS75]
G0 ARF R G Canning, Green Ln Cottage, Eardisland, Leominster, Herefordshire, HR6 9BN [IO82NF, SO45]
G0 ARG T R Ellinor Ariel Radio Grp, 5 Parkwood Rd, Banstead, SM7 1JJ [IO91VH, TQ25][Correspondence to Ariel Radio Gp (BBC Club), Kingswood Warren, Tadworth, Surrey, KT20 6NP.]
GM0 ARH A R Haxton, Lanimar, Dickson Ave, Hillside, Montrose, Angus, DD10 9EJ [IO86SR, NO76]
GW0 ARK K Hudspeth, 67 Bloomfield Rd, Blackwood, NP2 1LX [IO81JQ, ST19]
G0 ARL P E Shorland, 25 Henfor Cl, Marazion, TR17 0BU [IO70GC, SW53]
G0 ARP J Price, The Ct House, 11 Mount St., Shrewsbury, SY3 8QJ [IO82OR, SJ41]
G0 ARQ M Burchmore, 49 School Ln, Horton Kirby, Dartford, DA4 9DQ [JO01DJ, TQ56]
GM0 ART A Gray, 191 Greengairs Rd, Greengairs, Airdrie, ML6 7SZ [IO85AV, NS77]
G0 ARU J T Lumb, 2 Briarwood Ave, Bury St. Edmunds, IP33 3QF [JO02IF, TL86]
G0 ARV A Gates, 26 Ringway Rd, Gateacre, Liverpool, L25 3QS [IO83NJ, SJ48]
G0 ARW Details withheld at licensee's request by SSL.
GM0 ARY N S Hamilton, Craig Mhor, 39 Highfield Rd, Buckie, AB56 1BE [IO87MQ, NJ46]
G0 ARZ A Everard, 19 Roker Park Rd, Sunderland, SR6 9PF [IO94HW, NZ45]
G0 ASA D Michael, 110 Exeter Rd, Edmonton, London, N9 0LL [IO91XO, TQ39]
G0 ASB M J Wakeford, 11 Pendine Ave, Worthing, BN11 2NA [IO90TT, TQ10]
G0 ASC A S Curtis Winchmore Am RA, 1 Blagdens Cl, Southgate, London, N14 6DE [IO91WP, TQ29]
G0 ASD G F Simcox, 11 Tyrol Cl, Wollaston, Stourbridge, DY8 3NL [IO82WL, SO88]
G0 ASF R O Geis, 67 Battersby Rd, Catford, London, SE6 1SA [JO01AK, TQ37]
G0 ASG P H Fautley, The Old Vicarage, Thornton-le-Street, Thirsk, N Yorks, YO7 4DR [IO94HG, SE48]
G0 ASH J J Coupe, 65 Irongate, Bamber Bridge, Preston, PR5 6UY [IO83PR, SD52]
G0 ASI K J Rodden, 12 Bonis Cres, Stockport, SK2 7HH [IO83WJ, SJ98]
G0 ASK S Gray, 29 Verity Walk, Wordsley, Stourbridge, DY8 4XS [IO82WL, SO88]
G0 ASL J M Gray, 29 Verity Walk, Wordsley, Stourbridge, DY8 4XS [IO82WL, SO88]
G0 ASM N E Marston, 14 Greystoke Ave, Sunderland, SR2 9DX [IO94HV, NZ35]
G0 ASN S J Hendry, 48 Downer Rd North, Benfleet, SS7 3EG [JO01GN, TQ78]
G0 ASP E N Mason, 57 Hampton Hill, Wellington, Telford, TF1 2ER [IO82RQ, SJ61]
G0 ASQ L M Williams, 24 Alston Dr, Bare, Morecambe, LA4 6QR [IO84OB, SD46]
G0 AST J C Willis, Milverton, 89 Earlswood Common, Earlswood, Solihull, B94 5SJ [IO92CI, SP17]
G0 ASV A Taylor, 17 Egton Cl, Redcar, TS10 4PG [IO94LO, NZ62]
G0 ASX N V Ward, 104 Kingsley Rd, Bishops Tachbrook, Leamington Spa, CV33 9RZ [IO92FF, SP36]
GM0 ASY L Lambert, Netherfield, Ardnadamby Sandbank, Dunoon, Argyll, PA23 8QH [IO75MX, NS18]
G0 ASZ E J Gamble, 87 Silverdale Dr, Waterlooville, PO7 6DP [IO90LV, SU61]
GM0 ATA Dr R G Mulholland, 1 Larch Gr, Milton of Campsie, Glasgow, G65 8HG [IO75WX, NS67]
G0 ATB V V Herbert, 98 Blithdale Rd, Abbey Wood, London, SE2 9HL [JO01BL, TQ47]
G0 ATC Details withheld at licensee's request by SSL.
G0 ATD D E J Welch, 51 Verbena Way, Worle, Weston Super Mare, BS22 0RL [IO81MI, ST36]
G0 ATG T Goodyer, 23 Mundays Row, Horndean, Waterlooville, PO8 0HF [IO90UW, SU71]
G0 ATJ S E Cook, 301 Brodie Ave, Liverpool, L19 7ND [IO83NI, SJ48]
G0 ATK G T P Atkins, 97 South St., Tillingham, Southminster, CM0 7TH [JO01KQ, TL90]
GM0 ATL P J Smith, 29 Rowan Dr, Bearsden, Glasgow, G61 3HQ [IO75UW, NS57]
G0 ATN C A Waters, Sycamore Cottage, Millhill, Capel St Mary, Suffolk, IP9 2JE [JO02MA, TM03]
G0 ATO Details withheld at licensee's request by SSL.
G0 ATP C Abela, 14 Warren Rd, Barkingside, Ilford, IG6 1BJ [JO01BN, TQ48]
GM0 ATQ L Morgan, 25 Glamis Dr, Greenock, PA16 7NA [IO75OW, NS27]
G0 ATR Details withheld at licensee's request by SSL.
G0 ATS E W Green, Chylean, Tintagel, Cornwall, PL34 0HH [IO70PP, SX08]
G0 ATT J W Attlee, Rods Bygata 1, S42300 Torslanda, Sweden
G0 ATV A F G Willis, 23 Hockney Ave, Barton Seagrave, Kettering, NN15 5UF [IO92PJ, SP87]
G0 ATW J A Ferrier, 30 Grimsby Rd, Laceby, Grimsby, DN37 7DB [IO93WM, TA20]
G0 ATX A R Clark, Vanninecourt, Church Rd, Cam, Dursley, GL11 5PQ [IO81TQ, ST79]
G0 ATY F I Arnold, 6 Staden Park, Trimingham, Norwich, NR11 8HX [JO02QV, TG23]
G0 ATZ C A Best, 34 Julius Hill, Warfield, Bracknell, RG42 3UN [IO91PK, SU87]
G0 AUB F G Groves, 14 Edenfield Rd, Mobberley, Knutsford, WA16 7HE [IO83UH, SJ77]
G0 AUF R D G Griffiths, 26 Hamilton Rd, Morecambe, LA4 6QG [IO84OB, SD46]
G0 AUG T A Hopkinson, Whitbarrow Hall, Caravan Park, Berrier, Penrith, Cumbria, CA11 0XB [IO84NP, NY42]
G0 AUH M Hopkinson, Whitbarrow Hall, Caravan Park, Berrier, Penrith, Cumbria, CA11 0XB [IO84NP, NY42]
G0 AUI C L Stiller, 6 Barn Cottage Ln, Haywards Heath, RH16 3QW [IO91WA, TQ32]
G0 AUJ M D O'Dell, 22 Deer Cl, Hertford, SG13 7NR [IO91XS, TL31]
G0 AUK R J C Broadbent Amsat-Uk, 94 Herongate Rd, Wanstead Park, London, E12 5EQ [JO01AN, TQ48]
GM0 AUL R M McKenzie, 26 Gladstone Pl, Woodside, Aberdeen, AB2 2RP [IO87TH, NJ72]
G0 AUM J D Castree, 46 Chapel Ct, Brierley Hill, DY5 2UT [IO82WL, SO98]
G0 AUN Details withheld at licensee's request by SSL.
G0 AUO D A S Drybrough, Dolphin Cottage, Mounts Ln, Newnham, Daventry, NN11 3ES [IO92KF, SP55]
G0 AUQ Details withheld at licensee's request by SSL.[Op: G Parsons. Station located near Nuneaton.]
G0 AUR A Hogg, 1 Champions Way, South Woodham Ferrers, Chelmsford, CM3 5NJ [JO01HP, TQ89]
G0 AUT A B Utting, 9 Sydney Rd, Spixworth, Norwich, NR10 3PG [JO02PQ, TG21]
G0 AUU R A W D Stockbridge, Jaro, Berrys Hill, Berrys Green, Westerham, TN16 3AG [JO01AH, TQ45]
G0 AUV A E Johnson, 125 Charles St., Sileby, Loughborough, LE12 7SH [IO92KR, SK61]
G0 AUW R G Philpott, 4 Dukeswood, Radfall Rd, Chestfield, Whitstable, CT5 3PJ [JO01MI, TR16]
G0 AUX M K Daly, Granarogue, Carrickmacross, Co Managhan, Eire
G0 AUZ Details withheld at licensee's request by SSL.
GM0 AVB K G Graham, 106 Lochside Rd, Lochside, Dumfries, DG2 0LU [IO85EB, NX97]
GW0 AVD P Richards, Gilce, Llanfawr Rd, Holyhead, LL65 2HE [IO73QH, SH28]
G0 AVE D Brannon, 3 Barlow Rd, Chapel En-le-Frith, Stockport, Ches, SK12 6NH [IO83XI, SJ98]
GD0 AVF H Robins, 1 Mona St., Peel, IM5 1HJ
G0 AVI P Gellion, 6 Poplars Dr, Castle Bromwich, Birmingham, B36 9DR [IO92CM, SP18]
G0 AVJ M G Searley, 75 St. Dominic Park, Harrowbarrow, Callington, Cornwall, PL17 8BW [IO70VL, SX46]
G0 AVO Details withheld at licensee's request by SSL.
G0 AVP P A Raxworthy, 327 Fareham Rd, Gosport, PO13 0AB [IO90JT, SU50]
GM0 AVR Details withheld at licensee's request by SSL.[Op: C Roberts, 4 Ladieside, Brae, (North) Shetland Islands, UK, ZE2 9SX.]
G0 AVS H G Hough, 19 Valentia Rd, Hoylake, Wirral, L47 2AN [IO83JJ, SJ28]
G0 AVU G H Logan, Fenton Hill Farm, Wooler, Northd., NE71 6JL [IO85XO, NT93]
GW0 AVW T Steele, 92 Commercial St., Senghenydd, Caerphilly, CF8 4GZ [IO81HN, ST08]
G0 AVY M D Thatcher, 23 Fairfield Rd, Petts Wood, Orpington, BR5 1JR [JO01AJ, TQ46]
G0 AWA M W Hall, 5 Sandringham Way, Ponteland, Newcastle upon Tyne, NE20 9AE [IO95CA, NZ17]
G0 AWB Details withheld at licensee's request by SSL.[Op: A W Bradley.]
G0 AWC K Chew, 6 West Dr, Nafferton, Driffield, YO25 0QD [IO94TA, TA05]
G0 AWD Details withheld at licensee's request by SSL.
G0 AWE K Boon, 137 Chantry Rd, Romanby, Northallerton, DL7 8JJ [IO94GI, SE39]
G0 AWG W H McNamara, 22 Cissbury Ave, Peacehaven, BN10 8TJ [JO00AS, TQ40]
G0 AWH N Bergstrom-Allen, 35 Acorn Ave, Bar Hill, Cambridge, CB3 8DT [JO02AG, TL36]
GI0 AWK D Payne, 10 Grovemount Ct, Waterside, Altnagelvin, Londonderry, BT47 1JP [IO64IX, C41]
G0 AWL Details withheld at licensee's request by SSL.[Op: R E Hope. Station located near Lichfield, Staffs.]
G0 AWM C R Warr, 29 Barton Rd, Lancaster, LA1 4ER [IO84OA, SD45]
GW0 AWN H Hampson, Argoed, 3 Sychnant View, Pemaenmawr, Gwynedd, LL34 6TN [IO83BG, SH77]
G0 AWP J Wootton, 76 Woodland Dr, Anlaby, Hull, HU10 7HX [IO93SR, TA02]
G0 AWQ Details withheld at licensee's request by SSL.
G0 AWR D Proud, Willow Cottage, Tresevern, Stithians, Truro, TR3 7AT [IO70JE, SW73]
G0 AWS A W Sharrock, 373 Leach Ln, Sutton Leach, St. Helens, WA9 4LZ [IO83PK, SJ59]
GW0 AWT S E Richardson, 3 Bronfena, Caio, Llanwrda, SA19 8RF [IO82AB, SN64]
G0 AWV Details withheld at licensee's request by SSL.
G0 AWX J Fletcher, 5 Highridge Green, Bishopsworth, Bristol, BS13 8AT [IO81QK, ST56]
G0 AWY R K Titmuss, 70 Mallards Rise, Church Langley, Harlow, CM17 9PL [JO01BS, TL40]
G0 AWZ D J Richardson, Becksyde, Cow Moor Bridge, Stockton Ln, York, YO3 9UA [IO93LX, SE65]
G0 AXA A G Hargrave, 36 Fromondes Rd, Cheam, Sutton, SM3 8QR [IO91VI, TQ26]
G0 AXB J R Moore, 53 Thatchers Ct, Westlands, Droitwich, WR9 9EG [IO82WG, SO86]
G0 AXC R E Hamson Sidmouth ARS, 43 Arcot Park, Sidmouth, EX10 9HU [IO80JQ, SY18]
G0 AXD Dr S S Waters, 27 Mill Ln, Shepherdswell, Dover, CT15 7LJ [JO01OE, TR24]
G0 AXE M T Davenport, 119 Gravel Ln, Wilmslow, SK9 6EG [IO83UH, SJ87]
G0 AXF A W Ainsworth, 19 Huntsman Gr, Blakelands, Milton Keynes, MK14 5HS [IO92PB, SP84]
G0 AXH H Orchard, 3 Chapel Hill, Luxulyan, near Bodmin, Cornwall, PL30 5EN [IO70OJ, SX05]
G0 AXI Dr N J H Davies, Wood End House, Ridgeway Ln, Lymington, SO41 8AA [IO90FR, SZ39]
G0 AXJ Details withheld at licensee's request by SSL.[Correspondence to: 4/1 Meisenweg, 3208 Giesen, nr Hildesheim, Germany.]
G0 AXK B H Taylor, Charleston Cottage, Polcoverack, Coverack, Helston, TR12 6SW [IO70KA, SW71]
G0 AXM J M M Thomson, 7 Hatchgate Gdns, Burnham, Slough, SL1 8DD [IO91QM, SU98]
G0 AXN Details withheld at licensee's request by SSL.[Station located near Shaftesbury, Dorset, ST82]
G0 AXO Details withheld at licensee's request by SSL.
G0 AXP R Knight, 14 Redwood Dr, Burntwood, WS7 8AS [IO92AQ, SK00]
G0 AXQ P E Austin, 29 Salisbury Cl, Murston, Sittingbourne, ME10 3BP [JO01JI, TQ96]
G0 AXR R L Ashton, 42 Greenslade Gr, Hednesford, Cannock, WS12 5QR [IO92AR, SK01]
G0 AXS S C Birch, 29 Manners Rd, Southsea, PO4 0BA [IO90LT, SZ69]
G0 AXU G Woods, Leyland House, Farleton, Lancaster, Lancs, LA2 9LF [IO84QC, SD56]
G0 AXV S J Rigby, 55 Wood Ln, Streetly, Sutton Coldfield, B74 3LS [IO92BN, SP09]
GM0 AXX F A Gilhooly, 129 Main St., Davidsons Mains, Edinburgh, EH4 5AG [IO85IX, NT27]
GM0 AXY E K Dons, 37 Ashley Dr, Edinburgh, EH11 1RP [IO85JW, NT27]
G0 AXZ W D Johnson, 29 Wentworth Park, Allendale, Hexham, NE47 9DR [IO84VV, NY85]
G0 AYA M L Chown, 15 Hambleden Walk, Poundfield Est, Maidenhead, SL6 7UH [IO91PM, SU88]
GI0 AYB J O Thone, 23 Caldwell Park, Portrush, BT56 8PJ [IO65QE, C83]
G0 AYC C J Porter, 10 Cotefield Dr, Leighton Buzzard, LU7 8DQ [IO91QW, SP92]
G0 AYD D W Dixon, 3 Towns End, Wylye, Warminster, Wilts, BA12 0RN [IO91AD, SU03]
G0 AYE Details withheld at licensee's request by SSL.
G0 AYF W Collier, 38 Swan Ln, Hindley Green, Hindley, Wigan, WN2 4HF [IO83RM, SD60]
GI0 AYG M A Evans, 12 Tullymore Park, Ballymena, BT42 2AU [IO64UU, D10]
G0 AYI B Spencer, 161 West Ln, Hayling Island, PO11 0JW [IO90MT, SU70]
G0 AYM K C Luxton, 71 Woodland Cl, Barnstaple, EX32 0EG [IO71XB, SS52]
G0 AYN Details withheld at licensee's request by SSL.
GW0 AYP D P Graves, 185 Rhyl Coast Rd, Rhyl, LL18 3US [IO83GH, SJ08]
GW0 AYQ R A Smith, Craig y Wylan, 4 Glan Ysgethin, Talybont, LL43 2BB [IO72WS, SH52]
GM0 AYR M W Dalrymple Ayr ARG, 11 Shawfield Ave, Ayr, KA7 4RE [IO75QK, NS31]
GM0 AYT A H McDougall, Ceol Na Mara, 3 Bowfield Rd, West Kilbride, KA23 9LB [IO75NQ, NS24]
G0 AYU Details withheld at licensee's request by SSL.
G0 AYV D E Burtenshaw, Rivendell, Yapton Rd, Barnham, Bognor Regis, PO22 0BA [IO90QT, SU90]
GM0 AYW R W Evans, 13 Carlin Craig, Kinghorn, Burntisland, KY3 9RX [IO86JB, NT28]
G0 AYX P J Towell, 43 High St., Cranford, Kettering, NN14 4AA [IO92QJ, SP97]
G0 AYY A P Perry, Chala, Woodhouse Hill, Uplyme, Lyme Regis, Dorset, DT7 3SX [IO80MR, SY39]
G0 AYZ Details withheld at licensee's request by SSL.
GI0 AZA E J M Harper, 404 Foreglen Rd, Dungiven, Londonderry, BT47 4PN [IO64MW, C60]
GI0 AZB H I Evans, Ovlle House, 404 Foreglen Rd, Dungiven, Londonderry, BT47 4PN [IO64MW, C60]
GM0 AZC J Sherry, 26 Grahamshill Terr, Fankerton, Denny, FK6 5HX [IO86AA, NS78]
G0 AZD G R Hayter, 22 Golden Farm Rd, Beeches Est, Cirencester, GL7 1BX [IO91AR, SP00]
G0 AZE L Owen, 68 Clevedon Rd, Tickenham, Clevedon, BS21 6RD [IO81OK, ST47]
G0 AZF B F Gibbard, 58 Norwich Ave, Southend on Sea, SS2 4DG [JO01IN, TQ88]
G0 AZG T J Wharton, Bashall Eaves, nr Clitheroe, Lancs, BB7 3DA [IO83SV, SD64]
G0 AZH J G Wharton, 66 Hayhurst St., Clitheroe, BB7 1ND [IO83TU, SD74]
G0 AZI A A Siviter, Cherry Tree Cottage, New Rd, Summercourt, Newquay, TR8 5BX [IO70MI, SW95]
G0 AZJ J J Sales, 6 Ennerdale Cl, Ridge, Lancaster, LA1 3NB [IO84OB, SD46]
GI0 AZL M O R O'Rourke, Derrycammagh, Castlebellingham, Dundalk, Co Louth, XXXX XXX
G0 AZM M A Walton, 5 Home Farm Ct, Home Farm Cl, Leicester, LE4 0SU [IO92KP, SK50]
G0 AZP S R Tricker, 30 Boswell Rd, Cowley, Oxford, OX4 3HN [IO91JR, SP50]
G0 AZQ A J Pearce, 31 Oxford Rd, Owlsmoor, Camberley, Surrey, GU15 4TR [IO91PI, SU86]
G0 AZR J Norman, The End Peg, The Street, Matlaske, Norwich, Norfolk, NR11 7AQ [JO02OU, TG13]
G0 AZS M C Tinkler, 10 Carnation Cl, Crowthorne, RG45 6UX [IO91OJ, SU86]
G0 AZT E J Schneider, 1826 Van Ness, San Pablo, Ca 94806, USA
GM0 AZU J C Aiken, 48 Kirkwall Ave, Blantyre, Glasgow, G72 9NX [IO75WT, NS65]
GW0 AZW R Dooley, 98 Gelli Aur, Treboeth, Swansea, SA5 9DG [IO81AP, SS69]
G0 AZX R J Pritchard, Whitehall Cottage, Rushwick, Worcester, WR2 5TA [IO82UE, SO85]
G0 AZY E Bjart, 38 Parkhill Cl, Blackwater, Camberley, GU17 0LZ [IO91OH, SU85]
G0 AZZ Details withheld at licensee's request by SSL.[Station located near Chessington.]
G0 BAA P J Kirsop Nth Cheshire AR, 8 Warburton Cl, Lymm, WA13 9QE [IO83SJ, SJ68]
G0 BAB B J Daniel, 91 Beltinge Rd, Herne Bay, CT6 6HW [JO01NI, TR16]
G0 BAD P Lindley, 26 Fairfields Rd, Holmfirth, Huddersfield, HD7 1NP [IO93CN, SE10]

G0 BAF N J Quinn, 15 Newham Ln, Steyning, BN44 3LR [IO90TV, TQ11]
G0 BAG Details withheld at licensee's request by SSL.[Op: R C P Cox. Station located near Waterlooville.]
GW0 BAH P F Bateman, 9 Clos y Deri, Locks Ln, Nottage, Porthcawl, CF36 3PR [IO81DL, SS87]
G0 BAI P J Goodger, 2 Pembroke Ave, South Wigston, Syston, Leicester, LE7 2BZ [IO92LQ, SK61]
G0 BAJ R J Edinborough, Glendevon, 3 Manor Rd, Paignton, TQ3 2HT [IO80FK, SX86]
G0 BAK W E Schofield, 57 Woodhill Rd, Cookbridge, Leeds, LS16 7BZ [IO93EU, SE23]
G0 BAL B W Hardingham, 156 Middletons Ln, Hellesdon, Norwich, NR6 5SQ [JO02PP, TG21]
G0 BAM E A Smith, 166 Tudor Way, Dines Green, Worcester, WR2 5QY [IO82UE, SO85]
G0 BAN J Emerson, 26 Cardwell St., Roker, Sunderland, SR6 0JP [IO94HW, NZ45]
G0 BAO M C B Heath, 89 Manton Rd, Hamworthy, Poole, BH15 4PN [IO80XR, SY99]
G0 BAP R Harman, 42 Newlyn Cl, Midmere Ave, Bransholme, Hull, HU7 4PQ [IO93US, TA13]
G0 BAQ I A Irving, Fourwinds, Woodburn Dr, Leyburn, DL8 5HU [IO94CH, SE19]
G0 BAR A D Goodhew Brickfields Amateur Radio Soci, Strathwell Manor, Strathwell Park, Whitwell, Isle of Wight, PO38 2QU [IO90IO, SZ57]
G0 BAU S Craggs, 34 Allensgreen, Hall Cl Green, Cramlington, NE23 6SQ [IO95FC, NZ27]
G0 BAW L C W Haynes, 9 Heather Gdns, Belton, Great Yarmouth, NR31 9PP [JO02TN, TG40]
G0 BAY S Parker, 1 John St., Glossop, SK13 8DZ [IO93AK, SK09]
GW0 BAZ I J Bateman, 90 Maidenhall, Highnam, Gloucester, GL2 8DL [IO81UV, SO72]
G0 BBA C D B Stoakes, Skidden House Hotel, Skidden Hill, St Ives, Cornwall, TR26 2DU [IO70GF, SW54]
G0 BBB U Grunewald, Nuptown Orchard, Nuptown, Warfield, Berks, RG12 6HU [IO91PJ, SU86]
GW0 BBC A G Williams BBC Cardiff Club, 33 Heol Llanishen Fach, Cardiff, CF4 6LA [IO81JM, ST18]
G0 BBD E V Benou, 10 Culverlands Cl, Stanmore, HA7 3AG [IO91UO, TQ19]
G0 BBE L D B Conlon, 76 Deanwood Cres, Allerton, Bradford, BD15 9BL [IO93CT, SE13]
GW0 BBF D C Skinner, 28 Knoll Gdns, Carmarthen, SA31 3EJ [IO71UU, SN42]
GW0 BBG E J Sawkins, 126 Loughor Rd, Gorseinon, Swansea, SA4 6QY [IO71XP, SS59]
G0 BBH G Howard, 33 The Highway, Peacehaven, BN10 8XL [JO00AS, TQ40]
G0 BBI G Thornton, Park House, Hallgate Ln, Stalmine, Poulton-le-Flyde, Lancs, FY6 0LA [IO83MV, SD34]
G0 BBL J Verduyn, 14 Ragleth Gr, Trowbridge, BA14 7LE [IO81VH, ST85]
G0 BBN K J Hendry, 48 Downer Rd North, Benfleet, SS7 3EG [JO01GN, TQ78]
GW0 BBO J C Stanbury, 23 Crymlyn Parc, Skewen, Neath, SA10 6DG [IO81BP, SS79]
GW0 BBQ C Loosmore, 82 Craiglas, Llangeinor, Bridgend, Mid Glam, CF32 8PT [IO81FN, SS98]
GM0 BBR W L Kemp, 67 Brahan Terr, Letham, Perth, PH1 2LN [IO86GJ, NO02]
G0 BBS G Saxton, 16 South View, Scapegoat Hill, Golcar, Huddersfield, HD7 4NU [IO93BP, SE01]
G0 BBT E Snape, 1 Stephen Cres, Humberston, Grimsby, DN36 4DS [IO93XM, TA30]
GM0 BBU J Bell, 29 Parkwinding, Parkmains, Erskine, Renfrewshire, PA8 7AS [IO75SV, NS46]
G0 BBV H P J Watts, 44 Laurel Rd, Thorpe St. Andrew, Norwich, NR7 9LL [JO02QP, TG21]
G0 BBZ Details withheld at licensee's request by SSL.
GM0 BCA D Swanson, 67 Redford Loan, Edinburgh, EH13 0AU [IO85JV, NT26]
G0 BCE R Twiddy, 6-7 Victoria Pl, Boston, PE21 8UL [IO92XX, TF34]
G0 BCF R J Pickering, 14 Dalestorth Gdns, Skegby, Sutton in Ashfield, NG17 3FT [IO93ID, SK46]
G0 BCH P F Guppy, 37 Park Rd, Aldershot, GU11 3PX [IO91OF, SU84]
GM0 BCI R B Liedberg, 29 Clincarthill Rd, Rutherglen, Glasgow, G73 2LF [IO75VT, NS66]
GD0 BCJ P J McGrath, 87 Clagh Vane, Ballasalla, IM9 2HP
G0 BCK D A Ibbotson, 1 Axholme Ave, Crowle, Scunthorpe, DN17 4EG [IO93NO, SE71]
GW0 BCL Details withheld at licensee's request by SSL.
GD0 BCM W G Livsey, 22 Alberta Dr, Onchan, Douglas, IM3 1LS
G0 BCO A G Duffield, 32 Mount Cl, Honiton, EX14 8QZ [IO80JT, ST10]
GM0 BCP M C Jones, 44 Ainsdale Dr, Priorslee, Telford, Shropshire, TF2 9DJ [IO82SQ, SJ71]
GM0 BCQ D Morrison, Waverley Cottage, Home Farm, Ladykirk, Berwick-upon-Tweed, Berwickshire, TD15 1SU [IO85VR, NT84]
GW0 BCR Details withheld at licensee's request by SSL.
G0 BCS P F Rose, 53 South St., Pennington, Lymington, SO41 8DY [IO90FS, SZ39]
G0 BCT R J Sayer, 16 Hodds Hill, Peatmoor, Swindon, SN5 5BJ [IO91BN, SU18]
G0 BCU D Charnock, 44 Bramshill Cl, Birchwood, Warrington, WA3 6TZ [IO83RK, SJ69]
G0 BCW D Grevett, 45 East Bridge Rd, South Woodham Ferrers, Chelmsford, CM3 5SB [JO01HP, TQ89]
G0 BCX G Eden, 123 Windle Hall Dr, St. Helens, WA10 6QA [IO83PL, SJ59]
GM0 BCY A Carslaw, 51 Stonefield Dr, Lochfield, Paisley, PA2 7QY [IO75TT, NS46]
G0 BCZ R T Johnson, 89 Arlington Gdns, Romford, RM3 0EB [JO01CO, TQ59]
G0 BDA R F Hills, Horncop Bungalow, Heversham, Milnthorpe, Cumbria, LA7 7EB [IO84OF, SD48]
G0 BDB P Stephens, 45 Wyoming Cl, Plymouth, PL3 6SU [IO70WJ, SX55]
G0 BDD P Hulatt, 484 Hern Rd, Ramsey St. Marys, Ramsey, Huntingdon, PE17 1TJ [IO92WL, TL29]
G0 BDE B A Spencer, 146 Knighton Fields Rd East, Leicester, LE2 6DR [IO92KO, SK50]
G0 BDF P L Storey, 16 Spencer Rd, Lutterworth, LE17 4PG [IO92JL, SP58]
G0 BDG P Randall, 75 Brookmead Dr, Wallingford, OX10 9BH [IO91KO, SU68]
GU0 BDI D M Prosser, Rustlings, Les Friquets, St. Andrew, Guernsey, GY6 8SJ
G0 BDJ K J Dower, Springcote, Carlton Rd, Kingsdown, Deal, CT14 8DE [JO01QE, TR34]
G0 BDK B G Steele, 82 Margetts Rd, Kempston, Bedford, MK42 8DT [IO92SC, TL04]
G0 BDL D S S Fraser, 87 Main St., Milton, Derby, DE65 6EF [IO92FU, SK32]
G0 BDM D G Briggs, Ti Mor, 3 Danube Cl, Droitwich, Worcs, WR9 8DD [IO82WG, SO86]
G0 BDN A N Slater, Wood Nook, High Bradley Ln, Bradley, Keighley, West Yorks, BD20 9ES [IO93AW, SE04]
G0 BDP R H Wills, 8 Owlswood, Ridings Mead, Salisbury, SP2 8DN [IO91CB, SU12]
G0 BDR M Gee, 100 Plantation Hill, Kilton, Worksop, S81 0QN [IO93KH, SK57]
G0 BDS B D Scroggs, 10 Lyon Cl, Chelmsford, CM2 8NY [JO01FQ, TL70]
GI0 BDT W Boyd, 54 Farlough Rd, Dungannon, BT71 4DU [IO64PN, H86]
GI0 BDU V Fortune, 11 Enniskeen Ave, Rathcoole Est, Newtownabbey, BT37 9HN [IO74AQ, J38]
G0 BDW R G Kelsall, Keepers Cottage, Breach Ln, Sudbury, Ashbourne, DE6 5HH [IO92CV, SK13]
GI0 BDZ D D Pavis, 269 Lower Braniel Rd, Belfast, BT5 7NR [IO74BN, J37]
G0 BEA P R Vaughan, 102 Whitethorn Ave, West Drayton, UB7 8LB [IO91SM, TQ08]
GI0 BEB T Magee, 10 Abernethy Park, Glengormley, Newtownabbey, BT36 6QQ [IO74AQ, J38]
G0 BEC S J Christmas, 4 The Close, Little Weighton, Cottingham, HU20 3XA [IO93RS, SE93]
G0 BED J G Minshall, 58 Bold St., Leigh, WN7 1LT [IO83RL, SD60]
G0 BEE Details withheld at licensee's request by SSL.
GI0 BEG Dr B G Lindsay, 8 Glenville Parade, Newtownabbey, BT37 0TS [IO74BQ, J38]
G0 BEH R G Madder, 5 The Close, Kington St. Michael, Chippenham, SN14 6LE [IO81WL, ST97]
G0 BEJ Details withheld at licensee's request by SSL.
GM0 BEL G C Lindsay, 6 Netherhouse Ave, Lenzie, Kirkintilloch, Glasgow, G66 5NG [IO75WW, NS67]
G0 BEM R H Moores, 5 Love Ln, Petersfield, GU31 4BU [IO91MA, SU72]
G0 BEN D W Pykett, 13 West Bank, Saxilby, Lincoln, LN1 2LU [IO93QG, SK87]
G0 BEO L W Coley, The Bungalow, Stafford Rd, Great Wyrley, Walsall, WS6 6AX [IO82XP, SJ90]
G0 BEP D R A Boalch, Mill House, Mill Ln, Lyme Regis, DT7 3PU [IO80MR, SY39]
G0 BEQ Details withheld at licensee's request by SSL.
G0 BES S Illsley, 37 Pemerton Rd, Weeke Est, Winchester, SO22 6EU [IO91IB, SU43]
G0 BET R Gough, 149 Blackbrook Way, Moseley Green, Wolverhampton, WV10 8TD [IO82WP, SJ90]
G0 BEU M D L Williams, White Lodge, 21 Brook Ln, Felixstowe, IP11 7EG [JO01QX, TM33]
G0 BEV Dr M N S Hill, Windrush, Jesmond Gdns, Newcastle upon Tyne, NE2 2JN [IO94EX, NZ26]
GW0 BEW C Bluck, 3 Gelli Rd, Ton Pentre, Pentre, CF41 7AS [IO81GP, SS99]
G0 BEX A Penfold, 8 Lavender Cl, Chestfield, Whitstable, CT5 3QL [JO01MI, TR16]
GI0 BEY N S Turkington, 14 Ards Dr, Monkstown Est, Newtownabbey, BT37 0JN [IO74BQ, J38]
G0 BEZ M Stamps, 148 Hargham Rd, Attleborough, NR17 2JP [JO02MM, TM09]
GI0 BFA R I A McKersie, 57 Beechgrove Ave, Belfast, BT6 0ND [IO74BN, J37]
GI0 BFD A R Magee, 15 Glenwood Gdns, Carrigans, Enniskillen, Co Fermanagh, BT74 5LT [IO64EI, H24]
GU0 BFE H B Hodkinson, Arama, Les Baissieres, St Peter Port, Guernsey, GY1 2UD
GJ0 BFF D M Wilden, Yamaska, Cedar Gr, St. Saviour, Jersey, JE2 7GN
G0 BFG S G Perry, 29 Bowmans Dr, Battle, TN33 0LT [JO00FW, TQ71]
G0 BFH P D Thompson, 37 Queens Rd, Bishopsworth, Bristol, BS13 8LF [IO81QJ, ST56]
G0 BFI V E Heard, 12 St. Johns Ct, Tollesbury, Maldon, CM9 8QS [JO01JS, TL91]
G0 BFJ J B Stocks, 96 North St., Lockwood, Huddersfield, HD1 3SL [IO93CP, SE11]
G0 BFK K F James, Yew Tree Cottage, Whiston, Penkridge, Stafford, ST19 5QH [IO82WR, SJ81]
G0 BFM Dr A J Anderson, 16 Whitwell Rd, Norwich, NR1 4HB [JO02PP, TG20]
GD0 BFN J D Kneale, 25 Upper Dukes Rd, Douglas, IM2 4AZ
GI0 BFO R J Dixon, 16 Glenview Ave, Belfast, BT5 7LZ [IO74BN, J37]
G0 BFQ S J Gladwin, 120 Moor End Rd, Halifax, HX2 0JB [IO93BR, SE02]
GM0 BFS D F J Smith, 58 Tullibody Rd, Alloa, FK10 2LY [IO86CC, NS89]
G0 BFU L G Turpin, 14 Maddocks Hill, Warminster, BA12 8DJ [IO81VE, ST84]
GM0 BFW A McClelland, 30/4 Shandon Cres, Haldane, Balloch, Alexandria, G83 8EX [IO76RA, NS38]
G0 BFZ A L Nock, 57 Mushroom Green, Dudley, DY2 0EE [IO82WK, SO98]
G0 BGA G W D Allan, 24 Leadbetter Dr, Bromsgrove, B61 7JG [IO82XI, SO97]
G0 BGB M Davis, 7 Silver Poplars, Holyhead Rd, Albrighton, Wolverhampton, WV7 3AP [IO82VO, SJ80][Operator in Cadiz, Spain.]
G0 BGC Details withheld at licensee's request by SSL.
G0 BGD Details withheld at licensee's request by SSL.
GI0 BGE D Robinson, 53 Toome Rd, Ballee, Ballymena, BT42 2BT [IO64UU, D10]
G0 BGF Details withheld at licensee's request by SSL.
G0 BGG D V Rawlings, High St Garage, Burton, Chippenham, Wilts, SN14 7LT [IO81UM, ST87]
G0 BGH I M Bamford, 60 Coast Dr, Greatstone on Sea, New Romney, Kent, TN28 8NX [JO00LX, TR02]
G0 BGI S J Knight, 14A Manor Rd, Upton Lovel, Warminster, Wilts, BA12 6HR [IO81UC, ST83]
G0 BGK A W Brindle, 27 Cottam Ave, Ingol, Preston, PR2 3XE [IO83PS, SD53]
G0 BGL R Bignell, 31 Western Ave, Appledore, Bideford, EX39 1SD [IO71VB, SS43]
GU0 BGP G M Petit, -le-Vauquiedor, St. Andrews, St. Andrew, Guernsey, Channel Islands, GY6 8TT
G0 BGQ Details withheld at licensee's request by SSL.
G0 BGR R M Jones, Glenroy, 5 School Rd, Marton Moss, Blackpool, FY4 5DS [IO83LS, SD33]
G0 BGV H Eastwood, 3 The Brambles, Thorpe Willoughby, Selby, YO8 9LL [IO93KS, SE53]
G0 BGW Details withheld at licensee's request by SSL.[Station located near Mansfield, Notts.]
G0 BGX O R Pauley, 5 Clare Cl, Waterbeach, Cambridge, CB5 9PS [JO02CG, TL46]
G0 BGY J F Moult, 29 Mount Rd, New Malden, KT3 3JX [IO91UJ, TQ26]
G0 BHA P G White, 8 Audrey Gdns, Wembley, HA0 3TG [IO91UN, TQ18]
G0 BHB H C Varns, 6 Cutthorpe Rd, Newbold, Chesterfield, S42 7AE [IO93GG, SK37]
G0 BHD Details withheld at licensee's request by SSL.
G0 BHG M Rigby, 8 Irvine Rd, Largs, KA30 8JT [IO75NT, NS25]
G0 BHH P J Gregory, 26 Standish Dr, Rainford, St. Helens, WA11 8JY [IO83OM, SD40]
G0 BHI P W Davies, 34 Caernarvon Rd, Up Hatherley, Cheltenham, GL51 5JP [IO81WV, SO92]
GW0 BHJ D R Thomas, 28 Gwerthonor Rd, Gilfach, Bargoed, CF8 8JS [IO81HN, ST08]
G0 BHK E T Stiles, 20 Montford Rd, Whitenap, Romsey, Hants, SO51 5SS [IO90GX, SU32]
G0 BHO S G Robinson, 70 Melford Rd, Bilborough, Nottingham, NG8 4AP [IO92JX, SK54]
G0 BHP P Fambely, 126 Ashton Ln, Sale, M33 5QJ [IO83UK, SJ79]
G0 BHT W Hand, 46 Marlborough Rd, Ipswich, IP4 5AX [JO02OB, TM14]
G0 BHU M M Jamieson, 3 Rowan Way, Bourne, PE10 9SB [IO92TS, TF02]
G0 BHV H S Newns, 739 High St., Kingswinford, DY6 8AA [IO82WL, SO88]
G0 BHW G E Brown, Rose Cottage, Finch Hill, Bulmer, Sudbury, CO10 7EX [JO02IA, TL84]
G0 BHX C S Penton, 7 Oaks Dr, Cannock, WS11 1ET [IO82XQ, SJ91]
G0 BIA R I Armstrong, 64 Churchill Dr, Marske By The Sea, Redcar, TS11 6BE [IO94LO, NZ62]
G0 BID D M Arnold, 27 Stonewell Park Rd, Congresbury, Bristol, BS19 5DP [IO81OI, ST46]
G0 BIE D Lucas, 62 St. Austell Ave, Macclesfield, SK10 3NN [IO83WG, SJ87]
G0 BIF S C Dunn, 7 Deacons Way, Hitchin, SG5 2UF [IO91UX, TL13]
G0 BIG Details withheld at licensee's request by SSL.
G0 BIH R J Tetley, Beacon Croft, Shaw Ln, Lichfield, WS13 7AG [IO92CQ, SK10]
G0 BII B R Abrams, 12 The Tennis, Cassington, Witney, OX8 1EL [IO91HT, SP41]
G0 BIN E P Ashley, 2 School Rd, Bulkington, Nuneaton, CV12 9JB [IO92GL, SP38]
G0 BIP J G I Moore, Highbury House, Berkswell Rd, Meriden, Coventry, CV7 7LB [IO92EK, SP28]
G0 BIQ R Bellamy, 1 Campers Rd, Letchworth, SG6 3QT [IO91VX, TL23]
G0 BIR A Skinner, Halfway Lock Cottage, Upper Gambolds Ln, Stoke Prior, Bromsgrove, B60 3HB [IO82XH, SO96]
G0 BIS Details withheld at licensee's request by SSL.
G0 BIT F E Mance, 2 Shortheath Crest, Farnham, GU9 8SA [IO91OE, SU84]
G0 BIV J E Young, 3 Kensington Cl, Kings Sutton, Banbury, OX17 3XB [IO92IA, SP43]
G0 BIW M C Smith, 394 Longridge Rd, Barking, Essex, IG11 9EG [JO01BN, TQ48]
G0 BIX T J Dansey, Woodlands, 19 Hill Chase, Walderslade, Chatham, ME5 9HE [JO01GI, TQ76]
G0 BIY Details withheld at licensee's request by SSL.
G0 BJA G Vincent-Squibb, 7 Chudleigh Rd, Henley Green, Coventry, CV2 1AF [IO92GK, SP38]
G0 BJC R Wood, Stonnis, 227 The Hill, Cromford, Matlock, DE4 3QL [IO93FC, SK25]
G0 BJD S C Wood, 55 Megdale, Sheriff Fields, Matlock, DE4 3TE [IO93FD, SK26]
G0 BJF Details withheld at licensee's request by SSL.
GI0 BJH Details withheld at licensee's request by SSL.
G0 BJI D E Peacock, 15 Farmfield Rd, Banbury, OX16 9AP [IO92HB, SP43]
G0 BJJ P K Sargent, 14 Swanbourne Rd, Wick, Littlehampton, BN17 6HS [IO90RT, TQ00]
G0 BJK D R Thomas, 33 Chatsworth Rd, Stretford, Manchester, M32 9QF [IO83UK, SJ79]
G0 BJN A G Wright, 23 Rushdon Cl, Grays, RM17 5QN [JO01DL, TQ67]
G0 BJO M W Hampton, 2 Hardie Rd, Stanford-le-Hope, SS17 0PB [JO01FM, TQ68]
G0 BJR G Oliver, 158 High Barn St., Royton, Oldham, OL2 6RW [IO83WN, SD90]
G0 BJS M D J Redwin, 22 Peterborough Rd, Leyton, London, E10 6DL [IO91XN, TQ38]
G0 BJT K P Selby, 37 Omar Rd, Stoke, Coventry, CV2 5JX [IO92GJ, SP37]
G0 BJV Details withheld at licensee's request by SSL.
G0 BJW Details withheld at licensee's request by SSL.
G0 BJZ E Pritchard, 8 Stourmore Cl, Short Heath, Willenhall, WV12 5RF [IO82XO, SJ90]
G0 BKA J Leedham, 27 St. Andrews Cres, Stratford upon Avon, CV37 9QL [IO92DE, SP15]
G0 BKB V C Leedham, 27 St. Andrews Cres, Stratford upon Avon, CV37 9QL [IO92DE, SP15]
G0 BKC B F Glasper, 5 Saunton Rd, Billingham, TS23 3HN [IO94IO, NZ42]
G0 BKD B R Mawn, 46 Newstead Walk, Carshalton, SM5 1AW [IO91VJ, TQ26]
G0 BKE F A Barker, 17 Walders Ave, Sheffield, S6 4AY [IO93FJ, SK39]
G0 BKH B Minton, 49 Runcorn Rd, Barnton, Northwich, CW8 4ES [IO83RG, SJ67]
GW0 BKJ M Glover, 4 New Hospital Villa, Hospital Rd, Talgarth, Brecon, LD3 0DU [IO81JX, SO13]
GW0 BKK J H Webber, 21 Danygrug, Crickhowell, NP8 1DD [IO81KU, SO21]
G0 BKL E B Smith, 40 Grays End Cl, Grays, RM17 5QR [JO01DL, TQ67]
G0 BKN I M McIver, 31 Hartshill, Bedford, MK41 9AL [IO92SD, TL05]
G0 BKO Details withheld at licensee's request by SSL.
G0 BKP J H Dadswell, 30 Barlow Rd, Wendover, Aylesbury, HP22 6HS [IO91PS, SP80]
G0 BKR Dr G G Clark, 3 Southway, Beaconsfield, HP9 1DE [IO91PO, SU98]
GM0 BKS G Christison, 51 Springfield Park, Kinross, KY13 7QT [IO86GF, NO10]
G0 BKU G D Coles, 11 St. Peters Terr, Shepton Mallet, BA4 5BH [IO81RE, ST64]
GM0 BKX T Stewart, 104 Barrhill Rd, Cumnock, KA18 1PU [IO75UK, NS52]
G0 BLA Details withheld at licensee's request by SSL.
G0 BLB R Baker, Holmelea, Upper Bristol Rd, Clutton, Bristol, BS18 4RJ [IO81RI, ST65]
G0 BLD Details withheld at licensee's request by SSL.
G0 BLF J P Goode, 235 Church Hill Rd, Cheam, Sutton, SM3 8LB [IO91VI, TQ26]
G0 BLJ Details withheld at licensee's request by SSL.
G0 BLL A A Dorrill, Woodcote, Wood End Ln, Moulsoe, Newport Pagnell, Bucks, MK16 0HD [IO92PB, SP94]
G0 BLM P Mathews, 25 Shore Mount, Littleborough, OL15 8EN [IO83WP, SD91]
G0 BLN Details withheld at licensee's request by SSL.
G0 BLO B L Osborn, 4 Highfield Cl, Romford, RM5 3RX [JO01CO, TQ59]
G0 BLP Details withheld at licensee's request by SSL.
G0 BLQ H G Cooper, 47 Coppice Way, Aylesbury, Bucks, HP20 1XG [IO91OT, SP81]
G0 BLR J T Booth, 37 Budworth Walk, Handforth Rd, Wilmslow, SK9 2HR [IO83VI, SJ88]
G0 BLS A H Wilkie, 7 Willow Dr, Droitwich, WR9 7QE [IO82WF, SO96]
G0 BLT G J Blackmoor, 4 St. Godwalds Cres, Aston Fields, Bromsgrove, B60 2EB [IO82XH, SO96]
G0 BLU E Mustard, 108 Allandale, Highfield, Hemel Hempstead, HP2 5AT [IO91SS, TL00]
G0 BLV M Puncer, 17 St. Michaels Walk, Eye, Peterborough, PE6 7XG [IO92VO, TF20]
G0 BLW A T Crimlisk, 14 Long Ln, Aughton, Ormskirk, L39 5AT [IO83NN, SD40]
GM0 BLX J Sellars, Sel Crag, 30 Craw Rd, Paisley, PA2 6AD [IO75SU, NS46]
GW0 BLZ H Phillips, Rhos Cottage, Gorsefawr, Grovesend, Swansea, SA4 4WJ [IO71XQ, SN50]
G0 BMB R H Morgan, 41 Juniper Dr, Wombridge Common, Telford, TF2 6SJ [IO82SQ, SJ61]
G0 BMD S A Fromm, 2 Crofts Ln, Newton Longville, Milton Keynes, MK17 0BT [IO91OX, SP83]
G0 BME J M Evans, 212 Grand Dr, Raynes Park, London, SW20 9NB [IO91VJ, TQ26]
G0 BMF R M French, 78 Gilbert Ave, Rugby, CV22 7BZ [IO92II, SP47]
G0 BMG M R Green, 65 Rosamond Rd, Bedford, MK40 3UG [IO92SD, TL05]
G0 BMH J I McLean, 80 Whalley Rd, Clayton-le-Moors, Accrington, BB5 5DT [IO83TS, SD73]
GW0 BMI W K Edmonds, 1 Clas y Deri, Waunarlwydd, Swansea, SA5 4TP [IO71XP, SS69]
G0 BMJ G E Barnard, 12 Mortain Rd, Westham, Pevensey, BN24 5HL [JO00DT, TQ60]
G0 BMK R Mapstone, 11 Kemsley Cl, Northfleet, Gravesend, DA11 8PH [JO01EK, TQ67]
G0 BML T P Garvey, 162 Birchfields Rd, Manchester, M14 6PE [IO83VK, SJ89]
G0 BMN K S Hutchins, 23 Salisbury Rd, Tunbridge Wells, TN4 9DJ [JO01DD, TQ54]
G0 BMP L Gyurgyak, 1 Elmbank Rd, Paignton, TQ4 5NG [IO80FK, SX86]
G0 BMQ M Armstrong, 4 Medway Dr, Preston, Weymouth, DT3 6LF [IO80SP, SY68]
GI0 BMR R M Cherry, 25 Grangewood Rd, Dundonald, Belfast, BT16 0GW [IO74CO, J47]
G0 BMS I J Cooper, 1 Waterside, Columbia Way, Kings Lynn, PE30 2NA [JO02ES, TF62]
G0 BMT J C Walkley, 10 Exton, Dunster Cres, Weston Super Mare, BS24 9EH [IO81MH, ST35]
G0 BMU T C Drewitt, 137 Linden Rd, Reading, RG2 7EJ [IO91MK, SU77]
G0 BMX Details withheld at licensee's request by SSL.
G0 BMZ B E A Lindgren, Vestra, Rosenlundsvegen 22, S-66341, Hammaro, Sweden
G0 BNA Details withheld at licensee's request by SSL.
G0 BNC Details withheld at licensee's request by SSL.[Station located near Witney. Op: R Fleming.]
GW0 BND K N Smith, 29 Wessex Cl, Shavington, Crewe, CW2 5HX [IO83SB, SJ75]
G0 BNE Dr A J Knell, 13 Northd. Rd, Leamington Spa, CV32 6HE [IO92FH, SP36]
G0 BNG L D Britton, Widbrook House, Widbrook, Bradford on Avon, BA15 1UD [IO81VI, ST85]
GW0 BNH R W Bevan, 55 Heol y Gors, Cwmgors, Ammanford, SA18 1PT [IO81BS, SN71]
G0 BNI Details withheld at licensee's request by SSL.
G0 BNJ B J Northway, 3 Kingston Cl, Kingskerswell, Newton Abbot, TQ12 5EW [IO80FL, SX86]
G0 BNK Details withheld at licensee's request by SSL.
GW0 BNN J Bulpin, 81 Merthyr Rd, Pontypridd, CF37 4DD [IO81IO, ST09]
GW0 BNO D J Lindley, 29 Belvedere Cl, Kittle, Swansea, SA3 3LA [IO71XN, SS58]
GW0 BNP R C Johnson, 2 Ruskin Cl, Fairwater, Cwmbran, NP44 4QX [IO81LP, ST29]
GM0 BNQ D Macdonald, Greenbrae Cottage, Auchterless, Turriff, Aberdeenshire, AB53 8HD [IO87TL, NJ74]
G0 BNR N Keightley, Daintree Rd, Ramsey St. Marys, Ramsey, Huntingdon, Cambs, PE17 1TF [IO92WL, TL28]
G0 BNS Details withheld at licensee's request by SSL.[Station located near Orpington.]

G0 BNT M T Collicott, 16 Worcester Rd, Chichester, PO19 4DJ [IO90OU, SU80]
G0 BNU D G Wright, c/o Fco (Lagos), King Charles St, London, SW1A 2AH [IO91WM, TQ37]
G0 BNW W E M Wheeler, 201 Topsham Rd, Exeter, EX2 6AN [IO80FR, SX99]
G0 BNY C R Lee, 12 Clifton Green, Sunnybrow, Crook, DL15 0NP [IO94DQ, NZ13]
G0 BNZ J C Hewitt, 51 Three Oaks Rd, Wythall, Birmingham, B47 6HG [IO92BJ, SP07]
G0 BOB Details withheld at licensee's request by SSL.
G0 BOC E L Pitman, 35 Brackley Way, Totton, Southampton, SO40 3HP [IO90FW, SU31]
GW0 BOD B G Morris, School House, Heol Dyfed, Fishguard, Dyfed, SA65 9DS [IO71MX, SM93]
G0 BOE W Piotrowski, 63 Headley Dr, Epsom Downs, Epsom, KT18 5RP [IO91VH, TQ25]
G0 BOF W F Ardley, 353 Rush Green Rd, Romford, RM7 0NJ [JO01CN, TQ58]
G0 BOH D J Bowman, 6 Linksfield, Denton, Manchester, M34 3TE [IO83WL, SJ99]
G0 BOI D S J Gray, 58 Dumont Ave, St. Osyth, Clacton on Sea, CO16 8JP [JO01MS, TM01]
GW0 BOJ H N Wilmington, 10 Lon yr Awel, Pontyclun, CF7 9AW [IO81HN, ST08]
G0 BOK H C Davies, Inland Cottage, The Vauld, Marden, Herefordshire, HR1 3HA [IO82PD, SO54]
G0 BOL A H Hart, 54 Greenpark Rd, Exmouth, EX8 4JT [IO80HP, SY08]
G0 BOM A J Samouelle, 6 Poulsen Cl, Warminster, BA12 9QD [IO81VE, ST84]
G0 BON I P Rogers, 26 Milton Rd, Slough, SL2 1PF [IO91QM, SU98]
G0 BOO B H Gilbert, Sliver Micha, Hunts Corner, Banham, Norfolk, NR16 2HL [JO02MK, TM08]
G0 BOP F U Hook, Delta, Bracken Ln, Storrington, Pulborough, RH20 3HS [IO90SW, TQ11]
G0 BOQ P Macdonald, 77 Ousebank Way, Stony Stratford, Milton Keynes, MK11 1LD [IO92NB, SP74]
G0 BOR J D Goodall, 287 Newbrook Rd, Atherton, Manchester, M46 9QZ [IO83SM, SD60]
G0 BOT D L Ashby, 22 Middle Field Rd, Birmingham, B31 3EH [IO92AJ, SP07]
G0 BOW R E Bowgen, 2 Pelham Cl, Old Basing, Basingstoke, RG24 7HU [IO91LG, SU65]
G0 BOX C T Dobbs, 160 Leeds Rd, Barwick in Elmet, Leeds, LS15 4HS [IO93HT, SE33]
G0 BOY W J Fitzgerald, 37A St Domingo Gr, Everton, Liverpool, L5 6RP [IO83MK, SJ39]
G0 BPA B J Lyford, 37 Ivor Cl, Holbury, Southampton, SO45 2NY [IO90HT, SU40]
G0 BPB F Roche, Flat G, 8 Clerkenwell Cl, London, EC1R 0DY [IO91WM, TQ38]
G0 BPC Details withheld at licensee's request by SSL.[Correspondence to BPC, BCM 5114, London WC1N 3XX.]
G0 BPF W Johnstone, Stonehouse Farm Barn, Scales, Aspatria, Carlisle, CA5 3NW [IO84IT, NY14]
GM0 BPH J D Hogarth, 24 Broomlea, Kelso, TD5 7RB [IO85SO, NT73]
G0 BPJ J Patterson, 22 Nook Rd, Scholes, Leeds, LS15 4AU [IO93GU, SE33]
G0 BPK N P Ferguson, Royd Moor, Badsworth, Pontefract, Yorks, WF9 1AZ [IO93IP, SE41]
G0 BPL D F Farnham, 24 Downham Rd, Watlington, Kings Lynn, PE33 0HS [JO02EP, TF61]
G0 BPM Details withheld at licensee's request by SSL.
G0 BPN F Blades, 1 Rectory Cl, Garforth, Leeds, LS25 1HH [IO93HT, SE43]
GM0 BPO E C D Brown, Walterstead, Ladykirk, Berwick upon Tweed, Berwickshire, TD15 1XW [IO85VR, NT84]
G0 BPQ S Hall, Sunningdale, Whalley Rd, Langho, Blackburn, BB6 8EF [IO83ST, SD73]
G0 BPR S Whitnear, 55 Brier Cres, Nelson, BB9 0QD [IO83VT, SD83]
G0 BPS R A Pascoe, Seaview House, Crete Rd East, Folkestone, Kent, CT18 7EG [JO01OC, TR23]
GM0 BPT A T L Murray, 1 Gordon Rd, Edinburgh, EH12 6NB [IO85IW, NT27]
G0 BPU M R Johnson, 23 Camden Rd, Ipswich, IP3 8JW [JO02OB, TM14]
GW0 BPV F R Hastings, Guileen, Felinwynt, Aberporth, Cardigan, Dyfed, SA43 1RW [IO72QD, SN25]
G0 BPW M G Howes, 2 Kenworthy Rise, Leeds, LS16 7QW [IO93EU, SE24]
G0 BPX D A Norridge, 125 Ferry Rd, Marston, Oxford, OX3 0EX [IO91JS, SP50]
GM0 BPY A M McCreadie, 16 Fancove Pl, Eyemouth, TD14 5JQ [IO85WU, NT96]
G0 BPZ P A Maple, 8 Cavendish Dr, Ashbourne, DE6 1SR [IO93DA, SK14]
G0 BQA R C Mogford, 4 The Green, New Malden, KT3 3LD [IO91UJ, TQ26]
G0 BQB P J Smith, 36 Childs Rd, Alverthorpe, Wakefield, WF2 0BS [IO93FQ, SE32]
G0 BQD S R Meddick, 53 Reforne, Easton, Portland, DT5 2AW [IO80SN, SY67]
G0 BQE J E Chown, 15 Hambleden Walk, Maidenhead, SL6 7UH [IO91PM, SU88]
G0 BQF P Collett, 26 Hertford Rd, Hoddesdon, EN11 9JR [IO91XS, TL30]
G0 BQH D Cusick, Chapel Cottage, Holmans Pl, Porthleven, Helston, Cornwall, TR13 9HZ [IO70IC, SW62]
G0 BQI M A Chapman, 6A Rees St, Islington, London, N1 7AR [IO91WM, TQ38]
G0 BQK M F Hall, 31 Winchester Cl, Stratton St. Margare, Swindon, SN3 4HB [IO91DN, SU18]
G0 BQO J W D Rawson, 3 Cooks Cl, Kesgrave, Ipswich, IP5 7YT [JO02OB, TM24]
G0 BQP Dr J B Simpson, 18 Southdean Dr, Hemlington, Middlesbrough, TS8 9HH [IO94JM, NZ41]
GM0 BQQ I Laczko, 34 Airds Dr, Lochvale, Dumfries, DG1 4EW [IO85FB, NX97]
G0 BQT C G Atkinson, 126 High Wood Rd, Hoddesdon, EN11 9AU [IO91XS, TL31]
G0 BQU A Symons, 2 St. Michaels St., Penzance, TR18 2DG [IO70FC, SW43]
G0 BQV M Ashdown, 42 Alpine Ave, Tolworth, Surbiton, KT5 9RJ [IO91UJ, TQ26]
G0 BQW S L F Robinson, 114 Hopefield Ave, Frecheville, Sheffield, S12 4XE [IO93HI, SK38]
GI0 BQX D A Maguire, 16 Kilmacormick Dr, Enniskillen, BT74 6EP [IO64EI, H24]
G0 BQY P F Hawes, 20 Leverstock Green Rd, Hemel Hempstead, HP2 4HQ [IO91SS, TL00]
G0 BQZ S A Smith, 18 Stratford Dr, Eynsham, Witney, OX8 1QJ [IO91HS, SP40]
G0 BRA Details withheld at licensee's request by SSL.
G0 BRB L N Bailey, RAF Mildenhall, Box 3484, Bury St Edmunds, Suffolk, XX99 1AA
G0 BRC M J Pearson Bredhurst Receiving & Transmit, 56 Parkwood Green, Parkwood, Gillingham, ME8 9PP [JO01HI, TQ86]
G0 BRE P M Page, 11 Laburnum Rd, Weston Super Mare, BS23 3LL [IO81MI, ST36]
G0 BRG J G Butcher, 135 Gladstone St., West Bromwich, B71 1EW [IO92AM, SP09]
G0 BRH C Waters, 468 Buckfield Rd, Leominster, HR6 8SD [IO82OF, SO45]
G0 BRI J P Billingham Brighton College Amateur Radio, 22 Lawday Pl Ln, Farnham, GU9 0BT [IO91OF, SU84]
GM0 BRJ D S Wilson, Linnburn, Tak Ma Doon Rd, Kilsyth, Glasgow, G65 0RS [IO75XX, NS77]
G0 BRL M W H Bevan, 22 Spring Cres, Brown Edge, Stoke on Trent, ST6 8QH [IO83WB, SJ95]
G0 BRM N J Entwistle, Park Garden House, Park Garden, West Row, Bury St Edmunds, IP28 8PB [JO02FI, TL67]
GI0 BRO R A Wilson, 504 Castlereagh Rd, Belfast, BT5 6QA [IO74BN, J37]
G0 BRP Details withheld at licensee's request by SSL.
G0 BRQ A H Morris, 108 Lytchett Dr, Broadstone, BH18 9NR [IO80XS, SY99]
G0 BRR A J Turner Restormel Raynet, 19 Trelawney Rd, St. Austell, PL25 4JA [IO70OI, SX05]
GM0 BRS E C Brown The Border ARS, Waltersteads, Ladykirk, Berwick on Tweed, TD15 1XW [IO85VR, NT84]
G0 BRV Details withheld at licensee's request by SSL.
G0 BRW B J Whatling, 6 Rock Rd, Cam, Dursley, GL11 6LF [IO81TQ, ST79]
G0 BRY Details withheld at licensee's request by SSL.
G0 BRZ L T Rimmer, 29 Cheviot Rd, Hazel Gr, Stockport, SK7 5BH [IO83WI, SJ98]
G0 BSA M G Embling, 23 Sinodun Rd, Didcot, OX11 8HP [IO91JO, SU58]
G0 BSB M D Eccleston, Cadney Bank, The Cadney, Bettisfield, Whitchurch, Shropshire, SY13 2LP [IO82OV, SJ43]
G0 BSD D N Tringham, 417 Eastwood, Windmill Hill, Runcorn, WA7 6LJ [IO83PI, SJ58]
G0 BSF H F Rolfe, 46 Great Gdns Rd, Hornchurch, RM11 2BA [JO01CN, TQ58]
G0 BSG Details withheld at licensee's request by SSL.
G0 BSH R C Harman, 22 Ridgebrook Rd, Kidbrooke, London, SE3 9QN [JO01AL, TQ47]
G0 BSJ E R New, 1 Brocklands, Bedhampton, Havant, PO9 1NS [IO90MU, SU70]
G0 BSK K Bryant, Homestead Church Rd, Mabe Burthouse, Penryn, Cornwall, TR10 9HL [IO70KD, SW73]
G0 BSM D A Baker, 9 Croft Cres, Brownhills, Walsall, WS8 7DN [IO92AP, SK00]
G0 BSN R K Vincent-Squibb, 1 Alexander Ave, Earl Shilton, Leicester, LE9 7AF [IO92IN, SP49]
GI0 BSO D McGilloway, 17 Kingsfort Park, Londonderry, BT48 7SY [IO65IA, C41]
G0 BSP R A Snow, 73 Boxtree Rd, Harrow Weald, Harrow, HA3 6TN [IO91TO, TQ19]
G0 BSQ A Osborne, 8 Victory Cttgs, West Newton, Kings Lynn, PE31 6AT [JO02GT, TF62]
G0 BSS Details withheld at licensee's request by SSL.
G0 BST J C R I'Anson-Holton, 2 Winchester Rd, Northwood, HA6 1JE [IO91TO, TQ18]
G0 BSU R S Thawley, 1 Ridgeway, Adlington Rd, Wilmslow, Ches, SK9 2BP [IO83VH, SJ88]
G0 BSW F Brennan, 6 Winterburn Rd, Blackburn, BB2 4NJ [IO83SR, SD62]
G0 BSX Dr P D Meiring, 18 Slayleigh Ln, Sheffield, S10 3RF [IO93FI, SK38]
G0 BSY J G Rennie, Viejer Lodge, Covenham Rd, Yarburgh, nr Louth, Lincs, LN11 0PL [JO03AK, TF39]
G0 BTA D M E Round, 20 Colders Green, Meltham, Huddersfield, HD7 3JH [IO93BO, SE01]
GW0 BTB B T Botham, The Cherries, Benllech, Anglesey, Gwynedd, LL74 8SR [IO73VH, SH58]
G0 BTD Details withheld at licensee's request by SSL.
G0 BTH P Clark, 180 Roselands Dr, Paignton, TQ4 7RW [IO80FJ, SX85]
GM0 BTK Dr W D Rossmann, Burnbrae Cottage, Kirriemuir, Angus, DD8 5JH [IO86JQ, NO25]
GM0 BTL J Ritchie, Sunart, Stormyhill Rd, Portree, Isle of Skye, IV51 9DT [IO67VJ, NG44]
G0 BTN Details withheld at licensee's request by SSL.
G0 BTO Details withheld at licensee's request by SSL.
G0 BTP Details withheld at licensee's request by SSL.
G0 BTQ N W J Burge, 43 Bourn Rise, Pinhoe, Exeter, EX4 8QD [IO80GR, SX99]
G0 BTS L J Jenkins, 134 Manor Rd, Benfleet, SS7 4HU [JO01GN, TQ78]
G0 BTT E J Clay, 23 New St., Sleaford, NG34 7HG [IO92TX, TF04]
G0 BTU K D Tupman, 23 Moggs Mead, Petersfield, GU31 4NX [IO91MA, SU72]
G0 BTV S Coleman, 42 Regent St., Sutton in Ashfield, NG17 2EH [IO93IC, SK45]
GW0 BTW J F R Jackson, 23 The Green, Pembroke, SA71 4NU [IO71MQ, SM90]
G0 BTX D A Norman, Meadowbank, 9 Station Cttgs, Brampton Rd, Huntingdon, Cambs, PE18 6BW [IO92VH, TL27]
G0 BTZ A G Pollard, 8 Norman Pl, Leeds, LS8 2AW [IO93FU, SE33]
GW0 BUA E J Rees, 461 Carmarthen Rd, Cwmdu, Swansea, SA5 8LN [IO81AP, SS69]
G0 BUC K W Simpson, 5 Cedar Cl, The Elms, Torksey, Lincoln, LN1 2NH [IO93PH, SK87]
G0 BUD J J Spearing, 107 Bourne Way, Bromley, BR2 7EX [JO01AI, TQ36]
GM0 BUE A Stark, 12 Kelvin Way, Kilsyth, Glasgow, G65 9UL [IO75XX, NS77]
G0 BUF H Trueman, 12 Darville, Shrewsbury, SY1 2UG [IO82PR, SJ51]
GM0 BUH A Westwater, 22 Hill St., Dysart, Kirkcaldy, KY1 2XA [IO86KD, NT39]
GM0 BUI G Williamson, 2 Laburnum Gr, Burntisland, KY3 9EU [IO86JB, NT28]
G0 BUJ R F Pelling, The Spinney, Shirwell, Barnstaple, Devon, EX31 4JR [IO71XC, SS53]
G0 BUK I K Mitchell, Cornerfield, Five Ash Down, Uckfield, E Sussex, TN22 3AP [JO00BX, TQ42]
GM0 BUL A B Horton, Thornbank, Blairmore, Argyll, PA23 9TJ [IO75MW, NS17]
G0 BUP P Parish, 21 Cherry Tree Ln, Nettleham, Lincoln, LN2 2PR [IO93SG, TF07]
G0 BUQ H S Blake, 4 Blake Ave, Merrivale, Ross on Wye, HR9 5JP [IO81RV, SO62]
G0 BUR R A Burrows, 1 Browns Ave, Runwell, Wickford, SS11 7PT [JO01GO, TQ79]
G0 BUS D J Johnson, 34 Keats Gdns, Farmhill, Stroud, GL5 4DJ [IO81VS, SO80]
G0 BUV J A Howard, 130 Coventry Rd, Coleshill, Birmingham, B46 3EH [IO92DL, SP28]
G0 BUW P B R Martin, 39A The Grove, Bearsted, Maidstone, ME14 4JB [JO01GG, TQ75]
G0 BUX R H Clinton, 25 Butlers Gr, Great Linford, Milton Keynes, Bucks, MK14 5DT [IO92OB, SP84]
G0 BUY R A J Scheuber, 9 Southfield Rd, Hoddesdon, EN11 9EA [IO91XS, TL30]
G0 BUZ E J Piper, 44 Parsonage Est, Rogate, Petersfield, GU31 5HJ [IO91NA, SU82]
G0 BVB D A Keir, 3 Eastry Ave, Hayes, Bromley, BR2 7PE [JO01AJ, TQ36]
G0 BVC G G Broadhurst, Summerhayes, 10 Ottervale Cl, Rawridge, Honiton, EX14 9TA [IO80KU, ST20]
G0 BVD P J Oakley, Fenham Villa, 12 Bank St., Malvern, Worcs, WR14 2JN [IO82UC, SO74]
G0 BVH Details withheld at licensee's request by SSL.
G0 BVJ J W Atkinson, 2 Ct Ave, Halewood, Liverpool, L26 6LD [IO83OI, SJ48]
GM0 BVK E Evers, 34 Noddleburn Rd, Largs, KA30 8PZ [IO75NT, NS26]
G0 BVL V Lockhart, 17 Balfour St., Blyth, NE24 1NR [IO95FD, NZ38]
G0 BVM J M Speers, 187 Worsley Rd, Winton, Eccles, Manchester, M30 8BP [IO83TL, SJ79]
GW0 BVN D W Evans, 3 Heol Elfed, Cefn Caeau, Llanelli, SA14 9DY [IO71WQ, SN50]
G0 BVO J G Teasdale, 1 Newtown Bungalows, Newtown, Spennymoor, DL16 7QS [IO94EQ, NZ23]
G0 BVP K Lofthouse, 219 Southport Rd, Ulnes Walton, Leyland, Preston, PR5 3LP [IO83PP, SD51]
G0 BVQ C Hawkes, 12 Summer Hall Ing, Wyke, Bradford, BD12 8DN [IO93CR, SE12]
G0 BVS M J Arthur, 8 Hanbury Rd, Bedworth, Nuneaton, CV12 9BX [IO92GL, SP38]
G0 BVT A G Tonge, 30 Cardigan Ave, Morley, Leeds, LS27 0DP [IO93ER, SE22]
G0 BVU D R Davies, 30 Bullpit Rd, Balderton, Newark, NG24 3LY [IO93OB, SK85]
G0 BVV D J Clench, 4 Creston Way, Worcester Park, KT4 8PH [IO91VJ, TQ26]
G0 BVW Q Curzon, 154 Clophill Rd, Maulden, Bedford, MK45 2AE [IO92SA, TL03]
G0 BVX Details withheld at licensee's request by SSL.[Station located near Berkhamsted, Herts. Locator: IO91QS. WAB: SP90. Rateable district: Dacorum. Op: Alan Fincher. QSL via RSGB.]
GW0 BVY D J Lloyd, 6 Neyland Path, Fairwater, Cwmbran, NP44 4PX [IO81LP, ST29]
G0 BVZ Details withheld at licensee's request by SSL.
G0 BWB J D Chappell, 49 Midway, South Crosland, Huddersfield, HD4 7DA [IO93CO, SE11]
G0 BWE D Allen, 19 Stoneyhill, Abbotskerswell, Newton Abbot, TQ12 5LH [IO80EL, SX86]
G0 BWF W Frakes, 59 Wasdale Park, Seascale, CA20 1PD [IO84GJ, NY00]
G0 BWG J M Raynes, 267 Pelham Rd, Immingham, Grimsby, DN40 1JU [IO93VO, TA11]
G0 BWJ K E Miller, 8 Horsham Gdns, Sunderland, SR3 1UJ [IO94HV, NZ35]
G0 BWK R A Lee, 57 Hart Ln, Luton, LU2 0JF [IO91TV, TL12]
G0 BWL P H Carr, 8 Southfield Cl, Horbury, Wakefield, WF4 5AZ [IO93FP, SE21]
G0 BWN Details withheld at licensee's request by SSL.
G0 BWO D Wilkinson, 139 Grosvenor Rd, Dalton, Huddersfield, HD5 9HX [IO93DP, SE11]
G0 BWP B W Passmore, 364 Franklin Rd, Kings Norton, Birmingham, B30 1NG [IO92AK, SP08]
G0 BWQ K M Kitson, 278 Cowcliffe Hill Rd, Huddersfield, HD2 2NE [IO93CQ, SE11]
GM0 BWR B Coan, 60 Elmbank, Menstrie, FK11 7AR [IO86BD, NS89]
GM0 BWS V A Pennycook, 11 Highfields, Dunblane, FK15 9EQ [IO86AE, NN70]
GM0 BWT R A Lapsley, 51 Coney Park, Stirling, FK7 9LU [IO86AC, NS79]
GM0 BWU R Smith, 9 Queen Elizabeth Dr, Castle Douglas, DG7 1HH [IO84AW, NX76]
G0 BWV J L Puttock, 53 Alexandra Ave, Sutton, SM1 2PA [IO91VI, TQ26]
GW0 BWX R Mutch, 6 School Terr, Village Rd, Nercwys, Mold, CH7 4EX [IO83KD, SJ26]
G0 BWY G I Fearnside, 16 Lee Ct, Thwaites Brow, Keighley, BD21 4TL [IO93BU, SE04]
G0 BWZ Details withheld at licensee's request by SSL.
G0 BXA Dr A Haylock, 5 Severn Dr, Thornbury, Bristol, BS12 1EU [IO81RO, ST69]
G0 BXC P W V Hughes, 123 Garth Rd, Morden, SM4 4LF [IO91VJ, TQ26]
G0 BXD A J Wootton, 14 The Wold, Claverley, Wolverhampton, WV5 7BD [IO82UM, SO89]
G0 BXE N Clark, Hollow Rd, Forty Foot, Ramsey, Huntingdon, Cambs, PE17 1YA [IO92XL, TL38]
G0 BXF R L Tilley, 84 Taverners Rd, Rainham, Gillingham, ME8 9AQ [JO01HI, TQ86]
G0 BXG R A Dring, 22 Castle St., Eastwood, Nottingham, NG16 3GW [IO93IA, SK44]
G0 BXH B G Hansell, 7 Fry Rd, Stevenage, SG2 0QG [IO91VV, TL22]
G0 BXI D J Hitchens, 16 Skipton Rd, Chandlers Ford, Eastleigh, SO53 3BN [IO90HX, SU41]
G0 BXJ A W Peachey, 60 Upwell Rd, March, PE15 9EA [JO02BN, TL49]
G0 BXK A E C Wilkie, Norscot, Alstone, Tewkesbury, Glos, GL20 8JD [IO81XX, SO93]
G0 BXL T M White, 79 Elmbridge, Old Harlow, Harlow, CM17 0JY [JO01BS, TL41]
G0 BXM M D Smyth, 4 Bourton Gdns, Chaseside, Bournemouth, BH7 7HZ [IO90CR, SZ19]
G0 BXN B J Hall, 53 Lower Rd, Hednesford, Cannock, WS12 5ND [IO82XQ, SJ91]
G0 BXO J Leak, Flat 7, 56 Free School Ln, Halifax, HX1 2PW [IO93BR, SE02]
G0 BXP R C Coleman, 16 Mouse Ln, Rougham, Bury St. Edmunds, IP30 9JB [JO02JF, TL96]
G0 BXQ R Colvin, 46 Beechwood Ave, Woodley, Reading, RG5 3DG [IO91NK, SU77]
G0 BXR Details withheld at licensee's request by SSL.
G0 BXT A P Dyson, 96 Fourth Ave, Teignmouth, TQ14 9DR [IO80FN, SX97]
G0 BXV D A Dixon, 7 Mason Rd, Southgate, Crawley, RH10 6DN [IO91VC, TQ23]
G0 BXX J F Dyson, 44 Dawlish Cl, Hucknall, Nottingham, NG15 6NY [IO93JA, SK54]
GW0 BXZ L E Jefferies, Threeacres, Waterfal Rd, Llanrhaeadr-Ym-Mochnant, Oswestry, Shrops, SY10 0JX [IO82IT, SJ12]
G0 BYA S T Houlding, 90 Wordsworth Ave, Castleridge, Stafford, ST17 9UE [IO82WS, SJ92]
G0 BYF R G Aucote, 24 Windsor Dr, Winsford, CW7 1PJ [IO83RE, SJ66]
G0 BYG J P Gear, 4 Welwyn Ave, Mansfield Woodhouse, Mansfield, NG19 9DR [IO93JE, SK56]
G0 BYH G R Stokes, 22 Landrock Rd, Crouch End, London, N8 9HL [IO91WN, TQ38]
G0 BYI P P Chapman, 2 Duke Rd, Burntwood, WS7 8PY [IO92AQ, SK00]
G0 BYJ J P Lovelock, 3 Byron Cl, Pound Hill, Crawley, RH10 3BW [IO91WC, TQ23]
G0 BYK M P Jackson, Stockswood, Stocks Ln, Walberswick, Southwold, IP18 6UJ [JO02TH, TM47]
G0 BYM J R Madeley, Mill Cottage, Higher Wych, Malpas, Ches, SY14 7JR [IO82OX, SJ44]
G0 BYO F Hayhurst, 7 Kings Rd, Accrington, BB5 6BS [IO83TS, SD72]
G0 BYP G Webster, 27 Barkby Rd, Sheffield, S9 1JX [IO93GK, SK39]
G0 BYQ J M Buckmaster, 173 Whinney Ln, New Ollerton, Newark, NG22 9TJ [IO93ME, SK66]
G0 BYR A D Neal, 26 Homestead Dr, Wigston, LE18 2HN [IO92KN, SP69]
G0 BYS G Thompson, Cumstoun House, Burnhope, Durham, Co. Durham, DH7 0AG [IO94DT, NZ14]
G0 BYT W W Birrell, 33 Simpkin Cl, Eaton Socon, St. Neots, Huntingdon, PE19 3PD [IO92UF, TL15]
G0 BYU F G Cooper, 23 York Cl, Gillow Heath, Stoke on Trent, ST8 6SE [IO83VD, SJ85]
G0 BYV R G Rogers, 7 Rose Acre Rd, Littlebourne, Canterbury, CT3 1SY [JO01NG, TR25]
G0 BYX C Pavier, 12A Friend Ln, Edwinstowe, Mansfield, NG21 9QZ [IO93LE, SK66]
G0 BYY D G Hannaford, 7 Priors Ave, Bury St. Edmunds, IP33 3LT [JO02IF, TL86]
GW0 BYZ R A Jones, 9 Upper Row, Dowlais, Merthyr Tydfil, CF48 3NU [IO81HS, SO00]
GW0 BZA K Duckfield, Ellesmere, 29 Esplanade Ave, Porthcawl, CF36 3YS [IO81DL, SS87]
G0 BZB A T Volpe, 31 Cleveland Gdns, Newcastle upon Tyne, NE7 7QE [IO94EX, NZ26]
G0 BZC G Smith, 9 Blackett Ave, Stockton on Tees, TS20 2EX [IO94IN, NZ42]
GW0 BZE R T Jones, 12 Maes Machraeth, Llanfachraeth, Holyhead, LL65 4UF [IO73RH, SH38]
G0 BZF D A Reid, 5 Bridge Ct, Bridge Rd, Chertsey, KT16 8LX [IO91SJ, TQ06]
G0 BZG A G Hepworth, 50 Edmonton Rd, Clipstone Village, Mansfield, NG21 9AJ [IO93KD, SK56]
G0 BZH G Hodgson, 16 Dockroyd, Oakworth, Keighley, BD22 7RH [IO93AU, SE03]
G0 BZI Details withheld at licensee's request by SSL.[Op: D T Price. Station located at Marlbrook, Bromsgrove.]
G0 BZJ J R Newton, 98 Beaulieu Cl, Toothill, Swindon, SN5 8AJ [IO91CN, SU18]
G0 BZK R A Marsden, 26 Paxton Rd, Fareham, PO14 1AB [IO90JU, SU50]
GI0 BZM C J Colhoun, 85 Whitehill Park, Limavady, BT49 0QF [IO65MA, C62]
G0 BZN C P Johnson, 78 Ridgeway Rd, Chesham, HP5 2EW [IO91QR, SP90]
G0 BZO G W Maple, Carlyle House, 4 Carlyle Ave, Blackpool, FY4 1PF [IO83LS, SD33]
G0 BZP J Bates, 28 Westbourne Rd, West Bromwich, B70 8LD [IO82XM, SO99]
G0 BZQ Mt T Curwen, Flat 2, 140 Conway Rd, Llandudno Junction, LL31 9ND [IO83CG, SH77]
G0 BZS A M Ince, Burnside, Braefield, Glenurguhart, Inverness, IV3 6TN [IO77QI, NH43]
G0 BZT R Targonski, 4 Woodville Gdns, Sedgley, Dudley, DY3 1LB [IO82WN, SO99]
G0 BZU K Barlow, 105 Buller St., Bury, BL8 2BQ [IO83UO, SD71]
G0 BZV E Trett, 52 Hambledon Ave, Bierley, Bradford, BD4 6BA [IO93DS, SE12]
G0 BZW A J Porter, 1 Bloomfield Terr, Easton, Portland, DT5 2AB [IO80SN, SY67]
G0 BZX P R Smith, 8 Avalon Cl, Orpington, BR6 9BS [JO01BI, TQ46]
G0 CAA Details withheld at licensee's request by SSL.
G0 CAC Details withheld at licensee's request by SSL.
G0 CAD D G Hewitt, 1 Stadhampton Rd, Drayton St. Leonard, Wallingford, OX10 7AR [IO91KP, SU59]
G0 CAE K G Walsh, 73 Verne Common Rd, Portland, DT5 1EJ [IO80SN, SY67]
G0 CAF Details withheld at licensee's request by SSL.[Station located near Bishops Waltham.]
G0 CAG D J Talaber, 54 Southfield Park, North Harrow, Harrow, HA2 6HE [IO91TO, TQ18]
GI0 CAH P C Leonard, Clonurson Macken, Enniskillen, Co Fermanagh, BT92 3BU [IO64EG, H23]
G0 CAJ T J Morgan, 15 Windsor Rd, Somerset Rd, 7130 Cape, South Africa
G0 CAK G M Russell, 57 Silchester Rd, Pamber Heath, Tadley, RG26 3ED [IO91KI, SU66]
G0 CAL R P Faulkner, 10 Fell Wilson St., Warsop, Mansfield, NG20 0PT [IO93KE, SK56]
G0 CAM C Chislett, Penponds, Tregavethan, Truro, Cornwall, TR4 9EJ [IO70KG, SW74]

G0 CAN P C Shepperd, 19 Spencer Rd, Cobham, KT11 2AF [IO91TH, TQ16]

G0 CAO Details withheld at licensee's request by SSL.

G0 CAP J Burton, Bungalow, 22 Balance Hill, Uttoxeter, ST14 8BT [IO92BV, SK03]

GM0 CAQ Details withheld at licensee's request by SSL.

G0 CAR Details withheld at licensee's request by SSL.

G0 CAS N A Clarke, 39 Acacia Rd, Doncaster, DN4 6NR [IO93LM, SE60]

G0 CAT V M Trelease, 15 Springfield Park, Barripper, Camborne, TR14 0QZ [IO70IE, SW63]

G0 CAU Details withheld at licensee's request by SSL.

G0 CAW Details withheld at licensee's request by SSL.

G0 CAX I H Moore, 25 Granby Cl, Winyates East, Redditch, B98 0PJ [IO92BH, SP06]

G0 CAY K M Arnold, Portland Cottage, Pensilva, Liskeard, Cornwall, PL14 5PS [IO70TM, SX26]

G0 CAZ S C Hill, Fairview, Rosses Ln, Wichenford, Worcester, WR6 6YU [IO82TG, SO76]

GM0 CBA B A L Aitken, 48 Kenilworth Rise, Livingston, EH54 6JJ [IO85FV, NT06]

G0 CBB M J Phillips, 52 Rivington Dr, Burscough, Ormskirk, L40 7RP [IO83NO, SD41]

GM0 CBC J T Johnstone, 12 Peacock Pl, Ecclefechan, Lockerbie, DG11 3EQ [IO85IB, NY17]

G0 CBD G N Roby, 7 Lime Diglands Est, New Mills, Stockport, Ches, SK12 4HT [IO83XI, SJ98]

G0 CBE Details withheld at licensee's request by SSL.

G0 CBF M T Davies, 76 St. James Rd, Oldbury, Warley, B69 2DX [IO82XM, SO98]

G0 CBH N Kershaw, 2 Lidgate Cl, Batley Carr, Dewsbury, WF13 2DP [IO93EQ, SE22]

G0 CBI S McArthur, 63 Bradshaw Ln, Bradshaw, Halifax, HX2 9XD [IO93BS, SE03]

G0 CBJ A Whittingham, 43 Mossgrove Rd, Timperley, Altrincham, WA15 6LF [IO83UJ, SJ78]

G0 CBK N J Priestley, 7 Tabernacle Rd, Wotton under Edge, GL12 7EF [IO81TP, ST79]

GW0 CBL J Follant, 76 Wern Rd, Skewen, Neath, SA10 6DL [IO81BP, SS79]

G0 CBM C R C Wilkie, Grange Rd, Sutton on Sea, Sandilands, Mablethorpe, Lincs, LN12 2RE [JO03DH, TF58]

G0 CBN P Forster, 34 North Park Brook Rd, Callands, Warrington, WA5 5SX [IO83QJ, SJ59]

G0 CBO R R Hoffman, 14 Watling Rd, Attleborough, NR17 2DH [JO02MM, TM09]

G0 CBP R R McMahon, 34 Denmark Rd, Poole, BH15 2DB [IO90AR, SZ09]

GM0 CBQ J Macleod, 61 Fox Covert Ave, Edinburgh, EH12 6UH [IO85IW, NT27]

G0 CBR S S Yim, 308 Bolton Rd, Ewood, Blackburn, BB2 4HY [IO83SR, SD62]

G0 CBT S J Peat, 64 Grange Rd, Heaton Grange, Romford, RM3 7DX [JO01CO, TQ59]

G0 CBU W P Drea, 146 Slewins Ln, Hornchurch, RM11 2BS [JO01CN, TQ58]

G0 CBV C J F Deakin, 48 Draycott Rd, Tean, Stoke on Trent, ST10 4JF [IO92AW, SK03]

G0 CBW M K Bentley, Nickers Hill Farm, Falhouse Ln, Whitley, Dewsbury, WF12 0NL [IO93DP, SE21]

GM0 CBX R W Kerr, 35 Thomson Cres, Port Seton, Prestonpans, EH32 0AN [IO85MX, NT47]

G0 CBY B E Pearson, 13 Barnet Ct, Melbourne Ave, Ramsgate, CT12 6LP [JO01QI, TR36]

G0 CCA G G Coles, Mainstay, Woollard Ln, Whitchurch, Bristol, BS14 0QS [IO81RJ, ST66]

G0 CCB C J Gill, 52 Southfield Rd, Nailsea, Bristol, BS19 1JD [IO81PK, ST47]

G0 CCC C F H Young Caversham Con G, 18 Wincroft Rd, Caversham, Reading, RG4 7HH [IO91ML, SU77]

G0 CCF J D Broome, 38 Westfield Rd, Smethwick, Warley, B67 6AN [IO92AL, SP08]

G0 CCG H H Toh, 3 Priory Rd, Dover, CT17 9RQ [JO01PD, TR34]

G0 CCI Details withheld at licensee's request by SSL.[Op: Angelika Voss (DF2XV, ex-G5CCI) PO Box 49, Manningtree, Essex CO11 2SZ.]

G0 CCJ J Weinstock, 24 Hamilton Rd, Tiddington, Stratford upon Avon, CV37 7DD [IO92DE, SP25]

G0 CCK M W N Ward, Holly House, King St., Broseley, TF12 5NA [IO82SO, SJ60]

G0 CCL P Howarth CCL ARC, Cambs Consultants, Science Park, Cambridge, Cambs, CB4 4DW [JO02BF, TL46]

G0 CCM M Hurst, 2 Poplar Rd, Oughtibridge, Sheffield, S30 3HR [IO93GJ, SK38]

G0 CCN P L Hurst, 23 Cantilupe Cres, Aston, Sheffield, S31 0AS [IO93GJ, SK38]

G0 CCO Details withheld at licensee's request by SSL.[Op: L B Holderness. Station located near Weston-Super-Mare.]

G0 CCP Details withheld at licensee's request by SSL.

G0 CCQ J M J Smith, 24 The Dr, Checketts Ln, Worcester, WR3 7JS [IO82VF, SO85]

GW0 CCR R Cardwell Clwyd County RA, 23 Russell Ave, Colwyn Bay, Clwyd, LL29 7TR [IO83DG, SH87]

G0 CCT Details withheld at licensee's request by SSL.

G0 CCU L W Whitelegg, 30 Chatsworth Rd, Arnos Vale, Bristol, BS4 3EY [IO81RK, ST67]

G0 CCV J W Gaut, 18 First Ave, Clipstone Village, Mansfield, NG21 9EA [IO93KD, SK56]

G0 CCW Details withheld at licensee's request by SSL.

G0 CCX Details withheld at licensee's request by SSL.[Op: Andy Gilbert. Station located at Newhaven. JO00AT. QSL via PO Box 25, Newhaven, Sussex, BN9 0NW.]

G0 CDA M J Ryder, 9 Lincoln Cl, Woolston, Warrington, WA1 4LU [IO83RJ, SJ68]

G0 CDB J S May, 6 Hodson Cl, Paignton, TQ3 3NU [IO80FK, SX86]

GM0 CDC F Shaw, 105 Castle St., Forfar, DD8 3AH [IO86NP, NO45]

G0 CDD Details withheld at licensee's request by SSL.

G0 CDF J Morris, Charnwood, 31 Allen House Park, Woking, GU22 0DB [IO91RH, SU95]

GW0 CDG R T Williams, Flat 8, 38 Oakfield St., Cardiff, CF2 3RE [IO81KL, ST17]

G0 CDH A Hetherington, 47 Elm Park, Filton, Bristol, BS12 7PS [IO81RM, ST67]

G0 CDJ D C Bundle, 118 Crossdale Rd, Hindley, Wigan, WN2 4RQ [IO83RM, SD60]

G0 CDK P J Fallon, 5 Welbeck Rise, Bradford, BD7 4BU [IO93CS, SE13]

GI0 CDM J Neill, 16 Rathlin St., Woodvale Rd, Belfast, BT13 3DZ [IO74AO, J37]

G0 CDO J D Faulkner-Court, 4 The Old Pound, Halford, Shipston on Stour, CV36 5DB [IO92EC, SP24]

G0 CDP G C Rudge, 18 Herbert Rd, Smethwick, Warley, B67 5DD [IO92AL, SP08]

G0 CDQ M D Murphy, 10 Bayham Rd, Sevenoaks, TN13 3XA [JO01CG, TQ55]

G0 CDR J Warwick, 40 Kingsway, Newby, Scarborough, YO12 6SG [IO94SH, TA09]

G0 CDS C R Brind, 8 Pezenas Dr, Market Drayton, TF9 3UJ [IO82SV, SJ63]

GM0 CDV R Evans, 9 Low St., New Aberdour, Fraserburgh, AB43 6LL [IO87VP, NJ86]

G0 CDY A M B Hulme, 71 Victoria Gdns, Ferndown, BH22 9JQ [IO90BT, SU00]

G0 CDZ E A Hicks-Arnold, Ingleside, Junction Rd, Alderbury, Salisbury, SP5 3AZ [IO91DB, SU12]

GM0 CEA R Robertson, 183 High St., Newburgh, Cupar, KY14 6DY [IO86JI, NO21]

G0 CEB E Hutchinson, 32 South Parade, Worksop, S81 0BW [IO93KH, SK57]

G0 CEC P S Coenraats, 54 Falstaff Ave, Earley, Reading, RG6 5TG [IO91MK, SU77]

G0 CED Details withheld at licensee's request by SSL.

G0 CEE J A Gorwill, 37 Bristol Rd, Whitchurch, Bristol, BS14 0PF [IO81RJ, ST66]

G0 CEF A G Sheard, 1 Low Ln, Clayton, Bradford, BD14 6BQ [IO93CS, SE13]

G0 CEG P Worsdale, 10 Manton Rd, Stores Park, Lincoln, LN2 2JL [IO93RF, SK97]

G0 CEI P N Olliffe, Bitham Cottage, Lockings, Oxford, Oxon, OX12 8QW [IO91HN, SU48]

G0 CEJ C Baguley, 44 Royds Cres, Rhodesia, Worksop, S80 3HG [IO93KH, SK58]

G0 CEK F W Webb, 17 Low Downs Rd, Hetton-le-Hole, Houghton-le-Spring, DH5 9AW [IO94GT, NZ34]

G0 CEL N K Evely, Brookview, 3 Meadowbrook Cl, Exwick, Exeter, EX4 2NN [IO80FR, SX99]

G0 CEM T T Barry, 26 Gatcombe Rd, Hartcliffe, Bristol, BS13 9RB [IO81QJ, ST56]

G0 CEN F J Field, 19 The Maples, Nailsea, Bristol, BS19 2RT [IO81OK, ST47]

G0 CEO M Ohta, 5-38-1, Hachimandai, Kisarazu City, Chiba, Japan 292, X X

G0 CEP D J Craker, 7 Burnt House Ln, Stubbington, Fareham, PO14 2LF [IO90JT, SU50]

G0 CEQ D Griffiths, 66 Gallants Farm Rd, Barnet, EN4 8ER [IO91WP, TQ29]

G0 CER D K Harris, 36 Sunniside, Coalbrookdale, Shropshire, TF8 7ET [IO82SP, SJ60]

GW0 CES M G Smith, 8 Ridgeway Ave, Marford, Wrexham, LL12 8ST [IO83MC, SJ35]

G0 CET Details withheld at licensee's request by SSL.

G0 CEU C D Hawkins, 3 Offord Cl, Tottenham, London, N17 0TE [IO91XO, TQ39]

G0 CEV K D Towers, 7 Copeland Rd, Hucknall, Nottingham, NG15 8EB [IO93JA, SK54]

G0 CEW P M Allanson, Burnside, 16 Woodhouse Ln, Kirkhamgate, Wakefield, WF2 0SE [IO93FR, SE22]

G0 CEX K J Rennison, 23 Maplewood Dr, Thornton Cleveleys, FY5 1PN [IO83LU, SD34]

G0 CEY G Cadey, The Millstone, 45 St. Mildreds Rd, Westgate on Sea, CT8 8RJ [JO01QJ, TR37]

G0 CEZ C Tate, 25 Peak Ln, Fareham, PO14 1RX [IO90JU, SU50]

G0 CFA F A Richards, 30 Wilbert Gr, Beverley, HU17 0AN [IO93SU, TA03]

G0 CFB R J J Tyler, The Firs, Laundry Ln, Huntingfield, Halesworth, IP19 0PY [JO02RH, TM37]

GM0 CFC G M Percival, Lavenir, Carlyle Cres, Buckhaven, Leven, KY8 1DW [IO86LE, NT39]

G0 CFD F A Dimmock, 28 Glyndebourne Gdns, Corby, NN18 0PZ [IO92PL, SP88]

G0 CFE F L Whitehead, Mackworth College Derby, 18 Bath Rd, Mickleover, Derby, DE3 5BW [IO92FV, SK33]

G0 CFF A S Morency, Shepherds Farmhouse, South Perrot, Beaminster, Dorset, DT8 3HU [IO80PU, ST40]

G0 CFG B H Horner, 2 Ascot Gr, Bebington, Wirral, L63 2QT [IO83LI, SJ38]

G0 CFH R Lancaster, 219 Moor Ln, Chessington, KT9 2AB [IO91UI, TQ16]

G0 CFI R D Deans, 23 Aldringham Park, Aldringham, Leiston, IP16 4QZ [JO02SE, TM46]

GM0 CFK C S Knight, 19 Brown Cres, Thornton, Kirkcaldy, KY1 4AA [IO86KD, NT29]

G0 CFM B P Robbins, 46 Purton Cl, Kingswood, Bristol, BS15 2ZE [IO81SK, ST67]

G0 CFN T J Hastings, 1 Pottle Cl, Eynsham Rd, Botley, Oxford, OX2 9SN [IO91IR, SP40]

G0 CFQ R C Goodger, 17 Irlam St., Wigston, LE18 4QA [IO92KN, SP59]

G0 CFR D J Smith, 36 South Lawn, Witney, OX8 7HX [IO91GS, SP30]

G0 CFS E A Smith, 36 South Lawn, Witney, OX8 7HX [IO91GS, SP30]

G0 CFT T P Feaviour, 9 Holbrook Cres, Felixstowe, Suffolk, IP11 8NE [JO01PX, TM23]

G0 CFV Details withheld at licensee's request by SSL.

GM0 CFW K L Moffat, 4 Crosswood Ave, Balerno, EH14 7HT [IO85HV, NT16]

G0 CFY Details withheld at licensee's request by SSL.

G0 CFZ J A Wood, 3 Admirals Hard, Stonehouse, Plymouth, PL1 3RJ [IO70WI, SX45]

G0 CGA W F Roberts, 10 The Meadows, Station Rd, Hodnet, Shropshire, TF9 3QF [IO82RV, SJ63]

G0 CGB K A Crittenden, 8 Denny Ct, Bow Arrow Ln, Dartford, DA2 6PA [JO01CK, TQ57]

G0 CGC E G Fountaine, 19 Metcalfe Gr, Blakelands, Milton Keynes, MK14 5JY [IO92PB, SP84]

G0 CGD C J Oneill, 1 Links Ave, Little Sutton, South Wirral, L66 1QS [IO83MG, SJ37]

G0 CGE B J Griffin, 75 Greenham Wood, North Lake, Bracknell, RG12 7WH [IO91PJ, SU86]

G0 CGG B E Thompson Novartis Grimsby ARS, 45 Meadowbank, Great Coates, Grimsby, DN37 9PL [IO93WN, TA21]

G0 CGH R P Hurley, 59 Kimberley Rd, Chingford, London, E4 6DQ [JO01AP, TQ39]

G0 CGI B J Redfern, The Olde Blacksmith Forge, Newton, Cravenarms, Shropshire, SY7 9PG [IO82OK, SO48]

G0 CGJ D J Bateman, 43 Monk Rd, Bishopston, Bristol, BS7 8LE [IO81QL, ST57]

G0 CGL E D Carling, 1 Seaforth Lodge, Victoria Ave, Swanage, BH19 1AN [IO90AO, SZ07]

G0 CGM A G Cutcliffe, 9 Dixon Cl, Paignton, TQ3 3NA [IO80EK, SX86]

G0 CGO H R Jenman, 31 Brockington Rd, Bodenham, Hereford, HR1 3LR [IO82PD, SO55]

G0 CGQ J R Hazlewood, 12 Chatham Way, Haslington, Crewe, CW1 5NU [IO83TC, SJ75]

G0 CGS D H Morris, 2 Willow Cl, Brinsworth, Rotherham, S60 5JU [IO93HJ, SK48]

G0 CGT B T Birch, 59 Shepperson Rd, Hillsborough, Sheffield, S6 4FG [IO93FJ, SK39]

G0 CGU L B P Pinder, 15 Virley Cl, Heybridge, Maldon, CM9 4YS [JO01IR, TL80]

G0 CGV J J Lee, 29 Tewkesbury Dr, Lytham, Lytham St. Annes, FY8 4LN [IO83MR, SD32][Op: James J Lee (Jim)]

G0 CGW G Ashcroft, 49 Tantallon, Birtley, Chester-le-Street, DH3 2JG [IO94FV, NZ25]

G0 CGY A French-St-George, 31 Anson Rd, Exmouth, EX8 4NY [IO80HP, SY08]

G0 CGZ G Allison, 24 Southfield Rd, Scartho, Grimsby, DN33 2PL [IO93WM, TA20]

G0 CHB B M Small, 40 Queens Rd, Skegness, PE25 2ET [JO03DD, TF56]

G0 CHC M J Smith, 12 High St., West Wickham, Cambridge, CB1 6RY [JO02EC, TL64]

GW0 CHD Details withheld at licensee's request by SSL.

G0 CHF J P Carter, 36 Farm View, Yateley, Camberley, Surrey, GU17 7JA [IO91OH, SU85]

G0 CHG R F Moate, Garth House, Redbrooke St, Woodchurch, Ashford, Kent, TN26 3QS [JO01JC, TQ93]

G0 CHI K G Hodges, 15 Gravits Ln, Bognor Regis, PO21 5LT [IO90PS, SZ99]

G0 CHJ R G Poulter, 19 Homestead Way, Winscombe, BS25 1HL [IO81OH, ST45]

G0 CHK R Kerby, 56 Blackboy Ln, Fishbourne, Chichester, PO18 8BE [IO90OU, SU80]

G0 CHL K Dewhurst, 82 New St., Biddulph Moor, Stoke on Trent, ST8 7NW [IO83WC, SJ85]

GM0 CHM J Stephen, Whyntie Lodge, Wester Whyntie, Boyndie, Banff, AB45 2LJ [IO87QQ, NJ66]

G0 CHN J A Sutton, 40 Dane Valley Rd, Margate, CT9 3RX [JO01QJ, TR36]

G0 CHO C H Ousbey, Stores Cottage, Newbold on Stour, Stratford upon Avon, Warks, CV37 8TS [IO92EC, SP24]

G0 CHP R Angus, 25 Jellicoe House, Capstan Rd, Hull, HU6 7AS [IO93TS, TA03]

G0 CHQ J E Pepper, 7 Byron Way, Hayes, UB4 8AT [IO91SM, TQ08]

G0 CHR A E Steward, 94 Whittington Ave, Hayes, UB4 0AE [IO91TM, TQ08]

G0 CHT Details withheld at licensee's request by SSL.

G0 CHV T M Sherriff, 5 Pembroke Ave, Morecambe, LA4 6EJ [IO84NB, SD46]

G0 CHX FtLt A S Leigh, 15 Windsor Park, East Dereham, Norfolk, NR19 2SU [JO02LQ, TF91]

G0 CHY P A Moulton, 4 The Ridge, Withyham Rd, Groombridge, Tunbridge Wells, TN3 9QU [JO01CC, TQ53]

G0 CHZ Details withheld at licensee's request by SSL.

GM0 CIA D D Jones, 7 Field Cl, Chessington, KT9 2QD [IO91UI, TQ16]

G0 CIB Details withheld at licensee's request by SSL.

G0 CIC Details withheld at licensee's request by SSL.[Station located in London N11.]

G0 CIE Details withheld at licensee's request by SSL.[Station loacted in London N11.]

G0 CIF Details withheld at licensee's request by SSL.

G0 CIG A W Aldridge, 6 Vyvyans Terr, Praze, Camborne, TR14 0LD [IO70IE, SW63]

GM0 CII W H A Rogers, 170 Boswall Parkway, Edinburgh, EH5 2JJ [IO85JX, NT27]

G0 CIJ M J Littlewood, 75 Crouch Caravan Park, Pooles Ln, Hullbridge, Hockley, SS5 6PY [JO01HP, TQ89]

G0 CIL C E W Wallis, 182 Warren Rd, Woodingdean, Brighton, BN2 6DD [IO90XU, TQ30]

G0 CIM S J Dodd, 20 Tongdean Ave, Hovehton, E Sussex, BN1 1NW [IO90WT, TQ30]

G0 CIO M Hook, Skidden House Hotel, Skidden Hill, St Ives, Cornwall, TR26 2DU [IO70GF, SW54]

G0 CIP P Low, 6 Jevington Cl, Durrington, Worthing, BN13 3QF [IO90ST, TQ10]

G0 CIR R Jasper, 84 Rose Green Rd, Bognor Regis, PO21 3EQ [IO90PS, SZ99]

G0 CIS R A Bowman, 4 Church Walk, Raynes Park, London, SW20 9DL [IO91VJ, TQ26]

G0 CIV A J Civil, 4 Prescott St, Farrer, A.C.T. 2607, Australia, X X

G0 CIW D E Dawson, 13 Chetwode Dr, Epsom Downs, Epsom, KT18 5TL [IO91VH, TQ25]

G0 CIX R M Wilton, 39 Osborne Rd, Broadstairs, CT10 2AF [JO01RI, TR36]

G0 CJA L E Harper, 23 Beech Ave, Hazel Gr, Stockport, SK7 4QP [IO83WJ, SJ98]

G0 CJC C J Clode, 174 Kenn Rd, Clevedon, BS21 6LH [IO81NK, ST47]

G0 CJD K Spratley, 92 Plantation Hill, Worksop, S81 0QN [IO93KH, SK57]

G0 CJE Details withheld at licensee's request by SSL.

G0 CJG D N Slatter, 13 Hill Burn, Henleaze, Bristol, BS9 4RH [IO81QL, ST57]

GM0 CJK W Runcie, 15 Scylla Dr, Cove, Aberdeen, AB12 3EG [IO87WC, NJ90]

G0 CJM Details withheld at licensee's request by SSL.

G0 CJN J Price, 27 Wheatcroft, Wick, Littlehampton, BN17 7NY [IO90RT, TQ00]

G0 CJO R Nelson, Flat 1, 17 Ashburnham Rd, Hastings, TN35 5JN [JO00HU, TQ81]

G0 CJP C J Petty, 6868 Castlerock Dr, San Jose, Ca 95120, USA, X X

G0 CJQ R J Saw, 33 Rectory Ln, Southoe, St. Neots, Huntingdon, PE18 9YA [IO92UG, TL16]

G0 CJS F R Thompson, Greenacre, Holmeswood Rd, Holmeswood, Ormskirk, Lancs, L40 1TX [IO83NP, SD41]

GM0 CJT R R Macdonell, Taylor Hill, Easter Kinkell, Cononbridge, Dingwall, Ross Shire, IV7 8HY [IO77TN, NH55]

G0 CJV C A Scholey, 11A Guildford St., Grimsby, DN32 7PL [IO93XN, TA21]

G0 CJW S J B Butler, 113 Arley Dr, Widnes, WA8 4XU [IO83OI, SJ48]

G0 CJX J P Stott, Old Police House, Main Rd, Benhall, Saxmundham, Suffolk, IP17 1JY [JO02RE, TM36]

GM0 CJY G Potts, 1 Onslow Pl, Ballantrae, Girvan, KA26 0NW [IO75LC, NX08]

G0 CJZ R G Waller, 29 Valley Dr, Withdean, Brighton, BN1 5FA [IO90WU, TQ20]

G0 CKA G K Waller, The Pines, 18 Old London Rd, Brighton, BN1 8XQ [IO90WU, TQ30]

G0 CKB A J G Baker, 54 All Saints Ave, Prettygate, Colchester, CO3 4PA [JO01KV, TL92]

G0 CKC Details withheld at licensee's request by SSL.[Op: E J Horsey.]

G0 CKD F Wall, 23 Cecil House, Chingford Rd, Walthamstow, London, E17 5AQ [IO91XO, TQ39]

G0 CKE J D Centanni, 5 Mickle Meadow, Water Orton, Birmingham, B46 1SN [IO92DM, SP19]

G0 CKG Details withheld at licensee's request by SSL.

G0 CKH A R Affolter, 12 Newfound Dr, Norwich, NR4 7RY [JO02OO, TG10]

G0 CKI K W Tarbett, 20 Leeholme, Warden Gr, Houghton-le-Spring, DH5 8HR [IO94GU, NZ34]

GW0 CKK D H Robinson, Island View Caravan Park, Beach Rd, Swanbridge, Penarth, CF64 5UG [IO81JJ, ST16]

GW0 CKL I Gulyas, 2 Eglwysilan Way, Abertridwr, Caerphilly, CF83 4EQ [IO81IO, ST18]

G0 CKM C W D Gee, 6 Canterbury Cl, Dukinfield, SK16 5RT [IO83XL, SJ99]

G0 CKP S R Knell, 11 Cheviot Cl, Tonbridge, TN9 1NH [JO01DE, TQ54]

G0 CKQ D S Brown, 3 The Avenue, St. Andrew, St. Andrews, Bristol, BS7 9AF [IO81QL, ST57]

G0 CKT K E Ludlam, 434 PO Box, Ascot, SL5 0QY [IO91QJ, SU96]

G0 CKU J F Lewis, 34 Maple Cres, Shaw Est, Newbury, RG14 1LL [IO91IJ, SU46]

G0 CKV O I Lundberg, Rowan House, Cavendish Rd, Weybridge, KT13 0JW [IO91SI, TQ06]

GW0 CKX D J Matthews, 130 Parc Ave, Morriston, Swansea, SA6 8HU [IO81AP, SS69]

G0 CKY C H Kirk, 4 Kingswear Garth, Whitkirk, Leeds, LS15 8LS [IO93GT, SE33]

G0 CKZ D H G Fritsch, 6 Stanton Rd, Thelwall, Warrington, WA4 2HS [IO83RJ, SJ68]

G0 CLB Details withheld at licensee's request by SSL.

G0 CLC E W Rogers, Cross Foxes, Venn Green, Milton Damerel, Holsworthy, Devon, EX22 7DL [IO70UU, SS31]

G0 CLD H R Davey, 6 Cambridge Gr, Otley, LS21 1DH [IO93DV, SE24]

G0 CLF R H Moon, 22 Meagill Rise, Otley, LS21 2EJ [IO93DV, SE14]

G0 CLG S F Haynes, 9 Heather Gdns, Belton, Great Yarmouth, NR31 9PP [JO02TN, TG40]

G0 CLH D L J Lingard, 10 Barkways, Burwell, Cambridge, Cambs, CB5 0RG [JO02DF, TL56]

G0 CLJ Details withheld at licensee's request by SSL.

G0 CLK Details withheld at licensee's request by SSL.

G0 CLL I Wilson, 333 Rawling Rd, Gateshead, NE8 4UH [IO94EW, NZ26]

G0 CLM A S Greig, 47 The Park, Hewell Grange, Redditch, B97 6QF [IO92AH, SP06]

GM0 CLN C R Smith, 15 Seton Wynd, Port Seton, Prestonpans, EH32 0TY [IO85MX, NT47]

G0 CLO V B de Rose, 4 Briars Walk, Broadstairs, CT10 2XR [JO01RI, TR36]

G0 CLP C L Partington, 34 Ash Gr, Chinley, High Peak, SK23 6BQ [IO93AI, SK08]

G0 CLR R A Hunt, 146 North Walsham Rd, Sprowston, Norwich, NR6 7QQ [JO02PP, TG21]

G0 CLT P G Hodgson, 14 Catherine Howard Cl, Thetford, IP24 1TQ [JO02JK, TL88]

G0 CLV S T Barr, 11 Mallard Way, Moreton, Wirral, L46 7SJ [IO83KJ, SJ29]

G0 CLX G A J Glotham, 6 Slim Rd, Bentley, Walsall, WS2 0EG [IO82XO, SO99]

G0 CMB R G Burling, 28 Croydon Rd, Arrington, Royston, SG8 0DJ [IO92XD, TL34]

G0 CME Details withheld at licensee's request by SSL.

G0 CMF Details withheld at licensee's request by SSL.

G0 CMH C M Heward, 61 Grosvenor Ave, Streetly, Sutton Coldfield, B74 3PE [IO92BN, SP09]

GW0 CMI K J Voller, 5 Hillfield Rd, Parcllyn, Cardigan, SA43 2DH [IO72RD, SN25]

GI0 CMJ P A Campbell, 8 Woodvale Rd, Portstewart, BT55 7HY [IO65PE, C83]

G0 CML B Middleton, 170 Cadeby Rd, Sprotbrough, Doncaster, DN5 7SG [IO93JM, SE50]

G0 CMM Details withheld at licensee's request by SSL.[Operator: John Bell, 28 Stiles Avenue, Marple, Stockport, Cheshire, SK6 6LR. Tel and fax: (0161) 427 6094. RNARS: 1238. FOC: 1511.]

G0 CMN A R Woodhouse, 24 Taylor Cl, Hampton Hill, Hampton, TW12 1LE [IO91TK, TQ17]

GM0 CMO M A K Murray, 1 Gordon Rd, Edinburgh, EH12 6NB [IO85IW, NT27]

G0 CMQ E Ward, 5 Ashley Gr, Hebden Bridge, HX7 5NF [IO93AR, SE02]

G0 CMR R S Melia, 14 Friar Park Rd, Wednesbury, WS10 0TB [IO92AN, SP09]

G0 CMT E Gadeberg, Hojmarksvej 35, PO Box 56, 8700 Horsens, Denmark, XX99 9AA

G0 CMU C R Beecham, Moorview, 2 Endor Cres, Burley in Wharfedale, Ilkley, LS29 7QH [IO93DV, SE14]

G0 CMW J C Groom, Windyridge, High Rd, Cookham, Maidenhead, SL6 9JF [IO91PN, SU88]

G0 CNA M R Wyatt, 32 Stafford Rd, Bridgwater, TA6 5PH [IO81MC, ST33]

GM0 CNB I Hill, 9 Inchbrae Rd, Aberdeen, AB1 7AD [IO87TH, NJ72]

G0 CND I E Sharrott, 31 The Fleet, Stoney Stanton, Leicester, LE9 4DZ [IO92IN, SP49]

G0 CNE Details withheld at licensee's request by SSL.

G0 CNF D W Heftel, 19 Normanton Way, Histon, Cambridge, CB4 4XS [JO02BG, TL46]

G0 CNG C J Roberts, 72 Nairn Rd, Walsall, WS3 3XB [IO82XP, SJ90]
G0 CNH Details withheld at licensee's request by SSL.
GI0 CNI J G Young, 3 Beechland Rd, Magherafelt, BT45 6BG [IO64QS, H99]
GW0 CNJ S F Edwards, 2 Allen Ct, Boverton, Llantwit Major, CF61 2LN [IO81GJ, SS96]
GW0 CNK G Orchard, 8 Glyn Ave, Prestatyn, LL19 9NN [IO83HI, SJ08]
G0 CNL C R Rogers, 34 Martin Ct, Werrington, Peterborough, PE4 6JS [IO92UO, TF10]
G0 CNR Details withheld at licensee's request by SSL.[Op: H J E Davies.]
G0 CNS Details withheld at licensee's request by SSL.
G0 CNU J S McKenzie, 4 Hadrian Ct, Chollerford, Hexham, Northd., NE46 4DD [IO85VA, NY87]
G0 CNV A McManus, 84 Beverley Rd, Hessle, HU13 9BP [IO93SR, TA02]
GM0 CNW H Taylor, 20 Woodhaven Ave, Wormit, Newport on Tay, DD6 8LF [IO86MK, NO42]
G0 COA G P Coates, 1 Ash Brow, Flockton, Wakefield, WF4 4TE [IO93EP, SE21]
G0 COC R J Vallis, 38 Shepherds Cl, Bartley, Southampton, SO40 2LJ [IO90FV, SU31]
GM0 COD L Lambert Dunoon & Dst.Ar, Netherfield, Ardnadam, Dunoon, Argyll, PA23 8QH [IO75MX, NS18]
G0 COE E J Coe, 15 Matching Ln, Bishops Stortford, CM23 2PP [JO01BU, TL42]
G0 COG K C Stringer, 33 Brookes Rd, Flitwick, Bedford, MK45 1BU [IO92SA, TL03]
G0 COH J C Rixon, 10 Hartwell Rd, Hanslope, Milton Keynes, MK19 7BZ [IO92NC, SP74]
G0 COI J Atkinson, 90 Priors Rd, Oakley, Cheltenham, GL52 5AN [IO81XV, SO92]
G0 COJ B W Ellery, 384 Sutton Way, Great Sutton, South Wirral, L66 3LL [IO83NG, SJ37]
G0 COL C Shepherd, 11A Porritt Ln, Irton, Scarborough, YO12 4RL [IO94SF, TA08]
G0 COM A R Redding, 28 Warwick Ave, Bridgwater, TA6 5PF [IO81MC, ST33]
G0 COP Details withheld at licensee's request by SSL.
G0 COQ W A Kelsey-Stead, 44 Shelley Rd, Luton, LU4 0JA [IO91SV, TL02]
G0 COT N Geary, 12 Alton Way, Ashby de La Zouch, LE65 1ER [IO92GR, SK31]
GW0 COU S A Jones, 635 Clydach Rd, Ynystawe, Swansea, SA6 5AX [IO81BQ, SN60]
GW0 COV A A Morris, 18 Glan y Mor, Llanstephan, Carmarthen, SA33 5LL [IO71TS, SN31]
G0 COX N A Currey, 263 Frederick St., Coppice, Oldham, OL8 4HX [IO83WM, SD90]
G0 COY T W Frearson, 31 Paradise St., Rugby, CV21 3SZ [IO92JI, SP57]
G0 COZ J L Wayman, 62 Richards Cl, Tonedale, Wellington, TA21 0BE [IO80JX, ST12]
G0 CPA A D Prichard, 1 Polton Dale, Swindon, SN3 5BN [IO91DN, SU18]
G0 CPD G Forster, 34 North Park Brook Rd, Callands, Warrington, WA5 5SX [IO83QJ, SJ59]
G0 CPE L A J Gallant, Ridgeway, Hogs Back, Seale, Farnham, GU10 1EU [IO91PF, SU84]
G0 CPF I Whyte, 3 Mandeville Rd, Isleworth, TW7 6AD [IO91UL, TQ17]
G0 CPG J C Allen, Laredo, 5 Gipsy Ln, Staplegrove, Taunton, TA2 6LP [IO81KA, ST22]
GM0 CPI J W T Davidson, 27 Laws Rd, Kincorth, Aberdeen, AB1 5JT [IO87TH, NJ72]
G0 CPJ P Walker, 46 Ribble Ave, Freckleton, Preston, PR4 1RX [IO83NS, SD42]
G0 CPK Details withheld at licensee's request by SSL.
G0 CPN I R G Gurton, Grey Gable, 28 Bloomfield Rd, Harpenden, AL5 4DB [IO91TT, TL11]
G0 CPO G Childe S Norm Alfreton, 13 Meerbrook Cl, Oakwood, Derby, DE21 2BE [IO92GW, SK33]
G0 CPP J S Linfoot, Flat, 10 Pembroke Ct, Rectory Rd, Oxford, OX4 1BY [IO91JR, SP50]
GM0 CPQ A Mitchell, 14 Montrose Dr, Aberdeen, AB1 7DA [IO87TH, NJ72]
G0 CPR J F Stewart, 44 Salisbury Rd, Seaford, BN25 2DB [JO00BS, TV49]
G0 CPS N D Eagles, 14 Lincoln Rd, Kempston, Bedford, MK42 7HE [IO92SC, TL04]
G0 CPT A A Fielder, 269 Deeds Gr, High Wycombe, HP12 3PF [IO91OP, SU89]
G0 CPU M P Cracknell, 17 Windmill Fields, Harlow, CM17 0LQ [JO01BS, TL41]
G0 CPV S J Pocock, 3 Penshurst Rd, Ramsgate, CT11 8EG [JO01RI, TR36]
G0 CPW D K Pattison, 1 Heather Ct, Rushden, NN10 9SQ [IO92QG, SP96]
G0 CPX P R Storr, 988 Bradford Rd, East Bierley, Bradford, BD4 6PB [IO93DS, SE12]
G0 CPY P Harvatt, Hillside, Aldern Way, Bakewell, DE45 1AJ [IO93DF, SK26]
G0 CPZ B W Adams, 21 Calthorpe Gdns, Sutton, SM1 3DF [IO91VI, TQ26]
G0 CQB G T Cant, 4 The Mount, Docking, Kings Lynn, PE31 8LN [JO02HV, TF73]
GM0 CQC Dr N B Wilding, 2Fl, 68 Montpelier Park, Edinburgh, Midlothian, EH10 4NQ [IO85JW, NT27]
G0 CQE D C Attreed, 109 Gordon Rd, Ilford, IG1 2XT [JO01BN, TQ48]
G0 CQG W S Harkness, 2 Mill Field, Trusthorpe, Mablethorpe, LN12 2PG [JO03DH, TF58]
G0 CQH T B Neal, 34 Dene View, Ashington, NE63 8JF [IO95EE, NZ28]
G0 CQI G D Baker, 144 Beridge Rd, Halstead, CO9 1JU [JO01HW, TL83]
G0 CQJ J Kilmister, Parc Brause House, Penmenner Rd, The Lizard, Helston, TR12 7NR [IN79JX, SW71]
G0 CQK J E Coombes, 22 Chollerford Cl, Kingsmere, Gosforth, Newcastle upon Tyne, NE3 4RN [IO95EA, NZ26]
GM0 CQL P D Rudd, 41 Broadford Terr, Broughty Ferry, Dundee, DD5 3EF [IO86NL, NO43]
G0 CQO K R Clark, 14 Backfield Rise, Chapeltown, Sheffield, S30 4YR [IO93GJ, SK38]
G0 CQP Dr R A Berrisford, Hollington End Farm, Thorpe, Ashbourne, Derbyshire, DE6 2AU [IO93CB, SK15]
GM0 CQQ A N Herring, Mountpleasant, 8 Linlithgow Rd, Boness, EH51 0DD [IO86EA, NS98]
G0 CQR P M Smith, 93 Nottingham Rd, Long Eaton, Nottingham, NG10 2BY [IO92IV, SK43]
G0 CQS S J Faulkner, 96 Ashby Rd East, Bretby, Burton on Trent, DE15 0PT [IO92ET, SK22]
G0 CQT J H Aulsebrook, 38 South Rd, Beeston, Nottingham, NG9 1LY [IO92JV, SK53]
G0 CQU M Dean, Cranford, 3 Buxton Rd West, Disley, Stockport, SK12 2AE [IO83XI, SJ98]
GM0 CQV B Hynes, 31 Redwood Cres, Cove, Cove Bay, Aberdeen, AB1 3NZ [IO87TH, NJ72]
G0 CQY V Tharp, 14 Cumberland Rd, Congleton, CW12 4PH [IO83VE, SJ86]
G0 CRB A Paddock, 58 Hill St., Warwick, CV34 5PA [IO92FG, SP26]
G0 CRC Details withheld at licensee's request by SSL.
G0 CRD M G Wallis, Quernmore, Hammer Ln, Cowbeech, Hailsham, BN27 4JL [JO00DV, TQ61]
G0 CRE A R Green, 32 Boston Rd, Sleaford, NG34 7ET [IO92TX, TF04]
G0 CRF T Bailey, 65 Edge Ln, Chorlton Cum Hardy, Manchester, M21 9JU [IO83UK, SJ89]
G0 CRG I M Page Sussex Raynet, 127 Whyke Ln, Chichester, PO19 2AU [IO90OT, SU80]
G0 CRH Details withheld at licensee's request by SSL.
G0 CRI Details withheld at licensee's request by SSL.
G0 CRJ A Reader, The Pastures, 8 Daddyhole Rd, Torquay, TQ1 2ED [IO80FK, SX96]
G0 CRK D J Bowles, 30 Highfield Rd, March, PE15 8PE [JO02BN, TL49]
G0 CRL K A Moate, 32 Bournewood, Hamstreet, Ashford, TN26 2HL [JO01KB, TR03]
G0 CRM D T Danenberg, 146 Elms Vale Rd, Dover, CT17 9PN [JO01PC, TR34]
G0 CRN B Hopper, 189 Western Rd, Mickleover, Derby, DE3 5GT [IO92FV, SK33]
G0 CRO Details withheld at licensee's request by SSL.
G0 CRQ Details withheld at licensee's request by SSL.
G0 CRT N D Jelley, 64 Leicester Rd, Broughton Astley, Leicester, LE9 6QE [IO92JM, SP59]
G0 CRU K M Tham, c/o The Hongkong Bn, Gpo Box 64, Hong Kong
G0 CRV A K Balson, 10 Queens Rd, Ashley Down, Bristol, BS7 9HY [IO81RL, ST57]
G0 CRW C E Tucker Crowborough A R C, 6 Rosehill Gdns, Crowborough Hill, Crowborough, TN6 2ED [JO01CB, TQ53]
G0 CRX K R Pearson, 14 Kalima, Bishampton Rd, Flyford Flavell, Worcester, WR7 4BT [IO82XE, SO95]
G0 CRY T M Sparrey, 16 Rosemary Rd, Parkstone, Poole, BH12 3HB [IO90AR, SZ09]
G0 CRZ J A Wall, 7 Brook View Dr, Keyworth, Nottingham, NG12 5JN [IO92KU, SK63]
G0 CSA T A Tickner, -le-Bourg, Verteillac, 24320, France
G0 CSD K Perkins, 9 Southwood Gdns, Ramsgate, CT11 0BG [JO01QI, TR36]
G0 CSF F C Stewart, Shingles, Ingleborough Ln, St Mary's Platt, Sevenoaks, Kent, TN15 8JU [JO01DG, TQ65]
G0 CSH Details withheld at licensee's request by SSL.
G0 CSI Details withheld at licensee's request by SSL.
G0 CSJ A R Williams, 61 East Ave, Donnington, Telford, TF2 8BX [IO82SR, SJ71]
G0 CSK P Senior, 13 St. Michaels Ave, Swinton, Mexborough, S64 8NX [IO93IL, SK49]
G0 CSM R M Widdowson, 5 Kirton Park, Main St., Kirton, Newark, NG22 9LR [IO93MF, SK66]
GM0 CSN R W Trussler, 19 Royellen Ave, Earnock, Hamilton, ML3 8QH [IO75XS, NS75]
GW0 CSR D Roberts, 16 Rhedyw Rd, Llanllyfni, Caernarvon, LL54 6SG [IO73UB, SH45]
G0 CSS H Shillitto, 25 Commonside, Selston, Nottingham, NG16 6FN [IO93IB, SK45]
G0 CSU T Chadwick, 9 Ernest St., Prestwich, Manchester, M25 3HZ [IO83UM, SD80]
G0 CSV J W Capindale, 2 Rivan Gr, Grimsby, DN33 3BL [IO93WM, TA20]
G0 CSW W D Hudson, 9 Nethergate, Upper Gornal, Dudley, DY3 1XW [IO82WM, SO99]
G0 CSX O T Aplin, 2 Knight St., Knights Brow, Macclesfield, SK11 7AT [IO83WG, SJ97]
G0 CSY W H Dunstan, 5 Twemlow Ln, Holmes Chapel, Crewe, CW4 8DT [IO83UF, SJ76]
GM0 CSZ F Dinger, Shore-Acre, Inver By Tain, Ross-shire, IV20 1RX [IO87AT, NH88]
G0 CTE D W Smith, 42 Lynwood Gdns, Croydon, CR0 4QH [IO91WI, TQ36]
G0 CTF W C Sargent, Likoma, 32 Seaton Down Rd, Seaton, EX12 2SB [IO80LQ, SY29]
GW0 CTG G J Darrell, 11 Church St., Presteigne, LD8 2BU [IO82LG, SO36]
G0 CTH H T Churchman, Westcroft, Church Rd, Dodleston, Chester, CH4 9NG [IO83MD, SJ36]
GI0 CTI M Murphy, 14 Carnmore Dr, Daisy Hill, Newry, Co. Down, BT35 8PN [IO64TE, J02]
GW0 CTK J King, Rycote, Bryn Awel Rd, Benllech, Anglesey, Gwynedd, LL74 8TS [IO73VH, SH58]
G0 CTL Details withheld at licensee's request by SSL.
G0 CTM E C Graham, 64 Stretford Rd, Urmston, Manchester, M41 9LS [IO83TK, SJ79]
G0 CTN C T Noon, 59 Taylor Rd, Hindley, Wigan, WN2 4TL [IO83RM, SD60]
G0 CTO Details withheld at licensee's request by SSL.
G0 CTP J D Smart, 4 Sycamore Cl, Holmes Chapel, Crewe, CW4 7BT [IO83TE, SJ76]
G0 CTQ I A Whiffin, 42 Canute Rd, Minnis Bay, Birchington, CT7 9QH [JO01PJ, TR26]
G0 CTR P J Martin, 32 Warkworth Ct, Ellesmere Port, South Wirral, L65 9EN [IO83NG, SJ47]
G0 CTS R E Bradshaw, 19 Castle Ln, Bolsover, Chesterfield, S44 6PS [IO93IF, SK47]
G0 CTT Details withheld at licensee's request by SSL.
G0 CTU P H Comer, 1 Thornleigh Ave, Eastham, Wirral, L62 9AZ [IO83MH, SJ37]
G0 CTX R N Ludlam, 434 PO Box, Ascot, SL5 0QY [IO91QJ, SU96]
GM0 CTY J W Grant, 117 Main St., Carnwath, Lanark, ML11 8HP [IO85EQ, NS94]
G0 CTZ E H Gray, 4 Shortheath Crest, Farnham, GU9 8SA [IO91OE, SU84]
G0 CUA M B Dissanayake, 9 Sweyn Pl, Blackheath Park, London, SE3 0EZ [JO01AL, TQ47]
G0 CUB G T Smith, 55 Countess Way, Euxton, Chorley, Lancs, PR7 6PT [IO83QP, SD51]
G0 CUC R G Hackett, PO Box 39153, Moreleta Park, Pretoria, Rep of Sth Afr 0044, ZZ3 9PO
G0 CUD P Howes, 8 Balliol Rd, Daventry, NN11 4RE [IO92JF, SP56]
G0 CUG J T E Stephens, 19 Fane Cl, Stamford, PE9 1HG [IO92SP, TF00]
G0 CUH S Crane, Linden Cottage, Alma Farm, 70 Highertown, Truro, Cornwall, TR1 3QD [IO70LG, SW84]
G0 CUI L Hobson, 240 Hough Ln, Wombwell, Barnsley, South Yorks, S73 0LL [IO93HM, SE30]
G0 CUJ O Taylor, 4 Bransby Ave, Blackley, Manchester, M9 6JN [IO83VM, SD80]
G0 CUK Details withheld at licensee's request by SSL.
G0 CUL B Chilvers, 99 Links Ave, Hellesdon, Norwich, NR6 5PQ [JO02PP, TG21]
GW0 CUM C J G Laws, 7 Seys Cl, Cowbridge, CF7 7BW [IO81HN, ST08]
G0 CUN P M Connolly, 21 Hartwood Green, Hartwood Park, Chorley, PR6 7BJ [IO83QQ, SD51]
G0 CUO J Hewitt, 44 Penhill Cl, Ouston, Chester-le-Street, DH2 1SG [IO94EV, NZ25]
G0 CUP N W Steele Davies, 27 Vera Ave, Grange Park, London, N21 1RH [IO91WP, TQ39]
G0 CUQ G D Milton, 227 Eton Rd, Ilford, IG1 2UH [JO01BN, TQ48]
GI0 CUR G R Johnston, 10 Church Rd, Newnham, GL14 1AP [IO81ST, SO61]
G0 CUS Details withheld at licensee's request by SSL.
G0 CUT Details withheld at licensee's request by SSL.[Op: K G Westbury, tel: Aylesbury 86916.]
G0 CUU R V Privett, Trecollas, 12 Lloyd Park Ave, Croydon, CR0 5SA [IO91XI, TQ36]
G0 CUV J P Crompton, 14 Fastnet Cl, Wilsey, Haverhill, CB9 0LL [JO02FB, TL64]
G0 CUX R Bedford, 26 Devon Ave, Fleetwood, FY7 7EA [IO83LV, SD34]
GM0 CUY R J Dixon, 18 Mortonhall Park Green, Edinburgh, EH17 8SP [IO85JV, NT26]
G0 CUZ C D Morris, 12 Turners Hill Rd, Loewr Gornal, Dudley, DY3 2JU [IO82WM, SO99]
G0 CVA G Yates, 14 Mason St., Sutton in Ashfield, NG17 4HP [IO93JD, SK55]
G0 CVB K G Albon, 65 Belmont Rd, Kirkby in Ashfield, Nottingham, NG17 9DY [IO93JC, SK55]
G0 CVC E Buckley, 34 Newstead Terr, Newstead, Halifax, HX1 4TA [IO93BR, SE02]
G0 CVD F Nicholl, 17 Pendle Cres, Billingham, TS23 2RA [IO94IO, NZ42]
G0 CVG Details withheld at licensee's request by SSL.[Station located near Richmond, Surrey. Op: R Langtry.]
G0 CVH V Hughes, 27 Billy Ln, Clifton, Swinton, Manchester, M27 8FS [IO83UM, SD70]
G0 CVJ K Wortley, 41 Common End Ln, Lepton, Huddersfield, HD8 0AL [IO93DP, SE11]
G0 CVL C Hornsby, Oakerbank Farm, Skipton Rd, Killinghall, Harrogate, HG3 2AR [IO94FA, SE25]
G0 CVM W L Auty, 22 Hurley Cl, Great Sankey, Warrington, WA5 1XG [IO83QJ, SJ58]
G0 CVN I Howsham, West St, North Kelsey, Lincoln, Lincs, LN7 6EL [IO93SM, TA00]
G0 CVO P Jukes, 25 Oaklands Ave, West Wickham, BR4 9LF [IO91XI, TQ36]
GW0 CVP C V Phillips obo Rhosllanerchrugog Group, Aelwyd, Hill St., Rhosllanerchrugog, Wrexham, LL14 1LW [IO83LA, SJ24]
G0 CVS J Bernier, 27 Ladycross Rd, Hythe, Southampton, SO45 3JW [IO90HU, SU40]
G0 CVT C Dickinson, 23 Gisborne Rd, Sheffield, S11 7HA [IO93FI, SK38]
GW0 CVY D Edwards, 26 Russell Terr, Carmarthen, SA31 1SY [IO71UU, SN42]
G0 CVZ M J T Bowthorpe, 2 Chancery Ln, Eye, Peterborough, PE6 7YF [IO92VO, TF20]
G0 CWA N A Strong, 46 Malpas Dr, Great Sankey, Warrington, WA5 1HN [IO83QJ, SJ58]
G0 CWB M J Starkey, 23 Arthur St., Chadsmoor, Cannock, WS11 2HD [IO82XQ, SJ91]
G0 CWC A Hollyoake Sandwell ARS, 281 Causeway Green Rd, Oldbury, Warley, B68 8LT [IO82XL, SO98]
G0 CWD V Donachie, 26 Darwin Walk, Dereham, NR19 1BP [JO02LQ, TF91]
G0 CWF M Warr, Oak Lodge, 31 Steyn St, Barrydale, 6750 West Cape, South Africa, X X
GW0 CWG A Evans, 8 Rhos y Gad, Llanfairpwllgwyngyll, LL61 5JE [IO73VF, SH57]
G0 CWH Details withheld at licensee's request by SSL.[Op: J W Carden. Station located near Sittingbourne.]
G0 CWI Details withheld at licensee's request by SSL.[Op: K I Mathews. Station located near High Wycombe.]
G0 CWK J J Seymour, Mdina, Offwell, Honiton, Devon, EX14 9SA [IO80KS, SY19]
GM0 CWL D Chapaton, 1 Rue de L'Angile, 69005 Lyon, France, ZZ1 9RU
G0 CWM Details withheld at licensee's request by SSL.
G0 CWO L W Gough, 7 Congreve Rd, Blurton, Stoke on Trent, ST3 2HA [IO82WX, SJ84]
G0 CWP K Stather, 5 St. Margarets Rd, Bolton-le-Sands, Carnforth, LA5 8EN [IO84OC, SD46]
G0 CWQ A R Nesbitt, 42 Barnpark Rd, Teignmouth, TQ14 8PN [IO80GN, SX97]
GM0 CWR W D Pentland, 15 Stuart Cres, Edinburgh, EH12 8XR [IO85IW, NT17]
G0 CWS A F Hattersley, Top Ln, Townhead, Tideswell, Buxton, Derbyshire, SK17 8LP [IO93CG, SK17]
G0 CWT Details withheld at licensee's request by SSL.
G0 CWU F R Humphreys, 2 Kennedy Cl, Oakham, LE15 6LL [IO92PQ, SK80]
G0 CWW R C Gooden, 39 Heath Rd, Ipswich, IP4 5RZ [JO02OB, TM14]
G0 CWX M J Shotter, Peverley, Newport Rd, Somerton, Cowes, I O W, PO31 8PE [IO90IR, SZ49]
G0 CWZ T J H Cattley, Yew Tree Cottage, Overton Rd, Ifton Heath, St Martins, Oswestry, SY11 3DH [IO82LW, SJ33]
G0 CXD T J Moore, Highfields, Ashbourne Rd, Turnditch, Belper, DE56 2LH [IO93FA, SK24]
G0 CXE F R Rogers, 111 Nailers Dr, Burntwood, Staffs, WS7 0EY [IO92BQ, SK00]
G0 CXF Details withheld at licensee's request by SSL.
G0 CXG J L Lockhart, 17 Balfour St., Blyth, NE24 1NR [IO95FD, NZ38]
G0 CXJ A Beasley, 2 Ilmington Rd, Blackwell, Shipston on Stour, Warks, CV36 4PE [IO92EC, SP24]
GW0 CXK F D Johns, Kensington House, Kensington St., Fishguard, SA65 9LH [IO71MX, SM93]
G0 CXO S J Wellings, 11 Matlock Rd, Lower Farm Est, Bloxwich, Walsall, WS3 3QD [IO92AP, SK00]
G0 CXP Details withheld at licensee's request by SSL.
GM0 CXQ Details withheld at licensee's request by SSL.
GI0 CXR R Campbell, 55 Coolsythe Rd, Randalstown, Antrim, BT41 3HF [IO64TS, J09]
G0 CXS J K Wareham, Windrush, Ashbourne Rd, Turnditch, Belper, DE56 2LH [IO93FA, SK24]
G0 CXT Details withheld at licensee's request by SSL.
G0 CXU R Pitter, 57 Greenhill Way, Shortheath, Farnham, GU9 8TA [IO91OE, SU84]
G0 CXV R Clark, 34 Barnards Hill, Marlow, SL7 2NZ [IO91ON, SU88]
G0 CXX J A Jopling, 54 Redesdale Gdns, Dunston, Gateshead, NE11 9XH [IO94EW, NZ26]
G0 CYA Details withheld at licensee's request by SSL.
G0 CYB P I Kelsall, Red Croft, Rock Hill, Mansfield, NG18 2PG [IO93JD, SK56]
G0 CYC R E Curtis, 125 Handside Ln, Welwyn Garden City, AL8 6TA [IO91VT, TL21]
G0 CYD W C Snow, 4 St. Bernards Rd, Shirehampton, Bristol, BS11 9UR [IO81PL, ST57]
G0 CYE J A H Easton, 130 The Greenway, Epsom, KT18 7JB [IO91UH, TQ16]
G0 CYF D J Bennett, Franbie, Station Rd, Bere Ferrers, Yelverton, PL20 7JS [IO70VK, SX46]
GW0 CYG D J Davies, 40 Furzeland Dr, Bryncoch, Neath, SA10 7UG [IO81CQ, SN70]
GW0 CYI J R Knight, 4 Frankwell, Llanidloes, SY18 6HG [IO82FK, SN98]
GW0 CYK S I Radford, 1A South Parade, Maesteg, CF34 0AB [IO81EO, SS89]
G0 CYL M R Sefton, 8 Sandmoor Ave, Leeds, LS17 7DW [IO93FU, SE34]
G0 CYN R J Blackmore, 4 Haycocks Cl, Telford, TF1 3NN [IO82RQ, SJ61]
G0 CYO G Beddow, 12 Wulfruna Gdns, Finchfield, Wolverhampton, WV3 9HZ [IO82WN, SO89]
G0 CYP Details withheld at licensee's request by SSL.
G0 CYR D Bramley, 10 Thirlmere Cl, Huncoat, Accrington, BB5 6JQ [IO83TS, SD73]
G0 CYT Details withheld at licensee's request by SSL.
G0 CYU D G Hooper, 6 Kingswell Ave, Outwood, Wakefield, WF1 3DY [IO93GR, SE32]
G0 CYX J Faulkner, 31 Valley View, South Elmsall, Pontefract, WF9 2DD [IO93IO, SE41]
G0 CZB Details withheld at licensee's request by SSL.
G0 CZD M G Kinder, 12 Jessop Way, Haslington, Crewe, CW1 5FU [IO83TC, SJ75]
G0 CZE G P Drew, 2 Bridge Croft, Great Waltham, Chelmsford, CM3 1RE [JO01FT, TL61]
G0 CZG D F Middleton, 5 Wynne Cl, Broadstone, BH18 9HQ [IO80XS, SY99]
GW0 CZJ G A Philpot, 19 Upper Cliffe Cl, Penarth, S Glam, CF6 1BE [IO81HN, ST08]
G0 CZL W V Riley, 10 Hough, Halifax, HX3 7AP [IO93BR, SE12]
GM0 CZM W Taylor, 49 Pentland Ave, Port Glasgow, PA14 6LF [IO75QW, NS37]
G0 CZN B Roscoe, 6 Yew Tree Cl, Lutterworth, LE17 4TT [IO92JL, SP58]
G0 CZO Details withheld at licensee's request by SSL.[Station located near Worthing.]
G0 CZP J Nicholson, 7 Greenwood Sq, Grindon Village Est, Sunderland, SR4 8JT [IO94GV, NZ35]
G0 CZQ Details withheld at licensee's request by SSL.[Op: D G Saxton, 30 Belmont Park, Pensilva, nr Liskeard, Cornwall PL14 5QT]
G0 CZR K I Laws, 6 Crabbe Cl, Feltwell, Thetford, IP26 4BD [JO02GL, TL79]
G0 CZS J A Brown, 13 Cross Ln, Codnor, Ripley, DE5 9RZ [IO93HA, SK44]
G0 CZT Details withheld at licensee's request by SSL.
G0 CZU H W Richardson, 15 Sea View Pl, Cleator Moor, CA25 5LZ [IO84FM, NY01]
G0 CZV A G Harris, 16 West Hall Garth, South Cave, Brough, HU15 2HD [IO93QS, SE93]
G0 CZY P D Heath, The Cottage, Great Staughton Rd, Pertenhall, Bedford, MK44 2BA [IO92TG, TL06]
G0 CZZ A F Leader, Quintedeena, 67 Upper Park, Harlow, CM20 1TN [JO01AS, TL41]
G0 DAA Details withheld at licensee's request by SSL.
G0 DAB D A Buik, Firlands Lodge, 54 Buckshaft Rd, Ruspidge, Cinderford, GL14 3DZ [IO81RT, SO61]
G0 DAC D J Cowley, 81 Ashtree Rd, Walsall, WS3 4LS [IO92AP, SK00]
G0 DAD Details withheld at licensee's request by SSL.[Op: J Rogers. Station located near Woking.]
G0 DAE C W Tidwell, 66 Powerscourt Rd, Northend, Portsmouth, PO2 7JN [IO90LT, SU60]
G0 DAF J A Randall, 26 Marian Rd, Boston, PE21 9HA [IO92XX, TF34]
G0 DAG Rev P McArdie, St. Wilfred's Presbytery, 1 Winckley Sq, Preston, Lancs, PR1 3JJ [IO83PS, SD52]
G0 DAH B J Smith, 2 School Ln, Boxworth, Cambridge, CB3 8ND [IO92XG, TL36]
G0 DAI D A Isom, 36 Deerfold, Astey Village, Chorley, PR7 1UH [IO83QP, SD51]
G0 DAL G Daley, 5 Linden Rd, Forest Town, Mansfield, NG19 0EL [IO93JD, SK56]
G0 DAM R H Clayton, 171 Warning Tongue Ln, Doncaster, DN4 6TU [IO93LM, SE60]
GM0 DAN A C Templeton, Larch Ave, Lenzie, Kirkintilloch, Glasgow, Lanarkshire, G66 4HX [IO75WW, NS67]
G0 DAS D A Squire, 26 Claygate Ln, Hinchley Wood, Esher, KT10 0AQ [IO91UI, TQ16]
G0 DAT Details withheld at licensee's request by SSL.

G0 DAU M D Saunders, White Lodge, 22 Humphreys Cl, St. Cleer, Liskeard, PL14 5DP [IO70SL, SX26]
G0 DAV D P Paine, Woodland View, St Mellion, Cornwall, PL12 6RH [IO70UL, SX36]
G0 DAX D G Burt, 105 Cavalry Cres, Old Town, Eastbourne, BN20 8RL [JO00DS, TQ50]
G0 DAY K F Banks, 52 Hunter Ave, Burntwood, WS7 9AQ [IO92BQ, SK00]
G0 DBC J Hudson, 1 Linnet Way, Biddulph, Stoke on Trent, ST8 7UF [IO83WC, SJ85]
G0 DBD J J West, Stonecroft, 4 Trevella Rd, Bude, EX23 8NA [IO70RU, SS20]
G0 DBE L R Marsland, 154 Moss Ln, Litherland, Liverpool, L21 7NN [IO83ML, SJ39]
G0 DBG H G Ethell, 55 Lingdale Rd, Birkenhead, L43 8TF [IO83LJ, SJ28]
G0 DBH L C Molyneux, 7 Dunster Cl, Platt Bridge, Wigan, WN2 5HT [IO83QM, SD60]
G0 DBI K Danks, 5 Pittmore Rd, Burton, Christchurch, BH23 7EY [IO90CS, SZ19]
G0 DBJ G A Willetts, Waterside, 48 Stourton Cres, Stourbridge, DY7 6RR [IO82VL, SO88]
G0 DBK D B Kerr, 115 Willis Rd, Haddenham, Aylesbury, HP17 8HG [IO91MS, SP70]
G0 DBL Dr A Comfort, 2 Fitzwarren House, Hornsey Ln, London, N6 5LX [IO91WN, TQ28]
G0 DBM S R Lovesey, 22 Briarwood Cl, Leigh on Sea, SS9 4LE [JO01HN, TQ88]
G0 DBN R Haines, 24 Loyd St., First Ln, Anlaby, Hull, HU10 6UG [IO93SR, TA02]
G0 DBR M B Loakes, 1 Sandhills Cl, Belton, Loughborough, LE12 9TT [IO92HS, SK42]
G0 DBS P L Le Feuvre, 56 Greenfields Ave, Alton, GU34 2EE [IO91MD, SU73]
G0 DBT R E Malcolm, 9 Ennerdale Dr, Bradford, BD2 4JD [IO93DT, SE13]
G0 DBU E F Stewart, 24 Ingleton Rd, Newsome, Huddersfield, HD4 6QX [IO93CP, SE11]
GM0 DBW M I K Bolton, 18 Ferryhill Terr, Aberdeen, AB1 6SQ [IO87TH, NJ72]
G0 DBX D Beale, Kenwood, London Rd, Raithby Cum Maltby, Louth, LN11 8QH [IO93XI, TF38]
G0 DCB Details withheld at licensee's request by SSL.
G0 DCC T Carroll Emergency Plan, Co. Emergency Planning Unit, Fire Brigade Headquarters, Framwellgate More, Durham, DH1 5JR [IO94ET, NZ24]
G0 DCE C T Stevens, 9 Westwood Rd, Marlow, SL7 2AT [IO91ON, SU88]
G0 DCF R Green, Kenville, West Ln, Aughton, Sheffield, S31 0XS [IO93GJ, SK38]
G0 DCG M E Brown, 129 Tressillian Rd, Brockley, London, SE4 1XZ [IO91XL, TQ37]
G0 DCH J C E Proctor, 32 Cheviot Rd, Chester-le-Street, DH2 3AL [IO94FU, NZ25]
G0 DCI A W Merrylees, 90 Grangehill Rd, Eltham, London, SE9 1SE [JO01AL, TQ47]
G0 DCL R S F Cathmoir, 340 Straight Rd, Shrubend, Colchester, CO3 5DX [JO01KU, TL92]
G0 DCM A Pilling, 29 Northgate, Bispham, Blackpool, FY2 9NG [IO83LU, SD33]
G0 DCN G L Cheeseman, 19 Hazelwood Gr, Eastwood, Leigh on Sea, SS9 4DE [JO01HN, TQ88]
G0 DCO M Beirne, 50 Ullswater, Macclesfield, SK11 7YW [IO83WF, SJ97]
G0 DCP P Holdaway, 9 Harecourt Rd, Canonbury, London, N1 2LW [IO91WN, TQ38]
G0 DCR J A Frost, 36 York Gdns, Braintree, CM7 9NF [JO01GV, TL72]
G0 DCS P Ashton, 27 Dunsby Rd, Luton, LU3 2UA [IO91SV, TL02]
G0 DCU J H Faithfull, 3 Malting Mead, Endymion Rd, Hatfield, AL10 8AR [IO91VS, TL20]
G0 DCV A J Storer, 1A Carlisle Ave, Littleover, Derby, DE23 7ET [IO92FV, SK33]
G0 DCW A R Corallini, 8 Britannia, Puckeridge, Ware, SG11 1TG [JO01AV, TL32]
G0 DCY R H Swan, 23 Upperfield Rd, Maltby, Rotherham, S66 8BG [IO93JK, SK59]
G0 DCZ P J Richards, 24 Beech Tree Ave, Tile Hill, Coventry, CV4 9FG [IO92FJ, SP27]
G0 DDA P G White, 1 Horseshoe Lodge, Warsash, Southampton, SO31 9AY [IO90IU, SU50]
G0 DDB F B Craven, 170 Lindley Moor Rd, Mount, Huddersfield, HD3 3UE [IO93BP, SE11]
G0 DDD Details withheld at licensee's request by SSL.[Station located near Weoley Castle.]
G0 DDE B J Dignum, 16 Stirling Ct Rd, Burgess Hill, RH15 0PT [IO90WX, TQ31]
G0 DDF D T Fairchild, 89 Park Rd, Congresbury, Bristol, BS19 5HE [IO81OI, ST46]
G0 DDG P J Myhal, 14 Connawood Walk, Bray, Co Wicklow, Ireland
G0 DDI P Matkin, 1 Heath Rd, St. Leonards, Ringwood, BH24 2PZ [IO90BT, SU10]
G0 DDJ A E King, 35 Cranes Way, Borehamwood, WD6 2ET [IO91UP, TQ29]
GW0 DDK E I J Down, Silver Hill, Penycwm, Haverfordwest, Dyfed, SA62 6JZ [IO71KV, SM82]
GW0 DDL G Jones, 9 George St., New Quay, SA45 9QR [IO72TF, SN35]
G0 DDM Prof L J Audus, 38 Belmont Ln, Stanmore, HA7 2PT [IO91UO, TQ19]
G0 DDN C R Suarez, 13 Ferrers Rd, Weston, Stafford, ST18 0JN [IO82XU, SJ92]
G0 DDO A Stevenson, 1 Hercules Rd, Norwich, NR6 5HQ [JO02PP, TG21]
GW0 DDR Details withheld at licensee's request by SSL.
G0 DDS T R Solomon, 74 Sketty Rd, Enfield, EN1 3SF [IO91XP, TQ39]
G0 DDT J R Barnes, 262 King Henrys Dr, New Addington, Croydon, CR0 0AA [IO91XI, TQ36]
G0 DDU J J Norris, Aughton House, 15 Liverpool Rd North, Burscough, Ormskirk, L40 5TN [IO83NO, SD41]
G0 DDV N P Hewett, 49 Harrow Way, Carpenders Park, Watford, WD1 5EH [IO91TO, TQ19]
G0 DDW M D Hillier, 28 Meadow Walk, Bridgemary, Gosport, PO13 0YN [IO90JT, SU50]
G0 DDX Details withheld at licensee's request by SSL.
G0 DDY P R Piper, 44 Parsonage Est, Rogate, Petersfield, GU31 5HJ [IO91NA, SU82]
G0 DDZ M D Eastman, 23 Haughgate Cl, Woodbridge, IP12 1LQ [JO02PC, TM25]
G0 DEB D E Bannister, 60 St. Johns Ave, Bridlington, YO16 4NL [IO94VC, TA16]
G0 DEC D R Willicombe, 26 Falkland Ct, Braintree, CM7 9LL [JO01GV, TL72]
G0 DED W W Neal, St. Oswald, 17 Larkhill Ave, Upton, Wirral, L49 4PN [IO83KJ, SJ28]
G0 DEE R H Spilling, 19 Ketts Cl, Hethersett, Norwich, NR9 3JF [JO02OO, TG10]
G0 DEF M V Mutton, 39 Martin Rd, Kettering, NN15 6HF [IO92PJ, SP87]
G0 DEH P J Finbow, 6 Down Rd, Teddington, TW11 9HA [IO91UK, TQ17]
G0 DEI N Howorth, 42 Fairfield Ave, Rossendale, BB4 9TQ [IO83VQ, SD82]
G0 DEJ W S Rutt, 24 Coopers Ln, Verwood, BH31 7PG [IO90BV, SU00]
G0 DEK F J Underwood, Hobletts, Orsett Fen, Orsett, Essex, RM16 3LT [JO01EM, TQ68]
G0 DEN W H D Maynard, 11 Denham Rd, Canvey Island, SS8 9HB [JO01GM, TQ78]
G0 DEO W A Batey, 13 Cassiobury Ave, Feltham, TW14 9JE [IO91SK, TQ07][SysOp GB7DEO. 9H1IA]
G0 DEP D Jarrard, 26 Lingmell Ct, Tolladine Rd, Worcester, WR4 9YU [IO82VF, SO85]
GM0 DEQ R C A Alexander, 8 Grougar Dr, Kilmarnock, KA3 1UL [IO75SP, NS44]
G0 DER D J Gillmore, 4 Holly Ridge, Fenns Ln, West End, Woking, GU24 9QE [IO91QI, SU96]
GI0 DES M D McCann, Casa Haya, 26 Farmhill Rd, Balliamullan, Omagh, BT79 0PY [IO64IO, H47]
G0 DEU J C Tournant, 47 High St., Linton, Cambridge, CB1 6HS [JO02DC, TL54]
G0 DEW H M S Dew, 16 Longmoor Rd, Liphook, GU30 7NY [IO91OB, SU83]
GM0 DEX A M Goldie, 66 Meldrum Mains, Glenmavis, Airdrie, ML6 0QG [IO85AV, NS76]
G0 DEZ D W Watson, 30 Louis Way, Dunkeswell, Honiton, EX14 0XW [IO80JU, ST10]
G0 DFA D F Allsopp, 10 Chalfont Cl, Middleton on Sea, Bognor Regis, PO22 7SL [IO90QS, SU90]
G0 DFB B Nicholl, 19 St. Ives Rd, Skircoat Green, Halifax, HX3 0LT [IO93BQ, SE02]
G0 DFC L I Cropley, 215 Tring Rd, Aylesbury, HP20 1JH [IO91OT, SP81]
GI0 DFD R C McAlister, 78 Cairn Rd, Milebush, Carrickfergus, BT38 9AP [IO74BS, J39]
G0 DFE J P Stone, 12 Main Rd, Hawkwell, Hockley, SS5 4JN [JO01IO, TQ89]
G0 DFF J E Sidnell, 17 Barlings Rd, Harpenden, AL5 2AL [IO91TT, TL11]
G0 DFI D H Oakley, 6 Staplehurst Gdns, Cliftonville, Margate, CT9 3JB [JO01RJ, TR37]
G0 DFK P J Brookes, 5 Pistyll Terr, Pistyll, Pwllheli, LL53 6LS [IO72SW, SH34]
GM0 DFL R J Taylor, 77 Western Rd, Woodside, Aberdeen, AB24 4DR [IO87WE, NJ90][Op: R Taylor, School House, Kingswells, Aberdeen, AB1 8PC.]
G0 DFN C J Pithers, 21 Grenville Gdns, Dibden Purlieu, Southampton, SO45 4HG [IO90HU, SU40]
G0 DFO Details withheld at licensee's request by SSL.[Op: J Tomlinson, station located near Nelson, Lancs]
G0 DFP Details withheld at licensee's request by SSL.
G0 DFS R Doran DFS RC, Cont Rm Fire Serv H, Burton Rd, Littleover, Derby, DE3 6EH [IO92FV, SK33]
G0 DFT J S Maw, 4 Western Ave, West Denton, Newcastle upon Tyne, NE5 5BU [IO94DX, NZ16]
G0 DFU E W Spanner, 30 Lowtherville Rd, Ventnor, PO38 1AP [IO90JO, SZ57]
GM0 DFW P H Racionzer, 41 North St., Forfar, DD8 3BH [IO86NP, NO45]
GW0 DFY R J Anthony, 2 Barrfield Rd, Rhuddlan, Rhyl, LL18 2RY [IO83GH, SJ07]
G0 DFZ J Wardale, 32 Fortnam Rd, Upper Holloway, London, N19 3NR [IO91WN, TQ28]
G0 DGA D Gullick, Greenleas, Courthay Orchard, Pitney, Langport, Somerset, TA10 9AE [IO81OB, ST42]
G0 DGB L Connell, 24 Finchale Rd, Framwellgate Moor, Durham, DH1 5JN [IO94ET, NZ24]
G0 DGD B Herrett, 40 Hammersmith, Ripley, DE5 3RA [IO93HB, SK35]
G0 DGE J E Gullick, Greenleas, Courthay Orchard, Pitney, Langport, Somerset, TA10 9AE [IO81OB, ST42]
G0 DGF M D Beakhust, 63 Chadacre Rd, Stoneleigh, Epsom, KT17 2HD [IO91VI, TQ26]
G0 DGG F J S Ward, 12 Birkbeck Way, Greenford, UB6 8NY [IO91TM, TQ18]
G0 DGH G J Daniels, 86 Bordesley Rd, Morden, SM4 5LR [IO91VJ, TQ26]
G0 DGI G Atchinson, 71 Flamstead Ave, Loscoe, Heanor, DE7 7RP [IO92IX, SK44]
GM0 DGK A L Donaldson, 30 Jeanfield Cres, Forfar, DD8 1JR [IO86NP, NO44]
GM0 DGL D C Kempton, 54 Balgarvie Rd, Cupar, KY15 4AJ [IO86LH, NO31]
G0 DGN A Ryan, 16 Bradgate Ave, Heald Green, Cheadle, SK8 3AQ [IO83VI, SJ88]
GW0 DGO D G Owens, 427 Middle Rd, Gendros, Swansea, SA5 8EH [IO81AP, SS69]
G0 DGQ G M D Dudley, 95 Alfreton Rd, South Normanton, Alfreton, DE55 2BJ [IO93HC, SK45]
G0 DGU P D Brown, 17 Greyhound Way, Frettenham, Norwich, NR12 7NB [JO02PQ, TG21]
G0 DGV Details withheld at licensee's request by SSL.
G0 DGW D C Barham, 64 Gorran Ave, Rowner, Gosport, PO13 0NF [IO90JT, SU50]
G0 DGX Details withheld at licensee's request by SSL.
GW0 DHA R A Waller, 4 Rose Ct, Ty Canol, Cwmbran, NP44 6JH [IO81LP, ST29]
G0 DHB R E Evans, 20 Pulley Ave, Eaton Bishop, Hereford, HR2 9QN [IO82OB, SO43]
GM0 DHD A Lymer, 16 Gerson Park, Broxburn, West Lothian, EH52 6PL [IO85GW, NT07]
GW0 DHE G D Osborne, 26 Lawrence Cl, Bridgend, CF31 1JY [IO81FM, SS98]
GW0 DHG B Ristic, 168 Heather Rd, Newport, NP9 7QW [IO81MO, ST38]
G0 DHI A Rutherford, 4 Rockwood Gdns, Greenside, Ryton, NE40 4BB [IO94CW, NZ16]
G0 DHJ J Wraight, 59 Sandy Ln, Walton, Liverpool, L9 9AY [IO83ML, SJ39]
G0 DHL M N Mason, 57 Hampton Hill, Wellington, Telford, TF1 2ER [IO82RQ, SJ61]
G0 DHM D H K Moore, Stoke Hall Farm, Stoke on Tern, Market Drayton, Shropshire, TF9 2DU [IO82RU, SJ62]
GW0 DHQ D M Hutchinson, Plas Dyfi, Pennal, Machynlleth, Powys, SY20 9LB [IO82AN, SN69]
G0 DHR D H N Rutt, 3 Russell Pl, Highfield, Southampton, SO17 1NU [IO90HW, SU41]
G0 DHS A Kitching, 1 Borrowdale, Albany, Washington, NE37 1QD [IO94FV, NZ35]
G0 DHT K M Smith, Lower Carneggy Farm, Greenbottom, Chacewater, Truro, Cornwall, TR4 8QL [IO70KG, SW74]
G0 DHU J M Collins, 11 Rosewood Cl, Plymstock, Plymouth, PL9 9JB [IO70XI, SX55]
G0 DHV C P Miller, 20 Durham St., Scarborough, YO12 7PT [IO94TG, TA08]
GI0 DHW C Cleland, 19 Sheskin Way, Cregagh, Belfast, BT6 0ER [IO74BN, J37]
G0 DHZ H A Johnsen, 17 Thruxton Rd, Havant, PO9 4DG [IO90MU, SU70]
GW0 DIB G P McConnell, Paris House, 9 Bell St., Talgarth, Brecon, LD3 0BP [IO81JX, SO13]
G0 DIC Details withheld at licensee's request by SSL.[Station located near Boston.]
G0 DID J A Last, 154 Bucklesham Rd, Ipswich, IP3 8TZ [JO02OA, TM24]
G0 DIE J O Garr, 16 Tappers Cl, Topsham, Exeter, EX3 0DG [IO80GQ, SX98]
G0 DIG P Johnson, 5 Brook Bank, Whitehaven, CA28 8PZ [IO84FM, NX91]
G0 DIH P M Strong, 2 Jasper Cttgs, Cornworthy, Totnes, TQ9 7EY [IO80EJ, SX85]
G0 DIJ M A W Brook, 2 Melrose Ct, Ashley Rd, New Milton, BH25 5BY [IO90ES, SZ29]
G0 DIL K R Cawood, 150 Tyne Cres, Brickhill, Bedford, MK41 7YB [IO92SD, TL05]
G0 DIM A C Latham, 49 Tithe Barn Rd, Wootton, Bedford, MK43 9EZ [IO92RC, TL04]
G0 DIN Details withheld at licensee's request by SSL.
G0 DIO D A Goodship, 89 Doddshill Rd, Dersingham, Kings Lynn, PE31 6LP [JO02GU, TF63]
G0 DIP J B Huggins, 7 Coniston Dr, Jarrow, NE32 4AE [IO94GX, NZ36]
GW0 DIQ M C Smith, 7 Clos Gorsfawr, Gorseinon, Grovesend, Swansea, SA4 4GZ [IO71XQ, SN50]
G0 DIR S Caslake, 18 Queen Margarets Ave, Byram, Brotherton, Knottingley, WF11 9HT [IO93IR, SE42]
G0 DIS P A Pearce, 6 Swift Way, Wakefield, WF2 6SQ [IO93GP, SE31]
G0 DIU T V Rumble, 1 Victoria St., Brighouse, HD6 1HH [IO93CQ, SE12]
GW0 DIV R Griffiths, 5 Heol y Sarn, Llantrisant, Pontyclun, CF7 8DA [IO81HN, ST08]
GW0 DIX R M Rees, 6 Heol yr Ysgol, Llangynwyd, Maesteg, CF34 9SS [IO81EO, SS88]
G0 DIY Details withheld at licensee's request by SSL.
G0 DIZ R E Quaintance, 18 Queens Ave, Ilfracombe, EX34 9LN [IO71WE, SS54]
G0 DJA D J Ackrill, 104 Haigh Moor Rd, Tingley, Wakefield, WF3 1EF [IO93FR, SE22]
G0 DJB L J Burchell, 9 Hawthorn Way, Rayleigh, SS6 8SP [JO01HN, TQ89]
G0 DJC D J Collins, 71 Trench Rd, Tonbridge, TN10 3HG [JO01DF, TQ54]
G0 DJF P A Griffin, The Clyst, 16 Vineyard Rd, Hampton Park, Hereford, HR1 1TT [IO82PB, SO53]
GM0 DJG J G Walker, 28 Lomond Pl, Castlepark, Irvine, KA12 9PG [IO75QP, NS34]
G0 DJH Details withheld at licensee's request by SSL.[Station located near Reigate.]
GM0 DJI D J Tulloch, Fairview, Aywick, East Yell, Shetland, ZE2 9AX [IP90LN, HU58]
G0 DJK D J Keates, 13 Willow Rise, Witham, CM8 2LL [JO01HT, TL81]
G0 DJL C J Blount, Kimberley, Ken's Meadow, Holywell Bay, Newquay, Cornwall, TR8 5PT [IO70KJ, SW75]
G0 DJM M Rawson, 23 Oakdene Dr, Shadwell Ln, Leeds, LS17 8XW [IO93FU, SE33]
GM0 DJN D J Noble, 26 Monteith Gdns, Clarkston, Glasgow, G76 8NU [IO75UT, NS55]
G0 DJO A Robson, 19 Barnard Cl, Bedlington, NE22 6NE [IO95ED, NZ28]
G0 DJQ A Webster, 91 Thirlmere, Macclesfield, SK11 7YJ [IO83WF, SJ97]
G0 DJR J H Mills, Penny Farthing, Chapel Ln, Wanborough, Swindon, SN4 0AJ [IO91DN, SU28]
G0 DJS H Stemp, 5 Depot Rd, Horsham, RH13 5HB [IO91UA, TQ22]
G0 DJT P S Odegaard, 19 Church Way, Sanderstead, South Croydon, CR2 0JT [IO91XI, TQ36]
GW0 DJU D Lee, 14 Ffordd Dinas, Cwmavon, Port Talbot, SA12 9BS [IO81CO, SS79]
G0 DJV B W F Joy, 1 Riverbourne Rd, Milford, Salisbury, SP1 1NU [IO91CB, SU12]
G0 DJW Details withheld at licensee's request by SSL.[Op: G I Sydenham, 41 Alexandra Road, Beccles, Suffolk NR34 9UD.]
GW0 DJX A Cullen, 8 Pembroke Cl, Undy, Magor, Newport, NP6 3QD [IO81ON, ST48]
G0 DKA Details withheld at licensee's request by SSL.[Op: D G MacKay, 16 Henley Drive, Timperley, Altrincham, Cheshire WA15 6RY.]
G0 DKB P Buckenham, Shepherds Fell, The Street, Scoulton, Norwich, Norfolk, NR9 4PE [JO02KN, TF90]
G0 DKE O R Anese, 13 Dennis Rd, Kempston, Bedford, MK42 7HF [IO92RC, TL04]
GW0 DKF R C Thomas, Chapel Cottage, The Cutting, Llanfoist, Abergavenny, Monmouthshire, NP7 9NX [IO81LT, SO21]
GW0 DKG R Malpas, 3 Millfield, Whitland, SA34 0QN [IO71QT, SN11]
G0 DKJ A Wardell, 7 Claremount Rd, Boothtown, Halifax, HX3 6NX [IO93BR, SE02]
GM0 DKK A D Mackenzie, 13 Castle Cres, Torwood, Larbert, FK5 4ST [IO86BB, NS88]
G0 DKL T Shaw, 176 Hollins Rd, Oldham, OL8 3DG [IO83WM, SD90]
G0 DKM S R Daniels, 27 Willow Dr, Hutton, Weston Super Mare, BS24 9TJ [IO81MH, ST35]
G0 DKN A W W McClelland, Chapel Cottage, Chapel Ln, Redlynch, Salisbury, SP5 2HN [IO90DX, SU22]
G0 DKO G H Maskort, 133 Borstal St., Rochester, ME1 3JU [JO01FI, TQ76]
G0 DKP D K Phillips, Halfshire House, 2 Ct Farm Way, Churchill, Kidderminster, DY10 3LZ [IO82VJ, SO87]
G0 DKQ G Pool, 41 Radford St., Manton, Worksop, S80 2NQ [IO93KH, SK57]
G0 DKR M G Cole, 54 Ribble Rd, Stoke, Coventry, CV3 1AU [IO92GJ, SP37]
G0 DKS R F Barnes, Pentwyn, Graeme Rd, Norton, Yarmouth, PO41 0RX [IO90FQ, SZ38]
G0 DKT A Berry, 15 Glendale Dr, Mellor, Blackburn, BB2 7HB [IO83RS, SD63]
G0 DKU P McQuillan, 15 Dyer Rd, Shirley, Southampton, SO15 3EH [IO90GV, SU41]
G0 DKV E W Peberdy, 18 Arden Rd, Kenilworth, CV8 2DU [IO92FI, SP27]
G0 DKW Details withheld at licensee's request by SSL.
G0 DKX G J Birk, 30 Maple Dr, Alvaston, Derby, DE24 0FT [IO92GV, SK33]
G0 DKY J J Bass, 96 The Meadow, Copthorne, Crawley, RH10 3RH [IO91WD, TQ33]
G0 DKZ B C Yates, 9 Cloister Walk, Whittington, Lichfield, WS14 9LN [IO92CQ, SK10]
GW0 DLA E J Stuckey, 32 Stanley Rd, Gelli, Pentre, CF41 7NJ [IO81GP, SS99]
G0 DLB R A Burdett, 11 Fisher Ave, Hillmorton, Rugby, CV22 5HN [IO92JI, SP57]
G0 DLC K J Symonds, The Beeches, 30 Fairlea Cres, Northam, Bideford, EX39 1BD [IO71VA, SS42]
G0 DLD Details withheld at licensee's request by SSL.
G0 DLF A J Keon, 72 Rochelle Way, Alsace Park, Duston, Northampton, NN5 6YW [IO92MG, SP76]
G0 DLG L B McGowan, 25 Haldon Gr, Longbridge, Birmingham, B31 4LN [IO92AJ, SP07]
G0 DLH P E Swingler, 7 School Cl, Wales, Sheffield, S31 8QF [IO93GJ, SK38]
G0 DLI C H Hutchison, 20 Crockhurst, Southwater, Horsham, RH13 7XA [IO91TA, TQ12]
G0 DLJ P Roe, 79 Searby Rd, Sutton in Ashfield, NG17 5JS [IO93JC, SK55]
G0 DLK G C Hulme, 11 Landing Ln, Riccall, York, North Yorks, YO4 6PW [IO93LT, SE63]
G0 DLL W A Jackson, 22 Cliff Gdns, Oswald Rd, Scunthorpe, DN15 7PJ [IO93QO, SE81]
G0 DLM D L Moss, Beechwood Lodge, Leeds Rd, Lightcliffe, Halifax, HX3 8NU [IO93CR, SE12]
G0 DLN D G B Arthur, 43 Church Rd, Willesborough, Ashford, TN24 0JZ [JO01KD, TR04]
G0 DLO Details withheld at licensee's request by SSL.
G0 DLP P J Lee, 101 Fairdene Rd, Coulsdon, CR5 1RH [IO91WH, TQ25]
G0 DLQ G W Orange, 27 Connery, Hucknall, Nottingham, NG15 7AH [IO93JA, SK54]
G0 DLR L N W Buck, 21 Willow Walk, Culverstone, Meopham, Gravesend, DA13 0QS [JO01EI, TQ66]
G0 DLS D L Sparey, 21 Buxton Rd, Ashbourne, DE6 1EX [IO93DA, SK14]
G0 DLT J R Aizlewood, The Chimes, 8 Westminster Cl, Simonstone, Burnley, BB12 7ST [IO83TT, SD73]
G0 DLU Details withheld at licensee's request by SSL.[Station located near Sherborne.]
GW0 DLW J B Goldsmith, Ivydene, Four Crosses, Llanymynech, Powys, SY22 6PS [IO82KS, SJ21]
G0 DLX E Emerson, 22 Belton Gr, Birchencliffe, Huddersfield, HD3 3RF [IO93CQ, SE11]
G0 DLY Details withheld at licensee's request by SSL.
G0 DMA J H Slaney, 50 Laburnum Rd, Langold, Worksop, S81 9RR [IO93KJ, SK58]
G0 DMB C A Cade, 6 Ct Cl, Kirby Muxloe, Leicester, LE9 2DD [IO92JP, SK50]
G0 DME J Hallam, 34 Danethorpe Vale, Sherwood, Nottingham, NG5 3DA [IO92KX, SK54]
G0 DMF D McFarlane, 16 Springhead Ln, Ely, CB7 4QY [JO02DJ, TL58]
G0 DMH B E Cox, 21 Shelthorpe Rd, Loughborough, LE11 2PB [IO92JS, SK51]
G0 DMI Details withheld at licensee's request by SSL.[Station located near Reigate.]
G0 DMJ N C Beck, 36 Gr St., Great Hale, Sleaford, NG34 9JZ [IO92UX, TF14]
G0 DMK D M King, 94 Western Rd, Mickleover, Derby, DE3 5GQ [IO92FV, SK33]
G0 DML Details withheld at licensee's request by SSL.
G0 DMN S T Langham, 43 Greenwood Dr, Kirkby in Ashfield, Nottingham, NG17 8JT [IO93IC, SK45]
G0 DMO Details withheld at licensee's request by SSL.
G0 DMP D M Potter, 102 Normandy Ave, Beverley, HU17 8PF [IO93ST, TA03]
G0 DMQ B Holding, 36 Risingsun Rd, Gawsworth, Macclesfield, Ches, SK11 7UZ [IO83WF, SJ97]
G0 DMR M H Ross, 18 Tytherington Park Rd, Macclesfield, SK10 2EL [IO83WG, SJ97]
G0 DMS F Spencer, 35 Askew Gr, Repton, Derby, DE65 6GR [IO92FU, SK32]
G0 DMU I Morison, 4 Arley Cl, Macclesfield, SK11 8QP [IO83WG, SJ97]
G0 DMV R C Parrish, 89 Delamere Dr, Macclesfield, SK10 2PS [IO83WF, SJ97]
G0 DMW F P Cosgrove, 8 Wandsworth Rd, Heaton, Newcastle upon Tyne, NE6 5AD [IO94EX, NZ26]
G0 DMY S J Edwards, 15 Raleigh Cres, Goring By Sea, Worthing, BN12 6EF [IO90TT, TQ10]
G0 DNB E P Englehard, 153 Hurdsfield Rd, Macclesfield, SK10 2QX [IO83WG, SJ97]
G0 DNC Details withheld at licensee's request by SSL.[Op: J Nelson.]
G0 DND N Downs, 10 Oak St., Northwich, CW9 5LJ [IO83RG, SJ67]
G0 DNE Details withheld at licensee's request by SSL.
G0 DNF D J Fisher, 61 Park Ln, Cowplain, Waterlooville, PO8 8BB [IO90LV, SU61]
GM0 DNG G W Wallace, 25 Somerville Park, Lawthorn, Irvine, Ayrshire, KA11 2EL [IO75QP, NS34]
GM0 DNH P L Moore, 105 Fintry Dr, Dundee, DD4 9HQ [IO86ML, NO43]
G0 DNI G B D Wann, Manor Barn, Lower Farm, Shotatton, Ruyton Xi Towns, Shrewsbury, SY4 1JH [IO82MT, SJ32]
G0 DNJ Details withheld at licensee's request by SSL.
G0 DNK S Shaw, 93 Hipswell Highway, Wyken, Coventry, CV2 5FL [IO92GJ, SP37]
G0 DNL K R Baines, 52 Wolfit Ave, Balderton, Newark, NG24 3PH [IO93OB, SK85]

G0 DNO P F Bradford, 7 Teme Cl, Bromyard, HR7 4TP [IO82RE, SO65]
G0 DNP M Crossley, 3 Ryedale, Huddersfield, HD5 0HT [IO93DP, SE11]
G0 DNQ C J Anderton, 77 Kestrel Rd, Moreton, Wirral, Merseyside, L46 6BW [IO83KJ, SJ29]
G0 DNU R Hamblett, 49 Kneller Rd, Twickenham, TW2 7DF [IO91TK, TQ17]
G0 DNV R J Hammett, 47 Sowden Park, Barnstaple, EX32 8EJ [IO71XB, SS53]
G0 DNX R Williams, 38 March Vale Rise, Conisborugh, Conisbrough, Doncaster, DN12 2EW [IO93JL, SK59]
G0 DOA C Dale, 11 Roman Cl, Newby, Scarborough, YO12 5RG [IO94SG, TA08]
G0 DOB R Gray, 17 Rose Gdns, Washdyke Ln, Immingham, Grimsby, DN40 2RG [IO93VP, TA11]
G0 DOC H Kiff, Rock Cottage, Red Ln, The Cloud, nr Congleton, Ches, CW12 3QG [IO83WE, SJ96]
G0 DOD M Dodds, West Walls, Nursery Ln, Stockton on Tees, TS18 4DR [IO94IN, NZ41]
G0 DOE T D Purcell, 18 Marston Ave, Chessington, KT9 2HF [IO91UI, TQ16]
G0 DOG C S Purcell, 18 Marston Ave, Chessington, KT9 2HF [IO91UI, TQ16]
G0 DOH D J Henson, 19 Gair St., Hyde, SK14 4SB [IO83XK, SJ99]
G0 DOI D B G Price, 8 Westford Cl, Westford, Wellington, TA21 0DY [IO80IX, ST12]
G0 DOJ Details withheld at licensee's request by SSL.
G0 DOK R D Cornish, 28 Warnford Cres, Leigh Park, Havant, PO9 4PZ [IO90MU, SU70]
G0 DOM D J Oskis, 10 Moultrie Way, Cranham, Upminster, RM14 1NB [JO01DN, TQ58]
G0 DOO J D Ingram, 2 Sylvan Cres, Sutton in Ashfield, NG17 3DL [IO93JD, SK56]
G0 DOQ P J Hawkins, 52 Windsor Rd, Dorchester, DT1 2JB [IO80SR, SY69]
G0 DOR P C Davidson, 5 Derby Gr, Maghull, Liverpool, L31 5JJ [IO83MM, SD30]
G0 DOU H Grandfield, 2 Bolshaw Rd, Heald Green, Cheadle, SK8 3PJ [IO83VI, SJ88]
G0 DOV G R Lintern, 33 Salisbury Rd, Dover, CT16 1EX [JO01PD, TR34]
G0 DOW Details withheld at licensee's request by SSL.
GW0 DOX M Morgan, 11 North Mead, Sarn, Bridgend, CF32 9SA [IO81FM, SS98]
G0 DOY Details withheld at licensee's request by SSL.[Op: L F J Rumens, PO Box 117, Stockport, SK5 7HF.]
G0 DPA G A Haywood, 9 Welton Park, Welton, Daventry, NN11 5JW [IO92KG, SP56]
G0 DPC J Cook, 85 Newington Ave, Southend on Sea, SS2 4SP [JO01IN, TQ98]
G0 DPE B J Aldersey, 4 Salterbeck Terr, Salterbeck, Workington, CA14 5HP [IO84FO, NX92]
GM0 DPH Details withheld at licensee's request by SSL.
G0 DPI D H Barkley, 39 Fulbeck Ave, Goose Green, Wigan, WN3 5QN [IO83QM, SD50]
G0 DPJ K B Ellis, 162 Wolverhampton Rd, Dudley, DY3 1RF [IO82WN, SO99]
G0 DPK P R E Mc Caldon, 47 Merritt Rd, Didcot, OX11 7DF [IO91IO, SU59]
GW0 DPM B C T Neathey, 18 Castle Dr, Dinas, Powys, S Glam, CF6 4NP [IO81HN, ST08]
G0 DPO K R Glazebrook, 86 Deveraux Dr, Wallasey, L44 4DL [IO83LJ, SJ39]
G0 DPP C S Poole, 2 Elm Cttgs, Lower Rd, Westerfield, Ipswich, IP6 9AU [JO02NC, TM14]
G0 DPQ A J Henning, 24 Garfield Ave, Draycott, Derby, DE72 3NP [IO92HV, SK43]
G0 DPS J M Fyrth, 2 Merton Gdns, Farsley, Pudsey, LS28 5DZ [IO93DT, SE23]
G0 DPT M C Smith, 22 Long Leys, London, E4 9LW [IO91XO, TQ39]
GM0 DPU L M Craig, Camus, Furnace, Inveraray, Argyll, PA32 8XU [IO76JD, NS09]
GI0 DPV J Mangan, St Patricks Trng.Sc, Glen Rd, Belfast, N Ireland, BT11 8BX [IO74AO, J27]
G0 DPX J Brown, 72 Whitcliffe Rd, Cleckheaton, BD19 3BY [IO93DR, SE12]
G0 DPZ Details withheld at licensee's request by SSL.[Station located near Towcester.]
G0 DQA Details withheld at licensee's request by SSL.
G0 DQB R A White, 27 Windsor Walk, Scawsby, Doncaster, DN5 8NQ [IO93KM, SE50]
GM0 DQC H P Meikle, 37 Turfholm, Lesmahagow, Lanark, ML11 0ED [IO85BP, NS83]
G0 DQD P E Bryan, 23 Millersdale Ave, Mansfield Rd, Mansfield, NG18 5HS [IO93JD, SK55]
G0 DQE C H D Heaps N Notts Raynet Group 201, 12 Oak Tree Cl, Eakring Rd, Mansfield, NG18 3EN [IO93KD, SK56]
G0 DQF W A Rustman, 27 King Edward Rd, Laindon, Basildon, SS15 6HL [JO01FN, TQ68]
G0 DQG J M Hardman, 140 Church Rd, Earley, Reading, RG6 1HR [IO91MK, SU77]
G0 DQH J L Colwill, 5 Stinchar Dr, Chandlers Ford, Eastleigh, SO53 4QJ [IO90HX, SU42]
G0 DQI D A Harding, High Peak, Hillcrest Rd, Kingsdown, Deal, CT14 8EB [JO01QE, TR34]
GI0 DQJ D Livingstone, 11 Cabragh Rd, Tandragee, Craigavon, BT62 2HJ [IO64SI, J04]
G0 DQL Details withheld at licensee's request by SSL.
G0 DQM J J Williams, Maes yr Awel, 7 Suttonfield Rd, Sutton, Doncaster, DN6 9JX [IO93JO, SE51]
G0 DQO W A Cartwright, 149 Heath Ln, West Bromwich, B71 2BL [IO92AM, SP09]
G0 DQP E G Rees, 26 Elsham Cl, Bramley, Rotherham, S66 0XZ [IO93IK, SK49]
G0 DQQ S W Power, Upperleigh, Upton Scudamore, Warminster, Wilts, BA12 0AQ [IO81VF, ST84]
GM0 DQR P Corrigan, 112 Clarkston Rd, Cathcart, Glasgow, G44 3DH [IO75UT, NS56]
G0 DQS M A Glen, 10 Field Ln, Dursley, Glos, GL11 6JE [IO81TQ, ST79]
GW0 DQT A J Duck, 15 Ambryn Rd, New Inn, Pontypool, NP4 0NJ [IO81LQ, ST39]
GM0 DQV G George, 13 Balmoral Terr, Bishopmill, Elgin, IV30 2JH [IO87IP, NJ26]
G0 DQW C Hughes, 46 Gillow Rd, Kirkham, Preston, PR4 2JS [IO83NS, SD43]
GW0 DQY G E Hughes, 39 Thornhill Cl, Upper Cwmbran, Cwmbran, NP44 5TQ [IO81LP, ST29]
G0 DRA D R A Love, 4 St. Chads Rd, Lichfield, WS13 7LZ [IO92CQ, SK11]
G0 DRC R J Joy obo Dartmoor Radio Club, The Old School House, Maristow, Roborough, Devon, PL6 7BY [IO70WK, SX46]
G0 DRD S Carvin, 173 Beech Park, Lucan, Co Dublin, Ireland, X X
G0 DRE J L Webster, 3 Badby Rd West, Daventry, NN11 4HJ [IO92JG, SP56]
G0 DRF P Trask, 23 Dove St, York, YO2 1AQ [IO93KW, SE55]
G0 DRG E M Cousins, 23 Hamtun Cres, Totton, Southampton, SO40 3PB [IO90GW, SU31]
G0 DRH B J Harris, 9 Woodlands Cl, Rayleigh, SS6 7RG [JO01HN, TQ88]
GW0 DRI J Smith, 7 Clos Gorsfawr, Gorseinon, Grovesend, Swansea, SA4 4GZ [IO71XQ, SN50]
G0 DRJ J Connell, 24 Finchale Rd, Framwellgate Moor, Durham, DH1 5JN [IO94ET, NZ24]
G0 DRL G Howarth, 5 Glen Ave, Kearsley, Bolton, BL4 8QW [IO83TM, SD70]
G0 DRM D A Cookson, 70 Rope Ln, Wistaston, Crewe, CW2 6RD [IO83SB, SJ65]
G0 DRN Details withheld at licensee's request by SSL.
G0 DRO D H Roberts, 15 Park Brook Rd, Macclesfield, SK11 8QH [IO83WG, SJ97]
G0 DRQ R P Hope, 3 Farm Cres, Sittingbourne, ME10 4QD [JO01IH, TQ96]
G0 DRR O A Perry, The Maples, Horsham Rd, Cranleigh, GU6 8DT [IO91SD, TQ03]
GW0 DRS R J Ford, 36 Circular Dr, Ewloe, Deeside, CH5 3DA [IO83LE, SJ26]
G0 DRT P W Quested, Nethercroft, Southsea Ave, Minster on Sea, Sheerness, ME12 2NH [JO01JK, TQ97]
GM0 DRU I M Maclennan, 70 Kenneth St., Stornoway, Isle of Lewis, HS1 2DS [IO68TF, NB43]
G0 DRV W D Hoyle, 34 Warwick Gdns, Rayleigh, SS6 8TQ [JO01HN, TQ89]
G0 DRW T D Wright, 73 West St., Ryde, PO33 2QQ [IO90KR, SZ59]
G0 DRX P M Hill, 34 Church Rd, Whitchurch, Bristol, BS14 0PP [IO81RJ, ST66]
G0 DSB T A Winship, 32 Lytes Cary Rd, Keynsham, Bristol, BS18 1XD [IO81SJ, ST66]
G0 DSC S J Charlesworth, 17 Clifton Ln, Rotherham, S65 2AA [IO93HK, SK49]
GI0 DSG W McKeever, 14 Carnhill, Londonderry, BT48 8BA [IO65IA, C41]
G0 DSH J R Kirkby, Church Ln, Willoughby, Alford, Lincs, LN13 9SU [JO03CF, TF47]
G0 DSI Dr J A Wszeborowski, 18 Shotley Gdns, Gateshead, NE9 5DP [IO94EW, NZ26]
GW0 DSJ E Shipton, 34 Argoed, Kimnel Bay, Kinmel Bay, Rhyl, LL18 5LN [IO83FH, SH97]
G0 DSK J I Williams, Skelmorlie, 1 Dover Rd, Sandwich, CT13 0BH [JO01QG, TR35]
GW0 DSL J E Carson, Clwyd House, 11 Bryneithin Ave, Prestatyn, LL19 9LS [IO83HI, SJ08]
G0 DSM E Sprogis, 4 Balmoral Cl, Earls Barton, Northampton, NN6 0LZ [IO92PG, SP86]
G0 DSN L C Nash, Four Furlongs, Wells Rd, Stiffkey, Wells Next The Sea, NR23 1QE [JO02LW, TF94]
G0 DSO R A Calvert, Shortley Cl, Robin Hoods Bay, nr Whitby, N Yorks, YO22 4PB [IO94RK, NZ90]
GW0 DSP M J Lamb, 4 Hadfield Cl, Connahs Quay, Deeside, CH5 4JP [IO83LE, SJ26]
G0 DSQ Dr K M Myint, 347 Dogsthorpe Rd, Peterborough, PE1 3PF [IO92VO, TF10]
G0 DSR D Jones, 20 Marsh Green, Wigan, WN5 0PU [IO83PN, SD50]
GW0 DST V J Bird, 3 Hafnant, Winch Wen, Swansea, SA1 7LG [IO81BP, SS69]
G0 DSU G S Fisher, 19 Orde Cl, Pound Hill, Crawley, RH10 3NG [IO91WD, TQ23]
G0 DSV Details withheld at licensee's request by SSL.
GW0 DSW T A G Orchard, 5 Farnsworth Ct, Old Golf Rd, Rhyl, Denbighshire, LL18 3PB [IO83GH, SJ08]
G0 DSX N Hanking, 7 Clayside House, Kenton Ct, South Shields, NE33 4HP [IO94GX, NZ36]
GM0 DTA W G J Gevers, The New Farm House, Mains of Cairnrobin, Portlethen, Aberdeen, AB1 4SB [IO87TH, NJ72]
G0 DTB C Proffitt, 17 Tempest St., Bolton, BL3 4HR [IO83SN, SD60]
G0 DTC Dr D T Coxon, Thurston, Gate Farm Rd, Shotley, Ipswich, IP9 1QH [JO01PX, TM23]
G0 DTF W E Williams, 2 Gayton Cl, Upton By Chester, Chester, CH2 2HS [IO83NF, SJ46]
G0 DTG D Tarbuck, 32 Church Rd, Catshill, Bromsgrove, B61 0JY [IO82XI, SO97]
G0 DTI K M Sumner, 7 Largs Rd, Shadsworth, Blackburn, BB1 2JQ [IO83SR, SD72]
GM0 DTJ W Cousins, 18 Earlswells Dr, Cults, Aberdeen, AB1 9NW [IO87TH, NJ72]
G0 DTK J Clarke, 66 Harefield Rd, Swaythling, Southampton, SO17 3TH [IO90HW, SU41]
G0 DTL Details withheld at licensee's request by SSL.
G0 DTN R Lord, 7 Harewood Rd, Norden, Rochdale, OL11 5TG [IO83VP, SD81]
G0 DTO J Duckworth, 201 Cox Green Rd, Egerton, Bolton, BL7 9UT [IO83SP, SD71]
G0 DTP L C Quantrill, Innisfree, 1 Ironwell Ln, Hawkwell, Hockley, SS5 4JY [JO01IO, TQ89]
G0 DTQ W N Goldstraw, 5 Council Houses, Wantage Rd, Great Shefford, Hungerford, RG17 7DG [IO91GM, SU37]
G0 DTR I C Pordum, 79 Lanercost Dr, Fenham, Newcastle upon Tyne, NE5 2DL [IO94DX, NZ26]
G0 DTT R Mycock, 5 Alstone Rd, Heaton Chapel, Stockport, SK4 5AH [IO83VK, SJ89]
G0 DTU Details withheld at licensee's request by SSL.
G0 DTW S I Bolton, 40 Claytonwood Rd, Stoke on Trent, ST4 6LD [IO82VX, SJ84]
G0 DTX Details withheld at licensee's request by SSL.
G0 DUA S J Linden, 4 Downing Dr, Great Barton, Bury St. Edmunds, IP31 2RP [JO02JG, TL86]
G0 DUB F G A Mossop, 14 Websters Ln, Great Sutton, South Wirral, L66 2LH [IO83MG, SJ37]
G0 DUE Details withheld at licensee's request by SSL.
G0 DUF R G Hoare, 7 Springfield Cl, Watlington, OX9 5RF [IO91MP, SU69]
G0 DUG D H Marsden, 31 Aberfoyle, Ouston, Chester-le-Street, DH2 1RH [IO94EV, NZ25]
G0 DUH P J Murrell, 10 Irving Cl, Braunton, EX33 1DH [IO71VC, SS43]
G0 DUK K Bennett, 78 Rectory Rd, Upper Deal, Deal, Kent, CT14 9NB [JO01QF, TR35]
G0 DUM D Gibbons, 17 Della Ave, Barnsley, S70 6LG [IO93GN, SE30]
G0 DUN D Wakeford, 2 Rooley House Cottage, Rooley Ln, Sowerby Bridge, HX6 1NS [IO93AQ, SE02]
GI0 DUP R Miskimmin, 15 Abbeydale Ave, Newtownards, BT23 8RT [IO74DN, J46]
G0 DUQ R Wilkes, 47 Greenwood Park, Hednesford, Cannock, WS12 4DQ [IO82XR, SJ91]
G0 DUT I F M Duthie, 5 Park Cl, Scotby, Carlisle, CA4 8AX [IO84NV, NY45]
G0 DUU Details withheld at licensee's request by SSL.
G0 DUV K T Howard, 22 Waterloo Rd, Norwich, NR3 1EY [JO02PP, TG21]
GM0 DUX R D McGowan, 35 Ochiltree, Dunblane, FK15 0DF [IO86AE, NN70]
G0 DUY K H Oates, 9 East Woodyates, Salisbury, Wilts, SP5 5QZ [IO90AX, SU01]
G0 DUZ J H W Washington, 76 Rockingham Rd, Uxbridge, UB8 2UA [IO91SN, TQ08]
G0 DVB J L Morris, 18 Ellingdon Rd, Wroughton, Swindon, SN4 9HY [IO91CM, SU18]
G0 DVC P S T J Robinson, 256 Victoria Rd, Ruislip, HA4 0DW [IO91TN, TQ18]
G0 DVD A C Strong, 8 Forrest Ave, Essington, Wolverhampton, WV11 2AJ [IO82XP, SJ90]
G0 DVE S A Hutchings, 59 Leigh Rd, Wimborne, BH21 1AE [IO90AT, SZ09]
G0 DVF S M Wadsley, 56 Hedgerley, Chinnor, OX9 4TJ [IO91NQ, SP70]
G0 DVG Details withheld at licensee's request by SSL.
GM0 DVH N McNulty, 6 Main Rd, Crookedholm, Kilmarnock, KA3 6JT [IO75SO, NS43]
G0 DVI R Brookes, 57 Fir Rd, Thetford, IP24 3EX [JO02IJ, TL88]
G0 DVJ J P Mitchener, 10 Pine Cl, Brantham, Manningtree, CO11 1TP [JO01MX, TM13]
G0 DVL R S Nash, 28 Squires Way, Wilmington, Dartford, DA2 7NW [JO01CK, TQ57]
G0 DVM H Rimmer, 44 The Heys, Coppull, Chorley, PR7 4NX [IO83QP, SD51]
GM0 DVO A Gemmell, 23 Busby Rd, Carmunnock, Clarkston, Glasgow, G76 9BN [IO75VS, NS55]
G0 DVP G A Gulliford, 29 Windsor Rd, Westlea, Seaham, SR7 8DG [IO94HU, NZ44]
G0 DVR Details withheld at licensee's request by SSL.
G0 DVS E Holding, 11 Dane Cres, Ramsgate, CT11 7JU [JO01RI, TR36]
G0 DVT J Brindle, 40 Huntingfield Rd, Bury St. Edmunds, IP33 2JA [JO02IF, TL86]
GI0 DVU J C Henry, 3 Kirkwoods Park, Lisburn, BT28 3RR [IO64XM, J26]
G0 DVV Details withheld at licensee's request by SSL.
G0 DVW A S Douglas, Bryansford, 29 Blackheath Gr, Wonersh, Guildford, GU5 0PU [IO91RE, TQ04]
G0 DVY R O Ibbotson, Fern Lea, Maltby-le-Marsh, Alford, Lincs, LN13 0JP [JO03CH, TF48]
G0 DVZ M G Barber, 9 Grange Cross Cl, West Kirby, Wirral, L48 8BR [IO83KI, SJ28]
G0 DWA C J V Chick, Norley Moat House, Little Bradley, Haverhill, Suffolk, CB9 7JN [JO02FD, TL75]
G0 DWB D Wathen-Blower, 61 Dykes End, Collingham, Newark, NG23 7LD [IO93OD, SK86]
G0 DWC S Beadle, 18 The Shrubberies, Cliffe, Selby, YO8 7PW [IO93MS, SE63]
G0 DWD J B Hawes, Cherry Lodge, Woodwaye, Woodley, Reading, RG5 3HA [IO91NK, SU77]
G0 DWF D W Fouche, 17 Burlington Gdns, Gillingham, ME8 8TA [JO01HI, TQ86]
G0 DWG P F Unterhorst, 240 Leeds Rd, Ilkley, LS29 8LN [IO93CW, SE14]
G0 DWI L N Fennelow N.Cambs Rptr Gr, 39 Clarence Rd, Wisbech, PE13 2ED [JO02CQ, TF41]
G0 DWJ N F Hall, 10 Newnham Rd, Leamington Spa, CV32 7SN [IO92FH, SP36]
G0 DWM C D Pearsons, 12 Abbey Way, Farnborough, GU14 7DA [IO91PH, SU85]
GI0 DWN M J Dougan, 97 Redrock Rd, Collone, Armagh, BT60 2BN [IO64RH, H93]
G0 DWO P Labron, 7 Boulmer Rd, Longhoughton, Alnwick, NE66 3AQ [IO95EK, NU21]
GW0 DWQ S M T Outen, 2 Heol Vaughan, Burry Port, SA16 0HF [IO71UQ, SN40]
G0 DWR P F Bell, 5 Sages Ln, Privett, Alton, GU34 3NP [IO91LB, SU62]
G0 DWS J W Tubbs, 19 Greenhill Rd, Northfleet, Gravesend, DA11 7EZ [JO01EK, TQ67]
G0 DWT Details withheld at licensee's request by SSL.
G0 DWV C M Danby, 299 Reepham Rd, Hellesdon, Norwich, NR6 5AD [JO02PQ, TG21]
G0 DWW J Barson, 1 Springwell Cl, Grange Meadows, Maltby, Rotherham, S66 7HG [IO93JK, SK59]
GM0 DWY R A Price, 80 Eastern Ave, Largs, KA30 9EQ [IO75NT, NS25]
G0 DWZ Details withheld at licensee's request by SSL.
G0 DXA Details withheld at licensee's request by SSL.
G0 DXB W McGill, 14 Farquhar Rd, Maltby, Rotherham, S66 7PD [IO93JK, SK59]
G0 DXC N A Bryan, 93 Middle Rd, Sholing, Southampton, SO19 8FT [IO90HV, SU41]
G0 DXD Details withheld at licensee's request by SSL.
GM0 DXE H A Quin, Post Office, Edinbane, Isle of Skye, IV51 9PW [IO67SL, NG35]
G0 DXF C W Emmanuel, 29 Guillemot Way, Liverpool, L26 7WG [IO83NI, SJ48]
G0 DXI H Lakhaney, 60 Border Gdns, Shirley, Croydon, CR0 8LP [IO91XI, TQ36]
G0 DXJ H R Green, 18 Burnetts Fields, Horton Heath, Eastleigh, SO50 7DH [IO90IW, SU41]
G0 DXK M Bedford, 17 Mount St., Cowlersley, Huddersfield, HD4 5TD [IO93CP, SE11]
G0 DXN R W Phillips, 7 Fell Gr, Leamington Spa, CV32 7UR [IO92FH, SP36]
GW0 DXO D R Oates, 86 Queens Ave, Maesgeirchen, Bangor, LL57 1NG [IO73WF, SH57]
G0 DXP R M Hall, 9 Ingswell Dr, Notton, Wakefield, WF4 2NF [IO93GO, SE31]
GW0 DXQ P M Kelson, 92 Christchurch Rd, Newport, NP9 7SJ [IO81MO, ST38]
G0 DXS C K Chan, Jet Propulsion Lab, M/S 264-114, Pasadena, California 91109, U S A, X X
G0 DXT T J Pearson, 45 Frobisher Cl, Daventry, NN11 4JL [IO92KG, SP56]
GU0 DXX P G Guilbert, Linden Lea Rue de La Boullerie, Four Cabot, St. Andrew, Guernsey, Channel Islands, GY6 8XE[Station located near Evesham, locator: IO92BB). QSL via Bureau or G0ENR QTHR.]
GW0 DXZ G Stephens, Ty Coch, Rhydwen Pl, Craig Cefn Parc, Clydach, Swansea, SA6 5RN [IO81BQ, SN60]
G0 DYB F R Hamilton Wigan Raynet Gr, 329 North Rd, Atherton, Manchester, M46 0RF [IO83RM, SD60]
G0 DYC D Anderton, 22 Barrington Rd, Poulton, Wallasey, L44 9BP [IO83LJ, SJ39]
GM0 DYD D Young, 4 Primrose Ave, Rosyth, Dunfermline, KY11 2SS [IO86GB, NT18]
GM0 DYF C J O'Hare, 14 Pentland Rd, Bellfield, Kilmarnock, KA1 3RS [IO75SO, NS43]
G0 DYG D J Doyle, 40 Howson St., Rock Ferry, Birkenhead, L42 2BR [IO83LI, SJ38]
GW0 DYH S W Fergusson, 37 Station Rd, Old Colwyn, Colwyn Bay, LL29 9EL [IO83DG, SH87]
G0 DYI Details withheld at licensee's request by SSL.
G0 DYL Details withheld at licensee's request by SSL.
G0 DYN J Denovan, 8 Pulley Ave, Eaton Bishop, Hereford, HR2 9QN [IO82OB, SO43]
G0 DYQ J D Morris, 28 Henley Ct, Buckingham Gdns, Lichfield, WS14 9AT [IO92CQ, SK10]
G0 DYR Details withheld at licensee's request by SSL.[Correspondence via PO Box 249, Poole, Dorset BH15 2LR.]
G0 DYS P Hutt, 7 Lynwood Ave, Woodlesford, Leeds, LS26 8LH [IO93GS, SE32]
GW0 DYT T Garner, Cae'R Sais Bach, Rhostryfan, Caernarfon, Gwynedd, LL54 7PA [IO73VC, SH55]
GM0 DYU M A McWhinnie, 14 Braehead Dr, Cruden Bay, Peterhead, AB42 0NW [IO97BK, NK03]
G0 DYW Details withheld at licensee's request by SSL.[Op: I Dowse.]
GW0 DYZ Details withheld at licensee's request by SSL.
G0 DZA P M Wentworth, 46 Woodside Ave, Cinderford, GL14 2DW [IO81ST, SO61]
G0 DZB P J Onion, 56 New Park St., New Town, Colchester, CO1 2NA [JO01LV, TM02]
G0 DZC C T Cosgrif, 32 Lower Mead Dr, Burnley, BB12 0ED [IO83VT, SD83]
GM0 DZE J Duffy, Lylestone Farm, Kilwinning, Ayrshire, KA13 7QP [IO75QQ, NS34]
G0 DZF C H Northrop, Buena Vista, Garstang Rd East, Singleton, Poulton-le-Fylde, FY6 7SX [IO83MU, SD33]
G0 DZG J H Pickersgill, 38 Sefton Ave, Newcastle upon Tyne, NE6 5QR [IO94FX, NZ26]
G0 DZH D M Gough, 20 Lawn Cl, Ruislip, HA4 6ED [IO91TN, TQ08]
G0 DZI C W Kuss, 20 Windermere Rd, Haydock, St. Helens, WA11 0ES [IO83PL, SJ59]
G0 DZJ Details withheld at licensee's request by SSL.[Op: D S Jelley.]
GW0 DZL J H Davies, 5 Talbot St., Llanelli, SA15 1DG [IO71WQ, SS59]
G0 DZM P R Gainey, Prencott, Harley Wood, Nailsworth, Stroud, GL6 0LD [IO81VQ, ST89]
G0 DZN Details withheld at licensee's request by SSL.
G0 DZO D F A Hambidge, 55 Flora Rd, Hay Mills, Birmingham, B25 8BH [IO92CL, SP18]
G0 DZQ B E Stevens, 24 Waverley Cres, Runwell, Wickford, SS11 7LN [JO01GP, TQ79]
G0 DZU P Barker, Peartree Cottage, Ash Hill Common, Sherfield English, Romsey, SO51 6FU [IO91EA, SU22]
G0 DZV K W Rose, 57 Cheriton Rd, Winchester, SO22 5AX [IO91IB, SU42]
GM0 DZW R Webster, Birchwood, Crathie, Aberdeenshire, Scotland, AB3 5TN [IO87TH, NJ72]
G0 DZX J Huggins, 30 Oakwood Cres, Rawmarsh, Rotherham, S62 7JG [IO93HL, SK49]
G0 DZY B R Sparrow, Glendowie, Denham, Bury St Edmunds, Suffolk, IP29 5EW [JO02GF, TL76]
G0 DZZ Details withheld at licensee's request by SSL.
G0 EAA J A Handley, 469 Willerby Rd, Hull, HU5 5JD [IO93TS, TA02]
G0 EAB E A T Bennett, 3 Ferndale Rd, Whiteshill, Stroud, GL6 6BA [IO81VS, SO80]
G0 EAC Details withheld at licensee's request by SSL.
G0 EAD E Green, 13 Northfield Rd, Rising Bridge, Accrington, BB5 2SR [IO83UR, SD72]
G0 EAE M G Taylor, 5 Curtis Mews, Wellingborough, NN8 5PG [IO92PH, SP86]
G0 EAG A L Sammons, 1 Poneys, The St., Wickham Bishops, Witham, CM8 3NN [JO01IS, TL81]
G0 EAM A Moore, 69 Renfrew Ave, Blackbrook, St. Helens, WA11 9RW [IO83PL, SJ59]
G0 EAN C E Bibb, 54 Dorsett Rd, Wednesbury, WS10 0JF [IO92AN, SP09]
G0 EAO Details withheld at licensee's request by SSL.
G0 EAP Details withheld at licensee's request by SSL.
G0 EAQ Details withheld at licensee's request by SSL.
G0 EAR Details withheld at licensee's request by SSL.
GM0 EAS J I Horne, Cowgan, Ruthwell, Dumfries, DG1 4NS [IO85HA, NY16]
G0 EAT S R Anderson, Sands Ln, Moor, Holme on Spalding Moor, York, North Yorks, YO4 4EX [IO93OU, SE73]
G0 EAU A A Surooprajally, 26 Walton Ave, North Cheam, Sutton, SM3 9UB [IO91VI, TQ26]
G0 EAV A F S Dickinson, 112 Worple Rd, Isleworth, TW7 7HU [IO91UL, TQ17]
GW0 EAW J W Humphries, 19 Tai Newydd, Llanfaelog, Ty Croes, LL63 5TW [IO73SF, SH37]
G0 EAZ N W Mason, 57 James St., Macclesfield, SK11 8BW [IO83WG, SJ97]
G0 EBA J P McNamara, 63 Vicarage Rd, Thetford, IP24 2LW [JO02JK, TL88]

G0 EBB Details withheld at licensee's request by SSL.
G0 EBD M Element, 9 Longbridge Cl, Shrewsbury, SY2 5YD [IO82PR, SJ51]
G0 EBF L J P Taylor, 18 Manifold Rd, Eastbourne, BN22 8EH [JO00DS, TV69]
G0 EBG R G Fuller, 41 Burnham Rd, Hullbridge, Hockley, SS5 6BG [JO01HO, TQ89]
G0 EBH Revd C G G Montgomery, Jasmine Cottage, Ash Cross, Bradworthy, Holsworthy, Devon, EX22 7SP [IO70TW, SS31]
G0 EBI J R Speller, Casa Mia, 62 Harwich Rd, Little Clacton, Clacton on Sea, CO16 9NE [JO01NU, TM12]
G0 EBK R F Bickley, 8 East Woodhay Rd, Harestock, Winchester, SO22 6JH [IO91IB, SU43]
G0 EBL K P Mayes, Overdale, Goathland, Whitby, N Yorks, YO22 5AN [IO94PL, NZ80]
G0 EBN J M Menhams, 19 Rowe Terr, Westfield, Workington, CA14 3SU [IO84FP, NX92]
G0 EBP A S P Bowmaker, 1 Hestham Dr, Morecambe, LA4 4QD [IO84NB, SD46]
G0 EBQ N C Flatman, 2 Deben Valley Dr, Kesgrave, Ipswich, IP5 7FB [JO02OB, TM24]
G0 EBR Details withheld at licensee's request by SSL.
G0 EBS S Artus, 29 Long Mead Way, Tonbridge, TN10 3TF [JO01DF, TQ54]
GI0 EBT W R Hadden, 21 Sleepy Valley, Richhill, Armagh, BT61 9QY [IO64RI, H94]
G0 EBU Details withheld at licensee's request by SSL.
G0 EBV A Vest, 42 Chillingham Dr, Chester-le-Street, DH2 3TJ [IO94EU, NZ24]
G0 EBW D W Agar, 122 Salisbury Rd, Moseley, Birmingham, B13 8JZ [IO92BK, SP08]
G0 EBZ M S W Whitfield, Camp Farm, Elberton, Olveston, Bristol, BS12 3AQ [IO81RO, ST68]
G0 ECA S V Coe, 15 Chestnut Gr, Crewe, CW1 4BD [IO83SC, SJ75]
G0 ECB M Hayhurst, 42 Lowthwaite Dr, Nelson, BB9 0SU [IO83VT, SD83]
G0 ECC R A Pears St Austell ARC, 24 Westbourne Dr, St. Austell, PL25 5EA [IO70OI, SX05]
G0 ECD D J Bosworth Nth N'Hants Ray, 144 Rothwell Rd, Kettering, NN16 8UP [IO92PJ, SP87]
G0 ECF L B Orengo, 16 Elton Ave, Wembley, HA0 3EE [IO91UN, TQ18]
G0 ECG M Higgin, 24 Tiverton Dr, Briercliffe, Burnley, BB10 2JT [IO83VT, SD83]
G0 ECI R J Matthews, 51 Mayfair Dr, Fazeley, Tamworth, B78 3TG [IO92DO, SK10]
G0 ECJ H Hughes, 212A Stroud Rd, Gloucester, GL1 5SD [IO81VU, SO81]
G0 ECK P Jenkins, 49 Ewell Park Way, Ewell, Epsom, KT17 2NW [IO91VI, TQ26]
G0 ECL K H W Clift, 28 Redgate Rd, Girton, Cambridge, CB3 0PP [JO02BF, TL46]
G0 ECM M D Bell, 18 Linnet Cl, Patchway, Bristol, BS12 5RN [IO81QM, ST58]
GW0 ECN K G Watkinson, 53 Sandringham Dr, Ashford, TW15 3JH [IO91SK, TQ07]
GM0 ECO Details withheld at licensee's request by SSL.
G0 ECP T Unsworth, 3 Conifer Dr, Stockton on Tees, TS19 0LU [IO94HN, NZ42]
G0 ECQ C W A Quinnin, 97 Cowpen Rd, Blyth, NE24 5TR [IO95FC, NZ28]
G0 ECR K H Kirby Nth Ferriby Unt, 100 Etherington Dr, Hull, HU6 7JT [IO93TS, TA03]
G0 ECS G C Westaby, 2 Goodwood, Bottesford, Scunthorpe, DN17 2TP [IO93PN, SE80]
GM0 ECU R C G Low, 2 Craigie Pl, Crosshouse, Kilmarnock, KA2 0JR [IO75RO, NS33]
G0 ECW E C Wilson, 20 Wivelsfield Rd, Saltdean, Brighton, BN2 8FQ [IO90XT, TQ30]
G0 ECX R B Mott, 27 Walpole St., Weymouth, DT4 7HJ [IO80SO, SY67]
G0 ECY S W Sutcliffe, Highlands Scarr, Bottom Ln, Greetland, Halifax, Yorks, HX4 8PG [IO93BQ, SE02]
G0 ECZ B Clarke, The Corner Stores, 66 Northload St., Glastonbury, BA6 9JR [IO81PD, ST43]
G0 EDB K Y Lee, 20 Amber Rd, Block C 11-01, Singapore 1543
GW0 EDC B Hall, 7 Ferndale Cl, Penyffordd, Chester, CH4 0NH [IO83LD, SJ36]
G0 EDD E J Pearce, 14 Uplands Ave, Willenhall, WV13 3PA [IO82XN, SO99]
G0 EDE D J Proctor, 7 Main Ave, Westhill, Torquay, TQ1 4HZ [IO80FL, SX96]
G0 EDF G H Lunt, 45 Malvern Rd, Liverpool, L6 6BN [IO83MJ, SJ39]
G0 EDG J D M Edgerton, 12 Strode Gdns, Alveston, Bristol, BS12 2PL [IO81RO, ST68]
G0 EDH S J Boundy, Sha-Viste, Westground Way, Bossiney, Tintagel, Cornwall, PL34 0BH [IO70PQ, SX08]
G0 EDI J V J Rohowsky, 146 Moorland Ave, Lincoln, LN6 7HX [IO93RE, SK96]
GM0 EDJ P T Temple, 71B Kilburchan Rd, Johnstone, Renfrewshire, PA5 8RJ [IO75RT, NS46]
G0 EDK A R Lee, 44 Lynn Rd, North Shields, NE29 8HS [IO95GA, NZ36]
G0 EDL D A J Evans, 86 Forest Rd, Kirkby in Ashfield, Nottingham, NG17 9HD [IO93JB, SK55]
G0 EDM E Dennett, 123 Drift Rd, Clanfield, Waterlooville, PO8 0PD [IO90LW, SU61]
G0 EDN W Robinson, 17 Hammond Cl, Off Oldfield Rd, Hampton, TW12 2DE [IO91TJ, TQ16]
G0 EDO R T Ormond, 75 Desford Rd, Newbold Verdon, Leicester, LE9 9LG [IO92IP, SK40]
GM0 EDQ J Shaw, 28 Drumcross Rd, Bathgate, EH48 4HG [IO85EV, NS96]
GM0 EDR J C Bell, 8 Westleigh, Clifton Rd, Stockport, SK4 4BU [IO83VK, SJ89]
G0 EDS G Grundy Eastbourne District Scouts ARC, Scouts Am Rd Grp, 47 Northiam Rd, Eastbourne, BN20 8LP [JO00DS, TV59]
G0 EDT J I Hopwood, 53 St. Marys Rd, Stratford upon Avon, CV37 6XG [IO92DE, SP25]
G0 EDV A J Massey, Flat 16, 72 Gr Ln, Camberwell, London, SE5 8TW [IO91WL, TQ37]
G0 EDY P D Gould, 53 Green Rd, Kidlington, OX5 2EU [IO91IT, SP41]
G0 EDZ M S Schmid, 10 Katchside, Sutton Courtenay, Abingdon, OX14 4BH [IO91IP, SU49]
G0 EEA G Reffell, 26 Barnwood Rd, Gloucester, GL2 0RX [IO81VU, SO81]
G0 EEB G W Barwell, 56 Sandringham Way, Brierley Hill, DY5 3JR [IO82WL, SO98]
GI0 EEC R J A McElroy, 39 Abercorn Park, Portadown, Craigavon, BT63 5JW [IO64TK, J05]
G0 EED Details withheld at licensee's request by SSL.
G0 EEF G K Legg, 12 Churchill Rd, Wimborne, BH21 2AU [IO90AT, SZ09]
GM0 EEH J Hunter, 47 Millburn Cres, Armadale, Bathgate, EH48 3RD [IO85DV, NS96]
G0 EEI A E H Jones, 21 Cumberland Ave, Nantwich, CW5 6HA [IO83RB, SJ65]
G0 EEJ A S E Mitchell, 18 Burnards Ct, Berrycombe Rd, Bodmin, PL31 2NU [IO70PL, SX06]
G0 EEK M Gibson, 42 Joseph Patterson Cres, Ferryhill, Co. Durham, DL17 8RN [IO94FQ, NZ23]
G0 EEL G L Kirby, 21 Harding Ave, Eastbourne, BN22 8PJ [JO00DS, TQ60]
GD0 EEM M D Buckler, Fairy Oak, Ardonan Ln, Regaby, Ramsey, IM7 3HN
G0 EEN R W Wright, 11 Newman Ave, Lanesfield, Wolverhampton, WV4 6DA [IO82WN, SO99]
GI0 EEO J R Ryan, 11 Carnesure Heights, Comber, Newtownards, BT23 5RN [IO74DN, J46]
G0 EER R J Harris, 28 Tickenor Dr, Finchampstead, Wokingham, RG40 4UD [IO91NJ, SU76]
G0 EES S N Henderson, 44 Tiffany Ct, Redcliff Mead Ln, Bristol, BS1 6FD [IO81QK, ST57]
G0 EET B L Henderson, 7 Brettingham Gate, Broome Manor, Swindon, SN3 1NH [IO91CM, SU18]
G0 EEW W G Burrell, Hillcrest, Pilton West, Shirwell Rd, Barnstaple, Devon, EX31 4JH [IO71XC, SS53]
GM0 EEY D P Smith, 6 Lismore Ave, Motherwell, ML1 3RA [IO75XT, NS75]
G0 EEZ C J J Wright, 60 Gr Cres, Hanworth, Feltham, TW13 6LZ [IO91TK, TQ17]
G0 EFA C L Mansfield, 10 Priory Dr, Abbey Wood, London, SE2 0PP [JO01BL, TQ47]
G0 EFB D W Prizeman, 82 Federation Rd, Abbey Wood, London, SE2 0JP [JO01BL, TQ47]
GM0 EFC G M Perry, 14B Meadowfoot Rd, West Kilbride, Ayrshire, KA23 9BX [IO75NQ, NS24]
GM0 EFD C A Perry, 14B Meadowfoot Rd, West Kilbride, Ayrshire, KA23 9BX [IO75NQ, NS24]
GM0 EFH A J Buchan, 14 Jordanhill Dr, Jordanhill, Glasgow, G13 1SA [IO75TV, NS56]
G0 EFI V H Newman, 583 Rayleigh Rd, Eastwood, Leigh on Sea, SS9 5HR [JO01HN, TQ88]
G0 EFL L J Cohen, 17 Ashbury Dr, Blackwater, Camberley, GU17 9HH [IO91OH, SU85]
G0 EFN H T Dunne, Barnacre, 33A Drayton Rd, Newton Longville, Milton Keynes, MK17 0BH [IO91OX, SP83]
G0 EFO M G Shortland, 4 Hillier Rd, Guildford, GU1 2JQ [IO91RF, TQ05]
G0 EFP N H Hughes, Appleton, Two Hedges Rd, Woodmancote, Cheltenham, GL52 4PT [IO81XW, SO92]
GM0 EFQ H N Fisher, Parsonage Cottage, 1 Millhill Ln, Musselburgh, EH21 7RD [IO85LW, NT37]
G0 EFR L R S Wheeler, 21 Ham Cl, Beacon Park, Plymouth, PL2 3PB [IO70WJ, SX45]
G0 EFS P A Hancock, 7 Carlton Ave, Hayes, UB3 4AD [IO91SL, TQ07]
GM0 EFT R Neilson, 54 Macdonald Smith Dr, Carnoustie, DD7 7TB [IO86PL, NO53]
G0 EFU Maj J F Chater, 19 Finsbury Ave, York, YO2 1LW [IO93KW, SE65]
G0 EFV W J Holland, 23 Tredarvah Rd, Penzance, TR18 4LD [IO70FC, SW43]
GI0 EFW J O'Hara, 284 Foreglen Rd, Dungiven, Londonderry, BT47 4PJ [IO64LW, C60]
G0 EFX Details withheld at licensee's request by SSL.
G0 EFY B S Warman, 11 Walker Dr, Winterton, Scunthorpe, DN15 9PW [IO93QP, SE91]
G0 EFZ I E Gerrard, 154 Tilehouse Green Ln, Knowle, Solihull, B93 9EJ [IO92CJ, SP17]
G0 EGA Dr M A Williamson, Appledore, 8 Penfold Rd, Maidenbower, Crawley, RH10 7HU [IO91WC, TQ23]
G0 EGB Details withheld at licensee's request by SSL.
G0 EGC W J Stormont, 3 Bridge Cttgs, Greenham, Crewkerne, TA18 8QE [IO80NU, ST40]
G0 EGD S Preston, 61 Wolverley Ct, Woodside, Telford, TF7 5QU [IO82SP, SJ60]
G0 EGE W G Trotter, Bungalow, Stoupe Cross Farm, Hawsker, Whitby, YO22 4JU [IO94QL, NZ91]
GW0 EGF T R A Burley, 43 Maes Bleddyn, Llanllechid, Bangor, LL57 3EG [IO73XE, SH66]
G0 EGG D B Jackson, 41 Colman Ave, Perry Hall, Wednesfield, Wolverhampton, WV11 3RT [IO82XO, SJ90]
GW0 EGH D J Taylor, 81 Goldfinch Cl, Caldicot, Newport, NP6 4BW [IO81OO, ST48]
GM0 EGI B G Devlin, Borrodale, Main St., Thornhill, Stirling, FK8 3PW [IO76WE, NN60]
G0 EGJ D C Breed, 50 Wentworth Gr, Hartlepool, TS27 3PP [IO94JR, NZ43]
G0 EGK F Bass, 53 Dene Ave, Lemington, Newcastle upon Tyne, NE15 8AL [IO94DX, NZ16]
G0 EGL Details withheld at licensee's request by SSL.
G0 EGP D L Williamson, Wrybourne Lodge, 200 Bushbury Rd, Wolverhampton, WV10 0NA [IO82WO, SJ90]
GW0 EGQ G G Peters, Mile House Farm, Chester Rd, Dobshill, Deeside, CH5 3LZ [IO83LD, SJ36]
G0 EGR C E Davis, 49 Brackendale Rd, Bournemouth, BH8 9HY [IO90BR, SZ19]
G0 EGS J W Hodgson, Ashcroft, 97 Calton Rd, Gloucester, GL1 5ER [IO81VU, SO81]
G0 EGT G I Pitts, 1 Sheppey Cl, Broadfield, Crawley, RH11 9HB [IO91VC, TQ23]
G0 EGX A K Mills, 28 Birkin Cl, Tiptree, Colchester, CO5 0PB [JO01IT, TL81]
GW0 EHA I R Pemberton, 26 Stanley Gr, Ruabon, Wrexham, LL14 6AH [IO82LX, SJ34]
GW0 EHB G E Foster, 32 Braeside Ave, Hawarden, Deeside, CH5 3HW [IO83LE, SJ36]
G0 EHF Details withheld at licensee's request by SSL.
G0 EHG Details withheld at licensee's request by SSL.[Station located near Greenhithe. Op: A G Hay.]
G0 EHH F E Byrne, 13 Kenny Walk, Newton Aycliffe, DL5 5AQ [IO94FO, NZ22]
G0 EHK G G Cheetham, 172A Hesketh Ln, Tarleton, Preston, Lancs, PR4 6UD [IO83NQ, SD42]
GM0 EHL T A B Armstrong, Skye, 6 Middleshot Rd, Gullane, EH31 2DG [IO86OA, NT48]
G0 EHM D D Miller, 22 Chapeldown Rd, Torpoint, PL11 2HT [IO70VI, SX45]
GW0 EHN N D W Manger, 23 Clos Vernon Watkins, Gorseinon, Swansea, SA4 4DH [IO71XQ, SS59]
G0 EHO R C Mellor, 1 The Sq, Lybury Ln, Redbourn, St. Albans, AL3 7JB [IO91TT, TL01]
G0 EHP Details withheld at licensee's request by SSL.
G0 EHQ F G Skinner, Halfway Lock Cottage, Upper Gambolds Ln, Stoke Prior, Bromsgrove, B60 3HB [IO82XH, SO96]
G0 EHR M J Clark, Ferndown, Five Houses, Calbourne, Isle of Wight, PO30 4JT [IO90HQ, SZ48]
GW0 EHS F M Fennah, 59 Lon Helyg, Newtown, SY16 1HY [IO82IM, SO09]
G0 EHT J M Tommey, The Birches, Upton Bishop, Ross on Wye, HR9 7UF [IO81RW, SO62]
G0 EHV E Ashburner, 20 The Cedars, Eighton Banks, Gateshead, NE9 7BW [IO94FW, NZ25]
G0 EHW B J Coles, 32 Victoria St., Lostock Hall, Preston, PR5 5RA [IO83PR, SD52]
G0 EHX T W Elliott, The Linds, 137 The Broadway, Sunderland, SR4 8HE [IO94GV, NZ35]
G0 EHY A J Drinkwater, 20 St. Johns Cres, Tyler Hill, Canterbury, CT2 9NB [JO01MH, TR16]
G0 EIA B Taylor, Namparra, Quarry Rd, Pensilva, Liskeard, PL14 5NP [IO70TM, SX26]
G0 EIB L Allen, 28 Clarence Pl, Maltby, Rotherham, S66 7HA [IO93JK, SK59]
G0 EID D W Harbour, 20 Durkins Rd, East Grinstead, W Sussex, RH19 2ER [IO91XD, TQ33]
G0 EIF P M Masheder, 5 Hazel Cl, Penwortham, Preston, PR1 0YE [IO83PR, SD52]
G0 EIG M Holtham, 8 Railway St., Leyland, Preston, PR5 2XB [IO83PQ, SD52]
G0 EIH T A Briley, 2 Usborne Cl, Staplehurst, Tonbridge, TN12 0LD [JO01GD, TQ74]
G0 EIJ C Cottrill, 38 Arundel Dr, West Monkseaton, Whitley Bay, NE25 9PZ [IO95GA, NZ37]
GW0 EIK Details withheld at licensee's request by SSL.
G0 EIM M R Heyes, 11 Beech Cl, Isleham, Ely, CB7 5UU [JO02EH, TL67]
GI0 EIO Details withheld at licensee's request by SSL.
G0 EIP J L Wiggins, 22 Ripley Ct, Princess Ave, Lancaster, LA1 4RZ [IO84OA, SD46]
G0 EIQ M S G Richards, 3 Derwent Cres, Whetstone, London, N20 0QN [IO91VO, TQ29]
G0 EIR G C Kemp, 27 Shady Gr, Alsager, Stoke on Trent, ST7 2NQ [IO83UC, SJ75]
GM0 EIT A George, 7 Mid St., Keith, AB55 5AG [IO87MM, NJ45]
G0 EIV Details withheld at licensee's request by SSL.
G0 EIX Details withheld at licensee's request by SSL.[Station located near Canvey Is.]
G0 EIY S G Pryce, Narberth, Welshpool Rd, Bicton Heath, Shrewsbury, SY3 5AH [IO82OR, SJ41]
G0 EIZ W P Kenyon, 22 Barons Way, Lower Darwen, Darwen, BB3 0RG [IO83SR, SD62]
G0 EJA P Lynas, c/o Cwao Civilian Wing, 9 Signals Regiment, BFPO 59
G0 EJB A Winter, 66 Beechfield, Newton Aycliffe, DL5 7AY [IO94EO, NZ22]
G0 EJC A J Young, 25 Thorney Rd, Crowland, Peterborough, PE6 0AL [IO92WQ, TF21]
G0 EJD M R Moss, 1 Calder Cl, Cheylesmore, Coventry, CV3 5PN [IO92GJ, SP37]
GW0 EJE P N Oneill Pembrokeshire, Pembrokeshire Radio Society, Mount Pleasant, Ambleston, Pembrokeshire, SA62 5DP [IO71NV, SM92]
G0 EJF A K Henderson, 93 Chosen Dr, Churchdown, Gloucester, GL3 2QS [IO81VV, SO82]
G0 EJG Details withheld at licensee's request by SSL.
G0 EJH I M Crampton, 36 Langbank Ave, Nottingham, Notts, NG5 5BU [IO93JA, SK54]
G0 EJI L Y Garden, 227 Oak Cres, Burlington, Ontario, L7L 1H3
G0 EJJ G H Gardiner, 118 Princess Dr, Grantham, NG31 9PY [IO92QW, SK93]
G0 EJL P J Boaler, 21 Harts Cl, Birmingham, B17 9LE [IO92AL, SP08]
G0 EJO G D Nock, 20 Chigwell Rd, Bournemouth, BH8 9HW [IO90BR, SZ19]
GW0 EJP G Glendinning, 44 Ynys y Mond Rd, Alltwen, Pontardawe, Swansea, SA8 3BA [IO81BR, SN70]
G0 EJR A M Love, 15 Mountain Ash, Weston Park East, Bath, BA1 2UU [IO81TJ, ST76]
GI0 EJT S Rafferty, 81 Mullaghmore Dr, Omagh, BT79 7PQ [IO64IO, H47]
GI0 EJU V A G Hutchinson, 10 Golan Rd, Knockmoyle, Omagh, BT79 7TJ [IO64IQ, H48]
G0 EJV L Hodges, 34 Wiseholme Rd, Skellingthorpe, Lincoln, LN6 5TF [IO93QF, SK97]
G0 EJW Details withheld at licensee's request by SSL.
G0 EJX G G Noble, 5 Horden Rd, Billingham, TS23 3BY [IO94IO, NZ42]
GM0 EJY A M Clark, Mount Pleasant, School Hendry St., Portsoy, Banff, AB45 2RS [IO87PQ, NJ56]
G0 EJZ E G Holford, 8 Crown Ln, Shorne, Gravesend, DA12 3DY [JO01FJ, TQ67]
G0 EKC L P White, 22 Seaton Park, Seaton, Workington, CA14 1EU [IO84FP, NY03]
G0 EKD A W Dickason, 15 Farsands, Oakley, Bedford, MK43 7SJ [IO92RE, TL05]
G0 EKG R R Tomlinson, Meadow View Cottage, Osbaston, Nuneaton, Warks, CV13 0DR [IO92HP, SK40]
G0 EKH K J Harper, 2 Vale Rd, Newton Abbot, TQ12 1DZ [IO80EM, SX87]
G0 EKJ Details withheld at licensee's request by SSL.[Op: H S Johnson, 2 Ashley Drive, Seasalter, nr Whitstable, Kent CT5 4TB.]
G0 EKK H R C Wright, 90 Woodside, Ashby de La Zouch, LE65 2NU [IO92GR, SK31]
GM0 EKL Details withheld at licensee's request by SSL.[Op: Mrs E Innes.]
GM0 EKM C M Duncan, Roadside Cottage, Hoswick, Sandwick, Shetland, ZE2 9HL [IO99IX, HU42]
G0 EKN C J Nye, 6 Harding Rd, Chadwell St. Mary, Grays, RM16 4XD [JO01EL, TQ67]
G0 EKP Details withheld at licensee's request by SSL.[Op: Mrs S C Nye, 6 Harding Road, Chadwell-St-Mary, nr Grays, Essex, RM16 4XD.]
G0 EKQ Details withheld at licensee's request by SSL.
G0 EKR P S Nicholson East Kent Rad S, 34 Cliff Ave, Herne Bay, CT6 6LZ [JO01NI, TR16]
GM0 EKS Details withheld at licensee's request by SSL.
G0 EKV A J Duke, 31 Burstead Cl, Hollingdean, Brighton, BN1 7HT [IO90WU, TQ30]
G0 EKX M S Dodgson, 2 Harrod Dr, Birkdale, Southport, PR8 2HA [IO83LP, SD31]
GM0 EKY B R Rourke, 77 Larkfield Rd, Lenzie, Kirkintilloch, Glasgow, G66 3AS [IO75WW, NS67]
G0 ELB B Bray, 34 Newlands Dr, Forest Town, Mansfield, NG19 0HZ [IO93KD, SK56]
G0 ELC P Hall, 28 Maria Dr, Fairfield, Stockton on Tees, TS19 7JL [IO94HN, NZ41]
G0 ELE Details withheld at licensee's request by SSL.[Station located near Colne. Op: S A Morris.]
GU0 ELF J T Gethings, Lindale Guest House, Elm Gr, St. Peter Port, Guernsey, GY1 1XE
G0 ELG R Zielinski, 154 Bensham Ln, Thornton Heath, CR7 7EN [IO91WJ, TQ36]
G0 ELH W R A Stainton, 45 Elmwood Way, Basingstoke, RG23 8LJ [IO91KG, SU65]
G0 ELI N F Macartney, 4 York Ave, Wallasey, L44 9ER [IO83LJ, SJ39]
G0 ELJ D P E Dawson, 19 Nightingale Ave, Chelmsley Wood, Birmingham, B36 0RT [IO92DM, SP18]
G0 ELK W F Baron, 14 Mount Rd, Middleton, Manchester, M24 1DZ [IO83VN, SD80]
GM0 ELL N D M Elliot, 37 Cranworth St., Glasgow, G12 8AF [IO75UV, NS56]
G0 ELM P R Green, 17 Columbine Cl, Hough Green, Widnes, WA8 4TR [IO83OJ, SJ48]
G0 ELN S N Macdonald, 61 Pavilion Rd, Worthing, BN14 7EE [IO90TT, TQ10]
G0 ELO J Ellis, 21 Coxway, Clevedon, BS21 5AQ [IO81NK, ST47]
GM0 ELP D C Maxwell, 7 Lilac Hill, Hamilton, ML3 7HG [IO75XS, NS75]
G0 ELQ Details withheld at licensee's request by SSL.
G0 ELR J Bradshaw East Lancs Raynet, Fernleigh Cottage, Yeomans Farm, Briercliff, Burnley, BB10 3QU [IO83VT, SD83]
G0 ELS Details withheld at licensee's request by SSL.[Op: E J Stannard, 84 Hawthorn Road, Gt Cornard, Suffolk, CO10 0LW.]
G0 ELT Details withheld at licensee's request by SSL.[Op: E Clarke, Alton House, Lake St., Lower Gornal, Dudley, W. Mids, DY3 2AL.]
G0 ELU Details withheld at licensee's request by SSL.[Op: K Kyriacou. Station located near Malvern.]
G0 ELV A Bailey, Top Flat, 15 Cissbury Rd, Worthing, BN14 9LD [IO90TT, TQ10]
GM0 ELW Details withheld at licensee's request by SSL.
G0 ELX Details withheld at licensee's request by SSL.[Op: P S Wilson.]
GD0 ELY Details withheld at licensee's request by SSL.[Op: Mrs J E Brown, Cleckheaton, Ballaragh, Laxey, Isle of Man.]
G0 ELZ W H Cross, 13 Carno St., Liverpool, L15 4LB [IO83MJ, SJ38]
GW0 EMB H G Blore, 17 Kendal Way, Wrexham, LL12 8AF [IO83MB, SJ35]
GM0 EMC K B McLaren, Dalriada, Fogwatt, near Elgin, Moray, IV30 3SL [IO87IO, NJ25]
G0 EMF D Brown, 36 Alexandra St., Pelton, Chester-le-Street, DH2 1NT [IO94EU, NZ25]
G0 EMG Details withheld at licensee's request by SSL.
G0 EMI Details withheld at licensee's request by SSL.
G0 EMK M G Kendall, 88 Coldnailhurst Ave, Braintree, CM7 5PY [JO01GV, TL72]
G0 EML R K Bullock, 40 Little Harlescott Ln, Shrewsbury, SY1 3PY [IO82PR, SJ41]
G0 EMM K Dockray, 54 Kelsick Park, Seaton, Workington, CA14 1PY [IO84FP, NY03]
G0 EMQ Details withheld at licensee's request by SSL.[Op: J J Vinton, 1 Gill-an-Creet, St. Ives, Cornwall TR26 2EW.]
G0 EMR P D Page, 25 Broomhill Rd, Cove, Farnborough, GU14 9PT [IO91OH, SU85]
G0 EMT M I Chapman, 6 Badgers Cl, Binscombe, Godalming, GU7 3RW [IO91QE, SU94]
G0 EMX E A Parr, 74 Stanley Rd, Earlsdon, Coventry, CV5 6FF [IO92FJ, SP37]
GM0 EMY J J Reid, 127 Drylie St., Cowdenbeath, KY4 9AH [IO86HC, NT19]
G0 ENA R F Procter, 20 Costells Edge, Scaynes Hill, Haywards Heath, RH17 7PY [IO90XX, TQ32]
G0 ENB R J Felton, Hall Farm, Snetterton, Norwich, Norfolk, NR16 2LF [JO02LL, TL99]
G0 ENC Details withheld at licensee's request by SSL.
G0 END R McGarvie, 5 Dairy Farm Rd, Rainford, St. Helens, WA11 7JQ [IO83OM, SD40]
G0 ENE Details withheld at licensee's request by SSL.
G0 ENF G D Crawshaw, 51 Templeway West, Lydney, GL15 5JD [IO81RR, SO60]
GM0 ENG Details withheld at licensee's request by SSL.
G0 ENH Details withheld at licensee's request by SSL.
G0 ENJ D R Buckingham, 1 Westfield Rise, Saltdean, Brighton, BN2 8HR [IO90XT, TQ30]
G0 ENM D James, 28 Durham Rd, Alumwell Est, Walsall, WS2 9TF [IO82XN, SO99]
G0 ENN B A Bowden, 49 Springfield Dr, Westcliff on Sea, SS0 0RA [JO01IN, TQ88]
G0 ENO K J Dempster, 10 St. Pauls Way, Tickton, Beverley, HU17 9RW [IO93TU, TA04]
G0 ENP J Moriarty, 105 Brendon Cres, Billingham, TS23 2QU [IO94IO, NZ42]
G0 ENR D J Newbury, 8 Mayfield Rd, Pershore, WR10 1NW [IO82XC, SO94]
G0 ENS Details withheld at licensee's request by SSL.

GW0 ENT J J W Comerford, Bod Elen, Bontnewydd, Caernarfon, Gwynedd, LL54 7YE [IO73UC, SH45]
GW0 ENU D J Walters, 132 Tan y Bryn, Valley, Holyhead, LL65 3ER [IO73RG, SH27]
G0 ENV A L Wood, 262 Egmanton Rd, Meden Vale, Mansfield, NG20 9PY [IO93KF, SK56]
G0 ENW G A Fingerhut, Wild Rose, Behind Hayes, South Cheriton, Templecombe, BA8 0BP [IO81SA, ST62]
G0 ENY R Ford, 27 Albert Rd, Allesley, Coventry, CV5 9AS [IO92EK, SP28]
G0 ENZ T J Trudgeon, 2 The Terrace, East Portholland, St. Austell, PL26 6NA [IO70NF, SW94]
G0 EOA Details withheld at licensee's request by SSL.
G0 EOB Details withheld at licensee's request by SSL.
G0 EOC Details withheld at licensee's request by SSL.
G0 EOE J W C Palmer, 251 Church St., Braintree, CM7 5LH [JO01GV, TL72]
G0 EOF G E Potter, 88 Highlands Cl, Pinewood, Kidderminster, DY11 6JU [IO82UJ, SO87]
G0 EOG J M Jennings, 81 Newgate St., Burntwood, WS7 8TX [IO92AQ, SK00]
G0 EOH D E Bristow, Herniss Bungalow, Longdowns, nr Penryn, Cornwall, TR10 9DT [IO70JD, SW73]
G0 EOI Details withheld at licensee's request by SSL.
G0 EOJ D Shore, 9 Hawthorn Cl, Clowne, Chesterfield, S43 4SX [IO93IG, SK47]
G0 EOK P Gibson, 77 Shevington Moor, Standish, Wigan, WN6 0SQ [IO83PO, SD51]
G0 EOL W A Prater, 12 Alford Way, Winsford, CW7 2HL [IO83RE, SJ66]
G0 EOM A Deakin, 15 Lyme Gr, Romiley, Stockport, SK6 4DH [IO83XJ, SJ99]
G0 EON F W Cloke, 9 Mill Cl, East Coker, Yeovil, BA22 9LF [IO80QV, ST51]
GM0 EOO D McMahon, Seabrae, Carnoustie, Angus, Scotland, DD7 6AY [IO86PM, NO53]
G0 EOP L Herf, Old Chapel, Fore St., North Molton, South Molton, EX36 3HL [IO81CB, SS72]
G0 EOQ Details withheld at licensee's request by SSL.[Station located near Barnstaple. Locator: IO71XB]]
G0 EOR W L Glover, 3 Gothic Rd, Newton Abbot, TQ12 1LD [IO80EM, SX87]
G0 EOS P R Smith, 35 Tanglewood Cl, Shard End, Birmingham, B34 7QX [IO92CL, SP18]
GD0 EOT T J P Wilson, 6 Cooil Farrane, Saddlestone Valley, Douglas, IM2 1NX
G0 EOU Details withheld at licensee's request by SSL.[Op: B R Holloway, 4 Tapsters, Cadbury Heath, Bristol, BS15 5HN.]
G0 EOV Details withheld at licensee's request by SSL.
G0 EOX I Hunton, 123 Huddersfield Rd, Diggle, Oldham, OL3 5NU [IO83XN, SD90]
G0 EOY S W Outterside, 9 Northcote Ave, West Denton, Newcastle upon Tyne, NE5 5AL [IO94DX, NZ16]
G0 EOZ P G Kerton, 11 North Filham Cot, Filham, Ivybridge, Devon, PL21 9DH [IO80AJ, SX65]
G0 EPA G Whitham, 55 Bisley Gr, Bransholme, Hull, HU7 4PY [IO93US, TA13]
G0 EPE D J Cargill, 41 Grosvenor Rd, Skegness, PE25 2DD [JO03ED, TF56]
GM0 EPG Details withheld at licensee's request by SSL.
G0 EPH Details withheld at licensee's request by SSL.[Op: M Jones.]
G0 EPI S M Cooper, 14 Greenview Dr, Towester, Northants, NN12 7DL [IO92MC, SP74]
G0 EPJ R Jones, 1 Belfry Ave, St. George, Bristol, BS5 7PL [IO81RL, ST67]
G0 EPK E Bristow, 11 Castle Dene, Maidstone, ME14 2NH [JO01GG, TQ75]
G0 EPL J P Love, 191 High St., Henley in Arden, Solihull, B95 5BA [IO92CG, SP16]
G0 EPM D Bainbridge, 8 Norley Ln, Studley, Calne, SN11 9LS [IO81XK, ST97]
GM0 EPO J J Shades, 12 Briar Cottage, Newlands, Glasgow, G43 2TF [IO75UT, NS56]
G0 EPP A P Przybyla, 18 Cherwell, Sulgrave, Washington, NE37 3LA [IO94FV, NZ35]
G0 EPQ G R Rippin, Siesta, 27 Cliffsend Gr, Cliffsend, Ramsgate, CT12 5JT [JO01QH, TR36]
G0 EPR P N Wardale, 2 Holt Ct, Horseferry Pl, Greenwich, London, SE10 9HG [IO91XL, TQ37]
G0 EPT Details withheld at licensee's request by SSL.
G0 EPU Details withheld at licensee's request by SSL.
G0 EPV J R Collins, 49 Alspath Rd, Meriden, Coventry, CV7 7LU [IO92EK, SP28]
G0 EPX Details withheld at licensee's request by SSL.
G0 EPY C Hirst, 18 Lazenby Ave, Larkholme, Fleetwood, FY7 8QH [IO83LV, SD34]
G0 EPZ Details withheld at licensee's request by SSL.
GM0 EQA Details withheld at licensee's request by SSL.
G0 EQB K E Curson, 3 Stiffkey Rd, Warham, Wells Next The Sea, NR23 1NP [JO02KW, TF94]
G0 EQC I Williamson, Westholme, 4 Edgewell Rd, Prudhoe, NE42 6JH [IO94BW, NZ06]
G0 EQD J Archer, 1 Suffolk Pl, Bishop Auckland, DL14 6UT [IO94DP, NZ22]
G0 EQE D P Cunningham, 2 Fairmead Rd, Moreton, Wirral, L46 8TX [IO83KJ, SJ29]
G0 EQF J C Tyler, 49 Grasleigh Way, Allerton, Bradford, BD15 9BD [IO93BT, SE03]
G0 EQH E Shackleton, 32 Pennine Rd, Simmondley, Glossop, SK13 9UJ [IO93AK, SK09]
G0 EQI R W Mallinson, 3 Captain Cooks Cres, Whitby, YO22 4HL [IO94QL, NZ91]
G0 EQJ B Glassford, Gosling Letch, Thorpe Rd, Easington, Peterlee, SR8 3UA [IO94HS, NZ44]
G0 EQM J H Cook, 40 Enys Rd, Eastbourne, BN21 2ED [JO00DS, TV69]
G0 EQN W Gibbons, 8 Darnet Rd, Tollesbury, Maldon, CM9 8XG [JO01KS, TL91]
G0 EQP Details withheld at licensee's request by SSL.
G0 EQS Can S C Palmer, The Vicarage, Portsdown Hill Rd, Cosham, Portsmouth, PO6 1BE [IO90LU, SU60]
G0 EQV D C Buckley, 35 Beamish Rd, Canford Heath, Poole, BH17 8SB [IO90AR, SZ09][Correspondence via PO Box 249, Poole, Dorset.]
GM0 EQW T H Olsen, Ardchuan, Taynuilt, Argyll, PA35 1HY [IO76JJ, NM92]
GM0 EQX Details withheld at licensee's request by SSL.
GM0 EQZ Details withheld at licensee's request by SSL.
G0 ERA A C Lindsay Vle of Evshm AR, 21 Willow Rd, Four Pools, Evesham, WR11 6YW [IO92AB, SP04]
GM0 ERB N Calder, 16 Camesky Rd, Caol, Fort William, PH33 7ER [IO76KU, NN17]
G0 ERC E Kershenbaum, 243 Lillie Rd, London, SW6 7LN [IO91VL, TQ27]
G0 ERE M K Jones, 36 Fosters Ln, Bradwell, Milton Keynes, MK13 9HZ [IO92OB, SP83]
G0 ERF K F Watkins, 10 Rowan Dr, Billingshurst, RH14 9NF [IO91SA, TQ02]
G0 ERH J A Harrington, 18 Beeches Croft, Fradley, Lichfield, WS13 8RX [IO92CR, SK11]
G0 ERI J M Shaw, 77 Satchell Ln, Hamble, Southampton, SO31 4HH [IO90IU, SU40]
G0 ERL W Marbus, Elm Farm House, Debenham Rd, Mickfield, Suffolk, IP14 5LW [JO02NF, TM16]
G0 ERN E Webster, 17 Chippendale Rise, Otley, LS21 2BL [IO93DV, SE24]
G0 ERR R E Hoverd, The Sycamores, 108 Ecton Ln, Sywell, Northampton, NN6 0BB [IO92OG, SP86]
G0 ERS R A Smith, 79 Froxfield Rd, Havant, PO9 5PW [IO90MU, SU70]
GM0 ERT R Tannahill, 13 Barr Cres, Largs, KA30 8PX [IO75NT, NS26]
GM0 ERU Details withheld at licensee's request by SSL.[Op: A McLennan, 6 Mull Terrace, Oban, Argyle, PA34 4YB.]
GM0 ERV Details withheld at licensee's request by SSL.[Op: Mrs S E McLennan, 6 Mull Terrace, Oban, Argyle, PA34 4YB.]
G0 ERW Details withheld at licensee's request by SSL.[Op: W R Spencer, 81 Churchill Avenue, Bourne, Lincs, PE10 9QA.]
G0 ERX T M Balment, 41 Orchard Rd, Barnstaple, EX32 9JQ [IO71XB, SS53]
G0 ERY R B Saunders, 24322 Augustin, Mission Viejo, Ca 92691, USA
G0 ESA W F Wilkinson, Dendrum Lodge, Racemoor Ln, Oakworth, Keighley, W Yorks, BD22 7JH [IO93AU, SE03]
G0 ESB Details withheld at licensee's request by SSL.[Op: P Bramidge, 91 Gaia Lane, Lichfield, Staffs, WS13 7LS.]
G0 ESC Details withheld at licensee's request by SSL.
G0 ESE Details withheld at licensee's request by SSL.[Op: C J Hill. Station located near Fareham.]
G0 ESF Details withheld at licensee's request by SSL.[Op: John R Southgate, 36 Porch Way, Whetstone, London, N20 0DS.]
G0 ESI D F S Williams, Miramare, Egremont Rd, St. Bees, CA27 0AS [IO84EL, NX91]
G0 ESJ G C Richardson, 89 Parsons Rd, Redditch, B98 7EJ [IO92AH, SP06]
GW0 ESK C E Williams, 3 Lodge Orchard, Mona St., Amlwch, LL68 9RX [IO73TJ, SH49]
G0 ESL M J Musgrave, South Fursdon Farm, Yealmpton, Devon, PL8 2EN [IO80AI, SX55]
G0 ESO S A Llewellyn, 110 South Ave, Southend on Sea, SS2 4HU [JO01IN, TQ88]
G0 ESP J Helliwell, 92 Rookery Rd, Wombourne, Wolverhampton, WV5 0JG [IO82VM, SO89]
G0 ESR D R Owen, Delamere, 12 Wolverley Ave, Wollaston, Stourbridge, DY8 3PJ [IO82VL, SO88]
GI0 EST Details withheld at licensee's request by SSL.
GW0 ESU W Lee, 8 Bronheulog, Bodffordd, Isle of Anglesey, LL77 7SU [IO73UG, SH47]
G0 ESV Details withheld at licensee's request by SSL.[Op: M R Medhurst, 4 William Close, Walkford, Christchurch, Dorset, BH23 5SB.]
G0 ESW R Graham, 454 Lobley Hill Rd, Lobley Hill, Gateshead, NE11 0BS [IO94EW, NZ26]
G0 ESZ A J Reed, 3 Warwick Cl, Feniton, Honiton, EX14 0DT [IO80IS, SY19]
G0 ETA G K Luhman, 17 Broadview, Stevenage, SG1 3TS [IO91VV, TL22]
G0 ETB Details withheld at licensee's request by SSL.[Op: J Thorpe. Station located near Barnsley]
GM0 ETC Details withheld at licensee's request by SSL.
G0 ETD S M Burton, 49 Aske Rd, Redcar, TS10 2BP [IO94LO, NZ62]
GW0 ETF S Rolfe, Tyn Lon, Minffordd, Bangor, Gwynedd, LL57 4DR [IO73WF, SH57]
G0 ETG J W Tory, Croft Farm House, Bridlington Rd, Flamborough, Humberside, YO15 1PW [IO94WC, TA27]
G0 ETH A Hadley, Wishford Ho Cottage, Gt Wishford, Salisbury, Wilts, SP2 0BQ [IO91BB, SU03]
G0 ETI Details withheld at licensee's request by SSL.
G0 ETL G L Tatterson, 2 Eden Rd, Kirkstall Hill, Leeds, LS4 2TT [IO93FT, SE23]
GW0 ETM J A Davies, 68 Edward St., Fairview, Blackwood, NP2 1NY [IO81JP, ST19]
GD0 ETO Details withheld at licensee's request by SSL.
G0 ETP T D Howe, 5 Stonehurst Cl, Hartwell, Northampton, NN7 2JD [IO92ND, SP75]
G0 ETQ J Curnow, 33 Kings Ave, Falmouth, TR11 2QH [IO70LD, SW73]
G0 ETU E A Cabban, 12 Gate Cttgs, Chorleywood Common, Chorleywood, Rickmansworth, WD3 5LW [IO91RP, TQ09]
G0 ETV L E McQuire, Springfield, Staynall Ln, Hambleton, Blackpool, Lancs, FY6 9DR [IO83MV, SD34]
G0 ETW Details withheld at licensee's request by SSL.
G0 ETX Details withheld at licensee's request by SSL.[Op: S Hassan, 27 Eton Road, Ilford, Essex, IG1 2UD.]
G0 ETY Details withheld at licensee's request by SSL.
G0 ETZ S F Gurney, The Common, Exmouth, Devon, EX8 5EE [IO80HP, SY08]
GM0 EUA G W Child, 19 Highcroft, Kelso, TD5 7NB [IO85SO, NT73]
G0 EUB Details withheld at licensee's request by SSL.
G0 EUC Details withheld at licensee's request by SSL.[Op: Capt R G Burnet.]
G0 EUD C F Jackson, Red House Farmhouse, Grange Rd, Terrington St. Clement, Kings Lynn, Norfolk, PE34 4HQ [JO02CS, TF52]
G0 EUE LtCr D S E Row, Bridge End, Elmgrove Rd, Topsham, Exeter, EX3 0EJ [IO80GQ, SX98]
G0 EUF Details withheld at licensee's request by SSL.[Op: I Ball, 13 Mayfield Road, Enfield, Middx EN3 7LS.]
GI0 EUG E B Hagan, 34 Coolshinney Rd, Magherafelt, BT45 5JF [IO64QR, H88]
G0 EUH Details withheld at licensee's request by SSL.
G0 EUI Details withheld at licensee's request by SSL.
G0 EUJ K Neville, 5 Coleville Ave, Fawley, Southampton, SO45 1DA [IO90HT, SU40]
GM0 EUK J Penny, 23 Don House, Kellands Rd, Inverurie, AB51 3YG [IO87TG, NJ72]
GM0 EUM J Mackinnon, 5 Octavia Terr, Greenock, PA16 7SP [IO75OX, NS27]
G0 EUN J Nichol, 58 Benson Cres, Doddington Park, Lincoln, LN6 3NU [IO93QE, SK96]
GM0 EUO E Scott, 5 Braehead Cres, Peterhead, AB42 1EG [IO97CM, NK14]
G0 EUP M R Rigg, 46 Borrowdale Dr, Meadway, Rochdale, OL11 3JZ [IO83VO, SD81]
G0 EUQ R J Fuller, 12 Plummers Ln, Haynes, Bedford, MK45 3PL [IO92TB, TL04]
G0 EUR B W Barber, 1 Shore Pl, Trowbridge, BA14 9TB [IO81VH, ST85]
G0 EUS G Reynard, 183 Wheathead Ln, Keighley, BD22 6NB [IO93AU, SE04]
G0 EUT Details withheld at licensee's request by SSL.
GM0 EUU A Pontiero, 1 Dalmeny Rd, Hamilton, ML3 6PP [IO75XS, NS75]
G0 EUV Details withheld at licensee's request by SSL.
GM0 EUY D S Macadie, 107 Treeswoodhead Rd, Kilmarnock, KA1 4PB [IO75SO, NS43]
G0 EUZ R M P Jones, 118 Ramwells Brow, Bromley Cross, Bolton, BL7 9LQ [IO83TO, SD71]
G0 EVA D O Evans, 79 Lockwood Scar, Newsome, Huddersfield, HD4 6DY [IO93CP, SE11]
G0 EVB A Barnes, 9 Sherwood Ave, Irby, Wirral, L61 4XB [IO83KI, SJ28]
G0 EVD J G Noble, 21 Cecil Cres, Hatfield, AL10 0HF [IO91VS, TL20]
GW0 EVE V E Berry, The Berries, 70 Yerburgh Ave, West End, Colwyn Bay, Clwyd, LL29 7NB [IO83DH, SH87]
G0 EVF J Mowbray, 44 Monkdale Ave, Cowpen Est, Blyth, NE24 4EB [IO95FC, NZ28]
GW0 EVG N M Berry, The Berries, 40 Yerburgh Ave, Colwyn Bay, LL29 7NB [IO83DH, SH87]
G0 EVH A W Ferneyhough, 30 Bedford Dr, Sutton Coldfield, B75 6AU [IO92CN, SP19]
G0 EVI S F Monk, 310 Hinckley Rd, Leicester, LE3 0TN [IO92KP, SK50]
G0 EVJ S W Evans, 181 Curborough Rd, Lichfield, WS13 7PW [IO92CQ, SK11]
G0 EVK R J Watson, 63 Church Rd, Catworth, Huntingdon, PE18 0PA [IO92TI, TL07]
G0 EVL L C Walshe, 9 St. Stephens Gdns, Northallerton, DL7 8XN [IO94GH, SE39]
G0 EVN S C Parkin, 13 Queens Dr, Nuthall, Nottingham, NG16 1EG [IO92JX, SK54]
G0 EVO F Robinson, 32 Claremont Terr, Gillygate, York, YO3 7EJ [IO93LX, SE65]
G0 EVP K R Griffiths, 67 Dickens Ln, Poynton, Stockport, SK12 1NN [IO83WI, SJ98]
G0 EVQ A T Sewell, 50 Wood Ln, Aylsham, Burgh, Norwich, NR11 6TS [JO02PS, TG22]
G0 EVR Details withheld at licensee's request by SSL.
G0 EVS H Angus, 111 Great Elms Rd, Hemel Hempstead, HP3 9UQ [IO91SR, TL00]
G0 EVT J M Hoban, 3 Lake Lock Gr, Stanley, Wakefield, WF3 4JJ [IO93GR, SE32]
G0 EVU F C Samet, 44 Honeydon Ave, Eaton Socon, St. Neots, Huntingdon, PE19 3PJ [IO92UF, TL15]
G0 EVV D I Stansfield, 22 Low Stobhill, Morpeth, NE61 2SG [IO95DD, NZ28]
G0 EVW G R Watts, 46 Links Rd, Weymouth, DT4 0PE [IO80SO, SY67]
G0 EVX A M A Cater, 52 Wellbrook Rd, Bishops Cleeve, Cheltenham, GL52 4BW [IO81XW, SO92]
G0 EVY D M Strobel, Dell Cottage, Copy-Holt Ln, Stoke Pound, nr Bromsgrove, B60 3AY [IO82XH, SO96]
G0 EVZ S Males, 6 Lammas Path, Stevenage, SG2 9RN [IO91VV, TL22]
G0 EWA D J B Cox, Green Meadows, 43 Edge Hill Rd, Sutton Coldfield, B74 4PD [IO92BO, SP19]
G0 EWC R G Harry, 29 Eastwood Rd, St. Annes Park, Bristol, BS4 4RN [IO81RK, ST67]
G0 EWD P B Kennedy, 262 Green Rd, Springvale, Penistone, Sheffield, S30 6BH [IO93GJ, SK38]
GI0 EWE P J Holland, 35 Ashfield Rd, Clogher, BT76 0HJ [IO64JJ, H54]
GM0 EWF D Pettigrew, 3 Loaninghill Park, Uphall, Broxburn, EH52 5EB [IO85GW, NT07]
G0 EWH R J Newton, 74 Walker Ave, Stourbridge, DY9 9EL [IO82WK, SO98]
G0 EWI J A Daramy, 6 Boulton Cl, Linacre Woods, Chesterfield, S40 4XJ [IO93GF, SK37]
GM0 EWJ N W Spencer, 34 Oxgangs Rd North, Edinburgh, EH13 9DR [IO85JV, NT26]
GM0 EWK H Gilchrist, 140 Marchmont Rd, Edinburgh, EH9 1AQ [IO85JW, NT27]
G0 EWL Details withheld at licensee's request by SSL.
G0 EWN Details withheld at licensee's request by SSL.
G0 EWO E W Ogden, 55 Cumberland Ave, Basingstoke, RG22 4BQ [IO91KF, SU65]
GI0 EWP J B Hartin, 46 Enagh Park, Limavady, BT49 0SA [IO65MA, C62]
GM0 EWQ T Lennox, 7 Dalrigh, Dunollie Rd, Oban, PA34 5PG [IO76FL, NM83]
G0 EWR D W C Wale, 16 St. Veronica Rd, Deepcar, Sheffield, S30 5TP [IO93GJ, SK38]
G0 EWS Details withheld at licensee's request by SSL.[Op: M R Grisenthwaite. Station located near West Kirby.]
G0 EWT A Coates, 11 Canterbury Rd, Brotton, Saltburn By The Sea, TS12 2XG [IO94MN, NZ61]
GM0 EWU C C Craig, Knipoch Hotel, By Oban, Argyllshire, Scotland, PA34 4QT [IO76GI, NM82]
G0 EWV T Buckle, 15 Gleaves Ave, Harwood, Bolton, BL2 4ET [IO83TO, SD71]
GM0 EWX C J Macpherson, 6, Borve, Skeabost Bridge, Portree, Isle of Skye, IV51 9PE [IO67VK, NG44]
GW0 EWY W A Woods, 27 Salop Pl, Penarth, CF64 1HP [IO81JK, ST17]
G0 EWZ I M Mason, 56 Evenlode Cres, Coundon, Coventry, CV6 1BY [IO92FK, SP38]
G0 EXA A F Bendall, 3 St Michaels Gate, Brimfield, Ludlow, SY8 4NE [IO82PH, SO56]
G0 EXB P J Langdon, Dahlia Cottage, Clows Top, Kidderminster, Worcs, DY14 9HP [IO82SI, SO77]
G0 EXC K C Pardoe, Pool House, Bell Ln, Broadheath, Tenbury Wells, WR15 8QX [IO82SG, SO66]
G0 EXD Dr C J Challinor, 44 West Vale, Neston, South Wirral, L64 9SF [IO83LG, SJ27]
G0 EXE E McGuinness, 29 Station Rd, Neston, South Wirral, L64 9QJ [IO83LG, SJ27]
G0 EXF P Havenhand, 69 Queen St., Mosborough, Sheffield, S19 5BP [IO93GJ, SK38]
G0 EXG V A Harris, 17 Sutton Gr, Sutton, SM1 4LR [IO91VI, TQ26]
G0 EXH Details withheld at licensee's request by SSL.
G0 EXI S J Brunton, 5 Mill Rise, Bourton, Gillingham, SP8 5DH [IO81UB, ST73]
G0 EXJ Details withheld at licensee's request by SSL.[Correspondence via PO Box 249, Poole, Dorset, BH15 2LR.]
GD0 EXK Details withheld at licensee's request by SSL.
GD0 EXM W A Parkinson, 11 Kirby Hill, Braddan, Douglas, IM2 1PA
GM0 EXN J Eden, Dunnet Head Tearoom, Brough, Caithness, KW14 8YE [IO88HP, ND27]
G0 EXO M C Rolfe, 9 Herbert St., Whitehall, Bristol, BS5 9BL [IO81RL, ST67]
G0 EXP H T Knight, 15 Kenninghall Mount, Sheffield, S2 3WA [IO93GI, SK38]
G0 EXS E B Foreman, 10 Wilmington Cl, Brighton, BN1 8JE [IO90WU, TQ30]
G0 EXU A Davies-Jones, 10 Ponsford Rd, Knowle, Bristol, BS4 2UP [IO81RK, ST67]
G0 EXX Details withheld at licensee's request by SSL.
G0 EXY C L James, 68 Pickering Rd, Hull, HU4 6TE [IO93TR, TA02]
G0 EXZ Details withheld at licensee's request by SSL.[Station loacted near Bedford.]
G0 EYB J Martindale, 15 Itchenside Cl, Southampton, SO18 2LZ [IO90HW, SU41]
G0 EYC P Barnett, 2 Avondale Bungalows, Mansfield, NG18 2NJ [IO93JD, SK56]
G0 EYE A C Lunn, 45 St. Anthonys Ave, Eastbourne, BN23 6LN [JO00DT, TQ60]
G0 EYF B T Ford, 15 Derby Rd, Frankmarsh, Barnstaple, EX32 7HW [IO71XC, SS53]
G0 EYG C E Tite, 29 Wingfield Rd, Bromham, Bedford, MK43 8JY [IO92RD, TL05]
GW0 EYH R J Dawson, 74 Dylan Ave, Cefn Fforest, Blackwood, NP2 1NG [IO81JQ, ST19]
G0 EYI Details withheld at licensee's request by SSL.
G0 EYJ Details withheld at licensee's request by SSL.[Op: F Beesley, 386 Fernhill Road, Farnborough, Hants, GU14 9EL.]
G0 EYK Details withheld at licensee's request by SSL.
G0 EYL Details withheld at licensee's request by SSL.[Op: W McAdam, 2 Manor Orchards, Knaresborough, N.Yorks, HG5 0BW.]
G0 EYM R A Rowlett, 30 Cotswold Cl, Putnoe, Bedford, MK41 9LR [IO92SD, TL05]
G0 EYN K P Foster, 71 Dalestorth St., Sutton in Ashfield, NG17 4EW [IO93ID, SK55]
G0 EYO C A Pettitt, 23 Dark Ln, Hollywood, Birmingham, B47 5BS [IO92BJ, SP07]
G0 EYP R T Francis, 42 Carmarthen Rd, Cheltenham, GL51 5LA [IO81WV, SO92]
G0 EYQ B B Clifford, 15 Eastwood Ave, Burntwood, Staffs, WS7 8DX [IO92AQ, SK00]
G0 EYR P G Robinson, 35 Stoke Rd, Taunton, TA1 3EH [IO81KA, ST22]
G0 EYT M P Fox, 49 Manor Dr, Esher, KT10 0AZ [IO91TJ, TQ16]
G0 EYU C Talbot, 59 Heywood Ave, Austerlands, Oldham, OL4 4AZ [IO83XN, SD90]
G0 EYV L Smith, 2 Springvale Ave, Walsall, WS5 3QB [IO92AN, SP09]
G0 EYW D L Jones, 4 Granville Crest, Kidderminster, DY10 3QS [IO82VJ, SO87]
G0 EYX D R J M Southey, 253 Sandon Rd, Stafford, ST16 3HQ [IO82WT, SJ92]
G0 EYZ P A Orchard, 68 Simpson Rd, Bletchley, Milton Keynes, MK1 1BA [IO92PA, SP83]
G0 EZA F M Dent, Woodvale Bussage, Stroud, Glos, GL6 8AU [IO81VR, SO80]
G0 EZB S J Cowie, 2 Longville Cl, Abbeymead, Gloucester, GL4 5SG [IO81VU, SO81]
GI0 EZD C C Marshall, 31 Killaire Park, Bangor, Co Down, BT19 1EG [IO74DP, J48]
G0 EZE M C Hart, 16 Downs Park East, Bristol, BS6 7QD [IO81QL, ST57]
G0 EZF A M J Kolker, Connemara, Big Stone, Middlewich Rd, Cranage, Ches, CW4 8HH [IO83TF, SJ76]
GM0 EZG R F Murison, 2 Abbots Walk, Kirkcaldy, KY2 5NL [IO86JC, NT29]
GM0 EZH C M Burgess, 6 St. Machar Dr, Aberdeen, Aberdeenshire, AB2 1SF [IO87TH, NJ72]
G0 EZJ D H Drake, 60 Jessopp Ave, Bridport, DT6 4ES [IO80OR, SY49]
G0 EZM S A Buttle, 156 Kathleen Rd, Sholing, Southampton, SO19 8LN [IO90HV, SU41]
G0 EZN Details withheld at licensee's request by SSL.[Op: Susan Morgan, 28 Newfield Rise, Dollis Hill, London NW2 6YH.]
G0 EZO J H Wilkinson, 4 Brook Rd, Cheslyn Hay, Walsall, WS6 7ES [IO82XQ, SJ90]
GM0 EZP Details withheld at licensee's request by SSL.
GW0 EZQ K L Williams Llanelli A.R.S, 12 Danybanc, Llanelli, SA15 4NS [IO71WQ, SN50]

GM0 EZR P W Newton, 37 The Henge, Glenrothes, KY7 6XU [IO86KF, NO20]
GI0 EZS D Neill, 11 Roscolban Ct, Kesh, Co Fermanagh, N Ireland, BT93 1TF [IO64DM, H16]
G0 EZT E W Humphries, 53 Birches Barn Rd, Bradmore, Wolverhampton, WV3 7BL [IO82WN, SO89]
G0 EZU A G Davies, 27 Wemberham Cres, Yatton, Bristol, BS19 4BD [IO81OJ, ST46]
G0 EZV Details withheld at licensee's request by SSL.
G0 EZW H Bent, 97 Nottingham Rd, Selston, Nottingham, NG16 6BU [IO93IB, SK45]
GM0 EZX C S J Wood, Edelweiss, 17 Portnalong, Isle of Skye, IV47 8SL [IO67TH, NG33]
G0 EZY T G Jeacock, San Clu, Chaul End Village, Caddington, Luton, LU1 4AX [IO91SU, TL01]
G0 FAA Comm W J Flindell, Pentley House, Marston Rd, Sherborne, DT9 4BJ [IO80RW, ST61]
G0 FAB H Kay, 51 Colin Cres, Colindale, London, NW9 6EU [IO91VO, TQ28]
G0 FAD J R Bowers, 22 Kilmiston Dr, Fareham, PO16 8DY [IO90KU, SU60]
G0 FAE R J Beer, 65 Bridgefield Rd, Whitstable, CT5 2PH [JO01MI, TR16]
GI0 FAF Details withheld at licensee's request by SSL.
G0 FAG J Pollitt, 2 Dukes Dr, Hoddlesden, Darwen, BB3 3RA [IO83SQ, SD72]
G0 FAH W R Wright, 46 Homestall Rd, East Dulwich, London, SE22 0SB [IO91XK, TQ37]
G0 FAI Details withheld at licensee's request by SSL.
G0 FAJ L A Barnes, 2 Viscount Rd, Weymouth, DT4 9EP [IO80SO, SY67]
G0 FAK K N Forward, 17 Cowper Rd, Deal, CT14 9TW [JO01QF, TR35]
GM0 FAL Details withheld at licensee's request by SSL.
G0 FAQ Details withheld at licensee's request by SSL.[Op: P Samson, c/o 19 Far Gosford Street, Coventry, CV1 5DT.]
G0 FAU A J Doyle, 59 Brookhead Dr, Cheadle, SK8 2JA [IO83VJ, SJ88]
G0 FAY Details withheld at licensee's request by SSL.
G0 FBA R H G Checketts, 30 Trafalgar Ct, Nelson Dr, Christchurch, BH23 3RZ [IO90DR, SZ19]
G0 FBC C Hyatt, 44 Barnes Ln, Sarisbury Green, Southampton, SO31 7BZ [IO90IU, SU40]
G0 FBG G Hands, 59 Lister St., Nuneaton, CV11 4NU [IO92GM, SP39]
G0 FBH G E Watson, 6 Woburn Dr, Bower Grange, Bedlington, NE22 5YB [IO95FD, NZ28]
G0 FBK P L Fawcett, 7 Albert Hill, Bishop Auckland, DL14 6EH [IO94DP, NZ22]
G0 FBL B Sharman, 64 Collingwood Dr, Shiney Row, Houghton-le-Spring, DH4 7LP [IO94GU, NZ35]
G0 FBM D F Lawson, 52 Ryefield Rd, Eastfield, Scarborough, YO11 3DR [IO94TF, TA08]
G0 FBN Details withheld at licensee's request by SSL.[Op: T Sorbie, 20 Gladstone Terrace, Sunniside, Tow Law, Bishop Auckland, Co. Durham, DL13 4LS.]
G0 FBO C L Roberts, 86 Peake Rd, Brownhills, Walsall, WS8 7BZ [IO92AP, SK00]
G0 FBQ W T Wootton, 94 Dyas Ave, Great Barr, Birmingham, B42 1HF [IO92AM, SP09]
G0 FBR D A Lockwood, 14 Clifton Gdns, Goole, DN14 6AS [IO93NQ, SE72]
G0 FBS G D B Nicolson, 34 Chalbury Cl, Weymouth, DT3 6LE [IO80SP, SY68]
GW0 FBT C E Hughes, 18 Westway, Rogiet, Newport, NP6 3SP [IO81OO, ST48]
G0 FBU W B Hedley, 74 Park Ln, Wensleydale, Middleham, Leyburn, DL8 4QU [IO94CG, SE18]
G0 FBV P R Wardman, 96 Box Hill, Scarborough, YO12 5NG [IO94SG, TA08]
G0 FBW A Armstrong, 1 Montfalcon Cl, Peterlee, SR8 1DD [IO94HS, NZ44]
G0 FBX A E H Leigh, 7 Field Fare, Abbeydale, Gloucester, Glos, GL4 8NH [IO81VT, SO81]
G0 FBY R Bailey, Springbank, 49 Spring Ln, Kenilworth, CV8 2HD [IO92FI, SP27]
G0 FBZ Details withheld at licensee's request by SSL.[Station located near Nelson.]
G0 FCA I C Groom, 10 Fallbarn Cres, Rawtenstall, Rossendale, BB4 6AZ [IO83UQ, SD82]
G0 FCB C H Tatlow, Frenton Farm, Park Rd, Whitemoor, St Austell, Cornwall, PL26 7XQ [IO70NJ, SW95]
GI0 FCC R D K Addy, 52 Alexandra Ave, Whitehead, Carrickfergus, BT38 9SB [IO74DS, J49]
G0 FCD Details withheld at licensee's request by SSL.[Op: B G Cooke, 49 Shepherds Croft, Woodlands Est. Slad Rd. Stroud, Glos, GL5 1US.]
G0 FCE Details withheld at licensee's request by SSL.[Op: R J Guest.]
G0 FCG P V O'Neill, 36 Grantley Gdns, Mannamead, Plymouth, Devon, PL3 5BP [IO70WJ, SX45]
G0 FCH R A Last, 39 Upham Rd, Swindon, SN3 1DJ [IO91CN, SU18]
GM0 FCI P Reid, 129 McKinlay Cres, Irvine, KA12 8DR [IO75PO, NS33]
G0 FCL D F Kitchen, Hazelbeech, Lenwood Park, Northam, Bideford, North Devon, EX39 3PD [IO71VA, SS42]
G0 FCM I J Sircombe, 6 Manor Pl, Priors Park, Tewkesbury, GL20 5HL [IO81WX, SO83]
GW0 FCN E J Thorne, 61 The Avenue, Ystrad Mynach, Hengoed, CF8 8AF [IO81HN, ST08]
G0 FCO B G Cooke, 49 Shepherds Croft, Slad Rd, Stroud, GL5 1US [IO81VS, SO80]
GM0 FCP R Dickinson, 3 Langton Gdns, East Calder, Livingston, EH53 0DZ [IO85GV, NT06]
G0 FCQ D W James, 15 Kington Gdns, Chelmsley Wood, Birmingham, B37 5HX [IO92CL, SP18]
G0 FCT I W N Pawson, 3 Orion, Roman Hill, Bracknell, RG12 7YX [IO91OJ, SU86]
G0 FCU S H Kennedy, 3 The Knoll, Crow Ln, Reed, Royston, Herts, SG8 8AD [IO92XA, TL33]
G0 FCV A J Woods, 9 Blenheim Dr, Launton, Bicester, OX6 0EA [IO91KV, SP62]
G0 FCX A Traynor, 2 Mansfield Rd, Mossley, Ashton under Lyne, OL5 9JN [IO83XM, SD90]
G0 FCZ L Standley, 26 Ullswater Dr, Middleton, Manchester, M24 5RL [IO83VN, SD80]
G0 FDA R Dresser, 6 Acacia Ave, Fencehouses, Houghton-le-Spring, DH4 6JG [IO94GU, NZ35]
GI0 FDB Details withheld at licensee's request by SSL.
G0 FDD C I Shalley, 4 Almond Walk, Lydney, GL15 5LP [IO81RR, SO60]
G0 FDE M N Jackson, 2 Sunnybank, Watledge, Nailsworth, Stroud, GL6 0AP [IO81VQ, SO80]
GI0 FDG Details withheld at licensee's request by SSL.
G0 FDH R N Wishart, 15 Plumer Ave, Tanghall Ln, York, YO3 0PX [IO93LX, SE65]
G0 FDI A I Horti, 48 River Dr, Upminster, RM14 1AS [JO01DN, TQ58]
G0 FDJ K B Castley, 64 Tantelen Rd, Canvey Island, SS8 9QG [JO01GM, TQ78]
G0 FDK P Hodgkins, 7 Mere View Gdns, Scarborough, YO12 4DF [IO94TG, TA08]
G0 FDM Details withheld at licensee's request by SSL.
G0 FDN Details withheld at licensee's request by SSL.[Op: W N Young. Station located in London N17.]
G0 FDO Details withheld at licensee's request by SSL.
G0 FDP F W J Parradine, 87 Oakways, Off Bexley Rd, Eltham, London, SE9 2NZ [JO01AK, TQ47]
G0 FDR E B Cachart, 4 Cottage Cl, Heage, Belper, DE56 2BS [IO93GA, SK34]
G0 FDS J D Johnson, 46 Costead Manor Rd, Brentwood, CM14 4YN [JO01DP, TQ59]
G0 FDT M P Flatman, 22 Yarrow Dr, Carlton Colville, Lowestoft, NR33 8NG [JO02UL, TM59]
G0 FDU R Barker, 12 Low Cl, Greenhithe, DA9 9PE [JO01DK, TQ57]
G0 FDV D M Jardine, 128 London Rd, Dunstable, LU6 3EE [IO91RV, TL02]
G0 FDW D Wood, 62 Pentire Rd, Lichfield, WS14 9SQ [IO92CQ, SK10]
G0 FDX J A Lawson Central Lancs Amateur Radio Cl, 14 Kentmere Ave, Leyland, Farington, Preston, PR5 1UH [IO83PQ, SD52][Correspondence to: Central Lancs ARC, Chairman Phil Catterall, Sundale, Buckholes La, Wheelton, Chorley PR6 8JL. Club formed 1986. We have a successful contest and dx group - new members welcome: see affiliated clubs section]
G0 FDY C E Tanner, 4 Cardinals Ct, Queens Rd, Felixstowe, IP11 7PG [JO01QX, TM33]
G0 FDZ C R Whitmarsh, 35 Dorchester Ave, Bexley, DA5 3AH [JO01BK, TQ47]
G0 FEA Details withheld at licensee's request by SSL.
G0 FEC B Robinson, 68 Langholm Dr, Heath Hayes, Cannock, WS12 5EZ [IO92AQ, SK01]
G0 FED J A Wentworth, 8 Oakleigh Rd, Great Clacton, Clacton on Sea, CO15 4PX [JO01NT, TM11]
G0 FEE Details withheld at licensee's request by SSL.[Op: Dr M L Hamilton, 73 Sutton Square, Heston, Middx, TW5 0JB.]
G0 FEF M W Westall, 29 Sussex Ave, Melton Mowbray, LE13 0AP [IO92NS, SK71]
G0 FEG T Halliday, 43 Hudson Rd, Maghull, Liverpool, L31 5PA [IO83MM, SD30]
G0 FEH D M Thickett, 8 Burns Rd, Lillington, Leamington Spa, CV32 7EL [IO92FH, SP36]
G0 FEI V Ward, Romayne, St. Johns Rd, Belton, Great Yarmouth, NR31 9JT [JO02UN, TG40]
G0 FEJ G Marshall, Birchlands, 3 Longbridge Cl, Sherfield on Loddon, Hook, RG27 0DQ [IO91LH, SU65]
G0 FEK R D Wilson, 34 Belfairs Dr, Chadwell Heath, Romford, RM6 4EB [JO01BN, TQ48]
GW0 FEM G W Felton, 10 Penbodeistedd, Llanfechell, Amlwch, LL68 0RE [IO73SJ, SH39]
G0 FEN Details withheld at licensee's request by SSL.
G0 FEO A J Quy, 17 Fircroft, Kingsbury, Tamworth, B78 2JU [IO92DN, SP29]
G0 FEP Dr H Maclean, 15 Keystone Cres, Kew East, Victoria, Australia 3102
G0 FEQ K Howard, Keitel, Station Rd, Ulceby, South Humberside, DN39 6UQ [IO93UO, TA11]
GM0 FET J A Williamson, Cairn Cottage, Denside of Durris, Kincardshire, AB31 3DT [IO87RB, NO69]
GW0 FEU D E Davies, Coed Fryn, Lower Halkyn, Halkyn, Holywell, CH8 8ES [IO83JF, SJ27]
G0 FEV E D Kittrick, 64 Hollin Dr, Durkar, Wakefield, WF4 3PR [IO93FP, SE31]
G0 FEX Details withheld at licensee's request by SSL.
G0 FEZ K M Wragg, 11A Fall Rd, Heanor, Derbyshire, DE75 7PQ [IO93HA, SK44]
G0 FFA N Painter, 75 Chaytor Rd, Polesworth, Tamworth, B78 1JS [IO92EO, SK20]
G0 FFB S Maughan, 17 Upper Dane, Desborough, Kettering, NN14 2LB [IO92OK, SP78]
G0 FFE S J Wainwright, Elleberta, Siddington Rd, Cirencester, GL7 1PE [IO91AQ, SP00]
G0 FFF Details withheld at licensee's request by SSL.
G0 FFG M Laming, 71 Noke Shot, Harpenden, AL5 5HS [IO91TT, TL11]
G0 FFH Details withheld at licensee's request by SSL.
G0 FFI T J Cleaver Daventry Amateur Radio Club, 2 Ruskin Way, Daventry, NN11 4TT [IO92KF, SP56]
G0 FFK R McKenzie, 40 Fairway Ave, West Drayton, UB7 7AN [IO91SM, TQ07]
G0 FFL R W O'Keeffe, 40 Edinburgh Rd, Maidenhead, SL6 7SH [IO91PM, SU88]
G0 FFM Details withheld at licensee's request by SSL.[Op: S F Simmonds.]
G0 FFN E C Fisher, 19 Keats Way, West Drayton, UB7 9DR [IO91SL, TQ07]
G0 FFP Details withheld at licensee's request by SSL.[Op: Mrs T Billett, 31 St Leonards Road, Claygate, Surrey, KT10 0EL.]
G0 FFQ R J Baldock, 1A Thorneywood Rd, Long Eaton, Nottingham, NG10 2DZ [IO92IV, SK43]
G0 FFU A Utteridge, 10 Burland Ct, Hillside Est, Branston, Lincoln, LN4 1NY [IO93SE, TF06]
G0 FGA A M Walker, 2 Chelwood Dr, Sandhurst, Camberley, Surrey, GU17 8HT [IO91OH, SU85]
G0 FGB D H E Wilde, Wildwood, Restwell Ave, Cranleigh, Surrey, GU6 8PQ [IO91RD, TQ04]
G0 FGC J Biggs, 1 Hetton Cl, Warwick, CV34 5XP [IO92FH, SP26]
G0 FGE B L Bolt, 106 Barley Farm Rd, Exeter, EX4 1NJ [IO80FR, SX99]
G0 FGF J G Phillips, Abbeyfield House, 18 Clarence Rd South, Weston Super Mare, BS23 4BN [IO81MI, ST36]
G0 FGG J J Thompson, 10 Allendale Terr, Annfield Plain, Stanley, DH9 8JT [IO94DU, NZ15]
G0 FGH Details withheld at licensee's request by SSL.[Op: S B Black, 1 Long Barrow Road, Curzon Pk, Caine, Wilts, SN11 0HE.]
G0 FGI H T H Cromack, Argyll Lodge, Ridge Rd, Maidenhead, Torquay, Devon, TQ1 4DT [IO80FL, SX96]
G0 FGJ D A Joyce, 35 Beale St., Dunstable, LU6 1LZ [IO91RV, TL02]
G0 FGN R J Whiting, 27 Chester St., Cirencester, GL7 1HG [IO91AR, SP00]
GW0 FGO W J Waldron, 100 Porthmawr Rd, Cwmbran, NP44 1NB [IO81LP, ST29]
G0 FGP R H Bradwell, East Barn, Mayfield Rd, Summerfields, Oxford, OX2 7EN [IO91IS, SP50]
GW0 FGQ Details withheld at licensee's request by SSL.
G0 FGR A S Perkins, 111 Broadmead, Corsham, SN13 9AP [IO81VK, ST86]
G0 FGS Dr M J Bosley, Two Mile Ln, Higham, Highnam, Gloucester, Glos, GL2 8DW [IO81UV, SO72]
G0 FGU Details withheld at licensee's request by SSL.
G0 FGW R C Clements, 28 Willow Gr, Chippenham, SN15 1AR [IO81WL, ST97]
G0 FGX R A H McCreadie, 42 High St., Chard, TA20 1QS [IO80MU, ST30]
G0 FGZ Dr C J P Newton, 22 Crewkerne Cl, Nailsea, Bristol, BS19 2SN [IO81PK, ST47]
G0 FHA G Parkin-Coates, 357 Bradley Rd, Bradley Bar, Huddersfield, HD2 1PR [IO93CQ, SE12]
G0 FHC B W Trimmer, Sydney Cottage, Salisbury Rd, Plaitford, Romsey, SO51 6EE [IO90EX, SU21]
GM0 FHD E W B Mottart, 2 Keilhill Cttgs, King Edward, Banff, Scotland, AB4 3LT [IO87TH, NJ72]
GM0 FHE S W Allen, Berea Cottage, 88 Kirk St., Peterhead, AB42 1RY [IO97CM, NK14]
GM0 FHF Details withheld at licensee's request by SSL.[Op: C Ashdown, Dunora, Pier Road, Fairlie, Ayrshire, KA29 0AP.]
GM0 FHJ R L Menzies, 105 Yokermill Rd, Knightswood, Glasgow, G13 4HL [IO75TV, NS56]
G0 FHK R H Peart, 33 Fieldfare, Abbeydale, Gloucester, GL4 4WH [IO81VU, SO81]
G0 FHL B Thomas, 2 Hengrove Ct, Hurst Rd, Bexley, Kent, DA5 3LT [JO01BK, TQ47]
G0 FHM R Doughty, 4 Trinity Rd, Wisbech, PE13 3UN [JO02CQ, TF41]
G0 FHO Details withheld at licensee's request by SSL.
G0 FHR Details withheld at licensee's request by SSL.
GM0 FHS S A Baird, Polfearn Cottage, Tanuilt, Argyll, Scotland, PA35 1JQ [IO76JK, NN03]
G0 FHT G A F Bate, 7 Albany Ct, Redruth, TR15 2NY [IO70JF, SW74]
G0 FHV Details withheld at licensee's request by SSL.
G0 FHW T D Opie, 3 Colborne Ave, Illogan, Redruth, TR16 4EB [IO70IG, SW64]
G0 FHX A B Hocking, 79 Cornish Cres, Truro, TR1 3PE [IO70LG, SW84]
G0 FHY J Hocking, 79 Cornish Cres, Truro, TR1 3PE [IO70LG, SW84]
GW0 FHZ Details withheld at licensee's request by SSL.[Station located near Carmarthen.]
G0 FIC K J Tarry, 38 Tresithney Rd, Carharrack, Redruth, Cornwall, TR16 5QZ [IO70JF, SW74]
G0 FIE F A Edwards, 1 Ash Cl, Winscombe, BS25 1HT [IO81OH, ST45]
G0 FIF A G Duncan, 9 South Ave, Whiteley Village, Walton on Thames, KT12 4ED [IO91SI, TQ06]
G0 FIG A D V Trusler, St. Marys Cottage, 42 Mill Hill, Shoreham By Sea, BN43 5TH [IO90UU, TQ20]
G0 FIH Details withheld at licensee's request by SSL.[Op: G Haselden. Station located 10 miles East-South-East of Chester. WAB: SJ56, Loc: IO83QB.]
GI0 FII Details withheld at licensee's request by SSL.
G0 FIJ C L C Jones, 46 Wilmington Cl, Woodley, Reading, RG5 4LR [IO91NK, SU77]
G0 FIK Details withheld at licensee's request by SSL.[Station located near Potters Bar.]
G0 FIL Details withheld at licensee's request by SSL.
G0 FIM Details withheld at licensee's request by SSL.[Station located near Chichester.]
G0 FIN A S Findlay, 218 Lower Hillmorton Rd, Rugby, CV21 3TS [IO92JI, SP57]
G0 FIP E M Tugwell, 67 Upper Kingston Ln, Shoreham By Sea, BN43 6TG [IO90VU, TQ20]
GM0 FIQ H P Bohan, 12 Loch Way, Kemnay, Inverurie, AB51 5QZ [IO87SF, NJ71]
G0 FIR Details withheld at licensee's request by SSL.
G0 FIS Details withheld at licensee's request by SSL.[Station located near Alton.]
G0 FIT K R Menzel, 8, Higher Bockhampton, Dorchester, Dorset, DT2 8QJ [IO80TR, SY79]
G0 FIU R J Edwards, Stanmore Cottage, Brockley Corner-Cul, Bury St Edmunds, Suffolk, IP28 6UA [JO02IH, TL87]
G0 FIV C D Baxter, 231 Stanley Rd, Bootle, Merseyside, L20 3DY [IO83MK, SJ39]
G0 FIW I F Osborne, Alacoo, Tan Ln, Little Clacton, Clacton on Sea, CO16 9PS [JO01NU, TM12]
G0 FIY C A Cudmore, 26 Higher Dr, Oulton Broad, Lowestoft, NR32 3DA [JO02UL, TM59]
G0 FIZ Details withheld at licensee's request by SSL.
G0 FJA B M Samuels, 63 Mill Rd, Okehampton, EX20 1PR [IO80AR, SX59]
G0 FJB J S Bailey, Powney Cottage, Powney St, Milden, Ipswich, IP7 7AL [JO02KB, TL94]
G0 FJD D Tyers, 18 Gardeners Cl, Marlpoole Gdns, Kidderminster, DY11 5DW [IO82UJ, SO87]
GW0 FJH C Lonsdale, 6 Oak Tree Cl, New Inn, Pontypool, NP4 0DG [IO81LQ, SO30]
G0 FJI E V Aldridge, 10 Ashtree Rd, Hamilton, Leicester, LE5 1TW [IO92LP, SK60]
G0 FJJ A J Winkler, 2 Beechwood Ave, Thornton Heath, CR7 7DY [IO91WJ, TQ36]
G0 FJK D P Bird, 1 Minster Cottage, St Cross, Harleston, Norfolk, IP20 0NY [JO02QJ, TM28]
G0 FJL E W G Rose, 8 Grasmere Ave, Felixstowe, IP11 9SG [JO01QX, TM33]
G0 FJM J E Roberts, Schapbach, 221 Clevedon Rd, Tickenham, Clevedon, BS21 6RX [IO81OK, ST47]
G0 FJN J F Bloice, 12 Gilpin Rd, Oulton Broad, Lowestoft, NR32 3NS [JO02UL, TM59]
GW0 FJP D J Stanley, 9 Haywain Ct, Brackla, Bridgend, CF31 2ED [IO81FM, SS97]
GW0 FJQ A Marshall, 26 Avondale Rd, Gelli, Pentre, CF41 7TW [IO81GP, SS99]
G0 FJR D S Paynter, 32 Broom Gr, Knebworth, SG3 6BQ [IO91VU, TL22]
G0 FJS Details withheld at licensee's request by SSL.
G0 FJT J Hanson, Rhoslwyd, Tally, Llandeilo, Dyfed, SA19 7EJ [IO81AX, SN63]
GW0 FJV Details withheld at licensee's request by SSL.
GW0 FJW Details withheld at licensee's request by SSL.
G0 FJY C S Kentch, 55 Melrose Ave, Worthing, BN13 1NZ [IO90TT, TQ10]
G0 FJZ G V Bradley, 59 Main Rd, Watnall, Nottingham, NG16 1HE [IO93JA, SK54]
G0 FKE P B H Jacob, Abbotts Ct, Victoria Ave, Brandon, IP27 0HZ [JO02HK, TL78]
G0 FKF J A Smith, Erw Las, The Crossroads, St Day, Redruth, Cornwall, TR16 5PN [IO70JF, SW74]
G0 FKG C G S Skillings, 11 Curtis Rd, Mile Cross Ln, Norwich, NR6 6RB [JO02PP, TG21]
G0 FKH Details withheld at licensee's request by SSL.
G0 FKI A S Brown, Panorama, Highway Ln, Mount Ambrose, Redruth, TR15 1SE [IO70JF, SW74]
G0 FKJ C Currey, 1 Newport Cl, Portishead, Bristol, BS20 8DD [IO81OL, ST47]
G0 FKK R M Parrish, 11 Pitt Rd, Maidstone, ME16 8PA [JO01FG, TQ75]
G0 FKL T M M Shea, 4 Sandown Cttgs, London Rd, Teynham, Sittingbourne, ME9 9JE [JO01JH, TQ96]
GI0 FKM Details withheld at licensee's request by SSL.
GM0 FKP J Tohill, 71 Campsie Rd, Bellfield, Kilmarnock, KA1 3RY [IO75SO, NS43]
G0 FKQ C D Parnell, 50 Bradwell Dr, Nottingham, NG5 9DW [IO93KA, SK54]
G0 FKR Details withheld at licensee's request by SSL.
G0 FKS K J Stancliffe, 3 Upper Lambricks, Rayleigh, SS6 8BP [JO01HO, TQ89]
G0 FKV J W Kidgell, 7 Lowland Cl, Barnsley, S71 2NU [IO93GN, SE30]
G0 FKW A R Timms, 1 Westfield Barn, School Ln, Priors Marston, Rugby, Warks, CV23 8RR [IO92IF, SP45]
G0 FKX D M K Warren, 1 Ruby Terr, Porkellis, Helston, TR13 0LD [IO70JD, SW63]
G0 FKY J D Merifield, 84 Wareham Rd, Corfe Mullen, Wimborne, BH21 3LG [IO80XS, SY99]
G0 FKZ M A Buck, 21 Willow Walk, Culverstone, Meopham, Gravesend, DA13 0QS [JO01EI, TQ66]
G0 FLA W E L Frewen, Talisman, Rye Rd, Northiam, Rye, TN31 6NJ [JO00HX, TQ82]
G0 FLB I J Gillingham, 29 High St., Long Wittenham, Abingdon, OX14 4QH [IO91JP, SU59]
G0 FLC D H Beckley, 1 Beech Gr, Donington, Spalding, PE11 4XQ [IO92VV, TF23]
G0 FLD K H Pitman, 129 Reevy Rd, Wibsey, Bradford, BD6 3QF [IO93CS, SE12]
G0 FLG B A Parker, 120 Brooks Ln, Whitwick, Coalville, LE67 5DF [IO92HR, SK41]
G0 FLH A G Goddard, 183 Chapelfields Rd, Acomb, York, YO2 5AD [IO93KW, SE55]
G0 FLI N F Penistone, 114 Long Ln, Worrall, Sheffield, S30 3AF [IO93GJ, SK38]
G0 FLJ F L James, 40 Romsley Hill Grange, Farley Ln, Romsley, Halesowen, B62 0LN [IO82XJ, SO97]
G0 FLL C H Whitaker, 7 Gatesbield, New Rd, Windermere, Cumbria, LA23 2LA [IO84NJ, SD49]
G0 FLM Details withheld at licensee's request by SSL.
G0 FLN Details withheld at licensee's request by SSL.
GM0 FLO Details withheld at licensee's request by SSL.[Op: A N Young.]
G0 FLP J T Hammond, Ashmond House, Queens St., March, PE15 8SN [JO02BN, TL49]
G0 FLQ R Cheetham, 65 Avondale Ave, Hazel Gr, Stockport, SK7 4QE [IO83WJ, SJ98]
G0 FLT A Auker-Howlett, 29 Shakespeare, Burns Rd, Royston, SG8 5PX [IO92XB, TL34]
G0 FLU M A Crane, 40 Dukes Way, Newquay, TR7 2RW [IO70LJ, SW86]
G0 FLV R M Pennock, 85 Cameron St., Heckington, Sleaford, NG34 9RP [IO92UX, TF14]
G0 FLW L A Williams, Astwood, 56 Meriden Ave, Wollaston, Stourbridge, DY8 4QS [IO82VL, SO88]
G0 FLX D J Gray, 10 Regency Gdns, Tingley, Wakefield, WF3 1JS [IO93ER, SE22]
G0 FLZ C W B Hartley, 26 Grange Ave, Scarborough, YO12 4AA [IO94TG, TA08]
G0 FMA B Burbridge, 50 The Wroe, Emneth, Wisbech, PE14 8AN [JO02CP, TF40]
G0 FMB B E Collins, 28 Marlborough Rd, South Woodford, London, E18 1AP [JO01AO, TQ48]
G0 FMD A F Kent, 34 Brushfield Rd, Linacre Woods, Chesterfield, S40 4XE [IO93GF, SK37]
G0 FMG J A Voss, 4 Chaucer Ave, Mablethorpe, LN12 1DA [JO03DI, TF58]
G0 FMH M E Matheny, 59 Purley Rd, Cirencester, GL7 1ER [IO91AR, SP00]
G0 FMI R J Friston, Savma, Bradenham Rd, Shipdham, Thetford, IP25 7PJ [JO02KP, TF90]
G0 FMJ R Bennett, 12 Fawkes Cl, Kingswood, Bristol, BS15 4LR [IO81SL, ST67]
G0 FML W R Auker-Howlett, 40 Worcester Rd, Titton, Stourport on Severn, DY13 9PD [IO82UH, SO87]
G0 FMM Details withheld at licensee's request by SSL.
G0 FMN A McQuarrie, 1 Helston Pl, Fishermead, Milton Keynes, MK6 2JR [IO92PA, SP83]
G0 FMO J S Smethurst, 38 Mill View, Ferrybridge, Knottingley, WF11 8SR [IO93IQ, SE42]
G0 FMP I Hodgkiss, 41 Buckingham Rise, Worksop, S81 7ED [IO93KH, SK58]
GW0 FMQ R A S Rees, 2 Bryn Poeth, Tregarth, Bangor, LL57 4PG [IO73WE, SH66]

G0 FMU A J Turner, 17 The Dell, Great Warley, Brentwood, CM13 3AL [JO01DO, TQ59]
GM0 FMW D Enderby, 45 Oxgangs Park, Edinburgh, EH13 9LF [IO85JV, NT26]
G0 FMX A D Roberts, 10 Broomhurst Way, Muxton, Telford, TF2 8RG [IO82SR, SJ71]
G0 FMY S Wilkinson, 30 Field Ln, Morton, Gainsborough, DN21 3BY [IO93OK, SK89]
G0 FMZ Details withheld at licensee's request by SSL.
G0 FNA C T Halliday, 5 Ovington Dr, Kew, Southport, PR8 6JW [IO83MP, SD31]
G0 FNB S J Mudd, 91 Chalkwell Ave, Westcliff on Sea, SS0 8NL [JO01IM, TQ88]
GM0 FNE T M Wilson, 20 Mace Ct, Stirling, FK7 7XA [IO86BC, NS89]
G0 FNF I Gilbert, 36 Tyne Park, Taunton, TA1 2RP [IO81LA, ST22]
G0 FNH J G Toon, 2 Marston Ln Park, Rolleston on Dove, Burton on Trent, Staffs, DE13 9BJ [IO92EU, SK22]
G0 FNI Details withheld at licensee's request by SSL.
G0 FNJ D G Earnshaw, 1 Simonstone Ln, Simonston, nr Burnley, Lancs, BB12 7NX [IO83TT, SD73]
G0 FNK Details withheld at licensee's request by SSL.
GM0 FNL Details withheld at licensee's request by SSL.[Op: W Simpson.]
G0 FNM E Walker, 216 Milnrow Rd, Rochdale, OL16 5BB [IO83WO, SD91]
G0 FNO R E Ward, 9 Town End, Hungarton, Leicester, LE7 9JT [IO92MP, SK60]
G0 FNP P F Radcliffe, Hill View, Church Ln, Wilton nr Pickerin, North Yorks, YO18 7JY [IO94PF, SE88]
G0 FNS J A Brayshaw, 26 Ashfield Ave, Malton, YO17 0LE [IO94OD, SE77]
G0 FNT R R Smith, 6 Bassey Rd, Branton, Doncaster, DN3 3NS [IO93LM, SE60]
G0 FNU V M B Carswell, 12 Nash Cl, Earley, Reading, RG6 5SL [IO91MK, SU77]
G0 FNW Details withheld at licensee's request by SSL.
G0 FNX Details withheld at licensee's request by SSL.
G0 FNY Details withheld at licensee's request by SSL.[Licencee I Sargent. Station located near Ringwood. Op: I Sargent)]
G0 FOA Details withheld at licensee's request by SSL.
G0 FOC M J Clowes, 1 Locking Croft, Castle Vale, Birmingham, B35 7LD [IO92CM, SP19]
G0 FOD Details withheld at licensee's request by SSL.[Op: N J Grant.]
G0 FOE P H Hume-Spry, 23 Appledore Ave, Wollaton, Nottingham, NG8 2RE [IO92JW, SK53]
G0 FOF Details withheld at licensee's request by SSL.
G0 FOG C P Singleton, 78 Valley Rd, West Bridgford, Nottingham, NG2 6HQ [IO92KW, SK53]
G0 FOH Details withheld at licensee's request by SSL.
G0 FOI J R Wilde, 2 Bottoms Ln, Birkenshaw, Bradford, BD11 2NN [IO93DR, SE22]
G0 FOJ Details withheld at licensee's request by SSL.
G0 FOK R J C Cox, 12 East St., Thame, OX9 3JS [IO91MR, SP70]
GW0 FOL Details withheld at licensee's request by SSL.
G0 FOO W J Taylor, 45 Lyndhurst Rd, Southport, PR8 4JT [IO83LO, SD31]
G0 FOP Details withheld at licensee's request by SSL.
G0 FOQ Details withheld at licensee's request by SSL.
G0 FOR J A Dodds, 81 Halton Rd, Upton, Chester, CH2 1SN [IO83NF, SJ46]
G0 FOS Details withheld at licensee's request by SSL.
G0 FOT R S Gibbons, 3 Fairfield, Gamlingay, Sandy, SG19 3LG [IO92VD, TL25]
G0 FOW B Kearsley, Flat, 18 Elswick Rd, Southport, PR9 9NU [IO83MQ, SD31]
G0 FOY Details withheld at licensee's request by SSL.[Station located near Wilmslow.]
G0 FOZ Details withheld at licensee's request by SSL.
G0 FPC Details withheld at licensee's request by SSL.
G0 FPD Details withheld at licensee's request by SSL.
G0 FPG A R Harrison, 84 Second Ave, Carlton, Nottingham, Notts, NG4 1PP [IO92KX, SK64]
G0 FPI J S Spence, 60 Railey Rd, Northgate, Crawley, RH10 2BZ [IO91VC, TQ23]
G0 FPL A J Cordell, 20 Stanley Rd, Benfleet, SS7 3EL [JO01GN, TQ78]
G0 FPM P M Fennell, Keepside, 45 Badby Rd West, Daventry, NN11 4HJ [IO92JG, SP56]
G0 FPN D R Waller, 16 Woolacombe Lodge Rd, Birmingham, B29 6PX [IO92AK, SP08]
G0 FPO W J Coulthard, 1 Lambton St., Eccles, Manchester, M30 8DD [IO83TL, SJ79]
G0 FPP P E Smith, 104 Brightside Ave, Laleham, Staines, TW18 1NQ [IO91SK, TQ07]
G0 FPS R Watson, 18 Sheridan Cl, Leighton, Crewe, CW1 4TJ [IO83SC, SJ65]
G0 FPT A J Pettigrew, Toad Hall, 12 Pensford Cl, Crowthorne, RG45 6QR [IO91OJ, SU86]
G0 FPV I J Marquis, Ye Old Radio House, 20 Hazelwood Gr, Eastwood, Leigh on Sea, SS9 4DE [JO01HN, TQ88]
G0 FPX Details withheld at licensee's request by SSL.
GW0 FPY J P Bortowski, 4 Bryn Deiniol, Valley Rd, Llanfairfechan, LL33 0SR [IO83AF, SH67]
G0 FPZ R A Wileman, 5 Ermine Rise, Great Casterton, Stamford, PE9 4AJ [IO92RQ, SK90]
G0 FQA A S Wallis, 33 Sywell Ave, Wellingborough, NN8 5BZ [IO92PH, SP86]
G0 FQB P J Watts, 94 Salisbury Ave, Newbold, Chesterfield, S41 8PN [IO93GG, SK37]
G0 FQC P D Wood, 26 Church Rd, Bamber Bridge, Preston, PR5 6EP [IO83QR, SD52]
G0 FQD R Harvey, 2 Burnham Walk, Rainham, Gillingham, ME8 8SJ [JO01HI, TQ86]
G0 FQE A G McKay, 13 Common Ln, Sawston, Cambridge, CB2 4HW [JO02CC, TL44]
G0 FQF J R Rae, Sunnybank House, Burnley Rd East, Rossendale, Lancs, BB4 9PX [IO83VR, SD82]
G0 FQG D Tomkinson, 8 Landswood Park, Hartford, Northwich, CW8 1NF [IO83RF, SJ67]
G0 FQI D J Breen, 32 Skipton Cl, East Hunsbury, Northampton, NN4 0RB [IO92NE, SP75]
G0 FQJ Dr J A F Barnes, 10 Parsons Hill, Colchester, CO3 4DT [JO01KV, TL92]
G0 FQL D B Lawton, 32 Mansion Gdns, Huddersfield West, Huddersfield, HD4 7RF [IO93CP, SE11]
G0 FQM M J Mills, 44 Bedford Rd, Great Barford, Bedford, MK44 3JF [IO92TD, TL15]
G0 FQN J L Procter, 168 St. Davids Rd, Leyland, Preston, PR5 2UX [IO83PQ, SD52]
G0 FQO D M Berry, The Bungalow, Basil Rd, West Dereham, Kings Lynn Norfolk, PE33 9RP [JO02FN, TF60]
G0 FQP G J Owens, 73 Edinburgh Rd, Widnes, WA8 8BG [IO83OI, SJ48]
GM0 FQQ B Campbell, 93 Treeswoodhead Rd, Kilmarnock, KA1 4PB [IO75SO, NS43]
G0 FQR W E Pendlebury, 6 Timsbury Cl, Brieghtmet, Bolton, BL2 6TH [IO83TN, SD70]
GM0 FQS A J Smith, 60 Gordon Ave, Poltonahll, Bonnyrigg, EH19 2PQ [IO85KU, NT26]
G0 FQT Details withheld at licensee's request by SSL.
G0 FQU A R W D Date, 1 Yew Tree Terr, Gold St., Hanslope, Milton Keynes, Bucks, MK19 7LX [IO92OC, SP84]
GM0 FQV J Black, Solway View, Carlisle Rd, Annan, DG12 6QX [IO84JX, NY26]
G0 FQW S M Humphreys, 11 Cedar Cl, Ashby Fields, Daventry, NN11 5TY [IO92KG, SP56]
G0 FQX L Key, 91 Dexter Ave, Oldbrook, Milton Keynes, MK6 2RZ [IO92PA, SP83]
G0 FQZ C F Emblen, 4 Rosemary Dr, Gee Cross, Hyde, SK14 5EX [IO83XK, SJ99]
G0 FRB S Gresty, 4 Palace Rd, Sale, M33 6WU [IO83TK, SJ79]
GM0 FRC S S Waterall Falkirk RC, 3 Wavell St., Grangemouth, FK3 8TG [IO86DA, NS98]
G0 FRD S A Baldwin, 86 Central Ave, Los Gatos, Ca 95032, USA
G0 FRE Details withheld at licensee's request by SSL.
GM0 FRH T H Currie, 11 Deveron Rd, Troon, KA10 7ED [IO75QN, NS33]
GM0 FRI J S Cooper, 5E Hanover Ct, North Frederick Path, Glasgow, Lanarkshire, G1 2BQ [IO75VU, NS56]
G0 FRK Details withheld at licensee's request by SSL.[Op: G W Turton. Station located near Widnes.]
G0 FRL R Lowe, 12 Cavenham Gr, Bolton, BL1 4UA [IO83SO, SD70]
G0 FRM H A White, 2 Honeycroft, Welwyn Garden City, AL8 6HR [IO91VT, TL21]
G0 FRN D N E Oakes, 56 Middle Way, Chinnor, OX9 4TP [IO91NQ, SP70]
G0 FRO A J Medcalf, The Old School, School Ln, Harwell, Oxon, OX11 0ES [IO91IO, SU48]
G0 FRQ P Charlesworth, 6 Oakfield Ave, Barmby on The Marsh, Goole, DN14 7HH [IO93MR, SE62]
G0 FRR M J Owen obo Flight Refuelling, 3 Canford View Dr, Wimborne, BH21 2UW [IO90AT, SU00]
G0 FRS I F Ireland Farnborough, Contest Group, 118 Mytchett Rd, Mytchett, Camberley, GU16 6ET [IO91PG, SU85]
GM0 FRT Details withheld at licensee's request by SSL.
G0 FRU C R Osbourn, Bourn Bungalow, Back Ln, Burrough Green, Newmarket, Suffolk, CB8 9NB [JO02EE, TL65]
G0 FRV S A Adams, 63 Turnbull Dr, Leicester, LE3 2JU [IO92JO, SK50]
G0 FRX G D Cowling, Laissez Faire, Reedness, Goole, DN14 8ET [IO93OQ, SE72]
G0 FRY J D Walker, Wildersley Farm, Holbrook Rd, Belper, DE56 1PD [IO93GA, SK34]
G0 FRZ R Swinney, 73 Station Ave North, Fence Houses, Fencehouses, Houghton-le-Spring, DH4 6HT [IO94FU, NZ35]
G0 FSA Details withheld at licensee's request by SSL.
G0 FSB S A Barton, 22 Chapel Hill, Cromford, Matlock, DE4 3QG [IO93FC, SK35]
G0 FSD Details withheld at licensee's request by SSL.
G0 FSF D C Cawser, 18 Croft Cl, Rolleston on Dove, Burton on Trent, DE13 9AF [IO92EU, SK22]
G0 FSG C Easton, Dallmore, Lockwood Beck Rd, High Stanghow, Saltburn Cleveland, TS12 3LE [IO94MM, NZ61]
GM0 FSH W Mackenzie, 4 Riverview Terr, Grahamsdyke Rd, Bo'Ness, W Lothian, EH51 9ED [IO86EA, NT08]
G0 FSJ Details withheld at licensee's request by SSL.
G0 FSL T D Fleet, 51 The Cres, Boney Hay, Walsall, WS1 2DA [IO92AN, SP09]
G0 FSO Details withheld at licensee's request by SSL.
G0 FSP J A Pears, Bramble Corner, East Green, Kelsale, Saxmundham, IP17 2PH [JO02SF, TM46]
G0 FSR K W Hamer, 18 Moor Hall Ln, Stourport on Severn, DY13 8RA [IO82UI, SO87]
GM0 FSW N McAllister, 36 Kinneff Cres, Dundee, DD3 9RG [IO86LL, NO33]
G0 FSX G R Barker Read, 170 Hopgarden Rd, Tonbridge, TN10 4QY [JO01DF, TQ54]
G0 FSY D L Macdonald, Lincoln Lodge, Chalfont Ln, Chorleywood, Rickmansworth Herts, WD3 5PR [IO91RP, TQ09]
GM0 FSZ E Sandilands, 12 Kerr Ct, Girvan, KA26 0BP [IO75NF, NX19]
G0 FTA L Pearce, 4 Long Crest, Carleton, Pontefract, WF8 2QT [IO93IQ, SE42]
G0 FTB L J Hubbert, 68 Leicester Ave, Cliftonville, Margate, CT9 3DB [JO01QJ, TR37]
G0 FTD Details withheld at licensee's request by SSL.
GM0 FTG R Y H Hill, 43 Gunn Rd, Grangemouth, FK3 8RF [IO86DA, NS98]
GM0 FTH H A Livingston, Flat 3, 40 Chalmers St., Dunfermline, KY12 8DF [IO86GB, NT08]
G0 FTI R W Hughes, 46 The Boundary, Oldbrook, Milton Keynes, MK6 2HT [IO92PA, SP83]
GM0 FTJ N Mitchinson, 85 Forest Rd, Selkirk, TD7 5DD [IO85NN, NT42]
GM0 FTL M P Livingston, 2 Primrose Ave, Rosyth, Dunfermline, KY11 2SS [IO86GB, NT18]
G0 FTM Details withheld at licensee's request by SSL.
G0 FTN Details withheld at licensee's request by SSL.
G0 FTP F J Castle, 51 Wildern Ln, East Hunsbury, Northampton, NN4 0SN [IO92NE, SP75]
G0 FTS Dr A F Gerrard 1st Timperley, 13 Wentworth Ave, Timperley, Altrincham, WA15 6NG [IO83UJ, SJ78]
G0 FTU C W Jones, Surrey Assembly Hall, Brickhouse Ln, South Godstone, Godstone, Surrey, RH9 8JW [IO91XF, TQ35]
G0 FTW J W Bullard, 17 Granville Ave, Barnstaple, EX32 7AH [IO71XC, SS53]
GM0 FTX I B Birkett, 25 Darnhall Cres, Craigie, Perth, PH2 0HH [IO86GJ, NO12]
G0 FTY M E Bannister, 31 Gr Rd, Lee on The Solent, PO13 9JA [IO90JT, SU50]
GJ0 FTZ M G S Thomson, 20 Rosemount Est, James Rd, St. Saviour, Jersey, JE2 7RR
GM0 FUA J K Rae, 45 St. Johns Way, Borrowstoun, Boness, EH51 9JD [IO86EA, NT08]
G0 FUD J T Carter, 34 Healdswood St., Skegby, Sutton in Ashfield, NG17 3FR [IO93ID, SK46]
G0 FUE D W Denton, 7 Uplands Ave, East Ayton, Scarborough, YO13 9EU [IO94SG, SE98]
G0 FUG T R Lund, The Cottage, Ruckland Corner, Burwell, Louth, LN11 8PR [JO03AG, TF37]
G0 FUH R G Douglas, 5 Portnalls Rd, Coulsdon, CR5 3DD [IO91WH, TQ25]
G0 FUI P C Abbott, 60 Slonk Hill Rd, Shoreham By Sea, BN43 6HY [IO90UU, TQ20]
G0 FUM A M Hardiman, 57 Roedean Rd, Worthing, BN13 2BT [IO90TT, TQ10]
G0 FUN Details withheld at licensee's request by SSL.[APAU Contest Group, c/o G4WVX.]
G0 FUQ Details withheld at licensee's request by SSL.
GW0 FUR D Buckle, 36 Yewberry Ln, Malpas, Newport, NP9 6WL [IO81MO, ST39]
G0 FUS P G Fry, 2A Priors Way, Olivers Battery, Winchester, Hants, SO22 4HJ [IO91HB, SU42]
G0 FUU P S Firmin, 25 The Heights, Hastings, TN35 5EP [JO00HU, TQ81]
G0 FUV J F Mills, Greenacre, Brittenden Ln, Waldron, Heathfield, TN21 0RL [JO00CX, TQ52]
G0 FUW S Hartley, 5 Sydenham Building, Bath, BA2 3BS [IO81TJ, ST76]
G0 FUY A Firth, 10 Holroyd Hill, Wibsey, Bradford, BD6 1PQ [IO93CS, SE13]
G0 FUZ C E Muten, 25 Lime Gr, Royston, SG8 7DJ [IO92XB, TL34]
G0 FVB G T Stokes, 27 Wetherden St., Leyton, London, E17 8EH [IO91XN, TQ38]
G0 FVC J A Newman, 161 Junction Rd, Archway, London, N19 5PZ [IO91WN, TQ28]
G0 FVD R W N Nichol, 10 Grangefield Terr, New Rossington, Doncaster, DN11 0LT [IO93LL, SK69]
G0 FVE M J Barnes, Gloribee, 20 Smithtyne Ave, Dereham, NR19 1HW [JO02LQ, TF91]
G0 FVF A Howman, 32 Dereham Rd, Pudding Norton, Fakenham, NR21 7NA [JO02KT, TF92]
G0 FVG P Wilson, Mill Lodge, Mill Ln, Brininghám, Melton Constable, NR24 2QA [JO02MU, TG03]
G0 FVH D S Dolling, 10 Ribbledale, London Colney, St. Albans, AL2 1TD [IO91UR, TL10]
G0 FVI A J Gilfillan, 15 Willow Rd, Carlton, Nottingham, NG4 3BH [IO92LX, SK64]
GM0 FVJ L L Martin, 17 King St., Perth, PH2 8HR [IO86GJ, NO12]
G0 FVK Details withheld at licensee's request by SSL.
G0 FVN P Johnston, 11 Rainsborough Ave, Knottingley, WF11 8RP [IO93IQ, SE42]
G0 FVO J Leach, 19 Fyfe Cres, Baildon, Shipley, BD17 6DR [IO93CU, SE13]
G0 FVS J A Jackson, 165 Hall St., Briston, Melton Constable, NR24 2LQ [JO02MU, TG03]
G0 FVU M A Smith, 31 Cromford Rd, Crich, Matlock, DE4 5DJ [IO93GC, SK35]
G0 FVV D A S Tideswell, 11 Eyam Cl, Oak Tree Ln, Mansfield, NG18 3QS [IO93KD, SK56]
G0 FVW A G Dent, 3A Chestnut Walk, Beachamwell, Swaffham, Norfolk, PE37 8AZ [JO02HO, TF70]
G0 FVX Details withheld at licensee's request by SSL.[Op: A Barnes. Station located near Burton-on-Trent.]
G0 FVY L H Underseth, 35 Lydiate Park, Liverpool, L23 1XL [IO83ML, SD30]
G0 FVZ B Beardsley, 40 Lakeside Ave, Long Eaton, Nottingham, NG10 3GJ [IO92IV, SK43]
G0 FWD J L Adcock, 102 Richmond Way, Newport Pagnell, MK16 0LH [IO92PB, SP84]
G0 FWF D G Stanton, 4 Churchill Ct, Connaught Rd, London, N4 4NU [IO91WN, TQ38]
G0 FWH W R Taylor, 5 Wilson Cl, Danesmoor, Chesterfield, S45 9RU [IO93HD, SK46]
G0 FWI Details withheld at licensee's request by SSL.
GM0 FWJ V Reilly, 16 Foot Pl, Rosyth, Dunfermline, KY11 2DF [IO86GA, NT18]
G0 FWK Details withheld at licensee's request by SSL.[Op: M W Cooper. Station located in Manchester.]
G0 FWL Details withheld at licensee's request by SSL.
G0 FWM D Foy, 49 Revidge Rd, Blackburn, BB2 6JH [IO83RS, SD62]
G0 FWO D E T Cocks, 20 Woodchurch Cl, Chatham, ME5 7LW [JO01GI, TQ76]
G0 FWP J F Purvess, 389 Otley Old Rd, Leeds, LS16 6BX [IO93EU, SE23]
G0 FWR E Jones Widnes Runcn AR, 40 Fenwick Ln, Halton Lodge, Runcorn, WA7 5YU [IO83PH, SJ58]
G0 FWU Details withheld at licensee's request by SSL.
GM0 FWV Details withheld at licensee's request by SSL.
GM0 FWY P D Gray, 71 Craigmount Ave North, Edinburgh, EH4 8DT [IO85IW, NT17]
GW0 FWZ E G Imianowski, 11 Durand Rd, Caldicot, Newport, NP6 4BZ [IO81OO, ST48]
GW0 FXA G D Lomax, 2 Glyndwr, Grosmont, Abergavenny, NP7 8EP [IO81NW, SO42]
GW0 FXC R D Rowland, 60 Ombersley Rd, Newport, NP9 3EE [IO81LN, ST28][Op: R D (Bob) Rowland.]
G0 FXD D C Harrison, 41 North End Ln, Malvern, WR14 2NG [IO82UC, SO74]
G0 FXI R J Oram, 4 Hardy Ave, Duckmoor Rd, Ashton Gate, Bristol, BS3 2BP [IO81QK, ST57]
GM0 FXJ J Corrigan, 20 Kennoway Pl, Broughty Ferry, Dundee, DD5 3HT [IO86NL, NO43]
G0 FXK J Paxton, 27 Holborn View, Leeds, LS6 2RD [IO93FT, SE23]
G0 FXL D J Garratt, 99 Chaytor Rd, Polesworth, Tamworth, B78 1JS [IO92EO, SK20]
G0 FXO R Warren, 2 Kirkstone Rd, Bedworth, Nuneaton, CV12 8SA [IO92GL, SP38]
G0 FXQ I M Drury, 14 James Way, Hucclecote, Gloucester, GL3 3TE [IO81VU, SO81]
G0 FXR R Clark, 9 Kensington Ave, Flatts Ln Normanb, Middlesbrough, TS6 0QQ [IO94KN, NZ51]
G0 FXS K Rutter, 31 Burlington Dr, Western Downs, Stafford, ST17 9UL [IO82WT, SJ92]
G0 FXT C Richardson, 47 Leighton Cl, Crossgates, Scarborough, YO12 4LA [IO94SF, TA08]
G0 FXU Details withheld at licensee's request by SSL.[Op: M J Rose. Station located in Stanmore.]
G0 FXV C Belshaw, 38 Cobmoor Ave, Billinge, Wigan, WN5 7EQ [IO83PM, SD50]
G0 FXW Details withheld at licensee's request by SSL.
G0 FXX A F Bray, 6 Devoke Ave, Worsley, Manchester, M28 7EW [IO83TM, SD70]
G0 FXY M O Smith, 10 Graham Rd, Kirk Sandall, Doncaster, DN3 1HY [IO93LN, SE60]
G0 FXZ Details withheld at licensee's request by SSL.
GJ0 FYB R E Le Jehan, Sundora, 4 St. Marys Village, St. Mary, Jersey, JE3 3BQ
G0 FYC D Allman, 6 Ash Way, Ash Bank, Bucknall, Stoke on Trent, ST2 9DZ [IO83WA, SJ94]
G0 FYD I R McCabe, 99 Edgeway Rd, South Shore, Blackpool, FY4 3NH [IO83LS, SD33]
G0 FYE B Moss, 22 Battersby St., Ince, Wigan, WN2 2NA [IO83QN, SD60]
GW0 FYF R W H Atkins, 7 Plas Ebbw, Northville, Cwmbran, NP44 1QJ [IO81LP, ST29]
G0 FYG C C Baier, 3 Howe Cl, Catmere End, Saffron Walden, CB11 4XH [JO02CA, TL43]
G0 FYH R C Butterworth, 2 Bare Ave, Morecambe, LA4 6BE [IO84NB, SD46]
G0 FYL M J Samuels, 63 Mill Rd, Okehampton, EX20 1PR [IO80AR, SX59]
G0 FYM R T Oxley, 60 Valley View, South Elmsall, Pontefract, WF9 2DD [IO93IO, SE41]
G0 FYN Details withheld at licensee's request by SSL.
GW0 FYO A T Williams, The Countryman, Scurlage, Swansea, W Glam, SA3 1AZ [IO71VN, SS48]
G0 FYP C G Holland, 44 Brightstowe Rd, Burnham on Sea, TA8 2HP [IO81MG, ST35]
G0 FYQ D Armstrong, The Bungalow, Hollin Hall, Hardcastle Crags, Hebden Br'Ge, W York, HX7 7AP [IO83XS, SD92]
GU0 FYR Details withheld at licensee's request by SSL.
G0 FYS J J Keeley, 3A St. Marks Rd, Huyton, Liverpool, L36 0XA [IO83OJ, SJ49]
G0 FYU T P Liggins, 55 Kirkland St., Pocklington, York, YO4 2BX [IO93OW, SE84]
G0 FYV T H G Cox, 105 Hollytrees, Bar Hill, Cambridge, CB3 8SG [JO02AG, TL36]
G0 FYW B S Morrin, 26 Anstable Rd, Bare, Morecambe, LA4 6TQ [IO84NB, SD46]
G0 FYX S W Swain, 35 Mavis Cres, Havant, PO9 2AE [IO90MU, SU70]
G0 FYY A E Heath, 20 Birch Ave, Thorton, Thornton Cleveleys, FY5 2HS [IO83LV, SD34]
G0 FYZ P L Sharpe, 10 Brentwood Ave, Poulton-le-Fylde, FY6 7EA [IO83MU, SD33]
G0 FZA N Dickinson, 27 Somerset Ave, Lancaster, LA1 4BW [IO84OA, SD46]
G0 FZB J T Atherfold, 42 Mansell Rd, Shoreham By Sea, BN43 6GP [IO90UU, TQ20]
G0 FZC G C Palmer, 29 Lindsay Rd, Leicester, LE3 2EJ [IO92KO, SK50]
G0 FZD C B Nichols, 39 Sandringham Ave, Whiston, Rotherham, S60 4DS [IO93IJ, SK49]
G0 FZE B P Gould, 2 Parkdale, Ibstock, LE67 6JW [IO92HQ, SK41]
G0 FZF C T Guinan, 20 Putney Cl, Oldham, OL1 2JS [IO83WN, SD90]
G0 FZH D R R Moore, The Moorings, 12 Newtown Park, Newtown, Langport, TA10 9TF [IO81OB, ST42]
G0 FZL T A Burns, 41 Salmesbury Ave, Blackpool, FY2 0PR [IO83LU, SD33]
GM0 FZM K A Blabey, 33 Deantown Ave, Whitecraig, Musselburgh, EH21 8NS [IO85LW, NT36]
G0 FZN F Rogers, 43 Beacon Rd, Billinge, Wigan, WN5 7HF [IO83PM, SD50]
G0 FZO C C Lee, 2 Thompson Dr, Strensall, York, YO3 5ZN [IO94LA, SE66]
G0 FZP C J Price, 15 Gores Park, High Littleton, Bristol, BS18 5YG [IO81RH, ST65]
G0 FZQ J Storey, 26 Orwell Dr, Weston Heath, Birmingham, B38 8HZ [IO92AJ, SP07]
GU0 FZS J Vaughan, -le-Phare, Bruce Ln, St. Peter Port, Guernsey, Channel Islands, GY1 2LU
GI0 FZT P E Frend, Tara, 49 Carlton Heights, Bangor, Co Down, BT19 6ZB [IO74EP, J58]
G0 FZU J Tinsley, 21 Peckforton View, Kidsgrove, Stoke on Trent, ST7 4TA [IO83VB, SJ85]
G0 FZV J M Gille, Qsl via On9Caa, Aster Straat 4, 3660 Opglabbeek, Belgium
G0 FZX R G Emmett, 3 Lodge Cl, Brighstone, Newport, PO30 4BX [IO90HP, SZ48]
GW0 FZY Dr J D Woolgar, 292 Van Riebeeck Ave, Verwoerdburg, 0157, South Africa
G0 FZZ J E Foster, 23 Shrewsbury St., Hartlepool, TS25 5RQ [IO94JQ, NZ53]
G0 GAB L S Chadwick, 47 Red Hall Ln, Leeds, LS14 1NT [IO93GU, SE33]
G0 GAC Details withheld at licensee's request by SSL.
G0 GAD Details withheld at licensee's request by SSL.
G0 GAG M Cowley, 46 Mapletoft Ave, Mansfield Woodhouse, Mansfield, NG19 8HT [IO93JE, SK56]
GW0 GAI M S Green, 7 Lon yr Afon, Llanbradach, Caerphilly, CF8 3NZ [IO81HN, ST08]
G0 GAJ F C D Bunce, 81 Royal Navy Ave, Plymouth, PL2 2AQ [IO70VJ, SX45]
GM0 GAK J C McFarlane, 40 Lennox St., Renton, Dumbarton, G82 4LF [IO75RX, NS37]
G0 GAL E Howells, 5 Bowland Cl, Telford, TF3 5HE [IO82SQ, SJ60]

G0 GAO K R Petter, 10 Holkam Cl, Tilehurst, Reading, RG3 6BZ [IO91LJ, SU66]
G0 GAP W N Meecham, 38 Douglas Rd West, Stafford, ST16 3NX [IO82WT, SJ92]
G0 GAQ B Johnson, 67 Nursery Ln, Kingsthorpe, Northampton, NN2 7PT [IO92NG, SP76]
G0 GAR A Robinson, 28 Briarsleigh, Wildwood, Stafford, ST17 4QP [IO82WS, SJ92]
GM0 GAT A H Thomasson, 1 Eastside Green, Westhill, Skene, Westhill, AB32 6XY [IO87UD, NJ80]
G0 GAU M J Tyson, 9 Whitehouse Garden, Millbrook, Southampton, Hants, SO1 4FD [IO90HW, SU41]
GM0 GAV G A Taylor, 27 Cairnie Cres, Glencarse, Perth, PH2 7ND [IO86II, NO12]
G0 GAW V Richards, Wayside, Penwithick Rd, Penwithick, St Austell, Cornwall, PL26 8UH [IO70OI, SX05]
G0 GAX W F L Lawson, The Wilderness, Forton, nr Chard, Somerset, TA20 2NB [IO80MU, ST30]
G0 GAZ Details withheld at licensee's request by SSL.
GD0 GBA J G Carroll, Lingague Cottage, Ballakilpheric, Colby, Isle of Man, IM9 4BR
G0 GBC P A Dempster, 10 St. Pauls Way, Tickton, Beverley, HU17 9RW [IO93TU, TA04]
G0 GBD M B Thomson, 14 Lanes End, Totland Bay, PO39 0AL [IO90FQ, SZ38]
G0 GBE C J Thompson, 135 Stafford Rd, Bloxwich, Walsall, WS3 3PG [IO82XP, SJ90]
G0 GBF Dr G T Knowlson, 22 Norris Rd, Sale, M33 3QR [IO83UJ, SJ79]
G0 GBG B H Wilkinson, 3 Friarage Ave, Northallerton, DL6 1DZ [IO94GI, SE39]
GM0 GBH P M Young, 4 Primrose Ave, Rosyth, Dunfermline, KY11 2SS [IO86GB, NT18]
G0 GBI G K Loake, 81 Duchess Rd, Bedford, MK42 0SE [IO92SC, TL04]
G0 GBK L Smith, 87 Penrose House, Penrose St., London, SE17 3EA [IO91WL, TQ37]
G0 GBL D H Taylor, 38 Seward Rd, Badsey, Evesham, WR11 5HQ [IO92BC, SP04]
G0 GBN J E Henshaw, 7 Gorsefield Cl, Bromborough, Wirral, L62 6BU [IO83MH, SJ38]
G0 GBO Details withheld at licensee's request by SSL.
G0 GBP D W Porter, Birchgreen, 146 Inskip, Skelmersdale, WN8 6JX [IO83ON, SD40]
G0 GBQ H Whitfield, 81 Tenter Balk Ln, Adwick-le-St., Doncaster, DN6 7EE [IO93JN, SE50]
G0 GBR Details withheld at licensee's request by SSL.
G0 GBS G T D Jarman, 212 High St., Clapham, Bedford, MK41 6BS [IO92SD, TL05]
G0 GBT E H Hougham, 44 Foxborough Hill, Woodnesborough, Sandwich, CT13 0NU [JO01PG, TR35]
G0 GBU M R Goddard, Wits End, Stillington, York, North Yorks, YO6 1LF [IO94KC, SE56]
G0 GBV D Hackett, 2A Lancaster Way, Market Deeping, Peterborough, Cambs, PE6 8LA [IO92UQ, TF11]
G0 GBW C B Oswald, 3 Belvedere Dr, Ganstead Ln, Bilton, Hull, HU11 4AX [IO93US, TA13]
G0 GBX W A Jones, 2 Northfield Rd, Hull, HU3 6TL [IO93TR, TA02]
G0 GBY M T Francis, 132 Rectory Ave, Rochford, Essex, SS4 3TB [JO01IO, TQ89]
G0 GBZ N Millar, 25 Copperfield Cl, Kettering, NN16 9EW [IO92PK, SP88]
G0 GCA M G Collins, 147 Cheyne Way, Farnborough, GU14 8SD [IO91OH, SU85]
G0 GCB Details withheld at licensee's request by SSL.
G0 GCD Details withheld at licensee's request by SSL.
GM0 GCF K Joynson, Roadside Cottage, Craiglemine, Whithorn, Newton Stewart, Wigtownshire, DG8 8NE [IO74SR, NX43]
G0 GCH Details withheld at licensee's request by SSL.
G0 GCI M P Wright, Randalls, Cryals Rd, Matfield, Tonbridge, TN12 7HL [JO01ED, TQ63]
G0 GCK M A Johnson, 2 de Gaunte Rd, Brompton, Northallerton, DL6 2QD [IO94GI, SE39]
G0 GCL Details withheld at licensee's request by SSL.
G0 GCM R H Morgan, 18 Bridge Meadow, Great Sutton, South Wirral, L66 2LF [IO83MG, SJ37]
G0 GCN E R Flower, 19 Hartington Rd, Salisbury, Wilts, SP2 7LG [IO91CB, SU13]
GM0 GCO B H Carson, 46 Tweed Dr, Bearsden, Glasgow, G61 1EJ [IO75TV, NS57]
G0 GCP Details withheld at licensee's request by SSL.
G0 GCQ J R Rivers, 8 Allan Rd, Seasalter, Whitstable, CT5 4AH [JO01LI, TR06]
GI0 GCR C G Robinson, 28 Rectory Rd, Doagh, Ballyclare, BT39 0PT [IO64XR, J29]
G0 GCS D Gent, 3 Brampton St., Brampton, Barnsley, S73 0XR [IO93HM, SE40]
G0 GCW Details withheld at licensee's request by SSL.
G0 GCX Details withheld at licensee's request by SSL.
G0 GCY L J Jackson, 36 Ramsden Ave, Langold, Worksop, S81 9PB [IO93KJ, SK58]
G0 GCZ Details withheld at licensee's request by SSL.
G0 GDA N D F Brown, 14 Madison Ave, Bispham, Blackpool, FY2 9HE [IO83LU, SD34]
GM0 GDD A McMillan, 33 Lachlan Cres, Linburn, Erskine, PA8 6HJ [IO75SV, NS47]
G0 GDG D A Ellson, 17 Central Ave, Northfield, Birmingham, B31 4HN [IO92AJ, SP07]
GM0 GDH N S Macgilp, 5 Ormonde Ave, Glasgow, G44 3QU [IO75UT, NS55]
GW0 GDI N C Erskine, 302 Caerphilly Rd, Birchgrove, Cardiff, CF4 4NS [IO81JM, ST18]
G0 GDJ J A Brooks, 2 Nettleton Rd, Caistor, Market Rasen, LN7 6NB [IO93UL, TA10]
GW0 GDK G J Reynolds, Summerfield, Padeswood Rd, Buckley, CH7 2JN [IO83LD, SJ26]
G0 GDL M Rodgers, 22 Harewood Rd, Allestree, Derby, DE22 2JN [IO92GW, SK33]
G0 GDM D D McNab, 11 Ardern Cl, Coombe Dingle, Bristol, BS9 2QT [IO81QL, ST57]
G0 GDN R Thorley, 326 Lapwing Ln, Manchester, M20 6UW [IO83VK, SJ89]
G0 GDO Details withheld at licensee's request by SSL.
GI0 GDP Details withheld at licensee's request by SSL.
G0 GDQ H L Wilson, 23 Wadham Rd, Woodthorpe, Nottingham, NG5 4JB [IO92KX, SK54]
G0 GDS J L Hyde, 7 Wandells View, Brantingham, Brough, HU15 1QL [IO93RS, SE92]
G0 GDT W H Clapp, 21 Dronsfield Rd, Fleetwood, FY7 7BW [IO83LW, SD34]
G0 GDU J D Crowder, Sunnycroft, Bleasby Rd, Thurgarton, Nottingham, NG14 7FW [IO93MA, SK64]
G0 GDW Details withheld at licensee's request by SSL.
G0 GDY Details withheld at licensee's request by SSL.
G0 GDZ Details withheld at licensee's request by SSL.
G0 GEA A C Wheater, The Bungalow, Green Ln, Carleton, Pontefract, WF8 3NW [IO93IQ, SE42]
G0 GEB R P East, 25 Bridgemont, Whaley Bridge, Stockport, Ches, SK12 7PB [IO83XI, SJ98]
G0 GEC R P Merrell GEC Marconi Club ARS, 40 Fanton Walk, Wickford, SS11 8QT [JO01GO, TQ79]
GM0 GEE L T Stapleton, 22 Ashie Rd, Inverness, IV2 4EN [IO77VK, NH64]
G0 GEF P W C Chipman, 15 Park Rd, Baddesley Ensor, Atherstone, CV9 2DS [IO92EN, SP29]
G0 GEH L J Howarth, 12 Thomas St., Hemsworth, Pontefract, WF9 4AY [IO93HO, SE41]
GW0 GEI S R Jones, Bron Heulog, Rhostrewea, Llangefni, Ynys Mon, North Wales, LL77 7AJ [IO73TG, SH47]
G0 GEK P J Blake, 30 St Cuthberts Ct, St Cuthberts Pl, Cleveland Ave, Darlington, DL3 7UX [IO94FM, NZ21]
G0 GEL J Pruden, 77 Browning Cres, Fordhouses, Wolverhampton, WV10 6BQ [IO82WO, SJ90]
GW0 GEM Details withheld at licensee's request by SSL.
G0 GEO S G Pruden, 77 Browning Cres, Fordhouses, Wolverhampton, WV10 6BQ [IO82WO, SJ90]
G0 GEP G E Perry, 11 The Woodlands, Upton upon Severn, Worcester, WR8 0PQ [IO82VB, SO84]
G0 GEQ J Rogerson, 15 South Edge, Shann Park, Keighley, BD20 6JZ [IO93AU, SE04]
G0 GER R Knighton, 262 Victoria Rd West, Thornton Cleveleys, FY5 3QB [IO83LV, SD34]
G0 GES L C Thatcher, 126 Waltham Ave, Hayes, UB3 1TE [IO91SM, TQ07]
G0 GEU S G Holt, 22 Sulby Gr, Morecambe, LA4 6HD [IO84NB, SD46]
GW0 GEV T R Hurst, Woodside, Parc Seymour, Penhow, Gwent, NP6 3AB [IO81OO, ST49]
G0 GEY Details withheld at licensee's request by SSL.
G0 GEZ D Creek, 4 Lodge Cl, Brighstone, Newport, PO30 4BX [IO90HP, SZ48]
G0 GFA C E Armitage, Copperfields, 19 Park Rd, Barlow, Selby, YO8 8ES [IO93LS, SE62]
G0 GFB Details withheld at licensee's request by SSL.
G0 GFC R S Johnson, 3 Lance Dr, Chase Terr, Burntwood, WS7 8FA [IO92AQ, SK01]
G0 GFD K Venn, 28 Streamleaze, Titchfield Common, Fareham, PO14 4NP [IO90IU, SU50]
G0 GFE B E Keechan, 24 Tolley Rd, Birchen Coppice, Kidderminster, DY11 7EW [IO82UI, SO87]
G0 GFF R Gibson, 6 Dryden Rd, Hartlepool, TS25 4HR [IO94JQ, NZ43]
G0 GFG M M Hill, 86 Commercial St., Willington, Crook, DL15 0AA [IO94DR, NZ23]
GW0 GFH R E Bennett, 75 Ashcroft Cres, Fairwater, Cardiff, CF5 3RL [IO81JL, ST17]
G0 GFI C R Burrows, 29 Hampden Rd, Malvern Link, Malvern, WR14 1NB [IO82UD, SO74]
G0 GFJ B Hill, 23 Honor Rd, Prestwood, Great Missenden, HP16 0NJ [IO91PQ, SP80]
G0 GFK J W G Denford, 10 Churchill Rd, Bideford, EX39 4HG [IO71VA, SS42]
G0 GFL A M M Marriott, 8 Melway Gdns, Child Okeford, Blandford Forum, DT11 8EP [IO80VV, ST81]
GW0 GFN D H A Anderson, Penrheol Farm, Meidrim, Carmarthen, Dyfed, SA33 5NX [IO71SU, SN32]
G0 GFP A J Fowler, 78 Beckingham Rd, Guildford, GU2 6BU [IO91QF, SU95]
G0 GFQ K J Martin, 21 All Saints Cl, Weybourne, Holt, NR25 7HH [JO02NW, TG14]
G0 GFR B Holloway, 10 Verulam Rd, Parkstone, Poole, BH14 0PP [IO90AR, SZ09]
G0 GFT M Hawkins, 5 Springfield Ave, Accrington, BB5 0EZ [IO83TR, SD72]
G0 GFU J A Dodsworth, 29 Arbor Ln, Winnersh, Wokingham, RG41 5JE [IO91NK, SU77]
GM0 GFV J M Angiolini, 58 Muirfield Rd, Cumbernauld, Glasgow, G68 0EX [IO85AX, NS77]
G0 GFX G E Carrington, 22 Abbey Rd, Ulceby, DN39 6TJ [IO93UO, TA11]
G0 GFY C F James, 180 Mitcham Rd, Croydon, CR0 3JF [IO91WJ, TQ36]
G0 GFZ P C Taylor, 18 Redriff Rd, Romford, RM7 8HD [JO01BO, TQ59]
G0 GGA R J Parkins, 12 Ashleigh Gdns, Barwell, Leicester, LE9 8LE [IO92HN, SP49]
G0 GGB A E Norman, Providence House, North Market Rd, Winterton on Sea, Great Yarmouth, NR29 4BH [JO02UR, TG41]
G0 GGE P W Burrow, 9 Minsmere Rd, Belton, Great Yarmouth, NR31 9NX [JO02TN, TG40]
G0 GGF C P Holloway, 6A New Rd, Reepham, Norwich, NR10 4LP [JO02NS, TG12]
G0 GGG N M Rogers, 34 Broadway, Warminster, BA12 8EB [IO81VE, ST84]
G0 GGH F L Sell, 17 Auriel Ave, Dagenham, Essex, RM10 8BS [JO01CN, TQ58]
G0 GGJ J Mortimer, 25 Leycester Dr, Scale Hall, Lancaster, LA1 5HW [IO84OB, SD46]
G0 GGL D P Ellins, 52 Littlewood, Stokenchurch, High Wycombe, HP14 3TF [IO91NP, SU79]
G0 GGM Dr B W Fitzsimmons, Gurney Vale, 64 Belle Vue Rd, Wivenhoe, Colchester, CO7 9LD [JO01LU, TM02][WAB book no 8211. Interested in packet radio, computers in radio generally and HF.]
G0 GGN R B Samways, 119 Station Rd, Quainton, Aylesbury, HP22 4BX [IO91MU, SP71]
G0 GGQ G W Redding, 50 Great Hill, Shefford, SG17 5EA [IO92UB, TL13]
G0 GGR G G Rackstraw, 11 Carbrooke Rd, Griston, Thetford, IP25 6QE [JO02KN, TL99]
G0 GGS J E Willis, Strassburger-Hof-St, 77620 Wolfach, Germany
G0 GGT B Hartley, 237 Barkerhouse Rd, Nelson, BB9 9NQ [IO83VU, SD83]
G0 GGU Details withheld at licensee's request by SSL.
G0 GGV B G Mulhall, 7 Drovers Pl, Westlands, Droitwich, WR9 9DB [IO82WG, SO86]
G0 GGX Details withheld at licensee's request by SSL.
G0 GHB J C Graham, Caxtonian, Brimbelow Rd, Hoveton, Norwich, NR12 8UJ [JO02QR, TG31]
G0 GHD N Houghton, 5 Heather Cl, Swanwick, Alfreton, DE55 1RU [IO93HB, SK45]
G0 GHE Details withheld at licensee's request by SSL.
GW0 GHF B D Williams, 10 Pant y Celyn Rd, Llandough, Penarth, CF64 2PG [IO81JK, ST17]
GW0 GHG D D Roberts, 16 Min y Mor, Ty Croes, Aberffraw, Ty Croes, LL63 5PQ [IO73SE, SH36]
G0 GHH P G Cross, Balls Farm Cottage, Musbury Rd, Axminster, EX13 5TT [IO80LS, SY29]
G0 GHK G Boothroyd Finningley Amateur Radio Soc., Finningley Ar Society, 38 Ascot Ave, Cantley, Doncaster, DN4 6HE [IO93KM, SE60]
G0 GHL T S Reeves, 50 Park Rd, Willaston, Nantwich, CW5 6PN [IO83SB, SJ65]
G0 GHM D Coxon, 7 Kingston Way, Nailsea, Bristol, BS19 2RA [IO81OK, ST47]
GM0 GHN T A Taylor, The Cottage, The Greens, Mosstowie, Elgin, Morayshire, IV30 3TU [IO87HP, NJ16]
G0 GHO W Small, 24 Barrowdale Cl, Exmouth, EX8 5PN [IO80HP, SY08]
G0 GHP A V Wood, 150 Hainault Ave, Westcliff on Sea, SS0 9EX [JO01IN, TQ88]
GW0 GHQ Details withheld at licensee's request by SSL.[Op: K G Mottley. Station located near Brecon. IO81HW.]
G0 GHR I G Beacham, 26 Great Parks, Holt, Trowbridge, BA14 6QP [IO81VI, ST86]
G0 GHT R Pearce, Waycott New Inn Cro, Shebbear Beaworthy, Devon, EX21 5RU [IO70VU, SS40]
G0 GHV V G Hodskinson, Maplecroft, 51 Waterswallows Rd, Fairfield, Buxton, SK17 7JJ [IO93BG, SK07]
G0 GHW G H Whitehead, 16 Leafield Pl, Trowbridge, BA14 9TH [IO81VH, ST85]
G0 GHX R M L Biles, 164 Lonnen Rd, Pilford, Wimborne, BH21 7AZ [IO90AT, SU00]
G0 GHY Details withheld at licensee's request by SSL.
G0 GIA R A Keeley-Osgood, 35 Norse Walk, Corby, NN18 9DG [IO92PL, SP88]
GM0 GIB M J Gibb, 6 Tillybrig, Skene, Dunecht, Westhill, AB32 7BE [IO87TE, NJ70]
G0 GIC Details withheld at licensee's request by SSL.
G0 GID G P Renggli, 102 Stewart Rd, Bournemouth, BH8 8NX [IO90BR, SZ19]
G0 GIE D B Adams, 12 Milton Cres, The Straits, Sedgley, Dudley, DY3 3DR [IO82WM, SO99]
G0 GIF J L Catterson, 26B Whiston Ln, Huyton, Liverpool, Merseyside, L36 1TY [IO83OK, SJ49]
G0 GIG Details withheld at licensee's request by SSL.
GW0 GIH P J Thomas, 41 The Uplands, Brecon, LD3 9HT [IO81HW, SO02]
G0 GII B E Jackson, 11 Bourchier Way, Halstead, CO9 1AY [JO01HW, TL82]
G0 GIL J G Carter, 112 Landor Rd, Whitnash, Leamington Spa, CV31 2JZ [IO92FG, SP36]
G0 GIN K L Mills, 2 Edwardian Cl, Wootton, Northampton, NN4 6LX [IO92NE, SP75]
GW0 GIQ W J Smith, 41 Rockleigh Ave, Aberbargeod, Bargeod, Mid Glam, CF81 9BQ [IO81JQ, SO10]
G0 GIR Details withheld at licensee's request by SSL.
G0 GIS Details withheld at licensee's request by SSL.
G0 GIT M W Jinks, Caixa Postal 003, Campina Grande de Sul, Parana, Brasil, Cep 83430-000, X X
G0 GIU Details withheld at licensee's request by SSL.
G0 GIV Details withheld at licensee's request by SSL.
G0 GIW Details withheld at licensee's request by SSL.
G0 GIY D M Fripp, 16 Norman Rd, Ashford, TW15 1QP [IO91SK, TQ07]
G0 GIZ R A Smith, Freshfields, 5 Robinson Ave, Alford, LN13 0PW [JO03CG, TF47]
G0 GJA D H Wright, 55 Homefield Rd, Hemel Hempstead, HP2 4BZ [IO91SS, TL00]
G0 GJB J A Brown, 4 Downs Cote Ave, Westbury on Trym, Bristol, BS9 3JX [IO81QL, ST57]
G0 GJC M J H Hole, 10 Evans Gr, Marshalswick, St. Albans, AL4 9PJ [IO91US, TL10]
GW0 GJD R M Radcliffe, 3 Bryncerdin Rd, Newton, Swansea, SA3 4UB [IO71XN, SS68]
G0 GJE R C Cleverley, 22 The Tinings, Monkton Park, Chippenham, SN15 3LX [IO81WL, ST97]
G0 GJF R T Hicks, 39 Wickham Ave, Shirley, Croydon, CR0 8TZ [IO91XI, TQ36]
G0 GJG B W Phillips, 6 Greenways, Penkridge, Stafford, ST19 5HD [IO82WR, SJ91]
G0 GJH G J Hughes, 51 Kingsmead, Seaford, BN25 2HA [JO00BS, TV49]
G0 GJK Details withheld at licensee's request by SSL.
G0 GJL S T Moyses, Alte Heerstr 134, 41564, Kaarst, Germany
G0 GJM J F Rattigan, 4 Grosvenor St., Barrow in Furness, LA14 4AH [IO84JD, SD17]
G0 GJN T Roberts, 14 Glenpark Gdns, St George, Bristol, BS5 7NG [IO81RL, ST67]
G0 GJO F R Guggiari, 31 Stratfield Rd, Basingstoke, RG21 5RS [IO91KG, SU65]
G0 GJP Details withheld at licensee's request by SSL.
G0 GJQ W F J Etchells, 12 Green Ln Gdns, Gateshead, NE10 0AE [IO94FW, NZ26]
G0 GJR A B Beard, 99 Station Rd, Surfleet, Spalding, PE11 4DB [IO92WU, TF22]
G0 GJS Details withheld at licensee's request by SSL.
G0 GJV M J Goodey, 62 Rose Hill, Binfield, Bracknell, RG42 5LG [IO91OK, SU87]
G0 GJW A C Sale, 52 The Grove, North Cray, Sidcup, DA14 5NQ [JO01BK, TQ47]
G0 GJX F J Norton, 147 Wells Rd, Glastonbury, BA6 9AN [IO81PD, ST54]
G0 GJZ Details withheld at licensee's request by SSL.
G0 GKB G P Keat, Lowenna, 25 Alexandra Way, Crediton, EX17 2EA [IO80DT, SS80]
G0 GKC Details withheld at licensee's request by SSL.
G0 GKD K P Dunne, 4 Campden Rd, Benhall, Cheltenham, GL51 6AA [IO81WV, SO92]
G0 GKE L J Roffey, 2 Orchard Terr, Barnstaple, EX32 9DY [IO71XB, SS53]
GM0 GKF Details withheld at licensee's request by SSL.[Op: Fiona Vincent Stapleton, Ravenscourt, 23 Strathkinness High Road, St Andrews, Fife KY16 9UA]
G0 GKH D Birch, 32 Union St., Trowbridge, BA14 8RY [IO81VH, ST85]
G0 GKI C Crawford, 68 Harborough Ave, Sidcup, DA15 8HP [JO01BK, TQ47]
G0 GKK A H Pickering, Copper Beach Cottage, 5 West Farm Ct, Manor Rd, Consett, DH8 6TL [IO94CV, NZ15]
G0 GKL M A Stevens, 33 Langham Rd, Hastings, TN34 2JE [JO00HU, TQ81]
G0 GKN J R J G Dowse, 46 Nantwich Rd, Middlewich, CW10 9HG [IO83SE, SJ66]
G0 GKO G Lumsden, Spencer Buildings, Front St, Hutton Henry, Hartlepool, Cleveland, TS27 4RT [IO94HR, NZ43]
G0 GKP J F Bonner, 40 Lyles Rd, Cottenham, Cambridge, CB4 4QR [JO02BG, TL46]
GM0 GKR J R Stapleton, Ravenscourt, 23 Strathkinness High Rd, St. Andrews, KY16 9UA [IO86OI, NO41]
G0 GKS P R Baker, 49 Cowper St., Northampton, NN1 3QR [IO92NF, SP76]
G0 GKU H I Brugsch, 17 Kent Cl, Kidderminster, DY10 1NS [IO82VI, SO87]
G0 GKW Details withheld at licensee's request by SSL.
G0 GKY D Burton, 49 Aske Rd, Redcar, TS10 2BP [IO94LO, NZ62]
G0 GLA M A Hoppe, 67 Belmont Rd, Maidenhead, SL6 6LG [IO91PM, SU88]
GM0 GLD N Lee, 3 Braikley Ave, Tarves, Ellon, AB41 7PU [IO87VI, NJ83]
G0 GLG H L Torunski, Hamlyn, 12 Holm Way, Southwold, Bicester, OX6 9YL [IO91KV, SP52]
GM0 GLH T Mitchell, 44A Main St., Stoneyburn, Bathgate, EH47 8BY [IO85EU, NS96]
GW0 GLI L M Edwards, 23 Tan y Bryn, Llanbedr Dc, Llanbedr Dyffryn Clwyd, Ruthin, LL15 1AQ [IO83IC, SJ15]
G0 GLJ Details withheld at licensee's request by SSL.
G0 GLK G E Scott, 17 Norma Cres, Whitley Bay, NE26 2PD [IO95GA, NZ37]
G0 GLL Details withheld at licensee's request by SSL.
G0 GLM Details withheld at licensee's request by SSL.[Licencee George A Waterfall. Station located near Crowborough. Locator: JO01CB, WAB: TQ53.]
G0 GLO G Sims Glossop & D ARG, 85 Surrey St., Glossop, SK13 9AJ [IO93AK, SK09]
G0 GLQ P A Davenport, Millwright Arms, Coten End, Warwick, CV34 4NU [IO92FG, SP26]
G0 GLS J W Smithurst, 103 Cavendish Rd, Worksop, S80 2SN [IO93KH, SK57]
GW0 GLT Details withheld at licensee's request by SSL.
G0 GLU M M Fry, 1 Hebden Ave, Warwick, CV34 5XD [IO92FH, SP26]
GM0 GLV E McWhinnie, 34 Irvine Dr, North Clippens, Linwood, Paisley, PA3 3TB [IO75RU, NS46]
G0 GLW G L White, 7 Oakengates, Hanworth, Bracknell, RG12 7QJ [IO91OJ, SU86]
GW0 GLX M S D Mc Farland, Min Afon Garreg Faw, Groeslon, Caernarvon, Gwynedd, LL54 7ED [IO73UB, SH45]
G0 GLY W Brough, 35 The Knoll, Brixworth, Northampton, NN6 9HY [IO92NH, SP77]
G0 GLZ R P Sugden, 15 Elsie St., Goole, DN14 6DU [IO93NR, SE72]
G0 GMA P L Sweeting, 69 Old Roselyon Rd, St. Blazey, Par, PL24 2LN [IO70PI, SX05]
G0 GMB M B Baker, 25 Pentlands, Stony Stratford, Fullers Slade, Milton Keynes, MK11 2AF [IO92NB, SP73]
G0 GMC C R Cook, 45 Stonepound Rd, Hassocks, BN6 8PR [IO90WW, TQ31]
GM0 GMD T Astbury, 8 Auchenlay Holdings, Auchenlay, Dunblane, FK15 9NA [IO86AE, NN70]
G0 GME R Turner, 8 Maddocks St., Saltaire, Shipley, BD18 3JL [IO93CU, SE13]
GI0 GMG Details withheld at licensee's request by SSL.
G0 GMH R D Snowden, 3B Conyers Dr, Aston, Sheffield, S31 0AY [IO93GJ, SK38]
GM0 GMI J G Bavin, Garvan, 10 Grampian Way, Barrhead, Glasgow, G78 2DH [IO75TT, NS55]
G0 GMJ J A Bowles, 38 Rydal Gr, Huddersfield Rd, Liversedge, WF15 7DN [IO93DQ, SE22]
G0 GMK D W Wheeler, 33 Larkspur Gdns, Holbury, Southampton, SO45 2QH [IO90HT, SU40]
G0 GML B P Ogden, 14 Fermandy Ln, Crawley Down, Crawley, RH10 4UB [IO91XC, TQ33]
G0 GMM G D Whalley, 52 Bowling Park Dr, Bradford, BD4 7ES [IO93DS, SE13]
GM0 GMN J Bertram, 47 Bournmouth Rd, Gourock, Renfrewshire, PA19 1HN [IO75OW, NS27]
GM0 GMO G M Wylie, 9 Friar Ave, Bishopbriggs, Glasgow, G64 2HP [IO75VW, NS67]
G0 GMQ Details withheld at licensee's request by SSL.
G0 GMR Details withheld at licensee's request by SSL.
G0 GMS A B Read, 15 Yarrow Cl, Horsham, RH12 5FP [IO91UB, TQ13]
G0 GMT A Owen, 6 Cromwell Cres, Lambley, Nottingham, NG4 4PJ [IO92LX, SK64]
G0 GMU Details withheld at licensee's request by SSL.
G0 GMV Details withheld at licensee's request by SSL.
G0 GMW Details withheld at licensee's request by SSL.

GW0 GMX G C Frayne, 15 Nash Ave, Bronwydd Rd, Carmarthen, SA31 2AX [IO71UU, SN42]
G0 GMY P W Whitelock, 2 Hayes Villas, Frimley Rd, Ash Vale, Aldershot, GU12 5NS [IO91PG, SU85]
G0 GMZ T H Wardle, Cres View, 504 Leasowe Rd, Moreton, Wirral, L46 3RD [IO83LK, SJ29]
G0 GNA J Weller, Pytchley, Chichester Cl, Dorking, RH4 1LP [IO91UF, TQ15]
G0 GND A Campion, 3 Masseycroft, Whitworth, Rochdale, OL12 8PZ [IO83VP, SD81]
G0 GNE R L Maddison, Tom Butt Cottage, Hope St, Elstead, Surrey, GU8 6DE [IO91PE, SU94]
G0 GNF G N Frykman, 8 Orchard Cl, Bishops Itchington, Leamington Spa, CV33 0QS [IO92GF, SP35]
G0 GNI M C D Anderson, 17 Orchard Rd, Seaview, PO34 5JE [IO90KR, SZ69]
G0 GNJ A H Robinson, 12 Atlantic Cres, Burnham on Sea, TA8 1NF [IO81MF, ST34]
G0 GNL K Pickering, 24 St. Augustines Rd, Camden Town, London, NW1 9RN [IO91WN, TQ28]
G0 GNM S J Goodfield, 47 The Martins, Stroud, GL5 4PG [IO81VR, SO80]
G0 GNO A D Stratta, 37 Oak Hall Park, Burgess Hill, RH15 0DH [IO90WW, TQ31]
G0 GNP B M Ackerman, 31 Melton Mill Ln, High Melton, Doncaster, DN5 7TE [IO93JM, SE40]
G0 GNQ L F West, The Bungalow, Courtwood Primary School, Courtwood Ln, Addington, Surrey, CR0 9HX [IO91XI, TQ36]
G0 GNR Details withheld at licensee's request by SSL.
G0 GNS Details withheld at licensee's request by SSL.[Station located near Tadworth.]
GM0 GNT W W Shackleton, Gairney Cottage, Gairney Bank, Kinross, Tayside, KY13 7JX [IO86HE, NT19]
G0 GNU J Norton, 7 Maudsley St., Bradford, BD3 9JT [IO93DT, SE13]
G0 GNV M Mundy, Welbeck House, 19 Manor Rd, Burgess Hill, RH15 0NW [IO90WX, TQ31]
G0 GNW G H Goddard, 113 Linden Walk, Louth, LN11 9HT [IO93XI, TF38]
G0 GNX Details withheld at licensee's request by SSL.
GM0 GNY L M Graupner, 11 James St., Lossiemouth, IV31 6AU [IO87IR, NJ27]
G0 GNZ L J Waddoups, Oulder Cottage, 4 Oulder Hill, Rochdale, OL11 5LF [IO83VO, SD81]
GM0 GOA Details withheld at licensee's request by SSL.
G0 GOB R G T Drage, 13 Manor Ride, Brent Knoll, Highbridge, TA9 4DY [IO81MG, ST35]
G0 GOC A G McGregor, 115 Shaldon Dr, Morden, SM4 4BQ [IO91VJ, TQ26]
G0 GOD J H Perry, 83 Water Orton Rd, Castle Bromwich, Birmingham, B36 9EY [IO92CM, SP18]
G0 GOE P B Bozac, Green Farm Villa, Church Ln, Shadoxhurst, Ashford, TN26 1LS [JO01JC, TQ93]
G0 GOF S Lightfoot, 7 Woburn Rd, Marston Moreteyne, Bedford, MK43 0NH [IO92RB, SP94]
G0 GOH R P Cornell, 81 Mercel Ave, Armthorpe, Doncaster, DN3 3HS [IO93LN, SE60]
G0 GOI J Macham, 9 Bankfield Gr, Scot Hay, Newcastle, ST5 6AR [IO83UA, SJ84]
G0 GOJ H M Oakley, 16 Mountfield Rd, Waterloo, Huddersfield, HD5 8RA [IO93DP, SE11]
G0 GOL Details withheld at licensee's request by SSL.
GM0 GON A Harrison, Moneen, High Askomil, Campbeltown, Argyll, Scotland, PA28 6EN [IO75EK, NR72]
G0 GOO B Evans, 17 Clarence Pl, Maltby, Rotherham, S66 7HA [IO93JK, SK59]
G0 GOQ B J W Wood, 4 Hunter Rd, Brookenby, Binbrook, Market Rasen, LN8 6EG [IO93VK, TF29]
G0 GOR W G Jones, 42 The Park, Kingswood, Bristol, BS15 4BL [IO81RL, ST67]
GW0 GOT Details withheld at licensee's request by SSL.
GM0 GOV R J Dinning, South Brae, Dunlop, Ayrshire, Scotland, KA3 4BP [IO75RQ, NS34]
G0 GOX F A Goodger, 66 Selkirk Cl, Merley, Wimborne, BH21 1TP [IO90AS, SZ09]
G0 GOZ G A L Reay, 53 Tithe Barn Rd, Stafford, ST16 3PL [IO82WT, SJ92]
G0 GPA G P Alban, 12 Ingram Cres, Dunscroft, Doncaster, DN7 4JG [IO93LN, SE60]
G0 GPB K G Love, 16 Tenscore Ave, Walsall, WS6 7BX [IO82XP, SJ90]
G0 GPC G P Conway, 53 Hobart Rd, Cannock, WS12 5SG [IO92AQ, SK01]
G0 GPD Details withheld at licensee's request by SSL.
G0 GPE D G Wells, Browtop, Old Ln, Crowborough, TN6 2AD [JO01CB, TQ53]
GI0 GPG D McKee, 38 Nursery Rd, Armagh, Co. Armagh, BT60 4BL [IO64PI, H84]
G0 GPH P Butterworth, 1 The Avenue, Seedfield, Bury, BL9 5DQ [IO83UO, SD81]
G0 GPK J F Burrows, Applefields, Wilmingham Ln, Thorley, Yarmouth, PO41 0SL [IO90GQ, SZ38]
G0 GPM J Q R Plummer, 17 Lonsdale Terr, Dearham, Maryport, CA15 7EW [IO84GQ, NY03]
G0 GPN S G Rees, 47 Roosevelt Ave, Leighton Buzzard, LU7 8EN [IO91QW, SP92]
G0 GPO A G Beer, 19 Island Rd, Sturry, Canterbury, CT2 0EA [JO01NH, TR16]
GW0 GPQ R P Rees, Bryn Isaf, 22 Pentremeurig Rd, Carmarthen, Carmarthenshire, SA31 3ER [IO71UU, SN32]
G0 GPR R Wordsworth, 59 Highgate Ln, Goldthorpe, Rotherham, S63 9BA [IO93IM, SE40]
G0 GPS I D Jones, 6 Sanderling Ct, Spennals Valley, Kidderminster, DY10 4TS [IO82VI, SO87]
G0 GPT S Carter, 22 Bear Rd, Hanworth, Feltham, TW13 6RA [IO91TK, TQ17]
G0 GPV J E Barnes, 24 Burleigh Pl, Oakley, Bedford, MK43 7SG [IO92RE, TL05]
G0 GPW E A Featherstone, 27 Ct Cl, Aylesbury, HP21 8BY [IO91OT, SP81]
G0 GPX M V Wise, 11 Coles Ln, Oakington, Cambridge, CB4 5BA [JO02AG, TL46]
G0 GPZ Details withheld at licensee's request by SSL.
GI0 GQA J A H Ogilby, 45 Ballynarry Rd, Portadown, Craigavon, BT62 1TX [IO64RL, H96]
GW0 GQC B A Morgan, 5 Brynmawr, Bettws, Bridgend, CF32 8SD [IO81EN, SS88]
G0 GQD Details withheld at licensee's request by SSL.
G0 GQF Details withheld at licensee's request by SSL.
GI0 GQG J T Mills, 6 Cloghan Rd, Drumnakilly, Omagh, BT79 0LF [IO64KO, H57]
G0 GQH A B Jenner, 53 The Leys, Woburn Sands, Milton Keynes, MK17 8QG [IO92QA, SP93]
G0 GQI A S Orton, 2 The Grove, Brampton Abbotts, Ross on Wye, HR9 7JH [IO81RW, SO62]
G0 GQJ T C Waters, 42 Tregundy Rd, Perranporth, TR6 0EF [IO70KI, SW75]
G0 GQK Details withheld at licensee's request by SSL.
G0 GQO S R Taylor, 68 Churncote, Stirchley, Telford, TF3 1YN [IO82SP, SJ70]
G0 GQP D R Jackson, 38 Chestnut Cres, Bletchley, Milton Keynes, MK2 2LA [IO91PX, SP83]
G0 GQQ D M Newton, The Boundaries, Thorpe Rd, Haddiscoe, Norwich, NR14 6PP [JO02TM, TM49]
G0 GQT M D Bishop, 52 Lingley Dr, Frindsbury, Wainscott, Rochester, ME2 4NE [JO01GJ, TQ77]
G0 GQU Details withheld at licensee's request by SSL.
G0 GQV P R Leech, Holly Tree Cottage, 1 Goat Ln, Strumpshaw, Norwich, NR13 4NF [JO02RO, TG30]
G0 GQW H G Jones, 171 Grange Ln, Gateacre, Liverpool, L25 5JY [IO83NJ, SJ48]
G0 GQX M J Chappell, 43 Aigburth Hall Ave, Grassendale, Liverpool, L19 9EA [IO83NI, SJ38]
G0 GQY D G Smith, 62 Beresford Rd, Dorking, RH4 2DG [IO91UF, TQ14]
G0 GQZ K T F Rutgers, 22 Marriotts Cl, Felmersham, Bedford, MK43 7HD [IO92RE, SP95]
G0 GRB P T Attew, 23 Kingsleigh Cl, Trunch, North Walsham, NR28 0QU [JO02QU, TG23]
G0 GRC F C Seddon Grantham Rad Cl, 13 Saltersford Rd, Grantham, NG31 7HH [IO92QV, SK93]
GM0 GRD R Kelly, 10 Barnshean Ave, Patna, Ayr, KA6 7PB [IO75SI, NS40]
G0 GRG A Lovell, 28 Frobisher Ave, Portishead, Bristol, BS20 9XB [IO81OL, ST47]
G0 GRI I L Carter, 12 Bobbin Ln, Westwood, Bradford on Avon, BA15 2DL [IO81UI, ST85]
GW0 GRJ Details withheld at licensee's request by SSL.
G0 GRK G P Stevens, 33 Langham Rd, Hastings, TN34 2JE [JO00HU, TQ81]
GM0 GRL D S Moore, 16 Keith Ave, Broomridge, Stirling, FK7 7UA [IO86AC, NS89]
G0 GRM G R G Meanley, The Hayloft, 53 Northd. Rd, Leamington Spa, CV32 6HF [IO92FH, SP36]
G0 GRO B J Phillips, 12 Fairview Ave, Weston, Crewe, CW2 5LX [IO83TB, SJ75]
G0 GRP Details withheld at licensee's request by SSL.
GW0 GRQ J G Mitchell, 44 Crossways St., Barry, CF63 4PQ [IO81IJ, ST16]
G0 GRR B Gray, Orchard End, 48 Smith House Ln, Brighouse, HD6 2LF [IO93CR, SE12]
G0 GRS G R Sawford, 17 Church Rd, Pytchley, Kettering, NN14 1EL [IO92PI, SP87]
G0 GRT B J Hill, 38 Abelwood Rd, Long Hanborough, Witney, OX8 8DD [IO91HT, SP41]
G0 GRU R Foster, 48 Empingham Rd, Ketton, Stamford, PE9 3RP [IO92RP, SK90]
G0 GRV Y O S I Katoh, 2-9-7-101 Minami-Aoyama, Minato-Ku, Tokyo, Japan, 107 X
GM0 GRW R C Young, 2 Quarryhead Cottage, Trabboch, Mauchline, KA5 5JE [IO75SL, NS42]
G0 GRX C P Ashlin Bolton Raynet G, 13 Brantfell Gr, Bolton, BL2 5LY [IO83TO, SD70]
G0 GRY P J Giles, 41 Shute Hill, Mawnan Smith, Falmouth, TR11 5HQ [IO70KC, SW72]
G0 GRZ F Pounder, 29 Read Ave, Beeston, Nottingham, NG9 2FJ [IO92JW, SK53]
G0 GSA R J Hall, The Cottage, Stretton under Fosse, Rugby, Warks, CV23 0PE [IO92IK, SP48]
G0 GSB C G Leonard, Bm/Trye, London, WC1N 3XX [IO91WM, TQ38]
G0 GSC B F Law, 47 Colindeep Ln, Sprowston, Norwich, NR7 8EG [JO02PP, TG21]
G0 GSD Details withheld at licensee's request by SSL.
G0 GSF Dr B A Austin, 110 Frankby Rd, West Kirby, Wirral, L48 9UX [IO83KJ, SJ28]
GM0 GSG H A Cameron, Garden Cottage, Balloch Alyth, Blairgowrie, Scotland, PH11 8JN [IO86JP, NO24]
G0 GSH C J Summers, 18 Hays Ln, Hinckley, LE10 0LA [IO92HM, SP49]
G0 GSJ D Howie, 22 Jason St., Walney, Barrow in Furness, LA14 3EJ [IO84JC, SD16]
G0 GSK W Blay, 50 Fir St., Cadishead, Manchester, M44 5AU [IO83SK, SJ79]
G0 GSL T M Ritchie, 122 Walpole Rd, South Woodford, London, E18 2LL [JO01AO, TQ39]
G0 GSM T E Pritchard, 21 Newlyn Dr, Bredbury, Stockport, SK6 1EF [IO83WK, SJ99]
G0 GSN N Pope, 1 Knowsley Rd West, Wilpshire, Blackburn, BB1 9PW [IO83SS, SD63]
G0 GSO Dr W Connor, 8 Staines Cl, Appleton, Warrington, WA4 5NP [IO83RI, SJ68]
G0 GSP R A B Wood, 2 Sea View Caravan Site, Bank Ln, Warton, Preston, PR4 1TD [IO83NR, SD42]
G0 GSQ G Stead, 131 Belfield Rd, Rochdale, OL16 2XL [IO83WO, SD91]
G0 GSR F D Johnson, 9 Manor Cl, Tavistock, PL19 0PN [IO70WN, SX47]
GW0 GST A Williams, 34 Gwydyr Rd, Llandudno, LL30 1HQ [IO83CH, SH78]
G0 GSU A Dixon, 13 Keats Gr, Penistone, Sheffield, S30 6GU [IO93GJ, SK38]
G0 GSV J Lawson, 42 Augustus Dr, Bedlington, NE22 6LE [IO95ED, NZ28]
GW0 GSW Rev V H Jones, 3 Ael y Glyn, Nant Rd, Harlech, LL46 2UJ [IO72WU, SH53]
G0 GSX C Guerrero, 30 Archbishop, Amigo House, Glacis Est, Gibraltar
G0 GSY B C Thomsen, 13 Oak Way, Cleethorpes, DN35 0RA [IO93XM, TA20]
G0 GSZ P Hunter, 2 Mayes Cl, Norwich, NR5 9AR [JO02OP, TG10]
G0 GTC D Nicholls, 5 Grains Rd, Delph, Oldham, OL3 5DS [IO83XN, SD90]
G0 GTD T E Davey, 11 Vicarage Ln, Harmston, Lincoln, LN5 9SL [IO93RD, SK96]
G0 GTE P Daly, 48 Lincoln Rd, St. Nicholas Est, Stevenage, SG1 4PJ [IO91VW, TL22]
G0 GTI A M W Dickinson, 6 Church Ln, Bessacarr, Doncaster, DN4 6QB [IO93LM, SE60]
GI0 GTJ T A Simpson, 5 Ballymaconnell Rd South, Bangor, BT19 6DG [IO74EP, J58]
GW0 GTK K M Thomas, Camrose Farm, Camrose, Haverfordwest, Pembrokeshire, SA62 6JA [IO71LU, SM92]
GM0 GTL T Lorimer, 443 Delgatie Ct, Pitteuchar, Glenrothes, KY7 4RW [IO86KE, NT29]
G0 GTM A W Baxter, 37 Newsham Rd, Blyth, NE24 5TJ [IO95FC, NZ38]
G0 GTN J R M Bumford, 19 Bewdley Ave, Telford Est, Monkmoor, Shrewsbury, SY2 5UQ [IO82PR, SJ51]
G0 GTO Details withheld at licensee's request by SSL.[Op: T N Stanley. Station located near central London.]
G0 GTQ Details withheld at licensee's request by SSL.
G0 GTR Details withheld at licensee's request by SSL.
G0 GTS Details withheld at licensee's request by SSL.
GM0 GTU Details withheld at licensee's request by SSL.
G0 GTV S P Moore, 2 Sheppards Cl, Heighington, Lincoln, LN4 1TU [IO93SF, TF06]
G0 GTX Details withheld at licensee's request by SSL.
GW0 GUA Details withheld at licensee's request by SSL.
GW0 GUB W D G Rees, 28 Great North Rd, Milford Haven, SA73 2LU [IO71LR, SM90]
G0 GUC A D Russon, 92 St. Georges Rd, Sledmere Est, Dudley, DY2 8ER [IO82XL, SO98]
G0 GUD D J Law, 208 Long St., Dordon, Tamworth, B78 1QA [IO92EO, SK20]
G0 GUE M C Tyou, Treweath, Trewennack, Helston, Cornwall, TR13 0PL [IO70JC, SW62]
G0 GUF J E Youde, 19 Charlotte Cl, Priory Fields, Little Haywood, Stafford, ST18 0QJ [IO92AT, SK02]
G0 GUG M Grove, 9 Foxcote Ln, Cradley, Halesowen, B63 2JJ [IO82WK, SO98]
G0 GUH Details withheld at licensee's request by SSL.
G0 GUI W S Cook, Minerva, Chattenden Ln, Chattenden, Rochester, Kent, ME3 8LF [JO01GK, TQ77]
GM0 GUJ J D Cumming, 9 Westfield, Kincardine, Alloa, FK10 4PN [IO86DB, NS98]
G0 GUK D J Beaver, 8 School Fields, North Petherton, Bridgwater, TA6 6QJ [IO81LC, ST23]
G0 GUL P J Solman, 73 Princethorpe Way, Binley, Coventry, CV3 2HG [IO92GJ, SP37]
GM0 GUM I O Edwards, 3 John St., Dunfermline, KY11 4TP [IO86GB, NT18]
G0 GUN C J Critchley, 3 Beaconsfield View, Robert Rd, Hedgerley, Slough, SL2 3XT [IO91QN, SU98]
G0 GUO J R Rosindale, Treworder Farm, Raun Minor, Helston, Cornwall, TR12 7JL [IN79JX, SW71]
G0 GUP Details withheld at licensee's request by SSL.
G0 GUR Details withheld at licensee's request by SSL.
G0 GUS Details withheld at licensee's request by SSL.
G0 GUT N G Reed, 30 Wrey Ave, Liskeard, PL14 3HX [IO70SL, SX26]
G0 GUW B L Standen, 43 Westover Gdns, St. Peters, Broadstairs, CT10 3EY [JO01RI, TR36]
GU0 GUX J Scheffer, Route de Carteret, Cobo, Castel, Guernsey, Channel Islands, GY5 7YS
GW0 GUY E E Hollowell, 54 Portfield, Haverfordwest, SA61 1BW [IO71MT, SM91]
G0 GVA J A Lawson, 14 Kentmere Ave, Leyland, Farington, Preston, PR5 1UH [IO83PQ, SD52]
G0 GVC Details withheld at licensee's request by SSL.
G0 GVD C A Houghton, 8 Wheatfield Rd, Cronton, Widnes, WA8 5BU [IO83OJ, SJ48]
G0 GVE S A Thomas, 14 Goodley, Oakworth, Keighley, BD22 7PD [IO93AU, SE03]
G0 GVF M Davidson, 18 Cerro Donoso, T.M. de Guaro, Malage, Spain 29108
G0 GVG A N Slark, 61 Chapeltown Rd, Bromley Cross, Bolton, BL7 9NB [IO83TO, SD71]
G0 GVI Details withheld at licensee's request by SSL.
G0 GVJ E Stals, North Lake Farm, Lapford, Crediton, Devon, EX17 6NA [IO80CV, SS71]
G0 GVL R J Lee, The Penthouse, 44 Amery St, Sliema, Malta, X X
G0 GVM Details withheld at licensee's request by SSL.
G0 GVN C G Wackett, 7 Newell Rd, Hemel Hempstead, HP3 9PD [IO91SR, TL00]
G0 GVO J G Anderson, 42 Old Slade Ln, Iver, SL0 9DR [IO91RM, TQ07]
GW0 GVQ Details withheld at licensee's request by SSL.[Station located in Barry, IO81IK, WAB: ST16. Mailbox: GW0GVQ @ GB7SIG. QSL via RSGB Bureau.]
G0 GVS N Clayton, 220 Milnrow Rd, Rochdale, OL16 5BB [IO83WO, SD91]
G0 GVT J R Gibb, 47 Longroyd Pl, Beeston, Leeds, LS11 5HD [IO93FS, SE33]
G0 GVU J M Gibb, 47 Longroyd Pl, Beeston, Leeds, LS11 5HD [IO93FS, SE33]
G0 GVX B S Sherwood, 363 Old Laira Rd, Laira, Plymouth, PL3 6DH [IO70WJ, SX55]
G0 GVZ A G R Grimes, 37 Cavendish Ave, Cambridge, CB1 4UR [JO02BE, TL45]
G0 GWA Dr S Browne, 3 High Barnes Cttgs, Battlegate Rd, Boxworth, Cambridge, CB3 8NJ [IO92XF, TL36]
G0 GWB B L Gentle, 4 Hare Park Terr, Upper Stondon, Henlow, SG16 6LH [IO92UA, TL13]
G0 GWC D G Roberts, 1 Fair Ln, The Sycamores, Shaftesbury, SP7 8RT [IO81VA, ST82]
G0 GWD A Brown, 6 Laing Sq, Wingate, TS28 5JE [IO94HR, NZ33]
GW0 GWE R M Edwards, 11 Trem y Eglwys, Coed y Glyn, Wrexham, LL13 7QE [IO83MA, SJ34]
G0 GWF L J Whitemore, 2 Netherton Wood Farm Cottage, West End, Nailsea, Bristol, BS19 2DG [IO81OJ, ST46]
G0 GWG C A Partridge, 44 Pine Cl, South Wonston, Winchester, SO21 3EB [IO91IC, SU43]
G0 GWI P W Bell, 6 Dorchester Cl, Hale, Altrincham, WA15 8PW [IO83UJ, SJ78]
G0 GWK J Ralph, 3 Long Hedge, Lambourn, Hungerford, RG17 7LY [IO91FM, SU37]
G0 GWL D D Baker, Dunbar Lodge, 11 High Rd, Middlestone Village, Bishop Auckland, Co Durham, DL14 8AE [IO94EQ, NZ23]
G0 GWM R E R Sawkins, 48 South Rd, Saffron Walden, CB11 3DN [JO02CA, TL53]
G0 GWN J H Faulconbridge, 32 Beridge Rd, Halstead, CO9 1LB [JO01HW, TL83]
G0 GWP A J Horton, 184 Mount Pleasant, Keyworth, Nottingham, NG12 5ET [IO92LU, SK63]
G0 GWS W Duff, Highfield Maundown, Wiveliscombe, Taunton, Somerset, TA4 2BU [IO81HB, ST02]
G0 GWV L Andrew, 40 Mill Hayes Rd, Knypersley, Stoke on Trent, ST8 7BU [IO83WC, SJ85]
G0 GWW W J Harvey, 158 Parkway, Welwyn Garden City, AL8 6HZ [IO91VT, TL21]
G0 GWX D J Anstiss, 7 Kipling Dr, Newport Pagnell, MK16 8EB [IO92PC, SP84]
G0 GWY G Birch, 92 Scotter Rd, Scunthorpe, DN15 8AT [IO93PO, SE81]
G0 GWZ D McCarty, 10 Glenmead Rd, Great Barr, Birmingham, B44 8UG [IO92BM, SP09]
G0 GXA P J Highley, The Lodge, Mansel Lacy, Hereford, HR4 7HQ [IO82NC, SO44]
G0 GXB J D Kirby, 33 Brinkinfield Rd, Chalgrove, Oxford, OX44 7QX [IO91LQ, SU69]
G0 GXD Details withheld at licensee's request by SSL.
G0 GXF P G Hirst, 57 Etherington Dr, Hull, HU6 7JT [IO93TS, TA03]
G0 GXH Details withheld at licensee's request by SSL.
G0 GXI S A Suter, 24 Keld Head Orchard, Kirkbymoorside, York, YO6 6EF [IO94MG, SE68]
G0 GXJ R R Bradshaw, 4 Upper Ley Dell, Chapeltown, Sheffield, S30 4AL [IO93GJ, SK38]
G0 GXM G R Munday, 33 Fisher Rd, Harrow, HA3 7JX [IO91UO, TQ19]
G0 GXO H Swaddle, Beech Hurst, 12 Belmont Gdns, Haydon Bridge, Hexham, NE47 6HG [IO84VX, NY86]
G0 GXP C Everett, 52 St. Pegas Rd, Peakirk, Peterborough, PE6 7NF [IO92UP, TF10]
GW0 GXQ J A Williams, 91 Mold Rd, Buckley, CH7 2JA [IO83LE, SJ26]
G0 GXR W S Rutter, 103 Allendale Rd, Walker, Newcastle upon Tyne, NE6 2SX [IO94FX, NZ26]
G0 GXS R M Wareham, 8 Meeting Ln, Needingworth, St. Ives, Huntingdon, PE17 3SN [IO92XH, TL37]
G0 GXT D J Mellor, Woodland View, Farm Tump, Ruardean Woodside, Glos, GL17 9AR [IO81RU, SO61]
G0 GXV R J Drewett, Churchyard Cottage, Church Ln, Ashbury, Swindon, SN6 8LZ [IO91EN, SU28]
G0 GXX M Smallwood, 12 Rowan Walk, Hornsea, HU18 1TT [IO93WV, TA24]
G0 GXZ M S Butler, 1 Springhead Ave, Willerby Rd, Hull, HU5 5HZ [IO93TS, TA02]
G0 GYA C M Taylor, 37 Manor Park Ave, Allerton Bywater, Castleford, WF10 2DN [IO93HR, SE42]
G0 GYB C Hunt, 1 Compton St., Holmewood, Chesterfield, S42 5RP [IO93HE, SK46]
G0 GYD Details withheld at licensee's request by SSL.
G0 GYG Details withheld at licensee's request by SSL.
G0 GYH D Goodall, 5 Coach Dr, Newthorpe Grange Es, Eastwood, Nottingham, NG16 3DR [IO93IA, SK44]
G0 GYI R P Ide, Wayside, Colchester Main Rd, Alresford, Colchester, CO7 8DH [JO01MU, TM02]
G0 GYJ M D R Dawson, 11 Eastholme Dr, York, YO3 6SU [IO93KX, SE55]
GM0 GYL M Duffy, Lylestone Farm, Kilwinning, Ayrshire, KA13 7QP [IO75QQ, NS34]
GM0 GYM T Quinn, 40 Drumry Rd, Clydebank, G81 2LL [IO75TV, NS57]
GM0 GYN P M Gibson, 7 Rogerhill Dr, Kirkmuirhill, Lanark, ML11 9XS [IO85AQ, NS74]
G0 GYO M Mc Pherson, 4 Highfield Pl, Brunswick Green, Wideopen, Newcastle upon Tyne, NE13 7HW [IO95EB, NZ27]
G0 GYP C Fiedler, 51 Fleet Ln, Tockwith, York, YO5 8QD [IO93IX, SE45]
GM0 GYQ Dr H Garmany, Ayrebrayde House, Forthill Rd, Broughty Ferry, Dundee, Angus, DD5 3DL [IO86NL, NO43]
GM0 GYT J M Ritchie, 24 Kirkton Cres, Knightswood, Glasgow, G13 3AQ [IO75TV, NS56]
G0 GYU P A Walters, Stonecroft, Main St, Knaresborough, North Yorkshire#, HG5 9LD [IO94GB, SE36]
G0 GYW A R L Thomas, 49 Byron Cl, Bletchlley, Bletchley, Milton Keynes, MK3 5BD [IO91PX, SP83]
G0 GYX P W Nixon, 17 Wallner Cres, Felpham, Bognor Regis, PO22 7QE [IO90QT, SU90]
G0 GYY P D Markham, 15 Victoria Rd, Walton on The Naze, CO14 8BU [JO01PU, TM22]
G0 GYZ Details withheld at licensee's request by SSL.
G0 GZA J M Heywood, 52 Springbank Rd, Cheltenham, Glos, GL51 0NH [IO81WV, SO92]
G0 GZB J Bartlett, 20 Thatchers Cl, Horsham, RH12 5TL [IO91UB, TQ13]
G0 GZC D C Bradley, 1 Audley Pl, Sutton, SM2 6RW [IO91VI, TQ26]
G0 GZE P Wylie, 15 Semley Rd, Hassocks, BN6 8PD [IO90WW, TQ31]
G0 GZF A D Pierce, 38 High Ridge, Godalming, GU7 1YF [IO91QE, SU94]
G0 GZI R P Jeffery, 167 Delamere St., Over, Winsford, CW7 2LY [IO83RE, SJ66]
G0 GZK Details withheld at licensee's request by SSL.
G0 GZL R B Daniels, Key House, South Hill, Rolleston on Dove, Burton on Trent, DE13 9AT [IO92EU, SK22]
G0 GZM M S Johns, 3 Burwood Ave, Kenley, CR8 5NT [IO91WH, TQ36]
G0 GZN L A Jasper, 27A Sea View Rd, Parkstone, Poole, BH12 3LP [IO90AR, SZ09]
G0 GZO K Walters, 18 Leabrooks Ave, Sutton in Ashfield, NG17 5HU [IO93JC, SK55]
G0 GZP Details withheld at licensee's request by SSL.
GW0 GZQ P J Davenport, Ty Cerrig, 1 Pant-y-Groes, Moylegrove, Dyfed, SA43 3BY [IO72PB, SN14]
GW0 GZR G E Foster Alyn & Deeside Amateur Radio S, 32 Braeside Ave, Hawarden, Deeside, CH5 3HW [IO83LE, SJ36]

G0 GZS Details withheld at licensee's request by SSL.
G0 GZT B Dixon, 36 Meadow Ln, Milehouse, Newcastle, ST5 9AJ [IO83VA, SJ84]
G0 GZU N R Harris, 16 Gibbs Field, Thorley Park, Bishops Stortford, CM23 4EY [JO01BU, TL42]
G0 GZV K G Bailey, 35 Edgehill Rd, Chislehurst, BR7 6LA [JO01AK, TQ47]
G0 GZW J E Spoard, 29 Quarry Rd, Alveston, Bristol, BS12 2JL [IO81RO, ST68]
GM0 GZX C H Brown, Flat 3/1, 3 Brisbane St, Glasgow, Lanarkshire, G42 9HL [IO75UT, NS56]
G0 GZZ G V Kinal, 11 Canonbury Park North, London, N1 2JZ [IO91WN, TQ38]
G0 HAA P M Hodgson, 23 Hensley Ct, The Glebe, Norton, Stockton on Tees, TS20 1TE [IO94IO, NZ42]
G0 HAB S Jones, 44 Springcroft, Parkgate, South Wirral, L64 6SE [IO83LH, SJ27]
G0 HAD A R Box, 21 St. Michaels Rd, Melksham, SN12 6HN [IO81WI, ST96]
G0 HAE R F R Isaac, 54 Bowcombe, Netley Abbey, Southampton, SO31 5GP [IO90HV, SU40]
G0 HAF Details withheld at licensee's request by SSL.
G0 HAH Details withheld at licensee's request by SSL.
G0 HAL A J Paterson, 1 Birch Gr, Manor Farm, Timperley, Altrincham, WA15 7YH [IO83UJ, SJ88]
G0 HAN Details withheld at licensee's request by SSL.
GM0 HAQ A Fraser, 37 Attadale Rd, Inverness, IV3 5QH [IO77VL, NH64]
G0 HAS A S Jordan, 10 Broadley Park, North Bradley, Trowbridge, BA14 0SS [IO81VH, ST85]
G0 HAT J C Bales, 49 High St., Pinner, HA5 5PJ [IO91TO, TQ18]
G0 HAU E J Hazell, 1 Heather Walk, Tonbridge, TN10 3NA [JO01DF, TQ54]
G0 HAV D I Wicker, 28 Lee Warner Ave, Fakenham, NR21 8ER [JO02KU, TF93]
G0 HAW C I Blackmoor, 4 St. Godwalds Cres, Aston Fields, Bromsgrove, B60 2EB [IO82XH, SO96]
G0 HAX F J W Perry, 9 Chestnut Ln, Matfield, Tonbridge, TN12 7JJ [JO01ED, TQ64]
G0 HAY D C J Thomas, 29 Kimberley, Wilnecote, Tamworth, B77 5LD [IO92EO, SK20]
G0 HBA R F Curzon, 24 Edwards Dr, Wellingborough, NN8 3JJ [IO92PH, SP86]
G0 HBB D H J Brown, 15 Osborne Ave, Tuffley, Gloucester, GL4 0QN [IO81UT, SO81]
G0 HBC A B Sammons, 53 Lillington Rd, Shirley, Solihull, B90 2RY [IO92CJ, SP17]
GW0 HBD A Kings, 13 Fairfield Rd, Bulwark, Chepstow, NP6 5JP [IO81PP, ST59]
G0 HBE R C Hillier, 24 The Croft, Urchfont, Devizes, SN10 4RT [IO91AH, SU05]
GM0 HBF C F Fraser, Kelp Cottage, Locheport, North Uist, Western Isles, HS6 5EU [IO67IN, NF86]
G0 HBG R A Marriner, 525B Ferndale Rd, Swindon, SN2 1DG [IO91CN, SU18]
GM0 HBI J G Cormack, Dunrobin Farm, Golspie, Sutherland, Scotland, KW10 6RH [IO87AX, NC80]
G0 HBJ R E Millett, 8 Sidestrand Rd, Newbury, RG14 6HP [IO91HJ, SU46]
GM0 HBK C Robertson, 3, Sasaig, Teangue, Sleat, Isle of Skye, IV44 8RD [IO77BC, NG60]
GM0 HBM B S Adam, Bogbain Cottage, Lochussie, Maryburg, Dingwall, Inverness, IV7 8HJ [IO77RN, NH55]
G0 HBN W Riley, 22 Boulsworth Dr, Trawden, Colne, BB8 8SJ [IO83WU, SD93]
G0 HBO S Holmes, 17 Portland Gdns, Low Fell, Gateshead, NE9 6UX [IO94EW, NZ25]
G0 HBQ D Underwood, Otter Bank, Clifton Rd, Blackpool, FY4 4QF [IO83LT, SD33]
G0 HBR R Rawsthorne, 15 Peebles Cl, Garswood, Ashton in Makerfield, Wigan, WN4 0SP [IO83PL, SJ59]
GM0 HBT C A Henery, 62 Douglas Park Cres, Bearsden, Glasgow, G61 3DN [IO75UW, NS57]
G0 HBU A Dwyer, 10 Shaftway Cl, Haydock, St. Helens, WA11 0YQ [IO83QL, SJ59]
G0 HBV T B Sandilands, 8 Cheviot View, Lowick, Berwick upon Tweed, TD15 2TY [IO95AP, NU03]
G0 HBW R Fitzgerald, 93 Finland Rd, Brockley, London, SE4 2JQ [IO91XL, TQ37]
G0 HBY S Grant, 9 Meadow Gdns, Hiveacres Est, Tweedmouth, Berwick upon Tweed, TD15 2FF [IO85XS, NT95]
GW0 HBZ D J Wright, Drws-y-Nant, St. Asaph Ave, Kinmel Bay, Rhyl, Clwyd, LL18 5EY [IO83FH, SH97]
G0 HCA R Horton Harrogate Col Rd, Clarence Dr, Harrogate, N Yorks, HG1 2QG [IO93FX, SE25]
GW0 HCB P R Matheson, 34 Pentwyn, Radyr, Cardiff, CF4 8RE [IO81IM, ST18]
G0 HCC T J Groves Herts County C/, Emergency Planning, Herts Co. Council, Co. Hall, Pegs Ln, Hertford, Herts, SG13 8DE [IO91XS, TL31]
G0 HCD T Carroll, 2 West Durham Cttgs, Billy Row, Roddymoor, Crook, DL15 9QX [IO94CR, NZ13]
G0 HCF K Faulconbridge, 66 Beridge Rd, Halstead, CO9 1LB [JO01HW, TL83]
G0 HCI M R Bodle, 40 Hewett Rd, North End, Portsmouth, PO2 0QP [IO90LT, SU60]
GI0 HCJ P Guy, 91 Ballyoran Park, Portadown, Craigavon, BT62 1JW [IO64SK, J05]
GW0 HCN T Hayden, 7 Attlee Cl, Garnlydan, Ebbw Vale, NP3 5ES [IO81JT, SO11]
G0 HCO M J Barnett, 3 Cartwright Dr, Shaw, Swindon, SN5 9QU [IO91CN, SU18]
GM0 HCQ M E P Gloistein, Rose Cottage, Kinlochleven, Argyll, PA70 7RE [IO66XI, NM52]
G0 HCR S M Turner, 39 Banister Way, Wymondham, NR18 0TY [JO02NN, TG10]
G0 HCS Details withheld at licensee's request by SSL.
G0 HCT Details withheld at licensee's request by SSL.
G0 HCU J D Burnett, 18 Carmen Ave, Shrewsbury, SY2 5NR [IO82PQ, SJ51]
GM0 HCW J Montgomery, 15 Wardlaw Rd, Bearsden, Glasgow, G61 1AL [IO75UV, NS57]
G0 HCX M J E Butler, 1 Deal Castle Rd, Deal, CT14 7BB [JO01QF, TR35]
G0 HDA Details withheld at licensee's request by SSL.
G0 HDB A M Davies, 11 Gravel Pits Cl, Bredon, Tewkesbury, GL20 7QL [IO82WA, SO93]
G0 HDC G E R Evans, 34 Coronation Dr, Donnington, Telford, TF2 8HY [IO82SR, SJ71]
G0 HDD A J Adey, 37 Cranmere Ave, Wolverhampton, WV6 8TR [IO82VO, SJ80]
G0 HDE Details withheld at licensee's request by SSL.
G0 HDF R N Bartlam, 34 Quarry Rd, Selly Oak, Birmingham, B29 5NX [IO92AK, SP08]
G0 HDG A S Edwards, 34 Manor Rd, Maltby, Rotherham, S66 7EG [IO93JK, SK59]
G0 HDH C W Worsfold, Osaka, 7 West View Rd, Gurnard, Cowes, PO31 8NR [IO90IR, SZ49]
G0 HDI B C Walker, Lantilla, Elmfield Ln, Calshot, Southampton, SO45 1BJ [IO90IT, SU40]
G0 HDJ A C Douglas, Threave House, Blind Ln, Barton St. David, Somerton, TA11 6BW [IO81QC, ST53]
GI0 HDK I C A Oakes, 8 McCormack Dr, Lurgan, Craigavon, BT66 8LF [IO64TK, J05]
G0 HDL C H Hardisty, 3 Oak Tree Ave, Wigan Rd, Leyland, Lancs, PR5 2PJ [IO83PQ, SD52]
G0 HDM D Muldoon, 68 Wroxham Dr, Wollaton, Nottingham, NG8 2QS [IO92JW, SK53]
G0 HDN Details withheld at licensee's request by SSL.
G0 HDP B Wainwright, 19 Durkar Low Ln, Durkar, Wakefield, WF4 3BL [IO93FP, SE31]
G0 HDS R C Hurt, 7 Chestnut Ave, Immingham, Grimsby, DN40 1BH [IO93VO, TA11]
G0 HDV K C Coxon, 29 Chapel Rd, Broughton, Brigg, DN20 0HW [IO93RN, SE90]
G0 HDX R S Hartley, 9 Gloucester Ave, Accrington, BB5 4BG [IO83TS, SD72]
GW0 HDY R P J Finnis, 6 Sharpes Way, Bulwark, Chepstow, NP6 5TG [IO81PP, ST59]
G0 HDZ I K Rose, 82 Little Brays, Harlow, CM18 6ES [JO01BS, TL40]
G0 HEA B J Griffin, 63 Thomas More Cl, Kearsley, Bolton, Lancs, BL4 8ND [IO83TM, SD70]
G0 HED D E Marshall, The Anvil, Stortford Rd, Clavering, Saffron Walden, CB11 4PE [JO01BX, TL43]
G0 HEE A Salt, 20 Freemantle Rd, Bagshot, GU19 5NF [IO91PI, SU96]
G0 HEF R McKeever, 38 Brookfield Ave, Runcorn, WA7 5RF [IO83PI, SJ58]
G0 HEG Details withheld at licensee's request by SSL.
G0 HEH Details withheld at licensee's request by SSL.
G0 HEJ E Lloyd, 43 Queensway, Newton, Chester, CH2 1PF [IO83NE, SJ46]
G0 HEL A C Nicholls, 22 Windsor Ave, Melksham, SN12 6BE [IO81WI, ST96]
G0 HEM G M Gardner, New House, Birdbush Ave, Saffron Walden, CB11 4DJ [JO02CA, TL53]
G0 HEN G A Henstock, 16 The Coppice, Enfield, EN2 7BY [IO91WP, TQ39]
G0 HEO T Hart, 23 Gregory Cres, Great Horton, Bradford, BD7 4PG [IO93CS, SE13]
G0 HER K C Kibblewhite, 53 Woodcote, Bedford, MK41 8EL [IO92SD, TL05]
G0 HES P Bowers, 57 Lapwing Gr, Palace Fields, Runcorn, WA7 2TJ [IO83PH, SJ58]
G0 HET P Nutkins, Higher Spence, Wooton Fitzpaine, Charmouth, Dorset, DT6 6DF [IO80MS, SY39]
G0 HEV P D Brindley, 2 Beech Park, School Rd, Great Barton, Bury St. Edmunds, IP31 2JL [JO02JH, TL97]
G0 HEW J G McHale, 21 Bonython Rd, Newquay, TR7 3AW [IO70LK, SW86]
G0 HEX D G Cheetham, 4 Battersbay Gr, Hazel Gr, Stockport, SK7 4QW [IO83WJ, SJ98]
G0 HFA Dr J D Lansdowne, 8 Nansloe Cl, Helston, TR13 8BP [IO70IC, SW62]
G0 HFB W A W Lankshear, 57 St. Georges Rd, East Looe, Looe, PL13 1ED [IO70SI, SX25]
G0 HFC F Chadwick, 59 Beech Ave, Greenfield, Oldham, OL3 7AW [IO83XM, SD90]
G0 HFE T M Cadman, 28 Denbigh Ct, Ellesmere Port, South Wirral, L65 5DX [IO83NG, SJ47]
G0 HFG N Spencer, 4 Roundell Rd, Barnoldswick, Colne, BB8 6EB [IO83VW, SD84]
G0 HFH Details withheld at licensee's request by SSL.[Station located near Dartmoor.]
G0 HFI Details withheld at licensee's request by SSL.
G0 HFK R W Fuller, 376 Lewis Trust Bld, Vanston Pl, Fulham, London, SW6 1AT [IO91VL, TQ27]
G0 HFL Details withheld at licensee's request by SSL.
G0 HFN Details withheld at licensee's request by SSL.
G0 HFO D P Hirst, 10 The Rogers, Shanklin, PO37 7HH [IO90JP, SZ58]
G0 HFR M J Hopkins, Hylton Cottage, Grafton Beckford, Tewksbury, Glos, GL20 7AT [IO82XA, SO93]
G0 HFT L H Smith, 16 Woodlands Ave, Cheadle Hulme, Cheadle, SK8 5DD [IO83VJ, SJ88]
G0 HFU N W Jaques, 39 Chaucer Cres, Newbury, RG14 1TP [IO91HJ, SU46]
G0 HFW Details withheld at licensee's request by SSL.
G0 HFX C Parnell, 29 Southfield, Southwick, Trowbridge, BA14 9PW [IO81VH, ST85]
G0 HFZ Details withheld at licensee's request by SSL.
G0 HGA A M Sitton, 29 Hudson Rd, Stevenage, SG2 0ER [IO91VV, TL22]
G0 HGB G B Howell, Wealdon, Wonston, Hazelbury Bryan, Sturminster Newton, DT10 2EE [IO80TU, ST70]
GW0 HGC B W Roberts, 3 Bay View Terr, Pwllheli, LL53 5HN [IO72TV, SH33]
G0 HGG G E Entwistle, 58 Westwood Park, Bashley Cross Rd, New Milton, BH25 5TB [IO90DS, SZ29]
G0 HGH J Scott, 25 Nash Rd, Margate, CT9 4BT [JO01QJ, TR36]
G0 HGI B W Boden, 71 Park Head Rd, Sheffield, S11 9RA [IO93FI, SK38]
G0 HGK Details withheld at licensee's request by SSL.
G0 HGL M Lindley, 85 Westfield Ave, Mirfield, WF14 9PL [IO93DQ, SE22]
G0 HGM J N Jenkinson, North Seven Ash, Combe Martin, N Devon, EX34 0PB [IO81AE, SS64]
GW0 HGN T E Jones, 11 Lon Ogwen, Bangor, LL57 2UD [IO73WF, SH57]
G0 HGO D N Cady, 45 The Hill, Wheathampstead, St. Albans, AL4 8PR [IO91UT, TL11]
GI0 HGP Dr L G M Magfhogartai, 11 Liskey Brae, Fintona, Omagh, BT78 2AU [IO64IL, H46]
G0 HGS E W Lamb, 32 Kingsway, Luton, LU4 8EH [IO91SV, TL02]
G0 HGV Details withheld at licensee's request by SSL.[Op: D M Burgess, 3 Bradwell Mews, Edmonton, London, N18 2QX.]
G0 HGZ G Vaughan, 42 Ridge Ln, Nuneaton, CV10 0RB [IO92FN, SP29]
G0 HHA P Dixon, 68 Chelsea Rd, Sheffield, S11 9BR [IO93FI, SK38]
G0 HHB W J Blewett, 12 Westover Rd, Callington, PL17 7EW [IO70UM, SX36]
G0 HHC A Gorton, 7 Sterling Cl, Lexden, Colchester, CO3 5DP [JO01KV, TL92]
GW0 HHD G A Williams, Gwastad Annas, Barmouth, Gwynedd, LL42 1DX [IO72XR, SH61]
G0 HHG J V M Fildes, 67 Cranbrook Dr, Esher, KT10 8DN [IO91TJ, TQ16]
G0 HHH W G Cradock, 7 Brambling Rise, Spennells, Kidderminster, DY10 4JQ [IO82VI, SO87]
G0 HHI A W Garlinge, 117 Kitchener Cres, Poole, BH17 7HZ [IO90AS, SZ09]
G0 HHJ R A Fuller, 147 Wolseley Rd, Rugeley, WS15 2QT [IO92AS, SK01]
G0 HHK A V Carder, 18 Dince Hill Cl, Whimple, Exeter, EX5 2TE [IO80HS, SY09]
G0 HHL P J Scrivens, 1 Walnut Cl, Milton, Derby, DE65 6WA [IO92FU, SK32]
G0 HHP D M Jones, 69 Elswick Rd, Kingstanding, Birmingham, B44 0JG [IO92BN, SP09]
G0 HHQ J A Nolan, 49 Watling St., Strood, Rochester, ME2 3JH [JO01FJ, TQ76]
G0 HHR C Sewart, 148 Hibbert Ln, Marple, Stockport, SK6 7NU [IO83XJ, SJ98]
G0 HHS Details withheld at licensee's request by SSL.
G0 HHU C S Griffiths, 6 Cedar Wood, Cuddington, Northwich, CW8 2XR [IO83QF, SJ57]
GI0 HHV S N Johnstone, 6 Gilbert Cres, Bangor, BT20 4PE [IO74EP, J58]
G0 HHX R V Self, 65 High St., Cinderford, GL14 2SU [IO81ST, SO61]
GI0 HHZ R Fitzsimons, 9 Ingledene Park, Newtownards, BT23 8QT [IO74DN, J46]
G0 HIA Details withheld at licensee's request by SSL.
G0 HIC J B East, 35 Preachers Vale, Coleford, Bath, BA3 5PT [IO81SF, ST64]
G0 HID Details withheld at licensee's request by SSL.
G0 HIF A J Edmonds, 44 Blea Tarn Rd, Kendal, LA9 7NA [IO84PH, SD59]
GM0 HIG C Cameron, Garden Cottage, Balloch Alyth, Blairgowrie, Tayside, PH11 8JN [IO86JP, NO24]
G0 HIH P A Brown, Tree Tops, Rugby Rd, Weston under Wetherley, Leamington Spa, CV33 9BW [IO92GH, SP36]
G0 HIJ W A Roberts, 87 Holmfield Rd, Lytham St. Annes, FY8 1JX [IO83LS, SD32]
G0 HIK N Gregory, 23 Princewood Dr, Barrow in Furness, LA13 0RX [IO84JC, SD27]
GM0 HIM J S Pert, 56 Lochiel Dr, Milton of Campsie, Glasgow, G65 8ET [IO75WX, NS67]
G0 HIN T Hamilton, 116 Upper Chobham Rd, Camberley, GU15 1EJ [IO91PH, SU85]
G0 HIO M F Warrington, 53 Berry Hedge Ln, Burton on Trent, DE15 0DP [IO92ET, SK22]
G0 HIP G H Dodd, 5 Teapot Row, Clocktower Dr, Southsea, PO4 9YA [IO90LS, SZ69]
G0 HIQ Details withheld at licensee's request by SSL.[Op: A E Betts. Station located near Orpington.]
GW0 HIR A Edwards, 29 Camrose Walk, St. Dials, Cwmbran, NP44 4NW [IO81LP, ST29]
G0 HIS Details withheld at licensee's request by SSL.
G0 HIU M Valentine, 19 Knowsley Cres, Weeton, Preston, PR4 3ND [IO83MT, SD33]
G0 HIW D B Ross, East Gubb Cottage, West Buckland, North Devon, EX32 0SF [IO81BB, SS63]
G0 HIX Details withheld at licensee's request by SSL.
G0 HIY K B Hinchliffe, 18 Campbell Cres, East Grinstead, RH19 1JR [IO91XD, TQ33]
G0 HIZ D Hughes, 64 Mablins Ln, Coppenhall, Crewe, CW1 3RF [IO83SC, SJ65]
G0 HJB F Little, 26 Hoghton Rd, Longridge, Preston, PR3 3UA [IO83QT, SD63]
G0 HJC Details withheld at licensee's request by SSL.[Op: Maurice Small, 8 Cherry Tree Rd, Chinnor, Oxon, OX9 4QY. Tel: (01844) 51694. RNARS: 3188, BARTG: 4336. G-QRP: 4056. WAB: SP70, Book 7417. ISWL: G-20024. RS87034, AMIPRE, Ex G1XBN. Interest award chasing.]
G0 HJD Details withheld at licensee's request by SSL.
G0 HJE Details withheld at licensee's request by SSL.
G0 HJJ J Johnson, 49 Pontac Rd, New Marske, Redcar, TS11 8AW [IO94LN, NZ62]
G0 HJK J L Everett, 92 Thackeray Rd, Ipswich, IP1 6JB [JO02NC, TM14]
G0 HJL J M Levesley, 96 Brookside Rd, Bransgore, Christchurch, BH23 8NA [IO90DS, SZ19]
G0 HJM R A Smith, 100 Braemar Rd, Billingham, TS23 2AN [IO94IO, NZ42]
G0 HJP Details withheld at licensee's request by SSL.
G0 HJQ Details withheld at licensee's request by SSL.
G0 HJR R J Filby, 5 Sleaford Rd, Newark, NG24 1NL [IO93OB, SK85]
GM0 HJS F J McLay, 53 Victoria Rd, Grangemouth, FK3 9JN [IO86DA, NS98]
GM0 HJU J A Mackenzie, 34 Blair Dr, Milton of Campsie, Glasgow, G65 8DS [IO75WX, NS67]
GM0 HJV C A McClure, 15 Station Ave, Inverkip, Greenock, PA16 0BB [IO75NV, NS27]
G0 HJW P C Webber, 111 Carlford Cl, Martlesham Heath, Ipswich, IP5 7TA [JO02OB, TM24]
G0 HJX P Devine, 23 Laing Sq, Wingate, TS28 5JE [IO94HR, NZ33]
G0 HJZ J R Purnell, 8 Woodcote Cttgs, Graffham, Petworth, GU28 0NY [IO90PW, SU91]
G0 HKB R H J Conneely, Harford House, Wells Rd, Chilcompton, Bath, BA3 4EX [IO81SG, ST65]
G0 HKC E K Chambers, Greenholme, 19 Marina Rd, Durrington, Salisbury, SP4 8DB [IO91CE, SU14]
G0 HKD R H Webster, Dom Pavela, Sproatley Rd, Flinton, Hull, HU11 4NE [IO93WT, TA23]
G0 HKE J A Boyton, 3 Wenny Est, Chatteris, PE16 6UX [JO02AK, TL38]
G0 HKG A Cross, Brick Kiln Farm, Great Tey, Colchester, Essex, CO6 1AR [JO01IV, TL82]
G0 HKH A L Pacitto, East View Cottage, Hovingham, York, YO6 4LA [IO94ME, SE67]
G0 HKI L Compton, 44 Overdown Rise, Mile Oak, Portslade, Brighton, BN41 2YG [IO90VU, TQ20]
G0 HKJ C Potgieter, Brackenfield, 15 Stonewell Park Rd, Congresbury, Bristol, BS19 5DP [IO81OI, ST46]
G0 HKK J H Hall, 36 Sandfield Rd, West Bromwich, Sandwell, West Bromwich, B71 3NF [IO92AN, SP09]
GW0 HKL Details withheld at licensee's request by SSL.
G0 HKN R J Leigh, 5 Parham Cl, Littlehampton, BN17 6HF [IO90RT, TQ00]
G0 HKO D A Douglas, 31 The Park, Northway, Tewkesbury, GL20 8RD [IO82WA, SO93]
G0 HKP Details withheld at licensee's request by SSL.
GW0 HKQ D Suddes, Allt y Coed, Carmel Rd, Carmel, Holywell, CH8 8NU [IO83IG, SJ17]
G0 HKR Rev D B Measures, 23 Green Leys, Church Crookham, Fleet, GU13 0PN [IO91NG, SU85]
GI0 HKS Details withheld at licensee's request by SSL.
G0 HKT S J Mayer, 7 Mellstock Rd, Poole, BH15 3DN [IO90AR, SZ09]
G0 HKZ S J Liptrott, 20 Formosa Way, Liverpool, L10 7NL [IO83ML, SJ39]
G0 HLA T M Barker, Pear Tree Cottage, Ashill Common, Sherfield English, Romsey, Hants, SO51 6FU [IO91EA, SU22]
G0 HLB Dr R A Evans, 50 Westby St., Lytham St. Annes, FY8 5JG [IO83MR, SD32]
G0 HLC K S Weight, 47 Ayelands, New Ash Green, Longfield, DA3 8JN [JO01DI, TQ66]
G0 HLD Details withheld at licensee's request by SSL.
G0 HLF D R Wainwright, 10 Hawkins Cres, Shoreham By Sea, BN43 6TP [IO90UV, TQ20]
G0 HLG G W Le Maistre, 8 Stocks Ln, Newland, Malvern, WR13 5AZ [IO82UD, SO74]
G0 HLH Details withheld at licensee's request by SSL.
G0 HLI A J Dudley, 7 St. Michaels Cl, Willington, Derby, DE65 6EB [IO92FU, SK22]
GM0 HLK M Borthwick, 5 Dalgety Ave, Edinburgh, EH7 5UF [IO85KX, NT27]
G0 HLL R A Steel, 43 Westfield Ave, Wigston, LE18 1HY [IO92KO, SP59]
G0 HLM Details withheld at licensee's request by SSL.
G0 HLN Details withheld at licensee's request by SSL.
GI0 HLO Details withheld at licensee's request by SSL.
GM0 HLP I Maclachlan, 9 Garvel Rd, Milngavie, Glasgow, G62 7JD [IO75TW, NS57]
GW0 HLR Details withheld at licensee's request by SSL.
G0 HLS S J Deacon, 32 Westfield Way, Charlton Heights, Wantage, OX12 7EW [IO91HO, SU48]
G0 HLT Details withheld at licensee's request by SSL.
GM0 HLV D W Gill, Slowbend Cottage, West Garty, Loth Helmsdale, Sutherland, KW8 6HP [IO88CB, NC91]
G0 HLW M P Neale, 126 Brookvale Rd, Olton, Solihull, B92 7JB [IO92CK, SP18]
G0 HLX Details withheld at licensee's request by SSL.
G0 HLY I M Castleton, 96 Blacksmiths Cres, Sompting, Lancing, BN15 0BY [IO90TU, TQ10]
GW0 HMC P Stevens, Noddfa, Dwrbach, Fishguard, Dyfed, SA65 9RL [IO71MW, SM93]
G0 HMD M Mills, 44 East Acridge, Barton upon Humber, DN18 5HH [IO93SQ, TA02]
G0 HME C J Statham, 24 St. Johns Cl, Heather, Coalville, LE67 2QL [IO92GQ, SK31]
G0 HMF J R Tite, 40 Dingledery, Olney, MK46 5ES [IO92PD, SP85]
G0 HMG S N Finch, 22 Sylvan Way, Redhill, RH1 4DE [IO91WF, TQ24]
G0 HMH B A C Dias, 36 St. Matthews Rd, Cotham, Bristol, BS6 5TU [IO81QL, ST57]
G0 HMI S W Sharrard, 9 Smalley Cl, Underwood, Nottingham, NG16 5GE [IO93IB, SK45]
G0 HMJ C M Baker, Clan Park Farm, Feiashill Rd, Trysull, Wolverhampton, WV5 7HT [IO82VM, SO89]
G0 HMK T J Angier, 29 Sunnyhill Rd, Herne Bay, CT6 8LT [JO01NI, TR16]
GM0 HML F Robertson-Mudie, 18, Portnaguran, Isle of Lewis, HS2 0HD [IO68WG, NB53]
G0 HMN S P Wilkinson, 15 Carden Ave, Sutton Trust Est, Hull, HU9 4RA [IO93US, TA13]
G0 HMO E J Cooke, 89 Colwick Woods Ct, Colwick Rd, Nottingham, NG2 4BB [IO92KW, SK53]
G0 HMP P Taylor Hm Prison Ser A, 22 Summerhill Dr, Maghull, Liverpool, L31 3DW [IO83MM, SD30]
G0 HMQ G G B Schofield, 93 Albion House, Grimsby, DN32 7HG [IO93XN, TA21]
GM0 HMR S K Martin, 57 Kilmuir Pl, Invergordon, IV18 0QA [IO77VQ, NH76]
G0 HMS F C Rafferty, 9 Mortlake Cl, Hull, HU8 0QH [IO93US, TA13]
G0 HMT L E Lockwood, 8 Wood Dr, Stevenage, SG2 8PA [IO91VV, TL22]
G0 HMU K M Jones, 17 Houldsworth Cres, Bolsover, Chesterfield, S44 6SQ [IO93IF, SK47]
G0 HMV Details withheld at licensee's request by SSL.
G0 HMW Details withheld at licensee's request by SSL.
G0 HMX M Wragg, 60 Kingsway, Worksop, S81 0AG [IO93KH, SK57]
G0 HMY T Pearce, 25 Central Ave, Kirkby in Ashfield, Nottingham, NG17 7FQ [IO93JC, SK55]
G0 HMZ J C Kelly, 7 Collingwood Cres, Matlock, DE4 3TB [IO93FD, SK26]
G0 HNA A J Hammon, 15 Abbey Way, Farnborough, GU14 7DA [IO91PH, SU85]

G0 HNB Details withheld at licensee's request by SSL.
G0 HND A G Simmonds, Fern Way, Purssells Meadow, Naphill, High Wycombe, HP14 4SG [IO91OP, SU89]
GW0 HNE K Luke, 19 Heol y Gors, Cwmgors, Ammanford, SA18 1PE [IO81BS, SN71]
G0 HNF Details withheld at licensee's request by SSL.
G0 HNG B A Horton, 3 Cromer Rd, Finedon, Wellingborough, NN9 5LP [IO92QI, SP97]
G0 HNI J M Buckler, 19 Craighill Rd, Leicester, LE2 3FD [IO92KO, SK50]
GM0 HNJ J I Cowan, 23 Camore Cres, Camore, Dornoch, IV25 3HU [IO77XV, NH79]
GW0 HNK J L Roberts, 47 Pyle Inn Way, Pyle, Bridgend, CF33 6LP [IO81DM, SS88]
G0 HNL T P Cannon, 67 Wren Way, Farnborough, GU14 8TA [IO91OH, SU85]
G0 HNO T C Callanan, 39 Greenlands Way, Henbury, Bristol, BS10 7PH [IO81QM, ST57]
GM0 HNP E J Bottomley, Lion Lodge North, Coldstream, Berwickshire, TD12 4HE [IO85VQ, NT84]
GW0 HNS S S Yates, 46 y Berllan, Killay, Dunvant, Swansea, West Glam, SA2 8RE [IO71XO, SS59]
GW0 HNT W G South, Kimberlee, 45 Dan y Bryn, Tonna, Neath, SA11 3PJ [IO81CQ, SS79]
G0 HNU R Martin, 65 Moor Dr, Crosby, Liverpool, L23 2UT [IO83LL, SD30]
GM0 HNV P J Mackenzie, 16 Cedar Dr, Milton of Campsie, Glasgow, G65 8AY [IO75WX, NS67]
G0 HNW P F Widger, Notre Revie, Cinderhills Rd, Holmfirth, Huddersfield, HD7 1EH [IO93CN, SE10]
G0 HNX A J Scott, 9 Treaty Park, Birgham, Coldstream, TD12 4NG [IO85TP, NT73][Op: Alex J Scott AMIEE.]
G0 HNZ R Disney, 4 Maris Cl, Barton Green, Clifton, Nottingham, NG11 8SH [IO92JV, SK53]
G0 HOA T Goodwin, 63 Lynwood Dr, Merley, Wimborne, BH21 1UT [IO90AS, SZ09]
G0 HOB J McGlynn, 1 Primrose Cres, Leeds, LS15 7QW [IO93GT, SE33]
G0 HOC M J Harris, PO Box 14-778, Panmure, Auckland, New Zealand, X X
G0 HOD N T Colbourn, 14 Eastway Sq, Nailsea, Bristol, BS19 1EX [IO81OK, ST47]
G0 HOE Details withheld at licensee's request by SSL.
G0 HOF K L Barnett, 5 Morborne Rd, Folksworth, Peterborough, PE7 3SS [IO92UL, TL19]
G0 HOG J M Wilson, Flat, 2 Peter Lyell Ct, Flemming Ave, Ruislip, HA4 9BL [IO91TN, TQ18]
G0 HOI A C Mudd, 67 Alexandra Rd, Grimsby, DN31 1RE [IO93WN, TA20]
G0 HOJ J W Hoyland, 67 Coronation Rd, Wroughton, Swindon, SN4 9AT [IO91CM, SU18]
G0 HOK Details withheld at licensee's request by SSL.
GW0 HOL Details withheld at licensee's request by SSL.
G0 HON Details withheld at licensee's request by SSL.
G0 HOO Details withheld at licensee's request by SSL.
G0 HOP G Evans, 12 Marlowe Cl, Stevenage, SG2 0JJ [IO91VV, TL22]
G0 HOQ Details withheld at licensee's request by SSL.
G0 HOR B L Tufnail, 197 Portland Rd, Wyke Regis, Weymouth, DT4 9BH [IO80SO, SY67]
G0 HOS M A Reeds, 6 Leach Way, Riddlesden, Keighley, BD20 5DB [IO93BV, SE04]
G0 HOU R O R Smith, 6 Bassey Rd, Branton, Doncaster, DN3 3NS [IO93LM, SE60]
G0 HOV A T J Howes, 11 Stretton Rd, Wolston, Coventry, CV8 3FR [IO92HI, SP47]
GI0 HOW Details withheld at licensee's request by SSL.
G0 HOX I P Atkins, 6 Cayenne Park, Swindon, SN2 2SF [IO91CO, SU18]
G0 HOY T Rampton, Chalemar, Eddeys Ln, Headley Down, Bordon, Hants, GU35 8HU [IO91OC, SU83]
G0 HPA G W Smith, 3 Park Ln, Featherstone, Pontefract, WF7 6BL [IO93HQ, SE42]
GW0 HPC G Evans, 49 Pen yr Ally Ave, Neath, SA10 6DS [IO81BP, SS79]
G0 HPD G L Tugwell, 67 Upper Kingston Ln, Shoreham By Sea, BN43 6TG [IO90VU, TQ20]
G0 HPG L E Brookes, 177 Charnwood Cl, Rubery, Rednal, Birmingham, B45 0JY [IO82XJ, SO97]
G0 HPH P G Brookes, 177 Charnwood Cl, Rubery, Rednal, Birmingham, B45 0JY [IO82XJ, SO97]
G0 HPJ D W Mason, 40 Washingley Rd, Folksworth, Peterborough, PE7 3SY [IO92UL, TL18]
GM0 HPK A E Gaston, Ellena, Lochans Mill, Lochans, Stranraer, Dumfries and Galloway, DG9 9BA [IO74LU, NX05]
GM0 HPL T L McCutcheon, 25 Millburn Ct, Sheuchan St., Stranraer, DG9 0DX [IO74LV, NX06]
G0 HPM R Waddingham, Blandford Hithe, Brimpton Rd, Reading, Berks, RG7 4TD [IO91JJ, SU56]
G0 HPN L M Denyer, Siesta, Enborne Row, Wash Water, Newbury, RG20 0LX [IO91HI, SU46]
G0 HPO Details withheld at licensee's request by SSL.
GW0 HPQ P A Delaney, 85 Beatty Ave, Gillingham, ME7 2DA [JO01GJ, TQ76]
G0 HPR Details withheld at licensee's request by SSL.
G0 HPT H P Tills, 10 East Causeway Vale, Leeds, LS16 8LG [IO93FU, SE23]
G0 HPU N J Powell, 2 Walton Ave, Twyford, Banbury, OX17 3LB [IO92IA, SP43]
G0 HPV K Hardy, 12 Liquorpond St., Boston, PE21 8UF [IO92XX, TF34]
G0 HPX J Porter, 96 Sandleford Rise, Newbury, RG14 7ET [IO91IJ, SU46]
G0 HPZ Details withheld at licensee's request by SSL.
G0 HQB Details withheld at licensee's request by SSL.
G0 HQC Details withheld at licensee's request by SSL.
GW0 HQD Details withheld at licensee's request by SSL.
G0 HQE Details withheld at licensee's request by SSL.
GM0 HQG G W Flett, Stenadale, Scorradale Rd, Orphir, Orkney, KW17 2RF [IO88JW, HY30]
G0 HQH S S Mayer, 7 Wright Ave, Chesterton, Newcastle, ST5 7PB [IO83VA, SJ84]
G0 HQJ Details withheld at licensee's request by SSL.
G0 HQK J S J Jones, 89 Doddington, Hollinswood, Telford, TF3 2DQ [IO82SQ, SJ70]
G0 HQN C R Wilson, 27 Worsley Ave, Little Hulton, Worsley, Manchester, M28 0HQ [IO83TM, SD70]
G0 HQO J Bate, 1 Ferndown Dr, Irlam, Manchester, M44 6JP [IO83TK, SJ79]
G0 HQP T Leek, 6 Bowden Rd, Swinton, Manchester, M27 5FP [IO83UM, SD70]
G0 HQQ J S Baker, 17 Bifield, Orton Goldhay, Peterborough, PE2 5SN [IO92UN, TL19]
G0 HQS F W Hollebon, 30 Diana Cl, Falmouth, TR11 4HY [IO70KD, SW73]
GM0 HQT B D Bell, 4 Broadlee Bank, Tweedbank, Galashiels, TD1 3RF [IO85OO, NT53]
G0 HQU S E Broadhead, 1 Second Ave, Rawdon, Leeds, LS19 6NH [IO93DU, SE24]
G0 HQX R H Rea, 11 Wissage Ln, Lichfield, WS13 6DQ [IO92CQ, SK11]
G0 HRA Details withheld at licensee's request by SSL.
G0 HRB H R Barnes, 1 Park View, Lingards Wood, Marsden, Huddersfield, HD7 6LR [IO93BO, SE01]
G0 HRD L W Lawrence, 119 Ladywood Rd, Lanes End, Dartford, DA2 7LP [JO01DK, TQ57]
G0 HRF B S Barwick, 100 Westwood, Golcar, Huddersfield, HD7 4JY [IO93BP, SE01]
GW0 HRG J W Washington obo Holywell Raynet Group (H.A, 15 Sundawn Ave, Holywell, CH8 7BH [IO83JG, SJ17]
G0 HRH G B Vaughton, Higher Woodhayne, Hampton Ln, Whitford, nr Axminster, East Devon, EX13 7PB [IO80LS, SY29]
G0 HRJ C C Mason, 22 Eskdale Ave, Off Belle Vue Rd, Halifax, HX3 7NH [IO93CS, SE12]
G0 HRL G W Ward, 67 Sebright Rd, Wolverley, Kidderminster, DY11 5UA [IO82UJ, SO87]
G0 HRM Details withheld at licensee's request by SSL.
G0 HRQ C R Davies, 23 Angel Cl, Dukinfield, SK16 4XA [IO83WL, SJ99]
G0 HRR K W Mott, 191 Joyners Field, Harlow, CM18 7QD [JO01BR, TL40]
G0 HRS A S Fennell Hiderstone RS, 12 Vale Rd, Ramsgate, CT11 9LU [JO01QH, TR36]
GM0 HRT R E Harwood, Milliken Park Rd, Kilbarchan, Milliken Park, Johnstone, Renfrewshire, PA10 2DB [IO75RT, NS46]
G0 HRW D H Barkley Wigan & Dist AR, 39 Fulbeck Ave, Goose Greeen, Wigan, WN3 5QN [IO83QM, SD50]
G0 HRX C Deakin, 10 Farnol Rd, Yardley, Birmingham, B26 2AF [IO92CL, SP18]
GU0 HRY D Henry, La Mare Hailla, La Rue Des Salines, St Pierre Du Bois, Guernsey, Channel Islands, GY7 9JD
G0 HRZ M J G Vasey, 317 High Rd, Chadwell Heath, Romford, RM6 6AX [JO01BN, TQ48]
G0 HSA Details withheld at licensee's request by SSL.[Op: A P Bennett.]
GI0 HSB W H Dickson, 8 Drumglass Ave, Bangor, BT20 3HA [IO74DP, J48]
GM0 HSC H S Cumming, 55 Barrmill Rd, Mansewood, Glasgow, G43 1EQ [IO75UT, NS56]
G0 HSE Dr I F Jones Earlsheaton Hig, Earlsheaton High School Rd C, 51 Springfield Park, Mirfield, WF14 9PE [IO93DQ, SE22]
GM0 HSG T Mehta Glasgow High AR, High School of Glasgow, 637 Crow Rd, Glasgow, G13 1PL [IO75UV, NS56]
G0 HSH D M Denford, 20 Shaxton Cres, New Addington, Croydon, CR0 0NU [IO91XI, TQ36]
G0 HSI H F Simmons, 4 Clay Hills, Halesworth, IP19 8TH [JO02SI, TM37]
G0 HSK J Hakes, 86 Station Cres, Rayleigh, SS6 8AR [JO01HO, TQ89]
G0 HSL G F Harman, 32 Orchard Gr, Brixham, TQ5 9RH [IO80FJ, SX95]
G0 HSM T J Morgan, 21 Apfel Ln, Chadderton, Oldham, OL9 9TA [IO83WN, SD90]
G0 HSN P J Broadley, 12 Strawberry Cl, Braintree, CM7 1EG [JO01GU, TL72]
G0 HSO Details withheld at licensee's request by SSL.
G0 HSR T J Venn Hntngdnshire AR, 22 Eaton Cl, Hartford, Huntingdon, PE18 7SR [IO92WI, TL27]
G0 HSS Details withheld at licensee's request by SSL.
G0 HSV S G Jacob, 81 Skeena Hill, South Shields, London, SW18 5PW [IO91VK, TQ27]
G0 HSW B Statham, 11 Old Woods Hill, Torquay, TQ2 7NR [IO80FL, SX96]
G0 HSZ Details withheld at licensee's request by SSL.
G0 HTD H Todd, 105 Brownhill Rd, Blackburn, BB1 9QY [IO83SS, SD63]
G0 HTE A W Stevens, 29 Denbigh Cl, Hornchurch, RM11 3EA [JO01CN, TQ58]
G0 HTF D J Crucefix, 5 Gainsborough Mews, Kidderminster, DY11 6PZ [IO82UJ, SO87]
GM0 HTG P Hague, Kitticks, Rendall, Orkney, KW17 2PB [IO89LB, HY41]
GM0 HTH J S Grieve, Langamo, Harray, Orkney, KW17 2JU [IO89KA, HY31]
G0 HTJ D Body, 53 Gr Rd, Wollescote, Stourbridge, DY9 9AE [IO82WK, SO98]
G0 HTK T D St John-Murphy, Sherbourne, Sherbourne Dr, Windsor, Berks, SL4 4AE [IO91QL, SU97]
G0 HTL B Sargent, 25 Jordans Way, Bricket Wood, St. Albans, AL2 3SJ [IO91TR, TL10]
G0 HTM Dr R G Bushell, 121 Rickmansworth Rd, Watford, WD1 7JD [IO91TP, TQ09]
G0 HTO V M Morris, 16 Wentworth Cl, Longlevens, Gloucester, GL2 9RB [IO81VV, SO82]
G0 HTP S Ainsworth, 494 Overpool Rd, Ellesmere Port, Whitby, South Wirral, L66 2JJ [IO83MG, SJ37]
G0 HTS S Alder, 135 Whitefield Rd, St. George, Bristol, BS5 7UB [IO81RL, ST67]
GM0 HTT A W Flett, Shannon, Greeny, Dounby, Orkney, KW17 2HR [IO89JC, HY22]
G0 HTV Details withheld at licensee's request by SSL.
G0 HTX A R Coyle, 8 Farm Pl, Kensington, London, W8 7SX [IO91VM, TQ28]
G0 HTY P O Owen, 20 Ringway, Waverton, Chester, CH3 7NP [IO83OE, SJ46]
G0 HTZ Details withheld at licensee's request by SSL.
G0 HUA L W Mann, 7 Homefield Cl, Swanley, BR8 7JH [JO01CJ, TQ56]
G0 HUB A W Hubbard, 39 Forester Rd, Thorneywood, Nottingham, NG3 6LP [IO92KX, SK54]
G0 HUC D A Anderson, 101 Knox Ave, Harrogate, HG1 3JF [IO94FA, SE35]
G0 HUD D E Withers, 141 Broadway, Walsall, WS1 3HB [IO92AN, SP09]
G0 HUE J Bowes, 22 Vale View, Durnhope, Burnhope, Durham, DH7 0EA [IO94DT, NZ14]
G0 HUF P I W Peterson, 13 Orchard Rd, Smallfield, Horley, RH6 9QP [IO91WE, TQ34]
G0 HUG P R Hemphill, 6 Stour Gdns, Great Cornard, Sudbury, CO10 0JN [JO02IA, TL83]
G0 HUI G A Potter, Windrush, Millers Green, Wirksworth, Derby, DE4 4BL [IO93FB, SK25]
G0 HUK G Roberts, 11 Moor Park Dr, Addingham, Ilkley, LS29 0PU [IO93BW, SE04]
GW0 HUN M W R Roberts, 19 Bro Geirionydd, Trefriw, LL27 0JE [IO83CD, SH76]
GM0 HUO A H Pardoe, 120 North St., St. Andrews, KY16 9AF [IO86OI, NO51]
G0 HUP Details withheld at licensee's request by SSL.
G0 HUQ R Monk, 43 Lichfield Dr, G, Bury, BL8 1BJ [IO83UO, SD71]
G0 HUR J J Farrell, 85 Chatsworth Rd, Stretford, Manchester, M32 9QD [IO83UK, SJ79]
GW0 HUS G Blomley, 9 Park View, Llaneurgain, Northop, Mold, CH7 6DD [IO83KE, SJ26]
GW0 HUT T H L Hutton, 14 Lancaster Way, Osbaston, Monmouth, NP5 4DA [IO81PT, SO51]
GM0 HUU S T Macdonald, 46 Underwood Ln, Paisley, PA1 2SL [IO75SU, NS46]
G0 HUV R Fawkes, 9 Crickley Dr, Warndon, Worcester, WR4 9LB [IO82VF, SO85]
G0 HUW A I Dyson, 24 Newborough Cl, Austrey, Atherstone, CV9 3EX [IO92FP, SK20]
G0 HUZ Details withheld at licensee's request by SSL.[Licencee T J Cadney. Station located near Surrey Docks, London SE16 1HN.]
G0 HVA J F Curtis, Glebe Farm, West Knighton, Dorchester, Dorset, DT2 8PE [IO80TQ, SY78]
G0 HVB D A Clark, Maple End, West Well Ln, Theale, Wedmore, BS28 4SW [IO81OF, ST44]
G0 HVC W E Hutchins, 25 Manor Rd, Herne Bay, CT6 6RF [JO01OI, TR26]
GM0 HVD A Doran, 8 Windhill Cres, Mansewood, Glasgow, G43 2UP [IO75UT, NS56]
G0 HVF E G Fogg, 189 Upton Rd, Bexley, DA5 1HG [JO01BK, TQ47]
G0 HVG W C Phillips, 216 Long St., Easingwold, York, YO6 3JD [IO94JC, SE56]
G0 HVH G N Watson, 14 Burnholme Ave, York, YO3 0LU [IO93LX, SE65]
GI0 HVJ R Cunliffe, 82 Strabane Rd, Newtownstewart, Omagh, BT78 4JZ [IO64HR, H38]
G0 HVL Details withheld at licensee's request by SSL.
G0 HVM Details withheld at licensee's request by SSL.
G0 HVN D S Cottam, The School of Public Policy, J G Smith Building, University of Birmingham, Edgbaston, Birmingham, B15 2TT [IO92AK, SP08]
G0 HVO R F Minter, 15 Harold Cl, Pevensey Bay, Pevensey, E Sussex, BN24 6SL [JO00ET, TQ60]
G0 HVP D A W Cooper, 1 Bempstead Cttgs, Lenacre St., Eastwell, Ashford, TN26 1JD [JO01KE, TR04]
G0 HVQ D R Moody, 6 Woodcote, Longford, Gloucester, GL2 9RX [IO81VV, SO82]
G0 HVS D E Kearns, 228 Turncroft Ln, Offerton, Stockport, SK1 4AX [IO83WJ, SJ99]
G0 HVT A Harding, 17 St. Anns Rd South, Heald Green, Cheadle, SK8 3DZ [IO83VI, SJ88]
G0 HVU Details withheld at licensee's request by SSL.
G0 HVV Details withheld at licensee's request by SSL.
G0 HVW Details withheld at licensee's request by SSL.
G0 HVX R S McQuillan, 9 Sandpit Rd, Welwyn Garden City, AL7 3TW [IO91VS, TL21]
G0 HVZ L Hindle, 48 Pine Cres, Walton on The Hill, Stafford, ST17 0NF [IO82XS, SJ92]
GD0 HWA C P Howard, 5 Ballure Gr, Ramsey, IM8 1NF
GM0 HWB P B Lafferty, 18 Chesters Cres, Motherwell, ML1 3QU [IO75XT, NS75]
G0 HWC P A Young, 519 Harlestone Rd, Northampton, NN5 6NX [IO92MG, SP76]
G0 HWE W H Smith, 15 Hillside, Appleby Magna, Swadlincote, DE12 7AB [IO92FQ, SK30]
G0 HWH J B Altoft, 33 Manor Park, Borrowash, Derby, DE72 3LP [IO92HV, SK43]
G0 HWI P R Mardle, 25 Stanley Rd, Canvey Island, SS8 8HD [JO01HM, TQ88]
G0 HWJ Details withheld at licensee's request by SSL.
G0 HWK M P Drew, 347 Ashcroft Rd, Stopsley, Luton, LU2 9AE [IO91TV, TL12]
G0 HWL A O Browne, 140 Tongham Rd, Aldershot, GU12 4AT [IO91PF, SU84]
G0 HWN Details withheld at licensee's request by SSL.
G0 HWO J R Crawford-Baker, 2 Alfred Cl, Worth, Crawley, RH10 7SD [IO91WC, TQ23]
G0 HWP A F Wilson, 41 Bentley Dr, Walsall, WS2 8RX [IO82XO, SO99]
G0 HWQ P R Wood, 10 Erdige Dr, Crowborough, TN6 2TJ [JO01CB, TQ53]
G0 HWS P H Dowsett, Strath Arun, Tower Cl, Tower Hill, Horsham, RH13 7AF [IO91TB, TQ12]
G0 HWT P E Littlechild, 2 Coombe Pine, Bracknell, RG12 0TJ [IO91PJ, SU86]
G0 HWU A N Edwards, Creca, The Ridgeway, Shorne, Gravesend, DA12 3LW [JO01FJ, TQ67]
G0 HWV B W Lawton, 9 Byron Ct, Kidsgrove, Stoke on Trent, ST7 4JF [IO83VB, SJ85]
G0 HWX S J Unsworth, 26 Rupert Brooke Rd, Rugby, CV22 6HQ [IO92II, SP47]
G0 HWY P E Adams, 229 Upper Selsdon Rd, South Croydon, CR2 0DZ [IO91XI, TQ36]
GW0 HXB R Evans, 17 Broniestyn Terr, Trecynon, Aberdare, CF44 8EG [IO81GR, SN90]
G0 HXC F Pegg, 20 Swaledale Ave, Cowpen Est, Blyth, NE24 4DT [IO95FC, NZ28]
G0 HXD P Halsall, 22 Northway, Northwich, CW8 4DF [IO83RG, SJ67]
G0 HXF A Etheridge, Wyngra, Haywards Heath Rd, Balcombe, Haywards Heath, RH17 6NJ [IO91WB, TQ33]
GI0 HXH A D McCaldin, 7 Mount Ida Rd, Banbridge, BT32 4HF [IO64VJ, J14]
G0 HXL E Calthorpe, Rowles Cottage, School Ln, Rothwell, Lincs, LN7 6BB [IO93UL, TF19]
G0 HXM A N N Delves, 11 Willoughby Rd, Langley, Slough, SL3 8JH [IO91RM, TQ07]
G0 HXN D J Mann, 106 South Meadow, Crowthorne, RG45 7HP [IO91OI, SU86]
G0 HXP A J Cousins, 28 Azalea Ave, Wickford, SS12 0BQ [JO01GO, TQ79]
G0 HXQ C Baron, 75 Princess Rd, Firgrove, Milnrow, Rochdale, OL16 4BA [IO83WO, SD91]
G0 HXR G W Martin, 54 Royston Ave, Chingford, London, E4 9DF [IO91XO, TQ39]
GW0 HXS T L Williams, Bryndewi, Pencae, Llanarth, Dyfed, SA47 0QN [IO72UE, SN45]
G0 HXT Details withheld at licensee's request by SSL.
G0 HXU S Dawson, 44 Avondale Rd, Darwen, BB3 1NS [IO83SQ, SD62]
G0 HXX J N Mackinnon, 49 Cookson Rd, Parsons Cross, Sheffield, S5 8LQ [IO93GK, SK39]
G0 HXY Details withheld at licensee's request by SSL.
G0 HXZ Details withheld at licensee's request by SSL.
G0 HYF Details withheld at licensee's request by SSL.
G0 HYG Details withheld at licensee's request by SSL.
G0 HYH V R Grayson, 48 Churchfield Rd, Wolverhampton, WV10 6TL [IO82WO, SJ90]
G0 HYI Details withheld at licensee's request by SSL.
GW0 HYJ V M Cross, 4 Maes Hywel, Meinciau, Kidwelly, SA17 5LU [IO71VS, SN41]
GW0 HYL V Underwood, Rock Hill, Llanarthney, Carmarthen, SA32 8LJ [IO71VU, SN51]
G0 HYN D Robertson, 5 Meadowbarn Cl, Cottam, Preston, PR4 0AG [IO83PT, SD53]
G0 HYR R E Deakin, 12 Henley Cl, Perrycrofts, Tamworth, B79 8TQ [IO92DP, SK20]
G0 HYS S Hutchinson, 55 Richmond Ave, Handsworth, Sheffield, S13 8TH [IO93HI, SK48]
G0 HYT P J Gray, 2 Bryan Cl, Sunbury on Thames, TW16 7UA [IO91SK, TQ17]
GW0 HYU K J Richards, 1 Clos Caer Wern, Caerphilly, CF83 1SQ [IO81JN, ST18]
G0 HYV R Sanderson, 15 The Close, Rainworth, Mansfield, NG21 0DF [IO93KC, SK55]
GM0 HYW R M Carmichael, 8 Winifred St., Kirkcaldy, KY2 5SR [IO86JC, NT29]
G0 HYX C W Hutchinson, Cloakham House, Axminster, Devon, EX13 5RP [IO80MS, SY29]
GM0 HYY Details withheld at licensee's request by SSL.
G0 HYZ R D Finch, 18 Barr Common Rd, Walsall, WS9 0SY [IO92AO, SP09]
G0 HZB M E Roberts, 4 Elba Cl, Goodrington, Paignton, TQ4 7LW [IO80FJ, SX85]
G0 HZC G N Hutt, 2 Gull Cl, Gosport, Hants, PO13 0RT [IO90JT, SU50]
G0 HZD W D Brown, Hillside, Greensplatt, St Austell, Cornwall, PL26 8XX [IO70OI, SW95]
G0 HZE R A Howell, 161 Coneygree Rd, Stanground, Peterborough, PE2 8LH [IO92VN, TL29]
G0 HZF L Barnwell, 6 Barnwell Cl, Dunchurch, Rugby, CV22 6QH [IO92II, SP47]
G0 HZG P Sturgess, 11 Laburnum Ave, Newbold Verdon, Leicester, LE9 9LQ [IO92HP, SK40]
GW0 HZJ Dr J A V Pritchard, 13 Cefn Graig, Cardiff, CF4 6SW [IO81JM, ST18]
G0 HZK R F Muggleton, 23 Randolph Rd, Langley, Slough, SL3 7QF [IO91RM, TQ07]
G0 HZL M Hawkins, 14 Dunes Ave, Blackpool, FY4 1PY [IO83LS, SD33]
GM0 HZM J T Aitken, Shawhead Rd, Dumfries, Galloway, DG2 9SJ [IO85CB, NX87]
G0 HZN J C T Corallo, 74 Thorn Gr, Cheadle Hulme, Cheadle, SK8 7LP [IO83VI, SJ88]
GM0 HZO T A Leckie, Dykehead Cottage, Kippen By Stirling, Scotland, FK8 3JY [IO76UD, NS59]
G0 HZP Details withheld at licensee's request by SSL.
G0 HZQ R P Eldrett, Widnell Farm, Hampden Bottom, Great Missenden, Bucks, HP16 9PT [IO91OR, SP80]
G0 HZR Details withheld at licensee's request by SSL.
G0 HZS C E Gray, 42 Riber Ave, Athersley South, Barnsley, S71 3PU [IO93GN, SE30]
G0 HZT M C Darwood, 24 Devon House, Lower Beeches Rd, Northfield, Birmingham, B31 5JU [IO92AJ, SP07]
G0 HZU Details withheld at licensee's request by SSL.
G0 HZV Details withheld at licensee's request by SSL.
GW0 HZW Details withheld at licensee's request by SSL.
G0 HZX Details withheld at licensee's request by SSL.[Station located near Egham.]
G0 HZY H Harding, 29 Brighton Ave, Elson, Gosport, PO12 4BU [IO90KT, SU50]
G0 HZZ N F Coathup, 1 Heath Gr, Little Sutton, South Wirral, L66 5PD [IO83MG, SJ37]
G0 IAA H R Singer, Neptune Gap, Whitstable, Kent, CT5 1EL [JO01MI, TR16]
G0 IAD A R Dobbyn, Selwick Dr, Flamborough Head, Flamborough, Bridlington, North Humberside, YO15 1AP [IO94WC, TA27]
G0 IAE D Yeo, 356 Radcliffe Rd, Fleetwood, FY7 7NH [IO83LV, SD34]
G0 IAF A Fernandez, 2 Silverston Ave, Bognor Regis, West Sussex, PO21 2RB [IO90PS, SZ99]
G0 IAG A King, 2 Easton Manor, Park Rd, Little Easton, Dunmow, CM6 2JN [JO01DV, TL62]

G0 IAH T F Preece, White House, Pentre Ln, Bredwardine, Hereford, HR3 6BY [IO82MC, SO34]
G0 IAI P G Dean, Down Farm, Lovaton, Yelverton, Devon, PL20 6PT [IO70XL, SX56]
G0 IAJ A J Powell, 3 Vinings Rd, Sandown, PO36 8DU [IO90KP, SZ58]
G0 IAK M J Watts, Hainault, Coronation Ave, Bradford on Avon, BA15 1AX [IO81VI, ST86]
G0 IAL R C S Scott, 46 St. Albans Rd, Hemel Hempstead, HP2 4BA [IO91SR, TL00]
G0 IAM Details withheld at licensee's request by SSL.
G0 IAN Details withheld at licensee's request by SSL.[Correspondence via PO Box 25, Hereford, HR1 1YL.]
G0 IAP P H Allen, 25 Wayside Ave, Hornchurch, RM12 4LL [JO01CN, TQ58]
G0 IAQ Details withheld at licensee's request by SSL.
G0 IAS A R Hickman, The Conifers, High St., Elkesley, Retford, DN22 8AJ [IO93MG, SK67]
G0 IAT Details withheld at licensee's request by SSL.[Op: A J Swainsbury.]
GW0 IAU J M Evans, 3 Penrhgn Ave, Bangor, Gwynedd, LL57 1TH [IO73WF, SH57]
G0 IAV F G Dove, 24 Beresford Ave, Lower Bebington, Wirral, L63 7LS [IO83LI, SJ38]
G0 IAW S C Davis, 26 Sherwood Rd, Meols, Hoylake, Wirral, L47 9RT [IO83KJ, SJ28]
G0 IAX R A B Bailey, 10 St. Michaels Cl, Lyneham, Chippenham, SN15 4NZ [IO91AM, SU07][Station located near Cirencester.]
G0 IAY L S Wiltshire, 4 Clover Field, Lychpit, Basingstoke, RG24 8SR [IO91LG, SU65]
G0 IBB M G Brynes, 4 Rose Cttgs, Ecchinswell, Newbury, RG20 4TZ [IO91II, SU56]
GI0 IBC P J Hallam Radio Amateur Invalid and Blin, 95 Belfast Rd, Carrickfergus, BT38 8BX [IO74CR, J48]
G0 IBD Details withheld at licensee's request by SSL.
G0 IBF H M Nevey, 25 Folkestone Croft, Bromford Est, Birmingham, B36 8QY [IO92CM, SP18]
G0 IBH T Phillips, 32 Kingsway Ave, Paignton, TQ4 7AD [IO80FJ, SX85]
G0 IBJ R P Bishop, 25 Hennings Park Rd, Oakdale, Poole, BH15 3QU [IO90AR, SZ09]
G0 IBK J Wilkins, 13 Shaftesbury Ct, Buckthorn Ave, Stevenage, SG1 1TX [IO91VV, TL22]
G0 IBL Details withheld at licensee's request by SSL.
GM0 IBM M M McCreery IBM Larkfield Rd, 1 Duart Dr, Newton Mearns, Glasgow, G77 5DS [IO75US, NS55]
G0 IBN A Kersey, 35 Sceptre Cl, Tollesbury, Maldon, CM9 8XB [JO01KS, TL91]
G0 IBP Details withheld at licensee's request by SSL.
G0 IBQ D A Blain, 30 Somerville Rd, Sutton Coldfield, B73 6JA [IO92CN, SP19]
G0 IBT H Percy, 13 Cherry Tree Walk, Astley Cross, Stourport on Severn, DY13 0JT [IO82UH, SO86]
G0 IBU E Howdon, 36 Moseley Wood Green, Cookridge, Leeds, LS16 7HB [IO93EU, SE24]
G0 IBV R G Theyer, 4 Albion Rd, Cinderford, GL14 2TA [IO81ST, SO61]
G0 IBW D Jones, 20 Duhallow Cl, Hunters Hill, Guisborough, TS14 7PF [IO94LM, NZ61]
G0 IBX F A Griffin, 21 Holden Way, Great Gonerby, Grantham, NG31 8LA [IO92QW, SK83]
G0 IBY F Hubbard, 187 Standhill Rd, Carlton, Nottingham, NG4 1LE [IO92KX, SK54]
G0 IBZ M J Hackford, Snuggles Rockalls Rd, Polstead, Colchester, Essex, CO6 5AR [JO02KA, TL93]
G0 ICA C Smith, Asklund, Drayton Ln, Drayton Bassett, Tamworth, B78 3TS [IO92DO, SK10]
G0 ICB J D Howell, 21 Peaslands Rd, Saffron Walden, CB11 3ED [JO02DA, TL53]
G0 ICC R L Shirvington, 3 The Sq, Preston Bissett, Buckingham, MK18 4LP [IO91LX, SP62]
G0 ICD D G Matthews, 84 Roscoe Cres, Weston Point, Runcorn, WA7 4ER [IO83PH, SJ58]
G0 ICE M P Knott, 24 Walsingham Way, Ely, CB6 3AL [JO02DJ, TL58]
GM0 ICF Details withheld at licensee's request by SSL.
G0 ICG Sir T E Lees, Post Green, Lytchett Minster, Poole, Dorset, BH16 6AP [IO80XR, SY99]
G0 ICH F W Norton, 24 Greenfields Dr, Little Neston, South Wirral, L64 0TX [IO83LG, SJ27]
G0 ICI Details withheld at licensee's request by SSL.
G0 ICJ D G Dawkes, 83 Alcester Rd, Hollywood, Birmingham, B47 5NR [IO92BJ, SP07]
G0 ICK I J Kraven Sth Tottenhm AR, 55 Cranfield Cres, Cuffley, Potters Bar, EN6 4DZ [IO91WR, TL30]
G0 ICM P T Foxall, 48 Chawn Hill, Stourbridge, DY9 7JA [IO82WK, SO98]
GM0 ICP J Ker, 1 Kerr Pl, Peebles, EH45 8DA [IO85JP, NT24]
G0 ICQ J Corney, 21 Otterwood Ln, Acomb, York, YO2 3JP [IO93KW, SE55]
G0 ICT J D M Morrison, 52 Kimberley Cl, Dover, CT16 2JW [JO01PD, TR34]
GW0 ICU J E Weller, 14 St. Davids Cres, Penarth, CF64 3LZ [IO81JK, ST17]
G0 ICW M A Bagnall, 37 Heath Gap Rd, Cannock, WS11 2DY [IO82XQ, SJ91]
G0 ICX Details withheld at licensee's request by SSL.
G0 ICY E Sledge, 9 Main St., Failsworth, Manchester, M35 9PD [IO83WM, SD80]
G0 ICZ Details withheld at licensee's request by SSL.
G0 IDB S J L Jefferys, Broadmark House, 26 Meadway, Esher, KT10 9HF [IO91TI, TQ16]
G0 IDC R F W Wright, 16 Oxford St., Watford, WD1 8ES [IO91TP, TQ19]
G0 IDD D Clarke, 29 Birchfield Rd, Stapenhill, Burton on Trent, DE15 9PT [IO92ES, SK22]
G0 IDF D W Fletcher, 9 Hazel End, Garsington, Oxford, OX44 9AW [IO91KR, SP50]
G0 IDH M R Frost, Kernyk, Trevanion Terr, Carn Brea Village, Redruth, TR15 3BP [IO70JF, SW64]
G0 IDI M G Cooper, 6 Royal Cres, Scarborough, YO11 2RN [IO94TG, TA08]
GM0 IDJ J A Low, 2 Craigie Pl, Crosshouse, Kilmarnock, KA2 0JR [IO75RO, NS33]
G0 IDK Details withheld at licensee's request by SSL.
G0 IDL M D Bedell, 1 Pheasant Field, Spondon, Derbyshire, DE2 7LR [IO92FW, SK33]
G0 IDN Details withheld at licensee's request by SSL.
G0 IDP D E Ford, 2 Bedelands Cl, Burgess Hill, RH15 8BL [IO90WX, TQ32]
G0 IDR L R Heath, 158 Warwick Ave, Derby, DE23 6HL [IO92GV, SK33]
G0 IDS S McMaster, 24 Garburn Pl, Newton Aycliffe, DL5 7DE [IO94FO, NZ22]
GM0 IDV A Price, Upper Airsdale, Costa Evie, Orkney Isles, KW17 2NN [IO89JD, HY32]
GM0 IDY R Milne, 1 Catriona Terr, Penicuik, EH26 0LX [IO85JU, NT26]
G0 IDZ L Bailey, 22 Coventry Gr, Wheatley, Doncaster, Yorks, DN2 4QA [IO93KM, SE50]
G0 IEB R A Neal, 2 Ellis Gdns, Scalby, Scarborough, YO13 0ST [IO94SH, TA09]
G0 IEC W G Jennings, 21 Foxon Ln, Caterham, CR3 5SG [IO91WG, TQ35]
G0 IED R L Lawrence, 46 The Grange, Woodham, Newton Aycliffe, DL5 4SZ [IO94FP, NZ22]
G0 IEE K C Harris, 14 Dunstall Cl, St. Marys Bay, Romney Marsh, TN29 0QX [JO01LA, TR02]
G0 IEF A Thompson, 59 Leyland Rd, Harrogate, HG1 4RU [IO93FX, SE35]
G0 IEG S E Williams, 304 The Wells Rd, Mapperley, Nottingham, NG3 3AA [IO92KX, SK54]
G0 IEH T Lister, 6 Fordlands Cres, Fulford, York, YO1 4QQ [IO93LW, SE64]
G0 IEI D M Waddington, 56 Alderton Bank, Moortown, Leeds, LS17 5LG [IO93FU, SE23]
G0 IEJ J Riley, 51 Nuttall Ln, Ramsbottom, Bury, BL0 9JX [IO83UP, SD71]
G0 IEN Dr R P Wharton, 1407 Briar Bayou Dr, Houston, Texas, U.S.A 77077
G0 IEO R E Woodberry, 35 Whybridge Cl, Rainham, RM13 8BB [JO01CM, TQ58]
G0 IEP P P Cuffe, 115 Abercrombie Rd, Fleetwood, FY7 7AY [IO83LW, SD34]
G0 IEQ G C Kennedy, 11 Nantwich Rd, Great Sutton, South Wirral, L66 2SJ [IO83MG, SJ37]
G0 IER B B Smith, 48 Colchester Cl, Toothill, Swindon, SN5 8AG [IO91CN, SU18]
G0 IES D R Johnson, 32 Middleton Rd, North Reddish, Stockport, SK5 6SH [IO83WK, SJ89]
G0 IET M J Shell, 25 Western Ave, Easton on The Hill, Stamford, PE9 3NB [IO92RO, TF00]
GW0 IEU R W Prior, 3 Hampden Rise, Royston, SG8 9UB [IO92XB, TL34]
G0 IEV M D Rippington, 15 Queens Ave, Canterbury, CT2 8AY [JO01MG, TR15]
G0 IEW J A Rose, 1 Nelson Pl, Whiston, Prescot, L35 3PP [IO83OJ, SJ49]
G0 IEY S J B Tribe, 17 Shaftesbury Ave, Purbrook, Waterlooville, PO7 5HW [IO90LU, SU60]
G0 IEZ Details withheld at licensee's request by SSL.
G0 IFA K P Keitch, 10 Sycamore Cl, Willand, Cullompton, EX15 2SH [IO80HV, ST01]
G0 IFC A W Burnett, 1 Hornbeam Cl, Gornhay View, Tiverton, EX16 6EY [IO80GV, SS91]
G0 IFD T Street, Spring Manor, 323 Green Lanes, Manor House, London, N4 2ES [IO91WN, TQ38]
G0 IFF G M Spinney, 28 Savoy Rd, Brislington, Bristol, BS4 3SY [IO81RK, ST67]
G0 IFG Details withheld at licensee's request by SSL.[Op: Alan Harvey. Station located near Farnborough, Hants. Correspondence c/o GB3SN, PO Box 6, Alton, Hants, GU34 2SP. 24hr packet PMS 144.675 MHz.]
G0 IFH H J Joyce, Flat, 6 Ln End Ct, Ln End Rd, Bembridge, PO35 5UE [IO90LQ, SZ68]
G0 IFI P J Mountstevens, 11 Deacon Rd, Bridgwater, TA6 4QJ [IO81MD, ST33]
G0 IFJ M E Harris, 47 Sherwood Dr, Hull, HU4 7RG [IO93TR, TA02]
G0 IFK M J Faulkner, 17 Lynnbank Rd, Liverpool, L18 3HE [IO83NJ, SJ48]
G0 IFL R J Finch, 16 Kemperleye Way, Bradley Stoke, Bristol, BS12 8EB [IO81RM, ST68]
G0 IFM Details withheld at licensee's request by SSL.
G0 IFN R L Moriarty, 105 Brendon Cres, Billingham, TS23 2QU [IO94IO, NZ42]
G0 IFO R B Greenwood, 19 Robindale Ave, Etobicoke, Ontario, Canada M8W 4A8
G0 IFP Details withheld at licensee's request by SSL.
G0 IFQ P E Lait, Glenside, 8 Kingston Ln, Shoreham By Sea, BN43 6YB [IO90VU, TQ20]
G0 IFR D Furniss, 11 Mere Ct, Sandbach Rd North, Alsager, Stoke on Trent, ST7 2AP [IO83UC, SJ75]
G0 IFS N K Clayton, Ringinglow, Fair Lawn, Chestfield, Whitstable, CT5 3JZ [JO01MI, TR16]
G0 IFT P Nurse, 67 Grasleigh Way, Allerton, Bradford, BD15 9BD [IO93BT, SE13]
GD0 IFU W C Corkish, 17 Ballachrink Dr, Onchan, Douglas, IM3 4NU
G0 IFV G Bate, 1 Ferndown Dr, Irlam, Manchester, M44 6JP [IO83TK, SJ79]
G0 IFW M A McNamara, 44 King Down Rd, Blandford Camp, Blandford Forum, DT11 8BL [IO80WV, ST90]
G0 IFX B R Galloway, 17 Clifton Ave, Fallowfield, Manchester, M14 6UD [IO83VK, SJ89]
G0 IFY P H Barton, 15 Renoir Ct, Fairlands, Bognor Regis, PO22 9AY [IO90PT, SU90]
G0 IGA R S Clark, 9 Windsor Cl, Hove, BN3 6WQ [IO90VU, TQ20]
G0 IGB V Elliott, 22 Kirkstead Rd, Cheadle Hulme, Cheadle, SK8 7PZ [IO83VI, SJ88]
G0 IGC J F Golightly, 123 Littlefield Ln, Great, Grimsby, DN34 4PN [IO93WN, TA20]
G0 IGD S E Liddicoat, 4 Rosevear Terr, Roseveur, Bugle, St. Austell, PL26 8RJ [IO70OJ, SX05]
GM0 IGF J C Davidson, 1 Dowlaw Cttgs, Coldsmith, Eyemouth, Berwickshire, TD14 5TY [IO85VW, NT87]
G0 IGH G G Golightly, 123 Littlefield Ln, Grimsby, DN34 4PN [IO93WN, TA20]
G0 IGI H F Overington, 75 Palatine Rd, Goring By Sea, Worthing, BN12 6JR [IO90TT, TQ10]
GM0 IGJ J J Dickson, Eilrig, By Roberton, Hawick, TD9 7PR [IO85MI, NT30]
G0 IGK G F Knox, 33 St. Lukes Rd, Aller Park, Newton Abbot, TQ12 4NE [IO80EM, SX87]
G0 IGM M R Manning, 130 Underdale Rd, Shrewsbury, SY2 5EF [IO82PR, SJ51]
G0 IGN V P Cambery, 285 Hitchin Rd, Luton, LU2 7SL [IO91TV, TL02]
G0 IGO J C Ashcombe, 11 Bourne Hill, London, N13 4LJ [IO91WP, TQ39]
G0 IGP F A Hacking, 59 Malvern Gdns, Kenton, Harrow, HA3 9PA [IO91UO, TQ18]
G0 IGR G F Harkness, 30 Fountains Rd, Bramhall, Stockport, SK7 1ET [IO83VI, SJ88]
G0 IGS R Evans, 102 Blackburn Rd, Darwen, BB3 1QJ [IO83SQ, SD62]
G0 IGT Details withheld at licensee's request by SSL.
G0 IGZ L Mayfield, 1 Poplar Ave, New Mills, Stockport, Ches, SK12 4HR [IO83XI, SJ98]
G0 IHA G S Goodwin, 16 St. Catherines Rd, Winchester, SO23 0PP [IO91IB, SU42]
G0 IHB G E Walker, 36 Vale Royal River Park, Moulton, Northwich, CW7 2PF [IO83RF, SJ66]
G0 IHC M A J Nash, 49 Oakfield Way, Sharpness, Berkeley, GL13 9UT [IO81SR, SO60]
G0 IHE Q L Reed, 3 Carre Gdns, Worle, Weston Super Mare, BS22 0YB [IO81NI, ST36]
G0 IHF W J Hough, Furrocks Ln, Ness, South Wirral, L64 4EH [IO83LG, SJ27]
G0 IHH F A Barksfield OIC Pri Sch of, Elect Eng Arcade, 10 Corsica Rd, Seaford, BN25 1BB [JO00BS, TV49]
G0 IHI M J Eyers, 190 Greenhill Rd, Herne Bay, CT6 7RS [JO01NI, TR16]
G0 IHK I S Tough, 15 Headlands Way, Whittlesey, Peterborough, PE7 1RL [IO92WN, TL29]
G0 IHM U M Sadler, 10 Hunt Hall Ln, Welford on Avon, Stratford upon Avon, CV37 8HF [IO92CD, SP15]
G0 IHO R E Priestley, 11 South Ave, Ullesthorpe, Lutterworth, LE17 5DG [IO92IL, SP58]
G0 IHP S Hanslow, 62 Habberley Rd, Kidderminster, DY11 5PE [IO82UJ, SO87]
G0 IHS P Antonelos, 61 Woodvale Rd, Woolton, Liverpool, L25 8RY [IO83NJ, SJ48]
G0 IHU C G Smith, 66 Bronte Farm Rd, Shirley, Solihull, B90 3DF [IO92CJ, SP17]
G0 IHY J Redmond, 8 Oakdene Ave, Woolston, Warrington, WA1 4NU [IO83RJ, SJ68]
G0 IIA J Stokes, 46 Florin Ct, Charterhouse Sq, London, EC1M 6EU [IO91WM, TQ38]
G0 IIB B A Daniels, 18 Oakley Rd, Chinnor, OX9 4HB [IO91NQ, SP70]
G0 IIC G F Light, 4 Oak Dr, Crownhill, Plymouth, PL6 5TZ [IO70WJ, SX45]
G0 IID J I Batley, 3 Folldon Ave, Sunderland, SR6 9HP [IO94HW, NZ35]
G0 IIE J E Colliton, PO Box 17, Ramsgate 2219, Sydney, Australia
G0 IIF J C Green, 24 Woodcross Ave, Manor, Doncaster, DN4 6RU [IO93LM, SE60]
G0 IIG G Fleming, 168 Blythway, Welwyn Garden City, AL7 1DU [IO91VT, TL21]
G0 III G R Wratten, 1A Muriel Rd, Waterlooville, PO7 7TE [IO90LV, SU60]
G0 IIJ L G Wardale, 224 Greenhill Rd, Mossley Hill, Liverpool, L18 7HW [IO83NI, SJ48]
G0 IIK N W J A Ackland, 138 Harlech Gdns, Heston, Hounslow, TW5 9PT [IO91TL, TQ17]
G0 IIL J L Iceton, 5 Dorlcote Pl, Norton, Stockton on Tees, TS20 2PP [IO94IN, NZ42]
G0 IIM J R Giles, 1 Parklands Dr, Sale, M33 4NU [IO83TJ, SJ79]
G0 IIN R D Irving, 7 Frogmore Rd, Huntley, Gloucester, GL19 3EP [IO81TU, SO71]
GM0 IIO Details withheld at licensee's request by SSL.
G0 IIP R K Chambers, 113 Kneller Rd, Whitton, Twickenham, TW2 7DY [IO91TK, TQ17]
G0 IIQ Details withheld at licensee's request by SSL.
GW0 IIR Details withheld at licensee's request by SSL.
G0 IIS Details withheld at licensee's request by SSL.
G0 IIU W Jackson, 4 Langdale Pl, Lancaster, LA1 3NS [IO84OB, SD46]
GW0 IIV Details withheld at licensee's request by SSL.
GW0 IIW T Hayden Ebbw Vale Col Rd, 7 Attlee Cl, Garnlydan, Ebbw Vale, NP3 5ES [IO81JT, SO11]
GM0 IIX Details withheld at licensee's request by SSL.
G0 IIY H R Miller, 3 Blakenham Cl, Oulton Broad, Lowestoft, NR32 4RW [JO02UL, TM59]
GI0 IIZ Rev R T Davey, 15 Heathfield, Culmore Rd, Londonderry, BT48 8JD [IO65IA, C42]
GM0 IJA B C Skinner, Lynwood, Duffshill Rd, Portlethen, Aberdeen, AB1 4RX [IO87TH, NJ72]
GI0 IJB J Stevenson, 12 Glenisland Terr, Carrickfergus, BT38 8RD [IO74BQ, J38]
GW0 IJC E G D Murphy, Sancta Maria, 108 Oaklands, Swiss Valley Park, Llanelli, SA14 8DL [IO71WQ, SN50]
GM0 IJD P Taggerty, 132 Forres Dr, Glenrothes, KY6 2JY [IO86JE, NO20]
G0 IJE D R King, 131 Poole Ln, Bournemouth, BH11 9DZ [IO90BS, SZ09]
G0 IJI A W Muir, Nyauga, 5 New St., Stainland, Halifax, HX4 9QL [IO93BQ, SE01]
G0 IJJ S A P Johnson, Springfield Cottage, 16 Sandbeds Rd, Willenhall, WV12 4EY [IO82XO, SO99]
G0 IJK B M Puncher, Danbys Oast, Shooters Hill, Grafty Green, Maidstone, Kent, ME17 2AX [JO01IE, TQ84]
G0 IJM Details withheld at licensee's request by SSL.
G0 IJN A J Slade, Skerries, Summerhill, Althorne, Chelmsford, Essex, CM3 6BY [JO01JP, TQ99]
G0 IJO A S Russell, 69 Moortower, Harlow, Essex, CM18 6BE [JO01BS, TL40]
G0 IJP P H Taggerty, 20 Haycombe, Durweston, Blandford Forum, DT11 0PZ [IO80VU, ST80]
GM0 IJR I J Ross, 14 Kilmundy Dr, Burntisland, KY3 0JW [IO86JB, NT28]
G0 IJS T E Maskell, 85 Leggatt Dr, Bramford, Ipswich, IP8 4EY [JO02NB, TM14]
G0 IJU P R Elms, 29 Mill Ln, Gaywood, Kings Lynn, PE30 3DT [JO02FS, TF62]
GM0 IJV K F W Doughty, Quoylanks, Deerness, Orkney, KW17 2QQ [IO88OW, HY50]
GW0 IJY I W Roberts, 3 Isallt Est, Trearddur Bay, Holyhead, Gwynedd, LL65 2UP [IO73QG, SH27]
G0 IJZ Dr M C Walden, 13 St. Andrews Ave, Harrogate, HG2 7RH [IO93FX, SE35]
G0 IKB K Brown, 3 Orchard Way, Horsmonden, Tonbridge, TN12 8JX [JO01FD, TQ74]
G0 IKC R S Hole, 3 Holywell Park, Halwill, Beaworthy, EX21 5UD [IO70VS, SS40]
G0 IKD A E Galvin, 23 Poppleton Rd, Tingley, Wakefield, WF3 1UX [IO93FR, SE22]
G0 IKE L R Dodson, 1 Limmer Ln, Booker, High Wycombe, HP12 4QR [IO91OO, SU89]
G0 IKG Details withheld at licensee's request by SSL.
G0 IKH Details withheld at licensee's request by SSL.
G0 IKI M F H Holbrough, 21 Malyns Cl, Chinnor, OX9 4EW [IO91NQ, SP70]
G0 IKL G Hornsby, 16 Fanshaws Ln, Brickendon, Hertford, SG13 8PF [IO91WS, TL30]
G0 IKN S O McDonald, 152 Stony Ln, Burton, Christchurch, BH23 7LD [IO90CR, SZ19]
G0 IKO G R C Dryer, Parklands, 46 Bowthorpe Rd, Wisbech, PE13 2DX [JO02CP, TF40]
G0 IKP T G Clements, 72 Stanbridge Rd, Haddenham, Aylesbury, HP17 8HN [IO91NS, SP70]
G0 IKR M C Best, 11 Howe Cres, Corby, NN17 2RY [IO92PL, SP88]
G0 IKS Details withheld at licensee's request by SSL.
G0 IKT Details withheld at licensee's request by SSL.
G0 IKV D J Southworth, 65 Zetland St., Southport, PR9 9DL [IO83MP, SD31]
G0 IKX B F Parker, 15 Bede Terr, Bowburn, Durham, DH6 5DT [IO94FR, NZ33]
GM0 IKY A G Diamond, 136 James Brown Ave, Ayr, KA8 9SQ [IO75QL, NS32]
G0 IKZ J Pugh, 1 The Lawns, Everton, Sandy, SG19 2LB [IO92VD, TL25]
G0 ILA N B C Smith, 165 Upper Deacon Rd, Bitterne, Southampton, SO19 5LN [IO90HV, SU41]
GM0 ILB I L Brown, Vaddel, Busta, Brae, Shetland Isles, ZE2 9QN [IP90HJ, HU36]
G0 ILC J R Price, 30 Pottery Cl, Whiston, Prescot, L35 3RW [IO83OJ, SJ49]
G0 ILD B D R Harrison, 61 Foyle Rd, Blackheath, London, SE3 7RQ [JO01AL, TQ37]
G0 ILE E H West, 89 Farnham Rd, Welling, DA16 1LG [JO01BL, TQ47]
G0 ILF Details withheld at licensee's request by SSL.
G0 ILH J Collins, 15 Hilltop Rd, Dudley, DY2 7EY [IO82XM, SO98]
G0 ILI G J Pomroy, 49 Green End, Landbeach, Cambridge, CB4 4ED [JO02BG, TL46]
G0 ILJ Details withheld at licensee's request by SSL.
G0 ILK Details withheld at licensee's request by SSL.
G0 ILL A V Butcher, Kirkdale, 34 Highfields Rd, Marlpit Hill, Edenbridge, TN8 6JW [JO01AF, TQ44]
G0 ILM R W Honnor, 67 Carr View Ave, Balby, Doncaster, DN4 8AX [IO93KM, SE50]
G0 ILN R J Putnam, Hedgerows, 95 Martyns Way, Bexhill on Sea, TN40 2SH [JO00GU, TQ70]
G0 ILO P S Taylor, Top Flat, 20 Gordon Rd, Herne Bay, CT6 5QT [JO01NI, TR16]
GM0 ILQ I B F Ferguson, Treetops, Gowkhouse Rd, Kilmacolm, PA13 4DJ [IO75QV, NS36]
G0 ILR J Bradley, 31 Exeter Cl, Ashington, NE63 9RZ [IO95FE, NZ28]
G0 ILS R I Bird, 13 Cuffelle Cl, Chineham, Basingstoke, RG24 8RH [IO91LH, SU65]
G0 ILT T Drummond, 10 Delamere Rd, Earley, Reading, RG6 1AP [IO91MK, SU77]
G0 ILW L R Wilbraham, 17 Lennard Ave, West Wickham, BR4 9AZ [JO01AI, TQ36]
G0 ILZ M Hanraads, 6 Oak Hill, Hollesley, Woodbridge, IP12 3JY [JO02RB, TM34]
G0 IMA P R Pearce, 7 Otter Rd, Southway Park, Clevedon, BS21 6LQ [IO81NK, ST47]
G0 IMB R Battersby, 9 Montreal St., Hathershaw, Oldham, OL8 1LW [IO83WM, SD90]
G0 IMC A P Kerr, The Leys, 21 Watton Green, Watton, Norfolk, IP25 6RB [JO02KN, TF90]
G0 IMD R A Clamp, 276 The Parade, Greatstone, New Romney, TN28 8UL [JO00LW, TR02]
G0 IMF A Dudley, 81 Parklands Way, Poynton, Stockport, SK12 1AT [IO83WI, SJ98]
G0 IMG M S Gotch, 44 Audley Rd, Saffron Walden, CB11 3HD [JO02CA, TL53]
GM0 IMH C G D Taylor, 10 East Lennox Dr, Helensburgh, G84 9DG [IO76PA, NS38]
G0 IMI G H Parkes, 4 Denewood Cres, Bilborough Est, Nottingham, NG8 3DA [IO92JX, SK54]
G0 IMJ D G Clarke, Meabo, 7 Water Ln, Tiverton, EX16 6RA [IO80GV, SS91]
G0 IMK N Sparrey, The Ashes, Clows Top, Kidderminster, DY14 9HX [IO82SH, SO77]
G0 IMM L A E Wealthy, 247 Upton Rd, Ryde, PO33 3JG [IO90JR, SZ59]
G0 IMN J L Hudson, 65 Sorrel, Amington, Tamworth, B77 4HA [IO92EP, SK20]
G0 IMP N L Wheeldon, 28 Constance Ave, Lincoln, LN6 8SN [IO93QE, SK96]
G0 IMT Details withheld at licensee's request by SSL.[Op: B J Parker. Station located near Bournemouth. Locator IO90CR. Correspondence to P O Box 249, Poole, Dorset BH15 2LR.]
G0 IMV R L Hill, Marclecote, Ledbury Rd, Ross on Wye, HR9 7BE [IO81RW, SO62]
GM0 IMW A T McBride, 16 Kingshill Dr, Kings Park, Glasgow, G44 4QY [IO75VT, NS56]
G0 IMX C Lacey, 69 Field Ln, Pelsall, Walsall, WS4 1DQ [IO92AO, SK00]
GI0 IMY F McCann, 91 Derryloughan Rd, Coalisland, Dungannon, BT71 4QS [IO64QM, H96]
GM0 IMZ I McEwan, 52 Murrayfield Gdns, Edinburgh, EH12 6DF [IO85JW, NT27]
G0 INA E P W Cairns, 2 Stockhill Circus, Nottingham, NG6 0LS [IO92JX, SK54]
G0 IND R S Bhamra, Apantesis, 13 Saxon Walk, Foots Cray, Sidcup, DA14 6SJ [JO01BK, TQ47]
G0 INE Details withheld at licensee's request by SSL.
G0 INF D J Briggs Sheffield ARC, 365 PO Box, Sheffield, S1 1BY [IO93GJ, SK38]

G0 ING L Parker Nthants Expd Gr, 128 Northampton Rd, Wellingborough, NN8 3PJ [IO92PH, SP86]
G0 INH P J Sheldrake, 76 Severn Rd, Ipswich, IP3 0PT [JO02OA, TM14]
G0 INJ P Shuttlewood, 10 Church Ln, Costock, Loughborough, LE12 6UZ [IO92KT, SK52]
G0 INK S M Taylor, 37 Crestfield Dr, Pye Nest, Halifax, HX2 7HG [IO93BR, SE02]
GW0 INN P R Garner, 33 St. Albans Rd, Brynmill, Swansea, SA2 0BP [IO81AO, SS69]
G0 INO D G Elvin, Broadmere, 122 Mereside, Soham, Ely, CB7 5EG [JO02EI, TL57]
G0 INP D J C Stephens, Garden Flat, 21A Ford Park Rd, Plymouth, PL4 6RB [IO70WJ, SX45]
G0 INQ C W Hannell, Toat Lodge, Toat, Pulborough, West Sussex, RH20 1BZ [IO90RX, TQ02]
G0 INR Details withheld at licensee's request by SSL.
G0 INS P M Yardley, 13 Blackthorn Rd, Kenilworth, CV8 2DS [IO92FI, SP27]
G0 INT J M Lee, 2 York Terr, Birchington, CT7 9AZ [JO01PJ, TR36]
G0 INU L W Symons, 14 Maidenwell Rd, Plympton, Plymouth, PL7 1RB [IO70XJ, SX55]
G0 INV N J Wileman, 5 Ermine Rise, Great Casterton, Stamford, PE9 4AJ [IO92RQ, SK90]
G0 INX R T H Jones, 40 Lambs Farm Rd, Horsham, RH12 4DQ [IO91UB, TQ13]
G0 INY R Chainey, 5 Hovendens, Sissinghurst, Cranbrook, TN17 2LA [JO01GC, TQ73]
G0 INZ R A Ball, 40 Money Ln, West Drayton, UB7 7NX [IO91SM, TQ07]
GM0 IOA S Aitken, 44 Huntly Dr, Tanshall, Glenrothes, KY6 2HS [IO86JE, NO20]
G0 IOD T J Cleaver, 2 Ruskin Way, Daventry, NN11 4TT [IO92KF, SP56]
G0 IOE B Haynes, 6 Epping Walk, Furnace Green, Crawley, RH10 6LX [IO91VC, TQ23]
G0 IOF G R Moore, 64 Walmer Rd, Seaford, BN25 3TN [JO00BS, TV59]
G0 IOG J H Minns, Glenrosa, Barracks Bridge, Silloth, Carlisle, CA5 4NR [IO84HU, NY15]
G0 IOH P C Thomas, 21 The Greenway, Daventry, NN11 4EE [IO92KF, SP56]
G0 IOI T D Maunder, 19 St. Monica Rd, Sholing, Southampton, SO19 8FF [IO90HV, SU41]
G0 IOK G Markeson, 67 Fisher Rd, Harrow, HA3 7JX [IO91UO, TQ19]
G0 IOL M J Ranby, 194 Shelford Rd, Gedling, Nottingham, NG4 4JJ [IO92KX, SK64]
GD0 IOM R Ferguson, Moaney Moar House, Corlea Rd, Ballasalla, IM9 3BA
GW0 ION B O Cooke, 5 Squires Gate, Rogerstone, Newport, NP1 0BP [IO81LO, ST28]
G0 IOO A C Willis, 46 Maryland Rd, Thornton Heath, CR7 8DF [IO91WJ, TQ36]
G0 IOP Dr D H Melville, Little Buckden, Milberry Ln, Stoughton, Chichester, PO18 9JJ [IO90NV, SU81]
G0 IOQ F P Cosgrove Denton Park Middle School Radi, 8 Wandsworth Rd, Heaton, Newcastle upon Tyne, NE6 5AD [IO94EX, NZ26]
G0 IOR M J Robertson, Jopanida, 12 Southfield Rd, Scartho, Grimsby, DN33 2PL [IO93WM, TA20]
GI0 IOT M W Rabbett, 41 Richill Park, Waterside, Londonderry, BT47 1QY [IO64IX, C41]
G0 IOU B F Gleed, 19 Silver Birch Caravan Site, Walters Ash, High Wycombe, HP14 4UY [IO91OQ, SU89]
G0 IOV W G Ellis, 1 Wood View, Airmyn, Goole, East Riding of Yorks, DN14 8LA [IO93NR, SE72]
G0 IOX Details withheld at licensee's request by SSL.
GM0 IOY E Watt, 43 Larkfield Rd, Gourock, PA19 1YA [IO75OW, NS27]
G0 IOZ R J Southerington, 58 Brighton Ave, Syston, Leicester, LE7 2EB [IO92LQ, SK61]
G0 IPB L R Collins, 32 Parkstone Rd, Syston, Leicester, LE7 1LY [IO92LQ, SK61]
G0 IPC J A Hilton, 99 Kirby Rd, Stone, Dartford, DA2 6HD [JO01DK, TQ57]
G0 IPE P A Gardner, 29 Minstrel Gdns, Surbiton, KT5 8DD [IO91UJ, TQ16]
G0 IPF E A Malpas, 89 St. Marys Rd, Stratford upon Avon, CV37 6TL [IO92DE, SP25]
G0 IPG R F Tinkler, 38 Barrett Rd, Fetcham, Leatherhead, KT22 9HL [IO91TG, TQ15]
G0 IPH D Sutton, 1 Bullfinch Cl, Riverhead, Sevenoaks, TN13 2BB [JO01BG, TQ55]
G0 IPJ Details withheld at licensee's request by SSL.
G0 IPK J Marsh, 44 Richmond Gdns, Harrow Weald, Harrow, HA3 6AJ [IO91UO, TQ19]
G0 IPL N Pike, 42 Stanley Rd, Chingford, London, E4 7DB [JO01AP, TQ39]
G0 IPN A Greenwood, 4 Roach Pl, Rochdale, OL16 2DD [IO83WO, SD91]
G0 IPO E C Landor, 1 Redwood Park, Five Oak Green, Tonbridge, TN12 6WB [JO01EE, TQ64]
G0 IPS Details withheld at licensee's request by SSL.
G0 IPT D A Griggs, 5 Collingwood Ave, Muswell Hill, London, N10 3EH [IO91WO, TQ28]
G0 IPU Details withheld at licensee's request by SSL.
GM0 IPV B Stephenson, 11 Bishop Forbes Cres, Kinellar, Aberdeen, AB2 0TW [IO87TH, NJ72]
GM0 IPW Dr R J Dickie, Taynish, 11 Churchill Dr, Stornoway, Isle of Lewis, PA87 2NP [IO76JB, NS08]
G0 IPX E Longden Fists CW Club, 119 Cemetery Rd, Darwen, BB3 2LZ [IO83SQ, SD62]
G0 IPY Details withheld at licensee's request by SSL.[Station located near Crayford, Kent. Correspondence via: PO Box 127, Dartford, Kent, DA1 4LH.]
GI0 IQA Dr S Ruff, 9 Cooleen Park, Newtownabbey, BT37 0RR [IO74BQ, J38]
GM0 IQD D Ross, 1 Forbes Pl, Smithton, Inverness, IV1 2NN [IO77WL, NH74]
G0 IQF Details withheld at licensee's request by SSL.
G0 IQG D A Shaw, 11 Farm Cl, Southall, UB1 3JF [IO91TM, TQ18]
G0 IQH F Manning, 180 Priestley Terr, Wibsey, Bradford, BD6 1QU [IO93CS, SE13]
G0 IQI P D James, 29 Old Oaks, Waltham Abbey, EN9 1TJ [JO01AQ, TL30]
G0 IQK W A Chewter, 93 Eton Rd, Ilford, IG1 2UF [JO01AN, TQ48]
G0 IQL D R Loveridge, Hazeldene, 23 Bridge Rd, Worthing, BN14 7BX [IO90TT, TQ10]
G0 IQM S Atkinson, 26 Skipton Rd, Trawden, Colne, BB8 8QS [IO83WU, SD93]
G0 IQN C J Jones, 189 Moor Ln, Cranham, Upminster, RM14 1HN [JO01DN, TQ58]
G0 IQO R H Montgomery, 27 Wyngate Rd, Cheadle Hulme, Cheadle, SK8 6ER [IO83VI, SJ88]
GW0 IQP P R Williams, 27 Lon Helyg, Newtown, SY16 1HY [IO82IM, SO09]
G0 IQQ G Bradley, Fairhaven, 29 Ashbourne Ave, Cleckheaton, BD19 5JH [IO93DR, SE12]
G0 IQS Details withheld at licensee's request by SSL.
GW0 IQT C J Bossino, 7 Viola Ave, Rhyl, LL18 2NE [IO83GH, SJ08]
G0 IQV J Mellor, 2 Livesey Ln, Heywood, OL10 3HY [IO83VO, SD81]
G0 IQW G E Martin, 20 Clover Rise, Whitstable, CT5 3EY [JO01MI, TR16]
G0 IQX Details withheld at licensee's request by SSL.
G0 IQY J R Gibbons, 35 Bassett Green Cl, Bassett, Southampton, SO16 3QQ [IO90HW, SU41]
GW0 IQZ W Williams, 31 Stad Ty Croes, Llanfairpwllgwyngyll, LL61 5JR [IO73VF, SH57]
G0 IRA P P Fox, 5 Llandovery Cl, Winsford, CW7 1NA [IO83RE, SJ66]
GW0 IRC D Knibbs, 9 Lon y Deri, Caerphilly, CF83 1DS [IO81JN, ST18]
G0 IRD E A H Hayward, 69 Lichfield Ave, Tupsley, Hereford, HR1 2RL [IO82PB, SO53]
GM0 IRG Details withheld at licensee's request by SSL.
G0 IRH D M G Pond, 31 Quintilis, Roman Hill, Bracknell, RG12 7QQ [IO91OJ, SU86]
G0 IRI J L Wade, 12 Kendal Rd, Harlescott, Shrewsbury, SY1 4ER [IO82PR, SJ51]
G0 IRJ A C Wilson, 42 Bacheler St., Hull, HU3 2TZ [IO93TR, TA02]
G0 IRK N B Porter, 23 Calder Ct, Britannia Rd, Surbiton, KT5 8TS [IO91UJ, TQ16]
G0 IRL A K King, 131 Poole Ln, Kinson, Bournemouth, BH11 9DZ [IO90BS, SZ09]
G0 IRM N F Chambers, 78 Durley Ave, Pinner, HA5 1JH [IO91TN, TQ18]
G0 IRO Details withheld at licensee's request by SSL.
GW0 IRP C D Evans, 19 Bryn Terr, Caerau, Maesteg, CF34 0UR [IO81EP, SS89]
G0 IRQ F T Wagner, 11 Ives Rd, Fifers Ln, Norwich, NR6 6DY [JO02PP, TG21]
G0 IRR G E Robbins, 6 Oakwood Dr, Grimsby, DN37 9RN [IO93WN, TA20]
G0 IRS I R Scott, Fern Cottage, High St, Streatley, nr Reading Berks, RG8 9JB [IO91KM, SU58]
G0 IRU J McKenzie, 12 The Avenue, Wolfreton Rd, Anlaby, Hull, HU10 6QR [IO93SR, TA02]
G0 IRV C G Lockyer, 153 Norman Rd, West Malling, ME19 6RW [JO01EH, TQ65]
G0 IRW N C Hearne, 12 Woodland Cl, West Malling, ME19 6RR [JO01EH, TQ65]
G0 IRX P G Minnis, 3 Werneth Hollow, Woodley, Stockport, SK6 1PW [IO83WK, SJ99]
G0 IRY W J Gallacher, 9 Glenside Cl, Edgerton, Huddersfield, HD3 3AP [IO93CP, SE11]
GM0 IRZ Details withheld at licensee's request by SSL.
GM0 ISA D Arcari, 184 Fintry Dr, Fintry, Dundee, DD4 9LP [IO86ML, NO43]
G0 ISB R G Griffiths, 29 Dubbers, Godshill, Ventnor, PO38 3HX [IO90IP, SZ58]
G0 ISC R E Slyfield, 359 Ringwood Rd, Parkstone, Poole, BH12 4LT [IO90AR, SZ09]
G0 ISD Details withheld at licensee's request by SSL.
G0 ISE G P Carter, 44 Capenhurst Ln, Ellesmere Port, Whitby, South Wirral, L65 7AH [IO83NG, SJ37]
G0 ISF A D Goodhew, Strathwell Manor, Strathwell Park, Whitwell, Isle of Wight, PO38 2QU [IO90IO, SZ57]
G0 ISG R Hale, 5 Land Oak Dr, Kidderminster, DY10 2ST [IO82VJ, SO87]
G0 ISH R J Pownall, 99 Fairwater Dr, Woodley, Reading, RG5 3JQ [IO91NK, SU77]
G0 ISI R J Christopher, 33 Grange Park, Albrighton, Wolverhampton, WV7 3EN [IO82UP, SJ80]
G0 ISJ G F Parkin, 15 Oakwood, Horbury Rd, Wakefield, WF2 8JG [IO93FQ, SE31]
G0 ISK M W Glover, 50 Broadway, Brinsworth, Rotherham, S60 5ES [IO93HJ, SK48]
G0 ISL Rev B A Shersby, Bapchild Rectory, School Ln, Sittingbourne, Kent, ME9 9NL [JO01JH, TQ96]
G0 ISM R M Faversham, 19 Pelwood Rd, Camber, Rye, TN31 7RU [JO00JW, TQ91]
G0 ISO M V Edmunds, 27A Sea View Rd, Parkstone, Poole, BH12 3LP [IO90AR, SZ09]
GI0 ISQ D W Christie, 8 Ballytober Rd, Bushmills, BT57 8UX [IO65RE, C93]
GM0 IST J M Purtell, 31 Daleally Cres, Errol, Perth, PH2 7QA [IO86JJ, NO22]
G0 ISU Details withheld at licensee's request by SSL.
G0 ISV J A Wilson, Mill Lodge, Mill Ln, Brinigham, Melton Constable, NR24 2QA [JO02MU, TG03]
G0 ISW Details withheld at licensee's request by SSL.[Operator: Philip. Town: Penrith. WAB: NY52/7283/Eden. Mbx: GB7PEN. County: Cumbria (CBA). RSARS: 2384. QSL: via RSGB. Locator: IO84PP. QRV: HF/50/144/432.]
G0 ISX M J Williams, 22 New Ave, Kirkheaton, Huddersfield, HD5 0JD [IO93DP, SE11]
G0 ISY J E Davies, 68 Hollywood Gdns, Hayes, UB4 0DY [IO91TM, TQ18]
G0 ITB Details withheld at licensee's request by SSL.
G0 ITF W J Faulconbridge, 25 Langlodge Rd, Holbrooks, Coventry, CV6 4EG [IO92FK, SP38]
G0 ITH Details withheld at licensee's request by SSL.
G0 ITI K Tyler, 43 Nest Est, Mytholmroyd, Hebden Bridge, HX7 5BH [IO93AR, SE02]
GI0 ITJ Dr D Linton, 4 Elmwood, Cullybackey, Ballymena, Co. Antrim, BT43 5PY [IO64TV, D00]
G0 ITK H A Johnson, 1 Pickering Ct, Grimsby, DN37 9LA [IO93WN, TA20]
G0 ITL J E Dunkley, 55 Windmill Ave, Kilburn, Belper, DE56 0PQ [IO93GA, SK34]
G0 ITM J J Dutton, 16 Briarfield, Fatfield, Washington, NE38 8RX [IO94FV, NZ35]
G0 ITO A F OShaughnessy, Southby, Buckland, Faringdon, Oxon, SN7 8QR [IO91FQ, SU39]
G0 ITP I R Dexter, 15 Whitethorn Ave, Burnage, Manchester, M19 1EU [IO83VK, SJ89]
G0 ITQ R J Le Sbirel, 7 Filmer Ln, Sevenoaks, TN14 5AG [JO01CG, TQ55]
G0 ITR J E C Fanner, 132 Morland Rd, Croydon, CR0 6NE [IO91XJ, TQ36]
G0 ITS R C Wells, 35 Creswell Farm Dr, Stafford, ST16 1PG [IO82WT, SJ92]
G0 ITT Details withheld at licensee's request by SSL.
G0 ITU M G S Wood, 78 Haycliffe Rd, Bradford, BD5 9HB [IO93CS, SE13]
G0 ITZ E P Pieroni, 395 Huddersfield Rd, Millbrook, Stalybridge, SK15 3HU [IO83XL, SJ99]
G0 IUA S A Ratcliffe, 20 Merrion St., Farnworth, Bolton, BL4 7LG [IO83TN, SD70]
G0 IUB G F Cooke, 17 Macklin Cl, Hungerford, RG17 0BY [IO91FJ, SU36]
G0 IUC E Hendy, 28 George St., Weston Super Mare, BS23 3AS [IO81MI, ST36]
G0 IUD A E Wakeman, 133 Stanshawe Cres, Yate, Bristol, BS17 4EG [IO81TM, ST78]
G0 IUE J S Wheeler, Tiddleywinks, 6 Severn Rd, Melksham, SN12 8BQ [IO81WJ, ST96]
G0 IUG E Yallop, 19 Manor Gdns, Buckden, St. Neots, Huntingdon, PE18 9TN [IO92VH, TL16]
G0 IUH P S Battershill, 45 Winkworth Rd, Banstead, SM7 2QJ [IO91VH, TQ26]
G0 IUI K A Jones, 12 Delamere Dr, Macclesfield, SK10 2PW [IO83WF, SJ97]
G0 IUK E Pickerill, 6 Pitmore Walk, Moston, Manchester, M40 0GB [IO83VM, SD80]
G0 IUL C W Watson, 92 Cromwell Rd, Saffron Walden, CB11 4BE [JO02CA, TL53]
G0 IUM J D Mc Culloch, 39 Beech Dr, St. Ives, Huntingdon, PE17 4UB [IO92XI, TL37]
G0 IUN C W Stain, 6 Sutton Cl, Sutton in Ashfield, NG17 3DP [IO93JD, SK56]
G0 IUO Details withheld at licensee's request by SSL.
GI0 IUP G A Lyle, 40 Enagh Cres, Maydown, Londonderry, BT47 1UG [IO65IA, C41]
G0 IUQ Details withheld at licensee's request by SSL.[Station located near Hailsham.]
G0 IUR B J Neal, 36 Lulworth Rd, Keynsham, Bristol, BS18 2PX [IO81RJ, ST66]
G0 IUT Details withheld at licensee's request by SSL.
G0 IUV Details withheld at licensee's request by SSL.
G0 IUW R J Hill, 7 Berkeley Cl, Cashes Green, Stroud, GL5 4SA [IO81VR, SO80]
G0 IUY J B Tribe, 17 Shaftesbury Ave, Purbrook, Waterlooville, PO7 5HW [IO90LU, SU60]
G0 IVA Details withheld at licensee's request by SSL.
G0 IVB M V Llewellyn, 8 Pencroft Ln, Danesmoor, Chesterfield, S45 9DF [IO93HD, SK46]
G0 IVD P A McGarvey, 125 Esme Rd, Sparkhill, Birmingham, B11 4NJ [IO92BK, SP08]
G0 IVE D J Hornsey, Daleside, 6 Kings Ct, Hemsby, Great Yarmouth, NR29 4EW [JO02UQ, TG41]
G0 IVF B Smith, Ellenglaze, School Ln, Budock Water, Falmouth, TR11 5DJ [IO70KD, SW73]
GW0 IVG W J E Jones, 30 Maesglas, Pontiets, Pontyates, Llanelli, SA15 5SG [IO71VS, SN40]
G0 IVI K Edwards, 39 King Edward St., Sandiacre, Nottingham, NG10 5BS [IO92IW, SK43]
G0 IVM A F Thompson, 7 Chesley Cl, Durrington, Worthing, BN13 2TN [IO90ST, TQ10]
G0 IVP S J Parrish, Ambleside, Carlton Ave, Hornsea, HU18 1JG [IO93WW, TA24]
GM0 IVQ G J McKinlay, 68 Hillend Rd, Clarkston, Glasgow, G76 7XT [IO75US, NS55]
G0 IVR L J Kennard obo Itchen Valley ARC, 116 Ringwood Dr, North Baddesley, Southampton, SO52 9GT [IO90GX, SU31]
G0 IVS E Toombs, 46 Fords Ave, Healing, Grimsby, DN37 7RP [IO93VN, TA20]
GW0 IVT T Jones, Heathbrook, Maesmawr Cl, Talybont-on-Usk, Brecon, Powys, LD3 7JF [IO81IV, SO12]
G0 IVV J P Sheldrake, 76 Severn Rd, Ipswich, IP3 0PT [JO02OA, TM14]
G0 IVW G S Crawley, 4 Ardington Rise, Purbrook, Waterlooville, PO7 5QP [IO90LU, SU60]
G0 IVX H H Harrison, 37 Parkside, Tanfield Lea, Stanley, DH9 9NP [IO94DV, NZ15]
G0 IVY J I Broom, 13 Westgate, Oakham, LE15 6BH [IO92PQ, SK80]
G0 IVZ J P Fisher, Farland, Rillaton, Rilla Mill, Callington, Cornwall, PL17 7PA [IO70TM, SX27]
G0 IWA Details withheld at licensee's request by SSL.
G0 IWB J A Wells, 12 Church Rd, Grafham, Huntingdon, PE18 0BB [IO92UH, TL16]
G0 IWC M E Burroughs, 34 Littlestone Rd, Littlestone, New Romney, TN28 8LR [JO00LX, TR02]
GW0 IWD A Aston, Ty Newydd, y Ffor, Pwllheli, LL53 6UY [IO72TW, SH33]
G0 IWF F R Oakton, 180 Wragley Way, Stenson Fields, Sinfin, Derby, DE24 3DZ [IO92GU, SK33]
G0 IWG D W Blundell, 6 Cavell Rd, Hylton Castle Est, Sunderland, SR5 3QW [IO94GW, NZ35]
G0 IWH G E Stapleton, 28 Gload Cres, Orpington, BR5 4PR [JO01BI, TQ46]
G0 IWI Dr D L Ranson, Vic Inst Fornsc Pat, 57-83 Kavanagh St, Sth Melbourne, Victoria, Australia 3205
G0 IWJ G A Slingsby, 46 Bathurst Rd, Staplehurst, Tonbridge, TN12 0LQ [JO01GD, TQ74]
G0 IWN D A Greenhalgh, 8 Pleasant Rd, Milton, Portsmouth, Hants, PO4 8JU [IO90LT, SZ69]
G0 IWP W H Pipping, 13 Bushy Park, Bristol, BS4 2EG [IO81RK, ST57]
G0 IWQ Details withheld at licensee's request by SSL.
G0 IWR D G Chubb Iow Raynet Grou, Walls End Cottage, Heathfield Cl, Bembridge, PO35 5UG [IO90KQ, SZ68]
G0 IWT Details withheld at licensee's request by SSL.
G0 IWU R Smith, 4 Dawes East Rd, Burnham, Slough, SL1 8BT [IO91QM, SU98]
G0 IWW R J Weedon, 100 Berkeley Vale Park, Berkeley, GL13 9TQ [IO81SQ, ST69]
G0 IWY C Taylor, 37 New Meadow, Woodlands, Ivybridge, PL21 9PT [IO80AJ, SX65]
G0 IWZ Details withheld at licensee's request by SSL.[Correspondence via PO Box 249, Poole, Dorset, BH15 2LR.]
G0 IXA M A Dodd, 26 Washington Gr, Bentley, Doncaster, DN5 9RL [IO93KM, SE50]
G0 IXB S F Lumbard, 9 Bamborough Cl, Pipers Copse, Southwater, Horsham, RH13 7XF [IO91TA, TQ12]
G0 IXC J H Martin, 27 Firs Cres, Harrogate, HG2 9HF [IO93FX, SE35]
G0 IXD J V Masters, 9 Denshire Ct, Baston, Peterborough, PE6 9QL [IO92TR, TF11]
G0 IXE I Hogarth, 64 Winterburn Pl, Newton Aycliffe, DL5 7ET [IO94EO, NZ22]
G0 IXF R C Godwin, 24 Hobbs Rd, Broadfield, Crawley, RH11 9SA [IO91VC, TQ23]
G0 IXH J H L Royal, 22 Leatherhead Rd, Leatherhead, KT22 8TL [IO91UH, TQ15]
G0 IXI Details withheld at licensee's request by SSL.
GW0 IXJ D E Thomas, 11 Courtlands Park, Carmarthen, SA31 1EH [IO71UU, SN42]
GW0 IXK R O Griffiths, 71 Elder Gr, Llangunnor, Carmarthen, SA31 2LH [IO71UU, SN41]
G0 IXL D J Bussey, Jonanjo, Prince Andrews Rd, Norwich, Norfolk, NR6 6XJ [JO02PP, TG21]
GW0 IXM O E Williams, 39 Heol Cennen, Ffairfach, Llandeilo, SA19 6UL [IO81AU, SN62]
GM0 IXN G R Steedman, 41 Dunster Rd, Stirling, FK9 5HY [IO86AD, NS89]
GM0 IXO G D Michie, 2 Moncur St., Townhill, Dunfermline, KY12 0HN [IO86GC, NT18]
GW0 IXP L R Startup, 24 Parc Plas, Blackwood, NP2 1SJ [IO81JP, ST19]
G0 IXR T J Hackworth, 37 Churchill Dr, Ruddington, Nottingham, NG11 6DH [IO92KV, SK53]
G0 IXT G Lambert, Church House, Chapel Gate, East Parley, Christchurch Dorse, BH23 6BE [IO90BS, SZ19]
G0 IXU A McVety, 6 Foley Walk, Woodhouse Park, Manchester, M22 0LL [IO83UI, SJ88]
G0 IXV P J W Biscombe, The Red House, 32 Turketel Rd, Folkestone, CT20 2NZ [JO01NB, TR23]
G0 IXW S A Saywell, 4 Spur Cl, Abridge, Romford, RM4 1DS [JO01BP, TQ49]
G0 IXX M H Rowe, 15 Woodward Rd, Pershore, WR10 1LW [IO82XC, SO94]
G0 IXZ D E Fleetwood, Lynton House, Station Rd, Bolsover, Chesterfield, S44 6BH [IO93IF, SK47]
GM0 IYA W I Brown, 29 Bankhead Cres, Dennyloanhead, Bonnybridge, FK4 1RY [IO86BA, NS88]
G0 IYB A B Squire, Liege, Sunnyside, Combe-Martin, North Devon, EX34 0JH [IO71XE, SS54]
G0 IYC T D Walsh, 13 Charnwood Rd, Burton on Trent, DE13 0PN [IO92ET, SK22]
G0 IYD P Bates, 29 Juler Cl, North Walsham, NR28 0SY [JO02RU, TG33]
G0 IYE D J H Chalmers, 42 Thornbury Dr, Uphill, Weston Super Mare, BS23 4YH [IO81MH, ST35]
G0 IYG J Hatch, 57 Limes Ave, Darwen, BB3 2SG [IO83SQ, SD62]
G0 IYH A J Healless, 30 Ouseburn Rd, Livesey, Blackburn, BB2 4NL [IO83SR, SD62]
G0 IYJ R H Turley, 30 Ave Grimaldi, Luton, LU3 1TJ [IO91SV, TL02]
G0 IYK M D L Harrison, 12 Richmond Rise, Reepham, Norwich, NR10 4LS [JO02NS, TG12]
GM0 IYL Details withheld at licensee's request by SSL.
G0 IYM M Linton, 134 Eccles Old Rd, Salford, M6 8QQ [IO83UL, SJ79]
G0 IYO K Stanmore, 48 St. Michaels Ave, Bishops Cleeve, Cheltenham, GL52 4NX [IO81XW, SO92]
GM0 IYP J G Cormack Sutherland District A.R.C, Dunrobin Farm, Golspie, Sutherland, Scotland, KW10 6RH [IO87AX, NC80]
G0 IYQ N R Mitchelson, Trevena, Winskill, Penrith, Cumbria, CA10 1PD [IO84QQ, NY53]
G0 IYS P R Willis, 5 Binbrook Walk, Corby, NN18 9HH [IO92PL, SP88]
G0 IYT C S Savin, Jenalri, Union Ln, Pilling, Preston, PR3 6SS [IO83NV, SD44]
G0 IYU P J Kirk, 16 Eliot Rd, Royston, SG8 5AT [IO92XB, TL34]
G0 IYV T Jones, 159 Cobden View Rd, Crookes, Sheffield, S10 1HT [IO93FJ, SK38]
G0 IYW G Farndon, 28 Willow Cl, Bedworth, Nuneaton, CV12 8BE [IO92GL, SP38]
G0 IYX I W Roberts, 2 Samuels Fold, Pendlebury Ln, Haigh, Wigan, Lancs, WH2 1LT
G0 IYY M J A Lindsay, 10 High Green, Brooke, Norwich, NR15 1HR [JO02QM, TM29]
G0 IYZ Details withheld at licensee's request by SSL.
G0 IZA Details withheld at licensee's request by SSL.
G0 IZB H C Smith, 69 Chapel View, South Croydon, CR2 7LJ [IO91XI, TQ36]
GM0 IZC J D Carver, 181 Rutland Ave, High Wycombe, HP12 3LL [IO91OO, SU89]
G0 IZD H W Oliver, Staithe House, Main Rd, Brancaster Staithe, Kings Lynn, PE31 8BP [JO02HX, TF74]
G0 IZE R D Bates, 36 Maple Cres, Alveley, Bridgnorth, WV15 6LT [IO82TL, SO78]
G0 IZF A H Hill, 26 Cres Rd, Netherton, Dudley, DY2 0NW [IO82WL, SO98]
G0 IZG G J Bennett, 21 Porlock Cl, Platt Bridge, Wigan, WN2 5HY [IO83QM, SD60]
G0 IZH R L Timings, Penrhyn, 14A Moor Hall Dr, Sutton Coldfield, B75 6LP [IO92CN, SP19]
G0 IZI A W J Warwick, 58 Longworth Ave, Tilehurst, Reading, RG31 5JY [IO91LK, SU67]
G0 IZJ E Whittaker, 17 Packer St., Bolton, BL1 3LD [IO83SO, SD71]
G0 IZK M S Dennehy, 45 Vine Rd, Tiptree, Colchester, CO5 0LR [JO01IT, TL81]
G0 IZL N C Procter, 2 Primrose Ave, Underwood, Nottingham, NG16 5FY [IO93IB, SK45]
G0 IZN A J Halliday, 12 Fowley Common Ln, Culcheth, Warrington, WA3 5JJ [IO83SL, SJ69]
G0 IZO J Cunliffe, 21 Eldon Pl, Eccles, Manchester, M30 8QE [IO83TL, SJ79]

G0 IZP R A McDonald, 11 Lakeside, South Cerney, Cirencester, GL7 5XE [IO91AQ, SU09]
G0 IZQ J Bradnock, 5 Milverton Rd, Knowle, Solihull, B93 0HX [IO92DJ, SP17]
GW0 IZR A E Thompson, 19 Bryn Eithin, Pentre Halkyn, nr Holywell, Clywd, CH8 8HZ [IO83JF, SJ27]
G0 IZT C C Shedden, 29 Hooton Rd, Willaston, South Wirral, L64 1SE [IO83MH, SJ37]
G0 IZU Details withheld at licensee's request by SSL.
G0 IZV D A Gowers, 23 Glebe Ln, Sonning, Reading, RG4 6XH [IO91NL, SU77]
G0 IZW D E Brades, 3 Coldnailhurst Ave, Braintree, CM7 2SL [JO01GV, TL72]
G0 IZX Details withheld at licensee's request by SSL.
G0 IZY D A B Hicks, 137 Ash St., Ash, Aldershot, GU12 6LJ [IO91PF, SU85]
G0 IZZ H C V K Vannerley, 58 Rouncil Ln, Kenilworth, CV8 1FQ [IO92FH, SP27]
G0 JAA T S Watson, 27 Parkhill Rd, Smethwick, Warley, B67 6AS [IO92AL, SP08]
G0 JAB A H Greenhalgh, 17 Station Rd, Rushall, Walsall, WS4 1EP [IO92AO, SK00]
G0 JAC M J Hill, 21 Somersby Rd, Mapperley, Nottingham, NG3 5QB [IO92KX, SK54]
G0 JAF J A Foy, 23 Lee Rd, Nelson, BB9 8SD [IO83VU, SD83]
G0 JAG R Duval, 171 Bull Cl Rd, Norwich, NR3 1NY [JO02PP, TG20]
G0 JAH A M Horton, 45 Clobells, South Brent, TQ10 9JW [IO80CK, SX65]
GW0 JAI T G Jones, Sunnybank, Tregynon, Newtown, Powys, SY16 3EH [IO82HO, SJ00]
G0 JAJ G W West, 39 Ct Farm Rd, Eltham, London, SE9 4JL [JO01AK, TQ47]
G0 JAL W T Bell, 244 Westbourne, Woodside, Telford, TF7 5QR [IO82SP, SJ60]
G0 JAM J A M Morrison, 107 Crownmeadow, Colnbrook, Slough, Berks, SL3 0LU [IO91RL, TQ07]
G0 JAN J F Ilston, 6 Dovedale, Canvey Island, SS8 8HX [JO01HM, TQ88]
G0 JAO L A White, 56 Grange Rd, Leigh on Sea, SS9 2HT [JO01HN, TQ88]
G0 JAP C R T Collins, 9 Wivell Dr, Keelby, Grimsby, DN37 8HF [IO93UN, TA10]
G0 JAQ G Blackwood, 63 Illingworth Ave, Halifax, HX2 9JH [IO93BS, SE02]
G0 JAR R P Offord, 116 Townsend Rd, Snodland, ME6 5RL [JO01FH, TQ66]
GW0 JAS Details withheld at licensee's request by SSL.
GM0 JAV J W Craigie, 15 Garrioch St., Kirkwall, KW15 1PL [IO88MX, HY41]
G0 JAY Details withheld at licensee's request by SSL.[Station located near Bexleyheath. Op: A P Jay, PO Box 127, Dartford.]
G0 JAZ G Mountney, 3 Cranford Rd, Wilmslow, SK9 4DU [IO83VI, SJ88]
G0 JBA P E Boorman, 47 Diligent Dr, Milton Regis, Sittingbourne, ME10 2LQ [JO01II, TQ96]
G0 JBB A P Garnett, 21 Kirkstall Ave, Tottenham, London, N17 6PH [IO91WO, TQ38]
G0 JBC C E Collins, 12 Pavement Ln, Halifax, HX2 9JJ [IO93BS, SE02]
G0 JBD Details withheld at licensee's request by SSL.
GM0 JBE R McCart, Rovelea, Cumnock Rd, Dalmellington, Ayr, KA6 7PS [IO75TI, NS40]
G0 JBF Details withheld at licensee's request by SSL.
GW0 JBH A L Smith, 6 Collier St., Newport, NP9 7AT [IO81MO, ST38]
G0 JBI M L Schofield, 13 Vicar Park Rd, Norton Tower, Halifax, HX2 0NL [IO93BR, SE02]
G0 JBJ A R Tungate, 171 Marlborough Gdns, Faringdon, SN7 7DG [IO91EP, SU29]
G0 JBK Details withheld at licensee's request by SSL.
GD0 JBL W M Kelly, 14 Meadow Ct, Ballasalla, IM9 2DW
G0 JBM L B Humphreys, 19 Clinch Green Ave, The Highlands, Bexhill on Sea, TN39 5HN [JO00FU, TQ70]
GW0 JBN D E Duggan, 8 Buxton Cl, Pont Ebbw, Newport, NP9 3BN [IO81LN, ST28]
G0 JBO Eur Ing B D Lees, Preston Hall, Preston upon The Weald Moors, Telford, Shropshire, TF6 6DH [IO82SR, SJ61]
G0 JBP R A Lipscomb, Redmoor, Bickley Rd, Bickley, Bromley, BR1 2NF [JO01AJ, TQ46]
G0 JBQ S N Lipscomb, Montaza, Church Rd, Hartley, Longfield, DA3 8DR [JO01DJ, TQ66]
G0 JBR Details withheld at licensee's request by SSL.
G0 JBS M A Notman, 3 Pilling Ave, Lytham St. Annes, FY8 3QF [IO83LS, SD32]
G0 JBT D J Roberts, 100 Sherwood Park Ave, Sidcup, DA15 9JJ [JO01BK, TQ47]
G0 JBV A J McIntosh, 17 The Chase, Abbeydale, Gloucester, GL4 4WP [IO81VU, SO81]
G0 JBW Details withheld at licensee's request by SSL.
G0 JBX Details withheld at licensee's request by SSL.
G0 JBZ R C Stevens, 51 Beacon Park Cres, Upton, Poole, BH16 5PB [IO80XR, SY99]
G0 JCA J C Andress, 94 Norfolk Rd, Weymouth, DT4 0PP [IO80SO, SY67]
GW0 JCB Details withheld at licensee's request by SSL.
G0 JCD P W Busby, 6 Rimmers Ave, Southport, PR8 1HW [IO83LP, SD31]
G0 JCF D G Smith, 14 College Dr, Ruislip, HA4 8SB [IO91TN, TQ18]
G0 JCG A J Butler, 2 Haycroft Cl, Great Sutton, South Wirral, L66 2GY [IO83MG, SJ37]
G0 JCH J S Head, 21 Besley Cl, Tiverton, EX16 4JF [IO80GV, SS91]
GU0 JCI P J Horsepool, Top Flat Vista Du Guet, Route de Cobo, Castel, Guernsey, Channel Islands, GY5 7HD
G0 JCJ P Webster, 38 Springdale Rd, Broadstone, BH18 9BU [IO80XS, SY99]
G0 JCK W Pattinson, 2 The Green, Ticknall, Derby, Derbyshire, DE73 1JJ [IO92GT, SK32]
G0 JCM Details withheld at licensee's request by SSL.
GM0 JCN M Lovatt, 21 Crathes Gdns, Livingston, EH54 9EN [IO85FU, NT06]
G0 JCP D R Grey, 7 Cemetery Ln, Tweedmouth, Berwick upon Tweed, TD15 2BS [IO85XS, NT95]
G0 JCQ B R Rimmer, 170 Roe Ln, Southport, PR9 7PN [IO83MP, SD31]
GM0 JCR G P Currie, 40 Queens Dr, Burnfoot, Hawick, TD9 8EP [IO85OK, NT51]
GM0 JCS R Scott, Glenmayne, 8 Traquair Rd, Innerleithen, EH44 6PD [IO85LO, NT33]
G0 JCT P J Jones, Bronallt, Cenarth, Newcastle Emlyn, Dyfed, Wales, SA38 9JS [IO72RB, SN24]
GM0 JCX J E Smith, 50 Gala Park, Galashiels, TD1 1EZ [IO85OO, NT43]
G0 JCY P H Stuart, Skylark Corner, Seaborough Hill, Crewkerne, TA18 8PL [IO80OU, ST40]
G0 JCZ M P J Martin, 12519, Huntington, Venture Dri Houston, Texas USA 77099
G0 JDA E Riley, 28 Randolph Rd, Kearsley, Bolton, BL4 8EB [IO83TN, SD70]
G0 JDD I R Douce, 67 Glenbervie Dr, Leigh on Sea, SS9 3JT [JO01HN, TQ88]
G0 JDE R A Doyle, 61 St. Peters Rd, West Mersea, Colchester, CO5 8LN [JO01KS, TM01]
G0 JDF Details withheld at licensee's request by SSL.
G0 JDG A D Purseglove, 122 Chesterfield Rd, Huthwaite, Sutton in Ashfield, NG17 2QF [IO93ID, SK45]
G0 JDK A Crampton, 27 Oakland Ave, Huthwaite, Sutton in Ashfield, NG17 2JE [IO93ID, SK45]
G0 JDL J R Clarke, 4 Swallowfields, Carlton Colville, Lowestoft, NR33 8TP [JO02UK, TM59]
G0 JDM T S D Kewell, The Annexe, Parkway, Brittenden Ln, Waldron, Heathfield, TN21 0RH [JO00CX, TQ52]
G0 JDP Details withheld at licensee's request by SSL.
G0 JDQ M V Cozens, P O Box St 650, Harare, Zimbabwe, X X
G0 JDR F J Lees, Coban, Lewes Rd, Ringmer, Lewes, BN8 5QD [JO00AV, TQ41]
GW0 JDS J D Stonehouse, 49 Heol y Gelynen, Upper Brynamman, Ammanford, SA18 1SB [IO81BT, SN71]
G0 JDU Details withheld at licensee's request by SSL.
G0 JDV E P Burr, Lorne Croft, Wellpond Green, Standon, Ware, SG11 1NJ [JO01AV, TL42]
GW0 JDW D C Willis, 5 Dan Lan Rd, Llanelli, Dyfed, SA16 0NF [IO71VQ, SN40]
G0 JDX C A Mellors, 16 The Oval, Sutton in Ashfield, NG17 2FQ [IO93ID, SK45]
G0 JDY M S Lewis, 27 Plymouth Dr, Stubbington, Fareham, PO14 3RF [IO90JT, SU50]
G0 JDZ A Hill, 18 Garendon Rd, Shepshed, Loughborough, LE12 9NX [IO92IS, SK41]
G0 JEA R Kaye, 63 Coronation Dr, Birdwell, Barnsley, S70 5RL [IO93GM, SE30]
G0 JEC D Naylor, Marrick House, Coulton Terr, Kirk Merrington, Spennymoor Co Durha, DL16 7HN [IO94EQ, NZ23]
G0 JED Details withheld at licensee's request by SSL.
G0 JEE B Greer, 235 Burton Rd, Branston, Burton on Trent, DE14 3DR [IO92ES, SK22]
GM0 JEF J Fysh, 7 Chestnut Pl, Eilean Rise, Ellon, AB41 9HF [IO87XI, NJ92]
G0 JEG J Mackenzie, 5 York Rd, Wallasey, L44 9EN [IO83LJ, SJ39]
G0 JEI Details withheld at licensee's request by SSL.
G0 JEK C L Kelland, 40A Kingfishers, Gr, Wantage, OX12 7JN [IO91GO, SU48]
G0 JEL B J Beauchamp, 1 Beanfield Lns Cot, Uppingham Rd., Corby, Northants., NN17 2UJ [IO92PL, SP88]
G0 JEO R Robinson, 94 Overdale Rd, Wombwell, Barnsley, S73 0RN [IO93HM, SE40]
G0 JEP A P G Date, 15 Kings Rd, Cambridge, CB3 9DY [JO02BE, TL45]
GW0 JEQ Details withheld at licensee's request by SSL.[/P from John Baddoes School, Presteigne, Powys, LD8 2AY. Op: R T Wicks, Llangunilo, Knighton, Powys LD7 1SW.]
G0 JER R F Hughes, 12 Spencer St., Reddish, Stockport, SK5 6UH [IO83WK, SJ89]
GI0 JEV H Kernohan, 40 Lisnafillon Rd, Gracehill, Ballymena, BT42 1JA [IO64TU, D00]
G0 JEW P S Wells, 15 Gilmorton Rd, Lutterworth, LE17 4DY [IO92JK, SP58]
G0 JEX M C Koester, 136 Barnhorn Rd, Bexhill on Sea, TN39 4QG [JO00FU, TQ70]
G0 JEZ B W Wells, 37 Elder Rd, Denvilles, Havant, PO9 2UW [IO90MU, SU70]
G0 JFA A F Jones, Highercombe West, Highercombe, Dulverton, Somerset, TA22 9PT [IO81FB, SS93]
GM0 JFB F Bower, Gateside Cottage, Castlemain, Duns, Scotland, TD1 3TP [IO85PO, NT53]
G0 JFC J H Aithison, 14 Claymore Rise, Silsden, Keighley, BD20 0QQ [IO93AV, SE04]
G0 JFD G J Butcher, 12 Radley Cl, Hedge End, Southampton, SO30 2UW [IO90IW, SU41]
G0 JFE P Sixsmith, 10 Wisbeck Rd, Bolton, BL2 2TA [IO83TO, SD70]
G0 JFG Details withheld at licensee's request by SSL.
G0 JFJ G Berry, 59 Hollows Terr, Hayle, TR27 4EF [IO70HE, SW53]
GM0 JFK C T Harper, The School House, Arpafeelie, North Kessock, Inverness, IV1 1XD [IO77UM, NH65]
GM0 JFL B C Harper, 2 Ryefield Cttgs, Conon Bridge, Dingwall, IV7 8EY [IO77TN, NH55]
G0 JFM S Nicholls, 117 Derby Rd, Eastleigh, SO50 5GT [IO90HX, SU41]
G0 JFN Details withheld at licensee's request by SSL.
G0 JFP J A Scott, 11 Edenbridge Rd, Manchester, M40 2UP [IO83VL, SJ89]
GW0 JFQ N D Williams, 29 William Ct, Swansea, SA1 3QZ [IO81AO, SS69]
G0 JFT D E White, 12 Timperley Est, Colkirk, Fakenham, NR21 7NN [JO02KT, TF92]
G0 JFU L F Piercy, 19 Green St., Brockworth, Gloucester, GL3 4LT [IO81WU, SO81]
G0 JFV C A Antill, 30 Sheard Ave, Ashton under Lyne, OL6 8DS [IO83XM, SD90]
G0 JFW J F Wood, Fenrother, Cresswell Rd, Ellington, Morpeth, NE61 5HS [IO95FF, NZ29]
G0 JFX R T Riley, Higher Ford, Wembury, Plymouth, Devon, PL9 0DZ [IO70XH, SX54]
G0 JGB J G Barnett, 37 Sheppard St., Stoke on Trent, ST4 5AE [IO82VX, SJ84]
G0 JGD R Tunstall, 3 Grindon Cl, Westmonk Seaton, Whitley Bay, NE25 9EB [IO95GA, NZ37]
G0 JGE G Alexander, 6 The Cres, Eaglescliffe, Stockton on Tees, TS16 0JB [IO94HM, NZ41]
G0 JGF S J Cox, 60 Leawood Rd, Trent Vale, Stoke on Trent, ST4 6LA [IO82VX, SJ84]
G0 JGH A Loveridge, 7 Walnut Way, Northfield, Birmingham, B31 4ES [IO92AJ, SP07]
G0 JGM G H Meares, 41 Tor Cl, Worle, Weston Super Mare, BS22 0BZ [IO81MI, ST36]
G0 JGO E J Pilfold, 20 Mayes Cl, Warlingham, CR6 9LB [IO91XH, TQ35]
G0 JGP Details withheld at licensee's request by SSL.
GI0 JGQ F Finnigan, 55 Cloughreagh Park, Bessbrook, Newry, BT35 7EH [IO64TE, J02]
GW0 JGR Details withheld at licensee's request by SSL.
G0 JGS Details withheld at licensee's request by SSL.
G0 JGT W G Wilbraham, 17 Lennard Ave, West Wickham, BR4 9AZ [JO01AI, TQ36]
G0 JGU H W Bolwell, 5 Rosewood Gdns, Waterlooville, PO8 0LT [IO90MW, SU71]
G0 JGV K C Bewley, 19 High St., Watton At Stone, Hertford, SG14 3SX [IO91WU, TL21]
G0 JGX D W Ginsberg, Rayside, Birthwaite Rd, Windermere, LA23 1DF [IO84NJ, SD49]
G0 JGY F Clayton, 46 Carr Ln, Birkdale, Southport, PR8 3EF [IO83LO, SD31]
GW0 JHA Details withheld at licensee's request by SSL.
G0 JHB Details withheld at licensee's request by SSL.
G0 JHC N M Carr, 15 Westlands, Leyland, Preston, PR5 3XT [IO83PQ, SD52]
G0 JHD R P Jennings, 11 Blake St., Burslem, Stoke on Trent, ST6 4BE [IO83VB, SJ84]
GM0 JHE K A Hunter, Flat 2/2, 206 Skirsa St, Cadder, Glasgow, G23 5BA [IO75UV, NS56]
GM0 JHF T M Mitchell, 7 Hillocks Pl, Barassie, Troon, KA10 6TU [IO75QN, NS33]
G0 JHG C A Holmes, 12 Cavendish St., Lancaster, LA1 5QA [IO84OB, SD46]
GW0 JHH L J Ireland, 109 Dan y Cribyn, Ynysybwl, Pontypridd, CF37 3EU [IO81HP, ST09]
G0 JHJ W C Fowler, 1 East Orchard, Sileby, Loughborough, LE12 7SX [IO92KR, SK51]
G0 JHK M S Hedges, 19 Claverham Way, Battle, TN33 0JE [JO00FV, TQ71]
G0 JHL R J Wilmot, The Holmes, Holmes Rd, Horsington, Woodhall Spa, LN1 5HS [IO93RG, SK97]
G0 JHM R I Wilmot, The Holmes, Holmes Rd, Horsington, Woodhall Spa, LN1 5HS [IO93RG, SK97]
GI0 JHR H Oxtoby, 13 Castle Ct, Cookstown, BT80 8QJ [IO64PP, H87]
G0 JHT D N T Talbot, Southways, Tichborne Down, Alresford, SO24 9PL [IO91KB, SU53]
G0 JHU M Stephenson, 38 St. Helens Cres, Low Fell, Gateshead, NE9 6DH [IO94EW, NZ26]
G0 JHV Details withheld at licensee's request by SSL.
G0 JHW J H Waterhouse, 81 Barkham Rd, Wokingham, RG41 2RJ [IO91NJ, SU86]
G0 JIA A W Sharp, 138 Chausee D'Ixelles, 1050 Ixelles, Brussele, Belgium, X X
G0 JIB P Moran, 12 Sapphire Dr, Kirkby, Liverpool, L33 1UW [IO83NL, SD40]
G0 JIC Details withheld at licensee's request by SSL.[Op Gerald; RSGB, RAIBC, Locator: IO90BT, WAB: SU00. Correspondence to: PO Box 703, West Moors, Ferndown, Dorset, BH22 0YB.]
G0 JID W A Barnet, 4 Lune St., Crosby, Liverpool, L23 5TU [IO83LL, SJ39]
G0 JIF A S Fennell, 12 Vale Rd, Ramsgate, CT11 9LU [JO01QH, TR36]
G0 JIL G M Hampson, 7 Merryfield Cl, Bransgore, Christchurch, BH23 8BS [IO90DS, SZ19]
G0 JIM J W King, 4 Glenhurst Ave, Ruislip, HA4 7LZ [IO91SN, TQ08]
GM0 JIN Cmdr J Steven, 95 Craigentinny Ave, Edinburgh, EH7 6RH [IO85KX, NT27]
G0 JIO J A Sandum, 4 Martins Cl, Ramsgate, CT12 6TD [JO01QI, TR36]
G0 JIQ C R Crompton, 2 Graham Rd, Blacon, Chester, CH1 5LQ [IO83ME, SJ36]
G0 JIR A Potter, 8 Oaklands, The St., Mersham, Ashford, TN25 6NE [JO01LC, TR03]
G0 JIS A H Myland, 8 Burseldon Ct, East Cliff Rd, Dawlish, Devon, EX7 0BP [IO80GN, SX97]
G0 JIT R S Atherton, Frensham, Grange Ln, Whitegate, Northwich, CW8 2BQ [IO83RF, SJ66]
G0 JIU J D Leather, 3281 Spruce Ridge Rd, Quesnel, British Columbia, V2J 4R9, Canada
G0 JIV F A Gurney-Smith, Levens, Surlingham Rd, Bramerton, Norwich, NR14 7DN [JO02QO, TG20]
G0 JIW M E Crowsley, 1 Cranborne Cl, Bedford, MK41 0BQ [IO92SD, TL05]
G0 JIX T G Davies, 20 Kirkwood Ct, Shrewsbury, SY1 3SX [IO82PR, SJ41]
G0 JIY K H Bilton, 6 Park Vale Dr, Thrybergh, Rotherham, S65 4HZ [IO93IK, SK49]
G0 JIZ S V Ward, 21 Whinchat Way, Bradwell, Great Yarmouth, NR31 8SD [JO02UN, TG50]
G0 JJA Details withheld at licensee's request by SSL.
G0 JJD G H Thorne, 4 Barronwood Ct, Tarleton, Preston, PR4 6TR [IO83OQ, SD42]
G0 JJE H N Spratt, 9 Kennedy Cl, Halesworth, IP19 8EG [JO02SH, TM37]
GW0 JJF I N McCormick, Vale View, Railway Side, Clydach, nr Abergavenny, Gwent, NP7 0RD [IO81KT, SO21]
G0 JJG J A B Butt, 24 Lowry Way, Stowmarket, IP14 1UF [JO02LE, TM05]
G0 JJH E F Pearson, 7 Bullington End Rd, Castlethorpe, Milton Keynes, MK19 7ER [IO92OC, SP84]
G0 JJI P D Forshaw, 73 Galloway Rd, Hamworthy, Poole, BH15 4JS [IO80XR, SY99]
G0 JJK T F A Atkins, 46 Fallowfield, Ampthill, Bedford, MK45 2TP [IO92SA, TL03]
G0 JJL J Livesey, 9 Croft Bank, Penwortham, Preston, PR1 9BH [IO83PR, SD52]
G0 JJM A Eves, 136 Thistle Gr, Welwyn Garden City, AL7 4AQ [IO91VS, TL21]
G0 JJO M P Wheatley, 25 Sheringham Dr, Etching Hill, Rugeley, WS15 2YG [IO92AS, SK01]
G0 JJP R H Fowler, 50 Milton Rd, Cannock, WS11 2PJ [IO82XQ, SJ91]
G0 JJQ Details withheld at licensee's request by SSL.
G0 JJR Details withheld at licensee's request by SSL.
G0 JJS J Smith, 48 Oakleigh Gdns, Oldland Common, Bristol, BS15 6RH [IO81SK, ST67]
G0 JJU J J Upsher, Berwyn House, Edgerley, Oswestry, Shropshire, SY10 8EN [IO82MS, SJ31]
G0 JJW Details withheld at licensee's request by SSL.
G0 JJX Details withheld at licensee's request by SSL.
G0 JJY I J Bowen, 169 Clopton Rd, Stratford upon Avon, CV37 6TF [IO92DE, SP25]
G0 JJZ W D M Marks, 48 Bodiam Ave, Tuffley, Gloucester, GL4 0TJ [IO81UT, SO81]
G0 JKA A D Brook, 41 Cross Ln, Stocksbridge, Sheffield, S10 1WL [IO93FJ, SK38]
GW0 JKB J A Jenkins, 22 Highfield, Kingsbridge, Gorseinon, Swansea, SA4 6SZ [IO71XP, SS59]
G0 JKC Details withheld at licensee's request by SSL.
G0 JKE S Ward, 27 Greenock St., Sheffield, S6 4NB [IO93FJ, SK38]
GM0 JKF M J Thorogood, Learig, Woodside Terr, Banchory, AB31 5XJ [IO87SB, NO79]
G0 JKG F R Shinn, 77 Nether Edge Rd, Sheffield, S7 1RW [IO93GI, SK38]
G0 JKH D C Hinson, 6 Nethergate, Stannington, Sheffield, S6 6DJ [IO93FJ, SK38]
G0 JKI M W Edmunds, 2 Jubilee Rd, Bungay, NR35 1RE [JO02RK, TM38]
G0 JKJ S Marshall, 31 Postbridge Rd, Styvechale, Coventry, CV3 5AG [IO92FJ, SP37]
G0 JKM P J Renvoize, 18 Clifton Terr, Finsbury Park, London, N4 3JP [IO91WN, TQ38]
G0 JKO Details withheld at licensee's request by SSL.
G0 JKP A E Whibley, 33 Swanborough Dr, Brighton, BN2 5PJ [IO90WT, TQ30]
GM0 JKS G L Davidson, The Struan House Hotel, Carrbridge, Inverness Shire, PH23 3AS [IO87CG, NH92]
G0 JKU A C Bowering, 137 Knole Ln, Brentry, Bristol, BS10 6JN [IO81QM, ST57]
G0 JKV Details withheld at licensee's request by SSL.
G0 JKW P M Lewin, 24 Brookfields Rd, Wyke, Bradford, BD12 9LU [IO93CR, SE12]
G0 JKX Details withheld at licensee's request by SSL.[Op: W F Kelly, Groom House Farm, Chilton Trinity, Bridgwater, Somerset, TA5 2BG.]
G0 JKY B G W Hadley, 60 Chapel St., Pensnett, Brierley Hill, DY5 4EF [IO82WL, SO98]
G0 JKZ G K Christofi, 12 St. Martins Cl, Muirfield Rd, South Oxhey, Watford, WD1 6NJ [IO91TO, TQ19]
G0 JLA D R Larcombe, 6 Moorlands Rd, Fishponds, Bristol, BS16 3LF [IO81RL, ST67]
G0 JLD W T Horscroft, 2 Blenheim Cttgs, Blenheim Rd, Minehead, TA24 5QB [IO81GF, SS94]
G0 JLE R F Adams, 18 Dundridge Gdns, Bristol, BS5 8SZ [IO81RK, ST67]
G0 JLF J L Flowers, 11 Maple Gr, Spalding, PE11 2LE [IO92WS, TF22]
G0 JLH W A Newman, 27 Southwood Ave, Southbourne, Bournemouth, BH6 3QB [IO90CR, SZ19]
G0 JLI T J Davies, 35 Archenfield, Madley, Hereford, HR2 9NS [IO82NB, SO43]
GM0 JLJ E J Mottart, 2 Keilhill Cttgs, King Edward, Banff, Scotland, AB4 3LT [IO87TH, NJ72]
G0 JLL N Sheen, 84 Moyse Ave, Walshaw, Bury, BL8 3BL [IO83TO, SD71]
G0 JLO Details withheld at licensee's request by SSL.
G0 JLP M S Bray, 205 Woodlands Rd, Gillingham, ME7 2SW [JO01GJ, TQ76]
G0 JLQ C J Myler, 6 Lark Cres, Hartford, Huntingdon, PE18 7YN [IO92WI, TL27]
G0 JLS M J Brown, 4 River Gdns, Shawbury, Shrewsbury, SY4 4LA [IO82QS, SJ52]
G0 JLU E J F Wood, 65 Walford Rd, Rolleston on Dove, Burton on Trent, DE13 9AR [IO92EU, SK22]
G0 JLV D L H Vaughan, 23 Beckmeadow Way, Mundesley, Norwich, NR11 8LP [JO02RU, TG33]
G0 JLW Details withheld at licensee's request by SSL.
G0 JLX A Digby, 12 Albany Dr, Bishops Waltham, Southampton, SO32 1GE [IO90JW, SU51]
G0 JLY E P Smith, 6 Baily Cl, Glastonbury, BA6 9AU [IO81PD, ST54]
G0 JLZ F J Stewart, 6 Grants Rd, Enford, Pewsey, SN9 6DB [IO91CG, SU15]
G0 JMB M J Boon, 3 The Betchworth, Reigate Rd, Betchworth, RH3 7ET [IO91UF, TQ15]
G0 JMC Details withheld at licensee's request by SSL.
G0 JMD J M Davis, 62 Kingscote, Yate, Bristol, BS17 4YE [IO81SM, ST78]
G0 JME J P Pither, 74 Bucklands Rd, Teddington, TW11 9QS [IO91UK, TQ17]
GW0 JMF G P Pierce, 76 Prince Philip Ave, Garnlydan, Ebbw Vale, NP3 5DY [IO81JT, SO11]
G0 JMG D G Salt, 134 Kingsley Rd, Southsea, PO4 8HN [IO90LS, SZ69]
G0 JMH M D Morgan, Willow End, Whitestone, Hereford, HR1 3RY [IO82QB, SO54]
G0 JMI M C Parkin, 17 Bolle Rd, Alton, GU34 1PW [IO91MD, SU73]
GW0 JMJ W G Williams, 16 Chapel Cl, Elim Way, Pontllanfraith, Blackwood, NP2 2AD [IO81JP, ST19]
G0 JMK M Renshaw, 74 Teg Down Meads, Winchester, SO22 5ND [IO91HB, SU43]
GM0 JML M A Pivac, 36 Hume Cres, Bridge of Allan, Stirling, FK9 4SN [IO86AD, NS79]
G0 JMM M N Mathur, 310 Eastern Ave, Ilford, IG4 5AA [JO01AN, TQ48]
G0 JMN H Marshall, 23 Cranbourne Dr, Chorley, PR6 0LJ [IO83QP, SD51]
GM0 JMO J Bell, 5 Louisa Dr, Girvan, KA26 9AH [IO75NF, NX19]
G0 JMP Details withheld at licensee's request by SSL.

G0 JMR D A Williams, 4 Chapter Rd, Darwen, BB3 3PU [IO83SQ, SD72]
G0 JMS M Y Standen, 11 Hazel Gdns, Sonning Common, Reading, RG4 9TF [IO91MM, SU78]
G0 JMT Details withheld at licensee's request by SSL.
G0 JMU H Berry, Sandall House, School Rd, Laughton, Sheffield, S31 7YP [IO93GJ, SK38]
G0 JMW M J Williams, 51 Crackley Hill, Coventry Rd, Kenilworth, CV8 2EE [IO92FI, SP37]
G0 JMX A B Haynes, Holbeache Lodge, Trimpley, Bewdley, Worcs, DY12 1PB [IO82UJ, SO77]
G0 JMY S M Mason, 367 Norwich Rd, Ipswich, IP1 4HA [JO02NB, TM14]
G0 JMZ P J Farrar, 20 Cleveland Gr, Lupset Park, Wakefield, WF2 8LD [IO93FP, SE31]
G0 JNA R A Janes, 37 Valley View, Market Drayton, TF9 1EA [IO82SV, SJ63]
GM0 JNB W Bell, Picketlaw Farm, Harelaw Rd, Neilston, Glasgow, G78 3DE [IO75SR, NS45]
G0 JND A E Briggs, 49 Milldale Ave, Buxton, SK17 9BG [IO93AG, SK07]
G0 JNE T E Royle, 28 Byron Cl, Ettiley Heath, Sandbach, CW11 3GF [IO83TD, SJ76]
G0 JNG A P Stothard, Lanshaw Farm, Otley Rd, Beckwithshaw, Harrogate, N Yorks, HG3 1QX [IO93EW, SE25]
G0 JNH K J Lloyd, 24 Reservoir Rd, Edgeley Park, Stockport, SK3 9QJ [IO83WJ, SJ88]
G0 JNI C R Bagshaw, 64 Nunsfield Rd, Buxton, SK17 7BW [IO93BG, SK07]
G0 JNJ A Denny, 85 Delamere Dr, Macclesfield, SK10 2PS [IO83WF, SJ97]
G0 JNK A J Powers, 42 Newbridge Rd, Ambergate, Belper, DE56 2GS [IO93GB, SK35]
G0 JNN Details withheld at licensee's request by SSL.
G0 JNP Details withheld at licensee's request by SSL.
G0 JNQ S Mitchell, 78 Bellasize Park, Gilberdyke, Brough, HU15 2XU [IO93PS, SE83]
G0 JNR S Doveton, 2 Red Scar Dr, Scarborough, YO12 5RQ [IO94SG, TA08]
G0 JNS Details withheld at licensee's request by SSL.
G0 JNT L Keeton, 66 Worlaby Rd, Scartho, Grimsby, DN33 3JP [IO93WM, TA20]
G0 JNU P J Wheeler, 6 Oakhurst Rise, Carshalton Beeches, Carshalton, SM5 4AG [IO91VI, TQ26]
G0 JNZ T J Bray, 135 Fort Austin Ave, Chrownhill, Plymouth, PL6 5NR [IO70WJ, SX45]
G0 JOA Details withheld at licensee's request by SSL.
G0 JOD R Degg, 28 The Spinneys, Welton, Lincoln, Lincs, LN2 3TU [IO93SH, TF08]
G0 JOE Dr T M Artingstoll, 1 Whitefriars, High St., Chesterton, Cambridge, CB4 1NN [JO02BF, TL46]
GI0 JOF Details withheld at licensee's request by SSL.
G0 JOG B V Wilson, 41 Palmerston Cl, Ramsbottom, Bury, BL0 9YN [IO83UP, SD71]
G0 JOH Details withheld at licensee's request by SSL.
G0 JOJ J E Shirley, 50 West Bank Dr, South Anston, Anston, Sheffield, S31 7JG [IO93GJ, SK38]
G0 JOK J Willis, 2 Brooksby Way, Stratton St. Margaret, Swindon, SN3 4DX [IO91DN, SU18]
GM0 JOL Rev J Lincoln, The Manse, Killin, Perthshire, FK21 8TN [IO76UL, NN53]
G0 JOM J M Goddard, 28 Belfit Dr, Wingerworth, Chesterfield, S42 6UP [IO93GE, SK36]
G0 JON J H Swain, 1 Ganstead Way, Low Grange, Billingham, TS23 3SY [IO94IO, NZ42]
G0 JOO Details withheld at licensee's request by SSL.
G0 JOP J F Bryder, 110 Georgelands, Ripley, Woking, GU23 6DQ [IO91SH, TQ05]
G0 JOR Details withheld at licensee's request by SSL.
G0 JOS E J Christmas, 15 Norton Ave, Surbiton, KT5 9DX [IO91UJ, TQ16]
G0 JOT Details withheld at licensee's request by SSL.
G0 JOU E Gresham, 95 Squires Ave, Bulwell, Nottingham, NG6 8GL [IO93JA, SK54]
GM0 JOV A M Farquhar, 464 Lee Cres North, Bridge of Don, Aberdeen, AB22 8GJ [IO87WE, NJ91]
G0 JOW N Johnson, 62 Stanley Cres, Gravesend, DA12 5SY [JO01EJ, TQ67]
G0 JOX D M Sykes, 449 Westdale Ln, Mapperley, Nottingham, NG3 6DH [IO92KX, SK64]
G0 JOY Details withheld at licensee's request by SSL.
G0 JOZ R C Ebden, Flat 1, The Planes, 3 Overbury Ave, Beckingham, Kent, BR3 6PZ [IO91XJ, TQ36]
G0 JPA Details withheld at licensee's request by SSL.
G0 JPC K R Dale, 31 Cadshaw Cl, Birchwood, Warrington, WA3 7LR [IO83RK, SJ69]
G0 JPD R Thompson-Pettitt, 34 Pheasant Way, Kingsthorpe, Northampton, NN2 8BJ [IO92NG, SP76]
G0 JPE P E J Roberts, 98 Pinkneys Rd, Maidenhead, SL6 5DN [IO91OM, SU88]
G0 JPF S J Lee, 64 Reynard Way, Boughton Green, Northampton, NN2 8QX [IO92NG, SP76]
GM0 JPG D J Arnold, 1 Knockenhair, Knockenhair Rd, Dunbar, EH42 1BA [IO86RA, NT67]
GM0 JPH B P Sergeant, 73 Coniston Cres, Humberston, Grimsby, DN36 4AB [IO93XM, TA30]
G0 JPI S B Martin, 54 Royston Ave, Chingford, London, E4 9DF [IO91XO, TQ39]
G0 JPJ P K Smith, Austins, Layer Rd, Abberton, Colchester, CO5 7NH [JO01KU, TM01]
G0 JPL D J Smith, 32 Nest Ln, Wellingborough, NN8 4AU [IO92PH, SP86]
G0 JPM J P Mitchell, 8 Eldred Dr, Orpington, BR5 4PF [JO01BI, TQ46]
G0 JPQ C S Barber, 31 Greenfield Rd, Herringthorpe, Rotherham, S65 3NX [IO93IK, SK49]
GI0 JPR I T McKenzie, 11 Rockfergus Mews, Carrickfergus, BT38 7SL [IO74CR, J48]
G0 JPS C G Caton, 12 Ivor Rd, Bristol, BS5 9BW [IO81RL, ST67]
G0 JPU Dr M T Ferguson, 85 The Woodlands, Melbourne, Derby, DE73 1DQ [IO92GT, SK32]
G0 JPV S A Shaw, 15 Greenfield Ave, Rock, Marlbrook, Bromsgrove, B60 1HE [IO82XI, SO97]
GI0 JPW M A Easterbrook, 8 Bradford Heights, Carrickfergus, BT38 9EB [IO74CR, J49]
G0 JPX Details withheld at licensee's request by SSL.
G0 JPY D A Polley, 42 Lindfield Rd, Hampden Park, Eastbourne, BN22 0AJ [JO00DT, TQ50]
G0 JPZ M A Kirk, 5 The Paddock, Kirkby in Ashfield, Nottingham, NG17 8BT [IO93ID, SK46]
G0 JQA B Alderson Hambleton ARC, 43 Brompton Rd, Northallerton, DL6 1ED [IO94GI, SE39]
GM0 JQE I S Templeton, 39 Cairngorm Ct, Middleton Park, Irvine, KA11 1PN [IO75QP, NS34]
GI0 JQH G E Gilchrist, 71 Victoria Rd, Bangor, BT20 5ER [IO74EP, J58]
G0 JQI C A Johnson, 52 Tideswell Rd, Hazel Gr, Stockport, SK7 6JG [IO83WI, SJ98]
G0 JQJ Details withheld at licensee's request by SSL.
G0 JQK J Hughes, 18 Monmouth Rd, Wallasey, L44 3ED [IO83LK, SJ39]
G0 JQL Details withheld at licensee's request by SSL.
G0 JQM Details withheld at licensee's request by SSL.
G0 JQN Details withheld at licensee's request by SSL.
G0 JQO Details withheld at licensee's request by SSL.
G0 JQS A Downing, 1 Raglans, Alphington, Exeter, EX2 8XN [IO80FQ, SX98]
G0 JQU P A Dainty, 53 Macclesfield Rd, Hazel Gr, Stockport, SK7 6BG [IO83WI, SJ98]
G0 JQW H G Hudson, 113 Birchwood Rd, St. Annes, Bristol, BS4 4QS [IO81RK, ST67]
G0 JQX C Ditchfield, 22 Holly End, Quedgeley, Gloucester, GL2 4UY [IO81UT, SO81]
G0 JQZ S Chamberlain, 54 Henray Ave, Glen Parva, Leicester, LE2 9QJ [IO92KO, SP59]
G0 JRB J A Reed, 28 Kealholme Rd, Messingham, Scunthorpe, DN17 3ST [IO93QM, SE80]
G0 JRC J R Clayton, 49 Bramble Ln, Mansfield, NG18 3NP [IO93KD, SK56]
GI0 JRD B P McAnespie, 23 Ashton Park, Belfast, BT10 0JQ [IO74AN, J37]
G0 JRE I D Perks, Percuil Cottage, Oaklands Cl, Shalford, Guildford, Surrey, GU4 8JL [IO91RE, TQ04]
GW0 JRF F J Rees, Caerleon, Picton Rd, Tenby, SA70 7DP [IO71PQ, SN10]
G0 JRG Details withheld at licensee's request by SSL.
G0 JRH P F Scott-Dickinson, Ninicsu, 18 Pennington Dr, Weybridge, KT13 9RU [IO91SI, TQ06]
GI0 JRI K Murray, 67 Sicily Park, Belfast, BT10 0AN [IO74AN, J37]
G0 JRM C T Brown, 8 The Elms, Horringer, Bury St. Edmunds, IP29 5SE [JO02IF, TL86]
G0 JRN A J Hansley, 238 Milton Rd, Cowplain, Waterlooville, PO8 8SE [IO90LV, SU61]
G0 JRO C O J Sparkes, 13 Swathing, Cranworth, Thetford, IP25 7SJ [JO02LO, TF90]
G0 JRP Details withheld at licensee's request by SSL.
G0 JRQ Details withheld at licensee's request by SSL.[Station located near Walderstone, between Maidstone and Chatham; 500 feet asl on the North Downs. Grid ref: TQ758627. Op: Brian W Baker.]
G0 JRR C Curtis, 1 Westover Dr, Burton on Stather, Burton upon Stather, Scunthorpe, DN15 9HH [IO93PP, SE81]
G0 JRT T Hanratty, 12 Clarendon St., Consett, DH8 5LS [IO94CU, NZ15]
G0 JRW M R S Cave, High Weetslade Farm, Dudley, Cramlington, Northd., NE23 7LZ [IO95EB, NZ27]
G0 JRX T R Cave, High Weetslade Farm, Dudley, Cramlington, Northd., NE23 7LZ [IO95EB, NZ27]
G0 JRY C M Mulvany, 11 Erconwald St., East Acton, London, W12 0BP [IO91VM, TQ28]
G0 JRZ T W Brookes, 33 Water Park Rd, Bideford, EX39 3RB [IO71VA, SS42]
G0 JSA R Taylor, 22 Hood St., Accrington, BB5 1BW [IO83TS, SD72]
G0 JSB Details withheld at licensee's request by SSL.
G0 JSC J H Wheatley, 13 Elstead Gdns, Purbrook, Waterlooville, PO7 5EX [IO90LU, SU60]
G0 JSE J S Edwards, 49 The Fleet, Stoney Stanton, Leicester, LE9 4DZ [IO92IN, SP49]
G0 JSF B H Halmshaw, 7 Gerrard St., Rochdale, OL11 2EB [IO83WO, SD91]
G0 JSG P J Holden, 19 Briar Cl, Lowestoft, NR32 4SU [JO02UL, TM59]
G0 JSH Details withheld at licensee's request by SSL.[Station located near Weybridge. Op: C W Selby.]
G0 JSJ R J Jackett, Bourne House, 105 Moor Rd, Croston, Preston, PR5 7HP [IO83OQ, SD41]
G0 JSK A B Wakefield, 3 Blythe Pl, Sea Mill Ln, St. Bees, CA27 0BE [IO84EL, NX91]
G0 JSL G S Brown, 9 Western Dr, Leyland, Preston, PR5 3JB [IO83PQ, SD52]
G0 JSM J A Brown, 9 Western Dr, Leyland, Preston, PR5 3JB [IO83PQ, SD52]
G0 JSP P W C Fauchon, 114 Petersfield Ave, Staines, TW18 1DJ [IO91SK, TQ07]
G0 JSR S A C Rickman, 35 Cedar Way, Basingstoke, RG23 8NG [IO91KG, SU65]
GW0 JSS Details withheld at licensee's request by SSL.
G0 JST J H Wheatley obo Jubilee Sailing Trust (Ars, 13 Elstead Gdns, Waterlooville, PO7 5EX [IO90LU, SU60]
G0 JSU Details withheld at licensee's request by SSL.
G0 JSV T A Tierney, 2 Adam Ct, Opal St., London, SE11 4HP [IO91WL, TQ37]
GM0 JSW R Downs, 44 Holehouse Rd, Largs, KA30 9EL [IO75NT, NS25]
GW0 JSX R K Davies, 8 Princes Park, Rhuddlan, Rhyl, LL18 5RW [IO83GG, SJ07]
GJ0 JSY S W Smith-Gauvin, 31-le-Jardin A Pommiers, La Rue de Patier, St. Saviour, Jersey, JE2 7LT
G0 JSZ J K Crellin, 89 Wapshare Rd, West Derby, Liverpool, L11 8LR [IO83MK, SJ39]
G0 JTA D H Ellison, 2 Wheatlands Way, Harrogate, HG2 8PZ [IO93FX, SE35]
G0 JTB Details withheld at licensee's request by SSL.[Station located near Stourbridge.]
G0 JTC Details withheld at licensee's request by SSL.
G0 JTD R H Lyne, 32 Davenwood, Upper Stratton, Swindon, SN2 6LL [IO91CO, SU18]
GW0 JTE L W Horne, 29 Station Terr, Dowlais, Merthyr Tydfil, CF48 3PU [IO81HS, SO00]
GW0 JTF J R Hosking, 14 School Terr, Cwm, Ebbw Vale, NP3 6QY [IO81JR, SO10]
GU0 JTG J T Gethings, Lindale Guest House, Elm Gr, St. Peter Port, Guernsey, GY1 1XE
GW0 JTJ T H Watkins, Ty Unig, Forest Rd, Treharris, CF46 5HG [IO81IQ, ST09]
GW0 JTK P N Burton, 2 Edward Cl, Penyard, Merthyr Tydfil, CF47 0LA [IO81HR, SO00]
G0 JTL J H Hinchliffe, 19 The Terrace, Honley, Huddersfield, HD7 2DS [IO93CO, SE11]
G0 JTM P Smith, 999 Manchester Rd, Linthwaite, Huddersfield, HD7 5LS [IO93BO, SE01]
G0 JTN C A Smith, 35 Allendale Rd, Earley, Reading, RG6 7PD [IO91MK, SU77]
G0 JTO J W T Oxley, 37 Buckminster Gdns, Grantham, NG31 7SJ [IO92QV, SK93]
G0 JTP A J Hill, 159 Sandford Rd, Bradford, BD3 9NU [IO93DT, SE13]
G0 JTR B W Thatcher, 18 Harescombe, Yate, Bristol, BS17 4UA [IO81TM, ST78]
G0 JTT S P Hemsworth, 4 Spoonhill Rd, Stannington, Sheffield, S6 5PA [IO93FJ, SK38]
GW0 JTU A J A Lewis, 96 Roundhouse Cl, Nantyglo, Brynmawr, NP3 4QY [IO81JS, SO11]
G0 JTX L F Huffer, 3-le-Marchant Ave, Lindley, Huddersfield, HD3 3DF [IO93CP, SE11]
G0 JUA J Hardcastle, 37 Caithness Rd, Liverpool, L18 9SJ [IO83NI, SJ38]
GM0 JUB C Healy, 14 Inch Cres, Bathgate, EH48 1EU [IO85EV, NS96]
G0 JUD Details withheld at licensee's request by SSL.
G0 JUE T P Cruse, Watch Tower House, The Ridgeway, Mill Hill, London, NW7 1RS [IO91VO, TQ29]
G0 JUG Details withheld at licensee's request by SSL.
G0 JUI E M Gibson, 107 Church Ave, Meanwood, Leeds, LS6 4JT [IO93FT, SE23]
GW0 JUJ D Griffiths, 8 Barry Rd, Pontypridd, CF37 1HY [IO81HO, ST09]
G0 JUK N B Mayes, Cliffe Cottage, Rotherham Rd, Monkbretton, Barnsley, South Yorks, S71 5QX [IO93GN, SE30]
G0 JUL G S Hogben, 36 King St., Yeadon, Leeds, LS19 7QA [IO93DU, SE24]
G0 JUM S R Barker, 11 Pennington Cl, Copplestone, Crediton, EX17 5NA [IO80DT, SS70]
G0 JUN R Softley, 14 Topps Dr, Bedworth, Nuneaton, CV12 0DE [IO92GL, SP38]
G0 JUQ M J Eborall, Little Cadbold, Pillerton Priors, Warks, CV35 0PQ [IO92FD, SP24]
G0 JUR N F J Barnett, 60 Commercial Rd, Spalding, PE11 2HE [IO92WT, TF22]
G0 JUS A G Swanborough, 121 Sutton Rd, Maidstone, ME15 9AA [JO01GG, TQ75]
G0 JUT D R Horne, 24 Ringwood Dr, Leeds, LS14 1AP [IO93GU, SE33]
G0 JUU R Davies, 49 Truleigh Dr, Mile Oak, Portslade, Brighton, BN41 2YQ [IO90VU, TQ20]
G0 JUW T R Batchelor, 9 Somerset Cl, Sittingbourne, ME10 1JU [JO01II, TQ86]
G0 JUX Details withheld at licensee's request by SSL.
G0 JUY P R Barden, 38 Silver Cl, Tonbridge, TN9 2UY [JO01DE, TQ54]
G0 JUZ R Harrigan, 7 Torrington Gdns, London, N11 2AB [IO91WO, TQ29]
GM0 JVC Dr M A Grant, 43 Churchill Rd, Kilmacolm, PA13 4NA [IO75QV, NS36]
G0 JVD J A Harris, The Old Bakehouse, Thornicks, Winfrith Newburgh, Dorchester, DT2 8JZ [IO80UP, SY88]
G0 JVE R T McLean, 20 Beaconsfield Rd, Fareham, PO16 0QB [IO90JU, SU50]
G0 JVF D R Cleaver, Stovey Croft, The Common, Rowde, Devizes, Wilts, SN10 1SY [IO81XI, ST96]
G0 JVG M A Woodford, Holm Wood, 5 Orchard Cl, North St., South Petherton, TA13 5DX [IO80OW, ST41]
G0 JVI A Mawson, 38 Springbank Rd, Gildersome, Morley, Leeds, LS27 7DJ [IO93ES, SE22]
G0 JVK R Cook, 142 Mansfield Rd, Clipstone, Clipstone Village, Mansfield, NG21 9AL [IO93KE, SK66]
G0 JVN R J Horne, 7 Alexander Rd, Bentley, Walsall, WS2 0HJ [IO82XO, SO99]
G0 JVR J S Rule, The Workshop, Meaver Rd, Mullion, Helston, TR12 7DN [IO70JA, SW61]
G0 JVT D J Marjoram, 459 Norwich Rd, Ipswich, IP1 5DR [JO02NB, TM14]
G0 JVU N G Pattinson, 10 High Hall Cl, Trimley St. Martin, Trimley, Ipswich, IP10 0TJ [JO02PA, TM24]
GM0 JVV J S Stevenson, 52 Fernbrae Ave, Rutherglen, Glasgow, G73 4AE [IO75VT, NS65]
G0 JVW A G Thornton, 1 Primula Cl, Barton Green, Clifton, Nottingham, NG11 8SL [IO92JV, SK53]
G0 JVX M C Linehan, 17 Bolton St., Swanwick, Alfreton, DE55 1BU [IO93HB, SK35]
G0 JVY Details withheld at licensee's request by SSL.[Op: D J Tait, 54 Dickens Ave, Hillingdon, Middx, UB8 3DL.]
G0 JVZ A E Capewell, 25 Causeway, Darley Abbey, Derby, DE22 2BW [IO92GW, SK33]
G0 JWA B A Higton, 3 New St., Little Eaton, Derby, DE21 5AF [IO92GX, SK34]
G0 JWB W J Brady, 15 Bamford Cres, Accrington, BB5 2PQ [IO83TR, SD72]
G0 JWD J M Brambley, 25 Temple Dr, Nuthall, Nottingham, NG16 1BE [IO92JX, SK54]
G0 JWE D H Cliffe, 72 Pares Way, Ockbrook, Derby, DE72 3TL [IO92HW, SK43]
GW0 JWF B Matthews, 25 Manor Park, Newbridge, Newport, NP1 4RS [IO81KQ, ST29]
G0 JWG J W Gaunt, Fenton House, Church Rd, Watlington, Kings Lynn, PE33 0HE [JO02EQ, TF61]
G0 JWJ T R Bridgeland-Taylor, 1 Overbury Ct, Venns Ln, Hereford, HR1 1DG [IO82PB, SO54]
GM0 JWK D Arcari, 40 Glenmoy Ave, Dundee, DD3 8EX [IO86ML, NO33]
G0 JWL K Lindsay, 11A Pyrford Cl, Waterlooville, PO7 6BT [IO90LV, SU61]
G0 JWN E M Buckland, 5A Charles St, Prebbleton, Christchurch, New Zealand, X X
G0 JWO T W Forbes, 63 Wardle Dr, Annitsford, Cramlington, NE23 7DE [IO95FB, NZ27]
G0 JWQ G W Francis, 6 Henderson Cl, Bramford, Ipswich, IP8 4EZ [JO02NB, TM14]
GD0 JWR H S Richardson, Pitcairn, Quarter Bridge Rd, Douglas, Isle of Man, IM2 3RQ
G0 JWS A H Stokes, 3 Graham Cl, Hockley, SS5 5HE [JO01HO, TQ89]
G0 JWT C J Leslie, 19 Hornbeam Cres, Harrogate, HG2 8QA [IO93FX, SE35]
G0 JWV P F Bryant, Crugsillick Cottage, Ruan High Lanes, Truro, Cornwall, TR2 5JP [IO70MF, SW93]
GW0 JWW Details withheld at licensee's request by SSL.
G0 JWX M J Reed, Larks Rise, Great Hewas, Grampound Rd, Truro, TR2 4EP [IO70MI, SW95]
G0 JWY J G Curtis, Conway, 27 Southgate, Hornsea, HU18 1RE [IO93VV, TA24]
G0 JWZ H Williams, 88 Pool Rise, Springfield Est, Shrewsbury, SY2 6EL [IO82PQ, SJ51]
G0 JXC P A D Humphreys, 96 Newcastle Rd, Blyth, NE24 4AS [IO95FC, NZ37]
G0 JXD P R C Turner, Westernlea, 2 Beach Rd, Westward Ho, Bideford, EX39 1HQ [IO71VA, SS42]
G0 JXF Details withheld at licensee's request by SSL.
GW0 JXG M T Price, 4 Vale View, Pontllanfraith, Woodfieldside, Blackwood, NP2 0DB [IO81JP, ST19]
G0 JXJ D J Copeland, Wales Bank, Elm, Wisbech, Cambs, PE14 0AY [JO02BP, TF40]
G0 JXN J E Brightman, 35 Perrysfield Rd, Cheshunt, Waltham Cross, EN8 0TQ [IO91XR, TL30]
G0 JXO R Smith, 24 Kirkstead Rd, Carlisle, CA2 7RD [IO84MV, NY35]
G0 JXP K N Pile, 321 Staines Rd West, Ashford, TW15 1RS [IO91SK, TQ07]
G0 JXQ D H Davis, 17 Welbourne Cl, Raunds, Wellingborough, NN9 6HE [IO92RI, SP97]
G0 JXR P J Keasley, 92 Parrots Field, Hoddesdon, EN11 0QU [IO91XS, TL30]
G0 JXS C Ridgway, 19 Hollin Hill Rd, Clowne, Chesterfield, S43 4AX [IO93JG, SK57]
GW0 JXW Details withheld at licensee's request by SSL.
G0 JXX M G Hoddy, 52 Hayling Rise, High Salvington, Worthing, BN13 3AG [IO90TU, TQ10]
G0 JXY T N Bartholomew, 4 Humberstone Villas, North Somercotes, Louth, LN11 7NJ [JO03BK, TF49]
G0 JXZ I James, 21 St. Blaise Ave, Water Orton, Birmingham, B46 1RT [IO92DM, SP19]
G0 JYB Details withheld at licensee's request by SSL.
G0 JYC P Hodgson, Wenigerbach Str 1, 53819 Neunkirchen, Seelscheid, Germany, X X
G0 JYD J Drinkwater, Springfield, Mudhurst Ln, Disley, Stockport, SK12 2BY [IO83XI, SJ98]
G0 JYF S G Deakin, 25 Hungerford Ave, Westfield Park, Trowbridge, BA14 9ES [IO81VH, ST85]
G0 JYG M Wilson, The Old Chapel, Poulshot Rd, Poulshot, Devizes, SN10 1RW [IO81XI, ST96]
G0 JYH H J Ryan, Fairview, Imperial Ave, Minster on Sea, Sheerness, ME12 2HG [JO01JK, TQ97]
G0 JYI A W R Street, 11 Leigh Gdns, Leigh on Sea, SS9 2PX [JO01HN, TQ88]
G0 JYJ C W R Hodgson, 113 Roman Rd, East Ham, London, E6 3RY [JO01AM, TQ48]
G0 JYK J Sharp, 22 Boat Ln, Irlam, Manchester, M44 6EN [IO83TK, SJ79]
G0 JYL J A Bartram, 4 Calshot Cl, Calshot Rd, Calshot, Southampton, SO45 1BL [IO90IT, SU40]
G0 JYM Details withheld at licensee's request by SSL.
G0 JYN Details withheld at licensee's request by SSL.
G0 JYO P J Walker, 35 Wainbridge Cres, Pilning, Bristol, BS12 3LH [IO81QN, ST58]
G0 JYP Details withheld at licensee's request by SSL.
G0 JYQ M P G Gregory, 22 Yarrow Way, Locks Heath, Southampton, SO31 6TH [IO90IU, SU50]
G0 JYR J J Jepson, 21 Didcot Rd, Manchester, M22 1NJ [IO83UI, SJ88]
G0 JYT Details withheld at licensee's request by SSL.[Station located near Cirencester.]
G0 JYU D J Smith, 104 Hanley Rd, London, N4 3DW [IO91WN, TQ38]
G0 JYV J Rowlands, 67 Woodside, Gosport, PO13 0YX [IO90JU, SU50]
G0 JYW I P Pratt, 2 Clarence Pl, Didcot, OX11 8NT [IO91JO, SU58]
G0 JYX T G Cloke, 14 Bickley Cl, Hanham Green, Bristol, BS15 3TB [IO81RK, ST67]
G0 JYZ I K Broomhall, 49 Funtley Hill, Fareham, PO16 7XA [IO90JU, SU50]
G0 JZA N D Cox, 284 Charlton Rd, Westbury on Trym, Bristol, BS10 6JU [IO81QM, ST57]
G0 JZE A McFadyen, 26 Lewis Rd, Chipping Norton, OX7 5JS [IO91FW, SP32]
G0 JZF N K W Rogers, 21 Maisemore, Yate, Bristol, BS17 4UR [IO81TM, ST78]
G0 JZG Details withheld at licensee's request by SSL.
G0 JZH R W Morris, 96 Chandag Rd, Keynsham, Bristol, BS18 1QE [IO81SJ, ST66]
G0 JZJ F Russell, 37 Overpool Rd, Ellesmere Port, South Wirral, L66 1JW [IO83MG, SJ37]
G0 JZL G N Galley, 1 St. James Ave, Anston, Sheffield, S31 7DR [IO93GJ, SK38]
G0 JZM C H Barber, 471 Kings Rd, Stretford, Manchester, M32 8QN [IO83UK, SJ89]
GW0 JZN M W Randall, 15 Erw Wen, Pencoed, Bridgend, CF35 6YF [IO81FM, SS98]
G0 JZP Details withheld at licensee's request by SSL.[Op: P A Robinson. Station located near the Wirral.]
GW0 JZQ D Hughes, 16 Clyngwyn Rd, Ystalyfera, Swansea, SA9 2AE [IO81CS, SN70]
GW0 JZR R J Sellek, 12 Norseman Cl, Rhoose, Barry, CF62 3FY [IO81HJ, ST06]
G0 JZS G T Corbett, 359 London Rd, West End, Stoke on Trent, ST4 5AN [IO82VX, SJ84]
G0 JZT J F Chappell, 2 Wayside, Knott End on Sea, Poulton-le-Fylde, FY6 0DD [IO83MW, SD34]
G0 JZU W R Etherington, 15 East Bank, North End Rd, Yapton, Arundel, BN18 0DJ [IO90QT, SU90]
GM0 JZV J S Warden, 15 Smalls Sq, Brechin, DD9 7EH [IO86QR, NO65]
G0 JZW A T Elford, 10 Meadowlands, Lymington, SO41 9LB [IO90FS, SZ39]
G0 JZX B A Sanders, 6 Belmont Rd, Coventry, CV6 5HF [IO92GK, SP38]

G0 JZY R C Pocock, 32 Highfield Dr, Portishead, Bristol, BS20 8JD [IO81OL, ST47]
G0 JZZ A Daniels-Galey, 93 Leadwell Ln, Rothwell, Leeds, LS26 0SR [IO93GR, SE32]
G0 KAB A Pilkington, 40 Melbourne Rd, Deane, Bolton, BL3 5RN [IO83SN, SD70]
G0 KAC Details withheld at licensee's request by SSL.
GM0 KAE D A Paterson, 82 Stirling St., Alva, FK12 5EA [IO86CD, NS89]
G0 KAF Details withheld at licensee's request by SSL.
G0 KAG Details withheld at licensee's request by SSL.
G0 KAH K A Heard, 1 Lealand Gr, Drayton, Portsmouth, PO6 1ND [IO90LU, SU60]
GM0 KAI Details withheld at licensee's request by SSL.
GM0 KAJ Details withheld at licensee's request by SSL.
G0 KAK I J Walker, 11 Laburnum Ave, Lutterworth, LE17 4TZ [IO92JK, SP58]
GW0 KAM G B Jones, 14 Plantation Dr, Croesyceiliog, Cwmbran, NP44 2AN [IO81LP, ST39]
GI0 KAN W L Clarke, 2 Glencregagh Park, Belfast, BT6 0NT [IO74BN, J37]
G0 KAO P M Helliwell, 31 Lord St., Darwen, BB3 0HD [IO83SQ, SD62]
G0 KAQ E D Walker, 8 Primrose Hill, Warwick, CV34 5HW [IO92FH, SP26]
G0 KAR C M Winstanley, 27 Sandpiper Walk, Eastbourne, BN23 7SD [JO00DT, TQ60]
G0 KAS M D Stevens, 20 Melton Pl, Epsom, KT19 9EE [IO91UI, TQ26]
G0 KAT V W Chapman, 20 St. Chad, Barrow upon Humber, DN19 7AU [IO93TQ, TA02]
G0 KAU R Crocker, Little London Caravan Park, Lincoln Rd, Torksey Lock, Lincoln, LN1 2EL [IO93PG, SK87]
G0 KAV Details withheld at licensee's request by SSL.
G0 KAW Details withheld at licensee's request by SSL.
GW0 KAX P J Owen, 13 Highland Cl, Sarn, Bridgend, CF32 9SB [IO81FM, SS98]
GM0 KAZ A F White, 65 Orchard St., Galston, KA4 8EJ [IO75TO, NS43]
G0 KBA P Langtree, 243 Devonshire Rd, Atherton, Manchester, M46 9QB [IO83SM, SD60]
G0 KBB Details withheld at licensee's request by SSL.
GM0 KBC N McCormack, 4 Miller Ave, Innellan, Dunoon, PA23 7SU [IO75MV, NS16]
G0 KBD B Dickinson, 23 Gisborne Rd, Sheffield, S11 7HA [IO93FI, SK38]
G0 KBE Details withheld at licensee's request by SSL.
G0 KBF M W Wiggins, 1 Plough Cottage, Shustoke, Coleshill, Birmingham, B46 2AN [IO92EM, SP29]
G0 KBH Details withheld at licensee's request by SSL.
G0 KBI W T Burndred, 52 Everest Rd, Whitehill, Kidsgrove, Stoke on Trent, ST7 4DY [IO83VC, SJ85]
G0 KBJ E F Burndred, 52 Everest Rd, Kidsgrove, Stoke on Trent, ST7 4DY [IO83VC, SJ85]
G0 KBK R J Sleigh, 15 Tildesley Dr, Short Heath, Willenhall, WV12 4JD [IO82XO, SO99]
G0 KBL S Rudcenko, 7 Banstead Rd South, Sutton, SM2 5LF [IO91VI, TQ26]
G0 KBM D C Manning, 43 King St., Walton, Felixstowe, IP11 9DX [JO01QX, TM23]
G0 KBN G M Beech, 15 Fisher Ct, Cotmanhay, Ilkeston, DE7 8PZ [IO92IX, SK44]
G0 KBO V Kravchenko, 16 Birchfield House, Birchfield St., London, E14 8EY [IO91XM, TQ38]
G0 KBP M C A G Dearing, 1 Woodbine Villas, New Village Rd, Cottingham, HU16 4NF [IO93TS, TA03]
G0 KBQ J E Pryor, 69 West Drayton Rd, Uxbridge, UB8 3LD [IO91SM, TQ08]
G0 KBS L G Kay, 2 Childwall Cres, Childwall, Liverpool, L16 7PQ [IO83NJ, SJ48]
GM0 KBU S S Waterall, 3 Wavell St., Grangemouth, FK3 8TG [IO86DA, NS98]
G0 KBV L Andrew, 82 Green Ln, Boarshaw, Middleton, Manchester, M24 2WF [IO83VN, SD80]
G0 KBX D C V Hayward, Beech Cottage, South Rd, Tetford, Horncastle, LN9 6QB [IO93XF, TF37]
G0 KBY M R Andrews, 10 Riseway Cl, Norwich, NR1 4NJ [JO02PP, TG20]
G0 KBZ K Harrison, 6 Staveley Rd, Alford, LN13 0PN [JO03CG, TF47]
G0 KCA I W Walder-Davis, 93 Church St., St. Peters, Broadstairs, CT10 2TX [JO01RI, TR36]
G0 KCB D A E Beckley, 14 Lockington Rd, Stowmarket, IP14 1BQ [JO02LE, TM05]
G0 KCC Dr W L Randolph, 13 Links Rd, Poole, BH14 9QP [IO90AR, SZ09]
G0 KCD M A L Leech, 192 Windsor Rd, Ilford, IG1 1HE [JO01AN, TQ48]
G0 KCE V E Harding, 17 St. Anns Rd South, Heald Green, Cheadle, SK8 3DZ [IO83VI, SJ88]
G0 KCF C J Fosbrook, Cuckoo Farm, 4 Yew Tree Rd, Hayling Island, PO11 0QE [IO90MT, SU70]
G0 KCG D M Hall, 47 Sunningdale Rd, Portchester, Fareham, PO16 9PA [IO90KU, SU60]
G0 KCH M P McCarthy, 75 Taynton Dr, Merstham, Redhill, RH1 3PX [IO91WG, TQ25]
G0 KCI B F Parkes, 8 Longlands, Dawlish, EX7 9NE [IO80GN, SX97]
G0 KCJ A L R Johnson, 25 Seafield Rd, Tankerton, Whitstable, CT5 2LW [JO01MI, TR16]
G0 KCL Details withheld at licensee's request by SSL.
G0 KCM J R Fenton, 54 Norman Rd, Penkridge, Stafford, ST19 5EX [IO82WR, SJ91]
GM0 KCN D Smith, 12 Cannon St., Selkirk, TD7 5BP [IO85NN, NT42]
G0 KCP R H Beech, 120 Norcliffe Rd, Bispham, Blackpool, FY2 9EW [IO83LU, SD34]
G0 KCQ Details withheld at licensee's request by SSL.
GM0 KCV J A Tayler, Oakland, 17 Mountstuart Rd, Rothesay, PA20 9DY [IO75LU, NS06]
G0 KCW Details withheld at licensee's request by SSL.
GM0 KCY D A Michael, Butt of Lewis Lighthouse, Port of Ness, Isle of Lewis, PA86 0XH [IO76JB, NS08]
G0 KCZ W G C Bowles, Willow Gr, Little Common, North Bradley, Trowbridge, BA14 0TX [IO81VH, ST85]
G0 KDA P D Cooper, 53 Sea View Dr, Scarborough, YO11 3JS [IO94TG, TA08]
G0 KDB D H Greenhalgh, Hillcroft, Colby, Appleby in Westmorland, Cumbria, CA16 6BD [IO84RN, NY62]
GM0 KDC R K Smith, 21 Glen View Cres, Gorebridge, EH23 4BT [IO85LU, NT36]
G0 KDD B D Woodward, 58 Marine Dr, Bishopstone, Seaford, BN25 2RU [JO00AS, TQ40]
GM0 KDF R L Thomson, 25 Cheviot Rd, Hamilton, ML3 7HB [IO75XS, NS75]
G0 KDG R M Simpson, 25 Hoe View Rd, Cropwell Bishop, Nottingham, NG12 3DE [IO92MV, SK63]
G0 KDI R F Steans, 302 Walton Rd, West Molesey, KT8 2HY [IO91TJ, TQ16]
G0 KDJ J W McGarry, 22 Dunsford, Widnes, WA8 4NP [IO83OJ, SJ48]
G0 KDK D W Bradbury, 17 Dibble Cl, Lodge Farm Est, Willenhall, WV12 4EE [IO82XO, SJ90]
G0 KDL W M C K Cooper, 24 Ambleside Rd, Lightwater, GU18 5TA [IO91QI, SU96]
G0 KDN P W E Roake, The Mint, 5 Tyler Hill Rd, Blean, Canterbury, CT2 9HP [JO01MH, TR16]
GM0 KDO G C Kirkland, 2 St. Serfs Rd, Crook of Devon, Kinross, KY13 7PQ [IO86FE, NO00]
GM0 KDP I Dunbar, Hillcrest, 25 Kinord Dr, Aboyne, AB34 5JZ [IO87OB, NO59]
G0 KDQ M M Ward, 45 The Gills, Otley, LS21 2BY [IO93DV, SE24]
G0 KDR R Lintott, Upper Gr Farm, Rendham, Saxmundham, Suffolk, IP17 2AS [JO02RF, TM36]
G0 KDS S W Lindsay, 27 Bagnell Rd, Bristol, BS14 8PZ [IO81RJ, ST66]
G0 KDT P G Cracknell, 19 Fairlea Cl, Brookside, Dawlish, EX7 0NN [IO80GO, SX97]
G0 KDU J Hillerby, The Freehold, 20 Marvillion Ct, East Peckham, Tonbridge, Kent, TN12 5AW [JO01EF, TQ64]
G0 KDV R Barker Darenth Vall RS, 1 Hazel End, Swanley, BR8 8NU [JO01CJ, TQ56]
G0 KDX Details withheld at licensee's request by SSL.[Op: B E Ashton. Station located near Wigan.]
G0 KDY A J S Spry, 38 Fairfield Rd, Bude, EX23 8DJ [IO70RT, SS20]
G0 KEA M A Baxter, 29 Northfield, Bridgwater, TA6 7HA [IO81LD, ST23]
G0 KEB C L Frost, 61 Selbourne Ave, Tolworth, Surbiton, KT6 7NR [IO91UJ, TQ16]
G0 KEC H Opitz, 17 Oakland Rd, Newton Abbot, TQ12 4EA [IO80EM, SX87]
G0 KED J Bower, Linwood, Stain Ln, Theddlethorpe, Mablethorpe, LN12 1QB [JO03CI, TF48]
G0 KEE C S Simons, Logos 2, Postfach 1565, 74819 Mosbach, Germany, X X
G0 KEH Details withheld at licensee's request by SSL.
G0 KEJ A F Sanders, 122 Eastfield Rd, Wellingborough, NN8 1PS [IO92PH, SP96]
G0 KEK B C Curtis, Beggars Roost, Rea Barn Rd, Brixham, TQ5 9EE [IO80FJ, SX95]
GM0 KEL I H Wilson, 30 Howdenburn Ct, Jedburgh, TD8 6NP [IO85RL, NT62]
G0 KEM P M Reed, Larks Rise, Great Hewas, Grampound Rd, Truro, TR2 4EP [IO70MI, SW95]
G0 KEN K Wightman, 38 Bradford Ave, Bolton, BL3 2PF [IO83TN, SD70]
GD0 KEO A S C Birchenough, Glenouff, Lezayre, Ramsey, Isle of Man, IM7 2AT
G0 KEP P Griffiths, 103 Connaught Rd, Brookwood, Woking, GU24 0EU [IO91QH, SU95]
G0 KEQ R C Crawford, Moreton Grange, Moreton St., Prees, Whitchurch, SY13 2EF [IO82PV, SJ53]
GM0 KET Details withheld at licensee's request by SSL.
G0 KEU K G Clayton, 23 Hackness Dr, Newby, Scarborough, YO12 5SB [IO94SH, TA08]
G0 KEV K Gallagher, 24 Lakenheath Dr, Sharples, Bolton, BL1 7RJ [IO83SO, SD71]
G0 KEX A C OHara, 26 Thompson Ave, Bolton, BL2 5RJ [IO83TO, SD71]
G0 KEY S J Cole, 37 Vaux Cres, Hersham, Walton on Thames, KT12 4HE [IO91TI, TQ16]
G0 KEZ Details withheld at licensee's request by SSL.
G0 KFA K A Hadlow, 49 Stanley Gdns, Herne Bay, CT6 5SQ [JO01NI, TR16]
G0 KFB P C Bell, The Ferns, 11 East St., Alford, LN13 9EQ [JO03CG, TF47]
G0 KFE D Fish, 29 Winter Gr, Parr, St. Helens, WA9 2JT [IO83PK, SJ59]
G0 KFF K F Field, 1 Hebden, Wilnecote, Tamworth, B77 4HP [IO92EO, SK20]
G0 KFG A Gauld, 1 Hirstead Rd, Scarborough, YO12 6TW [IO94SH, TA08]
G0 KFH Details withheld at licensee's request by SSL.
G0 KFI R W Bonney, 3 Aragon Dr, Ruislip, HA4 9PR [IO91TN, TQ18]
G0 KFJ L Morley-Joel, Whitespar, Queen Hoo Ln, Tewin, Welwyn, AL6 0LT [IO91WU, TL21]
G0 KFN D F H Darke, The Old Fosse Forge, Stone, East Pennard, Shepton Mallet, BA4 6RY [IO81QC, ST53]
G0 KFO J E Turner, 36 The Grove, Greenhill, Herne Bay, CT6 7QD [JO01NI, TR16]
G0 KFP K A Wooller, 30 Ellesmere Rd, Eccles, Manchester, M30 9FD [IO83TL, SJ79]
G0 KFQ B I Wilson, 14 Lovell Rd, Cambridge, CB4 2QR [JO02BF, TL46]
G0 KFR T J Morgan, 4 Seldon Cl, Westcliff on Sea, SS0 0AD [JO01IN, TQ88]
G0 KFS A E Purcell, 33 Fishley Cl, Bloxwich, Walsall, WS3 3QA [IO92AP, SK00]
G0 KFT C H Dickerson, 1 Park Farm Ln, Nuthampstead, Royston, SG8 8LT [JO01AX, TL43]
G0 KFV M W Evans, 72 Ambrose St., Fulford Rd, York, YO1 4DR [IO93LW, SE65]
G0 KFW W C Cole, 5A Park Ln, Kemsing, Sevenoaks, TN15 6NU [JO01CH, TQ55]
G0 KFX Details withheld at licensee's request by SSL.
G0 KFY P J Elliot-West, 135 Tunstall Rd, Sunderland, SR2 9BB [IO94HV, NZ35]
G0 KGA A Danby, 53 Whitfield Ave, Pickering, YO18 7HX [IO94OF, SE88]
G0 KGB Details withheld at licensee's request by SSL.
G0 KGC A S Tabelin, Rainbow Farm, Halesworth Rd, Redisham, Beccles, NR34 8NE [JO02SJ, TM48]
G0 KGD Dr B Zacharov, The Firs, Old Castle, Malpas, Ches, SY14 7NE [IO82OX, SJ44]
G0 KGE H P Johnson, 2 Thirlmere, Kennington, Ashford, TN24 9BD [JO01KD, TR04]
G0 KGH Details withheld at licensee's request by SSL.
G0 KGI J C Coleman, 80 Ormston Ave, Horwich, Bolton, BL6 7ED [IO83RO, SD61]
G0 KGJ Details withheld at licensee's request by SSL.
G0 KGK Details withheld at licensee's request by SSL.
G0 KGL G Lindsay, 66 Jubilee Cres, Mangotsfield, Bristol, BS17 3AZ [IO81SL, ST67]
G0 KGM K Williams, 55 Sherwood Cres, Worle, Weston Super Mare, BS22 9NG [IO81MI, ST36]
G0 KGO Details withheld at licensee's request by SSL.
G0 KGP Details withheld at licensee's request by SSL.
G0 KGQ G A Armatage, 39 Priors Grange, Pittington, High Pittington, Durham, DH6 1DA [IO94FS, NZ34]
G0 KGR R G Beadle, 2 Edward Cttgs, Great Munden, Ware, Herts, SG11 1HT [IO91XV, TL32]
G0 KGT B A Albers, The Hawthorns, 14 Hartwith Green, Summerbridge, Harrogate, HG3 4HX [IO94DB, SE26]
G0 KGU Details withheld at licensee's request by SSL.
G0 KGV Details withheld at licensee's request by SSL.
G0 KGW G R Duckett, 4 The Orchard, High St., Pensford, Bristol, BS18 4BG [IO81RI, ST66]
G0 KGX S J Stewart, 56 Rydal St., Leigh, WN7 4DR [IO83RL, SD60]
G0 KGY M F Mavin, 52 Bywell Rd, Ashington, NE63 0LE [IO95FE, NZ28]
G0 KHA K S Seddon, 17 Dunmail Dr, Heron Hill, Kendal, LA9 7JG [IO84PH, SD59]
G0 KHB M S Davis, Wensum House, 18 Southampton Rd, Romsey, SO51 8AF [IO90GX, SU32]
G0 KHC Details withheld at licensee's request by SSL.
G0 KHD Details withheld at licensee's request by SSL.
G0 KHF P C M Witley, 18 Seagate Rd, Hunstanton, PE36 5BD [JO02FW, TF64]
G0 KHG Details withheld at licensee's request by SSL.
G0 KHH A Rogers, 3 Ripley Dr, Highfield, Wigan, WN3 6AJ [IO83PM, SD50]
G0 KHI W McEntee, 100 Broadway, Royton, Oldham, OL2 5BP [IO83WN, SD90]
G0 KHJ J Warburton, 92 Worsley Rd, Farnworth, Bolton, BL4 9LX [IO83TN, SD70]
G0 KHK P N Shaw, 15 Greenfield Ave, Rock, Marlbrook, Bromsgrove, B60 1HE [IO82XI, SO97]
G0 KHL R E Steadman, 716 Lincoln Rd, Peterborough, PE1 3HH [IO92UO, TF10]
G0 KHM M D Greenwood, 21 Dobb Top Rd, Holmbridge, Holmfirth, Huddersfield, HD7 1PQ [IO93CN, SE10]
G0 KHN S T Dunn, 24 Gatwick Cl, Lords Hill, Southampton, SO16 8HL [IO90GW, SU31]
GM0 KHP E M C Wylie, Dunedin, Dunnet, Thurso, Caithness, KW14 8YD [IO88HP, ND27]
G0 KHQ P E Hughes, 2 Butterbur Way, Killinghall, Harrogate, HG3 2XH [IO94FA, SE25]
G0 KHR E N Forsyth, 11 Brooklyn Rd, Dialstone Ln, Stockport, SK2 6BX [IO83WJ, SJ98]
G0 KHV Details withheld at licensee's request by SSL.
G0 KHW Details withheld at licensee's request by SSL.
G0 KHX Details withheld at licensee's request by SSL.
G0 KHY J R Jenkins, 3 Gosslan Cl, St. Ives, Huntingdon, PE17 6YZ [IO92XI, TL37]
G0 KHZ M Crane, 5 Harland Rd, Elloughton, Brough, HU15 1JT [IO93RR, SE92]
G0 KIA R N Harris, 33 Culver Ln, Earley, Reading, RG6 1DX [IO91MK, SU77]
G0 KIC B J Hayward, 183 Weyhill Rd, Andover, SP10 3LJ [IO91FF, SU34]
G0 KIE M O Cherrington, 11 Whiteoaks Ln, Greenford, UB6 8XE [IO91TM, TQ18]
GW0 KIG K P O'Reilly, 14 St. Fagans Rd, Cardiff, CF5 3AJ [IO81JL, ST17]
G0 KIH C S H Frater, 56 Lynch Rd, Farnham, GU9 8BX [IO91OF, SU84]
G0 KIJ C Witson, 66 Saddleback, Albany, Washington, NE37 1QB [IO94FV, NZ35]
G0 KIK S C Berry, 26 Westcott Cl, Widley, Plymouth, PL6 5YB [IO70WJ, SX45]
G0 KIM J A West, 242 Grane Rd, Haslingden, Rossendale, BB4 4PB [IO83UQ, SD72]
G0 KIN T W Harper, 5 Chesworth Rd, Harwood Park, Bromsgrove, B60 2HF [IO82XH, SO96]
G0 KIQ D S Johnson, 31 Craig Rd, Moss Ln Est, Macclesfield, SK11 7XN [IO83WF, SJ97]
G0 KIR Details withheld at licensee's request by SSL.
G0 KIT Details withheld at licensee's request by SSL.
G0 KIU D J Godfrey, 46 Park View Way, Mansfield, NG18 2RN [IO93JD, SK56]
G0 KIW P D Walker, 5 The Malverns, Abbeymead, Abbeydale, Gloucester, GL4 4WN [IO81VU, SO81]
G0 KIY N Brook, Regent House, 2 Back Regent Pl, Starbeck, Harrogate, HG1 4QR [IO94GA, SE35]
G0 KIZ Details withheld at licensee's request by SSL.
G0 KJA K S Welsh, 14 Ribbesford Dr, Stourport on Severn, DY13 8TG [IO82UI, SO87]
G0 KJC J Clayton, 49 Bramble Ln, Mansfield, NG18 3NP [IO93KD, SK56]
G0 KJF R G F Warner, 32 Main St., Marston Trussell, Market Harborough, LE16 9TY [IO92ML, SP68]
G0 KJG B C Bevington, 12 Buckingham Rd, Rowley Regis, Warley, B65 9JN [IO82XL, SO98]
G0 KJJ W J Ritchie, The Ptarmigan, Buttery Rd, Little Combe, Honiton Devon, XX99 1AA
G0 KJK K B Ranger, 28 Charter Rd, Altrincham, WA15 9RL [IO83TJ, SJ78]
G0 KJL Details withheld at licensee's request by SSL.
G0 KJM J Richards, 14 Southwood Dr East, Bristol, BS9 2QP [IO81QL, ST57]
G0 KJP D A Scott, 19 The Fillybrooks, Walton, Stone, ST15 0DH [IO82WV, SJ83]
G0 KJQ G Lovell, 217 Moorside Rd, Flixton, Manchester, Lancs, M41 5SJ [IO83TK, SJ79]
G0 KJS Details withheld at licensee's request by SSL.
G0 KJU J A Robertson, 28 Frith Rd, Bognor Regis, PO21 5LL [IO90PS, SZ99]
G0 KJV Details withheld at licensee's request by SSL.
G0 KJW J E Brown, 66 Abbey Meadows, Morpeth, NE61 2YA [IO95DD, NZ18]
GM0 KJX W H Brownhill, 25 Shore Rd, Invergordon, IV18 0ER [IO77VQ, NH76]
G0 KJY Details withheld at licensee's request by SSL.
GW0 KJZ J R Jones, 64 Cleviston Park, Llangennech, Llanelli, SA14 9UP [IO71WQ, SN50]
GJ0 KKB K Kirk-Bayley, 4 Fairfield Ave, La Pouquelaye, St. Helier, Jersey, JE2 3FT
G0 KKD A J Parker, Derwent House, Roslyn Rd, Hathersage, Sheffield, S30 1BY [IO93GJ, SK38]
GM0 KKE I Coulson, 11 Redcliffs, Kingoodie, Invergowrie, Dundee, DD2 5DL [IO86LK, NO32]
G0 KKF B R Hough, 2 Litherland Rd, Sale Moor, Sale, M33 2PE [IO83UK, SJ89]
G0 KKH J A Henderson, 64 Central Ave, Corringham, Stanford-le-Hope, SS17 7NG [JO01FM, TQ78]
G0 KKK S M Tomlinson, 403 Red Lees Rd, Burnley, BB10 4TF [IO83VS, SD83]
G0 KKL P W Mayer, 16 Haig Ave, Poole, BH13 7AJ [IO90AR, SZ09]
G0 KKM D A McAtee, 25 Firwood Dr, Tuffley, Gloucester, GL4 0AB [IO81VU, SO81]
GW0 KKN J Harris, 7 Smith St., Gelli, Pentre, CF41 7NG [IO81GP, SS99]
G0 KKO J Langan, 58 Lowther Rd, Millom, LA18 4PQ [IO84IE, SD17]
G0 KKR B G M Chapman, Millbrooke Cottage, Covenham, St Bartholomew, Louth, Lincs, LN11 0PB [JO03AK, TF39]
G0 KKS A G Nance, 33 Oak Cl, Copthorne, Crawley, RH10 3QT [IO91WD, TQ33]
G0 KKT I L D Osbourne, 22 Pentlands, Fullers Slade, Milton Keynes, MK11 2AG [IO92NB, SP73]
G0 KKU J Howard, 29 Station Rd, Penketh, Warrington, WA5 2PH [IO83QJ, SJ58]
G0 KKX C D Chinn, 28A Cavendish Rd, Henleaze, Bristol, BS9 4EA [IO81QL, ST57]
G0 KKY Details withheld at licensee's request by SSL.
G0 KLB L Platten, 59 Royds Ln, Rothwell, Leeds, West Yorks, LS26 0BJ [IO93GR, SE32]
GW0 KLC S O Roberts, Bryngwyn, Bryngran, Holyhead, Anglesey Gwynedd, LL65 3RB [IO73SG, SH37]
G0 KLF N Anderton, 29 Cliftonville Dr, Swinton, Manchester, M27 5NA [IO83UM, SD70]
G0 KLG R H Hodds, 17 Oaklands Dr, Willerby, Hull, HU10 6BJ [IO93SS, TA03]
G0 KLH R J Griffin, Midland Gliding Club, The Long Mynd, Church Stretton, Shropshire, SY6 6TA [IO82NM, SO49]
G0 KLI Details withheld at licensee's request by SSL.
G0 KLJ J E Leader, 9 Southerwicks, Corsham, SN13 9NH [IO81VK, ST86]
G0 KLK A A James, 14 Randle Dr, Four Oaks, Sutton Coldfield, B75 5LH [IO92CO, SP19]
G0 KLM Details withheld at licensee's request by SSL.
GW0 KLN C T J Ryalls, 3 Bryn Terr, Blaenclydach, Tonypandy, CF40 2RY [IO81GP, SS99]
GM0 KLO C D Grossart, 11 Woodlands Dr, Brightons, Falkirk, FK2 0TF [IO85DX, NS97]
GM0 KLP J S Pentland, 2 Glenniston Cttgs, Auchtertool, Fife, KY5 0AX [IO86IC, NT29]
G0 KLQ D A Cross, 32 Harvest Bank, Hyde Heath, Amersham, HP6 5RD [IO91QQ, SU99]
G0 KLR P J Irwin, 17 The Front, Buxton, SK17 7EQ [IO93BG, SK07]
G0 KLS R S Pangborn, 1 Waterworks Cttgs, Hampton Loade, Bridgnorth, Salop, WV15 6HD [IO82TL, SO78]
G0 KLT D Rogers Jones, 20 Birchwood, Leyland, Preston, Lancs, PR5 3QJ [IO83PQ, SD52]
G0 KLU P A Fairhurst, 6 Audwick Cl, Cheshunt, Waltham Cross, EN8 0RF [IO91XR, TL30]
G0 KLW K H Barker, 26 Orchard Way, Orchard Park, Trenches Ln, Langley, Bucks, SL3 6DH [IO91RM, TQ08]
GW0 KLY R M Jones, 57 Charles St., Porth, CF39 9YD [IO81HO, ST09]
G0 KLZ G A Peake, 49 Bredon Ave, Binley, Coventry, CV3 2AA [IO92GJ, SP37]
GI0 KMA Dr M G Rainey, 12 Marna Brae, Lisburn, BT27 4LD [IO64XM, J26]
G0 KMB K Bowdler, 18 Cavendish St., Leigh, WN7 1SG [IO83RM, SD60]
G0 KMC A J Slaughter, 42 Goss Ave, Waddesdon, Aylesbury, HP18 0LY [IO91NU, SP71]
GM0 KMD N Drysdale, 10 Newton Ave, Skinflats, Falkirk, FK2 8NP [IO86DA, NS98]
G0 KME Details withheld at licensee's request by SSL.[Station located near Bury St Edmunds.]
G0 KMF R K Holmshaw, 142 Oakleigh Park Dr, Leigh on Sea, SS9 1RU [JO01HN, TQ88]
GM0 KMG W S M Gibson, 180 Castlemilk Rd, Glasgow, G44 4NS [IO75VT, NS66]
GM0 KMJ P G Johnstone, 3 Bailie Fyfe Way, Overtown, Wishaw, ML2 0EH [IO85BS, NS75]
G0 KMK M Aslam, 38 Grey St., Burnley, BB10 1BA [IO83VT, SD83]
G0 KMM P Moir, 106 Borrowdale Rd, Moreton, Wirral, L46 0RQ [IO83KJ, SJ28]
G0 KMN S G Hepworth, 9 College View, Ackworth, Pontefract, WF7 7LA [IO93HP, SE41]
G0 KMP G P Aungiers, 6 Woodlands Cres, Barton, Preston, PR3 5HB [IO83PT, SD53]
G0 KMR E W Nichols, 12 Woodlands Rd, Haywards Heath, RH16 3JY [IO90WX, TQ32]
G0 KMS N Rees, 136 East Pines Dr, Thornton Cleveleys, FY5 3DS [IO83LU, SD34]
G0 KMT I D Broadbent, 15 Mowbray Rd, Fleetwood, FY7 7JB [IO83LW, SD34]
G0 KMU J Birkby, 10 Rankin Ave, Hesketh Bank, Preston, PR4 6PA [IO83NQ, SD42]
G0 KMV H King, 53 Highfields, Stanton Drew, Bristol, BS18 4DH [IO81RI, ST56]

G0 KMW H H Shepherd, Whydown, 3 White House Cl, Shippon, Abingdon, OX13 6LP [IO91IQ, SU49]
G0 KMX Details withheld at licensee's request by SSL.
G0 KMZ Details withheld at licensee's request by SSL.
G0 KNB A Mead, 14 Cliffe Park, Seaburn, Sunderland, SR6 9NS [IO94HW, NZ45]
GW0 KNC T W A Wilcox, Gellypystyll Farm, New Bungalow, Tranch, Pontypool, Gwent, NP4 6BP [IO81LQ, SO20]
G0 KNF J M Suarez-Fernandez, 15 Western Ave, New Milton, Barton on Sea, New Milton, BH25 7PY [IO90DR, SZ29]
G0 KNG Rev T Wilkinson, 16 Rose Mount Dr, Wallasey, Merseyside, L45 5JA [IO83LK, SJ39]
G0 KNH J D Greenwood, 43 Townend Ave, Low Akworth, Ackworth, Pontefract, WF7 7HE [IO93IP, SE41]
G0 KNJ R W Bygrave, 38 Foxhall Rd, Didcot, OX11 7AA [IO91JO, SU59]
G0 KNL W Christlo, 20 Trickett Rd, High Green, Sheffield, S30 4FN [IO93GJ, SK38]
G0 KNM G A Woodford, The Lodge, Brunswick Park, Woodgreen Rd, Wednesbury, WS10 9AU [IO82XN, SO99]
G0 KNN M Gregg, 22 Mayfields, Spennymoor, DL16 6RN [IO94EQ, NZ23]
G0 KNQ R Gallop, The Cottage, Church Cl, West Runton, Cromer, NR27 9QY [JO02OW, TG14]
G0 KNR R C Sterry, 1 Wavell Garth, Sandal Magna, Wakefield, WF2 6JP [IO93GP, SE31]
GM0 KNT Dr A Bates, Phorp Farmhouse, Dunphail By Forres, Morayshire, IV36 0QR [IO87EM, NJ04]
G0 KNU D P Clear, 1 Milestone Cttgs, Main Rd, Fishbourne, Chichester, PO18 8AU [IO90OU, SU80]
G0 KNV R H Crocker, 8 Hampden Rd, Cowley, Oxford, OX4 3LW [IO91JR, SP50]
G0 KNW C K Wiles, 5 Burnside, Morpeth, NE61 1TB [IO95DE, NZ28]
G0 KNX G L A Allen, 3 Ryton Cl, Canley, Coventry, CV4 8HF [IO92FJ, SP27]
G0 KNZ A Melia, 17 Grange Rd, Saltford, Bristol, BS18 3AH [IO81SJ, ST66]
G0 KOA I M Tyler, 12 Ramerick Gdns, Arlesey, SG15 6XZ [IO91UX, TL13]
G0 KOC A A Kinson, 29 Foxhall Rd, Didcot, OX11 7AQ [IO91IO, SU58]
GW0 KOD I A C Borland, 12 Bro Derfel, Tregarth, Bangor, LL57 4RR [IO73WE, SH56]
G0 KOE T Foxton, Dunbar, Dam Ln, Leavening, Malton, YO17 9SJ [IO94OB, SE76]
G0 KOF D F Henretty, Westerne Farm, Blackborough, Cullompton, Devon, EX15 2HQ [IO80IU, ST00]
G0 KOG G Todd, 52 Trevor Cres, St. James, Northampton, NN5 5PF [IO92MF, SP76]
G0 KOH J F Dover, 11 Roman Rd, Barton-le-Clay, Bedford, MK45 4QJ [IO91SX, TL03]
G0 KOI M J Cooper, 15 Woodleigh Ave, Harborne, Birmingham, B17 0NW [IO92AK, SP08]
G0 KOJ B E R Thomas, Torcottage, 12 The Dip, Newmarket, CB8 8AH [JO02FF, TL66]
G0 KOK P J Love, 9 Argyle Ave, Westbrook, Margate, CT9 5RW [JO01QJ, TR37]
G0 KOO A McDowell, Fern Cottage, The Gride, Old Leake, Boston, PE22 9LS [JO03AA, TF34]
G0 KOP Details withheld at licensee's request by SSL.
G0 KOQ Rev C D Garrick, 266 Wellington Rd North, Stockport, SK4 2QR [IO83VK, SJ89]
G0 KOR Details withheld at licensee's request by SSL.
G0 KOS A J Yeates, 22 Hampstead Rd, Brislington, Bristol, BS4 3HJ [IO81RK, ST67]
G0 KOU B A Arrowsmith, 25 Watchouse Rd, Chelmsford, CM2 8PT [JO01FQ, TL70]
GI0 KOV P J Brown, 64 Templegrove, Londonderry, BT48 0QN [IO65HA, C41]
GI0 KOW R W C Cummings, 19 Bachelors Walk, Keady, Armagh, BT60 2NA [IO64PG, H83]
GI0 KOX Details withheld at licensee's request by SSL.
G0 KOY B J Clues, 8 Acland Ave, Colchester, CO3 3RS [JO01KV, TL92]
G0 KPA J Ellison, 66 Fisher St., Paignton, TQ4 5ES [IO80FK, SX86]
G0 KPB I A Ridgway, 19 Hollin Hill Rd, Clowne, Chesterfield, S43 4AX [IO93JG, SK57]
GW0 KPD J James, 1 Pellau Rd, Margam, Port Talbot, SA13 2LF [IO81CN, SS78]
G0 KPE C F McGowan, Belvedere, Tylers Ln, Bucklebury, Reading, RG7 6TN [IO91JK, SU57]
GI0 KPF B J McCausland, 1 Bernagh Gdns, Dungannon, BT71 4AP [IO64OM, H76]
G0 KPG R K Moore, 34 Fishponds Rd, Kenilworth, CV8 1EZ [IO92EI, SP27]
G0 KPH P J Keighley, 42 Rawlinson Rd, Lillington, Leamington Spa, CV32 7QS [IO92FH, SP36]
GM0 KPK Details withheld at licensee's request by SSL.
GW0 KPM M G Priddy, Esquimalt, 25 The Highway, New Inn, Pontypool, NP4 0PW [IO81LQ, ST39]
G0 KPQ J U Morgan, Kosciol Chrzescijan Baptystow, Ul.Dabrowskiego 11, 80-153 Gdansk, Poland, 80 153
G0 KPT J C Kelly Derbyshire Dales Radio Commu C, 7 Collingwood Cres, Sheriff Fields, Matlock, DE4 3TB [IO93FD, SK26]
GW0 KPU R J Harper, Chain Cottage, 4 Gresford Rd, Llay, Wrexham, LL12 0NW [IO83MC, SJ35]
GW0 KPV G G Owen, 2 Ffordd Beibio, Holyhead, LL65 2EF [IO73QH, SH28]
G0 KPW Details withheld at licensee's request by SSL.
G0 KPY A Baker Munton, 15 Stockwell Rd, Leicester, LE2 3PN [IO92KO, SK60]
G0 KPZ D A Portch, 37 Rowley Ave, Sidcup, DA15 9LF [JO01BK, TQ47]
G0 KQA P A Davies, Silver Birches, Orchard Rd, Basingstoke, RG22 6NU [IO91KG, SU65]
G0 KQD A E Smith, 38 Hartley St., Boston, PE21 9BS [IO92XX, TF34]
G0 KQG A Hoyle, 6 Cambridge Terr, Otley, LS21 1JS [IO93DV, SE24]
G0 KQH B ODonoghue, 28 Emerson Rd, Poole, BH15 1QT [IO90AR, SZ09]
G0 KQI L A Painter, 185 Albion St., St. Helens, WA10 2HA [IO83PK, SJ59]
G0 KQJ I T Burgess, 99 Dugdell Cl, Ferndown, BH22 8BJ [IO90BT, SU00]
G0 KQK T A Chambers, Autumn, Castle Bytham, Grantham, Lincs, NG33 4RT [IO92RS, SK91]
G0 KQL Details withheld at licensee's request by SSL.
G0 KQM R Beck, 6 Hatters Cl, Copmanthorpe, York, YO2 3XQ [IO93KW, SE54]
G0 KQO G Elliott, 32 Chapel St., Newport, PO30 1PZ [IO90IQ, SZ48]
G0 KQP J Carroll, 5 Montagu View, Oakwood, Leeds, LS8 2RH [IO93FT, SE33]
G0 KQR R A Bundell, 24 Sylvan Ave, East Cowes, PO32 6PS [IO90IS, SZ59]
G0 KQS G B Griffiths, Alderley, 170 Storeton Rd, Prenton, Birkenhead, L42 8NB [IO83LI, SJ38]
G0 KQT A R Holdway, 18 The Quantocks, Thatcham, RG19 3SF [IO91IJ, SU56]
GW0 KQU G E F Slatter, 6 Glannant St., Penygraig, Tonypandy, CF40 1JT [IO81GO, SS99]
G0 KQV D L Clark, 7 Hills Rd, Sible Hedingham, Halstead, CO9 3JH [JO01HX, TL73]
GW0 KQX W A Cook, Llwyn-Benglog, Pant-y-Dwr, Rhayader, Powys, LD6 5LW [IO82GJ, SN97]
G0 KQY D Murrell, 56 Russell Rd, Horsell, Woking, GU21 4UY [IO91RH, SU95]
GW0 KQZ Details withheld at licensee's request by SSL.
G0 KRA Details withheld at licensee's request by SSL.
G0 KRB G M Phillips, 57 Hollytrees, Bar Hill, Cambridge, CB3 8SF [JO02AG, TL36]
G0 KRC G A F Philpotts, 62 Emeley Cl, Stourport on Severn, DY13 0AH [IO82UH, SO87]
G0 KRD D J Downes, 7 Sandy Ln, Fakenham, NR21 9ES [JO02KU, TF93]
G0 KRE K E Wilson, 4 Amberley Dr, Hove, BN3 8JS [IO90VU, TQ20]
G0 KRF E T Capps, Lawley House, The Fold, Dorrington, nr Shrewsbury, SY5 7JD [IO82OO, SJ40]
G0 KRG L D B Conlon Keighley Raynet Group, 76 Deanwood Cres, Allerton, Bradford, BD15 9BL [IO93CT, SE13]
G0 KRH C P Tarrant, 91 Dunes Rd, Greatstone, New Romney, TN28 8SW [JO00LX, TR02]
G0 KRI A W H Waller, 16 Bracey Ave, Norwich, NR6 7LB [JO02PP, TG21]
G0 KRK B G Cockfield, 47 Aston Rd, Willenhall, WV13 3DG [IO82XO, SO99]
G0 KRL I J Capon, Windon, The Green, Beyton, Bury St. Edmunds, IP30 9AJ [JO02KF, TL96]
G0 KRM S G Reid, 1 Laleham Gdns, Cliftonville, Margate, CT9 3PN [JO01QJ, TR37]
GW0 KRQ J N Cartwright, 20 Castlefield Pl, Mynachdy, Cardiff, CF4 3DU [IO81JL, ST17]
G0 KRR J R Hough, 1 Rock Ln, Linslade, Leighton Buzzard, LU7 7QQ [IO91PV, SP92]
G0 KRS G W Fuller Keighley ARS, 76 Deanwood Cres, Allerton, Bradford, BD15 9BL [IO93CT, SE13]
G0 KRT E L Masters, 91 Mayfair Ave, Worcester Park, KT4 7SJ [IO91VJ, TQ26]
G0 KRU A J Wright, 35 Blofield Rd, Brundall, Norwich, NR13 5NU [JO02RO, TG30]
G0 KRX P C Ruder, 34 Chelmsford Rd, South Woodford, London, E18 2PL [JO01AO, TQ39]
G0 KRY D G Sanders, 10 Withywood Cl, Willenhall, WV12 5DZ [IO82XO, SJ90]
G0 KSB Details withheld at licensee's request by SSL.
G0 KSC J L Johnson, 9 Bradley Cl, Canvey Island, SS8 9RS [JO01GM, TQ78]
G0 KSD R Allgood, 7 The Chase, Blofield, Norwich, NR13 4LZ [JO02RP, TG30]
G0 KSE Details withheld at licensee's request by SSL.
G0 KSF L E Ashworth, Five Burrow House, 39 Trevithick Rd, Tremorvah, Truro, TR1 1RX [IO70LG, SW84]
G0 KSG D G Jones, Croesaw, Canterbury Rd, Bilting, Ashford, TN25 4HE [JO01LF, TR04]
G0 KSH V H Sandiford, 69 Irongate, Bamber Bridge, Preston, PR5 6UY [IO83PR, SD52]
G0 KSJ J L Graves, 172 Hall Ln, Upminster, RM14 1AT [JO01DN, TQ58]
G0 KSK T Sheppard, 4 Lindrick Ave, Swinton, Mexborough, S64 8TE [IO93IL, SK49]
G0 KSL R C Torr, 68 East Towers, Pinner, HA5 1TL [IO91TO, TQ18]
G0 KSN H T Dabhi, 23 Shalgrove Field, Fulwood, Preston, PR2 3SX [IO83PT, SD53]
G0 KSP Details withheld at licensee's request by SSL.
G0 KSS K Nobbs, 50 Aynam Rd, Kendal, LA9 7DW [IO84PH, SD59]
G0 KSV R H Collett, 3 St. Nicholas Gr, Ingrave, Brentwood, CM13 3RA [JO01EO, TQ69]
G0 KSW K A Barth, F5Vae, 4 Porte Du Bois Robert, 78000 Versailles Epi D'Or, France, X X
G0 KSX E Y Hewitt, 23 Sherwood Dr, Clacton on Sea, CO15 4EB [JO01NT, TM11]
G0 KSY H L Howard, 23 The Everglades, Hempstead, Gillingham, ME7 3PY [JO01GI, TQ76]
GW0 KSZ T H Gill, 70 Caeglas Rd, Rumney, Cardiff, CF3 8JW [IO81KM, ST27]
G0 KTB G C Spencer, 19 High St., Swayfield, Grantham, NG33 4LL [IO92RT, SK92]
G0 KTC C M Ayres, 219 Ashingdon Rd, Rochford, SS4 1RS [JO01IO, TQ89]
G0 KTD A K Bonney, 6 Mitchell Rd, St. Austell, PL25 3AU [IO70OI, SX05]
GW0 KTE E G Kenyon, 16 Abbey Rd, Port Talbot, SA13 1HA [IO81CO, SS78]
G0 KTF W H Atkins, 51 Shipton Rd, Sutton Coldfield, B72 1NR [IO92CN, SP19]
G0 KTH Details withheld at licensee's request by SSL.
GM0 KTJ W K Ward, 65 Durrockstock Rd, Forbar, Paisley, PA2 0AR [IO75ST, NS46]
GW0 KTL G Edmunds, 14 Avon Cl, Bettws, Newport, NP9 6BZ [IO81LO, ST29]
G0 KTN T W Smithers, 9 Pennine Cl, Foresters Park, Melksham, SN12 7RX [IO81WI, ST96]
GM0 KTO J P Power, 43 Marwick St., Dennistoun, Glasgow, G31 3NE [IO75VU, NS66]
G0 KTP R Cockbill, 45 Mills Rd, Melksham, SN12 7DT [IO81WI, ST96]
G0 KTQ M Raines, 11 Kingsway Cl, Werneth, Oldham, OL8 1BE [IO83WM, SD90]
G0 KTR R Farnley, 6 Cardigan Rd, Hollinwood, Oldham, OL8 4SF [IO83WM, SD90]
G0 KTS J W Wright, Coombes, Ridgeway Cl, Heathfield, TN21 8NS [JO00DX, TQ52]
G0 KTT P L C Riley, Lyndene, 20 Arthog Rd, Hale, Altrincham, WA15 0LY [IO83UI, SJ78]
G0 KTU A E Miller, 11 Blackbrook Ave, Broadsands Park, Paignton, TQ4 7ND [IO80FJ, SX85]
G0 KTV J E Edgington, 83 Woking Rd, Guildford, GU1 1QL [IO91RG, SU95]
G0 KTW J E A Moss, Rudland House, Hill Rd, Kirkby in Cleveland, Middlesbrough, TS9 7AN [IO94JK, NZ50]
G0 KTX A R Hornsby, 328 Pelham Rd, Immingham, Grimsby, DN40 1PT [IO93VO, TA11]
G0 KTY G Salisbury, 1 Stad-y-Garnedd, Star Llanfair Pg, Anglesey, L60 6BB [IO83KI, SJ28]
G0 KTZ P G Pearce, 17 Lymewood Dr, Disley, Stockport, SK12 2LD [IO83XI, SJ98]
G0 KUA V P Kathuria, 2 Bevan Rd, Lovedean, Waterlooville, PO8 9QH [IO90LV, SU61]
G0 KUB S Stavrinides, Keble College, Parks Rd, Oxford, OX1 3PG [IO91IS, SP50]
G0 KUC D J Bloomfield, 8 Sunningdale Dr, Boston, PE21 8HZ [IO92XX, TF34]
G0 KUD P Haith, Warrenden, 17 Lime Tree Ave, Grimsby, DN33 2BB [IO93WN, TA20]
G0 KUE P E Webb, 181 Franciscan Rd, Tooting, London, SW17 8HP [IO91WK, TQ27]
G0 KUF N W L Buchanan, Meadowside, Jacobs Well Rd, Guildford, GU4 7PD [IO91RG, TQ05]
G0 KUI J G Wood, 6 West Terr, Stakeford, Choppington, NE62 5UL [IO95FD, NZ28]
GM0 KUJ J W McGifford, 52 Gartons Rd, Glasgow, G21 3HY [IO75VV, NS66]
G0 KUM Details withheld at licensee's request by SSL.[Op: A J Woodroffe. Station located near East Grinstead.]
G0 KUN Details withheld at licensee's request by SSL.[Op: A T Ainsworth. Station located near Swadlincote.]
G0 KUO Details withheld at licensee's request by SSL.
GM0 KUP F Mann, 26 Balmoral St., Falkirk, FK1 5HE [IO85CX, NS87]
G0 KUQ F L Hills, 20 Elderwood Cl, Eastbourne, E Sussex, BN22 0TL [JO00DT, TQ50]
G0 KUR F/Lt B Thompson, Bungal House, Main Rd, Maltby-le-Marsh, Alford, LN13 0JP [JO03CH, TF48]
G0 KUS D A Williams, 8 Pickwick Cl, Swindon, SN2 6TH [IO91CO, SU18]
G0 KUT Details withheld at licensee's request by SSL.
G0 KUU F W H Gillham, 260 Summerhouse Dr, Wilmington, Dartford, DA2 7PB [JO01CK, TQ57]
G0 KUV L C Yeldham, St. Josephs, Grange Rd, Buckfast, Buckfastleigh, TQ11 0EH [IO80CL, SX76]
G0 KUW C A Bennett, 67 King St., Clowne, Chesterfield, S43 4BS [IO93IG, SK47]
G0 KUX P N Kay, 90 Ave Rd, Southgate, London, N14 4EA [IO91WP, TQ29]
G0 KUY S Crane, 3 Hawkshead Dr, Royton, Oldham, OL2 6TW [IO83WN, SD90]
G0 KVA A L J Sargent, 25 Jordans Way, Bricket Wood, St. Albans, AL2 3SJ [IO91TR, TL10]
G0 KVB R Edmonds, 87 Burton Rd, Castle Gresley, Swadlincote, DE11 9EW [IO92ES, SK21]
G0 KVC H J Crouch, 21A Victoria Gdns, Horsforth, Leeds, LS18 4PJ [IO93ET, SE23]
GM0 KVD C Mackay, 99 Avenuepark St., Glasgow, G20 8LL [IO75UV, NS56]
GM0 KVE A M Dickson, South Lodge, Ave Rd, Kinross, KY13 7ES [IO86GE, NO10]
G0 KVF R S Croucher, 25 The Highway, Peacehaven, BN10 8XL [JO00AS, TQ40]
G0 KVG R A Neal, 144 Netherton Rd, Worksop, S80 2SB [IO93KH, SK57]
GM0 KVI E M Barclay, 46 Shortlees Rd, Shortlees, Kilmarnock, KA1 4RG [IO75SN, NS43]
G0 KVJ N Ashman Peterlee ARC, 11 Burdon Pl, Peterlee, SR8 5QZ [IO94IS, NZ44]
G0 KVK G W Cooper, 33 Lawnswood Rd, Wordsley, Stourbridge, DY8 5PH [IO82WL, SO88]
G0 KVL Details withheld at licensee's request by SSL.
G0 KVM W G Gravenor, 3 Foxhill Gr, Queensbury, Bradford, BD13 2JN [IO93BS, SE03]
G0 KVN D E M Winterburn, 28 Montagu Rd, Sprotbrough, Doncaster, DN5 8DH [IO93KM, SE50]
G0 KVO D H Pallister, 9 Curtis Hayward Dr, Quedgeley, Gloucester, GL2 4WJ [IO81UT, SO81]
G0 KVP P S Stokes, 3/4 Tregenna Hill, St. Ives, TR26 1SE [IO70GF, SW54]
GI0 KVQ G S Millar, 23 Ashgrove Rd, Newry, BT34 1QN [IO64UE, J02]
G0 KVR C F Mayo, 106 Burden Rd, Beverley, HU17 9LH [IO93TU, TA04]
G0 KVT Details withheld at licensee's request by SSL.[Op: Greg, c/o GB3SN, PO Box 6, Alton, Hants, GU34 2SP.]
G0 KVU H Sheratte, 26 Mansfield Rd, Wanstead, London, E11 2JN [JO01AN, TQ48]
G0 KVW Details withheld at licensee's request by SSL.
G0 KVX C D Shearing, 8 Windmill Cl, Beeding, Upper Beeding, Steyning, BN44 3JP [IO90UV, TQ11]
G0 KVY Details withheld at licensee's request by SSL.
G0 KVZ S J Adams, 39 Denys Dr, Fryerns, Basildon, SS14 3LP [JO01FN, TQ79]
GW0 KWA D J Clark, 37 Rotherslade Rd, Langland, Swansea, SA3 4QW [IO71XN, SS68]
G0 KWB K W Bell, 5 Jaysmith Cl, Carlisle, CA3 0QH [IO84MV, NY35]
G0 KWD K W Dyer, 5 Acacia Villa, Blackman Way, Hunston, Chichester, West Sussex, PO20 6NZ [IO90OT, SU80]
G0 KWE J Knowles, 10 Gr Hill, Hessle, HU13 0RT [IO93SR, TA02]
G0 KWF W F Taylor, 14 Rossiters Ln, St. George, Bristol, BS5 8TW [IO81RK, ST67]
G0 KWG S Lodge, 6 Craven Mount, Lister Ln, Halifax, HX1 5JL [IO93BR, SE02]
GW0 KWK Details withheld at licensee's request by SSL.
GM0 KWL B B Mulleady, 9 Elizabeth Cres, Falkirk, FK1 4JF [IO86CA, NS88]
GD0 KWM G Brown, 46 Waterloo Rd, Ramsey, IM8 1DZ
GW0 KWO K R Williams, 39 Lewis Dr, Churchill Park, Caerphilly, CF8 3FT [IO81HN, ST08]
G0 KWQ P C Bates, 46 Kingsley Ave, Lakeside, Redditch, B98 8PL [IO92AH, SP06]
GM0 KWR W Rose, 10 Dalnottar Dr, Old Kilpatrick, Glasgow, G60 5DP [IO75SW, NS47]
G0 KWS K W Scott, 38 The Gdns, Monkseaton, Whitley Bay, NE25 8BG [IO95GA, NZ37]
G0 KWT Details withheld at licensee's request by SSL.
GW0 KWU A V Thurlow, 11 Cefn Carnau Rd, Heath, Cardiff, CF4 4LZ [IO81JM, ST18]
GW0 KWV Details withheld at licensee's request by SSL.
GM0 KWW J Alexander, Shore Cottage, Curragh, Girvan, Ayrshire, KA26 9JH [IO75NG, NS10]
G0 KWX G J Brumby, 43 Oak St., Whitworth, Rochdale, OL12 8NP [IO83VQ, SD81]
G0 KXA Details withheld at licensee's request by SSL.
G0 KXB N A Charlton, 44 Passingham Walk, Cowplain, Waterlooville, PO8 8JH [IO90LV, SU61]
G0 KXC Details withheld at licensee's request by SSL.
G0 KXD H H Worden, 3 Tower View, Marsh House Ln, Darwen, BB3 3JB [IO83SQ, SD72]
GM0 KXF I W Hughes, 57 Mill Rd, Irvine, KA12 0JP [IO75QO, NS33]
G0 KXG J E Nicholls, 93 Swan Rd, Hanworth, Feltham, TW13 6PE [IO91TK, TQ17]
G0 KXH K D E Clayden, 7 The Bower, Alverstone Rd, Queen Bower, Sandown, PO36 0LD [IO90JP, SZ58]
GM0 KXJ J Bain, 50 Hareshaw Cres, Muirkirk, Cumnock, KA18 3PY [IO75WM, NS62]
G0 KXK Details withheld at licensee's request by SSL.
G0 KXL S Maton, 117 Woodchurch Rd, Prenton, Birkenhead, L42 9LJ [IO83LJ, SJ38]
G0 KXN L F Bailey, 14 Willow Ave, Fordingbridge, SP6 1LH [IO90CW, SU11]
G0 KXQ M P Ryan, 9031-203B St, Langley B.C, Canada, Vim 2C7
G0 KXS J F Robinson, 15 Frobisher Rd, Rose Green, Bognor Regis, PO21 3LT [IO90PS, SZ99]
G0 KXT R C S Withers Raycom ARC, 73 PO Box, Alcester, B49 5JB [IO92BF, SP05]
G0 KXV G M Meredith, 24 St. Thomas Rd, Luton, Beds, LU2 7UY [IO91TV, TL12]
G0 KXW A T Fitzmaurice, 14 Welwyn Cl, Thelwall, Warrington, WA4 2HE [IO83RJ, SJ68]
G0 KXX P C Shepherd, 7 Park Ave, Clitheroe, BB7 2HP [IO83TV, SD74]
G0 KXY P D Wroe, 44 Hillberry Cres, Warrington, WA4 6AF [IO83QJ, SJ68]
G0 KXZ A A Sockett, 35 Whernside Rd, Woodthorpe, Nottingham, NG5 4LB [IO92KX, SK54]
G0 KYA S W Nichols, 4 Suffield Cl, Cringleford, Norwich, NR4 6UB [JO02OO, TG10]
G0 KYB M R Kinder, 194 Whiteacre Rd, Ashton under Lyne, OL6 9PZ [IO83XL, SJ99]
G0 KYC Details withheld at licensee's request by SSL.
G0 KYE L H Landricombe, 19 Crackston Cl, Mayflower Green, Eggbuckland, Plymouth, PL6 5SN [IO70WJ, SX45]
G0 KYG P R Willetts, 197 Norwich Rd, Fakenham, NR21 8LH [JO02KT, TF92]
G0 KYH J M Elliott, Tregerrick, Martinstown, Dorcester, Dorset, DT2 9JN [IO80SQ, SY68]
G0 KYI S H Stevens, 13 Tregarthen, Treverbyn Rd, St. Ives, TR26 1HA [IO70GF, SW54]
G0 KYJ G S Liversidge, 16 Baywood Ct, Thornhill, Ontario, Canada, L3T 5W3
G0 KYK R Beardsmore, 2 Fitzmaurice Rd, Wednesfield, Wolverhampton, WV11 3EG [IO82XO, SJ90]
G0 KYL J E K Lawrence, Mallaw House, Murray Terr, Dipton, Co Durham, DH9 9HB [IO94CV, NZ15]
G0 KYM Details withheld at licensee's request by SSL.
G0 KYN R A Markham, 132 Whitethorn Ave, Yiewsley, West Drayton, UB7 8LB [IO91SM, TQ08]
G0 KYO J C Ryder, 313 Blackburn Rd, Darwen, BB3 0AB [IO83SR, SD62]
G0 KYP D Murdoch, 160 Thorpe Rd, Kirby Cross, Kirby Cross, Frinton on Sea, CO13 0NQ [JO01OU, TM22]
G0 KYQ F F E Rogister, 100 Ramillies Rd, Blackfen, Sidcup, DA15 9HZ [JO01BK, TQ47]
G0 KYR D W Cooper, 22 Kettering Pl, Cramlington, NE23 9XP [IO95FC, NZ27]
G0 KYS Nr R A Edgar, Bridge House, Stratton Strawless, Norwich, Norfolk, NR10 5LP [JO02PR, TG22]
GW0 KYT M C Rees, 12 Colemere St., Wrexham, LL13 7PD [IO83MA, SJ34]
GM0 KYU J Robertson, 134 Antigua Ct, Greenock, PA15 4QJ [IO75OW, NS27]
G0 KYV R W Hammett, Brixington, College Ln, Longdown, Exeter Devon, EX6 7SS [IO80ER, SX89]
G0 KYX Rev. P N Morgan, 12 Victoria Cl, Burgess Hill, RH15 9QS [IO90WW, TQ31]
GW0 KYY M J Warner, 76 Rhodfar Eos, Morriston, Cwmrhydyceirw, Swansea, SA6 6SW [IO81AQ, SN60]
GJ0 KYZ P R Mahrer, Prospect Pl, Beaumont, Jersey, Channel Islands, JE3 7BB
G0 KZA E A F Bishop, 21 Mandalay St., Basford, Nottingham, NG6 0BH [IO92JX, SK54]
G0 KZC Details withheld at licensee's request by SSL.
G0 KZD M R Withey, 9 Marnhull Rd, Longfleet, Poole, BH15 2EX [IO90AR, SZ09]
GW0 KZE C F R Dublon, Tyn-y-Waun Farm, Dare Rd, Cwmdare, Aberdare, Mid Glam, CF44 8UB [IO81GR, SN90]
GW0 KZF J K McMullan, 56 Vulcan St., Holyhead, LL65 1TN [IO73QH, SH28]
GW0 KZG A P Adams, 9 Dryden Rd, Penarth, CF64 2RT [IO81JK, ST17]
G0 KZH E Clayton, 220 Milnrow Rd, Rochdale, OL16 5BB [IO83WO, SD91]
G0 KZI J F Williams, 18 St. Andrews Cl, Holme Hale, Thetford, IP25 7EH [JO02JP, TF80]
GM0 KZJ G L Strang, 1 Grange Cres, Edinburgh, EH9 2EH [IO85JW, NT27]
GW0 KZK B W Parsons, 31 Crymlyn Parc, Neath, SA10 6DG [IO81BP, SS79]
G0 KZM D Egan, 56 Walker Ave, Wollescote, Stourbridge, DY9 9EL [IO82WK, SO98]
G0 KZN A C Sargeant, The Cottage, Station Rd, Commonside Old Leak, Boston, Lincs, PE22 9QR [JO03BA, TF45]
G0 KZO E Lomas, 2 Linney Rd, Bramhall, Stockport, SK7 3JW [IO83WJ, SJ88]
G0 KZQ R E Gardner, 3 Foxwood Gdns, Tamerton Foliot Rd, Plymouth, PL6 5ET [IO70WK, SX46]

G0 KZS F Dunn, 24 Ashfield Cres, Blacon, Chester, CH1 5AU [IO83ME, SJ36]
G0 KZT A Briers, 67A Waddington Ave, Coulsdon, Surrey, CR5 1QJ [IO91WH, TQ35]
GM0 KZU T G Brodie, 35 Riccarton Rd, Linlithgow, EH49 6HX [IO85EX, NT07]
G0 KZV R J Brown, Sandwell, Hector Stones, Woolavington, Bridgewater, Somerset, TA7 8EG [IO81ME, ST34]
GW0 KZW W S A Jones, Tanglewood, 2 Bryntirion Ave, Prestatyn, LL19 9PB [IO83HH, SJ08]
GM0 KZX B P Spink, 9 St. Andrews Cres, Mansewood Est, Dumbarton, G82 3ER [IO75RW, NS47]
G0 KZZ Details withheld at licensee's request by SSL.
G0 LAA E G Martin, 90 Grand Dr, Herne Bay, CT6 8LS [JO01NI, TR16]
G0 LAB Details withheld at licensee's request by SSL.
G0 LAD J R Parfett, 65 Brompton Ln, Strood, Rochester, ME2 3BA [JO01FJ, TQ76]
G0 LAE W G S Thompson, 158 Beech Rd, St. Albans, AL3 5AX [IO91US, TL10]
G0 LAG J J T Penney, 3 Wainfleet Rd, Thorpe St. Peter, Skegness, PE25 3QT [JO03ED, TF56]
G0 LAH Details withheld at licensee's request by SSL.
G0 LAI Details withheld at licensee's request by SSL.
G0 LAK J Rogers, 186 Beavers Ln, Birleywood, Skelmersdale, WN8 9BP [IO83OM, SD40]
GW0 LAL A Penlington, 7 Thornley Ave, Rhyl, LL18 4HS [IO83GH, SJ08]
G0 LAM Details withheld at licensee's request by SSL.[Station located near Staines. IARU locator: IO91SK. WAB: TQ07. R/D: Spelthorne. Op: G A Lamb, tel no: (01784) 456555. Member of Echelford ARS. QSL via RSGB.]
G0 LAN A Taylor, Brigantia, Sessay, Thirsk, North Yorks, YO7 3NL [IO94IE, SE47]
G0 LAP S W Jeffery, 9 Elkington Rd, Yelvertoft, Northampton, NN6 6LU [IO92KJ, SP67]
G0 LAQ M D Dowthwaite, 13 Leamington Rd, Morecambe, LA4 4RL [IO84NB, SD46]
G0 LAR R Hole Launceston ARS, 3 Holywell Park, Halwill, Beaworthy, EX21 5UD [IO70VS, SS40]
G0 LAT Details withheld at licensee's request by SSL.
G0 LAU J A Barber, 66 Whitburn Rd, Beeston, Nottingham, NG9 6HR [IO92IV, SK43]
G0 LAW P G Lewis Hants.Police RC, 49 Estridge Cl, Bursledon, Southampton, SO31 8FN [IO90IV, SU41]
G0 LAX A Duggan, Fern Cottage, Higher Rads End, Eversholt, Beds, MK17 9ED [IO91RX, SP93]
GW0 LAY A G Williams, 33 Heol Llanishen Fach, Cardiff, CF4 6LA [IO81JM, ST18]
G0 LAZ J B A Pryer, Rothesay, 43 Abbey Rd, Sandbach, CW11 3HA [IO83TD, SJ76]
GW0 LBA A S Hughes, Derwen Las, Valley Rd, Llanfairfechan, LL33 0SS [IO83AF, SH67]
G0 LBB R Batty, 31 Spring Ln, Balderton, Newark, NG24 3NZ [IO93OB, SK85]
G0 LBC Details withheld at licensee's request by SSL.
G0 LBF J M Mealey, 32 Cambridge Rd, Bootle, L20 9LG [IO83MK, SJ39]
G0 LBG L A Hancox, 3 Hillfoot Rd, Woolton, Liverpool, L25 7UJ [IO83NI, SJ48]
G0 LBH A J Grant, 3 Cragg Wood Cl, Horsforth, Leeds, LS18 4RL [IO93ET, SE23]
GW0 LBI L E Smart, 33 Nantgwyn, Trelewis, Treharris, CF46 6DB [IO81IP, ST19]
GW0 LBJ C Lake, 16 Minehead Ave, Sully, Penarth, CF64 5TH [IO81JJ, ST16]
G0 LBK D J Law, 86 Bateman Rd, Hellaby, Rotherham, S66 8HB [IO93JK, SK59]
G0 LBL Dr M A Williamson, Appledore, 8 Penfold Rd, Maidenbower, Crawley, RH10 7HU [IO91WC, TQ23]
GM0 LBM D G A Muirhead, 151 Baldwin Ave, Knightswood, Glasgow, G13 2JX [IO75TV, NS57]
GM0 LBN Dr J S Clark, 35 Jedburgh Ave, Rutherglen, Glasgow, G73 3EN [IO75VT, NS66]
G0 LBO J D Ross, The Gables, Jack Ln, Moulton, Northwich, CW9 8QA [IO83RF, SJ66]
G0 LBQ R A Perks, 120 Cranes Park Rd, Sheldon, Birmingham, B26 3ST [IO92CK, SP18]
GM0 LBR B F Gourlay, Muirhead House, Chryston, Glasgow, G69 9ND [IO75WV, NS66]
G0 LBT K Tromans, 6 The Asshawes, Heath Charnock, Chorley, PR6 9JW [IO83QO, SD51]
G0 LBV R Womack, 3 Everest Dr, Melton Mowbray, LE13 0SH [IO92NS, SK72]
G0 LBW J A Barlow, 18 Colston Rd, Bulwell, Nottingham, NG6 9JY [IO93JA, SK54]
G0 LBY A B Hick, 15 Lowood Ln, Birstall, Batley, WF17 9DJ [IO93ER, SE22]
G0 LBZ P C Cahill, 56 Dene Rd, Headington, Oxford, OX3 7EE [IO91JR, SP50]
G0 LCA Details withheld at licensee's request by SSL.
G0 LCB A I Cleaver, 19 Newlands Dr, Gr, Wantage, OX12 0NY [IO91GO, SU39]
G0 LCE K Robinson, 33 Mirlaw Rd, Whitelea Chase, Cramlington, NE23 6UB [IO95EB, NZ27]
G0 LCG Details withheld at licensee's request by SSL.
G0 LCH M J S Nash, 22 Northleigh Cl, Maidstone, ME15 9RP [JO01GF, TQ75]
G0 LCI B J Hollis, 112 Lichfield Rd, Stafford, ST17 4LJ [IO82WT, SJ92]
G0 LCJ B A Lucock, 15 Mayfield Rd, Newquay, TR7 2DG [IO70LJ, SW86]
G0 LCL I P Link, 27 Woodleigh Ave, Harborne, Birmingham, B17 0NW [IO92AK, SP08]
G0 LCN D J P Prendiville, 40 Caerleon Dr, Bitterne, Southampton, SO19 5LF [IO90HV, SU41]
G0 LCP B J Davey, 30 Long Down Gdns, Plymouth, PL6 8SB [IO70WJ, SX55]
G0 LCQ G T Greed, 18 Nursteed Park, Devizes, SN10 3AN [IO91AI, SU06]
G0 LCR F N Charnley Lancashire County Raynet, 30 Dunkirk Ave, Fulwood, Preston, PR2 3RY [IO83PS, SD53]
G0 LCS K J Rochester, 22 Langford Rd, Barnet, EN4 9DS [IO91WP, TQ29]
G0 LCT G W Moss, 15 Coppice Ave, Hatfield, Doncaster, DN7 6AH [IO93LN, SE60]
G0 LCU B Walker, 70 King George Rd, Loughborough, LE11 2PA [IO92JS, SK51]
G0 LCV J A Fidoe, 85 Sedgemoor Rd, Bridgwater, TA6 5NS [IO81MC, ST33]
G0 LCW W E Cowling, 14 Russell Cl, Kensworth, Dunstable, LU6 3RW [IO91RU, TL01]
G0 LCX D G Weatherill, The Old School House, West Harptree, Bristol, BS18 6EB [IO81QH, ST55]
G0 LCY I H Jervis, 291 West End Rd, South Ruislip, Ruislip, HA4 6QS [IO91TN, TQ18]
GI0 LCZ J S Nixon, 85 Cahard Rd, Ballynahinch, BT24 8YD [IO74BJ, J45]
G0 LDB M S Mallinson, 25 The Fairway, Banbury, OX16 0RR [IO92HB, SP44]
GM0 LDC E W Newton, 43 Napier Rd, Glenrothes, KY6 1DS [IO86JE, NO20]
G0 LDD Details withheld at licensee's request by SSL.
GM0 LDE L McCormick, 9 School Ln, Kirkcaldy, KY1 3HQ [IO86KD, NT29]
GM0 LDF C J Mahady, 290 Strathmore Ave, Dundee, DD3 6SJ [IO86ML, NO33]
GI0 LDI D G Keys, 71 Madison Ave, Eglinton, Londonderry, BT47 [IO64JX, C51]
G0 LDJ D J Cansfield, 275 Salisbury Rd, Totton, Southampton, SO40 3LZ [IO90FW, SU31]
G0 LDK G S Hulcoop, 22 Jubilee Rd, Broughton Astley, Leicester, LE9 6PL [IO92JM, SP59]
G0 LDM M Bidwell, 206 Anns Hill Rd, Gosport, PO12 3RE [IO90KT, SZ69]
G0 LDO R A Summerfield, 64 Station Rd, Broughton Astley, Leicester, LE9 6PT [IO92JM, SP59]
G0 LDP K Starkey, 4 Berwyn Way, Stockingford, Nuneaton, CV10 8QN [IO92FM, SP39]
GW0 LDQ G W J Kift, 23 Field Cl, Morriston, Swansea, SA6 6QD [IO81AQ, SS69]
G0 LDR J F Marlow, 21 Thames Rise, Kettering, NN16 9JL [IO92PJ, SP88]
G0 LDS J E Wiles Moroni AR Ass U, 38 Northwood Ln, Clayton, Newcastle, ST5 4BN [IO82VX, SJ84]
GM0 LDT J K Cutt, Gerbo, North Ronaldsay, Orkney, KW17 2BE [IO89SI, HY75]
G0 LDU K Allies, 6 Alston Cl, Hazel Gr, Stockport, SK7 5LR [IO83WI, SJ98]
G0 LDV E E Snell, 139 Clifton Common, Brighouse, HD6 4JF [IO93CQ, SE12]
GI0 LDW Details withheld at licensee's request by SSL.
GM0 LDX K Strathdee, 7 Baxter Pl, Lhanbryde, Elgin, IV30 3QE [IO87JP, NJ26]
G0 LDY K J Jenkinson, 2 Madeira Ave, Codsall, Wolverhampton, WV8 2DS [IO82VO, SJ80]
GW0 LDZ B W Garland, 20 Bryn Ave, Upper Brynamman, Ammanford, SA18 1BD [IO81BT, SN71]
G0 LEA D J Gray, 63 Queen St., Grange Villa, Chester-le-Street, DH2 3LU [IO94EU, NZ25]
G0 LEB C Finch, 16 Kemperleye Way, Bradley Stoke, Bristol, BS12 8EB [IO81RM, ST68]
GI0 LEC J J Maguire Lough Erne ARC, 4 Lawnakilla Park, Enniskillen, BT74 7JN [IO64EI, H24]
G0 LEE R G Lee, 41 Liverpool Old Rd, Much Hoole, Preston, PR4 4RB [IO83OR, SD42]
G0 LEF T Bell, 16 North Seaton Rd, Newbiggin By The Sea, NE64 6XT [IO95FE, NZ38]
GM0 LEG W Findlay, 58 Northfield Ave, Shotts, ML7 5HR [IO85CT, NS85]
GW0 LEH Details withheld at licensee's request by SSL.
G0 LEI Details withheld at licensee's request by SSL.
G0 LEJ M W Huggett, Rosslyn, Station Rd, Brampton, CA8 1EX [IO84PW, NY56]
GI0 LEK G G Cooley, 22 Bloomfield Park, Galliagh, Londonderry, BT48 8HA [IO65IA, C42]
G0 LEL F C Sunley, 39 Winton Rd, Northallerton, DL6 1QQ [IO94GI, SE39]
G0 LEM G W Howard, 9 Bestwood Park, Clay Cross, Chesterfield, S45 9LD [IO93HD, SK36]
G0 LEN P Worsdale West Lincs Ry G, Emergency Planning Department, Fire Brigade Department, South Park Ave, Lincoln, LN5 8EL [IO93RF, SK96]
G0 LEO J G Frizzell, 4 Maple Rd, Rubery, Rednal, Birmingham, B45 9EA [IO82XJ, SO97]
G0 LEP D J Stewart, Buckskin, 16 Prescelly Cl, Basingstoke, RG22 5DN [IO91KG, SU65]
G0 LEQ Details withheld at licensee's request by SSL.
G0 LES E L Simpson, Poplar Farm, New Rd, Old Snydale, Pontefract, WF7 6EZ [IO93HQ, SE42]
G0 LET Details withheld at licensee's request by SSL.
G0 LEU P Johnson, 31 Rockingham Dr, Scarborough, YO12 5PG [IO94SG, TA08]
G0 LEV D J Painter, 45 St. Katherines Ave, Bridport, DT6 3DE [IO80PR, SY49]
GM0 LEW D J Archibald, 113 Sutherland Way, Knightsridge, Livingston, EH54 8HY [IO85FV, NT06]
G0 LEX R A C Robinson, 21 Elms Rd, Fleet, GU13 9EG [IO91OG, SU85]
G0 LEZ Details withheld at licensee's request by SSL.
G0 LFA N F S Swallow, 178 Barcroft St., Cleethorpes, DN35 7DX [IO93XN, TA20]
G0 LFD M Collinson, Blue House Farm, Chilton, Ferryhill, Co Durham, DL17 0HP [IO94FP, NZ23]
G0 LFE Rev K P Gray, The Manse, 21 Anderson Cl, Wisbech, PE13 1SA [JO02BP, TF40]
G0 LFF R E Hide, 74 Maple Dr, Burgess Hill, RH15 8DL [IO90WX, TQ32]
G0 LFH P R Mustchin, 66 Spinney North, Pulborough, RH20 2AT [IO90SX, TQ01]
G0 LFI F Cotton, 49 Cornwall Rd, Fratton, Portsmouth, PO1 5AR [IO90LT, SU60]
G0 LFJ G M Pearce, 199 Hoblands, Haywards Heath, RH16 3NA [IO90XX, TQ32]
G0 LFK Details withheld at licensee's request by SSL.
G0 LFL Details withheld at licensee's request by SSL.
G0 LFM V A Sancto, Meadowbank, 15A Spratling St, Ramsgate, Kent, CT12 5AW [JO01QI, TR36]
G0 LFN S W Southwell, Sullys, 12 Somerset Rd, Southsea, PO5 2NL [IO90LS, SZ69]
G0 LFR H C H Merewether, Simpson, 5 Sycamore Gdns, Dymchurch, Romney Marsh, TN29 0LA [JO01MA, TR12]
G0 LFS P Gledhill, 8 Broadway, Brinsworth, Rotherham, S60 5ES [IO93HJ, SK48]
G0 LFU Details withheld at licensee's request by SSL.
G0 LFV P Fisher, Chevalier, Marks Corner, Newport, Isle of Wight, PO30 5VH
G0 LFW Details withheld at licensee's request by SSL.
G0 LFX M R Harrison, 3 Mill House, London Rd, Maresfield, TN22 2ED [JO00BX, TQ42]
G0 LFY D G Recardo, 1 Heronfield Cl, Church Hill South, Redditch, B98 8QL [IO92BH, SP06]
G0 LFZ A M Recardo, 1 Heronfield Cl, Redditch, B98 8QL [IO92BH, SP06]
G0 LGA R Letts, 28 Catlin Cres, Shepperton, TW17 8EU [IO91SJ, TQ06]
G0 LGB J G Walker, 35 Chelwood Dr, Allerton, Bradford, BD15 7YD [IO93CT, SE13]
G0 LGC L G Culshaw, 15 Naunton Ave, Leigh, WN7 4SX [IO83RL, SD60]
G0 LGD A D Ackroyd, 5 Lidget Ave, Scholemoor, Bradford, BD7 2PJ [IO93CS, SE13]
G0 LGE M G Brooman, 19 Victor Ave, Cliftonville, Margate, CT9 3DY [JO01RJ, TR37]
G0 LGF T A Evennett, Homestead, Poundgreen, Shipdham, Thetford, Norfolk, IP25 7LS [JO02KP, TF90]
G0 LGG N J Challacombe, 17 Tanners Ln, Chalk House Green, Reading, RG4 9AD [IO91ML, SU77]
G0 LGI Details withheld at licensee's request by SSL.
G0 LGJ M Taylor, 6 Welden Rd, Scarning, Dereham, NR19 2UB [JO02LQ, TF91]
G0 LGK E W Wall, Shrubbery Cottage, Felderland Ln, Worth, Deal, CT14 0BT [JO01PG, TR35]
G0 LGM A J Cook, 9 Stonechat Ave, Abbeydale, Gloucester, GL4 4XD [IO81VU, SO81]
G0 LGO A J Turton, 58 Highfield Ln, Quinton, Birmingham, B32 1QT [IO82XK, SO98]
G0 LGQ W T McClelland, 1 Wells Cttgs, Ravenglass, CA18 1SP [IO84HI, SD09]
G0 LGS Details withheld at licensee's request by SSL.
G0 LGT Details withheld at licensee's request by SSL.[Op: A C E Germaney.]
G0 LGV H C Magill, 2 The Knolls, Beeston, Sandy, SG19 1PL [IO92UC, TL14]
G0 LGW R S Caine, 148 Dumpton Park Dr, Broadstairs, CT10 1RP [JO01RI, TR36]
G0 LGX A E Milton, 106 New Dover Rd, Capel-le-Ferne, Folkestone, CT18 7JN [JO01OC, TR23]
G0 LGY S Mattei, Almabrook House, Thynnes Ln, Mattishall, Dereham, NR20 3PN [JO02MP, TG01]
G0 LGZ B R Grimes, 2 South View, Newport Rd, Ventnor, PO38 1AN [IO90JO, SZ57]
G0 LHB A Okubo, 1427-9-608 Yamazaki-Cho, Machida-City, Tokyo 195, Japan
G0 LHC M B Oswald, 8 Chudleigh Way, Ruislip Manor, Ruislip, HA4 8TP [IO91TN, TQ18]
G0 LHD R Caton, 13 Goss Barton, Nailsea, Bristol, BS19 2XD [IO81OK, ST47]
G0 LHE R W Wilmot, Elm Tres, Drayson Ln, Crick, Northampton, NN6 7SR [IO92KI, SP57]
G0 LHF Details withheld at licensee's request by SSL.
G0 LHH G E Coole, 1 Pound Ln, Stottesdon, Kidderminster, Worcs, DY14 8UJ [IO82SK, SO68]
GM0 LHK Details withheld at licensee's request by SSL.
G0 LHL M Humphreys, 60 The Walkway, Ladybridge, Bolton, BL3 4NT [IO83SN, SD60]
G0 LHM B Tuffrey, 53 Sheffield Rd, Warmsworth, Doncaster, DN4 9QR [IO93JL, SE50]
G0 LHN J L Butterworth, 38 Stuart Ave, Moreton, Wirral, L46 9PF [IO83KJ, SJ29]
GW0 LHO C Mullen, 4 Marshfield Ct, Tonyrefail, Porth, CF39 8NG [IO81GO, ST08]
G0 LHQ W A Williams, 4 The Cres, Littleham, Exmouth, EX8 2PE [IO80HP, SY08]
G0 LHR L H Robinson, 82 Grassholme, Wilnecote, Tamworth, B77 4BZ [IO92EO, SK20]
G0 LHT Details withheld at licensee's request by SSL.
G0 LHU J M Lawton, 37 Southway, Horsforth, Leeds, LS18 5RN [IO93EU, SE23]
G0 LHV R Kay, 10 Meadow Ln, Newport, Brough, HU15 2QN [IO93PS, SE83]
G0 LHW T Lyon, 24 Hockenhull Ave, Tarvin, Chester, CH3 8LP [IO83OE, SJ46]
G0 LHX H L P Passmore, 142 Monks Dale, Yeovil, BA21 3HS [IO80QW, ST51]
G0 LHY A J Holden, 169 Clifford Bridge Rd, Binley, Coventry, CV3 2DX [IO92GJ, SP37]
G0 LHZ J M Carter, 22 Orchard Combe, Whitchurch Hill, Reading, Berks, RG8 7QL [IO91LM, SU67]
G0 LIA R James, 12, Norton, Craven Arms, Salop, SY7 9LT [IO82OK, SO48]
G0 LIB R B Weston, 38 Church Rd, Peasedown St. John, Bath, BA2 8AF [IO81SH, ST65]
G0 LIE Details withheld at licensee's request by SSL.
G0 LIF Details withheld at licensee's request by SSL.
G0 LII S P Hodgson, Mill Ln Farmhouse, Mill Ln, Irby in The Marsh, Skegness, PE24 5BB [JO03CD, TF46]
G0 LIJ Details withheld at licensee's request by SSL.
GW0 LIK C E Raymond, 23 Castle Pill Cres, Steynton, Milford Haven, SA73 1HD [IO71LR, SM90]
G0 LIL Details withheld at licensee's request by SSL.
GM0 LIM J P Duffy, 39 Kylerhea Rd, Thornliebank, Glasgow, G46 8AB [IO75UT, NS55]
G0 LIN C J Smith, 20 Endsleigh Ct, Lexden, Colchester, CO3 3QN [JO01KV, TL92]
G0 LIO Clr K F Brook, 23 Knowlys Ave, Heysham, Morecambe, LA3 2PA [IO84NB, SD46]
G0 LIP M H Turner Notts Cont Grp, 15 Witley Green, Luton, LU2 8TR [IO91TV, TL12]
G0 LIQ J G Cunningham, 11 Hawthorne Rise, Awsworth, Nottingham, NG16 2RG [IO92IX, SK44]
GM0 LIR P M Woods, 12 Brannock Pl, Newarthill, Motherwell, Lanarkshire, ML1 5DX [IO85AT, NS75]
GW0 LIS A P Wright, 8 Bryn Mor Terr, Holyhead, LL65 1EU [IO73QH, SH28]
G0 LIU G F Armstrong, 1 Vale Cl, Allerdale Gr, Cockermouth, CA13 0BW [IO84HP, NY12]
G0 LIW P A Hopkins, 42 Long Meadows, Burley in Wharfedale, Ilkley, LS29 7RY [IO93DV, SE14]
GI0 LIX I McKenzie Carrickfergus ARC, 11 Rockfergus Mews, Carrickfergus, BT38 7SL [IO74CR, J48]
G0 LIY P C Smit, 51 Quaker Ln, Northallerton, DL6 1EE [IO94GI, SE39]
G0 LJB P T Williams, 44 Meadow Rd, Mirehouse, Whitehaven, CA28 8EP [IO84FM, NX91]
G0 LJC C J Livesey, 14 Dene Dr, Longfield, DA3 7JR [JO01DJ, TQ66]
G0 LJD B J Howard, 15 Cambridge Rd, Strood, Rochester, ME2 3HW [JO01FJ, TQ76]
G0 LJE S C Thain, 2 Adelaide Gdns, Stonehouse, GL10 2PZ [IO81US, SO80]
G0 LJF M S Binks, 24 Mill Ln, Reddish, Stockport, SK5 6UU [IO83WK, SJ99]
G0 LJG D F Green, The Archways, St. Georges Rd, Semington, Trowbridge, BA14 6JQ [IO81WI, ST86]
G0 LJH C Holmes, 4 Dovefields, Rocester, Uttoxeter, ST14 5LT [IO92BW, SK13]
G0 LJJ D R Mackenny, 21 Chilton Way, Hungerford, RG17 0JR [IO91FJ, SU36]
G0 LJK W D Dancock, 11 St. Davids Cl, Stourport on Severn, DY13 8RZ [IO82UI, SO87]
G0 LJM D Roebuck, 85 Crook Farm Caravan Park, Glen Rd, Baildon, Shipley, West Yorks, BD17 5ED [IO93CU, SE13]
G0 LJN K C Alderson, 18 Trinity Gorse, Stafford, ST16 1SL [IO82WT, SJ92]
G0 LJP P A McLeod, Fair Lawns, Llangrove, Ross on Wye, Herefordshire, HR9 6EZ [IO81PU, SO51]
G0 LJQ D L Williams, 6 Oaktree Ave, Crookham Common, Thatcham, RG19 8DX [IO91JJ, SU56]
G0 LJS H J Simms, Pinewood Lodge, 276 Sandridge Ln, Bromham, Chippenham, SN15 2JW [IO81XJ, ST96]
G0 LJT J Atherton, 22 Ulleries Rd, Solihull, B92 8EF [IO92CK, SP18]
G0 LJV S R Swinbourne, 11 Stapleton Rd, Warmsworth, Doncaster, DN4 9LA [IO93JL, SE50]
GW0 LJW D C Goodwin, 25 Bevan Cres, Cefn Fforest, Blackwood, NP2 1EW [IO81JQ, ST19]
G0 LJX Details withheld at licensee's request by SSL.
GW0 LKA C R G Cruddas, Weedon House, Darran Rd, Risca, Newport, Gwent, NP1 6HA [IO81KO, ST29]
G0 LKB H Wright, 6 Ashcroft Dr, Old Whittington, Chesterfield, S41 9NU [IO93GG, SK37]
G0 LKC Details withheld at licensee's request by SSL.
G0 LKF P F Cottrell Ebrnw&Wlden RG, 6 The Dr, Ersham Park, Hailsham, BN27 3HP [JO00DU, TQ50]
G0 LKG A Horton, 80 Church Farm Cl, Dibden, Southampton, SO45 5TG [IO90GV, SU30]
GW0 LKH B F Wakefield, 14 Cadwaladr St., Mountain Ash, CF45 3RD [IO81HQ, ST09]
G0 LKI W G Cockerell, 3 Churchford Rd, Knowle, Braunton, EX33 2LT [IO71WD, SS43]
GW0 LKJ W K Halliwell, 20 Llwynon Rd, Oakdale, Blackwood, NP2 0LX [IO81JQ, ST19]
G0 LKK T Hardy, 2 Stableyard, Hardwick Hall, Doelea, Chesterfield, Derbys, S44 5QJ [IO93IE, SK46]
G0 LKM M M Tasker, Westview, Bank End, North Somercotes, Louth, Lincs, LN11 7LN [JO03BK, TF49]
G0 LKN G C Weller, 20 Cromwell Cl, Hopton, Stafford, ST18 0AT [IO82XU, SJ92]
G0 LKO R Baker, 8 Mays Ln, Stubbington, Fareham, PO14 2ER [IO90JT, SU50]
G0 LKP G Rhodes, 17 Gr St., Ossett, WF5 8LP [IO93FQ, SE21]
G0 LKR P Moran Kirkby ARC, 12 Sapphire Dr, Liverpool, L33 1UW [IO83NL, SD40]
GM0 LKT A P Ferris, 41 Longbraes Gdns, Kirkcaldy, KY2 5YJ [IO86JC, NT29]
G0 LKU R W Bowskill, 14 Rusland Park, Kendal, LA9 6AJ [IO84PH, SD59]
G0 LKW M W Wheaton Essex Repeater, 2 Crouch View, Rettendon, Rettendon Common, Chelmsford, CM3 8DS [JO01GP, TQ79]
G0 LKX K E Fisher, 27 Darcy Way, Tolleshunt Darcy, Maldon, CM9 8UD [JO01JS, TL91]
G0 LKY G I Civil, Whitehouse Farm, Magpie Ln, Little Warley, Brentwood, CM13 3DZ [JO01DO, TQ69]
G0 LKZ D J Bull, 5 Clive Hall Dr, Longstanton, Cambridge, CB4 5DT [JO02AG, TL36]
G0 LLA D Morrison, 35 Kilworth Ave, Shenfield, Brentwood, CM15 8PT [JO01DP, TQ69]
G0 LLB R M Smith, 34 Churchill Rise, Chelmsford, CM1 6FD [JO01FS, TL70]
G0 LLC M J Bridges, 7 Sun Rd, Woodland, Copley, Bishop Auckland, DL13 5NF [IO94BP, NZ02]
G0 LLD H Jones, 1 Doods Park Rd, Reigate, RH2 0PZ [IO91VF, TQ25]
G0 LLE P R Ferris, 116 Capel Rd, Forest Gate, London, E7 0JS [JO01AN, TQ48]
G0 LLG Details withheld at licensee's request by SSL.
G0 LLH Details withheld at licensee's request by SSL.
GM0 LLJ B Borrows, 27 Craigdimas Gr, Dalgety Bay, Dunfermline, KY11 5XR [IO86HA, NT18]
G0 LLK L Proctor, 57 Buchanan Rd, Arncott, Upper Arncott, Bicester, OX6 0PE [IO91KU, SP61]
G0 LLL J Roberts, 69 Barnoldswick Rd, Barrowfield, Barrowford, Nelson, BB9 6BQ [IO83VU, SD84]
G0 LLM S F Garrbutt, Whorlton Dene, Emmerson Cl, Swainby, Northallerton, DL6 3EL [IO94IJ, NZ40]
G0 LLN Details withheld at licensee's request by SSL.
G0 LLO H F Endersby, 145 Lindsay Ave, Lupset, Wakefield, WF2 8AP [IO93FQ, SE32]
G0 LLP L L Proud, 26 Drayton Ct, The Green, Hartshill, Nuneaton, CV10 0SL [IO92FN, SP39]
G0 LLR A W F Braisher, 50 Widford Rd, Hunsdon, Ware, SG12 8NW [JO01AT, TL41]
G0 LLT Details withheld at licensee's request by SSL.
G0 LLW Details withheld at licensee's request by SSL.
G0 LLX A J Bassett, 125 Stonyhill Ave, South Shore, Blackpool, FY4 1PW [IO83LS, SD33]
G0 LMA S J Crooks, 10 Mere Cl, Mountsorrel, Loughborough, LE12 7BP [IO92KR, SK51]
G0 LMB Details withheld at licensee's request by SSL.
G0 LMC C B Sibley, 11 Newport Cres, Waddington, Lincoln, LN5 9LZ [IO93RE, SK96]
G0 LMD M R Butler, 44 East Stratton, nr Winchester, Hants, SO21 3DU [IO91JD, SU53]

G0 LME R M Coaker, 12 Fen Rd, Little Hale, Sleaford, NG34 9BD [IO92UW, TF14]
G0 LMF Details withheld at licensee's request by SSL.
G0 LMG P Burchett, 62 Birchwood Ave, Wallington, SM6 7EN [IO91WI, TQ26]
G0 LMH Details withheld at licensee's request by SSL.
G0 LMI M J Dodding, Dore & Totley Golf Club, Bradway Rd, Bradway, Sheffield, S17 4QR [IO93GH, SK38]
G0 LMJ E T Garrott, Lynden, Clappers Ln, Earnley, Chichester, PO20 7JJ [IO90NS, SZ89]
GW0 LML D H Griffiths, 3 Hendre Ladus, Ystradgynlais, Swansea, SA9 1SE [IO81DS, SN71]
G0 LMO A Everitt, 58 Eastwood Rd, Aylestone, Leicester, LE2 8DB [IO92KO, SK50]
G0 LMQ B W Hill, 16 Fairland House, Masons Hill, Bromley, BR2 9JJ [JO01AJ, TQ46]
GI0 LMR W J Redmond, 6 Hazelwood Cres, Craigywarren, Ballymena, BT43 6TA [IO64UV, D10]
G0 LMT Details withheld at licensee's request by SSL.
G0 LMU Details withheld at licensee's request by SSL.
G0 LMX V Denecker, Kernanderry, Faringdon Rd, Frilford Heath, Abingdon, Oxon, OX13 6QJ [IO91HQ, SU49]
G0 LNA R W W Henderson, 65 Rowarth Rd, Manchester, M23 2UL [IO83UJ, SJ88]
G0 LNB G W Goodwin, 16 Hucklow Ave, Newall Green, Wythenshawe, Manchester, M23 2YX [IO83UJ, SJ88]
G0 LND Details withheld at licensee's request by SSL.
G0 LNE T Stokes, 22 Armada Cl, Erdington, Birmingham, B23 7PB [IO92BM, SP09]
G0 LNH Details withheld at licensee's request by SSL.
G0 LNI J H Stringer, 45 Balmoral Rd, Yeovil, BA21 5JQ [IO80QW, ST51]
G0 LNK P W Bower, 87 Blackbrook Rd, Loughborough, LE11 4PY [IO92JS, SK51]
G0 LNL G M Murphy, 15 Rose Ave, Bayworth Mead, Abingdon, OX14 1XX [IO91IQ, SU49]
GW0 LNM P D Pentecost, 2 Tir Llan, Llanarthney, Carmarthen, SA32 8JE [IO71WU, SN52]
G0 LNN D S Draycott, Grantchester Cottage, 3 Sycamore Gdns, Dymchurch, Romney Marsh, TN29 0LA [JO01MA, TR12]
G0 LNP A J Knowler, 33 Cherry Tree Rd, Rainham, Gillingham, ME8 8JY [JO01HI, TQ86]
GM0 LNQ T Nicholson, Mill of Schivas Cot, Ythanbank Ellon, Aberdeenshire, Scotland, AB4 0TN [IO87TH, NJ72]
G0 LNR Details withheld at licensee's request by SSL.
G0 LNS G W Robinson, 34 Willowfield Ave, Fawdon, Newcastle upon Tyne, NE3 3NH [IO95EA, NZ26]
GM0 LNT M G A Millward, Bendoran Boat Yard, Bunessanct, Isle of Mull, Argyll, X X
G0 LNV Dr T N Appleyard, 78 Chelsea Rd, Netheredge, Sheffield, S11 9BR [IO93FI, SK38]
G0 LNW T W Horabin, 69 Birchwood Ave, North Gosforth, Newcastle upon Tyne, NE13 6QB [IO95EB, NZ27]
G0 LNX I J Davison, 1 Bourner Cttgs; Digdog Ln, Frittenden, Cranbrook, TN17 2AX [JO01GD, TQ83]
G0 LNY P W Slegg, 8 Connaught Dr, Newton-le-Willows, WA12 8NE [IO83QK, SJ59]
GW0 LNZ R A Wincott, 44 Llandennis Rd, Roath Park, Cardiff, CF2 6EG [IO81JM, ST18]
GM0 LOA Details withheld at licensee's request by SSL.
GM0 LOD G L Collier, Mellbreak, Pool of Muckhart, Dollar, Clackmannanshire, FK14 7JW [IO86EE, NO00]
G0 LOE S M Phillips, 26 Belvedere Dr, Dukinfield, SK16 5NW [IO83XL, SJ99]
G0 LOF F H James, 6 Pinewood Cl, East Preston, Littlehampton, BN16 1HF [IO90ST, TQ00]
G0 LOG Details withheld at licensee's request by SSL.
G0 LOH K Dutson, 121 Maple Dr, Burgess Hill, RH15 8DE [IO90WX, TQ32]
GW0 LOI R B Jones, 31 Three Arches Ave, Heath, Cardiff, CF4 5NU [IO81JM, ST18]
G0 LOJ Dr C Budd, 10 Stanley Mead, Bradley Stoke, Bristol, BS12 0EG [IO81RN, ST68]
GM0 LOK J F Leggat, Ailach, St. Aethans Rd, Burghead, Elgin, IV30 2YR [IO87GQ, NJ16]
G0 LOM D G Cox, Tythings, Bluntington, Chaddesley Corbett, Kidderminster, DY10 4NR [IO82WJ, SO87]
GM0 LOO H L Hunter, 25 Braehead Rd, Kirkcaldy, KY2 6XP [IO86JD, NT29]
G0 LOP G Tweedy, 8 Greencliffe Dr, Clifton, York, YO3 6NA [IO93KX, SE55]
G0 LOQ S C Chappell, 49 Midway, South Crosland, Huddersfield, HD4 7DA [IO93CO, SE11]
G0 LOS J G Elliott, 37 Cornhill, Allestree Village, Allestree, Derby, DE22 2FS [IO92GW, SK33]
GM0 LOT R N Clasper, 32 Murieston Park, Livingston, EH54 9DT [IO85FU, NT06]
G0 LOU Details withheld at licensee's request by SSL.
G0 LOV Details withheld at licensee's request by SSL.
G0 LOW C C Riggs The Shortwave Shop Club, 18 Fairmile Rd, Christchurch, BH23 2LJ [IO90CR, SZ19]
G0 LOX C R Lansley, 15 Oak Way, Aldershot, GU12 4BB [IO91PF, SU84]
G0 LOZ I T Tomson, Valley View, Bewdley, Worcs, DY12 2JX [IO82UI, SO77]
GM0 LPB J Gault, 25 Beech Brae, Bishopmill, Elgin, IV30 2NS [IO87IP, NJ26]
G0 LPC Details withheld at licensee's request by SSL.
G0 LPD R P Dignum, 54 Westfield Dr, Greetwell, Lincoln, LN2 4RB [IO93SG, TF07]
G0 LPF R H Willkins, 20 Fairholme Dr, Yapton, Arundel, BN18 0JH [IO90QT, SU90]
G0 LPG B P M A Gaunt, Berkeley Gaunt, Oldfields Hotel, 102 Wells Rd, Bath, BA2 3AL [IO81TJ, ST76]
G0 LPI M J Green, 68 Butts Hill Rd, Woodley, Reading, RG5 4NP [IO91NK, SU77]
GM0 LPK P L Kidd, 234 Newbattle Abbey Cres, Dalkeith, EH22 3LU [IO85LV, NT36]
G0 LPN B Alperowicz, Three Pines, 14 Leith Rd, Beare Green, Dorking, RH5 4RQ [IO91UE, TQ14]
G0 LPO Details withheld at licensee's request by SSL.
G0 LPP Details withheld at licensee's request by SSL.
G0 LPQ J Hagen, 47 Rawcliffe Rd, Walton, Liverpool, L9 1AN [IO83MK, SJ39]
G0 LPS Details withheld at licensee's request by SSL.
G0 LPT G W Wegg, 23 Kerdane, Orchard Park Est, Hull, HU6 9EB [IO93TS, TA03]
G0 LPU A V Newton, 10 Rowan Ct, Greasby, Wirral, L49 3QH [IO83KI, SJ28]
G0 LPV B J Froggett, 28 Wood St., Longwood, Huddersfield, HD3 4RF [IO93CP, SE11]
G0 LPW B G Dix, 13 Sol y Vista, Frith Hill Rd, Godalming, GU7 2EF [IO91QE, SU94]
G0 LPX B S Garbutt, 35 Westfield Rd, Tockwith, York, YO5 8PY [IO93IX, SE45]
G0 LPY L C Tarver, 31 Park Dr, Swallownest, Sheffield, S31 0UL [IO93GJ, SK38]
G0 LPZ N A Gortmans, 38 Crane Dr, Verwood, BH31 6QB [IO90BV, SU00]
GD0 LQA P A Deakin, 10 School Rd, Onchan, Douglas, IM3 4LA
G0 LQB F W G Sampson, 89 Newcome Rd, Fratton, Portsmouth, PO1 5DR [IO90LT, SU60]
G0 LQC D F Briggs, 57 Charlton Dr, Charlton Brook, High Green, Sheffield, S30 4PA [IO93GJ, SK38]
G0 LQD P Valleley, 9 Lavender Rd, Basingstoke, RG22 5NN [IO91KF, SU64]
GD0 LQE G R Warburton, Cape Lodge, South Cape, Laxey, Isle of Man, IM4 7BY
G0 LQF R W Welch, 26 Straits Rd, Lower Gornal, Dudley, DY3 2UN [IO82WM, SO99]
G0 LQI M F Murphy, 133 Preston Rd, Preston, Weymouth, DT3 6BG [IO80SP, SY68]
G0 LQK M H Hinchliffe, 2 Ash Gr, New Longton, Preston, PR4 4XJ [IO83PR, SD52]
G0 LQM J A Hipwell, 5 Dolphin Cres, Paignton, TQ3 1AE [IO80FK, SX86]
G0 LQN L N Fish, 14 Cypress Ave, Thornton Cleveleys, FY5 2JA [IO83LV, SD34]
G0 LQO R H Taylor, 17 York Cl, Clayton-le-Moors, Accrington, BB5 5RB [IO83TS, SD73]
G0 LQP S D Davies, 42 Town End Field, Witham, CM8 1EU [JO01HS, TL81]
G0 LQQ J Barlow, 15 Lionel St., Burnley, BB12 6RA [IO83US, SD83]
G0 LQS Details withheld at licensee's request by SSL.
G0 LQT H W Smith, 66 The Avenue, Great Clacton, Clacton on Sea, CO15 4ND [JO01OT, TM11]
G0 LQU P J Fardell, 90 Beechwood Ave, St. Albans, AL1 4XZ [IO91US, TL10]
G0 LQV M D Fordham, 24A Main St., Prickwillow, Ely, Cambs, CB7 4UN [JO02EJ, TL58]
G0 LQW P Macolive, 6 Pembroke Way, Hayes, UB3 1PZ [IO91SL, TQ07]
G0 LQX T J Newstead, 5 Farnlea Dr, Bare, Morecambe, LA4 6JU [IO84NB, SD46]
G0 LQZ C R L Walkup, Darley Hall, Luton, Beds, LU2 8PP [IO91TV, TL12]
GM0 LRA G R Henderson Lorn Radio Amateurs, Tigh An Drochaid, Kilchrenan, By Taynuilt, Argyll, Scotland, PA35 1HD [IO76JJ, NN02]
GI0 LRB P O Strawbridge, 98 Moyola Dr, Shantallow, Londonderry, BT48 8EF [IO65IA, C42]
G0 LRD D R Tegerdine, 28 West Rising, Northampton, NN4 0TR [IO92NE, SP75]
G0 LRE J L Norman, 15 Thirlmere Rd, Chorley, PR7 2JH [IO83QP, SD51]
G0 LRG Details withheld at licensee's request by SSL.
G0 LRH R James, Solitaire, Low St., Ilketshall St. Margaret, Bungay, NR35 1QZ [JO02RJ, TM38]
G0 LRJ P A J Daymond, Radford House, Plymstock, Plymouth, Devon, PL9 9NH [IO70WI, SX55]
G0 LRK K Wall, 27 Broomfield Rd, Fleetwood, FY7 7HA [IO83LV, SD34]
G0 LRM A Littler, 17 Neuchatel Rd, London, SE6 4EH [IO91XK, TQ37]
G0 LRN B Robinson, 122 Harpenden Dr, Hatfield, Dunscroft, Doncaster, DN7 4HP [IO93LN, SE60]
G0 LRO D B Watmough, 54 Kingston Rd, Thackley, Bradford, BD10 8PD [IO93DU, SE13]
G0 LRQ Details withheld at licensee's request by SSL.
G0 LRR S Greenwood Rssdale Rayt Gr, 39 Rydal Rd, Hassingden, Haslingden, Rossendale, BB4 4EF [IO83UQ, SD72]
G0 LRS G V Allis, 117 Chessington Rd, West Ewell, Ewell, Epsom, KT19 9XB [IO91UI, TQ26]
G0 LRU F D Alderson, Old School House, Tattersett, Kings Lynn, Norfolk, PE31 8RS [JO02IT, TF82]
GJ0 LRV F H Poingdestre, Glenside, La Grande Route Des Sablons, Grouville, Jersey, JE3 9BB
G0 LRW Details withheld at licensee's request by SSL.
G0 LRX Details withheld at licensee's request by SSL.
G0 LRY Details withheld at licensee's request by SSL.
GI0 LRZ Dr N S J Mitchell, 6 Brae Rd, Ardaragh, Newry, BT34 1NZ [IO64VF, J13]
G0 LSA Details withheld at licensee's request by SSL.
GI0 LSB Details withheld at licensee's request by SSL.
G0 LSC F H Stothard, 1 St. Johns Way, Harrogate, HG1 3AL [IO94FA, SE35]
G0 LSD D R Strickland, 85 North Marine Rd, Scarborough, YO12 7HT [IO94TG, TA08]
G0 LSE P C G Harris, 208 South Rd, Hanworth, Feltham, TW13 6UH [IO91TK, TQ17]
G0 LSF D C Hall, 35 Molesworth Rd, Plympton, Plymouth, PL7 4NT [IO70XJ, SX55]
G0 LSG R J Joy, 61 Grenville Dr, Tavistock, PL19 8DP [IO70WM, SX47]
G0 LSH Details withheld at licensee's request by SSL.
G0 LSI D Peachey, Thornbury Cottage, Ashmill, Ashwater, Beaworthy, EX21 5HA [IO70UR, SX39]
G0 LSJ C M Jones, 179 Blandford Rd, Elford, Plymouth, PL3 6JZ [IO70WJ, SX55]
G0 LSK D F Taylor, 22 Meon Rd, Campden, Mickleton, Chipping Campden, GL55 6TD [IO92CC, SP14]
G0 LSM S Leak, 56 Greg St., Stockport, SK5 7LB [IO83WK, SJ89]
G0 LSN N C Billings, 235 Orchard Way, Beckenham, BR3 3EL [IO91XJ, TQ36]
G0 LSP L S Pawlik, 2 Woodcock Cl, Bamford, Rochdale, OL11 5QA [IO83VO, SD81]
G0 LSQ D E B Williams, 41 Ravensgate Rd, Charlton Kings, Cheltenham, GL53 8NS [IO81XV, SO91]
G0 LSS Details withheld at licensee's request by SSL.
GW0 LST S D Richards, Crydon, St. Andrews Rd, Wenvoe, Vale of Glam B, CF5 6AF [IO81IK, ST17]
G0 LSU J Hair, 84 Oxford St., Barrow in Furness, LA14 5QQ [IO84JD, SD27]
G0 LSX D Barber, 15 Roman Dr, Chesterton, Newcastle, ST5 7QB [IO83UA, SJ84]
G0 LSY J A Dodson, 4 The Larches, Butterfield Rd, Boreham, Chelmsford, CM3 3DP [JO01GS, TL70]
G0 LTA Details withheld at licensee's request by SSL.
GW0 LTC G W Edwards, 25 Herbert March Cl, Llandaff, Cardiff, CF5 2TD [IO81JM, ST17]
G0 LTD B L Tugwell, 67 Upper Kingston Ln, Shoreham By Sea, BN43 6TG [IO90VU, TQ20]
G0 LTE D C Prout, 8 Ferenberge Cl, Farmborough, Bath, BA3 1DH [IO81SI, ST66]
GI0 LTF H J Irwin, 9 Edward St., Armagh, BT61 7QU [IO64QI, H84]
G0 LTG L J Jenkins, 11 Randall Rd, Clifton Wood, Bristol, BS8 4TP [IO81QK, ST57]
GW0 LTH L J Hancock, 47 Brookfield Rd, Pontllanfraith, Blackwood, NP2 2AF [IO81JP, ST19]
GM0 LTJ E D Goodwin, 17 Springhill St., Douglas, Lanark, ML11 0PJ [IO85BN, NS83]
G0 LTN Details withheld at licensee's request by SSL.
G0 LTO R W Summers, 18A Rose Rd, Canvey Island, SS8 0BP [JO01HM, TQ78]
G0 LTP D R Freeman, 71 Longleaze, Wootton Bassett, Swindon, SN4 8AS [IO91BN, SU08]
GM0 LTQ N W Barrowman, 33 Burntbroom Gdns, Baillieston, Glasgow, G69 7NB [IO75WU, NS66]
G0 LTR K F Field Tamworth Raynet Group, 1 Hebden, Wilnecote, Tamworth, B77 4HP [IO92EO, SK20]
GI0 LTT S Beattie, 12 Dorwood Park, Newtownards, BT23 7BE [IO74DN, J46]
G0 LTU Details withheld at licensee's request by SSL.
G0 LTV R V Pearce, 199 Hoblands, Haywards Heath, RH16 3NA [IO90XX, TQ32]
G0 LTW Details withheld at licensee's request by SSL.
G0 LTX S Painting, Claytons, Inkpen, Newbury, Berks, RG17 9QE [IO91GJ, SU36]
G0 LTY D H Samuels, Braidwood, Enborne Row, Wash Water, Newbury, RG20 0NA [IO91HI, SU46]
G0 LTZ G W Wood, 170 Hall Rd, Hull, HU6 8SE [IO93TS, TA03]
G0 LUA D S Jewell, 1A St. Johns Rd, Wallingford, OX10 9AD [IO91KO, SU68]
G0 LUC P R Dyke, 41 High St., Puckeridge, Ware, SG11 1RX [JO01AV, TL32]
G0 LUD R M Spacey, 18 Longdale Ave, Ravenshead, Nottingham, NG15 9EA [IO93KB, SK55]
G0 LUE A L Coleman, 11 Eden St., Coventry, CV6 5HH [IO92GK, SP38]
GM0 LUF T R Traill, 30 Strathesk Rd, Penicuik, EH26 8EF [IO85JU, NT26]
G0 LUG S E M Thomas, 41 Meadowvale, Scawby, Brigg, DN20 9EW [IO93RM, SE90]
G0 LUI P R Draper, 265 Nottingham Rd, Ilkeston, DE7 5AT [IO92IX, SK44]
G0 LUJ G C Hughes, Wickets, Quarry Rd, Beckwithshaw, Harrogate, HG3 1QH [IO93FX, SE25]
G0 LUK D R Palmer, 1 Lewis Walk, Washcommon, Newbury, RG14 6TB [IO91HJ, SU46]
G0 LUL P J Clune, 50 St. Marks Rd, Mitcham, CR4 2LF [IO91WJ, TQ26]
G0 LUM W P Mitchell/Watson, 144 Shakespeare Cres, Dronfield, Sheffield, S18 6ND [IO93GH, SK37]
G0 LUN C W Stayt, 100 Cromwell Way, Kilington, Kidlington, OX5 2LL [IO91IT, SP51]
G0 LUO G F Collyer, 30 Field Ave, Blackbird Leys, Oxford, OX4 5PJ [IO91JR, SP50]
G0 LUP K L Chambers, 1 Green Ln, Tickton, Beverley, HU17 9RH [IO93TU, TA04]
G0 LUQ T Vale, Grange Farm, Launton, Bicester, OX6 0DX [IO91KV, SP62]
G0 LUR C A Holmes Univ Lanc ARS, Amateur Radio Society, 12 Cavendish St., Lancaster, LA1 5QA [IO84OB, SD46]
G0 LUT J D Lutterot, 63 Kennington Ave, Bishopston, Bristol, BS7 9EX [IO81QL, ST57]
G0 LUU K C Blackburn, 34 Horse Bank Dr, Lockwood, Huddersfield, HD4 5HN [IO93CP, SE11]
G0 LUW C R Fells, 21 Garthorne Ave, Mowden Park, Darlington, DL3 9XL [IO94EM, NZ21]
G0 LUY G Oneill, 17 Roker Cl, Darlington, DL1 2SL [IO94FM, NZ31]
G0 LUZ Details withheld at licensee's request by SSL.
G0 LVA T Aanestad, 14 Overdale Gdns, Sheffield, S17 3HE [IO93FH, SK38]
G0 LVB Details withheld at licensee's request by SSL.
GW0 LVC Details withheld at licensee's request by SSL.
G0 LVD W H Carter, Longacre, Harbury Ln, Heathcote, Warwick, CV34 6SL [IO92FG, SP36]
G0 LVF F J Talmage, 18 The Causeway, East Hanney, Wantage, OX12 0JN [IO91HP, SU49]
G0 LVG P Nilan, 15 Broomhall Rd, Pendlebury, Swinton, Manchester, M27 8XP [IO83UM, SD70]
G0 LVH J R C E Wimpenny, 15 Kildare Rd, Swinton, Manchester, M27 0YA [IO83TM, SD70]
GM0 LVI D J Warburton, Law Vista High St, Errol, Perth, PH2 7QQ [IO86JJ, NO22]
G0 LVJ B P Bradshaw, 18 Burley Ave, Lowton, Warrington, WA3 2ES [IO83RL, SJ69]
GM0 LVK L R Alexander, 97 Land St., Keith, AB55 5AP [IO87MN, NJ45]
G0 LVN D Byrne, 29 Holly St., Tottington, Bury, BL8 3EZ [IO83TO, SD71]
G0 LVO G A Thornton, 229 Queen St., Withernsea, HU19 2HH [JO03AR, TA32]
G0 LVP G W Pittam, 30 Westone Ave, Weston Favell, Northampton, NN3 3JJ [IO92NG, SP76]
G0 LVQ Details withheld at licensee's request by SSL.
G0 LVR J C Russell, 14 Highgate Ln, Farnborough, GU14 8AF [IO91PH, SU85]
G0 LVS M Y Petit, 29 Nares Rd, Rainham, Gillingham, ME8 9RG [JO01HI, TQ86]
G0 LVT B P Thornber, 78 Skipton Rd, Silsden, Keighley, BD20 9LL [IO93AV, SE04]
G0 LVV P Froggett, 28 Wood St., Longwood, Huddersfield, HD3 4RF [IO93CP, SE11]
G0 LVX T S Burns, 52 Somerset Dr, Bury, BL9 9DQ [IO83UO, SD80]
G0 LVY J Littler, 39 Wigan Rd, Golborne, Warrington, WA3 3TZ [IO83QL, SJ69]
G0 LVZ Details withheld at licensee's request by SSL.
G0 LWA C J Charles, 8 The Birches, Cheadle, Stoke on Trent, ST10 1EJ [IO92AX, SK04]
G0 LWC P G Timlett, 10 Reynolds Gdns, Moulton, Spalding, PE12 6PT [IO92XT, TF32]
GM0 LWD L A McWilliams, 40 Greenhead, Alva, FK12 5HG [IO86CD, NS89]
G0 LWE W E Short, 6 Kensington Ave, Normanby, Middlesbrough, TS6 0QQ [IO94KN, NZ51]
G0 LWF R E Pedley, 65 Mount Pleasant Rd, Chigwell, IG7 5EP [JO01BO, TQ49]
G0 LWH Details withheld at licensee's request by SSL.
G0 LWI F Butler, 8 Bradwell Rd, Buckhurst Hill, IG9 6BY [JO01AP, TQ49]
G0 LWL D P Spooner, 21 Fountayne Rd, Stoke Newington, London, N16 7EA [IO91XN, TQ38]
G0 LWM A P Mothew, 7 Ashfields, Loughton, IG10 1SB [JO01AP, TQ49]
G0 LWN D H Parsons, 107 Larkswood Rd, Chingford, London, E4 9DU [IO91XO, TQ39]
GI0 LWO Dr S H S Magill, 51 Dunboyne Park, Eglinton, Londonderry, BT47 3YJ [IO65JA, C52]
G0 LWP P A Wootton, 16 Wood Mead, Epping, CM16 6TD [JO01BQ, TL40]
G0 LWS Details withheld at licensee's request by SSL.
G0 LWU A C Scarr, 15 Church Park, Overton, Morecambe, Lancs, LA3 3RA [IO84NA, SD45]
G0 LWV H C Tulley, 41 Petts Wood Rd, Petts Wood, Orpington, BR5 1JT [JO01BJ, TQ46]
GM0 LWW L W Schofield, 645 Wellesley Rd, Methil, Leven, KY8 3PG [IO86LE, NT39]
G0 LWX Details withheld at licensee's request by SSL.
G0 LWY Details withheld at licensee's request by SSL.
G0 LWZ L M Williams, 4 Ridge Cres, Hawk Green, Marple, Stockport, SK6 7JA [IO83XJ, SJ98]
G0 LXA Details withheld at licensee's request by SSL.
GW0 LXD J B Duggan, 23 Heol Castell, Cefn Cribbwr, Bridgend, CF32 0BH [IO81EM, SS88]
G0 LXE A J Brown, Sandy Acre, Sandy Bank Rd, New York, Lincoln, LN4 4YE [IO93WB, TF25]
G0 LXF R J G Turner, 5 Darenth Ct, Quilter Rd, Orpington, BR5 4NS [JO01BI, TQ46]
G0 LXG B D Clulee, 11 Ascot Ride, Lillington, Leamington Spa, CV32 7TT [IO92FH, SP36]
G0 LXH W E M Arundel, 18 Paddock Cl, Castle Donington, Derby, DE74 2JW [IO92HU, SK42]
G0 LXI C P Sidney, 25 John McGuire Cres, Binley, Coventry, CV3 2QG [IO92GJ, SP37]
G0 LXJ Details withheld at licensee's request by SSL.
G0 LXK S Snarey, 49A Gr Rd, Hardway, Gosport, PO12 4JH [IO90KT, SU60]
G0 LXL G P Truckel, 44 Kestrel Cl, Chipping Sodbury, Bristol, BS17 6XD [IO81TM, ST78]
GI0 LXN W R Black, 14 Killyliss Rd, Fintona, Co. Tyrone, N Ireland, BT78 2DL [IO64HM, H46]
G0 LXP R C Gant, 25 Worcester Ave, Garstang, Preston, PR3 1FJ [IO83OV, SD44]
G0 LXS Details withheld at licensee's request by SSL.
G0 LXU Details withheld at licensee's request by SSL.
G0 LXV M W Lee, 23 Lyndale Rd, Redhill, RH1 2HA [IO91WG, TQ25]
G0 LXW A J Pollard, 65 Gr Rd, Bladon, Woodstock, OX20 1RG [IO91HU, SP41]
G0 LXX J P Mayfield, 9 Middlefell Way, Clifton Est, Nottingham, NG11 9JN [IO92JV, SK53]
G0 LXY Details withheld at licensee's request by SSL.
G0 LXZ G W J Harris, 133 Back Rd, Linton, Cambridge, CB1 6UJ [JO02DC, TL54]
G0 LYA Details withheld at licensee's request by SSL.
G0 LYC P J Hindle, 27 Wildwood Lawns, Wildwood, Stafford, ST17 4SE [IO82WS, SJ92]
G0 LYD Dr C T Ankcorn, 21A Church Ln, Ormesby, Middlesbrough, Cleveland, TS7 9AX [IO94JN, NZ51]
GW0 LYF D J Hobbs, 9 Llwynypia Terr, Llwynypia, Tonypandy, CF40 2JD [IO81GP, SS99]
G0 LYG F A Aris, 5 Horley Rd, Mottingham, London, SE9 4LF [JO01AK, TQ47]
GM0 LYH H Cochrane, 15 Hoylake Sq, Kilwinning, KA13 6RE [IO75PP, NS24]
G0 LYI W F Stevens, 97 Kelvin Gr, Portchester, Fareham, PO16 8LF [IO90KU, SU60]
G0 LYJ W Hughes, 26 Cambridge Cttgs, Richmond, TW9 3AY [IO91UL, TQ17]
G0 LYK S L Parker, 40 Boultons Ln, Crabbs Cross, Redditch, B97 5NY [IO92AG, SP06]
G0 LYL Details withheld at licensee's request by SSL.
GM0 LYM A G Donaldson, 14 Jamieson Dr, East Kilbride, Glasgow, G74 3EA [IO75WS, NS65]
G0 LYN L Roper, 85 Hart Ln, Luton, LU2 0JG [IO91TV, TL12]
GM0 LYO J C Fletcher, Riverside, Stair, Mauchline, Ayrshire, KA5 5PA [IO75SL, NS42]
G0 LYQ L H Selman, 156 Bradley Dr, Santa Cruz, California, 95060
G0 LYR P Spencer, 137 Sunrising, Looe, PL13 1NL [IO70SI, SX25]
GW0 LYS Details withheld at licensee's request by SSL.
GM0 LYT A M C L Fegen, Acharn, Losset Rd, Alyth, Blairgowrie, PH11 8BU [IO86JO, NO24]

G0 LYU Details withheld at licensee's request by SSL.[Correspondence via ZF1HJ.]
G0 LYV Details withheld at licensee's request by SSL.
G0 LYX D S Brown, 50 Philmead Rd, South Benfleet, Benfleet, SS7 5DW [JO01GN, TQ78]
G0 LYY J H G Allison, 154 Elm Terr, Tividale, Warley, B69 1TQ [IO82XM, SO99]
G0 LYZ C M Knaggs, 29 Wansford Rd, Driffield, YO25 7NB [IO94SA, TA05]
G0 LZA P D Dunne, 33 Tournament Rd, Glenfield, Leicester, LE3 8FQ [IO92JP, SK50]
GM0 LZC A G Young, 3 Broom Park, Gargunnock, Stirling, FK8 3BY [IO76XC, NS79]
G0 LZD R K Wood, 4 Burns Rd, Royston, SG8 5PT [IO92XC, TL34]
GM0 LZE D J Morrison, 27B Benside, Newmarket, Stornoway, Isle of Lewis, Scotland, PA86 0DZ [IO76JB, NS08]
G0 LZF J S Rivers, 211 Upper Wickham Ln, Welling, DA16 3AW [JO01BL, TQ47]
G0 LZG W A Miles-Williams, 8 Napier Rd, Maidenhead, SL6 5AW [IO91OM, SU88]
G0 LZI M J E Tudor, 32 Ringwood, Oxton, Birkenhead, L43 2LZ [IO83LJ, SJ38]
G0 LZJ C Price, 4 Greenway Cl, Helsby, Warrington, WA6 0QX [IO83OG, SJ47]
G0 LZL D Bates, 92 Thirlmere Rd, Partington, Manchester, M31 4PT [IO83SK, SJ79]
G0 LZM K A Abberstein, Winghale Priory, Waddingham Rd, South Kelsey, Lincs, LN7 6PN [IO93SL, TF09]
G0 LZN Details withheld at licensee's request by SSL.
G0 LZO Details withheld at licensee's request by SSL.
G0 LZP H R Pepper, 41 Edgar Rd, Yiewsley, West Drayton, UB7 8HN [IO91SM, TQ08]
G0 LZQ S Wilkinson, 4 Applesike, Longton, Preston, PR4 5BL [IO83OR, SD42]
G0 LZR D G Griffiths, 2 The Avenue, Mortimer, Mortimer Common, Reading, RG7 3QY [IO91LI, SU66]
G0 LZS T H Wright, 8 Glentham Cl, Lincoln, LN6 8BX [IO93RE, SK96]
G0 LZU L S Klatzko, 2 Ct Lodge Cottage, Glory Ln, Hastingleigh, Ashford, TN25 5HN [JO01MD, TR14]
G0 LZV C S J Nicholas, 19 Spring Crofts, Bushey, Watford, WD2 3AR [IO91TP, TQ19]
G0 LZW D J Riddick, 289 Hatfield Rd, St. Albans, AL4 0DH [IO91US, TL10]
G0 LZX R Knowles, 3 Avis Walk, Fazakerley, Liverpool, L10 4YT [IO83NL, SJ39]
G0 LZY R P Smith, Pleasanton, Church St., Gestingthorpe, Halstead, CO9 3AZ [JO02HA, TL83]
G0 MAA S Rawson, 4 Stafford Cl, Shirebrook Park, Glossop, SK13 8SG [IO93AK, SK09]
G0 MAB Details withheld at licensee's request by SSL.
GM0 MAC A W C Macfarlane, 34 Huntly Ave, Giffnock, Glasgow, G46 6LW [IO75UT, NS55]
G0 MAF M Plaskitt, 139 Willingham St., Grimsby, DN32 9PT [IO93XN, TA20]
G0 MAH G W Humphrey, 57 Haig Ave, Leyland, Preston, PR5 1WD [IO83PQ, SD52]
G0 MAI G H Grigg, 9 Washbrook Dr, Stretford, Manchester, M32 9DP [IO83UK, SJ79]
G0 MAL W M Bowden, 106 Manor Ln, Lapal, Halesowen, B62 8QW [IO82XK, SO98]
G0 MAM Details withheld at licensee's request by SSL.
GD0 MAN A J Wills Browne Manx Sthn DX G, 1 Holmes Ct, Rowany Dr, Port Erin, IM9 6LW
G0 MAO P M Denning, 13 Grey St., Harrogate, HG2 8DL [IO93FX, SE35]
GW0 MAP R J Wainwright, Brynfforest, Rhandirmwyn Rd, Gilyawm, Llandovery, Dyfed, SA20 0UL [IO82CB, SN73]
G0 MAR N J Buchan, Acorns, 37 Forge Rise, Uckfield, TN22 5BU [JO00BX, TQ42]
G0 MAS A H Sedgbeer, 5 Butterleigh Dr, Tiverton, EX16 4PN [IO80GV, SS91]
G0 MAT R Johnson, 30 Wheatlands, Titchfield Common, Fareham, PO14 4SL [IO90IU, SU50]
GW0 MAV R J Truran, 19 School St., Llanbradach, Caerphilly, CF8 3LB [IO81HN, ST08]
GW0 MAW N Davies, 2 The Alders, Wuan Vale Est, Oakdale, Blackwood, NP2 0LQ [IO81JQ, ST19]
G0 MAX M F Houghting, 76 Hayes Ln, Wimborne, BH21 2JG [IO90AT, SU00]
G0 MAY P Holmes, 11 Bingham Rd, Cotgrave, Nottingham, NG12 3JS [IO92LV, SK63]
G0 MAZ M A Gurr, Elan, Sandown Rd, Sandwich, CT13 9NY [JO01QG, TR35]
G0 MBA A R Horsman, 149 Clacton Rd, St. Osyth, Clacton on Sea, CO16 8PT [JO01NT, TM11]
G0 MBB A C Cutter, Hundred House, Pink Rd, Lacey Green, Princes Risborough, HP27 0PG [IO91OQ, SP80]
G0 MBC L J Rowley Mowmacre Boys C, 13 Ruddington Walk, Abbey Rise West, Leicester, LE4 2FH [IO92KP, SK50]
G0 MBE B Haworth, 9 Cedar Ave, Rossendale, BB4 6RR [IO83UQ, SD72]
G0 MBF D H F Blowers, 39 Prior Way, Colchester, CO4 5DH [JO01KV, TL92]
G0 MBG B G Drew, 59 Coventry St., Kidderminster, DY10 2BZ [IO82VJ, SO87]
G0 MBH M C Harper, 9 Birchwood, Thorpe St. Andrew, Norwich, NR7 0RL [JO02QP, TG20]
G0 MBI Details withheld at licensee's request by SSL.
G0 MBJ R Ellis, 8 Forest Rd, Market Drayton, TF9 3HX [IO82RV, SJ63]
G0 MBK Details withheld at licensee's request by SSL.
G0 MBL A P Plant, 28 Hawes Ave, Ramsgate, CT11 0RN [JO01QI, TR36]
G0 MBM A L Burley, 8 Peakirk Rd, Glinton, Peterborough, PE6 7LT [IO92UP, TF10]
G0 MBP B W Vanson, 25 St. Helens Rd, Southend, Westcliff on Sea, SS0 7LA [JO01IM, TQ88]
G0 MBQ S F Withers, 14 Rushes Rd, Petersfield, GU32 3BW [IO91MA, SU72]
G0 MBR I M McIver Mid-Beds Raynet Group, 37 Hartshill, Bedford, MK41 9AL [IO92SD, TL05]
G0 MBS B Sinclair, 48 East Cres, Duckmanton, Chesterfield, S44 5ET [IO93HF, SK47]
G0 MBT Details withheld at licensee's request by SSL.
G0 MBV R L Buckwell, 3 Newcomen Gr, Redcar, Cleveland, TS10 1BB [IO94LO, NZ62]
G0 MBX J A Ives-Whitaker, 17 Bedford Rd, Wells, BA5 3NH [IO81QE, ST54]
G0 MBY Details withheld at licensee's request by SSL.
G0 MBZ M Phillips, 14 Kingsclere Dr, Bishops Cleeve, Cheltenham, GL52 4TG [IO81XW, SO92]
G0 MCA R L Hollows, 1 Low Hill, Top Rd, Pontesbury, Shrewsbury, SY5 0YE [IO82NP, SJ40]
G0 MCB G P Hewitt, Blacksmiths Cottage, Rasen Rd, Tealby, Lincoln, LN8 3XL [IO93UJ, TF19]
G0 MCC A P McCabe, 58 St. Marys Rd, Doncaster, DN1 2NP [IO93KM, SE50]
G0 MCE R M Daw, 7 Solent Cl, Pendeford, Wolverhampton, WV9 5QF [IO82WO, SJ80]
GM0 MCJ H R Munro, 21 Ashwood Ave, Bridge of Don, Aberdeen, AB22 8XH [IO87WE, NJ91]
G0 MCK P Nicholson, 7 Hedge Hill Rd, East Challow, Wantage, OX12 9SD [IO91GO, SU38]
G0 MCM M A Sniezko-Blocki, 18 Lawnswood Ave, Chasetown, Walsall, West Midlands, WS7 8YD [IO92AQ, SK00]
G0 MCN J D McCann, 6 Stonebridge Rd, Steventon, Abingdon, OX13 6AS [IO91IP, SU49]
G0 MCO D F J Belcher, 7 Bower End, Chalgrove, Oxford, OX44 7YN [IO91LP, SU69]
G0 MCP K R Scott, 81 Churchill Dr, Newark, NG24 4LU [IO93OB, SK75]
G0 MCQ M C D Quicke, 53 Newfield Ave, Farnborough, GU14 9PJ [IO91OH, SU85]
G0 MCV S M Morley, 12 Elm Cl, Mountsorrel, Loughborough, LE12 7JJ [IO92KR, SK51]
G0 MCY J Smith, 38 Cherry Garden Ave, Folkestone, CT19 5LF [JO01NC, TR23][Op: Jeffrey Smith.]
G0 MDA Details withheld at licensee's request by SSL.
GM0 MDB D P North, 12 Kirkhill View, Kinellar, Aberdeen, AB21 0XX [IO87UE, NJ81]
G0 MDC D G Bunker, 14 High St., Mepal, Ely, CB6 2AW [JO02BJ, TL48]
GM0 MDD J E Clough, 84A Main Rd, Fairlie, Largs, KA29 0AD [IO75NS, NS25]
G0 MDF Details withheld at licensee's request by SSL.
G0 MDG Details withheld at licensee's request by SSL.
G0 MDI Details withheld at licensee's request by SSL.
G0 MDJ K G R Smithyes, 25 Green Ln, Radnage, High Wycombe, Bucks, HP14 4DJ [IO91NP, SU79]
G0 MDK C A Hobson, 1 Martindale Ave, Wimborne, BH21 2LE [IO90AT, SU00]
G0 MDL Details withheld at licensee's request by SSL.
G0 MDM R J Robbins, 3 North Approach, Kingswood, Watford, WD2 6EH [IO91TQ, TL10]
G0 MDN B Millward, 5 Regency Cl, Weddington, Nuneaton, CV10 0DF [IO92GM, SP39]
G0 MDO D Ward, 9 Little Ln, East Morton, Keighley, BD20 5UQ [IO93BU, SE14]
GW0 MDQ P J Firmstone, 26 Melwood Cl, Penyffordd, Chester, CH4 0NB [IO83LD, SJ36]
G0 MDR F Lupton, 10 Laurel Fields, Potters Bar, EN6 2BB [IO91VQ, TL20]
G0 MDS J M Coates Mansfld Scout Rd, 30 Abbott Rd, Mansfield, NG19 6DD [IO93JD, SK56]
G0 MDT M A Beal, 15 Conway Dr, Barnburgh, Doncaster, DN5 7JJ [IO93IM, SE40]
G0 MDU R V Leedham, 5 Reeves Cl, Totnes, TQ9 5WG [IO80DK, SX85]
G0 MDV M W Bellas, Millstone Cottage, Lazonby, Penrith, Cumbria, CA10 1AJ [IO84PR, NY53]
GM0 MDX W M Dempster, 124 Chatelherault C, Hamilton, Lanarkshire, ML3 7PW [IO75XS, NS75]
G0 MDZ Details withheld at licensee's request by SSL.[Op: A R Morton.]
G0 MEA P M Smith, 7 Prospect Dr, Fell Ln, Keighley, BD22 6DD [IO93AU, SE04]
G0 MEC M H G Spurgeon, 11 Homestead Rd, Bodicote Chase, Banbury, OX16 9TW [IO92IB, SP43]
G0 MEE B L Shelton, 31 Heathcote Ave, Hatfield, AL10 0RQ [IO91VS, TL20]
G0 MEF M E Frear, 18 Boulsworth Rd, Preston Grange, North Shields, NE29 9EN [IO95GA, NZ37]
G0 MEG Details withheld at licensee's request by SSL.[Op: Mrs M K Smith. Station located near Orpington.]
G0 MEH Details withheld at licensee's request by SSL.
G0 MEI J H Lockwood, 29 Henry Ave, Bowburn, Durham, DH6 5EL [IO94FR, NZ33]
G0 MEJ Details withheld at licensee's request by SSL.
G0 MEL Details withheld at licensee's request by SSL.
G0 MEN A N Fitzgerald, Pins 3, Charriere Blanche, 69130 Ecully, France, X X
G0 MEO R A Davis, 17 Welbourne Cl, Raunds, Wellingborough, NN9 6HE [IO92RI, SP97]
G0 MEP J R Maiden, 20 Hondwith Cl, Bradshaw, Bolton, BL2 3FQ [IO83TO, SD71]
G0 MEQ H A Rigby, 33 Herne Rise, Ilminster, TA19 0HH [IO80NW, ST31]
G0 MER L D Gomer, 61 Tamar Cl, Ferndown, BH22 8XE [IO90BT, SU00]
G0 MES Details withheld at licensee's request by SSL.
G0 MEU Details withheld at licensee's request by SSL.[Op: W G Abrahams, QSL via ON9CGB, PO Box 38, B-8510 Marke, Belgium.]
G0 MEV J Thorndyke, 23 Fordhams Cl, Stanton upon Hine Heath, Stanton, Bury St. Edmunds, IP31 2EE [JO02KH, TL97]
G0 MEW L T Whiteside, 9 Nutfield Gdns, Seven Kings, Ilford, IG3 9TB [JO01BN, TQ48]
G0 MEX H Horne, 29 Ashworth St., Waterfoot, Rossendale, BB4 7AY [IO83VQ, SD82]
G0 MEY M A L Coulter, 52 Pine Cl, Brant Rd, Waddington, Lincoln, LN5 9UT [IO93RE, SK96]
G0 MEZ C Thorndyke, 23 Fordhams Cl, Stanton upon Hine Heath, Stanton, Bury St. Edmunds, IP31 2EE [JO02KH, TL97]
G0 MFB J Keeley, Edge View Cl, Grindleford, via Sheffield, S30 1HF [IO93GJ, SK38]
GM0 MFD E R Wilson, 3 Kintail Pl, Broughty Ferry, Dundee, DD5 3TA [IO86NL, NO43]
GM0 MFE J R Nicholson, 19 Fairfield Rd, Dundee, DD3 8HR [IO86ML, NO33]
G0 MFH P Lee, 8 Escor Rd, Childwall, Liverpool, L25 4SH [IO83NJ, SJ48]
G0 MFI J D Moore, 46 Yarbury Way, Weston Super Mare, Avon, BS22 0DB [IO81NI, ST36]
G0 MFJ J R Flewitt, The Cottage, The Pitts, Pitts Ln, Ryde, Isle of Wight, PO33 3SU [IO90JR, SZ59]
G0 MFK R A Demain, 400 Radipole Ln, Weymouth, DT4 0SN [IO80SO, SY68]
G0 MFL Details withheld at licensee's request by SSL.
G0 MFM Details withheld at licensee's request by SSL.
GW0 MFN J Nap, Irenestraat 3, Breukelen U.T., Netherlands, 3621 Er
G0 MFO R Dievendorff, Carramore, Coldharbour Rd, Penshurst, Tonbridge, TN11 8EX [JO01CD, TQ54]
G0 MFP Details withheld at licensee's request by SSL.
G0 MFQ Capt G J Dunster, 3 Aldingbourne House, Aldingbourne Dr, Crockerhill, Chichester, PO18 0LG [IO90PU, SU90]
G0 MFR G J Ayre, 95 Quantock Rd, Bridgwater, TA6 7EJ [IO81LD, ST23]
G0 MFS H F Shepherd, 53 Milton Rd, Eastbourne, BN21 1SH [JO00DS, TV59]
G0 MFT G L Drake, The Bungalow, Church Ln, Tydd St. Giles, Wisbech, Cambs, PE13 5LG [JO02BR, TF41]
GM0 MFU J B H Black, 63 Westwood Rd, Glenrothes, KY7 5BB [IO86KE, NO20]
G0 MFV T W Shelley, 4 Richens Dr, Carterton, OX18 3XT [IO91ES, SP20]
G0 MFW W A Coulson, 26 Grayingham Rd, Kirton Lindsey, Gainsborough, DN21 4EL [IO93QL, SK99]
G0 MFY L S Choong, 14 School Cl, Stevenage, SG2 9TY [IO91VV, TL22]
G0 MGA S J Pryce, 62 Beach Rd, Eastbourne, BN22 7HA [JO00DS, TQ60]
G0 MGB Details withheld at licensee's request by SSL.
G0 MGC G Clark, 28 Manor Park, Tockington, Bristol, BS12 4NS [IO81RN, ST68]
G0 MGD D Lovell, 4 Grisedale Cl, Beechwood, Runcorn, WA7 2RL [IO83PH, SJ58]
GM0 MGE Details withheld at licensee's request by SSL.
G0 MGF Details withheld at licensee's request by SSL.
G0 MGG S W Smith, 60 Grange Rd, Tuffley, Gloucester, GL4 0PG [IO81UU, SO81]
G0 MGH A V Strevens, 14 Larchfield Way, Horndean, Waterlooville, PO8 9HE [IO90LV, SU71]
G0 MGI M J Goodall, 2 Meadow Ct, Littleport, Ely, CB6 1JW [JO02DL, TL58]
G0 MGJ K Hancock, 12 Westmorland Cl, Stoke on Trent, ST6 6UR [IO83VB, SJ85]
G0 MGK Details withheld at licensee's request by SSL.
G0 MGL G Lees, 22 Grendon Ave, Oldham, OL8 4HT [IO83WM, SD90]
G0 MGM R F Dunne, 77 Shawfield Rd, Ash, Aldershot, GU12 6RB [IO91PF, SU85]
G0 MGN J Brand, 38 Canterbury Gdns, Hadleigh, Ipswich, IP7 5BS [JO02LB, TM04]
GM0 MGO D J Macdonald, 2 Backmuir Cttgs, Keith, AB55 5PE [IO87MM, NJ44]
G0 MGP J M Tonks Rother Valley, College R C, 72 Muskoka Dr, Sheffield, S11 7RJ [IO93FI, SK38]
GW0 MGQ G P Budge, Windsor School, Bfpo 40, Forces Mail
G0 MGR Details withheld at licensee's request by SSL.
G0 MGS W C Baker, 3 Hillsboro Rd, Bognor Regis, PO21 2DX [IO90PS, SZ99]
G0 MGT D K Carruthers, 19 Creek View Ave, Hullbridge, Hockley, SS5 6LU [JO01HP, TQ89]
G0 MGU C W Brown, Tylers Rd, Roydon Hamlet, Roydon, Harlow, Essex, CM19 5LJ [JO01AR, TL40]
GM0 MGV L R Alexander Keith Gr Sc RC, 97 Land St., Keith, AB55 5AP [IO87MN, NJ45]
G0 MGW C D Wilson, 17 Thornsett Ct, Sharrow Ln, Sheffield, S11 8AQ [IO93GI, SK38]
G0 MGX H M Jones, 1 Kingsmuir Rd, Mickleover, Derby, DE3 5PY [IO92FV, SK33]
G0 MGZ Details withheld at licensee's request by SSL.
G0 MHA P Bakrania, Bina, 31 South Priors Ct, Lings, Northampton, NN3 8LD [IO92OG, SP86]
GI0 MHB P A McKee, 168 Ballynamoney Rd, Lurgan, Craigavon, BT66 6LD [IO64TL, J05]
G0 MHC G Ford, 11 Sandbanks Dr, Hart Station, Hartlepool, TS24 9RP [IO94JR, NZ43]
G0 MHD P C Overton, 34 Featherstone, Blindley Heath, Lingfield, RH7 6JY [IO91XE, TQ34]
G0 MHE B Hyde, 54 The Byway, Darlington, DL1 1EQ [IO94FM, NZ31]
G0 MHF J L Bisson, 14 Howbeck Dr, Oxton, Birkenhead, L43 6UY [IO83LJ, SJ38]
G0 MHG Details withheld at licensee's request by SSL.
G0 MHH M H Holloway, Hallgate Nurseries, Moulton, Spalding, Lincs, PE12 6QG [IO92XS, TF32]
GW0 MHK P L Lee, 22 Bronygraig, Bodedern, Holyhead, Gwynedd, LL65 3SY [IO73SH, SH38]
G0 MHN D Tebay, 8 Howards Way, Rustington, Littlehampton, BN16 2LT [IO90RT, TQ00]
GW0 MHO A Quayle, Ty Ar y Gornel, Tafarn y Grisiau Rd, Port Dinorwic, Gwynedd, LL56 4NZ [IO73VE, SH56]
G0 MHQ S E Meadows, 134 Medeswell, Orton Malborne, Peterborough, PE2 5PD [IO92UN, TL19]
G0 MHR K G Pollard Mid-Herts Ray G, 27 Far End, Hatfield, AL10 8TG [IO91VR, TL20]
GM0 MHS D Rendall, Aranthrue, 17 Scapa Cres, Kirkwall, KW15 1RL [IO88MX, HY41]
G0 MHU O M Cowling, 21 St. Margarets Ave, Plainmoor, Torquay, TQ1 4LW [IO80FL, SX96]
GM0 MHV J B Knox, Southport, Selkirk, Scotland, TD7 4AR [IO85NN, NT42]
GW0 MHY R A Jones, Flat 1, 24 Heol Cae Gurwen, Gwaun Cae Gurwen, Ammanford, Dyfed, SA18 1HB [IO81BS, SN71]
G0 MHZ R C Pardoe, 138 Fowler Rd, Prebendal Farm Est, Aylesbury, HP21 8QJ [IO91OT, SP81]
G0 MIA C W Murray, 21 Canterbury Rd, Ash, Aldershot, GU12 6SP [IO91PG, SU85]
G0 MIB D R Hussey, 15 The Ridings, Telscombe Cliffs, Peacehaven, BN10 7EF [IO90XT, TQ30]
G0 MIC M I Constantine, 19 Elmcroft, Fairview Ave, Woking, GU22 7NX [IO91RH, TQ05]
G0 MID R Jeffery, 3 New Rd, Paddock Wood, Tonbridge, TN12 6HP [JO01EE, TQ64]
G0 MIF I Buckle, 25 Portsmouth Cl, Rochester, ME2 2QY [JO01FJ, TQ76]
G0 MIG N A May, Park House, Old Down, Tockington, Bristol, BS12 4PT [IO81RN, ST68]
G0 MIH P R Swift, 26 Hazelwood Dr, Allington, Maidstone, ME16 0EA [JO01FG, TQ75]
G0 MIJ C Nolan, 95 Strodes Cres, Staines, TW18 1DG [IO91SK, TQ07]
G0 MIK M Ritson, 24 Chapel Rd, Pawlett, Bridgwater, TA6 4SH [IO81LE, ST24]
G0 MIL W B Walters, 6 Netherfield Rd, Somersall, Chesterfield, S40 3LS [IO93GF, SK36]
G0 MIM Details withheld at licensee's request by SSL.
G0 MIN A G Fisher Whitton ARG, 108 Heston Grange, Heston, Hounslow, TW5 0HD [IO91TL, TQ17]
G0 MIQ Details withheld at licensee's request by SSL.
GM0 MIS Details withheld at licensee's request by SSL.
G0 MIT G Bromfield, 63 Herondale Rd, Mossley Hill, Liverpool, L18 1JZ [IO83NJ, SJ38]
G0 MIU D C Bowen, 1 Osborne Ct, The Parade, Cowes, PO31 7QS [IO90IS, SZ49]
G0 MIV D J Woolnough, 38 High Leas, Beccles, NR34 9LF [JO02SK, TM48]
GM0 MIW A S McNicol, The Glebe House, Arbirlot, Arbroath, Angus, DD11 2NX [IO86QN, NO64]
G0 MIX M J Jones, 16 Navigation Cl, Murdishaw, Runcorn, WA7 6DD [IO83QH, SJ58]
G0 MIY Details withheld at licensee's request by SSL.
G0 MIZ J M B Munro, Flat 3/4, 24 The Strand, Ryde, PO33 1JD [IO90KR, SZ59]
G0 MJA G H Taylor, 33 Young Cl, Clacton on Sea, CO16 8UQ [JO01NT, TM11]
G0 MJB R C Daynes, 25 Redwood Cl, Long Lee, Keighley, BD21 4YG [IO93BU, SE04]
G0 MJC A P Keeble, 52 Garrick Green, Old Catton, Norwich, NR6 7AN [JO02PP, TG21]
G0 MJD Details withheld at licensee's request by SSL.
G0 MJE B K Widdows, 54 Knowland Gr, Norwich, NR5 8YF [JO02OP, TG10]
G0 MJF M R Weaver, 91 Mantle St., Wellington, TA21 8BB [IO80JX, ST12]
G0 MJG S N Cartlidge, 19 Thornfield Rd, Thornton, Crosby, Liverpool, L23 9XY [IO83LL, SD30]
G0 MJH B H Wilkinson N Yorks Ray Grp, 3 Friarage Ave, Northallerton, DL6 1DZ [IO94GI, SE39]
G0 MJI Details withheld at licensee's request by SSL.
G0 MJK D J Linnell, 19 Beech Ave, Northampton, NN3 2HE [IO92NF, SP76]
G0 MJL G T Jennings, 22 Yew Tree Cl, Cheltenham, GL50 4RQ [IO81XV, SO92]
G0 MJM Details withheld at licensee's request by SSL.
G0 MJN Details withheld at licensee's request by SSL.
G0 MJO G F Lucas, 49 Albany Ct, 56 Vincent Sq, London, SW1P 2NE [IO91WL, TQ27]
G0 MJP R T Davies, 59 Gaunts Way, Letchworth, SG6 4PL [IO91VX, TL23]
G0 MJQ Details withheld at licensee's request by SSL.
GM0 MJR E K Kane, 18 Glaskhill Terr, Penicuik, EH26 0EL [IO85JU, NT26]
G0 MJS Details withheld at licensee's request by SSL.
G0 MJT M J Tandy, 10 Palace Cl, Rowley Regis, Warley, B65 9LG [IO82XL, SO98]
G0 MJU A E Moss, 67 Henacre Rd, Kingsbridge, TQ7 1DP [IO80CG, SX74]
G0 MJV A S Williams, 16 Hoy Cres, Seaham, SR7 0JT [IO94HU, NZ45]
G0 MJW Details withheld at licensee's request by SSL.
G0 MJX J W Harper, 109 Baxter Ave, Kidderminster, DY10 2HB [IO82VJ, SO87]
G0 MJY D J Gourley, 86 Upton Rd, Kidderminster, DY10 2YB [IO82VJ, SO87]
G0 MJZ J Edwards, 42 Tenterfield Rd, Ossett, WF5 0RU [IO93FQ, SE22]
G0 MKA T Chapman, 17 Trevor Rd, Swinton, Manchester, M27 0YH [IO83TM, SD70]
G0 MKB A T Trend, 5 Chester St., Leigh, WN7 1LS [IO83RL, SD60]
G0 MKC W V Dunn, 1 Watling Ave, Seaham, SR7 8HZ [IO94HU, NZ44]
G0 MKF E W G Southcombe, Torview, Old Hill, Bickington, Newton Abbot, TQ12 6JU [IO80DM, SX77]
G0 MKG R D Watson, 37 Queensbury Dr, North Walbottle, Newcastle upon Tyne, NE15 9XF [IO94DX, NZ16]
G0 MKH Details withheld at licensee's request by SSL.
G0 MKI Details withheld at licensee's request by SSL.
G0 MKJ Details withheld at licensee's request by SSL.
G0 MKK M J Stockdale, First Floor, 22 Commercial St., Harrogate, HG1 1TY [IO93FX, SE35]
G0 MKL R Chell, 3 Elderberry Cl, Stourport on Severn, DY13 8TF [IO82UI, SO87]
GM0 MKM Details withheld at licensee's request by SSL.
G0 MKN K F Brady, 19B Furzefield Rd, Welwyn Garden City, AL7 3RL [IO91VT, TL21]
G0 MKO T C Liberson, Yacht Boon, c/o 6 South View Tc, St Judes, Plymouth, Devon, PL4 9DQ [IO70WI, SX45]
GW0 MKP N Grice, 21 Stad Ty Croes, Llanfairpwllgwyngyll, LL61 5JR [IO73VF, SH57]
G0 MKR E G Fountaine M.Keynes Raynet, 19 Metcalfe Gr, Blakelands, Milton Keynes, MK14 5JY [IO92PB, SP84]
G0 MKU G W Langford, 11 Nearhill Rd, Kings Norton, Birmingham, B38 8LB [IO92AJ, SP07]
G0 MKV S A Jones, 26 Clarendon St, Bloxwich, Walsall, West Midlands, WS3 2HT [IO82XO, SJ90]
G0 MKW A R Jones, 26 Clarendon St, Bloxwich, Walsall, West Midlands, WS3 2HT [IO82XO, SJ90]

G0 MKX Details withheld at licensee's request by SSL.
G0 MKY Dr J W Herries, Elmfold, East End, Witney, Oxon, OX8 6PZ [IO91GT, SP31]
G0 MKZ T D Pougher, 8 Borrowdale, Hull, HU7 6DR [IO93TS, TA03]
G0 MLA E S Walters, 12 Aspen Rd, Eaglescliffe, Stockton on Tees, TS16 0LN [IO94HM, NZ41]
G0 MLB K S Walters, 26 California Cl, Stockton on Tees, TS18 1PQ [IO94IN, NZ41]
G0 MLC W D Lowe, 54 St. Lesmo Rd, Edgeley, Stockport, SK3 0TX [IO83VJ, SJ88]
G0 MLD O H Buga, 10 Hazel Cres, Shinfield, Reading, RG2 7ND [IO91MK, SU77]
G0 MLE D J Sabin, Brookside, Walton Ln, Pillerton Priors, Warwick, CV35 0PJ [IO92FD, SP35]
G0 MLF P Marshall, 39 Pout Rd, Snodland, ME6 5EX [JO01FH, TQ66]
G0 MLG Details withheld at licensee's request by SSL.
G0 MLH T A Stilgoe, 37 Marlene Croft, Chelmsley Wood, Birmingham, B37 7JJ [IO92DL, SP18]
G0 MLI R I Weston, 94 Hanslope Cres, Bilborough Est, Nottingham, NG8 4BB [IO92JX, SK54]
G0 MLJ M L Jones, 8 College Cl, Hamble, Southampton, SO31 4QU [IO90IU, SU40]
G0 MLK D R Hardware, 59 Baulk Ln, Harworth, Doncaster, DN11 8PF [IO93LK, SK69]
G0 MLM T R Leeman, 5 Serlby Rise, Off Gordon Rd, Nottingham, NG3 2LS [IO92KX, SK54]
GW0 MLN E W Jones, 55 Blackoak Rd, Cyncoed, Cardiff, CF2 6QU [IO81KM, ST18]
G0 MLO K J Packard, 19 Elm Dr, Rayleigh, SS6 8AB [JO01HO, TQ89]
G0 MLP K Ritzema, Wildon Grange, Romaldkirk, Barnard Castle, Co. Durham, DL12 9EW [IO84XO, NY92]
G0 MLR J F Reid, 2 Bleng View Cottage, Wellington, Seascale, Cumbria, CA20 1BH [IO84GK, NY00]
G0 MLS Details withheld at licensee's request by SSL.
G0 MLT W T Stearn, Half Acre, Old Barn Ln, Hatch Beauchamp, Taunton, Somerset, TA3 6TN [IO80MX, ST21]
G0 MLV A H Pearson, 11 Peabody Cl, Devonshire Dr, London, SE10 8LB [IO91XL, TQ37]
G0 MLW B Harwood, 17 Stewart St., Seaham, SR7 7LQ [IO94IT, NZ44]
G0 MLX P E M Conant, Rudgwick, Bishopstone Rd, Coobe Bissett, Salisbury, Wilts, SP5 4LE [IO91BA, SU12]
G0 MLY C P Rolinson, 14 Waters View, Lichfield Rd, Pelsall, Walsall, WS3 4HJ [IO92AP, SK00]
G0 MLZ Details withheld at licensee's request by SSL.
G0 MMA K F Plumridge, 34 Three Counties Park, Sledge Green, Malvern, Worcs, WR13 6JW [IO82UA, SO83]
GW0 MMB P B Evans, Hirst Back, Tregynon, Newtown, Powys, SY16 3EW [IO82HO, SJ00]
G0 MMC J Cuthill, 17 Elmwood Dr, Igrow, Keighley, BD22 7DN [IO93AU, SE03]
G0 MMD Details withheld at licensee's request by SSL.
G0 MMG M M Golding, 125 Castle Sq, Backworth, Newcastle upon Tyne, NE27 0AZ [IO95FA, NZ27]
G0 MMH P Walker, 16 Whitby Cl, Coppenhall, Crewe, CW1 3XB [IO83SC, SJ65]
G0 MMI C A Underhill, 5 Gr Way, Waddesdon, Aylesbury, HP18 0LH [IO91MU, SP71]
G0 MMJ D J Wilkins, 18 Garendon Rd, Loughborough, LE11 4QD [IO92JS, SK51]
G0 MML R Kempster, 34 Crawford Gdns, Palmers Green, London, N13 5TD [IO91WO, TQ39]
GM0 MMN M Greig, 53 Dunbar Pl, Raithwood, Kirkcaldy, KY2 5SE [IO86JC, NT29]
G0 MMO N B Laud, Mulbrook, Pittywood Rd, Wirksworth, Matlock, DE4 4ED [IO93FB, SK25]
G0 MMP Details withheld at licensee's request by SSL.
G0 MMQ H E P Dadak, 13 Stour Ave, Norwood Green, Southall, UB2 4HL [IO91TL, TQ17]
G0 MMR Details withheld at licensee's request by SSL.
G0 MMT L C Catherall, Lawnswood, Ring Rd, Backford, Chester, CH1 6PG [IO83NG, SJ37]
G0 MMV J West, 2 Barrington St., Toronto, Bishop Auckland, DL14 7SA [IO94DQ, NZ23]
GW0 MMW L E Roberts, Ty-Coch, Tan-y-Bwlch, Maentwrog, Gwynedd, LL41 3AQ [IO82AW, SH64]
G0 MMX D Hebden, 128 George St., Shaw, Oldham, OL2 8DR [IO83WO, SD90]
GW0 MMY W R Ellis, Caerfa, Bryn Sannon, Brynford, Holywell, CH8 8AX [IO83JG, SJ17]
G0 MNA A Munir, 39 Gulberg V, Lahore, Pakistan
G0 MNB N A Munir, 39 Gulberg V, Lahore, Pakistan
G0 MNC J Williams, 423 Main Rd, Darnall, Sheffield, S9 4QJ [IO93HJ, SK38]
G0 MND T E Rogers, 7 Chaucer Dr, Biggleswade, SG18 8QG [IO92UB, TL14]
G0 MNE Details withheld at licensee's request by SSL.
G0 MNH M Brown, 15 Hamilton Row, Waterhouses, Durham, DH7 9AU [IO94DS, NZ14]
G0 MNI A K Carlile, 6 Marlborough Cl, Grimsby, DN32 7JD [IO93XN, TA21]
G0 MNJ E B Hancox, 35 St. Marys Rd, Stratford upon Avon, CV37 6XG [IO92DE, SP25]
G0 MNK J P Cooper, 15 Asford Gr, Bishopstoke, Eastleigh, SO50 6BG [IO90HX, SU41]
G0 MNL Details withheld at licensee's request by SSL.[Op: R J Grindley. Station located near Waterlooville.]
GW0 MNM Details withheld at licensee's request by SSL.
GW0 MNO N T R Bufton, 7 Laburnum Cl, Rassau, Ebbw Vale, NP3 5TS [IO81JT, SO11]
GW0 MNP M G Butler, Oakfield Villa, Bryncoch Rd, Sarn, Bridgend, Mid Glam, CF32 9PA [IO81FM, SS98]
GU0 MNQ J S Palmeri, Dronfield, Summerfield Rd, Vale, Guernsey, GY3 5UG
G0 MNS D Lisbona, 109 Hillfield Ct, Belsize Ave, London, NW3 4BE [IO91WN, TQ28]
GM0 MNV R G Gandy, 67 Alexander Rd, Glenrothes, KY7 4JG [IO86JE, NO20]
GM0 MNW K M Carmichael, 8 Colonsay Terr, Soroba, Oban, PA34 4YL [IO76GJ, NM82]
G0 MNY A K Dagnall, 10 Rosebury Ave, Leigh, WN7 1JZ [IO83RM, SD60]
GM0 MOC J A Mackenzie Mil.Of Camps.Ar, 16 Cedar Dr, Milton of Campsie, Glasgow, G65 8AY [IO75WX, NS67]
G0 MOD Details withheld at licensee's request by SSL.
GW0 MOF G Greenhalgh, 6 Clifton Gr, Rhyl, LL18 4AF [IO83GH, SJ08]
G0 MOH R F Greaves, 13 Jubilee Ave, Warboys, Huntingdon, PE17 2RT [IO92WJ, TL28]
GW0 MOI C L Brigstocke, Pant-y-Saer, Bwlch, Tynygongl, Anglesey, Gwynedd, LL74 8RG [IO73VH, SH58]
GW0 MOJ N V F Bray, The Conifers, Jubilee Terr, Conwy, Gwynedd, LL32 8SA [IO83BG, SH77]
G0 MOK R Hamer, 125 New St, Blackrod, Bolton, BL6 5AG [IO83RO, SD61]
G0 MOL J H Mehaffey, 6980 Ac Smith Rd, Dawsonville, Georgia, USA, 3054
G0 MOM S Kendall, 220 Marsh St., Barrow in Furness, LA14 1BQ [IO84JC, SD26]
G0 MON C A Mehaffey, 6980 Ac Smith Rd, Dawsonville, Georgia, USA, 3054
GM0 MOP Y H So, 11 Caledonian Ct, Falkirk, FK2 7FL [IO86CA, NS88]
GW0 MOQ N J Brush, 25 Heol y Ffynon, Efail Isaf, Pontypridd, Mid Glam, CF38 1AU [IO81IN, ST08]
G0 MOR Dr S M Morrey, 45 Melloway Rd, Rushden, NN10 6XX [IO92QH, SP96]
G0 MOU R C Clark, 5 St. Marys Way, Littlehampton, BN17 5QG [IO90RT, TQ00]
G0 MOV Details withheld at licensee's request by SSL.
GW0 MOW R J Harris, 25 Twynyffald Rd, Blackwood, NP2 1HQ [IO81JQ, ST19]
G0 MOX D A Gleek, 5 Highview Ave, Edgware, HA8 9TX [IO91UO, TQ29]
G0 MOZ Details withheld at licensee's request by SSL.
G0 MPB Details withheld at licensee's request by SSL.
GM0 MPC Details withheld at licensee's request by SSL.
G0 MPD Details withheld at licensee's request by SSL.
G0 MPE Details withheld at licensee's request by SSL.
G0 MPG Details withheld at licensee's request by SSL.[Op: J Ryan, 34 Watson Road, Killiney, Co Dublin. Ireland.]
G0 MPH C G Bale, 110 Wellsway, Keynsham, Bristol, BS18 1JB [IO81SJ, ST66]
G0 MPI S R Sutcliffe, 142 Sandy Ln, Cove, Farnborough, GU14 9JQ [IO91OH, SU85]
G0 MPJ B T Osborne, 12 Arminers Cl, Gosport, PO12 2HB [IO90KS, SZ69]
G0 MPK D J Knights, 11 King Edward St., Kirton Lindsey, Gainsborough, DN21 4NF [IO93QL, SK99]
G0 MPL Capt L A F Hughes, 12 Goodacres, Barnham, Bognor Regis, PO22 0JF [IO90QT, SU90]
G0 MPM J L Claughton, 12 St. Johns Cl, Donhead St. Mary, Shaftesbury, SP7 9NB [IO81WA, ST92]
G0 MPO A L Neenan, 50 Middleton Rd, Brownhills, Walsall, WS8 6JF [IO92AP, SK00]
G0 MPP J C Anderson, 15 Clifton Rd, Runcorn, WA7 4SX [IO83PH, SJ58]
G0 MPQ J I Wood, Garthmere, 4 Hunters Ln, Tattershall, Lincoln, LN4 4PB [IO93VC, TF25]
G0 MPR F Gibbons, 26 Tenbury Cl, Bentley, Walsall, WS2 0NH [IO82XO, SO99]
G0 MPS C W Wilson, 26 Cavendish Rd, Bournemouth, BH1 1RG [IO90BR, SZ09]
G0 MPU Details withheld at licensee's request by SSL.
G0 MPW J V Woods, 26 Compton Rd, Birkdale, Southport, PR8 4HA [IO83LP, SD31]
G0 MPY E Burton, 60 Langdale Rd, Runcorn, WA7 5PT [IO83PH, SJ58]
G0 MPZ B J Gifford, 20 Lewisham Rd, Gloucester, GL1 5EL [IO81VU, SO81]
G0 MQA B Southwell, Thornham, Moors Ln, St. Martins Moor, Oswestry, Shropshire, SY10 7BQ [IO82LV, SJ33]
G0 MQB J W K Lockyer, 19A Queens Ave, Minnis Bay, Birchington, CT7 9QN [JO01PJ, TR26]
G0 MQC P Capewell, 191 Monyhull Hall Rd, Birmingham, B30 3QN [IO92BJ, SP07]
G0 MQD T J Field, 128 Merafield Rd, Plympton, Plymouth, PL7 1SJ [IO70XJ, SX55]
G0 MQE J N M Peirce, 600 Highland Ave, Ottawa, Ontario, Canada, K2A 2K3
G0 MQF Details withheld at licensee's request by SSL.
G0 MQG Details withheld at licensee's request by SSL.
G0 MQH D Robinson, 4 Glastonbury Cl, Runcorn, WA7 1QW [IO83PI, SJ58]
G0 MQI R M Ingle, 16 Lintock Rd, Norwich, NR3 3NU [JO02PP, TG21]
G0 MQJ P J Robinson, 92 Greasby Rd, Greasby, Wirral, L49 3NG [IO83KJ, SJ28]
G0 MQK V J Murton, 4 Cross Park Rd, Wembury, Plymouth, PL9 0EU [IO70XH, SX54]
G0 MQL D J Franklin, Laurel Farm, 7 Holly Cl, West Winch, Kings Lynn, PE33 0PW [JO02EQ, TF61]
G0 MQM M C Hillier, 5 Sinodun Rd, Didcot, OX11 8HP [IO91JO, SU58]
GI0 MQN R Browning, 53 Caulside Park, Antrim, BT41 2DR [IO64VR, J18]
G0 MQP Details withheld at licensee's request by SSL.
G0 MQQ M H Edwards, 16 Ford Way, Armitage, Rugeley, WS15 4BX [IO92BR, SK01]
G0 MQR I R Tuson, 34 Brooklet Rd, Heswall, Wirral, L60 1UL [IO83LH, SJ28]
G0 MQS Details withheld at licensee's request by SSL.
G0 MQT R E Chisholm, 3 Yew Tree Cl, Derrington, Stafford, ST18 9ND [IO82WT, SJ82]
G0 MQU P J Smyth, 4 Dereham Rd, Pudding Norton, Fakenham, NR21 7NA [JO02KT, TF92]
G0 MQV N D Cook, 17 Moorside Rd, Richmond, DL10 5DJ [IO94DJ, NZ10]
G0 MQW C J R McWhinnie, 32 Horse Cl, Caversham, Reading, Berks, RG4 8TT [IO91ML, SU77]
G0 MQX R C G Bowers, 54 Buxton Rd, Dawley, Telford, TF4 2EW [IO82SP, SJ60]
G0 MRA E F Southon, 20 Edinburgh Cres, Kirton, Boston, PE20 1JT [IO92XW, TF33]
G0 MRB R W Broughton, 6 Lumley Pl, Lincoln, LN5 7UT [IO93RF, SK97]
G0 MRD D Gordon, 38 Hoppet Ln, Droylsden, Manchester, M43 7HX [IO83WL, SJ99]
G0 MRE G S Wells, 20 Parkfield Rd, Ruskington, Sleaford, NG34 9HS [IO93TB, TF05]
G0 MRF D Bowman, 31 Benson Cl, Hounslow, TW3 3QX [IO91TL, TQ17]
G0 MRH Details withheld at licensee's request by SSL.
GI0 MRI R Mc Ateer Rathlin Isle ARC, 23 Main St., Garvagh, Coleraine, BT51 5AA [IO64PX, C81]
GM0 MRJ M R Johnston, 27 Denholm Ct, Glenrothes, KY6 1JP [IO86JE, NO20]
G0 MRK J W Kelly, 14 Arden Walk, Sale, M33 5NY [IO83TK, SJ79]
G0 MRL L G F Bradshaw, 342 Manchester Rd, Blackrod, Bolton, BL6 5BG [IO83RO, SD61]
G0 MRM E Caligari, 209 Ormskirk Rd, Upholland, Skelmersdale, WN8 0AA [IO83PM, SD50]
G0 MRO R L Oakley, 4 Cross Keys Ln, Low Fell, Gateshead, NE9 6DA [IO94EW, NZ26]
G0 MRQ W H Heath, Larkspur, South Cres, Ripon, HG4 1SN [IO94FD, SE37]
G0 MRR C Denton-Powell, 1 Windrush Cl, Gr, Wantage, OX12 0NL [IO91GO, SU39]
G0 MRS Details withheld at licensee's request by SSL.
G0 MRU D B Thompson, West Fairhaven, Sandhills Rd, Salcombe, TQ8 8JP [IO80CF, SX73]
G0 MRV M R Viney, 13 Hill End Dr, Henbury, Bristol, BS10 7XL [IO81QM, ST57]
G0 MRX D J Wood Gt Mnchstr Red, G7 Manchester Branch, 439 Lower Broughton Rd, Salford, M7 2FX [IO83UM, SD80]
G0 MRY M Hazzledine, 66 Springfield Rd, Repton, Derby, DE65 6GP [IO92FU, SK32]
G0 MRZ B J Rowell, 73 Halsteads Rd, Barton, Torquay, TQ2 8HB [IO80FL, SX96]
G0 MSA C A P Fenton Sussex Cont Grp, Oakwell, Newtons Hill, Hartfield, E Sussex, TN7 4DH [JO01BC, TQ43]
G0 MSE R C Smith, 45 Kingsman Dr, Clacton on Sea, CO16 8UR [JO01NT, TM11]
G0 MSF G R Obey, 51 Chichester Cl, Murdishaw, Runcorn, WA7 6DQ [IO83QH, SJ58]
GI0 MSG T O'Shea, 29 Drumcairn Rd, Armagh, BT61 8DQ [IO64PI, H84]
GI0 MSH D P McElroy, 81 Keady Rd, Armagh, BT60 3AA [IO64QH, H84]
GI0 MSI S R E Nesbitt, 47 Mossfield, Glenanne, Armagh, BT60 2JF [IO64RF, H93]
GI0 MSJ T E Hall, 1 Hamiltonsbawn Rd, Armagh, BT60 1DL [IO64QI, H84]
GI0 MSK H G Rattray, 20 Charlemont Gdns, Armagh, BT61 9BB [IO64QI, H84]
G0 MSL J A Bobbett, 20 Fairfield Gdns, Glastonbury, BA6 9NH [IO81PD, ST43]
G0 MSM R H Cooper, 36 Long Meadow Cl, Plympton, Plymouth, PL7 4JG [IO70XJ, SX55]
G0 MSN Details withheld at licensee's request by SSL.
G0 MSO A E Webb, 12 Forthlin Rd, Allerton, Liverpool, L18 9TN [IO83NI, SJ48]
G0 MSP A P Guest, 15 Risemoor Rd, Bridgwater, TA6 6LB [IO81LC, ST23]
G0 MSR S J Rutt, 3 Russell Pl, Highfield, Southampton, SO17 1NU [IO90HW, SU41]
G0 MSS J M Taft, 6 Fairbourne Dr, Mickleover, Derby, DE3 5SA [IO92FV, SK33]
G0 MST J R Scotter, 7 Puffin Gdns, Peel Common, Gosport, PO13 0RF [IO90JT, SU50]
G0 MSU Dr D J Robertson, 23 St. Chads Rise, Leeds, LS6 3QE [IO93FT, SE23]
G0 MSV S R Miles, 98 Jersey Rd, Wolverton, Milton Keynes, MK12 5BH [IO92OB, SP84]
GW0 MSW E S J Goodwin, Tremayne, 11 Duchess Rd, Osbaston, Monmouth, NP5 3HT [IO81PT, SO51]
G0 MSX J Frank, 2 Selworthy Green, Ingleby Barwick, Stockton on Tees, TS17 0QT [IO94IM, NZ41]
GW0 MSY H E Duggan, 41 Maesglas Rd, Newport, NP9 3DE [IO81LN, ST28]
G0 MSZ J E Lycett, 14 Stanhope Rd North, Darlington, DL3 7AR [IO94FM, NZ21]
G0 MTA Rev F G Bligh, 6 Woodlands Rd, Bickley, Bromley, BR1 2AF [JO01AJ, TQ46]
G0 MTB P Pearson, 18 Ingoldsby Ave, Ingoldisthorpe, Kings Lynn, Norfolk, PE31 6NH [JO02GU, TF63]
G0 MTD S W Topping, 7 Beckstone Cl, Harrington, Workington, CA14 5QR [IO84FO, NX92]
GI0 MTE P R Robinson, 8 Annaboe Rd, Kilmore, Armagh, BT61 8NP [IO64RJ, H95]
G0 MTF G Sanders, 13 Bridevale Rd, Leicester, LE2 8DA [IO92KO, SK50]
G0 MTG A T Lavey, 12 Trenchard Cl, Stanmore, HA7 3SS [IO91UO, TQ19]
GW0 MTI M White, 52 James St., Trethomas, Newport, NP1 8FY [IO81JO, ST18]
G0 MTJ J Boothroyd, Quince Cottage, Church Ln, Shadoxhurst, Ashford, Kent, TN26 1LS [JO01JC, TQ93]
G0 MTK I Chapman, Mobile Systems Int Ltd, 1755 N.Collins Boulevard, Suite 400, Richardson, Texas 75080, XXXX XXX
G0 MTL D E Barnes, 8 Gourney Gr, Grays, RM16 2DA [JO01DM, TQ68]
G0 MTM G W Sweetenham, Silvertops, Mill Rd, Winfarthing, Diss, IP22 2DZ [JO02NK, TM18]
G0 MTN L J Volante, 200 Longmore Rd, Shirley, Solihull, B90 3EX [IO92CJ, SP17]
G0 MTO D Walton, 36 The Coppice, Beardwood Manor, Blackburn, Lancs, BB2 7BQ [IO83RS, SD62]
G0 MTP A D Owen, 26 Gresham St., Stoke, Coventry, CV2 4EU [IO92GJ, SP37]
G0 MTQ J B Baker, Moffat House, Church Rd, Broughton Moor, Maryport, CA15 7SS [IO84GQ, NY03]
G0 MTR C L Dunn Mid-Thames Raynet, 163A Farnham Rd, Slough, Berks, SL1 4XP [IO91QM, SU98]
G0 MTS S J Field, 172 Carter Ln East, South Normanton, Alfreton, DE55 2DZ [IO93IC, SK45]
G0 MTT R Williamson, 47 Ochre Dike Walk, Rockingham, Rotherham, S61 4DL [IO93HK, SK49]
G0 MTU J O H Lewarne, 53 Columbine Gdns, Walton on The Naze, CO14 8NN [JO01PU, TM22]
G0 MTV D Wright, 1 Wendy Ave, Ripon, HG4 1TD [IO94FD, SE37]
G0 MTW F J Puffett, 54 Two Hedges Rd, Bishops Cleeve, Cheltenham, GL52 4AB [IO81XW, SO92]
G0 MTX Details withheld at licensee's request by SSL.
G0 MTY P Stunden, 36 Meadow Ave, Preesall, Poulton-le-Fylde, FY6 0HA [IO83MW, SD34]
G0 MTZ Details withheld at licensee's request by SSL.
G0 MUA L F Hill, 3 Pyewipe Bungalows, Pyewipe Rd, Grimsby, DN31 2QW [IO93WN, TA21]
G0 MUB C Hartley, 3 Thornfield Ave, Longridge, Preston, PR3 3HL [IO83QT, SD63]
G0 MUD D I Layne Siemens Plessey, Siemens Plessey, Christchurch Ars, Grange Rd, Christchurch, Dorset, BH23 4JE [IO90DR, SZ19]
G0 MUH S Riley, 7 Crow Wood Ave, Burnley, BB12 0JG [IO83UT, SD83]
G0 MUK R J Sanders, 12 Warren Cl, Ringwood, BH24 2AJ [IO90CU, SU10]
G0 MUL S Behan, 219 Derbyshire Ln West, Stretford, Manchester, M32 9LJ [IO83UK, SJ79]
G0 MUN A M Collins, Parkhead, Drummuir, By Keith, Banffshire, AB55 3PQ [IO87LM, NJ34]
GM0 MUO Details withheld at licensee's request by SSL.
G0 MUP BR E J Coupe, 55 Parr St, Liverpool, Merseyside, L1 4JN [IO83MJ, SJ38]
G0 MUQ R L G Surrage, 80 Birch Rd, Farncombe, Godalming, GU7 3NU [IO91QE, SU94]
G0 MUR R Garrett, 27 Victoria Park Rd, Buxton, SK17 7PU [IO93BG, SK07]
G0 MUT Details withheld at licensee's request by SSL.
G0 MUU Details withheld at licensee's request by SSL.
GM0 MUV Details withheld at licensee's request by SSL.
G0 MUX P S Taylor, 2 Fairfield, Rawcliffe Bridge, Goole, DN14 8PB [IO93MQ, SE62]
G0 MUY M K Woodroffe, 14 Legsby Ave, Grimsby, DN32 0NP [IO93XN, TA20]
G0 MUZ J K Lockyer, 1 Rectory Cottage, The Common, Ewelme, Wallingford, Oxon, OX10 6HP [IO91LO, SU69]
G0 MVA J L Bailes, 239 Towngate, Ossett, WF5 0QE [IO93FQ, SE22]
G0 MVB L J Goddard, 4 Church Meadow, Milton under Wychwood, Chipping Norton, OX7 6JG [IO91EU, SP21]
G0 MVC C J Neil, 208 Stonelow Rd, Dronfield, Sheffield, S18 6ER [IO93GH, SK37]
G0 MVD I R Nash, 29 Cook Rd, Aldbourne, Marlborough, SN8 2EG [IO91EL, SU27]
G0 MVE M J Storkey, 9 Snatchup, Redbourn, St. Albans, AL3 7HD [IO91TT, TL11]
G0 MVF B Woodfine, 31 Woodlands Dr, Skelmanthorpe, Huddersfield, HD8 9DB [IO93EO, SE21]
G0 MVG H Gorman, 14 Cross Gr, Wigton, CA7 9DQ [IO84KT, NY24]
G0 MVH B E Wakefield, 84 Springfield Ave, Morley, Leeds, LS27 9PW [IO93ES, SE22]
GW0 MVI R C Taylor Porthmadog & Di, Bwlch Glas, Penrhyndeudraeth, Gwynedd, LL48 6RU [IO72XW, SH63]
G0 MVK E R Sadler, 21 Castle Cl, Killinghall, Harrogate, HG3 2DX [IO94FA, SE25]
G0 MVM A A Frost, Gavel House, Old Rd, Notton Lacock, near Chippenham, Wilts, SN15 2NF [IO81WK, ST96]
G0 MVN T P McDonnell, 44 Monmouth Rd, Yeovil, BA21 5NN [IO80QW, ST51]
G0 MVO P Taylor, 8 Ludlow Gdns, Quadring, Spalding, PE11 4QH [IO92VV, TF23]
G0 MVP Details withheld at licensee's request by SSL.
G0 MVQ Details withheld at licensee's request by SSL.
GW0 MVS D L Clarke, 146 Clydach Rd, Morriston, Swansea, SA6 6QB [IO81AQ, SS69]
G0 MVT W D Brindley, 41 Boon Hill Rd, Bignall End, Stoke on Trent, ST7 8LA [IO83UB, SJ85]
G0 MVU J A Ellis, 41 Derby Rd, Talke, Stoke on Trent, ST7 1SG [IO83UB, SJ85]
G0 MVV C Howes, 8 Alder Way, Hazel Slade, Cannock, WS12 5SX [IO92AR, SK01]
G0 MVW W S Barker, Fieldhead, School Ln, Tiddington, Thame, OX9 2NE [IO91LR, SP60]
G0 MVX J D Reeves, 5 Arrows Cres, Boroughbridge, York, YO5 9LP [IO94HC, SE36]
GM0 MVY J G Donnett, Dept of Anatomy, University College London, Gower St, London, WC1E 6BT [IO91WM, TQ28]
G0 MWA D Sadler, Maritime Mobile, Ross St, Opua, New Zealand, X X
G0 MWE R D Woodward, 22 Maryport Rd, Dearham, Maryport, CA15 7EG [IO84GQ, NY03]
G0 MWF W Foulsham, 52 South Cliff, Bexhill on Sea, TN39 3EE [JO00FU, TQ70]
G0 MWH R J W Atkins, 35 Spring Park Ave, Croydon, CR0 5EJ [IO91XI, TQ36]
G0 MWI J N Haden, 21 St Ives Cl, Copes Dr, Staffs, B79 8HL [IO92DP, SK20]
GM0 MWJ D S Robertson, 73 Kettilstoun Mains, Linlithgow, EH49 6SH [IO85EX, NS97]
GD0 MWL A Crowther, 3 Lime St., Port St. Mary, IM9 5ED
G0 MWM E R Bailey, 8 Blackthorn Cl, Thornton Cleveleys, FY5 2ZA [IO83LV, SD34]
G0 MWO A Gaffin, 509 Kenton Rd, Kenton, Harrow, HA3 0UL [IO91UO, TQ18]
G0 MWP A G Taylor, 1A Coolgardie Rd, Ashford, Middx, TW15 1EN [IO91SK, TQ07]
G0 MWQ B H N Koenen, 4 Oak Cl, Allestree, Derby, DE22 2JE [IO92GW, SK33]
GM0 MWR W S Meikle, Pardovan Farm, Philipstoun, Linlithgow, West Lothian, EH49 7RU [IO85FX, NT07]
G0 MWS S W McPhee, 12 West St., Stourbridge, DY8 1XN [IO82WK, SO88]
G0 MWT J R D Martyr Chelmsford ARS, 1 High Houses, Mashbury Rd, Great Waltham, Chelmsford, CM3 1EL [JO01FT, TL61]
G0 MWU B M Pebody, 30 High St., Oakfield, Ryde, PO33 1EL [IO90KR, SZ59]
G0 MWV D V Campbell, 2 Roebuck Green, Slough, SL1 5QY [IO91QM, SU98]
G0 MWW C J Murt, 17 Drake Rd, Padstow, PL28 8ES [IO70MM, SW97]
G0 MWX C H Hanks, 29 Pitsham Wood, Midhurst, GU29 9QZ [IO90PX, SU81]
G0 MWY A Morgan, 31 Brindley Ave, Latchford, Warrington, WA4 1RU [IO83RJ, SJ68]

G0 MWZ R N Rutherford, 26 St.Golder Rd, Newlyn Coombe, Penzance, Cornwall, TR18 5QW [IO70FC, SW42]
G0 MXA S N Black Border Comms Ltd, Border Communications Ltd, Durham Communications Centre, Drum Industrial Est, Chester-le-Street, Co Durham, DH2 1SX [IO94EV, NZ25]
G0 MXB D J Burnett, Cloonmaghaura, Williamstown, Co. Galway 0907, Eire
G0 MXD D P Wroe, 31 Abbots Way, Newcastle, ST5 2EX [IO83VA, SJ84]
G0 MXE F G Jennings, 182C Priory Rd, Hastings, TN34 3NL [JO00HU, TQ80]
GW0 MXG P R Taylor, 2 Pen y Dre, Highfields, Caerphilly, CF83 2NZ [IO81JO, ST18]
G0 MXH D E Kay, 12 Browning St., Hoddlesden, Darwen, BB3 3NE [IO83SQ, SD72]
G0 MXI D A Hopkins, 8 West Green, Cottingham, HU16 4BH [IO93SS, TA03]
G0 MXK R E Demchak, 23 Fordhams Cl, Stanton upon Hine Heath, Stanton, Bury St. Edmunds, IP31 2EE [JO02KH, TL97]
G0 MXM D V Rowles, 24 Brimsome Meadow, Highnam, Gloucester, GL2 8EW [IO81UV, SO82]
G0 MXN Details withheld at licensee's request by SSL.
GM0 MXP R M L Park, The Loft, Front St., Braco, Dunblane, FK15 9PX [IO86BG, NN80]
G0 MXQ F W W Gardiner, Copse View, Cirencester Rd, Northleach, Cheltenham Glos, GL54 3JJ [IO91BT, SP11]
G0 MXR G G Bell, 23 Park Garage, Agden Brow, Lymm, Ches, WA13 0UA [IO83SI, SJ78]
GI0 MXT G Browne, 43 Northwood Dr, Belfast, BT15 3QP [IO74AO, J37]
GW0 MXV D H Bell, 8 Dolphin Ct, Rhos on Sea, Colwyn Bay, LL28 4AW [IO83DH, SH87]
G0 MXW D Houghton, 77 Carr Ln, Norris Green, West Derby, Liverpool, L11 2UE [IO83NK, SJ39]
G0 MXX Prof B C Clarke, Linden Cottage, School Ln, Colston Bassett, Nottingham, NG12 3FD [IO92MV, SK63]
G0 MXY C D Erratt, 60 Allen Ct, Ridding Ln, Greenford, UB6 0JZ [IO91UN, TQ18]
GM0 MXZ T S Goody, Lambs' Park, Forgandenny, Perth, Scotland, PH2 9HS [IO86GI, NO01]
G0 MYA A Gray, The Manse, 21 Anderson Cl, Wisbech, PE13 1SA [JO02BP, TF40]
G0 MYB Details withheld at licensee's request by SSL.
G0 MYC R S Clifton, Heathwood, Thrigby Rd, Filby, Great Yarmouth, NR29 3HJ [JO02TP, TG41]
G0 MYD G A Hughes, 165 Mount Pleasant, Southcrest, Redditch, B97 4JJ [IO92AH, SP06]
G0 MYE Dr R P Watkins, Sycamore, 33 Aylesbury Rd, Wing, Leighton Buzzard, LU7 0PD [IO91PV, SP82]
G0 MYH J V Foster, 5 Jacobs Cl, Windmill Hill, Glastonbury, BA6 8EJ [IO81PD, ST53]
G0 MYJ D A Sutherland, 37 The Bentleys, Southend on Sea, SS2 6UJ [JO01IN, TQ88]
GW0 MYK S O Frost, Rhoslwyn, 4 Maes Tegid, Bala, Gwynedd, LL23 7BN [IO82EV, SH93]
G0 MYL Sir H Pigott, Brook Farm, Shobley, Ringwood, Hants, BH24 3HT [IO90DU, SU10]
G0 MYM P B Harrison, 47 Elmete Way, Leeds, LS8 2NA [IO93GT, SE33]
G0 MYN W T R Hughes, 60 Pineways, Appleton, Warrington, WA4 5EJ [IO83RI, SJ68]
G0 MYP D M Wordsworth, 15 Tetbury Dr, Bolton, BL2 5NP [IO83TO, SD70]
GM0 MYQ J B Frati, 10 Benbecula Rd, Aberdeen, AB1 6FU [IO87TH, NJ72]
G0 MYR R C Worsley, Omaru Pennance Rd, Lanner, Redruth, Cornwall, TR16 5TQ [IO70JF, SW74]
GM0 MYS S J McHenry, 19 Balmoral Gdns, Blantyre, Glasgow, G72 9NP [IO75WT, NS65]
G0 MYT Maj L J Colyton-Smith, 48 Shotesham Rd, Poringland, Norwich, NR14 7LN [JO02QN, TG20]
G0 MYU Details withheld at licensee's request by SSL.
GM0 MYV I A C Reid, Mdina, Auchentoul, Alford, Aberdeenshire, AB33 8NN [IO87OF, NJ51]
G0 MYW Details withheld at licensee's request by SSL.[Op: J R Hall. Station located near Marlow.]
G0 MYX R C Stephen, 97 Hunters Field, Stanford in The Vale, Faringdon, SN7 8ND [IO91FP, SU39]
GW0 MYY Details withheld at licensee's request by SSL.
G0 MYZ M B Kirk, 5 The Paddock, Kirkby in Ashfield, Nottingham, NG17 8BT [IO93ID, SK46]
G0 MZA Details withheld at licensee's request by SSL.
GM0 MZB W McTaggart, 30 Don St., Grangemouth, FK3 8HD [IO86DA, NS98]
GM0 MZD A G Coutts, 17 Sandy Loch Dr, Lerwick, ZE1 0SR [IP90JD, HU44]
G0 MZE W Castling, 2 Mitford Pl, Gosforth, Newcastle upon Tyne, NE3 3PQ [IO95EA, NZ26]
G0 MZF B A Nicholson, 75 Portway, Wells, BA5 2BJ [IO81QF, ST54]
G0 MZI S A Fox, Castle Cottage, Stratford Sub Castl, Salisbury, Wilts, SP1 3LB [IO91CC, SU13]
G0 MZJ K E Earp, 18 Main St., Kings Newton, Melbourne, Derby, DE73 1BX [IO92GT, SK32]
G0 MZK R T Little, Mythe House, 129 Slad Rd, Stroud, GL5 1RD [IO81VR, SO80]
G0 MZL D Fletcher, 129 North Bank Rd, Batley, WF17 8EX [IO93ER, SE22]
G0 MZN J C Nunn, 20 Somerton Gdns, Earley, Reading, RG6 5XG [IO91NK, SU77]
G0 MZO W Unsworth, 8 Thurlby Cres, Ermine Est East, Lincoln, LN2 2HU [IO93RF, SK97]
G0 MZP R S Kay, 7 Alderson Rd, Worksop, S80 1UZ [IO93KH, SK57]
G0 MZQ W G Greed, 5 West View, Creech St. Michael, Taunton, TA3 5QP [IO81LA, ST22]
G0 MZR T D Fletcher, 57 Hawthorne Way, Carlton in Lindrick, Worksop, S81 9HN [IO93KI, SK58]
G0 MZT Details withheld at licensee's request by SSL.
G0 MZU Rev J P Carroll, 31 George St., Balsall Heath, Birmingham, B12 9RG [IO92BK, SP08]
G0 MZV S J Barnfather, 8 Willow Dr, Bracknell, RG12 2HX [IO91PK, SU86]
G0 MZW Details withheld at licensee's request by SSL.
G0 MZX Details withheld at licensee's request by SSL.
G0 MZY F Woodhall, 13 Whitegate Dr, Clifton, Swinton, Manchester, M27 8RE [IO83UM, SD70]
G0 MZZ A Benson, 2 Fieldhead Ln, Birstall, Batley, West Yorks, WF17 9BQ [IO93ER, SE22]
G0 NAA A P Leake, Thorpe Garth, East Newton, Aldbrough, Hull, N.Humberside, HU11 4SD [IO93WT, TA23]
G0 NAB Details withheld at licensee's request by SSL.
GJ0 NAC L A Gray, 37 Poonah Rd, St. Helier, Jersey, JE2 3XJ
GM0 NAE J D Carlin, 24 Hillcrest Ave, Glenbburn, Paisley, PA2 8QW [IO75ST, NS46]
G0 NAG M J Proudler, 182 Acklam Rd, Middlesbrough, TS5 4HA [IO94IN, NZ41]
G0 NAH R Newton, 33 Farrowdene Rd, Reading, RG2 8SD [IO91MK, SU76]
GM0 NAI J N Fisher, 23 Ranfurly Rd, Bridge of Weir, PA11 3EL [IO75RU, NS36]
G0 NAJ J A Neary, 266 Yew Tree Ln, Dukinfield, SK16 5DN [IO83XL, SJ99]
G0 NAL T E F Carr, 23 Fitzharrys Rd, Abingdon, OX14 1EL [IO91IQ, SU49]
G0 NAN Details withheld at licensee's request by SSL.
G0 NAO N A Oakley, 42 Harewood Cres, Old Tupton, Chesterfield, S42 6HS [IO93GE, SK36]
G0 NAP P N Howell, 21 Clare Pl, Plymouth, PL4 0JW [IO70WI, SX45]
G0 NAQ G Furmage, 92 Weaverham Rd, Norton, Stockton on Tees, TS20 1QL [IO94IO, NZ42]
G0 NAR C A P Fenton, Oakwell, Newtons Hill, Hartfield, E Sussex, TN7 4DH [JO01BC, TQ43]
G0 NAS L R Aykroyd, 32 Coach Rd, Astley, Tyldesley, Manchester, M29 7ER [IO83SM, SD70]
G0 NAT Dr H M Good, 6671 W, Indiantown Rd 56-169, Jupiter, Fl 33458, USA
G0 NAU M W Best, 81 Maybury Rd, Holderness Rd, Kingston upon Hull, Hull, HU9 3LB [IO93US, TA13]
G0 NAV M P Runciman, 36 Kingsmead, Barnet, EN5 5AY [IO91VP, TQ29]
G0 NAX D J Edmonds, 15 Youngs Rd, Barkingside, Ilford, IG2 7LF [JO01BN, TQ48]
GM0 NAZ A G Heggie, 75 Doon Walk, Craigshill, Livingston, EH54 5AD [IO85GV, NT06]
GM0 NBA T Adam, 33 Lambie Cres, Newton Mearns, Glasgow, G77 6JU [IO75TS, NS55]
G0 NBB M R Watkins, 9 Benacre Rd, Whitstable, CT5 4NY [JO01MI, TR16]
G0 NBC L Steenvoorden, 1 Thornbury Rd, Immingham, Grimsby, DN40 1HH [IO93VO, TA11]
G0 NBD A G Brown, 20 Sheen Rd, Wallasey, L45 1HA [IO83LK, SJ39]
G0 NBE R K Allen, 2 Hilltop Ave, Basford, Newcastle, ST5 0QF [IO83VA, SJ84]
G0 NBF Details withheld at licensee's request by SSL.
GM0 NBG J G McVittie, 19 Beech Way, Girvan, KA26 0BX [IO75NF, NX19]
G0 NBH J W Goodwin, Hankelow Ct, Hall Ln, Hankelow, Crewe, CW3 0JB [IO83SA, SJ64]
G0 NBJ N R Foster, 12 Chequer Ln, Upholland, Skelmersdale, WN8 0DE [IO83PM, SD50]
G0 NBK S Golding, 86 Trevelyan Ct, Longbenton, Newcastle upon Tyne, NE12 8TD [IO95EA, NZ26]
GM0 NBM J R Hayes, 11 Nairn Way, Cumbernauld, Glasgow, G68 0HX [IO85AX, NS77]
G0 NBN D Williams, 49 Riverside, Combwich, Bridgwater, TA5 2RB [IO81LE, ST24]
GM0 NBO M P Hughes, 14 Combfoot Cttgs, Mid Calder, Livingston, EH53 0AD [IO85GV, NT06]
G0 NBQ H W Stevens, Ashlea, 26 Grange Ave, Cheadle Hulme, Cheadle, SK8 5EN [IO83VJ, SJ88]
G0 NBW B Nolan, 4 Shetland Rd, South Shore, Blackpool, FY1 6LP [IO83LT, SD33]
G0 NBY S S Nixon, PO Box 29, 5120 Manger, Radoy, Norway
GI0 NCA R Pinkerton, 9 Cloghole Rd, Campsie, Londonderry, BT47 3JW [IO65JA, C42]
G0 NCC D Wicker No.R.A.C.C., 28 Lee Warner Ave, Fakenham, NR21 8ER [JO02KU, TF93]
G0 NCE O T Wheeler, 7 Aveling Cl, Hoo, Rochester, ME3 9BZ [JO01GK, TQ77]
G0 NCG A F Broadbent, 128 Ringway, Thornton Cleveleys, FY5 2NW [IO83LV, SD34]
G0 NCH R Magri, 13 Roebuck Rd, Chessington, KT9 1JY [IO91UI, TQ16]
G0 NCJ Details withheld at licensee's request by SSL.
G0 NCK Details withheld at licensee's request by SSL.
G0 NCL R J Harris, Berghers Hill, Wooburn Common, Wooburn Green, High Wycombe, Bucks, HP10 0JP [IO91PN, SU98]
G0 NCM Details withheld at licensee's request by SSL.
GW0 NCN Details withheld at licensee's request by SSL.
G0 NCO D M Murray, 14 Littlestead Cl, Caversham, Reading, RG4 6UA [IO91ML, SU77]
G0 NCQ M J Hemmings, 2 Holly Walk, Whitestone, Nuneaton, CV11 6UU [IO92GM, SP39]
G0 NCR Details withheld at licensee's request by SSL.
G0 NCS C R J Healey, 22 Stirling Rd, St. Budeaux, Plymouth, PL5 1PD [IO70VJ, SX45]
G0 NCT F L Norman, 37 Third Ave, Canvey Island, SS8 9SU [JO01GM, TQ78]
GW0 NCU S E David, 142 Robert St., Manselton, Swansea, SA5 9NH [IO81AP, SS69]
G0 NCV W Chappell, 5 Willow Dyke, Corbridge, NE45 5JR [IO84XX, NY96]
G0 NCW A C Judge, 106 Bicknor Rd, Parkwood, Maidstone, ME15 9PD [JO01GF, TQ75]
G0 NCX R Hughesdon, 3 Lyndhurst Rd, Gosport, PO12 3QY [IO90KT, SZ69]
G0 NCY R J Hartley, 99 Carr Rd, Fleetwood, FY7 6QQ [IO83LW, SD34]
G0 NCZ W E F Wilson, 11 Crofthead Dr, Cramlington, NE23 6LG [IO95FB, NZ27]
GW0 NDA J R Frost, Rhoslwyn, Maes Tegid, Bala, LL23 7BN [IO82EV, SH93]
G0 NDB R Evans, Tonkins Quay House, Mixtow, Lanteglos By Fowey, Cornwall, PL23 1NB [IO70QI, SX15]
G0 NDC R Little, Maranatha, Higher Moresk Rd, Truro, Cornwall, TR1 1BW [IO70LG, SW84]
G0 NDD J Jackson, 26 Wadham Cl, Peterlee, SR8 2NN [IO94HS, NZ44]
G0 NDE G Hutchinson, Pear Tree Barn, Long Hyde Rd, South Littleton, Evesham, WR11 5TH [IO92BC, SP04]
G0 NDF C R Peters, Arundel Lodge, Arundel Rd, Brighton, BN2 5TD [IO90WT, TQ30]
G0 NDG G W Lamb, 18 East View, Bedlington, NE22 7HD [IO95FD, NZ28]
G0 NDH R W Hopley, 26 Mannington Way, West Moors, Ferndown, BH22 0JE [IO90BT, SU00]
G0 NDI A H Izzard, 230 Alwold Rd, Weoley Castle, Birmingham, B29 5JS [IO92AK, SP08]
G0 NDJ A Izzard, 230 Alwold Rd, Weoley Castle, Birmingham, B29 5JS [IO92AK, SP08]
G0 NDK A L Sinclair, 60 Church Ave, West Sleekburn, Choppington, NE62 5XG [IO95FD, NZ28]
G0 NDL P Crook, 73 Wansunt Rd, Bexley, DA5 2DJ [JO01BK, TQ57]
G0 NDM A Aspinal, 44 Meadow Cres, Woodchurch, Wirral, L49 8HY [IO83KJ, SJ28]
G0 NDN Details withheld at licensee's request by SSL.
GW0 NDP K Lasham, Llys Afan, 13 Ffordd y Felin, Dolgellau, LL40 1HU [IO82BR, SH71]
G0 NDQ Details withheld at licensee's request by SSL.
G0 NDR Details withheld at licensee's request by SSL.
G0 NDS J B Johnson Northampton ARG, 30 Millside Cl, Kingsthorpe, Northampton, NN2 7TR [IO92NG, SP76]
G0 NDT Details withheld at licensee's request by SSL.
G0 NDU J R Dykes, 33 Mill House Dr, Cheltenham, GL50 4RG [IO81XV, SO92]
G0 NDV P M Timmins, 21 Cox Rd, Sheffield, S6 4SX [IO93FJ, SK39]
G0 NDW C Harradine, 48 High St., Hale Village, Woolton, Liverpool, L25 7TF [IO83NI, SJ48]
G0 NDY R S T B Wayne, Colkirk House, Manor House St., Horncastle, LN9 5HF [IO93WE, TF26]
GW0 NDZ G Davies, 4 Crichton St., Treorchy, CF42 6DF [IO81FP, SS99]
G0 NEA K C Bewley Group of National Experimental, 19 High St., Watton At Stone, Hertford, SG14 3SX [IO91WU, TL21]
G0 NEB J A Dorning, 51 OSullivan Cres, Blackbrook, St. Helens, WA11 9RE [IO83PL, SJ59]
GW0 NEC V E Fletcher, 23 Lon Cymru, Llandudno, LL30 1SJ [IO83CH, SH78]
G0 NED E J Dudley, 95 Hilderstone Rd, Stoke on Trent, ST3 7NS [IO82WW, SJ93]
G0 NEE M Stott, Wellview, 12 Castle View, Ovingham, Prudhoe, NE42 6AT [IO94BX, NZ06]
GM0 NEG R J Young, 1 Spey St., Garmouth, Fochabers, IV32 7NJ [IO87KQ, NJ36]
G0 NEH Details withheld at licensee's request by SSL.
G0 NEI B Birkby, 10 Rankin Ave, Hesketh Bank, Preston, PR4 6PA [IO83NQ, SD42]
GW0 NEJ R G Sourbutts, Swn-yr-Afon Forge, Machynlleth, Powys, SY20 8RZ [IO82CO, SH70]
G0 NEL G Kirby-Parkinson, 32 Keats Ave, Bolton-le-Sands, Carnforth, LA5 8HH [IO84OC, SD46]
G0 NEM M R Purser, 17 Firecrest Rd, Chelmsford, CM2 9SN [JO01FR, TL70]
G0 NEN G P Lewin, The Hawthorns, Hawthorn Dr, Wheaton Aston, Stafford, ST19 9NQ [IO82VR, SJ81]
G0 NEO J H Boland, 28 Vicarage Rd, Orrell, Wigan, WN5 7AX [IO83PM, SD50]
G0 NEP P R Whitling, 17 Balcomb Cres, Margate, CT9 3XJ [JO01QI, TR36]
G0 NEQ R H Massey Barr Beacon ARC, 48 Quicksand Ln, Aldridge, Walsall, WS9 0BA [IO92AO, SP09]
G0 NER S M Stones, Maison Neuve, 47210, St Martin de Villereal, Lot Et Garonne, France, X X
G0 NES D G Bryant, 35 Truemans Heath Ln, Hollywood, Birmingham, B47 5QE [IO92BJ, SP07]
GM0 NET T Stewart Ayrshire Raynet Group, 104 Barrhill Rd, Cumnock, KA18 1PU [IO75UK, NS52]
G0 NEU A A Dawson, 182 Ladysmith Rd, Enfield, EN1 3AE [IO91XP, TQ39]
G0 NEV M A Carter, 1 Littlemoor Rd, Preston, Weymouth, DT3 6LA [IO80SP, SY68]
G0 NEW Details withheld at licensee's request by SSL.
G0 NEY Details withheld at licensee's request by SSL.
G0 NEZ D J Gosling, 31 Sempill Rd, Hemel Hempstead, HP3 9PF [IO91SR, TL00]
G0 NFA D Gilbert, 2 Greenfield Cttgs, Bentley, Farnham, GU10 5HZ [IO91NE, SU74]
G0 NFC K R Nunn, Oak House, New Rd, Fritton, Great Yarmouth, NR31 9HP [JO02TN, TG40]
G0 NFD J Whitehand, 1 Harbour Terr, Boscastle, PL35 0AE [IO70PQ, SX09]
G0 NFE R B Ransome, 66 Spencer Way, Stowmarket, Suffolk, IP14 1UB [JO02LE, TM05]
G0 NFG D A Hopper, 28 Western Ave, Herne Bay, CT6 8TU [JO01NI, TR16]
G0 NFH J D Acton, 63 Bevington Cl, Patchway, Bristol, BS12 5NP [IO81QM, ST58]
G0 NFI P Edwards, 5 Broomhall Cl, Oswestry, SY10 7HF [IO82LU, SJ22]
G0 NFJ H D Garland, 9 Field Cl, Abridge, Romford, RM4 1DL [JO01BP, TQ49]
G0 NFL M L George, 2 Jubilee Terr, Middle St., Isham, Kettering, NN14 1HG [IO92PI, SP87]
GW0 NFN J Butler, 178 Squirrel Walk, Fforest, Pontarddulais, Swansea, SA4 1UG [IO71XR, SN50]
G0 NFO R R Charteris, 7 Kennedy Cl, Kidderminster, DY10 1LR [IO82VJ, SO87]
G0 NFP Details withheld at licensee's request by SSL.
G0 NFQ Details withheld at licensee's request by SSL.
G0 NFR R Glynn, 118 Pelham Rd, Birmingham, B8 2PD [IO92CL, SP18]
G0 NFT T M J Jones, The Cottage, 99 Uxbridge Rd, Hampton Hill, Hampton, TW12 1SL [IO91TK, TQ17]
G0 NFV A J Hunt, 146 North Walsham Rd, Sprowston, Norwich, NR6 7QQ [JO02PP, TG21]
G0 NFW O L A F Skutsch, Margareten Str 16, 26446 Friedeburg, West Germany
G0 NFY A Cordwell, 71 Tadcaster Rd, Woodseats, Sheffield, S8 0RA [IO93GI, SK38]
G0 NFZ R S Hodges, 33 Harnall Cl, Shirley, Solihull, B90 4QR [IO92CJ, SP17]
G0 NGB L Taylor, Homewood, 40 Stotfold Rd, Church End, Arlesey, SG15 6XT [IO92UA, TL13]
G0 NGC R J Allwood, 27 Whitehall Gdns, Acton, London, W3 9RD [IO91UM, TQ18]
G0 NGD C A H Stenbacka, 11 Mount View, Billericay, CM11 1HB [JO01FP, TQ69]
G0 NGE W M Clarke, 6 Whernside Way, Leyland, Preston, PR5 2ZN [IO83PQ, SD52]
G0 NGF J M Rhodes, Bank Top Farm, Greenhead, Carlisle, Cumbria, CA6 7HA [IO84RX, NY66]
G0 NGG R A Brown, 20 King Edward Rd, Stanford-le-Hope, SS17 0EF [JO01FM, TQ68]
G0 NGH A S Pemberton, 22 Borough Way, Potters Bar, EN6 3HB [IO91VQ, TL20]
G0 NGI D R Johnson, 15 St. Michaels Rd, Sheerwater Est, Woking, GU21 5PY [IO91RH, TQ06]
GM0 NGJ A B Caldwell, 7 Gladstone Terr, New Deer, Turriff, AB53 6TE [IO87VM, NJ84]
G0 NGK P J Walmsley, Valley View, Longworth Ave, Coppull, Lancs, PR7 4PJ [IO83QP, SD51]
G0 NGL Details withheld at licensee's request by SSL.
G0 NGP P Wilson, 74 Coleraine Rd, Blackheath, London, SE3 7PE [JO01AL, TQ37]
G0 NGQ Details withheld at licensee's request by SSL.
GW0 NGR Details withheld at licensee's request by SSL.
G0 NGS J G Hetherington Gravesham & Dis, Scout Rd Fellowshi, 44 Brookside Rd, Istead Rise, Gravesend, DA13 9JJ [JO01EJ, TQ67]
G0 NGT Details withheld at licensee's request by SSL.
GW0 NGU Details withheld at licensee's request by SSL.
G0 NGV Details withheld at licensee's request by SSL.[Op: Raymond J Lee - Ray. QSL to Smiths Bakery, 277-281 Lytham Road, South Shore, Blackpool, Lancashire, FY4 1DP. WAB sq: SD33. RNARS: 3549.]
G0 NGW R W Ramplin, 17 Cross St., Langold, Worksop, S81 9SL [IO93KJ, SK58]
G0 NGX Details withheld at licensee's request by SSL.
G0 NGZ J C Skidmore, 6 Moray Cl, Halesowen, B62 9PP [IO82XL, SO98]
G0 NHA F Bray, 178 Thornley St., Burton on Trent, DE14 2QR [IO92ET, SK22]
G0 NHB Dr W A Stallard, 28 Wheatfield Rd, Stanway, Colchester, CO3 5YJ [JO01KV, TL92]
G0 NHC J J Kearney, The Dales, North Rd, Chester-le-Street, Co Durham, DH3 4AQ [IO94FU, NZ25]
GU0 NHD K F Benton, Keukenhof, Route de Carteret, Castel, Guernsey, GY5 7YS
GW0 NHE A Smith, 7 Clos Gors Fawr, Grovesend, Gorseinon, Swansea, SA4 2GZ [IO71XQ, SS59]
G0 NHF A L Slater, 25 Statham Ave, New Tupton, Chesterfield, S42 6YE [IO93HE, SK36]
G0 NHG J P Godwin, 22 Stonebeck Ave, Harrogate, HG1 2BW [IO94FA, SE25]
G0 NHJ J C Robson, 35 Melling Rd, Cramlington, NE23 6AS [IO95EB, NZ27]
G0 NHK K Robson, 35 Melling Rd, Cramlington, NE23 6AS [IO95EB, NZ27]
G0 NHM N H Robertshaw, 9 Chapel Ln, Anwick, Sleaford, NG34 9SX [IO93TA, TF15]
G0 NHN N G Cherry, 803 Manchester Rd, Stocksbridge, Sheffield, S30 5DR [IO93GJ, SK38]
G0 NHO K D Crookes, 64 Heron Dr, Audenshaw, Manchester, M34 5QX [IO83WL, SJ99]
G0 NHP T L Maguire, 1 Gosford St., Balsall Heath, Birmingham, B12 9ER [IO92BL, SP08]
G0 NHQ J Ford, Norland, 3 Kenwood Ave, Hale, Altrincham, WA15 9DE [IO83UJ, SJ78]
G0 NHR D Sommerfield Nunsfield House, 51 Spindletree Dr, Oakwood, Derby, DE21 2DG [IO92GW, SK33]
G0 NHS A Parr, 25 Sherwood Ave, Droylsden, Manchester, M43 7JJ [IO83WL, SJ99]
GM0 NHT H Cherrie, 2 Cearn Chilleagraidh, Stornoway, Isle of Lewis, PA87 2UJ [IO76JB, NS08]
G0 NHW Details withheld at licensee's request by SSL.
G0 NHX E R Holloway, 6 Downesway, Alderley Edge, SK9 7XB [IO83VH, SJ87]
G0 NHZ A D Pollard, 2 Forest Cl, Cowplain, Waterlooville, PO8 8JE [IO90LV, SU61]
G0 NIA G Sleigh, 7 Windsor Park Rd, Buxton, SK17 7NP [IO93BG, SK07]
G0 NIB R R Lucking, 62 Ember Farm Way, East Molesey, KT8 0BL [IO91TJ, TQ16]
G0 NIC C R Riddell, 98 Mount Rd, High Barnes, Sunderland, SR4 7NN [IO94HV, NZ35]
G0 NID E Page, 10 Balmer Dr, Newall Green, Manchester, M23 2YQ [IO83UJ, SJ88]
G0 NIE J R Allen, Brentor, Tavistock, Devon, PL19 0LR [IO70WO, SX48]
G0 NIF S D W Wilkins, 14 Arnold Rd, Laleham, Staines, TW18 1LX [IO91SK, TQ07]
G0 NIG N R Smith, 45 The Gills, Otley, LS21 2BY [IO93DV, SE24]
G0 NIH S T P Costelloe, 15 Budock Terr, Falmouth, TR11 3ND [IO70LD, SW83]
G0 NII Details withheld at licensee's request by SSL.
G0 NIJ S A Belcher, 52 Kynaston Rd, Didcot, OX11 8HD [IO91JO, SU58]
G0 NIK Details withheld at licensee's request by SSL.
G0 NIL D Woods, 30 Longridge Ave, Stalybridge, SK15 1HG [IO83XL, SJ99]
G0 NIN N H Beer, Holkham, Headley Fields, Headley, Bordon, GU35 8PU [IO91OC, SU83]
G0 NIP T R Taylor, 14 Meadow Way, Jaywick, Clacton on Sea, CO15 2SQ [JO01NS, TM11]
G0 NIQ J S Naylor, 16 Dorchester Cl, Basingstoke, RG23 8EX [IO91KG, SU65]
G0 NIS Details withheld at licensee's request by SSL.
G0 NIT J D Sharples, 2 Windsor Gr, Bodmin, PL31 2BP [IO70PL, SX06]
G0 NIV Details withheld at licensee's request by SSL.
G0 NIX W G Jones, 13 Kilrush Terr, Woking, GU21 5EG [IO91RH, TQ05]

GW0 NIY R W Milton, 49 Heol y Deri, Rhiwbina, Cardiff, CF4 6HD [IO81JM, ST18]
G0 NIZ H E Hill, The Acorns, Prince Cres, Staunton, Gloucester, GL19 3RF [IO81UX, SO72]
G0 NJB F G Sargent, 3 Crossings Cl, Mary Tavy, Tavistock, PL19 9QP [IO70WO, SX57]
G0 NJD R Mallett, 71 Olivet Rd, Woodseats, Sheffield, S8 8QR [IO93GI, SK38]
G0 NJE Details withheld at licensee's request by SSL.
G0 NJF A R J Binns, 25 Dines Way, Hermitage, Thatcham, RG18 9TF [IO91IK, SU57]
G0 NJG K L Card, 25 Dines Way, Hermitage, Thatcham, RG18 9TF [IO91IK, SU57]
G0 NJJ N J Jones, 19 Foxhollow, Bar Hill, Cambridge, CB3 8EP [JO02AG, TL36]
G0 NJK R Wilson, 62 Ventnor Ave, Hartlepool, TS25 5NA [IO94JQ, NZ53]
GM0 NJL R Watson, 24 Hillock Ave, Redding, Falkirk, FK2 9UT [IO85DX, NS97]
G0 NJM Details withheld at licensee's request by SSL.
G0 NJO J Howard, 171 Nutbeem Rd, Eastleigh, SO50 5JS [IO90HX, SU41]
GM0 NJP M Murakami, 5-8 Takanidai, Takatsuki, Osaka 569, Japan, XX XX
G0 NJQ P J Schlatter, Churchgate, Sutton Rd, Cookham, Maidenhead, SL6 9SN [IO91PN, SU88]
G0 NJS M W M Davie, First House, Hyde Valley, Welwyn Garden City, Herts, AL7 4ND [IO91VS, TL21]
G0 NJT Details withheld at licensee's request by SSL.
G0 NJY Details withheld at licensee's request by SSL.
G0 NJZ T Dodds, 33 Westgate, Oldbury, Warley, B69 1BA [IO82XL, SO98]
G0 NKA I L McAvoy, 6 Sidney St., Lincoln, LN5 8DB [IO93RF, SK96]
G0 NKC M J Connolly, 1 Clifton Cl, Yeovil, BA21 5LB [IO80QW, ST51]
GW0 NKF E O E James, 1 Pellau Rd, Margam, Port Talbot, SA13 2LF [IO81CN, SS78]
GW0 NKG M E York, 26 Penallta Rd, Ystrad Mynach, Hengoed, CF82 7AN [IO81JP, ST19]
GW0 NKI A J Tanner, 24 Parkhill Terr, Treboeth, Swansea, SA5 7DJ [IO81AP, SS69]
GW0 NKJ D T Davies, 6 Dulais Fach Rd, Tonna, Neath, SA11 3JW [IO81CQ, SS79]
G0 NKK Details withheld at licensee's request by SSL.
G0 NKL B T Osborne Submarine A.R.C, 12 Arminers Cl, Gosport, PO12 2HB [IO90KS, SZ69]
G0 NKM H B Monks, 100 Crossefield Rd, Cheadle Hulme, Cheadle, SK8 5PF [IO83VJ, SJ88]
G0 NKN S J Harrison, St Ives Manor, St. Ives Est, Harden Rd, Bingley, West Yorks, BD16 1AT [IO93BU, SE03]
G0 NKO K J OBrien, 2 Temple Dr, Nuthall, Nottingham, NG16 1BE [IO92JX, SK54]
G0 NKQ V J Nunns, 8 Trevithick Rd, Tregurra Parc, Truro, Cornwall, TR1 1RU [IO70LG, SW84]
G0 NKR Details withheld at licensee's request by SSL.
G0 NKU S C Fowler, 6 Atherton St., Edgeley, Stockport, SK3 9JN [IO83VJ, SJ88]
G0 NKV M E Power, 18 Chevington Dr, Heaton Mersey, Stockport, SK4 3RG [IO83VJ, SJ89]
GM0 NKX P Logan, 28 Hartrigge Cres, Jedburgh, TD8 6HT [IO85RL, NT62]
G0 NKY N E Feakes, 45 Langley Rd, Staines, TW18 2EH [IO91RK, TQ07]
G0 NKZ K L Everard, Woodside, Staple Ln, West Quantoxhead, Taunton, TA4 4DE [IO81IE, ST14]
G0 NLA R M A Bryan, 23 Quarry Ln, Halesowen, B63 4PB [IO82XK, SO98]
GW0 NLB W S Rees, 51 Heol Capel Ifan, Pontyberem, Llanelli, SA15 5HF [IO71VS, SN41]
G0 NLF Details withheld at licensee's request by SSL.
G0 NLG R J Chapman, 50 Graeme Rd, Enfield, EN1 3UT [IO91XP, TQ39]
G0 NLI C Macham, 9 Bankfield Gr, Scot Hay, Newcastle, ST5 6AR [IO83UA, SJ84]
G0 NLJ S A Oliver, 24 Sixty Acres Rd, Prestwood, Great Missenden, HP16 0PE [IO91PQ, SP80]
G0 NLK S Burwell, 28 Manor Ln, Shipton Rd, York, YO3 6TX [IO93KX, SE55]
G0 NLL J Bowers, 73 Buxton Rd, New Mills, High Peak, SK22 3JT [IO83XI, SJ98]
G0 NLM C J Ridley, 205 Leeson Dr, Ferndown, BH22 9TL [IO90BT, SU00]
G0 NLO G B Wharton, Onanole, Clitheroe Rd, Bashall Eaves, Clitheroe, BB7 3DA [IO83SV, SD64]
G0 NLP W Worswick, Ivy Cottage, Dunsop Bridge, Nr.Clitheroe, Lancs, BB7 3BB [IO83RW, SD65]
G0 NLQ P L Dunn, 10 Endsleigh Cl, Upton, Chester, CH2 1LX [IO83NF, SJ46]
G0 NLR Details withheld at licensee's request by SSL.
G0 NLS Details withheld at licensee's request by SSL.
G0 NLT D R J Wentworth, 7 Gilbeys Cl, Wordsley, Stourbridge, DY8 4XU [IO82WL, SO88]
GM0 NLU N Harvey, The Shieling, Tealing, By Dundee, DD4 0QU [IO86MM, NO43]
G0 NLV F J Fairman, 26 Marina Gdns, Cheshunt, Waltham Cross, EN8 9QY [IO91XQ, TL30]
G0 NLW F Schnell, 4 Windmill Rd, Sunbury on Thames, TW16 7HX [IO91SK, TQ07]
G0 NLX S E Pearson, 41 Broad St., Chesham, HP5 3EA [IO91QR, SP90]
GW0 NLY I G Batchelor, 14 Samuel Cres, Gendros, Swansea, SA5 8DW [IO81AP, SS69]
G0 NMA M J Coburn Marconi ARG, Marconi Instruments Ltd, Longacres House, Norton Green Rd, Stevenage, SG1 2BA [IO91VV, TL22]
G0 NMB A Haberman, 102 Garden Rd, Walton on The Naze, CO14 8SJ [JO01PU, TM22]
G0 NMC T M Neal, 11 Avondale Rd, Palmers Green, London, N13 4DX [IO91WO, TQ39]
G0 NMD Revd L E Austin, The Rectory, 2 Barnfield, Mill Ln, Bratton Fleming, Barnstaple, North Devon, EX31 4RT [IO81AC, SS63]
G0 NMG R R Gordon, 15 The Layne, Bognor Regis, PO22 6JL [IO90QT, SU90]
G0 NMH B Markey, 5 Hulbert Croft, Almondbury, Huddersfield, HD5 8SD [IO93DP, SE11]
G0 NMI Details withheld at licensee's request by SSL.
G0 NMJ J W Denniss, 61 Checkstone Ave, Bessacarr, Doncaster, DN4 7JY [IO93LL, SE60]
G0 NMK Details withheld at licensee's request by SSL.
G0 NML Details withheld at licensee's request by SSL.
G0 NMN L D James, 8 Opal Way, Woosehill, Wokingham, RG41 3UL [IO91NK, SU76]
G0 NMO Details withheld at licensee's request by SSL.
G0 NMP P J Chapman, 4 Windsor Ct, Watton, Thetford, IP25 6XB [JO02JN, TF90]
G0 NMQ Details withheld at licensee's request by SSL.
G0 NMR Details withheld at licensee's request by SSL.
G0 NMS J L Howes, 39 Pound Hill, Bacton, Stowmarket, IP14 4LP [JO02MG, TM06]
G0 NMT Details withheld at licensee's request by SSL.
GI0 NMV S M McClurg, 14 Larch Gr, Mossley, Newtownabbey, BT36 5NQ [IO74AQ, J38]
G0 NMW R R Johns, 5 Wychwood Dr, Meyrick Park, Bournemouth, BH2 6JG [IO90BR, SZ09]
G0 NMX Details withheld at licensee's request by SSL.
G0 NMY M Longson, 54 Beresford St., Shelton, Stoke on Trent, ST4 2EX [IO83VA, SJ84]
G0 NNA Details withheld at licensee's request by SSL.
GW0 NNB D B Jones, 45 Parc Penrhiw, Bettws, Betws, Ammanford, SA18 2SP [IO81AS, SN61]
G0 NNE R W Hart, 6 Chatsworth Rd, Halesowen, B62 8TA [IO82XL, SO98]
G0 NNF J A Yates, 87 Princess Rd, Oldbury, Warley, B68 9PW [IO92AL, SP08]
G0 NNG R H Westley, 91 Lincoln Way, Stefan Hill, Daventry, Northants, NN11 4SU [IO92JG, SP56]
G0 NNI R J Gosling, 86 Chambersbury Ln, Hemel Hempstead, HP3 8BB [IO91SR, TL00]
G0 NNJ S C French, 1 Marcus Ave, Thorpe Bay, Southend on Sea, SS1 3LB [JO01JM, TQ98]
GI0 NNM P O'Neill, 7 Ballyoran Park, Portadown, Craigavon, BT62 1JN [IO64SK, J05]
G0 NNN W S Forbes, 11 Moss Side, Allonby, Maryport, CA15 6QW [IO84GS, NY04]
G0 NNO M J Shore, 12 Boscoppa Rd, Bethel, St. Austell, PL25 3DR [IO70OI, SX05]
G0 NNP D A King, Iden, Rye, E Sussex, TN31 7PT [JO00JX, TQ92]
G0 NNR B C Thomas, Creekside, Greenbank Rd, Devoran, Truro Cornwall, TR3 6PQ [IO70KF, SW73]
G0 NNS C M Hendry Norwich North Scouts Fellowshi, 16 Levishaw Cl, Buxton, Norwich, NR10 5HQ [JO02PS, TG22]
G0 NNT V C Martinelli, 62 Angelo St, Sliema Slm 13, Malta, X X
G0 NNU L Payne, 147 Upper Marehay, Ripley, DE5 8JG [IO93HA, SK34]
GW0 NNY M A Griffiths, Glenmore Main Rd, Tonteg Pontypridd, Mid Glam, South Wales, CF38 1LS [IO81IN, ST08]
G0 NNZ J E Belfield, 17 Burtondale Rd, Crossgates, Scarborough, YO12 4JR [IO94SF, TA08]
G0 NOA J Clark, Cherry Garden, Windsor Rd, Crowborough, TN6 2HR [JO01CB, TQ52]
G0 NOB L Leek, Pine Cottage, 12 Thatcher Ave, Torquay, TQ1 2PD [IO80GL, SX96]
G0 NOC R E Hamson, 43 Arcot Park, Sidmouth, EX10 9HU [IO80JQ, SY18]
G0 NOE Details withheld at licensee's request by SSL.[Station located near Halesowen.]
G0 NOG B A Stockwell, Stockwell (Hwds) Lt, Ibstone Rd, Stokenchurch, Bucks, HP14 3TW [IO91NP, SU79]
G0 NOH J W K Reed, 23 Morehall Ave, Folkestone, CT19 4EQ [JO01NC, TR23]
G0 NOK Details withheld at licensee's request by SSL.
G0 NOL C J Baker, 19 Elizabeth Rd, Walsall, WS5 3PF [IO92AN, SP09]
G0 NOM K E P Fowler, 171 Queens Rd, Cheadle Hulme, Cheadle, SK8 5HX [IO83VJ, SJ88]
G0 NON T C Behan, Maytree Cottage, Marley Ln, Haslemere, GU27 3RG [IO91PB, SU83]
GW0 NOO J S Coburn, 54 Queensway, Hope, Wrexham, LL12 9PE [IO83LC, SJ35]
GW0 NOP P Coburn, 54 Queensway, Hope, Wrexham, LL12 9PE [IO83LC, SJ35]
G0 NOQ R V Skuse, Higher Centry, Totnes Rd, Kingsbridge, S.Devon, TQ7 2HF [IO80DG, SX74]
G0 NOR C E Jones, 40 Ambleside Cres, Warrington, WA2 9NE [IO83RJ, SJ69]
G0 NOS Details withheld at licensee's request by SSL.
G0 NOT Details withheld at licensee's request by SSL.
G0 NOU Dr W H Ayers, Peach Lodge, Bakewell Rd, Eyam, Sheffield, S30 1QA [IO93GJ, SK38]
G0 NOV P E R Adams, 25 Appleton Ave, Great Barr, Birmingham, B43 5LY [IO92AN, SP09]
GI0 NOX S F McAteer, 23 Highfield Park, Highfield, Craigavon, BT64 3AF [IO64TK, J05]
G0 NOY Details withheld at licensee's request by SSL.
GM0 NOZ S I Bremner, 32 Exeter Dr, Glasgow, G11 7XB [IO75UU, NS56]
G0 NPA R M Stephens, 50 Windrush Way, Abingdon, OX14 3SX [IO91IQ, SU59]
G0 NPC G K H Hill, 1 Gleneagles Ct, Edwalton, Nottingham, NG12 4DN [IO92KV, SK63]
G0 NPE P G Hulme, 196 Fromond Rd, Winchester, SO22 6ED [IO91HB, SU43]
G0 NPF D L Delacassa, 154 Midhurst Gdns, Hillingdon, Uxbridge, UB10 9DP [IO91SN, TQ08]
G0 NPG K O Heaviside, 7 Hertford Chase, Colton, Leeds, LS15 9EP [IO93GS, SE33]
G0 NPH Details withheld at licensee's request by SSL.
G0 NPI J Podvoiskis, 54 Stanwell Rd, Swinton, Manchester, M27 5TD [IO83TM, SD70]
G0 NPJ L Jackson, 60 East Park Ave, Darwen, BB3 2SQ [IO83SQ, SD62]
G0 NPK D V E Goulbourne, Widows Croft Farm, Hollingworth, Hyde, Ches, SK14 8LE [IO93AL, SK09]
GW0 NPL S W J Instone, 61 Llanfach Rd, Newbridge, Abercarn, Newport, NP1 5LA [IO81KP, ST29]
GW0 NPM H Thomas, 34 Upland Rd, Pontllanfraith, Blackwood, NP2 2ND [IO81JP, ST19]
G0 NPN B P Harris, 23 Pound Rd, Highworth, Swindon, SN6 7LA [IO91DP, SU29]
G0 NPO I D Brown, Egremont, Arterial Rd, Nevendon, Basildon, SS14 3JN [JO01FO, TQ79]
G0 NPP T E Watson, 22 Hollydene, Kibblesworth, Gateshead, NE11 0NR [IO94EW, NZ26]
G0 NPQ H Carruthers, 55 Inskip Terr, Gateshead, NE8 4AJ [IO94EW, NZ26]
GM0 NPS R Rankin, 276 Woodhall Ave, Coatbridge, ML5 5DF [IO75XU, NS76]
G0 NPU J E Marshall, 28 The Dovecote, Horsley, Derby, DE21 5BS [IO92GX, SK34]
G0 NPV B T Ham, 15 Laburnum Dr, Barnstaple, EX32 8PX [IO71XB, SS53]
G0 NPY P Yates, 31 Wallpark Cl, Brixham, Devon, TQ5 9UN [IO80FJ, SX95]
G0 NQA A R Gurbutt, 66C Watts Ln, Louth, LN11 9DG [JO03AI, TF38]
G0 NQB L Rogers, 18 Home Hill, Hextable, Swanley, BR8 7RR [JO01CJ, TQ57]
GI0 NQC A D Smith, 69 Antrim St., Lisburn, BT28 1AU [IO64XM, J26]
G0 NQE C B Wilkinson, 8 Westfield Ave, Knottingley, WF11 0JH [IO93JQ, SE42]
G0 NQF R F A de Almeida, 67 Ching Way, London, E4 8YE [IO91XO, TQ39]
G0 NQG S R Latham, 27 Rockside Gdns, Frampton Cotterell, Bristol, BS17 2HL [IO81SM, ST68]
G0 NQI J W Shepherd, 34 Lapwing Cl, Bradley Stoke, Bristol, BS12 0BJ [IO81RN, ST68]
G0 NQJ D Scaplehorn, 9 Stockwell Ave, Mangotsfield, Bristol, BS17 3DR [IO81SL, ST67]
G0 NQK R A Edwards, Manor Cottage, Manor Rd, Elmsett, Ipswich, IP7 6PN [JO02LC, TM04]
G0 NQN T D Fricker, 57 Wentworth Dr, Bishops Stortford, CM23 2PD [JO01BU, TL42]
G0 NQO Details withheld at licensee's request by SSL.
GM0 NQP C J Jenkins, 30 Cumbrae Dr, Tamfourhill, Falkirk, FK1 4AH [IO85CX, NS87]
GW0 NQQ S A Hudson, 87 Wordsworth Ave, Penarth, CF64 2RP [IO81JK, ST17]
G0 NQU K R Saagi, 1 Woodbridge Walk, Hollesley, Woodbridge, IP12 3LA [JO02RB, TM34]
G0 NQV D M Litchfield, 37 Graeme Rd, Enfield, EN1 3UU [IO91XP, TQ39]
G0 NQW D F Marshall, 15 Whisby Ct, Holton-le-Clay, Grimsby, DN36 5BG [IO93XM, TA20]
G0 NQX I P Carter, The School House, School Ln, Budock Water, Falmouth, Cornwall, TR11 5DT [IO70KD, SW73]
G0 NQY S L Seggar, 145 Mount View Rd, Norton Lees, Sheffield, S8 8PJ [IO93GI, SK38]
G0 NQZ Details withheld at licensee's request by SSL.
G0 NRA G W Lowe, 25 Manor House Ct, Kirkby in Ashfield, Nottingham, NG17 8LH [IO93IC, SK45]
G0 NRB R W Bellamy, 4 Wimbourne Walk, Corby, NN18 0BN [IO92PL, SP88]
G0 NRD R Bowmaker, 50 Balcaskie Rd, Eltham, London, SE9 1HQ [JO01AK, TQ47]
G0 NRE K Ellison, 4 Gainsborough Rd, Blackpool, FY1 4DZ [IO83LT, SD33]
G0 NRF G P Stilgoe, 74 McCarthy Way, Finchampstead, Wokingham, RG40 4UA [IO91NJ, SU76]
G0 NRG E C Young Northampton Ryn, 1 Ashton Rd, Roade, Northampton, NN7 2LF [IO92ND, SP75]
G0 NRI W J Hilton, 68 Roxborough Rd, Harrow, HA1 1PB [IO91TO, TQ18]
G0 NRJ R S M Croucher, 66 Loop Rd, Westfield, Woking, GU22 9BQ [IO91RH, TQ05]
G0 NRK J M Butler, 14 Fairfield Rd, Barnard Castle, DL12 8EB [IO94AN, NZ01]
G0 NRL Details withheld at licensee's request by SSL.
G0 NRM R F Stout, 7 Thornbridge Dr, Frecheville, Sheffield, S12 4YF [IO93HI, SK38]
G0 NRN G R Harrison, 14 Hardy Ave, South Ruislip, Ruislip, HA4 6SX [IO91TN, TQ18]
G0 NRO W R C Dimmock, 14 Alder Cl, Ash Vale, Aldershot, GU12 5QS [IO91PG, SU85]
G0 NRQ A W Dean, 9 School Hill, Chickerell, Weymouth, DT3 4BA [IO80SO, SY68]
G0 NRR D Fox, 20 Littlemead Ln, Exmouth, EX8 4RF [IO80HP, SY08]
GM0 NRT W D Cardno, 52 Salisbury Terr, Aberdeen, AB1 6QH [IO87TH, NJ72]
G0 NRU Dr R C Smith, 28 Thorney Green Rd, Stowupland, Stowmarket, IP14 4AB [JO02ME, TM05]
G0 NRX S A Godbold, 13 Dawn Cres, Beeding, Upper Beeding, Steyning, BN44 3WH [IO90UV, TQ11]
G0 NRZ A H Pill, 5 St. Leonards Cl, Upton St. Leonards, Gloucester, GL4 8AL [IO81VT, SO81]
G0 NSA T Brown, 5 St. Valentines Cl, Kettering, NN15 5EG [IO92PJ, SP87]
G0 NSB L S Bailey, 38 Gordon Hill, Enfield, EN2 0QP [IO91WP, TQ39]
G0 NSC W Thornton, 527 Rotherham Rd, Smithies, Barnsley, S71 1XB [IO93GN, SE30]
G0 NSE W Smith, 38 West Park Ave, Newby, Scarborough, YO12 6HH [IO94SG, TA08]
G0 NSF R J Payne, 117 Melody Rd, Biggin Hill, Westerham, TN16 3PL [JO01AH, TQ45]
GJ0 NSG P Crespel, 704 PO Box, Jersey, JE4 0PH
G0 NSH B R Piggott, 135 Daniells, Welwyn Garden City, AL7 1QT [IO91VT, TL21]
G0 NSI J H Man, 1 Holmbush Way, Southwick, Brighton, BN42 4YA [IO90VU, TQ20]
G0 NSL C B Russell, 163 Halton Rd, Runcorn, WA7 5RJ [IO83PI, SJ58]
G0 NSN M N Harrold, 37 Eastfield Rd, Dunston, Duston, Northampton, NN5 6TG [IO92MF, SP76]
G0 NSO T A Barfield, 91 Ollerton Rd, New Southgate, London, N11 2JY [IO91WO, TQ29]
G0 NSP B Teasdale, 18 Valley Forge, Washington, NE38 7JN [IO94FV, NZ35]
GW0 NSQ W Sawbridge, 4 Gordon Ave, Prestatyn, LL19 8RY [IO83HH, SJ08]
GW0 NSR A P Tuite, 44 Gorlan, Conwy, LL32 8RS [IO83BG, SH77]
G0 NST T A Wainwright, Dolphin Barn, Weatheroak Hill, Alvechurch, Birmingham, B48 7EA [IO92BI, SP07]
G0 NSU Details withheld at licensee's request by SSL.[Op: W G Reeve. Station located near Huntingdon.]
G0 NSW B C E Byrne, 83 Archer Rd, Redditch, B98 8DJ [IO92AH, SP06]
G0 NSX Details withheld at licensee's request by SSL.
G0 NSY Details withheld at licensee's request by SSL.
GW0 NSZ J Swinden, 8 Lon y Berllan, Abergele, LL22 7JF [IO83EG, SH97]
G0 NTA A H Jarvis, Willowmead, Nugents Park, Hatch End, Pinner, HA5 4RA [IO91TO, TQ19]
G0 NTB B C Jarvis, Willowmead, Nugents Park, Hatch End, Pinner, HA5 4RA [IO91TO, TQ19]
G0 NTC A Rybalka, 191 Sion Ave, Kidderminster, DY10 2YJ [IO82VJ, SO87]
G0 NTE Details withheld at licensee's request by SSL.
G0 NTF M J Nixon, Little Theobald, Sandy Cross, Heathfield, E Sussex, TN21 8BT [JO00DX, TQ52]
G0 NTG D R A Chawner, The Croft, 49 St. Anns Rd, Middlewich, CW10 9BY [IO83SE, SJ76]
G0 NTH A H P Gardner, 137A Castle Rd, Newport, PO30 1DP [IO90IQ, SZ48]
GM0 NTI L J Grieve, Elhanan, Myrtlefield Ln, Westhill, Inverness, IV1 2UE [IO77WL, NH74]
G0 NTJ A Williams, 48 Eskmont Ridge, London, SE19 3PZ [IO91WJ, TQ37]
G0 NTK Details withheld at licensee's request by SSL.
GM0 NTL R R B G Fraser, Hopefield Cottage, Gladsmuir, Tranent, East Lothian, EH33 2AL [IO85NW, NT47]
G0 NTM R A Matthews, 100 Horseferry Rd, London, E14 8DY [IO91XM, TQ38]
GW0 NTO Details withheld at licensee's request by SSL.
G0 NTP Details withheld at licensee's request by SSL.
G0 NTQ Details withheld at licensee's request by SSL.
G0 NTR J Harrison, 43 Churchfield Ct, Peterborough, PE4 6GB [IO92UO, TF10]
G0 NTT L D Lloyd, 8 Coastal Rise, Hest Bank, Lancaster, LA2 6HJ [IO84OC, SD46]
G0 NTV Details withheld at licensee's request by SSL.
GM0 NTW S J S Gray, 16 Barkerland Ave, Larchfield, Dumfries, DG1 4HR [IO85EB, NX97]
GW0 NTX Details withheld at licensee's request by SSL.
G0 NTZ A Burton, 20 Manor St., Evenwood, Bishop Auckland, DL14 9QB [IO94CO, NZ12]
G0 NUC Details withheld at licensee's request by SSL.
G0 NUD B S Bell, 74 Henderson Rd, Currock, Carlisle, CA2 4PZ [IO84MU, NY45]
G0 NUE P Hewitt, 26 Highfield Rd, North Thoresby, Grimsby, DN36 5RT [IO93XL, TF29]
G0 NUF A R Holroyd, 46 Montrose Ave, Lillington, Leamington Spa, CV32 7DY [IO92FH, SP36]
G0 NUG J V Hills, 20 Elderwood Cl, Eastbourne, E Sussex, BN22 0TL [JO00DT, TQ50]
G0 NUH M W M Darling, 132 Knowland, Highworth, Swindon, SN6 7NE [IO91DP, SU29]
GM0 NUI R W Honeyman, 81 Glen Ave, Largs, KA30 8RH [IO75NT, NS26]
G0 NUJ E R Usher, Crooms Hill Farm, Medstead, Alton, Hants, GU34 5LZ [IO91LD, SU63]
G0 NUL D S Forster, 15 Marlborough Park, Havant, PO9 2PP [IO90MU, SU70]
G0 NUM G A Rock, Old Paddocks, 17 Keswick Way, Verwood, BH31 6HP [IO90BU, SU00]
G0 NUN R S Barker, 5 Wickridge Cl, Uplands, Stroud, GL5 1ST [IO81VS, SO80]
G0 NUO G P Du Feu, 17 Oak Rd, Tavistock, PL19 9LJ [IO70WM, SX47]
G0 NUP K J Prince, 59 Chantry Rd, East Ayton, Scarborough, YO13 9ER [IO94SG, SE98]
GM0 NUQ R Handyside, 2 Davidson Quadrant, Hardgate, Clydebank, G81 6JL [IO75SW, NS47]
G0 NUR A Bushell, 121 Rickmansworth Rd, Watford, WD1 7JD [IO91TP, TQ09]
GW0 NUS G W Dyer, 15 Park Rd, Newbridge, Newport, NP1 4RE [IO81KQ, ST29]
G0 NUT D McKay, 43 Mordales Dr, Marske By The Sea, Redcar, TS11 7HT [IO94LO, NZ62]
G0 NUU R G Browne, Fox Cottage, Bukehorn Rd, Thorney, Peterborough, Cambs, PE6 0QG [IO92WP, TF20]
GW0 NUV A A Brigstocke, Pant-y-Saer, Bwlch, Tynygongl, Gwynedd, LL74 8RG [IO73VH, SH58]
G0 NUX C C Dakin, 40 Welland Croft, Bicester, OX6 8GD [IO91JV, SP52]
G0 NUZ L Wildman, 22 Berrys Wood, Newton Abbot, TQ12 1UP [IO80EM, SX87]
G0 NVA F A Stainsby, 11 Stonehouse Park, Thursby, Carlisle, CA5 6NS [IO84LU, NY35]
G0 NVC D A Hoppe, 354A Bourne Rd, Pode Hole, Spalding, Lincs, PE11 3LL [IO92VS, TF22]
G0 NVD J P Nothard, Ashmount, Fockerby, Garthorpe, Scunthorpe, DN17 4RZ [IO93PP, SE81]
G0 NVJ S K Winter, 203 Arnold Est, Druid St., London, SE1 2XR [IO91XL, TQ37]
G0 NVL Details withheld at licensee's request by SSL.
G0 NVM J K Chandler, 31 Henry Rd, Chelmsford, Essex, CM1 1RG [JO01FR, TL70]
GW0 NVN Details withheld at licensee's request by SSL.
G0 NVO P Oldham, 59 Wellspring Dale, Stapleford, Nottingham, NG9 7ET [IO92IW, SK43]
G0 NVP Details withheld at licensee's request by SSL.
GM0 NVQ I M Palmer, 72 Bellevue St., Edinburgh, EH7 4BY [IO85JX, NT27]
G0 NVR J D Morrison, 347 Heywood Rd, Prestwich, Manchester, M25 2RN [IO83UN, SD80]
G0 NVS M K Fletcher, 51 Greasley St., Bulwell, Nottingham, NG6 8NG [IO92JX, SK54]
G0 NVT P Boyle, 99 Heath Rd, Penketh, Warrington, WA5 2BY [IO83QJ, SJ58]
G0 NVU A McGrady, 68 Muston Rd, Filey, YO14 0AL [IO94UF, TA18]
G0 NVV G G W Price, 96 Moorpark Rd, Northfield, Birmingham, B31 4HE [IO92AJ, SP07]
G0 NVX C S Watts, 41 Salter St., Berkeley, GL13 9BU [IO81SQ, ST69]

G0 NVY P Hanson, 10 Parkfield Rd, Ruskington, Sleaford, NG34 9HS [IO93TB, TF05]
G0 NVZ J Welsh, 17 New House Park, St. Albans, AL1 1UA [IO91UR, TL10]
G0 NWB Details withheld at licensee's request by SSL.
G0 NWC Details withheld at licensee's request by SSL.
G0 NWE G Egan, 11 Shepherds Row, Castlefields, Runcorn, WA7 2LG [IO83PI, SJ58]
G0 NWF S A Webster, 31 Park Est, Shavington, Crewe, CW2 5AW [IO83SB, SJ75]
GI0 NWG A R Williamson, 23 Iskymeadow Rd, Armagh, BT60 3JS [IO64PG, H83]
G0 NWH E W Ogden North West Hants Raynet Group, 55 Cumberland Ave, Basingstoke, RG22 4BQ [IO91KF, SU65]
GM0 NWI A J Cunningham, 33 Broom Ct, St. Ninians, Stirling, FK7 7UL [IO86AC, NS89]
G0 NWJ G Blomeley, 13 Edale Gr, Sale, M33 4RG [IO83TJ, SJ79]
G0 NWM R W Armstrong Tynemouth ARC, 6 Barnstaple Rd, North Shields, NE29 8QA [IO95GA, NZ37]
GI0 NWN M Coyle, 67 Glen Rd, Londonderry, BT48 0BY [IO65HA, C41]
G0 NWQ S G Smith, South View, 29 Exeter Rd, Crediton, EX17 3BW [IO80ES, SX89]
GW0 NWR E K Shipton N.W.R.R.C., 34 Argoed Chester A, Kinmel Bay, Rhyl Clywd, LL18 5AY [IO83FH, SH98]
G0 NWS A Edwards, 5 Gr Cl, Watchet, TA23 0HN [IO81IE, ST04]
G0 NWT L C Nash Nnarg, Four Furlongs, Wells Rd, Stiffkey, Wells Next The Sea, NR23 1QE [JO02LW, TF94]
G0 NWU R R Fletcher, 50 Sleaford Rd, Boston, PE21 8EU [IO92XX, TF34]
G0 NWV D C Brown, 65 Warstones Dr, Penn, Wolverhampton, WV4 4PF [IO82VN, SO89]
GW0 NWW Details withheld at licensee's request by SSL.
G0 NWX D C Shiels, 66 Lime Tree Rd, Birmingham, B8 2XQ [IO92BL, SP18]
G0 NWY I D Peters, 52 Carnaby Rd, Darlington, DL1 4NS [IO94FM, NZ31]
G0 NXA Details withheld at licensee's request by SSL.
G0 NXB P E Flint, 53 Crowhurst Rd, Longbridge, Birmingham, B31 4PB [IO92AJ, SP07]
G0 NXC R Sillito, 25 Naisbett Ave, Peterlee, SR8 4BW [IO94IS, NZ44]
G0 NXD P R Brazenall, 118 Pr of Wles Ct, Eve Hill, Dudley, West Midlands, DY1 2TA [IO82WM, SO99]
G0 NXE F A Rogers, 45 Roundways, Ruislip, HA4 6EA [IO91TN, TQ08]
G0 NXF D Robinson, 5 Hazel Gr, Welton, Lincoln, LN2 3JX [IO93SH, TF07]
G0 NXG G A R Ballantyne, 1 Church Ln, Scredington, Sleaford, NG34 0AG [IO92TW, TF04]
G0 NXH P M Cunningham, 2 The Park, Mistley, Manningtree, CO11 2AL [JO01MW, TM13]
G0 NXI L W Edgecumbe, 51 Aller Park Rd, Aller Park, Newton Abbot, TQ12 4NH [IO80EM, SX86]
G0 NXJ D K Sullivan, 14737 Pickets Post Rd, Centreville Va 22020, U S A, X X
G0 NXK S Scott, 7 Lonsdale Pl, East Ayton, Scarborough, YO13 9HS [IO94SF, SE98]
G0 NXL B L Ewald, 61 Verney Rd, Dagenham, RM9 5LP [JO01BN, TQ48]
G0 NXM R A P Najman, 9 Bevin House, Alfred St., Bow, London, E3 2BB [IO91XM, TQ38]
G0 NXN B W Mitchell, 2 Mariners Ct, Great Wakering, Southend on Sea, SS3 0DR [JO01JN, TQ98]
GM0 NXO G Fyall, 105 St. Kilda Cres, Kirkcaldy, KY2 6DR [IO86JD, NT29]
G0 NXP R H L Thomson, 11 Cobham Gr, Whiteley, Fareham, PO15 7JQ [IO90IV, SU50]
G0 NXQ W J Love, 2 Longmead Cttgs, Milborne St. Andrew, Blandford Forum, DT11 0HU [IO80US, SY89]
G0 NXR Details withheld at licensee's request by SSL.
G0 NXS A K Ellis, Alston, Cumbria, CA9 3LZ [IO84TT, NY74]
G0 NXT S Platts, 15 Holywell Ave, Smisby Rd, Ashby de La Zouch, LE65 2HL [IO92GS, SK31]
G0 NXU R T Sudbury, 16 Beechwood Dr, Thornton Cleveleys, FY5 5EH [IO83LU, SD34]
GW0 NXW G Scanlin, 40 Rutland St., Grangetown, Cardiff, CF1 7TD [IO81JL, ST17]
G0 NXX J A Lynch, 14 The Pastures, Cayton, Scarborough, YO11 3UU [IO94TF, TA08]
GM0 NYB B E Byrne, 83 High St., Kirkcudbright, DG6 4JW [IO74XU, NX65]
G0 NYD J D Burrow, 36 Longfield Dr, Crag Bank, Carnforth, LA5 9EJ [IO84OC, SD47]
G0 NYE Details withheld at licensee's request by SSL.
G0 NYF W Anderson, 6 Hulton Dist Centr, Little Hulton, Worsley, Manchester, M28 6AU [IO83TM, SD70]
GJ0 NYG Details withheld at licensee's request by SSL.
G0 NYH J B Moseley, 42 Burford Rd, Chipping Norton, OX7 5DZ [IO91FW, SP32]
GI0 NYI W E J Benson, 6 Woodford Cl, Armagh, BT60 2DZ [IO64QH, H84]
G0 NYJ S Y Au, Lot 1374 Fuk Hi St, Yuen Long Indust Es, Yuen Long Nt, Hong Kong
G0 NYK J A Hoose, 91 Brevere Rd, Hedon, Hull, HU12 8LX [IO93VR, TA12]
G0 NYL B Jackson, 7 Ferriby Rd, Scunthorpe, DN17 2EQ [IO93QN, SE80]
G0 NYM M P Borer, 37 Broadway, Ripley, DE5 3LJ [IO93HB, SK45]
G0 NYN C J Price, 9 The Pastures, Repton, Derby, DE65 6GG [IO92FU, SK32]
GI0 NYO E Lowry, 10 The Gables, Moneynick Rd, Randalstown, Antrim, BT41 3JY [IO64UR, J08]
GM0 NYP N B Purtell, 31 Daleally Cres, Errol, Perth, PH2 7QA [IO86JJ, NO22]
G0 NYQ J Pape, Long Cast Cottage, 12 High Wiend, Appleby in Westmorland, CA16 6RD [IO84SN, NY62]
G0 NYR R J Cheetham, 8 Fairway, Huyton, Liverpool, L36 1UD [IO83OK, SJ49]
G0 NYS D H R Cox, 10 Calder Cl, Bollington, Macclesfield, SK10 5LJ [IO83WH, SJ97]
G0 NYT K W Bicknell, 12 Oak Dr, Kidlington, OX5 2HL [IO91IT, SP41]
G0 NYV P F Stanford, Camping Villasol, 03500 Benidorm, Alicante, Spain
G0 NYX Dr C G P J Wahlgren, 35 Prebend Mns S, Chiswick High Rd, London, W4 2LU [IO91VL, TQ27]
G0 NYY S A Nicholas, 144 Warwick Rd, Upper Edmonton, London, N18 1RT [IO91XO, TQ39]
G0 NYZ S Maloney, 34 Keswick Rd, Normanby, Middlesbrough, TS6 0BN [IO94KN, NZ51]
G0 NZA M D Lowe, 25 Manor House Ct, Kirkby in Ashfield, Nottingham, NG17 8LH [IO93IC, SK45]
G0 NZC Details withheld at licensee's request by SSL.
G0 NZE A G Benfield, 12 St. Marys Ct, Weald, Bampton, OX18 2HX [IO91FR, SP30]
G0 NZF L F Andrews, West Libbear, Shebbear, Beaworthy, Devon, EX21 5SZ [IO70VU, SS40]
G0 NZH G Edwards, 26 Kestrel Ave, Meir Heath, Stoke on Trent, ST3 7RD [IO82WX, SJ94]
G0 NZI C D Peake, 25 Omega Pl, Railway Terr, Rugby, CV21 3HW [IO92II, SP57]
G0 NZJ Rev P Forbes, 14 East St., Lilley, Luton, LU2 8LW [IO91TW, TL12]
G0 NZL N J Harding, 22 Knypersley Ave, Offerton, Stockport, SK2 5SR [IO83WJ, SJ98]
GM0 NZM C Spence, 108 Whinhall Ave, Airdrie, ML6 0HB [IO85AU, NS76]
GW0 NZN J H Smith, 6 Cherry Gr, Croespenmaen, Crumlin, Newport, NP1 4DF [IO81KQ, ST19]
G0 NZQ S J Brown, 20 Abbey Wood Rd, London, SE2 9NP [JO01BL, TQ47]
G0 NZR D A Catterall, 86 Broomfield Rd, Swanscombe, DA10 0LT [JO01DK, TQ67]
G0 NZS J F Ryan, 14 The Paddock, Upton, Wirral, L49 6NP [IO83KJ, SJ28]
G0 NZT D R Nash, 27 Sandling Ave, Horfield, Bristol, BS7 0HS [IO81RL, ST57]
G0 NZU R Blanning, 38 Northville Rd, Northville, Bristol, BS7 0RG [IO81RM, ST67]
G0 NZV D Jennings, 25 Bloom St., Edgeley, Stockport, SK3 9LA [IO83VJ, SJ88]
G0 NZX W G Griffith, 105 Kingston Hill, Kingston upon Thame, Surrey, KT2 7PZ [IO91UK, TQ17]
G0 OAB D G Griffith, 5 Upthorpe Dr, Wantage, OX12 7DF [IO91GO, SU48]
GM0 OAD D A Croft-Smith, 22Burns Cres, Irvine, Ayrshire, KA11 1AQ [IO75QP, NS34]
G0 OAE B A Franklin, 38 Sherwood Dr, Melton Mowbray, LE13 0LL [IO92NS, SK71]
G0 OAF J F Hackett, 67 Shropshire Ave, Chaddesden, Derby, DE21 6EW [IO92GW, SK33]
G0 OAG Details withheld at licensee's request by SSL.
G0 OAI E P Laing, 11 Leonard Ave, Nottingham, NG5 2LW [IO92KX, SK54]
G0 OAJ J C Morrice, 2 Pembroke Rd, Newquay, TR7 3HW [IO70LK, SW86]
G0 OAK Details withheld at licensee's request by SSL.
G0 OAM Details withheld at licensee's request by SSL.
G0 OAN Details withheld at licensee's request by SSL.
G0 OAO E M Jackson, Thule Farmhouse, Gidleigh, Chagford, Newton Abbot, Devon, TQ13 8HT [IO80BQ, SX68]
G0 OAP D K Coulson, 107 Fairfield Cres, Newhall, Swadlincote, DE11 0TB [IO92FS, SK22]
G0 OAQ Details withheld at licensee's request by SSL.
G0 OAR M M Ellis 1st Headley Sct, The Squirrels, Tower Rd, Hindhead, GU26 6SN [IO91PC, SU83]
G0 OAS N L Goddard, 15 Canada Rd, Cobham, KT11 2BB [IO91TH, TQ16]
G0 OAT R E G Petri, Tarnwood, Denesway, Meopham, Gravesend, DA13 0EA [JO01EJ, TQ66]
G0 OAU T J Pendleton S.N.A.R.C, 53 Ashby Rd, Kegworth, Leics, DE7 2DJ [IO92IX, SK44]
GM0 OAV T Roberts, Mosshill, Brora, Sutherland, Scotland, KW9 6NG [IO88BA, NC90]
G0 OAW W D Waldron, Redstone Farm, Germans Week, Beaworthy, Devon, EX21 5BQ [IO70VR, SX49]
G0 OAY J D McKenzie, 67 Crane Way, Whitton, Twickenham, TW2 7NH [IO91TK, TQ17]
G0 OAZ J C Fox, 3 The Plovers, Brighton Rd, Lancing, BN15 8LN [IO90UT, TQ10]
G0 OBA K C Barker, 4 Fort Widley Cttgs, Southwick Hill Rd, Cosham, Portsmouth, PO6 3EU [IO90LU, SU60]
GW0 OBB W I Evans, Brynawel, Cross Inn, Llanon, Dyfed, SY23 5NB [IO72WG, SN56]
G0 OBC C J Crabb, 19 Earls Dr, Clayton, Newcastle, ST5 3QR [IO82VX, SJ84]
G0 OBD E D Howarth, 30 Born Ct, New St., Ledbury, HR8 2DX [IO82SA, SO73]
G0 OBE J E Clarke, 16 Silver Birch Ave, Bedworth, Nuneaton, CV12 0AZ [IO92GL, SP38]
G0 OBF T E Hamilton, 21 Heath Rd, Middlestone Moor, Spennymoor, DL16 7DT [IO94EQ, NZ23]
G0 OBG A J Cox, 8 Spence Ave, Byfleet, West Byfleet, KT14 7TG [IO91SH, TQ06]
G0 OBH H W Cox, 30 Woodward Terr, Horns Cross, Greenhithe, DA9 9DD [JO01DK, TQ57]
G0 OBI A E V Wells, 6 Chichester Cl, Peacehaven, BN10 8TS [JO00AS, TQ40]
G0 OBJ G K Pratley, 7 Wytham Cl, Eynsham, Witney, OX8 1NS [IO91HS, SP40]
G0 OBK K M Warnes, 3 Bluebell Cl, Underwood, Hucknall, Nottingham, NG15 6TX [IO93JA, SK54]
G0 OBL D D Wellman, Mount Path Cottage, Pilgrims Way East, Otford, Sevenoaks, Kent, TN14 5RX [JO01CH, TQ55]
G0 OBM W F Noyce, 4 Norton Dr, Fareham, PO16 7PY [IO90JU, SU50]
G0 OBN C Hodgson, 12 Princess Rd, Seaham, SR7 7TB [IO94HU, NZ44]
G0 OBO K Weeks, 11 Sandwich Rd, Preston Grange, North Shields, NE29 9HT [IO95GA, NZ37]
G0 OBP K B Hudson, 2 Colwill Walk, Plymouth, PL6 8XF [IO70WJ, SX55]
G0 OBQ G M Lang, 63 Grosvenor Dr, Whitley Bay, NE26 2JR [IO95GA, NZ37]
G0 OBR Rev P McArdle Barlboro Hall ARC, St. Wilfred's Presbytery, 1 Winckley Sq, Preston, Lancs, PR1 3JJ [IO83PS, SD52]
G0 OBS Details withheld at licensee's request by SSL.
G0 OBT S Fort, 59 Coombe Dale, Seamills, Bristol, BS9 2JF [IO81QL, ST57]
G0 OBU J K Fallon, 32 Hunters Lodge, Preston Old Rd, Cherry Tree, Blackburn, BB2 5LX [IO83RR, SD62]
G0 OBW C A Bird, 3 Highlands Dr, Offerton, Stockport, SK2 5HX [IO83WJ, SJ98]
G0 OCB R J Dingle, 29 Castle View, Witton-le-Wear, Bishop Auckland, DL14 0DH [IO94DP, NZ12]
G0 OCC G Allen, 2 Haworth Dr, Orrell, Bootle, L20 6EJ [IO83ML, SJ39]
G0 OCD R Carroll, 7 Lambert Ct, Lamberts Beach, Mackay Queensland, Australia 4740
G0 OCG M Rodgers Ochil Hills Cn, 22 Harewood Rd, Allestree, Derby, DE22 2JN [IO92GW, SK33]
GM0 OCH Details withheld at licensee's request by SSL.
G0 OCI Details withheld at licensee's request by SSL.
G0 OCJ Details withheld at licensee's request by SSL.
G0 OCK B Pilkington, 3 Stockholm St., Burnley, BB11 5EB [IO83US, SD83]
G0 OCL F E Pilkington, 3 Stockholm St., Burnley, BB11 5EB [IO83US, SD83]
G0 OCP P J Hill, 54 Singleton, Sutton Hill, Telford, TF7 4JH [IO82SP, SJ70]
G0 OCR F D Batkin, 43 Belmont Rd, Penn, Wolverhampton, WV4 5UD [IO82WN, SO99]
G0 OCV R Smith, 289 New Rd, Staincross, Mapplewell, Barnsley, S75 6EP [IO93FO, SE31]
G0 OCW P A Brazier, 1 Ravenshore Cttgs, Holcombe Rd, Rossendale, BB4 4AN [IO83UQ, SD72]
G0 OCY P C Beeston, 100 Suffield Rd, High Wycombe, HP11 2JL [IO91OP, SU89]
GM0 ODB J G Kane, 24 Locherburn Pl, Houston, Johnstone, PA6 7NH [IO75RU, NS46]
G0 ODD L G Hutton Torbay (Raibc), Livewire Rg, 46 Penwill Way, Paignton, TQ4 5JQ [IO80FK, SX85]
G0 ODE D C Williams, 36 Greensted, Sawbridgeworth, Herts, CM2 9NY [JO01FR, TL70]
G0 ODF D R C Hayward, 172 Great Gregorie, Lee Chapel South, Basildon, SS16 5QF [JO01FN, TQ68]
G0 ODH D Hibberd, 24 Staveley Cl, Bucknall, Stoke on Trent, ST2 9PU [IO83WA, SJ94]
G0 ODI R J Sutton, 87 Downs Valley Rd, Woodingdean, Brighton, BN2 6RG [IO90XT, TQ30]
G0 ODJ Dr D J Tivey, 6 Newport Dr, Chichester, PO19 3QQ [IO90OU, SU80]
G0 ODK W C Everett, 120 Wantage Rd, Reading, RG3 2SF [IO91LJ, SU66]
G0 ODL Details withheld at licensee's request by SSL.
G0 ODM J Chomer, 9 Szold St, Ramat Hasharon, 47225, Israel
G0 ODN M C Hall, 31 Meendhurst Rd, Cinderford, GL14 2EF [IO81ST, SO61]
G0 ODO J G Keating, 26 Shelley Ave, Torquay, TQ1 4PF [IO80FL, SX96]
G0 ODQ J Hall, 1 Church Ln, Chinnor, OX9 4PW [IO91NQ, SP70]
G0 ODR C M Hendry, 16 Levishaw Cl, Buxton, Norwich, NR10 5HQ [JO02PS, TG22]
G0 ODS M J Treacher, 7 High Gr, Welwyn Garden City, AL8 7DW [IO91VT, TL21]
G0 ODT B Knight, 13 Southcote Rise, Ruislip, HA4 7LN [IO91SN, TQ08]
G0 ODU K E Petherick, 17 Castle Cl, Totternhoe, Dunstable, LU6 1QJ [IO91RV, SP92]
G0 ODV Details withheld at licensee's request by SSL.
GM0 ODW D Whitelaw, 12 Langour, Devonside, Tillicoultry, FK13 6JG [IO86DD, NS99]
G0 ODY G Fleming, 27 Crawthorne Cres, Deighton, Huddersfield, HD2 1LB [IO93CQ, SE11]
G0 OEA A E Moggridge, Outer Bailey, Kingsland, Leominster, Herefordshire, HR6 9QN [IO82OF, SO46]
G0 OEB C B Donald, Dexter House, 8 Greenway, Aldridge, Walsall, WS9 8XE [IO92BO, SK00]
G0 OED A C Mardo, 10 Meadow View, Uffculme, Cullompton, EX15 3DS [IO80IV, ST01]
G0 OEF Details withheld at licensee's request by SSL.
G0 OEG A J Shacklock, 14 Paparoa Rd, Howick, Aukland, New Zealand
GI0 OEH K Patterson, 8 Beechwood Gdns, Moira, Craigavon, BT67 0LB [IO64VL, J16]
G0 OEI M B Hopkins, 30 Commonside, Brownhills, Walsall, WS8 7AY [IO92AP, SK00]
G0 OEJ M Garbutt, 92 Owlet Rd, Windhill, Shipley, BD18 2LT [IO93CT, SE13]
G0 OEK D R Spooner, 60 St. Pauls Rd, Staines, TW18 3HH [IO91RK, TQ07]
G0 OEL T G Benton, 49 High St., Longstowe, Cambridge, CB3 7UN [IO92WE, TL35]
G0 OEN L R Davies Castle Hse Schl, Electronics Common Cl, 22 Pinewoods, Church Aston, Newport, TF10 9LN [IO82TS, SJ71]
GM0 OEO Details withheld at licensee's request by SSL.
G0 OEP Details withheld at licensee's request by SSL.
G0 OEQ J F Blichfeldt, Beacon Hall Farm Cottage, Beneden, Kent, TN17 4BT [JO01HB, TQ83]
G0 OER Revd P J Roberts, 13 Henleaze Ave, Henleaze, Bristol, BS9 4EU [IO81QL, ST57]
G0 OES D L Owen, 9 Cornfield Dr, Boley Park, Lichfield, WS14 9UG [IO92CQ, SK10]
GW0 OET F G Jacob, 5 Princess Louise Rd, Llwynypia, Tonypandy, CF40 2LY [IO81GP, SS99]
G0 OEU J E Whorton, 14 Burntwood Cres, Treeton, Rotherham, S60 5QF [IO93HJ, SK48]
G0 OEV J K Horrocks, 21 Vallian Croft, Birmingham, B36 8NH [IO92CM, SP18]
G0 OEW D Rooke, The Grange, 107 Wybunbury Rd, Willaston, Nantwich, CW5 7ER [IO83SB, SJ65]
G0 OEY A R Kerrison, 5 Seaview Ave, Little Oakley, Harwich, CO12 5JB [JO01OW, TM22]
G0 OEZ H E Chorley, 3 Ashmead Dr, Gotherington, Cheltenham, GL52 4ES [IO81XX, SO92]
G0 OFA R H E Dennis, 35 Woodleigh Rd, Barton, Newton Abbot, TQ12 1PN [IO80EM, SX87]
G0 OFC D J Townsend, 16 Church St., Long Buckby, Northampton, NN6 7QH [IO92LH, SP66]
G0 OFD J M Gilbert, 37 Riversway, Kings Lynn, PE30 2EE [JO02ES, TF62]
G0 OFE J R Smith, 61 Mallard Rd, Colehill, Wimborne, BH21 2NL [IO90AT, SU00]
G0 OFF S J Hipkin, 51 Kirby Rd, Walton on The Naze, CO14 8QZ [JO01PU, TM22]
GW0 OFH S M Williams, 5 Brynmelyn Ave, Llanerch, Llanelli, SA15 3RU [IO71WQ, SN50]
G0 OFJ Details withheld at licensee's request by SSL.[Station located near Farringdon. Locator: IO91EP. WAB: SU29. Op: Gerry]
GM0 OFL J Wilkie, Rosebank, Ladybank Rd, Pitlessie, Cupar, KY1 7SP [IO86KD, NT29]
GM0 OFM J D Park, Cramond, 12 Monkstown, Ladybank, Cupar, KY1 7JX [IO86KD, NT29]
G0 OFN I R Clabon, 14 Melrose Ave, Twickenham, TW2 7JE [IO91TK, TQ17]
G0 OFO F J W Deakin, 8 Chapel Ct, Sherburn, Durham, DH6 1HS [IO94FS, NZ34]
G0 OFP D J Keane, 68 The Warren, Holbury, Southampton, SO45 2QD [IO90HU, SU40]
G0 OFR G Borrowdale, 30 Barton View, Penrith, CA11 8AX [IO84PQ, NY53]
G0 OFS J R Adams, 96 Henley Rd, Springbank, Cheltenham, GL51 0PD [IO81WV, SO92]
G0 OFT S E Duncan, 10 Huntingdon Rise, Bradford on Avon, BA15 1RJ [IO81UI, ST86]
G0 OFW P A Lightfoot, 18 Fields Cl, Alsager, Stoke on Trent, ST7 2ND [IO83UC, SJ85]
G0 OFX F A Rawlins, 12 Arundel Rd, Eastleigh, SO50 4PQ [IO90HX, SU42]
G0 OFY J E Wane, 26 Coniston Ave, Euxton, Chorley, PR7 6NY [IO83PP, SD51]
G0 OGA A J Bevis, 16 Chapel River Cl, Weyhill Gdns, Andover, SP10 3UE [IO91GE, SU34]
G0 OGB J Wallis, 10 Middlewood Rd, Lanchester, Durham, DH7 0HL [IO94DT, NZ14]
GM0 OGC H A W Nelson, 39 Millfield, Cupar, KY15 5UU [IO86LH, NO31]
G0 OGD G A Nattrass, 24 Ritsons Rd, Black Hill, Consett, DH8 0AW [IO94BU, NZ05]
GW0 OGE M Free, 38 Waterside Cl, Quedgeley, Gloucester, Glos, GL2 4LF [IO81UU, SO81]
G0 OGG J Brown, 1 Jackson Rd, Houghton, Carlisle, CA3 0NW [IO84MW, NY45]
GW0 OGI D T Keely, Pensarn Cottage, Bryn Du, Ty Croes, Anglesey, LL63 5SH [IO73SF, SH37]
G0 OGJ K E Marshall, 5 Wilson Cres, Lostock Gralam, Northwich, CW9 7QH [IO83SG, SJ67]
GW0 OGK N Shaw, 9 Llys Edward, Towyn, Abergele, LL22 9NY [IO83FH, SH97]
G0 OGL S J Glanville, 3 Seneschal Rd, Cheylesmore, Coventry, CV3 5LF [IO92GJ, SP37]
G0 OGM S Bowerman, 24 Colston Bassett, Emerson Valley, Milton Keynes, MK4 2BA [IO92OA, SP83]
G0 OGN R G Hall, 10 Chapel St., Stratford upon Avon, CV37 6EP [IO92DE, SP25]
G0 OGP Y J Powell, 18 Carrington Rd, Stockport, SK1 2QE [IO83WJ, SJ99]
G0 OGS S J Malpass, 27 Vale St., Upper Gornal, Dudley, DY3 3XD [IO82WM, SO99]
G0 OGU Details withheld at licensee's request by SSL.
G0 OGV S R Moore, 34 Fishponds Rd, Kenilworth, CV8 1EZ [IO92EI, SP27]
G0 OGW C B Douglas, 12 Greenglades, West Hunsbury, Northampton, NN4 9YW [IO92MF, SP75]
G0 OGX J N Pennington, Flat, 5 Craddock Ct, Bodenham Rd, Hereford, HR1 2TS [IO82PB, SO54]
G0 OGY J R Evans, Rose Cottage, Slade Ln, Thornton Hough, Wirral, L63 4LB [IO83LI, SJ38]
GM0 OGZ R F Goodall, 3 Croft Crunie Cttgs, Muir of Ord, Ross-shire, IV6 7SB [IO77UN, NH65]
G0 OHA A White, 42 Rosedale Ave, Black Hill, Consett, DH8 0DZ [IO94BU, NZ05]
G0 OHC Details withheld at licensee's request by SSL.
G0 OHE Details withheld at licensee's request by SSL.
G0 OHF F H Wilson, 3 Foundry Mews, Burgh-le-Marsh, Skegness, PE24 5HQ [JO03DD, TF56]
GI0 OHG E Bennett, 31 Fernisky Park, Kells, Ballymena, BT42 3LL [IO64VT, J19]
GM0 OHH Details withheld at licensee's request by SSL.
G0 OHI Details withheld at licensee's request by SSL.
GW0 OHJ D J Workman, 4 Rhuddland Rd, St Mathews Park, Buckley, CH7 3QA [IO83LE, SJ26]
G0 OHK N D King, 7 Fountains Cl, Washington, NE38 7TA [IO94FV, NZ35]
G0 OHL R S Dale, 80 Lee Rd, Dovercourt, Harwich, CO12 3SB [JO01PW, TM23]
G0 OHN M L Lawson, Catchwater Meadow, Orby Rd, Burgh-le-Marsh, Skegness, PE24 5JD [JO03CE, TF46]
G0 OHQ R J Bunyan, 34 Stafford Rd, Sidcup, DA14 6PU [JO01BK, TQ47]
G0 OHR J T Armitage, 2 Sheffield Rd, Birdwell, Barnsley, S70 5UZ [IO93GM, SE30]
GI0 OHT W C Stanley, 95 Bangor Rd, Newtownards, BT23 7BZ [IO74DN, J46]
G0 OHV F G Shubert, Hill Top House, Manor Rd, Hagworthingham, Spilsby, PE23 4LL [JO03AE, TF36]
G0 OHW J Vasek, 20 Westhall Rd, Kew, Richmond, Surrey, TW9 4EE [IO91UL, TQ17]
G0 OHY A Mather, 15 Stanley Rd, Walkden, Worsley, Manchester, M28 3DT [IO83TM, SD70]
G0 OHZ Details withheld at licensee's request by SSL.
G0 OIA H A Taylor, 5 Lea Vale, Crayford, Dartford, DA1 4DL [JO01CK, TQ57][PO Box 127, Dartford, Kent, DA2 7LU.]
G0 OIB T C Harris, 10 Firle Rd, Peacehaven, BN10 8DD [JO00AT, TQ40]
G0 OID F J Tett, 36 Farhalls Cres, Horsham, RH12 4DA [IO91UB, TQ13]
G0 OIE M C Gathergood, 54 Robin Ln, Bentham, Lancaster, LA2 7AG [IO84RC, SD66]
G0 OIF D V Read, 12 York Rd, Winchmore Hill, London, N21 2JL [IO91XP, TQ39]
G0 OII R A R Pullen, No1 Ridings Ct, Crown Cres, South Cliff, Scarborough, North Yorks, YO11 2BJ [IO94TG, TA08]
G0 OIK P G King, Chad Ln Farm, Flamstead, St Albans, Herts, AL3 8HW [IO91ST, TL01]
G0 OIL Details withheld at licensee's request by SSL.
G0 OIM B Turnbull, 56 Bryans Cl Rd, Calne, SN11 9AB [IO81XK, ST97]
G0 OIN A G Fairey, Four Ways, St. Johns Rd, New Romney, TN28 8EW [JO00LX, TR02]
G0 OIO J N Fuller, 13 Lucastes Ln, Haywards Heath, RH16 1LB [IO91WA, TQ32]

G0 OIP L Crosby, 2 Hurford Ave, Great Sutton, South Wirral, L65 7AY [IO83MG, SJ37]
G0 OIQ A L Welland, Sarnia, Stroud Farm Rd, Holyport, Maidenhead, SL6 2LH [IO91PL, SU87]
G0 OIR G J Wicks, 28 Old School Ln, Milton, Cambridge, CB4 6BS [JO02BF, TL46]
G0 OIS T Smith, Hembury House, Sellars Rd, Hardwicke, Gloucester, GL2 4QD [IO81UT, SO71]
G0 OIT S Parry, 71 Claremont Rd, Wavertree, Liverpool, L15 3HJ [IO83MJ, SJ38]
G0 OIU A Smith Jones, 16 Armley Rd, Liverpool, L4 2UN [IO83MK, SJ39]
G0 OIV A Sait, 124 Dicksons Dr, Newton, Chester, CH2 2BX [IO83NE, SJ46]
G0 OIW M P T Palmer, 28 Westfield Rd, Caversham, Reading, RG4 8HH [IO91ML, SU77]
G0 OIY J C Smith, 59 Charlecote Dr, Dudley, DY1 2GG [IO82WL, SO98]
G0 OIZ J G Purcell, 5 Mallard Way, Hickling, Norwich, NR12 0YU [JO02SS, TG42]
G0 OJA Details withheld at licensee's request by SSL.
G0 OJB L M Chadwick, 97 Althorpe Rd, Luton, Beds, LU3 1JX [IO91SV, TL02]
GW0 OJC Details withheld at licensee's request by SSL.
GI0 OJD Details withheld at licensee's request by SSL.
G0 OJE Details withheld at licensee's request by SSL.
G0 OJH Details withheld at licensee's request by SSL.
GW0 OJI Details withheld at licensee's request by SSL.
G0 OJJ A W Green, 23 Cobbs Pl, Great Yarmouth, NR30 2EE [JO02UO, TG50]
GW0 OJK J M Kerslake, 44 York Rd, Weybridge, KT13 9DX [IO91SI, TQ06]
G0 OJN J D Smith, 34 Rappart Rd, Wallasey, L44 6QE [IO83LJ, SJ39]
G0 OJP R J Melton, 5 Nursery Way, Pott Row, Grimston, Kings Lynn, PE32 1DQ [JO02GS, TF72]
G0 OJR B J Fox, 156 Hawksford Cres, Low Hill, Bushbury, Wolverhampton, WV10 9SN [IO82WO, SJ90]
G0 OJS S I John, 40 Elizabeth Ave, Brixham, TQ5 0AY [IO80FJ, SX95]
G0 OJT W L Jenkinson, 7 Moortown Rd, Watford, WD1 6JH [IO91TO, TQ19]
G0 OJU M Goddard, Selsted Garage, Selsted, near Dover, Kent, CT15 7HJ [JO01OD, TR24]
G0 OJV Details withheld at licensee's request by SSL.
G0 OJW J Taylor, c/o Glenifer, Homer Park, Hooe, Plymouth, PL9 9NN [IO70WI, SX55]
G0 OJX P D Williams, 7 The Greebys, Paignton, TQ3 3DN [IO80FK, SX86]
G0 OJY A C Hurt, 8 Lime Cl, Ware, SG12 7ND [IO91XT, TL31]
G0 OJZ Details withheld at licensee's request by SSL.
G0 OKA D P Martin, 67 Mill St., Torrington, EX38 8AL [IO70WW, SS41]
G0 OKB Details withheld at licensee's request by SSL.
G0 OKC J M Venton, 15 Trehaverne Terr, Truro, Cornwall, TR1 3SE [IO70LG, SW84]
G0 OKD R M Bradley, 42 The Croft, South Normanton, Alfreton, DE55 2BU [IO93HC, SK45]
G0 OKE Details withheld at licensee's request by SSL.
G0 OKF S J Bolam, 100 Bushfield Rd, Scunthorpe, DN16 1NA [IO93QN, SE81]
G0 OKH E F Bridgeman, 8 Rue de La Chaussee, Penze 29670, Taule, France
G0 OKI R J Morris, 4 Greenway Gdns, Kings Norton, Birmingham, B38 9RY [IO92AJ, SP07]
GM0 OKJ J Fraser, 2 Barra Pl, Stenhousemuir, Larbert, FK5 4UF [IO86CA, NS88]
G0 OKK B E G Crowe-Haylett, 13 Lynton Cl, Ely, CB6 1DJ [JO02DJ, TL58]
G0 OKL J Collins, Delver House, Reach, Cambridge, CB5 0JF [JO02DG, TL56]
G0 OKN R A Maloney, Rosewell Kents, Jacobstow, Bude, Cornwall, EX23 0BN [IO70RR, SX19]
G0 OKO H A Jones, Fair View, The Green, Lower Brailes, Banbury, OX15 5HZ [IO92FB, SP33]
GI0 OKQ S Ferguson, 4 Riverdale, Tamnamore Rd, Dungannon, BT71 6PZ [IO64QL, H86]
G0 OKR D R Edisbury, 6 Orchard Dr, Handforth, Wilmslow, SK9 3BL [IO83VI, SJ88]
GM0 OKS Details withheld at licensee's request by SSL.
G0 OKT D B White, 122 Griffiths Dr, Ashmore Park, Wednesfield, Wolverhampton, WV11 2JW [IO82XO, SJ90]
GI0 OKU C V Foote, 4 Bushfield Rd, Moira, Craigavon, BT67 0JB [IO64WM, J16]
G0 OKV K Cowell, 111 Fullingdale Rd, The Headlands, Northampton, NN3 2PZ [IO92NG, SP76]
G0 OKW Details withheld at licensee's request by SSL.
G0 OKX Details withheld at licensee's request by SSL.
G0 OKY I D Wye, New House, Hook Rd, Amcotts, Scunthorpe, DN17 4AZ [IO93PO, SE81]
G0 OKZ J T Thorpe, Four Jays, 46A High St., Misterton, Doncaster, DN10 4BU [IO93NK, SK79]
G0 OLB Details withheld at licensee's request by SSL.
G0 OLC Details withheld at licensee's request by SSL.
GM0 OLD D M McLaren, 53 Alder Rd, Milton of Campsie, Glasgow, G65 8JA [IO75WW, NS67]
G0 OLE D A Lockwood obo Boothferry ARC, 14 Clifton Gdns, Goole, DN14 6AS [IO93NQ, SE72]
GM0 OLF D M Phillips, East Grange Steading House, Inverarity, By Forfar, DD8 2JN [IO86NO, NO44]
G0 OLG E T Lambourne, 206 Brownedge Rd, Lostock Hall, Preston, PR5 5AJ [IO83PR, SD52]
G0 OLH Details withheld at licensee's request by SSL.
G0 OLI Details withheld at licensee's request by SSL.
G0 OLK Details withheld at licensee's request by SSL.
G0 OLL E Platts, 38 Swanbourne Rd, Sheffield, S5 7TL [IO93GK, SK39]
G0 OLM G S Gardiner, 8 Elm Gr, Nayland, Colchester, CO6 4LL [JO01KX, TL93]
GW0 OLN G A Clement, 16 Oakwood Rd, Brynmill, Swansea, SA2 0DN [IO81AO, SS69]
G0 OLO D Collinson, 20 Carlisle Cres, Penshaw, Houghton-le-Spring, DH4 7RD [IO94GU, NZ35]
GW0 OLP C P Roberts, Brookside, 59 Bertha Rd, Margam, Port Talbot, SA13 2AP [IO81CN, SS78]
G0 OLR L S Roberts, Rose Cottage, Middleham, Leyburn, North Yorks, DL8 4QN [IO94CG, SE18]
G0 OLS T M Humphries, 23 Sycamore Dr, Lutterworth, LE17 4TR [IO92JL, SP58]
G0 OLT L Tringale, 19 Lysander Rd, Kings Hill, West Malling, Kent, ME19 4TT [JO01EG, TQ65]
G0 OLX D H Stanton, Tropi Wotsit, 8 Thurnham Way, Tadworth, KT20 5PR [IO91VH, TQ25]
G0 OLY Details withheld at licensee's request by SSL.
GW0 OLZ G L Smith, 51 St. Cadocs Rd, Trevethin, Pontypool, NP4 8JW [IO81LR, SO20]
G0 OMB B Walker, 46 Frostoms Rd, Workington, CA14 3UR [IO84FP, NX92]
GM0 OMC C E Cook, Briarwood, 95 Old Edinburgh Rd, Inverness, IV2 3HT [IO77VL, NH64]
G0 OMD A G Gilbert, 19 Farrs Ave, Andover, SP10 2AH [IO91GE, SU34]
G0 OME D E Rawlinson, Southdene, Holbeach Drove, South Lincs, PE12 0PS [IO92XQ, TF31]
G0 OMF D Hupton, 90 Warwick Rd, Atherton, Manchester, M46 9PQ [IO83SM, SD60]
G0 OMG N Moody, 34 Burns Ln, Warsop, Mansfield, NG20 0PG [IO93KE, SK56]
G0 OMH P Burbeck, 5 Wouldham Terr, Saxville Rd, St. Pauls Cray, Orpington, BR5 3AT [JO01BJ, TQ46]
G0 OMI J D Richards, 35 Briar Edge, Burn Cottage, Forest Hall, Newcastle upon Tyne, NE12 0JN [IO95FA, NZ26]
G0 OMJ Dr P S Schein, Us Bioscience, 12 The Courtyards, Hatters Ln, Watford, WD1 8YH [IO91SP, TQ09]
G0 OMK Details withheld at licensee's request by SSL.
G0 OMM S T Adams, 8A Tovil Green, Maidstone, ME15 6RJ [JO01GG, TQ75]
G0 OMN G I Charman, 19 Welland Rd, Worthing, BN13 3LN [IO90TU, TQ10]
G0 OMP Details withheld at licensee's request by SSL.
G0 OMR Details withheld at licensee's request by SSL.
G0 OMS Details withheld at licensee's request by SSL.
G0 OMT T Bailey, 21 Gargrave Pl, Lupset, Wakefield, WF2 8AR [IO93FQ, SE32]
G0 OMU R J Dack, Poplar View Cottage, Low Moor Side Ln, New Farnley, Leeds, West Yorks, LS12 5HU [IO93ES, SE23]
GM0 OMV Details withheld at licensee's request by SSL.
G0 OMX J W Boddy, 1 Graffham Cl, Chichester, PO19 4AW [IO90OU, SU80]
G0 OMY J H Maclagan-Wedderburn Harpenden ARC, Aldwickbury School, Wheathampstead Rd, Harpenden, AL5 1AD [IO91UT, TL11]
G0 OMZ R T Lomas, 7 Chaunterell Way, Abingdon, OX14 5PP [IO91IP, SU49]
G0 ONA P Nicholls, 101 Rochester Ave, Feltham, TW13 4EF [IO91SK, TQ07]
GM0 ONB N D Shaxted Csmt Group, Viewbank Cottage, Shieldhill Rd, Reddingmuirhead, Falkirk, FK2 0DU [IO85DX, NS97]
G0 ONC Details withheld at licensee's request by SSL.
GI0 OND J Lappin, 46 Grange Rd, Ballytrue, Kilmore, Armagh, BT61 8NX [IO64RK, H95]
G0 ONE F R E D Marland, 116 Boyds Walk, Dukinfield, SK16 4AU [IO83XL, SJ99]
G0 ONF V B Szendzielarz, 5 Granville Rd, Urmston, Manchester, M41 0XY [IO83TK, SJ79]
G0 ONG J R Mobbs, 5 Distaff Rd, Poynton, Stockport, SK12 1HN [IO83WI, SJ98]
G0 ONH B Fellows, 40 Highfield Cres, Colley Gate, Halesowen, B63 2BE [IO82XL, SO98]
G0 ONJ E R Howell, Guildford Magnetic Imaging, Egerton Rd, Guildford, Surrey, GU2 5RG [IO91QF, SU94]
G0 ONK Details withheld at licensee's request by SSL.
G0 ONM P N Morgan Northease Manor, School Rc, 12 Victoria Cl, Burgess Hill, RH15 9QS [IO90WW, TQ31]
GM0 ONN I D Barnetson, 18 Ardivot Pl, Coulardbank, Lossiemouth, IV31 6TE [IO87IR, NJ27]
G0 ONO R Horsfield, 223 Middlewich Rd, Northwich, CW9 7DN [IO83SG, SJ67]
G0 ONQ C N Foster, 2 Marlow Ave, Upton, Chester, CH2 1NQ [IO83NF, SJ46]
G0 ONR S C Smith, 9 Jennifer Rd, Bromley, BR1 5LP [JO01AK, TQ37]
G0 ONS J E P Chinnery, 31 Kingsway, Kingsthorpe, Northampton, NN2 8HD [IO92NG, SP76]
G0 ONT R J Broom, 379 Exeter Rd, Exmouth, EX8 3NS [IO80HP, SY08]
GW0 ONU D P Harris, 2 Sheppard St., Pwllgwaun, Pontypridd, CF37 1HT [IO81HO, ST09]
GM0 ONV D J Macliver, 60 Neilshill Caravan Park, Mossblown, Ayr, KA6 5AU [IO75RM, NS42]
G0 ONW N F Lees, 49 Flansham Park, Bognor Regis, PO22 6QH [IO90QT, SU90]
GM0 ONX L J Paget, 40 Davaar Dr, Kilmarnock, KA3 2JG [IO75SP, NS44]
GW0 ONY J Edwards, Hafan Deg, Bryngwran, Holyhead, Gwynedd, LL65 3PL [IO73SG, SH37]
G0 OOB D E Walpole, Lahore, Damgate Ln, Acle, Norwich Norfolk, NR13 3DH [JO02SP, TG40]
G0 OOC D G Still, 25 Brooks Green Park, Emms Ln, Brooks Green, Horsham, RH13 8QR [IO91TA, TQ12]
G0 OOD T C Chapman, 13 Bolingbroke Rd, Norwich, NR3 2SJ [JO02PP, TG21]
G0 OOF R J S Williams, Dyffryn Coed, 25 Peghouse Rise, Stroud, GL5 1RU [IO81VS, SO80]
G0 OOG D D Gorman, Flat 6, 12 Queens Gate Terr, South Kensington, London, SW7 5PF [IO91VL, TQ27]
G0 OOH Details withheld at licensee's request by SSL.
G0 OOI W S Humphries, 76 Mortlake Rd, Kew, Richmond, TW9 4AS [IO91UL, TQ17]
G0 OOJ R D Bennett, Thorncroft, 8 Eddeys Ln, Headley Down, Bordon, GU35 8HU [IO91OC, SU83]
G0 OON P Healey, 10 Wroxham Rd, Great Sankey, Warrington, WA5 3EE [IO83QJ, SJ58]
G0 OOO R Clayton Scarborough SEG, 9 Green Island, Irton, Scarborough, YO12 4RN [IO94SF, TA08][Op: Roy Clayton, obo Scarborough Special Events Group. ILA: 740, FISTS: 1000, WAB: 11000, TA08. QSLs for special GB stations forwarded direct, or via bureau. SWL reports welcomed.]
G0 OOP T Kier, 9 Newbridge Way, Truro, TR1 3LX [IO70KG, SW84]
G0 OOR A B Jex, 7 Wilby Rd, Norwich, NR1 2NJ [JO02PO, TG20]
G0 OOS L Marobin, 60 Tudor Ct, King Henrys Walk, London, N1 4NU [IO91XN, TQ38]
G0 OOT J K Zervas, 8 Priory Cl, Abbots Park, Chester, CH1 4BX [IO83NE, SJ46]
G0 OOU R E Field, 34 Piltdown Cl, Hastings, TN34 1UU [JO00GU, TQ81]
G0 OOV R P Johnson, 17 Cedar Rd, Kettering, NN16 9PU [IO92PJ, SP87]
G0 OOW Dr D I W Phillips, 48 Welbeck Ave, Highfield, Southampton, SO17 1SS [IO90HW, SU41]
G0 OOX N V Davies, 38 Towerscroft Ave, St. Leonards on Sea, TN37 7JB [JO00GV, TQ81]
G0 OOY Details withheld at licensee's request by SSL.
G0 OPA Details withheld at licensee's request by SSL.
G0 OPB A J Canning, 261 Loddon Bridge Rd, Woodley, Reading, RG5 4BL [IO91NK, SU77]
G0 OPC M Marriott, Greenfield View, March Rd, Friday Bridge, Wisbech, PE14 0HA [JO02BO, TF40]
G0 OPD A F Clark, 20 Middle Mead, Fareham, PO14 3EG [IO90JU, SU50]
G0 OPE C Wright, 60 Brindley Cl, Sheffield, S8 8PX [IO93GI, SK38]
G0 OPF B W Akehurst, 5 Charltons Way, Tunbridge Wells, TN4 8JS [JO01CC, TQ53]
G0 OPG C R V Knowlson, 28 Hill Dr, Handforth, Wilmslow, SK9 3AR [IO83VI, SJ88]
G0 OPH I P Heptinstall, 16 Fulmar Rd, Stockton on Tees, TS20 1SL [IO94IO, NZ42]
G0 OPI A C Bennett, 32 Gainsborough Rd, Bournemouth, BH7 7BD [IO90BR, SZ19]
GM0 OPJ S Weir Macpac, 19 Ellismuir Rd, Baillieston, Glasgow, G69 7HW [IO75WU, NS66]
G0 OPL W S Cowell, 21 Elm Way, Trench, Telford, TF2 6RS [IO82SR, SJ61]
G0 OPM G B Melia, 1 Marlborough Ave, Warton, Preston, PR4 1BP [IO83NS, SD42]
GW0 OPP R H Owens, 62 Ty Llwyd Parc Est, Quakers Yard, Treharris, CF46 5LB [IO81IP, ST19]
G0 OPQ A Stocks, 3 Limestone Way, Burniston, Scarborough, YO13 0DQ [IO94SH, TA09]
G0 OPR A C Stone, 22 Meadowvale, Norwich, NR5 0NJ [JO02OP, TG11]
GM0 OPS J M Dundas, 103 Rockmount Ave, Thornliebank, Glasgow, G46 7DP [IO75UT, NS55]
G0 OPU A J R Heatley, 25 Watson Gr, Norwich, NR2 4LF [JO02PP, TG20]
G0 OPV R K Heatley, 23 The Waterside, Hellesdon, Norwich, Norfolk, NR6 5QN [JO02PQ, TG21]
G0 OPW Details withheld at licensee's request by SSL.
GM0 OPX D G McFerran, Ardlair, Beanshill, Milltimber, Aberdeen, AB1 0ER [IO87TH, NJ72]
GW0 OPY D Griffith, Craig Artro, Llanbedr, Gwynedd, LL45 2LU [IO72XT, SH52]
G0 OPZ D Nicholson Cent Yorks Rsf, 25B Glenholme Rd, Farsley, Pudsey, LS28 5BY [IO93DT, SE23]
G0 OQB Details withheld at licensee's request by SSL.
G0 OQC M J Deeley, PO Box 172, Black Rock, Victoria, Australia 3193
G0 OQD A J Porter, Kinross, 12 Brooklands Rd, Bletchley, Milton Keynes, MK2 2RN [IO91PX, SP83]
G0 OQE F J Porter, Kinross, 12 Brooklands Rd, Bletchley, Milton Keynes, MK2 2RN [IO91PX, SP83]
G0 OQG Dr H O Hughes, 4 Wood Ln, Scarcliffe, Chesterfield, S44 6TF [IO93IF, SK46]
GW0 OQH W E Davies, 17 Mount Pleasant Sq, Ebbw Vale, NP3 6LF [IO81JS, SO11]
G0 OQI K W Zak, 33 Greenfield Rd, Spinney Hill, Northampton, NN3 2LJ [IO92NG, SP76]
G0 OQJ B R Allt, 19 Watkiss Dr, Rugeley, WS15 2PN [IO92AS, SK01]
G0 OQK N J Garrod, 23 Pynchester Cl, Uxbridge, Middx, UB10 8JY [IO91SN, TQ08]
G0 OQM Details withheld at licensee's request by SSL.
GW0 OQN G L Nash, 95 Newport Rd, Cwmcarn, Cross Keys, Newport, NP1 7LY [IO81KP, ST29]
G0 OQO Details withheld at licensee's request by SSL.
G0 OQP A Caton, 20 Lower Oxford Rd, Basford, Newcastle, ST5 0PB [IO83VA, SJ84]
G0 OQQ B Wood, 52 Ashfield Ave, Rylands, Beeston, Nottingham, NG9 1PY [IO92JW, SK53]
G0 OQR A W Glen, 70 Moscow Rd East, Edgeley, Stockport, SK3 9QL [IO83WJ, SJ88]
G0 OQS N S Dean, 13 St. Marys Ave, Billinge, Wigan, WN5 7QL [IO83PL, SJ59]
G0 OQT M W Jones, 20 Winchester Rd, Burnham on Sea, TA8 1HY [IO81MF, ST34]
G0 OQU Details withheld at licensee's request by SSL.
G0 OQW Details withheld at licensee's request by SSL.
G0 OQX J East, 29 Hiskins, Wantage, OX12 9HU [IO91GO, SU38]
G0 OQY J H Hyde, 27 Malvern Ct, Whetstone Gr, Bushbury, Wolverhampton, WV10 9TL [IO82WO, SJ90]
G0 OQZ H G Dawson, 6 Maer Top Way, Barnstaple, EX31 1RZ [IO71XC, SS53]
G0 ORA Details withheld at licensee's request by SSL.
G0 ORC V L Shirley, 160 Over Ln, Belper, DE56 0HN [IO93GA, SK34]
G0 ORD Dr E N Chantler, 45 Longhurst Ln, Marple Bridge, Stockport, SK6 5AE [IO83XJ, SJ98]
G0 ORE N A Reddish, 15 Drakes Cl, Redditch, B97 5NG [IO92AG, SP06]
G0 ORG N F Robertson, 21 Battles Ln, Kesgrave, Ipswich, IP5 7XF [JO02OB, TM24]
G0 ORH K J Chandler, 4 Park Ave, Thatcham, RG18 4NP [IO91IJ, SU56]
G0 ORI Details withheld at licensee's request by SSL.
G0 ORJ Details withheld at licensee's request by SSL.
G0 ORK S Humberstone, Farley Hill, Matlock, Derbyshire, DE4 3LL [IO93FD, SK26]
G0 ORL C E Rowley, 31 Keepers Croft, East Goscote, Leicester, LE7 3ZJ [IO92LR, SK61]
G0 ORM D Birch, 31 Grasmere Terr, Maryport, CA15 7QN [IO84GQ, NY03]
G0 ORN P W Mortimer, 60 Lower Farnham Rd, Aldershot, GU12 4EA [IO91PF, SU84]
G0 ORO D Martin, Bayshore, Gilcrux, Carlisle, Cumbria, CA5 2QD [IO84HR, NY13]
G0 ORP M C Simpson, 3 Front St., Barnby in The Willo, Barnby, Newark, NG24 2SA [IO93PB, SK85]
G0 ORR Details withheld at licensee's request by SSL.[Op: C Frost. Correspondence via 6269 Elm St., Vancouver BC, V6N 1B2, Canada.]
G0 ORS C B Lee, 51 Moor Grange Ct, West Park, Leeds, LS16 5EB [IO93EU, SE23]
G0 ORT D R Leonard, Three Ashes Cottage, 442 Outwood Common Rd, Billericay, CM11 1ET [JO01FP, TQ69]
G0 ORU J J Rose, 4 Rugby Rd, Acton, London, W4 1AT [IO91UM, TQ27]
G0 ORV V F Wilton, Fairthorn Trotts Ln, Pooks Green, Marchwood, Southampton, Hants, SO4 4WQ [IO90HW, SU41]
G0 ORW G N Gee, Gaynesfords, Undershore, Walhampton, Lymington, Hants, SO41 5SA [IO90FS, SZ39]
G0 ORX J D Melton, 4 Charlwoods Cl, Copthorne Bank, Crawley, West Sussex, RH10 3QZ [IO91WD, TQ33]
G0 ORY A G J Moss, 40 Westcotes Dr, Leicester, LE3 0QR [IO92KP, SK50]
G0 OSA C Wilkinson, Dendrum Lodge, Race Moor Ln, Oakworth, Keighley, West Yorks, BD22 7JH [IO93AU, SE03]
GW0 OSB I G Price, 16 Carmarthen Ct, Caerphilly, CF8 2TX [IO81HN, ST08]
G0 OSC G H Mason, 18 Nithsdale Rd, Liverpool, L15 5AX [IO83MJ, SJ38]
G0 OSD G Alexander, 15 Brackley Way, Totton, Southampton, SO40 3HP [IO90FW, SU31]
G0 OSE D A R Powell, 17 Ledger Dr, Addlestone, KT15 1AS [IO91RI, TQ06]
G0 OSF G Biggs, 16 Maple Dr, Newport, PO30 5QP [IO90IQ, SZ48]
G0 OSG R G Brazier, 9 Wheelers Walk, Blackfield, Southampton, SO45 1WX [IO90HT, SU40]
G0 OSH H Davies Nortel (Paignton) Amateur Radi, 33 Sandown Rd, Ocean Heights, Paignton, TQ4 7RL [IO80FJ, SX85]
G0 OSK C G Saggers, 49 Revels Rd, Bengeo, Hertford, SG14 3JU [IO91XT, TL31]
G0 OSM Details withheld at licensee's request by SSL.
G0 OSN Details withheld at licensee's request by SSL.
G0 OSO P J J Markham, 13 Massey Rd, Lincoln, LN2 4BN [IO93RF, SK97]
G0 OSP Dr P A G Leach, 17 The Wicket, Hythe, Southampton, SO45 5AU [IO90HU, SU40]
GW0 OSQ B J West, 83 Blaendare Rd, Pontypool, NP4 5RU [IO81LQ, SO20]
G0 OSR Dr H O Middleton, Gordon House, 30 Parsons Heath, Colchester, CO4 3HX [JO01LV, TM02]
G0 OSS J D Porter, 30 Fermor Way, Crowborough, TN6 3BD [JO01CB, TQ52]
G0 OSU J R Collier, 27 Birdham Cl, North Bersted, Bognor Regis, PO21 5TD [IO90PT, SU90]
G0 OSV P L Foster, 7 Frobisher Terr, Falmouth, TR11 2NB [IO70LD, SW83]
G0 OSW R P Sainsbury, Bridge Farmhouse, Southampton Rd, Landford, Salisbury, SP5 2ED [IO90EX, SU21]
G0 OSX N J Shackley, 20 Pear Tree Rd, Ashford, TW15 1PW [IO91SK, TQ07]
G0 OTA Details withheld at licensee's request by SSL.
GM0 OTB R D Pugh, 28 Pladda Rd, Saltcoats, KA21 6AQ [IO75OP, NS24]
GI0 OTC T A Doherty, 37 Magheramenagh Dr, Portrush, BT56 8SP [IO65QE, C83]
G0 OTD D M McCarthy, 29 Rosslyn Dr, Moreton, Wirral, L46 0SU [IO83KJ, SJ28]
G0 OTE E R Bowell, 7 Bede House Bank, Bourne, PE10 9JX [IO92TS, TF11]
G0 OTF G S George, 211 Bromford Rd, Hodge Hill, Birmingham, B36 8HA [IO92CM, SP18]
G0 OTH R P Topliss, 10 Wilford Gr, Skegness, PE25 3EZ [JO03ED, TF56]
GM0 OTI Dr J S Grieve, Elhanan, Myrtlefield Ln, Westhill, Inverness, IV1 2UE [IO77WL, NH74]
G0 OTJ J R Cummins, Tarr House, Lumb Ln, Darley Dale, Matlock, DE4 2HP [IO93EE, SK26]
GI0 OTL M A Simcox, 51 Mullaghboy Rd, Islandmagee, Larne, BT40 3TR [IO74DU, D40]
G0 OTM Details withheld at licensee's request by SSL.
G0 OTN Details withheld at licensee's request by SSL.
G0 OTO Details withheld at licensee's request by SSL.
GM0 OTP G A Sim, Liathach, 29 Bogie St., Huntly, AB54 8DX [IO87OK, NJ53]
G0 OTQ E Hunter, 45 Westway, Newcastle upon Tyne, NE15 9HL [IO94DX, NZ16]
GI0 OTR Details withheld at licensee's request by SSL.
GM0 OTS W A McIntosh, Firth View, 63 Harbour St., Hopeman, Elgin, IV30 2RU [IO87GR, NJ16]
G0 OTT Details withheld at licensee's request by SSL.
GM0 OTU A King, 31 Pendreich Gr, Bonnyrigg, EH19 2EH [IO85KV, NT36]
G0 OTW Details withheld at licensee's request by SSL.
GM0 OTX T Scott, 19 Marchburn Dr, Penicuik, EH26 9HE [IO85JT, NT26]
GW0 OTY W C Cooper, 50 Tennyson Rd, Penarth, CF64 2SA [IO81JK, ST17]
G0 OTZ D G Monte, 30 Chestnut Hill, Eaton, Norwich, NR4 6NL [JO02PO, TG20]
G0 OUB Details withheld at licensee's request by SSL.

G0 OUD S B Hill, 4 Tennyson Cl, Penistone, Sheffield, S30 6GY [IO93GJ, SK38]
G0 OUE Details withheld at licensee's request by SSL.
G0 OUF Details withheld at licensee's request by SSL.
GW0 OUH H Griffiths, 45 Jubilee Rd, Godreaman, Aberdare, CF44 6DD [IO81GQ, SO00]
G0 OUI Details withheld at licensee's request by SSL.
G0 OUJ R Walker, 270 Stourbridge Rd, Halesowen, B63 3QR [IO82XK, SO98]
G0 OUK J G Hinton, 6 Petworth Dr, Whirlow Dale Park, Sheffield, S11 9QU [IO93FI, SK38]
G0 OUL Details withheld at licensee's request by SSL.
GI0 OUM R T Ferris, 3 Kingsland Dr, Belfast, BT5 7EY [IO74BO, J37]
G0 OUO S W Palk, 2 Lydeard Mead, Bishops Lydeard, Taunton, Somerset, TA4 3UD [IO81JB, ST12]
GW0 OUP Details withheld at licensee's request by SSL.
G0 OUQ Details withheld at licensee's request by SSL.
G0 OUR E G Fountaine OU ARC, Open University, Walton Hall, Milton Keynes, MK7 6AA [IO92PA, SP83]
G0 OUT D M Basford Dronfield & District A.R.C, 91 Hollins Spring Ave, Dronfield, Sheffield, S18 6RP [IO93GH, SK37]
GW0 OUV M Williams, 8 Caiach Terr, Trelewis, Treharris, CF46 6DH [IO81IP, ST19]
GI0 OUX J D Skelly, Bredagh Glen, Moville, Co Donegal, Ireland
GM0 OUY Details withheld at licensee's request by SSL.
G0 OVA Details withheld at licensee's request by SSL.
G0 OVB G W Linnegar, 20 Durham Rd, Dagenham, RM10 8AN [JO01CN, TQ58]
G0 OVC B J Godfrey, 824 Lea Bridge Rd, London, E17 9DN [IO91XN, TQ38]
GM0 OVD R W O Darroch, 36 Tweed St., Dunfermline, KY11 4NA [IO86GB, NT18]
G0 OVE K L Mohammed, 63 Shirley Gdns, Barking, IG11 9XB [JO01BM, TQ48]
G0 OVG G A Webb, 3 Scott Ave, Weddington, Nuneaton, CV10 0DP [IO92GM, SP39]
G0 OVI Details withheld at licensee's request by SSL.
G0 OVJ Details withheld at licensee's request by SSL.
G0 OVK R T J Mansell, 16 Parkes St., Willenhall, WV13 2LP [IO82XN, SO99]
G0 OVM Details withheld at licensee's request by SSL.
G0 OVN Details withheld at licensee's request by SSL.
G0 OVO A D Wiltshire, 81 Whomerley Rd, Stevenage, SG1 1SS [IO91VV, TL22]
G0 OVP R J Layzell, 37 Alpraham Cres, Chester, CH2 1QX [IO83NF, SJ46]
G0 OVQ A Bannister, 34 Morningside Dr, East Didsbury, Manchester, M20 5PL [IO83VJ, SJ88]
G0 OVR A G E Chappelle, 130 Baring Rd, Lee, London, SE12 0PU [JO01AK, TQ47]
G0 OVS H W Smith, 17 Gannahs Farm Cl, Walmley, Sutton Coldfield, B76 2TF [IO92CN, SP19]
G0 OVT B L Navier, 12 Brooklyn Ave, Brooklyn St., Hull, HU5 1ND [IO93TS, TA03]
G0 OVU M Hajdukiewicz, Shamrock Cottage, Church Ln, Sapperton, nr Cirencester, Glos, GL7 6LQ [IO81WR, SO90]
G0 OVV M E Bolton, 85 Oak Park Rd, Wordsley, Stourbridge, DY8 5YJ [IO82WL, SO88]
G0 OVW D M Hall, 13 Franklands Dr, Addlestone, KT15 1EQ [IO91RI, TQ06]
G0 OVX Details withheld at licensee's request by SSL.
G0 OVY P E Maggs, 33 Springs Bridge Rd, Manchester, M16 8PW [IO83VK, SJ89]
G0 OWA J Wright, 10 Whalley Rd, Heskin, Chorley, PR7 5NY [IO83PP, SD51]
G0 OWB J W F Fox, 11 West Gr, Royston, Barnsley, S71 4RY [IO93GO, SE31]
G0 OWD D Weston, 91 Hill Top Ln, Rotherham, S61 2EQ [IO93HK, SK39]
G0 OWE D T Matthews, 54 The Wynding, Willow Grange, Bedlington, NE22 6HW [IO95ED, NZ28]
G0 OWF Details withheld at licensee's request by SSL.
G0 OWG H W Marriott, 13 Station St., Lymington, SO41 3BA [IO90FS, SZ39]
G0 OWH J L Dobbs, St. Aideans Vicarage, 498 Manchester Rd, Rochdale, OL11 3HE [IO83VO, SD81]
G0 OWI A T Hawkridge, Thorntrees, 109 Allerton Rd, Bradford, BD8 0AA [IO93CT, SE13]
G0 OWJ A R Cooper, 28 Belmont Rd, Pensnett, Brierley Hill, DY5 4EX [IO82WL, SO98]
G0 OWK S J Searle, 14 Edison Gdns, Colchester, CO4 4AJ [JO01LV, TM02]
GM0 OWM A W Wright Orkney Wrlss Ms, Crosslea, Berstane Rd, St. Ola, Kirkwall, KW15 1SZ [IO88MX, HY41]
G0 OWN Details withheld at licensee's request by SSL.
G0 OWO M Levy, Eur Ing, 223 Woodcock Hill, Kenton, Harrow, HA3 0PG [IO91UN, TQ18]
G0 OWP D F Edwards, 9 Mark Rd, Hightown, Liverpool, L38 0BG [IO83LM, SD20]
G0 OWR C J Howard, 75 Westbury Park, Wootton Bassett, Swindon, SN4 7DN [IO91BM, SU08]
G0 OWS J Davis, 61 Somers Rd, Malvern, WR14 1JA [IO82UD, SO74]
G0 OWT J F C Marks, 61 Sebright Rd, Wolverley, Kidderminster, DY11 5UA [IO82UJ, SO87]
G0 OWU R W J Wilkes, 39 Hillside Rd, Wrens Nest, Dudley, DY1 3LE [IO82WM, SO99]
G0 OWV J W Harbottle, 42 Littlemede, Eltham, London, SE9 3EB [JO01AK, TQ47]
G0 OWW Details withheld at licensee's request by SSL.
G0 OWX W T Horscroft W Somerset ARC, 2 Blenheim Cttgs, Blenheim Rd, Minehead, TA24 5QB [IO81GF, SS94]
G0 OWY A McAsey, 24 Keswick Ave, Bromborough, Wirral, L63 0NP [IO83MH, SJ38]
G0 OWZ Details withheld at licensee's request by SSL.
G0 OXA G L Landen-Turner, 59 Mill Rd, Higher Bebington, Bebington, Wirral, L63 5PA [IO83LI, SJ38]
G0 OXB P R Draycott, 77 Old Taunton Rd, Bridgwater, TA6 3NU [IO81MD, ST33]
G0 OXE C J Tapping The Morse Club, 56 Wolfe Cres, Charlton, London, SE7 8TR [JO01AL, TQ47]
GI0 OXG J F Baker, 509 Enniskeen, Drumgor, Craigavon, BT65 4AB [IO64TK, J05]
G0 OXH G F Morley, 19 Palgrave House, Fleet Rd, London, NW3 2QJ [IO91WN, TQ28]
G0 OXI Details withheld at licensee's request by SSL.
G0 OXJ Details withheld at licensee's request by SSL.
G0 OXL Details withheld at licensee's request by SSL.
G0 OXM Details withheld at licensee's request by SSL.
G0 OXN J R Hart, 15 Heather Ln, Thistle Flat Est, Crook, DL15 9TW [IO94DR, NZ13]
G0 OXO Details withheld at licensee's request by SSL.
G0 OXQ Details withheld at licensee's request by SSL.
G0 OXR C Franks, 11 Orchard Cl, Crook, DL15 8QU [IO94DR, NZ13]
GM0 OXS M A Beith, 30 Raith Rd, Fenwick, Kilmarnock, KA3 6DB [IO75SP, NS44]
G0 OXT P M Hutchinson, Chippings, Coles Ln, Kingskerswell, Newton Abbot, TQ12 5BA [IO80FM, SX86]
G0 OXV K R Mahood, Brooklands Lodge, Heskin Ln, Ormskirk, L39 1LR [IO83NN, SD40]
G0 OXW V Soutter, 2 Hyde Barton, Churchill Way, Northam, Bideford, EX39 1NX [IO71VA, SS42]
G0 OXX Dr J C Berridge, Bracklyn, St. Clare Rd, Walmer, Deal, CT14 7QB [JO01QE, TR35]
G0 OXY M F Gray, 268 Redwood Gr, Bedford, Beds, MK42 9NQ [IO92SD, TL04]
G0 OYA M F J Clapperton, Rookery Cottage, 23 Waterloo, Puriton, Bridgwater, TA7 8BB [IO81ME, ST34]
G0 OYB B C Monk, 22 Carshalton Ave, Cosham, Portsmouth, PO6 2JT [IO90LU, SU60]
G0 OYC K Saunders, 1 Chesham Way, Watford, WD1 8NX [IO91TP, TQ09]
GW0 OYD R A Fray, Tan y Bryn, Llaneilian, Amlwch, Anglesey, LL68 9LS [IO73UJ, SH49]
GM0 OYE A C Gourlay, 46 Fonab Cres, Pitlochry, PH16 5SR [IO86DQ, NN95]
G0 OYF S Harvey, 68 Stuart Rd, Rowley Regis, Warley, B65 9HZ [IO82XL, SO98]
G0 OYH B Taylor, 3 Anne Cl, Stoke Hill, Exeter, EX4 7DL [IO80FR, SX99]
G0 OYI G Holden, 4 Oak Rd, Carleton, Penrith, CA11 8TS [IO84PP, NY53]
G0 OYJ T H Gonsalves, 30 Cunnington St., Chiswick, London, W4 5EN [IO91UL, TQ27]
G0 OYK K D Brettell, 38 Delves Cres, Walsall, WS5 4LR [IO92AN, SP09]
G0 OYL W M Waring, 11 Sycamore Gr, Prudhoe, NE42 6QA [IO94BX, NZ06]
G0 OYM M E Trahearn, 16 Grange Ln, Lichfield, WS13 7ED [IO92BQ, SK11]
G0 OYN D J Hedley, 124 North End Ave, Portsmouth, PO2 8NH [IO90LT, SU60]
G0 OYO D I James, 33 Mallet Rd, Ivybridge, PL21 9TD [IO80AJ, SX65]
G0 OYP B Barber, 3 Catherine Ave, Mansfield Woodhouse, Mansfield, NG19 9AZ [IO93JD, SK56]
G0 OYQ S Lowe, 258 North Rd, Hull, HU4 6LB [IO93TR, TA02]
G0 OYR N J Ashfield, 167 Greville Rd, Warwick, CV34 5PU [IO92FH, SP26]
G0 OYS D R Temple, 4 Cameron Ave, Abingdon, OX14 3SR [IO91IQ, SU59]
GM0 OYT J Hutchison, 13 Windford Rd, Mastrick, Aberdeen, AB1 6NQ [IO87TH, NJ72]
GM0 OYU M Chesters, Dunessa, Fintray, Dyce, Aberdeen, AB2 0HY [IO87TH, NJ72]
GM0 OYV C N L Cowper, 9 Oxgangs Rd, Edinburgh, EH10 7BG [IO85JV, NT26]
G0 OYX D Medley, 9 Northolme Cres, Hessle, HU13 9HU [IO93SR, TA02]
G0 OYY G J Mantle, 12 Poplar Cl, Tividale, Warley, B69 1RP [IO82XM, SO99]
G0 OYZ L W Barnes, 2A Hallcroft Rd, Whittlesey, Peterborough, PE7 1LP [IO92WN, TL29]
G0 OZA J B Whitehead, 120 Broomhill Rd, Bulwell, Nottingham, NG6 9GJ [IO92JX, SK54]
GW0 OZB A Gardner, 28 Usk Ct, Thornhill, Cwmbran, NP44 5UN [IO81LP, ST29]
G0 OZD D J Tanner, Home Field House, Dunbridge Ln, Awbridge, Romsey, SO51 0GQ [IO91FA, SU32]
G0 OZE J A Maddocks, 41 Oakdale Dr, Shipley, BD18 1PD [IO93CT, SE13]
GI0 OZF D A Tilson, 64 Main St., Ballycarry, Carrickfergus, BT38 9HH [IO74DS, J49]
G0 OZG D J Turner, 27 Aylesbury Ave, Langley Point, Eastbourne, BN23 6AB [JO00DS, TQ60]
G0 OZH S Squibb, 36 Frognal Gdns, Teynham, Sittingbourne, ME9 9HU [JO01JH, TQ96]
G0 OZI E J May, 10 Clyde Cres, Wharton, Winsford, CW7 3LA [IO83SE, SJ66]
G0 OZJ G D Gourley, 6A Longsight Ln, Cheadle Hulme, Cheadle, SK8 6PW [IO83VI, SJ88]
G0 OZK K S Burrows, 10 Basil St., Heaton Norris, Stockport, SK4 1QL [IO83WK, SJ89]
G0 OZL B J Smith, 19 Fieldstone Ct, Northpark Howick, New Zealand 1705
G0 OZM C D Rapson, Kaloma, Northiam Rd, Broad Oak, Rye, TN31 6EP [JO00HW, TQ81]
G0 OZN Details withheld at licensee's request by SSL.
G0 OZO J G Harris, 44 Boston Rd, Heckington, Sleaford, NG34 9JE [IO92UX, TF14]
G0 OZP B Salt, 9 Ashville Gdns, Moore End Rd, Pellon, Halifax, HX2 0PJ [IO93BR, SE02]
GI0 OZQ D A Gillespie, 81 Lisfannon Park, Londonderry, BT48 9DU [IO64IX, C41]
G0 OZR G C Markham, 2 Edwin Ave, Woodbridge, IP12 1JS [JO02PC, TM25]
G0 OZS I G Moffat, 30 Daimler Rd, Ipswich, IP1 5PQ [JO02NB, TM14]
G0 OZT E A Shephard, 70 Cliff Rd, Hornsea, HU18 1LZ [IO93WW, TA24]
G0 OZV D J Taylor, 41 Gr Rd, Woodbridge, IP12 4LG [JO02PC, TM24]
G0 OZW Details withheld at licensee's request by SSL.
G0 OZX Details withheld at licensee's request by SSL.
G0 OZY J S Tucker, Oakhurst Farm, Kings Hill, Burwash, Etchingham, TN19 7DP [JO00EX, TQ62]
G0 OZZ Details withheld at licensee's request by SSL.
G0 PAA Dr B E Keiser, 2046 Carrhill Rd, Vienna, Va 22181, USA
G0 PAB P A Betts, 68 Manor Rd, Scunthorpe, DN16 3PA [IO93QN, SE80]
G0 PAD A D Jacobs, 38 Cottesmore Ave, Barton Seagrave, Kettering, NN15 6QX [IO92PJ, SP87]
G0 PAE C R Hewitt, 28 Amersham Ave, Langdon Hills, Basildon, SS16 6SJ [JO01EN, TQ68]
G0 PAF C R French, Cresta, 27 Chapel Rd, Sarisbury Green, Southampton, SO31 7FB [IO90IV, SU50]
G0 PAG N C Page, 158 Manchester Rd, Worsley, Manchester, M28 3LU [IO83TM, SD70]
G0 PAH Details withheld at licensee's request by SSL.
G0 PAI I T Leitch, 70 Hanover Rd, Rowley Regis, Warley, B65 9DZ [IO82XL, SO98]
G0 PAJ E H Binns, Crabtree House, Springfield, Bentham, Lancaster, LA2 7BA [IO84SC, SD66]
G0 PAK W D Barnett, 94 Park Ave, Penistone, Sheffield, S30 6DN [IO93GJ, SK38]
G0 PAN D F Elkington, 45 Heathfield, Adel, Leeds, LS16 7AB [IO93EU, SE23]
G0 PAO C S Muddimer, 41 Brandeston Cl, Great Waldingfield, Sudbury, CO10 0XY [JO02JB, TL94]
G0 PAQ E Simons, 14 Gurth Ave, Edenthorpe, Doncaster, DN3 2LW [IO93LN, SE60]
G0 PAR D I How, 25 Lovelace Rd, West Dulwich, London, SE21 8JY [IO91WK, TQ37]
G0 PAS M P Lord, 5 Wasdale Green, Cottingham, HU16 4HN [IO93TS, TA03]
G0 PAU J I Forsyth, 102 Langley Rd, Watford, WD1 3PJ [IO91TQ, TQ19]
G0 PAV A Dyson, 80 Mansell Rd, Greenford, UB6 9EW [IO91TM, TQ18]
G0 PAW P A Weaving, 7 Reigate Rd, Brighton, BN1 5AJ [IO90WU, TQ30]
G0 PAX B S Inch, 17 Mount Pleasant, Harefield, Uxbridge, UB9 6BE [IO91SO, TQ09]
G0 PAY R J Swift, 5 Greenwood Cl, Westhuntspill, West Huntspill, Highbridge, TA9 3SF [IO81MF, ST34]
G0 PAZ D I Utley, 7 Pavilion Cl, Brierley, Barnsley, S72 9LR [IO93HO, SE41]
G0 PBA K Garbutt, 92 Owlet Rd, Windhill, Shipley, BD18 2LT [IO93CT, SE13]
G0 PBB W J Bonser Forest of Dean Amateur Radio S, 24 Meend Garden Terr, Cinderford, GL14 2EB [IO81ST, SO61]
G0 PBC A Smith, 22 Goodwood Park Rd, Northam, Bideford, EX39 2RR [IO71VA, SS42]
G0 PBE D A Yates, 12 Walnut Cl, Woolston, Warrington, WA1 4HA [IO83RJ, SJ68]
G0 PBF J N Brown, 20 Broad St., Hoyland, Barnsley, S74 9DY [IO93GM, SE30]
G0 PBH D B Mason, 18 Rodman St., Woodhouse Mill, Sheffield, S13 9WT [IO93HI, SK48]
GW0 PBJ L B Wright, 32 Blantern Rd, Higher Kinnerton, Chester, CH4 9DA [IO83MD, SJ36]
G0 PBK G T S Clarke, 18 Lilac Ave, Beverley, HU17 9UT [IO93TU, TA04]
G0 PBL P M Davies, 85 Church Rd, Byfleet, West Byfleet, KT14 7NG [IO91SI, TQ06]
G0 PBM A G Razzell, 96 Weston Rd, Aston on Trent, Derby, DE72 2BA [IO92HU, SK42]
G0 PBN A A Moulder, 10 Parsonage Rd, Rainham, RM13 9LW [JO01CM, TQ58]
G0 PBO A Coleman, 16 Tennyson Rd, Ruskington, Sleaford, NG34 9HF [IO93TB, TF05]
G0 PBP A J Evans, 24 Oakleigh Ave, Glen Parva, Leicester, LE2 9TH [IO92KN, SP59]
G0 PBQ Details withheld at licensee's request by SSL.
G0 PBR R Clark, 4 Haigh St., Cleethorpes, DN35 8QN [IO93XN, TA30]
G0 PBS D J Webber, 37 Woodhurst Dr, Denham, Uxbridge, UB9 5LL [IO91RO, TQ08]
G0 PBT J T Green, 7 Walsh Cl, Priorslee, Telford, TF2 9RY [IO82SQ, SJ71]
G0 PBU D Bradley, Stoneybanks Cottage, 28 High St., Bozeat, Wellingborough, NN29 7NF [IO92PF, SP95]
G0 PBV N J Plumb, 12 Ferringham Ln, Ferring, Worthing, BN12 5NQ [IO90ST, TQ00]
G0 PBW R E Brown, 26 Lynnes Cl, Blidworth, Mansfield, NG21 0TU [IO93KC, SK55]
G0 PBY R A Freer, 117 Devitt Way, Broughton Astley, Leicester, LE9 6NQ [IO92JM, SP59]
G0 PBZ M R Wustrau, 39 Bedford Rd, Houghton Conquest, Bedford, MK45 3LS [IO92SB, TL04]
G0 PCA K T Godwin, 11 St. Lukes Way, Allhallows, Rochester, ME3 9PR [JO01HL, TQ87]
G0 PCB E J Godwin, 11 St. Lukes Way, Allhallows, Rochester, ME3 9PR [JO01HL, TQ87]
G0 PCD S C Farrow, 65 The Hide, Netherfield, Milton Keynes, MK6 4HQ [IO92PA, SP83]
G0 PCE R D F Barnes, 113 Queens Quay, Upper Thames St., London, EC4V 3EJ [IO91WM, TQ38]
G0 PCF B R Foxall, Foxglove Cottage, 9 Truslers Hill Ln, Albourne, Hassocks, BN6 9DU [IO90VW, TQ21]
GM0 PCH I R Drummond, PO Box 8640, Pc22057, Salmiya, Kuwait, X X
GW0 PCJ C L Watson, 4 Brookland Cl, Maesycwmmer, Hengoed, CF82 7RH [IO81JP, ST19]
G0 PCK A M Lord, 16 Shakespeare Ave, Lichfield, WS14 9BE [IO92CQ, SK10]
G0 PCM I D Calvert, 16 Nabwood Dr, Shipley, West Yorks, BD1 4EJ [IO93DT, SE13]
G0 PCN H R McAlroy, 75 Rounthorn Rd, Bagely, Manchester, M23 1EP [IO83UJ, SJ88]
G0 PCP Details withheld at licensee's request by SSL.
G0 PCQ I N Yeo, Chyventon, Smithams Hill, East Harptree, Bristol, BS18 6BZ [IO81QG, ST55]
G0 PCR Details withheld at licensee's request by SSL.
G0 PCS A C Douglas Preston Comunit, School Arc, Preston School, Monks Dale, Yeovil, BA21 3JD [IO80QW, ST51]
GI0 PCU Details withheld at licensee's request by SSL.
G0 PCV M J Hiendl, Trelawney Farm, Towntanna, Ponsanooth, Cornwall, TR3 7HT [IO70KE, SW73]
G0 PCW J C Budden, 20 Gordon Ave, Portswood, Southampton, SO14 6WD [IO90HW, SU41]
G0 PCX P Russell Erewash Valley Amateur Radio G, 9 Baggot St., West Hallam, Ilkeston, Derbyshire, DE7 6HA [IO92HX, SK44]
G0 PCY J J Radford, 93 Hook Rd, Surbiton, KT6 5AF [IO91UJ, TQ16]
G0 PCZ B R Lody, 41 Galsworthy Rd, Chertsey, KT16 8EP [IO91RJ, TQ06]
GW0 PDA W E Cole, 34 Erw Goch, Waunfawr, Aberystwyth, Ceredigion, SY23 3AZ [IO72XJ, SN68]
GW0 PDB G P Griffiths, Dolcoed, Llanddysul, Dyfed, SA44 4AE [IO72UA, SN44]
G0 PDD K A Howard, Meadow View, Coverham Rd, Berry Hill, Coleford, GL16 7AU [IO81QT, SO51]
G0 PDE D A P Livingstone, 8 Alkerton Terr, Eastington, Stonehouse, GL10 3AU [IO81UR, SO70]
G0 PDF P Watkins, Fourwinds, 44 Hyde Ln, Kinver, Stourbridge, DY7 6AF [IO82VK, SO88]
G0 PDH D Hyde, 9 Empress Ave, Marple, Stockport, SK6 7BG [IO83XJ, SJ98]
GJ0 PDJ M J Turner, 4-le-Clos Sara, St. Lawrence, Jersey, JE3 1GT
G0 PDK W Marsden, 8 Albert Rd, Eston, Middlesbrough, TS6 9QW [IO94KN, NZ51]
G0 PDL Details withheld at licensee's request by SSL.
G0 PDM M J Glover, 124 St. James Ave, Southend on Sea, SS1 3LN [JO01JM, TQ98]
G0 PDN D Beedan, 1 Stonehill Croft, Shirley, Solihull, B90 4TD [IO92CJ, SP17]
G0 PDO W J Gardner, 78 Stratford Rd, Warwick, CV34 6AT [IO92EG, SP26]
G0 PDP A P Farmer, 43 Barn Platt, Ashford, TN23 7UH [JO01KD, TR04]
GM0 PDQ M E Kusin, East Overhill Farm, Stewarton, Ayrshire, KA3 5JP [IO75SQ, NS44]
G0 PDS Details withheld at licensee's request by SSL.
G0 PDT B Tear, 36 Ashbourne Ave, Cleckheaton, BD19 5JJ [IO93DR, SE12]
G0 PDU Details withheld at licensee's request by SSL.
G0 PDV R T Netherway, 28 Snowdon Rd, Fishponds, Bristol, BS16 2EJ [IO81RL, ST67]
G0 PDW Details withheld at licensee's request by SSL.
G0 PDY Details withheld at licensee's request by SSL.
G0 PDZ Details withheld at licensee's request by SSL.[Op: I M Lowe. Station located near Welling, Kent.]
GW0 PEA J L Pearson, 6 Stag Terr, Abergarwed, Neath, SA11 4DH [IO81DQ, SN80]
GW0 PEB R G Williams, Erw Las, 10 Ffordd Iago, Groeslon, Caernarvon, LL54 7DH [IO73UB, SH45]
G0 PEC I G Tutt, 7 Mitchells Rd, Haylands, Ryde, PO33 3JA [IO90JR, SZ59]
GI0 PED P J Serridge, 21 Lassara Heights, Warrenpoint, Newry, BT34 3PG [IO64UC, J11]
G0 PEF I M Williams, Glenroy, Newport Rd, Godshill, Ventnor, PO38 3HR [IO90IP, SZ58]
G0 PEG J M Jenner, 73 Firs Ln, Folkestone, CT19 4QF [JO01NC, TR13]
G0 PEH A M Lifton, 70 Scrapsgate Rd, Minster on Sea, Sheerness, ME12 2DJ [JO01JK, TQ97]
GM0 PEI A N Pollock, 113 Gartmorn Rd, Sauchie, Alloa, FK10 3PD [IO86CD, NS99]
G0 PEJ G Ford, 5 Rosslyn Cl, Hockley, SS5 5BP [JO01HO, TQ89]
G0 PEK K M Richardson, 21 Jillian Way, Ashford, TN23 5DT [JO01KD, TQ94]
G0 PEM P C Oliver, 12 Tudor End, Kennington, Ashford, TN24 9DR [JO01KD, TR04]
G0 PEN Details withheld at licensee's request by SSL.
GM0 PEO C Andersz, Windrush, School Rd, Port Elphinstone, Inverurie, AB51 3XJ [IO87TG, NJ72]
G0 PEP P W Waters Waters & Stanton Electronics, 9 Tudor Way, Hockley, SS5 4EY [JO01HO, TQ89]
G0 PEQ P J Cook, 88 Sprowston Rd, Norwich, NR3 4QW [JO02PP, TG21]
G0 PER K D O Kreuchen, 5 Burnham Gdns, Cranford, Hounslow, TW4 6LS [IO91TL, TQ17]
G0 PES J M Cornall, 45 Plumpton Ave, Blackpool, FY4 3RB [IO83LS, SD33]
G0 PET A J Cunningham, 2 Orchard Cl, Normandy, Guildford, Surrey, GU3 2EU [IO91PF, SU95]
G0 PEV R Dawson, 6 Oxton Ln, Tadcaster, LS24 8AG [IO93IV, SE44]
G0 PEW J R Lyne, 157 Westwick Rd, Sheffield, S8 7BW [IO93GH, SK38]
GM0 PEX P Bendermacher, 1 Cedar Dr, Milton of Campsie, Glasgow, G65 8AY [IO75WX, NS67]
G0 PEY R Pearson, 145 Woodhall Rd, Chelmsford, CM1 4AF [JO01FS, TL70]
GI0 PEZ A M Willis, 685 Lisburn Rd, Belfast, BT9 7GU [IO74AN, J37]
G0 PFA M Sole, 44 Chestnut Ave, Ewell, Epsom, KT19 0SZ [IO91UI, TQ26]
GI0 PFB R McKeown, 2 Green Cres, Knock, Belfast, BT5 6JE [IO74BO, J37]
G0 PFC D Smith, 23 Coleridge Dr, Baxenden, Accrington, BB5 2PU [IO83TR, SD72]
G0 PFD A E R Davison, 31 Clifford St., Hornsea, HU18 1HZ [IO93WW, TA24]
G0 PFE R N Lees, Lyndric, 23 New Queen St., Scarborough, YO12 7HL [IO94TG, TA08]
G0 PFF P J Downs, 60 Primrose Ridge, Godalming, GU7 2NX [IO91QE, SU94]
G0 PFI E A Ball, 120 Inver Rd, Bispham, Blackpool, FY2 0RP [IO83LU, SD33]
G0 PFJ F J Poynter, 7 Howards Way, Cawston, Norwich, NR10 4AZ [JO02OS, TG12]
GI0 PFK N Liggett, 14 Kilmaine Rd, Bangor, BT19 6DT [IO74EP, J58]
GI0 PFL S S McClean, 86 Circular Rd, Newtownards, BT23 4BW [IO74DO, J47]
G0 PFM E J Ashworth, 88 Hawthorn Ave, Colchester, CO4 3JP [JO01LV, TM02]
G0 PFN D J Catchpole, 15 Vera Rd, Hellesdon, Norwich, NR6 5HU [JO02PP, TG21]
G0 PFO D A Butler, 1901 Dean Ave, Holt, Michigan, USA 48842
G0 PFQ S J Streluk, 11 Ninefoot Ln, Belgrave, Tamworth, B77 2NA [IO92DO, SK20]
G0 PFT M D Farrell, Hobberley House, Hobberley Ln, Shadwell, Leeds, LS17 8LX [IO93GU, SE33]

G0 PFU K Wignall, 4 Weavers Fold, South Rd, Bretherton, Preston, PR5 7AP [IO83OQ, SD42]
G0 PFV G A F W Newby-Robson, Hillrise, Lamarsh, Bures, Suffolk, CO8 5EP [JO01JX, TL83]
G0 PFX Details withheld at licensee's request by SSL.
G0 PFY R W Marshall, 66 Oakwood Hill, Loughton, IG10 3EP [JO01AP, TQ49]
GW0 PFZ A R Powell, Rich Lyn, Carmel Rd, Carmel, Clwyd, CH8 7DF [IO83JG, SJ17]
G0 PGA C A Smith, 186 Mansfield Rd, Warsop, Mansfield, NG20 0DG [IO93KE, SK56]
G0 PGB C Hosking, 32 Queen St., Penzance, TR18 4BH [IO70FC, SW42]
GI0 PGC J T Forsythe, 1 Coulson Ave, Lisburn, BT28 1YJ [IO64XM, J26]
GM0 PGD A Paterson, 21 Kirkwood Ave, Redding, Falkirk, FK2 9UF [IO85DX, NS97]
GW0 PGE Details withheld at licensee's request by SSL.
G0 PGF P V Campbell, 105 Carnarvon Rd, South Woodford, London, E18 2NT [JO01AO, TQ39]
GM0 PGG L Taylor, Rogarth, Milton of Balhall, Menmuir nr Brechin, Angus, Tayside, DD9 6SE [IO86OR, NO56]
G0 PGI Dr D E Beckly, Knighton Buckland, Honachorum, Yelverton, Devon, PL20 7LH [IO70WL, SX46]
G0 PGJ G S Smith, Ormonde, 9 Parc Ave, St. Ives, TR26 2DN [IO70GF, SW54]
G0 PGK D Lawrence, 7 Richmond Rd, Appledore, Bideford, EX39 1PE [IO71VB, SS43]
G0 PGL D J Blight, 31 Priorswood Rd, Taunton, TA2 7PS [IO81KA, ST22]
G0 PGM P C Maskell, 17 Harcourt St., Market Harborough, LE16 9AG [IO92ML, SP78]
G0 PGO E J Couzens, 30 Thorncroft, Hornchurch, RM11 1EU [JO01CN, TQ58]
G0 PGQ M D Molloy, 20 The Lawn, Whittlesford, Cambridge, CB2 4NG [JO02BC, TL44]
G0 PGS P G Slater, 1 Greyhound Rd, Glemsford, Sudbury, CO10 7SJ [JO02IC, TL84]
G0 PGT J P C Newman, Bank Farm, West St, Chickerell, Weymouth, Dorset, DT3 4DY [IO80RO, SY68]
GI0 PGU H Irvine, 19 Hydepark Manor, Newtownabbey, BT36 4PA [IO74AQ, J38]
GM0 PGV A F Aitkenhead, 54 Swanston Gdns, Edinburgh, EH10 7DE [IO85JV, NT26]
G0 PGW G Dunn, 6 Rosewood Ave, Haslingden, Rossendale, BB4 5NG [IO83UQ, SD72]
G0 PGX S A Thomas, Creekside, Greenbank Rd, Devoran, Truro, Cornwall, TR3 6PQ [IO70KF, SW73]
G0 PGY J G Underwood, 56 Bassenhally Rd, Whittlesey, Peterborough, PE7 1RR [IO92WN, TL29]
G0 PHB A Wainwright, 39 Caldwell Rd, Widnes, WA8 7JT [IO83PI, SJ58]
G0 PHD C E Whitehead, 27-28 St. Nicholas St., Scarborough, YO11 2HF [IO94TG, TA08]
G0 PHE P J Long, 40D Curborough Rd, Lichfield, WS13 7NQ [IO92CQ, SK11]
G0 PHF E Ineson, 29 Pingate Ln, Cheadle Hulme, Cheadle, SK8 7LX [IO83VI, SJ88]
GM0 PHG D J McLaughlin, 96 Craighlaw Ave, Eaglesham, Glasgow, G76 0HA [IO75US, NS55]
G0 PHH Details withheld at licensee's request by SSL.
G0 PHI P A Hirst, 4 Brook House Ln, Shelley, Huddersfield, HD8 8LX [IO93DO, SE21]
GM0 PHM A J McCafferty, 44 Hillend Cres, Clarkston, Glasgow, G76 7XX [IO75US, NS55]
G0 PHN M V Kelly, 66 Sixteenth Ave, Hull, HU6 9JP [IO93TS, TA03]
G0 PHO C B Wilson, 448 Hythe Rd, Willesborough, Ashford, TN24 0JH [JO01KD, TR04]
G0 PHP K J Green, 39 Fleetgate, Barton upon Humber, DN18 5QA [IO93SQ, TA02]
G0 PHQ G K Wilcox, 13 Grizedale, Sutton Park Est, Hull, HU7 4AY [IO93TS, TA03]
G0 PHR M J Andrews, 9 Irving Rd, Solihull, B92 9DQ [IO92CK, SP18]
G0 PHS G W Dodsworth, 9 South St., Normanton, WF6 1EE [IO93GQ, SE32]
G0 PHT A P Hemmings, 35 Ravensthorpe Dr, Loughborough, LE11 4WA [IO92JS, SK51]
G0 PHU P J E Pleydell, 30 Rahn Rd, Epping, CM16 4JB [JO01BQ, TL40]
G0 PHY O W Williams, 30 Franklin Rd, Biggleswade, SG18 8DX [IO92UB, TL14]
G0 PHZ J R Louca, 40 Orpen Gdns, Bristol, BS7 9UA [IO81RL, ST67]
G0 PIA J M Brown, 14 St. Georges Ave, Hornchurch, RM11 3PD [JO01CN, TQ58]
G0 PIB K C Simmonds, 7 Sandringham Ave, West Bridgford, Nottingham, NG2 7QS [IO92KW, SK53]
G0 PID B F Thomas, 112 Pen Park Rd, Southmead, Bristol, BS10 6BP [IO81QM, ST57]
G0 PIE Details withheld at licensee's request by SSL.
G0 PIF Details withheld at licensee's request by SSL.
G0 PIH Details withheld at licensee's request by SSL.
G0 PIJ Details withheld at licensee's request by SSL.
G0 PIK A B Clements, 37 Sun St., Isleham, Ely, CB7 5RU [JO02EI, TL67]
G0 PIL T O Brien, 15 Wilwick Ln, Macclesfield, SK11 8RS [IO83WG, SJ97]
G0 PIM Details withheld at licensee's request by SSL.
G0 PIN A G Pinnock, 1 Rutland Gdns, Ealing, London, W13 0ED [IO91UM, TQ18]
G0 PIO C Jones, 55 Church Rd, Saxilby, Lincoln, LN1 2HH [IO93QG, SK87]
G0 PIP Details withheld at licensee's request by SSL.[Op:]
GM0 PIQ L E A Patey, 17B Chapel St., Airdrie, ML6 6LG [IO85AU, NS76]
G0 PIS J L Bird, 12 Beresford Gdns, Romford, RM6 6RX [JO01BN, TQ48]
G0 PIT A L Freeman, 22 Cromer Dr, Wallasey, L45 4RS [IO83LK, SJ39]
G0 PIU G Papadopoulos, 28 Herbert Rd, Emerson Park, Hornchurch, RM11 3LD [JO01CN, TQ58]
GM0 PIV M T P Black, Drumtochty, 37 Clepington Rd, Maryfield, Dundee, DD4 7EL [IO86ML, NO43]
G0 PIX Details withheld at licensee's request by SSL.
GM0 PIY C Pollock, Rhiannan, Tannadice, Forfar, Angus, DD8 3PZ [IO86NQ, NO45]
G0 PIZ Details withheld at licensee's request by SSL.
GW0 PJA P Baston, 7 Allerton Cl, Pen y Ffordd, Clwyd, CH4 0NJ [IO83LD, SJ36]
G0 PJC A M Jones, High Gables, Moss Ln, Brereton Heath, Congleton, CW12 4SX [IO83UE, SJ86]
GM0 PJD Details withheld at licensee's request by SSL.
GW0 PJF B Ross, 25 Ty Hen, Rhostrehwfa, Llangefni, LL77 7EZ [IO73UG, SH47]
G0 PJG J J Geraghty, 25 Circuit Cl, Willenhall, WV13 1EB [IO82XO, SO99]
GI0 PJH W J Stewart, 23 Sandy Gr, Magherafelt, BT45 6PU [IO64PT, H89]
G0 PJI J V McBride, 2 Hemming Way, Hutton, Weston Super Mare, BS24 9RT [IO81MH, ST35]
G0 PJK Details withheld at licensee's request by SSL.
G0 PJM M V Hughes, 2 Glasshampton Lodge, Shrawley, Worcester, WR6 6TQ [IO82UH, SO76]
G0 PJO M D Waller, Olive Cottage, 6 Church Rd, Chelmondiston, Ipswich, IP9 1HS [JO01OX, TM23]
GM0 PJP K A Scott, 26 Strathern Rd, West Ferry, Broughty Ferry, Dundee, DD5 1PN [IO86NL, NO43]
G0 PJQ E G Pratt, 24 Pier Rd, Gillingham, Kent, ME7 1RJ [JO01GJ, TQ76]
G0 PJR P J Ruffle, 7 Meadway St., Chasetown, Walsall, West Midlands, WS7 8TW [IO92AQ, SK00]
G0 PJS P J Spicer, 86 Main St., Wilsford, Grantham, NG32 3NR [IO92RX, TF04]
G0 PJU J Brown, 20 Stamford Ave, Seaton Delaval, Whitley Bay, NE25 0PA [IO95FB, NZ37]
G0 PJV J W Robertson, 18 Council Rd, Ashington, NE63 8RZ [IO95FE, NZ28]
G0 PJW C J Wormald, 78 Waterloo Rd, Bramhall, Stockport, SK7 2NU [IO83WI, SJ88]
GW0 PJX C W Seward, 10 Caer Gog, Pantymwyn, Mold, CH7 5EX [IO83JE, SJ26]
G0 PJY P Graham, 97 Parklands, Little Sutton, South Wirral, L66 3QH [IO83MG, SJ37]
G0 PJZ R A A Dorling, 1 Horner Pl, The Grove, Witham, CM8 2UG [JO01HT, TL81]
GW0 PKA H A Teaney, Pendre Cottage, Pwll Glas, Mold, CH7 1RA [IO83KE, SJ26]
GM0 PKB W A McLean, Willow Cottage, Auchteraw Rd, Fort Augustus, PH32 4BW [IO77PD, NH30]
G0 PKD P E Maggs 123rd Manchester Scout Group, 33 Springs Bridge Rd, Manchester, M16 8PW [IO83VK, SJ89]
G0 PKE C R Tremble, 14 Newby St., Battersea, London, SW8 3BG [IO91WL, TQ27]
G0 PKF P R French, 15 Popes Cres, Pitsea, Basildon, SS13 3AD [JO01GN, TQ78]
G0 PKG R W E Matthews, 12 Brookfield Ave, Syston, Leicester, LE7 2AB [IO92LQ, SK61]
G0 PKH Details withheld at licensee's request by SSL.
G0 PKI A G Burrell, 39 Benfleet Park Rd, South Benfleet, Benfleet, SS7 5HG [JO01GN, TQ78]
G0 PKJ D A Stallon, Osbourne Villa, London Rd, Brimscombe, Stroud, GL5 2QF [IO81VR, SO80]
G0 PKL Details withheld at licensee's request by SSL.
G0 PKN T F Finneran, 23 Longdales Rd, Lincoln, LN2 2JR [IO93RF, SK97]
GM0 PKP W J Carroll, 20 Pinewood Rd, Mayfield, Dalkeith, EH22 5HX [IO85LU, NT36]
GM0 PKQ F Grant, Silverknowes, Arbeadie Rd, Banchory, AB31 5XA [IO87SB, NO79]
G0 PKR K Ritson, 14 Dunsdale Rd, Holywell, Whitley Bay, NE25 0NG [IO95FB, NZ37]
G0 PKS Details withheld at licensee's request by SSL.[Op: P F Norman. Correspondence via Wellington School, Wellington, Somerset, TA21 8NT.]
G0 PKT A R Horsman C L P K, 12 Hanwell Cl, Clacton on Sea, CO16 7HF [JO01NT, TM11]
G0 PKU Details withheld at licensee's request by SSL.
G0 PKV E L Bennett, 27 Lapwing Cl, Bradley Stoke, Bristol, BS12 0BJ [IO81RN, ST68]
GM0 PKW O J Fairgrieve, 8 Aird, Point, Isle of Lewis, H52 0EU
GM0 PKX E D Michael, 32 Collieston Way, Bridge of Don, Aberdeen, AB22 8SL [IO87WE, NJ91]
G0 PLA T B Reddish, 72 Edgmond Cl, Redditch, B98 0JQ [IO92BH, SP06]
G0 PLB K C Murray, Viamory, Wistanswick, Market Drayton, Shropshire, TF9 2BD [IO82SU, SJ62]
G0 PLC P C Gosnell, 230 Rowley Gdns, London, N4 1HN [IO91WN, TQ38]
G0 PLD T Pogson, 64 New North Rd, Crimble, Slaithwaite, Huddersfield, HD7 5BW [IO93BP, SE01]
G0 PLG W Hargreaves, 24 Moorside Rd, Honley, Huddersfield, HD7 2ER [IO93CO, SE11]
GM0 PLH Details withheld at licensee's request by SSL.
GI0 PLI Details withheld at licensee's request by SSL.
G0 PLK T A Kennedy, 1 Nursery Cttgs, Woodgreen, Fordingbridge, SP6 2AL [IO90CX, SU11]
G0 PLL M Gill, 32 Derbyshire Rd, Barrow in Furness, LA14 5NB [IO84JD, SD27]
G0 PLM W A U Titze, Dr.-Doerfler-Str.20, D-91781 Weissenburg/Bay, Germany, X X
GW0 PLN P Moore, 83 Poplar Rd, Fairwater, Cardiff, CF5 3PT [IO81JL, ST17]
GW0 PLP D Kirby, Arosfa, 7 Heol Enlli, Tanygroes, Cardigan, SA43 2JE [IO72RC, SN24]
GD0 PLQ Details withheld at licensee's request by SSL.
GD0 PLR W J Smith, 1 High View Rd, Douglas, IM2 5BQ
GD0 PLT R M Evans, 46 Howe Rd, Onchan, Douglas, IM3 2AZ
G0 PLV Details withheld at licensee's request by SSL.
G0 PLW B Jones, 80 Foxearth Ave, Clifton, Nottingham, NG11 8JS [IO92KV, SK53]
G0 PLX J I Parker, 24 Egmont St., Salford, M6 7LA [IO83UM, SD80]
G0 PLZ D J Lindsay, 2 Birches Cl, Selsey, Chichester, PO20 9EP [IO90OR, SZ89]
G0 PMA P J Macphail, 128 Westwood Rd, Tilehurst, Reading, RG3 5PZ [IO91LJ, SU66]
G0 PMB G J Banks, 10 Gregory Rd, Glasshoughton, Castleford, WF10 4PH [IO93HR, SE42]
G0 PMD Details withheld at licensee's request by SSL.
G0 PMF G R Dellbridge, 19 Cleeve Cl, Astley Cross, Stourport on Severn, DY13 0NY [IO82UH, SO86]
G0 PMG R G Dellbridge, 24A Calder Rd, Stourport on Severn, Worcs, DY13 8QD [IO82UI, SO87]
G0 PMH D F T Tabberer, 2 Fort Royal Hill, Worcester, WR5 1BT [IO82VE, SO85]
G0 PMI R S Spencer, 4 Barstow Ave, Hull Rd, York, YO1 3HE [IO93LW, SE65]
G0 PMJ D W Hoggart, 15 McMullen Rd, Darlington, DL1 1BW [IO94FM, NZ31]
G0 PMK Capt G Tedd, 1 The Spinney, 42 West Common, Harpenden, AL5 2JW [IO91TT, TL11]
G0 PMM D G Carrott, Cedamyd Dawn Rise, Copthorne, Crawley, West Sussex, RH10 3RL [IO91WD, TQ33]
G0 PMN F B Greenall, The Coach House, Woodlands, Lelant, St. Ives, TR26 3EB [IO70GE, SW53]
GI0 PMO A N Fawcett, Waird, Marwick, Birsay, Orkney, KW17 2ND [IO89IC, HY22]
G0 PMP M Overend, 4 Highfield Mount, Thornhill, Dewsbury, WF12 0QU [IO93EP, SE21]
G0 PMS R H Sweeney, 12 Ash Gr Cl, Bodenham, Hereford, HR1 3LT [IO82PD, SO55]
G0 PMT G D Robinson, 48 St. Andrews Gdns, Lincoln, LN6 7UQ [IO93RF, SK96]
G0 PMU R E Nolson, 50 Shelf Hall Ln, Shelf, Halifax, HX3 7NA [IO93CS, SE12]
GM0 PMW A C Renwick, 11 Victoria Rd, Annan, DG12 6BD [IO84IX, NY16]
G0 PMX J A Garnham, 1 Ashbourne Dr, Coxhoe, Durham, DH6 4SE [IO94GR, NZ33]
G0 PMY C L Rushton, Ashcroft, Market St., Hambleton, Poulton-le-Fylde, FY6 9AP [IO83MV, SD34]
G0 PMZ I Brydon, 12 Pearce Rd, Maidenhead, SL6 7LF [IO91PM, SU88]
G0 PNA M P Cranwell, 7 The Orchard, Bishopsteignton, Teignmouth, TQ14 9RB [IO80FN, SX97]
G0 PNB R M Hope, 7 Irwell Green, Taunton, TA1 2TA [IO81LA, ST22]
GW0 PNC H S Hartwell, Ffosyffin, Llanfair Clydogau, Lampeter, Dyfed, SA48 8LL [IO82AD, SN65]
GW0 PND J K Jones, 4 Lletai Ave, Pencoed, Bridgend, CF35 5PW [IO81GM, SS98]
GW0 PNE D H Hutson, Sandalwood, 60 Glyndwr Rd, Llysfaen, Colwyn Bay, LL29 8TA [IO83EG, SH87]
G0 PNF W S Warren, 38 Stoneyhurst Dr, Curry Rivel, Langport, TA10 0JH [IO81NA, ST32]
G0 PNG G J Buckley, 4 Beach Rd, Preesall, Poulton-le-Fylde, Lancs, FY6 0HQ [IO83MW, SD34]
GW0 PNI D J Pitkin, Charsfield Dental Practice, Priory St., Cardigan, Dyfed, SA43 1BU [IO72QB, SN14]
G0 PNJ W Curtis, 1 Chase End, Hurworth, Darlington, DL2 2JH [IO94FL, NZ31]
GD0 PNK L C Ellison, 1 Kerrocoar Dr, Onchan, Douglas, IM3 1LR
G0 PNM P W Sobye, 34 Elizabeth Cl, Bodmin, PL31 1HY [IO70PL, SX06]
G0 PNN R C Waight, 41 Annalee Rd, South Ockendon, RM15 5BZ [JO01DM, TQ58]
G0 PNO P K Studdart, 656 Rayleigh Rd, Hutton, Brentwood, CM13 1SJ [JO01EP, TQ69]
GM0 PNS J D Harris Pabay Radio Club, Isle of Pabay, Broadford, Isle of Skye, IV49 9BP [IO77BG, NG62]
G0 PNT S F Poulter, 119 Aragon Rd, Morden, SM4 4QG [IO91VJ, TQ26]
G0 PNV K C Price, Sabena, Beeches Rd, Crowborough, E Sussex, TN6 2BN [JO01CB, TQ53]
G0 PNY R M Pearce, 28 Wood St., Longwood, Huddersfield, HD3 4RF [IO93CP, SE11]
G0 PNZ Dr D R Jenkin, Hope Cottage, Yarnscombe, Barnstaple, Devon, EX31 3LW [IO70XX, SS52]
GW0 POA M A Hale, 5 Marchwood Cl, Rumney, Cardiff, South Glam, CF3 8LZ [IO81KM, ST28]
GI0 POB G R Eldridge, 59 Beechwood Gdns, Bangor, BT20 3JD [IO74DP, J48]
G0 POC Dr P E Elwood, Earlswood, 37 Cliff Ave, Nettleham, Lincoln, LN2 2PU [IO93SG, TF07]
GM0 POD W E McCallum, 212 Middleton St., Alexandria, G83 0DJ [IO75RX, NS37]
G0 POE S Kendrick, 11 Winnipeg Rd, Bentley, Doncaster, DN5 0ED [IO93KN, SE50]
GM0 POF S Dempster, 124 Chatelherault Cres, Hamilton, ML3 7PW [IO75XS, NS75]
G0 POH D L Byrer, 49 Covent Garden Rd, Caister on Sea, Great Yarmouth, NR30 5SB [JO02UP, TG51]
G0 POJ J G M M Hirst, 10 Curtois Cl, Branston, Lincoln, LN4 1LJ [IO93SE, TF06]
G0 POK D M Quinnear, 5 Heath Dr, Chelmsford, CM2 9HA [JO01FR, TL70]
G0 POM P R Harris, 44 Boston Rd, Heckington, Sleaford, NG34 9JE [IO92UX, TF14]
G0 POQ D E Kemp, 7 St. Nicholas Ave, Hessle High Rd, Hull, HU4 7AH [IO93TR, TA02]
G0 POR B M Cross, 3 Bargus Cl, Steventon, Abingdon, OX13 6SU [IO91IO, SU49]
G0 POS Dr C Sumner, 7 Barncroft Dr, Hempstead, Gillingham, ME7 3TJ [JO01GI, TQ76]
G0 POT M Sansom, 38 Romsey Rd, Shirley, Southampton, SO16 4DA [IO90GW, SU31]
G0 POU J M Crosby, 9 Hermitage Cl, North Mundham, Chichester, PO20 6JZ [IO90OT, SU80]
G0 POW J Collins, 1019 Byron S South, Whitby, Ontario, Canada L1N 4Sa
G0 POY A Eskelson, 90 Charlton Cres, Barking, IG11 0NL [JO01BM, TQ48]
GW0 POZ D J Morgan, Coedybryn, Synod Inn, Llandysul, Dyfed, Wales, SA44 6JE [IO72TD, SN35]
G0 PPD R Stewart, 63 Calway Rd, Taunton, TA1 3EG [IO81KA, ST22]
GM0 PPE T F Frame, 28 Cedar Dr, Port Seton, Prestonpans, EH32 0SN [IO85MX, NT47]
G0 PPF Details withheld at licensee's request by SSL.
GW0 PPG S D Richards, Crydon, St. Andrews Rd, Wenvoe, Vale of Glam, CF5 6AF [IO81IK, ST17]
G0 PPH W S Blythe, 25 Bartlett Cl, Fareham, PO15 6BQ [IO90JU, SU50]
G0 PPI D Chenery, 18 Barwick Rd, Dover, CT17 0LL [JO01PD, TR24]
G0 PPJ P Johnson, 20 Bearmore Rd, Cradley Heath, Warley, B64 6DU [IO82XL, SO98]
G0 PPK W Gill, 21 Flockton Ave, Standish Lower Ground, Wigan, WN6 8LH [IO83PN, SD50]
G0 PPL G P P Lattka, South Fen Lodge, 9 The Row, Sutton, Ely, CB6 2PD [JO02BJ, TL47]
G0 PPM K M G Powell, 6 Badgers Way, Nailsworth, Stroud, GL6 0HE [IO81VQ, SO80]
G0 PPN Details withheld at licensee's request by SSL.
G0 PPO Details withheld at licensee's request by SSL.
G0 PPP Details withheld at licensee's request by SSL.
G0 PPQ P Jackson, 26 Bempton Gr, Birstall, Batley, WF17 9QZ [IO93ER, SE22]
G0 PPR G Whaling, 32 The Croft, Little Snoring, Fakenham, NR21 0JS [JO02KU, TF93]
G0 PPS D K Egan Prudential ARS, 19 Sycamore Cl, Longmeadow, Dinas Powys, Vale of Glam, CF64 4TG [IO81JK, ST17]
G0 PPT Details withheld at licensee's request by SSL.
G0 PPU Details withheld at licensee's request by SSL.
G0 PPX J J OMara, 18 Tarrant Gr, Quinton, Birmingham, B32 2NW [IO92AK, SP08]
G0 PPY N A Turner, 31 Shamrock Ave, Whitstable, CT5 4EL [JO01MI, TR06]
G0 PPZ M E Adams, 75 Yateley Cres, Great Barr, Birmingham, B42 1JH [IO92AM, SP09]
G0 PQA A S Heather, 4 Ridgeway Heights, Ridgeway Rd, Torquay, TQ1 2ND [IO80FL, SX96]
G0 PQB S A Slater, 24 Lullington Garth, Borehamwood, WD6 2HE [IO91UP, TQ19]
G0 PQC D A W Challis, 22 Gainsborough Dr, Beltinge, Herne Bay, CT6 6QH [JO01NI, TR26]
G0 PQD K J Skuse, 99 Maple Dr, Burnham on Sea, TA8 1DH [IO81MF, ST34]
G0 PQE R Gee, 73 Chatsworth Rd, Pudsey, LS28 8JX [IO93DT, SE23]
G0 PQF A J Judge, 44 Thorley Ln, Bishops Stortford, CM23 4AD [JO01BU, TL41]
G0 PQG A M Harper, 83 High St., Great Houghton, Barnsley, S72 0AU [IO93HN, SE40]
G0 PQH C B Derrick, 7 St. Patricks Rd, Taunton, TA2 7JE [IO81KA, ST22]
GW0 PQI W H Winters, 20 y Berllan, Penmaenamawr, Gwynedd, LL34 6HB [IO83AG, SH77]
G0 PQK Details withheld at licensee's request by SSL.
G0 PQN E C S Bradley-Feary, South Woodcote, Waldens Rd, Horsell, Woking, GU21 4RH [IO91RH, SU95]
G0 PQO K L Martin, 8 Taylors Cl, Meppershall, Shefford, SG17 5NH [IO92UA, TL13]
GW0 PQP Details withheld at licensee's request by SSL.
G0 PQQ Details withheld at licensee's request by SSL.
G0 PQR C F Wardle, P O Box N 3189, Nassau, Bahamas, X X
G0 PQS Details withheld at licensee's request by SSL.
G0 PQT Details withheld at licensee's request by SSL.
G0 PQU Details withheld at licensee's request by SSL.
GM0 PQV J Maguire, 64 High St., Loanhead, EH20 9RR [IO85KV, NT26]
G0 PQW P R Bartholomew, 29 Beatrice Ave, East Cowes, PO32 6HR [IO90IR, SZ59]
G0 PQX S Shipley, 102 Jackson St., Goole, DN14 6DH [IO93NR, SE72]
G0 PQY A S Langford, 53 Cambridge Ave, Bottesford, Scunthorpe, DN16 3PH [IO93QN, SE80]
G0 PQZ W W Barnes, 12 Brockley Hall Rd, London, SE4 1RH [IO91XK, TQ37]
G0 PRB R K Kingston, Small Ashes Farm, Marden, Hereford, Herefordshire, HR1 3DA [IO82PD, SO54]
G0 PRE K Daniels, 14 Windsor Wharf, Hackney Wick, London, E9 5NY [IO91XN, TQ38]
G0 PRF J A Goodwin, 15 Deer Croft Rd, Salendine Nook, Huddersfield, HD3 3SN [IO93BP, SE11]
G0 PRH M Grassi, 10 Oaks Dr, St. Leonards, Ringwood, BH24 2QT [IO90BT, SU10]
G0 PRI L J Ward, 20 The Green, Newby, Scarborough, YO12 5JA [IO94SH, TA08]
G0 PRJ W D Long, 29 Mitchells Rd, Haylands, Ryde, PO33 3JA [IO90JR, SZ59]
G0 PRK Details withheld at licensee's request by SSL.
GW0 PRM B D Goodier, 14 Meadowbank, Old Colwyn, Colwyn Bay, LL29 8EX [IO83DH, SH87]
G0 PRN T A C Bruin, Seaford, 38 Kirkley Cliff Rd, Lowestoft, NR33 0DB [JO02UL, TM59]
GM0 PRO P Greenway, 5 Java Pl, Craignure, Isle of Mull, PA65 6BG [IO76DL, NM73]
G0 PRQ M J Shield, Lilac Cottage, 1 High Rd, Stanley, Crook, DL15 9SN [IO94DR, NZ13]
G0 PRR P A Roper, 31 Fern Way, Ilfracombe, EX34 8JS [IO71WE, SS54]
G0 PRS C A Baverstock Poole Radio Sct, 43 Tatnam Rd, Longfleet, Poole, BH15 2DW [IO90AR, SZ09]
G0 PRT Details withheld at licensee's request by SSL.
G0 PRU D D Dyer Prudential ARS, High Bank Cottage, Underhill, Moulsford, Wallingford, OX10 9JH [IO91KN, SU58]
G0 PRV E Stokes, Grosvenor Rd PO, Cheadle Hulme, Cheadle, Ches, SK8 5QJ [IO83VJ, SJ88]
G0 PRW Details withheld at licensee's request by SSL.
G0 PRY D McNab, 10 Rainham Gdns, Alvaston, Derby, DE24 0DJ [IO92GV, SK33]
G0 PRZ D C Symonds, 29 Pine Cres, Chandlers Ford, Eastleigh, SO53 1LN [IO90HX, SU42]
G0 PSB I T Wright, 222 Barton Rd, Barton Seagrave, Kettering, NN15 6RZ [IO92PJ, SP87]
G0 PSC M Suzuki, 3-5-13 Hamadayama, Suginamiku, Tokyo 168, Japan, X X
G0 PSD Details withheld at licensee's request by SSL.
G0 PSE Details withheld at licensee's request by SSL.
G0 PSF P N Yeatman, 110 Sunnymead Dr, Waterlooville, PO7 6BX [IO90LV, SU61]
G0 PSG R Carvell, 26 Greenfield Ave, Kettering, NN15 7LL [IO92PJ, SP87]
G0 PSI J A Wood, 18 Kennedy Ave, Long Eaton, Nottingham, NG10 3GF [IO92IV, SK43]

G0 PSJ S Jacques, 37 St. Francis Rd, Gosport, PO12 2UG [IO90KS, SZ69]
G0 PSK G W A Hawkins, 8 Broughton Rd, West Ayton, Scarborough, YO13 9JW [IO94SF, SE98]
G0 PSL P W Daddy, 52 Seafield Ave, Holdernessrd, Hull, HU9 3JQ [IO93US, TA13]
G0 PSM Details withheld at licensee's request by SSL.
G0 PSN Details withheld at licensee's request by SSL.
G0 PSO P S O'Nion, 7 Ettington Cl, Cheltenham, GL51 0NY [IO81WV, SO92]
GU0 PSP M J Dowding, Lancrage, Les Marais, St. Pierre Du Bois, Guernsey, GY7 9LD
GM0 PSQ W M Ireland, 78 Torridon Rd, Broughty Ferry, Dundee, DD5 3JH [IO86NL, NO43]
GM0 PSS T S Goody Ssrc, Strathallan School, Forgandenny, Perth, Perthshire, PH2 9EG [IO86GI, NO01]
G0 PST C Smelt, Poplar House, Endyke Ln, Cottingham, Hull, N H'Side, HU6 8TG [IO93TS, TA03]
G0 PSU Dr T B Smith, 4 York St., Bedford, MK40 3RJ [IO92SD, TL04]
GW0 PSV G A Wardman, 129 Shingrig Rd, Nelson, Treharris, CF46 6DU [IO81IP, ST19]
G0 PSW K E Brown, 131 Bramble Rd, Hatfield, AL10 9SD [IO91US, TL20]
G0 PSY S J Brodie, Top of The Hill, Hospital Rd, Wicklewood, Wymondham Norfolk, NR18 9PR [JO02MN, TG00]
G0 PSZ L G Banaszak, 17 Stoney Piece Cl, Bozeat, Wellingborough, NN29 7NS [IO92PF, SP95]
G0 PTA R F Attwood, 2 Elizabeth Rd, Basingstoke, RG22 6AX [IO91KG, SU65]
G0 PTB J B Fallon, 132 Belthorn Rd, Belthorn, Blackburn, BB1 2NN [IO83SR, SD72]
G0 PTC J Bailey, 21 Mayfair Ave, Mansfield, NG18 4EQ [IO93JD, SK56]
G0 PTD A Washington, 22 Elm Tree Dr, Bignall End, Stoke on Trent, ST7 8NG [IO83UB, SJ85]
G0 PTG J L M Mattison, Aylmers Farm, Sheering Lower Rd, Old Harlow, Harlow, CM17 0NE [JO01BS, TL41]
G0 PTI H P Aigeldinger, 14 Peregrine Ave, Morley, Leeds, LS27 8TD [IO93FR, SE22]
G0 PTJ Details withheld at licensee's request by SSL.
G0 PTK D J Dunford, 15 Lower Rea Rd, Brixham, TQ5 9UD [IO80FJ, SX95]
G0 PTL D Caley, 5 Crosswood Cl, Lothian Way, Bransholme, Hull, HU7 5BU [IO93UT, TA13]
G0 PTM A M Baird, 65 Waterpump Ct, Thorplands, Northampton, NN3 8UR [IO92NG, SP76]
G0 PTN N C Heggerty, Abersoch, Main Rd, Ansty, Coventry, CV7 9JA [IO92HK, SP38]
G0 PTO G Widdop, 9 Coventry Ave, Cheadle Heath, Stockport, SK3 0QS [IO83VJ, SJ88]
GM0 PTP R Bennett, 20 Kettins Terr, Downfield, Dundee, DD3 9RJ [IO86LL, NO33]
GI0 PTQ P A T R Keenan, Drumbadreeuagh, Belleek, Co Fermanagh, N Ireland, BT93 3FT [IO54WL, G95]
G0 PTR A Ryland, 9 Asquith Rd, Leckhampton, Cheltenham, GL53 7EJ [IO81XV, SO92]
G0 PTS Details withheld at licensee's request by SSL.
G0 PTT K Caunce, Northmead, Main Rd, Bouldnor, Yarmouth, Isle of Wight, PO41 0UX [IO90GQ, SZ38]
G0 PTU J N Davies, Siebenburgenweg 9, Hinrichssegen, 8206 Bruckmuehl, Germany
G0 PTW J B Sinkinson, 11 Lowlands Rd, Morecambe, LA4 5SB [IO84NB, SD46]
GW0 PTX J M Jones, 48 Morfar Garreg, Pwllheli, LL53 5AU [IO72TV, SH33]
GM0 PTY A Higgins, 37 Knock Point, Stornoway, Isle of Lewis, PA86 0BW [IO76JB, NS08]
G0 PTZ A J Wedgwood Military Wireless Amateur Radi, 10 Milner Pl, London, N1 1TN [IO91WM, TQ38]
G0 PUA J Bennett, 18 Riviere Towans, Phillack, Hayle, TR27 5AF [IO70HE, SW53]
G0 PUB P R Swynford, 219 Wykeham Rd, Earley, Reading, RG6 1PL [IO91MK, SU77]
G0 PUD D F Shaw, 50 Elmpark Way, Rodley Moor, Rochdale, OL12 7JQ [IO83VP, SD81]
G0 PUF Details withheld at licensee's request by SSL.
G0 PUG M P Sables obo South Yorks Packet Users G, 54 Harvey St., Deepcar, Sheffield, S30 5QB [IO93GJ, SK38]
GW0 PUH K T Larcombe, Tudor Bungalow, Osborne Rd, Pontypool, NP4 6NR [IO81LR, SO20]
G0 PUK A Johnson, 3 Plantation Ave, Swalwell, Newcastle upon Tyne, NE16 3JN [IO94DW, NZ26]
G0 PUL M T Day, 23 Queens Gdns, Cranham, Upminster, RM14 1NG [JO01DN, TQ58]
GW0 PUM D P Jenkins, Gwalia House, 143A Priory St., Carmarthen, SA31 1LR [IO71UU, SN42]
GM0 PUN H J Heritage, 6 Newton Pl, Rosyth, Dunfermline, KY11 2LX [IO86HA, NT18]
G0 PUO C F O Sullivan, 5 Donoghue Cttgs, Maroon St., London, E14 7SH [IO91XM, TQ38]
GW0 PUP G B Brown, 17 High St., Senghenydd, Caerphilly, CF8 4GG [IO81HN, ST08]
G0 PUQ H M OHare, 4 Clandon Ct, Albion Rd, Sutton, SM2 5TA [IO91VI, TQ26]
G0 PUR J C M Andrews, 27 Edward Way, Ashford, TW15 3AY [IO91SK, TQ07]
G0 PUT N A Richardson, 7 Wellfield Cl, Tilehurst, Reading, RG3 5HP [IO91LJ, SU66]
G0 PUU M L Scarr, 2A Rectory Rd, Piddlehinton, Dorchester, Dorset, DT2 7TE [IO80TS, SY79]
G0 PUV R J Maskill 115 Peterborough Squadron ARS, 21 Clayton, Orton Goldhay, Peterborough, PE2 5SB [IO92UN, TL19]
G0 PUW G W Taylor, 25 Chalford Rd, Erdington, Birmingham, B23 5DE [IO92BM, SP09]
G0 PUX C A Collins, 1 Nina Villas, Goudhurst Rd, Marden, Tonbridge, TN12 9JQ [JO01FE, TQ74]
G0 PUY C S Duckworth, 121 Mill Gate, Newark, NG24 4UA [IO93OB, SK75]
GI0 PUZ I Crawford, 101 Ardowen, Monbrief, Craigavon, BT65 5EB [IO64TK, J05]
G0 PVA O P Haffenden, 6 Berrystead, Hartford, Northwich, CW8 1NG [IO83RF, SJ67]
G0 PVB B Sketcher, 84 The Oval, Gilstead, Bingley, BD16 4RQ [IO93CU, SE13]
G0 PVE K Greaves, 22 Marsden Ave, Queniborough, Leicester, LE7 3FL [IO92LQ, SK61]
G0 PVF P A Benson, 21 Farleigh Rd, New Haw, Addlestone, KT15 3HS [IO91RI, TQ06]
GI0 PVG T Lyons, 3 Clanbrassil Gdns, Portadown, Craigavon, Co. Armagh, BT63 5YD [IO64SK, J05]
G0 PVH Details withheld at licensee's request by SSL.
G0 PVI R M J Loveland, 6 Juniper Cl, Whiteshill, Whitehill, Bordon, GU35 9EZ [IO91NC, SU73]
G0 PVJ E R Hewitt, 8 Embleton Rd, Headley Down, Bordon, GU35 8AJ [IO91OC, SU83]
G0 PVK Details withheld at licensee's request by SSL.
G0 PVL Details withheld at licensee's request by SSL.
G0 PVN C J Fleet, 8 Gurney Ct, Eaton Bray, Dunstable, LU6 2DZ [IO91RU, SP92]
G0 PVO L Hewitt, Sunny Nook, Grains Rd, Shaw, Oldham, OL2 8JF [IO83XN, SD90]
G0 PVP C F Duffy, 590 Chorley Old Rd, Bolton, BL1 6AA [IO83SO, SD61]
G0 PVQ P T E Fuller, 19 Greenwood Ct, Webb Cl, Broadfield, Crawley, RH11 9JH [IO91VC, TQ23]
G0 PVR J D Davies, 54 Buckingham Rd, Stalybridge, SK15 1BL [IO83XL, SJ99]
G0 PVS R W Phillips, 23 Hammonds Croft, Hixon, Stafford, ST18 0PQ [IO82XT, SJ92]
G0 PVT D C Henderson, 7 Love Ave, Fordley, Dudley, Cramlington, NE23 7BH [IO95EB, NZ27]
G0 PVU Rev J Roberts, 31 Seaton Way, Marshside, Southport, PR9 9GJ [IO83MQ, SD32]
G0 PVV W J Collins, 61 Woodrush Rd, Purdis Farm, Ipswich, IP3 8RB [JO02OA, TM24]
G0 PVW Details withheld at licensee's request by SSL.
G0 PVX A C Wilson, Wateredge, Crosthwaite, Kendal, Cumbria, LA8 8HX [IO84NH, SD49]
G0 PVY A Heward, 22 Ross Ave, Leasowe, Wirral, L46 2SB [IO83LK, SJ29]
G0 PVZ T Cantwell, 57 Oriel Dr, Old Roan, Liverpool, L10 3JN [IO83ML, SJ39]
G0 PWA D Williams, 31 Piper Hill Ave, Manchester, M22 4DZ [IO83UJ, SJ89]
G0 PWB B D Barrett, 26 Rosamund Ave, Pickering, YO18 7HF [IO94OF, SE88]
G0 PWC B A Dawe, 6 Ullswater Ave, Stourport on Severn, DY13 8QP [IO82UI, SO87]
G0 PWD R F Daniels, 18 Coniston Ave, Berwick, Victoria, Australia 3806
G0 PWE J C Lloyd, 203 Chestnut St., Ashington, NE63 0BT [IO95FE, NZ28]
G0 PWF F W G Parkman, 35 Beatrice Ave, East Cowes, PO32 6HR [IO90IR, SZ59]
G0 PWG A H G Woods, 14 Anstead Dr, Rainham, RM13 7QS [JO01CM, TQ58]
G0 PWH P W Hughes, 51 Highfield Cres, Brighton, BN1 8JD [IO90WU, TQ30]
G0 PWI B Davies, Llais Ceirw, 5 Maesmor Cttgs, Maerdy, Corwen, LL21 0NS [IO82GX, SJ04]
G0 PWJ J L Harwood, 171 West Coker Rd, Yeovil, BA20 2HE [IO80QW, ST51]
G0 PWK S Alder, 1 Oakdene Terr, Middlestone Moor, Spennymoor, DL16 7BA [IO94EQ, NZ23]
G0 PWL S J Wright, 33 Virginia Ave, Lydiate, Liverpool, L31 2NN [IO83MM, SD30]
G0 PWM H A S Bond, 36 Bachelor Hill, York, YO2 3BD [IO93KW, SE55]
G0 PWN N Lee, 27 Coach Rd, Brotton, Saltburn By The Sea, TS12 2RB [IO94MN, NZ61]
G0 PWO A S Boyes, 7 Thornwood Covert, Foxwood Hill, York, YO2 3LF [IO93KW, SE55]
G0 PWP W Jones, 78 Chantry Rd, East Ayton, Scarborough, YO13 9ER [IO94SG, SE98]
G0 PWQ W T Tonks, 295 Quinton Rd West, Quinton, Birmingham, B32 1PG [IO92AK, SP08]
G0 PWR K E Staley, 69 Tedder Ave, Henlow Camp, Henlow, SG16 6HW [IO92UA, TL13]
G0 PWS Details withheld at licensee's request by SSL.
G0 PWU G F Brown, 21 Armada Dr, Teignmouth, TQ14 9NF [IO80FN, SX97]
G0 PWV H J Chorley, 19 Cleeve Rd, Priorswood, Taunton, TA2 8DX [IO81KA, ST22]
G0 PWW Rev T W Edwards, P O Box 8966, Camp Lejeune, North Carolina 28547, USA, X X
G0 PWX G A Richards, 87 Woodlands Rd, Ditton, Aylesford, ME20 6EF [JO01FH, TQ75]
G0 PWZ P Waite, 75 Stockton Rd, Darlington, DL1 2RZ [IO94FM, NZ31]
G0 PXA G Petri, Mount Holly, Castledon Rd, Downham, Billericay, CM11 1LH [JO01GO, TQ79]
G0 PXB C C Marsh, 4 South Parade, South Kyme, Lincoln, LN4 4AQ [IO93VA, TF14]
G0 PXC G F Reeve, 17 Prior Rd, Thorpe St. Andrew, Norwich, NR7 0LX [JO02QP, TG21]
G0 PXD A Harrison, 25 Lansbury Ave, Rossington, New Rossington, Doncaster, DN11 0AA [IO93LL, SK69]
G0 PXE M W Cook, 41 Royston Ave, Doncaster, DN5 9RB [IO93KM, SE50]
G0 PXF K T Linsley, 132 Rein Rd, Tingley, Wakefield, WF3 1JB [IO93ER, SE22]
G0 PXG M J Hardman, 10 Lansdale Gdns, Manchester, M19 1QX [IO83VK, SJ89]
G0 PXH B Wilkinson, 20 Portree Cres, Blackburn, BB1 2HB [IO83SR, SD72]
G0 PXI P W Rigby, 41 St. Huberts Rd, Great Harwood, Blackburn, BB6 7AS [IO83TS, SD73]
G0 PXJ P McVey, 18 Worlebury Hill Rd, Weston Super Mare, BS22 9SP [IO81MI, ST36]
G0 PXK N J Pratt, 23 Hall Ln, Whitwick, Coalville, LE67 5FD [IO92HR, SK41]
G0 PXL D V Martland, 6 Omega Way, Trentham, Stoke on Trent, ST4 8TF [IO82VX, SJ84]
G0 PXM G M Kirby, Tralee, Brookside, Brighstone, Isle of Wight, Hants, PO30 4DJ [IO90HP, SZ48]
G0 PXN P G Corkin, Whinney Moor Cottage, Wynyard Est, Wolviston, Billingham, Cleveland, TS22 5NH [IO94HP, NZ42]
G0 PXO J R W Morgan, 5 Sealy Cl, Wirral, L63 9LP [IO83MH, SJ38]
G0 PXP T Cox, 60 Seven Oaks Cres, Beeston, Nottingham, NG9 3FP [IO92JW, SK53]
G0 PXQ C A Bell, 17 Jubilee Sq, South Hetton, Durham, DH6 2TR [IO94HT, NZ34]
GM0 PXR F W Coghill, Largiemore, Otter Ferry, By Tighnabruaich, PA21 2DH [IO76IA, NR98]
GI0 PXS J K Madden, The Cottage, 53 Clarendon St., Londonderry, BT48 7ER [IO65IA, C41]
G0 PXT E T Denman, 15 Clare Way, Bexleyheath, DA7 5JU [JO01BL, TQ47]
GM0 PXV P J Barclay, 87 Silverknowes Gdns, Edinburgh, EH4 5NF [IO85IX, NT27]
G0 PXX E Mason, 59 Radburn Rd, New Rossington, Doncaster, DN11 0RP [IO93LL, SK69]
G0 PXY D C Smith, 43 Totteridge Rd, Enfield, EN3 6NF [IO91XQ, TQ39]
G0 PXZ G D Walker, 54 Burnage Ln, Burnage, Manchester, M19 2NL [IO83VK, SJ89]
G0 PYB Details withheld at licensee's request by SSL.
GM0 PYC A Duncan Banff & Dist AR, Drumduan, Bellevue Rd, Banff, AB45 1BJ [IO87RP, NJ66]
G0 PYD M N Hague, Brookview, Millburn Grange, Coventry Rd, Kenilworth, CV8 2FE [IO92FI, SP27]
G0 PYF A H de Buriatte, Tanglewood, East End, North Leigh, Oxon, OX8 6PZ [IO91GT, SP31]
G0 PYI G L W Bodaly, 41 Robert St., Northampton, NN1 3BL [IO92NF, SP76]
G0 PYJ J E Swatton, 22 Hallsland, Crawley Down, Crawley, RH10 4XZ [IO91XD, TQ33]
G0 PYK P J Brettell, 26 Beverston Rd, Tipton, DY4 0DF [IO82XN, SO99]
G0 PYL Details withheld at licensee's request by SSL.
GM0 PYM J Quigley Paisley & D ARS, 90 George St., Paisley, PA1 2JR [IO75SU, NS46]
GW0 PYN C L Rogers, 6 Netley Rd, Rhyl, LL18 2AN [IO83GH, SJ08]
GU0 PYO C J A Petit, Homedale-le-Vauquiedor, St. Andrew, Guernsey, GY6 8TT
G0 PYQ P J Moran, 40 Uldale Rd, Upperby, Carlisle, CA2 4PP [IO84MU, NY45]
G0 PYR Details withheld at licensee's request by SSL.
G0 PYS D T Rose, 6 Newtown Ln, Kimbolton, Huntingdon, PE18 0HT [IO92TH, TL06]
GW0 PYT Details withheld at licensee's request by SSL.
GW0 PYU H E J Clarke, 3 Tynrallt Ave, Bridgend, CF31 1PQ [IO81FM, SS98]
G0 PYV M S H Hainesborough, 47 Elmwood, Sawbridgeworth, CM21 9NN [JO01BT, TL41]
G0 PYW A Haworth, 16 Anson Way, Braintree, CM7 9TN [JO01GV, TL72]
G0 PYX Details withheld at licensee's request by SSL.
G0 PYY Details withheld at licensee's request by SSL.
GW0 PYZ Details withheld at licensee's request by SSL.
G0 PZA Details withheld at licensee's request by SSL.
G0 PZC D D M Flitterman, Flat 7, 1 Rutland Gate, London, SW7 1BL [IO91WM, TQ27]
G0 PZD G Holmes, 6 Darleydale Dr, Eastham, Wirral, L62 8EX [IO83MH, SJ38]
G0 PZE J S Rule Poldhu ARC, The Workshop, Meaver Rd, Mullion, Helston, TR12 7DN [IO70JA, SW61]
G0 PZF J P OConnell, 11 Suffolk Ave, Shirley, Southampton, SO15 5EF [IO90GV, SU41]
G0 PZG K H Bremerman, 7 Dunfane, Billericay, CM11 1EL [JO01FP, TQ69]
G0 PZH C A Lazenby, 28 Castle Way, Hessle, HU13 0DU [IO93SR, TA02]
G0 PZI N B Campbell, Beeleigh, Chalfont Ave, Little Chalfont, Amersham, HP6 6RF [IO91RQ, SU99]
G0 PZJ R L Pope Duxford RS, 95 Northolt Ave, Bishops Stortford, CM23 5DS [JO01CV, TL42]
G0 PZK W J F Cocker, 9 Spinacre, Becton Green, Barton on Sea, New Milton, BH25 7DF [IO90ER, SZ29]
G0 PZM N H Byron, 2 St. Aidans View, Boosbeck, Saltburn By The Sea, TS12 3LS [IO94MN, NZ61]
G0 PZN A M D Jones, Edorene, Tincleton, Dorchester, Dorset, DT2 8QR [IO80UR, SY79]
G0 PZO C K Jordan, 31 Rocky Bank Rd, Tranmere, Birkenhead, L42 7LB [IO83LJ, SJ38]
G0 PZP W Rabbitt, 21 Barnfield Rd, Woolston, Warrington, WA1 4NW [IO83RJ, SJ68]
GW0 PZQ R Burdon, 25 Derrie Ave, Abergele, LL22 7TF [IO83EG, SH97]
G0 PZR S M Cooper Penzance Radio Club, 6 Polmeere Rd, Treneere Est, Penzance, TR18 3PD [IO70FD, SW43]
GW0 PZS T Edwards, 6 Cottage Home, Newborough, Llanfairpwllgwyngyll, LL61 6SY [IO73TD, SH46]
GW0 PZT E J Allely, Dwyfor, Rhiw, Pwllheli, Gwynedd, LL53 8AE [IO72QT, SH22]
GW0 PZU A Ward, 158 Mold Rd, Alltami, Mynydd Isa, Mold, CH7 6TF [IO83KD, SJ26]
GM0 PZV G F Paterson, 39 Underwood Pl, Balloch, Inverness, IV1 2RF [IO77WL, NH74]
G0 PZW Dr J A Birch, 5 Breamish Dr, Washington, NE38 9HS [IO94FV, NZ25]
G0 PZX A C Dennis, 211 Beacon Rd, Luton, Chatham, ME5 7BU [JO01GI, TQ76]
G0 PZY Details withheld at licensee's request by SSL.
GW0 PZZ M I Owen, 90 Shakespeare Ave, Penarth, CF64 2RX [IO81JK, ST17]
GM0 RAB Details withheld at licensee's request by SSL.
GW0 RAD J L Lewis, White Ox Cottage, 6 High St., Abergwili, Carmarthen, SA31 2JA [IO71UU, SN42]
G0 RAE R Walker, 39 Ashbourne Dr, Pontefract, WF8 3RA [IO93IQ, SE42]
G0 RAF R Pickles RAF Waddington ARC, The Bell Inn, Far Ln, Coleby, Lincoln, LN5 0AH [IO93RD, SK96]
GU0 RAG K G de La Haye, Flat 6, 65 Victoria Rd, St. Peter Port, Guernsey, GY1 1JB
G0 RAH Details withheld at licensee's request by SSL.
G0 RAI Details withheld at licensee's request by SSL.
G0 RAJ A L Heathcote, 25 Commonhall Ln, Hadleigh, Benfleet, SS7 2RN [JO01HN, TQ88]
G0 RAL P J Vallis, Gryphon Dirtham Ln, Effingham, Leatherhead, Surrey, KT24 5SD [IO91TG, TQ15]
G0 RAM M P King, 17 Green Acres, Long Meadow, Stevenage, SG2 8ND [IO91VV, TL22]
G0 RAN M L Jamil, 29 Harrow Cl, Blackford Bridge, Bury, BL9 9UD [IO83UN, SD80]
GM0 RAO A G Williamson, Cairn Cottage, Denside of Durris, nr Banchory, Kincardineshire, AB31 3DT [IO87RB, NO69]
G0 RAP Details withheld at licensee's request by SSL.
G0 RAQ Details withheld at licensee's request by SSL.
G0 RAR Details withheld at licensee's request by SSL.
G0 RAS V H Maddex, Walnut Cottage The, Vines Shabbington, Aylesbury, Bucks, HP18 9HH [IO91LS, SP60]
G0 RAT W B Barnes, 17 Saxon Way, Bradley Stoke, Bristol, BS12 9AR [IO81RM, ST68]
G0 RAU D W Woodnutt, 17 Hill Farm Rd, Chalfont St. Peter, Gerrards Cross, SL9 0DD [IO91RO, TQ09]
G0 RAV R W Ravenscroft, 4 The Paddock, Lidlington, Bedford, MK43 0RW [IO92RA, SP93]
G0 RAX R G Preston, 45 Long Meadow, Skipton, BD23 1BP [IO83XX, SD95]
GM0 RAY M M McCreery East Renfrews Rd, 1 Duart Dr, Newton Mearns, Glasgow, G77 5DS [IO75US, NS55]
G0 RAZ Details withheld at licensee's request by SSL.[Station located near Ely, Cambridgeshire. Op: A R Last. QSL via G4NSY.]
G0 RBA E W Bannister, 62 St. Georges Rd, Winsford, CW7 1BY [IO83RE, SJ66]
G0 RBB M D Batchelor, 16 Clementi Ave, Holmer Green, High Wycombe, HP15 6TN [IO91PP, SU99]
GI0 RBC J M C B Thompson, 3 Strandburn Park, Sydenham, Belfast, BT4 1ND [IO74BO, J37]
G0 RBD D P Kiely, 45 Redland, Chippenham, SN14 0JB [IO81WL, ST97]
G0 RBE Details withheld at licensee's request by SSL.
G0 RBG H W Holmes Broadland RG, 7 Parkland Cres, Old Catton, Norwich, NR6 7RQ [JO02PP, TG21]
GW0 RBH Dr R B Hughes, 17 Pentrosfa Rd, Llandrindod Wells, LD1 5NL [IO82HF, SO05]
G0 RBI S Ward, 43 Farmleigh Gdns, Great Sankey, Warrington, WA5 3FA [IO83QJ, SJ58]
G0 RBJ P B Evans, 8B Royden St., Liverpool, L8 4UN [IO83MJ, SJ38]
G0 RBK J F Tattersall, 34 Shellfield Rd, Southport, PR9 9UR [IO83MQ, SD31]
G0 RBL M D Adams J Hanson Sch AR, J Hanson School, Croye Cl, Andover, Hants, SP10 3AB [IO91GE, SU34]
G0 RBM C F Boland, 13 Rushfield Cres, Brookvale, Runcorn, WA7 6BN [IO83PH, SJ58]
G0 RBN P G Dolling RAF Brize Norton ARC, Time House, 30 The Cres, Carterton, OX18 3SJ [IO91ER, SP20]
GI0 RBO J N Kernohan, 15 Tullygrawley Rd, Teeshan, Ballymena, BT43 5NP [IO64UV, D00]
G0 RBQ R N Gibbs, 33 Winterbourne, Horsham, West Sussex, RH12 5JW [IO91UC, TQ13]
G0 RBR J D Brown, Larchdale, Church Rd, Charing, Ashford, TN27 0HE [JO01JF, TQ95]
GI0 RBS Rev R C Rainey, 14 Mount Eden, Limavady, BT49 0RP [IO65MA, C62]
G0 RBT J H Donaldson, 7 Lakeside, South Shields, NE34 7HA [IO94HX, NZ36]
G0 RBV P S Brunton, 276 Maidstone Rd, Rochester, ME1 3PJ [JO01GI, TQ76]
G0 RBW T B Jones, 4 Lowerfold Cres, Rochdale, OL12 7HZ [IO83VP, SD81]
G0 RBX A H Brooks, 22 High St., Alderney, Guernsey, GY9 3UG
G0 RBY Details withheld at licensee's request by SSL.
GW0 RBZ M F Allen, Maelgwyn House, Bow St, Dyfed, SY24 5BE [IO72XK, SN68]
G0 RCB T E Baker, 99 Greenland Rd, Durrington on Sea, Worthing, BN13 2RN [IO90TT, TQ10]
G0 RCC Details withheld at licensee's request by SSL.
G0 RCE K P Preston, The Cottage, Banks Green, Bentley, nr Redditch Worcs, B97 5SU [IO82XH, SO96]
G0 RCF E J Carrington, 4 Lancaster Dr, East Grinstead, West Sussex, RH19 3XF [JO01AD, TQ43]
GW0 RCG R C Giddings, 28 Ashgrove, Port Talbot, SA12 8PP [IO81CO, SS79]
G0 RCH T W Cullup, 13 London St., Whittlesey, Peterborough, PE7 1BP [IO92WN, TL29]
G0 RCI A Gibson, 1 Oakleigh Rd, Grantham, NG31 7NN [IO92PV, SK83]
G0 RCJ J O Topham, 23 St. Nicholas View, West Boldon, East Boldon, NE36 0RF [IO94GW, NZ36]
G0 RCK Details withheld at licensee's request by SSL.
G0 RCL O D Baldwin, 7 Gildercliffe, Scarborough, YO12 6NT [IO94SG, TA08]
G0 RCM Details withheld at licensee's request by SSL.
G0 RCN N T Allen, 78 Bargates, Christchurch, BH23 1QL [IO90CR, SZ19]
G0 RCO A F Cundy, 31 Alexandra Rd, St. Austell, PL25 4QW [IO70OI, SX05]
G0 RCP P J Mellors, 11 Nursery Rd, Colerne, Chippenham, SN14 8BZ [IO81UK, ST87]
G0 RCS G P Bateman R Sig Scarb ARC, Apartment, 31-33 Filey Rd, Scarborough, YO11 2TP [IO94TG, TA08]
G0 RCU R E W Thomas, 12 Kings Walk, Bishopsworth, Bristol, BS13 8AX [IO81QK, ST56]
G0 RCW F G A Mossop W Cheshire Ray, 14 Websters Ln, Great Sutton, South Wirral, L66 2LH [IO83MG, SJ37]
G0 RCX C A Garbett, 49 Wheatley St., West Bromwich, B70 9TL [IO82XM, SO99]
G0 RCY M J Crimes, 27 Dunmore Rd, Little Sutton, South Wirral, L66 4PD [IO83MG, SJ37]
GW0 RCZ Details withheld at licensee's request by SSL.
GM0 RDA G H Adamson, 31 Kirkgate, Burntisland, KY3 9DL [IO86JB, NT28]
G0 RDB C E Fernie, 19 Arbury Rd, Cambridge, CB4 2JB [JO02BF, TL46]
GM0 RDC C M Robinson, Greenhouse Farm, Lilliesleaf, Melrose, Roxburghshire, TD6 9EP [IO85PM, NT52]
G0 RDD M J Prendergast, 21 Westmorland Rise, Peterlee, SR8 2EX [IO94HS, NZ44]
G0 RDF L Wolstenholme, The Hollies, Avondale Rd, Chesterfield, Derbyshire, S40 4TF [IO93GF, SK37]
G0 RDG G J Kowalski, 47 Graveney Pl, Springfield, Milton Keynes, MK6 3LU [IO92PA, SP83]
G0 RDH B A Watson, 7 Branksome Dr, Morecambe, LA4 5UJ [IO84NB, SD46]
G0 RDI I Philipps, 24 Acres End, Amersham, HP7 9DZ [IO91QP, SU99]
GI0 RDJ I Mc Mullan, 35 Howard Pl, Park, Lisburn, BT28 1EX [IO64XM, J26]
G0 RDL M R Smith, 25 Ewart Rd, Walsall, WS2 0EU [IO82XO, SO99]

GI0 RDM K M McGuckin, 20 Lisnahull Park, Dungannon, BT70 1UH [IO64OM, H76]
G0 RDN G L Johnston, 11 Granville St., Deal, CT14 7EZ [JO01QF, TR35]
G0 RDO J A Snell, 5 Waverley Rd, Newton Abbot, TQ12 2ND [IO80EM, SX87]
G0 RDP D Peat, Jeswyn, Brookland Ave, Mansfield, NG18 5NB [IO93JD, SK56]
G0 RDQ Details withheld at licensee's request by SSL.
G0 RDR B Lowe, 48 Northwood Falls, Woodlesford, Woodlesford, Leeds, LS26 8PD [IO93GS, SE32]
G0 RDS A Williams, 30 Swan Cl, Talke, Stoke on Trent, ST7 1TA [IO83UB, SJ85]
G0 RDT D Treen, 13 Peveril Rd, Old Duston, Northampton, NN5 6JW [IO92MF, SP76]
G0 RDU S W A Emms, 33 Whitworth Ave, Stokealdermoor, Coventry, CV3 1EQ [IO92GJ, SP37]
G0 RDV L J L Davies, 2 Kettonby Gdns, Headlands, Kettering, NN15 6BT [IO92PJ, SP87]
G0 RDW Details withheld at licensee's request by SSL.
G0 RDX P M Walker, Moze Cross Cottage, Beaumont Rd, Great Oakley, Harwich, CO12 5BQ [JO01OV, TM12]
G0 RDY G Steel, Long Cl, 82 Whatton Rd, Kegworth, Derby, DE74 2DT [IO92IT, SK42]
GM0 RDZ S Smith, 12 Home Ave, Duns, TD11 3HQ [IO85US, NT75]
G0 REA R E A James, Woodpeckers, Freshwater Ln, St. Mawes, Truro, TR2 5AR [IO70LD, SW83]
G0 REB C G Salmon, 20 Benham Rd, Greens Norton, Towcester, NN12 8DB [IO92LD, SP64]
G0 REC P L Hunt, 358 James Reckitt Ave, Hull, HU8 0JA [IO93US, TA13]
GM0 RED J M Murdoch obo East Dunbartonshire Raynet, 4 Cedar Dr, Milton of Campsie, Glasgow, G65 8AY [IO75WX, NS67]
G0 REE D N Jones, Mill Cottage, 120 Heathfield Rd, Keston, BR2 6BA [JO01AI, TQ46]
G0 REF J W Andrews Epp For Dist RG, 85 Little Cattins, Harlow, CM19 5RN [JO01AR, TL40]
G0 REG A J Wright Hewtt Reg, Hewett School, Cecil Rd, Norwich, NR1 2PL [JO02PO, TG20]
G0 REI H D Harrop, 79 Swanland Rd, Hessle, HU13 0NN [IO93SR, TA02]
G0 REJ Details withheld at licensee's request by SSL.
G0 REK D N Nicklin, 59 Laurel Rd, Armthorpe, Doncaster, DN3 2ES [IO93LM, SE60]
G0 REL D R Gaskell, 18 Woodcroft, Kennington, Oxford, OX1 5NH [IO91JR, SP50]
G0 REM T A Melnyczuk, 58 Northfield Rd, Milfield, Peterborough, PE1 3QJ [IO92VO, TF10]
G0 REN C R Wienrich, Cookhamdene, Manorpark, Chislehurst, Kent, BR7 5QD [JO01AJ, TQ46]
G0 REO J S Jones, 57 Cannon Hill Rd, Coventry, CV4 7BT [IO92FJ, SP37]
G0 REP A H Blackburn, Otterburn, 2 Blackthorne Rd, Stratford upon Avon, CV37 6TD [IO92DE, SP15]
G0 REQ D Hibberd, 11 Borrowdale, Brownsover, Rugby, CV21 1NH [IO92JJ, SP57]
G0 RER G M Hampson Dorset Raynet, 7 Merryfield Cl, Bransgore, Christchurch, BH23 8BS [IO90DS, SZ19]
GM0 RES Details withheld at licensee's request by SSL.
G0 RET W J G Wilson, 8 Kenilworth Gdns, Hayes, UB4 0AY [IO91SM, TQ08]
G0 REU T P Lam, Flat 2, Wing Fat House 3/F, Ma Miu Rd Yuen Long, Hong Kong
G0 REV A Bowmaker, Post Cottage, Ardley Rd, Somerton, Bicester, OX6 4LP [IO91IW, SP42]
GM0 REW J A Smith, 157 Elgin Dr, Glenrothes, KY6 2JT [IO86JE, NO20]
G0 REY J R Mackie, Field House, Birgham, Coldstream, Berwickshire, TD12 4NE [IO85UP, NT73]
GM0 REZ A J Dailey, 40 Croft Pl, Eliburn, Livingston, EH54 6RJ [IO85FV, NT06]
G0 RFA S P McGill, 2 Rombalds Pl, Armley, Leeds, LS12 2BD [IO93ET, SE23]
G0 RFB C A Topliss, 61 Woodstock Rd, Kingswood, Bristol, BS15 2UE [IO81SL, ST67]
G0 RFC D W Blackford Corn BR RAF ARS, 11 Hendras Parc, Carbis Bay, St. Ives, TR26 2TT [IO70GE, SW53]
G0 RFD M K Smith, 197 Conygre Gr, Bristol, BS12 7HZ [IO81RM, ST67]
G0 RFE A Moore, 139 Argyle St., Heywood, OL10 3RS [IO83VO, SD81]
G0 RFF C J Bourne, Essams, 11 The Grove, Ersham Park, Hailsham, BN27 3HU [JO00DU, TQ50]
G0 RFG E Hyde, 63 Newlyn Dr, Sale, M33 3LH [IO83UJ, SJ79]
G0 RFI S J Brackley, 28 Elwyn Dr, Liverpool, L26 0UY [IO83OI, SJ48]
G0 RFJ S Butler, Salesnook, Cocker Bar Rd, Leyland, Preston, PR5 3TA [IO83OQ, SD52]
GM0 RFK J Leitch, 42 St. Michaels Dr, Cupar, KY15 5BS [IO86MH, NO31]
G0 RFL T P Pooley, 133 Hardie Rd, Dagenham, RM10 7BT [JO01BN, TQ58]
G0 RFM J H Copplestone, 21 Birks Ave, Lees, Oldham, OL4 3PR [IO83XN, SD90]
G0 RFP Details withheld at licensee's request by SSL.
G0 RFQ G Buck, 3 James St., Colne, BB8 0HN [IO83WU, SD83]
G0 RFS Dr C W W Bradley, Reuterweg 52, Frankfurt /Main, Germany
G0 RFT R Lagar, 25 Neville Ave, Orford, Warrington, WA2 9BQ [IO83RJ, SJ69]
G0 RFV K Goodworth, Cantley, 12 Ewood Dr, Doncaster, DN4 6AU [IO93LM, SE60]
G0 RFW A McLean, 146 Station Rd, Earls Barton, Northampton, NN6 0NX [IO92OG, SP86]
G0 RFX P G Walford, Suite 184, 2 Old Brompton Rd, London, SW7 3DQ [IO91VL, TQ27]
G0 RFY D R Horton, 118 Accrington Rd, Burnley, BB11 5AE [IO83US, SD83]
G0 RFZ P Davis, 23 Castle Dr, Berwick upon Tweed, TD15 1NS [IO85XS, NT95]
G0 RGB Details withheld at licensee's request by SSL.
G0 RGC J H C Bridge, Oak Haven, Priory Rd, Dawlish, EX7 9JG [IO80GN, SX97]
G0 RGE M J Jenkinson, 25 Porchester Cl, Hucknall, Nottingham, NG15 7UB [IO93JA, SK54]
G0 RGF Details withheld at licensee's request by SSL.
G0 RGG Details withheld at licensee's request by SSL.
G0 RGH L A Crane Harig, Goodwill, Rectory Rd, Wrabness, Manningtree, CO11 2TR [JO01OW, TM13]
G0 RGJ R Provins, 42 Forest View Rd, Tuffley, Gloucester, GL4 0BX [IO81VT, SO81]
G0 RGL D J Edmondson, 64 Raleigh Ave, Hayes, UB4 0EF [IO91TM, TQ18]
G0 RGM J M Trice, Camfield, Langley Ln, Ifield, Crawley, West Sussex, RH11 0NB [IO91VC, TQ23]
G0 RGN B A Woodhead, 16 Dow St., Hyde, SK14 4BS [IO83XL, SJ99]
G0 RGO Revd J M Drummond, 14 Bulls Head Cotts, Turton, Bolton, Lancs, BL7 0HS [IO83TP, SD71]
G0 RGP A E Gibbs, 17 Manor Bend, Galmpton, Brixham, TQ5 0PB [IO80FJ, SX85]
G0 RGQ N Johnson, 12 Mulgrave Rd, Roe Green, Worsley, Manchester, M28 2RW [IO83TM, SD70]
G0 RGR Details withheld at licensee's request by SSL.
G0 RGS J R Masterson, Economic Skips, Tunnel Ave, Greenwich, London, XX99 1AA
G0 RGT Xx J A C Q Bellinger, Penthouse, 108 Hulbert Rd, Bedhampton, Havant, PO9 3TG [IO90LU, SU70]
G0 RGU Details withheld at licensee's request by SSL.
G0 RGV Details withheld at licensee's request by SSL.
G0 RGX J T Sandys, Tarn Cottage, The Maultway, Camberley, Surrey, GU15 1PS [IO91PI, SU96]
G0 RGY Details withheld at licensee's request by SSL.
G0 RHA R Archer, 44 Herringthorpe Ln, Rotherham, S65 3AS [IO93IK, SK49]
G0 RHB L F Mulford, 55 Mill Farm Cres, Hounslow, TW4 5PF [IO91TK, TQ17]
GW0 RHC K G Dyer, 34 Lundy Dr, West Cross, Swansea, SA3 5QL [IO71XN, SS68]
G0 RHD M T Hirst, 196 Parkhouse Farm Way, Havant, PO9 4DS [IO90LU, SU70]
GW0 RHE S Jones, 12 Danybanc, Felinfoel, Llanelli, SA15 4NS [IO71WQ, SN50]
G0 RHF P Ellwood, 10 Ingol Gr, Hambleton, Poulton-le-Fylde, FY6 9DN [IO83MV, SD34]
G0 RHG D J Stewart, 36 Monson Rd, Redhill, RH1 2EZ [IO91WG, TQ25]
G0 RHH D A Barnett, 3 Cartwright Dr, Shaw, Swindon, SN5 9QU [IO91CN, SU18]
G0 RHI B E Dooks, 7 Manor Dr, Kirby Hill, Boroughbridge, York, YO5 9DY [IO94HC, SE36]
G0 RHJ Details withheld at licensee's request by SSL.
G0 RHK P C Ford, 31 Basin Rd, Chichester, PO19 2PY [IO90OT, SU80]
G0 RHL D M Cardell, 22 Millview Rd, Heckington, Sleaford, NG34 9JP [IO92UX, TF14]
G0 RHM J S Cardell, 22 Millview Rd, Heckington, Sleaford, NG34 9JP [IO92UX, TF14]
G0 RHO J H Belling, 77 Chantry Rd, Marden, Tonbridge, TN12 9JD [JO01FE, TQ74]
GM0 RHP D N Crooke, 2 Main St., Carnock, Dunfermline, KY12 9JQ [IO86FC, NT08]
G0 RHQ Details withheld at licensee's request by SSL.
G0 RHR A E Pickles, 46 Lynton Rd, Southport, PR8 3AW [IO83LO, SD31]
G0 RHS Details withheld at licensee's request by SSL.
G0 RHT Details withheld at licensee's request by SSL.
G0 RHU M R Parish, 83 Harold Rd, Stubbington, Fareham, PO14 2QS [IO90JT, SU50]
G0 RHV J N Parish, 83 Harold Rd, Stubbington, Fareham, PO14 2QS [IO90JT, SU50]
G0 RHW J N Gee, 27 Farnley Cl, Norton, Windmill Hill, Runcorn, WA7 6NN [IO83QI, SJ58]
G0 RHY A Howarth, 12 Welbeck Rd, Worsley, Manchester, M28 2SL [IO83TM, SD70]
G0 RHZ C J G Purcell, Southerhay, The St., Felthorpe, Norwich, NR10 4AB [JO02OR, TG11]
G0 RIB A M Shaw, Offshore, Salterns Cl, Haylins Island, Hants, PO11 9PL [IO90MS, SZ79]
G0 RIC R Cannell, 284 Archway Rd, Highgate, London, N6 5AU [IO91WN, TQ28]
G0 RIE D T Reilly, 15 Shutewater Cl, Bishops Hull, Taunton, TA1 5EH [IO81KA, ST22]
G0 RIF D A Barnes, 6 Beacon Dr, Loughborough, LE11 2BD [IO92JS, SK51]
G0 RIH Details withheld at licensee's request by SSL.
G0 RII J R Spacey, 43 Woodlands Rd, Allestree, Derby, DE22 2HG [IO92GW, SK33]
G0 RIK N G G Stockwell, 20 Sutton Rd, Sutton Poyntz, Preston, Weymouth, DT3 6BX [IO80SP, SY78]
GW0 RIL L R Rees, Dunblair, Tremont Rd, Llandrindod Wells, LD1 5EB [IO82HF, SO06]
GW0 RIM Details withheld at licensee's request by SSL.
G0 RIN Details withheld at licensee's request by SSL.
G0 RIO Details withheld at licensee's request by SSL.
GM0 RIP J Austwick, Lower Flat, Holm Farm, Beattock, Dumfriesshire, DG10 9PG [IO85GH, NT00]
G0 RIQ D H Wisbey, 22 Rutland Dr, Hornchurch, RM11 3EN [JO01CN, TQ58]
G0 RIR W S Lewis, 59 Edmund Rd, Hastings, TN35 5LE [JO00HU, TQ81]
G0 RIS Dr J W E Fellows, Green Tye, 20 St. Peters Rd, West Mersea, Colchester, CO5 8LJ [JO01KS, TM01]
G0 RIT A D Nunneley, The Potter's Wheel, Mullion Cove, nr Helston, Cornwall, TR12 7ET [IO70IA, SW61]
G0 RIU P J Davis, 95 Buxton Rd, Stratford, London, E15 1QX [JO01AN, TQ38]
GM0 RIV N D Baird Raynet Inverness, 23 Scorguie Ave, Inverness, IV3 6SD [IO77UL, NH64]
G0 RIW A L Male, 68 Redhall Rd, Lower Gornal, Dudley, DY3 2NL [IO82WM, SO99]
G0 RIX B A Cook, 7 Rosewood Gdns, New Milton, BH25 5NA [IO90ES, SZ29]
G0 RIY G F Watson, 24 Richards Cl, Exmouth, EX8 4LQ [IO80HP, SY08]
G0 RIZ B H Body, Penolver, Scarcewater Vean, St. Clement, Truro, TR1 1TA [IO70LG, SW84]
G0 RJA K A Jones, 10 Westway, Fairfield, Droylsden, Manchester, M43 6FH [IO83WL, SJ99]
G0 RJB J W Burt, 62 Garland Sq, Tangmere, Chichester, PO20 6JF [IO90PU, SU90]
G0 RJC V C Turner, 7 Highfield Cres, Baildon, Shipley, BD17 5NR [IO93CU, SE13]
G0 RJD A R Patrick, 29 Tower Rd, Boston, PE21 9AH [IO92XX, TF34]
G0 RJE R J Enright, 45 Gorham Dr, Tonbridge, TN9 2DU [JO01DE, TQ64]
G0 RJF D S Taylor, 31 Southdale Rd, Ossett, WF5 8BA [IO93FQ, SE22]
GM0 RJG E Kelly, Durness, Newbridge, Dumfries, DG2 0QX [IO85EC, NX97]
G0 RJH E J White, Hillbrook, 1 Brook Cl, Helston, TR13 8NY [IO70IC, SW62]
G0 RJI N A Rapson, Flat A, Gr House, St. Austell Rd, St. Blazey Gate, Par, Cornwall, PL24 2EF [IO70PI, SX05]
G0 RJJ E P Foord, 65 Dane Ct Gdns, Broadstairs, CT10 2SD [JO01RI, TR36]
G0 RJK Details withheld at licensee's request by SSL.
G0 RJM A Jeffs, Inglenook, Sutton Rd, Huttoft, Alford, LN13 9RL [JO03DG, TF57]
G0 RJN H M Vicary, The Brambles, Wrotham Rd, Meopham Green, Gravesend, Kent, DA13 0QA [JO01EI, TQ66]
GI0 RJO L C Douglas, Mullans Hill, Drumsurn, Limavady, Derry, N Ireland, BT49 0PP [IO64NX, C71]
G0 RJP G A F Philpotts, 62 Erneley Cl, Stourport on Severn, DY13 0AH [IO82UH, SO87]
G0 RJQ R E A McCulloch, Welbeck, Pebble Ln, Brackley, NN13 7DA [IO92KA, SP53]
G0 RJT Rev H D Leak, 15 Sutherland Rd, Tittensor, Stoke on Trent, ST12 9JQ [IO82VW, SJ83]
GI0 RJU J M M Tosh, 452 Doagh Rd, Newtownabbey, BT36 6AW [IO74AQ, J38]
GW0 RJV G T Rogers, Maesgwersyl, Garthmyl, Montgomery, Powys, SY15 6RS [IO82JN, SO19]
G0 RJX E A Gaskell, 18 Woodcroft, Kennington, Oxford, OX1 5NH [IO91JR, SP50]
G0 RJY H Klose, Webereistrasse 4, D-86842 Tuerkheim, W Germany
G0 RKA O C Selden, 72 Emlyn Rd, London, W12 9TD [IO91VL, TQ27]
G0 RKB D P Roberts, 20 Beech Gr, Trowbridge, BA14 0HG [IO81VH, ST85]
G0 RKC A Alecio, 15 Camperdown Terr, Exmouth, EX8 1EH [IO80GO, SX98]
G0 RKD Details withheld at licensee's request by SSL.
G0 RKE C S Burgess, 12 Middleway, Grotton, Oldham, OL4 5SH [IO83XM, SD90]
G0 RKF Details withheld at licensee's request by SSL.
G0 RKG R K Gaskell, 18 Woodcroft, Kennington, Oxford, OX1 5NH [IO91JR, SP50]
GW0 RKH K Lawson, Dan-y-Deri, Aberedw, Builth Wells, Powys, LD2 3AR [IO82HD, SO05]
G0 RKJ R G Weiss, Tudor Hall, Wykham Park, Banbury, Oxon, OX16 9UR [IO92HB, SP43]
G0 RKK D M Gooding, 51 The Spinney, Downend, Fareham, PO16 8QD [IO90KU, SU60]
GJ0 RKM R A Hutt, 1 Milestone Cttgs, Main Rd, Fishbourne, Chichester, PO18 8AU [IO90OU, SU80]
G0 RKN H Burn, Ashleigh, Leek Rd, Werrington, Stoke on Trent, ST9 0DG [IO83XA, SJ94]
G0 RKO Details withheld at licensee's request by SSL.
G0 RKP J Aubin, 24 Beechwood Ave, Burnley, BB11 2PL [IO83US, SD83]
G0 RKQ R Plumtree, 80 Dewsbury Ave, Scunthorpe, DN15 8BP [IO93PO, SE81]
G0 RKS G R Goss, Slade Farm, Village Ln, Hedgerley, Slough, SL2 3XD [IO91QN, SU98]
G0 RKT D A Dukesell, 29 Gretton Rd, Fairfield, Buxton, SK17 7PW [IO93BG, SK07]
GM0 RKU P M Craft, 2 Luke Pl, Broughtyferry, Broughty Ferry, Dundee, DD5 3BN [IO86NL, NO43]
G0 RKV V C Webley, 1 Bates Cl, Willen, Milton Keynes, MK15 9HZ [IO92PB, SP84]
G0 RKY D J Coldwell, 3 Poplars Gr, Maidenhead, SL6 8HD [IO91PM, SU88]
G0 RLA P Harvey, 6 St. Johns Cl, Slitting Mill, Rugeley, WS15 2TG [IO92AS, SK01]
G0 RLB B Stoneley, 44 Ilthorpe, Orchard Park Est, Hull, HU6 9ER [IO93TS, TA03]
G0 RLE A N McGuire, 28 Haggerston Ct, Etal Park, Newcastle upon Tyne, NE5 4TQ [IO95DA, NZ26]
G0 RLF G E Low, 40 Lilac Cres, Runcorn, WA7 5JX [IO83PH, SJ58]
G0 RLH E G Miles, 31 Winnipeg Rd, Bentley, Doncaster, DN5 0ED [IO93KN, SE50]
G0 RLI J P Thomas, 204 Watchouse Rd, Galleywood, Chelmsford, CM2 8NF [JO01FQ, TL70]
G0 RLJ P H Tyson, 44 Windmill Ave, Kilburn, Belper, DE56 0PQ [IO93GA, SK34]
G0 RLK N Voisey, 10 Cricket Cl, Kirkby in Ashfield, Nottingham, NG17 9FQ [IO93IB, SK45]
G0 RLL T J Dyson, 27 Throstle Nest, Batley, WF17 7SN [IO93ER, SE22]
G0 RLM L K Moreton, 10 Looseleigh Ln, Crownhill, Plymouth, PL6 5EX [IO70WK, SX46]
G0 RLN K Taylor, 29 School Rd, Chequerfield, Pontefract, WF8 2AJ [IO93IQ, SE42]
G0 RLO K A Conlon, Braunfels, 76 Deanwood Cres, Allerton, Bradford, BD15 9BL [IO93CT, SE13]
GW0 RLQ J F T Ellwood, 5 Smallwood Rd, Baglan, Port Talbot, SA12 8AP [IO81CO, SS79]
G0 RLR Details withheld at licensee's request by SSL.
G0 RLS P D U V Ashcroft, 5 Rocklands, Gordon Rd, Finchley, London, N3 1EN [IO91VO, TQ29]
G0 RLT R L Taylor, 75 Newton St., Southport, PR9 7AS [IO83MP, SD31]
G0 RLV E D Jones, 16 Fisher Ave, Rugby, CV22 5HN [IO92JI, SP57]
G0 RLW A A Robinson, 1 Bowes Gr, Bishop Auckland, DL14 6LQ [IO94DP, NZ22]
G0 RLX R A Cope, 6 Magnolia Rise, Lyme Rd, Axminster, Devon, EX13 5BH [IO80MS, SY39]
G0 RLY J P Karkoszka, 5 Wood St., Coldshaw, Haworth, Keighley, BD22 8BJ [IO93AT, SE03]
GM0 RLZ C J Brown, 9 Newton Cres, Rosyth, Dunfermline, KY11 2QW [IO86HA, NT18]
G0 RMA J O Sims, 9 The Beeches, Hilperton Rd, Trowbridge, BA14 7HG [IO81VH, ST85]
GW0 RMB S Ferris, Arfryn, y Ffor, Pwllheli, Gwynedd N Wales, LL53 6UB [IO72TV, SH33]
G0 RMC M Charlton, 11 Holcombe Dr, Plymstock, Plymouth, PL9 9JD [IO70WI, SX55]
G0 RMD P R Calter, 8 Exeter Rd, Scunthorpe, DN15 7AT [IO93QO, SE81]
G0 RME F G Wood, Grangefields, 213 Crewe Rd, Alsager, Stoke on Trent, ST7 2JJ [IO83UC, SJ75]
G0 RMF R W Homden, 67 Pier Rd, Northfleet, Gravesend, DA11 9NA [JO01EK, TQ67]
G0 RMG R S Jones, 6 Wychwood Dr, Crabbs Cross, Redditch, B97 5NW [IO92AG, SP06]
G0 RMI G L Fogell, Rippleway, Paddock Cl, Napton, Rugby, Warks, CV23 8JA [IO92IF, SP46]
G0 RMJ S M Hogg, 20 West St., Ventnor, PO38 1NQ [IO90JO, SZ57]
G0 RMK M Levy Menorah Prim Sc, Arcade, 223 Woodcock Hill, Kenton, Harrow, HA3 0PG [IO91UN, TQ18]
GM0 RML A G Smart, 6 Alton Bank, Tradespark, Nairn, IV12 5PJ [IO87BN, NH85]
GM0 RMM Details withheld at licensee's request by SSL.
G0 RMN A W Younger, 25 Stanley Rd, West Bridgford, Nottingham, NG2 6DF [IO92KW, SK53]
G0 RMO M G Miller, 8 Pilton Walk, Westerhope, Newcastle upon Tyne, NE5 4PQ [IO94DX, NZ16]
G0 RMP R M P Seal, 2 Shaftesbury Rd, Bridlington, YO15 3NP [IO94VB, TA16]
G0 RMQ Details withheld at licensee's request by SSL.
G0 RMR C J Rabey, 26 Opal Way, Wokingham, RG41 3UL [IO91NK, SU76]
GM0 RMT Details withheld at licensee's request by SSL.
G0 RMU R E Clover, Teffont, 42 Warren Rd, New Haw, Addlestone, KT15 3UA [IO91SI, TQ06]
GM0 RMV M Verity, 19 Vivian Terr, Edinburgh, EH4 5AW [IO85IX, NT27]
GM0 RMW K Houston, Waterstonehill, California, By Falkirk, Stirlingshire, FK1 2DG [IO85DX, NS97]
G0 RMX D J Esdale, New House, Kings Rd, Malvern, WR14 4HL [IO82TB, SO74]
GI0 RMZ Details withheld at licensee's request by SSL.
G0 RNA T J Rawlinson, Vancouver Hotel, 343/7 South Prom, Blackpool, Lancs, FY1 6BJ [IO83LT, SD33]
G0 RNC Details withheld at licensee's request by SSL.
G0 RND R Noakes, 10 Ashfield, Chineham, Basingstoke, RG24 8UF [IO91LG, SU65]
G0 RNE Details withheld at licensee's request by SSL.
G0 RNF I N Hunnisett, 69 Cornwall Rd, Ruislip, HA4 6AJ [IO91TN, TQ08]
G0 RNG Details withheld at licensee's request by SSL.
G0 RNH M Ahmed, 27 Sandpiper Bridge, Swindon, SN3 5DY [IO91DN, SU18]
G0 RNI D Thorpe Luton Rep Grp, 70 Willow Way, Ampthill, Bedford, MK45 2SP [IO92SA, TL03]
G0 RNJ P A Habib, 8 Hamilton Cl, Chertsey, KT16 9LP [IO91RJ, TQ06]
GW0 RNK K L Williams, 12 Danybanc, Felinfoel, Llanelli, SA15 4NS [IO71WQ, SN50]
G0 RNL B A Stilgoe, 140 Kelynmead Rd, Kitts Green, Birmingham, B33 8LE [IO92CL, SP18]
G0 RNM L R Smith, 1 Perring Cl, Sharnbrook, Bedford, MK44 1JE [IO92RF, SP95]
G0 RNN I T Tutt, 6 Dunster Cl, Brighton, BN1 7ED [IO90WU, TQ30]
G0 RNO E D Prothero, 5 Home Meadow Dr, Flackwell Heath, High Wycombe, HP10 9JY [IO91PO, SU88]
G0 RNP D G Eves, 64 Hillingdon Rd, Gravesend, DA11 7LG [JO01EK, TQ67]
G0 RNQ B R Willson, 4 Caldew Gr, Sittingbourne, ME10 4SL [JO01II, TQ96]
GM0 RNR R L Thomson RN ARS Scotland, 25 Cheviot Rd, Silvertonhill, Hamilton, ML3 7HB [IO75XS, NS75]
G0 RNS J A White, 5 The Avenue, Brighton, BN2 4GF [IO90WU, TQ30]
G0 RNT T R Cooper, 22 Wilton Gdns, Chatsworth Park, New Milton, BH25 5UT [IO90ES, SZ29]
G0 RNU N J Illing, 35 Park Ave, Barking, IG11 8QU [JO01BN, TQ48]
G0 RNV B Sherriff, 6 St. Polycarps Dr, Holbeach Drove, Spalding, PE12 0SF [IO92XR, TF31]
G0 RNW R Williams, 6C Linden Gdns, Chiswick, London, W4 2EG [IO91UL, TQ27]
G0 RNX S Onions, 78 Mill Ln, Albrighton, Wolverhampton, WV7 3ND [IO82UP, SJ80]
G0 RNY A Attle, 25 Moorside, Middlestone Moor, Spennymoor, DL16 7DY [IO94EQ, NZ23]
G0 RNZ P E Attwood, 100 Botley Rd, Oxford, OX2 0HH [IO91IS, SP40]
G0 ROA H Seidner, 401 East 80th St, 25A New York, New York 10021, USA
G0 ROB R C Gilbert, 35 Lower Rd, West Malvern, Malvern, WR14 4BX [IO82TC, SO74]
G0 ROC Details withheld at licensee's request by SSL.
G0 ROD C Reaney, 109 Holbrook Rd, Belper, DE56 1PB [IO93GA, SK34]
G0 ROE R W F J Swann, 8 Lawrence Rd, Ham, Richmond, TW10 7LR [IO91UK, TQ17]
G0 ROF C J Radley, 280 Burley Rd, Thorney Hill, Bransgore, Christchurch, BH23 8DQ [IO90DT, SZ19]
G0 ROG Details withheld at licensee's request by SSL.
G0 ROH G T Lines, 141 Blackborough Rd, Reigate, RH2 7DA [IO91VF, TQ25]
G0 ROI Details withheld at licensee's request by SSL.
G0 ROJ Details withheld at licensee's request by SSL.
GW0 ROL D E Brown, 11 Dowland Rd, Penarth, CF64 3QX [IO81JK, ST17]
G0 RON R J McNeil, 3 Hatchell Dr, Bessacarr, Doncaster, DN4 6SH [IO93LM, SE60]
G0 ROO R A Pascoe The Kanga Gang, Seaview House, Crete Rd East, Folkestone, Kent, CT18 7EG [JO01OC, TR23]
G0 ROQ V J Bromley, La Petit Maison, Kergroas, Tregoman, Glomel, 22110,France, X X
G0 ROR S Lidster, 3 Clarkfield Dr, Morecambe, LA4 6UG [IO84NB, SD46]
G0 ROS R J Kent, 40 Waxes Cl, Abingdon, OX14 2NG [IO91IQ, SU59]

G0 ROT M J Davis, 3 Carlton Rd, Southport, PR8 2PG [IO83LO, SD31]
G0 ROU A D Butcher, 7 Crummock Pl, Mereside, Blackpool, FY4 4TP [IO83MT, SD33]
G0 ROV A E Hall, 30 Faraday Cl, Arborfield, Reading, RG2 9NR [IO91NJ, SU76]
G0 ROW A G Gurnhill, 53 Millbrook Ave, Denton, Manchester, M34 2DQ [IO83WK, SJ99]
G0 ROX D P Lee, 131 Abbotsbury Rd, Weymouth, DT4 0JX [IO80SO, SY67]
G0 ROY R R Biddle, 21 Kingsway West, Newton, Chester, CH2 2LA [IO83NE, SJ46]
G0 ROZ C Hardy Dorset Police AR, Dorset Police Station, Ashley Rd, Parkstone, Poole, Dorset, BH14 0BD [IO90AR, SZ09]
G0 RPA I McAvoy, 5 Lytchett Way, Upton, Poole, BH16 5LS [IO80XR, SY99]
G0 RPC R W Steele, 185 Northdown Park Rd, Margate, CT9 3UJ [JO01RJ, TR37]
G0 RPD J E Barton, 183 Windy Arbor Rd, Whiston, Prescot, L35 3SF [IO83OJ, SJ49]
G0 RPF L Smith, 28 Chester Rd, Stockton Heath, Warrington, WA4 2RX [IO83RI, SJ68]
G0 RPG J A Riley, 1 Chatsworth Ave, Culcheth, Warrington, WA3 4LD [IO83RK, SJ69]
G0 RPH Details withheld at licensee's request by SSL.
G0 RPI L Williams, 106 Redgate, Ormskirk, L39 3NY [IO83NN, SD40]
G0 RPJ D H Wesil, 30 Old Shoreham Rd, Shoreham By Sea, BN43 5TD [IO90UU, TQ20]
G0 RPK H F McGuinness, 20 Eastfield Ave, Fareham, PO14 1EG [IO90JU, SU50]
G0 RPL N Alison, 152 Gossops Dr, Gossops Green, Crawley, RH11 8HE [IO91VC, TQ23]
G0 RPM N H Williams, 11 Berkeley Gdns, London, N21 2BE [IO91XP, TQ39]
G0 RPO R Dowd, Belgrano, 1 Watson Ave, Golborne, Warrington, WA3 3QX [IO83QL, SJ59]
G0 RPP Details withheld at licensee's request by SSL.
G0 RPQ M G Jones, 160 Oakbrook Rd, Sheffield, S11 7ED [IO93FI, SK38]
G0 RPR Details withheld at licensee's request by SSL.
GI0 RPS W G McHugh Harps, 47 Main St., Hamiltonsbawn, Armagh, BT60 1LP [IO64RI, H94]
G0 RPT W Bannister, 81 Ormskirk Rd, Chapel House, Skelmersdale, WN8 8TR [IO83ON, SD40]
G0 RPU J C Symonds, La Cumbre, 35 Byward Dr, Crossgates, Scarborough, YO12 4JE [IO94TF, TA08]
G0 RPV W P Till, 97 Haslar Cres, Waterlooville, PO7 6DD [IO90LV, SU61]
G0 RPW D A Wilson, 6 The Chase, Calcot, Reading, RG3 7DN [IO91LJ, SU66]
G0 RPX R D Evans, 426 Hawthorn Cres, Cosham, Portsmouth, PO6 2TX [IO90LU, SU60]
G0 RPY C A Button, 8 Heywood Rd, Diss, IP22 3DJ [JO02NJ, TM18]
G0 RPZ G R Porter, The Bungalow, Holy Island, Northd., TD15 2SE [IO95CQ, NU14]
G0 RQE J Gray, 34 Thornton St., Kimberworth, Rotherham, S61 2LE [IO93HK, SK49]
G0 RQF K G Hales, Brook View, 15 Meadow Bank Rd, Hereford, HR1 2ST [IO82PB, SO53]
G0 RQG J F Gill, 4 Banbury, Burwarton, Bridgnorth, WV16 6QN [IO82RK, SO68]
G0 RQH D Hughes, 28 The Pines, Yapton, Arundel, BN18 0EG [IO90QT, SU90]
G0 RQJ Details withheld at licensee's request by SSL.
GI0 RQK Details withheld at licensee's request by SSL.
G0 RQL D G Roomes, Shop Cross, Milton Damerel, Holsworthy Devon, EX22 7NY [IO70UV, SS31]
GW0 RQM L L Ogden, Coed Celyn, Barmouth Rd, Dolgellau, LL40 2EW [IO82BR, SH71]
G0 RQN P G Robertson, 9 Newfield Rd, Liss, GU33 7BW [IO91NB, SU72]
G0 RQO Details withheld at licensee's request by SSL.
GW0 RQP Details withheld at licensee's request by SSL.
G0 RQQ Details withheld at licensee's request by SSL.
GW0 RQS L A Pritchard, 86 Bryn Rd, Markham, Blackwood, NP2 0QE [IO81JQ, SO10]
G0 RQT D E Presley, 16 Lillington Rd, Radstock, Bath, BA3 3NR [IO81SG, ST65]
G0 RQU H J Parkinson, 114 Ainsworth Ln, Bolton, BL2 2PY [IO83TO, SD71]
G0 RQV F G W Nelson, 5 Chester Grange, Glebe Rd, Grimsby, DN33 2HW [IO93WM, TA20]
G0 RQW W E Naylor, 80 Burnside, Parbold, Wigan, WN8 7PE [IO83OO, SD41]
G0 RQX D M Townend, 3 Ladycroft Cl, Radbrook Green, Shrewsbury, SY3 6BB [IO82OQ, SJ41]
G0 RQY J F Cuthbertson, Wollow Wands, Brimpton, Reading, RG7 4SP [IO91JJ, SU56]
G0 RQZ R L Lawrence, 4 Dale Park Rise, Leeds, LS16 7PP [IO93EU, SE23]
G0 RRA Details withheld at licensee's request by SSL.
G0 RRB A Stanners, 33 Netley Cl, New Addington, Croydon, CR0 0QR [IO91XI, TQ36]
G0 RRC R Smith, Lykkebo, The Street, Burstall, Ipswich, Suffolk, IP8 3DN [JO02MB, TM14]
G0 RRD D R Daws, 20 Tynings Rd, Forest Green, Nailsworth, Stroud, GL6 0EQ [IO81VQ, ST89]
G0 RRE B A Wheway, 7 Chevin Ave, Mickleover, Derby, DE3 5GW [IO92FV, SK33]
G0 RRF P B Smith, 2 Swale Dr, Bramble Rise, Wellingborough, NN8 5ZL [IO92PH, SP86]
G0 RRG A D Bettley Ridgeway Repeater Group, 1 Dovetrees, Covingham, Swindon, SN3 5AX [IO91DN, SU18]
G0 RRI I P Burden, 79 Tonning St., Lowestoft, NR32 2AL [JO02VL, TM59]
G0 RRJ D V Cox, 19 Exbury Way, Andover, SP10 3UH [IO91GE, SU34]
GM0 RRK M Boyce, 93 Ledi Dr, Bearsden, Glasgow, G61 4JP [IO75TW, NS57]
G0 RRL Details withheld at licensee's request by SSL.
G0 RRM P Brumby, 69 Gilbert Walk, Netherstowe Est, Lichfield, WS13 6AU [IO92CQ, SK10]
G0 RRO J R Breingan, 44 Farmstead Rd, Corby, NN18 0LG [IO92PL, SP88]
G0 RRP J K Brash, Sea Mews, Knott End on Sea, Poulton-le-Fylde, Lancs, FY6 0DX [IO83MW, SD34]
G0 RRQ G Blair, Elgon, Castle Hills, Northallerton, North Yorks, DL7 8UR [IO94GI, SE39]
G0 RRR Details withheld at licensee's request by SSL.
G0 RRS Details withheld at licensee's request by SSL.
G0 RRT P J Ambrose, 12 Welford Gdns, Abingdon, OX14 2BW [IO91IQ, SU59]
GM0 RRU J W Balfour, North Lodge, Ethie Barns Farm, Inverkeilor, Arbroath, Tayside, DD11 5SP [IO86RO, NO64]
G0 RRV A J Tomson, Valley View, Bewdley, Worcs, DY12 2JX [IO82UI, SO77]
G0 RRW Details withheld at licensee's request by SSL.
G0 RRX P R Bethell, 6 Givendale Dr, Higher Crumpsall, Manchester, M8 4PY [IO83VM, SD80]
G0 RRY M T Dabhi, 14 Emerson Rd, Preston, PR1 5SN [IO83PS, SD53]
G0 RRZ R K Carrington, 3 Lords Cl, Bolsover, Chesterfield, S44 6TU [IO93IF, SK47]
G0 RSA J A King, 39 Nursery Gdns, St. Ives, Huntingdon, PE17 6NL [IO92XH, TL37]
GM0 RSE S Spence Morse Enthusiasts Group, 90 Simshill Rd, Glasgow, G44 5EN [IO75VT, NS55]
GM0 RSF J Carr, 2 Martin Brae, Livingston, EH54 6UR [IO85FV, NT06]
GI0 RSH Details withheld at licensee's request by SSL.
GM0 RSI J G Ritchie, 36 James Mitchell Pl, Mintlaw, Peterhead, AB42 5ES [IO97AM, NK04]
G0 RSK A Dunster, 50 Ludlow Rd, Alum Rock, Birmingham, B8 3BY [IO92BL, SP18]
G0 RSL K A White, 99 Low Ln, Bare, Morecambe, LA4 6PS [IO84OB, SD46]
G0 RSM K White, 14 Hardwick Rd East, Manton, Worksop, S80 2NS [IO93KH, SK57]
G0 RSN Details withheld at licensee's request by SSL.[WAB SZ09, locator: IO90BS. Correspondence c/o 9 Periton Lane, Minehead, Somserset, TA24 8AQ.]
G0 RSO Details withheld at licensee's request by SSL.
G0 RSP Details withheld at licensee's request by SSL.
G0 RSQ P L Walker, 114 Elswick, Tanhouse, Skelmersdale, WN8 6BT [IO83ON, SD40]
G0 RSS R Simmonds, 4 Corys Cl, Norwich Rd, Bramerton, Norwich, NR14 7DP [JO02QO, TG20]
G0 RST E Bardsley, 30 Blue Bell Cl, Hyde, SK14 4HU [IO83XL, SJ99]
G0 RSU G P Weston, 51 Rufford Ave, Bramcote, Beeston, Nottingham, NG9 3JG [IO92IW, SK53]
G0 RSV W B Webster, 4 Horsham Rd, Owlsmoor, Camberley, Surrey, GU15 4YY [IO91PI, SU86]
G0 RSW R C Waters, 56 Elmsleigh Dr, Leigh on Sea, SS9 3DN [JO01HN, TQ88]
G0 RSX P Vagars, 40 Jervison St., Longton, Stoke on Trent, ST3 5DD [IO82WX, SJ94]
G0 RSY A D Gibbs, 35 Queens Rd, Cowes, PO31 8BW [IO90IS, SZ49]
GW0 RTA T Arakawa, 21 Trem y Foel, Sychdyn, Mold, CH7 6HA [IO83KE, SJ26]
G0 RTC T Chisholm, 316 Birchfield Rd East, Northampton, NN3 2SY [IO92NG, SP76]
G0 RTD Details withheld at licensee's request by SSL.
G0 RTF I M Clarke, Chelmwood, 7 Howard Cres, Seer Green, Beaconsfield, HP9 2XR [IO91QO, SU99]
GM0 RTG Details withheld at licensee's request by SSL.
G0 RTH A Elcoate, 9 Parsonage Ln, Laindon, Basildon, SS15 5YN [JO01FN, TQ68]
G0 RTI S A Harriss, 6 Redland Rd, Leamington Spa, CV31 2PB [IO92FG, SP36]
G0 RTJ I B G Harrison, 2 Ivy Cttgs, Long Crichel, Wimborne, BH21 5LA [IO80XV, ST91]
G0 RTK Details withheld at licensee's request by SSL.
G0 RTL J L Flowers Radio-Tele Lincolnshire Group, Joanne, Belnie Ln, Gosberton, Spalding, PE11 4HN [IO92WU, TF23]
G0 RTM P McKnight, 39 Dunmail Dr, Kendal, LA9 7JG [IO84PH, SD59]
GW0 RTP C N Llewellyn, 107 Margam St., Cymmer, Port Talbot, SA13 3EF [IO81EP, SS89]
G0 RTQ D J Lawrence, 75 Church St., Ilkeston, DE7 8QP [IO92IX, SK44]
GW0 RTR R T Rees, 45 Sandy Rd, Llanelli, SA15 4BR [IO71VQ, SN40]
GM0 RTS R T Sewell, 5 Willow Dr, Girvan, KA26 0DE [IO75NF, NX19]
G0 RTT D A H Laister, 23 Barrowby Rd, Broom, Rotherham, S60 3HF [IO93IK, SK49]
G0 RTU P N Kirkup, 337 Wheatley Ln Rd, Fence, Burnley, BB12 9QA [IO83VU, SD83]
G0 RTV P R King, Cornish Cottage, Priors Ct Rd, Hermitage, Thatcham, RG18 9TG [IO91IK, SU57]
G0 RTW A J Jeeves, 27 Ash Way, Newton Abbot, TQ12 4LW [IO80EM, SX87]
G0 RTX K J Eaton, 22 Honingham Rd, Ilkeston, DE7 9JZ [IO92IX, SK44]
GM0 RTY D J Inns, 19 Drumfork Rd, Helensburgh, G84 7TN [IO75PX, NS38]
G0 RTZ G M Hurst, 4 Lytton Strachey Path, London, SE28 8DU [JO01BM, TQ48]
G0 RUA G G Hooton, 9 Marlborough Ave, Wellingborough, NN8 5YN [IO92PH, SP86]
GI0 RUC R E Kerr, 194 Shore Rd, Greenisland, Carrickfergus, BT38 8TX [IO74BQ, J38]
G0 RUD P G Marriott, 59 Oaklands Way, Fareham, PO14 4LF [IO90IU, SU50]
G0 RUE I P D Bowden, 30 Albany Rd, Brentford, TW8 0NF [IO91UL, TQ17]
G0 RUF N W Taylor, 19 Castle Cl, Leconfield, Beverley, HU17 7NX [IO93SV, TA04]
G0 RUH M J Roberts, 82 Glover Rd, Scunthorpe, DN17 1AS [IO93QN, SE80]
G0 RUJ D M Rowlands, 2 Springhill Rd, Dawley, Telford, TF4 3DF [IO82SP, SJ60]
G0 RUK A S Caley, 16 Burdale Cl, Driffield, YO25 7SG [IO93SX, TA05]
G0 RUM Details withheld at licensee's request by SSL.
G0 RUN I Braddock, 1C, Bollington, Macclesfield, Ches, SK10 5LN [IO83WH, SJ97]
G0 RUP Details withheld at licensee's request by SSL.
G0 RUR P Simpson, 231 Caxton St., Sunnyhill, Derby, DE23 7RB [IO92GV, SK33]
G0 RUS R C Lenthall, 182 Chelmsford Ave, Grimsby, DN34 5DB [IO93WN, TA20]
G0 RUT R Russell, 4 Hinton Rd, Carisbrooke, Newport, PO30 5QZ [IO90IQ, SZ48]
G0 RUV M J Gent, 111 Portland St., Clowne, Chesterfield, S43 4SA [IO93JG, SK57]
GM0 RUW J B Coughtrie, 61 Bells Burn Ave, Linlithgow, EH49 7LD [IO85FX, NT07]
G0 RUX W T Taylor, 21 Summerdale Rd, Cudworth, Barnsley, South Yorks, S72 8XG [IO93GN, SE30]
G0 RUY A C Pritchard, 29 Brockley Rd, Leonard Stanley, Stonehouse, GL10 3NB [IO81UR, SO80]
G0 RVA V H Booth, 32 Stockdale Cl, Arnold, Nottingham, NG5 9QB [IO93KA, SK54]
GW0 RVB Details withheld at licensee's request by SSL.
GW0 RVC H C Taylor Rhymney Valley Amateur Radio, 2 Pen y Dre, Highfields, Caerphilly, CF83 2NZ [IO81JO, ST18]
G0 RVD R S Bibb, 98 Saunders Cl, Kettering, NN16 0AU [IO92PJ, SP87]
G0 RVE A P Pierce, 34 Church Cl, Shawbury, Shrewsbury, SY4 4JX [IO82QS, SJ52]
G0 RVF Details withheld at licensee's request by SSL.
GW0 RVG Details withheld at licensee's request by SSL.
G0 RVH K M Dailey, 55 Chesterton Ave, Harpenden, AL5 5SU [IO91TT, TL11]
G0 RVI J J Davis, 1 Oak Farm Cl, Blackwater, Camberley, GU17 0JU [IO91OI, SU86]
G0 RVJ R M Elliott, 3 Stratford Dr, The Willows, Aylesbury, HP21 8PL [IO91OT, SP81]
G0 RVK M A Fogg, 15 Elm Gr, Bisley, Woking, GU24 9DG [IO91QH, SU95]
G0 RVL D B Clift, 30 Lynfield Rd, Lichfield, WS13 7BU [IO92BQ, SK11]
G0 RVM A S Gawthrope, 27 Coriander Dr, Bradley Stoke, Bristol, BS12 0DJ [IO81RN, ST68]
G0 RVN Details withheld at licensee's request by SSL.
G0 RVQ Details withheld at licensee's request by SSL.
GW0 RVR Details withheld at licensee's request by SSL.
G0 RVS Details withheld at licensee's request by SSL.
G0 RVT Details withheld at licensee's request by SSL.
G0 RVU Details withheld at licensee's request by SSL.
G0 RVV P J Barker, 1 Whitriggs Cl, Haverigg Rd, Millom, LA18 4EL [IO84IE, SD17]
G0 RVW D L Hughes, 7 Mellor Cl, Windmill Hill, Runcorn, WA7 6QB [IO83PI, SJ58]
G0 RVX Details withheld at licensee's request by SSL.
G0 RVY Details withheld at licensee's request by SSL.
G0 RVZ Details withheld at licensee's request by SSL.
G0 RWA B Chorley, 19 Cleeve Rd, Taunton, TA2 8DX [IO81KA, ST22]
G0 RWB Details withheld at licensee's request by SSL.
G0 RWC Details withheld at licensee's request by SSL.
G0 RWD Details withheld at licensee's request by SSL.
G0 RWF Details withheld at licensee's request by SSL.
G0 RWG Details withheld at licensee's request by SSL.
GW0 RWH Details withheld at licensee's request by SSL.
G0 RWI E Johns, 3 The Rowans, Portishead, Bristol, BS20 8QR [IO81OL, ST47]
G0 RWJ D R King, 78 Andersey Way, Abingdon, OX14 5NW [IO91IP, SU49]
G0 RWK Details withheld at licensee's request by SSL.
G0 RWL Details withheld at licensee's request by SSL.
G0 RWM R W A Martin, 52 Leacroft, Staines, Middx, TW18 4NN [IO91SK, TQ07]
G0 RWQ N Monument, Highlands, 32 Rosecroft Way, Cloverfields, Thetford, Norfolk, IP24 2XR [JO02JK, TL88]
G0 RWR Details withheld at licensee's request by SSL.
G0 RWS W J Scott, 28 Kingsbury Ave, St. Albans, AL3 4TA [IO91TS, TL10]
G0 RWT R D Pine, Rhodanna, Tennis Ct Rd, Paulton, Bristol, BS18 5LU [IO81SH, ST65]
GM0 RWU Dr J W Ponton, 3 High St., Edinburgh, EH1 1SR [IO85JW, NT27]
G0 RWV F Marshall, 9 Bucklers Ct, Anchorage Way, Lymington, SO41 8JN [IO90FS, SZ39]
G0 RWW Details withheld at licensee's request by SSL.
G0 RWX J Butcher, Stonehurst, 48 Deepdene Ave, Dorking, RH5 4AE [IO91UF, TQ14]
G0 RWY Dr D W C Ramsay, 2 Old Church Rd, Colwall, Malvern, WR13 6ET [IO82TB, SO74]
G0 RWZ Details withheld at licensee's request by SSL.
G0 RXA Details withheld at licensee's request by SSL.
G0 RXB M L Rowe, 9 Runford Ct, Shenley Lodge, Milton Keynes, MK5 7BB [IO92OA, SP83]
G0 RXE Details withheld at licensee's request by SSL.
G0 RXG Details withheld at licensee's request by SSL.
G0 RXH R A Wells, 33 Sandholme, Steeple Claydon, Buckingham, MK18 2QE [IO91MW, SP72]
G0 RXI R C Weston, 10 Graham Rd, West Kirby, Wirral, L48 5DW [IO83JJ, SJ28]
G0 RXJ M T Wilderspin, 59 Underwood Pl, Oldbrook, Milton Keynes, MK6 2NU [IO92OA, SP83]
GW0 RXL Details withheld at licensee's request by SSL.
G0 RXN Details withheld at licensee's request by SSL.
G0 RXO F W Collins, 22 Paxton Rd, Wollescote, Stourbridge, DY9 8YD [IO82WK, SO98]
GM0 RXP M Bottomley, Lion Lodge North, Coldstream, Berwickshire, TD12 4HE [IO85VQ, NT84]
G0 RXQ F Lockey, 15 Mark Anthony Ct, Hayling Island, Hamts, PO11 0AE [IO90MS, SZ79]
G0 RXR Details withheld at licensee's request by SSL.
G0 RXT Details withheld at licensee's request by SSL.
G0 RXU F J Nethercott, 6 Laking Ave, Broadstairs, CT10 3NE [JO01RI, TR36]
G0 RXV C Eves, 2 Prospect Ave, Sherburn in Elmet, Leeds, LS25 6LR [IO93JT, SE43]
G0 RXW Details withheld at licensee's request by SSL.
G0 RXX Details withheld at licensee's request by SSL.
G0 RXY Details withheld at licensee's request by SSL.
G0 RXZ Details withheld at licensee's request by SSL.
G0 RYA G Roberts, Highbank Farm, London Rd, Clanfield, Waterlooville, Hants, PO8 0QD [IO90MW, SU71]
G0 RYC Details withheld at licensee's request by SSL.
GM0 RYD J R Van Dyke, 112 Alexander Ave, Largs, KA30 9EX [IO75NT, NS26]
G0 RYF Details withheld at licensee's request by SSL.
G0 RYG Details withheld at licensee's request by SSL.
G0 RYH Details withheld at licensee's request by SSL.
GI0 RYK R W White, 1 Woodland Park, Lisburn, BT28 1LD [IO64XM, J26]
G0 RYL R B Hodges, 1A Clements Ln, Chiswell, Portland, DT5 1AS [IO80SN, SY67]
G0 RYM Details withheld at licensee's request by SSL.
GI0 RYN Details withheld at licensee's request by SSL.
G0 RYO J P Webb, Littlejohns Cottage, Milton Damerel, Holsworthy, Devon, EX22 7DL [IO70UU, SS31]
G0 RYP C Martin, 13 St Lawrence St, B'Kara Bkro9, Malta, X X
G0 RYQ Details withheld at licensee's request by SSL.
G0 RYR T J Ballinger, 9 Somerville Ct, Cirencester, GL7 1TG [IO91AQ, SP00]
G0 RYS M S Vann Richmond School Amateur, Richmond School, Darlington Rd, Richmond, DL10 7BQ [IO94DJ, NZ10]
GW0 RYT Details withheld at licensee's request by SSL.
GI0 RYU B Millar, 312 Churchill Park, Portadown, Craigavon, BT62 1EY [IO64SK, J05]
G0 RYV A R Smith, Crown Cottage, Stone, Berkeley, Glos, GL13 9LE [IO81SP, ST69]
G0 RYW Details withheld at licensee's request by SSL.
G0 RYY Details withheld at licensee's request by SSL.
G0 RZA G Waterfield, 51 Vancouver Dr, Winskill, Burton on Trent, DE15 0EY [IO92ET, SK22]
G0 RZB D McDonnell, Glencoe, The Ridge, Redlynch, Salisbury, SP5 2LN [IO90DX, SU11]
G0 RZC J Richardson, 85 Oakshaw Dr, Rochdale, OL12 7PF [IO83VO, SD81]
G0 RZD W Bongers, Whitewebs, 9 Yeovilton Cl, Everton, Lymington, SO41 0JS [IO90ER, SZ29]
G0 RZE Details withheld at licensee's request by SSL.
G0 RZF Details withheld at licensee's request by SSL.
G0 RZG Details withheld at licensee's request by SSL.
G0 RZI B Easdon, 20 Winder Gate, Frizington, CA26 3QS [IO84GM, NY01]
G0 RZJ Details withheld at licensee's request by SSL.
G0 RZM B P Judd, 61 Kensington Rd, Reading, RG3 2SZ [IO91LJ, SU66]
G0 RZO W D Francis, 21 White Hart Cl, Buntingford, SG9 9DG [IO91XW, TL32]
G0 RZP D W Allan, 283 Cliffe Ln, Gomersal, Cleckheaton, BD19 4SB [IO93DR, SE22]
G0 RZR P Ward Lincoln & District Amateur, 13 Keddington Ave, Ermine West, Lincoln, LN1 3SU [IO93RG, SK97]
G0 RZS Details withheld at licensee's request by SSL.
G0 RZT D S Irvine, 8 Susan Cl, Romford, RM7 8ET [JO01BO, TQ58]
G0 RZV Details withheld at licensee's request by SSL.
GM0 RZY J Sutherland, 'Woodlands', Braidwood, Carluke, Lanarkshire, ML8 5NE [IO85BR, NS84]
G0 SAA B C Hicks, 28 Hope St., Halesowen, B62 8LU [IO82XL, SO98]
G0 SAC A D Cross Sutton Area Contest Group, 31 Mountcombe Cl, Surbiton, KT6 6LJ [IO91UJ, TQ16]
G0 SAD Details withheld at licensee's request by SSL.
G0 SAE Details withheld at licensee's request by SSL.
G0 SAG Details withheld at licensee's request by SSL.
G0 SAH A Holdsworth, 26 Chelveston, Welwyn Garden City, AL7 2PW [IO91WT, TL21]
GW0 SAJ Dr H V Jones, 28 Penmaen Terr, Swansea, SA1 6HZ [IO81AO, SS69]
G0 SAM Details withheld at licensee's request by SSL.
G0 SAN Details withheld at licensee's request by SSL.
GI0 SAP Details withheld at licensee's request by SSL.
G0 SAQ Details withheld at licensee's request by SSL.
G0 SAR M J Watson Suffolk Raynet, The Tubbery, Henley, Ipswich, Suffolk, IP6 0BR [JO02NC, TM15]

G0 SAT R M Lloyd Madley Amateur Radio Group, Madley Comms Centre, BT Madley, Madley, Hereford, HR2 9NH [IO82NA, SO43]
GW0 SAU A F Skellern, 35 Madoc St., Porthmadog, LL49 9BU [IO72WW, SH53]
G0 SAV R C Cotsford, 31 Nevyll Ct, Station Rd, Southend on Sea, SS1 3UE [JO01JM, TQ98]
G0 SAW Details withheld at licensee's request by SSL.
G0 SAY C A Thorpe, 78 Bowland Rd, Wythenshawe, Manchester, M23 1JX [IO83UJ, SJ88]
G0 SBA Details withheld at licensee's request by SSL.
G0 SBC R Harris, 142 St Nicolas Park, Dr, Nuneaton, Warks, CV11 6EE [IO92GM, SP39]
G0 SBD Details withheld at licensee's request by SSL.
G0 SBE Details withheld at licensee's request by SSL.
G0 SBH T D Wernham, 5 The Hill, Wangford, Beccles, NR34 8AT [JO02TI, TM47]
G0 SBI S A Bell, 56 Nortonwood Ln, Windmill Hill, Runcorn, WA7 6QG [IO83PI, SJ58]
GM0 SBJ H F Sweeney, 139 Juniper Ave, Greenhills, East Kilbride, Glasgow, G75 9JP [IO75VR, NS65]
G0 SBK M R Jenkins, 9 Tothill Rd, Swaffham Prior, Cambridge, CB5 0JX [JO02DG, TL56]
G0 SBL Details withheld at licensee's request by SSL.
G0 SBM C J Coker South Devon Raynet Group, 46 Clarendon Rd, Ipplepen, Newton Abbot, TQ12 5QS [IO80EL, SX86]
G0 SBN Details withheld at licensee's request by SSL.
G0 SBO E A Hodgson, 21 Royd Ave, Mapplewell, Barnsley, S75 6HH [IO93FO, SE31]
G0 SBP F G Parkinson, 106 Westmoreland St., Darlington, DL3 0NU [IO94FM, NZ21]
G0 SBQ J P O'Riordan, 12 Hulton Cl, Boreham, Chelmsford, CM3 3BU [JO01GS, TL70]
G0 SBR I M Lee, Clayton Lodge, Sunnyside, Edgerton, Huddersfield, HD3 3AD [IO93CP, SE11]
G0 SBS Details withheld at licensee's request by SSL.
G0 SBT Details withheld at licensee's request by SSL.
G0 SBU B J Wedgwood, 40 Ford St., Delves Ln, Consett, DH8 7AE [IO94CU, NZ14]
G0 SBV R H Talbot, 11 Whitefield Rd, Holbury, Southampton, SO45 2HP [IO90HT, SU40]
G0 SBW Details withheld at licensee's request by SSL.
G0 SBX Details withheld at licensee's request by SSL.
G0 SBY J Thompson, 78 de Lacy St., Ashton on Ribble, Preston, PR2 2AP [IO83PS, SD53]
G0 SBZ W Sandle, Flat B, 507 Harrogate Rd, Leeds, LS17 7DU [IO93FU, SE34]
G0 SCA Details withheld at licensee's request by SSL.
G0 SCB J Langridge, 6 The Leys, Berrycroft, Berkeley, GL13 9AF [IO81SQ, ST69]
G0 SCC C Keeler, Pipers Hay, Rutters Ln, Ilminster, TA19 9AN [IO80NW, ST31]
G0 SCD V G Clarke, 12 Howe Hill Cl, York, YO2 4SN [IO93KX, SE55]
G0 SCF Details withheld at licensee's request by SSL.
G0 SCG A A Leavey, 14 Cherry Cl, Ealing, London, W5 4JW [IO91UL, TQ17]
G0 SCK D C Britton, 31 Clay Bottom, Bristol, BS5 7EJ [IO81RL, ST67]
G0 SCL S C Lawrence, 4 Dale Park Rise, Leeds, LS16 7PP [IO93EU, SE23]
G0 SCM F J Binnington, 7 Webbs Cl, Combs, Stowmarket, Suffolk, IP14 2NZ [JO02LE, TM05]
GW0 SCN A N Jones, 12 Danybanc, Felinfoel, Llanelli, SA15 4NS [IO71WQ, SN50]
GM0 SCO J C Lefever Scottish Office Amateur, Scottish Office Am Rd Club, 38 Woodburn Terr, Edinburgh, EH10 4ST [IO85JW, NT27]
G0 SCP T Hughes, 1 Southwold Way, Clacton on Sea, CO16 8BY [JO01NT, TM11]
G0 SCQ D J Brusch, 378 Bricknell Ave, Hull, HU5 4QD [IO93TS, TA03]
G0 SCR P N F A Lewis Caterham Radio Group, Sky Waves, 20 Annes Walk, Caterham, CR3 5EL [IO91WH, TQ35]
G0 SCS P N Paterson Division of Science and Techno, Division Science & Technology, North Oxon College, Broughton Rd, Banbury, OX16 9QA [IO92HB, SP44]
G0 SCT R D R Bricknell, 165 Eastwood Old Rd, Leigh on Sea, SS9 4RZ [JO01HN, TQ88]
G0 SCV G N Belt, 5 Allerton Hill, Chapelallerton, Chapel Allerton, Leeds, LS7 3QB [IO93FT, SE33]
GM0 SCW R J Anderson, 34 Endrick Dr, Paisley, PA1 3TX [IO75TU, NS46]
G0 SCY W E Best, 61 Gainsborough, North Lake, Bracknell, RG12 7WL [IO91PJ, SU86]
G0 SCZ D E Williams, 48 Penarth Rise, Sherwood Vale, Nottingham, NG5 4EE [IO92KX, SK54]
G0 SDA C F Bird, 63 Brackenwoods, Necton, Swaffham, PE37 8EX [JO02JP, TF80]
G0 SDB Details withheld at licensee's request by SSL.
G0 SDC P A Robinson South Downs College Radio & El, 11 The Avenue, Hambrook, Chichester, PO18 8TZ [IO90NU, SU70]
G0 SDD C B James, 7 St. James Park, Lower Milkwall, Coleford, Glos, GL16 7LG [IO81QS, SO50]
G0 SDE B C Jupp, 7 Abbots Cl, Rainham, RM13 9LA [JO01CM, TQ58]
G0 SDF J R Atkins, 2 Ches Gdns, Chessington, KT9 2PR [IO91UI, TQ16]
G0 SDG P T Knight, 29 Isbury Rd, Marlborough, SN8 4AJ [IO91DK, SU16]
G0 SDH Details withheld at licensee's request by SSL.
G0 SDI N J Stamp, 162 Hawthorn Ave, Anlaby Rd, Hull, HU3 5PY [IO93TR, TA02]
G0 SDJ A S McMullon, Hothersall Ln, Longridge, Hothersall, Preston, Lancs, PR3 2XB [IO83RT, SD63]
G0 SDK N Darwin, Brindle Nook, Denaby Ln, Old Denaby, Doncaster, DN12 4LE [IO93IL, SK49]
G0 SDL J Wilson, Appletrees, Stoke Rd, Combeinteignhead, Newton Abbot, Devon, TQ12 4RE [IO80FM, SX97]
G0 SDN Details withheld at licensee's request by SSL.
G0 SDO Details withheld at licensee's request by SSL.
G0 SDP Details withheld at licensee's request by SSL.
G0 SDQ L A Talkowski, 83 Church Rd, Peasedown St. John, Bath, BA2 8AB [IO81SH, ST65]
G0 SDR Details withheld at licensee's request by SSL.
GM0 SDS Details withheld at licensee's request by SSL.
G0 SDT J S Sparkes, 87 Chyvelah Ope, Gloweth, Truro, TR1 3YB [IO70KG, SW74]
GM0 SDV Details withheld at licensee's request by SSL.
G0 SDW M A Pattman, 4 Branscombe Rd, Bristol, BS9 1SN [IO81QL, ST57]
G0 SDX J D Faulkner-Court Willpower Contest Group, 4 The Old Pound, Halford, Shipston on Stour, CV36 5DB [IO92EC, SP24]
G0 SDY Details withheld at licensee's request by SSL.
G0 SDZ J E Thomas, La Clapotis, 3 Denis Semeria, 06230 St Jean, Cap Ferrat, France, X X
G0 SEA T A Melnyczuk The Far Canal Contest Group, 58 Northfield Rd, Milfield, Peterborough, PE1 3QJ [IO92VO, TF10]
G0 SEB J M Shepherd, 25 Station Rd, St. Helens, Ryde, PO33 1YF [IO90KQ, SZ68]
G0 SEC J E Curtis, 66 Rockhampton Cl, Littlemoor, Weymouth, DT3 6NG [IO80SP, SY68]
G0 SED G G Bullock, 9 Springwood, Haxby, York, YO3 3YN [IO94LA, SE65]
G0 SEF D N Stolting, 80 Carr Ln, York, YO2 5HY [IO93KX, SE55]
G0 SEG R V Cuff, 16 Carleton Glen, Pontefract, WF8 2RT [IO93IQ, SE42]
GW0 SEH D B Richards, 53 Lletty Rd, Upper Tumble, Llanelli, SA14 6BN [IO71WS, SN51]
GM0 SEI R K Vennard, 4 Braehead, Girdle Toll, Irvine, KA11 1BD [IO75QP, NS34]
GW0 SEJ Details withheld at licensee's request by SSL.
G0 SEK Details withheld at licensee's request by SSL.
G0 SEL R G Kemp, 7 The Mews, Trinity Green, Gosport, PO12 1EZ [IO90KT, SZ69]
G0 SEM Details withheld at licensee's request by SSL.
G0 SEN S J Thompson, Elmtree Cottage, Woodrow, Fifehead Neville, Sturminster Newton, Dorset, DT10 2AQ [IO80TV, ST71]
GW0 SEO G T Jones, Brithdir, Llanbedrog, Pwllheli, Gwynedd, Wales, LL53 7PA [IO72SU, SH33]
GM0 SEP R M Cowan Strathclyde Emergency Planning, 85 Eastwoodmains Rd, Clarkston, Glasgow, G76 7HG [IO75UT, NS55]
G0 SEQ J Weiss, 11 Buttery Well Ln, Kendal, LA9 4HZ [IO84PH, SD59]
G0 SER G R Brookes, 73 Chadwell Ave, Sholing, Southampton, SO19 8GE [IO90HV, SU41]
G0 SET H Pearson, 55 Cowper Rd, River, Dover, CT17 0PL [JO01PD, TR24]
G0 SEU L B Payas, 36 Tintern Cl, Popley, Basingstoke, RG24 9HE [IO91KG, SU65]
G0 SEV E T Denman Sevenoaks & District Amateur, Sevenoaks District Council, Argyle Rd, Sevenoaks, Kent, TN13 1HG [JO01CG, TQ55]
G0 SEW K A Green, 13 Knowle Rd, Sheffield, S5 9GA [IO93GK, SK39]
G0 SEY E Russell, 60 Icknield Way, Tring, HP23 4HZ [IO91PT, SP91]
G0 SFA B V Hyde, 108 St. Bedes Cres, Cambridge, CB1 3UB [JO02CE, TL45]
GW0 SFB Details withheld at licensee's request by SSL.
G0 SFC C L Roberts, 16 Derby Rd, Maidstone, ME15 7JB [JO01GG, TQ75]
G0 SFD M Gudonis, 25 South View Park, Woodford, Plympton, Plymouth, PL7 4JE [IO70XJ, SX55]
G0 SFE K R Spring, 18 Greenway, Woodmancote, Cheltenham, GL52 4HU [IO81XW, SO92]
G0 SFF Details withheld at licensee's request by SSL.
G0 SFH I Glossop, 31 Earl Marshal Dr, Sheffield, S4 8JZ [IO93GJ, SK39]
GW0 SFI B C Hull, 13 Shirley Dr, Heolgerrig, Merthyr Tydfil, CF48 1SE [IO81HR, SO00]
G0 SFJ A C Thomas, 21 Great Bowden Rd, Market Harborough, LE16 7DE [IO92NL, SP78]
G0 SFL Details withheld at licensee's request by SSL.
G0 SFN J M Stephens, 4 The Fairway, Portchester, Fareham, PO16 8NS [IO90KU, SU60]
GW0 SFO K Moore, Llynfell, Oakford, Llanarth, Dyfed, SA47 0RW [IO72UE, SN45]
GW0 SFP B M Rish, 33 Oaklands Rd, Chirk Bank, Wrexham, LL14 5DP [IO82LW, SJ23]
G0 SFQ J Stirling, 16 Sandringham Ave, Wisbech, PE13 3ED [JO02CQ, TF41]
G0 SFR B P Burke, 4 Gaskell St., Pendlebury, Swinton, Manchester, M27 6QB [IO83TM, SD70]
G0 SFS M Best, 260 Kenyon Way, Little Hulton, Manchester, M38 0PU [IO83SM, SD70]
G0 SFT P P McDonald, 13 Heathfield, Culmore Rd, Londonderry, BT48 8JD [IO65IA, C42]
G0 SFV D J Burton, 100 Carden Hill, Hollingbury, Brighton, BN1 8DB [IO90WU, TQ30]
GI0 SFX D McCorkell, 27 Dalriada Walk, Ballymena, BT42 4DY [IO64UU, D10]
G0 SGA Dr J G Brockis, Moorcroft Flats, 23 Hamilton Dr East, Holgate, York, YO2 4DW [IO93KW, SE55]
G0 SGB S G Bryan Sgb Contest & Expedition Group, 91 Kilnhurst Rd, Rawmarsh, Rotherham, South Yorks, S62 5QQ [IO93HL, SK49]
G0 SGE D M Bell obo East Midlands Communica, Spindrift, The Green, Westborough, Newark, Notts, NG23 4HQ [IO93OC, SK85]
G0 SGF J R Barrett, Round Cottage, Philleigh, Truro, Cornwall, TR2 5NB [IO70MF, SW83]
GW0 SGG W J Holt, 14 Heather Cres, Sketty, Swansea, SA2 8HE [IO81AO, SS69]
GM0 SGH M C Brown, 3 Arnott Rd, Blackford, Auchterarder, PH4 1QE [IO86CG, NN80]
G0 SGI J M Sankey, 9 Hawthorne Ave, Newton, Preston, PR4 3TB [IO83NS, SD43]
G0 SGJ C T Birch, 5 Raynel Dr, Leeds, LS16 6BS [IO93EU, SE23]
GW0 SGL S L Locke, 216 Bryntirion, Ynysboeth Mathewstown, Ynysboeth, Mountain Ash, CF45 4EJ [IO81HP, ST09]
G0 SGM C E Read, 2 Castle View, Gosport, PO12 4LS [IO90KT, SU60]
G0 SGN Details withheld at licensee's request by SSL.
G0 SGP A Danton, 3 Cliffe Cl, Ruskington, Sleaford, NG34 9AT [IO93TB, TF05]
G0 SGQ M L Green, 15 Lodge Cl, Holt, Norfolk, NR25 6SN [JO02NV, TG03]
G0 SGR Details withheld at licensee's request by SSL.
G0 SGS A Rigby Skegness Gramar School, Skegness Grammar School, Vernon Rd, Skegness, PE25 2QS [JO03ED, TF56]
G0 SGT D C Huddleston, 162 Manor Rd, Newton St. Faith, Norwich, NR10 3LG [JO02PQ, TG21]
G0 SGU E Jones, 40 Fenwick Ln, Halton Lodge, Runcorn, WA7 5YU [IO83PH, SJ58]
G0 SGV J C W Allen, 86 Station Rd, Kings Langley, WD4 8LB [IO91SR, TL00]
G0 SGW Details withheld at licensee's request by SSL.
G0 SGX F B Dingwall, 20 Whitehills Rd, Loughton, IG10 1TS [JO01AP, TQ49]
G0 SGY Details withheld at licensee's request by SSL.
G0 SHA E R Churchman, 139 Cambridge Rd, Great Shelford, Cambridge, CB2 5JJ [JO02BD, TL45]
G0 SHB E H Lambourn, 7 Broadlands Ave, Enfield, EN3 5AH [IO91XP, TQ39]
G0 SHC M Lane, 10 Two Sisters Cl, Sutton Bridge, Spalding, PE12 9XP [JO02CS, TF42]
GM0 SHD G R Balfour, 6 Kirkden St., Friockheim, Arbroath, DD11 4SX [IO86QP, NO55]
G0 SHE Details withheld at licensee's request by SSL.
G0 SHF Details withheld at licensee's request by SSL.
G0 SHG M Hinken, 104 Barberwood Rd, Blackburn, Rotherham, S61 2DD [IO93GK, SK39]
G0 SHH J P Travers, 10 Lodge Cl, Little Oakley, Harwich, CO12 5EF [JO01OW, TM22]
G0 SHJ R Harrison, 8 Middlestone Cl, Gorleston, Great Yarmouth, NR31 6JB [JO02UN, TG50]
G0 SHM B Coates, 55 Maple Dr, Northstead, Scarborough, YO12 6LW [IO94SG, TA08]
G0 SHN Details withheld at licensee's request by SSL.
G0 SHO B C Lawrence, 70 Beacon Rd, Rolleston on Dove, Burton on Trent, DE13 9EG [IO92EU, SK22]
G0 SHP A M Pratt, 4 Chestnut Cl, Braunton, EX33 2EH [IO71WC, SS43]
G0 SHQ Details withheld at licensee's request by SSL.
G0 SHR H A Watson, Heather Watson, Manor Barn, Lower Hill, Tockholes, Lancs, BB3 0NF [IO83RQ, SD62]
G0 SHS C I Duff, Woodlea, Highfield Rd, Croston, Preston, Lancs, PR5 7HH [IO83OP, SD41]
G0 SHT S Rossi, via Teognide 154, Oo125 Roma, Italy, X X
G0 SHU G Bennett, 57 Princess Way, Euxton, Chorley, PR7 6PL [IO83QP, SD51]
G0 SHW Details withheld at licensee's request by SSL.
G0 SHX P W Johnson, PO Box 403, Wonthaggi, Victoria, Austrailia, X X
G0 SHY M D Bamber, 1 Penair Cres, Truro, TR1 1YS [IO70LG, SW84]
G0 SHZ R Bennett, 3 Hogarth Cl, Plymouth, PL9 8EX [IO70XI, SX55]
GM0 SIA P M Brooks, 3 Jamiesons Ct, Kelso, TD5 7EU [IO85SO, NT73]
G0 SIB Details withheld at licensee's request by SSL.
GM0 SIC Details withheld at licensee's request by SSL.
G0 SID Details withheld at licensee's request by SSL.
G0 SIE A R Swingler, 9 Princess Dr, Wistaston, Crewe, CW2 8HP [IO83SB, SJ65]
G0 SIG K J Prince Signallers Interest Group, 59 Chantry Rd, East Ayton, Scarborough, YO13 9ER [IO94SG, SE98]
G0 SIH J Rupp, 35 Courtland Rd, Shiphay, Torquay, TQ2 6JU [IO80FL, SX86]
G0 SII T Richards, 142 Princes Mews, Royston, SG8 9BN [IO92XB, TL34]
G0 SIK N W G Coram, Haytor, 56 Oakland Park, South Barnstable, Devon, EX31 2HX [IO71WB, SS53]
G0 SIL Details withheld at licensee's request by SSL.
GM0 SIM Details withheld at licensee's request by SSL.
G0 SIO G Goad, 2 Westholme, Orpington, BR6 0AN [JO01BJ, TQ46]
G0 SIP Details withheld at licensee's request by SSL.
G0 SIQ Revd R N Myerscough, 10 Kelling Rd, Holt, NR25 6RT [JO02NV, TG03]
G0 SIR Details withheld at licensee's request by SSL.
GW0 SIS K Barrett, 36 Priory Ave, Bridgend, CF31 3LR [IO81FL, SS97]
G0 SIT Details withheld at licensee's request by SSL.
G0 SIU B M Durrant, 31 The Dr, Shoreham By Sea, BN43 5GB [IO90UU, TQ20]
G0 SIV J E Wilcox, 31 Soane St., Basildon, SS13 1QU [JO01GN, TQ78]
G0 SIW B F Ellison, 6 Eskdale Rd, Ashton in Makerfield, Wigan, WN4 8QT [IO83QL, SJ59]
G0 SIX P J S Turner ., Flat 6, 132 Marine Parade, Brighton, BN2 1DE [IO90WT, TQ30]
G0 SIY A S Hopkinson, 26 Partridge Way, Chadderton, Oldham, OL9 0NS [IO83WN, SD80]
GM0 SIZ Details withheld at licensee's request by SSL.
G0 SJA Details withheld at licensee's request by SSL.
G0 SJB S J Barraclough, 46 Western Ave, Birstall, Batley, WF17 0PF [IO93ER, SE22]
G0 SJC C D Orman, 6 Berry Park Cl, Plymouth, PL9 9AQ [IO70WI, SX55]
G0 SJF J Tumber, 9 Eldred Ave, Brighton, BN1 5EB [IO90WU, TQ20]
G0 SJH S J Harris, 19 Mundays Boro, Puttenham, Guildford, Surrey, GU3 1AZ [IO91PF, SU94]
G0 SJI Details withheld at licensee's request by SSL.
G0 SJK Details withheld at licensee's request by SSL.
G0 SJP M Windle, 37 Gleton Ave, Hove, BN3 8LN [IO90VU, TQ20]
G0 SJQ Details withheld at licensee's request by SSL.
G0 SJR FNO R A Brand, 17 Brougham Rd, Acton, London, W3 6JD [IO91UM, TQ28][Op: Flight Nursing Officer R A Brand, RGN, Tech TNC, REMT(P), MIFPA. RAFARS 2695, RAOTA 913.]
G0 SJS D Pugh, 33 Brocstedes Ave, Ashton in Makerfiel, Ashton in Makerfield, Wigan, WN4 0NJ [IO83QM, SD50]
G0 SJT R N Kendrick, 68 Nansen Ave, Oakdale, Poole, BH15 3DD [IO90AR, SZ09]
G0 SJU T S Bousfield, 8 Harpington View, Mordon, Sedgefield, Co Durham, TS21 2EZ [IO94GP, NZ32]
G0 SJV P F Gostick, 25 Cashmere Ln, Cashmere, Queenslands, Australia 4500, X X
G0 SJW W H Cross HMS Plymouth Radio Club, 13 Carno St., Liverpool, L15 4LB [IO83MJ, SJ38]
G0 SJY K D Oliver, 50 Everingham Cres, Sheffield, S5 7LL [IO93GJ, SK39]
G0 SJZ J A Dobson, 7 Wallace Gdns, Lofthouse Gate, Wakefield, WF3 3SL [IO93FR, SE32]
G0 SKA C R Mitchell, Old Tiles, Beaconsfield Rd, Farnham Common, Slough, SL2 3LZ [IO91QN, SU98]
G0 SKB Details withheld at licensee's request by SSL.
G0 SKC Details withheld at licensee's request by SSL.
G0 SKD Details withheld at licensee's request by SSL.
G0 SKE Details withheld at licensee's request by SSL.
G0 SKG A H Faulkner, 105 Corbyn Rd, Russells Hall Est, Dudley, DY1 2JZ [IO82WM, SO98]
G0 SKI M J Foy, 335 South Ave, Southend on Sea, SS2 4HR [JO01IN, TQ88]
G0 SKJ K J Cockburn, 11 Highlands Ave, Barrow in Furness, LA13 0AU [IO84JC, SD27]
GM0 SKL Details withheld at licensee's request by SSL.
G0 SKM M R Tunstall, 24 Barbrook Ave, Weston Park, Longton, Stoke on Trent, ST3 5UG [IO82WX, SJ94]
G0 SKN P W Hartley, 18 Gravel Hill, Merley, Wimborne, BH21 1RR [IO90AS, SZ09]
GW0 SKO R Hale, 3 Wood Cl, Lisvane, Cardiff, CF4 5TT [IO81JM, ST18]
G0 SKP Details withheld at licensee's request by SSL.
G0 SKQ C Haines, 29 Woodlands Cl, Walton, Stone, ST15 0DX [IO82WV, SJ83]
G0 SKR J Goodall, 37 Woodfield Rd, Bear Cross, Bournemouth, BH11 9EU [IO90BS, SZ09]
G0 SKT Details withheld at licensee's request by SSL.
G0 SKU Details withheld at licensee's request by SSL.
G0 SKV R A Beattie, 25 Springfield Cl, Buckden, St. Neots, Huntingdon, PE18 9UR [IO92UH, TL16]
G0 SKW K B Walker, 37 Normanton Dr, Mansfield, NG18 3AQ [IO93JD, SK56]
G0 SKX R V Ardern, 43 Coppice Rd, Poynton, Stockport, SK12 1SL [IO83WI, SJ98]
G0 SKZ C J Hogan, 35 Pike Purse Ln, Richmond, DL10 4PS [IO94DJ, NZ10]
G0 SLB C F K Mattison, Aylmers Farm, Sheering Lower Rd, Old Harlow, Harlow, CM17 0NE [JO01BS, TL41]
GW0 SLC R J Thomas, 6 Grovers Cl, Glyncoch, Pontypridd, CF37 3DF [IO81IO, ST09]
G0 SLD P F Westripp, 88 Grange Rd, Tunbridge Wells, TN4 8QB [JO01CD, TQ53]
G0 SLE J D Hutchison, 20 Harbour Heights Ln, St Catherines, Ontario L2N 4K3, Canada, X X
G0 SLF S Collins, 1019 Byron St South, Whitby, Ontario L1N 4S3, Canada, X X
G0 SLG S L Gough, Applegate, Newtown, Langport, TA10 9SE [IO81OB, ST42]
G0 SLH C R Shoesmith, 2 Caravelle Gdns, Northolt, UB5 6EU [IO91TM, TQ18]
G0 SLI T W Day, 21 Mowbray Rd, Ham, Richmond, TW10 7NQ [IO91UK, TQ17]
G0 SLJ Dr D S Pepper, 11 Earls Ave, Folkestone, CT20 2HW [JO01NB, TR23]
G0 SLK E Patterson, 45 Sandhurst Rd, Rainhill, Prescot, L35 8NE [IO83OK, SJ49]
G0 SLL Details withheld at licensee's request by SSL.
GW0 SLM E Catherall, Tyn y Lon, Gwredog, Rhosgoch, LL66 0AX [IO73SJ, SH39]
G0 SLN P Grainger, 14 Menteith Cl, Lambton, Washington, Tyne & Wear, NE38 0PJ [IO94FV, NZ25]
G0 SLO Details withheld at licensee's request by SSL.
G0 SLP M A Coultas, 4 Suffolk Gdns, Marsden, South Shields, NE34 7JF [IO94HX, NZ36]
G0 SLQ S L Quinn, 48 Aldsworth Cl, Springwell Village, Gateshead, NE9 7PG [IO94FW, NZ25]
G0 SLR R J Lisle, 21 Porlock Cl, Penketh, Warrington, WA5 2QE [IO83QJ, SJ58]
G0 SLU C J Barr, 17 Knighton Rd, Otford, Sevenoaks, TN14 5LD [JO01CH, TQ55]

G0 SLV D B Gregory, 29 Ludlow Gr, Blackpool, FY2 0PZ [IO83LU, SD33]
G0 SLW J P Waite, 28 Overdown Rise, Mile Oak, Portslade, Brighton, BN41 2YG [IO90VU, TQ20]
G0 SLY C G Kratzer, Rb Thompson, 2 Oakfield Ln, Keston, BR2 6BY [JO01AI, TQ46]
G0 SMA Details withheld at licensee's request by SSL.
G0 SMD Details withheld at licensee's request by SSL.
G0 SMF M W Farrey, 38 Kingston Ave, Bearpark, Durham, DH7 7DL [IO94ES, NZ24]
G0 SMG Details withheld at licensee's request by SSL.
G0 SMH B C Marchant, 20 Wrench Rd, Norwich, NR5 8AS [JO02PP, TG10]
G0 SMI Details withheld at licensee's request by SSL.
G0 SMJ M F Jackson, 44 Dulwich Rd, Holland on Sea, Clacton on Sea, CO15 5NA [JO01OT, TM11]
G0 SMK P A Orchard obo Milton Keynes Scout, 68 Simpson Rd, Bletchley, Milton Keynes, MK1 1BA [IO92PA, SP83]
G0 SML Details withheld at licensee's request by SSL.
G0 SMM J R O'Nion, 7 Ettington Cl, Cheltenham, GL51 0NY [IO81WV, SO92]
G0 SMN A B McKenzie, 311 Weston Rd, Weston Coyney, Stoke on Trent, ST3 6HA [IO82WX, SJ94]
G0 SMO C R Cash, 7 Park Ln, Park Village, Wolverhampton, WV10 9QE [IO82WO, SJ90]
G0 SMP S M Pountain, 33 Milldale Ave, Buxton, SK17 9BE [IO93AG, SK07]
G0 SMQ F S Brazier, 12 Golden Riddy, Linslade, Leighton Buzzard, LU7 7RJ [IO91PW, SP92]
G0 SMR J Ballard, 30 Brainerd St., Liverpool, L13 7EH [IO83MK, SJ39]
G0 SMS Details withheld at licensee's request by SSL.
GM0 SMT Details withheld at licensee's request by SSL.
GI0 SMU A G Hanna, 39 Dalton Cres, Comber, Newtownards, BT23 5HE [IO74CN, J46]
G0 SMV I C Murray, 62 Gedney Rd, Long Sutton, Spalding, PE12 9JN [JO02BS, TF42]
G0 SMW Details withheld at licensee's request by SSL.
G0 SMX Details withheld at licensee's request by SSL.
G0 SMY L G Silvester, 22 Nordseter Lodge, Sea Ln, Rustington, Littlehampton, BN16 2RE [IO90RT, TQ00]
G0 SMZ R G Clews, 99 Kilbury Dr, Worcester, WR5 2NG [IO82VE, SO85]
G0 SNB W J Bonser, 24 Meend Garden Terr, Cinderford, GL14 2EB [IO81ST, SO61]
G0 SNC Details withheld at licensee's request by SSL.
G0 SND B Forster, Flat, 5 Stanley Rd, Stocksbridge, Sheffield, S8 9JB [IO93GI, SK38]
G0 SNE Details withheld at licensee's request by SSL.
G0 SNF J C Culling, 12 The Retreat, Princes Risborough, HP27 0JQ [IO91OR, SP80]
GM0 SNG G L Collier obo Dollar Academy A.R.C., Dollar Academy, Dollar, Scotland, FK14 7DU [IO86DD, NS99]
G0 SNJ Details withheld at licensee's request by SSL.
G0 SNK A J Gill, Bradgate, Kings Ln, Sway, Lymington, SO41 6BQ [IO90ES, SZ29]
G0 SNM K D Killick, 15 Popplechurch Dr, Lyncroft, Swindon, SN3 5DE [IO91DN, SU18]
G0 SNO D M Lauder, 20 Sutherland Cl, Barnet, EN5 2JL [IO91VP, TQ29]
G0 SNP J C Du Heaume, 10 Water Ln, Pill, Bristol, BS20 0EQ [IO81PL, ST57]
G0 SNQ H Davis, 44 Kenyon St., Ashton under Lyne, OL6 7DU [IO83WL, SJ99]
G0 SNR J Bagley South Norfolk Raynet, South Norfolk District Council, Swan Ln, Long Stratton, Norwich, NR15 2XE [JO02OL, TM19]
G0 SNS B E Harrison, 8 Elm Park, Pontefract, WF8 4LG [IO93IQ, SE42]
GM0 SNT A W Carpenter, 9 Glenbervie Rd, Kirkcaldy, KY2 6HR [IO86JD, NT29]
G0 SNU I K Gray, 27 Meadow Cl, Lavenham, Sudbury, CO10 9RU [JO02JC, TL94]
G0 SNV J Worsnop, 35 Westwood Ave, Eccleshill, Bradford, BD2 2NJ [IO93DT, SE13]
G0 SNW P L P Rogers, 126 Bradford Rd, Otley, LS21 3LE [IO93DV, SE14]
G0 SNX N W Johnson, 12 Bleach Mill Ln, Menston, Ilkley, LS29 6HE [IO93DV, SE14]
G0 SNY E S Wagner, 191 London Rd South, Poynton, Stockport, SK12 1LQ [IO83WI, SJ98]
G0 SNZ A Flood, 3 Ongar Walk, Sherwell Rd, Blackley, Manchester, M9 8JD [IO83VM, SD80]
G0 SOA Details withheld at licensee's request by SSL.
G0 SOB Details withheld at licensee's request by SSL.
G0 SOC P Hardiman Scouts of Croydon, 7 Osborne Rd, Thornton Heath, CR7 8PD [IO91WJ, TQ36]
G0 SOE A Yeomans, 61 Rushton Rd, Desborough, Kettering, NN14 2RR [IO92OK, SP88]
G0 SOF G Hedley, 1 Lisle Ln, Ely, CB7 4AS [JO02DJ, TL58]
G0 SOG F Mellings, 7 Todds Cl, Horley, RH6 8LB [IO91VE, TQ24]
G0 SOH O M Hazell, 13 Garland Rd, Poole, BH15 2LA [IO90AR, SZ09]
G0 SOI R Newman, Flat, 38 Eagle Cl, Russells Hall, Dudley, DY1 2JX [IO82WM, SO99]
G0 SOJ Details withheld at licensee's request by SSL.
G0 SOL Details withheld at licensee's request by SSL.
G0 SOM R M Roulstone, 5 Havenwood Rise, Clifton Est, Nottingham, NG11 9HD [IO92JV, SK53]
G0 SON Details withheld at licensee's request by SSL.
G0 SOO P J Antliff, 76 Albert Dr, Woking, GU21 5QZ [IO91RH, TQ06]
G0 SOP Details withheld at licensee's request by SSL.
G0 SOQ B D Summers, 1 St. Clements Hill, Norwich, NR3 4DE [JO02PP, TG21]
G0 SOR Details withheld at licensee's request by SSL.
G0 SOT Details withheld at licensee's request by SSL.
G0 SOU J E Brown, 53 Ashby Rd, Kegworth, Derby, DE74 2DJ [IO92IU, SK42]
G0 SOV D J Mason, 2 Lawrence Cl, Charlton Kings, Cheltenham, GL52 6NN [IO81XV, SO92]
G0 SOX P A Chapman, 28 Planton Way, Brightlingsea, Colchester, CO7 0LB [JO01MT, TM01]
G0 SPA P A Benson, 7 Crofton Cl, Attenborough, Beeston, Nottingham, NG9 5HX [IO92JV, SK53]
G0 SPB G W C Rusby, 12 Park Meadow, Princes Risborough, HP27 0EB [IO91NR, SP80]
G0 SPC M A Chawner, Navas House, 4 The Paddock, Kingsclere, Newbury, Berks, RG15 8SP [IO91LJ, SU66]
G0 SPE Details withheld at licensee's request by SSL.
G0 SPF T P Hill, 67 Uxbridge Rd, Hampton Hill, Hampton, TW12 1SL [IO91TK, TQ17]
G0 SPG D B Irwin, 124 Hawcoat Ln, Barrow in Furness, LA14 4HS [IO84JD, SD27]
G0 SPH K Brooks, 15 Tarn Cl, Winsford, CW7 2SA [IO83RE, SJ66]
G0 SPJ F Cangir, Turkish Embassy, 43 Belgrave Sq, London, SW1X 8PA [IO91WL, TQ27]
G0 SPK D G Neal, 490 Aureole Walk, Newmarket, CB8 7BQ [JO02EG, TL66]
G0 SPL Details withheld at licensee's request by SSL.
G0 SPM A H Hill, Hillcrest School & Comm Col., Simms Ln, Netherton, Dudley, West Midlands, DY3 0PB [IO82WM, SO99]
G0 SPO Dr J F Jefferys, 4 Long Dolver Drove, Soham, Ely, CB7 5UP [JO02EI, TL67]
G0 SPP A A Main, 6 West Meadows Dr, Cleadon, Sunderland, SR6 7TZ [IO94HW, NZ36]
G0 SPQ I R Wilson, 3 Caring Ln, Bearsted, Maidstone, ME14 4NJ [JO01HG, TQ85]
G0 SPR Details withheld at licensee's request by SSL.
G0 SPS M J Forder, 157 Kennington Rd, Kennington, Oxford, OX1 5PE [IO91JR, SP50]
G0 SPT Details withheld at licensee's request by SSL.
G0 SPV Details withheld at licensee's request by SSL.
G0 SPW R Simpson, 63 South Dene, South Shields, NE34 0HB [IO94GX, NZ36]
G0 SPX J M Sparks, 34 Green Park Ave, Skircoat Green, Halifax, HX3 0SR [IO93BQ, SE02]
G0 SPY J Thorley, 18 Bates Cl, Castleton, Rochdale, OL11 2TU [IO83WO, SD81]
G0 SPZ D M Rickard-Worth, 4 Sherwood Rd, Knaphill, Woking, GU21 2DE [IO91QH, SU95]
G0 SQA A E Vinters, 106 Halifax Rd, Ripponden, Sowerby Bridge, HX6 4AG [IO93AQ, SE02]
G0 SQC Details withheld at licensee's request by SSL.
G0 SQD T W McCulloch, 1 Wells Rd, Wheatley, Doncaster, DN2 4HQ [IO93KM, SE50]
G0 SQE E Young, 44 Joicey Rd, Low Fell, Gateshead, NE9 5HN [IO94EW, NZ26]
G0 SQF J T Bubez, 4 Southway, Burgess Hill, RH15 9ST [IO90WX, TQ31]
G0 SQH D Higbee, 12 Shelley Cl, Ashley Heath, Ringwood, BH24 2JA [IO90BU, SU10]
G0 SQI N A Blythe, 63 Humphrey Ave, Bromsgrove, B60 3JE [IO82XH, SO96]
G0 SQJ J J Kelly, 13 Dean St., Darwen, BB3 1HH [IO83SQ, SD62]
G0 SQK M R Stuckey, 162 Stockingstone Rd, Luton, LU2 7NJ [IO91TV, TL02]
G0 SQL R E Bishop, 99 St. Pauls Ave, Kenton, Harrow, HA3 9PR [IO91UO, TQ18]
G0 SQO R W Musicer, 2349 Crestcliff Dr, Tucker Georgia, 30084 USA
G0 SQP F A Ott, 16 Hornbeam Cl, Chelmsford, CM2 9LW [JO01FR, TL70]
G0 SQR Details withheld at licensee's request by SSL.
G0 SQS M A Hewitt, 1 Harpswell Hill Park, Hemswell, Gainsborough, Lincs, DN21 5UT [IO93QJ, SK99]
GW0 SQT A L Mackay, 23 Dock St., Cogan, Penarth, CF64 2LA [IO81JK, ST17]
G0 SQU R Farnley Tameside Amateur Radio Soc, 6 Cardigan Rd, Hollinwood, Oldham, OL8 4SF [IO83WM, SD90]
GI0 SQV J D Lyttle, Annagh Lodge, 9 Annaghanoon Rd, Warringstown, Co. Down, BT66 7RZ [IO64UK, J15]
G0 SQW Details withheld at licensee's request by SSL.
G0 SQX T Donley, 21 Elmridge, Leigh, WN7 1HN [IO83RM, SD60]
GW0 SQY S D Morgan, 79 Coedpenmaen Rd, Trallwng, Pontypridd, CF37 4LR [IO81IO, ST09]
G0 SQZ Details withheld at licensee's request by SSL.
G0 SRA B C Hanson, 19 Romsley Rd, Bartley Green, Birmingham, B32 3PR [IO82XK, SO98]
G0 SRB R E Boittier, 52 Tithelands, Harlow, CM19 5NB [JO01AS, TL40]
G0 SRC L Kirby Sth Derbs & Ashby Wlds A.R.G, 41 Woodville Rd, Overseal, Swadlincote, DE12 6LU [IO92FR, SK21]
GM0 SRD R A J Donaldson, 49 Arkaig Dr, Crossford, Dunfermline, KY12 8YW [IO86GB, NT08]
GW0 SRE D W Price, 1 Rhas Cottage, Pontyates, Llanelli, SA15 5SF [IO71VS, SN40]
GW0 SRF D M Daniels, 73 Bethania Rd, Upper Tumble, Llanelli, SA14 6DT [IO71WS, SN51]
G0 SRG S Green Sunderland Raynet Group, 133 Sevenoaks Dr, Sunderland, SR4 9NQ [IO94GV, NZ35]
G0 SRJ N J Ransom, 14 Azalea Cl, London, W7 3QA [IO91TM, TQ18]
GI0 SRL A S Harbison, 26 Ballymartin Rd, Templepatrick, Ballyclare, BT39 0BW [IO64XR, J28]
GI0 SRN E A McElroy, 39 Abercorn Park, Portadown, Craigavon, BT63 5JW [IO64TK, J05]
G0 SRO I A Maclean, 18 Westview Terr, Stornoway, Isle of Lewis, PA87 2HP [IO78JB, NO08]
GI0 SRP N Averill, 3 Edmund Ct, Tobermore, Magherafelt, BT45 5QA [IO64PT, H89]
GM0 SRQ D Beacher, 17 Cairn Gr, Crossford, Dunfermline, KY12 8YD [IO86GB, NT08]
G0 SRR M F Baugh, 97 Wilson Ave, Deal, CT14 9NJ [JO01QF, TR35]
G0 SRS Details withheld at licensee's request by SSL.
G0 SRT R Thompson, 11 Derwent Cl, Seaham, SR7 7BS [IO94HU, NZ44]
GI0 SRU R Hamilton, 126 Belmont Rd, Belfast, BT4 2AQ [IO74BO, J37]
G0 SRV B S Otty, 103 Fifers Ln, Hellesdon, Norwich, NR6 6EF [JO02PP, TG21]
G0 SRX J E D Loader, 9 Fernlea Ave, Ferndown, BH22 8HG [IO90BT, SZ09]
G0 SRY J B Smart, 60 Blaze Park, Wall Heath, Kingswinford, DY6 0LN [IO82VM, SO88]
G0 SRZ E G Hart, 75 The St., Wareham, Warham, Wells Next The Sea, NR23 1NL [JO02KW, TF94]
GI0 SSA J K Stevenson, 1 Newtownards Rd, Donaghadee, BT21 0DY [IO74FP, J58]
GW0 SSB Details withheld at licensee's request by SSL.
G0 SSC G Mills, 36 Ivanhoe Rd, Conisborough, Conisbrough, Doncaster, DN12 3JT [IO93JL, SK59]
GW0 SSD T Derbyshire, Carreg-Landeg, Pentraeth, Anglesey, Gwynedd, LL75 8YH [IO73VG, SH57]
G0 SSE N J Morton, Mont Rose, Lake Ln, Barnham, Bognor Regis, PO22 0AE [IO90QU, SU90]
G0 SSF C H Wileman, 414 Havant Rd, Farlington, Portsmouth, PO6 1NF [IO90LU, SU60]
G0 SSG R J Andre, 54 Covertside, Wirral, L48 9UL [IO83KI, SJ28]
GW0 SSI Details withheld at licensee's request by SSL.
G0 SSK G A Johnson, 13 Wilkinson St., Leigh, WN7 4DQ [IO83RL, SD60]
G0 SSL A C McEvoy, 12 Fountains Ave, Haydock, St. Helens, WA11 0RS [IO83QL, SJ59]
G0 SSM H G H Oswick, 57 Belfield, Digmoor, Skelmersdale, WN8 9HQ [IO83OM, SD40]
G0 SSN C E Ireland, 14 Castlefields, Gravesend, DA13 9EJ [JO01EJ, TQ66]
G0 SSP Details withheld at licensee's request by SSL.
GM0 SSQ A N Winchester, 23 Craigmount Ave North, Edinburgh, EH12 8DL [IO85IW, NT17]
G0 SSR G R Dellbridge Stourport & District Scout A.R, 19 Cleeve Cl, Astley Cross, Stourport on Severn, DY13 0NY [IO82UH, SO86]
G0 SST S M Blumfield, 43 Stapleton Rd, Meole, Shrewsbury, SY3 9LU [IO82OQ, SJ41]
G0 SSU P G Newbold, 9 Laurel Dr, Newport, TF10 7LY [IO82TS, SJ71]
G0 SSV Details withheld at licensee's request by SSL.
GI0 SSW S Caughey, 54 Oakwood Rd, Carrickfergus, BT38 8EU [IO74CR, J48]
G0 SSX P S Ellis, 4 Rostwold Way, Norwich, NR3 3NN [JO02PP, TG21]
G0 SSY D G Webb, Cams Cottage, 3 Cams Ln, Hambledon, Waterlooville, Hants, PO7 4SP [IO90KW, SU61]
G0 SSZ R W L Stanley, 10 Beechroyd, Radclife Ln, Pudsey, LS28 8BH [IO93ET, SE23]
G0 STA L G Robinson, The Wheatsheaf, Higher Ln, Broomedge, Lymm, WA13 0TR [IO83SI, SJ78]
GM0 STB R J Aitkenhead Scottish Tourist Board Radio, 7 Waterside Gdns, Hamilton, ML3 7PY [IO75XS, NS75]
GI0 STC P Dellett, 35 Oakfield Pl, Kilrea, Coleraine, BT51 5SA [IO64RW, C91]
G0 STE Details withheld at licensee's request by SSL.
G0 STF T R Clements, 1 Sandown Ln, Wavertree, Liverpool, L15 8HY [IO83MJ, SJ38]
G0 STG A P Sutherland, 42 Grange Rd, Bearley, Stratford upon Avon, CV37 0SE [IO92DF, SP16]
G0 STH R T Vaughan obo St Helens and District, 6 Dellside Gr, St. Helens, WA9 5AR [IO83OK, SJ59]
G0 STJ Details withheld at licensee's request by SSL.
G0 STK A G Hardcastle, 3 St. Johns Gr, Kirk Hammerton, York, YO5 8DE [IO93IX, SE45]
GI0 STM K P Murray, 17 Glebe Ct, Mullaghconnor, Dungannon, BT70 3PU [IO64NO, H77]
G0 STN Details withheld at licensee's request by SSL.
G0 STP Details withheld at licensee's request by SSL.
G0 STQ Details withheld at licensee's request by SSL.
G0 STR W J Shaw, 161 Springwood Cres, Edgware, HA8 8SH [IO91UP, TQ29]
GI0 STS R Todd, 73 Lakeview Park, Drumgor, Craigavon, BT65 4AL [IO64TK, J05]
G0 STT D A Rutter, 2 Laburnum Cl, Balderton, New Balderton, Newark, NG24 3AF [IO93OB, SK85]
G0 STU Details withheld at licensee's request by SSL.
G0 STV Details withheld at licensee's request by SSL.
G0 STW C R Kendrick, 18 Ainger Rd, Upper Dovercourt, Harwich, CO12 4TS [JO01PW, TM23]
G0 STX Details withheld at licensee's request by SSL.
G0 STY A V Jackson, 88 Chamberlain Cres, Shirley, Solihull, B90 2DJ [IO92BJ, SP17]
G0 SUA A Edwards, 49 Griggs Meadow, Dunsfold, Godalming, GU8 4ND [IO91RC, TQ03]
G0 SUB G R Thomas, Glendalough, Upper Kitesnest Ln, Whiteshill, Stroud, Gloucester, GL6 6BH [IO81VS, SO80]
G0 SUC B R Smith, 2 Hermitage Way, Sleights, Whitby, YO22 5HG [IO94QK, NZ80]
G0 SUE Details withheld at licensee's request by SSL.
GM0 SUF H J Butcher, 14 Newbattle Rd, Dalkeith, EH22 3DB [IO85LV, NT36]
G0 SUG Details withheld at licensee's request by SSL.
GM0 SUH J Y C Montgomery, 36 Bonaly Cres, Colinton, Edinburgh, EH13 0ER [IO85IV, NT26]
G0 SUI G T Scarlett, 14 Warren Dr, Eldwick, Bingley, BD16 3BX [IO93CU, SE13]
G0 SUJ Details withheld at licensee's request by SSL.
G0 SUK D P Lee West Dorset VHF Group, 131 Abbotsbury Rd, Weymouth, DT4 0JX [IO80SO, SY67]
G0 SUL J W Ward, 8 Wilderness Rd, Guildford, GU2 5QN [IO91QF, SU94]
G0 SUM Details withheld at licensee's request by SSL.
G0 SUN Details withheld at licensee's request by SSL.
G0 SUO D J R de Castillo, 52 Bronte Cres, Woodhall Farm, Hemel Hempstead, HP2 7PR [IO91SS, TL01]
GU0 SUP P F H Cooper, 1 Clos Au Pre, La Route de La Hougu, Castel, Guernsey, GY5 7FQ
G0 SUQ I L Johnson, 181 Broad St., Sidemoor, Bromsgrove, B61 8NQ [IO82XI, SO97]
G0 SUS B B Hall, 20A Leach Green Ln, Rubery, Rednal, Birmingham, B45 9BL [IO82XJ, SO97]
G0 SUT J Burdett, 68 Rangemore St., Burton on Trent, DE14 2EE [IO92ET, SK22]
G0 SUU J D Ogier, 37 Hill Park Rd, Torquay, TQ1 4LD [IO80FL, SX96]
G0 SUV Details withheld at licensee's request by SSL.
G0 SUX D Brown, The Meadows, Charlecote Garden Centre, Charlecote, Warks, CV35 9ER [IO92EE, SP25]
GM0 SUY C L Auld, 148 Echline Dr, South Queensferry, EH30 9XG [IO85HX, NT17]
G0 SUZ G Quick, 264 Rolleston Rd, Burton on Trent, DE13 0AY [IO92EU, SK22]
G0 SVA D J Nock, 431 Locking Rd, Weston Super Mare, BS22 8QN [IO81MI, ST36]
G0 SVB P J Herrmann, 5992 Royal Ct, Lockport N.Y., USA, 14094-9530, X X
G0 SVD M F Rhodes, 15 Manor Rd, Garden Village, Hatfield, AL10 9LJ [IO91VS, TL20]
G0 SVG Details withheld at licensee's request by SSL.
G0 SVH E G Wright, 3 Lindadale Cl, Fern Gore, Accrington, BB5 0NQ [IO83TR, SD72]
G0 SVI A J Beaver, 6 Mayland Green, Mayland, Chelmsford, CM3 6BD [JO01JQ, TL90]
G0 SVJ Details withheld at licensee's request by SSL.
G0 SVK J Hosfield, 76 Santon Way, Seascale, CA20 1NF [IO84GJ, NY00]
G0 SVL J P Alldred, 4 Hawthorne Rd, Rochdale, OL11 5JG [IO83VO, SD81]
G0 SVM A Mills, 23 Bilston Rd, Liverpool, L17 6AS [IO83MI, SJ38]
G0 SVN N Savin, 7 Bannard Rd, Maidenhead, SL6 4NG [IO91OM, SU88]
G0 SVO Details withheld at licensee's request by SSL.
G0 SVP G G Broadhurst, 9 Sharples St., Accrington, BB5 0HQ [IO83TR, SD72]
G0 SVQ A R Haydock, 8 Corbridge Cl, Blackpool, FY4 5EZ [IO83LT, SD33]
GM0 SVS Dr M Whiteley, 9 Pathfoot Ave, Bridge of Allan, Stirling, FK9 4SA [IO86AD, NS89]
G0 SVT Details withheld at licensee's request by SSL.
G0 SVU R W Morris, 146 Archer Rd, Stevenage, SG1 5HH [IO91VV, TL22]
G0 SVV S Winfield, 3 Garbett St., Accrington, BB5 0QJ [IO83TR, SD72]
G0 SVW P R Eaton, 118 Fulmar Ln, Wellingborough, NN8 4AG [IO92PH, SP86]
G0 SVX I Roebuck, 3 Lodge Hill Dr, Kiveton Park, Sheffield, S31 8RU [IO93GJ, SK38]
GJ0 SVZ M J Cooper, Flat 2 Mount Aubin, 67 Don Rd, St. Helier, Jersey, Channel Islands, JE2 4QD
G0 SWB Details withheld at licensee's request by SSL.
G0 SWC R A Eeles, 50 Nightingale Rd, Guildford, GU1 1EP [IO91RF, SU95]
GW0 SWD M A C Meehan South Wales Data Group, 50 Gwalia Cres, Gorseinon, Swansea, SA4 4DN [IO71XQ, SS59]
G0 SWE S A Whitbourn, 50 Nightingale Rd, Guildford, GU1 1EP [IO91RF, SU95]
G0 SWF J W Durrant, 16 Bugdens Ln, Verwood, BH31 6EY [IO90BV, SU00]
G0 SWG S Earle, 6 Tavistock Ave, Didcot, OX11 8NA [IO91JO, SU58]
G0 SWH B R Fletcher, 58 Broomfield Ave, Worthing, BN14 7SB [IO90TT, TQ10]
G0 SWI M J Marsh, 21 Stour Gdns, Great Cornard, Sudbury, CO10 0JN [JO02IA, TL83]
G0 SWK K W Bishop, 16 Parkers Hill, Tetsworth, Thame, OX9 7AJ [IO91LR, SP60]
G0 SWL C P Niles, 23 Randsfield Ave, Brough, HU15 1BE [IO93RR, SE92]
G0 SWM R Maude, 14 Crystal Cl, Goodrington, Paignton, Devon, TQ4 7LF [IO80FJ, SX85]
G0 SWN J C Fogg obo Gilwell Park Scout Rad, 16 Cotefield Dr, Leighton Buzzard, LU7 8DQ [IO91QW, SP92]
G0 SWO B Atkinson, 165 Alliance Ave, Anlaby Rd, Hull, HU3 6QY [IO93TR, TA02]
G0 SWP Details withheld at licensee's request by SSL.
G0 SWQ Details withheld at licensee's request by SSL.
G0 SWS T J Stow, 43 Kings Dr, Hassocks, BN6 8DY [IO90WW, TQ31]
G0 SWT L G Tandy, 8 Matthew Ct, Shenley Church End, Milton Keynes, MK5 6HJ [IO92OA, SP83]
G0 SWU P G Broad, 78 Lenham Rd, Sutton, SM1 4BG [IO91VI, TQ26]
G0 SWV Details withheld at licensee's request by SSL.
G0 SWW M E Stevens, 22 Lytham Dr, Waltham, Grimsby, DN37 0DG [IO93WM, TA20]
GM0 SWX J R Meredith, 22 Ryan Gdns, Innermessan, Stranraer, Wigtonshire, DG9 PQD
G0 SWY M I Humphrey, 4 Bluebell Rd, Bassett, Southampton, SO16 3LQ [IO90HW, SU41]
G0 SWZ J A Sharp, 10 Vale View, Aldridge, Walsall, WS9 0HW [IO92BO, SP09]
G0 SXA W Daly, 85 Lordens Rd, Huyton, Liverpool, L14 9PA [IO83NK, SJ49]
G0 SXC Details withheld at licensee's request by SSL.
GM0 SXD F Doyle, 1 Hawthorn Rd, Strathaven, ML10 6HA [IO75XQ, NS64]
GW0 SXE J G Williams, Alltwen, 44 Mayfield Dr, Buckley, CH7 2PN [IO83KE, SJ26]
G0 SXF Details withheld at licensee's request by SSL.
G0 SXG K Stammers, 102 Eaton Rd, Appleton, Abingdon, OX13 5JJ [IO91HR, SP40]
G0 SXH Details withheld at licensee's request by SSL.
G0 SXK A S Dodd, 20 Braemar Ave, Chelmsford, CM2 9PW [JO01FR, TL70]

G0 SXM D J Smith, The Bungalow, Langford Hele Farm, Marchamchurch, Bude, Cornwall, EX23 0HR [IO70RS, SS20]
G0 SXN B W Watts, The Wattsits, 32 Delta Rd, Chobham, Woking, GU24 8PY [IO91QI, SU96]
G0 SXO A R Segar, 21 Woodcroft, Kennington, Oxford, OX1 5NH [IO91JR, SP50]
GM0 SXP R W Cliff, 32 Lochardil Rd, Inverness, IV2 4LD [IO77VK, NH64]
GM0 SXQ M W Hepburn, Toll House, Corse, Lumphanan, Banchory, Kincardineshire, AB31 4RY [IO87PD, NJ50]
G0 SXS K J Wheeler, 11 Deacon Way, Rugeley, WS15 3JZ [IO92AS, SK01]
G0 SXU C F P Bocock, 1 Charles Lister Ct, Lister Cl, Dover, CT17 0TP [JO01PD, TR34]
G0 SXW P R Cressey, Skidby Hill Farm, Beverley Rd, Skidby, Cottingham, HU16 5TF [IO93ST, TA03]
G0 SXY P G N Davies, 26 Campion Hall Dr, Millbrook, Didcot, OX11 9RN [IO91JO, SU58]
G0 SXZ J A Green, 23 Litchard Park, Bridgend, CF31 1PF [IO81FM, SS98]
G0 SYA Dr D I Tock, Stratford Dr, Woodburn Green, Wooburn Green, High Wycombe, Bucks, HP10 0QH [IO91PN, SU98]
G0 SYB D S Church, 57A Brighton Rd, Newhaven, BN9 9NG [JO00AT, TQ40]
G0 SYE J N Coates, 17 Stanshawes Dr, Yate, Bristol, BS17 4ET [IO81SM, ST78]
G0 SYF R Pearce, 14 Shepherds Leaze, Wotton under Edge, GL12 7LQ [IO81TP, ST79]
GW0 SYG T J Perry The Cleddau Amateur Radio Soci, 70 Lawrenny St., Neyland, Milford Haven, SA73 1TB [IO71MR, SM90]
G0 SYH A M P Dolby, Payrole, 82100 Castelferrus, Tarn Et Garonne, France, X X
G0 SYI K Treasure, 30 Grace Park Rd, Brislington, Bristol, BS4 5JA [IO81RK, ST67]
G0 SYJ Details withheld at licensee's request by SSL.
G0 SYK S D Oram, 1 Lydiard Way, Trowbridge, BA14 0UG [IO81VH, ST85]
GM0 SYL E M C Dons obo Scottish Young Ladies, 37 Ashley Dr, Edinburgh, EH11 1RP [IO85JW, NT27]
G0 SYM Details withheld at licensee's request by SSL.
GW0 SYN J C Cooke, Hope and Anchor, 94 Vale St., Denbigh, Clwyd, LL16 3BW [IO83HE, SJ06]
G0 SYO Details withheld at licensee's request by SSL.
G0 SYQ Details withheld at licensee's request by SSL.
G0 SYR B G Petifer, 14 Wood Ln, Caterham, CR3 5RT [IO91WG, TQ35]
G0 SYS Details withheld at licensee's request by SSL.
G0 SYT D R Firks, Bryn Garth Cottage, Much Birch, Hereford, HR2 8HJ [IO81PX, SO43]
GM0 SYU G H Anderson, 22 Springvale St., Saltcoats, KA21 5LP [IO75OP, NS24]
GM0 SYV J Kelly, 28 Riverbank Dr, Mossend, Bellshill, ML4 2PR [IO75XT, NS75]
GM0 SYW A R Smith, 51 Ivanhoe Dr, Saltcoats, KA21 6LX [IO75OP, NS24]
GM0 SYY J B Hutchens, 2 Provost Gate, Larkhall, ML9 1DN [IO85AR, NS75]
GW0 SYZ I Beals, Maes y Coed, Maenygroes, New Quay, Dyfed, SA45 9RL [IO72TE, SN35]
GM0 SZA I Stones, 176 North Deeside Rd, Peterculter, AB1 0UD [IO87TH, NJ72]
GW0 SZB A J Plumbley, 2 Mount Pleasant, Lon Uchaf, Morfa Nefyn, Pwllheli, LL53 6AD [IO72RW, SH23]
G0 SZC W R Duce, 16 Gillmans Rd, Orpington, BR5 4LA [JO01BJ, TQ46]
G0 SZD S K White, 14 Hardwick Rd East, Manton, Worksop, S80 2NS [IO93KH, SK57]
G0 SZE C C Andrews, 34 Forest Rd, Athersley North, Barnsley, S71 3BG [IO93GN, SE30]
GM0 SZF M Volkert, 37 Forest Ave, Aberdeen, AB1 4TU [IO87TH, NJ72]
G0 SZG J O Greene, 308 Cedar Rd, Camphill, Nuneaton, CV10 9DY [IO92GM, SP39]
GI0 SZH W J I McEwen, 234 Legahory Ct, Legahory, Craigavon, BT65 5DH [IO64TK, J05]
G0 SZI D J Green, 18 Lea Oak Gdns, Fareham, PO15 6TA [IO90JU, SU50]
G0 SZJ S H Fletcher, 'Rainbow Dancer', Chichester Yacht Basin, Birdham, Chichester, West Sussex, PO20 7EJ [IO90OT, SU80]
G0 SZK L J Joyce, 51 Woodlands Rd, Northmeads, Bognor Regis, PO22 9EE [IO90PT, SU90]
GM0 SZL Dr G M Clarke, 59 Hillpark Ave, Edinburgh, EH4 7AL [IO85IX, NT27]
GW0 SZN M Lawrence, 1 Greenwood Cttgs, Gelli Groes, Pontllanfraith, Blackwood, NP2 2JB [IO81JP, ST19]
G0 SZO J C Everard, Woodside, Staple Ln, West Quantoxhead, Taunton, TA4 4DE [IO81IE, ST14]
GM0 SZQ M Murphy, Roadinghead, Bardarroch, Ochiltree, Cumnock Ayrshire, KA18 2RR [IO75TK, NS41]
G0 SZR Details withheld at licensee's request by SSL.
G0 SZS S D Bone, 18 Pullmans Pl, Staines, TW18 4LD [IO91RK, TQ07]
G0 SZT D J Allibone, Virginia, North St., Langport, TA10 9RH [IO81OA, ST42]
GW0 SZU B E Osborne, 5 Litchard Cross, Bridgend, CF31 1NZ [IO81FM, SS98]
G0 SZV J Milnes, 381 Staincliffe Rd, Dewsbury, WF13 4RB [IO93EQ, SE22]
GW0 SZW D W Evans Coleshill Radio Invalid & Blin, 3 Heol Elfed, Cefn Caeau, Llanelli, SA14 9DY [IO71WQ, SN50]
G0 SZX W E Hunt, 8 Denmark Rd, Exeter, EX1 1SL [IO80FR, SX99]
G0 SZY R B Herrick, 9 Tiberius Rd, Luton, LU3 3QJ [IO91SV, TL02]
G0 SZZ V A Andreyev, 75 Warley Ave, Dagenham, RM8 1JS [JO01BN, TQ48]
G0 TAA W G J Chandler, 32 Suffolk Dr, Chandlers Ford, Eastleigh, SO53 3HW [IO90HX, SU41]
G0 TAB E H Tabor, 5 Taylors Ave, Cleethorpes, DN35 0LF [IO93XN, TA30]
G0 TAC P Chadwick, 2 Auden Pl, Longton, Stoke on Trent, ST3 1SJ [IO82WX, SJ94]
G0 TAD G A Jewell, Les Bardots, 47120 Savignac de Duras, France, X X
GM0 TAE C Porter, 20 Baird Ave, Kilwinning, KA13 7AR [IO75PP, NS34]
GW0 TAF Details withheld at licensee's request by SSL.
G0 TAG D Beane, 3 Mary Proud Ct, Oak Piece, Welwyn, AL6 0XG [IO91VU, TL21]
G0 TAH K F C Aldus, 77 Springvale Rd, Kings Worthy, Winchester, SO23 7ND [IO91IC, SU43]
G0 TAJ T A J Millington, 11 Polwhaveral Terr, Falmouth, TR11 2LR [IO70LD, SW83]
G0 TAK R Walker, 3 Elderberry Cl, Thornton Cleveleys, FY5 2ZB [IO83LV, SD34]
G0 TAL S N Walsworth, 4 The Homestead, Heckmondwike, WF16 9JL [IO93ER, SE22]
G0 TAM A Farrow, 18 The Green, Trimingham, Norwich, Norfolk, NR11 8ED [JO02QV, TG23]
G0 TAN S B Venner, Tanglewood, 4 Cuckoo Ln, West End, Woking, GU24 9NG [IO91QI, SU96]
G0 TAO R M N Lokuge, 11 Porchester Cl, Southwater, Horsham, RH13 7XR [IO91TA, TQ12]
G0 TAP Details withheld at licensee's request by SSL.
G0 TAR B D Lucas, 8 Gilbert Cl, Hempstead, Gillingham, ME7 3QQ [JO01GI, TQ76]
G0 TAS J W Taylor, 19 Castle Cl, Leconfield, Beverley, HU17 7NX [IO93SV, TA04]
G0 TAT M K Wilmot, Southview, Roman Rd, Bleadon, Weston Super Mare, BS24 0AB [IO81MH, ST35]
GW0 TAU Details withheld at licensee's request by SSL.
G0 TAV T V Arumugam, 19 Donnington Rd, Kenton, Harrow, HA3 0NB [IO91UN, TQ18]
G0 TAW I Patrick, 31 Courtfields, Swaffham, PE37 7ET [JO02IP, TF80]
G0 TAX T Welch, 63 Vicarage Cl, New Silksworth, Sunderland, SR3 1JF [IO94HU, NZ35]
GM0 TAY I S Strachan obo Tayside Raynet, Hope Cottage, 238 Coupar Angus Rd, Muirhead, Dundee, DD2 5QN [IO86LL, NO33]
G0 TAZ J E Burgess, Combeside House, Symonsburrow, Hemyock, Cullompton, Devon, EX15 3XA [IO80JW, ST11]
GW0 TBA Details withheld at licensee's request by SSL.
G0 TBB R P Bullough, 33 Booth Rd, Little Lever, Bolton, BL3 1JR [IO83TN, SD70]
G0 TBC S Lawson, 27 Broadlands Ave, Boyatt Wood, Eastleigh, SO50 4PP [IO90HX, SU42]
G0 TBD Details withheld at licensee's request by SSL.[Op: T. Barclay. Station located in the Ilford postcode area. WAB square: TQ48. Book no: 13433. Fists: 1163. SWL reports welcome.]
G0 TBF T G Chadwick, 38 Galloway Rd, Bentilee, Stoke on Trent, ST2 0QH [IO83WA, SJ94]
G0 TBG Details withheld at licensee's request by SSL.
GM0 TBH J D McMaster, 96 Cunningham Cres, Ayr, KA7 3JB [IO75QK, NS32]
G0 TBI S J McKinnon, 145 Enville Rd, Kinver, Stourbridge, DY7 6BN [IO82VK, SO88]
G0 TBJ J D Cardiff, 32 High St., Markyate, St. Albans, AL3 8PB [IO91SU, TL01]
G0 TBK D A Fleet, 14 Flint Rd, Upper Hatherley, Cheltenham, GL51 5JE [IO81WV, SO92]
G0 TBM J S Goulden, 5 Mayfield Rd, Tunbridge Wells, TN4 8ES [JO01DD, TQ53]
G0 TBN Details withheld at licensee's request by SSL.
G0 TBO P N Clark, 7 Shirley Ave, Ramsgate, CT11 7AT [JO01RI, TR36]
G0 TBR Details withheld at licensee's request by SSL.
G0 TBS W J Lucas, 8 North St., Bletchley, Milton Keynes, MK2 2PY [IO91PX, SP83]
GW0 TBT J C Roberts, 10 Coronation St., Cefn Mawr, Wrexham, LL14 3PN [IO82LX, SJ24]
G0 TBU R Brightwell, 4 Buckingham Rd, Margate, CT9 5SS [JO01QJ, TR37][Correspondence to PO Box 97, Sandwich, Kent.]
G0 TBV M Cossins, 259 Lynchford Rd, Farnborough, GU14 6HX [IO91PG, SU85]
G0 TBW G Davis, 7 Broxton Ave, Bolton, BL3 3TG [IO83SN, SD60]
G0 TBX Details withheld at licensee's request by SSL.
G0 TBY R Harvey, 12 Hill Crest, Altofts, Wakefield, West Yorks, WF6 2NT [IO93GR, SE32]
G0 TBZ J Hamilton, 1 Phillips Cl, Haswell, Durham, DH6 2BW [IO94GS, NZ34]
G0 TCA R K Seaward, 13 Blythe Cl, Catford, London, SE6 4UW [IO91XK, TQ37]
GM0 TCC K A Walker, 2 Station View, Kirtlebridge, Lockerbie, Dumfriesshire, DG11 3LR [IO85JB, NY27]
G0 TCD I C Hamilton, 139 Elstree Rd, Woodhall Farm, Hemel Hempstead, HP2 7QW [IO91SS, TL01]
G0 TCE P N Taylor, 67 Rectory Park, Sanderstead, South Croydon, CR2 9JR [IO91XI, TQ36]
G0 TCF P Loch, 25 Leeholme, Houghton-le-Spring, DH5 8HR [IO94GU, NZ34]
G0 TCG Details withheld at licensee's request by SSL.
G0 TCH T E Noszkay, 127 Lonsdale Ave, Weston Super Mare, BS23 3SQ [IO81MH, ST35]
G0 TCI B Hughes, 29 East View Cl, Radwinter, Saffron Walden, CB10 2TZ [JO02EA, TL63]
G0 TCJ T H Worrall, 9 Barnstaple Cl, Wigston, LE18 2QX [IO92KN, SP69]
G0 TCK F Fuller, 29 Lannock Rd, Hayes, UB3 2NG [IO91SM, TQ08]
G0 TCL D R Parsons, 22 Shawbury Ave, Higher Bebbington, Wirral, L63 8LR [IO83LI, SJ38]
G0 TCM F G Poulter, 2 Leighfield Cl, Bedford, MK41 0AG [IO92SD, TL05]
GW0 TCN E J Evans, 26 Frederick Pl, Llansamlet, Swansea, SA7 9RY [IO81BP, SS69]
G0 TCO M C Roper, 1 The Cttgs, Norwich Rd, Colton, Norwich, NR9 5BY [JO02NP, TG10]
G0 TCP P C Maynard, Seaton House, Lower Rd, Westerfield, Ipswich, IP6 9AR [JO02OC, TM14]
G0 TCQ Details withheld at licensee's request by SSL.
G0 TCR Sqn Ldr C E Savage, 27 Old Church Green, Kirk Hammerton, York, YO5 8DL [IO93IX, SE45]
G0 TCS D Smith, 43 Ridgeway, Nettleham, Lincoln, LN2 2TL [IO93SG, TF07]
GM0 TCU C G Pirie, 10 Annesley Park, Torphins, Banchory, AB31 4HG [IO87QC, NJ60]
GW0 TCV A Thomas, 4 Gordon Terr, King St., Mold, CH7 1LD [IO83KE, SJ26]
G0 TCW A B Furmston, 46 Twydall Ln, Gillingham, ME8 6JE [JO01GI, TQ76]
G0 TCX Details withheld at licensee's request by SSL.
G0 TCY M J Templeman, 14 Hengist St., Tonge Fold, Bolton, BL2 6BX [IO83TN, SD70]
G0 TCZ Details withheld at licensee's request by SSL.
GW0 TDA P G Price, 16 Heol Dyddgen, Crwbin, Kidwelly, SA17 5DG [IO71US, SN40]
G0 TDB P C Batson, 56 Sqn, RAF Coningsby, Lincs, LN4 4SY [IO93WC, TF25]
G0 TDC S Tinsley, 31 Havisham Cl, Locking Stumps, Birchwood, Warrington, WA3 7NB [IO83RK, SJ69]
G0 TDD T D Dubourg, 50 Wellington Rd, Oxton, Birkenhead, L43 2JF [IO83LJ, SJ38]
G0 TDE R Barrett, 140 Berries Ave, Bude, EX23 8QP [IO70RT, SS20]
G0 TDF Details withheld at licensee's request by SSL.[However station located near Lichfield. Correspondence to R D Briggs, PO Box 11, Rugeley, Staffordshire, WS15 4YR.]
G0 TDG N R Hotson, 9A Merlin Rd, Scunthorpe, DN17 1ND [IO93QN, SE80]
G0 TDI Details withheld at licensee's request by SSL.
G0 TDJ S R Smith, 124 Parkside Ave, Barnehurst, Bexleyheath, DA7 6NL [JO01CL, TQ57]
G0 TDK J Anderson, 104 Mab Ln, West Derby, Liverpool, L12 6RL [IO83NK, SJ49]
G0 TDL N A Yates, 11 Beverley Cl, Balsall Common, Coventry, CV7 7GA [IO92EJ, SP27]
G0 TDM J G Sutton, 15 Lowther St., Penrith, CA11 7UW [IO84OQ, NY53]
G0 TDN S T Maclennan, Shalam, Rising Sun, Callington, Cornwall, PL17 8JE [IO70UM, SX37]
GM0 TDO A I Macadam, 18 Loch Pl, South Queensferry, EH30 9NG [IO85HX, NT17]
GI0 TDP J G Driscoll, 67 Whinney Hill, Holywood, BT18 0HG [IO74CP, J47]
G0 TDQ P B Luscombe, 25 Brookdale Ct, Brookdale Cl, Brixham, TQ5 9JW [IO80FJ, SX95]
G0 TDR T D Round, 26 Castle St., Kinver, Stourbridge, DY7 6EL [IO82VK, SO88]
G0 TDS K A Shaddick, 32 Kings Head Ln, Bishopsworth, Bristol, BS13 7DD [IO81QK, ST56]
G0 TDT M D P Nokes, 10 Elsworth Pl, Cambridge, CB2 2RG [JO02BE, TL45]
G0 TDU P D Galsworthy, 112 Watchfield Ct, Sutton Ct Rd, Chiswick, London, W4 4ND [IO91UL, TQ27]
G0 TDV R J King, 10 Lansdown Rd, Soundwell, Kingswood, Bristol, BS15 1XB [IO81RL, ST67]
G0 TDW W Pettinger, 24 Pye Nest Gdns, Pye Nest, Halifax, HX2 7JX [IO93BR, SE02]
G0 TDX Y Kimoto, 48 Birchall Ave, Culcheth, Warrington, WA3 4DD [IO83RK, SJ69]
G0 TDY A S Frank, 15 Home Ground, Cricklade, Swindon, SN6 6JG [IO91BP, SU09]
G0 TDZ H Jones, 391 Chester Rd, Woodford, Stockport, SK7 1QQ [IO83WI, SJ88]
GM0 TEA A R Cherry, 39 Clark Rd, Edinburgh, EH5 3AR [IO85JX, NT27]
G0 TEB C P Sexton, Lapswater, Marsh, Honiton, Devon, EX14 9AL [IO80LV, ST21]
G0 TEC A J Mee, 18 Forest Rise, Thurnby, Leicester, LE7 9PF [IO92LP, SK60]
G0 TED E C Thane, 19 Churchill Rd, Walton, Stone, ST15 0EB [IO82WV, SJ83]
G0 TEE T L Speight, 1 Lyndene Ave, Roe Green, Worsley, Manchester, M28 2RJ [IO83TM, SD70]
G0 TEF S Morecroft, 4 Arran Cl, Bolton, BL3 4PP [IO83SN, SD60]
G0 TEH S R Lumb, 21 Edward St., Sowerby Bridge, HX6 2NJ [IO93BR, SE02]
G0 TEI R J Sparks, 10 Glebe Ct, Southampton, SO17 1RH [IO90HW, SU41]
G0 TEK D Seddon, 2 Hurstlea Ct, Alderley Edge, SK9 7QF [IO83VH, SJ87]
G0 TEL R P Bellenot, 113 Melfort Rd, Thornton Heath, CR7 7RX [IO91WJ, TQ36]
G0 TEM A Merrix, 56 Morris St., West Bromwich, B70 7SP [IO92AM, SP09]
G0 TEN G E Walker, 36 Vale Royal River Park, Moulton, Northwich, CW7 2PF [IO83RF, SJ66]
G0 TEO L A Baker, 13 Strathville Rd, South Shields, London, SW18 4QX [IO91VK, TQ27]
GD0 TEP A C Kissack, 30 High View Rd, Douglas, IM2 5BH[IO74SD. QSL Manager for GD4IOM, direct cards require SAE or IRC. E-mail: Andy_Kissick@compuserve.com.]
G0 TEQ S H Jennings, Flat 2, 143 Merritts Brook Ln, Northfield, Birmingham, B31 1UH [IO92AK, SP08]
G0 TER Details withheld at licensee's request by SSL.
G0 TES R W Dicks, 148 Finedon Rd, Irthlingborough, Wellingborough, NN9 5UB [IO92QH, SP97]
GM0 TET S Conner, 9 Winton Dr, Kelvinside, Glasgow, G12 0PZ [IO75UV, NS56]
G0 TEV J A Mullin, Prom Office, Bradbury Bks, Krefeld, Bfpo 35, X X
GU0 TEW R C L Sample, Nantucket, La Villiaze Rd, Forest, Guernsey, GY8 0HQ
GM0 TEX A D Forrest, White Lodge, 1 Laggan Rd, Newton Mearns, Glasgow, G77 6LP [IO75US, NS55]
G0 TEY Details withheld at licensee's request by SSL.
G0 TEZ Details withheld at licensee's request by SSL.
G0 TFB L J Wallen Wallen Antennae Radio Club, Unit, 1 Trinity Pl, Ramsgate, CT11 7HJ [JO01RI, TR36]
G0 TFC A B Stanley, Knock Cross, Knock, Appleby, Cumbria, CA16 6DT [IO84RP, NY62]
G0 TFD D J Eyre, 29 Old Acre Ln, Brocton, Stafford, ST17 0TW [IO82XS, SJ91]
GM0 TFE C C M Stewart, Leffnoll Cottage, Cairnryan, DG9 8QU [IO74LW, NX06]
GM0 TFF A McGhie, 16 Boyach Cres, Isle of Whithorn, Newton Stewart, DG8 8LD [IO74TQ, NX43]
GD0 TFG J G Dowling, 2 Ballabridson Park, Ballasalla, IM9 2ES
G0 TFI D Roberts, 1 Whitehead Cl, Staining, Blackpool, FY3 0DZ [IO83MT, SD33]
G0 TFJ M J Chance, 19 St. Stephens Rd, Selly Park, Selly Oak, Birmingham, B29 7RR [IO92AK, SP08]
G0 TFK J Sutcliffe, 4 Lancaster Gate, Nelson, BB9 0AP [IO83VT, SD83]
G0 TFL J G Williams, 107 Clay Ln, Rochdale, OL11 5QW [IO83VO, SD81]
G0 TFM K J Milsom, 1 Amber Villas, Sutton St. Nicholas, Hereford, HR1 3DF [IO82QD, SO54]
G0 TFN R H Perkins, 15 Potters Cl, Brinklow, Rugby, CV23 0NS [IO92HJ, SP47]
GD0 TFO J Bellis, 2 Cedar Walk, Douglas, IM2 5NG
G0 TFP J A Brett, 11 Manor Rd, Astley, Tyldesley, Manchester, M29 7PH [IO83SM, SD60]
GM0 TFQ H R Wignall, 7 Windyedge, Upperboat, Inverurie, AB51 3WJ [IO87TG, NJ72]
G0 TFR A Vining, The Cedars, Thorney Rd, Crowland, Peterborough, Cambs, PE6 0LH [IO92WP, TF20]
G0 TFS Details withheld at licensee's request by SSL.
G0 TFT A Gibson, 28 Finchale Terr, Jarrow, NE32 3TX [IO94GX, NZ36]
G0 TFU M P D Wilson, 6 The Chase, Calcot, Reading, RG3 7DN [IO91LJ, SU66]
G0 TFV T G Carvell, 18 Park View, School Ln, Peasmarsh, Rye, TN31 6UR [JO00IX, TQ82]
GM0 TFW Details withheld at licensee's request by SSL.
G0 TFX S E Roberts, 20 Beech Gr, Trowbridge, BA14 0HG [IO81VH, ST85]
GD0 TFY Details withheld at licensee's request by SSL.
G0 TGA Details withheld at licensee's request by SSL.
G0 TGB T G Blore, Glendhoon, Laneham St., Rampton, Retford, DN22 0JX [IO93OH, SK87]
G0 TGC Details withheld at licensee's request by SSL.
G0 TGD Details withheld at licensee's request by SSL.
GM0 TGE I J J Ross, South Bogside, St Katherines, Inverurie, Aberdeenshire, AB51 8SY [IO87UJ, NJ83]
G0 TGF J McGoff, 55 Knights End Rd, March, PE15 9QA [JO02AM, TL49]
GM0 TGG L A Mackenzie, 90 Tay St., Newport on Tay, DD6 8AP [IO86MK, NO42]
G0 TGH S D Leak, 29 Red Hall Chase, Whinmoor, Leeds, LS14 1NR [IO93GU, SE33]
G0 TGI Details withheld at licensee's request by SSL.
G0 TGJ Details withheld at licensee's request by SSL.
G0 TGK P M Lambert, 232 Jesmond Dene Rd, Newcastle upon Tyne, NE2 2JU [IO94EX, NZ26]
G0 TGL Details withheld at licensee's request by SSL.
G0 TGM S D Fowler, 38 Meadowcroft, Aylesbury, HP19 3LN [IO91OT, SP81]
G0 TGN J W Lee, 24 Thirlwall Ave, Conisbrough, Doncaster, DN12 3JZ [IO93JL, SK59]
G0 TGO A M Riddell, 114 Ventnor Ct, Leaside Way, Southampton, Hants, SO16 3EE [IO90HW, SU41]
G0 TGP W J Ford, 3 Cauldron Barn Rd, Swanage, BH19 1QF [IO90AO, SZ07]
G0 TGQ O L Cubitt, 97 Sutton Ln, Langley, Slough, SL3 8AU [IO91RL, TQ07]
G0 TGR D Greenacre, 38 Toon Cres, Bury, BL8 1JB [IO83UO, SD71]
G0 TGS M F Miller, 73 Macclesfield Rd, Buxton, SK17 9AG [IO93AG, SK07]
G0 TGT Details withheld at licensee's request by SSL.
G0 TGU W R Collier, 20 Wainers Croft, Greenleys, Milton Keynes, MK12 6AL [IO92OB, SP83]
G0 TGV L C Wilson-Dutton, 59 Mount Pleasant, Tadley, RG26 4BN [IO91KI, SU56]
G0 TGW D Hay, 19 Munks Cl, West Harnham, Salisbury, SP2 8PB [IO91CB, SU12]
G0 TGX P D E Ireland, 2 The Ridings, Willerby Rd, Hull, HU5 5HW [IO93TS, TA02]
G0 TGY Details withheld at licensee's request by SSL.
G0 THA P T Cole, 1 New Cttgs, Udimore, Rye, TN31 6AN [JO00IW, TQ81]
G0 THB G E Batty, 13 Mercury Gdns, Hamble, Southampton, SO31 4NZ [IO90IU, SU40]
G0 THD P D Hart, 32 Harvard Rd, Ringmer, Lewes, BN8 5HW [JO00AV, TQ41]
G0 THE R F Boniface, 7 Coombes Cl, Beckley, Rye, TN31 6TR [JO00HX, TQ82]
G0 THF K Greatorex, 54 Lilac Gr, Glapwell, Chesterfield, S44 5NG [IO93IE, SK46]
G0 THG Details withheld at licensee's request by SSL.
G0 THH H Hudders, 37 Chearsley Rd, Long Crendon, Aylesbury, HP18 9BT [IO91MS, SP60]
G0 THI W E S Davey, 15 Park Ave, Histon, Cambridge, CB4 4JU [JO02BG, TL46]
G0 THJ J S J Stowell, 10 Manley Cl, Antrobus, Northwich, CW9 8BB [IO83RF, SJ67]
G0 THK T H Koeze, 43 Upper Selsdon Rd, South Croydon, CR2 8DG [IO91XI, TQ36]
G0 THL G Redman, 28 Landsdowne Ave, Audenshaw, Manchester, M34 5SZ [IO83WL, SJ99]
G0 THM Details withheld at licensee's request by SSL.
G0 THN J D Mercer, 20 Lovel Way, Speke, Liverpool, L24 3XE [IO83NI, SJ48]
GI0 THO Details withheld at licensee's request by SSL.
G0 THP Details withheld at licensee's request by SSL.
G0 THQ R J King, 6 Mayon Green Cres, Sennen, Penzance, TR19 7BS [IO70DB, SW32]
GI0 THR J McDonald, 54 Bettys Hill Rd, Newry, BT34 2ND [IO64UD, J12]
G0 THS S J Graham, Church Farm Cottage, Sunk Island Rd, Sunk Island, East Yorks, nr Hull, HU12 0DY [IO93XQ, TA22]
G0 THT A P L Hall obo Lowe Electronics Bristo, 3 Vicarage Ct, Hanham, Bristol, BS15 3BL [IO81RK, ST67]
G0 THV J A C Coote, 8 St. Francis Chase, Bexhill on Sea, TN39 4HZ [JO00FU, TQ70]
G0 THW A M Brentnall, Kastelli, Fewcott, Bicester, Oxon, OX6 9NX [IO91JW, SP52]

G0 THX V Robins, 8 Lansdowne Rd, Hailsham, BN27 1LJ [JO00DV, TQ51]
G0 THY M J Preston, 15 Poplar Cl, Kidlington, OX5 1HH [IO91IT, SP41]
G0 THZ Details withheld at licensee's request by SSL.
G0 TIA P A Lodge, 46 Onibury Rd, Midanbury, Southampton, SO18 2DD [IO90HW, SU41]
G0 TIB L J Everitt, 6 Belmont Rd, Rednal, Birmingham, B45 9LW [IO82XJ, SO97]
GI0 TIC Details withheld at licensee's request by SSL.
G0 TID C I Alexander, 25 Diamedes Ave, Stanwell, Staines, TW19 7JE [IO91SK, TQ07]
GI0 TIE D A Coulter, 23 Oakdale, Ballygowan, Newtownards, BT23 5TS [IO74CM, J46]
G0 TIG H C Janes, 91 Thorpe Bay Gdns, Southend on Sea, SS1 3NW [JO01JM, TQ98]
G0 TIH A R Denney, 19 Lodge Hall, Harlow, Essex, CM18 7SU [JO01BR, TL40]
G0 TII E J Baxter, 6 Wampool St., Silloth, Carlisle, CA5 4AA [IO84HU, NY15]
G0 TIJ D R Page, 11 Valley Cl, Yeovil, BA21 4RS [IO80QW, ST51]
G0 TIK Details withheld at licensee's request by SSL.
G0 TIL J R Parmenter, 48 Honey Way, Royston, SG8 7EU [IO92XB, TL34]
G0 TIM T J Daly, 2 Carter Rd, Ivybridge, PL21 0RX [IO80AJ, SX65]
G0 TIN Details withheld at licensee's request by SSL.
G0 TIP G A Thorne, 19 Lapwing Ln, Brinnington, Stockport, SK5 8JY [IO83WK, SJ99]
G0 TIS Details withheld at licensee's request by SSL.
GI0 TIU R Farrier, 14 Sepon Park, Lisburn, BT28 3BQ [IO64XM, J26]
G0 TIV Details withheld at licensee's request by SSL.
G0 TIW T G Parker, The Bungalow, 178 Green End Ln, Hemel Hempstead, HP1 2BQ [IO91SS, TL00]
G0 TIX P A Wilson, 2 Staley Cl, Stalybridge, SK15 3HJ [IO83XL, SJ99]
G0 TIY Details withheld at licensee's request by SSL.
G0 TIZ E S A Tometzki, 11 Southey Cl, Narborough, Enderby, Leicester, LE9 5QZ [IO92JO, SP59]
G0 TJA T Jackson, 20 Canon Hoare Rd, Aylsham, Norwich, Norfolk, NR11 6UU [JO02OS, TG12]
G0 TJB Details withheld at licensee's request by SSL.
G0 TJC L Taylor, 35 Leafield Rd, Darlington, DL1 5DF [IO94FM, NZ21]
G0 TJD A J Wedgwood, 10 Milner Pl, London, N1 1TN [IO91WM, TQ38]
G0 TJE S R Sullivan, 7 Gosfield Rd, Dagenham, RM8 1JY [JO01BN, TQ48]
G0 TJF A W Roe, 15 Chelmorton Pl, Chaddesden, Derby, DE21 4QL [IO92GW, SK33]
G0 TJG M Spafford, 8 Netherleigh, Dark Ln, Calow, Chesterfield, Derbyshire, S44 5AD [IO93HF, SK47]
G0 TJH Details withheld at licensee's request by SSL.
G0 TJI G H Woods, 126 Luddenham Cl, Stanhope Est, Ashford, TN23 5SA [JO01KD, TQ94]
GI0 TJJ A Hamilton, 60 Parkland Ave, Lisburn, BT28 3JP [IO64XM, J26]
G0 TJL Details withheld at licensee's request by SSL.
G0 TJN T J Newland, 80 Burnway, Hornchurch, RM11 3SG [JO01CN, TQ58]
G0 TJP A D Person, 21 Bath View, Stratton on The Fosse, Bath, BA3 4RE [IO81SG, ST65]
G0 TJQ C M Dowell, 19 Field Top, Bailiffe Bridge, Brighouse, HD6 4EQ [IO93CR, SE12]
G0 TJR L M Ellams, 131 Broadway, Dunscroft, Doncaster, DN7 4HB [IO93LN, SE60]
G0 TJS J J Sheils, 35 Quarry Ln, Halesowen, B63 4PB [IO82XK, SO98]
G0 TJT R Francis, 28 Amber St., Allenton, Derby, DE24 8FT [IO92GV, SK33]
GI0 TJU N B Buchanan, 30 North Circular Rd, Lisburn, BT28 3AH [IO64XM, J26]
GI0 TJV A G Gibson, 111 Manor Park, Lisburn, BT28 1EY [IO64XM, J26]
G0 TJW E E Waddoups, 53 Highlands Rd, Basildon, SS13 2HX [JO01GN, TQ78]
G0 TJX P A Kunzler, 32 Dolphin Ct, Dolphin Way, Rustington, Littlehampton, BN16 2EW [IO90ST, TQ00]
G0 TJY T Herd, The Old Dairy, High St., Sproughton, Ipswich, IP8 3AP [JO02NB, TM14]
G0 TJZ A Senior, St Helena, Egremont, Cumbria, CA22 2EL [IO84FL, NY01]
G0 TKA M A Charleston, 22 Bure Homage Gdns, Christchurch, BH23 4DR [IO90DR, SZ19]
GM0 TKB L W Thomas, Greengeo, Scarfskerry, Caithness, KW14 8XN [IO88IP, ND27]
GM0 TKC T Maxwell, 28 Cedar Gr, Dunfermline, KY11 5BH [IO86GB, NT18]
G0 TKD S J Devine, 51 Victoria Hall Gdns, Matlock, DE4 3SQ [IO93FD, SK25]
GM0 TKE J C McLuckie, 20 Woodlands Dr, Brightons, Falkirk, FK2 0TF [IO85DX, NS97]
G0 TKF W Stuart, The Corner House, 56 Occupation Rd, Harley, Rotherham, S62 7UF [IO93GL, SK39]
G0 TKG D J Mawson, 14 Windermere Cl, Worksop, S81 7QE [IO93KH, SK58]
G0 TKH B A Whalley, 3 Thirtle Cl, Clacton on Sea, CO16 8YH [JO01NT, TM11]
G0 TKI A E Adams, 91 Leslie St, Waitara, New Zealand, X X
G0 TKJ T Clayton, 40 Morrison Rd, Darfield, Barnsley, S73 9ED [IO93HM, SE40]
G0 TKL E Roberts, 43 Ashbourne Cres, Sale, M33 3LQ [IO83UJ, SJ79]
GM0 TKM W M McDonald, 1 Sprotwell Terr, Sauchie, Alloa, FK10 3LD [IO86CD, NS89]
G0 TKN H E Ritchie, Cayuga, Gayton Parkway, Heswall, Merseyside, L60 3SS [IO83LH, SJ28]
G0 TKO Details withheld at licensee's request by SSL.
G0 TKP Details withheld at licensee's request by SSL.
GM0 TKQ J J Teague, 1 Crawford St., Motherwell, ML1 3AD [IO85AS, NS75]
G0 TKR Details withheld at licensee's request by SSL.
G0 TKT P K Hamblett, 13 Ironside Cl, Bewdley, DY12 2HX [IO82UI, SO77]
G0 TKU D M Wilkins obo Trio-Kenwood A.R.C., 802 Kenton Ln, Harrow Weald, Harrow, HA3 6AG [IO91UO, TQ19]
G0 TKV Details withheld at licensee's request by SSL.
G0 TKW G Gee, 23 Moorside Cres, Droylsden, Manchester, M43 7HT [IO83WL, SJ99]
GW0 TKX A F Mason, 101 Aneurin Bevan Ave, Gelligaer, Hengoed, CF8 8ET [IO81HN, ST08]
G0 TKZ R J Hamblin, 5 Lingfield Rd, Edenbridge, TN8 5DR [JO01AE, TQ44]
G0 TLA R Lythall, 71 Bennett St., Kimberworth, Rotherham, S61 2JZ [IO93HK, SK49]
G0 TLC Details withheld at licensee's request by SSL.
GM0 TLD Details withheld at licensee's request by SSL.
G0 TLE P J W Grigson, 20 Wilton Rd, Malvern, WR14 3RH [IO82UC, SO74]
G0 TLF J F Hackett Landau Forte College ARS, 67 Shropshire Ave, Chaddesden, Derby, DE21 6EW [IO92GW, SK33]
G0 TLG P D Duell, 3 Treeside Rd, Shirley, Southampton, SO15 5FY [IO90GW, SU41]
G0 TLH A R East, 117 Hazelton Rd, Colchester, CO4 3DY [JO01LV, TM02]
G0 TLI C L Vince, 8 Kent Rd, Swindon, SN1 3NJ [IO91CN, SU18]
GW0 TLJ A C Mathias, 11 Great North Rd, Milford Haven, SA73 2LH [IO71LR, SM90]
G0 TLK Details withheld at licensee's request by SSL.
G0 TLL Details withheld at licensee's request by SSL.
G0 TLM Details withheld at licensee's request by SSL.
G0 TLN G J Sim, Flat 3, 55 Western Rd, Brighton, BN1 2EB [IO90WT, TQ30]
G0 TLP A J Whitwam, 6 Winston Ct, Churchill, Kidderminster, Worcs, DY10 3JA [IO82VJ, SO87]
G0 TLQ R A Parsons, 10 Waterside, Isleham, Ely, CB7 5SH [JO02FI, TL67]
G0 TLR Details withheld at licensee's request by SSL.
G0 TLS R C Upton, 1 Magpie Ln, Swindon, SN3 5DR [IO91DN, SU18]
G0 TLT D H Davies, 9 Kirton Cl, Meden Vale, Mansfield, NG20 9QZ [IO93KF, SK56]
G0 TLU P F Thompson, Flat 3, 29 Cannon Pl, Brighton, BN1 2FB [IO90WT, TQ30]
G0 TLV H W Lambert, 78 Old Station Rd, Halesworth, IP19 8JQ [JO02SI, TM37]
G0 TLW S Parker, 28 Hornby Dr, Bolton, BL3 4RP [IO83SN, SD60]
GM0 TLX A O Wright, Crosslea, Berstane Rd, St. Ola, Kirkwall, KW15 1SZ [IO88MX, HY41]
G0 TLZ J J Trefry, 67 Axminster Cl, Eastfield Lea, Cramlington, NE23 9UE [IO95FC, NZ27]
G0 TMA Details withheld at licensee's request by SSL.
G0 TMB Details withheld at licensee's request by SSL.
G0 TMC Details withheld at licensee's request by SSL.
G0 TMD Details withheld at licensee's request by SSL.
G0 TME B M Park, 39 Tenth Ave, Northville, Bristol, BS7 0QJ [IO81RM, ST67]
G0 TMF M F Fleetwood, 9 Reynolds Cl, Swindon, Dudley, DY3 4NQ [IO82VM, SO89]
G0 TMG G S Ravenscroft, 40 Cedar Rd, Willenhall, WV13 3BZ [IO82XO, SO99]
G0 TMH C M Neary, 58 The Hive, Northfleet, Gravesend, DA11 9DF [JO01DK, TQ67]
G0 TMI G E French, 56 Ringstead Rd, Catford, London, SE6 2BP [IO91XK, TQ37]
G0 TMJ E W Jones, 1 Ivel View, Sandy, SG19 1AU [IO92UD, TL14]
G0 TMK W S Howell, 41 Chestnut Ave, Shavington, Crewe, CW2 5BJ [IO83SB, SJ75]
G0 TML A H Rowley, 32 Spring Ln, Flore, Northampton, NN7 4LS [IO92LF, SP66]
G0 TMN J H Allcroft, 37 Vicars Rd, Chorlton Cum Hardy, Manchester, M21 9JB [IO83UK, SJ89]
G0 TMP M A Lane, 1 Chasely Cres, Up Hatherley, Cheltenham, GL51 5RY [IO81WV, SO92]
GW0 TMQ Details withheld at licensee's request by SSL.
G0 TMR M R Ravenscroft, 40 Cedar Rd, Willenhall, WV13 3BZ [IO82XO, SO99]
GI0 TMS M J Smyth, 41 Coolaghy Rd, Newtownstewart, Omagh, BT78 4LG [IO64GR, H38]
G0 TMT M J Tuttle, Millstone, 7 Mill Ln, Horsford, Norwich, NR10 3ES [JO02OQ, TG11]
GW0 TMU D N Edwards, 68 Heol y Meinciau, Pontyates, Llanelli, SA15 5RT [IO71VS, SN40]
GW0 TMV T Vlismas, Maes yr Awel, Newport Rd, Crymych, Pembrokeshire, SA41 3RR [IO71QX, SN13]
G0 TMW G E Boswell, 7 Chestnut Ave, Wootton, Northampton, NN4 6LA [IO92NE, SP75]
G0 TMX Details withheld at licensee's request by SSL.
G0 TMY Details withheld at licensee's request by SSL.
G0 TMZ N J Lilley, 33 Downton Ln, Milford on Sea, Downton, Lymington, SO41 0LG [IO90ER, SZ29]
G0 TNA Capt R K Nicholls, 6 Elmslie Cl, Epsom, KT18 7JT [IO91UH, TQ25]
G0 TNB B F G Chaplin, 1 Southfields, Kings Lynn, PE30 4BA [JO02FS, TF62]
G0 TNC G C Stephenson, 27 Colne Ct, Grantham, NG31 7QY [IO92QV, SK93]
G0 TND B L Ward, 16 Fishers Cl, Blandford, Blandford Forum, DT11 7EL [IO80WU, ST80]
G0 TNF M A Wills, 16 Mayburgh Cl, Eamont Bridge, Penrith, CA10 2BW [IO84PP, NY52]
G0 TNG P C Hayward, 63 Devereaux Cres, Ebley, Stroud, GL5 4PX [IO81UR, SO80]
G0 TNH L J Crow, 181 Foxlydiate Cres, Batchley, Redditch, B97 6NS [IO92AH, SO06]
G0 TNI Details withheld at licensee's request by SSL.
G0 TNJ N H Durrant, 13 River Ln, Kings Lynn, PE30 4HD [JO02FS, TF62]
GM0 TNK D I Mackenzie, 33 Balnafettack Rd, Inverness, IV3 6TF [IO77UL, NH64]
G0 TNL A Beattie, Pine Croft, Blitterlees, Silloth, Carlisle, Cumbria, CA5 4JJ [IO84HU, NY15]
G0 TNM R W Guppy, 4 Church St., Helmdon, Brackley, NN13 5QJ [IO92KB, SP54]
G0 TNN D J Silverson, The Wickers, 63 Downside, Shoreham By Sea, BN43 6HF [IO90UU, TQ20]
G0 TNO C J Billington, 5 Lamers Rd, Ramridge Est, Stopsley, Luton, LU2 9BL [IO91TV, TL12]
G0 TNP P B Unstead, 29 Varna Rd, Bordon, GU35 0DG [IO91NC, SU83]
G0 TNQ A L J Lewis, 152 Crossfield Rd, Birmingham, B33 9QG [IO92CL, SP18]
G0 TNR T B Minett, 20 Thames Side, Staines, TW18 2HA [IO91RK, TQ07]
G0 TNS H D Hauton, 21 Chancel Rd, Scunthorpe, DN16 3LD [IO93QN, SE80]
G0 TNT Details withheld at licensee's request by SSL.
G0 TNU Details withheld at licensee's request by SSL.
G0 TNV K Coleman, 50 Main St., Long Lawford, Rugby, CV23 9AZ [IO92IJ, SP47]
G0 TNX M A Morrissey, 68 Dunnisher Rd, Wythenshawe, Manchester, M23 2YN [IO83UJ, SJ88]
G0 TNY A A Rose, Heronsgate, River Gdns, Bray, Maidenhead, Berks, SL6 2BJ [IO91PM, SU97]
G0 TOB Details withheld at licensee's request by SSL.
G0 TOC M Litchman, 22 Oak Tree Cl, Loughton, IG10 2RE [JO01AP, TQ49]
G0 TOD T W Northover, 13 Dagenham Ave, Dagenham, RM9 6LD [JO01BM, TQ48]
G0 TOE A J R Gallagher, 14 Wynsome St., Southwick, Trowbridge, BA14 9RB [IO81VH, ST85]
GM0 TOF M J Farnworth, 30 Roseangle, Dundee, DD1 4LY [IO86MK, NO32]
G0 TOG Details withheld at licensee's request by SSL.
G0 TOH H Punshon, 23 Amersham Cres, Peterlee, SR8 5JJ [IO94HS, NZ44]
GW0 TOI Details withheld at licensee's request by SSL.
G0 TOJ J E Townend, 44 Liphill Bank Rd, Holmfirth, Huddersfield, HD7 1LQ [IO93CN, SE10]
G0 TOK B Dixon, 97 Sunny Blunts, Peterlee, SR8 1LN [IO94HR, NZ43]
GW0 TOM T H Beedle, 2 Chestnut Gr, Maesteg, CF34 0NT [IO81EO, SS89]
G0 TON P N Greenhalgh obo HMS Bronington Radio Cl, 13 Primrose Ave, Urmston, Manchester, M41 0TY [IO83TK, SJ79]
G0 TOQ S W Harrison, 2 Oak Lea Cl, Mapplewell, Barnsley, S75 6LY [IO93GO, SE31]
G0 TOR Details withheld at licensee's request by SSL.
G0 TOS B D Harper, 51 Cross Ln, Newby, Scarborough, YO12 6DQ [IO94SH, TA08]
G0 TOT R E Claxton, 15 Heatherside, Studland, Swanage, BH19 3DA [IO90AP, SZ08]
G0 TOU S T Chamley, 3 Linden Dr, Clitheroe, BB7 1JL [IO83TU, SD74]
G0 TOV Details withheld at licensee's request by SSL.
GM0 TOW Details withheld at licensee's request by SSL.
G0 TOX C H Wheeler, 190 Mount Pleasant, Redditch, B97 4JL [IO92AH, SP06]
G0 TOY K C Li, 16 Garthland Dr, Arkley, Barnet, EN5 3BB [IO91VP, TQ29]
G0 TOZ R Patchitt, 1 St. Andrews Ct, Church Ln, Immingham, Grimsby, DN40 2HE [IO93VO, TA11]
G0 TPA A C Taylor, The Grey House, Paxford, Chipping Campden, Glos, GL55 6XP [IO92DA, SP13]
G0 TPB B W Smith, 34 Wheatlands, Fareham, PO14 4SL [IO90IU, SU50]
G0 TPC Details withheld at licensee's request by SSL.
G0 TPD J F Gayther, 33 Greenways, Winchcombe, Cheltenham, GL54 5LQ [IO91AW, SP02]
G0 TPE A J Davis, Tryfan, 320 Preston Old Rd, Blackburn, BB2 2TX [IO83RR, SD62]
G0 TPG B W Taylor, 3 Stonepits Ln, Hunt End, Redditch, B97 5LX [IO92AG, SP06]
G0 TPH A R Horne, 7 Williams Cl, Littlethorpe, Leicester, LE9 5JG [IO92JN, SP59]
GM0 TPI D Harris, 15 Glen Garry, East Kilbride, Glasgow, G74 2BN [IO75WS, NS65]
G0 TPJ S A Cross, 41 Whitewell Dr, Upton, Wirral, L49 4PF [IO83KJ, SJ28]
G0 TPK A J Mitchell, 87 Bridgewater Rd, Berkhamsted, HP4 1JN [IO91RS, SP90]
GW0 TPL D R Blundell, 138 Heritage Park, St. Mellons, Cardiff, CF3 0DS [IO81KM, ST28]
G0 TPM P Mercer, 10 Holmcliffe Ave, Bankfield Park, Huddersfield, HD4 7RJ [IO93CP, SE11]
G0 TPN M Padgett, 20 South View, Rothwell, Leeds, LS26 0NT [IO93GS, SE32]
G0 TPO M G Cook, 46A Brooksbys Walk, Homerton, London, E9 6DA [IO91XN, TQ38]
G0 TPP P Pimblett, 5 Edgeside, Great Harwood, Blackburn, BB6 7JS [IO83TS, SD73]
GW0 TPR B A Carter, Rhyd-y-Mwyn, Cilgwyn St, Llanerchymedd, LL71 8EA [IO73TH, SH48]
G0 TPS Details withheld at licensee's request by SSL.
G0 TPT Details withheld at licensee's request by SSL.
G0 TPU Details withheld at licensee's request by SSL.
G0 TPV Details withheld at licensee's request by SSL.
G0 TPX Details withheld at licensee's request by SSL.
G0 TPY J W Brunt, 1 Dane Gr, Cheadle, Stoke on Trent, ST10 1QS [IO92AX, SK04]
G0 TPZ Details withheld at licensee's request by SSL.
GM0 TQA J Quigley, 90 George St., Paisley, PA1 2JR [IO75SU, NS46]
GM0 TQB M J Devries, 10 Bruce Rd, Glasgow, G41 5EJ [IO75UU, NS56]
G0 TQC K J Sharman, 21 Ferrers Green, Churston Ferrers, Brixham, TQ5 0LF [IO80FJ, SX85]
GI0 TQD J G Gough, 50 Culmore Point, Londonderry, BT48 8JW [IO65IB, C42]
G0 TQE M G Jones obo Eckington School Radio, 160 Oakbrook Rd, Sheffield, S11 7ED [IO93FI, SK38]
G0 TQF J P Fenton-Coopland obo Noel-Baker Community Schoo, Bracknell Dr, Alveston, Derby, DE24 0BR [IO92GV, SK33]
G0 TQG D E Houghton, 2 Hereford Ave, Golborne, Warrington, WA3 3NA [IO83QL, SJ69]
G0 TQJ C M Vernon, 57 Parker Rd, Wittering, Peterborough, PE8 6AN [IO92SO, TF00]
GM0 TQK K W Traill, 2 Third St., Newtongrange, Dalkeith, EH22 4PU [IO85LU, NT36]
GW0 TQM Dr C S Littlejohns, 18 Church Meadow, Rhydymwyn, Mold, CH7 5HX [IO83JE, SJ26]
G0 TQN Details withheld at licensee's request by SSL.
G0 TQO Details withheld at licensee's request by SSL.
G0 TQP J M Smith, High Tree, Radford Ln, Lower Penn, Wolverhampton, WV3 8JT [IO82VN, SO89]
G0 TQR A N Baker, 23 Trematon Dr, Ivybridge, PL21 0HT [IO80BJ, SX65]
G0 TQS C M Forsyth, 8 Oriole Dr, Exeter, EX4 4SJ [IO80FR, SX99]
G0 TQT J W Joll, 16 Jephson Rd, St. Judes, Plymouth, PL4 9ET [IO70WJ, SX45]
G0 TQU G B Morris, 725 Calle Del Ficus, El Tosalet, Javea, Alicante, Spain, X X
G0 TQV K P Taylor, 1 Chapel Cl, Reepham, Lincoln, LN3 4EJ [IO93SG, TF07]
G0 TQX R A Stelp, Solveigs House, Penn, Bucks, HP10 8NX [IO91PP, SU99]
G0 TQY Details withheld at licensee's request by SSL.
G0 TQZ M J Emm, Highwood Cottage, Daggons Rd, Alderholt, Fordingbridge, SP6 3DJ [IO90BV, SU11]
G0 TRB R L C Betts, 15 Cleasby, The Shires, Wilnecote, Tamworth, B77 4JL [IO92EO, SK20]
G0 TRC P Fambely obo Trafford Amateur Radio, 126 Ashton Ln, Sale, M33 5QJ [IO83UK, SJ79]
G0 TRD T E B Thorman, 39 Temple Sheen Rd, East Sheen, London, SW14 7QF [IO91UL, TQ27]
G0 TRE N J Dimbleby, 4 Rossetti Pl, Holmer Green, High Wycombe, HP15 6XA [IO91PQ, SU99]
G0 TRF Details withheld at licensee's request by SSL.
G0 TRG C H Field Trafford Radio Group, 54 Kingston Dr, Urmston, Manchester, M41 9FG [IO83TK, SJ79]
G0 TRH D R Brooke, 24 Bedford St., Scarborough, YO11 1DB [IO94TG, TA08]
G0 TRI L A G Pritchard, The Granary, Greenway Farm, Whitchurch, Ross-on-Wye, Herefordshire, HR9 6DH [IO81PU, SO51]
G0 TRJ R L Makepeace, 39 Brendon, Laindon, Basildon, SS15 5XJ [JO01FN, TQ68]
G0 TRK J Ogden, 65 Elm St., Middleton, Manchester, M24 2EQ [IO83VN, SD80]
G0 TRM C H Page, 1 The Leeway, Danbury, Chelmsford, CM3 4PS [JO01HR, TL70]
G0 TRN J de Frece, West Side, Wilkin Hill, Barlow, Sheffield, S18 5TE [IO93GH, SK37]
G0 TRO Details withheld at licensee's request by SSL.
G0 TRP M J Williams, Immanuel, 3 Hill End, Withiel, Bodmin, PL30 5NJ [IO70OK, SW96]
G0 TRQ C N Omeje, Directors Office, Works Services Department, University of Nigeria, Nsukka,Enuga State, Nigeria,West Africa
G0 TRR J E Schiefer, 7 Cotham Gr, Cotham, Bristol, BS6 6AL [IO81QL, ST57]
G0 TRT Details withheld at licensee's request by SSL.
G0 TRU Details withheld at licensee's request by SSL.
G0 TRV Details withheld at licensee's request by SSL.
G0 TRW R G Cross, 7 The Island, Anthorn, Carlisle, CA5 5AN [IO84JV, NY15]
G0 TRX Details withheld at licensee's request by SSL.
G0 TRY G Lyon, 33 Barlborough Rd, Pemberton, Wigan, WN5 9HZ [IO83PM, SD50]
GW0 TRZ Details withheld at licensee's request by SSL.
GI0 TSA D S Moore, 3 Knightsbridge, Cres Link, Londonderry, BT47 1FE [IO64IX, C41]
G0 TSB B J Snell, 30 Queens Cres, Brixham, TQ5 9PJ [IO80FJ, SX95]
G0 TSD M Hampson, 39 Brook Hill, Clowne, Chesterfield, S43 4RP [IO93IG, SK47]
GW0 TSE L J Owen, Cartref, Llangain, Carmarthen, Dyfed, SA33 5AH [IO71TT, SN31]
GW0 TSF P R Owen, Cartref Llangain, Carmarthen, Dyfed, SA33 5AH [IO71TT, SN31]
G0 TSG P E Ryder, 35 Stotfield Rd, Bilborough, Nottingham, NG8 4DB [IO92JW, SK54]
G0 TSH K J Hutt, The Crossing House, Fenwick Ln, Fenwick, Doncaster, DN6 0EZ [IO93KP, SE51]
G0 TSI H W Norton N Sefton Amateur Radio Club, 57 Cornwall Way, Ainsdale, Southport, PR8 3SG [IO83LO, SD31]
G0 TSJ S Ruud, 5 Wood St., Haworth, Keighley, BD22 8BJ [IO93AT, SE03]
G0 TSK G R Wilkins, 28 Byron Way, Bicester, OX6 8YR [IO91KV, SP52]
GW0 TSL H Chapman, 302 Holton Rd, Barry, CF63 4HW [IO81IJ, ST16]
G0 TSM D J Collins, 10 Swift Cl, Eastleigh, SO50 9LD [IO90HX, SU41]
G0 TSN R A Charlesworth obo Hoddesdon Radio Club, 6 Curzon Ave, Enfield, EN3 4UD [IO91XP, TQ39]
G0 TSP Details withheld at licensee's request by SSL.
G0 TSR N W Depledge, 166 Farleigh Rd, Pershore, WR10 1LY [IO82WC, SO94]
GI0 TSS C T Tait, 33 Riverside Dr, Harmony Heights, Lisburn, BT27 4HE [IO64XM, J26]
G0 TST Details withheld at licensee's request by SSL.
GW0 TSW P A Redman, 87 Wepre Park, Connahs Quay, Deeside, CH5 4HL [IO83LF, SJ26]
GW0 TSX E Jenkins, 5 Waunscil Ave, Bridgend, CF31 1TX [IO81FM, SS97]

G0 TSY R E Sharp, The Ridge, Woodfalls, Redlynch, Salisbury, Wilts, SP5 2LN [IO90DX, SU11]
G0 TTA M P Dunkley, 30 Roseangle, Dundee, DD1 4LY [IO86MK, NO32]
G0 TTD M J Smith, 176 Nettleham Rd, Lincoln, LN2 4DQ [IO93RF, SK97]
G0 TTE Details withheld at licensee's request by SSL.
GW0 TTF W G Price, Tramore, 67 High St., Laleston, Bridgend, CF32 0HL [IO81EM, SS87]
G0 TTG M Warriner, 34 Cabrera Ave, Virginia Water, GU25 4EZ [IO91RJ, SU96]
G0 TTH N E Taylor, 27 Heron Ln, Crossgates, Scarborough, YO12 4TW [IO94SF, TA08]
G0 TTI R J A Rayment, 145 Feeches Rd, Southend on Sea, SS2 6TF [JO01IN, TQ88]
G0 TTJ Details withheld at licensee's request by SSL.
G0 TTK Details withheld at licensee's request by SSL.
G0 TTL R A Booth, 32 Buttermere Cl, Anston, Sheffield, S31 7GA [IO93GJ, SK38]
G0 TTM A P Radley, 16 Kingsley Ln, Thundersley, Benfleet, SS7 3TU [JO01HN, TQ78]
GW0 TTN P M Brzenczek, 39 Oaklands Rd, Sebastopol, Pontypool, NP4 5DB [IO81LQ, ST29]
G0 TTO L E Chadwick, 2 Auden Pl, Longton, Stoke on Trent, ST3 1SJ [IO82WX, SJ94]
G0 TTP Details withheld at licensee's request by SSL.
G0 TTQ B W Trivett, Moltkestrasse 42, 42551 Velbert, Germany
G0 TTR N J Parkinson, 42 West St., Winterton, Scunthorpe, DN15 9QF [IO93QP, SE91]
G0 TTS P D Bulmer, 61 Middleham Ave, Huntingdon Rd, York, YO3 9BD [IO93LX, SE65]
G0 TTT Details withheld at licensee's request by SSL.
G0 TTV M D Benton, 22 Hillcrest, Chedgrave, Norwich, NR14 6HX [JO02RN, TM39]
G0 TTW K C Yeates, Newlands, Ashby Ln, Bitteswell, Lutterworth, LE17 4SQ [IO92JL, SP58]
GM0 TTY W McBurney, 6 Hill St., Tillicoultry, FK13 6HF [IO86DD, NS99]
G0 TTZ Details withheld at licensee's request by SSL.
G0 TUA Details withheld at licensee's request by SSL.
G0 TUB R M Wilson, Garth Cottage, Garstang Rd, St. Michaels, Preston, PR3 0TD [IO83OU, SD44]
G0 TUC D F Wilkes, 63 Runnymede Ave, Bearcross, Bournemouth, BH11 9SG [IO90AS, SZ09]
G0 TUD Details withheld at licensee's request by SSL.
G0 TUE Details withheld at licensee's request by SSL.
G0 TUH Details withheld at licensee's request by SSL.
G0 TUI B W Hudson, 5 Rylands Rd, Southend on Sea, SS2 4LW [JO01IN, TQ88]
G0 TUJ W Spencer, 111 Rosmead St., Newbridge Rd, Hull, HU9 2TE [IO93US, TA12]
G0 TUK Details withheld at licensee's request by SSL.
G0 TUL R J Woollard, 12 Swallow Way, Colehill, Wimborne, BH21 2NH [IO90AT, SU00]
G0 TUM B P Cooper, 43 Wych Elm Cl, Hornchurch, RM11 3AJ [JO01CN, TQ58]
G0 TUN A J Powney, 16 Westbrook Way, Wombourne, Wolverhampton, WV5 0EA [IO82VM, SO89]
G0 TUO Details withheld at licensee's request by SSL.
G0 TUP N Callow, 3 Maunleigh, Forest Town, Mansfield, NG19 0PP [IO93JD, SK56]
G0 TUQ D M Ferguson obo Leiston Amateur Radio, 3 Aldeburgh Rd, Leiston, IP16 4JY [JO02SE, TM46]
GM0 TUR D A Turnbull, 13 Almond Cres, Paisley, PA2 0NQ [IO75ST, NS46]
GM0 TUS R G Young, 50 Berrywell Dr, Duns, TD11 3HG [IO85US, NT75]
G0 TUT B Haresign, 1 Meynell Mount, Rothwell, Leeds, LS26 0LQ [IO93GS, SE32]
G0 TUU R C Hudson, Norton House, 27 Tome View, Auckley, Doncaster, DN9 3PQ [IO93LM, SE60]
G0 TUV Details withheld at licensee's request by SSL.
G0 TUW G Woolfenden, 11 Chesshire Cl, Areley Kings, Stourport on Severn, DY13 0EB [IO82UH, SO86]
G0 TUX J W Fossey, 12 Hitchin Rd, Arlesey, SG15 6RP [IO92UA, TL13]
G0 TUY Details withheld at licensee's request by SSL.
G0 TVA D J L Howard, 3 Dalby Ave, Swinton, Manchester, M27 0HZ [IO83TM, SD70]
G0 TVB P F Rigg, Birchwood House, Walsden, Todmorden, Lancs, OL14 6QX [IO83WQ, SD92]
G0 TVC A E Lanham, 7 College Ln, Stratford upon Avon, CV37 6DD [IO92DE, SP25]
G0 TVD Details withheld at licensee's request by SSL.
G0 TVE D J Seton, 92 Baslow Dr, Heald Green, Cheadle, SK8 3HP [IO83VI, SJ88]
G0 TVF Details withheld at licensee's request by SSL.
G0 TVH Details withheld at licensee's request by SSL.
G0 TVI Details withheld at licensee's request by SSL.
G0 TVJ C N Morris, 29 Bartholomew Rd, Bishops Stortford, CM23 3TP [JO01BU, TL42]
GW0 TVK N R Juby, 64 Rhodfa Wen, Llysfaen, Colwyn Bay, LL29 8LE [IO83DH, SH87]
G0 TVL S M Birkenshaw, 19A Vale Head Gr, Knottingley, WF11 8JL [IO93IQ, SE42]
G0 TVM A M Bashir, Hillcrest, 70 Smith Ln, Bradford, BD9 6DQ [IO93CT, SE13]
G0 TVN B J Wright, 11 Bracken Cl, Feniscowles, Blackburn, BB2 5AH [IO83RR, SD62]
G0 TVO B Hewson, 14 Boardman St., Todmorden, OL14 5JG [IO83WR, SD92]
G0 TVQ L M Mansfield obo Nottingham Amateur Repe, 26 Dovecote Ln, Beeston, Nottingham, NG9 1HU [IO92JW, SK53]
G0 TVR C J Binnell, 146 Hales Cres, Smethwick, Warley, B67 6QX [IO92AL, SP08]
G0 TVS C D Saunders, 17 Bure Rd, Friars Cliff, Christchurch, BH23 4ED [IO90DR, SZ19]
G0 TVT R K Hemmings, 39 Ridgmont Rd, Seabridge, Newcastle, ST5 3LD [IO82VX, SJ84]
G0 TVU M S Binns, 49 Fairview, Pontefract, WF8 3NT [IO93IQ, SE42]
G0 TVV Details withheld at licensee's request by SSL.
G0 TVW D S Rodman, 23 High Trees, Dore, Sheffield, S17 3GF [IO93FH, SK38]
GW0 TVX R L Elms, 7 Manor Daf Gdns, St. Clears, Carmarthen, SA33 4ES [IO71ST, SN21]
GW0 TVY Details withheld at licensee's request by SSL.
G0 TVZ C M Sher, 13 Lynton Mead, Totteridge, London, N20 8DG [IO91VO, TQ29]
G0 TWA R Smith, 29 Sedley Ave, Nuthall, Nottingham, NG16 1EN [IO92JX, SK54]
GM0 TWB I F Lindsay, Fallady Cottage, Letham, Angus, DD8 2SP [IO86PP, NO54]
G0 TWD Revd J M Drummond obo Wigan Deanery High Scho, 14 Bulls Head Cttgs, Turton, Bolton, Lancs, BL7 0HS [IO83TP, SD71]
G0 TWE Revd T J Walker, The Rectory, Louth Rd, Binbrook, Market Rasen, LN8 6BJ [IO93VK, TF29]
GW0 TWF M J Patterson, 6 Devonshire Pl, Port Talbot, SA13 1SG [IO81CO, SS78]
G0 TWH M V Jewkes, 11 Maple Gr, Crewe, CW1 4DY [IO83SC, SJ75]
G0 TWI Details withheld at licensee's request by SSL.
G0 TWJ J Rickard, 40 Coppice Rd, Kingsclere, Newbury, RG20 5RS [IO91JH, SU55]
GM0 TWK D McKenzie, 11 Fulmar Cres, Ardersier, Inverness, IV1 2SY [IO77XN, NH75]
G0 TWL M J Lewis, 5 Hayway Farm Cttgs, Willersey, nr Broadway, Worcs, WR12 7PB [IO92BB, SP03]
G0 TWM R Hughes, 211 Brantingham Rd, Chorlton Cum Hardy, Manchester, M21 0TT [IO83UK, SJ89]
G0 TWN Details withheld at licensee's request by SSL.
GW0 TWO P R W Murdoch, Llys Helig, 19 Penhelig Rd, Aberdovey, LL35 0PT [IO72XN, SN69]
G0 TWP K A Moakes, 70 Berry Hill Ln, Mansfield, NG18 4BW [IO93JD, SK55]
GW0 TWQ T D Lloyd, Chardonnay, Cwmffrwd, Carmarthen, Dyfed, SA31 2LP [IO71UU, SN41]
GW0 TWR C D Harrison, 28 Brynau Wood, Cimla, Neath, SA11 3YQ [IO81CP, SS79]
GW0 TWS J P C Phillips, 1 Bryngwastad Rd, Gorseinon, Swansea, SA4 4XQ [IO71XQ, SS59]
G0 TWT B J Lee, The Larches, Thirlstane Rd, Malvern, WR14 3PL [IO82UC, SO74]
G0 TWV K Stewart, 25 Millfield, Seaton Sluice, Whitley Bay, NE26 4DD [IO95GB, NZ37]
G0 TWW Details withheld at licensee's request by SSL.
GI0 TWX Dr I K C Chin, 7 Donegall Gdns, Whitehead, Carrickfergus, BT38 9LP [IO74DS, J49]
G0 TWY W E Aceves Ii, Box 1206, Mhs, Harrogate, North Yorks, HG3 2RF [IO94DA, SE25]
G0 TWZ Details withheld at licensee's request by SSL.
G0 TXA B A Carter, 16 Hollow St., Chislet, Canterbury, CT3 4DS [JO01OH, TR26]
G0 TXD Details withheld at licensee's request by SSL.
G0 TXE Details withheld at licensee's request by SSL.
G0 TXF E Daly, 38 Wood Ln, Greasby, Wirral, L49 2PU [IO83KJ, SJ28]
GM0 TXG E Young, 24 Birnam Cres, Bearsden, Glasgow, G61 2AU [IO75UW, NS57]
G0 TXI Details withheld at licensee's request by SSL.
G0 TXJ N D Service, Craigpark, Maddiston Rd, Rumford, Falkirk, FK2 0SB [IO85DX, NS97]
G0 TXK R A McKinnon, 30 John Smale Rd, Sticklepath, Barnstaple, EX31 2HR [IO71XB, SS53]
G0 TXL P C Elliott, 22 Seymour Rd, Mitcham Junction, Mitcham, CR4 4JX [IO91WJ, TQ26]
GW0 TXM R E Thomas, 49 Morfar Garreg, Pwllheli, LL53 5AU [IO72TV, SH33]
G0 TXN M V Hines, 2 Lockerley Cl, Lymington, SO41 8ER [IO90FS, SZ39]
G0 TXO Details withheld at licensee's request by SSL.
GW0 TXP A R Smith, Delfan Stores, Gwyddgrug, Pencader, Dyfed, SA39 9AX [IO71VX, SN43]
G0 TXQ J P Llewellyn-Jones, 148 Dudley Rd, Cot Hill, Plympton, Plymouth, PL7 1SA [IO70XJ, SX55]
GW0 TXS D F Warner, 58 Solva Rd, Clase, Swansea, SA6 7NU [IO81AP, SS69]
G0 TXU Details withheld at licensee's request by SSL.
G0 TXV J D Bacon Anglia TV Amateur Radio Soc, Weather Department, Anglia Television, Norwich, NR1 3JG [JO02PP, TG20]
G0 TXW J Kerrigan, 53 Cromarty Ave, Crosland Moor, Huddersfield, HD4 5LG [IO93CP, SE11]
G0 TXY A J Hicks, 29 Oak Tree Ln, Strensall, York, YO3 5TE [IO94LA, SE65]
G0 TYA M R Trotman, 44 Armscroft Rd, Barnwood, Gloucester, GL2 0SJ [IO81VU, SO81]
G0 TYB R Wright, 17 Romney Walk, Dereham, NR19 1BL [JO02LQ, TF91]
G0 TYC W S Y McGuire, 68 Boyd Ave, Dereham, NR19 1ND [JO02LP, TF91]
G0 TYD B Shepherdson, 30 Upper St., Quainton, Aylesbury, HP22 4AY [IO91MV, SP72]
G0 TYE P Geduldig, Ulmenstrasse 11/2, 75397 Simmozheim, Germany, X X
G0 TYG J A K Reid, 4 Harles Acres, Hickling, Melton Mowbray, LE14 3AF [IO92MU, SK62]
G0 TYH P Boyes, Lingmoor Keighley Rd, Cowling, Keighley, West Yorks, BD22 0LA [IO83XV, SD94]
G0 TYI F A Thornton, 82 Longfield Rd, Bolton, BL3 3SZ [IO83SN, SD60]
G0 TYJ O H Ireland, 58 Wrestwood Ave, Eastbourne, BN22 0ES [JO00DT, TQ50]
G0 TYK L H Worrall, 106 Kingsway, Widnes, WA8 7QR [IO83PI, SJ58]
G0 TYL R L Phillips, 57 Hollytrees, Bar Hill, Cambridge, CB3 8SF [JO02AG, TL36]
G0 TYM T J Allison, 2 Westlands, Stokesley, Middlesbrough, TS9 5BU [IO94JL, NZ50]
G0 TYN M O Holmes, 231A Hoylake Rd, Wirral, L46 0SL [IO83KJ, SJ28]
G0 TYO Details withheld at licensee's request by SSL.
G0 TYP A Fowler, 44 Fifth Ave, Edwinstowe, Mansfield, NG21 9NR [IO93LE, SK66]
G0 TYQ G M Clark, 2 Keith Rd, Swanton Morley, Dereham, NR20 4NQ [JO02LQ, TG01]
G0 TYS S Allanson, 6 St. Georges Ct, Havercroft, Wakefield, WF4 2EH [IO93HP, SE31]
G0 TYT Details withheld at licensee's request by SSL.
G0 TYV W E Jones, 4 St. Johns Church Rd, Folkestone, CT19 5BQ [JO01OC, TR23]
G0 TYW P D Cocker, 6 Grange Park Cl, Penwortham, Preston, PR1 0JS [IO83PS, SD52]
G0 TYX Details withheld at licensee's request by SSL.
G0 TYY Details withheld at licensee's request by SSL.
G0 TYZ T Turley, 29 Friars Ave, Great Sankey, Warrington, WA5 2AR [IO83QJ, SJ58]
G0 TZA Details withheld at licensee's request by SSL.
G0 TZC P A Sables, 54 Harvey St., Deepcar, Sheffield, S30 5QB [IO93GJ, SK38]
G0 TZD R K Leah, 56 King St West, Apartment 7, Stoney Creek, Ontario, Canada, L8G 1H8
G0 TZE M Grant, 38 Crusader Rd, Hedge End, Southampton, SO30 0PE [IO90IV, SU41]
G0 TZF E Marshall, Fern Bank, Fountain St., Nelson, BB9 7XU [IO83VU, SD83]
GW0 TZG M W M Chandler, 15 Homerton St., Matthewstown, Mountain Ash, CF45 4YP [IO81HP, ST09]
G0 TZH M A Bushnell, Rose Cottage, Revel, St. Ashton, Rugby, CV23 0PH [IO92IK, SP48]
G0 TZI N Stork, 79 Constable Rd, Flamborough, Bridlington, YO15 1NN [IO94WC, TA27]
G0 TZJ T L Malson, Pinfold, Chapel Ln, North Scarle, Lincoln, LN6 9EX [IO93PE, SK86]
G0 TZK Details withheld at licensee's request by SSL.
G0 TZL M S Evans, 5 Beehive Yard, Denmark St., Diss, IP22 3LQ [JO02NJ, TM17]
G0 TZM K R Winfield, Russettwalls, 58 Bretby Ln, Bretby, Burton on Trent, DE15 0QW [IO92ET, SK22]
GM0 TZN Details withheld at licensee's request by SSL.
G0 TZO R J Nelson, Elgon Lodge, French Drove, Thorney, Peterborough, PE6 0PF [IO92WP, TF20]
G0 TZP A E Seals, 372A London Rd, Westcliff on Sea, SS0 7HZ [JO01IN, TQ88]
G0 TZQ G C Mallion, 64 Horningsea Rd, Fen Ditton, Cambridge, CB5 8SZ [JO02CF, TL46]
G0 TZR J Macknish, Robins Croft, 21A Knoll Rise, Orpington, BR6 0EJ [JO01BI, TQ46]
G0 TZS Details withheld at licensee's request by SSL.
G0 TZT S J Emmett, Moorlands, Main Rd, Brighstone, Newport, PO30 4DJ [IO90HP, SZ48]
G0 TZU P Fox-Roberts, 2 Cannobie Cl, Darlington, DL3 8RU [IO94FM, NZ21]
G0 TZV G H Bryant, 54 Drew Rd, Wollescote, Stourbridge, DY9 0UP [IO82WK, SO98]
G0 TZW Details withheld at licensee's request by SSL.
G0 TZX Dr J S Katz, Pine Tree, Kingstonridge, Kingston, Lewes, E Sussex, BN7 3JU [IO90XU, TQ30]
G0 TZY A Davies, 20 Peel Park Cl, Accrington, Lancs, BB5 6PL [IO83TS, SD72]
G0 TZZ C G Soames, 10 Stigands Gate, Dereham, NR19 2HF [JO02LQ, TF91]
G0 UAA I Fallows, 34 Hargher St., Burnley, BB11 4EG [IO83US, SD83]
G0 UAB R Milsom, 18 Montreal Cl, Worcester, WR2 4DY [IO82VE, SO85]
G0 UAC C P Martin, 21 Wharfedale Pl, Newcastle upon Tyne, NE6 4LQ [IO94FX, NZ26]
G0 UAD G T Rowe, 8 Dove Cl, Bolton upon Dearne, Rotherham, S63 8JL [IO93IM, SE40]
G0 UAF D Upton, 10 Gelder Clough, Ashworth Rd, Heywood, OL10 4BD [IO83VO, SD81]
GI0 UAG R H Anderson, Derry Lodge, 2 Tullymally Rd, Portaferry, Newtownards, BT22 1JX [IO74FJ, J65]
G0 UAH Details withheld at licensee's request by SSL.
G0 UAI S D W Marshall, 68 Parkfield Ave, Hampden Park, Eastbourne, BN22 9SF [JO00DT, TQ60]
G0 UAK R W Thompson, 38 Buckingham Rd, Margate, CT9 5SS [JO01QJ, TR37]
G0 UAL Details withheld at licensee's request by SSL.
G0 UAN Details withheld at licensee's request by SSL.
G0 UAO J Smith, 124 Parkside Ave, Barnehurst, Bexleyheath, DA7 6NL [JO01CL, TQ57]
G0 UAP P J Westbury, 5 Smithfield Rd, Winton, Bournemouth, BH9 2QJ [IO90BR, SZ09]
G0 UAQ Details withheld at licensee's request by SSL.
G0 UAS A R G Smith, 3 Woodcourt Cl, Sittingbourne, ME10 1QT [JO01IH, TQ96]
GM0 UAT Details withheld at licensee's request by SSL.
G0 UAU S B Ashmore, 1 Highfield Rd, Great Barr, Birmingham, B43 5AW [IO92AM, SP09]
G0 UAW R S Wood, 104 Hainault Ave, Westcliff on Sea, SS0 9EY [JO01IN, TQ88]
GI0 UAX Details withheld at licensee's request by SSL.
G0 UAY E Fletcher, 5 Butt Hill Ct, Bury New Rd, Prestwich, Manchester, M25 9NT [IO83UM, SD80]
G0 UAZ Details withheld at licensee's request by SSL.
G0 UBA R J Gibbs, 357 Downham Way, Bromley, BR1 5EW [JO01AK, TQ47]
G0 UBC Details withheld at licensee's request by SSL.
G0 UBD Details withheld at licensee's request by SSL.
G0 UBE Details withheld at licensee's request by SSL.
G0 UBF D A Rowland, 3 Parklands Cl, South Molton, EX36 4EU [IO81BA, SS72]
G0 UBG D S Endean, 6 Higher Westonfields, Totnes, TQ9 5QY [IO80DK, SX86]
G0 UBJ K J Beach, Taylor Cove, Harwich Rd, Beaumont, Clacton on Sea, CO16 0AX [JO01OV, TM12]
G0 UBK C J Carvell, 6 Field Cl, Whitby, YO21 3LR [IO94QL, NZ81]
G0 UBL M A Stracey, 84 Doeshill Dr, Wickford, SS12 9RD [JO01GO, TQ79]
G0 UBM J M Horsfield, 17 Osberton Pl, Sheffield, South Yorks, S11 8XL [IO93GI, SK38]
G0 UBO E C Martin, 9 The Valley Green, Welwyn Garden City, AL8 7DQ [IO91VT, TL21]
G0 UBQ Details withheld at licensee's request by SSL.
G0 UBU M N Page, 43 Kingsway, Mildenhall, Bury St. Edmunds, IP28 7HP [JO02GI, TL77]
G0 UBV Details withheld at licensee's request by SSL.
G0 UBX A J Quince, 96 Mallard Hill, Brickhill, Bedford, MK41 7QT [IO92SD, TL05]
G0 UBY D H Duffill, 45 Delius Cl, Anlaby Park Rd Nt, Hull, HU4 7NR [IO93TR, TA02]
G0 UBZ J W Flack, 40 Rosedale, Worksop, S81 0TB [IO93KH, SK58]
G0 UCA Dr T L Ogden, Coed Celyn, Barmouth Rd, Dolgellau, LL40 2EW [IO82BR, SH71]
GM0 UCB R F de Ath, Aros Park Lodge, Tobermory, Isle of Mull, Argyll, PA75 6QA [IO66XO, NM55]
G0 UCC M Sayegh, 1 Hoylake Rd, East Acton, London, W3 7NP [IO91UM, TQ28]
G0 UCD M F West, 12 Jenny Gill Cres, Skipton, BD23 2RR [IO83XW, SD95]
G0 UCE J L Swartz, 1208 Leland Ave, Springfield, Illinois 62704, USA, X X
G0 UCF Details withheld at licensee's request by SSL.
G0 UCH C H Hawes, 6 Dallin Rd, Bexleyheath, DA6 8EJ [JO01BK, TQ47]
G0 UCI R A O Jones, Gorswen, 29 Avon Dale, Newport, TF10 7LS [IO82TS, SJ71]
G0 UCK B K R Linehan, 20 Torridon Ct, Bletchley, Milton Keynes, MK2 3PP [IO91PX, SP83]
G0 UCN P F Turner, 59 Waverley, Woodside, Telford, TF7 5LT [IO82SP, SJ60]
G0 UCP Dr J Seager, 2 Waterford Rd, Oxton, Birkenhead, L43 6UT [IO83LJ, SJ28]
G0 UCQ P Gater, 42 Oxford Rd, Orrell, Wigan, WN5 8PQ [IO83PN, SD50]
G0 UCR Details withheld at licensee's request by SSL.
G0 UCS W J Dingley, 63 Hurst Green, Mawdesley, Ormskirk, L40 2QS [IO83OP, SD41]
G0 UCT B M OBrien, 47 Harscroft, Linton Glade, Croydon, CR0 9LB [IO91XI, TQ36]
G0 UCU T J Fisher, 62 Hamberts Rd, South Woodham Ferrers, Chelmsford, CM3 5TU [JO01HP, TQ89]
G0 UCV J K Baxendale, 6 Woodstock Dr, Tottington, Bury, BL8 4BW [IO83TO, SD71]
G0 UCX T M Taylor, 1 Cooke Gdns, Branksome, Poole, BH12 1QE [IO90BR, SZ09]
G0 UCY G I Young, c/o Adrian Southern, University Computing Service, University of Leeds, Leeds, LS2 9JT [IO93FT, SE23]
G0 UCZ E R Vaughan, 10 Water Tower Pl, St. Albans Pl, London, N1 0YW [IO91WM, TQ38]
G0 UDA Details withheld at licensee's request by SSL.
G0 UDB D N Birch, 29 Parc An Maen, Porthleven, Helston, TR13 9AU [IO70IC, SW62]
G0 UDC Details withheld at licensee's request by SSL.
G0 UDD C M Knaggs H U C R A, Emergency Planning Serv, 39 Meaux Rd, Wawne, Hull, HU7 5XD [IO93TT, TA03]
G0 UDE C R Woodlock, 10D Carmina Pl, 7-9 Deep Water Bay Dr, Hong Kong, X X
G0 UDF E M Wyman, 1 Bridle Rd, Kings Acre, Hereford, HR4 0PP [IO82PB, SO44]
G0 UDH Details withheld at licensee's request by SSL.
G0 UDI J A Murphy, 8 Spencer Ave, Wribbenhall, Bewdley, DY12 1DB [IO82UJ, SO77]
GW0 UDJ D W Jones, Bryn Awelon, Bryn Sannan, Brynford, Holywell, Flintshire, CH8 8AX [IO83JG, SJ17]
G0 UDK A R Girling, 107 Chapel Ln, Sands, High Wycombe, HP12 4BY [IO91OP, SU89]
GM0 UDL A W Cowan, House of Shannon, Wester Templand, Fortrose, Scotland, IV10 8RA [IO77WN, NH75]
GW0 UDM J D Reed, Coed Mor House, 3 The Avenue, Prestatyn, LL19 9RD [IO83HH, SJ08]
G0 UDN Details withheld at licensee's request by SSL.
G0 UDO R Dodd, 2 Brinscall Green, Stoke on Trent, ST6 6RN [IO83VB, SJ85]
G0 UDP K N Fairbotham, 93 Station Rd, Upper Poppleton, York, YO2 6PZ [IO93KX, SE55]
G0 UDQ Details withheld at licensee's request by SSL.
GW0 UDR J R Greenall obo Bods Radio Club, Llys Gwyn, 3 Norton Ave, Prestatyn, LL19 7NL [IO83HI, SJ08]
G0 UDS C Clifton, 8606 Basswood #6, Pierrefonds, Quebec, Canada, H8Y 1S7
G0 UDT M A Pawlowski, Box 576, Menwith Hill Station, North Yorks, HG3 2RF [IO94DA, SE25]
G0 UDX Details withheld at licensee's request by SSL.
GM0 UDY A Hyslop, 9 Shalloch Sq, Girvan, KA26 0EA [IO75NF, NX19]
G0 UDZ M Baister, 28 Simonside Terr, Newbiggin, Newbiggin By The Sea, NE64 6PU [IO95FE, NZ38]
G0 UEA J Heald, 94 Haylings Rd, Leiston, IP16 4DT [JO02SE, TM46]
G0 UEB R E Fisher, 14 Colindeep Ln, Sprowston, Norwich, NR7 8EG [JO02PP, TG21]
G0 UEC Details withheld at licensee's request by SSL.
G0 UED I Harkness, 21 Gr End Gdns, Gr End Rd, St. Johns Wood, London, NW8 9LL [IO91VM, TQ28]
G0 UEE P Dyson, 32 Carr Field Ln, Bolton upon Deame, Rotherham, S63 8AP [IO93IM, SE40]
G0 UEF A Brooke, Rock House, 17 Wolseley Rd, Gloucester, GL2 0PJ [IO81VU, SO81]
GI0 UEG R A Wilson Radio Amateur Special Event Gr, 504 Castlereagh Rd, Belfast, BT5 6QA [IO74BN, J37]
G0 UEH R Hislop, 79 Norwood Ave, Hasland, Chesterfield, S41 0NJ [IO93HF, SK46]

G0 UEI M J Myhill, Mayes Farm, Melton Rd, Wymondham, NR18 0SE [JO02NO, TG10]
GM0 UEJ Details withheld at licensee's request by SSL.
G0 UEK S M Payne, 55 Binstead Lodge Rd, Binstead, Ryde, PO33 3TL [IO90JR, SZ59]
GM0 UEL D Hill, 47 Sheardale Dr, Coalsnaughton, Tillicoultry, FK13 6LN [IO86DD, NS99]
G0 UEM S W Copley, 28 Kiln Cl, Old Catton, Norwich, NR6 7HZ [JO02PQ, TG21]
G0 UEN Details withheld at licensee's request by SSL.
GW0 UEO J Ewan, 71 Kingston Dr, Connahs Quay, Deeside, CH5 4TN [IO83LF, SJ26]
GM0 UEQ W G Brown, 65 Coronation Cres, Larkhall, ML9 1PU [IO85AR, NS74]
G0 UER Details withheld at licensee's request by SSL.
G0 UES H Rushton, 30 Kent Ave, East Cowes, PO32 6QN [IO90IS, SZ59]
GM0 UET R G Henderson, 22 Bowmont Pl, Gardenhall, East Kilbride, Glasgow, G75 8YG [IO75VS, NS65]
G0 UEU A D Jackson, 38 Ross Cl, Saffron Walden, CB11 4AY [JO02CA, TL53]
G0 UEW R A Jones St Christophers School ARC, 36 Cloisters Rd, Letchworth, SG6 3JS [IO91VX, TL23]
G0 UEX Details withheld at licensee's request by SSL.
G0 UEY R A Y M Muggeridge, 29 Groomsland Dr, Billingshurst, RH14 9HB [IO91SA, TQ02]
G0 UFB P Short, 23 Barn Cl, Hartford, Huntingdon, PE18 7XF [IO92WI, TL27]
G0 UFC Dr D J Meacock, The Limes, Davids Ln, Benington, Boston, PE22 0BZ [JO02BX, TF44]
G0 UFD Details withheld at licensee's request by SSL.
G0 UFE S J Bird, 15 Ludlow Dr, Stirchley, Telford, TF3 1EG [IO82SP, SJ60]
G0 UFF R J Reynolds, 4 Linley Cl, East Tilbury, Grays, Essex, RM18 8PT [JO01FL, TQ67]
G0 UFG H L Earnshaw, 69 Maldon Rd, Bitterne, Southampton, SO19 7AF [IO90HV, SU41]
G0 UFH A Oren, 22 Stonebridge Park, London, NW15 5PF [IO91VN, TQ28]
G0 UFI G M Brady, Bordercot, Ripon Rd, Baldersby, Thirsk, YO7 4PS [IO94GE, SE37]
G0 UFM H Thorpe, 6 Schofield Ct, Rowsley, Matlock, Derbyshire, DE4 2GZ [IO93EE, SK26]
G0 UFN P H Dean, 80 Escallond Dr, Seaham, SR7 8JZ [IO94HT, NZ44]
G0 UFO Prof V G Civita, 1 Crossway, Lewes, E Sussex, BN7 1NE [IO90XV, TQ41]
G0 UFP C R Reynolds, 64 Cecil Rd, Lancing, BN15 8HP [IO90UT, TQ10]
G0 UFQ T D Churchard, 18 Ruthven Rd, Bristol, Avon, BS4 1ST [IO81QK, ST56]
G0 UFS D H Barber, 43 Margetson Rd, Sheffield, S5 9LS [IO93GK, SK39]
G0 UFT K Lees, 23 Renfrew St., Derby, DE21 6GB [IO92GW, SK33]
G0 UFU C J Jameson, 35 Bilberry Gr, Taunton, TA1 3XN [IO80LX, ST22]
G0 UFV P B Shaw, 33 Market Ct, Market St., Launceston, PL15 8XA [IO70TP, SX38]
G0 UFW J K Marvill, 242 Hillmorton Rd, Rugby, CV22 5BG [IO92JI, SP57]
G0 UFX N R Harvey, 28 Southern Rd, Ward End, Birmingham, B8 2EH [IO92CL, SP18]
G0 UFY J D Brown, 3 Slipper Mill, Slipper Rd, Emsworth, Hants, PO10 8BS [IO90MU, SU70]
G0 UFZ Dr J S Howard, Clifton Villa, 26 South St., Cottingham, HU16 4AS [IO93TS, TA03]
G0 UGA B Moody, 38 Bromwich Rd, Willerby, Hull, HU10 6SF [IO93SS, TA02]
G0 UGD Details withheld at licensee's request by SSL.
G0 UGE Details withheld at licensee's request by SSL.
G0 UGF J A Foad, Greenfields, Crumps Ln, Ulcombe, Maidstone, Kent, ME17 1EX [JO01HE, TQ84]
GM0 UGG A M Aird, 3 Graystones, Kilwinning, KA13 7DT [IO75PP, NS34]
GM0 UGH M R Westland, Ardlea, 65 Grange Rd, Alloa, FK10 1LU [IO86CC, NS89]
G0 UGI H G Argument, 9 Oxley Cl, Shepshed, Loughborough, LE12 9LS [IO92IS, SK41]
G0 UGJ D M Basford, 91 Hollins Spring Ave, Dronfield, Sheffield, S18 6RP [IO93GH, SK37]
G0 UGK Details withheld at licensee's request by SSL.
G0 UGM E Peasey, 31 Elm St., Colne, BB8 0RQ [IO83WU, SD84]
GM0 UGO A D Wilson, 59 Baker Rd, Abingdon, OX14 5LJ [IO91IP, SU49]
G0 UGP A E J Monk, Dinton Cottage, Bull Ln, Chalfont St. Peter, Gerrards Cross, SL9 8RL [IO91RO, SU98]
GW0 UGQ M Webb, 9 Bryn Awelon, Gronant, Prestatyn, LL19 9UG [IO83HI, SJ08]
G0 UGR M J Chaloner, Barnhay, Berkeley Rd, Newport, Berkeley, Glos, GL13 9PY [IO81SQ, ST69]
G0 UGS F T Taberner, 51 Canford View Dr, Colehill, Wimborne, BH21 2UW [IO90AT, SU00]
GW0 UGT Details withheld at licensee's request by SSL.
G0 UGU R P Ward, 64 Astral Way, Sutton on Hull, Hull, HU7 4YA [IO93US, TA13]
GI0 UGW Details withheld at licensee's request by SSL.
G0 UGX J M Kinsella, 85 Lowther St., Coventry, CV2 4GL [IO92GJ, SP37]
G0 UGY S P Jackson, The Bungalow, Alcester Rd, Studley, B80 7PD [IO92BG, SP06]
G0 UGZ L H Willford, 15 Clarence Pl, Stonehouse, Plymouth, PL1 3JW [IO70WI, SX45]
GW0 UHA L Ross, 25 Ty Hen, Rhostrehwfa, Llangefni, LL77 7EZ [IO73UG, SH47]
GM0 UHC I D Ropper, 2 Deerhill, Dechmont, Broxburn, EH52 6LY [IO85FW, NT07]
G0 UHD K J Hore, 15 Heriot Way, Great Totham, Maldon, CM9 8BW [JO01IS, TL81]
GI0 UHE Details withheld at licensee's request by SSL.
G0 UHF R R Darwent, 139 The Oval, Firth Park, Sheffield, S5 6SQ [IO93GK, SK39]
G0 UHG R H Warne, 18 Martindale Rd, Churchdown, Gloucester, GL3 2DW [IO81VV, SO82]
G0 UHH Details withheld at licensee's request by SSL.
G0 UHI D K Norton, 52 Letchworth Rd, Leicester, LE3 6FG [IO92KP, SK50]
GW0 UHJ W Griffiths, 147 High St., Tonyrefail, Porth, CF39 8PL [IO81GO, ST08]
G0 UHK Prof M A Peiperl, 45 High St., Harrow, HA1 3HT [IO91TN, TQ18]
G0 UHM L M Ruddock, 2 Cross Ln, Waterlooville, PO8 9TJ [IO90LV, SU61]
G0 UHN R Seabourne, Farleighs, 158 High St., Ventnor, PO38 1QN [IO90JO, SZ57]
GW0 UHO P Sage, 4 Gladstone Terr, Miskin, Mountain Ash, CF45 3BS [IO81HQ, ST09]
G0 UHP D J Garratt, S/Y Norfolk Quetzal, Larnaca Marina, Larnaca, Cyprus, X X
G0 UHQ M J Minihane, 60 Wolesey Dr, Walton on Thames, Surrey, KT12 3BA [IO91TJ, TQ16]
G0 UHR R M Smith, Millpond View, 17 Marina Rd, Mundesley, Norwich, NR11 8BJ [JO02RV, TG33]
G0 UHS R J Hatch, 10 Carisbrooke Ave, Southshore, Blackpool, FY4 5DA [IO83LS, SD33]
G0 UHT J B Bulteel Clifton Amateur Radio Club, 15 Elmdale Rd, Clifton, Tyndalls Park, Bristol, BS8 1SF [IO81QK, ST57]
G0 UHU G M Bloyce, 8 Stanwyn Ave, Clacton on Sea, CO15 3AR [JO01NT, TM11]
G0 UHW D A Robins, 45 Colwyn Dr, Knypersley, Stoke on Trent, ST8 7BJ [IO83WC, SJ85]
G0 UHY Details withheld at licensee's request by SSL.
G0 UID J D Parker, Meadowsweet, Church End, Syresham, Brackley, NN13 5HU [IO92LB, SP64]
G0 UIF Details withheld at licensee's request by SSL.
GM0 UIG C A Cowan, 85 Eastwoodmains Rd, Clarkston, Glasgow, G76 7HG [IO75UT, NS55]
G0 UIH S P Lawman, 44, Barnwell, Peterborough, Cambs, PE8 5PS [IO92SK, TL08]
G0 UII J M Stewart, 26 Speak Cl, Heath, Wakefield, WF1 4TG [IO93GQ, SE32]
G0 UIJ H F Wyatt, 19 Cornwallis Ave, Worle, Weston Super Mare, BS22 9PF [IO81MI, ST36]
G0 UIK A G Taylor, 18 Tilleycombe Rd, Portland, DT5 1LG [IO80SN, SY67]
G0 UIL D F T Brice, 38 St. Pauls Rd, Honiton, EX14 8BR [IO80JT, ST10]
GM0 UIN Details withheld at licensee's request by SSL.
G0 UIO T Bickell, 161 High St., Chesterton, Cambridge, CB4 1NL [JO02BF, TL45]
GW0 UIP N T Wallace, Tan-y-Bryn, Bryn Rd, Flint, Clwyd, CH6 5HU [IO83KF, SJ27]
G0 UIQ W L Furze, 2 Lynwood Rd, Cromer, NR27 0EE [JO02PW, TG24]
G0 UIR P A Peasey, 31 Elm St., Colne, BB8 0RQ [IO83WU, SD84]
G0 UIS R L Darby, 38 Abbotsbury, Great Hollands, Bracknell, RG12 8QU [IO91OJ, SU86]
G0 UIT P Heywood, Bryher, Easton, Somerset, BA5 1DU [IO81PF, ST54]
G0 UIU J H Clifton, 21 Park Rd, Hilton Est, Featherstone, Wolverhampton, WV10 7HS [IO82XP, SJ90]
G0 UIW A S Jones, 57 Crab Ln, Harrogate, HG1 3BQ [IO94FA, SE35]
G0 UIX A F Lambert, 66 North Home Rd, Cirencester, GL7 1DR [IO91AR, SP00]
GW0 UIZ B E Galsworthy, 30 Pen yr Yrfa, Morriston, Swansea, SA6 6BA [IO81AQ, SS69]
G0 UJA Details withheld at licensee's request by SSL.
GU0 UJC R L Renouf, Rondor, Mare de Carteret, Grandes Rocques, Castel Guernsey, GY5 7XE
G0 UJD M R Bartle, 14 Litton Ave, Skegby, Sutton in Ashfield, NG17 3AB [IO93JD, SK56]
G0 UJE P C Mullineaux, Dept of Engineering, University of Lancaster, Bailrigg, Lancaster, LA1 4YR [IO84OA, SD45]
G0 UJF B Mellor, 24 Denbigh Dr, Shaw, Oldham, OL2 7EQ [IO83WN, SD90]
GI0 UJG R J Stinson, 51 Cloncarrish Rd, Birches, Portadown, Craigavon, BT62 1RN [IO64RL, H95]
G0 UJH Details withheld at licensee's request by SSL.
G0 UJI D A Sparkes, 17 St. Marys Dr, Bedfont, Feltham, TW14 8JT [IO91SK, TQ07]
GW0 UJJ G B Evans, C.P.P.U., Hill House, Picton Terr, Carmarthen, Dyfed, SA31 3BS [IO71UU, SN41]
G0 UJK Details withheld at licensee's request by SSL.
G0 UJP J D Fleetwood, 11 Chichester Rd, Bitterne, Southampton, SO18 6BB [IO90HW, SU41]
G0 UJR Dr J C Berridge, Bracklyn, St. Clare Rd, Walmer, Deal, CT14 7QB [JO01QE, TR35]
G0 UJS Details withheld at licensee's request by SSL.
G0 UJU D Barlow, c/o C.W.A.O., 9 Signal Regiment, BFPO 59
G0 UJW J G W Wintle, 140 Camelot Cl, King Arthurs Way, Andover, SP10 4BQ [IO91GF, SU34]
G0 UJY T S Eyers, 25 North Rd, Haywards Heath, RH16 3NJ [IO90WX, TQ32]
G0 UKA J C Black, 8 Cornwood Cl, Finchley, London, N2 0HP [IO91VO, TQ28]
G0 UKB B E Jones, 47 Pine Cres, Chandlers Ford, Eastleigh, SO53 1LN [IO90HX, SU42]
GW0 UKC M D Price, 50 Llangorse Rd, Cwmbach, Aberdare, CF44 0HR [IO81HQ, SO00]
GM0 UKD S W Munro, 17 Berwick Cres, Linwood, Paisley, PA3 3TF [IO75RU, NS46]
GW0 UKE D Lewis, 11 Heol Parc Glas, Merthyr Tydfil, CF48 1HG [IO81HS, SO00]
GW0 UKF J M Griffith, Craig Artro, Llanbedr, Gwynedd, Wales, LL45 2LU [IO72XT, SH52]
GW0 UKG A G Powell, 80 St. Andrews Cres, Abergavenny, NP7 6HN [IO81LT, SO31]
G0 UKH K W Hallam, Main Rd, Knockholt Pound, Knockholt, Sevenoaks, Kent, TN14 7JE [JO01BH, TQ45]
G0 UKI Details withheld at licensee's request by SSL.
G0 UKJ W A Shrubsall, 19 Dean Rd, Sittingbourne, ME10 2DG [JO01II, TQ96]
G0 UKK K Stanyer, 15 Wilbrahams Way, Alsager, Stoke on Trent, ST7 2NR [IO83UC, SJ75]
G0 UKM M Russell, 24 Conway Ave, Whitefield, Manchester, M45 7AZ [IO83UM, SD80]
G0 UKN Details withheld at licensee's request by SSL.
G0 UKO R Theakston, 8 Heslington Rd, York, YO1 5AT [IO93LW, SE65]
G0 UKP B J Jopson, 8 Brampton Cl, Westcliff on Sea, SS0 0DY [JO01IN, TQ88]
G0 UKQ Details withheld at licensee's request by SSL.
G0 UKR Details withheld at licensee's request by SSL.
G0 UKS R Towler, 77 Glebe Rd, Hull, HU7 0DU [IO93US, TA13]
GW0 UKT W Waldron, c/o New St Post Office, Pontnewydd, Cwmbran, Gwent, NP44 1EE [IO81LP, ST29]
G0 UKV R E Daly, 51 Queen St., Southminster, CM0 7BB [JO01JP, TQ99]
G0 UKW Details withheld at licensee's request by SSL.
G0 UKX R Wilkin, 117 Barn Meads Rd, Wellington, TA21 9BD [IO80JX, ST11]
GM0 UKZ C J F Gibson, 45 Tiree Pl, Greenfarm, Newton Mearns, Glasgow, G77 6UJ [IO75TS, NS55]
G0 ULA N R Foster, 5 Lees Gr, Leesbrook, Oldham, OL4 5LG [IO83WM, SD90]
G0 ULB D W Curtis, Pinchbeck Cottage, Moorhen Ave, St. Lawrence, Southminster, CM0 7LU [JO01JR, TL90]
GW0 ULC L Kirby, Arosfa, 7 Heol Enlli, Tanygroes, Cardigan, SA43 2JE [IO72RC, SN24]
G0 ULF S R Adams, 11 Pield Heath Ave, Hillingdon, Uxbridge, UB8 3PB [IO91SM, TQ08]
G0 ULG K J Dewing, Flat 3, 23 St. Marys Rd, Cromer, NR27 9DJ [JO02PW, TG24]
G0 ULH L C Harris, 183A Painswick Rd, Gloucester, GL4 4AG [IO81VU, SO81]
G0 ULI M P Kaliski, 93 Abbeydale Cl, Church Langley, Harlow, Essex, CM17 9QB [JO01BS, TL40]
G0 ULJ C J Darker, Urb Buena Vista, Casa Blanca Villas 320, 03194 La Marina, Alicante, Spain, X X
GM0 ULK S P Gould, 22 Cromlet Pl, Oldmeldrum, Inverurie, AB51 0DW [IO87UI, NJ82]
G0 ULL E W Williams, 50 Broad Lawn, New Eltham, London, SE9 3XD [JO01AK, TQ47]
G0 ULM P Wilson, Pear Tree Cottage, 56 Highfield Rd, Blacon, Chester, CH1 5AZ [IO83ME, SJ36]
G0 ULN L Fuller, 68 Huntingdon St., Hull, HU4 6QJ [IO93TR, TA02]
G0 ULO P A Ravenscroft, 35 Conway Rd, Ashton in Makerfield, Wigan, WN4 8UQ [IO83QL, SD50]
G0 ULP L Parsons, 28A Hope Farm Rd, Great Sutton, South Wirral, L66 2TN [IO83MG, SJ37]
G0 ULQ W C Bucknell, 119 Fossway, Dodsworth Ave, Heworth, York, YO3 7SQ [IO93LX, SE65]
G0 ULS F J Mortimer, 115 Dell Rd, Oulton Broad, Lowestoft, NR33 9NX [JO02UL, TM59]
G0 ULU R House, Norrells, Steeple Rd, Mayland, Chelmsford, CM3 6BB [JO01JQ, TL90]
G0 ULV Details withheld at licensee's request by SSL.
GW0 ULX T E Griffiths, 77 Gwili Terr, Mayhill, Swansea, SA1 6TN [IO81AP, SS69]
G0 ULY J C Stokes, 44 Caldecott Rd, Leicester, LE3 1GJ [IO92JO, SK50]
G0 ULZ G P Toop, 5 Hunt Rd, Blandford Forum, Dorset, DT11 7LZ [IO80WU, ST80]
G0 UMA P R Hutty, 38 Sunningdale Rd, Hessle, HU13 9BN [IO93SR, TA02]
G0 UMB J A Anderson, 104 Mab Ln, West Derby, Liverpool, L12 6RL [IO83NK, SJ49]
GW0 UMC G C Bowen, Charalis, Picton Pl, Carmarthen, SA31 3BZ [IO71UU, SN42]
G0 UMD D M Russell-Smith, 6 Braeside Cl, Winchester, SO22 4JL [IO91HB, SU42]
G0 UME P A Hutty, 38 Sunningdale Rd, Hessle, HU13 9BN [IO93SR, TA02]
G0 UMF Details withheld at licensee's request by SSL.
G0 UMG Details withheld at licensee's request by SSL.
G0 UMH M R Vaslet, Heatherlea, Adbury Holt, Newtown, Newbury, RG20 9BN [IO91II, SU46]
G0 UMI E T Latter, 76 Crossway, Plympton, Plymouth, PL7 4HY [IO70XJ, SX55]
GM0 UMJ Dr S F C Heerma Van Voss, Blar A Chaoruinn, Fort William, Inveness-shire, PH33 6SZ [IO76KS, NN06]
G0 UMK R J Adams, 4 Clifford Rd, Poynton, Stockport, SK12 1HY [IO83WI, SJ98]
G0 UML N P Watling, 36 All Saints Walk, Mattishall, Dereham, NR20 3RF [JO02MP, TG01]
G0 UMM N Roberson, 6 Long Ln, West Winch, Kings Lynn, PE33 0PG [JO02ER, TF61]
G0 UMN J P Massett, 8 Rospeath Cres, Manadon, Plymouth, PL2 3SY [IO70WJ, SX45]
GW0 UMO J L Joseph, 21 St. Nicholas Rd, Bridgend, CF31 1RT [IO81FM, SS98]
G0 UMP M R Parker, 90 Cavendish Rd, Patchway, Bristol, BS12 5HH [IO81RM, ST58]
G0 UMQ S Brierley, 23 Spodden Fold, Whitworth, Rochdale, OL12 8TP [IO83VP, SD81]
G0 UMR F Langley, 6 Hutchins Cl, Cledford, Middlewich, CW10 0EX [IO83SE, SJ76]
G0 UMS R S Petrie, 11 St. James Cl, Yeovil, BA21 3AH [IO80QW, ST51]
G0 UMT D M Williams, 5 Little Meadow Way, Bideford, EX39 3QZ [IO71VA, SS42]
G0 UMU H M Crawford, 15 Dukes Meadow, Woolsington, Newcastle upon Tyne, NE13 8AU [IO95DA, NZ16]
G0 UMV P B Johnson, 52 Evesham Rd, Cookhill, Alcester, B49 5LJ [IO92AF, SP05]
G0 UMX C M J Pickersgill, 6 The Dr, Bardsey, Leeds, LS17 9AE [IO93GV, SE34]
G0 UMY C Lowe, 5 Broxton Ave, Orrell, Wigan, WN5 8NP [IO83PM, SD50]
G0 UNA Dr J S J Craig, Partridge Cottage, Redpale, Dallington, Heathfield, TN21 9NR [JO00EW, TQ61]
G0 UNB J Jeffers, 11 Polywell, Appledore, Bideford, EX39 1SG [IO71VB, SS43]
G0 UNC G A Hancock, 3C Richmond St., Marlborough Ave, Hull, HU5 3JY [IO93TS, TA03]
G0 UND D B Scargill, 36 Meadow Ln, North Hykeham, Lincoln, LN6 9RE [IO93RE, SK96]
G0 UNE J Spencer, 6 Redcar Rd, Sunderland, SR5 5QA [IO94GW, NZ35]
G0 UNF G A Taylor, Bramley Grange Hotel, Horsham Rd, Bramley, Guildford, GU5 0BL [IO91RE, TQ04]
G0 UNG B Massey, 40 East St., Ashton in Makerfield, Wigan, WN4 8ST [IO83QL, SJ59]
G0 UNJ C Brown, 5 Dunsley Ave, New Moston, Manchester, M40 3NB [IO83VM, SD80]
G0 UNK T R Wright, 2 Regent Rd, Church, Accrington, BB5 4AR [IO83TS, SD72]
G0 UNL Details withheld at licensee's request by SSL.
G0 UNM J Kirkland, 15 The Cres, Breaston, Derby, DE72 3DE [IO92IV, SK43]
G0 UNN W R P Leggatt, 6 Rye Cl, Polegate, BN26 6LT [JO00DT, TQ50]
G0 UNO Details withheld at licensee's request by SSL.
G0 UNP F B S Fitchett, 11 Ellenbrook Rd, Worsley, Manchester, M28 1FX [IO83TM, SD70]
G0 UNQ N P Chetwood, Cerasus Lodge, Chain House Ln, Whitestake, Preston, PR4 4LE [IO83PR, SD52]
G0 UNR N J Rushton, 7 Sandringham Dr, Aldridge, Walsall, WS9 8HD [IO92BO, SK00]
G0 UNT S A Ray, 19 Furnival St., Cobridge, Stoke on Trent, ST6 2PD [IO83VA, SJ84]
G0 UNU Details withheld at licensee's request by SSL.
GI0 UNV Details withheld at licensee's request by SSL.
G0 UNW R E Martin, 44 Threadneedle Cres, Willowdale, Ontario, Canada, M2H 1Z6
G0 UNX T Gamble, 357 Downham Way, Bromley, BR1 5EW [JO01AK, TQ47]
G0 UNY M H Lindley, 23 Townend Ln, Deepcar, Sheffield, S30 5TN [IO93GJ, SK38]
G0 UNZ I A Harrison, 51 Sutton Park, Blunsdon, Swindon, SN2 4BB [IO91CO, SU19]
G0 UOA P C Nottle, 16 Mayflower Cl, Bere Alston, Yelverton, PL20 7DA [IO70VL, SX46]
G0 UOB A Chilinski, 7 Goldenash Ct, Northampton, NN3 8JE [IO92OG, SP76]
G0 UOC F Oakes, 18 Lime Ave, Northwich, CW9 8DS [IO83RF, SJ67]
G0 UOD E W Sheather, Clare Cottage, North End, Motcombe, Shaftesbury, SP7 9HX [IO81VB, ST82]
G0 UOE K C Tanner, 6 South Meadway, High Ln, Stockport, SK6 8EJ [IO83XI, SJ98]
G0 UOF B Shade, 1 Ridge Ln, Sleights, Whitby, North Yorks, YO21 1SA [IO94QL, NZ80]
G0 UOG D A Brunsch, 69 Milton Rd, Holway, Taunton, TA1 2JQ [IO81LA, ST22]
G0 UOH M P Fallon, 2 Abbots Way, Cressex, High Wycombe, HP12 4NR [IO91OO, SU89]
G0 UOI R J Fox, 14 Malcolm Gdns, Polegate, BN26 6PN [JO00CT, TQ50]
G0 UOJ R E Bradley, 214 Woodcock St., Hull, HU3 5DP [IO93TR, TA02]
G0 UOL D G Hards, 38 Halsbury Rd, Tiverton, EX16 4AE [IO80GV, SS91]
G0 UOM N F Taylor, 16 Josephine Rd, Holmes, Rotherham, S61 1BJ [IO93HK, SK49]
G0 UON Details withheld at licensee's request by SSL.
G0 UOO Details withheld at licensee's request by SSL.
G0 UOP R S Linford obo University of Plymouth, School of Electronic, Communication Engineering, University of Plymouth, Drake Circus Plymouth, Devon, PL4 8AA [IO70WJ, SX45]
G0 UOQ P Creissen, 143 Hawthorn Bank, Spalding, PE11 2UN [IO92WS, TF22]
G0 UOR Details withheld at licensee's request by SSL.
G0 UOS J K Butterworth, 48 Greenside, Mapplewell, Barnsley, S75 6AY [IO93FO, SE31]
GM0 UOU C G Muir, 23 Main Rd, Castlehead, Paisley, PA2 6AN [IO75SU, NS46]
G0 UOV S C Porter, 20 Newbridge Rd, Ambergate, Belper, DE56 2GR [IO93GB, SK35]
G0 UOW F J Cornish, 29 Ridgewood Dr, Burton upon Stather, Scunthorpe, DN15 9YE [IO93PP, SE81]
G0 UOY A Strong obo Dept of Electronics, Department of Electronics, University of York, Heslington, York, YO1 5DD [IO93LW, SE65]
G0 UOZ A J Morris, 4 Pleasant Terr, Lincoln, Lincs, LN5 8DA [IO93RF, SK96]
G0 UPA P Vukasinovic, 3 Stanhope St., Ashton under Lyne, OL6 9QY [IO83XL, SJ99]
G0 UPB J F Barker, Riverside House, Homecroft Dr, Uckington, Cheltenham, GL51 9SN [IO81WW, SO92]
G0 UPC Details withheld at licensee's request by SSL.
G0 UPD R C Brinkley, Fishermans Haul Cot, Felixstowe Ferry, Felixstowe, Suffolk, IP11 9RZ [JO01QX, TM33]
GM0 UPE Dr G R Sutherland, 22 Montrose Dr, Bearsden, Glasgow, G61 3LG [IO75UW, NS57]
G0 UPF O Pajo, 10 Moorhouse Ave, Stanley, Wakefield, WF3 4BD [IO93GR, SE32]
G0 UPG J M Dunne, 40 Egmont Rd, Hamworthy, Poole, BH16 5BZ [IO80XR, SY99]
G0 UPH C I Kneller, 26 Sedgley Rd, Bournemouth, BH9 2JW [IO90BR, SZ09]
G0 UPJ D R Clarence, 11 Melton Dr, New Hartley, Northd., NE25 0RD [IO95FC, NZ37]
G0 UPK K M Harmer, 1 West Meadow Rd, Braunton, EX33 1EB [IO71VC, SS43]
G0 UPL H J A Summers, 10 Falconer St., Bishops Stortford, CM23 4FE [JO01BU, TL42]
G0 UPM A E Blake, 137 Newhouse Rd, Marston, Blackpool, FY4 4JP [IO83LT, SD33]
G0 UPN Dr A M Segar, 15 Marlborough Cres, Woodstock, OX20 1YJ [IO91HU, SP41]
G0 UPO M I May, 1 Green Ln, Hadfield via, Hadfield, Hyde, SK14 8DT [IO93AK, SK09]
G0 UPP S J Bennett, 2 Willis Vean, Mullion, Helston, TR12 7DF [IO70JA, SW61]
G0 UPQ Details withheld at licensee's request by SSL.
G0 UPR Details withheld at licensee's request by SSL.
G0 UPS P A Zimmermann, 85 Wimborne Rd West, Wimborne, BH21 2DH [IO90AT, SZ09]
G0 UPU R L Kendall, Lynch Ln Farm, Greenway Ln, Gretton, Cheltenham, GL54 5ER [IO91AX, SP03]
G0 UPV T Berrisford, 126 Star & Garter Rd, Longton, Stoke on Trent, ST3 7HN [IO82WX, SJ94]
G0 UPW P T Johnson, 26 Grasmere Ave, Hullbridge, Hockley, Essex, SS5 6LF [JO01HO, TQ89]
G0 UPX S R A McFetridge, 76 Park Ln, Baildon, Shipley, BD17 7LQ [IO93DU, SE13]

G0 UPY M J Rogers, 136 Goddard Way, Saffron Walden, CB10 2ED [JO02CA, TL53]
G0 UPZ C F Hibbs, 5 Edgcumbe Gdns, Newquay, TR7 2QD [IO70LJ, SW86]
G0 UQA A S Levy Oulder Hill Radio Soc, 31 Ashbrook Cres, Rochdale, OL12 9AJ [IO83WP, SD91]
G0 UQB D K Brittain, 9 Highfield Rd, Cookley, Kidderminster, DY10 3UB [IO82VK, SO88]
G0 UQC R Birkett, 14 Greta St., Keswick, CA12 4HS [IO84KO, NY22]
G0 UQD G Emmerson, 72 The Gables, Widdrington, Morpeth, NE61 5RB [IO95EF, NZ29]
G0 UQE B H J Jones, 42 Ryde Park Rd, Rednal, Birmingham, B45 8RE [IO92AJ, SP07]
G0 UQF G Merrils, 2 East St., Darfield, Barnsley, S73 9AE [IO93HM, SE40]
GW0 UQH S T Provan, Shamrolee, Marloes, Haverfordwest, Dyfed, SA62 3BE [IO71JR, SM70]
G0 UQI L J Snowden, 25 Brixham Rd, Paignton, TQ4 7HG [IO80EK, SX85]
G0 UQJ N D Smith, 36 Maple Rd, Sutton Coldfield, B72 1JP [IO92CN, SP19]
GI0 UQK P A Sinclair, 87 Rosepark, Belfast, BT5 7RH [IO74CO, J47]
G0 UQL S Jackson, 10 Colindeep Ln, Hendon, London, NW4 4SG [IO91VO, TQ28]
G0 UQN Details withheld at licensee's request by SSL.
G0 UQO G J A Rekers, 219 Woodcote Rd, Purley, CR8 3PB [IO91WI, TQ26]
G0 UQP F Waters, 96 Stockley Rd, Barmston Village, Washington, NE38 8DR [IO94FV, NZ35]
G0 UQQ N P Baskerville, 83 Crusader Dr, Sprotbrough, Doncaster, DN5 7RS [IO93KM, SE50]
G0 UQR Details withheld at licensee's request by SSL.
G0 UQS R Eke, 106 Kings Rd, Wallsend, NE28 9JQ [IO95FA, NZ26]
G0 UQT S G Nash, 27 Clevedon Rd, Failand, Bristol, BS8 3UG [IO81PK, ST57]
G0 UQU J J Squires, 44 St. Marys Rd, Doncaster, DN1 2NP [IO93KM, SE50]
G0 UQV M J Parkin, 109 Armthorpe Rd, Wheatley Hills, Doncaster, DN2 5NR [IO93KM, SE50]
G0 UQW M Singh-Gill, 30 King Edwards Gdns, London, W3 9RQ [IO91UM, TQ18]
G0 UQX Details withheld at licensee's request by SSL.
G0 UQY P J Cox, 17 Hyde Ln, Upper Beeding, Steyning, BN44 3WJ [IO90UV, TQ11]
G0 UQZ D B Eyre, 41 Lindsay Rd, Parson Cross, Sheffield, S5 7WE [IO93GK, SK39]
G0 URA C A Myhan, 1 Camelia Gr, Fair Oak, Eastleigh, SO50 7GZ [IO90IX, SU51]
G0 URB Details withheld at licensee's request by SSL.
G0 URC P W Ball, 12 Warren Way, Digswell, Welwyn, AL6 0DH [IO91VT, TL21]
GM0 URD I M Thomson, 33 Aytoun Gr, Dunfermline, KY12 9YA [IO86GB, NT08]
G0 URE Details withheld at licensee's request by SSL.
G0 URF M H W Dodsworth, 359 Upper Town St, Bramley, Leeds, LS13 3JX [IO93ET, SE23]
G0 URG Details withheld at licensee's request by SSL.
G0 URH J R Hodges, 16340 NE 83Ro St, Apt A101, Redmond Wa 98052, USA, X X
GI0 URI S G Robinson, 19 Pattonville, Cunninghams Ln, Dungannon, BT71 6DD [IO64OL, H86]
G0 URK Details withheld at licensee's request by SSL.
G0 URL P Hayler, 21 Harbour Ln, Edgworth, Turton, Bolton, BL7 0PA [IO83TP, SD71]
G0 URM Details withheld at licensee's request by SSL.
GI0 URN D W Beattie, 14 Joanmount Gdns, Belfast, BT14 6NX [IO74AO, J37]
G0 URO D R Fosh, 23 Itchenor Rd, Hayling Island, PO11 9SN [IO90MS, SZ79]
G0 URP Details withheld at licensee's request by SSL.
G0 URR R M Webb, 9 Bingham Cl, Verwood, BH31 6TS [IO90BU, SU00]
G0 URT S M Bradbury, 37 Beech Ln, Romiley, Stockport, SK6 4AF [IO83WJ, SJ99]
G0 URU J A Cartlidge, 19 Thornfield Rd, Thornton, Crosby, Liverpool, L23 9XY [IO83LL, SD30]
G0 URW Details withheld at licensee's request by SSL.
G0 URX Dr D C Craig, Pear Tree Cottage, Cripps Corner, Staplecross, E Sussex, TN32 5QS [JO00GX, TQ72]
G0 URY D J Carver, 46 Chalkwell Park Dr, Leigh on Sea, SS9 1NJ [JO01HN, TQ88]
GM0 URZ A F Edwards, Allandhu, 12 Corvisel Rd, Newton Stewart, DG8 6LN [IO74SW, NX46]
G0 USA L K A Civita, 1 Cross Way, Lewes, BN7 1NE [IO90XV, TQ41]
GI0 USB B R Cumberland, 1 Ravelstone Ave, Bangor, BT19 1EQ [IO74EP, J57]
GI0 USC J R Smith, 15 The Horse Park, Boneybefore, Carrickfergus, Co Antrim, BT38 7ED [IO74CR, J48]
G0 USE J Davy-Jones, 76 Kyoto Ct, Bognor Regis, PO21 2UL [IO90PS, SZ99]
G0 USF Details withheld at licensee's request by SSL.
G0 USH Details withheld at licensee's request by SSL.
GM0 USI A J Dimmick, 78B Fergus Dr, Glasgow, Lanarkshire, G20 6AP [IO75UV, NS56]
G0 USJ S Johnson, 36 Langar Woods, Langar, Nottingham, NG13 9HZ [IO92MV, SK73]
G0 USK P V P Perera, 13 Dalcross Rd, Hounslow, TW4 7RA [IO91TL, TQ17]
G0 USL Details withheld at licensee's request by SSL.
G0 USM B Parker, 24 Mayfield Rd, Chorley, PR6 0DG [IO83QP, SD51]
G0 USP Details withheld at licensee's request by SSL.
GI0 USQ P A Fox-Roberts, 8 Lynwood Park, Holywood, BT18 9EU [IO74CP, J47]
G0 USR M R A Murray, 50 Telegraph Ln East, Norwich, NR1 4AR [JO02PP, TG20]
GI0 USS K G Chambers, 28 Gr Meadows, Banbridge, BT32 3EJ [IO64VK, J15]
G0 UST Details withheld at licensee's request by SSL.
G0 USU Details withheld at licensee's request by SSL.
G0 USV B H Neep, Forest View Farm, Peckleton Ln, Desford, Leicester, LE9 9JU [IO92IO, SK40]
GI0 USW P G McDonald, 11 Graymount Cres, Belfast, N Ireland, BT36 7DZ [IO74AP, J37]
GI0 USX Details withheld at licensee's request by SSL.
G0 USY K Davies, Oakdene, Caravan Site, 1A Golden Bank, Falmouth, Cornwall, TR11 5BE [IO70LD, SW83]
G0 USZ P S Schwartz, 3291 NW 63 St, Fort Lauderdale, Florida, USA, 33309, X X
G0 UTA C W Gurney, 10 Sandford Walk, Exeter, EX1 2ER [IO80FR, SX99]
G0 UTB P W Cordrey, C\O David Abbott, 18 Strawberry Hill, Berrydale, Northampton, NN3 5HL [IO92OG, SP86]
GW0 UTC J S Lomas, 47 Bryn Eglwys, Rhos on Sea, Colwyn Bay, LL28 4YY [IO83DH, SH88]
GM0 UTD H R Urquhart, The Cless, Stobo, Peebles, Tweeddale, EH45 8NU [IO85IO, NT13]
GI0 UTE D Auld, 9 Chilton Rd, Carrickfergus, BT38 7JT [IO74CR, J48]
G0 UTF D R Procter, 15 Church Cl, Tollerton, York, YO6 2ES [IO94JB, SE56]
G0 UTG W T Little, 133 Potters Field, Harlow, CM17 9DD [JO01BS, TL40]
G0 UTH Details withheld at licensee's request by SSL.
G0 UTI C B Jobson, Northmead, Main Rd, Bouldnor, Yarmouth, Isle of Wight, PO41 0UX [IO90GQ, SZ38]
G0 UTJ Details withheld at licensee's request by SSL.
G0 UTK Details withheld at licensee's request by SSL.
G0 UTL Details withheld at licensee's request by SSL.
G0 UTM B Watson, 6 Shakespeare Ave, Scunthorpe, DN17 1SA [IO93PN, SE80]
G0 UTN S J Harrison, 101 Dalestorth Rd, Skegby, Sutton in Ashfield, NG17 3AG [IO93JD, SK56]
G0 UTO Details withheld at licensee's request by SSL.
G0 UTP R C Walker, 96 Broadwell Rd, Middlesbrough, TS4 3NJ [IO94JN, NZ51]
G0 UTR D R N Coleman, Honeybourne, 1 Kerstin Cl, Wymans Brook, Cheltenham, GL50 4SA [IO81XV, SO92]
GI0 UTS E A Smyth, 4 Ballykeel Rd, Lurgan, Craigavon, BT67 9JU [IO64UM, J16]
G0 UTT A J Slade Dengie Hundred Amateur Radio S, Skerries, Summerhill, Althorne, Chelmsford, Essex, CM3 6BY [JO01JP, TQ99]
G0 UTU P Humphreys, 19 Langden Cl, The Paddock, Culcheth, Warrington, WA3 4DR [IO83RK, SJ69]
GI0 UTV I E Ross, 47 Castlereagh Pl, Belfast, BT5 4NN [IO74BO, J37]
G0 UTW T K Albee, 1 Belgrave Cres, Harrogate, HG2 8HZ [IO93FX, SE35]
G0 UTX P Curtis, 31 Dorking Cl, Ings Rd Est, Hull, HU8 9DG [IO93US, TA13]
G0 UTY Details withheld at licensee's request by SSL.
G0 UTZ R T E Davies, 36 Woodside Ave, Brown Edge, Stoke on Trent, ST6 8RX [IO83WB, SJ95]
G0 UUA Dr W Hutchings, Bitternsdale Farm, Bustomley Ln, Leigh, Stoke on Trent, ST10 4PE [IO82XW, SJ93]
GM0 UUB G M Matthews, 7 Glenochil Park, Tullibody, Alloa, FK10 3AG [IO86CD, NS89]
G0 UUC W A Slater, 44 Hope St., Brampton, Barnsley, S75 2AY [IO93GN, SE30]
G0 UUD R C Paramor, Wendover, 8 Mersea Ave, West Mersea, Colchester, CO5 8JL [JO01KS, TM01]
G0 UUE J R Allinson, 296 Derbyshire Ln, Sheffield, S8 8SF [IO93GI, SK38]
G0 UUF S C Errington, 23 Pinewood Dr, Bletchley, Milton Keynes, MK2 2HT [IO91PX, SP83]
G0 UUG M W E G Harris, 82 Wyckham Rd, Castle Bromwich, Birmingham, B36 0HS [IO92CM, SP18]
G0 UUH L M S Ashby, Willowside, Ebbs Ln, East Hanney, Wantage, OX12 0HL [IO91HP, SU49]
G0 UUI R C Newport, The Old Mill, Bucks Mills, Bideford, North Devon, EX39 5DY [IO70TX, SS32]
G0 UUL K Cox, 55 Reddicap Heath Rd, Sutton Coldfield, B75 7DX [IO92CN, SP19]
G0 UUM J F Lewis, Burley Cottage, Shortwood, Standon, Stafford, ST21 6RG [IO82UW, SJ73]
G0 UUN J Lewis, Burley Cottage, Shortwood, Standon, Stafford, ST21 6RG [IO82UW, SJ73]
G0 UUO Prof J L H O'Riordan, 14 Northampton Park, London, N1 2PJ [IO91WN, TQ38]
G0 UUP M Stevens, 24 Oakroyd Cl, Burgess Hill, RH15 0QN [IO90WX, TQ32]
G0 UUQ P D May, Maple Leaf Cottage, Rescorla, St Austell, Cornwall, PL26 8YT [IO70OJ, SX05]
G0 UUR C D Branch, 10 Queens Rd, Thame, OX9 3NQ [IO91MR, SP70]
G0 UUS G R Hanson, 7 Bradley Mews, Saffron Walden, CB10 2BG [JO02DA, TL53]
G0 UUT I Sadeh, 107 Hillside Ave, Borehamwood, WD6 1HH [IO91UP, TQ19]
G0 UUU P J Earnshaw, Dunelm, Ayton Rd, Irton, Scarborough, YO12 4RQ [IO94SF, TA08]
G0 UUV E Harley, 40 Beechfield Rd, Corsham, SN13 9DW [IO81VK, ST87]
G0 UUW Maj J Daw BEM, Scillonia, 5 Linda Rd, Fawley, Southampton, SO45 1DJ [IO90HT, SU40]
G0 UUX W B Ellis, 23 Cliff Boulevard, Kimberley, Nottingham, NG16 2JJ [IO93IA, SK44]
G0 UUZ A J Breeze, 1 Park Rd West, Wollaston, Stourbridge, DY8 3NG [IO82VK, SO88]
G0 UVA Details withheld at licensee's request by SSL.
G0 UVB I C Templeton, 41 Beattie Rise, Hedge End, Southampton, SO30 2RF [IO90IW, SU41]
G0 UVC T H Clark, 30 Deer Orchard Cl, Cockermouth, CA13 9JH [IO84HP, NY13]
GI0 UVD Details withheld at licensee's request by SSL.
GM0 UVE Details withheld at licensee's request by SSL.
G0 UVG P R Ellis, 9 Matilda Gdns, Shenley Church End, Milton Keynes, MK5 6HT [IO92OA, SP83]
G0 UVH T M Bosher, The Coach House, Long Orchard, Cobham, Surrey, KT11 1EL [IO91SH, TQ06]
G0 UVJ I McLuskie, 9 Holly Ln, Alsager, Stoke on Trent, ST7 2RS [IO83UC, SJ85]
G0 UVK M Saunders, 8 Sackville Gdns, Hove, BN3 4GH [IO90VT, TQ20]
G0 UVL L Cartwright, 115 Selkirk Gr, Norley Hall, Wigan, WN5 9XY [IO83PN, SD50]
G0 UVM S H Barthorpe, 4 Wensleydale Dr, Brinsworth, Rotherham, S60 5JY [IO93HJ, SK48]
G0 UVN B C Goolding, 10 Oakwell Cl, Stevenage, SG2 8UG [IO91WV, TL22]
G0 UVP R J Jermy, 4 Ramsdean Ave, Wigston, LE18 1DX [IO92KO, SP69]
G0 UVQ A J Davidson, 23 Mansfield Rd, Basingstoke, RG22 6DX [IO91KG, SU65]
G0 UVR G H Akse, Dr Cottage, High St, Ebberston, Scarborough, YO13 9PA [IO94QF, SE88]
G0 UVS Details withheld at licensee's request by SSL.
G0 UVT J Goldie, The Old Rectory, Dodleston, Chester, Ches, CH4 9JR [IO83MD, SJ36]
G0 UVU Details withheld at licensee's request by SSL.
G0 UVX G K Valentine, Lansbury, Railway Terr, Carharrack, Redruth, TR16 5RL [IO70JF, SW74]
G0 UVY Details withheld at licensee's request by SSL.
G0 UWA E J M Shanklin, The Old Rectory, Dodleston, Chester, Ches, CH4 9JR [IO83MD, SJ36]
G0 UWB R R Moyle, Aitchill House, Lower Brailes, Banbury, Oxon, OX15 5AP [IO92FB, SP33]
GW0 UWD Details withheld at licensee's request by SSL.
G0 UWF S M Bowles, 59 Longmead Rd, Paignton, TQ3 1AX [IO80FK, SX86]
G0 UWG Details withheld at licensee's request by SSL.
G0 UWH J J Hartigan, 83 Parsons Rd, Southcrest, Redditch, B98 7EJ [IO92AH, SP06]
G0 UWI H R Chipper, 36 Newbridge Way, Truro, TR1 3LX [IO70KG, SW84]
G0 UWK I Goodier, 59 Gill Bank Rd, Kidsgrove, Stoke on Trent, ST7 4HJ [IO83VB, SJ85]
GW0 UWM Details withheld at licensee's request by SSL.
G0 UWO N J Winfield, Warm Waters, 57 Knights Way, Mount Ambrose, Redruth, TR15 1PA [IO70JF, SW74]
G0 UWQ L S Wright, 29 Albert Rd, Benfleet, SS7 4DJ [JO01GN, TQ78]
G0 UWS A M Sharman, 21 Ferrers Green, Churston Ferrers, Brixham, TQ5 0LF [IO80FJ, SX85]
G0 UWU M J Claridge, 88 Sandyleaze, Elmbridge, Gloucester, GL2 0PX [IO81VU, SO81]
G0 UWV Details withheld at licensee's request by SSL.
G0 UWW S H Fisk, St Peters House, Stow Rd, Magdalen, Kings Lynn Norfolk, PE34 3BX [JO02EQ, TF51]
G0 UWX D Pollard, 191 High Rd, Halton, Lancaster, LA2 6QB [IO84OB, SD56]
G0 UWY Details withheld at licensee's request by SSL.
G0 UXA Details withheld at licensee's request by SSL.
G0 UXB Details withheld at licensee's request by SSL.
GI0 UXD D A Burns, 207 Rathfriland Rd, Dromara, Dromore, BT25 2EQ [IO64XI, J24]
GI0 UXE Details withheld at licensee's request by SSL.
G0 UXF C Whittaker, 55 Mallett Cres, Bolton, BL1 5TQ [IO83SO, SD61]
G0 UXG P J Blunt, 17 Offens Dr, Staplehurst, Tonbridge, TN12 0LR [JO01GD, TQ74]
G0 UXH A Mawson, 93 Glenridding Dr, Barrow in Furness, LA14 4PA [IO84JD, SD27]
G0 UXI M T J Whitehead, 17 Thorncliffe Rd, Sharples, Bolton, BL1 7ER [IO83SO, SD71]
G0 UXK P B Cooney, 181 Brennan Rd, Tilbury, RM18 8BA [JO01EL, TQ67]
G0 UXM P J Dwyer, 2 Thicket Terr, Anerley Rd, Anerley, London, SE20 8DH [IO91XJ, TQ37]
G0 UXN T J Cahill, 64 Kursaal Way, Southend on Sea, SS1 2UZ [JO01IM, TQ88]
G0 UXO B Hillman, 2 Holmes Chapel Rd, Congleton, CW12 4NE [IO83VD, SJ86]
G0 UXP M V Bushnell obo Tmc Amateur Radio Club, Rose Cottage, Revel, St. Ashton, Rugby, CV23 0PH [IO92IK, SP48]
G0 UXQ C Lees, 11 Stuart St., Clayton, Manchester, M11 4DQ [IO83VL, SJ89]
G0 UXR T M Gilmore, 48 Ash Ln, Hale, Altrincham, WA15 8PD [IO83UJ, SJ78]
G0 UXS E Lowman, 17 Lower Millhayes, Hemyock, Cullompton, EX15 3SL [IO80JW, ST11]
G0 UXT Details withheld at licensee's request by SSL.
GW0 UXU G R Hannan, 9 Horrocks Rd, Chester, CH2 1HE [IO83NF, SJ46]
GW0 UXX D Cumiskey, 23 Pemba Dr, Buckley, CH7 2HQ [IO83LE, SJ26]
G0 UXZ A Walmsley, 41 Nettlecombe, Shaftsbury, Dorset, SP7 8PR [IO81VA, ST82]
G0 UYA S B Chamberlin, 6 Yew Tree Ct, Hockering, Dereham, NR20 3JR [JO02MQ, TG01]
G0 UYB Details withheld at licensee's request by SSL.
G0 UYC D Rolph, 202 Dereham Rd, New Costessey, Norwich, NR5 0SW [JO02OP, TG11]
G0 UYD A J Colvin, 22 Circular Dr, Bebington, Port Sunlight, Wirral, L62 5EP [IO83MI, SJ38]
G0 UYE A Colton, 9 Pineway, Lodge Farm, Bridgnorth, WV15 5DS [IO82TM, SO79]
G0 UYF P Moss, 7 Kingsway, Houghton-le-Spring, DH5 8DD [IO94GU, NZ34]
G0 UYG A J Forster, 59 Roseberry Ave, Stokesley, Middlesbrough, TS9 5HF [IO94JL, NZ50]
G0 UYH A E Smart, Nine Hamelin St, Pye Green Rd, Cannock, Staffs, WS11 2SE [IO82XQ, SJ91]
G0 UYI N A F Hawkesford, 17 Owen Rd, Bilston, WV14 6QH [IO82XN, SO99]
G0 UYJ P D Lord, 2 Edgedale Cl, Crowthorne, RG45 7QH [IO91OI, SU86]
G0 UYL Details withheld at licensee's request by SSL.
G0 UYM J R Hall obo Thames Amateur Radio, Pump Cottage, Pump Ln, Marlow, SL7 3RB [IO91OO, SU88]
G0 UYN Details withheld at licensee's request by SSL.
G0 UYP R S Arkell, 33 Chatsworth Cres, Rushall, Walsall, WS4 1QU [IO92AO, SK00]
G0 UYQ M P Melbourne, 32 Lake Farm Rd, Rainworth, Mansfield, NG21 0ED [IO93KC, SK55]
G0 UYR R Johnson, Flat 13, Ida's Ct, Princes Rd, Hull, HU5 2RD [IO93TS, TA03]
G0 UYS S Jeffrey, 19 Ruswarp Ln, Whitby, YO21 1NB [IO94QL, NZ81]
G0 UYT J A Hemming, 10 Brook Dr, Bartley Green, Birmingham, West Midlands, B32 3LQ [IO92AK, SP08]
G0 UYU W Plant, 17 Lawson Terr, Knutton, Newcastle, ST5 6DS [IO83VA, SJ84]
G0 UYV B Smith, 80A Bramcote Ln, Chilwell, Beeston, Nottingham, NG9 4ES [IO92JW, SK53]
G0 UYW E F Powell, 63 Princess St., Chadsmoor, Cannock, WS11 2JT [IO82XR, SJ91]
G0 UYX L D Morris, 29 Partridge Cl, Washington, NE38 0ES [IO94FV, NZ25]
GI0 UYY L M O'Flaherty, 1 Ravensdale Villas, Newry, BT34 2PG [IO64UD, J02]
GM0 UYZ J F Wheeler, 6 Whiteash Pl, Fochabers, IV32 7HS [IO87KO, NJ35]
G0 UZA R F Ayling obo Detling Amateur Radio Club, 25 Nash Ct Rd, Margate, CT9 4DH [JO01QJ, TR36]
G0 UZB A W Denby, Amberwood, Brindley Brae, Kinver, West Mids, DY7 6LR [IO82VK, SO88]
GI0 UZC A J Allen, 49 Grahams Bridge Rd, Dundonald, Belfast, BT16 0DB [IO74CO, J47]
G0 UZD A M Gisby, 6 Kingston Cl, Hove, BN3 8LE [IO90VU, TQ20]
G0 UZE K H Brookes, 39 Kingsmead, Chorley, PR7 3JY [IO83QP, SD51]
GI0 UZG Details withheld at licensee's request by SSL.
G0 UZH K S Barber, 5 Park Highatt Dr, Shipdham, Thetford, IP25 7LG [JO02KP, TF90]
G0 UZI P J Carvell, 18 Little Ridge Ave, St. Leonards on Sea, TN37 7LS [JO00GV, TQ71]
G0 UZJ K M Hogg, 30 Dale Park Ave, Winterton, Scunthorpe, DN15 9UY [IO93QP, SE91]
GW0 UZK A J Rushton, 29 Queen St., Blaengarw, Pontycymer, Bridgend, CF32 8AH [IO81FP, SS99]
G0 UZL Details withheld at licensee's request by SSL.
GM0 UZM B M C Walker, 12 Cumbrae Terr, Kirkcaldy, KY2 6SF [IO86JD, NT29]
GW0 UZN R J Newman obo Barry College, Barry College, Faculty of Technology, School of Engineering (M5), Colcot Rd, Barry, CF6 8YJ [IO81HN, ST08]
G0 UZO I Worswick, 53 Mavis Dr, Coppull, Chorley, PR7 5AE [IO83QO, SD51]
G0 UZP P Zrobok, 12 Victoria Ave, Kidsgrove, Stoke on Trent, ST7 1HB [IO83VC, SJ85]
G0 UZQ P A Saunders, Helena, Orchard Ln, Goathland, Whitby, YO22 5JT [IO94PJ, NZ80]
G0 UZS G E Gallagher-Daggitt, Hayfield House, 9 Tullis Cl, Sutton Courtenay, Abingdon, OX14 4BD [IO91IP, SU49]
G0 UZT Details withheld at licensee's request by SSL.
G0 UZU A Kearney, 18 Wayside Mews, Maidenhead, SL6 7EJ [IO91PM, SU88]
GM0 UZV W R Cargill, 23 Ceres Rd, Craigrothie, Cupar, KY15 5QB [IO86LG, NO31]
G0 UZW N K G Asquith, 39 Cambourne Cl, Adwick-le-St., Doncaster, DN6 7DB [IO93JN, SE50]
GW0 UZX F Thompson, 15 Aneurin Cres, Twynyrodyn, Merthyr Tydfil, CF47 0TB [IO81HR, SO00]
G0 UZY J J Shotter, 41 Summerheath Rd, Hailsham, BN27 3DR [JO00DU, TQ50]
G0 UZZ Details withheld at licensee's request by SSL.
GI0 VAB P Moore, 59 Belmont Ave, Strandtown, Belfast, BT4 3DE [IO74BO, J37]
G0 VAC M B Jemmison, 28 Trehill Rd, Ivybridge, PL21 0AZ [IO80BJ, SX65]
G0 VAD J Whitehall, 29 Melrose Terr, Newbiggin By The Sea, NE64 6XN [IO95FE, NZ38]
G0 VAE C L Smith, 33 Normandy Ave, Barnhill Est, Colchester, CO2 8SD [JO01KU, TM02]
G0 VAF R S Mason, 24A Lady Ediths Park, Scarborough, YO12 5PD [IO94SG, TA08]
G0 VAG P G Mason, 91 Hampton Rd, Scarborough, YO12 5PX [IO94SG, TA08]
G0 VAH A Heyworth, Caldecott Farm, Caldecott, Farndon, Chester, CH3 6PE [IO83NB, SJ45]
G0 VAI R D Frow, Oakleigh Farm, High Halden, Ashford, Kent, TN26 3HX [JO01IC, TQ83]
G0 VAJ S A Hobden, 71 Old Shoreham Rd, Southwick, Brighton, BN42 4RD [IO90VU, TQ20]
G0 VAK Details withheld at licensee's request by SSL.
G0 VAL D E Wyatt, 37 Southfield Rd, Hoddesdon, EN11 9EA [IO91XS, TL30]
G0 VAM A Witter, 99 Grafton St., St. Helens, WA10 4HJ [IO83OK, SJ59]
G0 VAO P F Smith, 58 Catisfield Rd, Fareham, PO15 5LY [IO90JU, SU50]
G0 VAP G J Withers, 5 Whorlton Cl, Kemplah Park, Guisborough, TS14 8LW [IO94LM, NZ61]
G0 VAQ D W Green, 8 Borrowdale Dr, South Croydon, CR2 9JS [IO91XH, TQ36]
G0 VAR C Wilson, 15 Biddenden Cl, Bearsted, Maidstone, ME15 8JP [JO01GG, TQ75]
G0 VAS V Ikonomou, 39 Amberley Rd, Bush Hill Park, Enfield, EN1 2QY [IO91XP, TQ39]
G0 VAT A D Croydon, 1 Cheriton Dr, Ravenshead, Nottingham, NG15 9DG [IO93KC, SK55]
G0 VAU T C James, 114 Broadway, Loughborough, LE11 2JG [IO92JS, SK51]
G0 VAV J R Farrington, 6 The Gravel, Mere Brow, Preston, PR4 6JX [IO83NP, SD41]
GW0 VAW R Chatwin, New St Post Office, Pontnewydd, Cwmbran, Gwent, NP44 1EE [IO81LP, ST29]
G0 VAX B R Bowers, 23 Rake Cl, Wirral, L49 0XD [IO83KJ, SJ28]
G0 VAY G A Gower, 14 Beck Garth, Hedon, Hull, HU12 8LH [IO93VR, TA12]
G0 VAZ R P A Barker, 56 Southend Rd, Weston Super Mare, BS23 4JZ [IO81MH, ST35]
G0 VBA Details withheld at licensee's request by SSL.
G0 VBB D G Goulborn, 31 Ellingham Rd, Hemel Hempstead, HP2 5LE [IO91SS, TL00]

GW0 VBC	M P Tidball, 7 Church Walk, Lower Green, Tettenhall, Wolverhampton, WV6 9LS [IO82WO, SJ80]
G0 VBD	A E Barr, 8 Hornby Park, Liverpool, L18 3LL [IO83NJ, SJ48]
GM0 VBE	B Higton, The Straith, Priestland, Darvel, KA17 0LP [IO75UO, NS53]
GW0 VBF	Details withheld at licensee's request by SSL.
G0 VBG	W L Chandler, 6 St. Thomas Cl, Hinton Waldrist, Faringdon, SN7 8RP [IO91GQ, SU39]
G0 VBH	W H Chandler, 6 St. Thomas Cl, Hinton Waldrist, Faringdon, SN7 8RP [IO91GQ, SU39]
G0 VBI	P Christodoulou, 27 Citizen House, Harvist Est, London, N7 7ND [IO91WN, TQ38]
G0 VBJ	Details withheld at licensee's request by SSL.
G0 VBK	M G Blackmore, 9 Barnwood Cl, Kingswood, Bristol, BS15 4JA [IO81SL, ST67]
G0 VBL	A Myhan, 1 Camelia Gr, Fair Oak, Eastleigh, SO50 7GZ [IO90IX, SU51]
G0 VBM	R A Caddy, 7 Thompsons Rd, Keresley End, Coventry, CV7 8JU [IO92FL, SP38]
G0 VBN	J M Cressey, 32 Ballifield Rd, Sheffield, S13 9HX [IO93HI, SK48]
G0 VBO	L P D Vaughan, Karnack, Kearsney Ct, Temple Ewell, Dover, CT16 3EB [JO01PD, TR24]
G0 VBP	F Madeley, 138 Coningsby Dr, Kidderminster, DY11 5LZ [IO82UJ, SO87]
G0 VBQ	F Webster, 14 Redbank Ave, Erdington, Birmingham, B23 7JR [IO92BM, SP09]
G0 VBR	Details withheld at licensee's request by SSL.
G0 VBS	I R Poynter, 1 Park Hill, Awsworth, Nottingham, NG16 2RD [IO92IX, SK44]
G0 VBT	G W H Dawson, 22 High St., Tean, Stoke on Trent, ST10 4DZ [IO92AW, SK03]
GW0 VBU	A R Willis, 5 Danlan Rd, Pembrey, Burry Port, SA16 0UF [IO71UQ, SN40]
G0 VBV	J S Loader, 21 Canford View Dr, Colehill, Wimborne, BH21 2UW [IO90AT, SU00]
G0 VBX	J Collingwood, 41 Preston Ave, Alfreton, DE55 7JY [IO93HC, SK45]
G0 VBZ	D G Stinton, 43 The Meadows, Bidford on Avon, Alcester, B50 4AP [IO92BD, SP05]
G0 VCA	P D Wilson, 59 Oakfield Rd, Bishops Cleeve, Cheltenham, Glos, GL52 4LA [IO81XW, SO92]
G0 VCB	D J McCormick, 2 Littleworth Cottage, Duncote, Towcester, NN12 8AL [IO92LD, SP65]
G0 VCC	S J Large, 15 Miles Cl, Yapton, Ford, Arundel, BN18 0TB [IO90QT, SU90]
G0 VCD	L W Browne, 236 Glynswood, Chard, TA20 1BG [IO80MV, ST30]
G0 VCE	J C Everitt, 55 Risborough Rd, Bedford, MK41 9QR [IO92SD, TL05]
G0 VCF	G Dainotto, Flat 7, 65 Holland Park, London, W11 3SJ [IO91VM, TQ28]
G0 VCG	Details withheld at licensee's request by SSL.
G0 VCH	Details withheld at licensee's request by SSL.
G0 VCI	Details withheld at licensee's request by SSL.
G0 VCJ	K A Emblen, 33 Havelock Rd, Bognor Regis, PO21 2HB [IO90PS, SZ99]
GW0 VCK	N A Love, 9 Tredegar Rd, New Tredegar, NP2 6AL [IO81JR, SO10]
G0 VCL	Details withheld at licensee's request by SSL.
GM0 VCN	W M Long, 52 Stirling Rd, Milnathort, Kinross, KY13 7XG [IO86GF, NO10]
G0 VCO	M McCracken, Avoca, Barton Stacey, Winchester, Hants, SO21 3RL [IO91HD, SU44]
G0 VCP	Details withheld at licensee's request by SSL.
GM0 VCQ	Details withheld at licensee's request by SSL.
G0 VCR	M A Lawrence, 110 Rathbone Rd, Smethwick, Warley, B67 5JE [IO92AL, SP08]
G0 VCS	Details withheld at licensee's request by SSL.
G0 VCT	Details withheld at licensee's request by SSL.
G0 VCU	Details withheld at licensee's request by SSL.
G0 VCV	J K Partridge, 27 Leigh Rd, Penhill, Swindon, SN2 5DE [IO91CO, SU18]
G0 VCW	R W Evans, 18 Studley Rd, Wootton, Bedford, MK43 9DL [IO92RC, TL04]
G0 VCY	P E Clark, 62 Ashburnham Rd, Southend on Sea, SS1 1QE [JO01IM, TQ88]
G0 VCZ	G N Smith, 49 Grenville Cl, Highgrove Park, Churchdown, Gloucester, GL3 1LY [IO81VV, SO82]
G0 VDC	A S Lockton, Bridge Green House, Gissing Rd, Burston, Diss, IP22 3UD [JO02NJ, TM18]
G0 VDE	W J Rothwell, Blackburn Cottage, Saxtead Rd, Framlingham, Woodbridge, Suffolk, IP13 9PU [JO02PF, TM26]
G0 VDG	Details withheld at licensee's request by SSL.
G0 VDI	Details withheld at licensee's request by SSL.
G0 VDJ	E E Webster, 33 Cherry Gdns, Bitton, Bristol, BS15 6JA [IO81SK, ST67]
G0 VDK	K W Adams, 2 Brook Cttgs, Smalls Hill Rd, Leigh, Reigate, RH2 8PE [IO91VE, TQ24]
G0 VDL	C Stangroom, 35 Parker Rd, Grays, RM17 5YW [JO01DL, TQ67]
G0 VDM	K Coventry, 317 Witchards, Basildon, SS16 5BN [JO01FN, TQ78]
G0 VDN	Dr B Logan, 22 Chiltern House, Hillcrest Rd, Ealing, London, W5 1HL [IO91UM, TQ18]
G0 VDO	C R Pond, 316 St. Faiths Rd, Old Catton, Norwich, NR6 7BL [JO02PQ, TG21]
G0 VDP	K L J Hutley, Three Ways, 1 Walden House Rd, Great Totham, Maldon, CM9 8PJ [JO01IS, TL81]
G0 VDQ	A Ramsden, 60 The St., Lound, Lowestoft, NR32 5LR [JO02UM, TM59]
G0 VDR	L C Goffin, The Hollies, Belaugh Green, Coltishall, Norwich, Norfolk, NR12 7AJ [JO02QR, TG21]
G0 VDS	Details withheld at licensee's request by SSL.
G0 VDT	P Clark, 85 Uplands, Welwyn Garden City, AL8 7EH [IO91VT, TL21]
G0 VDU	J E Newman, 11 Brockstone Rd, St. Austell, PL25 3DW [IO70OI, SX05]
G0 VDV	D H Steele, Tanglewood, Beckingham St., Tolleshunt Major, Maldon, CM9 8LL [JO01JS, TL91]
G0 VDW	Details withheld at licensee's request by SSL.
G0 VDX	D A Sparkes obo Kingston College Radio Clu, 17 St. Marys Dr, Bedfont, Feltham, TW14 8JT [IO91SK, TQ07]
G0 VDZ	N C Newby, Ashbend, 167 Watersplash Rd, Shepperton, TW17 0EN [IO91SJ, TQ06]
G0 VEA	J R Mellor, 1 Vision Hill Rd, Budleigh Salterton, EX9 6EE [IO80IP, SY08]
G0 VEB	R B Verrall, 17 Hadley Hall, Lynwood Gr, Winchmore Hill, London, N21 3JP [IO91WP, TQ39]
G0 VEC	B E Hoare, 14 Beacon Oak Rd, Tenterden, TN30 6RY [JO01IB, TQ83]
G0 VEF	P J M Borley, 18 Greenside Walk, The Dales, Nottingham, NG3 7HJ [IO92KX, SK64]
G0 VEG	J G Nevison, Wanlass, 24 Brow Cres, Windermere, LA23 2EZ [IO84NI, SD49]
G0 VEH	J E Mulye, 83 Forest Dr, East London, E11 1JX [JO01AN, TQ38]
G0 VEI	B J Davison, Monks Farm, Coles Oak Ln, Dedham, Colchester, CO7 6DR [JO01LW, TM03]
G0 VEJ	L Toy, 10 Trevella Vean, St. Erme, Truro, TR4 9BS [IO70LH, SW85]
GM0 VEK	P Davie, 331 Wallacewell Rd, Balornock, Glasgow, G21 3RP [IO75VV, NS66]
GW0 VEM	A J McCleverty, 27 The Dr, Bardsey, Leeds, LS17 9AE [IO93GV, SE34]
GW0 VEN	G A Bingham, 43 Bryncoed Park, Rhyl, LL18 4SD [IO83GH, SJ08]
G0 VEO	R Myatt, 48 Adaston Ave, Eastham, Wirral, L62 8BS [IO83MH, SJ37]
G0 VEP	Details withheld at licensee's request by SSL.
GM0 VEQ	T A Brejnak, Szubinska 6/93, 01-958, Warszawa, Poland
G0 VER	Details withheld at licensee's request by SSL.
G0 VES	R G Brade Venus Electronic Software ARC, 26 Pevensey Way, Frimley Green, Frimley, Camberley, GU16 5YJ [IO91PH, SU85]
G0 VET	P B Burnand, 34 Northgate, Hornsea, HU18 1ES [IO93VW, TA24]
G0 VEU	G A Chantry, Brook House, Weston, Oswestry, Shropshire, SY10 9ES [IO82LU, SJ22]
GW0 VEW	D R Davies, 27 Twyniago, Pontarddulais, Swansea, SA4 1HX [IO71XR, SN50]
G0 VEX	P Dickinson, 34 Marshall Ave, Bridlington, YO15 2DS [IO94VC, TA16]
GM0 VFA	I L Brown obo Shurton Hill Radio Club, Vaddel, Busta, Brae, Shetland Isles, ZE2 9QN [IP90HJ, HU36]
G0 VFB	D F Banks, 6 Kirkstall Cl, Walsall, WS3 2SS [IO82XO, SJ90]
G0 VFC	Details withheld at licensee's request by SSL.
GM0 VFD	A S Adam, 10 Greenmount Rd North, Burntisland, KY3 9JQ [IO86JB, NT28]
G0 VFE	W I Vernon-Jones, 13 Erleigh Dene, Newbury, RG14 6JG [IO91IJ, SU46]
GW0 VFF	S Emanuel, 98 Moorland Rd, Cimla, Neath, SA11 1JL [IO81CP, SS79]
G0 VFG	Details withheld at licensee's request by SSL.
G0 VFH	D F Carter, 1 Falcon Green, Portsmouth, PO6 1LW [IO90LU, SU60]
G0 VFI	M I Towlson, 82 Chelynch, Doulting, nr Shepton Mallet, Somerset, BA4 4PY [IO81RE, ST64]
G0 VFJ	K C Hubbard, 2 Stulpfield Rd, Grantchester, Cambridge, CB3 9NL [JO02BE, TL45]
G0 VFK	Details withheld at licensee's request by SSL.
G0 VFL	K E Gardner, Treetops, Cemetry Rd, Weston, nr Crewe, Ches, CW2 5LJ [IO83TB, SJ75]
G0 VFM	A J Whitlock, 50 Greenfield Ave, Spinney Hill, Northampton, NN3 2AF [IO92NG, SP76]
G0 VFN	Details withheld at licensee's request by SSL.
GW0 VFQ	D M Landin, 26 Elm Gr, Rhyl, LL18 3PE [IO83GH, SJ08]
G0 VFR	Details withheld at licensee's request by SSL.
G0 VFS	R G Bailey, 13 Whiteland Rise, Westbury, Wilts, BA13 3HP [IO81VG, ST85]
GI0 VFT	K S J Cree, 93 Owenroe Dr, Kilcoolly, Bangor, BT19 1QJ [IO74DP, J48]
G0 VFU	F Cokayne, 101 Neston Dr, Bulwell, Nottingham, NG6 8QY [IO92JX, SK54]
G0 VFV	G R Marley, Whitestone Farm Cottage, Staintendale, Scarborough, YO13 0EZ [IO94SI, SE99]
G0 VFW	T J Thirlwell, Barden House, 58 Chesham Rd, Bovingdon, Hemel Hempstead, HP3 0EA [IO91RR, TL00]
G0 VFX	W R L Fowler, Bradley Lodge, Droitwich Rd, Bradley Green, Redditch, B96 6QU [IO82XG, SO96]
GM0 VFY	G J Stuart, Easter Ardoe Cottage, Ardoe, Aberdeen, Aberdeenshire, AB12 5XT [IO87VC, NJ80]
G0 VFZ	R G S Barnes, 18 Battle Rd, Tewkesbury Park, Tewkesbury, GL20 5TZ [IO81WX, SO83]
G0 VGA	E A Nottingham, 44 Sycamore Rd, Northway, Ashchurch, Tewkesbury, GL20 8PY [IO82WA, SO93]
G0 VGB	D J London, 12 Kestrel Cl, Norton, Stockton on Tees, TS20 1SF [IO94IO, NZ42]
G0 VGC	Details withheld at licensee's request by SSL.
G0 VGD	J V W Constance, 22 Gorse Cres, Ditton, Aylesford, ME20 6EU [JO01FG, TQ75]
G0 VGF	Details withheld at licensee's request by SSL.
G0 VGH	M T Lewis Bridlington Raynet Group, Melmara, 1 Barnard Way, Hedon, Hull, HU12 8QB [IO93VR, TA12]
GM0 VGI	G A Anderson, 4 Middleton Terr, Bridge of Don, Aberdeen, AB22 8HW [IO87WE, NJ91]
G0 VGJ	Prof D G Graham Cbe, Gilmour Mews, Battlebarrow, Appleby in Westmorland, CA16 6XT [IO84SN, NY62]
G0 VGK	P Jenkinson, 4 Taplin Cl, Stafford, ST16 1NW [IO82WT, SJ92]
GI0 VGL	W G Warnock, 98 Skerriff Rd, Newtownhamilton, Altnamachin, Newry, BT35 0PJ [IO64QD, H82]
G0 VGM	Details withheld at licensee's request by SSL.
G0 VGN	A Fawcett, 24 Durham Dr, Oswaldtwistle, Accrington, BB5 3AT [IO83TR, SD72]
G0 VGP	B M Davies, 18 Wyresdale Gdns, Lancaster, LA1 3FA [IO84OB, SD46]
GI0 VGQ	J P A Sherlock, Woodbank, 451 Shore Rd, Whiteabbey, Newtownabbey, BT37 9SE [IO74BQ, J38]
G0 VGR	P J Tamplin, 1 Tyler Dr, Rainham, Gillingham, ME8 9LT [JO01HI, TQ86]
G0 VGS	Details withheld at licensee's request by SSL.
G0 VGT	A Trent, 87 Woodman Rd, Coulsdon, CR5 3HQ [IO91WH, TQ25]
G0 VGU	P M Ostcliffe, 7 Cobcar St., Elsecar, Barnsley, S74 8DA [IO93GM, SE30]
GI0 VGV	A R Wright, 36 Florida Rd, Killinchy, Newtownards, BT23 6SD [IO74DL, J46]
GW0 VGW	D A T George, 2 Bryn Hefin, Mynyddygarreg, Kidwelly, SA17 4RE [IO71UR, SN40]
G0 VGX	D K Austin, 72 Romney Rd, Willesborough, Ashford, TN24 0RR [JO01KD, TR04]
G0 VGY	P Keen, 89 Raleigh Ave, Hayes, UB4 0EF [IO91TM, TQ18]
G0 VGZ	Details withheld at licensee's request by SSL.
G0 VHB	I M Herron, 19 Swinburne Terr, Dipton, Stanley, DH9 9EH [IO94CU, NZ15]
GM0 VHC	J G Cutt, 35 Simpson Rd, Bridge of Don, Aberdeen, AB23 8EP [IO87WE, NJ90]
G0 VHD	J B Atkinson obo Bridlington & District ARS, 8 Woodcock Rd, Flamborough, Bridlington, YO15 1LJ [IO94WC, TA27]
G0 VHE	S D Esler, 10 Welholme Rd, Grimsby, DN32 0DU [IO93WN, TA20]
G0 VHF	D J Bartlett The W and H Contest Group, 80 Burnway, Hornchurch, RM11 3SG [JO01CN, TQ58]
GI0 VHG	P Hughes, 17 Cardinal Dalton Park, Keady, Armagh, BT60 3TS [IO64PG, H83]
G0 VHH	G J Mann, 286 Wootton Rd, Kings Lynn, PE30 3BJ [JO02FS, TF62]
G0 VHI	W E Stainforth, 2 Grangefield Terr, New Rossington, Doncaster, DN11 0LT [IO93LL, SK69]
G0 VHJ	G J Holland, 29 Plantation Dr, Ellesmere Port, South Wirral, L66 1JT [IO83MG, SJ37]
G0 VHK	D Godding, 46 Brockley Cl, Tilehurst, Reading, RG3 4YP [IO91LJ, SU66]
G0 VHL	Details withheld at licensee's request by SSL.
G0 VHM	Details withheld at licensee's request by SSL.
G0 VHO	Dr D Ross, Leyland Ln, Ulnes Walton, Leyland, Preston, Lancs, PR5 3LB [IO83PQ, SD52]
G0 VHP	Details withheld at licensee's request by SSL.
G0 VHQ	A P Harding, 6 Avocet Way, Banbury, OX16 9YA [IO92IB, SP43]
GM0 VHR	C G Cook, 71 Hartrigge Cres, Jedburgh, TD8 6HT [IO85RL, NT62]
G0 VHS	T J Simons, 1 Studland Way, Redlands, Weymouth, DT3 5RJ [IO80SP, SY68]
G0 VHT	P W Morrison, 42 Stanton Dr, Pegswood, Morpeth, NE61 6YW [IO95EE, NZ28]
G0 VHU	R Maude, 85 Watkinson Rd, Halifax, HX2 9DA [IO93BR, SE02]
G0 VHW	A G Sparks, 125 Merlin Rd, Scunthorpe, DN17 1LN [IO93PN, SE80]
G0 VHX	D McGill, 7 Stapleford Ave, Ermine East, Lincoln, LN2 2DR [IO93RF, SK97]
G0 VHY	R Wilson, 7 Scarratt Cl, Forsbrook, Stoke on Trent, ST11 9AP [IO82XX, SJ94]
G0 VHZ	R E Jenkins, 27 Roman Rd, Salisbury, SP2 9BH [IO91CB, SU13]
GI0 VIB	A J Smith, 42 Ballycullen Rd, Charlemont, Moy, Dungannon, BT71 7HT [IO64PK, H85]
G0 VIC	Details withheld at licensee's request by SSL.
G0 VID	D J Morris, 117 Lonsdale Ave, Intake, Doncaster, DN2 6HF [IO93KM, SE60]
G0 VIE	D M Brooks, 10 Meadow Cl, Whaley Bridge, Woodley, Stockport, SK6 1QZ [IO83WK, SJ99]
G0 VIF	D D Canning, 3 Mill Gr, Whissendine, Oakham, LE15 7EY [IO92OR, SK81]
G0 VIG	L Merrick, 4 Berryfield Glade, Churchdown, Gloucester, GL3 2BT [IO81VV, SO82]
G0 VIH	Details withheld at licensee's request by SSL.
G0 VII	G T Boundey, 7 Palm Terr, Tantobie, Stanley, DH9 9PS [IO94DV, NZ15]
G0 VIJ	J D Johnson, 1 Rosa Vella Dr, Norwich Rd, Dereham, NR20 3SB [JO02LQ, TG01]
G0 VIK	D G Wood, 22 Cleveland Way, Shelley, Huddersfield, HD8 8NQ [IO93DO, SE11]
G0 VIL	Details withheld at licensee's request by SSL.
G0 VIM	M J Rivers, 153 Ashford Rd, Bearsted, Maidstone, ME14 4NE [JO01GG, TQ85]
G0 VIN	Details withheld at licensee's request by SSL.
G0 VIP	Details withheld at licensee's request by SSL.
G0 VIQ	E M Sully, 10 The Paddock, Pound Hill, Crawley, RH10 7RQ [IO91WC, TQ23]
G0 VIR	Details withheld at licensee's request by SSL.
G0 VIS	D M Imber, The Firs, Lamb Ln, Sible Hedingham, Halstead, CO9 3RS [JO01HX, TL73]
GM0 VIT	W J Henderson, 1 Drumglen, Bridge of Cally, Blairgowrie, PH10 7JL [IO86HP, NO15]
G0 VIU	Details withheld at licensee's request by SSL.
GM0 VIV	J Dunn, 50 Duddingston Rd, Edinburgh, EH15 1SG [IO85KW, NT27]
G0 VIW	Details withheld at licensee's request by SSL.
G0 VIX	M I Rutland, Roselea, 28 Eastfield Ave, Fareham, PO14 1EG [IO90JU, SU50]
GM0 VIY	Dr J D L Oates, 14 Craighlaw Ave, Eaglesham, Glasgow, G76 0EU [IO75US, NS55]
G0 VIZ	Details withheld at licensee's request by SSL.
G0 VJB	B Vaughan, 22 Hill Foot, Nabwood, Shipley, BD18 4EP [IO93CT, SE13]
G0 VJC	C S Smith, 19 Wiscombe Ave, Penkridge, Stafford, ST19 5EH [IO82WR, SJ91]
G0 VJD	J S Davidson, Hyde Vicarge, Fording Bridge, Hants, SP6 2QJ [IO90CV, SU11]
GI0 VJE	C Hannigan, 41 Lisnafin Park, Strabane, BT82 9DF [IO64GT, H39]
G0 VJF	Details withheld at licensee's request by SSL.
G0 VJG	Details withheld at licensee's request by SSL.
G0 VJH	J A Herrington, 84 Glenn Rd, Poringland, Norwich, NR14 7LU [JO02QN, TG20]
G0 VJI	R A Clay, 38 Hubbards Rd, Chorleywood, Rickmansworth, WD3 5JJ [IO91RP, TQ09]
G0 VJJ	S S Gould, 87 Wentworth Dr, Bedford, MK41 8QD [IO92SD, TL05]
G0 VJK	M Stanton, 11 Eldean Rd, Duston, Northampton, NN5 6RF [IO92MG, SP76]
G0 VJL	A L Jackson, 14 Earls View, Sutton in Craven, Keighley, BD20 7PR [IO93AV, SE04]
G0 VJM	A D Howell, 5 Moorland Rd, Weston Super Mare, BS23 4HW [IO81MH, ST35]
G0 VJN	P D Andres, 29A Langhill Ave, Knowle, Bristol, BS4 1TN [IO81QK, ST56]
GW0 VJO	F J Rees obo Tenby Amateur Radio Club, Caerleon, Picton Rd, Tenby, SA70 7DP [IO71PQ, SN10]
GJ0 VJP	N R Collier-Webb, La Bruyere, Clos Des Landes, St Brelade, Jersey, Channel Islands, JE3 8DJ
G0 VJR	R A Henshall, 9 Murrayfield Dr, Willaston, Nantwich, CW5 6QF [IO83SB, SJ65]
GW0 VJS	M R Carey obo Mid Glam Amateur Radio Gro, 47 Heol Ty Gwyn, Maesteg, CF34 0BD [IO81EO, SS89]
G0 VJT	Details withheld at licensee's request by SSL.
G0 VJU	J Saunders, 12 Honiton Cl, Northfield, Birmingham, B31 1TH [IO92AJ, SP07]
GI0 VJV	Details withheld at licensee's request by SSL.
G0 VJW	Details withheld at licensee's request by SSL.
G0 VJX	M F Thackray, 11 Long Rd, Tydd Gote, Tydd, Wisbech, PE13 5RB [JO02BR, TF41]
G0 VJY	R Welbourn, 30 Bower Rd, Swinton, Mexborough, S64 8NU [IO93IL, SK49]
G0 VJZ	H B Jeffrey, 11 Linden Rd, Stalybridge, SK15 2SL [IO83XL, SJ99]
G0 VKA	M Keating, 511 Kings Rd, Kingstanding, Great Barr, Birmingham, B44 9HL [IO92BN, SP09]
G0 VKB	P A Fisher, 49 Brixham Dr, Wyken, Coventry, CV2 3LA [IO92GK, SP38]
G0 VKC	Details withheld at licensee's request by SSL.
G0 VKE	A S Willis, 13 Horseguards Dr, Maidenhead, SL6 1XL [IO91PM, SU88]
G0 VKF	M A I George, 3 Oak Cres, Cherry Willingham, Lincoln, LN3 4AX [IO93SF, TF07]
GM0 VKG	J E Clough obo Largs & District ARS, obo Largs & District Ars, 84 Main Rd, Fairlie, Largs, KA29 0AD [IO75NS, NS25]
G0 VKH	H G Venus, 45 St. Albans Rd, Seven Kings, Ilford, IG3 8NN [JO01BN, TQ48]
G0 VKI	A Hopkinson, 34 Welby St., Fenton, Stoke on Trent, ST4 4PL [IO82WX, SJ84]
G0 VKJ	T F Bootyman, 6 Stable Ct, Welbeck, Worksop, S80 3LP [IO93JG, SK57]
G0 VKK	H Greenwood, 4 Bardney Rd, Hunmanby, Filey, YO14 0LX [IO94UE, TA07]
G0 VKL	R A Butler, 34 Commissioners Rd, Strood, Rochester, Kent, ME2 4EB [JO01GJ, TQ76]
G0 VKM	R A Gould, 37 The Avenue, Welwyn, AL6 0PW [IO91VU, TL21]
G0 VKN	W F Hull, 22 Buena Vista Gdns, Glenholt, Plymouth, PL6 7JG [IO70WK, SX56]
G0 VKO	Details withheld at licensee's request by SSL.
GI0 VKP	Details withheld at licensee's request by SSL.
GM0 VKQ	Details withheld at licensee's request by SSL.
GD0 VKS	H Klein, Odenwaldstrasse 30, 60528 Frankfurt/Main, Germany, X X
G0 VKT	P Sayer, 90 Pitcroft Ave, Earley, Reading, RG6 1NN [IO91MK, SU77]
G0 VKU	P W Mooney, 80 Puxley Rd, Deanshanger, Milton Keynes, MK19 6LW [IO92NB, SP74]
G0 VKV	J C Martin, 2 Bishopton Lodge, 346 Birmingham Rd, Stratford upon Avon, Warks, CV37 0RE [IO92DF, SP15]
G0 VKW	K Wilkinson, Rannoch, Prescott Fields, Baschurch, Shrewsbury, SY4 2EL [IO82NS, SJ42]
G0 VKX	Dr R J G Smith, 5 Carisbrooke Cl, Havant, PO9 2QW [IO90MU, SU70]
G0 VKY	Details withheld at licensee's request by SSL.
G0 VKZ	C A Brough, 103 Spalding Ave, Clifton, York, YO3 6JJ [IO93KX, SE55]
G0 VLA	Details withheld at licensee's request by SSL.
G0 VLC	Dr A J Betts, Ivy Nook, Jarvis Ln, Steyning, West Sussex, BN44 3GL [IO90UV, TQ11]
GI0 VLE	W Dalton, 16 Junction Rd, Milltown, Randalstown, Antrim, BT41 4NP [IO64UR, J18]
G0 VLF	R J McAleer, 10 Rosedale Ave, Shotley Bridge, Consett, DH8 0DY [IO94BU, NZ05]
G0 VLG	Details withheld at licensee's request by SSL.
G0 VLH	Details withheld at licensee's request by SSL.
G0 VLI	D E Taylor, 24 Ladder Hill, Wheatley, Oxford, OX33 1SX [IO91KR, SP50]
G0 VLJ	G J Heald, 5 Greenway Cl, Weymouth, DT3 5BQ [IO80SP, SY68]
G0 VLK	A E Palfreeman, 29 Boulby Rd, Redcar, TS10 5EB [IO94KO, NZ52]
G0 VLM	P S Milan, 24 Marth Gunn Rd, Bevendean, Brighton, E Sussex, BN2 4NX [IO90WU, TQ30]
G0 VLN	Details withheld at licensee's request by SSL.
G0 VLO	Details withheld at licensee's request by SSL.
G0 VLP	Capt J W Kleijn, 23 Waterloo Mns s, Waterloo Cres, Dover, CT17 9BT [JO01PC, TR34]
G0 VLQ	B Close, 74 Heston Ave, Hounslow, TW5 9EX [IO91TL, TQ17]
G0 VLR	P N A Stickland obo Leicester Raynet Group, PO Box 4, Cosby, Leics, LE9 1ZX [IO92JN, SP59]
G0 VLV	D V Hardman, 15 Wordsworth Ave, Bolton-le-Sands, Carnforth, LA5 8HJ [IO84OC, SD46]
G0 VLW	C Watson, 11 Little Dene Copse, Lymington, SO41 8EW [IO90FR, SZ39]
G0 VLX	R W J Wellham, 10 Millers Cl, Goring, Reading, RG8 9BS [IO91KM, SU58]
G0 VLY	Details withheld at licensee's request by SSL.
G0 VLZ	J C Redpath, 22 Holmes Cres, Wokingham, RG41 2SD [IO91NJ, SU86]
G0 VMA	G Skupski, 57 Three Nooks, Bamber Bridge, Preston, PR5 8EN [IO83QR, SD52]

G0 VMB B Hayward, 58 East Hill, Colchester, CO1 2QZ [JO01KV, TM02]
G0 VMC J M Williams, 34 Maskelyne Cl, Battersea, London, SW11 4AA [IO91WL, TQ27]
GW0 VMD K Peacey, 2 Robin Cl, Coed y Gloriau, Cyncoed, Cardiff, CF2 7HN [IO81KM, ST18]
G0 VME M D Macdonald, 45 Eastdale Cl, Kempston, Bedford, MK42 8LY [IO92SC, TL04]
G0 VMF P E Quirk, 70 Sands Ln, Oulton Broad, Oulton, Lowestoft, NR32 3HS [JO02UL, TM59]
G0 VMG Details withheld at licensee's request by SSL.
G0 VMH H Underwood, 3 Dunlin Cl, Thornton Cleveleys, FY5 2RG [IO83LV, SD34]
G0 VMI B J Osborne, 56 Paradise Ln, Hall Green, Birmingham, B28 0DU [IO92BK, SP18]
G0 VMJ S Hahn, 26 Watling St., Gillingham, ME7 2YH [JO01GI, TQ76]
G0 VMK P C Nicholls, 11 Station Rd North, Belton, Great Yarmouth, NR31 9NF [JO02TN, TG40]
GW0 VML D I Wright, 91 High St., Rhosllanerchrugog, Wrexham, LL14 1AN [IO83LA, SJ24]
G0 VMO W W Wright, 8 Laithes Ln, New Lodge, Barnsley, S71 3AB [IO93GN, SE30]
G0 VMP J Roze, 9 Ralfland View, Shap, Penrith, CA10 3PF [IO84PM, NY51]
G0 VMQ P I Ellis, 104 Gravesend Rd, Strood, Rochester, ME2 3PN [JO01FJ, TQ76]
GW0 VMR P A S Smith, Bron Awel, Bryn Isa Rd, Brynteg, Wrexham, Clwyd, LL11 6NS [IO83LB, SJ35]
GW0 VMS L Beedle, 2 Chestnut Gr, Maesteg, CF34 0NT [IO81EO, SS89]
G0 VMT Dr T F Moorhead, 53 Childwall Priory Rd, Liverpool, L16 7PA [IO83NJ, SJ48]
GM0 VMV E D Kennedy, 12 Morningside Dr, Edinburgh, EH10 5LY [IO85JW, NT27]
GW0 VMW R H Price, 42 Ffordd Naddyn, Glan Conwy, Colwyn Bay, LL28 5NH [IO83CG, SH87]
GW0 VMX Details withheld at licensee's request by SSL.
G0 VMY J S Harvey, 59 Southgate, Sutton Hill, Telford, TF7 4HE [IO82SP, SJ60]
GW0 VMZ J A Davies, 75 Brondeg, Heolgerrig, Merthyr Tydfil, CF48 1TP [IO81HR, SO00]
G0 VNA G T G Kent, Plum Tree Cottage, Colesbrook Ln, Gillingham, Dorset, SP8 4HH [IO81UB, ST82]
G0 VNB M Brain, Coldharbour, Low Rd, Little Cheverell, Devizes, SN10 4JZ [IO81XG, ST95]
GW0 VND M H Goodridge, 17 Charles St., Neyland, Milford Haven, SA73 1SA [IO71MR, SM90]
G0 VNE H B Stokes, 9 Causeway Glade, Dore, Sheffield, S17 3EZ [IO93FH, SK38]
GU0 VNF J G Fraser, Rue Maze Dental Surgery, Lindfield, Rue Maze, St Martins, Guernsey, GY4 6LJ
G0 VNG Details withheld at licensee's request by SSL.
G0 VNH C N Langdon, 652 Hotham Rd South, Hull, HU5 5LE [IO93TS, TA03]
G0 VNI S D Williams, The Croft, Ringwood Rd, Bartley, Southampton, SO40 7LA [IO90FV, SU31]
G0 VNJ M J Guy, 38 Sandy Ln, Charlton Kings, Cheltenham, GL53 9DQ [IO81XU, SO91]
G0 VNK J B C Harders, Kalckreuthweg 17, 22607 Hamburg 52, Germany
G0 VNN A Glenister, 24 Masons Dr, Necton, Swaffham, PE37 8EE [JO02JP, TF80]
G0 VNO D J Johns, 8 Hill Fold, Dawley Bank, Telford, TF4 2QE [IO82SQ, SJ60]
G0 VNP Details withheld at licensee's request by SSL.
G0 VNQ W B Askam, 10 Staunton Cl, Castle Donington, Derby, DE74 2XA [IO92HU, SK42]
G0 VNR Details withheld at licensee's request by SSL.
G0 VNS Details withheld at licensee's request by SSL.
G0 VNT M D Whitney, 39 Bereweeke Rd, Bognor Regis, PO22 7EG [IO90QS, SZ99]
G0 VNU Capt B J Podmore, 9 Redhill, Trinity Fields, Stafford, ST16 1LG [IO82WT, SJ92]
G0 VNV E Troughton, 31 Leyburn Rd, Livesey, Blackburn, BB2 4NQ [IO83SR, SD62]
G0 VNW J G Carrington, 101 Papplewick Ln, Hucknall, Nottingham, NG15 8BG [IO93JB, SK55]
G0 VNX D R Stuart, 293 Burgess Rd, Bassett, Southampton, SO16 3BA [IO90HW, SU41]
G0 VNY J Crawford, 19 Bury Cl, Colchester, CO1 2YR [JO01KV, TM02]
G0 VNZ Details withheld at licensee's request by SSL.
G0 VOA S J Lamb, 126 Ringway, Waverton, Chester, CH3 7NR [IO83OE, SJ46]
G0 VOB K R Fuller, 28 Bradshaw Way, Irchester, Wellingborough, NN29 7DP [IO92QG, SP96]
G0 VOC Details withheld at licensee's request by SSL.
G0 VOD A M R Baker, 10 Knights Rd, Bearwood, Bournemouth, BH11 9ST [IO90AS, SZ09]
G0 VOE C A Iles, 83 Wood Rd, Chaddesden, Derby, DE21 4LZ [IO92GW, SK33]
G0 VOF M Walmsley, 121 Roe Lee Park, Blackburn, BB1 9SA [IO83SS, SD63]
GW0 VOG D G Roberts, Moss Bank, Bwlch, Tynygongl, Gwynedd, LL74 8RH [IO73VH, SH58]
G0 VOH R J Aley, 39 Westwood Ave, March, PE15 8AX [JO02BN, TL49]
G0 VOJ S C Williams, 6 Oak Rd, Clanfield, Waterlooville, PO8 0LJ [IO90MW, SU71]
G0 VOK N P Reilly, 7 Parliament St., Castle, Northwich, CW8 1HJ [IO83RG, SJ67]
GM0 VOL V Nerurkar, 26 Fothringham Dr, Monifieth, Dundee, Tayside, DD5 4SW [IO86OL, NO43]
GM0 VOM Details withheld at licensee's request by SSL.
G0 VON K Cummings, 56 Red Lion St., Earby, Colne, BB8 6RD [IO83WW, SD94]
G0 VOP M J Haddon obo Portland ARC, 1 Victoria Pl, Easton, Portland, DT5 2AA [IO80SN, SY67]
G0 VOQ T J Morgan, 73 Kingrosia Park, Clydach, Swansea, SA6 5PL [IO81BQ, SN60]
G0 VOR H A Moore, Langdale, Highgate Rd, Hayfield, Highpeak, Ches, SK22 2JL [IO93AJ, SK08]
G0 VOT C J Templeton, 59 Oaklands Way, Fareham, PO14 4LF [IO90IU, SU50]
GM0 VOU P Scott, 30 Main St., Newmills, Dunfermline, KY12 8SS [IO86FB, NT08]
G0 VOV D S Hartwell, 57 Wyatts Covert, Denham, Uxbridge, UB9 5DJ [IO91RO, TQ08]
G0 VOW Details withheld at licensee's request by SSL.
G0 VOX P H Stanley, 12 Beech Ln, Romiley, Stockport, SK6 4AF [IO83WJ, SJ99]
G0 VOY Details withheld at licensee's request by SSL.
G0 VOZ Details withheld at licensee's request by SSL.
GU0 VPA Details withheld at licensee's request by SSL.
GW0 VPB B T Lewis, 36 George St., Brynmawr, NP3 4TW [IO81JT, SO11]
G0 VPC M G Davies, 1 Marina Heights, West Hill Rd, St. Leonards on Sea, TN38 0NF [JO00GU, TQ70]
G0 VPE D B Pibworth obo Reading & W Berkshire Ray, 20 Marathon Cl, Woodley, Reading, RG5 4UN [IO91NK, SU77]
G0 VPF Details withheld at licensee's request by SSL.
GM0 VPG N D Thackrey, Leachkin Lodge, Upper Leachkin, Inverness, IV3 6PN [IO77UL, NH64]
G0 VPH M Austin, 107 Spicer Cl, London, SW9 7UE [IO91WL, TQ37]
G0 VPI Details withheld at licensee's request by SSL.
G0 VPJ J H Stacey, 16 Crane Dr, Verwood, BH31 6QB [IO90BV, SU00]
G0 VPK A B Dixon obo Chce Community School RC, Chase Community School, Churchbury Ln, Enfield, EN1 3HQ [IO91XP, TQ39]
G0 VPL Details withheld at licensee's request by SSL.
GM0 VPM M Patel, East Aryburn Cottage, Aryburn Farm, Parkhill Dyce, Aberdeen, Scotland, AB2 0AN [IO87TH, NJ72]
G0 VPN P Evans, 23 Dorning Rd, Swinton, Manchester, M27 5UX [IO83UM, SD70]
G0 VPO A P Robinson, 49 Mitchell Terr, Myrtle Park, Bingley, BD16 1ER [IO93BU, SE13]
GW0 VPR E W S Meredith obo St Tybie ARC, 5 Woodfield Rd, Llandybie, Ammanford, SA18 3UR [IO71XT, SN61]
G0 VPS D B Bennett, 3 Sivilla Rd, Kilnhurst, Rotherham, S62 5TY [IO93HL, SK49]
G0 VPT L T Smith, Hillside, Kings Mill Ln, Painswick, Stroud, GL6 6SA [IO81VS, SO80]
G0 VPU M J Burbridge, 3 Kirklands, Hest Bank, Lancaster, LA2 6ER [IO84OC, SD46]
G0 VPV G Pesarini, 94 Herongate Rd, Wanstead, London, E12 5EQ [JO01AN, TQ48]
G0 VPW M J Rhodes, 102 Malvern Cres, Little Dawley, Telford, TF4 3JF [IO82SP, SJ60]
G0 VPX M Worsfold, 20 Park Dr, Rustington, Littlehampton, BN16 3DY [IO90ST, TQ00]
G0 VPY E E Weston, 33 William St., St. Johns, Tunbridge Wells, TN4 9RP [JO01DD, TQ54]
G0 VPZ J E Greenfield, 36 Barttelot Rd, Horsham, RH12 1DQ [IO91UB, TQ13]
G0 VQA J R Groves, 19 Elm Rd, Southborough, Tunbridge Wells, TN4 0HD [JO01DD, TQ54]
G0 VQB M A Grainger, 174 Woodlands Rd, Gillingham, ME7 2SX [JO01GJ, TQ76]
G0 VQC N L Edis, 32 Brooklands Rd, Bakersfield, Nottingham, NG3 7AL [IO92KX, SK54]
G0 VQD P A Newberry, Thetford, 25 Wilton Rd, Redhill, RH1 6QR [IO91VF, TQ24]
G0 VQE C V H Fenn, 26 Prideaux Brune Ave, Bridgemary, Gosport, PO13 0UE [IO90JT, SU50]
G0 VQG M A Thomas, 154 Jedburgh Dr, Darlington, DL3 9UW [IO94EM, NZ21]
G0 VQH Details withheld at licensee's request by SSL.
G0 VQI F J Farrell, 23 Seaton Ave, Thornton Cleveleys, FY5 2NS [IO83LV, SD34]
G0 VQJ J Metcalfe, 158 Barrowford Rd, Colne, BB8 9QR [IO83VU, SD84]
G0 VQK S A Haigh, 2 Locker Ave, Longford, Warrington, WA2 9PS [IO83QJ, SJ69]
G0 VQL G M Guild, 7 Coronation Rd, Preston Brook, Runcorn, WA7 3AR [IO83QH, SJ58]
G0 VQM C J Reid, 54 Montacute Rd, New Addington, Croydon, CR0 0JE [IO91XI, TQ36]
G0 VQN N Edgeler, 16 Trevelva Rd, Truro, TR1 1QW [IO70LG, SW84]
G0 VQO N D Haynes, 139 Hull Rd, Anlaby, Hull, HU10 6ST [IO93SR, TA02]
G0 VQQ R J Lee, Flat B, 2 Haugh Shaw Rd, Halifax, West Yorks, HX1 3LE [IO93BR, SE02]
G0 VQR T Cannon, 35 Loddon Bridge Rd, Woodley, Reading, RG5 4AP [IO91NK, SU77]
G0 VQS Details withheld at licensee's request by SSL.
G0 VQT F J Seabourne, 11 Heyford Ave, Merton, London, SW20 9JT [IO91VJ, TQ26]
GW0 VQU Details withheld at licensee's request by SSL.
G0 VQV A Falamerzi, 2 Ventnor Way, Fareham, PO16 8RU [IO90KU, SU50]
G0 VQW A G Jack, 59 Affleck Cl, Toothill, Swindon, SN5 8DG [IO91CN, SU18]
G0 VQX D Thomas, 30 Droitwich Cl, Harmans Water, Bracknell, RG12 9EQ [IO91PJ, SU86]
G0 VQY P R Wooding, 31 Douglas Ave, Brixham, TQ5 9EL [IO80FJ, SX95]
GW0 VQZ N Jones, 88 Park St., Penrhiwceiber, Mountain Ash, CF45 3YL [IO81HQ, ST09]
G0 VRA R A Mackay, 27 Parkland Dr, Elton, Chester, CH2 4PG [IO83OG, SJ47]
G0 VRB A C Everett, 6 Western Ave, Epping, CM16 4JR [JO01BQ, TL40]
G0 VRC J F France obo Ridge College Radio Club, Cobham Cottage, 34 Ladythorn Rd, Bramhall, Stockport, SK7 2ER [IO83WI, SJ88]
G0 VRD Details withheld at licensee's request by SSL.
G0 VRE A D Challis, 12 Moorland Cres, Boultham Moor, Lincoln, LN6 7NL [IO93RE, SK96]
G0 VRF M H Waples, 19 James Rd, Wellingborough, NN8 2LR [IO92PG, SP86]
G0 VRI Details withheld at licensee's request by SSL.
G0 VRK C E Seabridge, 13 Hillside Ave, Forsbrook, Stoke on Trent, ST11 9BH [IO82XX, SJ94]
GW0 VRL C P Saunders, 14 Portway, Bishopston, Swansea, SA3 3JR [IO71XO, SS58]
G0 VRM R C Andreang obo Raywell Park Scout ARS, 6 Beech Ave, Bilton, Hull, HU11 4EN [IO93VS, TA13]
GM0 VRP R Phillips, 18 Broomridge Rd, St. Ninians, Stirling, FK7 0DT [IO86AC, NS89]
G0 VRR Details withheld at licensee's request by SSL.
G0 VRS K Gillen, 46 Cottingley Green, Leeds, LS11 0JD [IO93FS, SE23]
G0 VRT M Ritson, 14 Dunsdale Rd, Holywell, Whitley Bay, NE25 0NG [IO95FB, NZ37]
G0 VRU B W Gee, Barracks Cottage, New Main Rd, Scamblesby, Louth, LN11 9XQ [IO93WG, TF27]
G0 VRV S A Philipps, 24 Acres End, Amersham, HP7 9DZ [IO91QP, SU99]
G0 VRW P L C Wadhams, Brickwood, Bigberry Rd, Chartham Hatch, Canterbury, Kent, CT4 7ND [JO01MG, TR15]
G0 VRX C Jenkins, 31 Ashbrook Cres, Smallbridge, Rochdale, OL12 9AJ [IO83WP, SD91]
G0 VRY P J Maccormick, 7 Poplar Cl, Horsford, Norwich, NR10 3SE [JO02OQ, TG11]
G0 VRZ K H Tuer, Broad Ing, Tirril, Penrith, Cumbria, CA10 2LL [IO84OP, NY42]
G0 VSA T B Finch, 58 Cranborne Rd, Potters Bar, EN6 3AJ [IO91VQ, TL20]
G0 VSB A M Gibbs, 18 Grange Cl, Ludham, Great Yarmouth, NR29 5PZ [JO02SR, TG31]
GW0 VSC T Page, 12 Ffordd Cadfan, Tywyn, LL36 9EE [IO72XO, SH50]
G0 VSD Details withheld at licensee's request by SSL.
G0 VSE D Milner, 8 Quarrie Dene Ct, Leeds, LS7 3PH [IO93FT, SE33]
G0 VSF R Keers, 26 Coquet Terr, Heaton, Newcastle upon Tyne, NE6 5LE [IO94FX, NZ26]
GI0 VSG I M Smyth, 26 Ballytyrone Rd, Loughgall, Armagh, BT61 8QA [IO64QJ, H95]
GW0 VSH M Wright, 6 Cwm Eithin, Wrexham, LL12 8JY [IO83MB, SJ35]
G0 VSI Dr A J Betts obo Lancing College ARC, Lancing College, Lancing, West Sussex, BN15 0RW [IO90UU, TQ10]
G0 VSJ P C Taylor, 133 Beech Dr, Shifnal, TF11 8HZ [IO82TP, SJ70]
G0 VSK A E Thomas, 3 Barnfield Way, Cannock, WS12 5PR [IO92AR, SK01]
G0 VSL K P Hill, 15 Carlton Cres, Chase Terr, Walsall, West Midlands, WS7 8EP [IO92AQ, SK00]
G0 VSM T S Day, 46 Beatrice Ave, Saltash, PL12 4NG [IO70VJ, SX45]
G0 VSN Details withheld at licensee's request by SSL.
GW0 VSO P Burchell, 4 Brentwood Pl, Ebbw Vale, NP3 6JR [IO81JS, SO11]
G0 VSP Details withheld at licensee's request by SSL.
G0 VSQ Details withheld at licensee's request by SSL.
G0 VSR E Scott, 18 Manor Gdns, Killinghall, Harrogate, HG3 2DS [IO94FA, SE25]
G0 VSS R R King, 19 Greenhayes, Cheddar, BS27 3HZ [IO81OG, ST45]
GW0 VST P H Sandham, Gorwelion, Llanddarog, Caerfyrddin, Dyfed, SA32 8BJ [IO71WT, SN51]
GW0 VSW J C Mason, 12 Llwyn y Bryn, Crymlyn Parc, Skewen, Neath, SA10 6DZ [IO81BP, SS79]
G0 VSX N Jordan, 53 Longford Ct, Belle Vue Est, Hendon, London, NW4 2BX [IO91VO, TQ28]
G0 VSY G C Forster, 1 Apple Orchard, Prestbury, Cheltenham, GL52 3EH [IO81XV, SO92]
G0 VSZ D L Forster, 1 Apple Orchard, Prestbury, Cheltenham, GL52 3EH [IO81XV, SO92]
G0 VTA R D Collins, 85 Malvern Rd, Swindon, SN2 1AU [IO91CN, SU18]
G0 VTB F R Bourne, 4 Three Oaks Cttgs, Butchers Ln, Three Oaks, Hastings, TN35 4NG [JO00HV, TQ81]
G0 VTC L King, 4 Glenhurst Ave, Ruislip, HA4 7LZ [IO91SN, TQ08]
G0 VTD S Towler, 77 Glebe Rd, Stoneferry, Hull, HU7 0DU [IO93US, TA13]
G0 VTE C L Thomas, The Croft, 23 Park Ln, Knebworth, SG3 6PH [IO91VU, TL22]
GM0 VTF Details withheld at licensee's request by SSL.
G0 VTH Details withheld at licensee's request by SSL.
G0 VTI T Ibbitson, 36 Knoll Park, East Ardsley, Wakefield, WF3 2AX [IO93FR, SE32]
G0 VTK J M Leithead, 4 Ladybank Rise, Arnold, Nottingham, NG5 8QG [IO92KX, SK54]
G0 VTL H G Bradfield, West Burton View, 41 Station Rd, Walkeringham, Doncaster, DN10 4JL [IO93OK, SK79]
G0 VTM J A Pearson, 23 Glebe Ave, Mitcham, CR4 3DZ [IO91VJ, TQ26]
G0 VTN R E Shaw, 58 Albert Dr, Sheerwater, Woking, GU21 5QZ [IO91RH, TQ06]
G0 VTP R J Goss, 39 Silverdale Rd, Tadley, RG26 4JL [IO91KI, SU56]
G0 VTQ R B Glassey, 33 Saddleback Rd, Camberley, GU15 4BT [IO91PI, SU86]
G0 VTR Details withheld at licensee's request by SSL.
GI0 VTS R H M Boyle, 13 Sherwood Rd, Ballymaconnell, Bangor, BT19 6DJ [IO74EP, J58]
G0 VTT B R Richards, 17 Ironstone Rd, Chase Terr, Burntwood, Walsall, West Midlands, WS7 8NB [IO92AQ, SK00]
G0 VTU D I Stocks, 215 Watling St, Wilnecote, Tamworth, Staffs, B77 5BB [IO92DO, SK20]
G0 VTV K K Groom, 47 Hawthorne Rd, Sittingbourne, ME10 1BB [JO01II, TQ96]
G0 VTW Details withheld at licensee's request by SSL.
G0 VTX Details withheld at licensee's request by SSL.[Op David Walden, Paignton, Devon. SX85.]
G0 VUA T Barrow, 6 York Pl, Coalville, LE67 4TH [IO92IR, SK41]
G0 VUB T Cartwright, 13 Holland Ct, Evelyn Rd, Walthamstow, London, E17 9HB [IO91XO, TQ38]
G0 VUC C G Boughton, 12 Harwell Cl, Tamworth, B79 8SA [IO92DP, SK20]
G0 VUD B L Partridge, 32 Windmill Cl, Brixham, TQ5 9SH [IO80FJ, SX95]
G0 VUE Details withheld at licensee's request by SSL.
G0 VUF F R Garcia, 7 Cowper Rd, Worthing, BN11 4PD [IO90TT, TQ10]
G0 VUG K R J Townsend, 36 Dark St. Ln, Plympton, Plymouth, PL7 1PN [IO70XJ, SX55]
G0 VUH A France, 45 Wincobank Ave, Sheffield, S5 6AZ [IO93GK, SK39]
G0 VUI Details withheld at licensee's request by SSL.
GM0 VUJ R C Clarke, Nesella, Sandhead, Stranraer, Wigtonshire, DG9 9JR [IO74MT, NX05]
G0 VUL A W Hardie, Tana Merah, Church Ln, Clarborough, Retford, DN22 9NQ [IO93NI, SK78]
G0 VUM P R Cooper, 51 Acre Gate, Angram Gate, High Green, Sheffield, S30 4FT [IO93GJ, SK38]
G0 VUN T J Wooding, 101 Park Farm Rd, Ryarsh, West Malling, ME19 5JX [JO01EH, TQ66]
G0 VUO Details withheld at licensee's request by SSL.
GW0 VUP F M Muraca, 44 Gronant Rd, Prestatyn, LL19 9NB [IO83HI, SJ08]
G0 VUR Details withheld at licensee's request by SSL.
G0 VUS G Austin, 8 Aldershot Sq, Faringdon, Sunderland, SR3 3ES [IO94GU, NZ35]
G0 VUT C P A Turner, 28 Reading St., Broadstairs, CT10 3AZ [JO01RI, TR36]
G0 VUU B W Hooper, 14 Victoria Gdns, Fordingbridge, SP6 1DJ [IO90CW, SU11]
G0 VUV Details withheld at licensee's request by SSL.
G0 VUW Details withheld at licensee's request by SSL.
G0 VUX C J Walker obo Bolton School Amateur RA, Bolton School Boys Division, Chorley New Rd, Bolton, BL1 4PA [IO83SN, SD70]
GM0 VUY R K Turnbull, 23 Spartleton Pl, Dundee, DD4 0UJ [IO86NL, NO43]
G0 VUZ I C Washington, 34 Boxwood Cl, Kingston Ln, West Drayton, UB7 9PD [IO91SM, TQ07]
G0 VVA M Newbold, 10 Shaw St., Derby, DE22 3AS [IO92GW, SK33]
GI0 VVB J Serridge, 21 Lassara Heights, Warrenpoint, Newry, BT34 3PG [IO64UC, J11]
GI0 VVC J L Serridge, 21 Lassara Heights, Warrenpoint, Newry, BT34 3PG [IO64UC, J11]
G0 VVE Details withheld at licensee's request by SSL.
G0 VVF D E Turner, 54 Wood St., Ashby de La Zouch, LE65 1EG [IO92GR, SK31]
G0 VVG A Lomas, 23 Lever St., Radcliffe, Manchester, M26 4PB [IO83TN, SD70]
G0 VVH F K Garstang, 58 Anson St., Barrow in Furness, LA14 1NW [IO84JC, SD16]
GI0 VVJ J Donnelly, 37 Ballymaginaghy Rd, Castlewellan, BT31 9BH [IO74AG, J33]
G0 VVK P J Neale, 15 Kenmore Walk, Wibsey, Bradford, BD6 3JQ [IO93CS, SE13]
G0 VVM P N Cresswell, 6 Branton Cl, Basingstoke, RG22 6UW [IO91KG, SU65]
G0 VVN J W Flynn, 86 Cross Ln, Primrose Hill, Huddersfield, HD4 6DL [IO93CP, SE11]
G0 VVO D P McNamara, 18 Ayleswade Rd, Salisbury, SP2 8DR [IO91CB, SU12]
G0 VVP R Hatton, 140 Glasgow St., St. Georges Rd, Hull, HU3 3PS [IO93TR, TA02]
G0 VVQ J K Mahon, 2 Rob Ln, Newton-le-Willows, WA12 0DR [IO83QK, SJ59]
G0 VVR C V Leigh, 4 Corrick Cl, Draycott, Cheddar, BS27 3UB [IO81OG, ST45]
G0 VVS Details withheld at licensee's request by SSL.
G0 VVT Details withheld at licensee's request by SSL.
G0 VVU C E Parish, 60 Lower Cres, Linford, Stanford-le-Hope, SS17 0QP [JO01EL, TQ67]
G0 VVV M C Headey, 20 Hawthylands Cres, Hailsham, BN27 1HG [JO00DU, TQ51]
G0 VVW D N Partridge, Cherry Dene, Evesham Rd, Broadway, WR12 7DG [IO92BB, SP03]
G0 VVX M R L Samuel Croham Callers, 71 Brighton Rd, South Croydon, CR2 6EE [IO91WI, TQ36]
G0 VVY M L Soper, 11 Daisy Links, Exwick, Exeter, EX4 2PQ [IO80FR, SX99]
G0 VVZ Revd D Matthiae, Tunstall Rectory, Sittingbourne, Kent, ME9 9DU [JO01JH, TQ96]
GM0 VWA Details withheld at licensee's request by SSL.
G0 VWB M M Davies, 23 Star Ln, Folkestone, CT19 4QH [JO01NC, TR13]
G0 VWC Details withheld at licensee's request by SSL.
GW0 VWD A R Ball, 156 Merlin Cres, Bridgend, CF31 4QJ [IO81EM, SS88]
G0 VWE P W Whitfield, 47 Ranelagh Rd, Felixstowe, IP11 7HA [JO01QX, TM33]
G0 VWF C A Howell, 24 Bellring Cl, Belvedere, DA17 6LP [JO01BL, TQ47]
GM0 VWG Details withheld at licensee's request by SSL.
G0 VWH M S Wright, 27 Ellesmere Cl, Hucclecote, Gloucester, GL3 3DH [IO81VU, SO81]
G0 VWI Details withheld at licensee's request by SSL.
GI0 VWK J N McParland, 30 Ballymoyer Rd, Newtownhamilton, Newry, BT35 0AL [IO64RF, H93]
G0 VWL J F Coles, 33 Radley Gdns, Kenton, Harrow, HA3 9NZ [IO91UO, TQ18]
G0 VWP T F Sayner, 59 Horner St., York, YO3 6DZ [IO93KX, SE55]
GM0 VWQ D M Cockburn, Langholm, Seafield St, Whitemills, Bannf, Grampian, AB45 2NA [IO87RQ, NJ66]
G0 VWR Details withheld at licensee's request by SSL.
G0 VWS G D Dickin, 23 Dunmaster Way, Stirchley, Telford, TF3 1DR [IO82SP, SJ60]
G0 VWT T Holland, 135 Newcastle Rd, Stone, ST15 8LF [IO82WV, SJ83]
GI0 VWU A McCabe, 40 Rydalmere St., Belfast, BT12 6GF [IO74AO, J37]
G0 VWV R Giles, 9 Bowergreen, Lordswood, Chatham, Kent, ME5 8TN [JO01GH, TQ76]
G0 VWW T Robson, 58 Burton Rd, Lincoln, LN1 3LB [IO93RF, SK97]
G0 VWX E Collinson, 40 Cock Robin Ln, Catterall, Preston, PR3 1YL [IO83OV, SD44]
GM0 VWZ S E Lawrie, 4 Glenavon Dr, Airdrie, ML6 8QG [IO85AU, NS76]
GM0 VXA P A Marriott, 15 Innermanse Quadrant, Newarthill, Motherwell, ML1 5TD [IO85AT, NS76]

GM0	VXB	S Cockburn, Langholm, Seafield St, Whitehills, Banff, Grampian, AB45 2NA [IO87RQ, NJ66]
G0	VXC	M Coles, 133 Highthorne Rd, Kilnhurst, Mexborough, S Yorks, S64 5UU [IO93IL, SK49]
G0	VXD	G C Clayton, 27 Cotterill Rd, Knottingley, WF11 0HB [IO93IQ, SE42]
G0	VXE	D Herbert, 50 St. Leonards Cres, Scarborough, YO12 6SP [IO94SG, TA08]
G0	VXF	M J Field, 29 Welton Old Rd, Welton, Brough, HU15 1NU [IO93RR, SE92]
G0	VXG	R L Wilkinson, 139 Church Rd, Jackfield, Telford, TF8 7ND [IO82SO, SJ60]
G0	VXH	J Hampson, Marloes, Borrowby, Thirsk, N Yorks, YO7 4QP [IO94HH, SE48]
G0	VXI	F M Lawson, 6 Kirkdale Ave, New Moston, Manchester, M40 3LB [IO83VM, SD80]
G0	VXJ	M J Finch, 73 Cordingley Way, Donnington, Telford, TF2 7LJ [IO82SR, SJ71]
G0	VXK	J W Davies, 26 Beverley Cl, Wylde Green, Sutton Coldfield, B72 1YF [IO92CM, SP19]
G0	VXL	Details withheld at licensee's request by SSL.
G0	VXM	G D Simmons, 90 Pollards Fields, Ferrybridge, Knottingley, WF11 8TD [IO93IR, SE42]
G0	VXN	Details withheld at licensee's request by SSL.
G0	VXO	Details withheld at licensee's request by SSL.
GM0	VXQ	P Mulheron, 78 South Commonhead Ave, Airdrie, ML6 6PA [IO85AV, NS76]
G0	VXR	A L Radley, 10 Woodbridge Ct, Vicarage Rd, Woodford Green, IG8 8NL [JO01AO, TQ49]
G0	VXS	C G Newman, 38 Kingfisher, Regent Leisure Park, Regent Rd, Morecambe, UK, LA3 3DF [IO84NB, SD46]
G0	VXU	Details withheld at licensee's request by SSL.
G0	VXV	P D Nicholls, 11 Westfield Mount, Yeadon, Leeds, LS19 7NL [IO93DU, SE14]
G0	VXW	M Rendall, 15 Alminstone Cl, Newton Heath, Manchester, M40 1PR [IO83VL, SD80]
G0	VXX	A F Fry, 96 Westcroft Gdns, Morden, SM4 4DL [IO91VJ, TQ26]
G0	VXY	G D Forde, 11 Horsley House, St. Norbert Rd, London, SE4 2LG [IO91XL, TQ37]
GM0	VXZ	I B D Terris, 1 Linden Ave, Wishaw, ML2 8SE [IO85BT, NS85]
G0	VYA	P A Hollands, 48 Waterside, Kings Langley, WD4 8HH [IO91SR, TL00]
G0	VYB	T J Cleghorn, Leonard House, Rosemary Ln, Whitehaven, CA28 9AB [IO84EN, NX91]
G0	VYC	M W Dawson, 11 Owls Retreat, Longridge, Colchester, CO4 3FE [JO01LV, TM02]
G0	VYD	C W Sedgwick, 28 Woodall Ave, Scarborough, YO12 7TH [IO94TG, TA08]
GW0	VYF	P A Simmons, 135 Castle Way, Dale, Haverfordwest, SA62 3RN [IO71JR, SM80]
GW0	VYG	T W Jones Anglesey Cluster Support Group, Penrhiw Bach, Brungwran, Anglesey, Gwynedd, LL65 3RD [IO73SG, SH37]
G0	VYH	M D Kowalsky, Box1167, Menwith Hill Station, Harrogate, North Yorks, HG3 2RF [IO94DA, SE25]
GM0	VYI	R Johnson, 22 Torwood Ave, Grangemouth, FK3 0DN [IO86DA, NS98]
G0	VYJ	P Baxter Chapel Green Radio Soc, 16 Mortomley Cl, High Green, Sheffield, S30 4HZ [IO93GJ, SK38]
G0	VYK	R A Heesom, Ardings, Lindfield Rd, Ardingly, Haywards Heath, RH17 6TR [IO91XB, TQ32]
GM0	VYL	P Maver, 69 Mayfield Cres, Musselburgh, EH21 6EX [IO85LW, NT37]
GM0	VYM	D M Campbell, 83 Broomhall Rd, Edinburgh, EH12 7PS [IO85IW, NT17]
G0	VYN	M G Guest, 5 Blackwater Ln, Pound Hill, Crawley, RH10 7RL [IO91WC, TQ23]
G0	VYO	P D Myerscough, 6 Walton Cl, Huby, York, YO6 1HA [IO94KC, SE56]
G0	VYP	W Southworth, 58 Moyse Ave, Walshaw, Bury, BL8 3BL [IO83TO, SD71]
G0	VYQ	T J McInerney, 41 Newton Way, Tongham, Farnham, GU10 1BY [IO91PF, SU84]
G0	VYR	N H Vickerstaff, 3 Caradon Cl, Woking, GU21 3DU [IO91RH, SU95]
G0	VYS	A Rennie, 40 Viking Rd, Acomb, York, YO2 5EZ [IO93KX, SE55]
G0	VYT	M J McGarry, 14 Adrians Cl, Mansfield, NG18 4HG [IO93JD, SK55]
G0	VYU	R G Raynor-Smith, 10 Marsh Rd, Trowbridge, BA14 7PR [IO81VI, ST86]
G0	VYV	E J Smith, Magnolia House, Main Rd, West Huntspill, Highbridge Somerset, TA9 3QZ [IO81MF, ST34]
G0	VYW	Details withheld at licensee's request by SSL.
G0	VYX	Details withheld at licensee's request by SSL.
GM0	VYY	D G Macconnell, 4 Johnstone Ln, Carluke, ML8 4NR [IO85CR, NS85]
G0	VYZ	Details withheld at licensee's request by SSL.
G0	VZA	B Howlett, Bae Ltd, PO Box 98, Dhahran 31932, Saudi Arabia
G0	VZB	R A R Ferris, 6 Dozmere Cl, Feock, Truro, TR3 6RL [IO70LE, SW83]
G0	VZC	J R Terry, 72 Schofield Ave, Witney, OX8 5JW [IO91GT, SP31]
G0	VZD	J M Wadman, 11 Ash Cl, Wymondham, NR18 0HR [JO02NN, TG10]
G0	VZE	E Cook, 1 Glandore Rd, Weston Coyney, Stoke on Trent, ST3 5QW [IO82WX, SJ94]
G0	VZF	E J Walker, 30 Juniper Way, Tilehurst, Reading, RG3 6NB [IO91LJ, SU66]
G0	VZG	Details withheld at licensee's request by SSL.
G0	VZH	R Morris, 16 Stephens Cl, Luton, LU2 9AN [IO91TV, TL12]
G0	VZI	P A Robinson, 5 Coppice Cl, Haxby, York, YO3 3RR [IO94LA, SE65]
G0	VZJ	Details withheld at licensee's request by SSL.
G0	VZK	C Sampson, 1 Warton Ln, Austrey, Atherstone, CV9 3EJ [IO92FP, SK20]
G0	VZL	E H Levring, 3 Evelyn Croft, Wylde Green, Sutton Coldfield, B73 5LF [IO92BN, SP19]
G0	VZM	I A Nightingale, 42 Spilsby Rd, Horncastle, LN9 6AW [IO93WE, TF26]
G0	VZN	K J Voller, 20 Browns Ln, Uckfield, TN22 1RY [JO00BX, TQ42]
G0	VZO	J B Koops, 64 Winchester Ave, Nuneaton, CV10 0DW [IO92GM, SP39]
G0	VZP	S G Jones, 12 Burggracht, 1930 Zaventem, Belgium
G0	VZQ	Details withheld at licensee's request by SSL.
G0	VZR	R C K Leah, 9 Edward Rd, Biggin Hill, Westerham, TN16 3HN [JO01AH, TQ45]
G0	VZS	D Gilbert Bentley Amateur Radio Club, 2 Greenfield Cttgs, Bentley, Farnham, GU10 5HZ [IO91NE, SU74]
G0	VZT	S R Hopton, 5 Wellington Cl, Marske By Sea, Marske By The Sea, Redcar, TS11 6NW [IO94LO, NZ62]
GM0	VZU	Details withheld at licensee's request by SSL.
G0	VZV	Details withheld at licensee's request by SSL.
G0	VZW	Details withheld at licensee's request by SSL.
G0	VZX	P M Dart, 208 Elburton Rd, Plymouth, PL9 8HU [IO70XI, SX55]
G0	VZY	Details withheld at licensee's request by SSL.
G0	VZZ	D Taylor, 6 The Foreland, Canterbury, CT1 3NT [JO01NG, TR15]
GI0	WAA	T R Carlisle, 29 Dundrod Rd, Nutts Corner, Crumlin, BT29 4SR [IO64WP, J27]
G0	WAB	Details withheld at licensee's request by SSL.
G0	WAC	R Ray AR Training Unit, 54 Gladstone Rd, Chesham, HP5 3AD [IO91QQ, SP90]
G0	WAD	C Dyson, 21 Highmoor, Kirkhill, Morpeth, NE61 2AS [IO95DD, NZ18]
G0	WAE	D E F Phillips, 14 Seymour Rd, Newton Abbot, TQ12 2PU [IO80EM, SX87]
G0	WAF	Details withheld at licensee's request by SSL.
G0	WAG	Details withheld at licensee's request by SSL.
GI0	WAH	W A Hutchman, 35 Carlingford Park, Newry, BT34 2NY [IO64UD, J02]
GI0	WAI	Details withheld at licensee's request by SSL.
G0	WAJ	Details withheld at licensee's request by SSL.
G0	WAK	Details withheld at licensee's request by SSL.
G0	WAL	T W Reed, 10 Ashmeads Way, Wimborne, BH21 2NZ [IO90AT, SU00]
G0	WAM	S Stevens, Quality Box, 45 Tregonissey Rd, St. Austell, Cornwall, PL25 4DH [IO70OI, SX05]
G0	WAN	M J Gill, 23 Walmers Ave, Higham, Rochester, ME3 7EH [JO01FJ, TQ77]
G0	WAO	P Davies, 57 Bonfire Hill Rd, Crawshawbooth, Rossendale, BB4 8PG [IO83UR, SD82]
G0	WAP	Details withheld at licensee's request by SSL.
G0	WAQ	Details withheld at licensee's request by SSL.
G0	WAR	P F Hulse, Eastling, Stock Ln, Wilmington, Dartford, DA2 7BY [JO01CK, TQ57]
G0	WAS	A D Smith, Moorcroft, 9 Moor Ln, Maulden, Bedford, MK45 2DJ [IO92SA, TL03]
G0	WAT	P N Brice-Stevens, 31 Lodge Field, Welwyn Garden City, Herts, AL7 1SD [IO91VT, TL21]
G0	WAU	Details withheld at licensee's request by SSL.
G0	WAV	Details withheld at licensee's request by SSL.
G0	WAW	I J Oura, Quoins, Gloucester Rd, Upper Swainswick, Bath, BA1 8AD [IO81TK, ST76]
G0	WAX	L G Merrin, 15 Hornsland Rd, Canvey Island, SS8 8LX [JO01HM, TQ88]
G0	WAY	R J Slimmon, 70 Long Knowle Ln, Wednesfield, Wolverhampton, WV11 1JH [IO82WO, SJ90]
G0	WAZ	H Asmussen, Park Farm, Gorcott Hill, Beoley, Redditch, B98 9EN [IO92BH, SP06]
G0	WBA	K Fradgley, 84 Church Rd, Wordsley, Stourbridge, DY8 5AU [IO82WL, SO88]
G0	WBB	R E Anderson, 3565 N 950W, Pleasant View, Utah 84414, USA
G0	WBC	Dr C E Mortimer, 19 Greenland Dr, Leicester, LE5 1AB [IO92LP, SK60]
G0	WBE	Details withheld at licensee's request by SSL.
G0	WBF	T B McCarthy, 13 Hobart Gdns, Thornton Heath, CR7 8LR [IO91WJ, TQ36]
G0	WBG	J H Day, The Apartment, Whitegates, Homefield Paddock, Beccles Suffolk, NR34 9NE [JO02SK, TM49]
G0	WBH	M R Knights, 11 The Island, Steeple Claydon, Buckingham, MK18 2NU [IO91MW, SP62]
G0	WBI	M de Lannoy, 54 Spring Ave, Keighley, BD21 4UJ [IO93BU, SE04]
G0	WBJ	G R Nakisa, 11 Springbank, 86 Graham Rd, Malvern, Worcs, WR14 2HX [IO82UC, SO74]
G0	WBK	J Pascoe, 1 Latrobe St, Droylsden, Manchester, M43 6HU [IO83WL, SJ99]
G0	WBL	S R Hughes, 43 The Cloisters, Rickmansworth, WD3 1HL [IO91SP, TQ09]
G0	WBM	Details withheld at licensee's request by SSL.
G0	WBO	Details withheld at licensee's request by SSL.
GW0	WBP	P R Durdin, 12 St. Nicholas Cres, Penally, Tenby, SA70 7PF [IO71PP, SS19]
GW0	WBQ	D V Howells, 3 Giltar Terr, Penally, Tenby, SA70 7QD [IO71PP, SS19]
G0	WBR	T M Johnson, 17 Bridge Rd, Chichester, PO19 4NW [IO90OU, SU80]
GW0	WBS	P M Wentworth Monmouth School Amateur School, 46 Woodside Ave, Cinderford, GL14 2DW [IO81ST, SO61]
G0	WBT	J Woodcock, 54 Longworth Rd, Horwich, Bolton, BL6 7BE [IO83RO, SD61]
G0	WBU	T R Hartley, 62 Ostman Rd, Acomb, York, YO2 5QQ [IO93KX, SE55]
G0	WBV	R P Burchell, 1 Rankins Farm Cttgs, Linton, Maidstone, Kent, ME17 4AU [JO01GF, TQ74]
GM0	WBW	R P Reid, 14 Ash Hill Pl, Aberdeen, AB16 5JA [IO87WD, NJ90]
G0	WBX	Details withheld at licensee's request by SSL.
G0	WBY	Details withheld at licensee's request by SSL.
G0	WBZ	P E Webb Brdford Qlr Group, 181 Franciscan Rd, Tooting, London, SW17 8HP [IO91WK, TQ27]
G0	WCA	Details withheld at licensee's request by SSL.
G0	WCB	A W Bathurst, 81 Heatherstone Ave, Dibden Purlieu, Southampton, SO45 4LE [IO90HU, SU40]
G0	WCC	W C A Cowell, 15 Higher Copythorne, Brixham, TQ5 8QB [IO80FJ, SX95]
GI0	WCE	D Cromie, 11 Cherryvalley Park West, Belfast, BT5 6PU [IO74BO, J37]
G0	WCF	S P Adams, 79 Longthornton Rd, Streatham, London, SW16 5QF [IO91WJ, TQ26]
G0	WCG	Details withheld at licensee's request by SSL.
G0	WCH	B C P Prestage, 5 Balls Ln, Thursford, Fakenham, NR21 0BX [JO02LU, TF93]
G0	WCI	M S Hand, 150 Curtin Dr, Moxley, Wednesbury, WS10 8RN [IO82XN, SO99]
G0	WCJ	J S Slater, 25 Croft Ln, Diss, IP22 3NA [JO02NJ, TM18]
G0	WCK	D Hornby, 7 Milton St., West Bromwich, B71 1NJ [IO82XM, SO99]
G0	WCM	A R Lowe, 34 Bridgwood, Telford, TF3 1LY [IO82SP, SJ70]
G0	WCN	Details withheld at licensee's request by SSL.
G0	WCO	P E Anness, 33 William Mead Garden, Norwich, Norfolk, NR1 4PD [JO02PP, TG20]
G0	WCP	R J Swindells, 129 Stamford St., Stalybridge, SK15 1LU [IO83XL, SJ99]
G0	WCR	M J Knott, 76 New Barns Ave, Mitcham, CR4 1LF [IO91WJ, TQ26]
G0	WCS	S Ormerod, 6 New Wokingham Rd, Crowthorne, RG45 7NR [IO91OI, SU86]
G0	WCT	T J Law, Golden Lodge, Wood Ln, Fordham Heath, Colchester, CO3 5TR [JO01JV, TL92]
G0	WCU	R Roberts, 10 The Queens Dr, Millend, Rickmansworth, WD3 2LL [IO91SP, TQ09]
G0	WCW	Details withheld at licensee's request by SSL.
G0	WCZ	G D K Sutherland, 25 Prospero Rd, London, N19 3QX [IO91WN, TQ28]
G0	WDA	Details withheld at licensee's request by SSL.
GM0	WDB	Details withheld at licensee's request by SSL.
G0	WDC	D C Clark, Meadowcroft Bungalow, Ugthorpe, Whitby, North Yorks, YO21 2BL [IO94OL, NZ71]
GM0	WDD	G C Stirling, 12 Swanston Pl, Edinburgh, EH10 7DD [IO85JV, NT26]
G0	WDE	Details withheld at licensee's request by SSL.
GM0	WDF	J J Dunlop, West Dougliehill Fr, Port Glasgow, Renfrewshire, PA14 5XF [IO75PW, NS37]
G0	WDG	L A Morgan, Aldouran, Nether Compton, Sherborne, Dorset, DT9 4PZ [IO80RW, ST61]
G0	WDH	B C Ballam, Eastwood, Daneshill, Dalwood, Axminster, Devon, EX13 7HH [IO80LT, ST20]
G0	WDJ	A Pearson, 30 The Maples, Harlow, CM19 4QY [JO01BR, TL40]
G0	WDK	D M Hackett, Habitat, Station Rd, Walpole St. Andrew, Wisbech, PE14 7LZ [JO02DR, TF51]
G0	WDL	T R Rennie, 47 The Causeway, Potters Bar, EN6 5HF [IO91VQ, TL20]
G0	WDM	Details withheld at licensee's request by SSL.
G0	WDO	E Brown, 128 Old Nazeing Rd, Broxbourne, EN10 6QY [IO91XR, TL30]
G0	WDP	P D Hanson, Hydon Hill, Clock Barn Ln, Hydon Heath, Godalming Surrey, GU8 4BA [IO91QD, SU94]
G0	WDQ	S Collins, 85 Malvern Rd, Swindon, SN2 1AU [IO91CN, SU18]
G0	WDS	Details withheld at licensee's request by SSL.
G0	WDT	R E Emery, 10 Penarth Pl, Newcastle, ST5 2JL [IO83VA, SJ84]
G0	WDU	Details withheld at licensee's request by SSL.
G0	WDV	S Steddy, 1 Springfield Cl, Wenvoe, Cardiff, CF5 6DA [IO81IK, ST17]
G0	WDW	W D A Williamson, Monfa, Walford Heath, Shrewsbury, Shropshire, SY4 2HT [IO82OS, SJ41]
G0	WDX	R J T Morgan World DX Radio Group, 56 The Meadows, Ennerdale, Hull, HU7 6EE [IO93TS, TA03]
G0	WDY	Details withheld at licensee's request by SSL.
G0	WDZ	Details withheld at licensee's request by SSL.
G0	WEA	Details withheld at licensee's request by SSL.
G0	WEB	R A Webb, 25 Greenfield Croft, Bilston, WV14 8XD [IO82XN, SO99]
G0	WEC	J S Virdee, 2 Ember Rd, Slough, SL3 8ED [IO91RM, TQ07]
GM0	WED	E Holt, Ashwell, Cannigall Park, St Ola, Orkney, KW15 1SX [IO88MW, HY40]
G0	WEE	Details withheld at licensee's request by SSL.
G0	WEF	N J Kenworthy, 10 Byron Ave, Prestwich, Manchester, M25 9LT [IO83UM, SD80]
G0	WEH	J C Hill, 22 Sandy Point Rd, Hayling Island, PO11 9RP [IO90MS, SZ79]
G0	WEI	S R Peddie, 246 Uxbridge Rd, Hatch End, Pinner, HA5 4HS [IO91TO, TQ19]
G0	WEJ	A T Buswell, 12 Park Rd, Denmead, Waterlooville, PO7 6NE [IO90LV, SU61]
G0	WEK	A M Robinson, 43 Fremantle Rd, High Wycombe, HP13 7PQ [IO91PP, SU89]
G0	WEN	Details withheld at licensee's request by SSL.
G0	WEO	D A Waterfield, 21 Cedar Rd, Mickleton, Chipping Campden, GL55 6SZ [IO92CC, SP14]
G0	WEP	W E Pauline, Bell Cottage, 52 Oxford St., Ramsbury, Marlborough, SN8 2PG [IO91EK, SU27]
G0	WEQ	J Trotter Callington School Radio Club, 27 Maynards Park, Bere Alston, Yelverton, PL20 7AR [IO70VL, SX46]
GW0	WER	P J F Moran, 1 Plas Issa, Brook St, Rhosymedre, Wrexham, Clwyd, LL14 3EE [IO82LX, SJ24]
G0	WES	Details withheld at licensee's request by SSL.
G0	WEU	J P C Gwinnett, 9 Old Fold Ln, Hadley, Barnet, EN5 4QN [IO91VP, TQ29]
G0	WEV	S J McMullen, 70 Sylvan Ave, Timperley, Altrincham, WA15 6AB [IO83UJ, SJ78]
G0	WEX	S J Elliott, The Gables, 1 Oddfellows Rd, Hathersage, Sheffield, Yorks, S30 1DU [IO93GJ, SK38]
GW0	WEY	J W Doores, 14 Parc Tyddyn, Red Wharf Bay, Pentraeth, LL75 8NQ [IO73VH, SH58]
G0	WEZ	Dr P A Ewing, 23 Cherry Orton Rd, Orton, Peterborough, PE2 5EQ [IO92UN, TL19]
GM0	WFA	T Clark, 43 Meadowbank, Kirkwall, KW15 1QJ [IO88MX, HY41]
GM0	WFB	J Keenan, 53 Clermiston Cres, Edinburgh, EH4 7DF [IO85IX, NT27]
G0	WFC	R G Simmons, 29 Pinewood Dr, Bletchley, Milton Keynes, MK2 2HT [IO91PX, SP83]
G0	WFE	S J Kirwan, 9 Broomhouse Cl, Denby Dale, Huddersfield, HD8 8UX [IO93EN, SE20]
G0	WFF	C Whelan, 50 Garrick Cl, Ings Rd, Hull, HU8 0ST [IO93US, TA13]
G0	WFG	A E Vinters, 106 Halifax Rd, Ripponden, Sowerby Bridge, HX6 4AG [IO93AQ, SE02]
G0	WFH	C J Gresswell, 121 Granby Ct, Granby, Bletchley, Milton Keynes, MK1 1NG [IO92PA, SP83]
GM0	WFI	C M Young, Park Neuk, Banchory Devenick, Aberdeen, AB12 5XN [IO87WC, NJ90]
G0	WFJ	A Wills, 27 Fawns Cl, Ermington, Ivybridge, PL21 9NB [IO80BI, SX65]
G0	WFK	H M Bamford, Upper Twynings Farm, Pumphouse Ln, Hanbury, Droitwich, Worcs, WR9 7EB [IO82XG, SO96]
G0	WFL	M W Street, 262 Carter Knowle Rd, Sheffield, S7 2EB [IO93GI, SK38]
G0	WFM	P W Wallace, 8 Wemmick Cl, Rochester, ME1 2DL [JO01GI, TQ76]
G0	WFN	M J Rivers Swadelands School Radio Club, 153 Ashford Rd, Bearsted, Maidstone, ME14 4NE [JO01GG, TQ85]
G0	WFO	D J Tarry, 18 Harles Acres, Hickling, Melton Mowbray, LE14 3AF [IO92MU, SK62]
G0	WFP	W F Painz, Langcliffe House, Badger Ln, Blackshawhead, Hebden Bridge, HX7 7JX [IO83XR, SD92]
G0	WFQ	M L Troy, 90 Evenlode Rd, Millbrook, Southampton, SO16 9EH [IO90GW, SU31]
G0	WFR	R I Cooper East Anglia VHF DX Group, Post Office Stores, Little Oakley, Harwich, Essex, CO12 5JF [JO01OV, TM22]
G0	WFS	Details withheld at licensee's request by SSL.
G0	WFT	D F Saunders, 14 Shelton Ave, Toddington, Dunstable, LU5 6EL [IO91RW, TL02]
G0	WFU	M K Schroepfer, 3A Union Cres, Margate, CT9 1NR [JO01QJ, TR37]
G0	WFV	A J Corbett, 5 Mill Rise, Robertsbridge, TN32 5EF [JO00FX, TQ72]
G0	WFW	Details withheld at licensee's request by SSL.
G0	WFY	A J Robinson, 14 Parkancreeg, Carnon Downs, Truro, TR3 6HN [IO70KF, SW74]
G0	WGA	M A Merrin, 15 Hornsland Rd, Canvey Island, SS8 8LX [JO01HM, TQ88]
G0	WGB	J Atkins, 6 Carolina Gdns, Plymouth, PL2 2ER [IO70VJ, SX45]
G0	WGC	Details withheld at licensee's request by SSL.
G0	WGD	H E Ware, 77 Bucklands Rd, Teddington, TW11 9QS [IO91UK, TQ17]
GW0	WGE	D W J Thomas, 85 Tan y Bryn, Burry Port, SA16 0LD [IO71UQ, SN40]
G0	WGF	Details withheld at licensee's request by SSL.
GW0	WGG	I G Cripwell, Grassbanks, West Ln, Templeton, Narberth, Dyfed, SA67 8SX [IO71PS, SN11]
G0	WGH	J G Edwards, 21 Ridgeway, Ottery St. Mary, EX11 1DT [IO80IS, SY19]
G0	WGI	K Sherwin, 8 Meadow Way, Badbury, Swindon, SN4 0ET [IO91DM, SU18]
G0	WGJ	G M Jarvis, Rulow House, Buxton Old Rd, Macclesfield, SK11 0AG [IO83WG, SJ97]
GW0	WGK	Details withheld at licensee's request by SSL.
G0	WGL	Details withheld at licensee's request by SSL.
GW0	WGM	Details withheld at licensee's request by SSL.
GW0	WGN	Details withheld at licensee's request by SSL.
G0	WGO	C M Sharp, Peppermill Studio, 27 Pulham Ln, Wetwang, Driffield, YO25 9XT [IO94RA, SE95]
G0	WGP	R A Glover, 5 The Vineries, Burgess Hill, RH15 0ND [IO90WX, TQ31]
G0	WGQ	D P Snaith, 10 Greenwood Ave, Ilkeston, DE7 5PL [IO92IX, SK44]
G0	WGR	J J Orbell, Sundial Barn, The St., Bunwell, Norwich, NR16 1NA [JO02NM, TM19]
GI0	WGS	M J Rabbett, 41 Richill Park, Londonderry, BT47 1QY [IO64IX, C41]
G0	WGT	Details withheld at licensee's request by SSL.
G0	WGU	Details withheld at licensee's request by SSL.
G0	WGV	I H French, Little Shuffle, Tolgus Mount, Redruth, Cornwall, TR15 3TA [IO70JF, SW64]
GW0	WGW	B H James, 18 Duffryn St., Aberbargoed, Bargoed, CF8 9ET [IO81HN, ST08]
G0	WGX	Details withheld at licensee's request by SSL.
G0	WGY	D T Carr, 46 Homelea Cres, Lingwood, Norwich, NR13 4BW [JO02RO, TG30]
G0	WGZ	W G Metcalfe, 170 Ashlands Rd, Northallerton, DL6 1HD [IO94GI, SE39]
GI0	WHA	Details withheld at licensee's request by SSL.
G0	WHB	Details withheld at licensee's request by SSL.
G0	WHC	W H Chesterton, 61 Butt Ln, Blackfordby, Swadlincote, DE11 8BG [IO92FS, SK31]
G0	WHD	G Hassall, 47 Shadowmoss Rd, Manchester, M22 0LH [IO83UI, SJ88]
G0	WHE	Details withheld at licensee's request by SSL.
GM0	WHF	J A Butterworth, 9 McKenzie Rd, Buckie, AB56 1DH [IO87MQ, NJ46]
G0	WHG	Details withheld at licensee's request by SSL.
G0	WHH	Dr I G Burnside, 30 Lime Cres, Sandal, Wakefield, WF2 6RY [IO93GP, SE31]
GI0	WHI	Details withheld at licensee's request by SSL.
G0	WHJ	Details withheld at licensee's request by SSL.
G0	WHK	Details withheld at licensee's request by SSL.

G0 WHL E Barnes, 10 Cranbourne Rd, Rochdale, OL11 5JD [IO83VO, SD81]
GM0 WHM K Malone, 30 Komarom Pl, Dalkeith, EH22 2LT [IO85LV, NT36]
G0 WHN J C Edwardes, St. Ewe, Yawl Hill Ln, Uplyme, Lyme Regis, DT7 3XF [IO80MR, SY39]
G0 WHO R M Clutson, 151 Stepney Rd, Scarborough, North Yorks, YO12 5NJ [IO94SG, TA08]
G0 WHP Details withheld at licensee's request by SSL.
G0 WHQ K J Wilson, The Bedford, 34 St. Johns Rd, Buxton, SK17 6XL [IO93AG, SK07]
G0 WHS P Wood, Willow Garth, Sutton Ln, Barmby Moor, York, YO4 5HX [IO93NW, SE74]
GW0 WHT T A Roberts, 18 Ash Tree Cl, Radyr, Cardiff, CF4 8RX [IO81IM, ST18]
GW0 WHU J C Dart, 55 Caerau Rd, Caerau, Maesteg, CF34 0PB [IO81EP, SS89]
G0 WHV R A Bessell, 6 Bayford Lodge, Wellington Rd, Hatch End, Pinner, HA5 4NJ [IO91TO, TQ19]
G0 WHW M Freedman, 22 Stanhope Gdns, Ilford, IG1 3LQ [JO01AN, TQ48]
G0 WHX Details withheld at licensee's request by SSL.
G0 WHY P P White, Orchard Farm, 29 Romford Rd, Pembury, Tunbridge Wells, Kent, TN2 4JB [JO01DD, TQ64]
G0 WHZ Details withheld at licensee's request by SSL.
G0 WIA C J P J Cullington, 7 Cloudstock Gr, Little Hulton, Manchester, M38 0DU [IO83SM, SD70]
GM0 WIB M A McDermott, 69 High St., New Aberdour, Fraserburgh, AB43 6LD [IO87VP, NJ86]
G0 WIC E Clark, 8 Rose Cttgs, Shotton Colliery, Durham, DH6 2NF [IO94HS, NZ44]
G0 WID P J Gill, 1 Harbour View, Truro, TR1 1XJ [IO70LG, SW84]
G0 WIE P C Swan, 37 Heathfield Rd, London, W3 8EJ [IO91UM, TQ17]
G0 WIF T B Evans, Padside Green Barn, Padside, Summerbridge, Harrogate, HG3 4AL [IO94CA, SE15]
G0 WIG P Ellison, 28 Kingscote Rd East, Cheltenham, GL51 6JS [IO81WV, SO92]
G0 WIH T G Parker obo Dacorum AR Transmitting So, The Bungalow, 178 Green End Ln, Hemel Hempstead, HP1 2BQ [IO91SS, TL00]
G0 WII R Wolfe, 56 High St., Langford, Biggleswade, SG18 9RU [IO92UB, TL14]
G0 WIJ J L Gray, 4 Church Terr, Church Rd, East Harling, Norwich, NR16 2NA [JO02LK, TL98]
G0 WIL M J Williams, The Croft, Ringwood Rd, Bartley, Southampton, SO40 7LA [IO90FV, SU31]
GW0 WIM D G Crick, 29 Merlin Cres, Bridgend, CF31 4QW [IO81EM, SS88]
G0 WIN A H Jubb obo West Cambs Contest Group, 30 West St., Great Gransden, Sandy, SG19 3AU [IO92WE, TL25]
G0 WIO G S Monteiro, 123 Gadebridge Rd, Hemel Hempstead, HP1 3EN [IO91SS, TL00]
G0 WIP J M A Sheppard John Cabot Radio Club, 37 Oakfield Rd, Kingswood, Bristol, BS15 2NT [IO81RK, ST67]
G0 WIQ K E Stanhope, 19 Kirkstone Rd, Harrogate, North Yorks, HG1 4SJ [IO93FX, SE35]
G0 WIR G Egar, 24 Fords Ave, Healing, Grimsby, DN37 7RR [IO93VN, TA20]
G0 WIS G E Woodbury, 2 Park Way, Droitwich, WR9 9HE [IO82VG, SO86]
G0 WIT M E Perry, 52 Somerset Ave, Chessington, KT9 1PN [IO91UI, TQ16]
G0 WIU D P Ditchman, 88 Mongeham Rd, Great Mongeham, Deal, CT14 9PD [JO01QF, TR35]
G0 WIV T Y Zaim, 17A Blossom Ct, Discovery Bay, Hong Kong
G0 WIW N J Appleby, 25 Home Farm, Withcall, Louth, LN11 9RL [IO93XI, TF28]
G0 WIX A J Beeching, 174 Gr Rd, Rayleigh, SS6 8UA [JO01HN, TQ89]
G0 WIY W J King, 67 Greenacres Park, Meysey Hampton, Cirencester, GL7 5JH [IO91CQ, SP10]
G0 WIZ I C Waugh, 20 Heaton Way, Tiptree, Colchester, CO5 0DZ [JO01JT, TL81]
G0 WJA L Coleman, 1 Kerstin Cl, Cheltenham, GL50 4SA [IO81XV, SO92]
G0 WJB E Tait, 25 Stepney Dr, Scarborough, YO12 5DP [IO94SG, TA08]
G0 WJC M Briscoe, 34 Winterton Dr, Low Moor, Bradford, BD12 0UX [IO93CR, SE12]
G0 WJD D M Jackson, 44 Dulwich Rd, Holland on Sea, Clacton on Sea, CO15 5NA [JO01OT, TM11]
G0 WJE Details withheld at licensee's request by SSL.
G0 WJF J N Price, 7 Maresfield Gdns, London, NW3 5SJ [IO91VN, TQ28]
G0 WJG Details withheld at licensee's request by SSL.
G0 WJH J A Kemp, 85 St. Andrews Way, Church Aston, Newport, TF10 9JQ [IO82TS, SJ71]
GI0 WJI R Bicker, 62 Spa Rd, Ballynahinch, BT24 8PT [IO74BJ, J34]
G0 WJJ D P Mullaney, 62 Darby Rd, Wednesbury, WS10 0PN [IO82XN, SO99]
G0 WJK A Cobb, 37 Nordham, North Cave, Brough, HU15 2LT [IO93QS, SE83]
G0 WJM J W Cave, 14A Tillington Rd, Hereford, HR4 9QJ [IO82PB, SO44]
G0 WJN P I Howland, 21 Camp Hill Rd, Worcester, WR5 2HE [IO82VE, SO85]
G0 WJO P Bishop, Old Post House, East Bergholt, nr Colchester, Essex, CO7 6SE [JO01MX, TM03]
G0 WJP P W Swire, 2 Pimlott Gr, Prestwich, Manchester, M25 9TR [IO83UM, SD80]
G0 WJQ S Brackenbury, 66 Airmyn Rd, Goole, DN14 6XD [IO93NR, SE72]
G0 WJR R J Wilkinson, 84 Park Rd, Bolton, BL1 4RQ [IO83SN, SD70]
G0 WJS D Mellings, 36 Hillwood Dr, Glossop, SK13 8RJ [IO93AK, SK09]
G0 WJT Details withheld at licensee's request by SSL.
G0 WJU R Crewe, 11 Osbert Cl, Norwich, NR1 2NL [JO02PO, TG20]
G0 WJV C H Page, 73 Two Saints Cl, Hoveton, Norwich, NR12 8QR [JO02QR, TG31]
G0 WJW T J Newstead Bay Amateur Radio Group, 5 Farnlea Dr, Bare, Morecambe, LA4 6JU [IO84NB, SD46]
G0 WJX R A Davies, 84 Hob Hey Ln, Culcheth, Ches, WA3 4NW [IO83RK, SJ69]
GM0 WJY T Callaway, 10 Netherlaw, Grange Rd, North Berwick, EH39 4RF [IO86PB, NT58]
G0 WJZ I B Donachie, 72 Gresham Rd, Norwich, Norfolk, NR3 2QP [JO02PP, TG21]
G0 WKA M R Drever, 66 Milton Rd, Branton, Doncaster, DN3 3PB [IO93LM, SE60]
G0 WKB Details withheld at licensee's request by SSL.
G0 WKC Details withheld at licensee's request by SSL.
G0 WKE N M Goodkin, 27 Hermitage Way, Madeley, Telford, TF7 5SY [IO82SP, SJ60]
G0 WKF N T Ryder, 48 Standlake Ave, Hodge Hill, Birmingham, B36 8JS [IO92CM, SP18]
G0 WKG R A Tyrell, 28 St. Martins Gdns, Leeds, LS7 3LE [IO93FT, SE33]
G0 WKH M Thomas, 63 Corfe Way, Broadstone, BH18 9ND [IO80XS, SY99]
G0 WKI H Yearl, 191 Tamworth Rd, Kettlebrook, Tamworth, B77 1BT [IO92DO, SK20]
G0 WKJ C Towle, 53 Brampton Gr, Harrow, HA3 8LE [IO91UO, TQ18]
G0 WKK Details withheld at licensee's request by SSL.
G0 WKL R L Martin, Terpolders, School Ln, Denmead, Waterlooville, PO7 6NA [IO90KV, SU61]
G0 WKM N D Purchon, 54 Gondar Gdns, London, NW6 1HG [IO91VN, TQ28]
G0 WKN K R Whitmore, Amosford, Lutton Gowts, Lutton, Spalding, PE12 9LQ [JO02BT, TF42]
G0 WKP W Hunter, 36 Harvey Rd, Bishopstoke, Eastleigh, SO50 6GT [IO90IX, SU41]
G0 WKQ B Burdis The South Tyneside Amateur Rad, 10 Johnston Ave, Hebburn, NE31 2LJ [IO94FX, NZ36]
GW0 WKR Details withheld at licensee's request by SSL.
G0 WKS E Eaton, 7 St. Martins Cl, Poulton-le-Fylde, FY6 7NT [IO83LU, SD34]
G0 WKT M D Pugh, 4 Winster Ave, Dorridge, Solihull, B93 8ST [IO92CJ, SP17]
G0 WKU P D Yea, 89 Laxton Rd, Taunton, TA1 2XF [IO81LA, ST22]
G0 WKV Details withheld at licensee's request by SSL.
G0 WKW V Pleshkevich, 22 Chestnut Dr, Claverham, Bristol, BS19 4LN [IO81OJ, ST46]
G0 WKY D Hunt, 10 Glebe Way, Oakham, LE15 6LX [IO92PP, SK80]
G0 WKZ B Pybus, 29 Howick Park, Monkwearmouth, Sunderland, SR6 0AQ [IO94HV, NZ35]
G0 WLA Dr W Laidler, 8 Sunningdale, Orton Waterville, Peterborough, PE2 5UB [IO92UM, TL19]
G0 WLB Details withheld at licensee's request by SSL.
G0 WLC B Cole, 499 Lightwood Rd, Stoke on Trent, ST3 7EN [IO82WX, SJ94]
G0 WLD M C Russell, 62 Lonsdale Rd, London, SW13 9JS [IO91VL, TQ27]
GM0 WLE J Stewart, 118 Glen Tennet, St. Leonards, East Kilbride, Glasgow, G74 3UY [IO75WS, NS65]
G0 WLF W W Storace-Rutter, 46 Norbury Ct Rd, London, SW16 4HT [IO91WJ, TQ36]
G0 WLG P Blizzard, 107 Pickersleigh Gr, Malvern, WR14 2LU [IO82UC, SO74]
GW0 WLH Dr D Shewring, 12A Hollubush Rd, Cyncoed, Cardiff, South Glam, CF2 6TA [IO81KM, ST18]
GW0 WLI J G Ruddle, 80 Woolaston Ave, Lakeside, Cardiff, CF2 6HA [IO81JM, ST18]
G0 WLJ J M Lees, 30 Clowes Ave, Alsager, Stoke on Trent, ST7 2RL [IO83UC, SJ85]
G0 WLK D Dunn, 27 Alexandra Rd, Fullbrook, Walsall, WS1 4DX [IO92AN, SP09]
G0 WLL Details withheld at licensee's request by SSL.
GW0 WLN C Purcell, 99 Norton Bridge, Pontypridd, Mid Glam, CF37 4ND [IO81IO, ST09]
G0 WLO R W McKay, 14 Clarefield Ct, North End Ln, Ascot, SL5 0EA [IO91QJ, SU96]
G0 WLP E Pearce, 6 Rotherham Rd, Clowne, Chesterfield, S43 4PS [IO93IG, SK47]
GW0 WLQ R Evans, 4 Llantrisant Rd, Tonyrefail, Porth, CF39 8PP [IO81GN, ST08]
G0 WLR Dr M W Pettigrew, 41 Wostenholm Rd, Sheffield, S7 1LB [IO93GI, SK38]
G0 WLS B K Chamberlain, 2 Smiths Walk, Oulton Broad, Lowestoft, NR33 8QN [JO02UL, TM59]
G0 WLT Details withheld at licensee's request by SSL.
G0 WLU Details withheld at licensee's request by SSL.
G0 WLV L M J Turton, 16 Norton Park Dr, Norton, Sheffield, S8 8GP [IO93GH, SK38]
GI0 WLW Rev H A Quinn, Parochial House, 20 The Diamond, Pomeroy, Dungannon, BT70 2QX [IO64MO, H67]
G0 WLX A Suckling, 20 Palmerston Cres, Palmers Green, London, N13 4UA [IO91WO, TQ39]
GW0 WLZ Cpt D J Melhuish, 1323 SE 17th St, #329, Fort Lauderdale, Florida 33316, USA
G0 WMA L J Andrews, 19 The Becks, Alvechurch, Worcs, B48 7NE [IO92AI, SP07]
G0 WMB A E Houghton, 2 Beanhill Cres, Alveston, Bristol, BS12 2JG [IO81RO, ST68]
G0 WMC P J Norman, 18 Everest Pl, Swanley, BR8 7BX [JO01CJ, TQ56]
G0 WMD M A de Silva, 40 PO Box, Hounslow, TW4 7JB [IO91TL, TQ17]
G0 WME Details withheld at licensee's request by SSL.
G0 WMF Details withheld at licensee's request by SSL.
G0 WMG G J Studd, 34 The Broadway, Lancing, BN15 8NY [IO90UT, TQ10]
G0 WMH A E Hughes, Flat 2, 43 Church Rd, Ramsgate, Kent, CT11 8RD [JO01RI, TR36]
G0 WMJ J Walker, 249 Mosley St., Blackburn, BB2 3RX [IO83SR, SD62]
GM0 WMK Details withheld at licensee's request by SSL.
G0 WML D S Tyler, 7 Stephens Cres, Horndon on The Hill, Stanford-le-Hope, SS17 8LZ [JO01EM, TQ68]
G0 WMM M A Pickstock, 35 Portsdown Ave, Drayton, Portsmouth, PO6 1EL [IO90LU, SU60]
G0 WMN W McNab, 74 Parkhouse Rd, Minehead, TA24 8AF [IO81GE, SS94]
G0 WMO S G Goulding, 117 Greystone Rd, Broadgreen, Liverpool, L14 6UF [IO83NJ, SJ49]
G0 WMP M J Mullins, Ivy Cottage, 113 Endlebury Rd, London, E4 6PX [IO91XP, TQ39]
G0 WMQ E R Williams, 24 Oxton Rd, Wallasey, L44 4EU [IO83LJ, SJ39]
G0 WMR R Stringfellow Red Rose Amateur Radio Group, 48 Richmond Rd, Eccleston, Chorley, PR7 5SR [IO83PP, SD51]
G0 WMS S M Williams, 11 Cedar Ave, Sleaford, NG34 8BW [IO93TA, TF04]
GW0 WMT S J Robinson, 341 Caerphilly Rd, Birchgrove, Cardiff, CF4 4QF [IO81JM, ST18]
G0 WMU R Davis, 49 Goodway Rd, Great Barr, Birmingham, B44 8RL [IO92BM, SP09]
G0 WMV M P Grubb, Tregenver House, Flamouth, Cornwall, TR11 2QS [IO70KD, SW73]
G0 WMX G G Giuliani, 32 Davison St., Newburn, Newcastle upon Tyne, NE15 8NB [IO94DX, NZ16]
GW0 WMY P J Hanmer, 14 Maes Geraint, Pentraeth, LL75 8UR [IO73VG, SH57]
G0 WMZ R Duckworth, 79 Werneth Rd, Woodley, Stockport, SK6 1HR [IO83XK, SJ99]
G0 WNA Details withheld at licensee's request by SSL.
GW0 WNB S G Blumson, 15 Bryn Colwyn, Colwyn Bay, LL29 9LJ [IO83DG, SH87]
G0 WNC Details withheld at licensee's request by SSL.
G0 WND T Sunouchi, Panasonic Uk/Beg, Panasonic House, Willoughby Rd, Bracknell, Berks, RG12 8FP [IO91OJ, SU86]
G0 WNE K S G Allack Oakbank School Keighley, 10 Goose Eye, Oakworth, Keighley, BD22 0PD [IO93AU, SE04]
G0 WNF E Clark, 58A Church St., Whitby, YO22 4AS [IO94QL, NZ91]
G0 WNH Details withheld at licensee's request by SSL.
G0 WNI J C E Bones, Lower Ware, Ware Ln, Lyme Regis, DT7 3EL [IO80MR, SY39]
G0 WNJ M Bottomley, 21 Priory Park, Grosmont, Whitby, YO22 5QQ [IO94PK, NZ80]
G0 WNK J D Bowman, Darvell, Robertsbridge, E Sussex, TN32 5DR [JO00FX, TQ72]
G0 WNL M R Trotman White Noise Listeners Contest, 44 Armscroft Rd, Barnwood, Gloucester, GL2 0SJ [IO81VU, SO81]
GM0 WNM D M Livingston, Zalongou 9, Arta 47100, Greece
G0 WNN Details withheld at licensee's request by SSL.
G0 WNO Details withheld at licensee's request by SSL.
G0 WNP M D Wilson, 28 The Approach, Scholes, Leeds, LS15 4AN [IO93GT, SE33]
G0 WNQ Details withheld at licensee's request by SSL.
GM0 WNR A G Campbell, 17 Arran Rd, Motherwell, ML1 3NA [IO75XT, NS75]
GM0 WNS I A M Calder, 2 Nechtan Pl, Letham, Forfar, DD8 2LA [IO86OP, NO54]
G0 WNT Details withheld at licensee's request by SSL.
G0 WNU F J Havard Hollins High Sch Radio Club Ac, Whitehalgh Farm, Whitehalgh Ln, Langho, Blackburn, BB6 8ET [IO83ST, SD73]
G0 WNV Details withheld at licensee's request by SSL.
GW0 WNW Details withheld at licensee's request by SSL.
G0 WNX P S Dye, 17 Thorpes Ave, Denby Dale, Huddersfield, HD8 8SP [IO93EN, SE20]
G0 WNY R K Dressel, Flat 4, 62 Windsor Rd, Oldham, OL8 4AL [IO83WM, SD90]
G0 WNZ A N Sharp, 18 Pemerton Rd, Weeke, Winchester, SO22 6EU [IO91IB, SU43]
G0 WOA M Lyon, 182 Wilmslow Rd, Heald Green, Cheadle, SK8 3BG [IO83VI, SJ88]
GM0 WOB P Cox, 54 Queens Dr, Troon, KA10 6SE [IO75QN, NS33]
G0 WOC J Doxey, Armouth House, Tinderbox Ln, Burnaston, Etwell Derby, DE6 6LG [IO92DX, SK24]
G0 WOD Details withheld at licensee's request by SSL.
G0 WOE J S Model, 7No71 Willow St, Itasca, Illinois, USA, X X
G0 WOF T R Webster, 72 Thornley Rd, Eatontown, New Jersey, 07724, X X
G0 WOH D C Zulawski, 2808 Catnip St, El Paso, Texas, 79925, X X
G0 WOI B F H Wood Avon Scouts Amateur Radio Club, 193 Robin Way, Chipping Sodbury, Bristol, BS17 6JU [IO81TM, ST78]
G0 WOJ D H Wilkinson, 39 Boulsworth Cres, Nelson, BB9 8DF [IO83VU, SD83]
G0 WOK Details withheld at licensee's request by SSL.
G0 WOL W E Woodnoth, 171 Rosecroft Dr, Nottingham, NG5 6EL [IO92KX, SK54]
G0 WOM T M Renshaw, 1 Basford View, Cheddleton, Leek, ST13 7HJ [IO83XB, SJ95]
G0 WON M V G Kelly, 9 Thetford Rd, Brandon, IP27 0BS [JO02HK, TL78]
GI0 WOO J P A Sherlock Belfast Royal Academy A R C, Woodbank, 451 Shore Rd, Whiteabbey, Newtownabbey, BT37 9SE [IO74BQ, J38]
G0 WOP P Dabell, 21 Tatton Rd North, Heaton Moor, Stockport, SK4 4RL [IO83VK, SJ89]
G0 WOQ N R Goodkin, 27 Hermitage Way, Madeley, Telford, TF7 5SY [IO82SP, SJ60]
G0 WOS J Godfrey, Flat 2, 4 Kings Ave, Ealing, London, W5 2SH [IO91UM, TQ18]
G0 WOT Details withheld at licensee's request by SSL.
G0 WOU D T Atkinson, 596 Wolseley Rd, St. Budeaux, Plymouth, PL5 1UX [IO70VJ, SX45]
G0 WOV F G Barnes, 3 Unity St., Chippenham, SN14 0AR [IO81WK, ST97]
GI0 WOW R A Cooper, 69 Antrim St., Lisburn, BT28 1AU [IO64XM, J26]
G0 WOX P V P Perera Cardinal Wiseman RC School ARC, 13 Dalcross Rd, Hounslow, TW4 7RA [IO91TL, TQ17]
G0 WOY Details withheld at licensee's request by SSL.
G0 WOZ J J Hunter, 58 Ecos Ct, Frome, BA11 1HZ [IO81UF, ST74]
G0 WPA Details withheld at licensee's request by SSL.
G0 WPB R G Tee, 19 Westfield Terr, Westfield, Radstock, Bath, BA3 3UT [IO81SG, ST65]
G0 WPC N I Crossley, 78 Plumbley Hall Rd, Mosborough, Sheffield, S20 5BL [IO93HH, SK48]
G0 WPF J G Towle, 36 Old Orchard, Haxby, York, YO3 3DT [IO94LA, SE65]
GW0 WPG Details withheld at licensee's request by SSL.
G0 WPH P Howard, 162 High St., Hanham, Bristol, BS15 3HH [IO81RK, ST67]
GM0 WPI Details withheld at licensee's request by SSL.
G0 WPJ Details withheld at licensee's request by SSL.
G0 WPK W J Jones, 9 Ribston Cl, Worcester, WR2 6EH [IO82VE, SO85]
G0 WPL J R Wozniak, 40 Cockhill, Trowbridge, BA14 9BQ [IO81VH, ST85]
G0 WPM N R Gilboy, 6 Talisman Cl, Sherburn, Durham, DH6 1RJ [IO94GT, NZ34]
G0 WPN A Marshall, 16 Gr Rd, Chesterfield, S41 8LN [IO93GG, SK37]
G0 WPO N Griffiths, 14 Howarth Cross St., Rochdale, OL16 2PB [IO83WP, SD91]
G0 WPP Details withheld at licensee's request by SSL.
GM0 WPQ Details withheld at licensee's request by SSL.
GM0 WPR Details withheld at licensee's request by SSL.
G0 WPS R J Leeds Cromer High School Radio Club, Sunholme, The Green, Aldborough, Norwich, NR11 7AA [JO02OU, TG13]
GW0 WPT H Griffiths, 18 Maes Cynfaen, Brynford, Holywell, CH8 8LA [IO83JG, SJ17]
GM0 WPU K M Faloon, Moss-Side Croft, 6 Rothiemay, Huntley, Aberdeenshire, AB54 5NY [IO87PK, NJ54]
GI0 WPV O Price, 2 Meadow Ct, Newtownards, BT23 8YE [IO74DN, J46]
GM0 WPW T M Halligan, 7 Little Spott Cttgs, Dunbar, East Lothian, EH42 1RH [IO85RX, NT67]
G0 WPX S Telenius-Lowe Wpx Contest Group, Belvista, 27 Hertford Rd, Stevenage, SG2 8RZ [IO91VV, TL22]
GM0 WPY N Mohammed, Anadrill, Kirkton Ave, Dyce, Aberdeen, AB2 0UB [IO87TH, NJ72]
G0 WPZ Details withheld at licensee's request by SSL.
G0 WQA J A Ward, 49 Woodhall Dr, Ermine East, Lincoln, LN2 2AE [IO93RG, SK97]
G0 WQB P Dixon, 15 Church Cl, Sharow, Ripon, HG4 5BL [IO94FD, SE37]
G0 WQC J M Keeling, 31 Tudor Dr, Otford, Sevenoaks, TN14 5QP [JO01CH, TQ55]
G0 WQD Details withheld at licensee's request by SSL.
GW0 WQE S A Mason, 25 Hawarden Way, Mancot, Deeside, CH5 2EL [IO83LE, SJ36]
G0 WQH G E Pooler, 18 Johnstone Cl, Moss Rd, Wrockwardine Wood, Telford, TF2 7DA [IO82SR, SJ61]
G0 WQI Details withheld at licensee's request by SSL.
G0 WQJ Details withheld at licensee's request by SSL.
G0 WQK E Spencer, 8 Station Rd, Drayton, Portsmouth, PO6 1PH [IO90LU, SU60]
G0 WQL A C Bird, 115 Windsor Rd, Carlton in Lindrick, Worksop, S81 9DH [IO93KI, SK58]
G0 WQM M Atkinson, 26 Skipton Rd, Trawden, Colne, BB8 8QS [IO83WU, SD93]
G0 WQN C H Orchard, The Diver, Ashey Rd, Rhyde, Isle of Wight, PO33 4AY [IO90JQ, SZ58]
GW0 WQP D Taylor, Cilgwyn, Wellfield Rd, Marshfield, Cardiff, CF3 8UB [IO81LM, ST28]
G0 WQQ D E K Bennett, Shrove Furlong, Longwick Rd, Princes Risborough, HP27 9HE [IO91OR, SP80]
G0 WQR J Young, 19 Wycombe Rd, Princes Risborough, HP27 0EE [IO91NR, SP80]
G0 WQS Details withheld at licensee's request by SSL.
G0 WQT Details withheld at licensee's request by SSL.
G0 WQU C Costis, Apollo Court, Shop C, 232 Makarios Ave, Limassol, Cyprus
GW0 WQV Details withheld at licensee's request by SSL.
G0 WQW J R Thompson, 24 Shakespeare Rd, London, W7 1LR [IO91TM, TQ18]
GW0 WQX Details withheld at licensee's request by SSL.
G0 WQY L Mansfield, 25 Carlton Rd, Derby, DE23 6HB [IO92GV, SK33]
G0 WQZ S M Godbold, 81 Blenheim St., Princes Ave, Hull, HU5 3PR [IO93TS, TA02]
G0 WRA J Hatch Wincanton Amateur Radio Club, Peveril, 42 Bowden Rd, Templecombe, BA8 0LF [IO80TX, ST72]
G0 WRB J L Walmsley, PO Box 1637, Yeovil, Somerset, BA21 4YF [IO80QW, ST51]
G0 WRC L J Volante Wythall Contest Group, 200 Longmore Rd, Shirley, Solihull, B90 3EX [IO92CJ, SP17]
G0 WRD H Taylor, 70 Beecher St., Halesowen, B63 2DP [IO82XK, SO98]
G0 WRE P Scarratt, 339 Utting Ave East, Norris Green, Liverpool, L11 1DF [IO83NK, SJ39]
G0 WRF S Walford, 49 Dempster Ave, Goole, DN14 5RZ [IO93NQ, SE72]
G0 WRG B H L Lowe Walsall Raynet Group, 19 Wolverhampton Rd, Bloxwich, Walsall, WS3 2EZ [IO82XO, SJ90]
GM0 WRH E A Castle, 38 Davieland Rd, Giffnock, Glasgow, G46 7LU [IO75UT, NS55]
GW0 WRI R Lewis, Sycamore House, 26 Bryn Rd, Upper Brynamman, Ammanford, SA18 1AU [IO81BT, SN71]
GW0 WRJ Details withheld at licensee's request by SSL.
G0 WRK P M Taylor, Old Acres, Priory Rd, Ilchester, Yeovil, BA22 8NY [IO81PA, ST52]
G0 WRL Details withheld at licensee's request by SSL.
G0 WRM M R White Lichfield Raynet Group, 38 Tamworth Rd, Amington, Tamworth, B77 3BT [IO92DP, SK20]

G0 WRN J Hodges, 48 Beach Rd, Severn Beach, Bristol, BS12 3PF [IO81QN, ST58]
G0 WRO Details withheld at licensee's request by SSL.
G0 WRP Details withheld at licensee's request by SSL.
G0 WRQ O R Baxter, 5 Church Path, Bridgwater, TA6 7AJ [IO81LD, ST23]
GM0 WRR J Scott, 70 Montford Ave, Glasgow, G44 4PA [IO75VT, NS56]
G0 WRS M E Isherwood Warrington Amateur Radio Club, 52 St. Bridgets Cl, Cinnamon Brow, Fearnhead, Warrington, WA2 0EW [IO83RJ, SJ69]
G0 WRT S P Winfield, 150 Tinshill Rd, Cookridge, Leeds, LS16 7PN [IO93EU, SE23]
GM0 WRU R S Holmes, 50 Branchalmuir Cres, Newmains, Wishaw, ML2 9DY [IO85BS, NS85]
GM0 WRV R W Spink, 2 Cormorant Brae, Cove, Aberdeen, AB12 3WH [IO87WC, NJ90]
GW0 WRW Dr P Evans, Maes y Gr, Capel Dewi, Carmarthen, Dyfed, SA32 8AH [IO71VU, SN41]
G0 WRX J A Boyer, 33 Great Ellshams, Banstead, SM7 2BA [IO91VH, TQ25]
GM0 WRY S Koncz, Cayman, 8 Inglis Farm, Cockenzie, Prestonpans, EH32 0JT [IO85MX, NT47]
G0 WRZ S W Fry, 151 Rowney Ave, Wimbish, Saffron Walden, CB10 2YE [JO01DX, TL53]
G0 WSB M Brimley, 42 Grange Rd, Netley Abbey, Southampton, SO31 5FE [IO90HV, SU40]
G0 WSC R G Connett, 15 Channels Ln, Horton, Ilminster, TA19 9QL [IO80MW, ST31]
G0 WSD G E D Swann, 1 Beaver Cl, Fishbourne, Chichester, PO19 3QU [IO90OU, SU80]
GW0 WSE Details withheld at licensee's request by SSL.
G0 WSF G Schnapp, 8 Common Field, Wigginton, Tring, HP23 6EW [IO91QS, SP91]
GM0 WSG Details withheld at licensee's request by SSL.
G0 WSH Details withheld at licensee's request by SSL.
G0 WSI Details withheld at licensee's request by SSL.
G0 WSJ Details withheld at licensee's request by SSL.
G0 WSK Details withheld at licensee's request by SSL.
G0 WSL J W Sharp, 26 Longstock Cres, Totton, Southampton, SO40 8ED [IO90FV, SU31]
G0 WSM Details withheld at licensee's request by SSL.
GM0 WSN Details withheld at licensee's request by SSL.
G0 WSO B T Marshall, 24 Braemar Ave, Dunblane, FK15 9ED [IO86AE, NN70]
G0 WSP P Croft, 82 Granby Rd, Buxton, SK17 7TJ [IO93BF, SK07]
G0 WSQ Details withheld at licensee's request by SSL.
GM0 WSR D I Mackinnon Strathclyde Regional Raynet Gr, 5 Octavia Terr, Greenock, PA16 7SP [IO75OX, NS27]
G0 WSS Details withheld at licensee's request by SSL.
G0 WST Details withheld at licensee's request by SSL.
GW0 WSU H I Jones, 35 Davis Ave, Bryncethin, Bridgend, CF32 9JJ [IO81FN, SS98]
G0 WSV Details withheld at licensee's request by SSL.
G0 WSW Dr M Farrugia, 66 Clarence Mews, London, SE16 1GD [IO91XM, TQ37]
G0 WSX B M Winlow, 58 Postern Cres, Morpeth, NE61 2JN [IO95DD, NZ18]
G0 WSY J R Adams, 4 Avondale Rd, Farnworth, Bolton, BL4 0PA [IO83SN, SD70]
G0 WTA N C Midworth, 1 Highfields Dr, Loughborough, LE11 3JS [IO92JS, SK51]
G0 WTB C Hicks, 59 Elmsfield Ave, Norden, Rochdale, OL11 5XW [IO83VO, SD81]
G0 WTC W C T Clark, 41 Brook Cl, Jarvis Brook, Crowborough, TN6 2ET [JO01CB, TQ53]
G0 WTD T J H Stokes, 33 Talbot Dr, Euxton, Chorley, PR7 6PD [IO83QP, SD51]
G0 WTF A P Wright, 12 Hawbush Rise, Welwyn, AL6 9PN [IO91VT, TL21]
G0 WTG D P Way, 56 Fernside Rd, Poole, BH15 2JJ [IO90AR, SZ09]
G0 WTH Details withheld at licensee's request by SSL.
G0 WTK R A Jenkins, 13 Baulk Ln, Worksop, S81 7DF [IO93KH, SK58]
G0 WTL G Fisher, 6 Totternhoe Rd, Dunstable, LU6 2AG [IO91RV, TL02]
G0 WTM D R Sutton, 32 Queensway, Euxton, Chorley, PR7 6PW [IO83QP, SD51]
G0 WTN Details withheld at licensee's request by SSL.
G0 WTO E M Birch, 14 Duchy Cl, Chelveston, Wellingborough, NN9 6AW [IO92RH, SP96]
GM0 WTP J Whitecross, 3/5 Royston Mains Cres, Edinburgh, EH5 1NQ [IO85JX, NT27]
G0 WTQ W G Pearce, 52 Treveneague Gdns, Plymouth, PL2 3SX [IO70WJ, SX45]
G0 WTR M E Robertson, 21 Battles Ln, Kesgrave, Ipswich, IP5 7XF [JO02OB, TM24]
GW0 WTT Details withheld at licensee's request by SSL.
G0 WTU D F Baldwin, 1 Pound Pl, East St., Bovey Tracey, Newton Abbot, TQ13 9EJ [IO80DO, SX87]
G0 WTV P A Beeston, 2 Copners Way, Holmer Green, High Wycombe, HP15 6SQ [IO91PP, SU89]
G0 WTW Details withheld at licensee's request by SSL.
G0 WTX Details withheld at licensee's request by SSL.
G0 WTZ Details withheld at licensee's request by SSL.
G0 WUA P J Harris, 1 Newlands, Landkey, Barnstaple, EX32 0NJ [IO81AB, SS63]
G0 WUB T H Best, 3 Gordon Rd, Topsham, Exeter, EX3 0LJ [IO80GQ, SX98]
G0 WUC N D Dentamaro, PO Box 5274, Limassol, Cyprus, 3820
G0 WUD S T Smith, The Palm Tree, Beacon Ln, Woodnesborough, Sandwich, CT13 0PA [JO01PG, TR35]
G0 WUE S J Smith, The Palm Tree, Beacon Ln, Woodnesborough, Sandwich, CT13 0PA [JO01PG, TR35]
G0 WUG G M Bishop, 8 Bulstrode Pl, Kegworth, Derby, DE74 2DS [IO92IU, SK42]
G0 WUH H E White, 173 Centenary Ave, Marsden, South Shields, NE34 6SQ [IO94HX, NZ36]
G0 WUI W N Cassidy, 17 Catcheside Cl, Whickham, Newcastle upon Tyne, NE16 5RX [IO94DW, NZ26]
G0 WUJ Details withheld at licensee's request by SSL.
G0 WUK R P James Ping Jockey's Contest Group, 4 Pentland Pl, Bearsden, Glasgow, G61 4JU [IO75TW, NS57]
GW0 WUL C Minard, 3 Riverside Cl, Aberfan, Merthyr Tydfil, CF48 4RN [IO81HQ, SO00]
GW0 WUM E L Roobottom, 2 Ddol Cttgs, Abergwyngregyn, Llanfairfechan, Gwynedd, LL33 0LH [IO73XF, SH67]
G0 WUN C C Naldoken, 43 Belgrave Sq, London, SW1 8PA [IO91WL, TQ27]
G0 WUO R A Berkeley, 4 Raleigh Rd, Leasowe, Moreton, Wirral, L46 2QZ [IO83KK, SJ29]
G0 WUP W F A Steele, 77 Ashtree Rd, Frome, BA11 2SE [IO81UF, ST74]
G0 WUQ J Steele, 77 Ashtree Rd, Frome, BA11 2SE [IO81UF, ST74]
G0 WUR G W D Steele, 77 Ashtree Rd, Frome, BA11 2SE [IO81UF, ST74]
G0 WUT Details withheld at licensee's request by SSL.
G0 WUU J G Purcell, 22 Thelton Ave, Broadbridge Heath, Horsham, RH12 3LS [IO91TB, TQ13]
G0 WUV J R Dickinson, 112 Stoneleigh Ave, Newcastle upon Tyne, NE12 8XQ [IO95EA, NZ26]
G0 WUW V J Claridge, 88 Sandyleaze, Gloucester, GL2 0PX [IO81VU, SO81]
GM0 WUX J B Donnan, 41 Annick Dr, Dreghorn, Irvine, KA11 4ER [IO75QO, NS33]
G0 WUY A Williamson, 8 Farrar St., York, YO1 3BZ [IO93LW, SE65]
G0 WUZ R F Rous, 30 Cooks Spinney, Harlow, CM20 3BJ [JO01BS, TL41]
G0 WVA Details withheld at licensee's request by SSL.
G0 WVB D G Still AR Group TS Vindicatrix Associ, 25 Brooks Green Park, Emms Ln, Brooks Green, Horsham, RH13 8QR [IO91TA, TQ12]
G0 WVD T A Lishman, Applegarth, 13 Meadow Way, Sandown, PO36 8QE [IO90KP, SZ68]
G0 WVE B Richardson, 12 Stoney Ln, Barrow, Bury St. Edmunds, Suffolk, IP29 5DD [JO02HF, TL76]
GW0 WVF D J Collins, 3 Bryncoed Terr, Penpedairheol, Hengoed, CF8 8DD [IO81HN, ST08]
G0 WVG A J Stirling, 111 Elm Rd, Thetford, IP24 3HL [JO02IJ, TL88]
G0 WVI Details withheld at licensee's request by SSL.
G0 WVJ N D Purchon Kingsbury High School Amateur, 54 Gondar Gdns, London, NW6 1HG [IO91VN, TQ28]
G0 WVK Details withheld at licensee's request by SSL.
GW0 WVL M Jones, 61 Mount Pleasant Est, Brynithel, Abertillery, NP3 2HU [IO81KQ, SO20]
G0 WVM M G Brooker, 18 Yoells Ln, Waterlooville, PO8 9SP [IO90LV, SU61]
GI0 WVN R Finlay, 7 The Mews, 160 Groomsport Rd, Bangor, Co. Down, BT20 5QP [IO74EP, J58]
G0 WVO R Merritt, 126 Mullway, Letchworth, SG6 4BE [IO91VX, TL23]
G0 WVQ H Thomas, Jm Karkamp 30, 45259, Essen
G0 WVT W J Carwood, 57 Upton Rd, Kidderminster, DY10 2YB [IO82VJ, SO87]
G0 WVU Details withheld at licensee's request by SSL.
G0 WVV D I Jones, 4 Back Bower Ln, Gee Cross, Hyde, SK14 5NS [IO83XK, SJ99]
G0 WVW D I Webb, 55 Dudley Rd, Ellesmere Port, South Wirral, L65 8DQ [IO83NG, SJ37]
G0 WVX Details withheld at licensee's request by SSL.
G0 WVY F A Holt, 22 First Ave, Tottington, Bury, BL8 3JA [IO83TO, SD71]
G0 WVZ Details withheld at licensee's request by SSL.
G0 WWA D M Nicholas, 1 Windsor Rd, Millfields Est, West Bromwich, B71 2NX [IO82XN, SO99]
G0 WWC W McManus, 85 July Rd, Anfield, Liverpool, L6 4BS [IO83MK, SJ39]
G0 WWD D W Weston, 25 Ley Ln, Kingsteignton, Newton Abbot, TQ12 3JE [IO80EN, SX87]
G0 WWE S E Fitton, 14 Fairview Cl, Norden, Rochdale, OL12 7SR [IO83VP, SD81]
G0 WWF P Baron, 6 Silver Hill Rd, Hyde, SK14 5QA [IO83XK, SJ99]
G0 WWG J W Brett, Willow Farm, Wangfield Ln, Curdridge, Southampton, Hants, SO3 2DA [IO90HW, SU41]
G0 WWH A Houghton, 193 Skagen Ct, Bolton, BL1 2JF [IO83SO, SD71]
G0 WWJ Details withheld at licensee's request by SSL.
G0 WWK C E Parker, 10 Linden Ave, Sheffield, S8 0GA [IO93GI, SK38]
G0 WWL A D Babbage, 10 Heather Cl, Honiton, EX14 8YP [IO80JS, SY19]
G0 WWM Prof T Yukawa, #206-20-8-1-Chome, Fukumitsu-Nishi, Gifu-5O2, Japan, 50E
G0 WWO J B Otter, 97 Hilltop Rd, Dronfield, Sheffield, S18 6UN [IO93GH, SK37]
G0 WWP S J N Griffith, 12 Beech Gr, Cliffsend, Ramsgate, CT12 5LD [JO01QH, TR36]
G0 WWQ H R Burton, 137 Markland Rd, Dover, CT17 9NL [JO01PC, TR24]
G0 WWR Details withheld at licensee's request by SSL.
G0 WWS Details withheld at licensee's request by SSL.
G0 WWT J R F Smith, 54 Greenfield Ave, Kettering, NN15 7LL [IO92PJ, SP87]
G0 WWU R Pratt, 4 King John Ave, Gaywood, Kings Lynn, PE30 4QA [JO02FS, TF62]
G0 WWV Details withheld at licensee's request by SSL.
G0 WWW M Nicolaou, 15 Willow Rd, Carlton, Nottingham, Notts, NG4 3BH [IO92LX, SK64]
GM0 WWX R W Johnstone, 8 Harris Ct, Alloa, FK10 1DD [IO86CC, NS89]
GW0 WWY N D Hughes, 2 Brynheulog Terr, Aberaman, Aberdare, CF44 6EW [IO81GQ, SO00]
G0 WWZ M G Charlesworth, 6 Curzon Ave, Enfield, EN3 4UD [IO91XP, TQ39]
G0 WXA S Everitt, 125 Victoria Rd, Oldbury, Warley, B68 9UL [IO92AL, SP08]
G0 WXB Details withheld at licensee's request by SSL.
G0 WXC R Needham, 1 The Leas, Sedgefield, Stockton on Tees, TS21 2DS [IO94GP, NZ32]
G0 WXD P Atkinson, 27 Ranworth Rd, Bramley, Rotherham, S66 0SP [IO93IK, SK49]
G0 WXE A Kelleher, 144 Alibon Rd, Dagenham, RM10 8DE [JO01BN, TQ48]
G0 WXF P F Wright, 60 Farnborough Rd, Clifton Est, Nottingham, NG11 8GF [IO92JV, SK53]
G0 WXG J P Lamb, 45 Spen View, Heckmondwike Rd, Dewsbury Moor, Dewsbury, WF13 3PZ [IO93EQ, SE22]
G0 WXH S G Croot, 15 Barker Ave, Jacksdale, Nottingham, NG16 5JH [IO93IB, SK45]
G0 WXI G A Osler, 4 Mill Rd, Waterbeach, Cambridge, CB5 9RQ [JO02CG, TL46]
G0 WXJ P H Badham, 7 Bucklewood, Tintern Ave, Worcester, Worcs, WR3 8EJ [IO82VF, SO85]
G0 WXK P Holdway, 63 Russell Cres, Sleaford, Lincs, NG34 7JF [IO92TX, TF04]
G0 WXL E Harper, 160 Manwood Rd, Crofton Park, London, SE4 1SE [IO91XK, TQ37]
G0 WXM Details withheld at licensee's request by SSL.
G0 WXN M I Thomas, 1 Plantaganet Pl, Broomfield Rd, Romford, RM6 6JU [JO01BN, TQ48]
G0 WXP S D Parkes, 8 Woodlands Rd, Woodlands, Doncaster, DN6 7JX [IO93JN, SE50]
G0 WXQ G M Beech Lace Web Amateur Radio Club, 15 Fisher Ct, Cotmanhay, Ilkeston, DE7 8PZ [IO92IX, SK44]
G0 WXR Details withheld at licensee's request by SSL.
G0 WXS I E Taylor, 41 Sovereign Dr, Botley, Southampton, SO30 2SR [IO90IV, SU51]
G0 WXT Details withheld at licensee's request by SSL.
G0 WXU J M Cliff, 8 St. Johns Walk, Adwick upon Dearne, Mexborough, S64 0NS [IO93IM, SE40]
G0 WXV M Smith, Flat 1, 51 Fitzjohns Ave, London, NW3 6PH [IO91VN, TQ28]
GW0 WXW P B Higgs Marford& District Amateur Radi, Parkside, Rossett, Wrexham, Clwyd, LL12 0BP [IO83NC, SJ35]
GM0 WXX F A Mackay, Shalimar, Alligin, Achnasheen, Ross Shire, IV22 2HB [IO77EN, NG85]
G0 WXY H Davies, 41 St. Clements Ct, Pendle St., Barrowford, Nelson, BB9 8QU [IO83VU, SD83]
G0 WXZ D J Milne, 6 Sunnyhill Rd, Southbourne, Bournemouth, BH6 5HW [IO90CR, SZ19]
G0 WYA N R Taylor, 61 Oldbury Rd, Rowley Regis, Warley, B65 0NP [IO82XL, SO98]
GI0 WYB J Simpson, 3 Gordonville Park, Ballymoney, BT53 7EU [IO65SB, C92]
G0 WYC B Fleury, Little Croft, Bennington Rd, Walkern, Herts, SG2 7HX [IO91WV, TL22]
G0 WYD R S Coleman, 8 The White House, Bourne Rd, Tidworth, SP9 7RE [IO91EF, SU24]
G0 WYE Details withheld at licensee's request by SSL.
G0 WYF C R Rudge, 37 Kingsland Rd, Alton, GU34 1LA [IO91MD, SU73]
G0 WYG D J Biginton, 67 Capstone Rd, Downham, Bromley, BR1 5NA [JO01AK, TQ37]
G0 WYI C I Sharpe, 35 Fairmead Cl, Nottingham, NG3 3EQ [IO92KX, SK54]
GI0 WYJ O J P Donnan, 119 Belsize Rd, Lisburn, BT27 4BT [IO64XM, J26]
G0 WYK W J Carress, Manor Barton, North Curry, Taunton, Somerset, TA3 6LP [IO81MA, ST32]
G0 WYL Dr B G S Hardie, 2/44 Clifton Gdns, London, W9 1DT [IO91VM, TQ28]
G0 WYM A R Shields, 8 Thames Dr, Melton Mowbray, LE13 0DS [IO92NS, SK71]
G0 WYN A E Godwin, 27 Melbourne Ave, Dronfield Woodhouse, Sheffield, S18 5YW [IO93GH, SK37]
GI0 WYO R J Kilgore, 29 Duncastle Park, New Buildings, Londonderry, BT47 2QL [IO64HW, C41]
G0 WYP S J Barker, 64 Granada Rd, Denton, Manchester, M34 2LJ [IO83WK, SJ89]
G0 WYQ J A Richardson, 24 Brockhall Rd, Kingsthorpe, Northampton, NN2 7RY [IO92NG, SP76]
G0 WYR C M Fox, 31 Pierson Rd, Windsor, SL4 5RE [IO91QL, SU97]
G0 WYS K Matthew, 3 Marconi Cl, Helston, TR13 8PD [IO70IC, SW62]
G0 WYT I R L Seabright, 117 Ainsdale Rd, Leicester, LE3 0UE [IO92KP, SK50]
G0 WYU E G Martin, 74 Chariot Rd, Illogan Highway, Redruth, TR15 3LH [IO70IF, SW64]
G0 WYV P D Little, 24 Pickwick Cres, Rochester, ME1 2HZ [JO01GJ, TQ76]
G0 WYW D Austin, 26 Brookside, Witton Gilbert, Durham, DH7 6RT [IO94ET, NZ24]
G0 WYX L A Austin, 26 Brookside, Witton Gilbert, Durham, DH7 6RT [IO94ET, NZ24]
G0 WYY R Bell, 41 Ravenshill Rd, West Denton, Newcastle upon Tyne, NE5 5EA [IO94DX, NZ16]
G0 WYZ S Gillen, 46 Cottingley Green, Leeds, LS11 0JD [IO93FS, SE23]
G0 WZA D J Wood, 2 Buckingham Mews, Flitwick, Bedford, MK45 1TB [IO91RX, TL03]
G0 WZB B Burdis, 10 Johnston Ave, Hebburn, NE31 2LJ [IO94FX, NZ36]
G0 WZC S M Cooper, 6 Polmeere Rd, Treneere Est, Penzance, TR18 3PD [IO70FD, SW43]
G0 WZD Dr J R Paloschi, 1 Lander Cl, Milton, Cambridge, CB4 6EB [JO02BF, TL46]
G0 WZE E Robinson, 113 Pennine Way, Carlisle, CA1 3QH [IO84NV, NY45]
G0 WZG K Norris, 15 East View, Choppington, NE62 5UF [IO95FD, NZ28]
G0 WZH S P Frankum, 5 Lower End, Wingrave, Aylesbury, HP22 4PG [IO91PU, SP81]
G0 WZI N L Lewis, 2604 Read Ave, Belmont, California, USA, 9400 2
G0 WZJ J R Pickering, 10 The Maltings, Malton, YO17 0HJ [IO94OD, SE87]
G0 WZK N G Collis Bird, 1 Drayton Park, London, N5 1NU [IO91WN, TQ38]
G0 WZL J D Rushton, 54 Constable Ave, Burnley, BB11 2NU [IO83VS, SD83]
G0 WZM I Kitchen, 25 Christchurch Mount, Epsom, KT19 8LU [IO91UI, TQ16]
G0 WZN L J Wood, 318 Fort Austin Ave, Plymouth, PL6 5TQ [IO70WJ, SX55]
GM0 WZO I G Finlayson, 39 Paisley Ave, Edinburgh, EH8 7LG [IO85KW, NT27]
G0 WZQ Dr J Nieschalk, 15 Dominion Rd, Brandon, Durham, DH7 8AY [IO94ER, NZ23]
G0 WZR C A Scaffer, 8/9 Quay Terr, Riverside, Reedham, Norwich, NR13 3TG [JO02SN, TG40]
G0 WZS J M Harper, 4 Friston Downs, Friston, Eastbourne, BN20 0ET [JO00CS, TV59]
G0 WZT M F Hall, 33 Chandos Rd, Newbury, RG14 7EP [IO91IJ, SU46]
G0 WZU I A A Dorrell, Unsers, Culford Rd, Ingham, Bury St. Edmunds, IP31 1NP [JO02IH, TL87]
G0 WZV K F Aston, 10 Browning Cl, Lexden, Colchester, CO3 4JJ [JO01KV, TL92]
GI0 WZW R D Pollock, 9 Larchwood Mews, Ballygowan Rd, Banbridge, BT32 3XJ [IO64UI, J14]
G0 WZX B L Stevens, 53 Rivermead Rd, Camberley, GU15 2SD [IO91OH, SU85]
G0 WZY M Davies, New House Cottage, Cublington, Madley, Hereford, Herefordshire, HR2 9NX [IO82NB, SO43]
GW0 WZZ M J Bobby, Hafan, Church St., Penycae, Wrexham, LL14 2RL [IO83LA, SJ24]
G0 XAA K Evans Ansty Contest Club, Little Field House, Cuckfield Rd, Ansty, Haywards Heath, W Sussex, RH17 5AL [IO90WX, TQ22]
G0 XAB R Dawson, 28 Calf Cl, Haxby, York, YO3 3NS [IO94LA, SE65]
GI0 XAC Dr A K L Chin, Site 29 Meadowbrook, Mullaghboy Islandmagee, Larne, Co. Antrim, N Ireland, BT40 3UG [IO74CT, J49]
G0 XAD B P Wright, 96 Ellenborough Cl, Thorley, Bishops Stortford, CM23 4HU [JO01BU, TL42]
G0 XAE M J Buckland, The Homestead, Bollow, Westbury on Severn, Glos, GL14 1QX [IO81TT, SO71]
G0 XAF P Coles, 87 Badgworth, Yate, Bristol, BS17 4YJ [IO81SM, ST78]
G0 XAH D A Tecklenburg, 46 Highfield Walk, Yaxley, Peterborough, PE7 3ET [IO92VM, TL19]
G0 XAI Details withheld at licensee's request by SSL.
G0 XAJ D J Denney, 6 Hill St., Kettering, NN16 8EE [IO92PJ, SP87]
G0 XAK S M Curtis, 125 Mancroft Ave, Bristol, BS11 0HZ [IO81QL, ST57]
GM0 XAL A S Milligan, 96 Beveridge St., Dunfermline, KY11 4PY [IO86GB, NT18]
G0 XAM L G Camber, Sundown, Strawberry Gdns, Jackies Ln, Newick, E Sussex, BN8 4QT [JO00AX, TQ42]
G0 XAN G R Aylward, 53 Overdown Rise, Portslade, Brighton, BN41 2YF [IO90VU, TQ20]
G0 XAO P J Cole, 48 Trembel Rd, Mullion, Helston, TR12 7DY [IO70JA, SW61]
GW0 XAP B P Blake, 39 Heol Sant Gattwg, Llanspyddid, Brecon, LD3 8PD [IO81GW, SO02]
GW0 XAQ S P Hughes, 45 Dew St., Haverfordwest, SA61 1ST [IO71MT, SM91]
G0 XAR S J Farthing, 38 Duxford Cl, Bowerhill, Melksham, SN12 6XN [IO81WI, ST96]
G0 XAS J P G Gavin, 143 Station Rd, Impington, Cambridge, CB4 4NP [JO02BF, TL46]
G0 XAT R Wallbank, 32 Truro Pl, Heath Hayes, Cannock, WS12 5YJ [IO92AQ, SK01]
G0 XAU C R Martin, 7 Mayfair Cl, Dukinfield, SK16 5HR [IO83XL, SJ99]
GM0 XAV A E Main, 5 Campsie Ct, Kirkintilloch, Glasgow, G66 4QQ [IO75WW, NS67]
G0 XAW A Santillo, 34 Wearde Rd, Saltash, PL12 4PP [IO70VJ, SX45]
G0 XAX Details withheld at licensee's request by SSL.
G0 XAY R T Elford, Prospects, Tormarton Rd, Acton Turville, Badminton, GL9 1HP [IO81UM, ST88]
G0 XAZ H W Simmons, 96 Porlock Rd, Millbrook, Southampton, SO16 9JF [IO90GW, SU31]
G0 XBA A D Hill, 13 Sycamore Way, Winklebury, Basingstoke, RG23 8AD [IO91KG, SU65]
G0 XBB W F Brierley, 15 Holmestrand Ave, Burnley, BB11 5DW [IO83US, SD83]
G0 XBC M E Soane, 24 Nurseries Rd, Wheathampstead, St. Albans, AL4 8TP [IO91UT, TL11]
G0 XBD W M Coopman, 1711-55 St Pl, Moline, Il, USA, 612 65
G0 XBE P P Rushen, 7 Cress Croft, Cressing Rd, Braintree, CM7 3YR [JO01HU, TL72]
G0 XBF Details withheld at licensee's request by SSL.
G0 XBG A N Marinho, 13 Pipkin Way, Oxford, OX4 4AR [IO91JR, SP50]
GI0 XBH K T Morgan, 34 Locan St., Belfast, BT12 7NE [IO74AO, J37]
G0 XBI P Roberts, 29 Bidston Rd, Anfield, Liverpool, L4 7XJ [IO83MK, SJ39]
G0 XBJ J T Stockwell, 75 Medworth, Orton Goldhay, Peterborough, PE2 5RY [IO92UN, TL19]
GI0 XBK A E W Ardill, 2 Lonsdale Ct, Shore Rd, Newtownabbey, BT37 0FA [IO74BQ, J38]
G0 XBL H B Watts, 7 Hartwood Rd, Liverpool, L32 7QH [IO83NL, SJ49]
G0 XBM Details withheld at licensee's request by SSL.
G0 XBN K A Smith, 2 Brantwood Gr, Bradford, BD9 6QB [IO93CT, SE13]
G0 XBO E H Smith, 8 Nene Rd, Hunstanton, PE36 5BZ [JO02FW, TF64]
G0 XBP E W Smith, 7 Waverley Ave, Beeston, Nottingham, NG9 1HZ [IO92JW, SK53]
G0 XBQ C Levingston, 7 Shortheath Crest, Farnham, GU9 8SA [IO91OE, SU84]
G0 XBU I C B Morley, Forest Cottage, School Ln, Eakring, Newark, NG22 0DE [IO93MD, SK66]
G0 XBV Details withheld at licensee's request by SSL.
G0 XCF C W Foley, 12 Cross St., Northam, Bideford, EX39 1BS [IO71VA, SS42]
GM0 XCW R W Ferguson Gmdx Group, 24 Braemar Ave, Dunblane, FK15 9ED [IO86AE, NN70]
G0 XDI M D Cabban, 37 Ryman Ct, Stag Ln, Chorleywood, Rickmansworth, WD3 5HN [IO91RP, TQ09]
G0 XDL G J Edwards, 8 Clover Ct, Clover Hill, Skipton, BD23 1BD [IO83XX, SD95]

G0 XDX Details withheld at licensee's request by SSL.
G0 XEG D E Riches, 92 Barons Rd, Bury St. Edmunds, IP33 2LY [JO02IF, TL86]
G0 XEL Details withheld at licensee's request by SSL.
G0 XEM Details withheld at licensee's request by SSL.
GM0 XFK J Neary, 17 Harkins Ave, Blantyre, Glasgow, G72 0RQ [IO75WS, NS65]
G0 XFT J D Carp, 33Su Det Akrotiri, Bfpo 57, X X
G0 XGK G C Kinder, 2 Church Cl, Brookwood, Woking, GU24 0AB [IO91QH, SU95]
G0 XGL G R Lawrence, 44B Island View, Greyshott Ave, Fareham, Hants, PO14 3JD [IO90JU, SU50]
G0 XGM R P Burgess, 22 Lee Cl, Kidlington, OX5 2XZ [IO91IT, SP41]
G0 XIT B J Davis, 36 Ampthill Rd, Maulden, Bedford, MK45 2DH [IO92SA, TL03]
G0 XJB Details withheld at licensee's request by SSL.
G0 XJK J P Kemble, 88 Mayfield Rd, Ipswich, IP4 3NG [JO02OB, TM14]
G0 XJS J D R Sliman, 97 Bredon, Yate, Bristol, BS17 4TE [IO81SM, ST78]
G0 XKK K Keenan, 25 Harris Rd, Harpur Hill, Buxton, SK17 9JS [IO93BF, SK07]
G0 XKT K R Taylor, 7 Newton Gdns, Paddock Wood, Tonbridge, TN12 6AJ [JO01EE, TQ64]
G0 XLI Details withheld at licensee's request by SSL.
GW0 XLK M Meredith, 5 Woodfield Rd, Llandybie, Ammanford, SA18 3UR [IO71XT, SN61]
G0 XOX P M Thorndike, 56 Durham Rd, Southend on Sea, SS2 4LU [JO01IN, TQ88]
G0 XPD A M R Sutton, Karena Gweek, Helston, Cornwall, TR12 6UB [IO70JC, SW72]
GI0 XPE M Carson, 148 Mount Vernon Park, Off Shore Rd, Belfast, BT15 4BJ [IO74AP, J37]
G0 XRC D Fox Exmouth Amateur Radio Club, 20 Littlemead Ln, Exmouth, EX8 4RF [IO80HP, SY08]
G0 XRN J M Barrett, 38 St. Annes Gdns, Woolston, Southampton, SO19 9FJ [IO90HV, SU41]
G0 XRO K J Hoare, 119 Marlins Turn, Hemel Hempstead, HP1 3LW [IO91SS, TL00]
G0 XTA R C Skells, 95 Sutton Rd, Leverington, Wisbech, PE13 5DR [JO02BQ, TF41]
G0 XTC Details withheld at licensee's request by SSL.
G0 XTL T R Layphries, 33 Harvey Rd, Evesham, WR11 5BQ [IO92AC, SP04]
G0 XTM N A Marshall, 2 Coombe Ct, Station Approach Roa, Tadworth, KT20 5AL [IO91VG, TQ25]
G0 XTT H G S Nicol, 5 Russell Hill, Purley, CR8 2JB [IO91WI, TQ36]
G0 XTV Details withheld at licensee's request by SSL.
G0 XVC R W Nixon, 6 St. Aidans Pl, Black Hill, Consett, DH8 5SU [IO94BU, NZ15]
G0 XVL D J Veale, 5 Heathfield Cl, Dronfield, Sheffield, S18 6RJ [IO93GH, SK37]
G0 XVS J A Thompson, 22 Kendal Dr, Dronfield, Dronfield Woodhouse, Sheffield, S18 5NA [IO93GH, SK37]
G0 XXX D J Knowler 5Xx Group Daventry, Little Fawsley, Fawsley Park, Daventry, Northants, NN11 6BU [IO92KF, SP55]
GW0 XYL J I Hockley, 44 Brookfields, Crickhowell, NP8 1DJ [IO81KU, SO21]
G0 XYS Details withheld at licensee's request by SSL.
GI0 XYZ Details withheld at licensee's request by SSL.
G0 YAE D Phillips, 28 Prince of Wales Rd, Great Totham, Maldon, CM9 8PX [JO01IS, TL81]
G0 YAK Details withheld at licensee's request by SSL.
G0 YAP I A P Bevan, 14 Pretty Dr, Scole, Diss, IP21 4DG [JO02NI, TM17]
G0 YBU D Cuff, Lake Rising, Lake, Salisbury, Wilts, SP4 7BP [IO91CD, SU13]
G0 YCA C J Adams, 13 High Row, Washington, NE37 2LZ [IO94FW, NZ35]
GW0 YDX R M Bates, PO Box 923, Woodale, Illinois, USA, 6019 093
G0 YEF P W Mister, 407A The Spa, Melksham, SN12 6QL [IO81WI, ST96]
GI0 YES D McAlpine Hilltop Amateur Radio Club, 35 Carnamena Ave, Cregagh, Belfast, BT6 9PJ [IO74BN, J37]
GI0 YEW Details withheld at licensee's request by SSL.
GW0 YEY D O Westmoreland, 27 Palmerston Rd, Barry, CF63 2NR [IO81JJ, ST16]
G0 YKC G K Clampin, 10 Chapelgate, Sutton St. James, Spalding, PE12 0EE [JO02AR, TF31]
G0 YKK Details withheld at licensee's request by SSL.
G0 YLM Details withheld at licensee's request by SSL.
G0 YLO C D Monksummers Wincanton Ladies Contest Group, 29 Cloverfields, Peacemarsh, Gillingham, SP8 4UP [IO81UA, ST82]
G0 YNM M W Blackburn, 3 Brookside, Old Langho, Blackburn, BB6 8AP [IO83ST, SD73]
G0 YOU Details withheld at licensee's request by SSL.
G0 YOY Details withheld at licensee's request by SSL.
G0 YRT B A Howarth, 23 Yew Tree Rd, Denton, Manchester, M34 6JY [IO83WK, SJ99]
G0 YSS A P Jones, 122 Slater St., Latchford, Warrington, WA4 1DW [IO83RJ, SJ68]
G0 YTF J P Clarke, Chelmwood, 7 Howard Cres, Seer Green, Beaconsfield, HP9 2XR [IO91QO, SU99]
G0 YYY R J Konowicz, 12 Ambleside Cres, Farnham, GU9 0RZ [IO91OF, SU84]
G0 YZC P Newcombe, 6 Church Ln, Bessacarr, Doncaster, DN4 6QB [IO93LM, SE60]
G0 ZAA S A Balding, Beechedge, Church Cl, Banningham, Norwich, Norfolk, NR11 7DY [JO02PT, TG22]
GI0 ZAK F Tanner, 62 Disraeli St., Belfast, BT13 3HW [IO74AO, J37]
GM0 ZAM J Glennon, 68 Carronshore Rd, Carron, Falkirk, FK2 8EE [IO86CA, NS88]
G0 ZAP R D Crozier, 53 Parsonage Brow, Upholland, Skelmersdale, WN8 0JG [IO83ON, SD40]
G0 ZAT W G Skipper, 27 Central Ave, South Shields, NE34 6AY [IO94HX, NZ36]
G0 ZDL D Lister, Blue Firs, Speen Rd, North Dean, High Wycombe, HP14 4NN [IO91OQ, SU89]
G0 ZDX J L Walmsley South Coast Hilltoppers, PO Box 1637, Yeovil, Somerset, BA21 4YF [IO80QW, ST51]
G0 ZEC E Cooper, 10 Church Dr, Orton Waterville, Peterborough, PE2 5EX [IO92UN, TL19]
G0 ZEE C D Monksummers, 29 Cloverfields, Peacemarsh, Gillingham, SP8 4UP [IO81UA, ST82]
G0 ZEN Details withheld at licensee's request by SSL.
G0 ZEP Details withheld at licensee's request by SSL.
G0 ZER E Robinson, 59 Barrack Hill, Armagh, BT60 1BL [IO64QI, H84]
GI0 ZGB Details withheld at licensee's request by SSL.
G0 ZGN Details withheld at licensee's request by SSL.
G0 ZHP K Jasinski The Polish South ARC - London, 35B Friars Pl Ln, London, W3 7AQ [IO91UM, TQ28]
GW0 ZIG B J Lennox, 49 Beatty Ave, Roath Park, Cardiff, CF2 5QR [IO81JM, ST17]
G0 ZIP F W G Marston, 1 Weaver Rd, Leicester, LE5 2RL [IO92LP, SK60]
G0 ZMC M V Conlon, 3 Selside, Brownsover, Rugby, CV21 1PG [IO92JJ, SP57]
G0 ZMH M R Howell, 1 East Terr, Blennerhasset, Carlisle, CA5 3QY [IO84IS, NY14]
G0 ZPV J G Roberts, Garfield House, 64 Broughton Ln, Wistaston, Crewe, Ches, CW2 8JR [IO83SC, SJ65]
G0 ZZY Details withheld at licensee's request by SSL.
G0 ZZZ P D Measom, 47 Malyons Rd, London, SE13 7XD [IO91XK, TQ37]

G1

G1 AAD N M Youd, 30 Eastfield Dr, Hanslope, Milton Keynes, MK19 7NQ [IO92OC, SP84]
G1 AAE Details withheld at licensee's request by SSL.
G1 AAG A Withers, 23 Fernie Rd, Guisborough, TS14 7LZ [IO94LM, NZ61]
G1 AAH P Worledge, 8 Forest Edge Rd, Sandford, Wareham, BH20 7BX [IO80WQ, SY98]
G1 AAL A Woodward, 40 Berwood Farm Rd, Wylde Green, Sutton Coldfield, B72 1AG [IO92CM, SP19]
G1 AAP M N Wilson, 18 Briars Cl, Southwood, Farnborough, GU14 0PB [IO91OG, SU85]
G1 AAQ E E Writer, 9 Fir Tree Cl, Shanklin, PO37 7EX [IO90JP, SZ58]
G1 AAR M A West, The Lair, Piddinghoe, Newhaven, E Sussex, BN9 9AH [JO00AT, TQ40]
G1 AAS Details withheld at licensee's request by SSL.[Station located near Margate.]
G1 ABA D M Lambert, 72 Johnson Dr, Barrs Ct, Bristol, BS15 7BS [IO81SK, ST67]
GW1 ABB E V Nelmes, 7 Dunraven Terr, Treorchy, CF42 6EL [IO81FP, SS99]
G1 ABJ A B J Norton, Kerkira, Long Ln, Chapel En-le-Frith, Stockport, SK12 6TA [IO83XI, SJ98]
G1 ABM I Macdonald, 139 Elmway, South Pelaw, Chester-le-Street, DH2 2LG [IO94EU, NZ25]
G1 ABO J W Markwell, 18 Fenner Cl, Folkstone, Folkestone, CT20 3NH [JO01NB, TR23]
G1 ABQ F R Macdonald, Boothlands Farm, Newdigate, Dorking, Surrey, RH5 5BS [IO91TD, TQ13]
G1 ABW B Webber, Glendarvel, Wayfarers Park, Shootersway Ln, Berkhamsted, HP4 3UR [IO91QS, SP90]
G1 ABX C A Passey, Rylstone Hotel, 129 West Hill Rd, Bournemouth, BH2 5PH [IO90BR, SZ09]
G1 ABY Details withheld at licensee's request by SSL.
G1 ACB G M Gifford, 42 Green Park, Brinkley, Newmarket, CB8 0SQ [JO02EE, TL65]
G1 ACG D Holmes, 15 St. Catherines Ave, Market Bosworth, Nuneaton, CV13 0LX [IO92HO, SK40]
G1 ACH Details withheld at licensee's request by SSL.
G1 ACJ A Hawkes, 12 Summer Hall Ing, Wyke, Bradford, BD12 8DN [IO93CR, SE12]
GI1 ACN T N Hutton, 23 Enniscrone Park, Portadown, Craigavon, BT63 5DQ [IO64SK, J05]
GW1 ACV W F Harrison, Top Flat, Craiglwyd Hall, Penmaenmawr, Gwynedd, LL34 6ER [IO83AG, SH77]
G1 ACY M V Holley, 38 Brendon Rd, Bedminster, Bristol, BS3 4PL [IO81QK, ST57]
G1 ADB Details withheld at licensee's request by SSL.
G1 ADD Details withheld at licensee's request by SSL.
G1 ADE H Kirk, 34 Tilworth Rd, Hull, HU8 9BN [IO93US, TA13]
GM1 ADG Details withheld at licensee's request by SSL.
GM1 ADI R T Stevens, 10 Tiel Path, Glenrothes, KY7 5AX [IO86KE, NO20]
G1 ADP Details withheld at licensee's request by SSL.
G1 ADS A J Perkins, 11 Claygate Rd, Wimblebury, Cannock, WS12 5RN [IO92AQ, SK01]
G1 ADW A D Whiteman, 17 May Cl, Chessington, KT9 2AP [IO91UI, TQ16]
GW1 ADY L Rees, 40 High St., Abergwili, Carmarthen, SA31 2JB [IO71UU, SN42]
G1 AEA A J Read, 16 Western Cl, Penton Park, Chertsey, KT16 8QB [IO91SJ, TQ06]
G1 AEB J R Savage, 32 Balliol Rd, Bicester, OX6 7HP [IO91KV, SP52]
G1 AEF R S Smith, 130 Winchester Rd, Fordhouses, Wolverhampton, WV10 6EZ [IO82WP, SJ90]
G1 AEI P R Stevenson, 19 Triangle East, Oldfield Park, Bath, BA2 3HZ [IO81TJ, ST76]
G1 AEJ L N Smith, 4 Penhale Rd, Braunstone, Leicester, LE3 2UU [IO92JO, SK50]
G1 AEQ D Lewis, 468 Wigan Rd, Bolton, BL3 4QH [IO83SN, SD60]
G1 AES Details withheld at licensee's request by SSL.[Op: S J Lee. Station located near Dunstable.]
G1 AET F R Lawson, 10 Avebury Cl, Tuffley, Gloucester, GL4 0TS [IO81UT, SO81]
G1 AEU A Lott, 27 Queens Cres, Brixham, TQ5 9PJ [IO80FJ, SX95]
G1 AEV M K Morecroft, 10 Tintagel Gdns, Rochester, ME2 2RD [JO01FJ, TQ76]
G1 AEW A E Westney, 73 Ulster Ave, Shoeburyness, Southend on Sea, SS3 9HL [JO01JM, TQ98]
G1 AEX G R H Muggeridge, Brookers House, Saddlers Wall Ln, Brookland, Romney Marsh, Kent, TN29 9RT [JO01JA, TQ92]
G1 AEY D W Lee, 27 Lombardy Dr, Maidstone, ME14 5TA [JO01GG, TQ75]
GM1 AFD D Mathieson, 14 Hawksmuir, Kirkcaldy, Fife, KY1 2PW [IO86KC, NT29]
G1 AFI D Mills, 25 Lower Park Cres, Poynton, Stockport, SK12 1EF [IO83WI, SJ98]
G1 AFJ P F Keyte, 11 Woodward Rd, Pershore, WR10 1LW [IO82XC, SO94]
G1 AFK B Kelsey, 2 Brookbarn Farm Cttgs, Courtwick Ln, Wick, Littlehampton, BN17 7PE [IO90RT, TQ00]
G1 AFT J C Higgs, 46 Birley Rd, Whetstone, London, N20 0EZ [IO91VP, TQ29]
G1 AFW R J Harris, Ashington House, Rodmer Cl, Minster on Sea, Sheerness, ME12 2BS [JO01JK, TQ97]
G1 AGA A S Gee, 74 New St., Milnsbridge, Huddersfield, HD3 4LD [IO93CP, SE11]
G1 AGB D M Gee, 74 New St., Milnsbridge, Huddersfield, HD3 4LD [IO93CP, SE11]
G1 AGI Details withheld at licensee's request by SSL.
G1 AGK D R Woodruffe, 7 Orchard Cl, Melton, Woodbridge, IP12 1LD [JO02PC, TM25]
G1 AGM G Williams, 22 Moor Tarn Ln, Walney, Barrow in Furness, LA14 3LP [IO84IC, SD16]
G1 AGN Details withheld at licensee's request by SSL.
G1 AGQ T J Woods, 36 Stoney Brow, Roby Mill, Skelmersdale, WN8 0QE [IO83PN, SD50]
G1 AGT J S Woods, 15 Fossard Way, Scawthorpe, Doncaster, DN5 7XY [IO93JN, SE50]
G1 AGV E Yorke, 49 Ingleton Cl, Nuneaton, CV11 6WB [IO92GM, SP39]
G1 AGW D J Yeatman, 31 Gilchrist Ave, Greenhill, Herne Bay, CT6 7SG [JO01NI, TR16]
G1 AHA Details withheld at licensee's request by SSL.
GM1 AHC D R Lindsay, 83 St. Andrews Dr, Thurso, KW14 8QB [IO88FO, ND16]
GM1 AHF A P McCormack, 18 Harris Ct, Muirton, Perth, PH1 3DD [IO86GK, NO12]
GM1 AHG N M C G McCormack, 18 Harris Ct, Muirton, Perth, PH1 3DD [IO86GK, NO12]
G1 AHM A Martland, Knowleswood, Wrennals Ln, Eccleston, Chorley, Lancs, PR7 5PW [IO83PP, SD51]
G1 AHP Details withheld at licensee's request by SSL.
G1 AHQ D W Mobley, The Little Butts, Aynhoe, Banbury, Oxon, OX17 3AF [IO91IX, SP53]
G1 AHS A P Mitchell, 7 Cross Park St., Horbury, Wakefield, WF4 6AE [IO93FP, SE21]
G1 AHT Details withheld at licensee's request by SSL.
GW1 AHU R C Huckle, 13 Aldergrove Cres, Doddington Park, Lincoln, LN6 0SJ [IO93QF, SK96]
G1 AHW T J Redden, Westoe, Whickham Highway, Gateshead, NE11 9QH [IO94EW, NZ26]
G1 AHZ R Oliver, 10 Taylor Terr, West Allotment, Newcastle upon Tyne, NE27 0EF [IO95FA, NZ37]
G1 AIA E Oliver, 9 Taylor Terr, West Allotment, Newcastle upon Tyne, NE27 0EF [IO95FA, NZ37]
G1 AIB E W C Robinson, 5 Elmley Cl, Malvern, WR14 2QT [IO82UC, SO74]
G1 AIF R Robinson, 9 Prospect Terr, New Kyo, Stanley, DH9 7TR [IO94DU, NZ15]
G1 AIG S B Rimell, 1 Pullin Ct, Warmley, Bristol, BS15 5YL [IO81SK, ST67]
GM1 AIH M M Ogg, 1 Woodlea Gr, Udny Station, Ellon, Gordon, AB41 0RG [IO87WI, NJ93]
G1 AII J Rushton, 8 Keats Rd, Stonebroom, Alfreton, DE55 6JG [IO93HD, SK45]
G1 AIO A Peters, 4 Tollemache Cl, Manston, Ramsgate, CT12 5LX [JO01QI, TR36]
G1 AIP J Nolan, 16 Henry Wise House, Vauxhall Bridge Rd, London, SW1V 2SU [IO91WL, TQ27]
G1 AIT D R P Plyer, 20 Hawthorn Way, Royston, SG8 7JS [IO92XB, TL34]
G1 AIV R H R Gray, 279 Stockport Rd, Timperley, Altrincham, WA15 7SP [IO83UJ, SJ78]
G1 AJA A T Hopwood, 2 Rose Meadow, Dassels, Braughing, Ware, SG11 2RS [JO01AW, TL32]
G1 AJB D W Howard, Rame Common Farm, Carnkie, Wendron, Helston, Cornwall, TR13 0DY [IO70JD, SW73]
G1 AJC T J Howe, 4 Willow Cres, Broughton Gifford, Melksham, SN12 8NB [IO81VJ, ST86]
G1 AJD M Jacobsen, 3 Green Ln, Tickton, Beverley, HU17 9RH [IO93TU, TA04]
G1 AJE M Chambers, 1 Green Ln, Tickton, Beverley, HU17 9RH [IO93TU, TA04]
G1 AJG L K Hutton, 14 Folly Cl, Hitchin, SG4 9DG [IO91UW, TL12]
G1 AJI Details withheld at licensee's request by SSL.
G1 AJN G A Overy, 38 Otterden St., London, SE6 3SJ [IO91XK, TQ37]
G1 AJR Details withheld at licensee's request by SSL.
G1 AJS N P Parker, 67 Bustleholme Ln, West Bromwich, B71 3BD [IO92AN, SP09]
G1 AJU J A Parsons, 40 Tynings Cl, Kidderminster, DY11 5JP [IO82UJ, SO87]
G1 AJV E J Roberts, 4 Willow Way, Redditch, B97 6PH [IO92AH, SP06]
G1 AJW R A Phelps, 27 Newton Cl, Oakenshaw, Redditch, B98 7YR [IO92AG, SP06]
G1 AJX Details withheld at licensee's request by SSL.
G1 AJY A J Young, 4 Woodlea, Leybourne, West Malling, ME19 5QY [JO01FH, TQ65]
G1 AJZ P M Rayson, 1 Grange Gdns, Taunton, TA2 7EN [IO81KA, ST22]
G1 AKA H Rowley, Gelert, 44 Tytherington Dr, Tytherington, Macclesfield, SK10 2HJ [IO83WG, SJ97]
G1 AKC D J Reid, 215 Green Lanes, Wylde Green, Sutton Coldfield, B73 5LX [IO92CN, SP19]
G1 AKD A B Rideout, 7 Beech Rd, Martock, TA12 6DT [IO80OX, ST41]
G1 AKI K Rayner, -le-Chalet, 76 Coombe Vale Rd, Teignmouth, TQ14 9EW [IO80FN, SX97]
G1 AKN Details withheld at licensee's request by SSL.[Op: A Bloor. Station located near Woking.]
G1 AKR Details withheld at licensee's request by SSL.
GW1 AKT M Allington, 16 Llewelyns Est, Denbigh, LL16 3NR [IO83GE, SJ06]
G1 AKV T J Alexander, 8 Greenway, Eastbourne, BN20 8UG [JO00DS, TQ50]
G1 AKX D G Boast, 6 Astley Rd, Hemel Hempstead, HP1 1HU [IO91SS, TL00]
G1 ALA J F P Bird, 17 Sherrards Way, Barnet, EN5 2BW [IO91VP, TQ29]
G1 ALC P M Barker, 8 Burdon Cl, Willerby, Hull, HU10 6QZ [IO93SR, TA02]
G1 ALD A A Brown, 40 Sutherland Rd, Edmonton, London, N9 7QG [IO91XP, TQ39]
G1 ALF Details withheld at licensee's request by SSL.
G1 ALK R Batchelor, 20 Mallard Way, Lower Stoke, Rochester, ME3 9ST [JO01HK, TQ87]
G1 ALL M F C Bond, 25 Grangeway, Houghton Regis, Dunstable, LU5 5PR [IO91SV, TL02]
G1 ALN R S Swift, 56 Suffolk Dr, Whiteley, Fareham, PO15 7DJ [IO90IV, SU50]
G1 ALR A C Stafford, 24 Bourne St., Croydon, CR0 1XL [IO91WI, TQ36]
G1 ALT Details withheld at licensee's request by SSL.[Op: J K Smithson.]
G1 ALU K F Spiers, Robins Post, North Heath Ln, Horsham, RH12 5PJ [IO91UB, TQ13]
G1 ALW P R Smith, 9 King Edwards Rise, Ascot, SL5 8JZ [IO91PK, SU97]
G1 ALX T A Stead, 30 Grassmere Ave, Westone, Northampton, NN3 3DP [IO92NG, SP76]
G1 ALZ B D Smith, 8 Forester Way, The Poplars, Kidderminster, DY10 1NT [IO82VI, SO87]
GM1 AMC A Stirling, 178 Romford Rd, Aveley, South Ockendon, RM15 4PJ [JO01DM, TQ58]
G1 AMG A M Gould, 21 Radlet Cl, Taunton, TA2 8EB [IO81KA, ST22]
G1 AMI Details withheld at licensee's request by SSL.
G1 AML I W Tidey, 7 Victoria Cl, Horley, RH6 7AP [IO91VE, TQ24]
G1 AMM D J C Taylor, 5 Fairfield Gdns, Rothwell, Leeds, LS26 0GD [IO93GS, SE32]
G1 AMN N D Webb, 11 Coopers Cl, Thorpe Marriott, Taverham, Norwich, NR8 6QZ [JO02OQ, TG11]
G1 AMS J W Winter, 9 Standish Cl, Wyken, Coventry, CV2 5NN [IO92GJ, SP37]
G1 AMV Details withheld at licensee's request by SSL.
G1 AMW Details withheld at licensee's request by SSL.
G1 ANA J F Coles, 15 Martindale Cl, Eccleshill, Bradford, BD2 3SR [IO93DT, SE13]
G1 ANF T P Crowe, 15 Lambert Rd, Kendray, Barnsley, S70 3AA [IO93GN, SE30]
G1 ANI M Cooper, 33 Park View, Royston, Barnsley, S71 4AA [IO93GO, SE31]
G1 ANN A J Millner, 4 Monkreed Villas, Longfield Rd, Longfield, DA3 7AR [JO01EJ, TQ66]
G1 ANS K J Elvin, 97 Jeans Way, Dunstable, LU5 4PR [IO91SV, TL02]
G1 ANV R K Edgar, 19 Butt Hedge, Marston, York, YO5 8LW [IO93JW, SE55]
GW1 ANW D K Evans, 63 Teglan Park, Tycroes, Ammanford, SA18 3RA [IO71XS, SN61]
G1 ANY Details withheld at licensee's request by SSL.
G1 ANZ P A Thirst, The Haywain, Thirsts Farm, Happisburgh Rd, East Ruston, NR12 0RU [JO02ST, TG32]
G1 AOC R C Doughty, 10 Northdown Way, Margate, CT9 3QX [JO01QJ, TR37]
G1 AOE J R Darley, 44 Thompson Way, Rickmansworth, Herts, WD3 2GP [IO91RO, TQ09]
G1 AOF R Dean, 10 Livingstone Rd, Ellesmere Port, South Wirral, L65 2BE [IO83NG, SJ47]
G1 AOL A J Galpin, 2 Haywood, Birch Hill, Bracknell, RG12 7WG [IO91PJ, SU86]
G1 AOQ I Gorsuch, Elmstone Farm, Fosten Green, Biddenden, Ashford, TN27 8ER [JO01HC, TQ83]
G1 AOR P G Grayshon, 90 Park Lea, Bradley Grange, Bradley, Huddersfield, HD2 1QP [IO93DQ, SE12]
G1 AOV M J Handley, 71 Aversley Rd, Kings Norton, Birmingham, B38 8PD [IO92AJ, SP07]
G1 AOY Details withheld at licensee's request by SSL.
G1 APA C M Hibbert, 5 Ashness Gdns, Greenford, UB6 0RL [IO91UN, TQ18]
G1 APC P Hill, 50 Sycamore Rise, Newbury, RG14 2LZ [IO91IJ, SU46]
G1 APD A P Daw, 21 Southern Rd, West End, Southampton, SO30 3ES [IO90IW, SU41]
G1 APH R J Hoole, Grindleford, Crewkerne Rd, Raymonds Hill, Axminster, EX13 5SY [IO80MS, SY39]
G1 APL R F Hopkins, 39 Forty Ln, Wembley, HA9 9EU [IO91UN, TQ28]
G1 APT R Prior, 16 Kelso Cl, Great Horkesley, Colchester, CO6 4TS [JO01KW, TL92]
GW1 APU K J Pierson, Celynfa, Penybont Ll.E, Oswestry, Salop, SY10 9JQ [IO82KT, SJ22]
G1 APV Details withheld at licensee's request by SSL.
G1 APW Details withheld at licensee's request by SSL.
G1 AQF A P Matthews, 44 Essex Cl, Dines Green, Worcester, WR2 5RW [IO82UE, SO85]
G1 AQI J Aldersey, 88 Firth View Walk, Westerfield, Workington, CA14 3RA [IO84FP, NX92]
GW1 AQJ P J Arnold, 5 Cefn Graig, Rhiwbina, Cardiff, CF4 6SW [IO81JM, ST18]
G1 AQP A Bell, 1 Purbeck Dr, Lostock, Bolton, BL6 4JF [IO83SN, SD60]
G1 AQQ Details withheld at licensee's request by SSL.
GM1 AQV J C Blackman, Orchard House, Hawick, Borders, TD9 9ST [IO85OK, NT51]
G1 AQX K B Armstrong, 30 Cobholm Pl, Kings Hedges, Cambridge, CB4 2UN [JO02BF, TL46]

GD1 AQY S Broad, 50 Selborne Dr, Douglas, IM2 3NH
G1 AQZ Details withheld at licensee's request by SSL.
GW1 ARC G Orchard Rhyl&Dist ARC, 8 Glyn Ave, Prestatyn, LL19 9NN [IO83HI, SJ08]
G1 ARD G Ashton, 101 Wickenby Garth, Wawne Rd, Bransholme, Hull, HU7 4RF [IO93US, TA13]
G1 ARF J E Makin, 6 Cambridge House, Courtfield Gdns, Ealing, London, W13 0HP [IO91UM, TQ18]
GM1 ARG Details withheld at licensee's request by SSL.
G1 ARH R Lupton, 19 Ave Cl, Starbeck, Harrogate, HG2 7LJ [IO94GA, SE35]
G1 ARJ J E Lowe, 3 Lichfield, Biggleswade, Beds, SG18 8JF [IO92UC, TL14]
GM1 ARK S G Lewcock, 79 Redcraigs, Kirkcaldy, KY2 6TS [IO86JD, NT29]
G1 ARL C R Mandall, 11 Hazel Rd, Park St., St. Albans, AL2 2AH [IO91TR, TL10]
G1 ARN Details withheld at licensee's request by SSL.
G1 ARR C E Jones, 19 Lucan Dr, Laleham, Staines, TW18 1QS [IO91SK, TQ07]
G1 ARV Details withheld at licensee's request by SSL.
G1 ARZ H M Kidman, Padfield Cottage, Ruett Ln, Farrington Gurney, Bristol, BS18 5UP [IO81RH, ST65]
GM1 ASA H Kilpatrick, 80 Livingstone Terr, Irvine, KA12 9DN [IO75QP, NS34]
GD1 ASB G A Kissack, 31 Marathon Rd, Douglas, IM2 4HN
G1 ASD K Knight, 80 Winton Rd, Reading, RG2 8HJ [IO91MK, SU76]
G1 ASE J C Knowles, 6 Southview Rd, Dudley, DY3 3PG [IO82WN, SO99]
G1 ASG T Stokes, 24 Atlantic Way, Sheffield, S8 7FZ [IO93GH, SK38]
G1 ASN J B Stansfield, 12 Harty House, Church St., Eccles, Manchester, Lancs, M30 0LT [IO83TL, SJ79]
G1 AST Details withheld at licensee's request by SSL.
G1 ASU Details withheld at licensee's request by SSL.
G1 ASW T J Cooper, 5 Downham Ct, Long Lodge Dr, Walton on Thames, KT12 3BZ [IO91TJ, TQ16]
G1 ASX M J Croft, 46 Poplar Ave, Euxton, Chorley, PR7 6BE [IO83PQ, SD51]
GM1 ASY J D Christie, 99 Meadow Cres, New Elgin, Elgin, IV30 3ER [IO87IP, NJ26]
G1 ATA D W Cotton, 24 Stirling Rise, Stretton, Burton on Trent, DE13 0JP [IO92ET, SK22]
G1 ATC R Degg obo Air Cadet Radio Soc, 28 The Spinneys, Welton, Lincoln, Lincs, LN2 3TU [IO93SH, TF08]
G1 ATG C B Clarke, 29 Huyton Hey Rd, Huyton, Liverpool, L36 5SF [IO83OJ, SJ49]
G1 ATL K R Bishop, 8 Sandbanks Gr, Hailsham, BN27 3LS [JO00DU, TQ50]
G1 ATN D F Barker, 1079 Stratford Rd, Hall Green, Birmingham, B28 8AU [IO92BK, SP18]
G1 ATQ H J Rogers, 36 Catsbrook Rd, Runfold Est, Luton, LU3 2ET [IO91SV, TL02]
G1 ATU Details withheld at licensee's request by SSL.
GM1 ATW E M King, Marionville, nr Donisbristle, Cowdenbeath, Fife, KY4 8EU [IO86HC, NT18]
G1 ATY B F Jarratt, 2 Claydon Cl, Aylesbury, HP21 8EB [IO91OT, SP81]
GW1 ATZ G R Morris, 6 Kent Ave, Shotton, Deeside, CH5 1BE [IO83LE, SJ36]
G1 AUH R J Holland, 16 Coniston Ave, St. Annes on Sea, Lytham St. Annes, FY8 3DQ [IO83LS, SD33]
G1 AUI C M Hayes, Flat 5, 37 Great Ormond St., London, WC1N 3HZ [IO91WM, TQ38]
G1 AUM G Gumbrell, 48 Church St., Deeping St. James, Peterborough, PE6 8HD [IO92UQ, TF10]
GW1 AUO W F Griffiths, Powys, Ednyfed Hill, Amlwch Port, Gwynedd, LL68 9HW [IO73UJ, SH49]
G1 AUQ R Green, 2 Fleets Ln Cottage, Fleets Rd, Sturton By Stow, Lincoln, LN1 2DN [IO93QH, SK88]
GW1 AUT N D Gordon, 21 Picket Mead Rd, Newton, Swansea, SA3 4SA [IO71XN, SS68]
G1 AUU S D J Goan, 36 Wellers Gr, Cheshunt, Waltham Cross, EN7 6HU [IO91XR, TL30]
G1 AUX T A Coton, 47 Skidmore Ave, Dosthill, Tamworth, B77 1NJ [IO92DO, SK20]
GM1 AUZ I Crockford, 12 McGregor Cres, Peterhead, AB42 1GE [IO97CM, NK14]
G1 AVA B Carter, 10 Opal St., Ingrow, Keighley, BD22 7BP [IO93AU, SE03]
G1 AVB R R W Davidson, 3 Eastridge Dr, Highridge, Bishopsworth, Bristol, BS13 8HQ [IO81QJ, ST56]
G1 AVC J Dean, 27 Ionic Rd, Stoneycroft, Liverpool, L13 3DU [IO83NK, SJ39]
G1 AVD T J Eastham, 12 Fairhaven Rd, Blackburn, BB2 3EE [IO83SR, SD62]
G1 AVF B Eastick, 30 Farmcote Rd, Aldermans Green, Coventry, West Midlands, CV2 1SA [IO92GK, SP38]
G1 AVI Details withheld at licensee's request by SSL.
G1 AVJ G M A Flower, 5 The Scop, Almondsbury, Bristol, BS12 4DU [IO81RN, ST68]
G1 AVQ Details withheld at licensee's request by SSL.
G1 AVW H F Sterry, 23 Eardisley Cl, Matchborough East, Redditch, B98 0BX [IO92BH, SP06]
G1 AVZ J Toolan, 12 Stillington Rd, Huby, York, YO6 1HW [IO94KC, SE56]
GW1 AWA Details withheld at licensee's request by SSL.
G1 AWD T J Wells, 2 Stephens Cl, Mortimer, Mortimer Common, Reading, RG7 3TL [IO91LJ, SU66]
G1 AWF A T Wyspianski, 53 Alington Cres, Kingsbury, London, NW9 8JL [IO91UN, TQ28]
G1 AWI J E Reed, 89 Shaftesbury Ave, Montpelier, Bristol, BS6 5LU [IO81QL, ST57]
G1 AWJ Dr J T B Moyle, 35 Midland Rd, Olney, MK46 4BL [IO92PD, SP85]
G1 AWK D G Moulson, Newbridge Cottage, Newbridge Rd, Tiptree, Colchester, CO5 0JA [JO01JT, TL91]
GM1 AWL Details withheld at licensee's request by SSL.
G1 AWP W Scott, 4 Wooden Farm, Lesbury, Alnwick, Northd., NE66 2TW [IO95EJ, NU20]
G1 AWU A J Pickles, 23 St. Christophers Cl, Isleworth, TW7 4NP [IO91TL, TQ17]
G1 AXE I Broadbent, 93 Hollinhall St., Oldham, OL4 3EL [IO83XN, SD90]
G1 AXF L Storkey, 9 Snatchup, Redbourn, St. Albans, AL3 7HD [IO91TT, TL11]
GW1 AXG R J Beckinsale, 1 Llys Eirlys, Park View, Rhyl, LL18 4LX [IO83GH, SJ08]
GM1 AXI T Ball, 12 Collier St., Johnstone, PA5 8AR [IO75RU, NS46]
G1 AXK R W Bird, 119 Highfields Rd, Chasetown, Walsall, West Midlands, WS7 8QS [IO92AQ, SK00]
G1 AXR T Chapman, 118 Austin Rd, Luton, LU3 1UB [IO91SV, TL02]
G1 AXS C J Costigan, 52 Westfield Rd, Westtown, Backwell, Bristol, BS19 3ND [IO81PJ, ST46]
GW1 AXU J Cook, 22 Northlands Park, Northway, Bishopston, Swansea, SA3 3JW [IO71XO, SS58]
G1 AXW C Dann, 25 Magnolia Walk, Quedgeley, Gloucester, GL2 4GD [IO81UU, SO81]
GW1 AYA M E J Chick, 2 Magnolia Way, Llantwit Fardre, Pontypridd, CF38 2PQ [IO81HN, ST08]
GW1 AYB P J Chick, 2 Magnolia Way, Efail Isaf, Llantwit Fardre, Pontypridd, CF38 2PQ [IO81HN, ST08]
G1 AYC R Dodge, 16 Mill Rd, Maldon, CM9 5HZ [JO01IR, TL80]
G1 AYL J J Devenish, 10 Toronto Cl, Worthing, BN13 2TD [IO90TT, TQ10]
G1 AYM Details withheld at licensee's request by SSL.[Correspondence to: N Negus, G6AWT, obo Gloucester ARS, 41 Oxtalls Lane, Longlevens, Gloucester GL2 9HP]
G1 AYP K A Lawton, Meadowbank, Ridgeway Rd, Sutton St Nicholas, Hereford, HR1 3BJ [IO82PC, SO54]
GM1 AYT J D McAllister, Redwood Lodge, Coach Rd, Kilsyth, Glasgow, G65 0PR [IO75XX, NS77]
G1 AYU Details withheld at licensee's request by SSL.
G1 AZA J R E Cook, 72 Valebridge Rd, Burgess Hill, RH15 0RP [IO90WX, TQ32]
G1 AZD A J Drage, 51 Greenbank Ave, Kettering, NN15 7EF [IO92PJ, SP87]
G1 AZE B W Davies, 22 Hillside Rd, Four Oaks, Sutton Coldfield, B74 4DQ [IO92BO, SP19]
G1 AZJ R L Endersby, 3 Woodside, Vigo Village, Meopham, Gravesend, DA13 0SU [JO01EH, TQ66]
G1 AZM H E Everitt, 167 Alliance Ave, Anlaby Rd, Hull, HU3 6QY [IO93TR, TA02]
G1 BAB R L Wilson, Cleves Cottage, 32 Tudor Cl, Seaford, BN25 2LX [JO00BS, TV49]
G1 BAF Details withheld at licensee's request by SSL.
G1 BAI K P Lowe, 18 Seaton Rd, Felixstowe, IP11 9BP [JO01QX, TM23]
G1 BAL T J Mahoney, 9 Waveney Dr, Higham, Barnsley, S75 1PU [IO93FN, SE30]
GM1 BAN D R Morrison, Wirane, 4 West Murkle, Murkle, Thurso, KW14 8YT [IO88GO, ND16]
G1 BAQ A G Miller, 44 Spring Gdns, Newport Pagnell, MK16 0EE [IO92PC, SP84]
G1 BAR B A Norris, Iearleswood, Benfleet, Essex, SS7 1DN [JO01GN, TQ78]
GW1 BAV G W Pritchard, Gorngol Rhedyn, Cemaes Bay, Anglesey, LL67 0HY [IO73SJ, SH39]
G1 BAW G J B Phillips, Whixall Marina Ltd, Alders Ln, Whixall, Whitchurch, SY13 2QP [IO82PV, SJ53]
G1 BAX J H Peters, 81 Emmbrook Rd, Wokingham, RG41 1JN [IO91NK, SU76]
G1 BBA T W Tilley, 10 Elmhurst Cl, Haverhill, CB9 8EG [JO02FB, TL64]
GI1 BBG D J Scott, 1 Shalom Park, Castlereagh, Belfast, BT6 9RY [IO74BN, J37]
GW1 BBH J M Sharkey, 33 Ffordd Morfa, Llandudno, LL30 1ES [IO83CH, SH78]
G1 BBI K R Ford, 3 Swaledale, Wildridings, Bracknell, RG12 7ES [IO91OJ, SU86]
G1 BBJ T W Blezard, Newlands, Monkton Deverill, Warminster, Wilts, BA12 7EX [IO81VD, ST83]
G1 BBK B N Artingstall, 19 Town Ln, Denton, Manchester, M34 6AF [IO83WK, SJ99]
G1 BBR A W Fraser, 54 Thorndene Way, Westgate Hill, Bradford, BD4 0SW [IO93DS, SE22]
G1 BBT J Dackham, 4 Overbury Cl, Lanehouse, Weymouth, DT4 9UE [IO80SO, SY67]
G1 BBY W T Brown, 53 Drummonds Cl, Longhorsley, Morpeth, NE65 8UR [IO95CF, NZ19]
G1 BCB A K Blake, 17 Apsley Rd, Lower Weston, Bath, BA1 3LP [IO81TJ, ST76]
G1 BCC C G Barnes, 20 Appleton Ave, Pedmore, Stourbridge, DY8 2JZ [IO82WK, SO98]
G1 BCE B J Brough, 6 Higgs Rd, Ashmore Park, Wednesfield, Wolverhampton, WV11 2PD [IO82XO, SJ90]
GW1 BCI A L Gray, The Graylands, 92 Wenallt Rd, Rhiwbina, Cardiff, CF4 6TP [IO81JM, ST18]
G1 BCL Details withheld at licensee's request by SSL.
G1 BCQ J A Russell, 45 Hillside Cres, Walsall, WS3 4JL [IO92AO, SK00]
G1 BCU R J Tagg, The Pines, 38 Salhouse Rd, Rackheath, Norwich, NR13 6QH [JO02QP, TG21]
G1 BCX K E Waites, 49 Shoreditch Rd, Taunton, TA1 3DF [IO81KA, ST22]
GW1 BDF K Jones, 10 Trinity Rd, Tonypandy, CF40 1DQ [IO81GO, SS99]
GW1 BDG O Jones, 10 Trinity Rd, Tonypandy, CF40 1DQ [IO81GO, SS99]
GW1 BDH B Jones, 8 Maes Derw, Llandudno Junction, LL31 9AL [IO83CG, SH87]
G1 BDI B Jones, 56 Mount Grace Rd, Luton, LU2 8EP [IO91TV, TL12]
G1 BDK P G Burge, 1 Winnal Farm Cottage, Kinlet, Bewdley, DY12 3BJ [IO82TK, SO78]
G1 BDP M D Broad, 7 Steventon Rd, Drayton, Abingdon, OX14 4JX [IO91IP, SU49]
G1 BDS D J Butt, 30 Jessopp Ave, Bridport, DT6 4AN [IO80PR, SY49]
G1 BDT S W Bradbury, 74 Sutton Rd, Kidderminster, DY11 6QS [IO82UJ, SO87]
G1 BDU A J Bradshaw, Heron Lodge Farm, Shuttle Hillock Rd, Bickershaw, Wigan, WN2 4RP [IO83RM, SD60]
G1 BDZ C G Cull, 17 Shakespeare Rd, Addlestone, KT15 2SR [IO91SI, TQ06]
G1 BEB R I Cocking, 304 Scar Ln, Golcar, Huddersfield, HD7 4AU [IO93BP, SE11]
G1 BED Details withheld at licensee's request by SSL.
G1 BEG J W Ellsmore, 11 Douglas Rd, Halesowen, B62 9HS [IO82XL, SO98]
G1 BEJ I R Dixon, 60A Woodlands Rd, Allestree, Derby, DE22 2HF [IO92GW, SK33]
G1 BEK G M Death, 105 Belvedere Rd, Ipswich, IP4 4AD [JO02OB, TM14]
GI1 BES J W Gibbard, 2 Almond Ct, Speke Rd, Garston, Liverpool, L19 2QZ [IO83NI, SJ48]
GI1 BEU L F Gough, 78 Culmore Point, Londonderry, BT48 8JW [IO65IB, C42]
GI1 BEW V E Graham, Cavancarragh, Lisbellaw, Co Fermanagh, BT74 6BN [IO64EI, H24]
G1 BEZ Details withheld at licensee's request by SSL.
G1 BFF T W F Fishlock, 62 Red Barn Rd, Brightlingsea, Colchester, CO7 0SJ [JO01MT, TM01]
G1 BFG A J Harper, 144 Ashfield Rd, Blackpool, FY2 0EN [IO83LU, SD34]
G1 BFH Details withheld at licensee's request by SSL.
G1 BFK R G O Else, 19 Chatsworth Ave, Crich, Matlock, DE4 5DY [IO93GB, SK35]
G1 BFQ B C Green, 6 Graham Dr, Fairgreen, Middleton, Kings Lynn, PE32 1RL [JO02FR, TF61]
G1 BFS P J F Sainsbury, Battle Rd, Punnetts Town, Heathfield, TN21 9DS [JO00DX, TQ62]
G1 BFU Details withheld at licensee's request by SSL.
G1 BFX Details withheld at licensee's request by SSL.
G1 BFZ J A Salisbury, 2 Henning Cottage, Lindal in Furness, Ulverston, Cumbria, LA12 0LT [IO84KE, SD27]
G1 BGB Details withheld at licensee's request by SSL.
G1 BGC S Javes, 22 Brookdean Rd, Worthing, BN11 2PB [IO90TT, TQ10]
G1 BGF D A McLachlan, 71 Yeoman St., Bonsall, Matlock, DE4 2AA [IO93FC, SK25]
G1 BGH P N Nicholls, 331 Ditchfield Rd, Hough Green, Widnes, WA8 8JX [IO83OI, SJ48]
G1 BGJ E P Morgan, 79 Mayland Ave, Canvey Island, SS8 0BU [JO01HM, TQ78]
G1 BGK G Lang, Stancliffe, Woodland Rd, Ashburton, Newton Abbot Devon, TQ13 7DR [IO80CM, SX77]
G1 BGM Details withheld at licensee's request by SSL.
G1 BGQ P E Sampson, 23 Westfield Rd, Mirfield, WF14 9PW [IO93DQ, SE22]
G1 BHB S A Matthews, 222 Widney Ln, Solihull, B91 3JY [IO92CJ, SP17]
G1 BHF S R Oldfield, The Sycamores, Fulford Rd, Fulford, Stoke on Trent, ST11 9QT [IO82XW, SJ93]
G1 BHL K Palmer, 3 Woodside Ave, Boothville, Northampton, NN3 6JL [IO92NG, SP76]
G1 BHM T M Prouse, Ventula, Wodalott Cross, Thornbury, Holsworthy, Devon, EX22 7BT [IO70UU, SS30]
G1 BHO B A Palmer, 28 Westminster Cres, Burnbridge, Burn Bridge, Harrogate, HG3 1LY [IO93FW, SE35]
G1 BHQ R D Pritchard, 41 Greenland Ave, Maltby, Rotherham, S66 7EU [IO93JK, SK59]
G1 BHR I A E Rabbitt, 15 Hunsbury Green, Ladybridge, Northampton, NN4 9UL [IO92MF, SP75]
G1 BHS M E Rumbelow, The Chase, Knott Park, Oxshott, Leatherhead, Surrey, KT22 0HR [IO91TH, TQ15]
G1 BHT S P Rogers, 16 Norwich Rd, Northwood, HA6 1NB [IO91TO, TQ18]
G1 BHV R A Green, 20 Haygate Dr, Wellington, Telford, TF1 2BY [IO82RQ, SJ61]
G1 BHW C D Vaughan, 42 Middleton Rd, Moordown, Bournemouth, BH9 2SU [IO90BS, SZ09]
G1 BHX J F Sanders, 6 Epping Walk, Melksham, SN12 7HW [IO81WJ, ST96]
G1 BIA H J Wilmshurst, 54 Gleton Ave, Hove, BN3 8LL [IO90VU, TQ20]
GM1 BID A E Watkins, 36 Cullen Cres, Kirkcaldy, KY2 6EP [IO86JD, NT29]
G1 BIF T Whelan, 43 Martin Ave, Little Lever, Bolton, BL3 1NX [IO83TN, SD70]
G1 BIM R N Ross, 13 Rydal Cl, Rugby, CV21 1JP [IO92JJ, SP57]
G1 BIN B R Talbot, 27 Shuttleworth Rd, Clifton on Dunsmore, Clifton upon Dunsmore, Rugby, CV23 0DB [IO92JJ, SP57]
G1 BIQ Details withheld at licensee's request by SSL.
G1 BJA M A Seabrook, Lyndene, 5 Mill View, Gazeley, Newmarket, CB8 8RN [JO02GG, TL76]
G1 BJE A J Stimpson, 15 Windsor Cl, Kings Sutton, Banbury, OX17 3QT [IO92IA, SP43]
G1 BJJ D T Hodgetts, 39 Fairway, Shelfield, Walsall, WS4 1RP [IO92AO, SK00]
G1 BJK P J L Lough, 87 Finchley Rd, Kingstanding, Birmingham, B44 0LB [IO92BN, SP09]
G1 BJN C B S Stroud, 83 Damers Rd, Dorchester, DT1 2LB [IO80SR, SY69]
G1 BJW P C Smith, 26 Ripley Rd, Luton, LU4 0AT [IO91SV, TL02]
G1 BJY Details withheld at licensee's request by SSL.
GM1 BKF L Calder, 49 Falcon Ave, Edinburgh, EH10 4AN [IO85JW, NT27]
G1 BKI M J York, 3 Cam Cl, Corby, NN17 2LJ [IO92PM, SP89]
G1 BKJ J R Prior, 6 Emfield Gr, Grimsby, DN33 3BS [IO93WM, TA20]
G1 BKL D N Ambler, Corrig, 4 Old Main Rd, Pawlett, Bridgwater, TA6 4RY [IO81LE, ST34]
GM1 BKR J G Rankin, 3 Spalding Dr, Largs, KA30 9BZ [IO75NT, NS26]
G1 BKV N P Frey, 76 Glebelands, Pulborough, RH20 2JJ [IO90SW, TQ01]
G1 BLA J R White, 47 Sandpiper Rd, Bayview Park, Whitstable, CT5 4DP [JO01MI, TR16]
G1 BLB P Thurman, 11 Copperfield Dr, Langley, Maidstone, ME17 1SX [JO01HF, TQ85]
G1 BLJ S J Lovell, 10 Ash Cres, Kingswinford, DY6 8DJ [IO82WL, SO88]
G1 BLK C S Ridley, 14 Painswick Rd, Hall Green, Birmingham, B28 0HH [IO92BK, SP18]
GM1 BLM I A G Ross, 3 Kenbank Cres, Bridge of Weir, PA11 3AY [IO75RU, NS36]
G1 BLO N D Swan, 8 Tyrrells Ct, Bransgore, Christchurch, BH23 8BU [IO90DS, SZ19]
G1 BLQ P S Lloyd, 29 Lodge Cl, Hertford, SG14 3DH [IO91XT, TL31]
G1 BLT M S Courcoux, 14 Haydock Cl, Bletchley, Milton Keynes, MK3 5LL [IO91OX, SP83]
G1 BLV Details withheld at licensee's request by SSL.
GM1 BLX I M Dewar, 11 Abbotshall Rd, Kirkcaldy, KY2 5PH [IO86KC, NT29]
G1 BMB G J Fallows, 66 Ulverston Rd, Swarthmoor, Ulverston, LA12 0JF [IO84KE, SD27]
G1 BMC P Bowen, 27 Aintree Cl, Uxbridge, UB8 3HS [IO91SM, TQ08]
G1 BMI B Widdup, The Bungalow, Oxcliffe New Farm, Oxcliffe Rd, Heysham, Lancs, LA3 3EF [IO84NB, SD46]
G1 BMJ B D Wood, 40 Woodgreen Rd, Stopsley, Luton, LU2 8BU [IO91TV, TL12]
G1 BMK T J Lindsell, 58 Baron Gdns, Barkingside, Ilford, IG6 1PB [JO01BO, TQ48]
G1 BML Details withheld at licensee's request by SSL.
G1 BMN N Lamb, 106 St. Davids Rd, Leyland, Preston, PR5 2XY [IO83PQ, SD52]
G1 BMP S J Campbell, 35 St. Georges Rd, Saltash, PL12 6EH [IO70VJ, SX45]
G1 BMQ S V Taylor, 39 Millview Ct, Lever Ln, Rochford, SS4 1DH [JO01IN, TQ89]
GW1 BMS Details withheld at licensee's request by SSL.
G1 BMT P J Tuthill, 12 Herbert Rd, Salisbury, SP2 9LF [IO91CB, SU13]
G1 BMW P Winterton, 27 Cranbrook Cl, Hayes, Bromley, BR2 7QA [JO01AJ, TQ46]
GM1 BNA R A Main, 43 Maryknowe, Gauldry, Newport on Tay, DD6 8SL [IO86LJ, NO32]
G1 BNE Details withheld at licensee's request by SSL.[Station located near Porth. Op: A J Perkins, PO Box 55, Luton, Beds LU1 1XG.]
G1 BNG S M Marsh, 28 Orcheston Rd, Charminster, Bournemouth, BH8 8SR [IO90BR, SZ19]
G1 BNN S A Tilly, 16 Mitchell Dr, Milburn Park, Ashington, NE63 9JT [IO95FE, NZ28]
GM1 BNP 7R R J Holt, Whitlam Farmhouse, Newmachar, Aberdeenshire, AB5 0RS [IO87TH, NJ72]
G1 BNV S J Henderson, 7 Havering, Castlehaven Rd, London, NW1 8TH [IO91WN, TQ28]
G1 BNX R Huxley, 83 Gleneagles Rd, Wyken, Coventry, CV2 3BH [IO92GK, SP38]
G1 BOB C J Balsdon, 4 Queens Hayes, Willey Ln, Sticklepath, Okehampton, EX20 2NG [IO80AR, SX69]
G1 BOE Details withheld at licensee's request by SSL.
G1 BOO F A Crompton, 24 Alcester Rd, Sale, M33 3QP [IO83UJ, SJ79]
G1 BOP M A Carroll, 69 Selbourne St., Eastwood, Rotherham, S65 1RP [IO93HK, SK49]
GM1 BOT G D Ogg, 65 Downie Park, Dundee, DD3 8JW [IO86ML, NO43]
G1 BOX M R Platten, 48 Brier Rd, Borden, Sittingbourne, ME10 1YL [JO01II, TQ86]
G1 BOZ Details withheld at licensee's request by SSL.
G1 BPB P Carter, 120 Gifford Cl, Two Locks, Cwmbran, NP44 7NZ [IO81LP, ST29]
G1 BPE M Barwick, 32 St. Georges Rd, Harrogate, HG2 9BS [IO93FX, SE35]
G1 BPH Details withheld at licensee's request by SSL.
G1 BPI J A Bullwinkle, The Willows, Baumber, Horncastle, Lincs, LN9 5ND [IO93VG, TF27]
G1 BPM R A F Jenkins, Currypool Cottage, Cannington, Bridgwater, Somerset, TA5 2NH [IO81KD, ST23]
G1 BPR Details withheld at licensee's request by SSL.
G1 BPU L M Staal, 17 Hillfield Park, Winchmore Hill, London, N21 3QJ [IO91WP, TQ39]
G1 BQC J B Small, 88 Gervase Rd, Burn Oak, Edgware, HA8 0EP [IO91UO, TQ29]
G1 BQI G Smith, 92 Lime Rd, Accrington, BB5 6BJ [IO83TS, SD72]
G1 BQJ I W Tebby, 29 Fairbourne Ln, Caterham, CR3 5AZ [IO91WG, TQ35]
GM1 BQP W E Macrobbie, Clach Mhuilinn, 44 Moraypark Terr, Culloden, Inverness, Invernessshire, IV1 2RG [IO77WL, NH74]
G1 BQQ J Sowerbutts, 22 Worsley St., Accrington, BB5 2PA [IO83TR, SD72]
G1 BQR M Spinks, 26 Church Hill, Royston, Barnsley, S71 4NH [IO93GO, SE31]
G1 BQV M C Sunderland, 36 Moorlands Ave, Yeadon, Leeds, LS19 6AD [IO93DU, SE24]
G1 BQW P J Symonds, 93 Mead Way, Coulsdon, CR5 1PQ [IO91WH, TQ35]
G1 BQZ T M Jones, 8 Lime St., Harrogate, HG1 4BG [IO93FX, SE35]
G1 BRB A J Patton, 72 Sanctuary Way, Great, Grimsby, DN37 9RZ [IO93WN, TA20]
G1 BRD J T Smith, 127 Wolverhampton Rd, Cannock, WS11 1AR [IO82XQ, SJ90]
G1 BRE S N Walker, 52 Bell Hagg Rd, Walkley, Sheffield, S6 5BS [IO93FJ, SK38]
G1 BRN Details withheld at licensee's request by SSL.
G1 BRP K J Crawley, 19 Park Mount, Harpenden, AL5 3AS [IO91TT, TL11]
G1 BRR T Bailey, 21 Gargrave Pl, Lupset, Wakefield, WF2 8AR [IO93FQ, SE32]
G1 BRS I D Brotherton Bournemouth Radio Soc, 6 Cranfield Ave, Wimborne, BH21 1DE [IO90AT, SU00]
G1 BRV Details withheld at licensee's request by SSL.
GM1 BSG J A Ross, 8 Cairnoch Way, Bannockburn, Stirling, FK7 8PN [IO86BC, NS89]
GI1 BSJ J J P Cunningham, 4 Garvaghy Rd, Portglenone, Ballymena, BT44 8EF [IO64SU, C90]
G1 BSN C S Catton, 68 Ward Ave, Grays, RM17 5RW [JO01DL, TQ67]
G1 BSO Details withheld at licensee's request by SSL.
GM1 BSP Details withheld at licensee's request by SSL.[Station located near Tenterden.]
G1 BSQ Details withheld at licensee's request by SSL.
G1 BSX J Morrison, 53 Broad Hinton, Waltham Chase, Twyford, Reading, RG10 0LP [IO91NL, SU77]
G1 BSY K J Morris, 3 Moravian Cl, Dukinfield, SK16 4EW [IO83WL, SJ99]
G1 BSZ R S Nash, Roann, Pimlico, Hemel Hempstead, Herts, HP3 8SH [IO91TR, TL00]

G1 BTA Details withheld at licensee's request by SSL.
G1 BTF A H Hardy, 14 Parsonage Rd, Rainham, RM13 9LW [JO01CM, TQ58]
G1 BTG K Hardacre, 13 St. Johns St., Bridlington, YO16 5NL [IO94VC, TA16]
G1 BTI K Forrest, 61 Woodbury Rd, Bridgwater, TA6 7LJ [IO81LD, ST23]
GM1 BTL W D Erskine, 30 Market Rd, Kirkintilloch, Glasgow, G66 3JL [IO75WW, NS67]
G1 BTT G A Cusick, 4 Clough Ave, Marple Bridge, Stockport, SK6 5AQ [IO83XJ, SJ98]
G1 BTU J A J Corbu, 6 Ave F.Goby, 06460 St Vallier de Thiey, France, X X
G1 BTV D E Holt, 2 London Heights, Dudley, DY1 2QZ [IO82WM, SO99]
G1 BUB T W Dolden, 609 Mierscourt Rd, Rainham, Gillingham, ME8 8RD [JO01HI, TQ86]
G1 BUJ Details withheld at licensee's request by SSL.
GM1 BUL D G Briers, East Brucehill Cott, New Deer, Turriff, Aberdeen, Aberdeenshire, AB4 8YJ [IO87TH, NJ72]
G1 BUN J G Cherrington, 10 Stanifield Ln, Farington, Preston, PR5 2GA [IO83PQ, SD52]
G1 BUO Details withheld at licensee's request by SSL.
G1 BUQ J N Spink, 38 Hemlingford Rd, Sutton Coldfield, B76 1JQ [IO92CM, SP19]
G1 BUV M R Osborne, 5 Wells Rd, Riseley, Bedford, MK44 1DY [IO92SG, TL06]
G1 BUY C M Rose, 5 West Woods, Crewe Rd South, Edinburgh, EH4 1RA [IO85JX, NT27]
GM1 BVA R L V Nieto, 12 Cow Wynd, Falkirk, FK1 1PL [IO85CX, NS87]
G1 BVI Details withheld at licensee's request by SSL.
G1 BVN Details withheld at licensee's request by SSL.
GM1 BVT W D Pettett, 15 Strude Howe, Alva, FK12 5JU [IO86CD, NS89]
G1 BVV G M Pemberton, 8 Hotchin Rd, Sutton on Sea, Mablethorpe, LN12 2NP [JO03DH, TF58]
G1 BWE D J Cook, 15 Broadlands, Broadmeadows, South Normanton, Alfreton, DE55 3NW [IO93HC, SK45]
G1 BWG R E Dodd, 5 Halesworth Rd, Chandlers Keep, Pendeford, Wolverhampton, WV9 5PH [IO82WO, SJ90]
G1 BWH P R Eaton, 8 Chester Rd, Barnwood, Gloucester, GL4 3AX [IO81VU, SO81]
G1 BWI J Eastham, 81 Park Lee Rd, Blackburn, BB2 3NZ [IO83SR, SD62]
G1 BWJ K Falconer, 9 The Fairway, Morpeth, NE61 2DW [IO95DD, NZ18]
G1 BWO P J Stokes, 46 Meadow View Rd, Sudbury, CO10 7NY [JO02IA, TL84]
G1 BWP W J Webb, 47 Barclay Ct, Park View, Hoddesdon, EN11 8PY [IO91XS, TL30]
G1 BWT Details withheld at licensee's request by SSL.[Station located near Leighton Buzzard.]
G1 BWU M Cooper Barnsley & DARC, 33 Park View, Royston, Barnsley, S71 4AA [IO93GO, SE31]
GM1 BWV J S Adams, 14 Leys Dr, Crimond, Fraserburgh, AB43 8RW [IO97AO, NK05]
G1 BWW D F Ash, 33 The Avenue, Leighton Bromswold, Ramsey, Huntingdon, PE17 1AS [IO92WK, TL28]
G1 BWX P A Caton, 39 Farmerie Rd, Hundon, Sudbury, CO10 8HA [JO02GC, TL74]
G1 BWZ N T Fieldsend, 112 Charnock Dale Rd, Sheffield, S12 3HR [IO93GI, SK38]
G1 BXB Details withheld at licensee's request by SSL.
GM1 BXG R A Beech, 56 Tarmangie Dr, Dollar, FK14 7BP [IO86EE, NS99]
GM1 BXI R Clark, 5 Abbotsfield Terr, Auchterarder, PH3 1DD [IO86DH, NN91]
GW1 BXJ Details withheld at licensee's request by SSL.
G1 BXN G A McDonald, 17 Brookeville, Brighouse Rd, Hipperholme, Halifax, HX3 8EA [IO93CR, SE12]
G1 BXT I R Thomas, 33 Barchester Way, Tonbridge, TN10 4HR [JO01DF, TQ64]
G1 BYB D J Hatton, 54 Chelsfield Way, Crossgates, Leeds, LS15 8XE [IO93GT, SE33]
G1 BYI J L Moore, 68 Sudworth Rd, Wallasey, L45 5BX [IO83LK, SJ39]
G1 BYJ G S C Noon, 48 Sanderling Cl, Letchworth, SG6 4HY [IO91VX, TL23]
G1 BYK N Pratt, 16 Burley Cl, Desford, Leicester, LE9 9HX [IO92IO, SK40]
G1 BYN R F Simmons, 78 Mayfield Rd, Thornton Heath, CR7 6DJ [IO91WJ, TQ36]
G1 BYO S C Slaughter, 3 Solomons Passage, Peckham Rye, London, SE15 3UH [IO91XL, TQ37]
G1 BYP P A Snitch, 79 Albion Ave, York, YO2 5QZ [IO93KX, SE55]
G1 BYQ D C Hatton, 34 Avocet Way, Langford Village, Bicester, OX6 0YP [IO91KV, SP52]
G1 BYS A C Kempton, 14 Lower Gravel Rd, Bromley, BR2 8LT [JO01AJ, TQ46]
G1 BYT N E Kinselley, 4 Stancliffe Rd, Bedford, MK41 9AN [IO92SD, TL05]
G1 BYY Details withheld at licensee's request by SSL.
G1 BZB C R Wigmore, 47 Hollingbourne Cres, Crawley, RH11 9QJ [IO91VC, TQ23]
G1 BZD P J Ward, 3 Harrogate St., Bradford, BD3 0LG [IO93DT, SE13]
G1 BZE C A Weeds, 45 Reeds, Cricklade, Westmill Park, nr Swindon, Wilts, SN6 6JF [IO91BP, SU09]
G1 BZM Details withheld at licensee's request by SSL.[Station located near Brackley.]
GM1 BZR D M Cameron, 14 Queen St., Castle Douglas, DG7 1HX [IO84AW, NX76]
G1 BZU R Baker Royal Naval Amateur Radio Soci, Royal Naval Am Rad Society, HMS Collingwood, Fareham, Hants, PO14 1AS [IO90JU, SU50]
G1 BZW R J Kimber, 38 Greenmere, Brightwell Cum Sotwell, Wallingford, OX10 0QG [IO91KO, SU59]
G1 BZY Details withheld at licensee's request by SSL.
G1 CAA Details withheld at licensee's request by SSL.[Op: J A G Airlie, 56 Douglas Avenue, Whitstable, Kent, CT5 1RU.]
G1 CAE Details withheld at licensee's request by SSL.
GI1 CAI J J Mc Bride, Dunaree, 20 Oldcastle Rd, Newtownstewart, Omagh, BT78 4HX [IO64HR, H38]
G1 CAN P T Shonfield, 242 Chickerell Rd, Weymouth, DT4 0QY [IO80SO, SY67]
G1 CAQ A A Mills, 29 Carlyle Rd, Maltby, Rotherham, S66 7LP [IO93JK, SK59]
GW1 CAV R W Roberts, 3 Cae Fron, Upper Llandwrog, Caernarvon, LL54 7BB [IO73VB, SH55]
G1 CAX D Deakin, Restholme, Mosham Rd, Blaxton, Doncaster, DN9 3BA [IO93ML, SE60]
G1 CAY D A Shea, 38 Ranworth Ave, Hoddesdon, EN11 9NR [IO91XS, TL31]
G1 CBB L J Walker, 39 Severus Ave, Acomb, York, YO2 4LX [IO93KW, SE55]
G1 CBG K Rodgerson, 10 Leigh Rd, Clifton, Bristol, BS8 2DA [IO81QL, ST57]
G1 CBK S T Mole, 53 Parkfield Rd, Gillingham, ME8 7TA [JO01HI, TQ86]
G1 CBL L R Mason, 27 Mead Way, Malvern, WR14 1SB [IO82UD, SO74]
GM1 CBQ P E Gibb, 5 Sunnyside, Mid Yell, Shetland, ZE2 9BS [IP90LO, HU59]
G1 CBS J D Hatt, 35 Field Rd, Fox Ln Est, Farnborough, GU14 9DJ [IO91OH, SU85]
G1 CBX D E Eastwood, 152 Halifax Rd, Todmorden, OL14 5RE [IO83WR, SD92]
G1 CBY S E Fountaine, 19 Metcalfe Gr, Blakelands, Milton Keynes, MK14 5JY [IO92PB, SP84]
G1 CCC Details withheld at licensee's request by SSL.[Station located near Daventry.]
GM1 CCI C M Watson, 11 Ladybridge Houses, Banff, Banffshire, AB45 2JR [IO87QP, NJ66]
G1 CCM T McMillan, 47 Sandsend Rd, Eston, Middlesbrough, TS6 8AF [IO94KN, NZ51]
GM1 CCN C Orr, Easter, Cowden Farm, Dalkeith, Mid Lothian, EH22 2NS [IO85LV, NT36]
G1 CCW F C Haselden, 7 Chestnut Ave, Gosfield, Halstead, CO9 1TD [JO01HW, TL72]
G1 CCX P B Kennedy, 77 Swarcliffe Rd, Harrogate, HG1 4QZ [IO93FX, SE35]
G1 CDG Details withheld at licensee's request by SSL.
GW1 CDH D R Davies, 10 Bryn Castell, Abergele, LL22 8QA [IO83EG, SH97]
G1 CDJ R J Hagar, 11 Berkeley Cres, Stourport on Severn, DY13 0HJ [IO82UH, SO86]
G1 CDQ A A Sturman, 26 Dean Way, Aston Clinton, Aylesbury, HP22 5GB [IO91PT, SP81]
G1 CDV Details withheld at licensee's request by SSL.
G1 CDW M D Wende, Sunbeams, Green Rd, Wivelsfield Green, Haywards Heath, RH17 7QD [IO91XA, TQ32]
G1 CDY D Forth, 20 West Common Cres, Scunthorpe, DN17 1DJ [IO93QN, SE80]
GI1 CDZ L M Campbell, 2 Downshire Park, Carnreagh, Hillsborough, BT26 6HB [IO64XL, J25]
G1 CEB Details withheld at licensee's request by SSL.
GM1 CEJ R S Stout, 16 Ardoch Park, Balgeddie, Glenrothes, KY6 3PJ [IO86JF, NO20]
G1 CEO R G Day, 17 Barry Ave, Bicester, OX6 8DZ [IO91KV, SP52]
GI1 CET J S Barr, 13 Roosevelt Rise, Belfast, BT12 5RN [IO74AO, J37]
GW1 CEV A J Jones, 53 Central Dr, Shotton, Deeside, CH5 1LS [IO83LE, SJ36]
G1 CEY Details withheld at licensee's request by SSL.
G1 CFA P M Middleton, 5 Fieldhouse, Cinderhills, Holmfirth, Huddersfield, HD7 1EN [IO93CN, SE10]
G1 CFB K Rhodes, 795A Hessle High Rd, Hull, North Humberside, HU4 6QE [IO93TR, TA02]
G1 CFC D J Grime, 124 Queens Ave, Bromley Cross, Bolton, BL7 9BP [IO83TO, SD71]
G1 CFG C J Rigby, 3 Springs Rd, Longridge, Preston, PR3 3TE [IO83QU, SD63]
G1 CFJ G Gardner, 165 Brookhouse Rd, Brookhouse, Lancaster, LA2 9NY [IO84PB, SD56]
G1 CFK D C Bowles, 40 Beetons Way, Bury St. Edmunds, IP32 6RE [JO02IG, TL86]
GM1 CFL L Ingram, Calle La Barraca 20, Torrevieja, (Alicante), Spain, X X
G1 CFZ J Bottomley-Mason, 48 Leeds Rd, Eccleshill, Bradford, BD2 3AY [IO93DT, SE13]
G1 CGD Details withheld at licensee's request by SSL.
G1 CGH A D Rawlins, 1 Worth Farm Cttgs, The St., Worth, Deal, CT14 0DF [JO01QG, TR35]
G1 CGJ M D L Davies, Kuling, Bridgwater Rd, Winscombe, BS25 1NB [IO81OH, ST45]
G1 CGP A W Gillard, 28 Moor Tarn Ln, Walney Island, Walney, Barrow in Furness, LA14 3LP [IO84IC, SD16]
G1 CGQ J W Mantle, 27 Highley Cl, Winyates East, Redditch, B98 0PL [IO92BH, SP06]
G1 CGU G C Fitzpatrick, Malmshead, 26 Ponsford Rd, Minehead, TA24 5DY [IO81GE, SS94]
G1 CHA D C Rodger, 29 Cherington Cl, Manchester, M23 0FE [IO83UJ, SJ89]
G1 CHM C A Milburn, Field House, Copper Hill, Copperhouse, Hayle, TR27 4LY [IO70HE, SW53]
G1 CHN A T A James, 6 Harvey Cl, Lawford, Manningtree, CO11 2HW [JO01MW, TM13]
G1 CHQ M H Wells, 7 Vint Rise, Idle, Bradford, BD10 8PU [IO93DU, SE13]
GM1 CHT A C Hyde, The Studio, Biggar Rd, Edinburgh, EH10 7DX [IO85JV, NT26]
G1 CHV C A Compton, 21 Vange Riverview Centre, Vange, Basildon, SS16 4NE [JO01FN, TQ78]
G1 CIA M Ferentiuk, 74 Fallowfield Dr, Rochdale, OL12 6LZ [IO83VP, SD81]
G1 CIH C Bylo, 50 Farndale St., York, YO1 4BP [IO93LW, SE65]
G1 CIK J P Milne, 15 Small Cres, Linden Village, Buckingham, MK18 7DE [IO91MX, SP73]
G1 CIR J N Webberley, Hulmewalfield House, Hulme Walfield, Congleton, Ches, CW1 2JG [IO83SC, SJ75]
G1 CIT M Whalley, 6 Rookery Walk, Clifton, Shefford, SG17 5HW [IO92UA, TL13]
G1 CIV D M Owen, 8 Southey Cl, Widnes, Ches, WA8 7EU [IO83PI, SJ58]
G1 CIY G W Evans, Summerfeild, Beulah, Newcastle Emlyn, Dyfed, SA38 9QB [IO72RC, SN24]
G1 CJB R J S Smith, 36 Station Rd, Furnby, Flimby, Maryport, CA15 8QN [IO84FQ, NY03]
G1 CJC L Gilbert, 11 Park Ave, Darley Dale, Matlock, DE4 2FX [IO93ED, SK26]
G1 CJH J C Hawkins, 42 Radstock Ave, Hodge Hill, Birmingham, B36 8HD [IO92CL, SP18]
GW1 CJJ P L Williams, 16 Cwm Rd, Dyserth, Rhyl, LL18 6BB [IO83HH, SJ07]
G1 CJK R V Bailey, 318 Plumstead Common Rd, London, SE18 2RT [JO01BL, TQ47]
G1 CKB F E G Jenkins, 15 The Ct, Blanchmans Rd, Warlingham, CR6 9BT [IO91XH, TQ35]
G1 CKF Details withheld at licensee's request by SSL.
G1 CKJ T Martin, 201 Gloucester Rd, Kidsgrove, Stoke on Trent, ST7 4DQ [IO83VC, SJ85]
G1 CKK J W Petrie Baker, Flat 1, 24 Canfield Gdns, London, NW6 3LA [IO91VN, TQ28]
G1 CKL Details withheld at licensee's request by SSL.
G1 CKQ A E Hughes, 5 Newton Dr, West Kirby, Wirral, L48 9UP [IO83KI, SJ28]
G1 CKR T I Miller, 26 Kirkland Cl, Hereford, HR1 1XP [IO82PB, SO53]
G1 CKT R J Nelson, 6 Undertown, Ugborough, Ivybridge, PL21 0NH [IO80BJ, SX65]
GI1 CKU T N Gardiner, 22 Carnmoney Rd, Newtownabbey, BT36 6HW [IO74AQ, J38]
G1 CKV N Derbyshire, 54 Windy Arbor Rd, Whiston, Prescot, L35 3SG [IO83OJ, SJ49]
G1 CKW Details withheld at licensee's request by SSL.
G1 CKY P R Turner, 3 Juniper Cl, Leicester Forest East, Leicester, LE3 3JX [IO92JO, SK50]
GW1 CLA A F Fish, Blaentir Farm, Cwmcou, Newcastle Emlyn, Dyfed, SA38 9PH [IO72RB, SN24]
G1 CLD S Patterson, 28 The Parks, Sundorne Gr, Shrewsbury, SY1 4TJ [IO82PR, SJ51]
G1 CLG Details withheld at licensee's request by SSL.
G1 CLJ P S Kennedy, 12 Newbroke Rd, Rowner, Gosport, PO13 9UJ [IO90KT, SU50]
G1 CLL Details withheld at licensee's request by SSL.
G1 CLT R Bokor, 75 Attlee Rd, Middlesbrough, TS6 7NA [IO94KN, NZ51]
GW1 CLZ P D Brewerton, c/o 33 Bridge St, Upminster, Essex, RM14 2LX [JO01CN, TQ58]
GM1 CMF P A G Carnegie, 21B Pasteur Ln, Dundee, DD2 1UU [IO86LL, NO33]
G1 CMH J Norwood, Flat 28 The Manor, Church Rd, Churchdown, Gloucester, Glos, GL3 2HT [IO81VV, SO81]
G1 CMP Details withheld at licensee's request by SSL.
G1 CMT I I Oxley, 6 Vickers Ave, South Elmsall, Pontefract, WF9 2LN [IO93IO, SE41]
G1 CMW A E Smith, 67 Fabian Cres, Shirley, Solihull, B90 2AB [IO92CJ, SP17]
G1 CMZ S P Lewkowicz, Flat 2, 105 Moorland Rd, Weston Super Mare, BS23 4HU [IO81MI, ST35]
GM1 CNH N D Stewart, 12 Grigor Ct, 204 Telford Rd, Edinburgh, EH4 2PL [IO85JX, NT27]
G1 CNI S E Dwyer, PO Box 44, Tasmoor 2573, New South Wales, X X
G1 CNN P J Beeson, 6 Oxford Ct, Queens Dr, London, W3 0HH [IO91UM, TQ18]
G1 CNV T J Thornton, Orchard Walk, 23 Crookham Rd, Fleet, GU13 8DP [IO91NG, SU85]
G1 CNY T Evans, 10 Greenwood Cl, Upton, Pontefract, WF9 1NU [IO93IO, SE41]
G1 COB J P Biddlecombe, Forest Glen, Old Cross Roads, Cadnam, Southampton, Hants, SO4 2NL [IO90HW, SU41]
G1 COP I W Waite, 101 Badshot Lea Rd, Badshot Lea, Farnham, GU9 9LP [IO91OF, SU84]
G1 COR Details withheld at licensee's request by SSL.
G1 COT I F S Boxall, 14 Little Common Rd, Bexhill on Sea, TN39 4JB [JO00FU, TQ70]
G1 COV D Green, 67 Coombe Park Rd, Coventry, CV3 2NW [IO92GJ, SP37]
G1 COW R H W Penfold, 118 Holbeck, Great Hollands, Bracknell, RG12 8XF [IO91OJ, SU86]
G1 COX A R Berkeley, 42 Ringley Dr, Whitefield, Manchester, M45 7LR [IO83UN, SD70]
G1 CPC J J Arthur, St. Aubin, Plomer Green Ln, Downley, High Wycombe, HP13 5XN [IO91OP, SU89]
G1 CPD G A Ghetti, 7 Rue de Provence, 75009, Paris, France
G1 CPF S Cook, 60 First Ave, Grimsby, DN33 1AB [IO93WN, TA20]
G1 CPM E P Rose, 3 North Luffenham Rd, South Luffenham, Oakham, LE15 8NP [IO92QO, SK90]
G1 CPO C J Haygarth, Carlin, 3 Rew Cl, Ventnor, PO38 1BH [IO90JO, SZ57]
G1 CPQ Details withheld at licensee's request by SSL.
G1 CPU G J Milligan, 607B Ecclesall Rd, Sheffield, S11 8PT [IO93FI, SK38]
G1 CQA R J Chaney, 55 Bartlow Rd, Linton, Cambridge, CB1 6LY [JO02DC, TL54]
G1 CQB Details withheld at licensee's request by SSL.
GM1 CQC H J L Smith, 601 Ferry Rd, Edinburgh, EH4 2TT [IO85JX, NT27]
G1 CQF P Simms, 141 Brays Rd, Sheldon, Birmingham, B26 2UL [IO92CL, SP18]
G1 CQG D J Perry, 5 Beech Hill, Wellington, TA21 8ER [IO80JX, ST12]
G1 CQL Details withheld at licensee's request by SSL.
G1 CQN Details withheld at licensee's request by SSL.
G1 CQR D H M Fuller, 26 Longfields, Ely, CB6 3DN [JO02CJ, TL57]
G1 CQT P M Turley, 21 Middlewood Dr, Heaton Mersey, Stockport, SK4 2DF [IO83VJ, SJ89]
G1 CQV Details withheld at licensee's request by SSL.
G1 CQX P R Lemasonry, 7 Eastwood Rd, Sittingbourne, ME10 2LZ [JO01II, TQ86]
G1 CRB R G Smart, 54 Exminster Rd, Stivichall, Coventry, CV3 5NW [IO92GJ, SP37]
G1 CRD M L Sheridan, 21 Drayton Ave, Stratford upon Avon, CV37 9PF [IO92DE, SP15]
G1 CRP G C Hersee, 99 Wilton Rd, Shirley, Southampton, SO15 5JH [IO90GW, SU41]
G1 CRT W A Cambridge, 5 Grosvenor Ave, Richmond, Surrey, TW10 6PD [IO91UL, TQ17]
GD1 CRZ W J Ferris, 21 Vicarage Mews, Douglas, IM2 2NR
G1 CSA J P Walton, 23 Keighley Ave, Downhill, Sunderland, SR5 4BU [IO94GW, NZ35]
G1 CSC Details withheld at licensee's request by SSL.
G1 CSF Details withheld at licensee's request by SSL.
G1 CSL Details withheld at licensee's request by SSL.
G1 CSN R C Beasley, 27 Retford Cl, Harold Hill, Romford, RM3 9NA [JO01CO, TQ59]
G1 CSO J E Dent, 30 Longbridge Rd, Bramley, Tadley, RG26 5AN [IO91LH, SU65]
G1 CSR N Sanderson Civil Service Amateur Radio So, 54 Kelvedon Cl, Chelmsford, CM1 4DG [JO01FS, TL70]
G1 CSS M A Wilson, Ambergate, London Rd, Hook, Hants, RG27 9EQ [IO91MG, SU75]
G1 CSZ A Cockram, 70 Arlington Dr, Marston, Oxford, OX3 0SJ [IO91JS, SP50]
G1 CTF Details withheld at licensee's request by SSL.
G1 CTT J N Hobbs, 9 Bracken Bank, Lychpit, Basingstoke, RG24 8TQ [IO91LG, SU65]
G1 CUB K J Tarrant, 12 Juniper Cl, Whiteshill, Whitehill, Bordon, GU35 9EZ [IO91NC, SU73]
GM1 CUC H D Mattinson, 12 Linns View, Harelaw, Canonbie, DG14 0RR [IO85NC, NY47]
G1 CUG D A Laughton, 2 Stamford Rd, Essendine, nr Stamford, Lincs, PE9 4LR [IO92SQ, TF01]
G1 CUH B A Reid, 32 Arlington Dr, Alvaston, Derby, DE24 0AU [IO92GV, SK33]
G1 CUJ J A Jones, San-Melv, Wycombe Rd, Stokenchurch, High Wycombe, Bucks, HP14 3RP [IO91NP, SU79]
G1 CUW J C Flowers, 102 Northfield Rd, Harborne, Birmingham, B17 0TA [IO92AK, SP08]
G1 CUZ S W Seal, Crantock, Bellingdon, Chesham, Bucks, HP5 2XW [IO91QR, SP90]
G1 CVA Details withheld at licensee's request by SSL.
G1 CWD P Bullough, 11 Druids View, Crossflatts, Bingley, BD16 2DY [IO93BU, SE04]
G1 CWE Details withheld at licensee's request by SSL.
G1 CWI M J Kemp, 37 Berrylands Rd, Surbiton, KT5 8PA [IO91UJ, TQ16]
G1 CWL Details withheld at licensee's request by SSL.
G1 CWN Details withheld at licensee's request by SSL.
G1 CWO A J Stringer, 39 Hurst Rd, Longford, Coventry, CV6 6EL [IO92GK, SP38]
G1 CWW G Stone, 37 Canterbury Dr, Ashby de La Zouch, LE65 2QQ [IO92GS, SK31]
G1 CWZ D J Penrose, 7 Two Ashes, Bayston Hill, Shrewsbury, SY3 0QF [IO82OQ, SJ40]
GI1 CXA Details withheld at licensee's request by SSL.
G1 CXE J Palmer, 53 Southwood Rd, Greatmoor, Stockport, SK2 7DJ [IO83WJ, SJ98]
G1 CXG Details withheld at licensee's request by SSL.
G1 CXK P Brock, 21 Ladyside Cl, Bransholme Cl, Bransholme, Hull, HU7 5AB [IO93UT, TA13]
G1 CXQ C H Roberts, 11 Adel Wood Dr, Leeds, LS16 8JQ [IO93FU, SE23]
G1 CXS Details withheld at licensee's request by SSL.
GM1 CYB Details withheld at licensee's request by SSL.
G1 CYD A E B Brennan, 56 Claverdon Dr, Great Barr, Birmingham, B43 5HP [IO92AM, SP09]
G1 CYN N M Tweedy, 29 Shoals Walk, Oulton Broad, Lowestoft, NR33 9HG [JO02UL, TM59]
G1 CYQ B S Wheeldon, 27 Lawrence Walk, Newport Pagnell, MK16 8RF [IO92PC, SP84]
G1 CYR W J D French, 21 Dalelands West, Market Drayton, TF9 1DQ [IO82SV, SJ63]
G1 CYS Details withheld at licensee's request by SSL.
G1 CYY T P B Brien, 54 Central Ave, Fartown, Huddersfield, HD2 1DA [IO93CP, SE11]
G1 CZU M P Abram, 28 Langport Dr, Vicars Cross, Chester, CH3 5LY [IO83NE, SJ46]
G1 CZW R J Silcocks, 69 Kennaway Rd, Clevedon, BS21 6JJ [IO81NK, ST47]
G1 CZX Details withheld at licensee's request by SSL.
G1 DAE I M Rusby, 12 Park Meadow, Princes Risborough, HP27 0EB [IO91NR, SP80]
G1 DAK S A James, 107 Spendmore Ln, Coppull, Chorley, PR7 4PY [IO83QP, SD51]
G1 DAN D A Smith, Range House, Kingsbury, Tamworth, Staffs, B78 2DX [IO92DN, SP29]
GI1 DAO J G M Lester, 37 Glebe Rd, Annahilt, Hillsborough, BT26 6NE [IO74AK, J35]
G1 DAT P W Burnett, 8 High St., Eston, Middlesbrough, TS6 0QY [IO94KN, NZ51]
G1 DAU G H Speirs, 19 Poets Way, Winchester, SO22 5BX [IO91IB, SU42]
G1 DAV D J Forsey, 11 Abbey Rise, Wollaston, Wellingborough, NN29 7QA [IO92PG, SP96]
G1 DAX P P L Costigan, 10 The Paddock, Clevedon, BS21 6JU [IO81NK, ST47]
G1 DAZ S Burchell, 40 Daisy Cl, Cambridge, CB4 3XH [JO02BF, TL46]
G1 DBH B J Cobb, 39 Church Ln, Bedminster, Bristol, BS3 4NE [IO81QK, ST57]
G1 DBI G W Doig, 78 Plane Tree Dr, Crewe, CW1 4ES [IO83SC, SJ75]
G1 DBK J Dunning, Roslyn, 80 Station Rd, Winsford, CW7 3DD [IO83RE, SJ66]
G1 DBL L R G Owen, 27 Coniston Dr, Holmes Chapel, Crewe, CW4 7LA [IO83TE, SJ76]
G1 DBR D J Ross, 113 Nun House Dr, Winsford, CW7 3LE [IO83SE, SJ66]
G1 DBZ R D Cooper, 31 Erskine Cres, Arbourthorne, Sheffield, S2 3LQ [IO93GI, SK38]
G1 DCB M Senior, 28 Broom Cres, Tarvin, Chester, CH3 8HA [IO83OE, SJ46]
G1 DCG D W D Dixon, 23 Wemberham Cres, Yatton, Bristol, BS19 4BD [IO81OJ, ST46]
G1 DCU P B Gardner, 66 Belmont Rd, Wollescote, Stourbridge, DY9 8BE [IO82WK, SO98]
G1 DCX M C Race, 76 Lonsdale Rd, Stamford, PE9 2SG [IO92RP, TF00]
G1 DCY S J Richmond, 17 Heron Cl, Alvechurch, Birmingham, B48 7PT [IO92AI, SP07]
G1 DCZ J A Sandall, 7 St. Rd, Compton Dundon, Somerton, TA11 6PX [IO81PC, ST43]

G1 DDD P D Gaskell St.Helens Raynet Group, 131 Greenfield Rd, Dentons Green, St. Helens, WA10 6SH [IO83OL, SJ59]
G1 DDH Details withheld at licensee's request by SSL.
G1 DDI C I Eyre, 32 Breedon Hill Rd, Derby, DE23 6TG [IO92GV, SK33]
G1 DDK M J Abraham, Skywave Marine Services, Sq Sail Boatyard, Charlestown Harbour, St. Austell, Cornwall, PL25 3NJ [IO70OI, SX05]
G1 DDR R Oakley, 20 Halton Ln, Wendover, Aylesbury, HP22 6AR [IO91PS, SP80]
G1 DDS D Seccombe, 14 Millfield, Bedlington, NE22 5DZ [IO95ED, NZ28]
G1 DDV Details withheld at licensee's request by SSL.
G1 DDY B A Hall, Irene Rd, Stoke Dabernon, Cobham, Surrey, KT11 2SR [IO91TH, TQ16]
G1 DEA S J Barnett, 11 Ridge St., Wollaston, Stourbridge, DY8 4QF [IO82VL, SO88]
G1 DEN P D Edinburgh, 77 Westerley Ln, Shelley, Huddersfield, HD8 8HP [IO93DO, SE21]
G1 DEO B Davies, 122 Hall Ln, Willington, Crook, DL15 0QD [IO94DR, NZ13]
G1 DEP J J Dunhill, 8 Brentwood Ave, Thornton Cleveleys, FY5 3QR [IO83LU, SD34]
G1 DEQ D L J Gilbey, 7 Victory Way, Cottenham, Cambridge, CB4 4TG [JO02BG, TL46]
G1 DER J R Hacker, 83 Fernbank Rd, Ascot, SL5 8JS [IO91PK, SU96]
G1 DES D G Smith, 14 College Dr, Ruislip, HA4 8SB [IO91TN, TQ18]
G1 DET Details withheld at licensee's request by SSL.
G1 DEU D E Hagger, 7 Pecockes Cl, Great Cornard, Sudbury, CO10 0NQ [JO02JA, TL84]
G1 DEV E H Hardwick, 5 Seaview, Oakmere Park, Little Neston, Merseyside, L64 0XP [IO83LG, SJ27]
G1 DEW Details withheld at licensee's request by SSL.
G1 DEX H Irvin, 46 Northway Gdns, London Park Est, Mirfield, WF14 0LP [IO93DQ, SE12]
G1 DEY Details withheld at licensee's request by SSL.[Correspondence via B M Box 86, London WC1N 3XX.]
G1 DEZ P G Baxter, 27 Manor Cres, Brinsworth, Rotherham, S60 5HG [IO93HJ, SK48]
G1 DFF M D Smith, 22 Cedars Ave, Wombourne, Wolverhampton, WV5 0JX [IO82VM, SO89]
G1 DFI J M Swift-Hook, 12 Warwick Dr, Greenham, Newbury, RG14 7TT [IO91IJ, SU46]
G1 DFM A J Westlake, 47 Quarry Rd, Kingswood, Bristol, BS15 2NZ [IO81RK, ST67]
G1 DFP G R Fielding, 35 Amos Ave, Litherland, Liverpool, L21 7QH [IO83ML, SJ39]
G1 DFR A D Gemmill, 9 Caynham Village, Ludlow, Salop, SY8 3BJ [IO82QI, SO57]
G1 DFT I J Hampson, 57 Cornwall Way, Ainsdale, Southport, PR8 3SG [IO83LO, SD31]
G1 DFW D Hoare, 51 Hartington Rd, Dronfield, Sheffield, S18 6LE [IO93GH, SK37]
G1 DFZ R H Jobbins, 8 Newark Rd, Hartlepool, TS25 2LA [IO94JP, NZ42]
G1 DGG Details withheld at licensee's request by SSL.
G1 DGL Details withheld at licensee's request by SSL.[Op: Richard C Simpson, 16 Ashridge, Cove, Farnborough, Hants, GU14 9UY. IO91OH. Tel: (01276) 38905, aka Herbert. Packet @ GB7VIR. Member Amsat-UK and Surrey RAYNET. Other interests: Acorn Archimedes and drama.]
G1 DGM S A Parslow, Stedi Tani, 1 Willington Cl, Little Harlescott L, Shrewsbury, SY1 3RH [IO82PR, SJ51]
G1 DGN J W Skerritt, 4 Ingoldsby Rd, Bitchfield, Grantham, NG33 4DT [IO92RU, SK92]
G1 DGP J Smith, 160 Tyldesley Rd, Atherton, Manchester, M46 9AB [IO83SM, SD60]
G1 DGS P D Pauley, 5 Clare Cl, Waterbeach, Cambridge, CB5 9PS [JO02CG, TL46]
G1 DGV A E Davey, 36 Linden Cl, Dunstable, LU5 4PF [IO91SV, TL02]
G1 DGW I J Johnson, 24 York Rd, Maghull, Liverpool, L31 5NL [IO83MM, SD30]
G1 DGY A C Koch, 65 Collier Ln, Ockbrook, Derby, DE72 3RP [IO92HV, SK43]
G1 DHA S P Tolley, 80 Carrington Rd, Friar Park, Wednesbury, WS10 0HX [IO92AN, SP09]
G1 DHM G W Miller, 32 Belbroughton Cl, Lodge Park, Redditch, B98 7NH [IO92AH, SP06]
G1 DHQ D Palmer, 18 Brushfield Ave, Sileby, Loughborough, LE12 7NX [IO92KR, SK61]
G1 DHX Details withheld at licensee's request by SSL.
G1 DHY N J Roe, 41 Highfield Ln, Chaddesden, Derby, DE21 6PH [IO92GW, SK33]
G1 DIA P Rowe, 7 Arrowcroft Rd, Guilden Sutton, Chester, CH3 7ES [IO83OE, SJ46]
G1 DIG S C Cadman, 71 Gayfield Ave, Withymoor Village, Brierley Hill, DY5 2BU [IO82WL, SO98]
G1 DIK A Smith, Windycross, Newbourn Rd, Waldringfield, Woodbridge, IP12 4PT [JO02PB, TM24]
G1 DIL A J Witts, 16 Redstone Dr, Highley, Bridgnorth, WV16 6EQ [IO82TK, SO78]
G1 DIM C F Smith, 37 Ivory Cl, Tuffley, Gloucester, GL4 0QY [IO81UU, SO81]
G1 DIW Details withheld at licensee's request by SSL.
G1 DJA Details withheld at licensee's request by SSL.
G1 DJD L L Chivers, 8 Lovekyn Cl, Londod Rd, Kingston upon Thames, KT2 6RY [IO91UJ, TQ16]
G1 DJG Details withheld at licensee's request by SSL.
G1 DJI J E Short, 16 Tycehurst Hill, Loughton, IG10 1BU [JO01AP, TQ49]
G1 DJJ Details withheld at licensee's request by SSL.
G1 DJM D J Mansfield, 31 Holcombe Ave, Kings Lynn, PE30 5NY [JO02ER, TF61]
G1 DJQ N Lofthouse, Cambridge Park, 8 Abbott Clough Ave, Knuzden, Blackburn, Lancs, BB1 3LP [IO83SR, SD72]
G1 DJT Details withheld at licensee's request by SSL.
G1 DJU C J Whitby, 7 Wentworth Way, Stoke Bruerne, Towcester, NN12 7SA [IO92MD, SP74]
G1 DKA W J Harris, 17 Redacre Rd, Abbey Hey, Manchester, M18 8RU [IO83WL, SJ89]
G1 DKC J Roberts, 35 Walmsley Rd, Eccleston, St. Helens, WA10 5JR [IO83OL, SJ49]
G1 DKE M J Spry, Four Views, Old Winslade, Clyst-St-Mary, Exeter, Devon, EX5 1AS [IO80GR, SX99]
G1 DKF S G Willetts, 48 Stourton Cres, Stourton, Stourbridge, DY7 6RR [IO82VL, SO88]
G1 DKG P S McNaney, 7 Hillside Ave, Bridgnorth, WV15 6BS [IO82TM, SO79]
G1 DKI M F L Lindenberg, 26 Manston Dr, Perton, Wolverhampton, WV6 7LX [IO82VO, SJ80]
GW1 DKK J E Price, Wendon, 14 Garth Est, Pontllyfni, Caernarvon, LL54 5ET [IO73TB, SH45]
G1 DKP A Atkinson, 109 Hadrian Rd, Jarrow, NE32 3TS [IO94GX, NZ36]
G1 DKV A Charlton, 20 Bailey Cres, South Elmsall, Pontefract, WF9 2TL [IO93IO, SE41]
G1 DKW S R Hickman, 37 Rugeley Rd, Chase Terr, Walsall, West Midlands, WS7 8AG [IO92AQ, SK00]
G1 DKX A G Thomas, 92 Singleton Cres, Goring By Sea, Worthing, BN12 5DJ [IO90ST, TQ00]
G1 DKY J W Miller, 61 Sunray Ave, Bromley, BR2 8EL [JO01AJ, TQ46]
G1 DLA R H Deacon, 22 Islip Gdns, Northolt, UB5 5BX [IO91TN, TQ18]
G1 DLC J Swales, 7 Blaykeston Cl, Seaton, Seaham, SR7 0PJ [IO94HU, NZ45]
G1 DLH M Ogle, 8 Spring Cl, Daventry, NN11 4HG [IO92KG, SP56]
G1 DLJ G Hope, 3 Farm Cres, Sittingbourne, ME10 4QD [JO01IH, TQ96]
G1 DLL Details withheld at licensee's request by SSL.[Op: J Darley. Station located near Gillingham.]
G1 DLO C L Holdsworth, Squirrels Wood, Horseshoe Ridge, St George's Hill, Weybridge, Surrey, KT13 0NR [IO91SI, TQ06]
GW1 DLP W H Jones, 160 Christchurch Rd, Newport, NP9 7SA [IO81MO, ST38]
G1 DLU J W Challis, 1 Slades Cl, Chestfield, Whitstable, CT5 3NN [JO01MI, TR16]
G1 DMC D J Frith, Laindon, 135 Mellow Purgess, Basildon, SS15 5XA [JO01FN, TQ68]
G1 DME N E W Mendham, Cranbrook, 2A Wingate Rd, Folkstone, Kent, CT19 5QE [JO01OC, TR23]
G1 DMF M J Thorogood, 13 Pine Gr Park, Swavesey, Cambridge, CB4 5RG [IO92XH, TL36]
G1 DMH L J Lees, 59 Beresford Rd, Long Eaton, Nottingham, NG10 3EF [IO92IV, SK43]
G1 DMJ D M Jones, Shasta, 12 Portland Dr, Walton Hills, Much Wenlock, TF13 6EY [IO82RO, SO69]
G1 DMM L C Jackson, Eastlea, 43 Hazelhurst St, Hanley, Stoke on Trent, Staffs, ST1 3HD [IO83WA, SJ84]
G1 DMN P R Snow, 14 Bewholme Ave, Darlington, DL3 7HP [IO94FM, NZ21]
G1 DMR R K Manser, 53 Downs Barn Boulevard, Downs Barn, Milton Keynes, MK14 7LL [IO92PB, SP84]
G1 DMS D M Segal, 1 Mason House, 1/3 Valley Dr, Kingsbury, London, NW9 9NQ [IO91UN, TQ18]
G1 DMW F E Latham, Higher Ln, Parbold, Dalton, Wigan, Lancs, WN8 7RA [IO83ON, SD40]
G1 DMX Details withheld at licensee's request by SSL.
G1 DNA J A Birkmyre, 21 Magenta Cres, St. Johns, Newcastle upon Tyne, NE5 1YL [IO95DA, NZ16]
G1 DNI W G Darling, 2 Strathaird Ave, Walney, Barrow in Furness, LA14 3DE [IO84JC, SD16]
G1 DNK B S Cunningham, 1 Martins Dr, Ferndown, BH22 9SG [IO90BT, SU00]
G1 DNP R A Collins, 5 Sussex Gdns, Woodley, Reading, RG5 4JN [IO91NK, SU77]
G1 DNQ L V Collins, 12 Maple Gr, Roundswell, Barnstaple, EX31 3QP [IO71WB, SS53]
G1 DNX K Clegg, Wolverhampton Rd, Cookey, Cookley, Kidderminster, Worcs, DY10 3RX [IO82VK, SO88]
G1 DNY R A Clay, 38 Hubbards Rd, Chorleywood, Rickmansworth, WD3 5JJ [IO91RP, TQ09]
G1 DNZ G P Clarke, 24 Brisbane Dr, Stapleford, Nottingham, NG9 8ND [IO92IW, SK43]
G1 DOA K J Chappell, 17 Linton Cl, Winyates East, Redditch, B98 0NA [IO92BH, SP06]
G1 DOG H Foster, Ringstone, 30 Ivy Rd, Macclesfield, SK11 8QB [IO83WG, SJ97]
G1 DOG I R Cheeseman, 445 Uttoxeter Rd, Blythe Bridge, Stoke on Trent, ST11 9NT [IO82XX, SJ94]
G1 DOJ T D Brodrick, South Bank, Hallathrow Rd, Paulton, Bristol, BS18 5LJ [IO81RH, ST65]
G1 DOK A G Brixton, 5 Don Rd, Worcester, WR4 9ET [IO82VE, SO85]
G1 DOL R M Breakspear, Parkside, 78 Main Rd, Long Hanborough, Witney, OX8 8JY [IO91HT, SP41]
G1 DON D J Macnamara, 56 Macdonald St., Orrell, Wigan, WN5 0AJ [IO83PM, SD50]
G1 DOR G T Bennett, 6 Wordsworth Dr, Market Drayton, TF9 1ND [IO82SV, SJ63]
G1 DOS P R Barrett, 7 Mead Cl, Stoke St. Michael, Bath, BA3 5JB [IO81SF, ST64]
G1 DOT D J Barker, 60 Rolvenden Rd, Wainscott, Rochester, ME2 4PG [JO01GJ, TQ77]
G1 DOW Details withheld at licensee's request by SSL.
G1 DPB J E Ashby, 95 Southway, Horsforth, Leeds, LS18 5RW [IO93EU, SE23]
G1 DPH B W Barber, 15 Hedingham Rd, Dagenham, RM8 2NA [JO01BN, TQ48]
G1 DPI A A Barratt, 23 Wilberforce Rd, Anston, Sheffield, S31 7EG [IO93GJ, SK38]
G1 DPJ C D Beasley, 12 East Leys Ct, Moulton, Northampton, NN3 7TX [IO92NG, SP76]
G1 DPN D Bettany, 10 Redbrook Cres, Melton Mowbray, LE13 0EU [IO92NS, SK71]
G1 DPT T M Cairney, Hall Ln, Walton, Lutterworth, Leics, LE17 5RP [IO92KL, SP58]
G1 DPW S W Cmoch, 68 Blenheim Cres, London, W11 1NZ [IO91VM, TQ28]
G1 DPX R P Colley, 12 Glenfield Rd, Banstead, SM7 2DG [IO91VH, TQ25]
G1 DPY A Cooke-Sanderson, Flat 4, 65 Ellis Rd, Crowthorne, RG45 6PP [IO91OI, SU86]
G1 DQB D J Cowtan, 32 Finmere, Avon Park, Rugby, CV21 1RT [IO92JJ, SP57]
G1 DQD C A Anderson, 11 Swallowfield Dr, Hessle High Rd, Hull, HU4 6UG [IO93TR, TA02]
G1 DQL R I Bradley, 1 Audley Pl, Sutton, SM2 6RW [IO91VI, TQ26]
G1 DQQ D A Dwight, 19 The Highway, Stanmore, HA7 3PL [IO91UO, TQ19]
G1 DQR E C H Eccleston, 24 Milton Rd, Weston Super Mare, BS23 2SL [IO81MI, ST36]
G1 DQS A C Edmunds, 16 The Greenway, Hurst Green, Oxted, RH8 0JZ [JO01AF, TQ45]
G1 DQU M Elliott, 52 Wellfield Rd, Alrewas, Burton on Trent, DE13 7EZ [IO92DR, SK11]
GW1 DQV B H Emary, 2 The Paddocks, Penarth, CF64 5BW [IO81JK, ST17]
G1 DQX Details withheld at licensee's request by SSL.
G1 DQZ J T Rimmer, Church View, Mere Ln, Halsall, Ormskirk, L39 8RT [IO83MN, SD30]
G1 DRA Details withheld at licensee's request by SSL.
G1 DRF Details withheld at licensee's request by SSL.
G1 DRG G R Foster, 19 Asquith Ave, Burnholme, York, YO3 0PZ [IO93LX, SE65]
G1 DRI K J Gill, 33 Hazel Croft, Peterborough, PE4 5BJ [IO92UO, TF10]
G1 DRM D R Merry, 266 The Rowans, Milton, Cambridge, CB4 6ZL [JO02BF, TL46]
G1 DRP Details withheld at licensee's request by SSL.
GW1 DRQ C G Morrison, 5 The Beeches, Holywell, CH8 7SW [IO83JG, SJ17]
G1 DRR Details withheld at licensee's request by SSL.
G1 DRW D I Hart, 72 Wye Dean Rise, Hereford, HR2 7XZ [IO82PB, SO43]
G1 DRX P E D Hodson, 83 Princess Margaret Ave, Cliftonville, Margate, CT9 3EF [JO01RJ, TR37]
G1 DRY S Cox, 25 Church Cl, Stoke St. Gregory, Taunton, TA3 6HA [IO81MA, ST32]
G1 DSA J R Critchley, 3 Beaconsfield View, Robert Rd, Hedgerley, Slough, SL2 3XT [IO91QN, SU98]
G1 DSB J H Darling, 145 Hartlands, Bedlington, NE22 6JJ [IO95ED, NZ28]
G1 DSC G Davies, 30 Hillside Rd, Blidworth, Mansfield, NG21 0TR [IO93KC, SK55]
G1 DSE R L M Daw, 6 Hall Dr, Weston Coyney, Stoke on Trent, ST3 6PF [IO82WX, SJ94]
G1 DSF A M Daw, 19 Rowan Cl, Walton, Stone, ST15 0EP [IO82WV, SJ93]
G1 DSG M E Degerdon, 25 Rosslyn Rd, Billericay, CM12 9JN [JO01EO, TQ69]
G1 DSI W S Mockett, 23 Moor Ln, Maidenhead, SL6 7JX [IO91PM, SU88]
G1 DSJ P Q Morgan, 29 Brisbane Rd, Reading, RG3 2PE [IO91LJ, SU66]
GM1 DSK D S Keay, 34 King St., Stanley, Perth, PH1 4NA [IO86GL, NO13]
G1 DSM A W T Hicks, 5 Restwell Ave, Cranleigh, GU6 8PQ [IO91RD, TQ04]
G1 DSP D Hoult Spalding&Dis AR, Chespool House, Gosberton Risegate, Spalding, Lincs, PE11 4EU [IO92VU, TF23]
G1 DSQ V J Huggett, 20 Thornbury Ave, Weeley, Clacton-on-Sea, Essex, CO16 9HN [JO01NU, TM12]
G1 DSY B C Petch, 2 Bedwell Rd, Tottenham, London, N17 7AH [IO91XO, TQ39]
G1 DSZ J R Phillips, 20 The Meadows, Broomfield, Herne Bay, CT6 7XF [JO01NI, TR16]
GW1 DTA M J Pilot, 92 Llanllienwen Rd, Morriston, Cwmrhydyceirw, Swansea, SA6 6LU [IO81AQ, SS69]
G1 DTB R Powis, 85 Walsall Rd, West Bromwich, B71 3HH [IO92AM, SP09]
G1 DTC A K Quarterman, 30 Hope Cl, The Watergardens, Sutton, SM1 4AT [IO91VI, TQ26]
G1 DTE W Merz, 38 Lime Ave, Colchester, CO4 3NL [JO01LV, TM02]
G1 DTF A J Middleton, 2 Beccles Way, Bramley, Rotherham, S66 0SJ [IO93IK, SK49]
GI1 DTH H T Arnold, 4 Drumclay Rd, Enniskillen, N Ireland, BT74 6NG [IO64EI, H24]
GM1 DTJ R Main, 39 Balgove Ave, Gauldry, Newport on Tay, DD6 8SQ [IO86LJ, NO32]
G1 DTS E J Kier, 9 Newbridge Way, Truro, TR1 3LX [IO70KG, SW84]
G1 DTV Details withheld at licensee's request by SSL.
G1 DUB J Lisle, 213 Whitehall Rd, Gateshead, NE8 4PS [IO94EW, NZ26]
G1 DUE Details withheld at licensee's request by SSL.
G1 DUF Details withheld at licensee's request by SSL.
G1 DUH Details withheld at licensee's request by SSL.
G1 DUI P A J Norman, 3 Church View, Witchford, Ely, CB6 2HH [JO02CJ, TL57]
G1 DUJ B A Oakley, 6 Staplehurst Gdns, Cliftonville, Margate, CT9 3JB [JO01RJ, TR37]
G1 DUL A J Paddon, 107 Wolsey Ave, Intake, Doncaster, DN2 6EL [IO93KM, SE50]
G1 DUO P P Richards, 114 Northleach Cl, Church Hill, Redditch, B98 8RD [IO92BH, SP06]
G1 DUS D Roberts, Westpark, 296 Westleigh Ln, Leigh, WN7 5PW [IO83RM, SD60]
G1 DUT J W Robertson, Everslea, Congleton Rd, Kermincham, Congleton, CW12 2LL [IO83UF, SJ86]
G1 DUU W E Sage, Jewalls, 25 Springfield, Norton St. Philip, Bath, BA3 6NR [IO81UH, ST75]
G1 DUV Details withheld at licensee's request by SSL.
G1 DVA P T Middlehurst, 17 Greeba Ave, Warrington, WA4 6AP [IO83QJ, SJ68]
G1 DVD B R Marshall, 1 Anglers Way, Chesterton, Cambridge, CB4 1TZ [JO02BF, TL46]
G1 DVH J D Knighton, 90 Sherwood Cres, Longford Gdns, Market Drayton, TF9 1NP [IO82RV, SJ63]
GM1 DVO C Hepworth, 4 Northfield Farm Cttgs, St. Abbs, Eyemouth, Berwickshire, TD14 5QF [IO85WV, NT96]
G1 DVP R M Head, 12 Rookley Ct, Linnet Way, Purfleet, RM19 1TW [JO01DL, TQ57]
G1 DVT S Green, 61 Lees Rd, Mossley, Ashton under Lyne, OL5 0PG [IO83XM, SD90]
G1 DVU J M Green, 788 The Ridge, St. Leonards on Sea, TN37 7PS [JO00GV, TQ81]
G1 DVV D A Green, 44 Maple Rd, Boston, PE21 0BZ [IO92XX, TF34]
G1 DVW P A Gould, 40 Reynolds Cl, Flanderwell, Rotherham, S66 0XL [IO93IK, SK49]
G1 DVX D A Gibbons, 2 Marshcroft Dr, Cheshunt, Waltham Cross, EN8 8XA [IO91XQ, TL30]
G1 DWC R C Everett, First Floor Flat, 73 Fordwych Rd, London, NW2 3TL [IO91VN, TQ28]
G1 DWI D C Shaw, 34A Harding Ave, Upper Haugh, Rawmarsh, Rotherham, S62 7ED [IO93HL, SK49]
G1 DWL B Simpson, 9 Dunheved Rd South, Thornton Heath, CR7 6AD [IO91WJ, TQ36]
G1 DWN K M Smart, 2 St. Andrews Cl, Pedmore, Stourbridge, DY8 2LR [IO82WK, SO98]
GU1 DWO A M Smith, La Cambrette, Rue Des Reines, Forest, Guernsey, Channel Islands, GY8 0JB
G1 DWR D Soar, 8 Northcliffe Ave, Nottingham, NG3 6DA [IO92KX, SK64]
G1 DWT J N Dwight, 59 Highfield Rd, Bramley, Leeds, LS13 2BX [IO93ET, SE23]
G1 DWU A P Swales, 6 Conifer Gr, Great Sankey, Warrington, WA5 3BQ [IO83QJ, SJ58]
G1 DXD P D Darke, 18 Colchester Cl, Prittlewell, Southend on Sea, SS2 6HR [JO01IN, TQ88]
G1 DXG J W Cross, Stonehaven, 33 Silver St., Lyme Regis, DT7 3HS [IO80MR, SY39]
G1 DXH R P Crissell, 182 Lodge Ln, Grays, RM16 2TP [JO01DL, TQ67]
G1 DXQ P A Postle, 20 Courtenay Cl, Norwich, NR5 9LB [JO02OP, TG10]
G1 DXS Details withheld at licensee's request by SSL.
G1 DXU J D Welch, 76 Ten Acre Way, Valley Park, Rainham, Gillingham, ME8 8TL [JO01HI, TQ86]
G1 DYB A L S White, 5 New Rd, Fairoak, Eastleigh, SO50 8EN [IO90IX, SU41]
G1 DYC D A Winkley, 14 Paget Cl, Crofters Meade, Bromsgrove, B61 7JE [IO82XI, SO97]
G1 DYD Details withheld at licensee's request by SSL.
G1 DYE Details withheld at licensee's request by SSL.
G1 DYN Details withheld at licensee's request by SSL.[Station located near Runcorn. Locator: IO83PH. Op: J V Snowling. Prestel MBX 219996483. QRV 6m/2m/70cms.]
G1 DYP V Salem, 46 Chatsworth Rd, Ealing, London, W5 3DB [IO91UM, TQ18]
G1 DYQ N R Prosser, 35 Holmfirth Cl, Belmont, Hereford, HR2 7UG [IO82PA, SO43]
G1 DYR R A Munday, 12 Glisson Rd, Hillingdon, Uxbridge, UB10 0HH [IO91SM, TQ08]
G1 DYT P J Allen, 8 Ashmead Cres, Birstall, Leicester, LE4 4GS [IO92KQ, SK61]
G1 DYW R J Andrews, 262 Malden Way, New Malden, KT3 5QT [IO91UJ, TQ26]
G1 DZB N J Babbage, 248 Molesey Ave, West Molesey, KT8 2ET [IO91TJ, TQ16]
G1 DZD L D Ball, 14 St. Wilfrids Rd, Burgess Hill, RH15 8BD [IO90WX, TQ31]
G1 DZF G S Barlow, 6 Margaret House, Ancaster Ave, Chapel St. Leonards, Skegness, PE24 5SL [JO03EF, TF57]
G1 DZH A Fisher, 11 Rosedale Way, Forest Town, Mansfield, NG19 0QR [IO93KD, SK56]
G1 DZM Details withheld at licensee's request by SSL.
G1 DZO M P Bird, 194 Harwoods Rd, Watford, Herts, WD1 7RT [IO91TP, TQ19]
G1 DZP K Bishop, 31 Ravenslea, Ravenstone, Coalville, Leics, LE6 2AT [IO92JP, SK50]
G1 DZQ A J Boreham, 4 Walnut Rd, Kirton, Boston, PE20 1XG [IO92XW, TF33]
G1 DZR J Boon, 137 Chantry Rd, Northallerton, DL7 8JJ [IO94GI, SE39]
G1 DZY K Bricknall, 99 Dykes Way, Gateshead, NE10 8QN [IO94FW, NZ26]
G1 DZZ K E Bridle, 8 Hardy Ave, Dorchester, DT1 1LL [IO80SR, SY69]
G1 EAB A Bolton, 5 Willow Cres, Gedling, Nottingham, NG4 4BL [IO92LX, SK64]
G1 EAE A J Broughton, Cobwebs, The Fleet, Fittleworth, Pulborough, RH20 1HS [IO90RX, TQ01]
GM1 EAF R McMurray Perth & Dist AR, 36 Low Rd, Perth, PH2 0NF [IO86GJ, NO12]
GI1 EAG Details withheld at licensee's request by SSL.
GM1 EAH W S Buchanan, 167 Muirfield Dr, Newcastle Pcnt, Glenrothes, KY6 2PX [IO86JE, NO20]
G1 EAJ M D Bunting, 22 Ling Cl, Coltishall, Norwich, NR12 7HZ [JO02QR, TG22]
G1 EAM A S Bush, 15 Pelton Ave, Belmont, Sutton, SM2 5NN [IO91VI, TQ26]
G1 EAN A A Butler, Ty Ni, Hall Ln, Harbury, Leamington Spa, CV33 9HG [IO92GF, SP36]
G1 EAP L W Moody, 26 Sinodun View, Warborough, Wallingford, OX10 7DF [IO91KP, SU59]
GW1 EAV S M Davies, Laburnum House, Guilsfield, Welshpool, Powys, SY21 9PX [IO82KQ, SJ21]
G1 EAX R G Dawkins, 17 Dacer Cl, Stirchley, Birmingham, B30 3BZ [IO92AK, SP08]
G1 EAZ G B Denton, 11 Highland Rd, Amersham, Bucks, HP7 9AU [IO91QQ, SU99]
G1 EBA A D'Agostino, Nuts, 2 Kings Rd, Lancing, BN15 8EA [IO90UT, TQ10]
G1 EBB A A Di Duca, 15 Moray Cl, Off Firth Park Cres, Halesowen, B62 9PP [IO82XL, SO98]
G1 EBD Details withheld at licensee's request by SSL.
G1 EBL J Freer, 54A High Ln East, West Hallam, Ilkeston, DE7 6HW [IO92HX, SK44]
G1 EBP C E Jermany, 5 Lexington Cl, Hemsby, Great Yarmouth, NR29 4ES [JO02UQ, TG41]
G1 EBS A W Johnstone, 49 Moorland Rd, Pudsey, LS28 8EN [IO93DT, SE23]
G1 EBT A F Jones, Gelli Aur, 56 Copthorne Dr, Shrewsbury, SY3 8RX [IO82OR, SJ41]
G1 EBU L B Butters, The White House, 58 Glendale Ave, Edgware, HA8 8HH [IO91UO, TQ19]
G1 EBV S J Cade, 6 Ct Cl, Kirby Muxloe, Leicester, LE9 2DD [IO92JP, SK50]
G1 EBW P J Challen, 45 Reedswood Rd, St. Leonards on Sea, TN38 8DW [JO00GU, TQ70]
G1 EBX S M Challen, 21 Fletcher Ave, St. Leonards on Sea, TN37 7QU [JO00GV, TQ71]
G1 EBZ R I Charlton, Meer Booth Rd, Antons Gowt, Boston, Lincs, PE22 7AB [IO93WA, TF24]
G1 ECA I Cheek, 179 Herbert Rd, High Wycombe, HP13 7HR [IO91PP, SU89]
G1 ECC D R Chippendale, 19 East Park Ave, Darwen, BB3 2SQ [IO83SQ, SD62]
G1 ECD I F A Chubb, 2 Mount Pleasant, Perry St., Chard, TA20 2QG [IO80MU, ST30]

G1 ECE B P Clark, 9 Conigree, Chinnor, OX9 4JY [IO91MQ, SP70]
G1 ECI J A Christy, 1 Edinburgh Dr, Hindley Green, Hindley, Wigan, WN2 4HL [IO83RM, SD60]
G1 ECV J H Gardener, 32 Beckington Cres, Chard, TA20 2BU [IO80MU, ST30]
G1 ECZ Details withheld at licensee's request by SSL.
G1 EDA R F G Goff, 21 Findon Rd, Elson, Gosport, PO12 4EP [IO90KT, SU60]
G1 EDD R W Green, 18 Yardley Green, Aylesbury, HP20 2HE [IO91OT, SP81]
G1 EDE D C Graydon, 12 Lespic St., Northallerton, North Yorks, DL7 8QY [IO94GI, SE39]
G1 EDK P L Hammond, Flat 4, 82 Rowland Dr, Herne Bay, CT6 7SD [JO01NI, TR16]
G1 EDM J Hargreaves, 5 Nuttall Ave, Little Lever, Bolton, BL3 1PW [IO83TN, SD70]
G1 EDO T Harrison, 4 Hornby Ave, Blackley, Manchester, M9 6QZ [IO83VM, SD80]
G1 EDP M V Hazell, The Prospect, Lords Hill, Coleford, GL16 8BG [IO81QT, SO51]
G1 EDS J B Hedgecock, 11 Foxmoor Cl, Oakley, Basingstoke, RG23 7BQ [IO91JG, SU55]
G1 EDT W A Hewitt, 99 Derrydown Rd, Perry Barr, Birmingham, B42 1RY [IO92BM, SP09]
G1 EDU A G Hicks, 91 Warwick Rd, Broughton Astley, Leicester, LE9 6SA [IO92JM, SP59]
G1 EDX M Holtam, 16 Cowley Cl, Cheltenham, GL51 6NP [IO81WV, SO92]
G1 EEA S Howcroft, 23 Alderley Ave, Blackpool, FY4 1QG [IO83LS, SD33]
G1 EEC J E Hubbard, 3 Burder St., Loughborough, LE11 1JH [IO92JS, SK52]
G1 EEL D S King, 32 Hamlet St., Warfield, Bracknell, RG42 3EF [IO91PK, SU87]
G1 EEM P W King, 555 Reading Rd, Winnersh, Wokingham, RG41 5HJ [IO91NK, SU77]
G1 EEO M J Kirby, Church Cottage, Burrington, Umberleigh, Devon, EX37 9JG [IO80AW, SS61]
G1 EEQ S P Musto, 8B Millbeck Ct, Lawson Ave, Cottingham, North Humberside, HU16 4EY [IO93TS, TA03]
G1 EEY A A Levinas, 102 Warrenside, Huddersfield, HD2 1LG [IO93CQ, SE11]
GW1 EEZ J R Lewis, 80 Pencisely Rd, Llandaff, Cardiff, CF5 1DQ [IO81JL, ST17]
G1 EFA R D Liddiard, Yew Tree Cottage, Mill Rd, Wyverstone, Stowmarket, IP14 4SE [JO02LG, TM06]
G1 EFF A D Marriott, 75 St. Johns Rd, Cudworth, Barnsley, S72 8DE [IO93HN, SE30]
G1 EFG C V McAra, 95 Clock House Rd, Beckenham, BR3 4JU [IO91XJ, TQ36]
G1 EFK G J Means, Ferry Farm, Witham Bank, Chapel Hill, Lincoln, LN4 4QA [IO93VB, TF25]
G1 EFL M N Medcalf, 47 Paddock Dr, Chelmsford, CM1 6UX [JO01FS, TL70]
G1 EFM G Hunt, Ellsmere, St. Johns Rd, Belton, Great Yarmouth, NR31 9JT [JO02UN, TG40]
G1 EFO P Hyde, 212 King St., Cottingham, HU16 5QJ [IO93TS, TA03]
G1 EFS S A M Newell, 7 Edward Rd West, Walston St. Mary, Clevedon, BS21 7DY [IO81NK, ST47]
G1 EFT P W Nicholson, 24 Higher Ranscombe Rd, Brixham, TQ5 9HF [IO80FJ, SX95]
G1 EFU A Nixon, 14 Carlton Rd, Lowton St. Lukes, Lowton, Warrington, WA3 2EP [IO83RL, SJ69]
G1 EFX C R A Nutkins, Higher Spence, Wooton Fitzpaine, Charmouth, Dorset, DT6 6DF [IO80MS, SY39]
G1 EFY P K W Owers, 8 The Cres, Eastbourne, BN20 8PH [JO00DS, TQ50]
G1 EGB G H Page, 23 Maskelyne Cl, Battersea, London, SW11 4AA [IO91WL, TQ27]
G1 EGC Details withheld at licensee's request by SSL.[Op: M A F Page. Station located near Colchester.]
G1 EGE K J Pay, 11 Willow St., Congleton, CW12 1RL [IO83VD, SJ86]
G1 EGK Dr J M Preece, 9 Skippon Terr, Thorner, Leeds, LS14 3HA [IO93GU, SE34]
G1 EGL R G Preston, 45 Gaynor Cl, Wymondham, NR18 0EA [JO02NN, TG10]
G1 EGM D R Pullen, 20 Sussex Rd, Knaphill, Woking, GU21 2RA [IO91QH, SU95]
G1 EGO A S Redgate, Radcliffe Mobile Home Park, 38 Oak Ave, Radcliffe on Trent, Nottingham, NG12 2AP [IO92LW, SK63]
G1 EGU J E Morton, 1 Blandford Rd, Ipswich, IP3 8SL [JO02OB, TM14]
G1 EGZ A Adams, Radnor, Shorts Rd, Carshalton, Surrey, SM5 2PB [IO91VI, TQ26]
G1 EHA B Ainley-Smith, Galen House, Middleton-on-Sea, W Sussex, PO22 6DB [IO90QT, SU90]
G1 EHB P J Allcock, 15 Highlands Park, Seal, Sevenoaks, TN15 0AQ [JO01CG, TQ55]
G1 EHE M Appleton, 8 Headon View, West Meon, Petersfield, GU32 1LH [IO91KA, SU62]
G1 EHF Details withheld at licensee's request by SSL.[Op: D J Austen. Station located near Bracknell.]
GW1 EHI R Davies, 68 Edward St., Fairview, Blackwood, NP2 1NY [IO81JP, ST19]
GM1 EHK D K Birch, 52 Watters Cres, Lochgelly, KY5 9LD [IO86IC, NT19]
G1 EHM P A Bird, 4 Parkside Ave, Tilbury, RM18 8DT [JO01EL, TQ67]
G1 EHR V J Brooksbank, 43 Salcombe Dr, Glenfield, Leicester, LE3 8AG [IO92JP, SK50]
G1 EHS B C Brodribb, 1 Ponswood Rd, St. Leonards on Sea, TN38 9BU [JO00GU, TQ71]
G1 EHT G J Hobbs, 39 Winchester Way, Cheltenham, GL51 5EZ [IO81WV, SO92]
G1 EHU M Hostekens, 1 Ponswood Rd, St. Leonards on Sea, TN38 9BU [JO00GU, TQ71]
G1 EHX C A C Cameron, The Nook, Bridstow, nr Ross-on-Wye, Herefordshire, HR9 6QJ [IO81QW, SO52]
G1 EHY C Robinson, 56 Stanley Rd, Radcliffe, Manchester, M26 4HG [IO83TN, SD70]
G1 EIB N Purkins, 32 Astral Rd, Hessle, HU13 9DD [IO93SR, TA02]
G1 EIH D Samber, 102 Midsummer Ave, Hounslow, TW4 5BB [IO91TL, TQ17]
G1 EIL C E D Sharp, 3 Howard Rd, North Holmwood, Dorking, RH5 4HZ [IO91UF, TQ14]
G1 EIO B Smith, 43 Oak Ave, Hindley Green, Hindley, Wigan, WN2 4LZ [IO83RM, SD60]
G1 EIP G H Smith, 52 Penhill Cres, St. Johns, Worcester, WR2 5PX [IO82VE, SO85]
G1 EIR R C Smith, 29 Windmill Ln, Henbury, Bristol, BS10 7XE [IO81QM, ST57]
G1 EIT H J Stalley, Shootisham, Woodbridge, Suffolk, IP13 3EJ [JO02PE, TM26]
G1 EIV S F Stanley, 11 Mandeen Gr, Loxley Park, Mansfield, NG18 4FA [IO93JD, SK56]
G1 EIX H S Stephens, 23 Eland Way, Cherry Hinton, Cambridge, CB1 4XQ [JO02CE, TL45]
G1 EIZ M W Stewart, 2 Patmore Link Rd, Hemel Hempstead, HP2 4PX [IO91SS, TL00]
G1 EJA R A Stone, 1 Poplar Cl, Godington Park, Ashford, TN23 3DY [JO01KD, TQ94]
G1 EJB P M Struve, 3 Monkton Cl, Swindon, SN3 2EU [IO91DN, SU18]
G1 EJE Details withheld at licensee's request by SSL.
G1 EJH F Thomasson, 2, Kershaw Walk, Manchester, Lancs, M12 4AL [IO83VL, SJ89]
G1 EJK G M Tomlinson, 4 Werneth Cl, Denton, Manchester, M34 6LR [IO83WK, SJ99]
G1 EJX Details withheld at licensee's request by SSL.
G1 EKC M J Davis, 6 Fords Cl, Bledlow Ridge, High Wycombe, HP14 4AP [IO91NQ, SU79]
G1 EKH A J Doxey, 1 Crabtree Cl, Wirksworth, Matlock, DE4 4FP [IO93FB, SK25]
G1 EKM S J H Evans, 93 Park Meadow, Hatfield, AL9 5HE [IO91VS, TL20]
G1 EKP G D K Fowler, 18 Lossie Dr, Iver, SL0 0JS [IO91RM, TQ08]
G1 EKU N W Kernahan, 16 Melrose Ave, Churchtown, Southport, PR9 9UY [IO83MQ, SD32]
G1 EKX J T Green, Clearview, New Mills, Newtown, Powys, SY16 3NT [IO82HO, SJ00]
G1 ELE R S Watt, 88 Graham Cres, Mile Oak, Portslade, Brighton, BN41 2YB [IO90VU, TQ20]
G1 ELI R W Weymouth, 5 Blundell Pl, Blackwell, Carlisle, CA2 4SJ [IO84MU, NY45]
G1 ELJ R D Williams, 6 Jay Rd, Kingswinford, DY6 7RR [IO82WM, SO88]
G1 ELK P J Wilson, 16 Hamilton Rd, Cockfosters, Barnet, EN4 9HE [IO91WP, TQ29]
G1 ELQ B Bentley, Sandy Ridge, Church St., Rookery, Stoke on Trent, ST7 4RS [IO83VC, SJ85]
G1 ELX A J Challinor, Las Alondras, 12 Trussell Cl, Acton Trussell, Stafford, ST17 0RL [IO82WS, SJ91]
G1 ELZ M J Cook, 8 Thompson Cl, Salisbury, SP2 8QU [IO91CB, SU12]
G1 EME P L Crosland obo The Worcester Moonbounce S, Sprackets Orchard, Curry Rivel, Langport, Somerset, TA10 0PP [IO81NA, ST32]
G1 EML H A Hill, 3 Dale Cl, Wrecclesham, Farnham, GU10 4PQ [IO91OE, SU84]
G1 EMM K J Hill, 3 Dale Cl, Wrecclesham, Farnham, GU10 4PQ [IO91OE, SU84]
G1 EMR W J Harris East Manchester, 17 Redacre Rd, Manchester, M18 8RU [IO83WL, SJ89]
G1 EMT J T Hopkins, 70 Redhouse Ln, Aldridge, Walsall, WS9 0DQ [IO92AO, SK00]
G1 EMU C Howard, 8 Glenmore Rd, Exeter, EX2 5HB [IO80GR, SX99]
G1 EMW R G Dearsley, Prince William Farm, Main Rd, Wereham, Kings Lynn, Norfolk, PE33 9BD [JO02FO, TF60]
GW1 EMZ C E Digweed, Tyn yr Ardd, The Pant, Penrhyndeudraeth, Gwynedd, LL48 6NH [IO72XW, SH63]
G1 ENA G S Edwards, 22 Whalley Ln, Uplyme, Lyme Regis, DT7 3UR [IO80MR, SY39]
G1 END M R Friedman, 23 Northway Ct, Green Ave, Mill Hill, London, NW7 4PY [IO91UO, TQ29]
GW1 ENG P S Gibson, The Nook, Trimsaran Rd, Kidwelly, SA17 4EB [IO71UR, SN40]
GM1 ENI J J Goode, 55 Belwood Rd, Milton Bridge, Penicuik, EH26 0QN [IO85JU, NT26]
G1 ENN M J Hart, Ham Rd, Brean, Berrow, Burnham on Sea, Somerset, TA8 2RN [IO81MG, ST35]
G1 ENP P J Hewett, 117 Leopold Ave, Birmingham, B20 1EX [IO92AM, SP09]
G1 ENR A D Deakin, 53 High Haden Rd, Cradley Heath, Warley, B64 7PJ [IO82XL, SO98]
G1 ENS A E Jones, Station House, Station Rd, Gobowen, Oswestry, SY11 3JS [IO82LV, SJ33]
GM1 EOA J D Minaudo, Meadowside, Old Glasgow Rd, Newbridge, Dumfries, DG2 0QX [IO85EC, NX97]
G1 EOC D G Hurn, 26 Blakedown Rd, Linslade, Leighton Buzzard, LU7 7XJ [IO91PV, SP92]
G1 EOH S P Braybrooke, 6 Tubbenden Ln, Orpington, BR6 9PN [JO01BI, TQ46]
GW1 EOI G S John, 31 Gellifawr Rd, Garnlwyd, Morriston, Swansea, SA6 7PN [IO81AP, SS69]
G1 EOJ M P Kay, 9 Dove Tree Rd, Leighton Buzzard, LU7 8UP [IO91QW, SP92]
G1 EOK E Keeble, 17 Moat Ave, Green Ln, Coventry, CV3 6BT [IO92FJ, SP37]
G1 EOM H W Kinghorn, 22 Hillrise Ave, Sompting, Lancing, BN15 0LX [IO90UU, TQ10]
G1 EOO T G Lancaster, 219 Moor Ln, Chessington, KT9 2AB [IO91UI, TQ16]
G1 EOQ E V Lawrence, The Mendips, 10 Hunts Mead, Lenthay, Sherborne, DT9 6AL [IO80RW, ST61]
GI1 EOS P W Leitch, 212 Belfast Rd, Muckamore, Antrim, BT41 2EY [IO64WQ, J18]
G1 EOU N D Levin, 23 Sheerstock, Haddenham, Aylesbury, HP17 8EZ [IO91MS, SP70]
G1 EOV Details withheld at licensee's request by SSL.
G1 EOY R Lund, 2 Winter Hill Cotts, Darwen, Lancs, BB3 0LB [IO83SQ, SD62]
G1 EPD D Gardener, 960 Walker Rd, Walker, Newcastle upon Tyne, NE6 3JL [IO94FX, NZ26]
G1 EPF L Marshall, 75 Acacia Cres, Beech Hill, Wigan, WN6 8NJ [IO83QN, SD50]
G1 EPJ E Murphy, 64 Springfield Rd, Gatley, Cheadle, SK8 4PF [IO83VJ, SJ88]
G1 EPO J K Pragnell, Sundale, Northampton Rd, Whitfield, Brackley, NN13 7TY [IO92KB, SP53]
GW1 EPR R Rees, 16 Railway View, Caldicot, Newport, NP6 4GB [IO81PO, ST48]
G1 EPV Details withheld at licensee's request by SSL.
G1 EPZ H A Skippon, 3 Thornborough Cres, Leyburn, DL8 5DY [IO94CH, SE19]
G1 EQB R D Stringer, Mount Cottage, Lower Penkridge Rd, Acton Trussell, Stafford, ST17 0RJ [IO82WS, SJ91]
G1 EQF P R Uttridge, 11 Fiddington Clays, Market Lavington, Devizes, SN10 4BT [IO91AG, SU05]
G1 EQL P N J Wootton, 20 Oakhill Rd, Dronfield, Sheffield, S18 6EJ [IO93GH, SK37]
G1 EQM R N Agacy, 23 Highgate Ln, Bolton on Dearne, Bolton upon Dearne, Rotherham, S63 8HR [IO93IM, SE40]
G1 EQT Details withheld at licensee's request by SSL.
G1 EQU R H Percival, 23 Plumtree Rd, Church Ln, Thorngumbald, Hull, HU12 9QG [IO93VR, TA22]
GW1 ERA A E Price, 23 Dol Wen, Pencoed, Bridgend, CF35 6RS [IO81GM, SS98]
G1 ERB R D Pruitt, 5 Parrington Way, Lawford, Manningtree, CO11 2LZ [JO01MW, TM03]
G1 ERC D T Richardson, 13 Lime Cl, Langtoft, Peterborough, PE6 9RA [IO92TQ, TF11]
G1 ERF S J Rogers, 31 Morgan Rd, Southsea, PO4 8JS [IO90LT, SZ69]
G1 ERG J A Scofield, 4 Meadow Cttgs, Main St., Grandborough, Rugby, CV23 8DQ [IO92IH, SP46]
G1 ERJ Details withheld at licensee's request by SSL.
G1 ERM D J Salter, 81 Norfolk St., Cambridge, CB1 2LD [JO02BE, TL45]
G1 ERO M Scott, 36 Glebe Cres, Forest Hall, Newcastle upon Tyne, NE12 0JR [IO95FA, NZ27]
G1 ERQ R W Stevens, 172 Branksome Ave, Stanford-le-Hope, SS17 8DE [JO01FM, TQ68]
G1 ERS D J Strange, 90 Combe Rd, Farncombe, Godalming, GU7 3SL [IO91QE, SU94]
G1 ERU S M Mole, 16 Sharpley Dr, Seaton Farm Est, Seaham, SR7 0LE [IO94HU, NZ44]
G1 ERV A W Moseley, 15 Gillsway, Northampton, NN2 8HT [IO92NG, SP76]
G1 ERZ A J Moules, 5 Hill Rd, Borstal, Rochester, ME1 3NJ [JO01FI, TQ76]
G1 ESA S J Abrahams, 33 Marle Croft, Whitefield, Manchester, M45 7NB [IO83UN, SD80]
G1 ESC C O Mountain, 1 Bryony Way, Deeping St. James, Peterborough, PE6 8SZ [IO92UQ, TF11]
G1 ESE J Alvey, 43 Alnwick St., Newburn, Newcastle upon Tyne, NE15 8PT [IO94DX, NZ16]
G1 ESL G R Davies, 32 Umberslade Rd, Selly Oak, Birmingham, B29 7RZ [IO92AK, SP08]
G1 ESO R A Gane, 4A Capel Rd, Matson, Gloucester, Glos, GL4 6JP [IO81VU, SO81]
G1 ESS N J Hyland, 20 Clent View Rd, Bartley Green, Birmingham, B32 4LN [IO82XK, SO98]
GW1 ESU P M Jones, 34 Finchley Rd, Fairwater, Cardiff, CF5 3AX [IO81JL, ST17]
G1 ESW K M Laughton, 33 Tiverton Cl, Radcliffe, Manchester, M26 3UJ [IO83TN, SD70]
G1 ESX K Love, 63 Buxton Rd, Spixworth, Norwich, NR10 3PP [JO02PQ, TG21]
G1 ETA J Mc Chrystal, 10 Thirlmere Ave, Larkholme, Fleetwood, FY7 8NL [IO83LV, SD34]
G1 ETD J R Newland, 18 Thornhill Cres, London, N1 1BJ [IO91WM, TQ38]
GD1 ETE R H Nicholls, Storeton, 1 Alberta Dr, Onchan, Douglas, IM3 1LT
GW1 ETH M G Sparks, 15 Woodland Rd, Whitchurch, Cardiff, CF4 2BU [IO81JM, ST17]
G1 ETU C A Taylor, 269 Coventry Rd, Hinckley, LE10 0NE [IO92HM, SP49]
G1 ETW M S Thomas, 99 Olympic Way, Eastleigh, SO50 8QS [IO90IX, SU41]
G1 ETX C J Thomas, 4 Hemdean Gdns, West End, Southampton, SO30 3BB [IO90HW, SU41]
G1 ETZ C D Walker, 1 Shepherds Cl, Shepshed, Loughborough, LE12 9SQ [IO92IS, SK42]
G1 EUA B F Wall, 11 Coombe View, Coombe Valley, Teignmouth, TQ14 9UY [IO80FN, SX97]
G1 EUB R H Wallace, 2 Church Cl, Prestolee, Radcliffe, Manchester, M26 1HG [IO83TN, SD70]
G1 EUC C R Ward, 6 Hopping Jacks Ln, Danbury, Chelmsford, CM3 4PN [JO01HR, TL70]
G1 EUD D Wiles, 62 Taylor St., Tunbridge Wells, TN4 0DX [JO01DD, TQ54]
G1 EUF R G Wilson, St Farm, Henny, Sudbury, Suffolk, CO10 7LS [JO02IA, TL83]
G1 EUG D J Wolfe, 48 Wilby Ln, Great Doddington, Wellingborough, NN29 7TP [IO92PG, SP86]
G1 EUH J R Woods, 1 Mill Ln, Burscough Bridge, Burscough, Ormskirk, L40 5TJ [IO83NO, SD41]
G1 EUI M E J Wright, 71 Oakridge Rd, High Wycombe, HP11 2PL [IO91OP, SU89]
G1 EUJ 12841 N J Lock, 6 Delius Cl, Basingstoke, RG22 4DS [IO91KF, SU65]
G1 EUM S W Foote, 42 Fairview Ave, Stanford-le-Hope, SS17 0DT [JO01FM, TQ68]
G1 EUN A J Friend, 46 Pevensey Rd, Slough, SL2 1UQ [IO91QM, SU98]
G1 EUQ A P Gammon, 5 Sommerville Cl, Faversham, ME13 8HP [JO01KH, TR06]
G1 EUT M Coupe, 46A Williamthorpe Cl, North Wingfield, Chesterfield, S42 5NG [IO93HE, SK46]
G1 EUV M I Gill, The Cottage, Barrowell Green, Winchmore Hill, London, N21 3AU [IO91WP, TQ39]
G1 EUW L V Gladdis, 12 Hendy Rd, East Cowes, PO32 6QQ [IO90IS, SZ59]
G1 EUX E E Griffin, 16 Marigold Cl, Nettleham Rd, Lincoln, LN2 4SZ [IO93RG, SK97]
G1 EVA M E Hattersley, Top Ln, Townhead, Tideswell, Buxton, Derbyshire, SK17 8LP [IO93CG, SK17]
G1 EVC G D Hogg, 15 West End Rd, Silsoe, Bedford, MK45 4DU [IO92SA, TL03]
G1 EVD A C Holden, The Copse, Mayland Green, Mayland, Chelmsford, Essex, CM3 6BD [JO01JQ, TL90]
G1 EVG D L Holmes, 29 Grange Cttgs, Marsden, Huddersfield, HD7 6AJ [IO93AO, SE01]
G1 EVS J Lucas, 17 Cherry Hinton, Maygate, Oldham, OL1 2PU [IO83WN, SD90]
G1 EVV C Mylchreest, 8 Parren Ave, Whiston, Prescot, L35 3SB [IO83OJ, SJ49]
G1 EVW R Murphy, 33 Arundel Rd, Heatherside, Camberley, GU15 1DL [IO91PH, SU95]
G1 EWB E G Pearson, 15 Norman St., Middleton, Manchester, M24 2JP [IO83VN, SD80]
G1 EWC A P Webster, 49 Uplands Croft, Werrington, Stoke on Trent, ST9 0LF [IO83WA, SJ94]
G1 EWD Details withheld at licensee's request by SSL.
G1 EWE T J Williams, 29 Danson Rd, Bexleyheath, DA6 8HA [JO01BK, TQ47]
G1 EWH S R Bell, The Haven, Low St., East Drayton, Retford, DN22 0LN [IO93NG, SK77]
G1 EWI A C Boor, 24 Wheeldon St., Gainsborough, DN21 1BS [IO93OJ, SK88]
G1 EWP Details withheld at licensee's request by SSL.
GW1 EWQ R E Corrall, The Fir House, Trelyston, Leighton, Welshpool, Powys, SY21 8HZ [IO82LP, SJ20]
G1 EWR R A Coxshall, 41 Cassandra Gate, Cheshunt, Waltham Cross, EN8 0XE [IO91XR, TL30]
G1 EWT K D G Cullen, Old Police House, Firs Rd, Alderbury, Salisbury, SP5 3BD [IO91DA, SU12]
GW1 EWW K Edwards, 25 Woodland Rd, Neath, SA11 3AL [IO81CP, SS79]
GJ1 EXC M E Green, 1 Peel Terr, La Route Du Fort, St. Helier, Jersey, JE2 4PA
G1 EXG J P Hare, 19 Sea Ln, Goring By Sea, Worthing, BN12 4QB [IO90ST, TQ10]
G1 EXR W J Cosgrove, 62 Twyford Ave, Great Wakering, Southend on Sea, SS3 0EX [JO01JN, TQ98]
G1 EXU R J Cloke, 24 Cornflower Dr, Chelmsford, CM1 6XY [JO01GR, TL70]
G1 EXV D J Cooper, 27 Bracebridge St., Nuneaton, CV11 5PA [IO92GM, SP39]
G1 EXZ V J Philbrick, 5 Jennifer Way, East Cowes, PO32 6BS [IO90IS, SZ59]
G1 EYD K Rutter, 101 Allendale Rd, Newcastle upon Tyne, NE6 2SX [IO94FX, NZ26]
G1 EYG G D Shipperley, 72 Hithercroft Rd, Downley, High Wycombe, HP13 5RH [IO91OP, SU89]
G1 EYJ I D Smith, 216 Corbets Tey Rd, Upminster, RM14 2BL [JO01CN, TQ58]
G1 EYL M S Spurr, 116 Coniston Rd, Dronfield Woodhouse, Sheffield, S18 5PZ [IO93GH, SK37]
GU1 EYP J C Thompson, Mossbank, Les Nouettes, Forest, Guernsey, GY8 0ED
G1 EYS H N Stuart-Turner, 45 Ashdown Ave, Saltdean, Brighton, BN2 8AH [IO90XT, TQ30]
G1 EYT P J Vickers, 21 Blackwood Dr, Sutton Coldfield, B74 3QP [IO92BN, SP09]
G1 EYX J Ward, 55 Richter Rd, Jackson, New Jersey, USA 08527, ZZ9 9AA
G1 EZI S A Armstrong, 5 Dashpers, Brixham, TQ5 9LJ [IO80FJ, SX95]
G1 EZJ C J Barker, 52 Spode St., Stoke on Trent, ST4 4DY [IO82VX, SJ84]
G1 EZU D R Harpham, 37 Lytham Rd, Kirkby in Ashfield, Nottingham, NG17 8NQ [IO93IC, SK45]
G1 EZV Details withheld at licensee's request by SSL.
G1 EZW G Hartley, Broadfield House, Broadfield, Oswaldtwistle, Accrington, BB5 3RY [IO83TR, SD72]
G1 FAA S D Jeffery, 17 Bishop Butt Cl, Orpington, BR6 9UF [JO01BI, TQ46]
G1 FAD T P Kenney, 7 Hickin Cl, Charlton, London, SE7 8SH [JO01AL, TQ47]
GM1 FAF J M Marshall, Drummorlie, Wallyford Toll, Wallyford, Musselburgh, EH21 8JT [IO85LW, NT37]
GM1 FAI A K Miller, 21 Merker Terr, Linlithgow, EH49 6DD [IO85EX, NS97]
G1 FAJ W C Mock, 73 Oxford St., Burnham on Sea, TA8 1EW [IO81MF, ST34]
G1 FAL T Osborne, 9 Oxford Dr, Halewood, Liverpool, L26 0TN [IO83OI, SJ48]
G1 FAM Details withheld at licensee's request by SSL.
G1 FAN I H Reynolds, 12 Sherwood, North Stifford, Grays, RM16 5UD [JO01DL, TQ68]
G1 FAV Details withheld at licensee's request by SSL.
G1 FAZ C P Williams, 132 Weoley Park Rd, Selly Oak, Birmingham, B29 5HA [IO92AK, SP08]
G1 FBE D Telford, 16 Springfields, Wigton, CA7 9JS [IO84KT, NY24]
G1 FBH V J Bobin, 37 Carisbrooke Cl, Eastbourne, BN23 8EQ [JO00DT, TQ60]
GW1 FBI J G Hughes, 17 Pentrosfa Rd, Llandrindod Wells, LD1 5NL [IO82HF, SO05]
G1 FBK Details withheld at licensee's request by SSL.
GW1 FBL T D Boorman, 43 Ffordd Taliesin, Killay, Swansea, SA2 7DF [IO71XO, SS69]
GM1 FBM B M S Borland, Beechwood, Upper Muirhall Rd, Kinfauns, Perth, Perthshire, PH2 7LL [IO86HJ, NO12]
G1 FBO L W Browning, The Little Cottage, Fair Green, Glemsford, Sudbury, CO10 7PH [JO02IC, TL84]
G1 FBQ Dr S Fraser, Walnut Tree Cottage, Main Rd, Fyfield, Abingdon, OX13 5LN [IO91HQ, SU49]
G1 FBS P A C Gracie, 16 Antoneys Cl, Pinner, HA5 3LP [IO91TO, TQ19]
G1 FBU K H Goodchild, 115 Gloucester Rd, Newbury, RG14 5JJ [IO91IJ, SU46]
G1 FBW T N Howchen, 37 Central Ave, Corringham, Stanford-le-Hope, SS17 7NG [JO01FM, TQ78]
G1 FBY A Hynd-Smith, 9 Limecrag Ave, Durham, DH1 1DF [IO94FS, NZ24]
G1 FBZ W A E Hicks, 7 Meadow Cl, Thundersley, Benfleet, SS7 3RJ [JO01HN, TQ78]
G1 FCK R White, 5 Caer Urfa Cl, South Shields, NE33 2BY [IO95GA, NZ36]
G1 FCQ A E Perks, 96 School Ln, Chase Terr, Walsall, West Midlands, WS7 8LE [IO92AQ, SK00]
G1 FCR M I F Pell, 10 The Avenue, Hipperholme, Halifax, HX3 8NP [IO93CR, SE12]
G1 FCT C J Reade, 7 Wilmar Cl, Hayes, UB4 8ET [IO91SM, TQ08]
G1 FCY P Cook, Westway Boundary Hs, Shiney Row, Houghton-le-Spring, Tyne & Wear, DH4 4PZ [IO94FU, NZ35]
G1 FDB Details withheld at licensee's request by SSL.
G1 FDD N E Groeber, 113 Kings Rd, Kings Heath, Birmingham, B14 6TN [IO92BK, SP08]
G1 FDK Details withheld at licensee's request by SSL.[Station located near Brighton, postcode area: BN3, locator: IO90. Op: Henry Jaremko. Packet mail to G1FDK @ GB7VRB.]
G1 FDL V W Kelk, 7 Rowan Pl, Garforth, Leeds, LS25 2JR [IO93HS, SE43]
G1 FDO T R King, 32 Bagnall Ave, Arnold, Nottingham, NG5 6FT [IO92KX, SK54]
G1 FED D Rosewarn, 16 Charles Cres, Taunton, TA1 2XN [IO81LA, ST22]
G1 FEF C P Smith, 69 High St., Belton, Doncaster, DN9 1NR [IO93ON, SE70]
G1 FEH R G Seymour, 16 Rosemead South, Benfleet, Essex, SS7 3JQ [JO01GN, TQ78]
G1 FEJ J Saveall, Nascott Bungalow, Beach Rd, Woolacombe, Devon, EX34 7BT [IO71VE, SS44]
GM1 FEM I H Smith, 13 Newmills Gr, Balerno, EH14 5SY [IO85IV, NT16]

G1 FEO G S Jones, 8 Kenilworth Rd, Lighthorne Heath, Leamington Spa, CV33 9TH [IO92GE, SP35]
G1 FEP D R Twidale, 18 Kinnaird Rd, Wallasey, L45 5HN [IO83LK, SJ39]
G1 FER C W Tye, 2 Ellesmere Ct, Brackley, NN13 6BT [IO92KA, SP53]
G1 FET P M Taylor, 29 Dunstall Rd, Halesowen, B63 1BB [IO82XK, SO98]
G1 FEV D C Watson, 112 Keith Way, Southend on Sea, SS2 6SQ [JO01IN, TQ88]
G1 FFA R W Brown, 69 Moreton Ave, Great Barr, Birmingham, B43 7QR [IO92BN, SP09]
G1 FFB R A Littleboy, Hillcrest, Bromley Green Rd, Ruckinge, Ashford, TN26 2EG [JO01KC, TQ93]
G1 FFH T S Firth, 126 Tombridge Cres, Kinsley, Pontefract, WF9 5HE [IO93HO, SE41]
G1 FFL A T Robson, 12 Greetham Rd, Bedgrove Est, Aylesbury, HP21 9BS [IO91OT, SP81]
G1 FFO P J A Thomas, 9 Awefields Cres, Smethwick, Warley, B67 6PR [IO92AL, SP08]
G1 FFR S L Tucker, 28 Peregrine Rd, Offerton, Stockport, SK2 5UR [IO83WJ, SJ98]
G1 FFU D E Woolmer, 2 Muccleshell Cl, Havant, PO9 2HR [IO90MU, SU70]
G1 FFV C Barnett, 8 Wilmslow Cres, Thelwall, Warrington, WA4 2JE [IO83RJ, SJ68]
G1 FGI P A Escreet, 198 Front St., Sowerby, Thirsk, YO7 1JN [IO94HF, SE48]
G1 FGK W G Grint, 15 Ivythorn Rd, St, BA16 0TE [IO81PC, ST43]
G1 FGS A S Hasted, 9 Roscrea Dr, Bournemouth, BH6 4LU [IO90CR, SZ19]
G1 FGT A T Heal, Westfield Post Office, 8A Coronation Ave, Yeovil, BA21 3DX [IO80QW, ST51]
G1 FHH P Johnson, 30 Copplestone Gr, Stoke on Trent, ST3 5UD [IO82WX, SJ94]
G1 FHI S B Murray, 51 Huddersfield Rd, Milnrow, Rochdale, OL16 3QZ [IO83WO, SD91]
G1 FHK C Peart, 33 Fieldfare, Abbeydale, Gloucester, GL4 4WH [IO81VU, SO81]
G1 FHO B L A Rivers, 211 Upper Wickham Ln, Welling, DA16 3AW [JO01BL, TQ47]
G1 FHR R H Roots, 57 Cobdown Cl, Ditton, Aylesford, ME20 6SZ [JO01FH, TQ75]
G1 FHY S Wise, 11 Clive Rd, West Dulwich, London, SE21 8DA [IO91WK, TQ37]
G1 FHZ G Wragg, 224 Inchbonnie Rd, South Woodham Ferrers, Chelmsford, CM3 5WU [JO01HP, TQ89]
G1 FIM P M McDonnell, 55 Lodge Hall, Harlow, CM18 7SY [JO01BR, TL40]
G1 FIP S J Rowlandson, 48 Greville Rd, Warwick, CV34 5PB [IO92FG, SP36]
G1 FIU A P Stringer, 40 Penzance Way, Horeston Grange, Nuneaton, CV11 6FW [IO92GM, SP39]
GM1 FIX A D Turnbull, Flat A, 40 Ainslie Pl, Perth, Perthshire, PH1 5DD [IO86GJ, NO12]
G1 FIZ J M Pottinger, 30 Valley Dr, Yarm, TS15 9JQ [IO94IM, NZ41]
GW1 FJC S Williams, 8 Queensway, Deeside, CH5 1HT [IO83LF, SJ36]
G1 FJD A C Wood, 2 Towning Cl, Deeping St. James, Peterborough, PE6 8HR [IO92UQ, TF11]
G1 FJF A B Bawden, 67 Silo Dr, Farncombe, Godalming, GU7 3NZ [IO91QE, SU94]
G1 FJH P K Bruce, 14 Hine Ave, Newark, NG24 2LH [IO93OB, SK85]
GW1 FJI R Bullock, St Tewdrics Pl, Mathern, Chepstow, Gwent, NP6 6JW [IO81PO, ST59]
G1 FJJ M T Cammish, 20 Chantry Ave, Hartley, Longfield, DA3 8DD [JO01DJ, TQ66]
G1 FJM A R Edwards, Syston, Fulbrook, Burford, Oxford, Oxon, OX8 4BL [IO91GT, SP31]
GM1 FJQ Details withheld at licensee's request by SSL.
G1 FJS K A Davis, 7 Hebden Rd, Lower Westwood, Bradford on Avon, BA15 2BX [IO81UH, ST85]
G1 FKF A Miller, 1 Blackthorn Cres, Cannock, WS12 5SW [IO92AR, SK01]
G1 FKJ G A Portlock, Windmill Cottage, Oxford St, Aldbourne, Marlborough, Wilts, SN8 2DH [IO91EL, SU27]
GW1 FKL G D Howells, 70 Meadow St., Treforest, Pontypridd, CF37 1SS [IO81IO, ST08]
G1 FKP D T Le Vine, Anglecroft, Borough Rd, Tatsfield, Westerham, TN16 2LA [JO01AH, TQ45]
G1 FKQ Details withheld at licensee's request by SSL.
G1 FKS C J Macliesh, 33 River Way, Twickenham, TW2 5JP [IO91TK, TQ17]
G1 FKT Details withheld at licensee's request by SSL.
GW1 FKY K J Eaton, 21 Westminster Way, Cefn Glas, Bridgend, CF31 4QX [IO81EM, SS88]
G1 FLA G Doyle, 89 Victoria Park Rd, Tunstall, Stoke on Trent, ST6 6DX [IO83VB, SJ85]
G1 FLI C E Franklin, 3 Park Rd, Rugby, CV21 2QU [IO92IJ, SP57]
G1 FLL P J Gray, 31 Rudgard Ave, Cherry Willingham, Lincoln, LN3 4JQ [IO93SF, TF07]
GM1 FLO G Hardy, 46 The Beeches, Nantwich, CW5 5YP [IO83RB, SJ65]
GM1 FLQ R S Monahan, 26 Beechwood Dr, Renfrew, PA4 0PN [IO75TU, NS56]
G1 FLT G A Norman, 93 Beloe Ave, Bowthorpe, Norwich, NR5 9BL [JO02OP, TG10]
G1 FLV M S Parr, 5 Suffolk Gr, Leigh, WN7 4TA [IO83RL, SD60]
G1 FLW P Partridge, 18 Chaucers Dr, St Peters Field, Galley Common, Nuneaton, Warks, CV12 9SD [IO92GL, SP38]
G1 FLX S J Pickstone, 48 Oak Tree Dr, London, N20 8QH [IO91VP, TQ29]
G1 FLY A W Rosier, 3 Rowley Ct, Rowley Dr, Botley, Southampton, SO30 2LB [IO90IW, SU41]
G1 FMA A G Robinson, 5 Alford Rd, Heaton Chapel, Stockport, SK4 5AW [IO83VK, SJ89]
G1 FMC R A Rose, 44 Linden Rd, Newport, PO30 1RJ [IO90IQ, SZ48]
G1 FMD Details withheld at licensee's request by SSL.
G1 FMJ D Wakes, 6 Spencer St., Eldon Ln, Bishop Auckland, DL14 8TL [IO94EP, NZ22]
GM1 FML A Q Beale, Top Right, 1 Holyrood Cres, Glasgow, G20 6HJ [IO75UU, NS56]
G1 FMT C J Hardy, 110 Jubilee Rd, Waterlooville, PO7 7RG [IO90LV, SU61]
G1 FMU K C Harris, 8 Trelawney Rise, Callington, PL17 7PT [IO70UM, SX36]
GM1 FMV G G Hind, 21 Eildon View, Melrose, TD6 9RH [IO85PO, NT53]
G1 FMW D E Lewis, 4 Westwood Gr, Solihull, B91 1QB [IO92CJ, SP17]
GM1 FMX W McCandlish, Lingdowey, Stoneykirk Rd, Stranraer, DG9 7BX [IO74LV, NX05]
G1 FNA L R B Phillips, 2 Stratton Green, Bedgrove Est, Aylesbury, HP21 7EP [IO91OT, SP81]
G1 FND N J Stephens, 7 Quarry Rd, Alveston, Bristol, BS12 2JL [IO81RO, ST68]
G1 FNQ T J C Blight, 36 Fairview Way, Baswich, Stafford, ST17 0AX [IO82WT, SJ92]
GM1 FNT Details withheld at licensee's request by SSL.
G1 FNU J A Dodd, 38 The Quadrant, North Shields, NE29 7HP [IO95GA, NZ36]
GM1 FNX P E Ewing, 27 Boyd Orr Cres, Kilmaurs, Kilmarnock, KA3 2QB [IO75RP, NS44]
GM1 FNY D C Fisher, 2 Links View, St. Fergus, Peterhead, AB42 3HY [IO97BN, NK05]
G1 FOA P J Franklin, 84 Bodmin Rd, Springfield, Chelmsford, CM1 6LL [JO01FR, TL70]
G1 FOD J R Hamriding, Smithy Cottage, Weeton Ln, Weeton, Leeds, West Yorks, LS17 0AW [IO93FW, SE24]
G1 FOE P G Holt, 27 Sandown Cl, Blackwater, Camberley, GU17 0EN [IO91OI, SU86]
GW1 FOF M L James, 32 Kipling Dr, London, SW19 1TW [IO91VK, TQ27]
G1 FOG A Kay, 7 Manor Farm Cl, Bingley, BD16 1RX [IO93CT, SE13]
G1 FOM P V Lyttle, 3 Woodlands, East Ardsley, Wakefield, WF3 2JG [IO93FR, SE32]
G1 FON M M Mangan, 48 Emblett Dr, Bradley Park, Newton Abbot, TQ12 1YJ [IO80EM, SX87]
G1 FOO G A Nicholls, 45 River Ln, Gaywood, Kings Lynn, PE30 4HD [JO02FS, TF62]
G1 FOP B R Payne, 24 Danbury Vale, Danbury, Chelmsford, CM3 4LA [JO01HR, TL70]
GM1 FOS J A Skinner, 163 Milnafua, Alness, IV17 0YT [IO77VQ, NH67]
G1 FOU S J Tyler, 6 Brayford Cl, Northampton, NN3 3LU [IO92NF, SP76]
G1 FOW K Worsley, 102 Cabul Cl, Orford, Warrington, WA2 7SE [IO83RJ, SJ68]
G1 FPC A G Sands, 7 Kimberley Ave, Seymour St., Hull, HU3 5PP [IO93TR, TA02]
GM1 FPD D G King, Mo Dachaidh, Benderloch, By Oban, Argyll, PA37 1QP [IO76HL, NM93]
G1 FPF A T Comley, 10 Pellinore Rd, Exeter, EX4 9BJ [IO80GR, SX99]
G1 FPK R G Kerridge, 80 Melton Rd, Wymondham, NR18 0DE [JO02NN, TG10]
G1 FPS Details withheld at licensee's request by SSL.
G1 FPY A J Watson, 85 Old Birmingham Rd, Lickey End, Bromsgrove, B60 1DF [IO82XI, SO97]
GW1 FPZ Details withheld at licensee's request by SSL.
G1 FQI D J Jackson, 44 Millport Dr, Hessleh Rd, Hull, HU4 7DU [IO93TR, TA02]
G1 FQK E A Tutt, 14 Frome Ct, Longmead Way, Tonbridge, TN10 3TS [JO01DE, TQ54]
G1 FQV Details withheld at licensee's request by SSL.
G1 FQX J F Lamb, 52 Binstead Rd, Kingstanding, Birmingham, B44 0TL [IO92BN, SP09]
G1 FRD G H Bennett, 14 Thessaly Rd, Stratton, Cirencester, GL7 2NG [IO91AR, SP00]
G1 FRH D I Playfair, 68 Raymonds Dr, Thundersley, Benfleet, SS7 3PW [JO01GN, TQ78]
G1 FRL G L Mott, 191 Joyners Field, Harlow, CM18 7QD [JO01BR, TL40]
G1 FSH Details withheld at licensee's request by SSL.
GI1 FSJ N Colgan, 11 St. Johns Park, Moira, Craigavon, BT67 0NL [IO64VL, J16]
G1 FSM L J Insole, 22 Mapledene Ave, Hullbridge, Hockley, SS5 6JB [JO01HO, TQ89]
GM1 FSU I J Menzies, 104 Newburgh Circle, Bridge of Don, Aberdeen, AB22 8XB [IO87WE, NJ91]
G1 FSW R C Consolante, 19 Chestnut Gdns, Stamford, PE9 2JY [IO92RP, TF00]
GM1 FSZ K Hall, 12 McCash Pl, Parkview, Kirkintilloch, Glasgow, G66 4BE [IO75WW, NS67]
G1 FTD J H Hobbs, Fetchalls, The Green, Beyton, Bury St. Edmunds, Suffolk, IP30 9AF [JO02KF, TL96]
GM1 FTE A Jaconelli, 73 Springboig Rd, Glasgow, G32 0DB [IO75WU, NS66]
G1 FTH A Marston, 266 Greenmoor Rd, Nuneaton, CV10 7EP [IO92GM, SP39]
G1 FTK N C Apps, Brensett Mill Ln, Mill Corner, Northiam, E Sussex, TN31 6HU [JO00HX, TQ82]
G1 FTO Details withheld at licensee's request by SSL.
G1 FTR R A Farnworth, 65 Green Meadows, Westhoughton, Bolton, BL5 2BN [IO83RN, SD60]
G1 FTU J Pearson, 42 Chesterfield Rd, Barlborough, Chesterfield, S43 4TT [IO93IG, SK47]
G1 FTV A J Bryant, 27 Haydon St., Swindon, SN1 1DT [IO91CN, SU18]
G1 FTX E C Demeza, 2 Adam Cl, St. Leonards on Sea, TN38 9QW [JO00GV, TQ71]
GM1 FTZ H C Simpson, Clachan Farm Cotts, Rosneath, Helenburgh, Dumbartonshire, G84 0QR [IO75OX, NS28]
G1 FUA Details withheld at licensee's request by SSL.
GM1 FUD G S Drain, 166 Brownside Rd, Burnside, Rutherglen, Glasgow, G73 5AZ [IO75VT, NS66]
G1 FUG A M Sagar, 28 Rangoon Rd, Solihull, B92 9DB [IO92CK, SP18]
G1 FUI R A Earnshaw, Top Farm, Stocksmoor Rd, Midgley, nr Wakefield, West Yorks, WF4 4JQ [IO93EP, SE21]
G1 FUJ B G Jones, 4 Coldrick Cl, Kingswood, Bristol, BS15 4RT [IO81SL, ST67]
G1 FVA K H Irons, 14 Beech Gr, Houghton, Carlisle, CA3 0NU [IO84MW, NY45]
G1 FVC I G Batten, 17 Cornfield Rd, Birmingham, B31 2EB [IO92AJ, SP07]
G1 FVE D M Ward, 36 Croxby Ave, Scartho, Grimsby, North East Lincs, DN33 2NW [IO93WM, TA20]
G1 FVF F J Pover, Site Office, Sunland Holiday Bungalows, Tregea Hill, Portreath, Cornwall, TR16 4PE [IO70IG, SW64]
G1 FVH M V Jordan, 160 Beta Rd, Cove, Farnborough, GU14 8PH [IO91OH, SU85]
G1 FVN R P Foster, Glen Alva, Ewyas Harold, Hereford, HR2 0JB [IO81NW, SO32]
G1 FVP R D Slone, 55 Chilton Way, Hungerford, RG17 0JR [IO91FJ, SU36]
G1 FWA A W Musk, 44 Raedwald Dr, Moreton Hall Est, Bury St. Edmunds, IP32 7DD [JO02IF, TL86]
GW1 FWE J E Duggan-Keen, Bodlondeb, Chapel St., Caerwys, Mold, CH7 5AE [IO83IF, SJ17]
G1 FWF H R Morgan, 2 Mayfield Park South, Fishponds, Bristol, BS16 3NG [IO81RL, ST67]
GI1 FWK A Crowe, 1 Ardymagh Rd, Glenwherry, Ballyclare, BT39 9TJ [IO74AS, J39]
G1 FWR P Harman, Dingley Dell, 25 North St., Maldon, CM9 5HH [JO01IR, TL80]
G1 FWS F W Swaine, 52 Downlands, Chells Manor Village, Stevenage, SG2 7BH [IO91WV, TL22]
G1 FWU T C Norbury, 19 Charles Cope Rd, Orton Waterville, Peterborough, PE2 5ER [IO92UN, TL19]
G1 FWY C J Pitt, 44 Darcy Way, Tolleshunt Darcy, Maldon, CM9 8UD [JO01JS, TL91]
G1 FWZ B L Lakey, 45 Madam Ln, Weston Super Mare, BS22 9PW [IO81MI, ST36]
G1 FXB A J Turquand, 63 Sundown Ave, Dunstable, LU5 4AL [IO91RV, TL02]
G1 FXC R W Lambourne, Rosedale, Townsend, Marsh Gibbon, Bicester, OX6 0EY [IO91LV, SP62]
G1 FXD O W Rogers, Lower Polgrain, St Wenn, Bodmin, Cornwall, PL30 5PS [IO70NK, SW96]
G1 FXE P K Gorman, 10 Elizabeth Cl, Off Kingsway, Wellingborough, NN8 2JA [IO92PG, SP86]
G1 FXG I R Hawkins, Rocrisar Tut Hill, Fornham All Saints, Bury St Edmunds, Suffolk, IP28 6LD [JO02IG, TL86]
GM1 FXJ Details withheld at licensee's request by SSL.
G1 FXM T E Young, 10 The Acorns, Hockley, SS5 5AS [JO01HO, TQ89]
G1 FXS P C J Striplin, 19 Ebrington Rd, Malvern, WR14 4NL [IO82TC, SO74]
G1 FXT R S Robinson, 5 Lilac Cl, Newton Longville, Milton Keynes, MK17 0DQ [IO91OX, SP83]
G1 FXU I Andrew, The Willows, Neatmoor Hall Farm, Nordelph, Downham Market, Norfolk, PE38 0BY [JO02DO, TF50]
G1 FXW W Colclough, 12 Meadow Cl, Forsbrook, Stoke on Trent, ST11 9DW [IO82XX, SJ94]
G1 FYC M R Trotman, 44 Armscroft Rd, Barnwood, Gloucester, GL2 0SJ [IO81VU, SO81]
G1 FYE C B Sterland, 103 Main St., Distington, Workington, CA14 5UJ [IO84FO, NY02]
G1 FYF C J Bradley, 5 Derwent Ave, Biggleswade, SG18 8LY [IO92UB, TL14]
G1 FYN W King, 50 Elizabeth St., Widdrington, Morpeth, NE61 5NW [IO95EF, NZ29]
G1 FYQ R Greenhough, Pontefract Dist Ars, 36 Churchbalk Ln, Pontefract, WF8 2QQ [IO93IQ, SE42]
G1 FYS K Boothroyd, 16 Kelvin Ave, Dalton, Huddersfield, HD5 9HG [IO93CP, SE11]
GM1 FYW K W Chapman, 80 Glendinning Cres, Edinburgh, EH16 6DN [IO85KW, NT27]
G1 FZE H M Gould, 2 Parkdale, Ibstock, LE67 6JW [IO92HQ, SK41]
G1 FZL P A Dyer, 18 Christopher Cl, Yeovil, BA20 2EH [IO80QW, ST51]
G1 FZQ T W Brooker, 4 Poffley End, Hailey, Witney, OX8 5US [IO91GT, SP31]
G1 FZS R Fleet, 193 Cowley Dr, Woodingdean, Brighton, BN2 6TG [IO90XT, TQ30]
G1 FZV A J Ogden, 5 Lower Bristol Rd, Clutton, Bristol, BS18 4PB [IO81RH, ST65]
G1 FZY R M Turner, 72 Dale St., Chatham, ME4 6QG [JO01GJ, TQ76]
G1 GAD F J McLoughlin, 21 Darwin Cres, Montagu Est, Gosforth, Newcastle upon Tyne, NE3 4TT [IO94EX, NZ26]
G1 GAF Details withheld at licensee's request by SSL.
G1 GAI A Goodison, 29 Clerke St., Cleethorpes, DN35 7NE [IO93XN, TA20]
G1 GAM Details withheld at licensee's request by SSL.
G1 GAN P G Cartwright, Westlake House, West Coker, Yeovil, Somerset, BA22 9AH [IO80PW, ST51]
G1 GAO Details withheld at licensee's request by SSL.
G1 GAP D R Woods, 33 Bishops Ave, Hill, Worcester, WR3 8XA [IO82VE, SO85]
G1 GAR M A C Kipping, 46 Old Hardenwaye, Totteridge, High Wycombe, HP13 6TJ [IO91PP, SU89]
G1 GAS C I Kelley, 5 Russell Way, Wootton, Bedford, MK43 9EX [IO92RC, TL04]
G1 GAV A S Burton, 13 Oakwood Cres, Worrall, Sheffield, S30 3AW [IO93GJ, SK38]
G1 GAW R W Smith, 18 Curlew, Wilnecote, Tamworth, B77 5PL [IO92EO, SK20]
G1 GBC W T Boucher, 12 Highfield Terr, Ilfracombe, EX34 9LG [IO71WE, SS54]
G1 GBE F R Charlesworth, 24 Dunster Rd, Southport, PR8 3AQ [IO83LO, SD31]
G1 GBF J P Delaney, 97 Roose Rd, Barrow in Furness, LA13 9RJ [IO84JC, SD26]
G1 GBI M S Pearson, 34 Downside Rd, Sutton, SM2 5HP [IO91VI, TQ26]
G1 GBL Details withheld at licensee's request by SSL.
G1 GBO Details withheld at licensee's request by SSL.
G1 GBR A Ziemacki, 3 Wheatcroft Rd, Rawmarsh, Rotherham, S62 5JR [IO93IL, SK49]
G1 GBV D J Evans, 59 Watlington Rd, Benfleet, SS7 5DT [JO01GN, TQ78]
G1 GBX D E Vanbeck, 101 Upper St., Islington, London, N1 1QN [IO91WM, TQ38]
G1 GCF F Clough, 20 The Greenway, Haxby, York, YO3 3FE [IO94LA, SE65]
G1 GCJ G L Morris, 21 Orchard Pl, Deer Park, Ledbury, HR8 2XD [IO82SA, SO73]
G1 GCS W F Brown, 28 South View Gdns, Andover, SP10 2AG [IO91GE, SU34]
G1 GCT M C Thurlow, 20 Francis St., Brightlingsea, Colchester, CO7 0DG [JO01MT, TM01]
G1 GCU Details withheld at licensee's request by SSL.
G1 GCV P R Beevers Jnr, 4 Sherwood Pl, Kirkby in Ashfield, Nottingham, NG17 9ED [IO93JB, SK55]
G1 GCY G L Gowland, 7 Canewdon Hall Cl, Canewdon, Rochford, SS4 3PY [JO01IO, TQ89]
G1 GCZ J Vinson, Akhurst Cottage, Shepherds Hill, Selling, Faversham, ME13 9RS [JO01KG, TR05]
G1 GDA M B Austin, 10 Simon Pl, Wideopen, Newcastle upon Tyne, NE13 7HT [IO95EB, NZ27]
G1 GDB D R Thwaytes, 2 Harriston, Aspatria, Carlisle, CA5 2EF [IO84IS, NY14]
G1 GDD P M Chrismas, Chesworth House, Pound Green, Buxted, E Sussex, TN22 4JW [JO00BX, TQ52]
G1 GDH R A Bullen, Hay Wain, Kingswood Rd, Shortlands, Bromley, BR2 0HQ [JO01AJ, TQ36]
G1 GDJ C J Godward, 14 Goldstone Way, Hove, BN3 7PB [IO90VU, TQ20]
GM1 GDO J W Morton, 6 Deanpark Pl, Balerno, EH14 7ED [IO85HV, NT16]
G1 GDQ T R Twine, 1 Parkside, Fort William Rd, Vange, Basildon, SS16 5JX [JO01FN, TQ78]
G1 GDR B R Smith, 4 Planetree Cl, Bromsgrove, B60 1AW [IO82XI, SO97]
G1 GDT C M Groom, Woodstock Cottage, 2 Woodstock Terr, Uley, Dursley, GL11 5SW [IO81UQ, ST79]
G1 GDU M Thornton, 40 West Ave, Filey, YO14 9BE [IO94UE, TA18]
G1 GED J L Luscombe, 30 Brookdale Ct, Brookdale Cl, Brixham, TQ5 9JW [IO80FJ, SX95]
G1 GEI G T Hunt, Four Seasons, Westmarsh Ash, Canterbury, Kent, CT3 2LP [JO01PH, TR26]
G1 GEL K W H Harris, 66 Dinting Rd, Glossop, SK13 9DY [IO93AK, SK09]
GM1 GEQ T W G Menzies, 112 Buckstone Terr, Edinburgh, EH10 6QR [IO85JV, NT26]
G1 GER A S Goodings, 3 Maple Gdns, Bradwell, Great Yarmouth, NR31 8ND [JO02UN, TG50]
GM1 GES W J Barbour, 27 Drove Rd, Langholm, DG13 0JW [IO85MD, NY38]
G1 GET F H Wood, Cross Swords, The Sq, Skillington, Grantham, NG33 5HB [IO92QT, SK82]
GW1 GEX T Williams, Belmont, 9 Victoria St., Aberaeron, SA46 0DA [IO72UF, SN46]
G1 GEY D Stoker, 9 Moorlands, Flint Hill, Dipton, Stanley, DH9 9LW [IO94CV, NZ15]
G1 GFA N G Pitt, 37 Shelley Dr, Four Oaks, Sutton Coldfield, B74 4YD [IO92BO, SK10]
G1 GFC S L Bradley, 75 New Rd, Sawston, Cambridge, CB2 4BN [JO02BD, TL44]
G1 GFD A J Crook, 21 Treyew Rd, Truro, TR1 2BY [IO70LG, SW84]
G1 GFF V C Thomas, 17 St. Pauls Rd, St. Leonards on Sea, TN37 6RS [JO00GU, TQ81]
G1 GFQ E W Longstaffe, 3 Fenwick Ln, Halton Lodge, Runcorn, WA7 5YU [IO83PH, SJ58]
GW1 GFU T A H Woodcock, 21 Bryn Colwyn, Penmaenhead, Colwyn Bay, LL29 9LJ [IO83DG, SH87]
G1 GFW J Jacobs, 11 Delamere Cl, Birmingham, B36 9TW [IO92CM, SP19]
G1 GFZ R J Pelling, Spring Cottage, Main Rd, Westfield, Hastings, TN35 4SL [JO00HV, TQ81]
G1 GGB R M Phillips, 58 Baranscraig Ave, Patcham, Brighton, BN1 8RE [IO90WU, TQ30]
G1 GGC C F Ralph, 26B Warren Ave, Brighton, E Sussex, BN2 6BJ [IO90XU, TQ30]
G1 GGF T J Thomas, 236 Park Rd, Cowes, PO31 7NQ [IO90IS, SZ49]
G1 GGJ A Akrill, 102 Waveney Rd, Longhill Est, Hull, HU8 9LZ [IO93US, TA13]
G1 GGK F W Coldham, 24 Wagon Ln, Solihull, B92 7PW [IO92CL, SP18]
G1 GGN R A Barkley, 9 Eagle Cl, Erpingham, Norwich, NR11 7AW [JO02OT, TG13]
G1 GGT R E Sharp, Arosa, The Park, Harwell, Didcot, OX11 0HB [IO91IO, SU48]
G1 GGU R E Pickett, 40 Parkside, Shoreham By Sea, BN43 6HA [IO90UU, TQ20]
G1 GHD T L Major, 55 Wash Ln, Yardley, Birmingham, B25 8QZ [IO92CL, SP18]
G1 GHF Details withheld at licensee's request by SSL.
G1 GHG K E Knibbs, 8 Ferguson Way, Huntington, York, YO3 9YG [IO93LX, SE65]
GD1 GHK W D Corlett, Glen Mie, Kerrocruin, Kirk Michael, Isle of Man, IM6 1AF
G1 GHO S R Hartley, 80 Nottingham Rd, Somercotes, Alfreton, DE55 4LY [IO93HB, SK45]
G1 GHU T R Smith, Little Ct, 5 Marlborough Ave, Torquay, TQ1 1TT [IO80FL, SX96]
G1 GHY A F Laszkiewicz, 38 Langley Ln, Ifield, Crawley, RH11 0NA [IO91VC, TQ23]
GM1 GHZ P Thompson Backpackers, 5 Isabella Pl, Scone, Perth, PH2 6TE [IO86HK, NO12]
G1 GIA I R Sinclair, 27 Holders Hill Gdns, London, NW4 1NP [IO91VO, TQ29]
G1 GID G Watt, 48 Southdown Rd, Portslade, Brighton, BN41 2HN [IO90VU, TQ20]
G1 GIE R E Buckley, 7 Blackthorn Gr, Woburn Sands, Milton Keynes, MK17 8PY [IO92QA, SP93]
G1 GIF C A Chesters, The Old School, Hulland, Derbys, DE6 3EH [IO93EA, SK24]
G1 GIG M J Howells, 9 Alta Vista Cl, Teignmouth, TQ14 8UW [IO80GN, SX97]
G1 GIJ D Hadjidakis, 19 Eastfield Rd, Royston, SG8 7ED [IO92XB, TL34]
G1 GIT J Langford, Abpopa, 2 Hillside Rd, Belford, NE70 7NB [IO95CO, NU13]
GD1 GJB P Halsall, 5 Cronk Rd, Union Mills, Douglas, IM4 4NJ
G1 GJF Details withheld at licensee's request by SSL.
G1 GJK K G Seller, 1A Smallford Ln, Smallford, St. Albans, AL4 0SA [IO91UR, TL10]
G1 GJM F Astles, 26 Main St., Halton, Runcorn, WA7 2AN [IO83PH, SJ58]
G1 GJR A G Evans, Ditchling, Newton Purcell, Buckingham, Bucks, MK18 4BA [IO91KX, SP62]
GW1 GJS H R Boulter, 9 Dillwyn Rd, Penllergaer, Swansea, SA4 1BT [IO71XQ, SS69]
G1 GKA R T Mason, 32 Linden Dr, Evington, Leicester, LE5 6AH [IO92KO, SK60]
G1 GKC Details withheld at licensee's request by SSL.
G1 GKF R A Mann, Little Chysauster, Gulval, Penzance, TR20 8XA [IO70FD, SW43]
G1 GKH G H Hinds, 35 Lime Gr, Burntwood, Walsall, West Midlands, WS7 0HA [IO92BQ, SK00]
GI1 GKI Details withheld at licensee's request by SSL.
G1 GKK S Brooke, 14 Saxton Ave, Heanor, DE75 7PZ [IO93HA, SK44]

G1 GKN A J Tyler, 41 Beadle Way, Great Leighs, Chelmsford, Essex, CM3 1RT [JO01GT, TL71]
G1 GKR A J Barlow, 9 Clarence St., Royton, Oldham, OL2 6LR [IO83WN, SD90]
G1 GKU Details withheld at licensee's request by SSL.
G1 GKY Details withheld at licensee's request by SSL.
G1 GLA Details withheld at licensee's request by SSL.
G1 GLF F J Pavey, 125 Sussex Rd, Petersfield, GU31 4LB [IO90MX, SU72]
G1 GLS R Lees, 68 Princess St., Broadheath, Altrincham, WA14 5HA [IO83TJ, SJ78]
G1 GLZ B F Start, Dotland Grange, Hexham, Northd., NE46 2JY [IO84WW, NY95]
G1 GMF R L Sievert, 5 Sandmoor Rd, New Marske, Redcar, TS11 8BP [IO94LN, NZ62]
G1 GMG S M Gainswin, 1 Buckfast Rd, Buckfast, Buckfastleigh, TQ11 0EA [IO80CL, SX76]
G1 GMH D F Peat, 23 Hill Bottom Cl, Whitchurch Hill, Reading, RG8 7PX [IO91LM, SU67]
G1 GMP J S Beevers, 60 Seymour Rd South, Manchester, M11 4PR [IO83VL, SJ89]
G1 GMV A K Brewer, 25 Ackerman Rd, Dorchester, DT1 1NZ [IO80SR, SY79]
G1 GMX J J Mold, 2 Old Lime Ct, Hurst Rise, Matlock, DE4 3EP [IO93FD, SK36]
G1 GNA A C Green, 39 Glenfield Cres, Newbold, Chesterfield, S41 8SF [IO93GG, SK37]
G1 GNL J K Newbold, 100 Weston Rd, Aston on Trent, Derby, DE72 2BA [IO92HU, SK42]
G1 GNP D Roe, 9 The Orchard, Fairfield Rd, Horsley Woodhouse, Ilkeston, DE7 6DD [IO93HA, SK34]
G1 GNS D Johnson, 31 Coniston Ave, Penketh, Warrington, WA5 2QY [IO83QJ, SJ58][Op: Dave Johnson.]
G1 GNV Details withheld at licensee's request by SSL.
G1 GNX G R Leonard, Cross Heyes, 231 Hale Rd, Hale, Altrincham, WA15 8DN [IO83UJ, SJ78]
G1 GNY J H Holmes, 1 Alexander Terr, Hazlerigg, Newcastle upon Tyne, NE13 7BT [IO95EB, NZ27]
GW1 GOE Details withheld at licensee's request by SSL.
G1 GOG T J Liversidge, 19 Spruces Dr Mews, Netherton, Huddersfield, Yorks, HD4 7WB [IO93CO, SE11]
G1 GOH Details withheld at licensee's request by SSL.
G1 GOO M C Hawkins, 101 Tobyfield Rd, Bishops Cleeve, Cheltenham, GL52 4NZ [IO81XW, SO92]
G1 GOP A J Abbott, 5 Heathcote Gdns, Rudheath, Northwich, CW9 7JB [IO83SG, SJ67]
G1 GOQ S Abbott, 5 Heathcote Gdns, Rudheath, Northwich, CW9 7JB [IO83SG, SJ67]
G1 GOR S B Ackroyd, 15 Kimberley Rd, Parkstone, Poole, BH14 8SQ [IO90AR, SZ09]
G1 GOY P R Miles, 4 Grafton Cl, Hartwell, Northampton, NN7 2JE [IO92ND, SP75]
G1 GOZ S P Sims-Mindry, Basement Flat, 28 Wickham Rd, Fareham, PO16 7EY [IO90JU, SU50]
G1 GPE D Murray, 2 The Model Village, Long Itchington, Rugby, CV23 8RB [IO92HG, SP46]
G1 GPL Details withheld at licensee's request by SSL.
G1 GPM M I Stevenson, 6 Charnock Cres, Sheffield, S12 3HB [IO93GI, SK38]
G1 GPS P E Palmer, 16 Trulock Rd, Tottenham, London, N17 0PH [IO91XO, TQ39]
G1 GPT A E Parsons, 961 Leeds Rd, Deighton, Huddersfield, HD2 1UP [IO93DQ, SE11]
G1 GQB J M Bagshaw, 56 Davids Dr, Wingerworth, Chesterfield, S42 6TP [IO93GE, SE36]
G1 GQC G W Lowe, obo Mansfield Am Rd Society, 25 Manor House Ct, Kirkby in Ashfield, Nottingham, NG17 8LH [IO93IC, SK45]
G1 GQE T J G Preston, 3 Lundy Walk, Hailsham, BN27 3BJ [JO00DU, TQ51]
G1 GQH G Proudlove, 16 Darnhall School Ln, Over, Winsford, CW7 1JR [IO83RE, SJ66]
G1 GQI R J R Reynolds, 16 The Avenue, Southlands, Mansfield, NG18 4PN [IO93JD, SK55]
G1 GQJ C M Clark, 9 Conigre, Chinnor, OX9 4JY [IO91MQ, SP70]
G1 GQK N E Risbridger, 3 Weston Sq, Albury, Guildord, Surrey, GU5 9AF [IO91SF, TQ04]
G1 GQQ M A Rowbotham, 37 Crawford Rise, Arnold, Nottingham, NG5 8QF [IO92KX, SK54]
G1 GQW P H See, 1 The Titheway, Middle Littleton, Evesham, WR11 5LP [IO92BC, SP04]
G1 GQY A K Armstrong, 18 Flaxfield Way, Kirkham, Preston, PR4 2AY [IO83NS, SD43]
G1 GQZ C M Armstrong, 18 Flaxfield Way, Kirkham, Preston, PR4 2AY [IO83NS, SD43]
G1 GRB R W D Arnold, 147 Craigmoor Ave, Strouden Park, Bournemouth, BH8 9LT [IO90BR, SZ19]
G1 GRN L Skorupinski, 49 Pool Ln, Winterley, Sandbach, CW11 4RZ [IO83TC, SJ75]
G1 GRP P Slater, 32 Winthorpe Ave, Westgate, Morecambe, LA4 4RE [IO84NB, SD46]
G1 GRZ W M Baker, 144 Beridge Rd, Halstead, CO9 1JU [JO01HW, TL83]
G1 GSB P E Standley, Mahe, Church Ln, Bunwell, Norwich, NR16 1SL [JO02NL, TM19]
G1 GSD M J Stanley, Harby, Brightlingsea Rd, Thorrington, Colchester, CO7 8JH [JO01MU, TM02]
G1 GSF R J Still, 23 Drove Rd, Biggleswade, SG18 8HD [IO92UC, TL14]
G1 GSG M R Taylor, 7 Marshall Rd, Cropwell Bishop, Nottingham, NG12 3DP [IO92MV, SK63]
G1 GSJ W M Bateman, Waterloo, Whixall, Whitchurch, Shropshire, SY13 2PX [IO82PV, SJ43]
G1 GSK M E Baylis, 45 Florence Ave, Hove, BN3 7GX [IO90VU, TQ20]
G1 GSM J Blackburn, 26 Cherry Tree Cl, Romiley, Stockport, SK6 4HD [IO83XJ, SJ99]
G1 GSN P A Bradfield, 118 East Rd, Langford, Biggleswade, SG18 9QP [IO92UB, TL14]
G1 GSR D S Tetlow, 7 Northway, Newcastle upon Tyne, NE15 9EY [IO94DX, NZ16]
G1 GST J A Thomas, 59 Cross Ln, Dudley, DY3 1PD [IO82WN, SO99]
G1 GSW C L Tinker, 87 Keepers Ln, Weaverham, Northwich, CW8 3BN [IO83RG, SJ67]
G1 GSX S K Titherington, 3 Cherry Rd, Ainsdale, Southport, PR8 3RU [IO83LO, SD31]
G1 GSY G Bridle, 7 Constantius Ct, Brandon Rd, Church Crookham, Fleet, GU13 0YF [IO91NG, SU85]
G1 GTA P P Butler, 59 Buttermere Ave, Seacliffe, Whitehaven, CA28 9PX [IO84EM, NX91]
G1 GTF E Chilton, 33 Kersall Ct, Nottingham, NG6 9DT [IO92JX, SK54]
G1 GTG C Chilton, 7 Ramblers Dr, Oakwood, Derby, DE21 2XN [IO92HW, SK33]
G1 GTH C B Clark, 8 Kestrel Bank, Netherton, Huddersfield, HD4 7LD [IO93CO, SE11]
GW1 GTI Details withheld at licensee's request by SSL.
G1 GTK D J Towers, Rowans, 50 Westbeech Rd, Pattingham, Wolverhampton, WV6 7AQ [IO82UO, SO89]
G1 GTL A R Peaty, 30 Draper Rd, Christchurch, BH23 3AP [IO90CR, SZ19]
G1 GTM A S Turner, 5 Evans Cl, Ashley Heath, Ringwood, BH24 2JQ [IO90BU, SU10]
G1 GTO C B Saunders, 12208 Quail Run Row, Bayonet Point, Florida 34667, USA, X X
G1 GTP B C Warnaby, 9 Brechin Gr, Hartlepool, TS25 3DD [IO94JP, NZ52]
G1 GTQ A G Clarke, 18 Waterloo Rd, Brighouse, HD6 2AT [IO93CQ, SE12]
G1 GTR P A Clarke, 18 Waterloo Rd, Brighouse, HD6 2AT [IO93CQ, SE12]
G1 GTS R J Clarke, 47 Peartree Rd, Enfield, EN1 3DE [IO91XP, TQ39]
G1 GTU J M Coleman, Nameloc, Long Green, Wortham, Diss, IP22 1PU [JO02MI, TM07]
G1 GTV Details withheld at licensee's request by SSL.
G1 GTY B Johnson, 8 Park House Dr, Prestbury, Macclesfield, SK10 4HY [IO83WH, SJ97]
G1 GUB I C Weller, 52 Oakhill Cl, Ashtead, KT21 2JQ [IO91UH, TQ15]
G1 GUC C J Wheatley, Ailsa Craig, Hampstead Nurreys Rd, Hermitage, Newbury, Berks, RG18 9RS [IO91IK, SU57]
G1 GUE Details withheld at licensee's request by SSL.
G1 GUI S A Watts, 1 Denbeck Wood, Eastleaze, Swindon, SN5 7EJ [IO91CN, SU18]
G1 GUL M E Cowtan, 32 Finmere, Avon Park, Rugby, CV21 1RT [IO92JJ, SP57]
G1 GUT T J Worthington, 80 Greenwood Ln, Wallasey, L44 1DW [IO83LK, SJ39]
G1 GUW J E Lamb, 42 Yew Tree Rd, Bebington, Wirral, L63 2NL [IO83LI, SJ38]
GW1 GVG P R R Herrits, 113 Maes Glas, Caerphilly, CF8 1JW [IO81HN, ST08]
G1 GVJ P M Allen, The White House, Flintersill, Dent, Cumbria, LA10 5QR [IO84SG, SD78]
G1 GVM D K Gale, 33 Monks Way, Eastleigh, SO50 5BE [IO90HX, SU41]
G1 GVP J V Gibbon, 18 Eagle St., Penn Fields, Wolverhampton, WV3 7DN [IO82WN, SO99]
G1 GVT A C Dale, 2 Glenfield Cl, Hillfield, Solihull, B91 3XY [IO92CJ, SP17]
G1 GWE A Friel, 10 Marvejols Park, Cockermouth, CA13 0QR [IO84IQ, NY13]
G1 GWF B Maxwell, Hillcrest, Castle View, Egremont, Cumbria, CA22 2NA [IO84FL, NY01]
G1 GWG C H May, 10 Clyde Cres, Wharton, Winsford, CW7 3LA [IO83SE, SJ66]
G1 GWJ A S Gillon, 94 Pelham St., Ashton under Lyne, OL7 0DU [IO83WL, SJ99]
G1 GWO J A Green, 32 Elizabeth Cl, Highwoods, Colchester, CO4 4YU [JO01LV, TM02]
G1 GWR Details withheld at licensee's request by SSL.
G1 GWS J E Mossop, 14 Websters Ln, Great Sutton, South Wirral, L66 2LH [IO83MG, SJ37]
G1 GWX R Patrick, 9 Brant Ave, Illingworth, Halifax, HX2 8DL [IO93BR, SE02]
G1 GXB K L Ray, 4 Elm Rd, Bishops Waltham, Southampton, SO32 1JR [IO90JW, SU51]
G1 GXC S J Ray, 75 The Meads, Burnt Oak, Edgware, HA8 9HE [IO91UO, TQ29]
G1 GXF T W Scott, 9 Walker Dr, Leigh on Sea, SS9 3QS [JO01HN, TQ88]
G1 GXH I Sinclair, 59 Westbourne Park, Urmston, Manchester, M41 0XR [IO83TK, SJ79]
GW1 GXM T J Spain, 7 Heol y Deri, Cwmgwili, Llanelli, SA14 6PH [IO71XS, SN51]
G1 GXP R S Thatcher, 50 Hannel Rd, Fulham, London, SW6 7RB [IO91VL, TQ27]
GW1 GXQ R J Tulk, Home Farm Lodge, Penylan, Ruabon, Wrexham, Clwyd, LL14 6HS [IO82MX, SJ34]
G1 GXW B J Woodhouse, 14 Woodcutters Ave, Leigh on Sea, SS9 4PL [JO01HN, TQ88]
G1 GXX A C Mayes, 6 Chiltern Gdns, Hornchurch, RM12 4SQ [JO01CN, TQ58]
G1 GYA G F Haigh, 36 Pightle Way, Lyng, Norwich, NR9 5RL [JO02MR, TG01]
G1 GYC M J Hallsworth, 15 Stokesay Dr, Hazel Gr, Stockport, SK7 5PW [IO83WI, SJ98]
G1 GYF D C Harvey, 264 Rangefield Rd, Bromley, BR1 4QY [JO01AK, TQ47]
G1 GYH J N Hay, 23 Manor Cl, Wilmslow, SK9 5PX [IO83VH, SJ88]
G1 GYI M A Maidment, Gr Villa, Pious Drove, Upwell Wisbech, Cambs, PE14 9AL [JO02CO, TF50]
G1 GYJ F Mallows, 31 Booth Rd, Hartford, Northwich, CW8 1RD [IO83RF, SJ67]
G1 GYM P McEwen, 26 Walton Ave, North Shields, NE29 9BS [IO95GA, NZ36]
G1 GYN R McMullan, 42 Corringham Rd, Wembley, HA9 9PU [IO91UN, TQ18]
G1 GYQ A A Hayward, 1 Cleveland Rd, Basildon, SS14 1NF [JO01FN, TQ78]
G1 GYR D M Denison, 40 Merlin Way, Leckhampton, Cheltenham, GL53 0LU [IO81WV, SO92]
GW1 GZB V Mitchell, Tyn-y-Giat, Llandyfrydog, Llanerchymedd, Anglesey, LL71 8AL [IO73TI, SH48]
G1 GZG M J Newport, 9 Highbury Park, Exmouth, EX8 3EJ [IO80HP, SY08]
G1 GZI A N Farmar, Hawks Pl, Horslett Hill, Clawton, Holsworthy, EX22 6RS [IO70TS, SX39]
G1 GZJ C Featherstone, The Bungalow, Brougham Terr, Hartlepool, TS24 8ET [IO94JQ, NZ53]
G1 GZK K Feay, 19 Dorset Ave, Diggle, Oldham, OL3 5PL [IO93AN, SE00]
G1 GZL Details withheld at licensee's request by SSL.
G1 GZM I Ford, 8 Blakedon Rd, Wednesbury, WS10 7HY [IO82XN, SO99]
G1 GZW Details withheld at licensee's request by SSL.
G1 GZY Details withheld at licensee's request by SSL.
G1 HAB H A Birkmyre, 21 Magenta Cres, St. Johns, Newcastle upon Tyne, NE5 1YL [IO95DA, NZ16]
G1 HAC J Hilton, 32 Dowry St., Fitton Hill, Oldham, OL8 2LP [IO83WM, SD90]
G1 HAF A P Hill, The Acorns, Prince Cres, Staunton, Gloucester, GL19 3RF [IO81UX, SO72]
G1 HAG Details withheld at licensee's request by SSL.
GW1 HAN D Price, Hen Dwr, Nant y Gamar Rd, Llandudno, LL30 3BD [IO83CH, SH78]
G1 HAO T Piper, 19 The Cres, Tanfield Lea, Stanley, DH9 9NQ [IO94DV, NZ15]
G1 HAW S F Balon, 10 Woodlands Ave, Leigh, WN7 3HL [IO83RL, SJ69]
GW1 HAX N H Bevan, Gerann Ridge, Mwyn Fynydd, Newtown, Powys Wales, SY16 2LJ [IO82IM, SO19]
G1 HBC T Hopkins, 25 Dalewood, Welwyn Garden City, AL7 2JP [IO91WT, TL21]
G1 HBD A C Hornby, 2 Maple Cl, Winnersh, Wokingham, RG41 5PE [IO91NK, SU77]
G1 HBE A J Howlett, 43 Cheetham Hill Rd, Dukinfield, SK16 5JL [IO83XL, SJ99]
G1 HBF M Hughes, 2 Chaldon Rd, Canford Heath, Poole, BH17 8DB [IO90AS, SZ09]
G1 HBO B J Cole, 59 South Worple Way, Mortlake, London, SW14 8PB [IO91UL, TQ27]
G1 HBQ Details withheld at licensee's request by SSL.
G1 HBR K J Jackaman, 186 Charlton Rd, Charlton, London, SE7 7DW [JO01AL, TQ47]
GW1 HBU A E Jones, 6 Stanley St., Beaumaris, LL58 8ET [IO73WG, SH67]
G1 HBV E R B Jones, 63 Chester Gdns, Enfield, EN3 4BD [IO91XP, TQ39]
G1 HBW F M Jones, 184 Harwich Rd, Little Clacton, Clacton on Sea, CO16 9PU [JO01NU, TM12]
G1 HCC E Kent, 100 Waskerley Rd, Washington, NE38 8DS [IO94FV, NZ35]
G1 HCE A Kimmings, 18 Riley Park, Kirkburton, Huddersfield, HD8 0SA [IO93DO, SE11]
G1 HCM F W P Dawson, 3 Barnsbury Cl, New Malden, KT3 5BP [IO91UJ, TQ26]
G1 HCO S Durrant, 58 Tyrone Rd, Stockton on Tees, Fairfield, Stockton on Tees, TS19 7JW [IO94HN, NZ41]
G1 HCR J S Evans, 36 Wheatfields Rd, Shinfield, Reading, Berks, RG2 9DG [IO91MJ, SU76]
G1 HCU G Gratton, 5 Nursery Ave, Ovenden, Halifax, HX3 5SZ [IO93BR, SE02]
G1 HCV S J Graves, 135 Queen Anne Ave, Bromley, BR2 0SH [JO01AJ, TQ36]
GW1 HCW C P Griffiths, 16 Shelley Rd, Priory Park, Haverfordwest, SA61 1RX [IO71MT, SM91]
G1 HCX F J Hart, 27 Highfield Rd, Hale, Altrincham, WA15 8BX [IO83UJ, SJ78]
GM1 HDF J Johnston, 143 Arbroath Ave, Cardonald, Glasgow, G52 3HH [IO75TU, NS56]
G1 HDG P K Greed, 12 Bailey Cl, Windsor, Berks, SL4 3RD [IO91QL, SU97]
G1 HDK W Akhurst, Kent Lodge, 20 Newton Rd, Faversham, Kent, ME13 8DZ [JO01KH, TR06]
G1 HDM P V Aram, 4 Severn Rd, Chilton, Didcot, OX11 0PW [IO91IN, SU48]
G1 HDO A A Appleton, 14 Clarendon Rd, Bournemouth, BH4 8AL [IO90BR, SZ09]
G1 HDP R A Ashbee, Silver Wood, Skinners Ln, Ashtead, KT21 2LY [IO91UH, TQ15]
G1 HDQ K N Ravenhill, Jasmine Cottage, Alston, Axminster, Devon, EX13 7LG [IO80MT, ST30]
G1 HDR R A Stanford, Lime Tree Cottage, 1 South End, Bassingbourn, Royston, SG8 5NG [IO92XB, TL34]
G1 HDS Details withheld at licensee's request by SSL.
G1 HDX M E Robertson, Fishtoft Rd, Fishtoft, Boston, Lincs, PE21 0QS [JO02AX, TF34]
G1 HEA A G Steele, 92 Eelholme View St., Keighley, BD20 6AY [IO93BV, SE04]
G1 HED P E Smith, 274 Middle Park Way, Leigh Park, Havant, PO9 4NL [IO90MU, SU70]
G1 HEJ J M Alexander, 1 Locarno Rd, Swanage, BH19 1HY [IO90AO, SZ07]
GI1 HEK W J Barnes, 70 Disraeli St., Crumlin Rd, Belfast, BT13 3HW [IO74AO, J37]
G1 HEN D G Coates, 2 Penfold Dr, Countesthorpe, Leicester, LE8 5TP [IO92KN, SP59]
G1 HEP C J Heptonstall, 65 Ings Mill Ave, Clayton West, Huddersfield, HD8 9QG [IO93EO, SE21]
G1 HEQ K J Tucker, 507 New North Rd, Ilford, IG6 3TF [JO01BO, TQ49]
G1 HER G M A Dasilva-Hill, 12 St. Stephens Cres, Thornton Heath, CR7 7NP [IO91WJ, TQ36]
G1 HEU G B Tybora, 37 Nunsfield Dr, Alvaston, Derby, DE24 0GH [IO92GV, SK33]
G1 HEW P E Travers, 49 West Bank Dr, South Anston, Anston, Sheffield, S31 7JG [IO93GJ, SK38]
G1 HEX J W B Dunn, 8 Ettrick Terr North, Craghead, Stanley, DH9 6BE [IO94DU, NZ25]
G1 HEY Mt J B Todd, Atlast, 7 Marine Ave West, Sutton on Sea, Mablethorpe, Lincs, LN12 2TX [JO03DH, TF58]
G1 HEZ C Tindale, 4 Station Rd, Beamish, Stanley, DH9 0QU [IO94EV, NZ25]
G1 HFA R Winder, 176 Ambleside Rd, Lancaster, LA1 3ND [IO84OB, SD46]
G1 HFC Details withheld at licensee's request by SSL.[Station located near Sudbury.]
G1 HFD Details withheld at licensee's request by SSL.
G1 HFE S Wood, 18 Rosemellin, Camborne, TR14 8QF [IO70IF, SW64]
G1 HFH D J Ward, 10 Fulshaw Ave, Wilmslow, SK9 5JA [IO83VH, SJ88]
G1 HFI I M Wall, 11 Coombe View, Coombe Valley, Teignmouth, TQ14 9UY [IO80FN, SX97]
G1 HFJ L G Wooldridge, 1 Maidenhead Rd, Hartcliffe, Bristol, BS13 0PS [IO81QJ, ST56]
G1 HFK W Willoughby, 27 Foxwood Gr, Sheffield, S12 2FN [IO93HI, SK38]
G1 HFR R Butlin, 10 Salterford Rd, Hucknall, Nottingham, NG15 6GA [IO93JA, SK54]
G1 HFS K Burgess, 32 Hendon St., Leigh, WN7 1TS [IO83RM, SD60]
G1 HFT E M Bennett, 33 Station Rd, Barrow Hill, Chesterfield, S43 2NL [IO93HG, SK47]
GW1 HFW F Hewins, Hillcrest, Wern Rd, Rhosesmor, Mold, CH7 6PY [IO83JF, SJ26]
G1 HFY B J Watson, 20 St. Marys Gdns, Hilperton Marsh, Trowbridge, BA14 7PG [IO81VI, ST85]
G1 HGA K L Yates, 3 Flaxland Cres, Sileby, Loughborough, LE12 7SB [IO92KR, SK61]
G1 HGB J R Neville, 44 Thorpe House Ave, Sheffield, S8 9NG [IO93GI, SK38]
GM1 HGC B Newton, 12 Glascairn Ave, Portlethen, Aberdeen, AB12 4QF [IO87WB, NO99]
G1 HGD M F Newell, 90 Arthur St., Kenilworth, CV8 2HG [IO92FI, SP27]
G1 HGE R E Nason, 178 Wavell Cl, Stewartby, Bedford, MK43 9LW [IO92RB, TL04]
G1 HGF T A Standring, 52 Beechcroft Rd, Ipswich, IP1 6BD [JO02NB, TM14]
G1 HGG J M Newton, 51 Kaye Rd, Mansfield, NG19 6JP [IO93JD, SK56]
G1 HGQ R W Beadle, 10 Westlyn Cl, Rainham, RM13 9JP [JO01CM, TQ58]
G1 HGR R I Bagnall, 10 Chiltern Cl, Ashurst Bridge, Totton, Southampton, SO40 7PT [IO90FV, SU31]
G1 HGT A R Berkerey, 36 Erlesmere Gdns, London, W13 9TY [IO91UM, TQ17]
G1 HGW G E Bristow, 21 Dixon Ave, Grimsby, DN32 0AJ [IO93XN, TA20]
G1 HGY P J Parry, Forge House, Church Rd, Hargrave, Wellingborough, NN9 6BQ [IO92SH, TL07]
G1 HHC D T Bolt, Old School House, Maristow, Roborough, Plymouth, PL6 7BY [IO70WK, SX46]
G1 HHD E H Bolt, Old School House, Maristow, Roborough, Plymouth, PL6 7BY [IO70WK, SX46]
G1 HHH T H Ransom Hastings Elec Rd, 9 Lyndhurst Ave, Hastings, TN34 2BD [JO00HU, TQ81]
G1 HHL Details withheld at licensee's request by SSL.
GW1 HHM K R Roberts, 31 Cybi Pl, Holyhead, LL65 1DT [IO73QH, SH28]
G1 HHO G S Reeve, 10 Badgers Copse, New Milton, BH25 5PE [IO90ES, SZ29]
G1 HHP L M Silcock, 25 Beaver Dr, Sheffield, S13 9QL [IO93HI, SK48]
G1 HHS A J Burrows, 1 Browns Ave, Runwell, Wickford, SS11 7PT [JO01GO, TQ79]
G1 HHT A L Benstock, 17 Windermere Dr, Leeds, LS17 7UZ [IO93FU, SE33]
G1 HHU N T Ball, 50A Ashover Ave, Manchester, M12 5LT [IO83VL, SJ89]
G1 HHW W J Curtis, 34 Bentley Cl, Rectory Farm, Northampton, NN3 5JS [IO92OG, SP86]
G1 HIA P C Smith, 5 Western Rd, Horfield, Bristol, BS7 8UP [IO81QL, ST57]
G1 HIB M Standing, Flat, 9 Gressingham House, Gressingham Dr, Lancaster, LA1 4RG [IO84OA, SD45]
G1 HIC Details withheld at licensee's request by SSL.
G1 HID R Smith, Bank End Cottage, Main St., Kelfield, York, YO4 6RG [IO93KU, SE53]
G1 HIE B L R Scott, 67 Cowley Rd, Uxbridge, UB8 2AE [IO91SM, TQ08]
G1 HIG T L Ravelini, 14 Meath Cl, Orpington, BR5 2HF [JO01BJ, TQ46]
G1 HIH D Sledden, 7 Towneley Ave, Huncoat, Accrington, BB5 6LP [IO83TS, SD73]
G1 HIJ R M Dimmock, 67 Meadway, Dunstable, LU6 3JT [IO91RV, TL02]
G1 HIM F R Stephens, 22 Roedean Rd, Worthing, BN13 2BS [IO90TT, TQ10]
G1 HIO M Horsfield, 59 Queens Dr, Newton-le-Willows, WA12 0LY [IO83QL, SJ59]
G1 HIP K Horsfield, 59 Queens Dr, Newton-le-Willows, WA12 0LY [IO83QL, SJ59]
G1 HIR K S Craven, 669 Spring Bank Wes, Hull, HU3 6LG [IO93TS, TA02]
G1 HIU J B Clarke, 9 Park Ln, Bulmer, Sudbury, CO10 7EQ [JO02IA, TL83]
G1 HIZ Details withheld at licensee's request by SSL.
G1 HJA V I Trevitt, 17 Peterstow Cl, London, SW19 6JW [IO91VK, TQ27]
G1 HJD B J Taylor, 111 High St., Warboys, Huntingdon, PE17 2TB [IO92XJ, TL38]
G1 HJH J A Trusler, St. Marys Cottage, 42 Mill Hill, Shoreham By Sea, BN43 5TH [IO90UU, TQ20]
G1 HJJ D P Creek, 161 Willoughby Rd, Boston, PE21 9HR [IO92XX, TF34]
G1 HJL I M Copland, 69 Parkside Terr, Cullingworth, Bradford, BD13 5AD [IO93BT, SE03]
G1 HJO Details withheld at licensee's request by SSL.
G1 HJP T Carter, 7 St. Pauls Cl, Gorefield, Wisbech, PE13 4NL [JO02BQ, TF41]
G1 HJS M K Todd, 38 The Churchlands, New Romney, TN28 8LB [JO00LX, TR02]
G1 HJW A J Wayland, 31 King Henry Dr, Rochford, SS4 1HY [JO01IN, TQ88]
G1 HKD R A Denton, 8 Northolme, Gainsborough, DN21 2JD [IO93OJ, SK89]
G1 HKE Details withheld at licensee's request by SSL.
G1 HKF C J Maclennan, 66 Sandsfield Ln, Gainsborough, DN21 1DD [IO93OJ, SK88]
G1 HKM F S Woods, 275 Scotter Rd, Scunthorpe, DN15 7EH [IO93PO, SE81]
G1 HKP P A Wooldridge, 104 St. James Ln, Coventry, CV3 3GS [IO92GJ, SP37]
G1 HKQ A Woodward, 207 Avondale Dr, Widnes, WA8 7XB [IO83OI, SJ48]
G1 HKR T A Whittam, 27 Dimples Ln, Garstang, Preston, PR3 1RD [IO83OV, SD44]
G1 HKS D J Wilson, 34 Belfairs Dr, Chadwell Heath, Romford, RM6 4EB [JO01BN, TQ48]
G1 HKT R A Woodley, 1 Melton Dr, Didcot, OX11 7JP [IO91JO, SU59]
G1 HKU Details withheld at licensee's request by SSL.
GW1 HKY M S Williams, 88 Garth Owen, Newtown, SY16 1JW [IO82IM, SO19]
G1 HLF Details withheld at licensee's request by SSL.
G1 HLH F C Stewart, Shingles, Ingleborough Ln, St Mary's Platt, Sevenoaks, Kent, TN15 8JU [JO01DH, TQ65]
G1 HLK D Poole, Valency, 9 Hiliary Gdns, Stanmore, HA7 2NH [IO91UO, TQ19]
G1 HLO R P Penn, 12 Helens Gate, Thomas Rochford Way, Cheshunt, Waltham Cross, EN8 0SQ [IO91XR, TL30]

G1 HLP D J Plant, 15 Heathcombe Rd, Bridgwater, TA6 7PD [IO81LD, ST23]
G1 HLQ M Edwards, 6 Wansford Green, Goldsworth Park, Woking, GU21 3QH [IO91QH, SU95]
G1 HLS W F Etherton-Scott, 62 Spencer Rd, Walthamstow, London, E17 4BD [IO91XO, TQ39]
G1 HLT I D Fay, 7 Oakridge Cl, Forest Town, Mansfield, NG19 0EY [IO93KD, SK56]
G1 HLV J W Lee, Deighton Manor, Deighton, Northallerton, North Yorks, DL6 2SN [IO94GJ, NZ30]
G1 HLY R W Powell, 55 Lumley Rd, Horley, RH6 7JF [IO91WE, TQ24]
G1 HMA Details withheld at licensee's request by SSL.
G1 HMF B Rudkin, 24 Thrush Cl, Melton Mowbray, LE13 0QF [IO92NS, SK71]
G1 HMI T J Rock, More House, Haywood Dr, Tettenhall, Wolverhampton, WV6 8RF [IO82VO, SO89]
G1 HMK C D Selby, PO Box 3085, Onerahi, Whangarei, New Zealand
G1 HMO A J Steele, 4 Royville Pl, Stoke on Trent, ST6 1RP [IO83WB, SJ84]
G1 HMP P Shaw, 12 Croft Gdns, Grimscar Valley, Birkby, Huddersfield, HD2 2FL [IO93CP, SE11]
G1 HMT G W Gray, 107 Elliott Rd, March, PE15 8BT [JO02AN, TL49]
G1 HMW W E Gardner, 25 Prospect Pl, Wing, Leighton Buzzard, LU7 0NT [IO91PV, SP82]
G1 HMY P D Batty, 14 Woodville Rd, Penwortham, Preston, Lancs, PR1 9DR [IO83PR, SD52]
G1 HMZ K Breedon, 17 Emmanuel Ave, Arnold, Nottingham, NG5 9QN [IO93KA, SK54]
G1 HND M J Burling, 28 Croydon Rd, Arrington, Royston, SG8 0DJ [IO92XD, TL34]
G1 HNG P J Beesley, 73 Jubilee Rd, Aberdare, CF44 6DE [IO81GQ, SO00]
G1 HNH P E Bannister, 38 Regent St., Stowmarket, Suffolk, IP14 1RJ [JO02LE, TM05]
G1 HNI Details withheld at licensee's request by SSL.
G1 HNN R B Cockman, 31 Kensington Rd, Southend on Sea, SS1 2SX [JO01IM, TQ88]
G1 HNU M R Gray, 28 The Close, Bradwell, Great Yarmouth, NR31 8DR [JO02UN, TG50]
G1 HNW R F Harding, 55 Westwood Rd, Tilehurst, Reading, RG31 5PN [IO91LL, SU67]
G1 HNX M A Hulett, 34 Kingsley Rd, Allestree, Derby, DE22 2JH [IO92GW, SK33]
GM1 HNZ A K S Simmers, Loanside Cottage, Crossroads, Keith, Banffshire, AB55 6LP [IO87NN, NJ45]
G1 HOD A Smith, 2 Manvers Rd, West Bridgford, Nottingham, NG2 6DH [IO92KW, SK53]
G1 HOG J W Summers, 6 Bramble Cl, Copthorne, Crawley, RH10 3QB [IO91WD, TQ33]
G1 HOI Details withheld at licensee's request by SSL.
G1 HOJ B J Baylis, 118 Eastgate, Deeping St. James, Peterborough, PE6 8RD [IO92UP, TF10]
G1 HOK Details withheld at licensee's request by SSL.
G1 HOL S F Cook, 26 Sutton Cl, Winyates West, Redditch, B98 0JR [IO92BH, SP06]
G1 HOP R J Chaston, 11A Boswell Dr, Walsgrave, Coventry, CV2 2DL [IO92GK, SP38]
G1 HOQ Details withheld at licensee's request by SSL.
G1 HOU Details withheld at licensee's request by SSL.
G1 HPB R W S Wearing, 163 Birmingham Rd, Stratford upon Avon, CV37 0AP [IO92DE, SP15]
GM1 HPC Details withheld at licensee's request by SSL.
G1 HPD Details withheld at licensee's request by SSL.
G1 HPG J Orr, 13 Haldane Cl, Brierley, Barnsley, S72 9LL [IO93HO, SE41]
G1 HPI A Harness, 27 St. Andrews Cres, Black Hill, Consett, DH8 8PD [IO94BU, NZ05]
G1 HPJ D R Healey, 15 High St. North, West Mersea, Colchester, CO5 8JU [JO01KS, TM01]
GW1 HPP G Jones, Bodalaw Chwilog, Pwllheli, Gwynedd, LL53 6PS [IO72UW, SH43]
G1 HPQ K R Jones, 11 Westvale Rd, Timperley, Altrincham, WA15 7RL [IO83TJ, SJ78]
G1 HPS T G Jones, Willow Dale, Foxcotte Rd, Charlton, Andover, SP10 4AR [IO91GF, SU34]
G1 HPU P J James, 8 Higher Pipers Wood Cttgs, Missenden Rd, Little Missenden, Amersham, HP7 0RQ [IO91QQ, SU99]
G1 HPV T R Jones, 175 New Rd, Great Wakering, Southend on Sea, SS3 0AR [JO01JN, TQ98]
G1 HPZ I G Russell, 368 Whitehall Rd, St. George, Bristol, BS5 7BT [IO81RL, ST67]
G1 HQA Details withheld at licensee's request by SSL.
G1 HQC L Carline, 42 Storth Ln, Kiveton Park, Sheffield, S26 5QT [IO93II, SK48]
G1 HQD C Coughlan, 32 Higher Brockwell, Sowerby Bridge, HX6 1BT [IO93AQ, SE02]
G1 HQE C R Close, 3 Hay Green, Therfield, Royston, SG8 9QL [IO92XA, TL33]
G1 HQG A N Coley, Mayfield, Lower Draggons, Fordingbridge, Hants, SP6 3EE [IO90BW, SU01]
G1 HQH I T Church, 26 Usk Way, Didcot, OX11 7SQ [IO91JO, SU59]
G1 HQI P L Redman, 49 Elmdale Cl, Aspen Gdns, Warsash, Southampton, SO31 9RW [IO90IU, SU40]
G1 HQJ D J Robinson, 16 Green Ln, Lenham, Platts Heath, Maidstone, ME17 2NS [JO01IF, TQ85]
G1 HQK I C Richardson, 22 Mallard Way, Riverside, March, PE15 9HT [JO02BN, TL49]
G1 HQN J Rattenbury, 28 Abbotts Rd, New Barnet, Barnet, EN5 5DP [IO91VP, TQ29]
G1 HQO G P Spaven, 302 Church Rd, Redfield, St. George, Bristol, BS5 8AJ [IO81RL, ST67]
G1 HQQ F F Jensen, 79 The Drakes, Shoeburyness, Southend on Sea, SS3 9NY [JO01JM, TQ98]
GI1 HQU W A Kerr, 32 Kings Rd, Belfast, Co. Antrim, BT5 6JJ [IO74BO, J37]
G1 HQW J G Kierman, Elmgrove, 25 Sutton Rd, Leverington, Wisbech, PE13 5DN [JO02BQ, TF41]
G1 HRA D S Lloyd, 35 Charles Cl, Abbotts Barton, Winchester, SO23 7HT [IO91HB, SU43]
G1 HRD V P Allen, 23 Finchdean House, Tangley Gr, London, SW15 4EQ [IO91VK, TQ27]
G1 HRE S H Loveridge, 5 Aggborough Cres, Kidderminster, DY10 1LG [IO82VJ, SO87]
G1 HRF S G Lane, 29 Wellfield, Clayton-le-Moors, Accrington, BB5 5WA [IO83TS, SD73]
G1 HRH M R Gregory, 45 Larksfield Ave, Bournemouth, BH9 3LW [IO90BS, SZ19]
G1 HRI S R Davies, 20 Needwood Rd, Woodley, Stockport, SK6 1LQ [IO83WK, SJ99]
G1 HRL E H Dillow, 18 Laburnum Gr, Warwick, CV34 5TG [IO92FH, SP26]
G1 HRM T M Davenport, 36 Rydale Rd, Sherwood, Nottingham, NG5 3GS [IO92KX, SK54]
G1 HRQ B J Madore, 15 St. Johns Hill, Ryde, PO33 1EU [IO90KR, SZ59]
G1 HRU S J Hill, 26 Cres Rd, Netherton, Dudley, DY2 0NW [IO82WL, SO98]
G1 HRV K W Higgins, 22 Thatchers Ln, Cliffe, Rochester, ME3 7TN [JO01GL, TQ77]
G1 HRY A J Davis, 7 The Courtyard, Fisherwick Wood Ln, Fisherwick Wood, Lichfield, WS13 8QQ [IO92DQ, SK10]
G1 HSA S Arnold, 30 Pine Ave, Newton-le-Willows, WA12 8JE [IO83QK, SJ59]
G1 HSC B J Edwards, 4 Middlewich Rd, Holmes Chapel, Crewe, CW4 7EA [IO83TE, SJ76]
G1 HSE J H Emery, Pasmor, Trevowah Rd, Crantock, Newquay, TR8 5RU [IO70KJ, SW76]
G1 HSF M Evans, Fernlea, Moorlynch, Bridgwater, Somerset, TA7 9BT [IO81ND, ST33]
G1 HSG N D Evans, 25 Chetwyn Ave, Bromley Cross, Bolton, BL7 9BN [IO83SO, SD71]
G1 HSH M F Ellerby, 21 Ickworth Ct, Ingleby Barwick, Stockton on Tees, TS17 0PA [IO94IM, NZ41]
G1 HSI M S Glazier, 19 West Pl, Brookland, Romney Marsh, TN29 9RG [JO01JA, TQ92]
G1 HSJ L Godden, The Conifers, 14 Pirehill Ln, Walton, Stone, ST15 0JN [IO82WV, SJ83]
G1 HSL J L Girt, 22 Medway Rd, Ipswich, IP3 0QH [JO02OA, TM14]
G1 HSM L F Heller, Flat 2, 110 Henley Rd, Caversham, Reading, RG4 6DH [IO91ML, SU77]
G1 HSN Details withheld at licensee's request by SSL.
G1 HSS Details withheld at licensee's request by SSL.
G1 HST R Jones, 70 Devonport Rd, London, W12 8NU [IO91VM, TQ27]
G1 HSW J E Kennedy, 16 Back Ln, Congleton, CW12 4PP [IO83VD, SJ86]
G1 HSX P Kimber, 8 Jurys Gap Rd, Lydd, Romney Marsh, TN29 9BD [JO00KW, TR02]
G1 HTF H Heron, 43 Cheetham Hill Rd, Dukinfield, SK16 5JL [IO83XL, SJ99]
GM1 HTI S Gray, 64 Lockerbie Rd, Dumfries, DG1 3BL [IO85EB, NX97]
G1 HTL J M Foster, 25 Hunter Rd, Arnold, Nottingham, NG5 6QZ [IO92KX, SK64]
G1 HTN P A Farrow, 12 St. Georges Lodge, Queens Rd, Weybridge, KT13 0AB [IO91SI, TQ06]
G1 HTO R J Fortescue, 14 The Walk, Marine Parade, Lyme Regis, Dorset, DT7 3JE [IO80MR, SY39]
G1 HTT L W Gray, 87 London Rd, Coventry, CV1 2JQ [IO92GJ, SP37]
G1 HTW R Guest, 83 Worksop Rd, Swallownest, Sheffield, S31 0WB [IO93GJ, SK38]
GU1 HTY B P Ayres, Rousay, Bailiffs Cross Rd, St. Andrew, Guernsey, GY6 8RY
G1 HUE J T Lee, 2 South Lodge, Linslade Rd, Heath & Reach, Leighton Buzzard, LU7 0EB [IO91PW, SP92]
G1 HUL J G Andrews, 64 Bradgate Rd, Markfield, LE67 9SN [IO92IQ, SK41]
G1 HUM R Beech, 6 Law Cliff Rd, Birmingham, B42 1LP [IO92AM, SP09]
G1 HUP Details withheld at licensee's request by SSL.
G1 HUX D Mansfield, Little Ash, St. Ln, Lower Whitley, Warrington, WA4 4EN [IO83RH, SJ67]
G1 HVA M J Bugby, 99 Belle Vue Rd, Earl Shilton, Leicester, LE9 8AA [IO92IN, SP49]
G1 HVL P S Howarth, 63 Whitley Spring Cres, Ossett, WF5 0RF [IO93FQ, SE22]
G1 HVN Details withheld at licensee's request by SSL.
G1 HVZ K D Harper, 18 Bedlington Walk, Billingham, TS23 3XW [IO94IO, NZ42]
G1 HWH G S Jackson, Pathways, Downbarton Rd, St. Nicholas At Wade, Birchington, CT7 0PY [JO01PI, TR26]
G1 HWJ P S Milner, 3 Larne Ave, Cheadle Heath, Stockport, SK3 0UJ [IO83VJ, SJ88]
G1 HWK T E McCarthy, 25 Henley Ave, North Cheam, Sutton, SM3 9SG [IO91VJ, TQ26]
G1 HWM P A Milner, 3 Briggs Villas, Queensbury, Bradford, BD13 2EP [IO93BS, SE13]
G1 HWO T D W Miller, 27 Richmond Way, Oadby, Leicester, LE2 5TR [IO92LO, SP69]
G1 HWP R E Davies, 7 The Nook, Tupsley, Hereford, HR1 1NH [IO82PB, SO54]
G1 HWR L R N Mills, 54 Petters Rd, Ashtead, KT21 1NE [IO91UH, TQ15]
G1 HWY M R Jupp, 54 Shooting Field, Steyning, BN44 3RQ [IO90UV, TQ11]
G1 HWZ G W E Jukes, 12 Vine St., Kidderminster, DY10 2TS [IO82VJ, SO87]
G1 HXM Details withheld at licensee's request by SSL.
G1 HXN G E King, 1 Sudan Cottage, Frogge Ln, Coltishall, Norwich, NR12 7JU [JO02QR, TG22]
G1 HXO H E Kirby, 42 Tarnston Rd, Deer Park Est, Hartlepool, TS26 0PQ [IO94JQ, NZ43]
G1 HXP J C Lennard, 10 Orston Rd East, West Bridgford, Nottingham, NG2 5FU [IO92KW, SK53]
G1 HXR R P Davis, 6 Fairway Dr, Northmoor Park, Wareham, BH20 4SG [IO80WQ, SY98]
G1 HXT G R Eden, Heathend Cottage, Cromhall, Wotton under Edge, Glos, GL12 8AS [IO81SO, ST68]
G1 HXZ C A Cave, 20 Meadow View, Banbury, OX16 9SR [IO92IB, SP43]
G1 HYA N A Cramp, 19 Meadowgate Croft, Lofthouse, Wakefield, WF3 3SS [IO93FR, SE32]
G1 HYC D C Curson, 3 Stiffkey Rd, Warham, Wells Next The Sea, NR23 1NP [JO02KW, TF94]
G1 HYG B Crowther, 104 John St., Beamish, Stanley, DH9 0QP [IO94EU, NZ25]
GU1 HYN B Bolderston, 12 Hartlebury Est, Steam Mill Ln, St. Martin, Guernsey, GY4 6NH
G1 HYO M Green, 21 Thorfinn Terr, Thurso, KW14 7LL [IO88FO, ND16]
G1 HYQ D A Chenoweth, 20 Churchlands Rd, Bedminster, Bristol, BS3 3PW [IO81QK, ST57]
G1 HYT E D Cook, 129 Days Ln, Sidcup, DA15 8JT [JO01BK, TQ47]
G1 HYU K J Church, 63 Alice Fisher Cres, Kings Lynn, PE30 2PE [JO02ES, TF62]
G1 HYX A P Chance, 18 Egdon Glen, Crossways, Dorchester, DT2 8BQ [IO80UQ, SY78]
G1 HZA G Loosley, 48 Dormer Ave, Wing, Leighton Buzzard, LU7 0TF [IO91PV, SP82]
G1 HZD C J Legate, 5 Kemble Cl, Potters Bar, EN6 5EG [IO91VQ, TL20]
G1 HZG A Lovell, 28 Frobisher Ave, Portishead, Bristol, BS20 9XB [IO81OL, ST47]
G1 HZH T G Langley, 2 Glenbrook Dr, Lidget Green, Bradford, BD7 2QF [IO93CT, SE13]
G1 HZJ M W Devine, 8 Maple Rd, Gravesend, DA12 5JR [JO01EK, TQ67]
G1 HZL M C Donaldson, 54 Glebe St., Burnley, BB11 3LH [IO83VS, SD83]
G1 HZQ M Fisher, 27 Fairview Ave, Carlinghow, Batley, WF17 8EE [IO93ER, SE22]
G1 HZR K Farrar, 8 Ascot Ave, Cantley, Doncaster, DN4 6HE [IO93KM, SE60]
G1 HZW Details withheld at licensee's request by SSL.
GI1 HZX A J Gilbody, 5 The Plateau, Piney Hills, Belfast, BT9 5QP [IO74AN, J37]
G1 IAB S D M Matthews, 5 Green House Rd, Wheatley Hills, Doncaster, DN2 5NG [IO93KM, SE50]
G1 IAD D M Morton, 14 St. Georges Rd, Pakefield, Lowestoft, NR33 0JW [JO02UL, TM59]
G1 IAG P Morris, 18 Greenway Cl, Colindale, London, NW9 5AZ [IO91UO, TQ29]
G1 IAL A J Plant, 11 Halesbury Ct, Ombersley Rd, Halesowen, B63 4PE [IO82XK, SO98]
G1 IAP Details withheld at licensee's request by SSL.
G1 IAR M J M Marsh, 23 Simcoe Way, Dunkeswell, Honiton, EX14 0UR [IO80JU, ST10]
G1 IAV P J Costello, Newgrange, Poplar Rd, New Milton, Hants, BH25 5XP [IO90ES, SZ29]
GW1 IAW I A Woodward, Ty Canol, Heol Maelor, Coedpoeth, Wrexham, LL11 3LR [IO83LB, SJ25]
G1 IAY C D Starkey, 31 Almsford Oval, Harrogate, HG2 8EJ [IO93FX, SE35]
G1 IBF G Tullock, 4 Northfield Ave, Driffield, YO25 7EX [IO94SA, TA05]
G1 IBG G C Nicholson, 22 Stortford Hall Park, Bishops Stortford, CM23 5AL [JO01CU, TL42]
G1 IBJ C J Diaper, 41 Beechwood Ave, Chatham, ME5 7HJ [JO01GI, TQ76]
G1 IBM A Kearney IBM ARC Lond CG, 18 Wayside Mews, Maidenhead, SL6 7EJ [IO91PM, SU88]
G1 IBO D Buss, 33 Meon Cl, Chelmsford, CM1 7QG [JO01FR, TL70]
G1 IBP A G Heaysman, 325 Broomfield Rd, Chelmsford, CM1 4DU [JO01FS, TL70]
G1 IBS W A Chadwick, 73 Hengrove Cres, Ashford, TW15 3DF [IO91SK, TQ07]
G1 IBX C N H Drayton, 11 Ayresome Ave, Roundhay, Leeds, LS8 1BB [IO93FU, SE33]
G1 ICA D A Keable, 90 King Edward Rd, Rugby, CV21 2TE [IO92IJ, SP57]
G1 ICH M Adams, 61 Monks Park Rd, Northampton, NN1 4LU [IO92NF, SP76]
G1 ICK D Winton, 16 Lord Ave, Clayhall, Ilford, IG5 0HP [JO01AO, TQ48]
G1 ICO P L Whittle, 53 Slough Ln, Kingsbury, Middx, NW9 8YB [IO91UN, TQ28]
G1 ICP J B Young, 14 Margeholes, Watford, WD1 5AP [IO91TP, TQ19]
G1 ICQ A R Orgee, 54 Riverview Cl, Off Hallow Rd, Worcester, WR2 6DA [IO82VE, SO85]
G1 ICX D M Palmer, 14 Walcups Ln, Great Massingham, Kings Lynn, Norfolk, PE32 2HR [JO02HS, TF72]
G1 IDE Details withheld at licensee's request by SSL.[Station located near Ashford.]
G1 IDF A S Edwards, 51 Redrock Rd, Rotherham, S60 3JN [IO93IJ, SK49]
G1 IDJ G Perry, 61 Ollands Rd, Reepham, Norwich, NR10 4EL [JO02NS, TG12]
G1 IDN Details withheld at licensee's request by SSL.
GM1 IDP G D Russell, 7 Turner Park, Balerno, EH14 7BT [IO85HV, NT16]
G1 IDV G L Bain, 99 Longford Ln, Longlevens, Gloucester, GL2 9HB [IO81VV, SO82]
G1 IDZ D M Young, 12 Windrush Cl, Bramley, Guildford, GU5 0BB [IO91RE, TQ04]
GW1 IEB L C Tatham, Rhos y Grug, Llangwnadl, Pwllheli, Gwynedd, LL53 8NW [IO72QU, SH23]
G1 IEC P Walton, 2 Albert Rd, Hill, Bromsgrove, B61 7BE [IO82XH, SO96]
G1 IEJ G F Brown, 21 Armada Dr, Teignmouth, TQ14 9NF [IO80FN, SX97]
G1 IEK W J Bradford, 6 Lynndale Ave, Cowpen, Blyth, NE24 4DY [IO95FC, NZ28]
GM1 IEL J Brown, 24 Southgreen Dr, Airth By Falkirk, Stirlingshire, Scotland, FK2 8JP [IO86CB, NS98]
G1 IEM A R J Bird, 3 Cudham Gdns, Cliftonville, Margate, Kent, CT9 3HG [JO01RJ, TR37]
G1 IEO J E Turner, 2 Hilberry Rd, Canvey Island, Essex, SS8 7EQ [JO01HM, TQ88]
G1 IEP G B Tomkins, 73 Oakleigh Rd, Clayton, Bradford, BD14 6NP [IO93CS, SE13]
G1 IER J E Tupper, 185 Northmoor Way, Northmoor Park, Wareham, BH20 4SB [IO80WQ, SY98]
G1 IEV R J Usher, 7 Wych Elms, Park St., St. Albans, AL2 2AN [IO91TR, TL10]
G1 IEX D A Broughton, Quaintways Cottage, Clatterway Hill, Bonsall, Matlock, Derbyshire, DE4 2AH [IO93FC, SK25]
G1 IEY S C Reigate, 9 Effingham Rd, Croydon, CR0 3NF [IO91WJ, TQ36]
G1 IFA G D Richardson, 52 Hartley Down, Christchurch Rd, Bournemouth, BH1 3PJ [IO90BR, SZ19]
G1 IFF A Rose, 15 Elderwood Way, Tuffley, Gloucester, GL4 0RA [IO81UU, SO81]
G1 IFH G Reading, 22 Stubbin Cl, Rawmarsh, Rotherham, S62 7DQ [IO93HL, SK49]
G1 IFL T P Williams, Homewood, 40 Stotfold Rd, Church End, Arlesey, SG15 6XT [IO92UA, TL13]
G1 IFM Details withheld at licensee's request by SSL.
G1 IFS M A Ford, 3 Hill View, Baddeley Green, Stoke on Trent, ST2 7AR [IO83WB, SJ95]
G1 IFV N J J Ginger, 152 Hawks Rd, Hailsham, BN27 1NA [JO00DV, TQ51]
G1 IFW W A Gain, 14 Clarence Rd, St. Leonards on Sea, TN37 6SD [JO00GU, TQ81]
G1 IFX D H Garratt, 3 Fort Rd, Mountsorrel, Loughborough, LE12 7HB [IO92KR, SK51]
G1 IFY M B Wyatt, 4 Gritstone Rd, Matlock, DE4 3GB [IO93FD, SK36]
G1 IGA C R Brooks, 78 Leggatts Wood Ave, Watford, WD2 5RP [IO91TQ, TQ19]
G1 IGC S E Brookes, 52 Larch Gr, Kendal, LA9 6AU [IO84PH, SD59]
G1 IGE S G Bates, 6 The Green, Swanwick, Alfreton, DE55 1BL [IO93HB, SK45]
G1 IGP G C Spinks, 89 Uplands Rd, Oadby, Leicester, LE2 4NT [IO92LO, SK60]
G1 IGQ B B Chambers, 55 Palm Ave, Footscray, Sidcup, DA14 5JF [JO01BJ, TQ47]
G1 IGW D Cliff, 23 Grey Towers Dr, Nunthorpe, Middlesbrough, TS7 0LT [IO94JM, NZ51]
G1 IGX G K Brogden, 10 Sidmouth Cl, Bedford, MK40 3BS [IO92SD, TL05]
G1 IGZ R Shaw, 11 Gaunt Rd, Bramley, Rotherham, S66 0YL [IO93IK, SK49]
G1 IHA R A Stravens, 13 Chandos Rd, East Finchley, London, N2 9AR [IO91VO, TQ29]
GW1 IHB P Spashett, Llys-y-Coed, Trefriw, Gwynedd, LL27 0QA [IO83BD, SH76]
G1 IHE J S Smith, 65 Woods Ave, Hatfield, AL10 8QF [IO91VS, TL20]
G1 IHI M E Godsave, Redburn, 35 Furlong Cl, Midsomer Norton, Bath, BA3 2PR [IO81SG, ST65]
G1 IHJ A M Homer, 33 Ensall Dr, Wordsley, Stourbridge, DY8 4XX [IO82WL, SO88]
G1 IHK Details withheld at licensee's request by SSL.
G1 IHL S J Hopkins, 98 Ct Rd, Kingswood, Bristol, BS15 2QP [IO81RK, ST67]
G1 IHN R D Hargreaves, 58 Horsewell Ln, Wigston, LE18 2HQ [IO92KN, SP69]
G1 IHO D C W Johnson, 3 Harbour View, Combwich, Bridgwater, TA5 2QU [IO81LE, ST24]
G1 IHS C Currie, 33 Ashridge Dr, Bricket Wood, St. Albans, AL2 3SR [IO91TQ, TL10]
G1 IHT J A Clarke, 39 Soundwell Rd, Staple Hill, Bristol, BS16 4QQ [IO81RL, ST67]
G1 IHY J E Shilson, 3 Hereford Cl, Desborough, Kettering, NN14 2XA [IO92OK, SP88]
G1 IIC Details withheld at licensee's request by SSL.
G1 IIE Details withheld at licensee's request by SSL.
GI1 IIL G C J Scullion, 13 Orritor Cres, Cookstown, BT80 8BQ [IO64PP, H87]
G1 IIO B Thornton, 21 Valley Rd, Greenhills Est, Banbury, OX16 9BQ [IO92IB, SP43]
G1 IIX B M Lee, 31 Merton Ave, Farsley, Pudsey, LS28 5DX [IO93DT, SE23]
G1 IIY D D Turner, 131 Farmstead Rd, Corby, NN18 0LJ [IO92PL, SP88]
GW1 IIZ J B Underwood, Rock Hill, Llanarthney, Dyfed, SA32 8LJ [IO71VU, SN51]
G1 IJC M D Williamson, 9 East View, Raynham Rd, Hempton, Fakenham, NR21 7LW [JO02KT, TF92]
G1 IJJ J B Lainchbury, 17 Pearmain Ave, Wellingborough, NN8 4SF [IO92PH, SP86]
G1 IJK Details withheld at licensee's request by SSL.
G1 IJM M G Shoosmith, Four Winds, Little Kimble, Aylesbury, Bucks, HP17 0UE [IO91OS, SP80]
G1 IJQ D M Martin, 27 St. Andrews Rd, Stratton, Bude, EX23 9AG [IO70RT, SS20]
G1 IJT J Diver, 24 Tamworth Rd, Coventry, CV6 2EL [IO92FK, SP38]
G1 IJY R J Wise, 7 Patricia Ave, Goring By Sea, Worthing, BN12 4NE [IO90TT, TQ10]
G1 IJZ S J Worrall, 40 Pipers Ln, Oldbury, Nuneaton, CV10 0RH [IO92FN, SP39]
G1 IKD D G Meaton, 32 The Dell, Luton, LU2 8SX [IO91TV, TL12]
G1 IKE K Martin, 17 Warley Way, Frinton on Sea, CO13 9PA [JO01PU, TM22]
G1 IKF R Blakemore, 3 Lyndale Mews, Kilpin Hill, Dewsbury, WF13 4BU [IO93EQ, SE22]
G1 IKG P J McGahon, 520 Chessington Rd, West Ewell, Epsom, KT19 9HH [IO91UI, TQ16]
G1 IKI D R West, 60 Queen Anne Gdns, Falmouth, TR11 4SW [IO70KD, SW73]
G1 IKL P OSullivan, Quince Cottage, 46 Kings Rd, Shalford, Guildford, GU4 8JX [IO91RF, TQ04]
GW1 IKN H Penny, 184 Church Rd, Teddington, TW11 8QL [IO91TK, TQ17]
GM1 IKQ L Davies, 24 Achintore Rd, Caol, Fort William, PH33 7PQ [IO76KU, NN17]
G1 IKT S P Elliott, Manor House, Bewholme, Driffield, Nth Humberside, YO25 8DX [IO93VW, TA14]
G1 IKV B J Austin, 16 Heathlands, Moor Ln, Westfield, Hastings, TN35 4QZ [JO00GV, TQ81]
G1 ILA E P Dudley, 18 Brandon Mews, Barbican, London, EC2Y 8BE [IO91WM, TQ38]
G1 ILB Details withheld at licensee's request by SSL.
G1 ILC S Colley, 8 Tennyson Rd, Maltby, Rotherham, S66 7LU [IO93JK, SK59]
G1 ILF P G Ellis, 30B Skellow Rd, Carcroft, Doncaster, South Yorks, DN6 8HJ [IO93JN, SE50]
G1 ILG B L Evans, 45 Red Willow, Harlow, CM19 5PA [JO01AS, TL40]
G1 ILH G D Farr, 18 The Loont, Over, Winsford, CW7 1EU [IO83RE, SJ66]
G1 ILJ C Wood, Corner Cottage, Swaythorpe Farm, Thwing, Driffield, YO25 0ED [IO94TC, TA06]
G1 ILN C Bonfield, 14 The Furrows, Luton, LU3 2LF [IO91SV, TL02]
G1 ILO R A Bell, 30 Valley Ave, Halifax, HX3 8UD [IO93CR, SE12]
G1 ILS L Preece, 23 Ormonde Rd, Woking, GU21 4RZ [IO91RH, SU95]
G1 ILT S Pritchard, 26 Bryony Ct, Leeds, LS10 4SS [IO93FR, SE32]
G1 ILV A R Robinson, 16 Shaw Green, Storth, Milnthorpe, LA7 7JB [IO84OF, SD48]
G1 ILW Details withheld at licensee's request by SSL.[Op: E Rother.]
G1 ILY C P Sims, 10 Palm Cl, Exmouth, EX8 5NZ [IO80HP, SY08]
GM1 IMB Details withheld at licensee's request by SSL.

G1 IMD M D Hall, 6 Poplar Ave, New Mills, Stockport, Ches, SK12 4HR [IO83XI, SJ98]
G1 IME N E Hopkins, Hylton Cottage, Grafton, Beckford, Tewkesbury, Glos, GL20 7AT [IO82XA, SO93]
G1 IMF C L Hembery, 15 Wivenhoe Ct, Berkeley, Frome, BA11 2DF [IO81UF, ST74]
G1 IMG R E Harwood, 67 Longlands, Adeyfield, Hemel Hempstead, HP2 4DB [IO91SS, TL00]
G1 IMH Details withheld at licensee's request by SSL.
G1 IMI C M Foreman, Thornham Farm, Wansford, Driffield, East Yorks, YO25 8JJ [IO93TX, TA05]
G1 IMJ I M Johnson, 78 Croft Rd, Cosby, Leicester, Leics, LE9 1SE [IO92JN, SP59]
G1 IML M Fisher, 8 Telford Ave, Great Wyrley, Walsall
G1 IMM A J Gee, 24 Granhams Cl, Great Shelford, Cambridge, CB2 5LG [JO02BD, TL45]
G1 IMN Details withheld at licensee's request by SSL.
G1 IMQ Details withheld at licensee's request by SSL.
G1 IMY R Laycock, 24 Farmcroft Rd, Mansfield Woodhouse, Mansfield, NG19 8QT [IO93JE, SK56]
G1 INA A J Lowe, 47 Springfield Park Rd, Chelmsford, CM2 6EB [JO01FR, TL70]
G1 INB P J E Laker, 207 Columbia Rd, Ensbury Park, Bournemouth, BH10 4EE [IO90BS, SZ09]
G1 INC D T Linford, 75 Mercot Cl, Oakenshaw South, Redditch, B98 7YY [IO92AG, SP06]
G1 IND V G E Lowe, Cantley, 7 Castell Cres, Doncaster, DN4 6LG [IO93LM, SE60]
G1 INH D L Munro, Jacaranda, Aldham, Ipswich, Suffolk, IP7 6NH [JO02LB, TM04]
G1 INI B Ginsburg, 27 Park Cres, Elstree, Borehamwood, Herts, WD6 3PT [IO91UP, TQ19]
G1 INJ Dr R Ginsburg, 3 Basing Hill, London, NW11 8TE [IO91VN, TQ28]
G1 INK S J Green, 8 Granby Rd, Fairfield, Buxton, SK17 7TW [IO93BF, SK07]
G1 INL S Gore, 139 Spoondell, Dunstable, LU6 3JF [IO91RV, TL02]
G1 INQ F A Storer, 26 The Sycamores, Broadmeadows, South Normanton, Alfreton, DE55 3AE [IO93HC, SK45]
GM1 INS B Skakle, 26 Thompson Terr, Fraserburgh, AB43 9NY [IO87XQ, NJ96]
G1 INU M A S Sweet, 48 Howick Park Dr, Penwortham, Preston, PR1 0LU [IO83PR, SD52]
G1 INW D I Rea, 58 Chamberlain Cres, Shirley, Solihull, B90 2DQ [IO92BJ, SP17]
G1 INY R T Wojtuszek, 16 Macaulay Ave, White Cross, Hereford, HR4 0JJ [IO82PB, SO44]
G1 IOF R J Hodge, 69 Downview Rd, Felpham, Bognor Regis, PO22 8JA [IO90QT, SU90]
G1 IOH Details withheld at licensee's request by SSL.
G1 IOJ Details withheld at licensee's request by SSL.[Op: L G S Cole, 11 Silverdale Drive, Winlaton, Tyne & Wear, NE21 6EQ.]
G1 ION R S Cox, 62 Rochester Ave, Woodley, Reading, RG5 4NB [IO91NL, SU77]
G1 IOO D R Camac, 6 Wisbeck Rd, Tonge Fold, Bolton, BL2 2TA [IO83TO, SD70]
G1 IOP A Cheer, 15 Stibbs Way, Bransgore, Christchurch, BH23 8HG [IO90DS, SZ19]
G1 IOQ J P Holden, 16 Kershaw Way, Newton-le-Willows, WA12 0AZ [IO83QL, SJ59]
G1 IOR R Hebdige, 17 Tunstall Way, Walton, Chesterfield, S40 2RH [IO93GF, SK37]
GW1 IOT Details withheld at licensee's request by SSL.
G1 IOU G E Holland, Shanida, Waterloo, Puriton, Bridgwater, TA7 8BB [IO81ME, ST34]
G1 IOV Details withheld at licensee's request by SSL.
G1 IPA A J Fowler IPARC, 78 Beckingham Rd, Guildford, GU2 6BU [IO91QF, SU95]
G1 IPC J D Methven, 42 Alpha Rd, West Green, Crawley, RH11 7AZ [IO91VC, TQ23]
G1 IPD D A Mobbs, 64 Cranford Rd, Kingsthorpe, Northampton, NN2 7QX [IO92NG, SP76]
G1 IPE A Medcalf, 23 Allesborough Dr, Pershore, WR10 1JH [IO82WC, SO94]
G1 IPF Details withheld at licensee's request by SSL.
G1 IPK G R Wigham, 23 Beech Ave, Spennymoor, DL16 7ST [IO94EQ, NZ23]
G1 IPO R N Allen, 10 Scotter Rise, Sheringham, NR26 8YD [JO02OW, TG14]
G1 IPP M A Allen, 23 Waterloo Cres, Countesthorpe, Leicester, LE8 5SU [IO92KN, SP59]
G1 IPQ M E Cheeseman, 63 Ringwood Dr, North Baddesley, Southampton, SO52 9GN [IO90GX, SU32]
G1 IPU G W Coote, Springfield Hall Co, Lawn Ln, Chelmsford, Essex, CM1 5TJ [JO01FR, TL60]
G1 IPX R A Clarke, 185 Central Ave, Canvey Island, SS8 9QP [JO01GM, TQ78]
G1 IPY J H Rowlands, 2 Wellfield, Longton, Preston, PR4 5BX [IO83OR, SD42]
G1 IQA G Adkins, 117 Connolly Dr, Rothwell, Kettering, NN14 6TN [IO92OK, SP88]
G1 IQE F T Angwin, 15 Hatton Park Rd, Wellingborough, NN8 5BA [IO92PH, SP86]
G1 IQF M Ames, 7 Northgate, Leyland, Preston, PR5 2NR [IO83PQ, SD52]
G1 IQG A W Bloodworth, 79 Hands Rd, Heanor, DE75 7HB [IO93HA, SK44]
G1 IQK B Shaw, 23 Lodge Dr, Culcheth, Warrington, WA3 4ES [IO83RK, SJ69]
G1 IQN J F H Spicer, 33 Triandra Way, Yeading, Hayes, UB4 9PB [IO91TM, TQ18]
GW1 IQS I B Jones, 7 The Oaks, Quakers Yard, Treharris, CF46 5HQ [IO81IQ, ST09]
G1 IQU D V Jolley, 212 Eastern Esplanade, Southend on Sea, SS1 3AD [JO01IM, TQ98]
G1 IRC J Gee Ipswich RC, 11 Charlton Ave, Ipswich, IP1 6BH [JO02NB, TM14]
G1 IRG S J Manning, 11 Broomhill Cres, Southfields, Northampton, NN3 5BH [IO92OG, SP86]
GW1 IRL E Bowen, 46 Cwmdu Rd, Pontardawe, Swansea, SA8 4QU [IO81CR, SN70]
G1 IRQ N A Tansley, 31A Wignals Gate, Holbeach, Spalding, PE12 7HL [JO02AT, TF32]
G1 IRV M G L Wicks, 21 Weoley Ave, Selly Oak, Birmingham, B29 6PP [IO92AK, SP08]
G1 IRW E S Walker, 8 Mead Rd, Ham, Richmond, TW10 7LG [IO91UK, TQ17]
G1 IRX M H Weir, 17 Pasteur Dr, Leegomery, Telford, TF1 4PQ [IO82RR, SJ61]
G1 IRY Details withheld at licensee's request by SSL.
G1 ISA W L Manley, 6 Hassocks Cl, London, SE26 4BS [IO91XK, TQ37]
G1 ISB D Macken, 9 Lodge Bank Rd, Smithybridge, Littleborough, OL15 8QS [IO83WP, SD91]
G1 ISE K F Maskelyne, 25 Halstead Rd, Bitterne Park, Southampton, SO18 2PQ [IO90HW, SU41]
GI1 ISG Details withheld at licensee's request by SSL.
G1 ISJ D R Barker, 6 Chadwick Rd, Haresfinch, St. Helens, WA11 9AN [IO83PL, SJ59]
G1 ISP B Etherington, 24 Broomcroft Rd, Ossett, WF5 8LH [IO93FQ, SE21]
G1 ISR P B Kenington, Watleys Cottage, 74 North Rd, Winterbourne, Bristol, BS17 1PX [IO81SM, ST68]
G1 ISS B Lyons, 268 St. Johns Rd, Walthamstow, London, E17 4JN [IO91XO, TQ39]
G1 ISW R P Henderson, Veterinary Cottage, The Ridgeway, Crudwell, Malmesbury, Wilts, SN16 9EQ [IO81XP, ST99]
G1 ISX C B Hall, 6 Nairn Green, Watford, WD1 6NW [IO91TP, TQ19]
G1 ISY N L Morris, 15 Turners Cl, Highnam, Gloucester, GL2 8EH [IO81UV, SO72]
G1 ITE P J Hayler, 27 Birch Way, Heathfield, TN21 8BB [JO00DX, TQ52]
G1 ITJ K A Edmett, Youngs Paddock, Middleton Rd, Winterslow, Salisbury, Wilts, SP5 1RS [IO91EC, SU23]
G1 ITL D G Gilbey, 34 Farnhurst Rd, Barnham, Bognor Regis, PO22 0JN [IO90QT, SU90]
G1 ITN A L Guest, 53 St. Helens Rd, Solihull, B91 2DB [IO92CK, SP18]
G1 ITP P Glover, 4 Hogan Way, Kingston Hill, Stafford, ST16 3YN [IO82WT, SJ92]
G1 ITR R White, Yeritiz, Exbury Rd, Blackfield, Southampton, SO45 1XD [IO90HT, SU40]
G1 ITS T Williams, 86 Hillcrest Rd, Rochdale, OL11 2QB [IO83VO, SD81]
G1 ITY T R Hanman, 9 Heafield Dr, Kegworth, Derby, DE74 2GG [IO92IU, SK42]
GM1 IUA N R Harris, Deveron, Hill Rd, Ballingry, Lochgelly, KY5 8NP [IO86ID, NT19]
G1 IUB G A Horton, Arnside, Crete Rd West, Folkestone, CT18 7AA [JO01OC, TR23]
G1 IUD C Sermons, 17 Wellside, Marks Tey, Colchester, CO6 1XG [JO01JV, TL92]
G1 IUE S L Kelsall, Red Croft, Rock Hill, Mansfield, NG18 2PG [IO93JD, SK56]
G1 IUF M V Kilkenny, 138 Stanbury Rd, Haworth Park, Hull, HU6 7BW [IO93TS, TA03]
G1 IUM E R Johnson, 32 Barnes Dr, Lydiate, Liverpool, L31 2LW [IO83MM, SD30]
G1 IUN Details withheld at licensee's request by SSL.
G1 IUQ J F Heley, 52 Queens Cres, Stainforth, Doncaster, DN7 5PJ [IO93LO, SE61]
G1 IUT T R Christmas, 3 Mount Rd, Cosby, Leicester, LE9 1SX [IO92JN, SP59]
G1 IUW G R Diaper, 89 East St., Sudbury, CO10 6TP [JO02IA, TL84]
G1 IUY I L Glenwright, 67 Caldy Rd, Belvedere, DA17 6JT [JO01BL, TQ47]
G1 IVE Details withheld at licensee's request by SSL.
G1 IVF D Lowe, 21 Farndon Rd, Market Harborough, LE16 9NW [IO92ML, SP78]
G1 IVG C D Lowe, Paseo Del Mar, 14F 20 1A, O8350 Areyns de Mar, Barcelona, Spain
G1 IVH J Lee, 13 Crich Rd, Inkersall, Chesterfield, S43 3SN [IO93HG, SK47]
G1 IVI M J Grey, 7 Cemetery Ln, Tweedmouth, Berwick upon Tweed, TD15 2BS [IO85XS, NT95]
G1 IVJ Details withheld at licensee's request by SSL.[Op: G E Gloin.]
G1 IVK T A Garnham, 8 Pagisters Rd, Abingdon, OX14 2LY [IO91IQ, SU59]
G1 IVL R A Hudson, 103 McIntyre Rd, St. Johns, Worcester, WR2 5LQ [IO82VE, SO85]
G1 IVN P H P Loweth, Dogwood Cottage, Cackle St, Brede, Rye, E Sussex, TN31 6DY [JO00HX, TQ82]
G1 IVO L G Ladner, 7 Polventon Cl, Heamoor, Penzance, TR18 3LD [IO70FD, SW43]
G1 IVP J R Lamb, 5 Honeycroft, Loughton, IG10 3PR [JO01AP, TQ49]
G1 IVV G Merrington, Cartref Ball Ln, Kingsley, Ches, WA6 8HP [IO83QG, SJ57]
G1 IVY R A Cowell, 9 Redmond Cl, Etchinghill, Rugeley, WS15 2XG [IO92AS, SK01]
G1 IWA M Conley, 5 Orpington Ave, Walker, Newcastle upon Tyne, NE6 2RL [IO94FX, NZ26]
G1 IWC M R Cashmore, 2 Seaton Rise, Leicester, LE5 1SR [IO92LP, SK60]
G1 IWE T J Coombs, 114 Talbot St., Whitwick, Coalville, LE67 5AZ [IO92HR, SK41]
G1 IWG J Crossley, 11 Cintra Ave, Ashton on Ribble, Preston, PR2 2HR [IO83PS, SD53]
G1 IWR Details withheld at licensee's request by SSL.
G1 IWS A M Mattocks, The Dunstable Arms, 27 Cromer Rd, Sheringham, NR26 8AB [JO02OW, TG14]
G1 IWT R C J Moore, 9 Rowland St., Allenton, Derby, DE24 9BT [IO92GV, SK33]
G1 IWX P Lee, 287 Armshead Rd, Werrington, Stoke on Trent, ST9 0NB [IO83WA, SJ94]
G1 IXE V A Green, 50 Alcove Rd, Fishponds, Bristol, BS16 3DR [IO81RL, ST67]
G1 IXF I B Green, 50 Alcove Rd, Fishponds, Bristol, BS16 3DR [IO81RL, ST67]
G1 IXJ P Bolas, 4 Moat Bank, Bretby, Burton on Trent, DE15 0QJ [IO92ET, SK22]
G1 IXN M J Cochrane, 57 Goulden St., Seedley, Salford, M6 5PZ [IO83UL, SJ89]
G1 IXU R Hood, 47 Beechwood Rd, Bedworth, Nuneaton, CV12 9AQ [IO92GL, SP38]
G1 IXV C D Haver, 29A Edenham Rd, Hanthorpe, Bourne, PE10 0RB [IO92TT, TF02]
GM1 IXW R Izatt, Criffel, 33 Melrose Rd, Galashiels, TD1 2AT [IO85OO, NT53]
G1 IXY Details withheld at licensee's request by SSL.
G1 IYA A S Greenwood, 94 Wharncliffe Dr, Bradford, BD2 3SY [IO93DT, SE13]
G1 IYB B L Haines, 66 North Dr, Gr, Wantage, OX12 7PN [IO91HO, SU49]
G1 IYE I E Hawes, 129 Manor Rd, Ash, Aldershot, GU12 6QB [IO91PF, SU85]
G1 IYM Details withheld at licensee's request by SSL.[Op: Miss L C Rimmer. Station located near Sutton Colfield.]
G1 IYN M B Richardson, 138 Northd. Ave, Welling, DA16 2PY [JO01BL, TQ47]
G1 IYX J Turner, 33 Corbyn St., Hornsey, London, N4 3BY [IO91WN, TQ38]
G1 IYY W H Jackman, 38 Gosden Hill Rd, Burpham, Guildford, GU4 7JD [IO91RG, TQ05]
G1 IZA D H Lamb, 33 Cherston Rd, Loughton, IG10 3PL [JO01AP, TQ49]
G1 IZB F J Smith, 6 Mill Cl, Marshchapel, Grimsby, DN36 5TP [JO03AL, TF39]
G1 IZD C J Stubbs, 3 Cartmel Cl, Macclesfield, SK10 3PE [IO83WG, SJ97]
G1 IZH A M Trueman, Ragdale, Main St, Wilsthorpe, Stamford, Lincs, PE9 4PE [IO92TQ, TF01]
G1 IZN R S Mitchell, 45 Kent Cl, Mitcham, CR4 1XN [IO91WJ, TQ36]
G1 IZP M A Mackender, 28 New St., Doddington, March, PE15 0SP [JO02AL, TL39]
G1 IZU J Timperley, 413 Rossendale Rd, Burnley, BB11 5HJ [IO83US, SD83]
G1 JAA R Lees-Oakes, 15 Byron Ave, Droylsden, Manchester, M43 6QB [IO83WL, SJ99]
G1 JAB J A Burke, 48 Medina Rd, Portsmouth, PO6 3HD [IO90LU, SU60]
G1 JAC M R Osborne, 34 Burrstock Way, Rainham, Gillingham, ME8 8TR [JO01HI, TQ86]
G1 JAF A E Trigell, 8 Brownsea Cl, New Milton, BH25 5UG [IO90DS, SZ29]
G1 JAG K G Powell, 35 The Reddings, Red Rd, Borehamwood, WD6 4ST [IO91UP, TQ19]
G1 JAH J A Hagues, 80 Barnfield Gdns, Eastern Green, Penzance, Cornwall, TR18 3RH [IO70FD, SW43]
G1 JAJ R V Womersley, 180 Main St., Grenoside, Sheffield, S30 3PR [IO93GJ, SK38]
G1 JAL P S Westbury, 120 Bunbury Rd, Northfield, Birmingham, B31 2DN [IO92AJ, SP07]
G1 JAR Details withheld at licensee's request by SSL.
G1 JAS P F Ayles, 128A Tankerton Rd, Whitstable, CT5 2AN [JO01MI, TR16]
G1 JAU P Halifax, 5 Wyck Ct, St Aubyns Rd, Fishergate, W Sussex, BN4 1PW [IO90WT, TQ30]
G1 JAX T Hudson, 27 Ramsden Cres, Carlton in Lindrick, Worksop, S81 9BB [IO93KI, SK58]
G1 JAZ S R Richard, 14 Wilson Walk, Gilroyd, Dodworth, Barnsley, S75 3QJ [IO93FM, SE30]
G1 JBB C N Richards, Pentillie House, Cliff Rd, Mevagissey, nr St Austell, Cornwall, PL26 6TF [IO70OG, SX04]
G1 JBC Details withheld at licensee's request by SSL.
G1 JBD Details withheld at licensee's request by SSL.
G1 JBE D G Blackburn, 24 Reservoir St., Darwen, Lancs, BB3 1LQ [IO83SQ, SD62]
GW1 JBF J W Bircham, yr Hen Efail, Rhoscefnhir, Pentraeth, Anglesey, Gwynedd, LL75 8YU [IO73VG, SH57]
G1 JBG J Beacon, 11 Durrants Path, Chesham, HP5 2LH [IO91QR, SP90]
G1 JBI P J Buckingham, 1C Biddenham Turn, Biddenham, Bedford, Beds, MK40 4AT [IO92SD, TL05]
G1 JBJ S M Bartlett, Honeysuckle Farm, Jarvis Gate, Sutton St. James, Spalding, PE12 0EU [JO02AS, TF31]
G1 JBN R W Cooper, 30 Lambeth Dr, Stirchley, Telford, TF3 1QW [IO82SQ, SJ60]
G1 JBO D R Crawford, 17 Lavender Way, Bourne, PE10 9TT [IO92TS, TF02]
G1 JBT E S Davey, 19 Northfield Rd, Swaffham, PE37 7JB [JO02IP, TF80]
G1 JBV T Eden, 5 Waddow Gr, Waddington, Clitheroe, BB7 3JL [IO83TV, SD74]
G1 JBW B Ellison, 931 Burnley Rd, Todmorden, OL14 7ET [IO83WR, SD92]
G1 JBZ I S J Halsey, 1 Darnborough St., York, YO2 1AU [IO93LW, SE65]
G1 JCC I P Jefferson, 125 Telscombe Way, Stopsley, Luton, LU2 8QP [IO91TV, TL12]
G1 JCH A G Adderley, 43 The Heys, Coppull, Chorley, PR7 4NX [IO83QP, SD51]
G1 JCL M R Munn, 11 Foxley Rd, Queenborough, ME11 5AW [JO01IK, TQ97]
G1 JCP J C Pasfield, Fairlands, White Lodge Cres, Thorpe-le-Soken, Clacton on Sea, CO16 0HT [JO01OU, TM12]
G1 JCQ W M Croston, 13 Greencroft, Penwortham, Preston, PR1 9LA [IO83PR, SD52]
G1 JCW A F Duffy, 6 Camden Rd, Layton, Blackpool, FY3 8HN [IO83LT, SD33]
G1 JDE O P Graffham, 25 Southfield Ave, Edgbaston, Birmingham, B16 0JN [IO92AL, SP08]
G1 JDF D H K Gray, 38 Blunt St., Maybank, Newcastle, ST5 9NA [IO83VA, SJ84]
G1 JDM J D McKernan, 21 Hurley Rd, Durrington, Worthing, BN13 2PA [IO90TT, TQ10]
G1 JDO P Oliver, 67 High St., Great Houghton, Barnsley, S72 0AU [IO93HN, SE40]
G1 JDP M J B Overton, Eastrigg, Scorers Ln, Great Lumley, Chester-le-Street, DH3 4JH [IO94FU, NZ24]
G1 JDQ P J Paterson, Oak Lea, 11A Fletsand Rd, Wilmslow, SK9 2AD [IO83VH, SJ88]
G1 JDR M D K Phillips, Halfshire House, 2 Ct Farm Way, Churchill, Worcs, DY10 3LY [IO82VJ, SO87]
G1 JDT G B Palmer, 5 Dunstar Ave, Audenshaw, Manchester, M34 5LJ [IO83WL, SJ99]
G1 JDU D P Phillips, 23 Felskirk Rd, Manchester, M22 1PX [IO83UI, SJ88]
G1 JDW D Riley, 12 Booth St., Audley, Stoke on Trent, ST7 8EP [IO83UB, SJ75]
G1 JEA Revd D A Hart, The Vicarage, Thorndon Gate, Ingrave, Brentwood, CM13 3RG [JO01EO, TQ69]
G1 JEH K F Schneider, White Lodge, 65 Alpha Rd, Birchington, CT7 9ED [JO01PJ, TR36]
G1 JEM J Shadwell, 93 Fairway, Castleton, Rochdale, OL11 3BZ [IO83VO, SD81]
G1 JEO B J Strett, 30 Scott Rd, Lowton, Warrington, WA3 2HH [IO83RL, SJ69]
G1 JEP P W Sherwin, 433 Fox Hollies Rd, Acocks Green, Birmingham, B27 7QA [IO92CK, SP18]
G1 JER Dr J R Johnson, 5 Hunters Ride, Appleton Wiske, Northallerton, DL6 2BD [IO94HK, NZ30]
G1 JET Details withheld at licensee's request by SSL.
G1 JEZ S Taylor, 47 Heyes Dr, Wallasey, L45 8QL [IO83LK, SJ29]
GU1 JFA A J Rowsell, Sunkist, Springlea Rd, Vale, Guernsey, GY6 8EX
G1 JFE J T Warner, Nursery Farm, Bold Ln, Aughton, Lancs, L39 6SH [IO83MM, SD30]
GM1 JFF A Weddell, 10 High St., Eyemouth, TD14 5EU [IO85WU, NT96]
G1 JFL M Woolridge, 23 Marina Dr, Maybank, Newcastle, ST5 9NL [IO83VA, SJ84]
G1 JFN D S Weeks, 53 Burns Ave, Brake Farm, Plymouth, PL5 3LQ [IO70WJ, SX45]
G1 JFQ B A Acheson, 32 Lords Ln, Brighouse, HD6 3RF [IO93CQ, SE12]
G1 JFR A M Antmony, Holly Cottage, Crondall Rd, Crookham Village, Fleet, GU13 0SU [IO91NG, SU75]
GW1 JFT R C Beaugie, 32 Ct Gdns, Rogerstone, Newport, NP1 9FU [IO81LO, ST28]
G1 JFU Wg Cdr D D Bryant, 22 Highfield Park, Heaton Mersey, Stockport, SK4 3HD [IO83VJ, SJ89]
GW1 JFV R W D Brooks, 24 Erw Lon, Pen y Cwm, Haverfordwest, SA62 6AU [IO71KU, SM82]
G1 JFZ F J Bolton, 5 Turners Building, Barrow Rd, New Holland, North Lincs, DN19 7RB [IO93TQ, TA02]
G1 JGD N T C Cullis, 4 Elmswood, Dickens Park, Chigwell, IG7 5JQ [JO01BO, TQ49]
G1 JGE M Colley, 118 Devon Cres, Birtley, Chester-le-Street, DH3 1HP [IO94FV, NZ25]
G1 JGF L R Cox, 7 Timberdine Ave, Worcester, WR5 2BD [IO82VE, SO85]
G1 JGM B Easey, 4 Ash Trees, East Brent, Highbridge, TA9 4DQ [IO81MF, ST35]
G1 JGN Details withheld at licensee's request by SSL.
G1 JGQ Details withheld at licensee's request by SSL.
G1 JGR C A M Fortnum, 11 Ayr Cl, Stamford, PE9 2TS [IO92RP, TF00]
G1 JGS M K Garland, 12 Downsview Rd, Newport, Isle of Wight, PO30 2AT [IO90IQ, SZ58]
G1 JGT J Giller, 9 Alberta Cres, Huntingdon, PE18 7TL [IO92VI, TL27]
G1 JGY H Ketley, 1 Tewkesbury Ave, Mansfield Woodhouse, Mansfield, NG19 8LA [IO93JE, SK56]
G1 JHB C J R Lawrence, 43 Mountside Cres, Prestwich, Manchester, M25 3JF [IO83UM, SD80]
G1 JHD M Plant, Golf Cottage, Osborne House Est, East Cowes, PO32 6JZ [IO90IS, SZ59]
G1 JHG W D McCormick, 33 Bryanston Rd, Aigburth, Liverpool, L17 7AL [IO83MJ, SJ38]
G1 JHI Details withheld at licensee's request by SSL.
G1 JHK B T Hazell, 213 Prince Ave, Westcliff, Essex, SS0 0JU [JO01IN, TQ88]
G1 JHL M N Hubball, 34 Shipbrook Rd, Rudheath, Northwich, CW9 7EJ [IO83SG, SJ67]
G1 JHM A G W Harding, 11 Mallard Cl, Kempshott, Basingstoke, RG22 5JP [IO91KF, SU54]
G1 JHP H J G Hamer, 42 Long Cl, Westacres, Leyland, Preston, PR5 3WB [IO83PQ, SD52]
G1 JHR M C Jones, 5 Oakley Hill, Merley, Wimborne, BH21 1RJ [IO90AS, SZ09]
G1 JHS Details withheld at licensee's request by SSL.[Station located near Weybridge.]
GM1 JHU J P Adams, 3A Glenpatrick Rd, Elderslie, Johnstone, PA5 9BH [IO75SU, NS46]
G1 JHX A Bennett, 4 Hampson Rd, Stretford, Manchester, M32 9JH [IO83UK, SJ79]
G1 JHY A M Potter, 25 Robinsons Meadow, Ledbury, HR8 1SU [IO82SA, SO73]
G1 JHZ S M Potter, 5 Cotswold Rd, Malvern, Worcs, WR14 2QF [IO82UC, SO74]
GW1 JIE K W Robertson, Tyn y Pwll, Fachwen, nr Llanberis, Caernarfon, LL55 3HD [IO73WD, SH56]
G1 JIG S C Ridgard, 5 Orchard Way, Luton, LU4 9LT [IO91SV, TL02]
G1 JIH E Rowell, 80 Kings Delph, Whittlesey, Peterborough, PE7 2PD [IO92VN, TL29]
G1 JII A Probyn, 87 Westonfields Dr, Longton, Stoke on Trent, ST3 5JH [IO82WX, SJ94]
G1 JIJ J E Passfield, 2 Parker Rd, Chelmsford, CM2 0ES [JO01FR, TL70]
GM1 JIU J G Small, 167 Alloway Dr, Kirkintilloch, Glasgow, G66 2SB [IO75WW, NS67]
G1 JIW P J Spooner, Queen Anne Cottage, Tavistock Pl, Basford, Stoke on Trent, ST4 6HY [IO83VA, SJ84]
G1 JIY D W Batham, 1 Clarence Pl, Lower Weston, Bath, BA1 3EW [IO81TJ, ST76]
G1 JJA S Bessent, 407 Evesham Rd, Crabbs Cross, Redditch, B97 5JA [IO92AG, SP06]
GI1 JJC W J Brown, 19 Benbraddagh Ave, Limavady, BT49 0AP [IO65MB, C62]
G1 JJG A R Boulton, 41 Caithness Dr, Crosby, Liverpool, L23 0RG [IO83LL, SJ39]
GD1 JJH J J Hall, Fernleigh Hotel, Palace Rd, Douglas, Isle of Man, 1M2 4LB
GM1 JJJ Details withheld at licensee's request by SSL.
G1 JJK P A Naylor, 14 Wrockwardine Rd, Wellington, Telford, Salop, TF1 3DB [IO82RQ, SJ61]
G1 JJQ J P Schulz, 95 Baywell, Leybourne, West Malling, ME19 5QE [JO01FH, TQ65]
G1 JJR V A Smith, 15B Oakfield Rd, Harringay, London, N4 4NH [IO91WN, TQ38]
G1 JJS B E Sexton, 2 Wilmslowe, Canvey Island, SS8 8HU [JO01HM, TQ88]
G1 JJT Details withheld at licensee's request by SSL.
G1 JKB T D Walton, 149 Randall Ave, Cricklewood, London, NW2 7TA [IO91VN, TQ28]
G1 JKF N G Leaney, 3 Rubbra Cl, Browns Wood, Milton Keynes, MK7 8DP [IO92PA, SP93]
GM1 JKJ A J Britton, 15 Glenbrook, Balerno, EH14 7JE [IO85HV, NT16]
G1 JKL A W Crouch, 56 Lichfield Rd, Cambridge, CB1 3TP [JO02BE, TL45]
G1 JKN P J Cooper, 17 Crmwell Cl, Tutbury, Burton on Trent, Staffs, DE13 9HZ [IO92DU, SK22]
G1 JKP R A Coleman, 18 London Rd, Thatcham, RG18 4LQ [IO91JJ, SU56]
G1 JKR K L Wootton, 16 West St., Over, Cambridge, CB4 5PL [JO02AH, TL36]
G1 JKV T Whittaker, 25 Felix Rd, Walton on Thames, KT12 2LD [IO91TJ, TQ16]
G1 JKX J T West, 4 Coronation Terr, Longhorsley, Morpeth, NE65 8UN [IO95CF, NZ19]

GM1 JKY Details withheld at licensee's request by SSL.
G1 JLB M D Denison, 9 Derwent Rd, Harrogate, HG1 4SG [IO93FX, SE35]
G1 JLE D J Freeborough, 17 The Haven, Fulbourn, Cambridge, CB1 5BG [JO02CE, TL55]
G1 JLG B J Giddings, 71 Tyrone Rd, Southend on Sea, SS1 3HD [JO01JM, TQ98]
G1 JLM J M McCloskey, Southleigh, 71 Ashcombe Rd, Dorking, RH4 1LZ [IO91TF, TQ15]
GM1 JLP K G Robson, 13 Woodstock Ave, Galashiels, TD1 2EE [IO85OO, NT53]
G1 JLQ J K Yarnall, 22 Atherstone Cl, Matchborough East, Redditch, B98 0BD [IO92BH, SP06]
G1 JLS K L More, 77 Loveys Rd, Yapton, Arundel, BN18 0HQ [IO90QT, SU90]
GM1 JLU J R Ness, 22 Lovat St., Largs, KA30 9NE [IO75NT, NS25]
G1 JLZ W Horner, 28 Claremont Rd, Sedgley, Dudley, DY3 1HW [IO82WM, SO99]
G1 JMC B G Harris, 44 Boston Rd, Heckington, Sleaford, NG34 9JE [IO92UX, TF14]
G1 JMD P F Hall, 1 Rowberry Cttgs, Leys Rd, Harvington, Evesham, WR11 5NA [IO92AD, SP04]
G1 JMF A J Hooper, 5 Nine Elms Rd, Longlevens, Gloucester, GL2 0HA [IO81VV, SO81]
G1 JMG H Hawkins, 38 North View, Eastcote, Pinner, HA5 1PE [IO91TN, TQ18]
G1 JMK M E Justice, 6 Stanley Terr, Pans Ln, Devizes, SN10 5AJ [IO91AI, SU06]
G1 JMM A G Slinger, 10 Longlands View, Kendal, LA9 6HJ [IO84PI, SD59]
G1 JMN R Andrews, Mount View, Park Ln, Hallow, Worcester, WR2 6PQ [IO82UF, SO85]
G1 JMP R H Ainsworth, 95 Heysham Cl, Murdishaw, Runcorn, WA7 6DT [IO83PH, SJ58]
G1 JMS J M Stoddart, 113 Minehead Way, Stevenage, SG1 2JJ [IO91VV, TL22]
G1 JMV Details withheld at licensee's request by SSL.
G1 JMW W E Smith, 37 Peake Rd, Walsall, WS8 7BZ [IO92AP, SK00]
G1 JMX Details withheld at licensee's request by SSL.
G1 JMY F R Taylor, Wold Lodge, Pocklington Rd, Huggate, York, YO4 2YJ [IO93QX, SE85]
GM1 JNC A T Campbell, 17 Moulin Circus, Cardonald, Glasgow, G52 3JY [IO75TU, NS56]
G1 JND Details withheld at licensee's request by SSL.
G1 JNG G Eccles, 1 Bridge Pl, Amersham, HP6 6JF [IO91QQ, SU99]
GW1 JNI A J Fennah, 59 Lon Helyg, Newtown, SY16 1HY [IO82IM, SO09]
G1 JNM P Gething, 27 Bowlwell Ave, Nottingham, Notts, NG5 9HX [IO93JA, SK54]
G1 JNQ P R Auld, 80 Milestone Rd, Stone, Dartford, DA2 6DN [JO01CK, TQ57]
GW1 JNR P M Adcock, Bleake House, Cefn Coch, Welshpool, Powys, SY21 0AE [IO82HO, SJ00]
G1 JNT D Brentnall, 395 Queen Mary Rd, Sheffield, S2 1EB [IO93GI, SK38]
G1 JNX S T L Langston, 32 Chevening Cl, Broadfield, Crawley, RH11 9QU [IO91VC, TQ23]
G1 JNY A R Larkin, 16 Thetford Cl, Danesholme, Corby, NN18 9PH [IO92PL, SP88]
G1 JOA B J Marsh, 90 Ellingham Rd, Adeyfield, Hemel Hempstead, HP2 5LL [IO91SS, TL00]
G1 JOC Details withheld at licensee's request by SSL.
G1 JOD R J H Norton, 7 Moor Meadow, Shobdon, Leominster, HR6 9NT [IO82NG, SO46]
G1 JOG M L Ridden, 9 Woodlands Gdns, Romsey, SO51 7TE [IO90GX, SU32]
G1 JOI J S Shuttle, 5 Sheredan Rd, Highams Park, London, E4 9RW [JO01AO, TQ39]
G1 JOJ B D Smith, 146 Battram Rd, Ellistown, Coalville, LE6 1GB [IO92JP, SK50]
G1 JOL B J R Shane, 94 Winchester Rd, Highams Park, Chingford, London, E4 9JP [IO91XO, TQ39]
G1 JOO R W Seymour, 5 Clifton Pl, Easton, Bristol, BS5 0SE [IO81RL, ST67]
G1 JOR J D Ormsby-Rymer, 128 Harescombe, Yate, Bristol, BS17 4UE [IO81TM, ST78]
G1 JOT F J Light, 29 Silversea Dr, Westcliff on Sea, SS0 9XD [JO01IN, TQ88]
G1 JOW Details withheld at licensee's request by SSL.
G1 JOY D G Orton, 20 Goodacre Rd, Ullesthorpe, Lutterworth, LE17 5DL [IO92JL, SP58]
G1 JPB Details withheld at licensee's request by SSL.
G1 JPC M H Ward, Four Winds, 13 Westfield Ave, Raunds, Wellingborough, NN9 6DQ [IO92RI, SP97]
G1 JPD J Williams, 8 Cotswold Gate, Hendon Way, London, NW2 1QS [IO91VN, TQ28]
G1 JPE D C Graham, 106 Regent Terr, Billy Mill Ave, North Shields, NE29 0QJ [IO95GA, NZ36]
GW1 JPF D I Barton, 70 Sandilands Rd, Tywyn, LL36 9AT [IO72WO, SH50]
G1 JPI M V Taylor, 4A Seadown Parade, Bowness Ave, Sompting, Lancing, BN15 9TP [IO90TT, TQ10]
GM1 JPJ R Jamieson, 6 Salmon Ln, Stonehaven, AB3 2NZ [IO87TH, NJ72]
G1 JPK T G Jefferies, 98 St. Johns Rd, Frome, BA11 2BD [IO81UF, ST74]
G1 JPL Details withheld at licensee's request by SSL.
G1 JPP A F Hawes, 25 Folly Cl, Fleet, GU13 9LN [IO91OG, SU85]
G1 JPS G J Chilton, 5 Horner Ct, South Birkbeck Rd, London, E11 4HY [JO01AN, TQ38]
G1 JPT B M Gleave, 1 Fearnley Way, Newton-le-Gillows, Merseyside, WA12 9NR [IO83QK, SJ59]
G1 JQA B J Penney, 59 Snodhurst Ave, Chatham, ME5 0TB [JO01GI, TQ76]
G1 JQG N Strothard, 35 Milton Dr, Scholes, Leeds, LS15 4BS [IO93GT, SE33]
G1 JQH D L J Spalding, 171 Minster Rd, Minster on Sea, Sheerness, ME12 3LH [JO01JK, TQ97]
G1 JQI Details withheld at licensee's request by SSL.
G1 JQJ Details withheld at licensee's request by SSL.[Op: A J Graves.]
G1 JQK S Gibbs, 43 Reddish Vale Rd, Reddish, Stockport, SK5 7EU [IO83WK, SJ89]
GI1 JQP J R Innes, 141 Glenburn Rd, Dunmurray, Dunmurry, Belfast, BT17 9BB [IO74AN, J26]
G1 JQQ A Straker, 24 Leslie Dr, Amble, Morpeth, NE65 0PX [IO95FI, NU20]
G1 JQR D F Sell, 17 Auriel Ave., Dagenham, Essex, RM10 8BS [JO01CN, TQ58]
G1 JQT G L Smith, 20 Smailes St., Stanley, DH9 7NU [IO94DU, NZ15]
G1 JQW Details withheld at licensee's request by SSL.
G1 JRC B W Borrer, 125 Sackville Rd, Hove, BN3 3WF [IO90VT, TQ20]
G1 JRD J A Barber, Hylands, 4 Chestnut Path, Canewdon, Rochford, SS4 3QQ [JO01IO, TQ89]
G1 JRF D A Bruckshaw, 18 Old Moat Dr, Northfield, Birmingham, B31 2LY [IO92AJ, SP07]
G1 JRL H T Cook, 31 Butley Rd, Felixstowe, IP11 8NY [JO01QX, TM23]
GW1 JRM N W G Davies, 67 Trinant Terr, Pentwyn Crumlin, Newport, NP1 4JJ [IO81KQ, SO20]
G1 JRP C P Davis, 8 Spiers Cl, Tadley, RG26 3SF [IO91KI, SU66]
G1 JRR R Chalker, 89 Mount Rd, Chessington, KT9 1JH [IO91UI, TQ16]
G1 JRT Details withheld at licensee's request by SSL.
G1 JRU D J Evans, 63 Malwood Rd West, Hythe, Southampton, SO45 5DL [IO90HU, SU40]
G1 JRW D A Gilchrist, 204 Great West Rd, Heston, Hounslow, TW5 9AW [IO91TL, TQ17]
G1 JRX Dr M S Girgis, Rozel, Wilson Rd, Hartlebury, Kidderminster, DY11 7XU [IO82VH, SO87]
G1 JRZ G C Hobbs, 3 Glebe Cottage, Bremhill, Calne, Wilts, SN11 9LD [IO81XK, ST97]
G1 JSA Details withheld at licensee's request by SSL.
G1 JSK P D Lees, 2 Russet Cl, Braintree, Essex, CM7 7DR [JO01GV, TL72]
G1 JSP J R F Marston, 119 Devana Rd, Leicester, LE2 1PL [IO92KO, SK60]
G1 JSS R D Smith, 9 Lovelace Gdns, Walton on Thames, KT12 5HJ [IO91TI, TQ16]
G1 JST P A Johnson, 3 Lascelles Ln, Northallerton, DL6 1EP [IO94GI, SE39]
G1 JSY W E Whitehead, 42 Langton Ave, Chelmsford, CM1 2BS [JO01FR, TL60]
G1 JSZ Details withheld at licensee's request by SSL.
G1 JTC S Baskerville, Shalimar, Gr Ln, Headingley, Leeds, LS6 2AP [IO93FT, SE23]
G1 JTD R M Boyes, 89 Hazelhurst Rd, Daisy Hill, Bradford, BD9 6AB [IO93CT, SE13]
GJ1 JTF L H Jackson, Eskdale, 5 Sunshine Ave, Five Oaks St Saviou, Jersey C I Je4 7Ts, Jersey, JE99 1AA
GM1 JTJ D J R Dickson, Flat 0/1/35 Midlock St, Cessnock, Glasgow, G51 1SE [IO75UU, NS56]
GM1 JTK A Doig, 18 Gotterstone Dr, Broughty Ferry, Dundee, DD5 1QW [IO86NL, NO43]
G1 JTM M E Ferris, 11 Gannon Rd, Worthing, BN11 2DT [IO90TT, TQ10]
G1 JTN A M Whitehead, 8 Harwell Rd, Sutton Courtenay, Abingdon, OX14 4BN [IO91IP, SU59]
G1 JTO Details withheld at licensee's request by SSL.
G1 JTQ E M Gibson, 144 Wales Rd, Kiveton Park, Sheffield, S31 8RE [IO93GJ, SK38]
G1 JTR L Gibson, 144 Wales Rd, Kiveton Park, Sheffield, S31 8RE [IO93GJ, SK38]
G1 JTX L J Sharman, 14 Northlands Ave, Orpington, BR6 9LY [JO01BI, TQ46]
G1 JTY A K Grayson, 28 Wedge Ave, Haydock, St. Helens, WA11 0DY [IO83PL, SJ59]
G1 JTZ B Linn, 35 New St., Carcroft, Doncaster, DN6 8EH [IO93JO, SE51]
G1 JUD R Robinson, 24 Affleck Ave, Radcliffe, Manchester, M26 1HN [IO83TN, SD70]
G1 JUH R V Barwise, 439 Mill St., Dingle, Liverpool, L8 4RD [IO83MJ, SJ38]
G1 JUI M D Lister, Beaconfield, Middle Rd, Lytchett Matravers, Poole, BH16 6HJ [IO80XS, SY99]
G1 JUL R J F Metcalfe, 57 Newton Lodge Dr, Leeds, LS7 3DQ [IO93FT, SE33]
G1 JUO I Penney, 59 Snodhurst Ave, Walderslade, Chatham, ME5 0TB [JO01GI, TQ76]
G1 JUP M A Baker, 7 Reigate Rd, Brighton, BN1 5AJ [IO90WU, TQ30]
G1 JUR P F Bache, Flat 4, 256 Bills Ln, Shirley, Solihull, B90 2PP [IO92BJ, SP17]
G1 JUY Details withheld at licensee's request by SSL.
GW1 JVB G B Evans, Bwthyn Bach, 2 Old Village Rd, Barry, CF62 6RA [IO81IJ, ST16]
G1 JVF G D Hannan, 20 Arlington Dr, Woodsmoor, Stockport, SK2 7EB [IO83WJ, SJ98]
G1 JVH C P Kirkman, The Nant, Nantmawr, Oswestry, Salop, SY10 9HN [IO82KT, SJ22]
GM1 JVI G R Sore, 6 Langholm St., Newcastleton, TD9 0QX [IO85OE, NY48]
G1 JVL J V Leggett, 10 Home Park Rd, Nuneaton, CV11 5UB [IO92GM, SP39]
G1 JVM D Turton, 8 Lightwoods Rd, Smethwick, Warley, B67 5AY [IO92AL, SP08]
G1 JVN F A Ursell, 110 Watt Ln, Sheffield, S10 5RE [IO93FI, SK38]
G1 JVO C M Ursell, 110 Watt Ln, Sheffield, S10 5RE [IO93FI, SK38]
GM1 JVU R-I Aitken, 81 Rashgill, Locharbriggs, Dumfries, DG1 1QN [IO85FC, NX98]
G1 JVY A W Smith, Woodlands, Old School Ln, Stanford, Biggleswade, SG18 9JL [IO92UB, TL14]
GM1 JWC G M Reid, 20 Broomhall Loan, Edinburgh, EH12 7PY [IO85IW, NT17]
G1 JWD R K Osborne, 24 Brockington Rd, Bodenham, Hereford, HR1 3LR [IO82PD, SO55]
G1 JWG D J McKay, 15 Wellington Cres, Baughurst, Tadley, RG26 5PJ [IO91JI, SU56]
GM1 JWJ R G Male, 13 Briar Gr, Forfar, DD8 1DQ [IO86NP, NO45]
G1 JWL M B Warren, 81 Triumph Walk, Castle Bromwich, Birmingham, B36 9NU [IO92DM, SP19]
GW1 JWN W L Thomas, 17 Maesgrug, Stop & Call, Goodwick, SA64 0HB [IO72LA, SM93]
G1 JWO A F Stone, 28 Whithybed Ln, Alvechurch, nr Birmingham, West Midlands, B48 7NY [IO92AI, SP07]
G1 JWY T R Yorke, 12 Shanklin Dr, Weddington, Nuneaton, CV10 0BA [IO92GM, SP39]
G1 JXA R E Whatley, Bondleigh Moor House, North Tawton, Devon, EX20 2AQ [IO80AT, SS60]
GI1 JXE G T Murray, Ashgrove, 61 Monteith Rd, Katesbridge, Banbridge, BT32 5RD [IO64VH, J13]
G1 JXI Details withheld at licensee's request by SSL.
G1 JXR D J O'Sullivan, 23 Kenilworth Cres, Parkfields, Wolverhampton, WV4 6TA [IO82WN, SO99]
G1 JXS C B B Bunkum, 8 Trelawney Rd, Callington, PL17 7EE [IO70UM, SX36]
G1 JXU A E Kippax, 228 Manchester Rd, Burnley, BB11 4HG [IO83US, SD83]
G1 JXX H Williams, 24 Vaughan Cl, Four Oaks, Sutton Coldfield, B74 4XR [IO92BO, SK10]
G1 JXZ D E Critoph, 26 Pevensey Cl, Aylesbury, HP21 9UB [IO91OT, SP81]
G1 JYA G Coombs, 50 Fosseway Cl, Colerne, Chippenham, Wilts, SN14 8EF [IO81UK, ST87]
G1 JYD C H Longley, 45 California Ave, Scratby, Great Yarmouth, NR29 3NS [JO02UQ, TG51]
G1 JYE D Meffan, 53 St. Clements Rd, Boscombe, Bournemouth, BH1 4DX [IO90BR, SZ19]
G1 JYH M F Cherry, 12 Mount Caburn Cres, Peacehaven, BN10 8DW [JO00AT, TQ40]
G1 JYJ R J Dewhurst, 4 Fairfield Rd, North Shore, Blackpool, FY1 2RA [IO83LT, SD33]
G1 JYK S C Langdale, The Lodge, Low Bridges, Stocksfield, Northd., NE43 7SF [IO94BW, NZ05]
G1 JYR D M Fraley, 1334 Warwick Rd, Knowle, Solihull, B93 9LQ [IO92DJ, SP17]
G1 JYT K G B Gibbens, 13 Benenden Rd, Wainscott, Rochester, ME2 4NU [JO01GJ, TQ77]
GM1 JYV K Grehan, 1 Rosebank Ln, Forfar, DD8 2BG [IO86NP, NO45]
G1 JYW Dr G W Hadfield, Fleet Rise, Fleet Hill, Finchampstead, Wokingham, RG11 4LE [IO91LJ, SU66]
G1 JYZ A S Philpott, Ravenscourt, 2 Ocean View Rd, Ventnor, PO38 1AA [IO90JO, SZ57]
G1 JZG M B Wilmshurst, Chalklands, West Ashby, Horncastle, Lincs, LN9 5PT [IO93WF, TF27]
G1 JZJ R Andrews, 52 Lindridge Rd, Erdington, Birmingham, B23 7HX [IO92BM, SP09]
G1 JZK A C Austin, 44 Mendip Cres, Putnoed, Bedford, MK41 9EP [IO92SD, TL05]
G1 JZL S B Beckett, 15 Peaks Ave, New Waltham, Grimsby, DN36 4LJ [IO93XM, TA20]
GM1 JZM D Johnstone, 7 Gleneagles Ave, Glenrothes, KY6 2QA [IO86JE, NO20]
G1 JZN J C Jacklin, 26 Rockmill End, Willingham By Stow, Willingham, Cambridge, CB4 5HY [JO02AH, TL47]
G1 JZT R G Everitt, 55 Risborough Rd, Bedford, MK41 9QR [IO92SD, TL05]
G1 JZU D Goulden, 26 Derwent Walk, Greenacres, Oldham, OL4 2DJ [IO83WN, SD90]
G1 JZX J C Hesketh, 110 Molyneux Rd, Westhoughton, Bolton, BL5 3UJ [IO83RN, SD60]
G1 JZY T Mitchell, 22 Grundy Ave, Prestwich, Manchester, Lancs, M25 9TG [IO83UM, SD80]
G1 JZZ D E Porter, 21 Langdale Dr, Burscough, Ormskirk, L40 5SE [IO83NO, SD41]
G1 KAC A D Brooks, 45 Northfield Rd, Townhill Park, Southampton, SO18 2QE [IO90HW, SU41]
G1 KAG A P Watson, 6 Bell Ct, Kingston Rd, Tolworth, Surbiton, KT5 9NR [IO91UJ, TQ16]
G1 KAJ R G Bull, Newlyn, 4 Farnham Rd, Fleet, GU13 9JD [IO91OG, SU85]
G1 KAK C Buttery, Yew Tree Cottage, Chapel Ln, Merstone, Newport, PO30 3DD [IO90IP, SZ58]
G1 KAO A Hawxby, 73 Anthea Dr, Huntington, York, YO3 9DB [IO93LX, SE65]
G1 KAP S A Hall, 21 Edward Ave, Saltdean, Brighton, BN2 8QJ [IO90XT, TQ30]
G1 KAR A P Compton Southdown A R S, 25 Framfield Rd, Uckfield, TN22 5AH [JO00BX, TQ42]
G1 KAS M Hughes, 43 Coach Rd, Baildon, Shipley, BD17 5HS [IO93CU, SE13]
G1 KAT C J Lawrence, 23 Brutus Dr, Coleshill, Birmingham, B46 1UF [IO92DM, SP19]
G1 KAW K D Rampton, 20 Lower Neatham Mill Ln, Holybourne, Alton, GU34 4ET [IO91MD, SU74]
G1 KAX S Racz, 10 Slateacre Rd, Gee Cross, Hyde, SK14 5LB [IO83XK, SJ99]
G1 KBA J Wilson, 15 Rosslyn Cres East, Preesall, Poulton-le-Fylde, FY6 0QB [IO83MW, SD34]
G1 KBC S L Barrington, Fawley Cottage, Butt Ln, Normanton on Soar, Loughborough, LE12 5EE [IO92JT, SK52]
G1 KBH P F Bloy, 43 Post Mill, Fairstead Est, Kings Lynn, PE30 4QZ [JO02FS, TF61]
G1 KBJ A Wagstaff, 124 Langwith Rd, Langwith Junction, Mansfield, NG20 9RP [IO93JE, SK56]
G1 KBL M Rack, 212 Willingham St., Grimsby, DN32 9PY [IO93XN, TA20]
G1 KBN P Coombs, 4 Holcombe Cl, Whitwick, Coalville, LE67 5BR [IO92HR, SK41]
G1 KBO Details withheld at licensee's request by SSL.
GM1 KBZ S Crockford, 12 McGregor Cres, Peterhead, AB42 1GE [IO97CM, NK14]
GM1 KCH W Curran, 3 Hillbank Terr, Kirriemuir, DD8 4HR [IO86LQ, NO35]
G1 KCI Details withheld at licensee's request by SSL.
G1 KCJ T N Parkinson, 50 Argyll Rd, North Shore, Blackpool, FY2 9UE [IO83LU, SD33]
G1 KCR J A Smith, 12 Elmhirst Rd, Lutterworth, LE17 4QB [IO92JK, SP58]
G1 KCS A Scrutton, Ashleigh, Butt Hill, Napton, Rugby, CV23 8NE [IO92IG, SP46]
G1 KCU J D Warrington, 204 High St., Feltham, TW13 4HX [IO91TK, TQ17]
G1 KCW H J Elleray, 11 Cresthill Rd, Beacon Park, Plymouth, PL2 2RG [IO70WJ, SX45]
G1 KCY B J Broadfoot, 4 South Ave, Prescot, L34 1LY [IO83OK, SJ49]
G1 KDB Details withheld at licensee's request by SSL.
GW1 KDE S L Williams, 55 Romilly Park Rd, Barry, CF62 6RR [IO81IJ, ST16]
G1 KDG P M Beynon, 18 Rednal Rd, Kings Norton, Birmingham, B38 8DR [IO92AJ, SP07]
G1 KDH K J Nell, 16 Avern Rd, West Molesey, KT8 2JB [IO91TJ, TQ16]
G1 KDO D W Cattell, 29 Braemar Dr, Erdington, Birmingham, B23 7HW [IO92BM, SP09]
G1 KDQ Details withheld at licensee's request by SSL.
GI1 KDS K C Lewis, 763 Antrim Rd, Belfast, BT15 4EP [IO74AP, J37]
G1 KDX D J Ransome, 13 St. Annes Cres, Newtonhill, Stonehaven, Kincardineshire, AB39 3WX [IO87WA, NO99]
G1 KEA Details withheld at licensee's request by SSL.
G1 KEB R Farmer, 33 Skidmore Rd, Coseley, Bilston, WV14 8SE [IO82XN, SO99]
G1 KEH S Wilson, 2 Parsonage Oast, Monkton St., Monkton, Ramsgate, CT12 4JS [JO01PI, TR26]
G1 KEI P F Smith, 47 Bostock Rd, Abingdon, OX14 1DW [IO91IQ, SU49]
G1 KEJ Details withheld at licensee's request by SSL.
G1 KEM S J Morris, 4 Benning Way, Wokingham, RG40 1XX [IO91OK, SU86]
G1 KEO P A Foster-Jones, 9 Stowford Rd, Headington, Oxford, OX3 9PJ [IO91JS, SP50]
G1 KEP J G Houghton, Glenwood, Raby Rd, Thornton Hough, Wirral, L63 4JS [IO83LH, SJ38]
GW1 KEU W L Haynes, Ty'R Ysgol, y Fron, Upper Llandwrog, Caernarfon, Gwynedd, LL54 7BB [IO73VB, SH55]
G1 KEV H E Denton, 27 Melrose Gdns, Hersham, Walton on Thames, KT12 5HF [IO91TI, TQ16]
G1 KEY G R Ney, Basted Ln, Crouch, Borough Green, Sevenoaks, Kent, TN15 8PZ [JO01DG, TQ65]
G1 KFD Details withheld at licensee's request by SSL.
G1 KFF R Coates, 35 James Andrew Cres, Greenhill, Sheffield, S8 7RJ [IO93GH, SK38]
G1 KFG J V Feay, 19 Dorset Ave, Diggle, Oldham, OL3 5PL [IO93AN, SE00]
G1 KFH J L Richmond, 11 Elm Ave, Pennington, Lymington, SO41 8BD [IO90FR, SZ39]
G1 KFI J Griffiths, 62 Victoria Mount, Horsforth, Leeds, LS18 4PX [IO93ET, SE23]
G1 KFO A J Eadie, 55 Windsor Park, Musselburgh, EH21 7QH [IO85LW, NT37]
G1 KFQ P King, 10 Hockley Ln, Eastern Green, Coventry, CV5 7FR [IO92EJ, SP27]
G1 KFR K F Rutherford, 11 Saffron Cl, Padgate, Warrington, WA2 0QZ [IO83RJ, SJ69]
G1 KFT S D Archer, 35 Ashurst Rd, Walmley, Sutton Coldfield, B76 1JE [IO92CM, SP19]
G1 KGA O D Himmo, 13 Dinorben Ct, 79/81 Woodcote Rd, Wallington, Surrey, SM6 0PZ [IO91WI, TQ26]
G1 KGC P M Simpson, 45 Amwell Rd, Cambridge, CB4 2UH [JO02BF, TL46]
G1 KGE M A Phelps, 35 Belmont Ave, Hereford, HR2 7JQ [IO82PB, SO53]
G1 KGL S Dorrington, Bahnhof Strasse 19, 64347, Griesheim, Germany
G1 KGO S Coben, 106 Fleming Mead, Mitcham, CR4 3LW [IO91WK, TQ27]
G1 KGQ C A Buxton, 10 Burwood Ave, Mansfield, NG18 3DZ [IO93KD, SK56]
G1 KGU J L Amos, Mite View, Ravenglass, Cumbria, CA18 1SW [IO84HI, SD09]
G1 KGV C O Ashcroft, 86 Avondale Ave, North Finchley, London, N12 8EN [IO91VO, TQ29]
GI1 KGZ P S Knott, 47 Pretoria St., Belfast, BT9 5AQ [IO74AN, J37]
G1 KHH Details withheld at licensee's request by SSL.
G1 KHM K H Morgan, 91 Headlands, Fenstanton, Huntingdon, PE18 9LP [IO92XH, TL36]
G1 KHS D W Tucker, 5 Uplands Cl, Hawkwell, Hockley, SS5 4DN [JO01HO, TQ89]
G1 KHX D Brooking, 16 Greenhill Cl, Worle, Weston Super Mare, BS22 0PE [IO81MI, ST36]
G1 KIB J S Martin, 2 Sackville Ct, Gullicott Ln, Hanwell, Banbury, OX17 1HQ [IO92HC, SP44]
G1 KIC J W J Jennings, 19 The Cornfields, Wick St. Lawrence, Weston Super Mare, BS22 9DY [IO81MI, ST36]
G1 KID S J Kemp, 1 Rectory Cl, Stanwick, Wellingborough, NN9 6QR [IO92RH, SP97]
G1 KIE C Brown, 11 Newbolds Rd, Wolverhampton, West Midlands, WV10 0SA [IO82WO, SJ90]
G1 KIG J A Boot, 33 Orchard Rd, East Cowes, PO32 6DU [IO90IS, SZ59]
G1 KII D S Beale, 88 Long Innage, Cradley, Halesowen, B63 2UY [IO82WL, SO98]
G1 KIJ D A Marsters, 21 Cow Ln, Rampton, Cambridge, CB4 4QG [JO02BH, TL46]
G1 KIK W Thompson, 36 Cliffield Rd, Swinton, Mexborough, S64 8PX [IO93IL, SK49]
G1 KIL J I Rimmer, 16 Rookery Dr, Rainford, St. Helens, WA11 8BB [IO83OL, SD40]
G1 KIM G C Clift, Morva, 1 Brook Way, Friars Cliff, Christchurch, BH23 4HA [IO90DR, SZ19]
G1 KIV P R Bridger, Linden Lea, Frith Common, Eardiston, Tenbury Wells, WR15 8JX [IO82SH, SO66]
G1 KIW J D Moss, 42 Chantry Ln, Necton, Swaffham, PE37 8ET [JO02JP, TF80]
G1 KIZ T A Head, 36A Ashacre Ln, Worthing, BN13 2DH [IO90TU, TQ10]
GW1 KJE E G Collier, The Coach House, Pentreheyling, Churchstoke, Montgomery, Powys, SY15 6HU [IO82KM, SO29]
GM1 KJF C B Horton, 61 Dixon Ave, Kirn, Dunoon, PA23 8JE [IO75MX, NS17]
G1 KJG C A Robinson, 33 Wakering Ave, Shoeburyness, Southend on Sea, SS3 9BE [JO01JM, TQ98]
G1 KJH J Beach, 11 Highview Gdns, Edgware, HA8 9UB [IO91UO, TQ29]
G1 KJQ M E Haymes, 33 Cornmill Dr, Liversedge, WF15 7EE [IO93DQ, SE22]
G1 KJX B S Hobbs, 32 Blackhorse Ln, Swavesey, Cambridge, Cambs, CB4 5QR [JO02AH, TL36]
G1 KJY D A Haller, 6 Bondyke Cl, St. Margarets Ave, Cottingham, HU16 5ND [IO93SS, TA03]
G1 KKA P Montgomery, 7 Birchwood Cl, Tavistock, PL19 8DR [IO70WN, SX47]
G1 KKD T D Elcock, Little Grange, 33 Cromford Dr, Mickleover, Derby, DE3 5JT [IO92FV, SK33]
G1 KKE D K Rose, 47 Waltham Gdns, Banbury, OX16 8FB [IO92IB, SP44]
G1 KKF A J Lawes, 8 Lynchet Down, Brighton, E Sussex, BN1 7FR [IO90WU, TQ30]
G1 KKG M J White, 7 Tyneham Cl, Sandford, Wareham, BH20 7BE [IO80WQ, SY98]
G1 KKH P J Cunliffe, 37 Rectory Rd, Worthing, BN14 7PE [IO90TT, TQ10]
GM1 KKI K L Johnston, Innisfree, Gulberwick, Shetland, ZE2 9LL [IP90IE, HU34]
GW1 KKJ K D Taylor, 23 Vardre Ave, Deganwy, Conwy, LL31 9UT [IO83CH, SH77]

G1 KKS I F Gott, Tayman House, The St., Acton Turville, Badminton, GL9 1HH [IO81UM, ST88]
G1 KKZ A A Dixon, 40 Beckwith Rd, Rotherham, S65 3PD [IO93IK, SK49]
G1 KLC D J Boyd-Livingston, 8 Woodlands Rd, Baughurst, Tadley, RG26 5NZ [IO91KI, SU56]
G1 KLI M D Smith, 8 Central Ave, Peterborough, PE1 4JJ [IO92VO, TF10]
G1 KLK A Dutton, 111 St. Michaels Rd, Crosby, Liverpool, L23 7UL [IO83LL, SD30]
G1 KLO J Partington, 9 Tithebarn Hill, Glasson Dock, Lancaster, LA2 0BY [IO83NX, SD45]
G1 KLW P A L Golding, 80 Birdbrrok Rd, Kidbrook, London, SE3 9PQ [JO01AL, TQ47]
G1 KLZ Details withheld at licensee's request by SSL.
G1 KMA Details withheld at licensee's request by SSL.
G1 KMC B B Booth, 11 Warwick Ave, Scotforth, Lancaster, LA1 4EY [IO84OA, SD46]
GM1 KMH J T Bell, 34 Union Rd, Grangemouth, FK3 8AB [IO86DA, NS98]
G1 KMJ J D Couzins, 30 Camden Rd, St. Peters, Broadstairs, CT10 3DR [JO01RI, TR36]
G1 KMN N C Thompson, 10 Belmont Cres, Old Town, Swindon, SN1 4EY [IO91CN, SU18]
G1 KMS I C Millar, The Grange, 105 High St., Weston Favell, Northampton, NN3 3JX [IO92NF, SP76]
G1 KNA J M Burdett, Glencorse, 13 Fairfax Ave, Selby, YO8 0AZ [IO93LS, SE63]
G1 KNB L Blake, 55 Coniston Cl, London, SW20 9NJ [IO91VJ, TQ26]
G1 KNC J Casson, 1 The Cres, Bamber Bridge, Preston, PR5 6RJ [IO83QR, SD52]
G1 KNF S M Humes, 20 Ghyllside Rd, Northiam, Rye, TN31 6QG [JO00HX, TQ82]
G1 KNG E H Humes, 20 Ghyllside Rd, Northiam, Rye, TN31 6QG [JO00HX, TQ82]
G1 KNH I L Humes, 79 Lowry Cres, Mitcham, CR4 3NX [IO91WJ, TQ26]
G1 KNI S M Martin, 6 Prinsted Walk, Fareham, PO14 3AD [IO90JU, SU50]
G1 KNK G W Mellors, 14 St. Peters Ave, Church Warsop, Mansfield, NG20 0RZ [IO93KF, SK56]
G1 KNU P Sharp, Purbrook Cottage, Lyme Rd, Axminster, Devon, EX18 5BL [IO80BV, SS61]
G1 KNX S T Jones, 31 Church St., Tewkesbury, GL20 5PD [IO81WX, SO83]
G1 KNZ J M Washby, 83 School Ln, Dewsbury, WF13 4RY [IO93EQ, SE22]
G1 KOD J Rodgers, 5 Bridge Ave, Latchford, Warrington, WA4 1RJ [IO83RJ, SJ68]
G1 KOG E V Beir, 17 Deansway, Hemel Hempstead, HP3 9UE [IO91SR, TL00]
G1 KOH G M Hall, 14 Evington Parks Rd, Leicester, LE2 1PR [IO92KO, SK60]
G1 KOK I Button, 17 Monsall Dr, South Normanton, Alfreton, DE55 2BG [IO93HC, SK45]
G1 KOM M W Coombs, 49 Wansbeck Ave, Stanley, DH9 6HT [IO94DU, NZ15]
G1 KON L P McCoy, 56 Curate Rd, Anfield, Liverpool, L6 0BZ [IO83MK, SJ39]
G1 KOO Details withheld at licensee's request by SSL.
G1 KOP J W Gibbard, 2 Almond Ct, Liverpool, L19 2QZ [IO83NI, SJ48]
G1 KOT M F Lynn, 194 Great Gregorie, Basildon, SS16 5QG [JO01FN, TQ68]
G1 KOW L D Tamblin, 14 Copper Leaf Cl, Moulton, Northampton, NN3 7HS [IO92NG, SP76]
G1 KPC D Carter, 50 Helford Pl, Fishermead, Milton Keynes, MK6 2DQ [IO92PA, SP83]
G1 KPI T C Houghton, 22 Collingwood Rd, Paignton, TQ4 5PG [IO80FK, SX86]
G1 KPS Details withheld at licensee's request by SSL.[Station located near Canvey Is.]
G1 KPU D J Skilton, 42 Commercial Rd, Eastbourne, BN21 3XF [JO00DS, TV69]
G1 KPV A Rayner, 8 Rea Barn Rd, Brixham, TQ5 9DU [IO80FJ, SX95]
G1 KPY H S Patterson, 8 Evesham Cl, Stockton Heath, Warrington, WA4 6LJ [IO83RI, SJ68]
G1 KPZ M I Gillott, 2 Firthwood Ave, Coal Aston, Sheffield, S18 6BQ [IO93GH, SK37]
G1 KQE H W Sweet, 5 Dence Cl, Herne Bay, CT6 6BH [JO01NI, TR16]
G1 KQH S A Wigg, 45 Cambrian Ln, Rugeley, WS15 2XH [IO92AS, SK01]
GM1 KQK Dr R A August, Smiddyhill House, Stracathro, By Brechin, Angus, DD9 7QE [IO86QS, NO66]
GW1 KQN A M Bowyer, Letter F Farm, 51 Mildenhall Rd, Littleport, Ely, CB7 4SY [JO02DL, TL58]
G1 KQS T J Lockett, 21 Valley Rd, Spital, Chesterfield, Derbyshire, S41 0HB [IO93HF, SK37]
GW1 KQV C Caudy, 43 Graham Ave, Pen y Fai, Bridgend, CF31 4NR [IO81EM, SS88]
GW1 KQY B J Francis, 4 Heol Tir Coch, Efail Isaf, Pontypridd, CF38 1BW [IO81IN, ST08]
G1 KRB Dr K R Bentley, 46 Freeman Rd, Didcot, OX11 7DD [IO91IO, SU59]
G1 KRD N J Morton, Mont Rose, Lake Ln, Barnham, Bognor Regis, PO22 0AE [IO90QU, SU90]
G1 KRF T Howes, 136 Station St., Cheslyn Hay, Walsall, WS6 7EQ [IO82XP, SJ90]
G1 KRI M Townsend, 15 Chalk End, Pitsea, Basildon, SS13 3NN [JO01FN, TQ78]
G1 KRK E T Ersser, 19 Sandhurst Rd, Margate, CT9 3HR [JO01RJ, TR37]
G1 KRN R L H Mitchell, 15 Meadow Pl, Bondgate, Selby, YO8 0LU [IO93LS, SE63]
G1 KRR D A Athersmith, 55 Rakesmoor Ln, Barrow in Furness, LA14 4LQ [IO84JD, SD27]
G1 KRU A Graves, 49 Robin Ln, Edgmond, Newport, TF10 8JL [IO82TS, SJ71]
G1 KRX B A Piper, 26 Hare Law Gdns, Stanley, DH9 8DG [IO94CU, NZ15]
G1 KRY L A White, The Firs, Bennetts Ave, West Kingsdown, Sevenoaks, TN15 6AT [JO01DH, TQ56]
G1 KRZ J J Sugden, No 4 Cottage, Norham West Mains, Berwick upon Tweed, Northd., TD15 2JY [IO85WR, NT94]
GM1 KSB R P Stokes, Easter Buckieburn, Denny, Stirlingshire, Scotland, FK6 5JJ [IO76XB, NS78]
G1 KSC S A Harrison, 44 Rosslyn Rd, Whitwick, Coalville, LE67 5PT [IO92HR, SK41]
G1 KSE A C Robinson, 5 Hood Cl, Locks Heath, Southampton, SO31 6ST [IO90IU, SU50]
G1 KSH P J Sherwood, 3 Otham Cl, Canterbury, CT2 7QX [JO01NH, TR15]
G1 KSK E G Mullin, 26 Fearnhead Ln, Fearnhead, Warrington, WA2 0BE [IO83RJ, SJ69]
G1 KSN V J Tankard, 59 Warren Dr, Ifield, Crawley, RH11 0DT [IO91VC, TQ23]
G1 KSQ M O D Kennedy, 3 Webbs Cl, Bromham, Bedford, MK43 8NH [IO92RD, TL05]
G1 KSS D Williamson, 14 Parkstone Pl, Eaglescliffe, Stockton on Tees, TS16 9EP [IO94HM, NZ41]
G1 KSU Details withheld at licensee's request by SSL.
G1 KSV T W Nicholls, Hillside, 6 Panorama View, Dales View Park, Salterforth Colne, Lancs, BB8 5SH [IO83VV, SD84]
G1 KTE T Corsellis, 2 Maners Way, Cambridge, CB1 4SL [JO02BE, TL45]
G1 KTF D J Webb, 75 Barkham Ride, Finchampstead, Wokingham, RG40 4HB [IO91NJ, SU76]
G1 KTH Details withheld at licensee's request by SSL.
G1 KTJ Details withheld at licensee's request by SSL.
GW1 KTK Details withheld at licensee's request by SSL.
G1 KTN Details withheld at licensee's request by SSL.
G1 KTU P M Edwards, Elberton Forge, Elberton, Olveston, Bristol, BS12 3AE [IO81RO, ST68]
GW1 KTW C A Thomas, 18 Acrefield Ave, Guilsfield, Welshpool, SY21 9PN [IO82KQ, SJ21]
G1 KTY B K Rider, Rose Cottage, Coley Rd, East Harptree, Bristol, BS18 6AP [IO81QH, ST55]
G1 KTZ D S Robins, Ayala, Higher Rd, Pensilva, Liskeard, PL14 5NQ [IO70TM, SX26]
G1 KUG L Preece, 31 Pilling Cl, Walsgrave, Coventry, CV2 2HR [IO92GK, SP38]
GM1 KUI G G Smith, 38 Crown Cttgs, Stuartfield, Peterhead, AB42 5HR [IO87XM, NJ94]
G1 KUN T J Sexton, 41 St. Bedes Gdns, Cherry Hinton, Cambridge, CB1 3UF [JO02CE, TL45]
G1 KUQ P L Andrews, Physics Department, Univ of Birmingham, PO Box 363, Edgebaston, Birmingham, B15 2TT [IO92AK, SP08]
G1 KVC J A Norris, 3 St. Pauls Cl, Adlington, Chorley, PR6 9RS [IO83QO, SD61]
G1 KVD M A Webb, Peacross Cottage, Yarcombe, nr Honiton, Devon, EX14 9LX [IO80LV, ST20]
G1 KVI D F Bedson, 9 Rosemary Ln, Haskayne, Downholland, Ormskirk, L39 7JP [IO83MN, SD30]
G1 KVM Details withheld at licensee's request by SSL.
G1 KVN Details withheld at licensee's request by SSL.
G1 KVO M P Haydon, 10 Cherry Cl, Hockley, SS5 5BG [JO01HO, TQ89]
G1 KVR M D Wood, Kviabol, Sheep Pen Ln, Seaford, BN25 4QR [JO00BS, TV49]
G1 KVW I Wilson, 45 Meadway, Halstead, Sevenoaks, TN14 7EY [JO01BH, TQ46]
GM1 KWA B L Simpson, Glenan, Fassifern Rd, Fort William, PH33 6LJ [IO76KT, NN17]
G1 KWF S R Watson, 21 Lynn Cl, Leigh Sinton, Malvern, WR13 5DU [IO82UD, SO75]
GM1 KWG J Winterbourne, Birkenbush, Buckie, Banffshire, AB56 2AL [IO87MQ, NJ46]
G1 KWH S M Dorrington, 184 Manor Rd, South Benfleet, Benfleet, SS7 4HY [JO01GN, TQ78]
G1 KWJ Details withheld at licensee's request by SSL.
G1 KWK M D Duerden, 12 Masefield Ave, Bradford, BD9 6EX [IO93CT, SE13]
G1 KWP Details withheld at licensee's request by SSL.
G1 KWS D Godfrey, Lords Hill Cottage, Longbridge Deverill, Warminster, Wilts, BA12 7DY [IO81VD, ST84]
G1 KWX C W Carter, 22 Haddon Ave, Halifax, HX3 0NE [IO93BQ, SE02]
G1 KXF M A Williamson-Armsby, 97 Weeth Rd, Camborne, TR14 7NA [IO70IF, SW64]
G1 KXP S L Smith, 5 Andrew Gr, Macclesfield, SK10 1QR [IO83WG, SJ97]
G1 KXQ M K Bloxham, 34 Northcote Rd, Cove, Farnborough, GU14 9EA [IO91OH, SU85]
G1 KXW Details withheld at licensee's request by SSL.
G1 KXY G C Garrett, 25 Oakapple Rd, Southwick, Brighton, BN42 4YL [IO90VU, TQ20]
G1 KXZ A J B Moore, 76 Thompson Hill, High Green, Sheffield, S30 4JU [IO93GJ, SK38]
G1 KYN D Jackson, 68 Dorset Ave, Barley Mow Est, Birtley, Chester-le-Street, DH3 2DU [IO94FV, NZ25]
G1 KYV S J Youngs, 29 Great Eastern St., Cambridge, Cambs, CB1 3AB [JO02BE, TL45]
G1 KZA E Kilner, 3 Ruskin Cl, West Melton, Wath upon Dearne, Rotherham, S63 6NU [IO93HL, SE40]
G1 KZD T B Asker, 53 Rivermead, Stalham, Norwich, NR12 9PJ [JO02SS, TG32]
G1 KZI Details withheld at licensee's request by SSL.
G1 KZL Details withheld at licensee's request by SSL.
G1 LAJ Details withheld at licensee's request by SSL.
G1 LAO J C Flower, 12 Balmoral Cl, Alton, GU34 1QY [IO91MD, SU73]
G1 LAP J G Beecham, Newholme, Whitemill Ln, Walton, Stone, ST15 0EG [IO82WV, SJ83]
G1 LAR L A Rogers, 37 Gilbey Rd, Tooting, London, SW17 0QQ [IO91VK, TQ27]
G1 LAW E J Boyce, 214 London Rd, Benfleet, SS7 5SJ [JO01GN, TQ78]
G1 LBE Details withheld at licensee's request by SSL.
G1 LBH P Tomkins, 64 Glenwood Gdns, Bedworth, Nuneaton, CV12 8DA [IO92GL, SP38]
GI1 LBI Details withheld at licensee's request by SSL.
G1 LBK M A Kirk, Dunley Gdns, Areley Kings, Stourport-on-Severn, DY13 0LL [IO82UH, SO77]
G1 LBM C J Taylor, Freelands, Clacton Rd, Wix, Manningtree, CO11 2RU [JO01NV, TM12]
G1 LBQ C Taylor, 5 Gadbury Ave, Atherton, Manchester, M46 0LQ [IO83RM, SD60]
G1 LBR Details withheld at licensee's request by SSL.
G1 LCB A W T Bailey, 8 Spindlebury, Cullompton, EX15 1SY [IO80HU, ST00]
G1 LCC J W Edwards, 1 Herons Way, Runcorn, WA7 1UH [IO83QI, SJ58]
G1 LCE S M Turner, Rainow Villa, under Rainow Rd, Timbersbrook, Congleton, CW12 3PL [IO83WD, SJ86]
G1 LCH L Sparks, 9 Hawk Pl, Moresby Parks, Whitehaven, CA28 8YG [IO84FN, NX91]
G1 LCI D J Hurst, 31 Meadow Way, Dorney Reach, Maidenhead, SL6 0DR [IO91PM, SU97]
G1 LCN G G Clegg, 16 The Pastures, Lower Westwood, Bradford on Avon, BA15 2BH [IO81UH, ST85]
G1 LCR D M Harrison Leicester Raynet Grp, 6 Kirkland Cl, Castle Donington, Derby, DE74 2QY [IO92HU, SK42]
G1 LCS J M Burrows, 1 Browns Ave, Runwell, Wickford, SS11 7PT [JO01GO, TQ79]
G1 LCY R T Brown, 3 Penmare Cl, Hayle, TR27 4PJ [IO70HE, SW53]
G1 LDC P M Gibson, 1 Binney Rd, Northwich, CW9 5PZ [IO83SG, SJ67]
G1 LDG Details withheld at licensee's request by SSL.
G1 LDI Details withheld at licensee's request by SSL.
G1 LDJ Details withheld at licensee's request by SSL.
G1 LDM Details withheld at licensee's request by SSL.
G1 LDN J Y Burnet, 41 Douglas Cres, Thornhill, Southampton, SO19 5JP [IO90HV, SU41]
G1 LDP C B B F Farmer, 21 Fairbourne Ave, Wilmslow, SK9 6JQ [IO83VH, SJ87]
G1 LDQ Details withheld at licensee's request by SSL.
G1 LDR Details withheld at licensee's request by SSL.
G1 LDS J E Wiles Moroni ARA (Uk), 38 Northwood Ln, Clayton, Newcastle, ST5 4BN [IO82VX, SJ84]
G1 LDT Details withheld at licensee's request by SSL.
G1 LED K M Pinkard, 7 Mulberry Cl, Conwy, LL32 8GS [IO83BG, SH77]
G1 LEH M Nicholson, 7 Ellerbeck Cl, Workington, CA14 4HY [IO84FP, NY02]
GW1 LEL A Rowland, 8 Gosen Terr, Trevor, Trefor, Caernarvon, LL54 5HR [IO72SX, SH34]
G1 LEO J V G Brittain, 26 Saxby Cl, Eastbourne, BN23 7BH [JO00DT, TQ60]
G1 LEQ G L Adams Evening Study A, 2 Ash Gr, Knutsford, WA16 8BB [IO83TH, SJ77]
G1 LES J L Buckle, 14 Alder Cl, Mapplewell, Barnsley, S75 6JA [IO93FO, SE31]
G1 LEX J C Blything, 319 Gt Brickkiln S, Wolverhampton, West Midlands, WV3 0PZ [IO82WN, SO99]
G1 LFD V M Middleton, 5 Fieldhouse, Cinderhills, Holmfirth, Huddersfield, HD7 1EN [IO93CN, SE10]
G1 LFG L E Halsall, 133 Sultan Rd, Buckland, Portsmouth, PO2 7AT [IO90LT, SU60]
G1 LFI G O Hepworth, 3 College View, Ackworth, Pontefract, WF7 7LA [IO93HP, SE41]
GW1 LFN D A Rees, 97 Woodfield Rd, Llandybie, Ammanford, SA18 3UT [IO71XT, SN61]
GW1 LFO D Benson Highfields ARC, Highfields Ctre For, Physically Handcped, Heath Cardiff, CF4 3RB [IO81JM, ST17]
G1 LFR D Love, 17 Longstaff Ave, Rawnsley, Cannock, WS12 5QE [IO92AQ, SK01]
G1 LGB G P J Gundry, 51 Dorien Rd, Raynes Park, London, SW20 8EL [IO91VJ, TQ26]
G1 LGJ S B Stephens, 41 Elliotts Ln, Codsall, Wolverhampton, WV8 1PG [IO82VP, SJ80]
GI1 LGM C J Dowdall, 20 Gransha Dr, Glen Rd, Belfast, BT11 8AL [IO74AO, J37]
G1 LGO E W Coleman, 127 Allington Dr, Birstall, Leicester, LE4 4FF [IO92KQ, SK51]
G1 LGQ P White, 25 Witton Rd, Ferryhill, DL17 8QE [IO94FQ, NZ23]
G1 LGW D W Hamilton, 4 Bannister Dr, Abbey Gdns, Drypool, Hull, HU9 1EJ [IO93UR, TA12]
G1 LGX Details withheld at licensee's request by SSL.
G1 LGY B J Bishop, 26 Robin Gdns, Calmore, Totton, Southampton, SO40 8US [IO90FW, SU31]
G1 LHD M C Goodes, 17 Ashmead Cl, Lordswood, Chatham, ME5 8NY [JO01GI, TQ76]
G1 LHE G R D Courtney, 13 Purcell Rd, Crowthorne, RG45 6QN [IO91OJ, SU86]
G1 LHL B J Mayson, 2 Windmill Cttgs, Allhallows Rd, Lower Stoke, Rochester, ME3 9SP [JO01HK, TQ87]
G1 LHQ S L King, 2 Oakenholt Cottage, Eynsham Rd, Farmoor, Oxford, OX2 9NL [IO91IS, SP40]
G1 LHR R Taylor, 18 Redriff Rd, Collier Row, Romford, RM7 8HD [JO01BO, TQ59]
GW1 LHV J M O'Nions, Pant-Glas, Gegin Ln, Mineram, Wrexham, Clwyd, LL11 3YT [IO83LB, SJ25]
G1 LHW L T Shooter, 9 Albert Rd, Corfe Mullen, Wimborne, BH21 3QB [IO80XS, SY99]
G1 LHZ D Leach, 75 Overdale Rd, Romiley, Stockport, SK6 3JA [IO83WJ, SJ98]
G1 LIB A McKeever, 25 Allendale Dr, Copford, Colchester, CO6 1BP [JO01JV, TL92]
G1 LIE B J A Wickins, 40 Heath Rd, Pamber Heath, Tadley, RG26 3DS [IO91KI, SU66]
G1 LIF G S Smith, Whitewood, High St, Nutley, E Sussex, TN22 3NQ [JO01AA, TQ42]
G1 LIG M J Coker, 5 Penling Cl, Cookham, Maidenhead, SL6 9NF [IO91PN, SU88]
G1 LII Details withheld at licensee's request by SSL.
G1 LIJ P R Porter, 21 Farthings, Knaphill, Woking, GU21 2JS [IO91QH, SU95]
G1 LIK S N Church, 24 Lovel End, Chalfont St. Peter, Gerrards Cross, SL9 9PA [IO91RO, SU99]
G1 LIQ C M Wood, 195 Kimbolton Cres, Stevenage, SG2 8RW [IO91VV, TL22]
G1 LIS Details withheld at licensee's request by SSL.
G1 LIT J Southwell, 40 Downsview, Small Dole, Henfield, BN5 9YB [IO90UV, TQ21]
G1 LIZ Details withheld at licensee's request by SSL.
G1 LJA Details withheld at licensee's request by SSL.
G1 LJB Details withheld at licensee's request by SSL.
G1 LJJ A K Everett, 2 Downings View, Windmill Hill, Launceston, Cornwall, PL15 9AG [IO70TP, SX38]
G1 LJK Details withheld at licensee's request by SSL.
G1 LJL B S Duncan, 13 Westwick Gr, Sheffield, S8 7DP [IO93GH, SK38]
G1 LJM Details withheld at licensee's request by SSL.
G1 LJQ P M Johnson, 338 Creek Rd, March, PE15 8SD [JO02BN, TL49]
G1 LJR Details withheld at licensee's request by SSL.
G1 LJT R B Wheeler, 36 Brakspear Dr, Corsham, SN13 9NE [IO81VK, ST86]
G1 LJU Details withheld at licensee's request by SSL.
G1 LJY Details withheld at licensee's request by SSL.
GM1 LKD J R W Craib, 10 Cameron Rd, Bridge of Don, Aberdeen, AB23 8QN [IO87WE, NJ91]
GW1 LKG Details withheld at licensee's request by SSL.[Op: R J Cannon. Station located near Southport.]
G1 LKH R W Hilton, 8 Hogshill Ln, Cobham, KT11 2AQ [IO91TI, TQ16]
G1 LKJ P A D Manning, 39 Ashbury Cres, Merrow Park, Guildford, GU4 7HG [IO91RG, TQ05]
G1 LKK C W B Wardle, 16 Tedworth Ave, Sinfin, Derby, DE24 3BS [IO92FU, SK33]
G1 LKL Details withheld at licensee's request by SSL.
G1 LKP Details withheld at licensee's request by SSL.
G1 LLA C M Davis, 122 Herbert Ave, Parkstone, Poole, BH12 4HU [IO90AR, SZ09]
G1 LLJ M P Hutchings, 18 Orchard Rd, Paulton, Bristol, BS18 5YZ [IO81SH, ST65]
G1 LLM Details withheld at licensee's request by SSL.
G1 LLO G V Tanner, 116 Ascot Gdns, Southall, UB1 2SB [IO91TM, TQ18]
G1 LLQ P J Dollimore, 96 Hayes Bridge Ct, Uxbridge Rd, Hayes, Middx, UB4 0JH [IO91TM, TQ18]
G1 LLU I V Olver, 5 Church Meadow, Deal, CT14 9QZ [JO01QF, TR35]
G1 LLW D W J Rogers, 36 Guessens Rd, Welwyn Garden City, AL8 6RH [IO91VT, TL21]
G1 LLZ M G Jackson, 11 Church Farm Cl, Chapel St. Leonards, Skegness, PE24 5SQ [JO03EF, TF57]
G1 LMA K B Levitt, 87 Wolversdene Rd, Andover, SP10 2AU [IO91GE, SU34]
G1 LMC J N Trainer, 86 Plessey Rd, Blyth, NE24 3HX [IO95FC, NZ38]
G1 LMI J T Sanders, 35 Halkingcroft, Langley, Slough, SL3 7BB [IO91RM, SU97]
G1 LML J Thornton, 7 Queens Rd, Vicars Cross, Chester, CH3 5HA [IO83NE, SJ46]
G1 LMM S A Oliver, 155 Linton Rd, Loose, Maidstone, ME15 0AS [JO01GF, TQ75]
G1 LMN S R Froggatt, 255 Rushton Rd, Desborough, Kettering, NN14 2QB [IO92OK, SP88]
G1 LMQ G K Donachie, Church Lodge, Oxborough, Kings Lynn, Norfolk, PE33 9PS [JO02GN, TF70]
G1 LMR J B Brettell, 6 Beech Ct, Walsall, WS1 2NQ [IO92AN, SP09]
G1 LMU S A Hodgetts, 101 Ladyfield Rd, Chippenham, SN14 0AW [IO81WK, ST97]
G1 LMW D H Partridge, 44 Trumpet Terr, Cleator, CA23 3DY [IO84FM, NY01]
G1 LMX N Chapple, 1 St. Johns West, Bedlington, NE22 7DY [IO95FD, NZ28]
G1 LNA D L Wood, 18 Rosemellin, Camborne, TR14 8QF [IO70IF, SW64]
G1 LNQ P R Bury, 2 Manor Rise, Thornton in Craven, Skipton, BD23 3TP [IO83WW, SD94]
G1 LNR L J Button, 37 Abbots Way, Preston Farm, North Shields, NE29 8LU [IO95GA, NZ36]
G1 LNT Details withheld at licensee's request by SSL.
GW1 LNY Details withheld at licensee's request by SSL.
G1 LOA R E Day, 155 Bromford Ln, West Bromwich, B70 7HR [IO82XM, SO99]
GM1 LOB Details withheld at licensee's request by SSL.
G1 LOC R V Locke, 439 Valley Rd, Basford, Nottingham, NG5 1HX [IO93JA, SK54]
G1 LOE Details withheld at licensee's request by SSL.
G1 LOK P S Blyth, 6 Burney Rd, South Lynn, Kings Lynn, PE30 5LD [JO02ER, TF61]
G1 LOP R H Burrow, 9 Gheluvelt Ave, Hurcott Rd, Kidderminster, DY10 2QP [IO82VJ, SO87]
GW1 LOR Details withheld at licensee's request by SSL.
G1 LOU L C Vaisey, Esperanza, 5 Hudson Cl, Poulner, Ringwood, BH24 1XL [IO90CU, SU10]
G1 LOV P W Foster, 61 Cumpsty Rd, Liverpool, L21 9HX [IO83ML, SJ39]
G1 LPC R H Chapman, 37 Rutland Rd, Skegness, PE25 2AX [JO03ED, TF56]
G1 LPF D H Pike, Silver Trees, 28 Birchwood Dr, Wilmington, Dartford, DA2 7NE [JO01CK, TQ57]
G1 LPJ A L Smith, 5 Yaxley Cl, Thurnby, Leicester, LE7 9UU [IO92LP, SK60]
G1 LPL R Brown, The Lilacs, Scar Ln, West Barnby, Whitby, YO21 3SD [IO94PM, NZ81]
G1 LPN A R G Kroll, 196 Wordsworth Rd, Horfield, Bristol, BS7 0EH [IO81RL, ST67]
G1 LPQ D Thorpe, 216 Diamond Ave, Kirkby in Ashfield, Nottingham, NG17 7NA [IO93JC, SK55]
G1 LPS T Roxby, 20 Middleham Walk, Grange Est, Spennymoor, DL16 6LX [IO94EQ, NZ23]
G1 LQB L T Clarke, 10 Stonecliff Park, Prebend Ln, Welton, Lincoln, LN2 3JS [IO93SH, TF08]
G1 LQC P M Chambers, 26 Drummond Gdns, Epsom, KT19 8RP [IO91UI, TQ26]
G1 LQF S T Overton, 92 Bramhall Ln South, Bramhall, Stockport, SK7 2EA [IO83WI, SJ88]
G1 LQH M D Lewis, 35 Lower Mead, Herne Farm, Petersfield, GU31 4NR [IO91MA, SU72]
G1 LQM C Costello, Threeways, The Green, Stalham, Norwich, NR12 9PZ [JO02SS, TG32]
G1 LQT C A Matthews, 36 Minterne Waye, Hayes, UB4 0PD [IO91TM, TQ18]
G1 LQV G J Austin, 10 St. Peters Cl, Eastcote, Ruislip, HA4 9JT [IO91TN, TQ18]
G1 LQX R B Saunders, 3 Lancaster Rd, Cressex Industrial Estat, High Wycombe, HP12 3NN [IO91OO, SU89]

G1 LQZ P C Beaver, 6 Mayland Green, Mayland, Chelmsford, CM3 6BD [JO01JQ, TL90]
G1 LRJ J O Critchley, 28 Elmer Gdns, Edgware, HA8 9AR [IO91UO, TQ19]
G1 LRK C M Bassett, 11 Redcastle Rd, Thetford, IP24 3NF [JO02IJ, TL88]
G1 LRN R Metcalfe, 5 Curlcroft Rd, Neston, Corsham, SN13 9RR [IO81VJ, ST86]
G1 LRR L V Pinto, 43 Napoleon Rd, St. Margarets, Twickenham, TW1 3EW [IO91UK, TQ17]
G1 LRU R Jones, 8 Mullen Ave, Downs Barn, Milton Keynes, MK14 7LU [IO92PB, SP84]
G1 LSB P S Brockett, 146 Winsover Rd, Spalding, PE11 1HQ [IO92WS, TF22]
G1 LSF J Cranfield, 80 Dexter Ave, Oldbrook, Milton Keynes, MK6 2QH [IO92PA, SP83]
G1 LSJ S J Haynes, Yew Tree Farm, Yew Tree Ln, Astley Bridge, Bolton, BL1 8TZ [IO83SO, SD71]
G1 LSK I Wiseman, 13 Swift Gdns, St. Giles, Lincoln, LN2 4NA [IO93RF, SK97]
G1 LSL W H Gray, 11 Muswell Cl, Salthouse Rd, Hull, East Riding of Yorks, HU8 0RE [IO93US, TA13]
G1 LSN M J Felton, 1 Barnwell Cl, Wistaston, Crewe, CW2 6TG [IO83SB, SJ65]
G1 LSX J E Humphreys, 16 Church Croft, Madley, Hereford, HR2 9LT [IO82NB, SO43]
G1 LSZ P J Lambert, 22 Cullingworth Ave, Haworth Park, Hull, HU6 7DD [IO93TS, TA03]
G1 LTC E W Rodd, 28 Amherst Rd, Pennycomequick, Plymouth, PL3 4HH [IO70WJ, SX45]
G1 LTE G D Eades, 30 Oakenhayes Cres, Minworth, Sutton Coldfield, B76 9RP [IO92CM, SP19]
G1 LTH D I Williams, 40 Dorchester Waye, Hayes, UB4 0HX [IO91TM, TQ18]
G1 LTI P D Smith, 65 Oatlands, Gossops Green, Crawley, RH11 8EH [IO91VC, TQ23]
G1 LTK A G Sturgess, Goonhilly, 11 Keats Cl, Earl Shilton, Leicester, LE9 7DU [IO92IN, SP49]
G1 LTL D J Gardner, New House, Birdbush Ave, Saffron Walden, CB11 4DJ [JO02CA, TL53]
GM1 LTM U A Wallace, No 1 Holding Wester, Kincardie, Crieff, Tayside, PH7 3RP [IO86CI, NN82]
G1 LTQ R Baker, 2 Godwit Rd, Southsea, PO4 8YS [IO90LT, SU60]
G1 LTZ S P Durrant, 11 Aintree Dr, Waterlooville, PO7 8NE [IO90LV, SU61]
G1 LUC N J Spring, Sycamore Cottage, Darracott, Georgeham, Braunton, EX33 1JY [IO71VD, SS43]
G1 LUF J D Wright, 298 Field Rd, Bloxwich, Walsall, WS3 3NB [IO92AO, SK00]
G1 LUI V R Hoskins, 44 Vicarage Rd, Whitehall, Redfield, Bristol, BS5 9AF [IO81RL, ST67]
G1 LUM M S White, 592 Mappowder, Mappowder, Sturminster Newton, Dorset, DT10 2EH [IO80TU, ST70]
G1 LUN P J Baker, 17 Gr Farm Cl, Leeds, LS16 6DA [IO93EU, SE23]
G1 LUR W E Lambert, 64 Kimberley St., Hollinwood, Oldham, OL8 4NX [IO83WM, SD90]
G1 LUU J Burton, 262 Carlton Hill, Carlton, Nottingham, NG4 1FY [IO92KX, SK64]
G1 LUV Details withheld at licensee's request by SSL.
G1 LUX C J Deacon, 12 Russet Way, Burnham on Crouch, CM0 8RB [JO01JP, TQ99]
G1 LUY P Wareing, 63 Fields End, Huyton, Liverpool, L36 5YQ [IO83NJ, SJ48]
GM1 LUZ C Campbell, 18 Parkview Ave, Falkirk, FK1 5JX [IO85CX, NS87]
G1 LVH D R Barnett, The Beehive, 20 Staythorpe Rd, Rolleston, Newark, NG23 5SG [IO93NB, SK75]
G1 LVK C E Street, Russets, Isle-Brewers, Taunton, TA3 6QN [IO80NX, ST32]
G1 LVN G W Paley, 10 Gaddum Rd, Manchester, M20 6SZ [IO83VK, SJ89]
G1 LVR P R Cousins, 38 Braunston Dr, Willowtree Est, Yeading, Hayes, UB4 9RB [IO91TM, TQ18]
G1 LVV A K Rhodes, 9 Midland Terr, New Mills, Stockport, Ches, SK12 4NL [IO83XI, SJ98]
G1 LVW P J Parker, 43 Meadow Cl, Farmoor, Oxford, OX2 9PA [IO91HS, SP40]
G1 LVY J A Dorman, Glenfield, Main St., Cotesbach, Lutterworth, LE17 4HZ [IO92JK, SP58]
G1 LVZ F H West, 14 Ashley Dr, Whitton, Twickenham, TW2 6HW [IO91TK, TQ17]
G1 LWE J Etchells, 34 Link Ave, Urmston, Manchester, M41 9NJ [IO83TK, SJ79]
G1 LWF T M J Finch, 9 Halstead Rd, Bitterne Park, Southampton, SO18 2PQ [IO90HW, SU41]
G1 LWH B C Whitehouse, 3 Barratts Rd, Kings Norton, Birmingham, B38 9HU [IO92AJ, SP07]
G1 LWL M E Patrick, 39 Poplar Rd, Healing, Grimsby, DN37 7RE [IO93VN, TA20]
G1 LWP S B Davies, 3 Bluebellwood Cl, Walmley, Sutton Coldfield, B76 2UB [IO92CN, SP19]
GW1 LXD M Fisher, 40 Magor St., Newport, NP9 0GU [IO81MN, ST38]
G1 LXI Details withheld at licensee's request by SSL.
G1 LXM Details withheld at licensee's request by SSL.
GW1 LXN Details withheld at licensee's request by SSL.
G1 LXP Details withheld at licensee's request by SSL.
G1 LXQ Details withheld at licensee's request by SSL.
G1 LXR Details withheld at licensee's request by SSL.[Station located near Biggin Hill.]
G1 LXU T J Burke, 42 Albert Rd, Cleethorpes, DN35 8LX [IO93XN, TA30]
G1 LYI T E Wright, 38 Gregson Rd, Widnes, WA8 0BX [IO83PI, SJ58]
G1 LYO T Green, The Walkers, Rushall, Much Marcle, Ledbury Herefordshi, HR8 2PE [IO82RA, SO63]
G1 LYV D H Ison, 28 Brunwins Cl, Wickford, SS11 8EA [JO01GO, TQ79]
G1 LZF B W Dickinson, 178 Wycliffe Gdns, Shipley, BD18 3JB [IO93CU, SE13]
G1 LZH P D Taylor, 10 Pickenham Rd, Hollywood, Birmingham, B14 4TG [IO92BJ, SP07]
G1 LZK T Campbell, Irthing Vale, Caravan Park, Old Church Ln, Brampton, Cumbria, CA8 2AA [IO84PW, NY56]
G1 LZM F J Ormett, 44 Manor Dr, Hinchley Wood, Esher, KT10 0AX [IO91TI, TQ16]
G1 LZQ A C Sparkes, 13 Swathing, Cranworth, Thetford, IP25 7SJ [JO02LO, TF90]
G1 LZS S G Stallworthy, 131 Lower Glen Rd, St. Leonards on Sea, TN37 7AR [JO00GV, TQ71]
G1 LZT P J M Creed, 32 Osborne Rd, Penn, Wolverhampton, WV4 4AY [IO82WN, SO89]
G1 LZW G R Bahlke, 52 Adisham Green, Kemsley, Sittingbourne, ME10 2SR [JO01II, TQ96]
G1 LZY B C Pyke, 2 Vicarage Gdns, Milton Abbot, Tavistock, PL19 0NA [IO70UO, SX47]
G1 MAB S G Hutton, Awelon, Turners Ln, Llynclys Hill, Oswestry, Salop, SY10 8LL [IO82LT, SJ22]
G1 MAC M P Macbeth, Baytree Cottage, 58 The Common, Abberley, Worcester, WR6 6AY [IO82TH, SO76]
G1 MAD W Colclough Moorlands Dis Rd, 12 Meadow Cl, Forsbrook, Stoke on Trent, ST11 9DW [IO82XX, SJ94]
G1 MAK A K Macbeth, 26 Severn Terr, Worcester, WR1 3EH [IO82VE, SO85]
G1 MAL D M Simmons, 47 Lower St., Haslemere, GU27 2NY [IO91PC, SU93]
GD1 MAN J H Kneale, 23 Clypse Rd, Willaston, Douglas, IM2 6JE
G1 MAR N L Gutteridge Midland ARS, 68 Max Rd, Quinton, Birmingham, B32 1LB [IO92AK, SP08]
G1 MAV G W Seaman, Foxborrow, 22A Mount Rd, Bexleyheath, DA6 8JS [JO01BK, TQ47]
GW1 MAX R Pullen, 8 Carmarthen Ct, Hendre Denny Park, Caerphilly, CF8 2TX [IO81HN, ST08]
G1 MBA A L Smith, 1 Constable Rd, St. Ives, Huntingdon, PE17 6EP [IO92XI, TL37]
G1 MBE C J Batty, 32 The Warings, Heskin, Chorley, PR7 5NZ [IO83PP, SD51]
G1 MBG J A Brown, 3A Edinburgh Rd, Kings Worthy, Winchester, SO23 7NY [IO91IC, SU43]
G1 MBM M J Kitson, 278 Cowcliffe Hill, Rd Fixby, Huddersfield, West Yorks, HD2 2NE [IO93CQ, SE11]
G1 MBN P Doyle, 11 Clifford Ave, Longton, Preston, PR4 5BH [IO83OR, SD42]
GM1 MBT I G Finlayson, 2F3 16 Roseneath Pl, Edinburgh, Midlothian, EH9 1JB [IO85JW, NT27]
GW1 MBV J Follett, 136 Westbourne Rd, Penarth, CF64 3HH [IO81JK, ST17]
G1 MBW M C Smith, 22 Long Leys, London, E4 9LW [IO91XO, TQ39]
G1 MBX M J Fisher, 72 Park Ln, Norwich, NR2 3EF [JO02PP, TG20]
GM1 MBZ J T Smith, Chapel Park Cottage, Methlick, Ellon, Aberdeenshire, AB4 0EU [IO87TH, NJ72]
GM1 MCA O S Barr, 19 Cleveden Gdns, Glasgow, G12 0PU [IO75UV, NS56]
GW1 MCD Details withheld at licensee's request by SSL.
G1 MCG D E Minton, 49 Runcorn Rd, Barnton, Northwich, CW8 4ES [IO83RG, SJ67]
G1 MCI A Rawson, 43 Co. Rd North, Bricknell Ave, Hull, HU5 4HN [IO93TS, TA03]
G1 MCJ D White, 33 Sobraon House, Elm Rd, Kingston upon Thames, KT2 6JD [IO91UJ, TQ16]
GM1 MCN Rev C T Stanley, 70 Cairngorm Cres, Aberdeen, AB12 5BR [IO87WC, NJ90]
G1 MCO D J Furby, 7 Parry Cl, Portsmouth, PO6 4SL [IO90KU, SU60]
G1 MCR E M Hampson Merseyside Cty, Raynet Group, 21 Marlowe Rd, Wallasey, L44 3DA [IO83LK, SJ39]
G1 MCT G J Williams, 29 Coleridge Rd, Barnby Dun, Doncaster, DN3 1AN [IO93LN, SE60]
G1 MCW W J Higgins, 15 Redburn Cl, Dingle, Liverpool, L8 4XR [IO83MJ, SJ38]
G1 MCX M J Richards, 10 Blackthorn Way, Alcester, B49 6BW [IO92BF, SP05]
G1 MCY C S C Toogood, 16 Penlea Ave, Bridgwater, TA6 6JU [IO81CE, ST23]
G1 MCZ E F Bottomley, Green Ln, Glusburn, Keighley, West Yorks, BD20 8RU [IO83XV, SD94]
G1 MDA R G Tye, Pendle View, Ribchester Rd, Hothersall, Preston, PR3 3XA [IO83RT, SD63]
G1 MDC D G Tommey, 99 Fairfield Park Rd, Fairfield Park, Bath, BA1 6JR [IO81TJ, ST76]
G1 MDE D W Harrison, 26 Valley Rd, Barnoldswick, Colne, BB8 6AZ [IO83VW, SD84]
G1 MDF G K Richardson, 1 The Spinneys, Lewes, E Sussex, BN7 2RN [JO00AV, TQ41]
G1 MDG R Ray Chesham&Dist AR, 54 Gladstone Rd, Chesham, HP5 3AD [IO91QQ, SP90]
GM1 MDH E M Rodgers, 2 Norse Rd, Scotstoun, Glasgow, G14 9HP [IO75TV, NS56]
G1 MDJ J A Murch, Prinsted, Emsworth, Hants, PO10 8HS [IO90NU, SU70]
G1 MDK Details withheld at licensee's request by SSL.
G1 MDL N Walker, 70 Maple Ave, Beeston Rylands, Beeston, Nottingham, NG9 1PW [IO92JW, SK53]
GM1 MDO J Stewart, 104 Barrhill Rd, Cumnock, KA18 1PU [IO75UK, NS52]
G1 MDQ M H Mitchell, 4 Carlton Ave, Harlington, Hayes, UB3 4AD [IO91SL, TQ07]
G1 MDS M G Lewis, Westbank, 46 Weyside Rd, Guildford, GU1 1HX [IO91RF, SU95]
G1 MEG P E Cox, 25 Church Cl, Stoke St. Gregory, Taunton, TA3 6HA [IO81MA, ST32]
G1 MET Dr R R Heywood, 22 Catterall Cl, Blackpool, FY1 3RB [IO83LT, SD33]
G1 MEZ A E Shenton, 19 King Edward St., Darlaston, Wednesbury, WS10 8TN [IO82XN, SO99]
GM1 MFD A W T Laing, 7 Vandeleur Gr, Edinburgh, EH7 6UE [IO85KW, NT27]
GD1 MFF G H Waft, 4 Whitebridge Rd, Onchan, Douglas, IM3 4HS
G1 MFG G C Read, L Eglise, Durley St., Durley, Southampton, SO32 2AA [IO90IW, SU51]
G1 MFR R J Rackstraw, 24 Albion Rd, Chalfont St. Giles, HP8 4EW [IO91RP, SU99]
G1 MFW A Empringham, Paulzanne, Bank End, North Somercotes, Louth, LN11 7LN [JO03BK, TF49]
G1 MFX E P Dickinson, 7 Mount Barrow Rd, Ulverston, Cumbria, LA12 9NH [IO84KE, SD27]
GW1 MFY G A Williams, 2 Godrer Mynydd, Gwernymynydd, Mold, CH7 4AD [IO83KD, SJ26]
G1 MGF A C Hall, 172 Aldershot Rd, Guildford, GU2 6BL [IO91QG, SU95]
G1 MGH D W H Poole, 54 Willis Rd, Haddenham, Aylesbury, HP17 8HF [IO91MS, SP70]
GW1 MGI L K Long, 18 Pentre Poeth Rd, Bassaleg, Newport, NP1 9LL [IO81LN, ST28]
G1 MGN T J Jones, 35 Manta Rd, Dosthill, Tamworth, B77 1PE [IO92DO, SK20]
GW1 MGR H T John Mid Glam Ray Gr, 11 Penylan, Bridgend, CF31 1QW [IO81FM, SS98]
G1 MGT D Norman, 5 Virginia Park Rd, Gosport, PO12 3DZ [IO90KT, SU50]
G1 MGU J M Studd, 39 Wallington Rd, Copnor, Portsmouth, PO2 0HB [IO90LT, SU60]
G1 MGZ M D Brophy, 78 Foley Rd West, Streetly, Sutton Coldfield, B74 3NP [IO92BO, SP09]
G1 MHA C C Corner, 19 Swinbourne Gdns, Whitley Bay, NE26 3AZ [IO95GB, NZ37]
G1 MHF G M Fleming, 25 Waverton Ave, Prenton, Birkenhead, L43 0XB [IO83LI, SJ28]
G1 MHM W J Taylor, 44 Colesbourne Rd, Benhall, Cheltenham, GL51 6DL [IO81WV, SO92]
G1 MHP D B Gilbert, 5 Wymbush Gdns, Hartcliffe, Bristol, BS13 0AZ [IO81QJ, ST56]
G1 MHU Details withheld at licensee's request by SSL.
G1 MIB Details withheld at licensee's request by SSL.
GI1 MIC Details withheld at licensee's request by SSL.
G1 MIH J Moran, 27 Mellor Gr, Bolton, Lancs, BL1 6DA [IO83SO, SD61]
GM1 MII Details withheld at licensee's request by SSL.
G1 MIL P Thompson, 326 Broad St., Crewe, CW1 4JH [IO83SC, SJ75]
G1 MIN T Wilkins, 10 Dorking Gr, Birmingham, B15 2DS [IO92BL, SP08]
G1 MIR E Sherriff, 10 Wainfleet Rd, Great, Grimsby, DN33 1LD [IO93WN, TA20]
G1 MIY R F Lunnon, 9 Hennerton Way, High Wycombe, HP13 7UE [IO91PP, SU89]
G1 MJD Details withheld at licensee's request by SSL.
G1 MJE G S Jeckells, Hogals End, Mill St., Bradenham, Thetford, IP25 7QN [JO02KP, TF90]
G1 MJI D K Fry, 8 Montpelier Rd, Ilfracombe, EX34 9HP [IO71WE, SS54]
GI1 MJJ F M Gilliland, 48 Malone Heights, Belfast, BT9 5PG [IO74AN, J36]
GM1 MJK D Miller, 142 Cameron Dr, Newfarmloch, Kilmarnock, KA3 7PL [IO75SO, NS43]
G1 MJN M J Newbold, 124 Hillcross Ave, Morden, SM4 4EG [IO91VJ, TQ26]
G1 MJO A H Hemming, 64 Haslucks Green Rd, Shirley, Solihull, B90 2EJ [IO92CJ, SP17]
G1 MJT R Naylor, 31 Cross Flatts Dr, Leeds, LS11 7HY [IO93FS, SE23]
G1 MJV N A Liddiard, 84 Westward House, Leiston, IP16 4HU [JO02SF, TM46]
G1 MJX Details withheld at licensee's request by SSL.
GM1 MKC C J Strong, Little Couchercairn, St Katherines, Inverurie, Aberdeenshire, AB5 8TQ [IO87TH, NJ72]
G1 MKE R P Cox, Carlingford, Brimley Dr, Lower Brimley, Teignmouth, TQ14 8LE [IO80GN, SX97]
G1 MKJ Details withheld at licensee's request by SSL.
G1 MKP M K Pauley, 5 Clare Cl, Waterbeach, Cambridge, CB5 9PS [JO02CG, TL46]
G1 MKR R W Hughes Milton Keynes Raynet Group, 46 The Boundary, Oldbrook, Milton Keynes, MK6 2HT [IO92PA, SP83]
G1 MKT G M Jewers, 1 Sywell Leys, Rugby, CV22 5SD [IO92II, SP57]
GW1 MKV A J Coe, Woodbrook Cottage, Paddock Row, Ruabon, Clwyd, LL14 6DD [IO82LX, SJ34]
G1 MLC H M Taha, 53 Barbers Hill, Werrington, Peterborough, PE4 5ED [IO92UP, TF10]
GW1 MLE G Evans, Maes yr Onnen, 134 Waterloo Rd, Capelhendre, Ammanford, SA18 3RY [IO71XS, SN51]
G1 MLH J P Tuck, 18 Denholm Cl, Ringwood, BH24 1TF [IO90CU, SU10]
G1 MLJ D F Whelan, 6 Valestone Ave, Hamsworth, Hemsworth, Pontefract, WF9 4JQ [IO93HO, SE41]
G1 MLK A C Bull, 34 Bishops Gate, Birmingham, B31 4AJ [IO92AJ, SP07]
G1 MLO P D Pugh, 22 Boythorpe Rd, Chesterfield, S40 2NB [IO93GF, SK37]
G1 MLP B P Brown, The Warren, High Rd, Whaplode, Spalding, PE12 6TG [IO92XT, TF32]
G1 MLR P A Thomas, 49 Derwent Dr, Purley, CR8 1ER [IO91WH, TQ36]
GM1 MLS A B Pearson, 1 Bridgefield, Inverbervie, Montrose, DD10 0SR [IO86UU, NO87]
G1 MLV P Kelsey, 573 Stannington Rd, Stannington, Sheffield, S6 6AB [IO93FJ, SK38]
GM1 MLW C W Lindsay, 23 Ashley Rise, Bonhill, Alexandria, G83 9NL [IO75RX, NS37]
GM1 MLY M J Bull, 9 Grenitote, Sollas, North Uist, Western Isles, PA82 5BP [IO76JB, NS08]
G1 MMA S M Butcher, 155 Crow Ln East, Newton-le-Willows, WA12 9UD [IO83QK, SJ59]
G1 MMD K D Morris, 9 Vera Cres, Rainworth, Mansfield, NG21 0EU [IO93KC, SK55]
G1 MME A W Morgan, 4A Wolverton Gdns, London, W5 3LJ [IO91UM, TQ18]
G1 MMG I P A Scott Iverson, 35 Milverton Cres, Abington Park, Northampton, NN3 3AT [IO92NF, SP76]
G1 MMI T P D Ryder, 166 Vandyke Rd, Leighton Buzzard, LU7 8HS [IO91QW, SP92]
G1 MMN C J Lovett, 49 Tame St. East, Walsall, WS1 3LB [IO92AN, SP09]
G1 MMO R C Egerton, 68 Hawarden Rd, Pen y Ffordd, Penyffordd, Chester, CH4 0JE [IO83LD, SJ36]
G1 MMR M M Robson, 12 Greetham Rd, Bedgrove, Aylesbury, HP21 9BS [IO91OT, SP81]
G1 MMS Details withheld at licensee's request by SSL.
G1 MMX J A Kidd, 19 Montagu Dr, Oakwood, Leeds, LS8 2PD [IO93GT, SE33]
G1 MMZ A T Roffey, 32 Hertford Rd, Digswell, Welwyn, AL6 0DB [IO91VT, TL21]
G1 MNA P J Edwards, 94 St. Johns Rd, Colchester, CO4 4JE [JO01LV, TM02]
G1 MNB Details withheld at licensee's request by SSL.
GW1 MNC A D Dykes, 16 The Mercies, Porthcawl, CF36 5HN [IO81DL, SS87]
G1 MNP H D Gray, 6 Nore Farm Ave, Emsworth, PO10 7NA [IO90MU, SU70]
G1 MNU I R Jukes, 39 Ponds Rd, Chelmsford, CM2 8QP [JO01FQ, TL70]
G1 MNX P Lane, 1 St. Davids Cl, Lower Willingdon, Eastbourne, BN22 0UZ [JO00DT, TQ50]
G1 MNY R J Smith, 3 Florence Farm Park, London Rd, West Kingsdown, Sevenoaks, TN15 6BP [JO01DI, TQ56]
G1 MOB Dr N S Booth, 20 Springfield, Ovington, Prudhoe, NE42 6EH [IO94BX, NZ06]
G1 MOE A M Anderson, 7 Thirlmere, Liden, Swindon, SN3 6LA [IO91CN, SU18]
G1 MOF B E Britten, 115 Mayflower Dr, Coventry, CV2 5NL [IO92GJ, SP37]
GW1 MOJ Details withheld at licensee's request by SSL.
GM1 MON J McQueen, 1 Balcathie Farm Cttgs, Arbroath, DD11 2PD [IO86QN, NO63]
G1 MOS H Whitbread, Foresters, Main Rd, Martlesham, Woodbridge, IP12 4SL [JO02PB, TM24]
G1 MOV M A Ballantyne, 248 Calshot Rd, Great Barr, Birmingham, B42 2BX [IO92AN, SP09]
G1 MOW D Mallin, 60 Arundel Rd, Littlehampton, BN17 7DF [IO90RT, TQ00]
G1 MPC M J Human, 89 Links Way, Croxley Green, Rickmansworth, WD3 3RW [IO91SP, TQ09]
G1 MPD M C Champion, 5 Adcote Cl, Barwell, Leicester, LE9 8DT [IO92HN, SP49]
G1 MPG C R Alefs, 28 Spains Hall Pl, Basildon, SS16 5UR [JO01FN, TQ78]
G1 MPH Details withheld at licensee's request by SSL.
G1 MPJ T C Gage, Bolter End Ln, Ln End, Bolter End, High Wycombe, Bucks, HP14 3NB [IO91NP, SU89]
G1 MPL C J Hitcham, 27 Kirby Cane Walk, Lowestoft, NR32 3EL [JO02UL, TM59]
G1 MPP A W M Baxter, 94 Abbeyfield Dr, Fareham, PO15 5PF [IO90JU, SU50]
GW1 MPR G W Roberts, Berwynne, Heol Offa, Coedpoeth, Wrexham, LL11 3EN [IO83LB, SJ25]
G1 MPT D G Bowden, 4 Cornmill Cl, Bardsey, Leeds, LS17 9EG [IO93GV, SE34]
G1 MPU P Brolan, 2 Mount Rd, Barnet, EN4 9RL [IO91WP, TQ29]
G1 MPW J Cooke, 2 Alpine Cl, Croydon, CR0 5UN [IO91WI, TQ36]
G1 MPZ J H Moore, 11 Elmwood, Barton Rd, Worsley, Manchester, M28 2PF [IO83TL, SD70]
GM1 MQA R G Cooke, 1 Springfield Rd, Lytham St. Annes, FY8 1TW [IO83LS, SD32]
G1 MQB P A Barker, 7 Verbena Cl, St. Anns, Nottingham, NG3 4PZ [IO92KX, SK54]
G1 MQC G P Du Feu CEPO ARC, Co. Emergency Planning Dept, Devon Fire & Rescue Service, Clyst St George, Exeter, Devon, EX3 ONW
GM1 MQE A M Norrie, 1 Bellard Rd, West Kilbride, KA23 9JT [IO75NQ, NS24]
G1 MQF M Booker, 7 Beech Ave, Warton, Preston, PR4 1BY [IO83NS, SD42]
G1 MQQ G Stainton, 168 Slades Rd, Golcar, Huddersfield, HD7 4JR [IO93BP, SE01]
G1 MRD J F W B Boddington, Greenlands, South Ln, Thornton Dale, Pickering, YO18 7QU [IO94PF, SE88]
G1 MRE P L Langford, 33 Briscoe Rd, Hoddesdon, EN11 9DG [IO91XS, TL30]
G1 MRI P Gardiner, 13A Reading Rd, Northolt, UB5 4PG [IO91TN, TQ18]
G1 MRP H Powell, 5 Carter Rd, Great Barr, Birmingham, B43 6JR [IO92AN, SP09]
G1 MRQ J A Smith, 10 Poplar Cl, Huntingdon, PE18 7BP [IO92VI, TL27]
G1 MRT B C Beck, 29 Whitney Moor Cl, Retford, DN22 7AT [IO93MH, SK78]
G1 MRU B Houghton, 24 Moss Ln, Lydiate, Liverpool, L31 4DQ [IO83MM, SD30]
G1 MRX P A Young, 30 Badminton Rd, Maidenhead, SL6 4QT [IO91PM, SU88]
GM1 MRY W Andrew, 11 Eddington Gdns, Chryston, Glasgow, G69 0JW [IO75XW, NS67]
G1 MRZ P Gregory, Mrz Communctns R Cl, 25 Wye Rd, Newcastle, ST5 4AZ [IO82VX, SJ84]
G1 MSA A S Taylor, 215 Nuffield Rd, Coventry, CV6 7HZ [IO92GK, SP38]
G1 MSB E J Turner, 14 Lauderdale Gdns, Bushbury, Wolverhampton, WV10 8AY [IO82WP, SJ90]
G1 MSG M White, 5 Edendale, Oulton Broad, Lowestoft, NR32 3JZ [JO02UL, TM59]
G1 MSH A R Johns, 20 Grangefield Ave, Rossington, New Rossington, Doncaster, DN11 0LS [IO93LL, SK69]
G1 MSK J N Riley, Hillcrest, Norwich Rd, Chedgrave, Norwich, NR14 6BQ [JO02RN, TG30]
GM1 MSN W J Clark, 66 Winstanley Wynd, The Grange, Kilwinning, KA13 6EB [IO75PP, NS24]
GM1 MSO D A H Clark, 76 Winstanley Wynd, The Grange, Kilwinning, KA13 6EB [IO75PP, NS24]
G1 MSR J W C Cockroft, 8 Harris Rd, Standish, Wigan, WN6 0QR [IO83PO, SD51]
G1 MSS Dr I E Coates, The Nook, Lower Wokingham Rd, Crowthorne, RG45 6DB [IO91OJ, SU86]
G1 MSU Details withheld at licensee's request by SSL.
G1 MSX D M Hall, 53 Lower Rd, Hednesford, Cannock, WS12 5ND [IO82XQ, SJ91]
G1 MSY K Ludgate, 138 Halton Rd, Runcorn, WA7 5RW [IO83PI, SJ58]
G1 MTB S P White, 12 Morgan Cl, Sidley, Bexhill on Sea, TN39 5EQ [JO00FU, TQ70]
G1 MTC A E Heath, 68 Cromer Rd, Northwood, Hanley, Stoke on Trent, ST1 6QN [IO83WA, SJ84]
G1 MTG T K Gamble, Highfield Lawn, Longford Ln, Longford, Derby, DE6 3DT [IO92EW, SK23]
G1 MTJ S C Tickle, 14 Rothesay Dr, Crosby, Liverpool, L23 0RF [IO83LL, SJ39]
G1 MTK C J Moore, Turners View, 50 Knowlys Rd, Morecambe, LA3 2PG [IO84NB, SD46]
G1 MTP N A J McNeil, 3 Grosvenor Cl, Water Ln, York, YO3 6PR [IO93KX, SE55]
G1 MTU D T Ward, 27 Penzer St., Kingswinford, DY6 7AA [IO82WL, SO88]
G1 MTY Details withheld at licensee's request by SSL.
G1 MUC C J Shingles, 20 Spencer Cl, Lingwood, Norwich, NR13 4BB [JO02RO, TG30]
G1 MUK Details withheld at licensee's request by SSL.
G1 MUM H S Phillips, 6 Peaks Down, Peatmoor, Swindon, SN5 5BH [IO91BN, SU18]
GU1 MUP P K Rudd, Val Des Arquets, Les Arquets, St. Pierre Du Bois, Guernsey, GY7 9HE
G1 MUQ P Hillier, Whyle House, Pudleston, nr Leominster, Hereford & Worceste, HR6 0RE [IO82QF, SO56]

G1 MUT P Lawrence, 4 Digmire Ln, Thorpe, nr Ashbourne, Derbyshire, DE6 2AW [IO93CB, SK15]
GM1 MUY L A Coxon, 40 Hamilton St., Barnhill, Broughty Ferry, Dundee, DD5 2RE [IO86NL, NO43]
G1 MVE P A Tither, 32 Manor Ave, Marston, Northwich, CW9 6DS [IO83SG, SJ67]
G1 MVG Dr F G Marshall, Hartwell, Newgrounds, Godshill Fordingbrd, Havant Hants, SP6 2LJ [IO90DW, SU11]
G1 MVI P J Bowe, 197 Gloucester Ave, Chelmsford, CM2 9DX [JO01FR, TL70]
GM1 MVK J Kane, Lintibert Rd, Muthill Crieff, Perthshire, PH5 2AH [IO86CH, NN81]
GW1 MVL D I Wright, 91 High St., Rhosllanerchrugog, Wrexham, LL14 1AN [IO83LA, SJ24]
G1 MVQ S N Hamilton-Cooper, 32 Malvern Cres, Ashby de La Zouch, LE65 2JZ [IO92GS, SK31]
G1 MVT D P Driscoll, 11 Caernarvon Cl, Hemel Hempstead, HP2 4AN [IO91SS, TL00]
G1 MVV P W Scovell, Pentlands, 1 Greenfields Cl, Totton, Southampton, SO40 3NE [IO90GW, SU31]
GW1 MVZ D M Petrie, 48A Lower Quay Rd, Hook, Haverfordwest, SA62 4LR [IO71MS, SM91]
G1 MWF B Nicholson, 58 Wexford Ave, Greatfield Est, Hull, HU9 5DT [IO93VS, TA13]
G1 MWI R B Moakes, 41 Pennyroyal Ct, Reading, RG1 6HE [IO91MK, SU77]
GM1 MWK J R Grieve, Craigiefield Park, St Ola, Orkney, KW15 1TE [IO88MX, HY41]
G1 MWP Details withheld at licensee's request by SSL.
G1 MWQ Details withheld at licensee's request by SSL.
G1 MWS B E Clay Macclesfld&Dist, 64 Coppice Rd, Poynton, Stockport, SK12 1SN [IO83WI, SJ98]
G1 MWT M P Clancy, 4 Golf Cres, Norton Tower, Halifax, HX2 0LD [IO93BR, SE02]
G1 MXC K G Sampson, 40 Crisp Rd, Lewes, BN7 2TX [IO90XV, TQ41]
G1 MXD G A Waldron, 68 Jubilee Rd, Corfe Mullen, Wimborne, BH21 3TJ [IO80XS, SY99]
GM1 MXE G Schafers, Dahlsteven, Orphir, Orkney, KW17 2RD [IO88KW, HY30]
G1 MXM I M Hunt, Four Seasons, Westmarsh Ash, nr Canterbury, Kent, CT3 2LP [JO01PH, TR26]
G1 MXO C P Ratcliffe, 28 Vicarage Ln, Wilpshire, Blackburn, BB1 9HX [IO83SS, SD63]
G1 MXV J O Branigan, 81 Trinity St., Gainsborough, DN21 1JF [IO93OJ, SK88]
GM1 MXW W C Donaldson, 8 Kenneth Rd, Greenacres, Motherwell, ML1 3AN [IO75XS, NS75]
G1 MYB Details withheld at licensee's request by SSL.
G1 MYD F J Carsboult, 6 Carbonels, Great Waldingfield, Sudbury, CO10 0RQ [JO02JB, TL94]
G1 MYE L Chiappi, Gianicolo, Clitheroe Rd, Barrow Whalley, nr Blackburn, Lancs, BB6 9AS [IO83ST, SD73]
GM1 MYF R D Jones, Gordondale, 20 Paradise Rd, Kemnay, Inverurie, AB51 5NJ [IO87SF, NJ71]
G1 MYM E Emons, 18 Haig Rd, Stanmore, HA7 4EP [IO91UO, TQ19]
G1 MYO F S Ross, 2 Mount Pleasant, Steeple Claydon, Buckingham, MK18 2QS [IO91MW, SP72]
G1 MYQ P C England, Moonstones, Down Ampney, Cirencester, Glos, GL7 5QS [IO91BQ, SU19]
GM1 MYR R M Cook, Briarwood, 95 Old Edinburgh Rd, Inverness, IV2 3HT [IO77VL, NH64]
G1 MYX M C Macartney, 4 York Ave, Wallasey, L44 9ER [IO83LJ, SJ39]
G1 MZC A Squire, 16 Henwick Hall Ave, Ramsbottom, Bury, BL0 9YH [IO83UP, SD71]
G1 MZD D J Barlow, 47 Brookfield Rd, Haversham, Milton Keynes, MK19 7AF [IO92OB, SP84]
G1 MZG S E Banks, 18 Sheerstock, Haddenham, Aylesbury, HP17 8EU [IO91MS, SP70]
G1 MZH E R Schamp, 21 Beechwood Ave, Melton Mowbray, LE13 1RT [IO92NS, SK72]
GD1 MZJ J H Sutherland, Archallagan Park, Marown, Isle of Man, IO9 9SU
G1 MZM M A Bignell, 53 Rosebay Ave, Hawkesley, Birmingham, B38 9QT [IO92AJ, SP07]
G1 MZT T G Tipper, 114 Paddock Ln, Oakenshaw, Redditch, B98 7XT [IO92AG, SP06]
G1 MZW I D Gillson, 13 Beech Green, Southcourt, Aylesbury, HP21 8JG [IO91OT, SP81]
G1 NAA M P Crabtree, 13 Stamford Ave, Frimley, Camberley, GU16 5XA [IO91PH, SU85]
G1 NAB G Rainy Brown, 104 Meadow Way, Theale, Reading, RG7 5DG [IO91LK, SU67]
G1 NAF K M Vickers, 44 Earlesmere Ave, Balby, Doncaster, DN4 0QE [IO93KM, SE50]
G1 NAG Details withheld at licensee's request by SSL.
G1 NAK K H Green, Chylean, Penpethy, Tintagel, Cornwall, PL34 0HH [IO70PP, SX08]
G1 NAN A J Gateley, 2 Langmere Rd, Watton, Thetford, IP25 6LG [JO02JN, TF90]
G1 NAQ A Ashton, 6 Lansdowne Cres, Darton, Barnsley, S75 5PW [IO93FN, SE30]
G1 NAT D J McGowan, 7 Eccles Cl, Henley Green, Coventry, CV2 1EF [IO92GK, SP38]
GI1 NAV G H H Beattie, 4 Dromore Rd, Carrickfergus, BT38 7PJ [IO74CR, J48]
G1 NBG C R Coopland, 182 Long Ln, Dalton, Huddersfield, HD5 9SF [IO93CP, SE11]
G1 NBJ D T N Hughes, 128 Sigston Rd, Beverley, HU17 9PA [IO93TU, TA04]
G1 NBK W M Winning, Plump House, Main St, Terrington, York, North Yorks, YO6 4QB [IO94MD, SE67]
G1 NBP Dr P M Allan, 16 Farmstead Cl, Gr, Wantage, OX12 0BD [IO91GO, SU39]
G1 NBQ Details withheld at licensee's request by SSL.
G1 NBR M Losekoot, Dept of Elec Eng, University of Surre, Guildford, GU2 5XH [IO91QF, SU95]
G1 NBT A Bradley, 59 Main Rd, Watnall, Nottingham, NG16 1HE [IO93JA, SK54]
G1 NBU L D Wellbeloved, 8 Orchard Cl, South Wonston, Winchester, SO21 3EY [IO91HC, SU43]
G1 NBX H Kaminski, Nebraska, Southover Way, Hunston, Chichester, PO20 6NY [IO90OT, SU80]
G1 NBY Rev R F Roeschlaub, 20 Pannatt Hill, Millom, LA18 5DB [IO84IF, SD18]
G1 NCC J N Potter, 27 Alexandra Rd, Pudsey, LS28 8BX [IO93DT, SE23]
G1 NCE Details withheld at licensee's request by SSL.
G1 NCF K Andrews, 41 Emerson Ave, Stainforth, Doncaster, DN7 5QF [IO93LO, SE61]
G1 NCG K G Powell, 74 Barnes Cre, Bournemouth, BH10 5AW [IO90BS, SZ09]
G1 NCK D L R Bates, 71 Nicholas Cres, Fareham, PO15 5AJ [IO90JU, SU50]
G1 NCL C Holmes, 16 Industrial St., Pelton, Chester-le-Street, DH2 1NR [IO94EU, NZ25]
G1 NCN H W Jones, 8 Warren Cl, Old Catton, Norwich, NR6 7NL [JO02PP, TG21]
G1 NCO Details withheld at licensee's request by SSL.
G1 NCR P O Sambrook Nth Cheshire AR, 117 Conway Rd, Salemoor, Sale, M33 2TL [IO83UK, SJ79]
GW1 NCS A Snelgrove, Redesdale, 17 Richmond Terr, Tredegar, NP2 4LE [IO81JS, SO10]
G1 NDA Details withheld at licensee's request by SSL.
GM1 NDE M A Turnbull, 23 South Union St., Cupar, KY15 5BB [IO86LH, NO31]
G1 NDG Details withheld at licensee's request by SSL.
G1 NDK K A R Dunn, Marylands, Maidstone Rd, Staplehurst, Tonbridge, TN12 0RH [JO01GE, TQ74]
G1 NDL K H Harrison, 20 Springfield Ave, Ashbourne, DE6 1BJ [IO93DA, SK14]
G1 NDM Details withheld at licensee's request by SSL.
G1 NDQ M C Taylor, 29 Harewood Ave, Highroad Well Ln, Halifax, HX2 0LU [IO93BR, SE02]
G1 NDU J P Kelly, 1 Whittaker Dr, Smithybridge, Littleborough, OL15 8QR [IO83WP, SD91]
G1 NDV R Airey, 30 White Horse Cres, Gr, Wantage, OX12 0PY [IO91GO, SU38]
G1 NEB A J Dearman, 232 Birmingham Rd, Redditch, B97 6EL [IO92AH, SP06]
GW1 NED H F M Anderson, Penrheol Farm, Meidrim, Carmarthen, Dyfed, SA33 5NX [IO71SU, SN32]
G1 NEE R S Hindley, 10 Briar Ave, Hollins Green, Rixton, Warrington, WA3 6JH [IO83SJ, SJ69]
G1 NEG P E McGarry, 10 Douglas Ave, Soothill, Batley, WF17 6HG [IO93ER, SE22]
G1 NEJ L W Opit, 93 Chatburn Rd, Clitheroe, BB7 2AS [IO83TV, SD74]
G1 NEK N Gerolemou NE Kent Raynet, 63 Cobblers Bridge Rd, Herne Bay, CT6 8NT [JO01NI, TR16]
G1 NEN A J Hefford, 31 High St., Rushton, Kettering, NN14 1RQ [IO92OK, SP88]
GM1 NET R M Campbell Strathclyde R G, 83 Shawwood Cres, Newton Mearns, Glasgow, G77 5ND [IO75US, NS55]
G1 NEV D Dawson, 4 Hawkesworth Ln, Guiseley, Leeds, LS20 8HA [IO93DU, SE14]
GM1 NEW J Kerins, 30 Beech Ave, Newton Mearns, Glasgow, G77 5PP [IO75US, NS55]
G1 NEZ B H Stiff, 4 Timberlaine Rd, Pevensey Bay, Pevensey, BN24 6DE [JO00ET, TQ60]
G1 NFB D P Bagley, 49 Green Ln, Malvern Wells, Malvern, WR14 4HT [IO82UB, SO74]
G1 NFE C J Duffy, 189 Fir Ln, Royton, Oldham, OL2 6SS [IO83WN, SD90]
G1 NFN R K Hammond, 124 Maney Hill Rd, Sutton Coldfield, B72 1JU [IO92CN, SP19]
G1 NFQ R M Vivian, Flat 1, 26 Beer Rd, Seaton, EX12 2PD [IO80LQ, SY29]
G1 NFT Details withheld at licensee's request by SSL.
G1 NFV Details withheld at licensee's request by SSL.
GW1 NGD G R Friend, 33 Denham Ave, Llanelli, SA15 4DB [IO71VQ, SN40]
G1 NGE R A Nelson, Woodlands, Norwich Rd, Hevingham, Norfolk, NR10 5QX [JO02PS, TG22]
GM1 NGH Details withheld at licensee's request by SSL.
GW1 NGL K R Roberts, 331 Heol y Coleg, Vaynor, Newtown, SY16 1RA [IO82HM, SO09]
GW1 NGN P Johansson, 63 Grange Rd, Rhyl, LL18 4AD [IO83GH, SJ08]
G1 NGO D W Rookard, 15 Fairfield Rd, Brentwood, CM14 4LR [JO01DO, TQ59]
GW1 NGX A F Hodges, 47 Faenol Isaf, Tywyn, LL36 0DW [IO72WN, SH50]
G1 NGZ M Donoghue, 6 Harrison Dr, North Weald, Epping, CM16 6JD [JO01CR, TL40]
G1 NHB Details withheld at licensee's request by SSL.[Op: K J Price. Station located near Weybridge.]
G1 NHW Details withheld at licensee's request by SSL.
G1 NHX P R W Severn, 310 Worlds End Ln, Birmingham, B32 2SB [IO92AK, SP08]
G1 NIB Details withheld at licensee's request by SSL.
G1 NID Details withheld at licensee's request by SSL.
G1 NIG M P Thompson, 43 Manor Rd, Horsham St. Faith, Norwich, NR10 3LF [JO02PQ, TG21]
G1 NIO P A Radcliffe, 91 Linwood Dr, Potters Green, Coventry, CV2 2LZ [IO92GK, SP38]
G1 NIS Details withheld at licensee's request by SSL.
G1 NIT M L R Virtue, 50 Borthwick Park, Orton Wistow, Peterborough, PE2 6YY [IO92UN, TL19]
G1 NIV J V Young, 24 Cambridge Rd, Barking, IG11 8NW [JO01AM, TQ48]
G1 NIW P D Broder, 10 Birkett Cl, Sharples, Bolton, BL1 7DQ [IO83SO, SD71]
G1 NIY Details withheld at licensee's request by SSL.
G1 NIZ Details withheld at licensee's request by SSL.
G1 NJC K D Taylor, 241 Daventry Rd, Cheylesmore, Coventry, CV3 5HH [IO92FJ, SP37]
G1 NJG B J Asker, 34 Post Office Rd, Frettenham, Norwich, NR12 7AB [JO02PQ, TG21]
G1 NJI R C Foster, 35 Colin Rd, Barnwood, Gloucester, GL4 3JL [IO81VU, SO81]
G1 NJJ C P Gregory, 1 Three Oaks Rd, Wythall, Birmingham, B47 6HG [IO92BJ, SP07]
G1 NKF I C Spindler, 1 Spring Cttgs, Brewery Ln, Thrupp, Stroud, Glos, GL5 2EA [IO81VR, SO80]
G1 NKN Dr A G Mason, 25 Middle Rd, Leatherhead, KT22 7HN [IO91UH, TQ15]
G1 NKP D J F Leaver, 8 Limbourne Dr, Heybridge, Maldon, CM9 4YU [JO01IR, TL80]
G1 NKS P M Thom, Southern House, 9 Southern Rd, Cheltenham, GL53 9AW [IO81XU, SO91]
G1 NKU Details withheld at licensee's request by SSL.
G1 NKZ L C Hall, 115 Edgeley Rd, Stockport, SK3 9NG [IO83WJ, SJ88]
G1 NLB Details withheld at licensee's request by SSL.
GM1 NLD Details withheld at licensee's request by SSL.
G1 NLG W R Shelley, 5 Lower Putton Ln, Chickerell, Weymouth, DT3 4AN [IO80SO, SY68]
G1 NLJ R H S Sprake, 75 Upfield Rd, Hanwell, London, W7 1AW [IO91TM, TQ18]
G1 NLK G J Smith, 33 Redcar Cl, Northolt, UB5 4EJ [IO91TN, TQ18]
G1 NLQ D J Arter, 18 Essex Rd, Westgate on Sea, CT8 8AP [JO01QJ, TR36]
G1 NLS G G Borrett, 27 Sheepcote Rd, Eton Wick, Windsor, SL4 6JA [IO91QL, SU97]
G1 NLU D J Tulk, 8 Cleves Cl, Weymouth, DT4 9JU [IO80SO, SY67]
G1 NLV Details withheld at licensee's request by SSL.
G1 NLZ R E Brown, 15 Johnson Rd, Great Baddow, Chelmsford, CM2 7JL [JO01GR, TL70]
G1 NMF D Buttimore, 61 Burnham Rd, Leigh on Sea, SS9 2JR [JO01HN, TQ88]
G1 NMH N M G Hillman, 236 Wheelers Ln, Kings Heath, Birmingham, B13 0SR [IO92BK, SP08]
G1 NMK Details withheld at licensee's request by SSL.
G1 NMN M S Durey, 71 Orchard Rd, Maldon, CM9 6EW [JO01IR, TL80]
G1 NMO Details withheld at licensee's request by SSL.
G1 NMP N K Fenner, 22 Gowers Field, Aylesbury, HP19 3QB [IO91OT, SP81]
G1 NMQ L G Gibbs, 304 Wexham Rd, Slough, SL2 5QL [IO91RM, SU98]
G1 NMR M M Kelly, 18 Fitzmaurice Rd, Christchurch, BH23 2DY [IO90CR, SZ19]
G1 NMW A J Stone, 34 Masefield Rd, Warminster, BA12 8HR [IO81VE, ST84]
G1 NMY B M Hitchcock, 111 Kingsdown Cres, Dawlish, EX7 0HB [IO80GO, SX97]
G1 NNA B D Lloyd, 9 Hornbeam Walk, Witham, CM8 2SZ [JO01HT, TL81]
G1 NNB G B Lloyd, 9 Hornbeam Walk, Witham, CM8 2SZ [JO01HT, TL81]
G1 NNC G F T Huntley, Salterton, Church Rd, Kennington, Ashford, TN24 9DQ [JO01KD, TR04]
G1 NNF R A Kenny, 35 Broom Leys Rd, Coalville, LE67 4DD [IO92HR, SK41]
G1 NNL V J Linstead, 12 Birches Green Rd, Erdington, Birmingham, B24 9SR [IO92CM, SP19]
G1 NNN S J Moore, 104 Gloucester Ave, Chelmsford, CM2 9LF [JO01FR, TL70]
G1 NNP Details withheld at licensee's request by SSL.[Op: H E Morgan. Station located near Woodford Green.]
G1 NNQ S J Healey, 107 Church Rd, Shoeburyness, Southend on Sea, SS3 9EY [JO01JM, TQ98]
G1 NNU M J Lowe, 76 Kingskerswell Rd, Newton Abbot, TQ12 1DG [IO80EM, SX87]
G1 NNV D W Rees, Derwyn, 40 Chine Walk, West Parley, Ferndown, BH22 8PX [IO90BS, SZ09]
G1 NOC Details withheld at licensee's request by SSL.
G1 NOD Details withheld at licensee's request by SSL.[Op: Hazel Tether. Station located 5km S Towcester, Northants.]
G1 NOL T J Jones, 7 Dove Cl, Thorley, Bishops Stortford, CM23 4JD [JO01BU, TL41]
G1 NOM M C Fox, 29 Phillips Field Rd, Great Cornard, Sudbury, CO10 0JH [JO02IA, TL84]
G1 NON S K Fox, 18 Hartest Way, Great Cornard, Sudbury, CO10 0LA [JO02JA, TL84]
G1 NOO I F Allgood, 53 The Avenue, Leighton Bromswold, Huntingdon, PE18 0SH [IO92TI, TL17]
G1 NOR G Galbraith, 44 Parker Rd, Grays, RM17 5YN [JO01DL, TQ67]
G1 NOS K Brookes, 20 School Ave, Guidepost, Choppington, NE62 5DN [IO95ED, NZ28]
G1 NOX Details withheld at licensee's request by SSL.
G1 NOZ H Grant, 6 Felling Dene Gdns, Gateshead, NE10 0NA [IO94FW, NZ26]
G1 NPA P Allan, 214 Westwood Rd, Sutton Coldfield, B73 6UQ [IO92BN, SP09]
G1 NPI C M Badcock, 7 Heathfield Rd, Chandlers Ford, Eastleigh, SO53 5RP [IO91HA, SU42]
G1 NPJ K T Hyslop, 111 Swingate Ln, Plumstead, London, SE18 2DB [JO01BL, TQ47]
G1 NPM G P Batchelor, 14 Arterial Rd, Rayleigh, Essex, SS6 7TR [JO01HN, TQ78]
G1 NPN Details withheld at licensee's request by SSL.
G1 NPP P J Pritchard, The Grange, Gaddesby, Leics, LE7 4WN [IO92MR, SK61]
G1 NPT Details withheld at licensee's request by SSL.
G1 NPU C Marshall, 5 George St., Gainsborough, DN21 2PU [IO93OJ, SK89]
G1 NQB C Carpenter, 22 Falconwood Ave, Welling, DA16 2SQ [JO01BL, TQ47]
G1 NQG Details withheld at licensee's request by SSL.
G1 NQH M Cook, Brooksdie Cottage, Brook Ln, Great Easton, Market Harborough, LE16 8SJ [IO92OM, SP89]
G1 NQK Details withheld at licensee's request by SSL.
G1 NQM J Saunders, 125 Pursey Dr, Bradley Stoke, Bristol, BS12 8DP [IO81RM, ST68]
G1 NQN K M Benfold, 56 Cornwall Ave, Blackpool, FY2 9QW [IO83LU, SD33]
G1 NQO J L Jacques, 65 Daggers Hall Ln, Marston, Blackpool, FY4 4AX [IO83LT, SD33]
G1 NQQ F J Drake, 4 Kits Cl, Chudleigh, Newton Abbot, TQ13 0LG [IO80EO, SX87]
G1 NQS I C Elsworth, 88 Mungo Park Way, Orpington, BR5 4EQ [JO01BJ, TQ46]
G1 NQT Details withheld at licensee's request by SSL.
G1 NQU M Peacock, 38 Foxhill Dr, Queensbury, Bradford, BD13 2JH [IO93BS, SE03]
G1 NQW Details withheld at licensee's request by SSL.
G1 NQX N E Flisher, 3 Broadlands Ave, New Romney, TN28 8JE [JO00LX, TR02]
G1 NQY C J Butcher, 21 Bruce Rd, Writtle, Chelmsford, CM1 3EE [JO01FR, TL60]
G1 NRE D Paul, 22 Bruce Rd, Kempston, Bedford, MK42 7EU [IO92SC, TL04]
G1 NRG R W Steele Northampton Ryn, 98 Obelisk Rise, Boughton Green, Northampton, Northants, NN2 8QU [IO92NG, SP76]
G1 NRK P Slater, 12 Apsley Cl, Bishops Stortford, CM23 3PX [JO01BU, TL41]
G1 NRM A H Harrison, 34 Marsh Ln, Mill Hill, London, NW7 4QP [IO91VO, TQ29]
G1 NRN Details withheld at licensee's request by SSL.
GW1 NRS M A Hill Newport Amateur Radio Soc, 13 Maesglas Gr, Newport, NP9 3DJ [IO81LN, ST28]
G1 NRT T Palmer, 274 Findon Rd, Worthing, BN14 0HD [IO90TU, TQ10]
G1 NRU Details withheld at licensee's request by SSL.
G1 NRV Details withheld at licensee's request by SSL.
G1 NRX P J Hill, The Acorns, Prince Cres, Staunton, Glos, GL19 3RF [IO81UX, SO72]
G1 NRY I Leach, 36 Harrowden, Bradville, Milton Keynes, MK13 7DA [IO92OB, SP84]
G1 NSA Details withheld at licensee's request by SSL.
G1 NSB R R Winterburn, 3 Inglewood Cl, Larkholme, Fleetwood, FY7 8PE [IO83LV, SD34]
G1 NSD D I Young, 13 Crawshaw Park, Pudsey, LS28 7EP [IO93ET, SE23]
G1 NSG H S Goodwin, 91 Grange Ln, Four Oaks, Sutton Coldfield, B75 5LD [IO92CO, SP19]
G1 NSK F I H Godfrey, 50 Eskdale Dr, Worksop, S81 7QB [IO93KH, SK58]
G1 NSL L Thompson, 59 Leyland Rd, Harrogate, HG1 4RU [IO93FX, SE35]
G1 NSP J P E Hughes, 2 Butterbur Way, Killinghall Moor, Harrogate, HG3 2XH [IO94FA, SE25]
G1 NSQ C A Joyce, 70 Campbell Rd, Twickenham, TW2 5BY [IO91TK, TQ17]
G1 NSR T F Hurley, 11 Hewitt Ave, Fayre Oaks, Hereford, HR4 0QP [IO82PB, SO44]
G1 NST N F Dodd, Chedburgh House, Hall Ln, Knapton, North Walsham, NR28 0RZ [JO02RU, TG33]
GM1 NSU R Hardy, Laurelhill Cottage, Snowdon Pl Ln, Stirling, FK7 9JW [IO86AC, NS79]
G1 NSV D F Kemplen, Coriander, 2 Vicarage Cl, Menheniot, Liskeard, PL14 3QG [IO70TK, SX26]
G1 NSW P G Corley, Aegis, High Rd, Fobbing, Stanford-le-Hope, SS17 9HN [JO01FM, TQ78]
G1 NTB A Johnson, 8 Brampton Gdns, Eynesbury, St. Neots, Huntingdon, PE19 2DU [IO92UF, TL15]
G1 NTE C J E Askew, 78 West Gr, Woodford Green, IG8 7NR [JO01AO, TQ49]
G1 NTG C Mapp, 16 Parkside Dr, Arnside, Carnforth, LA5 0BU [IO84NE, SD47]
G1 NTH R A Stanley, 41 Shakespeare Rd, Eaton Socon, St. Neots, Huntingdon, PE19 3HG [IO92UF, TL15]
G1 NTI B Longstaff, 3 Railway Gdns, Stanley, DH9 8QB [IO94DU, NZ15]
G1 NTK P J Hull, 5 Richmond Park, Bishops Hull, Taunton, TA1 5LL [IO81KA, ST22]
G1 NTL J McShane, 12 Virginia Gdns, Middlesbrough, TS5 8BT [IO94JM, NZ41]
G1 NTN A C Edwards, 9 Lincoln Cl, Great Woodley Est, Romsey, SO51 7TJ [IO90GX, SU32]
G1 NTO C S Macleod, 7A Church Rd, West Kirby, Wirral, L48 0RL [IO83JI, SJ28]
G1 NTP B W Haden, 72 Charlton Rd, Blackheath, London, SE3 8TT [JO01AL, TQ47]
G1 NTV D A W Merry, 18 Tremabe Park, Dobwalls, Liskeard, PL14 6JS [IO70RK, SX26]
G1 NTW N J Marshall, 3 Sunnyside, Oxted Rd, Godstone, RH9 8BP [IO91XG, TQ35]
G1 NTX S N Taylor, 5 Collingwood, Farnborough, Hants, GU14 6LX [IO91PG, SU85]
G1 NUA G Crocker, 98 Pound Rd, Kingswood, Bristol, BS15 4QU [IO81SL, ST67]
G1 NUG D M Mydat, 14 Mapleleafe Gdns, Ilford, IG6 1LG [JO01AO, TQ48]
G1 NUH R Hardwick, 5 Seaview, Oakmere Park, Little Neston, Ches, L64 0XP [IO83LG, SJ27]
G1 NUJ Details withheld at licensee's request by SSL.
G1 NUN S Lomas, 62 Springfield Rd, Etwall, Derby, DE65 6LA [IO92EV, SK23]
G1 NUO E T Oram, 31 Nathaniel Walk, Tring, HP23 5DG [IO91QT, SP91]
G1 NUP B Johnson, 7 St. Davids Ln, Flamborough, Bridlington, YO15 1BE [IO94WD, TA27]
G1 NUR Details withheld at licensee's request by SSL.[Op: A Bewick. Station located near Whitley Bay.]
G1 NUS J R Thornley, 270 Hurdsfield Rd, Macclesfield, SK10 2PN [IO83WG, SJ97]
G1 NUU Details withheld at licensee's request by SSL.
G1 NVE V S Wood, 175 Windleshaw Rd, Dentons Green, St. Helens, WA10 6TP [IO83OL, SJ59]
G1 NVG F L Badman, Sarahs Cottage, Bridge Ln, Ladbroke, Leamington Spa, CV33 0DE [IO92HF, SP45]
G1 NVL G A Officer, 3 Kinloch St., Manchester, M11 4DL [IO83VL, SJ89]
G1 NVN A S Parkin, 1 Dunelm Walk, Leadgate, Consett, DH8 7QT [IO94CU, NZ15]
G1 NVO B W Kimber, 27 Ct Rd, Brockworth, Gloucester, GL3 4ES [IO81WU, SO81]
G1 NVQ Details withheld at licensee's request by SSL.
G1 NVS R J Pennington, 5 Park Cl, Northway, Tewkesbury, GL20 8RB [IO82WA, SO93]
G1 NWA C Rickerby, 113 Cliftonville Rd, Woolston, Warrington, WA1 4BJ [IO83RJ, SJ68]
GW1 NWF R D Gray, 36 Heol Pentre Felen, Morriston, Swansea, SA6 6BY [IO81AQ, SS69]
G1 NWG C H Scates, 17 Trecastle Way, Carleton Rd, London, N7 0EL [IO91WN, TQ28]
G1 NWH H Walker, 24 Castleton Ave, Riddings, Alfreton, DE55 4AG [IO93HB, SK45]
G1 NWM J Westwood, 9 Landbeach Rd, Milton, Cambridge, CB4 6DA [JO02BF, TL46]
G1 NWO J F Sharp, 3 Inkerman Rd, Eton Wick, Windsor, SL4 6LE [IO91QL, SU97]
G1 NWR Details withheld at licensee's request by SSL.

G1 NWT B Worviell, Cliddesden, The St., Motcombe, Shaftesbury, SP7 9PF [IO81VA, ST82]
G1 NWU B H Smith, 1 Kilworth Ct, Lutterworth Rd, North Kilworth, Lutterworth, LE17 6JE [IO92KK, SP68]
G1 NWZ M J Spacey, 54 Dovehouse Hill, Luton, LU2 9ES [IO91TV, TL12]
G1 NXB G R Fardoe, 3 Park Ave, Wallasey, L44 9DZ [IO83LJ, SJ39]
G1 NXR V Barrett, 10 Goldbrook Cl, Heywood, OL10 2QP [IO83VN, SD80]
G1 NXS B G Martlew, 38 Hob Hey Ln, Culcheth, Warrington, WA3 4NW [IO83RK, SJ69]
G1 NXT K Rushton, 9 Laburnum Ave, Woolston, Warrington, Ches, WA1 4NY [IO83RJ, SJ68]
G1 NXU S M Bishop, 31 Ravenslea, Ravenstone, Colville, Leics, LE6 2AT [IO92JP, SK50]
G1 NXV Dr R J Roycroft, Roadside House, Knutsford Rd, Chelford, Macclesfield, SK11 9AS [IO83UG, SJ87]
G1 NXX G P Fisher, 160 Netherton Rd, Worksop, S80 2SF [IO93KH, SK57]
G1 NYB B G Good, 29 Wharton St., Grimsby, DN31 2EG [IO93WN, TA20]
G1 NYG D Thompson, 200 Halifax Rd, Hove Edge, Brighouse, HD6 2QG [IO93CR, SE12]
G1 NYJ D J Hopton, 32 Braemar Ave, Flixton, Manchester, M31 3HP [IO83SK, SJ79]
GW1 NYO P D Dombrowski, 30 Hilary Rd, Newbridge, Newport, NP1 5DD [IO81KQ, ST29]
G1 NYS R A P Parkhurst, 28 Bahram Rd, Epsom, KT19 9DN [IO91UI, TQ26]
G1 NYZ D P Robinson, 4 Rushden Dr, Reading, RG2 8LJ [IO91MK, SU77]
G1 NZB Details withheld at licensee's request by SSL.
GM1 NZD P Bates, Craig Du, Church Terr, Newtonmore, Inverness, PH20 1DT [IO77WB, NN79]
GW1 NZF R J Stuckey, 8 Gelli Crossing, Gelli, Pentre, CF41 7UD [IO81GP, SS99]
G1 NZK A P Davis, 73A Milton Rd, Taunton, TA1 2JQ [IO81LA, ST22]
G1 NZL C Elliott, 7 Elizabeth Diamond Gdns, South Shields, NE33 5HX [IO94GX, NZ36]
G1 NZO Details withheld at licensee's request by SSL.
G1 NZP E Elliott, 7 Redhouse Rd, Hebburn, Tyne & Wear, NE31 2XS [IO94FX, NZ36]
G1 NZQ Details withheld at licensee's request by SSL.
G1 NZR D J Bond, 4 Alfred Rd, Haydock, St. Helens, WA11 0QD [IO83QL, SJ59]
G1 NZT Details withheld at licensee's request by SSL.
G1 NZZ R I Nicol, 37 Thicknall Dr, Stourbridge, DY9 0YH [IO82WK, SO98]
G1 OAC Details withheld at licensee's request by SSL.
G1 OAE R A L Steel, 7 Derwent Bank, Seaton, Workington, CA14 1EE [IO84FP, NY03]
G1 OAH Details withheld at licensee's request by SSL.
G1 OAJ A Mather, 11 Allenby Rd, Cadishead, Manchester, M44 5EA [IO83SK, SJ79]
G1 OAN D G Shaw, 29 Wyndham Ave, Clifton, Swinton, Manchester, M27 6PY [IO83TM, SD70]
G1 OAP A English, 53 St. Michaels Way, Childs Ercall, Market Drayton, TF9 2DE [IO82RT, SJ62]
G1 OAR P J Wallace, 7 Trinity View, Ketley Bank, Telford, TF2 0DX [IO82SQ, SJ61]
G1 OAS Details withheld at licensee's request by SSL.
G1 OAU V J English, 3 Colne Rd, Earith, Huntingdon, PE17 3PX [JO02AI, TL37]
G1 OAV Details withheld at licensee's request by SSL.
G1 OAW J R Smith, 62 Elson Ln, Elson, Gosport, PO12 4EU [IO90KT, SU60]
G1 OAY R W Hardingham, 69 Aylesbury Cl, Norwich, NR3 3LB [JO02PP, TG21]
G1 OBA I R Bpophy, 78 Foley Rd West, Streetly, Sutton Coldfield, B74 3NP [IO92BO, SP09]
G1 OBB Details withheld at licensee's request by SSL.
G1 OBC I L A M Evans, 6 Park End, Lichfield, WS14 9US [IO92CQ, SK10]
G1 OBD Details withheld at licensee's request by SSL.
G1 OBE M W Waddoups, 14 Home Mead, Barnard Rd, Galleywood, Chelmsford, CM2 8WA [JO01FQ, TL70]
G1 OBF Details withheld at licensee's request by SSL.
G1 OBM J A Miller, 5 Rubicon Ave, Wickford, SS11 8LL [JO01GO, TQ79]
G1 OBP Details withheld at licensee's request by SSL.
G1 OBQ A T Goacher, 59 Quincewood Gdns, Tonbridge, TN10 3LS [JO01DF, TQ54]
G1 OBR C W Burton, 13 Yew Tree Park, Congresbury, Bristol, BS19 5ER [IO81OI, ST46]
G1 OBU K A Haddington, 78 Gamble Hill Dr, Bramley, Leeds, LS13 4JL [IO93ET, SE23]
G1 OBY Details withheld at licensee's request by SSL.
G1 OCH C J Hillman, Mayjon, Crow, Ringwood, Hants, BH24 3ER [IO90CU, SU10]
G1 OCK R W Liepziger, 1 Moorfield Ave, Poulton-le-Fylde, FY6 7QE [IO83LU, SD34]
G1 OCN Details withheld at licensee's request by SSL.
G1 OCS A Heap, 35 Kingfisher Gr, Bradford, BD8 0NW [IO93CT, SE13]
G1 OCY K J Minihane, 60 Wolsey Dr, Walton on Thames, KT12 3BA [IO91TJ, TQ16]
G1 ODB D Price, 20 Fosse Way, Nailsea, Bristol, BS19 2BG [IO81OK, ST47]
G1 ODD B M Westlake, 47 Quarry Rd, Kingswood, Bristol, BS15 2NZ [IO81RK, ST67]
G1 ODJ T Boycott, 17 Brook St., Whitley Bay, NE26 1AF [IO95GB, NZ37]
G1 ODK Details withheld at licensee's request by SSL.
G1 ODQ C Bond, 6 Copse Cl, Hugglescote, Coalville, LE67 2GL [IO92HR, SK41]
G1 ODS J Rudd, 7 Garden Farm, West Mersea, Colchester, CO5 8DU [JO01LS, TM01]
G1 ODT A S Pargeter, Acres Cottage, Smallburn Rd, Longhorsley, Morpeth, Northd., NE65 8QH [IO95CF, NZ19]
G1 ODW G F Scantlebury, 15 Chellew Rd, Truro, TR1 1LR [IO70LG, SW84]
G1 ODX I J Ford, 16 Burley Rd, Harestock, Winchester, SO22 6LJ [IO91HB, SU43]
G1 ODZ G L Fountain, Tinkers Ln, Champneys, Wigginton, Tring, Herts, HP23 6JB [IO91QS, SP90]
G1 OEB A J Orchard, Flat No 3, 35 The High St, Bovingdon, Hemel Hempstead, Herts, HP3 0HG [IO91RR, TL00]
G1 OEF Dr E A Lowings, Great Halls, Aylesbeare, Exeter, Devon, EX5 2BY [IO80HR, SY09]
G1 OEH S A Whittaker, 42 Alanbrooke, Gravesend, DA12 1NA [JO01EK, TQ67]
G1 OEK G T Smith, 10 Pumphreys Ct, Pumphreys Rd, Charlton Kings, Cheltenham, GL53 8DD [IO81XV, SO92]
G1 OEM T D R Webb, 95 Deveraux Cres, Ebley, Stroud, Glos, GL4 5BX [IO81VU, SO81]
G1 OEN D A Richardson, 51 Sullivan Cres, Browns Wood, Milton Keynes, MK7 8DW [IO92PA, SP93]
G1 OEP J Harris, 109 Hook Rise South, Tolworth, Surbiton, KT6 7NA [IO91UI, TQ16]
G1 OEQ D W Tribute, 18 Molesey Park Cl, East Molesey, KT8 0NN [IO91TJ, TQ16]
G1 OER A G Parkin, 4 Waverley Rd, Farnborough, GU14 7EY [IO91PG, SU85]
G1 OET K R Doswell, 15A Queen St., Desborough, Kettering, NN14 2RE [IO92OK, SP88]
G1 OFA F H G Hearne, 150 Victoria St., Dunstable, LU6 3BB [IO91RV, TL02]
G1 OFF P A Bates, 122 London Rd, Crawley, RH10 2LD [IO91VD, TQ23]
G1 OFG P K Howard, The Cottage, Treffry, Lanhydrock, Bodmin, Cornwall, PL30 5AF [IO70PK, SX06]
G1 OFI N C Taylor, 3 Lyndhurst Ave, Bethel, St. Austell, PL25 3HJ [IO70OI, SX05]
G1 OFL R Hall-Osman, 32 Havelock Rd, Gravesend, DA11 0JG [JO01EK, TQ67]
G1 OFO D H Talbot, 8 Waterloo Cl, Puriton, Bridgwater, TA7 8BD [IO81ME, ST34]
G1 OFX W C Coates, 3 Graysmead, Sible Hedingham, Halstead, CO9 3NX [JO01HX, TL73]
G1 OFY Capt P Hendy, 23 Almsford Rd, Oatlands, Harrogate, HG2 8EQ [IO93FX, SE35]
G1 OFZ Details withheld at licensee's request by SSL.
G1 OGB Details withheld at licensee's request by SSL.
G1 OGC J C Kelly, 7 Collingwood Cres, Matlock, DE4 3TB [IO93FD, SK26]
G1 OGE A Wilson, 7 Hawthorne Cl, West Heath, Congleton, CW12 4UF [IO83VE, SJ86]
G1 OGH K Atkinson, 62 Fines Park, Stanley, DH9 8QY [IO94DU, NZ15]
G1 OGJ E W Spence, 23 Abbey Rd, Ulceby, DN39 6TJ [IO93UO, TA11]
G1 OGM D A Durell, 1 Chart Gdns, Dorking, RH5 4DP [IO91UF, TQ14]
G1 OGR R J Welch, 2 Broadlands Ave, Waterlooville, PO7 7JE [IO90LV, SU60]
G1 OGU Details withheld at licensee's request by SSL.
G1 OGV R J Bicknell-Thompson, 4 Linden Ct, Greenfrith Dr, Tonbridge, TN10 3LW [JO01DF, TQ54]
G1 OGY Details withheld at licensee's request by SSL.[Station located near Danbury, Essex, CM3.]
GM1 OGZ D Madden, 19 Waverley Gdns, Elderslie, Johnstone, PA5 9AJ [IO75SU, NS46]
G1 OHD B E Beswick, 2 Hollins Ave, Buxton, SK17 6LH [IO93BG, SK07]
G1 OHH S M Griffin, 6 Raygill Pl, Lancaster, LA1 2UQ [IO84OB, SD46]
G1 OHI P M Osborn, 122 Harrington St., Leicester, LE4 6ES [IO92KP, SK60]
G1 OHL Details withheld at licensee's request by SSL.
G1 OHM Details withheld at licensee's request by SSL.[Op: M Twyman, G6KOA, obo The South Birmingham RAYNET Group, c/o Hampstead House, Fairfax Road, West Heath, Birmingham, B31 3QY.]
G1 OHN Details withheld at licensee's request by SSL.
G1 OHQ Details withheld at licensee's request by SSL.
G1 OHS Details withheld at licensee's request by SSL.[Station located in London N21.]
G1 OHT Details withheld at licensee's request by SSL.
G1 OHU J R Freeman, 81 West Hill, Rotherham, S61 2EX [IO93HK, SK39]
G1 OHX V Pears, 10 Fremantle Rd, South Shields, NE34 7RF [IO94HX, NZ36]
G1 OIF R A Wright, 10 St. Michaels Ct, Slough, SL2 2NF [IO91QM, SU98]
GW1 OIK W C Jaggard, 6 Marston Rd, Rhos on Sea, Colwyn Bay, LL28 4SG [IO83CH, SH88]
GM1 OIN D M Ingram, 49 Henrietta St., Avoch, IV9 8QT [IO77VN, NH75]
G1 OIO W A Benton, 2 Regents Cl, Seaford, BN25 2EB [JO00BS, TQ40]
G1 OIP Details withheld at licensee's request by SSL.
G1 OIR Details withheld at licensee's request by SSL.
G1 OIS N Ellis, 1A Northcote Rd, Croydon, CR0 2HX [IO91WJ, TQ36]
G1 OIW S R Ludford, 9 Castle Mound, Barby, Rugby, CV23 8TN [IO92JH, SP57]
G1 OIZ R Young, 2 Jubilee Gdns, Bakestone Moor, Whitwell, Worksop, S80 4PW [IO93JG, SK57]
G1 OJB P J I Critchley, 4 Shandon Ave, Manchester, M22 4DP [IO83UJ, SJ89]
G1 OJD D A Facer, 7 Lowry Cl, Bedworth, Nuneaton, CV12 8DG [IO92GL, SP38]
G1 OJL S J Evenden, 11 Chapel St., Tavistock, PL19 8DX [IO70WN, SX47]
G1 OJO D R Pearson, 77 Hope Park, Bromley, Kent, BR1 3RG [JO01AJ, TQ37]
G1 OJP Details withheld at licensee's request by SSL.
G1 OJT R Barnish, 64 Braithwell Rd, Maltby, Rotherham, S66 8JU [IO93JK, SK59]
G1 OKB A Ibbotson, 62 Crag View Cres, Oughtibridge, Sheffield, S35 0GD [IO93FK, SK39]
G1 OKD R A Penfold, 7 Folkstone Cl, Chippenham, SN14 0XZ [IO81WK, ST97]
G1 OKF D I Cannon, 44 Grange Bottom, Royston, SG8 9UQ [IO92XB, TL34]
G1 OKI D W Hyde, 108 St. Bedes Cres, Cambridge, CB1 3UB [JO02CE, TL45]
G1 OKJ R A Pettican, 52 Shepherds Way, Saffron Walden, CB10 2AH [JO02DA, TL53]
G1 OKM D Speirs, 53 Tideswell Rd, Great Bar, Birmingham, B42 2DU [IO92BM, SP09]
G1 OKP R M Lloyd, 8 Kymin Lea, Wyesham, Monmouth, NP5 3TF [IO81PT, SO51]
G1 OKT M P Delaforce, Jeannsmeadow, Launcells, Bude, Cornwall, EX23 9NN [IO70ST, SS20]
G1 OKU E N Lee, 4 Gr Rd, Burnham on Sea, TA8 2HF [IO81MF, ST34]
G1 OKV G L Hendricks, 5 Jay Cl, Standalone Farm, Letchworth, SG6 4YH [IO91VX, TL23]
G1 OKY J R Cottrell, 33 Elton Dr, Spital, Wirral, L63 9HD [IO83MI, SJ38]
G1 OLD J F C Scott, 51 Water St., Accrington, BB5 6QU [IO83TS, SD72]
G1 OLE D C J Bates, 10 Upton Gdns, Worthing, BN13 1DA [IO90TT, TQ10]
G1 OLF A Heys, 151-153 Burnley Rd, Accrington, BB5 6DH [IO83TS, SD72]
G1 OLH R Purvess, 98 Birchwood Hill, Leeds, LS17 8NS [IO93FU, SE33]
G1 OLJ Details withheld at licensee's request by SSL.
G1 OLL M J Pinckston, 17 Trehaverne Terr, Truro, Cornwall, TR15 3SE [IO70IF, SW64]
G1 OLM J G Hesketh, 87 Condor Gr, Marston, Blackpool, FY1 5NA [IO83LT, SD33]
G1 OLT Details withheld at licensee's request by SSL.
G1 OLZ D G Turtle, 8 Coronation Ct, Stelling Rd, Erith, DA8 3JN [JO01CL, TQ57]
G1 OMG Details withheld at licensee's request by SSL.
G1 OMH Details withheld at licensee's request by SSL.
G1 OMI J R Canning, 130 Main Rd, Duston, Northampton, NN5 6RA [IO92MF, SP76]
G1 OMK Details withheld at licensee's request by SSL.
G1 OMN Details withheld at licensee's request by SSL.
G1 OMO Details withheld at licensee's request by SSL.
G1 OMP A E Woodford, 81 Harrold Rd, Rowley Regis, Warley, B65 0RL [IO82XL, SO98]
G1 OMQ Details withheld at licensee's request by SSL.
G1 OMT H Jeckells, Hogals End, Mill St., Bradenham, Thetford, IP25 7QN [JO02KP, TF90]
G1 OMU K J Pye, 5 Teme Ave, Brooklands Est, Wellington, Telford, TF1 3HU [IO82RR, SJ61]
G1 OMV W Huxley, 7 St. Michaels Rd, Newcastle, ST5 9LN [IO83VA, SJ84]
G1 OMX P J Cumiskey, 1 York Terr, Gateshead, NE10 9NB [IO94FW, NZ26]
G1 OMY D Ainscough, 11 Tressel Dr, Sutton Manor, St. Helens, WA9 4BS [IO83PJ, SJ59]
G1 OMZ E J H Rogers, 20 Rashleigh Ave, St. Stephens, Saltash, PL12 4NS [IO70VJ, SX45]
G1 ONC P M Maitland, 7 Spinners Ct, Stalham, Norwich, NR12 9EQ [JO02SS, TG32]
G1 OND B A Hastry, 56 Kilsyth Cl, Fearnhead, Warrington, WA2 0SQ [IO83RK, SJ69]
G1 ONF R G Creedy, 18 Panton Cl, Kings Lynn, PE30 5NB [JO02ER, TF61]
G1 ONG Details withheld at licensee's request by SSL.
G1 ONH F Slater, 32 Winthorpe Ave, Westgate, Morecambe, LA4 4RE [IO84NB, SD46]
G1 ONK L L Boston, 18 Athenaeum Ct, Highbury New Park, Highbury, London, N5 2DN [IO91WN, TQ38]
GI1 ONL Details withheld at licensee's request by SSL.
G1 ONQ F W Pearce, 59 Minerva Way, Wellingborough, NN8 3TP [IO92PH, SP86]
G1 ONY Details withheld at licensee's request by SSL.
G1 OOA Details withheld at licensee's request by SSL.
G1 OOB J Clark, Hindhayes, 25 Leigh Rd, St, BA16 0HB [IO81PC, ST43]
G1 OOE Details withheld at licensee's request by SSL.
G1 OOG A E Cooper, 16 Brindles Cl, Linford, Stanford-le-Hope, SS17 0RS [JO01EL, TQ67]
G1 OOJ M J Watson, 18 The Old Brewery, Newtown, Bradford on Avon, BA15 1NF [IO81UI, ST86]
G1 OOM B R Woodcock, 27 Main St., Cosby, Leicester, LE9 1UW [IO92JN, SP59]
G1 OOS E W F Rowthorn, 4 Woburn Ct, Rushden, NN10 9HL [IO92QG, SP96]
G1 OOU J F Pearce, 19 Cawthorne Rd, Kettlethorpe, Wakefield, WF2 7HW [IO93GP, SE31]
G1 OOW D W Street, 78 Stockfield Rd, Acocks Green, Birmingham, B27 6BB [IO92CK, SP18]
G1 OOX A T Harding, 6 Ullswater Gr, Alresford, SO24 9NP [IO91KB, SU53]
GD1 OOY S Wilson, Mistral, 18 Hilltop Rise, Douglas, IM2 2LE
G1 OOZ D Parsons, 1 Carlyle Rd, Rowley Regis, Warley, B65 9BQ [IO82XL, SO98]
G1 OPA D J Smith, 15 Billington Cl, Stoke Hill Est, Coventry, CV2 5NQ [IO92GJ, SP37]
G1 OPD P E Elsom, 12 Kettlebridge Ln, East Halton, Grimsby, DN40 3PR [IO93UP, TA11]
GW1 OPE S Rake, 9 Attlee Way, Cefn Golau, Tredegar, NP2 3TA [IO81JS, SO10]
G1 OPG K C Chappell, 21 Victoria St., Long Eaton, Nottingham, NG10 3EW [IO92IV, SK43]
G1 OPH Details withheld at licensee's request by SSL.
G1 OPJ A T Brown, 53 Middleton St., Blyth, NE24 2LS [IO95FD, NZ38]
G1 OPK J G H Hunter, Hunters Lodge, 21 Avondale Cl, Whitstable, CT5 3QA [JO01MI, TR16]
G1 OPL A W Roy, 16 Foljambe Rd, Brimington, Chesterfield, S43 1DD [IO93HG, SK47]
GM1 OPO G Askew, Glenormiston Farm, Innerleithen, Peebles, EH44 6RD [IO85KP, NT33]
G1 OPS A Johnson, 37 Wainwright Ave, Wombwell, Barnsley, S73 8LS [IO93HM, SE30]
G1 OPT P G Harvey, Swynford Cottage, Wagon Ln, Hook, RG27 9EJ [IO91MG, SU75]
G1 OPV P W Drew, 20 Russell St., Accrington, BB5 2NF [IO83TR, SD72]
G1 OPW F V Cox, Flat 5, 67 Great Pulteney St., Bath, BA2 4DL [IO81TJ, ST76]
G1 OPZ F Anderson, 2 Monmouth Paddock, Norton St. Philip, Bath, BA3 6LA [IO81UH, ST75]
G1 OQB R J Tams, 7 Hermitage Rd, Abingdon, OX14 5RN [IO91IQ, SU49]
G1 OQF C F Caines, 8 Abel Smith Garden, Branston, Lincoln, LN4 1NN [IO93SE, TF06]
G1 OQG D M Fryer, 44 Haig Rd, Aldershot, GU12 4PR [IO91PF, SU85]
G1 OQI C E Kill, 73 Aarons Hill, Godalming, GU7 2LH [IO91QE, SU94]
G1 OQK R W Collins, 27 Cavendish Rd, Chesham, HP5 1RW [IO91QQ, SP90]
G1 OQM M I Palmer, 250 Kinson Rd, East Howe, Bournemouth, BH10 5EP [IO90BS, SZ09]
G1 OQO M R Williamson, The Grange, Plumley Moor Rd, Plumley, Knutsford, WA16 9RS [IO83TG, SJ77]
GM1 OQT J W Watson, 3 Biggar Rd, Cleland, Motherwell, ML1 5PB [IO85BT, NS85]
G1 OQU A B Windsor, Buckskin, 28 Mendip Cl, Basingstoke, RG22 5BP [IO91KG, SU65]
G1 OQV J W Connor, 28 Church St., Hungerford, RG17 0JE [IO91FJ, SU36]
G1 OQW A J Boot, 63 Hunters Way, Stoke on Trent, ST4 5EF [IO82VX, SJ84]
G1 OQX M R Dunham, 5 King St., Wimblington, March, PE15 0QF [JO02BM, TL49]
G1 OQY R Farrow, Meadow Croft, Lady Drove, Downham Market, PE38 0AG [JO02DO, TF50]
G1 ORA M Ellis Oswestry&Dis AR, Eagle Communications, Unit E3 Bank Top Ind. Est., St. Martins, Oswestry, Salop, SY10 7BB [IO82LW, SJ33]
G1 ORB D R Tomsett, Porpoises, 58 East Front Rd, Pagham, Bognor Regis, PO21 4ST [IO90PS, SZ89]
G1 ORC G Oliver Oldham Am Rad C, 158 High Barn St., Royton, Oldham, OL2 6RW [IO83WN, SD90]
G1 ORD W J Roberts, 21 Denman Cl, Retford, DN22 7QG [IO93MH, SK68]
G1 ORG D J Stanley, 25 Kingsley Cres, Bulkington, Nuneaton, CV12 9PS [IO92GL, SP38]
G1 ORJ Details withheld at licensee's request by SSL.
G1 ORK D J Spicer, 35 Strood Rd, St. Leonards on Sea, TN37 6PN [JO00GU, TQ81]
G1 ORL C Barlow, 16 Fosseway South, Midsomer Norton, Bath, BA3 4AN [IO81SG, ST65]
G1 ORN F T Daniels, 6 Middlemead, Stratton on The Fosse, Bath, BA3 4QH [IO81SG, ST65]
GW1 ORP J H Marlow, West Bulthy Bulthy, Middletown, Welshpool, Powys, SY21 8ER [IO82LR, SJ31]
G1 ORS B M Williams, 3 Welton Cl, Wilmslow, SK9 6HD [IO83VH, SJ87]
G1 ORT S J Smale, 9 Tamworth Rd, Coventry, CV6 2JH [IO92FK, SP38]
G1 ORY Details withheld at licensee's request by SSL.
G1 OSA A J Sleigh, 2 Rock Terr, Bath Rd, Buxton, SK17 6HN [IO93BG, SK07]
G1 OSD J R Bradley, 64 Saffrondale, Anlaby, Hull, HU10 6QD [IO93SR, TA02]
G1 OSE R M Howlett, 29 Upper Park, Harlow, CM20 1TW [JO01BS, TL41]
G1 OSG W M Beilby, 119 Beaconsfield, Withernsea, HU19 2EW [JO03AR, TA32]
G1 OSH G E Slater, 12A Apsley Cl, Bishops Stortford, CM23 3PX [JO01BU, TL41]
G1 OSJ M W D Howard, 8 Abbotts Cres, St. Ives, Huntingdon, PE17 6YB [IO92XI, TL37]
G1 OSK P Alexander, 4 Standard Rd, Downe, Orpington, BR6 7HL [JO01AI, TQ46]
G1 OSL G T Hunter, 2 Dilloway St., Dentons Green, St. Helens, WA10 4LN [IO83OK, SJ59]
G1 OSM R B Newman, 35 Whitehall Walk, St. Neots, Huntingdon, PE19 2EE [IO92UF, TL15]
G1 OSN R Jaramillo, 27 Alterton Cl, Goldsworth Park, Woking, GU21 3DD [IO91QH, SU95]
G1 OSO A P Durbridge, 16 Nightingale Dr, Mytchett, Camberley, GU16 6BZ [IO91PG, SU85]
G1 OSP J C Woollons, 28 Columbus Ravine, Scarborough, YO12 7JT [IO94TG, TA08]
GW1 OSQ E O Evans, 6 Picton Cres, New Quay, SA45 9QB [IO72TF, SN35]
GM1 OSZ J J Smith, 41 Dickie Dr, Peterhead, AB42 1HB [IO97CM, NK14]
G1 OTA D Lee, 25 Elm View, Steeton, Keighley, BD20 6SZ [IO93AV, SE04]
G1 OTE A Copsey, 32 Coles Way, Riddlesden, Keighley, BD20 5DD [IO93BV, SE04]
GW1 OTI K Nickson, Robyn, 25 Burntwood Rd, Buckley, CH7 3EL [IO83LE, SJ26]
G1 OTJ N R Parr, 9 Manvers Rd, West Bridgford, Nottingham, NG2 6DJ [IO92KW, SK53]
G1 OTN R J Weight, 1 Crowland Rd, Thornton Heath, CR7 8RP [IO91WJ, TQ36]
G1 OTQ C E J Beardwell, 15 Silverdale, Stapleford, Nottingham, NG9 7EX [IO92IW, SK43]
G1 OTZ J G Macdonald, 42 Lion Ln, Haslemere, GU27 1JD [IO91PC, SU83]
G1 OUA H Saunders, 11 Abbey Dr, Luton, LU2 0LG [IO91TV, TL12]
G1 OUK W A Lynn, 54 Willesby Rd, Spalding, PE11 2AX [IO92WT, TF22]
G1 OUM B A T Woodage, 11 Reeves Way, Wokingham, RG41 2PS [IO91NJ, SU86]
GW1 OUP D E George, 24 Ty Fry Cl, Brynmenyn, Bridgend, CF32 8YB [IO81FN, SS98]
G1 OUR M D Batten, 5 Barton Rd, Berrow, Burnham on Sea, TA8 2LT [IO81LG, ST25]
G1 OUT C J Thorndyke, 23 Fordhams Cl, Stanton upon Hine Heath, Stanton, Bury St. Edmunds, IP31 2EE [JO02KH, TL97]
G1 OUX P Bolderson, 113 Kirkdale Cres, Leeds, West Yorks, LS12 6AY [IO93ES, SE23]
G1 OVA Details withheld at licensee's request by SSL.
G1 OVB J H Helm, 2 Alisan Rd, Carleton, Poulton-le-Fylde, FY6 7QF [IO83LU, SD33]
G1 OVG T R Powell, 11 Wymering Ln, Portsmouth, PO6 3QT [IO90FU, SU60]
G1 OVH Details withheld at licensee's request by SSL.

GM1 OVJ W Forsyth, Aldernaig, Avoch, Ross & Cromarty, Scotland, IV9 8QL [IO77VN, NH75]
G1 OVO R R Cox, 60 Prospect Cres, Whitton, Twickenham, TW2 7EA [IO91TK, TQ17]
G1 OVS M L Bullough, Nystad, Hague Bar Rd, New Mills, High Peak, Derbyshire, SK12 3EA [IO83XI, SJ98]
GM1 OVW R J Hetherington, 37 Brockwood Ave, Penicuik, EH26 9AN [IO85JT, NT26]
G1 OVY P McClelland, 7 Quayside, Little Neston, South Wirral, L64 0TB [IO83LG, SJ27]
G1 OVZ W G Hodgson, 8 Bolton House Rd, Bickershaw, Wigan, WN2 4AB [IO83RM, SD60]
G1 OWD M W J Forsyth, 13 Hillside Cl, Paulton, Bristol, BS18 5PN [IO81SH, ST65]
G1 OWI M L Branch, 38 Kynaston Rd, Didcot, OX11 8HD [IO91JO, SU58]
G1 OWJ A J Copsey, 13 Monro Ave, Crownhill, Milton Keynes, MK8 0BB [IO92OA, SP83]
G1 OWK G Garner, 8 Lansdowne Rd, Swadlincote, DE11 9DZ [IO92FS, SK21]
G1 OWM G R Woodley, 16 Albert St., St. Barnabas, Oxford, OX2 6AY [IO91IS, SP50]
G1 OWR J O Greene, 308 Cedar Rd, Camphill, Nuneaton, CV10 9DY [IO92GM, SP39]
GM1 OWV C D McBean, Kirkton Farm, Balblair, Dingwall, Ross-shire, IV7 8LG [IO77VQ, NH76]
G1 OWZ J T C Scott, 20 Foxton, Woughton Park, Milton Keynes, MK6 3AS [IO92PA, SP83]
G1 OXB D J M Owen, 6 The Garland, Leen Ct, Nottingham, NG7 2HR [IO92JW, SK53]
GM1 OXC C Greig, 12 Credon Dr, Airdrie, ML6 9RT [IO85AU, NS76]
G1 OXD Dr S W Anderson, 179 Rolleston Rd, Burton on Trent, DE13 0LD [IO92ET, SK22]
GM1 OXE A N Constance, 8 Ladeside Rd, Blackburn, Bathgate, EH47 7JW [IO85EV, NS96]
G1 OXF R G Lewis, 35 Lower Mead, Herne Farm, Petersfield, GU31 4NR [IO91MA, SU72]
GW1 OXJ I W Jones, 1 Tyn Llan, Dinas, Caernarvon, LL54 5UB [IO73UC, SH45]
G1 OXL R P Kitzmann, Copperfield, Main Rd, Arreton, Newport, PO30 3AL [IO90JQ, SZ58]
G1 OXM J G Powell, 7 Maori Dr, Frodsham, Warrington, WA6 7BS [IO83PH, SJ57]
G1 OXO J Ilston, 6 Dovedale, Canvey Island, SS8 8HX [JO01HM, TQ88]
GM1 OXQ I McKune, 16 Queensberry Ct, Dumfries, DG1 1BT [IO85EB, NX97]
G1 OXT R G Faulkner, 10 Fell Wilson St., Warsop, Mansfield, NG20 0PT [IO93KE, SK56]
G1 OXV D N J Neech, 208 Green End Rd, Cambridge, CB4 1RL [JO02BF, TL46]
G1 OXX Details withheld at licensee's request by SSL.
G1 OYF V B Shirley, 18 Crotch Cres, New Marston, Marston, Oxford, OX3 0JJ [IO91JS, SP50]
G1 OYG D M Crowe, 18 Bengairn Ave, Patcham, Brighton, BN1 8RH [IO90WU, TQ30]
G1 OYM A P Ingram, 78 Kenwood Gdns, Gants Hill, Ilford, IG2 6YG [JO01AN, TQ48]
G1 OYQ Details withheld at licensee's request by SSL.
G1 OYS Details withheld at licensee's request by SSL.
G1 OYT R H C Jelley, 30 Ashbury Cres, Merrow Park, Guildford, GU4 7HG [IO91RG, TQ05]
G1 OYU B Toon, 2 Marstonlane Park, Rolleston on Dove, Burton-on-Trent, Staffs, DE13 9BJ [IO92EU, SK22]
G1 OYW Details withheld at licensee's request by SSL.[Op: D G Baldwin. Station located near Godalming.]
G1 OYY Details withheld at licensee's request by SSL.
G1 OYZ P R Smith, 189 Rolleston Rd, Burton on Trent, DE13 0LD [IO92ET, SK22]
G1 OZB A W Saunders, Firtrees, Wheatcroft Ave, Bewdley, DY12 1DD [IO82UJ, SO77]
G1 OZD Dr J D Anderson, 179 Rolleston Rd, Burton on Trent, DE13 0LD [IO92ET, SK22]
G1 OZF Details withheld at licensee's request by SSL.
G1 OZO D K Fuller, The Doone, Fourteen Acre Ln, Three Oaks, Hastings, TN35 4NB [JO00HV, TQ81]
G1 OZP R Rook, 74 Stonegate, Hunmanby, Filey, YO14 0PU [IO94UE, TA17]
G1 OZR I G F Thomson, Morvarose, 143 Pottingfield Rd, Rye, TN31 7BW [JO00IW, TQ92]
G1 OZV M J Jolly, The Oaks, 6 Gwealhellis Warren, Helston, TR13 8PQ [IO70IC, SW62]
G1 OZZ Details withheld at licensee's request by SSL.
G1 PAA B A Oldbury, 4 Tithebarn Rd, Rugeley, WS15 2QW [IO92AS, SK01]
G1 PAB L N Creek, 382 Ripon Rd, Stevenage, SG1 4NQ [IO91VW, TL22]
G1 PAC Details withheld at licensee's request by SSL.
G1 PAD D A Reeves, Isla Blanca, 21 Falmouth Rd, Congleton, CW12 3BH [IO83VD, SJ86]
G1 PAF P A Foster, 67 Newdigate Rd, Harefield, Uxbridge, UB9 6EL [IO91SO, TQ09]
G1 PAG K W Tomkinson, 3 Heysham Cl, Weston Coyney, Stoke on Trent, ST3 6RG [IO82WX, SJ94]
G1 PAH P A Hughes, 13A St. Patricks Rd, Yeovil, BA21 3EX [IO80QW, ST51]
G1 PAJ J E Bugg, 37 Westmoreland Rd, Felixstowe, IP11 9TF [JO01QX, TM33]
G1 PAK C J Parker, 5 York Cl, Kings Langley, WD4 9HX [IO91SR, TL00]
G1 PAL A Armstrong, 3 Wardley Hall Rd, Wardley, Swinton, Manchester, M27 9QE [IO83TM, SD70]
G1 PAS H J Kelle, Rednend Farm, Station Rd, Pilning, Bristol, BS12 3JW [IO81QN, ST58]
G1 PAT P M Chapman, 24 Broad Ln, Moulton, Spalding, PE12 6PN [IO92XT, TF32]
G1 PAW P Wetherill, 16 Queensthorpe Rise, Leeds, LS13 4JR [IO93ET, SE23]
G1 PBB D J Smith, 71 Ashbourne Ave, New Springs, Aspull, Wigan, WN2 1HW [IO83QN, SD60]
G1 PBD R J Muddimer, 31 Lynmouth Dr, Wigston, LE18 1BP [IO92KO, SK50]
G1 PBF D W Pearce, 32 Marshall Rd, Willenhall, WV13 3PB [IO82XN, SO99]
G1 PBH Details withheld at licensee's request by SSL.
G1 PBU D R H Green, 39 Heritage Park, Hatch Warren, Hants, RG22 4XT [IO91KF, SU64]
G1 PBW Details withheld at licensee's request by SSL.
G1 PBX E L Churchill, 87 Bradley Cres, Shirehampton, Bristol, BS11 9SR [IO81PL, ST57]
G1 PBY A J Parrott, 54 Dockin Hill Rd, Doncaster, DN1 2QU [IO93KM, SE50]
G1 PCA R P Blandford, 7 Dryleaze, Keynsham, Bristol, BS18 2DA [IO81SK, ST66]
G1 PCD P C Dicken, 3 Mossfield, Cobham, KT11 1DF [IO91TI, TQ16]
G1 PCG J Dunwell, 8 Cochrane Cl, Thatcham, RG19 4QX [IO91IJ, SU56]
G1 PCH Details withheld at licensee's request by SSL.
G1 PCN E M M Benzie, 6 Priors Park, Emerson Valley, Milton Keynes, MK4 2BT [IO92OA, SP83]
G1 PCR Details withheld at licensee's request by SSL.
G1 PCS Details withheld at licensee's request by SSL.
G1 PCU C Mather, 5 Knolles Rd, Cowley, Oxford, OX4 3HT [IO91JR, SP50]
G1 PCW Details withheld at licensee's request by SSL.
G1 PCX Details withheld at licensee's request by SSL.
G1 PDA E I Evans, 14 Montbelle Rd, New Eltham, London, SE9 3PB [JO01AK, TQ47]
G1 PDC Details withheld at licensee's request by SSL.
G1 PDD A J Dale, 24 Nelson St., Wolstanton, Newcastle, ST5 8BW [IO83VA, SJ84]
G1 PDH B R Gummow, 24 Hendra Vean, Truro, TR1 3TU [IO70LG, SW84]
GM1 PDL A Whitehead, 24 Lochpark Pl, Denny, FK6 5AA [IO86BA, NS88]
G1 PDO R K Mitchell, 17 Church Ln Dr, Coulsdon, CR5 3RG [IO91WG, TQ25]
G1 PDS S Cliffe, 24 Dalehurst Rd, Bexhill on Sea, TN39 4BN [JO00FU, TQ70]
G1 PDY P J Cognet, 1 Little Heath, Astley, Stourport on Severn, Worcs, DY13 0RE [IO82UH, SO86]
G1 PED G J Solway, 31 Windermere Rd, Babbacombe, Torquay, TQ1 3RF [IO80FL, SX96]
G1 PEE S Jackson, 16 Wensley Dr, Lancaster, LA1 2JA [IO84OB, SD46]
G1 PEI L A Berridge, 33 Wesley Dr, Worle, Weston Super Mare, BS22 0TJ [IO81NI, ST36]
G1 PEK M G Sutton, 17 Barton Cl, Witchford, Ely, CB6 2HS [JO02CJ, TL47]
GM1 PEL B G Taynton, 32 Broomhall Rd, Edingburgh, EH12 7PD [IO85IW, NT17]
G1 PEN M M Dockray, 54 Kelsick Park, Seaton, Workington, CA14 1PY [IO84FP, NY03]
G1 PER M W Payne, 106 Blackbridge Ln, Needles Farm Est, Horsham, RH12 1SA [IO91TB, TQ13]
G1 PET J A Dowling, 38 Lyndsey Cl, Farnborough, GU14 9TG [IO91OH, SU85]
G1 PEU G P Gibbons, 43 Buckland Ave, Basingstoke, RG22 6JA [IO91KG, SU65]
G1 PEY E Parkes, 1 Silkstone View, Platts Common, Hoyland, Barnsley, S74 0QR [IO93GM, SE30]
GW1 PFK A F Romano, The Glen, Glen Rd, West Cross, Swansea, SA3 5QJ [IO71XN, SS68]
G1 PFP Details withheld at licensee's request by SSL.
G1 PFQ Details withheld at licensee's request by SSL.
GM1 PFU C G Hewlett, Viewforth, 24 Evelyn Terr, Perth, PH2 0BP [IO86GJ, NO12]
G1 PFY J P Gold, 94 Carbery Ave, Southbourne, Bournemouth, BH6 3LQ [IO90CR, SZ19]
G1 PFZ W Gill, Kittycroyd, Underwood Dr, Rawdon, Leeds, LS19 6LB [IO93DU, SE23]
G1 PGB J E Davies, 19 Elm Gdns, Lichfield, WS14 9AH [IO92CQ, SK10]
G1 PGF Details withheld at licensee's request by SSL.
G1 PGG Dr P M Forster, Gulland House, Atlantic Terr, New Polzeath, Wadebridge Cornwal, PL27 6UG [IO70NN, SW97]
G1 PGI C C Smith, 84 Woodcote Gr Rd, Coulsdon, CR5 2AD [IO91WH, TQ26]
G1 PGJ D H Castle, 8 Woodhall Ct, Welwyn Garden City, AL7 3TD [IO91VT, TL21]
GM1 PGL J Blackadder, 3 Auchincloch Dr, Banknock, Bonnybridge, FK4 1LA [IO85AX, NS77]
G1 PGN D E Brooks, 10 Elmtree Rd, Ruskington, Sleaford, NG34 9BT [IO93TB, TF05]
GM1 PGO G A P Bower, Gateside Cottage, Castlemains, Duns, Scotland, TD11 3TP [IO85TT, NT75]
GM1 PGQ R T Pennycook, 11 Highfields, Dunblane, FK15 9EQ [IO86AE, NN70]
G1 PGS D C Hands, 45 Croft Ave, West Wickham, BR4 0QH [IO91XJ, TQ36]
G1 PHA G W Day, 102 Meadlands Dr, Petersham, Richmond, TW10 7ED [IO91UK, TQ17]
G1 PHB R H Maudsley, 102 Meadow St., Preston, PR1 1TS [IO83PS, SD53]
GM1 PHD Dr N R Muir, 25 Drylaw House Gdns, Edinburgh, EH4 2UE [IO85JX, NT27]
GI1 PHF J F Campbell, 4 Woodford Gr, Newtownabbey, BT36 6TN [IO74AQ, J38]
G1 PHJ H R W Johnson, 27 Ridgeway Ave, Gravesend, DA12 5BD [JO01EK, TQ67]
G1 PHK R V Baines, 8 Richmond Ave, Sedgley Park, Prestwich, Manchester, M25 0NA [IO83UM, SD80]
G1 PHL Details withheld at licensee's request by SSL.
G1 PHN M L Morley, 7 Homemead Gr, Rubery, Rednal, Birmingham, B45 9RH [IO82XJ, SO97]
G1 PHQ Details withheld at licensee's request by SSL.
G1 PHS P H Street, 12 Ledston Ave, Garforth, Leeds, LS25 2BP [IO93HS, SE43]
G1 PHU B R Burton, Natson, Tedburn St Mary, Exeter, Devon, EX6 6ET [IO80DR, SX79]
G1 PHV C R Marsh, 85 Cromwell Cres, Market Harborough, LE16 9JW [IO92ML, SP78]
G1 PHZ N Silverthorne, 161 Newtown Rd, Malvern, WR14 1PJ [IO82UC, SO74]
G1 PIB M A Smith, 86 Slades Dr, Chislehurst, BR7 6JY [JO01AK, TQ47]
G1 PIC J P Parkinson, 3A Gleneagles Dr, Penwortham, Preston, PR1 0JT [IO83PS, SD52]
G1 PIF D F Rogers, 20 Chapel Cl, Acomb, Hexham, NE46 4RX [IO84WX, NY96]
GW1 PIH H B Owen, Llys Gwynedd, Bethel, Caernarfon, Gwynedd, LL55 1YB [IO73VD, SH56]
G1 PII G E Cooper, 39 Church Rd, Harlington, Dunstable, LU5 6LE [IO91SX, TL03]
G1 PIX R D Bibby, 40 Morval Cres, Runcorn, WA7 2QS [IO83PH, SJ58]
G1 PIY F Cholerton, Frans Cottage, 110 London Rd, Stockton Heath, Warrington, WA4 6LE [IO83RI, SJ68]
G1 PJC G W Siarey, 57 Egmont Rd, Sutton, SM2 5JR [IO91VI, TQ26]
G1 PJE D H Cant, Camp Ln, Edgehill, Warmington, Banbury, Oxon, OX17 1DH [IO92GD, SP34]
G1 PJH J Abram, 130 Larkfield Ln, Southport, PR9 8NP [IO83MQ, SD31]
G1 PJI S Scott, 48 Benthal Rd, London, N16 7DA [IO91XN, TQ38]
G1 PJJ P A Turvey, 165 Snargate St., Dover, CT17 9BZ [JO01PC, TR34]
G1 PJK J J Foster, 56 St. Matthews Ave, Litherland, Liverpool, L21 5JT [IO83ML, SJ39]
G1 PJL Dr D J Brookfield, 1 Holme Cottage, Leighton Rd, Neston, South Wirral, L64 3SQ [IO83LH, SJ27]
G1 PJM P J Mitchell, 11 Wingle Tye Rd, Burgess Hill, RH15 9HR [IO90WW, TQ31]
G1 PJO K Thompson, 216A Lincoln Rd, Peterborough, PE1 2NE [IO92VO, TF10]
GW1 PJP P J Probert, 7 Albany Rd, Blackwood, NP2 1DZ [IO81JP, ST19]
G1 PJQ Details withheld at licensee's request by SSL.
G1 PJT R E G Wood, Lynwood, Halley Rd, Broad Oak, Heathfield, TN21 8TG [JO00DX, TQ62]
G1 PJV N H Saunders, 9 Kestrel Wood Way, York, YO3 9EQ [IO93LX, SE65]
G1 PJX G H England, 17 Barnfield, Much Hoole, Preston, PR4 4GE [IO83OQ, SD42]
G1 PJZ J G Rogers, 55 York Rd, Driffield, YO25 7AY [IO94SA, TA05]
G1 PKA J Davey, 1 Bellamy Ave, Morecambe, LA4 4QS [IO84NB, SD46]
G1 PKG A M Stockton, 190 Sommerfield Rd, Woodgate Valley South, Birmingham, B32 3TA [IO92AK, SP08]
G1 PKM P K Miller, 81 Meadway, Woolavington, Bridgwater, TA7 8HA [IO81ME, ST34]
GM1 PKN D B Gillies, 10 Killeonan, Campbeltown, PA28 6PL [IO75EJ, NR61]
G1 PKO C J Jones, 52 The Dr, Bury, BL9 5DL [IO83UO, SD81]
G1 PKP M J Lloyd, 243 Stand Ln, Radcliffe, Manchester, M26 1JA [IO83UN, SD70]
G1 PKQ R Wright, 34 Park Rd, Stretford, Manchester, M32 8DQ [IO83UK, SJ79]
G1 PKR D J Bannister, 11 Keats Dr, Swadlincote, DE11 0DS [IO92FS, SK32]
G1 PKS V H Johnston, 14 Auckland Cl, London, SE19 2DA [IO91XJ, TQ36]
G1 PKU M D Watts, 242 Wellmeadow Rd, Catford, London, SE6 1HS [JO01AK, TQ37]
G1 PKV J A Sennitt, 44 Pear Tree Ave, Newhall, Swadlincote, DE11 0NB [IO92FS, SK22]
G1 PKW P C Kingsley-Williams, Back Rd, Llanarmon Yn Ial, Llanarmon Yn Ial, Mold, Clwyd, CH7 4QD [IO83JC, SJ15]
GW1 PKX A N Thrussell, 31 Bedlington Terr, Barry, CF62 7JA [IO81IJ, ST16]
GM1 PKY R J Deasington, Westhall, Milnathort, Kilnross-shire, KY13 7RP [IO86GG, NO00]
G1 PKZ A D Evans, 142 Oak Park Rd, Stourbridge, DY8 5YE [IO82WL, SO98]
G1 PLA M A Sleath, 1 Davos Way, Skegness, PE25 1EL [JO03ED, TF56]
G1 PLB P Robinson, 63 Droversdale Rd, Bircotes, Doncaster, South Yorks, DN11 8AZ [IO93LK, SK69]
G1 PLE C Griffith, 5 Park Cl, Yaxley, Peterborough, PE7 3JW [IO92UM, TL19]
GW1 PLJ J D Morgan, Holly Cottage, The Old Racecourse, Oswestry, SY10 7PQ [IO82KU, SJ23]
G1 PLP C Champ, 20 Lundy Cl, Broadoak, Broadfield, Crawley, RH11 9HF [IO91VC, TQ23]
G1 PLT P L Taylor, Brionys Patch, 13 Altona Rd, Loudwater, High Wycombe, HP10 9RW [IO91PO, SU99]
G1 PLU S A Goy, 352 Chanterlands Ave, Hull, HU5 4ED [IO93TS, TA03]
G1 PLV R P Cilia, 18 London Fields House, Kensington Rd, Crawley, RH11 9NS [IO91VC, TQ23]
GM1 PLY Details withheld at licensee's request by SSL.
G1 PMA D Harding, 37 Junction Cottage, Hardham, Pulborough, Sussex, RH20 1LA [IO90RW, TQ01]
G1 PMD I D Broom, 28 Shirburn Rd, Crownhill, Plymouth, PL6 5PG [IO70WJ, SX45]
G1 PMF P M Felton, 13 New St., Sudbury, CO10 6JB [JO02IA, TL84]
G1 PMG W F Kilburn, 3 Westroyd Ave, Pudsey, West Yorks, LS28 8JA [IO93DS, SE23]
G1 PMJ D C Loon, 18 Stourcliffe Rd, Wallasey, L44 3AF [IO83LK, SJ39]
G1 PMN Details withheld at licensee's request by SSL.
GW1 PMQ V M Davies, Coed Fryn, Lower Halkyn, Halkyn, Holywell, CH8 8ES [IO83JF, SJ27]
G1 PMR Details withheld at licensee's request by SSL.
G1 PNC T Collins, 6 Barnacre Ave, Newall Green, Wythenshawe, Manchester, M23 2TZ [IO83UJ, SJ88]
GW1 PND K P Hassall, 219B Robert St., Milford Haven, SA73 2HR [IO71LR, SM90]
G1 PNF R Broadberry, 9 Maple Ave, Fishponds, Bristol, BS16 4HJ [IO81RL, ST67]
G1 PNL D M Johnson, 27 Ridgeway Ave, Gravesend, DA12 5BD [JO01EK, TQ67]
GM1 PNP Details withheld at licensee's request by SSL.
G1 PNR Details withheld at licensee's request by SSL.
G1 PNU Details withheld at licensee's request by SSL.
G1 PNV Details withheld at licensee's request by SSL.
G1 PNX D Staples, 2 Bulcote Rd, Clifton, Nottingham, NG11 8FD [IO92JV, SK53]
GM1 POA J N Jamieson, 11 Binns Rd, Glasgow, G33 5HU [IO75WV, NS66]
G1 POC C J Elsom, 8 King Ave, Maltby, Rotherham, S66 7HX [IO93JK, SK59]
G1 POD E P Mitchell, 11 Wingle Tye Rd, Burgess Hill, RH15 9HR [IO90WW, TQ31]
G1 POG R J Letts, 12 Hawkshead Rd, Shaw, Oldham, OL2 7QY [IO83WO, SD90]
G1 POJ K S Phillips, Flat 3, Bridges Hall, White Knights Rd, Reading, RG6 6BG [IO91MK, SU77]
G1 POK D D Harris, 143 Collingwood Rd, Sutton, SM1 2QW [IO91VI, TQ26]
G1 POM E T Baker, 19 Ramsey Rd, Thornton Heath, CR7 6BX [IO91WJ, TQ36]
G1 POR A B Porter, 1125 Yardley Wood Rd, Warstock, Birmingham, B14 4LS [IO92BJ, SP07]
G1 POS J J Page, 3 Alexandra Rd, Hampton, Evesham, WR11 6QQ [IO92AC, SP04]
G1 POV R A Smith, 18 Hornby Ave, Westcliff on Sea, SS0 0LE [JO01IN, TQ88]
G1 PPB T P Roots, 11 Windermere Ave, Eastern Green, Coventry, CV5 7GP [IO92FJ, SP27]
G1 PPD A P Shons, 108 Southdown Rd, Catherington, Waterlooville, PO8 0NF [IO90LW, SU71]
G1 PPG G R F Joyner, 7 Blenkinsop Castle Home Park, Greenhead, Carlisle, CA6 7JS [IO84RX, NY66]
G1 PPH M A Hill, 18 Garendon Rd, Shepshed, Loughborough, LE12 9NX [IO92IS, SK41]
G1 PPK J H Knight, 183 Northd. Ave, Thornton Cleveleys, FY5 2JS [IO83LV, SD34]
G1 PPO A G Wade, 15 Almond Gr, Woodland Park, Scarborough, YO12 5QG [IO94SG, TA08]
G1 PPQ S Walker, 22 Ward Cl, Aylestone, Leicester, LE2 8NJ [IO92KO, SK50]
G1 PPU D R Walker, 11 Laburnum Ave, Lutterworth, LE17 4TZ [IO92JK, SP58]
G1 PPW P A J Elgar, 3 Bingham Dr, Laleham, Staines, TW18 1QX [IO91SK, TQ07]
G1 PPX G Nicholls, 2 Leybrook Croft, Hemsworth, Pontefract, WF9 4JA [IO93HO, SE41]
G1 PPZ A W Parton, 6 Wolverley Rd, Halesowen, B63 4LX [IO82XK, SO98]
GW1 PQE D Lewis, 7 High Terr, Holyhead, LL65 1SP [IO73QH, SH28]
G1 PQJ P J Wilson, 25 Danesbower Cl, Blofield, Norwich, NR13 4LR [JO02RP, TG30]
G1 PQK P Harrison, 9 Eleanor Gdns, Dagenham, RM8 3EJ [JO01BN, TQ48]
G1 PQO S J Stevens, The Catkins, 35 Charwood Rd, Wokingham, RG40 1RY [IO91OJ, SU86]
G1 PQT J R Mayes, 44 Foxwarren, Claygate, Esher, KT10 0JZ [IO91TI, TQ16]
G1 PQW K E Johnson, 20 Rolling Dales Cl, Maltby, Rotherham, S66 8EJ [IO93JK, SK59]
G1 PQX R F Moat, 27 Pioneer Rd, Dover, CT16 2AR [JO01PD, TR34]
G1 PQY G D Jones, 25 Myvod Rd, Wednesbury, WS10 9BT [IO82XN, SO99]
G1 PRA Details withheld at licensee's request by SSL.
G1 PRB Details withheld at licensee's request by SSL.
G1 PRE K J Moore, Silver Seas, Trerieve Est, Downderry, Torpoint, PL11 3LZ [IO70TI, SX35]
G1 PRL R K Williams, 54 Windways, Little Sutton, South Wirral, L66 1JF [IO83MG, SJ37]
G1 PRM A Corduroy, 60 Trowley Hill Rd, Flamstead, St. Albans, AL3 8EE [IO91ST, TL01]
G1 PRP A T Moore, Wyndrush, Northend Ln, Droxford, Southampton, SO32 3QN [IO90KX, SU61]
G1 PRS C E Ladley, 25 Laburnum Cres, Louth, LN11 8SG [JO03AI, TF38]
G1 PRW A W Swift, 29 Eldon St., Winskill, Burton on Trent, DE15 0LU [IO92ET, SK22]
G1 PRX P F Scovell, Four Winds, Lessland Ln, Sandford, Ventnor, PO38 3AR [IO90JO, SZ58]
G1 PRZ B L Saich, 65 Orchard Rise West, Sidcup, DA15 8TA [JO01BK, TQ47]
G1 PSJ M Crosby, 11 Loxley Ave, Shirley, Solihull, B90 2QE [IO92BJ, SP17]
G1 PSL C J Sparrow, 2 Gate St., Rochdale, OL11 1PN [IO83WO, SD81]
G1 PSS A L Laszkiewicz, 18 Wickham Way, Haywards Heath, RH16 1UQ [IO91WA, TQ32]
GM1 PST P Stanhope, The Roundal, Alva House, Alva, Clackmannanshire, FK12 5HU [IO86CD, NS99]
GM1 PSU I A Manson, 85 Bishops Park, Mid Calder, Livingston, EH53 0ST [IO85GV, NT06]
GM1 PSZ D W Liddle, 9 Rullion Rd, Penicuik, EH26 9HS [IO85JT, NT26]
G1 PTH A K Roberts, 60 Windsor Rd, Ipswich, IP1 4AN [JO02NB, TM14]
G1 PTO M K Brickwood, 7 The Fairway, Camberley, GU15 1EF [IO91PH, SU85]
G1 PTS Details withheld at licensee's request by SSL.
G1 PTT R A Neilson, 1 Bowness Ave, Sompting, Lancing, BN15 9TS [IO90TT, TQ10]
G1 PTY S M Atkinson, 17 High St., Great Hale, Sleaford, NG34 9LE [IO92UX, TF14]
G1 PTZ S N Atkinson, Brook Villa, Hinders Ln, Huntley, Gloucester, GL19 3EZ [IO81SU, SO71]
G1 PUC L R Burkett, 4 Selwyn Cres, Radley, Abingdon, OX14 3AW [IO91JQ, SU59]
GI1 PUM D S Benson, 121 Dungannon Rd, Derrykeevan, Portadown, Craigavon, BT62 1UG [IO64RL, H95]
G1 PUO D F West, 30 Farm Ave, Swanley, BR8 7JA [JO01BJ, TQ56]
G1 PUQ C J Tripp, 76 Chapel Ln, Fowlmere, Royston, SG8 7SD [JO02AC, TL44]
G1 PUV C R Wiseman, Whiterock, 14 Whiteridge Rd, Kidsgrove, Stoke on Trent, ST7 4TH [IO83VC, SJ85]
G1 PUY J M de Renzi, Wharfdale House, 4 Pinfold Cl, Bickerton, Wetherby, LS22 5JW [IO93IW, SE45]
G1 PUZ S Box, 103 Ilkeston Rd, Bramcote, Beeston, Nottingham, NG9 3JT [IO92IW, SK53]
GM1 PVD H A G Murray, 29F Hayfield, Edinburgh, EH12 8UJ [IO85IW, NT17]
G1 PVJ H W Harwood, 123 Northgate St., Great Yarmouth, NR30 1BP [JO02UO, TG50]
GI1 PVO Details withheld at licensee's request by SSL.[Op: Dr W I Campbell. Station located near Newtownards.]
G1 PVQ J R Johnson, 16 The Grove, Abingdon, OX14 2DQ [IO91IQ, SU59]
G1 PVS W Macleod, 129 Lichfield Rd, Dagenham, RM8 2AX [JO01BN, TQ48]
G1 PVT D M Broad, 14 Albion Rd, Westcliff on Sea, SS0 7DR [JO01IN, TQ88]
G1 PVU J S Allen, 23 Beech Rd, St, BA16 0RY [IO81PC, ST43]

G1 PVW M J Baxter, 23 Holly Rd, Wednesbury, WS10 9NX [IO82XN, SO99]
G1 PVY Details withheld at licensee's request by SSL.
G1 PVZ J A Vincent, Brookfield, North St., Crewkerne, TA18 7AW [IO80OV, ST41]
G1 PWD A M Moss, Rudland House, Hill Rd, Kirkby in Cleveland, Middlesbrough, TS9 7AN [IO94JK, NZ50]
G1 PWF D C King, 79 Wootton Dr, Hemel Hempstead, HP2 6LA [IO91SS, TL00]
G1 PWH P Walters S Norm Alfreton, Alfreton & District Arcade, 6 Victoria St., Alfreton, DE55 7GS [IO93HC, SK45]
G1 PWI S I Thurman, 11 Salisbury Terr, Akroydon, Halifax, HX3 6NB [IO93BR, SE02]
G1 PWJ R W Gascoyne, 2 Torpoint Cl, Wyken, Coventry, CV2 3LY [IO92GK, SP38]
GM1 PWL K C Robertson, Kilfinan View, Brenfield, Tarbert Rd, Ardrishaig, Argyll, PA30 8ER [IO75GX, NR88]
G1 PWM J Bulman, 3 South View, Littlethorpe, Ripon, HG4 3LL [IO94FC, SE36]
GM1 PWS R J Manson, Omahanui, Goddard Rd, Tasman Rd1, Upper Moutere 7152, New Zealand
G1 PWU D S Dwyer, 24 Alder Way, Melksham, SN12 6UL [IO81WI, ST96]
G1 PWW M J Reynolds, 110 Bilborough Rd, Bilborough Est, Nottingham, NG8 4DN [IO92JX, SK54]
G1 PWY P W Gardner, 31 Kingsway, Heysham, Morecambe, LA3 2EB [IO84NB, SD46]
G1 PXE W E J Jones, 16 Park Ave, Woodborough, Nottingham, NG14 6EB [IO93LA, SK64]
G1 PXG G C Kirkpatrick, 23 Hornhatch, East Shalford Ln, Chilworth, Guildford, GU4 8AY [IO91RF, TQ04]
G1 PXH C E Daily, 1 Railway Cttgs, Wendens Ambo, Saffron Walden, CB11 4LA [JO02CA, TL53]
G1 PXL R D Wood, 12 Warwick Rd, Ashford, TW15 3PG [IO91SK, TQ07]
G1 PXM Dr R Blakeway, Ashdene, 152 Shawfield Rd, Ash, Aldershot, GU12 6SG [IO91PG, SU85]
G1 PXQ A Scivetti, 10 Rippleside, Basildon, SS14 1UA [JO01FN, TQ78]
G1 PXW P M Hollands, 16 Hazel Cres, Thornbury, Bristol, BS12 1BX [IO81RO, ST69]
GI1 PXX S A Argue, 2 Canvy Manor, Portadown, Craigavon, BT63 5LP [IO64TJ, J05]
G1 PYA Details withheld at licensee's request by SSL.
G1 PYC S C Johnson, 115 Manchester Rd, Leigh, WN7 2LE [IO83SL, SD60]
G1 PYJ P G T Powell, 10 Selsdon Cl, Wythall, Birmingham, B47 6HP [IO92BJ, SP07]
G1 PYL M W Cammell, 10 Quarr Rd, Carshalton, SM5 1ER [IO91VJ, TQ26]
G1 PYQ K E P Rowe, 15 Woodward Rd, Pershore, WR10 1LW [IO82XC, SO94]
G1 PYR B Johnston, 45 Westcombe Ct, Westcombe Park Rd, Blackheath, London, SE3 7QB [JO01AL, TQ37]
G1 PYU R Green, 41 Keswick Rd, Heaton Chapel, Stockport, SK4 5JU [IO83VK, SJ89]
G1 PZA W J Grech-Cini, 13 Manor Garth Rd, Kippax, Leeds, LS25 7PD [IO93HS, SE43]
G1 PZB H J Grech-Cini, 13 Manor Garth Rd, Kippax, Leeds, LS25 7PD [IO93HS, SE43]
G1 PZD M Pattison, 13 Mixes Hill Rd, Stopsley, Luton, LU2 7TX [IO91TV, TL02]
G1 PZF Details withheld at licensee's request by SSL.
G1 PZH D C E Williams, 203 High St., Shirley, Solihull, B90 1JN [IO92BJ, SP07]
G1 PZK H G Kraus, 263 St. Michaels Ave, Yeovil, BA21 4NA [IO80QW, ST51]
G1 PZP C A Collett, Yaffle, 26 Hertford Rd, Hoddesdon, EN11 9JR [IO91XS, TL30]
GM1 PZT G Hardacre, 242 Sutherland Way, Knightsbridge, Livingston, EH54 8JB [IO85FV, NT06]
G1 PZX Details withheld at licensee's request by SSL.
GW1 PZZ F W Haines, 52 Pantydwr, Three Crosses, Swansea, SA4 3PG [IO71XP, SS59]
G1 RAB T E Hunt, 76 St. Pancras Ave, Plymouth, PL2 3RY [IO70WJ, SX45]
G1 RAD Details withheld at licensee's request by SSL.
G1 RAE Details withheld at licensee's request by SSL.
G1 RAG J C V Jones, 10 Huntington Cl, Redditch, B98 0NF [IO92BH, SP06]
G1 RAH Details withheld at licensee's request by SSL.
G1 RAM C A G Adams, 2 Ash Gr, Knutsford, WA16 8BB [IO83TH, SJ77]
G1 RAO K Hughes, 20 Pickering Cl, Bury, BL8 1UE [IO83UO, SD71]
G1 RAP R A Prosser, 27 Dorset Gdns, Rochford, SS4 3AH [JO01IO, TQ89]
G1 RAS P Owen Ras Club, 3 Farndon Green, Wollaton Park, Nottingham, NG8 1DU [IO92JW, SK53]
G1 RAX D F Bendall, Brambles, 17 Berryfield Rd, Hordle, Lymington, SO41 0HQ [IO90ES, SZ29]
G1 RAY R F Burton, 108 Wolverton Rd, Rednal, Birmingham, B45 8RN [IO92AJ, SP07]
G1 RBA A R Wheatley, 3 Woodsbank Terr, Wednesbury, WS10 7RQ [IO82XN, SO99]
GI1 RBI W H A McKeown, 62 Derrychara Dr, Enniskillen, BT74 6JH [IO64EI, H24]
G1 RBJ Details withheld at licensee's request by SSL.
G1 RBK Details withheld at licensee's request by SSL.
GM1 RBM Details withheld at licensee's request by SSL.
G1 RBX R K Baker, 3 Hazelton Cl, Solihull, B91 3GA [IO92CJ, SP17]
G1 RBY P J Hurp, 55 Brooklyn Gr, Cosely, Bilston, WV14 8YH [IO82XM, SO99]
G1 RBZ G A Norris, 26 Westwood Rd, Leyland, Preston, PR5 2NS [IO83PQ, SD52]
GW1 RCC R Cardwell Clwyd County RA, 23 Russell Ave, Colwyn Bay, LL29 7TR [IO83DG, SH87]
G1 RCD D T Bolt Dartmoor Rad Cl, c/o Old School Hse, Maristow Roborough, Plymouth, PL6 7BY [IO70WK, SX46]
G1 RCE A J R Hills, 1 Burwood Ave, Kenley, CR8 5NT [IO91WH, TQ36]
G1 RCI P L Mason, 34 Central Park Ave, Wallasey, L44 0AQ [IO83LK, SJ39]
G1 RCN P T Wilson, 146 Wilkinson St., Nottingham, NG8 5FJ [IO92JX, SK54]
GM1 RCP P Easton, Machrihanish, 45 Whiteloch Rd, Macmerry, Tranent, EH33 1PF [IO85NW, NT47]
G1 RCT Details withheld at licensee's request by SSL.
G1 RCV A R Burchmore Cray Valley ARS, obo Cray Valley Radio Society, 49 School Ln, Horton Kirby, Dartford, DA4 9DQ [JO01DJ, TQ56]
G1 RCW J Chetwynd, 35 Cordelia Cl, Dibden, Southampton, SO45 5UD [IO90GU, SU40]
G1 RCX P Tweney, 42 Spareacre Ln, Eynsham, Witney, OX8 1NP [IO91HS, SP40]
GM1 RDG J J Horsburgh, Don Villa, 66 Harlaw Rd, Inverurie, AB51 4TB [IO87TH, NJ72]
G1 RDI R J Thomas, 44 Valentine Ave, Bexley, DA5 3HE [JO01BK, TQ47]
G1 RDJ Details withheld at licensee's request by SSL.
G1 RDU P I E Fanning, 2 Waters Edge, Brighton Rd, Lancing, BN15 8LN [IO90UT, TQ10]
G1 RDX S T Newbold, 14 Robins Cl, Cressex, High Wycombe, HP12 4NY [IO91OO, SU89]
G1 REG Details withheld at licensee's request by SSL.
G1 REH D L Connolly, 87 Regent on The River, William Morris Way, London, SW6 2UU [IO91VL, TQ27]
G1 REL G P Hawker, 46 Southfield Dr, North Ferriby, HU14 3DX [IO93RR, SE92]
G1 REO J T Telford, 85 Medway, Great Lumley, Chester-le-Street, DH3 4HU [IO94FU, NZ24]
G1 REQ Details withheld at licensee's request by SSL.
G1 RET Details withheld at licensee's request by SSL.
G1 REU V Knight, 13 Tavistock Cl, Sittingbourne, ME10 1JY [JO01II, TQ86]
GM1 REY R Caine, 8 Church St., Eyemouth, TD14 5DH [IO85WU, NT96]
GM1 REZ N P C Caine, 8 Church St., Eyemouth, TD14 5DH [IO85WU, NT96]
G1 RFB R D Palmer, 75 Upper Yarborough Rd, East Cowes, PO32 6EE [IO90IS, SZ59]
G1 RFC R Armitage, 8 Northolmby St., Howden, Goole, DN14 7JL [IO93NR, SE72]
G1 RFH J W Slater Forest Heath Raynet, 47 Broom Rd, Lakenheath, Brandon, IP27 9EZ [JO02GJ, TL78]
G1 RFO D L Towse, 62 Bramley Rd, Broadwater, Worthing, BN14 9DS [IO90TT, TQ10]
G1 RFQ D J Jenks, 36 West Dene, Gaddesden Row, Hemel Hempstead, Herts, HP2 6HU [IO91ST, TL01]
G1 RFS T A W Niner, 281 Nightingale Rd, Edmonton, London, N9 8QL [IO91XP, TQ39]
G1 RFX K W Tonner, Bank Cottage, 6 Lower Wyche Rd, Malvern, WR14 4ET [IO82TC, SO74]
G1 RGG Details withheld at licensee's request by SSL.
G1 RGJ J M North, 13 Weaverthorpe Rd, Woodthorpe, Nottingham, NG5 4ND [IO92KX, SK54]
G1 RGK M J North, 19 Georgia Dr, Arnold, Nottingham, NG5 8HX [IO93KA, SK54]
GM1 RGM D Keddie, 26 Daleally Cres, Errol, Perth, PH2 7QA [IO86JJ, NO22]
G1 RGT G C Williams, 76 Eastern Ave, Pinner, HA5 1NJ [IO91TN, TQ18]
G1 RGV I A Huggins, 394 Beccles Rd, Carlton Colville, Lowestoft, NR33 8HN [JO02UK, TM59]
GM1 RHA K W McCormick, 179 Springfield Park, Johnstone, PA5 8JT [IO75RU, NS46]
G1 RHB K E Sears, 43 Collyer Rd, Calverton, Nottingham, NG14 6ND [IO93KA, SK64]
G1 RHE W W Berry, The Bungalow, Basil Rd, West Dereham, Kings Lynn, Norfolk, PE33 9RP [JO02FN, TF60]
G1 RHH D C McKay, 42 Wingfield Dr, Chaddesden, Derby, DE21 4PW [IO92GW, SK33]
GM1 RHL Details withheld at licensee's request by SSL.
G1 RHM Details withheld at licensee's request by SSL.
G1 RHO A Emmerson, 19 Whitley Rd, Benton Sq, Newcastle upon Tyne, NE12 9SU [IO95FA, NZ37]
GW1 RHQ J Drake, 8D Merton Pl, Pwllycrohan Ave, Colwyn Bay, Clwyd, LL29 7BU [IO83DH, SH87]
GD1 RHT E J Moore, 21 Slieau Whallian Park, St. Johns, Douglas, IM4 3JH
G1 RHV Details withheld at licensee's request by SSL.[Michael, Callendar. QRP only.]
G1 RHW T Ibbitson, 36 Knoll Park, East Ardsley, Wakefield, WF3 2AX [IO93FR, SE32]
GM1 RHX C A Bewley, Gorrenberry Farm, Hermitage Water, nr Hawick, Scotland, TD9 0LT [IO85OG, NY49]
G1 RIE M Paterson, 10 Pendeford Sm-Hol, Barnhurst Ln, Bilbrook Codsall, Wolverhampton, WV8 1RS [IO82WO, SJ80]
GM1 RIG I J McGowan, 144 Alloa Rd, Stenhousemuir, Larbert, FK5 4HQ [IO86CA, NS88]
G1 RIH W H Hough, 225 Fleetwood Rd, Thornton Cleveleys, FY5 1RA [IO83LU, SD34]
GM1 RII F H W Baylis, Tarn Hows, Kilchrenan, Taynuilt, Argyll, PA35 1HF [IO76JI, NN02]
G1 RIJ Details withheld at licensee's request by SSL.
GW1 RIK Details withheld at licensee's request by SSL.
G1 RIP F I Derry, 43 Biddle Rd, Littlethorpe, Leicester, LE9 5HE [IO92JN, SP59]
G1 RIR L F J Shears, 7 Lower Furlongs, Brading, Sandown, PO36 0DX [IO90KQ, SZ68]
G1 RIV T J Dyson, 27 Throstle Nest, Batley, WF17 7SN [IO93ER, SE22]
G1 RIY M A Gray, 3 Central Sq, Waterhouses, Stoke on Trent, ST10 3HP [IO93BB, SK05]
G1 RJA M S Johnson, 28 Bittles Green, Motcombe, Shaftesbury, SP7 9NX [IO81VA, ST82]
G1 RJD W A Blower, 129 Kingsway, Kirkby in Ashfield, Nottingham, NG17 7FH [IO93JC, SK55]
G1 RJE P Allen, 6 Helston Cl, Wigston, LE18 2JH [IO92KN, SP69]
G1 RJG G Owen Smith, 23 Cherry Brook Dr, Paignton, TQ4 7LZ [IO80FJ, SX85]
G1 RJN S J Hall, Eastwick Rd, Great Bookham, Bookham, Leatherhead, Surrey, KT23 4BA [IO91TG, TQ15]
GM1 RJS J W Hambrook, 32 Blackdales Ave, Largs, KA30 8HU [IO75NS, NS25]
GW1 RJU G A Parsons, 22 Elm Park, Crundale, Haverfordwest, SA62 4DN [IO71MT, SM91]
G1 RJW R J Wicks, 32 Shelley Cl, Northcourt, Abingdon, OX14 1PR [IO91IQ, SU59]
G1 RKD B Allport, 1 Percy Dr, Swarland, Morpeth, NE65 9JN [IO95DH, NU10]
G1 RKF Details withheld at licensee's request by SSL.
GM1 RKI A Morris, 32 Biggiesknowe, Peebles, EH45 8HS [IO85JP, NT24]
G1 RKM D G Roney, 5 Highcroft, Woolavington, Bridgwater, TA7 8EU [IO81ME, ST34]
G1 RKP D Seal, 19 Thirlmere Way, Rawtenstall, Rossendale, BB4 8QE [IO83UR, SD82]
G1 RKR R J Luker, 369 Stafford Rd, Caterham, CR3 6NP [IO91XH, TQ35]
G1 RLB R Lawson, 24 Newby Gr, Beacons End, Birmingham, B37 6QR [IO92DL, SP18]
G1 RLD D G R Beeton, 34 Midsummer Meadow, Inkberrow, Worcester, WR7 4HD [IO92AF, SP05]
G1 RLF R I M Walter, 10 Birch Meadow, Clehonger, Hereford, HR2 9RH [IO82OA, SO43]
G1 RLI P A Webb, 40 Links Rd, Penn, Wolverhampton, WV4 5RF [IO82WN, SO99]
G1 RLK I S Bradshaw, 35 Stanley Rd, Morecambe, LA3 1UR [IO84NB, SD46]
G1 RLR P A G Pedley, 24 Appledore Rd, Orchard Hills, Walsall, WS5 3DT [IO92AN, SP09]
G1 RLT P J Vipond, 16 Vale St., Sunderland, SR4 7NB [IO94HV, NZ35]
G1 RLZ J G Krom, 20 Eason Dr, Abingdon, OX14 3YD [IO91JQ, SU59]
G1 RMC I R Jackson SW London Rayne, 5 Vivien Cl, Chessington, KT9 2DE [IO91UI, TQ16]
G1 RMH Details withheld at licensee's request by SSL.
G1 RMN M J Richards, 20 Tas Combe Way, Willingdon, Eastbourne, BN20 9JA [JO00DT, TQ50]
G1 RMW P J Danagher, 6 Brou Cl, East Preston, Littlehampton, BN16 1DB [IO90ST, TQ00]
G1 RNA Details withheld at licensee's request by SSL.
G1 RND R W Marden, 9 Meadow View, Goldsithney, Penzance, TR20 9HB [IO70GC, SW53]
G1 RNH Details withheld at licensee's request by SSL.
G1 RNL M P Kinsella, 93 Locket Rd, Harrow, HA3 7NP [IO91UO, TQ19]
G1 RNP M J Saunders, 63 Ferry Rd, Rye, TN31 7DJ [JO00IW, TQ92]
G1 RNU G M McCullough, 35 Ringlow Park Rd, Swinton, Manchester, M27 0HA [IO83TM, SD70]
G1 RNV B E Purse, 28 Holford Rd, Guildford, GU1 2QF [IO91RF, TQ05]
G1 RNZ G E Saville, 4 Shannon Ct, Downs Barn, Milton Keynes, MK14 7PP [IO92PB, SP84]
G1 ROD I A Lupton, 19 Ave Cl, Starbeck, Harrogate, HG2 7LJ [IO94GA, SE35]
G1 ROH D C Gower, 68 Wood Common, Hatfield, AL10 0UB [IO91VS, TL20]
G1 ROK P G Court, 29 Chaffinch Rd, Beckenham, BR3 4LT [IO91XJ, TQ36]
G1 ROL S O'Connor, 37 Acre Moss Ln, Morecambe, Lancs, LA4 4NA [IO84NB, SD46]
G1 ROM D A O'Connor, 37 Acre Moss Ln, Morecambe, Lancs, LA4 4NA [IO84NB, SD46]
G1 RON F R Donnachie, 60 Leeds Rd, Kippax, Leeds, LS25 7HQ [IO93HS, SE43]
G1 ROO E R East, 70 Woodcote Gr Rd, Coulsdon, CR5 2AD [IO91WH, TQ26]
G1 ROX A K Donald, Three Greens, Primrose Ln, Stockbury, Sittingbourne, Kent, ME9 7QR [JO01HH, TQ86]
G1 RPA R F Miller, 1 Hamelsham Ct, Hailsham, BN27 3EL [JO00DU, TQ50]
G1 RPG A T A Oakes, Manor House 'C', Hunts Common, Hartley Wintney, Hook, Hants, RG27 8AA [IO91NH, SU75]
GM1 RPJ D G Norrie, 14 Strathmore Ave, Kirriemuir, DD8 4DJ [IO86MQ, NO35]
G1 RPN Details withheld at licensee's request by SSL.
G1 RPP C G Broadbent, 8 Leafe Cl, Chilwell, Beeston, Nottingham, NG9 6NR [IO92IV, SK43]
G1 RPQ P C Cervini, 342 Linnet Dr, Chelmsford, CM2 8AL [JO01FR, TL70]
G1 RPR G J Palastanga, Corner House, Commister Ln, Ixworth, Bury St. Edmunds, IP31 2HE [JO02JH, TL97]
G1 RPT M J Tribe, 11 Heathlands Cl, Crossways, Dorchester, DT2 8TS [IO80UQ, SY78]
G1 RPU Details withheld at licensee's request by SSL.
G1 RPV R D Hardiman, 27 Staithe Rd, Martham, Great Yarmouth, NR29 4PT [JO02TR, TG41]
G1 RPX L Bland, 20 Castleton Ave, Middlesbrough, TS5 5JA [IO94IN, NZ41]
G1 RPY I A Kerr, 21 Lingdale, Cheveley Park Estat, Belmont, Durham, DH1 2AN [IO94FS, NZ34]
GM1 RQD D S Marwick, 17 Laverock Rd, Kirkwall, KW15 1EE [IO88MX, HY41]
GW1 RQF I E Davies, Benallt, 11 Stryd Fawr, Penygroes, Caernarfon, Gwynedd, LL54 6PL [IO73UB, SH45]
G1 RQH J B Pollard, 12 Little Twitten, Cooden, Bexhill on Sea, TN39 4SS [JO00FU, TQ70]
G1 RQI P R Quirk, 24 Walton Gdns, Folkestone, CT19 5PR [JO01OC, TR23]
G1 RQM M F C Aquilina, 2 Yardley, Letchworth, SG6 2SR [IO91VX, TL23]
G1 RQZ C R Selwyn-Smith, Miranda, Ash Island, East Molesey, Surrey, KT8 9AX [IO91TJ, TQ16]
G1 RRE J A Eckersley, 88 New Heys Way, Bradshaw, Bolton, BL2 4AQ [IO83TO, SD71]
G1 RRG H S Johnstone, 16 Riverside Cres, Otley, LS21 2RS [IO93DV, SE24]
GM1 RRJ B Elliott, 195 Braehead Rd, Cumbernauld, Glasgow, G67 2BL [IO85AW, NS77]
G1 RRM Details withheld at licensee's request by SSL.
G1 RRR K Bareham, 30 Orchard Rd, Horsham, RH13 5NF [IO91UA, TQ22]
G1 RRS Details withheld at licensee's request by SSL.
G1 RRW A M Henderson, The Holt, Ditchling Common, Ditchling, Hassocks, BN6 8SG [IO90WW, TQ31]
G1 RRX M L Fletcher, Country House, Main Rd, Harlaston, Tamworth, B79 9HT [IO92DQ, SK21]
G1 RSB Details withheld at licensee's request by SSL.
G1 RSE J Rod, 28 Haywards Ave, Weymouth, DT3 5JU [IO80SP, SY68]
G1 RSF A A Dean, 2 Maggs Mead, Shalford Ln, Charlton Musgrove, Wincanton, BA9 8HF [IO81TB, ST73]
G1 RSH Details withheld at licensee's request by SSL.
G1 RSK C K Broughton, 78 Peaksfield Ave, Grimsby, DN32 9QG [IO93XN, TA20]
G1 RSL Details withheld at licensee's request by SSL.
G1 RSO W E R Eastick, 7 Hemmant Way, Gillingham, Beccles, NR34 0LF [JO02SL, TM49]
G1 RSP H Allan, 14 South Rd, Prenton, Birkenhead, L42 7JW [IO83LJ, SJ38]
GI1 RSR C J Fogarty, 30 Moeran Park, Portadown, Craigavon, BT62 3QN [IO64SJ, H95]
G1 RSZ Details withheld at licensee's request by SSL.
G1 RTB Details withheld at licensee's request by SSL.
G1 RTD Details withheld at licensee's request by SSL.
G1 RTF G A Osler, 4 Mill Rd, Waterbeach, Cambridge, CB5 9RQ [JO02CG, TL46]
G1 RTG Details withheld at licensee's request by SSL.
G1 RTJ Details withheld at licensee's request by SSL.
G1 RTS Details withheld at licensee's request by SSL.
G1 RTV A Bottrill, 19 Durnford Way, Cambridge, CB4 2DP [JO02BF, TL46]
G1 RTX I Poole, 8 Bates Cl, Higham Ferrers, Rushden, NN1 8HF [IO92NF, SP76]
G1 RTY G R Stacey, 6 Highfield Rd, Petersfield, GU32 2HN [IO91MA, SU72]
G1 RUG A A Kay, Pear Tree Cottage, Hale House Ln, Churt, Farnham, GU10 2JG [IO91OD, SU83]
G1 RUR R F Hudson, 5 Loman Rd, Mytchett, Camberley, GU16 6BS [IO91PG, SU85]
G1 RUV Details withheld at licensee's request by SSL.
G1 RUY T Campbell, 5 Farne Rd, Shiremoor, Newcastle upon Tyne, NE27 0PQ [IO95FA, NZ37]
G1 RUZ Dr E J H Byrne, 25 South Rd, Aigburth, Grassendale Park, Liverpool, L19 0LS [IO83NI, SJ38]
G1 RVA J K Taylor, 1 Wildmoor Rd, Shirley, Solihull, B90 3PT [IO92CK, SP18]
G1 RVE Details withheld at licensee's request by SSL.
G1 RVF B Dempster, 43 Nest Farm Cres, Wellingborough, NN8 4TQ [IO92PH, SP86]
G1 RVH C Stancer, 20 Overton Ave, Willerby, Hull, HU10 6AR [IO93SS, TA03]
G1 RVK I N Brooks, 16 Buttercup Ct, Deeping St. James, Peterborough, PE6 8TF [IO92UQ, TF11]
G1 RVP P F Scott, 30 The Junipers, Barkham, Wokingham, RG41 4UX [IO91NJ, SU76]
G1 RVS R V Souter, 4 Risby Garth, Skidby, Cottingham, HU16 5UE [IO93SS, TA03]
G1 RVT B C Kavanagh, 110 Top Rd, Calow, Chesterfield, S44 5SY [IO93HF, SK47]
G1 RVV D A Fyles, 3 Chapel Ln, Lathom, Burscough, Ormskirk, L40 7RA [IO83NO, SD41]
GI1 RWD J Watson, 3 St. Judes Ave, Belfast, BT7 2GZ [IO74BN, J37]
G1 RWI G Hawkins, 34 Acacia Dr, Lower Pilsley, Chesterfield, S45 8DY [IO93HD, SK46]
G1 RWK Details withheld at licensee's request by SSL.
G1 RWR K A Lee, 33 North Rd, Eastdene, Rotherham, S65 2RR [IO93IK, SK49]
G1 RWS R W Smith, 30 Woodland Ave, Kidderminster, DY11 5AW [IO82UJ, SO87]
G1 RWT K V Whitton, 11 Dursley Rd, Shirehampton, Bristol, BS11 9XB [IO81PL, ST57]
G1 RWU Details withheld at licensee's request by SSL.
G1 RWX K A Ikin, 15 Broadway, Farnworth, Bolton, BL4 0HQ [IO83SN, SD70]
G1 RXB S A Jarvis, 11 Charnwood Way, Blackfield, Southampton, SO45 1ZL [IO90HT, SU40]
GI1 RXM A A Murphy, 8 Grange Cres, Saintfield, Ballynahinch, BT24 7NP [IO74BK, J45]
G1 RXQ Details withheld at licensee's request by SSL.
G1 RXR J Jones, 5 Arlington Rd, Lipson Vale, Plymouth, PL4 7ER [IO70WJ, SX45]
G1 RXV S B Mugele, 63 Holland Pines, Bracknell, RG12 8UY [IO91OJ, SU86]
G1 RYB M G Challis, Journey's End, Bagpath, Tetbury, Glos, GL8 8YG [IO81UP, ST89]
G1 RYE C J Turk, 7 Elmsmead Cttgs, Iden, Rye, TN31 7PU [JO00IX, TQ92]
G1 RYF M R G O'Callaghan, 1 Lawrence Rd, Biggleswade, SG18 0LS [IO92UC, TL14]
G1 RYK J L Weaver, Sonning, 7 Woodley Rd, Orpington, BR6 9BN [JO01BI, TQ46]
G1 RYO C A Bartnik, 51 French St., Sunbury on Thames, TW16 5JL [IO91TJ, TQ16]
G1 RYQ S G Marshall, 43 Naseby Rd, Belper, DE56 0ER [IO93GA, SK34]
G1 RYS I McCulloch, 5 Knighthead Point, The Quarterdeck, London, E14 8SR [IO91XL, TQ37]
GM1 RZB J T Kinnell, 13 Chattan Ave, Stirling, FK9 5RD [IO86AD, NS79]
G1 RZJ Details withheld at licensee's request by SSL.
G1 RZL Details withheld at licensee's request by SSL.
G1 RZR S Barlow, Oak Cottage, Bullock Ln, Ironville, Nottingham, NG16 5NP [IO93HB, SK45]
G1 RZT Details withheld at licensee's request by SSL.
G1 RZU D Straw, 286 Derby Rd, Beeston, Bramcote, Nottingham, NG9 3JN [IO92IW, SK53]
G1 RZV Details withheld at licensee's request by SSL.
G1 RZZ R J Wood, 40 Ashville Gdns, Pellon, Halifax, HX2 0PL [IO93BR, SE02]
G1 SAA R J Warner, 26 Shirley Rd, Histon, Cambridge, CB4 4JR [JO02BG, TL46]
G1 SAJ R J White, 61 Bournemead Ave, Northolt, UB5 6PX [IO91TM, TQ18]
G1 SAK M J Weatherley, 95 Cambalt Rd, Puney Hill, London, SW15 6EX [IO91VL, TQ27]
G1 SAM A G Hodgkinson, 7 Grassmere Terr, Stoke on Trent, ST6 7EU [IO83VB, SJ85]

G1 SAN A Hollyoake Sandwell ARS, 281 Causeway Green Rd, Oldbury, Warley, B68 8LT [IO82XL, SO98]
G1 SAQ I R P Collyer, 3 St. Farm Bungalows, The St., Brantham, Manningtree, CO11 1PL [JO01NX, TM13]
G1 SAR M J Watson Suffolk Raynet, The Tubbery, Henley, Ipswich, Suffolk, IP6 0BR [JO02NC, TM15]
G1 SAU Details withheld at licensee's request by SSL.
G1 SBB Details withheld at licensee's request by SSL.
GM1 SBD M Angiolini, Innis Chonain, Mill Rd, Cambusbarron, Stirling, FK7 9LP [IO86AC, NS79]
G1 SBG A R Page, 7 Selsdon Cl, Kidderminster, DY11 6BD [IO82UJ, SO87]
G1 SBI E A Wagoner, 21 Wycliffe Rd West, Wyken, Coventry, CV2 3DX [IO92GJ, SP37]
G1 SBJ J R McKinlay, 5 Woodway, Horsforth, Leeds, LS18 4HY [IO93ET, SE23]
G1 SBK J E Spink, Highfields, Church Ln, Adel, Leeds, LS16 8DE [IO93EU, SE23]
G1 SBN J M Davison, 29 Glenfield Ave, Wetherby, LS22 6RN [IO93HW, SE44]
GW1 SBO W M Edwards, 59 Danycoed, Aberystwyth, SY23 2HD [IO72XK, SN58]
G1 SBP F Mason, 22 Lewis Cl, St. Anns, Nottingham, NG3 1NR [IO92KX, SK54]
G1 SBT C G Storey, 12 Vereker Dr, Sunbury on Thames, TW16 6HF [IO91TJ, TQ16]
G1 SBZ T H Lumley, 32 Downland Rd, Woodingddan, Brighton, BN2 6DJ [IO90WU, TQ30]
G1 SCA S C Allen, 28 Neville Rd, Luton, LU3 2JJ [IO91SV, TL02]
G1 SCL N Stackhouse, 16 Tintern Ave, Flixton, Manchester, M31 3FJ [IO83SK, SJ79]
G1 SCN C G N Taylor, Wycoller, 1 Bridle Rd, Claygate, Esher, KT10 0ET [IO91UI, TQ16]
G1 SCO R T Brown, 52 Challenger Dr, Sprotbrough, Doncaster, DN5 7RY [IO93KM, SE50]
G1 SCP Details withheld at licensee's request by SSL.
G1 SCQ K M Kent, 5 Jubilee Rd, Heacham, Kings Lynn, PE31 7AR [JO02FV, TF63]
G1 SCR G M Evans Shropshire County Raynet, 16 Kynaston Dr, Wem, Shrewsbury, SY4 5DE [IO82PU, SJ52]
G1 SCV A J Faulkner, 17 Ashcroft Cl, St. Leonards Park, Matson, Gloucester, Glos, GL4 6JX [IO81VT, SO81]
G1 SCX C Kirk, Jacobs Gutter Ln, Hounsdown, Totton, Southampton, Hants, SO40 9FU [IO90GV, SU31]
G1 SDA F A Shore, 3 Deansway Ct, 29 Dean Park Rd, Bournemouth, Dorset, BH1 1HY [IO90BR, SZ09]
G1 SDD Details withheld at licensee's request by SSL.
G1 SDH J H S Shepherd, 140 The Broadway, Herne Bay, CT6 8HY [JO01NI, TR16]
G1 SDJ C N Cooper, 167 Valley Rd, Ipswich, IP1 4PQ [JO02NB, TM14]
G1 SDN R Mordue, 29 Sycamore Cl, Witham, CM8 2PE [JO01HT, TL81]
G1 SDO S E Ashdown, 217 Kings Parade, Holland on Sea, Clacton on Sea, CO15 5TR [JO01OT, TM21]
G1 SDP P G H Nicholls, 86 Lower Hanham Rd, Hanham, Bristol, BS15 3BZ [IO81RK, ST67]
G1 SDX G Taylor, 3 Lyndhurst Ave, Bethel, St. Austell, PL25 3HJ [IO70OI, SX05]
G1 SEA D M Tucker, 35 Brockhall Rise, Kings Haven, Heanor, DE75 7TL [IO93HA, SK44]
G1 SEC J S Barlow, 14 Dartington Ave, Woodley, Reading, RG5 3PD [IO91NK, SU77]
G1 SED P N Hedicker, Hilvista, 26 Malden Rd, Sidmouth, EX10 9LS [IO80JQ, SY18]
G1 SEF A C Fearnley, 1 Dover Rd, Wanstead, London, E12 5DZ [JO01AN, TQ48]
G1 SEG P F Sawyer, 102 St. James Rd, Bridlington, YO15 3NJ [IO94VB, TA16]
G1 SEH Details withheld at licensee's request by SSL.
G1 SEI D H Seaman, 154 Franklin Ave, Tadley, Basingstoke, Hants, RG26 4EU [IO91KI, SU56]
G1 SEJ Details withheld at licensee's request by SSL.
G1 SEM L F Acott, 60 The Gdns, Southwick, Brighton, BN42 4AN [IO90VT, TQ20]
G1 SEN Details withheld at licensee's request by SSL.
G1 SEO M J Lines, 22 Dowding Cl, Woodley, Reading, RG5 4NL [IO91NK, SU77]
G1 SES M P Halden, 19 Fenwick Ln, Halton Lodge, Runcorn, WA7 5YU [IO83PH, SJ58]
G1 SEU J E Billett, 31 St. Leonards Rd, Claygate, Esher, KT10 0EL [IO91TI, TQ16]
G1 SFA B V Hyde, 108 St. Bedes Cres, Cambridge, CB1 3UB [JO02CE, TL45]
G1 SFG Details withheld at licensee's request by SSL.
G1 SFI Details withheld at licensee's request by SSL.
G1 SFN Details withheld at licensee's request by SSL.[Op: M L A Saunders. Station located near Great Missenden]
G1 SFO Details withheld at licensee's request by SSL.
G1 SFU D Coffey, 14 Shawbridge, Harlow, CM19 4NJ [JO01BS, TL40]
G1 SGA R I Blunt, 8 The Cres, Wolverhampton, WV6 8LA [IO82VO, SO89]
GW1 SGE A J Morgan, 3 Gelli Newydd, Golden Gr, Carmarthen, SA32 8LP [IO71XU, SN51]
GW1 SGG M Jones, 48 Maes Alltwen, Dwygyfylchi, Penmaenmawr, LL34 6UA [IO83BG, SH77]
GW1 SGH R J Thorne, 6 Cromwell Ave, Rhyddings, Neath, West Glam, SA10 8DW [IO81CQ, SS79]
G1 SGM R K Parkin, Craigside, 15 Holly Dr, Tinshill Ln, Leeds, LS16 6EF [IO93EU, SE23]
G1 SGN Details withheld at licensee's request by SSL.[Station located near Ormskirk.]
G1 SGO L C Johnson, 31 Edward Ave, Jacksdale, Nottingham, NG16 5LB [IO93IB, SK45]
G1 SGP N R Barnes, Hillside, Derby Rd, Wirksworth, Matlock, DE4 4AR [IO93FB, SK25]
G1 SGR N J Marsh, 25 Langney Green, Eastbourne, BN23 6HY [JO00DT, TQ60]
G1 SGS B L M Kimber, 51 Ct Lodge Rd, Gillingham, ME7 2QX [JO01GJ, TQ76]
G1 SGT B R Turner, 13 Chestfield Cl, Rainham, Gillingham, ME8 7DR [JO01HI, TQ86]
G1 SGU P T B Hill, 35 Clavering Ave, Barnes, London, SW13 9DX [IO91VL, TQ27]
G1 SGW J Ashurst, 40 Denton Ln, Chadderton, Oldham, OL9 8PU [IO83WM, SD90]
G1 SGX Details withheld at licensee's request by SSL.
G1 SGZ P R Gamble, 9 Windmill Cl, Ockbrook, Derby, DE72 3TE [IO92HW, SK43]
G1 SHC I N Barnes, Trelowarth, Watergate Ln, St. Mabyn, Bodmin, Cornwall, PL30 3BJ [IO70OM, SX07]
G1 SHG E A Rice, 21 Hollabury Rd, Bude, EX23 8JA [IO70RU, SS20]
G1 SHH A P Compton, Fairlight, 25 Framfield Rd, Uckfield, TN22 5AH [JO00BX, TQ42]
G1 SHI M Kuik, 196 Prestbury Rd, Macclesfield, Ches, SK10 3BS [IO83WG, SJ97]
G1 SHJ D N Weston, 130 Stonebank Rd, Kidsgrove, Stoke on Trent, ST7 4HL [IO83VB, SJ85]
G1 SHM A G S Wilkinson, 27 Barbridge Rd, Cheltenham, GL51 0BP [IO81WV, SO92]
G1 SHN G B Richardson, 16 Courtlands, Haywards Heath, RH16 4JD [IO90WX, TQ32]
G1 SHQ I M Woodford, The Lodge, Brunswick Park, Woodgreen Rd, Wednesbury, WS10 9AU [IO82XN, SO99]
G1 SHT D Wooster, 4 Shepherds Way, Chesham, HP5 1RH [IO91QQ, SP90]
G1 SHU N R Carr, 8 Mob Ln, High Heath, Pelsall, Walsall, WS4 1BB [IO92AO, SK00]
G1 SHV Details withheld at licensee's request by SSL.[Station located near Hampton.]
G1 SIB E P Fraser, 47 Longdown Rd, Congleton, CW12 4QH [IO83VE, SJ86]
G1 SID C R Siddons, 35 Stephenson Ave, Tilbury, RM18 8XB [JO01EL, TQ67]
G1 SIG B G Scholte, 14 St. Michaels Dr, Cannock, WS12 5LB [IO92AQ, SK01]
G1 SIO J N Robson, Ealands, The Stanners, Corbridge, NE45 5BA [IO84XX, NY96]
G1 SIP R J Webb, 54 Ashby Ave, Chessington, KT9 2BU [IO91UI, TQ16]
G1 SIR Details withheld at licensee's request by SSL.
G1 SIU A Bingley, 51 Kirkby Folly Rd, Sutton in Ashfield, NG17 5HP [IO93JC, SK55]
G1 SIV S A ODell, 41 Pevensey Park Rd, Westham, Pevensey, BN24 5HW [JO00DT, TQ60]
G1 SIW R J ODell, 41 Pevensey Park Rd, Westham, Pevensey, BN24 5HW [JO00DT, TQ60]
G1 SJB C M Thompson, 135 Stafford Rd, Bloxwich, Walsall, WS3 3PG [IO82XP, SJ90]
G1 SJF Details withheld at licensee's request by SSL.
G1 SJG E G Boydon, 56 Oliver Leese Ct, Ten Butts Cres, Stafford, ST17 9HP [IO82WS, SJ92]
G1 SJJ E M Oldbury, 4 Tithebarn Rd, Rugeley, WS15 2QW [IO92AS, SK01]
G1 SJO A M Banthorpe, 32 Long Cl, Station Rd, Lower Stondon, Henlow, SG16 6JS [IO91UX, TL13]
G1 SJR C E Parker, 180 Girtin House, Academy Gdns, Northolt, UB5 5PJ [IO91TM, TQ18]
G1 SJT A D Long, 3 Windsor Hill, Princes Risborough, HP27 9HZ [IO91OR, SP80]
G1 SJU D Maciver, 176 Burges Rd, London, E6 2BS [JO01AM, TQ48]
G1 SJZ P Randall, Calle Debla, 4 Arroyo de-le-Miel, Malaga 29630, Spain, X X
G1 SKA J A Thompson, 25 Cutsyke Crest, Cutsyke, Castleford, WF10 5HU [IO93HR, SE42]
G1 SKI K J Weaver, 28 Sawyers Cl, Burgess Hill, RH15 0QB [IO90WW, TQ31]
G1 SKL J Bayliss, 79 Priors Croft, Old Woking, Woking, GU22 9EY [IO91RH, TQ05]
G1 SKM W Haden, 33 Poplar Ave, Chelmsley Wood, Birmingham, B37 7RD [IO92DL, SP18]
G1 SKQ J G Forster, 46 Gowland Ave, Fenham, Newcastle upon Tyne, NE4 9NH [IO94EX, NZ26]
G1 SKR Me E A Smith, 22 Fleece Rd, Surbiton, KT6 5JN [IO91UJ, TQ16]
G1 SKV B Abell, 44 Quarry Ave, Nottingham, NG6 8GZ [IO92JX, SK54]
G1 SKW R B Sidwell, 81 Oakengates Rd, Donnington, Telford, TF2 7LQ [IO82SR, SJ71]
G1 SKY G W A Fordham, 66 Cromer Rd, Mundesley, Norwich, NR11 8DD [JO02RV, TG33]
G1 SLA C Baker, 17 Dawlish Ave, Chadderton, Oldham, OL9 0RF [IO83WN, SD80]
G1 SLE R Drabble, 68 St. Lawrence Ave, Bolsover, Chesterfield, S44 6HT [IO93IF, SK46]
G1 SLF S Davies, 102 Wold Rd, Pocklington, Pocklington, York, YO4 2QG [IO93OW, SE84]
G1 SLG M T Butcher, 32, Grafton Underwood, Kettering, Northants, NN14 3AA [IO92QJ, SP98]
G1 SLI B Elliott, 51 Allerhope, Hall Cl Grange, Cramlington, NE23 6SX [IO95FB, NZ27]
G1 SLO K Turner, 44B Foxgrove Rd, Beckenham, BR3 5DB [IO91XJ, TQ36]
G1 SLQ J C Peerless, 101 Greenside, Aycliffe Rd, Borehamwood, WD6 4JD [IO91UQ, TQ19]
G1 SLU K G Ford, 2 Jersey Ave, St. Annes, Bristol, BS4 4RA [IO81RK, ST67]
GM1 SLW F A McLaren, 17 Troon Ct, Greenhills, East Kilbride, Glasgow, G75 8TA [IO75VR, NS65]
GI1 SLZ J J Kellagher, 20 Fort Lee, Derrylin, Enniskillen, BT92 9GE [IO64ED, H22]
G1 SMA Details withheld at licensee's request by SSL.[Station located near Weybridge.]
G1 SMB M D Chitty, Timbercroft, Faris Ln, Woodham, Addlestone, KT15 3DL [IO91RI, TQ06]
G1 SMD I A N Galpin, 19 Palmer Rd, Oakdale, Poole, BH15 3AR [IO90AR, SZ09]
GM1 SMF R R Rodgers, 19 Carn Dearg Rd, Claggan, Fort William, PH33 6QA [IO76KT, NN17]
G1 SMI R W Waters, 87 Castlehey, Skelmersdale, WN8 9DU [IO83PM, SD50]
GW1 SMJ F Beavan, Uplands, Bronllys, Brecon, Pwys, LD3 0HN [IO82JA, SO13]
G1 SMT C S Manning, 406 Fishponds Rd, Fishponds, Eastville, Bristol, BS5 6RQ [IO81RL, ST67]
G1 SMX P J McCracken, 265 Wigman Rd, Bilborough, Nottingham, NG8 4AG [IO92JX, SK54]
G1 SMY S M Fisher, 23A Regent St., Shanklin, PO37 7AF [IO90JP, SZ58]
GI1 SMZ M R McKersie, 57 Beechgrove Ave, Belfast, BT6 0ND [IO74BN, J37]
G1 SND Details withheld at licensee's request by SSL.
G1 SNG Details withheld at licensee's request by SSL.[Op: J E Bryce. Station located near Oldham.]
G1 SNH Details withheld at licensee's request by SSL.
G1 SNI Details withheld at licensee's request by SSL.
G1 SNJ Details withheld at licensee's request by SSL.
G1 SNK Details withheld at licensee's request by SSL.
GM1 SNL Details withheld at licensee's request by SSL.
G1 SNO D R Tubb, 42 Hill Farm Rd, Marlow, SL7 3LU [IO91OO, SU88]
G1 SNQ I Boss, 64 Chiltern Rd, Swadlincote, DE11 9SJ [IO92FS, SK21]
G1 SNT S N Twigger, Manbrook Cottage, Church St, Willersby, Worcestershiree, WR12 7PN [IO92BB, SP13]
G1 SNU D J Tunbridge, 12 Burnham Rd, Latchingdon, Chelmsford, CM3 6EU [JO01IQ, TL80]
GM1 SNW Details withheld at licensee's request by SSL.
G1 SOA Details withheld at licensee's request by SSL.
G1 SOB T A Yetton, 7 Warwick Cl, Canvey Island, SS8 9YB [JO01HM, TQ78]
G1 SOG R W Steam, 18 Kings Ave, Chippenham, SN14 0UJ [IO81WK, ST97]
G1 SOM M G Daniels Somerset Raynet, 6 Middlemead, Stratton on The Fosse, Bath, BA3 4QH [IO81SG, ST65]
G1 SOT S M Webb, 2 Gordon Rd, Shenfield, Brentwood, CM15 8LR [JO01DO, TQ69]
G1 SOU Details withheld at licensee's request by SSL.[Op: R Fitzpatrick 78a Maidstone Road, Rainham, Gillingham, Kent, ME8 0DR.]
G1 SOX W Stennett, 9 Spice Ave, Wyberton, Boston, PE21 7BQ [IO92XW, TF34]
G1 SOY J Moseley, 886 Manchester Rd, Linthwaite, Huddersfield, HD7 5QS [IO93BP, SE01]
G1 SPA D M Harrison, 6 Kirkland Cl, Off The Common, Castle Donnington, Derbyshire, DE7 2QY [IO92IX, SK44]
G1 SPE A Stevens, 7 Wigmore Rd, Aylesbury, HP19 3HU [IO91NT, SP81]
G1 SPJ A H Gibbs, 86 Broadmark Rd, Upton Lea, Slough, SL2 5PN [IO91RM, SU98]
G1 SPK B P Johnson, 71 Glover Pl, Bootle, L20 4QR [IO83MK, SJ39]
G1 SPO A R Dixon, Clifton House, 63 High St., Skellingthorpe, Lincoln, LN6 5TS [IO93QF, SK97]
G1 SPQ Details withheld at licensee's request by SSL.
G1 SPT S P Tatem, 55 Chelwood Rd, Chellaston, Derby, DE73 1SJ [IO92GU, SK33]
G1 SPU A J Burnett, 14 Waterford Ct, Elworthy Cl, Stafford, ST16 3QT [IO82WT, SJ92]
G1 SPW B J Saunders, 1 Castle Cl, Tarring, Worthing, BN13 1BT [IO90TT, TQ10]
G1 SQA J M Yates, 37 Flaxpiece Rd, Clay Cross, Chesterfield, S45 9HB [IO93HD, SK36]
G1 SQB A H Ross, 84 Benhurst Ave, Elm Park, Hornchurch, RM12 4QT [JO01CN, TQ58]
G1 SQC K Boote, Link View, Fowlers Ln, Light Oaks, Stoke on Trent, ST2 7NB [IO83WB, SJ95]
G1 SQG P M Dennis, 18 West View Rd, Bere Alston, Yelverton, PL20 7DD [IO70VL, SX46]
G1 SQH S J Chubb, 28 Hollymoor Garden, Beaminster, Dorset, DT8 3NH [IO80PT, ST40]
G1 SQI J W Bewley, 21 Duloe Gdns, Pennycross, Plymouth, PL2 3RS [IO70WJ, SX45]
G1 SQM D Baily, 10 Trent Cl, Sompting, Lancing, BN15 0EJ [IO90TT, TQ10]
GW1 SQT D T Jones, 41 Penrhys Rd, Ystrad Rhondda, Ystrad, Pentre, CF41 7SJ [IO81GP, SS99]
G1 SQW R Rawson, 43 Co. Rd North, Bricknell Ave, Hull, HU5 4HN [IO93TS, TA03]
G1 SQY Details withheld at licensee's request by SSL.
GM1 SQZ G J W Pocock, 1 Pitcairn Gr, Hairmyres, East Kilbride, Glasgow, G75 8TN [IO75VS, NS65]
G1 SRA T Binns, Cross Farm Cottage, Oxenhope, Keighley, West Yorks, BD22 9LE [IO93AT, SE03]
GW1 SRB K M Gough, 2 Church Rd, Abertridwr, Caerphilly, CF8 4DL [IO81HN, ST08]
G1 SRD P M Foulds, 41 Hanbury Rd, Bedworth, Nuneaton, CV12 9BX [IO92GL, SP38]
G1 SRJ J E Smith, 4 Townend Villas, Humbleton, Hull, HU11 4NR [IO93WS, TA23]
G1 SRN R Pettinger, 38 St. Andrews Cres, Hoyland, Barnsley, S74 9HE [IO93GM, SE30]
GM1 SRP J C McCulloch, 53 Milverton Ave, Bearsden, Glasgow, G61 4BG [IO75TW, NS57]
GM1 SRR M A Christmas, 2 Birch Rd, Killearn, Glasgow, G63 9SQ [IO76TB, NS58]
G1 SRZ Details withheld at licensee's request by SSL.
G1 SSB Details withheld at licensee's request by SSL.[Op: G V Stewart. Station located near Crystal Palace.]
G1 SSF Details withheld at licensee's request by SSL.
G1 SSL M M J Belcher, 52 Kynaston Rd, Didcot, OX11 8HD [IO91JO, SU58]
G1 SSN Details withheld at licensee's request by SSL.
G1 SSO Details withheld at licensee's request by SSL.
G1 SSR Details withheld at licensee's request by SSL.
G1 SSS J M Banfield, 2 Laleham Cl, Eastbourne, BN21 2LQ [JO00DS, TV69]
G1 SST Details withheld at licensee's request by SSL.
G1 SSZ F R Bowhill, Broadfield Farm, 78 East Gomeldon Rd, Gomeldon, Salisbury, SP4 6NB [IO91DC, SU13]
G1 STA J W Buntin, 50 Commercial Rd, Hayle, TR27 4DH [IO70HE, SW53]
G1 STE Details withheld at licensee's request by SSL.
G1 STK A M Goddard, 65 Langley Hall Rd, Solihull, B92 7HE [IO92CK, SP18]
G1 STL J B Gleed, 83 Gr Ln, Harborne, Birmingham, B17 0QT [IO92AK, SP08]
G1 STO W A Bamford, 7 Poole Ave, Milton, Stoke on Trent, ST2 7JJ [IO83WB, SJ95]
G1 STP V A Trolan, 12 Broadwell Rd, Solihull, B92 8QH [IO92CK, SP18]
G1 STQ J L Taylor, 9 Hanover St., Newcastle, ST5 1AU [IO83VA, SJ84]
G1 SUE S A Pryke, 22 Sycamore Rd, Stowupland, Stowmarket, IP14 4DR [JO02ME, TM05]
G1 SUH J K Speakman, 130 Dicconson St., Wigan, WN1 2BA [IO83QN, SD50]
G1 SUI M McGibbon, 4 Bournemouth Parade, Hebburn, NE31 2AU [IO94GX, NZ36]
G1 SUJ Details withheld at licensee's request by SSL.
G1 SUK A D Dimmock, 26 Foxford Walk, Wythenshawe, Manchester, M22 5QN [IO83UJ, SJ88]
G1 SUM T Cull, 25 Queensway, Darras Hall, Ponteland, Newcastle upon Tyne, NE20 9RZ [IO95CA, NZ17]
G1 SUT Details withheld at licensee's request by SSL.
G1 SUV Details withheld at licensee's request by SSL.
G1 SUW K Draper, 115 Haynes Rd, Hornchurch, RM11 2HX [JO01CN, TQ58]
G1 SVD R G Roper, 85 Hart Ln, Luton, LU2 0JG [IO91TV, TL12]
G1 SVF S Fletcher, 54 Heights Ave, Rochdale, OL12 6JH [IO83WP, SD81]
GW1 SVG R P Hope Swansea ARS, 75 Priors Way, Dunvant, Swansea, SA2 7UH [IO71XO, SS59]
G1 SVI T A Metcalfe, 17 Honeydon Ave, Eaton Socon, St. Neots, Huntingdon, PE19 3PJ [IO92UF, TL15]
G1 SVJ C R Murphy, Mortimer House, Brays Cl, Hyde Heath, Amersham, HP6 5RZ [IO91QQ, SU99]
G1 SVL K J Stocker, 53 Westmorland Rd, North Harrow, Harrow, HA1 4PL [IO91TO, TQ18]
G1 SVN M A Churchman, Westcroft, Church Rd, Dodleston, Chester, CH4 9NG [IO83MD, SJ36]
G1 SVP B A Howard, Rame Common Farm, Carnkie, Wendron, Helston Cornwall, TR13 0DY [IO70JD, SW73]
GM1 SVQ C Pringle, Carnslloch, Ches Home, Dumfries, Scotland, DG1 1SN [IO85EC, NX98]
G1 SVR E G Churchyard Severn Valley Rd, 11 Greenfields Dr, Bridgnorth, WV16 4JW [IO82SM, SO79]
G1 SVV D Osborne, 8 Bromley Rd, Frating, Elmstead, Colchester, CO7 7BT [JO01LV, TM02]
G1 SVW H N B Bostock, 2 Denham Cl, Wivenhoe, Colchester, CO7 9NS [JO01LU, TM02]
G1 SWE M E Foster, 56 St. Matthews Ave, Litherland, Liverpool, L21 5JT [IO83ML, SJ39]
G1 SWF S J Roberts, 13 Plough Rd, Epsom, KT19 9RA [IO91UI, TQ26]
G1 SWH G H Schoof, 5 Canal Row, Haigh, Wigan, WN2 1NA [IO83QO, SD50]
G1 SWI B J Gillett, 18 Rookery Cl, Fenny Drayton, Nuneaton, CV13 6BB [IO92GN, SP39]
GW1 SWN Details withheld at licensee's request by SSL.[Op: Graham.]
G1 SWP E Gordon, Greenways, 3 East Oakwood, Oakwood, Hexham, NE46 4LG [IO84XX, NY96]
G1 SWR Details withheld at licensee's request by SSL.
G1 SWS Details withheld at licensee's request by SSL.
G1 SWU G J Denham, 36 Redstone Farm Rd, Hall Green, Birmingham, B28 9NT [IO92CK, SP18]
G1 SWX C P Harrap, 16 Lichfield Rd, Fareham, PO14 4QN [IO90IU, SU50]
G1 SWY A J Graffham, 29 New Rd, Milford, Godalming, GU8 5BE [IO91QE, SU94]
G1 SWZ R A Wright, 61 Quarry Rd, Hurtmore, Godalming, GU7 2RW [IO91QE, SU94]
G1 SXA P J White, 163 Mount Rd, Chessington, KT9 1JJ [IO91UI, TQ16]
G1 SXB G A Whetstone, 60 Worple Rd, Staines, TW18 1EE [IO91SK, TQ07]
G1 SXF P R Woodhouse, 8 Greenhill Rd, Halesowen, B62 8EZ [IO82XL, SO98]
G1 SXJ W E Coggin, 27 West Hall Garth, South Cave, Brough, HU15 2HA [IO93QS, SE93]
GW1 SXN P L J O'Brien, 12 Church St., Caernarvon, LL55 1SW [IO73UD, SH46]
G1 SXO M D Wright, 14 Beeston Rd, Broughton, Chester, CH4 0SB [IO83MD, SJ36]
GW1 SXP G Wright, 22 Wepre Park, Deeside, CH5 4HN [IO83LF, SJ26]
GW1 SXT M J Kerry, 40 Oaklands Rd, Sebastopol, Pontypool, NP4 5BZ [IO81LQ, ST29]
GW1 SXU P A Janes, 19 Fair View, Chepstow, NP6 5BX [IO81PP, ST59]
GM1 SXW R D Murray, 91 Stockiemuir Ave, Bearsden, Glasgow, G61 3LL [IO75UW, NS57]
GM1 SXX A I Copland, 99 Bruce Rd, Paisley, PA3 4SQ [IO75TU, NS46]
G1 SXY Y D Entwistle, Park Garden House, Park Garden, West Row, Bury St Edmunds, IP28 8PB [JO02FI, TL67]
GM1 SXZ W Hunter, 31 Hamilton Dr, New Cumnock, Cumnock, KA18 4JP [IO75VJ, NS61]
GM1 SYC W C Graham, 7 Brunt Pl, Dunbar, EH42 1RT [IO85RX, NT67]
G1 SYG Details withheld at licensee's request by SSL.
GI1 SYM G Thompson, 57 Rosepark, Donaghadee, BT21 0BN [IO74FP, J57]
G1 SYR J M Guy, Jackdaw Cottage, Burmarsh, Romney Marsh, Kent, TN29 0JR [JO01MB, TR13]
G1 SYS Details withheld at licensee's request by SSL.
G1 SYT M D Tatum, 11 Kempton Cl, Ipswich, IP1 6QZ [JO02NB, TM14]
G1 SYU M R Oubridge, 12 Wilmin Gr, Loughton, Milton Keynes, MK5 8EU [IO92OA, SP83]
G1 SYV B Goodier, 56 Clively Ave, Clifton, Swinton, Manchester, M27 8RU [IO83UM, SD70]
G1 SYZ D E Setterfield, 3 Waldon Cl, Hillcrest View, Plympton, Plymouth, PL7 2ZA [IO70XJ, SX55]
GI1 SZC Dr D K McManus, 38 Deanfield, Bangor, BT19 6NX [IO74EP, J58]
G1 SZD T C F Trengove, 8 Kemp Cl, Truro, TR1 1EF [IO70LG, SW84]
GM1 SZH D Boyd, West Main St, Blackburn, West Lothian, EH47 7LS [IO85EU, NS96]
G1 SZK Details withheld at licensee's request by SSL.
GM1 SZM W B Robertson, 28 Dewars Ave, Kelty, KY4 0BG [IO86HD, NT19]
GM1 SZN L F Morrison, 92 Gr Rd, Broughty Ferry, Dundee, DD5 1LB [IO86NL, NO43]
G1 SZT W M Dowkes, Woodlea, Gillamoor Rd, Kirkbymoorside, York, YO6 6EL [IO94MG, SE68]
G1 TAG J P Gaunt, 25 Church Ln, Tittleshall, Kings Lynn, PE32 2QD [JO02JS, TF82]

G1 TAH Details withheld at licensee's request by SSL.
G1 TAI P Gabel, 30 Showsley Rd, Shutlanger, Towcester, NN12 7RW [IO92MD, SP75]
G1 TAL P I Kerr, 115 Willis Rd, Haddenham, Aylesbury, HP17 8HG [IO91MS, SP70]
G1 TAN P W Evans, 19 Carlton Hill, Hernebay, Herne Bay, CT6 8HN [JO01NI, TR16]
G1 TAQ Details withheld at licensee's request by SSL.
G1 TAR C Anderton, 5 Leyland Ave, Hindley, Wigan, WN2 3SB [IO83RM, SD60]
G1 TAU J R Drewry, 32 Mersey Rd, Heaton Mersey, Stockport, SK4 3DJ [IO83VJ, SJ89]
G1 TAZ F G Perkin, 2 Pullington Cottage, Sweetshouse, Bodmin, PL30 5AW [IO70PK, SX06]
G1 TBB C W Stynes, 41 Richland Rd, Stoneycroft, Liverpool, L13 7BN [IO83MK, SJ39]
G1 TBI W R Monk, Brook House, River View, Litton Mill, near Buxton, Derbyshire, SK17 8SW [IO93CG, SK17]
G1 TBK D C Watson, 72 Dawes Ave, West Bromwich, B70 7LS [IO92AM, SP09]
G1 TBL A J Stock, 16 Starkie Dr, Oldbury, Warley, B68 9NX [IO92AL, SP08]
G1 TBN A C Malhi, 5 Beech Ave, Warton, Preston, PR4 1BY [IO83NS, SD42]
G1 TBO C J J Bittan, 176 Ninfield Rd, Bexhill on Sea, TN39 5DA [JO00FU, TQ70]
G1 TBR P M McGough, 78 Upper Rd, Madeley, Telford, TF7 5DJ [IO82SP, SJ60]
G1 TBV G Morgan, 13 Deal St., Great Lever, Bolton, BL3 2DA [IO83SN, SD70]
GM1 TBW A K Napier, Jehrada Cottage, Longhaven, Peterhead, Aberdeenshire, AB4 7NY [IO87TH, NJ72]
G1 TBX G Williams, 16 Coppice Rd, Talke, Stoke on Trent, ST7 1UB [IO83UB, SJ85]
G1 TCE Details withheld at licensee's request by SSL.
G1 TCF G J King, 2 Towers Rd, Heybridge, Maldon, Essex, CM9 7AP [JO01IR, TL80]
G1 TCH Details withheld at licensee's request by SSL.[Operator: Clyde. Tel: (01773) 719839. IO93HA. WAB: SK44, Derbyshire, Amber Valley dist., 7034, 7819. Station sometimes at Burgess Hill, IO90WW, TQ31, West Sussex, Mid-Sussex dist.]
G1 TCM Details withheld at licensee's request by SSL.
GM1 TCN T Curran, 15A Tillyloss, Kirriemuir, DD8 4DB [IO86MQ, NO35]
GM1 TCP J R Campbell, 15 Birchwood Rd, Cumnock, KA18 1NG [IO75VK, NS51]
GW1 TCZ Details withheld at licensee's request by SSL.
G1 TDB Details withheld at licensee's request by SSL.
G1 TDG Details withheld at licensee's request by SSL.
G1 TDK A J Brewer, 3 Walton Ct, Sheen Park, Richmond, TW9 1UL [IO91UL, TQ17]
G1 TDO F Sanchez-Garci, 74 Gorthorpe, Hull, HU6 9EZ [IO93TS, TA03]
G1 TDP J R Spry, 21 Christchurch Gdns, Waterlooville, PO7 5BT [IO90LU, SU60]
G1 TDQ F G Shepherd, 20 Frances Rd, Purbrook, Waterlooville, PO7 5HH [IO90LU, SU60]
G1 TDR A Crosby, 3 Fern Gr, Middlestone Moor, Spennymoor, DL16 7DR [IO94EQ, NZ23]
GM1 TDT D A Robertson, 131 Foxbar Rd, Paisley, PA2 0BD [IO75ST, NS46]
GM1 TDU J M Rooney, 2 Slains Rd, Bridge of Don, Aberdeen, AB22 8TT [IO87WE, NJ91]
GW1 TDV A J Potts, 19 Northfield Cl, Caerleon, Newport, NP6 1EZ [IO81MO, ST39]
GW1 TDW E Powell, 9 Railway Terr, Penygraig, Tonypandy, CF40 1LE [IO81GO, SS99]
GW1 TDX E Evans, 4 Crown Ave, Ynyswen, Treorchy, CF42 6DY [IO81FP, SS99]
G1 TEJ Details withheld at licensee's request by SSL.
G1 TES W L Roberts, 28 Ventnor St., Rochdale, OL11 1QD [IO83WO, SD81]
G1 TEW J Gadsby, 25 Sefton Cl, Stapenhill, Burton on Trent, DE15 9BU [IO92ET, SK22]
G1 TEX N G Swann, 9 Alexandra Rd, Parkstone, Poole, BH14 9EL [IO90AR, SZ09]
G1 TEY R E Thomas, 14 Montgomery St., Hove, BN3 5BF [IO90VT, TQ20]
GW1 TFB A L Hughes, 52 Bishops Walk, St. Asaph, LL17 0SZ [IO83GG, SJ07]
GI1 TFC R A McMaster, 43 Craigs Rd, Carrickfergus, BT38 9RL [IO74CR, J49]
GM1 TFG Details withheld at licensee's request by SSL.
G1 TFK Dr M D Addlesee, 36A The Limes, Harston, Cambridge, CB2 5QT [JO02BD, TL45]
GW1 TFL Details withheld at licensee's request by SSL.
G1 TFM J K P Bishop, 299 Summerwood Rd, Isleworth, TW7 7QP [IO91UK, TQ17]
G1 TFN Details withheld at licensee's request by SSL.
GW1 TFU D Thomas, 10 Woodside Ave, Litchard, Bridgend, CF31 1QF [IO81FM, SS98]
G1 TFV Details withheld at licensee's request by SSL.[Op: C R Cassey. Station located near Winchester.]
G1 TFY D M Mackinnon, 12 Stannells Cl, Luddington Rd, Stratford upon Avon, CV37 9SA [IO92DE, SP15]
GM1 TFZ S L Bavin, Garvan, 10 Grampian Way, Barrhead, Glasgow, G78 2DH [IO75TT, NS55]
GM1 TGA D G Cooper, 43 Hamilton Cres, Bishopton, PA7 5JT [IO75RV, NS47]
G1 TGC R S Halsall, 53 Stancliffe Rd, Sharston, Wythenshawe, Manchester, M22 4PG [IO83UJ, SJ88]
G1 TGI Details withheld at licensee's request by SSL.
G1 TGJ Details withheld at licensee's request by SSL.
G1 TGM Details withheld at licensee's request by SSL.
GM1 TGS J Cooper, Lonan Dr No.1, Oban, Argyll, Scotland, PA34 4NN [IO76GJ, NM82]
G1 TGZ R W McLintock, 10 The Close, Riverhead, Sevenoaks, TN13 2HE [JO01CG, TQ55]
G1 THA V F Collins, 29 Hillcrest Ave, Chandlers Ford, Eastleigh, SO53 2JS [IO90HX, SU42]
G1 THD A M Simmons, 22 Willow Way, Princes Risborough, HP27 9AY [IO91PO, SU89]
G1 THE C P Houghton, 30 Blakemore, Brookside, Telford, TF3 1PR [IO82SP, SJ70]
G1 THF Details withheld at licensee's request by SSL.
G1 THG D Moore, East View, The Common, East Stour, Gillingham, SP8 5NB [IO81UA, ST82]
G1 THO D G Moreton, 25 Carr Green Ln, Mapplewell, Barnsley, South Yorks, S75 6DY [IO93FN, SE30]
G1 THP M W G Fudge, 8 West Hill Park, Winchester, SO22 5DY [IO91IB, SU42]
GM1 THS G R Slessor, 27 Scurdie Ness, Aberdeen, AB1 3NG [IO87TH, NJ72]
G1 THV Details withheld at licensee's request by SSL.
G1 THW A P Taylor, 9 Barleymow Ct, Betchworth, RH3 7HF [IO91UF, TQ24]
G1 THY D D Waine, Mv Sea Priestess, PO Box 380, Richmond, Surrey, TW9 1GU [IO91UL, TQ17]
G1 TIE A F Green, 20 Mordern House, Harewood Ave, London, NW1 6NR [IO91WM, TQ28]
G1 TIF J W Batey, The Hemmel, Fell Ln, Barrasford, Hexham, Northd., NE48 4BD [IO85WB, NY97]
G1 TIH F Bell, 143 Peter St., Blackpool, FY1 3NN [IO83LT, SD33]
G1 TII Details withheld at licensee's request by SSL.
G1 TIJ P R Seaman, 18 Rectory Rd, Mellis, Eye, Suffolk, IP23 8DZ [JO02MH, TM07]
G1 TIL M Watson, 6 Woburn Dr, Bower Grange, Bedlington, NE22 5YB [IO95FD, NZ28]
G1 TIQ S Crellin, 89 Wapshare Rd, West Derby, Liverpool, L11 8LR [IO83MK, SJ39]
GW1 TIU E Griffiths, 19 Heol Dylan, Pontardulais Rd, Gorseinon, Swansea, SA4 4LR [IO71XQ, SS59]
G1 TJH C J Barfoot, 6 Maldon Cl, Bishopstoke, Eastleigh, SO50 6BD [IO90HX, SU41]
GW1 TJK A J Evans, Maes yr Onnen, 134 Waterloo Rd, Capel Hendre, Ammanford, SA18 3RY [IO71XS, SN51]
G1 TJL R Ross, 101 Gorse Cres, Stretford, Manchester, M32 0UQ [IO83UK, SJ89]
GJ1 TJP J G Poole, Jardin Du Puits, La Longue Rue, St. Martin, Jersey, JE3 6ED
G1 TJR R C Bromley, 33 Bromley Rd, Lytham St. Annes, FY8 1PQ [IO83LR, SD32]
G1 TJT W L Adam, 76 Lantern House, Alpha Gr, London, E14 8LL [IO91XM, TQ37]
G1 TJW B T Smith, 25 The Ferns, Tetbury, GL8 8JE [IO81WP, ST89]
G1 TJX P J Harper, 2 Quarry Building, Horbury, Wakefield, WF4 5NQ [IO93FP, SE21]
G1 TKA L A Pemberton, 29 Boothmeadow Cour, Thorplandsi, Northampton, NN3 1YH [IO92NG, SP76]
G1 TKH A J Smith, 24 Packington Ln, Coleshill, Birmingham, B46 3EL [IO92DL, SP28]
GM1 TKI G Higgins, 19 Cedar Rd, Banknock, Stirlingshire, Scotland, FK4 1JP [IO85AX, NS77]
G1 TKN S M Jackson, 2 Holme View Park, Upperthong Ln, Holmfirth, Huddersfield, HD7 2UZ [IO93CN, SE10]
GW1 TKO P H Dotchon, Kenilworth, Crosshill, The Knap, Barry, CF62 6SR [IO81IJ, ST16]
G1 TKQ I R T Burdon, 2 Dairy Cttgs, Mutterton, Cullompton, Devon, EX15 1RN [IO80HU, ST00]
G1 TKT A B Bowers, 83 Bishopdale, Telford, TF3 1SE [IO82SP, SJ70]
G1 TKX Details withheld at licensee's request by SSL.
G1 TKY P R Draper, 4 Woodmans Croft, Hatton, Derby, DE65 5QQ [IO92DU, SK23]
G1 TLA B Pattenden, Inshallah, 3 Copthorne Dr, Lightwater, GU18 5TE [IO91QI, SU96]
G1 TLB Details withheld at licensee's request by SSL.
G1 TLC A Myers, 7 Hillside, Chelveston, Wellingborough, NN9 6AQ [IO92RH, SP96]
G1 TLE M Rowe, 11 Parkfield, Stillington, York, YO6 1JR [IO94KC, SE56]
G1 TLH D J Koopman, 24 Cemetery Rd, Dereham, NR19 2ET [JO02LQ, TF91]
G1 TLI S A Wilkinson, Dalton House, Percy St. West, Thornley, Durham, DH6 3AP [IO94GS, NZ33]
G1 TLQ Details withheld at licensee's request by SSL.
G1 TLW A J Lloyd, 96 Fairdene Rd, Coulsdon, CR5 1RF [IO91WH, TQ25]
G1 TLY Details withheld at licensee's request by SSL.
G1 TMD P J Carpenter, 162 Wickham Way, Beckenham, BR3 3AS [IO91XJ, TQ36]
G1 TMF J F Firth, 29 Churston Cl, Clayton, Newcastle, ST5 4LP [IO82VX, SJ84]
GW1 TMJ Details withheld at licensee's request by SSL.
GW1 TMK Details withheld at licensee's request by SSL.
G1 TMQ H L King, 18 Helliers Rd, Chard, TA20 1LL [IO80MU, ST30]
GW1 TMR Details withheld at licensee's request by SSL.
G1 TMW A B Caspersz, 25 Cheltenham Pl, Kenton, Harrow, HA3 9NB [IO91UO, TQ18]
G1 TNE Details withheld at licensee's request by SSL.
G1 TNG Details withheld at licensee's request by SSL.
G1 TNK J Flattley, 53 The Dr, Bredbury, Stockport, SK6 2ED [IO83WK, SJ99]
G1 TNL Details withheld at licensee's request by SSL.
G1 TNN Details withheld at licensee's request by SSL.
G1 TNP A N Karande, 30 Ravenhill Ave, Knowle, Bristol, BS3 5DU [IO81RK, ST57]
G1 TNR D J Jordan, 1 Cranesbill Cl, Featherstone, Wolverhampton, WV10 7TY [IO82WP, SJ90]
G1 TOA Details withheld at licensee's request by SSL.
G1 TOB T G R Hall, 3 Saville Rd, Twickenham, TW1 4BQ [IO91TK, TQ17]
G1 TOF Details withheld at licensee's request by SSL.
G1 TOK R Mayou, 6 Anglesey St., Hednesford, Cannock, WS12 5AB [IO82XR, SJ91]
G1 TOL S J Talbot, 42 Birdcombe Rd, Westlea, Swindon, SN5 7BL [IO91CN, SU18]
G1 TOP Details withheld at licensee's request by SSL.
G1 TOQ Details withheld at licensee's request by SSL.
G1 TOT M E Horton, 105 Cedar Cres, North Baddesley, Southampton, SO52 9FX [IO90GX, SU32]
GI1 TOU Details withheld at licensee's request by SSL.[Op: E B E Smyth, 53 Church Road, Newtownbreda, Belfast, Co Down, N Ireland, BT8 4AL.]
G1 TOW P J Chapman, 462 Cherry Hinton Rd, Cambridge, CB1 4EA [JO02BE, TL45]
G1 TOX Details withheld at licensee's request by SSL.
G1 TOZ Details withheld at licensee's request by SSL.
G1 TPC M G Bellamy, 2 Nelson Dr, Rothwell, Kettering, NN14 6DZ [IO92OK, SP88]
G1 TPK P Moysey, 1 Tilly Cl, Plymouth, PL9 9DD [IO70WI, SX55]
G1 TPN R J S I Whateley, 14 Eastfield Rd, Delapre, Northampton, NN4 8PE [IO92NF, SP75]
G1 TPO Details withheld at licensee's request by SSL.[Op: R A Steel, 33 Greenlaw Road, Southfield Green, Cramlington, Northumberland, NE23 6NP.]
G1 TPS R Smith, Bron Castell, Twtil, Harlech, Gwynedd, LL46 2UA [IO72WU, SH53]
G1 TPU Details withheld at licensee's request by SSL.
G1 TPV D A Juett, 10 Leys Rd, Cambridge, CB4 2AU [JO02BF, TL46]
G1 TQH K P Scroggins, 44 Hillcroft Rd, Herne, Herne Bay, CT6 7EW [JO01NI, TR16]
G1 TQN A J Hobbs, 31 Ashley Pl, Warminster, BA12 9QJ [IO81VE, ST84]
G1 TQP Details withheld at licensee's request by SSL.
G1 TQR J W Harris, 48 Beech Cl, Corby, NN17 2AF [IO92PM, SP89]
G1 TQT M G Costello, Flat L5, 3 Gulston Walk, Chelsea, London, SW3 4UJ [IO91WL, TQ27]
G1 TQU C W Chambers, 8 Dagtail Ln, Redditch, B97 5QT [IO92AG, SP06]
G1 TQW Details withheld at licensee's request by SSL.
G1 TQX Details withheld at licensee's request by SSL.
G1 TQY P J Bishop, 2 Spruce Ave, Whiteshill, Whitehill, Bordon, GU35 9TA [IO91NC, SU73]
G1 TRB Details withheld at licensee's request by SSL.
G1 TRC G C Oldfield Trafford A.R.C., 6 Newcroft Rd, Urmston, Manchester, M41 9NN [IO83UK, SJ79]
G1 TRD Details withheld at licensee's request by SSL.
G1 TRF K Nicholson, 11 Latton Cl, Chilton, Didcot, OX11 0SU [IO91IN, SU48]
G1 TRL J R Deacon, 28 Dollicott, Haddenham, Aylesbury, HP17 8JG [IO91MS, SP70]
G1 TRN A R Higgins, Meadow End, Portway, St, BA16 0SE [IO81PC, ST43]
G1 TRV J J Davies, 40 Malvern House, Pickersleigh Cl, Halesowen, B63 4TH [IO82XK, SO98]
G1 TRZ W Hamilton-Sturdy, 2 Strawberry Hill, Northampton, NN3 5HL [IO92OG, SP86]
GM1 TSE Details withheld at licensee's request by SSL.
G1 TSL J D Riddoch, 12 Nelson Dr, Washingborough, Lincoln, LN4 1HN [IO93SF, TF07]
G1 TSM Details withheld at licensee's request by SSL.
G1 TSQ Details withheld at licensee's request by SSL.
G1 TSV S V Corrigan, 163 Blackburn Rd, Heapey, Chorley, PR6 8EJ [IO83QQ, SD52]
G1 TSY S D Grant, Broomhill, Hackford Vale, Reepham, Norfolk, NR10 4QU [JO02NS, TG02]
G1 TTB G T L Birkby, 44 Lady Bay Rd, West Bridgford, Nottingham, NG2 5DS [IO92KW, SK53]
G1 TTC Details withheld at licensee's request by SSL.[Op: K A Howard.]
G1 TTG J M Brickwood, 7 The Fairway, Camberley, GU15 1EF [IO91PH, SU85]
G1 TTJ Details withheld at licensee's request by SSL.
G1 TTK G E Lewis, 57 Edgecumbe Rd, Roche, St. Austell, PL26 8JH [IO70NJ, SW96]
G1 TTM Details withheld at licensee's request by SSL.
G1 TTR Details withheld at licensee's request by SSL.
G1 TTY Details withheld at licensee's request by SSL.
G1 TUI J P Tracy, 18 Preston St., Kirkham, Preston, PR4 2ZA [IO83NS, SD43]
G1 TUJ Details withheld at licensee's request by SSL.[Op: G A Riocreux. Station located near Wokingham.]
G1 TUN P H Beaven, 18 Leamington Rd, Urmston, Manchester, M41 7AZ [IO83TK, SJ79]
G1 TUS R G Rodley, Meadow Cottage, Cold Ashby Rd, Guilsborough, Northants, NN6 8QP [IO92LI, SP67]
G1 TUX R W Ogle, 38 St. Marys Walk, Harrogate, HG2 0LS [IO93FX, SE25]
GI1 TVH W S Brown, The Warren, 38 Springvale Rd, Ballywalter, Newtownards, BT22 2RS [IO74GM, J66]
G1 TVJ Details withheld at licensee's request by SSL.[Op: J Turnbull, tel (01322) 21364.]
G1 TWF L F Barker, 8 Wellesley Rd, Colchester, CO3 3HF [JO01KV, TL92]
G1 TWG A J D'Cruz, 4 Bure Dr, Witham, CM8 1UB [JO01HT, TL81]
G1 TWH S M A Greenfield, Byways, Brightlingsea Rd, Thorrington, Colchester, CO7 8JH [JO01MU, TM02]
G1 TWN P H Matthews, 41 Vivian Rd, Bodmin, PL31 1QX [IO70PL, SX06]
G1 TWQ A Hart, 10 Petterson Dale, Coxhoe, Durham, DH6 4HA [IO94GR, NZ33]
G1 TWS M R Dench, 110 Eastwood Rd, Rayleigh, SS6 7JR [JO01HN, TQ89]
G1 TWT A E Scott, 21 Hexham, Oxclose, Washington, NE38 0NR [IO94FV, NZ25]
G1 TWW R M Reeves, 40 Kennett Rd, Halterworth Park, Romsey, SO51 5PQ [IO90GX, SU32]
G1 TWX M G Foreman, 14 Townlane Rd, Bury St. Edmunds, IP33 2TE [JO02IF, TL86]
G1 TWY B J Panton, Lavers, Preston Rd, Lavenham, Sudbury, Suffolk, CO10 9QD [JO02JC, TL94]
G1 TXC Details withheld at licensee's request by SSL.
G1 TXD Details withheld at licensee's request by SSL.
GM1 TXE W Grieve, 11 Turnberry Dr, Kilmarnock, KA1 4LJ [IO75SO, NS43]
G1 TXG Details withheld at licensee's request by SSL.
G1 TXO M S Riches, 32 Wyncham Ave, Sidcup, DA15 8ER [JO01BK, TQ47]
G1 TXQ E G Roberts, 37 Buccleuch House, Clapton Common, London, E5 9AN [IO91XN, TQ38]
G1 TXR C G Stubbs, 290 Upminster Rd North, Rainham, RM13 9JR [JO01CM, TQ58]
G1 TXV H P Killick, 17 Harewood Cl, Poulton-le-Fylde, FY6 7RB [IO83MU, SD34]
G1 TXW D Defries, 31 Farleigh Rd, Stoke Newington, London, N16 7TB [IO91XN, TQ38]
G1 TXY Details withheld at licensee's request by SSL.
G1 TYA Details withheld at licensee's request by SSL.
G1 TYB Details withheld at licensee's request by SSL.
G1 TYH Details withheld at licensee's request by SSL.
G1 TYK D A W Lyons, 150 Fouracre Cres, Bristol, BS16 6PZ [IO81SM, ST67]
G1 TYN Details withheld at licensee's request by SSL.
G1 TYP E Summers, 262 Huddersfield Rd, Stalybridge, SK15 3DZ [IO83XL, SJ99]
G1 TYQ Details withheld at licensee's request by SSL.
G1 TYU R Ward, 88 Little Barn Ln, Mansfield, NG18 3JJ [IO93JD, SK56]
G1 TYY Details withheld at licensee's request by SSL.
G1 TZC Details withheld at licensee's request by SSL.
G1 TZF C J Buck, 44 Este Rd, Birmingham, B26 2ES [IO92CL, SP18]
G1 TZJ Details withheld at licensee's request by SSL.[Op: M G Collins. Station located near Clacton-on-Sea.]
G1 UAE D C Norman, 19 Eastcote Gr, Southend on Sea, SS2 4QA [JO01IN, TQ88]
G1 UAF A S Webster, 20 Green Park, Brinkley, Newmarket, CB8 0SQ [JO02EE, TL65]
G1 UAH Details withheld at licensee's request by SSL.
G1 UAI Details withheld at licensee's request by SSL.[Station located near Denham. Correspondence via PO Box 137, Uxbridge, Middx, UB9 5QT.]
G1 UAJ Details withheld at licensee's request by SSL.[Op: A P Jones, 8 Maygoods View, Cowley, near Uxbridge, Middx, UB8 2HG.]
G1 UAN Details withheld at licensee's request by SSL.
G1 UAO Details withheld at licensee's request by SSL.
G1 UAR C E W Mayhew, 80 St. David Rd, Eastham, Wirral, L62 0BT [IO83MH, SJ38]
G1 UAS Details withheld at licensee's request by SSL.
G1 UAU Details withheld at licensee's request by SSL.[Op: P F Evans.]
G1 UAW Details withheld at licensee's request by SSL.
G1 UAY K D Varnals, Regent Hotel, Fern Lea Terr, St. Ives, TR26 2BH [IO70GF, SW54]
G1 UAZ C J Musson, 14 Alfreton Rd, South Normanton, Alfreton, DE55 2AS [IO93HC, SK45]
G1 UBA T W Last, 85 Westwood Rd, Salisbury, SP2 9HR [IO91CC, SU13]
G1 UBC J M Pedley, 24 Appledore Rd, Walsall, WS5 3DT [IO92AN, SP09]
G1 UBH M A Howell, 4 Chattisham Cl, Church Meadow, Stowmarket, IP14 2RE [JO02ME, TM05]
G1 UBN S A Knight, 8 Longmeadow Garden, Hythe, Southampton, SO4 6BQ [IO90HW, SU41]
G1 UBO Details withheld at licensee's request by SSL.
G1 UBT E Dunn, 12 Streete Ct, Westgate on Sea, CT8 8BT [JO01QJ, TR36]
G1 UBV N Brigden, 121 Brocks Dr, North Cheam, Sutton, SM3 9UP [IO91VI, TQ26]
G1 UBW R A Honeysett, Joskins, Old Wives Lees, Canterbury, Kent, CT4 8AS [JO01LG, TR05]
G1 UBY Details withheld at licensee's request by SSL.
G1 UCB D J Dawson, 68 Rivington Ave, Blackpool, FY2 9DG [IO83LU, SD34]
G1 UCC D A Boot, 13 Westland St., Penkhull, Stoke on Trent, ST4 7HE [IO83VA, SJ84]
G1 UCD N K Rogers, 2 Great Cliffe Rd, Eastbourne, BN23 7AY [JO00DT, TQ60]
G1 UCH Details withheld at licensee's request by SSL.
G1 UCI D D Roberts, 56 Meadow Ln, Ainsdale, Southport, PR8 3RS [IO83LO, SD31]
G1 UCJ D Booth, The Hawthorns, Parkgate Rd, Newdigate, Dorking, RH5 5AH [IO91UE, TQ14]
G1 UCN M Davis, 37 South St., Ellistown, Coalville, LE67 1EJ [IO92HQ, SK41]
G1 UCO D Ralph, 55 St. Marys Rd, Smethwick, Warley, B67 5DH [IO92AL, SP08]
G1 UCP J Bridgewood, 17 Church Ln, Scredington, Sleaford, NG34 0AG [IO92TW, TF04]
G1 UCZ D H Dewar, 224 Seaside Rd, Aldbrough, Hull, HU11 4RY [IO93WU, TA23]
G1 UDA L A Hardy, 17 Buccaneer Way, Hethersett, Norwich, NR9 3QJ [JO02OO, TG10]
G1 UDB G J Wardle, 53 Braine Rd, Wetherby, LS22 6NP [IO93HW, SE44]
G1 UDE D Grayson, 28 Chesterfield Rd, Swallownest, Sheffield, S31 0TL [IO93GJ, SK38]
G1 UDH Details withheld at licensee's request by SSL.
GW1 UDK Details withheld at licensee's request by SSL.
G1 UDL P Kinghorn, 36 Blueburn Dr, Killingworth, Newcastle upon Tyne, NE12 0GA [IO95FA, NZ27]

G1 UDN S E Webb, 13 Kingston Rd, Taunton, TA2 7SA [IO81KA, ST22]
G1 UDP N W Jones, 8 Downing Ave, Basforde, Newcastle, ST5 0JY [IO83VA, SJ84]
G1 UDR J G M Scott, 15 Oakwood Rd, Chandlers Ford, Eastleigh, SO53 1LW [IO90HX, SU42]
G1 UDS P Spence, 4 Masefield Dr, Stafford, ST17 9UR [IO82WT, SJ92]
G1 UDT M G Boydon, 56 Oliver Leese Ct, Ten Butts Cres, Stafford, ST17 9HP [IO82WS, SJ92]
G1 UDW Details withheld at licensee's request by SSL.
G1 UDX M V Stasuik, 58 Ingram Rd, Nottingham, NG6 9GS [IO92JX, SK54]
G1 UEA D J Mills, 56 Canterbury Rd, Birchington, CT7 9AS [JO01PJ, TR36]
G1 UEB Details withheld at licensee's request by SSL.
G1 UEG Details withheld at licensee's request by SSL.
G1 UEO J R Aldred, 1E North End Rd, Steeple Claydon, Buckingham, MK18 2PF [IO91MW, SP62]
G1 UEQ J M W Sims, 345 Blandford Rd, Hamworthy, Poole, BH15 4HP [IO80XR, SY99]
G1 UES L G Cannard, 3 Victoria Terr, Quaves Ln, Bungay, NR35 1DF [JO02RK, TM38]
G1 UEV S J Brown, Oak Ridge, 26 Saxon Rd, Blackfield, Southampton, SO45 1WY [IO90HT, SU40]
G1 UFA B V Bailey, 10 Milton Rd, Waterlooville, PO7 6AA [IO90LV, SU61]
G1 UFD J G Owen, 85 Tennyson Ave, New Malden, KT3 6NA [IO91VJ, TQ26]
G1 UFH Details withheld at licensee's request by SSL.
G1 UFJ R M Davis, 192 Greenhill Rd, Herne Bay, CT6 7RS [JO01NI, TR16]
G1 UFL Details withheld at licensee's request by SSL.
G1 UFM S C Pugh, 18 Styal Ave, Stretford, Manchester, M32 9SJ [IO83UK, SJ79]
G1 UFN Details withheld at licensee's request by SSL.
G1 UFQ Details withheld at licensee's request by SSL.
G1 UFS N E Chandler, 7 Sherlock Ave, Parklands, Chichester, PO19 3AE [IO90OU, SU80]
G1 UFT P R J Zara, 17 Yarwell Dr, The Meadows, Wigston, LE18 3QF [IO92KO, SP69]
G1 UFU W A McKay, 8 Westbury Rd, Leicester, LE2 6AG [IO92KO, SK50]
G1 UFV D Hodgekins, 165 Newmarket, Louth, LN11 9EJ [JO03AI, TF38]
G1 UFX A Hern, Block H Flat 7, Peabody Sq, Blackfriars Rd, London, SE1 8JJ [IO91WL, TQ37]
G1 UGG K R Blagg, 19 Lyndhurst Ave, Blackpool, FY4 3AX [IO83LT, SD33]
G1 UGH T G Chaplin, 21 Shillitoe Cl, Bury St. Edmunds, IP33 3DU [JO02IF, TL86]
G1 UGJ B A Lowe, Little Maltings, Bradfield St Clare, Bury St.Edmunds, Suffolk, IP30 0ED [JO02JE, TL95]
G1 UGK J C Wilkinson, Mentor House, Chare Rd, Stanton, Bury St. Edmunds, IP31 2DX [JO02KH, TL97]
G1 UGL P D Beardshaw, 12 Halesworth Cl, Walton, Chesterfield, S40 3LW [IO93GF, SK36]
G1 UGM Details withheld at licensee's request by SSL.
G1 UGO S M Lexton, 84 Highway Rd, Evington, Leicester, LE5 5RF [IO92KO, SK60]
G1 UGV D J Bodman, 34 Churchlands, North Bradley, Trowbridge, BA14 0TD [IO81VH, ST85]
G1 UHB S Tomkins, 73 Oakleigh Rd, Clayton, Bradford, BD14 6NP [IO93CS, SE13]
G1 UHD Details withheld at licensee's request by SSL.
G1 UHE Details withheld at licensee's request by SSL.
G1 UHI Details withheld at licensee's request by SSL.
G1 UHJ R Morris, 16 Stephens Cl, Luton, LU2 9AN [IO91TV, TL12]
G1 UHL Details withheld at licensee's request by SSL.
G1 UHO Details withheld at licensee's request by SSL.
G1 UHQ Details withheld at licensee's request by SSL.[Op: B Bowker, High Lane, Stockport, Cheshire.]
G1 UHU Details withheld at licensee's request by SSL.
G1 UHW Details withheld at licensee's request by SSL.
G1 UHY Details withheld at licensee's request by SSL.
G1 UHZ Details withheld at licensee's request by SSL.
G1 UIB R D Moxon, 16 Kielder Oval, Harrogate, HG2 7HQ [IO93GX, SE35]
G1 UID H D Kentfield, 1 Torquay Ave, Gosport, PO12 4NS [IO90KT, SU60]
G1 UIG Details withheld at licensee's request by SSL.
GW1 UIK V J Sellek, 12 Norseman Cl, Rhoose, Barry, CF62 3FY [IO81HJ, ST06]
G1 UIO M McVittie, 19 Alder Cres, Parkstone, Poole, BH12 4BD [IO90BR, SZ09]
G1 UIU Details withheld at licensee's request by SSL.
G1 UJB Details withheld at licensee's request by SSL.
G1 UJR Details withheld at licensee's request by SSL.
G1 UJU Details withheld at licensee's request by SSL.
G1 UJV J Griffths, 6 Cedar Wood, Cuddington, Northwich, CW8 2XR [IO83QF, SJ57]
G1 UJX J A Huddlestone, 8 Wilmot Ave, Chaddesden, Derby, DE21 6PL [IO92GW, SK33]
G1 UKA M D Wilkie, 1 Celandine Cl, Billericay, CM12 0SU [JO01FP, TQ69]
G1 UKC Details withheld at licensee's request by SSL.
G1 UKE Details withheld at licensee's request by SSL.[Op: W Dykes.]
G1 UKG C W Brown, 42 Mandeville Cl, Vanbrugh Park, Blackheath, London, SE3 7AH [JO01AL, TQ47]
G1 UKH R Rodgers, Carmel, Chelsfield Ln, Orpington, BR6 7RS [JO01BI, TQ46]
G1 UKI Details withheld at licensee's request by SSL.
G1 UKL Details withheld at licensee's request by SSL.
G1 UKQ Details withheld at licensee's request by SSL.
G1 UKR Details withheld at licensee's request by SSL.
G1 UKW C A Crane, 3 Hawkshead Dr, Royton, Oldham, OL2 6TW [IO83WN, SD90]
G1 UKX Details withheld at licensee's request by SSL.
G1 ULB Details withheld at licensee's request by SSL.[Op: G J Walker, 6 Cannon Street, Eccles, Manchester, M30 0FT.]
G1 ULE A A Lee, 91 Old Vicarage Park, Narborough, Kings Lynn, PE32 1TG [JO02HQ, TF71]
G1 ULN Details withheld at licensee's request by SSL.[Op: J Holdford. Station located near Corby.]
G1 ULP T Garner, Woolhara, Gorefield Rd, Leverington, Wisbech, PE13 5AS [JO02BQ, TF41]
G1 ULQ P J Rowsell, Thomley Hall Farm, Worminghall, Aylesbury, Bucks, HP18 9JZ [IO91LS, SP60]
G1 ULR D A Ross, 2 Jemmetts Cl, Dorchester on Thame, Oxon, OX9 8RA [IO91MQ, SP70]
G1 ULS Details withheld at licensee's request by SSL.
G1 ULZ L G Dalby, 19 Braeside Rd, West Moors, Ferndown, BH22 0JS [IO90BT, SU00]
G1 UMA C Nelson, 31 Fairburn Cl, Fairfield, Stockton on Tees, TS19 7SN [IO94HN, NZ41]
G1 UME Details withheld at licensee's request by SSL.
G1 UMF K B Morpeth, 9 St. Michaels Ave, Gedling, Nottingham, NG4 3NN [IO92KX, SK64]
G1 UMH Details withheld at licensee's request by SSL.
G1 UMJ P A Carslake, 38 Loppets Rd, Tilgate, Crawley, RH10 5DW [IO91VC, TQ23]
G1 UMK B W Drury, 14 Wheatsheaf Ln, Long Bennington, Newark, NG23 5DU [IO92OX, SK84]
G1 UML Details withheld at licensee's request by SSL.
G1 UMM Details withheld at licensee's request by SSL.
G1 UMR Details withheld at licensee's request by SSL.
G1 UMS J C M Preece, White House, Pentre Ln, Bredwardine, Hereford, HR3 6BY [IO82MC, SO34]
G1 UMU Details withheld at licensee's request by SSL.
G1 UMV Details withheld at licensee's request by SSL.[Op: T Jones. Station located near Tamworth.]
G1 UMX Details withheld at licensee's request by SSL.[Op: M A Hirst, 21 Keswick Avenue, Kingston Vale, London, SW15 3QH.]
G1 UMY R Wiltshire, Danesboro, Stonehill Rd, Ottershaw, Chertsey, KT16 0ER [IO91RJ, TQ06]
G1 UNB P T G Durrant, Wagtails, 138 Tilt Rd, Cobham, KT11 3HR [IO91TH, TQ15]
G1 UND Details withheld at licensee's request by SSL.
G1 UNF Details withheld at licensee's request by SSL.
G1 UNG Details withheld at licensee's request by SSL.
G1 UNL Details withheld at licensee's request by SSL.
G1 UNP Details withheld at licensee's request by SSL.
G1 UNU Details withheld at licensee's request by SSL.[Op: K W A Etwell. Station located near Kidderminster.]
G1 UNW Details withheld at licensee's request by SSL.
G1 UOD Details withheld at licensee's request by SSL.
G1 UOL A Muster, 38 Vaughan Gdns, Ilford, IG1 3NZ [JO01AN, TQ48]
G1 UOM Details withheld at licensee's request by SSL.[Op: S D Hewitt, 22 Trafalgar Rd, Ilkley, W Yorks, LS29 8HH.]
G1 UOR W S Rodgers, 9 Hillcrest, Skelmersdale, WN8 9JZ [IO83ON, SD40]
GW1 UOV A J Ham, 133 Maple Dr, Brackla, Bridgend, CF31 2PR [IO81FM, SS97]
G1 UOZ Details withheld at licensee's request by SSL.[Op: A W Knight.]
G1 UPH Details withheld at licensee's request by SSL.
G1 UPK Details withheld at licensee's request by SSL.[Op: T Haslam.]
G1 UPL Details withheld at licensee's request by SSL.
G1 UPR Details withheld at licensee's request by SSL.[Station located near Woking.]
G1 UPT J P Ravelini, 14 Meath Cl, Orpington, BR5 2HF [JO01BJ, TQ46]
G1 UPU Details withheld at licensee's request by SSL.
G1 UPX Details withheld at licensee's request by SSL.
G1 UPY Details withheld at licensee's request by SSL.
G1 UQA Details withheld at licensee's request by SSL.
G1 UQC Details withheld at licensee's request by SSL.
G1 UQF Details withheld at licensee's request by SSL.
G1 UQG Details withheld at licensee's request by SSL.
G1 UQI Details withheld at licensee's request by SSL.
G1 UQK Details withheld at licensee's request by SSL.
G1 UQO Details withheld at licensee's request by SSL.
G1 UQT Details withheld at licensee's request by SSL.
GW1 URD R Ameson, 32 Godrer Gaer, Llwyngwril, LL37 2JZ [IO72XP, SH50]
GW1 URF Details withheld at licensee's request by SSL.
G1 URG K N Forward South Kent Raynet, 17 Cowper Rd, Deal, CT14 9TW [JO01QF, TR35]
G1 URH Details withheld at licensee's request by SSL.
G1 URJ N D Capon, 13 Westcroft, Berinsfield, Oxford, OX10 7NL [IO91JP, SU59]
G1 URL J W Haywood, 2 Elm Rd, Shoeburyness, Southend on Sea, SS3 9PB [JO01JM, TQ98]
G1 URO T Villiers, 31 Eaton Ave, High Wycombe, HP12 3BP [IO91OP, SU89]
G1 URQ D A Traynor, 13 Elgar Cl, Great Sutton, South Wirral, L65 7AZ [IO83MG, SJ37]
G1 URR C P Gough, 67 Pickmere Ln, Wincham, Northwich, CW9 6EB [IO83SG, SJ67]
G1 URW J Carman, 5 Melbourne Rd, Blacon, Chester, CH1 5JQ [IO83ME, SJ36]
G1 URZ T E Bennett, 16 Montgomery Ave, Hemel Hempstead, HP2 4HE [IO91SS, TL00]
G1 USF P M Napp, 23 Harriot Dr, Newcastle upon Tyne, NE12 0EU [IO95EA, NZ27]
G1 USH K Mountford, 22 Hollington Dr, Oxford Tunstalll, Stoke on Trent, ST6 6TZ [IO83VB, SJ85]
G1 USI A F Tawney, Waterside, Riverside, Shaldon, Teignmouth, Devon, TQ14 0DJ [IO80FM, SX97]
GM1 USN J O Challis, Bay Villa, Strachur, Argyll, PA27 8DD [IO76LE, NN00]
GW1 USQ G J Bolster, 52 St. Josephs Ct, Llanelli, SA15 1NR [IO71WQ, SN50]
G1 USV R York, 10 Severn View Rd, Thornbury, Bristol, BS12 1AY [IO81RO, ST69]
G1 USW P R Daniels, 29 Station Rd, Wickwar, Wotton under Edge, GL12 8NB [IO81TO, ST78]
GW1 USX C A Erskine, 302 Caerphilly Rd, Birchgrove, Cardiff, CF4 4NS [IO81JM, ST18]
G1 UTA S H Wilson, 16 Meadow Ct, Wimborne Rd, Bournemouth, BH9 2BU [IO90BS, SZ09]
G1 UTC C R Thomas, 48 Rushfield Rd, Liss, GU33 7LP [IO91NA, SU72]
G1 UTF R E F Beer, 3 Rectory Rd, St. Mary in The Marsh, Romney Marsh, TN29 0BU [JO01LA, TR02]
G1 UTG R A H Gash, 23 Baldwin Rd, Greatstone, New Romney, TN28 8SY [JO00LX, TR02]
G1 UTI Details withheld at licensee's request by SSL.
G1 UTJ A C D Lees, 35 Meadow Cl, Hockley Heath, Solihull, B94 6PF [IO92CI, SP17]
G1 UTM P H Yearsley, 25 Dinmor Rd, Manchester, M22 1NN [IO83UI, SJ88]
G1 UTP S J Darlington, 17 Eleanor Rd, Royton, Oldham, OL2 6BH [IO83WN, SD90]
G1 UTR T McKown, 46 Reins Lee Ave, Filton Hill Est, Oldham, OL8 2QG [IO83WM, SD90]
G1 UTS G May, 95 Moorfield Ave, Denton, Manchester, M34 7TX [IO83WK, SJ99]
G1 UTW P J Keavey, 4 Brandlesholme Rd, Bury, BL8 1AS [IO83UO, SD71]
G1 UTY P D Mackley, 105 Kingsway North, Braunstone, Leicester, LE3 3BE [IO92JO, SK50]
G1 UTZ A Peters, 10 Hill View Cl, Grantham, NG31 7PH [IO92QV, SK93]
G1 UUF I Grounsell, 9 Waverton Cl, Southfield Lea, Cramlington, NE23 6PF [IO95FB, NZ27]
G1 UUK G A Perry, 6 Morgan Cl, Arley, Coventry, CV7 8PR [IO92FM, SP28]
G1 UUL S D Brough, Flat 6, 47 Wolverhampton Rd, Cannock, WS11 1AP [IO82XQ, SJ90]
G1 UUP A Robins, 38 Eastbourne Rd, Willingdon, Eastbourne, BN20 9NS [JO00DT, TQ50]
G1 UUS M J Carter, 30 Hornbeam Dr, Tile Hill, Coventry, CV4 9UJ [IO92FJ, SP37]
G1 UUT R Bygate, 91 Woodcote Ave, Kenilworth, CV8 1BE [IO92EI, SP27]
G1 UUX P A Moss, 26 Bath St., Market Harborough, LE16 9EL [IO92NL, SP78]
G1 UUZ M S Davis, 20 Pavilion Ave, Smethwick, Warley, B67 6LA [IO92AL, SP08]
G1 UVB J T Hencher, 36 Glenthorne Ave, Brickfields, Worcester, WR4 9TS [IO82VE, SO85]
G1 UVD R Rothery, 2 Highcroft, Mount Pleasant, Batley, WF17 7NT [IO93EQ, SE22]
G1 UVE A D Shackleton, 31 Ashton St., Harehills, Leeds, LS8 5BY [IO93FT, SE33]
G1 UVI M T Stanley, 35 Moorgate Rd, Kippax, Leeds, LS25 7ET [IO93HS, SE43]
G1 UVK A P Linfoot, 7 St. Marys Gr, Osbaldwick, York, YO1 3PZ [IO93LW, SE65]
G1 UVL D A Bedford, 10 Hambleton Ave, Osbaldwick, York, YO1 3PP [IO93LX, SE65]
GW1 UVM B M Gigg, 29 Fairways Cres, Fairwater, Cardiff, CF5 3DZ [IO81JL, ST17]
GW1 UVN J F Jones, Silver Springs, Church Rd, Gilwern, Abergavenny, Gwent, NP7 0EL [IO81KT, SO21]
G1 UVQ Details withheld at licensee's request by SSL.
G1 UVR T Roscoe, 11 Tennyson St., Derker, Oldham, OL1 4NH [IO83WN, SD90]
G1 UVU Details withheld at licensee's request by SSL.
G1 UVW Details withheld at licensee's request by SSL.
G1 UVZ Details withheld at licensee's request by SSL.
G1 UWD D Peachey, 28 Broad St., Truro, TR1 1JD [IO70LG, SW84]
G1 UWE M D Peachey, 28 Broad St., Truro, TR1 1JD [IO70LG, SW84]
G1 UWF Details withheld at licensee's request by SSL.
G1 UWG Details withheld at licensee's request by SSL.
G1 UWH M C Robins, 43 Fairfield, Gamlingay, Sandy, SG19 3LG [IO92VD, TL25]
G1 UWL M L Troy, 90 Evenlode Rd, Millbrook, Southampton, SO16 9EH [IO90GW, SU31]
G1 UWM G J Moreton, 12 Barn Rd, Shifnal, TF11 8EL [IO82TQ, SJ70]
G1 UWN A L Jones, 89 Doddington, Hollinswood, Telford, TF3 2DQ [IO82SQ, SJ70]
G1 UWP Details withheld at licensee's request by SSL.
G1 UWR P Dixon, 56 Vale View, Nuneaton, CV10 8AP [IO92GM, SP39]
G1 UWS Details withheld at licensee's request by SSL.
G1 UWU Details withheld at licensee's request by SSL.[Op: R J C Burt.]
G1 UWV B Dixon, Terridene, Park Rd, Barton on Sea, New Milton, BH25 6QE [IO90ER, SZ29][Op: Brian Dixon.]
G1 UWW Details withheld at licensee's request by SSL.
G1 UWX Details withheld at licensee's request by SSL.
G1 UXA Details withheld at licensee's request by SSL.
G1 UXB Details withheld at licensee's request by SSL.[Op: J S Booth.]
G1 UXC Details withheld at licensee's request by SSL.[Op: Dr I S Ross. Station located near Keele.]
G1 UXG Details withheld at licensee's request by SSL.
G1 UXH Details withheld at licensee's request by SSL.
G1 UXJ P A Howett, 37 Imperial Ave, Kidderminster, DY10 2RA [IO82VJ, SO87]
G1 UXP Details withheld at licensee's request by SSL.[Op: A Woodthorpe. Station located near Harrogate.]
G1 UXQ Details withheld at licensee's request by SSL.[Op: B Cherry, 138 Cemetery Road, Barnsley, S Yorks, S70 1XH.]
G1 UXR Details withheld at licensee's request by SSL.
GW1 UXW Details withheld at licensee's request by SSL.[Op: M L Lewis. Station located near Cwmbran.]
G1 UXZ P E Sage, 40 Wentworth Dr, Cliffe Woods, Cliffe, Rochester, ME3 8UL [JO01GK, TQ77]
G1 UYA Details withheld at licensee's request by SSL.
G1 UYE Details withheld at licensee's request by SSL.
G1 UYH Details withheld at licensee's request by SSL.
G1 UYK Details withheld at licensee's request by SSL.
G1 UYL D J Thurtell, 41 Percy Gdns, Isleworth, TW7 6BX [IO91UL, TQ17]
G1 UYM Details withheld at licensee's request by SSL.
G1 UYP Details withheld at licensee's request by SSL.
G1 UYT S Tomkinson, 3 Heysham Cl, Weston Coyney, Stoke on Trent, ST3 6RG [IO82WX, SJ94]
GW1 UYW K R Wallis, Cartref Newydd, Plot 3, Porth y Cwm, Llanarmon D.C., LL20 X [IO83BD, SH76]
G1 UYZ D J Payne, 8 Gascoigne Dr, Spondon, Derby, DE21 7GL [IO92HW, SK33]
G1 UZC P S Jarrett, 17 Wolmers Hey, Great Waltham, Chelmsford, CM3 1DA [JO01FT, TL61]
G1 UZD J R Yates, 8 Holt Dr, Wickham Bishops, Witham, CM8 3JR [JO01IS, TL81]
G1 UZR K Small, 55 Windmill Ave, Birstall, Leicester, LE4 4JN [IO92KQ, SK50]
G1 UZS E C Lees, 11 Edale Ave, Mickleover, Derby, DE3 5FY [IO92FW, SK33]
G1 UZT C I Tuckley, 28 Hume Brae, Immingham, Grimsby, DN40 1PD [IO93VO, TA11]
G1 UZW Details withheld at licensee's request by SSL.
G1 VAB D W Goodwill, 94 Palmerston St., Derby, DE23 6PF [IO92GV, SK33]
G1 VAC D Allsebrook, 12 Portman Chase, Sinfin Moor, Sinfin, Derby, DE24 3BQ [IO92GV, SK33]
GM1 VAD S Scanlain, Crossraguel, Old Newton, Cawdor, Nairnshire, IV12 5RA [IO87AM, NH85]
G1 VAE Details withheld at licensee's request by SSL.
G1 VAF Details withheld at licensee's request by SSL.
G1 VAG A Grant, 26 Fountains Ave, Boston Spa, Wetherby, LS23 6PX [IO93HV, SE44]
G1 VAJ A J Leach, 8 Eskdale, Brownsover, Rugby, CV21 1NJ [IO92JJ, SP57]
G1 VAL M N Pearson, 5 Windgate, Silsden, Keighley, BD20 0LG [IO93AV, SE04]
G1 VAN P A Van Falier, 572 Stafford Rd, Fordhouses, Wolverhampton, WV10 6NN [IO82WO, SJ90]
G1 VAO A Andrews, 12 Kings Lea, Ossett, WF5 8RY [IO93EQ, SE22]
GW1 VAW B Williams, 10 Tynybedw Terr, Treorchy, CF42 6RL [IO81FP, SS99]
GM1 VAX R Burns, Holmhill, Silverhillock, Cornhill, Banffshire, AB45 2ES [IO87PO, NJ55]
G1 VAY D Hearn, 90 Princes Dr, Valley Dip, Seaford, E Sussex, BN25 2TX [JO00BS, TQ40]
GI1 VAZ G A Richardson, 6 Cedarhurst Rise, Beechill Rd, Belfast, BT8 4RJ [IO74BN, J36]
GW1 VBA P M Buckle, 36 Yewberry Ln, Malpas, Newport, NP9 6WL [IO81MO, ST39]
G1 VBB R J Bedwell, 36 Spindleberry Gr, Nailsea, Bristol, BS19 1QF [IO81PK, ST47]
GM1 VBD R Scott, Beechview, Enzie Slackhead, Buckie, Moray, AB5 2BR [IO87TH, NJ72]
GM1 VBE S Clink, Southsyde, Woodhead Ave, Bothwell, Lanarkshire, G71 8AR [IO75XT, NS75]
GM1 VBG G R Ogilvie, 54 North Rd, Johnstone, PA5 8NF [IO75RT, NS46]
G1 VBL R J Kelsall, 144 Newland, Witney, OX8 6JH [IO91GS, SP30]
G1 VBP P A Smith, 26 Thomas Ave, Radcliffe on Trent, Nottingham, NG12 2HT [IO92LW, SK63]
G1 VBQ P L Wright, 81 High St., Syston, Leicester, LE7 1GQ [IO92LQ, SK61]
G1 VBR Details withheld at licensee's request by SSL.
G1 VBT Details withheld at licensee's request by SSL.
G1 VBX Details withheld at licensee's request by SSL.
G1 VCA Details withheld at licensee's request by SSL.
G1 VCB Details withheld at licensee's request by SSL.
G1 VCD Details withheld at licensee's request by SSL.
G1 VCE J T Thorpe, 20 Fernhill Way, Wolvey, Hinckley, LE10 3LP [IO92HL, SP48]
GW1 VCN Details withheld at licensee's request by SSL.
GW1 VCP Details withheld at licensee's request by SSL.
G1 VCT Details withheld at licensee's request by SSL.[Station located near Burton-on-Trent.]
G1 VCY Dr G Brown, 1 Langford Cres, Benfleet, SS7 3JP [JO01GN, TQ78]
G1 VCZ A H E Hewes, 89 Fronks Rd, Dovercourt, Harwich, CO12 4EQ [JO01PW, TM23]
G1 VDA Details withheld at licensee's request by SSL.

G1 VDB Details withheld at licensee's request by SSL.
G1 VDC M Jeffers, 20 Three Crowns Rd, Colchester, CO4 5AD [JO01KV, TL92]
G1 VDE D A Taylor, 21 Munday Cl, Bussage, Stroud, GL6 8DG [IO81WR, SO80]
G1 VDF K D Griffiths, 47 Swiss Walk, Fayre Oaks, Kings Acre Rd, Hereford, HR4 05X
G1 VDI Dr J H Wood, 24 The Villas, Stoke on Trent, ST4 5AQ [IO82VX, SJ84]
G1 VDO S Preston, 50 Milton Ave, Peasey Hill, Malton, YO17 0LB [IO94OD, SE77]
G1 VDS W Potter, 34 Marshall Dr, Pickering, YO18 7JT [IO94OF, SE88]
GW1 VDT P Whatley, 28 Treetops, Portskewett, Newport, NP6 4SQ [IO81PO, ST48]
G1 VDY J S D Dale, 317 Waterside Dr, Blurton, Stoke on Trent, ST3 3LG [IO82WX, SJ84]
GM1 VDZ G R Neil, 48 Dalfarson Ave, Dalmellington, Ayr, KA6 7TY [IO75TH, NS40]
G1 VEB Details withheld at licensee's request by SSL.
G1 VEQ Details withheld at licensee's request by SSL.
G1 VEU Details withheld at licensee's request by SSL.
G1 VFE Details withheld at licensee's request by SSL.
G1 VFG Details withheld at licensee's request by SSL.
G1 VFI Details withheld at licensee's request by SSL.[Op: B S Smith, 23 Bradfield Close, Burpham, Guildford, Surrey GU4 7XT.]
GM1 VFQ J Murdoch, 8 Primpton Ave, Dalrymple, Ayr, KA6 6EL [IO75QJ, NS31]
GM1 VFR Details withheld at licensee's request by SSL.
GM1 VFS Details withheld at licensee's request by SSL.
G1 VGA K G Crane, 92 Dimond Rd, Bitterne Park, Southampton, SO18 1JS [IO90HW, SU41]
G1 VGC W N Aldcroft, 4 Allington St., Liverpool, L17 7AD [IO83MJ, SJ38]
G1 VGE L Lee, 7 Wyedale Rd, Haydock, St. Helens, WA11 0HN [IO83PL, SJ59]
G1 VGF Details withheld at licensee's request by SSL.
G1 VGH A R Harding, 39 North Denes Rd, Great Yarmouth, NR30 4LU [JO02UO, TG50]
G1 VGI K J Waterson, 20 Cadogan Rd, Bury St. Edmunds, IP33 3QJ [JO02IF, TL86]
G1 VGM S Ball, 7 Leavington House, East Hill Rd, Ryde, Isle of Wight, PO33 1LU [IO90KR, SZ69]
G1 VGN A J Peach, The Old Smithy, Seaford Ln, Naunton Beauchamp, Pershore, WR10 2LL [IO82XE, SO95]
G1 VGO D Holdsworth, 62 Pontefract Rd, Wombwell, Barnsley, S73 0YG [IO93HM, SE40]
G1 VGP G Anderson, 1 White Rose Mead, Garforth, Leeds, LS25 2EG [IO93HT, SE43]
G1 VGW Details withheld at licensee's request by SSL.
G1 VGX Details withheld at licensee's request by SSL.
GM1 VGZ J Crowden, Brigga, Main St., Castletown, Thurso, KW14 8TU [IO88HO, ND16]
G1 VHC Details withheld at licensee's request by SSL.
G1 VHG Details withheld at licensee's request by SSL.
G1 VHL Details withheld at licensee's request by SSL.[Op: G E Brinkman.]
G1 VHM Details withheld at licensee's request by SSL.[Op: T V Warren, 2 Central Avenue, Sandiacre, Notts, NG10 5FN.]
G1 VHN Details withheld at licensee's request by SSL.
G1 VHQ Details withheld at licensee's request by SSL.
GI1 VHU Details withheld at licensee's request by SSL.
G1 VHW W Killeen, 12 Meriden Gr, Lostock, Bolton, BL6 4RQ [IO83RN, SD60]
G1 VHX C L P M Melia, Melia, 21 Boggart Hill Gdns, Leeds, LS14 1LJ [IO93GT, SE33]
G1 VHY D Symonds, 79 Kingsway, Kirkby in Ashfield, Nottingham, NG17 7EH [IO93JC, SK55]
G1 VIB Details withheld at licensee's request by SSL.
G1 VID T Howe, 7 The Vale, Broadstairs, CT10 1RB [JO01RI, TR36]
G1 VIE Details withheld at licensee's request by SSL.[Op: T J Pickett. Station located near Rochester.]
G1 VIG D Carrick, 7 May Cl, Owlsmoor, Camberley, Surrey, GU15 4UG [IO91PI, SU86]
G1 VII N Pooley, 64 Lynwood Gr, Orpington, BR6 0BH [JO01BJ, TQ46]
G1 VIK Details withheld at licensee's request by SSL.
G1 VIM Details withheld at licensee's request by SSL.
G1 VIN C F Kirkland, 29 Shelley Rd, Enderby, Narborough, Leicester, LE9 5QX [IO92JO, SP59]
G1 VIO A G Eades, 41 Woodhall Gate, Pinner, HA5 4TX [IO91TO, TQ19]
G1 VIP M S Walker, 'Bessbrook', Cogdean, Corfe Mullen, Dorset, BH21 3DS [IO80XS, SY99]
G1 VIR I R Berry, 2 Pottery Cttgs, Trefonen, Oswestry, SY10 9DF [IO82KU, SJ22]
G1 VIS S I Grimes, Orchard End, 73 Ryston Rd, Denver, Downham Market, PE38 0DP [JO02EO, TF60]
G1 VIT N H Cliff, 12 Newchurch Rd, Wellington, Telford, TF1 1JH [IO82RQ, SJ61]
G1 VIW R H G Paterson, 27 Copt Heath Dr, Knowle, Solihull, B93 9PA [IO92DJ, SP17]
G1 VIY T Depledge, 13 Peel Dr, Astbury, Congleton, CW12 4RF [IO83VD, SJ86]
G1 VIZ A J Hemenway, 69 Sixth Ave, Heworth, York, YO3 0UR [IO93LX, SE65]
GW1 VJB B R Flounders, 76 Springfield Rd, Sebastopol, Pontypool, NP4 5BX [IO81LQ, ST29]
G1 VJE M B Sayers, 21 Chesham Rd North, Weston Super Mare, BS22 8AD [IO81MI, ST36]
G1 VJF F M Rappolt, Gold Tops, 49 Farm Rd, Maidenhead, SL6 5JA [IO91OM, SU88]
G1 VJG C M Rhenius, 5 Water St., Cambridge, CB4 1NZ [JO02BF, TL45]
G1 VJH C S Harper, 2 Lyme Green Park, Sutton, Lyme Green, Macclesfield, SK11 0LD [IO83WF, SJ97]
G1 VJJ C J Silvey, 43 Suffolk Ave, Leigh on Sea, SS9 3HF [JO01IN, TQ88]
G1 VJM R W Vandepeer, 148 Laburnum Rd, Strood, Rochester, ME2 2LD [JO01FJ, TQ76]
G1 VJN B M Jones, 73 Tonge Rd, Murston, Sittingbourne, ME10 3NR [JO01JI, TQ96]
G1 VJQ T Monaghan, 15 Mulgrave Rd, Roe Green, Worsley, Manchester, M28 2RW [IO83TM, SD70]
G1 VJS Details withheld at licensee's request by SSL.
G1 VJT D M Groom, 19 Fern Way, Horsham, RH12 5XE [IO91UB, TQ13]
G1 VKA B F A Gould, 20 Kings Coughton Ln, Kings Coughton, Alcester, B49 5QE [IO92BF, SP05]
G1 VKB A E Morris, 140 Astwood Rd, Worcester, WR3 8EZ [IO82VE, SO85]
G1 VKC D Close, 12 Maple Walk, Lead Ln, Ripon, HG4 2ND [IO94FC, SE36]
GM1 VKG J H Howarth, Cruachan, 25 The Loan, Torphichen, Bathgate, EH48 4NF [IO85EW, NS97]
GM1 VKI J W R Kelly, 15 Morlich Rd, Dalgety Bay, Dunfermline, KY11 5UF [IO86HA, NT18]
G1 VKJ B M Duffy, 26 Cherry Blossom Cl, Billing Leys, Northampton, NN3 9DN [IO92OG, SP86]
G1 VKL S Bullers, 150 Eckington Rd, Coal Aston, Sheffield, S18 6AZ [IO93GH, SK37]
G1 VKN S A Matthews, 54 Belper Ave, Carlton, Nottingham, NG4 3SD [IO92KX, SK64]
G1 VKP Details withheld at licensee's request by SSL.
G1 VKT E Dale, 29 Hulme Rd, Leigh, WN7 5BT [IO83RM, SD60]
G1 VKY D M Musgrove, 27 Main St., Eastwood, Nottingham, NG16 3JJ [IO93IA, SK44]
G1 VKZ Details withheld at licensee's request by SSL.
GM1 VLA A Lee, Sandana, Chapelknowe Rd, Kirkpatrick Fleming, Lockerbie, Dumfriesshire, DG11 3BA [IO85KA, NY27]
G1 VLB J D Stannard, 10 Hornbeam Cl, Farnborough, GU14 9TR [IO91OH, SU85]
G1 VLC Details withheld at licensee's request by SSL.
G1 VLD A R Burkitt, 6 Chewells Cl, Haddenham, Ely, CB6 3XE [JO02BI, TL47]
G1 VLG Details withheld at licensee's request by SSL.
G1 VLM Details withheld at licensee's request by SSL.
G1 VLR Details withheld at licensee's request by SSL.
G1 VLS G D Turner, 121 Kingshurst Rd, Northfield, Birmingham, B31 2LJ [IO92AJ, SP07]
G1 VLT Details withheld at licensee's request by SSL.
GW1 VLW Details withheld at licensee's request by SSL.
GI1 VLZ Details withheld at licensee's request by SSL.
GW1 VMA M J S Hills, Ty Newydd Rhos, Llandyssul, Dyfed, SA44 5HE [IO71TX, SN33]
GI1 VMF J Oliver, 29 Callan Bridge Park, Armagh, BT60 4BU [IO64PI, H84]
G1 VMG Details withheld at licensee's request by SSL.
G1 VMH Details withheld at licensee's request by SSL.
G1 VMI Details withheld at licensee's request by SSL.
GW1 VMK Details withheld at licensee's request by SSL.
G1 VML Details withheld at licensee's request by SSL.
G1 VMP Details withheld at licensee's request by SSL.
G1 VMR Details withheld at licensee's request by SSL.
G1 VMS Details withheld at licensee's request by SSL.
G1 VMT Details withheld at licensee's request by SSL.[Op: R G Andrews. Station located near Broadstairs.]
G1 VMU Details withheld at licensee's request by SSL.
G1 VMW Details withheld at licensee's request by SSL.
G1 VMX E V Driver, 39 Witham Rd, Woodhall Spa, LN10 6RW [IO93VD, TF16]
G1 VMY Details withheld at licensee's request by SSL.
G1 VNE S Nocera, Strada Provinciale, Mulazzano 156, Lesigano de Bagni, Parma, Italy, 430 10
G1 VNK D M Reed, 89 Shaftesbury Ave, Montpelier, Bristol, BS6 5LU [IO81QL, ST57]
G1 VNL A J Smith, 112 Manor Ln, Charfield, Wotton under Edge, GL12 8TN [IO81TO, ST79]
G1 VNM S L Eyers, 190 Greenhill Rd, Herne Bay, CT6 7RS [JO01NI, TR16]
G1 VNR R Leeson, 178 Tudor Rd, Hinckley, LE10 0EH [IO92HN, SP49]
G1 VNS K G Baum, 11 St. Ives Rd, Wigston, LE18 2JB [IO92KN, SP69]
G1 VNU S G Exell, 5 Longfield Rd, Dorking, Surrey, RH4 3DE [IO91TF, TQ14]
G1 VNV Details withheld at licensee's request by SSL.
G1 VNY A L James, 173 Hillside Rd, Hastings, TN34 2QJ [JO00GV, TQ81]
G1 VNZ A S Atkins, 1 Angela Cl, Pebsham, Bexhill on Sea, TN40 2RH [JO00GU, TQ70]
G1 VOB M E Prior, 36 Bassnage Rd, Halesowen, B63 4HQ [IO82XK, SO98]
G1 VOC A James, 16 Wilson St., Wombwell, Barnsley, S73 8LP [IO93HM, SE30]
G1 VOD E T Silcock, 82 Woolgreaves Dr, Sandal, Wakefield, WF2 6DT [IO93GP, SE31]
G1 VOJ A Duce, 16 Gillmans Rd, Orpington, BR5 4LA [JO01BJ, TQ46]
G1 VON S J Howard, 91 Greenbarn Way, Blackrod, Bolton, BL6 5TE [IO83RO, SD61]
G1 VOP Dr D Hettiarratchi, 2 Carham Cl, Gosforth, Newcastle upon Tyne, NE3 5DX [IO95EA, NZ26]
G1 VOQ P N Tandy, Old Channel Hill Farmhouse, Damerham, Fordingbridge, Hants, SP6 3HA [IO90BW, SU11]
G1 VOX M S Cowgill, 25 Long Causeway, Stanley, Wakefield, WF3 4JA [IO93GR, SE32]
G1 VOY R Ward, Overdale, Egton, Whitby, North Yorks, YO21 1UE [IO94PK, NZ80]
GI1 VPA D W Smythe, 57 Ballymacormick Ave, Bangor, BT19 6AY [IO74EP, J58]
G1 VPC M Wigley, Ground Floor Flat, 23 Rutland Rd, Hove, BN3 5FF [IO90VT, TQ20]
G1 VPE R J Wilcockson, 11 Highland Rd, New Whittington, Chesterfield, S43 2EZ [IO93HG, SK37]
G1 VPH S J Currie, 45 Orford Ave, Warrington, WA2 7QH [IO83RJ, SJ68]
GM1 VPK Details withheld at licensee's request by SSL.
GM1 VPL Details withheld at licensee's request by SSL.
G1 VPM Details withheld at licensee's request by SSL.
G1 VPP Details withheld at licensee's request by SSL.
G1 VPR Details withheld at licensee's request by SSL.
G1 VPS Details withheld at licensee's request by SSL.
G1 VPU J Newman, 27 Southwood Ave, Southbourne, Bournemouth, BH6 3QB [IO90CR, SZ19]
G1 VPV Details withheld at licensee's request by SSL.
G1 VPW Details withheld at licensee's request by SSL.
G1 VPX Details withheld at licensee's request by SSL.
G1 VQH G A King, Mill Cottage, 48 Mill St., Coton in The Elms, Swadlincote, DE12 8ES [IO92ER, SK21]
G1 VQI Details withheld at licensee's request by SSL.
G1 VQK E J Wright, 94 Bachelor Gdns, Harrogate, HG1 3EA [IO94FA, SE35]
G1 VQO Details withheld at licensee's request by SSL.
G1 VQQ Details withheld at licensee's request by SSL.
G1 VQV Details withheld at licensee's request by SSL.
G1 VQW Details withheld at licensee's request by SSL.
G1 VRA E A Jones, 3 The Green, Rampton, Cambridge, CB4 4QB [JO02BG, TL46]
G1 VRC T E Nicholson, 8 East St., High Spen, Rowlands Gill, NE39 2HD [IO94CW, NZ15]
G1 VRF W T Smith, 136 Wayside Green, Woodcote, Reading, RG8 0QJ [IO91LM, SU68]
G1 VRG Details withheld at licensee's request by SSL.
G1 VRH Details withheld at licensee's request by SSL.
G1 VRJ J G Ager, 20 Kirktonhill Rd, Westlea, Swindon, SN5 7AF [IO91CN, SU18]
G1 VRP M E Carter, 16 St. Wulstan Way, Southam, Leamington Spa, CV33 0TT [IO92HF, SP45]
GW1 VRR J H Williams, 31 Syr Davids Ave, Pencisely, Llandaff, Cardiff, CF5 1GH [IO81JL, ST17]
GW1 VRW C K King, 48 Severn Ave, Barry, CF62 7PW [IO81IJ, ST16]
G1 VSD W Bennett, 90 Garwood Rd, Yardley, Birmingham, B26 2AW [IO92CL, SP18]
G1 VSE D J Herridge, 8 Kelvinside, Stanford-le-Hope, SS17 8BP [JO01FM, TQ68]
G1 VSH G T Pettit, 7 Dunster Cres, Hornchurch, RM11 3QD [JO01CN, TQ58]
G1 VSJ G I Conway, 25 Hazelwood, Firwood Park, Chadderton, Oldham, OL9 9TB [IO83WN, SD80]
G1 VSK A Heyes, 2A Margaret Ave, Woolston, Warrington, WA1 3UN [IO83RJ, SJ68]
G1 VSL Details withheld at licensee's request by SSL.
G1 VSM D W Pratt, 17 Worcester Gdns, Greenford, UB6 0BH [IO91TN, TQ18]
G1 VSO C J Champ, 34 Kennett Cl, Gr, Wantage, OX12 0NJ [IO91GO, SU39]
G1 VSQ J V Ellis, 1 Delbury Cttgs, East Grinstead Rd, North Chailey, Lewes, BN8 4DD [IO90XX, TQ32]
GM1 VSR A Bain, 4 Bawdley Head, Fraserburgh, Aberdeenshire, AB4 5SE [IO87TH, NJ72]
G1 VSU Details withheld at licensee's request by SSL.
G1 VSV Details withheld at licensee's request by SSL.
G1 VTA Details withheld at licensee's request by SSL.
G1 VTC Details withheld at licensee's request by SSL.
G1 VTD Details withheld at licensee's request by SSL.
G1 VTE Details withheld at licensee's request by SSL.
G1 VTI Details withheld at licensee's request by SSL.
G1 VTK Details withheld at licensee's request by SSL.
G1 VTN C E Peacock, 64 Cleveland, Tunbridge Wells, TN2 3NH [JO01DD, TQ54]
G1 VTO B B C Kemp, 193 Cavalry Park, March, PE15 9DL [JO02BM, TL49]
G1 VTQ B Gorman, 40 Maudland Bank, Preston, PR1 2YL [IO83PS, SD52]
G1 VTS J Smith, 9 Birchway, Hayes, UB3 3PA [IO91TM, TQ18]
G1 VTU R E Arnold, The Ridge, Hickman Rd, Galley Common, Nuneaton, CV10 9NG [IO92FM, SP39]
G1 VTV A Cowley, 30 New Rd, Bromham, Chippenham, SN15 2JB [IO81XJ, ST96]
G1 VTW N K Smith, 23 Haddon Cl, Macclesfield, SK11 7YG [IO83WF, SJ97]
GM1 VTY D W M Muir, 13 Greenend Gr, Edinburgh, EH17 7QE [IO85KV, NT26]
G1 VUA Details withheld at licensee's request by SSL.
G1 VUE M P Rayner, 13 St. Bartholomews Cl, Cam, Dursley, GL11 5US [IO81TQ, SO70]
G1 VUG M B Cooper, Osborne House, Main St., Fulstow, Louth, LN11 0XF [IO93XK, TF39]
GM1 VUH H Hume, 5 South Middleton Cottage, Middleton, Gorebridge, EH23 4RF [IO85LT, NT35]
G1 VUI Details withheld at licensee's request by SSL.
G1 VUK S Hunt, 37 Tanza Rd, Hampstead, London, NW3 2UA [IO91WN, TQ28]
GM1 VUL Details withheld at licensee's request by SSL.[Op: F A S Keighren.]
G1 VUQ Details withheld at licensee's request by SSL.[Op: E F W Brewer, 85 Mead Way, Bushey, Watford, Herts, WD2 2DJ.]
G1 VUS Details withheld at licensee's request by SSL.
GM1 VUT Details withheld at licensee's request by SSL.
GM1 VUU Details withheld at licensee's request by SSL.
G1 VUX Details withheld at licensee's request by SSL.
G1 VUY D White, 14 Waggoners Way, Bugbrooke, Northampton, NN7 3QT [IO92MF, SP65]
G1 VVA K W Orme, 676 High St., Tunstall, Stoke on Trent, ST6 5PJ [IO83VB, SJ85]
G1 VVB S D Patrick, 9 Brant Ave, Illingworth, Halifax, HX2 8DL [IO93BR, SE02]
G1 VVE G J Bindon, 70 Gill Cres, Taunton, TA1 4NS [IO81KA, ST22]
G1 VVF M J Clews, 137 Wyatt Rd, Sutton Coldfield, B75 7ND [IO92CN, SP19]
G1 VVJ S W Long, 56 Harriseahead Ln, Harriseahead, Stoke on Trent, ST7 4RB [IO83VC, SJ85]
GW1 VVK D C Phillips, 34 Graig Terr, Graig, Pontypridd, CF37 1NH [IO81HO, ST08]
G1 VVN E Gouldsbrough, 33 Lingwood Rd, Great Sankey, Warrington, WA5 3EN [IO83QJ, SJ58]
G1 VVT H T Mawson, 112 Paget St., Loughborough, LE11 5DU [IO92JS, SK51]
G1 VVU J Stephens, 34 King St., Seahouses, NE68 7XR [IO95EN, NU23]
G1 VVX I Andronov, 3 Thurloe Pl Mews, London, SW7 2HL [IO91VL, TQ27]
G1 VVY J E Waters, Bryher, 27 Hamilton Cl, Epsom, KT19 8RG [IO91UI, TQ26]
GM1 VWA J Gilruth, 88 Fintry Cres, Dundee, DD4 9EX [IO86ML, NO43]
G1 VWB S Cross, 27 Arun, East Tilbury, Grays, Essex, RM18 8SX [JO01FL, TQ67]
G1 VWH W Kirk, 3 Villiers Dr, Arbourthorne, Sheffield, S2 2AW [IO93GI, SK38]
G1 VWM K F Bell, 19 Goodwin Dr, Hogsthorpe, Skegness, PE24 5NY [JO03DF, TF57]
G1 VWP S M Goodwin, 41 Mount Rd, Prestwich, Manchester, M25 2GP [IO83UM, SD80]
G1 VWU A L Driver, Grace Barn, Pencarrow, Advent, Camelford, Cornwall, PL32 9RZ [IO70PO, SX18]
G1 VWZ R C L Huntley, The Old Granary, 49 Main St., Wetwang, Driffield, YO25 9XL [IO94RA, SE95]
G1 VXI Details withheld at licensee's request by SSL.
G1 VXJ Details withheld at licensee's request by SSL.
G1 VXK Details withheld at licensee's request by SSL.
G1 VXL Details withheld at licensee's request by SSL.
G1 VXN Details withheld at licensee's request by SSL.
G1 VXQ Details withheld at licensee's request by SSL.
G1 VXT G Round, 14 Bridge Garth, Clifford, Wetherby, LS23 6HF [IO93HV, SE44]
G1 VXU P J Sterry, 23 Eardisley Cl, Matchborough East, Redditch, B98 0BX [IO92BH, SP06]
G1 VXV Details withheld at licensee's request by SSL.
G1 VXX A C Powell, Bei Gani, 5 St. Peters Cl, Headley, Thatcham, Berks, RG19 8AE [IO91II, SU56]
G1 VYA H J Seddon, 15 Westfield Gr, Wigan, WN1 2QJ [IO83QN, SD50]
G1 VYB T A Smith, 16 Trafalgar Terr, Scarborough, North Yorks, YO12 7QG [IO94TG, TA08]
G1 VYC E A Walker, North Walk Farm Cottage, Louth Rd, Hainton, Market Rasen, LN8 6LB [IO93VI, TF28]
GM1 VYG J McDonald, 4 Braeside, Bowfield Rd, Howwood, Johnstone, Renfrewshire, PA9 1BP [IO75RT, NS36]
G1 VYK Details withheld at licensee's request by SSL.
G1 VYM D Eveleigh, 180 Belswains Ln, Hemel Hempstead, HP3 9XA [IO91SR, TL00]
G1 VYS A T Walker, 49 Selworthy Rd, Norton Green, Stoke on Trent, ST6 8PL [IO83WB, SJ95]
GW1 VYT T M H Randall, Plas Pitcairn, 36 Newlands Park, Valley, Holyhead, Gwynedd, LL65 3AR [IO73RG, SH28]
G1 VYW Details withheld at licensee's request by SSL.
GI1 VYZ Details withheld at licensee's request by SSL.
G1 VZC D J Phelps, 35 Belmont Ave, Hereford, HR2 7JQ [IO82PB, SO53]
GM1 VZG T C D Gilmour, Fiold, Rope Walk, Kirkwall, Orkney, KW15 1XJ [IO88MX, HY40]
G1 VZJ Details withheld at licensee's request by SSL.
G1 VZK Details withheld at licensee's request by SSL.
G1 VZM Details withheld at licensee's request by SSL.
G1 VZT B S Rayner, 6 Budds Ln, Wittersham, Kent, TN30 7EL [JO01IA, TQ92]
G1 VZW B Hitchen, 40 Methuen Ave, Fulwood, Preston, PR2 9QX [IO83PS, SD53]
G1 WAB J Wainwright Derbys Workd All Britain G.Arc, 8 Common Ln, Cutthorpe, Chesterfield, S42 7AN [IO93GG, SK37]
G1 WAC D G Dawkes Wythall Rad Club, 83 Alcester Rd, Hollywood, Birmingham, B47 5NR [IO92BJ, SP07]
G1 WAE C W Rogers, 55 Moore Ave, St. Helens, Merseyside, WA9 2PP [IO83PK, SJ59]
G1 WAI Details withheld at licensee's request by SSL.
G1 WAL Details withheld at licensee's request by SSL.[Op: K G Swinburn.]
G1 WAP B Stott, 35 Sheridan Rd, Laneshawbridge, Colne, BB8 7HW [IO83WU, SD94]
G1 WAR S R Parker Wigston Rad Club, 22 Lincoln Dr, Syston, Leicester, LE7 2JW [IO92LQ, SK61]
G1 WAS Details withheld at licensee's request by SSL.
G1 WAT Details withheld at licensee's request by SSL.

G1 WAW P W Mitchell Wessex Aw Club, 99 Pine Rd, Winton, Bournemouth, BH9 1LU [IO90BR, SZ09]
G1 WBG F H Deverell, 502 Sutton Rd, Walsall, WS5 3AY [IO92AN, SP09]
G1 WBL A J Palmer, 16A Ramsey Rd, Whittlesey, Peterborough, PE7 1DR [IO92WN, TL29]
G1 WBM R E Henson, 42 Yew Tree Dr, Shirebrook, Mansfield, NG20 8QJ [IO93JE, SK56]
G1 WBN Details withheld at licensee's request by SSL.[Op: S G Mitchell. Station located near Whitchurch.]
G1 WBO Details withheld at licensee's request by SSL.
G1 WBP W B Povey, Baxters, 16 Worfield, Bridgnorth, Shropshire, WV15 5LF [IO82TN, SO79]
GM1 WBT P L Dingle, Fordale, Mill Rd, Old Meldrum, Aberdeenshire, AB5 0BD [IO87TH, NJ72]
G1 WBV Details withheld at licensee's request by SSL.
G1 WCB M J P Whateley, 9 Millside Cl, Royal Park, Kingsthorpe, Northampton, NN2 7TR [IO92NG, SP76]
G1 WCO Details withheld at licensee's request by SSL.
G1 WCQ E Eastwood, 56 The Mede, Freckleton, Preston, PR4 1JB [IO83NS, SD43]
G1 WCR Details withheld at licensee's request by SSL.
GW1 WCS Details withheld at licensee's request by SSL.
G1 WCT Details withheld at licensee's request by SSL.
G1 WCY J D Reeves, 5 Arrows Cres, Boroughbridge, York, YO5 9LP [IO94HC, SE36]
G1 WCZ Details withheld at licensee's request by SSL.
G1 WDC Details withheld at licensee's request by SSL.
G1 WDE Details withheld at licensee's request by SSL.
G1 WDF R Bird, 21 Aster Cl, Clacton on Sea, CO16 7DA [JO01NT, TM11]
G1 WDJ J K Banks, 76 Chestnut Dr South, Leigh, WN7 3JX [IO83RL, SJ69]
G1 WDL Details withheld at licensee's request by SSL.
G1 WDM Details withheld at licensee's request by SSL.[Op: W E Daniell, Easterly House, 29 Easterly Rd, Leeds, LS8 2TN.]
G1 WDO J G Marlett, 6 Delamere Ave, Sutton Manor, St. Helens, WA9 4AP [IO83PJ, SJ59]
G1 WDQ T Hutton, 23 Dines Cl, Wilstead, Bedford, MK45 3BU [IO92SB, TL04]
G1 WDV Details withheld at licensee's request by SSL.
G1 WDW Details withheld at licensee's request by SSL.
G1 WDX Details withheld at licensee's request by SSL.
G1 WDZ Details withheld at licensee's request by SSL.
G1 WEA Details withheld at licensee's request by SSL.
G1 WEB Details withheld at licensee's request by SSL.[Op: J V Webb.]
G1 WEF R H Cornwell, 13 Milford Rd, Little Thurrock, Grays, RM16 2QL [JO01EL, TQ68]
G1 WEG J W D Scherrer, 5 Claymore Dr, Ickleford, Hitchin, SG5 3UB [IO91UX, TL13]
G1 WEQ J F G Merriday, 5 Elm Cl, Alresford, Colchester, CO7 8EE [JO01MU, TM02]
G1 WER G Williams, 22 Wyburns Ave, Rayleigh, SS6 7QU [JO01HN, TQ88]
G1 WEX R S Taylor, 70 Beecher St., Halesowen, B63 2DP [IO82XK, SO98]
GI1 WFA Dr C P Ward, 30 Ringsend Rd, Limavady, BT49 0QJ [IO65NB, C72]
G1 WFD N I Edwards, 6 Whitewell Rd, Frome, BA11 4EL [IO81UF, ST74]
G1 WFG S G Lake, 42 Haling Park Rd, South Croydon, CR2 6NE [IO91WI, TQ36]
G1 WFJ A Woodhouse, 5 Dudley Rd, Kingswinford, DY6 8BT [IO82WL, SO88]
G1 WFK Details withheld at licensee's request by SSL.
G1 WFM A Pounder, 18 Totts Ln, Walkern, Stevenage, SG2 7PL [IO91WW, TL22]
G1 WFO P R Burden, 110 Westbury Leigh, Westbury, BA13 3SH [IO81VF, ST84]
GI1 WFP Details withheld at licensee's request by SSL.
G1 WFR Details withheld at licensee's request by SSL.
G1 WFU R F Dickson, 49 Ashgrove, Peasedown St. John, Bath, BA2 8EF [IO81SH, ST75]
G1 WFY Details withheld at licensee's request by SSL.
G1 WGF J S McCullough, 35 Ringlow Park Rd, Swinton, Manchester, M27 0HA [IO83TM, SD70]
G1 WGH Details withheld at licensee's request by SSL.
G1 WGI Details withheld at licensee's request by SSL.
G1 WGJ Details withheld at licensee's request by SSL.
GI1 WGK W W A Steele, 230 Abbey Rd, Millisle, Newtownards, BT22 2JG [IO74FO, J57]
G1 WGM K M Perry, 25 Hillary Dr, Crowthorne, RG45 6QF [IO91OJ, SU86]
GW1 WGR M Jenkins West Glam Raynet, 25 Stepney Rd, Cockett, Swansea, SA2 0FZ [IO81AP, SS69]
GW1 WGV Details withheld at licensee's request by SSL.
G1 WGW W G Watson, 24 High St., Wall Heath, Kingswinford, DY6 0HB [IO82VM, SO88]
G1 WHC Details withheld at licensee's request by SSL.
G1 WHH Details withheld at licensee's request by SSL.
G1 WHN Details withheld at licensee's request by SSL.
G1 WHP P Gregson, 14 Tarragon Dr, Stoke on Trent, ST3 7YE [IO82WX, SJ94]
G1 WHV D H C Paton, 1 Higher Penponds Rd, Penponds, Camborne, Cornwall, TR14 0QG [IO70IE, SW63]
G1 WHY J T Lowe, 23 Hoylake Dr, Tividale, Warley, B69 1QA [IO82XM, SO98]
G1 WIA Details withheld at licensee's request by SSL.
GM1 WIB R S Wylie, Dummet, Thurso, Caithness, KW14 8YD [IO88HP, ND27]
G1 WIH Details withheld at licensee's request by SSL.
G1 WIK K J Morris, 44 Leamington Rd, Lanehouse Est, Weymouth, DT4 0EZ [IO80SO, SY67]
GW1 WIO Details withheld at licensee's request by SSL.
G1 WIR J Monaghan, 20 The Cheethams, Blackrod, Bolton, BL6 5RR [IO83RN, SD60]
G1 WIS D J Mountain, 45 Westway Gdns, Redhill, RH1 2JB [IO91WG, TQ25]
G1 WIW R E Dowdeswell, 42 Dovecote Way, Barwell, Leicester, LE9 8EX [IO92HN, SP49]
G1 WIY P J Smith, 31 Essex Gdns, Market Harborough, LE16 9JS [IO92ML, SP78]
G1 WIZ Details withheld at licensee's request by SSL.[Op: R H Horsley, 14 Manse Crescent, Burley in Wharfedale, Ilkley, W Yorks, LS22 7LA.]
GU1 WJA W J Ayres, Rousay, Baliffs Cross Rd, St Andrews, Guernsey C I, Guernsey, GY99 1AA
G1 WJC G R Tootell, 10 Leen Dr, Hucknall, Nottingham, NG15 8BW [IO93JB, SK54]
G1 WJG J A Coates, 35 School Ln, Lubenham, Market Harborough, LE16 9TW [IO92ML, SP78]
G1 WJK G E Richards, 3 Pleasant Cl, Kingswinford, DY6 9TQ [IO82VL, SO88]
G1 WJO A Blackwell, 1 Gladstone Terr, Hinckley, LE10 1HE [IO92HM, SP49]
G1 WJP P Denning, 8 Hastings Terr, Marshfields, Bradford, BD5 9PL [IO93CS, SE13]
G1 WJQ A R Collinson, 367 Filey Rd, Scarborough, YO11 3JG [IO94TG, TA08]
G1 WJR W J Rollins, 127 Clacton Rd, St. Osyth, Clacton on Sea, CO16 8PR [JO01NT, TM11]
GW1 WJT Details withheld at licensee's request by SSL.
G1 WJZ Details withheld at licensee's request by SSL.
G1 WKE A G Parker, 4 Portway, Baughurst, Tadley, RG26 5PD [IO91KI, SU56]
G1 WKH N P Graves, 26 Ratton Dr, Eastbourne, BN20 9BS [JO00DS, TQ50]
G1 WKK J Arnott, 86 Constantine Way, Hatch Warren, Basingstoke, RG22 4UR [IO91KF, SU64]
G1 WKP Details withheld at licensee's request by SSL.
GM1 WKR Details withheld at licensee's request by SSL.
G1 WKS S Fenwick West Kent ARS, 28 Gimble Way, Pembury, Tunbridge Wells, TN2 4BX [JO01DD, TQ64]
G1 WKT R Reed, 159 Roper Ave, Marlpool, Heanor, DE75 7DB [IO93HA, SK44]
G1 WKV R S Ellis, 20 Pine Glen Ave, Ferndown, BH22 9QP [IO90BT, SU00]
G1 WKW Details withheld at licensee's request by SSL.
G1 WKZ G J Evans, 4 The Mallards, Fareham, PO16 7XR [IO90JU, SU50]
G1 WLD S K Evans, 4 The Mallards, Fareham, PO16 7XR [IO90JU, SU50]
G1 WLE R C Bavister, Michaelmas House, Hollow St., Great Somerford, Chippenham, SN15 5JD [IO81XN, ST98]
GI1 WLJ G C McCutcheon, 73 Tullynagardy Rd, Newtownards, BT23 4TB [IO74DO, J47]
G1 WLM Details withheld at licensee's request by SSL.
G1 WLO M J Head, 15 Blackfields Ave, Bexhill on Sea, TN39 4JL [JO00FU, TQ70]
G1 WLQ A C Bennett, 57 Grovebury Ct, Wootton, Bedford, MK43 9HZ [IO92RC, TL04]
G1 WLU S M Brookes, Ivy Cottage, Haselor, Alcester, Warwicks, B49 6LX [IO92CF, SP15]
G1 WLW A A Fordyce, 41 Benscliffe Dr, Loughborough, LE11 3JP [IO92JS, SK51]
G1 WMD Details withheld at licensee's request by SSL.
G1 WMN K A Harvey, 61 Westfield Rd, Northchurch, Berkhamsted, HP4 3PW [IO91QS, SP90]
GM1 WMO A B Shields, 7 St. Colms Pl, School St., Largs, KA30 8DN [IO75NT, NS25]
G1 WMP Details withheld at licensee's request by SSL.
GM1 WMS R M C Cadwallader, Station Cottage, Forsinard, Sutherland, Scotland, KW13 6YT [IO88BJ, NC94]
G1 WMV B P Catchpoole, 23 Abraham Cl, Willen Park, Milton Keynes, MK15 9JA [IO92PB, SP84]
G1 WNK Details withheld at licensee's request by SSL.
G1 WNL F P Thomas, 4H Green End, Newtownabbey, BT37 9NQ [IO74BP, J38]
G1 WNS Details withheld at licensee's request by SSL.
G1 WNW T A Smith, 9 Brackenwood Rd, St. Johns, Woking, GU21 1XF [IO91QH, SU95]
G1 WNY A G Jones, 2 Ashbank Villas, via, Bentham, Lancaster, LA2 7HX [IO84RC, SD66]
G1 WOJ R A McWilliams, 35 Humphrey Cres, Urmston, Manchester, M41 9PU [IO83UK, SJ79]
G1 WOR R M Bray Worthing & Dist, Vhf Contest Group, 14 Hadlow Way, Lancing, BN15 9DE [IO90UT, TQ10]
G1 WOS J Berry, 22 Walton St., Easton, Bristol, BS5 0JG [IO81RL, ST67]
G1 WPD Details withheld at licensee's request by SSL.
G1 WPF M D Cabban, 37 Ryman Ct, Stag Ln, Chorleywood, Rickmansworth, WD3 5HN [IO91RP, TQ09]
G1 WPG B L Groome, 21 Barndale Dr, Ridge, Wareham, BH20 5BX [IO80WQ, SY98]
G1 WPH L A Eden, 23 Elm Green Cl, Bath Rd, Worcester, WR5 3HD [IO82VE, SO85]
G1 WPL R Balkwell, 2 Franklyn Rd, Droylsden, Manchester, M43 6DS [IO83WL, SJ99]
G1 WPR T Bromley, 7 Brookside, Desborough, Kettering, NN14 2UD [IO92OK, SP88]
G1 WQA M D J Cressey, 51 Hedge Field Rd, Barrowby, Grantham, NG32 1TA [IO92PV, SK83]
G1 WQC R R Pratt, 11 Park Rd, Ryde, PO33 2BG [IO90KR, SZ59]
G1 WQE A J Norris, 6 Bawtry Walk, Carlton Rd, Nottingham, NG3 2GF [IO92KX, SK54]
G1 WQH R F Booth, 25 Railswood Dr, Pelsall, Walsall, WS3 4BD [IO92AP, SK00]
G1 WQI D E Kozma, 11 Pyenot Dr, Pyenot Hall Ln, Cleckheaton, BD19 5AX [IO93DR, SE12]
G1 WQL P K Brodie, 73 Summergangs Rd, Hull, HU8 8JX [IO93US, TA13]
GW1 WQM P J Maris, Barnlake House, Burton, Milford Haven, Dyfed, SA73 1PA [IO71MR, SM90]
G1 WQN S M Mangan, 48 Emblett Dr, Newton Abbot, TQ12 1YJ [IO80EM, SX87]
G1 WQQ J F Button, 1 Ross Cottage, Southey Green, Sible Hedingham, Halstead Essex, CO9 3RN [JO01GX, TL73]
G1 WQU T F Gregg, Carino, Old Gore Ln, Emborough, Bath, BA3 4SJ [IO81RG, ST65]
G1 WQY K A T Webber, 2 Henniker Rd, Ipswich, IP1 5HD [JO02NB, TM14]
G1 WRC L N Fennelow N.Cambs Rptr Gr, 39 Clarence Rd, Wisbech, PE13 2ED [JO02CQ, TF41]
G1 WRD M J Simpson, 49 Elliott Dr, Inkersall, Chesterfield, S43 3DZ [IO93HF, SK47]
G1 WRF Details withheld at licensee's request by SSL.
G1 WRG W R Garrett, Holmleigh, 37 Pollard Ln, Bradford, BD2 4RN [IO93DT, SE13]
G1 WRH G Marshall, 1 Portland Cl, Braintree, CM7 9NJ [JO01GV, TL72]
G1 WRK Details withheld at licensee's request by SSL.
G1 WRM Details withheld at licensee's request by SSL.
G1 WRN T R Yorke North Warks Raynet Group, 12 Shanklin Dr, Weddington, Nuneaton, CV10 0BA [IO92GM, SP39]
G1 WRO M A Smith, 40 Diamond Ave, Greenacres, Rainworth, Mansfield, NG21 0FF [IO93KC, SK55]
G1 WRP L M Neeson, 24 Braeside, Tweedmouth, Berwick upon Tweed, TD15 2BZ [IO85XS, NT95]
G1 WRQ T Ellis, 18 Manor St., Sneinton, Nottingham, NG2 4JP [IO92KW, SK53]
G1 WRS Details withheld at licensee's request by SSL.
G1 WRU J V Jinks, 27 Taryn Dr, Primrose Gdns, Darlaston, Wednesbury, WS10 8XY [IO82XN, SO99]
GW1 WRV R Marston, 4 Ivor Terr, Dowlais, Merthyr Tydfil, CF48 3SW [IO81HS, SO00]
G1 WRY A White, 34 Pains Way, Amesbury, Salisbury, SP4 7RG [IO91CE, SU14]
G1 WSA Details withheld at licensee's request by SSL.
G1 WSC J D Bland, 7 Cres St., Grimsby, DN31 2HB [IO93WN, TA20]
G1 WSD D J Garratt, 87 Garden Rd, Eastwood, Nottingham, NG16 3FY [IO93IA, SK44]
G1 WSE J W Frizell, 17 St. Johns Terr, Lewes, BN7 2DL [JO00AU, TQ41]
G1 WSF D A Pettican, 52 Shepherds Way, Saffron Walden, CB10 2AH [JO02DA, TL53]
G1 WSH J W Johnstone, 8 Law View, Overtown, Wishaw, ML2 0RQ [IO85BS, NS85]
G1 WSL B G Manton, 23 Rydal Rd, Gosport, PO12 4ES [IO90KT, SU60]
G1 WSM M N Franklin, 15 Palatine Rd, Durrington, Goring By Sea, Worthing, BN12 6JR [IO90TT, TQ10]
G1 WSN J M Spillett, Mockbeggar Cottage, Cross Lanes, Mockbeggar, nr Ringwood, Hants, BH24 3NQ [IO90CV, SU10]
G1 WSR M J Watson St Edmundsbury Raynet, The Tubbery, Henley, Ipswich, Suffolk, IP6 0BR [JO02NC, TM15]
G1 WSS P J Gallagher, Holly Cottage, Meldon Park Corner, Meldon, Morpeth, Northd., NE61 3SL [IO95CE, NZ18]
G1 WSW M H Flewitt, 38 Laburnum Ave, Newbold Verdon, Leicester, LE9 9LQ [IO92HP, SK40]
G1 WSZ V T Tosney, 126 Norburn Park, Witton Gilbert, Durham, DH7 6SQ [IO94ET, NZ24]
G1 WTB E G P Musson, 110 Marples Ave, Mansfield Woodhouse, Mansfield, NG19 9DW [IO93JE, SK56]
G1 WTH C R Piddock, 118 Howley Grange Rd, Halesowen, B62 0HU [IO82XK, SO98]
G1 WTN R Wroe, 11 Poplars Rd, Kendray Est, Barnsley, S70 3ND [IO93GN, SE30]
G1 WTS M P Roper, 19 Normay Rise, Newbury, RG14 6RY [IO91HI, SU46]
G1 WTT F Hancock-Baker, 1 St. Oswalds, Newtown, Chester, CH1 3HN [IO83NE, SJ46]
G1 WTU R S Rockett, 34 Whitelake Rd, Tonbridge, TN10 3TJ [JO01DF, TQ54]
G1 WTW C I Underwood, Rosebury, 4 Hawthorn Rd, Godalming, GU7 2NE [IO91QE, SU94]
G1 WTX P T Egan, 105 Macclesfield Rd, Buxton, SK17 9AD [IO93AG, SK07]
G1 WTY B J Parkes, 5 Elm Dr, Brackley, NN13 6ES [IO92KA, SP53]
GW1 WTZ C Green, 63 Tan y Lan Terr, Morriston, Swansea, SA6 7DU [IO81AP, SS69]
G1 WUA M L Ansell, 203 Willowfield, Harlow, CM18 6RZ [JO01BS, TL40]
G1 WUC M J Garner, 40 Studley Rd, Harrogate, HG1 5JU [IO93FX, SE35]
G1 WUH K A Carr, 41 Surrey Rd, Dagenham, RM10 8ES [JO01BN, TQ58]
G1 WUJ Details withheld at licensee's request by SSL.
G1 WUK M J Stirland, 19 Nightingale Ave, Hathern, Loughborough, LE12 5JE [IO92IT, SK52]
G1 WUM R E Miles, Haseley Lodge, Birmingham Rd, Haseley, Warwick, CV35 7HF [IO92EH, SP26]
G1 WUP T R Tozer, Glenview Cottage, Grenville Rd, Lostwithiel, Cornwall, PL22 0EP [IO70QJ, SX15]
G1 WUR Details withheld at licensee's request by SSL.
G1 WUS S B Burns, 1 The Bungalows, Neville Rd, Peterlee, SR8 4PQ [IO94IS, NZ44]
G1 WUU J A Neate, 30 Berry Ave, Paignton, TQ3 3QN [IO80FK, SX86]
G1 WUY Details withheld at licensee's request by SSL.
G1 WVD M J Shrago, 12 Oakwood Rd, Bricket Wood, St. Albans, AL2 3PU [IO91TR, TL10]
G1 WVF Details withheld at licensee's request by SSL.
G1 WVK J J Power, 1 The Colt House, Shrublands Ave, Berkhamsted, Herts, HP4 3JH [IO91RS, SP90]
G1 WVL B S Sharp, 42 Hazelhurst Dr, Garstang, Preston, PR3 1WB [IO83OV, SD44]
G1 WVM R W Vowles, 47 Tyndale Ave, Yate, Bristol, BS17 5EX [IO81SN, ST78]
G1 WVO D J Willey, 17 Bridge Pl, Saxilby, Lincoln, LN1 2QA [IO93QG, SK87]
G1 WVR S H Bradshaw, Welland Valley Am Rd Society, 82 Arden Way, Market Harborough, LE16 7DD [IO92NL, SP78]
G1 WVS P F Gibson, 14 Nene Side Cl, Badby, Daventry, NN11 3AD [IO92JF, SP55]
G1 WVV R Sutton, 28 Shrubbery Gdns, Wem, Shrewsbury, SY4 5BX [IO82PU, SJ52]
G1 WVW R C Jones, 8 Downing Ave, Basford, Newcastle, ST5 0JY [IO83VA, SJ84]
G1 WVZ R P McCutcheon, 11 Mayflower Dr, Rugeley, WS15 2SW [IO92AS, SK01]
G1 WWB R I Eeles, 23 Elgin Ave, Ashford, TW15 1QE [IO91SK, TQ07]
GM1 WWD K A Penny, 24 Don House, Kellands Rd, Inverurie, Aberdeenshire, AB51 9YG [IO87SG, NJ72]
GW1 WWE J B Peake, 70 Higher Ln, Mumbles, Langland, Swansea, SA3 4PD [IO81AN, SS68]
GM1 WWG Details withheld at licensee's request by SSL.
G1 WWH A Benn, Stores & P O, Burneston Village, Bedale, North Yorks, DL8 2HT [IO94FG, SE38]
G1 WWI M W Dronfield, White Lodge Farm, High Bradfield, Sheffield, South Yorks, S6 6JL [IO93EK, SK29]
G1 WWK J A Vernon, 15 Hambridge Rd, Bishops Itchington, Leamington Spa, CV33 0RH [IO92GF, SP35]
G1 WWP J D Sharpe, 10 Stocking Green Cl, Hanslope, Milton Keynes, MK19 7NH [IO92OC, SP84]
G1 WWR S J Cockshoot, 72 Princess Margaret Ave, Margate, CT9 3EF [JO01RJ, TR37]
G1 WWT Details withheld at licensee's request by SSL.
G1 WWV Details withheld at licensee's request by SSL.
G1 WWX Details withheld at licensee's request by SSL.
G1 WWY L Donald, Sundale, Northampton Rd, Whitfield, Brackley, NN13 7TY [IO92KB, SP53]
G1 WWZ B Baker, 32 Emmanuel Rd, Burntwood, Walsall, West Midlands, WS7 9AD [IO92BQ, SK00]
G1 WXA Details withheld at licensee's request by SSL.
G1 WXB Details withheld at licensee's request by SSL.
G1 WXC Details withheld at licensee's request by SSL.[Op: D Wells, 35 Western Rd North, Somping, W Sussex, BN15 9UX.]
G1 WXD M Russell, 6 All Saints Rd, Wimbledon, London, SW19 1BX [IO91VK, TQ27]
G1 WXF J R Pearson, 17 Hebden Ave, Woodloes Park, Warwick, CV34 5XD [IO92FH, SP26]
G1 WXI Details withheld at licensee's request by SSL.
G1 WXK M Bell, 151 Towngate, Ossett, WF5 0PP [IO93FQ, SE22]
G1 WXM K W J Tuddenham, 109 Grenville Rd, Balby, Doncaster, DN4 9JJ [IO93KL, SE50]
G1 WXO J N Storer, 33 Thorpe Hill Dr, Heanor, DE75 7DN [IO93HA, SK44]
G1 WXR W J Hawkesworth, 27 Bedford Rd, Holland on Sea, Clacton on Sea, CO15 5LF [JO01OT, TM11]
G1 WXS P A J Springall, 31 The Orchards, Epping, CM16 7BB [JO01BQ, TL40]
G1 WXT M C Moorecroft, 4 St. Davids Rd, Locksheath, Locks Heath, Southampton, SO31 6EP [IO90IU, SU50]
G1 WXU G Parsons, 73 Worthing Ave, Elson, Gosport, PO12 4DB [IO90KT, SU50]
G1 WXW P K Prescott, 13 The Boltons, Waterlooville, PO7 5QR [IO90LU, SU60]
G1 WXZ Details withheld at licensee's request by SSL.
G1 WYA R M Webb, Hill Side Cottage, 59 Thorpe Ln, Austerlands, Oldham, OL4 3QW [IO83XN, SD90]
G1 WYB E Coupe, 28 Wellington Rd, Blackburn, Lancs, BB2 2NQ [IO83SR, SD62]
G1 WYC S Smith, 58 Fairfields, Holbeach, Spalding, PE12 7JE [JO02AT, TF32]
G1 WYE L J Ibbotson, 19 Highgate, Cleethorpes, DN35 8NR [IO93XN, TA30]
G1 WYG D J Biginton, 67 Capstone Rd, Downham, Bromley, BR1 5NA [JO01AK, TQ37]
G1 WYK A Woodcock, 3 Simmonds Pl, Wednesbury, WS10 8BN [IO82XN, SO99]
G1 WYP H J Milsom, 50 Sams Ln, Blunsdon, Swindon, SN2 4AZ [IO91CO, SU19]
GI1 WYZ R A Kennedy, 3 St. Annes Cres, Glengormley, Newtownabbey, Co. Antrim, BT36 5JZ [IO74AQ, J38]
GI1 WZA B J Davison, 29 Glendale Park, Newtownbreda, Belfast, BT8 4HT [IO74BN, J36]
G1 WZB I D Ross, 46 Fordbridge Rd, Ashford, TW15 2SJ [IO91SK, TQ07]
G1 WZG J E Endicott, 28 Whiteley Ave, Totnes, TQ9 5FQ [IO80DK, SX76]
GW1 WZI J E Cartwright, 20 Castlefield Pl, Mynachdy, Cardiff, CF4 3DU [IO81JL, ST17]
G1 WZK D C Heaton, 39 Bridgewater Rd, Hrold Hill, Romford, RM3 7UB [JO01CO, TQ59]
G1 WZO L C Leach, Leyland, The St., Frampton on Severn, Gloucester, GL2 7ED [IO81TS, SO70]
G1 WZQ A D Utting, 20 Davenport Rd, Goodwood, Leicester, LE5 6SA [IO92LP, SK60]
G1 WZR K Young Epc ARC, Co. Director Dep, Derbyshire CO.Counc, Co.Offices Matlock, DE4 3AG [IO93FD, SK26]
G1 WZV S P Benney, 3 Kirkstone Walk, The Glebe, Nuneaton, CV11 6EZ [IO92GM, SP39]
G1 WZX J S Jones, 10 Ferndale Cres, Gobowen, Oswestry, Salop, SY11 3PJ [IO82LV, SJ33]
G1 XAA H E Jacklin, 26 Rockmill End, Willingham By Stow, Willingham, Cambridge, CB4 5HY [JO02AH, TL47]
G1 XAD Details withheld at licensee's request by SSL.
G1 XAE Details withheld at licensee's request by SSL.
G1 XAF M D Smith, 51 Shakespeare Rd, London, W7 1LU [IO91UM, TQ18]
G1 XAG R G M Hart, 165 High St., Harston, Cambridge, CB2 5QD [JO02BD, TL45]
G1 XAJ J H Franklin, 17 Oak Cl, Thorpe-le-Soken, Clacton on Sea, CO16 0HU [JO01OU, TM12]
G1 XAK W H Daniels, 29 Station Rd, Wickwar, Wotton under Edge, GL12 8NB [IO81TO, ST78]
G1 XAL A A Dangerfield, 29 Passage Rd, Saul, Gloucester, GL2 7LB [IO81TS, SO70]
G1 XAM J A Bryant, 12 Dale Tree Rd, Barrow, Bury St. Edmunds, IP29 5AD [JO02HF, TL76]
G1 XAN Details withheld at licensee's request by SSL.
G1 XAP P Whittingham, 28 Wedge Ave, Haydock, St. Helens, WA11 0DY [IO83PL, SJ59]

G1 XAR P Banham, La Corbiere, 3 Ronaldsway Cl, Bacup, OL13 9PY [IO83VQ, SD82]
G1 XAS J F Hume, 39 Epsom Rd, Rugby, CV22 7PF [IO92II, SP47]
G1 XAV G N Smith, Japonica House, 2 Ollerton Rd, Edwinstowe, Mansfield, NG21 9QG [IO93LE, SK66]
G1 XAW K D Young, 85 Greenaway Ln, Hackney, Matlock, DE4 2QA [IO93FD, SK26]
G1 XAX N R Elliott, 3 Rockvale Terr, Matlock Bath, Matlock, DE4 3NW [IO93FC, SK25]
G1 XAZ Details withheld at licensee's request by SSL.
G1 XBE T Beecher, 77 Grime Ln, Sharlston Common, Wakefield, WF4 1EH [IO93HP, SE31]
GW1 XBF J W N Northam, Sully House, Station Rd, Mundesley, Norwich, NR11 8JH [JO02RV, TG33]
GW1 XBG P J Smith, 35 Terr Rd, Mount Pleasant, Swansea, SA1 6HN [IO81AO, SS69]
GI1 XBH Details withheld at licensee's request by SSL.[Op: M A Shaw. Station located near Comber, Co Down, BT23.]
GI1 XBI E McGuinness, 2 Benbraddagh Ave, Limavady, BT49 0AR [IO65MB, C62]
GM1 XBK K W McClure, 21 Hamilton Cres, Bishopton, PA7 5JT [IO75RV, NS47]
G1 XBL B Darby, Pippins, Green St., Kempsey, Worcester, WR5 3QB [IO82VD, SO84]
G1 XBO Dr D R Hutton, Westfields, Sharperton, near Rothbury, Morpeth, Northd., NE65 7AE [IO85XH, NT90]
G1 XBR S G Loney, 4 Mendip Rd, Millbrook, Southampton, SO16 4BN [IO90GW, SU31]
G1 XBU Details withheld at licensee's request by SSL.
G1 XBV N A Watson, 12 Abbey Cl, Pyrford, Woking, GU22 8RY [IO91RH, TQ05]
G1 XBX A E Manning, 4 Lewthorne Cttgs, Ilsington, Newton Abbot, TQ13 9RR [IO80DN, SX77]
G1 XBY D F J Henson, South Devon Hotel, St Marychurch, Torquay, Devon, TQ1 4NP [IO80FL, SX96]
G1 XCB P Davies, 91 Station Rd, Hadfield, Hyde, SK14 7AR [IO93AL, SK09]
G1 XCC Details withheld at licensee's request by SSL.
G1 XCK Details withheld at licensee's request by SSL.
G1 XCL B S Wilson, 25 Pennine Ln, Golborne, Warrington, WA3 3EZ [IO83RL, SJ69]
G1 XCQ Details withheld at licensee's request by SSL.
G1 XCW Details withheld at licensee's request by SSL.
G1 XCY Details withheld at licensee's request by SSL.
G1 XCZ Details withheld at licensee's request by SSL.
G1 XDC Details withheld at licensee's request by SSL.
G1 XDD Details withheld at licensee's request by SSL.
G1 XDF A I Millership, 72 Park Rd, Burgess Hill, RH15 8HG [IO90WW, TQ31]
G1 XDG R P W Smith, 28 Langley Heath Dr, Sutton Coldfield, B76 2XB [IO92CN, SP19]
GM1 XDH Details withheld at licensee's request by SSL.
G1 XDJ Details withheld at licensee's request by SSL.
G1 XDK Details withheld at licensee's request by SSL.
G1 XDO J A Chaldecott, 20 Haynes Ave, Poole, BH15 2ED [IO90AR, SZ09]
G1 XDS J Fyson, One Redlands Est, Ibstock, Leics, LE6 1HT [IO92JP, SK50]
G1 XDT R J Wharrie, 69 Tollgate, Spalding, PE11 1NJ [IO92WS, TF22]
G1 XDU M R Peacock, 28 Bodgara Way, Liskeard, PL14 3BL [IO70SK, SX26]
G1 XDW G Horner, 25 Wolseley Rd, Great Yarmouth, NR31 0EJ [JO02UO, TG50]
GW1 XDY C Bayliss-Blomley, The Cedars, 9 Park View, Northop, Mold, CH7 6DD [IO83KE, SJ26]
GM1 XEA P R Thomson, 13 Westwood Dr, Westhill, Skene, Westhill, AB32 6WW [IO87UD, NJ80]
GM1 XEB M McCulloch, 53 Milverton Ave, Bearsden, Glasgow, G61 4BG [IO75TW, NS57]
G1 XEF D J Osborne, 12 Brookfield Terr, Lustleigh, Newton Abbot, Devon, TQ13 9TA [IO80DO, SX78]
G1 XEH R A Burton, 18 Churchfield, Harpenden, AL5 1LL [IO91TT, TL11]
G1 XEL Details withheld at licensee's request by SSL.
G1 XEM Details withheld at licensee's request by SSL.
G1 XEN Details withheld at licensee's request by SSL.
G1 XEP L J Lambert, 37 Fulford Gr, South Oxhey, Watford, WD1 6QQ [IO91TP, TQ19]
G1 XES E J Turner, 1104 Wimborne Rd, Moordown, Bournemouth, BH10 7AA [IO90BS, SZ09]
G1 XET C R Burton, 6 Nicholas Rd, Henley on Thames, RG9 1RB [IO91MM, SU78]
G1 XEX D I J Packham, South View, Mill St., Iden Green, Cranbrook, TN17 4HH [JO01GB, TQ83]
GW1 XFB D B Evans, Bwthyn Bach, 2 Old Village Rd, Barry, CF62 6RA [IO81IJ, ST16]
G1 XFI Details withheld at licensee's request by SSL.
GM1 XFJ Details withheld at licensee's request by SSL.
G1 XFL K Lanham, 22 Ascot Cl, Ladywood, Birmingham, B16 9EY [IO92AL, SP08]
G1 XFM J B Shaw, 33 Park Farm Cl, Horsham, RH12 5EU [IO91UB, TQ13]
G1 XFO M J Price, 31 Wattis Rd, Smethwick, Warley, B67 5BB [IO92AL, SP08]
G1 XFP D L Broadley, 25 Greenfield Rd, Stonesfield, Witney, OX8 8EQ [IO91GU, SP31]
G1 XFQ Details withheld at licensee's request by SSL.
G1 XFR B R Bance, 36 Leafy Oak Rd, Gr Park, London, SE12 9RS [JO01AK, TQ47]
G1 XFX A Harvey, 26 Kidborough Rd, Gossops Green, Crawley, RH11 8HW [IO91VC, TQ23]
GI1 XGA T J Campbell, 27 Silverbrook Park, New Buildings, Londonderry, BT47 2RD [IO64HX, C41]
G1 XGE D J Hutchinson, 22 Brandon Way, Bransholme, Kingswood, Hull, HU7 3EL [IO93UT, TA03]
G1 XGG M S Ballard, 15 Cross St., Stockingford, Nuneaton, CV10 8HY [IO92FM, SP39]
G1 XGL J J Young, 69 Ferrers Ave, West Drayton, UB7 7AB [IO91SM, TQ07]
G1 XGM Details withheld at licensee's request by SSL.
G1 XGP S Blinkhorn, 102 Lord Roberts Ave, Leigh on Sea, SS9 1NE [JO01IN, TQ88]
G1 XGS P D Wilson, 65 Recreation Rd, Bourne, PE10 9HD [IO92TS, TF02]
G1 XGU Details withheld at licensee's request by SSL.
G1 XGW A G Gray, 35 Belgrave Rd, Oldham, OL8 1LT [IO83WM, SD90]
G1 XGZ D J Richards, 8 Thackeray Ct, Elystan Pl, London, SW3 3LB [IO91WL, TQ27]
G1 XHA J de Bank, 16 Henley Orchards, Ludlow, SY8 1TN [IO82PJ, SO57]
GW1 XHG D Austin, Erw Mau, Rhosesmor Mountain, Rhosesmor, Mold, CH7 6PP [IO83JF, SJ26]
G1 XHJ J C Phelps, 30 Ellerdene Cl, Headless Cross, Redditch, B98 7PW [IO92AH, SP06]
G1 XHL E A Rudgley, 38 Edgecomb Rd, Stowmarket, IP14 2DN [JO02LE, TM05]
G1 XHM B C Amor, 45 St. Stephens Rd, Selly Oak, Birmingham, B29 7RR [IO92AK, SP08]
G1 XHO T Williams, 145 Bulwell Ln, Old Basford, Nottingham, NG6 0BS [IO92JX, SK54]
G1 XHR E M Goodwin, Hankelow Ct, Hall Ln, Hankelow, Crewe, CW3 0JB [IO83SA, SJ64]
G1 XHT D R Birch, 1 Highfield Cres, Rock Ferry, Birkenhead, L42 2DP [IO83LI, SJ38]
G1 XHW W J Brewer, 7 Firs Way, Rattlers Rd, Brandon, IP27 0HE [JO02HK, TL78]
GM1 XHZ T Valentine, 12 Rossie Terr, Ferryden, Montrose, DD10 9RX [IO86SQ, NO75][Op: Tom Valentine. Tel: (01674) 76503. IARU Locator: IO86RQ. Member of RSGB.]
GM1 XIA J D Mackenzie, 117 Innes Park Rd, Skelmorlie, PA17 5BB [IO75NU, NS16]
GI1 XIB J G Wilkinson, 8 Belmont Heights, Antrim, BT41 1BD [IO64VR, J18]
G1 XIC K C Harris Callington School RC, Obo/Callington School R C, Department of Physics, Launceston Rd, Callington, Cornwall, PL17 7BT [IO70UM, SX36]
G1 XIE R J Dyer, 76 Sandy Ln, Farnborough, GU14 9HJ [IO91OH, SU85]
G1 XII T H Lovatt, 5 Acre Rise, Willenhall, WV12 4SL [IO82XO, SJ90]
G1 XIK Details withheld at licensee's request by SSL.
GM1 XIN W Allan, Corse Farm, Kininmonth, Peterhead, Aberdeenshire, AB4 8JU [IO87TH, NJ72]
G1 XIO A J Faram, 4 Wellington Rd, Gillingham, ME7 4NN [JO01GJ, TQ76]
G1 XIP Details withheld at licensee's request by SSL.
G1 XIQ Details withheld at licensee's request by SSL.
G1 XIV G J R Reynolds, The Thatched Cottage, St. Thomas Dr, Pagham, Bognor Regis, PO21 4TN [IO90PS, SZ89]
G1 XIX Details withheld at licensee's request by SSL.
G1 XIY Dr M R Nottingham, 13 Ottowa Cl, Worcester, WR2 4XN [IO82VE, SO85]
GM1 XJE Dr J Hopkins, 9 Pathfoot Ave, Bridge of Allan, Stirling, FK9 4SA [IO86AD, NS89]
GW1 XJJ H J Worgan, 29 Mayfield Ave, Laleston, Bridgend, CF32 0LH [IO81EM, SS87]
G1 XJK F W Tilley, 37 Fant Ln, Maidstone, ME16 8NP [JO01FG, TQ75]
G1 XJN D Jones, 258 Windmill Ave, Kettering, NN15 6PF [IO92PJ, SP87]
G1 XJP Details withheld at licensee's request by SSL.
G1 XJS Details withheld at licensee's request by SSL.
G1 XJT A G Nichols, Herland Bungalow, Godolphin Cross, Breage, Helston, TR13 9RL [IO70HD, SW63]
G1 XJV Details withheld at licensee's request by SSL.
G1 XJZ D H Layton, The Nook, 7 Turton St., Kidderminster, DY10 2TH [IO82VJ, SO87]
G1 XKD G Lawton, 23 Fiske Ct, Cavendish Rd, Sutton, SM2 5ER [IO91VI, TQ26]
G1 XKK Details withheld at licensee's request by SSL.
G1 XKL B Smith, Hirsts Cottage, Spa Ln, Lathom, Ormskirk, L40 6JG [IO83ON, SD40]
G1 XKM Details withheld at licensee's request by SSL.
G1 XKN A Chambers, 34 Haunchwood Dr, Sutton Coldfield, B76 1JR [IO92CM, SP19]
G1 XKQ B J Neate, 30 Berry Ave, Paignton, TQ3 3QN [IO80FK, SX86]
G1 XKR Details withheld at licensee's request by SSL.
G1 XKW M J Harvey, 49 Willow Ave, Bradwell, Great Yarmouth, NR31 8JH [JO02UN, TG50]
G1 XKY E Marsh, 15 Beacon Cl, Rubery, Rednal, Birmingham, B45 9DA [IO82XJ, SO97]
G1 XLG C J Proctor, 309 Chester Rd, New Oscott, Sutton Coldfield, B73 5BJ [IO92BN, SP19]
GM1 XLH C Cullingworth, Lochmoss, Ythanwells, Huntly, Aberdeenshire, AB5 6HA [IO87TH, NJ72]
GI1 XLK S J L Baird, 11 Laral Park, Monkstown, Newtownabbey, BT37 0LH [IO74BQ, J38]
G1 XLN J S Banks, 47 Kidsgrove Rd, Stoke on Trent, ST6 5SJ [IO83VB, SJ85]
G1 XLP Details withheld at licensee's request by SSL.
G1 XLT R J Perrat, 18 Petts Hill, Northolt, UB5 4NL [IO91TN, TQ18]
G1 XLW A Harrington, 90 St. Pauls Cres, Pelsall, Walsall, WS3 4ET [IO92AP, SK00]
G1 XLX D Hall, 50 Skeggles Cl, Huntingdon, PE18 6SN [IO92VI, TL27]
G1 XLZ K Shearman, 17 Newlands, North Allerton, N Yorks, OL6 1SJ [IO83WL, SJ99]
GD1 XMA M J Haley, Ballastrooan, Sound Rd, Glen Maye, Isle of Man, IM5 3BJ
G1 XMI J A Brown, 78 Park Way, St. Austell, PL25 4HR [IO70OI, SX05]
G1 XMJ Details withheld at licensee's request by SSL.
G1 XMK Details withheld at licensee's request by SSL.
G1 XMP C W Brookes, 3 Violet Cl, Chatham, ME5 9ND [JO01GH, TQ76]
G1 XMT Details withheld at licensee's request by SSL.
G1 XMV Dr E S Bale, 19 Shooters Hill, Pangbourne, Reading, RG8 7DZ [IO91KL, SU67]
GD1 XMW R Q Moughtin, Foxdale Rd, The Hope, St. Johns, Douglas, Isle of Man, IM4 3AU
G1 XNI N J Dingle, 29 Castle View, Witton-le-Wear, Bishop Auckland, DL14 0DH [IO94DP, NZ12]
G1 XNK R Rafter, 8 Bishops Walk, Ilchester, Yeovil, BA22 8NS [IO81PA, ST52]
G1 XNR A M Clinch, 90 Tong Rd, Little Lever, Bolton, BL3 1QG [IO83TN, SD70]
G1 XNS Details withheld at licensee's request by SSL.
G1 XNX M S A Cowell, 4B Broughman Rd, Marsden, Huddersfield, West Yorks, HD7 6BN [IO93AO, SE01]
G1 XOA R Reddington, 5 Beeston Cl, Birchwood, Warrington, WA3 6LU [IO83RK, SJ69]
GM1 XOG I Wilson, 41 Laburnum Rd, Cumbernauld, Glasgow, G67 3AA [IO85AW, NS77]
GM1 XOI T G Costford Mid Lanark ARS, 14 Redburn Ct, Whitelees, Cumbernauld, Glasgow, G67 3NL [IO85AX, NS77]
G1 XOL G M Shilson, 121 Ramwells Brow, Bromley Cross, Bolton, BL7 9LG [IO83TO, SD71]
G1 XOP E Hull, 43 Birch Ave, Newhall, Swadlincote, DE11 0NQ [IO92FS, SK22]
G1 XOS Details withheld at licensee's request by SSL.
GW1 XOT B M Blake, Brynamlwg, Devils Bridge, Aberystwyth, Dyfed, SY23 4RD [IO82BI, SN77]
G1 XOX Details withheld at licensee's request by SSL.
G1 XOZ C Harding, 24 Bryer Cl, Bridgwater, TA6 6UR [IO81LC, ST23]
G1 XPA P Tonge, 58 Smallbrook Ln, Leigh, WN7 5QA [IO83RM, SD60]
G1 XPB L G Rohrlach, 57 Forbes Ave, Potters Bar, EN6 5NB [IO91WQ, TL20]
G1 XPD L J Wheatley, 25 Hobbis House, Redditch Rd, Kings Norton, Birmingham, B38 8LS [IO92AJ, SP07]
GM1 XPE J R Graham, 18 Wheatriggs Ave, Milfield, Wooler, NE71 6HU [IO85WO, NT93]
G1 XPF A R Duell, 12 Temple Sheen, East Sheen, London, SW14 7RP [IO91UL, TQ27]
GW1 XPP J F Mathers, Brookfield, 16 Glanffynnon, Llanrug, Caernarvon, LL55 4PS [IO73VD, SH56]
G1 XPR Details withheld at licensee's request by SSL.
GI1 XPV K R Nesbitt, 7 Inchkeith Rd, Ballykeel, Ballymena, BT42 4AR [IO64UU, D10]
G1 XPW M W J King, 104 Green Ln, Vicars Cross, Chester, CH3 5LE [IO83NE, SJ46]
G1 XPY J T Wood, 93 Sunnymead Dr, Waterlooville, PO7 6BW [IO90LV, SU61]
G1 XQG D J Pepper, Haselmere, 67 Fairfield Dr, Dorking, RH4 1JG [IO91UF, TQ15]
G1 XQP J L Jackson, Greengable Upcott, Bishops Hull, Taunton, Somerset, TA4 1AQ [IO81KA, ST12]
GW1 XQT Details withheld at licensee's request by SSL.
GW1 XQX Details withheld at licensee's request by SSL.
G1 XRA I L Townsend, 284 Langer Ln, Wingerworth, Chesterfield, S42 6UD [IO93GE, SK36]
G1 XRC R F Maynard Exmouth ARC, 37 Marions Way, Exmouth, EX8 4LF [IO80HP, SY08]
G1 XRE S Staton, 6 Greenhowsyke Ln, Northallerton, DL6 1HP [IO94GI, SE39]
G1 XRM M J Braybrook, 877A London Rd, Westcliff on Sea, Essex, SS0 9SZ [JO01IN, TQ88]
G1 XRO C Frost, 5 Edgecombe Ave, Worle, Weston Super Mare, BS22 9AY [IO81MI, ST36]
G1 XRP C J Wright, 33 Meadowbrook, Lincoln Rd, Ruskington, Sleaford, NG34 9FJ [IO93TB, TF05]
G1 XRQ J T Koenig, 216 Bretch Hill, Banbury, Oxon, OX16 0LU [IO92HB, SP44]
G1 XRT R H Taylor, 24 Hoestock Rd, Sawbridgeworth, CM21 0DZ [JO01BT, TL41]
GW1 XRV R M Partridge, 7 Heol Esgyn, Rhigos, Aberdare, CF44 9BJ [IO81FR, SN90]
G1 XRY D Buckley, 37 Oval Rd, Croydon, CR0 6BJ [IO91WJ, TQ36]
G1 XSA S Cattle, 107 Ferndene Gr, High Heaton, Newcastle upon Tyne, NE7 7PL [IO95FA, NZ26]
G1 XSI A R Rickers, Watchtower House, The Ridgeway, Mill Hill, London, NW7 1RN [IO91VO, TQ29]
G1 XSO R J Rae, 60 Robin Way, Chipping Sodbury, Bristol, BS17 6JP [IO81TM, ST78]
GM1 XSS M E Morrison, 83 Kirk Brae, Cults, Aberdeen, AB1 9QX [IO87TH, NJ72]
G1 XSV M R Scarr, 15 Biddesden Ln, Faberstown, Ludgershall, Andover, SP11 9PG [IO91EG, SU25]
G1 XTA S Hodson, Flagstones, 12 Duns Tew, Bicester, Oxon, OX6 4JR [IO91HW, SP42]
G1 XTB Details withheld at licensee's request by SSL.
G1 XTD I Clarke, 66 Colchester Rd, Southend on Sea, SS2 6HP [JO01IN, TQ88]
GM1 XTF Details withheld at licensee's request by SSL.
G1 XTJ P R Jones, 10 Malhamdale Ave, Rainhill, Prescot, L35 4QF [IO83OJ, SJ49]
GI1 XTK B M Braniff, 5 Cintons Park, Downpatrick, Co. Down, BT30 6NS [IO74DH, J44]
G1 XTN T Milman, 46 Walnut Ave, Tickhill, Doncaster, DN11 9EJ [IO93KK, SK59]
G1 XTO Details withheld at licensee's request by SSL.
G1 XTP J R Grant, 17 Delius Cl, Lowestoft, NR33 9DN [JO02UL, TM59]
G1 XTX H S Bickerton, 57 Ruskin Rd, Crewe, CW2 7JR [IO83SC, SJ75]
G1 XUA W J Condliffe, 7 Marriotts Cl, Felmersham, Bedford, MK43 7HD [IO92RE, SP95]
G1 XUC K F Hughes, 21 Cedar Cl, Patchway, Bristol, BS12 5HD [IO81RM, ST58]
GW1 XUD R S Andrews, 270 Barry Rd, Barry, CF62 8BJ [IO81IJ, ST16]
G1 XUE G Henne, 57 Heaf Gdns, Bentley Cl, Royal British Legion Vil, Aylesford, ME20 7SF [JO01FH, TQ75]
G1 XUG B C M Knights, 2 Bryans Cres, North Crawley, Newport Pagnell, MK16 9LR [IO92QC, SP94]
G1 XUH M E Thornton, 527 Rotherham Rd, Smithies, Barnsley, S71 1XB [IO93GN, SE30]
G1 XUJ Details withheld at licensee's request by SSL.
G1 XUM M J Whiting, 13 Winsford Cl, Harwill Cres, Aspley, Nottingham, NG8 5JR [IO92JX, SK54]
G1 XUO L J Mackenzie, 21 Priory Mead, Doddinghurst, Brentwood, CM15 0NB [JO01DQ, TQ59]
G1 XUP R H Gower, 8 Steele Cl, Marks Tey, Colchester, CO6 1XD [JO01JU, TL92]
G1 XUQ Dr I Harrison, 18 Charlecote Dr, Wollaton, Nottingham, NG8 2SB [IO92JW, SK53]
G1 XUR A P Harrison, 18 Charlecote Dr, Wollaton, Nottingham, NG8 2SB [IO92JW, SK53]
G1 XUW D G W Austin, 17 Patricia Ave, Horstead, Norwich, NR12 7EW [JO02QR, TG21]
GW1 XVC S P Ward, 16 Turnpike Rd, Croesyceiliog, Cwmbran, NP44 2AH [IO81LP, ST39]
G1 XVD C M Snow, Gable End, Ipswich Rd, Nedging Tye, Ipswich, IP7 7BN [JO02LC, TM04]
G1 XVF J T Pottage, 18 Pennine Cl, Huthwaite, Sutton in Ashfield, NG17 2QD [IO93ID, SK45]
GM1 XVI Dr D H Outram, 2 Lawhead Rd West, St. Andrews, KY16 9NE [IO86OI, NO41]
G1 XVL Details withheld at licensee's request by SSL.
GW1 XVM J C Duggan, 42 Vivian Rd, Newport, NP9 0EQ [IO81MO, ST38]
G1 XVR H D Briggs, 14 Ombersley Cl, Redditch, B98 7UU [IO92BG, SP06]
G1 XVT P P Hannan, 151 Star Rd, Peterborough, PE1 5HG [IO92VN, TL29]
G1 XVW A Pogorzelski, 2 Caterham Ct, Beauchamp Rd, London, SE19 3DL [IO91WJ, TQ36]
G1 XVY R Sacharewicz, 15 Milford Cl, Walkwood, Redditch, B97 5PZ [IO92AG, SP06]
G1 XWD A Rhodes, Jubilee Hse Kent Ave, Theddlethorpe, St Helens, Mablethorpe, Lincs, LN12 1QE [JO03CI, TF48]
G1 XWJ Details withheld at licensee's request by SSL.
G1 XWK A J E Rich, 9 Hartford Cl, Harborne, Birmingham, B17 8AU [IO92AL, SP08]
G1 XWM M J B Cox, 23 Cecil Dr, Tividale, Warley, B69 3LA [IO82XM, SO99]
G1 XWN G L Andrews, 22 Arnhem Gr, Braintree, CM7 5UQ [JO01GV, TL72]
G1 XWO F F Williams, 15 Hartsbourne Way, Stafford, ST17 4NR [IO82WS, SJ92]
G1 XWR R Worton, 54C Dunns Bank, Brierley Hill, DY5 2ER [IO82WL, SO98]
G1 XWS H W D Seatory, Ivydene, The St., Hollesley, Woodbridge, IP12 3QU [JO02RB, TM34]
G1 XWT Details withheld at licensee's request by SSL.
G1 XWV A G Cant, 3 Woodmount, Crockenhill, Swanley, BR8 8ER [JO01BJ, TQ56]
G1 XWW Details withheld at licensee's request by SSL.
G1 XWX Details withheld at licensee's request by SSL.[Station located near Edmonton.]
G1 XWZ F R Millbank, 7 Avebury Cl, Burnham on Sea, TA8 2TU [IO81MF, ST34]
G1 XXA J C Elwood, 15 Church Ln, Holton-le-Clay, Grimsby, DN36 5AQ [IO93XM, TA20]
G1 XXE P J Yeates, 9 Arlington Rd, St. Annes, Bristol, BS4 4AF [IO81RK, ST67]
G1 XXF J D Ellison, 68 Rocket Way, Forest Hall, Newcastle upon Tyne, NE12 9RL [IO95FA, NZ26]
G1 XXH R F Tapp, 33 Station Rd, Earls Barton, Northampton, NN6 0NT [IO92OG, SP86]
G1 XXI T Russell, 48 Weymouth Ave, Parr, St. Helens, WA9 3QX [IO83PK, SJ59]
G1 XXJ D V Holland, 89 Dugdell Cl, Tricketts Crose, Ferndown, BH22 8BJ [IO90BT, SU00]
GW1 XXL R G Flynn, 9 Railway Terr, Blaenclydach, Tonypandy, CF40 2DA [IO81GO, SS99]
G1 XXP K S How, 11 Hawthorn Ave, Stopsley, Luton, LU2 8AW [IO91TV, TL12]
G1 XXR S Austin, 5 Mercia Rd, Baldock, SG7 6RZ [IO91VX, TL23]
G1 XXS Details withheld at licensee's request by SSL.
G1 XXV Details withheld at licensee's request by SSL.
G1 XXW P J King, 27 Church St., Witham, CM8 2JP [JO01HT, TL81]
G1 XYD A Bates, 29 Juler Cl, North Walsham, NR28 0SY [JO02RU, TG33]
G1 XYH Details withheld at licensee's request by SSL.[Op: T G Balderson, 99 Earl St, Grimsby, S Humberside, DN31 2PJ.]
GM1 XYL Details withheld at licensee's request by SSL.[Op: Mrs J C Henderson. Station located near Warwick.]
G1 XYM K M Low, Woodend, Bonner Ln, Calverton, Nottingham, NG14 6FX [IO93LA, SK64]
G1 XYN I D Pritchard, 8 Hoon Ave, Newcastle-U-Lyme, Staffordshire, ST5 9NY [IO83VA, SJ84]
G1 XYO S D Glazzard, 109 Highfields Rd, Chasetown, Burntwood, WS7 8QS [IO92AQ, SK00]
G1 XYP A P Griffiths, 94 Rosslyn Dr, Moreton, Wirral, L46 0SZ [IO83KJ, SJ28]
G1 XYR D J Searle, 33 Claypool Rd, Kingswood, Bristol, BS15 2QJ [IO81RK, ST67]
G1 XYS A Brown, 5 Somersby Dr, Newcastle upon Tyne, NE3 3TN [IO95EA, NZ26]
G1 XYT J E Roberts, 28 Airedale Dr, Garforth, Leeds, LS25 2JF [IO93HS, SE43]
G1 XYV M G J Minshull, 23 Littlemore Cl, Saughall Massie, Wirral, L49 4GS [IO83KJ, SJ28]
G1 XYX Details withheld at licensee's request by SSL.
G1 XYZ E Haskett Kings Lynn ARC, 23 Gloucester Rd, Gaywood, Kings Lynn, PE30 4AB [JO02FS, TF62]
G1 XZA I J Rawlingson, 1 Wadham St., Penkhull, Stoke on Trent, ST4 7HF [IO83VA, SJ84]
G1 XZB J E Rawlingson, 1 Wadham St., Penkhull, Stoke on Trent, ST4 7HF [IO83VA, SJ84]
G1 XZG D Collins, 5 Elmwood Cl, Birchwood, Lincoln, LN6 0LZ [IO93QF, SK97]
G1 XZH Details withheld at licensee's request by SSL.
GW1 XZI R A Magwood, 13 Inverness Pl, Roath, Cardiff, CF2 4RU [IO81JL, ST17]
G1 XZJ A G Bradley, 55 Dukes Dr, Newbold, Chesterfield, S41 8QB [IO93GG, SK37]
G1 XZP B A Devine, 17 Newcomen House, York Rise, London, NW5 1DT [IO91WN, TQ28]

G1 XZQ R Carville, 66 Ludlow Rd, Paulsgrove, Portsmouth, PO6 4AE [IO90KU, SU60]
G1 XZW R Hudson, 84 Roman Rd, Luton, LU4 9SE [IO91SV, TL02]
G1 XZX B J Strutt, 55 Coleman Rd, Dagenham, RM9 6JU [JO01BM, TQ48]
G1 YAA T Lilley, Spylaw Cotage, Bilton, Alnwick, Northd., NE66 2TA [IO95EJ, NU21]
G1 YAB C V R Rogers, 197A Bramford Ln, Ipswich, IP1 4DS [JO02NB, TM14]
G1 YAE B Thompson, 1 Littlehoughton, Longhoughton, Alnwick, Northd., NE66 3JZ [IO95EK, NU21]
G1 YAF A M Tyler, 16 Harridge Rd, Leigh on Sea, SS9 4HA [JO01HN, TQ88]
G1 YAH J S McSoley, 88 Rodings Ave, Stanford-le-Hope, SS17 8DT [JO01FM, TQ68]
G1 YAN A Bray, 178 Thornley St., Burton on Trent, DE14 2QR [IO92ET, SK22]
G1 YAO K Bromage, 13 Plover Gr, Kidderminster, Worcs, DY10 4TG [IO82VI, SO87]
G1 YAR Details withheld at licensee's request by SSL.
G1 YAS D Elliott, The Forge, 103 The Causeway, Burwell, Cambridge, CB5 0DU [JO02DG, TL56]
G1 YAV Details withheld at licensee's request by SSL.[Op: Dharm, The Lilacs, Moxley, Wednesbury, West Midlands, WS10 8SA.]
G1 YAW R G Coulson, 67 Huntly Gr, Peterborough, PE1 2QW [IO92VN, TL19]
G1 YAX S Banks, 93 Whitacre, Peterborough, PE1 4SX [IO92VO, TF20]
G1 YBA I P Hardaker, 7 Cotgrave Cl, Strelley, Nottingham, NG8 6QE [IO92JX, SK54]
G1 YBB S D Clements, 46 Brampton Rd, Newton Farm Est, Hereford, HR2 7DF [IO82PA, SO43]
G1 YBD Details withheld at licensee's request by SSL.
GW1 YBF L Ward, 11 Verlands Way, Pencoed, Bridgend, CF35 6TY [IO81FM, SS98]
G1 YBG Dr J H W Ballance, Ty Celyn, 5 Folly Ln, Hereford, HR1 1LY [IO82PB, SO54]
G1 YBI A T Jones, 43 Oakleigh Rd, Droitwich, WR9 0RP [IO82VG, SO86]
G1 YBK J H Fyson, 1 Redlands Est, Ibstock, Leicester, LE6 1HT [IO92JP, SK50]
G1 YBM J C Pedley, 92 Ashfield Dr, Moira, Swadlincote, DE12 6HQ [IO92FR, SK31]
G1 YBN Details withheld at licensee's request by SSL.
G1 YBQ Details withheld at licensee's request by SSL.
G1 YBT J A Bagshaw, 14 Saxelby Gdns, Nottingham, NG6 8JZ [IO93JA, SK54]
G1 YBU K Ogle, 153 Alnwick Rd, Intake, Sheffield, S12 2GG [IO93GI, SK38]
G1 YBZ P Barnett, 40 Thames Cl, Congleton, CW12 3RL [IO83VD, SJ86]
G1 YCB Details withheld at licensee's request by SSL.
G1 YCJ I M Day, 2 New Bungalow, Stowbridge Farm, Stretham, Ely, Cambs, CB6 3LF [JO02CH, TL57]
G1 YCM A N Laughlan, 8 Kempton Cl, Droylsden, Manchester, M43 7JL [IO83WL, SJ99]
G1 YCN D M Lewis, 81 Ashton Ave, Rainhill, Prescot, L35 0QR [IO83OJ, SJ49]
G1 YCR R W Lawrence, 82 Moseley St., Southend on Sea, SS2 4NN [JO01IN, TQ88]
G1 YCV D Brook, 41 Cross Ln, Stocksbridge, Sheffield, S10 1WL [IO93FJ, SK38]
G1 YCY Details withheld at licensee's request by SSL.
G1 YDA M J Davies, The Willows, Eye Rd, Hoxne, Eye, IP21 5BA [JO02OI, TM17]
G1 YDD A R Forster, 56 Tantobie Rd, Denton Burn, Newcastle upon Tyne, Tyne & Wear, NE15 7DQ [IO94DX, NZ26]
G1 YDG A C Miles, Yew Tree House, Main St., East Hanney, Wantage, OX12 0HT [IO91HP, SU49]
G1 YDH Details withheld at licensee's request by SSL.
G1 YDI C P Lambeth, 98 Hockmore Tower, Pound Way, Oxford, OX4 3YG [IO91JR, SP50]
G1 YDJ T Polley, 9 Otter Rd, Clevedon, BS21 6LQ [IO81NK, ST47]
G1 YDK M J P Worner, 8 Causeway View, Nailsea, Bristol, BS19 2XG [IO81OK, ST47]
GW1 YDN J F Day, 20 St. Johns Dr, Pencoed, Bridgend, CF35 5NF [IO81FM, SS98]
G1 YDQ J P Carpenter, 34 Carey Park, Killigarth, Looe, PL13 2JP [IO70RI, SX25]
G1 YDT M C Readings, 18 Hadow Way, Quedgeley, Gloucester, GL2 4YJ [IO81UT, SO81]
G1 YDW D Puttick, 21 Sandyfield Cres, Cowplain, Waterlooville, PO8 8SQ [IO90LV, SU61]
GI1 YEA L M O'Flaherty, 1 Ravensdale Villas, Newry, BT34 2PG [IO64UD, J02]
G1 YEB Details withheld at licensee's request by SSL.
G1 YED W Ross-Fraser, 47 Lichford Rd, Arbourthorne, Sheffield, S2 3LB [IO93GI, SK38]
G1 YEF Details withheld at licensee's request by SSL.
GW1 YEH Details withheld at licensee's request by SSL.
G1 YEK J P Devaney, 4 Nantwich Rd, Fallowfield, Manchester, M14 7AP [IO83VK, SJ89]
G1 YEL F J Whitehead, 15 Leagate, Urmston, Manchester, M41 9LD [IO83TK, SJ79]
G1 YEP F K Russell, 7 Glenmore Ave, Liverpool, L18 4QE [IO83NJ, SJ38]
G1 YES B L Underhay, 24 Rutland Rd, Southall, UB1 2UP [IO91TM, TQ18]
G1 YEU M Eales, 23/07 Victoria Centre Flats, Milton St, Nottingham, NG1 3PW [IO92KW, SK54]
G1 YEV Dr D Martin, 73 Summerfields Way, Ilkeston, DE7 9HE [IO92IX, SK44]
G1 YEX Details withheld at licensee's request by SSL.
G1 YEY P L K Newing, 50 Eagles Dr, Melton Mowbray, LE13 0BA [IO92NS, SK71]
G1 YEZ A S Lord, 66 Salcombe Dr, Glenfield, Leicester, LE3 8AF [IO92JP, SK50]
G1 YFC P W Neades, 84 Nicholson Ct, Bobblestock, Hereford, HR4 9TD [IO82PB, SO44]
G1 YFD S P Lycett, 27 Ropewalk, Alcester, B49 5DD [IO92BF, SP05]
G1 YFE J S Dent, 1 Lynwood Cttgs, Fodder Fen, Manea, Cambs, PE15 0HH [JO02CM, TL49]
G1 YFF J Bevan, 76 St. Philips Rd, Newmarket, CB8 0EN [JO02EF, TL66]
G1 YFH E H Cox, 41 Lister Rd, Beechdale Est, Walsall, WS2 7HN [IO82XO, SJ90]
G1 YFI J Simmonds, 94 Gravel Hill, Tile Hill South, Coventry, CV4 9JH [IO92FJ, SP27]
G1 YFK G Otero, 3 Fry Rd, Stevenage, SG2 0QG [IO91VV, TL22]
GM1 YFO P G R Mirtle, Obelisk Cottage, Hopetown Est, South Queensferry, EH20 9SL [IO85KV, NT26]
G1 YFQ D G Scothern, 1 Wold View, Mulberry Rd, Claxby, Market Rasen, Lincs, LN8 3YS [IO93UK, TF19]
G1 YFR Details withheld at licensee's request by SSL.
G1 YFT R M Allsopp, 271 Wigston Ln, Aylestone, Leicester, LE2 8DL [IO92KO, SK50]
G1 YFU Details withheld at licensee's request by SSL.
G1 YFV Details withheld at licensee's request by SSL.
G1 YFW Details withheld at licensee's request by SSL.
G1 YFX Details withheld at licensee's request by SSL.
G1 YFY Details withheld at licensee's request by SSL.
G1 YGA Details withheld at licensee's request by SSL.
G1 YGD Details withheld at licensee's request by SSL.
G1 YGG Details withheld at licensee's request by SSL.
G1 YGH L Walker, 21 Alexandra Rd, Sale, M33 3EF [IO83UK, SJ79]
G1 YGJ Details withheld at licensee's request by SSL.[Station located near Tamworth.]
G1 YGP S B Jarman, 55 The Meadows, Todwick, Sheffield, S31 0JG [IO93GJ, SK38]
G1 YGR Details withheld at licensee's request by SSL.
GM1 YGV R W E Johnstone, 10 Lundy Rd, Inverlochy, Fort William, PH33 6NX [IO76KT, NN17]
GM1 YGW Dr G R G Craib, 9/1 Echline Rigg, South Queensferry, EH30 9XN [IO85HX, NT17]
G1 YGY C D Weaver, 11 Thirlmere, Swindon, SN3 6LA [IO91CN, SU18]
GW1 YHA P George, 24 Ty Fry Cl, Brynmenyn, Bridgend, CF32 8YB [IO81FN, SS98]
G1 YHB J G Moggeridge, 67 Park Rd, Didcot, OX11 8QT [IO91IO, SU58]
G1 YHE D J G Coate, 17 Swallow Cl, Poole, BH17 7UW [IO80XR, SY99]
G1 YHF M B Drinkwater, Green Quarter House, Green Quarter, Kentmere, Cumbria, LA8 9JP [IO84OK, NY40]
G1 YHG M J H Kennedy, Milestones, Spetisbury, Dorset, DT11 9DW [IO80WT, ST90]
G1 YHI K R Davies, 2 Orchard Cl, Lytchett Minster, Poole, BH16 6JH [IO80XR, SY99]
G1 YHJ G N W Williams, 2 Cotton Cl, Broadstone, BH18 9AJ [IO90AS, SZ09]
G1 YHL D M Crawshaw, 60 Forest Rd, Torquay, Devon, TQ1 4JR [IO80FL, SX96]
G1 YHM G Tonge, 19 Longmeadow Rd, Saltash, PL12 6DW [IO70VJ, SX45]
G1 YHO B Parkinson, 15 The Croft, Sheriff Hutton, York, YO6 1PQ [IO94MC, SE66]
G1 YHP J M Hogg, 16 Moorston Cl, Naisberry Park, Hartlepool, TS26 0PJ [IO94JQ, NZ43]
G1 YHQ B Gadsden, 61 Scruton Ave, Sunderland, SR3 1SG [IO94HV, NZ35]
G1 YHR Details withheld at licensee's request by SSL.
G1 YHS W H Jardine, 17 Bridge Park, Gosforth, Newcastle upon Tyne, NE3 2DX [IO95EA, NZ26]
G1 YHT S Osborne, 45 Broomfield Ave, Battle Hill Est, Wallsend, NE28 9AE [IO95FA, NZ36]
G1 YHV S R Briscoe, Chimneys, 8 Corfe View Rd, Corfe Mullen, Wimborne, BH21 3LZ [IO80XS, SY99]
G1 YHX F E Lines, 97 Oakhurst Dr, Wickford, SS12 0NW [JO01GO, TQ79]
G1 YIB Details withheld at licensee's request by SSL.
G1 YIC R H Hamilton, 12 Eakring Rd, Bilsthorpe, Newark, NG22 8PY [IO93LD, SK66]
G1 YIK Details withheld at licensee's request by SSL.
G1 YIL C A Grellis, 96 Crock Ln, Bridport, DT6 4DQ [IO80OR, SY49]
G1 YIO Details withheld at licensee's request by SSL.
G1 YIQ J A Stafford, 28 Vale Way, Kings Worthy, Winchester, SO23 7LL [IO91IC, SU43]
G1 YIZ A J Ward, 42 Felstead Cres, Sunderland, SR4 0AB [IO94GV, NZ35]
G1 YJB G M Evans, 16 Kynaston Dr, Wem, Shrewsbury, SY4 5DE [IO82PU, SJ52]
G1 YJH M Blackman, 39 Lyndhurst Ave, Mill Hill, London, NW7 2AD [IO91VO, TQ29]
G1 YJI P N Kay, 90 Ave Rd, Southgate, London, N14 4EA [IO91WP, TQ29]
G1 YJJ R B Colman, 11A Oulton Hall, Marine Parade East, Clacton on Sea, CO15 6JU [JO01NS, TM11]
G1 YJK K W Boughton, Meade Cottage, 2 Raglan Rd, Frinton on Sea, CO13 9HH [JO01PT, TM21]
G1 YJL W A Pond, Wheatlands, Calais St., Boxford, Sudbury, CO10 5JA [JO02KA, TL94]
G1 YJQ Details withheld at licensee's request by SSL.
G1 YJR J F Davies, 70 Ash Rd, Cuddington, Northwich, CW8 2PB [IO83QF, SJ57]
G1 YJW S P Bates, 21 St. Peters Cl, Wilpshire, Blackburn, BB1 9HH [IO83ST, SD63]
GM1 YKE J A Campbell, 15 Birchwood Rd, Cumnock, KA18 1NG [IO75VK, NS51]
G1 YKI J C Heathfield, 82 Auriel Ave, Dagenham, RM10 8BT [JO01CN, TQ58]
G1 YKK S M OConnor, 32 Whitfield Cross, Glossop, SK13 8NW [IO93AK, SK09]
G1 YKL J M Brackenridge, 21 St. Mark Rd, Deepcar, Sheffield, S30 5TF [IO93GJ, SK38]
G1 YKO R O Broome, 145 Sprotborough Rd, Doncaster, DN5 8BW [IO93KM, SE50]
GW1 YKT W M John, 4 Heol y Bryn, Rhiwbina, Cardiff, CF4 6HY [IO81JM, ST18]
G1 YKV N Stewart, 6 Huntington Cl, Moreton, Wirral, L46 6HU [IO83KJ, SJ28]
G1 YKX M K Rouse, 105 Great Spenders, Basildon, SS14 2NS [JO01FN, TQ78]
GW1 YKY S G B Jones, 14 Plantation Dr, Croesyceiliog, Cwmbran, NP44 2AN [IO81LP, ST39]
G1 YKZ R Burt, 68 Gordon Rd, Basildon, SS14 1PQ [JO01FN, TQ78]
G1 YLA S T Stanger, 3 Stoney Ln, Springwell, Gateshead, NE9 7SJ [IO94FW, NZ25]
GM1 YLB S J Doyle, 1 Hawthorn Rd, Strathaven, ML10 6HA [IO75XQ, NS64]
G1 YLC G G Gisborne, 90 Mayfield Dr, Hucclecote, Gloucester, GL3 3DX [IO81VU, SO81]
G1 YLE P J Adams, 25 Main Rd, Kesgrave, Ipswich, IP5 7AQ [JO02OB, TM24]
G1 YLG A R Hodkin, 18 Habershon Dr, Chapeltown, Sheffield, S30 4ZT [IO93GJ, SK38]
G1 YLH R D K Fisher, 23 Barrie Gr, Hellaby, Rotherham, S66 8HG [IO93JK, SK59]
G1 YLJ A Hunt, 14 Sandalwood Cl, Shires Green, Willenhall, WV12 5YJ [IO82XO, SJ90]
G1 YLM E C P Bradshaw, 38 Whiteford Dr, Kettering, NN15 6HH [IO92PJ, SP87]
G1 YLN M G Swetman, Taunton Deane Dist.Emer.P Dept, Somerset Co. Council, Room A311 Co. Hall, Taunton, Somerset, TA1 4DY [IO81KA, ST22]
G1 YLQ C T Thomas, 152 Bristol Rd, Edgbaston, Birmingham, B5 7XH [IO92BL, SP08]
G1 YLU Details withheld at licensee's request by SSL.
G1 YLV M H A Bayliss, 27 Manor Cl, Tunstead, Norwich, NR12 8EP [JO02QR, TG22]
G1 YLX J H C Sutton, The Partch, Salthouse, Holt, Norfolk, NR25 7XG [JO02NW, TG04]
G1 YMA W K Scoles, The Old Rectory, Great Snoring, Fakenham, Norfolk, NR21 0HP [JO02KU, TF93]
G1 YMC S J Fitzpatrick, 32 Beechwood Rd, Kings Heath, Birmingham, B14 4AD [IO92BK, SP08]
GM1 YME J G B Hein, 78 Montgomery St., Edinburgh, EH7 5JA [IO85JX, NT27]
G1 YMH M R Homer, 86 Victoria Rd, Quarry Bank, Brierley Hill, DY5 1DB [IO82WL, SO98]
G1 YMJ A Smith, 6 Norton Cres, Towcester, NN12 6DN [IO92MD, SP64]
G1 YMN G A Brooks, 16 Cameron Cl, Lillington, Leamington Spa, CV32 7DZ [IO92FH, SP36]
G1 YMO Details withheld at licensee's request by SSL.
G1 YMP S Ellis, 6 Newark Rd, Hindley, Wigan, WN2 3HR [IO83QM, SD60]
GW1 YMQ Details withheld at licensee's request by SSL.
G1 YMU Details withheld at licensee's request by SSL.
G1 YMV H Johnson, 2 Greenbank Ave, Storth, Milnthorpe, Cumbria, LA7 7JP [IO84OF, SD47]
G1 YMW Details withheld at licensee's request by SSL.
G1 YMY A Hussain, 206 Hereford Way, Middleton, Manchester, M24 2NJ [IO83VN, SD80]
G1 YND Details withheld at licensee's request by SSL.
G1 YNF B Logan, 27 Ruth House, Otley Rd, Bradford, BD3 0DB [IO93DT, SE13]
G1 YNG Details withheld at licensee's request by SSL.
G1 YNH C J Arundel, 91 Willow Garth Ave, Leeds, LS14 2EA [IO93GU, SE33]
G1 YNJ M J R Rocke, Orchard House, 55 Tarvin Rd, Littleton, Chester, CH3 7DD [IO83OE, SJ46]
G1 YNL Details withheld at licensee's request by SSL.
G1 YNO B Surtees, 5 Haweswater Gr, West Auckland, Bishop Auckland, DL14 9LQ [IO94DP, NZ12]
G1 YNP M J Cowley, 30 New Rd, Bromham, Chippenham, SN15 2JB [IO81XJ, ST96]
G1 YNQ J Crow, 71 Stockshill Rd, Ashby, Scunthorpe, DN16 2LQ [IO93QN, SE90]
G1 YNU W N Leigh, 37 Brendjean Rd, Morecambe, LA4 5SE [IO84NB, SD46]
G1 YNW Details withheld at licensee's request by SSL.
G1 YNX B J Johnson, 14 Berkeley Rd, Bolton, Lancs, BL1 6PS [IO83SO, SD71]
G1 YNY Details withheld at licensee's request by SSL.
G1 YOA G J Hawkins, Llc International, Inc., 7925 Jones Branch Dr, McLean, Va, USA 22102
G1 YOB N J Power, 25 Mill Green, Wolverhampton, WV10 6LX [IO82WP, SJ90]
G1 YOD R A Chivers, 13 Padwick Cl, Basingstoke, RG21 8XS [IO91KG, SU65]
G1 YOF A L Lockwood, 1 Chudley Cl, Exmouth, EX8 4DY [IO80HP, SY08]
G1 YOH W Lomas, Durnford Gdns, Mount Pleasant, Langton Matravers, Swanage, BH19 3HH [IO80XO, SY97]
G1 YOI A P Briggs, 7 College St., Brighton, BN2 1JG [IO90WT, TQ30]
G1 YOK Details withheld at licensee's request by SSL.
G1 YOM I P Morgan, 316 Middle Rd, Sholing, Southampton, SO19 8NT [IO90HV, SU41]
G1 YOR Details withheld at licensee's request by SSL.
G1 YOS M J Avenell, Lime House, Worlds End, Beedon, Newbury, RG20 8SD [IO91IL, SU47]
GJ1 YOT N A Paisnel, St Clements Farm, St Clement, Jersey, Channel Islands, JE2 6QQ
G1 YOY J J Bowen, Garden Flat, 3 West Mall, Clifton, Bristol, BS8 4BH [IO81QK, ST57]
G1 YPD Details withheld at licensee's request by SSL.
G1 YPH A C Roberts, 18 Surtees Gr, Fenton, Stoke on Trent, ST4 3HH [IO82WX, SJ94]
GM1 YPJ L J Davies, 24 Ardgour Rd Cao, Fort William, Inverness-shire, PH33 7HN [IO76KU, NN17]
G1 YPL B A Collins, 5 Randalls Row, High Banks, Loose, Maidstone, ME15 0EG [JO01GF, TQ75]
G1 YPM R J Northcott, 32 Lichfield Rd, Exwick, Exeter, EX4 2EU [IO80FR, SX99]
G1 YPR G R Dearden, 125 Campsall Field Rd, Wath upon Deame, Rotherham, S63 7ST [IO93HL, SK49]
G1 YPT G Hartshorn, Oxford House, 11 Lime Ave, Ripley, DE5 3HD [IO93HB, SK35]
G1 YPU E C Rowberry, 6 Rawlyn Cl, Cambridge, CB5 8NN [JO02BF, TL45]
G1 YPY Details withheld at licensee's request by SSL.
G1 YPZ L W Pritchett, 75 Dalehead, Harrington Sq, London, NW1 2JL [IO91WM, TQ28]
G1 YQG A R Higham, 3 Tonge Park Ave, Tonge Moor, Bolton, BL2 2QR [IO83TO, SD71]
G1 YQH Details withheld at licensee's request by SSL.[Station located near Herne.]
G1 YQI P E Bennett, 7 Woburn Ave, Firwood, Bolton, BL2 3AY [IO83TO, SD71]
G1 YQJ Details withheld at licensee's request by SSL.
G1 YQK Details withheld at licensee's request by SSL.
G1 YQL C E C Stagg, 559 Dividy Rd, Bucknall, Stoke on Trent, ST2 0BX [IO83WA, SJ94]
G1 YQN P B L McShea, Woodlands, Boxley Rd, Chatham, ME5 9JD [JO01GI, TQ76]
G1 YQO R A D Fletcher, 67 Furlong Ave, Arnold, Nottingham, NG5 7AS [IO93KA, SK54]
G1 YQP M J West, 27 Nidderdale Rd, The Meadows, Wigston, LE18 3XW [IO92KN, SP69]
G1 YQS D J M Johnson, 20 Hay Cl, Kidderminster, DY11 5DH [IO82UJ, SO87]
G1 YQU J A Stapleford, 7 Garfield Rd, Hugglescote, Coalville, LE67 2HU [IO92HR, SK41]
G1 YQW Details withheld at licensee's request by SSL.
G1 YQY W J Oakes, 2 Hillcrest, Scotton, Catterick Garrison, DL9 3NJ [IO94DI, SE19]
G1 YQZ D Webster, 4 Alwyne Gr, Shipton Rd, York, YO3 6RT [IO93KX, SE55]
G1 YRC R Morton York ARC, 19 Chapel Walk, Riccall, York, YO4 6NU [IO93LT, SE63]
G1 YRE S E Piper, Willow End, Ipswich Rd, Witnesham, Ipswich, IP6 9HT [JO02OC, TM14]
G1 YRF D G Buggs, 2 Archway Cttgs, Valley Rd, Leiston, IP16 4AR [JO02TF, TM46]
G1 YRK C R Bocking, 43 Winton Dr, Cheshunt, Waltham Cross, EN8 9JR [IO91XQ, TL30]
G1 YRM C W Mead, 32 Sandy Rd, Potton, Sandy, SG19 2QQ [IO92VD, TL24]
G1 YRP D A Bath, 34 Arkwright Rd, Sanderstead, South Croydon, CR2 0LL [IO91XI, TQ36]
G1 YRR W R P Fry, 227 London Rd North, Merstham, Redhill, RH1 3BN [IO91WG, TQ25]
G1 YRS R L Shaddick, 6 Haylands, Portland, DT5 2JZ [IO80SN, SY67]
G1 YRV R S Peters, 2 Heath Cl, Sussex Rd, Orpington, BR5 4JG [JO01BJ, TQ46]
G1 YRW Details withheld at licensee's request by SSL.
G1 YRX K J Stone, 1 Lewins Way, Cippenham, Slough, SL1 5JQ [IO91QM, SU98]
G1 YRY S C Roberts, 5 Troon Ct, Baird Ave, Southall, UB1 3LY [IO91TM, TQ18]
G1 YSA M J R Crick, 85 Ashurst Rd, Friern Barnet, London, N12 9AU [IO91WO, TQ29]
G1 YSD Details withheld at licensee's request by SSL.
GI1 YSG P J Kennedy, Oakwood, 29 Barnfield Rd, Lisburn, BT28 3TQ [IO64XN, J26]
GW1 YSM A T Hughes, 35 Queen St., Pontypridd, CF37 1RN [IO81IO, ST08]
G1 YSR P I A Shaw, 18 Knotts Pl, Granville Rd, Sevenoaks, TN13 1HD [JO01CG, TQ55]
G1 YSX D T Taylor, 103 Southend, Garsington, Oxford, OX44 9DL [IO91KR, SP50]
G1 YSY R S Marsden, 25 High Acres, Banbury, OX16 9SL [IO92IB, SP43]
G1 YSZ Details withheld at licensee's request by SSL.
G1 YTC P Dighton, 98 Millfield, Sittingbourne, ME10 4TP [JO01II, TQ96]
G1 YTO K Wenman, 2 Hythe Rd, Sittingbourne, ME10 2LR [JO01II, TQ86]
G1 YTR J H V Batchelor, 26 Ascham Pl, Eastbourne, BN20 7QQ [JO00DS, TV69]
G1 YTU A Clayton, 2 Woodfield Cl, Darfield, Barnsley, S73 9EP [IO93HM, SE40]
G1 YTV S K Fisher, Farland, Rillaton, Rilla Mill, Callington, Cornwall, PL17 7PA [IO70TM, SX27]
G1 YUB B Harrison, 15 Helmington Terr, Hunwick, Crook, DL15 0LQ [IO94DQ, NZ13]
G1 YUD G G Williams, 26 Folkestone Rd, Lytham St. Annes, FY8 3EQ [IO83LS, SD33]
G1 YUE D J T Bessant, 71 Moorpark Rd, Northfield, Birmingham, B31 4HE [IO92AJ, SP07]
G1 YUG R E Pankhurst, 31 Harridge Rd, Leigh on Sea, SS9 4HA [JO01HN, TQ88]
GM1 YUH K B Jones, Hyne Isle of Coll, Inner Hebrides, Argyll, Scotland, PA78 6TB [IO66QO, NM15]
G1 YUI T S Dixon, 1 Wareham Ct, Barnstaple Rd, Scunthorpe, DN17 1YD [IO93PN, SE80]
G1 YUL J W Robson, 28 Eastfield St., Sunderland, SR4 7SA [IO94HV, NZ35]
G1 YUN K W Hetherington, 22 Hoyle Ave, Fenham, Newcastle upon Tyne, NE4 9QX [IO94EX, NZ26]
GM1 YUO Details withheld at licensee's request by SSL.
G1 YUP I W Donald, 32 Shorestone Ave, Cullercoats, North Shields, NE30 3NE [IO95GA, NZ37]
G1 YUQ G Beaumont, 29 Castle View, Sandal, Wakefield, WF2 7HZ [IO93GP, SE31]
G1 YUS A Lees, 692 Walmersley Rd, Bury, BL9 6RN [IO83UO, SD81]
G1 YUU A G Bagworth, 5 The Lindales, Pogmoor, Barnsley, S75 2DT [IO93FN, SE30]
G1 YUV W H Broadbent, 3 George St., Thurnscoe, Rotherham, S63 0DF [IO93IN, SE40]
G1 YUX J Garnett, 21 Vicarage Cl, Mossley Hill, Liverpool, L18 7HU [IO83NI, SJ38]
G1 YVF Details withheld at licensee's request by SSL.[Station located near Redditch.]
G1 YVI K W J Biddlecombe, 7 Jevington Ct, Jevington Cl, Worthing, BN13 3QF [IO90ST, TQ10]
G1 YVL Details withheld at licensee's request by SSL.
G1 YVN Details withheld at licensee's request by SSL.
G1 YVR Details withheld at licensee's request by SSL.
G1 YVS C F Newby Robson, 39 Towngate East, Market Deeping, Peterborough, PE6 8DP [IO92UQ, TF11]

G1 YVT Details withheld at licensee's request by SSL.
G1 YVV I J Coleman, 69 Glebelands, West Molesey, KT8 2PY [IO91TJ, TQ16]
G1 YVY H A Miller, 47 Engleheart Dr, East Bedfont, Feltham, TW14 9HL [IO91TK, TQ17]
G1 YVZ A P Bell, 89 Kings Ave, New Malden, KT3 4DU [IO91UJ, TQ26]
G1 YWF K D Alford, 17 Hinchley Way, Esher, KT10 0BD [IO91UJ, TQ16]
G1 YWI A S Williams, 26 Matlock Rd, Lower Farm Est, Bloxwich, Walsall, WS3 3QD [IO92AP, SK00]
G1 YWL T G Wellington, 32 Fifth Ave, Grantham, NG31 9SY [IO92QW, SK93]
G1 YWM M F Maddison, 47C High St. South, Rushden, NN10 0QZ [IO92QG, SP96]
G1 YWN A D Whitworth, 183 Logan St., Bulwell, Nottingham, NG6 9FX [IO92JX, SK54]
G1 YWT M E Blackley, 76 Wombwell Ln, Stairfoot, Barnsley, S70 3NX [IO93GN, SE30]
G1 YWU G S Abbey, 126 Gaisby Ln, Shipley, BD18 1AQ [IO93CT, SE13]
G1 YWX B Wainwright, 243 Long Ln, Aughton, Ormskirk, L39 5BY [IO83NN, SD40]
G1 YWY M Jones, 55 Moorfoot Way, Melling Mount, Liverpool, L33 1WY [IO83NM, SD40]
G1 YXA B G Dixon, 16 Dyrham Parade, Stoke Lodge, Patchway, Bristol, BS12 6EF [IO81RM, ST68]
G1 YXF J W Pallister, 1 Third Ave, Highfields, Dursley, GL11 4NT [IO81TQ, ST79]
G1 YXG B Davidson, 2 Tennyson View, Elm Ln, Calbourne, Newport, PO30 4JS [IO90HQ, SZ48]
G1 YXH R J Harrison, 14 St. Leonards Ave, Chatham, ME4 6HL [JO01GI, TQ76]
G1 YXJ A E Palmer, 14 Garibaldi Rd, Redhill, RH1 6PB [IO91VF, TQ24]
GW1 YXR N J H Williams, 31 Syr Davids Ave, Cardiff, CF5 1GH [IO81JL, ST17]
G1 YXT C D Wise, 28 Southlands, East Grinstead, RH19 4BZ [IO91XC, TQ33]
G1 YXY L E White, The Garden Flat, 47 Hamerton Rd, Northfleet, Gravesend, DA11 9DX [JO01DK, TQ67]
G1 YYD E A Brown, 25 Cork Rd, Bowerham, Lancaster, LA1 4BD [IO84OA, SD46]
G1 YYH J Heaton, 85 Morris Green Ln, Bolton, BL3 3JD [IO83SN, SD70]
G1 YYL T A Mulloy, 7 Roundhill, Kirby Muxloe, Leicester, LE9 2DY [IO92JP, SK50]
G1 YYP M A Illston, 4 The Sett, Whatcote Rd, Oxhill, Warwick, CV35 0RE [IO92FC, SP34]
G1 YYV Details withheld at licensee's request by SSL.
G1 YYY D J Penny Braintree Ray G, Thirty Nine Steps, 13 Newham Cl, Braintree, Essex, CM7 7PR [JO01GV, TL72]
G1 YZB Details withheld at licensee's request by SSL.
GW1 YZF D Edwards, 240 Berthin, Greenmeadow, Cwmbran, NP44 4LB [IO81LP, ST29]
G1 YZH R Baxter, 10 Eastover, Bredbury Green, Romiley, Stockport, SK6 3ES [IO83WJ, SJ99]
G1 YZI P Rennison, Foxhall Cottage, North End, Kelshall, Royston, SG8 9SE [IO92XA, TL33]
G1 YZJ R Rennison, Foxhall Cottage, North End, Kelshall, Royston, SG8 9SE [IO92XA, TL33]
G1 YZL D W Long, 21 Muzzle Patch, Tibberton, Gloucester, GL2 8EE [IO81TV, SO72]
G1 YZN Details withheld at licensee's request by SSL.
G1 YZO A J Stavert-Dobson, 34 Chapel St., Birdwell, Barnsley, S70 5UG [IO93GM, SE30]
G1 YZQ J S Dobson, 34 Chapel St., Birdwell, Barnsley, S70 5UG [IO93GM, SE30]
G1 YZR Details withheld at licensee's request by SSL.
G1 YZY Details withheld at licensee's request by SSL.
G1 ZAA Details withheld at licensee's request by SSL.
G1 ZAD Details withheld at licensee's request by SSL.
G1 ZAF Details withheld at licensee's request by SSL.
G1 ZAL Details withheld at licensee's request by SSL.
G1 ZAQ Details withheld at licensee's request by SSL.
G1 ZAR Details withheld at licensee's request by SSL.
G1 ZAT Details withheld at licensee's request by SSL.
G1 ZAY A G Robinson, 31 Heathview Rd, Sockets Heath, Grays, RM16 2RS [JO01EL, TQ67]
G1 ZBA P Leeman, 449 Montagu Rd, Edmonton, London, N9 0HR [IO91XO, TQ39]
G1 ZBB S J Brown, 6 Pathfinder Way, Ramsey, Huntingdon, PE17 1LX [IO92WK, TL28]
GM1 ZBD P R Tebbutt, 5 St. Colme Rd, Dalgety Bay, Dunfermline, KY11 5LH [IO86IA, NT18]
GW1 ZBE A C Cooper, Brynhyfryd, Llanfachreth, Dolgellau, Gwynedd, LL40 2EH [IO82BS, SH72]
G1 ZBF Details withheld at licensee's request by SSL.
G1 ZBG M G Bason, 52 Wroslyn Rd, Freeland, Witney, OX8 8HH [IO91HT, SP41]
G1 ZBH G F Brock, 148 Lonsdale Dr, Rainham, Gillingham, ME8 9HX [JO01HI, TQ86]
G1 ZBJ A J Manning, 12 Clifford Dr, Heathfield, Newton Abbot, TQ12 6GX [IO80EN, SX87]
G1 ZBK J C Hancock, 21 Cumberland Cl, Aylesbury, HP21 7HH [IO91OT, SP81]
G1 ZBL N J Marsh, 16 Laurel Cl, North Warnborough, Hook, RG29 1BH [IO91MG, SU75]
G1 ZBM G A Blinman, 30 Charlton Rd, Midsomer Norton, Bath, BA3 4AE [IO81SG, ST65]
G1 ZBN M E Rice, 73 Woodbine Terr, Blyth, NE24 3DP [IO95FC, NZ38]
G1 ZBO E G C McLusky, 11 Ripon Rd, Killinghall, Harrogate, HG3 2DG [IO94FA, SE25]
G1 ZBP A Whipp, 114 Lower Manor Ln, Burnley, BB12 0EF [IO83VT, SD83]
G1 ZBU G H Newby, 77 Darby Rd, Garston, Liverpool, L19 9AN [IO83NI, SJ38]
G1 ZBW W P Baker, 103 New Ln, Harwood, Bolton, BL2 5BY [IO83TO, SD71]
G1 ZBY Details withheld at licensee's request by SSL.
GM1 ZCD P G Hughes, Arndean, 26 Kingseat Rd, Dunfermline, KY12 0DD [IO86GC, NT18]
G1 ZCH Details withheld at licensee's request by SSL.
G1 ZCK Details withheld at licensee's request by SSL.
G1 ZCS E G Davis, 10 Fairfield Dr, Oulton Broad, Lowestoft, NR33 8QG [JO02UL, TM59]
G1 ZCV Details withheld at licensee's request by SSL.
G1 ZCW Details withheld at licensee's request by SSL.
G1 ZCY P B O'Connor, Flat 12, 39 Varden Croft, Edgbaston, Birmingham, B5 7LR [IO92BL, SP08]
G1 ZDA J A Harrison, 12 Oswin Pl, Walsall, WS3 1PU [IO92AO, SK00]
G1 ZDB Details withheld at licensee's request by SSL.
G1 ZDC C J Brown, 8 The Warren, Cotgrave, Nottingham, NG12 3TH [IO92LV, SK63]
G1 ZDD Details withheld at licensee's request by SSL.
G1 ZDE Details withheld at licensee's request by SSL.
G1 ZDF Major D Fleetwood, 9 Reynolds Cl, Swindon, Dudley, DY3 4NQ [IO82VM, SO89]
G1 ZDG P E Whittaker, 33 Abbotts Green, Croydon, Surrey, CR0 5BL [IO91XI, TQ36]
G1 ZDI Details withheld at licensee's request by SSL.
G1 ZDK Details withheld at licensee's request by SSL.
G1 ZDM Details withheld at licensee's request by SSL.
G1 ZDR J Angus, 8 Gravel Rd, Bromley, BR2 8PF [JO01AI, TQ46]
G1 ZDT A R Gregory, 9 Fordbridge Rd, Ashford, TW15 2TD [IO91SK, TQ07]
G1 ZDU B G Rowles, 91 Crowshott Ave, Stanmore, HA7 2PA [IO91UO, TQ19]
G1 ZDX M I Bodecott, 6 Seymour Pl, South Norwood, London, SE25 4XU [IO91XJ, TQ36]
G1 ZDY C W Tipp, 77 Kimberley Rd, Croydon, CR0 2PZ [IO91WJ, TQ36]
G1 ZEC G A Stevens, 25 Ave Rd, New Milton, BH25 5JP [IO90ES, SZ29]
G1 ZED B Haworth, 139 Manchester Rd, Accrington, BB5 2NY [IO83TR, SD72]
G1 ZEI J M Wyatt, 10 St. Georges Hill, Perranporth, TR6 0DZ [IO70KI, SW75]
G1 ZEK D J Ault, 68 Moira Dale, Castle Donington, Derby, DE74 2PJ [IO92IU, SK42]
GM1 ZEL C R Wilson, Parkview, Knock, Huntly, Aberdeenshire, AB54 5LJ [IO87PK, NJ54]
G1 ZES H W Stenhouse, 21 Cheviot Gr, Pegswood, Morpeth, NE61 6YJ [IO95EE, NZ28]
G1 ZEU S M Aspey, 23 Colwell Ct, Newton Aycliffe, DL5 7PS [IO94EP, NZ22]
G1 ZEX R Davies, 71 Higher Croft Rd, Lower Darwen, Darwen, BB3 0QT [IO83SR, SD62]
G1 ZFB K R W Barton, Hameau de Cassoulet, Miribel Lanchatre, 38450 Vif, France
G1 ZFD J Davies, 71 Higher Croft Rd, Lower Darwen, Darwen, BB3 0QT [IO83SR, SD62]
G1 ZFF T Voisey, 26 Gorlands Rd, Chipping Sodbury, Bristol, BS17 6LA [IO81TM, ST78]
G1 ZFG J C Stephenson, 5 Hunstrete, Pensford, Bristol, BS18 4NT [IO81RI, ST66]
GW1 ZFX J Milosevic, 38 Thornhill Cl, Upper Cwmbran, Cwmbran, NP44 5TQ [IO81LP, ST29]
G1 ZGC V C Hales, 42 Churchill Rd East, Wells, BA5 3HU [IO81QF, ST54]
G1 ZGF R P Jackson, 25 Meadow Dr, Knutsford, WA16 0DT [IO83TH, SJ77]
G1 ZGH J M Sharpe, 7 Dales, Featherstone Ln, North Featherstone, Pontefract, West Yorks, WF7 6AH [IO93HQ, SE42]
G1 ZGM R P Ross, 19 Glynleigh Dr, Polegate, BN26 6LU [JO00DT, TQ50]
G1 ZGO L P M Hopper, 6 Tealby Rd, Ridding Est, Scunthorpe, DN17 2HA [IO93PN, SE80]
G1 ZGT C J Moore, 88 Co. Rd, Swindon, SN1 2EP [IO91CN, SU18]
G1 ZGU Details withheld at licensee's request by SSL.
G1 ZGW Details withheld at licensee's request by SSL.
G1 ZGX Details withheld at licensee's request by SSL.
G1 ZGZ R D N Mee, 57 Ashby Rd, Kegworth, Derby, DE74 2DJ [IO92IU, SK42]
G1 ZHD Details withheld at licensee's request by SSL.
G1 ZHH J A Bond, 1 Manley Terr, Station Rd, Liskeard, PL14 4DW [IO70SK, SX26]
GW1 ZHI C F Owens, 7 Frondeg, Southsea, Wrexham, LL11 6RH [IO83LB, SJ35]
G1 ZHL M R Dharas, 225 Redmile Walk, Welland Est, Peterborough, PE1 4UR [IO92VO, TF20]
G1 ZHM P E Hunt, 198 Daws Heath Rd, Rayleigh, SS6 7NP [JO01HN, TQ88]
G1 ZHN M A Griffiths, 2 Muirway, South Benfleet, Benfleet, SS7 4LS [JO01GN, TQ78]
G1 ZHQ V V Da-Costa, 48A Elmwood Rd, London, SE24 9NR [IO91WK, TQ37]
G1 ZHS Details withheld at licensee's request by SSL.
GW1 ZHX G J Morris, 28 Ruabon Rd, Wrexham, LL13 7PB [IO83MA, SJ34]
G1 ZHZ J L Hall, 27 Quarry Hill Rd, Ilkeston, Derbyshire, DE7 4DA [IO92IX, SK44]
G1 ZIT Details withheld at licensee's request by SSL.
G1 ZIU D R Martin, Glenville, Shortswood, Ashford, Kent, TN27 8DW [JO01HC, TQ83]
GM1 ZIV J D Large, 9 Maitland Terr, Kildrochat, Stranraer, DG9 9EX [IO74MU, NX05]
G1 ZIY Details withheld at licensee's request by SSL.
GW1 ZJC P H Tulk, Home Farm Lodge, Penylan Ruabon, Wrexham, Clwyd, LL14 6HS [IO82MX, SJ34]
GI1 ZJF J J Lynch, 21 Beechill Park East, Belfast, BT8 4NY [IO74BN, J36]
GM1 ZJI J Fish, Craigbrae, Eskdale Muir, Langholm, Dumfriesshire, DG13 0QH [IO85JG, NY29]
G1 ZJJ W H Flavell, 4 Birchland Cttgs, Broughton, Claverley, Wolverhampton, WV5 7AS [IO82UM, SO89]
G1 ZJK D W J Ellard, 15 Highlands Dr, The Green, Daventry, NN11 5ST [IO92KG, SP56]
G1 ZJM M J O'Neill, 17 Sunnyside Rd, Beeston, Nottingham, NG9 4FH [IO92JW, SK53]
G1 ZJO R M Lillycrop, 40 Chestnut Way, Milford Haven, SA73 1BP [IO71LR, SM90]
G1 ZJP R E Offer, 30 Alexandra Rd, Spalding, PE11 2PX [IO92WS, TF22]
G1 ZJQ D Smith, 2 Hampton Cl, Eastfield Glade, Cramlington, NE23 9FD [IO95FC, NZ27]
G1 ZJR E Mansfield, 2 Pine Walk, Weybourne, Holt, NR25 7HJ [JO02NW, TG14]
G1 ZJU A V Ridler, 16 Dilston Gdns, Sunderland, SR4 7TD [IO94HV, NZ35]
G1 ZJW S J Sturgeon, 83 St. Annes Rd, Aylesbury, HP21 8RB [IO91OT, SP81]
G1 ZKB Details withheld at licensee's request by SSL.
G1 ZKC Details withheld at licensee's request by SSL.
G1 ZKD Details withheld at licensee's request by SSL.
GW1 ZKE M Grindle, 57 Islwyn St., Cwmfelinfach, Ynysddu, Newport, NP1 7HY [IO81JO, ST19]
G1 ZKJ Details withheld at licensee's request by SSL.
G1 ZKK M R Allen, 79 College Rd, Sutton Coldfield, B73 5DL [IO92BN, SP09]
G1 ZKL Details withheld at licensee's request by SSL.
GW1 ZKM C S C Cronin, 1 Goldsmith Cl, Cefn Glas, Bridgend, CF31 4RZ [IO81EM, SS88]
G1 ZKT A Hide, 1 PO Box, North Shields, NE30 1HD [IO95GA, NZ36]
G1 ZLA M J Roberts, 18 Craster Dr, Nottingham, NG6 7FJ [IO93JA, SK54]
G1 ZLB J M Kynaston, Smithy Cottage, Main St, Widmerpool, Nottingham, NG12 5PY [IO92LU, SK62]
G1 ZLC P S Ashby, 19 Burnett Rd, Streetly, Sutton Coldfield, B74 3EL [IO92BO, SP09]
G1 ZLD M J Bignell, 57 Ramsey Rd, Halstead, CO9 1AS [JO01HW, TL83]
G1 ZLF W W Johnson, 40 Northorpe Rd, Donington, Spalding, PE11 4XU [IO92VV, TF23]
G1 ZLG Details withheld at licensee's request by SSL.
G1 ZLH Details withheld at licensee's request by SSL.
G1 ZLL D L Ball, 3 Hull Rd, Lydd on Sea, Romney Marsh, TN29 9PQ [JO00LW, TR02]
G1 ZLO Details withheld at licensee's request by SSL.
G1 ZLQ K Ellis, 161 Stoney Ln, Spondon, Derby, DE21 7QE [IO92HV, SK43]
G1 ZLS J R Parker, 1 Beech Way, Ashby de La Zouch, LE65 2SR [IO92GS, SK31]
G1 ZLU Details withheld at licensee's request by SSL.
G1 ZME R A Coatman, 6 Penzance Rd, St. Buryan, Penzance, TR19 6DZ [IO70EC, SW42]
G1 ZMG R A Hoad, Broad Lea, Amsbury Rd, Coxheath, Maidstone, ME17 4DN [JO01FF, TQ75]
G1 ZMJ Rev D M Roberts, 31 Seaton Way, Marshside, Southport, PR9 9GJ [IO83MQ, SD32]
G1 ZMS C R Cook Mid Sussex ARS, 45 Stonepound Rd, Hassocks, BN6 8PR [IO90WW, TQ31]
G1 ZMV M R Laird, 4 Eversfield Cl, Andover, Hants, SP10 3EN [IO91GF, SU34]
G1 ZMW C J Tubey, 274 Burton Rd, Midway, Swadlincote, DE11 7LY [IO92FS, SK32]
G1 ZMY S D Copsey, 119 Blackbridge Ln, Horsham, RH12 1SD [IO91TB, TQ13]
GW1 ZNC S E Elworthy, 70 Maple Dr, Brackla, Bridgend, CF31 2PF [IO81FM, SS97]
G1 ZND A P Soble, 19 Curtis Hayward Dr, Quedgeley, Gloucester, GL2 4WJ [IO81UT, SO81]
GM1 ZNI Details withheld at licensee's request by SSL.
G1 ZNK A E Edwards, 68 Middlemarch Rd, Coventry, CV6 3GF [IO92FK, SP38]
G1 ZNL Details withheld at licensee's request by SSL.
GM1 ZNR V E Roberts, 4 Ladieside, Brae, Shetland, ZE2 9SX [IP90HJ, HU36]
G1 ZNT W H Barton, 27 Hornby Cres, Clock Face, St. Helens, WA9 4RY [IO83PK, SJ59]
G1 ZNX R E Agnew, 156 Goswell End Rd, Harlington, Dunstable, LU5 6NT [IO91SX, TL03]
G1 ZNZ A M Steele, 20 Ronson Ave, Trent Vale, Stoke on Trent, ST4 6PX [IO82VX, SJ84]
G1 ZOB R L Brown, 8 Vicarage Gate, St. Erth, Hayle, TR27 6JB [IO70GD, SW53]
G1 ZOD A H Daniels, 37 Burnet Ave, Burpham, Guildford, GU1 1YF [IO91RG, TQ05]
G1 ZOE K Edwards, 9 Europa Ave, Sandwell Valley, West Bromwich, B70 6TL [IO92AM, SP09]
GW1 ZOI Details withheld at licensee's request by SSL.
G1 ZOM Details withheld at licensee's request by SSL.
G1 ZOQ Details withheld at licensee's request by SSL.
G1 ZOV P M Bateman, 40 Hathaway Gdns, London, W13 0DH [IO91UM, TQ18]
G1 ZOY M P Knowles, 32 Keyes Cl, Birchwood, Warrington, WA3 6RU [IO83RK, SJ69]
G1 ZPA N Johanssen, Flat 1, 13 Kingsland Rd, Cheadle Heath, Stockport, SK3 0NB [IO83VJ, SJ88]
G1 ZPG J G Shaw, 4 Southfield Rd, Marske By The Sea, Redcar, TS11 7BP [IO94LO, NZ62]
G1 ZPJ P M R Read, 58 Godolphin Rd, Helston, TR13 8QJ [IO70IC, SW62]
G1 ZPQ J S Pitchford, 7 Firecrest Dr, Leegomery, Telford, TF1 4FZ [IO82RR, SJ61]
G1 ZPU R W Compton, Clay End Farm, Sutton, Sandy, Beds, SG19 2NE [IO92VC, TL24]
G1 ZPZ V W R Gilbertson, 36 Branksea Ave, Hamworthy, Poole, BH15 4DP [IO80XR, SY99]
G1 ZQC Details withheld at licensee's request by SSL.
G1 ZQE D Marsden, 94 Blackford Rd, Shirley, Solihull, B90 4BX [IO92CJ, SP17]
GM1 ZQF G M M Milne, 6 Alexandra St., Alyth, Blairgowrie, PH11 8AS [IO86JO, NO24]
G1 ZQN Details withheld at licensee's request by SSL.
G1 ZQQ Details withheld at licensee's request by SSL.
G1 ZQR R K Twyman, Farm End Cottage, Halls Ln, Waltham St Lawrence, Berks, RG10 0JB [IO91OL, SU87]
G1 ZQY Details withheld at licensee's request by SSL.
G1 ZQZ Details withheld at licensee's request by SSL.
G1 ZRE Details withheld at licensee's request by SSL.
G1 ZRF Details withheld at licensee's request by SSL.
G1 ZRJ J S A Nesbit, 3 Orchard Hill, Windlesham, GU20 6DB [IO91QI, SU96]
G1 ZRN M A Hynes, 62 Clifton St., Bideford, EX39 4EU [IO71VA, SS42]
G1 ZRP M Crook, 21 Treyew Rd, Truro, TR1 2BY [IO70LG, SW84]
G1 ZRQ D M P Barrett, 77 Trevillis Park, Liskeard, PL14 4EQ [IO70SK, SX26]
G1 ZRR N S Youngman-Smith, 12 Timber Way, Chinnor, OX9 4EU [IO91NQ, SP70]
G1 ZRS J A Cantwell, 10 Cathedral Dr, Fairfield, Stockton on Tees, TS19 7JT [IO94HN, NZ42]
G1 ZRT C A Guymer, 74 West Common Ln, Scunthorpe, DN17 1DU [IO93QN, SE80]
G1 ZRU Details withheld at licensee's request by SSL.
G1 ZRZ Details withheld at licensee's request by SSL.
G1 ZSA Details withheld at licensee's request by SSL.
G1 ZSF A Lewis, 76 Reading Rd, Finchampstead, Wokingham, RG40 4RA [IO91NI, SU76]
G1 ZSG C J Bell, 41 Handel Rd, Canvey Island, SS8 7HL [JO01HM, TQ88]
G1 ZSK T C Adams, 10 Ivybridge Cl, Kingston Ln, Uxbridge, UB8 3TT [IO91SM, TQ08]
G1 ZSO I Shepherd, 89B High St., Yeadon, Leeds, LS19 7TA [IO93DU, SE24]
G1 ZSP W A Nutt, 7 Wolsey Way, Empress Rd, Loughborough, LE11 1PR [IO92JS, SK51]
G1 ZSR M J Salzman, 89 Delamere Rd, Bedworth, Nuneaton, CV12 8SG [IO92GL, SP38]
G1 ZST L H Sherwood, 50 Thornton Rd, Manchester, M14 7WT [IO83VK, SJ89]
G1 ZSU E E Baskerville, 83 Crusader Dr, Sprotbrough, Doncaster, DN5 7RS [IO93KM, SE50]
G1 ZSX C F Quantrell, Brentor, 19 North Hill Park, St. Austell, PL25 4BJ [IO70OI, SX05]
G1 ZSY A N Hughes, 7 Harland Way, Tunbridge Wells, TN4 0TQ [JO01DE, TQ54]
GM1 ZTA Dr J B Lloyd, 38 Riverside Dr, Haddington, EH41 3QN [IO85OX, NT57]
GM1 ZTB W W Bell, 77 Bongate, Jedburgh, TD8 6DU [IO85RL, NT62]
G1 ZTG P C Wilsdon, 53 Highways Ave, Euxton, Chorley, PR7 6QD [IO83PP, SD51]
G1 ZTJ G J Bailey, 34 Newton Rd, Bideford, EX39 2LL [IO71VA, SS42]
G1 ZTK W E D Causer, 47 Sandringham Rd, Wombourne, Wolverhampton, WV5 8EF [IO82VM, SO89]
G1 ZTM A W Corp, 16 Underhill Rd, St, BA16 0NS [IO81OC, ST43]
GW1 ZTP G Jones, Dolarfon, 21 Roumania Dr, Craig y Don, Llandudno, LL30 1UY [IO83CH, SH78]
G1 ZUB P C Thompson, Berry Brow, Wetherby Rd, Scarcroft, Leeds, LS14 3AU [IO93GU, SE34]
G1 ZUC Dr R D Johnson, Mile House, Lansdown Rd, Bath, BA1 5SY [IO81TJ, ST76]
G1 ZUF J Gibson, 114 Burscough St., Ormskirk, L39 2EY [IO83NN, SD40]
G1 ZUG N Adams, 117 Cottingham Rd, Hull, HU5 2DH [IO93TS, TA03]
G1 ZUH M Pomroy, 21 Nook Farm Ave, Syke, Rochdale, OL12 0SH [IO83WP, SD81]
G1 ZUI Details withheld at licensee's request by SSL.
G1 ZUK Details withheld at licensee's request by SSL.
GM1 ZUN S C Campbell, 9 Arbuthnott Pl, Stonehaven, AB3 2JA [IO87TH, NJ72]
G1 ZUS N J Watts, 23 St. Marys Cl, Taddiport, Torrington, EX38 8AS [IO70WW, SS41]
G1 ZUU J C V Jones Avon Valley ARS, 10 Huntington Cl, Redditch, B98 0NF [IO92BH, SP06]
G1 ZUV Details withheld at licensee's request by SSL.
G1 ZUW Details withheld at licensee's request by SSL.
G1 ZVC S J Beith, 18 Ave Rd, New Milton, BH25 5JP [IO90ES, SZ29]
G1 ZVE H C Barugh, 40 Ruden Way, Epsom Downs, Epsom, KT17 3LN [IO91VH, TQ26]
G1 ZVO W M J Scott, 16 Sweetbriar Ln, Holcombe, Dawlish, EX7 0JZ [IO80GN, SX97]
G1 ZVQ Details withheld at licensee's request by SSL.
G1 ZVZ K J Fish, Kj's Pl, Merrifield Rd, Wainfleet, Skegness, Lincs, PE24 4AE [JO03CC, TF55]
G1 ZWA Details withheld at licensee's request by SSL.
G1 ZWB B A Ward, 190 Jeans Way, Dunstable, LU5 4PR [IO91SV, TL02]
G1 ZWH T H Sanders, Lowena Vean, 6 Treganoon Rd, Mount Ambrose, Redruth, TR15 1NN [IO70JF, SW74]
G1 ZWJ N Mercer Lancs Amateur, Digital Society, 19 Sycamore Rd, Brookhouse, Lancaster, LA2 9PB [IO84PB, SD56]
G1 ZWK E R Tallett, 4 Roughlee Old Hall, Roughlee, Nelson, BB9 6NN [IO83VU, SD84]
G1 ZWQ J Bowker, 12 Turton Heights, Bolton, BL2 3DU [IO83TO, SD71]
G1 ZWW Details withheld at licensee's request by SSL.
G1 ZWX P Tryner, 130 Alfreton Rd, Newton, Alfreton, DE55 5TR [IO93HC, SK45]
G1 ZWY C E Roe, 69 Mercaston Cl, Holme Hall, Chesterfield, Derbyshire, S40 4UE [IO93GF, SK37]
G1 ZXA Details withheld at licensee's request by SSL.
G1 ZXF J M Ullett, 76 Rownhams Rd, Maybush, Southampton, Hants, SO1 6DY [IO90HW, SU41]
GI1 ZXM P J Smyth, 9 Glennor Cres West, Hillsborough Rd, Carryduff, Co. Down, BT8 8JF [IO74BM, J36]

GW1 ZXN V A Allen, 4 Beverley Cl, Oatlands Dr, Weybridge, KT13 9LW [IO91SJ, TQ06]
G1 ZXQ Details withheld at licensee's request by SSL.
GM1 ZXT Details withheld at licensee's request by SSL.
G1 ZYB A D Collins, 120 Plumberow Ave, Hockley, SS5 5AT [JO01HO, TQ89]
G1 ZYD B J Greene, 93 Bognor Rd, Chichester, PO19 2NW [IO90OT, SU80]
G1 ZYJ A W S Ainger, 16 Hillside Rd, Harpenden, AL5 4BT [IO91TT, TL11]
G1 ZYM P Chesworth, 3 The Hawthornes, Rufford, Ormskirk, L40 1UP [IO83OP, SD41]
G1 ZYN N E Suffolk, 2 Tamerton Rd, Eyres Monsell Est, Leicester, LE2 9DD [IO92KO, SK50]
G1 ZYS F H Woodland, Rodanum, Middle Rd, Sway, Lymington, SO41 6BB [IO90ES, SZ29]
G1 ZYW Details withheld at licensee's request by SSL.
GI1 ZYY W Gill, Riverside Farm, 51 Ballydrain Rd, Comber, Co Down, N Ireland, BT23 5SY [IO74DM, J46]
G1 ZZC P R Golds, 5 Highfield Rd, Worthing, BN13 1PX [IO90TT, TQ10]
G1 ZZX Details withheld at licensee's request by SSL.
G1 ZZY Details withheld at licensee's request by SSL.

G2

G2 AA A J Barratt, 20 Cross Ln, Burton Lazars, Melton Mowbray, LE14 2UH [IO92NR, SK71]
G2 AAN J H Clarke, Coach House, Willinghurst, Shamley Green, Guildford, Surrey, GU5 0SU [IO91SE, TQ04]
G2 AAY L Chappell, 107 Buxton Old Rd, Disley, Stockport, SK12 2BU [IO83XI, SJ98]
G2 ABC Details withheld at licensee's request by SSL.
GW2 ABJ G D Edwards, 2 Heol y Glo, Tonna, Neath, SA11 3NJ [IO81CQ, SS79]
G2 ABR C L Mayman, 6 Church Rd, Fordham, Colchester, CO6 3NA [JO01JW, TL92]
G2 ACG D Y Adalian, Herons, The Droveway, St Margarets Bay, Dover, Kent, CT15 6BZ [JO01QD, TR34]
G2 ACK M T Aitken, Dormers, Shovelstrode, East Grinstead, West Sussex, RH19 3PH [JO01AD, TQ43]
G2 ACN B J Bale, The Shiel, The Glade, Kingswood, Tadworth, KT20 6LL [IO91VH, TQ25]
GM2 ACY J C Carslaw, 22 Ardencaple Dr, Helensburgh, G84 8PS [IO76OA, NS28]
G2 ACZ G Whitehead, Funing, 30 Church Ln, Mablethorpe, LN12 2NU [JO03DH, TF58]
G2 ADA F Jones, 4 Stalbridge Cl, Stalbridge, Sturminster Newton, DT10 2ND [IO80TW, ST71]
G2 ADC Details withheld at licensee's request by SSL.
G2 ADM J F Stone, 3 Arndale House, South Clifton St., Lytham St. Annes, FY8 5HL [IO83MR, SD32]
G2 ADR E Parvin, Cherry Croft, Beechway Cl, Main St, Upper Poppleton, York, YO2 6JE [IO93KX, SE55]
GW2 ADZ H W Parker, 4 Nant y Dderwen, Drefach, Llanelli, Dyfed, SA14 7DD [IO71WT, SN51]
G2 AFV P Carbutt, 2 Penrhyn Walk, Barnsley, South Yorks, S71 5DR [IO93GN, SE30]
GI2 AFW H Shaw, 5 Fairview Gdns, Bangor, BT20 4QS [IO74EP, J58]
G2 AGG M Forster, 34 Marlborough Rd, Elmfield, Ryde, PO33 1AB [IO90KR, SZ69]
G2 AGH A G Hobson, 10 Bellomonte Cres, Drayton, Norwich, NR8 6EJ [JO02OQ, TG11]
G2 AGO J Moss, 11 Ascot Cl, Hainault, Ilford, IG6 3AE [JO01BO, TQ49]
G2 AGR W A Rice, 49 Colebrook Rd, Shirley, Solihull, B90 2JZ [IO92BJ, SP17]
G2 AHC R W Bishop, 442 Long Ln, Bexleyheath, DA7 5JW [JO01BL, TQ47]
G2 AHU R G Cracknell, 18 Green Ln Cres, Yarpole, Leominster, HR6 0BQ [IO82OG, SO46]
G2 AIS Details withheld at licensee's request by SSL.
G2 AIV M B Carson, Long Cl, 8 Ramley Rd, Lymington, SO41 8GQ [IO90FS, SZ39]
G2 AIW M E Lambeth, 11 Ellerman Ave, Twickenham, TW2 6AA [IO91TK, TQ17]
G2 AJS W N Maddock, The Green, Chulmleigh, N Devon, EX18 7DA [IO80BV, SS61]
G2 AJV Prof R C Jennison, Wildwood, Nackington Rd, Canterbury, CT4 7AY [JO01NG, TR15]
GM2 AJW J W Jack, Malindella, Main Rd, Locharbriggs, Dumfries, DG1 1RZ [IO85FC, NX98]
G2 AKK W Lishman, 28 Lightbown St., Darwen, BB3 0DY [IO83SQ, SD62]
G2 AKR D Barber, 16 Boxgrove Rd, Sale, M33 6QW [IO83TK, SJ79]
G2 AKY E J Williams, 35 Selangor Ave, Emsworth, PO10 7LR [IO90MU, SU70]
G2 ALM R F Wilkins, 36 Offington Gdns, Worthing, BN14 9AU [IO90TT, TQ10]
G2 ALN E W Taylor, 76 Sidney Rd, Blackley, Manchester, M9 8AT [IO83VM, SD80]
G2 ALO R P Munn, Little Downs, Sandgate Ln, Storrington, Pulborough, RH20 3HJ [IO90SW, TQ11]
G2 ALZ W H Heywood, 8 Talbot Cl, Shavington, Crewe, CW2 5EU [IO83SB, SJ65]
G2 AMG H W Mitchell, Stone Cottage, Yeovil Rd, Halstock, Yeovil, BA22 9RR [IO80QU, ST50]
G2 AMM T G Willey, 14 The Beeches, Ponteland, Newcastle upon Tyne, NE20 9SZ [IO95CB, NZ17]
G2 AMQ F G Cockerill, 3 Daen Ingas, Danbury, Chelmsford, CM3 4DB [JO01GR, TL70]
G2 AMV B O'Brien, Tanglewood, Anthony's Way, Heswall, Merseyside, L60 0BP [IO83KH, SJ28][RSGB Vice-President.]
G2 ANC J Bromiley, 28 Clive Rd, Westhoughton, Bolton, BL5 2HR [IO83RM, SD60]
G2 AND R H Broadbent, 23 Fenay Dr, Fenay Bridge, Huddersfield, HD8 0AB [IO93DP, SE11]
G2 ANS Details withheld at licensee's request by SSL.
GM2 AOL W S Hall, 21 Seabourne Gdns, Broughty Ferry, Dundee, DD5 2RT [IO86NL, NO43]
G2 AOY J R Muddell, 4 Spring Cl, Colwall, Malvern, WR13 6RE [IO82TB, SO74]
G2 AOZ G W F Ashford, 12 Langdale Way, Frodsham, Warrington, WA6 7LE [IO83PH, SJ57]
G2 AQJ R Collins, 33 Elm Cl, Laverstock, Salisbury, SP1 1SA [IO91CB, SU13]
G2 AQN C Renshaw, 51 Osgodby Cres, Scarborough, YO11 3JP [IO94TF, TA08]
G2 ART F H P Cawson, 43 Trafalgar Rd, Southport, PR8 2HF [IO83LP, SD31]
G2 ARU R A Loveland, Lashburn, Wandleys Ln, Walberton, Arundel, BN18 0QR [IO90QU, SU90]
G2 AS H V Booth, 45 Marston Rd, Crookes, Sheffield, S10 1HG [IO93FJ, SK38]
G2 ASF F A Noakes Coventry ARS, 4 Baronsfield Rd, Coventry, CV3 5GP [IO92GJ, SP37]
GM2 ASU C Clark, Pinewood, Mounthigh, Balblair, Dingwall, IV7 8LH [IO77VP, NH66]
G2 ASX E H Griffiths, The Stables, Hull Pl, Sholden, Deal, CT14 0AQ [JO01QF, TR35]
G2 ATM S Read, 21 Birkland Ave, Plains Rd, Mapperley, Nottingham, NG3 5LA [IO92KX, SK54]
G2 ATZ J Harris, 181 Eastbourne Rd, Willingdon, Eastbourne, BN20 9NB [JO00DT, TQ50]
G2 AUB N I Neame, 2 Eighth Ave, Lancing, BN15 9XD [IO90UU, TQ10]
G2 AUI G T Noakhes, 80 Starbeck Dr, Little Sutton, South Wirral, L66 4TS [IO83MG, SJ37]
G2 AUK S J Harris, 8 Gollands Cl, Brixham, TQ5 8JZ [IO80FJ, SX95]
G2 AVF A G Woofenden, 10 Shortridge Ln, Enderby, Leicester, LE9 5PA [IO92JO, SP59]
G2 AVI L R K Gregory, The Well House, The Downs, Herne Bay, Kent, CT6 6JP [JO01NI, TR16]
G2 AVV Details withheld at licensee's request by SSL.
G2 AXO W J Purser, Chimney End, Shutlanger Rd, Towcester, Northants, NN12 7SB [IO92MD, SP74]
G2 AXU K Mallett, 3 Flanders Cl, Marnhull, Sturminster Newton, DT10 1LH [IO80UX, ST71]
G2 AXY Dr J G Turner, 13 Promenade, Southport, Merseyside, PR8 1QY [IO83LP, SD31]
G2 AYG J G Openshaw, 516 The Mount, Walmersley Rd, Bury, Lancs, BL9 6QT [IO83UO, SD81]
G2 AZC E L Wills, 34 Hazeldown Rd, Teignmouth, TQ14 8QR [IO80FN, SX97]
G2 AZM E L Oakley, 67 South Rd, Northfield, Birmingham, B31 2QZ [IO92AJ, SP07]
G2 AZP P W Crowley, 136A Blackbrook Ln, Bromley, Kent, BR1 2HP [JO01AJ, TQ46]
G2 BAH L S Gumbrill, 53 Park St., Kendal, LA9 5QP [IO84PH, SD59]
G2 BAP W F Badcock, 33 Gaddesden Row, Hemel Hempstead, Herts, HP2 6HL [IO91ST, TL01]
G2 BBC D A Pick Ariel Rad Group, BBC Club Birmingham, Pebble Mill Rd, Birmingham, B5 7QQ [IO92BK, SP08]
G2 BBI L F Steel, 1B Trinity Ave, Westcliff on Sea, SS0 7PU [JO01IM, TQ88]
G2 BBN G A Banham, Dinard, 7 Abbots View, Kings Langley, WD4 8AW [IO91SR, TL00]
G2 BCB E A L Barrall, 10 Mewstone Ave, Wembury, Plymouth, PL9 0JY [IO70XH, SX54]
G2 BCI H M Tainton, 3 Woodstone Ave, Stoneleigh, Epsom, KT17 2JS [IO91VI, TQ26]
G2 BCY J E Corston, 23 Whitefield Terr, Newcastle upon Tyne, NE6 5DU [IO94FX, NZ26]
G2 BDL Details withheld at licensee's request by SSL.
G2 BDV I D Brotherton, The Moorings, 6 Cranfield Ave, Wimborne, BH21 1DE [IO90AT, SU00]
G2 BFC T N Forbes, 88B Winchester Rd, Shirley, Southampton, SO16 6US [IO90GW, SU31]
GW2 BFD C Pritchard, Wildwoods, 22 Brynffrwd Cl, Coychurch, Bridgend, CF35 5EP [IO81FM, SS97]
G2 BFO D D Silvester, Rookesbury, Uplands Rd, Denmead, Waterlooville, PO7 6HF [IO90LV, SU61]
GM2 BFV W L L Lowson, 13 Yeaman St., Forfar, DD8 2JH [IO86NP, NO45]
G2 BGG J S H Garner, Barbon, Aigburth Hall Rd, Liverpool, L19 9DG [IO83NI, SJ38]
G2 BGI Details withheld at licensee's request by SSL.
G2 BGU K S J Gasson, 31 Orchard Way, Horsmonden, Tonbridge, TN12 8LA [JO01FD, TQ74]
G2 BHG G A Harrison, 13 High View Park, Cromer, NR27 0HQ [JO02PW, TG24]
G2 BHQ A J Barnett, 45 St. James Ave, Sutton, SM1 2TQ [IO91VI, TQ26]
GW2 BHS L A Hensford, The Stables, Plas Llysyn, Carno, Caersws, Powys, SY17 5JT [IO82FN, SN99]
G2 BHY A P Bonner, The Dwellings, Turners Green, Heathfield, E Sussex, TN21 9RA [JO00DW, TQ61]
G2 BIM L W J Leask, 40 South Lawn, Sidmouth, EX10 9NT [IO80JQ, SY18]
G2 BJK G F Brown, Villa Westfield, Wells Rd, Draycott, Cheddar, Somerset, BS27 3SF [IO81PG, ST45]
G2 BJW R A W Lucas, 38 Junction Rd, Gillingham, ME7 4EQ [JO01GJ, TQ76]
G2 BKO N D N Belham, 7 Binyon Cl, Badsey, Evesham, WR11 5EY [IO92BC, SP04]
G2 BKZ R A McTait, 20 Rowland Rd, Stevenage, SG1 1TE [IO91VV, TL22]
G2 BLA M W A Pyle, Audale, 16 Dudley Hill Cl, Welwyn, AL6 0QQ [IO91VU, TL21]
GM2 BLC D B Black, Seapoint, Kippford, Dalbeattie, Kirkcudbrightshire, DG5 4LL [IO84CV, NX85]
G2 BLL W Layton, 14 Parc Monga, Constantine, Falmouth, TR11 5AR [IO70JC, SW72]
G2 BMI J Bramhill, 16 Cissbury Ct, Findon Rd, Worthing, BN14 0BF [IO90TU, TQ10]
GM2 BMJ T D Jardine, Malindella, Main Rd, Locharbriggs, Dumfries, DG1 1RZ [IO85FC, NX98]
G2 BNI D B Drage, 14 Acklands Ln, Long Bennington, Newark, NG23 5EW [IO92OX, SK84]
G2 BNY H Newman, 23 Bedefield Cromer St., London, WC1H 8DY [IO91WM, TQ38]
G2 BOI K V Draycott, 7 Raines Ln, Grassington, Skipton, BD23 5NJ [IO94AB, SD96]
G2 BOX G E Smith, 18 Roman Way, Desborough, Kettering, NN14 2QL [IO92OK, SP88]
G2 BOZ Details withheld at licensee's request by SSL.
G2 BPF Nr P F Ballard, 133 Sycamore Rd, Farnborough, GU14 6RE [IO91PG, SU85]
G2 BPW I W K Smith, Thalia Lodge, North Luffenham, Oakham, Rutland, LE15 8JZ [IO92QO, SK90]
G2 BQP P E Gully, 23 Lawrence Gr, Henleaze, Bristol, BS9 4EL [IO81QL, ST57]
G2 BQY R G Holland Trowbridge D AR, 37 Danvers Way, Westbury, BA13 3UF [IO81VG, ST85]
G2 BRR R G Rugg, 29 Milbourne Park, Milbourne, Malmesbury, SN16 9JE [IO81WN, ST98]
G2 BRS I D Brotherton Bournemouth Radio Soc, 6 Cranfield Ave, Wimborne, BH21 1DE [IO90AT, SU00]
G2 BSI D C Watson, 10 Ground Ln, Hatfield, AL10 0HH [IO91VS, TL20]
G2 BSJ R J Biltcliffe, 3 Church View, Steeple Claydon, Buckingham, MK18 2QR [IO91MW, SP72]
G2 BSW R J Ward, Apartado 330, 29680 Estepona, Spain
G2 BTJ D W L Robinson, 20 Halewood Cl, Gateacre, Liverpool, L25 3PJ [IO83NJ, SJ48]
G2 BTO G Openshaw, 7 Broadhead Rd, Turton, Bolton, BL7 0BG [IO83TP, SD71]
GM2 BUD D Tannock, 38 Sunnyside Cres, Mauchline, KA5 6DX [IO75TM, NS42]
G2 BUJ P H Greenwood, 32 Pound Ln, Pinehurst, Swindon, SN2 1PS [IO91CN, SU18]
G2 BUP H L Gibson, 8 Springfield, Norton St. Philip, Bath, BA3 6NR [IO81UH, ST75]
G2 BUV C E Teesdale, 6 Remigius Gr, Glebe Park, Lincoln, LN2 4QJ [IO93RF, SK97]
G2 BUY P L Stride, 4 Bramblemead, Balcombe, Haywards Heath, RH17 6HU [IO91WB, TQ33]
GM2 BWF E D Fleming, 6 Clachan, Ashfield, Dunblane, FK15 0JL [IO86AF, NN70]
G2 BWW A Barrett, 38 Haw Ln, Bledlow Ridge, High Wycombe, HP14 4JJ [IO91NQ, SU89]
G2 BXH F Perkins, 28 Marlborough Cl, Great Hayward, Stafford
G2 BXP M J Prestidge, 48 Parkfield Rd, Oldbury, Warley, B68 8PT [IO82XL, SO98]
G2 BXZ K Hinch, 14 Pinfold Cl, Bridlington, YO16 5GH [IO94VC, TA16]
G2 BY H E Whatley, 56 Stenbury View, Wroxall, Ventnor, PO38 3DD [IO90JO, SZ57]
G2 BYP J W M Mackay, 4 The Crest, Hillcrest, Whitehaven, CA28 6TJ [IO84FN, NX91]
G2 BZQ R Q Marris, 35 Kingswood House, Farnham Rd, Slough, SL2 1DA [IO91QM, SU98]
G2 BZR R K Bassford, 59 Watling St., Dorden, Hints, Tamworth, B78 3DF [IO92CP, SK10]
GI2 BZV R R B Cowden, 48 Ballyduff Rd, Newtownabbey, BT36 6PB [IO74AQ, J38]
G2 CAZ M S Ellis, Highfield, 15 Hartley Rd, Altrincham, WA14 4AZ [IO83TJ, SJ78]
G2 CBC W E G Smith, 47 Lea Gdns, Peterborough, PE3 6BY [IO92UN, TL19]
G2 CBH W W Turner, 24 Southdown Dr, Thornton Cleveleys, FY5 5BL [IO83MU, SD34]
G2 CCH R H Hawkes, 121 Eversley Ave, Barnehurst, Bexleyheath, DA7 6RQ [JO01CL, TQ57]
G2 CDT F H Martin, 22 Walders Ave, Wadsley, Sheffield, S6 4AY [IO93FJ, SK39]
G2 CFC R G Frisby, 25 Knighton Rise, Leicester, LE2 2RF [IO92KO, SK60]
G2 CFH F E Hamnett, 13 Greenway, Binstead, Ryde, PO33 3SD [IO90JR, SZ59]
G2 CGF S G Griffiths, 93 Brands Hill Ave, High Wycombe, HP13 5PX [IO91PP, SU89]
G2 CGL E C Grafton, 12 Dalesway, Kirk Ella, Hull, HU10 7NE [IO93SR, TA02]
G2 CH R B Newman Oundle Scl ARS, 20 Glapthorn Rd, Oundle, Peterborough, PE8 4JQ [IO92SL, TL08]
G2 CHI W G Bailey, 25 Lenham Rd East, Saltdean, Brighton, BN2 8AF [IO90XT, TQ30]
G2 CIL G A Hook, Cheam Cottage, 22 Felpham Way, Felpham, Bognor Regis, PO22 8QT [IO90QS, SU90]
G2 CIN T M Lott, 21 Morley St., Kettering, Northants, NN16 9LJ [IO92PJ, SP87]
G2 CIW J F Moseley, 74 Coverham Rd, Berry Hill, Coleford, GL16 7RD [IO81QT, SO51]
G2 CJK A Clarkson, 6 Mather Ave, Accrington, BB5 5AU [IO83TS, SD72]
G2 CJL J R Lane, Hanam Manor, Cheddar, Somerset, BS27 3AG [IO81OG, ST45]
G2 CJO Details withheld at licensee's request by SSL.
G2 CKB W Livens, 10 Cotton Dr, Hertford, SG13 7SU [IO91XT, TL31]
G2 CKQ Mjr R S Trevelyan, 2 Centry Ct, Higher Ranscombe Roa, Ranscombe, Brixham, TQ5 9HA [IO80FJ, SX95]
G2 CMW J F West, 10 Nien Oord, Clacton on Sea, CO16 8TT [JO01NT, TM11]
GJ2 CNC Dr E Banks, 21 Willows Ct, Green St., St. Helier, Jersey, JE2 4ZA
G2 CNN P M Branton, 38 Lynford Hall Park, Lynford, Thetford, IP26 5HW [JO02IM, TL89]
G2 CNO L G Blunden, 1 Mowbray Cl, Leominster, HR6 9AZ [IO82OF, SO45]
G2 CO F Cooknell, 65 Coombe Valley Rd, Preston, Weymouth, DT3 6NL [IO80SP, SY68]
G2 CP H P Wiggins, 50 Nares St., Scarborough, YO12 7RR [IO94TG, TA08]
GM2 CPC C Orr, Emahroo, 36 Main St., Invergowrie, Dundee, DD2 5AA [IO86LL, NO33]
G2 CPM W B Mansell, 8 Cochrane Cl, Thatcham, Berks, RG19 4QX [IO91IJ, SU56]
G2 CQJ Dr J C Harvey, Lancaffe, Devon Tors Rd, Yelverton, PL20 6DN [IO70WL, SX56]
G2 CQX P V Pugh, 8 Beech Cl, Hanwood, Shrewsbury, SY5 8RA [IO82OQ, SJ40]
G2 CR Dr S A O'Hagan, Sounion, Mill Ln, Thimbleby, Horncastle, LN9 5JS [IO93WF, TF26]
GM2 CRV G Cardoo, 9 Muirbrae Rd, Rutherglen, Glasgow, G73 4NE [IO75VT, NS65]
G2 CUJ J B Jones, 55 Sabine Rd, London, SW11 5LN [IO91WL, TQ27]
G2 CVA H G Collard, 13 Abbey Ccottages, Dereham Rd Hempton, Fakeham Norfolk, NR21 7JY [JO02KT, TF92]
G2 CVO F H Osborn, 28 Elmwood Dr, West Mersea, Colchester, CO5 8RD [JO01LS, TM01]
G2 CVV F C Ward, 5 Uplands Ave, Littleover, Derby, DE23 7GE [IO92FV, SK33]
G2 CVY W H J Yeo, Ebberly House, Landkey Rd, Barnstaple, EX32 9BW [IO71XB, SS53]
GM2 CWL C K Haswell, 6 Cameron Ave, Balloch By, Balloch, Inverness, IV1 2JT [IO77WL, NH74]
G2 CXH G Miles, Farthings, Appledram Ln South, Chichester, PO20 7PE [IO90OT, SU80]
G2 CXO E M Challons, 20 Windmill Balk Ln, Woodlands, Doncaster, DN6 7SE [IO93JN, SE50]
G2 CXT A R Richardson, 23 Montgomery Rd, Newbury, RG14 6HT [IO91HJ, SU46]
G2 CYN M D Hely, 25 High St., Olney, MK46 4EB [IO92PD, SP85]
G2 CYT H J Willis, 39 Nightingale Cl, Verwood, BH31 6NW [IO90BV, SU00]
G2 CZO V Spence, 9 Willows Ave, Maltby in Cleveland, Maltby, Middlesbrough, TS8 0BJ [IO94IM, NZ41]
G2 CZS R B Sachs, The Dell, Burnham Rd, Woodham Mortimer, Maldon Essex, CM9 6SP [JO01HQ, TL80]
G2 DAD C A Robinson, 12 St. Leonards Cl, Upton St. Leonards, Gloucester, GL4 8AL [IO81VT, SO81]
G2 DAN S E Whiteley, 142 Brisbane Rd, Mickleover, Derby, DE3 5JW [IO92FV, SK33]
G2 DAU E G Gamble, 9 Lords Mount, Berwick upon Tweed, TD15 1LY [IO85XS, NT95]
G2 DBA P M S Hedgeland, 16 Amersham Hill, Gdns, High Wycombe, Bucks, HP13 6QP [IO91PP, SU89]
G2 DBI J A Pateman, Bennett's Cottage, Whipsnade, Dunstable, Beds, LU6 2LG [IO91RU, TL01]
G2 DBP G E Brown, 146 Hollingbury Rd, Brighton, BN1 7JD [IO90WU, TQ30]
G2 DBT G L Sanderson, Owslebury, Winchester, Hants, SO21 1LN [IO91IA, SU52]
G2 DBW E Wood, 3 Royd St., Slaithwaite, Huddersfield, HD7 5EB [IO93BO, SE01]
G2 DCU Dr L Turgill, 11A West Heath Dr, London, NW11 7QG [IO91VN, TQ28]
G2 DDS W A Brooks, Marchfield, Chapel Gdns, Lindford, Bordon, GU35 0TA [IO91NC, SU83]
GW2 DDX F C Smith, 11 Llewellyn St., Barry, CF63 1BZ [IO81IJ, ST16]
G2 DFL D G B Knight, 29 Hay Lease Cres, Hereford, HR2 7AN [IO82PB, SO43]
G2 DFP J M Lowe, 72 Primrose Ln, Gilstead, Bingley, BD16 4QP [IO93CU, SE13]
G2 DFY K R Bunker, 5 Lord Cecil Ct, Pike Ln, Thetford, IP24 2DR [JO02JJ, TL88]
G2 DGB A G Short, 12 Grosvenor Cres, Dorchester, DT1 2BA [IO80SQ, SY68]
G2 DGF Details withheld at licensee's request by SSL.
GW2 DHM W D Andrews, 69 Fairwater Gr West, Cardiff, CF5 2JN [IO81JL, ST17]
G2 DJ R I Buckby Derby & District Amateur Radio, 20 Eden Bank, Ambergate, Belper, DE56 2GG [IO93GB, SK35]
G2 DJA J H Palmer, 23 Silverhill Ave, St. Leonards on Sea, TN37 7HQ [JO00GV, TQ81]
G2 DKI P N Ridout, 95 Barlows Ln, Andover, SP10 2HB [IO91GE, SU34]
GI2 DKN Details withheld at licensee's request by SSL.
G2 DLJ W R Chaffe, 53 Pinfold Cl, Repton, Derby, DE65 6FR [IO92FU, SK32]
GW2 DLK G V Williams, Adre, Holyhead Rd, Llanfair, Llanfairpwllgwyngyll, LL61 5YX [IO73VF, SH57]
G2 DLO Dr P W Green, 14 The Common, Evington, Leicester, LE5 6EA [IO92LO, SK60]
G2 DLX D R Mitchell, 1 Cottage, Denstroude Ln, Blean nr Canterbury, Kent, CT2 9JX [JO01MH, TR16]
G2 DML T Crossfield, 8 Nevill Rd, Stocksfield, NE43 7JX [IO94BW, NZ06]
G2 DNH Details withheld at licensee's request by SSL.[Op: R Brennand.]
GW2 DNJ N Brierley, Minera, 6 Trinity Cres, Llandudno, LL30 2PQ [IO83BH, SH78]
G2 DOH R F H Nicholson, 8 Pine Rd, Strood, Rochester, ME2 2HX [JO01FJ, TQ76]
G2 DOJ D Godwin, 37 Dollis Hill Ave, Cricklewood, London, NW2 6EU [IO91VN, TQ28]
G2 DOT K Clark, 21 Copland Meadows, Totnes, TQ9 6ER [IO80DK, SX76]
G2 DPL P E Smith, 52 Grantham Dr, Woodhill, Bury, BL8 1XW [IO83UO, SD71]
G2 DPQ K S Amos, 1 Byron Cl, Upper Caldecote, Biggleswade, Beds, SG18 9DF [IO92UC, TL14]
G2 DPY S G Mercer, 7 Seaside Ave, Lancing, BN15 8BY [IO90UT, TQ10]
G2 DQU Lord B N R Rix Cbe DL, 8 Ellerton Rd, Wimbledon Common, London, SW20 0EP [IO91VK, TQ27]
G2 DQW Details withheld at licensee's request by SSL.
G2 DQX R J Woodroffe, 19 Balmoral Dr, Southport, PR9 8QB [IO83MP, SD31]
GM2 DRB G H Heppel, 7 Finlayson Pl, Thrumster, Wick, KW1 5TT [IO88KJ, ND34]
G2 DSB H S Rawling, 4 Ln End Cl, Shinfield, Reading, RG2 9AS [IO91MJ, SU76]
G2 DSF N Booth, 49 Baggrave St., Leicester, LE5 3QW [IO92KP, SK60]
G2 DSP R Allen, 6 Arnhem Rd, Bognor Regis, PO21 5LB [IO90PT, SU90]
G2 DSY J L Dale, 17 Lansdowne Gdns, Romsey, SO51 8FN [IO90GX, SU32]
GM2 DTB R Graham, Carlingwark House, Castle Douglas, Kirkudbrightshire, DG7 1TH [IO84AW, NX76]
G2 DTD L W Limb, Newark Farm, Ozleworth, Wotton under Edge, Glos, GL12 7PZ [IO81UP, ST79]
G2 DTQ A Goode, Ivy Glen, 71 Church Rd, Shareshill, Wolverhampton, WV10 7LD [IO82XP, SJ90]
G2 DTS A D Monkhouse, Woodspring, 6 St. Marys Cl, Lower Swell, Cheltenham, GL54 1LJ [IO91DW, SP12]
GW2 DUR M N Lapper, 21 Heol Gwili, Llwynhendy, Llanelli, SA14 9HF [IO71WQ, SN50]
G2 DUS I B Howard, 40 Regent St., Stotfold, Hitchin, SG5 4EA [IO92VA, TL23]
G2 DVA D R Bradley, 3 Keswick Dr, Frodsham, Warrington, WA6 7LT [IO83PH, SJ57]
G2 DVP Details withheld at licensee's request by SSL.
G2 DW Dr B F Wickham, 35 Wraymead Pl, Wray Park Rd, Reigate, RH2 0EF [IO91VF, TQ25]
G2 DWB N W Webster, 50 Shaw Dr, Scartho, Grimsby, DN33 2JB [IO93XM, TA20]
GM2 DWW J C Ford, 87 Ardenslate Rd, Kirn, Dunoon, PA23 8NL [IO75MX, NS17]
G2 DWZ S Archer, 30 Westmorland Way, Jacksdale, Nottingham, NG16 5LZ [IO93IB, SK45]
G2 DXA Details withheld at licensee's request by SSL.
G2 DXH R H Hespley, 3 Cotsdale Rd, Penn Fields, Wolverhampton, WV4 5LF [IO82WN, SO89]
G2 DXK L Knight, 123 Baldock Rd, Letchworth, SG6 2EQ [IO91VX, TL23]
GW2 DXQ J Burton, 48 Bron-y-Craig, Llangefni, Anglesey, Gwynedd, LL77 7RE [IO73UG, SH47]

G2 DXU R C Lever, 11 Wharncliffe Gdns, Highcliffe on Sea, Christchurch, BH23 5DN [IO90DR, SZ29]
G2 DYF C C Butler, 205 Highlands Rd, Fareham, PO15 5PW [IO90JU, SU50]
G2 DYM Details withheld at licensee's request by SSL.[Op: Richard Benham-Holman, (also VK6YN, ZL3AAS), G2DYM Aerials, Cobhamden Castle, Beerdown Uplowman, nr Tiverton, Devon EX16 7PH. Tel: (013986) 215.]
G2 DYY Details withheld at licensee's request by SSL.
G2 DZF J H English, 7 Raleigh Cres, Goring By Sea, Worthing, BN12 6EF [IO90TT, TQ10]
G2 DZH N H Talbot, 105 Westwood Ln, Welling, DA16 2HJ [JO01BL, TQ47]
G2 FA Dr D S Pepper Folkstone & District ARS, 11 Earls Ave, Folkestone, CT20 2HW [JO01NB, TR23]
G2 FAB D H Garrad, Flate, 7 Macauley Rd, Clapham, London, SW4 0QP [IO91WL, TQ27]
G2 FBN Details withheld at licensee's request by SSL.
G2 FBU J C M McNeil Greig, Northwood House, Burndell Rd, Yapton, Arundel, BN18 0HR [IO90QT, SU90]
G2 FCA A E Burnard, 20 Kipling Dr, Newport Pagnell, MK16 8EB [IO92PC, SP84]
G2 FCP F L Varley, 39 Nettleton Rd, Mirfield, WF14 9AW [IO93DQ, SE12]
G2 FDF W F E Limehouse, 37 Cabin Ln, Oswestry, SY11 2LS [IO82LU, SJ32]
G2 FFD D Skipworth, West End Rd, Benington, Boston Lincs, PE22 0BU [JO02AX, TF34]
G2 FFK Dr G M Holme, 90 Coleshill Rd, Marston Green, Birmingham, B37 7HW [IO92DL, SP18]
G2 FFN S C Fisher, Te Anau, Main Rd, Maltby-le-Marsh, Alford, LN13 0JP [JO03CH, TF48]
G2 FFO R Johnson, 21 Leaverholme Cl, Cliviger, Burnley, BB10 4TT [IO83VS, SD83]
G2 FGB S R Minson, 52 Fairview Ave, Stanford-le-Hope, SS17 0DT [JO01FM, TQ68]
G2 FHF C D Didcott, Elmbridge, Rectory Ln, Appleby Magna, Swadlincote, DE12 7BQ [IO92FQ, SK31]
GM2 FHH L Hardie, 21 Inchbrae Dr, Aberdeen, AB10 7AJ [IO87WC, NJ90]
G2 FHK D A Smith, 18 Lucas Ave, Harrow, HA2 9UJ [IO91TN, TQ18]
G2 FHM J A Sadler, Abendaros, Barlings Ln, Langworth, Lincoln, Lincs, LN3 5DF [IO93TG, TF07]
GI2 FHN E R Sandys, 25 Moira Park, Bangor, BT20 4RJ [IO74EP, J58]
G2 FIF L W Bazley, 26 Cross St., Moulton, Northampton, NN3 7RZ [IO92NG, SP76]
G2 FIX A C A Newman, 74 Victoria Rd, Wilton, Salisbury, SP2 0DY [IO91BB, SU03]
G2 FJT T C Irving, 17 Uplands Rd, Northwood, Cowes, PO31 8AL [IO90IR, SZ49]
G2 FKO K J Symonds Appledore & Dar, The Beeches, 30 Fairlea Cres, Northam, Bideford, EX39 1BD [IO71VA, SS42]
G2 FKP J M Reid, 26 Minsterley Ave, Shepperton, TW17 8QT [IO91SJ, TQ06]
G2 FKT T J Parry, 17 Cliff Field Rd, Sheffield, S Yorks, S8 9EJ [IO93GI, SK38]
G2 FKZ C E Newton, 83 Hollingthorpe Rd, Hall Green, Wakefield, WF4 3NW [IO93FP, SE31]
G2 FLH T E Clarke, 453 Herries Rd, Sheffield, S5 8TJ [IO93GJ, SK39]
G2 FLY G Edwards, 71 Deakin Rd, Erdington, Birmingham, B24 9AL [IO92BM, SP19]
GW2 FLZ B H Green, 1 Clwyd Ct, Tan y Bryn Rd, Colwyn Bay, Conwy, LL28 4AH [IO83DH, SH88]
G2 FMU M Jackson, 5 Newton Dr, Accrington, BB5 2JT [IO83TR, SD72]
GJ2 FMV E S Chapman, Les Quatres Vents, Clos Des Gellettes, St. Lawrence, Jersey, JE3 1EN
G2 FMW E A Baker, Silver Leaves, Bridge Hill Bridge, Canterbury, Kent, CT4 5AX [JO01NF, TR15]
G2 FNK J H Ellis, 26 Cowleaze, Martinstown, Dorchester, DT2 9TD [IO80SQ, SY68]
G2 FNT C D Phillips, Oddacre, 68 Joydens Wood Rd, Bexley, DA5 2HT [JO01CK, TQ57]
GW2 FOF Details withheld at licensee's request by SSL.
G2 FPY J Harris, 9 Merryfield Cres, Angmering, Littlehampton, BN16 4DA [IO90ST, TQ00]
G2 FQD A L Rogers, 10 Spring Lodge, Elizabeth Way, Orpington, BR5 4BH [JO01BJ, TQ46]
G2 FQP L J Avory, 198 Station Rd, Kingswood, Bristol, BS15 4XR [IO81SL, ST67]
G2 FQR N W Austin, 20 Worcester Cl, Reading, RG3 3BN [IO91LJ, SU66]
G2 FQS Details withheld at licensee's request by SSL.
GW2 FRB E Naish, 46 Brook St., Taibach, Port Talbot, SA13 1TG [IO81CO, SS78]
G2 FRI T J Swain, 2 Monroe Dr, East Sheen, London, SW14 7AR [IO91UL, TQ17]
GU2 FRO E B H Woolley, -le-Pavillon, Sark, Guernsey, Channel Islands, GY9 OSA
G2 FRT W Croxall, 32 Wylde Green Rd, Sutton Coldfield, B72 1HD [IO92CN, SP19]
G2 FRY A Shillito, 123 Bramerton Rd, Nottingham, NG8 4NH [IO92JX, SK54]
G2 FRZ W N Handley, 89 Parrenthorn Rd, Prestwich, Manchester, M25 2RL [IO83UN, SD80]
G2 FS L K Winsor, 70 Station Rd, Hessle, HU13 0BG [IO93SR, TA02]
G2 FSA R L Harvey, 42 Broadhurst, Ashtead, KT21 1QD [IO91UH, TQ15]
G2 FSH A H Vaughan, 81 St. Marys Rd, Benfleet, SS7 1NL [JO01GN, TQ78]
G2 FSP J E Forde, Hillsboro, Shoplane, Cold Hatton, Telford, Salop, TF6 6PZ [IO82RS, SJ62]
G2 FSR J A Hunt, 4 Warmdene Rd, Brighton, BN1 8NL [IO90WU, TQ30]
G2 FSS J A Caley, 110 Shakespeare Dr, Kidderminster, DY10 3QY [IO82VJ, SO87]
G2 FTB F T Baker, 63 Newbury Gdns, Stoneleigh, Epsom, KT19 0NY [IO91VI, TQ26]
G2 FTK F A Noakes, 4 Baronsfield Rd, Cheylesmore, Coventry, CV3 5GP [IO92GJ, SP37]
G2 FTY G E Duffin, 20 Byron Rd, Redditch, B97 5EB [IO92AG, SP06]
G2 FUB K W Viles, 27 Cresta Gdns, Mapperley Rise, Nottingham, NG3 5GD [IO92KX, SK54]
G2 FUD A W Owen, Gwenarth, 184 Hale Rd, Hale, Altrincham, WA15 8SQ [IO83UI, SJ78]
G2 FUF R W Sheppard, 98 New Rd, Durrington, Worthing, BN13 3JR [IO90SU, TQ10]
G2 FUM H Hunt, 21 Highfield Ave, Melton Mowbray, LE13 0NQ [IO92NS, SK72]
G2 FUU T Knight, Homefield, Back Ln, Nazeing, Waltham Abbey, EN9 2DD [JO01AR, TL40]
GM2 FVV W Girvan, Barnsdale Villa, Fairhill Rd, Whins of Milton, Stirling, FK7 0LL [IO86AC, NS79]
G2 FVX S A Deverell, Cornerways, Old Post Office Rd, Chevington, Bury St. Edmunds, IP29 5RD [JO02HE, TL75]
GW2 FVZ D C Morris, 48 Pen y Cefn Rd, Caerwys, Mold, CH7 5BH [IO83JF, SJ17]
G2 FWZ S I Biggs, 27 Weythorne Dr, Birtle, Bury, BL9 7TX [IO83VO, SD81]
G2 FXD G T Eustace, Old Westlyan Chapel, The Entry, Wickham Skieth, Eye, Suffolk, IP23 8LY [JO02MG, TM06]
G2 FXJ S Moisy, 15 Charles St., Headless Cross, Redditch, B97 5AA [IO92AH, SP06]
G2 FXL K B Whittaker, 4 Greame Rd, Bridlington, YO16 5TQ [IO94VC, TA16]
G2 FXO E J Harris, 4 Maisemore Ave, Patchway, Bristol, BS12 6BT [IO81RM, ST68]
G2 FXQ S W Saddington, South Ridding, Sibson Rd, Sheepy Parva, Atherstone, CV9 3RE [IO92GO, SK30]
G2 FXS J H Knowles, 73 Glanton Rd, North Shields, NE29 8LQ [IO95GA, NZ36]
G2 FXV M Middleton, 7 Northd. St., Alnmouth, Alnwick, NE66 2RS [IO95EJ, NU21]
G2 FXZ J B Hodgetts, 59 Woodland Rd, Halesowen, B62 8JS [IO82XL, SO98]
G2 FYO H Terraneau, 2653 Nutmeg Circle, Simi Valley, California 93065, USA
GW2 FYV M S H Arthur, 33 Llys Teg, Dunvant, Swansea, SA2 7QQ [IO71XO, SS59]
G2 FYY M B Rowles, 28 Church St., Weldon, Corby, NN17 3JY [IO92QL, SP98]
GW2 FZ Details withheld at licensee's request by SSL.[Station located in Cheshire.]
G2 FZN J Harrison, 22 Bulkeley Rd, Cheadle, SK8 2AD [IO83VJ, SJ88]
G2 FZO C J Hine, White Ladies, Redesdale Pl, Moreton in Marsh, GL56 0EQ [IO91DX, SP23]
G2 FZU S Eyre, Kestrel, Halloughton, Southwell, Notts, NG25 0QP [IO93MB, SK65]
G2 GC G W R Field, 2 Lovaine Gr, Sandal, Wakefield, WF2 7NF [IO93GP, SE31]
G2 HAJ P C Trezise, 30 Norbury Cl, Chandlers Ford, Eastleigh, SO53 1PZ [IO90HX, SU42]
G2 HAO T H Salisbury, 17 Berrington Rd, Nuneaton, CV10 0LD [IO92FM, SP39]
G2 HAX S P Shackleford, 20 Coniston Dr, Tilehurst, Reading, RG30 6XS [IO91LL, SU67]
G2 HBA C H Spencer, 12 Sunny Bank, London, SE25 4TQ [IO91XJ, TQ36]
G2 HBC F D Webb, 4 Gordon Rd, Ramsgate, CT11 7HU [JO01QI, TR36]
G2 HBQ R I G St John, Hill Cottage, Tripp Hill, Fittleworth, Pulborough, RH20 1ER [IO90RW, TQ01]
G2 HCA L T Sanders, 18 Burbage Ave, Stratford upon Avon, CV37 0DU [IO92DE, SP15]
G2 HCG B Sykes, 52 Marine Dr, Barton on Sea, Hants, BH25 7DX [IO90ER, SZ29]
G2 HCH Details withheld at licensee's request by SSL.
GW2 HCJ R C Taylor, Bwlch Glas, Penrhyndeudraeth, Gwynedd, LL48 6RU [IO72XW, SH63]
G2 HCP C F Steeden, 2 Brighton Ave, Lytham St. Annes, FY8 1XQ [IO83LS, SD32]
GD2 HCX A Haycock, Hampton Croft, Clanna Rd, Santon, Douglas, IM4 2HP
GM2 HDH A W Turner, 31 Duddingston Park South, Edinburgh, EH15 3NZ [IO85KW, NT37]
G2 HDR C N Chapman, 12 Reedley Rd, Westbury on Trym, Bristol, BS9 3ST [IO81QL, ST57]
G2 HDU C W Cragg, 20 Cheltenham Rd, Sedgeberrow, Evesham, WR11 6UL [IO92AB, SP03]
G2 HFC N Lowe, 7 Bellingham Ave, Wigan Ln, Wigan, WN1 2NE [IO83QN, SD50]
G2 HFD Dr H S Reeve, Netherton, Leeks Hill, Melton, Woodbridge, IP12 1LW [JO02PC, TM25]
G2 HFP R N Higson, 1A Haydock Fold Cttgs, under Billinge Ln, Blackburn, BB2 6RN [IO83RR, SD62]
GW2 HFR A Ellis, Wern Las, Ala Rd, Pwllheli, LL53 5BL [IO72SV, SH33]
G2 HFW Details withheld at licensee's request by SSL.
G2 HGA C A Shutt, 25 Barons Way, Kingsthorpe, Northampton, NN2 8HP [IO92NG, SP76]
G2 HHH T H Bayliss, 55 Foxlydiate Cres, Redditch, B97 6NJ [IO92AH, SP06]
G2 HHV J Spivey, 72 Manor Farm Dr, Soothill, Batley, WF17 6HQ [IO93ER, SE22]
G2 HIO A Walmsley, Bridge Hill, 12 Brookside Rd, Breadsall, Derby, DE21 5LF [IO92GW, SK33]
G2 HIX G G P Holden OBE, Brook Lodge, 2 Westbrook Dr, Chesterfield, S40 3PQ [IO93GF, SK37]
GW2 HIY E M Davies, 3 Talton Ct, Prestatyn, LL19 9HF [IO83HH, SJ08]
G2 HJD Details withheld at licensee's request by SSL.
G2 HJT E J Wellman, Castleshaw House, Castleshaw, Delph, Oldham, OL3 5LZ [IO83XN, SD90]
G2 HKG R O Lowson, Moss House, Penton, Carlisle, Cumbria, CA6 5RT [IO85NB, NY47]
G2 HKK Details withheld at licensee's request by SSL.
G2 HKQ A R Knight, 17 Moorland Cres, Upton, Poole, BH16 5LA [IO80XR, SY99]
G2 HKS R P B Udall, Longfield, 20 Upper Way, Upper Longdon, Rugeley, WS15 1QA [IO92BR, SK01]
G2 HKU E H Trowell, Hamlyn, Saxon Ave, Minster on Sea, Sheerness, ME12 2RP [JO01JK, TQ97]
G2 HKW J A R Uphill, 37 Peverells Wood Ave, Chandlers Ford, Eastleigh, SO53 2BS [IO90HX, SU42]
G2 HLB C R Maltby, The Willows Farm, Stallingborough Rd, Immingham, Grimsby, DN40 1NR [IO93VO, TA11]
G2 HLL F H Pickard, 932 Scott Hall Rd, Leeds, LS17 6HL [IO93FU, SE33]
G2 HLP D R Hearsum, Sl-322 Locust Ln, Senecaville, Ohio, USA 43780
G2 HLU Dr H Owen, Arbutus, Durnford Drove, Langton Matravers, Swanage, BH19 3HG [IO80XO, SY97]
G2 HMB N F Tomlinson, 5 Orchard Terr, Rowlands Gill, NE39 1EG [IO94CW, NZ15]
G2 HMK T G Brown, 99 Brinkburn Dr, Darlington, DL3 0JY [IO94FM, NZ21]
G2 HMY G R Priday, 41 Mark Rd, Headington, Oxford, OX3 8PB [IO91JS, SP50]
G2 HNA J J L Weaver, 7 Cramer St., Stafford, ST17 4BX [IO82WT, SJ92]
G2 HNF F A Leach, 59 Well Ln, Galleywood, Chelmsford, CM2 8QZ [JO01FQ, TL70]
G2 HNI L J Hewitt, 8 Lawn Rd, Portswood, Southampton, SO17 2EY [IO90HW, SU41]
G2 HNU D M Byrne, 89 Ferry St., Stapenhill, Burton on Trent, DE15 9EZ [IO92ET, SK22]
G2 HOJ F Hunt, 217 Clee Rd, Cleethorpes, DN35 9HU [IO93XN, TA20]
G2 HOS J R Endall, 68 Bittell Rd, Barnt Green, Birmingham, Hereford & Worcester, B45 8LY [IO92AI, SP07]
G2 HOX F J T Tuckfield, Cynon House, 44 Willow Park Rd, Wilberfoss, York, YO4 5PS [IO93NW, SE75]
GW2 HPG Details withheld at licensee's request by SSL.
G2 HPH L G H Mobley, 29 Ladbrook Rd, Mount Nod, Coventry, CV5 7JW [IO92FJ, SP27]
G2 HR R J Chapman Silverthorn Rd, 50 Graeme Rd, Enfield, EN1 3UT [IO91XP, TQ39]
G2 HS P E Hale, 13 Seaton Dr, Ashford, TW15 3ET [IO91SK, TQ07]
G2 HV J B Alderman South Sussex Raynet, 38 Greenacres, Shoreham By Sea, BN43 5WY [IO90UU, TQ20]
G2 HW H Whalley, 15 Norman Rd, Sale, M33 3DF [IO83UK, SJ79]
G2 IC C P A Turner Radio Club of Thanet, 28 Reading St., Broadstairs, CT10 3AZ [JO01RI, TR36]
G2 JL R V A Allbright, 12 North Parade, Penzance, TR18 4SJ [IO70FC, SW43]
G2 JR H B Burton, 149 Longfellow Rd, Coventry, CV2 5HN [IO92GJ, SP37]
G2 JT P Jones, 81 Windsor Rd, Oldham, OL8 4AL [IO83WM, SD90]
GD2 KAU Details withheld at licensee's request by SSL.
G2 KF A E Robinson, 1 Nathans Cl, Tretherras, Newquay, Cornwall, TR7 2SP [IO70LJ, SW86]
G2 KG C G Hill, 7 Catherine Cl, Woodhall Farm, Hemel Hempstead, HP2 7LN [IO91SS, TL00]
G2 KI G A Spencer, Hollybank, Scotts Way, West Chinnock, Crewkerne, TA18 7PU [IO80OV, ST41]
G2 KU R M Herbert, 24 Norfolk Ave, Sanderstead, South Croydon, CR2 8BN [IO91XI, TQ36]
G2 KV J K Todd, Longacre, Dunns Ln, Iwerne Minster, Blandford Forum, Dorset, DT11 8NG [IO80VW, ST81]
G2 LL T H Ransom obo Hastings Electronics and, 9 Lyndhurst Ave, Hastings, TN34 2BD [JO00HU, TQ81]
G2 LO I P Jefferson Ariel Radio Grp, Cti Club, PO Box 98, Warwick, Warks, CV34 6TN [IO92FG, SP36]
G2 LV R J Leeves, 20 Tower Park, South Molton, Devon, EX36 4EP [IO81BA, SS72]
G2 LW F H Lawrence, Maple House, 23 Vancouver Rd, London, SE23 2AG [IO91XK, TQ37]
G2 MJ R T Hunt, 12 Risedale Dr, Longridge, Preston, PR3 3SB [IO83QT, SD63]
GM2 MP C W Tran North of Scotland Contest Grou, Achnacoille, Lamington, Invergordon, Ross Shire, IV18 0PE [IO77WS, NH77]
G2 MT J D F Peers Marconi Radio S, The Grove, Warren Ln, Stanmore, Middx, HA7 4LY [IO91UP, TQ19]
GW2 NF R H Wright, 89 Llandudno Rd, Rhos on Sea, Colwyn Bay, LL28 4PJ [IO83CH, SH88]
G2 NJ W Carter, 34 West Parade, Peterborough, PE3 6BD [IO92UN, TL19]
G2 NM D A Parsons, 27 St. Leodegars Way, Hunston, Chichester, PO20 6PE [IO90OT, SU80]
G2 OG J M Hogg, 9 Bwlch y Fedwen, Dwyran, Llanfairpwllgwyngyll, LL61 6LZ [IO73UE, SH46]
GW2 OP Details withheld at licensee's request by SSL.
G2 OR C H Ollett, 54 Rochford Garden Way, Rochford, SS4 1QJ [JO01IO, TQ89]
G2 OS Dr G A V Sowter, 73 Downs Hill, Beckenham, BR3 5HD [IO91XJ, TQ36]
G2 OT Details withheld at licensee's request by SSL.[QSL via the Secretary, G3HCQ.]
G2 PA R G Auckland, 60 High St., Sandridge, St. Albans, AL4 9BZ [IO91US, TL11]
G2 PB R K Clegg, Allandale, 347 Whalley Rd, Accrington, BB5 5DF [IO83TS, SD72]
G2 PK J C H Ellison, 6 Stanage Cl, Long Meadow, Worcester, WR4 0HQ [IO82VE, SO85]
G2 PS E A Parsons, 7 Hampton Ln, Winchester, SO22 5LF [IO91HB, SU43]
G2 PT J Piggott, 155 Northwood Way, Northwood, HA6 1RF [IO91TO, TQ19]
G2 PU S R R Kharbanda, Ivett Lodge, 39 London Rd, Harston, Cambridge, CB2 5QQ [JO02BD, TL45]
G2 QT F H Cooper, Woodleys, Sellindge, Ashford, Kent, TN25 6JX [JO01MC, TR13]
G2 RD R E T Dabbs, 17 Manor Ave, Caterham, CR3 6AP [IO91WG, TQ35]
G2 RO W A Roberts, Swallowfield, Beeson, Kingsbridge, Devon, TQ7 2HW [IO80DG, SX84]
GU2 RS F V Mourant, Ivy Mount, 51 Mount Durrand, St Peterport, Guernsey, GY9 9IV
G2 RSA P A King, 32 Millstream Way, Leegomery, Telford, TF1 4QR [IO82SR, SJ61]
G2 RT C C Partridge, 416 Henry Kendall, Village, Wyoming Nsw 2250, Australia
GW2 SB J W B Baker, 27 Highfield Ave, Myn Dd Isa, Mynydd Isa, Mold, CH7 6XY [IO83KE, SJ26]
G2 SH J N Shearme, Chevin, Penn St., Amersham, HP7 0PY [IO91QP, SU99]
G2 SP L M Mansfield, 26 Dovecote Ln, Beeston, Nottingham, NG9 1HU [IO92JW, SK53]
G2 SU A Robinson Northern Hgts Rd, 9 Illingworth Cl, Illingworth, Halifax, HX2 9JQ [IO93BS, SE02]
G2 TA Capt R C Ray, Springfield, Bushey Heath, Bushey, Watford, Herts, WD2 3JL [IO91TP, TQ19]
G2 TO P Brindley Bury St Eds ARC, 2 Beech Park, School Rd, Great Barton, Bury St. Edmunds, IP31 2JL [JO02JH, TL97]
G2 TS Details withheld at licensee's request by SSL.
G2 TV R M Herbert Baird Museum AR, 24 Norfolk Ave, South Croydon, CR2 8BN [IO91XI, TQ36]
G2 UD J M Cooper, Flat 18, 1 Grand Ave, Hove, BN3 2LA [IO90WT, TQ20]
G2 UG S P Ortmayer Halifax & District Amateur Rad, 14 The Cres, Hipperholme, Halifax, HX3 8NQ [IO93CR, SE12]
G2 UH D M Hayward, Hope, Churchwell St., Bradford Abbas, Sherborne, DT9 6RG [IO80QW, ST51]
G2 UK Dr A C Gee, East Keal, 21 Romany Rd, Oulton Broad, Lowestoft, NR32 3PJ [JO02UL, TM59]
G2 UT K M Reid, 4 Harles Acres, Hickling, Melton Mowbray, LE14 3AF [IO92MU, SK62]
G2 VH Details withheld at licensee's request by SSL.[Op: B G Byne. Correspondence via 5 Rosina Close, Waterlooville, nr Portsmouth, Hants, PO7 5SL.]
G2 VJ R A Wybrow, 111 Ferndown Rd, Solihull, B91 2AX [IO92CK, SP18]
G2 VO J J Platt, 24 Brackenley Dr, Embsay, Skipton, BD23 6QN [IO93AX, SE05]
G2 WQ A Brown, Oakwood, Lower Frankton, Oswestry, Shropshire, SY11 4PB [IO82MV, SJ33]
G2 XD D J Bruce, 23 Milton Dr, Newport Pagnell, MK16 9AS [IO92PC, SP84]
G2 XG E A Davie, 7 Cranworth Cres, Chingford, London, E4 7HN [JO01AP, TQ39]
G2 XK E Knowles, East View, Mewith, Bentham, Lancaster, LA2 7DH [IO84RC, SD66]
G2 XP J L Puttock Sutton & Cheam, Rs, 53 Alexandra Ave, Sutton, SM1 2PA [IO91VI, TQ26]
G2 XQ F E Marshall, 5 Shrubbery Ln, Wyke Regis, Weymouth, DT4 9LU [IO80SO, SY67]
G2 XV W Dunell Cambridge Dis Rd, 4 Orchard Rd, Haslingfield, Cambridge, CB3 7JT [JO02AD, TL45]
G2 YD G R A Wright, 62 Dornafield Dr, Ipplepen, Newton Abbot, Devon, TQ12 5RH [IO80EL, SX86]
G2 YK A E Markwick, 13 Dunstable Rd, Richmond, TW9 1UH [IO91UL, TQ17]
G2 YW E O J Woodward, 3345 University Woods, Victoria BC, Canada, V8P 5R2
G2 ZU F W Ellenger, 37 Harper Rd, Salisbury, SP2 7HG [IO91CB, SU13]

G3

G3 AAE J D Kay, 75 Roundmead Ave, Loughton, IG10 1PZ [JO01AP, TQ49]
G3 AAG Details withheld at licensee's request by SSL.
G3 AAH L Allen, 14 Frampton Cl, Bournville, Birmingham, B30 1QT [IO92AK, SP08]
G3 AAJ R J C Broadbent MBE, 94 Herongate Rd, Wanstead Park, London, E12 5EQ [JO01AN, TQ48]
G3 AAK K W C Bunston, 18 Daniells Walk, Lymington, SO41 3PN [IO90FS, SZ39]
G3 AAO W H Longhurst, 432 Ringwood Rd, Ferndown, BH22 9AY [IO90BT, SU00]
G3 AAS M D Glynn, 39 Moor Allerton Dr, Leeds, LS17 6RY [IO93FU, SE33]
G3 AAT Details withheld at licensee's request by SSL.
G3 AAV G N Glover, 4 Heath Royd, Halifax, HX3 0NW [IO93BR, SE02]
G3 AAZ G G Gibbs, Windward, 7 The Grove, Hartford, Huntingdon, PE18 7YD [IO92WI, TL27]
G3 ABA L J Kennard, 116 Ringwood Dr, North Baddesley, Southampton, SO52 9GT [IO90GX, SU31]
G3 ABH B E Crane, 6 Green Mead, Preston, Yeovil, BA21 3RJ [IO80QW, ST51]
G3 ABM H E Bull, 6 Rowton Dr, Ellesmere Port, Great Sutton, South Wirral, L65 7EG [IO83MG, SJ37]
G3 ABP S R C Dorman, Jax, East Cliff, East Looe, Cornwall, PL13 1DE [IO70SI, SX25]
G3 ABS W D Heath, 15 Hollybank Ave, Upper Cumberworth, Huddersfield, HD8 8NY [IO93DN, SE20]
G3 ACB K W King, Martinfield, Matching Green, Harlow, Essex, CM17 0PS [JO01CS, TL51]
GM3 ACL D Robb, 3 George Cres, Clydebank, G81 2EE [IO75TV, NS57]
G3 ACQ H J Harmsworth, 1 Staplehurst Ave, Broadstairs, CT10 1SH [JO01RI, TR36]
G3 ACR H C Harrison, 38 Baker St., Stapenhill, Burton on Trent, DE15 9LX [IO92ES, SK22]
G3 ACT F R Owen, 21 Aspley Ct, Woburn Rd, Woburn Sands, Milton Keynes, MK17 8PA [IO92QA, SP93]
G3 ADJ G L Fish, 44 Billing Ave, Finchampstead, Wokingham, RG40 4JE [IO91NJ, SU76]
G3 ADQ A W Walmsley, 18 Elmwood Ave, Barwick in Elmet, Leeds, LS15 4JT [IO93HT, SE33]
G3 ADR R L Threadingham, 223 Highland Rd, Southsea, PO4 9EZ [IO90LS, SZ69]
G3 ADS R Sawkins, 21 Newcombe Rd, Farnham, GU9 9DJ [IO91OF, SU84]
G3 ADV P Jackson, Fields View Cottage, Longhill Ln, Hankelow, Ches, CW3 0JG [IO83SA, SJ64]
G3 ADZ M J D W J Haylock, 49 Harvest House, Cobbold Rd, Felixstowe, IP11 7SP [JO01QX, TM33]
G3 AEF D J Sole, 6 Blackthorne Rd, Hyde, SK14 5EG [IO83XK, SJ99]
GM3 AEI J E Smallwood, Beulah, 140 Main St, Neilston, Glasgow, G78 3JX [IO75SS, NS45]
G3 AEO J Price, 13, Little London, Andover, Hants, SP11 6JE [IO91GF, SU34]
G3 AER G S Wright, 70 Gunton Dr, Lowestoft, NR32 4QB [JO02VL, TM59]
G3 AEU H S Young, Yellowstones, Heathside Rd, Woking, Surrey, GU22 7EY [IO91RH, TQ05]
G3 AEX G C Fox, 66 Homemead Rd, Bromley, BR2 8BA [JO01AJ, TQ46]
GM3 AEY H G Henderson, 54 Barassie Dr, Kirkcaldy, KY2 6HP [IO86JD, NT29]
G3 AEZ Details withheld at licensee's request by SSL.[Op: J E Greenwell, Eastfield, Beare Green, Dorking, Surrey, RH5 4RW.]
G3 AFF R Short, 84 Windmill Gr, Portchester, Fareham, PO16 9HH [IO90KU, SU60]
G3 AFK V A Bagnall, 60 Lincoln Rd, Ruskington, Sleaford, NG34 9AP [IO93TB, TF05]
G3 AFR H Bates, 1 Bulcote Dr, Burton Joyce, Nottingham, NG14 5AZ [IO92LX, SK64]
G3 AFT B C Bond The Grafton ARS, 86 Agar Gr, Camden Town, London, NW1 9TL [IO91WN, TQ28]
G3 AFV W E Atkinson, Sewage Works, Church Marshes, Milton, Sittingbourne, ME10 2QE [JO01II, TQ96]
G3 AFY V G Croucher, 39 Kirkstall Ave, Tottenham, London, N17 6PH [IO91WO, TQ38]
G3 AGA E L D Davey-Thomas, St James. Fore St, Goldsithney, Penzance, Cornwall, TR20 9JP [IO70GC, SW53]

GW3 AGB W Evans, 68 Heol Gleien, Lower Cwmtwrch, Swansea, SA9 2TT [IO81CS, SN71]
G3 AGC W A Curphey, 8 Emily Davison Ave, Springhill, Morpeth, NE61 2PL [IO95DD, NZ18]
G3 AGF R L Edginton, 9 Churchill Rd, Seaford, BN25 2UL [JO00BS, TQ40]
G3 AGO A M Bryant, Ridgewood, The Ridgeway, Cranleigh, GU6 7HR [IO91SD, TQ03]
G3 AGP F J B Barns, Avgda Joan Carles 1, No 9 Piso No 1, Blanes, Gerona 17300, Spain, X X
G3 AGR J R Dunne, 7 Fleetwood Cl, Minster on Sea, Sheerness, ME12 3LN [JO01JK, TQ97]
G3 AGW H A Edwards, 61 Birchfield Way, Yew Tree Est, Walsall, WS5 4EE [IO92AN, SP09]
G3 AGX L D Colley, Micasa, 13 Ferry Rd, Wawne, Hull, HU7 5XU [IO93TT, TA03]
G3 AGZ SLdr R A Evenett, 8 Grays Walk, Chesham, HP5 2EQ [IO91QR, SP90]
G3 AHB L G Coote, 2 Fairview, New Rd, High Littleton, Bristol, BS18 5JL [IO81RH, ST65]
G3 AHD D G Dean L'pool & Dis AR, 24 Hathaway Rd, Gateacre, Liverpool, L25 4ST [IO83NJ, SJ48]
G3 AHE R W James, 13 Windsor Park Rd, Hayes, UB3 5HZ [IO91SL, TQ07]
GW3 AHN T Higginson, 176 Countisbury Ave, Llanrumney, Cardiff, CF3 9RS [IO81KM, ST28]
G3 AHO C W Finch, 1 Rushall Green, Darley Heights, Luton, LU2 8TL [IO91TV, TL12]
GM3 AHR A R Thomson, Meadowrise, 4 Law View Gdns, Bonnybank, Leven, KY8 5SW [IO86LF, NO30]
G3 AHS D G Thompson, 17 Fair Oak Way, Tadley, Baughurst, Tadley, RG26 5NT [IO91JI, SU56]
GD3 AHV G W Ripley, Corlea Bungalow, Ronague Rd, Ballasalla, Isle of Man, IM9 3BA
G3 AHW Details withheld at licensee's request by SSL.
G3 AHX G H Banner, 49 Penponds Rd, Porthleven, Helston, TR13 9LP [IO70IC, SW62]
G3 AID J H G Davey, 73 Birchett Rd, West Heath, Farnborough, GU14 8RG [IO91OH, SU85]
G3 AIK K N Watkins, Bow House, Hurst, Martock, Somerset, TA12 6JU [IO80OX, ST41]
G3 AIN D Withers, 14 Redgates Ct, Calverton, Nottingham, NG14 6LR [IO93KA, SK64]
G3 AIO S Fenwick, 28 Gimble Way, Pembury, Tunbridge Wells, TN2 4BX [JO01DD, TQ64]
G3 AIU K A H Rogers, 17 Brook Meadow, Wroughton, Swindon, SN4 9LA [IO91CM, SU18]
G3 AIV I Beresford-Pym, Satsang, 30 Davenport Rd, Felpham, Bognor Regis, PO22 7JS [IO90QS, SZ99]
G3 AIZ C C Olley, 9 The Ridings, Market Rasen, LN8 3EE [IO93UJ, TF18]
G3 AJD T Moore, 58 Old Fold View, Barnet, EN5 4EB [IO91VP, TQ29]
G3 AJK R J A Earland, 26 The White House, Our Lady of Sorrows St, St Pauls Bay, Spb 09, Malta
G3 AJL D Campbell, 22 The Hawthorns, Eccleston, Chorley, PR7 5QW [IO83PP, SD51]
G3 AJP J D Baker, 23 Mostyn Cl, Sutton, Ely, CB6 2QJ [JO02BJ, TL47]
G3 AJS J Binning, 293 Perry St., Billericay, CM12 0RB [JO01FP, TQ69]
G3 AJW C P Eatwell, 4 Chaplin Cl, Metheringham, Lincoln, LN4 3HN [IO93TD, TF06]
G3 AJX G Stanton, 44 Lynford Way, Winchester, SO22 6BW [IO91IB, SU43]
G3 AKF B G W Taylor, 3 Hawthorn Cl, Woodford Halse, Daventry, NN11 3NY [IO92JE, SP55]
G3 AKI F Knowles, 1 Mayfield Cl, Bishops Cleeve, Cheltenham, GL52 4NA [IO81XW, SO92]
G3 AKJ D W E Wheele, 4 Mannings Way, Barnstaple, EX31 1QF [IO71XC, SS53]
GM3 AKK J B Rimmer, 7 Borebrae, Newmilns, KA16 9EJ [IO75UO, NS53]
GM3 AKM Details withheld at licensee's request by SSL.
G3 AKN R Milne, 19 Musgrave Rd, Chinnor, OX9 4PL [IO91NQ, SP70]
G3 AKU R A Harding, 11A Upper Packington Rd, Ashby de La Zouch, LE65 1ED [IO92GR, SK31]
G3 AKX R G Lascelles, 5 Middlewich Rd, Cranage, Holmes Chapel, Crewe, CW4 7EA [IO83TE, SJ76]
G3 ALA C Holt, Hob Hay Cottage, Elkstones, Longnor, Buxton, SK17 0LU [IO93AD, SK05]
GM3 ALB G D Gearing, North Logierieve, Udny Ellon, Aberdeenshire, AB4 0PT [IO87TH, NJ72]
G3 ALC P C Spence, 91 West Rd, Oakham, LE15 6LT [IO92PQ, SK80]
G3 ALG G S Starling, 207 Shirley Rd, Croydon, CR0 8SB [IO91XJ, TQ36]
G3 ALI R S Small, 13 Rydal Cl, Stowmarket, IP14 1QX [JO02LE, TM05]
G3 ALK E J Holmes, 7 Castle Dr, Ilford, IG4 5AE [JO01AN, TQ48]
G3 ALP H W O'Donnell, Hill View, Third Ave, Bucknall, Stoke on Trent, ST2 8NG [IO83WA, SJ94]
GM3 ALZ B McKdavidson, 42 Smithfield Dr, Aberdeen, AB1 7XN [IO87TH, NJ72]
GJ3 AME P M Landor, Lauge, Rue Des Raisies, St Martin, Jersey, JE3 6AT
G3 AMF K G Thompson, 11 Ten Bell Ln, Soham, Ely, CB7 5BJ [JO02EI, TL57]
G3 AMG G G Jessup, 19 Normanhurst Rd, Borough Green, Sevenoaks, TN15 8HT [JO01DG, TQ65]
G3 AMH H Green, 9 Robert Ave, Cundy Cross, Barnsley, S71 5RB [IO93GN, SE30]
G3 AMK B Littleproud, St. Michael, 28 Beccles Rd, Bradwell, Great Yarmouth, NR31 8DF [JO02UN, TG50]
G3 AMO W G Houghton, 25 Oak Cl, Dibden Purlieu, Southampton, SO45 4PJ [IO90GU, SU40]
G3 AMW R Towler Hull and District Amateur Radi, 77 Glebe Rd, Hull, HU7 0DU [IO93US, TA13]
GI3 AMY J T Collett, 3 Thornbury Dr, Mansfield, NG19 6NB [IO93JD, SK56]
G3 ANE A Corless, 20 Beech Ave, Huddersfield, HD5 8DZ [IO93DP, SE11]
G3 ANG J W Emmott, 6 Meadowcroft, Euxton, Chorley, PR7 6BU [IO83PP, SD51]
G3 ANH H F Dean, 32 Langdale Rd, Heaton Chapel, Stockport, SK4 5AR [IO83VK, SJ89]
G3 ANI J R Senior, Westering, Curbar Ln, Curbar, Calver, Hope Valley, S32 3YF [IO93EG, SK27]
G3 ANJ Details withheld at licensee's request by SSL.
G3 AOJ A J Clements, 57 Broad Walk, Heston, Hounslow, TW5 9AA [IO91TL, TQ17]
G3 AOK V H Cheeseman, 30 Homewater House, Upper High St., Epsom, KT17 4QJ [IO91UI, TQ26]
G3 AOP Details withheld at licensee's request by SSL.
G3 AOS J G Barnes, 14 Coal Pit Ln, Langley, Macclesfield, SK11 0DQ [IO83WF, SJ97]
G3 AOT T Pattinson, 16 Durham Ave, Washington, NE37 1AQ [IO94FV, NZ25]
G3 AOV R T Foster, 66 Willow Rd, Solihull, B91 1UF [IO92CJ, SP17]
G3 APL J Russon, 59 Ridge Rd, Kingswinford, DY6 9RE [IO82VL, SO88]
G3 APN B D Rabbage, 43 High St., Dawlish, EX7 9HF [IO80GN, SX97]
G3 APO I R Richard, 9 Havelock St., Wokingham, RG41 2XT [IO91NJ, SU86]
G3 APU J H Stanier, 2 Manor Ct, Parkham, Bideford, EX39 5PG [IO70UX, SS32]
G3 APV G R B Wilson, 20 Fell Croft, Dalton in Furness, Cumbria, LA15 8DD [IO84JD, SD27]
G3 APX C W Barrett, 22 Hollyfield Ave, London, N11 3BY [IO91WO, TQ29]
G3 APZ F H Worker, 78 Jayshaw Ave, Great Barr, Birmingham, B43 5RU [IO92AN, SP09]
G3 AQB W Stephenson, 20 Chapel Ct, Chapel Row, Seahouses, NE68 7TD [IO95EN, NU23]
G3 AQC L V Mayhead, Roanoke, Bosham Hoe, Bosham, Chichester, PO18 8EU [IO90NT, SU80]
G3 AQF H F P Weston, Gleneagles, Trevelmond, Liskeard, Cornwall, PL14 4LY [IO70RK, SX26]
G3 AQM F J Gregory, 89A Winslade Rd, Sidmouth, EX10 9EZ [IO80QQ, SY18]
G3 AQS R E Miller, 7 Queens Rd, Whitley Bay, Tyne & Wear, NE26 3AR [IO95GB, NZ37]
G3 AQX S Roberts, Cottage Farm, Slack Ln, Wessington, Derby, DE5 6BY [IO93HA, SK44]
G3 ARE F W Chubb, 2 Brook Cl, Plympton, Plymouth, PL7 1JR [IO70XJ, SX55]
G3 ARL H J Buckett, 16 The Mall, Brading, Sandown, Isle of Wight, PO36 0BU [IO90KQ, SZ68]
G3 ARO W J Marten, 52 Sherwood Ave, London, SW16 5EJ [IO91WJ, TQ26]
GW3 ARP E W Nield, 27 Bayswater Rd, Sketty, Swansea, SA2 9HA [IO81AO, SS69]
GW3 ARS J Sagar, 75 Hookland Rd, Newton, Porthcawl, CF36 5SG [IO81DL, SS87]
G3 ARU H J Smith, 13 Dover Rd, London, E12 5DZ [JO01AN, TQ48]
G3 ARZ C L Waywell, 4 Almers Cl, Houghton Conquest, Bedford, MK45 3LG [IO92SB, TL04]
G3 ASE H S King, 7 Needingworth Rd, St. Ives, Huntingdon, PE17 4JN [IO92XH, TL37]
G3 ASG R F Fautley, 7 Kingfisher Rd, Downham Market, PE38 9RQ [JO02EO, TF60]
G3 ASH R A Jackson, 43 Sparrows Herne, Basildon, SS16 5HW [JO01FN, TQ78]
G3 ASJ T G Kelsey, 23 All Saints Ct, Market Weighton, York, YO4 3NT [IO93QU, SE84]
G3 ASM S E Hincks, 2 Downholme Gr, Hartburn, Stockton on Tees, TS18 5HD [IO94HN, NZ41]
G3 ASQ P C W Ives, 21 Riverside Cl, Hellesdon, Norwich, NR6 5AU [JO02PP, TG11]
G3 ASR J W Bluff Edgware & D RC, 52 Winchester Rd, Kenton, Harrow, HA3 9PE [IO91UO, TQ18]
G3 AST J A Plowman, 17 Orchardleigh, East Chinnock, Yeovil, BA22 9EN [IO80PW, ST41]
G3 ASV G J Pope, 6 Brookland Rise, London, NW11 6DL [IO91VO, TQ28]
G3 ASX D R S Paine, 43 Wilton Rd, Muswell Hill, London, N10 1LX [IO91WO, TQ29]
G3 AT J H Ayre, 3416 Claremore Ave, Ca 90808, USA, X X
G3 ATC R Degg obo Air Cadet Radio Soc, 28 The Spinneys, Welton, Lincoln, Lincs, LN2 3TU [IO93SH, TF08]
G3 ATH H Pain, 15 Skipton Rd, Embsay, Skipton, BD23 6QT [IO93AX, SE05]
G3 ATI Details withheld at licensee's request by SSL.
G3 ATJ F W G Wells, 19 Lucena Ct, Brickfields, Stowmarket, IP14 1RN [JO02LE, TM05]
G3 ATK Dr E H P Young, Orchard House, Camel St., Marston Magna, Yeovil, BA22 8DB [IO81RA, ST52]
G3 ATM D Nasey, 17A Park Ct, St. Brannocks Rd, Ilfracombe, EX34 8PP [IO71WE, SS54]
G3 ATX A G Perry, Pantiles, 9 The Mount, Weybridge, KT13 9LT [IO91SJ, TQ06]
G3 AUA H B Morton, 120 Gravel Hill Cl, Bexleyheath, DA6 7PY [JO01BK, TQ47]
G3 AUB N R Paul, 8 Longden Ln, Macclesfield, SK11 7EN [IO83WG, SJ97]
GM3 AUE A M Mc Ghie, 1 Boyach Cres, Isle of Whithorn, Newton Stewart, DG8 8LD [IO74TQ, NX43]
G3 AUS Details withheld at licensee's request by SSL.
G3 AUU A J Hill, 33 Everest Rd, Cheltenham, GL53 9LL [IO81XV, SO91]
G3 AUX Dr K L Owen, 53 Merestones Dr, The Park, Cheltenham, GL50 2SU [IO81WV, SO92]
GM3 AVA W W W Peat, 61 Stirling Rd, Larbert, FK5 4SG [IO86BA, NS88]
G3 AVE F C P Flanner, 1 Ludford Cl, Sutton Coldfield, B75 6DW [IO92CN, SP19]
G3 AVJ E J Fox, 5 Belfield Cres, Huyton, Liverpool, L36 5TR [IO83NJ, SJ49]
G3 AVL R F S Reynolds, 12 Eastham Rake, Eatham, Wirral, L62 9AA [IO83MH, SJ37]
G3 AVN P W Parker, 18 Belvedere Ct, High St., Dawlish, EX7 9ST [IO80GN, SX97]
G3 AVO GCpt G K McKay, 4 Stevenson Ct, Eaton Ford, St. Neots, Huntingdon, PE19 3LF [IO92UF, TL16]
G3 AVV G Gunnill, 1 Kenmoor Cl, Preston, Weymouth, DT3 6JZ [IO80SP, SY68]
G3 AWA Dr A J Woiwod, 21 The Avenue, Beckenham, Kent, BR3 2DG [IO91XJ, TQ36]
GW3 AWC J F Thomas, 37 Clifton Rise, Abergele, LL22 7DL [IO83EG, SH97]
GM3 AWF D F Craig, 13 Clifford Rd, North Berwick, EH39 4PW [IO86PB, NT58]
G3 AWI C N Perry, 8 Rectory Rd, Church Warsop, Mansfield, NG20 0RX [IO93KF, SK56]
G3 AWK R F Gough, 3 Sycamore Cl, Cherry Willingham, Lincoln, LN3 4BJ [IO93SF, TF07]
G3 AWP P W Gifford, 21 Bengal Rd, Winton, Bournemouth, BH9 2ND [IO90BR, SZ09]
G3 AWQ D R Hill, 90 Clover Ridge, Dr West, Ajax Ontario, Canada L1S 3E8
G3 AWR C D Hammett, 48 Hadrian Rd, Newcastle upon Tyne, NE4 9QH [IO94EX, NZ26]
GM3 AWW W S Murray, 16 Broom Rd, Newton Mearns, Glasgow, G77 5DP [IO75US, NS55]
G3 AXF E C Brown, 64 Kingsdown Rd, Leytonstone, London, E11 3LP [JO01AN, TQ38]
G3 AXI R J Boal, 32 Park Cl, Linton, Swadlincote, DE12 6QB [IO92ER, SK21]
G3 AXK F O A Dawkins, 10 Oakham Cl, Moulton, Northampton, NN3 7DE [IO92NG, SP76]
G3 AXN C G Collop, Alcombe Cote, 19 Manor Rd, Alcombe, Minehead, TA24 6EH [IO81GE, SS94]
G3 AXS R A Hutcheson-Collins, 7 The Copse, Scarthoe, Grimsby, DN33 2LW [IO93XM, TA20]
GM3 AXX A Fraser, 58 Rigghead, Stewarton, Kilmarnock, KA3 3DQ [IO75RQ, NS44]
G3 AYC G D Eddowes Ariel Radio Grp, Flat 1, 47 The Avenue, Ealing, London, W13 8JR [IO91UM, TQ18][Correspondence to c/o Ariel Radio Group, Woodlands, London, W12.]
G3 AYS R A Watson, 23 Cunningham Park, Mabe Burnthouse, Penryn, TR10 9HB [IO70KD, SW73]
G3 AYY L Fletcher, 36 Scholes Ln, Prestwich, Manchester, M25 0AY [IO83UM, SD80]
G3 AYZ J F Turner, 7 Audley Cl, Burston, Diss, IP22 3XD [JO02MJ, TM18]
G3 AZ J J Hunter, 17 Jeffreys Way, Taunton, TA1 5JJ [IO81KA, ST12]
G3 AZI A McCann, 105 Todd Ln North, Lostock Hall, Preston, PR5 5UP [IO83PR, SD52]
G3 AZT C H Walker, Woodcote, Abingdon Rd, Tubney, Abingdon, OX13 5QQ [IO91HQ, SU49]
G3 AZW A S Bates, 68 Hill St., Hilperton, Trowbridge, BA14 7RS [IO81VI, ST85]
G3 AZY R M Cornish, 3 New Building, Greenham, Wellington, TA21 0JS [IO80IX, ST01]
G3 BAC R A Bastow, 15 Pitfield Dr, Meopham, Gravesend, DA13 0AY [JO01EI, TQ66]
G3 BAP R Cordingley, 61 Cleveleys Ave, Lancaster, LA1 5HE [IO84OB, SD46]
GW3 BAZ J Evans, 2 Clos Coedydafarn, Lisvane, Cardiff, CF4 5ER [IO81JM, ST18]
G3 BBC J M Eason Ariel Radio Grp, Lynwood, Holton, Oxon, OX9 1PU [IO91MQ, SP70][Correspondence via Ariel Radio Group, Film Studios, Ealing.]
G3 BBD J L Townend, 44 Liphill Bank Rd, Holmfirth, Huddersfield, HD7 1LQ [IO93CN, SE10]
G3 BBK J N Orrin, Greenacres, Church St., Old Heathfield, Heathfield, TN21 9AL [JO00DX, TQ52]
G3 BBR K J Wheatley, 36 Lynwood Rd, Redhill, RH1 1JS [IO91WF, TQ25]
G3 BBX D G Holloway, 101 St. Marks Rd, Henley on Thames, RG9 1LP [IO91NM, SU78]
G3 BCC D T Gerrard, 1 Vineyards Cl, Charlton Kings, Cheltenham, GL53 8NH [IO81XV, SO91]
G3 BCE D Nichols, West Hill, Milborne Port, Sherborne, Dorset, DT9 5LS [IO80SV, ST61]
G3 BCI V F Cotton, 45 Branksome Hill Rd, Bournemouth, BH4 9LF [IO90BR, SZ09]
GM3 BCL A G Anderson, West Balfour House, Durris, Banchory, AB3 3BJ [IO87TH, NJ72]
G3 BCM D Deacon, 17 Lonsdale Rd, South Norwood, London, SE25 4JJ [IO91XJ, TQ36]
GM3 BCX J M Sherriffs, Corrieview, Fort Augustus, Inverness Shire, PH32 4DP [IO77PD, NH30]
G3 BCY F W Unstead, 10 Lilac Cl, Bordon, Hants, GU35 0UY [IO91NC, SU73]
G3 BD M T Elvy, Flat 2, 15 Ellington Rd, Ramsgate, CT11 9SJ [JO01QI, TR36]
G3 BDD D Hudson, 8 Leaventhorpe Way, Fairweather Green, Bradford, BD8 0EQ [IO93CT, SE13]
G3 BDH R R Flaum, 135 The Dr, Ilford, IG1 3JF [JO01AN, TQ48]
G3 BDQ J D Heys, White Friars, Friars Hill, Guestling, Hastings, TN35 4EP [JO00HV, TQ81]
G3 BDT A L Searle, 30 Hawthorne Gr, Poulton-le-Fylde, FY6 7PN [IO83LU, SD34]
G3 BDU J F Washer, Brook Cottage, Sutton Valence Hill, Sutton Valence, Maidstone, ME17 3AS [JO01HE, TQ84]
G3 BDV P J Cott, 6 Hay Green, Danbury, Chelmsford, CM3 4NU [JO01HR, TL70]
G3 BDY R D Ball, West Croft, 20 Morant Rd, Ringwood, BH24 1SX [IO90CU, SU10]
G3 BEC B J Clark, 107 Eastland Rd, Yeovil, BA21 4EY [IO80QW, ST51]
G3 BEF Details withheld at licensee's request by SSL.
G3 BEG P C Bond, 5 The Fairway, Camberley, GU15 1EF [IO91PH, SU85]
G3 BEH A G Allan, Glendene, 8 Stonegallows, Taunton, TA1 5JN [IO81KA, ST22]
G3 BEJ A Cherrett, 4 Cormorant Cl, Ayton, Washington, NE38 0DE [IO94FV, NZ25]
G3 BEQ D J Rickard, 114 Dalmeny Ave, London, SW16 4RP [IO91WJ, TQ36]
G3 BEU Dr A R P Calder, Oldestairs, Kingsdown, Deal, Kent, CT14 8ES [JO01QE, TR34]
G3 BEX W J D Short, Highland Light, 26 Howard Cres, Seer Green, Beaconsfield, HP9 2XP [IO91QO, SU99]
G3 BEY R L Mailey, 66 Station Rd South, Belton, Great Yarmouth, NR31 9AA [JO02TN, TG40]
G3 BFC W H C Wheeler, Baycrest, 35 Marine Dr East, Barton on Sea, New Milton, BH25 7DU [IO90ER, SZ29]
G3 BFP J N Headland, 13 Tollers Ln, Coulsdon, CR5 1BE [IO91WH, TQ35]
G3 BG Details withheld at licensee's request by SSL.
G3 BGA Dr D Finlay-Maxwell, Prospect House, Prospect St, Huddersfield, HD1 2NU [IO93CP, SE11]
GM3 BGB S B Jagger, 65 Forsyth St., Greenock, PA16 8SX [IO75OW, NS27]
G3 BGF R Winkworth, 1 Collingwood Dr, Mundesley, Norwich, NR11 8JB [JO02RV, TG23]
G3 BGG C O Titley, 19 New Cl, Knebworth, SG3 6NU [IO91VU, TL22]
G3 BGL Rev P W Sollom, Douai Abbey, Upper Woolhampton, Reading, Berks, RG7 5TH [IO91JJ, SU56]
G3 BGM F D Shirreff, Wrightsbridge, Wanborough Marsh, Swindon, Wilts, SN4 0AR [IO91DN, SU28]
G3 BGO F R Piper, 14 Tylers Cl, Chelmsford, CM2 9DY [JO01FR, TL70]
G3 BGR S D Percival, 33 Arden Rd, Worcester, WR5 3BD [IO82VE, SO85]
G3 BGY A C White, 53 Ena Rd, London, SW16 4JE [IO91WJ, TQ36]
G3 BHA N S Taylor, 8 Aragon Way, Muscliffe Ln, Bournemouth, BH9 3SB [IO90BS, SZ19]
G3 BHF E C Hasted, 54 Plaxtol Rd, Erith, DA8 1NL [JO01BL, TQ47]
G3 BHK L R V Mitchell, Trehelig, New Rd, Zeals, Warminster, BA12 6NG [IO81UC, ST73]
G3 BHM H E Kempson, 8 Hounds Way, Hayes, Wimborne, BH21 2LD [IO90AT, SU00]
G3 BHT B G Meaden, 14 Aulton Rd, Four Oaks, Sutton Coldfield, B75 5PX [IO92CO, SP19]
G3 BHW E J Hancock, Little Hassocks, 32 Edmanson Ave, Westbrook, Margate, CT9 5EW [JO01QJ, TR37]
GM3 BHY R H Low, The Whins, Middle Rd, Rattray, Blairgowrie, PH10 7EP [IO86IO, NO14]
G3 BIF K H Bonnick, 77 Severn Dr, Burntwood, WS7 9JF [IO92BQ, SK00]
G3 BII A R Clark, 19 Lakes Ln, Beaconsfield, HP9 2LA [IO91QO, SU99]
G3 BIK E Chicken, 21 Townsend Cres, Kirkhill, Morpeth, NE61 2XP [IO95DD, NZ18]
G3 BIO J E Anderson, 110 Hadrian Ave, Dunstable, LU5 4SP [IO91SV, TL02]
G3 BIT G S Ellery, Penwerris, Higher Lincombe Rd, Wellswood, Torquay, TQ1 2HD [IO80FL, SX96]
G3 BIX J Collings, 17 Rowan Gr, St. Ippolyts, Hitchin, SG4 7SP [IO91UW, TL12]
G3 BJB E Dandy, 9 Lodge Dr, Malvern, WR14 4LS [IO82UC, SO74]
G3 BJC R E Sparry, 47 Victoria Rd, Trowbridge, BA14 7LD [IO81VH, ST85]
G3 BJD J L Colebrook, Green Gable, Springfield Rd, Bigrigg, Egremont, CA22 2TJ [IO84FM, NY01]
G3 BJJ J R Dawson, 72 Arbutus Dr, Coombe Dingle, Bristol, BS9 2PN [IO81QL, ST57]
G3 BJN T L Johnson, Hawthorn House, Bernard Ln, Green Hammerton, York, YO5 8BP [IO94IA, SE45]
G3 BJR D G Hopkins, 47 Bure Ln, Friars Cliff, Christchurch, BH23 4DJ [IO90DR, SZ19]
G3 BJS B C Seys, 59 Lansdowne Rd, Studley, B80 7RD [IO92BG, SP06]
G3 BJY H N Cantrill, 21 Hartshill Rd, Hartshorne, Swadlincote, DE11 7HN [IO92FS, SK31]
G3 BK Details withheld at licensee's request by SSL.
GM3 BKC A C Miller, 1 Cottage Cres, Camelon, Falkirk, FK1 4AX [IO86CA, NS88]
G3 BKG K C B Field, 28 Huntingdon Rise, Bradford on Avon, BA15 1RJ [IO81UI, ST86]
G3 BKJ H B Alderson, 31 Rumbold Rd, Edgerton, Huddersfield, HD3 3DB [IO93CP, SE11]
G3 BKL R Bland, 11 Great Croft, Firsdown, Salisbury, SP5 1SN [IO91DC, SU23]
G3 BKN E W Batten, 11 Webbs Cl, Ashley Heath, Ringwood, Hants, BH24 2EP [IO90BU, SU10]
G3 BLG E J Chatfield, Hanover Gdns, 156 Holland Rd, Clacton on Sea, CO15 6NF [JO01OT, TM11]
G3 BLH Details withheld at licensee's request by SSL.
G3 BLN P M Trowbridge, 4 Parkstone Heights, Poole, BH14 0QF [IO90AR, SZ09]
G3 BLO F G Sargent, 80 Grange Park, Bishopsteignton, Teignmouth, TQ14 9TS [IO80FN, SX97]
G3 BLR W W Fergusson, 71 Tabors Ave, Great Baddow, Chelmsford, CM2 7EL [JO01GR, TL70]
G3 BLS D A Walker, 32 South St., Osney, Oxford, OX2 0BE [IO91IR, SP50]
GM3 BMI A Bolton, 40 Bonaly Ave, Colinton, Edinburgh, EH13 0ET [IO85IV, NT26]
G3 BMM J W Lymer, 101 Arundel Dr, Beeston, Nottingham, NG9 3FQ [IO92JW, SK53]
G3 BMO H Speed, 45 Willow Glade, Huntington, York, YO3 9NJ [IO93LX, SE65]
G3 BMQ G C V Humphrey, 56 Park Ln, Wallington, SM6 0TN [IO91WI, TQ26]
G3 BMV A Shaw, 15 Revidge Rd, Blackburn, BB2 6JB [IO83RS, SD62]
G3 BMX G H Greenwood, 11 Greendown Cl, Valley Dr, Ilkley, LS29 8NQ [IO93CW, SE14]
G3 BNE G W Alderman, 35 Eynswood Dr, Sidcup, DA14 6JQ [JO01BK, TQ47]
G3 BNF A G Embleton, 16 Brandwood Park, Stacksteads, Bacup, OL13 0PA [IO83VQ, SD82]
G3 BNG R S Andrews, 101 Gr Ave, New Costessey, Norwich, NR5 0HZ [JO02OP, TG11]
G3 BNH Details withheld at licensee's request by SSL.
G3 BNU A D Allen, 36 Treelands Dr, Cheltenham, GL53 0DE [IO81XV, SO92]
G3 BNV F M J Exeter, 7 Wilberforce Rd, West Earlham, Norwich, NR5 8ND [JO02OP, TG10]
G3 BNW J E Bailey, 13 Heywood Rd, Alderley Edge, SK9 7PN [IO83VH, SJ87]
GM3 BNX J J C Shaw, 53 Lennel Mount, Coldstream, TD12 4NS [IO85VP, NT84]
G3 BOB G M Ward, Dovecote Barn, Weston Town, Evercreech, Somerset, BA4 6JG [IO81RD, ST63]
G3 BOC H M Synge, Lake Cottage, Llynclys Hill, Oswestry, Shropshire, SY10 8LL [IO82LT, SJ22]
G3 BOI A W Post, 17 Braintree Cl, Luton, LU4 0QX [IO91SV, TL02]
G3 BOK W G Rennison, Neasden, 18 Bucklesham Rd, Kirton, Ipswich, IP10 0PA [JO02PA, TM23]
G3 BON W J Rawlings, Manor Cottage, Church St., Winsham, Chard, TA20 4HU [IO80NU, ST30]
G3 BOR D W Hudson, Parkview, 37 Bridgehouse Ln, Haworth, Keighley, BD22 8QE [IO93AT, SE03]
G3 BOT A B Reeder, Castelo Do Repouso, Apartado 3015, Almancil, Algarve Portugal
G3 BP R M Garrett, Homestead, London Rd East, Amersham, HP7 9DT [IO91QP, SU99]
G3 BPE R G Holland, 37 Danvers Way, Westbury, BA13 3UF [IO81VG, ST85]
G3 BPF A R Painter, Gold Green, Upper Rochford, Tenbury Wells, Worcs, WR15 8SP [IO82RH, SO66]
G3 BPG J H C Richards, 64 Cres Gdns, Eastcote, Ruislip, HA4 8TA [IO91TN, TQ18]
G3 BPI J Hunter, 77 Kenmore Ave, Kenton, Harrow, HA3 8PA [IO91UO, TQ18]
G3 BPJ R S Shepherd, 59 Edgehill Cres, Leyland, Preston, PR5 1QU [IO83PQ, SD52]
G3 BPK D Snape Douglas Vall AR, 30 Culcross Ave, Highfield, Wigan, WN3 6AA [IO83PM, SD50]
G3 BPM P J H Matthews, Charnwood, Duckpool Ln, West Chinnock, Crewkerne, TA18 7QD [IO80OV, ST41]
G3 BPQ E Smith, 50 Ladysmith Rd, Ashton under Lyne, OL6 9BZ [IO83XM, SD90]

GM3 BQA J S McCaig, Woodlands, Wamphray Rd, North Berwick, East Lothian, EH39 5NR [IO86PB, NT58]
G3 BQE R J R Fussey, 9 Alicia Gdns, Kenton, Harrow, HA3 8JB [IO91UO, TQ18]
G3 BQL W Walker, 10 Gravel Hole Ln, Sowerby, Thirsk, YO7 1NS [IO94HF, SE48]
GM3 BQN J B S Gilfillan, Tigh Na Leven, By Tarbert, Argyll, PA29 6XZ [IO75FT, NR76]
G3 BQQ J C Tranter, 239 Holyhead Rd, Wellington, Telford, TF1 2EA [IO82RQ, SJ61]
G3 BQT E Hulme, 21 Brookside Cres, Greenmount, Bury, BL8 4BG [IO83TO, SD71]
G3 BQW Details withheld at licensee's request by SSL.[Op: J S Ellis.]
G3 BR S T Hall, 57 New St. Hill, Bromley, BR1 5AX [JO01AK, TQ47]
G3 BRA O C Doley, 25 North Rd, Berwick upon Tweed, TD15 1PW [IO85XS, NT95]
G3 BRD J D Lunn, 10 Southdown Rd, Seaford, BN25 4PB [JO00BS, TV49]
G3 BRQ K B Tackley, 1 Greenways, Fleet, GU13 9UG [IO91NG, SU85]
G3 BRR W J Leader, 45 Linden Gr, Chandlers Ford, Eastleigh, SO53 1LE [IO90HX, SU42]
G3 BRS P E Smith Bury Radio Soc, 52 Grantham Dr, Bury, BL8 1XW [IO83UO, SD71]
G3 BRT G O J Parfitt, 1 Branson Ct, Westfield Way, Bradley Stoke North, Bristol, BS12 0EN [IO81RN, ST68]
G3 BRU W Priestnall, 2 Corngrave Cl, Marske, Marske By The Sea, Redcar, TS11 7ER [IO94LO, NZ62]
G3 BRV R C Bennison, 49 Shorncliffe Cres, Folkestone, CT20 3PF [JO01NC, TR23]
G3 BRW Details withheld at licensee's request by SSL.
G3 BSA D Aitchison, 28 Avondale Dr, Astley, Tyldesley, Manchester, M29 7ES [IO83SM, SD70]
G3 BSF K D Faux, 11 The Dr, Potters Bar, EN6 2AP [IO91VQ, TL20]
G3 BSI S T Smith, Silver Birches, 23 Glenwood Rd, West Moors, Ferndown, Dorset, BH22 0EN [IO90BT, SU00]
G3 BSK M L W Prickett, 260 Haslucks Green Rd, Shirley, Solihull, B90 2LR [IO92BJ, SP17]
G3 BSL E B Hardy, 71 Scalpcliffe Rd, Burton on Trent, DE15 9AB [IO92ET, SK22]
G3 BSN P D Stanley, 1 Thames View, Cliffe Woods, Cliffe, Rochester, ME3 8LR [JO01GK, TQ77]
G3 BSO C L Turville, 10 Gynsill Ln, Anstey, Leicester, LE7 7AG [IO92JP, SK50]
GM3 BSQ M C Hately Aberdeen Amateur Radio Soc, 1 Kenfield Pl, Aberdeen, AB1 7UW [IO87TH, NJ72]
G3 BST J B Tuke, Spring Cottage, Pains Hill, Oxted, RH8 0RG [JO01AF, TQ45]
G3 BSU A F Cleall, 103 Alinora Cres, Goring By The Sea, Goring By Sea, Worthing, BN12 4HJ [IO90ST, TQ10]
G3 BTL J Tolman, 19 Back Rd, Linton, Cambridge, CB1 6LG [JO02DC, TL54]
G3 BTM N N Shires, Choice Hall, Duxford, Cambridge, Cambs, CB2 4QG [JO02BC, TL44]
G3 BTV R H Sumner, 73 Oxford Rd, Marston, Old Marston, Oxford, OX3 0PH [IO91JS, SP50]
G3 BUF B J Fost, Rest Harrow, The St., Swanton Novers, Melton Constable, NR24 2QY [JO02MU, TG03]
GI3 BUP F B Edwards, 4 Station Rd, Killough, Downpatrick, BT30 7QA [IO74EG, J53]
G3 BUR J F Collett, PO Box 651543, Benmore, 2010, S Africa, ZZ5 3SQ
G3 BVA E Digman, 75 Ramsden Rd, Orpington, BR5 4LU [JO01BJ, TQ46]
G3 BVB D R J Adair, 3 Belmont Cl, Shaftesbury, SP7 8NF [IO81VA, ST82]
G3 BVU C J Beanland, Co The Manager, Lloyds Bank Ltd, Witney, Oxon, OX8 6AE [IO91GS, SP30]
G3 BVW Flt Lt C P Townley, Exeter Rd, Moretonhampstad, Moretonhampstead, Newton Abbot, Devon, TQ13 8NW [IO80CP, SX78]
G3 BW M Gibbings Cumbria Contest Club, Glannaventa, 5 Meadowbank Ln, Grange Over Sands, LA11 6AT [IO84ME, SD37]
G3 BWI W R Timms, 22 Padway, Penwortham, Preston, PR1 9EL [IO83PR, SD52]
G3 BWN I R Fraser, 37 Chilwell Rd, Beeston, Nottingham, NG9 1EH [IO92JW, SK53]
G3 BWQ S H Iles, 29 River Bank, Winchmore Hill, London, N21 2AB [IO91WP, TQ39]
G3 BWR N B Greenall, 16 Shaw Ln, Prescot, L35 5BY [IO83OK, SJ49]
G3 BWV F E Springate, 150 Mackenzie Rd, Beckenham, BR3 4SD [IO91XJ, TQ36]
G3 BWX Maj A L Fayerman, 45 Copse Ave, West Wickham, BR4 9NN [IO91XI, TQ36]
G3 BWY W J Crossan, Highlands, Shernfold Park, Frant, Tunbridge Wells, TN3 9DL [JO01DC, TQ53]
G3 BXC J V Ashworth, Ash Corner, 2 Meadway, Oxshott, Leatherhead, KT22 0LZ [IO91TH, TQ16]
GM3 BXD G D Elliott, Torfness, 20 Glencaple Ave, Dumfries, DG1 4SJ [IO85EB, NX97]
G3 BXF L Austin Rugby Amateur Transmitting Soc, 12 Whittle Cl, Bawnmore Rd Bilto, Rugby, CV22 6JR [IO92II, SP47]
G3 BXS A G Stacey, 22 Montagu Rd, Datchet, Slough, SL3 9DJ [IO91RL, SU97]
GM3 BXW C Malcolm, 26 St. Clair Ave, Giffnock, Glasgow, G46 7QE [IO75UT, NS55]
GM3 BXX R Y A Robb, 82 Samson Ave, Kilmarnock, KA1 3ED [IO75SO, NS43]
G3 BXZ J T Parker, 9 Ashley Cl, Charlton Kings, Cheltenham, GL52 6LF [IO81XV, SO92]
G3 BYG N L H Williams, Chappel Lake Farm, Halwill, Beaworthy, Devon, EX21 5UF [IO70VS, SS40]
G3 BYN W Emery, 15 Spinney Rd, Chaddesden, Derby, DE21 6HW [IO92GW, SK33]
G3 BYV Details withheld at licensee's request by SSL.
G3 BYW W M Dunell, 4 Orchard Rd, Haslingfield, Cambridge, CB3 7JT [JO02AD, TL45]
G3 BYX D Dowson, 31 Dunelm Walk, Darlington, DL1 2DJ [IO94FM, NZ31]
G3 BYY E W Elliott, High Curley, 37 Staines Rd, Wraysbury, Staines, TW19 5BY [IO91RK, TQ07]
G3 BZB R T Cunliffe, 5 Silk Mill Ln, Tutbury, Burton on Trent, DE13 9LE [IO92DU, SK22]
G3 BZC J Bradshaw, Flat 8, 38/40 Grange Ave, Levenshulme, Manchester, M19 2FY [IO83VK, SJ89]
G3 BZQ Details withheld at licensee's request by SSL.
G3 BZR G W J Massey, 15 Ledmore Rd, Charlton Kings, Cheltenham, GL53 8RA [IO81XV, SO92]
G3 BZS C J Whistlecroft, 52 Shirley Rd, Droitwich, WR9 8NR [IO82WG, SO86]
G3 BZW Details withheld at licensee's request by SSL.[Op: R H Horner.]
G3 CAG R H Pearson, 5 Birchfield Gr, Bletchley, Milton Keynes, MK2 2RQ [IO91PX, SP83]
G3 CAJ R F D Prince, 52 Mafeking Rd, Southsea, PO4 9BG [IO90LS, SZ69]
G3 CAQ W C L Moorwood, 44 Station Rd, Codsall, Wolverhampton, WV8 1DA [IO82VP, SJ80]
G3 CAR P J Perkins Chiltern ARC, 26 Colne Rd, High Wycombe, HP13 7XN [IO91PP, SU89]
G3 CAZ J A Shaw, 128 Perth Rd, Ilford, IG2 6AS [JO01AN, TQ48]
GW3 CBA H S C J Kellaway, 34 Winston Rd, Barry, CF62 9SW [IO81IK, ST16]
G3 CBF G F Kelly, Linden, Dashpers, Brixham, TQ5 9LJ [IO80FJ, SX95]
G3 CBW H Walker, 20 Birchfield Dr, Eaglescliffe, Stockton on Tees, TS16 0ER [IO94HM, NZ41]
G3 CCA R J Maskill 51 (Evington) Sqn ATC A.R.C, 21 Clayton, Orton Goldhay, Peterborough, PE2 5SB [IO92UN, TL19]
G3 CCC H Barnes, 43 Waring Dr, Thornton Cleveleys, FY5 2SW [IO83LV, SD34]
GW3 CCF R Clarke, 16 Blaen Wern, Gwernymynydd, Mold, CH7 4AP [IO83KD, SJ26]
G3 CCH J H Stace, 38 Skippingdale Rd, Scunthorpe, DN15 8NU [IO93QO, SE81]
G3 CCL G H Ireland, 20 St. Chads Rd, Withington, Manchester, M20 4WH [IO83VK, SJ89]
G3 CCM W R Harris, 25 Cotton Rd, Potters Bar, EN6 5JT [IO91VQ, TL20]
G3 CCO D A V Williams, 27 Dene Way, Caldecote, Biggleswade, SG18 9DL [IO92UC, TL14]
G3 CCX Details withheld at licensee's request by SSL.[Op: P Craw. Station located near Littlehampton.]
G3 CDE Dr E G A Jackson, 12 Oak Tree Cl, Burpham, Guildford, GU4 7JQ [IO91RG, TQ05]
G3 CDJ R R Adams, 102 Carver Hill Rd, High Wycombe, HP11 2UD [IO91OO, SU89]
G3 CDM I W Gardner, 30 Pierremont Cres, Darlington, Co. Durham, DL3 9PB [IO94FM, NZ21]
GW3 CDP W D Evans, 71 Crymlyn Rd, Skewen, Neath, SA10 6EG [IO81BP, SS79]
G3 CDR Details withheld at licensee's request by SSL.
G3 CDY R W Roberts, 143 Meadow Way, Norwich, NR6 6XU [JO02PP, TG21]
G3 CEG C B King, Rhodas Cottage, Whiteway, Stroud, Glos, GL6 7EP [IO81WS, SO91]
G3 CEI Dr C W W Brown, Downlands, Off Hackwood Ln, Cliddesden, Basingstoke, Hants, RG25 2NH [IO91LF, SU64]
G3 CEL A Stafford, 29 Sandringham Rd, Hyde, SK14 5JA [IO83XK, SJ99]
G3 CET Details withheld at licensee's request by SSL.
GW3 CF F G Jones, Silverlyn, 15 Gronant Rd, Prestatyn, LL19 9DT [IO83HH, SJ08]
G3 CFA D J Vellacott, 31 Cleveland Rd, Bitterne Park, Southampton, SO18 2AP [IO90HW, SU41]
G3 CFG R S Lancaster, 30 Staines Way, Louth, LN11 0DF [JO03AI, TF38]
GI3 CFH D I Fulton Nth West Irelan, 120 Dunnalong Rd, Bready, Strabane, Co. Tyrone, BT82 0DP [IO64GW, C30]
GM3 CFK P J C Harrison, 62 Credon Dr, Airdrie, ML6 9RT [IO85AU, NS76]
G3 CFQ J N Howat, Cantley, 28 Lilac Gr, Doncaster, DN4 6PF [IO93LM, SE60]
G3 CFR J H Jowett, Ashleigh, Shute Rd, Kilmington, Axminster, Devon, EX13 7ST [IO80LS, SY29]
GM3 CFS J M Robson, Kirkfield, Olrig, Caithness, KW14 8HN [IO88HN, ND16]
G3 CFV F T Gay, 61 Abbey Rd, Yeovil, BA21 3EY [IO80QW, ST51]
G3 CFW N C H Burton, 46 Blacklands Dr, Hayes, UB4 8EX [IO91SM, TQ08]
G3 CGB R H Eaton-Williams, The Stewards, The Old Racecourse, Lewes, BN7 1UR [IO90XV, TQ31]
G3 CGD J J Yeend, 30 St. Lukes Rd, Cheltenham, GL53 7JJ [IO81XV, SO92]
G3 CGE R W C Gardner, 62 Rosewall Rd, Maybush, Southampton, SO16 5DW [IO90GW, SU31]
G3 CGP Details withheld at licensee's request by SSL.[Op: H S Sayer. Station located in London N20.]
G3 CGQ F W Tyler, 94 Alexandra Ave, Luton, LU3 1HJ [IO91SV, TL02]
G3 CHD S R Barker, Merston, Green Farm Ln, Shorne, Gravesend, DA12 3HL [JO01FK, TQ67]
G3 CHU A D Hunt, 110 Olton Boulevard East, Birmingham, B27 7ND [IO92CK, SP18]
G3 CIF J F Rogers, 52 Bosvean Gdns, Truro, TR1 3NQ [IO70LG, SW84]
GM3 CIG J E Priddy, 39 Hillfield Cres, Inverkeithing, KY11 1AH [IO86HA, NT18]
G3 CIK H D Romer, 96 Mortlake Rd, Kew, Richmond, TW9 4AS [IO91UL, TQ17]
G3 CIL M A Holley, 6586 196th St, Langley B.C., V2Y 1R3, Canada, X X
G3 CIM S A Denney, 52A Intwood Rd, Cringleford, Norwich, NR4 6AA [JO02OO, TG10]
G3 CIO Details withheld at licensee's request by SSL.[Correspondence to: Catterick ARC, HQ RSARS, Vimy Barracks, Catterick Garrison, N Yorks, DL9 3PS.]
GM3 CIX L J McDougall, 5 Cumnock Dr, Barrhead, Glasgow, G78 2HT [IO75TS, NS55]
G3 CJ E H Heaton-Jones, The Gateway, Farm Ln, Leckhampton, Cheltenham, GL53 0NN [IO81WV, SO92]
G3 CJD L F L Allen, 21 Inghead Rd, Slaithwaite, Huddersfield, HD7 5DS [IO93BP, SE01]
G3 CJG Maj J R Farr, Erma Cottage, 8 Julian Rd, Cadleigh Park, Ivybridge, PL21 9BU [IO80AJ, SX65]
G3 CJH P A H Smith, 37 Cobham Rd, Kingston upon Thames, KT1 3AE [IO91UJ, TQ16]
G3 CJI R B Miller, 47 Havering Cl, Clacton on Sea, CO15 4UX [JO01NT, TM11]
G3 CJJ P E Leventhall, 11A Chaldon Common Rd, Chaldon, Caterham, CR3 5DE [IO91WG, TQ35]
G3 CJP D E J Halfhide, Orchard Cottage, Five Lanes Rd, Marldon, Paignton, TQ3 1NQ [IO80EK, SX86]
G3 CJR A K Kerton, 8 Fabian Dr, Stoke Gifford, Bristol, BS12 6XN [IO81RM, ST68]
GW3 CKB R F Cashmore R.A.F.A.R.S. South Wales Group, 65 Michaelston Rd, Culverhouse Cross, Cardiff, CF5 4SX [IO81IL, ST17]
GI3 CKF G G R Mason, Crannagael Cranagil, 43 Ardress Rd, Portadown, Craigavon, BT62 1SE [IO64RK, H95]
GD3 CKO Capt I Morris, Ashley House, The Castleward Gree, Douglas, Isle of Man, IM2 5PS
G3 CKR Details withheld at licensee's request by SSL.[Correspondence to Warrington CG, PO Box 141, Warrington, Cheshire.]
G3 CLA J L Matthews, Mimosa, 53 Ham Shades Ln, Whitstable, CT5 1PA [JO01MI, TR16]
G3 CLC Details withheld at licensee's request by SSL.
G3 CLE J B Kennedy, Ashlea, 13 Highfield Rd, Malvern, WR14 1HR [IO82UD, SO74]
GM3 CLI Details withheld at licensee's request by SSL.
G3 CLK K J Vickery, 82 Copse Ave, West Wickham, BR4 9NP [IO91XI, TQ36]
G3 CLL J Willy, 13 Dennyview Rd, Abbots Leigh, Bristol, BS8 3RD [IO81QL, ST57]
G3 CLW W J Macinnes, 111 Primley Park, Paignton, TQ3 3JX [IO80FK, SX86]
G3 CMC G A Holland, 19 Peppercorn Caravan Park, The Baulk, Clapham, Bedford, MK41 6HD [IO92SD, TL05]
G3 CMH G W Davis Yeovil ARC, Broadview, East Lanes, Mudford, Yeovil, BA21 5SP [IO80QX, ST51]
G3 CMI J A Scott, The Old School House, Chalkpit Ln, Candlesby, Spilsby, Lincs, PE23 5SE [JO03CE, TF46]
G3 CMJ R A Titt, Royston, Roman Rond, Winterslow, Salisbury, Wilts, SP5 1QR [IO91EC, SU23]
G3 CML L H King, 48 Parc Du Bugnon, St Genis/Pouilly, Ain, France 01630
G3 CMT A H Hooke, 8 Warborough Rd, Churston Ferrers, Brixham, TQ5 0JY [IO80FJ, SX85]
G3 CMU H A Meyers, Cornerways, 2 Old Mill Ln, Wannock, Polegate, BN26 5NS [JO00CT, TQ50]
G3 CMY R W Rowsell, 20 Crawshay Dr, Emmer Green, Reading, RG4 8SX [IO91ML, SU77]
G3 CNA F C Potter, 18 Petworth Gdns, Raynes Park, London, SW20 0UH [IO91VJ, TQ26]
G3 CNC R E Molland, 62 Priory Rd, South Park, Reigate, RH2 8JB [IO91VF, TQ24]
G3 CNG R F P Luckman, 72 Binley Rd, Coventry, CV3 1FQ [IO92GJ, SP37]
GW3 CNM Details withheld at licensee's request by SSL.
G3 CNO A C Cake, 7 Wheatstone Rd, Southsea, PO4 0LJ [IO90LS, SZ69]
G3 CNW T L Fletcher, 1 Glencairn Ct, Lansdown Rd, Cheltenham, GL51 6QN [IO81WV, SO92]
G3 CNX G J Smith Grimsby ARS, 6 Fenby Cl, Great, Grimsby, DN37 9QJ [IO93WN, TA20]
G3 CO F R Howe Colchester Radio Amateurs, 29 Kingswood Rd, Colchester, CO4 5JX [JO01KV, TL92]
G3 COA M McKavney, 13 Dales Ln, Whitefield, Manchester, M45 7JN [IO83UN, SD80]
GM3 COB J Paterson, 5 Stornoway Cres, Cambusnethan, Wishaw, ML2 8XS [IO85BS, NS85]
G3 COF L F Sutton, 44 Kinnaird Ave, Bromley, BR1 4HQ [JO01AK, TQ37]
GW3 COI J Worthington, Penrhyn Bach, Bwlch Tocyn, Pwllheli, Gwynedd, LL53 7BU [IO72ST, SH32]
G3 COJ A H B Bower, 19 Chapel Rd, Flackwell Heath, High Wycombe, HP10 9AB [IO91PO, SU88]
G3 CON L W Crabbe, 15 Naseby House, Cromwell Rd, Cheltenham, GL52 5DT [IO81XV, SO92]
G3 COO F W Charrett, 8 Mavis Cres, Havant, PO9 2AE [IO90MU, SU70]
GM3 COQ D Oswald, 8 Redfield Rd, Montrose, DD10 8TW [IO86SR, NO75]
G3 COY V J Reynolds, 25 Yoxall Ave, Hartshill, Stoke on Trent, ST4 7JJ [IO83VA, SJ84]
G3 CPC R D Charlton, 7 St. Margarets Dr, Twickenham, TW1 1QL [IO91UL, TQ17]
G3 CPD A R C Durling, 46 Valley Dr, Great Sutton, South Wirral, L66 3PZ [IO83MG, SJ37]
G3 CPG L Damon, 1 Layton Ave, Malvern, WR14 2ND [IO82UC, SO74]
G3 CPH F H Hatt, 23 Ravens Cl, Enfield, EN1 3UR [IO91XP, TQ39]
G3 CPK Details withheld at licensee's request by SSL.
G3 CPN M J Stevens, 16 Golf Links Rd, Ferndown, BH22 8BY [IO90BT, SU00]
G3 CPS E C Gray, 23 Burrow Down, Eastbourne, BN20 8ST [JO00CS, TQ50]
G3 CPT D A Capp, 46 Stoke Rd, Bletchley, Milton Keynes, MK2 3AD [IO91PX, SP83]
G3 CQE W M Brennan, La Casa, Broughtons Dr, Misterton, Crewkerne, TA18 8LW [IO80OU, ST40]
G3 CQK J B Dodson, 74 Shelton Ln, Halesowen, B63 2XF [IO82XK, SO98]
G3 CQL M S Clarke, 9 Park Ln, Bulner Tye, Bulmer, Sudbury, CO10 7EQ [JO02IA, TL83]
G3 CQO G R Burch, 13 Moorhouse Est, Ashington, NE63 9LL [IO95FE, NZ28]
G3 CQQ Details withheld at licensee's request by SSL.
G3 CQR P R Burridge, 9 Quarr Dr, Sherborne, Dorset, DT9 4HZ [IO80RW, ST61]
G3 CQU K G Raffield, 113 Waddington Ave, Coulsdon, CR5 1QP [IO91WH, TQ35]
G3 CRC M Jackson Clacton ARC, 44 Dulwich Rd, Holland on Sea, Clacton on Sea, CO15 5NA [JO01OT, TM11]
G3 CRF A W Jones, 39 Campleshon Rd, Rainham, Gillingham, ME8 9LF [JO01HI, TQ86]
G3 CRH H H A Sanders, Mill Cottage, Mill Ln, Hammerwich, Walsall, West Midlands, WS7 0JR [IO92BP, SK00]
G3 CRJ B J Shaw, The Green, Staveley, Kendal, Cumbria, LA8 9NS [IO84OJ, SD49]
G3 CRP R J Pigou, 52 Rowan Rd, Bexleyheath, DA7 4BW [JO01BL, TQ47]
G3 CRR A E Glozier, 2 Cheddar Ct, Station Rd, Cheddar, BS27 3DT [IO81OG, ST45]
G3 CRS R Baker Royal Naval Radio Amateur Soci, Royal Naval Am Rad Society, HMS Collingwood, Fareham, Hants, PO14 1AS [IO90JU, SU50]
G3 CSA T B Saggerson Ellesmre P&D AR, 18 Ploughmans Way, Great Sutton, South Wirral, L66 2YJ [IO83MG, SJ37]
G3 CSC S J Roddan, 18 Muncaster Dr, Rainford, St. Helens, WA11 8NR [IO83OM, SD40]
G3 CSE C W Smith, 128 The Commons, Welwyn Garden City, AL7 4SB [IO91VS, TL21]
GI3 CSK J Sturrock, 4 Coutrai Park, Strabane, BT82 8HQ [IO64GT, H39]
G3 CSL W W Jones, 17 Spawell Cl, Lowton, Warrington, WA3 2TF [IO83RL, SJ69]
GM3 CSO A C Fulton, 43 Mount Charles Cres, Ayr, KA7 4PA [IO75QK, NS31]
G3 CSP E Brown, 117 Beck Rd, Sheffield, S5 0GE [IO93GK, SK39]
G3 CSR Details withheld at licensee's request by SSL.[Correspondence to: Civil Service ARS, Recreation Centre, Monck St, London, SW1P 2BL.]
G3 CSS W K Owen, Greenways, Burghill, Hereford, HR4 7RW [IO82OC, SO44]
G3 CST J I Moore, 6 Dartmouth St., Stafford, ST16 3TU [IO82WT, SJ92]
G3 CSY K A Hill, 30 Hestham Ave, Morecambe, LA4 4PZ [IO84NB, SD46]
GM3 CTG J E Hemphill, 31 Dundonald Rd, Kilmarnock, KA1 1RU [IO75SO, NS43]
G3 CTI J H Burr, 37 Layhams Rd, West Wickham, BR4 9HD [JO01AI, TQ36]
G3 CTP J W Swift, 20 Leighlands, Crawley, RH10 3DW [IO91WC, TQ23]
G3 CTQ H Westwell, 224 Dickson Rd, Blackpool, FY1 2JS [IO83LT, SD33]
G3 CTR R L Whorwell, 65 John Kennedy House, Rotherhithe Old Rd, London, SE16 2QF [IO91XL, TQ37]
G3 CTZ A Jones, 17 Oakley Way, Old Tupton, Chesterfield, S42 6JD [IO93GE, SK36]
G3 CU H F Knott, 15 Hampden Rd, Wantage, OX12 7DP [IO91GO, SU48]
G3 CUC Details withheld at licensee's request by SSL.[Op: W H Moore, Manderley, Church Brough, Kirkby Stephen, Cumbria, CA17 4EJ.]
G3 CUF H Ashworth, Ridgecroft, 97 Winchcombe Rd, Sedgeberrow, Evesham, WR11 6UA [IO92AB, SP03]
G3 CUI F R Ellory, Cleeve House, Norwood Ln, Iver, SL0 0EW [IO91RM, TQ08]
G3 CUN J Leonard, 122 Glenavon Rd, Highters Heath, Birmingham, B14 5BS [IO92BJ, SP07]
G3 CUR R Collette, 8A Woolwich Rd, Upper Belvedere, Kent, DA17 5EW [JO01BL, TQ47]
G3 CUY E F Paul, 91 Windmill Dr, Brighton, BN1 5HH [IO90WU, TQ20]
G3 CVF D Pidgeon, 6A Verlands Rd, Preston, Weymouth, DT3 6BY [IO80TP, SY78]
G3 CVG S Jackson, 22 Brook Rd, Morecambe, LA3 1AY [IO84NB, SD46]
G3 CVI B H Thwaites, 118 Baddow Hall Cres, Great Baddow, Chelmsford, CM2 7BU [JO01GR, TL70]
G3 CVK P Bolton, 50 Meadow Rd, Malvern Link, Malvern, WR14 2SD [IO82UC, SO74]
G3 CVM D N Gibson, View Point, Down Ln, Braunton, EX33 2LE [IO71WC, SS43]
G3 CVW R F Saunders, Virginia Cottage, 1 Churchfields, Sandiway, Northwich, CW8 2JS [IO83QF, SJ67]
G3 CVX J B Taylor, Sherbrook, Primrose Ln, Old Fallings, Wolverhampton, WV10 8RS [IO82WO, SJ90]
G3 CWC E C Lark, 10 Colman Ave, Stoke Holy Cross, Norwich, NR14 8NA [JO02PN, TG20]
G3 CWH R B Rogers, 107 Rotherham Rd, Coventry, CV6 4FH [IO92FK, SP38]
G3 CWI R Newstead, 27 Gleave Ave, Bollington, Macclesfield, SK10 5LX [IO83WH, SJ97]
G3 CWT F W Vale, 40 Ferry St., Stapenhill, Burton on Trent, DE15 9EY [IO92ET, SK22]
G3 CWV C Wallis, Wychwood, Snailswell Ln, Ickleford, Hitchin, SG5 3TP [IO91UX, TL13]
G3 CWW A W W Timme, 287 Gillroyd Ln, Linthwaite, Huddersfield, HD7 5SY [IO93BO, SE01]
GI3 CWY E S Wilson, Althammond, 57 Belfast Rd, Whitehead, Carrickfergus, BT38 9SP [IO74DR, J49]
G3 CXI P J Cooper, 11 Hardy Rd, Bishops Cleeve, Cheltenham, GL52 4BN [IO81XW, SO92]
G3 CXP R A Gill, 45 Biggin Ln, Ramsey, Huntingdon, PE17 1NB [IO92WK, TL28]
G3 CXR E H Clegg, 13 Selby, Abbey Park, Illingworth, Halifax, HX2 9LQ [IO93BS, SE02]
G3 CXT G H Clarke, 75 Paradise Ln, Hall Green, Birmingham, B28 0DZ [IO92BK, SP18]
G3 CYH L Austin, 12 Whittle Cl, Bawnmore Rd Bilto, Rugby, CV22 6JR [IO92II, SP47]
G3 CYI C A Green, 7 Denford Way, Wellingborough, NN8 5UB [IO92PH, SP86]
G3 CYL G J Bennett, 16 Coxheath Rd, Church Crookham, Fleet, GU13 0QJ [IO91NG, SU85]
G3 CYU J D Wilson, 1 Beeches Farm Rd, Crowborough, TN6 2NY [JO01CB, TQ53]
G3 CYX P F P Lambert, 11 Marlborough Cl, Musbury, Axminster, EX13 6AP [IO80LR, SY29]
G3 CZL R N Buckman, Heathfield, Hang Hill Rd, Bream, Lydney, GL15 6LQ [IO81RS, SO60]
G3 CZM G W K Chalwin, 11 Cormorant Pl, College Town, Sandhurst, Berks, GU47 0XY [IO91OI, SU86]
G3 CZO H Buckley, The Moss, 48 Park Mount Dr, Macclesfield, SK11 8NT [IO83WG, SJ97]
G3 CZU G Skinner Dorking & Dis Rd, Woodend, Oaks Ln, Mid Holmwood, Dorking, Surrey, RH5 4ES [IO91UE, TQ14]
G3 DAC A R Edwards, 55 Wordsworth Dr, Crewe, CW1 5JJ [IO83SC, SJ75]
G3 DAM Details withheld at licensee's request by SSL.
GM3 DAP A W Adam, 4A Scroggiehill, Almondbank, Perth, PH1 3NL [IO86FK, NO02]
G3 DAQ R Braithwaite, 32 Rupert Cres, Queniborough, Leicester, LE7 3TU [IO92LQ, SK61]
G3 DAV J D Waller, 26 Hermitage Way, Sleights, Whitby, YO22 5HG [IO94QK, NZ80]
G3 DBM Dr J A Blundell, Windmill, 7 Swanbourne Rd, Mursley, Milton Keynes, MK17 0JA [IO91OW, SP82]
G3 DBY J S Hallatt, 10 Queens Gr, Chorley, PR7 1JX [IO83QP, SD51]
G3 DCC E J King, 109 Marlborough Park Ave, Sidcup, DA15 9DY [JO01BK, TQ47]
G3 DCE F W F Humphries, Little Hayes, 1 Meadway, Sidmouth, EX10 9JA [IO80JQ, SY18]
G3 DCN F A Barrell, 4 Ethelwine Pl, The Cres, Abbots Langley, Herts, WD5 0DS [IO91TQ, TL00]
G3 DCO B D Coyne, 3 Brocks Way, Shiplake, Henley on Thames, RG9 3JG [IO91NM, SU77]
G3 DCS E H Chaudri, South Bank, 215A Woodbridge Rd, Ipswich, IP4 2QS [JO02OB, TM14]

G3 DCU Details withheld at licensee's request by SSL.[Op: W Shreuer. Correspondence via K1YZW.]
G3 DCV Details withheld at licensee's request by SSL.
G3 DCZ R G McDonald, 60 Dudley Dr, Morden, SM4 4RJ [IO91VJ, TQ26]
G3 DDA K W Dyson, 10 The Ridings, Cliftonville, Margate, CT9 3EJ [JO01RJ, TR37]
G3 DDH A Townend, 180 Anlaby Park Rd South, Hull, HU4 7BU [IO93TR, TA02]
G3 DDK E J Hartley, 85 Sands Ln, Lowestoft, NR32 3ET [JO02UL, TM59]
GM3 DDL J Jackson, 74 Cairngorm Cres, Paisley, PA2 8AW [IO75ST, NS46]
G3 DDN B N Tait, 23 Chymedden, Trebarwith Cres, Newquay, Cornwall, TR7 1DX [IO70LJ, SW86]
G3 DDS F Bergelin, 1 Church Green, Stanford in The Vale, Faringdon, SN7 8LQ [IO91FP, SU39]
G3 DDX G A Woodhouse, Sanderstead Cottage, 12 Sandows Ln, St. Ives, TR26 1QW [IO70GF, SW54]
GW3 DDY J E Sketch, 30 Dan yr Heol, Cyncoed, Cardiff, CF2 6JU [IO81JM, ST18]
G3 DEB T A Bennett, 4 West Way, Lymington, SO41 8DZ [IO90FS, SZ39]
GM3 DEE R P Russell, 11 Sunnylaw Dr, Brediland, Paisley, Renfrewshire, PA2 9NU [IO75SU, NS46]
G3 DEF B J Gealer, 27 Mns S Gdns, Evesham, WR11 6BX [IO92AC, SP04]
G3 DEJ T E Wiseman, 70 Dove House Ln, Solihull, B91 2EG [IO92CK, SP18]
G3 DEM J Page, Pendragon, 3 Farm Corner, Middleton on Sea, Bognor Regis, PO22 6LX [IO90QT, SU90]
G3 DEP H T Parsons, Little Croft, One Footways, Wootton Bridge, Ryde, Isle of Wight, PO33 4NQ [IO90JR, SZ59]
G3 DEQ H N Woodnutt, 128 Redlands Ln, Fareham, PO14 1HF [IO90JU, SU50]
GW3 DEX F N Howard, 7 John Lewis St., Hakin, Milford Haven, SA73 3HT [IO71LR, SM80]
G3 DEY E T Ford, 177 Latters Orchard, Old Rd, Wateringbury, Maidstone, ME18 5PR [JO01EG, TQ65]
G3 DFA D B Gaggs, 16 Handforth Gr, Old Hall Ln, Manchester, M13 0UH [IO83VK, SJ89]
G3 DFD M C Farley, Tandridgr, 82 Leas Rd, Warlingham, CR6 9LL [IO91XH, TQ35]
G3 DFE Details withheld at licensee's request by SSL.
G3 DFH S G Crow, Friarnin, Church Ln, Orleton, Ludlow, SY8 4HU [IO82PH, SO46]
G3 DFL G C Hill, 18 Lightwoods Hill, Bearwood, Smethwick, Warley, B67 5EA [IO92AL, SP08]
GM3 DGD J Dickie, 20 Murdiston Ave, Callander, FK17 8AY [IO76VF, NN60]
G3 DGH D G Hardcastle, 829 Carrol, Harlingen, Texas 78550, United States of America
G3 DGR B F Goodger, Cam-Or-Nant, Nant y Caws, Morda, Oswestry, Shropsire, SY10 9AP [IO82LU, SJ22]
GW3 DGT Details withheld at licensee's request by SSL.
G3 DGW D Early, The White House, Potterspury Lodge, Towcester, Northants, NN12 7LL [IO92MC, SP74]
G3 DHB Brig D H Baynham, 2 Byrons Cl, Bishops Waltham, Hants, SO3 1RS [IO90HW, SU41]
G3 DHE N Nelson, Tooley Park, Peckleton Ln, Earlshilton, Leics, LE9 7GH [IO92IN, SP49]
G3 DHU R Smith, Grasmere, Eddyfield Rd, Oxspring, Sheffield, South Yorks, S30 6YH [IO93GJ, SK38]
G3 DHY E J Edwards, 9 Bradworth Cl, Osgodby, Scarborough, YO11 3PZ [IO94TF, TA08]
G3 DIC C H Bullivant, 7 Rackclose Park, Chard, TA20 1RD [IO80MV, ST30]
G3 DID J H L Doyle, 16 Stow Gr, Hodge Hill, Birmingham, B36 8AY [IO92CM, SP18]
GM3 DIE T W Dickson, 91 Milton Rd West, Edinburgh, EH15 1RA [IO85KW, NT27]
G3 DIF P P J Butler, 9 Greencroft Gdns, Cayton, Scarborough, YO11 3SE [IO94TF, TA08]
G3 DII J Bell, Ilex, Tattershall Rd, Woodhall Spa, LN10 6TL [IO93VD, TF16]
GM3 DIN A A Clark, 11 Regent Park Sq, Glasgow, G41 2AF [IO75UU, NS56]
G3 DIT Details withheld at licensee's request by SSL.[Details c/o G3JZV QTHR.]
GW3 DJA A P J Mould, Llwyn-y-Fesen, Maesquarrie Rd, Betws, Ammanford, Dyfed, SA18 2PE [IO81AS, SN61]
G3 DJE G E Lumley, Guildy Hall, Muker, Richmond, North Yorks, DL11 6QG [IO84WJ, SD99]
G3 DJJ E T Whetton, 38 Beechcroft Cres, Streetly, Sutton Coldfield, B74 3SH [IO92BN, SP09]
G3 DJK K E Rosier, 31 Croindene Rd, London, SW16 5RE [IO91WJ, TQ36]
GM3 DJS J F McCreight, 40 Auchenharvie Rd, Saltcoats, KA21 5RL [IO75OP, NS24]
GM3 DJT J M Mitchell, 38 Marchbank Gdns, Balerno, EH14 7ET [IO85HV, NT16]
G3 DJY Maj P C Akass, 21 Ashley Rd, Farnborough, GU14 7EZ [IO91PG, SU85]
G3 DKB T A W Vallard, 10 Calder Ave, Greenford, UB6 8JQ [IO91TM, TQ18]
G3 DKD G V Taylor, 2 Hutton Terr, Bradford, BD2 2DY [IO93DT, SE13]
GW3 DKE W J Saunderson, 54 Heol y Waun, Pontlliw, Swansea, SA4 1EW [IO71XQ, SN60]
G3 DKH E H Mortimore, 4 Newington Cl, Coundon, Coventry, CV6 1PP [IO92FK, SP38]
G3 DKJ Details withheld at licensee's request by SSL.
G3 DKN Details withheld at licensee's request by SSL.
G3 DKO J W Stevenson, 27 Priory Cl, Sporle, Kings Lynn, PE32 2DU [JO02IP, TF81]
G3 DKR K E Roberts, The Bungalow, 1A Eversley Cres, Winchmore Hill, London, N21 1EL [IO91WP, TQ39]
GM3 DKW Dr J Hossack, 2 Panton Green, Livingston, EH54 8RY [IO85FV, NT06]
G3 DLG H G Curtis, 8 Greenhill Ct, Melcombe Ave, Weymouth, DT4 7TE [IO80SO, SY68]
G3 DLH P Evans, Sunlea, Wheal Speed, Carbis Bay, St. Ives, TR26 2PT [IO70GE, SW53]
G3 DLO J E Whitington, Syke House, Church St., Broughton in Furness, LA20 6ER [IO84JG, SD28]
G3 DMC A J Birkinshaw, Brincliffe, 219 Teagues Cres, Trench, Telford, TF2 6RA [IO82SR, SJ61]
G3 DMO C Earnshaw, 35 Rogersfield, Langho, Blackburn, BB6 8HB [IO83ST, SD73]
G3 DMQ Dr A S Curry, 24 Lima Ct, Bath Rd, Reading, RG1 6NG [IO91MK, SU77]
GW3 DMV M R Hewitt, 54 Abbey Rd, Rhos on Sea, Colwyn Bay, LL28 4NU [IO83DH, SH88]
G3 DND Details withheld at licensee's request by SSL.[Station located near Leiston.]
G3 DNE Details withheld at licensee's request by SSL.
G3 DNF Dr G J Bennett, 52 Whinmoor Cres, Leeds, LS14 1EW [IO93GU, SE33]
G3 DNH J A Spicer, 291 Green Ln, Coventry, CV3 6EH [IO92FJ, SP37]
G3 DNJ G F Weller, 54 Tadorne Rd, Tadworth, KT20 5TF [IO91VH, TQ25]
G3 DNK Details withheld at licensee's request by SSL.
G3 DNN G H Saville, 2 Gaskell Cl, Littleborough, OL15 8EB [IO83WP, SD91]
G3 DNQ Details withheld at licensee's request by SSL.[Op: D H Maclean. Station located near Rickmansworth.]
G3 DNR P O'Brien, 17 Rumfields Rd, Broadstairs, CT10 2PJ [JO01RI, TR36]
G3 DNS N C King, 31 Great Norwood St., Cheltenham, GL50 2AW [IO81XV, SO92]
GM3 DNV M J George, Hillhead Cookney, Netherley, Stonehaven, Kincardineshire, AB39 3SA [IO87VA, NO89]
GM3 DOD A M Murray, 50 Castlepark Dr, Fairlie, Largs, KA29 0DG [IO75NS, NS25]
G3 DOI G E Turvey, Gr House, 85 Park Gr, Barnsley, S70 1QB [IO93GN, SE30]
G3 DOJ W J Omer, 81 Eastfield Rd, Burnham, Slough, SL1 7EL [IO91PM, SU98]
G3 DON D W Cock, 6 Dover Rd East, Northfleet, Gravesend, DA11 0RG [JO01EK, TQ67]
G3 DOT J S Allen, 4 Philip Ave, Waltham, Grimsby, DN37 0QD [IO93WM, TA20]
G3 DOV D C Dove, 3 Walnut Gr, Watton, Thetford, IP25 6EY [JO02JN, TF90]
G3 DOX Details withheld at licensee's request by SSL.
G3 DPG Details withheld at licensee's request by SSL.
GM3 DPL E G Morgan, 16 Norval Pl, Rosyth, Dunfermline, KY11 2RJ [IO86GA, NT18]
G3 DPM D A Cooknell, 23 The Hyde, Winchcombe, Cheltenham, GL54 5QR [IO91AW, SP02]
G3 DPR L K Ayre, 15 Sea Rd, Barton on Sea, New Milton, BH25 7NA [IO90DR, SZ29]
G3 DPS J Cooper, 1 Kings Pond Bungalows, Ashdell Rd, Alton, Hants, GU34 2TA [IO91MD, SU73]
G3 DPW R L Knight, 35 Sussex Rd, South Croydon, CR2 7DB [IO91WI, TQ36]
G3 DPX C E Pollard, 3 Hillside, Sidbury, Sidmouth, Devon, EX10 0QZ [IO80JR, SY19]
G3 DQC J L Salter, 5 Anson Cl, Wheatley, Oxford, OX33 1YD [IO91KR, SP60]
G3 DQG G T Mortimer, Stoneleigh, 122 Roker Ln, Pudsey, LS28 9ND [IO93ES, SE23]
G3 DQL G E Sumption, Private Mailbag 274, Serekunda, The Gambia, West Africa
G3 DQQ D Winterburn, 47 Hilda Ave, Tottington, Bury, BL8 3JE [IO83TO, SD71]
G3 DQT J Ayres, 8 Cornfield Rd, Seaford, BN25 1SW [JO00BS, TV49]
G3 DQW B W Vaughan Peterborough Radio and Electro, 368 Fulbridge Rd, Paston, Peterborough, PE4 6SJ [IO92UO, TF10]
G3 DQY J Vaughan, 20 The Thatchings, Polegate, BN26 5DT [JO00CT, TQ50]
G3 DRB R S Trickey, 31 Pensby Ave, Chester, CH2 2DD [IO83NE, SJ46]
GW3 DRK W J Smale, 23 John St., Porth, CF39 9SD [IO81HO, ST09]
G3 DRL J M Willies, 10 Thirlby Rd, North Walsham, NR28 9BA [JO02QT, TG22]
G3 DRN E G Allen, 30 Bodnant Gdns, Wimbledon, London, SW20 0UD [IO91VJ, TQ26]
G3 DRP W Fletcher, 23 Wesley Pl, Stapleford, Nottingham, NG9 8DP [IO92IW, SK43]
GW3 DRV O D Jones, 4 Chalybeate Gdns, Aberaeron, SA46 0DL [IO72UF, SN46]
G3 DSC Details withheld at licensee's request by SSL.
GM3 DSD A R Trayler, 52 Albert Ave, Glasgow, G42 8RD [IO75UU, NS56]
G3 DSK R A Lord, 22 Elizabeth Cres, East Grinstead, RH19 3JA [IO91XD, TQ33]
G3 DSS Maj G S S Symons, Melbury View, Manston, Sturminster Newton, Dorset, DT10 1HB [IO80UW, ST81]
G3 DSV R W P Wilson, 4 Brookside, Pathfinder Village, Exeter, EX6 6DE [IO80ER, SX89]
G3 DSX Details withheld at licensee's request by SSL.
G3 DSZ A F C Kent, 23 Pagehall Cl, Scartho, Grimsby, DN33 2HF [IO93WM, TA20]
G3 DTA G F Nottingham, Carmel House, 59 Stockton Ln, York, YO3 0BP [IO93LX, SE65]
G3 DTG E W Clary, Flat 41, Bromford Ct, Houldey Rd, Birmingham, B31 3HJ [IO92AJ, SP07]
G3 DTJ C H Ellison, 8 Penketh Rd, Great Sankey, Warrington, WA5 2TA [IO83QJ, SJ58]
G3 DTP E Jackson, 125 Bolton Rd, Rochdale, OL11 3LP [IO83VO, SD81]
G3 DTU Details withheld at licensee's request by SSL.
G3 DTX I A Duck, Chenies, Loudhams Wood Ln, Chalfont St. Giles, HP8 4AR [IO91RP, SU99]
G3 DUH C H Belsham, 48 Clent View Rd, Norton, Stourbridge, DY8 3JJ [IO82VK, SO88]
G3 DUI H Osbaldeston, 126 Trafalgar St., Ashton under Lyne, OL7 0HD [IO83WL, SJ99]
G3 DUL H H Pickering, 52 Springwood Gdns, Woodthorpe, Nottingham, NG5 4HE [IO92KX, SK54]
GM3 DUM G Ireland, 4 Bush Terr, Musselburgh, EH21 6DF [IO85LW, NT37]
G3 DUN K F L Lansdowne, 44 Weatherhill Rd, Smallfield, Horley, RH6 9NQ [IO91WE, TQ34]
GM3 DUS W B Storry, Tigh Na Cladaich, Muasdale By Tarbert, Argyll, PA29 6XD [IO75DO, NR64]
G3 DUW R Hodgson, The Shealing, Forest Moor Dr, Knaresborough, HG5 8JT [IO93GX, SE35]
G3 DUZ B Froggatt, 1 Plantation Rd, West Wellow, Romsey, Hants, SO51 6DF [IO90EW, SU21]
G3 DVA J H Cancellor, 59 London Rd, Halesworth, IP19 8LS [JO02SI, TM37]
GJ3 DVC N R Collier Webb Jersey A.R.S., La Bruyere-le-Clos Des Landes, St. Brelade, Jersey, JE3 8DY
G3 DVF A H Cain, 18 Oaky Balks, Alnwick, NE66 2QE [IO95DJ, NU11]
G3 DVK F F Oldfield, 42 Symonds Ave, Rawmarsh, Rotherham, S62 7LP [IO93HL, SK49]
G3 DVL F Harrop, 15 Keymer Rd, Hollingbury, Brighton, BN1 8FB [IO90WU, TQ30]
G3 DVQ R H Pounder, Fair Wind, 44 Hartley Hill, Purley, CR8 4EN [IO91WH, TQ35]
G3 DVV J O Brown, Boulters Barn Cottage, Churchill Rd, near Chipping, Norton, Oxon, OX7 5UT [IO91FW, SP22]
G3 DVY L Baty, 5 Dipton Cl, Eastwood Grange Est, Hexham, NE46 1UG [IO84XX, NY96]
G3 DWI G Lusty, The Martins, High St., Chipping Campden, GL55 6AG [IO92CB, SP13]
G3 DWQ G Lancefield, 191 Higher Walton Rd, Walton-le-Dale, Preston, PR5 4HS [IO83QR, SD52]
G3 DWS W E Massingham, 173 Black Haynes Rd, Birmingham, B29 4RE [IO92AK, SP08]
G3 DWW Details withheld at licensee's request by SSL.[Station located in London SW20.]
G3 DXB R Gladwell, 14 Perne Rd, Cambridge, CB1 3RT [JO02BE, TL45]
G3 DXD D D Rolph, 2 Victoria Ct, Victoria Rd, Marlow, SL7 1DR [IO91ON, SU88]
G3 DXJ T H Holbert, 92 Conway Dr, Shepshed, Loughborough, LE12 9PP [IO92IS, SK41]
G3 DXQ A J Adams, Brooklands, Chestnut Garth, Roos Hall North, Humberside, HU12 0LE [IO93XS, TA23]
G3 DXS L H T Butler, 10 Harris Cl, Denton, Manchester, M34 2PU [IO83WK, SJ99]
GI3 DXU H E Richards, 3 Kinnegar Rd, Finaghy, Belfast, BT10 0DP [IO74AN, J36]
G3 DXY B O Leach, 50 Merevale Rd, Gloucester, GL2 0QY [IO81VU, SO81]
G3 DXZ C F Fletcher, 12 Park Cres, Retford, DN22 6UF [IO93MH, SK78]
G3 DYH S J Reynolds, Merrydown, Cottage Rd, Wigston Magna, Leics, LE18 3SA [IO92KN, SP69]
G3 DYO N G Alder, Greenwoods, Eastfield Rd, Ross on Wye, HR9 5JY [IO81RV, SO52]
GM3 DZB A Duncan, Drumduan, Bellevue Rd, Banff, AB45 1BJ [IO87RP, NJ66]
GW3 DZJ F F R Pardy, 5 y Bryn, Glan Conwy, Colwyn Bay, Clwyd, LL28 5NJ [IO83CG, SH87]
G3 DZS H W Fudge, 6 Mounters Cl, Marnhull, Sturminster Newton, DT10 1NT [IO80UX, ST71]
G3 DZT Details withheld at licensee's request by SSL.[Op: J H Beamand. Station located near Lichfield.]
G3 DZV G Greaves, Ivy House Farm, Fairfield Rd, Poulton-le-Fylde, FY6 8DN [IO83MT, SD33]
G3 DZW S T Chrees, 44 Oakleigh Rd St, New Southgate, London, N11 1LA [IO91WO, TQ29]
G3 EAE G A Billington, 75 Mount Vernon Rd, Barnsley, S70 4DW [IO93GM, SE30]
GM3 EAK R Macfarlane, PO Box 472, Blantyre, Malawi
G3 EAO V D Bullett, 38 Sidney Rd, Walton on Thames, KT12 2LY [IO91TJ, TQ16]
G3 EAT W H C Burden, 102 Westhill Rd, Wyke Regis, Weymouth, DT4 9NF [IO80SO, SY67]
G3 EAY D M J Wood, 45 Pilgrim Cl, Great Chesterford, Saffron Walden, CB10 1QG [JO02CB, TL54]
G3 EBE Details withheld at licensee's request by SSL.
G3 EBG W H Reckitt, Fairview, 32 Hardwicke Fields, Haddenham, Ely, CB6 3TW [JO02BI, TL47]
G3 EBH G C Newby, St. Minver, 25 Sudbrooke Ln, Nettleham, Lincoln, LN2 2RW [IO93SG, TF07]
G3 EBJ T W Chandler, Temerloh, South Green, Mattishall, Dereham, NR20 3JT [JO02MP, TG01]
G3 EBK B B Gale, 13 Hill Way, Ashley Heath, Ringwood, BH24 2HZ [IO90BU, SU10]
G3 EBL H G Baker, 51 Oakwood, Leam Ln Est, Gateshead, NE10 8LU [IO94FW, NZ25]
G3 EBO R G Garland, 3 Spy Cl, Lytchett Matravers, Poole, BH16 6DQ [IO80XS, SY99]
G3 EBP P E R Courcoux, 75 Parkside Dr, Watford, WD1 3AU [IO91TP, TQ09]
G3 EBV S B Squire, Leefield, 4 Little Green Ln, Croxley Green, Rickmansworth, WD3 3JQ [IO91SP, TQ09]
GJ3 ECC A G Martin, La Cachette, Les Fourneaux, Rue de La Corbiere, St Brelade, Jersey, JE3 8HP
GW3 ECH R J Price, Llanwern, Brecon, Powys, LD3 7UW [IO81IW, SO12]
G3 ECI D W McKay, 120 Dell Rd, Oulton Broad, Lowestoft, NR33 9NT [JO02UL, TM59]
G3 ECM P W Bowles, 29 Coleman Ave, Hove, BN3 5ND [IO90VT, TQ20]
G3 ECP J E Brown, Manor Cottage, 2 The Maltings, Alconbury, Huntingdon, PE17 5DZ [IO92UI, TL17]
GI3 ECQ G F McGarry, 19 Largy Rd, Crumlin, BT29 4RN [IO64UP, J17]
G3 ECS K J Ottrey, 51 Priory Ct Rd, Westbury on Trym, Bristol, BS9 4DB [IO81QL, ST57]
G3 EDD B D A Armstrong, 39 Angle End, Great Wilbraham, Cambridge, CB1 5JG [JO02DE, TL55]
G3 EDK J Taylor, 15 Butlers Cl, Aston-le-Walls, Daventry, NN11 6UH [IO92ID, SP45]
G3 EDM Details withheld at licensee's request by SSL.
G3 EDS K G Perkins, Sarnen, Croft Rd, Evesham, Worcs, WR11 4NE [IO92AC, SP04]
G3 EDT J E Rickaby, 19 Hall Dr, Cropwell Bishop, Nottingham, NG12 3DT [IO92MV, SK63]
G3 EDV J Smale, 87 Farmcombe Rd, Tunbridge Wells, TN2 5DQ [JO01DC, TQ53]
G3 EDW P R Golledge, Greystones, Shyners Terr, Merriott, TA16 5NS [IO80OV, ST41]
G3 EDX M N Perrins, 28 Rylands Ave, Gilstead, Bingley, BD16 3NJ [IO93CU, SE13]
GM3 EDZ T P Hughes, 8 Ossian Ave, Ralston, Paisley, PA1 3AY [IO75TU, NS56]
G3 EEH Dr J G Watkinson, The Moorings, 63 Ruffa Ln, Pickering, YO18 7HN [IO94OF, SE88]
G3 EEL L Critchley, Park House, 27 Park Cres, Peterborough, Cambs, PE1 4DX [IO92VO, TF10]
G3 EEO G L Jackson Nunsfield Hse Rd, 2 Franklyn Dr, Alvaston, Derby, DE24 0FR [IO92GV, SK33]
G3 EEQ K C Gill, 27 Sea View Gdns, Sunderland, SR6 9PN [IO94HW, NZ45]
G3 EEW F J Heasman, 1 St. Johns Church Rd, Folkestone, CT19 5BQ [JO01OC, TR23]
G3 EEZ A Wakeman, 1 Kendal Cl, Aldersley, Wolverhampton, WV6 9LD [IO82WO, SJ80]
G3 EFB Details withheld at licensee's request by SSL.
GD3 EFD M A Thompson, Thie Shielin, School House Rd, Dreemskerry, Ramsey, IM7 1BJ
G3 EFE A R Bryant, 14 Cherry Tree Cl, Hughenden Valley, High Wycombe, HP14 4LP [IO91OQ, SU89]
GM3 EFH G K Syme, 9 Maitland St., Tayport, DD6 9DL [IO86NK, NO42]
G3 EFI P J Powell, 16 Whowell Fold, Halliwell, Bolton, BL1 8BT [IO83SO, SD71]
G3 EFK W T Clegg, Riverside, 38 Thornicks, Winfrith Newburgh, Dorchester, DT2 8JZ [IO80UP, SY88]
G3 EFL W H Preston, 8 Pencraig View, Greytree, Ross on Wye, HR9 7JR [IO81QW, SO52]
G3 EFP J C Pennell, 43 Harlyn Dr, Northwood Hills, Pinner, HA5 2DF [IO91TO, TQ18]
G3 EFR Details withheld at licensee's request by SSL.
G3 EFS W H Borland, 25 Broadoaks Way, Bromley, BR2 0UA [JO01AJ, TQ36]
G3 EFX C D Friel Rad Soc Harrow, 5 Windmill Hill, Ruislip, HA4 8QF [IO91TN, TQ08]
G3 EFY T W A Smith, Trevose, 1 Sussex Cl, Exeter, EX4 1LP [IO80FR, SX99]
G3 EFZ O Postle, 19 Laburnum Gr, Whitby, South Wirral, L66 2PD [IO83NG, SJ37]
G3 EGC J V Hoban, 13 Druids Cl, Egerton, Bolton, BL7 9RF [IO83SP, SD71]
G3 EGF T Kellett, Braville, St. Ives Rd, Leadgate, Consett, DH8 7SJ [IO94CU, NZ15]
G3 EGQ D J Griffen, 37 Trelissick Fields, Hayle, TR27 6HZ [IO70GE, SW53]
G3 EGS R W Collett, 121 Bournville Ln, Birmingham, B30 1LH [IO92AK, SP08]
GM3 EGU A Mercer, 183 Townhill Rd, Dunfermline, KY12 0DQ [IO86GC, NT18]
G3 EGV R Staniforth, 26 Winslow Rd, Preston, Weymouth, DT3 6NE [IO80TP, SY78]
G3 EGX L Roberts, 18 Croxteth Ave, Wallasey, L44 5UL [IO83LK, SJ39]
G3 EGY Details withheld at licensee's request by SSL.
G3 EHA G F Hendriksen, 2 Highcroft Cres, Leamington Spa, CV32 6BN [IO92FH, SP36]
G3 EHE K J Mather, 10 The Martlets, Rustington, Littlehampton, BN16 2TY [IO90RT, TQ00]
G3 EHG Details withheld at licensee's request by SSL.[Op: R V Jordan, Dulverton, Nurton Park, Pattingham, Wolverhampton, WV6 7AB.]
G3 EHM K H Parkes, 41 Golborn Ave, Meirheath, Stoke on Trent, ST3 7JQ [IO82WW, SJ93]
GW3 EHN J O Thomas, 76 Waun Rd, Loughor, Swansea, SA4 6QN [IO71XP, SS59]
G3 EHP J W Wilmot, Mixtow Farm House, Mixtow, Lanteglos-By-Fowley, Cornwall, PL23 1NB [IO70QI, SX15]
G3 EHQ H E Bone, 2 Waterville Gdns, Orton Waterville, Peterborough, PE2 5LG [IO92UN, TL19]
G3 EHT J Tremain, High Lanes, St Issey, Waderbridge, Cornwall, PL27 7RY [IO70MM, SW97]
G3 EHV F R Tipping, 6 Wealdview Rd, Heathfield, TN21 0XA [JO00CX, TQ52]
G3 EHW A S Watkins, 19 Barrow Gr, Sittingbourne, ME10 1LB [JO01II, TQ86]
G3 EHZ Details withheld at licensee's request by SSL.
G3 EID T J Stephenson, 23 Devonshire St., Stockton on Tees, TS18 3QQ [IO94IN, NZ41]
G3 EIE L Bergna, 23 Cranley Dr, Ruislip, HA4 6BZ [IO91TN, TQ08]
G3 EIG Details withheld at licensee's request by SSL.[Op: C S C Colson. Station located near Ringwood.]
G3 EIW R L Halls, 10 Brookmead Way, Langstone, Havant, PO9 1RT [IO90MU, SU70]
G3 EIX P J Naish, 25 Arkena Ave, Epping, Nsw 2121, Australia
GW3 EIZ C S S Lyon, Ardraeth, Malltraeth, Bodorgan, Gwynedd, LL62 5AW [IO73TE, SH36]
G3 EJC D E Flowers, 3 Seymour Cl, East Molesey, KT8 0JY [IO91TJ, TQ16]
G3 EJD Details withheld at licensee's request by SSL.
G3 EJF J E Hodgkins, Bridge House, Hunton, near Bedale, North Yorks, DL8 1PX [IO94DH, SE19]
GU3 EJL S Green, Valongis, 9 PO Box, Alderney, Guernsey, GY9 3AF
G3 EJP G B Osborn, 136 All Saints Rd, Kings Heath, Birmingham, B14 6AT [IO92BK, SP08]
GW3 EJR J B Armstrong, Mirianog, 1 Brynbedw, Llechryd, Cardigan, SA43 2NJ [IO72QB, SN24]
G3 EJV R Hargreaves, 4 Tower Hill, Clitheroe, BB7 1PD [IO83TV, SD74]
G3 EKD A A H Sparrow, Craiglynne, Far Westrip, Stroud, Glos, GL6 6HE [IO81US, SO80]
G3 EKE L A F Stockley, PO Box 621, Sisters, Oregon 97759, U.S.A., X X
G3 EKG Details withheld at licensee's request by SSL.
G3 EKI R D Hunt, 9 Cemmaes Ct Rd, Hemel Hempstead, HP1 1ST [IO91SS, TL00]
G3 EKJ H F Mattacks, Fieldfare, Eastbourne Rd, Halland, Lewes, BN8 6PS [JO00BW, TQ51]
G3 EKL R A Webb, 3 Hillcrest, Scotton, Catterick Garrison, DL9 3NJ [IO94DI, SE19]
G3 EKT J H Saynor obo Royal Air Force ARS Co Du, 28 Lune Rd, Norton, Stockton on Tees, TS20 1AZ [IO94IO, NZ42]
G3 EKW S E Williams ARC of Nottm, 304 The Wells Rd, Mapperley, Nottingham, NG3 3AA [IO92KX, SK54]
G3 EKX N J Birkett, 42 Halvarras Rd, Playing Pl Village, Playing Pl, Truro, TR3 6HD [IO70LF, SW84]
G3 ELF F W Malpass, 16 Lower Dorrington Terr, London Rd, Stroud, GL5 2AR [IO81VR, SO80]
G3 ELH P R A Dolphin, 3 Buckmore Ave, Petersfield, GU32 2EF [IO91MA, SU72]
G3 ELI C Granger, 8 Bury End, Pirton, Hitchin, SG5 3QB [IO91UX, TL13]
G3 ELQ T G Warburton, 24 Esplanade, Sterte, Polle, Dorset, BH15 2BA [IO90AR, SZ09]
G3 ELS B Rudd, Orchard Bungalow, The Holloway, Droitwich, WR9 7AH [IO82WG, SO96]
G3 ELV G A Vallely RAF Henlow AR & E Club, 2 Kayser Ct, Biggleswade, SG18 8BG [IO92UC, TL14]
G3 ELW W J Roscrow, 195 Western Rd, Leigh on Sea, SS9 2PQ [JO01HN, TQ88]
G3 ELY D Parker, 147 Moorside St., Droylsden, Manchester, M43 7HQ [IO83WL, SJ99]
G3 ELZ F R Peterson, 129 Welholme Ave, Great, Grimsby, DN32 0BP [IO93WN, TA20]

G3 EMF P G Lewis, 49 Estridge Cl, Lowford, Bursledon, Southampton, SO31 8FN [IO90IV, SU41]
G3 EMG Col J T Rogers, Field Farm House, The Drove, Sleaford, NG34 8JQ [IO92SX, TF04]
GW3 EMI M Price Hopkins, Hafan Deg, 2 Waterfall Rd, Dyserth, Rhyl, LL18 6ET [IO83GH, SJ07]
G3 EMJ A J Smith, 9 Lindford Cl, Oakwood, Derby, DE21 4TA [IO92GW, SK33]
G3 EMK T E Price, Caorle, 26 Low Habberley, Kidderminster, Worcs, DY11 5RA [IO82UJ, SO87]
GJ3 EML J H E Watson, 6 Portelet Dr, La Route de Noirmont, St. Brelade, Jersey, JE3 8JY
G3 EMY R S Moreton, 91 Umberslade Rd, Selly Oak, Birmingham, B29 7SB [IO92AK, SP08]
G3 ENB W E Gates, 16 High Mill Dr, Scarborough, YO12 6RN [IO94SH, TA09]
GM3 END K Callow, 40 Glenmavis Dr, Bathgate, EH48 4DQ [IO85EV, NS96]
G3 ENG Details withheld at licensee's request by SSL.
G3 ENI Cdr A J R Pegler, Brook House, Forest Cl, East Horsley, Leatherhead, KT24 5BU [IO91SG, TQ15]
GM3 ENJ K Street, Foinaven, 3 York Pl, Dunfermline, KY12 0DA [IO86GB, NT18]
G3 ENL S May, Flat11 Egerton Ct, Egerton Rd, Davenport, Stockport, Ches, SK3 8SR [IO83WJ, SJ98]
GW3 ENN G W H King, 29 Albany Ct, Beach Rd, Penarth, CF64 1JU [IO81JK, ST17]
G3 ENO R Green, Curlews, Curlew Dr, West Charleton, Kingsbridge, TQ7 2AA [IO80CG, SX74]
G3 ENR J F Hucklebridge, Pound Cottage, Poor Hill, Farmborough, Bath, BA3 1AP [IO81SI, ST66]
G3 ENV P H Poole, 2C The Avenue, Hatch End, Pinner, HA5 4EP [IO91TO, TQ19]
G3 ENX D M Webber, 3 Shirburn Rd, Crownhill, Plymouth, PL6 5PG [IO70WJ, SX45]
G3 ENZ A Johnson, 9 Main Rd, Smalley, Ilkeston, DE7 6EE [IO93HA, SK44]
GM3 EOB C H Merrilees, 6 Spoutwells Dr, Scone, Perth, PH2 6RR [IO86HK, NO12]
G3 EON D G King, 16 Bartholomew Way, Chester, CH4 7RJ [IO83NE, SJ46]
G3 EOO J Hamlett, 23 Riddings Rd, Timperley, Altrincham, WA15 6BW [IO83UJ, SJ78]
GW3 EOP S Roberts Port Talbot ARC, 70 Cimla Rd, Neath, SA11 3TR [IO81CP, SS79]
GW3 EPF P J Curtis, 19 Heol y Nant, Whitchurch, Cardiff, CF4 6BS [IO81JM, ST18]
G3 EPG M N Fletcher, 13 Park Ave, Cheadle, Stoke on Trent, ST10 1LZ [IO92AX, SK04]
G3 EPK R L S Harrison, The Red House, Staines Green, Hertford, Herts, SG14 2LN [IO91WS, TL21]
GW3 EPN Details withheld at licensee's request by SSL.
G3 EPO K I Procter, 11 Boucher Rd, Budleigh Salterton, EX9 6JF [IO80IP, SY08]
G3 EPP E W Trevitt, Vine Ct, Portsmouth Rd, Milford, Godalming, GU8 5HJ [IO91QE, SU94]
G3 EPV R D Emes, Lower Winslow, Portway, Burghill, Hereford, Herefordshire, HR4 8NG [IO82PC, SO44]
G3 EQF P E Templeman, 78 Little Barn Ln, Mansfield, NG18 3JJ [IO93JD, SK56]
G3 EQJ J F Acquier, The Shepherds Cottage, 38 Chapel St., Stoke By Clare, Sudbury, CO10 8HS [JO02GB, TL74]
GW3 EQL G Haring, Uplands, Oak Ln, Machen, Newport, Gwent, NP1 8SS [IO81KO, ST28]
G3 EQM Dr J A Theobald, 34 Barnfield Rd, Exeter, EX1 1RX [IO80FR, SX99]
G3 EQX Capt J L Rowe, Barling Rd, Barling, Great Wakering, Southend on Sea, Essex, SS3 0LZ [JO01JN, TQ98]
GM3 EQZ J Solly, Fortissat View, Salsburgh, Shotts, Lanarkshire, ML7 4NS [IO85BU, NS86]
G3 ERA J Wood, 42 High St., Steyning, BN44 3YE [IO90UV, TQ11]
G3 ERD J Anthony Derby Dist ARS, 77 Brayfield Rd, Littleover, Derby, DE23 6GT [IO92FV, SK33]
G3 ERO S E C Fryer, 10 Boldre Cl, New Milton, BH25 7TH [IO90DR, SZ29]
G3 ERR J A W Edwards, 1 Braemar Gdns, Colindale, London, NW9 5LA [IO91UO, TQ29]
G3 ESA J H Oakes, 21 Sandiway, Knutsford, WA16 8BU [IO83TH, SJ77]
G3 ESB A D Hitchcock, 38 West Rd, Spondon, Derby, DE21 7AB [IO92HW, SK33]
G3 ESF A R Harrower, The Magnolias, 30 Barn Cl, Crewkerne, TA18 8BL [IO80OV, ST40]
G3 ESK L A Potter, Hive End, 2 Linden Dr, Chatteris, Cambs, PE16 6DZ [JO02AK, TL38]
G3 ESL R E Playle, 4 Theobalds Ct, Theobalds Rd, Leigh on Sea, SS9 2ND [JO01HN, TQ88]
G3 ESP W Farrar, 1 Barnsley Rd, Ackworth, Pontefract, WF7 7BS [IO93HP, SE41]
GD3 ESV Rev F Ness, Green Tops, Westmoreland Rd, Douglas, IM1 4AW
G3 ESW B Insull, 24 Hartland Ave, Weeping Cross, Stafford, ST17 0EJ [IO82XS, SJ92]
G3 ESY P W F Jones, 13 Blenheim Cl, Hereford, HR1 2TY [IO82PB, SO54]
G3 ETA C E Hall, Oakfield, Church Ln, Buckland Ripers, Weymouth Dorset, DT3 4BT [IO80SP, SY68]
G3 ETH J L Goldberg, 3 Haslin Cres, Christleton, Chester, CH3 6AN [IO83NE, SJ46]
G3 ETJ J E Jenner, 251 Marlpool Ln, Kidderminster, DY11 5DD [IO82UJ, SO87]
G3 ETK A J Cowley Ariel Radio Grp, BBC Club, Oxford Rd, Manchester, M60 1SJ [IO83VL, SJ89]
G3 ETQ G B Whitfield, 3 Goldsborough Rd, Doncaster, DN2 5HW [IO93KM, SE50]
G3 ETU L Toke, 34 Park Ln Ct, Bury New Rd, Salford, M7 4LP [IO83UM, SD80]
G3 ETX F R Wilson, 30 Glencoe Ave, Ilford, IG2 7AN [JO01BN, TQ48]
G3 ETY G H Lang, 14 Plover Cl, Bamford, Rochdale, OL11 5PU [IO83VO, SD81]
G3 ETZ J C Mackellar, Sparrows House, Chapel Rd, Cockfield, Bury St. Edmunds, IP30 0HE [JO02JD, TL95]
G3 EUE E F Jones, White Lodge, Crofters Wood, The Street, Bramber Sussex, BN4 3WE [IO90WT, TQ30]
GD3 EUI B S Pover, Orry Cottage, Ramsey Rd, Knocksharry, Peel, IM5 2AF
G3 EUK R W Curtis, 6 St. Christophers Cl, Bath, BA2 6RG [IO81TJ, ST76]
G3 EUS J G Fitzgerald, 98 Wymondley Rd, Hitchin, SG4 9PX [IO91UW, TL12]
G3 EVC D J Pye, 93 The Broadway, Perry Barr, Birmingham, B20 3ED [IO92BM, SP09]
G3 EVK E A Baker, 7 Southernhay Rd, Verwood, BH31 7AN [IO90BV, SU00]
G3 EVT R J Mutton, Summer Hayes, Mill Ln, Oversley Green, Alcester, B49 6LF [IO92BE, SP05]
GI3 EVU H M Humphreys, 10 Mount Eden Park, Malone Rd, Belfast, BT9 6RA [IO74AN, J37]
G3 EVV F J Hill, 328 Thong Ln, River View Park, Gravesend, DA12 4LQ [JO01EK, TQ67]
G3 EVX A Hibberson, Top End, 9 Hunsbury Cl, Hunsbury Hill, Northampton, NN4 9UE [IO92MF, SP75]
GM3 EWC R B Irvine, 8 Woodend Rd, Aberdeen, AB1 6YH [IO87TH, NJ72]
G3 EWF A C Harris, 5 Wickham Ct, Wickham Hill, Stapleton, Bristol, BS16 1DQ [IO81RL, ST67]
G3 EWH T Hewitson, 2 East Gr Rd, Exeter, EX2 4LX [IO80FR, SX99]
G3 EWJ E R B Harmer, 32 Abinger Ave, Cheam, Sutton, SM2 7LJ [IO91VI, TQ26]
G3 EWM P F Green, 23 Tilton Rd, Borough Green, Sevenoaks, TN15 8RS [JO01DG, TQ65]
G3 EWT C Tamkin, 4 Stanmer Villas, Brighton, BN1 7HP [IO90WU, TQ30]
G3 EWY P F Walder, 10 Whyke Rd, Chichester, PO19 2HL [IO90OU, SU80]
G3 EXG Details withheld at licensee's request by SSL.
G3 EXL D C Derham, 3 Riverside House, Riverbank, Saltash, Cornwall, PL12 6AZ [IO70UL, SX46]
G3 EXP A J Bassett, 24 St. Georges Dr, Ferndown, BH22 9EF [IO90BT, SZ09]
GM3 EXS A W Clark, Rowallan, 112 Cradlehall Park, Westhill, Inverness, IV1 2DB [IO77WL, NH74]
G3 EXU J W Cunliffe, 12 Blenheim Cl, Lostock Hall, Preston, PR5 5YX [IO83PR, SD52]
GM3 EXX A Olone, 2081 Great Western Rd, Glasgow, G13 2XX [IO75TV, NS56]
G3 EYB Details withheld at licensee's request by SSL.
G3 EYC H C Kutscherauer, 72 George V Ave, Pinner, HA5 5SW [IO91TO, TQ18]
G3 EYH M T Pullen, High Banks, 213 Northwick Rd, Bevere, Worcester, WR3 7EJ [IO82VF, SO85]
G3 EYO E J Duesbury, 3 Smithson Cl, Moulton, Richmond, DL10 6QP [IO94EK, NZ20]
G3 EYU J W Ringrose, 21 Longleaze, Wootton Bassett, Swindon, SN4 8AX [IO91BN, SU08]
G3 EYY N J Rowe, 48 Hollingbourne Rd, Herne Hill, London, SE24 9ND [IO91WK, TQ37]
G3 EYZ G R Minson, 52 Fairview Ave, Stanford-le-Hope, SS17 0DT [JO01FM, TQ68]
GM3 EZA R T S Johnstone, 16 Stirling Ave, Bearsden, Glasgow, G61 1PE [IO75TV, NS57]
G3 EZB J S Rackett, Little Vectis, Folgate Ln, Old Costessey, Norwich, NR8 5DP [JO02OP, TG11]
G3 EZE G H Standing, Windrush, Stonebarrow Ln, Charmouth, Bridport, DT6 6RA [IO80NR, SY39]
GM3 EZI C B G Staples, 202 Riverside Rd, Kirkfieldbank, Lanark, ML11 9JJ [IO85CQ, NS84]
G3 EZX S Wood, 70 Derwent Rd, Warrington, WA4 6AZ [IO83QJ, SJ68]
G3 EZZ J Eaton, 109 St. James Rd, Bridlington, YO15 3NJ [IO94VB, TA16]
GM3 FAH J C McCulloch, 41 Loch St., Townhill, Dunfermline, KY12 0HQ [IO86GC, NT18]
G3 FAM H W Brooker, Westwinds, 16 Avalon Cl, Orpington, BR6 9BS [JO01BI, TQ46]
GM3 FAO A F Davidson, Monymusk, 25 Doonholm Rd, Alloway, Ayr, KA7 4QQ [IO75QK, NS31]
G3 FAU V C Cundall, 311 Archer Rd, Stevenage, SG1 5HF [IO91VV, TL22]
G3 FBH A Bell, 1 Micklegate, Murdishaw, Runcorn, WA7 6HT [IO83QH, SJ58]
G3 FBI Capt C Dunkerley, 71 Cecil Rd, Hale, Altrincham, WA15 9NT [IO83TI, SJ78]
G3 FBN W J E Bolton, 16 Trotsworth Ave, Virginia Water, GU25 4AL [IO91RJ, TQ06]
G3 FBR J F Lewis, 13 Woodland Mews, Elm Way, Heathfield, TN21 8YD [JO00DX, TQ52]
G3 FBU W C Brown, 79 Mill Hill, Deal, CT14 9EW [JO01QF, TR35]
G3 FCB E D Melville, 59 Valley Prospect, Newark, NG24 4QN [IO93OB, SK75]
G3 FCD N Clegg, 22 Bent Lathes Ave, Rotherham, S60 4BN [IO93IJ, SK49]
G3 FCK A W McNeill, 40 Turnpike Rd, Newbury, RG14 2NF [IO91IJ, SU46]
G3 FCM A Cowley, 13 Steward Cl, Stuntney, Ely, CB7 5TW [JO02DJ, TL57]
G3 FCS D W Greenwood, 14 Mill Hill, Fartown, Pudsey, LS28 8NR [IO93ES, SE23]
G3 FCT S J Coe, 8 Priory Rd, Faversham, ME13 7EJ [JO01KH, TR06]
G3 FCV E L Bartholomew, 42 Audley Ave, Gillingham, ME7 3AY [JO01GI, TQ76]
G3 FCW G Denby, 31 Scotland Way, Horsforth, Leeds, LS18 5SQ [IO93EU, SE23]
G3 FD H T Brock, 56 Chapel Ln, Hadleigh, Benfleet, SS7 2PP [JO01HN, TQ88]
G3 FDC Revd H Makin, 46 Upper Highfield, Gibb Ln, Mount Tabor, Halifax West Yorks, HX2 0UG [IO93BR, SE02]
G3 FDG Details withheld at licensee's request by SSL.[Op: R G Morris.]
GM3 FDN J C Petrie, 4 Thistle Pl, Grangemouth, FK3 0BU [IO86DA, NS98]
G3 FDS C F Ford, 1C Tower Rd, Epping, CM16 5EL [JO01BQ, TL40]
G3 FDU Details withheld at licensee's request by SSL.
G3 FDW M Gibbings, Glannaventa, 5 Meadowbank Ln, Grange Over Sands, LA11 6AT [IO84ME, SD37]
GW3 FDZ D G Whitehead, Tyddyn Bach, Dyffryn Ardudwy, Gwynedd, LL44 2RQ [IO72WT, SH52]
G3 FEC I D McCarthy Swindon Dis ARC, 76 High St., Purton, Swindon, SN5 9AD [IO91BO, SU08]
G3 FED R O Watts, Dial Cottage, Bannut Tree Ln, Bridstow, Ross on Wye, HR9 6AJ [IO81QW, SO52]
G3 FET L F Rawlings, 7 Gillridge Green, Crowborough, TN6 2UN [JO01BB, TQ53]
G3 FEV J R Platt, Springfield House, Newchurch Rd, Rossendale, Lancs, BB4 7QX [IO83UQ, SD82]
G3 FEW E A Rule, 15 Norwich Rd, Lenwade, Norwich, NR9 5SH [JO02NR, TG11]
G3 FEX B C Oddy, Three Corners, Merryfield Way, Storrington, Pulborough, RH20 4NS [IO90SW, TQ01]
GI3 FFF J P Clarke Ballymena ARC, 154 Galgorm Rd, Ballymena, BT42 1DE [IO64UU, D00]
G3 FFH J A C K Frings, Hill Farm, Gore Ln, Uplyme, Lyme Regis, DT7 3RJ [IO80MR, SY39]
G3 FFL J H O Parker, Linskill House, 15 Blackburns Yard, Church St., Whitby, YO22 4DS [IO94QL, NZ91]
GM3 FFQ W Donaldson, 3 Inverallan Dr, Bridge of Allan, Stirling, FK9 4JR [IO86AD, NS79]
G3 FFR W A Darbyshire, 93 Heyhouses Ln, Lytham St. Annes, FY8 3RN [IO83LS, SD32]
G3 FFY M H Stedman, 63 Cranston Park, Ave, Upminster, Essex, RM14 3XD [JO01DN, TQ58]
G3 FGB Details withheld at licensee's request by SSL.
G3 FGC Details withheld at licensee's request by SSL.
G3 FGD Details withheld at licensee's request by SSL.
G3 FGH D R Leah, 16 St. Marys Cl, Bath, BA2 6BR [IO81TJ, ST76]
G3 FGP R L Brooks, 10 The Oval, New Barn, Longfield, DA3 7HD [JO01EJ, TQ66]
G3 FGT L F Crosby, 11 Loxley Ave, Shirley, Solihull, B90 2QE [IO92BJ, SP17]
G3 FGW M O Denny, 69 Ryan Ct, Bryanston St., Blandford Forum, DT11 7XE [IO80WU, ST80]
G3 FHG K S Martin, 11 Hockley Rise, Hockley, SS5 4QE [JO01HO, TQ89]
G3 FHK Details withheld at licensee's request by SSL.
G3 FHL G C Bagley, 49 Green Ln, Malvern Wells, Malvern, WR14 4HT [IO82UB, SO74]
G3 FHM J R Brannigan, 11 Clifton St., Bury, BL9 5DY [IO83UO, SD81]
G3 FHN E W B Aldworth, Denbigh, 26 Meadow Way, Fairlight, Hastings, TN35 4BN [JO00HV, TQ81]
G3 FHW N Ratcliffe, 55 Queensway, Fenham, Newcastle upon Tyne, NE4 9TA [IO94EX, NZ26]
G3 FIA A D Lowden, 3 Boscobel Rd, Great Barr, Birmingham, B43 6BB [IO92AN, SP09]
G3 FIB G A Livesey, 14 Dene Dr, Longfield, DA3 7JR [JO01DJ, TQ66]
G3 FIC J Glover, 53 Swanpool Ln, Aughton, Ormskirk, L39 5AY [IO83NN, SD40]
G3 FIJ F R Howe, 29 Kingswood Rd, Colchester, CO4 5JX [JO01KV, TL92]
G3 FIK K W Perfect, Littleton House, Pipe Ridware, Rugeley, Staffs, WS15 3QL [IO92BS, SK02]
G3 FIR B S Farrow, Gardencourt, 135 Tally House Rd, Shadoxhurst, Ashford, TN26 1HW [JO01JC, TQ93]
G3 FIT N E Ashman, 1 East Dr, East Brent, Highbridge, TA9 4JQ [IO81MG, ST35]
G3 FIY P L Spencer, Rose Orchard, Lower St., Dittisham, Dartmouth, TQ6 0HY [IO80EI, SX85]
GM3 FIZ D M Sangster, 32 Mortimer Ct, Dalgety Bay, Dunfermline, KY11 5UQ [IO86HA, NT18]
G3 FJ J Sharples, 11 The Lodge, Lavender Rd, Waterlooville, PO7 8BX [IO90LU, SU60]
GM3 FJA Dr W E D Sleat, 9 Doocot Rd, St. Andrews, KY16 8QP [IO86OH, NO41]
G3 FJE B T Elliott Shefford Dist Rd, 4 Ivel View, Sandy, SG19 1AU [IO92UD, TL14]
G3 FJF Details withheld at licensee's request by SSL.
GW3 FJI E L Jones, 8 Merllyn Rd, Rhyl, LL18 4HH [IO83GH, SJ08]
G3 FJL J Hall, 250 Scraptoft Ln, Leicester, LE5 1PA [IO92LP, SK60]
G3 FJN J A Barson, 30 Coxon St., Spondon, Derby, DE21 7JG [IO92HW, SK43]
G3 FJO A O Ellefsen, 121 The Furlongs, Ingatestone, CM4 0AL [JO01EQ, TQ69]
G3 FJQ Details withheld at licensee's request by SSL.[Station located near Southport.]
G3 FJT Details withheld at licensee's request by SSL.
G3 FJV G F Dutton, 1 The Cttgs, Fleets Rd, Sturton By Stow, Lincoln, Lincs, LN1 2DN [IO93QH, SK88]
GI3 FJX J Davidson, 7 Keel Point, Dundrum, Newcastle, BT33 0NQ [IO74BF, J43]
G3 FJY M A Pollard, Orchard Bungalow, Upper Up, South Cerney, Gloucester, GL7 5UR [IO91AQ, SU09]
G3 FK E W Taylor, Dawn Cottage, 9 Forest Rd, West Moors, Ferndown, BH22 0EU [IO90BT, SU00]
G3 FKB Details withheld at licensee's request by SSL.
G3 FKF Details withheld at licensee's request by SSL.
G3 FKI E C Lambert, 6 Abercorn Gdns, Kenton, Harrow, HA3 0PB [IO91UN, TQ18]
G3 FKJ W F Jeffery, 94 Gubbins Ln, Harold Wood, Romford, RM3 0BL [JO01CO, TQ59]
G3 FKM Dr E J Allaway, 10 Knightlow Rd, Birmingham, B17 8QB [IO92AL, SP08]
G3 FKU D Barlow, 15 Kinnerley St., Walsall, WS1 2LD [IO92AN, SP09]
G3 FKV H Collinson, 27 Lincoln Cl, Swanton Morley, Dereham, NR20 4NB [JO02LQ, TG01]
GJ3 FKW K S Ball, Oxenford Cottage, St Lawrence, Jersey, Channel Islands, Jersey, JE9 9OX
G3 FKY J Parker, 36 North Ave, Leek, ST13 8DP [IO83XC, SJ95]
G3 FLB G W Nailor, 4 Buckingham Rd, Swindon, SN3 1JA [IO91CN, SU18]
G3 FLG P Harvey, 11 Kinloss Rd, Wirral, L49 3PS [IO83KJ, SJ28]
GD3 FLH A K Sinclair Iom ARS, 1 Marathon Dr, Douglas, IM2 4BP
G3 FLJ H McIntyre, 42 Dunvegan Dr, Southampton, SO16 8DD [IO90GW, SU41]
G3 FLN N L Draper, 115 Kingsdown Way, Townhill Park, Southampton, SO18 2GQ [IO90HW, SU41]
G3 FLQ F Robinson, 35 Avon Cl, Higham, Barnsley, S75 1PD [IO93FN, SE30]
G3 FLR H Priestley, 879 Oldham Rd, Rochdale, OL16 4RY [IO83WO, SD91]
G3 FLV L Keighley, 24 St. Annes Rd, Headingley, Leeds, LS6 3NX [IO93FT, SE23]
G3 FM J Duckworth, The Long Barn, Brough, Kirkby Stephen, Cumbria, CA17 4BZ [IO84UM, NY71]
G3 FML H J Finch, 10 The Homeyards, Shaldon, Teignmouth, TQ14 0EQ [IO80FM, SX97]
G3 FMO G Elliott, Oatlands, Southend Rd, Howe Green, Chelmsford, CM2 7TD [JO01GQ, TL70]
G3 FMR C L T Dwyer, 8 Glebe Rd, Wells Next The Sea, NR23 1AZ [JO02KW, TF94]
G3 FMT D W Robinson, Bray Cottage, Dane Hill, North Aston, Bicester, OX6 4JE [IO91IX, SP42]
G3 FMU D McDiarmid, 102 Shalloak Rd, Broadoak, Broad Oak, Canterbury, CT2 0QH [JO01NH, TR16]
G3 FMW J Stockley, 22 Manor Gdns, Killinghall, Harrogate, HG3 2DS [IO94FA, SE25]
G3 FMZ B R Brown, 40 Coatham Rd, Redcar, Cleveland, TS10 1RS [IO94LO, NZ62]
G3 FNJ N F Joly, 28 Oakington Ave, Harrow, HA2 7JJ [IO91TN, TQ18]
G3 FNK C Drinkwater, 29 Eaton Ave, Bletchley, Milton Keynes, MK2 2HN [IO91PX, SP83]
G3 FNL R N Grubb, 7762 Brockway Dr, Boulder, Colorado 80303, USA
G3 FNM W R Parkinson, 141 Norris Rd, Sale, M33 3GS [IO83UJ, SJ79]
G3 FNO G W Morgan, 27 Kestrel Cl, Downley, High Wycombe, HP13 5JN [IO91OP, SU89]
G3 FNQ D T Bagshaw, 2 Harewood Ave, Ainsdale, Southport, PR8 2PH [IO83LO, SD31]
G3 FNT P Dean, 17 Pineheath Rd, High Kelling, Holt, NR25 6QF [JO02NV, TG14]
G3 FNU Details withheld at licensee's request by SSL.
G3 FNY R E Wand, Greyfell, Bendish, Hitchin, Herts, SG4 8JH [IO91UV, TL12]
G3 FNZ J A ' Lambert, 49 Rede Ct Rd, Strood, Rochester, ME2 3SP [JO01FJ, TQ76]
GD3 FOC L A Higgins, 56 Victoria Rd, Castletown, IM9 1ED
G3 FOD H Moxon, 51 Redoak Ave, Barrow in Furness, LA13 0LJ [IO84JC, SD26]
G3 FOE A Royle, Rosehearty, 19 Beaufort Cl, Alderley Edge, SK9 7HU [IO83VH, SJ87]
G3 FOO A Seed, 31 Withert Ave, Bebington, Wirral, L63 5NE [IO83LI, SJ38]
G3 FOP Details withheld at licensee's request by SSL.
G3 FOQ D B L Delanoy, Martindale, Halls Ln, Norton, Bury St. Edmunds, IP31 3LG [JO02KG, TL96]
G3 FOR R J Corps, Haldon, Hillside Rd, Aldershot, Hants, GU11 3LX [IO91OF, SU84]
G3 FOZ J D Slater, Grange Farm House, Grange Rd, Geddington, Kettering, NN14 1AL [IO92PK, SP88]
G3 FP B R Arnold, 5 Salcott Rd, Beddington, Croydon, CR0 4PS [IO91WI, TQ36]
G3 FPB J Wooller, 48 Fleckers Dr, Up Hatherley, Cheltenham, GL51 5BD [IO81WV, SO92]
GW3 FPC D Stephenson, Washpool Cottage, Velindre Farchog, Crymych, Dyfed, SA41 3UY [IO72OA, SN03]
GW3 FPF P F Jones, The Flat, 123 Wellington Rd, Rhyl, LL18 1LE [IO83GH, SJ08]
GW3 FPH J W Hayes, 4 St. Marys Dr, Village Rd, Northop Hall, Mold, CH7 6JF [IO83KE, SJ26]
G3 FPI W B Hopkins, 12 Hawkesbury Rd, Shirley, Solihull, B90 2QR [IO92BJ, SP17]
G3 FPK N A S Fitch, 40 Eskdale Gdns, Purley, CR8 1EZ [IO91WH, TQ36]
G3 FPN J R Davey, 19 Southey St., Keswick, CA12 4EF [IO84KO, NY22]
G3 FPQ D L Courtier-Dutton, Markham Oak Cottage, Dockenfield Rd, Bucks Horn Oak, Farnham, Surrey, GU10 4LP [IO91NE, SU84]
G3 FPY J E Dew, 62 Monks Park Ave, Horfield, Bristol, BS7 0UH [IO81QL, ST57]
G3 FQC W G Edwards-Hanham, Fircroft, 11 Lower Howsell Rd, Malvern Link, Malvern, WR14 1EQ [IO82UD, SO74]
G3 FQY Details withheld at licensee's request by SSL.
G3 FRE W H Frith, 56 Ringleas, Cotgrave, Nottingham, NG12 3NE [IO92LV, SK63]
GM3 FRI W A Mitchell, 32 Gordon Rd, Aberdeen, AB1 7RL [IO87TH, NJ72]
GW3 FRK V C Morgan, Hirael, Blaenplwyf, Aberystwyth, Dyfed, SY23 4DH [IO72WI, SN57]
G3 FRN G N Myatt, 10 Ship Ln, Combwich, Bridgwater, TA5 2QT [IO81LE, ST24]
GM3 FRU D Wark, 24 Dirleton Gdns, Alloa, FK10 1NL [IO86CC, NS89]
G3 FRW Details withheld at licensee's request by SSL.
G3 FRX J A Wilkes, 1 Luccombe Pl, Upper Shirley, Southampton, SO15 7RL [IO90GW, SU41]
GM3 FRZ G B Esslemont, 7 Hazledene Rd, Aberdeen, AB15 8LB [IO87WD, NJ90]
G3 FSA A R Davis, Willow Cottage, Hedging, North Newton, Bridgwater, TA7 0DE [IO81MB, ST32]
G3 FSN A C Butcher, 70 Hughenden Ave, High Wycombe, HP13 5SN [IO91OP, SU89]
G3 FSO Details withheld at licensee's request by SSL.
GW3 FSW Details withheld at licensee's request by SSL.
G3 FSX R J Ellis, Laura House, 79 Sunte Ave, Haywards Heath, RH16 2AB [IO91WA, TQ32]
G3 FTE LtCm K D McInnes, 116 Applegarth Caravan Park, Seasalter Ln, Seasalter, Whitstable, CT5 4BZ [JO01MI, TR06]
G3 FTH J S Hale, 136 Bush Rd, Cuxton, Rochester, ME2 1HB [JO01FJ, TQ76]
G3 FTK L C Gray, 109 Foxholes Rd, Parkstone, Poole, BH15 3NE [IO90AR, SZ09]
G3 FTP E N Davis, 379 Kings Rd, Ashton under Lyne, OL6 9EW [IO83XM, SD90]
G3 FTQ A Frost, 11 Ingleboro Dr, Purley, CR8 1ED [IO91WH, TQ36]
GI3 FTT W Brennan, 10 Dunhugh Park, Londonderry, BT47 2NL [IO64HX, C41]
G3 FTU J R Jones, 76 Chantry Rd, Romanby, Northallerton, DL7 8JL [IO94GH, SE39]
G3 FUH Details withheld at licensee's request by SSL.[Op: M Taylor, 23a Albany Road, St. Leonards-on-Sea, E Sussex, TN38 0LP.]
G3 FUJ W K Scott, 10 Pavilion Rd, Littleover, Derby, DE23 6XL [IO92GV, SK33]
G3 FUN R E Kemsley, 1 St. Marys Rd, Faversham, ME13 8EH [JO01KH, TR06]
GM3 FUT F J Hawke, Meadowcroft, Roy Bridge, Inverness Shire, PH31 4AQ [IO76NV, NN28]
G3 FVA D J Armitage Sth Manchestr Rd, 12 Loughborough Cl, Sale, M33 5UF [IO83TK, SJ79]
G3 FVC E C Palmer, 1 Highbank, Watchet, TA23 0DG [IO81IE, ST04]
G3 FVD R K Mildren, Ken Berry, 13 Queens Cres, Bodmin, PL31 1QP [IO70PL, SX06]

G3 FVL H J Hudson, 77 Cres Rd, Wood Green, London, N22 4RU [IO91WO, TQ29]

G3 FVO Details withheld at licensee's request by SSL.

G3 FVV R C Fagg, 46 Captains Hill, Alcester, B49 6QN [IO92BF, SP05]

G3 FWA J Bennett, 14 Chinnor Cl, Goldington, Bedford, MK41 9TQ [IO92SD, TL05]

G3 FWB P L Hunt, Bridgend, Dukes Dr, Calver, Sheffield, S30 1YP [IO93GJ, SK38]

G3 FWD B Purchase, 126 Renton Rd, Wolverhampton, WV10 6XH [IO82WO, SJ90]

G3 FWG R T G Tremelling, Rose Eglos Bungalow, Treveryn Parc, Budock Water, Falmouth, TR11 5EH [IO70KD, SW73]

G3 FWI W E Sutton, Old Orchards, Romsey Rd, Kings Somborne, Stockbridge, SO20 6PN [IO91GB, SU33]

G3 FWN A Hall, 11 Sandford Ave, Long Eaton, Nottingham, NG10 1BQ [IO92IV, SK43]

G3 FWU L O Richardson, Primrose Farm, Hunger Hill, East Stour, Gillingham, SP8 5JR [IO81UA, ST82]

G3 FXG A Benyon, Carrer Mas Andreaus 46, Benicasim 12560, Spain, 1250 X

GW3 FXI P H Cardwell, 3 Old Talbot, Llanwnog, Caersws, SY17 5JG [IO82GM, SO09]

GD3 FXN A D Radcliffe, Rose Cottage, Ballafreer Ln, Union Mills, Douglas, IM4 4AS

G3 FXV S H D Golding, 23 Ramsbury Cl, Blandford Forum, DT11 7UF [IO80WU, ST80]

G3 FYF P R Acke, Kinghurst Farm, Holne, Devon, TQ13 7RU [IO80CM, SX76]

G3 FYP P S Robson, 51 Potter Hill, Pickering, YO18 8AF [IO94OF, SE78]

G3 FYQ C B Wilkinson Pontefract Dis, 8 Westfield Ave, Knottingley, WF11 0JH [IO93JQ, SE42]

G3 FYR W E Gardner, 31 Priory Ave, Petts Wood, Orpington, BR5 1JE [JO01AJ, TQ46]

G3 FYS J L Hooper, 28 Cunningham Cl, Ringwood, BH24 1XW [IO90CU, SU10]

G3 FYX R W Emery, 30 Station Rd, Winterbourne Down, Bristol, BS17 1EP [IO81RM, ST67]

G3 FZL G M C Stone, 11 Liphook Cres, Forest Hill, London, SE23 3BN [IO91XK, TQ37]

G3 FZR M W Capewell, 1 Parnell Rd, Wirral, L63 9JR [IO83LI, SJ38]

G3 FZS J H Brent, 26 Redhill Dr, Fishponds, Bristol, BS16 2AQ [IO81RL, ST67]

G3 FZW E A Matthews, 2 The Parchments, Lichfield, WS13 7NA [IO92CQ, SK11]

GW3 GA S Holmes, Havenholm, Maenygroes, New Quay, Dyfed, SA45 9RL [IO72TE, SN35]

G3 GAA W P Jeans, 1 Lancaster Ride, Penn, High Wycombe, HP10 8DU [IO91PP, SU89]

G3 GAD G A Day, 57 Westfield Rd, Backwell, Bristol, BS19 3ND [IO81PJ, ST46]

G3 GAF Dr C T Dollery, 101 Corringham Rd, London, NW11 7DL [IO91VN, TQ28]

G3 GAG W Eckersley, 50 Chaddock Ln, Worsley, Manchester, M28 1DD [IO83SM, SD70]

GW3 GAH A W Foster, Moryn Wen, 18 Nant Bychan, Moelfre, LL72 8HE [IO73VI, SH58]

G3 GAI R K Burbidge, 11 Third Ave, Worthing, BN14 9NZ [IO90TU, TQ10]

G3 GAO L J Avery, Garden Cl, Greenway Rd, St. Marychurch, Torquay, TQ1 4NJ [IO80FL, SX96]

G3 GAQ D D Bottomley, 24 Midhope Rd, Woking, GU22 7UE [IO91RH, TQ05]

G3 GAR C R Dickenson, 5545 Natoma Dr, Fort Myers, Florida 33919, USA, X X

G3 GAW D J Redshaw, 75 Lowndes Park, Driffield, YO25 7BE [IO94SA, TA05]

G3 GAZ G H Phillips, 3 Hazling Dane, Shepherdswell, Dover, CT15 7LS [JO01OE, TR24]

G3 GBB A J Munro, Manor Farm House, Mill St., Gislingham, Eye, IP23 8JR [JO02MH, TM07]

G3 GBD S R Hancock, 53 Friary Grange Park, Winterbourne, Bristol, BS17 1NA [IO81RM, ST68]

GD3 GBG A W Moore, 114 Ballabrooie Dr, Douglas, IM1 4HQ

G3 GBI A C Elliott, 119 Rusper Rd, Ifield, Crawley, RH11 0HW [IO91VC, TQ23]

G3 GBN S H Feldman, 4 Beechholm Mews, Waltham Cross, EN8 0DH [IO91XR, TL30]

G3 GBS M L A Sandoz, Edelweiss, Broad Ln, Tanworth in Arden, Solihull, B94 5DP [IO92BI, SP17]

G3 GBU A Allen Stoke on Trent Amateur Radio S, 3 Wayfield Gr, Stoke on Trent, ST4 6DB [IO83VA, SJ84]

GM3 GBX G Dawson, 32 Morningside Pl, Edinburgh, EH10 5EY [IO85JW, NT27]

GM3 GBY J Bryce, Easter Highfield, Dalry, Ayrshire, KA24 4HT [IO75PR, NS35]

GM3 GBZ G R Balfour obo Strathmore ARC, 6 Kirkden St., Friockheim, Arbroath, DD11 4SX [IO86QP, NO55]

G3 GC E H Godfrey, Dorset Reach, 60 Chilton Gr, Yeovil, BA21 4AW [IO80QW, ST51]

GD3 GCE P T Gordon, Dormer House, Walpole Dr, Ramsey, Isle of Man, IM8 1NA

G3 GCI H Griffiths, 10 Whitelock Rd, Abingdon, OX14 1NZ [IO91IQ, SU49]

G3 GCW B G A Jones, 44 Winner Hill Rd, Paignton, TQ3 3BT [IO80FK, SX86]

G3 GDA J E Armstrong, 64 Colwell Dr, Witney, OX8 7NQ [IO91GS, SP30]

G3 GDB G A Bird, 16 Simnel Rd, London, SE12 9BG [JO01AK, TQ47]

G3 GDH D M Silveston, 192 Rosemary Ave, Minster on Sea, Sheerness, ME12 3HX [JO01JK, TQ97]

GM3 GDS W J Graham, 7 Braehead, Douglas, Lanark, ML11 0PT [IO85BN, NS83]

G3 GDU W B Kendal, 12 Weald Dr, Furnace Green, Crawley, RH10 6JU [IO91VC, TQ23]

G3 GDY W F Ord, 7 Girvan Cl, Stanley, DH9 6UY [IO94DU, NZ25]

G3 GEA D C Callender, Rose Cottage, Snaisgill Rd, Middleton in Teesdale, Barnard Castle, DL12 0RP [IO84XP, NY92]

G3 GED B Bracewell, 108 St. Martins Rd, Blackpool, FY4 2EA [IO83LS, SD33]

G3 GEF J T A Andrews, Dents Barn, Hutton Roof, Kirkby Lonsdale, Carnforth Lancs, LA6 2PG [IO84QE, SD57]

G3 GEG E C Cooper, Ciren, 19 Ventnor Rd, Apse Heath, Sandown, PO36 0JT [IO90JP, SZ58]

G3 GEI Dr R A Hancock Solihull Amateur Radio Soc, 80 Ulleries Rd, Solihull, B92 8EE [IO92CK, SP18]

G3 GEJ Details withheld at licensee's request by SSL.[Op: L M Airey, 32 Brookside Close, Bedale, N Yorks, DL8 2DR.]

G3 GET P J Coppins, 79 Westerham Rd, Sittingbourne, ME10 1XF [JO01II, TQ86]

G3 GEV S C Hollingshurst, 4 Ward Cl, Erith, DA8 3EH [JO01CL, TQ57]

G3 GEX P L Burton, The Old School, Sandridgebury Ln, Sandridge, St. Albans, AL3 6JB [IO91US, TL11]

G3 GFC N A F Williams, Offwell Barton, Offwell, Nr.Honiton, Devon, EX14 9SA [IO80KS, SY19]

G3 GFG Details withheld at licensee's request by SSL.

GW3 GFM A N Lawes, 33 Cheriton House, Cardiff Rd, Llandaff, Cardiff, CF5 2DL [IO81JL, ST17]

G3 GFR R J Isbill, 27 Queenhythe Rd, Guildford, GU4 7NU [IO91RG, SU95]

G3 GFT G F Oldfield, Maitland, 13 Westbourne Ave, Wrea Green, Preston, PR4 2PL [IO83MS, SD33]

GM3 GG G Mortimer, 10 St. Combs Ct, Banff, AB45 1GA [IO87RP, NJ66]

G3 GGG R A Bishop, 31 Blenheim Cl, Didcot, OX11 7JQ [IO91JO, SU59]

G3 GGH P S Horn, Darfield, 50 Barrack Rd, Bexhill on Sea, TN40 2AZ [JO00FU, TQ70]

G3 GGI A A Laurence, 70 Firs Ave, London, N11 3NQ [IO91WO, TQ29]

G3 GGK P J Simpson, The Beagles, 109 Highfields, Caldecote, Cambridge, CB3 7NX [IO92XF, TL35]

G3 GGL A W G Wormald, Sabrina Lodge, 15 Sabrina Dr, Bewdley, DY12 2RJ [IO82UJ, SO77]

G3 GGN D Shute, 100 Wick St., Wick, Littlehampton, BN17 7JS [IO90RT, TQ00]

G3 GGO C N Wridgway, 11 St. Andrews Ave, Windsor, SL4 4EH [IO91QL, SU97]

G3 GGP G L Silburn, 38 Springbank Cres, Leeds, LS6 1AB [IO93FT, SE23]

G3 GGR J H Sykes, 49 Chapel St., Pelsall, Walsall, WS3 4LW [IO92AP, SK00]

G3 GGS W E Waring, 51 Church Rd, Leyland, Preston, PR5 2AA [IO83PQ, SD52]

G3 GGU G W Smith, Greenacres, Top Rd, Calow, Chesterfield, S44 5AE [IO93HF, SK47]

GI3 GGY J A Porter, 237 Culmore Rd, Derry City, Co. Derry, BT48 8JL [IO65IB, C42]

G3 GHB A T Eley, Tanglewood, Kington, Flyford Flavell, Worcester, WR7 4DH [IO82XE, SO95]

G3 GHI D A R Naylor, 4 Cullesden Rd, Kenley, CR8 5LR [IO91WH, TQ35]

G3 GHN A J Gould Clifton ARS, 25 Clarendon Rise, Lewisham, London, SE13 5ES [IO91XL, TQ37]

G3 GHS J G Holland, Tanglewood, Portheast Way, Gorran Haven, St. Austell, PL26 6JA [IO70OF, SX04]

G3 GHY C Smith, Sunnycot, Main Rd, East Boldre, Brockenhurst, SO42 7WL [IO90GT, SU30]

G3 GIB A D Wake, 42 Charles Ave, Watton, Thetford, IP25 6BZ [JO02JN, TF90]

G3 GIE Details withheld at licensee's request by SSL.

GM3 GIG J J Maconochie, 14 Craiglockhart Rd, Edinburgh, EH14 1HL [IO85JW, NT27]

G3 GIH J C Bird, Grange Farm, Euston, Thetford, Norfolk, IP24 2QG [JO02JJ, TL87]

G3 GII J P Clark, Hunters Moon, Botley Rd, Horton Heath, Eastleigh, SO50 7DN [IO90IW, SU41]

G3 GIL P Forrest, 341 Catcote Rd, Hartlepool, TS25 3EB [IO94JP, NZ42]

G3 GIQ H F Lewis, 271 Popes Ln, Ealing, London, W5 4NH [IO91UM, TQ17]

G3 GIW D W Birt, 99 Stoddens Rd, Burnham on Sea, TA8 2DD [IO81MF, ST35]

G3 GIY H Gregory, 31 Kempsford Cl, Oakenshaw, Redditch, B98 7YS [IO92AG, SP06]

G3 GIZ D G C Hicks Chester&Dist AR, 12 Toll Bar Rd, Christleton, Chester, CH3 5QX [IO83NE, SJ46]

GM3 GJB W A Macfarlan, 49 Shannon Dr, Falkirk, FK1 5HU [IO85CX, NS87]

G3 GJJ P B Watson, 5 High Garth, Winston, Darlington, DL2 3RY [IO94CN, NZ11]

GW3 GJQ S/Lr R Handley, 16 y Bryn, Glan Conwy, Colwyn Bay, Clwyd, LL28 5NJ [IO83CG, SH87]

G3 GJU L Rivers-Bland, Ailsa Dene, Kiln Ln, Hambleton, Poulton-le-Fylde, FY6 9BH [IO83MV, SD34]

G3 GJV H Brooke, Thistledown, Church Hill, Spofforth, Harrogate, HG3 1AG [IO93GW, SE35]

G3 GJW T I Lundegard, Saxby, Botsom Ln, West Kingsdown, Sevenoaks, TN15 6BL [JO01DI, TQ56]

G3 GJX E B Grist, Holmbury, Wheeler Ln, Witley, Godalming, GU8 5QU [IO91QD, SU94]

G3 GJY J O Yarker, Fieldway, Whitby Rd, Pickering, YO18 7HQ [IO94OF, SE88]

G3 GJZ H D Rodman, 6 Edinburgh Rd, Newmarket, CB8 0QF [JO02EF, TL66]

G3 GKC I Rosevear, 20 Christchurch Rd, Bradford on Avon, BA15 1TB [IO81VI, ST86]

G3 GKF E R Honeywood, 105 Whytecliffe Rd, Purley, Surrey, CR8 2AZ [IO91WI, TQ36]

G3 GKG G B Horsfall, 183 Chester Rd, Macclesfield, SK11 8QA [IO83WG, SJ97]

G3 GKI Me V F Kershaw, 78 West Garth, Cayton, Scarborough, YO11 3SD [IO94TF, TA08]

GM3 GKJ J Cockburn, Havenhoe Farm, Longcliffe, Brassington, Matlock, DE4 4HN [IO93EC, SK25]

G3 GKK D Bowers, 31 Rawlings Ln, Fowey, PL23 1DT [IO70QI, SX15]

G3 GKS R G Christian, 27 Howey Rise, Frodsham, Warrington, WA6 6DN [IO83PG, SJ57]

GW3 GKZ M D Fowler, Ty Gwyn Abergwynant, Penmaenpool, Dolgellau, Gwynedd, LL40 1YF [IO82AR, SH61]

G3 GLA B J Mase, 18 Norton Dr, Norwich, NR4 6JD [JO02PO, TG20]

G3 GLB J E Lacey, 50 Petersham Ave, Byfleet, West Byfleet, KT14 7HY [IO91SI, TQ06]

G3 GLE C J A Stewart, Box 63363, Muthaiga, Nairobi, Kenya

G3 GLK P R Graham, The Tors, Whitwell Rd, Ventnor, PO38 1LJ [IO90JO, SZ57]

G3 GLL T N Green, 6 Woodrolfe Rd, Tollesbury, Maldon, CM9 8SB [JO01KS, TL91]

G3 GLO K G Cass, 53 Maple Dr, Burnham on Sea, TA8 1DQ [IO81MF, ST34]

G3 GLQ W V Sutton, 57 Ashfurlong Cres, Sutton Coldfield, B75 6EN [IO92CN, SP19]

G3 GLV D A Burns, 3B Fitzroy Lodge, The Grove, Highgate, London, N6 6LH [IO91WN, TQ28]

G3 GLW P B E Willis, 23 Douglas Cres, Thornhill, Southampton, SO19 5JP [IO90HV, SU41]

G3 GLX J Simmonds, 99 Foljambe Ave, Walton, Chesterfield, S40 3EY [IO93GF, SK36]

GW3 GLY I C Williams, Croft Cottage, 129 Newton Rd, Newton, Swansea, SA3 4ST [IO71XN, SS68]

G3 GMC R N McVey, 18 Worlebury Hill Rd, Weston Super Mare, BS22 9SP [IO81MI, ST36]

G3 GMK K W May, 42 Millbrook Tower, Windermere Ave, Southampton, SO16 9FX [IO90GW, SU31]

G3 GMM E McFarland, 13 St. Oswalds Cres, Brereton Green, Brereton, Sandbach, CW11 1RW [IO83UE, SJ76]

G3 GMN H W Elsworthy, 90 Oxstalls Dr, Longlevens, Gloucester, GL2 9DE [IO81VV, SO82]

G3 GMS M K Thayne, 14 Tynedale Ave, Monkseaton, Whitley Bay, NE26 3BA [IO95GB, NZ37]

G3 GMT J W Knox, 2 Derwent Way, Little Neston, South Wirral, L64 9RX [IO83LG, SJ27]

G3 GMV N H Rhodes, 7 Moseley Wood Dr, Cookridge, Leeds, LS16 7HD [IO93EU, SE24]

G3 GMW L J Nichols, 5 Middle Pasture, Peterborough, PE4 5AU [IO92UO, TF10]

G3 GMY F E A Green, 5 Silvercliffe Gdns, New Barnet, Barnet, EN4 9QT [IO91WP, TQ29]

G3 GMZ Details withheld at licensee's request by SSL.

G3 GN R W Cavill, 189 Lower Rd, Great Bookham, Bookham, Leatherhead, KT23 4AU [IO91TG, TQ15]

G3 GNA D Macmillan, Brook Farm, Broadwas on Teme, Worcester, WR6 5NE [IO82TE, SO75]

G3 GNB K G Reid, 37 Priest Ave, Wokingham, RG40 2LT [IO91OJ, SU86]

GM3 GNE D A J Menzies, 7 Rysland Ave, Newton Mearns, Glasgow, G77 6EA [IO75US, NS55]

G3 GNK J Walker, 39 Lyngate Ave, Lowestoft, NR33 9JD [JO02UL, TM59]

GM3 GNM A C W Biddell, Eastferry, By Dunkeld, Perthshire, PH8 0HY [IO86FN, NO04]

G3 GNQ G C Cutting, Langtons, 35 The St., Galleywood, Chelmsford, CM2 8QN [JO01FQ, TL70]

G3 GNR Details withheld at licensee's request by SSL.[Op: R E Short, North Trew Farm, Highampton, Beaworthy, Devon, EX21 5JG.]

GM3 GNX Details withheld at licensee's request by SSL.[Op: J L Fraser, 37 Witchhill Road, Fraserburgh, Aberdeenshire, AB4 5NR.]

G3 GOM P W Thurlow, 20 Doncaster Rd, Braithwell, Rotherham, S66 7BB [IO93JK, SK59]

G3 GON R M Sharp, 217 Stockwood Ln, Bristol, BS14 8NF [IO81RJ, ST66]

G3 GOS P L Peach, The Firs, Goldsmith Ln, All Saints, Axminster, EX13 7LU [IO80MT, ST30]

G3 GOT B W Le Grys, 8 Kitchener Way, Shotley Gate, Ipswich, IP9 1RW [JO01PW, TM23]

G3 GOV W W Smith, 8 Hill Crest, Mannamead, Plymouth, PL3 4RW [IO70WJ, SX45]

G3 GOX A B Crane, 566 Hanworth Rd, Hounslow, TW4 5LH [IO91TK, TQ17]

G3 GPB R J Radford, 21 Farm Cl, Ringwood, BH24 1RZ [IO90CU, SU10]

G3 GPE K Smethurst, 1 Ham Lane(Bungalow, Aston, Bampton, Oxon, OX18 2DE [IO91FR, SP30]

G3 GPG H Y Strain, 129 Bradbourne Vale Rd, Sevenoaks, TN13 3DJ [JO01CG, TQ55]

G3 GPQ B J Killick, 22 St. Catherines, Lincoln, LN5 8LY [IO93RF, SK96]

G3 GPX P J Bartram, 83 Gowing Rd, Hellesdon, Norwich, NR6 6UH [JO02PQ, TG21]

G3 GPZ P C Probert, 61 Oxford St., Northwood, Cowes, PO31 8PT [IO90IR, SZ49]

G3 GQC G W Lowe obo Manfield Am Radio Soc, 25 Manor House Ct, Kirkby in Ashfield, Nottingham, NG17 8LH [IO93IC, SK45]

G3 GQF W Bartle, 78 Dolcoath Rd, Camborne, TR14 8RP [IO70IF, SW64]

G3 GQO M F D Steed, 5 Swallow Hill, Thurlby, Bourne, PE10 0JB [IO92TR, TF01]

G3 GQR G A Burton, 124 Curzon Ln, Alvaston, Derby, DE24 8RG [IO92GV, SK33]

G3 GR J J Campbell, 105 Whitecroft Rd, Sheldon, Birmingham, B26 3RL [IO92CK, SP18]

G3 GRB N H J Cliff, 155 St. Marks Rd, Wolverhampton, WV3 0QN [IO82WN, SO99]

G3 GRC R J T Athey, PO Box 17, Chelsea, Quebec, Canada Jox 1No

GI3 GRD W J C Curtis, 8 Station Rd, Kesh Post Office, Enniskillen, Co Fermanagh, N Ireland, BT93 1UN [IO64DM, H16]

G3 GRF G R Foggin, 16 Maclagan Rd, Bishopthorpe, York, YO2 1QW [IO93KW, SE54]

GM3 GRG D R Rollo, 25 Beaufort Dr, Kirkintilloch, Glasgow, G66 1AX [IO75WW, NS67]

G3 GRM Details withheld at licensee's request by SSL.

G3 GRO D Atter, 1 Little Crabtree, West Green, Crawley, RH11 7HW [IO91VC, TQ23]

G3 GRQ C S Hebden, 12 Abbeygate, Thetford, IP24 1AY [JO02IJ, TL88]

G3 GRS D W Blakeley Gravesend ARS, 338 Thong Ln, Gravesend, DA12 4LQ [JO01EK, TQ67]

G3 GRT E W E Taylor, 1 Windsor Rd, Wrenthorpe, Wakefield, WF1 2BT [IO93FQ, SE32]

G3 GRU M H Jones, Forest View, Poulton Hill, Marlborough, SN8 1AZ [IO91DK, SU16]

G3 GRV G L Halse, 14 Green End Gdns, Hemel Hempstead, HP1 1SN [IO91SS, TL00]

G3 GRW E H Goldsmith, 42 Copenhagen Tower, International Way, Weston, Southampton, SO19 9NU [IO90HV, SU40]

G3 GRX E L Simpson, Everdene, Fell Ln, Penrith, CA11 8AW [IO84PQ, NY53]

GW3 GRY F L Wiseman, Hillcrest, Llanvaches, Newport, Gwent, NP6 3BA [IO81OO, ST49]

GI3 GSB W J Galloway, 65A Connor Rd, Templepatrick, Ballyclare, BT39 0EA [IO64WR, J28]

G3 GSC J C Johnson, Woodthorpe, 55 Spetchley Rd, Worcester, WR5 2LR [IO82VE, SO85]

G3 GSI B S Atkinson, Barklye, Swife Ln, Broad Oak, Heathfield, TN21 8UR [JO00DX, TQ62]

GW3 GSJ E E Hewins, Hillcrest, Wern Rd, Rhosesmor, Mold, CH7 6PY [IO83JF, SJ26]

G3 GSL A N Rennison, 14 Brentwood Rd, Anderton, Chorley, PR6 9PL [IO83QO, SD61]

G3 GSO T W Bryan, Quarndon Lodge Residential Hom, 210 Burley Ln, Quarndon, Derby, DE22 5JS [IO92GX, SK34]

G3 GSR Details withheld at licensee's request by SSL.[Station located near Wimborne.]

G3 GSY K C A Terry, 162 Church Path, Deal, CT14 9TU [JO01QF, TR35]

G3 GTA J E Shute, 23 Brae Rd, Winscombe, BS25 1LJ [IO81OH, ST45]

G3 GTF B W N Harris, Burgess Croft, Croft Rd, Crowborough, TN6 1HA [JO01BB, TQ53]

G3 GTJ J Teague, Perrotts, Lydford on Fosse, Somerton, Somerset, TA11 7HA [IO81QB, ST53]

G3 GTN T J Navin, 43 Catholic Ln, Sedgley, Dudley, DY3 3UF [IO82WM, SO99]

GM3 GTQ A I McPhedran, 3 Argyll Rd, Bearsden, Glasgow, G61 3JX [IO75UW, NS57]

GI3 GTR R B McKinty, 3 Rhanbuoy Park, Craigavad, Holywood, BT18 0DX [IO74CP, J48]

G3 GTW D Kirk, 1 Townsend Ave, Sedgley, Dudley, DY3 3SJ [IO82WN, SO99]

G3 GUD A Bosworth, Arosfa, Watts Green, Chearsley, Aylesbury, Bucks, HP18 0DD [IO91MS, SP71]

G3 GUE A F Dowling, Church Cottage, Frittenden, Cranbrook, Kent, TN17 2DD [JO01HD, TQ84][Op: A F Dowling, MVO, MBE. TD.]

G3 GUL Details withheld at licensee's request by SSL.

G3 GUN P S L Lansley, 6 Youngwoods Way, Alverstone Garden Villag, Sandown, Isle of Wight, PO36 0HE [IO90JP, SZ58]

G3 GUP E T Howell, 164 Beeches Rd, Chelmsford, CM1 2RZ [JO01FR, TL60]

G3 GUR J W Scully, 1 Wyde Feld, Bognor Regis, PO21 3DH [IO90PS, SZ99]

G3 GVC D F Childs, 18 Glam D, Catherington, Waterlooville, PO8 0TR [IO90LW, SU61]

GM3 GVD J A Dunlop, 5 Nutberry Pl, Strathaven, ML10 6HW [IO75XQ, NS64]

G3 GVH W H C Greenwood, Parklands, 70 London Rd North, Poynton, Stockport, SK12 1BY [IO83WI, SJ98]

G3 GVM F L A Robins, 59 Titchfield Rd, Stubbington, Fareham, PO14 2JF [IO90JT, SU50]

G3 GVV R J Hughes, 10 Farm Ln, Tonbridge, TN10 3DG [JO01DE, TQ54]

G3 GVY M G James, 2 Hurstleigh Terr, Harrogate, HG1 4TF [IO93FX, SE35]

GW3 GWA R G Goulding, 10 Earle St, Wrexham, Clwyd, LL13 7DH [IO83MA, SJ34]

G3 GWB R W Guppy Northampton ARC, 4A Church St, Helmdon, Brackley, Northants, NN13 5QJ [IO92KB, SP54]

G3 GWC E J Ramsdale, 8 May Cttgs, Monkswell Ln, Coulsdon, CR5 3SX [IO91VG, TQ25]

G3 GWD M C Pavely, 52 Maidstone Rd, Pembury, Tunbridge Wells, TN2 4DE [JO01DD, TQ64]

G3 GWE Details withheld at licensee's request by SSL.

G3 GWF A Gale, 76 Forest Rd, Ajax, Ont, Canada L152N3

G3 GWH G E Martin, 1 Hedingham Gdns, Roborough, Plymouth, PL6 7DX [IO70WK, SX46]

G3 GWI Details withheld at licensee's request by SSL.[Op: N Spivey.]

GM3 GWL Details withheld at licensee's request by SSL.[Station located near Ayr.]

G3 GWR A G Stormont, Meadow View, 59 Ings Ln, Kellington, Goole, DN14 0NS [IO93KR, SE52]

G3 GWU A Stenhouse, 2 Clarks Ln, Aston on Trent, Derby, DE72 2AB [IO92HU, SK42]

G3 GWY E W Hancock, 51 Fairfield Rd, Epping, CM16 6ST [JO01BQ, TL40]

G3 GWZ Details withheld at licensee's request by SSL.

G3 GXG C M Lee, 5 Haywood Ct, Waltham Abbey, EN9 3DP [JO01AQ, TL30]

G3 GXI A A H Moss Eccles&Dist ARS, 53 Peveril Cl, Whitefield, Manchester, M45 6NS [IO83UM, SD80]

G3 GXN R Mapplebeck, 39 The Osiers, Leicester, LE3 2XN [IO92JO, SK50]

G3 GXQ W E Roberts, 24 Leeds Rd, Barwick in Elmet, Leeds, LS15 4JD [IO93HT, SE33]

G3 GXW J G Lamb, Royal Hospital, Chelsea, London, SW3 4SR [IO91WL, TQ27]

G3 GXX W S Horsfall, The Grange, Ravenstonedale, Kirkby Stephen, Cumbria, CA17 4NG [IO84SK, NY70]

G3 GYC P J Ingram, 28 John Amery Dr, Stafford, ST17 9NA [IO82WS, SJ92]

G3 GYD Details withheld at licensee's request by SSL.[Op: G Robertson. Station located near Camberley.]

G3 GYE P T Pitts, Westmoors, Trezelah, Gulval, Penzance, TR20 8XD [IO70FD, SW43]

G3 GYF A J F Powell, 19 Woodside, Stroud, GL5 1PL [IO81VR, SO80]

G3 GYQ C J Spackman, Highcroft, Hatchers Cres, Blunsdon, Swindon, SN2 4AQ [IO91CO, SU19]

G3 GYU J Wild, 11 Merlewood, Ramsbottom, Bury, BL0 0HE [IO83UP, SD71]

G3 GYW S E Stevenson, 98A Snakes Ln, Southend on Sea, Essex, SS2 6UA [JO01IN, TQ88]

G3 GYZ W G Wooller, Stella Harris, Fyrsway, Fairlight, Hastings, TN35 4BG [JO00HU, TQ81]

G3 GZH R E Brown, 16 Whipsnade Caravan Park, Whipsnade, Dunstable, LU6 2LP [IO91RU, TL01]

G3 GZI J G S Clarke, 37 Denny View, Portishead, Bristol, BS20 8BT [IO81OL, ST47]

G3 GZJ F J Crisp, Boskenwyn Gdns, Gweek, Helston, TR12 7AB [IO70JC, SW62]

G3 GZQ W J Roberts, 34 Barn Park, Buckfastleigh, TQ11 0AS [IO80CL, SX76]

G3 GZT R Moores, 117 Horton Rd, Brighton, BN1 7EG [IO90WU, TQ30]

G3 GZX A E Bladon, 7 Linksview, Wallasey, L45 0NQ [IO83LK, SJ29]

G3 GZZ A A Bevan, 14 Parsonage Rd, Berrow, Burnham on Sea, TA8 2NL [IO81LG, ST25]

G3 HAA J R Morgan, 10 Bamber Gdns, Southport, PR9 7PQ [IO83MP, SD31]

G3 HAB D J Black, Alameda, 1 Portnalls Rise, Coulsdon, CR5 3DA [IO91WH, TQ25]

G3 HAC H G Kimber, 30 Princes Boulevar, Bebington, Wirral, L63 5LN [IO83LI, SJ38]

G3 HAG R P Hughes, 148 Meadowhead, Sheffield, S8 7UF [IO93GI, SK38]

GW3 HAI C G Maxfield, Bryn Goleu, Henbre, Conwy, Gwynedd, LL32 8RX [IO83BG, SH77]

G3 HAL R A Parrott, 3 Ash Gr, Chard, TA20 1BZ [IO80MV, ST30]

GM3 HAM P J Bates Lothians Radio Soc, 9 Winton Terr, Edinburgh, EH10 7AP [IO85JV, NT26]

G3 HAN M J Hitchman, 12 Briar Walk, Oadby, Leicester, LE2 5UE [IO92LO, SP69]
G3 HAO E H Webster, 6 Grange Ave, Hastings, TN34 2QE [JO00GV, TQ81]
GM3 HAT M C Hately, 1 Kenfield Pl, Aberdeen, AB1 7UW [IO87TH, NJ72]
G3 HAV A E Gee, 274 Thorne Rd, Doncaster, DN2 5AJ [IO93KM, SE50]
G3 HAZ R Rew, 38 Manor Ln, Halesowen, B62 8QB [IO82XK, SO98]
G3 HB G L Benbow, 81 Anglesmede Cres, Pinner, HA5 5ST [IO91TO, TQ18]
G3 HBI R J Brooker, 5 Bear Hill, Kingsclere, Newbury, RG20 5QA [IO91IH, SU55]
G3 HBN J R Bolton, Flat A, 40 Queens Gate Terr, London, SW7 5PH [IO91VL, TQ27]
G3 HBR B Hummerstone, 71 St. Leonards Rd, Chesham Bois, Amersham, HP6 6DR [IO91QQ, SU99]
GM3 HBT T Hall, Rosewood, 50 Hamilton St., Larkhall, ML9 2AU [IO85AR, NS75]
G3 HBV D T Jennings, 8 Rushers Cl, Pershore, WR10 1HF [IO82XC, SO94]
G3 HBW A L Mynett, 10 Prior Gr, Chesham, HP5 3AZ [IO91QR, SP90]
G3 HBZ N E A Rush, 77 Maryland Way, Sunbury on Thames, TW16 6HW [IO91TJ, TQ16]
G3 HCH F J Church, 10 Bigstone Gr, Tutshill, Chepstow, NP6 7EN [IO81QP, ST59]
G3 HCJ S E Davies, 7 Campden Rd, Benhall, Cheltenham, GL51 6AA [IO81WV, SO92]
GW3 HCL D E C Lockyer, Gwenfro, 4 Gerddi Menai, Bangor Rd, Caernarvon, LL55 1LN [IO73UD, SH46]
G3 HCM Details withheld at licensee's request by SSL.[Op: D Dumbleton, Highthorn, Melbourne, York, YO4 4QQ.]
G3 HCN Details withheld at licensee's request by SSL.
G3 HCO G A Errock, 28 Winstanley Rd, Sale, M33 2AR [IO83UK, SJ79]
GI3 HCP R D Buckley, 56 Gransha Rd, Bangor, BT20 4TL [IO74EP, J58]
G3 HCQ S Gabriel, Millbrook House, 3 Mill Drove, Bourne, PE10 9BX [IO92TS, TF02]
G3 HCT J Bazley, Brooklands, Henley Rd, Ullenhall, Solihull, B95 5NW [IO92BG, SP16]
G3 HCU A J Martin, Green Dragon, High St., Chipping Campden, GL55 6AL [IO92CB, SP13]
G3 HCW A E Ashby, 22 Rossiter Dr, Knottingley, WF11 0EX [IO93IQ, SE42]
G3 HCX J Arundel, Kia Ora, 1 Crest Dr, Carleton, Pontefract, WF8 2RA [IO93IQ, SE42]
G3 HCY H W Cross, 5 Chippenham Cl, Pinner, HA5 2NF [IO91TO, TQ08]
G3 HCZ B Edmondson, 1 Harbour Ln, Edgeworth, Turton, Bolton, BL7 0PA [IO83TP, SD71]
GW3 HDF K Groves, 6 Overleigh Dr, Buckley, CH7 2PA [IO83KE, SJ26]
GW3 HDH E C Taylor, 8 Conduit St., Port Talbot, SA13 1TA [IO81CO, SS78]
G3 HDJ L J Smith, 118 Charnwood Ave, Westone, Northampton, NN3 3DY [IO92NG, SP76]
GD3 HDL Dr S E Kelly, Wayside, Ballaquane Rd, Peel, Isle of Man, IM5 1PS
G3 HDM S G Campbell, 34 North St., Maldon, CM9 5HL [JO01IR, TL80]
G3 HDO B W Arnold, 159 Oldfield Rd, Coventry, CV5 8FQ [IO92FJ, SP37]
G3 HDQ W Baker, 148 Redditch Rd, Alvechurch, Birmingham, B48 7RX [IO92AI, SP07]
GW3 HDR R J Gilbert, 4 Llwyn yr Eos, Morriston, Swansea, SA6 6AT [IO81AQ, SS69]
G3 HDS Details withheld at licensee's request by SSL.
G3 HDT J D Graham, 18 Wheatriggs Ave, Milfield, Wooler, NE71 6HU [IO85WO, NT93]
G3 HEA J U Burke, Malvern, 11 Ham Meadow, Marnhull, Sturminster Newton, DT10 1LR [IO80UW, ST71]
G3 HEE Details withheld at licensee's request by SSL.[Station located near Stamford.]
G3 HEH P L E Bennett, 4 St Mathews Rd, Eversdal 7550, Cape Town, Rsa
G3 HEJ D V M Stanners, Tanglewood, Samarkand Cl, Prior Rd, Camberley, GU15 1DG [IO91PH, SU85]
G3 HEL Details withheld at licensee's request by SSL.
GM3 HEN A C White, Byeways, Middle Rd, Whiting Bay, Isle of Arran, KA27 8QH [IO75KL, NS02]
G3 HEO D P Hobbs, 5 Highfield Rd, Drayton, Norwich, NR8 6ER [JO02OQ, TG11]
G3 HEQ J H Lomas, 2 Fenay Bank Side, Fenay Bridge, Huddersfield, HD8 0BN [IO93DP, SE11]
G3 HER Details withheld at licensee's request by SSL.
G3 HES K G Pugh, 115 Ryhall Rd, Stamford, PE9 1UJ [IO92SP, TF00]
GW3 HEU D Rickers, 16 Bryn Eglwys Rd, Wrexham, LL13 9LA [IO83MB, SJ35]
G3 HFA R Cairns, 71 Springfield Ave, West Kirby, Wirral, L48 9XB [IO83KJ, SJ28]
GD3 HFC F B Arrowsmith, The Evergreens, South Cape, Laxey, Isle of Man, IM4 7JB
G3 HFM A R Vickers, Foxcroft, 4 Woodlands End, Chelford, Macclesfield, SK11 9BF [IO83UG, SJ87]
GU3 HFN P J Bannier The Guernsey AR, 10-le-Bouet, Longstore, St Peter Port, Guernsey, GY1 2BA[Op: S Henry, obo Guernsey ARS, PO Box 100 , Guernsey.]
G3 HFO N A Smith, 7 The Byeways, Surbiton, KT5 8HT [IO91UJ, TQ16]
G3 HFS H J Ridge, 2 Beramic Cl, Connor Downs, Hayle, TR27 5DP [IO70HE, SW53]
G3 HFW E F Brooks, Leafield, Killerby, Cayton, Scarborough, YO11 3TW [IO94TF, TA08]
G3 HFX P J Wilson, 4 Hampden Cl, Middleton on Sea, Bognor Regis, PO22 7SN [IO90QS, SU90]
G3 HFY J W D Hobbs South London College Radio Soc, 111 Kilmartin Ave, Norbury, London, SW16 4RA [IO91WJ, TQ36]
G3 HFZ J H G Yardley, 30 Ulwell Rd, Swanage, BH19 1LL [IO90AO, SZ07]
GM3 HGA J McCall, 1 Pinewood Pl, Aberdeen, AB1 8LT [IO87TH, NJ72]
G3 HGD Details withheld at licensee's request by SSL.
G3 HGE T H A Withers, Woodpeckers, West Stow, Bury St Edmunds, Suffolk, IP28 4DN [JO02GH, TL77]
G3 HGI I J Soars, 84 Ridge Rd, Kingswinford, DY6 9RG [IO82VL, SO88]
GW3 HGJ D M Foster, Pentwyn House, Newchurch, nr Chepstow, Gwent, NP6 6DD [IO81OQ, ST49]
GW3 HGL B Clark, 97 Rhos Rd, Rhos on Sea, Colwyn Bay, LL28 4TT [IO83DH, SH88]
G3 HGM J A Ewen, 21 The Cres, Caddington, Luton, LU1 4HZ [IO91SU, TL01]
G3 HGQ F Fennell, 48 Brangwyn Ave, Patcham, Brighton, BN1 8XG [IO90WU, TQ30]
G3 HGR Details withheld at licensee's request by SSL.[Op: P Knight. Station located near Sevenoaks.]
G3 HGW D M Bradshaw, 4 Northampton Rd, Chapel Brampton, Northampton, NN6 8AE [IO92MG, SP76]
G3 HHD T J Hayward, Skirt Bank, Nether Silton, Thirsk, North Yorks, YO7 2LL [IO94IH, SE49]
GW3 HHF S Jones, Tadworth, 11 Lawson Rd, Wrexham, LL12 7BA [IO83MB, SJ35]
G3 HHM F J S Chandler, 15 Icknield St., Beoley, Church Hill North, Redditch, B98 9AD [IO92BH, SP06]
GI3 HHN R J Armstrong, 29 Darkfort Dr, Portballintrae, Bushmills, BT57 8TT [IO65RF, C94]
G3 HHR A L Thwaites, Westagarth, Rushley Mount, Host Bank, Lancaster, LA2 6EE [IO84OC, SD46]
G3 HHT J A Bassford, 14 Constable Cres, Whittlesey, Peterborough, PE7 1YY [IO92WN, TL29]
G3 HHU Dr J C W Ickringill, 28 Deena Cl, Queens Dr, London, W3 0HR [IO91UM, TQ18]
G3 HHV A M Hunt, 9 Pengelly, Callington, PL17 7DZ [IO70UM, SX36]
G3 HIA H C Young, 1 Derwent Ave, Droylsden, Manchester, M43 6HJ [IO83WL, SJ89]
GD3 HIC Details withheld at licensee's request by SSL.
G3 HIF A Reid, 205 Mortimer Rd, South Shields, NE34 0RT [IO94GX, NZ36]
G3 HIJ Details withheld at licensee's request by SSL.
G3 HIU V C Webley, 817 PO Box, Milton Keynes, MK6 3LE [IO92PA, SP83]
GI3 HJA P J Flanagan, 87 Main St., Gortin, Omagh, BT79 8NH [IO64JR, H48]
G3 HJC J T Coulman, 11A Lilac Ave, Willerby, Hull, HU10 6AE [IO93SS, TA03]
G3 HJD D H Careless, 33 Manor Bend, Churston, Galmpton, Brixham, TQ5 0PB [IO80FJ, SX85]
G3 HJF L J Smith, Baram, 64 Galley Ln, Arkley, Barnet, EN5 4AL [IO91VP, TQ29]
GI3 HJH R McBurney, 8 Main Rd, Ballymartin, Newry, BT34 4NU [IO74AB, J31]
G3 HJK B J Mitchell, 98 Queensway, Heald Green, Cheadle, SK8 3ET [IO83VI, SJ88]
G3 HJP G Cooper, 25 Plantation Ave, Leeds, LS15 0LL [IO93GT, SE33]
G3 HJS R V Woodford, 19 Fairlie Rd, Littlemore, Oxford, OX4 3SW [IO91JR, SP50]
G3 HJY R A Houtby, 19 Shrewsbury Ct, Shrewsbury Ave, Orton Longueville, Peterborough, PE2 7HT [IO92UN, TL19]
G3 HKA C W Booth, 10 Oldmead Walk, Uplands, Bristol, BS13 7BL [IO81QK, ST56]
G3 HKD D C Money, 125 Wroxham Rd, Norwich, NR7 8AD [JO02PP, TG21]
GM3 HKF G R Singleton, The Bungalow, Portling Farm, Dalbeattie, Scotland, DG5 4PZ [IO84DU, NX85]
G3 HKH M J F Harrison, 3 Stert St., Abingdon, OX14 3JF [IO91IQ, SU49]
G3 HKJ C F Page, 1 Denchfield Rd, Banbury, OX16 9EB [IO92IB, SP43]
G3 HKN F L Shakespeare, Fairways, 53 Ashby Rd East, Bretby, Burton on Trent, DE15 0PS [IO92ET, SK22]
G3 HKO D A Wood, 28 Hillcrest Ave, Scarborough, YO12 6RQ [IO94SH, TA09]
G3 HKQ L V Westmoreland, 5 Gill Green Walk, Clarborough, Retford, DN22 9JP [IO93NI, SK78]
G3 HKT A R Partner, 10 The Tanners, Titchfield Common, Fareham, PO14 4BH [IO90JU, SU50]
G3 HKU F Ratcliffe, 73 Crawford Ave, Leyland, Preston, PR5 2JP [IO83PQ, SD52]
GU3 HKV E H Page, Clos Du Murier, Rue de Bas, St. Sampson, Guernsey, Channel Islands, GY2 4HJ
G3 HKZ J F Hegerty, 20 Edith Rd, Maidenhead, SL6 5DY [IO91OM, SU88]
G3 HLG D E Johnson, Robins, Station Rd, Collingham, Newark Notts, NG23 7RA [IO93OD, SK86]
G3 HLI M S Bradford, 101 Oxendon Way, Ernsford Grange, Binley, Coventry, CV3 2HA [IO92GJ, SP37]
G3 HLM W E Harris, 48 Princes Dr, Sale, M33 3JB [IO83UK, SJ79]
G3 HLN P B Woods, 145 Hollybush Ln, Welwyn Garden City, AL7 4JT [IO91VS, TL21]
G3 HLP G A Brown, 56 Pipers Ln, Hoole, Chester, CH2 3LS [IO83NE, SJ46]
G3 HLR A D Dickins, Copperfield, Chapel Ln, Broughton, Brigg, DN20 0HP [IO93RN, SE90]
G3 HMB I E Elliot, Grange House, Manningtree Rd, Stutton, Ipswich, IP9 2SW [JO01NX, TM13]
G3 HMD Details withheld at licensee's request by SSL.
G3 HMF G G Kenyon, 76 Ashbrook Ave, Dane Bank, Denton, Manchester, M34 2GF [IO83WK, SJ99]
G3 HMG A G Macgregor, 14 Quantock Gr, Williton, Taunton, TA4 4PD [IO81ID, ST04]
G3 HMO J M Osborne, 141 Chadwick Rd, London, SE15 4PY [IO91XL, TQ37]
G3 HMQ J A W Robson, 32 St. Stephens Rd, Cold Norton, Chelmsford, CM3 6JE [JO01IQ, TL80]
G3 HMR G B Moser, 77 Valley Dr, Kendal, LA9 7AQ [IO84PH, SD59]
G3 HMV N A M Bolton, 2 Selborne Villas, Clayton, Bradford, BD14 6JZ [IO93CS, SE13]
G3 HN Col J W W Cock, Downside, 21 Chyngton Rd, Seaford, BN25 4HL [JO00BS, TV49]
G3 HNB L E Maund, 56 Exchange Rd, Stevenage, SG1 1PZ [IO91VV, TL22]
GW3 HNC B Dyer, 5 The Avenue, Woodland Park, Prestatyn, Clwyd, LL19 9RD [IO83HH, SJ08]
GM3 HNE G Campbell, 17 Roseburn Terr, Edinburgh, EH12 5NG [IO85JW, NT27]
GI3 HNM C H Davies, 1 Inisharoan Ct, Greenwell St., Newtownards, BT23 8LZ [IO74DN, J46]
G3 HNP A G Edwards, 8 Linnet Cl, Bradwell, Great Yarmouth, NR31 8JF [JO02UN, TG50]
G3 HNY C K Ashton, High Lodge, Longhouse Ln, Poulton-le-Fylde, Lancs, FY6 8DE [IO83MT, SD33]
G3 HOH F J Longman, 22 Queens Ct, High St. North, Dunstable, LU6 1LD [IO91RV, TL02]
G3 HOI H B Heath, 36 Fernlea Rd, Weston Super Mare, BS22 8NE [IO81MI, ST36]
GW3 HOJ A R Holbrook, Swn y Gwylan, 95 Pencaerfenni Park, Crofty, Swansea, SA4 3SG [IO71WP, SS59]
G3 HOK G E R Crapper, 56 Darby Cres, Sunbury on Thames, TW16 5LA [IO91TJ, TQ16]
GM3 HOM J Reilly, 30 Park Cres, Bishopbriggs, Glasgow, G64 2NS [IO75VV, NS67]
G3 HOO J R Pechey, 17 Orchard Way, Sandiacre, Nottingham, NG10 5NF [IO92IV, SK43]
GM3 HOQ D D Stobie, 32 Corstorphine Hill Ave, Edinburgh, EH12 6LE [IO85IW, NT27]
G3 HOT Details withheld at licensee's request by SSL.
G3 HOU G E West, Flat 1, 1 Somerville Gdns, Tunbridge Wells, TN4 8EP [JO01DD, TQ53]
G3 HOX H Osbaldeston The Manchester DX Radio Club, 126 Trafalgar St., Ashton under Lyne, OL7 0HD [IO83WL, SJ99]
G3 HOY J Parkin, 78 Hayburn Ave, National Ave, Hull, HU5 4LX [IO93TS, TA03]
G3 HPB F J Tooley, 58 Salvington Hill, Worthing, BN13 3BB [IO90TU, TQ10]
G3 HPC W A Stonehouse, Carmelite Lodge, Stott Cl, Efford, Plymouth, PL3 6HA [IO70WJ, SX55]
G3 HPD F Dews, Croft House Farm, 65 Northorpe Ln, Mirfield, WF14 0QN [IO93DQ, SE22]
G3 HPJ T Shepherd, 59 Pantain Rd, Loughborough, LE11 3LZ [IO92JS, SK51]
G3 HPM P J Mullock, 2 Rayners Cl, Fowlmere, Royston, SG8 7TF [JO02AC, TL44]
G3 HPO R C Griffiths, 9 Gordon Cl, Chertsey, KT16 9PR [IO91RJ, TQ06]
G3 HPZ D W G Boast, 70 Lakedale Rd, Plumstead, London, SE18 1PS [JO01BL, TQ47]
GM3 HQC Details withheld at licensee's request by SSL.
G3 HQG G Atkins, 20 Mansfield Rd, Killamarsh, Sheffield, S31 8BX [IO93GJ, SK38]
G3 HQH H Froggatt, Moncrieff, Hague Bar Rd, New Mills, High Peak, SK22 3EA [IO83XI, SJ98]
GD3 HQR A W Anderson, 7 Howstrake Dr, Onchan, Douglas, IM3 1BP
G3 HQS C J Baker, Refinesis, 16 Boulderside Cl, Thorpe St. Andrew, Norwich, NR7 0JJ [JO02QP, TG20]
G3 HQT P J Ball, 68 Brook Ln, Warsash, Southampton, SO31 9FG [IO90IU, SU40]
G3 HQX J Brodzky, 27 Stavedown Rd, South Wonston, Winchester, SO21 3HA [IO91HC, SU43]
G3 HRB J Coatsworth, 72 Lisle Rd, Co, South Shields, NE34 6DH [IO94HX, NZ36]
G3 HRD J Ellis, 9 Boscaswell Terr, Pendeen, Penzance, TR19 7DS [IO70ED, SW33]
G3 HRE F Watson, 54 Tavistock Rd, Cambridge, CB4 3ND [JO02BF, TL46]
G3 HRF V Wood, 113 Maldon Rd, Burnham on Crouch, CM0 8DB [JO01JP, TQ99]
G3 HRH R C Hills, 2 The Dell, Otterbourne Rd, Shawford, Winchester, SO21 2DE [IO91HA, SU42]
G3 HRK D F Willies, 17 Campion Way, Sheringham, NR26 8UN [JO02OW, TG14]
G3 HRN D L Wright, 29 Pinewoods, Church Aston, Newport, TF10 9LN [IO82TS, SJ71]
G3 HRP T J Wright, 8 Beaconsfield, Carrs Meadow, Withernsea, HU19 2EP [JO03AR, TA32]
G3 HRQ Details withheld at licensee's request by SSL.
G3 HRR G R Taylor, 7 Hill St., Cannock, WS12 5DJ [IO92AQ, SK01]
G3 HRU G T Senior, 8A York Ln, Knaresborough, HG5 0AJ [IO94GA, SE35]
G3 HRX J C Hilling, 24 Gloucester Rd, Gaywood, Kings Lynn, PE30 4AB [JO02FS, TF62]
GM3 HRZ Details withheld at licensee's request by SSL.
GM3 HSF W H Hier, Hazeldene, North Ailey Rd, Cove, Helensburgh, G84 0ND [IO75NX, NS28]
G3 HSG F Peirson, Elmslac, Church Wynd, Burneston, Bedale, North Yorks, DL8 2JB [IO94FG, SE38]
G3 HSL F B Peppert, 173 King Oswy Dr, Hartlepool, TS24 9SA [IO94JR, NZ43]
G3 HSP A F Ward, 29 Sandy Ln, Cromer, NR27 9JT [JO02PW, TG24]
G3 HSR Dr J B Smith, 8 Heathcote Pl, Hursley, Winchester, SO21 2LH [IO91HA, SU42]
G3 HSS H J Smith, Villa Cervino, Botley Rd, Horton Heath, Eastleigh, SO50 7DT [IO90IW, SU41]
G3 HST G K Allen, Moor Farm Cottage, East Portlemouth, nr Salcombe, South Devon, TQ8 8PW [IO80CF, SX73]
G3 HSU K J Richards, 25 Weir Rd, Hemingford Grey, Huntingdon, PE18 9EH [IO92WH, TL27]
G3 HSV D E Alesbury, 23 Cullerne Rd, Swindon, SN3 4HU [IO91DN, SU18]
G3 HSW J Cassidy, 31 Woodvale Gdns, Wylam, NE41 8ES [IO94CX, NZ16]
G3 HTA J D Forward, Sunrays, Barnstaple Cross, Crediton, Devon, EX17 2EP [IO80DT, SS80]
G3 HTB M P Squance, 20 Mayfair, West Cliff Rd, Bournemouth, BH4 8BG [IO90BR, SZ09]
G3 HTC G E Storey, 12 Vereker Dr, Sunbury on Thames, TW16 6HF [IO91TJ, TQ16]
G3 HTD R Storey, 56 Southlands Gr, Scarborough, YO12 5PH [IO94SG, TA08]
G3 HTF L W Barclay, Pendragon, 12 St. Stephens Rd, Cold Norton, Chelmsford, CM3 6JE [JO01IQ, TL80]
G3 HTJ W N Walker, 53 Wolfridge Ride, Alveston, Bristol, BS12 2PR [IO81RO, ST68]
G3 HTO R H Dolton, 62 Dilkoosh Rd, Northdene, Natal 4093, South Africa, X X
G3 HTP E G Drackley, 32 Windsor Rd, Chobham, Woking, GU24 8LA [IO91QI, SU96]
G3 HTX W F Hipwell, 289 Kings Ash Rd, Paignton, TQ3 3XG [IO80EK, SX86]
G3 HU T B Fox, 60 Carpenters Wood Dr, Chorleywood, Rickmansworth, WD3 5RJ [IO91RP, TQ09]
G3 HUA R E Holloway, Sturrow Cottage, Lewes Rd, Lindfield, Haywards Heath, RH16 2LQ [IO91XA, TQ32]
G3 HUB M E J Harrison, Rolling Hills, Brandy Ln, Lerryn, Lostwithiel, Cornwall, PL22 0QH [IO70QJ, SX15]
G3 HUD M R Brown, South Lodge, Great North Rd, Bawtry, Doncaster, DN10 6AA [IO93LK, SK69]
G3 HUI J O Kastner-Walmsley, 413 Clifton Dr North, Lytham St. Annes, FY8 2NW [IO83LS, SD32]
GW3 HUJ R Russell, 6 Bryn Terr, Gyffin, Conwy, LL32 8LU [IO83BG, SH77]
G3 HUK M P Morrissey, 1 Hamilton Rd, Church Crookham, Fleet, GU13 0AS [IO91OG, SU85]
G3 HUL D M Mallett, 45 Crown Rd, New Costessey, Norwich, NR5 0ES [JO02OP, TG11]
GM3 HUN W F Hunter, 4 Baird Gr, Edinburgh, EH12 5RP [IO85JW, NT27]
G3 HUO K J J Young, 80 Darbys Ln, Oakdale, Poole, BH15 3ET [IO90AR, SZ09]
G3 HUR D W Brough, 18 Lark Hall Rd, Macclesfield, SK10 1QP [IO83WG, SJ97]
G3 HUT Details withheld at licensee's request by SSL.[Op: Flt Lt M Doubleday, 832 East Rochester Way, Sidcup, Kent, DA15 8PD.]
G3 HUX J V Matthews, 4 Berrington Gr, Ashton in Makerfield, Wigan, WN4 9LD [IO83QL, SJ59]
G3 HVA D G Pinnock, 2 Oak Cl, Oakley, Basingstoke, RG23 7DD [IO91JF, SU55]
G3 HVE A E Broadbent, 5 Fernhill Walk, Blackwater, Camberley, GU17 9HB [IO91OH, SU85]
G3 HVH E Basilio, 111 Vale Rd, Portslade, Brighton, BN41 1GE [IO90VU, TQ20]
G3 HVI S Baskeyfield, 46 Golborn Ave, Meir Heath, Stoke on Trent, ST3 7JQ [IO82WW, SJ93]
G3 HVJ A E Chappell, 22206 Del Valle St, Woodland Hills, California USA 91364, ZZ2 2DE
GM3 HVK J Craig, 152 Avon Rd, Larkhall, ML9 1QB [IO85AR, NS74]
GM3 HVT J K Smith, 7 Cedar Ave, Methil, Leven, KY8 2AY [IO86LE, NO30]
G3 HVX W H Wells, Vyrnwy, Dilwyn, Hereford, HR4 8HS [IO82NE, SO45]
G3 HVY R M Murcott, 94 Norton Ln, Hammerwich, Burntwood, WS7 0HW [IO92BQ, SK00]
G3 HWD A E Jeffrey, 42 Dennis Rd, Padstow, PL28 8DF [IO70MM, SW97]
G3 HWE R F Perrett, 26 Second Ave, Heworth, York, YO3 0RX [IO93LX, SE65]
G3 HWF N W H Perch W.Y.G. RAF ARS, 40 Highfield Ave, Bailiff Bridge, Brighouse, HD6 4EB [IO93CR, SE12]
G3 HWL W L Hitchings, Pine Hollow, 5 North View Rd, Budleigh Salterton, EX9 6BY [IO80IP, SY08]
G3 HWM J F Cowling, Noon Sight, 19 The Dr, Hullbridge, Hockley, SS5 6LZ [JO01HP, TQ89]
GM3 HWN L R Turnbull, 1 Central Dr, Stenhousemuir, Larbert, FK5 4DA [IO86CA, NS88]
G3 HWQ J Hawkins, 5 Dirker Cttgs, Marsden, Huddersfield, HD7 6AU [IO93AO, SE01]
G3 HWS G R Marshall, 39 Kew Rd, Birkdale, Southport, PR8 4HH [IO83LP, SD31]
G3 HWW K R Cass York Amateur Radio Soc, 4 Heworth Village, York, YO3 0AF [IO93LX, SE65]
G3 HWX B J Whitty, Fourways Morris Ln, Halsall, Ormskirk, Lancs, L39 8SX [IO83MO, SD31]
G3 HWY W R Thomas, 30 Alexandra Park, Paulton, Bristol, BS18 5QT [IO81SH, ST65]
G3 HWZ J Ellis, 210A Chester Rd North, Kidderminster, Worcs, DY10 1TN [IO82VJ, SO87]
G3 HXG F Davies, 2 Telford Rd, Wellington, Telford, TF1 2EL [IO82RQ, SJ61]
GI3 HXH Dr J J Cosgrove, 22 Culmore Rd, Londonderry, BT48 7RS [IO65IA, C41]
GI3 HXV R R Parsons, 27 Mandeville Ave, Stratheden Heights, Newtownards, BT23 8XA [IO74DN, J46]
GW3 HXX L Jenkins, Chung SU, Erw-Hir, Llantristant, Mid Glam, CF7 8BY [IO81HN, ST08]
G3 HXZ Details withheld at licensee's request by SSL.
G3 HYG D R P Topping, Bentley, Middle St., Nazeing, Waltham Abbey, EN9 2LB [JO01AR, TL30]
G3 HYH S P Hay, 27 Acres Rd, Leicester Forest East, Leicester, LE3 3HB [IO92JO, SK50]
G3 HYJ O F Simkin, 6 Riseway Cl, Valley Dr Est, Norwich, NR1 4NJ [JO02PP, TG20]
G3 HYR B V Lockey, 67 Boulton Ln, Alvaston, Derby, DE24 0FF [IO92GV, SK33]
G3 HYV D A V Palmer, 2 Yew Tree Rd, Charlwood, Horley, RH6 0DE [IO91VD, TQ24]
GM3 HYX C B Rattray, Bracklyn, 58 Aberdour Rd, Dunfermline, KY11 4PE [IO86GB, NT18]
G3 HZE R M C D Cobb, 57 The Avenues, Norwich, NR2 3QR [JO02PP, TG20]
G3 HZI C L Hatfull, 2 Claverham Cl, Battle, TN33 0JF [JO00FV, TQ71]
G3 HZJ W J Walsh, 4 Meadowbrook Rd, Dorking, RH4 1DH [IO91UF, TQ14]
G3 HZL D F J Walmsley, 2 St. Margarets Ct, Uttoxeter Rd, Draycott, Stoke on Trent, Staffs, ST11 9SL [IO82XX, SJ94]
G3 HZM M Barnsley, Greenways, 11 Cemetery Rd, Denton, Manchester, M34 6FG [IO83WK, SJ99]
G3 HZP H D James, 23 Rampton Rd, Willingham By Stow, Willingham, Cambridge, CB4 5JG [JO02AH, TL46]
G3 HZR B Harris, 1 Newbridge Cttgs, Midgehole, Hebden Bridge, West Yorks, HX7 7AL [IO83XS, SD92]
G3 HZT P S Fraser, 45 The Martlet, The Upper Dr, Hove, BN3 6NT [IO90WU, TQ20]
G3 HZW D C Mainhood, The Gables, Banwell Rd, Hutton, Weston Super Mare, BS24 9TZ [IO81MH, ST35]
GM3 HZX Details withheld at licensee's request by SSL.
G3 IAB M R Curtis, 15 Selbourne St., South Shields, NE33 2TB [IO94GX, NZ36]
G3 IAF M J Marlow, Heather Patch, Tilford Rd, Tilford, Farnham, GU10 2EP [IO91OD, SU84]
G3 IAI C C Robinson, Boughton Mill Farm, Welford Rd, Chapel Brampton, Northampton, NN6 8AB [IO92MG, SP76]
G3 IAL Details withheld at licensee's request by SSL.
G3 IAR M Crowther-Watson, Highfield, St. Clere Hill Rd, West Kingsdown, Sevenoaks, TN15 6AH [JO01DH, TQ56]
G3 IAV R Bayfield, 25 Rowan Cl, Portslade, Brighton, BN41 2PT [IO90VU, TQ20]
G3 IBE L B Landon, 26 Trafford Cl, Leek, ST13 5BG [IO83XC, SJ95]
G3 IBI P A Scutt, 60 Bryony Rd, Stubbington, Fareham, PO14 2QS [IO90JT, SU50]
G3 IBK T R W Trowbridge, 17 Crabtree Crvn Pk, 3 East St, Cannington, Bridgwater, TA5 2HH [IO81LD, ST23]
G3 IBN Dr A R Bailey, 7 Nelson Rd, Ilkley, LS29 8HN [IO93CW, SE14]
GD3 IBQ K Holt, 61 Millford Ave, Nepean, Ontario K2G-1C4, Canada, X X
GM3 IBU A W Wright, Crosslea, Berstane Rd, St. Ola, Kirkwall, KW15 1SZ [IO88MX, HY31]
G3 IBY Dr T H Wilmshurst, 4 Eastern Rd, West End, Southampton, SO30 3EQ [IO90IW, SU41]

G3 ICA G H Adams, Sue-Marey, Selsley Hill Rd, Stroud, Glos, GL5 5JS [IO81VR, SO80]
G3 ICB A P Bull, 91 Lower Way, Thatcham, RG19 3RS [IO91IJ, SU56]
G3 ICG K S P McFarlane, Clifton, 18 Needham Rd, Harleston, IP20 9JY [JO02PJ, TM28]
G3 ICH P N Pitt, 52A Ringwood Rd, St. Ives, Ringwood, BH24 2NY [IO90CU, SU10]
G3 ICK C C Dobson, 16 Lowick Rd, Thrapston, Islip, Kettering, NN14 3JY [IO92RJ, SP97]
G3 ICO G W Davis, Broadview, East Lanes, Mudford, Yeovil, BA21 5SP [IO80QX, ST51]
G3 ICQ D Wigington, 21 Shiphay Ave, Torquay, TQ2 7ED [IO80FL, SX86]
G3 ICX Details withheld at licensee's request by SSL.[Op: E O Wright. Station located near Ledbury.]
G3 ICZ W J Clowes, 144 Norton Ln, Norton-le-Moors, Stoke on Trent, ST6 8BZ [IO83WB, SJ85]
G3 ID A E Tupman, Deanaity, 16 Fairlea Rd, Dawlish, EX7 0LR [IO80GO, SX97]
G3 IDB A D Brooks, 45 Northfield Rd, Townhill Park, Southampton, SO18 2QE [IO90HW, SU41]
G3 IDC F E Johnstone, Honeymead, Bath Rd, Sturminster Newton, DT10 1DR [IO80UW, ST71]
G3 IDG F A Herridge, 96 George St., Basingstoke, RG21 7RW [IO91KG, SU65]
G3 IDI S J V Cakebread, 94 Friars Rd, London, E6 1LL [JO01AM, TQ48]
G3 IDQ J D Booth, 11 Barley Rd, Ipstones, Stoke on Trent, ST10 2QF [IO93AA, SK04]
GM3 IDS J D Gentles Dunfermline ARS, Culra, 19 Clufflat Brae, South Queensferry, EH30 9YQ [IO85HX, NT17]
G3 IDT D I Thompson, 156 Gilbert Rd, Cambridge, CB4 3PB [JO02BF, TL46]
G3 IDW R Reynolds, 6 Church Way, Stratton St. Margare, Swindon, SN3 4NF [IO91DN, SU18]
G3 IDY R Robson, 66 Tilstock Cres, Sutton Farm, Shrewsbury, SY2 6HQ [IO82PQ, SJ51]
G3 IDZ G J E Neville, 33 North Rd, St. Albans, AL3 4NA [IO91TR, TL10]
G3 IED G H M Yule, 2 Dulwich Cl, Newport Pagnell, MK16 0PA [IO92PB, SP84]
G3 IEG R Stringer, 14 Lampits Ln, Great Barrow, Chester, CH3 7JJ [IO83OF, SJ46]
GW3 IEQ P H Hudson, Heulwen, Dinas Dinlle, Caernarfon, LL54 5TW [IO73TC, SH45]
G3 IER D G Martin, 88 Tennyson Rd, St. Marks, Cheltenham, GL51 7DB [IO81WV, SO92]
G3 IEW Details withheld at licensee's request by SSL.[Op: Stan J Heard, Park Beavers, Lympstone, Devon, EX8 5HQ.]
G3 IEZ S Bates, 101 Polstain Rd, Threemilestone, Truro, TR3 6DB [IO70KG, SW74]
G3 IFA F Allsopp, 13 Larch Cl, Allestree, Derby, DE22 2JA [IO92GW, SK33]
G3 IFB F H Bliss, Coppulex, North Rd, The Reddings, Cheltenham, GL51 6RE [IO81WV, SO92]
G3 IFC A Benstead, 24 Amble Rd, Chelmsford, Ma 01824-1907, U S A, X X
G3 IFF R A Coley, 3 Thruxton Rd, Leigh Park, Havant, PO9 4DG [IO90MU, SU70]
G3 IFN K F Norvall, 24 Ryedene, Basildon, SS16 4SY [JO01FN, TQ78]
G3 IFX R R Cooke, 9 Lee Cres, Ilkeston, DE7 5EF [IO92IX, SK44]
G3 IGC A Garforth, 110 Foxdenton Ln, Chadderton, Oldham, OL9 9QR [IO83WM, SD80]
GW3 IGG J P G Jones, Heywood, 40 Lower Quay Rd, Hook, Haverfordwest, SA62 4LR [IO71MS, SM91]
G3 IGH H R C Sanders, 8 Caldicot Cl, Aylesbury, HP21 9UF [IO91OT, SP81]
G3 IGI L E R Hall, 24 Calthorpe Rd, Walsall, WS5 3LX [IO92AN, SP09]
G3 IGM R G Hindes, 57 The Ridgeway, Acton, London, W3 8LW [IO91UM, TQ17][Op: Robert G Hindes. RSGB Member.]
G3 IGN L C Marshall, 11 Coopers Ct, Church St., Charlton Kings, Cheltenham, GL53 8AP [IO81XV, SO92]
G3 IGP J G H Pearce, 16 Old French Horn Ln, Hatfield, AL10 8AJ [IO91VS, TL20]
G3 IGQ M Blewett, Univ of Surrey E & Ar, Electronic Eng Dept, University of Surrey, Guildford, Surrey, GU2 5XH [IO91QF, SU95]
G3 IGR Details withheld at licensee's request by SSL.
G3 IGU K H Coates, 76 Copley Cres, Scawsby, Doncaster, DN5 8QP [IO93JM, SE50]
G3 IGV J W Birkbeck, 4 Tregullan View, Bodmin, PL31 1BH [IO70PL, SX06]
G3 IGW M G Whitaker, Rosedene, Wood Ln, Hipperholme, Halifax, HX3 8HB [IO93CR, SE12]
G3 IGX L F Cowling, Lindeth, 6 Sarabeth Dr, Tunley, Bath, BA3 1EA [IO81SH, ST65]
G3 IGZ D W Bruce, 22 Brownspring Dr, New Eltham, London, SE9 3JX [JO01AK, TQ47]
G3 IHB J Russell, 10 Elm Ct, Hyde Lea, Stafford, ST18 9BJ [IO82WS, SJ92]
G3 IHH G L Fish Arborfield ARC, c/o OIC Pri See Rem, Arborfield, Reading, RG2 9NH [IO91MJ, SU76]
GW3 IHN R Brace, 79 Culfor Rd, Loughor, Swansea, SA4 6UA [IO71XP, SS59]
G3 IHR H R Henly, Stanmer, 99 Moredon Rd, Swindon, SN2 2JG [IO91CO, SU18]
G3 IHX N J Bond, 333 Hillandale Dr, Charlotte Nc 28270, USA, X X
G3 IIA Details withheld at licensee's request by SSL.
G3 III G P B S Lovelock, Shambles, Whatcote, Shipston on Stour, Warks, CV36 5EF [IO92FC, SP34]
G3 IIN M J Griffin, Michaelmas, Southdown Rd, Freshwater Bay, Freshwater, PO40 9UA [IO90FQ, SZ38]
G3 IIO D R Harriott, 23 Hamsey Cres, Lewes, BN7 1NP [IO90XV, TQ31]
G3 IIV A Davies, Paarl, 129 Cotwall End Rd, Sedgley, Dudley, DY3 3YQ [IO82WM, SO99]
G3 IIW M Sands, Beech Lea, St. Marks Rd, Tunbridge Wells, TN2 5LU [JO01DC, TQ53]
G3 IIX H E R Burfield, Fox Corner, Priory Ln, Snape, Suffolk, IP17 1SA [JO02SE, TM45]
G3 IIY E C Clayson, Pevensey, 137 High St., Harrold, Bedford, MK43 7ED [IO92QE, SP95]
G3 IJA J Allan, 29 Aberdour Rd, Goodmayes, Ilford, IG3 9SA [JO01BN, TQ48]
G3 IJE M J Powell, 10 Park Ln, Pickmere, Knutsford, WA16 0JX [IO83GG, SJ67]
G3 IJH W P Packham, 8 Winsford Rd, Oakland Manor, Bury St. Edmunds, IP32 7JJ [JO02IG, TL86]
G3 IJI Details withheld at licensee's request by SSL.
G3 IJL A F Sephton, 16 Bloemfontein Ave, Shepherds Bush, London, W12 7BL [IO91VM, TQ28]
G3 IJN Details withheld at licensee's request by SSL.
G3 IJS J F Stratfull, 55 Craigweil Ln, Aldwick, Bognor Regis, PO21 4XN [IO90PS, SZ99]
GM3 IJT I J C Taylor, Soroba House, Ardfern, Lochgilphead, Argyll, PA31 8QR [IO76FE, NM80]
G3 IJU E Briggs, 32 Lethbridge Rd, Wookey Heights, Wells, BA5 2FN [IO81QF, ST54]
G3 IJV R D Harvey, 16 Gatesgarth Cl, Bakersmead, Hartlepool, TS24 8RB [IO94JQ, NZ53]
G3 IJW G S Garrett, 226 Rydal Dr, Bexleyheath, DA7 5DG [JO01BL, TQ47]
G3 IKA A B Hutchence, 79 Winchester Rd, Chandlers Ford, Eastleigh, SO53 2GG [IO90HX, SU42]
G3 IKB D A Giddens, 89 Pollards Oak Rd, Oxted, RH8 0JE [JO01AF, TQ45]
G3 IKG L D J Corsi, 16 Orchard Cl, Ruislip, HA4 7LS [IO91SN, TQ08]
G3 IKL R T Craxton, 103 Clifton Rd, Rugby, CV21 3QH [IO92JI, SP57]
G3 IKN V A Stagg, 7 The Hermitage, Warfield St., Warfield, Bracknell, RG42 6AS [IO91PK, SU87]
G3 IKQ R H Chilton, 80 Plantation Rd, Hextable, Swanley, BR8 7SB [JO01CJ, TQ57]
G3 IKR J P Moore, The White House, Earls Common Rd, Stock Green, Redditch, B96 6TB [IO82XF, SO95]
G3 IKS Details withheld at licensee's request by SSL.
G3 IKW Details withheld at licensee's request by SSL.
G3 ILB D S Kendall, 2130 Plaza Del Amo, Unit 144, Torrance, Calif 90501 USA, ZZ2 1PL
G3 ILD Details withheld at licensee's request by SSL.
G3 ILE E Marsh, 63 Willows Ln, Accrington, BB5 0SQ [IO83TR, SD72]
G3 ILO Details withheld at licensee's request by SSL.[Station located near Stroud.]
GI3 ILV Details withheld at licensee's request by SSL.[Op: Major (Rtd) James Thompson MBE. Station located in Co Armagh.]
G3 IMC B E Clark, 6 Shalcross Dr, Cheshunt, Waltham Cross, EN8 8UX [IO91XQ, TL30]
G3 IMF P J Faulkner, The Mount, Causeway Head Rd, Sheffield, S17 3DY [IO93FH, SK38]
G3 IMG R Turner, 55 High St., Pensnett, Brierley Hill, DY5 4RP [IO82WM, SO98]
G3 IMK S C Walters, 61 Sussex Gdns, Chessington, KT9 2PU [IO91UI, TQ16]
G3 IMN F E Perrisset, The Retreat, Old Roman Rd, Martin Mill, Dover, CT15 5JY [JO01QD, TR34]
G3 IMV J Hunter, 28 Whiteley Cres, Bletchley, Milton Keynes, MK3 5DG [IO91OX, SP83]
G3 IMW S J M Whitfield, The Corn Loft, Shillingford Abbot, Exeter, EX2 9QH [IO80FQ, SX98]
G3 IMX E G Jolliffe, 96 Cowes Rd, Newport, PO30 5TP [IO90IR, SZ49]
G3 INA Details withheld at licensee's request by SSL.[Station located near Alford.]
G3 INB P A Niblock, 52 Main St., Netherseal, Swadlincote, DE12 8DA [IO92FR, SK21]
G3 ING A J Gillham, 33 Masefield Ave, Southall, UB1 2NE [IO91TM, TQ18]
G3 INI D A Import, The Old Rectory, Dufton, Appleby, Cumbria, CA16 6DA [IO84SP, NY62]
G3 INL A A Chisholm, 95A Salthill Rd, Clitheroe, BB7 1PE [IO83TV, SD74]
G3 INN N S Lilley, Treloen, 62 Queens Rd, Thame, OX9 3NQ [IO91MR, SP70]
G3 INP G W Stanway, 2 Hawthorn Rd, Redcar, TS10 3NU [IO94LO, NZ62]
G3 INR P B B Buchan, 79 Cavendish Ave, Cambridge, CB1 4UR [JO02BE, TL45]
G3 INU R J Appleby, 27 Harrow Ct, Stevenage, SG1 1JS [IO91VV, TL22]
GW3 INW A Davies, 29 Sketty Park Cl, Sketty, Swansea, West Glam, SA2 8LR [IO81AO, SS69]
G3 INY E R Tudor, 133 High St., Harrold, Bedford, MK43 7ED [IO92QE, SP95]
G3 INZ J Tournier, Avalon, 13 Greenlands, Flackwell Heath, High Wycombe, HP10 9PL [IO91PO, SU88]
G3 IOA A B Langfield, 44 Romney St., Moston, Manchester, M40 9LD [IO83VM, SD80]
G3 IOB P Revell, 54 Lytham Rd, Perton, Wolverhampton, WV6 7YY [IO82VO, SO89]
G3 IOE A H Edgar, 10 Western Ave, West Denton, Newcastle upon Tyne, NE5 5BU [IO94DX, NZ16]
G3 IOI N R Pascoe, 118 London Rd, Wickford, SS12 0AR [JO01GO, TQ79]
G3 IOJ B H A Rixon, 5 The Ridgeway, Hertford, SG14 2JE [IO91WT, TL31]
G3 IOM Details withheld at licensee's request by SSL.
G3 ION G A Allcock, 71 Bassett Green Cl, Southampton, SO16 3QX [IO90HW, SU41]
G3 IOR P J A Gowen, 17 Heath Cres, Hellesdon, Norwich, NR6 6XD [JO02PP, TG21]
G3 IOW Details withheld at licensee's request by SSL.
G3 IOZ Details withheld at licensee's request by SSL.
G3 IPC J W Pike, 42 Staithe Rd, Wisbech, PE13 3TF [JO02CQ, TF40]
G3 IPD C W Oakley, 4 Cross Keys Ln, Low Fell, Gateshead, NE9 6DA [IO94EW, NZ26]
G3 IPG G H Phipps, 12 Mill Cl, Pulham Market, Diss, IP21 4TQ [JO02OK, TM18]
G3 IPJ E F Woodward, 58 Marine Dr, Bishopstone, Seaford, BN25 2RU [JO00AS, TQ40]
G3 IPL D A Winters, 43 Manor Cl, Harpole, Northampton, NN7 4BX [IO92MF, SP66]
G3 IPO H Henshaw, 10 Delamere Rd, Gatley, Cheadle, SK8 4PH [IO83VJ, SJ88]
G3 IPP M L Dance, Golf Cottage, 8 St. Johns Rd, West Green, Crawley, RH11 7BD [IO91VC, TQ23]
G3 IPV P W Haylett, Vancover House, Kimberley Rd, Bacton, Norwich, NR12 0EN [JO02RU, TG33]
G3 IPY C W Hope, 37 Britten Dr, Malvern, WR14 3LG [IO82UC, SO74]
G3 IPZ Details withheld at licensee's request by SSL.
G3 IQE J E Fuller, Rm 1 Beechwood Lodge, Priestley Rd, Basingstoke, Hants, RG24 9NR [IO91KG, SU65]
G3 IQF R A Fowler, 49 Westhorpe Park Caravan Site, Westhorpe, Marlow, SL7 3RH [IO91ON, SU88][Op: R A Fowler (Bob). Tel: (016284) 6421. G-QRP Club 024. RAFARS 840]
G3 IQI R Davenport, 5 Greenbank, Connor Downs, Hayle, TR27 5DA [IO70HE, SW53]
GM3 IQL A Lawrence, 2 Castleblair Ln, Dunfermline, KY12 9DR [IO86GB, NT08]
G3 IQM Details withheld at licensee's request by SSL.
G3 IQX E W Popplewell, 71 Thornbury Rd, Southbourne, Bournemouth, BH6 4HU [IO90CR, SZ19]
G3 IQY A B Rees, 105 Salisbury Rd, Worcester Park, KT4 7BZ [IO91UI, TQ26]
G3 IRA J P Wren, Windrush, 29 Carisbrook Trc, Chiseldon, Swindon, Wils, SN4 0LW [IO91DM, SU17]
GW3 IRK A C Whitehill, Cey-Fra-Cy, Penrhyncoch, Aberystwyth, Dyfed, SY23 3EG [IO82AK, SN68]
G3 IRM P Lumb, 2 Briarwood Ave, Bury St. Edmunds, IP33 3QF [JO02IF, TL86]
G3 IRP R W Plumb, 25 Love Ln, Morden, SM4 6LQ [IO91VJ, TQ26]
G3 IRQ P M Rackham, Upyonda, Otley Bottom, Otley, Ipswich, IP6 9NG [JO02OD, TM25]
GM3 IRV Dr E L Godfrey, 1 Strathearn Pl, Edinburgh, EH9 2AJ [IO85JW, NT27]
G3 IRW R A Wade, Hillside, 51 Main Rd, Hundleby, Spilsby, PE23 5LZ [JO03AE, TF36]
G3 IRX Details withheld at licensee's request by SSL.
G3 ISB C J Brock, Keusgasse 21, 52159 Roetgen, West Germany, XXX XXX
G3 ISD E J Hatch, 147 Borden Ln, Sittingbourne, ME10 1BY [JO01II, TQ86]
G3 ISG S E Green, Oakwood Lodge, Corston Fields, Corston, Bath, BA2 9EZ [IO81SJ, ST66]
GW3 ISJ J J Caulfield, 34 Greenmeadow Dr, Tongwynlais, Cardiff, CF4 7LU [IO81JM, ST18]
G3 ISK K Easter, 44 Rainsborowe Rd, Colchester, CO2 7JR [JO01KV, TL92]
G3 ISL G B Davis, 4 Castle Mount Ave, Scalby, Scarborough, YO13 0PJ [IO94SH, TA09]
G3 ISP P Cairns, 237 Victoria Rd West, Hebburn, NE31 1UH [IO94FX, NZ36]
G3 ISQ R H C White, 43, Hinton Parva, Wimborne, Dorset, BH21 4JG [IO80XU, ST90]
G3 IST S H Turner, 8001 Bayshore Dr, Seminole 34646, Florida, USA
G3 ISU J F Lucas, Briarwood, Barton-le-Street, Malton, North Yorks, YO17 0PL [IO94ND, SE77]
G3 ISX C J Leal, 61 Light Oaks Ave, Light Oaks, Stoke on Trent, ST2 7NF [IO83WB, SJ95]
G3 ISZ P D Morris, 206A Great West Rd, Hounslow, TW5 9AW [IO91TL, TQ17]
GJ3 IT Capt R C S Reid, 62 Elizabeth Ave, La Route Orange, St. Brelade, Jersey, JE3 8GR
G3 ITB T H Bartlett, Yew Tree Farm, 19 The Street, Hardley, Norwich, Norfolk, NR14 6BY [JO02SN, TG30]
GW3 ITD M R Davies, Hafan Wen, Bettws Ifan, Beulah, Newcastle Emlyn, SA38 9QL [IO72SC, SN24]
GM3 ITE R W G Costford, 22 Haldane Pl, Murray Three, East Kilbride, Glasgow, G75 0LN [IO75VS, NS65]
G3 ITF B S Freeman, 47 Gorham Ave, Rottingdean, Brighton, BN2 7DP [IO90XT, TQ30]
G3 ITH R D Franklin, 2 Berkeley Dr, Kingswinford, DY6 9DX [IO82VL, SO88]
G3 ITK P Haddock, 19 Aldersyde St., Bolton, BL3 3BB [IO83SN, SD70]
G3 ITL J E Humpoletz, 76 Marlborough Rd, Braintree, CM7 9LR [JO01GV, TL72]
GM3 ITN L Hamilton, Halls Land, Hardgate, Clydebank, Glasgow, G81 6NR [IO75TW, NS47]
G3 ITP J T Parker, 38 Beverley Way, Malvern, WR14 1LA [IO82TD, SO74]
GW3 ITT J S Cairns, 2 Ffordd Tirion, Sychdyn, Mold, CH7 6DY [IO83KE, SJ26]
G3 IUB D S Cottam Birmingham A R S, The School of Public Policy, J G Smith Building, University of Birmingham, Edgbaston, Birmingham, B15 2TT [IO92AK, SP08]
G3 IUC R D McMillan, 23 Francis Cl, Horndon on The Hill, Stanford-le-Hope, SS17 8NT [JO01EM, TQ68]
G3 IUE M W Newell, Southerly Rosudgeon, Penzance, Cornwall, TR20 9PA [IO70GC, SW52]
G3 IUJ R A Rogerson, 19 Martins Rd, Shortlands, Bromley, BR2 0EE [JO01AJ, TQ36]
G3 IUK C M Swift, Four Winds, 25 Charnwood Ave, Belper, DE56 1EA [IO93GA, SK34]
G3 IUL G H Taylor, 4 Edward Rd, Feltham, TW14 9RF [IO91SL, TQ07]
G3 IUO G W Allen, Rear Flat, 160 West St., Bedminster, Bristol, BS3 3LY [IO81QK, ST57]
G3 IUS V H Emms, 21 Badgers Hollow, Checkley, Stoke on Trent, ST10 4NW [IO92AW, SK03]
G3 IUV G S Loveday, 2 St. Aldwyns Cl, Horfield, Bristol, BS7 0UQ [IO81RL, ST57]
G3 IUW L E Pritchard, Green Horizons, Send Hill, Send, Woking, GU23 7HR [IO91RG, TQ05]
G3 IUY J Presland, 14 Townsend Cl, Barkway, Royston, SG8 8ER [JO01AX, TL33]
G3 IUZ Details withheld at licensee's request by SSL.
G3 IVA H I Wright, Rest Harrow, Welborne Rd, Mattishall, Dereham, NR20 3LJ [JO02MP, TG01]
G3 IVB Details withheld at licensee's request by SSL.
G3 IVC A Sycamore, The Clock House Flat, The Street, Sedlescombe, Battle, E Sussex, TN33 0QE [JO00GW, TQ71]
G3 IVF H E Smith, Greenacres, The Green, Kirk Langley, Ashbourne, DE6 4NN [IO92EW, SK23]
G3 IVG F Stocks, 252 Edgeside Ln, Waterfoot, Rossendale, BB4 9TY [IO83VQ, SD82]
G3 IVH E J Younge, 11 Charlottes, Washbrook, Ipswich, IP8 3HZ [JO02NA, TM14]
GI3 IVJ C J Rourke, Box 86, Porto Santo Island, Madeira, Portugal P9400
GW3 IVK D P T Evans, 11 Hill View, Bryn y Baal, Mold, CH7 6SL [IO83KE, SJ26]
G3 IVM B Bush, Woodstock, Burys Bank Rd, Greenham, Thatcham, RG19 8DB [IO91IJ, SU56]
G3 IVQ E J Brickstock, 3 Vauxhall Rd, Stourbridge, DY8 1EX [IO82WK, SO98]
GW3 IVR N H Bromley, 176 Westbourne Rd, Penarth, CF64 5BR [IO81JK, ST17]
G3 IVZ W E Stephen, 44 Pettys Brook Rd, Chineham, Basingstoke, RG24 8RW [IO91LH, SU65]
G3 IW Details withheld at licensee's request by SSL.[obo Vectis Wireless Group.]
G3 IWA A J Worrall, 63 Thrupp Ln, Thrupp, Stroud, GL5 2DF [IO81VR, SO80]
G3 IWB M C W Fozzard, 6 Petworth Gdns, Southend on Sea, SS2 4TG [JO01JN, TQ98]
G3 IWC W F Cox, 12 Cliff Cl, Reedham, Norwich, NR13 3TS [JO02SN, TG40]
G3 IWE A M H Wyse, 29 Tregainlands Park, Washaway, Bodmin, PL30 3AU [IO70OL, SX06]
G3 IWH I J Hall, 46 Bushmead Rd, Luton, LU2 7EU [IO91TV, TL02]
G3 IWP A Parr, 43 Argyle Rd, Poulton-le-Fylde, FY6 7EW [IO83MU, SD33]
G3 IWT J P Hewitt, 1 Cherry Tree Ct, Wadham Park, Crewkerne, TA18 7DH [IO80OV, ST41]
G3 IWV J Parker, 11 Castle Rd, Winton, Bournemouth, BH9 1PH [IO90BR, SZ09]
G3 IWW R M T Hopkins, 34 Shelley Cl, Abingdon, OX14 1PR [IO91IQ, SU59]
GM3 IWX W J Ritchie, 8 Cheviot Pl, Grangemouth, FK3 0DE [IO86DA, NS98]
G3 IXA Dr F D M Livingstone, 15 Chaff Cl, Whiston, Rotherham, S60 4JH [IO93IJ, SK49]
G3 IXB G M Cook, 30 Nicholas Ave, Whitburn, Sunderland, SR6 7DG [IO94HW, NZ46]
G3 IXC L G Cratchley, 2 Stratford Ave, Walmersley, Bury, BL9 5LB [IO83UO, SD81]
G3 IXG G R Cobb, Western House, 75 Ampthill Rd, Shefford, SG17 5AZ [IO92TA, TL13]
G3 IXI K H Landon, Leedons Park, Childswickham Rd, Broadway, WR12 7HB [IO92BA, SP03]
G3 IXN M M Lovejoy, 73 Stoneham Ln, Swaythling, Southampton, SO16 2NZ [IO90HW, SU41]
G3 IXO H J Tyson, Hillfield, 13 Beech Rd, Shipham, Winscombe, BS25 1SA [IO81OH, ST45]
GM3 IXW R D Jack, 30 Hill Terr, Markinch, Glenrothes, KY7 6EN [IO86KE, NO20]
G3 IXZ R T Bowden, Madley, 41 Brockington Rd, Bodenham, Hereford, HR1 3LP [IO82PD, SO55]
G3 IY J Pollard, 161 Woodgrove Rd, Burnley, BB11 3EQ [IO83VS, SD83]
G3 IYB J M Bentley-Beard, 1 The Pastures, Dog Ln, Napton, Rugby, CV23 8LT [IO92IF, SP46]
G3 IYF D H E Baker, Long Haul, 3 Chapel Ln, North Scarle, Lincoln, LN6 9EX [IO93PE, SK86]
G3 IYQ Details withheld at licensee's request by SSL.
G3 IYT S R Walker, 71 Humberston Ave, Humberston, Grimsby, DN36 4SR [IO93XM, TA20]
G3 IYU K B Smith, Ginbac, Upper Minety, Malmesbury, Wilts, SN16 9PT [IO91AO, SU09]
G3 IYX G J Leask, 17 Queen Anne St., New Bradwell, Milton Keynes, MK13 0BB [IO92OB, SP84]
G3 IZA D S Allison, 71 South Hill Rd, Bromley, BR2 0RW [JO01AJ, TQ36]
G3 IZD I E Davies, 7 Hawcoat Ln, Barrow in Furness, LA14 4HE [IO84JD, SD27]
G3 IZF D B Taylor, 24 Woodville Ave, Crosby, Liverpool, L23 3BZ [IO83LL, SJ39]
G3 IZG J W Wells, 23 Gainsborough Rd, Blackpool, FY1 4DZ [IO83LT, SD33]
G3 IZJ M J Faulkner, 35 Abbey Way, Farnborough, GU14 7DD [IO91PH, SU85]
G3 IZM J S Harper Bill, 1 Shepherds Cl, Staple Hill, Bristol, BS16 5LE [IO81SL, ST67]
G3 IZQ H Hyman, 19 Black Horse Dr, Acton, Massachusetts 01720, USA, ZZ1 9BL
G3 IZU D Aveling, 9 Springcroft, Hartley, Dartford, Kent, DA3 8AR [JO01DJ, TQ66]
G3 IZV Details withheld at licensee's request by SSL.
G3 IZW D J Weaver, 7 Rolfe Dr, Burgess Hill, RH15 0LA [IO90WW, TQ31]
G3 JAA P J Pitt, 52A Ringwood Rd, St. Ives, Ringwood, BH24 2NY [IO90CU, SU10]
G3 JAG J A Crux, 4 Sandyway, Prestwich, Manchester, M25 8PG [IO83UM, SD80]
G3 JAH E D Evans, 23 Hazelmount Dr, Warton, Carnforth, LA5 9HR [IO84OD, SD47]
G3 JAL R C Taylor, 304 Brigstock Rd, Thornton Heath, CR7 7JE [IO91WJ, TQ36]
G3 JAM B J P Howlett, 1 Cheltenham Garden, Loughton, Essex, IG10 3AW [JO01AP, TQ49]
G3 JAS K Riley, 15 Pinfold Way, Weaverham, Northwich, CW8 3NL [IO83RG, SJ67]
G3 JAU C R Davies, 107 Talbot Rd, Bournemouth, BH9 2JE [IO90BR, SZ09]
G3 JAX R S Gibson, 18 Fairfield Rd, Bosham, Chichester, PO18 8JH [IO90NU, SU80]
GW3 JAZ B M Poole, 57 Annefield Park, Gresford, Clwyd, LL12 8NR [IO83MC, SJ35]
G3 JBF L P Brown, Ladygate, St. Michaels Rd, Penkridge, Stafford, ST19 5AH [IO82WR, SJ91]
GW3 JBH J S Hammond, Lynwood, Crick, Chepstow, Gwent, NP6 4UW [IO81PO, ST49]
GW3 JBN F N Mathers, 17 Penlon, Menai Bridge, LL59 5LR [IO73VF, SH57]
G3 JBP J Claydon, Glentara, Albury Rd, Little Hadham, Ware, SG11 2DQ [JO01BV, TL42]
G3 JBQ Details withheld at licensee's request by SSL.
G3 JBR D P Tipper, 10 Lowdale Ave, Scarborough, YO12 6JW [IO94TG, TA08]
G3 JBT J Irlam, 13 Ashmeads Ct, Colehill, Wimborne, BH21 2LG [IO90AT, SU00]
G3 JBV D W A Smith, 11 Draycote Cres, Darlington, DL3 8DX [IO94FM, NZ21]
GW3 JBZ J Brace, 12 Heol Gwili, Gorseinon, Swansea, SA4 4GE [IO71XQ, SS59]
G3 JCB Details withheld at licensee's request by SSL.
GM3 JCC Details withheld at licensee's request by SSL.
GI3 JCD G Graham, 24 Lawnbrook Dr, Lisburn, BT27 4UB [IO64XM, J26]
G3 JCJ C L Antrobus, 10 Rodger Rd, Woodhouse, Sheffield, S13 7RH [IO93HI, SK48]
G3 JCK F G Chilvers, Hughley House, Twyford Ln, Foulsham, Dereham, NR20 5SE [JO02MS, TG02]
G3 JCL C K Lawson, 23 Alington Gr, Wallington, SM6 9NH [IO91WI, TQ26]
G3 JCR K J Smith, 20 Manor House Gdns, Abbots Langley, WD5 0DH [IO91SQ, TL00]

G3 JCU J Craven, 5 Uplands Cl, Clayton Heights, Queensbury, Bradford, BD13 1ET [IO93CS, SE13]
G3 JCW B E Greville, 11 Venns Acre, Symn Ln, Wotton under Edge, GL12 7BE [IO81TP, ST79]
G3 JCZ W E Rymer, Ebury House, Great Ln, Hackleton, Northampton, NN7 2AN [IO92OE, SP85]
G3 JDA C C Price, 3 Hornshurst Rd, Rotherfield, Crowborough, TN6 3ND [JO01CB, TQ52]
G3 JDC G Metcalfe, Hungate Lodge, Bishop Monkton, Harrogate, North Yorks, HG3 3QL [IO94GC, SE36]
G3 JDD R G R Dobson, 16 Howden Rd, Fulham, South Australia, 5024
G3 JDG D L Gibson, 2 Bede Ct, Little Gaddesden, Berkhamsted, HP4 1NY [IO91RT, SP91]
GW3 JDJ Details withheld at licensee's request by SSL.[Op: C J Haycock.]
G3 JDK H N Kirk, 54 Allendale Rd, Rotherham, S65 3BY [IO93IK, SK49]
G3 JDM P J Wright, 10 Hillcrest, Stafford, ST17 9YA [IO82WT, SJ92]
G3 JDO H Martin, 7 Nairn St., Jarrow, NE32 4HX [IO94GX, NZ36]
G3 JDP A B Altschul, 136 Pennine Dr, London, NW2 1NL [IO91VN, TQ28]
GM3 JDR D Robertson, Aukengill, Wick, Caithness, Scotland, KW1 4XP [IO88KN, ND36]
G3 JDT B J Read, Glenside, 4 Hatton Ln, Hatton, Warrington, WA4 4BY [IO83QI, SJ68]
GM3 JDX W M Roger, 9 Fairmile Ave, Edinburgh, EH10 6RJ [IO85JV, NT26]
G3 JDY B Atkinson obo Royal Air Force ARS E Ridi, 165 Alliance Ave, Anlaby Rd. Hull, HU3 6QY [IO93TR, TA02]
G3 JEB G E Bartle, 30 Victoria Rd, Capel-le-Ferne, Folkestone, CT18 7LR [JO01OC, TR23]
G3 JEP Rev J E Penney, 28 Oakleigh Rd, Exmouth, EX8 2LN [IO80HO, SY08]
G3 JES I Bolton, 168 Downs Rd, Canterbury, CT2 7TW [JO01MH, TR15]
GW3 JET C W C Richards, 73 Jalan Pantai, 71000 Port Dickson, Malaysia
GW3 JEZ R R Bastin, 86 Christchurch Rd, Newport, Gwent, NP9 7SP [IO81MO, ST38]
G3 JFC B M Stone, 12 Robertson Ave, Leasingham, Sleaford, NG34 8NJ [IO93SA, TF04]
G3 JFD B J C Brown, 130 Ashland Rd West, Sutton in Ashfield, NG17 2HS [IO93ID, SK45]
GM3 JFG Rev I W T D McHardy, Beech Tree Cottage, Navity, Cromarty, Ross-shire, IV11 8XY [IO77XP, NH76]
G3 JFH T A Russell, 15 Orchards Rd, Bishops Cleeve, Cheltenham, Glos, GL52 4LX [IO81XW, SO92]
G3 JFL Details withheld at licensee's request by SSL.
G3 JFP J F Proctor, 35 Westmeston Ave, Rottingdean, Saltdean, Brighton, BN2 8AL [IO90XT, TQ30]
G3 JFQ G Parlett, 74 Queens Rd, Wisbech, PE13 2PH [JO02BP, TF40]
G3 JFR N B Cottrell, 28 Colley Wood, Kennington, Oxford, OX1 5NF [IO91JR, SP50]
G3 JFS Details withheld at licensee's request by SSL.
G3 JFU V A W Bennington, Sandbank, Wisbech St Mary, nr Wisbech, Cambs, PE13 4SE [JO02AP, TF40]
G3 JFW F S White, Sejenane, 82 Nottingham Rd, Hucknall, Nottingham, NG15 7QE [IO93JA, SK54]
G3 JFY M J I Lillington, 23 Stapleford Cl, Romsey, SO51 7HU [IO91GA, SU32]
GW3 JGA J E T Lawrence, 40 Aberconwy Rd, Prestatyn, LL19 9HL [IO83HI, SJ08]
G3 JGB Details withheld at licensee's request by SSL.[Op: D W Laverack. Station located near Scunthorpe.]
G3 JGC Details withheld at licensee's request by SSL.[Op: R Bliss. Station located near Poole.]
G3 JGE V Owen, 5 Ness Gr, Cheadle, Stoke on Trent, ST10 1TA [IO92AX, SK04]
G3 JGH C Merrett, Thameside Hfe, Beldham Gdns, West Molesey, Surrey, KT8 1TF [IO91TJ, TQ16]
GM3 JGS W G Shand, 4 Sunnyhill Pl, Turriff, AB53 4EU [IO87SN, NJ75]
GJ3 JGY T Wood, 90 Les Quennevais Park, St. Brelade, Jersey, JE3 8GD
G3 JHC B Jenkinson, 22 Calderbrook Dr, Cheadle Hulme, Cheadle, SK8 5RT [IO83VJ, SJ88]
G3 JHI R L S Hathaway, 30 Berkeley Dr, Hornchurch, RM11 3PY [JO01CN, TQ58]
G3 JHK R Kilminster, 32 Shaw Rd, Heaton Moor, Stockport, SK4 4AE [IO83VK, SJ89]
G3 JHL J H Lepper, Turlington, Salisbury Rd, Shootash, Romsey, Hants, SO51 6GA [IO90FX, SU32]
G3 JHM D T Hayter, High Peak, Telegraph Ln, Four Marks, Alton, Hants, GU34 5AW [IO91LC, SU63]
G3 JHP E W G Allen, 11 Newlands Cl, Horley, RH6 8JR [IO91VE, TQ24]
G3 JHS A Wilson, Bardon, 7 Main St., Rempstone, Loughborough, LE12 6RH [IO92KT, SK52]
G3 JHU C W E Pavey, 3 Field Cl, Chatham, ME5 9TD [JO01GI, TQ76]
G3 JID A K Lord, 24 Cowling Gdns, Menheniot, Liskeard, PL14 3QJ [IO70TK, SX26]
G3 JIE D C Youngs, 12 Fox Gr, East Harling, Norwich, NR16 2PS [JO02LK, TL98]
GM3 JIG K R Hodge, Oakburn, Ardrossan Rd, Seamill, West Kilbride, KA23 9LX [IO75NQ, NS24]
GM3 JIH A M M Speed, 14 Nevis Pl, Broughty Ferry, Dundee, DD5 3EL [IO86NL, NO43]
G3 JII A J Isaac, Wave Crest, Marine Dr, Bigbury on Sea, Kingsbridge, TQ7 4AS [IO80BG, SX64]
GM3 JIJ J D Hague, 73 Lower Bayable Pnt, Stornoway, Isle of Lewis Wi, HS6 0QB [IO67IO, NF86]
G3 JIK P Eccleston, 26 Sowdley Green, Wheaton Aston, Stafford, ST19 9QB [IO82VR, SJ81]
G3 JIP J W Hill, 15 Kingsway, Chalfont St. Peter, Gerrards Cross, SL9 8NS [IO91RO, TQ08]
G3 JIR J A Hardcastle, 8 Norwood Gr, Rainford, St. Helens, WA11 8AT [IO83OM, SD40]
G3 JIS R V Heaton, 20 Tewkesbury Ave, Urmston, Manchester, M41 0RJ [IO83TK, SJ79]
GD3 JIU M R Thompson, 3 Cl Cam, Port Erin, IM9 6NB
G3 JIX Dr K L Smith, Staple Farm House, Durlock Rd, Staple, Canterbury, CT3 1JX [JO01PG, TR25]
G3 JIZ J M Read, The Manse, 46 East St., Alresford, SO24 9EQ [IO91KC, SU53]
G3 JJA E F Steventon, 72 Sandylands Park, Wistaston, Crewe, CW2 8HD [IO83SB, SJ65]
G3 JJG G F Gearing, 35 Merestones Dr, The Park, Cheltenham, GL50 2SU [IO81WV, SO92]
G3 JJJ J J Johnson, 12 Ln Head, Windermere, LA23 2DW [IO84NI, SD49]
G3 JJM Details withheld at licensee's request by SSL.
GM3 JJQ D M Millar, 51 Tiree Cres, Polmont, Falkirk, FK2 0UX [IO85DX, NS97]
G3 JJR J Rickwood, 852 Stafford Rd, Fordhouses, Wolverhampton, WV10 6NU [IO82WP, SJ90]
G3 JJU R E Hurst, 31 Avondale Rd, Fleet, GU13 9BP [IO91OG, SU85]
G3 JJW J B Johnson, 44 Castle Ave, Duston, Northampton, NN5 6LF [IO92MG, SP76]
G3 JJZ D J S Newton, 4 Denmark Rd, Bromley, BR1 3AB [JO01AJ, TQ46]
G3 JKB Details withheld at licensee's request by SSL.[Op: D E Simmonds, Parsonage Farm, Binbrook, Lincs, LN3 6BN.]
GM3 JKC C Cooper, 28 Kippford St., Glasgow, G32 9BW [IO75WU, NS66]
G3 JKD J Davison, 6 Eden Cl, Heighington Village, Newton Aycliffe, DL5 6RU [IO94EO, NZ22]
G3 JKE G W Thomas, 13 Essex Dr, Galmington, Taunton, TA1 4JX [IO81KA, ST22]
G3 JKF K V Frankl1N, 10 Weald Dr, Furnace Green, Crawley, RH10 6JU [IO91VC, TQ23]
G3 JKK R M Allen, 44 Clifton Rd, Henlow, SG16 6BJ [IO92UA, TL13]
G3 JKM D Buckland, Moorings, Wellingham, Kings Lynn, PE32 2TH [JO02JS, TF82]
G3 JKN Details withheld at licensee's request by SSL.[Station located near Spalding, Lincs.]
GM3 JKS F Claytonsmith, Knockycoid Cottage, Barrhill, South Ayrshire, KA26 0QY [IO75PB, NX27]
G3 JKT J Huggett, 20 Thornbury Ave, Weeley, Clacton on Sea, Essex, CO16 9HN [JO01NU, TM12]
G3 JKU J J Forbes, Gilgory, Lingla Bank, Frizington, Cumbria, CA26 3TD [IO84GM, NY01]
G3 JKV W F Blanchard, The Trundle, Tower Hill, Dorking, RH4 2AN [IO91UF, TQ14]
G3 JKX M J Street, 12 Ullswater Cl, Priorslee, Telford, TF2 9RB [IO82SQ, SJ71][Tel: (01952) 299677. RAFARS.]
G3 JKY A J Gould, 25 Clarendon Rise, Lewisham, London, SE13 5ES [IO91XL, TQ37]
G3 JKZ F W Lynes, 2 Sunnyside, Field Assarts, Witney, OX8 5NQ [IO91FT, SP31]
G3 JLB L W A Belger, 103 Whitehill Rd, Gravesend, DA12 5PL [JO01EK, TQ67]
G3 JLH I L Hampton, 14 Hill View Rd, Bournemouth, BH10 5BE [IO90BS, SZ09]
G3 JLN F G Blain, High Ridge, Howgate Ln, Bembridge, Isle of Wight, PO35 5QW [IO90LQ, SZ68]
G3 JLQ B S Thomas, 34 Barton Rd, Market Bosworth, Nuneaton, CV13 0LQ [IO92HP, SK40]
G3 JLS L H F Southwell, 56 Mountfield, Hythe, Southampton, SO45 5AQ [IO90HU, SU40]
G3 JLZ V J Ludlow, 6 Raleigh Cres, Chells, Stevenage, SG2 0EQ [IO91VV, TL22]
G3 JMA J M Appleyard, 194 The Briars, Harlow, Essex, CM18 7JJ [JO01BR, TL40]
G3 JMB J Brooker, 8 Barrowfield, Cuckfield, Haywards Heath, RH17 5ER [IO91WA, TQ32]
G3 JME M Watson, 38 The Paddock, Boroughbridge Rd, York, YO2 6AW [IO93KX, SE55]
G3 JMG J M Gale, 33 Island Cl, Hayling Island, PO11 0NJ [IO90MT, SU70]
G3 JMH V Callaghan, 18 Dean Cl, High Wycombe, HP12 3NS [IO91OP, SU89]
G3 JMJ D E Nunn, Oak Lea, Crouch House Rd, Edenbridge, TN8 5EL [JO01AE, TQ44]
G3 JMK D L Hurrell, 18 Chaucer Cl, Fareham, PO16 7PD [IO90JU, SU50]
GM3 JMM J M Murdoch, 4 Cedar Dr, Milton of Campsie, Glasgow, G65 8AY [IO75WX, NS67]
G3 JMO A L Taylor, 42 Runswick Ave, Redcar, TS10 5EL [IO94KO, NZ52]
G3 JMR L G Barlow, 15 Kinnerley St., Walsall, WS1 2LD [IO92AN, SP09]
G3 JMX P C Hayward, 31 Pinewood Ave, Lowestoft, NR33 9AQ [JO02UL, TM59]
G3 JMY E C Halliday, 4 Parkside Ave, Winterbourne, Bristol, BS17 1LU [IO81RM, ST68]
G3 JMZ J Hilton, Windsor House, Preston Rd, Charnock Richard, Chorley, PR7 5HH [IO83PP, SD51]
G3 JNB V E Brand, West Barn, Low Common, Bunwell, Norwich, NR16 1SY [JO02NL, TM19]
G3 JNI J A K Pitcher, 63 The Chase, Holland on Sea, Clacton on Sea, CO15 5PZ [JO01OT, TM21]
G3 JNJ D A Platt, 22 Charcroft Gdns, Enfield, EN3 7HA [IO91XP, TQ39]
G3 JNM T R Whittaker, 16 Acresdale, Lostock, Bolton, BL6 4PJ [IO83RO, SD60]
G3 JNO F D Buck, 31 Cordillera St, Metro Montanna, Montalban, Rizal Philippines
G3 JNP J H Halman, 131 Reservoir Rd, Gloucester, GL4 6SX [IO81VU, SO81]
G3 JNW H L Fleming, Flat 1, 1 Chubb Hill Rd, Whitby, YO21 1JP [IO94QL, NZ81]
G3 JNY S Ellis, 30 Grangefield Ct, Garforth, Leeds, LS25 1LQ [IO93HS, SE33]
G3 JNZ V D Knibbs, 38, Mixbury, Brackley, Northants, NN13 5RR [IO91KX, SP63]
GM3 JOA H E Stanway, 30 Durham Ave, Edinburgh, EH15 1PA [IO85KW, NT27]
GM3 JOB G R Bryce, 3 West Bowhouse Way, Girdle Toll, Irvine, KA11 1NJ [IO75QP, NS34]
G3 JOC O S Chilvers, 26 Wensum Valley Cl, Hellesdon, Norwich, NR6 5DJ [JO02OP, TG11]
G3 JOE J Brown, South Lodge, Great North Rd, Bawtry, Doncaster, DN10 6AA [IO93LK, SK69]
G3 JOF D B J Hicks, 4 Earnley Rd, Hayling Island, PO11 9SU [IO90MS, SZ79]
G3 JON J Bell, 30 Alms Hill Rd, Sheffield, S11 9RS [IO93FI, SK38]
G3 JOR V E Capell, Endways, 15 Copse Rd, Bexhill on Sea, TN39 3UA [JO00FU, TQ70]
G3 JOT F G Whatley, 1 Mill Cl, Wroughton, Swindon, SN4 9AR [IO91CM, SU18]
G3 JOX A B Greaves, Jacobswell, Woodhill Rd, Sandon, Chelmsford Essex, CM2 7SF [JO01GR, TL70]
GI3 JOZ J E Williamson, 40 Main St., Castlerock, Coleraine, BT51 4RA [IO65OE, C73]
G3 JPB C H Noden, Brownhills Farm, Brownhills, Market Drayton, Shropshire, TF9 4BE [IO82SW, SJ63]
G3 JPE P E Johnson, Camels, Annscroft, Shrewsbury, Shrops, SY5 8AN [IO82OP, SJ40]
GM3 JPF R Smith, 21 Garvel Rd, Milngavie, Glasgow, G62 7JD [IO75TW, NS57]
G3 JPG R R Parker, 6 Cambridge Rd, Chingford, London, E4 7BP [JO01AP, TQ39]
G3 JPJ Details withheld at licensee's request by SSL.[Station located in HA postcode area.]
G3 JPL Details withheld at licensee's request by SSL.
G3 JPM B F Grainge, Grainge House, Little Thurlow Green, Haverhill, Suffolk, CB9 7JH [JO02FD, TL65]
G3 JPO M E Fielding, 68 Mitford Rd, South Shields, NE34 0EQ [IO94GX, NZ36]
G3 JPP E H Price, 7 Lawrence Cl, Shurdington, Cheltenham, GL51 5SZ [IO81WU, SO91]
GW3 JPT C R Reynolds, Beacon View, Bronwylfa Rd, Welshpool, SY21 7RD [IO82KP, SJ20]
G3 JPU D G Plant, 5 Wilfred Owen Ave, Oswestry, SY11 2NB [IO82LU, SJ32]
G3 JPX F W Sherlock, 8 Tudor Rd, Canvey Island, SS8 0ND [JO01GM, TQ78]
G3 JPZ I C Denney, 5 Howard Cl, Harleston, IP20 9HY [JO02PJ, TM28]
G3 JQ Dr E T Webster, 91 Thirlmere, Macclesfield, SK11 7YJ [IO83WF, SJ97]
G3 JQC G W Hawksworth, 16 Birkhead St., Heckmondwike, WF16 0BE [IO93EQ, SE22]
G3 JQI A N Barton, 6 Lloyd Rd, Taverham, Norwich, NR8 6LL [JO02OQ, TG11]
G3 JQJ G Moore, 59 Royds Ln, Rothwell, Leeds, LS26 0BJ [IO93GR, SE32]
G3 JQK E V Jones, Appleton Thorne, Lower Broad Ln, Illogan, Redruth, TR15 3HJ [IO70IF, SW64]
G3 JQL J S Haggart, 22 Alnwick Rd, Newton Hall, Durham, DH1 5NL [IO94FT, NZ24]
G3 JQN J W D Hobbs, 111 Kilmartin Ave, Norbury, London, SW16 4RA [IO91WJ, TQ36]
G3 JQQ D A Bundey, 217 Bloomfield Rd, Bath, BA2 2AY [IO81TI, ST76]
G3 JQS J Guttridge, Victoria Cottage, The Common, West Wratting, Cambridge, CB1 5LR [JO02ED, TL65]
G3 JQV R D Sexton, 16B Beauford Rd, Horam, Heathfield, E Sussex, TN21 0EB [JO00CW, TQ51]
G3 JRC R C Brown, 53 Weldon Way, Merstham, Redhill, RH1 3QA [IO91WG, TQ25]
G3 JRD R G Dancy, 1 Ladds Corner, Eastcourt Ln, Gillingham, ME7 2UW [JO01HJ, TQ76]
G3 JRH P R Horne, 31 Ashley Gdns, Chandlers Ford, Eastleigh, SO53 2JH [IO90HX, SU42]
GW3 JRJ F Buckley, Penysarn, Anglesey, Gwynedd, LL69 9AQ [IO73UJ, SH48]
G3 JRK J R Knight, 10 Lynton Dr, Burnage, Manchester, M19 2LQ [IO83VK, SJ89]
G3 JRL F J Armstrong, 4 Medway Dr, Preston, Weymouth, DT3 6LF [IO80SP, SY68]
G3 JRM J Elsdon Pye Lowestoft Rd, 15 Union Rd, Lowestoft, NR32 2BZ [JO02UL, TM59]
G3 JRS J R Simpson, Maple Croft, Kegworth Rd, Kingston-on-Soar, Nottingham, NG11 0DB [IO92IU, SK52]
GI3 JRW R W Semple, 2 Rathmoyle Park, Carrickfergus, BT38 7NF [IO74CR, J48]
G3 JRX K Boddy, 4 Colwall Ave, Priory Rd, Hull, HU5 5SN [IO93TS, TA03]
G3 JRY S Auty, 3 Rochford Cres, Boston, PE21 9AE [IO92XX, TF34]
G3 JSA D A Wilcox, 13 Richards Dr, Dartmouth, Nova Scotia, Canada, B3A 2P1
G3 JSB S B Jeffrey, 5 Almsford Pl, Harrogate, HG2 8EH [IO93FX, SE35]
G3 JSF A W Baker, Red Gables, Pett Rd, Pett, Hastings, E Sussex, TN35 4HE [JO00HV, TQ81]
G3 JSG J D Shenington - Gun, 72 Fitzroy Cres, Woodley, Reading, RG5 4EX [IO91NK, SU77]
G3 JSJ D V Pritchard, 6 Sundew Rd, Broadstone, BH18 9NX [IO80XS, SY99]
G3 JSK D J Dean, 8 Bradford Rd, Corsham, SN13 0QR [IO81VK, ST87]
GM3 JSO Details withheld at licensee's request by SSL.
G3 JSP P D Rowe, 3 Elton Rd North, Nottingham, NG5 1BU [IO92KX, SK54]
G3 JSR P L Chapman, 27 Ilfracombe Gdns, Romford, RM6 4RL [JO01BN, TQ48]
G3 JSU L W Sampson, 57 Milford Ct, Brighton Rd, Lancing, BN15 8RN [IO90UT, TQ10]
GW3 JSV Details withheld at licensee's request by SSL.[Op: D A S Holmes, Berriew, Powys, SY21 8AU.]
G3 JSW D K Clarke, Friars Oak, Walberton Green, Walberton, Arundel, BN18 0AT [IO90QU, SU90]
GM3 JSX J A J Dalrymple, Homelea, Woodside Rd, Gretna, Dumfrieshire, CA6 5AW [IO85MA, NY36]
G3 JSY C E Nicholson, 19 St. Stephen Rd, Sticker, St. Austell, PL26 7HA [IO70NH, SW95]
G3 JTG E G Gibbins, 45 Church Cl, Locks Heath, Southampton, SO31 6LR [IO90IU, SU50]
G3 JTH W J G Hector, 11 Swan Hill, Shrewsbury, SY1 1NL [IO82OQ, SJ41]
G3 JTI F W J Broomfield, Flat, 27 Yewbarrow Lodge, Main St., Grange Over Sands, LA11 6EB [IO84NE, SD47]
G3 JTJ J T J Jones, Westerland, 1 Roborough Cl, Derriford, Plymouth, PL6 6AH [IO70WK, SX46]
G3 JTK G T Allen, 119 Haymoor Rd, Parkstone, Poole, BH15 3NR [IO90AR, SZ09]
G3 JTO F E Gell, 93 Pasture Rd, Stapleford, Nottingham, NG9 8HR [IO92IW, SK43]
G3 JTQ R F Griffiths, 7 Dever Way, Oakley, Basingstoke, RG23 7AQ [IO91JF, SU55]
G3 JTS T L Stoakes, 62 Dudley Rd, Ilford, IG1 1ET [JO01AN, TQ48]
G3 JTT P K Thompson, 30 Farnol Rd, Yardley, Birmingham, B26 2AF [IO92CL, SP18]
G3 JTX E G Wiffen, 28 Fitzroy Cl, Bassett, Southampton, SO16 7LW [IO90HW, SU41]
G3 JUB S Turner, 5 Balfe St., Seaforth, Liverpool, L21 4NR [IO83LL, SJ39]
G3 JUC R G Timms, Redcroft, Llangrove, Ross-on-Wye, Herefordshire, HR9 6EY [IO81PU, SO51]
GM3 JUD C Urquhart, 36 Mercat Green, Kinross, Perth, PH2 6HT [IO86IL, NO13]
GM3 JUF A McEachen, 40 Lamont Ave, Bishopton, PA7 5LJ [IO75SV, NS47]
G3 JUL G C Voller, 56 Marlborough Rd, Ashford, TW15 3QA [IO91SK, TQ07]
GW3 JUN J O D Jenkins, 5 Woodland Ave, Margam, Port Talbot, SA13 2LP [IO81CN, SS78]
G3 JUU D A Adams, 23 Arlington Gdns, Attleborough, NR17 2NH [JO02MM, TM09]
GW3 JUV J B Kitchin, 25 West Rd, Bridgend, CF31 4HD [IO81FM, SS97]
G3 JUW C L Lovell, 5 Montpelier Rd, Ilfracombe, EX34 9HP [IO71WE, SS54]
G3 JUX J McFarlane, 141 Tyler Gr, Walton, Stone, ST15 0JA [IO82WV, SJ83]
G3 JUY Flt A Mallinder, 234 Nottingham Rd, Mansfield, NG18 4SH [IO93JD, SK55]
G3 JVC J E Cleeve, 44 Ditton Hill Rd, Long Ditton, Surbiton, KT6 5JD [IO91UJ, TQ16]
G3 JVF Details withheld at licensee's request by SSL.
G3 JVL M H Walters, 26 Fernhurst Cl, Hayling Island, PO11 0DT [IO90MS, SZ79]
G3 JVM R H Medcraft, 134 Dulverton Rd, Ruislip Manor, Ruislip, HA4 9AG [IO91TN, TQ18]
G3 JVN D S Keen, 14 Penina Ave, Newquay, TR7 2LE [IO70LJ, SW86]
G3 JVO F F Lee, 32B Station Rd, Harpenden, AL5 4SE [IO91TT, TL11]
G3 JVP V J Purdy, 99 Belmont Rd, Uxbridge, UB8 1QX [IO91SN, TQ08]
G3 JVR D J Nokes, Wakehurst, Highgate Ln, Sutton on Sea, Mablethorpe, LN12 2LJ [JO03DH, TF58]
GW3 JVW J D Davies, Glynteg, 6 Caemawr Gdns, Porth, CF39 9DB [IO81HO, ST09]
GM3 JVX J M Allan, 18 Glebe Rd, Beith, KA15 1EY [IO75QR, NS35]
G3 JWC E T Ward, 21 Rangemore St., Burton on Trent, DE14 2ED [IO92ET, SK22]
G3 JWF W F Foster, 13 Leasway, Upminster, RM14 3AJ [JO01DN, TQ58]
G3 JWG L Stuart, 11 Tutbury Cl, Ashby de La Zouch, LE65 1XD [IO92GR, SK31]
G3 JWH J W Harber, 7 Balmoral Rd, Hornchurch, RM12 4NR [JO01CN, TQ58]
G3 JWI R M Page-Jones, 34 Edwards Way, Hutton, Brentwood, CM13 1BT [JO01EP, TQ69]
G3 JWK K W Starnes, Willow Lodge, Over Hall Dr, Winsford, Ches, CW7 1EY [IO83RE, SJ66]
G3 JWN F D Walker, 1 Shannon Cl, Rastrick, Brighouse, HD6 3LG [IO93CQ, SE12]
G3 JWP B C Leighton, 225 Bramford Rd, Ipswich, IP1 4AG [JO02NB, TM14]
G3 JWQ B A Maycock, Hill House, Bullock Ln, Riddings, Alfreton, DE55 4BP [IO93HB, SK45]
GM3 JWV Flt. Lt J J Snee, Bridgend Annan Water, Moffat, Dumfriesshire, Scotland, DG10 9LS [IO85GI, NT00]
G3 JWW W Walker, 11 Highfield, Harlow, CM18 6HE [JO01BS, TL40]
G3 JWY B Reddington, 1 Grange Cl, Outlane, Huddersfield, HD3 3FU [IO93BP, SE01]
G3 JXA G W Alderman, 30 Ellerton Rd, Tolworth, Surbiton, KT6 7TX [IO91UJ, TQ16]
G3 JXB Details withheld at licensee's request by SSL.
G3 JXC Details withheld at licensee's request by SSL.
G3 JXG F T Hodgson, 23 Cave Cres, Castle Park, Cottingham, HU16 5LA [IO93SS, TA03]
GW3 JXN Dr J E Tindle, Gwernant, Tan y Groes, nr Cardigan, Dyfed, Wales, SA43 2JS [IO72SC, SN24]
G3 JXR P P Haughey, 7 Pulborough Cl, Bletchley, Milton Keynes, MK3 7TU [IO91OX, SP83]
G3 JXT A E Deakin, 6 Hampton Cl, Newport, TF10 7RB [IO82TS, SJ71]
G3 JXU D G Chatfield, Ashlea, 41 Ecton Ln, Sywell, Northampton, NN6 0BA [IO92OH, SP86]
G3 JXV R W Peters, 12 Chase Dr, South Woodham Ferrers, Chelmsford, CM3 5PR [JO01HP, TQ89]
G3 JXW W J Wills, 17 Heron House, Barrow Hill Est, St. Johns Wood, London, NW8 7AJ [IO91WM, TQ28]
G3 JYA E A Smith, 21 Chapel Cl, Finningley, Doncaster, DN9 3DL [IO93ML, SK69]
G3 JYF B Bellringer, 36 Green Ln, Redruth, TR15 1JU [IO70JF, SW64]
G3 JYG J Kirby, 14 Grovelands Rd, Hailsham, BN27 3BZ [JO00CU, TQ50]
G3 JYK J R Dockerill, 14 Elizabeth Ct, Sutton, Ely, CB6 2QW [JO02BJ, TL47]
G3 JYL R M Woodman, 31 Crawley Rd, Horsham, RH12 4DS [IO91UB, TQ13]
G3 JYO W J Grainger, 61 Loxley Ave, Shirley, Solihull, B90 2QF [IO92BJ, SP17]
G3 JYP W B Capstick, Condenser Gap, Bongate, Morland, Appleby in Westmorland, CA16 6HW [IO84SN, NY61]
G3 JYS R G Finch, 8 Chalfont Cl, Allesley Park, Coventry, CV5 9HL [IO92FJ, SP27]
G3 JZE M P Scanlon, Little Pl, Pump Ln, Framfield, Uckfield, TN22 5RG [JO00BW, TQ51]
G3 JZF J E Smith, 17 Mulroy Rd, Sutton Coldfield, B74 2QA [IO92CN, SP19]
G3 JZG R J Riding, 5 Victoria Rd, Bridgnorth, WV16 4LA [IO82SN, SO79]
G3 JZI W D Wildigg, 31 Goldhurst Dr, Tean, Stoke on Trent, ST10 4LS [IO92AW, SK03]
G3 JZJ J Noble, 18 Coronation Dr, Birdwell, Barnsley, S70 5RJ [IO93GM, SE30]
G3 JZL W K Montford, 3 The Close, Main St, Brandon, Coventry, CV8 3JF [IO92HI, SP47]
G3 JZP J Hodgkins, Bridge House, Hunton, Bedale, North Yorks, DL8 1PX [IO94DH, SE19]
G3 JZT R A Cheetham, 7 Parkway, Stockport, SK3 0PX [IO83VJ, SJ88]
G3 JZU G Tomlinson, 11 Glenfield Rd, Stockton on Tees, TS19 7PW [IO94HN, NZ41]
G3 JZV Details withheld at licensee's request by SSL.[Op: Rev T. R. Mortimer, 59 First Avenue, Farlington, Portsmouth, PO6 1JL.]
G3 JZW H W Gadsden, 8 Kingsmead, Edlesborough, Dunstable, LU6 2JN [IO91QU, SP91]
G3 JZX H G Lassman, 5 Fircroft Cl, Woking, GU22 7LZ [IO91RH, TQ05]
G3 KAA L S Cutting, 43 Nappsbury Rd, Luton, LU4 9AL [IO91SV, TL02]
G3 KAC R J Wilkinson University of Bristol Amateur, 84 Park Rd, Bolton, BL1 4RQ [IO83SN, SD70]
G3 KAE J A Rowley, 36 Carr Ln, East Ayton, Scarborough, YO13 9HW [IO94SF, SE98]
G3 KAF J F France, Cobham Cottage, 34 Ladythorn Rd, Bramhall, Stockport, SK7 2ER [IO83WI, SJ88]
G3 KAG A Parker, Hillside, Roston, Ashbourne, Derbyshire, DE6 2EH [IO92CX, SK14]
G3 KAH D Hall, 212 Adswood Rd, Stockport, SK3 8PB [IO83VJ, SJ88]
GM3 KAI J C Bain, Mill House, Main St, Reston, Eyemouth, Berwickshire, TD14 5JN [IO85VU, NT86]

GW3 KAJ D K Jagger, Briarside, Gorn Rd, Llanidloes, SY18 6DQ [IO82FK, SN98]
GM3 KAK J P Hunter, 14 Curlew Rise, Gretna, DG16 5LB [IO84LX, NY36]
GM3 KAM D C Mather, Lochwood Smithy Hse, Beattock, Moffat, DG10 9PS [IO85GG, NY09]
G3 KAN A T Shrewsbury, 36 Winchester Rd, Delapre, Northampton, NN4 8AY [IO92NF, SP75]
G3 KAP R H Taylor, Ledbury House, Charing, Ashford, Kent, TN27 0LS [JO01JF, TQ94]
G3 KAR D G Hammond, Christen Mares, Willersey Hill, Willersey, Broadway, WR12 7PF [IO92BB, SP13]
G3 KAU L S Laszkiewicz, 38 Langley Ln, Ifield, Crawley, RH11 0NA [IO91VC, TQ23]
G3 KAW J W Maddison, 18 Larmour Rd, Great, Grimsby, DN37 9HH [IO93WN, TA20]
G3 KAX G E Mackrell, 1 Sidlow Cttgs, Doversgreen Rd, Reigate, Surrey, RH2 8PN [IO91VF, TQ24]
G3 KAY R J Lang, 18 Willow Cl, Saxilby, Lincoln, LN1 2QL [IO93QG, SK87]
G3 KAZ J E Saunders, 7 Queens Rd, Bradford Abbas, Sherborne, DT9 6RR [IO80QW, ST51]
G3 KBH Dr M P Hughes, Northdean, Brimstone Ln, Meopham, Gravesend, DA13 0BW [JO01EI, TQ66]
G3 KBI T S Waller, 12 Skelton Rd, Brotton, Saltburn By The Sea, TS12 2TJ [IO94MN, NZ61]
GM3 KBP A G Kerr, 47 Hillpark Ave, Edinburgh, EH4 7AH [IO85IX, NT27]
G3 KBQ P H Huntsman, 16 Cres Ave, Hexham, NE46 3DP [IO84WX, NY96]
G3 KBR R A Huntsman, 23 Worts Causeway, Cambridge, CB1 4RJ [JO02BE, TL45]
G3 KBS N Coupe, 19 Hillary Dr, Crowthorne, RG45 6QF [IO91OJ, SU86]
G3 KBX E J O'Brien, 62 Benfieldside Rd, Consett, DH8 0SD [IO94BU, NZ05]
GM3 KBZ J A Dunlop, 23 Cloberfield Gdns, Milngavie, Glasgow, G62 7LH [IO75UW, NS57]
GM3 KC G Stevenson, 129 Murray St., Montrose, DD10 8JQ [IO86SR, NO75]
G3 KCB B N Green, 18 Kenilworth Rd, Sale, M33 5FB [IO83TK, SJ79]
G3 KCD P Bedwell, 3 Roman Way, Dibden Purlieu, Southampton, SO45 4RP [IO90GU, SU40]
G3 KCE R I Rimmer, 45 Glencoe Rd, Parkstone, Poole, BH12 2DW [IO90AR, SZ09]
G3 KCF R M Kent, 8 Woodfield Ln, Stowmarket, IP14 1BN [JO02LE, TM05]
G3 KCG D J Tyerman, 20 Grace Gdns, Bishops Stortford, CM23 3EX [JO01BU, TL41]
G3 KCJ A H Webb, 286 Ashcroft Rd, Luton, LU2 9AE [IO91TV, TL12]
G3 KCN A G G Wills, 12 Rainsford Way, Hornchurch, RM12 4BJ [JO01CN, TQ58]
G3 KCR D W Payne, The Bramley, Eridge Rd, Crowborough, TN6 2SL [JO01BB, TQ53]
G3 KCT D W Blythe, 24 Connaught St., London, W2 2AF [IO91WM, TQ28]
G3 KCV J W F Saunders, 7 Stone Ln, Yeovil, BA21 4NN [IO80QW, ST51]
GM3 KCY G I Buchanan, 30 Gilmour Ave, Clydebank, G81 6AW [IO75TW, NS47]
G3 KCZ W J Siertsema, 21 Rowles Cl, Kennington, Oxford, OX1 5LX [IO91JR, SP50]
G3 KDA M G Rimmer, 9 Hazeldown Ave, Preston, Weymouth, DT3 6HT [IO80SP, SY68]
GW3 KDB P A Miles, y Gorlan, Cross Inn, Llandysul, Ceredigion, SA44 6NP [IO72TE, SN35]
G3 KDD V P Barrett, 18 Treza Rd, Porthleven, Helston, TR13 9NB [IO70IC, SW62]
G3 KDE P R Cheeseman, Laroch Post Office, Ln North Mundham, Chichester, W Sussex, PO20 6JY [IO90OT, SU80]
G3 KDH T H Price, 7 Mount St, Darlington, Wa 6070, Western Australia
G3 KDN D L Roberts, 64 Ledsham Rd, Little Sutton, South Wirral, L66 4QJ [IO83MG, SJ37]
G3 KDO A Mills, 18 Sandyacres, Rothwell, Leeds, LS26 0LY [IO93GS, SE32]
G3 KDP A G Bounds, 18 Harwood Ave, Tamerton Foliot, Plymouth, PL5 4NX [IO70WK, SX46]
G3 KDQ J C D Brock, 147 Wollaton Vale, Wollaton, Nottingham, NG8 2PE [IO92JW, SK53]
GI3 KDR J A Stringer, 16 Glebe Manor, Hillsborough, BT26 6NS [IO74AK, J25]
G3 KDU M Crawford, 95 Victoria Ave, Princes Ave, Hull, HU5 3DW [IO93TS, TA03]
G3 KDV R Coleman, 10 Lytes Rd, Brixham, TQ5 9SN [IO80FJ, SX95]
G3 KEB Details withheld at licensee's request by SSL.
G3 KEC J M Garner, 29 Sennen Cl, Torpoint, PL11 2JJ [IO70VJ, SX45]
G3 KED Details withheld at licensee's request by SSL.
G3 KEF Details withheld at licensee's request by SSL.[Station located in Harlow, TL40.]
G3 KEG C C Rogers, 100 Sparth Rd, Clayton-le-Moors, Accrington, BB5 5QD [IO83TS, SD73]
G3 KEK G G Carr, 88 Woodrow Cres, Knowle, Solihull, B93 9EQ [IO92DJ, SP17]
G3 KEL R Bray, Croft House, Blencogo, Wigton, Cumbria, CA7 0BZ [IO84IT, NY14]
G3 KEQ J P Mitchell, Chellow Dene, Viewlands Ave, Westerham Hill, Westerham, TN16 2JE [JO01BH, TQ45]
G3 KES D G O White, 69 Easton St., Portland, DT5 1BS [IO80SN, SY67]
G3 KEU T Leighfield, 123 Avonmead, Greenmeadow Est, Swindon, SN2 3PA [IO91CO, SU18]
G3 KEV M C Hamilton, 473 Scalby Rd, Scarborough, YO12 6UA [IO94SH, TA09]
GM3 KEZ J W Little, 33 Manor Ct, Manor St., Forfar, DD8 1BR [IO86NP, NO45]
GW3 KFA L Lindley, 14 Carlines Ave, Deeside, CH5 3RD [IO83LE, SJ36]
G3 KFC F W Clasby, 103 Stanley Park, Liverpool, L21 9JS [IO83ML, SJ39]
G3 KFD D J Billingham, Omega, 216 Standhills Rd, Kingswinford, DY6 8JR [IO82WL, SO88]
GW3 KFE Details withheld at licensee's request by SSL.[Op: E P Essery.]
G3 KFG H H Taylor, Kildare, Morton Rd, Brading, Sandown, PO36 0BJ [IO90KQ, SZ68]
G3 KFK K F Kippen, 36 Cressing Rd, Braintree, CM7 3PP [JO01GV, TL72]
G3 KFN A R Baker, 37 Boulter Cl, Roborough, Plymouth, PL6 7AY [IO70WK, SX56]
G3 KFS D V Preston, Ashfield, Tanworth Ln, Beoley, Redditch, B98 9EH [IO92BH, SP06]
G3 KFT J Readings, Flat 4 Malvern Hill House, East Approach Dr, Pittville, Cheltenham, Glos, GL52 3JE [IO81XV, SO92]
G3 KFU P T Barry, 21 Old Pasture Rd, Frimley, Camberley, GU16 5SA [IO91PH, SU85]
G3 KFZ Details withheld at licensee's request by SSL.
G3 KGA A R Morrison, 30 Sullington Gdns, Worthing, BN14 0HR [IO90TU, TQ10]
G3 KGB Details withheld at licensee's request by SSL.
G3 KGC R J B Morgan, El Nido Del Aguila, Canillas de Aceituno, 29716 Malaga, Spain
G3 KGF J S Foster, 1 Granville Gr, Stockton on Tees, TS20 2EP [IO94IN, NZ42]
GW3 KGI M T Bowen, 24 Parklands View, Sketty, Swansea, SA2 8LX [IO81AO, SS69]
G3 KGM D Maclennan, 66 Old Farm Ave, Sidcup, DA15 8AH [JO01BK, TQ47]
G3 KGN A C Edwards, 48 Fillebrook Ave, Leigh on Sea, SS9 3NT [JO01IN, TQ88]
G3 KGP M A Palmer, Fairways, 8 Gwealdues, Helston, TR13 8JZ [IO70IC, SW62]
GM3 KGT J Nicolson, 24 Pottersfield Rd, Woodmancote, Cheltenham, GL52 4PY [IO81XW, SO92]
GW3 KGV K A Bates, The Bridge, 5 Ffordd Nant Coch, Llangadfan, Welshpool Powys, SY21 0PW [IO82GQ, SJ01]
G3 KGW J D Smith, 8 Church Cl, Gnosall, Stafford, ST20 0DD [IO82US, SJ82]
G3 KGX B H Lawrence, 26 Browning Rd, Enfield, EN2 0EL [IO91XP, TQ39]
G3 KH A H Parker, 133 Station Rd, Cropston, Leicester, LE7 7HH [IO92JQ, SK51]
G3 KHC J S Cushing, 20 Kildowan Rd, Goodmayes, Ilford, IG3 9XW [JO01BN, TQ48]
G3 KHF F Parr, 140 Hall St., Briston, Melton Constable, NR24 2LQ [JO02MU, TG03]
GM3 KHH W G Cecil, Innes House, Oran, Buckie, AB56 5EP [IO87MP, NJ46]
G3 KHK D P Connolly, 78 Cromer Rd, Mundesley, Norwich, NR11 8DD [JO02RV, TG33]
G3 KHQ A A Langley, 58 Dumbarton Rd, Brixton Hill, London, SW2 5LU [IO91WK, TQ37]
G3 KHR J W Fox, 25 Langdale Cres, Bexleyheath, DA7 5DZ [JO01BL, TQ47]
G3 KHS T B Cutmore, Hawthorne Cottage, Brigg Rd, South Kelsey, Market Rasen, LN7 6PH [IO93SL, TF09]
G3 KHU R W Gabbitas, 12 Thornyville Dr, Oreston, Plymouth, PL9 7LF [IO70WI, SX55]
G3 KHZ D Cox, 18 Station Rd, Castle Bytham, Grantham, NG33 4SB [IO92RS, SK91]
G3 KIC A J Hudson, Thornham, Clacton Rd, Weeley Heath, Clacton on Sea, CO16 9DR [JO01NU, TM12]
GM3 KIG W J H Eaton, 100 Craigleith Hill Cres, Edinburgh, EH4 2JP [IO85JX, NT27]
G3 KIH F G Unwin, 32 Woodland Dr, Worksop, S81 7JU [IO93KH, SK58]
G3 KII G D Lively, 9 Wilson Rd, Shurdington, Cheltenham, GL51 5SN [IO81WU, SO91]
G3 KIJ E F Lugmayer, 17 Borough End, Beccles, NR34 9YW [JO02SK, TM48]
G3 KIL R L Messer, The Shambles, Swinbrook Rd, Carterton, Oxford, OX8 3DX [IO91GT, SP31]
G3 KIN M Ashdown Kingston and District A.R.C, 42 Alpine Ave, Tolworth, Surbiton, KT5 9RJ [IO91UJ, TQ26]
G3 KIP K G Grover, 1 Powdermill Cl, Tunbridge Wells, TN4 9DR [JO01DD, TQ54]
G3 KIQ J A Elliot, 2 Pennine Cl, Higher Blackley, Manchester, M9 6HR [IO83VM, SD80]
G3 KIW G W Jenner, Paradise Ln, Bucklebury, Chapel Row, Reading, Berks, RG7 6NU [IO91JK, SU56]
G3 KJ K D Jackson, 2 Little Toller, Roman Rd, Chilworth, Southampton, SO16 7HG [IO90HW, SU41]
GM3 KJA R W T Horne, 38 Howden Cres, Jedburgh, TD8 6JY [IO85RL, NT61]
G3 KJC R A Church, Three Birches, Sandy Cl, Hermitage, Thatcham, RG18 9QP [IO91IK, SU57]
GM3 KJE J P Scott, 5 Garthdee Terr, Aberdeen, AB10 7JE [IO87WD, NJ90]
GM3 KJF J Wilson, 5 Braeside, Annbank, Ayr, KA6 5EW [IO75RL, NS42]
GM3 KJI E W Pollard, 265 Eldon St., Greenock, PA16 7QE [IO75OX, NS27]
G3 KJK L E Wilkes, Parc Crane, Penmenner Rd, The Lizard, Helston, TR12 7NN [IN79JX, SW71]
G3 KJM K J Monk, 143 Hampton Rd, Southport, PR8 5DJ [IO83MP, SD31]
GW3 KJN I G Winter, 5 Uwch y Nant, Mynydd Isa, Mold, CH7 6YP [IO83KE, SJ26]
G3 KJP K B Peace, 44 The Pines, Honiton, EX14 8JG [IO80JT, ST10]
G3 KJS Details withheld at licensee's request by SSL.[Op: W T Smith, 32 Lumley Road, Chester, Cheshire, CH2 2AQ.]
G3 KJT J R Edgington, 5 Oldhall Ct, High St., Whitwell, Hitchin, SG4 8AL [IO91UV, TL12]
GW3 KJW P E W Allely, Rhiw, Pwllheli, Gwynedd, LL53 8AE [IO72QT, SH22]
G3 KJX B Alderson, 43 Brompton Rd, Northallerton, DL6 1ED [IO94GI, SE39]
G3 KJY J A York, 13 Melville Ave, Barnoldswick, Colne, BB8 5JS [IO83VW, SD84]
GM3 KJZ G Paterson, 3 Ferry Barns Ct, North Queensferry, Inverkeithing, KY11 1ET [IO86HA, NT18]
G3 KKB K D Hallam, 276 Harborne Park Rd, Harborne, Birmingham, B17 0BL [IO92AK, SP08]
G3 KKC A R Rumbelow, 7 Hoof Cl, Littleport, Ely, CB6 1HU [JO02DK, TL58]
G3 KKD I M Waters, 39 Stow Rd, Stow Cum Quy, Quy, Cambridge, CB5 9AD [JO02CF, TL56]
GW3 KKG F C Charlton, Crogan, Llandegfan, Anglesey, LL59 5TH [IO73WF, SH57]
G3 KKJ A Shannon, Tideway, Coast Rd, Roosebeck, Ulverston, LA12 0RG [IO84KC, SD26]
G3 KKM K R Barton, 24 Oriole Way, Larkfield, Aylesford, ME20 6LW [JO01FH, TQ65]
G3 KKN N G Armstrong, No 1 Alm Cttgs, The Gorse, COLeford, Glos, GL16 8QE [IO81QT, SO51]
G3 KKP J Burgess, Moorend, Main St., Guiseley, Leeds, LS20 8NX [IO93DU, SE14]
G3 KKQ D S Booty, 1 Georgian Cl, Leacroft, Staines, TW18 4NR [IO91SK, TQ07]
G3 KKT Details withheld at licensee's request by SSL.
G3 KKZ P Champion, The Lodge, Tydcombe Rd, Warlingham, Surrey, CR3 9LU [IO91XG, TQ35]
GM3 KLA W A Sinclair, Ark, Haroldswick, Unst, Shetland, ZE2 9ED [IP90OT, HP61]
G3 KLB Details withheld at licensee's request by SSL.
G3 KLC J S Bennett, Koivula, Station Rd, Hubberts Bridge, Boston, PE20 3QT [IO92WX, TF24]
G3 KLD R E B Russell, 43 Ingestre Rd, Hall Green, Birmingham, B28 9EQ [IO92BK, SP18]
G3 KLF I H Crowther, 3 Glenelg, Fareham, PO15 6JU [IO90JU, SU50]
G3 KLH D G Alexander, Staddle Barn, Homington Rd, Coombe Bissett, Salisbury, Wilts, SP5 4LR [IO91BA, SU12]
G3 KLJ W E Thornton, 40 Stanley Ave, Harborne, Birmingham, B32 2HA [IO92AL, SP08]
G3 KLK P E C Page, 7 Marconi Way, Southall, UB1 3JP [IO91TM, TQ18]
G3 KLL B Mercer, Hilltop House, Knott Hill Ln, Delph, Oldham, OL3 5RJ [IO83XN, SD90]
G3 KLO E P Barlow, 5 Birchview Ct, Parker St., Derby, DE1 3HH [IO92GW, SK33]
G3 KLP J R Young, Woodglades, 34 The Demesne, North Seaton, Ashington, NE63 9TP [IO95FE, NZ28]
G3 KLQ P F Yeates, 6 Wheatfield Rd, Lincoln, LN6 0PS [IO93QF, SK97]
G3 KLT I J Eamus, 10A Vale Rd, Aylesbury, HP20 1JA [IO91OT, SP81]
G3 KLV G Vine, Primrosa, 62 Overstone Rd, Sywell, Northampton, NN6 0AW [IO92OH, SP86]
G3 KLY M J Lloyd, 6 Berwick Cl, Seaford, BN25 2NU [JO00BS, TV49]
G3 KLZ D G Enoch, 7A Mount Rd, Evesham, WR11 6BE [IO92AC, SP04]
G3 KMA R Balister, La Quinta, Mimbridge, Chobham, Woking, Surrey, GU24 8AR [IO91RI, SU96]
G3 KMC R T Coggin, 12 Hunters Way, Sawtry, Huntingdon, PE17 5SJ [IO92UK, TL18]
G3 KMD T W Bass, 32 Whitmore Ave, Grays, RM16 2HX [JO01DL, TQ68]
G3 KMG D H Plumridge, Rose Cottage, Castleside, Consett, DH8 9AP [IO94BT, NZ04]
G3 KMH W H Ferguson, 9 Woodbine Terr, Hexham, NE46 3LE [IO84WX, NY96]
G3 KMI C E Thompson Southampton Uni, Radio Club, Students Union, Highfield Southampt, SO9 5NH [IO90HW, SU41]
G3 KMM J L Crowther, Vine Cottage, 26 Church Rd, Whitchurch, Bristol, BS14 0PP [IO81RJ, ST66]
GM3 KMN I McKenzie, 17 Beechwood Rd, Cumbernauld, Glasgow, G67 2NN [IO85AW, NS77]
G3 KMO M A Birch, Willow Lodge, The Heath, Hevingham, Norwich, NR10 5QW [JO02OR, TG12]
G3 KMP C A May, 7 Parkwood Rd, Hastings, TN34 2RN [JO00GV, TQ81]
G3 KMQ R G Heslop, Fairways, Meadow Dr, Bude, EX23 8HZ [IO70RU, SS20]
G3 KMS D H Swain, 3 Nevy Fold Ave, Horwich, Bolton, BL6 6QG [IO83RO, SD61]
G3 KMV R N Birchall, Poole Farm, Poole, Nantwich, Ches, CW5 6AL [IO83RC, SJ65]
G3 KMY D G Radford, Glenburn, Brassington Ln, Old Tupton, Chesterfield, S42 6LB [IO93HE, SK36]
G3 KNB K Ballance, 18 Rambleford Way, Parkside, Stafford, ST16 1TW [IO82WT, SJ92]
G3 KND J S P Hardy, Vogelenzang, 1B Roberts Rd, Aldershot, GU12 4RD [IO91PF, SU85]
G3 KNG A F Embrey, Elliotts Piece, 63 Elliotts Ln, Codsall, Wolverhampton, WV8 1PG [IO82VP, SJ80]
G3 KNI F W Pickard, 14 Oatfield Cl, Cranbrook, TN17 3NH [JO01GC, TQ73]
G3 KNJ R J Wyatt, 8 Millbrook Rd, Bushey, Watford, WD2 2BU [IO91TP, TQ19]
G3 KNM Details withheld at licensee's request by SSL.
G3 KNU P Jackson, 7 Ferriby Rd, Scunthorpe, DN17 2EQ [IO93QN, SE80]
G3 KNZ A W Eccles, 78 Uplands Ave, Connahs Quay, Deeside, CH5 4LG [IO83LE, SJ26]
G3 KOA T P Robinson, 32 Campbell Cres, East Grinstead, RH19 1JR [IO91XD, TQ33]
G3 KOB R V Goodman, 49 Holly Hedge Rd, Frimley, Camberley, GU16 5ST [IO91PH, SU85]
G3 KOC J D Pearson, 10 Woodgarth Villas, Oxmarsh Ln, New Holland, Barrow upon Humber, DN19 7RF [IO93TQ, TA02]
G3 KOD P J D Kay, 34 Woodland Rise, Muswell Hill, London, N10 3UG [IO91WO, TQ28]
G3 KOG W J Blanchard, 28 Church Ave, North Ferriby, HU14 3BY [IO93RR, SE92]
G3 KOJ R J Ezra, 52 Highfield Ave, Waterlooville, PO7 7PX [IO90LV, SU61]
G3 KOM F C Foulkes, Bankside Kings Toll, Matfield, Tonbridge, Kent, TN12 7HA [JO01ED, TQ64]
G3 KOP C B Jones, 36 Greenfields Rd, Shelfield, Walsall, WS4 1RS [IO92AO, SK00]
G3 KOQ B Parker, 9 Yewdale Ave, Heysham, Morecambe, LA3 2LR [IO84NA, SD46]
G3 KOS B A Faithfull, 68 Lampton Rd, Blowell Gdns, Long Ashton, Bristol, BS18 9AQ [IO81PK, ST57]
GI3 KOT G A Meaney, 15 Sarajac Cres, Cavehill Rd, Belfast, BT14 6SD [IO74AP, J37]
G3 KOU Details withheld at licensee's request by SSL.
G3 KOX N J Waite, 7 Lancercost Cl, Welwyn, AL6 0RW [IO91VU, TL21]
G3 KOZ W D Henderson, 9 Chiselbury Gr, Salisbury, SP2 8EP [IO91CB, SU12]
G3 KPB S A Moore, 1 Uplands, Canterbury, CT2 7BL [JO01NH, TR15]
GM3 KPD A M Coutts, 77 Barnton Park View, Barnton, Edinburgh, EH4 6EL [IO85IX, NT17]
G3 KPJ A W Butcher, 160 Lupin Dr, Chelmsford, CM1 6FJ [JO01FR, TL70]
G3 KPM V E Oliva, 66 Friars Walk, Southgate, London, N14 5LP [IO91WP, TQ29]
G3 KPO D Byrne, Lynwood, 52 West Hill Rd, Ryde, PO33 1LN [IO90KR, SZ69]
G3 KPT G V Farrance, 51 Amberley Green, Great Barr, Birmingham, B43 5TJ [IO92AM, SP09]
G3 KPU E Prince, Cubert, Bawtry Rd, Hatfield Woodhouse, Doncaster, DN7 6PQ [IO93LN, SE60]
G3 KPV J R Killeen, Tobago Lodge, White Cross, East Hill, Ottery St Mary, East Devon, EX11 1QB [IO80IR, SY19]
G3 KPZ C Nagle, 507 South End Rd, Elm Park, Hornchurch, RM12 5NX [JO01CN, TQ58]
G3 KQB Details withheld at licensee's request by SSL.
G3 KQD Details withheld at licensee's request by SSL.
G3 KQE D Heathcote, 22 Argyle St., Atherton, Manchester, M46 0AW [IO83SM, SD60]
G3 KQF J Anthony, 77 Brayfield Rd, Littleover, Derby, DE23 6GT [IO92FV, SK33]
G3 KQG Dr E D James, Kennford, Exeter, Devon, EX6 7TZ [IO80FP, SX98]
G3 KQI J L Howard, Hurst Farm, Newton Harcourt, Leics, LE8 9FH [IO92LN, SP69]
G3 KQJ M J Sparrow, White Orchard, 64 Showell Ln, Penn, Wolverhampton, WV4 4TT [IO82VN, SO89]
G3 KQL J L Weatherley, 116 Mercury Ct, Indialantic, Florida 32903, USA, ZZ1 1ME
G3 KQN J R Walton, 17 Wilmire Rd, Wolviston Ct, Billingham, TS22 5EN [IO94IO, NZ42]
G3 KQP Dr J B Poulter, Hove To, Ln End, Instow, Bideford, EX39 4LB [IO71VB, SS43]
G3 KQQ Details withheld at licensee's request by SSL.[Op: C A Mattacks, 68 Middlesex Drive, Bletchley, Milton Keynes, MK3 7EU.]
G3 KQR Dr D V Foster, 56 Elmbridge Ave, Tolworth, Surbiton, KT5 9HA [IO91UJ, TQ26]
G3 KQS E G Simpson, 5 Cottingham Gr, Bletchley, Milton Keynes, MK3 5AA [IO91PX, SP83]
G3 KQU A Thompson, 11 Caernarvon Cl, Shepshed, Loughborough, LE12 9QB [IO92IS, SK41]
G3 KQV J E M Ryley, 30 St. Helens Dr, Leicester, LE4 0GS [IO92KP, SK50]
G3 KQW Dr R F Williams, St. Marys, High St., Bures, CO8 5HZ [JO01JX, TL93]
G3 KQY R J Disley, 6 St. Margarets Rd, Farington, Leyland, Preston, PR5 2XT [IO83PQ, SD52]
GW3 KRD W J Northcott, 6 Craig Las, Letterston, Haverfordwest, SA62 5SQ [IO71MW, SM92]
G3 KRG E Yard, 45 Westfields, Stanley, DH9 7DB [IO94DU, NZ15]
G3 KRL J Schofield, 92 High St., Harlton, Cambridge, CB3 7ES [JO02AD, TL35]
G3 KRS H Hinde, 97 Burton Rd, Withington, Manchester, M20 1HZ [IO83VK, SJ89]
G3 KRT G L D Hodges, 102 Torrington Rd, Ruislip, HA4 0AU [IO91TN, TQ18]
G3 KRW K R Whelan, Killiney, Longsplatt, Kingsdown, Corsham, SN13 8DF [IO81VJ, ST86]
G3 KRX W T Addy, 14 Crestor Rd, Liverpool, L25 6DW [IO83NJ, SJ48]
G3 KRY G R Byles, Springtime, Heywood Rd, Shelfanger, Diss, IP22 2DJ [JO02NJ, TM18]
G3 KRZ J Greenwood, Paradise Farm, Wigtoft Bank, Wigtoft Boston, Lincs, PE20 2QE [IO92WV, TF23]
GM3 KSD G R B Robertson, Lavalette, Manse Rd, Moulin, Pitlochry, PH16 5EP [IO86DR, NN95]
G3 KSF R E D Harper, 53 Limekiln Ln Es, Limekiln Ln, Holbury, Southampton Hants, SO4 1HF [IO90HW, SU41]
G3 KSG F L Bones, 7707 Briarwood Dr, Port Richey, Florida 34668, USA
G3 KSH A R Gilding, 34 Ashley Way, Brighstone, Newport, PO30 4HH [IO90HP, SZ48]
G3 KSK J J Phillips, Shaston, Winterhay Ln, Ilminster, TA19 9BB [IO80MW, ST31]
G3 KSP P O Hooper, 1 Victoria Mews, Morecambe, LA4 5QD [IO84NB, SD46]
G3 KSU A R Williams, 7 Chandler Cl, Jump Farm, Devizes, SN10 3DS [IO91AI, SU06]
G3 KSW L A Wild, One Dig Dag Hill, Cheshunt, Waltham Cross, Herts, EN7 6NS [IO91XR, TL30]
GI3 KSY C A Friend, 39 Phillip St., Strand Rd, Londonderry, BT48 7PN [IO65IA, C41]
G3 KTA P G Munt, 130 Chipstead Way, Woodmansterne, Banstead, SM7 3JR [IO91VH, TQ25]
G3 KTC R E Tucker, 56 Highfields Cl, Ashby de La Zouch, LE65 2FN [IO92GR, SK31]
GM3 KTD W K Wylie, 2 Lovat Rd, Kinlochleven, PA40 4RQ [IO76MR, NN16]
G3 KTF R D May, Chelmer, Chelmer Cl, Little Totham, Maldon, CM9 8JN [JO01IS, TL81]
G3 KTH M J Darkin, 3 Adrian Cl, Tagwell Rd, Droitwich, WR9 7AY [IO82WG, SO96]
G3 KTI M V Rees, Blue Pillars, 6 Gr Cres, Coombs Park, Coleford, GL16 8AZ [IO81QT, SO51]
G3 KTJ G P Rigby, 30A Pimbo Ln, Upholland, Skelmersdale, WN8 9QQ [IO83PM, SD50]
G3 KTK Details withheld at licensee's request by SSL.
G3 KTL Details withheld at licensee's request by SSL.
G3 KTN J F Brown, 223 Middle Park Way, Havant, PO9 4NQ [IO90MU, SU70]
G3 KTO Details withheld at licensee's request by SSL.
G3 KTP E E West, 79 St. Wilfrids Rd, West Hallam, Ilkeston, DE7 6HG [IO92HX, SK44]
G3 KTR A D J Rock, 4 Causeway St, Hudson, Mass, USA 01749, ZZ9 9HU
G3 KTT F A Hunter, 8 St. Hildas Ave, Holy Cross, Wallsend, NE28 7AB [IO94FX, NZ36]
G3 KTU J T Ault, Bradgate, 89 Southbourne Coast Rd, Southbourne, Bournemouth, BH6 4DX [IO90CR, SZ19]
G3 KTX J V Tomlinson, 88 Linby Rd, Hucknall, Nottingham, NG15 7TW [IO93JB, SK54]
G3 KTZ C Lindsay, 27 Hertford Rd, Enfield, EN3 5JD [IO91XP, TQ39]
G3 KUD J R Duncan, 9 Springhill Cl, Westlea, Swindon, SN5 7BG [IO91CN, SU18]
G3 KUE G Lancefield Preston ARS, 191 Higher Walton Rd, Walton-le-Dale, Preston, PR5 4HS [IO83QR, SD52]
G3 KUF H T Falstein, 21 Beechwood Rd, Easton in Gordano, Bristol, BS20 0NA [IO81PL, ST57]
G3 KUG H Peabody, 182 Cavendish Rd, Walsall, WS2 7HY [IO82XO, SJ90]
G3 KUL D Stephenson, 16 Chard Ct, Fortfield Rd, Bristol, BS14 9NL [IO81RJ, ST66]
G3 KUS Details withheld at licensee's request by SSL.
GW3 KUY C T Edwards, 10 Golden Ave, Sandfields Est, Port Talbot, SA12 7RP [IO81CO, SS79]
G3 KVA J C Hall, Corfe Lodge, Ipswich Rd, Long Stratton, Norwich, NR15 2TA [JO02OL, TM19]
GI3 KVD D M Jones, 5 Whitehill Park, Limavady, BT49 0QF [IO65MA, C62]

G3 KVE T K Wright, 24 Stuart Rd, Bootle, L20 9ET [IO83MK, SJ39]
G3 KVF Details withheld at licensee's request by SSL.[Op: V A W Frisbee. Station located near Lowestoft.]
G3 KVG J S Charles, 87 Lees Hall Rd, Sheffield, S8 9JL [IO93GI, SK38]
G3 KVH J B Barnes, 20 Appleton Ave, Pedmore, Stourbridge, DY8 2JZ [IO82WK, SO98]
G3 KVJ S Tomlinson, 31 The Quarry, Alwoodley, Leeds, LS17 7NH [IO93FU, SE24]
G3 KVM P R Cavanagh, Carrfield House, Carrfield Ln, Sheffield, S8 9HU [IO93GI, SK38]
G3 KVP D Kitchen, Folkingham Pl, Folkingham, near Sleaford, Lincs, NG34 0SE [IO92TV, TF03]
G3 KVR S F J Davis, 3 Coronation Rd, Banwell, Weston Super Mare, BS24 6AZ [IO81NH, ST35]
G3 KVT A G Smith, Winston House, Felthorpe Rd, Attlebridge, Norwich, NR9 5TF [JO02NQ, TG11]
GW3 KVX R Pattinson, 15 Maes yr Eglwys, Llansantffraid, SY22 6BE [IO82KS, SJ22]
G3 KWA J L Parry, 7 The Green, Harrold, Bedford, MK43 7DB [IO92QE, SP95]
GW3 KWB R H Neville, 35 Beechcroft Rd, Newport, NP9 8AG [IO81MO, ST38]
G3 KWC D R Page, 100 Empingham Rd, Stamford, PE9 2SU [IO92SP, TF00]
G3 KWE M B Aburrow, 6 Blake Dene Rd, Lilliput, Poole, BH14 8HQ [IO90AR, SZ09]
G3 KWG J A H Spratt, 196 Shaftesbury Ave, Southend on Sea, SS1 3AJ [JO01IM, TQ98]
G3 KWH J F A G Vaux, 62 Saddleback Rd, Shaw, Swindon, SN5 9RN [IO91BN, SU18]
G3 KWJ N B Valentine, Bicknell Farm, Butcombe, Blagdon, BS18 6XG [IO81PI, ST56]
G3 KWK R W Nolan, 6 Plymouth Cl, Redditch, B97 4NP [IO92AH, SP06]
G3 KWN LtCl W C Nicoll, Yonder, Milldown Rd, Blandford Forum, DT11 7DE [IO80VU, ST80]
G3 KWO K W Dawson, 44 Avondale Rd, Darwen, BB3 1NS [IO83SQ, SD62]
G3 KWP P G Lynch, 9 Somerville Rd, Eton, Windsor, SL4 6PB [IO91QL, SU97]
G3 KWT Details withheld at licensee's request by SSL.
G3 KWW Dr R W Wilkinson, 9 Granary Wharf, Commercial Rd, Weymouth, DT4 8AL [IO80SO, SY67]
G3 KWY A Swain, 18 Romway Cl, Shepshed, Loughborough, LE12 9DT [IO92IS, SK41]
G3 KXB D E Pantony, 6 Longtye Dr, Chestfield, Whitstable, CT5 3NG [JO01MI, TR16]
GW3 KXC J H Rowntree, Talgarth, 4 Maes Llydan, Benllech, Tyn y Gongl, LL74 8RD [IO73VH, SH58]
G3 KXE E W Bettles, 15 St. Francis Ave, Bitterne, Southampton, SO18 5QL [IO90HW, SU41]
G3 KXF D S Roden, Skip End, 52 Cokeham Rd, Sompting, Lancing, BN15 0AE [IO90TT, TQ10]
G3 KXI D E Keeler, 15 Elm Rise, Witham, CM8 2LE [JO01HT, TL81]
G3 KXP G H Tillett, 42 Park Ln, Hornchurch, RM11 1BD [JO01CN, TQ58]
GM3 KXQ S A Floyd, 3 Crarae Pl, Newton Mearns, Glasgow, G77 6XX [IO75TS, NS55]
G3 KXT R I Richardson, 104 Cologne Rd, Bovington, Wareham, BH20 6NP [IO80VQ, SY88]
G3 KXU T A Cousens, 117 Morris St., Swindon, SN2 2HS [IO91CN, SU18]
G3 KXV V Johnston, 9 Holbeck Ave, Brookfield, Middlesbrough, TS5 8DR [IO94JM, NZ41]
G3 KXW T D Casey, 16 Norbury Gdns, Hamble, Southampton, SO31 4LX [IO90IU, SU40]
GW3 KXX R H G Weaver, 59 Broad St., Leckwith, Cardiff, CF1 8BZ [IO81JL, ST17]
G3 KXY R S Williams, 24 Sunny Bank Ave, Blackpool, FY2 9EQ [IO83LU, SD33]
GW3 KYA R B Davies, 16 Vancouver Dr, Penmaen, Blackwood, NP2 0UQ [IO81JQ, ST19]
G3 KYE J Orr, 102 Manor House Ln, Yardley, Birmingham, B26 1PR [IO92CL, SP18]
G3 KYF K G Sullivan, 14 Wigston Rd, Blaby, Leics, LE8 4FU [IO92KN, SP59]
G3 KYL J E Holt, 56 Mawbey House, Old Kent Rd, Southwark, London, SE1 5PQ [IO91XL, TQ37]
G3 KYM H Stamper, Far View Farm, Lanjeth, St Austell, Cornwall, PL26 7UX [IO70NI, SW95]
GI3 KYP A D Patterson, 24 Cyprus Ave, Belfast, BT5 5NT [IO74BO, J37]
G3 KYS R Kendall, 1 Charlottes, Washbrook, Ipswich, IP8 3HZ [JO02NA, TM14]
G3 KYV S W H Harrison, 50 Allenby Rd, Biggin Hill, Westerham, TN16 3LG [JO01AH, TQ45]
G3 KYZ D J Clarke, Primrose Mount, Old Neighbourhood, Chalford Hill, Stroud Glos, GL6 8AA [IO81WR, SO80]
G3 KZB M Ward, Hartley Green Ln, Milford, Godalming, Surrey, GU8 5BG [IO91QE, SU94]
G3 KZC R F Harknett, 28 Woodyleaze Dr, Hanham, Bristol, BS15 3BY [IO81RK, ST67]
G3 KZE J D G Davies, 25 Bakers Ave, Potton, Sandy, SG19 2PJ [IO92VD, TL24]
G3 KZG A J Bills, 46 High St, Kinver, Stourfridge, West Midlands, DY7 6HF [IO82VK, SO88]
G3 KZJ J M Dart, 2 Orchard Gr, Brixham, TQ5 9RH [IO80FJ, SX95]
G3 KZN D W Blakeley, 338 Thong Ln, Gravesend, DA12 4LQ [JO01EK, TQ67]
G3 KZR I S Davies, Lusty Hill Farm, Bruton, Somerset, BA10 0BS [IO81SC, ST63]
GW3 KZT A J P James, 143 Gaer Park Dr, Newport, NP9 3NS [IO81LN, ST28]
G3 KZU M B Dolan, 15 Ringwood Rd, Headington, Oxford, OX3 8JB [IO91JS, SP50]
GW3 KZW R B Ratcliffe, Llwynberis, Rhydargaeau Rd, Carmarthen, SA32 7AH [IO71UV, SN42]
G3 KZX L J Loveland, 21 Roseland Cl, Keyworth, Nottingham, NG12 5LQ [IO92KU, SK63]
G3 KZY J Rathbone, 4 Sandy Bank, Whixall, Whitchurch, SY13 2NS [IO82PW, SJ53]
G3 KZZ D I Forster, 281 Mortimer Rd, South Shields, NE34 0DR [IO94GX, NZ36]
G3 LAA Details withheld at licensee's request by SSL.
G3 LAC P B Inglis-Smith, 25 High St., Long Crendon, Aylesbury, HP18 9AL [IO91MS, SP60]
GW3 LAD E G White, 41 St. Albans Ave, Cardiff, CF4 4AS [IO81JM, ST17]
G3 LAG H B F Gow, 43 Ringstead Cres, Overcombe, Weymouth, DT3 6PT [IO80SP, SY68]
GW3 LAI G E Livingston, Rose Lodge, Cilcain Rd, Pantymwyn, Mold, CH7 5NJ [IO83JE, SJ16]
G3 LAS J B Butcher, Westlands, Westland Green, Little Hadham, Ware, SG11 2AJ [JO01AV, TL42]
GM3 LAU F W Adkin, 5 Cosway Mns S, Shroton St., London, NW1 6UE [IO91WM, TQ28]
GM3 LAW W A Walker, 15 Ashgrove Ave, Maybole, KA19 8BJ [IO75PI, NS21]
G3 LAZ R L Gerrard, 46 Hadrian Ave, Dunstable, LU5 4SP [IO91SV, TL02]
G3 LBA R T Greenwood, Campion House, Rectory Ln, Longworth, Abingdon, OX13 5DZ [IO91GQ, SU39]
G3 LBC M McGill, Garden View, Harras Moor, Whitehaven, Cumbria, CA28 6SG [IO84FN, NX91]
G3 LBL P Chapman, 2 Woolston Dr, Tyldesley, Manchester, M29 8WL [IO83SM, SD70]
G3 LBM A H Mulcahy, 14 Mulberry Ct, Holmer Green, High Wycombe, HP15 6TF [IO91PP, SU99]
G3 LBO P Laughton, 3 Newlay Wood Cl, Horsforth, Leeds, LS18 4SL [IO93ET, SE23]
G3 LBS Dr G Cleeton, 42 The Covert, University of Keele, Keele, Newcastle, ST5 5AZ [IO83UA, SJ84]
GM3 LBX Revd J L R Crawley, Cove, Campbeltown Rd, Tarbert, Argyll, PA29 6SX [IO75HU, NR86]
G3 LBZ L F Jones, 1 Brattswood Dr, Church Lawton, Stoke on Trent, ST7 3EJ [IO83UC, SJ85]
G3 LCA C F Hutchings, 151 Barn Meads Rd, Wellington, TA21 9AP [IO80JX, ST11]
G3 LCB R E Wolpers, 20 Langdon Shaw, Sidcup, DA14 6AU [JO01BK, TQ47]
G3 LCF P C Baldwin, 117 Tarring Rd, Worthing, BN11 4HE [IO90TT, TQ10]
G3 LCG G P Bateman, Hartford Ct, 33 Filey Rd, Scarborough, YO11 2TP [IO94TG, TA08]
G3 LCH M Pharaoh, 49 Streathbourne Rd, London, SW17 8QZ [IO91WK, TQ27]
G3 LCI H V Young, 345 Leasowe Rd, Moreton, Wirral, L46 2RE [IO83KK, SJ29]
G3 LCK D J Bradford, Crooked End House, 8 The Pippins, Wilton, Ross on Wye, Herefordshire, HR9 6BQ [IO81QV, SO52]
G3 LCL Sqld A D Baylis, Arntrees, 25 Westbury Rd, Warminster, BA12 0AW [IO81VF, ST84]
GM3 LCP J A R Hughes, 6 Claypotts Terr, Broughty Ferry, Dundee, DD5 1LE [IO86NL, NO43]
GW3 LCQ M Williams, 12 Pierhens Ave, Llandudno Junction, Gwynedd North Wales, LL31 9EL [IO83CG, SH77]
G3 LCR D Garlick, 20 Western Fields, Ruddington, Nottingham, NG11 6JE [IO92KV, SK53]
G3 LCS Details withheld at licensee's request by SSL.
G3 LCT Details withheld at licensee's request by SSL.
G3 LCW L A Hood, Little Cheverelle, Cherry Ln, Great Mongeham, Deal, CT14 0HF [JO01QF, TR35]
G3 LCX K H Lander, 196 Spencers Croft, Harlow, CM18 6JP [JO01BS, TL40]
G3 LCY J L F Tamlin, 53 Hele Gdns, Plympton St. Maurice, Plymouth, PL7 1JY [IO70XJ, SX55]
G3 LCZ T W Hickinbottom, 13 Almond Gr, Fairfield, Stockton on Tees, TS19 7DJ [IO94HN, NZ41]
G3 LD F W Foster, 16 Meadow Way, Rickmansworth, WD3 2NQ [IO91SP, TQ09]
G3 LDB Details withheld at licensee's request by SSL.
GW3 LDC J T Phillips, Derwen Fawr, 9 Trelawny Cl, Usk, NP5 1SP [IO81NQ, SO30]
G3 LDG B E Gee, Daisy Bank, Carlton Rd, Felmersham, Bedford, MK43 7JL [IO92RE, SP95]
GW3 LDH A C Wright, 6 Cwm Eithin, Wrexham, LL12 8JY [IO83MB, SJ35]
G3 LDI R J Cooke, The Old Nursery, The Drift, Swardeston Common, Lower East Carleton, NR14 8LQ [JO02ON, TG10]
G3 LDJ K Day, 45 Thick Hollins Dr, Meltham, Huddersfield, HD7 3DR [IO93BO, SE11]
G3 LDO P G Dodd, 37 The Ridings, East Preston, Littlehampton, BN16 2TW [IO90ST, TQ00]
G3 LDR R C Rule, 30 Forest Ave, Newcastle upon Tyne, NE12 9AH [IO95FA, NZ26]
G3 LDT L Bond, Knavesmire Clarke L, Bollington, Macclesfield, Ches, SK10 5AH [IO83WG, SJ97]
G3 LDU R Ballantyne, 61 Copse Ave, Weybourne, Farnham, GU9 9DZ [IO91OF, SU84]
G3 LDW Details withheld at licensee's request by SSL.[Station located near Halesowen.]
G3 LDY R D Taylor, 105 Sea Rd, Carlyon Bay, St. Austell, PL25 3SJ [IO70PI, SX05]
GI3 LEG D A Wilson, 189 Cregagh St., Belfast, BT6 8NL [IO74BO, J37]
G3 LEI D E Mills, 13 Primrose Terr, Shrubbery Rd, Gravesend, DA12 1JN [JO01EK, TQ67]
G3 LEJ M G Hudson, 6 Dewsbury Rd, Gomersal, Cleckheaton, BD19 4LD [IO93DR, SE22]
G3 LEK L Kitching, 18 Upton Pl, Western Springs, Rugeley, WS15 2PS [IO92AS, SK01]
G3 LEO G N L Brigham, The Manor, Carthorpe, Bedale, North Yorks, DL8 2LP [IO94FF, SE38]
G3 LEQ G L Adams, 2 Ash Gr, Knutsford, WA16 8BB [IO83TH, SJ77]
GM3 LER J G Stewart, 10 Pinewood Ave, Aberdeen, AB15 8NB [IO87WD, NJ90]
G3 LET P A F Hobbs, Middle House, Tilgate Forest Lodge, Pease Pottage, West Sussex, RH11 9AF [IO91VB, TQ23]
G3 LEW G Weale, 23 St. Michaels Way, Partridge Green, Horsham, RH13 8LA [IO91UB, TQ13]
GW3 LFC R R Copestake, 19 Glan y Pwll, Nefyn, Pwllheli, LL53 6EH [IO72RW, SH34]
G3 LFD R Widders, 82 Azalea Walk, Eastcote, Pinner, HA5 2EH [IO91TN, TQ18]
G3 LFE R T Clark, 2 Robartes Ct, Redannick, Truro, TR1 2XX [IO70LG, SW84]
G3 LFF R W C Chapman, 1 Green Ln, Hadzor, Droitwich, WR9 7DP [IO82WG, SO96]
GI3 LFH S Allan, Woodside, 8 Comber Rd, Saintfield, Ballynahinch, BT24 7BB [IO74CL, J45]
GJ3 LFJ H R Mesny, La Trigale, Route de Leglise, St. Lawrence, Jersey, JE3 1LA
G3 LFR I A Bunney, The Nook, 27 Spring Gdns, Newport Pagnell, MK16 0EE [IO92PC, SP84]
G3 LFV Details withheld at licensee's request by SSL.[Op: R G Manser (Ron). QSL via RSGB.]
G3 LFX D A G Pedder, 37 Hersham Rd, Walton on Thames, KT12 1LE [IO91TJ, TQ16]
G3 LFY K J Salter, 1 The Green, Culmington, Ludlow, SY8 2DA [IO82OK, SO48]
G3 LFZ Details withheld at licensee's request by SSL.
G3 LGA M G Hayward, Brindle, Romsey Rd, Broughton, Stockbridge, SO20 8DB [IO91FC, SU33]
G3 LGF G B Falding, 16 Delabere Rd, Bishops Cleeve, Cheltenham, GL52 4AJ [IO81XW, SO92]
G3 LGK B M Sandall, Amber Croft, Main Rd, Higham, Alfreton, DE55 6EH [IO93GD, SK35]
G3 LGL J E French, 75 Osmaston Rd, Norton, Stourbridge, DY8 2AN [IO82WK, SO88]
GM3 LGM W McGill, 13 Livingstone Cres, St. Andrews, KY16 8JL [IO86OI, NO51]
G3 LGN M A Niman, 9 Montgomery Dr, Unsworth, Bury, BL9 8PL [IO83UN, SD80]
G3 LGQ P D Marsden, 49 Southfield Park, North Harrow, Harrow, HA2 6HF [IO91TO, TQ18]
G3 LGR Details withheld at licensee's request by SSL.[Op: M A Hooles, 465 Whitton Avenue West, Greenford, Middx, UB6 0DX.]
G3 LGS C B C Hill, 7 South Dr, Tranmer Park, Guiseley, Leeds, LS20 8JF [IO93DU, SE14]
G3 LGT J Tate, Pine Holt, 34 Queens Rd, Fleet, GU13 9LE [IO91OG, SU85]
GM3 LGU R I Pryde, March Cottage, Toward, Dunoon, Argyll, PA23 7UB [IO75MU, NS16]
G3 LGV R E Hardman, 80 Green Ln, Garden Suburb, Oldham, OL8 3BA [IO83WM, SD90]
G3 LGW D G Spencer, Paladin, 89 Watling St., Hints, Tamworth, B78 3DE [IO92CP, SK10]
G3 LGX Details withheld at licensee's request by SSL.
G3 LGY J J Mulroy, Strandgatan 50B, S216 12 Malmo, Sweden, X X
G3 LHA R L Bastin, 40 Stamford Ave, Stivichall, Coventry, CV3 5BX [IO92FJ, SP37]
G3 LHB W G H Blanchard, 46 King George Ave, Loughborough, LE11 2NU [IO92JS, SK51]
G3 LHC B L Bonehill, 29 Charleton Way, West Charleton, Kingsbridge, TQ7 2AN [IO80CG, SX74]
G3 LHG E W Smith, 3 Meadow Ave, Wetley Rocks, Stoke on Trent, ST9 0BD [IO83XA, SJ94]
G3 LHH V T Walkley, 4 Shaw Ln, Holmfirth, Huddersfield, HD7 1PY [IO93CN, SE10]
G3 LHJ D Webber, 43 Lime Tree Walk, Milber, Newton Abbot, TQ12 4LF [IO80EM, SX87]
GW3 LHK G T Griffiths, Glyndwr, Lampeter Rd, Aberaeron, Ceredigion, SA46 0ED [IO72UF, SN46]
G3 LHN R D Muir, 19 Eastwick Dr, Bookham, Leatherhead, KT23 3PY [IO91TG, TQ15]
G3 LHP D Earnshaw, Ashford Lodge, 85 Alkington Rd, Whitchurch, SY13 1SU [IO82PX, SJ54]
G3 LHQ R T Burns, Wyedale, Havikil Ln, Scotton, Knaresborough, HG5 9HN [IO94FA, SE35]
G3 LHS L N Matthews, 16 Shirley Ave, Ramsgate, CT11 7AT [JO01RI, TR36]
G3 LHU C W Gubbins, Lower Kitesnest Ln, Whiteshill, nr Stroud, Glos, GL6 6BL [IO81VS, SO80]
GM3 LHV J Ellerby, 6 Clackmae Gr, Edinburgh, EH16 6PD [IO85JV, NT26]
G3 LHZ Dr M J Underhill, Hatchgate, Tandridge Ln, Lingfield, RH7 6LL [IO91XE, TQ34]
G3 LIA R J Rogers, 17 Elm Cl, Mulbarton, Norwich, NR14 8AU [JO02ON, TG10]
G3 LIK M Puttick, 21 Sandyfield Cres, Cowplain, Waterlooville, PO8 8SQ [IO90LV, SU61]
G3 LIO J A L Gibbs, 13 Bromley Rd, Macclesfield, SK10 3LN [IO83WG, SJ87]
G3 LIQ E Fell, 45 Delius Cl, Analby Park Rd Nt, Hull, HU4 7NR [IO93TR, TA02]
G3 LIT K Worrall, 44 The High St., Two Mile Ash, Milton Keynes, MK8 8HD [IO92OA, SP83]
G3 LIV J Melvin, 2 Salters Ct, Newcastle upon Tyne, NE3 5BH [IO95EA, NZ26]
GM3 LIW A C Wood, 2 Palmer Pl, Birkhill, Dundee, DD2 5RB [IO86LL, NO33]
G3 LIX Details withheld at licensee's request by SSL.
G3 LIZ E S Davies, 54 Rollason Rd, Erdington, Birmingham, B24 9BH [IO92BM, SP19]
G3 LJD J J H Davies, 57 Madeira Ct, Knightstone Rd, Weston Super Mare, BS23 2BH [IO81MI, ST36]
G3 LJF T H Ardern, Flat 3, 1 Stanley Mount, Sale, M33 4AF [IO83UK, SJ79]
G3 LJO R Smalley, 16 Ridge Gr, Heysham, Morecambe, LA3 2JN [IO84NB, SD46]
GW3 LJP A Bagley Mid Wales ARC, Ty Heulog, Ysfa Rd, Nantmel, Llandrindod Wells, Powys, LD2 6EW [IO82HD, SO05]
G3 LJQ C Leader, 847 Eastern Ave, Newbury Park, Ilford, IG2 7RZ [JO01BN, TQ48]
G3 LJR T E Saxton, Rock Farm House, 3 Acres Rd, Bebington, Wirral, L63 7QD [IO83LI, SJ38]
GW3 LJS Dr T W Bloxam, Maes y Bidiau, Abergorlech Rd, Brechfa, Carmarthen, SA32 7BH [IO71WX, SN53]
G3 LKV D D Locke, The Malthouse, Malthouse Ln, Foston, Derby, DE6 5PU [IO92CV, SK13]
GM3 LKY P Cohen, 2 Cedarwood Ave, Newton Mearns, Glasgow, G77 5QD [IO75US, NS55]
G3 LKZ Details withheld at licensee's request by SSL.
GM3 LLB A S Nelson, 44 McLean Gdns, Stonehouse, Larkhall, ML9 3LU [IO85AQ, NS74]
G3 LLD S W P Collier, 64 Slonk Hill Rd, Shoreham By Sea, BN43 6HY [IO90UU, TQ20]
G3 LLE K Webster, 47 Ashley Ln, Killamarsh, Sheffield, S31 8AB [IO93GJ, SK38]
G3 LLG R C H Loveday, 42 Bridle Path, Woodcote, Reading, RG8 0SE [IO91LM, SU68]
G3 LLJ Dr A J Hodgkinson, 27 Camborne Cl, Congleton, CW12 3BG [IO83VD, SJ86]
G3 LLK J A Gale, 66 Burys Bank Rd, Crookham Common, Thatcham, RG19 8DD [IO91IJ, SU56]
G3 LLL Details withheld at licensee's request by SSL.[Op: H Leeming, c/o Amateur Electronic UK Holdings, 45 Johnston St, Blackburn, BB2 1EF.]
G3 LLN M C Richardson, 4 Jessop Ave, Almondbury, Huddersfield, HD5 8UW [IO93DP, SE11]
GM3 LLP B M M Watson, Torloisk, 4 Caldwell Rd, West Kilbride, KA23 9LE [IO75NQ, NS24]
GI3 LLQ Details withheld at licensee's request by SSL.
G3 LLS H Stockley, 6 Priory Rd, Alcester, B49 5DY [IO92BF, SP05]
G3 LLV J A McElvenney, Airwork Limited, P O Box 223, Muscat Pc 111, Sultanate of Oman, X X
G3 LLX L R G H Reeve, 32 Cray Rd, Foots Cray, Sidcup, DA14 5BZ [JO01BJ, TQ47]
G3 LLZ D J Goacher, 27 Glevum Rd, Swindon, SN3 4AA [IO91DN, SU18]
G3 LMB A A Campbell, 43 Meersbrook Ave, Sheffield, S8 9EB [IO93GI, SK38]
G3 LMD Details withheld at licensee's request by SSL.[Op: J C Craig, PO Box 51, St Johns, Newfoundland, Canada, A1C 5H5.]
G3 LME K E Taylor, 7 Bowen Cl, Cheltenham, GL52 5EG [IO81XV, SO92]
G3 LMH R Wellbeloved, 8 Orchard Cl, South Wonston, Winchester, SO21 3EY [IO91HC, SU43]
G3 LMM G H Morgan, Dolphins, 53 The Ridgeway, Down End, Fareham, PO16 8RE [IO90KU, SU50]
G3 LMO Details withheld at licensee's request by SSL.
G3 LMP B Page, 12 Beechwood Ave, St. Albans, AL1 4YA [IO91US, TL10]
G3 LMQ J T Hamer, 7 Arundel Rd, Coventry, CV3 5JT [IO92FJ, SP37]
G3 LMR J K Eley, 112 Groby Rd, Glenfield, Leicester, LE3 8GL [IO92JP, SK50]
G3 LMS W D Bennett, 18 Earls Rd, Tunbridge Wells, TN4 8EA [JO01DD, TQ53]
G3 LMX T W Mitchell, 27 Hanmer Rd, Simpson, Milton Keynes, MK6 3AY [IO92PA, SP83]
G3 LNC J A Batham, 32 Ridgeway Ave, Dunstable, LU5 4QW [IO91RV, TL02]
G3 LNF M J Furness, Woodend, Sandy Ln, Cobham, KT11 2EG [IO91TI, TQ16]
G3 LNG G H Robbins, 35 Sunlight St., Anfield, Liverpool, L6 4AG [IO83MK, SJ39]
GM3 LNI H J Towns, 44 Glen Almond, East Kilbride, Glasgow, Lanarkshire, G74 2JU [IO75WS, NS65]
G3 LNK C J Bourne, 28 Roe Ln, Westlands, Newcastle, ST5 3PJ [IO82VX, SJ84]
G3 LNL P Lovelady, 14 Maunders Ct, Liverpool, Merseyside, L23 9YU [IO83LL, SD30]
G3 LNM R Scrivens, 17 Cragside View, Rothbury, Morpeth, NE65 7YU [IO95BH, NU00]
G3 LNN J E Symes, 20 Plants Brook Rd, Sutton Coldfield, B76 1EX [IO92CM, SP19]
G3 LNP A R Preedy, 7 Station Rd, Tring, HP23 5NG [IO91QT, SP91]
GW3 LNR A E Gwynne, 77 Edward St., Pant, Merthyr Tydfil, CF48 2BB [IO81HS, SO00]
G3 LNS G Beasley, Oaklands, Penn Ln, Tanworth in Arden, Solihull, B94 5HL [IO92BI, SP07]
G3 LNT J R Ambrose, 20 Cranborne Ave, Maidstone, ME15 7EB [JO01GG, TQ75]
G3 LNW J McGuire, 5 Primrose Way, Trevadlock Hall Park, Congdons Shop, Launceston, Cornwall, PL15 7PW [IO70SN, SX27]
G3 LOA G R Hartstone, 5 Woodcote Green, Downley, High Wycombe, HP13 5UN [IO91OP, SU89]
G3 LOC K J Jolly, 55 Storrington Way, Werrington, Peterborough, PE4 6QP [IO92UO, TF10]
GW3 LOD D M Rowse, 20 Peel Park Cl, Accrington, Lancs, BB5 6PL [IO83TS, SD72]
G3 LOE W Roberts, 13 Brean Rd, Stafford, ST17 0PA [IO82XS, SJ92]
G3 LOF Dr G D Peskett, 13 Warneford Rd, Oxford, OX4 1LT [IO91JR, SP50]
G3 LOJ A A Blythe, 3 Ripley Cl, Kirkbymoorside, York, YO6 6BS [IO94MG, SE68]
G3 LON G F Neal, 37 Fairlaine Rd, Eastbourne, E Sussex, BN21 1XF [JO00DS, TQ50]
G3 LOO M J Francis, Willow End, Fosters Ln, Tintagel, PL34 0BT [IO70PP, SX08]
G3 LOX B M Johnson, Rivendell Cottage, Manor Farm, Cattistock, Dorchester, Dorset, DT2 0JJ [IO80RT, SY59]
G3 LPA C R Coombe, 27 Garden Cl, Watton, Thetford, IP25 6DP [JO02KN, TF90]
G3 LPL P Sherdley, 2 Stable Yard, Taylors Ln, Pilling, Preston, PR3 6AB [IO83NW, SD44]
G3 LPN J P Hunt, 5 Mop Hale, Station Rd, Blockley, Moreton in Marsh, GL56 9EQ [IO92CA, SP13]
G3 LPO A B Burgess, 60 Alpraham Cres, Chester, CH2 1QX [IO83NF, SJ46]
G3 LPQ H W Wooding, 4 Millers Rise, Weston Super Mare, BS22 0SS [IO81NI, ST36]
G3 LPS E Pickering, 7 Hob Green, Mellor, Blackburn, BB2 7EP [IO83RS, SD63]
G3 LPT G Woods, Bamburgh House, Hunston, Bury St Edmunds, Suffolk, IP31 3EN [JO02KG, TL96]
GU3 LPV A H J Catts, 7 Little St., Alderney, Guernsey, GY9 3TT
G3 LQC R C Evans, 30 Chandler Cl, Bampton, OX18 2NW [IO91FR, SP30]
GW3 LQE A M Ernest, 6 Kymin Terr, Penarth, CF64 1WW [IO81KK, ST17]
G3 LQG J McCaig, 16 Leewood Rd, Weston Super Mare, BS23 2PB [IO81MI, ST36]
GM3 LQH J W Hudson, 37 The Green, Bathgate, EH48 4DA [IO85EV, NS96]
G3 LQI S G Williams, 58 Grinstead Ln, Lancing, BN15 9DZ [IO90UU, TQ10]
G3 LQJ R V Cox, 12 Kelling Cl, Holt, NR25 6RU [JO02NV, TG03]
G3 LQM Details withheld at licensee's request by SSL.
G3 LQN Details withheld at licensee's request by SSL.
G3 LQO E L Harris, 10 Girdle Rd, Walsworth, Hitchin, SG4 0AN [IO91UX, TL13]
G3 LQP R Brown, 32 Albert Rd, Sutton, SM1 4RX [IO91VI, TQ26]
G3 LQR S J W Freeman, West Farm, Cransforde, Woodbridge, Suffolk, IP13 9PQ [JO02QF, TM36]
G3 LQS Details withheld at licensee's request by SSL.
G3 LQT D S Buller, 39 Glebe Rd, Hemingford Grey, Huntingdon, PE18 9DS [IO92WH, TL27]
G3 LQU F R Vosper, 23 Lyons Rd, Holmbush, St. Austell, PL25 3HX [IO70OI, SX05]
G3 LQW K Wallace, 55 Lambourne Rd, West Knighton, Leicester, LE2 6HQ [IO92KO, SK50]
G3 LQX M A Nicholls, 63 Norwich Rd, Poringland, Norwich, Norfolk, NR14 7QX [JO02QN, TG20]

GI3 LQY J H Stronach, 1 Kinedale Cttgs, Lisburn Rd, Ballynahinch, BT24 8 [IO74BJ, J35]
G3 LRA C A E Eley, 30 North Down, Staplehurst, Tonbridge, TN12 0PQ [JO01GD, TQ74]
G3 LRB Details withheld at licensee's request by SSL.
GM3 LRG J F Gray, 47 South St., Greenock, PA16 8QG [IO75OW, NS27]
G3 LRH Details withheld at licensee's request by SSL.
G3 LRI J Blakey, 10 Wilson Terr, Newcastle upon Tyne, NE12 0JP [IO95FA, NZ26]
G3 LRL R Bowell, 16 Margarite Way, Wickford, SS12 0ER [JO01GO, TQ79]
G3 LRM S H W Tanner, 35 Apple Tree Cl, Woodmancote, Cheltenham, GL52 4UA [IO81XW, SO92]
G3 LRO Details withheld at licensee's request by SSL.
G3 LRP P N Ackley, Camelot, Greenside, Havercroft, Wakefield, WF4 2BG [IO93HO, SE31]
G3 LRQ M J Humphries, 2 South View Cl, Twyford, Reading, RG10 9AY [IO91NL, SU77]
G3 LRS R E Talbott obo Leicester Radio Soc, Thornfield, 33 Highfield St., Anstey, Leicester, LE7 7DU [IO92JQ, SK50]
G3 LRU J P Miller, Basement Flat, 63 Hova Villas, Hove, BN3 3DJ [IO90VT, TQ20]
G3 LRX R M Durell, Middleton Farm, Hubbards Hill, Warren St, Lenham, Kent, ME17 2EJ [JO01JF, TQ95]
G3 LSA D R Moore, 5 Seahaven Springs Est, Seaholme Rd, Mablethorpe, LN12 2QS [JO03DH, TF58]
GD3 LSF E S Ellis, Ballahams, 43 Governors Hill, Douglas, IM2 7AT
G3 LSJ C S Gerrard, 10 Bridle Cl, Southfields, Sleaford, NG34 7TD [IO92TX, TF04]
G3 LSL Details withheld at licensee's request by SSL.[Op: Dave Lunn. Station located near Andover.]
G3 LSQ P J Aitchison, Bolton House, Windmill Hill, London, NW3 6SJ [IO91VN, TQ28]
G3 LSS Details withheld at licensee's request by SSL.
G3 LST P F L Clarke, Corner House, Lower St., Stratford St. Mary, Colchester, CO7 6JR [JO01LX, TM03]
G3 LSW K L Willis, 151 Columbus Ravine, Scarborough, YO12 7QZ [IO94TG, TA08]
G3 LSX G M M Townsend, 18 Allenswood Rd, Eltham, London, SE9 6RP [JO01AL, TQ47]
G3 LSY D Morris, 16 Locket Rd, Wealdstone, Harrow, HA3 7LZ [IO91UO, TQ18]
G3 LSZ A Seldon, 67 Rural Vale, Northfleet, Gravesend, DA11 9JL [JO01EK, TQ67]
G3 LTB A Barnes, 33 Hawthorn Gr, Southport, PR9 7AA [IO83MP, SD31]
GM3 LTD J W Smyth, 222 Balunie Ave, Dundee, DD4 8TN [IO86NL, NO43]
G3 LTF P K Blair, Woodleigh, Upper Wyke, St. Mary Bourne, Andover, SP11 6EA [IO91GG, SU45]
G3 LTG J H Matthews, Silverwood, 8 Filbert Rd, Loddon, Norwich, NR14 6LW [JO02RM, TM39]
G3 LTM B E Moyler, 1 Bay Walk, Aldwick Bay Est, Bognor Regis, PO21 4ET [IO90PS, SZ99]
G3 LTN R Marriott, 28 Astrop Rd, Middleton Cheney, Banbury, OX17 2PQ [IO92IB, SP44]
G3 LTP R G Flavell, 174 Finchampstead Rd, Wokingham, RG40 3EY [IO91NJ, SU86]
G3 LTS J Stelfox, 41 Astley Gdns, Robert St., Dukinfield, SK16 4QE [IO83WL, SJ99]
G3 LTT H A Gray, 267 New Parks Boulevard, Braunstone Frith, Leicester, LE3 6NQ [IO92JP, SK50]
G3 LTV W E Robinson, 68 Wicks Ln, Formby, Liverpool, L37 1PX [IO83LN, SD20]
GM3 LTW A F Hunter, 2 Cargill Ave, Maybole, KA19 8AD [IO75PI, NS31]
G3 LTX R L Savage, Plas Gwyntog, Rhoslefain, Tywyn, Gwynedd, LL36 9ND [IO72WP, SH50]
G3 LUA A G Knowles, 73 Kingslea Rd, Solihull, B91 1TJ [IO92CJ, SP17]
G3 LUB D R Bowman, Blea Rigg, Little Ln, Loosley Row, Princes Risborough, HP27 0NX [IO91OQ, SP80]
G3 LUC E W Bate, 5 Elm Rd, Shildon, DL4 1BH [IO94EP, NZ22]
G3 LUF W L Chick, 28 Albion Rd, Christchurch, BH23 2JQ [IO90CR, SZ19]
G3 LUH K E Reader, 36 Balliol Rd, Welling, DA16 1PG [JO01BL, TQ47]
G3 LUI R E Hunter, 81 Waxwell Rd, Hullbridge, Hockley, SS5 6HG [JO01HO, TQ89]
G3 LUK R M Pearce 59 Squadron Air, 28 Wood St., Longwood, Huddersfield, HD3 4RF [IO93CP, SE11]
G3 LUL C D Harrington, Ons Tuis, 4A Maple Ave, Maidstone, ME16 0DD [JO01GG, TQ75]
G3 LUO C H Evans, Cucumber Hall Farm, Cucumber Ln, Essendon, Hatfield, AL9 6JB [IO91WR, TL20]
G3 LUP D Rhodes, 3 Sycamore Way, Huntington, Cannock, WS12 4QP [IO82XR, SJ91]
G3 LUW B C Whittaker, Burthorne, 2 New Town, Uckfield, TN22 5DB [JO00BX, TQ42]
G3 LUY Details withheld at licensee's request by SSL.
G3 LUZ F Machin, 70 Poors Ln, Hadleigh, Benfleet, SS7 2LN [JO01HN, TQ88]
GM3 LVA D J Simpson, Larchwood, Station Rd, Tomatin, Inverness, Scotland, IV13 7YR [IO87AI, NH72]
G3 LVB G R Brooks, 39 St. Catharines Way, Houghton on The Hill, Leicester, LE7 9HE [IO92MP, SK60]
G3 LVN D R Sibbald, 69 St. Georges Rd, Ilford, IG1 3PG [JO01AN, TQ48]
G3 LVP K F Eastty, 7 The Grange, Reddings Rd, Cheltenham, Glos, GL51 6RL [IO81WV, SO92]
G3 LVW R W B Smith, 40 Highwoods Dr, Marlow Bottom, Marlow, SL7 3PY [IO91OO, SU88]
G3 LWD P Stone, Bramley, Stone St., Lympne, Hythe, CT21 4JP [JO01MB, TR13]
G3 LWF L R Franklin, 32 Neeld Cres, Chippenham, SN14 0HT [IO81WL, ST97]
G3 LWI Capt J D F Francis, 3 Nightingale Cl, Bembridge, PO35 5YP [IO90LQ, SZ68]
G3 LWJ C F Way, 8 Stratford Pl, Eaton Socon, St. Neots, Huntingdon, PE19 3HY [IO92UF, TL15]
G3 LWK Details withheld at licensee's request by SSL.[Op: Harold Taylor, Banks, Southport, Lancs, PR9 8HD.]
G3 LWM J D Harris, 44 Fourth Ave, Frinton on Sea, CO13 9DX [JO01OT, TM21]
GM3 LWS E H Ross, 24 Ettrick Way, Glenrothes, KY6 1JL [IO86JE, NO20]
G3 LWT P W Buck, 17 Sanden Cl, Hungerford, RG17 0LA [IO91FJ, SU36]
GW3 LWU G P Brisbar, 97 Chambers Ln, Mynydd Isa, Mold, CH7 6UZ [IO83KE, SJ26]
G3 LWX E J Gane, 44 South View Ave, Swindon, SN3 1DZ [IO91CN, SU18]
G3 LX H P Arnfield, 7 Hurst Lea Rd, New Mills, Stockport, Ches, SK12 3HP [IO83XI, SJ98]
G3 LXB S W Jones, 43 New St., Chase Terr, Walsall, West Midlands, WS7 8BT [IO92AQ, SK00]
G3 LXD J L Hawkins, 17 Shasta Rd, Lesmurdie, Wa 6076, Australia
GW3 LXI G H Price, 9 Essex Rd, Pembroke Dock, SA72 6ED [IO71MQ, SM90]
G3 LXJ F J Fisher, 48 Nelmes Cres, Emerson Park, Hornchurch, RM11 2PR [JO01CN, TQ58]
G3 LXN B Negri, 18 Tollerton Ln, Tollerton, Nottingham, NG12 4FQ [IO92KV, SK63]
G3 LXP D E E Purchese, 68 Hughenden Rd, Marshalswick, St. Albans, AL4 9QS [IO91US, TL10]
G3 LXQ D L Gallop, 4 Volunteer Rd, Theale, Reading, RG7 5DN [IO91LK, SU67]
G3 LXW H A A Graves, 21 New Rd, Orpington, BR6 0DX [JO01BJ, TQ46]
G3 LXY W Rogerson, Hill Top Rd, Thornton, Bradford, W Yorks, BD13 3QX [IO93BT, SE03]
GM3 LYA J Morgan, Mayfield, Chapel Brae, Braemar, Ballater, AB35 5YT [IO87HA, NO19]
GU3 LYC T J de Putron, Shieling Cottage, La Rue Marquand, St. Andrew, Guernsey, GY6 8RB
G3 LYD Dr E E Henderson, The Homestead, High St., Godshill, Ventnor, PO38 3HZ [IO90IP, SZ58]
G3 LYE Details withheld at licensee's request by SSL.
GW3 LYF Details withheld at licensee's request by SSL.[Op: D L Jones, Corbetts, Queens Road, Ilkley, W Yorks, LS29 9QJ.]
G3 LYG Dr A E Macgregor, 10 Balroy Ct, Forest Hall, Newcastle upon Tyne, NE12 9AW [IO95FA, NZ26]
G3 LYK Details withheld at licensee's request by SSL.
G3 LYN R J Amblin, 21 Englishcombe Way, Bath, BA2 2EU [IO81TI, ST76]
G3 LYP Dr M D Scott, The Magnolias, Marlow Rd, Ln End, High Wycombe, HP14 3JW [IO91OO, SU89]
G3 LYT W J C Fennell, East View, Engine Dyke, Gedney Dyke, Spalding, PE12 0BE [JO02BT, TF42]
G3 LYU D T Price, 16 Dorset Ave, Glenfield, Leicester, LE3 8BB [IO92JP, SK50]
G3 LYV C R Rogers, 27 St. Martins Rd, Finham, Coventry, CV3 6ET [IO92FJ, SP37]
G3 LYW J F R Weston, Chenery Lodge, 44 Old Newbridge Hill, Bath, BA1 3LU [IO81TJ, ST76]
G3 LYX J R Crellin, 35 Meadow Way, Carlton Colville, Lowestoft, NR33 8LF [JO02UK, TM58]
GM3 LYY J T A Johnston, The Dolphins, Montgomerie Dr, Fairlie, Ayrshire, Scotland, KA29 0DZ [IO75NR, NS25]
G3 LYZ B Currey, Lagupie, Marmande 47180, France, X X
G3 LZC A E Stirland, 98 Aldreds Ln, Heanor, DE75 7HG [IO93HA, SK44]
G3 LZE M J Henry, 10 Robert St., Williton, Taunton, TA4 4PG [IO81ID, ST04]
G3 LZG Details withheld at licensee's request by SSL.
G3 LZI J A Oates, Cherry Tree Cottage, Green Moor, Wortley, Sheffield, S30 7DQ [IO93GJ, SK38]
G3 LZK J C B Steele, The Red House, Twyning Green, Twyning, Tewkesbury, GL20 6DF [IO82WA, SO93]
G3 LZM M A Bush, 52 Barrs Ct Rd, Hereford, HR1 1EQ [IO82PB, SO54]
G3 LZN G J R Ellison, Little Flushing, St. Peters Rd, Flushing, Falmouth, TR11 5TJ [IO70LE, SW83]
G3 LZO J Thomas, Great Harwood Lodge, Edward St., Great Harwood, Blackburn, BB6 7JB [IO83TS, SD73]
G3 LZQ J Dunnington, 73 West Hall Garth, South Cave, Brough, East Yorks, HU15 2HA [IO93QS, SE93]
G3 LZR E W Speller, 19 Hilltop Cres, Weeley, Clacton on Sea, CO16 9HZ [JO01NU, TM12]
G3 LZT H G Hall, The Mount, Habberley Rd, Trimpley, Bewdley, DY12 1NL [IO82UJ, SO77]
G3 LZV C Evans, 12 Stocks Hill, Bawburgh, Norwich, NR9 3LL [JO02OP, TG10][Operator: Clive Evans. Station located in Komlo 7300 Hungary.]
G3 LZY F D B McNamara, Sassoon, Andersons Ln, Bridgnorth, WV16 4PU [IO82SM, SO79]
G3 LZZ A M Pomfret, 2 Ingwell House, Main St., Grange Over Sands, LA11 6DP [IO84NE, SD47]
G3 MA E A Perkins, 40 Calton Rd, Gloucester, GL1 5DY [IO81VU, SO81]
G3 MAE Dr A E Wilson, 8 The Paddock, Appleton Wiske, Northallerton, DL6 2BE [IO94HK, NZ30]
G3 MAH B E Bailey, Herbs and Honey, Summer Hill, Burnham Rd, Althorne, Essex, CM3 6BX [JO01JP, TQ99]
G3 MAI R W Stevens, 138 Grange Dr, Stratton St. Margare, Swindon, SN3 4LA [IO91DN, SU18]
G3 MAJ E Holden, 10 Rowan Tree Cl, Greasby, Wirral, L49 3AW [IO83KJ, SJ28]
G3 MAK R Roberts, 28 Brockley Cres, Bleadon Hill, Weston Super Mare, BS24 9LL [IO81MH, ST35]
G3 MAL Details withheld at licensee's request by SSL.
G3 MAM Dr D B Sugden, 7 Church St., Alfreton, DE55 7AH [IO93HC, SK45]
G3 MAN D Hornsey, 183 Belton Ln, Grantham, NG31 9PL [IO92QW, SK93]
G3 MAO Details withheld at licensee's request by SSL.
G3 MAR N Gutteridge Midland ARS, 68 Max Rd, Quinton, Birmingham, B32 1LB [IO92AK, SP08]
GM3 MAS A B Pringle, 1 Falloch Rd, Milngavie, Glasgow, G62 7RR [IO75TW, NS57]
GU3 MAT Dr L H A Pilkington, Coppice End, Colborne Rd, St. Peter Port, Guernsey, GY1 1EP
G3 MAU J G Wardle, 17 Frederick Neal Ave, Coventry, CV5 7EH [IO92EJ, SP27]
G3 MAV J Bradley, 17 Talboys Walk, Tetbury, GL8 8YU [IO81WP, ST89]
G3 MAW D Noble, 30 Holland Rd, Westcliff on Sea, SS0 7SG [JO01IM, TQ88]
G3 MAX F N Nicholls, Montrose, Earles Ln, Marston, Northwich, CW9 6DX [IO83SG, SJ67]
GM3 MAY H F Stenhouse, Hazel Cottage, 8 West Donington St., Darvel, KA17 0AP [IO75UO, NS53]
G3 MAZ H V Bell, 35 Elm Trees, Long Crendon, Aylesbury, HP18 9DG [IO91MS, SP60]
GI3 MBB A Mc Murtry, 20 Towerview Cres, Bangor, BT19 6BA [IO74EP, J58]
GD3 MBC J H A L Churchill, Fair Isle, Lhoobs Rd, Foxdale, Douglas, IM4 3JB
G3 MBJ M Acton, 32 Hillcrest Ave, Winskill, Burton on Trent, DE15 0TZ [IO92ET, SK22]
G3 MBK D W Underdown, 26 Birch Rd, Farncombe, Godalming, GU7 3NT [IO91QE, SU94]
G3 MBL A G Edwards, 4 Rokewood Pl, Stanningfield, Bury St. Edmunds, IP29 4RF [JO02JE, TL85]
G3 MBM J D Masters, 67 High St., Landbeach, Cambridge, CB4 4DR [JO02BG, TL46]
G3 MBN B C Gibbs, 15 Moor Barton, Neston, Corsham, SN13 9SH [IO81VJ, ST86]
G3 MBO B A Aspinwall, 33 Clipstone Cres, Leighton Buzzard, LU7 8LU [IO91QW, SP92]
GU3 MBS S H Gibbs, 44 Les Prins, Vale, Guernsey, GY6 8HB
G3 MBU M Standige, 7 Hill Crest Ave, Burnley, BB10 4JA [IO83VS, SD83]
G3 MBW J E Collins, 15 Hillside Ave, Guiseley, Leeds, LS20 9DH [IO93DV, SE14]
G3 MCA D F Owen, 1 Mosslea Rd, Orpington, BR6 8HP [JO01AI, TQ46]
G3 MCB A V Williams, 1 Wyvern Rd, Sutton Coldfield, B74 2PS [IO92CN, SP19]
G3 MCC K W Worrall, 12 Lower Heath Ave, Congleton, CW12 2HJ [IO83VE, SJ86]
G3 MCD K O Holland, 40 Brookland Rise, London, NW11 6DS [IO91VO, TQ28]
G3 MCE L M Lee, 34 Westby Way, Poulton-le-Fylde, FY6 8AD [IO83MU, SD33]
G3 MCF J Wilson, 4 St. Augustine Gr, Bridlington, YO16 5DB [IO94VC, TA16]
G3 MCI H Hoyle, 373 Devonshire Rd, Blackpool, FY2 0RE [IO83LU, SD33]
G3 MCK G P Stancey, 14 Cherry Orchard, Staines, TW18 2DF [IO91RK, TQ07]
G3 MCL C P T Simpkins, 6 Compton Way, Olivers Battery, Winchester, SO22 4EY [IO91HB, SU42]
G3 MCN H James, The Ridge, Eddisbury Hill, Delamere, Northwich, Ches, CW8 2HX [IO83QF, SJ56]
G3 MCO Details withheld at licensee's request by SSL.
G3 MCP P G Goadby, 535 Welford Rd, Leicester, LE2 6FN [IO92KO, SK50]
G3 MCS W R Hawthorne, The Willows, Moortown Rd, Nettleton, Market Rasen, LN7 6HY [IO93UL, TA10]
G3 MCV B A Vaughan, 17 Richmond Cl, West Town, Hayling Island, PO11 0ER [IO90MS, SZ79]
G3 MCW R A E Fronius, 5 Hospital Cttgs, Cres Rd, Warley, Brentwood, CM14 5JA [JO01DO, TQ59]
G3 MCX W J Kennedy, 22 Croham Park Ave, South Croydon, CR2 7HH [IO91XI, TQ36]
G3 MDC Details withheld at licensee's request by SSL.[Op: G H Taylor.]
G3 MDD B S Mudge, 3 Meadow Way, Bracknell, RG42 1UE [IO91OK, SU87]
G3 MDG R Ray, 54 Gladstone Rd, Chesham, HP5 3AD [IO91QQ, SP90]
G3 MDH P A L Shoosmith, 39 Lakeside, Hightown, Ringwood, BH24 3DX [IO90CU, SU10]
G3 MDI M F Plummer, 17 Cadman Sq, Shenley Lodge, Milton Keynes, MK5 7DN [IO92OA, SP83]
GW3 MDK R Jones, Woodcote, 37 Coed Pella Rd, Colwyn Bay, LL29 7BB [IO83DH, SH87]
G3 MDL P Cunningham, 3 Clayfield Ave, Mexborough, S64 0HY [IO93IL, SE40]
G3 MDM G J McGee, 2 Ilynton Ave, Firsdown, Salisbury, SP5 1SH [IO91DC, SU23]
G3 MDN H D Jackson, Lilacs, Elms Green, Leominster, Herefordshire, HR6 0NS [IO82PE, SO55]
G3 MDP L S Owen, 53 Applegarth Dr, Newbury Park, Ilford, IG2 7TQ [JO01BN, TQ48]
G3 MDQ Dr G A H Heaney, 23A Wylde Green Rd, Sutton Coldfield, B72 1HD [IO92CN, SP19]
G3 MDR M H Hallet, Tresco, 38 Beechwood Cres, Chandlers Ford, Eastleigh, SO53 5PD [IO90HX, SU42]
G3 MDT R Langston, 59 Merchants Way, Canterbury, CT2 8PN [JO01MG, TR15]
G3 MEA S Harle, 24 Front St., Quebec, Durham, DH7 9DF [IO94DS, NZ14]
G3 MEC J S E Pearce, 86 Sopers Ln, Poole, BH17 7EU [IO90AR, SZ09]
G3 MED F A Griffiths, 105 Hillcroft Cres, Oxhey, Watford, WD1 4PA [IO91TP, TQ19]
G3 MEH R E Piper, 8 Osborne Way, Wigginton, Tring, HP23 6EN [IO91QS, SP91]
G3 MEJ Details withheld at licensee's request by SSL.
G3 MEK N Gaunt, 13 Whiteoaks Ln, Greenford, UB6 8XE [IO91TM, TQ18]
GW3 MEO J E Cronk, 2 Mostyn Ave, Prestatyn, LL19 9NF [IO83HI, SJ08]
G3 MEP R D Luscombe, 6 Barradon Cl, Barton, Torquay, TQ2 8QE [IO80FM, SX96]
G3 MER J D Davis, 16 Newmarket Rd, Furnace Green, Crawley, RH10 6NB [IO91VC, TQ23]
G3 MES A V Tillin, 11 Great Ellshams, Holly Ln, Banstead, SM7 2BA [IO91VH, TQ25]
G3 MEV C N Cory, Skylite, Chapel Ln, Ashford Hill, Newbury, RG15 8BE [IO91LJ, SU66]
G3 MEY J Lawrence, 2 Canterbury St., Chippenham, SN14 0EB [IO81WL, ST97]
G3 MEZ D J Earnshaw, 12 Kelsey Cl, Hunstanton, PE36 6HL [JO02GW, TF64]
G3 MFE D W Aird, 25 Milford Ave, Stony Stratford, Milton Keynes, MK11 1EY [IO92NB, SP73]
G3 MFG D A Close, 27 High St., Collyweston, Stamford, PE9 3PW [IO92RO, SK90]
G3 MFH G Dale, 20 Blythe Ave, Meir Heath, Stoke on Trent, ST3 7JY [IO82WW, SJ94]
G3 MFJ G F Firth, 13 Wynmore Dr, Bramhope, Leeds, LS16 9DQ [IO93EV, SE24]
G3 MFK M Camp, 82 Leicester Rd, Hinckley, LE10 1LT [IO92HN, SP49]
G3 MFL A J Russell, Peartree Cotttage, Stibb Green, Burbage, Marlborough, Wilts, SN8 3AS [IO91DI, SU26]
G3 MFO P J Elliot, 58 Westminster Ct, St. Albans, AL1 2DX [IO91TR, TL10]
G3 MFQ S A Kerrison, 32 Tiercel Ave, Sprowston, Norwich, NR7 8JN [JO02PP, TG21]
G3 MFU R W R Findlay, 43 Walsingham Ct, Leverington, Wisbech, PE13 5AQ [JO02BQ, TF41]
G3 MFW H G Woodhouse, Trenoweth, Porthpean Beach Rd, St. Austell, PL26 6AU [IO70OH, SX05]
G3 MFX H R Hodson, 105 Elm Ln, Sheffield, S5 7TX [IO93GK, SK39]
GW3 MFY W M Lee, Wallas Farm House, Wick Rd, Ewenny Bridgend, Mid Glam, CF35 5AE [IO81FL, SS97]
G3 MGB D B Smart, 202 London Rd, Knowle, Solihull, B93 9HU [IO92DJ, SP17]
G3 MGF S H Heaven, Melbourne House, Tickmorend, Horsley, Stroud, GL6 0PE [IO81VQ, ST89]
G3 MGH P J Clegg, 1 Frogmore Cl, Hughenden Valley, High Wycombe, HP14 4LN [IO91OQ, SU89]
G3 MGI D Binns, 5 Wentworth Rd, York, YO2 1DG [IO93KW, SE55]
G3 MGJ P Robinson, 4 St. Johns Ct, Stratford upon Avon, CV37 9AD [IO92DE, SP15]
G3 MGK . I A Kemp, 5 Oakdale Ave, Frodsham, Warrington, WA6 6PY [IO83PG, SJ57]
G3 MGL A V H Davis, 41 Gainsborough Rd, Crawley, RH10 5LD [IO91VC, TQ23]
G3 MGM Details withheld at licensee's request by SSL.
G3 MGP Details withheld at licensee's request by SSL.
G3 MGS C D Stephens, 12 Berks Rd, Bishopston, Bristol, BS7 8EX [IO81QL, ST57]
GM3 MGT A W Hope, 5 Collins Cres, Dalgety Bay, Dunfermline, KY11 5FG [IO86HA, NT18]
G3 MGU A T Dodson, 53 Simons Ln, Wokingham, RG41 3HG [IO91NK, SU76]
G3 MGV Details withheld at licensee's request by SSL.
G3 MGW R Wheeler, 26 Ladysmith Ave, Brightlingsea, Colchester, CO7 0JD [JO01MT, TM01]
G3 MGX J M Tomlinson, Silver Garth, Moor Ln, Roughton, Woodhall Spa, LN10 6YH [IO93WD, TF26]
G3 MGZ D R Revell, 18 Cumberland Ave, Basingstoke, RG22 4BG [IO91KF, SU65]
G3 MHC V A J Cole, Evergreen, Rawreth Ln, Rayleigh, Essex, SS6 9PZ [JO01HO, TQ89]
G3 MHD A E W Williams, 9 Bracken Ln, Fern Tree, Tasmania 7054, Australia, X X
G3 MHF M S Ockenden, 6 Selwyn Rd, Eastbourne, BN21 2LE [JO00DS, TV69]
G3 MHH G L Blake, 8 Sempringham Fen, Billingborough, Sleaford, Lincs, NG34 0NH [IO92UU, TF13]
G3 MHM S F Wheeler, 58 Scribers Ln, Hall Green, Birmingham, B28 0PA [IO92BK, SP17]
G3 MHN B J Hitchens, 74 Christ Church Ln, Lichfield, WS13 8AL [IO92BQ, SK10]
G3 MHQ E W Holt, 34 Lee Rd, Greenford, UB6 7DD [IO91UM, TQ18]
G3 MHR W E Lee, 6 Highfield Rd, Swanwick, Alfreton, DE55 1BW [IO93HB, SK35]
G3 MHT E J Landon, Meadows, Smithy Ln, Bigby, Barnetby, DN38 6ER [IO93TN, TA00]
G3 MHV Dr T G Langdon, 58 Upper Marsh Rd, Warminster, BA12 9PN [IO81VE, ST84]
GW3 MHW Details withheld at licensee's request by SSL.
G3 MHX M C Tate, 48 Crossgates, Bedwell Plash, Stevenage, SG1 1LS [IO91VV, TL22]
G3 MHY R Morris, Manfield, 3 Wrockwardine Rd, Wellington, Telford, TF1 3DA [IO82RQ, SJ61]
G3 MID R Harkin, 7 Troutbeck Ave, Leamington Spa, CV32 6NE [IO92FH, SP36]
GM3 MIE R Brunskill, 24 Templeton Cres, Prestwick, KA9 1JA [IO75QL, NS32]
G3 MIH B W Sutton, 117 Utting Ave East, Liverpool, L11 5AB [IO83MK, SJ39]
G3 MII P A Roper, Flat 1, 28 Cobham Rd, Westcliff on Sea, SS0 8EA [JO01IM, TQ88]
G3 MIP S Heilbron, 8 Beechwood Dr, Formby, Liverpool, L37 2DG [IO83LN, SD20]
G3 MIQ M Akehurst, 73 Gerda Rd, New Eltham, London, SE9 3SJ [JO01AK, TQ47]
G3 MIR D H Parr, Silverdale, Well St., Starcross, Exeter, EX6 8QH [IO80GO, SX98]
G3 MIZ Maj S A Bevan, 7 Patwell St., Bruton, BA10 0EQ [IO81SC, ST63]
G3 MJH A R Haggarty, 86 Albion House, Common Rd, Slough, SL3 8TF [IO91RL, TQ07]
G3 MJK Brig J C Clinch, The Pippins, Dummer Rd, Axford, Basingstoke, RG25 2ED [IO91KE, SU64]
G3 MJM A D Marshall, 16 Beverley Dr, Beverley, HU17 9GG [IO93SU, TA03]
G3 MJN L A Harvey, 755A London Rd, Westcliff on Sea, SS0 9SU [JO01IN, TQ88]
G3 MJO Details withheld at licensee's request by SSL.
G3 MJP M M B Philpott, 20 Cottes Way, Hill Head, Fareham, PO14 3NE [IO90JT, SU50]
G3 MJS E C Long, 30 Hunts Mead, Sherborne, DT9 6AJ [IO80RW, ST61]
G3 MJW D Edmunds, 15 Hope St., Bozeat, Wellingborough, NN29 7LU [IO92PF, SP95]
G3 MJX A Bird, 28 Moorfield Rd, Duxford, Cambridge, CB2 4PS [JO02BC, TL44]
G3 MKB A C W Watts, 31 Albert Rd, Ledbury, HR8 2DN [IO82SA, SO73]
G3 MKH G Rooney, 270 Spital Rd, Bromborough, Wirral, L62 2AW [IO83MI, SJ38]
GW3 MKT M M Hooks, 1 Llwyn Castan, Pentwyn, Cardiff, CF2 7DA [IO81KM, ST28]
G3 MKU A F Bower, 82 Anson Rd, Shepshed, Loughborough, LE12 9PU [IO92IS, SK41]
G3 MKV C Curtis, 2 Girton Cres, Hartford, Huntingdon, PE18 7QH [IO92WI, TL27]
G3 MKW W J Rider, 174 Oak Tree Ln, Bournville, Birmingham, B30 1TX [IO92AK, SP08]
G3 MLA J C Woodhouse, 369 Blackpool Old Rd, Highfurlong, Blackpool, FY3 7LX [IO83LU, SD33]
G3 MLD K W Darby, Dormer Cottage, 56 Pennyacre Rd, Teignmouth, TQ14 8LB [IO80GN, SX97]
G3 MLH D Yeoell, 31 Westbury Gdns, Higher Odcombe, Yeovil, BA22 8UR [IO80PW, ST51]
G3 MLO P W Weatherall, Woodside 6 Mile Grg, Stone St, Stelling Minnis, Canterbury Kent, CT4 6DN [JO01MD, TR14]
G3 MLP B C J Poole, 139 Lincoln Rd, Peterborough, PE1 2PW [IO92VN, TL19]
GU3 MLR S A Faulkner, Sentosa, Sous L'Englise, St Saviours, Guernsey, Guernsey, GY9 9SE
G3 MLS D Nappin, New Edge Farm, Colden Heptonstal, Hebden Bridge, West Yorks, HX7 7PG [IO83XS, SD92]
GM3 MLW W R Cook, 4 Liberton Brae, Edinburgh, EH16 6AE [IO85KW, NT27]
G3 MLX J Bourne, 42 Legarde Ave, Hull, HU4 6AP [IO93TR, TA02]

G3 MLY Details withheld at licensee's request by SSL.
G3 MMA D W Mayes, 5 Halstead Rd, Gosfield, Halstead, CO9 1PQ [JO01HW, TL73]
G3 MME P A Whitford, Vernon Ln, Kelstedge, Ashover, Chesterfield, Derbyshire, S45 0EA [IO93GE, SK36]
GI3 MMF B W McAleer, 90 Gortin Park, Belfast, BT5 7EQ [IO74BO, J37]
GI3 MMG D Noon, 34 Rodney Park, Bangor, BT19 6FN [IO74EP, J58]
G3 MMH Details withheld at licensee's request by SSL.[Correspondence to RAF Wyton ARC, Huntingdon, Cambridgeshire.]
G3 MMJ G R Browne, 20 The Cliff, Brighton, BN2 5RE [IO90WT, TQ30]
G3 MMK M Firth, 18 Shelf Moor, Shelf, Halifax, HX3 7PW [IO93CS, SE12]
G3 MML E G Augood, 32 Victoria Rd, Salisbury, SP1 3NG [IO91CB, SU13]
G3 MMM S Marriott, The Cottage, 28 The St., Kennington, Ashford, TN24 9HB [JO01KE, TR04]
G3 MMN B J Newman, Springfield, 101 Tally House Rd, Shadoxhurst, Ashford, TN26 1HW [JO01JC, TQ93]
G3 MMP Details withheld at licensee's request by SSL.
G3 MMQ J A Hedges, 35 Ferrymead Ave, Greenford, UB6 9TL [IO91TM, TQ18]
G3 MMS G A Whiting, 25 Obthorpe Ln, Thurlby, Bourne, PE10 0ES [IO92TR, TF01]
GW3 MMT L John, 15 Cedar Cl, Gowerton, Swansea, SA4 3AR [IO71XP, SS59]
GW3 MMU P M Fulton, 36 Sunnybank Rd, Blackwood, NP2 1HZ [IO81JQ, ST19]
G3 MNB H J Benjamin, 57 The Lea, Fleet, GU13 8AT [IO91NG, SU75]
G3 MND C B Wells, 184 Somersall Ln, Chesterfield, S40 3NA [IO93GF, SK36]
G3 MNJ J C Yates, 13 Nickleby Rd, Poynton, Stockport, SK12 1LE [IO83WI, SJ98]
G3 MNK E D D Turpin, 2 The Rye, Eaton Bray, Dunstable, LU6 2BQ [IO91QV, SP92]
G3 MNN T G Kelly, 78 Alameda House, Red Sands Rd, Gibraltar
G3 MNO D L Lisney, 119 Draycott Ave, Kenton, Harrow, HA3 0DA [IO91UN, TQ18]
G3 MNR Details withheld at licensee's request by SSL.
G3 MNS I Swan, 27 Dymoke Rd, Mablethorpe, LN12 2BY [JO03DH, TF58]
G3 MNT G A Farrall, 31 Springfield Rd, Gatley, Cheadle, SK8 4PE [IO83VJ, SJ88]
G3 MNV P W F Darragh, 48 Goodwood Park Rd, Northam, Bideford, EX39 2RR [IO71VA, SS42]
G3 MO S J Geary, 117 Taynton Dr, Merstham, Redhill, RH1 3PS [IO91WG, TQ25]
G3 MOA J H Ruff, High Bank, 17 Harts Cl, Teignmouth, TQ14 9HG [IO80FN, SX97]
G3 MOB Details withheld at licensee's request by SSL.
G3 MOE J H Moxey, 11 Westbury Rd, Cheltenham, GL53 9EN [IO81XV, SO91]
G3 MOJ Details withheld at licensee's request by SSL.
G3 MOK T Kelly, 47 Cutthorpe Rd, Newbold, Chesterfield, S42 7AD [IO93GG, SK37]
G3 MOL J Lixenberg, Orchard House, 77 Pembroke Cres, Hove, BN3 5DF [IO90VT, TQ20]
GW3 MOM C K Davies, 19A Heol y Dre, Cefneithin, Llanelli, SA14 7DR [IO71WT, SN51]
GW3 MOP L D Watts, 2 Trinity Terr, Burton, Milford Haven, SA73 1PA [IO71MR, SM90]
GM3 MOR Dr R Webster, Meric, 7 Woodmuir Cres, Newport on Tay, DD6 8HL [IO86MK, NO42]
GM3 MOU Details withheld at licensee's request by SSL.
GW3 MOV Dr C L Smith, Awelon, 29 Heol y Parc, Efail Isaf, Pontypridd, CF38 1AN [IO81IN, ST08]
G3 MOZ R G Dearsley, 2 Wisborough Lodge, Billingshurst Rd, Wisborough Green, Billingshurst, RH14 0DZ [IO91SA, TQ02]
G3 MP D J G Legge, 62 Fernleigh Ave, Nottingham, NG3 6FL [IO92KX, SK54]
G3 MPB A R Smith, 10 Goodwood Rd, Redhill, RH1 2HH [IO91VF, TQ25]
G3 MPD E L D Davey-Thomas obo Poldhu ARC, St. James, Fore St., Goldsithney, Penzance, TR20 9JP [IO70GC, SW53]
G3 MPF C F Smith, 29 Cloisters, Tarleton, Preston, PR4 6UL [IO83OQ, SD42]
G3 MPN D E Johnson, 54 Norwich Rd, Wymondham, NR18 0NT [JO02NN, TG10]
G3 MPO L J Robinson, 10 Greenways, Highcliffe, Christchurch, BH23 5AZ [IO90DR, SZ29]
GW3 MPP G C Price, 17 Celtic Cl, Undy, Magor, Newport, NP6 3PB [IO81ON, ST48]
G3 MPU Details withheld at licensee's request by SSL.
G3 MPW A S Walker, 14 St. Joans Dr, Scawby, Brigg, DN20 9BE [IO93RM, SE90]
G3 MPX J H F Wilshaw, 180 The Avenue, West Wickham, BR4 0EA [IO91XJ, TQ36]
G3 MPY W S Carruthers, 21 Leinster Ave, London, SW14 7JW [IO91UL, TQ27]
G3 MPZ C J Brabbins, 102 Talbot Rd, Northampton, NN1 4JB [IO92NF, SP76]
G3 MQD P T Greed, 18 Nursteed Park, Devizes, SN10 3AN [IO91AI, SU06]
G3 MQI R A Gill, 45 Biggin Ln, Ramsey, Huntingdon, PE17 1NB [IO92WK, TL28]
G3 MQK J T Hough, 1 Calder Ave, Freckleton, Preston, PR4 1DN [IO83NR, SD42]
GM3 MQO G K Olesen, 8 Rowallan Cres, Prestwick, KA9 2HE [IO75QL, NS32]
G3 MQR D J Robinson, 32 Bullock Wood Cl, Colchester, CO4 4HX [JO01LV, TM02]
G3 MQU R M W Rash, Beech Tree Farm, Wortham, Diss, Norfolk, IP22 7SS [JO02MI, TM07]
G3 MQX P Lane, 13 Wayside, Brixham, TQ5 8PY [IO80FJ, SX95]
G3 MQY Details withheld at licensee's request by SSL.
G3 MRA M G Campbell, 50 Chessel Ave, Bitterne, Southampton, SO19 4DX [IO90HV, SU41]
G3 MRB T L W Puryer, 118 Wolverton Rd, Newport Pagnell, MK16 8JQ [IO92PB, SP84]
G3 MRC B J Poole, 18 Grosvenor Ave, Kidderminster, DY10 1SS [IO82VJ, SO87]
G3 MRD S P Brett Hull Clge Fe AR, Dept Engineering, Queens Gdns, Hull, HU1 3DG [IO93UR, TA12]
G3 MRJ A Dean, 6 Hayward Cres, Verwood, BH31 6JT [IO90BU, SU00]
G3 MRP S J Butlin, 53 Vicarage Rd, Yardley, Birmingham, B33 8PH [IO92CL, SP18]
G3 MRQ D L Byne, Storm Bay, Church St., Charwelton, Daventry, NN11 3YT [IO92JE, SP55]
G3 MRS D R A Pontet, 4 Elsted Rd, Bexhill on Sea, TN39 3BG [JO00FU, TQ70]
G3 MRT R A Strafford, Chy Lowarth, Sparnock Farm, Kea, Truro, Cornwall, TR3 6EB [IO70KF, SW74]
G3 MRU Details withheld at licensee's request by SSL.
G3 MRV G W Carrick, Low Wood, Linstock, Carlisle, Cumbria, CA6 4PZ [IO84NV, NY45]
G3 MRX Dr P Robinson, 9 Barton Cl, Cambridge, CB3 9LQ [JO02BE, TL45]
G3 MRZ M A Crutchley, 40 Ufton Cres, Shirley, Solihull, B90 3SA [IO92CJ, SP17]
GM3 MSG J C Carmichael, Rowanbank, 7 Greystone Loaning, Dumfries, DG1 1PL [IO85EB, NX97]
G3 MSL R J Ives, 11 Coombe Dr, Fleet, GU13 9DY [IO91OG, SU85]
G3 MSO E K Tunstall, 11 The Broadway, Charlton on Otmoor, Kidlington, OX5 2UB [IO91JU, SP51]
G3 MSQ H J M Warren, 82 Meadowfield, Gosforth, Seascale, CA20 1HU [IO84GJ, NY00]
G3 MSS J Savage, 38 Bury St., Ruislip, HA4 7SU [IO91SO, TQ08]
G3 MSV A D Bishop, Wayside, Exmouth Rd, Colaton Raleigh, Sidmouth, EX10 0LE [IO80IQ, SY08]
G3 MSW K Ashcroft, Fendley Corner, Sauncy Wood, Harpenden, Herts, AL5 5DW [IO91UT, TL11]
GW3 MSY A C Davies, 7 Holly Lodge Gdns, Croesy Ceiliog, Croesyceiliog, Cwmbran, NP44 2NB [IO81LP, ST39]
G3 MTD B V Kissack, The Old Saddlers, Church St, Braunton, Devon, EX33 2EL [IO71WC, SS43]
G3 MTF Details withheld at licensee's request by SSL.
G3 MTG R A Prior, Woodlands, Ruishton, Taunton, Somerset, TA3 5LU [IO81LA, ST22]
GM3 MTH J McGill, 9 Ramsay Pl, Coatbridge, ML5 5RE [IO75XU, NS76]
G3 MTJ R V Skoyles, 2 Hay Cl, Great Oakley, Corby, NN18 8HX [IO92PL, SP88]
G3 MTK J F Knight, 54 Brize Norton Rd, Carterton, OX18 3JF [IO91ES, SP20]
G3 MTM Details withheld at licensee's request by SSL.
G3 MTP P G Gadsden, Rose Cottage, Salwayashs, Bridport, Dorset, DT6 5HX [IO80OS, SY49]
G3 MTQ B H Price, 324 Pershore Rd, Birmingham, B5 7QY [IO92BK, SP08]
G3 MTR B S Wolfe, 1 Valley Cl, Cheadle, SK8 1HZ [IO83VJ, SJ88]
GM3 MTS A B Wylie, 1 Willow Ct, Kirkwall, KW15 1PE [IO88MX, HY41]
G3 MTX A C Pointon, 164 Arundel Rd, Peacehaven, BN10 8HH [JO00AT, TQ40]
G3 MUA P J Lawlor, Woodside, Kilmuir, Kessock, Rosshire, IV1 1XG [IO77VM, NH65]
G3 MUF J J Owens, 12 Silva Cl, Pebsham, Bexhill on Sea, TN40 2SY [JO00GU, TQ70]
G3 MUI Details withheld at licensee's request by SSL.[Op: D J Durrant. Station located near Royston.]
G3 MUJ H R Morris, 47 Waldron Thorns, Heathfield, TN21 0AD [JO00DX, TQ52]
G3 MUL F C Lathwood, 6 Green Gate, Syerston, Newark, NG23 5NF [IO93NA, SK74]
GM3 MUM P S ODell, 109 East Trinity Rd, Edinburgh, Midlothian, EH5 3PT [IO85JX, NT27]
G3 MUN E B Ellam, 1 Laburnum Bungalows, Hinton, Chippenham, SN14 8HG [IO81LT, ST77]
G3 MUO Dr G F Gott, 10 Churchill Cres, Marple, Stockport, SK6 6HJ [IO83XJ, SJ98]
GI3 MUS W A R Bell, 200 Upper Newtownards Rd, Belfast, BT4 3ET [IO74BO, J37]
G3 MUW Details withheld at licensee's request by SSL.
G3 MUX C E H Benson, Orchard Croft, 85 Runcorn Rd, Moore, Warrington, WA4 6UA [IO83QI, SJ58]
GM3 MUZ J M Morrison, 5 Skigersta, Port of Ness, Isle of Lewis, PA86 0TX [IO76JB, NS08]
G3 MVD A Redfern, 26 Lancaster St., Dalton in Furness, LA15 8SD [IO84JD, SD27]
G3 MVK L R Davies, 22 Pinewoods, Church Aston, Newport, TF10 9LN [IO82TS, SJ71]
G3 MVM P B K Pierson, 7 Beehive Rd, Cheshunt, Goffs Oak, Waltham Cross, EN7 5NL [IO91WR, TL30]
G3 MVU A J W Adkins, 80 Marine Parade, Leigh on Sea, SS9 2NL [JO01HN, TQ88]
G3 MVV N O Miller, Avon, Gardiners Ln, Crays Hill, Billericay, Essex, CM11 2XA [JO01FO, TQ79]
G3 MVX J Burke, 120 Seabourne Rd, Bexhill on Sea, TN40 2SD [JO00GU, TQ70]
G3 MVZ F E Garrett, Little Acre, East Ashling, Chichester, West Sussex, PO18 9AR [IO90NU, SU80]
G3 MWB W C Povey, 35 Valley Rd, Newbury, RG14 6ET [IO91HJ, SU46]
G3 MWF R G Marden, 46 Highfield Rd, Winchmore Hill, London, N21 3HL [IO91WP, TQ39]
G3 MWG D E Bootman, Brookmans House, Putsborough Rd, Georgeham, Braunton, EX33 1JU [IO71VD, SS43]
G3 MWH G W Jennings, 32 Buckfield Rd, Leominster, HR6 8SF [IO82OF, SO45]
G3 MWK W Uttley, 11 Fortis Way, Salendine Nook, Huddersfield, HD3 3WW [IO93BP, SE11]
G3 MWL W M Lane, 117 High St., Yatton, Bristol, BS19 4DR [IO81OJ, ST46]
G3 MWM Dr D W Murden, 2 Mill Brow Cl, St. Helens, WA9 4JR [IO83PK, SJ59]
G3 MWN E D Rogers, 7 Buckleigh Rd, Wath upon Dearne, Rotherham, S63 7JB [IO93HL, SK49]
G3 MWO D A Beales, High View, Wood Cl, Tostock, Bury St Edmunds, IP30 9PX [JO02KF, TL96]
G3 MWP N S Beckett, 67 Sebastian Ave, Shenfield, Brentwood, CM15 8PP [JO01DP, TQ69]
G3 MWQ P J Groves, 26 Christine Ave, Wellington, Telford, TF1 2DX [IO82RQ, SJ61]
G3 MWS J T C Sladden, 5 Knavewood Rd, Kemsing, Sevenoaks, TN15 6RH [JO01CH, TQ55]
G3 MWV Details withheld at licensee's request by SSL.
GM3 MWX A C M Winton, 6 Windsor Dr, Falkirk, FK1 5QN [IO85CX, NS87]
G3 MWZ J D Casling, 19 Orchard Cl, Tavistock, PL19 8HA [IO70WN, SX47]
G3 MXA B S Collins, 64 Park Ave, Sittingbourne, ME10 1QY [JO01IH, TQ96]
G3 MXE K Darlington, 32549 Bad Oeynhausen, Elbinger Strasse 1, Fed. Rep Germany, X X
G3 MXF P J Cutler, 42 Hamble Rd, Oakdale, Poole, BH15 3NL [IO90AR, SZ09]
GD3 MXG Details withheld at licensee's request by SSL.
G3 MXH T E Downing, Archways, 94 Shipston Rd, Stratford upon Avon, CV37 7LR [IO92DE, SP25]
G3 MXJ D J Andrews, 18 Downsview Cres, Uckfield, TN22 1UB [JO00BX, TQ42]
G3 MXK D R Paice, 19 Laburnum Gr, Banbury, OX16 9DP [IO92IB, SP43]
GM3 MXN T Sorbie, Tamaur, 7 High Pleasance, Larkhall, ML9 2HJ [IO85AR, NS75]
G3 MXP J A Palfrey, Caprice, 4 Laverstock Park West, Laverstock, Salisbury, SP1 1QL [IO91CB, SU13]
G3 MXR J Wood, 22 Coppice Cl, Malvern, WR14 1LE [IO82UD, SO74]
G3 MXT G V Entwisle, 12 Booths Hall Gr, Worsley, Manchester, M28 1LQ [IO83TM, SD70]
G3 MXV H V Pierson, 65 Station Rd, Countesthorpe, Leicester, LE8 5TB [IO92KN, SP59]
G3 MXZ R Wesson, 58 Rochester Rd, Linthorpe, Middlesbrough, TS5 6QE [IO94JN, NZ41]
G3 MY Dr G M King, Hill Crest, Thornhill, Bamford, Sheffield, S30 2BR [IO93GJ, SK38]
G3 MYA A Martindale, 1 Dinsdale Rd, Leiston, IP16 4EX [JO02SE, TM46]
G3 MYC C J Cheatle, 56 Ashfurlong Cres, Sutton Coldfield, B75 6EN [IO92CN, SP19]
G3 MYE F A H Keen, Westways, Upper Tadmarton, Banbury, Oxon, OX15 5TB [IO92GA, SP33]
G3 MYF Details withheld at licensee's request by SSL.
G3 MYG R M F Inman, 60 Abercorn Rd, Mill Hill, London, NW7 1JL [IO91VO, TQ29]
G3 MYI Details withheld at licensee's request by SSL.
G3 MYM R W Micklewright, 5 Sandringham Rd, Yeovil, BA21 5JE [IO80QW, ST51]
G3 MYN R J Smith, The Lodge, Church Rd, Pimperne, Blandford Forum, DT11 8UB [IO80WV, ST90]
G3 MYT Details withheld at licensee's request by SSL.
G3 MYU H G J R Taylor, 56 Stapleton Cl, Marlow, SL7 1TZ [IO91ON, SU88]
G3 MYV Details withheld at licensee's request by SSL.[Op: E Yates. Station located in Surrey.]
GM3 MYW W Waugh, 4/29 Moncrieff Terr, Edinburgh, EH9 1LZ [IO85JW, NT27]
G3 MYY S J Boston, Beavers, Mill Ln, Bramford, Ipswich, IP8 4AU [JO02NB, TM14]
G3 MYZ P Nicholson, 2A Clarke Cres, Bempton, Bridlington, YO15 1JJ [IO94VD, TA17]
G3 MZA E D Hamblen, 64 Tollers Ln, Coulsdon, CR5 1BB [IO91WH, TQ35]
G3 MZB G A Ward, 35 Highfield Dr, Garforth, Leeds, LS25 1JY [IO93HS, SE43]
G3 MZC C B Sutcliffe, 1 Tollgate Rd, Culham, Abingdon, Oxon, OX14 4NL [IO91IP, SU59]
G3 MZF Details withheld at licensee's request by SSL.
G3 MZI J H Hood, 89 Freemens Way, Deal, CT14 9DQ [JO01QF, TR35]
G3 MZK Details withheld at licensee's request by SSL.
GW3 MZN R W Lightfoot, The Paddock, Oakford, Llanarth, Dyfed, SA47 0RN [IO72UE, SN45]
G3 MZP D F Alldrick, 261 Boldmere Rd, Sutton Coldfield, B73 5LL [IO92BM, SP19]
G3 MZU F B Breedon, 52 Barkers Ln, Wythall, Birmingham, B47 6BU [IO92BI, SP07]
G3 MZV G Brown, Rivelands Rd, Swindon, Cheltenham, Glos, GL51 9RF [IO81WW, SO92]
GM3 MZX M Pedreschi, Carse of Clary, Newton Stewart, Wigtownshire, DG8 6BH [IO74SV, NX46]
GW3 MZY Dr J D Last, The Orchard House, Gorddinog, Llanfairfechan, Gwynedd, LL33 0EG [IO83AF, SH67]
G3 MZZ A H Kightley, 29 The Parkway, Gosport, PO13 0PT [IO90JT, SU50]
G3 NAA R C Polley, 25 Tower Ave, Chelmsford, CM1 2PW [JO01FR, TL60]
G3 NAC Details withheld at licensee's request by SSL.
G3 NAE C K Richardson, 10 Fielders Way, East Wellow, Romsey, SO51 6EX [IO90FX, SU31]
G3 NAI R E Norman, 50 Bloxcidge St., Oldbury, Warley, B68 8QH [IO82XL, SO98]
G3 NAK G Mallinson, 145 Huddersfield Rd, Meltham, Huddersfield, HD7 3AJ [IO93BO, SE11]
G3 NAL R E Burgess-Lee, 3 Paddock Ln, Aldridge, Walsall, WS9 0BP [IO92AO, SK00]
G3 NAN R W Henderson, 5 West Ave, Pinner, HA5 5BZ [IO91TN, TQ18]
G3 NAP B C Sowter, 56 Alderminster Rd, Mount Nod, Coventry, CV5 7JU [IO92FJ, SP27]
G3 NAQ Dr G H Grayer, Bagatelle, 3 Southend, Brightwalton, Newbury, RG20 7BE [IO91HL, SU47]
G3 NAR B E Monk, 30 West St., Warwick, CV34 6AN [IO92EG, SP26]
G3 NAS Details withheld at licensee's request by SSL.
G3 NAT A D Brooker London Raynet G, 36 Pope Rd, Bromley, BR2 9QB [JO01AJ, TQ46]
G3 NAU Rev P R Heath, 22 Cobham Rd, Halesowen, B63 3JZ [IO82XK, SO98]
G3 NAV E R Cook, 57 Raeburn Rd, Great Barr, Birmingham, B43 7LQ [IO92BN, SP09]
G3 NAW J P Ryan, Garden Flat, 20 Lower Oldfield Park, Bath, BA2 3HL [IO81TJ, ST76]
G3 NAY S G Whithorn, 53 Torbay Rd, Allesley Park, Coventry, CV5 9JY [IO92FJ, SP37]
G3 NBB R D Lyder, High Meadow Cottage, Hawksdale, Dalston, Carlisle, CA5 7BJ [IO84MT, NY34]
G3 NBC K A V Hurrell, The Anchorage, Stour Row, Shaftesbury, Dorset, SP7 0QF [IO80UX, ST82]
G3 NBI R Smith, 2 Hill Crest, Rishworth, Sowerby Bridge, HX6 4QJ [IO93AP, SE01]
G3 NBL Dr J E Larson, Nyhem, Whitton Village, Stockton on Tees, Cleveland, TS21 1LQ [IO94HO, NZ32]
G3 NBN R J Weaving, 27 Beech Dr, Nailsea, Bristol, BS19 1QA [IO81PK, ST47]
G3 NBQ P T Burt, 245 Pilling Ln, Preesall, Poulton-le-Fylde, FY6 0HJ [IO83MW, SD34]
G3 NBR T R Ashby, Foye Lodge, 21 Albion Hill, Exmouth, EX8 1JS [IO80HO, SY08]
G3 NBS A Bairstow, 27 Williams Way, Longwick, Princes Risborough, HP27 9RP [IO91NR, SP70]
G3 NBU P J S Bendall, Kallieser Stieg 8, 24568 Kaltenkirchen, West Germany
G3 NBX A W Phillips, 68 Fort Austin Ave, Crownhill, Plymouth, PL6 5JW [IO70WJ, SX45]
G3 NBZ K R Thorne, 49 Sigdon Rd, Dalston, London, E8 1AP [IO91XN, TQ38]
G3 NC E J Rose, 84 Burnside, Withycombe, Exmouth, EX8 3AL [IO80HP, SY08]
G3 NCA P L Jenner, 37 Hillsboro Rd, Bognor Regis, PO21 2DX [IO90PS, SZ99]
G3 NCB H A A Bourner, Woodpeckers, Richborough Rd, Sandwich, CT13 9JE [JO01QG, TR35]
G3 NCC D J Cousins, 6936 Meadow Song Tr, Rockford, Illinois 61109, USA, ZZ6 9ME
G3 NCG R G Garman, 60 Fishers Field, Buckingham, MK18 1SN [IO92MA, SP73]
GJ3 NCJ O P Bradley, 34 Elizabeth Ave, La Route Orange, St. Brelade, Jersey, JE3 8GR
G3 NCL R Ray, Flat 4 Victoria Villas, Gladstone Rd, Chesham, Bucks, HP5 3AD [IO91QQ, SP90]
G3 NCM Details withheld at licensee's request by SSL.
G3 NCN J A Ellerton, 7 Cotterell Cl, Bracknell, RG42 2HL [IO91OK, SU87]
GM3 NCO A Mustard, 30 North Gyle Gr, Edinburgh, EH12 8JZ [IO85IW, NT17]
G3 NCR Details withheld at licensee's request by SSL.
GM3 NCS D H M Noble, Stanes, Cummingston, Burghead, Elgin, IV30 2XY [IO87GQ, NJ16]
G3 NCX J H W Broomhead, 62 Upper Holland Rd, Sutton Coldfield, B72 1ST [IO92CN, SP19]
G3 NDC C E Deamer, Gatehouse, The Grove, Warren Ln, Stanmore, Middx, HA7 4LD [IO91UP, TQ19]
G3 NDE G C Driver, 9 Teak Dr, Kearsley, Bolton, BL4 8RR [IO83TM, SD70]
G3 NDI C R Fry, 7 Thornbury Cl, Wokingham, Berks, RG44 6PE [IO91LJ, SU66]
G3 NDJ Mar C R Delhaye, 35 Cokeham Ln, Sompting, Lancing, BN15 9UP [IO90TT, TQ10]
G3 NDK R K Webb, 142 Penrose Ave, Carpenters Park, Watford, WD1 5AA [IO91TP, TQ19]
G3 NDM Details withheld at licensee's request by SSL.
G3 NDN D G Newey, 15 Clent View Rd, Norton, Stourbridge, DY8 3JE [IO82VK, SO88]
G3 NDO P Sorab, Woodgaston Cottage, Woodgaston Ln, Northney, Hayling Island, Hants, PO11 0RL [IO90MT, SU70]
G3 NDQ T W Byrne, Bongate, 19 Prestwich Ave, Worcester, WR5 1QA [IO82VE, SO85]
GW3 NDR D Harris, 5 Christopher Rd, Skewen, Neath, SA10 6LD [IO81BP, SS79]
G3 NDS R B Oliver, Sayonara, 6 Willis Cl, Great Bedwyn, Marlborough, Wilts, SN8 3NP [IO91EJ, SU26]
G3 NDV A Audsley, 15 Farne Ave, Wakefield, WF2 9EE [IO93FQ, SE32]
GU3 NDX W J Allisett, Springbank, Les Ozouets Rd, St Peter Port, Guernsey, Channel Islands, GY1 2UD
GI3 NEB J E Wilson, 76 Castlemore Ave, Belfast, BT6 9RG [IO74BN, J37]
GM3 NEC C S Miller, 19 Craighill Dr, Clarkston, Glasgow, G76 7TG [IO75US, NS55]
G3 NEF R E Barber, 12 Park Ct, Sandy, SG19 1NP [IO92UD, TL14]
G3 NEH J L Isles, 3 Drovers Croft, Greenleys, Milton Keynes, MK12 6AN [IO92OB, SP83]
G3 NEI P W Smith, Kintail, 6 Ct Rd, Godstone, RH9 8BT [IO91XF, TQ35]
G3 NEO P Bagshaw, 48 Kiveton Ln, Todwick, Sheffield, S31 0HL [IO93GJ, SK38]
G3 NEP P Bradley, Hollingarth, Highfield Rd, Grange Over Sands, LA11 7JA [IO84NE, SD47]
GM3 NEQ A D Finlay, 19 Fraser Ave, Newton Mearns, Glasgow, G77 6HP [IO75US, NS55]
G3 NEU T L Painter, 5 High Hatters Cl, Downham Market, PE38 9RP [JO02EO, TF60]
G3 NEV Details withheld at licensee's request by SSL.
GM3 NEX Rev N O Grady, St Gabriels, 74 West Loan, Prestonpans, EH32 9JX [IO85MW, NT37]
G3 NEZ J A Sole, 82 Stanbury Rd, Beverley High Rd, Hull, HU6 7BU [IO93TS, TA03]
G3 NFB J T Leviston, 9 Barnes Ave, Fearnhead, Warrington, WA2 0BL [IO83RJ, SJ69]
G3 NFC M W Cotton Burton & District Radio Societ, 113 Belvedere Rd, Burton on Trent, DE13 0RF [IO92ET, SK22]
G3 NFJ M A Coward, High Bank, 51 High St., Sutton Veny, Warminster, BA12 7AP [IO81WE, ST94]
GI3 NFM K McElhatton, 2A Orpheus Dr, Dungannon, BT71 6DR [IO64OM, H86]
G3 NFP Details withheld at licensee's request by SSL.[Op: L R Beckwith. Station located near Chippenham.]
G3 NFQ J A Cawley, 6 Willow Pl, Shrewsbury, SY3 8YQ [IO82OR, SJ41]
G3 NFS G E Lewis, 63 Mount Rd, Canterbury, CT1 1YF [JO01NG, TR15]
G3 NFT P M E Pavey, P O Box 7, Tooradin, Victoria 3980, Australia, X X
G3 NFV R Sykes, 16 The Ridgeway, Fetcham, Leatherhead, KT22 9AZ [IO91TG, TQ15]
G3 NFW J A Carroll, White Lodge, Hunston, Chichester, West Sussex, PO20 6PA [IO90OT, SU80]
G3 NFY B W Thist, 11 Church St, Minehead, Somerset, TA24 5JU [IO81GF, SS94]
G3 NGA M D G Breeze, 15 Scotts Ln, Wilbarston, Market Harborough, LE16 8QW [IO92OL, SP88]
G3 NGD Details withheld at licensee's request by SSL.
G3 NGI G W Davey, Ropes End, Newberry Hill, Berrynarbor, Ilfracombe, EX34 9SS [IO71XE, SS54]
G3 NGK D C Chapman, 6 Pickhurst Green, Hayes, Bromley, BR2 7QT [JO01AJ, TQ36]
G3 NGT J W Philips, 17 Jerram Cl, Gosport, PO12 2QH [IO90KS, SZ59]
GM3 NGW W Webb, 5 Thornlea Dr, Giffnock, Glasgow, G46 6DB [IO75UT, NS56]

G3 NGX H M Hogg, Crossways, Ferry Rd, South Stoke, Reading, RG8 0JL [IO91KN, SU68]
G3 NHB Dr D E Bowyer, 41 High St., Trumpington, Cambridge, CB2 2HR [JO02BE, TL45]
G3 NHE M Dann, 61 Alms Hill Rd, Parkhead, Sheffield, S11 9RR [IO93FI, SK38]
G3 NHF J A S Noble, Bolle Hall, Hoffleet Rd, Bicker, Boston, PE20 3AJ [IO92WW, TF23]
G3 NHG C D Gammon, 20 Belmont Dr, Failand, Bristol, BS8 3UU [IO81PK, ST57]
GU3 NHL C D H Lewis, Gwel-An-Mor, Mylor Churchtown, Falmouth, Cornwall, TR11 5UF [IO70LE, SW83]
G3 NHO Details withheld at licensee's request by SSL.
G3 NHP G A Peacock, Hallowsgate House, Flat Ln, Kelsall, Tarporley, CW6 0PU [IO83PE, SJ56]
GM3 NHQ T Harrison, 7 Cults Gdns, Broughty Ferry, Dundee, DD5 1QT [IO86NL, NO43]
G3 NHR H T Rogers, Aughavore, Birchfield Rd, Nordelph, Downham Market, Norfolk, PE38 0BT [JO02DO, TF50]
G3 NHU A D Besford, Conifers, 2A Halt Rd, Caister on Sea, Great Yarmouth, NR30 5NZ [JO02UP, TG51]
G3 NHV D A Hare, Salmon Tails, Shamrock Quay, Northam, Southampton, SO1 1QL [IO90HW, SU41]
GM3 NHW W K Heggie, 7 Davidson Park, Edinburgh, EH4 2PF [IO85JX, NT27]
G3 NHX G L Quarterman, 2 Milton Ave, Sutton, SM1 3QB [IO91VI, TQ26]
G3 NIC K Plant, Rose Cottage, Lincoln Rd, Nettleham, Lincoln, LN2 2NE [IO93RG, SK97]
G3 NID I A M Douglas, 6 Ansley Rd, Houghton, Huntingdon, PE17 2DQ [IO92WI, TL27]
G3 NIE C R Bell, Rosemere, 85 Waterbeach Rd, Landbeach, Cambridge, CB4 4EA [JO02BG, TL46]
GM3 NIG D Cram, Treeside Cottage, Ayr Rd, Newton Mearns, Glasgow, G77 6RT [IO75TS, NS55]
G3 NII R A Porter, 6 Clifton Rd, Shefford, SG17 5AA [IO92UA, TL13]
G3 NIJ B C Barker, 4 Glantlees, West Denton Est, Newcastle upon Tyne, NE5 2PJ [IO94DX, NZ26]
G3 NIL G C Munden, 126 Stanley Green Rd, Poole, BH15 3AQ [IO90AR, SZ09]
GW3 NIN J C Brogan, 38 Graig Park Circle, Newport, NP9 6HE [IO81LO, ST39]
G3 NIQ R L Gorton, 2 Clyde Cl, Clyde Ct, Redhill, RH1 4AF [IO91WF, TQ25]
G3 NIR G E Miles, 37 Henley Deane, Northfleet, Gravesend, DA11 8SU [JO01EK, TQ67]
G3 NIW P S Ives, Allt Na Crioch, Ockham Rd South, East Horsley, Leatherhead, KT24 6QJ [IO91SG, TQ05]
G3 NJA D Webber Torbay ARS, 43 Lime Tree Walk, Milber, Newton Abbot, TQ12 4LF [IO80EM, SX87]
G3 NJB S Nicholls World Assoc of Christian Ra's, 117 Derby Rd, Eastleigh, SO50 5GT [IO90HX, SU41]
G3 NJF M Knights, 255 Addington Rd, Irthlingborough, Wellingborough, NN9 5US [IO92QI, SP97]
GW3 NJG T M George, Windyridge, Ridgeway, Wyesham, Monmouth, NP5 3JX [IO81PT, SO51]
G3 NJK Dr V J Debono, St. Cross, 1 Glenside, Castle Bytham, Grantham, NG33 4SS [IO92RS, SK91]
G3 NJM J E P Philp, 61 Ewell Park Way, Ewell, Epsom, KT17 2NW [IO91VI, TQ26]
G3 NJP M T Phillips, Rosedene, Treworga, Ruan High Lanes, Truro, TR2 5NP [IO70MF, SW84]
G3 NJQ J D Simpson, 588 Dereham Rd, Norwich, NR5 8TE [JO02OP, TG10]
G3 NJU N Harrison, 45 Ullswater Rd, Flixton, Manchester, M31 2SY [IO83SK, SJ79]
G3 NJV P F Randall, Myresyke, Ruan Minor, Helston, Cornwall, TR12 7LU [IO70JA, SW71]
GW3 NJW C Whelan, 9 St. Margarets Pl, Cardiff, CF4 7AD [IO81JM, ST18]
G3 NJX R D Geeson, The Grove, Main Rd, Pentrich, Ripley, DE5 3RE [IO93GB, SK35]
GW3 NJY Dr M M Bibby, 19 The Avenue, Carlton, Poulton-le-Fylde, FY6 7NA [IO83MU, SD33]
G3 NJZ Details withheld at licensee's request by SSL.
G3 NKA A J Lingwood, 17 Briar Rd, St. Albans, AL4 9TH [IO91US, TL10]
G3 NKC D L Sharred, 246 Beacon St., Lichfield, WS13 7BH [IO92BQ, SK11]
GM3 NKG A Campbell, 22 Saltire Cres, Larkhall, ML9 2LG [IO85AR, NS75]
G3 NKH R Dowling, Orchard House, Oughtrington Ln, Lymm, Ches, WA13 0RD [IO83SJ, SJ68]
G3 NKJ R A Gill, 45 Biggin Ln, Ramsey, Huntingdon, PE17 1NB [IO92WK, TL28]
G3 NKL R M Jones, 5 Apsley Fold, Longridge, Preston, PR3 3TY [IO83QT, SD63]
GW3 NKM C H Jones, 77 Margam Rd, Port Talbot, SA13 2LB [IO81CN, SS78]
G3 NKN T W Cox, 2 Pityme Farm Rd, St. Minver, Wadebridge, PL27 6PL [IO70NN, SW97]
G3 NKO R S Ford, 11 Chaplains Ave, Cowplain, Waterlooville, PO8 8QL [IO90LV, SU61]
G3 NKQ C R Burchell, 4 Bakers Way, Perry, Huntingdon, PE18 0BS [IO92UG, TL16]
G3 NKR M A Rowlands, 204 Latchford Rd, Ottawa, Ontario, Canada, KIZ 5W2
G3 NKS D Thom, 9 Southern Rd, Cheltenham, GL53 9AW [IO81XV, SO91]
G3 NKW Details withheld at licensee's request by SSL.[Station located near Lymm.]
G3 NKY J T Christie, 3 Bosbury Rd, Malvern Link, Malvern, WR14 1TR [IO82UD, SO74]
GW3 NKZ A L Williams, Wren House, 3 Portland Pl, Lisvane, Cardiff, CF4 5EQ [IO81JM, ST18]
G3 NL G W Parkes, 43 Oldbury Rd, Worcester, WR2 6AA [IO82VE, SO85]
GM3 NLB Dr F Inglis, 3 Fleming Rd, Bishopton, PA7 5HW [IO75SV, NS47]
GW3 NLN R E Andrews, Berwyn, y Ffor, Pwllheli, Gwynedd, LL53 6UH [IO72TW, SH33]
G3 NLR O Heggs, 8 Kingsley Ave, Wilmslow, SK9 4EN [IO83VI, SJ88]
G3 NLW M Rosenthal, 279 Addiscombe Rd, Croydon, CR0 7HY [IO91XI, TQ36]
G3 NLX R J Hopkins, 47 Offington Ln, Worthing, BN14 9RG [IO90TT, TQ10]
G3 NLY R Smethers, 46 Church Rd, Burntwood, Staffs, WS7 9EA [IO92BQ, SK00]
GM3 NMA W T Mains, 12 Dryburgh Rd, Bearsden, Glasgow, G61 4DH [IO75TW, NS57]
G3 NMD J F Aungles Houghton Amateur Radio Club, 158 Burn Park Rd, Houghton-le-Spring, DH4 5DH [IO94GU, NZ34]
GW3 NMH H Perkins, Aptdo 384, 03730 Javea, Alicante, Spain
G3 NMI R Purdom, Croft Cottage, Berrow Rd, Burnham on Sea, TA8 2JJ [IO81MG, ST35]
G3 NMJ G C C Knapp, 4 Venture Cl, Bexhill on Sea, TN40 1TU [JO00FU, TQ70]
G3 NML M E Slater, 46 Ladywood, Boyatt Wood, Eastleigh, SO50 4RW [IO90HX, SU42]
GM3 NMN R H Dunlop, 39 Braid Dr, Glenrothes, KY7 4ES [IO86KE, NO20]
G3 NMT R G Fernandez, 52 Windermere Rd, Noctorum, Birkenhead, L43 9SW [IO83LJ, SJ28]
G3 NMX D Wills, Astronomy Departmen, University of Texas, Austin, Texas 78712 USA
G3 NMY Details withheld at licensee's request by SSL.
G3 NMZ G N Bath, 87 Stanmore Cres, Luton, LU3 2RJ [IO91SV, TL02]
G3 NN C Bolt Bradford A R S, 33 Manscombe Rd, Allerton, Bradford, BD15 7AQ [IO93CT, SE13]
G3 NNA M J Codd, 71A Higher Rd, Longridge, Preston, PR3 3SY [IO83QU, SD63]
GW3 NNB R J Evans, Cemlyn, Nefyn, Pwllhel, Gwynedd, LL53 6EG [IO72RW, SH34]
G3 NND H J Everest, 32 Lea Rd, Harpenden, AL5 4PG [IO91TT, TL11]
G3 NNG C L Desborough, 22 Westland Rd, Faringdon, SN7 7EY [IO91FP, SU29]
G3 NNK A J Reynolds, 11 Malvern Park, Beltinge, Herne Bay, CT6 6LW [JO01NI, TR16]
G3 NNQ A D Mason, 67 Boston Rd, Heckington, Sleaford, NG34 9JD [IO92UX, TF14]
G3 NNR J G McLoughlin, 41 Croft Holm, Moreton in Marsh, GL56 0JH [IO91DX, SP23]
G3 NNT S J Pilkington, The Quarries, Quarry Dr, Aughton, Ormskirk, L39 5BG [IO83NN, SD40]
G3 NNV P A Swanson, 1 Gordon Ave, Bromborough, Wirral, L62 6AL [IO83MH, SJ38]
G3 NNW K Taylor, 34 Shore Rd, Warsash, Southampton, SO31 9FU [IO90IU, SU40]
G3 NNY J H W Bolter, 32 Rotherham Ave, Luton, LU1 5PN [IO91SU, TL02]
GM3 NNZ Dr B W East, 26 Hyndford Rd, Lanark, ML11 9AE [IO85CQ, NS84]
G3 NOA P N Reynolds, Brook Bushes, Bramshaw, Lyndhurst, Hants, SO43 7JB [IO90EW, SU21]
G3 NOB R G Shepherd, 59 Pantain Rd, Loughborough, LE11 3LZ [IO92JS, SK51]
G3 NOC A J Waldie, Gwyn Lyn, 85 Park Rd, Berry Hill, Coleford, GL16 7AG [IO81QT, SO51]
G3 NOD R J Burton, 16 Lowcross Ave, Hutton Lowcross, Guisborough, TS14 8BP [IO94LM, NZ61]
G3 NOF D L McLean, 9 Cedar Gr, Yeovil, BA21 3JR [IO80QW, ST51]
G3 NOG R S Emmerson, 8 Harlsey Gr, Stockton on Tees, TS18 5DF [IO94HN, NZ41]
G3 NOH G D Eddowes, Flat 1, 47 The Avenue, Ealing, London, W13 8JR [IO91UM, TQ18]
G3 NOI R Cumming, 21 Britannia Way, Woodmancote, Cheltenham, GL52 4QP [IO81XW, SO92]
G3 NOK R A Knight, 187 Weedon Rd, Northampton, NN5 5DA [IO92MF, SP76]
G3 NOM R Gerrard, 37 Godward Rd, New Mills, Stockport, Ches, SK12 3BU [IO83XI, SJ98]
G3 NOP D J Peacock, Robin Hill, 62 Castle Rd, Cottingham, HU16 5JG [IO93SS, TA03]
G3 NOQ A G P Boswell, 27 Seabrook Rd, Great Baddow, Chelmsford, CM2 7JG [JO01GR, TL70]
G3 NOW R W Tomkys, 132 Victoria Rd, Bradmore, Wolverhampton, WV3 7HA [IO82WN, SO89]
G3 NOX J R T Royle, Keepers Cottage, Duddenhoe End, Saffron Walden, Essex, CB11 4UU [JO02BA, TL43]
G3 NPA G W Anderson, 46 Bearcroft, Weobley, Hereford, HR4 8TA [IO82ND, SO45]
G3 NPB D W Blackford, 11 Hendras Parc, Carbis Bay, St. Ives, TR26 2TT [IO70GE, SW53]
G3 NPC Dr J G Swanson, 23 Oatlands Rd, Burgh Heath, Tadworth, KT20 6BS [IO91VH, TQ25]
G3 NPF A C Wadsworth, 39 Church Rd, Broadbridge Heath, Horsham, RH12 3LD [IO91TB, TQ13]
G3 NPI G C Suggate, 26 Highlands Rd, Buckingham, MK18 1PL [IO92MA, SP73]
G3 NPJ J A Jones, 47 Rhodesway, Wirral, L60 2UA [IO83KH, SJ28]
G3 NPL E H Matthews, 20 Stockwell Furlong, Haddenham, Aylesbury, HP17 8HD [IO91MS, SP70]
G3 NPM A D Macdonald, 5 Arlington Cl, Swindon, SN3 3NB [IO91DN, SU18]
G3 NPO Details withheld at licensee's request by SSL.
GI3 NPP R Gibson, 109 Bush Rd, Dungannon, BT71 6QG [IO64PM, H86]
G3 NPT G Bell, 1 Fountain Cl, Hessle, HU13 0LB [IO93SR, TA02]
G3 NPY J Joslin, 150 Roman Bank, Skegness, PE25 1SE [JO03ED, TF56]
G3 NPZ T J Griffiths, 4 Morshead Cres, Fareham, PO16 7QP [IO90JU, SU50]
G3 NQA S W Hall, 76 Cheltenham Dr, Bromford, Birmingham, B36 8QG [IO92CM, SP18]
G3 NQE E G Jones, 9 Dorchester Rd, Morden, SM4 6QE [IO91VJ, TQ26]
G3 NQF R C Fenton, Harmins Green, France Lynch, Stroud, Glos, GL6 8LZ [IO81WR, SO90]
GI3 NQH J Beattie, 170 Lower Braniel Rd, Belfast, BT5 7NG [IO74BN, J37]
G3 NQJ J R Selkirk, 1714 Sprucewoodcl, Victoria V8N 1H3, British Columbia, Canada, ZZ1 7SP
G3 NQK J G Beddows, 49 Pinewood Park, New Haw, Addlestone, KT15 3BS [IO91SI, TQ06]
GW3 NQP H T Jones, 4 Uwch y Dre, Gwernymynydd, Mold, CH7 4AB [IO83KD, SJ26]
G3 NQR A G Goddard, 42 Pinner Park Ave, Harrow, HA2 6LF [IO91TO, TQ19]
G3 NQT R H J Levi, 24 Stanmore Way, Loughton, IG10 2SA [JO01AP, TQ49]
G3 NQV Details withheld at licensee's request by SSL.
G3 NQX W H Brown, 73 Church Ave, Preston, PR1 4UD [IO83QS, SD53]
G3 NQZ Dr G B Lockhart, 46 Gainsborough Dr, Leeds, LS16 7PE [IO93EU, SE24]
G3 NR A W Birt, 37 Horsell Park Cl, Woking, GU21 4LZ [IO91RH, SU95]
G3 NRB Details withheld at licensee's request by SSL.
G3 NRC C A G Miles, 8 Epsom Rd, Furnace Green, Crawley, RH10 6LU [IO91VC, TQ23]
G3 NRD J E P Packer, Butts Bank Farm, Gulval Churchtown, Penzance, Cornwall, TR18 3BB [IO70FD, SW43]
G3 NRH B N Perrin, Farm Gates, 10 Riverside Rd, West Moors, Wimborne, Dorset, BH22 0LQ [IO90BT, SU00]
G3 NRM M R Moore, 127 Adel Ln, Leeds, LS16 8BL [IO93EU, SE23]
GM3 NRP Details withheld at licensee's request by SSL.
G3 NRQ C D Higgins, 16 St. Leonards Ave, Grimsby, DN31 2BW [IO93WN, TA20]
G3 NRT Details withheld at licensee's request by SSL.
G3 NRU D Brook Foster, Green Ln, Old Wives Lees, Canterbury, CT4 8BJ [JO01LG, TR05]
G3 NRW A I H Wade, 7 Daubeney Cl, Harlington, Dunstable, LU5 6NF [IO91SX, TL03]
G3 NRX R E Murphy, 3 Lady Leasow, Radbrookgreen, Shrewsbury, SY3 6AB [IO82OQ, SJ41]
G3 NRZ C A Hogg, 7 Elm Gr, Lesney Park, Erith, DA8 3BL [JO01CL, TQ57]
G3 NSB R P Bray, 14 Hammonds Ln, Totton, Southampton, SO40 3LG [IO90GW, SU31]
G3 NSF T L Simpson, 184 Boston Rd, Holbeach, Spalding, PE12 8AG [JO02AT, TF32]
G3 NSG J Tyas, 2 Craven St., Barnoldswick, Colne, BB8 6AY [IO83VW, SD84]
G3 NSH R Benham, Bracken Vale, Herberts Way, Oldcroft, Lydney, Glos, GL15 4NS [IO81RS, SO60]
G3 NSI N Ashman, 11 Burdon Pl, Peterlee, SR8 5QZ [IO94IS, NZ44]
G3 NSM Dr R E McHenry, 26 Charlbury Rd, Oxford, OX2 6UU [IO91IS, SP50]
G3 NSN J G Waring, Hillgarth, West Buckland, Kingsbridge, Devon, TQ7 3AF [IO80BG, SX64]
G3 NSO G B Brookes, 27 Pineside Ave, Rugeley, WS15 4RG [IO92AQ, SK01]
GW3 NSP Dr J C Lennox, 49 Beatty Ave, Roath Park, Cardiff, CF2 5QR [IO81JM, ST17]
G3 NSS T R Spain, Manor View, Shotatton, Ruyton Xi Towns, Shrewsbury, SY4 1JD [IO82MT, SJ32]
G3 NSU B G Ellis, 140 Pudsey Rd, Greenthorpe, Leeds, LS12 3TZ [IO93ET, SE23]
GI3 NSV M C Donnelly, Tirgarve, Allistragh, Armagh, N Ireland, BT61 8EZ [IO64PJ, H85]
G3 NSW R E Kay, 7 Lea Dr, Blackley, Manchester, M9 7AR [IO83VM, SD80]
G3 NSY F J Hall, Radiohm, David Ave, Pontesbury, Shrewsbury, SY5 0QB [IO82NP, SJ30]
G3 NSZ D Roberts, 10 Queenswood Ave, Bebington, Wirral, L63 8NZ [IO83LI, SJ38]
G3 NTA G A Couzens, 47 Holmstead Ave, Whitby, YO21 1NA [IO94QL, NZ81]
G3 NTD A Marsden, 15 Northfield Way, Hall Croft, Retford, DN22 7LJ [IO93MH, SK78]
G3 NTG C E Griffin, 5 Range View, College Town, Sandhurst, GU47 0RH [IO91OI, SU86]
G3 NTI R Blain, 11 Mill Bank, Ness, South Wirral, L64 4BJ [IO83LG, SJ37]
GW3 NTK Details withheld at licensee's request by SSL.
GM3 NTL G W Watt, 17 Hopetoun Grange, Bucksburn, Aberdeen, AB2 9RA [IO87TH, NJ72]
G3 NTM W T Brown, 18 Georgian Cl, Staines, TW18 4NR [IO91SK, TQ07]
GW3 NTR C Bowman, 9 Prince St., Newport, NP9 8DS [IO81MO, ST38]
G3 NTV Details withheld at licensee's request by SSL.
GM3 NTX R Burt, 11 Saughton Mains Pl, Edinburgh, EH11 3PW [IO85IW, NT27]
G3 NUA J Hogg, 16 Moorston Cl, Naisberry Park, Hartlepool, TS26 0PJ [IO94JQ, NZ43]
G3 NUB M Bursnall, 17 Beech Cl, Buckingham, MK18 1PG [IO92MA, SP63]
G3 NUE Details withheld at licensee's request by SSL.
GM3 NUF Details withheld at licensee's request by SSL.
G3 NUG E N Cheadle, Further Felden, Longcroft Ln, Felden, Hemel Hempstead, HP3 0BN [IO91SR, TL00]
G3 NUL V M Johnston, 119 High St., Cheveley, Newmarket, CB8 9DG [JO02FF, TL66]
GI3 NUM J S McKinley, 4A Chestnut Hill Rd, Moira, Craigavon, BT67 0LW [IO64VL, J16]
GW3 NUO P M Williams, Crud y Gwynt, 27 Mynydd Garnllwyd Rd, Morriston, Swansea, SA6 7PB [IO81AP, SS69]
G3 NUQ I D Macarthur, 2 Bramley Cl, Bramhall, Stockport, SK7 2DT [IO83WI, SJ88]
GM3 NUU Dr J S Reid, Rochelle, Old in Rd, Findon, Aberdeen, Aberdeenshire, AB12 3RL [IO87WB, NO99]
G3 NVC T Howard, 103 Westley Rd, Acocks Green, Birmingham, B27 7UW [IO92CK, SP18]
G3 NVJ G W Hubber, 3 Antron Way, Mabe, Mabe Burnthouse, Penryn, TR10 9HS [IO70KD, SW73]
G3 NVK R Winters, 8 Epping Dr, Melton Mowbray, LE13 1UH [IO92NS, SK72]
G3 NVL R F Allen, 692 Hitchin Rd, Luton, LU2 7UH [IO91TV, TL12]
G3 NVM D G Arigho, 87 Westover Rd, Fleet, GU13 9DE [IO91OG, SU85]
G3 NVO N Vincent, Woodville, The Ridge, Cold Ash, Thatcham, RG18 9HX [IO91IK, SU56]
G3 NVP B K Mapp, 33 Cotswold Dr, Redcar, TS10 4AG [IO94LO, NZ52]
GM3 NVQ G Martin, 6 Harriebrae Park, Dunfermline, KY12 9EA [IO86GB, NT08]
GM3 NVU G A Maclauchlan, 16 Wellpark Terr, Bonnybridge, FK4 1DE [IO86BA, NS88]
G3 NVV M E Kinder, 12 Bilton Ln, Dunchurch, Rugby, CV22 6PZ [IO92II, SP47]
GI3 NVW W R Pollock, 155 Doogary Rd, Omagh, BT79 0HF [IO64JN, H46]
G3 NVX R H Davison, 10 Beckwith Cl, Harrogate, HG2 0BJ [IO93FX, SE25]
G3 NW L G Shaw, 17 Howard Ave, West Wittering, Chichester, PO20 8EX [IO90NS, SZ79]
GW3 NWC A C Fry, 6 Rose Ct, Pantsite, Newbridge, Newport, NP1 5LR [IO81KQ, ST29]
G3 NWD Details withheld at licensee's request by SSL.
G3 NWG Details withheld at licensee's request by SSL.
G3 NWH A W Collis, 9 The Langlands, Hampton Lucy, Warwick, CV35 8BN [IO92EF, SP25]
G3 NWL A D Lock, 7 Heather Cl, St. Leonards, Ringwood, BH24 2QJ [IO90BT, SU10]
G3 NWO Details withheld at licensee's request by SSL.
G3 NWP J Day, 51 Orchards Way, Walton, Chesterfield, S40 3DA [IO93GF, SK37]
G3 NWR A Seed Wirral ARS, 31 Withert Ave, Wirral, L63 5NE [IO83LI, SJ38]
GW3 NWS F R Clare, Glenview, Newport Rd, Magnor, Gwent, NP6 3BZ [IO81NO, ST48]
G3 NWU J W Thompson, 73 Eamont Gdns, Hartlepool, TS26 9JE [IO94JQ, NZ53]
G3 NWW M C Wakely, Orchardfarmhouse, Hill End, Longdon, Upton on Severn Wor, WR8 0RN [IO82UB, SO83]
G3 NWX K A Morgan, 97 Elmwood, Sawbridgeworth, CM21 9NN [JO01BT, TL41]
G3 NWY D Forster, 79 Westbrooke Ave, Hartlepool, TS25 5HX [IO94JQ, NZ53]
G3 NXC A B Plant, 178 Clay Ln, Yardley, Birmingham, B26 1DY [IO92CL, SP18]
G3 NXD R C Shuck, Tregarron,Lowe Ln, Wolverley, Kidderminster, Worcs, DY11 5QR [IO82UJ, SO87]
G3 NXI K H Schau, Clementine Cottage, Bristol Rd, Falfield, Wotton under Edge, GL12 8DE [IO81SP, ST69]
G3 NXJ B James, 2 Sheepscombe Dr, Worcester, WR4 9JX [IO82VF, SO85]
G3 NXK O Diplock, North Lodge Cottage, Messing Park, Colchester, Essex, CO5 9TD [JO01IT, TL81]
G3 NXL P A Lamming, 25 Leconfield Garth, Follifoot, Harrogate, HG3 1NF [IO93GX, SE35]
G3 NXN F R Wickens, 32 Kenilworth Ave, Wimbledon Park, London, SW19 7LW [IO91VK, TQ27]
G3 NXO A D Watt, 5 Singleton, Chichester, PO18 0HA [IO90OV, SU81]
G3 NXQ A K Barker, 52 Tunnel Hill, Worcester, WR4 9SD [IO82VE, SO85]
GW3 NXR T D J Miles, West Uplands Lodge, Upland Arms, Carmarthen, Dyfed, SA32 8DX [IO71UT, SN41]
G3 NXS F A Shaw, 69 Finedon Rd, Irthlingborough, Wellingborough, NN9 5TY [IO92QH, SP97]
G3 NXT W H Fletcher, 3 Orchard Cl, Metheringham, Lincoln, LN4 3DT [IO93TD, TF06]
G3 NXU B Booth, 5 The Labbott, Keynsham, Bristol, BS18 1BD [IO81SJ, ST66]
G3 NXV R H Jennings, Edensor, Grendon Hall, Atherstone, Warks, CV9 3DP [IO92FO, SK20]
G3 NXX I Miller, 59 Andrew Ln, High Ln, Stockport, SK6 8HY [IO83XI, SJ98]
G3 NXZ J W Howe, 18 Laburnum Gr, Conisbrough, Doncaster, DN12 2JW [IO93JL, SK59]
G3 NYA L D Strange, 19 Firhill Croft, Druids Heath, Birmingham, B14 5UP [IO92BJ, SP07]
G3 NYB W L Bingham, 7 Bolton Hill Rd, Doncaster, DN4 6DQ [IO93LM, SE60]
G3 NYD D E J Coles, 113 Berrow Rd, Burnham on Sea, TA8 2PH [IO81MG, ST35]
G3 NYE A J Taylor, 25 Burnside Rd, Gatley, Cheadle, SK8 4NA [IO83VJ, SJ88]
GM3 NYG J G Fish, 31 Oaklands Ave, Irvine, KA12 0SE [IO75QO, NS33]
G3 NYI W H Brownson, 20 Upland Ct Rd, Harold Wood, Romford, RM3 0TT [JO01CO, TQ59]
GI3 NYJ S A W Currie, 122 Belfast Rd, Comber, Newtownards, BT23 5QP [IO74CM, J46]
G3 NYK A J Melia, 67 Deben Ave, Martlesham Heath, Ipswich, IP5 7QR [JO02OB, TM24]
G3 NYR D K Rayner, 42 Canford Dr, Allerton, Bradford, BD15 7AU [IO93CT, SE13]
G3 NYS C J Whiteley, 30 Lynch Hill Park, Whitchurch, RG28 7NF [IO91IF, SU44]
G3 NYX J W Heaviside, Grisedale, 110 Cuckfield Rd, Hurstpierpoint, Hassocks, BN6 9RZ [IO90VW, TQ21]
G3 NYY Details withheld at licensee's request by SSL.[Op: Walter A F Davidson, PO Box 423, Tewkesbury, Glos, GL20 5RN.]
G3 NYZ A F Stafford, Blakefield, Jawbone Ln, Melbourne, Derbyshire, DE7 1BW [IO92IX, SK44]
GM3 NZJ J C W Driscoll, 47 Strathcona Pl, East Kilbride, Glasgow, G75 0HA [IO75VS, NS65]
G3 NZK Dr P G Robson, Glenross, Whiteshoot, Redlynch, Salisbury, SP5 2PR [IO90DX, SU21]
G3 NZL H S Chapman, 57 Athelstan Rd, Bitterne, Southampton, SO19 4DE [IO90HV, SU41]
G3 NZO G D Kidder, 60 College Rd, Bexhill on Sea, TN40 1TW [JO00FU, TQ70]
G3 NZP M W G Harman, 19 Hill House Cl, Turners Hill, Crawley, RH10 4YY [IO91WC, TQ33]
G3 NZR W T H Young, 5 Grasmere Gr, Frindsbury, Rochester, ME2 4PN [JO01GJ, TQ77]
G3 NZS H W Parkes, 35 Dovey Rd, Tividale, Warley, B69 1NT [IO82XM, SO98]
G3 NZU B Gilbert, 13 Baslow Dr, Heald Green, Cheadle, SK8 3HW [IO83VI, SJ88]
G3 NZV A J C Park, Waterside Cottage, Bowden Ln, Chapel-En-le-Frith, Stockport, Ches, SK12 6QF [IO83XI, SJ98]
G3 NZW S I W James, Beresford, Latchmoor Ave, Gerrards Cross, Bucks, SL9 8LJ [IO91RO, SU98]
G3 OA H G Cottis, 19 Kingswood Chase, Leigh on Sea, SS9 3BB [JO01HN, TQ88]
G3 OAB H Reeves, 138 Blandford Ave, Castle Bromwich, Birmingham, B36 9JE [IO92CM, SP19]
G3 OAD T C Haydu Jones, 1 Beggars Roost, Golf Course Rd, Painswick, Gloucester, GL6 6TJ [IO81VT, SO81]
G3 OAF W A Jeffs, Silver Jay, Colehill Ln, Wimborne, BH21 7AN [IO90AT, SU00]
G3 OAG Details withheld at licensee's request by SSL.
G3 OAH Dr P R M Whittlestone, Welcove, Croft Farm Dr, West Malvern, Malvern, WR14 4DT [IO82TC, SO74]
G3 OAK C J Dempster, 48 Western Ave, Woodley, Reading, RG5 3BH [IO91NL, SU77]
G3 OAL E W Lincoln, Lynholme, Millbank, Heighington Village, Newton Aycliffe, DL5 6RF [IO94EO, NZ22]
G3 OAN Details withheld at licensee's request by SSL.
G3 OAP Details withheld at licensee's request by SSL.
G3 OAR G Greenwood, 1 Maltkiln Ln, Castleford, WF10 4LF [IO93HR, SE42]
GI3 OAU D B McCutcheon, 38 Killyglen Rd, Larne, BT40 2HR [IO74BU, D30]
GM3 OAV D G Varney, 27 Trinley Rd, Knightswood, Glasgow, G13 2JB [IO75TV, NS57]

G3 OAX I F Anderson, 102 Hopefold Dr, Worsley, Manchester, M28 3PW [IO83TM, SD70]
G3 OAY R N Graham, 5 The Langlands, Hampton Lucy, Warwick, CV35 8BN [IO92EF, SP25]
G3 OAZ J Randall, 243 Paddock Rd, Basingstoke, RG22 6QP [IO91KG, SU65]
GM3 OBC R Thomson, Knowehead, Star, Glenrothes, Fife, KY7 6LA [IO86KF, NO30]
G3 OBD P V Dutfield, 16 Talbot Dr, Poole, BH12 5ED [IO90BR, SZ09]
GM3 OBG P B Bridges, 29 Kirkbank, Auchmithie, Arbroath, DD11 5SY [IO86RO, NO64]
G3 OBJ R Coulson, 106 Salterford Rd, Hucknall, Nottingham, NG15 6GD [IO93JA, SK54]
G3 OBL J R Tyrrell, 2 Briar Cl, Lyde Rd, Yeovil, BA21 5XA [IO80QX, ST51]
G3 OBV P H Harris, 15 Ratliffe Rd, Rugby, CV22 6HB [IO92II, SP47]
G3 OBX Details withheld at licensee's request by SSL.
G3 OBY C M Smith, 79 Hartwell Rd, Hanslope, Milton Keynes, MK19 7BY [IO92NC, SP74]
G3 OBZ M R Birkett, Orchard End, 185 Old Main Rd, Bulcote, Nottingham, NG14 5GS [IO92LX, SK64]
G3 OCA Details withheld at licensee's request by SSL.[QSL via G3EEO]
G3 OCB C Bowden, Tregwyn, Tregonning Rd, Stithians, Truro, Cornwall, TR3 7AA [IO70JE, SW73]
GW3 OCD V A Davies, 11 Cefn Ct, High Cross, Rogerstone, Newport, NP1 9AH [IO81LO, ST28]
G3 OCF J C Norman, 31 Derry Hill Rd, Arnold, Nottingham, NG5 8HQ [IO93KA, SK54]
G3 OCH J Hulett, 21 Exmoor Ave, Leicester, LE4 0BJ [IO92KP, SK50]
G3 OCI D W Hayter, 31A High View Rise, Crays Hill, Billericay, CM11 2XU [JO01FO, TQ79]
G3 OCK L W Hodgetts, Troutbeck, 164 Probert Rd, Oxley, Wolverhampton, WV10 6UA [IO82WO, SJ90]
G3 OCL M Gay, 35 Chandos Rd, Rodborough, Stroud, GL5 3QT [IO81VR, SO80]
G3 OCP D A R Wallace, 11 Station Rd, Haddenham, Aylesbury, HP17 8AN [IO91MS, SP70]
G3 OCQ O Diplock Colchester Inst, Radio Club, Sheepen Rd, Colchester, Essex, CO3 3LL [JO01KV, TL92]
G3 OCS O H Kennedy, 63 Warren Cres, East Preston, Littlehampton, BN16 1BL [IO90ST, TQ00]
G3 OCW SLdr F J Cubberley, The Cedars, Callow End, Worcester, WR2 4TE [IO82VD, SO84]
G3 ODB A E Pritchard, 8 de Lapre Cl, Orpington, BR5 4HR [JO01BJ, TQ46]
G3 ODC D A G Martin, 7 Seaview Ave, Eastham, Wirral, L62 0BD [IO83MH, SJ38]
G3 ODD E M Stables, Manor Croft, Hemingbrough, Selby, North Yorks, YO8 7QL [IO93MS, SE63]
G3 ODH S B Smythe, 46 Green Acres, Skipton, BD23 1BU [IO83XX, SD95]
G3 ODK N K Mort, 55 Manor Ln, Penwortham, Preston, PR1 0TA [IO83PR, SD52]
G3 ODL B Everard, 65 Croslands Park, Barrow in Furness, LA13 9LB [IO84JD, SD27]
G3 ODO W Buckett, 401 Colehill Ln, Wimborne, BH21 7AW [IO90AT, SU00]
GM3 ODP Dr T M N Salvesen, Easter Catter, Croftamie, Glasgow, G63 0EX [IO76SB, NS48]
G3 ODQ Details withheld at licensee's request by SSL.
G3 ODR Details withheld at licensee's request by SSL.
G3 ODU B P Carter, 4 Woodbastwick Rd, Sydenham, London, SE26 5LQ [IO91XK, TQ37]
G3 ODY R L Field, 7 Hill Top Walk, Woldingham, Caterham, CR3 7LJ [IO91XH, TQ35]
G3 OEA Details withheld at licensee's request by SSL.
G3 OEB R D Downs, 8 Chapel Ln, Benson, Wallingford, OX10 6LU [IO91KO, SU69]
G3 OEC Prof C J Isham, 2 Lime Gr, Ruislip, HA4 8RY [IO91TN, TQ18]
G3 OEF R M G Maule, International Wireless Comms, 12Th Floor Sun Hung Kai Centre, 30 Harbour Rd, Wanchai, Hong Kong, X X
G3 OEG E F Harverson, 179 London Rd, Staines, TW18 4HR [IO91SK, TQ07]
G3 OEI F W Fairclough, 28 Rimmer Green, Scarisbrick, Southport, PR8 5LP [IO83MO, SD31]
G3 OEM R A McCarty, 1 Baden Rd, Brighton, BN2 4DP [IO90WU, TQ30]
G3 OEP D J M Buddery, 33 Addison Rd, Gorleston, Great Yarmouth, NR31 0PA [JO02UO, TG50]
G3 OEW D L Saunders, Sutton Poyntz, Weymouth, Dorset, DT3 6LL [IO80TP, SY78]
G3 OFA Details withheld at licensee's request by SSL.
G3 OFB I L Whitworth, Milverton, Ave Rd, Lyme Regis, DT7 3AE [IO80MR, SY39]
G3 OFF P C Hunter, 48 Hallmead Rd, Sutton, SM1 1RD [IO91VI, TQ26]
G3 OFI B A Bisley, 281 James St, Parksville, British Columbia, Canada, V9P 2R9
G3 OFJ G P Mitchell, Chelwood, Furze Hill Rd, Headley Down, Bordon, GU35 8EZ [IO91OC, SU83]
G3 OFK N P Henry, 334 Nine Mile Ride, Finchampstead, Wokingham, RG40 3NJ [IO91NJ, SU86]
G3 OFL Details withheld at licensee's request by SSL.
G3 OFP G D Cunnah, 225 Springwell Ln, Balby, Doncaster, DN4 9AJ [IO93KL, SE50]
G3 OFQ R Lambert, 167 Hardhorn Rd, Poulton-le-Fylde, FY6 8ES [IO83MU, SD33]
GM3 OFT P G Bower, An Cluain, Ballplay Rd, Moffat, DG10 9JU [IO85GH, NT00]
G3 OFW H A C Blake, 19 Segsbury Gr, Harmans Water, Bracknell, RG12 9JL [IO91PJ, SU86]
G3 OFX R P Welch, 112 Copsewood Rd, Bitterne Park, Southampton, SO18 1QR [IO90HW, SU41]
G3 OFZ R J G Jones, The Spinney, 67 Higher Dr, Banstead, SM7 1PW [IO91VH, TQ26]
G3 OGA C C Dumbrille, Fairmount, 34 Gilbert Hill, Smith's Parish, Bermuda Flo5
GW3 OGC C W Parsons, Lower House Farm, Woodstock, Haverfordwest, Dyfed, SA63 4TE [IO71NV, SN02]
G3 OGE J Rose, Town Pit House, Woodbastwick Rd, Blofield, Norwich, NR13 4AB [JO02RP, TG31]
G3 OGF Details withheld at licensee's request by SSL.
G3 OGH A Brooker-Carey, 29 Byron St., Amble, Morpeth, NE65 0ER [IO95FH, NU20]
G3 OGJ A M Cameron, 7 Benedict Cl, Romsey, SO51 8PN [IO90GX, SU32]
G3 OGK Dr G R Kennedy, Thayers Wood Ln, Edwyn Ralph, nr Bromyard, Herefordshire, HR7 4LY [IO82RF, SO65]
G3 OGL F C A Cobbett, 97 Mayfair Ave, Worcester Park, KT4 7SJ [IO91VJ, TQ26]
G3 OGM J C Moore, 1 The Spinney, Fleet, GU13 8EP [IO91NG, SU75]
G3 OGN N B Lomas, 78 Durnford St., Middleton, Manchester, M24 5TZ [IO83VN, SD80]
G3 OGO J M Nisbet, 25 Elizabeth Dr, Wantage, OX12 9YA [IO91GO, SU38]
G3 OGP R J Powell, Garlands Farm, The Haven, Billingshurst, RH14 9BH [IO91SB, TQ03]
G3 OGQ G N Fare, 1 Old Hall Cl, Higher Walton, Warrington, WA4 6SZ [IO83QI, SJ68]
G3 OGU R Hodgkinson, 39 Clifton Rd, Brierfield, Nelson, BB9 5EX [IO83VT, SD83]
G3 OGW H Wilson, 9 Dale Rd, Darlington, DL3 8LX [IO94FM, NZ21]
G3 OGX J H G Allsop, 17 Hambro Hill, Rayleigh, SS6 8BN [JO01HO, TQ89]
G3 OGY Details withheld at licensee's request by SSL.
G3 OGZ M S Beer, 24 Byron Ct, Beech Gr, Harrogate, HG2 0LL [IO93FX, SE35]
G3 OHA D Calvert, 15 Stancombe View, Winchcombe, Cheltenham, GL54 5LE [IO91AW, SP02]
G3 OHC Cmr G C Badger, 11 Escrick Park Gdns, Escrick, York, YO4 6LZ [IO93LU, SE64]
G3 OHH R A Hargreaves, 46 Castle Rd, Mow Cop, Stoke on Trent, ST7 3PH [IO83VC, SJ85]
G3 OHK H M Cole, 25 Causeway Rd, Seaton, Workington, CA14 1PL [IO84FP, NY03]
G3 OHL D S White, 32 Cawflands, Durdar, Carlisle, CA2 4UH [IO84MU, NY45]
G3 OHM Details withheld at licensee's request by SSL.[Correspondence to R Whitwell, G4EBL, obo The South Birmingham Radio Society, c/o Hampstead House, Fairfax Road, West Heath, Birmingham, B31 3QY.]
G3 OHN Dr K G Whitehouse, 27 Howdles Ln, Brownhills, Walsall, WS8 7PL [IO92AP, SK00]
G3 OHP M J Winter, The Chimes, 9 Higham Rd, Cliffe, Rochester, ME3 7SH [JO01FK, TQ77]
GM3 OHQ B Willcox, 11 St. Peters Rd, Newtonhill, Stonehaven, AB3 3RG [IO87TH, NJ72]
G3 OHS J C Perry, 517 Longbridge Rd, Barking, IG11 9DD [JO01BN, TQ48]
G3 OHT E Thackeray, 5 Eastgate, Patrington, Hull, HU12 0RG [IO93XQ, TA32]
G3 OHV R J Taylor, Eagles Rest, 9 Eefies Way, Crowborough, TN6 2UH [JO01BB, TQ53]
G3 OHW J M Smith, 309 Norton Way South, Letchworth, SG6 1SX [IO91VX, TL23]
G3 OHX I Jackson, Brattle House, Manor Rd, Seer Green, Beaconsfield, HP9 2QU [IO91QO, SU99]
GM3 OIB K A J Younger, 183 Main St., Pathhead, EH37 5SQ [IO85MU, NT36]
G3 OIC I L Croxford, 16 Chesterwood, Hollywood, Birmingham, B47 5EN [IO92BJ, SP07]
G3 OIF P C R Squires, 191 Station Rd, Knowle, Solihull, B93 0PT [IO92DJ, SP17]
G3 OIH B Shields, 24 Churchfield, Fulwood, Preston, PR2 8GT [IO83PS, SD53]
G3 OIL M Wills, 23 Falcons Way, Salisbury, SP2 8NR [IO91CB, SU12]
GW3 OIM Details withheld at licensee's request by SSL.[Op: Cyril Brown,]
GW3 OIN J G Nicholas, 28 Hardy Ave, Rhyl, LL18 3BG [IO83GH, SJ08]
G3 OIP P M Holker, Hillfield Cottage, Scragged Oak Rd, Detling, Maidstone, ME14 3HB [JO01GH, TQ75]
G3 OIT K P Jillings, The White House, 348 The Chase, Thundersley, Benfleet, SS7 3DN [JO01HN, TQ78]
GM3 OIV W J Anderson, Norwood, Woodend, Winchburch, West Lothian, EH52 6QB [IO85GX, NT07]
G3 OJ J E Hobin, 14 St. Martins Green, Trimley St. Mary, Trimley, Ipswich, IP10 0UU [JO02PA, TM24]
G3 OJA P Allan, 27 Maywood Ave, East Didsbury, Manchester, M20 5GR [IO83VJ, SJ88]
GM3 OJC W Whyte, 488 North Anderson Dr, Aberdeen, AB1 7GJ [IO87TH, NJ72]
G3 OJE M D Bass, 84 Mount Park Ave, South Croydon, CR2 6DJ [IO91WI, TQ36]
G3 OJG Dr P F Gale, Garden Cottage, Sacombe Green, Ware, Herts, SG12 0JQ [IO91XU, TL31]
G3 OJI J H Sleight, Orchard House, School Hill, Napton, Rugby, CV23 8NN [IO92IF, SP46]
G3 OJJ J E Hutchinson, 1 Southside, Woodcote Park, Wilmcote, Stratford-on-Avon, CV37 9XJ [IO92CF, SP15]
G3 OJK J R Bates, 8 Spaxton Rd, Durleigh, Bridgwater, TA5 2AP [IO81LC, ST23]
G3 OJL M W Plaster, Coombe House, Wookey Hole, Wells, Somerset, BA5 1DG [IO81QF, ST54]
GI3 OJO Dr J F Breach, 1 Massey Park, Belfast, BT4 2JX [IO74BO, J37]
G3 OJQ E D Wilson, 2 Wharf Rd, Stamford, PE9 2DU [IO92SP, TF00]
G3 OJS H L Braham, 10 Glebe Way, Frinton on Sea, CO13 9HR [JO01OU, TM22]
G3 OJV P W Waters, 9 Tudor Way, Hawkwell, Hockley, SS5 4EY [JO01HO, TQ89]
G3 OJW J M Crossan, Highlands, Shernfold Park, Frant, Tunbridge Wells, TN3 9DL [JO01DC, TQ53]
G3 OJX A J Hobbs, 65 Spurfield, West Molesey, KT8 1RR [IO91TJ, TQ16]
G3 OJZ B R Todd White, 3 Alexandra Rd, Capel-le-Ferne, Folkestone, CT18 7LB [JO01OC, TR23]
G3 OKA J A Share, Aureol House, 82 Birkenhead Rd, Meols, Wirral, L47 0LB [IO83KJ, SJ28]
G3 OKB M T Ireson, 15 Digby Dr, North Luffenham, Oakham, LE15 8JS [IO92QO, SK90]
G3 OKD Z J Nilski, The Paddocks, Wistanswick, Market Drayton, Salop, TF9 2BA [IO82SU, SJ62]
G3 OKF J S Fitz-Patrick, 79 Sparks Ln, Heswall, Wirral, L61 7XF [IO83KI, SJ28]
G3 OKH G E Hillman, 504 Chester Rd, Kingshurst, Birmingham, B36 0LG [IO92DL, SP18]
GW3 OKM J R Mitchell, 18 Camden Cres, Brecon, LD3 7BY [IO81HW, SO02]
G3 OKQ J H R Russell, Greenfingers, 136 Oyster Ln, Byfleet, West Byfleet, KT14 7JQ [IO91SI, TQ06]
G3 OKS S R Smithies, Moorcroft, Fernhill Rd, Horley, Surrey, RH6 9SY [IO91WD, TQ24]
G3 OKT Dr J E Thompson, Airys Cottage, Church Ln, Playford, Ipswich, IP6 9DS [JO02OC, TM24]
G3 OKU M W Cross, 6 Brackendale Cl, Camberley, GU15 1HP [IO91PH, SU85]
G3 OKX J W Roberts, 21 Addison Dr, Alfreton, DE55 7LB [IO93HC, SK45]
G3 OKY D A G Vincent, 10 Leaveland Cl, Beckenham, BR3 3PL [IO91XJ, TQ36]
G3 OLB T Boucher, Sunnyside, Little Vigo, Yateley, Camberley, Surrey, GU17 7ES [IO91OH, SU85]
GM3 OLG Details withheld at licensee's request by SSL.
G3 OLH A A Remsbury, Nodali, 16 Little Green Ln, Chertsey, KT16 9PH [IO91RJ, TQ06]
G3 OLK D W Dalrymple, 7 Havisham Way, Chelmsford, CM1 4UY [JO01FS, TL60]
G3 OLM D J Walker, 82A Grosvenor Rd, Epsom Downs, Epsom, KT18 6JB [IO91UH, TQ25]
G3 OLP B Wadsworth, 5 Birch Ave, Todmorden, OL14 5NX [IO83WR, SD92]
G3 OLU J T Saunders, 26 Wasdale Park, Seascale, CA20 1PB [IO84GJ, NY00]
G3 OLW J G Burnett, Wenrisc, Chapel Ln, Kinsham, Tewkesbury, Glos, GL20 8HS [IO82WA, SO93]
G3 OLX J C G Parker, Palfreys, Picquets Way, Banstead, SM7 1AJ [IO91VH, TQ25]
G3 OLY J E Boylett, 6 Wheat Hill, Letchworth, SG6 4HJ [IO91VX, TL23]
G3 OLZ A Caley, 2 Grinkle Ln, Easington, Saltburn By The Sea, TS13 4NT [IO94NN, NZ71]
G3 OMA S Kay, 5 Chevalier Cl, Middleleaze, Swindon, SN5 9TS [IO91BN, SU18]
G3 OMB E C Partner, Orchard House, The Straightway, Birch, Colchester, CO2 0NR [JO01JT, TL91]
G3 OMC A E Jenkinson, 40 Walker Rd, Chadderton, Oldham, OL9 8DB [IO83WM, SD80]
G3 OMD A G Callegari, Danebridge Nursery, Much Hadham, Herts, SG10 6JG [JO01BU, TL41]
G3 OMH D F S Hayward, 6 Larkhill Rd, Yeovil, BA21 3HF [IO80QW, ST51]
G3 OMJ E Judkins, 19 Feversham Dr, Kirkbymoorside, York, YO6 6DH [IO94MG, SE68]
G3 OMK T Kirk, 54 Highfields Dr, Loughborough, LE11 3JT [IO92JS, SK51]
G3 OML C H McLewee, 111 Camborne Rd, Morden, SM4 4JN [IO91VJ, TQ26]
GW3 OMN M Jenkins, 25 Stepney Rd, Cockett, Swansea, SA2 0FZ [IO81AP, SS69]
GI3 OMQ Details withheld at licensee's request by SSL.
G3 OMR M Russoff, 54 Cissbury Ring North, London, N12 7AH [IO91VO, TQ29]
G3 OMS Dr R A Simpson, 23 Larkhill, Rushden, NN10 6BG [IO92QH, SP96]
G3 OMT A B T Russell, 76 Chadwick Rd, Sutton Coldfield, B75 7RA [IO92CN, SP19]
G3 OMU A Bradbury, 2 Lower Chestnut Dr, Basingstoke, RG21 8YN [IO91KG, SU65]
G3 OMX A S Edwards, 67 Central Ave, Church Stretton, SY6 6EF [IO82OM, SO49]
G3 OMY D A Hancock, 17 Forestlake Ave, Hightown, Ringwood, BH24 1QU [IO90CU, SU10]
G3 ONB K Hall-Brooks, 24 Tall Trees Park, Homes Old Mill Ln, Forest Town, Mansfield, Notts, NG19 0JP [IO93JD, SK56]
G3 OND J C Denman, 62 Parkside Way, Harrow, HA2 6DG [IO91TO, TQ18]
G3 ONE E L Groom, 17 Evelyn Rd, Great Leighs, Chelmsford, CM3 1QQ [JO01FU, TL71]
GI3 ONF R F S Sinton, 35 Rose Garden, Tandragee, Craigavon, BT62 2NJ [IO64SI, J04]
G3 ONG G C J Miller, 74B Upper Northam Rd, Hedge End, Southampton, SO30 4EB [IO90IV, SU41]
G3 ONI D P Woods, 38 Welton Dr, Wilmslow, SK9 6HE [IO83VH, SJ87]
G3 ONK D B J Keen, Keriel, 22530, Mur de Bretagne, France
G3 ONL P B Brodribb, 16 Ipswich Rd, Debenham, Stowmarket, IP14 6LB [JO02OF, TM16]
GW3 ONN G C Griffiths, 68 Poplars Rd, Mardy, Abergavenny, NP7 6LJ [IO81LU, SO31]
G3 ONP D G Lovesey, 11 Watson Rd, Oxley, Wolverhampton, WV10 6SB [IO82WO, SJ90]
G3 ONQ R Goodall, Hazelmere, 21 Church Ln, Halifax, HX2 0EF [IO93BR, SE02]
G3 ONR B J Reynolds, 17 Cresswells Mead, Holyport, Maidenhead, SL6 2YP [IO91PL, SU97]
G3 ONT L W Harvey, 99 St. Marys Rd, Kettering, NN15 7BN [IO92PJ, SP87]
G3 ONU D A Barry, 2 Catherine Cl, Shrivenham, Swindon, SN6 8ER [IO91EO, SU28]
G3 ONV J H Verity, Tall Pine, Station Rd, Kirby Muxloe, Leicester, LE9 2EN [IO92JP, SK50]
G3 ONW S A N Magill, 11 Werstan Cl, Malvern, WR14 3NH [IO82UC, SO74]
GI3 ONZ W H Chambers, 66 Hopefield Ave, Portrush, BT56 8HE [IO65QE, C83]
G3 OOE A E Palmer, 11 Links Rd, Hollywood, Birmingham, B14 4TW [IO92BJ, SP07]
G3 OOH Details withheld at licensee's request by SSL.
G3 OOK J Plenderleith, Deroga, Whitwell Rd, Sparham, Norwich, NR9 5PN [JO02MR, TG01]
G3 OOL J Hatch, Peveril, 42 Bowden Rd, Templecombe, BA8 0LF [IO80TX, ST72]
G3 OOP B G Havenhand, 11 The Coppice, High Wycombe, HP12 4SA [IO91OO, SU89]
G3 OOQ M J W Webb, 14 Townsend Rd, Tiddington, Stratford upon Avon, CV37 7DE [IO92DE, SP25]
G3 OOS J A Robinson, 39 Springfield Rd, Stoneygate, Leicester, LE2 3BB [IO92KO, SK60]
G3 OOU R F Burns, 84 Portnalls Rd, Coulsdon, CR5 3DE [IO91WH, TQ25]
G3 OOW M F Docker, Butterfly Cottage, 2 Frederick Rd, Malvern, WR14 1RS [IO82UD, SO74]
G3 OOZ C F Simpson, 17 Sandygate Rd, Marlow, SL7 3BD [IO91ON, SU88]
GU3 OPC N F Ward, Lilyvale Hotel, Hougue Du Pommier, Castel, Guernsey, Channel Islands, GY5 7YG
G3 OPD S J Newton, 93 Bowleaze Coveway, Preston, Weymouth, DT3 6PW [IO80SP, SY78]
G3 OPE C Urwin, Clifton House, 1 Nelson Terr, Chopwell, Newcastle upon Tyne, NE17 7JR [IO94CW, NZ15]
G3 OPH R C Atkinson, 7 Stirling Cl, Carterton, OX18 3PH [IO91ES, SP20]
G3 OPJ C K Harrisson, 28 Manor Cl, Bradford Abbas, Sherborne, DT9 6RN [IO80QW, ST51]
G3 OPK T A Proud, 11 Chesters Gdns, Ryton, NE40 4PG [IO94CX, NZ16]
G3 OPL A A Milham, 83 St. James Ave, Ramsgate, CT12 6DZ [JO01QI, TR36]
G3 OPM T Marshall, 52 Spen Burn, High Spen, Rowlands Gill, NE39 2DN [IO94CW, NZ15]
G3 OPT Dr D Taylor, 3 Snaefell Rise, Appleton, Warrington, WA4 5BW [IO83RI, SJ68]
GW3 OPU I M Watson, 11 Seaward Ave, Port Talbot, SA12 7LT [IO81CO, SS79]
G3 OPW J Cook, 13 Mill St., Somercotes, Alfreton, DE55 4JF [IO93HB, SK45]
G3 OQB J J Pink, 6 Spencer Walk, Rickmansworth, WD3 4EE [IO91SP, TQ09]
G3 OQC J W Woods, 1 Dean Rd, Cosham, Portsmouth, PO6 3DG [IO90LU, SU60]
G3 OQD M H Emmerson, 6 Mounthurst Rd, Hayes, Bromley, BR2 7QN [JO01AJ, TQ36]
G3 OQF R Kay, Residence, Grand Champs, Ch 1279 Bogis-Bossey, Switzerland, ZZ9 9RE
GM3 OQI J C Ramsay, 78 Wheatlands Ave, Bonnybridge, FK4 1PL [IO86BA, NS88]
G3 OQL F J Holdaway, 12 Kings Rd, Clevedon, BS21 7HA [IO81NK, ST47]
G3 OQM Details withheld at licensee's request by SSL.
G3 OQN Details withheld at licensee's request by SSL.
G3 OQO D M Henley, 36 Main St., Newbold, Rugby, CV21 1HW [IO92IJ, SP47]
GI3 OQR D Gibson, Suvorov, 93 Cavan Rd, Dungannon, BT71 6QN [IO64PM, H86]
G3 OQT Details withheld at licensee's request by SSL.
G3 OQU J D Powell, The Croft, 86 White Hill, Kinver, Stourbridge, DY7 6AU [IO82VK, SO88]
G3 OQV C E Fisher, 79 Pickering Rd, Hull, HU4 6TB [IO93TR, TA02]
G3 OQX J Lathrope, 1 Carr Ln, Wildsworth, Gainsborough, DN21 3ED [IO93OL, SK89]
G3 ORB S S Bosley, 67 Holly Ave, New Haw, Addlestone, KT15 3UD [IO91SI, TQ06]
G3 ORC R R J Caines, The Squirrels, Priestwood Rd, Meopham, Gravesend, DA13 0DA [JO01EI, TQ66]
G3 ORD D Munton, 576 Huntington Rd, Huntington, York, YO3 9QA [IO93LX, SE65]
G3 ORE P S Burson, 64 Wordsworth Dr, Cheam, Sutton, SM3 8HF [IO91VI, TQ26]
G3 ORG I K Taylor, Westfield End, 10 Westfield Rd, Henlow, SG16 6BN [IO92UA, TL13]
G3 ORH C E Harris, Cheysing, Shepherds Way, Langley, Maidstone, ME17 3LJ [JO01HF, TQ85]
G3 ORI J R Vickers, 45 Willow Park Dr, Oldswinford, Stourbridge, DY8 2HL [IO82WK, SO98]
G3 ORJ Details withheld at licensee's request by SSL.
G3 ORK R A Talbot, 9 Bracebridge Dr, Kew Meadows, Southport, PR8 6XH [IO83MP, SD31]
GW3 ORL D A G Williams, Penllyn, 14 Seymour Ave, Penhow, Newport, NP6 3AG [IO81NO, ST49]
G3 ORN W Thomas, 20 Vinnicombes Rd, Stoke Canon, Exeter, EX5 4BB [IO80FS, SX99]
G3 ORP P J Pickering, 21 Palmar Rd, Maidstone, ME16 0DL [JO01GG, TQ75]
G3 ORV M A Saunders, 40 Archfield Rd, Cotham, Bristol, BS6 6BE [IO81QL, ST57]
G3 ORW E J Gregory, 1 Singleton Cl, Bognor Regis, PO21 4JY [IO90PS, SZ99]
G3 ORX A G Rumbold, 31 Springfield Cl, Hawthorn, Corsham, SN13 0JR [IO81VK, ST87]
G3 ORY R G Titterington, Wiclif House, St Mary's Rd, Lutterworth, Leicester, LE17 4PS [IO92JK, SP58]
G3 OS F Green, 2 Fleets Ln Cottage, Fleets Rd, Sturton By Stow, Lincoln, LN1 2DN [IO93QH, SK88]
G3 OSH A W Haines, 22 St. Peters Cl, Horton, Ilminster, TA19 9RW [IO80MW, ST31]
G3 OSI D J Swanson, 48 Moscow Dr, Liverpool, L13 7DJ [IO83MK, SJ39]
G3 OSJ L H Parsons, 19 Pendragon Park, Glastonbury, BA6 9PG [IO81PD, ST53]
G3 OSP S E Plumtree, 55 Croxden Way, Eastbourne, BN22 0UH [JO00DT, TQ60]
G3 OSQ D J Beakhust, 3 St. Blaize Rd, Romsey, SO51 7JY [IO90GX, SU32]
G3 OSR P F Hughes, Glan y Mor, 14 Stockdove Way, Thornton Cleveleys, FY5 2AR [IO83LV, SD34]
G3 OSS A A McKenzie, 57 Fitzalan Rd, Finchley, London, N3 3PG [IO91VO, TQ28]
G3 OST D E J Wilson, Nocturum, Horseshoe Ln, Ash Vale, Aldershot, GU12 5LJ [IO91PG, SU85]
GW3 OSV J P O Bushell, Park East Farm, New Moat, Clarbeston Rd, Dyfed, SA63 4SA [IO71OV, SN02]
GM3 OSW Details withheld at licensee's request by SSL.
G3 OSY R H Joll, 9 Hewett Cl, College Way, Taunton, TA1 4YQ [IO81KA, ST22]
G3 OTA Details withheld at licensee's request by SSL.
G3 OTD W J Spilman, 73 Wainfleet Rd, Skegness, PE25 3RZ [JO03DD, TF56]
G3 OTE R Binns, 22 Rydings Dr, Brighouse, HD6 2DA [IO93CQ, SE12]
G3 OTH C A R Cook, Swiss Cottage, Netherton Ln, Bedlington, Northd., NE22 6DR [IO95ED, NZ28]
G3 OTJ E C J Whitworth, Turn Lee House, Midgley, Halifax, Yorks, HX2 6TT [IO93AR, SE02]
G3 OTK R J Harris, Badgers Cottage, Lasham, Alton, Hants, GU34 5RY [IO91LE, SU64]
G3 OTN P J Seaman, The Gabled Lodge, 25 Gringer Hill, Maidenhead, SL6 7LY [IO91PM, SU88]
G3 OTO E G Barker, 12 Hollin Head, Baildon, Shipley, BD17 7LJ [IO93DU, SE13]
G3 OTR M J Beckley, Mallards, Albury Rd, Little Hadham, Ware, SG11 2DN [JO01BV, TL42]
GI3 OTU A J Burge, 38 Bayview Rd, Bangor, BT19 6AR [IO74EP, J58]
GI3 OTV P G O Kane, 36 Coolkill, Sandyford, Dublin 18, Ireland[Paul O'Kane EI5DI, author of Super-Duper (SD), the contest logging program. E-mail: okanep@iol.ie. Web: http://www.ioe.ie/~okanep. Tel: 00 353 1295 3668.]
G3 OTW W C Miller, 418 Old Chester Rd, Rock Ferry, Birkenhead, L42 4PD [IO83LI, SJ38]
G3 OTX J Roberts, 8 Mayalls Cl, Tirley, Gloucester, GL19 4HW [IO81UW, SO82]

G3 OTY Capt R F G Cogzell, 242 Clydesdale Tower, Holloway Head, Birmingham, B1 1UJ [IO92BL, SP08]
G3 OUC P Painting, 15 Turnpike Rd, Shaw, Newbury, RG14 2ND [IO91IJ, SU46]
G3 OUF Details withheld at licensee's request by SSL.[Op: David Evans, PO Box 599, Hemel Hempstead, Herts, HP3 0SR.]
G3 OUI I H Dickinson, 33 Broadshard Ln, Ringwood, BH24 1RP [IO90CU, SU10]
G3 OUK Details withheld at licensee's request by SSL.
G3 OUP J C Knight, 40 Mountfield, Hythe, Southampton, SO45 5AQ [IO90HU, SU40]
G3 OUQ H B Bird, The Hollies, 66 Main Rd, Sheepy Magna, Atherstone, CV9 3QU [IO92FO, SK30]
G3 OUT A H Walker, 45 Horncastle Rd, Woodhall Spa, LN10 6UY [IO93VD, TF26]
GM3 OUU G W J Rennie, 60 Woodend Pl, Aberdeen, AB1 6AN [IO87TH, NJ72]
G3 OVC D J Mauchel, 15 Lyon Ct, Ayshe Ct Dr, Horsham, RH13 5RN [IO91UA, TQ22]
GW3 OVD R J Baker, Up Yonder, Pant yr Hesg Rd, West End Abercarn, Newport, Gwent, NP1 4TB [IO81KP, ST29]
G3 OVE Dr M C Brown, 23 Carpenters Ln, West Kirby, Wirral, L48 7EX [IO83JI, SJ28]
G3 OVH A F Abbey, 40 Laureston Dr, Stoneygate, Leicester, LE2 2AQ [IO92KO, SK60]
G3 OVI D A Jordan, 17 Colenutts Rd, Ryde, PO33 3HS [IO90JR, SZ59]
G3 OVK J A Frearson, 5 Caxton Cl, New Whittington, Chesterfield, S43 2EA [IO93HG, SK37]
G3 OVL M Hubbard, 7 Creake Rd, Syderstone, Kings Lynn, PE31 8SF [JO02IU, TF83]
G3 OVM Details withheld at licensee's request by SSL.
G3 OVQ Details withheld at licensee's request by SSL.
G3 OVR H Redfern, 22 Rosslyn Rd, Moston, Manchester, M40 9PE [IO83VM, SD80]
G3 OVS S A Isaac, 20 Read Rd, Ashtead, KT21 2HS [IO91UH, TQ15]
G3 OVT F Collett, 56 Walsworth Rd, Hitchin, SG4 9SX [IO91UW, TL12]
G3 OVX H W Hammett, 27 Courtman Rd, Tottenham, London, N17 7HT [IO91WO, TQ39]
G3 OVZ A C Woodroffe, 9 Spring Rd, Riddings, Alfreton, DE55 4BS [IO93HB, SK45]
G3 OWA P J Wooden, 31 Meadow View Cl, Ryde, PO33 3EY [IO90JR, SZ59]
G3 OWB J N Holland Carter, 37 Highfield Ave, Cambridge, CB4 2AJ [JO02BF, TL46]
G3 OWC G A Boyle, 46 Butlers Gr, Great Linford, Milton Keynes, MK14 5DT [IO92OB, SP84]
G3 OWE D E Saunders, 4A Ullswater Cres, Radipole, Weymouth, DT3 5HE [IO80SP, SY68]
G3 OWF Details withheld at licensee's request by SSL.
G3 OWJ P N Jarvis, 34 Kennedy Rd, Shrewsbury, SY3 7AB [IO82OQ, SJ41]
G3 OWO J E Taylor, 283 Garstang Rd, Fulwood, Preston, PR2 9XH [IO83PS, SD53]
G3 OWQ J R Clarke, 29 Long Brackland, Bury St. Edmunds, IP33 1JH [JO02IG, TL86]
GM3 OWU V W Stewart, 9 Juniper Ave, Juniper Green, EH14 5EG [IO85IV, NT16]
GM3 OXA A S Foster, Tower House, 111 Sinclair St, Helensburgh, Dunbartonshire, G84 9QD [IO76PA, NS28]
G3 OXG D W Thompson, 34 Sandy Rd, Potton, Sandy, SG19 2QQ [IO92VD, TL24]
G3 OXH K W Brooker, Utopia, Surf Cres, Eastchurch, Sheerness, ME12 4JU [JO01KK, TQ97]
GM3 OXK J Carson, 23 Whinny Rig, Heathhall, Dumfries, DG1 3RJ [IO85FC, NX97]
G3 OXL D R Westbury, 120 Bunbury Rd, Northfield, Birmingham, B31 2DN [IO92AJ, SP07]
G3 OXN D Swainson, 4 Grasmere Ave, Spondon, Derby, DE21 7JZ [IO92HW, SK43]
G3 OXO P E Morrison, 59 Castle Rd, Whitstable, CT5 2ED [JO01MI, TR16]
G3 OXQ D W Morley, Garth Cottage, Main St, Moor Monkton, York, YO5 8JA [IO94JA, SE55]
G3 OXR P S Garthwaite, 16 Newtown Ave, Royston, Barnsley, S71 4HF [IO93GO, SE31]
G3 OXS N B Rivett, 42 Sunningvale Ave, Biggin Hill, Westerham, TN16 3BX [JO01AH, TQ45]
GM3 OXU J Wright, Eilein Buidhe, Craobh Haven, Lochgilphead, Argyll, PA31 8UA [IO76FF, NM70]
G3 OXV C A Earl, 1 Mayfield Dr, Daventry, NN11 5QB [IO92KG, SP56]
GM3 OXX G R W Burt, 6 Glenside Ct, Armadale, Bathgate, EH48 3RX [IO85DV, NS96]
G3 OY Details withheld at licensee's request by SSL.
G3 OYB W Waters, 4 Calartha Rd, Pendeen, Penzance, Cornwall, TR19 7DZ [IO70ED, SW33]
G3 OYE Details withheld at licensee's request by SSL.
G3 OYF J G Wilcox, 22 Woodland Ave, Kingswood, Bristol, BS15 1PZ [IO81RL, ST67]
GI3 OYG J Semple, 5 Tullaghgore Rd, Forttown, Ballymoney, BT53 6QF [IO65RC, C92]
GM3 OYH D P Baird, 20 Main Rd, Crookedholm, Kilmarnock, KA3 6JT [IO75SO, NS43]
G3 OYI M T Littlewood, Springwood, 10 Springfield Ave, Honley, Huddersfield, HD7 2ED [IO93CO, SE11]
GW3 OYL D J Gilbert, 2 Tal y Fan, Glan Conwy, Colwyn Bay, LL28 5NG [IO83CG, SH87]
G3 OYN G M J Saunders, 17 Chester St., Caversham, Reading, RG4 8JH [IO91ML, SU77]
GM3 OYO Details withheld at licensee's request by SSL.
G3 OYQ W G Carpenter, 32 Shophouse Rd, Bath, BA2 1ED [IO81TJ, ST76]
G3 OYS E Bright, 2 Heathcote Rd, Whitnash, Leamington Spa, CV31 2NF [IO92FG, SP36]
G3 OYT G D Clinton, 2 Greenways, Abbots Langley, WD5 0EU [IO91SQ, TL00]
G3 OYU B J R Davies, Red Roofs, Crowhurst Rd, Crowhurst, Lingfield, RH7 6DG [IO91XE, TQ34]
GM3 OYV P R Ritchie, 35 Castle Ave, Edinburgh, EH12 7LB [IO85IW, NT17]
G3 OYX M W Rignall, Ashdown, The Street, Horsley, Nailsworth, Gloucester, GL6 0PU [IO81VQ, ST89]
G3 OZ G L Jackson, 2 Franklyn Dr, Alvaston, Derby, DE24 0FR [IO92GV, SK33]
GM3 OZB A J McKay, 2 Osprey Dr, Kilmarnock, KA1 3LQ [IO75SO, NS43]
G3 OZC J Holstead, 72 Woodlands Ave, Feniscowles, Blackburn, BB2 5NN [IO83RR, SD62]
G3 OZD P M Cross, 5 Lings Ln, Hatfield, Doncaster, DN7 6AB [IO93LN, SE60]
G3 OZE J H Grainger, 6 Fulford Cross, Fulford Rd, York, YO1 4PB [IO93LW, SE65]
G3 OZF D F Beattie, Mayerin, Church Way, Stone, Aylesbury, Bucks, HP17 8RG [IO91NT, SP71]
GM3 OZI Details withheld at licensee's request by SSL.
GM3 OZJ I G M Morgan, 43 Dalgety Gdns, Dalgety Bay, Dunfermline, KY11 5LF [IO86HB, NT18]
G3 OZK M D James, 36 Iver Ln, Iver, SL0 9LF [IO91SM, TQ08]
G3 OZL Dr A P Jeavons, Lyne House, 18 Lyne Rd, Kidlington, OX5 1AD [IO91IT, SP41]
G3 OZN E W Badger, 20 Tennyson Dr, Worksop, S81 0EE [IO93KH, SK67]
G3 OZQ Details withheld at licensee's request by SSL.
G3 OZS Details withheld at licensee's request by SSL.
G3 OZT R A E German, 10 Beverley Rd, Dibden Purlieu, Southampton, SO45 4HS [IO90HU, SU40]
G3 OZV J H Croysdale, 14 Malwood Rd, Hythe, Southampton, SO45 5FB [IO90HU, SU40]
GI3 OZW P L Dynes, 2 Mullaghmore Park, Dungannon, BT70 1UL [IO64OM, H76]
G3 OZY G R Sweet, 154 South St., Andover, SP10 2BS [IO91GE, SU34]
G3 OZZ J C E Ramsay, Fairview, Briar Cl, Fairlight, Hastings, TN35 4DP [JO00IV, TQ81]
GM3 PAB D Mitchell, 19 Airbles Farm Rd, Motherwell, ML1 3AZ [IO75XS, NS75]
GM3 PAE N V Clarke, Pathhead House, Cockburnspath, Berwickshire, TD13 5XB [IO85TW, NT77]
G3 PAF E Brailsford, 78 Stoops Ln, Bessacarr, Doncaster, DN4 7RY [IO93KM, SE50]
G3 PAG J J Davies, Cedar Croft, School Ln, Trottiscliffe, West Malling, ME19 5EH [JO01EH, TQ66]
G3 PAI J Rabson, Limes Farm House, Eyke, Woodbridge, Suffolk, IP12 2QG [JO02QC, TM35]
G3 PAK Dr M G Senior, 28 Broom Cres, Tarvin, Chester, CH3 8HA [IO83OE, SJ46]
G3 PAQ J D Davis, 62A Allfarthing Ln, Wandsworth, London, SW18 2AJ [IO91VK, TQ27]
G3 PAX J P Barker, 2 Worcester Ct, Pevensey Gdns, Worthing, BN11 5PB [IO90TT, TQ10]
G3 PBF J R Orford, 63 Flowerhill Way, Northfleet, Gravesend, DA13 9DS [JO01EJ, TQ67]
G3 PBI A J Davies, 69 Sycamore Rd, Chalfont St. Giles, HP8 4LG [IO91RP, SU99]
G3 PBK G Seward, 53 Noddington Ln, Whittington, Lichfield, WS14 9PA [IO92CQ, SK10]
G3 PBQ M G Keen, 71 Deakin Rd, Erdington, Birmingham, B24 9AL [IO92BM, SP19]
G3 PBT R W Hilsley, 1 Chelmerton Ave, Great Baddow, Chelmsford, CM2 9RE [JO01FR, TL70]
G3 PBU Details withheld at licensee's request by SSL.
G3 PBX Details withheld at licensee's request by SSL.
G3 PBY P J E Carey GEC Avionics, Bp Kent Club Ars, 99 Bells Ln Hoo, St Werburgh Rochest, ME3 9JD [JO01GK, TQ77]
G3 PCA J R Hooper, 50 Mortlake Rd, Ilford, IG1 2SX [JO01AN, TQ48]
G3 PCC P S G Cadman, 7 Sutton Rd, Witchford, Ely, CB6 2HX [JO02CJ, TL47]
G3 PCG D M E Askew, Eden House, Spanish Hill, North Newton, Bridgwater, TA6 6NB [IO81LB, ST23]
G3 PCJ T R N Walford, Upton Bridge Farm, Long Sutton, Langport, Somerset, TA10 9NJ [IO81OA, ST42]
G3 PCN Details withheld at licensee's request by SSL.[Op: R P Brown. Station located near Pinner.]
GM3 PCQ Details withheld at licensee's request by SSL.
G3 PCS C Stockdale, 23 Mountfield Ave, Waterloo, Huddersfield, HD5 8RD [IO93DP, SE11]
G3 PCT Details withheld at licensee's request by SSL.
G3 PCW M G Watling, 8 Preetz Way, Blandford Forum, DT11 7XG [IO80WU, ST80]
G3 PCX B J Dodge, 34 Downs Rd, Penenden Heath, Maidstone, ME14 2JN [JO01GG, TQ75]
G3 PCY Details withheld at licensee's request by SSL.
G3 PD Details withheld at licensee's request by SSL.
G3 PDC R H Curwen, 53 Karslake Rd, Liverpool, L18 1EY [IO83MJ, SJ38]
G3 PDD J Dolby, Oaklea, School Ln, Heage, Belper, DE56 2AL [IO93GB, SK35]
G3 PDE D M Paterson, 19 Allison House, Westward Rd, Hedge End, Southampton, Hants, SO3 4NR [IO90HW, SU41]
G3 PDG Details withheld at licensee's request by SSL.
G3 PDH M H Prestwood, Salatiga, Bell Ln, Salhouse, Norwich, NR13 6RR [JO02QQ, TG31]
G3 PDK Details withheld at licensee's request by SSL.[Op: G J Newnham. Station located in London SE13.]
G3 PDL P F Linsley, 12 Cambridge Cres, Brookenby, Binbrook, Market Rasen, LN8 6HB [IO93VK, TF29]
GI3 PDN R B Bell, 26 Ballymartin Rd, Templepatrick, Ballyclare, BT39 0BW [IO64XR, J28]
G3 PDP A W Ralls, 12 Oakhill Cl, Bursledon, Southampton, SO31 1AP [IO90IV, SU41]
GW3 PDW D L Warnett, Black Tar Cottage, Black Tar Hill, Llangwm, Haverfordwest, SA62 4JD [IO71NR, SM90]
GM3 PDX J R Barker, 44 Priory Rd, Linlithgow, EH49 6BS [IO85EX, NS97]
G3 PEC N J Carter, 107 Cranford Ln, Heston, Hounslow, TW5 9HQ [IO91TL, TQ17]
G3 PED L A Crane, Goodwill Rectory Rd, Wrabness, Manningtree, Essex, CO11 2TR [JO01OW, TM13]
G3 PEJ P Watson, 37 Chestnut Bank, Scarborough, YO12 5QJ [IO94SG, TA08]
G3 PEK B D Simpson, 20 Monterey St, St Ives, Nsw 2075, Australia, X X
G3 PEM C J W Thomson, 109 Hillside Gr, Chelmsford, CM2 9DD [JO01FR, TL70]
G3 PEN D J Penny, Thirtynine Steps, 13 Newnham Cl, Braintree, CM7 2PR [JO01GV, TL72]
G3 PET A G Widdowson, 34 Highfields Rd, Chasetown, Burntwood, WS7 8QU [IO92AQ, SK00]
G3 PEW Dr J E Hudson, 68 Lower St., Stansted, CM24 8LR [JO01CV, TL52]
GW3 PEX L France, 8 Conway Dr, Cwmbach, Aberdare, CF44 0LL [IO81HR, SO00]
G3 PEZ J M Gutteridge, 66 Croft Dr, Moreton, Wirral, L46 0QT [IO83KJ, SJ28]
G3 PFE G W Spriggs, Willingham, 5 Penlington Ct, Nantwich, CW5 6SA [IO83RB, SJ65]
G3 PFF J H Parham, 10 Berwick Cl, Taunton, TA1 4JW [IO81KA, ST22]
G3 PFH M Blunden, Linsley Cottage, Dynes Hall Rd, Great Maplestead, Halstead, CO9 2QS [JO01HX, TL83]
G3 PFJ J D Harris, 3 Chimney Mills, West Stow, Bury St. Edmunds, IP28 6ES [JO02IH, TL87]
G3 PFM A J Baker, Deans Drove, Lychett Matravers, Lytchett Matravers, Poole, Dorset, BH16 6EQ [IO80WR, SY99]
G3 PFO C D Barr, Riders Way, Collum Green Rd, Stoke Poges, Slough, SL2 4AX [IO91QN, SU98]
G3 PFP S A Greenfield, Byways, Brightlingsea Rd, Thorrington, Colchester, CO7 8JH [JO01MU, TM02]
GM3 PFQ Details withheld at licensee's request by SSL.[Op: J Balfour, 17 Bandon Avenue, Kirkcaldy, Fife, KY1 3BS.]
G3 PFR Dr M W Dixon, Woodstock, Gad Bank, Norley, Warrington, WA6 8LL [IO83QF, SJ57]
G3 PFS D G N King, Hartwell House, Hartwell Rd, Wroxham, Norwich, Norfolk, NR12 8TL [JO02QQ, TG31]
G3 PFT A N Heeley, 108 Valley Ln, Lichfield, WS13 6ST [IO92CQ, SK11]
GW3 PFV K Robbins, 1 Rhiw Parc Rd, Abertillery, NP3 1BS [IO81KR, SO20]
GM3 PFX C Small, Strathedin, Langbank Dr, Kilmacoln, Kilmacolm, PA13 4PL [IO75QV, NS37]
G3 PFY J G Watt, 55 Beech Way, Clarke Est, Basingstoke, RG23 8LS [IO91KG, SU65]
G3 PFZ E H Porter, 1 Waverley Ct, Winstanley, Wigan, WN3 6EJ [IO83PM, SD50]
G3 PGA A M J Hammond, 23 St. Andrews Rd, Fremington, Barnstaple, EX31 3BS [IO71WB, SS53]
G3 PGC R W Armstrong, 6 Barnstaple Rd, North Shields, NE29 8QA [IO95GA, NZ37]
GI3 PGD Details withheld at licensee's request by SSL.
G3 PGG T A Wilson, 3 Chrysanthemon St, Cy-4632 Kolossi, Cyprus, X X
G3 PGI W A Phillimore, 37 Perth Rd, Plaistow, London, E13 9DS [JO01AM, TQ48]
G3 PGK C A Pearless, 4 Newberry Rd, Weymouth, DT4 8LW [IO80SO, SY67]
G3 PGM E Davies, 11 Tape Ln, Hurst, Reading, RG10 0DP [IO91NK, SU77]
G3 PGN H A Buckenham, Larks Lodge, Larks Ln, Great Waltham, Chelmsford, CM3 1AD [JO01FS, TL61]
G3 PGQ D C Yates, 35 Greenhill, Evesham, WR11 4LX [IO92AC, SP04]
G3 PGT D G G Tilling, Woodhay, Woodhouse Ln, Uplyme, Lyme Regis, DT7 3SX [IO80MR, SY39]
G3 PGV Details withheld at licensee's request by SSL.
G3 PGW H H Goodman, Nortonbury House, Edward Pl, Tewkesbury, GL20 5HF [IO81WX, SO83]
G3 PGX Details withheld at licensee's request by SSL.
GM3 PGY A McEwen, 4 Reef Terr, Crossapol, Scarinish, PA77 6UT [IO66NL, NL94]
G3 PGZ F J Brookes, Sowerby Halt, Spring Hill, Arley, Coventry, CV7 8FE [IO92FM, SP28]
G3 PH C R H Broadhurst, 65 Church Walk, Atherstone, CV9 1PS [IO92FN, SP39]
G3 PHA Details withheld at licensee's request by SSL.[Op: J J Morris.]
G3 PHD I M Gardiner, 189 Brennan Rd, Tilbury, RM18 8BA [JO01EL, TQ67]
G3 PHJ J G Johnston, 9 School Ln, Buckden, St. Neots, Huntingdon, PE18 9TT [IO92VH, TL16]
G3 PHL B F J Davies, 17 Linksway, Leigh on Sea, SS9 4QY [JO01HN, TQ88]
G3 PHM J E Macarthy, 30 Richmond Ave, Merton Park, London, SW20 8LA [IO91VJ, TQ26]
G3 PHO P E H Day, 146 Springvale Rd, Sheffield, S6 3NU [IO93GJ, SK38]
G3 PHR J H Carter, PO Box 30329 Smb, Grand Cayman, Cayman Islands, B W I
G3 PHS Details withheld at licensee's request by SSL.
G3 PHT K C Brown, Breckland, Park Rd, Aldeburgh, IP15 5EL [JO02TD, TM45]
G3 PHU B S D Clark, 50 Brondesbury Park, London, NW6 7AT [IO91VN, TQ28]
G3 PHW B J Todd, 3 Icepits Cl, Great Barton, Bury St. Edmunds, IP31 2PB [JO02JG, TL96]
G3 PHX J Moore, 129 High Cross Rd, Poulton-le-Fylde, FY6 8BX [IO83MU, SD33]
G3 PHZ J C Fogg, 16 Cotefield Dr, Leighton Buzzard, LU7 8DQ [IO91QW, SP92]
G3 PIA C L Desborough Harwell Amateur Radio Soc, 22 Westland Rd, Faringdon, SN7 7EY [IO91FP, SU29]
G3 PID P R Chandler, 528 Goffs Ln, Cheshunt, Goffs Oak, Waltham Cross, EN7 5EW [IO91XQ, TL30]
G3 PIH F R Kent, 5 Juniper Ct, College Hill Rd, Harrow, HA3 7JF [IO91UO, TQ19]
G3 PIJ P E Mellett, 16 Tutton Hill, Colerne, Chippenham, SN14 8DN [IO81UK, ST87]
G3 PIK J Taylor, 21 Norton Hall Cl, Norton, Stoke on Trent, ST6 8LZ [IO83WB, SJ85]
GM3 PIL A Chisholm, Modnachaidh, Lochloy Rd, Nairn, IV12 5AF [IO87BO, NH85]
G3 PIN J Patten, 8 Leacroft Rd, Penkridge, Stafford, ST19 5BX [IO82WR, SJ91]
GW3 PIO C W Owen, 13 Brynffynnon, Star, Gaerwen, LL60 6BA [IO73VF, SH57]
GM3 PIP P I Park, Kranji, 23 Longside Rd, Mintlaw, Peterhead, AB42 5EJ [IO97AM, NK04]
G3 PIT Details withheld at licensee's request by SSL.
G3 PIY C A Isaacs, Holmeview,Brick Ln, Thorney Hill, Bransgrove, Christchurch, Dorset, BH23 8DU [IO90DT, SZ29]
G3 PIZ T G Watts, 15 Ringmore Dr, Guildford, GU4 7DQ [IO91RG, TQ05]
G3 PJB P J Bailey, 34 Pinks Hill, High Firs, Swanley, BR8 8AQ [JO01CJ, TQ56]
G3 PJC C J Arnold, 47 Peartree Ln, Danbury, Chelmsford, CM3 4LS [JO01HQ, TL70]
G3 PJD Details withheld at licensee's request by SSL.
G3 PJI Details withheld at licensee's request by SSL.
G3 PJK J V Mee, 205\207 Grimshaw Ln, Middleton, Manchester, M24 2BW [IO83VN, SD80]
G3 PJL J H Hampson, 1 Fawborough Rd, Northern Moor, Manchester, M23 9BU [IO83UJ, SJ89]
G3 PJN R Hattersley, Hill Top, Hill Ln, Holymoorside, Chesterfield, S42 7EP [IO93FE, SK36]
G3 PJQ A E G Aldridge, The White House, Buttermilk Ln, Rudford, Gloucester, GL2 8DY [IO81UV, SO72]
G3 PJR A Ellis, East Leigh, Low Rd, Barrowby, Grantham, NG32 1DJ [IO92PV, SK83]
G3 PJT Dr R C Whelan, 36 Green End, Comberton, Cambridge, CB3 7DY [JO02AE, TL35]
G3 PJW R S Unsworth, 8 Coleridge Rd, Billinge, Wigan, WN5 7EB [IO83PM, SD50]
G3 PJX R A C Doe, 1 Blackbrook Cttgs, Blackbrook, Dorking, RH5 4DS [IO91UE, TQ14]
G3 PJY R G H Millman, 38 Fowlmere Rd, Birmingham, B42 2EA [IO92BM, SP09]
G3 PKA E C Harris, 23 Southlea Ave, Southbourne, Bournemouth, BH6 3AB [IO90CR, SZ19]
G3 PKC J M Tinker, 72 Jackson Ave, Leeds, LS8 1NS [IO93FT, SE33]
G3 PKD R L Sharples, 40 Greetham Rd, Cottesmore, Oakham, LE15 7DB [IO92QR, SK91]
G3 PKH Details withheld at licensee's request by SSL.
G3 PKL C A Fox, 2 Mill Cttgs, Wareham Rd, Organford, Poole, BH16 6ET [IO80WR, SY99]
GW3 PKO H R Poole, Pengwaunydd, Cilgwyn Rd, Newport, Dyfed, SA42 0QG [IO72OA, SN03]
GI3 PKP S Wishart, 14 Cumberland Cl, Dundonald, Belfast, BT16 0AW [IO74CO, J47]
G3 PKQ J R Holmes, 36 Hillside Gdns, Walthamstow, London, E17 3RJ [JO01AO, TQ38]
G3 PKR K E Parker, 263 High St., Hayes, UB3 5ET [IO91SL, TQ07]
GM3 PKV H R Thornton, 25 Bay St., Fairlie, Largs, KA29 0AL [IO75NS, NS25]
G3 PKW Details withheld at licensee's request by SSL.
G3 PKX D H McGredy, Tudor Cottage, Stratford Bridge, Ripple, Tewkesbury, GL20 6HE [IO82VB, SO83]
GI3 PKY Rev P J OKelly, Parochial House, Tullyallen, Drogheda, Co Louth, Eire, X X
G3 PKZ R R Baker, 7 Rosewood Dr, Crews Hill, Enfield, EN2 9BT [IO91WQ, TQ39]
G3 PLA E B Ullathorne, 20 Raven Ave, Tibshelf, Alfreton, DE55 5NR [IO93HD, SK46]
G3 PLB R W Howe, 28 Egbert Gdns, Wickford, SS11 7BH [JO01GO, TQ79]
G3 PLC Details withheld at licensee's request by SSL.
G3 PLE D H Barlow, Pine, Churchtown, Cury Cross Lanes, Helston, TR12 7BW [IO70JB, SW62]
G3 PLH Details withheld at licensee's request by SSL.
G3 PLJ P Fairnington, 30 Orchard St., Weston Super Mare, BS23 1RQ [IO81MI, ST36]
G3 PLL FtLt R P Moore, 11 Burley Rd, Cottesmore, Oakham, LE15 7BZ [IO92PR, SK81]
G3 PLN G Smith, Roseway Cottage, Willerton Rd, North Somercotes, Louth, LN11 7NH [JO03BK, TF49]
GM3 PLO J Gray, Norland, Stromness, Orkney, Scotland, KW16 3DJ [IO88IW, HY20]
G3 PLP R W Cox, Catbells, West Hill Rd, West Hill, Ottery St. Mary, EX11 1UZ [IO80IR, SY09]
G3 PLR D A Skye, 112 Eastmoor Park, Harpenden, AL5 1BP [IO91TT, TL11]
G3 PLS Dr R G Fenby, Fir Tree Cottage, Warrington Rd, Mere, Knutsford, WA16 0TE [IO83TH, SJ78]
G3 PLT G H F Lawes, Holmewood, 10 Birch Ave, Bleadon, Weston Super Mare, BS24 0PA [IO81MH, ST35]
G3 PLW J E Norton, Bramley House, 29 Shrewsbury Rd, Cockshutt, Ellesmere, SY12 0JH [IO82NU, SJ42]
G3 PLX J P Martinez, High Blakebank, Underbarrow, Kendal, Cumbria, LA8 8BN [IO84NH, SD49]
G3 PLY G J McNeil, Titlarks Cottage, 168 Chobham Rd, Sunningdale, Ascot, SL5 0HU [IO91QJ, SU96]
G3 PMA C K Reaney, 11 Freeboard Rd, Millfield Farm, Leicester, LE3 2UN [IO92JO, SK50]
GM3 PMB W B Miller, Whiteleys, Alloways, Ayr, KA7 4EG [IO75QJ, NS31]
G3 PMC Details withheld at licensee's request by SSL.
G3 PMD A Tranter, 12 Orchard Rd, Burpham, Guildford, GU4 7JH [IO91RG, TQ05]
G3 PMF W F Craine, 5 The Cres, Abbots Langley, WD5 0DR [IO91TQ, TL00]
G3 PMH J B W Braithwaite March & Dist AR, 34 Lawn Ln, Little Downham, Ely, CB6 2TS [JO02DK, TL58]
G3 PMI B F Green, 99 Southfield Rd, Cowley Rd, Oxford, OX4 1NY [IO91JR, SP50]
G3 PMJ S Revell, 11 Mere Fold, Little Hulton, Worsley, Manchester, M28 0SX [IO83TM, SD70]
GM3 PML Dr D B Smith, East Neuk, Netherley, Stonehaven, Grampian, AB3 2NH [IO87TH, NJ72]
G3 PMM Details withheld at licensee's request by SSL.
G3 PMO A Spencer, 297 Liverpool Rd, Walmer Bridge, Preston, PR4 5QD [IO83OR, SD42]
G3 PMR A H Jubb, 30 West St., Great Gransden, Sandy, SG19 3AU [IO92WE, TL25]
G3 PMT J S Russell, 2 Robin Cl, Mulbarton, Norwich, NR14 8EF [JO02ON, TG10]
G3 PMV A J Feist, 1 Lowry Dr, Marple Bridge, Stockport, SK6 5BR [IO83XJ, SJ98]
G3 PMW K W Dews, 14 Baddow Pl Ave, Great Baddow, Chelmsford, CM2 7JN [JO01GR, TL70]
G3 PMX J R D Martyr, 1 High Houses, Mashbury Rd, Great Waltham, Chelmsford, CM3 1EL [JO01FT, TL61]
G3 PNB E G Mason, 154 White Lion Rd, Little Chalfont, Amersham, HP7 9NQ [IO91RQ, SU99]
G3 PNC R L Jack, 2 Rookfield Cl, Muswell Hill, London, N10 3TR [IO91WO, TQ28]
G3 PND S F Appleyard, 45 Shakespeare Rd, Colchester, CO3 4HZ [JO01KV, TL92]
G3 PNF Capt D A Bowden, 29 Knightsbridge Way, Bridgwater, TA6 4XR [IO81MD, ST33]
G3 PNH J B Wilson, 6 Ridings Mead, Chippenham, SN15 1PG [IO81WL, ST97]
G3 PNJ J W G Pethard, St-Die-Strasse 13, D7990, Friedrichshafen, Germany
G3 PNK G Milner, Sunnyside, Stanford Ln, Clifton, Shefford, SG17 5EU [IO92UA, TL13]

G3 PNO I J Hawkins, Victoria House, Victoria St., Totnes, TQ9 5EF [IO80DK, SX86]
G3 PNP J J Ward, 44 Butterworth Path, North St., Luton, LU2 0TP [IO91TV, TL02]
G3 PNQ A Floyd, 27 Beechfield, Parbold, Wigan, WN8 7AR [IO83OO, SD41]
G3 PNR W Higgins, 65 Hayden Ct, Eleanor Rd, Norwich, NR1 2RG [JO02PO, TG20]
G3 PNT C A Durell, 17 Ryders Ave, Westgate on Sea, CT8 8LW [JO01PJ, TR36]
G3 PNV G M E Andrews, 33 Highmoor, Amersham, HP7 9BU [IO91QQ, SU99]
G3 PNW D W Wyatt, 34 Offington Ln, Worthing, BN14 9RT [IO90TT, TQ10]
G3 PNX E H Wyatt, 34 Offington Ln, Worthing, BN14 9RT [IO90TT, TQ10]
G3 POC P O Cartwright, 37 Priory Cres, Bridlington, YO16 5SE [IO94VC, TA16]
G3 POD G H Foster, 6 Forestry Houses, Chawleigh, Chulmleigh, EX18 7LD [IO80BV, SS61]
G3 POF R E Whiting, 752 Jalan 17/34, 46400 Petaling Jaya, Selangor, Malaysia
G3 POG D V Mawdsley, 32 Watchyard Ln, Formby, Liverpool, L37 3JU [IO83LN, SD30]
GM3 POI C S Penna, North Windbreck, Deerness, Orkney, KW17 2QL [IO88OW, HY50]
G3 POJ A Perkins, 1 Spring Gdns Terr, Nuthall, Nottingham, NG16 1DL [IO92JX, SK54]
GW3 POM G W L Morgan, 8 Coed yr Esgob, Llantrisant, Pontyclun, CF7 8EL [IO81HN, ST08]
G3 POQ P D Hayes, 16 Melton Dr, Storrington, Pulborough, RH20 4LU [IO90SW, TQ01]
GI3 POS A G Smyth, 91A Gilford Rd, Lurgan, Craigavon, BT66 7EB [IO64UK, J05]
GM3 POT J G Walford, Chorcaill, Harbour Rd, Reay, Thurso, Caithness, KW14 7RG [IO88CN, NC96]
G3 POX G D Griffiths, 22 Glebe Ln, Buckden, St. Neots, Huntingdon, PE18 9TG [IO92VH, TL16]
G3 POY H E Smith, Gowles, Long Mill Ln, Crouch, Borough Green, Sevenoaks, Kent, TN15 8QF [JO01DG, TQ65]
G3 POZ D Lane, The Banks, Waterfield Dr, Warlingham, Surrey, CR6 9HP [IO91XH, TQ35]
G3 PPB J M Perkins, 23 Simmonds Cl, Bracknell, RG42 1FL [IO91OK, SU86]
G3 PPC D J O Taylerson, 18 The Grove, Teddington, TW11 8AS [IO91UK, TQ17]
G3 PPD R S Dodson, The Haven, Lound Rd, Blundeston, Lowestoft, NR32 5AT [JO02UM, TM59]
GM3 PPE M J Eccles, Sunnyside, Ballencrieff Toll, Bathgate, W Lothian, EH48 4LD [IO85EW, NS97]
GW3 PPF P A Schorah, 11 Dan y Graig, Pant Mawr, Cardiff, CF4 7HJ [IO81JM, ST18]
G3 PPG Details withheld at licensee's request by SSL.[Correspondence to Ariel R G, c/o B J Gealer, BBC Engr Training Centre, Wood Norton, Evesham, Worcs WR11 4TF.]
G3 PPI J C Huntley, 68 Lamorna Gr, Stanmore, HA7 1PG [IO91UO, TQ19]
GM3 PPJ H Hogg, 12 Valley Ct, Patna, Ayr, KA6 7LQ [IO75RI, NS41]
G3 PPN D A Rees-Jones, Sandfield House, Runnymede Cl, Gateacre, Liverpool, L25 5JU [IO83NJ, SJ48]
G3 PPO L R Hook, 79 Whiteley Cres, Bletchley, Milton Keynes, MK3 5DQ [IO91OX, SP83]
G3 PPR Dr J R G Beavon, 38 Abbots Way, Sherborne, DT9 6DT [IO80RW, ST61]
G3 PPT L G Sear, 4 Mount Pleasant Rd, Threemile Stone, Truro, Cornwall, TR3 6BB [IO70KG, SW74]
G3 PPU P J Smith, 56 Alphington Ave, Frimley, Camberley, GU16 5LR [IO91PH, SU85]
G3 PPX E W Curtis, 23 Daisy Dormer Ct, Trinity Gdns, London, SW9 8DW [IO91WL, TQ37]
G3 PQ M P Bayliss, 5 Arden Rd, Kenilworth, CV8 2DU [IO92FI, SP27]
G3 PQA J P F O Rogers, Dromore, Strand Ln, Cookham, Berks, SL6 9DN [IO91PN, SU88]
G3 PQC P D G Turk, 13 The Cres, Farnborough, GU14 7AR [IO91PG, SU85]
G3 PQD D E A St John, 26 Henry St., Rainham, Gillingham, ME8 8HE [JO01HI, TQ86]
G3 PQE J Thorn, 49 Ashcombe Rd, Weston Super Mare, BS23 3DU [IO81MI, ST36]
G3 PQF D H Dell, 7 Blunden Rd, Cove, Farnborough, GU14 8QJ [IO91OH, SU85]
G3 PQH Details withheld at licensee's request by SSL.
G3 PQI R F Bright, 77 Church Rd, Tiptree, Colchester, CO5 0HB [JO01JT, TL81]
G3 PQJ B D Cole, 5 Daneshill, Victoria Gr, East Cowes, PO32 6BJ [IO90IR, SZ59]
G3 PQK J Cameron, 14 Greenholm Rd, London, SE9 1UH [JO01AK, TQ47]
G3 PQM M A Thorp, Cecil, Thorrington Rd, Little Clacton, Clacton on Sea, Essex, CO16 9ES [JO01NT, TM11]
G3 PQP T H Foster, 136 Sladepool Farm Rd, Kings Heath, Birmingham, B14 5EF [IO92BJ, SP07]
G3 PQT M J Jones, 49 Gr Rd, Hoylake, Wirral, L47 2DS [IO83JJ, SJ28]
GM3 PQU Details withheld at licensee's request by SSL.
GI3 PQW W Campbell, 7 Silverstream Park, Bangor, BT20 3LU [IO74DP, J48]
G3 PQX J Butterworth, 61 Netherton Grange, Bootle, L30 8RF [IO83ML, SJ39]
G3 PQY J Lawrence, 2 Hall Rd, Hull, HU6 8SA [IO93TS, TA03]
GW3 PRA H J M Phillips, 11 Field Cl, Flint, CH6 5RQ [IO83KF, SJ27]
G3 PRC D T Hind Plymouth Radio Club, 4 Thornyville Villas, Plymouth, PL9 7LA [IO70WI, SX55]
G3 PRD W L Nilan, 16 Dunstan Cres, Worksop, S80 1AF [IO93KH, SK57]
G3 PRE W Armstrong, 24 Newbury St., South Shields, NE33 4UE [IO94GX, NZ36]
G3 PRF R Findlay, Nahlin, Trent Lock, Long Eaton, Nottingham, NG10 2FY [IO92IV, SK43]
G3 PRH M R Coward, Dormer Cottage, 51 Farleigh Rd, Backwell, Bristol, BS19 3PB [IO81PJ, ST46]
G3 PRI D J Quigley, 16 Crossways Rd, East Cowes, PO32 6JA [IO90IR, SZ59]
G3 PRK A H Yilmaz, 2 Cameron Cl, Southgate St., Long Melford, Sudbury, CO10 9TS [JO02IB, TL84]
GW3 PRL D M Snow, Rhwng-y-Ddwydre, Brynsiencyn, Anglesey, Gwynedd, LL61 6TZ [IO73UE, SH46]
G3 PRN L F Cox, 26 Little Pynchons, Harlow, CM18 7DD [JO01BS, TL40]
G3 PRO R D Evans, West Winds, Main St, Appleton Roebuck, York, YO5 7DA [IO93KU, SE54]
G3 PRQ E G Wooden, Herne Cottage, Burton, East Coker, Yeovil, BA22 9LR [IO80QW, ST51]
G3 PRR I S Partridge, The Rectory, Torrington Ln, East Barkwith, Market Rasen, Lincs, LN8 5RY [IO93UH, TF18]
G3 PRS Details withheld at licensee's request by SSL.
G3 PRU J C Nicholas-Letch, 3 Orchard Walk, Lavendon, Olney, MK46 4HE [IO92PE, SP95]
G3 PS A McCann, 5 Arrowsmith Dr, Hoghton, Preston, PR5 0DT [IO83QR, SD52]
G3 PSC J G J Holton, 1204 Greenford Rd, Greenford, UB6 0HQ [IO91TN, TQ18]
G3 PSG R I Armstrong North Riding Rafars, obo North Riding Rafars, 64 Churchill Dr, Marske By The Sea, Redcar, TS11 6BE [IO94LO, NZ62]
GM3 PSJ R B Harkness, 23 Heatherlie Park, Selkirk, TD7 5AL [IO85NN, NT42]
G3 PSM C J Thomas, 83 Salisbury Rd, Totton, Southampton, SO40 3HY [IO90GW, SU31]
G3 PSP Dr A J Masson, 407 Pacific Circle, Newbury Park, Ca 91320, USA
GI3 PSQ C G Bristow, 58 Bristow Park, Belfast, BT9 6TJ [IO74AN, J37]
G3 PSR M J Gibbs, 19 Hall Rd, Lordswood, Chatham, ME5 8PL [JO01GI, TQ76]
G3 PSS M J Kent, 99 London Rd, Newington, Sittingbourne, ME9 7NU [JO01HI, TQ86]
G3 PSU P R Martin, 15 St. Lukes Cl, Cannock, WS11 1BB [IO82XQ, SJ90]
G3 PSV D A Park, 3 Wingfield Gdns, Frimley, Camberley, GU16 5QD [IO91PH, SU95]
G3 PSW P A Spencer, 32 Harfield Rd, Sunbury on Thames, TW16 5PT [IO91TJ, TQ16]
G3 PSY J G Cottrell, Duxbury, Church Hill, High Halden, Ashford, Kent, TN26 3JB [JO01IC, TQ93]
G3 PSZ K G Jones, Ockton House, 24 Station Rd, Okehampton, EX20 1EA [IO70XR, SX59]
G3 PTB A W Tomalin, Chapel St, Barford, Norwich, Norfolk, NR9 4AB [JO02NP, TG10]
G3 PTD Details withheld at licensee's request by SSL.
G3 PTG R H Gealy, 6 Mill End Cl, Eaton Bray, Dunstable, LU6 2FH [IO91RU, SP92]
GM3 PTI K F Atter, 60 Hough Rd, Barkston, Grantham, NG32 2NS [IO92QX, SK94]
G3 PTJ J Robinson, 35 Farrow Rd, Leeds, LS12 3TB [IO93ET, SE23]
G3 PTM A Griffiths, 63 Wylde Green Rd, Sutton Coldfield, B72 1HD [IO92CN, SP19]
G3 PTN Z T Chowaniec, 33 Elmete Gr, Leeds, LS8 2LA [IO93GT, SE33]
G3 PTO J A Reynolds, 124 Merlin Way, Chipping Sodbury, Bristol, BS17 6XT [IO81TM, ST78]
G3 PTQ T J Chapman, 5 Maple Cl, Bottisham, Cambridge, CB5 9BQ [JO02DF, TL56]
G3 PTS Dr G M Holt, 7 Beech Cl, Olivers Battery, Winchester, SO22 4JY [IO91HA, SU42]
G3 PTT S H Goodall, Park Farm, Brailsford, Ashbourne, Derbyshire, DE6 3BE [IO92EX, SK24]
G3 PTU Details withheld at licensee's request by SSL.
G3 PTV D Critchlow, 7 Willow Rd, Armthorpe, Doncaster, DN3 3HD [IO93LM, SE60]
G3 PTX L Buckley, 188 Compstall Rd, Romiley, Stockport, SK6 4JF [IO83XJ, SJ99]
G3 PTZ A Bensley, 13 Lime Gr, Cherry Willingham, Lincoln, LN3 4BE [IO93SF, TF07]
G3 PUC A H Strange, 26 Gannet House, Hartington Pl, Eastbourne, BN21 3BL [JO00DS, TV69]
G3 PUO L D Rooks, 17 The Close, Clayton-le-Moors, Accrington, BB5 5RX [IO83TS, SD73]
G3 PUP L J Walker, 7 Bansons Way, Chipping, Ongar, CM5 9AS [JO01CQ, TL50]
G3 PUQ N S Semmens, 4 South Park, Redruth, TR15 3AW [IO70JF, SW64]
G3 PUR R J Tarr, Oakfield, 37 Warwick Ave, Earlsdon, Coventry, CV5 6DJ [IO92FJ, SP37]
G3 PUT C J Heath, 15 Canberra Gdns, Luton, LU3 2EU [IO91SV, TL02]
G3 PUU Details withheld at licensee's request by SSL.
G3 PUV Details withheld at licensee's request by SSL.
G3 PUW S Pennington, 6 The Limes, Bletchley, Milton Keynes, MK2 2JN [IO91PX, SP83]
G3 PUX I S Champion, Mill Bungalow, Wisborough Green, Billinghurst, Sussex, RH14 0DY [IO91RA, TQ02]
G3 PUZ D Hogan, 8 Beech Cl, Alderholt, Fordingbridge, SP6 3DG [IO90CV, SU11]
G3 PV E H Rickett, Wynnsted, 29 North Rd, Berkhamsted, HP4 3DU [IO91RS, SP90]
G3 PVA P J King, 10 Holne Chase, Morden, SM4 5QB [IO91VJ, TQ26]
G3 PVB L J Santer, 3 Barn Park, Liverton, Newton Abbot, TQ12 6HE [IO80DN, SX87]
G3 PVD G Y Loades, Brooklands, 32 Fox Ln, Hoghton, Preston, PR5 0JQ [IO83QR, SD52]
G3 PVG J Bennett, 11 Enderby Rd, Thurlaston, Blaby, Leicester, LE8 4GD [IO92KN, SP59]
G3 PVH D J Sumner, 20 Woodlands Way, Southwater, Horsham, RH13 7HZ [IO91TA, TQ12]
G3 PVJ Dr H D Coltman, 50 Baronsmead Rd, High Wycombe, HP12 3PG [IO91OP, SU89]
G3 PVR T H Burdon, 22 Shirley Walk, Coton Green, Tamworth, B79 8LD [IO92DP, SK10]
G3 PVS D J Hawkins, Fairview Farm, Stapley Ln, Ropley, Hants, SO24 0EL [IO91LB, SU63]
G3 PVT K J E Field, 42 Minworth Rd, Water Orton, Birmingham, B46 1NH [IO92DM, SP19]
G3 PVU J R Hunt, 28 Harris Rd, Lincoln, LN6 7PN [IO93RE, SK96]
G3 PVW Details withheld at licensee's request by SSL.
G3 PVX Details withheld at licensee's request by SSL.[Letter 18 Dec 1995]
GW3 PWA R R Hughes, 34 Albert St., Pentre, CF41 7JR [IO81GP, SS99]
G3 PWB I G Dufour, 3 Western Cl, Rushmere, Ipswich, IP4 5UU [JO02OB, TM24]
GW3 PWH Details withheld at licensee's request by SSL.
G3 PWJ R W Fisher, 34 Doctors Hill, Pedmore, Stourbridge, DY9 0YE [IO82WK, SO98]
G3 PWK J B W Braithwaite, Old Number 10, 34 Lawn Ln, Little Downham, Ely, Cambs, CB6 2TS [JO02DK, TL58]
G3 PWN G T Grimshaw, 1 Sandsacre Dr, Bridlington, YO16 5UA [IO94VC, TA16]
G3 PWQ A Marven, Green Hazel, Periton Rd, Minehead, TA24 8DR [IO81GE, SS94]
G3 PWS R Dalton, 23 Muswell Rd, Mackworth, Derby, DE22 4HN [IO92FW, SK33]
G3 PWU M W Cleland, 5 Northcourt Ave, Reading, RG2 7HE [IO91MK, SU77]
G3 PWW G P Seaman, 47 Careys Wood, Smallfield, Horley, RH6 9PA [IO91WE, TQ34]
G3 PWY D M Gresswell, 10 Cherrywood Gdns, Flackwell Heath, High Wycombe, HP10 9AX [IO91PO, SU88]
G3 PXC D H Caplan, 23 Princes Cres, Basingstoke, RG22 6DR [IO91KG, SU65]
G3 PXF A Petts, 88 Northfield Ln, Horbury, Wakefield, WF4 5JF [IO93FP, SE31]
GM3 PXG J M Donacie, 19 George Terr, Balfron, Glasgow, G63 0PL [IO76UB, NS58]
G3 PXH M J Bartlett, 3 Jessopp Ave, Bridport, DT6 4AN [IO80PR, SY49]
G3 PXI A J Evans, 57 Bowes Hill, Rowlands Castle, PO9 6BS [IO90MV, SU71]
GM3 PXK R McEwan Mid Lanarks ARS, Oakwood, 12 Valleyfield Dr, Cumbernauld, Glasgow, G68 9NW [IO75XW, NS77]
G3 PXL A J Hickin, 4B Laburnum Row, Torre, Torquay, TQ2 5QX [IO80FL, SX96]
G3 PXM R A K Pavey, The Rest, Fair View Ln, Colyford Colyton, East Devon, EX13 6QX [IO80LR, SY29]
G3 PXO Revd R N Myerscough Gresham's School Amateur Radio, 10 Kelling Rd, Holt, NR25 6RT [JO02NV, TG03]
G3 PXQ Details withheld at licensee's request by SSL.
G3 PXU G H Grove, 15 Ward Gr, Warwick, CV34 6QL [IO92FG, SP36]
G3 PXV R E Wiseman, 147 Oilmills Rd, Mereside, Ramsey, Huntingdon, PE17 1UA [IO92WL, TL28]
G3 PXX J J M Phillips, 18 Rock Farm Dr, Little Neston, South Wirral, L64 4DZ [IO83LG, SJ37]
GW3 PXY E Thomas, Swn y Nant, Lon Glanfred, Llandre, Bow St, Dyfed, SY24 5BY [IO72XL, SN68]
G3 PYA Details withheld at licensee's request by SSL.
G3 PYB P Blakeborough, Papine, 33 Cloisters Rd, Letchworth, SG6 3JR [IO91VX, TL23]
G3 PYC R J Polley, 88 Merryfield Dr, Horsham, RH12 2AX [IO91TB, TQ13]
GW3 PYD D C Stephens, 1 Awelfryn Terr, Merthyr Tydfil, CF47 9YP [IO81HS, SO00]
G3 PYE J F Wilson Philips Tele Lt, 24 PO Box, St. Andrews Rd, Cambridge, CB4 1DP [JO02BF, TL46]
G3 PYF J L Green, 68 Magdalen Ln, Wingfield, Trowbridge, BA14 9LQ [IO81UH, ST85]
G3 PYH A Broadbent, 52 Norman St., Failsworth, Manchester, M35 9EJ [IO83WM, SD90]
G3 PYI D N Coy, 26 Hardy Rd, Bishops Cleeve, Cheltenham, GL52 4BN [IO81XW, SO92]
G3 PYM A S Barlow, 29 Watergate St., Ellesmere, SY12 0EX [IO82NV, SJ43]
G3 PYN J A Birley, 4 Gr Ln, Holt, NR25 6EG [JO02NV, TG03]
G3 PYO J R Dann, 1 Finch Cl, Ditton, Aylesford, ME20 6ET [JO01FG, TQ75]
G3 PYP G F Hibberd, 33 Mead Park, Atworth, Melksham, SN12 8JS [IO81VJ, ST86]
G3 PYU P Jones, 49 Gr Rd, Hoylake, Wirral, L47 2DS [IO83JJ, SJ28]
G3 PYV R G Coleman, 4 Pembroke Cl, St. Ives, Huntingdon, PE17 4WA [IO92WI, TL37]
G3 PYW Rev A Speight, Glebe Cottage, Hollow Ln, Badingham, Woodbridge, IP13 8LZ [JO02QG, TM36]
GW3 PYX J Chetcuti, 3 Beechwood Dr, Penarth, CF64 3RB [IO81JK, ST17]
G3 PZB A Ash, 34 Coronation Ave, Cowes, PO31 8PN [IO90IR, SZ49]
G3 PZC F P J E Coles, 156 Stanley Rd, Teddington, TW11 8UD [IO91TK, TQ17]
G3 PZE C J Burkitt, The Old Wheelwright, 17 Oxford Rd, Breachwood Green, Hitchin, SG4 8NP [IO91UV, TL12]
GM3 PZG J Lauder, 43 Lochpots Rd, Fraserburgh, AB43 9NH [IO87XQ, NJ96]
G3 PZH C J Shakeshaft, 7 Merton Dr, Lache Ln, Chester, CH4 7PG [IO83NE, SJ36]
G3 PZK S G Collyer, 28 Barnham Rd, Greenford, UB6 9LR [IO91TM, TQ18]
G3 PZL P A Brown, Little Langford, Newlands Ln, Stoke Row, Henley on Thames, RG9 5PS [IO91LN, SU68]
G3 PZN C H Wood, 24 Talveneth, Pendeen, Penzance, TR19 7UT [IO70ED, SW33]
G3 PZQ D J Smith, 47 Ashtree Rd, Costessey, Norwich, NR5 0LR [JO02OP, TG11]
G3 PZS F B Stanbridge, 119 High St., Earith, Huntingdon, PE17 3PN [JO02AI, TL37]
G3 PZV P T Greed, 18 Nursteed Park, Devizes, SN10 3AN [IO91AI, SU06]
G3 PZX A A Ward, 20 Tower Cl, Costessey, Norwich, NR8 5AU [JO02OP, TG11]
G3 PZZ P D Smith, 38 Leasway, Wickford, SS12 0HE [JO01GO, TQ79]
G3 QD J G Treece, 32 Broughton St., Beeston, Nottingham, NG9 1BD [IO92JW, SK53]
GW3 QN J S Owen, 44 Kings Ave, Llandudno, LL30 2BQ [IO83CH, SH78]
G3 QX T R Barlow, 3 Heathway, Heswall, Wirral, L60 2TL [IO83KH, SJ28]
G3 RAC R J Pownall, Racal Comms Ltd Ard, Western Rd, Bracknell, Berks, RG12 1RG [IO91OK, SU86]
G3 RAD R T Trull, 1 Approach Rd, Broadstairs, CT10 1QT [JO01RI, TR36]
G3 RAE R A Eldridge, 82 Lowestoft Rd, Worlingham, Beccles, NR34 7RD [JO02TK, TM48]
G3 RAF R Finch RAF ARS, 1 Cherry Tree Cottage, Church Rd, Tylers Green, Bucks, HP10 8LN [IO91PP, SU99]
G3 RAL Details withheld at licensee's request by SSL.[Correspondence to: Loughborough & DARC, Hind Leys Community College, Forest St, Shepshed, Loughborough, Leics LE12 9DA.]
G3 RAM F C Langmaid, 12 Lower Faircox, Henfield, BN5 9UT [IO90UW, TQ21]
G3 RAN P E Lavender, Woodland Ct, 128 Springdale Rd, Corfe Mullen, Wimborne, BH21 3QL [IO80XS, SY99]
G3 RAR E D Hodgson, 12 Cornmoor Rd, Whickham, Newcastle upon Tyne, NE16 4PU [IO94DW, NZ26]
G3 RAU D S Moffatt, Mill House, Glentworth Cliff, Gainsborough, Lincs, DN21 5BZ [IO93RJ, SK98]
G3 RAW Details withheld at licensee's request by SSL.
GJ3 RAX Details withheld at licensee's request by SSL.
G3 RB K N Smith, 78 Thorntree Dr, West Monkseaton, Whitley Bay, NE25 9NW [IO95GA, NZ37]
GW3 RBA Details withheld at licensee's request by SSL.
G3 RBD F I Hanson, 207 Grant Rd, Liverpool, L14 0LG [IO83NJ, SJ49]
G3 RBG Details withheld at licensee's request by SSL.
G3 RBJ A N Payne, Laurel Bank, Sand Rd, Wedmore, BS28 4BZ [IO81OF, ST44]
G3 RBL G J Lloyd, 4 Firs Cl, Marlbrook, Bromsgrove, B60 1DR [IO82XI, SO97]
GW3 RBM Details withheld at licensee's request by SSL.[Op: J F Roberts, Tan y Cefn, Llanarmon yn Ial, Mold, Clwyd, CH7 4QU.]
G3 RBP R J Parsons, Nutcombe, Hanney Rd, Southmoor, Abingdon, Oxon, OX13 5HT [IO91GQ, SU39]
G3 RBQ J H Hollom, 27 Peace Rd, Stanway, Colchester, CO3 5HL [JO01KV, TL92]
G3 RBR F B S Rogerson, Rua Felizardo de, Lima 149, 4100 Porto, Portugal
G3 RBW K F Broome, 28 Templemead House, Homerton Rd, London, E9 5PT [IO91XN, TQ38]
GI3 RBX Details withheld at licensee's request by SSL.
G3 RBY A F Stagles, 8 Goodwood Cl, Cowplain, Waterlooville, PO8 8BG [IO90LV, SU61]
G3 RCA T O Austin, 45 Hallbridge Gdns, Upholland, Skelmersdale, WN8 0EP [IO83PN, SD50]
G3 RCB N Kingsley, 1 Wensleydale Gdns, Hampton, TW12 2LU [IO91TK, TQ17]
G3 RCD C J Brockbank, 31 Park Hill, Church Crookham, Fleet, GU13 0PW [IO91NG, SU85]
G3 RCE R J S Allbright, 221 Hayling Ave, Copnor, Portsmouth, PO3 6DZ [IO90LT, SU60]
G3 RCO L C de La Bertauche, Westleigh, Fore St., Beer, Seaton, EX12 3EQ [IO80KQ, SY28]
G3 RCQ D L Cole, 9 Troopers Dr, Carters Hill, Romford, RM3 9DE [JO01CO, TQ59]
G3 RCU Details withheld at licensee's request by SSL.
G3 RCV A R Burchmore Cray Valley Radio Soc Ltd, 49 School Ln, Horton Kirby, Dartford, DA4 9DQ [JO01DJ, TQ56]
G3 RCW G Pool Worksop ARS, 89 Norwood Ave, Hasland, Chesterfield, S41 0NJ [IO93HF, SK46]
G3 RCX L H Gibson, 7 Heycroft Rd, Eastwood, Leigh on Sea, SS9 5SW [JO01HN, TQ88]
G3 RCZ G T Thompson, 22 Warton Ave, Heysham, Morecambe, LA3 2LX [IO84NA, SD46]
G3 RD Details withheld at licensee's request by SSL.
G3 RDA S J Whitehead, 98 Oak Rd, Fareham, PO15 5HP [IO90JU, SU50]
GW3 RDB T S George, 80 Yew St., Troedyrhiw, Merthyr Tydfil, CF48 4EE [IO81HR, SO00]
G3 RDC A F H Wood, 28 Watts Ln, Hillmorton, Rugby, CV21 4PE [IO92JI, SP57]
G3 RDD J K Shaw, 58 Elizabeth Ave, North Hykeham, Lincoln, LN6 9RR [IO93RE, SK96]
G3 RDF J D Jeffrey, Old Church Cottage, Ipsden, Oxford, OX10 6AE [IO91KN, SU68]
G3 RDG K B Michaelson, 40 The Vale, Golders Green, London, NW11 8SG [IO91VN, TQ28]
G3 RDH J H Barnes, Southwinds, 4 Deepdene Dr, Dorking, RH5 4AD [IO91UF, TQ14]
G3 RDI Dr J R Kirkman, 58 Nursery Rd, Silksworth Ln, Sunderland, SR3 1NT [IO94HV, NZ35]
G3 RDK Dr D Doyle, 4 Wricklemarsh Rd, Blackheath, London, SE3 0NF [JO01AL, TQ47]
G3 RDN E J Sharp, 11 Forge Way, Billingshurst, RH14 9LJ [IO91SA, TQ02]
G3 RDO B J H Matson, 36 Windmill Rise, Woodhouse Eaves, Loughborough, LE12 8SG [IO92JR, SK51]
G3 RDP H B Cutts, 50 Cropton Rd, Hull, HU5 4LP [IO93TS, TA03]
G3 RDQ D C Griffiths, Upcote Cottage, Chilbolton, Stockbridge, Hants, SO20 6BA [IO91GD, SU33]
G3 RDR P L Rudwick, 3 Carberry Dr, Portchester, Fareham, PO16 9JB [IO90KU, SU60]
G3 RDS D Moore, 129 High Cross Rd, Poulton-le-Fylde, FY6 8BX [IO83MU, SD33]
G3 RDU R B R Palk, 14 Hillview Rd, Minehead, TA24 8EG [IO81GE, SS94]
G3 RDV S P Brindle, 206 Blackburn Rd, Oswaldtwistle, Accrington, BB5 4NZ [IO83TS, SD72]
G3 RDW A W Kendrick, Long Dr, Roman Rd, Sutton Coldfield, B74 3AA [IO92BO, SP09]
G3 RDZ J A Walker, 38 Ash Cl, Peterborough, PE1 4PG [IO92VO, TF20]
G3 RE M Bernard Royal Engineers Assoc Radio BR, 16 Mountbatten, Ave, Chatham, Kent, ME5 0JY [JO01GI, TQ76]
G3 REA C F Peers, 5 Heath Dr, Worcester, WR5 3TD [IO82VE, SO85]
G3 REB R A Cole, Lyndale, Brimscombe Ln, Brimscombe, Stroud, GL5 2RF [IO81VR, SO80]
G3 RED D C Sylvester, 10 Ivy Gr, Gunthorpe, Peterborough, PE4 7TW [IO92UO, TF10]
G3 REH H Neale, Thornlea, Fishersgate, Sutton St James, Spalding, PE12 0EZ [JO02AS, TF31]
G3 REL B W Woodfield, 49 Oakfield Rd, Blackwater, Camberley, GU17 9DZ [IO91OH, SU85]
G3 REM M Austin, 20 Dimple Park, Egerton, Bolton, BL7 9QE [IO83SP, SD71]
G3 REP R E Parkes, 2 Saxon Rd, Steyning, BN44 3FP [IO90UV, TQ11]
G3 REU C F Hearn, 70 Cranmer St., Long Eaton, Nottingham, NG10 1NL [IO92IV, SK43]
G3 REV R E Pulling, 5 Station Rd, Prescot, L34 5SN [IO83OK, SJ49]
G3 REW D R Morris, 5 Eden Park, Brixham, TQ5 9LS [IO80FJ, SX95]
GW3 REY E Lynn, 3 Trehwfa, Coed Mawr, Bangor, LL57 4TE [IO73WF, SH57]
G3 REZ S R Lane, 51 Lindenthorpe Rd, Broadstairs, CT10 1BQ [JO01RI, TR36]
GM3 RFA D E Garrington, 3 Sutherland Ave, Fort William, PH33 6JS [IO76KT, NN07]
G3 RFE T H Bell, 36 South View, Barrow in Furness, LA14 5NN [IO84JD, SD27]

G3 RFF Details withheld at licensee's request by SSL.
GD3 RFH Details withheld at licensee's request by SSL.
G3 RFJ W F Williams, Four Corners, Reigate Rd, Hookwood, Horley, RH6 0AR [IO91VE, TQ24]
GD3 RFK D C Dodd, Ellan Geay, Ballyockey Ln, Regaby Kirk Andreas, Isle of Man, IM7 3HP
G3 RFL Details withheld at licensee's request by SSL.
G3 RFN G Wild, 17 New Church Cl, Clayton-le-Moors, Accrington, BB5 5GH [IO83TS, SD73]
G3 RFO G A Simmons, 31 Cliff Rd, Worlebury, Weston Super Mare, BS22 9SE [IO81MI, ST36]
G3 RFP F C D Taylor, 4 Cross Keys Ct, Cottenham, Cambridge, CB4 4UW [JO02BG, TL46]
GM3 RFQ A J Nadauld, 12 Lower Broomieknowe, Lasswade, EH18 1LW [IO85KV, NT36]
G3 RFT Details withheld at licensee's request by SSL.
G3 RFX M C Phillips, 17 Richmond Hill, Clifton, Bristol, BS8 1BA [IO81QK, ST57]
G3 RGA P J H Toynton, 25 Greenways, Saffron Walden, CB11 3EZ [JO02CA, TL53]
G3 RGB A Moon, 6 Troon Cl, Saltersgill, Middlesbrough, TS4 3HX [IO94JN, NZ41]
G3 RGC T Matthews, 38 Foxhill, Wybers Wood, Grimsby, DN37 9QL [IO93WN, TA20]
G3 RGD R G Dobdinson, 73 Watwood Rd, Hall Green, Birmingham, B28 0TW [IO92BJ, SP17]
G3 RGE K G King, 18 Alexandria Ct, Glenmoor Rd, West Parley, Ferndown, BH22 8PW [IO90BS, SZ09]
G3 RGG B Johnson-Roberts, 45 Bean Oak Rd, Wokingham, RG40 1RH [IO91OJ, SU86]
G3 RGJ R J Weston, 43 Pearce Ave, Parkstone, Poole, BH14 8EG [IO90AR, SZ09]
GW3 RGL E A Hays, 23 Edgemoor Dr, Upper Killay, Swansea, SA2 7HH [IO71XO, SS59]
G3 RGM D R Mullins, 23 Kennington Palace Ct, Sancroft St., London, SE11 5UL [IO91WL, TQ37]
G3 RGN L Binns, Leamar, 707 Halifax Rd, Hartshead Moor, Cleckheaton, BD19 6LJ [IO93CR, SE12]
G3 RGO G H Dobbs, 14C St. Andrews Cres, Leasingham, Sleaford, NG34 8LS [IO92SX, TF04]
G3 RGP R G Pratt, 1 Colebrooke Ave, Ealing, London, W13 8JZ [IO91UM, TQ18]
G3 RGQ T T Harding, 73 Balland Field, Willingham By Stow, Willingham, Cambridge, CB4 5JT [JO02AH, TL47]
G3 RGR M C Niblett, 4 Hillcrest Gdns, Ramsgate, CT11 0DS [JO01QH, TR36]
G3 RGS D J Thomson, The Barn, Evenden Farm, Meopham, Gravesend, DA13 0JE [JO01EJ, TQ66]
GM3 RGU J J Connelly, 9 Glenhead Cres, Hardgate, Clydebank, G81 6LW [IO75TW, NS47]
G3 RGV G A Heapy, Rose Cottage, York Pl, Knaresborough, HG5 0AA [IO94GA, SE35]
GW3 RHC H Vowles, 19 Glannant Pl, Cwmgwrach, Neath, SA11 5TE [IO81ER, SN80]
G3 RHG D J Marshall, 52 Wyvern Ave, Long Eaton, Nottingham, NG10 1AG [IO92IV, SK43]
G3 RHH Dr P H Masterman, 5 Grundys Ln, Malvern Wells, Malvern, WR14 4HS [IO82UB, SO74]
G3 RHM G D Clarkson, 22 The Burlands, Feniton, Honiton, EX14 0UN [IO80JT, ST10]
G3 RHN Details withheld at licensee's request by SSL.
G3 RHP J Garrett, Shrubbery Farm, Otley, Ipswich, Suffolk, IP6 9PD [JO02PE, TM25]
G3 RHQ K Vickers, Hillview, Horkstow Rd, Barton upon Humber, South Humberside, DN18 5DZ [IO93SQ, TA02]
G3 RHR K Drinkwater, Brearton Lodge, Brearton, Harrogate, North Yorks, HG3 3BX [IO94FB, SE36]
G3 RHU M F Stanbridge, 183 Charlton Park, Midsomer Norton, Bath, BA3 4BR [IO81SG, ST65]
G3 RHW C C Cushion, 3 The Copse, Bridgwater, TA6 4DW [IO81MD, ST33]
G3 RHZ A E Wilkinson, 18 Tansey Cres, Stoney Stanton, Leicester, LE9 4BT [IO92IN, SP49]
G3 RIC F A Stanier, 192 Trent Valley Rd, Penkhull, Stoke on Trent, ST4 5HL [IO82VX, SJ84]
G3 RID D J Nancarrow, 6 Trythogga Rd, Gulval, Penzance, TR18 3NA [IO70FD, SW43]
G3 RIF H Harrison, 77 Moat Ave, Green Ln, Coventry, CV3 6BT [IO92FJ, SP37]
GW3 RIH W J Elton, Wayside, 15 Main Ave, Peterston Super Ely, Cardiff, CF5 6LQ [IO81IL, ST07]
G3 RII H Armstrong, 2 Trevance, Ln, Penryn, TR10 8RD [IO70KE, SW73]
GM3 RIJ M Hughes BBC Blackhill C, BBC Transmitting St, Black Hill, Salsburgh Lanarks, ML7 4NZ [IO85BU, NS86]
G3 RIK D Carden, 9 Wood Hey Gr, Rochdale, OL12 9TY [IO83WP, SD81]
G3 RIM T Emeney, 10 Kilnside, Claygate, Esher, KT10 0HS [IO91UI, TQ16]
G3 RIO J E Wade, St Peters, New Rd, Brading, Isle of Wight, PO36 0AG [IO90KQ, SZ68]
G3 RIQ F O Cooper, 75 Wood Ln, Cotton End, Bedford, MK45 3AP [IO92TC, TL04]
G3 RIR N Ackerley, 24 Macaulay Rd, Lutterworth, LE17 4XB [IO92JL, SP58]
G3 RIS F W Nash, 39 Bernard Rd, Cromer, NR27 9AW [JO02PW, TG24]
G3 RIT A Ball, 11 Fairview Ave, Weston, Crewe, CW2 5LX [IO83TB, SJ75]
G3 RIV G Simpson, 209 Barton Rd, Stretford, Manchester, M32 9RB [IO83UK, SJ79]
G3 RIX M R Tetley, 87 Main St., Irton, Scarborough, YO12 4RJ [IO94SF, TA08]
G3 RIY A J Chapman, 19 Andrews Way, Marlow Bottom, Marlow, SL7 3QJ [IO91OO, SU88]
G3 RJB B R Edwards, Whitegates, 41 Winston Rd, Putson, Hereford, HR2 6DJ [IO82PA, SO53]
G3 RJD G A Partridge, 14 Belinda Cl, The Manor, Willenhall, WV13 1BZ [IO82XO, SO99]
G3 RJE J D Hunt, 33 Rainhill Rd, Rainhill, Prescot, L35 4PA [IO83OK, SJ49]
G3 RJF I Walker, 28 Norrington Rd, Loose, Maidstone, ME15 9RA [JO01GF, TQ75]
G3 RJH Dr R J Harding, High Trees, Arrowsmith Rd, Canford Magna, Wimborne, BH21 3BG [IO90AS, SZ09]
G3 RJI A H Paul, 38 Avon Rd, Upminster, RM14 1QU [JO01DN, TQ58]
G3 RJK R L J Kissick, Kingleigh, Hooke Hill, Freshwater, PO40 9BH [IO90FQ, SZ38]
G3 RJM R Cutts, 60 Holmpton Rd, Withernsea, HU19 2QD [JO03AR, TA32]
G3 RJQ Details withheld at licensee's request by SSL.
GW3 RJR R J Rothery, 267 Fairbourne Village, Fairbourne, Gwynedd, LL38 2RQ [IO72XQ, SH61]
G3 RJS P F Barry, 16 Boleyn Dr, Bognor Regis, PO21 3LG [IO90PS, SZ89]
G3 RJT C M Garland, 48 Underbank End Rd, Holmfirth, Huddersfield, HD7 1ES [IO93CN, SE10]
G3 RJU P J Atkins, 26 Temple Sheen Rd, East Sheen, London, SW14 7QG [IO91UL, TQ27]
G3 RJV Rev G C Dobbs, St. Aidans Vicarage, 498 Manchester Rd, Rochdale, OL11 3HE [IO83VO, SD81]
G3 RJW G F Marshall, Osmond House, 115 Bute Rd, Wallington, SM6 8AE [IO91WI, TQ26]
G3 RJX B G Elcocks, 213 Perrywood Rd, Birmingham, B42 2BL [IO92AM, SP09]
G3 RK H A Spashett, The Glen, 61 Mill Rd, Ellingham, Bungay, NR35 2PY [JO02RL, TM39]
GW3 RKD Details withheld at licensee's request by SSL.
GI3 RKE P S Valentine, 4 Carnesure Heights, Comber, Newtownards, BT23 5RN [IO74DN, J46]
G3 RKH Revd J L Marshall, The Vicarage, Back Clough, Northowram, Halifax, HX3 7HH [IO93CR, SE12]
G3 RKI E J Young, Flat, 28 Friesian Cl, Lewsey Farm Est, Luton, LU4 0PU [IO91SV, TL02]
G3 RKJ N A Summers, 126 Kestrel House, Alma Rd, Enfield, EN3 4QE [IO91PI, XP, TQ39]
G3 RKK Dr A J Shepherd, 59 Lime Ave, Camberley, GU15 2BH [IO91PI, SU86]
G3 RKL Dr A J T Whitaker, 160 Derbyshire Ln, Sheffield, S8 8SE [IO93GI, SK38]
G3 RKM J B Meaker, 47 New Earth St., Mossley, Ashton under Lyne, OL5 0SL [IO83XM, SD90]
G3 RKN D Leese, 4 Harefield, Harlow, CM20 3EF [JO01BS, TL41]
GM3 RKO N J McIntosh, Riverside Cottage, Kirkton Pl, Kingoldrum, Kirriemuir, Angus, DD8 5HW [IO86LQ, NO35]
G3 RKP D A N Evetts, 35 Wood End Rd, Kempston, Bedford, MK43 9BB [IO92RC, TL04]
G3 RKQ A J Balmforth, 2 Hemper Gr, Sheffield, S8 7FF [IO93GH, SK38]
GW3 RKV R M Volck, Maes y Bryn, Rosebush, Maenclochog, Dyfed West Wales, SA66 7QS [IO71OW, SN02]
G3 RKX D M Thompson, 233 Ruskin Rd, Crewe, CW2 7JY [IO83SC, SJ75]
G3 RKZ B R Tibbert, 66 Horsley Rd, Kilburn, Belper, DE56 0NE [IO93GA, SK34]
G3 RLA C Phillips, Bella Vista, The Moorings, Heswall, Wirral, L60 9JT [IO83JI, SJ28]
G3 RLB D A Powell, 25 Cumberwall Dr, Enderby, Leicester, L49 5LB [IO83LI, SJ28]
G3 RLD R A Ramshaw, 132 Main Rd, Duston, Northampton, NN5 6RA [IO92MF, SP76]
G3 RLE B H Turner, 56 Bamford Way, Rochdale, OL11 5NB [IO83VO, SD81]
G3 RLF D J W Price, 8 Newland Rd, Droitwich, WR9 7AF [IO82WG, SO86]
G3 RLH R C Mathews, 4 Brazenhill Ln, Haughton, Stafford, ST18 9HS [IO82VS, SJ82]
G3 RLJ J E Harper, 12 Henley Cres, Westcliff on Sea, SS0 0NT [JO01IN, TQ88]
G3 RLL D W Grindell, 23 Park Hall Ave, Walton, Chesterfield, S42 7LR [IO93GF, SK36]
G3 RLN D J K Chapman, 11 Eagle Way, Abbeydale, Gloucester, GL4 4WS [IO81VU, SO81]
G3 RLO R A Davis, 39 Boxley Dr, West Bridgford, Nottingham, NG2 7GQ [IO92KV, SK53]
G3 RLS R Dale, 69 Glebe Ave, Pinxton, Nottingham, NG16 6HR [IO93IC, SK45]
G3 RLV M S Vann, 3 Mile Planting, Richmond, DL10 5DB [IO94DK, NZ10]
G3 RLW D G Crutchley, 23 Orchard Way, Cogenhoe, Northampton, NN7 1LZ [IO92OF, SP86]
G3 RLY N E Rock, 56 Tansey Green Rd, Pensnett, Brierley Hill, DY5 4TE [IO82WM, SO98]
G3 RMA J P Olway, 22 Langridge Rd, Paignton, TQ3 3PT [IO80EK, SX86]
G3 RMC J O Minks, 6 Daines Way, Thorpe Bay, Southend on Sea, SS1 3PF [JO01JM, TQ98]
G3 RMD F Regan, 7 Hilltop Rd, Cheltenham, GL50 4NW [IO81XV, SO92]
GW3 RME Details withheld at licensee's request by SSL.
G3 RMF B N R Magill, Yew Tree Cottage, Winnall Common, Allensmore, Herefordshire, HR2 9BS [IO82OA, SO43]
GW3 RMJ P L Jennings, Cwmfrain, Llanbister Rd, Llandrindod Wells, Powys, LD1 6UE [IO82JI, SO17]
G3 RMK R R Ruaux, Park View, Wallage Ln, Rowfant, Crawley, RH10 4NG [IO91WC, TQ33]
G3 RML D W Trowell, 10 Homestead Ln, Welwyn Garden City, AL7 4LU [IO91VS, TL21]
G3 RMN M J Smith, 121 Shirley Way, Shirley, Croydon, CR0 8PN [IO91XI, TQ36]
G3 RMQ J T Ingham, 6 Prospect St., Rawdon, Leeds, LS19 6DP [IO93DU, SE23]
G3 RMS Details withheld at licensee's request by SSL.
G3 RMW A K Sumesar-Rai, 16 Coppice Green, Bracknell, RG42 1TL [IO91OK, SU87]
G3 RMX W H Hall, 52 Barley Gate, Leven, Beverley, HU17 5NU [IO93UV, TA14]
G3 RMY J M Andrews, 12 Gerald Cl, Burgess Hill, RH15 0NB [IO90WW, TQ31]
G3 RMZ A S Pink, 37 Shute Park Rd, Plymstock, Plymouth, PL9 8RB [IO70XI, SX55]
G3 RNB S J Saines, 2 The Almshouses, Woolston, Williton, Taunton, TA4 4LN [IO81ID, ST03]
GW3 RNC J A Brown, Chantelle, 1 Bryony Cl, Twynyrodyn, Merthyr Tydfil, CF47 0LL [IO81HR, SO00]
GW3 RNE Details withheld at licensee's request by SSL.
GW3 RNH K T Larcombe Pontypool Amateur Radio Societ, Tudor Bungalow, Osborne Rd, Pontypool, NP4 6NR [IO81LR, SO20]
G3 RNM F W Pallant, Calves Croft, Nightingale Ln, Storrington, Pulborough, RH20 4NU [IO90SW, TQ01]
G3 RNN L K E Pleass, 7 Dashwood Cl, Slough, SL3 7NB [IO91RL, SU97]
GI3 RNO P P Greenan, 9 Ashville Park, Antrim, BT41 1HH [IO64VR, J18]
G3 RNP J D Price, 125 Oakfield Rd, Malvern, WR14 1DT [IO82UD, SO74]
G3 RNR G F Williams, Carniola, Whins Ln, Simonstone, Burnley, BB12 7QT [IO83TT, SD73]
G3 RNT D Broadbridge, 9 Burnt Acre, Chelford, Macclesfield, SK11 9SS [IO83UG, SJ87]
G3 RNV C A Galloway, 105 Dumbarton Rd, South Reddish, Stockport, SK5 7EX [IO83WK, SJ89]
G3 RNX W I B Walker, 44 South Rd, Weston Super Mare, BS23 2HE [IO81MI, ST36]
GM3 RNZ J Cowie, 12 Crawford Pl, Ladybank, Cupar, KY15 7NX [IO86KG, NO21]
G3 ROC R E Collins, Thorn Acacia, Northiam, Rye, E Sussex, TN31 6NJ [JO00HX, TQ82]
G3 ROD R Davenport, 7 Nether Cl, Duffield, Belper, DE56 4DR [IO92GX, SK34]
G3 ROG G J Morgan, 1 St. Johns Mead, Beggars Ln, Winchester, SO23 0HE [IO91IB, SU42]
G3 ROI Details withheld at licensee's request by SSL.
G3 ROK L A J Hull, 35 Oakmere Ln, Potters Bar, EN6 5LT [IO91VQ, TL20]
G3 ROM B V Sweetman, 25 Keyham Cl, Leicester, LE5 1FW [IO92LP, SK60]
G3 ROO I H Keyser, Rosemount, Church Whitfield, Dover, Kent, CT16 3HZ [JO01PD, TR34]
G3 ROQ R A Gill, 45 Biggin Ln, Ramsey, Huntingdon, PE17 1NB [IO92WK, TL28]
G3 ROS H Williams, Roslyn, Whalley Rd, Simonstone, Burnley, BB12 7HT [IO83TT, SD73]
G3 ROV Details withheld at licensee's request by SSL.
G3 ROW S W G C Wheaton Smith, 1828 Palmcroft Dr NE, Phoenix, Az 85007-1732, X X
G3 ROZ S Dyke, 13 Abbey Gr, Sandy, SG19 1QP [IO92UD, TL14]
G3 RPA J D Knowles, Springhill, Gilpins Ride, Berkhamstead, Herts, HP4 2PD [IO91RS, TL00]
G3 RPB K Spicer, Gr Cottage, Dallinghoo, Woodbridge, Suffolk, IP13 0LR [JO02PD, TM25]
G3 RPC M Goodey, Hilbre, Telegraph, St. Marys, Isles of Scilly, TR21 0NS [IN69UW, SV91]
G3 RPD G Clinch, Brook Cottage, South Brook Ln, Bovey Tracey, Devon, TQ13 9NB [IO80DO, SX87]
G3 RPI Details withheld at licensee's request by SSL.
G3 RPL T A Neyland, 22 Pax Hill, Hillfields, Bedford, MK41 8BT [IO92SD, TL05]
GM3 RPM J McAvoy, 120 Donaldswood Rd, Glenburn, Paisley, PA2 8EB [IO75ST, NS46]
G3 RPO F Seddon, 23 Countessway, Euxton, Chorley, PR7 6PT [IO83QP, SD51]
G3 RPQ C W Halling, 39 Overbrook Dr, Cheltenham, GL52 3HR [IO81XV, SO92]
G3 RPU Details withheld at licensee's request by SSL.
G3 RPV T J Venn, 22 Eaton Cl, Hartford, Huntingdon, PE18 7SR [IO92WI, TL27]
G3 RPY J Newman, 31 Beeson St., Grimsby, DN31 2QH [IO93WN, TA21]
G3 RPZ H J F Trunley, Bijou House, 75 Belgrave Rd, Eastwood, Leigh on Sea, SS9 5EL [JO01HN, TQ88]
G3 RQF D C Keith, 108 Lower Northern, Hedge End, Southampton, SO3 4FT [IO90HW, SU41]
G3 RQG Details withheld at licensee's request by SSL.[Op: D E Vaughan, High Beeches, Coppice Way, Haywards Heath, W Sussex, RH16 4NN.]
G3 RQJ P Woollett, How Green Farm, How Green Ln, Hever, Edenbridge, TN8 7NN [JO01BE, TQ44]
G3 RQP Dr D J Skyrme, 18 Gorselands, Newbury, RG14 6PX [IO91HJ, SU46]
GM3 RQQ H Robertson, 102 Orchy Cres, Bearsden, Glasgow, G61 1RE [IO75TV, NS57]
G3 RQR N L Kirtley, 14 Byron Ave, Winchester, SO22 5AT [IO91IB, SU42]
G3 RQS R A Rimmer, 15 Ellenborough Rd, Bishops Cleeve, Cheltenham, GL52 4AQ [IO81XW, SO92]
GW3 RQT Details withheld at licensee's request by SSL.
GI3 RQU Dr S J Laverty, 572 Antrim Rd, Belfast, BT15 5GL [IO74AP, J37]
G3 RQX P R Lewis, 20 Osborne Rd, Penn, Wolverhampton, WV4 4AY [IO82WN, SO89]
G3 RQY K D Brickham, 15 Tiercel Ave, Norwich, NR7 8JN [JO02PP, TG21]
G3 RQZ P M Madagan, 40 Lagham Park, South Godstone, Godstone, RH9 8ER [IO91XF, TQ34]
G3 RR Details withheld at licensee's request by SSL.[Correspondence to Rolls Royce ARC, c/o L G Logan, 19 Fenton Ave, Barnoldswick, Colne, Lancs BB8 6HB.]
G3 RRA I R Matheson, 17 The Ridings, Frimley, Camberley, GU16 5RA [IO91PH, SU85]
G3 RRD D W Marmont, Woodcot, St. Chloe, Amberley, Stroud, GL5 5AP [IO81VR, SO80]
G3 RRG P C Taylor, 44 Leegate Rd, Heaton Moor, Stockport, SK4 4AX [IO83VK, SJ89]
GW3 RRI R C Wilmot, Glyn Gwynedd, Rhostryfan, Caernarfon, Gwynedd, LL54 7PD [IO73VC, SH45]
G3 RRK A D W Atkins, 81 The Marles, Exmouth, EX8 4NU [IO80HP, SY08]
G3 RRL W S Teanby, Elmwood, Main Rd, Ealand, Scunthorpe, DN17 4JG [IO93OO, SE71]
G3 RRM J Hughes, 41 Highfield Ave, Great Sankey, Warrington, WA5 2TW [IO83QJ, SJ58]
G3 RRN Dr K E Jones, Field House, Wragby Rd, Sudbrooke, Lincoln, LN2 2QU [IO93SG, TF07]
G3 RRP R C Pine, 29 Sunnydale Gdns, London, NW7 3PD [IO91UO, TQ29]
G3 RRS R C Church R.A Lab A.R.S, Rutherford Appleton, Recreation Society, Chilton, Didcot, Oxon, OX11 0QX [IO91IN, SU48]
G3 RRY J A Russell, 95 Kingsbridge Rd, Morden, SM4 4PU [IO91VJ, TQ26]
G3 RSB R Scaife, 7 Woodgates Cl, North Ferriby, HU14 3JS [IO93RR, SE92]
G3 RSC J E Symes Sutton Coldfiel, Rs.20 Plantsbrook Rd, Walmley, Sutton Coldfield, B76 8EX [IO92CN, SP19]
G3 RSD J R Reynolds, 6 Fairfield Ct, Cleethorpes, DN35 0QW [IO93XM, TA30]
G3 RSE C J Cheney, 35 Metcalfe Rd, Cambridge, CB4 2DB [JO02BF, TL46]
G3 RSF A F Notschild, 46 Cunningham Dr, Lutterworth, LE17 4YR [IO92JK, SP58]
G3 RSI F M A J McKeracher, Highland Lodge, Church Ln, Ewshot, Farnham, GU10 5BD [IO91OF, SU84]
G3 RSJ Details withheld at licensee's request by SSL.[Station located in West Wiltshire.]
G3 RSK C Stringer, 11 Hawthorn Rd, Bishopsmead, Tavistock, PL19 9DL [IO70WM, SX47]
G3 RSM F S C Burnett, 4 Woodlands Dr, Fulwood, Preston, PR2 9SQ [IO83PT, SD53]
G3 RSP A J Pampling, 47 Altham Gr, Harlow, CM20 2PQ [JO01BS, TL41]
G3 RSS N L Hodgson Scout Rad Grp, 42 Tofts Gr, Rastrick, Brighouse, HD6 3NP [IO93CQ, SE12]
G3 RST R V Southern, 30 Barnfield, Crowborough, TN6 2RY [JO01CB, TQ53]
G3 RSU D P A Bindon, Forth House, Water St., Curry Rivel, Langport, TA10 0HH [IO81NA, ST32]
G3 RSW W J Mullarkey, The Willows, 7 Harper St., Hindley, Wigan, WN2 3HL [IO83RM, SD60]
G3 RSX G M Preece, 46 Windsor Gdns, Castlecroft, Wolverhampton, WV3 8LY [IO82VN, SO89]
G3 RSY Details withheld at licensee's request by SSL.
GW3 RTA W J Lewis, 5 Galon Uchaf Rd, Merthyr Tydfil, CF47 9TP [IO81HS, SO00]
G3 RTB Details withheld at licensee's request by SSL.
G3 RTE G J Kellaway, 110 Mimms Hall Rd, Potters Bar, EN6 3DY [IO91VQ, TL20]
GM3 RTJ G R Henderson, Tigh An Drochaid, Kilchrenan, By Taynuilt, Argyll, Scotland, PA35 1HD [IO76JJ, NN02]
G3 RTO N W F Pratt, 3 Grey St., Newthorpe, Nottingham, NG16 2EF [IO93IA, SK44]
G3 RTP J Pennington, Brambling, Forest Rd, Denmead, Waterlooville, PO7 6UE [IO90LV, SU61]
G3 RTR R T Rowse, Park Ln, Penyffordd, Chester, Ches, CH4 0HN [IO83LD, SJ36]
G3 RTU Details withheld at licensee's request by SSL.
G3 RTY H R Meers, 10 Lawnswood Ave, Chasetown, Walsall, West Midlands, WS7 8YD [IO92AQ, SK00]
GW3 RTZ A W Bennett, 43 Mill St., Usk, NP5 1AP [IO81NQ, SO30]
G3 RUB S O Hine, 7 Castleton Ave, Riddings, Alfreton, DE55 4AG [IO93HB, SK45]
G3 RUD E Workman, Sunset, 2 Burnham Dr, Bleadon Hill, Weston Super Mare, BS24 9LW [IO81MH, ST35]
GW3 RUE E R Edwards, Ceris, Ruthin Rd, Denbigh, LL16 3EU [IO83HE, SJ06]
G3 RUG G E Twiss, 9 Brae Head, Eaglescliffe, Stockton on Tees, TS16 9HP [IO94HM, NZ41]
G3 RUH J R Miller, 3 Bennys Way, Coton, Cambridge, CB3 7PS [JO02AE, TL45]
G3 RUJ R Powell, 1 Chepstow Park, Avonmead, Downend, Bristol, BS16 6SQ [IO81SM, ST67]
GW3 RUL Maj W Y Sturdy, 19 Cleggars Park, Lamphey, Pembroke, SA71 5NP [IO71NQ, SN00]
G3 RUO W Williamson, 84 Atfield Dr, Whetstone, Leicester, LE8 3NE [IO92KN, SP69]
GM3 RUP C R Morton, 295 Byres Rd, Hillhead, Glasgow, G12 8TL [IO75UV, NS56]
G3 RUQ M H Harris, 7 Church Rd, Abbots Leigh, Bristol, BS8 3QP [IO81QL, ST57]
G3 RUR A E Roberts, 62 Oak Way, Feltham, TW14 8AS [IO91SK, TQ07]
G3 RUV A T James, 37 Stratford Ave, Whipton, Exeter, EX4 8ES [IO80GR, SX99]
G3 RUX B V Marshall, 12 Park Ln, Pinhoe, Exeter, EX4 9HL [IO80GR, SX99]
G3 RUZ D H Martin, Castle Acre, Stockbridge Rd, Winchester, SO22 5JH [IO91HB, SU43]
G3 RVA R H Crowe, 37 Huccaby Cl, Brixham, TQ5 0RJ [IO80FJ, SX95]
G3 RVC Dr P Cochrane, 11 Forest Ln, Martlesham Heath, Ipswich, IP5 7ST [JO02OB, TM24]
G3 RVD I D Macey, 14 Hurst Park Rd, Twyford, Reading, RG10 0EY [IO91NL, SU77]
GW3 RVG S D C Sedgebeer, 50 Minffrwd Rd, Pencoed, Bridgend, CF35 6SD [IO81GM, SS98]
G3 RVI J M Walch, 52 Marsh House Rd, Ecclesall, Sheffield, S11 9SP [IO93FI, SK38]
GM3 RVL Dr H M Brash, 5 Hillview Dr, Edinburgh, EH12 8QW [IO85IW, NT17]
G3 RVM C I B Trusson, 27A Roman Way, Thatcham, RG18 3BP [IO91IJ, SU56]
G3 RVN F Jones, Mount Cottage, Flash Ln, Bollington, Macclesfield, SK10 5AQ [IO83WH, SJ97]
G3 RVP A A Mothersdale, 9 Millers Cl, Offord Darcy, St. Neots, Huntingdon, PE18 9SB [IO92VG, TL26]
GD3 RVQ D R Jackson, Lapithos, Ballaragh, Laxey, Isle of Man, IM4 7PL
G3 RVS Details withheld at licensee's request by SSL.
GW3 RVT T G Makinson, 69 Ffordd Derwen, Rhyl, LL18 2LU [IO83GH, SJ07]
G3 RVX J Colegate, 1 Oldmere Cttgs, High St., Bathford, Bath, BA1 7TJ [IO81UJ, ST76]
G3 RVY P Colegate, 65 Forest Rd, Melksham, SN12 7AB [IO81WJ, ST96]
G3 RWC J Harris, 5 Bradford Cl, The Grange, Comeytrowe, Taunton, TA1 4YH [IO81KA, ST22]
G3 RWE T B Yates, 3 Sycamore Cres, Macclesfield, SK11 8LL [IO83WG, SJ97]
G3 RWF P N Henwood, Conifers House, Church Rd, Littlebourne, Canterbury, CT3 1AU [JO01OG, TR25]
G3 RWG R W Gibson, 12 Park Rd, Chandlers Ford, Eastleigh, SO53 2EU [IO90HX, SU42]
G3 RWI Dr P H Cross, Home Farm House, Icomb, Stow-on-the-Wold, Glos, GL54 1JD [IO91DV, SP22]
G3 RWL R W L Limebear, 60 Willow Rd, Enfield, EN1 3NQ [IO91XP, TQ39]
G3 RWN R Harrod, 3 Glen Cl, Newton, Alfreton, DE55 5TU [IO93HD, SK45]
G3 RWO R L White, 7 Mill Cres, Kingsbury, Tamworth, B78 2LX [IO92DN, SP29]
GM3 RWP C D Bell East Midlands Communications, 18 Castle Ct, Stirling, FK8 1EL [IO86AC, NS79]
G3 RWR M C Cox, Scurf Barn, Bugbrooke, Northampton, NN7 3QA [IO92ME, SP65]
G3 RWU R N Francis, 4 Essex St, Te Atatu South, Auckland, New Zealand, X X
G3 RWV M A Sanders, 7 Netherby Cl, Gr Rd, Tring, HP23 5PJ [IO91QT, SP91]
G3 RWW G D Southern, 27 Eldred Rd, Liverpool, L16 8NZ [IO83NJ, SJ48]
GW3 RWX D M Thomas, 88 Cefn Graig, Rhiwbina, Cardiff, CF4 6JZ [IO81JM, ST18]
GW3 RXD G Llewelyn, Bayano, 33 Awelfryn, Amlwch, LL68 9DG [IO73TJ, SH49]
G3 RXF W J C Storeton-West, 45 Somerleyton Rd, Oulton, Lowestoft, NR32 4RB [JO02UL, TM59]

G3

G3 RXG R W G Burgess, 11 Beech Rd, Shipham, Winscombe, BS25 1SA [IO81OH, ST45]
G3 RXH H A Aspinall, Adare, Raikeswood Cres, Skipton, BD23 1ND [IO83XX, SD95]
G3 RXI E C Blundell, 29 Garden Cl, Hook, RG27 9QZ [IO91MG, SU75]
G3 RXL Details withheld at licensee's request by SSL.[Op: P L Avery. Station located near Fleet.]
G3 RXM J B Hope, 1 Drummond Cl, Bracknell, RG12 2QG [IO91PK, SU86]
G3 RXN C W Jacob, 21 Coquet Gr, Throckley, Newcastle upon Tyne, NE15 9JU [IO94CX, NZ16]
G3 RXP D J Mason, 5 Spa Top, Caistor, Market Rasen, LN7 6RB [IO93UL, TA10]
G3 RXQ S J Baker, 9 Swan Cl, Ivinghoe Aston, Leighton Buzzard, LU7 9DN [IO91QU, SP91]
G3 RXS W G Scarlett, 14 Warren Dr, Bingley, BD16 3BX [IO93CU, SE13]
GM3 RXU Prof I A Macpherson, Kirklands, Balcomie Rd, Crail, Fife, KY10 3XL [IO86QG, NO60]
GI3 RXV N M Graham, 47 Main St., Castle Dawson, Castledawson, Magherafelt, BT45 8AA [IO64RS, H99]
GM3 RXZ R K Marshall, Braehead, Balbardie Rd, Bathgate, EH48 1AP [IO85EV, NS96]
G3 RYC B R M Hanson, 1 Tower Rd, Portishead, Bristol, BS20 8RD [IO81OL, ST47]
GW3 RYE J Harris, 20 Penywern Rd, Ystalyfera, Swansea, SA9 2NJ [IO81CS, SN70]
G3 RYH J F Brodie, Moorhall Farm, Whitbourne, Worcester, WR6 5SF [IO82SE, SO75]
G3 RYK I J Grayson, 156 Little Brays, Harlow, CM18 6EY [JO01BS, TL40]
G3 RYN M S Barratt, 116 Evington Ln, Leicester, LE5 6DG [IO92KO, SK60]
G3 RYP D R Craggs, New House, Dacre Banks, Harrogate, North Yorks, HG3 4EW [IO94DB, SE16]
G3 RYQ J G Eason, 24 Cranborne Cres, Potters Bar, EN6 3AG [IO91VQ, TL20]
GW3 RYR Dr C Morgan, 33 West Gr, Merthyr Tydfil, CF47 8HJ [IO81HS, SO00]
G3 RYV P R Cox, Pippin Cottage, Church St., Moreton in Marsh, Glos, GL56 0LN [IO91DX, SP23]
G3 RYW Details withheld at licensee's request by SSL.
G3 RYX D McAndrew, 10 Burrill Dr, Wigginton, York, YO3 3ST [IO94KA, SE55]
G3 RYY N Penketh, Greenways Bolton Rd, Duxbury, Chorley, Lancs, PR7 4AJ [IO83QP, SD51]
G3 RYZ M G Byrne, 16 Downham Gdns, Tamerton Foliot, Plymouth, PL5 4QF [IO70WK, SX46]
G3 RZ Dr T A Appleby, Walnut Tree Bungalow, The St., Bodham, Holt, NR25 6NW [JO02OV, TG13]
G3 RZA Details withheld at licensee's request by SSL.
G3 RZC R D Pellett, Maple House, Marlpits Ln, Ninfield, Battle, TN33 9LD [JO00FV, TQ71]
G3 RZF D A Horton, 26 The Cres, Slough, SL1 2LQ [IO91QM, SU97]
G3 RZG M S Box, 18 Stottingway St., Upwey, Weymouth, DT3 5QA [IO80SP, SY68]
G3 RZI M P Moss, 1082 Evesham Rd, Astwood Bank, Redditch, B96 6ED [IO92AG, SP06]
G3 RZJ G F S Hall, 185 Dialstone Ln, Stockport, SK2 7LQ [IO83WJ, SJ98]
G3 RZM M F McDonald, PO Box 11155, Bloubergrant 7443, Capetown, Rsa
G3 RZP P E Chadwick, Three Oaks, Braydon, Swindon, Wilts, SN5 0AD [IO91BO, SU08]
G3 RZS N W Wooderson, 15 Cheney St., Pinner, HA5 2TF [IO91TO, TQ18]
G3 RZV A A Lawrance, 97 Dorchester Rd, Oakdale, Poole, BH15 3QZ [IO90AR, SZ09]
G3 RZX E Wilby, 21 Hall Cliffe Rd, Horbury, Wakefield, WF4 6BX [IO93FP, SE21]
G3 RZY C E Abrey, 77 Blackmore, Letchworth, SG6 2SZ [IO91VX, TL23]
G3 SAD A Wiltshire Stevenage & Dis, 81 Whomerley Rd, Stevenage, SG1 1SS [IO91VV, TL22]
GM3 SAE R McMillan, 54 Birchwood, Invergordon, Ross-shire, IV18 0BG [IO77VQ, NH66]
G3 SAF Details withheld at licensee's request by SSL.
G3 SAH R J Matthews, 354 Birchfield Rd, Webheath, Redditch, B97 4NQ [IO92AH, SP06]
GM3 SAN S Weir, 19 Ellismuir Rd, Baillieston, Glasgow, G69 7HW [IO75WU, NS66]
G3 SAO Details withheld at licensee's request by SSL.
G3 SAQ Rev A Coles, 6 Eden Park, Lancaster, LA1 4SJ [IO84OA, SD45]
G3 SAR R M Warner, Little Foxes, Rycroft Ln, Sevenoaks, TN14 6HT [JO01CF, TQ55]
G3 SAZ J Barker, 10 Frenchay Rd, Weston Super Mare, BS23 4JL [IO81MH, ST35]
GW3 SB T C Bryant, Maes y Crynwyr, Llwyngwril, LL37 2JQ [IO72XQ, SH51]
G3 SBA R C Marshall, The Dappled House, 30 Ox Ln, Harpenden, AL5 4HE [IO91TT, TL11]
GM3 SBC E T J Murphy, 65 Silverknowes Cres, Edinburgh, EH4 5JA [IO85IX, NT27]
GM3 SBE J C Sinclair, Am Fasgadh, Rathad Na Roinne, Port Charlotte, Islay, PA48 7TF [IO65TR, NR25]
G3 SBF B V Eames, 46 Ratby Ln, Markfield, LE67 9RJ [IO92IQ, SK40]
G3 SBI C W Horrabin, 17 Denbury Ave, Stockton Heath, Warrington, WA4 2BL [IO83RJ, SJ68]
G3 SBJ D J Millard, 20 Green Walk, Northgate, Crawley, RH10 2HX [IO91VC, TQ23]
G3 SBL P J Hindle GEC Measurement, Amateur Radio Soc, St Leonards Works, Stafford, ST17 4LX [IO82WT, SJ92]
G3 SBM D G Turner, 50 Hardings, Chalgrove, Oxford, OX44 7TJ [IO91LP, SU69]
G3 SBP R D Gynn, Tremonde, Kelly, Lifton, Devon, PL16 0HH [IO70UO, SX38]
G3 SBR M J Smith, 32 Albert Rd, Erdington, Birmingham, B23 7LT [IO92BM, SP09]
G3 SBU M J Reeds, 22 Elms Cross Dr, Bradford on Avon, BA15 2EL [IO81UI, ST86]
G3 SBV H L W Bellfield, The Little House, Stone St, Lympne, nr Hythe, Kent, CT21 4JP [JO01MB, TR13]
G3 SBW M S Chambers, 62 Whinney Ln, New Ollerton, Newark, NG22 9TH [IO93ME, SK66]
G3 SCD D B Dunn, Millstones, Scamblesby, Louth, Lincs, LN11 9XG [IO93XH, TF27]
G3 SCE K G A Gair, 30 Farrance Rd, Chadwell Heath, Romford, RM6 6EB [JO01BN, TQ48]
G3 SCH D J Becket, 65 Grenville Ave, Chelston, Torquay, TQ2 6DS [IO80FL, SX86]
G3 SCJ D W Power, Chapel Green, Fillongley, Coventry, CV7 8DX [IO92EL, SP28]
G3 SCL R G Houghton, 7 Bramshill Mansion, Dartmouth Park Hill, London, NW5 IJG
GI3 SCM T McCullough, 16 McCormack Gdns, Lurgan, Craigavon, BT66 8LE [IO64TK, J05]
G3 SCT S G Lydiate obo Thurrock Sea Cadet Corps, 11 Avondale Rd, Pitsea, Basildon, SS16 4TT [JO01FN, TQ78]
G3 SCV Rev G A Stanton, 8 Kennett Cl, Norwich, NR4 7JA [JO02PO, TG20]
GW3 SCX J Naylor, 32 Graig y Tewgoed, Cwmavon, Port Talbot, SA12 9YE [IO81CO, SS79]
G3 SCY C Seldon, 97 Gunners Rd, Shoeburyness, Southend on Sea, SS3 9SB [JO01JM, TQ98]
G3 SCZ R A Brown, 22 Lordswood, Silchester, Reading, RG7 2PZ [IO91KI, SU66]
G3 SDC R G Titterington de Montfort Uni Leicester Amat, de Montfort University Ars, Dept of Electronic Engineering, de Montfort University, The Gateway, Leicester, LE1 9BH [IO92KP, SK50]
G3 SDG J J Bottom, 48 Chesterton Ave, Harpenden, AL5 5SU [IO91TT, TL11]
G3 SDH P D Kelly, Martyndale, Main St., Compton Martin, Bristol, BS18 6JE [IO81QH, ST55]
GW3 SDK M R Kidman, 53 Nant y Glyn, Llanrug, Caernarvon, Gwynedd, LL55 4AH [IO73VD, SH56]
G3 SDL D I Court, 98 Andover Rd, Orpington, BR6 8BN [JO01BJ, TQ46]
G3 SDO K A Heathfield, 2 Georgian Cl, Broadway, Weymouth, DT3 5PF [IO80SP, SY68]
G3 SDQ Details withheld at licensee's request by SSL.
G3 SDS G R Watts South Dorset RS, obo South Dorset Radio Society, 46 Links Rd, Weymouth, DT4 0PE [IO80SO, SY67]
G3 SDT D Allen, Chelsea Cttgs, The Turnpike, Carleton Rode, Norwich, NR16 1RS [JO02NL, TM19]
G3 SDW K J Underwood, 44 Burleigh Rd, Shiphay, Torquay, TQ2 6JX [IO80FL, SX86]
G3 SDY G Edinburgh, 77 Westerley Ln, Shelley, Huddersfield, HD8 8HP [IO93DO, SE21]
G3 SED E M Devereux, 191 Botley Rd, Burridge, Southampton, SO31 1BJ [IO90IV, SU50]
G3 SEF R I Frew, Sawley House, 82 Wormholt Rd, London, W12 0LP [IO91VM, TQ28]
G3 SEG W A Gordon, 55 Trajan Ave, South Shields, NE33 2AN [IO95GA, NZ36]
G3 SEJ E C John, 52 Broadway Ave, Wallasey, L45 6TD [IO83LK, SJ39]
G3 SEK Dr I F White, 52 Abingdon Rd, Drayton, Abingdon, OX14 4HP [IO91IP, SU49]
G3 SEL F Powell, New Tregenver, Off Venton Rd, Falmouth, Cornwall, TR11 4JY [IO70KD, SW73]
G3 SEM P J Cort-Wright, 32 Brian Ave, Norwich, NR1 2PH [JO02PO, TG20]
G3 SEN R Dawes, Sarnia, 18 Sutherland Rd, Nottingham, NG3 7AP [IO92KX, SK54]
G3 SEP D L Buddery, 18 Upper Gordon Rd, Camberley, GU15 2HN [IO91PI, SU86]
G3 SEQ J E Crossfield, Forest Lodge, Chopwell Wood, Rowlands Gill, Tyne & Wear, NE39 1LT [IO94CV, NZ15]
GM3 SER H J Bremner, 17 Heriot Rd, Lenzie, Kirkintilloch, Glasgow, G66 5AX [IO75WW, NS67]
G3 SES P L Stevens, 20 Abbots Park, Chester, CH1 4AN [IO83NE, SJ46]
G3 SET G D Aram, 31 North Cl, Medmenham, Marlow, SL7 2EL [IO91ON, SU88]
G3 SEY R Mackey, The Tudor, 44 South St., Ossett, WF5 8LF [IO93FQ, SE21]
G3 SEZ P M A Luft, 37 Henwood Rd, Compton, Wolverhampton, WV6 8PQ [IO82VO, SO89]
G3 SFA T A Plant, 59 Eastbrook Dr, Romford, RM7 0YT [JO01CN, TQ58]
G3 SFB C T Hale, 16 Windmill Ct, Northern Cres, East Wittering, Chichester, PO20 8RJ [IO90NS, SZ79]
GW3 SFC A Richards, Hill Crest, 30 Well Pl, Cwmbach, Aberdare, CF44 0PB [IO81GQ, SO00]
G3 SFE P C P Everett, 339 Chichester Rd, North Bersted, Bognor Regis, PO21 5AN [IO90PT, SU90]
G3 SFG D F Berry Southgate ARC, 4 Holly Hill, Winchmore Hill, London, N21 1NP [IO91WP, TQ39]
GM3 SFH A J Oliphant, 11, Bridge of Westfield, Thurso, Caithness, KW14 7QN [IO88EN, ND06]
G3 SFK P L Kerry, 251 Upper Rainham Rd, Hornchurch, RM12 4EY [JO01CN, TQ58]
G3 SFL F Harrison, 42 Woodlands Rd, Cleadon, Sunderland, SR6 7UD [IO94HW, NZ36]
G3 SFM A H W Bailey, 12 Nelson Rd, Rainham, RM13 8AL [JO01CM, TQ58]
G3 SFO R H Jones, 140 Station Rd, Bawtry, Doncaster, DN10 6QD [IO93LK, SK69]
G3 SFP R H Troughton, 4 Owletts, Crabbett Park, Worth, Crawley, RH10 7SQ [IO91WC, TQ23]
GW3 SFQ R M Mugford, 27 Highfield Cl, Dinas Powys, CF64 4LR [IO81JK, ST17]
G3 SFT E J Bailey, 100 Throne Rd, Rowley Regis, Warley, B65 9JX [IO82XL, SO98]
G3 SFU P Woodfield, 49 Oakfield Rd, Hawley, Blackwater, Camberley, GU17 9DZ [IO91OH, SU85]
G3 SFV E E Meachen, 46 Rainsborough Gdns, Market Harborough, LE16 9LW [IO92ML, SP78]
G3 SFX Dr S F Pugh, 10 Coniger Rd, London, SW6 3TA [IO91VL, TQ27]
G3 SFY D W Shaw, 7 Yew Tree Ln, Slaithwaite, Huddersfield, HD7 5HU [IO93BO, SE01]
G3 SFZ J Carter, 224 Cavendish Ave, Ealing, London, W13 0JW [IO91UM, TQ18]
G3 SGA A D Jones, 45 Rowan Dr, Highcliffe, Christchurch, BH23 4QR [IO90DR, SZ19]
G3 SGB E W Paddock, 30 West St., Brant Broughton, Lincoln, LN5 0SF [IO93QB, SK95]
G3 SGC G W Morris, 35 Abbots Rise, Kings Langley, WD4 8AR [IO91SR, TL00]
G3 SGF P J Casemore, 9 Wellcroft Cttgs, Church Ln, Albourne, Hassocks, BN6 9BZ [IO90VW, TQ21]
G3 SGJ K D Smith, 58 Cemetery Rd, Ipswich, IP4 2HZ [JO02NB, TM14]
G3 SGK Dr B R H King, Watbridge Farm Hous, Ashendon, Aylesbury, Bucks, HP18 0HA [IO91MT, SP71]
G3 SGL A M Isaacs, Holme View, Brick Ln, Bransgore, Christchurch, BH23 8DU [IO90DT, SZ29]
G3 SGM R D Rose, 170 Winsley Rd, Bradford on Avon, BA15 1NZ [IO81UI, ST86]
G3 SGN C E Stone, 11 Liphook Cres, Forest Hill, London, SE23 3BN [IO91XK, TQ37]
G3 SGQ . R C Hill Esq B.E.M., 47 Eastway, Maghull, Liverpool, L31 6BS [IO83MM, SD30]
G3 SGR Dr J S J Craig, Partridge Cottage, Redpale Dallington, Heathfield, E Sussex, TN21 9NR [JO00EW, TQ61]
G3 SGS J H Clements, 147 Luton Rd, Dunstable, LU5 4LP [IO91SV, TL02]
G3 SGT A P Teale, 53 Hartland Dr, South Ruislip, Ruislip, HA4 0TH [IO91TN, TQ18]
G3 SGV J P Fallon, 16 College View, Mutley, Plymouth, PL3 4JB [IO70WJ, SX45]
G3 SGX R E Bona, 54 Syke Ings, Richings Park, Iver, SL0 9EU [IO91RM, TQ07]
G3 SGY A R Nesbitt, 43 Oaktree Cl, Middleton St. George, Darlington, DL2 1HJ [IO94GM, NZ31]
G3 SGZ T A M Chapple, 39 Maynards Park, Bere Alston, Yelverton, PL20 7AR [IO70VL, SX46]
G3 SHD L R Dray, 1 Chalfont Cl, Bradville, Milton Keynes, MK13 7HS [IO92OB, SP84]
G3 SHF B Naylor, 47 Chester Rd, Poynton, Stockport, SK12 1HA [IO83WI, SJ98]
GI3 SHI S H McKaig, 52 Beverley Gdns, Bangor, BT20 4NQ [IO74EP, J58]
G3 SHJ D H Heald, 20 Moor Field, Whalley, Clitheroe, BB7 9SA [IO83TT, SD73]
G3 SHK R G Pett, 5 Kingford Cl, Woodfalls, Salisbury, SP5 2NQ [IO90DX, SU12]
G3 SHL J Harlow, 7 The Harridge, Rochdale, OL12 7UX [IO83VP, SD81]
GM3 SHR J H Coster, 17 Glamis Pl, The Green, Dalgety Bay, Dunfermline, KY11 5UA [IO86HA, NT18]
G3 SHS R W Perrin, 297 Mutton Ln, Potters Bar, EN6 2AT [IO91VQ, TL20]
G3 SHU H D Critchlow, 63 Gattison Ln, Rossington, New Rossington, Doncaster, DN11 0NH [IO93LL, SK69]
G3 SHW W Shaw, 131 Woodsmoor Ln, Davenport, Stockport, SK3 8TJ [IO83WJ, SJ98]
G3 SHX R G West, 45 Friends Ave, Margate, CT9 3XE [JO01QJ, TR37]
G3 SHY R C Cottrell, Roan Cottage, 157 Ridge Ln, Watford, WD1 3SU [IO91SQ, TQ09]
G3 SHZ Dr J R Whittington, Twyford Manor, Twyford, Buckingham, Bucks, MK18 4EL [IO91LW, SP62]
G3 SIA B H Keyte, 9 Swanns Meadow, Bookham, Leatherhead, KT23 4JX [IO91TG, TQ15]
G3 SIB J Reed, 53 Malvern Rd, Balsall Common, Coventry, CV7 7DU [IO92EJ, SP27]
G3 SID M P Fox, 31 Pierson Rd, Windsor, SL4 5RE [IO91QL, SU97]
G3 SIH R F C Bennett, 14 Southleigh, Bradford on Avon, BA15 2EQ [IO81UI, ST86]
G3 SIK K R Pugh, Pipers Ash, 34 Highland Rd, Northwood, HA6 1JT [IO91TO, TQ19]
G3 SIO J E M Beardsmore, 67 Beachwood Ave, Kingswinford, DY6 0HL [IO82VM, SO89]
G3 SIP R A Merriman, Chantry, Edlington, Horncastle, Lincs, LN9 5RJ [IO93VF, TF27]
G3 SIQ A J Greenwood, 83 Ash Rd, Cuddington, Northwich, CW8 2PB [IO83QF, SJ57]
G3 SIR D R Durham, 29 Waverley Rd, Stratton St. Margaret, Swindon, SN3 4AY [IO91DN, SU18]
G3 SIT R I Kressman, 12 School Ln, Fenstanton, Huntingdon, PE18 9JR [IO92XH, TL36]
G3 SIU P Hearson, 14 Osgood Gdns, Orpington, BR6 6JU [JO01BI, TQ46]
G3 SIV D Amer, 77 Woodland Ave, Overstone, Northampton, NN6 0AH [IO92OG, SP86]
GW3 SIY G R Steele, 5 Golden Cl, West Cross, Swansea, SA3 5PE [IO71XO, SS68]
G3 SJD G Drake, Ridgeways, Grimescar Rd, Fixby, Huddersfield, HD2 2EB [IO93CQ, SE11]
G3 SJE J W Bluff, 52 Winchester Rd, Kenton, Harrow, HA3 9PE [IO91UO, TQ18]
G3 SJG Details withheld at licensee's request by SSL.
G3 SJH Dr C J Eyles, 9 St. Peters Rd, Harborne, Birmingham, B17 0AT [IO92AK, SP08]
G3 SJI M S Batt, 9 Grange Park, Henleaze, Westbury on Trym, Bristol, BS9 4BU [IO81QL, ST57]
G3 SJJ J C Burbanks, 16 Cotgrave Rd, Plumtree, Nottingham, NG12 5NX [IO92LV, SK63]
G3 SJK S M Cherry, Aquila, 4 Westhill Rd South, South Wonston, Winchester, SO21 3HP [IO91IC, SU43]
G3 SJM Details withheld at licensee's request by SSL.
G3 SJR W Tynan, 91 Westhorpe Rd, Gosberton, Spalding, PE11 4EN [IO92VU, TF23]
G3 SJV D J Viney, 117 St. Philips Ave, Eastbourne, BN22 8NA [JO00DS, TQ60]
G3 SJW S C Haigh, 8 Speldhurst Rd, Hackney, London, E9 7EH [IO91XM, TQ38]
G3 SJX P J Hart, The Willows, Paice Ln, Medstead, Alton, GU34 5PR [IO91LC, SU63]
GM3 SJY C M Lawrenson, Hollyburn, West Port, Falkland, Cupar, KY1 7BW [IO86KD, NT29]
G3 SKD N A Smith, Sunningdale, Green End Rd, Radnage, High Wycombe, HP14 4BY [IO91NP, SU79]
G3 SKF T C M Wigg, Setters, Hyde, Fordingbridge, Hants, SP6 2QB [IO90CV, SU11]
GD3 SKH C C Black, 33 Claughbane Dr, Ramsey, IM8 2BH
G3 SKI R A Bravery, 7 Copse Hill, Withdean, Brighton, BN1 5GA [IO90WU, TQ20]
G3 SKK W D James, 34 Nicholson Dr, Beccles, NR34 9UX [JO02SK, TM48]
G3 SKN D Naylor, Valleyview, Dale of Walls, Shetland, ZE2 9PE [IP90EG, HU15]
GW3 SKP Details withheld at licensee's request by SSL.
G3 SKR A R Gold, The Chalet, Aston Hill, Halton, Aylesbury, HP22 5NQ [IO91PS, SP81]
G3 SKT P K Cripps, 1 Gregory Cl, Harlaxton, Grantham, NG32 1JG [IO92PV, SK83]
G3 SKV S J Hobday, 31 Sackville Cres, Harold Wood, Romford, RM3 0EJ [JO01CO, TQ59]
G3 SKY A Ash low Rad Soc, 34 Coronation Ave, Cowes, PO31 8PN [IO90IR, SZ49]
GD3 SKZ K G Manktelow, Tramman House, Ballabeg, Isle of Man, IM9 4HA
G3 SLD R P Daish, 18 Underdown Ave, Waterlooville, PO7 5DH [IO90LU, SU60]
GI3 SLE Details withheld at licensee's request by SSL.
G3 SLG R E Simpson, The Brackens, Sycamore Rd, Harrow Hill, Drybrook, GL17 9JZ [IO81RU, SO61]
G3 SLI A C Osborne, 29 Carlton Rd, Caversham Heights, Caversham, Reading, RG4 7NT [IO91ML, SU67]
G3 SLJ D W Parsons, Bergweg 5, 83627 Osterwarngau, Germany
G3 SLK R Pickering, 69 School Hill, Chapel End, Nuneaton, CV10 0NF [IO92FM, SP39]
G3 SLL H R Tyreman, 37 Lavinia St, Seven Hills, Nsw 2147, Australia, ZZ3 7LA
G3 SLN F G Sawyer, 3 Addison Dr, Middleton, Manchester, M24 2PL [IO83VN, SD80]
G3 SLR Details withheld at licensee's request by SSL.
G3 SLT D R Ormerod, Writtle Lodge, 21 Valletta Cl, Chelmsford, CM1 2PT [JO01FR, TL70]
G3 SLU V Stimpson, 67 Alliance Ave, Hull, HU3 6QU [IO93TR, TA02]
G3 SLX J Q Smith, 256 Stone Rd, Hanford, Stoke on Trent, ST4 8NJ [IO82VW, SJ83]
G3 SLZ M E Butcher, Omega, Bodrugan Hill, Mevagissey, St. Austell, PL26 6PS [IO70OG, SX04]
G3 SMD R L Turner, 7 Paddocks Ln, Cheltenham, GL50 4NU [IO81XV, SO92]
G3 SMF I B Hamill, 74 Lampits Hill, Corringham, Stanford-le-Hope, SS17 9AJ [JO01FM, TQ78]
G3 SMH Details withheld at licensee's request by SSL.
G3 SMK G E Eaton, 288 Norton Ln, Earlswood, Solihull, B94 5LP [IO92BI, SP17]
G3 SMM W M Furness, 32 Westmorland Rd, Sale, Ches, M33 3QU [IO83UJ, SJ79]
G3 SMP Details withheld at licensee's request by SSL.
G3 SMR R J Tinkler, 11 Westhill Dr, Kettering, NN15 7LG [IO92PJ, SP87]
G3 SMT P G Torry, 26 Moss Ln, Bramhall, Stockport, SK7 1EH [IO83VI, SJ88]
G3 SMU B J Pugh, 1608 Scant Row, Chorley Old Rd, Horwich, Bolton, BL6 6PZ [IO83RO, SD61]
G3 SMV J E Smith, 18 Hounslow Rd, Mackworth Est, Derby, DE22 4BW [IO92FW, SK33]
G3 SMW B J C Spencer, 10 Jarry Ct, Gunthorpe Rd, Marlow, SL7 1UJ [IO91ON, SU88]
G3 SMX Details withheld at licensee's request by SSL.
GW3 SMY R Livsey, 39 Brompton Park, Rhos on Sea, Colwyn Bay, LL28 4TW [IO83DH, SH88]
G3 SMZ R T A Hill, 68 Chestnut St., Chadderton, Oldham, OL9 8HH [IO83WM, SD90]
G3 SN R P Ellis, 12 Hillside Rd, Saltash, PL12 6EX [IO70VJ, SX45]
G3 SNA S J Andrew, 6A Clydesdale Rise, Diggle, Oldham, OL3 5PX [IO93AN, SE00]
GJ3 SND B J Walster, -le-Ponterrin Cottage, La Rue Du Ponterrin, St. Saviour, Jersey, JE2 7HP
G3 SNE A R King, 117 Torrington St., Grimsby, DN32 9QJ [IO93XN, TA20]
G3 SNG A E Ambler, 12 Oakdene Rd, Marple, Stockport, SK6 6PJ [IO83XJ, SJ98]
G3 SNH W A Harrison, 94 West Park Dr, Blackpool, FY3 9HU [IO83LT, SD33]
G3 SNN A B Woolford, 28 Upper Hasfield, Hasfield, Gloucester, GL19 4LL [IO81UW, SO82]
G3 SNO G M Smith, Stoneycroft, Godsons Ln, Napton, Rugby, CV23 8LX [IO92IF, SP46]
G3 SNP M J Pitcher, Sandycot, Cadsden Rd, Princes Risborough, HP27 0NB [IO91OR, SP80]
G3 SNR G Morgan, Eaton House, Eaton Bank, Duffield, Belper, DE56 4BH [IO92GX, SK34]
G3 SNT R N Dixon, Copper Beeches, Parkhouse, Witton Gilbert, Co Durham, DH7 6TW [IO94DT, NZ24]
G3 SNV G Gallagher, 14 Ponting St., Swindon, SN1 2BL [IO91CN, SU18]
G3 SNX Details withheld at licensee's request by SSL.
G3 SNY J V Pearson, West End, 79 Bellevue, Stourbridge, DY8 5DB [IO82WL, SO88]
G3 SOA W A McCartney, Lychgate House, Uffington, Shrewsbury, SY4 4SN [IO82PR, SJ51]
G3 SOE R H Jennings, 31 Copper Beech Dr, Wombourne, Wolverhampton, WV5 0LH [IO82VM, SO89]
G3 SOI R S Pace, 22 The Dr, Woodlands Rd, Shotley Bridge, Consett, DH8 0DL [IO94BU, NZ05]
G3 SOL J B G Parker, 48 Cheltenham Dr, Leigh on Sea, SS9 3EH [JO01HN, TQ88]
GM3 SOM A B Miller, 23 Greenknowe Ave, Annan, DG12 6ER [IO84IX, NY16]
GW3 SON L A France, 5 Heol y Garth, Penparcau, Aberystwyth, SY23 1TE [IO72XJ, SN58]
GI3 SOO M Foley, 5 Woodland Dr, Cookstown, BT80 8PL [IO64PP, H87]
G3 SOP Details withheld at licensee's request by SSL.
G3 SOR Details withheld at licensee's request by SSL.
G3 SOU H McIntyre obo Southampton ARC, 90 Evenlode Rd, Millbrook, Southampton, SO16 9EH [IO90GW, SU31]
G3 SOX Details withheld at licensee's request by SSL.
GW3 SPA R F C Alban, 73 Plymouth Rd, Penarth, CF64 3DD [IO81JK, ST17]
G3 SPB P J Levay, 76 Woodgate Rd, Liskeard, PL14 6DY [IO70SK, SX26]
GW3 SPC T R Thomas South Powys ARC, Tymawr Farm, Llanwern, Brecon Powys, LD3 7UW [IO81IW, SO12]
G3 SPE Details withheld at licensee's request by SSL.[Op: J H Winnard. Station located near Bude.]
G3 SPI I D Dawe, 10 Selsden Cl, Elburton, Plymouth, PL9 8UR [IO70XI, SX55]
G3 SPJ C Wooff, 55 Bostall Hill, Abbey Wood, London, SE2 0QX [JO01BL, TQ47]
G3 SPK P W Corner, 6 Millfield Cl, Chatteris, PE16 6SS [JO02AK, TL38]
G3 SPL P D Lee, 1 Town End Ave, Holmfirth, Huddersfield, HD7 1YW [IO93CN, SE10]
G3 SPN N J Collins, 50 Exmoor Dr, Worthing, BN13 2PH [IO90TU, TQ10]
G3 SPO P R Oneill, Cross Cottage, Cowley, Cheltenham, Glos, GL53 9NN [IO81XT, SO91]
G3 SPP A Minett, 45 Patterdale Dr, Worcester, WR4 9HS [IO82VE, SO85]
GM3 SPT G McKay, 152 Inveresk St., Greenfield, Glasgow, G32 6TA [IO75WU, NS66]
G3 SPU R C Moore, 15 Waverley Gdns, Melksham, SN12 6AL [IO81WI, ST96]
G3 SPV K D Richardson, Brookfield Gr, The Dukes Dr, Ashford in The Water, Bakewell, DE45 1QQ [IO93DF, SK16]

G3 SPX Details withheld at licensee's request by SSL.
G3 SPY R G Harris Gpt(Coventry)Amateur Radio Soc, 44 Burbages Ln, Longford, Coventry, CV6 6AY [IO92FK, SP38]
G3 SPZ D F Miles, 45 Ashburton Rd, Ruislip Manor, Ruislip, HA4 6AA [IO91TN, TQ18]
G3 SQA P B Moss, 30 Coningsby Rd, Woodthorpe, Nottingham, NG5 4LH [IO92KX, SK54]
G3 SQC Details withheld at licensee's request by SSL.
G3 SQE E B Longstaffe, 24 High St., Mansfield Woodhouse, Mansfield, NG19 8AN [IO93JD, SK56]
G3 SQH A D Reffold, Whitewalls, 12 Woodlands Rise, North Ferriby, HU14 3JT [IO93RR, SE92]
G3 SQK A G Howell, 25 Thornhill Rd, Hednesford, Cannock, WS12 4LR [IO82XR, SJ91]
G3 SQM A Smith, 5 Varnells Terr, Hambleden, Henley on Thames, RG9 6SA [IO91NN, SU78]
G3 SQN J W R Grant, 8 Thornhill Way, Mannamead, Plymouth, Devon, PL3 5NP [IO70WJ, SX45]
G3 SQO D Best, Lonmore, Ribchester Rd, Wilpshire, Blackburn, BB1 9EE [IO83ST, SD63]
G3 SQQ J W Franks, 11 Thoresby Ave, Kirkby in Ashfield, Nottingham, NG17 7LY [IO93JC, SK55]
G3 SQR P Friend, 198 Devonshire Rd, Belmont, Durham, DH1 2BN [IO94FS, NZ34]
G3 SQU C S Clarke, 73 Larchside Cl, Spencers Wood, Reading, RG7 1DS [IO91MJ, SU76]
G3 SQV J S Davis, 17 Huston Cl, Barrow upon Soar, Barrow on Soar, Loughborough, LE12 8NB [IO92KR, SK51]
G3 SQX E F Taylor, PO Box 261304, Denver, Colorado 80226, USA
G3 SR Details withheld at licensee's request by SSL.
G3 SRA R J Chapman Silverthorn RC, 50 Graeme Rd, Enfield, EN1 3UT [IO91XP, TQ39]
G3 SRC M N Fagg Surrey RA Con C, 113 Bute Rd, Wallington, SM6 8AE [IO91WI, TQ26]
G3 SRE Dr K L Smith Thanet Elect Cl, Staple Farm House, Durlock Rd, Staple, Canterbury, CT3 1JX [JO01PG, TR25]
GW3 SRF D J Woolen, 17 Kensington Dr, Bridgend, CF31 4QS [IO81EM, SS88]
GW3 SRG A E Peake, 70 Higher Ln, Mumbles, Langland, Swansea, SA3 4PD [IO81AN, SS68]
G3 SRJ G R Carlisle, 3 Grimms Meadow, Walters Ash, High Wycombe, HP14 4UH [IO91OQ, SU89]
G3 SRM S Hulme, 7 Archery Cl, Countesthorpe, Leicester, LE8 5QB [IO92KN, SP59]
G3 SRO Details withheld at licensee's request by SSL.
G3 SRQ Details withheld at licensee's request by SSL.
G3 SRT R N Golding Salop A.R.S., 7 Belvidere Ave, Shrewsbury, SY2 5PF [IO82PQ, SJ51]
G3 SRU F Morrell, 18 Cross Rd, Birchington, CT7 9HN [JO01QJ, TR36]
GM3 SRV R M Tatton, 44 Craiglea Dr, Edinburgh, EH10 5PF [IO85JW, NT27]
G3 SRX N E Down, Amber Hill House, Sutterton Drove, Amber Hill, Boston, PE20 3RQ [IO93WA, TF24]
G3 SRZ K Rodgers, 30 Fore St., St. Blazey, Par, PL24 2NJ [IO70PI, SX05]
G3 SSA F Longson, 4 Chester Rd, Hull, HU5 5QE [IO93TS, TA03]
G3 SSC B R Evill, 54 Copsey Gr, Farlington, Portsmouth, PO6 1NB [IO90LU, SU60]
G3 SSE R Trevitt, 53 The Brow, Woodingdean, Brighton, BN2 6LP [IO90XT, TQ30]
GW3 SSG Details withheld at licensee's request by SSL.
G3 SSH D H Garside, 5 Arthur St., Prestwich, Manchester, M25 3HE [IO83UM, SD80]
GW3 SSK J E Williams, 62 Turberville St., Maesteg, CF34 0LU [IO81EO, SS89]
G3 SSM N A Currey, 31 Glebe Rd, Ashtead, KT21 2NT [IO91UH, TQ15]
G3 SSN J W Brand, 40 Mowbray Ave, Blackburn, BB2 3ET [IO83SR, SD62]
G3 SSO Details withheld at licensee's request by SSL.[Station located in Gloucester post code area.]
G3 SSQ C Spear, 25 Robert Ave, St. Albans, AL1 2QW [IO91TR, TL10]
GI3 SSR H C Lyttle, 7 Marmont Cres, Holywood Rd, Belfast, BT4 2GQ [IO74BO, J37]
G3 SSU D W Ryan, 62 Woodplumpton Ln, Woodplumpton, Preston, PR4 0AQ [IO83PT, SD53]
G3 SSZ L S Lavelle, 49 Jones Rd, Cheshunt, Goffs Oak, Waltham Cross, EN7 5JT [IO91WR, TL30]
G3 STB J W Jolly, 4 Renshaw Dr, Walton-le-Dale, Preston, PR5 4RA [IO83QR, SD52]
G3 STD E C John, obo St. Dunstans Ars, 52 Broadway Ave, Wallasey, L45 6TD [IO83LK, SJ39]
G3 STF P Sandiford, 4E Camden Hill, Tunbridge Wells, TN2 4TJ [JO01DD, TQ53]
G3 STG G A Griffiths, 11 The Grove, Asfordby, Melton Mowbray, LE14 3UF [IO92MS, SK71]
G3 STP Details withheld at licensee's request by SSL.
G3 STS Details withheld at licensee's request by SSL.
G3 STT W Haynes, 37 Hawthorn Gr, Southport, PR9 7AA [IO83MP, SD31]
GM3 STU C Auty Baltasound Js, School, Valsgarth, Haroldswick, Shetland, ZE2 9EF [IP90OT, HP61]
G3 STZ C R Thorn, 4 River Edge, Framilode, Gloucester, Glos, GL2 7LH [IO81TS, SO71]
G3 SUA E H Winstanley, Padua, 1 Drews Ct, Churchdown, Gloucester, GL3 2LD [IO81VU, SO81]
G3 SUG J J Jarvis, 56 Upper Churnside, The Beeches, Cirencester, GL7 1AP [IO91AR, SP00]
GW3 SUH K Hughes, 2 Graig Terr, Ferndale, CF43 4EU [IO81GP, ST09]
G3 SUI J W Burrows, 68 Grosvenor Rd, Sale, M33 6NW [IO83TK, SJ79]
G3 SUL D E Waller, 11 Lowndes Lodge, Hadley Rd, Barnet, EN5 5QW [IO91VP, TQ29]
GI3 SUM J Gould, 13 Maralin Ave, Bangor, BT20 4RQ [IO74EP, J58]
G3 SUN G Hodgkinson, Units 8/9/10, Adler Industrial Est, Betam Rd, Hayes, UB3 1ST [IO91SM, TQ08]
G3 SUR R J Hodges, 34 Newbridge Rd, Bath, BA1 3JZ [IO81TJ, ST76]
G3 SUS S B Jacobs, 16 Mayfield Park, Thorley Park, Bishops Stortford, CM23 4JL [JO01BU, TL41]
G3 SUV D E Ashby, Insteps Farm, White Colne, Colchester, Essex, CO6 2QB [JO01IW, TL83]
G3 SUX D J Bradshaw, 25 Meare Cl, Tadworth, KT20 5RZ [IO91VG, TQ25]
G3 SUY P J Bridgeman, 4 Dockwra Ln, Danbury, Chelmsford, CM3 4RQ [JO01HR, TL70]
GM3 SUZ D McLean, Whitecroft Farm, Barrs Brae, Port Glasgow, Renfrewshire, PA14 5QG [IO75PW, NS37]
G3 SVC H C Foster Spen Valley Amateur Radio Soci, 23 Ghyllroyd Dr, Birkenshaw, Bradford, BD11 2ET [IO93DR, SE22]
G3 SVD A P Hewitt, 3621 PO Box, Western Rd, Bracknell, RG12 1WJ [IO91OK, SU86]
GM3 SVE R R J Gibbs, Col Dene, Shieldhill Rd, Reddingmuirhead, Falkirk, FK2 0DU [IO85DX, NS97]
G3 SVI D H Davis, 303 Rayleigh Rd, Eastwood, Leigh on Sea, Essex, SS9 5PX [JO01HN, TQ88]
G3 SVK F L Curtis, 32 Elgin Ave, Harold Park, Romford, RM3 0YT [JO01CO, TQ59]
G3 SVL C Duckling, Many Oaks, Collington Ln West, Bexhill-on- Sea, E Sussex, TN39 3TD [JO00FU, TQ70]
G3 SVQ A H Yallop, Whitehill, 16 High St, Carlton, Bedford, Beds, MK43 7JX [IO92QE, SP95]
G3 SVR E G Churchyard Severn Valley Rd, 11 Greenfields Dr, Bridgnorth, WV16 4JW [IO82SM, SO79]
G3 SVS J D Garford, 31 Gravel Hill Ln, West Winch, Kings Lynn, PE33 0QG [JO02EQ, TF61]
G3 SVT A M Robins, 20 Featherston Dr, Burbage, Hinckley, LE10 2PN [IO92HM, SP49]
G3 SVW R P Smith, 16 Coniston Ave, Sale, M33 3GT [IO83UJ, SJ79]
GW3 SVY D Scourfield, Llwyncelyn, Bancyfelin, Carmarthen, SA33 5NJ [IO71SU, SN32]
G3 SWC B Tinton, Farthings, 1 Bridge Rd, Rudgwick, Horsham, RH12 3HD [IO91SC, TQ03]
GM3 SWF Details withheld at licensee's request by SSL.
G3 SWH P A Whitchurch, 21 Dickensons Gr, Congresbury, Bristol, BS19 5HQ [IO81OI, ST46]
GM3 SWK A J Shearer, 12 Coolin Dr, Portree, Isle of Skye, IV51 9DN [IO67VJ, NG44]
G3 SWM R Mannion Short Wave Mag, Arrowsmith Ct, Station Approach, Broadstone, Dorset, BH18 8PW [IO90AS, SZ09]
G3 SWO G S Thomas, 65 Queslett Rd, Great Barr, Birmingham, B43 6DR [IO92AN, SP09]
G3 SWP W A Boothman, 5 Millwood Rd, Balby, Doncaster, DN4 9DA [IO93KL, SE50]
G3 SWT A C Morris, 13 Eastfields, Pinner, HA5 2SR [IO91TO, TQ18]
G3 SWU T Heeley, 34 Worlaby Rd, Scartho, Grimsby, DN33 3JT [IO93WM, TA20]
G3 SWW H A Cooper, 9 Fortyfoot, Bridlington, YO16 5SA [IO94VC, TA16]
G3 SXA J R Croft, 14 Stanstead Rd, Forest Hill, London, SE23 1BW [IO91XK, TQ37]
G3 SXC A Critchley, 39 Westcliffe, Great Harwood, Blackburn, BB6 7PH [IO83TT, SD73]
G3 SXE L P J Lethbridge, 24 Furze Rd, High Salvington, Worthing, BN13 3BH [IO90TU, TQ10]
G3 SXH A P Henderson, 37 Barnardo Rd, St. Leonards, Exeter, EX2 4ND [IO80FR, SX99]
G3 SXI D M Ashmore, Forest End, Forest Rd, Pyrford, Woking, GU22 8LS [IO91RH, TQ05]
G3 SXK B J Doel, 11 North Dell, Chelmsford, CM1 6UP [JO01FS, TL70]
G3 SXL R G Anderson, 12 Fairy Dell, Marton in Cleveland, Middlesbrough, TS7 8LF [IO94JM, NZ51]
G3 SXP J Redford, Alma House, The St., Great Hockham, Thetford, IP24 1NH [JO02KL, TL99]
G3 SXQ E P Rockett, Devoran, The Causeway, Mark, Highbridge, TA9 4QT [IO81MF, ST34]
G3 SXR A S M Read, Readymoney Cove, Fowey, Cornwall, PL23 1JH [IO70QH, SX15]
G3 SXS N E Stoneman, Hilton Fir, 22A Keyberry Rd, Newton Abbot, TQ12 1BX [IO80EM, SX87]
G3 SXV B F Vincent, 18 Rowanhayes Cl, Ipswich, IP2 9SX [JO02NB, TM14]
G3 SXW R K Western, 7 Field Cl, Chessington, KT9 2QD [IO91UI, TQ16]
G3 SYA D Ashworth, 31 Belmont Ave, Ribbleton, Preston, PR2 6DH [IO83PS, SD53]
G3 SYB H H Barker, Azul Avion, Main Rd, Maltby-le-Marsh, Alford, LN13 0JP [JO03CH, TF48]
G3 SYC B K Booth, 39 Park Ln, Pontefract, WF8 4QH [IO93IQ, SE42]
G3 SYD S H Beauchamp, 1 Gosden Cl, Furnace Green, Crawley, RH10 6SE [IO91WC, TQ23]
G3 SYF Details withheld at licensee's request by SSL.
G3 SYG J T Marsh, 17 Fish Ln, Aldwick, Bognor Regis, PO21 3AH [IO90PS, SZ99]
G3 SYJ R I Taylor, 4 Ashdale Rd, Kesgrave, Ipswich, IP5 7PA [JO02OB, TM24]
G3 SYK A J Parker, 23 Cheviot Way, Hopton, Mirfield, WF14 8HW [IO93DP, SE11]
GW3 SYL R M Price, 49 Pant Hirwaun, Heol y Cyw, Bridgend, Mid Glam, CF35 6HH [IO81FN, SS98]
G3 SYM D R Coltart, Tregerry Farm, Treneglos, Launceston, Cornwall, PL15 8UF [IO70RQ, SX28]
GM3 SYO D J Mackay, 37 Willowbank, Wick, KW1 4NY [IO88KK, ND35]
G3 SYS D T Emerson, 3555 E, Thimble Peak Pl, Tucson, Arizona 85718, USA
G3 SYW I D Scott, 4 Hartmoor Cl, Stokenchurch, High Wycombe, HP14 3QL [IO91NP, SU79]
G3 SYX R N A Irving, Firglen, Beesby Rd, Maltby-le-Marsh, Alford, LN13 0JJ [JO03CH, TF48]
G3 SYZ A W G Rogers, Draycott, Primrose Hill, Fairlight, Hastings, TN35 4DN [JO00IV, TQ81]
G3 SZ A Chilvers, 3 Thurlin Rd, Kings Lynn, PE30 4PG [JO02FS, TF61]
G3 SZA D R Wilson, 18 Pelican Mead, Hightown, Ringwood, BH24 3RG [IO90CU, SU10]
G3 SZC I G West, 85 Priest Ave, Canterbury, CT2 8PP [JO01MG, TR15]
G3 SZF R R C Frost, 24 Mount Pleasant, Hertford Heath, Hertford, SG13 7QU [IO91XS, TL31]
G3 SZG J I Wright, Redwood, Hillside, Rothbury, Morpeth, NE65 7YG [IO95AH, NU00]
G3 SZH B Heape, 49 Templar Rd, Yate, Bristol, BS17 5TG [IO81TN, ST78]
G3 SZJ M Shardlow, 19 Portreath Dr, Allestree, Derby, DE22 2BJ [IO92GW, SK33]
G3 SZM J R Wuille, 45 Keymer Cres, Goring By Sea, Worthing, BN12 4LD [IO90TT, TQ10]
G3 SZO Details withheld at licensee's request by SSL.
GM3 SZP Details withheld at licensee's request by SSL.
G3 SZR C A Davis, 148 Birkbeck Rd, Beckenham, BR3 4SS [IO91XJ, TQ36]
G3 SZS R J Bee, 19 Hazelcroft, Churchdown, Gloucester, GL3 2DS [IO81VV, SO82]
G3 SZT Details withheld at licensee's request by SSL.
G3 SZU K Radford, 24 Broad Way, Wilburton, Ely, CB6 3RT [JO02CI, TL47]
G3 SZV B C Ward, 138 Co. Rd, Ormskirk, L39 1NN [IO83NN, SD40]
G3 SZY G J Douglas, 169 High St., Cheveley, Newmarket, CB8 9DG [JO02FF, TL66]
G3 TA C J Lambert, Stonecroft, Wsinstone, Cirencester, Glos, GL7 7JU [IO81XS, SO90]
G3 TAA K F Jessop, 26 Larchwood Rd, New Eltham, London, SE9 3SF [JO01AK, TQ47]
G3 TAD A B Williams Bristol ARC, 38 Seneca St., Bristol, BS5 8DX [IO81RK, ST67]
G3 TAF D A Cassere, Claremont, 20 Church Rd, Horsforth, Leeds, LS18 5LG [IO93EU, SE23]
G3 TAG R H J Gouldstone, Old Orchards, 11 School Ln, Toft, Cambridge, CB3 7RE [IO92XE, TL35]
G3 TAH Details withheld at licensee's request by SSL.
G3 TAI C F J Ward, 50 Lakeside, Bracknell, RG42 2LE [IO91PK, SU87]
G3 TAJ R T Marchant, Staple Farmhouse, Staple, Canterbury Kent, CT3 1JX [JO01PG, TR25]
GM3 TAL M J W Hamilton, 3 Charles Ct, Limekilns, Dunfermline, KY11 3LG [IO86GA, NT08]
G3 TAN Details withheld at licensee's request by SSL.
G3 TAO W D Eaton, 8 St. Aubyns Rd, London, SE19 3AD [IO91XK, TQ37]
G3 TAP A C Krarup, 8 Coniston Rd, Beeston, Nottingham, NG9 3AD [IO92JW, SK53]
G3 TAQ N H Bullock, 29 St. Marys Rd, Stowmarket, IP14 1LP [JO02LE, TM05]
G3 TAR R E Roberts, 15 Pineside Ave, Cannock Wood, Rugeley, WS15 4RG [IO92AQ, SK01]
G3 TAS W Smith, 13 Honbeam Rd, Stowupland, Stowmarket, IP14 4DJ [JO02ME, TM05]
G3 TAV G E Evans, Barn Farm, Pocknedge Ln, Holymoorside, Chesterfield, S42 7HL [IO93GF, SK37]
G3 TAW C Wood, 22 Habberley Rd, Kidderminster, DY11 6AA [IO82UJ, SO87]
G3 TAX J C Boydell, Abbotsford, 13 Lynch Rd, Farnham, GU9 8BZ [IO91OF, SU84]
G3 TAY A B Yarker, 6 Moor Top Rd, Halifax, HX2 0NP [IO93BR, SE02]
G3 TAZ R G Davies, 69 Stopsley Way, Luton, LU2 7UU [IO91TV, TL12]
G3 TB E T Burkitt, 41 Nickerwood Dr, Aston, Sheffield, S31 0BX [IO93GJ, SK38]
G3 TBA Details withheld at licensee's request by SSL.[Op: A Barsby, 6 Cromford Drive, Staveley, Chesterfield, Derbys, S43 3TB:]
G3 TBF H K Wilkins, 18 Borough Cl, Kingstanley, Kings Stanley, Stonehouse, GL10 3LJ [IO81UR, SO80]
G3 TBG Details withheld at licensee's request by SSL.
G3 TBH F M Gray, 52 Burton Manor Rd, Rising Brook, Stafford, ST17 9QQ [IO82WS, SJ92]
G3 TBJ C J Webster, 20 Piggotts Rd, Caversham, Reading, RG4 8EN [IO91ML, SU77]
G3 TBK J D Cree, 24 Old Lincoln Rd, Caythorpe, Grantham, NG32 3DF [IO93QA, SK94]
G3 TBQ T Quail, 51 Meadowbrook Rd, Moreton, Wirral, L46 0RR [IO83KJ, SJ28]
G3 TBU F J Gibbons, 6 Steeple Cl, Cleobury Mortimer, Kidderminster, DY14 8PD [IO82SJ, SO67]
G3 TBW T H Westbury, 1299 Evesham Rd, Astwood Bank, Redditch, B96 6AY [IO92AG, SP06]
G3 TBX J K Robinson, 4 Shaftesbury Ave, Bradford, BD9 6AJ [IO93CT, SE13]
G3 TCA M J C Burns, 21 Tewkesbury Dr, Sedgley Park, Prestwich, Manchester, M25 0HR [IO83UM, SD80]
G3 TCG M A Trundle, 9 Upper Mill, Wateringbury, Maidstone, ME18 5PD [JO01FG, TQ65]
G3 TCI A S B Bye, 7 Larkfield Ave, Gillingham, ME7 2LN [JO01GJ, TQ76]
G3 TCL M J G Dawson, 51 Spring Gr, Loughton, IG10 4QD [JO01AP, TQ49]
GM3 TCM D J Munro, 3 Sinclair Terr, Wick, KW1 5AD [IO88KK, ND35]
G3 TCO Dr A W Preece, 12 South Dene, Stoke Bishop, Bristol, BS9 2BW [IO81QL, ST57]
G3 TCQ R Hyde RAF North Luffenham ARC, 25 The Pastures, Cottesmore, Oakham, LE15 7DZ [IO92QR, SK91]
G3 TCR C Mott-Gotobed Basingstoke Amateur Radio Club, Cherry Trees, 17 Reading Rd, Chineham, Basingstoke, RG24 8LN [IO91LG, SU65]
G3 TCT G F Kimbell, 39 Downs Way, Bookham, Leatherhead, KT23 4BL [IO91TG, TQ15]
G3 TCU P R Guttridge, 33 Franklyn Rd, Godalming, GU7 2LD [IO91QE, SU94]
GW3 TCV J H Edwards, Penymaes, Trehelig, Welshpool, Powys, SY21 8SG [IO82KO, SJ20]
GM3 TCW J F Kelly, 144A Manse Rd, Newmains, Wishaw, ML2 9BL [IO85BS, NS85]
G3 TCY J C Lewis, 10 Sheringham Dr, Etchinghill, Rugeley, WS15 2YG [IO92AS, SK01]
G3 TCZ R W Freeman, 65 Higher Dr, Banstead, SM7 1PW [IO91VH, TQ26]
G3 TDC J Yates, Ferncliffe, Forest Dr Kinver, Stourbridge, DY7 6DX [IO82VK, SO88]
G3 TDF P Farley, 11 Arthur Rd, Erdington, Birmingham, B24 9EX [IO92CM, SP19]
G3 TDH R W Stevens, 19 Canberra Rd, Bramhall, Stockport, SK7 1LG [IO83WI, SJ88]
GM3 TDI Dr R J Teperek Rgit Amat RA So, Sch of Elect & E En, Robert Gordns Ins T, Schoolhill Aberdeen, AB9 1FR [IO87TH, NJ72]
G3 TDJ Details withheld at licensee's request by SSL.
G3 TDL R C Davis, 21 Denton Dr, Brighton, BN1 8LR [IO90WU, TQ30]
G3 TDM R D Mason, 28 Shrubbery Gdns, Winchmore Hill, London, N21 2QT [IO91WP, TQ39]
G3 TDR R S Hewes, 24 Brightside Ave, Laleham, Staines, TW18 1NG [IO91SK, TQ07]
GM3 TDS J A Shelton, 70 Mount Harriet Dr, Stepps, Glasgow, G33 6DG [IO75WV, NS66]
G3 TDT D A H Hollingsbee, 24 East Hatley, Hatley St. George, East Hatley, Sandy, SG19 3HZ [IO92WD, TL25]
G3 TDU F L Judges, 10 Mayday Gdns, Blackheath, London, SE3 8NN [JO01AL, TQ47]
G3 TDV G Fidler, 169 Ennerdale Rd, Walker Dene, Newcastle upon Tyne, NE6 4LJ [IO94FX, NZ26]
G3 TDW W G Western, Sandye Pl, 181 Topsham Rd, Exeter, EX2 4SQ [IO80FR, SX99]
G3 TDX E H Ingram, 36 Kenwood Rd, Knighton, Leicester, LE2 3PJ [IO92KO, SK60]
GI3 TDY R J Grange, Fairways, 13 Newlands Crscent, Portstewart, Co Londonderry, BT55 7JJ [IO65PE, C83]
G3 TDZ J R Hey, 8 Armley Grange Cres, Leeds, LS12 3QL [IO93ET, SE23]
G3 TEB G R J Addis, 34 Ryhill Way, Lower Earley, Reading, RG6 4AZ [IO91MK, SU76]
G3 TEC T C Rutherford, 12 Dorothy Sayers Dr, Witham, CM8 2LX [JO01HT, TL81]
G3 TEE F Stork, 20 Gay Meadows, Stockton on The Forest, York, YO3 9UJ [IO93MX, SE65]
G3 TEG R G Wainwright, 37 High Meadow, Hathern, Loughborough, LE12 5HW [IO92IT, SK52]
G3 TEH A R Storey, 23 Foster St., Stairfoot, Barnsley, S70 3EW [IO93GN, SE30]
G3 TEI R S Grant Thorn EMI Amateur Radio Club, Bentley, Middle St., Nazeing, Waltham Abbey, EN9 2LB [JO01AR, TL30]
G3 TEJ M Dighton, 7 The Close, Fairey Ave, Godmanchester, Huntingdon, PE18 8DU [IO92WH, TL27]
G3 TEK H E Newland, 22A Cromwell Rd, Basingstoke, RG21 5NR [IO91KG, SU65]
G3 TEL P H McPherson, 49 St. Johns Rd, Gr, Wantage, OX12 7NP [IO91GO, SU49]
GI3 TEN Dr D Linton Dept.Of Electl, 4 Elmwood, Cullybackey, Ballymena, Co. Antrim, BT43 5PY [IO64TV, D00]
G3 TEP B Atkinson, 6 Argyle St., Alnmouth, Alnwick, NE66 2SB [IO95EJ, NU21]
G3 TEQ Details withheld at licensee's request by SSL.
G3 TEU A Sherer, 35 Beverley Rd, Willerby, Hull, HU10 6AW [IO93SS, TA03]
G3 TEV M J Mills, Shepton, 3 Tylers Way, Chalford Hill, Stroud, GL6 8ND [IO81WR, SO80]
G3 TEX P Painter, 28 Burlington Rd, Manchester, M20 4QA [IO83VK, SJ89]
G3 TEY P Hargreaves, 46 Castle Rd, Mow Cop, Stoke on Trent, ST7 3PH [IO83VC, SJ85]
G3 TEZ R E Spencer, 263 Beckfield Ln, York, YO2 5PG [IO93KX, SE55]
G3 TFA G A Whenham, Hogs Hollow, Welsh Rd East, Southam, Leamington Spa, CV33 0NF [IO92HF, SP46]
G3 TFC J F Coggins, Hamelin Coventry Rd, Baginton Coventry, Warks, CV8 3AP [IO92GI, SP37]
G3 TFF G W Fuller, 17 Baden St., Haworth, Keighley, BD22 8HQ [IO93AU, SE03]
G3 TFG Capt E J Griffiths, (Overshorticombe), Fairlynch Ln, Braunton, Devon, EX33 1BT [IO71VC, SS43]
G3 TFI P G Stephens, Post Office Box 182, Fourways 2055, Johannesburg, Rep of South Africa
G3 TFL G E Rogers, 19 Manor Rd, Henley on Thames, RG9 1LT [IO91NM, SU78]
G3 TFM R Scadden, 29 Pinewood, Somerton, TA11 6JW [IO81OB, ST42]
G3 TFO J Auty, 64 Ainley Rd, Birchencliffe, Huddersfield, HD3 3QX [IO93CQ, SE11]
G3 TFP P R Bailey, Flat B 12Th Floor, Mountainville Ct, 7 Lok Fung Path, Fotan,New Territories, Hong Kong
G3 TFT Dr B H Posner, 39 Moor Cres, Gosforth, Newcastle upon Tyne, NE3 4AQ [IO94EX, NZ26]
G3 TFV E E Tokley, 14 Maple Way, Earl Shilton, Leicester, LE9 7HW [IO92IN, SP49]
GM3 TFY D H Guest, 31 Newmills Cres, Balerno, EH14 5SX [IO85IV, NT16]
G3 TFZ D T Legg, Dorina, Whitewayhead Ln, Knowbury, Ludlow Salops, SY8 3LF [IO82QJ, SO57]
G3 TGA Details withheld at licensee's request by SSL.
G3 TGB B W Ely, 375 Cressing Rd, Braintree, CM7 3PE [JO01GU, TL72]
G3 TGC D S Woods, 5 Grenville Rd, Padstow, PL28 8EX [IO70MM, SW97]
G3 TGD M Allenson, 4 The Orchard, Powick, Worcester, WR2 4SE [IO82UD, SO85]
G3 TGE D M Cahill, 3 Leys View, Sherington, Newport Pagnell, MK16 9NL [IO92PC, SP84]
G3 TGF C R Bonner, The Dwellings, Turners Green, Heathfield, E Sussex, TN21 9RA [JO00DW, TQ61]
GM3 TGG T G Gratton, 23 Culhorn Rd, Stranraer, DG9 8DB [IO74LV, NX06]
G3 TGK S D Parish, 6927 W Wagoner Rd, Glendale Az 85308, USA, ZZ1 2AS
G3 TGL A J Fantham, 52 Calverley Rd, Kings Norton, Birmingham, B38 8PW [IO92AJ, SP07]
G3 TGN Dr P G Collar, Queen Mary House, Brook Rd, Wormley, Godalming, GU8 5UA [IO91QD, SU93]
G3 TGO B W Vaughan, 368 Fulbridge Rd, Paston, Peterborough, PE4 6SJ [IO92UO, TF10]
G3 TGR J J Woods, 19 Furland Cl, Plymstock, Plymouth, PL9 9NG [IO70WI, SX55]
G3 TGT R G Clitheroe, 55 Privett Pl, Gosport, PO12 3SG [IO90KT, SZ59]
G3 TGW Details withheld at licensee's request by SSL.
G3 THA D W H Ashton, The Mistal, 16 Church St., Boston Spa, Wetherby, West Yorks, LS23 6DN [IO93HV, SE44]
G3 THC D R Stimson, Chestnut View, 94 Casterton Rd, Stamford, PE9 2UB [IO92SP, TF00]
G3 THD D Livesey, Calle Paises Bajos 5, 03186 Los Balcones, Torrevieja, Alicante, Spain, X X
G3 THF B McHugh, 283 Coppice Rd, Poynton, Stockport, SK12 1SP [IO83WI, SJ98]
G3 THG Wgcd A J Kent, Icentown House, Pitney, Langport, Somerset, TA10 9AJ [IO81OB, ST42]
GM3 THI Dr R D Harkess, Friarton Bank, Rhynd Rd, Perth, PH2 8PT [IO86HI, NO12]
G3 THK R Turner, 23 Grafton Rd, Reffley Est, Kings Lynn, PE30 3HA [JO02FS, TF62]
G3 THM L P Best, 3A Chipstead Ln, Sevenoaks, TN13 2AH [JO01CG, TQ55]
G3 THO H R Gelsthorpe, 223 Walton Rd, Walton, Chesterfield, S40 3BT [IO93GF, SK36]

G3 THQ B F Greenaway, 5 Lansdowne Gr, Neasden, London, NW10 1PL [IO91VN, TQ28]
G3 THS P A Last, 4 Hillside, Marham, Kings Lynn, PE33 9JJ [JO02GP, TF71]
G3 THU W A Robb, 19 Avon St., Clifton upon Dunsmore, Rugby, CV23 0DQ [IO92JJ, SP57]
G3 THV G J Swindells, 4 Fitzhenry Mews, Norwich, NR5 9BH [JO02OP, TG10]
G3 THW P J Walters, 32 Quail Green, Wightwick, Wolverhampton, WV6 8DF [IO82VO, SO89]
G3 THX C A Collins, 60 Alexandra Rd, Skegness, PE25 3RE [JO03DD, TF56]
G3 THY R E Wheeler, 143 Coventry Rd, Ilford, IG1 4QX [JO01AN, TQ48]
G3 THZ E F Spiers, 15 Hall Green Rd, West Bromwich, B71 3JS [IO92AN, SP09]
G3 TIB Details withheld at licensee's request by SSL.
G3 TIE A G F Dutton, 130 Wades Hill, Winchmore Hill, London, N21 1EH [IO91WP, TQ39]
G3 TIG P C Turner, 11 Manor Ct Rd, Hanwell, London, W7 3EJ [IO91TM, TQ18]
G3 TII J Burgon, Anvil Lodge, 11 Newport Dr, Winterton, Scunthorpe, DN15 9RG [IO93QP, SE91]
GI3 TIJ F C Eccles, Silverhill, Ballydawley, Moneymore, Co Londonderry, N Ireland, BT45 7NU [IO64QP, H88]
G3 TIK D R French, 37 Warner Rd, Ware, SG12 9JN [IO91XT, TL31]
G3 TIN B M Taylor, Perry House, 188 Walstead Rd, Walsall, WS5 4DN [IO92AN, SP09]
G3 TIP F H Gibbons, 11 Gold St., Barnsley, S70 1TT [IO93GN, SE30]
G3 TIQ Details withheld at licensee's request by SSL.
G3 TIR D A Stewart, 3 Grosvenor Terr, Teignmouth, TQ14 8NE [IO80FN, SX97]
G3 TIS J A Clarke, The Alders, Weekes Ln, West Brabourne, Ashford, TN25 5LZ [JO01LD, TR04]
G3 TIX R Hardy, 522 Halifax Rd, Bradford, BD6 2LP [IO93CS, SE12]
G3 TJA R G Street, Norcombe Farm Cottage, 16 Park Rd, Blockley, Moreton in Marsh, GL56 9BZ [IO92CA, SP13]
G3 TJB D J Norman, 11 Bernie Crossland Walk, Kidderminster, DY10 1XT [IO82VJ, SO87]
G3 TJC E Ross, 20 Briar Wood, Shipley, BD18 1NB [IO93CT, SE13]
G3 TJE P G Smith, Beggars Roost, 39 Brent St., Brent Knoll, Highbridge, TA9 4DT [IO81MF, ST35]
G3 TJH W B Bickham, 22 Ash Cres, Galmington, Taunton, TA1 5PW [IO81KA, ST22]
G3 TJI G A Roff, 47 Penshurst Rise, Frimley, Camberley, GU16 5XX [IO91PH, SU85]
GI3 TJJ J Boyce, 19 Dunvale Park, Londonderry, N Ireland, N Ireland, BT48 0AU [IO65IA, C41]
GI3 TJM R A H Miller, 17 Knockmore Park, Bangor, BT20 3SL [IO74DP, J48]
G3 TJR J D Burton, Flat, 11 Fulbourne House, Blackwater Rd, Eastbourne, BN20 7DN [JO00DS, TV69]
G3 TJS P D Goodenough, Llys Aderyn, 11 Guildford Rd, Lightwater, GU18 5RZ [IO91QI, SU96]
G3 TJT A D Bramall, 16 Linkswood Ave, Wheatley Hills, Doncaster, DN2 5QW [IO93KM, SE60]
G3 TJU L P Grant, 179 Hamstel Rd, Southend on Sea, Essex, SS2 4LA [JO01IN, TQ88]
G3 TJW J E Bright, Bewsley Croft, Copplestone, Crediton, Devon, EX17 5NX [IO80DT, SS70]
G3 TJX G Tillson, 95 Kelverlow St., Oldham, OL4 1LX [IO83WM, SD90]
G3 TJY D R H Jolly, Little Russel, Lytchett Minster, Poole, Dorset, BH16 6JD [IO80XR, SY99]
G3 TKA P S Duncan, 18 Pickering Rd, Hull, HU4 6TL [IO93TR, TA02]
G3 TKB J B Foster, Burnside, Hookergate Ln, Rowlands Gill, NE39 2AD [IO94CW, NZ15]
GW3 TKD A M Cooper, 4 The Beeches, Willow Bank, Hawarden, Deeside, CH5 3LJ [IO83LE, SJ26]
G3 TKF R W Thompson, 179 Newbridge Hill, Bath, BA1 3PY [IO81TJ, ST76]
GW3 TKG D J Locke, 201 Tyn y Tower, Baglan, Port Talbot, SA12 8YE [IO81CO, SS79]
GW3 TKH K J Winnard, 55 Lon y Celyn, Whitchurch, Cardiff, CF4 7BT [IO81JM, ST18]
G3 TKI D Bradshaw, 54 Plymouth Gr, Radcliffe, Manchester, M26 3WU [IO83TN, SD70]
G3 TKK Dr P R Doughty, Mallows, Ballam Rd, Westby, Preston, PR4 3PN [IO83MS, SD33]
G3 TKN V C Lear, 53 Chaplains Ave, Cowplain, Waterlooville, PO8 8QH [IO90LV, SU61]
G3 TKO Details withheld at licensee's request by SSL.[Op: E E Snow. Station loacted near Weston-Super-Mare.]
G3 TKQ G T Barrell, 28 Melton Rd, Wymondham, NR18 0DB [JO02NN, TG10]
G3 TKS J R Sanderson, 28 Finmere, North Lake, Bracknell, RG12 7WF [IO91PJ, SU86]
G3 TKW Details withheld at licensee's request by SSL.
GW3 TKZ W J Palmer, 42 Bloomfield Rd, Blackwood, NP2 1LX [IO81JQ, ST19]
GM3 TLA Dr D Pearson, 23 Binghill Rd West, Milltimber, AB13 0JB [IO87VC, NJ80]
G3 TLB K R Smith, Sheerland, Blackness Rd, Crowborough, TN6 2NB [JO01CB, TQ52]
G3 TLD M F J Selwyn, 50 Tufthorn Ave, Coleford, GL16 8PT [IO81QS, SO50]
G3 TLE Details withheld at licensee's request by SSL.
G3 TLF T F Adey, 7 St. Peters Cl, Orchard Land, Hutton, Driffield, YO25 9YZ [IO93SX, TA05]
G3 TLG J H Robley, 6 Pinewood Cl, Paddock Wood, Tonbridge, TN12 6JN [JO01EE, TQ64]
G3 TLH I D Brown, 45 Greenham Wood, Bracknell, RG12 7WJ [IO91PJ, SU86]
G3 TLI D F Heathershaw, The Old School, Cliff Ln, Mappleton, Hornsea, HU18 1XX [IO93WV, TA24]
G3 TLK A J Endacott, Red Acres, St Mary Church Rd, Coffinswell, Newton Abbot, TQ12 4SB [IO80FM, SX86]
GM3 TLN W McEwan, 19 Blair St., Kelty, KY4 0ET [IO86HD, NT19]
GW3 TLP I Jones, Tyddyn Brith, Gaerwen, Anglesey, Gwynedd, LL60 6HD [IO73VF, SH57]
GI3 TLT H M Irvine, Gransha House, 35 Rowreagh Rd, Kircubbin, Newtownards, BT22 1AS [IO74FL, J66]
G3 TLU J H Serlin, 45 Wychwood Ave, Canons Park, Edgware, HA8 6TQ [IO91UO, TQ19]
G3 TLV G C Wynes, Hill View, Wrenbury Wood, Wrenbury, Nantwich, CW5 8HH [IO83QA, SJ54]
G3 TLY S E Alexander, Pinetrees, Wilmslow Ave, Woodbridge, IP12 4HW [JO02PC, TM24]
G3 TLZ M J Walker, The Bield, Lower Blandford, St Mary, Blandford Forum, DT11 9ND [IO80WU, ST80]
G3 TMA I Buffham, Fir Tree House, Northgate, West Pinchbeck, Spalding, Lincs, PE11 3TB [IO92VT, TF22]
G3 TMB Details withheld at licensee's request by SSL.[Op: J M Baker, 29 Garstang Road, Southport, Lancs, PR9 - 9XW.]
G3 TMC R J Cross, 4 Overgreen Cl, Burniston, Scarborough, YO13 0JA [IO94SH, TA09]
G3 TMD E Parsons, 22 Colins Walk, Scotter, Gainsborough, DN21 3SR [IO93QM, SE80]
GI3 TME R A A Hargan, 17 Portlock Pl, Culmore Rd, Londonderry, BT48 8PR [IO65IB, C42]
G3 TMG Details withheld at licensee's request by SSL.
GW3 TMH K Williams, 20 Roland Ave, Kinmel Bay, Rhyl, LL18 5DL [IO83FH, SH98]
G3 TMI Details withheld at licensee's request by SSL.
GW3 TMJ A Taylor, 24 Emroch St., Goytre, Port Talbot, SA13 2YE [IO81CO, SS78]
GM3 TMK I M Ross, 70 Kenneth St., Inverness, IV3 5PZ [IO77VL, NH64]
G3 TMN Dr T M Newland, The Hamlet, Moorlands Ln, Tollerton, York, YO6 2ER [IO94JB, SE56]
G3 TMO A J Brown, Unit 17 Alan Walker Village, Dalmar Pl, Carlingford, Nsw 2118, Australia
GW3 TMP Details withheld at licensee's request by SSL.
G3 TMQ R J Harrison, 28 Briar Hill Rd, Delapre, Northampton, NN4 8LJ [IO92NF, SP75]
G3 TMR D A Emmett, PO Box 5307, Walmer, Port Elizabeth 6065, South Africa
GW3 TMS D C Smith, 2 Glan yr Afon Gdns, Sketty, Swansea, SA2 9HY [IO81AO, SS69]
G3 TMU C A Neale, 51 Oakfield Rd, Blackwater, Camberley, GU17 9DZ [IO91OH, SU85]
G3 TMX S G Bennett, 12 Angle Ln, Bury St Edmunds, Suffolk, IP33 1RF [JO02IF, TL86]
G3 TMZ Details withheld at licensee's request by SSL.
G3 TN T Noblet, Marget, Fleet Rd, Fleet, Weymouth, DT3 4EB [IO80RO, SY68]
G3 TNI J F Clingan, 41 Cranham Cl, Headless Cross, Redditch, B97 5AY [IO92AG, SP06]
GI3 TNK S Dornan, 9 Clonallon Gdns, Belfast, BT4 2BY [IO74BO, J37]
G3 TNM M P Steel, 9 Greenhill Main Rd, Sheffield, S8 7RA [IO93GH, SK38]
G3 TNO M T Healey, 56 Cissbury Rd, Ferring, Worthing, BN12 6QL [IO90ST, TQ00]
G3 TNQ C R Davis, 963 Manchester Rd, Bury, BL9 8DN [IO83UN, SD80]
GD3 TNS A K Sinclair, 1 Marathon Dr, Douglas, IM2 4BP
GM3 TNT D R McArthur, 74 Bathurst Dr, Alloway, Ayr, KA7 4UA [IO75QK, NS31]
G3 TNX V Allison, 24 Colston Gate, Cotgrave, Nottingham, NG12 3JY [IO92LV, SK63]
G3 TNY K J Spooner, Penncroft, 16 Booton Rd, Cawston, Norwich, NR10 4AH [JO02NS, TG12]
GW3 TOB A Coughlin, 37 Parc y Felin, Creigiau, Cardiff, CF4 8PB [IO81IM, ST08]
G3 TOC Details withheld at licensee's request by SSL.
G3 TOE Details withheld at licensee's request by SSL.
G3 TOF R T A Brown, 177 Radburn Cl, Harlow, CM18 7EH [JO01BR, TL40]
G3 TOI Details withheld at licensee's request by SSL.
G3 TOJ G R Steel, 10 Rossmere Ave, Rochdale, OL11 4BT [IO83VO, SD81]
G3 TOK J L Hall, 54 South Eden Park Rd, Beckenham, BR3 3BG [IO91XJ, TQ36]
G3 TOM F Forbes, 109 Matlock Cres, North Cheam, Sutton, SM3 9SY [IO91VI, TQ26]
G3 TON A M Fentham, 106 Elm Rd, New Malden, KT3 3HP [IO91UJ, TQ26]
G3 TOP A H Peperell, 16 Claremont Rd, Marlow, SL7 1BW [IO91ON, SU88]
G3 TOQ N C A Taylor, Rua Ministro Raul, F'Nandes 180 Apt1805, Botafogo R'D'Janeir, Brazil 22260
G3 TOR Details withheld at licensee's request by SSL.
G3 TOS J W Williams, 58 Springfield Rd, Sawston, Cambridge, CB2 4HX [JO02CC, TL44]
G3 TOV G P Miles, 200 Ladybank Rd, Mickleover, Derby, DE3 5RR [IO92FV, SK33]
G3 TOW A D Hirst, 10 Baristow Cl, Chester, CH2 2EA [IO83NE, SJ46]
G3 TOY R J Wright, High View Cottage, Tatenhill Common, Rangemore, Burton on Trent, DE13 9RT [IO92DT, SK12]
G3 TOZ G F Parkhurst, Fruit Farm, High St., Marsham, Norwich, NR10 5QD [JO02OS, TG12]
G3 TPB Dr J C Knight, 2120 North Pantops, Dr, Charlottesville, Va 22901 USA
G3 TPH R J Henville, 67 Salisbury Rd, Blandford Forum, DT11 7LW [IO80WU, ST80]
G3 TPI E M Wager, 29 York Rd, Loughborough, LE11 3DA [IO92JS, SK51]
G3 TPJ O S Tillett, 27 Cranbrook Dr, Gidea Park, Romford, RM2 6AP [JO01CN, TQ58]
G3 TPK R G Wells, 31 Hunting Gate, Birchington, CT7 9JA [JO01PJ, TR26]
G3 TPM M J Sharman, 2 Culford Rd, Fornham St. Martin, Bury St. Edmunds, IP28 6TN [JO02IG, TL86]
G3 TPO C S Ockendon, 29 Garlies Rd, Forest Hill, London, SE23 2RU [IO91XK, TQ37]
G3 TPP B E Eyre, 56A Ounsdale Rd, Wombourne, Wolverhampton, WV5 8BH [IO82VM, SO89]
G3 TPQ G J Harris, 12 Highridge Cl, Purton, Swindon, SN5 9BS [IO91BO, SU08]
G3 TPT R E Wills, 32 Albert Rd, Sutton, SM1 4RX [IO91VI, TQ26]
G3 TPV F A Robinson, 30 Manor Rd, Fawley, Holbury, Southampton, SO45 2NN [IO90HT, SU40]
G3 TPW S R Webb, 1 The Green, Swinton, Malton, YO17 0SY [IO94ND, SE77]
G3 TPX P Avill, 7 Moorland Cres, Mapplewell, Barnsley, S75 6NS [IO93FO, SE31]
G3 TPZ J S Adams, Silver Birches, 1 Meadow Way, Norwich, NR6 5NW [JO02PP, TG21]
G3 TQA A Robinson, 9 Illingworth Cl, Illingworth, Halifax, HX2 9JQ [IO93BS, SE02]
G3 TQC J C Sunderland, 7 Beavers Cl, Guildford, GU3 3BX [IO91QF, SU95]
G3 TQD R L Avery, 42 Lineholt Cl, Redditch, B98 7YU [IO92AG, SP06]
G3 TQE A J Booth, 81 Brockhurst Rd, Hodge Hill, Birmingham, B36 8JB [IO92CL, SP18]
G3 TQF G D Findon, 20 Flaxfield Cl, Groby, Leicester, LE6 0EZ [IO92JP, SK50]
G3 TQG Details withheld at licensee's request by SSL.
GM3 TQH W M Burke, 1 Ashton Rd, Glasgow, G12 8SP [IO75UU, NS56]
GW3 TQI D Benson, 16 Penrhos, Radyr, Cardiff, CF4 8RJ [IO81IM, ST18]
G3 TQL J H Jones, 97 Ravenscroft Rd, Ashmore Lake, Willenhall, WV12 4LW [IO82XO, SO99]
G3 TQM Details withheld at licensee's request by SSL.
G3 TQN D H E King, 4 Heath Rd, Thurston, Bury St. Edmunds, IP31 3PJ [JO02JG, TL96]
G3 TQO J Rowley, 22 Corner Park, Saffron Walden, CB10 2EF [JO02CA, TL53]
G3 TQP I E Davies, South View, Ashley Rd, St. Georges, Telford, TF2 9LF [IO82SQ, SJ71]
G3 TQQ J D Bottomley, 60 Trenance Gdns, Greetland, Halifax, HX4 8NN [IO93BQ, SE02]
G3 TQS P A Patrickson, 227 Weston Rd, Meir Heath, Stoke on Trent, ST3 6EF [IO82WX, SJ94]
G3 TQU E F Munroe, 83 Riverside, Leighton Buzzard, LU7 7HX [IO91QW, SP92]
G3 TQX G Grimshaw, 50 Rembrandt Way, Bury St. Edmunds, IP33 2LT [JO02IF, TL86]
G3 TQY M T Knights, Springside Farm, Tismans Common, Rudgwick Horsham, West Sussex, RH12 3DU [IO91SC, TQ03]
G3 TQZ R D Allan, 9 Christine Ave, Rushwick, Worcester, WR2 5SW [IO82UE, SO85]
G3 TRA J A Wilkinson, 25 St. Helens Dr, Leicester, LE4 0GS [IO92KP, SK50]
G3 TRB T A Barber, 48 Newland Rd, Droitwich, WR9 7AZ [IO82WG, SO86]
G3 TRC R T Collins, 8 Sylvan Way, Redhill, RH1 4DE [IO91WF, TQ24]
G3 TRD J Bellamy, 2A Devon Rd, Felixstowe, IP11 9AF [JO01QX, TM23]
G3 TRE K Rigby, 9 Batcliffe Dr, Leeds, LS6 3QB [IO93ET, SE23]
G3 TRF P J Pickering Maidstone YMCA, Ars, 21 Palmar Rd, Maidstone, ME16 0DL [JO01GG, TQ75]
G3 TRG R E Green, 2 Ragley Walk, Rowley Regis, Warley, B65 9NT [IO82XL, SO98]
G3 TRH R Farrance, 57 High Mead, Rayleigh, SS6 7DT [JO01HO, TQ89]
GM3 TRI A F Ferguson, 33 Unity Terr, Perth, PH1 2BG [IO86GJ, NO12]
G3 TRK D Kitson, 11 Deerstone Rd, Nelson, BB9 9LN [IO83VU, SD83]
G3 TRL A T Green, Dee Comms Ltd, Dutton Green, Stanney Mill, Chester, CH2 4RA [IO83NG, SJ47]
G3 TRP Details withheld at licensee's request by SSL.
G3 TRU H Bates, 20 Ashford Rd, Wellington, TA21 8QF [IO80JX, ST11]
G3 TRV M Smith, 161 Batley Rd, Kirkhamgate, Wakefield, WF2 0SP [IO93FQ, SE22]
G3 TRX C W R Bailey, 15 Seymour Ave, Thanet, Margate, CT9 5HT [JO01QJ, TR37]
G3 TRY W L Pechey, Jays Lodge, Crays Pond, Reading, RG8 7QG [IO91KM, SU68]
G3 TSA J S Denby, 107 Station Rd, Fenaybridge, Fenay Bridge, Huddersfield, HD8 0DE [IO93DP, SE11]
G3 TSE D J Brealy, 6 Acre Pl, Stoke, Plymouth, PL1 4QP [IO70VJ, SX45]
G3 TSF E M Glasscott, 28 Gunton Cliff, Lowestoft, Suffolk, NR32 4PF [JO02VL, TM59]
GW3 TSH Details withheld at licensee's request by SSL.[Op: Rob Wilcox.]
G3 TSI R B Flood, 147 City Way, Rochester, ME1 2BE [JO01GI, TQ76]
G3 TSJ G A Smith, 15 Bedford Ave, High Crompton, Shaw, Oldham, OL2 7DR [IO83WN, SD90]
G3 TSK Details withheld at licensee's request by SSL.[Station located in mid-Somerset.]
G3 TSM V J Mallows, 13 Greatfield Way, Rowlands Castle, PO9 6AG [IO90MV, SU71]
G3 TSO M J Grierson, Woodlands, 9 Coneygar Rd, Quenington, Cirencester, GL7 5BY [IO91CR, SP10]
GW3 TSQ J D Bowen, 33 Parklands View, Sketty Park, Sketty, Swansea, SA2 8LT [IO81AO, SS69]
G3 TSR Col P S Reader, HQ Cvww, Wentworth Barracks, Bfpo 15, ZZ9 9HQ
G3 TSS C A Waters, 1 Chantry Est, Corbridge, NE45 5JH [IO84XX, NY96]
G3 TSV T H Clay, 132 Underdale Rd, Shrewsbury, SY2 5EF [IO82PR, SJ51]
G3 TSW L E J Pierce, 141 Replingham Rd, Southfields, London, SW18 5LX [IO91VK, TQ27]
G3 TSZ A R Mac Walter, 142 Altrincham Rd, Wilmslow, SK9 5NQ [IO83UI, SJ88]
G3 TTB P B Clegg, 6 Ricketts Dr, Billericay, CM12 0HH [JO01EP, TQ69]
G3 TTC K M Orchard, 32 Myton Cres, Warwick, CV34 6QA [IO92FG, SP26]
G3 TTH T S Coltman, Home Farm, Normanton on Soar, Loughborough, LE12 5HB [IO92JT, SK52]
G3 TTI L F Meikle, 3 Hillcrest, West Woodburn, Hexham, NE48 2RZ [IO85VE, NY88]
G3 TTJ J G Barber, Paines Oast, East Hoathly, Lewes, E Sussex, BN8 6DT [JO00BW, TQ51]
G3 TTL E Turner, 38 Vine St., Whelley, Wigan, WN1 3PG [IO83QN, SD50]
G3 TTP Details withheld at licensee's request by SSL.
G3 TTU R Holt, 11 Morley Rd, Wheatley, Doncaster, DN1 2TW [IO93KM, SE50]
G3 TTW V Medici, Villa Medici, Callas, Bishop Burton, Beverley, HU17 8QL [IO93SU, SE93]
G3 TTY B J Field, 14 Glenalmond House, Manor Fields, Putney, London, SW15 3LP [IO91VK, TQ27]
G3 TTZ C A Watkins, 70 Calcott Rd, Knowle, Bristol, BS4 2HE [IO81RK, ST67]
G3 TUA Dr S Lazarus, 114 Beechwood Gdns, Clayhall, Ilford, IG5 0AQ [JO01AN, TQ48]
G3 TUC F J Lengyel, 31 Brook Rd, Epping, CM16 7BT [JO01BQ, TL40]
GW3 TUD J M Allen, Roseleigh, High St., Saundersfoot, SA69 9EJ [IO71PR, SN10]
G3 TUF F A Long, 37 St. Catherines Rd, Bitterne Park, Southampton, SO18 1LS [IO90HW, SU41]
GW3 TUG W C Burton, 9 Cendl Cres, Rassau Beafort, Rassau, Ebbw Vale, NP3 5PR [IO81JT, SO11]
G3 TUJ C Green, Bryden, Coolham Rd, West Chiltington, Pulborough, RH20 2LH [IO90TX, TQ12]
G3 TUK R Hall, 26 Frenchfield Gdns, Carleton, Penrith, CA11 8TX [IO84PP, NY52]
G3 TUL J M Copson, 101 Rampton Rd, Cottenham, Cambridge, CB4 4TJ [JO02BG, TL46]
G3 TUN J B Greenwood, 35 Cyprus Rd, Hatch Warren, Basingstoke, RG22 4UY [IO91KF, SU64]
G3 TUO G R Lambert, New Willow Cottage, Willow Ct, Hadnall, Shrewsbury Shrops, SY4 3DD [IO82PS, SJ51]
G3 TUQ G F Ward, Caversham, Halterworth Ln, Whitenap, Romsey, SO51 9AD [IO90GX, SU32]
G3 TUU C R Keeble, 86 Kirby Rd, Walton on The Naze, CO14 8RL [JO01PU, TM22]
G3 TUX C J Rees, PO Box 88, Haslemere, Surrey, GU27 2RF [IO91PD, SU93]
G3 TUY M G G Bruce, 3 Redlands Pl, Wokingham, RG41 4ED [IO91NJ, SU76]
G3 TVC L Rice, Beechwood, 11 Barnoldby Rd, Waltham, Grimsby, DN37 0JR [IO93WM, TA20]
G3 TVD J D Shersby, 29 Vale Sq, Ramsgate, CT11 9DE [JO01QH, TR36]
G3 TVF S Darbyshire, 41 Landswood Rd, Oldbury, Warley, B68 9QE [IO92AL, SP08]
G3 TVH J E Harknett, 60 Windmill Dr, Croxley Green, Rickmansworth, WD3 3FE [IO91SP, TQ09]
G3 TVI R A W Stevens, 64 Ferndale, Waterlooville, PO7 7PB [IO90LV, SU60]
G3 TVK E W P Jones, 59 Elm Dr, Hove, BN3 7JA [IO90VU, TQ20]
G3 TVL P M Hunt, 14 Walnut Cl, Epsom, KT18 5JL [IO91UH, TQ25]
G3 TVM H D Fletcher, 20 Westfield Rd, Great Shelford, Cambridge, CB2 5JW [JO02BD, TL45]
G3 TVN R J Williams, 23A Acacia Ave, Roby, Liverpool, L36 5TN [IO83NJ, SJ49]
G3 TVO T A Mellor, 2 Lesser Foxholes, Shoreham By Sea, BN43 5NT [IO90UU, TQ20]
G3 TVP Details withheld at licensee's request by SSL.
G3 TVR E G Churchyard, 11 Greenfields Dr, Bridgnorth, WV16 4JW [IO82SM, SO79]
G3 TVS A J R Pegler Thames Vall ART, Brook House, Forest Cl, East Horsley, Leatherhead, KT24 5BU [IO91SG, TQ15]
G3 TVT I M Fraser, 5 Blenmar Cl, Radcliffe, Manchester, M26 2XE [IO83UN, SD70]
G3 TVU I D Brown, 63 Peak View Dr, Ashbourne, DE6 1BR [IO93DA, SK14]
G3 TVV A C L Coates, 35 Mogg St., St. Werburghs, Bristol, BS2 9UB [IO81RL, ST67]
G3 TVW H M Davison, 18 Woodcock Rd, Flamborough, Bridlington, YO15 1LJ [IO94WC, TA27]
G3 TVX D G Ashwood, 114 Clifton Dr, Blackpool, FY4 1RR [IO83LS, SD33]
G3 TWB R A Ballard, 31 South Devon Ave, Nottingham, Notts, NG3 6FT [IO92KX, SK54]
G3 TWE Details withheld at licensee's request by SSL.
G3 TWG LtCm P J Patrick, Bedford Lodge, Camden Pl, Bourne End, SL8 5RW [IO91PN, SU88]
G3 TWJ M H Roach, 104 Old Lodge Ln, Purley, CR8 4DH [IO91WH, TQ35]
G3 TWN F Mason, 22 Brookhurst Rd, Bromborough, Wirral, L63 0EP [IO83MH, SJ38]
G3 TWO F Rhodes, 23 Quantock Ave, Bridgwater, TA6 7EB [IO81LD, ST23]
G3 TWQ A R A Major, Jamara, Aldeburgh Rd, Friston, Saxmundham, IP17 1NP [JO02SE, TM46]
G3 TWS J R G Corbett, Jacobs Ladder, Leys Hill, Walford, Ross on Wye Hfds, HR9 5QS [IO81QV, SO52]
G3 TWT E V Robinson, 176 Gill Ave, Fishponds, Bristol, BS16 2PH [IO81RL, ST67]
G3 TWX D C Woodhouse, Flat 3 Raylands, Rayes Ln, Newmarket, Suffolk, CB8 7AB [JO02EF, TL66]
G3 TWY G J Mills, 23 Harecroft Cres, Sapcote, Leicester, LE9 4FX [IO92IM, SP49]
G3 TWZ D J Farman, 4 Juniper Gr, Sheringham, NR26 8LX [JO02OW, TG14]
G3 TXA R E Wilkinson, 609 Green Lanes, Palmers Green, London, N13 4EP [IO91WP, TQ39]
G3 TXC N C Harris, April Cottage, Sheepcote Green, Clavering, Saffron Walden, CB11 4SJ [JO01BX, TL43]
G3 TXE A Parker, 11 Marlow Rd, Ipswich, IP1 5JJ [JO02NB, TM14]
G3 TXF N S Cawthorne, Falcons, St. Georges Ave, Weybridge, KT13 0BS [IO91SI, TQ06]
G3 TXH B G Levett, 198 Church St., Ellesmere Port, South Wirral, L65 2DT [IO83NG, SJ47]
G3 TXK C Moss, 2 Sutton Ln, Sutton House Farm Est, Adlington, Chorley, PR6 9PA [IO83QO, SD61]
G3 TXL J A Graham, Woodtown, Sampford Spiney, Yelverton, Devon, PL20 6LJ [IO70XM, SX57]
G3 TXN P J Tucker, 20 Bourne Mead, Bexley, Kent, DA5 1PJ [JO01CK, TQ57]
G3 TXO R H Sutton, 12A The Beeches, 2 Newmarket Rd, Royston, Herts, SG8 7DY [IO92XB, TL34]
G3 TXP L S Duffy, 60 Snatchup, Redbourn, St. Albans, AL3 7HB [IO91TT, TL11]
G3 TXQ S E Hunt, 21 Green St., Milton Malsor, Northampton, NN7 3AT [IO92ME, SP75]
G3 TXT R T Laing, 3 Lime Kiln Rd, Tackley, Kidlington, OX5 3BW [IO91IV, SP42]
G3 TXU B Peacock, 13 Station Rd, Mossley, Ashton under Lyne, OL5 9EA [IO83XM, SD90]
G3 TXX B Tiffany, 18 Fairfax Rd, Bingley, BD16 4DR [IO93BU, SE14]
G3 TXZ C E Tucker, 6 Rosehill Gdns, Crowborough Hill, Crowborough, TN6 2ED [JO01CB, TQ53]
G3 TYA J B Grant, Tanyanga, Wheal Leisure, Perranporth, TR6 0EY [IO70KI, SW75]
G3 TYB Details withheld at licensee's request by SSL.
G3 TYE I Barton, 3 Barnmeadow Rd, Liverpool, L25 4UG [IO83NJ, SJ48]
G3 TYG B D R Winslow, 10 Almond Walk, Hazlemere, High Wycombe, HP15 7RE [IO91PP, SU89]

G3 TYH M J Cooney, 9 Moorfield Ave, Ealing, London, W5 1LG [IO91UM, TQ18]
GW3 TYI D J West, 44 Glanmor Park Rd, Sketty, Swansea, SA2 0QE [IO81AO, SS69]
G3 TYO J Stringer, 1 Hazel Rd, Tavistock, PL19 9DN [IO70WM, SX47]
G3 TYP I A Jackson, 47 Stephen St., Rugby, CV21 2ES [IO92II, SP47]
GM3 TYQ Dr B P Simpson, 67 Partickhill Rd, Glasgow, G11 5AD [IO75UV, NS56]
G3 TYR K C Davis, 121 Weston Rd, Guildford, GU2 6AW [IO91QF, SU95]
GM3 TYS I G Drysdale, 82 Kirk Brae, Cults, Aberdeen, AB1 9QQ [IO87TH, NJ72]
G3 TYT W Maclacklan, The Lowlands, St. Michaels Rd, Penkridge, Stafford, ST19 5AH [IO82WR, SJ91]
G3 TYU M J Bredael, 6 Ollerton Gdns, Windy Nook, Gateshead, NE10 9RT [IO94FW, NZ26]
G3 TYY Details withheld at licensee's request by SSL.
G3 TZA J J A Riley, 3 Manor Way, Croxley Green, Rickmansworth, WD3 3LU [IO91SP, TQ09]
GI3 TZB W J M McKinney, 33 Heatherstone Rd, Bangor, BT19 6AE [IO74EP, J58]
GI3 TZC J J Canavan, 15 Barry St., Londonderry, BT48 7PJ [IO65IA, C41]
G3 TZD R Mansell, Rd1 Box 552A, Salt Point, New York, USA 12578
G3 TZE R G Armitage, 3 Holst Mead, Stowmarket, IP14 1TD [JO02LE, TM05]
GI3 TZF R A W Watson, 115 Locksley Park, Belfast, BT10 0AT [IO74AN, J36]
G3 TZG J H E Glanville, 3 Seneschal Rd, Coventry, CV3 5LF [IO92GJ, SP37]
G3 TZH A Dolby, Payrole, 82100 Castelferrus, Tarn Et Garonne, France, X X
G3 TZL P A Bowen, White House, Durleigh Marsh, Petersfield, GU31 5AX [IO91NA, SU72]
G3 TZM Details withheld at licensee's request by SSL.
G3 TZO P J Holland, Chatterton, Chapel Ln, Threapwood, Malpas, SY14 7AX [IO83OA, SJ44]
G3 TZP I E Rodwell, Wellfield, Park Ln, Ln End, High Wycombe, HP14 3NN [IO91OO, SU89]
G3 TZQ S G Ridgway, 8 Almeria Ct, Plympton, Plymouth, PL7 1TX [IO70XJ, SX55]
G3 TZS P G Jones, 6 Falcon Cl, Fareham, PO16 8PL [IO90KU, SU50]
G3 TZT M C Mead, 57 Great Melton Rd, Hethersett, Norwich, NR9 3HA [JO02OO, TG10]
G3 TZU J E Harding, Beech Tree Cottage, New Rd, Hanmer, Whitchurch, SY13 2JX [IO82OW, SJ43]
G3 TZV P Fry, 52 Spathfield Ct, Holmfield Cl, Stockport, SK4 2RR [IO83VK, SJ89]
GI3 TZX W H Nesbitt, 101 Belfast Rd, Bangor, Co. Down, BT20 3PP [IO74DP, J48]
G3 TZZ J W E Jackson, 18 Blakesware Gdns, Edmonton, London, N9 9HU [IO91XP, TQ39]
GM3 UA A S Pairman, Seabank, Largiebeg, Whiting Bay, Brodick, KA27 8RL [IO75LL, NS02]
G3 UAA D A Ramsey, The Orchard, Carmen Gr, Groby, Leics, LE6 0BA [IO92JP, SK50]
G3 UAC J D W Aitken, 135 Waddon Rd, Croydon, CR0 4JL [IO91WI, TQ36]
G3 UAE J O Gill, 22 Maddever Cres, Liskeard, PL14 3PT [IO70SK, SX26]
G3 UAF Dr M W Smith, 138 Market St., Clay Cross, Chesterfield, S45 9LY [IO93HE, SK36]
GM3 UAG Me J R Davidson, Cairntoul, Cassiegills, Ellon, Aberdeen, AB41 8QS [IO87XJ, NJ93]
G3 UAL J P Price, 6 St. Marys Terr, Leamington Spa, CV31 1JT [IO92FG, SP36]
G3 UAS T D W Morgan, 2 Park View, Hatch End, Pinner, HA5 4LN [IO91TO, TQ19]
G3 UAU Details withheld at licensee's request by SSL.
G3 UAX R H Stansfield, 3 Greencroft Gdns, Reading, RG3 3PL [IO91LJ, SU66]
GW3 UAY C D Butters, 105 Blackoak Rd, Cyncoed, Cardiff, CF2 6QW [IO81JM, ST18]
G3 UAZ E A Sweetman, 8 The Acorns, Wimborne Rd West, Wimborne, BH21 2EU [IO90AT, SZ09]
GI3 UBA R A Reid, 21 Ballymaconnell Rd, Bangor, BT20 5PN [IO74EP, J58]
G3 UBB D W Fill, 2 Brook Cl, Packington, Ashby de La Zouch, LE65 1WA [IO92GR, SK31]
G3 UBC W T Guilfoyle, 22 Dukeshill Rd, Bracknell, RG42 2DT [IO91OK, SU86]
G3 UBD G M Higgins, Lower Laithe Farm, Providence Ln, Oakworth, Keighley, BD22 7QS [IO93AU, SE03]
G3 UBI M J Fisher, Great House Barn, Lighthazels Rd, Ripponden, Sowerby Bridge York, HX6 4NP [IO93AQ, SE02]
GM3 UBJ Dr W S Hossack, Kincrig, 39 Skene St., Macduff, AB44 1RP [IO87SQ, NJ76]
G3 UBL C L K Ledger, 17 Orchard Cl, Bushey Heath, Bushey, Watford, WD2 1LW [IO91TP, TQ19]
G3 UBM J R E Marshall, 39 Gr Rd, Headingley, Leeds, LS6 2AQ [IO93FT, SE23]
G3 UBO J B Lowe, Greenway, Matlock Rd, Walton, Chesterfield, S42 7LD [IO93GF, SK36]
G3 UBP C Riches, 95 Hallford Way, Dartford, DA1 3AA [JO01CK, TQ57]
G3 UBR Details withheld at licensee's request by SSL.
G3 UBS B J Speakman, Merrydown, Burley Ln, Quarndon, Derby, DE6 4JS [IO92FW, SK23]
G3 UBU Details withheld at licensee's request by SSL.
G3 UBV D R Roberts, 8 Churnet Cl, Bedford, MK41 7ST [IO92SD, TL05]
G3 UBX J P H Burden, 2 Links Rd, Wolverhampton, WV4 5RF [IO82WN, SO99]
G3 UBY A M Clark, Sans Souci, Fairmead Rd, Saltash, PL12 4JH [IO70VJ, SX45]
G3 UBZ R McNair, 1079A Christchurch Rd, Bournemouth, BH7 6BQ [IO90CR, SZ19]
G3 UCA P G Sinclair, 32 Barn Meadow, Clayton Brook, Bamber Bridge, Preston, PR5 8DU [IO83QR, SD52]
G3 UCC R Hayward, 29 Golden Hill, Wiveliscombe, Taunton, TA4 2NT [IO81IB, ST02]
G3 UCD R W Pescod, 7 Brian Cl, Chelmsford, CM2 9DZ [JO01FR, TL70]
G3 UCE L B Uphill, 12 Rushley Mount, Hest Bank, Lancaster, LA2 6EE [IO84OC, SD46]
GM3 UCH W P Wright, 460 Main St., Stenhousemuir, Larbert, FK5 3JU [IO86CA, NS88]
GM3 UCI G D McCallum, 15 Quarry Rd, Law, Carluke, ML8 5HB [IO85BR, NS85]
GW3 UCJ M J P Evans, 31 Cilmaengwyn, Vnysmeudwy, Pontardawe, West Glam, SA8 4QL [IO81CR, SN70]
G3 UCK G W Downs, 2 Dyehouse, Wilsden Hill, Wilsden, Bradford, BD15 0BE [IO93BT, SE03]
G3 UCM Details withheld at licensee's request by SSL.
G3 UCR W R Byrom, 54 Chapel Rd, Tadworth, KT20 5SE [IO91VG, TQ25]
G3 UCT M G Taylor, Wychwood, 27 Glen Rd, Fleet, GU13 9QS [IO91NG, SU85]
G3 UCV Details withheld at licensee's request by SSL.
G3 UCW M D Pettit, 84 Western Rd, Sompting, Lancing, BN15 9UB [IO90TT, TQ10]
G3 UD G Bloor, 26 Leveson Rd, Hanford, Stoke on Trent, ST4 4QP [IO82VX, SJ84]
G3 UDA K A Linney, 57 Whitemere Rd, Shrewsbury, SY1 3BY [IO82PR, SJ41]
G3 UDC K N Bran, 9 Carnoustie Gr, Bletchley, Milton Keynes, MK3 7RP [IO91OX, SP83]
G3 UDD S A G Chandler, 94 Cannon Park Rd, Coventry, CV4 7AY [IO92FJ, SP37]
G3 UDH P A Butcher, 55 Offington Ln, Worthing, BN14 9RJ [IO90TT, TQ10]
G3 UDI Dr R J Butcher, Temple Lodge, Six Mile Bottom Rd, Little Wilbraham, Cambridge, CB1 5LD [JO02DE, TL55]
GM3 UDJ T M Grant, Laigh Hook Cottage, Strathaven, Lanarkshire, ML10 6RP [IO75WQ, NS64]
GM3 UDL Details withheld at licensee's request by SSL.
G3 UDN B D Clulee obo Mid Warwickshire A.R.S., 11 Ascot Ride, Lillington, Leamington Spa, CV32 7TT [IO92FH, SP36]
G3 UDP M J Brown, 4 Boyfields, Quadring, Spalding, PE11 4QQ [IO92VV, TF23]
G3 UDU Details withheld at licensee's request by SSL.
G3 UDV P E Lindsley, 5 Castlebar Mews, Ealing, London, W5 1RS [IO91UM, TQ18]
G3 UDW P Williams, 31 Falmouth Ave, Undercliffe, Bradford, BD3 0HL [IO93DT, SE13]
G3 UDZ G L Bolton, 8 Kings Ct, Kings Cl, Leyland, Preston, PR5 1SF [IO83PQ, SD52]
G3 UEC N Stanley, 9 Castle View, Sedgwick, Kendal, LA8 0JL [IO84PG, SD58]
G3 UED J M Jones, 12 Francis Groves Cl, Bedford, Beds, MK41 7DH [IO92SD, TL05]
G3 UEE D Diamond, 36 Darbys Ln, Oakdale, Poole, BH15 3ET [IO90AR, SZ09]
G3 UEG D I Gould, 2 Mayfield Cl, Old Harlow, Harlow, CM17 0LH [JO01BS, TL41]
G3 UEH D G Barson, 2 Pitreavie Dr, Hailsham, BN27 3XG [JO00CU, TQ51]
G3 UEK J M Whitehouse, PO Box 583, Laceys Spring, Alabama 35754-0583, USA, X X
G3 UEN R W Stiles, Outgate, Southsea Rd, Flamborough, Bridlington, YO15 1AE [IO94WC, TA27]
GW3 UEP R P V Plimmer, Tregroes, Llandysul, Dyfed, SA44 4NA [IO72UB, SN44]
G3 UEQ A H R Hearn, 53 Twyford Gdns, Worthing, BN13 2NT [IO90TT, TQ10]
G3 UER Details withheld at licensee's request by SSL.
G3 UES R S Hewes Echelford ARS, 24 Brightside Ave, Laleham, Staines, TW18 1NG [IO91SK, TQ07]
G3 UEU J Holmes, 9 Fir Trees Gr, Higham, Burnley, BB12 9BA [IO83UT, SD83]
G3 UEY D M Browning, 13 Beechcombe Cl, Pershore, WR10 1PW [IO82XC, SO94]
G3 UFA Rev T W Gladwin, 99 Warren Way, Digswell, Welwyn, AL6 0DL [IO91VT, TL21]
G3 UFB N G Brinkworth, 48 Cambridge Rd, St. Albans, AL1 5LD [IO91UR, TL10]
G3 UFC G D Hurst, 2 Sunny Bank Dr, Mirfield, WF14 0JN [IO93DQ, SE12]
G3 UFF P A Rodway, 141 Hazelwood Dr, St. Albans, AL4 0UY [IO91US, TL10]
G3 UFH T J Moss, 42 Chalfont Cl, Cherry Hinton, Cambridge, CB1 4NA [JO02CE, TL45]
G3 UFI P A Conway, 1 The Woodlands, The Ridge, Hastings, TN34 2SF [JO00GV, TQ81]
G3 UFJ L Symons, 31 Springfield Way, Threemilestone, Truro, TR3 6BJ [IO70KG, SW74]
G3 UFK Details withheld at licensee's request by SSL.[Op: M P Taylor, 48 Garfield Road, Enfield, Middx, EN3 4RR.]
G3 UFO Details withheld at licensee's request by SSL.
G3 UFP Details withheld at licensee's request by SSL.
G3 UFQ D W E Eckley, 27 Apsley Gr, Dorridge, Solihull, B93 8QP [IO92CI, SP17]
G3 UFS C J Smith, 50 Grand Ave, Lancing, BN15 9PZ [IO90UT, TQ10]
G3 UFV P Crawshaw, 35 Bishopton Ave, Stockton on Tees, TS19 0RA [IO94HN, NZ42]
G3 UFW J Stevens, Pendine, Romsey Rd, West Wellow, Romsey, SO51 6BG [IO90FX, SU21]
G3 UFX H G Julian, Brigantine, Market St., Penryn, TR10 8BH [IO70KE, SW73]
G3 UFY S V Knowles, 77 Bensham Manor Rd, Thornton Heath, CR7 7AF [IO91WJ, TQ36]
G3 UFZ C H Foulkes, 3 Manor Bend, Galmpton, Brixham, TQ5 0PB [IO80FJ, SX85]
G3 UGB A Woffenden, 89 Beechen Dr, Fishponds, Bristol, BS16 4BU [IO81RL, ST67]
G3 UGC J Smethurst, 81 Springside Rd, Bury, BL9 5JG [IO83UO, SD81]
G3 UGF R J Constantine, The Old Exchange, Burnley Rd, Mytholmroyd, Hebden Bridge, HX7 5PD [IO93AR, SE02]
G3 UGG Details withheld at licensee's request by SSL.
G3 UGJ D G Smith, 33 Rippington Dr, Old Marston, Marston, Oxford, OX3 0RJ [IO91JS, SP50]
G3 UGK Details withheld at licensee's request by SSL.
G3 UGL H R Quantick, 39 Springfield Way, Cranfield, Bedford, MK43 0JN [IO92QB, SP94]
G3 UGO E M Cooper, The Firs, Treskerby, Redruth, Cornwall, TR16 5AG [IO70JF, SW74]
G3 UGR C D Tabor, Banksia, High St., Queen Camel, Yeovil, BA22 7NH [IO81RA, ST52]
G3 UGT Details withheld at licensee's request by SSL.[Op: J Coote. Station located near Wallsend.]
G3 UGX R B Heaton, Flat 2, 95 Redington Rd, London, NW3 7RR [IO91VN, TQ28]
G3 UGY J S V Mortlock, 10 Wattendon Rd, Kenley, Surrey, CR8 5LU [IO91WH, TQ35]
G3 UHF C C Ward Sth Manchester, Arcade, 2 Arlington Dr, Stockport, SK2 7EB [IO83WJ, SJ98]
G3 UHJ R A Gordon, 77 Alwyn Rd, Darlington, DL3 0AH [IO94FN, NZ21]
G3 UHK J E C Baldwin, 41 Castle Dr, Maidenhead, SL6 6DB [IO91PM, SU88]
G3 UHN P A Neale, 98 Meadway, Harpenden, AL5 1JQ [IO91UT, TL11]
G3 UHQ C A Robinson, 1968 Coventry Rd, Sheldon, Birmingham, B26 3HJ [IO92CK, SP18]
G3 UHS C E Houltby, 9 Bayard St., Gainsborough, DN21 2JZ [IO93OJ, SK89]
GM3 UHT W Garner, Sarkshields Cottage, Eaglesfield, Lockerbie, Dumfriesshire, DG11 3AE [IO85KB, NY27]
G3 UHU D M Hampton, 30 Victoria Rd, Maldon, CM9 5HF [JO01IR, TL80]
G3 UHV C J Sutton, Braehead, Old Ln, Brown Edge, Stoke on Trent, ST6 8TG [IO83WC, SJ95]
G3 UHW H J Tomlinson, 32 Manor Rd, Farnborough, GU14 7EU [IO91PG, SU85]
G3 UHX A B Thorpe, 12 Newnham Ln, Ryde, PO33 4ED [IO90JR, SZ59]
G3 UHY L R D Ling, Cherrels, Mill Rd, Bethersden, Ashford, TN26 3DA [JO01JD, TQ94]
G3 UHZ R P Treneer, 15 Kensaway, Connordowns Hayle, Cornwall, TR27 5EQ [IO70HE, SW63]
G3 UI L L N Cobb, 27 Moorlands Cres, Halifax, HX2 8AA [IO93BR, SE02]
G3 UIB C C Hearn, 8 The Poles, Upchurch, Sittingbourne, ME9 7EX [JO01HJ, TQ86]
G3 UID K A P Baldock, 5321 92nd St NE, Marysville, Wa 98270, USA
G3 UIE Details withheld at licensee's request by SSL.
G3 UIF G N Thorne, Flagstaff House, Main St., Welwick, Hull, HU12 0RY [JO03AQ, TA32]
G3 UIJ M J Peake, 73 Jamieson House, Edgar Rd, Hounslow, TW4 5QH [IO91TK, TQ17]
G3 UIK J S Young, Shirley Lodge, 45 Graham Rd, Malvern, WR14 2HU [IO82UC, SO74]
G3 UIN Details withheld at licensee's request by SSL.[Op: M. McLeman. Station located near Perpignan, Dept 66, Pyrenees-Orientales.]
G3 UIP R Hill, 14 Norcliffe Rd, Blackpool, FY2 9AW [IO83LU, SD34]
G3 UIR K Craig, 26 Queens Rd, Black Hill, Consett, DH8 0BJ [IO94BU, NZ05]
G3 UIS A P Stone, West Lodge, The Downs, Poulton-le-Fylde, FY6 7EG [IO83MU, SD33]
G3 UJA B A McClory, 12 The Cres, Mottram St. Andrew, Macclesfield, SK10 4QW [IO83VH, SJ87]
G3 UJB B T Davis, 9 Old Rectory Cl, Instow, Bideford, EX39 4LY [IO71WB, SS43]
G3 UJC J N Read, Waldweg 15, Maitenbeth, 8091, Germany, X X
G3 UJE B D R Gale, Tall Trees, Noahs Ark Ln, Mobberley, Knutsford, WA16 7AX [IO83UH, SJ87]
G3 UJG J G Seal, Willow Cottage, 45 Doulting, Shepton Mallet, Somerset, BA4 4QE [IO81RE, ST64]
G3 UJI S R Turner, 9 Bullrush Cl, Dibden Purlieu, Southampton, SO45 4NN [IO90HU, SU40]
G3 UJK J W Burnham, 304 Desborough Ave, High Wycombe, HP11 2TJ [IO91OO, SU89]
G3 UJL Details withheld at licensee's request by SSL.
G3 UJO P Bradley, 60 Weyland Rd, Headington, Oxford, OX3 8PD [IO91JS, SP50]
G3 UJU M E F Haslam, Updene, The Dene, Hindon, Salisbury, SP3 6EE [IO81WC, ST93]
G3 UJV R M M Heath, 26 Lancaster Ave, Hadley Wood, Barnet, EN4 0EX [IO91VQ, TQ29]
G3 UJZ J G McNaught, Ryton House, Lechlaoe, Glos, GL7 3AR [IO91DQ, SU29]
G3 UK C V Whittaker, Fir Bank, Manor Rd, Whitchurch on Thames, Reading, RG8 7EW [IO91KL, SU67]
G3 UKB R A Cowdery, 89 Downlands, Chells Manor Villag, Stevenage, SG2 7BJ [IO91WV, TL22]
G3 UKC Dr K L Smith University of Kent A.R.S, Staple Farm House, Durlock Rd, Staple, Canterbury, CT3 1JX [JO01PG, TR25]
G3 UKD A Golding, 40 Unicorn Ln, Eastern Green, Coventry, CV5 7LJ [IO92FJ, SP27]
G3 UKE P W Adams, 34 Mount Pleasant Cl, Lightwater, GU18 5TP [IO91PI, SU96]
G3 UKF A G Mills, Farley, 20 Litcham Rd, Mileham, Kings Lynn, PE32 2PS [JO02JR, TF91]
GM3 UKG G M Grant, 35 Inward Rd, Buckie, AB56 1DD [IO87MQ, NJ46]
G3 UKH P A Hopwood, 58 Bolbec Rd, Newcastle upon Tyne, NE4 9EP [IO94EX, NZ26]
G3 UKI B J Curnow, 10 Maresfield Gdns, London, NW3 5SU [IO91VN, TQ28]
G3 UKK Details withheld at licensee's request by SSL.
G3 UKL M F Bennett, Shireley, Munns Ln, Hartlip, Sittingbourne Kent, ME9 7SY [JO01HI, TQ86]
G3 UKM M D Leighton, 85 Kemps Green Rd, Balsall Common, Coventry, CV7 7QF [IO92EJ, SP27]
G3 UKN C J Hewitt, 4 Hilltop Cl, Keens Park, Keens Ln, Worplesdon Guildfor, GU3 3HR [IO91QG, SU95]
G3 UKV M I Vincent, 9, Sleapford, Long Ln, Telford, Salop, TF6 6HQ [IO82RR, SJ61]
G3 UKW M J Newton, 11 Chestnut Cl, Rushmere St. Andrew, Rushmere, Ipswich, IP5 7ED [JO02OB, TM24]
G3 UKX L C Currie, 50 Crabtree Ave, Romford, RM6 5EX [JO01BO, TQ48]
G3 ULD G H Cawkwell, 50 Station Rd, Patrington, Hull, HU12 0NE [IO93XQ, TA32]
GM3 ULG J Haliburton Glenrothes and District Arc., 21 Bolam Dr, Burntisland, KY3 9HP [IO86JB, NT28]
G3 ULL D Walker, 9 Vectis Rd, Gosport, PO12 2QD [IO90KS, SZ59]
G3 ULN M Hibbitt, 123 Stanborough Rd, Plystock, Plymouth, PL9 8PJ [IO70XI, SX55]
G3 ULO I D Spencer, Fichtenweg loc, 53804 Much, West Germany, ZZ9 9IM
GM3 ULP G A Hunter, 12 Airbles Dr, Motherwell, ML1 3AS [IO85AS, NS75]
G3 ULT J C Nunn Reading & Dis Rd, 20 Somerton Gdns, Earley, Reading, RG6 5XG [IO91NK, SU77]
GM3 ULW Details withheld at licensee's request by SSL.
G3 ULY D Youngman, Hallowell, Culgaith, Penrith, Cumbria, CA10 1QW [IO84QP, NY62]
GM3 UM G P Millar, 5 Ettrick Gr, Edinburgh, EH10 5AW [IO85JW, NT27]
GW3 UMB J Taylor, 14 Ffordd Elias, Olo Colwyb, Colwyn Bay, LL29 9LA [IO83DG, SH87]
GW3 UMD N A Maxwell, 1 Nant Fawr Cres, Cardiff, CF2 6JN [IO81JM, ST18]
G3 UMF Dr A F Simpson, Forest Farmhouse, Old Rd, Shotover, Headington Oxford, OX3 8TA [IO91JS, SP50]
G3 UMH A H Dailey, 154 Harrogate Rd, Leeds, LS7 4NZ [IO93FT, SE33]
G3 UMI G S Milne, 142 Hayes Ln, Hayes, Bromley, BR2 9EL [JO01AJ, TQ46]
G3 UMK Details withheld at licensee's request by SSL.
G3 UML L S Margolis, 52 Park View Gdns, Hendon, London, NW4 2PN [IO91VN, TQ28]
G3 UMM P T Hudson, 105 Southlands, Weston, Bath, BA1 4DZ [IO81TJ, ST76]
G3 UMT B D Turvey, 90 Jenkinson Rd, Towcester, NN12 6AW [IO92MC, SP64]
G3 UMV P B Johnson, 52 Evesham Rd, Cookhill, Alcester, B49 5LJ [IO92AF, SP05]
GD3 UMW Details withheld at licensee's request by SSL.
GU3 UMX D J Ozanne, Eturs Lodge, Les Eturs, Castel, Guernsey, GY5 7DT
G3 UMZ S Tudor-Jones, 6 Green Ln, Freckleton, Preston, PR4 1RN [IO83NR, SD42]
G3 UNA D J Cutter, 34 Greengate Ln, Knaresborough, HG5 9EL [IO94GA, SE35]
G3 UNB R S Fairhead, 122 Oak Rd, Fareham, PO15 5HP [IO90JU, SU50]
G3 UNC Details withheld at licensee's request by SSL.
G3 UNF S W Yeomanson, 32 Gaynesford Rd, Forest Hill, London, SE23 2UQ [IO91XK, TQ37]
GW3 UNH N Halford, 567 Caerleon Rd, Newport, NP9 7LY [IO81MO, ST38]
G3 UNI T J Wood, 7 Wellfield Rd, Piddington, High Wycombe, HP14 3BP [IO91OP, SU89]
GM3 UNJ F E Taylor, 7 Letham Terr, Leven, KY8 4SQ [IO86ME, NO30]
G3 UNK A R Wakeman, 17 Galsworthy Rd, Goring By Sea, Worthing, BN12 6LN [IO90ST, TQ10]
G3 UNM A J Matthews, Thisledo, Dale Bank Rd, Cheadle, Stoke on Trent, ST10 1RE [IO92AX, SK04]
G3 UNP A W Ayres, 3 Swallow Cl, Bushey Heath, Bushey, Watford, WD2 1AU [IO91TP, TQ19]
G3 UNS T G Mills, 6 River Mead, Ifield, Crawley, RH11 0NS [IO91VD, TQ23]
G3 UNT B J Henman, 8 Woodlands Cl, Maidstone, ME14 2EX [JO01GG, TQ75]
G3 UNU Details withheld at licensee's request by SSL.
G3 UNV M Clift, 77 The Ridge, Great Doddington, Wellingborough, NN29 7TT [IO92PG, SP86]
G3 UNW F R Stoodley, Ragged House, Mundy Bois Rd, Egerton, Ashford, TN27 9ER [JO01IE, TQ94]
G3 UNY D E Moreman, 5 Vernon Ave, Woodingdean, Brighton, BN2 6BF [IO90XU, TQ30]
GW3 UO Details withheld at licensee's request by SSL.
G3 UOB A A Mayers, 3 Colby Rd, London, SE19 1HA [IO91XK, TQ37]
G3 UOC D R D Brown, Rexfield, Alcester Rd, Wootton Wawen, Solihull, B95 6BH [IO92CG, SP16]
G3 UOF M R Wadsworth, Reddings Cottage, Hawsley Hill, Lydbrook, Glos, GL17 9ST [IO81RU, SO61]
G3 UOG R E Jessemey, 3 Victoria Ave, Rayleigh, SS6 9DH [JO01HO, TQ79]
G3 UOI J C Firby, Beckdene, 4 Poplar Gr, Harden, Bingley, BD16 1LW [IO93BU, SE03]
G3 UOJ J Hartwell, Fulling Mill Oast, Caring Ln, Leeds, Maidstone, ME17 1TJ [JO01HG, TQ85]
G3 UOK Dr P K Grannell Univ of Keele, Dept of Physics, Keele, Staffs, ST5 5BG [IO83UA, SJ84]
G3 UOL W F Meinerts-Hahn, Hunters Farm, Garstang Rd, Pilling, Preston, PR3 6AL [IO83NW, SD44]
G3 UOM D I Horsburgh, 11 Delamare Way, Cumnor Hil, Oxford, OX2 9HZ [IO91IR, SP40]
G3 UON D H Geere, 3 Chapel Terr Mews, Brighton, BN2 1HU [IO90WT, TQ30]
GW3 UOO D Rogers, Green Tops, Megs Ln, Buckley, CH7 2AG [IO83LD, SJ26]
GU3 UOQ P P Le Boutillier, Fair Villa, Route de Leglise, Castel, Guernsey, GY5 7DL
G3 UOS Dr A J T Whitaker, Univer of Sheffield, Dept of Elec Eng, Mappin St, Sheffield, S1 3JD [IO93GJ, SK38]
GI3 UOY E P Clarke, 63 Barcroft Park, Newry, BT35 8EW [IO64TE, J02]
G3 UP A Leese, 2 Albany Pl, Louth, LN11 8EY [JO03AI, TF38]
G3 UPD H W Delmonte, 5 Scotts Cl, Colden Common, Winchester, SO21 1US [IO90IX, SU42]
G3 UPI T Codling, 21 Willow Cl, Saxilby, Lincoln, LN1 2QL [IO93QG, SK87]
G3 UPJ D A Trainer, 153 High St., Cherry Hinton, Cambridge, CB1 4LN [JO02CE, TL45]
G3 UPM T R Burke, 12 Worthing Rd, Laindon, Basildon, SS15 6AL [JO01FN, TQ68]
G3 UPN K R Snape, Delamere, Ryston End, Downham Market, Norfolk, PE38 9AX [JO02EO, TF60]
G3 UPT D Beare, 38 Morningside Rd, Aberdeen, AB10 7NT [IO87WD, NJ90]
G3 UPV M J Rutty, 55 Dartford, Corsley, Warminster, BA12 7NR [IO81UE, ST84]
G3 UPW P K Smith, 36 Ferncroft Ave, Eastcote, Ruislip, HA4 9JF [IO91TN, TQ18]
G3 UPX S P Higgins, 273 Breck Rd, Anfield, Everton, Liverpool, L5 6PU [IO83MK, SJ39]
G3 UPY D Houghton, 119 Welsby Rd, Leyland, Preston, PR5 1JD [IO83PQ, SD52]
G3 UPZ H W James, 12 Kingsdown Cl, Earley, Reading, RG6 5PN [IO91MK, SU77]
G3 UQD R W Whittington, 65 King Edward Ave, Worthing, BN14 8DG [IO90TT, TQ10]
G3 UQE R W White, Ferndale, Troubridge Rd, Helston, TR13 8DQ [IO70IC, SW62]
G3 UQH D Goddard, 4 Gravels Bank, Minsterley, Shrewsbury, SY5 0HG [IO82MO, SJ30]
GM3 UQI L R Simpson, 152 Caledonia Rd, Baillieston, Glasgow, G69 7AP [IO75WU, NS66]

G3 UQJ P Clifton, 89 Beech Rd, Roffey, Horsham, RH12 4NW [IO91UB, TQ13]
G3 UQK F Cummings, 23 Castle Hill Rd, Bury, BL9 7RN [IO83UO, SD81]
GJ3 UQM F N Jervis, 3 West Horton Cl, Eastleigh, SO50 8DA [IO90IX, SU41]
G3 UQR D A Robinson, 3 Marriott Cl, Irthlingborough, Wellingborough, NN9 5RB [IO92QH, SP97]
G3 UQS W J J Hewitt, 28 Beverley Rd, Sunbury on Thames, TW16 6NG [IO91SJ, TQ06]
G3 UQT K A Draycott, 28 Ladywood Rd, Kirk Hallam, Ilkeston, DE7 4NE [IO92IX, SK44]
G3 UQU J W Birtwistle, 29 Duchess Rd, Bedford, MK42 0SE [IO92SC, TL04]
G3 UQW A G Ball, Orchard House, Orchard Lea, Sherfield on Loddon, Hook, RG27 0ES [IO91LH, SU65]
G3 URA R G Whittering, The Coach House, 28 Flamstead End Rd, Cheshunt, Waltham Cross, EN8 0HT [IO91XR, TL30]
G3 URE J W Thexton, 4 Farrer Ct, Cambridge Park, Twickenham, TW1 2LB [IO91UK, TQ17]
G3 URG N Williamson, 30 Oldershaw Rd, East Leake, Loughborough, LE12 6NG [IO92JT, SK52]
G3 URI K S Franks, 251 Queens Rd, Smethwick, Warley, B67 6NX [IO92AL, SP08]
G3 URJ A J Moss, 17 Surrey Dr, Finchfield, Wolverhampton, WV3 9LW [IO82WN, SO89]
G3 URK I Campbell, 27 Lewis Cl, Adlington, Chorley, PR7 4JU [IO83QO, SD51]
G3 URL C Adams, 25 Avon Rd, Cannock, WS11 1LJ [IO82XQ, SJ90]
G3 URM I J Matthews, 3 Tower View Rd, Great Wyrlkey, Great Wyrley, Walsall, WS6 6HE [IO82XP, SJ90]
G3 URN Dr M E Jolley, 34469 N Circle Dr, Round Lake, Il 60073, USA
G3 URP J Gardner, 71 Ainslie Rd, Fulwood, Preston, PR2 3DE [IO83PS, SD53]
G3 URQ J B Letts, Bridgeways, Snows Ln, Keyham, Leicester, LE7 9JS [IO92LP, SK60]
G3 URR J L Weaver, 8 The Green, Aston on Trent, Derby, DE72 2AA [IO92HU, SK42]
G3 URU Dr R M Edworthy, 44 Middleton Ave, Littleover, Derby, DE23 6DL [IO92FV, SK33]
G3 URV F R Stevens, 38 Endhill Rd, Birmingham, B44 9RR [IO92BN, SP09]
G3 URW D A W White, 9 Lymefield Gr, Mile End, Stockport, SK2 6ER [IO83WJ, SJ98]
G3 URX J D Speake, 211 Milton Rd, Cambridge, CB4 1XG [JO02BF, TL46]
G3 URZ Dr B M Ewen-Smith, 57 Ludlow Rd, Church Stretton, SY6 6AD [IO82OM, SO49]
G3 US R Shadlock, Cleveland View, Ackworth, Pontefract, West Yorks, WF7 7DF [IO93HP, SE41]
G3 USA C W Taylor, 231 Robin Hood Ln, Hall Green, Birmingham, B28 0DH [IO92BK, SP18]
G3 USC M A Hall, The Bungalow, Upton Corner, Long Sutton, nr Langport, Somerset, TA10 9NL [IO81OA, ST42]
G3 USD D G Mason, 2A Devon Rd, Bedford, MK40 3DF [IO92SD, TL05]
G3 USE S J Down, 76 Falcon Ave, Bedford, MK41 7DX [IO92SD, TL05]
G3 USF Prof M Harrison, 1 Church Fields, Keele, Newcastle, ST5 5HP [IO83UA, SJ84]
G3 USH P F Allen, 144 Harecroft, Wilsden, Bradford, BD15 0BP [IO93BT, SE03]
GI3 USK H D Kernaghan, 1 Elizabeth Rd, Holywood, BT18 0PL [IO74CP, J47]
GM3 USL R K Vennard Cunningham & Dist ARC, 4 Braehead, Girdle Toll, Irvine, KA11 1BD [IO75QP, NS34]
G3 USO C J Walker, 30 Turnfield Rd, Cheadle, SK8 1JQ [IO83VJ, SJ88]
G3 USP Details withheld at licensee's request by SSL.
GI3 USS J T Barnes, White Gables, 95 Crawfordsburn Rd, Bangor, BT19 1BJ [IO74DP, J48]
G3 UST J R Turner, 20 Homemead Ave, Leicester, LE4 0LN [IO92KP, SK50]
G3 USW W Clough, 32 Jackson Cres, Rawmarsh, Rotherham, S62 7EN [IO93HL, SK49]
G3 USX M Robertson, Karibisha, Horsell Rise, Woking, GU21 4BG [IO91RH, TQ05]
G3 USZ R W Ingall, 5 The Rodings, Upminster, RM14 1RL [JO01DN, TQ58]
G3 UTA K J Smyth, 154 Scrub Ln, Benfleet, SS7 2JP [JO01HN, TQ88]
G3 UTC G Farr, 26 Burstead Dr, Billericay, CM11 2QN [JO01FO, TQ69]
GW3 UTE N W Williams, 8 Rannoch Dr, Cyncoed, Cardiff, CF2 6LQ [IO81KM, ST18]
GW3 UTG A Antley, Fairholme, 12 Fairfield Ave, Rhyl, LL18 3EE [IO83GH, SJ08]
GW3 UTL R N Barker, 51 Rockfield Dr, Llandudno, LL30 1PF [IO83CH, SH77]
G3 UTO P A Strand, Whiteoaks, 8 Brookwell Cl, Chippenham, SN15 1PJ [IO81WL, ST97]
GM3 UTQ I A Balloch, 28 Brentwood Dr, Glasgow, G53 7UJ [IO75TT, NS55]
G3 UTS T W R Belshaw, 20 Greencroft Rd, Delves Ln, Consett, DH8 7DY [IO94CU, NZ14]
G3 UTW G T Chaplin, 22 Waveney Rd, Harpenden, AL5 4QY [IO91TT, TL11]
G3 UTX R A Ridley, 9 Greenacre, Worlebury, Weston Super Mare, BS22 9SL [IO81MI, ST36]
GM3 UU A S McNicol, 12 Edgehill Terr, Aberdeen, AB1 5HA [IO87TH, NJ72]
G3 UUB N D Bateman, 10 Telford Cres, Woodley, Reading, RG5 4QT [IO91NL, SU77]
G3 UUC J E Nurse, Lisbue Farm, Newbridge, Penzance, Cornwall, TR20 8RG [IO70EC, SW42]
G3 UUG E Nightingale, 61 The Cockpit, Marden, Tonbridge, TN12 9TQ [JO01FE, TQ74]
G3 UUI M H Mapson, 253 Central Ave, Southend on Sea, SS2 4ED [JO01IN, TQ88]
G3 UUK V J Ingram, PO Box 14235, Farrarmere, Benoni, South Africa
G3 UUL T H Jones, 32 Oakwood Dr, Hucclecote, Gloucester, GL3 3JF [IO81VU, SO81]
G3 UUM A M Page, 19 Boulsworth Cres, Nelson, BB9 8DF [IO83VU, SD83]
G3 UUO J W Dudbridge, 42B Duncansby Rd, Stanmore Bay, Whangaparaoa 1463, Auckland, New Zealand
G3 UUP A P Teale Ealing & Dst AR, 53 Hartland Dr, South Ruislip, Ruislip, HA4 0TH [IO91TN, TQ18]
G3 UUQ A S Clelland, Rieschbogen 7, 85635 Hohenkirchen, Germany
G3 UUR Dr D Gordon-Smith, 2 Meadow View, Bourton Rd, Frankton, Rugby, CV23 9NX [IO92HH, SP47]
G3 UUS G B Packer, 7 Robinia Cl, Waterlooville, PO7 8HF [IO90LV, SU60]
G3 UUT J F Wilson, Westcroft Cottage, 20B High Green, Great Shelford, Cambridge, CB2 5EG [JO02BD, TL45]
G3 UUU L P V Newman, 13 Ascham Ln, Whittlesford, Cambridge, CB2 4NT [JO02BC, TL44]
G3 UUV R S Frost, Gravel House, Old Rd, Notton Lacock, Chippenham, SN15 2AF [IO81WK, ST97]
G3 UUW Details withheld at licensee's request by SSL.
G3 UUX E W Hibbert, 36 Heyhouses Ln, St. Annes, Lytham St. Annes, FY8 3RW [IO83LS, SD32]
G3 UUY D W Wright, St. Julians, 55 Old Rd, Harlow, CM17 0HD [JO01BS, TL41]
G3 UUZ H Bluer, 20 Trewellard Rd, Pendeen, Penzance, TR19 7ST [IO70DD, SW33]
GW3 UVA D E Knowles, The Clappers, Spon Green, Buckley, CH7 3BL [IO83LD, SJ26]
G3 UVB D F Barnes, 3 Quarry End, Begbroke, Kidlington, OX5 1SF [IO91IT, SP41]
G3 UVC Prof G A King, Systems Engineering Division, Southampton Institute of, Higher Education, East Park Terr, Southampton, SO14 0YN [IO90HV, SU41]
G3 UVH J T Purnell, 49A Oxford Rd, Rochford, SS4 1TG [JO01IO, TQ89]
G3 UVM M R G Simpson, 174 Welbeck Rd, Harrow, HA2 0RX [IO91TN, TQ18]
GM3 UVO T C Malin, Meadowbank, 3A Achgarve, Laide, Achnasheen, IV22 2NS [IO77FV, NG89]
G3 UVP G P Rive, 15 Edenfield Rd, Liverpool, L15 5BP [IO83NJ, SJ38]
G3 UVQ N Mercer, 19 Sycamore Rd, Brookhouse, Lancaster, LA2 9PB [IO84PB, SD56]
G3 UVR D Jones, Uplands, 39 Pensby Rd, Heswall, Wirral, L60 7RA [IO83KI, SJ28]
G3 UVS C A Mitchell, 30 Southway Ln, Roborough, Plymouth, PL6 7DJ [IO70WK, SX46]
G3 UVU J S Curry, Clonlea, New Ridley, Stocksfield, Northd., NE43 7RQ [IO94BW, NZ06]
G3 UVW R Harris Coventry Tech ARC, 44 Burbages Ln, Longford, Coventry, CV6 6AY [IO92FK, SP38]
G3 UVY L G Parkin, 8 Smithfield Cl, Ripon, HG4 2PG [IO94FC, SE36]
G3 UWA A Hall, 3 The Copse, Sandilands, Mablethorpe, LN12 2UB [JO03DH, TF58]
G3 UWB R D Hall, Crest Cottage, Chapel Ln, Grundisburgh, Woodbridge, Suffolk, IP13 6TS [JO02PC, TM25]
G3 UWE R B Simpson, 30 Heath Lawns, Fareham, PO15 5QB [IO90JU, SU50]
G3 UWH Cdr J W Endicott, The Mill House, Halse, Taunton, Somerset, TA4 3AQ [IO81JB, ST12]
G3 UWI T J Connell, 46 Redlands Ln, Fareham, PO14 1EZ [IO90JU, SU50]
G3 UWK Details withheld at licensee's request by SSL.
G3 UWL Dr D Grant, Blue Cedars, 40 Dukes Wood Dr, Gerrards Cross, SL9 7LR [IO91RN, SU98]
G3 UWM P J Marchant, 12 Laurel Way, Ickleford, Ickleford, Hitchin, SG5 3UP [IO91UX, TL13]
GM3 UWO E Dahle, 7 Nursery Ave, Kilmarnock, KA1 3DP [IO75SO, NS43]
G3 UWP R C Pickering, Copper Coins, 41 Maiden Greve, Malton, YO17 0BE [IO94OD, SE77]
G3 UWR C Bonsall, Parkside, Lodge Rd, Carcroft, Doncaster, DN6 8EB [IO93JO, SE51]
GW3 UWS M T Bowen Univ Coll of, Swansea A R South, 24 Parklands View, Sketty, Swansea, SA2 8LX [IO81AO, SS69]
G3 UWT P W Myers, 22 High St, Barnby Dun, Doncaster, DN3 1DS [IO93LN, SE60]
G3 UWU W A Mansfield, 94 Penberthy Rd, Helston, TR13 8AR [IO70IC, SW62]
GM3 UWX J Stirling, 25 Maxwell Rd, Bishopton, PA7 5HE [IO75SV, NS47]
G3 UWY J A Brighting, 27 Mounthurst Rd, Hayes, Bromley, BR2 7QW [JO01AJ, TQ36]
G3 UWZ M Newman, 26 Highbank, Westdene, Brighton, BN1 5GB [IO90WU, TQ20]
G3 UXF R J Rawlings, 10 Beech Ave, East Leake, Loughborough, LE12 6NU [IO92JT, SK52]
G3 UXG M Wales, 74 Australia Gr, South Shields, NE34 9DF [IO94GX, NZ36]
G3 UXH P J E Carey, 99 Bells Ln, Hoo St. Werburgh, Hoo, Rochester, ME3 9HX [JO01GK, TQ77]
G3 UXI G B Spinks, 18 St. Edmunds Way, Old Harlow, Harlow, CM17 0BN [JO01BS, TL41]
G3 UXJ B Donders, Gaydon Cottage, Stretton on Fosse, Moreton in Marsh, Glos, GL56 9SA [IO92DB, SP23]
G3 UXL K C Tam, 11 The Torre, Yeovil, BA21 3SL [IO80QW, ST51]
G3 UXM J Greaves, 23 Woodhouse Rd, Intake, Sheffield, S12 2AY [IO93HI, SK38]
G3 UXO Dr A J N Eardley, 43 Cranbourne Dr, Harpenden, AL5 1RJ [IO91TT, TL11]
G3 UXQ M R Smart, 12 Bromwich Ln, Pedmore, Stourbridge, DY9 0QZ [IO82WK, SO98]
G3 UXR N R Goddard, 1 Aston Mead, St. Catherines Hill, Christchurch, BH23 2SP [IO90CS, SZ19]
G3 UXX J E Baylis, 15 Adbolton Ave, Gedling, Nottingham, NG4 3NB [IO92KX, SK64]
G3 UXY A W E Baker, Kirkella, 1 Napier Rd, Maidenhead, SL6 5AR [IO91PM, SU88]
G3 UYB M E Shaw, Kingston, Hever Rd, Bough Beech, Edenbridge, TN8 7NU [JO01BE, TQ44]
G3 UYC J Peirson, Ashfield Farm, Ulting, Maldon, Essex, CM9 6QP [JO01HS, TL80]
G3 UYD E T Clarke, 65 Oakmount Rd, Chandlers Ford, Eastleigh, SO53 2LJ [IO90HX, SU42]
G3 UYE M H Richer, Dolphin Cottage, 1 Station Rd, Surfleet, Spalding, PE11 4DA [IO92WU, TF22]
G3 UYG J J Clegg, 11 South Park Rd, Gatley, Cheadle, SK8 4AL [IO83VJ, SJ88]
G3 UYH J M Broadbent, Kimberley, 7 James Cl, Llanon, SY23 5HP [IO72VG, SN56]
G3 UYK P R Kemble, 74 Teg Down Meads, Winchester, SO22 5ND [IO91HB, SU43]
G3 UYL D S Knott, Lymewood, Linden Cl, Prestbury, Cheltenham Glous, GL52 3DU [IO81XV, SO92]
G3 UYM H J Groves, 36 Grange Cl, Hitchin, SG4 9HD [IO91UW, TL12]
G3 UYN C E Malcolm, The Cottage, 25 Red Lion Ln, Overton, Basingstoke, RG25 3HH [IO91IF, SU54]
GM3 UYR Dr P Gamble, 21 St. Marys Dr, Perth, PH2 7BY [IO86HJ, NO12]
G3 UYV L H Navier, 19 Ormonde Ave, Hull, HU6 7LX [IO93TS, TA03]
G3 UYX J R Ball, 7 Moorfield Rd, Woodbridge, IP12 4JN [JO02PC, TM24]
GI3 UYZ Rev P J McGuigan, Parochial House, Feeny, Dungiven, Co.Derry, N Ireland, BT47 4TW [IO64LV, C60]
G3 UZB J R Shaw, 42 Stirling Rd, Redcar, TS10 2JZ [IO94LO, NZ62]
G3 UZD F Bilke, 2 Walcott Ave, Christchurch, BH23 2NG [IO90CR, SZ19]
G3 UZI I D Wollen, Douzaine, 12 Fort George, Guernsey, GY1 2TA
G3 UZJ D W Singleton, 55 Northwood Ln, Darley Dale, Matlock, DE4 2HR [IO93EE, SK26]
G3 UZK D W Bloomfield, 7 Larmer Cl, Fleet, GU13 8AY [IO91NG, SU75]
G3 UZL F A Cook, 120 Hulham Rd, Exmouth, EX8 3LD [IO80HP, SY08]
G3 UZM C P Haddock, 26 Featherbed Ln, Exmouth, EX8 3NE [IO80HP, SY08]
G3 UZN M P Rathbone, Cranmer House, 22 New St., Holt, NR25 6JH [JO02NV, TG03]
GW3 UZS J D Diplock, Cartref, 98 Pendwyallt Rd, Whitchurch, Cardiff, CF4 7EH [IO81JM, ST18]
G3 UZU A Upton, 36 Robin Way, Wirral, L49 7NA [IO83KI, SJ28]
G3 UZW R V Andrews, 44 The Park, Great Bookham, Bookham, Leatherhead, KT23 3LS [IO91TG, TQ15]
G3 UZX F Mitchell, 158 Cobham Rd, Fetcham, Leatherhead, KT22 9JR [IO91TH, TQ15]
G3 UZY Details withheld at licensee's request by SSL.
G3 UZZ G P Jones, 19 Windmill Ct, Windmill Rd, Ealing, London, W5 4DN [IO91UL, TQ17]
G3 VA J P Hawker, 37 Dovercourt Rd, Dulwich, London, SE22 8SS [IO91XK, TQ37]
GI3 VAF R C Best, 6 Knightsbridge Ct, Bangor, BT19 6SD [IO74EP, J58]
G3 VAI P A F Carter, 44 Birch Rd, Congleton, CW12 4NR [IO83VD, SJ86]
G3 VAJ I L M Gray, The Old Police House, Great Finborough, Stowmarket, Suffolk, IP14 3AA [JO02LE, TM05]
G3 VAK M J Sutcliffe, 26 Weald Rd, Burgess Hill, RH15 9SP [IO90WX, TQ31]
G3 VAL G A Talbot, Galebars, Kirkby Stephen, Cumbria, CA17 4HE [IO84TL, NY70]
G3 VAO Details withheld at licensee's request by SSL.
GM3 VAP C G Weston, 18 Kirkbrae Mews, Cults, Aberdeen, AB1 9QF [IO87TH, NJ72]
GM3 VAR Dr A J M Campbell, 2 Dunlin Cres, Crosslee, Johnstone, PA6 7JX [IO75RU, NS46]
G3 VAS G Jones, 294 Halling Hill, Harlow, CM20 3JU [JO01BS, TL41]
GI3 VAW R T Sherrard, 39 Shanreagh Park, Limavady, BT49 0SF [IO65MB, C62]
GW3 VBC D R Blanchard, 11 Daytona Dr, Northop Hall, Mold, CH7 6LP [IO83LE, SJ26]
G3 VBD D G Bingham, 30 Hazelmere Rd, Stevenage, SG2 8RX [IO91VV, TL22]
G3 VBE F G Miles, 65 Montgomery St., Hove, BN3 5BE [IO90VU, TQ20]
G3 VBG B A Morris, 80 Heathend Rd, Alsager, Stoke on Trent, ST7 2SH [IO83UC, SJ75]
G3 VBI H R Christopher, 19 Spa Hill, Kirton Lindsey, Gainsborough, DN21 4BA [IO93QL, SK99]
G3 VBL C Pedder, Thorncliffe, 5 Royalty Ln, New Longton, Preston, PR4 4JD [IO83PR, SD52]
G3 VBN Details withheld at licensee's request by SSL.
GW3 VBP D H Adams, 25 Timbers Sq, Cardiff, CF2 3SH [IO81KL, ST17]
G3 VBQ D K Wright, 5 Padin Cl, Chalford, Stroud, Glos, GL6 8FB [IO81WR, SO80]
G3 VBS Details withheld at licensee's request by SSL.
GM3 VBT T R Logan, 137 Buccleuch St., Garnethill, Glasgow, G3 6QN [IO75UU, NS56]
G3 VBV S T E A Boyce, 58 Woodbury Rd, Halesowen, B62 9AW [IO82XL, SO98]
G3 VBW E G Thomas, 4 Hemdean Gdns, West End, Southampton, SO30 3BB [IO90HW, SU41]
GM3 VBY F Hindley, The White House, 17 Main Rd, Alves, Elgin, IV30 3UR [IO87GP, NJ16]
G3 VC M H F Bridge, 6 Greenwich Cl, Bamford, Rochdale, OL11 5JN [IO83VO, SD81]
G3 VCA R Pickles, The Bell Inn, Far Ln, Coleby, Lincoln, LN5 0AH [IO93RD, SK96]
G3 VCE F C Ellery, 47 Newmorton Rd, Moordown, Bournemouth, BH9 3NX [IO90BS, SZ09]
G3 VCG D R Wilks, 36 Greenways, Chelmsford, CM1 4EF [JO01FS, TL70]
G3 VCH R P Philpott, Wilhlmstr 19, 7600 Offenburg, W Germany, ZZ7 6OF
GI3 VCI M H McFadden, 121 Greystown Ave, Belfast, BT9 6UH [IO74AN, J36]
G3 VCK J E Fenwick, 78 Loveridge Rd, London, NW6 2DT [IO91VN, TQ28]
G3 VCL B Clark, 60 Somerset Ave, Harefield, Southampton, SO18 5FS [IO90HW, SU41]
G3 VCM I H Anderson-Mochrie, 10214 Hunt Club Ln, Palm Beach Gdns, Fl33418, USA, X X
G3 VCN P A Kalas, Kerlins, 110A Underlane, Plympton, Plymouth, PL7 1QZ [IO70XJ, SX55]
GM3 VCO R Pallester, 36 McKenzie Dr, Balloch, Alexandria, G83 8HL [IO76RA, NS38]
G3 VCP R F Burns obo Crystal Palace Dist Rad, 84 Portnalls Rd, Coulsdon, CR5 3DE [IO91WH, TQ25]
G3 VCQ D J D Wilson, 6 Orton Rd, Barton Seagrave, Kettering, NN15 6UJ [IO92PI, SP87]
G3 VCR C J Rooney, 103 Hart Plain Ave, Cowplain, Waterlooville, PO8 8PN [IO90LV, SU61]
G3 VCT R K Hemmings, Wood View, Cryers Hill Rd, Cryers Hill, High Wycombe, HP15 6JR [IO91PP, SU89]
G3 VCU Details withheld at licensee's request by SSL.
G3 VCV D T Prout, Beechmount, St. Georges Ave, Kings Stanley, Stonehouse, GL10 3HN [IO81UR, SO80]
G3 VCW M A Baynham, 21 Copperfield Dr, Leeds, Langley, Maidstone, ME17 1SY [JO01HF, TQ85]
G3 VCY Dr C R I Clayton, Wildfield, West Flexford Ln, Wanborough, Guildford, GU3 2JW [IO91QF, SU94]
GM3 VCZ A McInnes, Morvendale, Newton Row, Wick, KW1 5SB [IO88KK, ND34]
G3 VD B A Pettit, 5 The Glades, Locks Heath, Southampton, SO31 6UX [IO90IU, SU50]
G3 VDB J T Evans, 7 Barncroft Cl, Chelford, Macclesfield, SK11 9SW [IO83UG, SJ87]
G3 VDC V C Cheesman, 142 Cuckfield Rd, Hurstpierpoint, Hassocks, BN6 9SD [IO90VW, TQ21]
G3 VDE J R Sellers, The Bungalow, Sandholme Rd, Eastrington, Goole, DN14 7QQ [IO93OS, SE73]
G3 VDF H Gregory, 44 Mowlands Cl, Sutton in Ashfield, NG17 5GH [IO93JC, SK55]
G3 VDH R L G Godwin, Hopworthy Moor Cottage, Pyworthy, Holsworthy, Devon, EX22 6XX [IO70TT, SS30]
G3 VDI L Millward, Oak Tree Cottage, Ashgate Rd, Ashgate, Chesterfield, S42 7JE [IO93GF, SK37]
G3 VDK S Bailey, 6 Minnie St., Keighley, BD21 1HY [IO93BU, SE04]
G3 VDL J A C St.Leger, Warmbrook, Throwleigh, nr Okehampton, Devon, EX20 2JF [IO80BQ, SX69]
G3 VDM M Clift, 49 Manor Ln, Halesowen, B62 8PZ [IO82XK, SO98]
G3 VDO I Hacking, 1 Pine Cres, Poulton-le-Fylde, Lancs, FY6 8EB [IO83MU, SD33]
G3 VDQ B E Heggs, 4 Woodlands Ct, Congleton Rd, Alderley Edge, SK9 7AB [IO83VH, SJ87]
G3 VDU P J Bennett, 56 Winchester Ave, Weddington, Nuneaton, CV10 0DW [IO92GM, SP39]
G3 VDV N R F Brinnen, 134 Victoria Rd, Mablethorpe, LN12 2AJ [JO03DI, TF58]
G3 VDW T F Underwood, 197 Highfield St., Coalville, LE67 3BT [IO92HR, SK41]
G3 VDX Details withheld at licensee's request by SSL.
G3 VEB R E Bridson, 14 Zig Zag Rd, Liscard, Wallasey, L45 7NZ [IO83LK, SJ39]
G3 VED G J J Wilkins, Forfrae House, Heronway, Hutton, Brentwood, CM13 2LX [JO01EO, TQ69]
G3 VEF M K Curran Fareham&Dist AR, 2 Bridges Ave, Paulsgrove, Portsmouth, PO6 4PA [IO90KU, SU60]
G3 VEH C J Morcom, 43 Kingsmill Rd, Basingstoke, RG21 3JU [IO91KG, SU65]
GM3 VEI I W Sheffield, Cottage Two, Sandys Mill, nr Haddington, East Lothian, EH41 3SB [IO85PX, NT57]
GD3 VEM F H Knight, 3 Queens Dr East, Ramsey, IM8 1EJ
GW3 VEN Mrf A H Dicker, 2 Jubilee Cttgs, Jubilee, Tenby, SA70 8DH [IO71PQ, SN10]
G3 VEO Dr M L Jeremy, Sheen Cottage, Thorpe Ln, Fylingthorpe, Whitby, YO22 4TH [IO94RK, NZ90]
GW3 VEP A E Evans, Woodland Vale, Ludchurch, Narberth, Dyfed, SA67 8JF [IO71PS, SN11]
G3 VER R M M Heath Verulam ARC, 26 Lancaster Ave, Hadley Wood, Barnet, EN4 0EX [IO91VQ, TQ29]
G3 VES H Martin, 1 Houghton Park Cttgs, Hazelwood Ln, Ampthill, Bedford, MK45 2EY [IO92SA, TL03]
G3 VET M C Langwade, Oakwood, 19 South Wootton Ln, Kings Lynn, PE30 3BS [JO02FS, TF62]
GW3 VEW K E Godfrey, Kimberley, Ludchurch, Narberth, Pembrokeshire, SA67 8JE [IO71PS, SN11]
GM3 VEY F Baxter, 24 Hillview Cres, Cults, Aberdeen, AB1 9RT [IO87TH, NJ72]
G3 VEZ A S N Blake, 65 Strathmore Rd, Moordown, Bournemouth, BH9 3NT [IO90BS, SZ09]
G3 VFB A R Matthews, Lavernock Cottage, The Street, Crookham Village, Aldershot Hants, GU13 0SH [IO91NG, SU75]
G3 VFC T T Chipperfield, 5 Lullingstone Cl, Hempstead, Gillingham, ME7 3TS [JO01GI, TQ76]
G3 VFD C W Westwood, 62 Blackbrook Ln, Bickley, Bromley, BR2 8AY [JO01AJ, TQ46]
G3 VFF D J Hine, 2 Richmond St., Great Bridge, West Bromwich, West Midlands, B70 0DD [IO82XM, SO99]
G3 VFH L R Moore, 15 Elmete Dr, Roundhay, Leeds, LS8 2LA [IO93GT, SE33]
G3 VFI J A Rampton, Apartado de Correos 112, 03740 Gata de Gorgos, Alicante, Spain, X X
GI3 VFK P Harvey, 254 High Donaghadee, Rd, Kilmaine, Bangor Co. Down, N Ireland, BT19 2NH [IO74EP, J58]
GW3 VFL A A Lightly, St. Anthony, 9 The Kymin, Monmouth, NP5 3SD [IO81PT, SO51]
G3 VFO T W Hart, Elmwood Pl, Tott Cl, Etchingham Rd, Burwash, E Sussex, TN19 7BE [JO01EA, TQ62]
G3 VFP Details withheld at licensee's request by SSL.
G3 VFQ R K Barlow, Kenwood, Higher Trehaverne, Truro, TR1 3RH [IO70LG, SW84]
G3 VFU S K Street, P O Box 26844, Adliya, Bahrain, Arabian Gulf, X X
G3 VFV Details withheld at licensee's request by SSL.
GI3 VFW D F Campbell Mid Ulster ARC, 109 Drumgor Park, Drumgor, Craigavon, BT65 4AH [IO64TK, J05]
G3 VFX D N Davison, 28 Treve Ave, Harrow, HA1 4AJ [IO91TN, TQ18]
GW3 VFZ M Hughes, Cefn Dinas, Bangor, Gwynedd, Wales, LL57 4DP [IO73WF, SH57]
G3 VG J Wood, 7 Sherring Cl, Wick Hill, Bracknell, RG42 2LD [IO91PK, SU87]
G3 VGB R Taylor, 6 Victoria Ave, Baxenden, Accrington, BB5 2XD [IO83TR, SD72]
G3 VGE M F Hickman, 75 Carlton Rd, Redhill, RH1 2BZ [IO91VF, TQ25]
G3 VGG J F Layton Brmsgve&Dist AR, Meadow View, Worcester Rd, Martley, Worcester, Worcs, WR6 6QA [IO82TF, SO75]
G3 VGH Dr B R G Hutchinson, 78 Strensall Rd, Huntington, York, YO3 9SH [IO94LA, SE65]
G3 VGI J S Webb, Stone Cottage, Lyons Rd, Slinfold, Horsham, RH13 7QT [IO91TB, TQ13]
GI3 VGL P B Strong, 9 Carnbrae Ave, Saintfield Rd, Belfast, BT8 4NH [IO74BN, J36]
G3 VGN W Easton, 24 Castle Rd, Whitby, YO21 3NQ [IO94QL, NZ81]
G3 VGO S T S Evans, Glengormley, Forth Coth, Carnon Downs, Truro, TR3 6JY [IO70KF, SW74]
G3 VGR D J Aldridge, 62 Roding View, Buckhurst Hill, IG9 6AQ [JO01AP, TQ49]
G3 VGU J C Outen, 31 Carlton Rd, Grays, RM16 2YB [JO01EL, TQ67]
G3 VGW R I Buckby, 20 Eden Bank, Ambergate, Belper, DE56 2GG [IO93GB, SK35]
G3 VGX Dr R S Orton, 38 Sandford Rd, Chelmsford, CM2 6DQ [JO01FR, TL70]
GW3 VGY R H Ricketts, 4 Waterston Rd, Llanstadwell, Milford Haven, SA73 1EH [IO71MQ, SM90]
G3 VGZ B Duffell, 7 Potto Cl, Levendale, Yarm, TS15 9RZ [IO94IM, NZ41]

G3 VHA M G D Skelding, 2 Church Ave, Clent, Stourbridge, DY9 9QT [IO82WJ, SO97]
G3 VHB C Swallow, 345 Walmley Rd, Sutton Coldfield, B76 1PG [IO92CN, SP19]
G3 VHC J K Phillips, 30 Long Ln, Aughton, Ormskirk, L39 5AT [IO83NN, SD40]
G3 VHE R A Evans, 23 Hardwell Cl, Gr, Wantage, OX12 0BN [IO91GO, SU39]
G3 VHF Details withheld at licensee's request by SSL.[Op: RSGB, Lambda House, Cranborne Road, Potters Bar, Herts EN6 3JE.]
GW3 VHG Details withheld at licensee's request by SSL.
G3 VHH J L Delves, 11 Willoughby Rd, Langley, Slough, SL3 8JH [IO91RM, TQ07]
G3 VHI G S Boultbee, 44 High St., Heckington, Sleaford, NG34 9QT [IO92UX, TF14]
G3 VHK Dr J M Robinson, 8 Lorraine Park, Harrow, HA3 6BX [IO91TO, TQ19]
G3 VHL H Buttress, 130 Elan Ave, Burlish Park, Stourport on Severn, DY13 8LR [IO82UI, SO87]
GI3 VHM V M Addidle, 228 Orby Dr, Belfast, BT5 6BE [IO74BN, J37]
G3 VHN J C Burge, 42 Chestnut Ave, Southborough, Tunbridge Wells, TN4 0BU [JO01DD, TQ54]
G3 VHS J H Cobb, Middle Cottage, Fyfield, Abingdon, Oxon, OX13 5LR [IO91HQ, SU49]
G3 VHT B L Norval, 50 Ashley Ct, Grand Ave, Hove, BN3 2NN [IO90VT, TQ20]
G3 VHW N R Humphrey, 10 Pembroke Cl, Boyatt Wood, Eastleigh, SO50 4QY [IO90HX, SU42]
GI3 VHY Details withheld at licensee's request by SSL.
G3 VHZ B Neary, Chao Phraya, 30 Laneham Cl, Bessacarr, Doncaster, DN4 7HU [IO93KL, SE60]
G3 VIC D Barney, 46 Holway Rd, Sheringham, NR26 8HR [JO02OW, TG14]
G3 VID T A Howe, 34 Princes Gdns, Cliftonville, Margate, CT9 3AR [JO01QJ, TR37]
G3 VIE P D de La Mothe, 35 Brookside, Wokingham, RG41 2ST [IO91NJ, SU76]
G3 VIG A F Goldsmith, 12 Orchard Cl, Roughton, Norwich, NR11 8SR [JO02PV, TG23]
G3 VII Details withheld at licensee's request by SSL.
G3 VIJ G W Perkins, Flat Holm, 6 Chapel Terr, Aspatria, Carlisle, CA5 2LU [IO84HS, NY14]
GM3 VIO G A C Currie, Burghfield Cottage, Cnoc An Lobht, Dornoch, IV25 3HN [IO77XV, NH78]
G3 VIP G Wood, 47 Church Ln, Holton-le-Clay, Grimsby, DN36 5AQ [IO93XM, TA20]
G3 VIR R G Brade, 26 Pevensey Way, Frimley Green, Frimley, Camberley, GU16 5YJ [IO91PH, SU85]
G3 VIS R W Cox, 1 Bridge Terr, Northallerton, DL7 8NH [IO94GI, SE39]
G3 VIV R H Hannaford, 3 Devonshire Rd, Bathampton, Bath, BA2 6UB [IO81UJ, ST76]
G3 VIX T S Stevens, 97 Broadacres, Hatfield, Herts, AL10 9LE [IO91VS, TL20]
G3 VIY R Vasper, 31 Oakland Rd, Forest Town, Mansfield, NG19 0EJ [IO93JD, SK56]
G3 VJB G W Cooper, The Firs, Treskerby, Redruth, Cornwall, TR16 5AG [IO70JF, SW74]
G3 VJE H A Cole, 3 Canberra Cres, Grantham, NG31 9RD [IO92QW, SK93]
G3 VJF P S Nicholson, 34 Cliff Ave, Herne Bay, CT6 6LZ [JO01NI, TR16]
G3 VJG M E Deutsch, 80 Windermere Rd, Kettering, NN16 8UF [IO92PJ, SP87]
G3 VJI J W Steel, 52 Rusland Park, Kendal, LA9 6AJ [IO84PH, SD59]
G3 VJJ F A Smedley, 13 Justice Ave, Saltford, Bristol, BS18 3DR [IO81SJ, ST66]
G3 VJK S C Cooke, 36 Park Ln, Castle Donington, Derby, DE74 2JF [IO92HU, SK42]
G3 VJM A G R Wood, Danehill, Brookhill Rd, Copthorne, Crawley, RH10 3PS [IO91WD, TQ33]
G3 VJN A Ryan, Isscc/Nasb, Builing 119, B-7010 Shape, Belgium
G3 VJO P G Hildebrand, 31 Crouch Hall Gden, Redbourn, Stalbans, Herts, AL3 7EL [IO91TT, TL11]
G3 VJP J T Eden, 14 Broadwell Rd, Solihull, B92 8QH [IO92CK, SP18]
G3 VJR J B Longstaff, 23 Harlington Rd, Adwick on Dearne, Adwick upon Dearne, Mexborough, S64 0NL [IO93IM, SE40]
G3 VJV C P Hartley, 16 Cyril Bell Cl, Off Pepper St., Lymm, WA13 0JS [IO83SJ, SJ68]
G3 VJX M H Gill, Perry Pear Barn, Bryans Green, Cutnall Green, Droitwich, Worcs, WR9 0ND [IO82VH, SO86]
GM3 VJY J L Evans, 64 Craigmount Ave, North, Edinborough, Lothians, EH12 8DL [IO85IW, NT17]
G3 VJZ R G N Soper, 50 Morritt Cl, York, YO3 9DY [IO93LX, SE65]
G3 VKB J Orr, 18 Randall Cl, Langley Slough, Slough, SL3 8RJ [IO91RL, TQ07]
G3 VKF K Kelly, 2 Longden Ln, Buxton Old Rd, Macclesfield, SK11 7EN [IO83WG, SJ97]
G3 VKH A Roberts, 10 Windmill Dr, Marchington, Uttoxeter, ST14 8JP [IO92CU, SK13]
G3 VKI F D Turner-Smith, St. Clair, 47 Park Rd, Camberley, GU15 2SP [IO91OH, SU85]
G3 VKK R Baines Chesterfield &, District Ars, 327 Langer Ln, Wingerworth, Chesterfield, S42 6TY [IO93GE, SK36]
GW3 VKL C S Beynon, 16 Hardy Cl, Woodfield Heights, Barry, CF62 9HJ [IO81IK, ST16]
G3 VKM R J Basford, Newgate, Thorpe Rd, Haddiscoe, Norwich, NR14 6PP [JO02TM, TM49]
G3 VKN P Mansell, 2 Highland Ave, Penwortham, Preston, PR1 0TJ [IO83PR, SD52]
G3 VKO L Emmett, Boxtreecottage, Whiteleaf, Princes/Risborough, Aylesbury Bucks, HP17 0LL [IO91OS, SP80]
G3 VKP M H Lay, 66 High St., Great Shelford, Cambridge, CB2 5EH [JO02BD, TL45]
G3 VKQ C D McEwen, 37 Malvern Way, Twyford, Reading, RG10 9PY [IO91NL, SU77]
G3 VKR Details withheld at licensee's request by SSL.
G3 VKT R A J Smith, 32 Wolseley Gdns, London, W4 3LR [IO91UL, TQ17]
G3 VKU D Hollingsworth, 4 Cairn View, Longframlington, Morpeth, NE65 8JT [IO95BG, NZ09]
G3 VKV G H S Jones, 32 The Grove, Hales Rd, Cheltenham, GL52 6SX [IO81XV, SO92]
G3 VKW K Evans, Little Field House, Cuckfield Rd, Ansty, Haywards Heath, W Sussex, RH17 5AL [IO90WX, TQ22]
G3 VKX S F Cummins, The Bungalow, Loton Park, Alberbury, Shropshire, SY5 9AG [IO82MR, SJ31]
GM3 VLB A H G Saunders, 6 Douglas Cres, Kelso, TD5 8BB [IO85SO, NT73]
G3 VLC C C Hawkins, 54 Benett Dr, Hove, BN3 6UQ [IO90VU, TQ20]
G3 VLD T A Denney, Spindrift-East-Terr, Walton on Naze, Essex, CO14 8PX [JO01PU, TM22]
G3 VLE Details withheld at licensee's request by SSL.
G3 VLF Dr T W Beamond, Park View, Middle Ln, Whatstandwell, Matlock, DE4 5EG [IO93GB, SK35]
G3 VLG A M Robins Hinckley Amateur Radio &, 20 Featherston Dr, Burbage, Hinckley, LE10 2PN [IO92HM, SP49]
G3 VLH J F Longhurst, 13 Hophurst Dr, Crawley Down, Crawley, RH10 4XA [IO91XC, TQ33]
G3 VLI A W Brooks, 10 Camfield Cl, Basingstoke, RG21 3AQ [IO91LG, SU65]
G3 VLJ Dr A D A Hansen, 1829 Francisco St, Berkeley, California, USA 94703, ZZ1 8FR
G3 VLL G R Gauntlett, 7 Riverside Dr, Sprotbrough, Doncaster, DN5 7LH [IO93JM, SE50]
G3 VLN J S Allin, 57 Burleigh Rd, West Bridgford, Nottingham, NG2 6FQ [IO92KW, SK53]
G3 VLP Details withheld at licensee's request by SSL.
G3 VLQ M A Tindal, 14 Higher Bullen, Barwick, Yeovil, BA22 9TZ [IO80QW, ST51]
G3 VLR B Rispin, 37 Ferry Rd, South Cave, Brough, HU15 2JG [IO93QS, SE93]
G3 VLT Details withheld at licensee's request by SSL.[Op: C J Linnell.]
GW3 VLU E S M Hotchkiss, 30 Beechwood Dr, Penarth, CF64 3QZ [IO81JK, ST17]
G3 VLW P C Martin, 17 Ash Hayes Dr, Nailsea, Bristol, BS19 2LG [IO81OK, ST47]
G3 VLX D Buckley, Little Oaks, Park Rd, Marden, Tonbridge, TN12 9LG [JO01GE, TQ74]
GM3 VMB Details withheld at licensee's request by SSL.
G3 VMD F C Reid, 466 New Bedford Rd, Luton, LU3 2BA [IO91SV, TL02]
G3 VMI D A Pike, 11 Cavalry Dr, March, PE15 9EQ [JO02BN, TL49]
G3 VMJ J D Peters, 93 Baddow Hall Cres, Chelmsford, CM2 7BU [JO01GR, TL70]
G3 VMK N D Chadwick, 67 Toton Ln, Stapleford, Nottingham, NG9 7HB [IO92IW, SK43]
G3 VML Details withheld at licensee's request by SSL.
G3 VMM Details withheld at licensee's request by SSL.
G3 VMO G E Fenner, 11 Melverley Gdns, Wimborne, BH21 1HJ [IO90AT, SU00]
G3 VMP B F D Mills, Highlands Cottage, Crow Ln, Weeley, Clacton on Sea, CO16 9AN [JO01NU, TM12]
G3 VMQ P A Tory, 79 Partridge Way, Cirencester, GL7 1BH [IO91AR, SP00]
G3 VMR R J Redding, September House, Cox Green Ln, Maidenhead, SL6 3EL [IO91PM, SU87]
G3 VMS Details withheld at licensee's request by SSL.
G3 VMT T J Poole, 64 Humber Cl, Thatcham, RG18 3DT [IO91IJ, SU56]
G3 VMU C J Davis, 23 Vernon Walk, St. Edmunds, Northampton, NN1 5ST [IO92NF, SP76]
G3 VMW S Wilson, 3 Crag Gdns, Bramham, Wetherby, LS23 6RP [IO93HV, SE44]
GW3 VMX G Thomas, 118 Southdown Rd, Port Talbot, SA12 7HS [IO81CO, SS79]
G3 VMY Dr E Searle, 203 Church Rd, Earley, Reading, RG6 1HW [IO91MK, SU77]
G3 VMZ D A Nicholls, Seekers Cove, Shaw, Melksham, Wilts, SN12 8EJ [IO81WJ, ST86]
G3 VN L F Hoskins, 9 Spencer Rd, Caterham, CR3 5LE [IO91WH, TQ35]
G3 VNA J S Fisher, Galena, 106 Dorchester Rd, Garstang, Preston, PR3 1EE [IO83OV, SD44]
G3 VNB R M Thomas, 1B Elfield Park Dve, Weymouth, DT4 9RB [IO80SO, SY67]
G3 VNC Details withheld at licensee's request by SSL.
G3 VND R Stevenson, 26 North Rocks Rd, Churston, Paignton, TQ4 6LF [IO80FJ, SX85]
G3 VNG D T Hind, 4 Thornyville Villas, Plymouth, PL9 7LA [IO70WI, SX55]
GM3 VNH P Hardy, 30 Straid A Cnoc, Barremman, Clynder, Helensburgh, G84 0QX [IO76OA, NS28]
G3 VNI S C Cammies, Hill Garth, Station Rd, Goathland, Whitby, North Yorks, YO22 5LY [IO94PJ, NZ80]
G3 VNN Maj T L Craze, 27 Francis Gdns, Abbotts Barton, Winchester, SO23 7HD [IO91IB, SU43]
GW3 VNO D L Hughes, Mountain Air, Began Rd, Old St. Mellons, Cardiff, CF3 9XJ [IO81KM, ST28]
G3 VNP P C F Dowles, 46 New Rd, Great Baddow, Chelmsford, CM2 7QT [JO01GR, TL70]
G3 VNQ M G Pritchard, 9 Tamarack Dr, Peekskill, Ny 10566, USA
G3 VNT L J Pearson, The St., Ashfield Cum Thorpe, Ashfield, Stowmarket, Suffolk, IP14 6LX [JO02OF, TM26]
G3 VNU J A Finch, 286 Sea Front, Hayling Island, PO11 0AZ [IO90MS, SZ79]
GM3 VNW J B Macphee, 24 Bourtreehall, Girvan, KA26 9EL [IO75NF, NX19]
G3 VNY I Walker, 45 Terry Dr, Walmley, Sutton Coldfield, B76 2PT [IO92CN, SP19]
GW3 VNZ D A Jacklin, 43 Pant y Celyn Rd, Llandough, Penarth, CF64 2PF [IO81JK, ST17]
G3 VO J R Brierley, 133 Moor End Rd, Mellor, Stockport, SK6 5NQ [IO83XJ, SJ98]
G3 VOD K R Denman, 1 Church Cottage, Sedgeford Rd, Docking, Kings Lynn, Norfolk, PE31 8LJ [JO02HV, TF73]
G3 VOF M G Foster, 17 Prospect Rd, Hornchurch, RM11 3TY [JO01CN, TQ58]
G3 VOI B W Beevers, 11 Mint St., Marsh, Huddersfield, HD1 4PE [IO93CP, SE11]
G3 VOJ R A J Wager, Robbins Manse Chase, Maldon, Essex, CM9 7EA [JO01IR, TL80]
GW3 VOL J G Phillips, 96 Maes y Sarn, Pentyrch, Cardiff, CF4 8QR [IO81IM, ST08]
G3 VOM D A Lane, 4 Bramley Cl, Swinton, Manchester, M27 0DR [IO83TM, SD70]
G3 VOO M R Barnett-Bone, Dorchester Hill, Blandford, Milborne St. Andrew, Blandford Forum, Dorset, DT11 0JG [IO80US, SY89]
G3 VOQ B A Meade, Pine Trees, Back Ln, Birdingbury, Rugby, Warks, CV23 8EN [IO92HH, SP46]
G3 VOR J H R Brooker, 19 Ambledale, Sarisbury Green, Southampton, SO31 7BR [IO90IU, SU50]
G3 VOS R B Cottrell, Larkhill, 47 Bullsland Ln, Chorleywood, Rickmansworth, WD3 5BD [IO91RP, TQ09]
G3 VOT G C Webster, Red House Farm, Ashford Ln, Bakewell, DE45 1NJ [IO93DF, SK17]
G3 VOU J S Barlow, 68 Willow Ave, Cheadle Hulme, Cheadle, SK8 6AX [IO83VJ, SJ88]
G3 VOV M A Lane, 56 Main St., Bushby, Leicester, LE7 9PP [IO92LP, SK60]
G3 VOW M J Fereday, Spindlewood, Stoney Ln, Ashmore Green, Thatcham, RG18 9HQ [IO91IK, SU46]
G3 VPA M D Rose, 59 Park Dr, Sittingbourne, ME10 1RD [JO01IH, TQ86]
G3 VPD Details withheld at licensee's request by SSL.
G3 VPE H S Pinchin, 61 Cole Bank Rd, Hall Green, Birmingham, B28 8EZ [IO92BK, SP18]
G3 VPF E J Harland, 3 Randall Cl, Chickerell, Weymouth, DT3 4AS [IO80RO, SY68]
G3 VPG C K Jacob, 18 Compton Way, Olivers Battery, Winchester, SO22 4HS [IO91HB, SU42]
G3 VPH J Mayall, 10 Manor Cl, Droitwich, WR9 8HG [IO82WG, SO86]
G3 VPK W J McClintock, Mapleleaf, Tiptree Rd, Great Braxted, Witham, CM8 3EJ [JO01IT, TL81]
GW3 VPL S L Bleaney, 18 Brooklands Cl, Dunvant, Swansea, SA2 7TS [IO71XP, SS59]
GM3 VPN J F J Gardner, Taringa, Edentown, Cupar, Fife, KY15 7UH [IO86KG, NO21]
G3 VPO G R Willey, 37 Loxwood Rd, Lovedean, Waterlooville, PO8 9TT [IO90LV, SU61]
G3 VPR R Harrison, 512 Broadgate, Weston Hills, Spalding, Lincs, PE12 6DA [IO92WS, TF22]
G3 VPS P J Lennard, 5 Parkside, East Grinstead, RH19 1JG [IO91XD, TQ33]
G3 VPT P Burgess, 26 William Peck Rd, Spixworth, Norwich, NR10 3QB [JO02PQ, TG21]
GI3 VPV R Aughey, 30 Glen Rd, Hillsborough, BT26 6ES [IO64WK, J25]
G3 VPW J S Wright, 5 Warmans Cl, Wantage, OX12 9XS [IO91GO, SU38]
G3 VPX I J Sumner, 132 Barrs Rd, Cradley Heath, Warley, B64 7EZ [IO82XL, SO98]
GI3 VQ J K Thompson, 3 Sandyhill Ave, Dunmurry, Belfast, BT17 9LT [IO74AM, J26]
G3 VQD F E W Welch, 6 Wheatley Cl, Bowhay Park, Exeter, EX4 1NT [IO80FR, SX99]
G3 VQF J A Moorhouse, 185 Aldermoor Rd, Lordswood, Southampton, SO16 5NQ [IO90GW, SU31]
G3 VQG R C Beadle, 5 Badgeney Rd, March, PE15 9AP [JO02BN, TL49]
G3 VQH B J Clegg, 8 Hillside, Leak Hall Ln, Denby Dale, Huddersfield, HD8 8QZ [IO93EN, SE20]
GM3 VQJ T L Darke, 46 Dumgoyne Dr, Bearsden, Glasgow, G61 3AW [IO75TW, NS57]
G3 VQM D M Harrington, 27 Bayview Rd, Peacehaven, BN10 8QD [JO00AS, TQ40]
G3 VQN S M Newbold, 1 Druids Cl, West Parley, Ferndown, BH22 8RU [IO90BS, SZ09]
G3 VQO L M Allwood, 9 Gorse End, Horsham, RH12 5XW [IO91UB, TQ13]
G3 VQP Details withheld at licensee's request by SSL.
G3 VQQ M R Hall, 5 Kings Lea, Kingsway, Ossett, WF5 8RY [IO93EQ, SE22]
G3 VQR A Henshaw, 3 Lewens Cl, Parkwood Rd, Wimborne, BH21 1JJ [IO90AT, SU00]
G3 VQS S Barraclough, 8 Park Way, Formby, Liverpool, L37 6ED [IO83LN, SD20]
G3 VQW B J Fawkes, 6 Oak Ave, Brickfields, Worcester, WR4 9UG [IO82VE, SO85]
G3 VQY J W Cumming, 30 Victoria Ave, Wickford, SS12 0DL [JO01GO, TQ79]
G3 VQZ R Chinn, 27 Riverbourne Rd, Milford, Salisbury, SP1 1NU [IO91CB, SU12]
G3 VRB J D Nias, 49 St. Margarets Rd, Bishopstoke, Eastleigh, SO50 6DG [IO90HX, SU41]
G3 VRD R Shaw, 74 Brant Rd, Lincoln, LN5 8SH [IO93RE, SK96]
G3 VRE D Birch Chippenham & District Amateur, 32 Union St., Trowbridge, BA14 8RY [IO81VH, ST85]
G3 VRF J Charlton, 57 Victoria Rd, Bidford on Avon, Alcester, B50 4AR [IO92BE, SP05]
G3 VRL Details withheld at licensee's request by SSL.
G3 VRO J Russell, Kirklands, Peareth Hall Rd, Washington, NE37 1NR [IO94FW, NZ25]
G3 VRP R Pearson, 6 Ruskin Cl, Stanley, DH9 6UF [IO94DU, NZ25]
G3 VRU P E Ford, 15 Doles Ln, Whitwell, Worksop, S80 4SN [IO93JG, SK57]
G3 VRV M A Huish, Becketts, Woodbury, Exeter, Devon, EX5 1JD [IO80HQ, SY08]
G3 VRW P R Lamb, 5 The Templars, Bridge End, Warwick, CV34 6PF [IO92FG, SP26]
G3 VRX R W Turner, 7 Highfield Cres, Baildon, Shipley, BD17 5NR [IO93CU, SE13]
G3 VRY J E Pitt, 30 Hillcroft Rd, Chesham, HP5 3DJ [IO91QR, SP90]
G3 VSB G O Jones, Braemar Alton Rd, Denstone, Uttoxeter, Staffs, ST14 5DH [IO92BX, SK04]
G3 VSE K C Thompson, 18 Gleneagles Dr, Morecambe, LA4 5BN [IO84NB, SD46]
G3 VSG N Sedgwick, 261 Walton Back Ln, Walton, Chesterfield, S42 7AA [IO93GF, SK36]
G3 VSH D Freedman, Rivermeade, Irwell Vale, Ramsbottom, Bury, BL0 0QA [IO83UQ, SD72]
G3 VSI N H Prince, 96 Foxglove Way, Springfield, Chelmsford, CM1 6QR [JO01GS, TL70]
G3 VSJ D Chaloner, 38 Barnfield Cl, Hoddesdon, EN11 9EP [IO91XS, TL30]
G3 VSK T W McCurry, 148 Moorgate Rd, Rotherham, S60 3AZ [IO93HK, SK49]
G3 VSL J A Arscott, 122 Woodlands Rd, Ashurst, Southampton, SO40 7AL [IO90FV, SU31]
G3 VSQ R S West, 10 Hawkshill Dr, Boxmoor, Felden, Hemel Hempstead, HP3 0BS [IO91RR, TL00]
G3 VSR T Barraclough, 25 Ennismore Rd, Crosby, Liverpool, L23 7UG [IO83LL, SD30]
G3 VSS Details withheld at licensee's request by SSL.
G3 VST F E J Moore, 13 Westfield Rd, Barnehurst, Bexleyheath, DA7 6LS [JO01CL, TQ57]
G3 VSU A R F Moore, The Belle Vue Tavern, Pegwell Rd, Ramsgate, Kent, CT11 0NJ [JO01QH, TR36]
G3 VSV D J Middleton, 8 Fulmar Cl, Bradwell, Great Yarmouth, NR31 8JG [JO02UN, TG50]
G3 VSX R J Scoffield, Fairfield, 11 Grantham Rd, Great Gonerby, Grantham, NG31 8JZ [IO92QW, SK93]
G3 VSY S E Black, 53 Calderbrook Dr, Cheadle Hulme, Cheadle, SK8 5RT [IO83VJ, SJ88]
GM3 VTB V T Budas, 20 Oak Ave, Bearsden, Glasgow, G61 3HD [IO75UW, NS57]
G3 VTC F A Elvis, The Lodge, Lyncroft House, Stafford Rd, Lichfield Staffs, WS13 8JB [IO92BQ, SK11]
G3 VTD R Price, 36 Hadleigh Rise, Pontefract, WF8 4SJ [IO93IQ, SE42]
G3 VTE R Swetmore, 18 Tideswell Rd, Sandford Hill, Longton, Stoke on Trent, ST3 5EG [IO82WX, SJ94]
G3 VTG D J Esplin, 8 Lyte Ln, West Charleton, Kingsbridge, TQ7 2BW [IO80DG, SX74]
GM3 VTH D S Coutts, 29 Barons Hill Ave, Linlithgow, EH49 7JU [IO85EX, NT07]
G3 VTI A R Burford, 138 Jessie Rd, Aldridge, Walsall, WS9 8HP [IO92BO, SK00]
G3 VTL J K Levett, 402 Cromwell Ln, Burton Green, Kenilworth, CV8 1PL [IO92EJ, SP27]
G3 VTN W F Burgess, 11 Cobden Ave, Portsmouth, PO3 6NA [IO90LT, SU60]
G3 VTO M P Coombs, 10 Horseshoe Walk, Widcombe, Bath, BA2 6DE [IO81TJ, ST76]
G3 VTP E S Deione, 6 Lisle Cl, Lymington, SO41 3NW [IO90FS, SZ39]
G3 VTR A W Davis, Fieldings, Bury Rd, Lawshall, Bury St.Edmunds, IP29 4PL [JO02ID, TL85]
G3 VTS C G Walker, 2 Georgian Cl, The Wheatridge, Abbeydale, Gloucester, GL4 5DG [IO81VU, SO81]
G3 VTT C L Turner, 105 The Everglades, Hempstead, Gillingham, ME7 3PZ [JO01GI, TQ76]
G3 VTY K R Robson, 8 Grange Rd, Burley in Wharfedal, Burley in Wharfedale, Ilkley, LS29 7NF [IO93DV, SE14]
G3 VUD P Bentley, 12 West Terr, Seaton Sluice, Whitley Bay, NE26 4RE [IO95GC, NZ37]
G3 VUE T R Mowbray, 18 King Georges Cl, Rollesby, Great Yarmouth, NR29 5HB [JO02TQ, TG41]
G3 VUF G Wright, The Old Chapel, Coanwood, Haltwhistle, Northd., NE49 0QS [IO84SW, NY65]
G3 VUH M J Blackwell, 43 Rushton Rd, Wilbarston, Market Harborough, LE16 8QL [IO92OL, SP88]
G3 VUI M R Harris, Box 226, Port Stanley, Falkland Islands, South Atlantic, ZZ2 2BO
G3 VUK R M Knight, 8 Narromine Dr, Calcot, Reading, RG3 7ZL [IO91LJ, SU66]
G3 VUL Dr J C Lotz, 29 Burton Manor Rd, Stafford, ST17 9QJ [IO82WS, SJ92]
G3 VUN G D Ackerley, 40 Linwood Ave, Strood, Rochester, ME2 3TR [JO01FJ, TQ76]
G3 VUS D W Latimer, 44 Lyndale Ave, Barrow in Furness, LA13 9AR [IO84JC, SD27]
G3 VUU B J Stork, 20 Oakdale Ave, Wibsey, Bradford, BD6 1RN [IO93CS, SE13]
G3 VUV E J Smith, 85 Boulton Grange, Telford, TF3 2LE [IO82SP, SJ70]
G3 VUY D E Bradley, 4 Felthorpe Cl, Upton, Wirral, L49 4GY [IO83KJ, SJ28]
G3 VVA Dr E J B McArthur, 20 Leige House, Manorside Cl, Upton, Wirral, Merseyside, L49 4PP [IO83KJ, SJ28]
GW3 VVC J E Parry, Ar Allt, Lon Hedydd, Llanfair, Anglesey, LL61 5JY [IO73VF, SH57]
G3 VVE H Robinson, 4 Cross St., Mansfield Woodhouse, Mansfield, NG19 9NA [IO93JD, SK56]
GM3 VVF A C Ross, 17 Tarvit Green, Glenrothes, KY7 4SJ [IO86KE, NT29]
G3 VVG B W Salt, Cruets, The Village, Buckland Monachorum, Yelverton, PL20 7NA [IO70WL, SX46]
G3 VVI P A Archer, 6 Kenton Howse, Elmore Rd, Patchway, Bristol, BS12 5NX [IO81RM, ST58]
G3 VVL K C Lax, The Nook, Rodington, Shrewsbury, SY4 4QN [IO82QR, SJ51]
GM3 VVM H Graham, 20 Stirling Dr, Linwood, Paisley, PA3 3LF [IO75RU, NS46]
G3 VVO J R B Brain, 68 Sheridan Rd, Bath, BA2 1RB [IO81TJ, ST76]
G3 VVP F W Porter, 45 Haddington Rd, Stoke, Plymouth, PL2 1RR [IO70VJ, SX45]
G3 VVR J Grace, Woodside, Easthorpe, Malton, N Yorks, YO17 0QX [IO94ND, SE77]
G3 VVT R Wilkinson, 18 Green Rd, Kendal, LA9 4QR [IO84OH, SD59]
G3 VVU I W King, 21 Chadwick Rd, Bobbers Mill, Nottingham, NG7 5NN [IO92JX, SK54]
G3 VVW J L Forrest, Little Orchard, 21 Bell Ln, Fetcham, Leatherhead, KT22 9ND [IO91TG, TQ15]
G3 VVZ R C Crines, 180 Brambles Chine, Monks Ln, Freshwater, PO40 9SY [IO90FQ, SZ38]
G3 VW R H Newland, Ham House, Lyme Rd, Uplyme, Lyme Regis, DT7 3XA [IO80MR, SY39]
G3 VWC A C Marriott, 28 Horseshoe Walk, Bath, BA2 6DF [IO81TJ, ST76]
G3 VWD C J Bean, 11 Nightingale Ln, Earlsdon, Coventry, CV5 6AY [IO92FJ, SP37]
G3 VWE J Phillips, 6 North Devon Rd, Fishponds, Bristol, BS16 2EU [IO81RL, ST67]
G3 VWF Details withheld at licensee's request by SSL.
G3 VWH B K Wilde, 34 Grangefields Rd, Shrewsbury, SY3 9DB [IO82OQ, SJ41]
G3 VWJ G W Westwood, 133 Torrisholme Rd, Lancaster, LA1 2TZ [IO84OB, SD46]
G3 VWK A H Hammett, Rosehill, Ladock, Truro, Cornwall, TR2 4PQ [IO70MH, SW85]
G3 VWL R E Kemp, 20 Bramble Cl, Worthing, BN13 3HZ [IO90TU, TQ10]
G3 VWP Details withheld at licensee's request by SSL.[Station located in Lodon SW1.]
G3 VWQ P W F Forster, 53 Alexandra Rd, Norwich, NR2 3EA [JO02PP, TG20]
GW3 VWT D J E Farraday, 21 Kenway Ave, Cimla, Neath, SA11 3TU [IO81CP, SS79]

G3

G3 VWX E H M Perks, The Oaklands, Bromfield Rd, Ludlow, SY8 1DW [IO82PJ, SO57]
GM3 VWY I M Malcolm, 54 Lawmill Gdns, St. Andrews, KY16 8QS [IO86OH, NO41]
G3 VXA M J Harrold, 26 Leys Cl, Harefield, Uxbridge, UB9 6QB [IO91SO, TQ09]
GW3 VXC M J Williams, Garnbrook Cottage, Glen Usk Rd, Llanhennock, Caerleon Newport, NP6 1LU [IO81MP, ST39]
G3 VXE G F Brindle, 8 Peckover Dr, Pudsey, LS28 8EF [IO93DT, SE13]
G3 VXF B E Ellis, Whitmore Lodge, Ridgemoor, Hindhead, Surrey, GU26 6QX [IO91PC, SU83]
G3 VXH J R Huffadine, 19 Orchard Way, Studley, B80 7NZ [IO92BG, SP06]
G3 VXI Details withheld at licensee's request by SSL.
G3 VXJ R F F Rylatt, 16 First Ave, Charmandean, Worthing, BN14 9NJ [IO90TU, TQ10]
G3 VXK R B Porter, 16 Millcroft, Crosby, Liverpool, L23 9XJ [IO83LL, SD30]
G3 VXM D T M Clemens, 66 Mayles Rd, Milton, Southsea, PO4 8NP [IO90LT, SZ69]
GW3 VXP Details withheld at licensee's request by SSL.
G3 VXS D M Peach, 56 Basford Park Rd, Newcastle, ST5 0PS [IO83VA, SJ84]
G3 VXX Maj G S S Symons Blanford Garr AR, Melbury View, Manston, Sturminster Newton, Dorset, DT10 1HB [IO80UW, ST81]
G3 VXZ M J Frey, 18 Rushington Ave, Maidenhead, SL6 1BZ [IO91PM, SU88]
G3 VYA B A Atkiss, 47 Russell Rd, Urmston, Partington, Manchester, M31 4DY [IO83SJ, SJ79]
G3 VYB R A Butterfield, 4 Barnhurst Cl, Childwall, Liverpool, L16 7QT [IO83NJ, SJ48]
G3 VYD J R Bourne, Tyndalls, 8 Kelvedon Rd, Wickham Bishops, Witham, CM8 3LZ [JO01IS, TL81]
G3 VYE J Doswell, 63 Heather Way, Harrogate, HG3 2SH [IO93FX, SE25]
G3 VYF M R Lee, 11 Sturrocks, Vange, Basildon, SS16 4PQ [JO01FN, TQ78]
G3 VYG R R Walpole, Woodfarm Cottage, Reymerston, Norwich, Norfolk, NR9 4QZ [JO02LO, TG00]
G3 VYI M A Franklin, 6 Tor Rd, Farnham, GU9 7BX [IO91OF, SU84]
G3 VYK P A Frost, 164 Newthorpe Common, Newthorpe, Nottingham, NG16 2EN [IO93IA, SK44]
G3 VYN M S Turner, Plumtree Cottage, Alburgh Rd, Hempnall, Norwich, NR15 2NT [JO02PL, TM29]
G3 VYP Details withheld at licensee's request by SSL.[Station located near Ludlow.]
G3 VYQ R A J Deane, 11 Shirwood Ave, Whickham, Newcastle upon Tyne, NE16 5JW [IO94DW, NZ26]
G3 VYS A W Nell, Ravens Oak, Landford, Salisbury, Wilts, SP2 2BE [IO91CB, SU13]
G3 VYU R G Chamberlain, 1 Thornmead, Werrington Meadows, Peterborough, PE4 7ZD [IO92UO, TF10]
G3 VYV D A Duff, Woodlea, Highfield Rd, Croston, Preston Lancs, PR5 7HH [IO83OP, SD41]
G3 VYW D A P Carter, 2 Wellswood Gdns, Rowlands Castle, PO9 6DN [IO90MV, SU71]
GI3 VYY B G Hamilton, 6 Trenchill Ave, Ballyclare, BT39 9JH [IO64XS, J29]
G3 VYZ L W Thompson, 44 Tillmouth Ave, Holywell, Whitley Bay, NE25 0NP [IO95FB, NZ37]
G3 VZE D Kennedy, 79 High St., Dunsville, Doncaster, DN7 4BS [IO93LN, SE60]
G3 VZF J H Adams, Chilterns, Bellingdon, Bucks, HP5 2XL [IO91QR, SP90]
G3 VZG R N Golding, 7 Belvidere Ave, Shrewsbury, SY2 5PF [IO82PQ, SJ51]
G3 VZH Dr C J Doran, 16 Wordsworth Rd, Penge, London, SE20 7JG [IO91XJ, TQ37]
G3 VZJ A R Clemmetsen, The Danes, Shelbridge Rd, Slindon Common, Arundel, W Sussex, BN18 0ND [IO90QU, SU90]
G3 VZL R B Newman, 20 Glapthorn Rd, Oundle, Peterborough, Cambs, PE8 4JQ [IO92SL, TL08]
G3 VZM F Houghton, 14 Windfield Gdns, Little Sutton, South Wirral, L66 1JJ [IO83MG, SJ37]
GW3 VZO V Hartshorn, Clwydfa, Rhyl Rd, Denbigh, LL16 3DP [IO83HE, SJ06]
G3 VZP R T Morrison, 10 Ballinode Cl, Cheltenham, GL50 4SQ [IO81WW, SO92]
G3 VZR E R Thompson, Meadowside, Bromberrow Heath, Ledbury, Herefordshire, HR8 1PF [IO81TX, SO73]
G3 VZT R K Johnson, The Hollies, Belaugh Green, Coltishall, Norwich, Norfolk, NR12 7AJ [JO02QR, TG21]
G3 VZU W Mooney, 538 Liverpool Rd, Great Sankey, Warrington, WA5 3LU [IO83QJ, SJ58]
G3 VZV G P Shirville, The Hill Farm, Potsgrove, Milton Keynes, Bucks, MK17 9HF [IO91QX, SP93]
G3 VZZ K C George, 194 Jeffcock Rd, Penn Fields, Wolverhampton, WV3 7AH [IO82WN, SO99]
G3 WA Details withheld at licensee's request by SSL.
G3 WAB P W Harrison, 71 Coachmans Ln, Baldock, SG7 5BG [IO91VX, TL23]
G3 WAC W A Coates, 21 Salcombe Ave, Weeping Cross, Stafford, ST17 0HN [IO82XT, SJ92]
G3 WAD Mt W Wade, 12 Berkswell Cl, Solihull, B91 2EH [IO92CK, SP18]
G3 WAE I Harris, Orchard Cottage, Bishops Cannings, Devizes, Wilts, SN10 2LD [IO91AJ, SU06]
G3 WAG D J Gillett, 135 Honey Ln, Waltham Abbey, EN9 3AX [JO01AQ, TL30]
G3 WAH N L Hodgson, 42 Tofts Gr, Rastrick, Brighouse, HD6 3NP [IO93CQ, SE12]
G3 WAL Details withheld at licensee's request by SSL.[Op: J W Barker. Station located near Rugby.]
G3 WAM M G Taplin, 88 Chartwell Ave, Wingerworth, Chesterfield, S42 6SP [IO93GE, SK36]
G3 WAO T F Biddlecombe, 3 Humber Cl, Stubbington, Fareham, PO14 3RH [IO90JT, SU50]
GM3 WAP A P Philp, Former Manse, High St, Alyth, Perthshire, PH11 8DW [IO86JP, NO24]
G3 WAR R North, 4 Cauldron Barn Rd, Swanage, BH19 1QF [IO90AO, SZ07]
G3 WAS R Smethers Lichfield A.R.S, 46 Church Rd, Burntwood, Staffs, WS7 9EA [IO92BQ, SK00]
G3 WAW A W Ellis, Churton Lodge, 35 Douglas Rd, Herne Bay, CT6 6AF [JO01NI, TR16]
G3 WBA I Currell, 47 Highdale Ave, Clevedon, BS21 7LU [IO81NK, ST47]
G3 WBB E D Avery, 10 Maple Gr, Boley Park, Lichfield, WS14 9XB [IO92CQ, SK10]
G3 WBC R E Bryant, 51 Wheatfield Rd, Luton, LU4 0TS [IO91SV, TL02]
G3 WBF A C Hunter, 43 Edgemoor Dr, Crosby, Liverpool, L23 9UE [IO83LL, SD30]
G3 WBG H J R Hindle, 6 Windsor Rd, Conisbrough, Conisbrough, Doncaster, DN12 3DF [IO93JL, SK59]
GW3 WBH F A Tonks, 8 Scandinavia Heights, Saundersfoot, SA69 9PE [IO71PQ, SN10]
G3 WBI P J Lewis, 27 St. Leonards Rd West, St. Annes, Lytham St. Annes, FY8 2PF [IO83LS, SD32]
G3 WBK Dr P S Tofts, 48 Rugby Rd, Brighton, BN1 6EB [IO90WU, TQ30]
G3 WBL K Weller, Charlbury House, Bayton Village, nr Kidderminster, Worcs, DY14 9LJ [IO82SI, SO67]
G3 WBM J E Durrant, 6 Greenacres Rd, Layer de la Haye, Colchester, Essex, CO2 OJP
G3 WBN A Thurlow, Chesnet House, Bishops Walk, Croydon, CR0 5BA [IO91XI, TQ36]
G3 WBP J D Broadley, 13 Portland Cl, Bedford, MK41 9NE [IO92SD, TL05]
G3 WBQ Details withheld at licensee's request by SSL.[Op: T Brook, White Cottage, Shophouse Lane, Farley Green, Guildford, Surrey, GU5 9EQ.]
GI3 WBR R E McCrea, Killynoogan, Tullyhommen, Co Fermanagh, N Ireland, BT93 8DF [IO64BJ, H15]
G3 WBS D Thomson, Pineview, Berrylands, Hartley, Longfield, DA3 8AP [JO01DJ, TQ66]
GW3 WBU B Vodden, 22 Heath Ave, Cowslip Est, Penarth, CF64 2QZ [IO81JK, ST17]
G3 WBW P R Heming, 68 Bedowan Meadows, Tretherras, Newquay, TR7 2SW [IO70LJ, SW86]
G3 WBX M V Kathrens, 102 New Rd, Hethersett, Norwich, NR9 3HQ [JO02OO, TG10]
G3 WBY A V Hoare, 26 Kingsway, Euxton, Chorley, PR7 6PP [IO83QP, SD51]
G3 WBZ D H Facer, 4 Lime Rd, Harrington, Workington, CA14 5QH [IO84FO, NX92]
GW3 WCA P R L Dunbar, Elms, Newcastle Emlyn, Dyfed, SA38 9RA [IO72SB, SN24]
G3 WCB D John, 41A Chequers Orchard, Iver, Bucks, SL0 9NJ [IO91RM, TQ08]
G3 WCD Dr C R Dillon, 63 High St., Toseland, Huntingdon, Cambs, PE19 4RX [IO92VF, TL26]
G3 WCE B L Edwards, 232 Earlham Rd, Norwich, NR2 3RH [JO02PP, TG20]
GM3 WCH W C Henry, Anderlea, Gulberwick, Lerwick, Shetland, ZE2 9JX [IP90JD, HU43]
G3 WCJ P F Hackett, Lot49, Bodeguero Way, Wooroloo, Australia 6558
G3 WCK R H Godley, 87 Clifden Rd, Clapton, London, E5 0LW [IO91XN, TQ38]
G3 WCL J J Croker, 29 Alexandra Rd, Bedminster Down, Bristol, BS13 7DF [IO81QK, ST56]
G3 WCM F Chidlow, 64 Mitchell Ave, Northside, Workington, CA14 1AA [IO84FP, NY02]
G3 WCO J A Rollason, Pavitts Cottage, Keers Green, Dunmow, CM6 1PQ [JO01DT, TL51]
G3 WCQ R G Bailey, 43 Earlsdon Ave South, Coventry, CV5 6DR [IO92FJ, SP37]
G3 WCS K J Wood, Hilbre, Warrington Rd, Antrobus, Northwich, CW9 6JB [IO83RH, SJ67]
G3 WCU J Pealing, 93 Fernside Rd, Poole, BH15 2JQ [IO90AR, SZ09]
GW3 WCV D Howell, 6 Douglas Cl, Danescourt, Llandaff, Cardiff, CF5 2QT [IO81JL, ST17]
G3 WCY B A Smith, 20 Bessingby Rd, Ruislip, HA4 9DA [IO91TN, TQ18]
G3 WDD T H Horrobin, 29 Ambleside Rd, Maghull, Liverpool, L31 6BU [IO83MM, SD30]
G3 WDG Dr C W Suckling, 314 Newton Rd, Rushden, NN10 0SY [IO92RG, SP96]
G3 WDI T F Weatherley, 16 Beverley Ct, Carlton Colville, Lowestoft, NR33 8JZ [JO02UK, TM58]
G3 WDK P Downey, Qrm Productions, Myrtle Cottage, Drewsteignton, Exeter, Devon, EX6 6QW [IO80CQ, SX79]
G3 WDL J S Boast, 118 Barnsley Rd, Moorends, Doncaster, DN8 4QR [IO93MP, SE61]
G3 WDM C M Care, 127 Brooklands Cres, Fulwood, Sheffield, S10 4GF [IO93FI, SK38]
G3 WDN E Fielding, 3 Orchard Rise, Worlingham, Beccles, NR34 7RZ [JO02TK, TM48]
G3 WDQ R A Millington, 2 Goodwood Green, Fernhill Heath, Worcester, WR3 7XG [IO82VF, SO85]
G3 WDR H J Kearsey, 6 Dukes Cl, Felixstowe, IP11 9NS [JO01QX, TM33]
G3 WDS D Spooner, 45 Otterburn Ave, Whitley Bay, NE25 9QR [IO95GA, NZ37]
G3 WDU I F Peterkin, 65 Old Town Mews, Old Town, Stratford upon Avon, CV37 6GR [IO92DE, SP15]
G3 WDV J B Gardner, 40 Esk Rd, Lowry Hill, Carlisle, CA3 0HW [IO84MV, NY35]
G3 WDW J T Webster, 13 The Bank, Eccleshill, Bradford, BD10 8BL [IO93DT, SE13]
G3 WDX P A Hickey, 16 Cross Rd, Oxhey, Watford, WD1 4DH [IO91TP, TQ19]
G3 WDY D Pike, 27 Cintra Park, Upper Norwood, London, SE19 2LQ [IO91XK, TQ37]
G3 WEA A R Cross, 58 Alston Rd, Barnet, EN5 4EY [IO91VP, TQ29]
G3 WEB G W Gardiner, 11 Langdale Ave, Ramsgate, CT11 0PQ [JO01QI, TR36]
G3 WEC A Yates, 28 Kingfisher Rd, Offerton, Stockport, SK2 5JR [IO83WJ, SJ98]
GM3 WED A J Rose, Craiglea, Culbokie By Dingwal, Ross-shire, IV7 8JH [IO77TO, NH65]
G3 WEF A L Beazley, 24 Tealsbrook, Covingham Park, Swindon, SN3 5AU [IO91DN, SU18]
G3 WEG P M J Webster, 7 Polesden Cl, North Millers Dale, Chandlers Ford, Eastleigh, SO53 1TW [IO90HX, SU42]
G3 WEI D C M Turner, Birchwood, Heath Top, Ashley Heath, Market Drayton, TF9 4QR [IO82TW, SJ73]
GI3 WEL R Knox, Knockboy House, 91 Banbridge Rd, Waringstown, Craigavon, BT66 7RU [IO64UK, J15]
GI3 WEM V I Gracey, 23 Cascum Rd, Banbridge, BT32 4LF [IO64UH, J14]
G3 WEP R M Dotchin, 2 The Cres, Shortstown, Bedford, MK42 0UJ [IO92SC, TL04]
GW3 WEQ C Collins, 21 Bron Wern, Llanddulas, Abergele, LL22 8JD [IO83EG, SH97]
G3 WER L C Hearn, 16 Vine Cl, Welwyn Garden City, AL8 7PS [IO91VT, TL21]
G3 WET J G Evans, 22 Harwood Rd, Lichfield, WS13 7PP [IO92CQ, SK11]
G3 WEU K H Gregory, Poldhu, 67 Clowne Rd, Barlborough, Chesterfield, S43 4EH [IO93IG, SK47]
G3 WEX A L Wragg, 29 Eastern Rd, Sutton Coldfield, B73 5PA [IO92CN, SP19]
GW3 WEZ J C Lawrence, 90 Pastoral Way, Ty Coch, Sketty, Swansea, SA2 9LY [IO81AO, SS69]
G3 WF D R Cockings, Elettra, 207 Birchfield Rd, Redditch, B97 4LX [IO92AH, SP06]
G3 WFB Details withheld at licensee's request by SSL.
G3 WFC M E Allard, 12 Greenway Park, Galmpton, Brixham, TQ5 0NA [IO80FJ, SX85]
G3 WFD T R Mallender, Kelwyn, 1 Chapel Ln, Lichfield, WS14 9BA [IO92CQ, SK10]
G3 WFF B G Tew, 96 Mill Ln, Sawston, Cambridge, CB2 4HZ [JO02BC, TL44]
G3 WFH D H Morris, 27 Albert Sq, Bowdon, Altrincham, WA14 2ND [IO83TJ, SJ78]
GM3 WFJ R S T J Andrew, The Old Manse, Kirkmichael, Blairgowrie, Perthshire, PH10 7NY [IO86FR, NO05]
G3 WFK Details withheld at licensee's request by SSL.
G3 WFM J J N Crabbe, 47 Torrington Dr, Potters Bar, EN6 5HU [IO91WQ, TL20]
GI3 WFP P McAlpine, 20 Gransha Rd South, Bangor, BT19 7QB [IO74EP, J57]
G3 WFT D C Holland, 32 Woodville Dr, Sale, M33 6NF [IO83UK, SJ79]
G3 WFW K M Hampson, 11 Gladstone Gr, Heaton Moor, Stockport, SK4 4BX [IO83VK, SJ89]
G3 WGB D R Taylor, Broken Hill, Lower Freystrop, Haverfordwest, Pembrokeshire, Wales, SA62 4ET [IO71MS, SM91]
G3 WGC K G Pollard Welwyn-Hatfield, Arcade, 27 Far End, Hatfield, AL10 8TG [IO91VR, TL20]
G3 WGE E K Law, 4 Brograve Cl, Galleywood, Chelmsford, CM2 8YA [JO01FQ, TL70]
G3 WGF G S Rose, 4 Saltdean Cl, Bexhill on Sea, TN39 3ST [JO00FU, TQ70]
G3 WGG K E Wright, Rozel Cottage, Maynards Green, Heathfield, E Sussex, TN21 0DE [JO00DW, TQ51]
G3 WGK B J Wormwell, 1 Towneley House, Longridge, Preston, Lancs, PR3 3AA [IO83QT, SD63]
G3 WGL N I Briggs, 27 St. Georges Dr, Caister on Sea, Great Yarmouth, NR30 5QW [JO02UP, TG51]
G3 WGM Details withheld at licensee's request by SSL.
G3 WGN D W Aslin, Parkside, Much Marcle, Ledbury, Herefordshire, HR8 2NW [IO81RX, SO63]
G3 WGQ J R Hartley, 2 Hall St., Cockbrook, Ashton under Lyne, OL6 6SD [IO83XL, SJ99]
G3 WGU S Williamson, 120 Warbreck Hill Rd, Blackpool, FY2 0TR [IO83LU, SD33]
G3 WGV J R Linford, Camberra Lodge, Heath Ride, Finchampstead, Wokingham, RG40 3QJ [IO91NI, SU86]
G3 WGY H V Ashford, 56 Guarlford Rd, Malvern, WR14 3QP [IO82UC, SO74]
G3 WGZ G W Sowden, Villa Clare, Lizard Point, Helston, Cornwall, TR12 7NU [IN79JX, SW71]
GI3 WHA L Hanna, 48 The Park, Millars Forge, Dundonald, Belfast, BT16 0QP [IO74CO, J47]
G3 WHB S Christie, Braywood Linn, Oakley Green Rd, Oakley Green, Windsor, SL4 4PZ [IO91PL, SU97]
G3 WHF D Richardson, 95 Brookfield Rd, Bedford, MK41 9LL [IO92SD, TL05]
G3 WHG M D Key, 12 Great Melton Rd, Hethersett, Norwich, NR9 3AB [JO02OO, TG10]
G3 WHJ A R Johnson, Cowslip Cottage, 121 Clevelode, Malvern, Worcs, WR13 6PD [IO82VC, SO84]
G3 WHK D A R Poulter, 119 Aragon Rd, Morden, SM4 4QG [IO91VJ, TQ26]
G3 WHL J Darrington, 10 Furnival Rd, Balby, Doncaster, DN4 0PJ [IO93KM, SE50]
G3 WHM A G Coker, 48 Charlock Way, Burpham, Guildford, GU1 1XZ [IO91RG, TQ05]
GU3 WHP M C Sloan, Portinfer Vale, Guernsey, GY6 8LN
G3 WHR R A W Brocks, 12 Blacksmiths Ln, Wickham Bishops, Witham, CM8 3NR [JO01IS, TL81]
GM3 WHT M J Smith, Vakterlee, Cumliewick, Sandwick, Shetland, ZE2 9HH [IP99JX, HU42]
GW3 WHU G C Parrott, Flat 5, 33 Princes Dr, Colwyn Bay, Clwyd, LL29 8PD [IO83DH, SH87]
G3 WHV K J Brackley, 14 Meadow Cres, Castle Donington, Derby, DE74 2LX [IO92HU, SK42]
G3 WHW J A Daykin, Old Railway Inn, Fancy Hill, Parkend, Lydney, GL15 4JN [IO81RS, SO60]
G3 WHZ Details withheld at licensee's request by SSL.
G3 WI F Hulme, 3 Garden Ave, Droylsden, Manchester, M43 7XA [IO83WL, SJ99]
G3 WIA R Ottley, 15 Polesteeple Hill, Biggin Hill, Westerham, TN16 3TE [JO01AH, TQ45]
G3 WIB P S Downham, 30 Southbury Ave, Enfield, EN1 1RL [IO91XP, TQ39]
G3 WIC K E Griffiths, 127 Utting Ave East, Liverpool, L11 5AB [IO83MK, SJ39]
G3 WIE Dr C J Bryant, 36 Blenheim Gdns, Southampton, SO17 3RQ [IO90HW, SU41]
GM3 WIG G Shankie, Appin, 3 Orchard St., Hawick, TD9 9JJ [IO85OK, NT51]
G3 WII F G E Clarke, 3 Juniper Cl, North Baddesley, Southampton, SO52 9FS [IO90GX, SU32]
GM3 WIJ N A Mackenzie, 57 Countesswells Terr, Aberdeen, AB1 8LQ [IO87TH, NJ72]
G3 WIK M Shorland, Lower End Cottage, Bredons Norton, Tewkesbury, Glos, GL20 7HB [IO82WB, SO93]
GM3 WIL D L Cossar, 52 Bentfield Dr, Prestwick, KA9 1TT [IO75QL, NS32]
G3 WIM J T Gale Wimbledon & District ARS, 55A Tattenham Way, Tadworth, KT20 5NE [IO91VH, TQ25]
G3 WIO E M OBrien, Tanglewood, Anthonys Way, Heswall, Merseyside, L60 0BP [IO83KH, SJ28]
G3 WIP Dr G V Bulger, 58 Newick Rd, Lower Clapton, London, E5 0RR [IO91XN, TQ38]
G3 WIR P Harman Burnham Beeches, Radio Club, 25 Pitts Rd, Slough, SL1 3XG [IO91QM, SU98]
G3 WIS B Day, 53 Widdows St., Leigh, WN7 2AE [IO83SL, SJ69]
G3 WIU W J F Bekenn, 35 Blackdown Ave, Rushmere St. Andrew, Rushmere, Ipswich, IP5 7AY [JO02OB, TM24]
GM3 WIV J McGowan, 2 Coila Ave, Prestwick, KA9 2BW [IO75QL, NS32]
G3 WIW A S Leach, 199 Braemor Rd, Calne, SN11 9EA [IO81XK, ST97]
GU3 WIY S K Sharman, Mont Morin, St. Sampsons, St. Sampson, Guernsey, Channel Islands, GY2 4JD
G3 WJA D S Farler, 107 Chandag Rd, Keynsham, Bristol, BS18 1QE [IO81SJ, ST66]
G3 WJC R Chilton, 9 Perry Ct, Hagley Rd West, Oldbury, Warley, B68 0BU [IO82XL, SO98]
GM3 WJE J S Thom, 64 Hawick Dr, Dundee, DD4 0TA [IO86NL, NO43]
G3 WJG G D Lean, 54 Blacketts Wood Dr, Chorleywood, Rickmansworth, WD3 5QH [IO91RP, TQ09]
G3 WJH W Wilkinson, Chiriqui, 15 Camerton Rd, Seaton, Workington, CA14 1LP [IO84FP, NY03]
G3 WJI P M White, Linden House, Willisham, Ipswich, Suffolk, IP8 4SP [JO02MC, TM05]
G3 WJJ D G Finnemore, 9 Beacon Rd, Broadstone, BH18 9JP [IO80XS, SY99]
G3 WJM B E Schoth, 21 Hampton Ln, Blackfield, Southampton, SO45 1ZA [IO90HT, SU40]
G3 WJN R Hassell Bennett, 30 Greenlands Ave, Redditch, B98 7QA [IO92AG, SP06]
G3 WJO L R Bryant, 9 Suncrest Est, Indian Queens, St. Columb, TR9 6PP [IO70MJ, SW95]
G3 WJP J H I Parnell, 40 Dolcoath Rd, Camborne, TR14 8RW [IO70IF, SW64]
G3 WJS J A Starling, 13 Grantham Cres, Ipswich, IP2 9PD [JO02NB, TM14]
G3 WJU C V Buckland, 101 Acorn Ave, Cowfold, Horsham, RH13 8RT [IO91UB, TQ12]
G3 WJY G A Tansley, 3 Parade Bank, Moulton, Northampton, NN3 7ST [IO92NG, SP76]
GM3 WKB D Topham, 15 Lower Burnmouth, Burnmouth, Eyemouth, TD14 5SP [IO85XU, NT96]
G3 WKD K B Attenborough, 28 Briony Ave, Hale, Altrincham, WA15 8QD [IO83UJ, SJ78]
G3 WKE Details withheld at licensee's request by SSL.[Op: Steve Braidwood.]
G3 WKF M J Richards, Wayside, Penwithick Rd, Penwithick, St. Austell, Cornwall, PL26 8UH [IO70OI, SX05]
G3 WKG E J Vesper, 53 Beechwood Dr, Woodford Green, IG8 9QY [JO01AO, TQ39]
G3 WKI M F W Hewins, 37 Ringwood Cl, Furnace Green, Crawley, RH10 6HQ [IO91VC, TQ23]
G3 WKL Details withheld at licensee's request by SSL.
GM3 WKM K G Melton, 9 Murray Pl, Smithton, Inverness, IV1 2PX [IO77WL, NH74]
G3 WKP P King, Nirvana, Comprigney Hill, Truro, TR1 3TX [IO70LG, SW84]
G3 WKR M B Goodwin, 6 Hobbs Hill, Rothwell, Kettering, NN14 6YG [IO92OK, SP88]
G3 WKS A Korda West Kent RC, 5 Windmill Ct, North St., Tunbridge Wells, TN2 4SU [JO01DD, TQ53]
G3 WKX P J S Bysshe Maidenhd&Dis AR, Orchard House, High Rd, Cookham, Maidenhead, SL6 9JT [IO91PN, SU88]
G3 WKZ Dr C R Bayliss, 3 Cleeve Ct, Streatley, Reading, RG8 9PS [IO91KM, SU58]
G3 WLA A C Macpherson, Esperance, 15A Monkstone Dr, Berrow, Burnham on Sea, Somerset, TA8 2NW [IO81LG, ST25]
G3 WLD J W Hall, 22 Haverhill Rd, Stapleford, Cambridge, CB2 5BX [JO02BD, TL45]
G3 WLG Dr M J Griffiths, The Oaklands, Holly Bush Ln, Clifton on Teme, Worcester, Worcs, WR6 6HQ [IO82SG, SO66]
G3 WLH Dr C Pell, 1 Glenville Gdns, Tower Rd, Hindhead, GU26 6SX [IO91PC, SU83]
G3 WLK P J O'Connor, 63 Mellstock Rd, Oakdale, Poole, BH15 3DW [IO90AR, SZ09]
G3 WLM R A Joyce, 20 Barking Cl, Luton, LU4 9HG [IO91SV, TL02]
GW3 WLN Dr A H Pritchard, Brynawelon, 15 Wingfield Rd, Whitchurch, Cardiff, CF4 1NJ [IO81JM, ST17]
G3 WLO E R Denton, 11 Highland Rd, Amersham, HP7 9AU [IO91QQ, SU99]
G3 WLP H P Stokoe, 15 Primley Park Walk, Leeds, LS17 7LB [IO93FU, SE34]
G3 WLS C S Town, 15 Knollbeck Cres, Wombwell, Brampton, Barnsley, S73 0TT [IO93HM, SE40]
G3 WLT Dr D R Firth, 3 School Ln, Shaldon, Teignmouth, TQ14 0DG [IO80FM, SX97]
G3 WLV J R Bushby, 14 Clayton Dr, Thurnscoe, Rotherham, S63 0RZ [IO93IN, SE40]
G3 WLW R T Millar, 1229 Leeds Rd, Bradley, Huddersfield, HD2 1UY [IO93DQ, SE12]
G3 WLX J L Green, 13 Ave Cesar Geoffray, Vaison La Romaine, Vaucluse 84110, France, X X
G3 WLY J H Harwood, 12 Longwood Ave, Cowplain, Waterlooville, PO8 8HX [IO90LV, SU61]
G3 WMA W P Shepperd, 28 Tyne Rd, Oakham, LE15 6SJ [IO92PP, SK80]
G3 WMB M K Bacon, The Retreat, Helham Green, Wareside, Ware Herts, SG12 7RE [JO01AT, TL41]
G3 WME M J D Groom, 409 Finchampstead Rd, California, Finchampstead, Wokingham, RG40 3RL [IO91NJ, SU76]
GM3 WML D G Smith, 5 Camaghael Rd, Caol, Fort William, PH33 7HU [IO76KU, NN17]
G3 WMM Details withheld at licensee's request by SSL.
G3 WMN J J Carter, 27 Old Hall Dr, Chapel Rd, Dersingham, Kings Lynn, PE31 6JT [JO02GU, TF63]
G3 WMO W H Davis, 23 Redlake Meadow, Bucknell, SY7 0AY [IO82MI, SO37]
G3 WMP Dr J Hopton, Greyfriars, 6 Wellfield Rd, Carmarthen, SA31 1DS [IO71UU, SN42]
G3 WMQ M D Watson, Chant House, Dark Ln, Nailsworth, Stroud, GL6 0DR [IO81VQ, ST89]
G3 WMR Details withheld at licensee's request by SSL.
G3 WMS I A W Vance, Larkfield, Debden Rd, Newport, Saffron Walden, CB11 3RU [JO01CX, TL53]
G3 WMT R M Dowling, 393 Blackfen Rd, Sidcup, DA15 9NJ [JO01BK, TQ47]
G3 WMX C Knott, Fox & Hounds, Broadway Rd, Charlton Adam, Somerton, TA11 7AU [IO81QB, ST52]
G3 WMY S P Smith, Five Oaks, Sandy Ln, Henfield, West Sussex, BN5 9UX [IO90UW, TQ21]
G3 WMZ L V Lawbury, 27 Dr Quiles Sotillos, Ciudad Quesada, Rojales 3170, Alicante, Spain
GM3 WNB J OHare, 208 Gilmartin Rd, Linwood, Paisley, PA3 3ST [IO75RU, NS46]
G3 WNC R K Todd, 17 Tudor Rd, West Bridgford, Nottingham, NG2 6EB [IO92KW, SK53]
G3 WND R P Aston, 'Glenroyd', St. Johns Rd, Mortimer, Berks, RG7 5TR [IO91JJ, SU56]
G3 WNG Maj J E Grigsby, 38 Rosewood Ave, Burnham on Sea, TA8 1HE [IO81MF, ST34]
G3 WNI W A Lindsay-Smith, Way Cl, Madford, Hemyock, Cullompton Devon, EX15 3QY [IO80JV, ST11]

G3 WNP T C Baker, 54 Hamilton Rd, Reading, RG1 5RD [IO91MK, SU77]
G3 WNQ E F Lingard, Tedulf Rottenrow, Theddlethorpe, Mablethorpe, Lincs, LN12 1NX [JO03CI, TF48]
G3 WNR K Grey, 15 Woodbourne Ave, Leeds, LS17 5PQ [IO93FU, SE33]
G3 WNS A D Willson, Hilltop, Cryers Hill Rd, Cryers Hill, High Wycombe, HP15 6LJ [IO91PP, SU89]
G3 WNU Details withheld at licensee's request by SSL.
G3 WNV D C Field, Rackhay, Prescott, Uffculme, Devon, EX15 3BA [IO80IW, ST01]
G3 WOD J Welford, 303 Scalby Rd, Scarborough, YO12 6TF [IO94SG, TA08]
G3 WOE M A White, Dentrys Schoolfield, Shiplake Cross, Henley on Thames, Oxon, RG9 4DH [IO91NL, SU77]
G3 WOF Details withheld at licensee's request by SSL.
G3 WOH E Grossmith, 4 Lincoln Way, Rainhill, Prescot, L35 6PJ [IO83OJ, SJ59]
G3 WOI C I B Trusson Newbury & District ARS, 27A Roman Way, Thatcham, RG18 3BP [IO91IJ, SU56]
GM3 WOJ C W Tran, Achnacoille, Lamington, Invergordon, Ross Shire, IV18 0PE [IO77WS, NH77]
G3 WOK D L B Clifton, 59 Grantham Rd, Bracebridge Heath, Lincoln, LN4 2LE [IO93RE, SK96]
G3 WOM M S Muir, 6 Broadstairs Ct, Sunderland, SR4 8NP [IO94GV, NZ35]
GM3 WON Details withheld at licensee's request by SSL.
G3 WOO R G Brace, 11 Cedar Cl, Sawbridgeworth, CM21 9NT [JO01BT, TL41]
G3 WOP M A Goldsbrough, 323 The Cedars, Abbey Foregate, Shrewsbury, SY2 6BY [IO82PQ, SJ51]
G3 WOQ D M Norman, 44 Allingham Rd, South Park, Reigate, RH2 8HX [IO91VF, TQ24]
G3 WOR B Ainsworth Wdarc, 23 Cokeham Rd, Sompting, Lancing, BN15 0AE [IO90TT, TQ10]
G3 WOS C S Gare, Old White Lodge, 183 Sycamore Rd, Farnborough, GU14 6RF [IO91PG, SU85]
G3 WOT M R Meads, 12 Burlington Way, Hemingford Grey, Huntingdon, PE18 9BS [IO92WH, TL27]
G3 WOV Details withheld at licensee's request by SSL.
GU3 WOW P N Hancock, La Breloque, Les Grandes Rues, St Pierre Du Bois, Guernsey, Guernsey, GY9 9LA
G3 WOX A S Hall, 24 The Martins, Crawley Down, Crawley, RH10 4XU [IO91XD, TQ33]
G3 WP J H Brazzill, 43 Forest Dr, Chelmsford, CM1 2TT [JO01FR, TL60]
GM3 WPA S M Hutchinson, 4 Wiston Pl, Dundee, DD2 3JR [IO86LL, NO33]
G3 WPB P R Smith, Holmleigh, 180 Victoria Rd, Ferndown, BH22 9JE [IO90BT, SU00]
G3 WPD A G Smith, 116 Bois Moor Rd, Chesham, HP5 1SS [IO91QQ, SP90]
GW3 WPE F Roberts, 54 South Dr, Rhyl, LL18 4ST [IO83GH, SJ07]
G3 WPF R S Unsworth, Spurs Lodge, Sagars Rd, Styal, Wilmslow, SK9 4HE [IO83VI, SJ88]
G3 WPG D C Dye, 10 Headington Cl, Bradwell, Great Yarmouth, NR31 8DN [JO02UN, TG50]
G3 WPH Dr M J Chamberlain, 10 Clifton Rise, Wargrave, Reading, RG10 8BN [IO91NL, SU77]
G3 WPI H P Shelvey, 16 Horsebridge Way, Rownhams, Southampton, SO16 8AZ [IO90GW, SU31]
G3 WPK R A McCowatt, 70 Heath Rd, Hounslow, TW3 2NW [IO91TL, TQ17]
G3 WPL A Warburton, Chai Hai, Fleet Rd, Holbeach, Spalding, PE12 7AX [JO02AT, TF32]
G3 WPM F R Bridges, 43 Neville Rd, Shirley, Solihull, B90 2QW [IO92BJ, SP17]
G3 WPN V R Bennellick, The Woverns, Castle Frome, Hereford, HR8 1HG [IO82SC, SO64]
G3 WPO Details withheld at licensee's request by SSL.
G3 WPP D C H Minett, 24 Goodwood Ave, Worcester, WR5 2HR [IO82VE, SO85]
G3 WPQ M A Kaye, Pucknell Lodge, Hollywell Ln, Bayton Common, Worcs, DY14 9NR [IO82SI, SO77]
G3 WPR C C S Richmond, 17 Hornbeam Rd, Buckhurst Hill, IG9 6JT [JO01AO, TQ49]
G3 WPS Details withheld at licensee's request by SSL.
G3 WPT R N Brown, 65 Staining Rise, Staining, Blackpool, FY3 0BU [IO83MT, SD33]
G3 WPV D Lamont, 7 Heather Dr, Lindford, Bordon, GU35 0RN [IO91NC, SU83]
G3 WPY E Hodgetts, 303 Birmingham Rd, Great Barr, Birmingham, B43 7AP [IO92AN, SP09]
G3 WPZ Dr J Squire, 19 Southfield Rd, Burley in Wharfedal, Burley in Wharfedale, Ilkley, LS29 7PA [IO93CV, SE14]
G3 WQ P B Jackson, 5 Rhodes Terr, Barlby, Osgodby, Selby, YO8 7HF [IO93LT, SE63]
G3 WQG D F Chalmers, Iona, 25 Willow Cl, Flackwell Heath, High Wycombe, HP10 9LH [IO91PO, SU98]
G3 WQJ J C Corson, 20 Turnberry Ave, Bunkers Hill 5241, East London, South Africa, ZZ2 0TU
G3 WQK B W G Gauntlett Southdown ARS, Heamoor, 4 Sandbanks Gdns, Hailsham, BN27 3TL [JO00DU, TQ50]
G3 WQL A C Conway, 17 Mountcastle Rd, Leicester, LE3 2BW [IO92KO, SK50]
G3 WQM J W Thompson, 80 Albion Ave, Acomb, York, YO2 5QY [IO93KX, SE55]
G3 WQO Details withheld at licensee's request by SSL.
G3 WQT Details withheld at licensee's request by SSL.
G3 WQU P L McKay, Untso, PO Box 490, Jerusalem, 91004, Israel
G3 WQW F G S Sims, 71 Lambley Ln, Burton Joyce, Nottingham, NG14 5BL [IO92LX, SK64]
G3 WQY T A Codrai, Coast Rd, Walcott, Norwich Norfolk, NR12 0PD [JO02ST, TG33]
G3 WRA S W Powell, 98 Kempton Ave, Hillcrest, Hereford, HR4 9TY [IO82PB, SO44]
G3 WRD R J Richardson, Common Crest, Drapery Common, Glemsford, Sudbury, CO10 7RW [JO02IC, TL84]
GW3 WRE B Jones, 6 Pentyla, Maesteg, CF34 0BB [IO81EO, SS89]
G3 WRI P A Brown, 30 Applerigg, Kendal, LA9 6EA [IO84PI, SD59]
G3 WRJ R R Bacon, 20 Cromwell Way, Pirton, Hitchin, SG5 3RD [IO91UX, TL13]
G3 WRL E J Northwood, 8 Derwood Gr, Werrington, Peterborough, PE4 5DD [IO92UO, TF10]
GM3 WRN C R McRae, 17 Valentine Dr, Danestone, Aberdeen, AB22 8YF [IO87WE, NJ91]
G3 WRO K L Haynes, 34 Pear Tree Mead, Harlow, CM18 7BY [JO01BS, TL40]
G3 WRR Q G Collier, 19 Grangecliffe Gdns, South Norwood, London, SE25 6SY [IO91WJ, TQ36]
G3 WRS I R Firth Wakefield&Dist, Am Radio Society, 6 Eastfield Dr, Woodlesford, Leeds, LS26 8SQ [IO93GS, SE32]
G3 WRT Dr I J Dilworth, Pound Ln, Capel St. Mary, Capel, Ipswich, Suffolk, IP9 2JB [JO02MA, TM03]
G3 WRU Details withheld at licensee's request by SSL.
G3 WRV Details withheld at licensee's request by SSL.
G3 WRX Details withheld at licensee's request by SSL.
G3 WSB K S Band, 11 Denewood Cl, Ridge Ln, Watford, WD1 3SZ [IO91TQ, TQ09]
G3 WSC D Atter Crawley ARC, 1 Little Crabtree, West Green, Crawley, RH11 7HW [IO91VC, TQ23]
G3 WSD A W Fisher, 44 Downside Gdns, Potton, Sandy, SG19 2RE [IO92VD, TL24]
G3 WSJ A Evans, 13 Tithe Barn Cres, Swindon, SN1 4JX [IO91CN, SU18]
G3 WSL S J Garner, 8 Heatherdene Rd, Chandlers Ford, Eastleigh, SO53 5BN [IO90HX, SU42]
G3 WSM B Storry, 508 Arleston Ln, Stenson Fields, Sinfin, Derby, DE24 3AA [IO92GU, SK33]
G3 WSN K A M Fisher, Clover Cottage, South Instow, Swanage, BH19 3DS [IO90AO, SZ07]
GM3 WSR V G Clark, 6 Parkhill Circle, Dyce, Aberdeen, AB2 7FN [IO87TH, NJ72]
GI3 WSS C A Billington, 33 Wood End, Holywood, BT18 9PN [IO74BP, J37]
GW3 WSU C S Beynon, 16 Hardy Cl, Woodfield Heights, Barry, CF62 9HJ [IO81IK, ST16]
G3 WSV J C Lawson, 1 Daffodil Way, Springfield, Chelmsford, CM1 6XB [JO01FS, TL70]
G3 WSZ P Gilson, 22 Carr Manor Pl, Moortown, Leeds, LS17 5DL [IO93FT, SE23]
G3 WTA Details withheld at licensee's request by SSL.[Op: Michael L Kinnersly-Taylor, 3 Town Farm Cottages, Stannington, nr Morpeth, Northumberland, NE61 6HP.]
G3 WTB R J T Baxter, 10 Windsor Gdns, Birkdale, Southport, Merseyside, PR8 2JR [IO83LP, SD31]
G3 WTD J R Davis, 71 Broughton Rd, Croft, Leicester, LE9 3EB [IO92JN, SP59]
G3 WTE Details withheld at licensee's request by SSL.
GI3 WTG Details withheld at licensee's request by SSL.
G3 WTL R Cooke, 13 Brighton St., Barrow in Furness, LA14 5HE [IO84JC, SD27]
G3 WTN R W Limehouse, 22 Begbroke Cres, Begbroke, Kidlington, Oxon, OX5 1RW [IO91IT, SP41]
G3 WTO Dr J Spencer, Hollybush, Southernby, Hesket Newmarket, Wigton, Cumbria, CA7 8JA [IO84MR, NY33]
G3 WTP S S Gould Bedford Dist RC, 87 Wentworth Dr, Bedford, MK41 8QD [IO92SD, TL05]
G3 WTQ P C Angold, 10 Hartford Ave, Wilmslow, SK9 6LP [IO83VH, SJ88]
G3 WTR D T Wright, 8 Calverley Park, Tunbridge Wells, TN1 2SH [JO01DD, TQ53]
G3 WTS J P Smith, Windycross, Newbourn Rd, Waldringfield, Woodbridge, IP12 4PT [JO02PB, TM24]
G3 WTU C G Parsons, 89 Staines Rd, Feltham, TW14 0JS [IO91TK, TQ17]
G3 WTV K J Baker, 33 Reading Rd, Woodley, Reading, RG5 3DA [IO91NK, SU77]
G3 WTY P Hodgkiss, 3 Tennyson Rd, Creswell, Worksop, Notts, S80 4DW [IO93JG, SK57]
GW3 WTZ M Jones, 55 Rowan Way, Malpas, Newport, NP9 6JN [IO81LO, ST39]
G3 WUA B R P Lindop, Melrose, 47 High St., Eccleshall, Stafford, ST21 6BW [IO82UU, SJ82]
G3 WUB Dr P Rice, 23 Christchurch Sq, Homerton, London, E9 7HU [IO91XM, TQ38]
G3 WUC Rev A E Measures, St. Josephs Brindle, Chapel Ford, Hoghton, Preston, Lancs, PR5 0DE [IO83QR, SD62]
G3 WUE D R Perrey, Middle Burnt Hills, Lanehead, Weardale, Bishop Auckland, Co Durham, DL13 1AJ [IO84VS, NY84]
G3 WUG I Elvins, 15 Welbeck Cl, Middlewich, CW10 9HX [IO83SE, SJ66]
G3 WUH W E Dufton, 22 Windsor Rd, Bexhill on Sea, TN39 3PB [JO00FU, TQ70]
G3 WUI G Spink, 60 Woodhouse Hill, Huddersfield, HD2 1DH [IO93CP, SE11]
G3 WUK J S Spencer Chapman, Tienda Pascal, Plaza Nueva S/N, Mojacar, Almeria Spain, ZZ2 3TP
G3 WUN D M Holden, 99 Sheerstock, Haddenham, Aylesbury, HP17 8EY [IO91MS, SP70]
GI3 WUO Dr L C Waring, 16 Belfast Rd, Holywood, BT18 9EL [IO74BP, J37]
G3 WUP H R Onn, 173 Woodfield Ave, Birchwood, Lincoln, LN6 0LU [IO93QF, SK97]
G3 WUW A R Papworth, Sandlewood, Cock Ln, Southend Bradfield, Berks, RG7 6HN [IO91KK, SU57]
G3 WUX T Robinson, 16 Nibthwaite Rd, Harrow, HA1 1TA [IO91UO, TQ18]
G3 WUZ P H Brown, The Briers, Brent Rd, Burnham on Sea, TA8 2JT [IO81LG, ST35]
G3 WVA T Mason, Admergill, Withnell Fold, Withnell, Chorley, PR6 8AZ [IO83RQ, SD62]
G3 WVG K I Pritchard, 9 Golf Cl, Pyrford, Woking, GU22 8PE [IO91RH, TQ06]
G3 WVJ V W Higgs, 20 St. Austell Rd, Park Hall, Walsall, WS5 3EF [IO92AN, SP09]
G3 WVO K R Cass, 4 Heworth Village, York, YO3 0AF [IO93LX, SE65]
G3 WVQ J F Barratt, 26 Johnstone Rd, Newent, GL18 1PZ [IO81TW, SO72]
G3 WVR J R Green, Honeysuckle Cottage, New Green, Bardfield Saling, Braintree, CM7 5EG [JO01FV, TL62]
G3 WVU N Harrison, 10 St. Vincents Way, Whitley Bay, NE26 1HS [IO95GB, NZ37]
GW3 WVV R J Barker, Maenygroes, New Quay, Dyfed, SA45 9RL [IO72TE, SN35]
G3 WVW R J Casemore, 11 Oakhurst Gdns, Chingford, London, E4 6BQ [JO01AP, TQ39]
G3 WVY Details withheld at licensee's request by SSL.[Op: P Beecroft. Station located near Scarborough.]
GW3 WWB W D W Brotherton, North Studdock Farm, Angle, Pembrokeshire, Dyfed, SA71 5BG [IO71KQ, SM80]
G3 WWC C J Carroll, 25 Glebe Cres, Broomfield, Chelmsford, CM1 7BH [JO01FS, TL71]
G3 WWD Dr S C Noble, 38 Priory Rd, Cottingham, HU16 4SA [IO93TS, TA03]
G3 WWF I R Firth, 6 Eastfield Dr, Woodlesford, Leeds, LS26 8SQ [IO93GS, SE32]
G3 WWG J A Ross, 24 Raby Rd, Stockton on Tees, TS18 4JA [IO94HN, NZ41]
G3 WWH R K Taylor, Gray's Honey Farm, Cross Drove, Warboys, Huntingdon Cambs, PE17 2UQ [IO92XK, TL38]
G3 WWI R E Oxley, 1 Elm Gr, Maidstone, ME15 7RT [JO01GG, TQ75]
G3 WWL B D Tipper, 271 Blackberry Ln, Four Oaks, Sutton Coldfield, B74 4JS [IO92BO, SK10]
GI3 WWM Details withheld at licensee's request by SSL.
GW3 WWN G W Evans, 18 Mount Pleasant, Tonna, Neath, SA11 3HX [IO81CQ, SS79]
G3 WWO B Rose, 9 Upton Rd, South Walsham, Norwich, NR13 6EL [JO02SP, TG31]
G3 WWS M Southall, 61 Grange Cl, Horam, Heathfield, TN21 0EF [JO00CW, TQ51]
G3 WWT J M Teed, 5 Island Cl, Staines, TW18 4YZ [IO91RK, TQ07]
GD3 WWW Details withheld at licensee's request by SSL.
G3 WWX P C Swann, 20 Roselea Ave, Welton, Lincoln, LN2 3RT [IO93SH, TF07]
GI3 WWY M Anderson, 5 Blacks Ln, Lisbane, Tandragee, Craigavon, BT62 2EF [IO64TI, J04]
GW3 WXA J Gough, Traleen, Rhydlewis, Llandyssul, Dyfed, SA44 5PN [IO72SC, SN34]
G3 WXB K W Watmough, Devonia Hotel, 73 Royal Parade, Eastbourne, BN22 7AQ [JO00DS, TV69]
G3 WXC P G Brooker, 28 Uplands Rd, Northwood, Cowes, PO31 8AL [IO90IR, SZ49]
G3 WXD M N L Zammit, Flat 6 Bovency Hous, Segsbury Gr, Harmans Water, Bracknell Berks, RG12 3JX [IO91PJ, SU86]
G3 WXF L W Brock, 72 Sparrows Herne, Basildon, SS16 5EN [JO01FN, TQ78]
G3 WXG I D Habens, 48 Carden Ave, Brighton, BN1 8NE [IO90WU, TQ30]
G3 WXH J L Arnold, 6 The Spinney, Bleadon Hill, Weston Super Mare, BS24 9LH [IO81MH, ST35]
G3 WXI Details withheld at licensee's request by SSL.
G3 WXJ G Sanby, 25 Norton Park Dr, Norton, Sheffield, S8 8GP [IO93GH, SK38]
G3 WXL Details withheld at licensee's request by SSL.
G3 WXM M S Smith, Linden, 86 Gr Rd, Tring, HP23 5PB [IO91QT, SP91]
G3 WXN L McKown, 310 Oldham Rd, Royton, Oldham, OL2 5AS [IO83WN, SD90]
G3 WXU S J Allbutt, 8 Langton Cl, Vinter Park, Maidstone, Kent, ME14 5PG [JO01GG, TQ75]
G3 WXW C R Traveller, 27 Silver St., Stansted, CM24 8HA [JO01CV, TL52]
G3 WXX Details withheld at licensee's request by SSL.
G3 WY R V Beekar, 5 Sandown Rd, Cheltenham Rd, Evesham, WR11 6XA [IO92AB, SP04]
G3 WYA Details withheld at licensee's request by SSL.
G3 WYB A J Tring, 22 Burnell Rd, Sutton, SM1 4EA [IO91VI, TQ26]
G3 WYD P W Patmore, 141 Cannons Cl, Bishops Stortford, CM23 2BL [JO01CV, TL42]
G3 WYG E P Gooding, 10 Gulpher Cttgs, Gulpher Rd, Felixstowe, IP11 9RQ [JO01QX, TM33]
G3 WYH R Hutton, 59 Draycott Rd, Long Eaton, Nottingham, NG10 3BB [IO92IV, SK43]
G3 WYI G J Simpson, 1 Marsh Terr, Darwen, BB3 0HF [IO83SQ, SD62]
G3 WYK P J S Bysshe, High Rd, Cookham Rise, Cookham, Maidenhead, Berks, SL6 9JT [IO91PN, SU88]
GM3 WYL A L Ritchie, 83 Larkfield Rd, Lenzie, Kirkintilloch, Glasgow, G66 3AS [IO75WW, NS67]
G3 WYN J K Gibson, Four Oaks, Tylers Green, Cuckfield, Haywards Heath, RH17 5DZ [IO91WA, TQ32]
G3 WYP D W Allan West Yorkshire Police ARC, 283 Cliffe Ln, Gomersal, Cleckheaton, BD19 4SB [IO93DR, SE22]
G3 WYT M W Edwards, 23 Burnside, Waterlooville, PO7 7QQ [IO90LV, SU61]
G3 WYU G R Smith, 35 The Cloisters, Ramsgate, CT11 9PL [JO01QH, TR36]
G3 WYV F Chatterton, 9 Somerset Gr, Rochdale, OL11 5YS [IO83VO, SD81]
G3 WYW P C J Bigwood, 18 The Martins, Kennet Lea, Thatcham, RG19 4FD [IO91JJ, SU56]
G3 WZ R J H Baldwin, 11 Meadow Ct, Whiteparish, Salisbury, SP5 2SE [IO91EA, SU22]
G3 WZA W Burnet, 56 Sleaford Rd, Boston, PE21 8EU [IO92XX, TF34]
G3 WZD Details withheld at licensee's request by SSL.
G3 WZE P G Cleary, 531 Diamond St, San Francisco, Ca 94114, U S A
G3 WZF A Butt, 15 Brownberrie Cres, Horsforth, Leeds, LS18 5PT [IO93EU, SE23]
G3 WZG P Murtha, 16 Approach Rd, Margate, CT9 2AN [JO01QJ, TR37]
G3 WZH N Ghani, 57 Fern Ave, Jesmond, Newcastle upon Tyne, NE2 2QU [IO94EX, NZ26]
G3 WZI K A Reeves, 9 Tibberton Cl, Solihull, B91 3UD [IO92CJ, SP17]
G3 WZJ A D Watt, Manor Farm, Eagle Hall, Swinderby, Lincoln, LN6 9HZ [IO93PE, SK86]
G3 WZK S J Beal, 27 Falcon Wood Rd, Croydon, CR0 9BE [IO91XI, TQ36]
G3 WZL D V McMurdo, 23 Meadow Walk, Maidstone, ME15 7RY [JO01GG, TQ75]
G3 WZN Details withheld at licensee's request by SSL.
G3 WZP G A Budden, 183 Leybourne Ave, Northbourne, Bournemouth, BH10 5NP [IO90BS, SZ09]
G3 WZR Dr R D Wright, 34 Broadleas Rd, Devizes, SN10 5DG [IO91AI, SU06]
G3 WZS H L Williams, PO Box 276, Tobermory, Ontario, Canada, N0H 2R0
G3 WZT J R Matthews, 46 Park Ln, West Grinstead, Horsham, RH13 8LT [IO91UB, TQ13]
GM3 WZV S C Hunter, Balnagown, By Muir of Ord, Ross Shire, IV6 7RS [IO77TM, NH54]
G3 WZW G W Laycock, 1 Campsall Cottage, Churchfield Rd, Campsall Doncaster, South Yorks, DN6 9BY [IO93JO, SE51]
G3 WZX R A D Mooney, Oakfield, Monastery, Enniskerry, Co Wicklow
G3 WZZ A J Huddleston, 53 Bolling Rd, Ilkley, LS29 8QA [IO93CW, SE14]
G3 XAB D Whittaker, 2 Stone Edge, Halifax Rd, Briercliffe, Burnley, BB10 3QH [IO83VT, SD83]
G3 XAC C J Whitehead, 10 Berkeley Dr, Read, Burnley, BB12 7QG [IO83TT, SD73]
G3 XAD Details withheld at licensee's request by SSL.
G3 XAE B W Hodkinson, 33 York Ct, Macclesfield, SK10 1PQ [IO83WG, SJ97]
G3 XAG J A Gibbon, The Bungalow, Manless Terr, Skelton in Cleveland, Saltburn By The Sea, TS12 2DQ [IO94MN, NZ61]
G3 XAI J E Temple, 4 Coquetdale Pl, Bedlington, NE22 5JD [IO95FD, NZ28]
G3 XAJ Details withheld at licensee's request by SSL.
G3 XAK L A Luff, 17 Campkin Rd, Cambridge, CB4 2NL [JO02BF, TL46]
G3 XAN W J D Forrester, 34 Keble Dr, Old Roan, Liverpool, L10 3LD [IO83ML, SJ39]
G3 XAO C R Harris, The Coach House, Monks Gate, Sproughton, Ipswich, IP8 3BS [JO02NB, TM14]
G3 XAQ A L Ibbetson, Katallin, Town Ln, Chartham Hatch, Canterbury, CT4 7NN [JO01MG, TR15]
G3 XAS C C Riggs, 12 Dene Walk, West Parley, Ferndown, BH22 8PQ [IO90BS, SZ09]
G3 XAU T J W Woodward, 33 Common Rd, Hemsby, Great Yarmouth, NR29 4LT [JO02UQ, TG41]
G3 XAV J D Rayment, 29 Middlefield Rd, Sawtry, Huntingdon, PE17 5SH [IO92UK, TL18]
G3 XAW M J Chouings, 32 Nunney Cl, Keynsham, Bristol, BS18 1XG [IO81SJ, ST66]
G3 XAX A S Paley, 19 Arbour Ln, Wickham Bishops, Witham, CM8 3NS [JO01IS, TL81]
G3 XAZ R G Stoppard, 4 Beaumaris Dr, Chilwell, Beeston, Nottingham, NG9 5PB [IO92JV, SK53]
G3 XBB D E Latimer, 59 Roecliffe Rd, Woodhouse Eaves, Loughborough, LE12 8TN [IO92JQ, SK51]
G3 XBE A F Walton, 40 Rooley Cres, Bradford, BD6 1BX [IO93CS, SE12]
G3 XBF S J Peat Barking Radio and Electronics, 64 Grange Rd, Heaton Grange, Romford, RM3 7DX [JO01CO, TQ59]
G3 XBH G R L Thompson, 25A Copleston Rd, London, SE15 4AN [IO91XL, TQ37]
G3 XBM R Lapthorn, 37 Spring Cl, Burwell, Cambridge, CB5 0HF [JO02DG, TL56]
G3 XBN F J Chamberlain, 43 Old Mill Cl, Patcham, Brighton, BN1 8WE [IO90WU, TQ30]
G3 XBQ A P Weseley, Loves House, Goudhurst Rd, Marden, Tonbridge, TN12 9NB [JO01FD, TQ74]
G3 XBW M S Wells, Smiths Orchard, Orchard St., Frome, BA11 3BX [IO81UF, ST74]
G3 XBY D F Harvey, 38 School Rd, Shirley, Solihull, B90 2BB [IO92CJ, SP17]
G3 XBZ P R Ciotti, 214 Rossmore Rd, Parkstone, Poole, BH12 2HN [IO90AR, SZ09]
G3 XCB K A R Edmonds, 32 Pear Tree Ave, Ditton, Aylesford, ME20 6EB [JO01FH, TQ75]
G3 XCD H Martin, 17 Vyner Rd, Wallasey, L45 6TE [IO83LK, SJ29]
G3 XCE E E Wells, 23 Briarfield Rd, Carleton, Poulton-le-Fylde, FY6 7PW [IO83LU, SD34]
G3 XCG R Ferguson, 29 Kinglass Rd, Spital, Bebington, Wirral, L63 9AJ [IO83MI, SJ38]
G3 XCI H T Ellis, Hough Fold, Hough Fold Way, Harwood, Bolton, BL2 3PU [IO83TO, SD71]
G3 XCJ W R Burden, 44 Spekehill, Eltham, London, SE9 3BW [JO01AK, TQ47]
G3 XCK J S Pegrum, 14 The Leys, Langford, Biggleswade, SG18 9RS [IO92UB, TL14]
G3 XCO Dr D S Meldrum, 34 Graham Rd, Ipswich, IP1 3QF [JO02NB, TM14]
G3 XCP J M Rennie, 75 Rosemont Rd, Aigburth, Liverpool, L17 6BY [IO83MJ, SJ38]
G3 XCQ K F Butcher, 5 Pennycross Park Rd, Peverell, Plymouth, PL2 3NP [IO70WJ, SX45]
GW3 XCR Details withheld at licensee's request by SSL.
G3 XCS C J Squires, 5 Frith Rd, Saltash, PL12 6EL [IO70VJ, SX45]
G3 XCT D Dade, 30 Hilden Park Rd, Hildenborough, Tonbridge, TN11 9BL [JO01DF, TQ54]
G3 XCW G G Winter, 14 Drakes Lea, Evesham, WR11 5BJ [IO92AC, SP04]
G3 XCX Details withheld at licensee's request by SSL.[Station located in Sedgefield, Cleveland, TS21.]
G3 XCY K Bristow, 2 Bittern Way, Verwood, Wimborne, Dorset, BH21 5NX [IO90AV, SU01]
GI3 XCZ G A Martin, 100 Drumconnelly Rd, Gortaclare, Sixmilecross, Omagh, BT79 0XS [IO64JM, H56]
G3 XDA R J Holderness, 16 Helmsley Way, Spalding, PE12 6BG [IO92WS, TF22]
G3 XDC H E Olley, 1 The Fairstead, Botesdale, Diss, IP22 1DG [JO02MI, TM07]
GI3 XDD S C Crampton, 135A Ballymena Rd, Doagh, Ballyclare, Co. Antrim, BT39 0TN [IO64WS, J29]
G3 XDE Details withheld at licensee's request by SSL.
G3 XDH T B Parkes, Penarlas Fach, Llanfyllin, Powys, SY22 5ND [IO82IR, SJ11]
G3 XDI C York, 18 Hebburn Rd, Salters Est, Stockton on Tees, TS19 8AL [IO94HN, NZ42]
G3 XDK A J Maris, 140 Edward St., Brighton, BN2 2JL [IO90WT, TQ30]
G3 XDL A E Long, 51 Beadon Rd, Bromley, BR2 9AS [JO01AJ, TQ46]
G3 XDM A H Benson, Claregate, 31 Oakhill Dr, Welwyn, AL6 9NW [IO91VU, TL21]
G3 XDO D H Stevens, 3 Bridge Ave, Woodley, Stockport, SK6 1QN [IO83WK, SJ99]
G3 XDP G Wilkinson, 509 Warrington Rd, Culcheth, Warrington, WA3 5QY [IO83SL, SJ69]
G3 XDQ Details withheld at licensee's request by SSL.
G3 XDR A F Goodwin, 2 Hamilton Ln, Exmouth, EX8 2JT [IO80HO, SY08]
G3 XDU K V Whitbread, Leighbank, Moor End Rd, Radwell, Bedford, MK43 7HY [IO92RF, TL05]

G3 XDV Details withheld at licensee's request by SSL.
GI3 XDX G G McDowell, 13 Redford Rd, Cullybackey, Ballymena, BT43 5PR [IO64TV, D00]
G3 XDY J H Quarmby, 12 Chestnut Cl, Rushmere St. Andrew, Rushmere, Ipswich, IP5 7ED [JO02OB, TM24]
G3 XDZ Z P Skrobanski, 1035 Pine Gr, Pointe Dr, Roswell, USA, Ga 30075
G3 XEB T R Baker, 39 Rookwood Park, Horsham, RH12 1UB [IO91TB, TQ13]
G3 XEC G Grundy, 29 Sheppey Ln, Hackleton, Northampton, NN7 2AL [IO92OE, SP85]
G3 XED C S Masters, 79 Kings Head Ln, Bristol, BS13 7DB [IO81QK, ST56]
G3 XEE N L Kinch, 17 Rose Carr Walk, Hornsea, HU18 1HN [IO93VV, TA24]
G3 XEG H E Livermore, 11 Roe Green Ln, Hatfield, AL10 0SH [IO91VS, TL20]
G3 XEI J C Hooper, More Hall, Brockton, Much Wenlock, Shropshire, TF13 6JU [IO82QM, SO59]
G3 XEK W R Prince, 54 Devonshire Ave, Ripley, DE5 3SS [IO93GB, SK35]
G3 XEM P K Booth, 39 Jenned Rd, Arnold, Nottingham, NG5 8FT [IO93KA, SK54]
G3 XEN P C Mullineaux, 27 Ashfield Ave, Lancaster, LA1 5EB [IO84OB, SD46]
G3 XEP A S Kessler White Rose ARS, 8 West Ct, West Ave, Roundhay, Leeds, LS8 2JP [IO93GT, SE33]
GI3 XEQ J A Bailie, 25 Upper Knockbreda, Rd, Belfast, N Ireland, N Ireland, BT6 0NA [IO74BN, J37]
G3 XER D Mannix, Vine Cottage, 34 Ashby Rd, Ticknall, Derby, DE73 1JJ [IO92GT, SK32]
GW3 XET V J Riley, Suncrest, Carreghofa Ln, Llanymnech, Powys, SY22 6LA [IO82KS, SJ22]
G3 XEV J J Cooper, 34 Arcal St., Sedgley, Dudley, DY3 1TG [IO82WM, SO99]
G3 XEW G L Childs, 115 Summerhouse Dr, Bexley, DA5 2ER [JO01CK, TQ57]
G3 XEY A J P Robinson, 48 Colton Rd, Shrivenham, Swindon, SN6 8AZ [IO91EO, SU28]
G3 XFA R R T Wilkins, 50 Brookside, Barnwell, Peterborough, PE8 5PS [IO92SK, TL08]
G3 XFB D F Jewson, 8 Johnsgate, Crestwood Park, Brewood, Stafford, ST19 9HZ [IO82WQ, SJ80]
GM3 XFC H R S Canale, 211A High St, Arbroath, Angus, DD11 1DZ [IO86RN, NO64]
G3 XFD R B Mannion, 48 Priory Rd, West Moors, Ferndown, BH22 0AY [IO90BT, SU00]
G3 XFE Details withheld at licensee's request by SSL.
G3 XFF E P Tuddenham, 36 King St., Felixstowe, IP11 9DX [JO01QX, TM23]
G3 XFH D E A Watts, 23 Maple Cl, Bristol, BS14 8HX [IO81RJ, ST66]
G3 XFL J K Harding, Whispers, 27 Northfield Dr, Truro, TR1 2BS [IO70LG, SW84]
G3 XFN G Coffin, 45 Egerton Rd, Streetly, Sutton Coldfield, B74 3PG [IO92BN, SP09]
G3 XFV Details withheld at licensee's request by SSL.
G3 XFW G L Parris, 124 Chelston Ave, Yeovil, BA21 4PR [IO80QW, ST51]
G3 XFZ Details withheld at licensee's request by SSL.
G3 XG A Robinson, 7 Lea Cl, Chelmer Rd, Braintree, CM7 3YP [JO01GV, TL72]
G3 XGC Details withheld at licensee's request by SSL.
G3 XGD G E Watson, 6 The Avenue, Lyneal, Ellesmere, SY12 0QJ [IO82OV, SJ43]
G3 XGE P N Greenhalgh, 13 Primrose Ave, Urmston, Manchester, M41 0TY [IO83TK, SJ79]
G3 XGH W A Jamison, Horseshoe Cottage, Town Fold, Marple Bridge, Stockport, Ches, SK6 5BT [IO83XJ, SJ98]
G3 XGK C J Langley, 5 Thurne Rd, Long Rd Est, Lowestoft, NR33 9DT [JO02UL, TM59]
G3 XGM G Marshall, 16 Inverness Rd, Southall, UB2 5QG [IO91TL, TQ17]
G3 XGQ Details withheld at licensee's request by SSL.
G3 XGS Details withheld at licensee's request by SSL.
G3 XGV G Fowles, Ruby House, Broad Marston, Stratford on Avon, CV37 8XY [IO92CC, SP14]
G3 XGW K Yates, Tibblestone Lodge, Ashton Rd, Beckford, Tewkesbury, GL20 7AU [IO82XA, SO93]
GM3 XGX J McEachran, 70 Stafford Dr, Greenock, PA16 0TE [IO75OW, NS27]
G3 XGY B A Harris, 4 Flamingo Cres, Worle, Weston Super Mare, BS22 8XH [IO81MI, ST36]
G3 XGZ M M Lambert, 16 Dundas Cl, Abingdon, OX14 3UZ [IO91IQ, SU59]
G3 XHC S G Mitchelmore, 177 Victoria Rd, Dartmouth, TQ6 9EG [IO80EI, SX85]
GW3 XHD B B Walters, 16 Broomhill, Port Talbot, SA13 2US [IO81CO, SS78]
GW3 XHG D L Griffiths, 7 Canning St., Ton Pentre, Pentre, CF41 7HF [IO81GP, SS99]
GW3 XHJ W Robinson, 54 Abergarw Rd, Brynmenyn, Bridgend, CF32 9LF [IO81FN, SS98]
G3 XHK A A Wickham, 1 The Laurels, 29 Sidney Rd, Walton on Thames, Surrey, KT12 2NA [IO91TJ, TQ16]
GI3 XHL W N Rawson, 79 Loopland Dr, Castlereagh Rd, Belfast, BT6 9DW [IO74BO, J37]
G3 XHM A C Lewis, Bradley Villa, 41 West St., Ryde, PO33 2UH [IO90KR, SZ59]
G3 XHP H W Pettit, 19 Cobbs Ln, Hough, Crewe, CW2 5JN [IO83SB, SJ75]
G3 XHQ D Graham, 4 Warleigh Rd, Brighton, BN1 4NT [IO90WU, TQ30]
G3 XHR H Butler, 20 Meadow View, Cattistock, Dorchester, DT2 0JF [IO80RT, SY59]
GW3 XHV W J Thomas, 98 Westlands, Baglan Moor, Port Talbot, SA12 7DE [IO81CO, SS79]
G3 XHW Dr J E Morris, 2 The Corniche, Sandgate, Folkestone, CT20 3TA [JO01NB, TR23]
G3 XHX J W Rudkin, Otago, Quethiock, Liskeard, Cornwall, PL14 3SQ [IO70TK, SX36]
G3 XHY Details withheld at licensee's request by SSL.[Op: Dr C C Tinline.]
G3 XHZ J Farrer, 221 Rye St., Bishops Stortford, CM23 2HE [JO01BV, TL42]
G3 XIB B A Johnson, 16 St. Margarets Ln, West Town, Backwell, Bristol, BS19 3JR [IO81PJ, ST46]
G3 XID D Mitchinson, 32 St. Aidans Ave, Grangetown, Sunderland, SR2 9SF [IO94HV, NZ45]
G3 XIE J B Williams, 4 Cardigan Terr, Wakefield, WF1 3DF [IO93GQ, SE32]
G3 XIG Details withheld at licensee's request by SSL.
G3 XIH W R Dixon, 17 Chestnut Bank, Scarborough, YO12 5QJ [IO94SG, TA08]
G3 XII F Harrison, 78 Lancaster Ln, Leyland, Preston, PR5 2SP [IO83QQ, SD52]
GM3 XIJ A J Binning, Mount Pleasant, Manse Brae, Lochgilphead, PA31 8RA [IO76GA, NR88]
G3 XIN T J Harris, Ssvc/Bfbs, Herford, Bfpo 15, X X
G3 XIP D J Aspinall, Osier Cottage, Bell Ln, Fenstanton, Huntingdon, PE18 9JX [IO92XH, TL36]
G3 XIQ K Finch, 7A High St., Sutton, Sandy, SG19 2NE [IO92VC, TL24]
G3 XIR Details withheld at licensee's request by SSL.
GW3 XIS Dr R A Belcher, 8 Bishops Gr, Derwen Fawr, Sketty, Swansea, SA2 8BE [IO81AO, SS69]
G3 XIT Details withheld at licensee's request by SSL.
G3 XIU W J Byers, 29 Newtown, Frizington, CA26 3QQ [IO84GM, NY01]
G3 XIV G G Bulleyment, 30 Brackley Ave, Fair Oak, Eastleigh, SO50 8FL [IO90IX, SU41]
G3 XIX J E Hobin, 14 St. Martins Green, Trimley St. Martin, Trimley, Ipswich, IP10 0UU [JO02PA, TM24]
G3 XIY R W Hall, 22 Cumbria Cl, Thornbury, Bristol, BS12 2YE [IO81RO, ST69]
G3 XIZ C J Osborn, 116 Holme Ct Ave, Biggleswade, SG18 8PB [IO92UB, TL14]
GW3 XJA D J Williams, 5 Coed Eithen St, Blaenavon, Gwent, NP4 9LQ [IO81LS, SO20]
G3 XJB Details withheld at licensee's request by SSL.[Op: J D Parkinson, 12 Jarmyns, Bishops Hull, Taunton, Somerset, TA1 5HG.]
GW3 XJC B S Luke, 33 Maiden St., Cwmfelin, Maesteg, CF34 9HP [IO81EO, SS88]
G3 XJE Dr P J Duffett-Smith, 41 Denmark Rd, Cottenham, Cambridge, CB4 4QS [JO02BG, TL46]
G3 XJG Dr J W Stenson, 49 Chesham Cl, Goring By Sea, Worthing, BN12 4BJ [IO90TT, TQ10]
G3 XJH A W Barry, 4 Templer Rd, Preston, Paignton, TQ3 1EL [IO80FK, SX86]
G3 XJI W A Wilkinson, 1 Scafell Dr, Kendal, LA9 7PE [IO84PH, SD59]
G3 XJK Details withheld at licensee's request by SSL.
G3 XJL D E Thomas, 19 Norfolk Rd, Maldon, CM9 6AZ [JO01IR, TL80]
G3 XJM J G Sawdy, 41 Ashbarn Cres, Winchester, SO22 4QH [IO91IB, SU42]
G3 XJN H Duncombe, Badgers Copse, Hampers Ln, Storrington, RH20 3HU [IO90SW, TQ11]
G3 XJO A L Jones, 2 Kenwell Ct, Woolstones, Woolstone, Milton Keynes, MK15 0BD [IO92PB, SP83]
G3 XJP P I Rhodes, 44 Manor Park Ave, Princes Risborough, HP27 9AS [IO91NR, SP80]
GW3 XJQ M A Shelley, Sunray, Pendine, Carmarthen, Dyfed, SA33 4PD [IO71RR, SN20]
GD3 XJR M S Dickinson, 33 PO Box, Athol St., Douglas, IM99 1BN
G3 XJS P Barville, 40 Watchet Ln, Holmer Green, High Wycombe, HP15 6UG [IO91PP, SU89]
G3 XJW L G Rix, 63 Edendale Rd, Melton Mowbray, LE13 0EW [IO92NS, SK71]
G3 XJY J R Jardine, Field Head Farm, Balance Hill, Uttoxeter, Staffs, ST14 8PU [IO92BV, SK03]
G3 XJZ C L Sykes, Racal Decca Tran St, Denhall Ln, Burton, South Wirral, L64 0TG [IO83LG, SJ27]
G3 XKA I J Cunningham, White Pines, St. Johns Hill Rd, Woking, GU21 1QY [IO91QH, SU95]
GW3 XKB K H Bevan, Renhold, 25 Bryn Gannock, Deganwy, Conwy, LL31 9UG [IO83CH, SH77]
G3 XKC R J Cole, Shiplaps, 20 Church Terr, London, NW4 4JU [IO91VO, TQ28]
G3 XKD M King, 15 Glebe Rd, Prestbury, Cheltenham, GL52 3DG [IO81XV, SO92]
G3 XKE C S Evans, 8 Blakelands Ave, Sydenham, Leamington Spa, CV31 1RJ [IO92FG, SP36]
G3 XKF J Sharratt, Woodleys Cottage, Marsh, Aylesbury, Bucks, HP17 8SP [IO91OS, SP80]
G3 XKG R J Stanton, Rostellan, Pilgrims Cl, Westhumble, Dorking, RH5 6AR [IO91UG, TQ15]
G3 XKH B R Ward, 12 Pagets Rd, Bishops Cleeve, Cheltenham, GL52 4AG [IO81XW, SO92]
G3 XKI W R D Nobbs, Grange Cottage, Brigg Rd, Howsham, Market Rasen, LN7 6LF [IO93SM, TA00]
G3 XKL C Gill, Little Acre, Marston, Devizes, Wilts, SN10 5SR [IO81XH, ST95]
G3 XKN J P Hind, 9 Dale Cl, Toddington, Dunstable, LU5 6EP [IO91RW, TL02]
G3 XKQ Details withheld at licensee's request by SSL.
G3 XKR Details withheld at licensee's request by SSL.
G3 XKS H O C Grattan, Wenford Bridge, Bodmin, Cornwall, PL30 3PN [IO70PN, SX07]
G3 XKU M E Rose, Canbury, 1A Goyfield Ave, Felixstowe, IP11 7RX [JO01QX, TM23]
G3 XKV R E Stratton, 60 Lateward Rd, Brentford, TW8 0PL [IO91UL, TQ17]
G3 XKW K R Hamilton-Wedgwood, Rosedale, Redmoor, Bodmin, PL30 5AR [IO70PJ, SX06]
G3 XKX D G Wills, 70 Hidcote Rd, Oadby, Leicester, LE2 5PF [IO92LO, SP69]
G3 XKY G R Schrager, 3 The Park, London, N6 4EU [IO91WN, TQ28]
GM3 XLB M R Giddings, 9 Lansdowne Cres, Glasgow, G20 6NQ [IO75UU, NS56]
G3 XLC F Cooke, 3 Laburnham Pl, Stoke on Trent, ST3 5NL [IO82WX, SJ94]
G3 XLF Details withheld at licensee's request by SSL.
G3 XLG R J Spreadbury, Holly House, Cross St., Drinkstone, Bury St. Edmunds, IP30 9TP [JO02KE, TL96]
G3 XLI P W Holland, 83 Gallows Hill Ln, Abbots Langley, WD5 0DD [IO91SQ, TL00]
GI3 XLK W R Magee, 6 Cherryvalley Park West, Belfast, BT5 6PU [IO74BO, J37]
G3 XLL J L Lockwood, 22 Egremont Rd, Diss, IP22 3NF [JO02NJ, TM18]
G3 XLM R N Lee, 5 Albany Rd, Earlsdon, Coventry, CV5 6JQ [IO92FJ, SP37]
G3 XLN Dr D D Russell, 1 Mountsfield Cl, Newport Pagnell, MK16 0JE [IO92PB, SP84]
G3 XLP I T Richardson, 57 Pelham Rd, Cowes, PO31 7DR [IO90IS, SZ49]
G3 XLQ B J Carling, 18016 Fertile Meadow Ct, Gaithersburg, Md 20877, USA, 20877
G3 XLR A S Bunyan, 87 Seymer Rd, Romford, RM1 4LA [JO01CO, TQ58]
G3 XLS T J Williams, 5 Greenwood Dr, Bolton-le-Sands, Carnforth, LA5 8AP [IO84OC, SD46]
G3 XLU R W S Hewett, 145 Old Ferry Rd, Saltash, PL12 6BN [IO70VJ, SX45]
G3 XLV Details withheld at licensee's request by SSL.
G3 XLW D G Powell, High Pines, Kingston, Kingsbridge, Devon, TQ7 4QD [IO80AH, SX64]
G3 XLX R K Littlewood, 24 Thornway, Mosley Common, Worsley, Manchester, M28 1YS [IO83SM, SD70]
G3 XLZ J J Tozer, 54 Ganges Rd, Stoke, Plymouth, PL2 3AZ [IO70WJ, SX45]
G3 XMB R Richardson, 42 King Edwards Rd, South Woodham Ferrers, Chelmsford, CM3 5PQ [JO01HP, TQ89]
G3 XMD D R Baker, 7 Broad Lawn, New Eltham, London, SE9 3XE [JO01AK, TQ47]
G3 XME C W Wright, 1 Tadley Hill, Tadley, RG26 3PJ [IO91KI, SU66]
G3 XMG M B Graham, 30 Moorlands Rd, Thornton, Liverpool, L23 1US [IO83ML, SD30]
G3 XMK A S Flather, 15 Porlock Gr, Leyfield Park, Trentham, Stoke on Trent, ST4 8TN [IO82VX, SJ84]
G3 XMM T W Morgan, 32 Grasmere Rd, Longlevens, Gloucester, GL2 0NQ [IO81VV, SO81]
G3 XMP A S Brasier, 56 Gatenby, Werrington, Peterborough, PE4 6JU [IO92UP, TF10]
G3 XMQ P Eggleton, Woodlands, 5 Rowhills, Farnham, GU9 9AT [IO91OF, SU84]
G3 XMR K W Hedges, 12 Wimborne Cl, Up Hatherley, Cheltenham, GL51 5QP [IO81WV, SO92]
GM3 XMS J M Macintosh, 6 David St., Coatbridge, ML5 3QL [IO75XU, NS76]
G3 XMU S F Jennings, 35 Halton Rd, Carterton, OX18 3SD [IO91ES, SP20]
G3 XMW M E B Brown, 93 Springfield Rd, Brighton, BN1 6DH [IO90WU, TQ30]
GM3 XMY D J Hobden, 6 Strathallan Cl, Glenrothes, KY7 4SW [IO86KE, NT29]
G3 XMZ C A Dodd, Saint Helen, Triq Il-Gross, Marsaskala, Zbr11, Malta
G3 XNB D Baldock, 107 St. Neots Rd, Eaton Ford, St. Neots, Huntingdon, PE19 3AE [IO92UF, TL16]
G3 XNE A F Smyth, Tamarisk, 62 Agnes Cl, Bude, EX23 8SB [IO70RT, SS20]
G3 XNG L D Grant, 14 Friars Gate, Kirkhill, Morpeth, NE61 2AY [IO95DD, NZ18]
G3 XNH Details withheld at licensee's request by SSL.
G3 XNI W J Hopkins, 15 Avon Walk, Riverdene, Basingstoke, RG21 4DJ [IO91LG, SU65]
GM3 XNJ C R Mc Anerney, 3 Hilltop Rd, Gourock, PA19 1YN [IO75OW, NS27]
G3 XNK D R Kidd, 22 Willow Way, Motcombe, Shaftesbury, SP7 9QH [IO81VA, ST82]
G3 XNN R E Jephcott, 3 Chatsworth Park, Thornbury, Bristol, BS12 1JF [IO81RO, ST69]
G3 XNO R Brown Otley AR Soc, 2 Layton Park Cl, Rawdon, Leeds, LS19 6PJ [IO93EU, SE23]
G3 XNP K F Arnold, 251 Appleby St., Cheshunt, Waltham Cross, EN7 6RB [IO91XR, TL30]
G3 XNT Details withheld at licensee's request by SSL.
G3 XNU J F Craine, 74 Beaufort Ave, Cubbington, Leamington Spa, CV32 7TD [IO92FH, SP36]
G3 XNX D C Chivers, 51 Alma Rd, Brixham, TQ5 8QR [IO80FJ, SX95]
G3 XOC M A Cooley, 21 Castle Rd, Newport, PO30 1DT [IO90IQ, SZ48]
G3 XOD R Horsman, 65 Pendennis Park, Brislington, Bristol, BS4 4JL [IO81RK, ST67]
G3 XOF A Dunford, Radbourne Lodge, 103 Radbourne St., Derby, DE22 3BW [IO92FW, SK33]
G3 XOI Eur Ing A M Gordon, 20 Hawkins Cres, Shoreham By Sea, BN43 6TP [IO90VU, TQ20]
GJ3 XOJ D F Gray, La Brecque-le-Hocq, St Clement, Jersey, JE2 6SD
G3 XOK R J Kearney, 32 Springfield Rd, Lower Somersham, Ipswich, IP8 4PQ [JO02MC, TM04]
G3 XON S G Casperd, 14 Dagden Rd, Shalford, Guildford, GU4 8DD [IO91RF, TQ04]
G3 XOP P E Featherstone, Laneside, Stewkley Rd, Cublington, Leighton Buzzard, LU7 0LR [IO91OV, SP82]
GM3 XOQ P C Weller, Mither Tap, Bridge Rd, Kemnay, Inverurie, AB51 5QT [IO87SF, NJ71]
G3 XOU D Wright, 15 Milton Cres, Down Park, Tavistock, PL19 9AL [IO70WN, SX47]
G3 XOV R M Johnson, 29 Hungary Hill, Stourbridge, DY9 7PS [IO82WK, SO98]
GD3 XPA R W Bevan, Ashbourne House, Ballacraine, St. Johns, Isle of Man, IM4 3NF
G3 XPB P R Butcher, Merrydown, Reigate Rd, Hookwood, Horley, RH6 0AR [IO91VE, TQ24]
G3 XPC R D Chapman, 22 Windsor Ride, Finchampstead, Wokingham, RG40 3LG [IO91NJ, SU76]
G3 XPD D W J Smith, 5 Peel St., Stafford, ST16 2DZ [IO82WT, SJ92]
G3 XPE Details withheld at licensee's request by SSL.
G3 XPH R S Grant, 43 Catisfield Ln, Fareham, PO15 5NT [IO90JU, SU50]
G3 XPI B Hallows, 3 Southdown Cl, Rochdale, OL11 4PP [IO83VO, SD81]
G3 XPJ K J George, 34 Third Ave, Northville, Bristol, BS7 0RT [IO81RM, ST67]
GW3 XPK J R Dore, Henfaes Isaf, Llangurig, Powys, SY18 6SN [IO82EJ, SN88]
G3 XPM R E Tinson, 29 Appleby Cres, Knaresborough, HG5 9LS [IO94GB, SE36]
G3 XPO K G Filmer, Foxhollow, 10 Canterbury Rd, Hawkinge, Folkestone, CT18 7BW [JO01OC, TR23]
GM3 XPQ G H Black, Sandsound Schoolhouse, Tresta, Shetland, ZE2 9LU [IP90HF, HU34]
G3 XPR I M Bassett-Smith, Avalon, 3 Greenhills Rd, Charlton Kings, Cheltenham, GL53 9ED [IO81XV, SO92]
G3 XPT G A Symonds, Arlington, 45 Westfield Rd, Dereham, NR19 1JB [JO02LP, TF91]
G3 XPU C Woodley, 170 Rugby Rd, Burbage, Hinckley, LE10 2ND [IO92HM, SP49]
G3 XPX G L K Crawford, 14 Yew Tree Rd, Southborough, Tunbridge Wells, TN4 0BA [JO01DD, TQ54]
G3 XPY A W Bagley, 4 The Alders, Romsley, Halesowen, B62 0PT [IO82XK, SO97]
G3 XPZ J Appleton, 53 Chapel St., Leigh, WN7 2PB [IO83RL, SD60]
G3 XQB N Whitelegg, 11 Rex Ave, Millhouse Green, Sheffield, S7 2GS [IO93FI, SK38]
G3 XQE K D Brown, 54 Wendover Rise, Allesley Park, Coventry, CV5 9JU [IO92FJ, SP37]
G3 XQJ G H Wren, 8 Aspin Park Dr, Knaresborough, HG5 8EY [IO94GA, SE35]
G3 XQL R Johnstone, The Villa, Withington, Shrewsbury, SY4 4PU [IO82QR, SJ51]
G3 XQM A Finch, 56 Valebridge Dr, Burgess Hill, RH15 0RW [IO90WX, TQ32]
GW3 XQO P M Salomon, 28 Ansell Rd, Plas Goulbourne, Wrexham, LL13 9NQ [IO83MB, SJ35]
G3 XQP Dr A J French, 35 Hanover House St, Johns Wood High St, London, NW8 7DY [IO91VM, TQ28]
G3 XQU Dr A G Gray, 2 Church Ln, Great Warley, Brentwood, CM13 3EP [JO01DN, TQ68]
G3 XQX W A Hardcastle, 13 Glorney Mead, Badshot Lea, Farnham, GU9 9NL [IO91OF, SU84]
G3 XQZ P A Simpson, Meadow Brook House, Hallaton Rd, Medbourne, Market Harborough, LE16 8DR [IO92OM, SP79]
G3 XRA J Gentzler, 450 Arterial Rd, Leigh on Sea, SS9 4DS [JO01HN, TQ88]
G3 XRC D C Carlsen, 57 Chignal Rd, Chelmsford, CM1 2JA [JO01FR, TL60]
G3 XRD G A Knight, 57 Oliver Rd, Kirk Hallam, Ilkeston, DE7 4JY [IO92IX, SK44]
G3 XRE D M Ferigan Royal Engnrs RS, Communications Training Wing, 3 Rsme Regiment Royal Engrs, Gibraltar Barracks, Camberley, Surrey, GU17 9LP [IO91OH, SU85]
G3 XRI P B W Williams, 33 Park Ave North, Newton-le-Willows, WA12 8HF [IO83QK, SJ59]
G3 XRJ S J Chappell, Trebehor, St Levan, Penzance, Cornwall, TR19 6LX [IO70EB, SW32]
G3 XRK D K Griffin, 12 Charles Rd, Whittlesey, Peterborough, PE7 2RG [IO92WN, TL29]
G3 XRL C N Scott, 18 Coronation Rd, Nuthall, Nottingham, NG16 1EP [IO92JX, SK54]
GW3 XRM D J Dunn, 9 Mill Bank, Llandegfan, Menai Bridge, Gwynedd, LL59 5RD [IO73VF, SH57]
G3 XRP R Croucher, 14 Orchard Way, Send, Woking, GU23 7HS [IO91RG, TQ05]
GI3 XRQ H Irvine Bangor&Dist ARS, Whitegables, 95 Crawfordsburn Rd, Bangor, BT19 1BJ [IO74DP, J48]
G3 XRT J R Hooper Ilford Grp RSGB, 50 Mortlake Rd, Ilford, IG1 2SX [JO01AN, TQ48]
G3 XSA D L F Standley, 5 East Mill Green, Bentley, Ipswich, IP9 2BW [JO01MX, TM13]
G3 XSB Details withheld at licensee's request by SSL.
G3 XSC K H Southgate, 44 Taylor Ave, Kew, Richmond, TW9 4ED [IO91UL, TQ17]
G3 XSE H J Allison, 89 Birchanger Ln, Birchanger, Bishops Stortford, CM23 5QF [JO01CV, TL52]
G3 XSF D W Martin, 32 Clifton Rd, Halifax, HX3 0BT [IO93BR, SE02]
G3 XSG E J Mortimer-Hampson, 16 Ernesettle Cres, Higher St. Budeaux, Plymouth, PL5 2EX [IO70VJ, SX45]
G3 XSH Dr W Redman-White, 5 Goddard Cl, West Wellow, Romsey, SO51 6RH [IO90FX, SU21]
G3 XSI T Haslam, 29 Backmoor Rd, Norton, Sheffield, S8 8LB [IO93GI, SK38]
G3 XSJ K R Brooks, Laycombe House, Coombe, Wotton-under-Edge, Glos, GL12 7NG [IO81UP, ST79]
G3 XSK K H E Dawson, 28 Springfield Gdna, Lowestoft, Suffolk, NR33 9EE [JO02UL, TM59]
G3 XSN B Donn, 7 Thurne Way, Liverpool, L25 4SQ [IO83NJ, SJ48]
G3 XSP J F Neville, 2222 Sandia Dr, Prescott, Arizona 86301, USA, X X
G3 XSR A F Sutton, 15 Gr Gdns, Littleton, Chester, CH3 7DL [IO83NE, SJ46]
G3 XSV A F Hydes, 1 Robinson Cl, Backwell, Bristol, BS19 3BT [IO81PJ, ST46]
G3 XSZ F H Mundy, Westlands, 2 Woodpeckers Dr, Winchester, SO22 5JJ [IO91HB, SU43]
G3 XTC P J R Borrett, 21 Kenley Walk, Cheam, Sutton, SM3 8ES [IO91VI, TQ26]
G3 XTD H F Speight, 15 Walney Ln, Aylestone Hill, Hereford, HR1 1JD [IO82PB, SO54]
G3 XTF H Webster, 16 Juniper Dr, Salisbury, SP1 3RA [IO91CC, SU13]
G3 XTG J E E Drury, 54 Clifton Rd, Henlow, SG16 6BL [IO92UA, TL13]
G3 XTH G R A King, 73 Grand Ave, Hassocks, Hassocks, BN6 8DD [IO90WW, TQ31]
G3 XTI J C Jarvie, 11 Guild Rd, Aston Cantlow, Solihull, B95 6JA [IO92CF, SP16]
G3 XTL C H Barlow, 6 Fell Wilson St., Warsop, Mansfield, NG20 0PT [IO93KE, SK56]
G3 XTM D J Ledger, 28 Moor View, Chudleigh, Newton Abbot, TQ13 0JB [IO80EO, SX87]
G3 XTN R Hough, Avondale, 27 Hill Wootton Rd, Leek Wootton, Warwick, CV35 7QL [IO92FH, SP26]
G3 XTP K J Lloyd, Harcourt, Selkirk Gdns, Cheltenham, GL52 5LY [IO81XV, SO92]
G3 XTQ M J Draycott, 119A High St. North, Stewkley, Leighton Buzzard, LU7 0EX [IO91OW, SP82]
GM3 XTR B P Dunn, 79 Battery Park Ave, Waterfront, Fort Matilda, Greenock, PA16 7JA [IO75OX, NS27]
G3 XTS B Payne, 45 Kellaway Ave, Westbury Park, Bristol, BS6 7XS [IO81QL, ST57]
G3 XTT D I Field, 105 Shiplake Bottom, Peppard Common, Henley on Thames, RG9 5HJ [IO91MM, SU78][Op: Donald (Don) I Field. Packet mail to GB7WOK. Also licensed as NK1G. Tel: (01734) 724192.]
G3 XTU H Lundie, 10 Northampton Rd, Bromham, Bedford, MK43 8PE [IO92RD, TL05]
G3 XTX Details withheld at licensee's request by SSL.
G3 XTY I Lewis, Hope Cottage, 18 Pitchers, Salway Ash, Bridport, DT6 5QS [IO80OS, SY49]
G3 XTZ G B Phillips, Age, 27 Station Rd, Ashford, TW15 2LP [IO91SK, TQ07]
G3 XUA Details withheld at licensee's request by SSL.
G3 XUB Rvdr F H Boardman, Woodside, Burtonwood Rd, Great Sankey, Warrington, WA5 3AN [IO83QJ, SJ58]

G3 XUC A Keohane, 6 Birchwood Fields, Tuffley, Gloucester, GL4 0AL [IO81VU, SO81]
G3 XUD P Kirby, 10 Junction Cl, Burgess Hill, RH15 0NZ [IO90WX, TQ31]
G3 XUE K Beesley, 73 Wenlock Rd, Shrewsbury, SY2 6JU [IO82PQ, SJ51]
G3 XUF A E Warner, 79 Kelvin Gr, Portchester, Fareham, PO16 8LF [IO90KU, SU60]
G3 XUH R E Pearson, 24 St. Saviours Cl, Bamber Bridge, Preston, PR5 6AH [IO83QR, SD52]
G3 XUL P J Carter, 21 Mayfield Ave, Worle, Weston Super Mare, BS22 0AA [IO81MI, ST36]
G3 XUM J W Moran, 30 Elsie St., Farnworth, Bolton, BL4 9HT [IO83TN, SD70]
G3 XUO K V Edwards, 99 Wilton Cres, Shirley, Southampton, SO15 7QF [IO90GW, SU41]
G3 XUP G R J Everest, 13 Noel Rise, Burgess Hill, RH15 8BW [IO90WX, TQ31]
G3 XUQ M J Lee, 95 Kings Rd, Hayling Island, PO11 0PE [IO90MT, SU70]
G3 XUS P D Simmons, 62 Lawes Ave, Newhaven, BN9 9SB [JO00AT, TQ40]
GM3 XUW R Johnston, 123 Craigmount Brae, Edinburgh, EH12 8XW [IO85IW, NT17]
G3 XUX E Fitzgerald, 4 Southwick Rd, Wickham, Fareham, PO17 6HS [IO90JV, SU51]
G3 XVA D A Pickles, 26 Padstow Ave, Fishermead, Milton Keynes, MK6 2ES [IO92PA, SP83]
G3 XVB A F Vizoso, Granby House, Brightwalton Holt, Newbury, Berks, RG16 0DA [IO91LJ, SU66]
G3 XVC M F Collopy, 3 Main Rd, Sotton At Hone, Sutton At Hone, Dartford, DA4 9HG [JO01CJ, TQ57]
G3 XVF Details withheld at licensee's request by SSL.[Op: R G Davy. Station located in Norwich, Norfolk, NR2 3QX. (JO02PP, TG20)]
G3 XVG T Barraclough, 52 Denshaw Ave, Denton, Manchester, M34 3NX [IO83WL, SJ99]
G3 XVH S M Franklin, 337 Hendon Way, London, NW4 3NB [IO91VN, TQ28]
G3 XVL C J McCarthy, 31 Philip Rd, Ipswich, IP2 8BQ [JO02NB, TM14]
G3 XVN P S Norris, 1 High St., Dunton, Biggleswade, SG18 8RN [IO92VB, TL24]
G3 XVO D J West, 1 Ct Orchard, Newton St. Cyres, Exeter, EX5 5BJ [IO80ES, SX89]
G3 XVP P V Pimblott, 40 Richmondfield Ln, Barwick in Elmet, Leeds, LS15 4EZ [IO93HT, SE43]
GW3 XVQ F A Jinks, 28 St. Anns Rd, Bonnie View, Blackwood, NP2 1PG [IO81JP, ST19]
G3 XVR D Higgins, Lyndenhurst, The Shrave, Four Marks, Alton, Hants, GU34 5BJ [IO91LC, SU63]
G3 XVS R C Kitching, 96 Golf Links Rd, Ferndown, BH22 8BZ [IO90BT, SU00]
G3 XVV M J Salmon, 54 Church Rd, Rivenhall, Witham, CM8 3PH [JO01HT, TL81]
G3 XVW K A Ball, 39 Spinney Cl, Northfield, Birmingham, B31 2JG [IO92AJ, SP07]
G3 XVY P Coull, 40 Wear Bay Cres, Folkestone, CT19 6BA [JO01OC, TR23]
G3 XVZ R F Lowe, 1 Lowlands Dr, Keyworth, Nottingham, NG12 5HG [IO92LV, SK63]
G3 XWA J A Ennis, 30 Hillcrest Ave, Carlisle, CA1 2QJ [IO84NV, NY45]
G3 XWB C K Cadogan, 8 Horncliffe Cl, Rawtenstall, Rossendale, BB4 6EE [IO83UQ, SD82]
G3 XWC Details withheld at licensee's request by SSL.
G3 XWD D C Watts, 40 Outlands Dr, Hinckley, LE10 0TW [IO92HN, SP49]
GM3 XWE Details withheld at licensee's request by SSL.
G3 XWH R Horton, 7 Carlton Rd, Harrogate, HG2 8DD [IO93FX, SE35]
G3 XWI E Tredgold, 17 New Windsor Dr, Rothwell, Leeds, LS26 0HU [IO93GS, SE32]
G3 XWJ Details withheld at licensee's request by SSL.
G3 XWK W J C Pinnell, 31 Nunhead Ln, London, SE15 3TR [IO91XL, TQ37]
G3 XWL J R Cripps, 3 Queens Ct, Queens Rd, Hawkhurst, Cranbrook, TN18 4JE [JO01GB, TQ73]
G3 XWM C J Harvey, 124 Essex Rd, Southsea, PO4 8DJ [IO90LT, SZ69]
G3 XWN G B Laycock, 48 Marina Terr, Golcar, Huddersfield, HD7 4RA [IO93BP, SE11]
G3 XWO T Lowe, 688 East J St, Chula Vista, California 92010, USA
G3 XWU R Pearson, 10 Eastleigh Cl, Harden Park, Boldon Colliery, NE35 9NG [IO94GW, NZ36]
G3 XWV C M Hamilton, Paddocks, 13 Grimpits Ln, Kings Norton, Birmingham, B38 9EY [IO92AJ, SP07]
G3 XWX P W Ward, 47 Radstock Ave, Ward End, Birmingham, B36 8HD [IO92CL, SP18]
G3 XWZ F J Clarke, 149 Somersall St., Mansfield, NG19 6EL [IO93JD, SK56]
GW3 XXB A E Evans, 74 Celyn Ave, Lakeside, Cardiff, CF2 6EQ [IO81JM, ST18]
G3 XXD K G Rigelsford, 14 Glebelands Ave, South Woodford, London, E18 2AB [JO01AO, TQ49]
G3 XXE P J Williams, 4 Plantation Cl, Aller Park, Newton Abbot, TQ12 4NS [IO80EM, SX86]
G3 XXF C M Vine, 14 Hamilton Rd, St. Albans, AL1 4PZ [IO91US, TL10]
G3 XXG P J Sharpen, 3 Western Rd, Flixton, Urmston, Manchester, M41 6LE [IO83TK, SJ79]
G3 XXI Prof P Curzen, 2 Bishops Dr, Old Blanford Rd, East Harnham, Salisbury, SP2 8NZ [IO91CB, SU12]
G3 XXJ R P Smith, 5 Morrison Cl, Marine Ave West, Sutton on Sea, Mablethorpe, LN12 2LU [JO03DH, TF58]
G3 XXK J F H Dark, Edelweiss, Main Rd, Rollesby, Norfolk, NR29 5ER [JO02TQ, TG41]
G3 XXM D P Richards, 14 Dorrells Rd, Longwick, Princes Risborough, HP27 9SL [IO91NR, SP70]
G3 XXN F A Pickersgill, 3 Church St., Langold, Worksop, S81 9NW [IO93KJ, SK58]
G3 XXO E G Birks, 46 Curzon Dr, Worksop, S81 0LP [IO93KH, SK58]
G3 XXQ Details withheld at licensee's request by SSL.
G3 XXR P R Higton, 13 Wilton Ave, Bradley, Huddersfield, HD2 1RN [IO93DQ, SE12]
G3 XXT Details withheld at licensee's request by SSL.
G3 XXX S E Bradnam, 7 Melvin Way, Histon, Cambridge, CB4 4HY [JO02BG, TL46]
G3 XXA S G Charles, 48 Elms Farm Rd, Elm Park, Hornchurch, RM12 5RD [JO01CN, TQ58]
G3 XYB G Maitland, Mousehole Cottage, 7 Battery Rd, Cowes, Isle of Wight, PO31 8DP [IO90IS, SZ49]
G3 XYC P W Crust, 16 London Ln, Wymeswold, Loughborough, LE12 6UB [IO92KT, SK52]
G3 XYD W A Gordon-Laycock, 12 Godmanston Cl, Poole, BH17 8BU [IO90AR, SZ09]
G3 XYE J S Clifton, Romford Cottage, Romford, Verwood, BH31 7LE [IO90BV, SU00]
G3 XYF J F Wresdell, Bracey Bridge Farm, Harpham, Driffield, East Yorks, YO25 0DE [IO94TA, TA06]
G3 XYG Dr M George, Northfield Cottage, Northfield Ln, Barnstaple, EX31 1QD [IO71XC, SS53]
G3 XYH J C Hill, 35 Windmill Ave, Marshalswick, St. Albans, AL4 9SJ [IO91US, TL10]
G3 XYI D C Fearnley, Chase Cottage, Stock Chase, Heybridge, Maldon Essex, CM9 7AA [JO01IR, TL80]
G3 XYJ R M Walker, 27 Archer Cl, Kings Langley, WD4 9HF [IO91SR, TL00]
G3 XYK A E H Douse, 2 Queens Rd, Felixstowe, IP11 7QT [JO01QX, TM33]
G3 XYL B J Janes, 2 Dagmar Rd, Tivoli, Cheltenham, GL50 2UG [IO81WV, SO92]
G3 XYO S J Line, Thistledown, Church Rd, Carleton Rode, Norwich, NR16 1RW [JO02NL, TM19]
G3 XYP D Rollitt, St. Peters, 29 High St., Navenby, Lincoln, LN5 0EE [IO93RC, SK95]
G3 XYS Details withheld at licensee's request by SSL.
G3 XYV I G Cooper, 118 Stagsden Rd, Bromham, Bedford, MK43 8QJ [IO92RD, TL05]
GW3 XYW D S Jones, 22 Alltiago Rd, Pontarddulais, Swansea, SA4 1HU [IO71XR, SN50]
G3 XYX Details withheld at licensee's request by SSL.
G3 XYZ E Haskett Kings Lynn ARC, 23 Gloucester Rd, Gaywood, Kings Lynn, PE30 4AB [JO02FS, TF62]
G3 XZB N M Edwards, 14 Churchill Cl, Cowes, PO31 8HQ [IO90IS, SZ49]
GJ3 XZE R F Allenet, Les Sablons, La Grande Route de La Cote, St. Clement, Jersey, JE2 6FW
G3 XZF W Felton, 31 Mountbatten Dr, Leverington, Wisbech, PE13 5AF [JO02BQ, TF41]
G3 XZG J R Browne, 82 Cresswell Rd, Chesham, HP5 1TA [IO91QQ, SP90]
G3 XZJ G A Maynard, 6 Brimblecombe Cl, Emmbrookon, Wokingham, RG41 1QH [IO91NK, SU86]
G3 XZK D B Gething, 31 Lower Lodge Ln, Hazlemere, High Wycombe, HP15 7AT [IO91PP, SU89]
GI3 XZM D H Vance, The Eaves, Reagh Island, Comber, Newtownards, BT23 6EN [IO74EM, J56]
G3 XZO M D O Rhodes, The Bungalow, 21 Halford Rd, Ettington, Stratford upon Avon, CV37 7TL [IO92ED, SP24]
G3 XZP D M Holbum, 19 Whitwell Way, Coton, Cambridge, CB3 7PW [JO02AF, TL45]
G3 XZQ R Hargreaves, 4 Thorntrees Dr, Thornhill, Egremont, CA22 2SU [IO84FL, NY00]
G3 XZR R J Lawton, San Miguel, Weston Ln, Totland Bay, PO39 0HE [IO90FQ, SZ38]
G3 XZS Details withheld at licensee's request by SSL.
G3 XZT Details withheld at licensee's request by SSL.
GW3 XZU T Stafford, 17 Woodland Gr, Froncysyllte, Llangollen, LL20 7RL [IO82LX, SJ24]
G3 XZV J M Sonley, Ravenscliffe, Lands Ln, Knaresborough, North Yorks, HG5 9DE [IO94GA, SE35]
G3 XZW G B Wills Taunton & DARS, Rainbow Crest, Oatens Ln, Churchstanton, Taunton, Somerset, TA3 7PU [IO80KV, ST11]
G3 XZX J A Lowe, St. Michaels, Littlemead Ln, Exmouth, EX8 3BU [IO80HP, SY08]
G3 XZY T R Garner, 122 Wainfleet Rd, Skegness, PE25 3RX [JO03DD, TF56]
G3 YA C Smith, 21 Copeland Rd, Wombwell, Barnsley, S73 8TF [IO93HM, SE30]
G3 YAA G S Dunn, 5 Eden Cl, Beverley, HU17 7HE [IO93SU, TA04]
G3 YAB M E Bath, 25 Barrington Rd, Southgate, Crawley, RH10 6DQ [IO91VC, TQ23]
G3 YAC P A Howarth, 1 Clay Close Ln, Impington, Cambridge, CB4 4NE [JO02BF, TL46]
G3 YAD M J Goodrich, 2 Highworth Cres, Yate, Bristol, BS17 4EY [IO81SM, ST78]
G3 YAE W Murphy, 11 Wheatley Rd, Wardley, Swinton, Manchester, M27 9RW [IO83TM, SD70]
GW3 YAF T D J Davies, Rhydwen, Heol Dwr, Llangyndeyrn, nr Carmarthen Dyfed, SA17 5HF [IO71UT, SN41]
G3 YAG W L Thompson, 2 Fern Cl, Frimley, Camberley, GU16 5QU [IO91PH, SU85]
G3 YAI T A Mills, 16 Hunts Hill, Glemsford, Sudbury, CO10 7RL [JO02IC, TL84]
G3 YAJ D S Sellen, Prospect House, Wignall St., Lawford, Manningtree, CO11 2HX [JO01MW, TM03]
G3 YAL B D Sheasby, 75 Amberley Ave, Bulkington, Nuneaton, CV12 9QY [IO92GL, SP38]
G3 YAM Details withheld at licensee's request by SSL.[Op: H L Spong. Station located near Farnborough.]
GM3 YAN Details withheld at licensee's request by SSL.
GM3 YAO F Offler, Ar Dachaidh, 3 King David Dr, Inverbervie, Montrose, DD10 0SW [IO86UU, NO87]
G3 YAP A M Yiacoumis, 1 Fountain Rd, Edgbaston, Birmingham, B17 8NJ [IO92AL, SP08]
G3 YAR I R Gildersleve, 7 Oak Park Rd, Newton Abbot, TQ12 1RQ [IO80EM, SX87]
G3 YAS P A Ellis, 152 Cambridge Rd, Hounslow, TW4 7BH [IO91TL, TQ17]
GM3 YAU J Hill, 67 Caledonian Rd, Stevenston, KA20 3LF [IO75PP, NS24]
GM3 YAV J M Boyd, Sannox, 14 Brookfield Cres, Stranraer, DG9 0HY [IO74LV, NX06]
G3 YAW M D Beales, 14 Goosen Green, Aylesbury, HP21 9BX [IO91OT, SP81]
G3 YBA E Cooper, Flat, 29 Redholme, Sandygate Rd, Sheffield, S10 5UA [IO93FJ, SK38]
G3 YBD D S Probin, 5 Cherry Tree Cl, Timperley, Altrincham, WA15 7QJ [IO83TJ, SJ78]
G3 YBE E J Gilbert, 2 Church Field, Stanford, Ashford, TN25 6UA [JO01LD, TR04]
G3 YBF E G Arnold, 46 Inhurst Ave, Waterlooville, PO7 7QR [IO90LV, SU60]
G3 YBG J G Rabjohns, Quarries Bungalow, Barley Ln, Exeter, EX4 1TA [IO80FR, SX89]
G3 YBH P H Storey, RR 1 Site 2 Box 233, Brechin, Ont Lok 1Bo, Canada
G3 YBK R J Donno, 6 Mincinglake Rd, Exeter, EX4 7EA [IO80FR, SX99]
G3 YBM R H Mitchell, 98 Marlborough Dr, Burgess Hill, RH15 0EU [IO90WW, TQ31]
GW3 YBN C Davies, 31 Park Prospect, Graigwen, Pontypridd, CF37 2HF [IO81HO, ST09]
G3 YBO R Baines, 327 Langer Ln, Wingerworth, Chesterfield, S42 6TY [IO93GE, SK36]
G3 YBP A F Young, Ford House, Kentisburyford, Barnstable, Devon, EX31 4NB [IO81AE, SS64]
GM3 YBQ K D Horne, 10 Blair Pl, Hygrove Park, Kirkcaldy, KY2 5SQ [IO86JC, NT29]
G3 YBR S G Cook, 6 Clarence Ct, Clarence Rd, Wotton under Edge, GL12 7EX [IO81TP, ST79]
G3 YBS R Lindsay Smith, 58 Chalgrove Rd, London, N17 0JD [IO91XO, TQ39]
G3 YBT Details withheld at licensee's request by SSL.[Op: C F Harvey. Station located near Chippenham]
G3 YBU B L Whittle, 7 Main Rd, Camerton, Hull, HU12 9NG [IO93WR, TA22]
G3 YBX E A Milton, 46 Chatsworth Rd, Torquay, TQ1 3BL [IO80FL, SX96]
G3 YBY I D McCarthy, 16 High St., Purton, Swindon, SN5 9AD [IO91BO, SU08]
GI3 YBZ J H McCann, 61 Glenpark Rd, Omagh, BT79 7SS [IO64IP, H47]
G3 YCA Details withheld at licensee's request by SSL.
GM3 YCB S Riddell, 16 Lewis Dr, Old Kilpatrick, Glasgow, G60 5LE [IO75SW, NS47]
G3 YCC F W Lee, 8 Westland Rd, Kirkella, Kirk Ella, Hull, HU10 7PJ [IO93SR, TA02]
GW3 YCD H A Griffiths, 39 Erw Faen, Tregarth, Bangor, LL57 4AT [IO73WE, SH66]
G3 YCE J W Peck, 7 Paddock Cl, Radcliffe on Trent, Nottingham, NG12 2BX [IO92LW, SK63]
GM3 YCG S Spence, 90 Smithfield Rd, Glasgow, G44 5EN [IO75VT, NS55]
G3 YCH J R Sharman, 21 Ferrers Green, Churston Village, Churston Ferrers, Brixham, TQ5 0LF [IO80FJ, SX85]
G3 YCJ P R Shead, Valley View, Hard Platts Ln, Holywell Green, Halifax, HX4 9HP [IO93BQ, SE01]
G3 YCN W E B Kent, Old Cottage, Hermitage Ln, Detling, Maidstone, ME14 3HP [JO01GH, TQ75]
G3 YCO R J Lewis, 115 Chester Rd, Ellesmere Port, Whitby, South Wirral, L65 6SB [IO83NG, SJ37]
G3 YCQ E J Horne, 1 Upper Halliford Rd, Shepperton, TW17 8RX [IO91SJ, TQ06]
G3 YCR Details withheld at licensee's request by SSL.
G3 YCV J Hibbert, 5 Cliff View Rd, Cliffs End, Cliffsend, Ramsgate, CT12 5ED [JO01QI, TR36]
G3 YCW R A Jackson Vange ARS, 43 Sparrows Herne, Basildon, SS16 5HW [JO01FN, TQ78]
G3 YCX A Cain, 42 Wood Ln, Prescot, L34 1LW [IO83OK, SJ49]
G3 YCY R E Barrett, 47 Marshals Dr, St. Albans, AL1 4RD [IO91US, TL10]
GD3 YDB J D Thorpe, 69 Waverley Ct, Laureston Ave, Douglas, IM2 3BZ
G3 YDC M E Brett, 39 Burgage Rd, Stogursey, Bridgwater, TA5 1RB [IO81KE, ST24]
G3 YDD R D Emes Hereford A.R.S., Lower Winslow, Portway, Burghill, Hereford, Herefordshire, HR4 8NG [IO82PC, SO44]
G3 YDE W J Bagwell, Mati, 93 Broadley Dr, Livermead, Torquay, TQ2 6UT [IO80FK, SX86]
GI3 YDH M G McIntyre, 36 Beechgrove Park, Belfast, BT6 0NR [IO74BN, J37]
GI3 YDM J I Dunlop, 34 Ballybentragh Rd, Muckamore, Dunadry, Antrim, BT41 2HJ [IO64WQ, J28]
GM3 YDN D J Nutt, Garden House, Kilkerran Est, Maybole, Ayrshire, KA19 7SG [IO75QH, NS30]
GI3 YDO W J McNally, 13 Meadowside, Dublin Rd, Antrim, BT41 4HD [IO64VQ, J18]
G3 YDU J H Peters, 43 Holtwood Rd, Glenholt, Plymouth, PL6 7HU [IO70WK, SX56]
GW3 YDX Details withheld at licensee's request by SSL.[Op: R G D Stone. Station located near Llanymynech.]
G3 YDY P L Selwood, 43 Keene Way, Galleywood, Chelmsford, CM2 8NT [JO01FQ, TL70]
G3 YDZ G P Radford, 28 Loxley Rd, Oulton Broad, Lowestoft, NR33 9PG [JO02UL, TM59]
G3 YEB H E Cox, 26 Little Pynchons, Harlow, CM18 7DD [JO01BS, TL40]
G3 YEC R H Edmondson, 16 Orchard Cl, Copford Green, Copford, Colchester, CO6 1DB [JO01JU, TL92]
G3 YED R J Nettleton, 48 Dale Ave, Stratford upon Avon, CV37 7EW [IO92DE, SP25]
G3 YEE R Short, 6 Cross Henley Rd, Bramley, Leeds, LS13 3AN [IO93ET, SE23]
G3 YEG N J Sears, 1 Courtenay Dr, Emmer Green, Reading, RG4 8XH [IO91ML, SU77]
GM3 YEH B J Beggs, 27 Burnawn Pl, Galston, KA4 8JY [IO75TO, NS53]
G3 YEI T H Anyon, 6 Hillylaid Rd, Thornton Cleveleys, FY5 4DX [IO83LV, SD34]
G3 YEN D P Kennard, 36 Worcester Rd, Chichester, PO19 4DZ [IO90OU, SU80]
GD3 YEO R C Rimmer, 27 Manor Ln, Farmhill, Douglas, IM2 2NP
G3 YEP R A Wakeley, 7 St. Augustines Cl, Hillcrest Rd, Portishead, Bristol, BS20 8JH [IO81OL, ST47]
G3 YEQ L E Miller, 66 Longwater Rd, Bracknell, RG12 7NU [IO91PJ, SU86]
G3 YER D Lowe, Haven Cottage, Linkside West, Hindhead, GU26 6PA [IO91OC, SU83]
G3 YET Dr W M Arnold, c/o Higher Woodhall, Exbourne, Okehampton, EX20 3QZ [IO70XT, SS50]
G3 YEU B J Short, 83 Rowanfield Rd, Cheltenham, GL51 8AF [IO81WV, SO92]
GM3 YEW D E Morris, Ash Cottage, Perth Rd, Abernethy, Perth, PH2 9LW [IO86IH, NO11]
G3 YEY D M Gair, 14 Lindens Dr, Sutton Coldfield, B74 2AQ [IO92BN, SP09]
G3 YEZ A H G King, Sunset Cottage, 16 Franklands Way, Burgess Hill, RH15 0AX [IO90WW, TQ31]
G3 YFD W T Hewitt, 22 Derby Rd, Heaton Moor, Stockport, SK4 4NE [IO83VK, SJ89]
G3 YFE J P Shaw, 57 London Rd, Amesbury, Salisbury, SP4 7EE [IO91CE, SU14]
G3 YFG S H Westell, 2 Whiteacre Ln, Barrow North, Barrow, Clitheroe, BB7 9BJ [IO83TU, SD73]
G3 YFI D H Wilkinson, Rosehurst, Margaret Ave, Bardsey, Leeds, LS17 9AU [IO93GV, SE34]
G3 YFK P J McAlister, Lower Winnington Fr, Halfway House, Shrewsbury, Shropshire, SY5 9DJ [IO82LQ, SJ31]
G3 YFL C L Wright, Holmwood, Brackley Ave, Hartley Wintney, Hook, RG27 8QX [IO91NH, SU75]
G3 YFM M J Smyth, 6 Healey Cl, Abingdon, OX14 5RL [IO91IQ, SU49]
G3 YFN Details withheld at licensee's request by SSL.
G3 YFO M G Bunce, 36 Burlington Rd, Burnham, Slough, SL1 7BQ [IO91QM, SU98]
G3 YFS D F Haver, 31 Edenham Rd, Hanthorpe, Bourne, PE10 0RB [IO92TT, TF02]
G3 YFT I A Willox, 29 Timperley Way, Upper Hatherley, Up Hatherley, Cheltenham, GL51 5RH [IO81WV, SO92]
G3 YFU E A Tomlin, Indalo, Magna Mile, Ludford, Market Rasen, LN8 6AJ [IO93VJ, TF18]
G3 YFV P W I'Anson, 5 Knightcote Dr, Leamington Spa, CV32 5FA [IO92FG, SP36]
G3 YFW P R Rosen, 5 Courtleigh Ave, Hadley Wood, Barnet, EN4 0HT [IO91VQ, TQ29]
G3 YFZ J W Shelley, Bridge View Cottage, Old Rd, Harbertonford, Totnes, TQ9 7TA [IO80DJ, SX75]
G3 YGA E K Warwick-Oliver, 21 Meautys, St. Stephens, St. Albans, AL3 4LU [IO91TR, TL10]
G3 YGB J C H Coleman, 18 Chester St., Coventry, CV1 4DJ [IO92FJ, SP37]
G3 YGC E A Elliott, 18 Bear St., Lowerhouse, Burnley, BB12 6NQ [IO83UT, SD83]
G3 YGD D H Brown, 7 Torver Cl, Burnley, BB12 8UH [IO83UT, SD83]
G3 YGE J Okas, Dipley Springs, Dipley Common, Hartley Wintney, Hook, RG27 8JS [IO91MH, SU75]
G3 YGF Dr J N Gannaway, Deanhill Barn, East Dean, Salisbury, Wilts, SP5 1HJ [IO91EA, SU22]
G3 YGG J E Kelly, 79 West Hill Ave, Epsom, KT19 8JX [IO91UI, TQ26]
GW3 YGH A W Hughes, 68 Higher Ln, Langland, Swansea, SA3 4PD [IO81AN, SS68]
G3 YGJ D K Brierley, 4 Waterloo Terr, Bideford, EX39 3DJ [IO71VA, SS42]
G3 YGL F Smith, 34 Bridle Rd, Eastham, Wirral, L62 8BR [IO83MH, SJ38]
GW3 YGM M A Osborne, 18 Norton Rd, Mumbles, Swansea, SA3 5TQ [IO71XN, SS68]
G3 YGR C Thomas, Oakdene, School Ln, Burghfield Common, Reading, Berks, RG7 3ES [IO91LJ, SU66]
GM3 YGS Details withheld at licensee's request by SSL.
G3 YGU H Matley, 5 Esk Ave, Fleetwood, FY7 8BZ [IO83LV, SD34]
G3 YGZ A Walsh, 12 Broomfield, Hullen Edge, Elland, HX5 0QX [IO93BQ, SE02]
G3 YHA T W Cooper, 28 Hopgrove Ln, York, YO3 9TG [IO93LX, SE65]
G3 YHB J I Baker, Thurlstone, 76 Seel Rd, Huyton, Liverpool, L36 6DJ [IO83OJ, SJ49]
G3 YHC W R Hermes, 22 Mallinson Cres, Harrogate, HG2 9HP [IO93FX, SE35]
G3 YHD T P Holroyd, 52 Bowers Ave, Davyhulme, Urmston, Manchester, M41 5TQ [IO83TK, SJ79]
G3 YHF C K Skelcher, 154 Addison Rd, Kings Heath, Birmingham, B14 7EP [IO92BK, SP08]
G3 YHG D R Harding, 17 Summerfield Cl, Wokingham, RG41 1PH [IO91NK, SU87]
G3 YHH J H Froud, Summer Park, New Rd, Teignmouth, TQ14 8UF [IO80GN, SX97]
G3 YHI R Vale, High House, Mounts Ln, Newnham, Daventry, NN11 3ES [IO92KF, SP55]
G3 YHJ P J Simmons, 99 Ticonderoga Gdns, Southampton, SO19 9HD [IO90HV, SU41]
G3 YHK J R Clemence, 97 Queens Rd, Felixstowe, IP11 7PG [JO01QX, TM33]
G3 YHM R A Harvey, 26 Birkdale Rd, Worthing, BN13 2QY [IO90TT, TQ10]
G3 YHN C J Pedley, 25 Fallowfield Rd, Orchard Hills, Walsall, WS5 3DH [IO92AN, SP09]
G3 YHO R A Yaxley, Ashness, Swaffham Rd, Wendling, Dereham, NR19 2LX [JO02KQ, TF91]
G3 YHP R B Rayner, Endways, Fairview Dr, Colkirk, Fakenham, NR21 7NT [JO02KT, TF92]
G3 YHQ D Mercer, 19 Kingsfield Dr, Didsbury, Manchester, M20 6JA [IO83VJ, SJ89]
G3 YHR C Briscoe, 8 Westlands Way, Leven, Beverley, HU17 5LG [IO93UV, TA14]
G3 YHS J S Cottingham, 192 Beech Rd, St. Albans, AL3 5AX [IO91US, TL10]
GJ3 YHU D R Robinson, La Chaleur, Clos de La Bruyere, St Lawrence, Jersey, Channel Islands, JE3 1FS
G3 YHV C R Chidgey, 46 Station Rd, Shirehampton, Bristol, BS11 9TX [IO81PL, ST57]
G3 YHY H Young, 93 Leaford Cres, Watford, WD2 5JQ [IO91TQ, TQ19]
G3 YIA M R Harris, 100 Chapel Ln, Wymondham, NR18 0DN [JO02NN, TG10]
G3 YIC V R Sedgley, 6 The Cres, Surbiton, KT6 4BN [IO91UJ, TQ16]
G3 YIE E D Lusty, Stanley End Farm, Bell Ln, Selsley, Stroud, GL5 5JY [IO81VR, SO80]
G3 YIF J H Weiner, 1 Chippendayle Dr, Harrietsham, Maidstone, ME17 1AD [JO01IF, TQ85]
G3 YIG R J Hathaway, Merrygill, Cranes Gate, Whaplode St Catherine, nr Holbeach, Lincs, PE12 6TA [IO92XT, TF32]
GW3 YIH F A Cobb, Mon Reve, Rhodfa Nant, Towyn, Abergele, LL22 9ND [IO83FH, SH97]
G3 YII N J D Smith, Fordby, Ingoldsby Ave, Ingoldisthorpe, Kings Lynn, PE31 6NH [JO02GU, TF63]
G3 YIJ G D Moore, Fairfield Lodge, 26-28 Townsend, Soham, Ely, CB7 5DB [JO02EI, TL57]
G3 YIN E I T Ellery, 14 Four Acres, Londonderry Farm, Bideford, EX39 3RW [IO71VA, SS42]
G3 YIP Details withheld at licensee's request by SSL.
G3 YIQ R W Jones, 22 Onslow Ave Mns s, Onslow Ave, Richmond, TW10 6QD [IO91UK, TQ17]
G3 YIR A A Ward Anglo Americn Rd, 20 Tower Cl, Costessey, Norwich, NR8 5AU [JO02OP, TG11]
G3 YIS Details withheld at licensee's request by SSL.
G3 YIT A Moss, 432 Bromford Rd, Birmingham, B36 8JH [IO92CM, SP18]
G3 YIU K R Bassett, 7 Mayhurst Rd, Hollywood, Birmingham, B47 5QG [IO92BJ, SP07]
G3 YIV Details withheld at licensee's request by SSL.

G3

G3 YIW J E Gallop, 55 Somervell Dr, Fareham, PO16 7QW [IO90JU, SU50]
G3 YIY R A Ingram, 11 Bank Terr, Mevagissey, St. Austell, PL26 6QZ [IO70OG, SX04]
GU3 YIZ J E Martin, Bonne Chance, Rue Des Marais, Vale, Guernsey, GY6 8BB
G3 YJD J H Davies, 25 Harkness Cl, Bletchley, Milton Keynes, MK2 3NB [IO91PX, SP83]
G3 YJE P R Merriman, Gaylors Cottage, Westmill, Buntingford, Herts, SG9 9LA [IO91XW, TL32]
G3 YJF G W Brandwood, 16 Hambleton Rd, Heald Green, Cheadle, SK8 3DW [IO83VI, SJ88]
G3 YJG G E Mason, 8 Leighton Rd, Sunderland, SR2 9HQ [IO94HV, NZ35]
G3 YJJ R T Palmer, 8 St. Francis Rd, Blackfield, Southampton, SO45 1XU [IO90HT, SU40]
GW3 YJL J L Lawrence, 18 Maesglas, Tredegar, NP2 3ET [IO81JS, SO10]
G3 YJN R E Hodge, 36 Binswood End, Harbury, Leamington Spa, CV33 9LN [IO92GF, SP35]
G3 YJO Details withheld at licensee's request by SSL.[Op: Prof. M N Sweeting.]
G3 YJP D J Benham, 200 Central Park South, 6D New York, New York 10019, USA, USA, ZZZ ZZZ
G3 YJQ F A G Bourne, 78 Normandy Way, Plymouth, PL5 1SR [IO70VJ, SX45]
G3 YJR G J Coyne, 188 Stannington View Rd, Sheffield, S10 1ST [IO93FJ, SK38]
G3 YJS M J Roche, 8 Northdown Cl, Maidstone, ME14 2ER [JO01GG, TQ75]
G3 YJU L G Newton, 23 Prebendal Ave, Aylesbury, HP21 8HZ [IO91OT, SP81]
G3 YJW R T Whitehouse, White Ct, Camp Hill, Chiddingstone Causeway, Tonbridge, TN11 8LE [JO01CE, TQ54]
G3 YJX A E Warne, Treryn, Trevanson Rd, Wadebridge, PL27 7HB [IO70NM, SW97]
G3 YJY J P Kealey, 40 Winchester Dr, Wallasey, L44 2AZ [IO83LK, SJ39]
G3 YJZ A P Mitchell, 89 Queen Annes Gr, Bush Hill Park, Enfield, EN1 2JU [IO91XP, TQ39]
GM3 YKA J Wiewiorka, 2 Cuillin Pl, Grangemouth, FK3 0DA [IO86DA, NS98]
G3 YKC D J Fayers, 1 Tismeads Cres, Swindon, SN1 4DP [IO91CN, SU18]
GM3 YKE R J English, 1 Berchem Pl, Saltcoats, KA21 5NX [IO75OP, NS24]
G3 YKI K A Vickers, Yew Cottage, 19 Combrook, Warwick, CV35 9HP [IO92FD, SP35]
G3 YKO D A Darwood, Briarwood Cottage, Packhorse Ln, Hadley Heath, Birmingham, West Midlands, B38 0DN [IO92BJ, SP07]
G3 YKP W J Sutton, 8 Hill Syke, Lowdham, Nottingham, NG14 7DE [IO93LA, SK64]
G3 YKS R G Butlin, 48 Roman Way, Market Harborough, LE16 7PQ [IO92ML, SP78]
GW3 YKU Details withheld at licensee's request by SSL.
G3 YKW Capt R J Walker, 17 Ballantyne South, Montreal West, Prov Quebec, Canada H4X 2Bi, ZZ1 7BA
GW3 YKZ M Biddiscombe, 20 Arlington Cl, Newport, NP9 6QF [IO81LO, ST39]
G3 YLA J D Bacon, Highways, Common Rd, East Tuddenham, Dereham, NR20 3AH [JO02MP, TG01]
G3 YLC G L Groom, 15 Grange Cl, Moreton Grange, Buckingham, MK18 1JJ [IO92MA, SP73]
GM3 YLD J F Frew, Queens Cottage, 87 Queen St., Dunoon, PA23 8AX [IO75MW, NS17]
G3 YLE Details withheld at licensee's request by SSL.
G3 YLG Details withheld at licensee's request by SSL.
GJ3 YLI A D Morrissey, Darlinghurst, Bagot Rd, St. Saviour, Jersey, JE2 7RG
G3 YLJ J Boraston, 15 Northfield Rd, Redcliffe Bay, Portishead, Bristol, BS20 8LE [IO81OL, ST47]
G3 YLL R F Brooks, 10 Leopold Cl, Bognor Regis, PO22 8JJ [IO90QT, SU90]
GJ3 YLN J R Speller, Lindau, Gorey Village, Jersey Cl, Jersey, JE9 9LI
G3 YLO Details withheld at licensee's request by SSL.
G3 YLP J B E Thorpe, 7 Pipers End, Wolvey, Hinckley, LE10 3LQ [IO92HL, SP48]
G3 YLR F R Blake, 59 Hillbury Park, Hillbury Rd, Alderholt, Fordingbridge, Hants, SP6 3BW [IO90CV, SU11]
G3 YLV P Jones, 4 Snowford Cl, Bramingham Park, Luton, LU3 3XU [IO91SV, TL02]
G3 YLW P S Lascelles, 25 Saxon Way, Dergingham, Dersingham, Kings Lynn, PE31 6LY [JO02GU, TF63]
G3 YLX J A C Brasier, Country Cottage, Scalp Rd, Fishtoft, Boston, PE21 0SH [JO02AW, TF34]
G3 YLY B A Hawes, 15 Bridge Ln, Wimblington, March, PE15 0RR [JO02BM, TL49]
G3 YMA R B Smith, 11 Chariot Rd, Illogan Highway, Redruth, TR15 3LG [IO70IF, SW64]
G3 YMC D W Sergeant, 8 Toll Gdns, 'racknell, Berks, RG12 9EX [IO91PJ, SU86]
G3 YMD B L Cuff Dover Radio Club, 51 Valley Rd, River, Dover, CT17 0QW [JO01PD, TR24]
G3 YMH R C Wainwright, The Olives, High St., Buxted, Uckfield, TN22 4LB [JO00BX, TQ42]
G3 YMJ J W G Pampling, 47 Altham Gr, Harlow, CM20 2PQ [JO01BS, TL41]
G3 YMK R W R Jones, Millfield House, Alton Ln, Four Marks, Alton, GU34 5AL [IO91LC, SU63]
G3 YMM T F Campbell Davis, 9 Cloister Rd, Acton, London, W3 0DE [IO91UM, TQ28]
G3 YMN J B Rhys, 64 Tregundy Rd, Perranporth, TR6 0LL [IO70KI, SW75]
G3 YMO J Bissett, 50 The Hollows, Bessacarr, Doncaster, DN4 7PP [IO93LL, SE60]
G3 YMP Details withheld at licensee's request by SSL.[Station located in Lodon N12.]
G3 YMS J H Ison, 12 Hunters Lodge, Catisfield, Fareham, PO15 5NE [IO90JU, SU50]
GI3 YMT M A Higgins, 1 Cairnshill Park, Cairnshill Rd, Belfast, BT8 4RG [IO74BN, J36]
G3 YMU J W Hibberd, Honeysuckle Cottage, Coxbank, Audlem, Ches, CW3 0EU [IO82RX, SJ64]
G3 YMV I G M Machardie, 33 Sandwich Cl, St. Ives, Huntingdon, PE17 6DQ [IO92XI, TL37]
G3 YMW D T Sapsworth, 16 Laxton Ave, Hardwick, Cambridge, CB3 7XL [JO02AF, TL35]
GM3 YMX D E Ferguson, 21 Pentland Dr, Edinburgh, Midlothian, EH10 6PU [IO85JV, NT26]
GI3 YMY N S Newell, 18 Kilmaine Ave, Bangor, BT19 6DU [IO74EP, J58]
G3 YNC C J Adams, 19 Hamlet Rd, Collier Row, Romford, RM5 2DS [JO01BO, TQ49]
GM3 YND I F Simpson, 14 Maccoll Rd, Cannich, Beauly, IV4 7LP [IO77OI, NH33]
G3 YNF B W Turner, 48 High St., Great Houghton, Northampton, NN4 7AF [IO92NF, SP75]
G3 YNG F Higgins, 27 School Rd, Thornton Cleveleys, FY5 5AW [IO83MU, SD34]
G3 YNH Details withheld at licensee's request by SSL.
G3 YNI A W Bailey, 85 Meadow St., Weston Super Mare, BS23 1QL [IO81MI, ST36]
G3 YNJ C J Powell, 38 Braeside Rd, St. Leonards, Ringwood, BH24 2PH [IO90BT, SU10]
G3 YNK D S Evans, Michigan Villa, Wall, Gwinear, Hayle, Cornwall, TR27 5HA [IO70HE, SW53]
GW3 YNM A E Hewitt, 6 Lon y Bedw, Bangor, LL57 4TN [IO73WF, SH57]
G3 YNO M D Booth, The Coach House, Humber Rd, North Ferriby, North Humberside, HU14 3DW [IO93RR, SE92]
G3 YNP D R Shepherd, Radnor, 52 Barnfield Rd, Northgate, Crawley, RH10 2DP [IO91VC, TQ23]
G3 YNR E Nash, 60 Peveril Rd, Chesterfield, S41 8RW [IO93GG, SK37]
G3 YNT Details withheld at licensee's request by SSL.
G3 YNU I J Stevenson, 18 Sittingbourne Rd, Wigan, WN1 2RR [IO83QN, SD50]
G3 YNY A R Richings, High Wold, The Hithe, Rodborough Common, Stroud, GL5 5BW [IO81VR, SO80]
G3 YOA A R Adams, The Gables, Chapel Rd, Trunch, North Walsham, NR28 0QG [JO02QU, TG23]
G3 YOB Details withheld at licensee's request by SSL.[Op: P M Dowdall.]
G3 YOC R Moore, 38/40 Sandygate, Wath on Dearne, Wath upon Dearne, Rotherham, S63 7LR [IO93HM, SE40]
G3 YOF Details withheld at licensee's request by SSL.
G3 YOG C H Crook, 19 Hatters Ln, Berwick upon Tweed, TD15 1BY [IO85XS, NT95]
G3 YOI Dr K J Falconer, Lumbo Farmhouse, St Andrews, Fife, KY16 8NS [IO86NH, NO41]
G3 YOK Details withheld at licensee's request by SSL.
G3 YOL S K Cole, Halebrook, Bridgwater Rd, Sidcot, Winscombe, BS25 1NH [IO81OH, ST45]
G3 YOM K S Beddoe, 30 Tamella Rd, Botley, Southampton, SO30 2NY [IO90IV, SU41]
G3 YON F R A N Webster, 16 Pembroke Rd, Dronfield, Sheffield, S18 6WH [IO93GH, SK37]
G3 YOO J D Webster, 18 Collaton Rd, Wigston, LE18 2GY [IO92KN, SP69]
G3 YOQ L A O Spinks, 2 Ventnor Ave, Grantham, NG31 7EA [IO92QV, SK93]
GM3 YOR A B Givens, 3 Murray Pl, Gourock, PA19 1TS [IO75OW, NS27]
G3 YOS C Gerrard, 10 Crofton Ln, Hill Head, Fareham, PO14 3LP [IO90JT, SU50]
G3 YOU Details withheld at licensee's request by SSL.
G3 YOV T J Gammage, 23 Artizan Rd, Northampton, NN1 4HU [IO92NF, SP76]
G3 YOY H D L Clark, Church House, 2 Chestnut Cl, Peakirk, Peterborough, PE6 7NW [IO92UP, TF10]
G3 YPD P S Chester, 44 Richmond Dr, Lichfield, WS14 9SZ [IO92CQ, SK10]
G3 YPE M D Greenwood, 21 Dobb Top Rd, Holmbridge, Holmfirth, Huddersfield, HD7 1PQ [IO93CN, SE10]
G3 YPK S N Wallis, La Floride, Route D'aubais, Sommieres 30250, France
G3 YPL D M Gray, 19 Westbury Gdns, Higher Odcombe, Yeovil, BA22 8UR [IO80PW, ST51]
G3 YPM Dr R K Moore, 20 Ebrington Rd, Malvern, WR14 4NL [IO82TC, SO74]
G3 YPN A R Turner, 7 St. Davids Way, Wickford, SS11 8EX [JO01GO, TQ79]
G3 YPP M E Costello, 26 Stockbridge Rd, Clifton, Shefford, SG17 5HA [IO92UA, TL13]
G3 YPQ P E Williams, Fuchsia Cottage, 50 Bell St., Swanage, BH19 2SB [IO90AO, SZ07]
G3 YPS S Atkinson, 13 Charles St., Gainsborough, DN21 2JA [IO93OJ, SK89]
G3 YPT P W Tomes, 1 Catalina Cl, Mudeford, Christchurch, BH23 4JG [IO90DR, SZ19]
G3 YPU P C Koker, 14 Mallard Ave, Kidderminster, DY10 4AE [IO82VJ, SO87]
G3 YPX Details withheld at licensee's request by SSL.
G3 YPY A J Head, 32 Weald View Rd, Tonbridge, TN9 2NQ [JO01DE, TQ54]
G3 YPZ J D Petters, Rose Cottage, Hannath Rd, Tydd Gote, Wisbech, PE13 5NA [JO02BR, TF41]
G3 YQ S Hunt, 33R Listers Camp, North Sea Ln, Humberston, Grimsby
G3 YQA F Wilson, 26 Humber Rd, North Ferriby, HU14 3DW [IO93RR, SE92]
G3 YQB D A Rankin, 6 Woodfield, Lacey Green, Princes Risborough, HP27 0QQ [IO91OQ, SP80]
G3 YQC J L Wood, 4 Lockhill, Upper Sapey, Worcester, WR6 6XR [IO82SG, SO66]
G3 YQD D J Taylor, 45 Norris Rd, Sale, M33 3GR [IO83UJ, SJ79]
G3 YQF R S Linford, Springtide, Church Rd, Webury, Devon, PL9 0LB [IO70XH, SX54]
G3 YQG N S A Waylett, Chilterns, 19 Kingshurst Dr, Preston, Paignton, TQ3 2LT [IO80FK, SX86]
G3 YQJ P Burnet, 14 Barnfield Cl, Sheffield, S10 5TF [IO93FJ, SK38]
GM3 YQK J T H Dillon, Steelbank Cottage, Dalry Rd, Kilwinning, KA13 6PL [IO75PQ, NS24]
G3 YQL P S Murfitt, 53 Codnor Denby Ln, Codnor, Ripley, DE5 9SP [IO93HA, SK44]
GW3 YQM Dr D Wynford-Thomas, 2 Tummel Cl, Cyncoed, Cardiff, CF2 6LR [IO81KM, ST18]
G3 YQO D M Kirkwood, 17 Woodside Way, Salfords, Redhill, RH1 5BD [IO91WF, TQ24]
GW3 YQP C B Hardie, 3 Berth-y-Glyd, Gyffin, Conwy, Gwynedd, LL32 8NP [IO83BG, SH77]
G3 YQQ J A Bibby, 167 The Green, Eccleston, Chorley, PR7 5SA [IO83PP, SD51]
G3 YQR S J Whiteman, 3 Woodside Cl, Kearsney, Dover, CT16 3BP [JO01PD, TR24]
G3 YQV C J Railton, Cadney, Highfield Rd, Whiteshill, Stroud, GL6 6AJ [IO81VS, SO80]
G3 YQW M E Funnell, 15 McIndoe Rd, East Grinstead, RH19 2DD [IO91XD, TQ33]
G3 YQZ M Johnson, 36 Coventry Rd, Bulkington, Nuneaton, CV12 9ND [IO92GL, SP38]
G3 YRB A T Simpson, 9 Dunheved Rd South, Thornton Heath, CR7 6AD [IO91WJ, TQ36]
G3 YRC A D Besford Great Yarmouth Radio Club, Conifers, 2A Halt Rd, Caister on Sea, Great Yarmouth, NR30 5NZ [JO02UP, TG51]
G3 YRE H A Scott, Cat Bells, 5 Plowden Park, Aston Rowant, Watlington, OX9 5SX [IO91MQ, SU79]
G3 YRF J N P Martin, 6 Bedford St., Brighton, BN2 1AN [IO90WT, TQ30]
G3 YRH B Dodds, 1 Croft View, Killingworth, Newcastle upon Tyne, NE12 0BT [IO95FA, NZ27]
G3 YRJ Details withheld at licensee's request by SSL.
GI3 YRL J H Branagh, 17 Rathmoyle Park West, Carrickfergus, BT38 7NG [IO74CR, J48]
G3 YRM C H Stanley, 142 Elm Tree Rd, Locking, Weston Super Mare, BS24 8EL [IO81MI, ST36]
G3 YRN C T Marshall, The Old Coach House, 81 Long Ashton Rd, Long Ashton, Bristol, BS18 9HY [IO81QK, ST57]
G3 YRO Details withheld at licensee's request by SSL.
G3 YRP I C Dudley, Oaklea, Gr Ln, Somersal Herbert, Ashbourne, Derbys, DE6 5PD [IO92CW, SK13]
G3 YRQ I J Parkinson, 61 Cinnamon Ln, Fernhead, Fearnhead, Warrington, WA2 0AG [IO83RJ, SJ69]
G3 YRT Details withheld at licensee's request by SSL.
G3 YRU P R Wilby, Green Farm, Main St., Huggate, York, YO4 2YQ [IO93QX, SE85]
G3 YRW Details withheld at licensee's request by SSL.[Station located in London N4.]
G3 YRX I C Elston, 11 Knowle Dr, Exwick, Exeter, EX4 2DF [IO80FR, SX99]
G3 YRZ Details withheld at licensee's request by SSL.
G3 YS G Troy, 9 Epworth Ct, Chapel St., Bentley, Doncaster, DN5 0DR [IO93KN, SE50]
GW3 YSA A Bartlett, 95 Penarth Rd, Cardiff, CF1 7JT [IO81JL, ST17]
G3 YSC K J Maskell Maidstone ARS, 19 Arethusa Rd, Rochester, ME1 2TZ [JO01GI, TQ76]
G3 YSD J Murdoch, 32 Scalegill Rd, Moor Row, CA24 3JL [IO84FM, NY01]
GW3 YSF C J Probert, 103 Bedwellty Rd, Cefn Fforest, Blackwood, NP2 1HB [IO81JQ, ST19]
G3 YSG M F W Taylor, 26 Switchback Rd, North, Maidenhead, Berks, SL6 7UE [IO91PM, SU88]
G3 YSH Details withheld at licensee's request by SSL.
G3 YSK A J Button, 13 Taplings Rd, Weeke, Winchester, SO22 6HE [IO91IB, SU43]
G3 YSL G B Davidson, 1 Carolyn Cres, Woodburn Est, Whitley Bay, NE26 3ED [IO95GB, NZ37]
G3 YSM M I Davidson, 24 Dingwall Dr, Greasby, Wirral, L49 1SG [IO83KJ, SJ28]
G3 YSN H H Smith, Carn View, Tregender Ln, Crowlas, Penzance Cornwall, TR20 8DJ [IO70GD, SW53]
GW3 YSP Details withheld at licensee's request by SSL.
G3 YSQ A P Pratt, 7 The Croft, West Hanney, Wantage, OX12 0LD [IO91HP, SU49]
G3 YSR C M Beattie, Mayerin, Churchway, Stone, Aylesbury, HP17 8RG [IO91NT, SP71]
G3 YST Details withheld at licensee's request by SSL.
G3 YSV D Heaton, 1 Jer Ln, Bradford, BD7 4NU [IO93CS, SE13]
G3 YSW N Thrower, 8 Upton Gdns, Tarring, Worthing, BN13 1DA [IO90TT, TQ10]
G3 YSX Dr S F Bryant, 3 Redstone Park, Redhill, RH1 4AS [IO91WF, TQ25]
G3 YSY R Hartshorn, 21 Hucklow Ave, Chesterfield, S40 2LT [IO93GF, SK37]
G3 YTE P C Gill, 43 The Westerings, Tye Green, Cressing, Braintree, CM7 8HQ [JO01HU, TL72]
G3 YTF K C Beale, 110 Castledon Rd, Wickford, SS12 0EJ [JO01GO, TQ79]
G3 YTH Details withheld at licensee's request by SSL.
G3 YTI S J Cooper, 24 Cambridge St., Darwen, BB3 3JH [IO83SQ, SD72]
GW3 YTJ Details withheld at licensee's request by SSL.
G3 YTK Details withheld at licensee's request by SSL.
GW3 YTL C J Lewis, 22 Erw Goch, Ruthin, LL15 1RR [IO83IC, SJ15]
G3 YTP E W Sillifant, Westview, Toll Rd, Bleadon, Weston Super Mare, BS23 4TZ [IO81MH, ST35]
G3 YTQ C J C Kidd, 20 Gorran Ave, Rowner, Gosport, PO13 0NF [IO90JT, SU50]
G3 YTR M G Hatt, 1 Larches Way, Crawley Down, Crawley, RH10 4UJ [IO91XC, TQ33]
GM3 YTS R W Ferguson, 24 Braemar Ave, Dunblane, FK15 9ED [IO86AE, NN70]
G3 YTT W R Taylor, 58 Collett, Glascote, Tamworth, B77 2DZ [IO92EO, SK20]
G3 YTU Details withheld at licensee's request by SSL.[Station located near Haywards Heath.]
G3 YTV D Greatbatch YTV ARS, Technical Support Manager, Yorks Television, Studio Centre, Leeds, LS3 1JS [IO93FT, SE23]
G3 YTW G H Clarke, 117 Bermuda, Nuneaton, CV10 7PW [IO92GM, SP38]
G3 YTX G Clamp, 9 Furse Cl, Camberley, GU15 1BF [IO91PH, SU95]
G3 YTY M S Edib, 84 Connaught Gdns, London, N13 5BT [IO91WO, TQ39]
GW3 YUC D Davies, 30 Wern Isaf, Penywern, Dowlais, Merthyr Tydfil, CF48 3NY [IO81HS, SO00]
G3 YUD P C Hawkins, 37 Alexandra Rd, Dorchester, DT1 2LZ [IO80SR, SY69]
G3 YUE Details withheld at licensee's request by SSL.
G3 YUF I Webster, 34 Armitage Rd, Balby, Doncaster, DN4 0TD [IO93KM, SE50]
G3 YUH R F Ayling, 25 Nash Ct Rd, Margate, CT9 4DH [JO01QJ, TR36]
G3 YUI P A D O'Dell, 92 Butterworth Path, Luton, LU2 0TR [IO91TV, TL02]
G3 YUJ R Steed, 53 Colchester Rd, Ipswich, IP4 3BT [JO02OB, TM14]
GJ3 YUL J A Botrel, 18 Willow Ct, Green St, St Helier, Jersey, Channel Islands, JE2 4UG
GD3 YUM J Parnell, Brookfield, 30 Mona St., Douglas, IM1 2QD
G3 YUQ E Elsley, 25 Elmsdale Rd, Wootton, Bedford, MK43 9JW [IO92RC, TL04]
G3 YUR A B Ryan, 4 Shepherds Hill, Harold Wood, Romford, RM3 0ND [JO01CO, TQ59]
G3 YUS D C Moule, 6 Thrushdale, Great Clacton, Clacton on Sea, CO15 4HX [JO01NT, TM11]
G3 YUT D F Molyneaux, 25 Crosby Rd, West Bridgford, Nottingham, NG2 5GG [IO92KW, SK53]
G3 YUU C J Lord, 40 Captains Cl, Sutton Valence, Maidstone, ME17 3BA [JO01HE, TQ84]
G3 YUV Details withheld at licensee's request by SSL.
G3 YUX R W Moore, 69 Ivatt, Tamworth, B77 2HQ [IO92EO, SK20]
G3 YUZ I C Wilson, Asprington, Garth Rd, Letchworth, SG6 3NG [IO91VX, TL23]
GW3 YVC O W A Wade, 1 Lomond Cres, Cyncoed, Cardiff, CF2 6ES [IO81KM, ST18]
G3 YVF Details withheld at licensee's request by SSL.
G3 YVH A C Boyne, 18 Crow Ln West, Newton-le-Willows, WA12 9YG [IO83QK, SJ59]
G3 YVI R A Gilbert, New Copse, Bentworth, Medstead, Alton, Hants, GU34 5NP [IO91LD, SU63]
G3 YVK H F Tabberer, 101 Broadclyst Gdns, Southend on Sea, SS1 3QU [JO01JN, TQ98]
GW3 YVN N J G Little, -le-Njg, 34 Lon Fawr, Caerphilly, CF83 1DA [IO81JN, ST18]
G3 YVO R J Hodges, 141 Appleby St., Cheshunt, Waltham Cross, EN7 6QU [IO91XR, TL30]
G3 YVP J Twiss, 16 Masefield Ave, Lamberhead Green, Orrell, Wigan, WN5 8HR [IO83PM, SD50]
G3 YVR G Bowden, 51 Leighlands, Pound Hill, Crawley, RH10 3DN [IO91WC, TQ23]
GU3 YVV R H M Outhwaite, Flat 2, 3 Choisi, Les Greaves, St Peter Port, Guernsey, GY1 1RP
GM3 YVX D W Coupar, 32 Gillies Pl, Broughty Ferry, Dundee, DD5 3LE [IO86NL, NO43]
G3 YVY D Hanley, 1 Darlington Ln, Norton, Stockton on Tees, TS20 1EW [IO94IO, NZ42]
G3 YVZ T Gardner, 64 Balmoral Terr, Heaton, Newcastle upon Tyne, NE6 5YA [IO94FX, NZ26]
G3 YWA E S Pepper, 30 Westfield Dr, Harpenden, AL5 4LP [IO91TT, TL11]
G3 YWB K A W Wickham, 33 John Gale Ct, Thorpe Marriat, Drayton, Norwich, NR8 6YW [JO02OQ, TG11]
G3 YWD E Darlington, Dawnwind, Mill Ln, Storrington, Pulborough, RH20 4NF [IO90SW, TQ01]
G3 YWF P P Smith, 76 Hackington Rd, Tyler Hill, Canterbury, CT2 9NQ [JO01MH, TR16]
G3 YWH F Hill, 24 Mount St. James, Knuzden, Blackburn, BB1 2DR [IO83SR, SD72]
G3 YWI G W H Grayson, One Elm, 58 Kaye Ln, Huddersfield, HD5 8XU [IO93CP, SE11]
G3 YWM P J Hubert, 575 Bramford Ln, Ipswich, IP1 5JX [JO02NB, TM14]
G3 YWO J M B Tripp, The Robbis, Manthorpe, Bourne, Lincs, PE10 0JE [IO92TR, TF01]
G3 YWP Details withheld at licensee's request by SSL.
G3 YWQ Details withheld at licensee's request by SSL.
G3 YWS J M Smith, 16 Woodlands, Winthorpe, Newark, NG24 2NL [IO93OC, SK85]
G3 YWT P Smith, Beechwood, Clarendon Rd, Alderbury, Salisbury, SP5 3AT [IO91DB, SU12]
G3 YWU S Fisher, Arkle, 31 Frith Ave, Delamere, Northwich, CW8 2JB [IO83QF, SJ56]
G3 YWV Details withheld at licensee's request by SSL.
G3 YWW A B Carpenter, 17 Victoria Ave, Upwey, Weymouth, DT3 5NG [IO80SP, SY68]
G3 YWX I D Poole, 5 Meadway, Staines, TW18 2PW [IO91RK, TQ07]
G3 YWZ K Buksh, 5 Broadfield Ave, Blackpool, FY4 3RA [IO83LS, SD33]
G3 YXB W A B Keay, 37 Leyborne Park, Kew, Richmond, TW9 3HB [IO91UL, TQ17]
G3 YXH S B Marshall, 14 Parkgrove Terr, Glasgow, G3 7SD [IO75UU, NS56]
G3 YXJ D Begg, 48 Eskhill, Penicuik, EH26 8DG [IO85JT, NT26]
G3 YXK Details withheld at licensee's request by SSL.
G3 YXM D A Pick, 178 Alcester Rd South, Kings Heath, Birmingham, B14 6DE [IO92BK, SP08]
G3 YXN P A Whalley, The Limes, School Ln, Warmingham, Sandbach, CW11 3QN [IO83SD, SJ76]
G3 YXO Dr D J Watson, Norton, Gote Ln, Ringmer, Lewes, BN8 5HX [JO00AV, TQ41]
G3 YXQ Dr R J Ireland, Box 609 Sackville, 23 St James St, New Brunswick, Canada, EOA 3CO
G3 YXR B Frogatt, IBM Laboratories Ar, 1 Plantation Rd, West Wellow, Romsey, Hants, SO51 6DF [IO90EW, SU21]
G3 YXS D T Naylor, 10 Meadow Hey, Bootle, L20 4PW [IO83LL, SJ39]
G3 YXW P J Dunford, 1 Charlock Way, Watford, WD1 8JY [IO91TP, TQ09]
G3 YXX D W Wood, 85 Epsom Rd, Guildford, GU1 3PA [IO91RF, TQ04]
GM3 YXY A H Thomson, Roselea, Ravenstruther, nr Larark, Scotland, ML11 7SE [IO85DQ, NS94]
G3 YXZ Details withheld at licensee's request by SSL.[Op: P J Marcham.]
G3 YY C T Fairchild, 4 Chalkland Rise, Woodingdean, Brighton, BN2 6RH [IO90XT, TQ30]
G3 YYC G P Sharples, 5 Bramlea Cl, Leominster, HR6 8RZ [IO82OF, SO45]
G3 YYD Details withheld at licensee's request by SSL.
G3 YYE H Lawrence, 14 Manor Ln, Throapham Dinnington, Dinnington, Sheffield, S31 7SW [IO93GJ, SK38]
G3 YYF R L G Kemp, 7 Forewood Rise, Crowhurst, Battle, TN33 9AH [JO00FV, TQ71]
G3 YYG J H Bolton, 3 Fyne Dr, Linslade, Leighton Buzzard, LU7 7YG [IO91PW, SP92]
G3 YYI P Dockerty, 5 Vicarage Ct, Heworth, Gateshead, NE10 8HG [IO94FW, NZ26]
G3 YYK R C North, 40 Swindon Rd, Swindon, SN1 3JJ [IO91CN, SU18]

G3 YYM G H Oxley, 13 Beech Dr, Ellesmere, SY12 0BU [IO82NV, SJ33]
G3 YYN P C Spurr, Windmill Farm Barn, High St., Great Linford, Milton Keynes, MK14 5AX [IO92OB, SP84]
G3 YYO Details withheld at licensee's request by SSL.
G3 YYP J Addy, 27 Linn St., Crumpsall, Manchester, M8 5SN [IO83VM, SD80]
G3 YYQ L R Balls, 2 Bure Cl, Great Yarmouth, NR30 1QU [JO02UO, TG50]
G3 YYR Dr A Parry, 56 Grimshaw Ln, Bollington, Macclesfield, SK10 5LY [IO83WH, SJ97]
G3 YYU M O Binns, 21 High Park, Morpeth, NE61 2TA [IO95DD, NZ18]
G3 YYZ J C Cuthbert, Fulbeck, 48 Mayes Ln, Ramsey, Harwich, CO12 5EJ [JO01OW, TM23]
G3 YZ D G Price, Calvados, 30 Celandine Bank, Woodmancote, Cheltenham, GL52 4HZ [IO81XW, SO92]
G3 YZA M Tompkins, 82 Henley Ave, Thornhill, Dewsbury, WF12 0LN [IO93EP, SE21]
G3 YZB S H Bassford, The Willows, 17 Sandhill, Littleport, Ely, CB6 1NT [JO02DK, TL58]
G3 YZG Details withheld at licensee's request by SSL.
G3 YZJ Details withheld at licensee's request by SSL.
G3 YZK Dr G M West, 6 Lammas Cl, Cowes, PO31 8DT [IO90IS, SZ49]
G3 YZN R E Awbery, Dashwood, Beacon View Rd, Elstead, Godalming, GU8 6DU [IO91PE, SU94]
G3 YZO Details withheld at licensee's request by SSL.
G3 YZP J R Birch, 8 St. Johns Cl, Swindon, Dudley, DY3 4PQ [IO82VM, SO89]
G3 YZR J P Porter, 94 Oaken Gr, Haxby, York, YO3 3QZ [IO94LA, SE65]
G3 YZS R T Jackson, 29 Olivia Dr, Leigh on Sea, SS9 3EF [JO01HN, TQ88]
G3 YZT A M Slaney, St. Austel, High St., Upper Beeding, Steyning, BN44 3TQ [IO90UV, TQ11]
G3 YZU T Smith, 8 Oakdean Pl, Nairn, IV12 4TU [IO87BN, NH85]
G3 YZW A S Armstrong, 423 Bideford Green, Linslade, Leighton Buzzard, LU7 7TY [IO91PW, SP92]
G3 YZX H Brindle, 3 Cliff Farm Approach, 94 Victoria Ave, Shanklin, Isle of Wight, PO37 6QW [IO90JP, SZ58]
G3 YZY H D Brindle, 7 The Peregrines, Birdwood Gr, Fareham, PO16 8QU [IO90KU, SU60]
G3 YZZ K G Beverstock, 16 Chaucer Cl, Emmer Green, Reading, RG4 8PA [IO91ML, SU77]
G3 ZAC R G Butler, 2 Grace Meadow, Whitfield, Dover, CT16 3HA [JO01PD, TR34]
G3 ZAE S J Benstead, 5 Town Farm Cttgs, Blackawton, Totnes, TQ9 7BU [IO80DI, SX85]
G3 ZAF Details withheld at licensee's request by SSL.
G3 ZAG B G Taylor, 27 Ridgeway, Wellingborough, NN8 4RU [IO92PH, SP86]
G3 ZAJ D M Sutton, Deer Wood, Canterbury Rd, Challock, Ashford, TN25 4DF [JO01KF, TR05]
G3 ZAL L G Huggett, 24 Hockers Ln, Detling, Maidstone, ME14 3JN [JO01GG, TQ75]
G3 ZAR P Rubinstein, 18 Westland Rd, Kirk Ella, Hull, HU10 7PJ [IO93SR, TA02]
GM3 ZAS S Norris, Glenleith, Shaw Rd, Prestwick, KA9 2LP [IO75QM, NS32]
G3 ZAT D Cox, 78 Church St. West, Pinxton, Nottingham, NG16 6PU [IO93IC, SK45]
G3 ZAU E M Lord, 41 Daven Rd, Congleton, CW12 3RB [IO83VD, SJ86]
G3 ZAV R J Mills, 45 High Price Cl, Murston, Sittingbourne, ME10 3AS [JO01JI, TQ96]
G3 ZAW K W Bird, Leys House, Main St., Upper Stowe, Northampton, NN7 4SH [IO92LE, SP65]
G3 ZAY M J Atherton, 41 Enniskillen Rd, Cambridge, CB4 1SQ [JO02BF, TL46]
G3 ZAZ F W Mills, 9 Lansdown Hill, Fulwood, Preston, PR2 3ND [IO83PT, SD53]
G3 ZBA Dr S Sefton, 8 Sandmoor Ave, Leeds, LS17 7DW [IO93FU, SE34]
GW3 ZBB S Jackson, 6 Anderson Cl, Epsom, KT19 8LY [IO91UI, TQ16]
G3 ZBC R H Conway, 40 Chiltern Ave, Northampton, NN5 6AP [IO92MF, SP76]
G3 ZBD D W E Owen, 29 Longley Rd, Sanford Hill, Longton, Stoke on Trent, ST3 1AT [IO82WX, SJ94]
G3 ZBE Details withheld at licensee's request by SSL.
G3 ZBF M J Matthews, 13 Bursill Cres, Ramsgate, CT12 6EZ [JO01QI, TR36]
G3 ZBG G A Moorfield, 17 Allendale Ave, Findon Valley, Worthing, BN14 0AH [IO90TU, TQ10]
G3 ZBI K Frankcom Nunsfield House, 1 Chesterton Rd, Spondon, Derby, DE21 7EN [IO92HW, SK43]
G3 ZBM W D C Worthington, 2 Hardy Cl, Wistaston Grange, Crewe, CW2 8DY [IO83SB, SJ65]
G3 ZBP M C Baker, 82 Folkestone Rd, Copnor, Portsmouth, PO3 6LR [IO90LT, SU60]
G3 ZBS J S McCall, 5 Sundew Cl, Wokingham, RG40 5YB [IO91OK, SU86]
G3 ZBU A P R Watt, 5 Brambling Rd, Horsham, RH13 6AX [IO91UB, TQ13]
G3 ZBW J M Ruscoe, Sandon, Stourbridge Rd, Wombourne, Wolverhampton, WV5 9BN [IO82VM, SO89]
G3 ZBZ B Cross, 176 Outwood Rd, Heald Green, Cheadle, SK8 3LL [IO83VI, SJ88]
G3 ZCA D Lake, 9 Grafton Cl, Kings Lynn, PE30 3EZ [JO02FS, TF62]
G3 ZCC Details withheld at licensee's request by SSL.
G3 ZCD R Fogg, Langton House, 46 Alma Rd, Windsor, SL4 3HA [IO91QL, SU97]
GW3 ZCF R J Trebilcock, 15 Broadmead Cres, Bishopston, Swansea, SA3 3BA [IO71XN, SS58]
G3 ZCG K J Young, 101 Thorncombe Cl, Canford Heath, Poole, BH17 9EF [IO90AS, SZ09]
G3 ZCH D W Hill, 11 Chapeltown Rd, Radcliffe, Manchester, M26 1YF [IO83UN, SD70]
G3 ZCI R E O'Brien, 9 Holmwood Garth, Hightown, Ringwood, BH24 3DT [IO90CU, SU10]
G3 ZCJ M R D Allerton Austin, 13 Kilpin Green, North Crawley, Newport Pagnell, MK16 9LZ [IO92QC, SP94]
GI3 ZCK G A Ward, 55 Claremont Ct, Belfast, BT9 6AP [IO74AO, J37]
G3 ZCL G E Hammersley, 74 Cammel Rd, West Parley, Ferndown, BH22 8SB [IO90BS, SZ09]
GD3 ZCM Details withheld at licensee's request by SSL.
GD3 ZCN A R Smyth, 20 Alberta Dr, Onchan, Douglas, IM3 1LS
G3 ZCR K J Winstanley, 33 Radnor Ave, Harrow, HA1 1SB [IO91TO, TQ18]
G3 ZCS M W R Fabb, 280 Horns Rd, Ilford, IG6 1BS [JO01BN, TQ48]
G3 ZCT J N P Beehlar, 12 Dulverton Rd, Leicester, LE3 0SA [IO92KP, SK50]
G3 ZCU Details withheld at licensee's request by SSL.[Station located near Orpington.]
G3 ZCV N F Harper, 42 Saxville Rd, St. Pauls Cray, Orpington, BR5 3AW [JO01BJ, TQ46]
G3 ZCW H J J Berridge, Wellington College, Crowthorne, Berks, RG11 7PU [IO91LJ, SU66]
G3 ZCX P T Fort, 1 Lowther Ln, Foulridge, Colne, BB8 7JY [IO83VV, SD84]
G3 ZCZ J E Kasser, POB 3419, Silver Spring, Md 20901, USA
G3 ZDB T N J Archard, 43 Pine Hill, Epsom, KT18 7BH [IO91UH, TQ25]
G3 ZDD D I Dewey, Ivordere, Catteshall Ln, Godalming, GU7 1LW [IO91QE, SU94]
G3 ZDF J J Kirk, 111 Stockbridge Rd, Donnington, Chichester, PO19 2QR [IO90OT, SU80]
G3 ZDG S R Cole, Saxon Croft, Cholderton Rd, Grateley, Andover, SP11 8LH [IO91EE, SU24]
GM3 ZDH R A Dixon, 2 Maidens, East Kilbride, Glasgow, G74 4RS [IO75VS, NS65]
G3 ZDM R C Muriel, 13 York Rd, Sale, M33 6EZ [IO83UK, SJ79]
G3 ZDO Details withheld at licensee's request by SSL.
G3 ZDP F Lee, 100 Mayswood Rd, Solihull, B92 9JE [IO92CK, SP18]
G3 ZDQ I Flemming, Rudderhams Cottage, Blandford Hill, Winterborne Whitechurch, Blandford Forum, DT11 0AA [IO80VT, ST80]
G3 ZDR W C Stampton, 19 Claremont Pl, Cutmore St., Gravesend, DA11 0PQ [JO01EK, TQ67]
G3 ZDT P L Morrison, Saddlers, Green Rd, Shipbourne, Tonbridge, TN11 9PL [JO01DF, TQ55]
G3 ZDU G C Marshall, Bassington, Hulne Park, Alnwick, NE66 3JE [IO95CK, NU11]
G3 ZDV Details withheld at licensee's request by SSL.
G3 ZDW R Hyde, 25 The Pastures, Cottesmore, Oakham, LE15 7DZ [IO92QR, SK91]
G3 ZDY D G Palmer, Acuda, S.L., Balmes,69, 3 2, 08007 Barcelona, Spain, X X
G3 ZDZ J M Halton, 8 Linden Way, High Ln, Stockport, SK6 8ET [IO83XI, SJ98]
GM3 ZEA W Dickie, Rayston, Carsphairn, Castle Douglas, Kirkcudbrightshire, DG7 3TQ [IO75UF, NX59]
G3 ZEB B J Robinson, The Gables, The St., West Somerton, Great Yarmouth, NR29 4EA [JO02TR, TG41]
G3 ZED A Rothwell, Brandon, Manor Brow, Keswick, CA12 4AP [IO84KO, NY22]
G3 ZEF R C Stephen, 97 Hunters Field, Stanford in The Vale, Faringdon, SN7 8ND [IO91FP, SU39]
G3 ZEG P R Bennett, 259 Leicester Rd, Mountsorrel, Loughborough, LE12 7DD [IO92KR, SK51]
G3 ZEH Details withheld at licensee's request by SSL.
G3 ZEJ R P Smith, Upcott Manor, Rackenford, Tiverton, Devon, EX16 8EA [IO80EW, SS81]
G3 ZEK M P Bailey, 12 Bridgers Mill, Haywards Heath, RH16 1TF [IO91WA, TQ32]
G3 ZEM R Henderson, Whitwell House, Whitwell on The Hill, York, YO6 7JJ [IO94NC, SE76]
G3 ZEN Dr A N Glaser, 155 Little Breach, Chichester, PO19 4UA [IO90OU, SU80]
G3 ZEO S J Wilders, The Old Post Office, Lasham, Alton, Hants, GU34 5SD [IO91LE, SU64]
G3 ZEP R H Tinning, 18 Ullswater Rd, Chester-le-Street, DH2 3HG [IO94EU, NZ25]
G3 ZEQ M J Brenig-Jones, Orchard House, Larters Ln, Middlewood Green, Stowmarket, IP14 5HB [JO02MF, TM06]
G3 ZER I C Mercer, 28 West Way, Rickmansworth, WD3 2EN [IO91SP, TQ09]
G3 ZES A Downing, 3 Green Meadows, Danbury, Chelmsford, CM3 4LD [JO01HR, TL70]
GM3 ZET T A Goodlad Lerwick Radio Club, 72 North Lochside, Lerwick, ZE1 0PJ [IP90KD, HU44]
GM3 ZEU C J Clarkson, 8 Moor Pl, Portlethen, Aberdeen, AB1 4TF [IO87TH, NJ72]
G3 ZEV Details withheld at licensee's request by SSL.
GD3 ZEX C G Douglas, Ballacowin Cottage, Glen Rd, Laxey, IM4 7AP
G3 ZEZ G N Coleman, 16 Kestrel Way, Clacton on Sea, CO15 4JE [JO01NT, TM11]
G3 ZFC C Davis, 3 Cross Rd, Haslington, Crewe, CW1 5SY [IO83TC, SJ75]
G3 ZFJ Details withheld at licensee's request by SSL.
G3 ZFP R Penberthy, Sunnyridge, Valley Rd, Studham, Dunstable, LU6 2NN [IO91RT, TL01]
G3 ZFQ J E Hall, 2 Sheridan Cl, Narborough, Leicester, LE9 5QW [IO92JO, SP59]
G3 ZFR R G Harris, 44 Burbages Ln, Longford, Coventry, CV6 6AY [IO92FK, SP38]
G3 ZFT A U Magnus, Woodland Cottage, Linkside East, Hindhead, GU26 6NY [IO91OD, SU83]
GM3 ZFU T S Rutherford, 15 Lade Ct, Lochwinnoch, PA12 4BT [IO75QT, NS35]
G3 ZFV J V Watts, Riverside, St. Georges Rd, Barnstaple, EX32 7AS [IO71XC, SS53]
G3 ZFX B R Cornwall, Hoadswood, Battle Rd, St. Leonards on Sea, TN37 7BS [JO00GV, TQ71]
G3 ZFZ G Gibson, 174 Roose Rd, Barrow in Furness, Cumbria, LA13 0EE [IO84JC, SD26]
G3 ZGA J E Hart, 146 St. Ln, Leeds, LS8 2AD [IO93FU, SE33]
G3 ZGC R T Jolliffe, 54 Glendale Ave, Wash Common, Newbury, RG14 6RU [IO91HI, SU46]
G3 ZGF B M Bailey, 5 Davenham Cl, Roborough, Plymouth, PL6 6BT [IO70WK, SX46]
G3 ZGG P M Storey, 9 Chadleigh Ln, Godmanchester, Huntingdon, PE18 8AL [IO92VH, TL27]
GM3 ZGH R W Yeoman, 162 Jamphlars Rd, Cardenden, Lochgelly, KY5 0ND [IO86ID, NT29]
G3 ZGI T A Oneill, 15 Ridgeway, West Parley, Ferndown, BH22 8TT [IO90BS, SZ09]
G3 ZGJ M S Wigg, 15 Garden View, Branksome Wood Rd, Bournemouth, BH4 9LA [IO90BR, SZ09]
G3 ZGN P J O Swarbrick, 1 Hill View, Charminster, Dorchester, DT2 9QX [IO80SR, SY69]
G3 ZGP R D Cridland, 13 Clarendon Ave, Redlands, Weymouth, DT3 5BG [IO80SP, SY68]
G3 ZGQ L H Mead, 7 Ferniefields, High Wycombe, HP12 4SP [IO91OO, SU89]
G3 ZGR H W Brownlee, 36 Bowness Ave, Battle Hill Est, Wallsend, NE28 9TB [IO95FA, NZ36]
G3 ZGT B Druce, Poplar Farm, Holme Rd, Spaldington, Goole, DN14 7NA [IO93OS, SE73]
G3 ZGU K G Richens, 12 Sambrook Cres, Market Drayton, TF9 1NG [IO82SV, SJ63]
G3 ZGY G C Paddock, 56 Clee View Rd, Wombourne, Wolverhampton, WV5 0BD [IO82VM, SO89]
G3 ZGZ D Woodhall, PO Box 8157, Edleen, South Africa 1625
G3 ZHA G F Gillam, 58 Downhall Rd, Rayleigh, SS6 9LY [JO01HO, TQ89]
G3 ZHB A C Stuart, 207 Saunders Ln, Mayford, Woking, GU22 0NT [IO91QH, SU95]
G3 ZHC N H Willmot, 2 Athlone Rd, Walsall, WS5 3QX [IO92AN, SP09]
G3 ZHE A Heyes, 20 Walsingham Rd, Penketh, Warrington, WA5 2AQ [IO83QJ, SJ58]
G3 ZHH R P Flowerday, 26 Wyckham Cl, Harborne, Birmingham, B17 0TB [IO92AK, SP08]
G3 ZHI I Abel, 52 Holly Tree Ave, Maltby, Rotherham, S66 8DY [IO93JK, SK59]
G3 ZHK T G Kellow, Glenvale, St. Dominick, Saltash, Cornwall, PL12 6TD [IO70UL, SX36]
G3 ZHL Dr J G Morgan, Cedars, Springhill, Longworth, Abingdon, OX13 5HL [IO91GQ, SU39]
G3 ZHO R F K Wilton, 8 Belgrave Terr, Addington, Liskeard, PL14 3EZ [IO70SL, SX26]
G3 ZHP D Marsden, 30 Sycamore Rise, Wooldale, Holmfirth, Huddersfield, HD7 2TJ [IO93CN, SE10]
GW3 ZHQ D M Rosser, 126 Clodien Ave, Heath, Cardiff, CF4 3NQ [IO81JM, ST17]
G3 ZHR K McGuckian, The Spinney, 2 Rock Ln, Warminster, Wilts, BA12 9JZ [IO81WE, ST84]
G3 ZHS R J Ray, 37 Doxey Fields, Stafford, ST16 1HJ [IO82WT, SJ82]
G3 ZHT B H Lundean, 13 Isis Cl, Lympne, Hythe, CT21 4JQ [JO01MB, TR13]
G3 ZHU G S Clark, Kenzie, Canterbury Rd, Swingfield, Dover, CT15 7HR [JO01OD, TR24]
G3 ZHV R S Bowler, 21 Pine Cl, South Wonston, Winchester, SO21 3EB [IO91IC, SU43]
G3 ZHW Details withheld at licensee's request by SSL.
G3 ZHX I E Somers, 4 Priory Cres, Aylesbury, HP19 3NU [IO91OT, SP81]
G3 ZHY M R Irving, 22 Wheatley Way, Chalfont St. Peter, Gerrards Cross, SL9 0JE [IO91RO, TQ09]
G3 ZHZ A D Macfadyen, 19 Oldfield Rd, Lower Willingdon, Eastbourne, BN20 9QD [JO00CT, TQ50]
G3 ZIB D F Tye, 21 Elmstone Dr, Tilehurst, Reading, RG31 5NS [IO91LL, SU67]
G3 ZIC J R Viney, Glebe Meadow, Hammersley Ln, Penn, High Wycombe, HP10 8HB [IO91PP, SU99]
G3 ZID A P Greathead, 20 Westland, Martlesham Heath, Ipswich, IP5 7SU [JO02OB, TM24]
G3 ZIE E M E Brown, 78 Summerhayes, Great Linford, Milton Keynes, MK14 5EX [IO92OB, SP84]
G3 ZIF H Wilson, 7 Westfield Dr, Skelmanthorpe, Huddersfield, HD8 9AN [IO93EO, SE21]
G3 ZIG R Reed, Oak Cottage, Dereham Rd, Bawdeswell, Dereham, NR20 4AA [JO02MR, TG02]
G3 ZII M Rathbone, 25 Halsall Rd, Birkdale, Southport, PR8 3DB [IO83LO, SD31]
G3 ZIJ J D Stables, 9 Milbanke Cl, Ouston, Chester-le-Street, DH2 1JJ [IO94EV, NZ25]
G3 ZIK A Mather, 15 Claughton Ave, Bolton, BL2 6US [IO83TN, SD70]
G3 ZIL G Griffiths, 14 Bassett Cl, Southampton, SO16 7PE [IO90HW, SU41]
G3 ZIN G S G Spencer, 16 West Lawn, Ipswich, IP4 3LJ [JO02OB, TM14]
G3 ZIO C H Harvey, Ham Cottage, 1A Elstan Way, Croydon, CR0 7PR [IO91XJ, TQ36]
G3 ZIV K J Nolan, West End Cottage, Woodhall, Selby, North Yorks, YO8 7TG [IO93MS, SE63]
G3 ZIY Details withheld at licensee's request by SSL.
G3 ZJB D J Brushwood, 142 Newbolt Rd, Paulsgrove, Portsmouth, PO6 4NT [IO90KU, SU60]
G3 ZJE K S Daniel, 49 Lidford Tor Ave, Paignton, TQ4 7ED [IO80EK, SX85]
G3 ZJF P N Broughton, 2 Arretine Cl, St. Albans, AL3 4JL [IO91TR, TL10]
G3 ZJG J D Garner, 50 Thorndale, Ibstock, Leicester, LE67 6JT [IO92HQ, SK41]
G3 ZJH P J Brouder, Aldebaran, 169 North Rd, Stoke Gifford, Bristol, BS12 6PH [IO81RM, ST68]
G3 ZJJ M J Peet, 31 White St., West Lavington, Devizes, SN10 4LP [IO91AG, SU05]
G3 ZJK C J L Milner, The Everglades, Sawbridge Rd, Grandborough, Rugby, CV23 8DN [IO92IH, SP46]
G3 ZJO E Bennett, 44 Central Ave, Whitehills, Northampton, NN2 8DZ [IO92NG, SP76]
G3 ZJP W N Fenton, 50 Orion Rd, Rochester, ME1 2UH [JO01GI, TQ76]
G3 ZJQ Dr R S Walker, 2 Chelwood Dr, Little Sandhurst, Sandhurst, Camberley, Surrey, GU17 8HT [IO91OH, SU85]
G3 ZJS J W Smith, 21 The Oval, Oadby, Leicester, LE2 5JB [IO92KO, SP69]
G3 ZJT R Clayton, c/o C.W.A.O, 9th Signals Regement, B.F.P.O 59
G3 ZJU D E W Melton, Winzer Lodge, Lynton Rd, Combe Martin, Ilfracombe, EX34 0NA [IO71XE, SS54]
G3 ZJV M J Firth, 63 Sycamore Rd, East Leake, Loughborough, LE12 6PP [IO92JT, SK52]
G3 ZJW Dr B H McCombe, 208 Thorpe Rd, Peterborough, PE3 6LB [IO92UN, TL19]
G3 ZJX B A Castle, Castles Nursery, 159 Elmers End Rd, Beckenham, BR3 4SZ [IO91XJ, TQ36]
G3 ZJY J F Greenwood, 91 Keyhaven Rd, Milford on Sea, Lymington, SO41 0TF [IO90FR, SZ29]
G3 ZJZ J P Mason, 35 Broad Way, Hockley, SS5 5EL [JO01IO, TQ89]
G3 ZKD W L Ball, 6 Coronation Dr, Penketh, Warrington, WA5 2DD [IO83QJ, SJ58]
G3 ZKE B C Bond, 86 Agar Gr, Camden Town, London, NW1 9TL [IO91WN, TQ28]
G3 ZKG J Riley, 41 Church Ave, Sleekburn, West Sleekburn, Choppington, NE62 5XF [IO95FD, NZ28]
G3 ZKH I Bateman, 1 Scots Dr, Wokingham, RG41 3XF [IO91NK, SU76]
G3 ZKI A B Williams, 38 Seneca St., St. George, Bristol, BS5 8DX [IO81RK, ST67]
G3 ZKL P J Scanlon, 174 Lostock Rd, Davyhulme, Urmston, Manchester, M41 0TB [IO83TK, SJ79]
G3 ZKM A Stewart, 39 Longfellow Rd, Maldon, CM9 6BD [JO01IR, TL80]
G3 ZKN D S Morgan, 9 Summerland Dr, Churchdown, Gloucester, GL3 2LZ [IO81VV, SO82]
G3 ZKO P F Lee, Hollyhock Cottage, Northmoor, Oxon, OX8 1SX [IO91HR, SP40]
G3 ZKQ A R Walton, 3 Fox Hill Cl, Selly Oak, Birmingham, B29 4AH [IO92AK, SP08]
G3 ZKS F W Webb, 37 Alwyne Gr, York, YO3 6RT [IO93KX, SE55]
GI3 ZKT T E Harding, 15 Rutherglen Park, Bangor, BT19 1DX [IO74DP, J48]
G3 ZKW A C Freer, 27 Atherstone Rd, Loughborough, LE11 2SH [IO92JS, SK51]
G3 ZKX Dr E Cronin, 4 Arundel Lodge, Salisbury Ave, Finchley, London, N3 3AL [IO91VO, TQ29]
G3 ZKZ J G Shaw, The Shieling, 13 Mill Ln, Barrow, Bury St. Edmunds, IP29 5BS [JO02HF, TL76]
G3 ZLB S H Holt, Flat 3, 3 Blenheim Ct, Marlborough Rd, Bournemouth, BH4 8DH [IO90BR, SZ09]
GM3 ZLC Details withheld at licensee's request by SSL.
G3 ZLD J M Barker, 1 Quay West, Ferry Rd, Teddington, Middx, TW11 9NH [IO91UK, TQ17]
G3 ZLE Me D J Ward, 54 Duncan Dr, Georgetown, Ontario, Canada, L7G 4MI
G3 ZLF R Nelson, 225 Walton Rd, Chesterfield, S40 3BT [IO93GF, SK36]
G3 ZLJ E D Dalton, 29 Windmill Ln, Castlecroft, Wolverhampton, WV3 8HJ [IO82VN, SO89]
G3 ZLM R N Hook, 35 Parkwood Cres, Hucclecote, Gloucester, GL3 3JH [IO81VU, SO81]
G3 ZLN D W N Thomas, 6 Preston Dr, Ipswich, IP1 6DR [JO02NB, TM14]
G3 ZLQ M J Adams, North Lodge, Wavering Ln, Gillingham, Dorset, SP8 5NH [IO81UA, ST72]
G3 ZLR A Ridley, 31 Racecourse Dr, Bridgnorth, WV16 4NR [IO82SM, SO79]
G3 ZLS S A Craske, Patent Law Chambers, 15 Queens Terr, Exeter, EX4 4HR [IO80FR, SX99]
G3 ZLX E G Jones, 94 Westbrook End, Newton Longville, Milton Keynes, MK17 0BX [IO91OX, SP83]
G3 ZLY V C Bird, 32 Benham Cl, Hope St., Battersea, London, SW11 2AY [IO91VL, TQ27]
GM3 ZMA J M Butler, Kirkholm, 11 Castle House, Stuartfield, Peterhead Aberdeens, AB4 8DE [IO87TH, NJ72]
G3 ZMB W L N Webster, 33-459 Carry Dr SE, Medicine Hat, Alberta Tib 3W1, Canada
G3 ZMC Details withheld at licensee's request by SSL.
G3 ZMD J S Reed, The Firs, Gloucester Rd, Upleadon, Newent, GL18 1EH [IO81TW, SO72]
G3 ZME M I Vincent Telford&Dist AR, 9 Sleapford, Long Ln, Telford Shropshire, TF6 6HQ [IO82RR, SJ61]
G3 ZMF Details withheld at licensee's request by SSL.
G3 ZMG J Maughan, 40 Windsor Dr, New Silks Worth, New Silksworth, Sunderland, SR3 1JQ [IO94HU, NZ35]
G3 ZMH D N McAuslan, Golden Sedge, St. End, Blagdon, Bristol, BS18 6TL [IO81PH, ST45]
G3 ZMK J W Askew, 29 Ivy Ave, South Shore, Blackpool, FY4 3QF [IO83LS, SD33]
G3 ZML M Owen, 10 Knightlow Rd, Birmingham, B17 8QB [IO92AL, SP08]
G3 ZMM R C Hodgkinson, 39 Oxford Rd, Carlton in Lindrick, Worksop, S81 9BD [IO93KI, SK58]
G3 ZMN Dr M J Turner, 14 Purley Bury Cl, Purley, CR8 1HU [IO91WI, TQ36]
G3 ZMO J A Callum, 31 Lambton Ct, High Rickleton, Washington, NE38 9HE [IO94FV, NZ25]
G3 ZMP R Beckerleg, 16 Lanuthnoe Est, St. Erth, Hayle, TR27 6HU [IO70GD, SW53]
G3 ZMQ J T Pickles, 38 Dean Beck Ave, Bradford, BD6 1DE [IO93DS, SE13]
G3 ZMS C R Cook Mid Sussex ARS, 45 Stonepound Rd, Hassocks, BN6 8PR [IO90WW, TQ31]
G3 ZMT J W Talboys, 33 Pilson Cl, Castle Bromwich, Birmingham, B36 8NB [IO92CM, SP18]
G3 ZMV L Wigglesworth, 23 Ravenswood Ave, Rockferry, Birkenhead, L42 4NY [IO83LI, SJ38]
G3 ZMX F E Jackson, 1 Harley Villas, Victoria Rd, Todmorden, OL14 5JB [IO83WR, SD92]
G3 ZMY Details withheld at licensee's request by SSL.[Op: J A James, 13 Hillside Gdns, Berkhamsted, HP4 2LE.]
G3 ZNB P D Hannam, Bogg Hall, Oulston, York, YO6 3RE [IO94KD, SE57]
GM3 ZNC J Mulheron, 10 Devonview Pl, Airdrie, ML6 9DF [IO85AU, NS76]
GD3 ZND J R Bartlett, Chickamauga, Deans Grange, Blackrock, Dublin, ZZ9 9CH
G3 ZNE N P Ingle, 81 Redmoor Cl, Tavistock, PL19 0ER [IO70WN, SX47]
G3 ZNF B M Whitford, 45 Chapel St., Shepshed, Loughborough, LE12 9AF [IO92IS, SK41]
G3 ZNG S G Birt, 82 Aintree Rd, Thornton Cleveleys, FY5 5HP [IO83LU, SD34]
G3 ZNH R J Coombes, 54 Kelsey Rd, Salisbury, SP1 1LA [IO91CB, SU13]
G3 ZNI M J Hawkins, Sandywood, Woodside Rd, Cobham, Surrey, KT11 2QR [IO91TH, TQ16]
G3 ZNK D Hainsworth, 48 Greenacre Park, Rawdon, Leeds, LS19 6AR [IO93DU, SE24]
GM3 ZNM S Henry, The Brakes, Upper Sound, Lerwick, ZE1 0SU [IP90JD, HU43]
GW3 ZNN W G Meredith, 11 Bryn y Glyn, Rhosddu, Wrexham, LL11 2HR [IO83LB, SJ35]
G3 ZNP A Mussell, 57 Bowland Ave, Childwall, Liverpool, L16 1JW [IO83NJ, SJ48]
G3 ZNQ S Valentine, 1 Ridings Cl, Doncaster, DN4 6UZ [IO93LM, SE60]
G3 ZNR D A Bailey, 4 Manor Park, Bradford, BD8 0LY [IO93CT, SE13]
G3 ZNT R Brown, 282 Luton Rd, Dunstable, LU5 4LF [IO91SV, TL02]
G3 ZNU M S Appleby, 6 Mandeville Rd, Prestwood, Great Missenden, HP16 9DS [IO91PQ, SP80]
G3 ZNV Capt J G Green, 23 Calton Gdns, Lyncombe Hill, Bath, BA2 4QG [IO81TJ, ST76]

G3 ZNW R J Blasdell, 32 Fulham Cl, Broadfield, Crawley, RH11 9NY [IO91VC, TQ23]
G3 ZNY F M Frisby, 42 Milford Ave, Stoney Stratford, Stony Stratford, Milton Keynes, MK11 1HE [IO92NB, SP73]
G3 ZOC J R Cunliffe, 12 Blenheim Cl, Lostock Hall, Preston, PR5 5YX [IO83PR, SD52]
G3 ZOD G J Smith, 10 Zurich Gdns, Bramhall, Stockport, SK7 3HZ [IO83WJ, SJ88]
G3 ZOE Details withheld at licensee's request by SSL.
G3 ZOF G Q Wheeler, 3 Holts Ln, Hilgay, Downham Market, PE38 0JG [JO02EN, TL69]
G3 ZOG A J Elliott, 15 Braemar Gdns, East Herrington, Sunderland, SR3 3PX [IO94GU, NZ35]
G3 ZOH B R George, 14 Pondfield Rd, Orpington, BR6 8HJ [JO01AI, TQ46]
G3 ZOI D J Deane, 10 Stephens Rd, Mortimer Common, Reading, RG7 3TU [IO91LJ, SU66]
G3 ZOL J R Powell, 156 Avon Way, Colchester, CO4 3YP [JO01LV, TM02]
G3 ZOM D R Pearson, 6 Fellows Ave, Wall Heath, Kingswinford, DY6 9ET [IO82VM, SO88]
G3 ZON K M Lacy, 9 Rhodes Way, Tilgate, Crawley, RH10 5DQ [IO91VC, TQ23]
G3 ZOP C R T Teasdale, 37 Woodland Rd, Hinckley, LE10 1JF [IO92HN, SP49]
G3 ZOQ G Foster, 3 Egerton Rd, Leyland, Preston, PR5 1YB [IO83PQ, SD52]
G3 ZOR J H Robertson, 173 Mongeham Rd, Great Mongeham, Deal, CT14 9LL [JO01QF, TR35]
GM3 ZOT J R Hewitt, 145 Queens Rd, Fraserburgh, AB43 9PU [IO87XQ, NJ96]
G3 ZOU R France, 25A Greenfield Rd, Middleton on The Wolds, Driffield, YO25 9UL [IO93RW, SE94]
G3 ZOW K W Clamp, 12 Cowlishaw Cl, Shardlow, Derby, DE72 2GS [IO92HU, SK43]
G3 ZOX P J Durham, 25 Tibbott Walk, Stockwood, Bristol, BS14 8DR [IO81RJ, ST66]
G3 ZOZ Dr D M H Cogman, 42 Parkside, London, NW7 2LP [IO91VO, TQ29]
G3 ZPA D O White, Rose Cottage, Whaddon Rd, Shenley Brook End, Bletchley Bucks, MK5 7AF [IO92OA, SP83]
G3 ZPB P L A Burton, Pound Cottage, 202 Coulsdon Rd, Coulsdon, CR5 2LF [IO91WH, TQ35]
G3 ZPD J E Rushforth, 23 Mountain View, Cockermouth, CA13 0DG [IO84HP, NY13]
G3 ZPE F J Raby, 20 Lime Tree Rd, Codsall, Wolverhampton, WV8 1NT [IO82VP, SJ80]
G3 ZPF Details withheld at licensee's request by SSL.
G3 ZPG R Shipman, 3 The Wranglands, Fleckney, Leicester, LE8 8TW [IO92LM, SP69]
G3 ZPI G S Braund, 184 Faversham Rd, Kennington, Ashford, TN24 9AE [JO01KD, TR04]
G3 ZPJ M Symons, 11 Tudor Lodge Parc, Crowlas, Penzance, Cornwall, TR20 9BW [IO70GD, SW53]
G3 ZPK H A Willis, 12 Combe View, Hungerford, RG17 0BZ [IO91FJ, SU36]
G3 ZPL Dr N Richardson, 1734 Webster St, Palo Alto, California 94301, USA, X X
G3 ZPM A H Stormont, The Hawthorns, Church Ln, Grainthorpe, Louth, LN11 7JR [JO03BK, TF39]
G3 ZPQ Details withheld at licensee's request by SSL.
G3 ZPR D I Mason, 26 Upton Rd, Fleetsbridge, Poole, BH17 7AH [IO90AR, SZ09]
G3 ZPW M D D Brown, Hope Farm, Trew, Breagell, nr Helston Cornwall, TR13 9NE [IO70IC, SW62]
G3 ZQA R O Rowntree, 78 Acorn Way, Woodthorpe, York, YO2 2RP [IO93KW, SE54]
G3 ZQB A R Seabrook, 63 St. Annes Rd, Willingdon, Eastbourne, BN20 9NJ [JO00DT, TQ50]
G3 ZQC J Smith, 2 Spindlewood Cl, Bassett, Southampton, SO16 3QD [IO90HW, SU41]
G3 ZQE D E Denny, 3 Sunnyway, Bosham, Chichester, PO18 8HQ [IO90NT, SU80]
G3 ZQF S V Carpenter, 4 Mount Ct, Hawes Ln, West Wickham, BR4 9AH [IO91XI, TQ36]
G3 ZQH Dr D A Barrett, 31 Kenilworth Rd, Beeston, Nottingham, NG9 2HR [IO92JW, SK53]
G3 ZQI B D Downer, 159 Howlands, Welwyn Garden City, AL7 4RL [IO91VS, TL21]
G3 ZQJ B A Stagg, 1 Naunton Way, Leckhampton, Cheltenham, GL53 7BQ [IO81XV, SO92]
G3 ZQL S Murray, Glenelg, 25 Prowses, Hemyock, Cullompton, EX15 3QG [IO80JV, ST11]
G3 ZQM Dr M N S Hill Tyneside ARS, Windrush, Jesmond Gdns, Newcastle upon Tyne, NE2 2JN [IO94EX, NZ26]
G3 ZQQ J A Peden Esq, 51A Bewdley Rd, Kidderminster, DY11 6RL [IO82UJ, SO87]
G3 ZQR B V Downton, Lickwith Cottage, Monkokehampton, Winkleigh, Devon, EX19 8SL [IO70XU, SS50]
G3 ZQS E Longden, 119 Cemetery Rd, Darwen, BB3 2LZ [IO83SQ, SD62]
G3 ZQT J Yu, 21 Langley Ave, Surbiton, KT6 6QN [IO91UJ, TQ16]
G3 ZQU M Goodrum, Church Ln, Stonham Parva, Little Stonham, Stowmarket, Suffolk, IP14 5JL [JO02NE, TM16]
G3 ZQV R Hague, 111 Mount Vernon Rd, Ward Green, Worsbrough, Barnsley, S70 4HH [IO93GM, SE30]
G3 ZQW B J Barrington, Pinlands Cottage, Bines Rd, Partridge Green, Horsham, RH13 8EQ [IO91UB, TQ12]
G3 ZQY N A K Clark, Chelsworth, Heaton Grange Rd, Gidea Park, Romford, RM2 5PP [JO01CO, TQ59]
G3 ZQZ R D Hartley, Holly Cottage, 43 Cromer Rd, Overstrand, Cromer, NR27 0NT [JO02QV, TG24]
GM3 ZRA Dr R D Elliot, 1321 East Bailey Rd, Naperville, Illinois 60565, USA, X X
G3 ZRB Details withheld at licensee's request by SSL.
GM3 ZRC J F Gray Greenock & DARC, 47 South St., Greenock, PA16 8QG [IO75OW, NS27]
G3 ZRE P J Ottewell, 65 The Martindales, Clayton-le-Woods, Chorley, PR6 7TJ [IO83QQ, SD52]
G3 ZRG I C Steward, 23 Orchard Dr, Watford, WD1 3DX [IO91TP, TQ09]
G3 ZRH A J Stokes, 34 Shenfield Cres, Brentwood, CM15 8BW [JO01DO, TQ69]
G3 ZRJ Details withheld at licensee's request by SSL.
G3 ZRK Details withheld at licensee's request by SSL.
G3 ZRL A McWatters, 38 Sutherland Mount, Harehills, Leeds, LS9 6DP [IO93FT, SE33]
G3 ZRM M Payne, 3 Waterside Cl, Bordon, GU35 0HB [IO91NC, SU83]
G3 ZRN D F Catherwood, Avondale, 14 Hatton Ln, Hatton, Warrington, WA4 4BY [IO83QI, SJ68]
G3 ZRP R C Perry, Little London Cottage, Little London, North Kelsey, Market Rasen, LN7 6JP [IO93SM, TA00]
G3 ZRQ D Maxfield, 40 Fegg Hayes Rd, Stoke on Trent, ST6 6RA [IO83VB, SJ85]
G3 ZRR Details withheld at licensee's request by SSL.
G3 ZRS P Rodmell, Field Head, Leconfield Rd, Leconfield, Beverley, HU17 7LU [IO93SU, TA04]
GM3 ZRT W Strachan, Sparnel, Cemetery Rd, Galston, KA4 8LL [IO75TO, NS53]
G3 ZRV Details withheld at licensee's request by SSL.
G3 ZRW C Whittaker, 76 Priory Dr, Darwen, BB3 3PT [IO83SQ, SD72]
G3 ZRX T I Lundegard The Raynet Association, Saxby, Botsom Ln, West Kingsdown, Sevenoaks, TN15 6BL [JO01DI, TQ56]
G3 ZRY G Stott, 8 Willow Rd, Chinnor, OX9 4RA [IO91NQ, SP70]
G3 ZRZ I M Cobbe, 24 Rossendale Ave North, Thornton Cleveleys, FY5 4NT [IO83LV, SD34]
G3 ZSA R D Armitage, 621 Halifax Rd, Bradford, BD6 2QS [IO93CS, SE12]
G3 ZSB V A Poore, 216 Powder Mill Ln, Twickenham, TW2 6EJ [IO91TK, TQ17]
GI3 ZSC K A McGonigal, 3 Glendale Gdns, Randalstown, Antrim, BT41 3EJ [IO64US, J09]
G3 ZSE Details withheld at licensee's request by SSL.
G3 ZSF A Houltby, 2 Sinderson Rd, Humberston, Grimsby, DN36 4UF [IO93XM, TA30]
GM3 ZSH J Donaldson, 13 Cromarty Dr, Milton, Invergordon, IV18 0PY [IO77XR, NH77]
G3 ZSI R Davey, 46 Swanborough Dr, Brighton, BN2 5PH [IO90WT, TQ30]
G3 ZSK J P Tysiorowski Merlin Skeleton Club, Merlin Skelton Club, Merlin Communications Inter, Skeleton Pastures, Penrith, Cumbria, CA11 9SY [IO84NR, NY43]
G3 ZSQ R Dunham, 42 Marsdale Dr, Stockingford, Nuneaton, CV10 7DE [IO92GM, SP39]
G3 ZSS P W Bacon, Highview, Monks Well, Farnham, Surrey, GU10 1RH [IO91OE, SU84]
G3 ZST T K Surgey, The Littlewood, Windmill Ln, Ladbroke, Leamington Spa, CV33 0BN [IO92HF, SP45]
G3 ZSU Details withheld at licensee's request by SSL.
G3 ZSX Dr K H Craig, 20 Alexander Cl, Abingdon, OX14 1XA [IO91IQ, SU59]
G3 ZSZ R James, 283 High St., New Whittington, Chesterfield, S43 2AP [IO93HG, SK47]
GM3 ZTA W Currie, 28B Watchmeal Cres, Faifley Clydebank, Clydebank, G81 5ED [IO75TW, NS57]
G3 ZTB Dr R G Ranson, 14 Eskdale Rd, Southbents, Sunderland, SR6 8AN [IO94HW, NZ46]
G3 ZTC Dr R B Pearce, 40 Northfield Rd, Headington, Oxford, OX3 9EW [IO91JS, SP50]
G3 ZTF G T Sparkes, 127 Redford Ave, Horsham, RH12 2HH [IO91TB, TQ13]
GI3 ZTG D McWhirter, 71 Main St., Sixmilecross, Omagh, BT79 9NL [IO64KN, H56]
GW3 ZTH J D V Ludlow, 44 Fox Hollows, Brackla, Bridgend, CF31 2NG [IO81FM, SS98]
G3 ZTI K W Marshall, 2 Keepers Mill, Woodmancote, Cheltenham, GL52 4QS [IO81XW, SO92]
G3 ZTJ C J A Morgan, The Villa, The Green, Wallsend, NE28 7PH [IO94FX, NZ36]
GI3 ZTL F Convery, 2 Coolagh Rd, Maghera, BT46 5JR [IO64PU, H89]
G3 ZTM L F W Lawes, 2 Ashley Dr, Walton on Thames, KT12 1JL [IO91TJ, TQ16]
G3 ZTP S A Elwell-Sutton, 30 Orchard Ct, Dundee, DD4 9DB [IO86ML, NO43]
G3 ZTR D Lockwood, 25 Thorntondale Dr, Bridlington, YO16 5GW [IO94VC, TA16]
G3 ZTT D A Bevan Mid Cheshire AR, 46 Park Ln, Hartford, Northwich, CW8 1PY [IO83RF, SJ67]
G3 ZTU J R Mace, Well Cottage, Tismans Common, Nr.Rudgwick, West Sussex, RH12 3DU [IO91SC, TQ03]
G3 ZTV P A S Webster, 2 Clabon Third Cl, Norwich, NR3 4HH [JO02PP, TG21]
GM3 ZTW R C Mitchell, 21 Loch Lea, Kirkmuirhill, Lanark, ML11 9ST [IO85AQ, NS74]
G3 ZTX P M Angell, Star Hill House, Star Hill, Nailsworth, Stroud, GL6 0NJ [IO81VQ, SO80]
G3 ZTY J Yale, 15 Rectory Ave, Corfe Mullen, Wimborne, BH21 3EZ [IO80XS, SY99]
G3 ZTZ P J Howell, 28 Buttermere Dr, Heatherside, Camberley, GU15 1RB [IO91PH, SU96]
G3 ZUA R E A Lawson, 163 Cole Green Ln, Welwyn Garden City, AL7 3JG [IO91VS, TL21]
G3 ZUB M J Davison, 43 Park Rd, Loughborough, LE11 2ED [IO92JS, SK51]
G3 ZUD L J Wicks, West Winds, The Fosse, Kinoulton, Nottingham, NG12 3ES [IO92LU, SK62]
G3 ZUE A E Nicholas, Verriotts Ln, Morecombelake, Morcombelake, Bridport, Dorset, DT6 6DU [IO80NR, SY49]
G3 ZUF R B Peters, 2 Pembroke Villas, The Green, Richmond, TW9 1QF [IO91UL, TQ17]
G3 ZUG C R Caunce, 12 Wallingford Rd, Handforth, Wilmslow, SK9 3JU [IO83VI, SJ88]
G3 ZUI M Johnson, Greentiles, Alford Rd, Maltby-le-Marsh, Alford, Lincs, LN13 0JW [JO03CH, TF48]
G3 ZUJ P J Felton, 3 Crediton Way, Claygate, Esher, KT10 0EB [IO91UI, TQ16]
G3 ZUK Dr R G Whitehead, Church End, Weston Colville, Cambridge, CB1 5PE [JO02ED, TL65]
G3 ZUL B Kennedy, 39 Redhill, Stourbridge, DY8 1NA [IO82WK, SO98]
G3 ZUM B L Lonnon, 5 Mickle Meadow, Water Orton, Birmingham, B46 1SN [IO92DM, SP19]
G3 ZUN D R Sharpe, 5 Scarletts Cl, Uckfield, TN22 2BA [JO00BX, TQ42]
G3 ZUO D E Ham, Holly Cottage, Spring Hill, Fordcombe, Tunbridge Wells, TN3 0SA [JO01CD, TQ54]
G3 ZUP J B Jobling, 14 Hawthorn Terr, Shilbottle, Alnwick, NE66 2XA [IO95DI, NU20]
G3 ZUQ R F Simms, 3 Saville Cres, Weston Super Mare, BS22 8PG [IO81MI, ST36]
G3 ZUS N Ewer, Springfields, 2 Astwick Barns, Croughton, Brackley, Northants, NN13 5LL [IO91JX, SP53]
G3 ZUT M E Thorne, Rossendale, Main Rd, Easter Compton, Bristol, BS12 3RE [IO81QM, ST58]
G3 ZUU J H Hammond, 1 Springhead Way, Crowborough, TN6 1LR [JO01BB, TQ52]
G3 ZUV B K Wilson, 29 Radcliffe Gdns, Carlton, Nottingham, NG4 1SB [IO92KX, SK64]
G3 ZUZ M E Fox, 41 Elgin Ave, Ashton in Makerfiel, Ashton in Makerfield, Wigan, WN4 0RH [IO83PL, SJ59]
G3 ZVD W K Allen, 1 Hollin Cl, Rossington, Doncaster, DN11 0XX [IO93LL, SK69]
GM3 ZVF Dr J L Swanston, 8 Townsend Cres, Kirkcaldy, KY1 1DN [IO86KC, NT29]
G3 ZVH D J Bedford, 64 St. James Ave, Upton, Chester, CH2 1NL [IO83NF, SJ46]
G3 ZVI P E Longhurst, 81 Collins Rd, Exeter, EX4 5DE [IO80FR, SX99]
G3 ZVJ Details withheld at licensee's request by SSL.
G3 ZVK J G Simons, 120 Bond Way, Pye Green, Cannock, WS12 4SN [IO82XR, SJ91]
G3 ZVM A Greenbank, Grahamsley, Westburn, Ryton, NE40 4EU [IO94CX, NZ16]
G3 ZVN G M Peck, 4 Koonowla Cl, Biggin Hill, Westerham, TN16 3BJ [JO01AH, TQ45]
G3 ZVO R Boyce, 63 Great Northern Rd, Dunstable, LU5 4BN [IO91RV, TL02]
G3 ZVQ J Bridge, 8 Highfield Gr, Lostockhall, Lostock Hall, Preston, PR5 5YB [IO83PR, SD52]
G3 ZVS E M K Park, Waterside Cottage, Bowden Ln, Chapel En-le-Frith, Stockport, Ches, SK12 6QF [IO83XI, SJ98]
G3 ZVT G M Rabstaff, Evesham, 26 Lichfield Dr, Prestwich, Manchester, M25 0HX [IO83UM, SD80]
G3 ZVU A Dawson, 30 Blackmoor Rd, Wellington, TA21 8ED [IO80JX, ST11]
G3 ZVV J C Gellatly, 11 Arches Dr, Bilsthorpe, Newark, Notts, NG22 8QF [IO93LD, SK66]
G3 ZVW S P White, 31 Amberley Rd, London, N13 4BH [IO91WP, TQ39]
GI3 ZVZ Dr D P Nicholls, 2 Printshop Rd, Templepatrick, Ballyclare, BT39 0HZ [IO64XQ, J28]
G3 ZWB J Jackson, 12 Clarence Chare, Newton Aycliffe, DL5 5HX [IO94FO, NZ22]
G3 ZWD P Flicos, 17 Gerald Rd, Bournemouth, BH3 7JZ [IO90BR, SZ09]
G3 ZWF P J Harpur, 62 Plumpton Ave, Hornchurch, RM12 6BD [JO01CN, TQ58]
GM3 ZWG W Frame, 13 Conon Ave, Bearsden, Glasgow, G61 1EN [IO75TV, NS57]
G3 ZWH D M Hill, 16 Hollow Ln, Snodland, ME6 5LS [JO01FH, TQ66]
G3 ZWI Details withheld at licensee's request by SSL.
G3 ZWK D T Raimbach, Silver Trees, 51 Old Wokingham Rd, Crowthorne, RG45 6SS [IO91OI, SU86]
G3 ZWL D Evans, Rock House, Woodlands End, Mells, Frome, BA11 3PF [IO81TF, ST74]
G3 ZWN A T Slingsby, 57 Regency Lodge, Albert Rd, Buckhurst Hill, IG9 6EF [JO01AO, TQ49]
G3 ZWR N A Hay, 19 Logan Rd, Walkerville, Newcastle upon Tyne, NE6 4SY [IO94FX, NZ26]
G3 ZWW M J Quee, 10 Estella Mead, Chelmsford, CM1 4XH [JO01FS, TL60]
G3 ZWY D N Ansell, 15 Stoney Stile Rd, Alveston, Bristol, BS12 2NG [IO81RO, ST68]
G3 ZWZ R Collinson, 1 Ullswater Rd, Lancaster, LA1 3PP [IO84OB, SD46]
G3 ZXA Dr A J M Wenham, 28 Pine Wood, Sunbury on Thames, TW16 6SG [IO91TK, TQ16]
GM3 ZXB A Robertson, 77 Cobden St., Dundee, DD3 6DD [IO86ML, NO33]
G3 ZXC G W Earnshaw, 12 Withy Parade, Fulwood, Preston, PR2 8JN [IO83PS, SD53]
G3 ZXD Details withheld at licensee's request by SSL.
GM3 ZXE M W Bannerman, 19 Inver Terr, Muirhead, Dundee, DD2 5LS [IO86LL, NO33]
G3 ZXF D W Corner, 122 Stortford Hall, Park, Bishops Stortford, Herts, CM23 5AP [JO01CU, TL42]
GM3 ZXG J Higgins, 8 Ardgowan Sq, Greenock, PA16 8ET [IO75OW, NS27]
GM3 ZXH L J Anderson, 18 Kirkton Park, Daviot, Inverurie, AB51 0HW [IO87TI, NJ72]
GW3 ZXI S J Brennan, 24 Gareth Cl, Thornhill, Cardiff, CF4 9AF [IO81JM, ST18]
GM3 ZXL Details withheld at licensee's request by SSL.[Op: Andy Allan, 2 Lanrig, Old Lindsaybeg Road, Chryston, Glasgow, G69 9HU.]
G3 ZXM M Brown, 5A The Woodlands, Rathfarnham Castle, Dublin 14, Eire
G3 ZXN Details withheld at licensee's request by SSL.
G3 ZXO J Burnie, 1 Chapel Meadow, Buckland Monachorum, Yelverton, PL20 7LR [IO70WL, SX46]
G3 ZXU Details withheld at licensee's request by SSL.
G3 ZXV P G Veale, 13 Lawford Gdns, Kenley, CR8 5JJ [IO91WH, TQ35]
G3 ZXW J G Midmore, 9 Whiteways, Wimborne, BH21 2PQ [IO90AT, SU00]
G3 ZXX Details withheld at licensee's request by SSL.
G3 ZXY P N Holtham, 50 Goolman St, Chapel Hill, Queensland, 4069, Australia, X X
G3 ZXZ M Stokes, 6 Deer Croft Cres, Huddersfield, HD3 3SG [IO93BP, SE11]
G3 ZYA Details withheld at licensee's request by SSL.[Op: S F Lakin, Torbryan, Newton Abbot, Devon, TQ12 5UR.]
G3 ZYC M I Sneap, Farm Cl, Pentrich, Ripley, Derby, DE5 3RR [IO93GB, SK35]
G3 ZYD R J Alton, 23 Cemetery Rd, Belper, DE56 1EJ [IO93GA, SK34]
G3 ZYE R Bellerby, Christs College, 4 St. Germans Pl, Blackheath, London, SE3 0NH [JO01AL, TQ47]
G3 ZYF K A Wilson, North View, Seapalling Rd, Ingham, Norwich, Norfolk, NR12 0TN [JO02SS, TG32]
G3 ZYN R A Gledhill, 38 Downlands, Waltham Abbey, EN9 1UH [JO01AQ, TL30]
G3 ZYO P A Connolly, 47 Norfolk Rd, Gravesend, DA12 2RX [JO01EK, TQ67]
G3 ZYP Details withheld at licensee's request by SSL.
G3 ZYQ A L Robinson, 264A Minster Rd, Minster on Sea, Sheerness, ME12 3LR [JO01JK, TQ97]
G3 ZYR A K Booer, Orchard House, 3 Lea Brooks Cl, Warboys, Huntingdon, PE17 2RA [IO92XJ, TL38]
G3 ZYS Details withheld at licensee's request by SSL.
G3 ZYT D L Taylor, 60 Kidderminster Rd, Bromsgrove, B61 7JY [IO82XI, SO97]
G3 ZYV D M Ferigan, 191 Gillingham Rd, Gillingham, ME7 4EP [JO01GJ, TQ76]
G3 ZYW C W Debney, Prospect Lodge, Station Rd, Kintbury, Hungerford, RG17 9UP [IO91GJ, SU36]
G3 ZYX R S Offord, Berberis Lodge, 10 Barberry Way, Blackwater, Camberley, GU17 9DX [IO91OH, SU85]
G3 ZYY T S Day, 46 Beatrice Ave, Saltash, PL12 4NG [IO70VJ, SX45]
G3 ZYZ Dr M C Joiner, 31 Kings Rd, Chalfont St. Giles, HP8 4HP [IO91RP, SU99]
G3 ZZA P V Rose, 21 Launceston Rd, Radcliffe, Manchester, M26 3UN [IO83TN, SD70]
G3 ZZB P J Bonfield, 90 Bucklesham Rd, Ipswich, IP3 8TS [JO02OA, TM14]
G3 ZZF D J Woolley, 21 Lulworth Ave, South Kenton, Wembley, HA9 8TP [IO91UN, TQ18]
G3 ZZG F D Vowles, 38 The Rowlands, Biggleswade, SG18 8NZ [IO92VB, TL24]
G3 ZZH M K Battersby, 80 Corte Yolande, Moraga, Ca 94556, California, USA, X X
G3 ZZI G Smith, The Orchard, Parkside Dr, Lowestoft, NR32 2QP [JO02UL, TM59]
G3 ZZJ Details withheld at licensee's request by SSL.
G3 ZZL Dr S J Keightley, The Old Parsonage, Steepleton Hill, Stockbridge, Hants, SO20 6JE [IO91GC, SU33]
G3 ZZM M G C Robinson, 7400 Old Bunch Rd, Wendell, North Carolina 27591, USA, X X
G3 ZZN Details withheld at licensee's request by SSL.
G3 ZZO Details withheld at licensee's request by SSL.[Op: M J Miller, 31 Coronation Road, Chatham, Kent, ME5 7DD.]
G3 ZZP D Matthews, 7 Boulsworth Ave, Haworth Park, Hull, HU6 7DZ [IO93TS, TA03]
G3 ZZQ R B Ludwell, 3 Bourne Honour, Tonwell, Ware, SG12 0HW [IO91XU, TL31]
G3 ZZS R M Wills, 21 Woodford Rd, Glenholt, Plymouth, PL6 7HX [IO70WK, SX56]
G3 ZZT R W Perrow, 17 Maidstone Dr, Marton, Marton in Cleveland, Middlesbrough, TS7 8QW [IO94JM, NZ51]
G3 ZZU C M Waldron, 22 Windermere Rd, Patchway, Bristol, BS12 5PW [IO81RM, ST68]
G3 ZZV G T Evans, 20 Creekside View, Tresillian, Truro, TR2 4BS [IO70MG, SW84]
G3 ZZW P S Chimber, 202 Wintersdale Rd, Leicester, LE5 2GP [IO92LP, SK60]
G3 ZZX A V Evans, Ashlea, Aston Munslow, Craven Arms, Shropshire, SY7 9ER [IO82PL, SO58]
G3 ZZY C C Cresswell, 1 Chapel Hill, Newquay, TR7 1ND [IO70KJ, SW86]
G3 ZZZ J Gibbs, 10 Waverley Rd, Westbrook, Margate, CT9 5QB [JO01QJ, TR37]

G4

GM4 AAF D Arcari Dundee ARC, 184 Fintry Dr, Dundee, DD4 9LP [IO86ML, NO43]
G4 AAH K Lawson, 233 Southwell Rd West, Mansfield, NG18 4HF [IO93JD, SK56]
G4 AAJ K B Runcorn, 10 Redland Green Rd, Redland, Bristol, BS6 7HE [IO81QL, ST57]
G4 AAL J F Layton, Meadow View, Worcester Rd, Martley, Worcester, Worcs, WR6 6QA [IO82TF, SO75]
GI4 AAM W H Langtry, 6 Brooklands Rd, Newtownards, BT23 4TL [IO74DO, J47]
G4 AAO Details withheld at licensee's request by SSL.
G4 AAQ P N Butterfield, 12 Lynwood Cres, Pontefract, WF8 3QT [IO93IQ, SE42]
G4 AAU W T Bowen, 126 Westfield Ln, Kippax, Leeds, LS25 7HU [IO93HS, SE43]
G4 AAW P L Miller, 104 Howard Dr, Allington, Maidstone, ME16 0QB [JO01FG, TQ75]
G4 AAX G Emmerson Northumbria ARC, 72 The Gables, Widdrington, Morpeth, NE61 5RB [IO95EF, NZ29]
G4 AAZ Details withheld at licensee's request by SSL.
G4 ABC P G K Cabban, Thornbury & District Arc, Thornbury Urc Hall, Rock St, Thornbury, Bristol, BS12 2BA [IO81RO, ST68]
G4 ABD Details withheld at licensee's request by SSL.
G4 ABE J Ellis, 4 Hazelmount Cres, Warton, Carnforth, LA5 9HS [IO84OD, SD47]
G4 ABF C J Watson, 96 West End Ave, Harrogate, HG2 9BY [IO93FX, SE25]
G4 ABG N M Kent, 12 Dean Cl, Weston Super Mare, BS22 0YN [IO81NI, ST36]
G4 ABI D S Radley, 12 South Cliff Tower, Bolsover Rd, Eastbourne, BN20 7JW [JO00DS, TV69]
G4 ABM C Todd, 7 Sandown Rd, Billingham, TS23 2BQ [IO94IO, NZ42]
G4 ABN T B J Atkins, 55 Havenbrook B'Var, Willowdale, Ontario, Canada M2J 1A7
G4 ABP Details withheld at licensee's request by SSL.
G4 ABQ J O Hudson, Greylag House, 46 High St., ODell, Bedford, MK43 7BB [IO92QF, SP95]
G4 ABS D Bedford, 2 Fairfields Walk, Stratford upon Avon, CV37 9PP [IO92DE, SP15]
G4 ABT P Oliver, 23 Knights Cl, West Bridgford, Nottingham, NG2 7HJ [IO92KV, SK53]
G4 ABU D J Gill, 32 Derbyshire Rd, Barrow in Furness, LA14 5NB [IO84JD, SD27]
G4 ABV K Frettsome, 375 Lode Ln, Solihull, B92 8NN [IO92CK, SP18]
G4 ABW L D Willey, 7 Oaklands Rd, Four Oaks, Sutton Coldfield, B74 2TB [IO92CN, SP19]
G4 ABX B M Macaulay, 38 Prospect Ln, Solihull, B91 1HN [IO92CK, SP18]
G4 ACF Capt R F G Cogzell ACF/CCF Int R N, 242 Clydesdale Towe, Holloway Head, Birmingham, B1 1UJ [IO92BL, SP08]
G4 ACI J Blackburn, 40 Carlton Ave, Upholland, Skelmersdale, WN8 0AE [IO83PM, SD50]
G4 ACJ H J R Reeve, 11 Heather Dr, Ferndown, BH22 9SD [IO90BT, SU00]
G4 ACK B H Scarisbrick, Shalfoi, 8 Abbey Cl, Wookey, Wells, BA5 1LF [IO81PF, ST54]

G4 ACL D Atkinson, 38 Hornbeam Rd, Theydon Bois, Epping, CM16 7JX [JO01BQ, TQ49]
GM4 ACM A C Miller, 38 Randolph Rd, Broomhill, Glasgow, G11 7LG [IO75UV, NS56]
GW4 ACO V B Kelly, 188 Llanelian Rd, Old Colwyn, Colwyn Bay, LL29 8UN [IO83DG, SH87]
G4 ACP J M Scherrer, 5 Claymore Dr, Ickleford, Hitchin, SG5 3UB [IO91UX, TL13]
G4 ACQ L W R Randall, 5 Linden Rd, Westgate on Sea, CT8 8BY [JO01QJ, TR36]
G4 ACS G C Weale, Nursery House, 4 Nursery Gr, Franche, Kidderminster, DY11 5BG [IO82UJ, SO87]
G4 ACU M E Levy, 5 Mangotsfield Rd, Magotsfield, Mangotsfield, Bristol, BS17 3JG [IO81SL, ST67]
G4 ACW N D Roe, 10 Ramsdean Rd, Stroud, Petersfield, GU32 3PJ [IO91MA, SU72]
G4 ACY R B Ratcliffe, 173 Montague Rd, Bilton, Rugby, CV22 6LG [IO92II, SP47]
G4 ACZ R J Mutton, Summer Hayes, Mill Ln, Oversley Green, Alcester, B49 6LF [IO92BE, SP05]
G4 ADA B A Coope, 39A High St., Nettleham, Lincoln, LN2 2PL [IO93SG, TF07]
G4 ADD W A Ricalton, 4 South Rd, Longhorsley, Morpeth, NE65 8UW [IO95CG, NZ19]
G4 ADE M Woollin, 26 Ashcourt Dr, Hornsea, HU18 1HE [IO93VV, TA24]
G4 ADJ Details withheld at licensee's request by SSL.
G4 ADK R Cutbush, 10 Western Rd, Upton Park, London, E13 9JF [JO01AM, TQ48]
GW4 ADL Dr T Davies, 44 Carnglas Rd, Tycoch, Sketty, Swansea, SA2 9BW [IO81AO, SS69]
G4 ADM A D Maish, 73 Edenfield Gdns, Worcester Park, KT4 7DX [IO91UI, TQ26]
G4 ADN Details withheld at licensee's request by SSL.
G4 ADP P McCurrie, Vinstone Cttgs, 212 Mannamead Rd, Hartley, Plymouth, PL3 5RF [IO70WJ, SX45]
G4 ADR N E Ayres, Glendevon, 68 Queens Rd, Thame, OX9 3NQ [IO91MR, SP70]
G4 ADS J C Chisman, 155 Ramsgate Rd, Broadstairs, CT10 2EP [JO01RI, TR36]
G4 ADT A Tibbett, 48 Park Cres, Darlington, DL1 5EF [IO94FM, NZ21]
G4 ADV P M Reed Newquay & D ARS, Larks Rise, Great Hewas, Grampound Rd, Truro, TR2 4EP [IO70MI, SW95]
G4 ADX Details withheld at licensee's request by SSL.
G4 AEB T J Baines, Highland House, Tendring Heath, Clacton, Essex, CO4 0BN [JO01LV, TM02]
G4 AEC S W Davies, 85 Moseley Wood Dr, Cookridge, Leeds, LS16 7HD [IO93EU, SE24]
G4 AED B Cator, 9 Saham Rd, Watton, Thetford, IP25 6EA [JO02JN, TF90]
G4 AEE M D Bedford, 4 Holme Houses, Oakworth, Keighley, BD22 0QY [IO93AU, SE04]
G4 AEG I P Kemp, 21 Rednal Rd, Kings Norton, Birmingham, B38 8DT [IO92AJ, SP07]
G4 AEH J V F Lee, 13 Redruth Cl, Sheraton Park, Nuneaton, CV11 6FG [IO92GM, SP39]
G4 AEI G V Prater, 5 Lutton Cl, Lower Earley, Reading, RG6 4AA [IO91MK, SU76]
G4 AEK P L Boswell, 3 Lahn Dr, Droitwich, WR9 8TQ [IO82WG, SO86]
G4 AEL R Cox, 34 Ratcliffe Dr, Stoke Gifford, Bristol, BS12 6UD [IO81RM, ST68]
G4 AEM P Ellis, 96 Whitelands Ave, Chorleywood, Rickmansworth, WD3 5RG [IO91RP, TQ09]
G4 AEO P Hunt, 93 Park Rd, Coalville, LE67 3AF [IO92HR, SK41]
G4 AEP W V Thomas, 58 Bearwood Rd, Wokingham, RG41 4SY [IO91NJ, SU76]
G4 AEQ Dr S W Redfern, 2 Church Cl, Church Ln, Eagle, Lincoln, LN6 9DJ [IO93PE, SK86]
G4 AER R J Sobey, 5 Fairway Cl, Liphook, GU30 7XD [IO91OB, SU83]
G4 AES K Walker, 3 Glen View, Shield Hall Ln, Sowerby, Sowerby Bridge, HX6 1NL [IO93AR, SE02]
G4 AEU M J Bassil, 55 Bedford Pl, Southampton, SO15 2DT [IO90HV, SU41]
G4 AEV R J Anderson, Trinafour, Abingdon Rd, Marcham, Abingdon, OX13 6NU [IO91HP, SU49]
G4 AEW R E Presswood, 89 Monks Dyke Rd, Louth, LN11 8DN [JO03AI, TF38]
G4 AEY R D Twist, 1 Birchwood Dr, Rushmere, Ipswich, IP5 7EB [JO02OB, TM24]
G4 AEZ B W Oughton, 176 South Lodge Dr, Southgate, London, N14 4XN [IO91WP, TQ39]
G4 AFA Dr N D Porter, 13 Charfield Cl, Winchester, SO22 4PZ [IO91IB, SU42]
G4 AFB J M Rickson, 34 Heyhouses Ln, St. Annes, Lytham St. Annes, FY8 3RW [IO83LS, SD32]
G4 AFE J R Tallentire, Oakdene Alston Rd, Middleton in Teesid, Co Durham, DL12 0UU [IO84WP, NY92]
GM4 AFF S G Cooper, 10 Cliff View, Newtonhill, Stonehaven, AB3 3GX [IO87TH, NJ72]
GI4 AFH G J Phillips, 38 Sketrick Ind Prk, Newtownards, Co Down, N Ireland, BT23 3BN [IO74DN, J46]
G4 AFI A H Cheetham, 39 Burns Ave, Church Crookham, Fleet, GU13 0BN [IO91OG, SU85]
G4 AFJ G W Dover, 31 Newbold Rd, Kirkby Mallory, Leicester, LE9 7QG [IO92HO, SK40]
G4 AFQ Details withheld at licensee's request by SSL.[Op: D P Warner. Station located near Ashford.]
G4 AFR F Nicholson, 5 Friars Terr, Barrow in Furness, LA13 0BX [IO84JC, SD26]
G4 AFS Details withheld at licensee's request by SSL.
G4 AFT D F Randles, 21 Whitcliffe Ave, Ripon, HG4 2JJ [IO94FD, SE37]
G4 AFU P G Rollin, Church Wynd, Burneston, Bedale, North Yorks, DL8 2JE [IO94FG, SE38]
G4 AFV T R Gaygan, 127 Imperial Dr, North Harrow, Harrow, HA2 7HW [IO91TN, TQ18]
G4 AFX A O Moore, Laundry Cottage, Coles Oak Ln, Dedham, Colchester, CO7 6DR [JO01LW, TM03]
G4 AFY R H Perrin, 8 Granville Crest, Offmore Farm, Kidderminster, DY10 3QS [IO82VJ, SO87]
G4 AFZ V M Bott, 25 Finkle St., Hensall, Goole, DN14 0QY [IO93KQ, SE52]
G4 AGA P E Carthey, 17 Kingston Rd, Trench, Telford, TF2 7HT [IO82SR, SJ61]
G4 AGB B Bull, 2 Maylands Gr, Barrow in Furness, LA13 0AN [IO84JD, SD27]
G4 AGC C R Wortham, 57 Cranleigh Dr, Swanley, BR8 8NZ [JO01CJ, TQ56]
G4 AGE R R Evans, Mansfield, Mooracre Ln, Bolsover, Chesterfield, Derbyshire, S44 6ER [IO93IF, SK47]
GM4 AGG A H Stewart W Scotland ARS, 599 PO Box, Glasgow, G3 6QH [IO75UU, NS56]
G4 AGH S W Pearson, 75 Gloucester Rd, Thornbury, Bristol, BS12 1JH [IO81RO, ST69]
G4 AGJ C P Price, 3 Mill View, Freckleton, Preston, PR4 1YQ [IO83NS, SD42]
GM4 AGL W R Ferguson, 72 High Parksail, Erskine, PA8 7HX [IO75SV, NS46]
G4 AGM Details withheld at licensee's request by SSL.
G4 AGN J F J Porter, 109 Heacham Dr, Leicester, LE4 0LL [IO92KP, SK50]
G4 AGQ Details withheld at licensee's request by SSL.
GM4 AGS J S Miller, Ingleby, 70 West Rd, Newport on Tay, DD6 8HP [IO86MK, NO42]
G4 AGT H P Thoennissen, Horseplot, Stockland, Honiton, Devon, EX14 9DL [IO80KT, ST20]
GW4 AGV A Beals, Maes-y-Coed, Maen-y-Groes, Newquay, Dyfed, SA45 9RL [IO72TE, SN35]
GM4 AGX J A Yates, Garths, Bressay, Shetland, Isle of Shetland, ZE2 9ER [IP90KD, HU43]
GM4 AHB Rev M J Gallagher, St Davids, Meadowhead Rd, Plains, Aidrie Lanarkshire, ML6 7JF [IO85AV, NS76]
G4 AHC T E O'Neill, 10 Dalehurst Cl, Wallasey, L44 8AE [IO83LK, SJ39]
GI4 AHD Dr F R Elder, 291 Glenshane Rd, Londonderry, BT47 3SW [IO64JW, C50]
G4 AHF R Ashall, 110 Waverley Cres, Droylsden, Manchester, M43 7WL [IO83WL, SJ99]
G4 AHG C R Chidgey Shirehampton AR, 46 Station Rd, Shirehampton, Bristol, BS11 9TX [IO81PL, ST57]
G4 AHH C Hayward, 5 Kennels Rd, Whittlebury, Towcester, NN12 8XW [IO92MC, SP64]
G4 AHI A Newton, Instow, St. Johns Rd, Wroxall, Ventnor, PO38 3EL [IO90JO, SZ58]
G4 AHJ M E Downey, 11 Woodlands Dr, Lepton, Huddersfield, HD8 0JB [IO93DP, SE11]
G4 AHK B J Palin, 11 Ashgrove Cl, Marlbrook, Bromsgrove, B60 1HW [IO82XI, SO97]
G4 AHM J W Stratton, 10 Brownshill, Maulden, Bedford, MK45 2BT [IO92SA, TL03]
G4 AHN D R Lax, 1 Gardeners Hill Rd, Boundstone, Wrecclesham, Farnham, GU10 4RL [IO91OE, SU84]
GW4 AHO K M Jones, 18 Caefelyn, Norton, Presteigne, LD8 2UB [IO82LH, SO36]
GI4 AHP T E Sloan, 26 Ballyknock Rd, Hillsborough, BT26 6EF [IO64WK, J15]
G4 AHQ J Dark, Civ Camping Ltd, Nrthmpton Vl Gon Ct, Newport Pagnell Roa, Northampton, NN4 0HP [IO92NE, SP75]
G4 AHS Details withheld at licensee's request by SSL.
G4 AHT M F Niven, 16 Treewall Gdns, Bromley, BR1 5BT [JO01AK, TQ47]
G4 AHX J G Clark, 9 Spath Walk, Cheadle Hulme, Cheadle, SK8 7NJ [IO83VI, SJ88]
G4 AHZ J A Kynaston, 12 Lamaleach Dr, Freckleton, Preston, PR4 1AJ [IO83NR, SD42]
G4 AIB P Holt, 41 Garden Ave, Ilkeston, DE7 4DF [IO92IX, SK44]
GI4 AID J M Mulholland, 75 Lurgan Rd, Glenavy, Crumlin, BT29 4QE [IO64VO, J17]
GM4 AIE W J Mackie, 8 Mossfield Ave, Oban, PA34 4EH [IO76GJ, NM82]
G4 AIF H J Horne, 20 Waveney Rd, Great Yarmouth, NR31 0LB [JO02UO, TG50]
G4 AIH F Weetman, 23 Kenmore Rd, Sale, M33 4LP [IO83TJ, SJ79]
G4 AIJ R G Jones, Sycamores, The Sheet, Ludlow, Shropshire, SY8 4JT [IO82PI, SO57]
G4 AIK K L Holburn, 34 Lydgate Ln, Wolsingham, Bishop Auckland, DL13 3LF [IO94BR, NZ03]
G4 AIQ G J Mitchell, Ashleigh Farm, Main St., Gayton-le-Marsh, Alford, LN13 0NW [JO03BI, TF48]
G4 AIR D A Bieber, Tonkins Quay House, Mixtow, Lanteglos By Fowey, Cornwall, PL23 1NB [IO70QI, SX15]
G4 AIT G Mason, 128 Bacup Rd, Todmorden, OL14 7HG [IO83WQ, SD92]
G4 AIU E A Morgan, Molescroft, 12 Kitts, Wellington, TA21 9AX [IO80JX, ST11]
G4 AIV D Scotney, 9 Spencer St., Rothwell, Kettering, NN14 6HD [IO92OK, SP88]
G4 AIW A Scarsbrook, 16 Greenbank Ave, Uppermill, Oldham, OL3 6EB [IO93AN, SE00]
G4 AIZ M B Houchen, 93 Enfield Chase, Hunters Hill, Guisborough, TS14 7LN [IO94LM, NZ61]
G4 AJ L J Ralli, 14 Oxford Sq, London, W2 2PB [IO91WM, TQ28]
G4 AJA C J Hoare, 16 Shrivenham Rd, Highworth, Swindon, SN6 7BZ [IO91DP, SU29]
G4 AJC Details withheld at licensee's request by SSL.
G4 AJE P W Brown, 33A March Rd, Wimblington, March, PE15 0RW [JO02BM, TL49]
G4 AJF N Ingman, 107 Derby Rd, Aston on Trent, Derby, DE72 2AE [IO92HU, SK43]
G4 AJG P E Perera, 43 Hillside Ave, Woodford Green, IG8 7QU [JO01AO, TQ49]
G4 AJI Details withheld at licensee's request by SSL.
G4 AJJ G R Smith, Greenacres, Main St., Sawdon, Scarborough, YO13 9DY [IO94RG, SE98][Op: Geoffrey R Smith, BSc MISTC, MBIM. Tel: (01723) 85212. XYL is G4LSS, QRA: ZQ58F. Packet facilities.]
G4 AJM Details withheld at licensee's request by SSL.
G4 AJN Details withheld at licensee's request by SSL.
G4 AJO R P Finch, 48 Allens Ln, Sprowston, Norwich, NR7 8EJ [JO02PP, TG21]
G4 AJQ Details withheld at licensee's request by SSL.
GM4 AJR L R M Donaldson, 11 Highgate Gdns, Aberdeen, AB1 7TZ [IO87TH, NJ72]
G4 AJS R J Hughes, Tonbridge School, High St., Tonbridge, TN9 1JP [JO01DE, TQ54]
G4 AJU I C Aldridge, 28 Robert St., Williton, Taunton, TA4 4PG [IO81ID, ST04]
GM4 AJV M J Mackinnon, 160 Carrick Knowe Dr, Edinburgh, EH12 7EW [IO85IW, NT27]
G4 AJW A J Wade, 139 Gilbert Rd, Cambridge, CB4 3PA [JO02BF, TL46]
G4 AJZ J J Harris, 10 Cooperage Cl, Liverpool, L8 6XH [IO83MJ, SJ38]
G4 AK J R Seager, Walnut Cottage, Sloothby Rd, Willoughby, Alford, LN13 9NW [JO03CF, TF47]
G4 AKA M K Diprose, 4 Russet Cl, Stanwell Moor, Staines, TW19 6AX [IO91RL, TQ07]
G4 AKB M L Court, 36 Elmsleigh Rd, Paignton, TQ4 5AU [IO80FK, SX86]
G4 AKC D S Starkie, 22 Acre Gate, South Shore, Blackpool, FY4 3LF [IO83LS, SD33]
G4 AKD I R Alexander, 46 Pettitts Ln, Dry Drayton, Cambridge, CB3 8BT [JO02AF, TL36]
G4 AKE C J Gent, 27 Walnut Ave, Alvaston, Derby, DE24 0PP [IO92GV, SK33]
G4 AKF R S Stott, 52 Fairfield Ave, Upminster, RM14 3AY [JO01DN, TQ58]
G4 AKG P C Fry, 11 Park Rd, Burgess Hill, RH15 8EU [IO90WX, TQ31]
G4 AKK P E Francis, 23 Ludlow Rd, The Poplars, Kidderminster, DY10 1NW [IO82VI, SO87]
GW4 AKO J Hogg, Meadow Cottage, Lords Meadow Farm, Pembroke, Pembrokeshire, SA71 4BB [IO71NQ, SM90]
G4 AKP J R Yates, 8 Marlborough Cl, Chichester, PO19 2XW [IO90OT, SU80]
G4 AKQ M Bernard, 16 Mountbatten, Ave, Chatham, Kent, ME5 0JY [JO01GI, TQ76]
G4 AKR G M Slack, 16 East Carr, Cayton, Scarborough, YO11 3TS [IO94TF, TA08]
G4 AKT S Fisher, 8 Beechcroft, Chestfield, Whitstable, CT5 3QF [JO01MI, TR16]
G4 AKU Details withheld at licensee's request by SSL.
G4 AKW G N Robinson, 2 Hasketon Rd, Woodbridge, IP12 4JR [JO02PC, TM24]
G4 AKY D P Hayes, 17 Telston Ln, Otford, Sevenoaks, TN14 5JX [JO01CH, TQ55]
GW4 AKZ Details withheld at licensee's request by SSL.
G4 AL G R Cox, 98 Whatton Rd, Kegworth, Derby, DE74 2DT [IO92IT, SK42]
G4 ALB N M Castledine, 1 Johns Cl, Burbage, Hinckley, LE10 2LY [IO92HM, SP49]
G4 ALC J B Balls, 48 Collingwood Rd, Great Yarmouth, NR30 4LR [JO02UO, TG50]
G4 ALD F P Donovan, 4 Rembrandt Dr, Northfleet, Gravesend, DA11 8NQ [JO01EK, TQ67]
G4 ALE Q G Collier Addiscombe ARC, 19 Grange Cliff Gdn, London, SE25 6SY [IO91WJ, TQ36]
G4 ALF K Law, 4 Ward Ct, Woodfield Park Rd, Emsworth, PO10 8BQ [IO90MU, SU70]
GW4 ALG S A Rawlings, 14 The Paddock, Chepstow, NP6 5BW [IO81PP, ST59]
G4 ALH Details withheld at licensee's request by SSL.
G4 ALI Details withheld at licensee's request by SSL.
GM4 ALK R Kinnes, 1 Cadham Villas, Markinch, Glenrothes, KY7 6PQ [IO86KE, NO20]
G4 ALL A W Palace, Flat, 15 Church St., Littlehampton, BN17 5EL [IO90RT, TQ00]
GI4 ALM J B Cairns, 24 Norwood Ave, Belfast, BT4 2EE [IO74BO, J37]
G4 ALN Details withheld at licensee's request by SSL.
G4 ALP W Stanway, 52 Bowerfield Ave, Hazel Gr, Stockport, SK7 6JA [IO83WI, SJ98]
G4 ALR M J Down, 95 High St., Henlow, SG16 6AB [IO92UA, TL13]
G4 ALT A L Taylor, 255 Hoo Rd, Kidderminster, DY10 1LY [IO82VJ, SO87]
G4 ALY R Bird, 6 The Cross, St Dominick Saltash, Cornwall, PL12 6TD [IO70UL, SX36]
G4 ALZ R G Bridgland, 20 Newling Way, Worthing, BN13 3DG [IO90TU, TQ10]
GD4 AM G L Danielson, 4 Merlins Ct, Queens Promenade, Ramsey, IM8 1ER
G4 AMD C G Heavens, 20928 NE 169th St, Woodinville, Wa 98072, USA, X X
G4 AMF J Cresswell, 7 Glinton Ave, Blackwell, Alfreton, DE55 5HD [IO93HC, SK45]
G4 AMI M Hearn, 63 Greswolde Rd, Solihull, B91 1DX [IO92CK, SP18]
G4 AMJ D R Evans, 7912 Fairview Rd, Boulder, Colorado, USA 80303
G4 AML L M G F Dumont, 78 Towan Ave, Fishermead, Milton Keynes, MK6 2JQ [IO92PA, SP83]
G4 AMN C J Wainwright, The Cooleen, Main St., Hoby, Melton Mowbray, LE14 3DT [IO92LS, SK61]
G4 AMP B M Flack, Ave Des Hospitaliers de, St Jean 7, 1410, Waterloo, Belgium
G4 AMT T R George, Sea Call, Cove Rd, Sennen, Penzance, Cornwall, TR19 7BT [IO70DB, SW32]
G4 AMW V M Goom, 48 Wayne Rd, Parkstone, Poole, BH12 3LF [IO90AR, SZ09]
GW4 AMX J S G Barrett, y Wern, Bryn Arthur, St Asaph, Denbighshire, LL17 0DP [IO83GG, SJ07]
G4 AMY R G Briggs, Lancaster New Rd, Garstang, Cabus, Preston, Lancs, PR3 1NL [IO83OV, SD44]
GW4 AMZ Details withheld at licensee's request by SSL.
G4 ANA Details withheld at licensee's request by SSL.[Op: S P Cook.]
GM4 ANB Dr J R Morris, 35 Main St., Hillend, Dunfermline, KY11 5ND [IO86HB, NT18]
G4 AND J A F King, 29 Tower Rd, Orpington, BR6 0SG [JO01BI, TQ46]
G4 ANE H Leach, 30 Taywood Rd, Thornton Cleveleys, FY5 2RT [IO83LV, SD34]
GW4 ANK R H Davenport, 14 Milward Rd, Barry, CF63 3QD [IO81IJ, ST16]
G4 ANL A Rodgers, 28 Allendale Rd, Stockton on Tees, TS18 4PW [IO94IN, NZ41]
G4 ANN R Hadfield, 45 Erica Way, Copthorne, Crawley, RH10 3XG [IO91WD, TQ33]
G4 ANP M J Valentine, 10 Thellusson Ave, Scawsby, Doncaster, DN5 8QN [IO93JM, SE50]
G4 ANQ P J Clayton, 90 Littleheath Rd, Selsdon, South Croydon, CR2 7SD [IO91XI, TQ36]
G4 ANS J S Redgate, 75 Kensington Gdns, Carlton, Nottingham, NG4 1DZ [IO92KX, SK64]
G4 ANT O S Chilvers E Anglian Cnt C, 26 Wensum Valley Cl, Hellesdon, Norwich, NR6 5DJ [JO02OP, TG11]
G4 ANU C Columbine, 5 Thornbury Dr, Mansfield, NG19 6NB [IO93JD, SK56]
G4 ANV P A Hudson, 3 Rowan Dr, Kilburn, Belper, DE56 0PG [IO93GA, SK34]
G4 ANW T J Slack, 34 Moggs Mead, Petersfield, GU31 4NX [IO91MA, SU72]
G4 ANY D G Stephens, Croeso Cottage, 31 Coton, Whitchurch, Shropshire, SY13 2RA [IO82PV, SJ53]
G4 AOA H T Mason, 9 Chatsworth Dr, Little Eaton, Derby, DE21 5AP [IO92GX, SK34]
G4 AOB M N Floyde, 6 Gilwell Park Cl, Prettygate, Colchester, CO3 4SP [JO01KV, TL92]
G4 AOC R Farmer, 4 Colebrook Cl, Worthing, BN11 2LA [IO90TT, TQ10]
G4 AOE G D Woodcock, 6 Meadow Cl, Belting, Herne Bay, CT6 6NN [JO01NI, TR16]
G4 AOG Details withheld at licensee's request by SSL.
G4 AOH T E Bryan, 28 Berwyn Ave, Keresley, Coventry, CV6 2FD [IO92FK, SP38]
G4 AOJ R P Horton, 31 Furze Ln, Purley, CR8 3EJ [IO91WI, TQ36]
G4 AOK T Winter, 6 Cunliffe Dr, Brooklands, Sale, M33 3WS [IO83UJ, SJ79]
G4 AOL D Harmer, 20 Binbrook Cl, Low Early, Lower Earley, Reading, RG6 3BW [IO91NK, SU77]
G4 AOM R Gorse, 40 Copthorne Dr, Breightmet, Bolton, BL2 6SB [IO83TN, SD70]
G4 AON D A J Johnson, 18 Blackthorn Cl, Newport, Brough, HU15 2QJ [IO93PS, SE83]
G4 AOO W A Winterburn, 28 Montagu Rd, Sprotbrough, Doncaster, DN5 8DH [IO93KM, SE50]
G4 AOP Flt Lt D J Hibbin, 165 Byron St., Loughborough, LE11 5JN [IO92JS, SK52]
G4 AOQ D R Ward, 60 New Rd, High Wycombe, HP12 4LG [IO91OP, SU89]
GM4 AOR Details withheld at licensee's request by SSL.[Op: K C Henderson.]
G4 AOS Dr J West, Horsley House, Rochester, Northd., NE19 1TA [IO85VG, NY89]
G4 AOV Details withheld at licensee's request by SSL.
G4 APA A P Ashcombe, 52 Windermere Rd, Handforth, Wilmslow, SK9 3NH [IO83VI, SJ88]
G4 APB K R May, 53 Shearwood Cres, Dartford, DA1 4SU [JO01CL, TQ57]
G4 APD P S Wells Rugby ATS, 15 Gilmorton Rd, Lutterworth, LE17 4DY [IO92JK, SP58]
G4 APF M Richards, 10 Dower Ct, Old Torquay Rd, Paignton, TQ3 2RT [IO80FK, SX86]
GM4 API D Hebenton, Craigmill Cttgs, Bridgefoot, Strathmartine, Dundee, Angus, DD3 0PH [IO86LM, NO33]
G4 APJ K R Punshon, 24 Newcombe Rd, Holcombe Brook, Ramsbottom, Bury, BL0 9UT [IO83UP, SD71]
GM4 APK E A G Davies, 17 Eskdale Rd, Bearsden, Glasgow, G61 1JY [IO75TV, NS57]
G4 APL P N F A Lewis, 20 Annes Walk, Caterham, CR3 5EL [IO91WH, TQ35]
G4 APM D W Turner, 16 Jersey St., Newark, NG24 4NE [IO93OB, SK75]
G4 APN G Ford Easington Amateur Radio Societ, 11 Sandbanks Dr, Hart Station, Hartlepool, TS24 9RP [IO94JR, NZ43]
G4 APO R I Hirst, 21 Manor Farm Ct, Thrybergh, Rotherham, S65 4NZ [IO93IK, SK49]
G4 APP W Grogan, 92 School Rd, Thornton Cleveleys, FY5 5AP [IO83MU, SD34]
G4 APS D Fiander, 34 Sheriff Ave, Canley, Coventry, CV4 8FD [IO92FJ, SP27]
G4 APU R C Kenward, 9 The Cres, Farnborough, GU14 7AR [IO91PG, SU85]
G4 APV R G Harris, 62 Whirlow Ct Rd, Sheffield, S11 9NT [IO93FI, SK38]
G4 AQ E G Filby, 1 Park Walk, Shaftesbury, Dorset, SP7 8JR [IO81VA, ST82]
G4 AQA P D Hall, 39 Mill Ln, Kirkella, Kirk Ella, Hull, HU10 7JE [IO93SR, TA02]
G4 AQB S J Macdonald, 58 Tarbet Dr, Breightmet, Bolton, BL2 6LT [IO83TN, SD70]
G4 AQF P J Saunders, Orchard Cottage, Vale Rd, Broadstairs, CT10 2JG [JO01RI, TR36]
G4 AQI J O N Spurling, 15 Tibbs Hill Rd, Abbots Langley, WD5 0EE [IO91TQ, TL00]
G4 AQJ K W Gordon, 96 Pear Tree Cres, Solihull Lodge, Shirley, Solihull, B90 1LF [IO92BJ, SP07]
G4 AQK D Davis, 23 Matley Moor, Liden, Swindon, SN3 6NL [IO91DN, SU18]
G4 AQL J W Welch, An Dara, North St., Hellingly, Hailsham, BN27 4DY [JO00CV, TQ51]
G4 AQN K G Cooper, 18 Brentford Ct, Saltburn, Brotton, Saltburn By The Sea, TS12 2XW [IO94MN, NZ61]
GM4 AQO J H Haliburton, 21 Bolam Dr, Burntisland, KY3 9HP [IO86JB, NT28]
G4 AQP W Ford, 6 Ormonde St., Newtown, Chester, CH1 3DD [IO83NE, SJ46]
G4 AQR I Cordingley, Orchard Cottage, Compton, Marldon, Paignton, TQ3 1TA [IO80EL, SX86]
G4 AQS M W Bliss, 53 Rowallan Dr, Bedford, MK41 8AS [IO92SD, TL05]
G4 AQT J Rowbotham, 18 Weldbank Cl, Chilwell, Beeston, Nottingham, NG9 5FU [IO92JW, SK53]
G4 AQU Details withheld at licensee's request by SSL.
G4 AQV R Saagi, 13 Braunstone Ave, Leicester, LE3 0JF [IO92KP, SK50]
G4 AQY Details withheld at licensee's request by SSL.
G4 AQZ Details withheld at licensee's request by SSL.
GW4 ARC A R Evans Rhyland District Amateur Radio, 4 Elm Gr, Rhyl, Denbighshire, LL18 3PE [IO83GH, SJ08]
G4 ARE R W Hammett Exeter ARS, Brixington, College Ln, Longdown, Exeter Devon, EX6 7SS [IO80ER, SX89]
G4 ARF B Bull Furness A.R.S., 2 Maylands Gr, Barrow in Furness, LA13 0AN [IO84JD, SD27]
G4 ARI T K Raven, 15 Preston Cl, Stanton under Bardon, Markfield, LE67 9TX [IO92IQ, SK41]
GM4 ARJ J W Ferguson, 26 Cleuch Ave, Tullibody, Alloa, FK10 2RX [IO86BD, NS89]
G4 ARK R A Bacon, 9 Daleside, Gerrards Cross, SL9 7JE [IO91RN, TQ08]
G4 ARL R Keighley, 129 Coleridge Cl, Hitchin, SG4 0QY [IO91UW, TL12]
G4 ARN Details withheld at licensee's request by SSL.[Op: D Manning, 97 Plumstead Road, Norwich, Norfolk, NR1 4JS.]
G4 ARO T C Covey, 88 Sunbury Ln, Walton on Thames, KT12 2HY [IO91TJ, TQ16]
G4 ARP Details withheld at licensee's request by SSL.

G4 ARS J A Ennis Carlis & Dis AR, 30 Hillcrest Ave, Carlisle, CA1 2QJ [IO84NV, NY45]
GM4 ARU J McIntyre, 12 Johnstone Ln, Carluke, ML8 4NR [IO85CR, NS85]
G4 ARX Details withheld at licensee's request by SSL.
G4 ARY A K Langford, 33 Briscoe Rd, Hoddesdon, EN11 9DG [IO91XS, TL30]
G4 ARZ J F Grieve, Freshfields, Crouch House Rd, Edenbridge, TN8 5EE [JO01AE, TQ44]
G4 ASA D Wright, 58 Elmbridge Parade, Greatfield Est, Hull, HU9 4JU [IO93US, TA13]
G4 ASB P G Wilson, 35 Jackson Rd, Newbourne, Woodbridge, IP12 4NR [JO02PA, TM24]
G4 ASE Details withheld at licensee's request by SSL.
G4 ASF R D McCurrach, Fco (Muscat), King Charles St, London, SW1A 2AH [IO91WM, TQ37]
G4 ASG P G Bayley, 125 Stoddens Rd, Burnham on Sea, TA8 2DE [IO81MF, ST35]
G4 ASH I Roberts, 1 Cauldwell Ct, Sandy, SG19 1DA [IO92UD, TL14]
G4 ASI F D S Emery, Room 10, Building 448, Biggin Hill Aiport, Kent, TN16 3BN [JO01AH, TQ46]
G4 ASJ B R Benbow, 1231 Pershore Rd, Stirchley, Birmingham, B30 2YT [IO92AK, SP08]
G4 ASK E G Rayland, 40 Sycamore Cl, Taunton, TA1 2QJ [IO81LA, ST22]
G4 ASL S G Ayling, Kitnocks, 89 Queens Rd, Alton, GU34 1JA [IO91MD, SU73]
G4 ASM A S Murphy, 8 Malibres Rd, Chandlers Ford, Eastleigh, SO53 5DT [IO90HX, SU42]
G4 ASN M W Baker, 3 Churnet Cl, Gr, Nottingham, NG11 8PH [IO92JV, SK53]
GU4 ASO R A Ayres, -le-Suquet. Marette de Bas, Capelles, St. Sampson, Guernsey, Channel Islands, GY2 4UX
G4 ASP J R Holding, 7 Doeford Cl, Newchurch, Culcheth, Warrington, WA3 4DL [IO83RK, SJ69]
G4 ASQ M P Jordan, 89 Stanhope Dr, Horseforth, Horsforth, Leeds, LS18 4EU [IO93EU, SE23]
G4 ASR D J Butler, Yew Tree Cottage, Lower Maescoed, Hereford, HR2 0HP [IO81MX, SO33]
G4 AST Details withheld at licensee's request by SSL.
G4 ASW M Yorke, 8 St John Pl, Port Washington, New York 11050, USA
G4 ASX O V Perry, Bay Cottage, Brownlow Rd, Redhill, RH1 6AW [IO91VF, TQ25]
G4 ASY D J Yeaman, 123 High St., Hinxton, Saffron Walden, CB10 1RF [JO02CC, TL44]
G4 ASZ M J Hurst, 7 Paget Pl, Newmarket, CB8 7DR [JO02EG, TL66]
G4 ATA J M Hotchin, 49 Woodlands Dr, Skelmanthorpe, Huddersfield, HD8 9DB [IO93EO, SE21]
G4 ATB R E Shapland, Aros, 39 Camus Cross, Isle Ornsay, Isle of Sky, IV43 8QS [IO77CD, NG61]
G4 ATC Flt V J Reynolds, Wing Radio Officer, 25 Yoxall Ave, Stoke on Trent, ST4 7JJ [IO83VA, SJ84]
G4 ATF Details withheld at licensee's request by SSL.
G4 ATG M J Kerry Bartg, 2 Beacon Cl, Seaford, BN25 2JZ [JO00BS, TQ40]
G4 ATH J E Duddington Thornton Clevel, 8 The Grove, Thornton Cleveleys, FY5 2JD [IO83LV, SD34]
G4 ATL D Bloomfield, Bitterns, Preston Crowmarsh, Oxon, OX9 6SL [IO91MQ, SP70]
G4 ATQ G R Hawkins, 18 Brook St., Leighton Buzzard, LU7 8LH [IO91QW, SP92]
G4 ATR J L Rogers, 24 Treza Rd, Porthleven, Helston, TR13 9NB [IO70IC, SW62]
G4 ATS G Oddy, 8 Pendas Way, Crossgates, Leeds, LS15 8HX [IO93GT, SE33]
G4 ATU S J Brown, 12 Cole Cres, Aughton, Ormskirk, L39 5AJ [IO83NN, SD40]
G4 ATX P L Pomeroy, 110 Millstrood Rd, Whitstable, CT5 1PT [JO01MI, TR16]
GW4 ATY R Jenkins, 35 William St., Twynyrodyn, Merthyr Tydfil, CF47 0RG [IO81HR, SO00]
G4 ATZ K M Cleary, Transener Sa, Pasco Colon 728, 60 Piso, (1063) Buenos Aires, Argentina
G4 AUB A J Smith, 7 Hobley Cl, Bilton, Rugby, CV22 7PU [IO92II, SP47]
G4 AUC S J Baugh, 70 Madingley, Bracknell, RG12 7TF [IO91OJ, SU86]
G4 AUD A Lacy, 58 Bilbrook Rd, Codsall, Wolverhampton, WV8 1ER [IO82VP, SJ80]
G4 AUF C D Friel, 5 Windmill Hill, Ruislip, HA4 8QF [IO91TN, TQ08]
G4 AUG R A Mortimer, 175 Englands Ln, Loughton, IG10 2NS [JO01AP, TQ49]
G4 AUH R C Walker, 215 Springfield Rd, Walmley, Sutton Coldfield, B76 2SZ [IO92CN, SP19]
G4 AUJ W F Shea, 32 High St., Brampton, Huntingdon, PE18 8TH [IO92VH, TL27]
G4 AUK Details withheld at licensee's request by SSL.
G4 AUL G K Mitchell, 10 Wealden Cl, Hildenborough, Tonbridge, TN11 9HB [JO01DF, TQ54]
G4 AUM N Nuttall, 1 Batts Park, Queens Dr, Taunton, TA1 4RE [IO81KA, ST22]
G4 AUN R A Collett, 70 Clifton Rd, Darlington, DL1 5DX [IO94FM, NZ21]
GM4 AUP I D Suart, 37 Meldrum Mains, Glenmavis, Airdrie, ML6 0QQ [IO85AV, NS76]
G4 AUQ F W Barker, 8 Burdon Cl, Willerby, Hull, HU10 6QZ [IO93SR, TA02]
G4 AUR J McBurney, Hilbre, 4 Fownhope Rd, Brooklands, Sale, M33 4RF [IO83TK, SJ79]
G4 AUS J Macdonald, 69 Marshall St., Leicester, LE3 5FA [IO92KP, SK50]
G4 AUV G Wing, 105 Moore Ave, Sprowston, Norwich, NR6 7LG [JO02PP, TG21]
G4 AUX P J Himsworth, 21 Lymmington Ave, Lymm, WA13 9NQ [IO83SJ, SJ68]
G4 AUY P J Sherwood, 43 Kingsland, Wellington, Arleston, Telford, TF1 2LE [IO82RQ, SJ61]
G4 AUZ M G Davies, Overdale, Sugden, Telford, Shropshire, TF6 6NB [IO82RR, SJ61]
G4 AV R H Forward, 5 Pickmere Cl, Thornton Cleveleys, FY5 2HP [IO83LV, SD34]
G4 AVB Details withheld at licensee's request by SSL.
GW4 AVC D W Bowers, 8 Maesydre Rd, Wrexham, LL12 7AS [IO83MB, SJ35]
G4 AVD R Berry, 7 Dundee Cl, Cinnamon Brow, Fearnhead, Warrington, WA2 0UJ [IO83RK, SJ69]
G4 AVE L V Cates, 45 Smoke Ln, Reigate, RH2 7HJ [IO91VF, TQ24]
G4 AVF A Fletcher, 7 Milton Dr, Chadderton, Oldham, OL9 9RA [IO83WM, SD90]
G4 AVJ G D Pople, 38 Killams Cres, Fullands Park, Taunton, TA1 3YB [IO80LX, ST22]
G4 AVK S P Ripley, 62 Palewell Park, East Sheen, London, SW14 8JH [IO91UL, TQ27]
G4 AVL P L Newby, 92 Long Rd, Lawford, Manningtree, CO11 2HS [JO01MW, TM03]
G4 AVN T N Thompson, 146 Hawthorn Rd, Ashington, NE63 9BG [IO95FE, NZ28]
G4 AVO R J Thompson, 30 Dene View, Ashington, NE63 8JF [IO95EE, NZ28]
G4 AVP P C Cook, Bagshott Cottage, Beaconhill, Sheldwich, Faversham Kent, ME13 0LA [JO01KG, TR05]
G4 AVS R C Wilson, Aerial House, 1 The Fields, Tunstall, Woodbridge, IP12 2HZ [JO02RD, TM35]
G4 AVT V S Evans, Beacon View, Parbold Hill, Parbold, Wigan, WN8 7TG [IO83PO, SD51]
G4 AVU Details withheld at licensee's request by SSL.
G4 AVV G Cluer, 12 Bingham Rd, Addiscombe, Croydon, CR0 7EB [IO91XJ, TQ36]
G4 AVW C A R Beyfus, 10 Denham Lodge, Oxford Rd, Denham, Uxbridge, UB9 4AA [IO91SN, TQ08]
G4 AVX A Newman, Lyndhurst, 111 Gladys Ave, North End, Portsmouth, PO2 9BD [IO90LT, SU60]
G4 AVY A Heckley, 8 Woodland Gr, Bembridge, PO35 5SG [IO90KQ, SZ68]
G4 AVZ Dr W E Arnold, Higher Woodhall, Exbourne, Okehampton, Devon, EX20 3QZ [IO70XT, SS50]
GM4 AWA R T Payne, Conifers, Lambourn, Wolfhill, Perth, PH2 6DX [IO86HM, NO13]
G4 AWB R A Macduff, 50 Kingsall Rd, Attadale, 6156, Western Australia
G4 AWD A J Young, 9 Larchfield House, Highbury Est, Highbury, London, N5 2DE [IO91WN, TQ38]
G4 AWF D V Wilson, Barnside, Straight Rd, Battisford, Stowmarket, IP14 2LZ [JO02LD, TM05]
G4 AWG G R Higgs, Firtree House, Perry Wood, Selling, Faversham, ME13 9SE [JO01LG, TR05]
G4 AWI G Cole, 19 Oxen Ave, Shoreham By Sea, BN43 5AF [IO90UU, TQ20]
G4 AWJ G R Thomas, Crofters Cottage, 9 Highcroft Cres, Heathfield, TN21 8HE [JO00DX, TQ52]
G4 AWM D W E Norfolk, 13 Oakwood Cres, Greenford, UB6 0RF [IO91UN, TQ18]
G4 AWO R R Gray, 5 Bellis House, Kilworth Cl, Welwyn Garden City, AL7 3HS [IO91VS, TL21]
GM4 AWP R H Parker, Rosemount, 77 Main St., Colmonell, Girvan, KA26 0RY [IO75ND, NX18]
G4 AWT G Boothroyd, 38 Ascot Ave, Cantley, Doncaster, DN4 6HE [IO93KM, SE60]
G4 AWU R Lane, Kelston, Doncaster Rd, Bawtry, Doncaster, DN10 6NQ [IO93LK, SK69]
G4 AWW N W Shepherd, Jokebi, Stonehall, Lydden, Dover, CT15 7JS [JO01OD, TR24]
G4 AWY R A Mekka, 57 St. Johns Rd, Caversham, Reading, RG4 5AL [IO91ML, SU77]
G4 AWZ P Matthews, 22 Rydens Rd, Walton on Thames, KT12 3DA [IO91TJ, TQ16]
G4 AXA N G Pope, 136 Ridgeway Dr, Bromley, BR1 5DD [JO01AK, TQ47]
G4 AXC C G Burden, Cedar Croft, Hengar Ln, St. Tudy, Bodmin, PL30 3PH [IO70PN, SX07]
G4 AXD G D Edy, 8 Farnborough Cl, Maidstone, ME16 8UE [JO01GG, TQ75]
G4 AXF J Jacques, 30 Centurian Way, Glebe Farm Est, Bedlington, NE22 6LD [IO95ED, NZ28]
G4 AXK J P Castell, 45 Layters Cl, Chalfont St. Peter, Gerrards Cross, SL9 9HS [IO91RO, SU99]
G4 AXL C Gerrard, 22 Kelso Dr, The Priorys, North Shields, NE29 9NS [IO95GA, NZ36]
G4 AXN A F Parker, 105 The St., Rockland St. Mary, Norwich, NR14 7HQ [JO02RO, TG30]
G4 AXO J H Wills, 48 Fairfield Rd, Winchester, SO22 6SG [IO91IB, SU43]
G4 AXP G K Clode, Haselbech, 20 Main Rd, Brookville, Thetford, IP26 4RB [JO02GM, TL79]
GM4 AXS P W C Wilberforce, Connel,By Oban, Argyll, PA37 1RA [IO76HL, NM93]
G4 AXU G D Parr, Chesil Coppice, West Bexington, Dorchester, Dorset, DT2 9DD [IO80QQ, SY58]
GI4 AXV J G Doherty, 172 Dunmore Rd, Guiness, Ballynahinch, BT24 8QQ [IO74BI, J34]
G4 AXW S R Jones, 40 Woodcot Park Dr, Wilmcote, Stratford upon Avon, CV37 9XT [IO92CF, SP15]
G4 AXX M Marsden, 38 Lambert Cross, Saffron Walden, CB10 2DP [JO02CA, TL53]
G4 AXY A M Mort, 86 Longfield Rd, Winnall Est, Winchester, SO23 0NU [IO91IB, SU42]
G4 AYA P H Jeffs, 41 The Downs, Blundellsands Rd West, Liverpool, L23 6XS [IO83LL, SJ39]
G4 AYB A H Kelle, Apt.18. Versailles, La Massana Park, La Massana, Andorra, XXXX XXX
G4 AYD T J Hodgetts, 15 Wiltons, Wrington, Bristol, BS18 7LS [IO81OI, ST46]
GW4 AYJ Details withheld at licensee's request by SSL.
G4 AYK P Perrins Mid Severn Vall, Raynet Group, 9 Merrick Cl, Halesowen, B63 1JY [IO82XK, SO98]
G4 AYL E P Lambert, 41 Brand Hill Dr, Crofton, Wakefield, WF4 1PF [IO93GP, SE31]
G4 AYM Details withheld at licensee's request by SSL.[Correspondence to: N Negus, G6AWT, obo Gloucester ARS, 41 Oxtalls Lane, Longlevens, Gloucester, GL2 9HP.]
G4 AYO M J Hewitt, 10 Blacka Moor View, Sheffield, S17 3GZ [IO93FH, SK38]
G4 AYP S S Kind, 7 Beckwith Rd, Harrogate, HG2 0BG [IO93FX, SE25]
GW4 AYQ J E Durrans, 87 The Links, Trevethin, Pontypool, NP4 8DQ [IO81LR, SO20]
G4 AYR T R R H Greenwood, 30 Ringwood Rd, Risinghurst, Headington, Oxford, OX3 8JA [IO91JS, SP50]
G4 AYS A E Crook, 153 Hawthorn Cottage, Shortheath, Moira, Swadlincote, Derbyshire, DE12 6BL [IO92FR, SK31]
G4 AYT Details withheld at licensee's request by SSL.
G4 AYV Details withheld at licensee's request by SSL.
G4 AYW W F Mills, 1 Alderwood Cl, Littlepark, Havant, PO9 3RG [IO90LU, SU60]
G4 AYY K Condliffe, 8 Conifer Gr, Blurton, Stoke on Trent, ST3 2EN [IO82WX, SJ84]
G4 AZA R F Winkworth, 13 Bagley Cl, Kennington, Oxford, OX1 5LS [IO91JR, SP50]
G4 AZC P I Martin, Wellow Top Rd, Wellow, Ningwood, Yarmouth, Isle of Wight, PO41 0TL [IO90GQ, SZ38]
G4 AZD A G Edgecock, Sunnydene, Station Rd, Thorrington, Colchester, CO7 8JA [JO01MU, TM02]
GW4 AZE R Curtis, 1 Avondale St., Ynysbeoth, Abercynon, Mountain Ash, CF45 4YU [IO81HP, ST09]
G4 AZF T L Flatt, 38 Beaulieu Gdns, Blackwater, Camberley, GU17 0LA [IO91OI, SU86]
G4 AZG Dr G H Macdonald, Pilgrims Cottage, Kingston, Canterbury, Kent, CT4 6HX [JO01NF, TR25]
G4 AZH M V Bushnell, Rose Cottage, Revel, St. Ashton, Rugby, CV23 0PH [IO92IK, SP48]
GW4 AZI D K Thomas, Sunnydale, Scurlage, Gower, Swansea, SA3 1BA [IO71VN, SS48]
G4 AZJ R Troughton, Hopton Grange, Short Gr Ln, Hopton, Diss, IP22 2RP [JO02LI, TL97]
G4 AZK D N Mason, 91 Buckingham Gdns, West Molesey, KT8 1TN [IO91TJ, TQ16]
G4 AZL P Justin, 57 Birchmead Ave, Pinner, HA5 2BQ [IO91TO, TQ18]
G4 AZM C Wilson, 16605 Cordoba St, Winter Garden, Florida 34787, USA
G4 AZN B M Crook, 41 Radley Rd, Abingdon, OX14 3PL [IO91IQ, SU59]
G4 AZO K J Dorrell, 1 Rue de Namur, L-2211, Luxembourg
G4 AZQ M R Sex, 17 Wealdon Cl, Southwater, Horsham, RH13 7HP [IO91TA, TQ12]
G4 AZR Details withheld at licensee's request by SSL.
G4 AZS A E Bayling, 55 Shelton Rd, Shrewsbury, SY3 8SU [IO82OQ, SJ41]
G4 AZT T J Barker, 4 Cres Cl, Cowley, Oxford, OX4 2NU [IO91JR, SP50]
G4 AZU J C Tiller, 21 Portal Rd, Winchester, SO23 0PX [IO91IB, SU42]
G4 AZV R P Millington, Laburnam Cottage, Asterley, Minsterley, Shropshire, SY5 0AW [IO82MP, SJ30]
GW4 AZW T W Lewis, Ty Gaeia, Llanidloes Rd, Newtown, SY16 1EZ [IO82IM, SO19]
G4 AZX J Robinson, 19 Sunnycroft Gdns, Cranham, Upminster, RM14 1HP [JO01DN, TQ58]
G4 AZZ Details withheld at licensee's request by SSL.[Magnum Contest Group. QSL (with IRCs) via GM3WTA.]
G4 BAB Details withheld at licensee's request by SSL.
GM4 BAE G D A Maxwell, Northbrae, Brookfield Cres, Stranraer, DG9 0HY [IO74LV, NX06]
G4 BAH R B P Carpenter, 33 Lark Rise, Martlesham Heath, Ipswich, IP5 7SA [JO02OB, TM24]
G4 BAI J T Roughton, Four Gables, Vicarage Cl, Hallow, Worcester, WR2 6PA [IO82VF, SO85]
G4 BAL M A Lee, 46 Harman Dr, Sidcup, DA15 8LY [JO01BK, TQ47]
G4 BAN P F Godfrey, 5 Parkway, Southgate, London, N14 6QU [IO91WP, TQ39]
G4 BAO J C Worsnop, 20 Lode Ave, Waterbeach, Cambridge, CB5 9PX [JO02CG, TL46]
G4 BAQ R J Chambers, 42 Beacon Way, Littlehampton, BN17 6QS [IO90RT, TQ00]
G4 BAS Details withheld at licensee's request by SSL.[Club of Friendship between UK and Russian radio amateurs. UK co-ordinator: Howard Ketley, 1 Tewkesbury Ave, Mansfield Woodhouse, Notts, NG19 8LA.]
G4 BAU R T Russell, 19 Welling Rd, Orsett, Grays, RM16 3DF [JO01EM, TQ68]
G4 BAV J Gee, 11 Charlton Ave, Ipswich, IP1 6BH [JO02NB, TM14]
G4 BAZ Details withheld at licensee's request by SSL.
G4 BB M Storey, 104 Stanfell Rd, Leicester, LE2 3GB [IO92KO, SK50]
G4 BBD M A Tooley, The Elms, 290 Spring Rd, Sholing, Southampton, SO19 2NX [IO90HV, SU41]
GI4 BBE R F Bolton, Ohmvilla, 69 Newcastle St., Kilkeel, Newry, BT34 4AQ [IO64XB, J31]
GM4 BBF D B Sleigh, 43 Murieston Rd, Livingston, EH54 9AX [IO85GU, NT06]
G4 BBH R C Ferryman, 2 Mayfield Gdns, Dover, CT16 2PQ [JO01PD, TR34]
G4 BBI P Nixon, 28 Lake View Ave, Chesterfield, S40 3DR [IO93GF, SK37]
G4 BBJ R L Ramsay, 1 Sapho Park, Gravesend, DA12 4NA [JO01EK, TQ67]
G4 BBL A R Thackery, Hayfield, 29 Moorside, Fulwood, Sheffield, S10 4LN [IO93FI, SK28]
G4 BBO J R Grice, 38 Crockford Rd, West Bromwich, B71 2ES [IO92AN, SP09]
G4 BBP Details withheld at licensee's request by SSL.
G4 BBQ D W King, 62 Ansley Rd, Stockingford, Nuneaton, CV10 8NU [IO92FM, SP39]
G4 BBR R C Harvey, 30 Wimborne Cl, Upper Hatherley, Up Hatherley, Cheltenham, GL51 5QP [IO81WV, SO92]
G4 BBS Details withheld at licensee's request by SSL.
G4 BBT Dr R A Hancock, 80 Ulleries Rd, Solihull, B92 8EE [IO92CK, SP18]
G4 BBU P Whittle, 20 Marlbrook Ln, Marlbrook, Bromsgrove, B60 1HN [IO82XI, SO97]
G4 BBW A P Smith, 9 Orchard Ct, Harestone Valley Roa, Caterham, CR3 6HE [IO91WG, TQ35]
G4 BBX F Chidlon, 64 Mitchell Ave, Northside, Workington, CA14 1AA [IO84FP, NY02]
G4 BBZ S P Ball, 1 Brindlegate, Kirkland St., Pocklington, York, YO4 2HB [IO93OW, SE84]
G4 BCA D L Tunnicliffe, Greycot, Mill Ln, Falfield, Wotton under Edge, Glos, GL12 8BU [IO81SP, ST69]
GW4 BCB Dr K R Johnston, 28 Heol Isaf, Radyr, Cardiff, CF4 8AL [IO81IM, ST18]
GW4 BCC R D Davies, 22 Lake Hill Dr, Cowbridge, CF7 7HR [IO81HN, ST08]
GW4 BCF R J Newman, 21 Catkin Dr, Cowslip Est, Penarth, CF64 2RD [IO81JK, ST17]
G4 BCG G W Wale, 2 The Jordans, Allesley Park, Coventry, CV5 9JT [IO92FJ, SP37]
G4 BCH P D Burgess, Bryher House, Howgate Rd, Bembridge, Isle of Wight, PO35 5UA [IO90KQ, SZ68]
G4 BCJ J S Wilson, 62 Wanstead Park Rd, Cranbrook, Ilford, IG1 3TQ [JO01AN, TQ48]
GM4 BCK Details withheld at licensee's request by SSL.
G4 BCM Details withheld at licensee's request by SSL.
G4 BCO B Houghton, 165 Hillside Rd, Hastings, TN34 2QJ [JO00GV, TQ81]
G4 BCP L A Graves, Woodpeckers, Hillside, Rothbury, Morpeth, NE65 7PT [IO95BH, NU00]
G4 BCS J L Buckingham, 1C Biddenham Turn, Bedford, MK40 4AT [IO92SD, TL05]
G4 BCT A Gordon, 34 Rose Ave, Wigan, WN6 8QX¶IO83QN, SD50]
G4 BCV J Binning Essex Raynet, 293 Perry St., Billericay, CM12 0RB [JO01FP, TQ69]
G4 BCW N P Hammersley, 1 Manor Gdns, Desford, Leicester, LE9 9QB [IO92IP, SK40]
G4 BCX A G B Helm, 38 Blandford Rd, Lower Compton, Plymouth, PL3 5DU [IO70WJ, SX45]
G4 BCY J H White, Heathcroft, Seymour Rd, Headley Down, Bordon, GU35 8JX [IO91OC, SU83]
G4 BCZ J White, 24 Derwent Cl, Bordon, GU35 0NR [IO91NC, SU73]
G4 BDB S G White, 19 Manilla Rd, Clifton, Bristol, BS8 4EB [IO81QK, ST57]
G4 BDC K W Collerton, Bp Exploration (Columbia) Ltd, Cra 9A No99-02, Bogota, Colombia
G4 BDD E F G Dowsett, 23 Everest Ln, Frindsbury, Rochester, ME2 3XA [JO01FJ, TQ77]
G4 BDE R Lister, 42 Cartmel Dr, Ulverston, LA12 9PF [IO84KE, SD27]
G4 BDG K Gray, 31 Broomfield, Adel, Leeds, LS16 6AE [IO93EU, SE23]
GM4 BDJ R B McCartney, Cairndhue, Walter St., Langholm, DG13 0AX [IO85LD, NY38]
G4 BDK A W Trowman, 155 Portland Rd, Edgbaston, Birmingham, B16 9TD [IO92AL, SP08]
GI4 BDL V Simpson, 26 Oak Grange, Waringstown, Craigavon, BT66 7SU [IO64UK, J15]
G4 BDN T W Kerr, 95 Aldwick Rd, Bognor Regis, PO21 2NY [IO90PS, SZ99]
G4 BDO Details withheld at licensee's request by SSL.[Op: G D Gambold.]
G4 BDP Details withheld at licensee's request by SSL.
GI4 BDR Dr N E Evans, 189 Gulladuff Rd, Bellaghy, Magherafelt, BT45 8LW [IO64RT, H99]
GW4 BDS R B Mowbray, 15 St. Margarets Dr, Llanelli, SA15 4EW [IO71VQ, SN40]
G4 BDU R W Ling, 13 Buttermere Cl, Kettering, NN16 8LZ [IO92PJ, SP87]
GW4 BDV J H Bird, Bryneynon, Tydraw Villas, Brynmenyn, Bridgend, CF32 9LR [IO81FN, SS98]
G4 BDW J Bagley, 109 Norwich Rd, Wymondham, NR18 0SJ [JO02NN, TG10]
G4 BDX M W Horoszko, Woodgarth Cottage, Reedness, Goole, South Humberside, DN14 8EX [IO93OQ, SE72]
G4 BDY D C Donnison, 23 Layston Park, Royston, SG8 9DS [IO92XB, TL34]
G4 BEB R C Browning, 6 Highwoods Dr, Marlow Bottom, Marlow, SL7 3PY [IO91OO, SU88]
G4 BEE Details withheld at licensee's request by SSL.[Station located near Chorley.]
GD4 BEG E M P Farrant, Skerrisdale, Michael, Isle of Man, Isle of Man, IM6 1MP
G4 BEJ E A Siggers, 15 Woodleigh Rd, Sutton Coldfield, B72 1ER [IO92CN, SP19]
G4 BEL R J Taylor, 12 The Rampart, Haddenham, Ely, CB6 3ST [JO02BI, TL47]
G4 BEM S Ford, 3 Hill View, Milton, Stoke on Trent, ST2 7AR [IO83WB, SJ95]
G4 BEO B K Hailstone, 6 Larkswood Rise, St. Albans, AL4 9JU [IO91US, TL10]
G4 BEQ G D Hotchkiss, 4 Erica Cl, Tudor Wood, Locks Heath, Southampton, SO31 6SD [IO90IU, SU50]
G4 BER A G Trend, 23 Bosley View, Congleton, CW12 3TU [IO83VD, SJ86]
GM4 BES G H Mercer, 5 Murray Villas, Heugh Rd, Portpatrick, Stranraer, DG9 8TE [IO74KU, NW95]
G4 BEU J G Small, 20 Hastings Rd, Birkdale, Southport, PR8 2LW [IO83LO, SD31]
G4 BEV R C Taylor, 6 Churchill Cres, Marple, Stockport, SK6 6HJ [IO83XJ, SJ98]
G4 BEZ J Phillipson, 51 Oaktree Dr, Hook, RG27 9RA [IO91MG, SU75]
G4 BFA Details withheld at licensee's request by SSL.
GM4 BFC A S Riddell, 140 Crewe Rd, Sandbach, CW11 4PX [IO83TD, SJ76]
G4 BFH J E Duddington, 8 The Grove, Thornton Cleveleys, FY5 2JD [IO83LV, SD34]
G4 BFK R Morgan, 25 Riggotts Way, Cutthorpe, Chesterfield, S42 7AW [IO93GG, SK37]
G4 BFM A N Kimble, Barrasmead, Staplehay, Trull, Taunton, TA3 7HF [IO80KX, ST22]
G4 BFP Details withheld at licensee's request by SSL.[Op: R S Wilson.]
G4 BFR D C Baldwin, 112 Moorland View Rd, Chesterfield, S40 3DF [IO93GF, SK37]
G4 BFS T C Sargent, 7 Riverview Flats, London Rd, Purfleet, RM19 1SL [JO01DL, TQ57]
G4 BFT C Johnson, 17 Verbena Way, Great Hay, Sutton Hill, Telford, TF7 4DX [IO82SP, SJ70]
G4 BFV D A Sinclair, 1 Waxwing Cl, Watermead, Aylesbury, HP19 3WT [IO91NT, SP81]
G4 BFW J R Matthews, 46 Park Ln, West Grinstead, Horsham, RH13 8LT [IO91UB, TQ13]
GW4 BFX Dr A B Milne, 11 Cromwell Ct, Forbesfield Rd, Aberdeen, AB15 4WB [IO87WD, NJ90]
G4 BFY Details withheld at licensee's request by SSL.
G4 BG A Duckworth, Ambergate, 2 Ashleigh Dr, Teignmouth, TQ14 8QX [IO80GN, SX97]
GI4 BGB P B Kelly, 30 Cahore Rd, Draperstown, Co Derry, N Ireland, N Ireland, BT45 7LY [IO64OS, H79]
G4 BGH A J Ruddell, 53 Steventon Rd, Drayton, Abingdon, OX14 4LA [IO91IP, SU49]
G4 BGL L Walsh, Nabs Head Ln, Sowerbutts Green, Samlesbury, Preston, Lancs, PR5 0UQ [IO83RS, SD63]
G4 BGM C Zeal, Yew Tree House, Church Ln, Rowledge, Farnham, GU10 4EN [IO91OE, SU84]
G4 BGQ Details withheld at licensee's request by SSL.
GM4 BGS S Liddell, 49 Inchbrae Rd, Cardonald, Glasgow, G52 3HA [IO75TU, NS56]
G4 BGT M A Staton, 65 Campbell Rd, Eastleigh, SO50 5AA [IO90HX, SU41]
G4 BGW I E Wilson, Whitethorn, Sandhurst Ln, Sandhurst, Gloucester, GL2 9NW [IO81VV, SO82]
G4 BGX Details withheld at licensee's request by SSL.

G4 BGZ J A Stanyon, 38 Edison Gr, Quinton, Birmingham, B32 2SQ [IO92AK, SP08]
G4 BHB P C Bellwood, 55 de Walden House, Allitsen Rd, St. Johns Wood, London, NW8 7BA [IO91VM, TQ28]
G4 BHC F A Stevens, 11 Hen Wythva, Camborne, TR14 7XN [IO70IF, SW63]
G4 BHD T H Goldsworthy, 24 Rose Cttgs, Camborne, TR14 8DD [IO70IF, SW64]
G4 BHE B W Macklin, 1 Cedar Tree Cl, Oakley, Basingstoke, RG23 7EG [IO91JF, SU54]
G4 BHJ M K Fochtmann, 1 Chapmans Way, Over, Cambridge, CB4 5PZ [JO02AH, TL37]
G4 BHO J H Robinson, 70 Trinity Rd, Stotfold, Hitchin, SG5 4EQ [IO92VA, TL23]
G4 BHP G T Benwell, The Did, Tunley, Bath, Avon, BA3 1DZ [IO81SH, ST65]
G4 BHQ L V Bright, 4 Dagley Farm Park, Shalford, Guildford, Surrey, GU4 8DE [IO91RF, SU94]
G4 BHT M R Hulands, Chy Carne, 14 Nene Rise, Cogenhoe, Northampton, NN7 1NT [IO92OF, SP86]
GM4 BHU D A Aitkenhead, 37/3 Cavalry Park Dr, Duddingston, Edinburgh, EH15 3QG [IO85KW, NT27]
G4 BHV A T Bernstein, 23 Barn Hall Ave, Colchester, CO2 8TE [JO01LU, TM02]
G4 BHX Dr J L Bowmer, Tillylodge Cottage, Lumphanan, Banchory, Kincardinshire, Scotland, AB31 4SD [IO87OD, NJ50]
G4 BHY H Kleeman, 41 Frognal, London, NW3 6YD [IO91VN, TQ28]
G4 BHZ M D Bull, 42 Church Leaze, Shirehampton, Bristol, BS11 9SZ [IO81PL, ST57]
G4 BI L J Philpott, 17 Tregonning Parc, St. Kevering, St. Keverne, Helston, TR12 6QF [IO70KB, SW72]
G4 BIA R A Hood, 21 Norwood Gdns, Ashford, TN23 1JP [JO01KD, TR04]
G4 BIB Details withheld at licensee's request by SSL.
G4 BIC J T Clegg, 75 Patch Ln, Bramhall, Stockport, SK7 1HR [IO83VI, SJ88]
G4 BID W Boyd, 17 Kipling Cl, Yateley, Camberley, Surrey, GU17 7YA [IO91OH, SU85]
G4 BIH H Butterworth, 57 St. Martins St., Castleton, Rochdale, OL11 2UW [IO83VO, SD81]
G4 BII D E Williams, 2 Poundon, Bicester, Oxon, OX6 0AZ [IO91LW, SP62]
G4 BIJ Details withheld at licensee's request by SSL.
G4 BIK P B Mellor, 10 Greenfields, Earith, Huntingdon, PE17 3QH [JO02AI, TL37]
G4 BIN N R W Long, Homedale, Bayford Hill, Wincanton, BA9 9LS [IO81TB, ST72]
G4 BIO A Chivers, 18 West Ave, Tylers Green, Penn, High Wycombe, HP10 8AE [IO91PP, SU99]
G4 BIP Details withheld at licensee's request by SSL.
GW4 BIS A Davies, 12 Church St., Troedyrhiw, Merthyr Tydfil, CF48 4HD [IO81HR, SO00]
GM4 BIT R Wilson, 5 Collins Dr, Loans, Troon, KA10 7HA [IO75QN, NS33]
G4 BIU W Elliot, 46A Senhouse St., Maryport, CA15 6BL [IO84GR, NY03]
G4 BIW Details withheld at licensee's request by SSL.
G4 BIX Dr D D Price, 34 Vanda Cres, St. Albans, AL1 5EX [IO91UR, TL10]
G4 BIZ A D H Paxton, 17 Queens Rd, Tichbourne Cottage, Lyndhurst, SO43 7BR [IO90FU, SU30]
G4 BJC R G Rugg I.S.W.L ARC, 29 Milbourne Park, Milbourne, Malmesbury, SN16 9JE [IO81WN, ST98]
G4 BJD G C Overton, Freens Ct House, Sutton St Nicholas, Hereford, HR1 3AY [IO82PC, SO54]
GW4 BJE Details withheld at licensee's request by SSL.
G4 BJF B R Marshall, 23 Sandgate Ave, Birstall, Leicester, LE4 3HQ [IO92KQ, SK50]
G4 BJG Details withheld at licensee's request by SSL.
G4 BJJ H Tickell, 26 Shear Brow, Blackburn, BB1 7EX [IO83SS, SD62]
GI4 BJK K J Patterson, 92 Castlereagh Rd, Belfast, BT5 5FR [IO74BO, J37]
G4 BJM Details withheld at licensee's request by SSL.
G4 BJN D J Harvey, 23 Lapwing Cl, Hemel Hempstead, HP2 6DS [IO91SS, TL00]
G4 BJO B W Greaves, 65 Stowupland Rd, Stowmarket, IP14 5AN [JO02ME, TM05]
G4 BJP S J N Popek, 42 Victoria Rd, Polegate, BN26 6DA [JO00CT, TQ50]
G4 BJQ I F Ireland, 118 Mytchett Rd, Mytchett, Camberley, GU16 6ET [IO91PG, SU85]
G4 BJS J C Loose, 43 Willows Cres, Birmingham, B12 9NE [IO92BK, SP08]
G4 BJT Dr M J Ware, 20 Bath Rd, Buxton, SK17 6HH [IO93BG, SK07]
G4 BJX W D Whatmore, 51 The Fairways, Sherford, Taunton, Somerset, TA1 3PA [IO81KA, ST22]
G4 BKA A J Neaves, 8 East House Dr, Hurley, Atherstone, CV9 2HB [IO92EN, SP29]
G4 BKB G C Jessup, 68 Danes Rd, Bicester, OX6 8LR [IO91JV, SP52]
G4 BKC C D Hopper, 10 Keats Dr, Towcester, NN12 6LT [IO92MD, SP64]
G4 BKE D H Wright, 4 Wynne Cl, Broadstone, BH18 9HQ [IO80XS, SY99]
G4 BKF T E Howarth, 71 Ford Rd, Upton, Wirral, L49 0TD [IO83KJ, SJ28]
GW4 BKG S Emlyn-Jones, 163 Llangewydd Rd, Cefn Glas, Bridgend, CF31 4JT [IO81EM, SS88]
G4 BKH A Chorley, 354 Denton Ln, Chadderton, Oldham, OL9 8QD [IO83WM, SD90]
G4 BKI P H Evans, Sunlea, Wheal Speed, Carbis Bay, St. Ives, TR26 2PT [IO70GE, SW53]
G4 BKM Details withheld at licensee's request by SSL.
G4 BKO J E Francis, Penrhyn, 204 Rykneld Rd, Littleover, Derby, DE23 7AN [IO92FV, SK33]
G4 BKP A G L Parker, The Nest, Beck St., Hepworth, Diss, IP22 2PN [JO02LI, TL97]
G4 BKQ R A Gubbins, 29 Meadow End, Gotham, Nottingham, NG11 0HP [IO92JU, SK53]
GI4 BKR W R Taggart, 35 Innishargie Gdns, Bangor, BT19 1SN [IO74DP, J48]
G4 BKS Details withheld at licensee's request by SSL.
GM4 BKV S Sutherland, 67 Greenfern Rd, Mastrick, Aberdeen, AB1 6TP [IO87TH, NJ72]
G4 BKW R Shears, 3 Windmill Park, Wrotham Heath, Sevenoaks, TN15 7SY [JO01EG, TQ65]
G4 BLD C , Croucher, 13 Magnolia Way, Pilgrims Hatch, Brentwood, CM15 9QS [JO01DP, TQ59]
GW4 BLE S R Cole, 101 Allt yr Yn Rd, Newport, NP9 5EF [IO81LO, ST28]
G4 BLF P Malon, 126 Barrow Hill, Goodworth Clatford, Andover, SP11 7RG [IO91GE, SU34]
G4 BLG A E Head, 19 Brodrick Gr, Abbey Wood, London, SE2 0SR [JO01BL, TQ47]
G4 BLH M Crawshaw, 50 Kibble Gr, Brierfield, Nelson, BB9 5EW [IO83VT, SD83]
G4 BLI A L Hart, 17 Minses Cl, Elburton, Plymouth, PL9 8DR [IO70XI, SX55]
G4 BLJ Details withheld at licensee's request by SSL.
G4 BLL P Burnett, 26 Stancliffe Way, Kirkheaton, Huddersfield, HD5 0JY [IO93DP, SE11]
G4 BLM N F Burton, 30 Holyrood, Great Holm, Milton Keynes, MK8 9AF [IO92OA, SP83]
G4 BLN Details withheld at licensee's request by SSL.
GM4 BLO Details withheld at licensee's request by SSL.
G4 BLQ Details withheld at licensee's request by SSL.
G4 BLS P W Appleby, 10 Downsview Way, Hailsham, BN27 3DH [JO00DU, TQ50]
G4 BLT R C Sterry, 1 Wavell Garth, Sandal Magna, Wakefield, WF2 6JP [IO93GP, SE31]
G4 BLU G F Young, 33 Mill Park Ave, Hornchurch, RM12 6HD [JO01CN, TQ58]
G4 BLX D E Austin, Limekiln, Upper Beacon Rd, Ditchling, Hassocks, BN6 8XD [IO90WV, TQ31]
G4 BMC D G Barrell, Woodlawn, Vaggs Ln, Hordle, Lymington, SO41 0FP [IO90ES, SZ29]
G4 BMD M B Hayes, 11 Denbigh Rd, Tunbridge Wells, TN4 9HS [JO01DD, TQ54]
G4 BME I E Farrow, 86 Byron Cres, Coppull, Chorley, PR7 5BD [IO83QO, SD51]
G4 BMK M J Kerry, 2 Beacon Cl, Seaford, BN25 2JZ [JO00BS, TQ40]
G4 BMM P Knight, 10 Collingtree, Luton, LU2 8HN [IO91TV, TL12]
G4 BMO D M Cloke, Church Cottage, Burton Cross, East Coker, Yeovil Somerset, BA22 9LY [IO80QW, ST51]
G4 BMP R L Sadler, 9 Meade King Gr, Woodmancote, Cheltenham, GL52 4UD [IO81XW, SO92]
G4 BMQ D S Harrop, 1 Edgecombe Cres, Rowner, Gosport, PO13 9RD [IO90JT, SU50]
G4 BMR D P Riddle, 26 Avonfield Ave, Bradford on Avon, BA15 1JE [IO81VI, ST86]
G4 BMU S East, Flat C, Linscott House, Russell Rd, Buckhurst Hill, Essex, IG9 5QE [JO01AP, TQ49]
G4 BMV W Bryan, 2 Easedale, Seaton Sluice, Whitley Bay, NE26 4HR [IO95GC, NZ37]
G4 BMW C R Pescod, 7 Brian Cl, Chelmsford, CM2 9DZ [JO01FR, TL70]
G4 BNB R A Wynn, 48 Darnley Rd, Woodford Green, IG8 9HY [JO01AO, TQ49]
G4 BNE Dr R N Herring, 96 St. Fabians Dr, Chelmsford, CM1 2PR [JO01FR, TL60]
G4 BNG G L Portrey, Bethelen, Grasmere, Ambleside, Cumbria, LA22 9RL [IO84LL, NY30]
G4 BNH F Hogg, 26 Dene Hill, Baildon, Shipley, BD17 5BA [IO93CU, SE13]
G4 BNI Details withheld at licensee's request by SSL.[Op: G H Taylor. Station located near Maidstone.]
GW4 BNJ D H Williams, 48 St. Hilary Dr, Killay, Swansea, SA2 7EH [IO71XO, SS59]
G4 BNK W H Wright, 27 St. Johns Rd, Cove, Farnborough, GU14 9RL [IO91OH, SU85]
G4 BNL R D Morley, Meadow View, 26 Holt Ln, Adel, Leeds, LS16 7NU [IO93EU, SE24]
GM4 BNM S W Homans, 4 Bucklerscroft, Kellas, Broughty Ferry, Dundee, DD5 3PQ [IO86NM, NO43]
G4 BNN P P Trott, Morialta, The Inner Down, Tockington, Bristol, BS12 4PR [IO81RO, ST68]
G4 BNO M C Ayling, 68 Littledown Ave, Queens Park, Bournemouth, BH7 7AS [IO90BR, SZ19]
G4 BNP J W Burgess, 11 Winters Ln, Ottery St. Mary, EX11 1AR [IO80IS, SY09]
G4 BNQ Details withheld at licensee's request by SSL.
G4 BNS A Collinson, Elmwood, 25 Thornton Rd, Pickering, YO18 7HZ [IO94OF, SE88]
G4 BNU Details withheld at licensee's request by SSL.
G4 BNV A S Le Good, Monkhams, Broad Oak Cl, West Hill, Ottery St Mary Devo, EX11 1XW [IO80IR, SY09]
G4 BNW M B Knight, 19 Friary Rd, Milton Green, Abbeymead, Gloucester, GL4 5FD [IO81VU, SO81]
G4 BNX I W Middleton, 6 Brentnall Ct, Kirk Cl, Beeston, Nottingham, NG9 5EZ [IO92JW, SK53]
G4 BNY L R Nayler, Wabey House, 156 Church St., Upwey, Weymouth, DT3 5QE [IO80SP, SY68]
GM4 BNZ E C Walker, 21 The Moorings, Paisley, PA2 9BD [IO75SU, NS46]
GM4 BOA E J Margetts, 17 Farndale, Stewartfield, East Kilbride, Glasgow, G74 4QS [IO75VS, NS65]
G4 BOB A H Wallwork, 12 Queensway, Brinscall, Chorley, PR6 8QQ [IO83RQ, SD62]
G4 BOC H Hodgson, 1 Harris St., Askam in Furness, LA16 7BY [IO84JE, SD27]
G4 BOF P H J Harry, Coppice, St Michaels Cl, Kingsland, Leominster, HR6 9QR [IO82OF, SO46]
G4 BOH C Cummings, Castle View, Childs Ln, Brownlow, Congleton, CW12 4TQ [IO83UD, SJ86]
G4 BOJ N E K Greenstreet, 78 Slayleigh Ln, Fulwood, Sheffield, S10 3RH [IO93FI, SK38]
G4 BOK J G Assenheim, 28 Athenaeum Rd, Whetstone, London, N20 9AE [IO91VP, TQ29]
G4 BOL R E Fineman, 4 Sherbourne Ave, Bradley Stoke, Bristol, BS12 8BB [IO81RM, ST68]
G4 BON J A Strutt, 7 Beechmere Rise, Etching Hill, Rugeley, WS15 2XR [IO92AS, SK01]
G4 BOO D J Rumens, 3 Flecker Cl, Thatcham, RG18 3BA [IO91IJ, SU56]
G4 BOP Dr P G Berwick, Easr End, Barton House, Morchard Bishop, Crediton, Devon, EX17 6RE [IO80DU, SS70]
G4 BOQ J A Hall, 8 Rushleydale, Springfield, Chelmsford, CM1 6JX [JO01FR, TL70]
G4 BOV A E Horton, Martletts, 52 Lower Cookham Rd, Maidenhead, SL6 8JZ [IO91PM, SU98]
G4 BOY A R Wilson, 11A Ansford Rd, Bromley, BR1 5QU [IO91XK, TQ37]
G4 BOZ A C Brock, Richard Wagner, Strasse 13, 37269 Eschwege, Germany, X X
G4 BP D P Tipper Scarborough ARS, 10 Lowdale Ave, Scarborough, YO12 6JW [IO94TG, TA08]
G4 BPC A Askew, Forrest House, Maltings Cl, Stewkley, Bucks, LU7 0HR [IO91OW, SP82]
G4 BPD P A Mustoe, Drapers House, Turkdean Northleach, Cheltenham, Glos, GL54 3NU [IO91BU, SP11]
G4 BPE A Evans, Fairfield Main St, Claypole, Newark, Notts, NG23 5BA [IO93PA, SK84]
G4 BPF Details withheld at licensee's request by SSL.
G4 BPG Details withheld at licensee's request by SSL.
G4 BPJ B Stone, 12 Forbes Rd, Newlyn, Penzance, TR18 5DQ [IO70FC, SW42]
G4 BPK R Chadwick, 4 The Green, Radcliffe on Trent, Nottingham, NG12 2LA [IO92LW, SK63]
G4 BPN N Kerstein, 80 Randall Ave, London, NW2 7SU [IO91VN, TQ28]
G4 BPO C J Hoare PO Research Cnt, Wheatstone Rd, Dorcan, Swindon, SN3 4RD [IO91DN, SU18]
G4 BPQ Details withheld at licensee's request by SSL.
G4 BPR P H Truran, 29 Queens Cres, Marshalswick, St. Albans, AL4 9QG [IO91US, TL10]
G4 BPV P R Barker, 2 Oriole Dr, Exeter, EX4 4SJ [IO80FR, SX99]
G4 BPW D F Reynolds, 21 Twentylands, Rolleston on Dove, Burton on Trent, DE13 9AJ [IO92EU, SK22]
G4 BPX Details withheld at licensee's request by SSL.
G4 BQB J Crocker, The Railway House, Stogumber, Taunton, Somerset, TA4 3TR [IO81IC, ST13]
G4 BQC B G Makeham, 64 Benomley Rd, Almondbury, Huddersfield, HD5 8LS [IO93CP, SE11]
GM4 BQD R N Muir, 9 Craigs Ct, Torphichen, Bathgate, EH48 4NU [IO85EW, NS97]
G4 BQF M Duce, 28 Thompson Ave, Canvey Island, SS8 7TS [JO01HM, TQ88]
G4 BQH D W Livsey, 18 Tollards Rd, Countess Wear, Exeter, EX2 6JJ [IO80GQ, SX99]
GI4 BQI W J McCullough, 92 Laurelvale Rd, Tandragee, Craigavon, BT62 2LW [IO64SI, H94]
G4 BQJ A E Hill, 3 Cambrai Ave, Warrington, WA4 6QU [IO83RJ, SJ68]
G4 BQM C D Daunt, 35 Runnymede, Great Lumley, Chester-le-Street, DH3 4LN [IO94FU, NZ24]
GI4 BQN N J Marsden, Arcadie, 32 Chard Rd, Drimpton, Beaminster, DT8 3RF [IO80OU, ST40]
G4 BQO M J Peach, 123 York Ave, East Cowes, PO32 6BB [IO90IS, SZ59]
G4 BQR W Carmichael, 47 Neath Dr, Ipswich, IP2 9TA [JO02NA, TM14]
G4 BQS B H Prichard, The Gables, Wootton Ln, Wootton, Canterbury, CT4 6RT [JO01OE, TR24]
G4 BQV R K L Mullard, 46 Green Ln, Clanfield, Waterlooville, PO8 0JX [IO90MW, SU71]
G4 BQW W B Glover, 27 Larchmere Dr, Hall Green, Birmingham, B28 8JB [IO92BK, SP18]
G4 BQX P A Wade, Gilton House, 5 Wath Wood Dr, Swinton, Mexborough, S64 8UW [IO93HL, SK49]
G4 BQY P J Aburrow, 25 Hill Cres, Worcester Park, KT4 8NB [IO91VJ, TQ26]
G4 BQZ Dr N O Thomson, Avalon, 43 Racecourt, Silverdale, Newcastle under Lyne, Staffs, ST5 6PD [IO83UA, SJ84]
G4 BRA M Goodey Bracknell ARC, 62 Rose Hill, Binfield, Bracknell, RG42 5LG [IO91OK, SU87]
G4 BRB A G Stewart, 7 King St., Driffield, YO25 7QW [IO94SA, TA05]
G4 BRC T I Lundegard Kent Raynet Group, Saxby, Botsom Ln, West Kingsdown, Sevenoaks, TN15 6BL [JO01DI, TQ56]
GM4 BRD J Clark, 1 Landemer Dr, Rutherglen, Glasgow, G73 2TA [IO75VT, NS66]
G4 BRE M A Stapleton, 69 Wakehurst Dr, Southgate, Crawley, RH10 6DU [IO91VC, TQ23]
G4 BRF R Mickleburgh, 85 Cary Park, Polperro, Cornwall, PL13 1JP [IO70SI, SX25]
G4 BRG E H Rogers, 21 Nevin Dr, Chingford, London, E4 7LL [IO91XP, TQ39]
G4 BRH J R Sniadowski, 42 Milesmere, Two Mile Ash, Milton Keynes, MK8 8DP [IO92OA, SP83]
G4 BRK N J Whiting, 33 The Dormers, Highworth, Swindon, SN6 7NZ [IO91DP, SU29]
GM4 BRM A Long, 34 Thornly Park Dr, Paisley, PA2 7RP [IO75TT, NS46]
G4 BRN A M Allan Midlands Contest & Constructor, 26 Gregory Ave, Green Ln, Coventry, CV3 6DL [IO92FJ, SP37]
G4 BRQ D L Andrews, 5 Northbourne Gdns, Northbourne, Bournemouth, BH10 6DJ [IO90BS, SZ09]
GW4 BRS C S Beynon, 16 Hardy Cl, Woodfield Heights, Barry, CF62 9HJ [IO81IK, ST16]
G4 BRT Details withheld at licensee's request by SSL.
G4 BRU A Lambe, Nimbus, Boscola, Truro, Cornwall, TR4 9ED [IO70LG, SW84]
G4 BRW M J Gordon, 57 Taunton Rd, Bridgwater, TA6 3LP [IO81MC, ST33]
G4 BRX P W Legg, 4 Linworth Rd, Bishops Cleeve, Cheltenham, GL52 4PB [IO81XW, SO92]
G4 BSA M Draper, Waverley, Bradfield St Clare, Bury St Edmunds, Suffolk, IP30 0EL [JO02JE, TL95]
G4 BSC J R Wells, Sunnyfield Ln, Up Hatherley, Upper Hatherley, Cheltenham, Glos, GL51 6JE [IO81WV, SO92]
G4 BSD D Hoose, Leonard Ches He, Oaklands, Dimples Ln, Garstang, PR3 1RD [IO83OV, SD44]
G4 BSH J F Swatman, Homelea Main Rd, Rollesby, Gt Yarmouth, Norfolk, NR29 5EQ [JO02TQ, TG41]
G4 BSK M A Rhind-Tutt, 121 Vale Rd, Windsor, SL4 5JR [IO91QL, SU97]
G4 BSM S P Grove, 31 Sheppard Way, Minchinhampton, Stroud, GL6 9BZ [IO81VQ, SO80]
G4 BSO R W Price, Little Belvedere, Canon Pyon Rd, Hereford, HR4 7RB [IO82PB, SO44]
G4 BSS J B Spence, Langford House, Langford Ln, Burley-in-Wharfedal, Ilkley, LS29 7NR [IO93DV, SE14]
G4 BSW N R Hadley, 323 Canterbury Rd, Westbrook, Margate, CT9 5JA [JO01QJ, TR36]
G4 BTE M R L Smith, 11 Finchfield Hill, Finchfield, Wolverhampton, WV3 9DQ [IO82VO, SO89]
GI4 BTG H W B Davidson, 106 Tudor Park, Newtownabbey, BT36 4WL [IO74AQ, J38]
GW4 BTH Details withheld at licensee's request by SSL.
G4 BTI D A Case, 18 Langborough Rd, Wokingham, RG40 2BT [IO91OJ, SU86]
G4 BTK A G Whitehouse, 690 Kingstanding Rd, Kingstanding, Birmingham, B44 9SS [IO92BN, SP09]
G4 BTN C J Brion, 4 North End, London, NW3 7HL [IO91VN, TQ28]
G4 BTO L J McWilliam, 5 Taunton Rd, Wallasey, L45 3JL [IO83LK, SJ29]
G4 BTS E D Rogers Mexbrgh Dist AR, 7 Buckleigh Rd, Wath upon Deame, Rotherham, S63 7JB [IO93HL, SK49]
G4 BTW I B Jolly, 1 Llewelyn Dr, Bryn y Baal, Mold, CH7 6SW [IO83KE, SJ26]
G4 BTX N R Monument, 6 Manor Rd, Martlesham, Martlesham Heath, Ipswich, IP5 7SY [JO02OB, TM24]
G4 BU R H Draper, 20 Prior St., Lincoln, LN5 7SW [IO93RF, SK97]
G4 BUB P W Cox, 53 Boleyn Ave, Enfield, EN1 4HR [IO91XP, TQ39]
G4 BUD B R Underwood, 3 Jackson Cl, Hampton Magna, Warwick, CV35 8SZ [IO92EG, SP26]
G4 BUE C J Page, Highcroft Farm, Gay St, Pulborough, West Sussex, RH20 2HJ [IO90SX, TQ01]
G4 BUF G E G Jolley, 77 Chestnut Way, Burton, Christchurch, BH23 7LR [IO90CS, SZ19]
G4 BUH M F Banahan, 27 Park Dr, Bradford, BD9 4DS [IO93CT, SE13]
G4 BUI J M Simpson, 19 Greenacres, Wetheral, Carlisle, CA4 8LD [IO84NV, NY45]
GI4 BUJ J H F Sander, 696 Doagh Rd, Newtownabbey, BT36 4TP [IO74AQ, J38]
G4 BUL D A Brudenell, 84 Porthcawl Green, Tattenhoe, Milton Keynes, MK4 3AL [IO91OX, SP83]
G4 BUN H Bunting, 8 Belvoir Ave, Hazel Gr, Stockport, SK7 6DL [IO83WI, SJ98]
G4 BUO D J Lawley, Carramore, Coldharbour Rd, Penshurst, Tonbridge, TN11 8EX [JO01CD, TQ54]
G4 BUP Dr P M Moss, Amalrie, Franklin Rd, North Fambridge, Essex, CM3 6NF [JO01IP, TQ89]
GW4 BUS R Prosser, 18 Danes Way, Leighton Buzzard, LU7 8LS [IO91QW, SP92]
G4 BUV P J Dyer, Mill Cottage, Mill Rd, Thompson, Thetford, IP24 1PH [JO02JM, TL99]
G4 BUW K G Lamb, The Haven, London Rd, Binfield, Bracknell, RG42 4BS [IO91OJ, SU86]
G4 BUX K Buxcey, Cold Wall Farm, Cobden Edge, Mellor Stockport, Ches, SK6 5NH [IO83XJ, SJ98]
GW4 BUZ J Howells, Bronllys, Vicarage Rd, Penygraig, Tonypandy, Mid Glam, CF40 1HR [IO81GO, ST09]
G4 BVA J A Edwards, 2 Chantry Cl, Teignmouth, TQ14 8FE [IO80GN, SX97]
G4 BVB R S Pridham, Victoria House, Chilsworthy, Gunnislake, Cornwall, PL18 9PB [IO70VM, SX47]
GM4 BVD A W Sampson, 47 Muirend Rd, Burghmuir, Perth, PH1 1JD [IO86GJ, NO02]
G4 BVE J D Clifford, Hobbiton, Holmes Chapel Rd, Lach Dennis, Northwich, CW9 7SZ [IO83SF, SJ77]
G4 BVF M J Sinclair, 28 Roker Park Ave, Ickenham, Uxbridge, UB10 8ED [IO91SN, TQ08]
G4 BVG A N Young, 90 Pine Ridge, Carshalton Beeches, Carshalton, SM5 4QH [IO91WI, TQ26]
G4 BVH P L Reed, 20 Greenfield Cres, Brighton, BN1 8HJ [IO90WU, TQ30]
G4 BVI G A Chenery, 44 Belstead Rd, Ipswich, IP2 8AZ [JO02NB, TM14]
GW4 BVJ R S Mortimore, Hendre, Colwinston, Cowbridge, CF71 7NL [IO81FL, SS97]
G4 BVK K Stevens, 20 Coberley, Hanham, Bristol, BS15 2ES [IO81RK, ST67]
GW4 BVN T R Davies, 26 Sandy Ridge, Aberavon, Port Talbot, SA12 6SU [IO81CO, SS79]
GM4 BVO C M Smith, Craigton Croft, Glenmoy, Kirriemuir, Angus, DD8 4NB [IO86MR, NO36]
G4 BVP M Noble, Harbet Shipley Rd, Southwater Horsham, West Sussex, RH13 7BG [IO91TA, TQ12]
G4 BVQ P J Kennedy, 18 Rushmere Ave, Levenshulme, Manchester, M19 3EH [IO83VK, SJ89]
G4 BVS Hhj S K Overend, Monticello Ct Rd, Newton Ferrers, Plymouth, Devon, PL8 1BZ [IO70XH, SX54]
GW4 BVT R N Osborne, 38 Penylan Cl, Bassaleg, Newport, NP1 9NW [IO81LN, ST28]
GM4 BVU N A Macdonald, 3 Townhill Rd, Earnock Est, Hamilton, ML3 9UX [IO75XS, NS65]
G4 BVV P Goben, 1 Petal Cl, Maltby, Rotherham, S66 7HJ [IO93JK, SK59]
G4 BVW A Reilly, 4 Moreton Dr, Poulton, Blackpool, FY3 0DR [IO83MT, SD33]
G4 BVY I R Dixon, 5 The Howsells, Lower Howsell Rd, Malvern, WR14 1AD [IO82UD, SO74]
G4 BVZ J L Davidson, Rosemount, Whiting Bay, Brodick, Isle of Arran, KA27 8PR [IO75KL, NS02]
G4 BWB R W Andrews, 80 Winchester Rd, Bristol, BS4 3NH [IO81RK, ST67]
G4 BWD A K Whillock Natsec ARC, Nats, Training and Conference Centre, Heckfield, nr Basingstoke, Hants, RG27 0LD [IO91MI, SU76]
G4 BWE S J Price, 9 Spurcroft Rd, Thatcham, RG19 3XX [IO91IJ, SU56]
G4 BWF R D Johnson, 29 Oakfield Ave, Markfield, LE67 9WH [IO92IQ, SK40]
G4 BWG S Marsh, 5 Ct Farm Cttgs, Tithepit Shaw Ln, Warlingham, Surrey, CR3 9AT [IO91XG, TQ35]
G4 BWH G G Sinclair, Azeleas, 5 Dornden Dr, Langton Green, Tunbridge Wells, TN3 0AA [JO01CD, TQ53]
G4 BWJ B T Cook, 31 Elm Dr, Hove, BN3 7JS [IO90VU, TQ20]
GI4 BWM J K McCullagh, 2 Holestone Rd, Doagh, Ballyclare, BT39 0SB [IO64XS, J29]
G4 BWN P J Funnell, 6 Bolero Cl, Wolaton, Nottingham, NG8 2BZ [IO92JW, SK54]
G4 BWP F C Handscombe, Sandholm, Bridge End Rd, Red Lodge, Bury St. Edmunds, IP28 8LQ [JO02FH, TL67]
G4 BWQ Details withheld at licensee's request by SSL.
G4 BWR M A Hildich, 22 Newlands Green, Clevedon, BS21 5BU [IO81NK, ST47]
G4 BWV A R L Burchmore, 49 School Ln, Horton Kirby, Dartford, DA4 9DQ [JO01DJ, TQ56]
G4 BWW Details withheld at licensee's request by SSL.
G4 BWX S E Egerton, 15 Hyde Rd, Torrisholme, Morecambe, LA4 6NU [IO84OB, SD46]
G4 BWY P W Willcocks, 27 Manor Rd, Barnet, EN5 2LE [IO91VP, TQ29]

GI4 BXB R J Brown, The Warren, 38 Springvale Rd, Ballywalter, Newtownards, BT22 2RS [IO74GM, J66]
G4 BXC H A Pearce, 32 Marshall Rd, Willenhall, WV13 3PB [IO82XN, SO99]
G4 BXD Details withheld at licensee's request by SSL.
GM4 BXG G Franklin, 43 Pinfold Ln, Holton-le-Clay, Grimsby, DN36 5DT [IO93XM, TA20]
G4 BXH D Hardy, Box 52831, Dubai, U A E, X X
G4 BXI C F Godden, 84 Cres Rd, Ramsgate, CT11 9QZ [JO01QI, TR36]
G4 BXM R W Spear, 93 Richmond Rd, Kingston upon Thames, KT2 5BP [IO91UJ, TQ16]
G4 BXO Details withheld at licensee's request by SSL.
G4 BXP Details withheld at licensee's request by SSL.
G4 BXQ A Pressley, 22 Springbank Ave, Farsley, Pudsey, LS28 5LW [IO93ET, SE23]
G4 BXR K A Chalkley, 105 Milford Ave, Stony Stratford, Milton Keynes, MK11 1EZ [IO92NB, SP73]
G4 BXS Rev J O Morris, 8 Great Mis Tor Cl, Yelverton, PL20 6DH [IO70WL, SX56]
G4 BXT P Homer, 81 Spring Vale South, Dartford, DA1 2LW [JO01CK, TQ57]
G4 BXU G W Butler, 18 Hobart Rd, Ramsgate, CT12 6NW [JO01QI, TR36]
G4 BXY H F Barker, 31 Briants Ave, Caversham, Reading, RG4 5AY [IO91ML, SU77]
G4 BXZ Dr J M Howell, 3 Gate Farm Rd, Shotley Gate, Ipswich, IP9 1QH [JO01PX, TM23]
G4 BY R W H Jennings, 15 New House Cl, Canterbury, CT4 7BQ [JO01MG, TR15]
GW4 BYA P A Braham, Maesawel, Black Lion Rd, Cross Hands, Llanelli, SA14 6SA [IO71XT, SN51]
G4 BYB R W Penman, 9 Southall Ave, Worcester, WR3 7LR [IO82VF, SO85]
GM4 BYC A J Pate, Ivy Cttgs, 55 Main St., Auchencairn, Castle Douglas, DG7 1QU [IO84BU, NX75]
G4 BYD A Atkinson, Knowle Cottage, Knowle Rd, Kirkheaton, Huddersfield, HD5 0DW [IO93DP, SE11]
G4 BYE T G Miller, 4 Jessop Rd, Stevenage, SG1 5NF [IO91VW, TL22]
GM4 BYF P J Bates, 9 Winton Terr, Edinburgh, EH10 7AP [IO85JV, NT26]
G4 BYG V A Lindgren, 143 Hull Rd, Anlaby, Hull, HU10 6ST [IO93SR, TA02]
G4 BYJ H L H Eesemann, 20 Kennington Ave, Bishopston, Bristol, BS7 9ET [IO81RL, ST57]
G4 BYL B A Preece, 82 Richmond Ave, Cliviger, Burnley, BB10 4JL [IO83VS, SD83]
G4 BYM B R Buzzing, 1 Westmead Cl, Heron Waters, Droitwich, WR9 9LG [IO82WG, SO86]
G4 BYO W H Tee, Springhill Cottage, 87 Higher Blandford Rd, Broadstone, BH18 9AE [IO80XS, SY99]
G4 BYP A G Scott, 70 Top Rd, Five Crosses, Frodsham, Warrington, WA6 6SN [IO83PG, SJ57]
G4 BYQ Details withheld at licensee's request by SSL.
G4 BYR I A Maslen, 40 Great Leylands, Harlow, CM18 6HR [JO01BS, TL40]
G4 BYS G Warren, 96 Parkside Dr, Cassiobury, Watford, WD1 3BB [IO91SP, TQ09]
GM4 BYT R H Cook, 132 Clachtoll, Lochinver, Lairg, IV27 4JD [IO78IE, NC02]
G4 BYV J R Tye, Inter-Nos, Swanton Morley, Dereham, Norfolk, NR20 4NU [JO02LQ, TG01]
G4 BYW J Lekesys, 4 Gleneagles Way, Fixby Park, Huddersfield, HD2 2NH [IO93CQ, SE11]
GW4 BYY K J Plumridge, Swn y Gwynt, High St., Llanberis, Caernarfon, Gwynedd, LL55 4EN [IO73WC, SH56]
G4 BYZ C J Mills, North Lodge, Margery Ln, Lower Kingswood, Tadworth, KT20 7BA [IO91VG, TQ25]
G4 BZA J Tyblewski, 120 St. Georges Ave, Northampton, NN2 6JF [IO92NG, SP76]
G4 BZB D A Parsons, 27 St. Leodegars Way, Hunston, Chichester, PO20 6PE [IO90OT, SU80]
GW4 BZD Details withheld at licensee's request by SSL.[Op: C Barnes. Station located near Menai Bridge.]
G4 BZE P J Bradley, Woodlands, Longdown, Exeter, Devon, EX6 7SR [IO80ER, SX89]
G4 BZF M B Reed, 66 Seymour Rd, Mannamead, Plymouth, PL3 5AY [IO70WJ, SX45]
G4 BZG R A Smith, 17 Styrrup Rd, Harworth, Doncaster, DN11 8LL [IO93LK, SK69]
GM4 BZI R H Bracey, 30 Sandpiper Rd, Lochwinnoch, PA12 4NB [IO75QT, NS35]
G4 BZJ A I Mitchell, 18 Malham Fell, Bracknell, RG12 7DU [IO91OJ, SU86]
G4 BZL D J Simpson, 8 Esholt Ave, Guiseley, Leeds, LS20 8AX [IO93DU, SE14]
G4 BZM M J Edwards, 13 Lechmere Cres, Malvern, WR14 1TY [IO82UD, SO74]
G4 BZO Details withheld at licensee's request by SSL.
G4 BZP F L Partington, 21 East Rd, Wymeswold, Loughborough, LE12 6ST [IO92KT, SK62]
G4 BZQ J F Richardson, 7 Oakfield Park, Much Wenlock, TF13 6HQ [IO82RO, SO69]
G4 BZR F A Jordan, 16 Elterwater Cres, Barrow in Furness, LA14 4PH [IO84JD, SD27]
G4 BZS M J Pasek, 11 Heath Ave, Halifax, HX3 0EA [IO93BR, SE02]
G4 BZU B O W Beaven, 7 Glam d, Up Hatherley, Cheltenham, GL51 5JF [IO81WV, SO92]
G4 BZV N Barton, 147 Whinney Ln, New Ollerton, Newark, NG22 9TJ [IO93ME, SK66]
G4 CA P A Wodehouse, 7 Grange Lodge, The Grange, London, SW19 4PR [IO91VK, TQ27]
G4 CAA B C Sheppard C A A A R S, Nats Ltd, Gate 2, Spectrum House, Gatwick Rd, Gatwick Airport South, RH6 0LG [IO91WD, TQ24]
GM4 CAB S Reynolds, 39 Panmure St., Broughty Ferry, Dundee, DD5 2EU [IO86NL, NO43]
GM4 CAC Dr G T Bryden, 6 Thistle Ave, Grangemouth, FK3 8YH [IO86DA, NS98]
G4 CAF D J Hogg, 70 Swanmore Rd, Littleover, Derby, DE23 7SY [IO92FV, SK33]
GM4 CAH P W Figures, 15 Curriehill Castl, Dr, Balerno, Midlothian, EH14 5TA [IO85IV, NT16]
GM4 CAI G Evans, 2 Mossend Pl, New Elgin, Elgin, IV30 3YB [IO87IP, NJ26]
G4 CAJ M B Farr, 23 Waterfall Way, Barwell, Leicester, LE9 8EH [IO92HN, SP49]
G4 CAK M N Scarlett, 44 Rowley Furrows, Linslade, Leighton Buzzard, LU7 7SH [IO91PW, SP92]
G4 CAL L Wade, Church View, Welbury, Northallerton, North Yorks, DL6 2SE [IO94HJ, NZ30]
GM4 CAM D Hamilton, 19 Abbots Cres, Doonfoot, Ayr, KA7 4JR [IO75QK, NS31]
G4 CAO Details withheld at licensee's request by SSL.
G4 CAP Details withheld at licensee's request by SSL.
GM4 CAQ R G Miles, 58 Fogralea, Lerwick, ZE1 0SE [IO90KD, HU44]
G4 CAR J Benson, 2 Saxon Walk, Lichfield, WS13 8AJ [IO92BQ, SK10]
GW4 CAT N L D Schofield, Maen Llwyd-Tan yr Allt, Llanllyfni, Caernarfon, Gwynedd, LL54 6RT [IO73VA, SH45]
GM4 CAU T C Wratten, 89 Hilton Rd, Aberdeen, AB24 4HX [IO87WE, NJ90]
G4 CAV A L Anderson, 1 Dewar Cl, Collingham, Wetherby, LS22 5JR [IO93HV, SE34]
G4 CAW Details withheld at licensee's request by SSL.
G4 CAX D R Borley, 95 Meadow Ln, Moulton, Northwich, CW9 8QQ [IO83RF, SJ66]
G4 CAY C J Parker, 25 Meadow Dale, Chilton, Ferryhill, DL17 0RW [IO94FP, NZ22]
GM4 CAZ J C Lefever, 38 Woodburn Terr, Edinburgh, EH10 4ST [IO85JW, NT27]
G4 CBA S Mulligan, 49 Springhead Ave, Willerby Rd, Hull, HU5 5HZ [IO93TS, TA02]
G4 CBC W F Shepherd, 12 Bryants Field, Beacon Gdns, Crowborough, TN6 1BH [JO01BB, TQ53]
G4 CBE P G Rolfe, 9 Westfields, St. Albans, AL3 4LR [IO91TR, TL10]
GI4 CBG R A Smyth, 58 Gilnahirk Rd, Belfast, BT5 7DH [IO74BO, J37]
G4 CBJ R A Stanbrook, 14 Blossom Cl, Botley, Southampton, SO30 2FR [IO90IV, SU51]
G4 CBK L H Weston, 52 Shireview Rd, Pelsall, Walsall, WS3 4EA [IO92AP, SK00]
G4 CBL P A Tomlinson, 55 Reldene Dr, Willerby Rd, Hull, HU5 5HS [IO93TS, TA02]
G4 CBM G Blakeley, Richmond, Old Woods, Shrewsbury, Shropshire, SY4 3AX [IO82OS, SJ42]
G4 CBN C L H Avis, Cirrus, Mill Rd, Chevington, Bury St. Edmunds, IP29 5QP [JO02HE, TL75]
G4 CBO D J W Aiken, 16 Woodland Gdns, North Wootton, Kings Lynn, Norfolk, PE30 3PX [JO02FS, TF62]
G4 CBQ P R Daniells, 21 Orchard Cl, Ockbrook, Derby, DE72 3RQ [IO92HV, SK43]
GW4 CBR M D Hunt, Deerbolt, Sageston, Tenby, Pembrokeshire, SA70 8SG [IO71OQ, SN00]
G4 CBS N A Swain, Hill Cottage, Camerton Hill, Camerton, Bath, BA3 1PS [IO81SH, ST65]
G4 CBT H J B Wall, 54 Little Harlescott Ln, Shrewsbury, SY1 3PZ [IO82PR, SJ41]
GM4 CBV H L C Mc Culloch, Flat 3, 6 Sir Michael St., Greenock, PA15 1PH [IO75OW, NS27]
G4 CBW A Horsfall, Eye of The Wind, 60 Talke Rd, Red St., Newcastle, ST5 7AH [IO83UB, SJ85]
G4 CBY T A Cooper, Lincs House, Brumby Wood Ln, Scunthorpe, South Humberside, DN17 1AF [IO93QO, SE81]
G4 CBZ A J Mepham, 1 Grand Cres, Rottingdean, Brighton, BN2 7GL [IO90XT, TQ30]
GW4 CC R Williams Swansea ARS, 5 Golden Cl, West Cross, Swansea, SA3 5PE [IO71XO, SS68]
G4 CCA M J L Fadil, 25 North Parade, Horsham, RH12 2DA [IO91UB, TQ13]
G4 CCB A W Brown, 14 Holly Rise, New Ollerton, Newark, NG22 9UZ [IO93LF, SK66]
G4 CCC C F H Young, 18 Wincroft Rd, Caversham, Reading, RG4 7HH [IO91ML, SU77]
G4 CCE R V Angell, 214 Lower Higham Rd, Chalk, Gravesend, DA12 2NN [JO01EK, TQ67]
G4 CCF Details withheld at licensee's request by SSL.
G4 CCH H A Ling, 8 Spa Hill, Kirton in Lindsey, Kirton Lindsey, Gainsborough, DN21 4NE [IO93QL, SK99]
G4 CCI J W Chapman, 7 Ravensthorpe Dr, Loughborough, LE11 4PU [IO92JS, SK51]
G4 CCJ E J Delmonte, Flat 1, 5 Hawkwood Mount, Clapton, London, E5 9EQ [IO91XN, TQ38]
G4 CCM H D U V Ashcroft, 86 Avondale Ave, North Finchley, London, N12 8EN [IO91VO, TQ29]
G4 CCN T R Keats, Bumbles, The St., Bredfield, Woodbridge, IP13 6AX [JO02PD, TM25]
G4 CCQ Details withheld at licensee's request by SSL.[Op: M P Stanton. Station located near Maidstone.]
G4 CCR A Campden, 41 Mountfield Rd, Spinney Hill, Northampton, NN3 6BE [IO92NG, SP76]
GW4 CCS Details withheld at licensee's request by SSL.
G4 CCT S N Hyman, 49 Southover, Woodside Park, London, N12 7JG [IO91VO, TQ29]
G4 CCU R R Wood, 1 Orchard Cl, Writtle, Chelmsford, CM1 3EG [JO01FR, TL60]
G4 CCV H Acomb, 27 Pickering Rd, West Ayton, Scarborough, YO13 9JE [IO94SF, SE98]
G4 CCW D R Sheen, 3 Foxearth Spur, Selsdon, South Croydon, CR2 8EP [IO91XI, TQ36]
G4 CCX Details withheld at licensee's request by SSL.
G4 CCY P M Fagg, 113 Bute Rd, Wallington, SM6 8AE [IO91WI, TQ26]
G4 CCZ P M Simons, Westwood, Faris Ln, Woodham, Addlestone, KT15 3DJ [IO91RI, TQ06]
G4 CDB G Lindsay, 11 Nuns Row, Kepier Est, Gilesgate, Durham, DH1 1HG [IO94FS, NZ24]
G4 CDC E H Morton, 6 Norfolk Ave, Burton upon Stather, Scunthorpe, DN15 9EW [IO93PP, SE81]
G4 CDD J Hinchcliffe Denby Dale ARC, 19 The Terrace, Honley, Huddersfield, HD7 2DS [IO93CO, SE11][Correspondence to the Secretary: Eric Stewart, G0DBU.]
G4 CDE Details withheld at licensee's request by SSL.
G4 CDG A J B Davidson, H.A & E Smith Limited, Brook St House, 47 Davies St, London, W17 1FJ [IO91VM, TQ28]
G4 CDH J T K Brade, 26 Pevensey Way, Frimley Green, Frimley, Camberley, GU16 5YJ [IO91PH, SU85]
G4 CDI G D Boardman, 9 Byron Rd, Weston Super Mare, BS23 3XQ [IO81MI, ST35]
G4 CDJ P E Jarrett, 2 Palmers Cttgs, Whitchurch Hill, Reading, RG8 7PP [IO91LM, SU67]
G4 CDK W M Barnes, Sibertshold, 12 Castle Rising Rd, South Wootton, Kings Lynn, PE30 3HR [JO02FS, TF62]
G4 CDL F E Mepham, 56 Primrose Hill Park, Charlton Adam, Somerton, Somerset, TA11 7AP [IO81QB, ST52]
G4 CDM C D Merry, 19 Faesten Way, Bexley, DA5 2JB [JO01CK, TQ57]
G4 CDO A F Raftery, 4 Harefield, Harlow, CM20 3EF [JO01BS, TL41]
G4 CDP J Everett, 9 Stoney Ln, Barrow, Bury St. Edmunds, IP29 5DD [JO02HF, TL76]
G4 CDQ N H Gordon, 16 Chosen Dr, Churchdown, Gloucester, GL3 2QT [IO81VV, SO82]
G4 CDT Dr A R Dexter, Conquest House, 11 Silsoe Rd, Maulden, Bedford, MK45 2AX [IO92SA, TL03]
G4 CDW G A Trickey, 3 Fairleigh Farm, Kington Langley, Chippenham, Wilts, SN15 5QF [IO81WL, ST97]
G4 CDY T G Giles, 37 Smitham Downs Rd, Purley, CR8 4NG [IO91WH, TQ36]
G4 CDZ J B Boden, 2 The Coppice, Whaley Bridge, High Peak, SK23 7LH [IO93AH, SK08]
GM4 CEA R D McCracken, 6 Binnie St., Gourock, PA19 1JS [IO75OX, NS27]
G4 CEC P V Knight, 26 Meadway, Harrold, Bedford, MK43 7DR [IO92QE, SP95]
G4 CEH R C G May, Tybaartstraat 50, B9170 Sint Gillis, Waas, Belgium
G4 CEI M P Baker, Willowcombe, Pangbourne Rd, Upper Basildon, Reading, RG8 8LN [IO91KL, SU67]
G4 CEJ R Moore, 17 Somme Ave, Flookburgh, Grange Over Sands, LA11 7LJ [IO84ME, SD37]
G4 CEK J R Bird, 21 Cambridge House, Courtfield Gdns, Ealing, London, W13 0HP [IO91UM, TQ18]
G4 CEM L A Reeves, 69 Whitehill Rd, Hitchin, SG4 9HP [IO91UW, TL12]
G4 CEN D C Davies, 35, Ruthellen Rd, Chelmsford, Massachusetts, U.S.A. 01824-4022
G4 CEO R I Leask, 80 Mill Rd, Sharnbrook, Bedford, MK44 1NP [IO92RF, TL05]
G4 CEP G R Morris, 7 Manor Rd, Sandy, SG19 1DT [IO92UD, TL14]
G4 CEQ E G Le Sueur, 14 Austendyke Rd, Weston Hills, Spalding, PE12 6BX [IO92WS, TF22]
G4 CEU D J Jarvis, 66 Leitrim Ave, Shoeburyness, Southend on Sea, SS3 9HF [JO01JM, TQ98]
G4 CEV K M Moorhouse, 78 Thorpe Ln, Almondbury, Huddersfield, HD5 8TS [IO93DP, SE11]
G4 CEW J W A Crooks, 32 Mount View Rd, Winchester, SO22 4JJ [IO91HB, SU42]
G4 CEX C Durant, 63 Ulleries Rd, Solihull, B92 8DX [IO92CK, SP18]
G4 CEY J E Ball, 8 Buntings Ln, Warmington, Peterborough, PE8 6TT [IO92TM, TL09]
G4 CF A Macfarlane, Braemar, Richmond Ave, Burscough, Ormskirk, L40 7RD [IO83NO, SD41]
G4 CFB K P Henry, 13 Guildford Rd, Brighton, BN1 3LU [IO90WT, TQ30]
GW4 CFC Dr L D Gruffydd, 45 Maes yr Hafod, Menai Bridge, LL59 5NB [IO73VF, SH57]
G4 CFF Details withheld at licensee's request by SSL.
G4 CFG P A Arnold, 14 George Birch Cl, Brinklow, Rugby, CV23 0NN [IO92HJ, SP47]
G4 CFH J R A Hill, 10 Albert Clarke Dr, Willenhall, WV12 5AU [IO82XO, SJ90]
G4 CFJ Details withheld at licensee's request by SSL.[Op: J E King. Station located near Maldon.]
G4 CFK Details withheld at licensee's request by SSL.
G4 CFN R J Kingshott, Heatherbrae, Woodend Rd, Deepcut, Camberley, GU16 6QH [IO91PH, SU95]
G4 CFO J J Carter, 9 North St., Cleethorpes, DN35 8NP [IO93XN, TA30]
G4 CFP W H Bones, 20 Thursford Gr, Blackrod, Bolton, BL6 5TW [IO83RO, SD61]
GI4 CFQ J G McSweeney, 109 Twaddell Ave, Belfast, BT13 3LG [IO74AO, J37]
G4 CFR C W Edmonds, 25 Ernest Clark Cl, Willenhall, WV12 4DY [IO82XO, SO99]
GM4 CFS G D Dodwell, 2 Muirton Pl, Kinloss, Forres, IV36 0UJ [IO87FP, NJ06]
G4 CFT M J Evans, White Lodge, Horton Rd, Horton, Slough, SL3 9NU [IO91RL, TQ07]
GI4 CFV R E Hall, Pinewood Lodge, 16 Tullyvarraga Hill, Shannon, Co Clare, Ireland, X X
G4 CFW R R Raven, 9 Southwood Cl, Ferndown, BH22 9HW [IO90BT, SU00]
G4 CFX J E Goodson, 7 Hestia Cl, Romsey, SO51 8PA [IO90GX, SU32]
G4 CFY A J Nailer, 12 Weatherbury Way, Dorchester, DT1 2EF [IO80SQ, SY68]
G4 CFZ M Stevens, 3 Rip Croft, Portland, DT5 2EE [IO80SM, SY67]
G4 CGA D E Sellwood, 47 Waterhall Ave, Highams Park, London, E4 6NA [JO01AO, TQ39]
G4 CGB D B Tromans, 68 Brook St., Wall Heath, Kingswinford, DY6 0JG [IO82VM, SO89]
G4 CGC B Otter, PO Box 34554, Lusaka, Zambia
G4 CGD A Richardson, 24 West House Cl, Princess Way, Wimbledon, London, SW19 6QU [IO91VK, TQ27]
G4 CGE W A Dore, Marden, Summerleys Rd, Princes Risborough, HP27 9QA [IO91NR, SP70]
G4 CGF W Badz, 36 Luckington Rd, Horfield, Bristol, BS7 0US [IO81QL, ST57]
G4 CGG R I'Anson, Croft Corner, 87 Tranby Ln, Anlaby, Hull, HU10 7DT [IO93SR, TA02]
G4 CGH M J Davies, 3 Monkstone Dr, Berrow, Burnham on Sea, TA8 2NW [IO81LG, ST25]
G4 CGK T Robson, 48 Annesley Rd, Newport Pagnell, MK16 0BG [IO92PB, SP84]
G4 CGM M C Duff, Clittaford Club, Moses Cl, Southway, Plymouth, PL6 6JP [IO70WK, SX46]
G4 CGP P M Wright, 4 Avill Way, Wickersley, Rotherham, S66 0DL [IO93IK, SK49]
G4 CGR K F Davies, Highview, Alcester Rd, Wootton Wawen, Solihull, B95 6BH [IO92CG, SP16]
G4 CGU R G Taylor, 23 Ridgacre Ln, Birmingham, B32 1EL [IO92AL, SP08]
G4 CGV C S Manklow, 39 Wigston Rd, Bury St. Edmunds, IP33 2HF [JO02IF, TL86]
G4 CGW J F Dunglinson, Blenheim, Willow Ln, Blackwater, Camberley, GU17 9DL [IO91OH, SU85]
G4 CGY G P Matthews Ciba-Geigy Sprt, &Social Arc, Wimblehurst Rd, Horsham W Sussex, RH12 4AA [IO91UB, TQ13]
G4 CGZ D R Newman, Flat B, 26 Stockfield Rd, Streatham, London, SW16 2LR [IO91WK, TQ37]
G4 CHD T E Adams, 3 Drake Cl, Innsworth Ln, Churchdown, Gloucester, GL3 1NE [IO81VV, SO82]
G4 CHE C S M Scott, 56 Western Way, Ponteland, Newcastle upon Tyne, NE20 9AP [IO95CA, NZ17]
G4 CHG P Ashton, 7 Conway Gr, Cheadle, Stoke on Trent, ST10 1QG [IO92AX, SK04]
G4 CHH J Heathershaw, The Old School, Cliff Ln, Mappleton, Hornsea, HU18 1XX [IO93WV, TA24]
G4 CHI P J Robinson, Longcroft House, Longcroft, Yoxall, Burton on Trent, DE13 8NT [IO92CS, SK12]
G4 CHJ A F Williams, 10 Olde Hall Rd, Featherstone, Wolverhampton, WV10 7BB [IO82WP, SJ90]
G4 CHL P J Howe, Moorcroft, The Chase, Woolaston, Lydney, GL15 6PT [IO81QQ, ST59]
G4 CHM R A McEwan, Fifth Acre, Carr Ln, South Normanton, Alfreton, DE55 2DN [IO93HC, SK45]
G4 CHO S F Jenkins, 31 Briar Cl, Horndean, Waterlooville, PO8 9ED [IO90LV, SU61]
GU4 CHY R W Allisette, Lilyvale House, Rue Des Houmets, Castel, Guernsey, Channel Islands, GY5 7XZ
G4 CHZ S G T Cripsey, 9 Highthorpe Cres, Cleethorpes, DN35 9PZ [IO93XN, TA20]
G4 CI D S Babbage, 63 Inveresk Gdns, Worcester Park, KT4 7BB [IO91VI, TQ26]
G4 CIA W E Cooper, 20 Planton Way, Brightlingsea, Colchester, CO7 0LB [JO01MT, TM01]
G4 CIB B M Woodcock, The Larches, Poolhay Cl, Corse Lawn, Gloucester, GL19 4NY [IO81VX, SO83]
G4 CIC S G Edmondson, 7 Browns Rd, Bradley Fold, Bolton, BL2 6RQ [IO83TN, SD70]
G4 CIF K Turner, 42 Norman Rd, Walsall, WS5 3QL [IO92AN, SP09]
G4 CIG A D Lincoln, 63 Cuckfield Rd, Hurstpierpoint, Hassocks, BN6 9RR [IO90VW, TQ21]
G4 CII A J Berriman, 32 Bethune Rd, Horsham, RH13 5JP [IO91UA, TQ22]
G4 CIJ J M Chennells, 10 Lower Cippenham Ln, Slough, SL1 5DF [IO91QM, SU98]
G4 CIL A Siese, PO Box Hm 1060, Hamilton, Bermuda Hmex, ZZ9 9HA
G4 CIM R S Little, Mont Apica, Thetford Rd, Ixworth, Bury St. Edmunds, IP31 2HB [JO02JH, TL97]
G4 CIN Details withheld at licensee's request by SSL.
G4 CIO Dr M E Phillips, Chapel House, The Cross, Nympsfield, Stonehouse, GL10 3TU [IO81UQ, SO80]
G4 CIR Details withheld at licensee's request by SSL.
G4 CIV Details withheld at licensee's request by SSL.
G4 CIZ A R Wallbank, 22 Oakfield Rd, Pamber Heath, Tadley, RG26 3DN [IO91KI, SU66]
G4 CJC Details withheld at licensee's request by SSL.
G4 CJF Details withheld at licensee's request by SSL.
G4 CJG J P J Goldsmith, Dolphins, Toll Rd, Porlock, Minehead, TA24 8JH [IO81EF, SS84]
G4 CJI Details withheld at licensee's request by SSL.
G4 CJJ M W Viner, 83 Kingscote Rd Wes, Up Hatherley, Cheltenham, Glos, GL51 6JP [IO81WV, SO92]
G4 CJK V Roney, 76 Hilton Ln, Great Wyrley, Walsall, WS6 6DT [IO82XP, SJ90]
G4 CJL F Ashcroft, Rose Cottage, Southmead Ln, Henstridge, Templecombe, BA8 0RJ [IO80TX, ST71]
G4 CJM J L Alcock, 1 Alma St., Fenton, Stoke on Trent, ST4 4PH [IO82VX, SJ84]
G4 CJO A Mountifield, 6 Sawyers Cl, Teg Down, Winchester, SO22 5JX [IO91HB, SU43]
G4 CJP V L Duffy, 2 Moor View Cl, High Harrington, Workington, CA14 4NX [IO84FO, NY02]
G4 CJR C A Crick, 19 The Dr, Coulsdon, CR5 2BL [IO91WH, TQ36]
G4 CJS Details withheld at licensee's request by SSL.
G4 CJT K J Hughes, 4 Epsom Pl, Cranleigh, GU6 7ET [IO91SD, TQ03]
G4 CJU G A Gayton, 2 Heyes Park, Hartford, Northwich, CW8 2AJ [IO83RF, SJ67]
G4 CJY B J Payne, 78 Carver Hill Rd, High Wycombe, HP11 2UA [IO91OO, SU89]
G4 CJZ Details withheld at licensee's request by SSL.
G4 CK R J Stellig, 8 Glebe Gdns, Motcombe, Shaftesbury, SP7 9QQ [IO81VA, ST82]
G4 CKA J D Burling, 83 Sele Rd, Hertford, SG14 1NQ [IO91WT, TL31]
G4 CKB J H Banester, Fairfield, Church Rd, Sutton, Norwich, NR12 9SA [JO02SS, TG32]
G4 CKF Details withheld at licensee's request by SSL.
G4 CKG R Price, 33 Coatsby Rd, Hollycroft, Kimberley, Nottingham, NG16 2TH [IO93IA, SK44]
G4 CKH G Jackson, 8 Southdown Cl, Rochdale, OL11 4PP [IO83VO, SD81]
G4 CKK P R Atkins, 60 Wentworth Way, Harborne, Birmingham, B32 2UX [IO92AK, SP08]
GM4 CKM P W Turnbull, 93 Granton Pl, Edinburgh, EH5 1AZ [IO85JX, NT27]
G4 CKN Details withheld at licensee's request by SSL.
G4 CKO D C Matheson, 18 Rosewood Cl, Burnham on Sea, TA8 1HG [IO81MF, ST34]
GM4 CKP W M C Macdonald, 74 Lochiel Dr, Milton of Campsie, Glasgow, G65 8ET [IO75WX, NS67]
G4 CKQ A J Horne, 1 Upper Halliford Rd, Shepperton, TW17 8RX [IO91SJ, TQ06]
G4 CKR S Scott, 1 Norfolk Pl, Blackpool, FY2 0BD [IO83LU, SD34]
G4 CKS D E Fitzgerald, 36 Vardens Rd, London, SW11 1RH [IO91VL, TQ27]
G4 CKT R Gwynne, 17 Dorrington Cl, Millfield Gdns, Milton, Stoke on Trent, ST2 7BZ [IO83WB, SJ95]
G4 CKW P R Hawes, 34 Sewell Cl, Aylesbury, HP19 3SB [IO91NT, SP71]
G4 CKX S W Taylor, 5 Chiltern Ave, Bishops Cleeve, Cheltenham, Glos, GL52 4XP [IO81XW, SO92]
G4 CLA P S Lindsay, Greystones, Stanford Cl, Cold Ashby, Northampton, NN6 6EW [IO92LJ, SP67]
G4 CLB C L Brown, 22 Warrender Way, Ruislip, HA4 8ED [IO91TN, TQ18]
G4 CLC D J Lewis, Brandywine, Coulston, Westbury, Wilts, BA13 4NY [IO81XG, ST95]
G4 CLD G Beaver, The Gables, Reading Rd, Woodcote, Reading, RG8 0QY [IO91LM, SU68]

G4 CLE T J Baker, 24 Hathaway Rd, Sutton Coldfield, B75 5JB [IO92BO, SP19]
G4 CLF J M Bryant, 60 West St., Newbury, RG14 1BD [IO91IJ, SU46]
G4 CLG S A Whittingham, 18 Northcroft, Shenley Lodge, Milton Keynes, MK5 7AJ [IO92OA, SP83]
G4 CLK Details withheld at licensee's request by SSL.
G4 CLL R Goodchild, Grey Cliffe House, Owmby Cliff Rd, Owmby By Spital, Market Rasen, LN8 2HL [IO93SI, TF08]
G4 CLN P G Redfern, 12 Wilbarn Rd, Preston, Paignton, TQ3 2BN [IO80FK, SX86]
G4 CLP J L Harrison, 18 Stamford St., Sale, M33 6LL [IO83UK, SJ79]
GM4 CLQ Details withheld at licensee's request by SSL.
G4 CLR I Hewer, 5 Windsor Dr, Tuffley, Gloucester, GL4 0QW [IO81UT, SO81]
G4 CLT Details withheld at licensee's request by SSL.
G4 CLV P F Cottrell, 6 The Dr, Ersham Park, Hailsham, BN27 3HP [JO00DU, TQ50]
G4 CLY N C Thompson, 6 Miena Way, Ashtead, KT21 2HU [IO91UH, TQ15]
G4 CLZ C A Todd, Beechwood, Grimpo, West Felton, Oswestry Shropshire, SY11 4SG [IO82MU, SJ32]
G4 CM R C C Middle, Cliff View, Church Ln, Aisthorpe, Lincoln, LN1 2SG [IO93RH, SK98]
G4 CMC P E Carr, 73 Mendip Rd, Yatton, Bristol, BS19 4HP [IO81OJ, ST46]
G4 CMF Details withheld at licensee's request by SSL.
G4 CMG T G Milne, Lynwood, Clovelly Rd, Beacon Hill, Hindhead, GU26 6RP [IO91OD, SU83]
G4 CMH D Spendlove, 22 Green Bank, Harwood, Bolton, BL2 3NG [IO83TO, SD71]
G4 CMI Details withheld at licensee's request by SSL.
G4 CMK R A Harker, 140 Victoria Rd, Beverley, HU17 8PJ [IO93ST, TA03]
G4 CMP P L Lennon, 53 Rycot Rd, Speke, Liverpool, L24 3TH [IO83NI, SJ48]
G4 CMR E F Beckett, Clifton, 3 Spring Gr, Fetcham, Leatherhead, KT22 9NN [IO91TG, TQ15]
G4 CMT R C Andreang, 6 Beech Ave, Bilton, Hull, HU11 4EN [IO93VS, TA13]
G4 CMU Details withheld at licensee's request by SSL.[Station located near Banstead.]
G4 CMV Details withheld at licensee's request by SSL.
GW4 CMW R M Powell, 16 Bulford Rd, Johnston, Haverfordwest, SA62 3EU [IO71LS, SM91]
G4 CMY A A Mann, 6 The Hedgerow, Longlevens, Gloucester, GL2 9JE [IO81VV, SO81]
G4 CMZ K A Archer, 24 Willson Rd, Littleover, Derby, DE23 7BZ [IO92FV, SK33]
G4 CND Details withheld at licensee's request by SSL.
G4 CNE Details withheld at licensee's request by SSL.
GM4 CNF J C Mathers, 57 Strathblane Rd, Milngavie, Glasgow, G62 8HA [IO75UW, NS57]
G4 CNG G J Yarnold, Longwood, The Paddock, Melmerby, Ripon, HG4 5HW [IO94GE, SE37]
G4 CNH L J Carpenter, 87 Kings Ave, Watford, WD1 7SB [IO91TP, TQ19]
G4 CNI P G Geiger, Lloyd Mount, Howard Dr, Hale, Altrincham, Ches, WA15 0LT [IO83UI, SJ78]
G4 CNJ F Pratt, 35 Graylands Rd, Bilborough, Nottingham, NG8 4FH [IO92JX, SK54]
G4 CNK W Johnson, 7 Burns Terr, Shotton Colliery, Durham, DH6 2PD [IO94HS, NZ34]
GW4 CNL G G Goodfield, 10 Lewis St., Church Village, Pontypridd, CF38 1BY [IO81IN, ST08]
GW4 CNM H G Roberts, 2 Cadnantcourt, Rating Row, Beaumaris, Anglesey Gwynedd, LL58 8AT [IO73WG, SH67]
G4 CNN J H Sexton, 24 Badgers Rise, Caversham, Reading, RG4 7QA [IO91ML, SU77]
G4 CNW R H Bold, 10 Haddon Cl, Stanground, Peterborough, PE2 8LS [IO92VN, TL29]
G4 CNZ D G Allen, 344 Coventry Rd, Hinckley, LE10 0NH [IO92HM, SP49]
G4 COA Details withheld at licensee's request by SSL.
G4 COE D Smith, 54 Warrington Rd, Leigh, WN7 3EB [IO83SL, SJ69]
G4 COF R Hooper, Kelly College, Parkwood Rd, Tavistock, PL19 0HZ [IO70WN, SX47]
GW4 COJ C J Roberts, 8 Oaklands Park Dr, Rhiwderin, Newport, NP1 9RB [IO81LN, ST28]
GM4 COK G O Szymanski, 56 Telford Rd, Edinburgh, EH4 2LX [IO85JX, NT27]
G4 COL I Braithwaite, 28 Oxford Ave, St. Albans, AL1 5NS [IO91UR, TL10]
G4 COM J R Compton, Aysgarth, Durley Brook Rd, Durley, Southampton, SO32 2AR [IO90IW, SU51]
G4 COQ J Holland, 18 Edinburgh Dr, Oswaldtwistle, Accrington, BB5 3AR [IO83TR, SD72]
G4 COR I D Harvey, 50 Callow Hill Way, Littleover, Derby, DE23 7RJ [IO92GV, SK33]
G4 COS J A Hansell, 210 Rutland Ave, High Wycombe, HP12 3LN [IO91OO, SU89]
G4 COT S P Brett, 8 Pinewood Gr, Nunburnholme Park, Hull, HU5 5YY [IO93TR, TA02]
G4 COU C J Bigger, 18 Wallace Rd, Loughborough, Leics, LE11 3NX [IO92JS, SK51]
G4 COV C D Cardwell, 11 Manor Cttgs, Heronsgate Rd, Chorleywood, Rickmansworth, WD3 5BJ [IO91RP, TQ09]
G4 COW R C Ridgwell, 186 New St., Great, Torrington, EX38 8BY [IO70WW, SS41]
GM4 COX J R Hood, 4 Murray Rd, Law, Carluke, ML8 5HR [IO85BR, NS85]
G4 CPA G S Hanson, 11 Churchill Way, Cross Hills, Keighley, BD20 7DN [IO93AV, SE04]
G4 CPB S E Walker, 1 Windyridge Rd, Walmley, Sutton Coldfield, B76 1HA [IO92CM, SP19]
G4 CPE A H Turner, 318 Sundon Park Rd, Luton, LU3 3AR [IO91SW, TL02]
G4 CPG M Howkins, 16 Beckett Ct, Gedling, Nottingham, NG4 4GS [IO92KX, SK64]
G4 CPI J Housden, 113 St. Bernards Rd, Whitwick, Coalville, LE67 5GY [IO92HR, SK41]
G4 CPJ J S Seymour, Lakam, Ferry Ln, North Muskham, Newark, Notts, NG23 6HB [IO93OC, SK75]
G4 CPL C G McGee, 10 Queensville Rd, London, SW12 0JJ [IO91WK, TQ27]
G4 CPM A P Fielding, 95 Hillcrest, Weybridge, KT13 8AS [IO91SI, TQ06]
G4 CPN Revd J P Bird, The Vicarage, Bridge St., Uffculme, Cullompton, EX15 3AX [IO80IV, ST01]
GI4 CPP W Adamson, 12 Greenland Walk, Ferris Park West, Larne, BT40 1NJ [IO74CU, D40]
G4 CPQ N P Scrogie, 1 The Knoll, Hayes, Bromley, BR2 7DD [JO01AJ, TQ46]
G4 CPS M Crawshaw, 0B0 Park High Sch Rd, Park High School, Venables Ave, Coln, Lancs, BB8 7DP [IO83WU, SD84]
G4 CPT E J Dyer, 6 Aldersbrook Ave, Enfield, EN1 3JB [IO91XP, TQ39]
G4 CPV R C Fisk, 16 Sterry Dr, Thames Ditton, KT7 0YN [IO91TJ, TQ16]
G4 CPW P F Wilson, 5 Pebble Cl, Lowestoft, NR32 4DR [JO02UL, TM59]
G4 CPY N W Grassby, 3, Lancaster Pl, Leicester, LE1 7HB [IO92KP, SK50]
G4 CQB Details withheld at licensee's request by SSL.
G4 CQC J J Calder, 59 Queens Cres, Gorleston, Great Yarmouth, NR31 7NJ [JO02UN, TG50]
G4 CQH J F Sperry, 50 Lochinver, Hanworth, Bracknell, RG12 7LD [IO91OJ, SU86]
G4 CQI A D Lanfear, 120 Charlton Rd, Kingswood, Bristol, BS15 1HF [IO81RL, ST67]
G4 CQJ Dr B Mellor, Silver Birches, Water Ln, Fewcott, Bicester, OX6 9NX [IO91JW, SP52]
GI4 CQL N R Kyle, 197 Aghafad Rd, Clogher, BT76 0XE [IO64JK, H55]
G4 CQM D Hilleard, Hazeldene, Bridgerule, Holsworthy, Devon, EX22 7EW [IO70ST, SS20]
G4 CQN M A Mawby, 129 Hook Rd, Goole, DN14 5XR [IO93NR, SE72]
G4 CQO S R Burgess, Bryher House, Howgate Rd, Bembridge, Isle of Wight, PO35 5UA [IO90KQ, SZ68]
G4 CQQ R W Taylor, 8 Park Ave, Markfield, LE67 9WA [IO92IQ, SK41]
G4 CQR D G Wood, 49 Wolsey Cres, Morden, SM4 4TD [IO91VJ, TQ26]
G4 CQS A Rowsby, 10 Echells Cl, Bromsgrove, B61 7EB [IO82XI, SO97]
G4 CQV P I Baldwin, 26 Ashford Rd, Fulshaw Park, Wilmslow, SK9 1QE [IO83VH, SJ87]
G4 CQX L E Palfrey, c/o PO Box 314, Paralimni, Cyprus, XX99 1AA
GW4 CQZ M P Doig, Helenfa, Ystrad Rd, Denbigh, Denbighshire, LL16 3HE [IO83HE, SJ06]
G4 CRB W J Oxley, Cloudshill, Well Lane, Stow on The Wold, Glos, GL54 1DB
G4 CRC G Bate Cornish ARC, 7 Albany Ct, Redruth, TR15 2NY [IO70JF, SW74]
G4 CRE D L Rush, 8 Sheaf Pl, Worksop, S81 7LE [IO93KH, SK58]
G4 CRF Details withheld at licensee's request by SSL.
G4 CRG K N Burgin, The Pike Lock House, Eastington, Stonehouse, Glos, GL10 3RT [IO81US, SO70]
G4 CRH M B Worvill, The Berwyns, Domgay Rd, Four Crosses, Llanymynech, SY22 6SL [IO82LR, SJ31]
G4 CRJ Details withheld at licensee's request by SSL.[Station located near High Wycombe.]
G4 CRK R J Sellman, 43 Mount Ave, Stone, ST15 8LW [IO82WV, SJ83]
GI4 CRL A E Henry, 23 Long Common, Country, Ballymena, BT42 2NU [IO64UU, D00]
G4 CRM J B Lennon, 130 Hambledon Rd, Waterlooville, PO7 6XA [IO90LV, SU61]
G4 CRN Dr A R Hall, Westhill,Bear Ln, Longdon, Tewkesbury, Glos, GL20 6BB [IO82VA, SO83]
G4 CRP K J Tyler, 86 Kingsway, Garforth, Leeds, LS25 1DQ [IO93HS, SE33]
GI4 CRQ R K Quigg, 101 Belvoir Dr, Belfast, BT8 4DN [IO74AN, J36]
G4 CRS A G Mackay E.C A.R.C., 2 Highcliffe Gr, New Marske, Redcar, TS11 8DU [IO94LN, NZ62]
G4 CRT L Kirby, 41 Woodville Rd, Overseal, Swadlincote, DE12 6LU [IO92FR, SK21]
G4 CRU M A Pearson, 5 Potterdale Dr, Little Weighton, Cottingham, HU20 3UU [IO93RT, SE93]
GM4 CRV J H Cregan, 15 Menzies Dr, Fintry, Glasgow, G63 0YG [IO76VB, NS68]
G4 CRW A J Holmes, 4 Castle Ave, Datchet, Slough, SL3 9BA [IO91QL, SU97]
G4 CRY P J Halls, 20 Tedder Rd, Acomb, York, YO2 3JB [IO93KW, SE55]
G4 CSD P A Hyde, High Trees, 8 Easton Cres, Billingshurst, West Sussex, RH14 9TU [IO91SA, TQ02]
G4 CSE M E Lewis, 10 Kenmore Dr, Filton, Bristol, BS7 0TT [IO81QM, ST57]
G4 CSF Details withheld at licensee's request by SSL.
G4 CSG B D Gibbs, Marigold Cottage, Dittons Rd, Polegate, BN26 6HU [JO00DT, TQ50]
G4 CSH N W Holding, 17 Mayfair Dr, Thornton Cleveleys, FY5 5BY [IO83LU, SD34]
G4 CSI C J Osborn-Jones, Hudnall House, Hudnall Ln, Little Gaddesden, Berkhamsted, HP4 1QQ [IO91RT, TL01]
G4 CSL R K Pegg, 401 Ratby Ln, Kirby Muxloe, Leicester, LE9 2AQ [IO92JP, SK50]
G4 CSM D Chaplin, 35 Lanes End, Totland Bay, PO39 0AL [IO90FQ, SZ38]
G4 CSN N Bylo, 7 The Croft, Sheriff Hutton, York, YO6 1PQ [IO94MC, SE66]
GI4 CSO J Mc Cormack, 12 Glengoland Cres, Dunmurry, Belfast, BT17 0JG [IO64XN, J27]
GI4 CSP G W J Robinson, 23 Old Moy Rd, Dungannon, BT71 6PS [IO64PL, H85]
G4 CST G C Hopkins, 83 Ermin St., Blunsdon, Swindon, SN2 4AA [IO91CO, SU19]
G4 CSV J Jackson Ba(Hons), 11 Sylvandale Ave, West Point, Manchester, M19 2FB [IO83VK, SJ89]
GW4 CSY B J Vickery, 6 Duffryn Cl, St. Nicholas, Cardiff, CF5 6SS [IO81IK, ST07]
G4 CSZ M R Riley, 5 Dunstarn Gdns, Leeds, LS16 8EJ [IO93FU, SE23]
G4 CTA A R Clewer, 32 Langlands Rd, Cottingley, Bingley, BD16 1QJ [IO93CT, SE13]
G4 CTC T Cann, Noahs Rough, Old Coach Rd, Wrotham, Sevenoaks, TN15 7NR [JO01DH, TQ66]
G4 CTE P J Bradshaw, The Brambles, 10 Hawksmede Way, Louth, LN11 0SR [JO03AJ, TF38]
G4 CTG Details withheld at licensee's request by SSL.
G4 CTI P H Ashcroft, 86 Avondale Ave, North Finchley, London, N12 8EN [IO91VO, TQ29]
G4 CTL Details withheld at licensee's request by SSL.
G4 CTM P L Barrett, 3 Bramshott Cl, Hitchin, SG4 9EP [IO91UW, TL12]
G4 CTO Details withheld at licensee's request by SSL.[Station located near Launceston.]
G4 CTP R S Whitworth, 26 Preston Cres, Inverkeithing, KY11 1DR [IO86HA, NT18]
G4 CTQ S T May, 77 Chaucer Dr, St. Giles, Lincoln, LN2 4LT [IO93RF, SK97]
G4 CTR R P Morris, Crichel House, Crichel Mount Rd, Evening Hill, Poole, Dorset, BH14 8LT [IO90AQ, SZ08]
G4 CTS B H Andrews, 52 Trenchard Cres, Springfield, Chelmsford, CM1 6FG [JO01FS, TL70]
G4 CTT Dr T J Thirst, The Haywain, Thirsts Farm, Happisburgh, Norfolk, NR12 0RU [JO02ST, TG32]
G4 CTU B P C Hitchins, 12 Parkland Ave, Kidderminster, DY11 6BX [IO82UJ, SO87]
GW4 CTV S S Mee, Cysgod y Gear, Cwmsymlog, Aberystwyth, Dyfed, SY23 3EZ [IO82AK, SN68]
G4 CTW F J Radwell, 25 Charles Ley Ct, Denny Cl, Fawley, Southampton, SO45 1FR [IO90HT, SU40]
G4 CTY A A Lightbody, 3 Elphicks Pl, Tunbridge Wells, TN2 5NB [JO01DC, TQ53]
G4 CTZ I M Cage, Rowan House, 5 Abbotts Gdns, Cawood, Selby, YO8 0TF [IO93KT, SE53]
GM4 CUB R H Paterson, Glenalla, 22 Doonfoot Rd, Ayr, KA7 4DN [IO75QK, NS32]
G4 CUC M P Cunningham, 29 Clark Ave, Edlington, Doncaster, DN12 1HN [IO93JL, SK59]
G4 CUD J F Fox, 19 St. Thomas Rd, Redbrooks, Barnsley, S75 1HW [IO93FN, SE30]
G4 CUE W L Pechey, Jays Lodge, Crays Pond, Reading, RG8 7QG [IO91KM, SU68]
G4 CUF A Byers, 2 Beechwood Ave, Ashton in Makerfield, Wigan, WN4 9LZ [IO83QL, SJ59]
G4 CUG R E Worsell, 8 Waterworks Cottage, S Friston, Eastbourne, Sussex, BN20 0AS [JO00CS, TV59]
G4 CUI Dr G G Cook, 1 St. Albans Rd, Fulwood, Sheffield, S10 4DN [IO93FI, SK38]
G4 CUJ Details withheld at licensee's request by SSL.
G4 CUK Details withheld at licensee's request by SSL.
G4 CUO D W Rowan, 13 Fleming Dr, Newark, NG24 2BA [IO93OC, SK85]
G4 CUQ B R Hughes, 40 Inverness Terr, London, W2 3JA [IO91VM, TQ28]
G4 CUR P H Burton, 76 Station Rd, Cholsey, Wallingford, OX10 9QB [IO91KN, SU58]
G4 CUS A P Turnbull, North Leas, Telham, Battle, E Sussex, TN33 0TP [JO00FW, TQ71]
GI4 CUV N Atkins, 38 Rosscoole Park, Belfast, BT14 8JX [IO74AP, J37]
GM4 CUX G R Winchester, 23 Craigmount Av Nt, Edinburgh, EH12 8DL [IO85IW, NT17]
G4 CUY Details withheld at licensee's request by SSL.[Op: Miss C Wade.]
GM4 CUZ S C Strachan, 4 Rhynie Rd, West Ferry, Broughty Ferry, Dundee, DD5 1RH [IO86NL, NO43]
G4 CVA Rev J A Wardle, 3 Vicars Ct, Southwell, NG25 0HP [IO93MB, SK75]
G4 CVC J W Everist, 15 West Harold, Swanley, BR8 7EJ [JO01CJ, TQ56]
G4 CVD P M Petty, 41 Hensley Rd, Bath, BA2 2DR [IO81TI, ST76]
G4 CVE J T Halford, Northfield, 5 Marryats Loke, Langham, Holt, NR25 7AE [JO02MW, TG04]
G4 CVF B C Sheppard, 21 Lambourne Ct, St. Johns Cl, Uxbridge, UB8 2UL [IO91SM, TQ08]
G4 CVG W D Bullock, 14 Saxon Dr, Rillington, Malton, YO17 8LZ [IO94PD, SE87]
G4 CVI Details withheld at licensee's request by SSL.
G4 CVK C S Williamson Old Swinford, Hospital School R South, 7 Hanbury Hill, Stourbridge, DY8 1BE [IO82WK, SO98]
G4 CVM R A Watson, 36 Abbots Cl, Knowle, Solihull, B93 9PP [IO92DJ, SP17]
G4 CVN D Williams, Half Acre, Rectory Ln, Bracknell, RG12 7AX [IO91OJ, SU86]
G4 CVO W L Wyer, 21 Chesterfield Ave, New Whittington, Chesterfield, S43 2BX [IO93HG, SK47]
G4 CVS Dr B R Pearson, 8 The Pastures, Edlesborough, Dunstable, LU6 2HL [IO91QU, SP91]
G4 CVU J Swingewood, 5 Blaze Park, Wall Heath, Kingswinford, DY6 0LL [IO82VM, SO88]
G4 CVW E W Law, 6 Blossom Hill, Erdington, Birmingham, B24 9DN [IO92CM, SP19]
G4 CVX Dr R J Sims, 345 Blandford Rd, Hamworthy, Poole, BH15 4HP [IO80XR, SY99]
G4 CVY Details withheld at licensee's request by SSL.
G4 CVZ A V Neilson, 16 Martlesham Cres, Wirral, L49 3PR [IO83KJ, SJ28]
G4 CW S R Smith North Kent RS, 124 Parkside Ave, Barnehurst, Bexleyheath, DA7 6NL [JO01CL, TQ57]
G4 CWA W Burton, 57A Cleveland Terr, Darlington, Co. Durham, DL3 8HN [IO94FM, NZ21]
G4 CWB D C Andrews, 100 Duchy Rd, Harrogate, HG1 2HA [IO93FX, SE25]
G4 CWC R Barrett, Lumina, Bridegate Ln, Hickling Pastures, Melton Mowbray, LE14 3QA [IO92MU, SK62]
G4 CWD A L Bagnall, 15 Ypres Rd, Allestree, Derby, DE22 2NA [IO92FW, SK33]
G4 CWE A J Humm, 32 Layton Rd, Hounslow, TW3 1YH [IO91TL, TQ17]
G4 CWG G H Crossland, 28 Grantley Dr, Harrogate, HG3 2XU [IO94FA, SE25]
G4 CWH Dr C R Smithers, 10 Grange Park, Bishops Stortford, CM23 2HX [JO01BV, TL42]
G4 CWJ J R Brindley, 20 Swan Cl, Talke, Stoke on Trent, ST7 1TA [IO83UB, SJ85]
G4 CWK Details withheld at licensee's request by SSL.
G4 CWL B Fletcher, 6 Burrows Cl, Penn, High Wycombe, HP10 8AR [IO91PP, SU99]
G4 CWM J H Pickles, 111 Linden Ave, Prestbury, Cheltenham, GL52 3DT [IO81XV, SO92]
G4 CWN J D Colclough, 16 Hackwood Cl, Barlaston, Stoke on Trent, ST12 9BE [IO82WW, SJ83]
G4 CWO B J Crowley, Egmont House, 101A Pound Ln, Sonning, Reading, Berks, RG4 0GG [IO91ML, SU77]
G4 CWP W R H Pevy, Brambletye, Ashtead Ln, Godalming, GU7 1SY [IO91QE, SU94]
G4 CWR P C Prior, 49 Woodbury Way, Axminster, EX13 5RE [IO80MS, SY29]
G4 CWS S C Wood, Hubbles, Bentworth, Alton, Hants, GU34 5RB [IO91LD, SU63]
G4 CWT G W Eves, Shirley Cottage, 112 Cooden Sea Rd, Little Common, Bexhill on Sea, TN39 4TA [JO00FU, TQ70]
GW4 CWU B Heppenstall, Gwelfor, Llanrhuddlad, Holyhead, Gwynedd, LL65 4BU [IO73RH, SH38]
G4 CWV F A Parr, 5 Benenden Rd, Frindsbury, Wainscott, Rochester, ME2 4NU [JO01GJ, TQ77]
G4 CWX Details withheld at licensee's request by SSL.
G4 CWY A R Sharp, 18 Vandyke Rd, Oadby, Leicester, LE2 5UB [IO92LO, SP69]
G4 CXA A J Cowley, 20 Broadfield Gr, Reddish, Stockport, SK5 6XN [IO83WK, SJ89]
G4 CXE P Bolton, 93 Westfields, Narborough, Kings Lynn, PE32 1SY [JO02HQ, TF71]
GM4 CXF J M Thomson, 31 Teviot Pl, Troon, KA10 7EE [IO75QN, NS33]
G4 CXG Details withheld at licensee's request by SSL.
GI4 CXH C R S Brennan, 58 Fortwilliam Park, Belfast, BT15 4AS [IO74AP, J37]
G4 CXJ B Mussell, Vine House, 73 Woodstock Rd, Witney, OX8 6ED [IO91GT, SP31]
GW4 CXK R W Evans, 60 Tan y Bryn, Rhymney, NP2 5LF [IO81IS, SO10]
G4 CXL R J Menday, Brindle Crest Camp, End Rd, St Georges Hill, Weybridge, KT13 0NR
GM4 CXM R P James, 4 Pentland Pl, Bearsden, Glasgow, G61 4JU [IO75TW, NS57]
GM4 CXP D Dance, 5 Wester Row, Greenlaw, Duns, TD10 6XE [IO85SQ, NT74]
G4 CXQ D L Dyer, Damir, 26 Locking Rd, Weston Super Mare, BS23 3DF [IO81MI, ST36]
G4 CXT M R C Bell, 50 Avocet Ln, Martlesham Heath, Ipswich, IP5 7SE [JO02OB, TM24]
G4 CXU C Gill, 34A Cannon Cl, Coventry, CV4 7AS [IO92FJ, SP37]
G4 CXW G E Spencer, Gatesgarth, Rockland Rd, Downend, Bristol, BS16 2SW [IO81RL, ST67]
G4 CXZ A C Thompson, 171 Boothferry Rd, Hull, HU4 6EX [IO93TR, TA02]
G4 CYA J C Otley, 13 Cruise Rd, Sheffield, S11 7EF [IO93FI, SK38]
G4 CYB F G Burnett, Herons Siege, Blundies Ln, Enville, Stourbridge, DY7 5HU [IO82UL, SO88]
G4 CYC K Green, Ao Te Aroa, 12 Hill Rd, Portchester, Fareham, PO16 8LB [IO90KU, SU60]
G4 CYE A P L Hall, 3 Vicarage Ct, Off Vicarage Rd, Hanham, Bristol, BS15 3BL [IO81RK, ST67]
G4 CYF C S Tully, Harmony, The Cres, Thorpe-le-Soken, Clacton on Sea, CO16 0EP [JO01NU, TM12]
G4 CYI J R Palfrey, The Barn, Station Rd, Congresbury, Bristol, BS19 5DX [IO81OI, ST46]
GI4 CYJ R G Skeats, 30 Clanconnel Gdns, Waringstown, Craigavon, BT66 7RP [IO64UK, J15]
GI4 CYK D J Fox, 71 Killymeal Rd, Dungannon, BT71 6LG [IO64PM, H86]
G4 CYM W J Stroud, 43 Wingfield Rd, Gravesend, DA12 1BS [JO01EK, TQ67]
G4 CYO K A Robinson, Woodhouse Cottage, 3 Patshull Rd, Pattingham, Wolverhampton, WV6 7DU [IO82UO, SO89]
G4 CYR S R Allen, Harriers, 96A Churchway, Haddenham, Aylesbury, HP17 8DT [IO91MS, SP70]
GI4 CYU H McIlroy, 28 Currans Brae, Moy, Dungannon, BT71 7SY [IO64PK, H85]
G4 CYW D F Borne, Roughways, Chub Tor, Yelverton, Devon, PL20 6HY [IO70WL, SX56]
G4 CYY C V Lewis, 54 Whelpley Hill Pk, Chesham, Bucks, HP5 3RJ [IO91RR, SP90]
G4 CYZ L H T Large, Captains House, Streat, Hassocks, Sussex, BN6 8SB [IO90XW, TQ31]
G4 CZA K J N Newman, 2 Skys Wood Rd, St. Albans, AL4 9NZ [IO91US, TL10]
G4 CZB J R Cockrill, 28 Northampton Rd, Harpole, Northampton, Northants, NN7 4DD [IO92MF, SP66]
G4 CZE A Mercer, Les Eversins Du Presle, 69870 Chambost-Allieres, France, X X
G4 CZH E T Brindley, 214 Sinfin Ave, Shelton Lock, Derby, DE24 9QA [IO92GV, SK33]
GW4 CZK A J Mercer, Tan Rallt, Brynteg, Isle of Anglesey, Gwynedd, LL78 7JL [IO73UH, SH48]
G4 CZL A J Rogers, 231 Village Way, Beckenham, BR3 3NN [IO91XJ, TQ36]
G4 CZP R M Crossley, 8 Heather Cl, Horsell, Woking, GU21 4JR [IO91RH, SU95]
G4 CZR C B Redfern, Boomstede 638, 3608 BR Maarssen, X X
G4 CZS Details withheld at licensee's request by SSL.
G4 CZT H C A Turner, 29 Burghley Rd, Wimbledon, London, SW19 5HL [IO91VK, TQ27]
G4 CZU P A Hadler, 30 Hill View Rd, Whitstable, CT5 4HX [JO01MI, TR16]
G4 CZV A R Wilson, 57 Ridgeway Rd, Willerby Rd, Hull, HU5 5HU [IO93TS, TA02]
GI4 CZW C J T Corderoy, 3 The Limes, Drumlyon, Enniskillen, BT74 5NQ [IO64EI, H24]
G4 CZX I G Godden, 163 Ringmer Rd, Worthing, BN13 1DZ [IO90TT, TQ10]
G4 CZZ R B Aggus, 68 Conifer Walk, Chells Manor Villag, Stevenage, SG2 7QS [IO91WV, TL22]
G4 DA J H Townshend, 11 Bloomfield Rd, Kingston upon Thames, KT1 2SF [IO91UJ, TQ16]
G4 DAC D H Squires, 91 Croham Valley Rd, South Croydon, CR2 7JJ [IO91XI, TQ36]
G4 DAD D A Denny, 144 Yarlside Rd, Barrow in Furness, LA13 0EX [IO84JC, SD26]
GM4 DAE W P McMillan, 149 Easterhill St., Tollcros, Glasgow, G32 8LE [IO75VU, NS66]
G4 DAF G A M Walker, 56 Goodwin Rd, Waddon, Croydon, CR0 4EG [IO91WI, TQ36]
GI4 DAH G Heaney, 49 Greystown Ave, Belfast, BT9 6UG [IO74AN, J36]
G4 DAJ T Aldridge, 53 Main Rd, East Morton, Keighley, BD20 5TE [IO93BU, SE04]
G4 DAL D R Alston, Flat 5, 65 The Ridgeway, Chingford, London, E4 6QW [IO91XP, TQ39]
G4 DAM R L Dence, 32 Hayeswood Rd, Stanley Common, Ilkeston, DE7 6GB [IO92HX, SK44]
G4 DAN R P Neave, 59 Harwich Rd, Mistley, Colchester, CO4 3BU [JO01LV, TM02]

G4

G4 DAP C R Ison, 19 Grays Cl, Chalgrove, Oxford, OX44 7TN [IO91KQ, SU69]
G4 DAQ W A Silvester, 2 Tudor Cl, Barton-le-Clay, Bedford, MK45 4NE [IO91SX, TL03]
G4 DAR N E Rock Dudley ARC, 56 Tansey Green Rd, Brierley Hill, DY5 4TE [IO82WM, SO98]
G4 DAT R W Davidson, Bramleys, Horsepond Rd, Gallowstree Common, Reading, RG4 9BT [IO91LM, SU68]
G4 DAU A E Pardy, 3 Riverway, Nailsea, Bristol, BS19 1HZ [IO81OK, ST47]
GI4 DAV D A Hart, 31 Downshire Rd, Carrickfergus, BT38 7QD [IO74CR, J48]
G4 DAX D S Smith, Red Roof, Goathland, Whitby, North Yorks, YO22 5AN [IO94PL, NZ80]
G4 DBA R H Percival, Orchard Cottage, The Green, Wetheral, Carlisle, CA4 8ET [IO84OV, NY45]
G4 DBC B G Capper, 1 Cumberland Cl, Hornchurch, RM12 6JJ [JO01CN, TQ58]
G4 DBD A M Borland, 50 Grebe Cl, Poynton, Stockport, SK12 1HU [IO83WI, SJ98]
G4 DBE Dr J H Clark, 15 Pond View Cl, Heswall Hills, Wirral, L60 1YH [IO83LH, SJ28]
G4 DBF Maj D J Freeston, 20 Coningham Rd, Whitley Wood, Reading, RG2 8QP [IO91MJ, SU76]
G4 DBG F E Kneale, 60 Summertrees Rd, Great Sutton, South Wirral, L66 2RP [IO83MG, SJ37]
G4 DBL T P Stanley, 72 Downs Rd, South Wonston, Winchester, SO21 3EW [IO91HC, SU43]
G4 DBM B P McGennity, 46 St. Andrews Rd, Boreham, Chelmsford, CM3 3BY [JO01GS, TL71]
G4 DBN N R Smith, Birch Tree House, Asselby, Goole, DN14 7HE [IO93NR, SE72]
G4 DBO R Primack, PO Box 18029, Delta, British Columbia, Canada, V4L 2M4
G4 DBP J B Anderson, West View Garage, Rudgate, Whixley, York, YO5 8AL [IO94IA, SE45]
G4 DBQ B A Roberts, 10 Sutcliffe Cl, London, NW11 6NT [IO91VN, TQ28]
G4 DBS Details withheld at licensee's request by SSL.
G4 DBU R J Edwards, 362 Garstang Rd, Fulwood, Preston, PR2 9RY [IO83PT, SD53]
G4 DBV G W Holtham, 36 Ellesboro Rd, Harborne, Birmingham, B17 8PT [IO92AL, SP08]
G4 DBW R W Hammond, 51 Poplar Dr, Greenhill, Herne Bay, CT6 7PY [JO01NI, TR16]
G4 DBX L J Stubbs, The Cottage, Middlewich Rd, Bradfield Green, Crewe, CW1 4RA [IO83SD, SJ65]
G4 DBY P T Walker, 48 Whitefields Dr, Richmond, DL10 7DL [IO94DJ, NZ10]
G4 DBZ D S Martin, 12 South Park, Redruth, TR15 3AW [IO70JF, SW64]
G4 DC P W Winsford, 16 Brisbane, Stonehouse, GL10 2PX [IO81US, SO80]
G4 DCB P Mortimer, 13A Elder Ave, Wickford, SS12 0LP [JO01GO, TQ79]
GI4 DCC W R Chesney, 52 Taylorstown Rd, Toomebridge, Antrim, BT41 3RT [IO64TT, J09]
G4 DCD C N Stephenson, 6 Livingstone Cl, Rothwell, Kettering, NN14 6HT [IO92OK, SP88]
G4 DCE J A Sketchley, 48 Coverdale, Thringstone, Whitwick, Coalville, LE67 5BP [IO92HR, SK41]
G4 DCF M W Booth, 15 Nether Royd View, Silkstone Common, Barnsley, S75 4QQ [IO93FM, SE20]
G4 DCG B H Licence, The Coach House, Lunecliffe, Ashton Rd, Lancaster, LA2 0AB [IO84OA, SD45]
G4 DCH C Tucker, 19 Holtwood Dr, Woodside, Ivybridge, PL21 9TH [IO80AJ, SX65]
G4 DCI P Hopewell, 15 Seatoller Cl, Edwalton Park, West Bridgford, Nottingham, NG2 6RB [IO92KW, SK63]
G4 DCJ D C Jarrett, 15 Groveside Est, East Rudham, Kings Lynn, PE31 8RL [JO02IT, TF82]
G4 DCK M R Holliday, Tosca, 1 Fairmile Gdns, Longford, Gloucester, GL2 9ED [IO81VV, SO82]
GM4 DCL T G Main, 15 Polton Rd, Lasswade, EH18 1AB [IO85KV, NT36]
G4 DCM P Rhodes, Parcela 1352, Calle de Zurbaran 8, Ciudad Quesada, 03170 Rojales,Alicante, Spain
G4 DCP P D Hull, Seymour Cottage, Forest Rd, Denmead, Waterlooville, PO7 6UA [IO90LV, SU61]
G4 DCQ Details withheld at licensee's request by SSL.
G4 DCR Details withheld at licensee's request by SSL.
G4 DCS M G Grant, 51 Manton, Bretton, Peterborough, PE3 9YT [IO92UN, TL19]
G4 DCV P E Whatton, 39 Wellington Rd, Deal, CT14 7AL [JO01QF, TR35]
G4 DCW D C Walker, Newlands Barn, 72 Tong Ln, Bradford, BD4 0RX [IO93ES, SE23]
G4 DCX E F Trickey, 53 Hollyguest Rd, Hanham, Bristol, BS15 3RN [IO81RK, ST67]
G4 DCY W Dransfield, Flat 6, Heathmounthall, Crossbeck Rd, Ilkley, LS29 9JN [IO93CW, SE14]
G4 DDA C D Lander, 4 Greenfield Rd, Preston, Paignton, TQ3 1DB [IO80FK, SX86]
G4 DDB E Connor, 17 The Strand, Walmer, Deal, CT14 7DY [JO01QF, TR35]
G4 DDC R A Joyce Dunst Dwns ARC, 20 Barking Cl, Luton, LU4 9HG [IO91SV, TL02]
G4 DDD R John, 32 Hundred Acre Rd, Streetly, Sutton Coldfield, B74 2LA [IO92BN, SP09]
G4 DDE L Rooke, 20 Fordwater Rd, Streetly, Sutton Coldfield, B74 2BQ [IO92BN, SP09]
G4 DDF B Price, Lower Bean Hall, Bungalow Church Rd, Bradley Green, Redditch, Worcs, B96 6SW [IO82XF, SO96]
G4 DDH R M B M Hodges, 11 Inlands Rise, Daventry, NN11 4DQ [IO92KG, SP56]
G4 DDI C J Guy, Hawthorn Folly, The-Cul-De-Sac, Stickford, Boston, Lincs, PE22 8EY [JO03AC, TF35]
G4 DDJ W B Gough, 13 Silver Ln, Billingshurst, RH14 9RP [IO91SA, TQ02]
G4 DDK S T Jewell, 56 Meadowlands, Kirton, Ipswich, IP10 0PP [JO02PA, TM23]
G4 DDL M J Pemberton, 37 Woodmancott Cl, Forest Park, Bracknell, RG12 0XU [IO91PJ, SU86]
G4 DDM R E Finch, 1 Cherrytree Cottage, Church Rd, Penn, High Wycombe, Bucks, HP10 8LN [IO91PP, SU99]
G4 DDN G Leonard, 65 Qualitas, Roman Hill, Bracknell, RG12 7QG [IO91OJ, SU86]
G4 DDP R E Clark, 41 Ave Rd, Bexleyheath, DA7 4EP [JO01BL, TQ47]
G4 DDS A Dalton-Kirby, 45 Sutton Rd, Kirk Sandall, Doncaster, DN3 1NY [IO93LN, SE60]
G4 DDT A F Ives, 24 Johnson Cres, Heacham, Kings Lynn, PE31 7LQ [JO02FV, TF63]
G4 DDU A J F Slade, 45 Ct Rd, Horfield, Bristol, BS7 0BU [IO81RL, ST57]
G4 DDV R G Bowman, 13 Wellington Rd, St. Albans, AL1 5NJ [IO91UR, TL10]
G4 DDW J D Berry, Lodge Farm, South Kilworth Rd, Walcote, Lutterworth, LE17 4LA [IO92KK, SP58]
G4 DDX R J L Pratt, 35 High St., Stevenage, SG1 3AU [IO91VV, TL22]
G4 DDY M N Fagg, 113 Bute Rd, Wallington, SM6 8AE [IO91WI, TQ26]
G4 DDZ N C Turner, 31 Wykeham Gr, Clee View Park, Perton, Wolverhampton, WV6 7TP [IO82VO, SO89]
G4 DEA P A Dunning, Cold Harbour, Bishop Burton, Beverley, N H'side, HU17 8QA [IO93SU, SE94]
G4 DED G K Day, 23 Old Parr Cl, Banbury, OX16 8HY [IO92HB, SP44]
G4 DEE Details withheld at licensee's request by SSL.
GM4 DEK I Sturrock, 23 Captains Rd, Edinburgh, EH17 8HR [IO85KV, NT26]
G4 DEL C W Cousins, Orchard Lodge, Green Glades, Emerson Park, Hornchurch, RM11 3BS [JO01CN, TQ58]
G4 DEM D E Walker, 16 Martins Rd, Hanham, Bristol, BS15 3EW [IO81RK, ST67]
G4 DEN A J Leigh, 6 Lynsey Cl, Halmerend, Stoke on Trent, ST7 8BT [IO83UA, SJ84]
G4 DEO A Wallis, 5 Nancevallon, Brea, Higher Brea, Camborne, TR14 9DE [IO70IF, SW64]
GW4 DEP D R Dabinett, Pentre Isaf, Llangyniew, Welshpool, Powys, SY21 0JT [IO82IQ, SJ10]
G4 DEQ A J Derrick, 4 Hillside Cttgs, Barrow St., Barrow Gurney, Bristol, BS19 3RX [IO81PJ, ST56]
G4 DES R K Clifft, 124 Jermayns, Basildon, SS15 5HB [JO01FN, TQ68]
G4 DEV S L Newport, 18 Chacewater, Barbourne, Worcester, WR3 7AN [IO82VF, SO85]
G4 DEW J J Males, 54 Cornelian Ave, Scarborough, YO11 3AW [IO94TG, TA08]
GM4 DEX J Sharp, Auchmuir Bridge, Leslie, Fife, KY6 3JD [IO86IE, NO20]
G4 DEY J E Balsdon, 31 Oakwood Ave, Borehamwood, WD6 1SP [IO91UP, TQ19]
G4 DEZ B C S Llewellyn, 110 South Ave, Southend on Sea, SS2 4HU [JO01IN, TQ88]
G4 DF B H Slatter, 87 Windy Arbour, Kenilworth, CV8 2BJ [IO92FI, SP27]
G4 DFA T R Ellinor, 5 Parkwood Rd, Banstead, SM7 1JJ [IO91VH, TQ25]
G4 DFB D F Berry, 4 Holly Hill, Winchmore Hill, London, N21 1NP [IO91WP, TQ39]
G4 DFC C J Goldingay, 71 Kingham Cl, Redditch, B98 0SB [IO92BH, SP06]
G4 DFD K N H Bailey, 6 Stebbings Cl, Hollesley, Woodbridge, IP12 3QY [JO02RB, TM34]
G4 DFE W O Raybould, Newlands, Bromyard Rd, Tenbury Wells, WR15 8BY [IO82QH, SO56]
G4 DFG P A G Gibbs, 4 Newgate, Pattingham, Wolverhampton, WV6 7AG [IO82UO, SO89]
G4 DFH J R Dales, 31 St. Pauls Way, Tickton, Beverley, HU17 9RW [IO93TU, TA04]
G4 DFI O L Cross, 28 Garden Ave, Bexleyheath, DA7 4LF [JO01BL, TQ47]
G4 DFN S A Widdett, 7 Burnside, West Coombe, Coventry, CV3 2RS [IO92GJ, SP37]
G4 DFO S J Wainwright, 14 Leasowes Rd, Kings Heath, Birmingham, B14 7AU [IO92BK, SP08]
G4 DFP A W Morecroft, 4 Arran Cl, Bolton, BL3 4PP [IO83SN, SD60]
G4 DFQ N C Dear, 3 Avon Ave, Ringwood, BH24 2BQ [IO90CT, SU10]
G4 DFS S W Booth, Oak Wood Lodge, Wellhill Rd, Wortley, Sheffield, S30 7DP [IO93GJ, SK38]
G4 DFT R W Perrin, 131 Acacia Ave, Ottawa, Ontario, Canada, K1M 0R2
G4 DFU F Skillington, 53 Temple Dr, Nuthall, Nottingham, NG16 1BE [IO92JX, SK54]
G4 DFV D J Walters, 11 King George V Ave, Mansfield, NG18 4ER [IO93JD, SK56]
G4 DFX J J Taylor, 20 Aylett Rd, Isleworth, TW7 6NP [IO91TL, TQ17]
G4 DFY R T Dedman, 2 Forest Villas, Long Mill Ln, Platt, Sevenoaks, TN15 8LQ [JO01DG, TQ65]
G4 DFZ K G Knight, 61 Westbourne Rd, Sutton in Ashfield, NG17 2FB [IO93ID, SK45]
G4 DGB T Crute, 52 Sevenoaks Dr, Hastings Hill, Sunderland, SR4 9NU [IO94GV, NZ35]
G4 DGF A D Matthews, 14 Hardy Green, Wellington Chase, Crowthorne, RG45 7QR [IO91OI, SU86]
G4 DGG Details withheld at licensee's request by SSL.
GI4 DGI Rev D C Coyle, 16 Northland Ave, Londonderry, BT48 7JN [IO65IA, C41]
G4 DGL Dr E L Mills, Rr#1, Rose Bay, Lunenburg Co, Nova Scotia, Canada, B0J 2X0
G4 DGM J G Meddings, 106 Goldthorn Hill, Wolverhampton, WV2 3HU [IO82WN, SO99]
G4 DGO J W Jones, The Shieling, 6A St Germains Ln, Marske-By-Sea, Redcar Cleveland, TS11 7LH [IO94LO, NZ62]
G4 DGQ J J C Dussart, Seagarth, Cresswell, Morpeth, Northd., NE61 5JU [IO95FF, NZ29]
GM4 DGT W D Stirling, 58 Tippet Knowes Park, Winchburgh, Broxburn, EH52 6UP [IO85GX, NT07]
G4 DGU Details withheld at licensee's request by SSL.
G4 DGW A C Gagnon, 60 Woodruff Ave, Hove, E Sussex, BN3 6PJ [IO90VU, TQ20]
G4 DGX J Stirrat, 106 St. Peters Rd, Uxbridge, UB8 3SB [IO91SM, TQ08]
G4 DGY D M W Boulton, 26 Lakewood Rd, Chandlers Ford, Eastleigh, SO53 1EW [IO90HX, SU42]
G4 DHA I Forse, Primrose Cottage, 66 Amberwood, Ferndown, BH22 9JT [IO90BT, SU00]
G4 DHC M H Smye, Heronsgate, Manorial Rd, Parkgate, South Wirral, L64 6QW [IO83LH, SJ27]
G4 DHE Me A Henry, 34 Queensway, Euxton, Chorley, PR7 6PW [IO83QP, SD51]
G4 DHF D Johnson, 65 West St., Bourne, PE10 9PA [IO92TS, TF02]
G4 DHH D H Hall, Greatfield Cottage, Trefusis Rd, Flushing, Falmouth Cornwall, TR11 5UB [IO70LD, SW83]
GM4 DHJ Details withheld at licensee's request by SSL.
G4 DHK R J Stanleigh, 26 Balmoral Rd, Bristol, BS7 9AZ [IO81RL, ST57]
G4 DHL C C Durnall, 120 Winchester Rd, Fordhouses, Wolverhampton, WV10 6EZ [IO82WP, SJ90]
GM4 DHN N F C Macleod, 54 Drum Brae South, Edinburgh, EH12 8TB [IO85IW, NT17]
G4 DHO Details withheld at licensee's request by SSL.
G4 DHP Details withheld at licensee's request by SSL.
G4 DHQ Details withheld at licensee's request by SSL.[Station located at Prestwood, near High Wycombe.]
G4 DHT R F Haverson, Kerri, Sunton, Collingbourne Ducis, Marlborough, SN8 3DZ [IO91EG, SU25]
G4 DHU D S Spender, Rose Cottage, Huntsmans Ln, Foxearth, Sudbury, CO10 7JX [JO02IB, TL84]
G4 DHV C F Jones, 49 Newport Pagnell Rd, Hardingstone, Northampton, NN4 6ER [IO92NE, SP75]
G4 DHX A L Trevitt, 17 Peterstow Cl, London, SW19 6JW [IO91VK, TQ27]
G4 DHY D H Bird, 50 Cambridge Rd, Crowthorne, RG45 7ER [IO91OI, SU86]
G4 DHZ S H Davis, 57 Sandy Ln, Wokingham, RG41 4SS [IO91NJ, SU76]
G4 DIA B Powell, 1 The Heights, Market Harborough, LE16 8BQ [IO92NL, SP78]
G4 DIC R Phipps, The Orchards, Northfield Ln, Wells Next The Sea, NR23 1LF [JO02KW, TF94]
G4 DIE I A R Dredge, 60 Springfield Cl, Rudloe, Corsham, SN13 0JR [IO81VK, ST87]
G4 DIG D M Hine, 5 Star Rd, Uxbridge, UB10 0QH [IO91SM, TQ08]
G4 DIH R F Coates, 57 Dalebrook Rd, Burton on Trent, DE15 0AB [IO92ET, SK22]
G4 DII A A Excell, 22 Malthouse Ct, Bishops Hull, Taunton, TA1 5HN [IO81KA, ST22]
GM4 DIJ J B Howie, 36 Clermiston Rd, Edinburgh, EH12 6XB [IO85IW, NT27]
GM4 DIN N F Burns, 24 Garioch Rd, Inverurie, AB51 4RQ [IO87TG, NJ72]
G4 DIP Details withheld at licensee's request by SSL.
G4 DIQ K R Arnold, 70 Queens Rd, Beeston, Radcliffe on Trent, Nottingham, NG1 1DJ [IO92KW, SK53]
G4 DIR W E Evans, 64 Grosvenor Dr, Debden, Loughton, IG10 2LG [JO01AP, TQ49]
G4 DIS K G Mills, 7 Montgomery Cl, Colchester, CO2 8SJ [JO01KU, TM02]
G4 DIT R G Siddall, 93 Rosebank Rd, Huyton, Liverpool, Merseyside, L36 3TY [IO83NK, SJ49]
G4 DIU A G Walker, 103 Torrington Rd, North End, Portsmouth, PO2 0TN [IO90LT, SU60]
G4 DIW Details withheld at licensee's request by SSL.
G4 DIY R T Bennett, 17 Truro Cl, Carr Mill, St. Helens, WA11 9EL [IO83PK, SJ59]
GM4 DIZ H Lydall, 27 Calder Rd, Edinburgh, EH11 3PF [IO85IW, NT27]
G4 DJA D W Spencer, 29 St. Helens Cres, Trowell, Nottingham, NG9 3PZ [IO92IW, SK43]
G4 DJB P J Roberts, 10 Tintagel Dr, Frimley, Camberley, GU16 5XQ [IO91PH, SU85]
G4 DJC R J Baker, 42 Rushleydale, Springfield, Chelmsford, CM1 6JX [JO01FR, TL70]
G4 DJD D J Carter, 43 Sturton Rd, Firvale, Sheffield, S4 7DE [IO93GJ, SK38]
G4 DJG V J W Coombs, 10 Maxwell Rd, Winton, Bournemouth, BH9 1DJ [IO90BR, SZ09]
G4 DJK D M Corkill, 1A Hardie Cres, Braunstone, Leicester, LE3 3DQ [IO92JO, SK50]
G4 DJL J J Craig, 113 Hull Rd, Woodmansey, Beverley, HU17 0TH [IO93TT, TA03]
G4 DJM J W Miller, 146 Southwell Rd, Rainworth, Mansfield, Notts, NG21 0EH [IO93KC, SK55]
G4 DJN J Kerr, 90 Mounthurst Rd, Hayes, Bromley, BR2 7PQ [JO01AJ, TQ36]
G4 DJP Dr J W H Chivers, Milestone Edge, 33 Hazelwood Rd, Duffield, Belper, DE56 4DP [IO92GX, SK34]
G4 DJR J Handley, 37 Money Rd, Caterham, CR3 5TF [IO91WG, TQ35]
GW4 DJW Details withheld at licensee's request by SSL.
G4 DJX A K Gray, 5 Meadow Cl, St. Albans, AL4 9TG [IO91US, TL10]
G4 DJY C P Steeden, Parklands, Chapel Rd, Marton Moss, Blackpool, FY4 5HT [IO83LS, SD33]
G4 DJZ A A Petrie, 3 Sharma Leas, Peterborough, PE4 6ZH [IO92UO, TF10]
G4 DKB P J Hubbard, 67 Knightbridge Walk, Billericay, CM12 0HL [JO01EP, TQ69]
G4 DKC T K A Smith, 9 Rubens St., Catford, London, SE6 4DH [IO91XK, TQ37]
G4 DKD E J Pascoe, 48 Bull Baulk, Middleton Cheney, Banbury, Oxon, OX17 2QQ [IO92IB, SP54]
G4 DKF Details withheld at licensee's request by SSL.
G4 DKG G Collins, 71 Washingborough Rd, Heighington, Lincoln, LN4 1QP [IO93SF, TF06]
G4 DKH K A Hastie, 3 The Woodlands, Kingsworthy, Kings Worthy, Winchester, SO23 7QQ [IO91IC, SU43]
G4 DKL G W A Pople, 11 Leonard Houlden Ct, Dorchester Rd, Taunton, TA2 7LN [IO81KA, ST22]
GM4 DKO Details withheld at licensee's request by SSL.
G4 DKP J G Etheridge, 10 Linden Ave, Halesowen, B62 9EL [IO82XL, SO98]
G4 DKQ J C Loughlin, 4 Cooper Ave South, Liverpool, L19 3PN [IO83NI, SJ38]
G4 DKR Details withheld at licensee's request by SSL.
G4 DKS I H Ford, The Spinney, Hazel Ave, Bristol, BS6 6UD [IO81QL, ST57]
G4 DKV M A L Pipes, Lower Hough Park, Jinglers Ln, Bradley, Ashbourne, Derbyshire, DE6 3EN [IO93EA, SK24]
G4 DKX N E Cartwright, Tithe House, Stutton Ln, Tattingstone, IP9 2NZ [JO01NX, TM13]
G4 DKZ D Dodd, 8 Lindal Cl, Dalton in Furness, LA15 8NL [IO84JD, SD27]
G4 DLA L E Turner, 160 Sandbach Rd, Lawton Heath End, Church Lawton, Stoke on Trent, ST7 3RB [IO83UC, SJ75]
G4 DLB Details withheld at licensee's request by SSL.
G4 DLC B Bourne, 15 Rhos Fawr, Maes y Mor, Abergele, LL22 9YH [IO83FH, SH97]
G4 DLD M R Garwood, 30 Rowan Cl, Maybush, Southampton, SO16 5DZ [IO90GW, SU31]
GM4 DLG R Bower, The Stables, Templeton Burn, Kilmarnock, KA3 6HP [IO75SO, NS43]
G4 DLK L H Jones, 7 Netherwood Rd, West Kensington, London, W14 0BL [IO91VL, TQ27]
G4 DLM R G Bishop, 12 Buttermere Dr, Warndon Est, Worcester, WR4 9HX [IO82VE, SO85]
G4 DLP R L Stoddon, 34 Cromwell Rd, Lancaster, LA1 5BD [IO84OB, SD46]
G4 DLQ J Milligan, 21 Charles Rd, Solihull, B91 1TS [IO92CJ, SP17]
G4 DLT R M Hill, Rose Lodge, Colne Fields, Somersham, Huntingdon, PE17 3DL [JO02AJ, TL37]
GM4 DLU A T McCudden, 9 Dryburgh Ln, West Mains, East Kilbride, Glasgow, G74 1BQ [IO75VS, NS65]
G4 DLW D W Plant, 16 Keats Cl, Ellesmere Port, Great Sutton, South Wirral, L66 2GA [IO83MG, SJ37]
G4 DLX Details withheld at licensee's request by SSL.
G4 DLY Details withheld at licensee's request by SSL.
G4 DLZ W G Taylor, 8 Park Ave, Markfield, LE67 9WA [IO92IQ, SK41]
GM4 DMA Details withheld at licensee's request by SSL.[Station located near Turriff, YR40, IO87VM./P Phillips Maureen Rig. AS69e, JO08UD.]
G4 DMB W E Green, Hemingford, Shawcross Rd, West Runton, Cromer, NR27 9NA [JO02PW, TG14]
G4 DMC R C Cleverley, 13 The Close, Melksham, SN12 6AG [IO81WI, ST96]
G4 DMD Details withheld at licensee's request by SSL.
G4 DME Details withheld at licensee's request by SSL.
G4 DMF J Wright, 10 Thorpes Rd, Heanor, DE75 7GQ [IO93HA, SK44]
G4 DMG D M Griffiths, 1 Shepherds Down, Alresford, SO24 9PP [IO91JB, SU53]
G4 DMH M F Horton, 79 Swinnow Gdns, Bramley, Leeds, LS13 4PH [IO93ET, SE23]
G4 DMI C J Armistead, 62 Landau Way, Turnford, Broxbourne, EN10 6LP [IO91XR, TL30]
GM4 DMK G Armstrong, 4 Ploughlands Cottage, Jedburgh, Roxburghshire, TD8 6TS [IO85RM, NT62]
G4 DML G R Moore, 61 Coggeshall Rd, Braintree, CM7 9EP [JO01GV, TL72]
G4 DMM I P Stinchcombe, 26 Plaistow Cres, Plymouth, PL5 2EA [IO70VJ, SX45]
G4 DMP D M Pratt, 11 Moorleigh Cl, Kippax, Leeds, LS25 7PB [IO93HS, SE43]
GM4 DMQ J L Pritchard, 36 Craigleith Hill Cres, Edinburgh, EH4 2JU [IO85JX, NT27]
GW4 DMR D G Bevan, 3 Trem y Foryd, Kinmel Bay, Rhyl, LL18 5JE [IO83FH, SH98]
G4 DMS P J Freeman, 1 Littleworth, Towcester, Northants, NN12 8AL [IO92LD, SP65]
G4 DMT J M Southall, 4 Tye Ln, Willisham, Ipswich, Suffolk, IP8 4SR [JO02MC, TM05]
G4 DMX J H Mitchell, Four Winds, 162 Manchester Rd, Wilmslow, SK9 2JW [IO83VH, SJ88]
GM4 DMZ C W Tran North of Scotland Vhf/Uhf, 'Achnacoille', Lanington, Kildary, Invergordon, Ross-shire, IV18 0PE [IO77WS, NH77]
G4 DNA S R C Green, 119 Oxford Rd, Abingdon, OX14 2AB [IO91IQ, SU59]
G4 DNB N W H Perch, 40 Highfield Ave, Bailiff Bridge, Brighouse, HD6 4EB [IO93CR, SE12]
G4 DNC G H Dodgshon, Larcombe Cottage, Diptford, Totnes, Devon, TQ9 7PD [IO80CJ, SX75]
G4 DND J S Kennedy, Trevanson, Gordon Pl, St Columb, Cornwall, TR9 6BG [IO70MK, SW96]
G4 DNE G F D Swaysland, 15 Andrew Ave, Rawtenstall, Rossendale, BB4 6EU [IO83UQ, SD82]
G4 DNG M J Whitaker, 332 Milton Rd, Cambridge, CB4 1LW [JO02BF, TL46]
G4 DNH J R Easteal, 6 Greenham Cl, Woodley, Reading, RG5 4EG [IO91NK, SU77]
G4 DNI P Weldon, 7 Coverdale, Heelands, Milton Keynes, MK13 7LZ [IO92OB, SP84]
G4 DNN T R O'Donnell, 4 Willian Pl, Hindhead, Surrey, GU26 6QZ [IO91PC, SU83]
G4 DNP R C Travis, via Passo Gardena, 66020 San Giovanni, Teatino (Ch), Italy, X X
GM4 DNS D Gallacher, 4 Hole Field, Kelso, Scotland, TD5 8BW [IO85UO, NT83]
G4 DNV T S G Makins, Coolmeen, Ballyfarnon, Co.Sligo, Eire, X X
G4 DNX D D Dyer, High Bank Cottage, Underhill, Moulsford, Wallingford, OX10 9JH [IO91KN, SU58]
G4 DOA A P Mead, Wingletang, 11 Yarnton Cl, Nine Elms, Swindon, SN5 9UQ [IO91CN, SU18]
G4 DOC D M James, 76 Gr Rd, Harpenden, AL5 1HD [IO91TT, TL11]
G4 DOE J C Alford, 26 Edmunds Ave, St. Pauls Cray, Orpington, BR5 3LF [JO01BJ, TQ46]
GM4 DOF R Davidson, 3 Hillcrest Ave, Kirkcaldy, KY2 5TU [IO86JC, NT29]
G4 DOG T H H Illingworth, 7 St. Anns Cres, Carlisle, CA3 9QA [IO84MV, NY35]
GM4 DOI Details withheld at licensee's request by SSL.
G4 DOJ Details withheld at licensee's request by SSL.
G4 DOK A R Eden, Torberrys House, Collins Ln, West Harting, Petersfield, GU31 5NZ [IO90NX, SU72]
G4 DOL P J Atkins, Harbour Lights, 12 Fairview Rd, Weymouth, DT4 9BN [IO80SO, SY67]
GI4 DOM D G Cafolla, 87 Stockmans Ln, Lisburn Rd, Belfast, BT9 7JD [IO74AN, J37]
GW4 DOO A V Kenyon, 6 Abbey Rd, Port Talbot, SA13 1HA [IO81CO, SS78]
G4 DOQ J G Willis, Kilncroft, Broadlayings, Woolton Hill, Newbury, RG20 9TS [IO91HI, SU46]
G4 DOW A C Chettle, Glenville, 4 Aspen Cl, Honiton, EX14 8YQ [IO80JS, SY19]
GM4 DOZ T Findlay, 37 Adamton Rd, Prestwick, Ayrshire, Scotland, KA9 2HY [IO75QL, NS32]
G4 DPA G E Austin, 16 Courtenay Rd, Wantage, OX12 7DN [IO91GO, SU48]
GM4 DPC T Wilson, School House, Boarhills, St Andrews, Fife, Scotland, KY16 8PP [IO86PH, NO51]
G4 DPD C J H Richardson, Brembridge Farm, Shillingford, Tiverton, Devon, EX16 9BT [IO81GA, ST02]
G4 DPF I S Ross, 12 Manor Cl, Buckden, St. Neots, Huntingdon, PE18 9XR [IO92UG, TL16]
G4 DPH G W F Jones, Alva, 7 The Avenue, Yatton, Bristol, BS19 4DA [IO81OJ, ST46]
G4 DPI J P Ball, 26 Royal Bank Rd, Marston, Blackpool, FY3 9PW [IO83LT, SD33]

G4 DPJ D Wear, 3 Westover Rise, Bristol, BS9 3LU [IO81QL, ST57]
GD4 DPK F Quayle, 1 Birch Hill Gdns, Onchan, Douglas, IM3 4ET
G4 DPN J Johnston, 13 Ct Dr, Waddon, Croydon, CR0 4QA [IO91WI, TQ36]
G4 DPO A M Nixon, 20 Stepney Rd, Falsgrave, Scarborough, YO12 5BN [IO94SG, TA08]
G4 DPP P J Slade, 2 Five Acres Ave, Bricket Wood, St. Albans, AL2 3PY [IO91TR, TL10]
G4 DPS O C Borsay, 15 Hyde Park Ave, North Petherton, Bridgwater, TA6 6SL [IO81LC, ST23]
G4 DPU J T Pilling, 223 Manchester Rd, Accrington, BB5 2PF [IO83TR, SD72]
G4 DPV S Ford, 3 Hill View, Milton, Stoke on Trent, ST2 7AR [IO83WB, SJ95]
G4 DPW P A Leslie-Reed, 43 Milehouse Ln, Newcastle, ST5 9JZ [IO83VA, SJ84]
G4 DPY Details withheld at licensee's request by SSL.
G4 DPZ D A Johnson, 96 Summerfields Ave, Halesowen, West Midlands, B62 9NR [IO82XL, SO98]
G4 DQB G Wallis, 61 Heath Rd, Market Bosworth, Nuneaton, CV13 0NX [IO92HO, SK30]
G4 DQC J P Deegan, 3 Plover Cl, Langley Green, Crawley, RH11 7RN [IO91VC, TQ23]
GM4 DQD J A Young, Zoar, Wadbister, Girlsta, Shetland, ZE2 9SQ [IP90JF, HU45]
G4 DQF K A Simpson, 17 Church Fold, Charlesworth, Broadbottom, Hyde, SK14 6EU [IO93AK, SK09]
G4 DQG A C Riley, 378 Hungerford Rd, Crewe, CW1 6HD [IO83SC, SJ75]
GM4 DQJ R M C P Grant, 31 Stormont Park, Scone, Perth, PH2 6SD [IO86HK, NO12]
GM4 DQK J T Kinghorn, 86 The Glebe, Kirkliston, EH29 9AT [IO85HW, NT17]
G4 DQL N C Hall, 5 Brooklyn Cres, Cheadle, SK8 1DX [IO83VJ, SJ88]
G4 DQN G G Spenceley, 168 Robin Hood Ln, Walderslade, Chatham, ME5 9LA [JO01GI, TQ76]
GI4 DQO H D J McLaughlin, 1A Taylor Park, Limavady, Co. Londonderry, BT49 0NT [IO65MB, C62]
G4 DQP V A Lewis, 355 Bolton Rd West, Holcombe Brook, Ramsbottom, Bury, BL0 9QZ [IO83UP, SD71]
G4 DQQ W K Thomas, 11 Murswell Ln, Silverstone, Towcester, NN12 8UT [IO92LC, SP64]
G4 DQR T C Forster, 149 Leicester Rd, Glen Parva, Leicester, LE2 9HP [IO92KO, SP59]
G4 DQT S G Taaffe, 56 San Marino, Muirhevna, Dublin Rd, Dundalk Eire
G4 DQW J Krzymuski, 13 Watlington Rd, Benson, Wallingford, OX10 6LT [IO91KO, SU69]
GM4 DQX Details withheld at licensee's request by SSL.
G4 DQY Details withheld at licensee's request by SSL.
G4 DQZ W P B Ellis, Gillhams House, Gillhams Ln, Linchmere, Haslemere, GU27 3ND [IO91OB, SU83]
G4 DR D P M Urquhart, 7 Padwell Ln, Bushby, Leicester, LE7 9PQ [IO92LP, SK60]
G4 DRA S Chester, 12 Paterson Rd, Aylesbury, HP21 8LL [IO91OT, SP81]
G4 DRB W Pickering, Byeways, Mill Ln, Crowborough, TN6 1DU [JO01BB, TQ53]
G4 DRD L J Mynett, 29 Sandbanks Way, Hailsham, BN27 3LW [JO00DU, TQ50]
G4 DRF G H Fuller, 4 Stanhope Ave, Woodhall Spa, LN10 6SP [IO93VD, TF16]
G4 DRH J Joyner, 36 Clifton St., Lytham, Lytham St. Annes, FY8 5EW [IO83MR, SD32]
G4 DRI I R Selby, 2 Ashley Cl, Welwyn Garden City, AL8 7LH [IO91VT, TL21]
G4 DRK A T J Thorpe, Corral, Snows Paddock, Windlesham, GU20 6LH [IO91QJ, SU96]
G4 DRO T P Brosnan, 8 Benedict Way, East Finchley, London, N2 0UR [IO91VO, TQ28]
GW4 DRR G Spencer, Tyn Cae, Llanfwrog, nr Holyhead, Anglesey, LL65 4YL [IO73RH, SH28]
G4 DRS J C Wayman, 22 Oakwood, Broadmayne, Dorchester, DT2 8UN [IO80TQ, SY78]
G4 DRU B Plastow, 185 Allesley Old Rd, Coventry, CV5 8FL [IO92FJ, SP37]
G4 DRW G L Smith, 18 Hamilton Way, Acomb, York, YO2 4LE [IO93KW, SE55]
G4 DRX D J McKone, 4 Danesway, Heath Charnock, Chorley, PR7 4EY [IO83QO, SD61]
GI4 DRY G Copeland, 9 Cherryville Park, Guildford Rd, Lurgan, Co Armaghn Ireland, N Ireland, BT66 7BA [IO64UK, J05]
G4 DRZ G C Carney, 94 Combe Ave, Portishead, Bristol, BS20 9JX [IO81OL, ST47]
G4 DSA G Kemp, 4 Chapter Way, Monk Bretton, Barnsley, S71 2HP [IO93GN, SE30]
G4 DSB Details withheld at licensee's request by SSL.
G4 DSC O Boniface, 11 Holmefield Rd, Ripon, HG4 1RZ [IO94FD, SE37]
G4 DSD R Woodman, 89 Western Way, Darras Hall, Ponteland, Newcastle upon Tyne, NE20 9AW [IO95CA, NZ17]
G4 DSE P M Zollman, 8 Blenheim Orchard, East Manney, East Hanney, Wantage, OX12 0JA [IO91HP, SU49]
G4 DSF S M Jones, 11 Alba Cl, Middleleaze, Swindon, SN5 9TL [IO91BN, SU18]
G4 DSG P C Draper, 115 Chesford Rd, Stopsley, Luton, LU2 8DP [IO91TV, TL12]
G4 DSH J E Lumkin, 137 Britannia Rd, Ipswich, IP4 5JX [JO02OB, TM14]
G4 DSI Dr I W McAndrew, South Winds, Outrigg, St. Bees, CA27 0AN [IO84EL, NX91]
G4 DSM P E Judkins, 18 St. Johns Sq, Wakefield, WF1 2RA [IO93FQ, SE32]
G4 DSN J L Dryden, 33 Old Station Rd, Newmarket, CB8 8DT [JO02EF, TL66]
GM4 DSO T Hughes, 15 Boreland Rd, Kirkcudbright, DG6 4HL [IO74XU, NX65]
G4 DSP D Hoult Spalding&Dis AR, Chespool House, Gosberton Risegate, Spalding, Lincs, PE11 4EU [IO92VU, TF23]
G4 DSR B Irwin, 97 Offerton Ln, Stockport, SK2 5BS [IO83WJ, SJ98]
GM4 DSS D S Smith, Strathcona, 48 Colinhill Rd, Strathaven, ML10 6HF [IO75XQ, NS64]
G4 DST Details withheld at licensee's request by SSL.
G4 DSX S D Williams, 3 Windham Cres, Wawne, Hull, HU7 5XW [IO93TT, TA03]
G4 DSY R C Miller, 21 Woodstock Ave, Sutton, SM3 9EG [IO91VJ, TQ26]
G4 DTA E R Endersby, 20 Wellesley Rd, Cliftonville, Margate, CT9 2UH [JO01QJ, TR37]
G4 DTB M Bryan, 127 Ledbury Rd, Hereford, HR1 1RQ [IO82PB, SO54]
G4 DTC R J Howgego, 31 Campbell Rd, Caterham, CR3 5JP [IO91WH, TQ35]
G4 DTD Details withheld at licensee's request by SSL.
G4 DTE A R Johnston, 1 Wheatlands Rd, Harrogate, HG2 8BB [IO93FX, SE35]
GM4 DTF Details withheld at licensee's request by SSL.
GM4 DTH P J Dick, Top Flat West, 21 West Maitland St., Edinburgh, EH12 5EA [IO85JW, NT27]
G4 DTI F Moffatt, 2 Oak Cl, Ceda Mount, Lyndhurst, SO43 7EF [IO90FU, SU20]
GM4 DTJ R Henderson, 2 Burdiehouse Ave, Edinburgh, EH17 8AW [IO85KV, NT26]
G4 DTK H S A Thayer, 45 Bradley Cres, Shirehampton, Bristol, BS11 9SP [IO81PL, ST57]
G4 DTL W L Young, 56 Lincoln Rd, Washingborough, Lincoln, LN4 1EG [IO93SF, TF07]
G4 DTO A Hindle, 43 Redhill Dr, Castleford, WF10 3EB [IO93IR, SE42]
G4 DTP D T Pells, 6 Clarence St., Stonebroom, Alfreton, DE55 6JW [IO93HD, SK45]
GW4 DTQ D Gibbon, 90 Grosvenor Rd, Prestatyn, LL19 7TS [IO83HI, SJ08]
G4 DTT W J Brooks, 11 Lowther Gr, Garforth, Leeds, LS25 1EN [IO93HS, SE43]
GW4 DTU A Roberts, Brynlludw, Van, Llanidloes, Powys, SY18 6NP [IO82FL, SN98]
G4 DTW S E Parsons, 54 Furze Cap, Kingsteignton, Newton Abbot, TQ12 3TE [IO80EN, SX87]
G4 DTZ R W Holland, 1 Station Rd, Castle Donington, Derby, DE74 2NJ [IO92HU, SK42]
G4 DUA R B Bearne, 30 Bouverie Ave, Salisbury, SP2 8DT [IO91CB, SU12]
G4 DUB R J Harden, 9 Caldew Gr, Trentham, Stoke on Trent, ST4 8UY [IO82VX, SJ84]
G4 DUC A G Johnson, 55 Magdalen Rd, Stoke on Trent, ST3 3HT [IO82WX, SJ84]
G4 DUF B W Phillips, Woody Nook, Petworth Rd, Wormley, Godalming, GU8 5TU [IO91QD, SU93]
G4 DUG Details withheld at licensee's request by SSL.
G4 DUH Details withheld at licensee's request by SSL.
G4 DUI P Wilson, 6 Hereford Rd, Colne, BB8 8JX [IO83VU, SD83]
G4 DUJ T R Morley, 34 Bickerton Point, South Woodham Ferrers, Chelmsford, CM3 5YG [JO01HP, TQ89]
G4 DUK S Addey, 110 All Saints Way, Aston, Sheffield, S31 0FD [IO93GJ, SK38]
G4 DUL M , Coburn, 16 Chapel Cl, Toddington, Dunstable, LU5 6AZ [IO91RW, TL02]
G4 DUM V G Long, 15 Barton Cl, Bexleyheath, DA6 8JP [JO01BK, TQ47]
G4 DUN R G Whitehead Dunn Nutrn Un Rd, Dunn Nutritional La, Downhams Ln, Milton Rd Cambridge, CB4 1XJ [JO02BF, TL46]
G4 DUO F J G Taylor, 7 Osterley Lodge, Church Rd, Osterley, Isleworth, TW7 4PQ [IO91TL, TQ17]
G4 DUP J D A Hall, Norton Lees, 43 Main St., Sewerby, Bridlington, YO15 1EH [IO94VC, TA16]
G4 DUQ P A Keane, 45 Bramblewood Rd, Worle, Weston Super Mare, BS22 9LW [IO81MI, ST36]
G4 DUS B H J Pickford, 78A Nightingale Rd, Rickmansworth, WD3 2BT [IO91SP, TQ09]
G4 DUT D L Elliott, 8 Curzon Rd, Maidstone, ME14 5BA [JO01GG, TQ75]
G4 DUU J Denton, 83 Bly Rd, Darfield, Barnsley, S73 9DP [IO93HM, SE40]
G4 DUV T J Hynes, 2 Deynes Rd, Debden, Saffron Walden, Essex, CB11 3LH [JO01DX, TL53]
G4 DUW J Goldbey, Waylands Gate, St. Johns Rd, Bashley, New Milton, BH25 5SD [IO90ES, SZ29]
GM4 DUX K J Hampson, 9 North Cres, Garlieston, Newton Stewart, DG8 8BA [IO74TS, NX44]
GW4 DUY D Chennell, 23 Cae Talcen, Penprisk, Pencoed, Bridgend, CF35 6RP [IO81GM, SS98]
G4 DUZ Details withheld at licensee's request by SSL.
G4 DV E B Gant, 13 St. Andrews Cres, Leasingham, Sleaford, NG34 8LS [IO92SX, TF04]
G4 DVA T R Stanway, 24 Fellbrook Ln, Bucknall, Stoke on Trent, ST2 8AQ [IO83WA, SJ94]
G4 DVB B D Price, 156 Parc Bryn Derwen, Llanharan, Pontyclun, CF72 9TX [IO81GN, SS98]
G4 DVE M Pugh, Tree Tops, Prospect Rd, Gornal Wood, Dudley, DY3 2TJ [IO82WM, SO99]
G4 DVF Details withheld at licensee's request by SSL.
GM4 DVG J N Douglas, 46 Hockers Ln, Detling, Maidstone, ME14 3JN [JO01GG, TQ75]
G4 DVH J Barnes, 6 Cross A Moor, Swarthmoor, Ulverston, LA12 0RT [IO84KE, SD27]
G4 DVI M Small, 1 Wingate Rd, Heaton Chapel, Stockport, SK4 2RJ [IO83VK, SJ89]
G4 DVJ R B Hall, 12 Britannia Gdns, Westcliff on Sea, SS0 8BN [JO01IM, TQ88]
G4 DVK M J Lang, 52 Gloucester Rd, Burnham on Sea, TA8 1JA [IO81MF, ST34]
G4 DVM M W Cartwright, Seacue, 8 Adelaide Ave, West Bromwich, B70 0SL [IO82XM, SO99]
G4 DVN S J Whalley, 1 Radley Way, Werrington, Stoke on Trent, ST9 0JN [IO83WA, SJ94]
G4 DVP P T Hicks, 1701 N Greenville Ave 304, Richardson, Texas 75081, USA
GW4 DVQ Details withheld at licensee's request by SSL.[Op: D P Kennedy. Station located near Llanfairpwllgwyngyll.]
G4 DVS E B Moore, 12 Brookland Ave, Wistastow, Crewe, CW2 8EJ [IO83SB, SJ65]
G4 DVV J E Thomas, 57 Bourton Ave, Stoke Lodge, Patchway, Bristol, BS12 6EB [IO81RM, ST68]
G4 DVW A E Lake, 7 Middleton Cl, Nuthall, Nottingham, NG16 1BX [IO92JX, SK54]
G4 DVY P E Linton, 21 Patchway, Chippenham, SN14 0HZ [IO81WL, ST97]
G4 DWA D R Throssell, Flat 3, 24 Birchwood Rd, Parkstone, Poole, BH14 9NP [IO90AR, SZ09]
G4 DWC D W Cannings, 5 Rowan Cl, Brackley, NN13 6PB [IO92KA, SP53]
G4 DWF Dr D W Faulkner, 1 Westland, Martlesham Heath, Ipswich, IP5 7SU [JO02OB, TM24]
G4 DWM T S Hunt, 33 Glenwood Cl, Swindon, SN1 4EB [IO91CN, SU18]
G4 DWO W B Ingham, Westfield Villa, Westfield Villas, Horbury, Wakefield, WF4 6EQ [IO93FP, SE21]
G4 DWP D J Mackinder, 44 Woodcote Way, Caversham Heights, Caversham, Reading, RG4 7HL [IO91ML, SU67]
G4 DWR M J Molloy, 153 Palmdale Dr, Scarborough, Ontario, Canada, MIT 1P2
GM4 DWS R Riddiough, Farnley, 30 Doonfoot Rd, Ayr, KA7 4DN [IO75QK, NS32]
G4 DWU J I H Blowers, 109 Scalby Rd, Scarborough, YO12 5QL [IO94SG, TA08]
G4 DWV Details withheld at licensee's request by SSL.
G4 DWW T Squires, 39 Blythe Ave, Congleton, CW12 4LQ [IO83VD, SJ86]
GW4 DWX M J Smith, Tonn Marr, Bronybuckley, Welshpool, SY21 7NQ [IO82KP, SJ20]
G4 DWZ P D Tucker, 8 Garland Cl, Hemel Hempstead, HP2 5HU [IO91SS, TL00]
G4 DXA Details withheld at licensee's request by SSL.
G4 DXB B F Chester, 69 Littlefield Ln, Grimsby, DN34 4NU [IO93WN, TA20]
G4 DXG Details withheld at licensee's request by SSL.
G4 DXH A F Pudsey, 528 Hotham Rd, South Wold Rd, Kingston upon Hull, Yorks, HU5 5RN [IO93TS, TA03]
G4 DXI J T Macrae, Park House, 1 Highsted Rd, Sittingbourne, ME10 4PS [JO01II, TQ96]
G4 DXJ L C Howe, 27 Pettits Boulevard, Romford, RM1 4PL [JO01CO, TQ59]
GI4 DXK W J Gordon, 17 Ballyheather Rd, Ballymagorry, Strabane, BT82 0BD [IO64HV, C30]
G4 DXL Dr L A W E Kemp, 189 Englishcombe Ln, Bath, BA2 2EW [IO81TI, ST76]
G4 DXN G J Williams, 11 Hadrian Cl, Lillington, Leamington Spa, CV32 7ED [IO92FH, SP36]
G4 DXO P D Jones, 102 Woodland Ave, Hove, BN3 6BN [IO90VU, TQ20]
G4 DXP C E P Howells, 11 West Garth, Carlton, Stockton on Tees, TS21 1DZ [IO94HO, NZ32]
G4 DXT T A Shaman, 9 Howard Cornish Rd, Marcham, Abingdon, Oxon, OX13 6PH [IO91HQ, SU49]
G4 DXW R Smith, 29 George St., Woodston, Peterborough, PE2 9PD [IO92VN, TL19]
G4 DXX D W Crompton, Hilltop, North Rd, Carnforth, LA5 9LU [IO84OD, SD47]
G4 DXY J W Spendlove, 15 Grammer St., Denby Village, Ripley, DE5 8PQ [IO93HA, SK44]
G4 DXZ J Garden, 42 Chatsworth Cres, Rushall, Walsall, WS4 1QU [IO92AO, SK00]
G4 DYA Details withheld at licensee's request by SSL.[Op: Richard J Lamont. Station located in Stone, Staffs. Tel: Stone (01785) 812784. Please QSL via bureau.]
G4 DYB H Mitchell, 50 Jermyn Cres, Hackenthorpe, Sheffield, S12 4QE [IO93HI, SK48]
G4 DYC M J Cooke, 4 Geddes Way, Mattishall, Dereham, NR20 3RE [JO02MP, TG01]
G4 DYG E P Rich, 392 Doncaster Rd, Stairfoot, Barnsley, S70 3RH [IO93GN, SE30]
G4 DYH G R Dunn, Thornton, 13 Orchid Way, Rugby, CV23 0SD [IO92JJ, SP57]
G4 DYI C Titheridge, 41 Church Walk, Worthing, BN11 2LT [IO90TT, TQ10]
G4 DYJ J W Cope, Main Rd, Keal Coates, Keal Cotes, Spilsby, Lincs, PE23 4AG [JO03AD, TF36]
G4 DYK J F Williams, 11 Heath Farm Rd, Norton, Stourbridge, DY8 3AX [IO82WK, SO88]
G4 DYM E Auty, Jesla, 5 Silverstone Way, Congresbury, Bristol, BS19 5ES [IO81OI, ST46]
G4 DYO B J McCartney, 123 Reading Rd, Finchampstead, Wokingham, RG40 4RD [IO91NI, SU76][Op: Brendan J McCartney. Tel + fax: (01734) 732393.]
G4 DYQ S A Voss, 5 Sandringham Ave, Earl Shilton, Leicester, LE9 7HY [IO92IN, SP49]
G4 DYT D J March, 86 Henley Ave, Norton, Sheffield, S8 8JJ [IO93GH, SK38]
G4 DYU B L Hazlewood, 62 Cooper Ave, Brierley Hill, DY5 3PE [IO82WL, SO98]
G4 DYV B E Whiting, Fellands Gate, Old Leake, Boston, Lincs, PE22 9QX [JO03AB, TF35]
G4 DYW M Crofts, 8 Rhugarve Gdns, Linton, Cambridge, CB1 6LX [JO02DC, TL54]
GW4 DYY R W Mander, 4 Withy Ave, Forden, Welshpool, SY21 8NJ [IO82KO, SJ20]
GM4 DYZ G F Bryson, 44 Isla Rd, Luncarty, Perth, PH1 3HN [IO86GK, NO02]
G4 DZA J A Lester, 42 Ardingly, Bracknell, RG12 8XR [IO91OJ, SU86]
G4 DZC Dr M W Bayes, Welby Lodge, 25 Welby Gdns, Grantham, NG31 8BN [IO92QW, SK93]
GW4 DZD D F Gwinnutt, Clomendy Wood, Usk Rd, Caerleon Gwent, NP6 1LR [IO81MP, ST39]
G4 DZE Details withheld at licensee's request by SSL.
G4 DZF Details withheld at licensee's request by SSL.
G4 DZH H Davies, 33 Sandown Rd, Ocean Heights, Paignton, TQ4 7RL [IO80FJ, SX85]
G4 DZK G Stocker, 8 Brook Dr, Astley, Tyldesley, Manchester, M29 7HS [IO83SM, SD70]
G4 DZL Details withheld at licensee's request by SSL.[Op: J S Smith, 91 Anson Road, Shepshed, Loughborough, Leics, LE12 9PT.]
GM4 DZM I Shewan, Distillery Rd, Old Meldrum, Oldmeldrum, Inverurie, Aberdeenshire, AB51 0ES [IO87UI, NJ82]
G4 DZN B J R Powell, Suncot, 2 Chapel St., Taliesin, Machynlleth, SY20 8JH [IO82AM, SN69]
G4 DZP M J Brimacombe, 79 Milehouse Rd, Milehouse, Plymouth, PL3 4AE [IO70WJ, SX45]
G4 DZQ S W Ascough, 42 Myton Dr, Solihull Lodge, Shirley, Solihull, B90 1HP [IO92BJ, SP07]
G4 DZS A D Watson, 59 Merdon Ave, Chandlers Ford, Eastleigh, SO53 1GD [IO90HX, SU42]
G4 DZT A Jennings, The White House, 29 Beverley Rd, Driffield, YO25 7RZ [IO94SA, TA05]
G4 DZU D Parker, Dalegarth, 50 Rein Rd, Tingley, Wakefield, WF3 1HZ [IO93ER, SE22]
G4 DZV F C Herod, 112 The Bramblings, Chingford Hatch, London, E4 6LU [JO01AO, TQ39]
G4 DZW A M Hockey, 2 Asney Rd, Walton, St, BA16 9RP [IO81OC, ST43]
GM4 DZX R F Macleod, Vesquoy, Rendall, Orkney, KW17 2EZ [IO89LB, HY42]
G4 EAA Details withheld at licensee's request by SSL.
G4 EAB J Blackburn, 9 Pitchford Rd, Albrighton, Wolverhampton, WV7 3LS [IO82UP, SJ80]
GM4 EAF A R Hutton Perth & Dist AR, 4 The Pheasantry, Linn Roadstanley, Perth Tayside, PH1 4QS [IO86GL, NO13]
G4 EAG S J Ruffle, Little Badgers, 39 Nightingale Ave, Cambridge, CB1 4SG [JO02BE, TL45]
G4 EAH N Jerram, 2 Leyside, Edgcumbe Park, Crowthorne, RG45 6EL [IO91OI, SU86]
GW4 EAI P W Whitburn, 33 Highmead, Penllwyn, Pontllanfraith, Blackwood, NP2 2PF [IO81JP, ST19]
G4 EAJ B D Wignall, 119 Ounsdale Rd, Wombourne nr, Wombourne, Wolverhampton, WV5 8BL [IO82VM, SO89]
G4 EAK M J Betts, 11 Hazelwood Dr, Pinner, HA5 3TU [IO91TO, TQ19]
G4 EAM E A M Minton, 17 Owen Ave, Long Eaton, Nottingham, NG10 2FR [IO92IV, SK53]
G4 EAN I R Brothwell, 56 Arnot Hill Rd, Arnold, Nottingham, NG5 6LQ [IO92KX, SK54]
G4 EAP A Birch, 5 Salisbury Cres, Hartshead, Ashton under Lyne, OL6 8DB [IO83XM, SD90]
G4 EAQ Dr A R Churchley, 46 Birchdale Rd, Appleton, Warrington, WA4 5AW [IO83RI, SJ68]
G4 EAS C J Ellery, 19 Wessex Way, Dorchester, DT1 2NR [IO80SR, SY69]
G4 EAT J R Wood, 20060 Rodrigues Ave#b, Cupertino, Ca 95014, USA, CM3 4ND [JO01HQ, TL70]
GM4 EAU C L Murray, 43 Malleny Ave, Balerno, EH14 7EJ [IO85HV, NT16]
GM4 EAW J R Mathers, 36 Alexander St., Dunoon, PA23 7EW [IO75MW, NS17]
G4 EAX J Gell, 30 Weston Cres, Sawley, Long Eaton, Nottingham, NG10 3BS [IO92IV, SK43]
G4 EAZ P A Holliman, 17 Arundel Rd, Tewkesbury, GL20 8AT [IO82WA, SO93]
GD4 EBA D H Kinrade, 1 Barrule View Terr, Higher, Foxdale, Douglas, IM4 3EW
GM4 EBB Details withheld at licensee's request by SSL.
G4 EBC Prof D Mattingly, 1 Moorview Cl, Exeter, EX4 6EZ [IO80FR, SX99]
G4 EBE G P Harcourt, Hard Farm House, Little Marsh Ln, Field Dalling, Holt, Norfolk, NR25 7LL [JO02LV, TG03]
G4 EBF G J Reason, 37 Park End, Croughton, Brackley, NN13 5LX [IO91JX, SP53]
G4 EBG B Meredith, 20 Kestrel Ave, Thorpe Hesley, Rotherham, S61 2TT [IO93GL, SK39]
G4 EBI A D Hamm, 166 Sylvan Rd, London, SE19 2SA [IO91XJ, TQ36]
G4 EBK G J Smith, 6 Fenby Cl, Grimsby, DN37 9QJ [IO93WN, TA20]
G4 EBL R Whitwell, 10 Westridge Rd, Kings Heath, Birmingham, B13 0DT [IO92BK, SP08]
G4 EBN M E S Valente, Glenville, Abbey Rd, Pity Me, Durham, DH1 5DQ [IO94ET, NZ24]
G4 EBO W E Gibbs, 25 Belvedere Rd, Exmouth, EX8 1QN [IO80HO, SY08]
G4 EBQ N Talbot, 59 Heywood Ave, Austerlands, Oldham, OL4 4AZ [IO83XN, SD90]
GI4 EBS J L McNerlin, 4 Castle Gdns, Limavady, BT49 0SD [IO65MB, C62]
G4 EBT D Taylor, 3 Crofters Dr, Cleminson Park, Cottingham, HU16 4SD [IO93TS, TA03]
G4 EBV A P Stainer, 23 The Batch, Batheaston, Bath, BA1 7DR [IO81UJ, ST76]
G4 EBW C W Hurley, 15 Buristead Rd, Great Shelford, Cambridge, CB2 5EJ [JO02BD, TL45]
G4 EBX P J Hopkinson, 11A Red Ln, South Normanton, Alfreton, DE55 3HA [IO93HC, SK45]
G4 EBY Details withheld at licensee's request by SSL.
G4 ECA Details withheld at licensee's request by SSL.
G4 ECC S OConnor, The East Byre, Kingstone Farm, nr Ilminster, Somerset, TA19 0NS [IO80NW, ST31]
G4 ECE Details withheld at licensee's request by SSL.
G4 ECF G Penney, 1 Hemsley Ct, Stoughton Rd, Guildford, GU2 6PW [IO91QG, SU95]
G4 ECL A Macconnachie, 155 Waveney Rd, St. Ives, Huntingdon, PE17 6FN [IO92XI, TL37]
G4 ECM Details withheld at licensee's request by SSL.
G4 ECO B Palmer, 35 Guilsborough Rd, West Haddon, Northampton, NN6 7AD [IO92LI, SP67]
G4 ECT J Crabbe Cheshunt Dis AR, 47 Torrington Dr, Potters Bar, EN6 5HU [IO91WQ, TL20]
G4 ECU D J Catling, 32 Brook Rd, Eaton Ford, St. Neots, Huntingdon, PE19 3AX [IO92UF, TL15]
G4 ECV Details withheld at licensee's request by SSL.
G4 ECY Details withheld at licensee's request by SSL.[Op: P Walker.]
G4 EDD J G Fletcher, 5 Hayeswood Rd, Stanley Common, Ilkeston, DE7 6GB [IO92HX, SK44]
G4 EDE Details withheld at licensee's request by SSL.
G4 EDG S P Taylor, 80 Nadder Park Rd, St. Thomas, Exeter, EX4 1NX [IO80FR, SX89]
G4 EDH G M Rose, 15 Pendennis Cl, Winklebury, Basingstoke, RG23 8JD [IO91KG, SU65]
G4 EDJ J S Cowsill, 21 Manor Cl, Bromham, Bedford, MK43 8JA [IO92RD, TL05]
G4 EDK A F P Ball, 31 Monmouth Rd, Dorchester, DT1 2DE [IO80SQ, SY68]
G4 EDL R J Crockett, 189 Ashburton Ave, Sevenkings, Ilford, IG3 9EN [JO01BN, TQ48]
G4 EDM W J Concannon, 155 Walton Rd, Sale, M33 4FS [IO83UJ, SJ79]
G4 EDN K M Currie, 71 Belle Vue Rd, Southend on Sea, SS2 4JE [JO01IM, TQ88]
G4 EDP P J Powell, 40 Carmarthen Rd, Up Hatherley, Cheltenham, GL51 5LA [IO81WV, SO92]
G4 EDQ Details withheld at licensee's request by SSL.
G4 EDR D E Mappin, 13 Willow Cl, Filey, YO14 9NY [IO94UF, TA18]
G4 EDW P Eaton, 29 Tilesford Cl, Monkspath, Shirley, Solihull, B90 4YF [IO92CJ, SP17]
G4 EDX J Fletcher, 69 Thackerays Ln, Woodthorpe, Nottingham, NG5 4HU [IO92KX, SK54]

G4 EDY M P Grindrod, 20 Castle Mead, Kings Stanley, Stonehouse, GL10 3LD [IO81UR, SO80]
G4 EDZ W H B Russell, 4 Colwyn Ave, Bare, Morecambe, LA4 6EQ [IO84NB, SD46]
GI4 EEB W M Fitzgerald, 1 The Maples, Gransha Rd, Bangor, BT19 7XY [IO74EP, J57]
G4 EED Details withheld at licensee's request by SSL.[Op: J Williams. Station located near Leamington Spa.]
G4 EEE A Wood, 16 Ramptons Meadow, Tadley, RG26 3UR [IO91KI, SU66]
G4 EEF S Foster, 6 Webster Cl, Hornchurch, RM12 6TF [JO01CN, TQ58]
G4 EEH D D Greer, 5 Potto Cl, Yarm, TS15 9RZ [IO94IM, NZ41]
G4 EEI Details withheld at licensee's request by SSL.
G4 EEJ R G Arak, 36 Roedean Cres, Brighton, BN2 5RH [IO90WT, TQ30]
G4 EEM Details withheld at licensee's request by SSL.
GW4 EEO D H Llewellyn, 13 Keynsham Rd, Whitchurch, Cardiff, CF4 1TS [IO81JM, ST18]
G4 EEQ Rev F Robinson, St Edwards Church, London Rd, Macclesfield, Ches, SK11 7RL [IO83WF, SJ97]
G4 EER Details withheld at licensee's request by SSL.
G4 EES P J R Smith, Forge House and Stables, Whistley Rd, Potterne, Devizes, Wilts, SN10 5TD [IO81XI, ST96]
G4 EEV D Warwick, Appletree Cottage, Colber Ln, Bishop Thornton, Harrogate, HG3 3JR [IO94EB, SE26]
G4 EEZ M J Bath, 55 Woodberry Rd, Wickford, SS11 8YE [JO01GO, TQ79]
G4 EFA R W Goad, 7 Chipstead House, Chipstead Rd, Cosham, Portsmouth, PO6 3JJ [IO90LU, SU60]
G4 EFB C J McCloud, 34 Stephen's Rd, Buckland, Portsmouth, Hants, PO2 7PG [IO90LT, SU60]
G4 EFD D Stubbs, 63 Moss Ln, Wardley, Swinton, Manchester, M27 9RD [IO83TM, SD70]
G4 EFE M L Peters, 11 Filbert Dr, Tilehurst, Reading, RG31 5DZ [IO91LL, SU67]
G4 EFG D W J Watton, 247 Bloxwich Rd, Walsall, WS2 7BB [IO92AO, SK00]
G4 EFH A E Johnson, Flat 1, Dean End, Lea Ross on Wye, Herefordshire, HR9 7LN [IO81SV, SO62]
G4 EFK C R Pearce, 71 Maple Dr, Burnham on Sea, TA8 1DH [IO81MF, ST34]
GM4 EFL J Crichton, 66 Oakbank Rd, East Calder, Livingston, EH53 0BY [IO85GV, NT06]
G4 EFO M F V Senior, 97 Littlehaven Ln, Roffey, Horsham, RH12 4JE [IO91UB, TQ13]
G4 EFP A Smith, 75 Bridgenorth Rd, Wirral, L61 8SH [IO83KI, SJ28]
GM4 EFR J L Moar, Hansel, Stangergill Cres, Castleton, Caithness, KW14 8UT [IO88HO, ND16]
G4 EFS P B Buzzing, 39 Kendlewood Rd, Kidderminster, DY10 2XG [IO82VJ, SO87]
G4 EFT Details withheld at licensee's request by SSL.
G4 EFU J Heptonstall, 42 Whitebirk Rd, Blackburn, BB1 3JD [IO83SR, SD72]
G4 EFV P B Revill, 8 Weston Park View, Otley, LS21 2DU [IO93DV, SE14]
G4 EFW Details withheld at licensee's request by SSL.
G4 EFX A Levitt, Strines Clough Farm, Blackshawhead, Hebden Bridge, W Yorks, HX7 7JA [IO83XR, SD92]
G4 EFY J B S Hurst, 12 Dukes Mead, Calthorpe Park, Fleet, GU13 8HA [IO91NG, SU75]
G4 EFZ B Nabb, 11 Fleet Ln, Woodlesford, Oulton, Leeds, LS26 8HT [IO93GS, SE32]
G4 EGB J E Fletcher, 114 Scholes Park Rd, Scarborough, YO12 6RA [IO94TH, TA09]
GM4 EGD I Brownlie, 16 Border St., Greenock, PA15 2EE [IO75PW, NS27]
G4 EGG W Higginson, 7 Arundale, Westhoughton, Bolton, BL5 3YB [IO83RN, SD60]
G4 EGH Details withheld at licensee's request by SSL.[Station located near Gillingham. Correspondence via PO Box CH18, Chatham, Kent.]
G4 EGM R Webster, 230 Huyton Ln, Huyton, Liverpool, L36 1TH [IO83OK, SJ49]
G4 EGN V B Coles, 205 Farmers Cl, Witney, OX8 6NS [IO91GT, SP31]
G4 EGQ P J Pennington, 6 Highland Cl, Folkestone, CT20 3SA [JO01NB, TR23]
G4 EGR D E Barwood, 41 Wingfield Rd, Knowle, Bristol, BS3 5EG [IO81QK, ST57]
G4 EGS M J Price, 24 Helvellyn Dr, Burnley, BB12 0TA [IO83UT, SD83]
G4 EGU P D Wolfe, 69 Alderney Rd, Slade Green, Erith, DA8 2JH [JO01CL, TQ57]
GM4 EGX R A Howard, 22 Kirkbrae Dr, Cults, Aberdeen, AB1 9RH [IO87TH, NJ72]
G4 EGY S J Liptrott, 40 Mapperley Orchard, Arnold, Nottingham, NG5 8AG [IO93KA, SK64]
G4 EHA Details withheld at licensee's request by SSL.
GM4 EHB E H Brockie, Ach-Na-Shee, Tobeqmory, Isle of Mull, PA75 6PX [IO66XO, NM55]
G4 EHD W B Tait, 51 Broadley Cres, Moorend Rd, Halifax, HX2 0RL [IO93BR, SE02]
G4 EHG C Bryan, 9 Brandy Hole Ln, Chichester, PO19 4RL [IO90OU, SU80]
G4 EHH Details withheld at licensee's request by SSL.
G4 EHI Details withheld at licensee's request by SSL.
G4 EHJ E A Wilby, 32 Caves Farm Cl, Sandhurst, GU47 8EA [IO91OI, SU86]
G4 EHK D J Goulbourne, 8 Hill Cl, Appley Bridge, Wigan, WN6 9JH [IO83PO, SD50]
G4 EHM J Dewe, 41 Rookery Rd, Swinton, Mexborough, S64 8HY [IO93IL, SK49]
G4 EHN Dr J C Axe, 5 Hillgate Pl, London, W8 7SL [IO91VM, TQ28]
GM4 EHO J W Sarjantson, 1 Veronica Cres, Kirkcaldy, KY1 2LH [IO86KD, NT29]
GM4 EHP I C Petrie, Ugie Cottage, Victoria Rd, Maud, Peterhead, AB42 4NL [IO87WM, NJ94]
G4 EHQ M J T Holley, 166 Hoo Marina Park, Vicarage Ln, Hoo, Rochester, ME3 9TH [JO01GJ, TQ77]
G4 EHR D G Kirton, 16 Silver Innage, Halesowen, B63 2PP [IO82WL, SO98]
G4 EHT W Watson, 12 Chadswell Heights, Lichfield, WS13 6BH [IO92CQ, SK11]
G4 EHU G F W Trenchard, 47 Holford Rd, Bridgwater, TA6 7NT [IO81LD, ST23]
G4 EHW D W Mason Greater Peterborough Amateur R, 40 Washingley Rd, Folksworth, Peterborough, PE7 3SY [IO92UL, TL18]
G4 EHX J R Fearn, 12 Plackett Cl, Breaston, Derby, DE72 3UG [IO92IV, SK43]
G4 EHY F Greenough, 7 Carnforth Ave, Hindley, Wigan, WN2 4LD [IO83RM, SD60]
G4 EHZ D G Worley, 16 Sussex Rd, Worthing, BN11 1DS [IO90TT, TQ10]
G4 EIA M W Wallis, 7 Bibury Cres, Hanham, Bristol, BS15 3EX [IO81RK, ST67]
G4 EIB J H Challenger, 42 Gibbons Hill Rd, Dudley, DY3 1QA [IO82WN, SO99]
G4 EIC E I Calvert, 163 Milner Rd, Heswall, Wirral, L60 5RY [IO83KH, SJ28]
G4 EID Details withheld at licensee's request by SSL.[Op: M R Howarth. Station located near Southport.]
G4 EIF R H O Francis, 27 Briarlyn Rd, Birchencliffe, Huddersfield, HD3 3NL [IO93CP, SE11]
G4 EIG J R Vickerstaff, 5 Luddington Rd, Solihull, B92 9QH [IO92CK, SP18]
G4 EII A Cunliffe, 35 Coultshead Ave, Billinge, Wigan, WN5 7HT [IO83PM, SD50]
G4 EIK R A Currell, 7 Bustlers Rise, Duxford, Cambridge, CB2 4QU [JO02BC, TL44]
G4 EIL G Oughtibridge, 75 Derby Rd, Thornbury, Bradford, BD3 8QA [IO93DT, SE13]
G4 EIM J Beaumont, 130 Hull Rd, Woodmansey, Beverley, HU17 0TH [IO93TT, TA03]
GW4 EIN D B Jones, Vine Tree Cottage, Mill Ln, Govilon, Abergavenny, NP7 9SA [IO81LT, SO21]
GD4 EIP C G Baillie-Searle, 2 Marguerite Pl, Foxdale, Douglas, IM4 3HE
G4 EIQ A P Brown, 567 Otley Rd, Adel, Leeds, LS16 7PH [IO93EU, SE24]
GW4 EIR J R Greenall, Llys Gwyn, 3 Norton Ave, Prestatyn, LL19 7NL [IO83HI, SJ08]
G4 EIS S Halfyard, 12 Nanjivey Terr, St. Ives, TR26 1BQ [IO70GF, SW54]
G4 EIT Details withheld at licensee's request by SSL.
G4 EIV J S Sondhis, 47 Emlyn Rd, Horley, RH6 8RX [IO91VE, TQ24]
GM4 EIW J P Dunnington, 4 Woodburn Way, Cumbernauld, Glasgow, G68 9BJ [IO75XW, NS77]
G4 EIX D Whalley, 1 Lees Farm Dr, Madeley, Telford, TF7 5SU [IO82SP, SJ60]
G4 EIY B J Thomas, 8 Whitehill Rd, Barton-le-Clay, Bedford, MK45 4PF [IO91SX, TL03]
GI4 EIZ W Stewart, 56 Ballysillan Park, Belfast, BT14 8HD [IO74AP, J37]
G4 EJA J Maynard, 32 Waldorf Heights, Hawley Hill, Blackwater, Camberley, GU17 9JH [IO91OH, SU85]
G4 EJD C S Bourne, 5 Brempton Croft, Hilderstone, Stone, ST15 8XL [IO82WV, SJ93]
G4 EJE J Brown, 127 Sandringham Ave, Upper Marham, Kings Lynn, PE33 9PJ [JO02GP, TF70]
GW4 EJG I W Adams, 4 Wyndham St., Troedyrhiw, Merthyr Tydfil, CF48 4JY [IO81HR, SO00]
G4 EJH K E Middleton, 92 South Rd, Portishead, Bristol, BS20 9DY [IO81OL, ST47]
GM4 EJI G Lucas, 20 Myreside Gdns, Kennoway, Leven, KY8 5TR [IO86LF, NO30]
G4 EJJ Details withheld at licensee's request by SSL.
G4 EJK D C Reardon, 291 Hemdean Rd, Caversham, Reading, RG4 7QP [IO91ML, SU77]
G4 EJL R Hart, 43 Moorcroft Rd, Sheffield, S10 4GS [IO93FI, SK28]
G4 EJM M S West, 43 Burrington Dr, Trentham, Stoke on Trent, ST4 8YD [IO82VX, SJ84]
G4 EJP P R Sheppard, 89 St. Catherines Dr, Leconfield, Beverley, HU17 7NY [IO93SV, TA04]
G4 EJQ R Clare-Noon, 91 Budshead Rd, Higher St. Budeaux, Plymouth, PL5 2PJ [IO70VJ, SX45]
G4 EJR Details withheld at licensee's request by SSL.
G4 EJS E J Stainer, 66 Herbert Rd, Bath, BA2 3PP [IO81TJ, ST76]
GW4 EJT A E Kilner, 1 Glandwr, Terr Rd, Aberdyfi, Gwynedd, LL35 0LP [IO72XN, SN69]
G4 EJU R W Hands, 19 Orwell Rd, Walsall, WS1 2PJ [IO92AN, SP09]
G4 EJW N Perkins, 231 Burnham Rd, Burnham on Sea, TA8 1LT [IO81MF, ST34]
GM4 EJX A R Murray, Almar, 67 Carronvale Rd, Larbert, FK5 3LH [IO86CA, NS88]
G4 EKB D J Epton, 61 Cartmel Dr, Dunstable, LU6 3PT [IO91RV, TL02]
GM4 EKC J M Mackinnon, 185 Deeside Gdns, Aberdeen, AB1 7QA [IO87TH, NJ72]
G4 EKD P J Spelman, 68 Hardwick St., Tibshelf, Alfreton, DE55 5QH [IO93HD, SK46]
GW4 EKE A Bagley, Ty Heulog, Ysfa Rd, Nantmel, Llandrindod Wells, Powys, LD2 6EW [IO82HD, SO05]
G4 EKF S J Sinclair, 29 Lealands, Lesbury, Alnwick, NE66 3QN [IO95EJ, NU21]
G4 EKG M V Tittensor, 16 Durcott Rd, Evesham, WR11 6EQ [IO92AC, SP04]
G4 EKJ C M Shaw, 10 St. Helens Rd, Harrogate, HG2 8LB [IO93FX, SE35]
G4 EKK Details withheld at licensee's request by SSL.
G4 EKL J A Lorton, 2 Charlotte Cl, Kirton, Newark, NG22 9LW [IO93ME, SK66]
G4 EKM S Green, 133 Sevenoaks Dr, Sunderland, SR4 9NQ [IO94GV, NZ35]
G4 EKO J B Harris, 383 Cemetary Rd, Rd9 Waunu, Whangarei, New Zealand
G4 EKQ D Baylis, 2 Greswolde Park Rd, Acocks Green, Birmingham, B27 6QD [IO92CK, SP18]
G4 EKS R E Holtham, 27 Peyton Cl, Eastbourne, BN23 6AF [JO00DS, TQ60]
G4 EKT D F Heathershaw Hornsea ARC, The Old School, Cliff Ln, Mappleton, Hornsea, HU18 1XX [IO93WV, TA24]
G4 EKV M T Lobb, 52 Ridge Park Ave, Mutley, Plymouth, PL4 6QA [IO70WJ, SX45]
G4 EKW M C Shaw, 50 White Rd, Nottingham, NG5 1JR [IO92JX, SK54]
G4 EKY L G Martin, 8 Oakleaf Cl, Bartley Green, Birmingham, B32 3DL [IO92AK, SP08]
G4 EKZ D S N Saul, 78 Ingleton Dr, Hala Carr, Lancaster, LA1 4QZ [IO84OA, SD45]
G4 ELA R O Dawson, Newhaven, 2 Bertram Dr, Meols, Wirral, L47 0LQ [IO83KJ, SJ28]
G4 ELB G T Bovenizer, 34 The Briary, Bexhill on Sea, TN40 2ET [JO00FU, TQ70]
G4 ELC G S Keay, 9 Buchanan Ave, Kings Park, Bournemouth, BH7 7AA [IO90BR, SZ19]
G4 ELE C F Elie, 90 Kynaston Ave, Thornton Heath, CR7 7BW [IO91WJ, TQ36]
G4 ELG D G Campbell, 12 Newton Cl, Newton Solney, Burton on Trent, DE15 0SL [IO92FT, SK22]
GD4 ELI S J Brown, Riverside, Glenauldyn, Ramsey, Isle of Man, IM7 2AD
G4 ELJ Dr D R Clark, Woodlands, Heatherdale Rd, Camberley, Surrey, GU15 2LT [IO91PH, SU85]
G4 ELK A Lewis, 1 Springcroft, Parkgate, South Wirral, L64 6SF [IO83LH, SJ27]
G4 ELL R E James-Robertson, The White Bungalow, 8 Whittington Rd, Worcester, WR5 2JU [IO82VE, SO85]
G4 ELM E R Jewell, 12 Patricks Copse Rd, Liss, GU33 7DL [IO91NB, SU72]
G4 ELN Details withheld at licensee's request by SSL.
G4 ELP D J Stockley, 10 The Leas, Chestfield, Whitstable, CT5 3QQ [JO01MI, TR16]
GI4 ELQ J A Cushnahan, 34 Cornakinnegar Rd, Lurgan, Craigavon, BT67 9JN [IO64UL, J06]
G4 ELR J Tiplady, 2 Redmayne Sq, Strensall, York, North Yorks, YO3 5YN [IO94LA, SE66]
G4 ELS J V Macleod, 4 Chiswick Ln, London, W4 2JE [IO91UL, TQ27]
G4 ELU G Beal, 10 Fairlea Cres, Northam, Bideford, EX39 1BD [IO71VA, SS42]
GM4 ELV Details withheld at licensee's request by SSL.
G4 ELY R C Panting, 124 Loddon Bridge Rd, Woodley, Reading, RG5 4AW [IO91NK, SU77]
G4 ELZ J J Pascoe, 3 Aller Brake Rd, Aller Park, Newton Abbot, TQ12 4NJ [IO80EM, SX86]
G4 EMA I A Welburn, 33 Bowland Way, Clifton, York, YO3 6PZ [IO93KX, SE55]
G4 EMB N J L Lockett, 18 Seagers, Hall Rd, Great Totham, Maldon, CM9 8PB [JO01IS, TL81]
G4 EMC J King, 35 Hornbeam Cl, Larkfield, Aylesford, ME20 6LY [JO01FH, TQ75]
G4 EMD R Edge, Laurel Cottage, Hopton Wafers, Kidderminster, Worcs, DY14 0NB [IO82RJ, SO67]
G4 EME D S Hall, Janderie, Crosemere Ln, Cockshutt, Ellesmere, SY12 0JR [IO82NU, SJ42]
G4 EMH G Parsley, 7 Rowan Rd, Martham, Great Yarmouth, NR29 4RY [JO02TQ, TG41]
G4 EMK G C L Parker, 6 Cedar Dr, Bourne, PE10 9SQ [IO92TS, TF02]
G4 EML C J Durbridge, 2 Send Villas, Sandy Ln, Send, Woking, GU23 7AP [IO91RG, TQ05]
G4 EMM Details withheld at licensee's request by SSL.
G4 EMO Details withheld at licensee's request by SSL.
G4 EMQ J A Purchon, 19 Warburton, Emley, Huddersfield, HD8 9QP [IO93EO, SE21]
G4 EMR Details withheld at licensee's request by SSL.
G4 EMT J Tawn, 11 Wyke Rd, Prescot, L35 5HL [IO83OK, SJ49]
G4 EMU D J Smith, 8 Hitherspring, Furze Hill, Corsham, SN13 9UT [IO81VK, ST86]
G4 EMV P I Johnson, 4 Chapel Ln, Hawley, Blackwater, Camberley, GU17 9ET [IO91OH, SU85]
G4 EMW Dr E M Warrington, 3 Long Meadow, Wigston, LE18 3TY [IO92KN, SP69]
GM4 EMX C J Hall, 18 Sumburgh Cres, Shedocksley, Aberdeen, AB1 6WF [IO87TH, NJ72]
G4 EMZ K F Maplesden, 55 The Heights, Northolt, UB5 4BP [IO91TN, TQ18]
G4 ENA P M Asquith, Well Cottage, The Green, Selsley, Stroud, GL5 5LN [IO81VR, SO80]
G4 ENB C A Asquith, 36 Sunningdale, Luton, LU2 7TE [IO91TV, TL02]
G4 ENF A F Fyffe, 5 Higher Ln, Upholland, Skelmersdale, WN8 0NL [IO83PM, SD50]
G4 ENG D R Horton, Streets Ln, Upper Landywood, Cheslyn Hay, Walsall, West Midlands, WS6 7NA [IO82XP, SJ90]
G4 ENJ K M Hunter, 2 Cavendish Ave, London, W13 0JG [IO91UM, TQ18]
GM4 ENK P Kelly, Glenfield, Dunrossness, Shetland Isles, ZE2 9JB [IO99IW, HU31]
G4 ENL P Jewitt, c/o Colenco Power Consulting, PO Box 130, Padang 130, Western Sumatra, Indonesia, X X
GM4 ENN A D Rae, 183 Campsie St., Glasgow, G21 4XY [IO75VV, NS66]
GM4 ENP Dr J P Johnston, 4 Lawhead Rd West, St. Andrews, KY16 9NE [IO86OI, NO41]
G4 ENQ Details withheld at licensee's request by SSL.[Op: F C Grant, 33 Morley Road, Chadwell Heath, Romford, Essex, RM6 6UX.]
G4 ENR K Brook, 154 Ridge Nethermoo, Liden, Swindon, Wilts, SN3 6NF [IO91DN, SU18]
G4 ENS A R Morris, Mentor House, Chare Rd, Stanton, Bury St. Edmunds, IP31 2DX [JO02KH, TL97]
G4 ENV Details withheld at licensee's request by SSL.
G4 ENW A R Gilbert, 47 Rayleigh Ave, Eastwood, Leigh on Sea, SS9 5DN [JO01HN, TQ88]
G4 ENZ M A Church, 2 Meadow Way, Churchdown, Gloucester, GL3 2AU [IO81VV, SO82]
G4 EOA T F Strickland, 1 Foxcote, St. Leonards on Sea, TN37 7HJ [JO00GV, TQ81]
G4 EOB G D Lawrance, 1 Swarthdale Ave, Ulverston, LA12 9HY [IO84KE, SD27]
G4 EOD D I King, 87 Staverton Rd, Werrington, Peterborough, PE4 6LY [IO92UO, TF10]
G4 EOF S R Lawrence, 207 Welland Park Rd, Market Harborough, LE16 9DW [IO92ML, SP78]
G4 EOG A V Heritage, 9 Torrington Cl, Springfield, Chelmsford, CM1 6HX [JO01FR, TL70]
G4 EOJ T K Ilott, 45 Parkside, Snettisham, Kings Lynn, PE31 7QF [JO02GU, TF63]
G4 EOL M J Coan, 4 Vera Rd, Hellesdon, Norwich, NR6 5HU [JO02PP, TG21]
GU4 EON M H Allisette, Les Amballes Lodge, Les Amballes, St. Peter Port, Guernsey, GY1 1WU
G4 EOO R W Greenwood, Elgon Lodge, French Drove, Thorney, Peterborough, PE6 0PF [IO92WP, TF20]
G4 EOR P R Stevens, 51 Beeches Rd, Chelmsford, CM1 2RX [JO01FR, TL60]
G4 EOT D Bussell, 26 Norbreck Cres, Wigan, WN6 7RF [IO83QN, SD50]
GM4 EOU J W Smith, 6 Rodger St., Cellardyke, Anstruther, KY10 3HU [IO86PF, NO50]
G4 EOV S Young, 96 Mereside Way, Olton, Solihull, B92 7AZ [IO92CK, SP18]
G4 EOW P J Baxter, 8 Birch Cl, Romsey, SO51 5SQ [IO90GX, SU32]
G4 EOZ Details withheld at licensee's request by SSL.
G4 EPA J B Pepper, 52 King Style Cl, Crick, Northampton, NN6 7ST [IO92KI, SP57]
G4 EPC Details withheld at licensee's request by SSL.
GW4 EPF J A Pile, 5 Western Cl, Thistle Boon, Mumbles, Swansea, SA3 4HF [IO81AN, SS68]
G4 EPG N A Mummery, Hvidhesten, Upper Dunsforth, Great Ousebourn, York, YO5 9RU [IO94IB, SE46]
G4 EPH J H Splaine, 43 Paulmont Rise, Temple Cloud, Bristol, BS18 5DZ [IO81RH, ST65]
GI4 EPK E J Coyle, 14 Colby Ave, Culmore Rd, Londonderry, BT48 8PF [IO65IA, C42]
G4 EPL L R Ward, 22 Coniston Rd, Hucknall, Nottingham, NG15 6NE [IO93JA, SK54]
G4 EPM N Lewis, 65 Brentwood Rd, Romford, RM1 2EU [JO01CN, TQ58]
G4 EPN A J Wright, 34 Webbs Way, Stoney Stanton, Leicester, LE9 4BW [IO92IN, SP49]
G4 EPS E P C S Skeggs, 57 Church Rd, Hatfield Peverel, Chelmsford, CM3 2LB [JO01HS, TL71]
G4 EPU M J Gray, 33 Claremont Dr, Pitsea, Basildon, SS16 4TL [JO01FN, TQ78]
G4 EPW L H Goulding, 24 Lancaster Dr, Hill, Lydney, GL15 5SL [IO81RR, SO60]
G4 EPX D J Chater-Lea, 71 Heathermount Dr, Crowthorne, RG45 6HJ [IO91OI, SU86]
GI4 EQA E C Mooney, 33 Piney Hill, Magherafelt, BT45 6PY [IO64PT, H89]
G4 EQC B K Smith, 11 Cheam Cl, Burntwood, WS7 9JS [IO92BQ, SK00]
G4 EQD N W Smith, 45 Manor Rd, Scunthorpe, DN16 3PB [IO93QN, SE80]
G4 EQE D G Smith, 7 Demesne Gdns, Martlesham Heath, Ipswich, IP5 7UA [JO02OB, TM24]
G4 EQG W G Lycett, 16 Spinney Cl, Walsall, WS3 4LB [IO92AO, SK00]
G4 EQH Details withheld at licensee's request by SSL.
G4 EQJ Lee, 12 Gainsborough Cl, Folkestone, CT19 5NB [JO01NC, TR23]
G4 EQK M Hale, 9 Cramer, Oreton, Kidderminster, Worcs, DY14 0UA [IO82RJ, SO67]
G4 EQL Dr M E Townsend, The Manor House, Main St., Fleckney, Leicester, LE8 8AP [IO92LM, SP69]
G4 EQM W R Evans, 9 Edwin Rd, Didcot, OX11 8LG [IO91IO, SU58]
GI4 EQN C G A Cupples, 2 Ballyhaskin Cttgs, Millisle, Newtownards, BT22 2JP [IO74FO, J67]
G4 EQP A D York, 1 Birdale Cl, Henbury, Bristol, BS10 7NU [IO81QM, ST57]
G4 EQQ P F Johnson, 58 Woodlands Rd, Allestree, Derby, DE22 2HF [IO92GW, SK33]
G4 EQR D J Evans, 103 Rhea Hall Est, Highley, Bridgnorth, WV16 6JY [IO82TK, SO78]
G4 EQT F Sharples, 209 Manchester Rd, Tyldesley, Manchester, M29 8WT [IO83SM, SD70]
G4 EQV P D Jones, 112 Norton Ln, Tidbury Green, Solihull, B90 1QT [IO92BJ, SP07]
GI4 EQW P J McCormack, 111 Duncreggan Rd, Londonderry, BT48 0AA [IO65IA, C41]
G4 EQX J A McIlroy, 17 Brownsfield Rd, Yardley Gobion, Towcester, NN12 7TY [IO92NC, SP74]
GM4 EQY J N Hately, 10 Crags Rd, Paisley, PA2 6RA [IO75ST, NS46]
G4 EQZ V K Faulkner, 18 Milton Cres, Talke, Stoke on Trent, ST7 1PF [IO83UB, SJ85]
G4 ERA D W Mepham, 8 The Close, Fairlight, Hastings, TN35 4AQ [JO00HU, TQ81]
G4 ERB B Skidmore, Rosedale, Folla Rule, Rothienorman, Inverurie, Aberdeenshire, AB51 8UN [IO87SJ, NJ73]
G4 ERC E R Chandler, 13 Kingsdown Ave, West Ealing, London, W13 9PS [IO91UM, TQ17]
G4 ERD A D G Hamilton, 1230 McLynn Ave N.E., Atlanta, Georgia 30306, USA, X X
G4 ERF P W Jones, Woodview, 10 Barrow Hill, Barrow, Bury St. Edmunds, IP29 5DX [JO02HF, TL76]
G4 ERG P Etheridge, 42 Wascana Cl, Anlaby Park Rd St., Hull, HU4 7BX [IO93TR, TA02]
G4 ERH J P Perry, c/o Wagons/Lits Apt168, San Antonio, Ibiza, Baleric Isles,Spain, ZZ9 9CO
G4 ERL E R E Lawley, 23 Briar Rigg, Keswick, CA12 4NN [IO84KO, NY22]
GI4 ERM K Bones, 42 Cloverdale Cres, Lisburn, BT27 4PS [IO64XM, J26]
G4 ERN E R Newsham, 31 Christopher Cl, Yeovil, BA20 2EH [IO80QW, ST51]
G4 ERO C S Leonard, 24 Lower Rd, Stuntney, Ely, CB7 5TN [JO02DJ, TL57]
G4 ERP R W Marshall, 40 Evesham Rd, Bishops Cleeve, Cheltenham, GL52 4SA [IO81XX, SO92]
G4 ERQ T Birchall, 11 Grig Pl, Alsager, Stoke on Trent, ST7 2SU [IO83UC, SJ75]
G4 ERR J Cummins, 5 Blenheim Orchard, Shurdington, Cheltenham, GL51 5TG [IO81WU, SO91]
G4 ERS J G Gamblen, Illfield, High Wych, Sawbridgeworth, Herts, CM21 0HX [JO01BT, TL41]
G4 ERT H V Marriott, 108 Leicester Rd, Quorn, Loughborough, LE12 8BB [IO92KR, SK51]
G4 ERU J P Taylor, 5 Luther Rd, Winton, Bournemouth, BH9 1LH [IO90BR, SZ09]
G4 ERV W C Coombes, 33 Clarence Park Rd, Boscombe, Bournemouth, BH7 6LF [IO90CR, SZ19]
G4 ERW D G Lurcook, 1 Norton Ave, Berrylands, Surbiton, KT5 9DX [IO91UJ, TQ16]
G4 ERX R A Elliott, 31 Marsland Rd, Sale, M33 3HP [IO83UK, SJ79]
G4 ERY D J Tyson, 114 Marshall Hill Dr, Nottingham, NG3 6HW [IO92KX, SK54]
G4 ERZ A Wells, 38 Sextant Rd, Beverley High Rd, Hull, HU6 7BA [IO93TS, TA03]
G4 ESC P N Browning, 68 Ashby Dr, Rushden, NN10 9HH [IO92QG, SP96]
G4 ESG D J Neal, 2 St. Margarets Ave, Ashford, TW15 1DR [IO91SK, TQ07]
GI4 ESI S T McClean, 14 Bamber Park, Ballymena, BT43 5HE [IO64UV, D00]

G4 ESJ M Stephenson, 3 Silverdale Rd, Beverley Rd, Hull, HU6 7HE [IO93TS, TA03]
G4 ESK G Benson, 2 Saxon Walk, Lichfield, WS13 8AJ [IO92BQ, SK10]
GW4 ESL P J Edwards, 14 Northfield Cl, Caerleon, Newport, NP6 1EZ [IO81MO, ST39]
G4 ESQ Details withheld at licensee's request by SSL.
G4 EST C Cartmel, 57 Stanley Ave, Rainford, St. Helens, WA11 8HU [IO83OM, SD40]
G4 ESU C R Rouse, 86 Melton Ave, Clifton, York, YO3 6QG [IO93KX, SE55]
G4 ESY D A Jackson, 16 Melrose Park, Keldgate, Beverley, HU17 8JL [IO93SU, TA03]
G4 ETA W Hopley, 41 Cherry Tree Rd, Brereton, Rugeley, WS15 1AY [IO92AR, SK01]
G4 ETC P J Keeble, 39 Bladen Dr, Rushmere St. Andrew, Ipswich, IP4 5UE [JO02OB, TM24]
G4 ETD A Firth, 49 Ravenswood Ave, Frindsbury, Rochester, ME2 3BY [JO01FJ, TQ76]
G4 ETG D B Humphries, 15 The Willows, Stevenage, SG2 8AN [IO91VV, TL22]
G4 ETI J R Shaw, 20 Castleton Gr, Jesmond, Newcastle upon Tyne, NE2 2HD [IO94EX, NZ26]
G4 ETJ R F Lyddon, 24 Gate Field Rd, Bideford, EX39 3QX [IO71VA, SS42]
G4 ETK C J Bourne, 10 Lansdowne Pl, London Rd, St. Albans, AL1 1LT [IO91TS, TL10]
G4 ETM J E Taylor, Brigantia, Sessay, Thirk, North Yorks, YO7 3NL [IO94IE, SE47]
G4 ETN B P Smith, 109 Chilton St., Bridgwater, TA6 3HY [IO81LD, ST23]
G4 ETO J M Roach, 33 Pound Ln, Topsham, Exeter, EX3 0NA [IO80GQ, SX98]
G4 ETP T Pinch, 1 Fernhill Cl, Ivybridge, PL21 9JE [IO80AJ, SX65]
G4 ETQ J M Davies, 12 St. Johns Cl, Claines, Worcester, WR3 7PT [IO82VF, SO85]
G4 ETU T M Allen, 2 Hillside, West Stoke, Chichester, W Sussex, PO18 9BL [IO90OU, SU80]
G4 ETV Details withheld at licensee's request by SSL.
G4 ETW B Cockfield Willenhall Dist, Amateur Radio Club, 47 Aston Rd, Willenhall, WV13 3DG [IO82XO, SO99]
G4 ETZ F S Webb, 166 Glastonbury Rd, Yardley Wood, Birmingham, B14 4DS [IO92BK, SP08]
GW4 EUA G J Smith, 96 Fonman Park Rd, Castle Park Est, Rhoose, Sth Glam, CF6 9BG [IO81HN, ST08]
G4 EUC G C Mendoza, 32 The Circuit, Cheadle Hulme, Cheadle, SK8 7LG [IO83VI, SJ88]
G4 EUE Details withheld at licensee's request by SSL.[Station located near Solihull.]
G4 EUF G E Mayo, Carlton House, Broad Ln, Markfield, Leicester, LE6 0TB [IO92JP, SK50]
G4 EUG G Payne, 28 Pollards Dr, Horsham, RH13 5HH [IO91UA, TQ22]
G4 EUI J M Hesman, 23 Beeches Rd, Great Barr, Birmingham, B42 2HH [IO92BM, SP09]
G4 EUJ R A Whiteley, Woodsend, Wood Ln, Gedling, Nottingham, NG4 4AD [IO92LX, SK64]
G4 EUK G G M Adcock, 2 Erringham Rd, Shoreham By Sea, BN43 5NQ [IO90UU, TQ20]
G4 EUL Details withheld at licensee's request by SSL.[Station located near Sandwich.]
G4 EUM Details withheld at licensee's request by SSL.
G4 EUN D T Freeman, Rhossili House, Malleson Rd, Gotherington, Cheltenham, GL52 4EY [IO81XX, SO92]
G4 EUQ Details withheld at licensee's request by SSL.[Op: W R Wynn.]
G4 EUR M J Tout, 34 Wysall Rd, Northampton, NN3 8TP [IO92OG, SP86]
G4 EUV W E Bidmead, 4 Pine Gr, Northville, Bristol, BS7 0SL [IO81RM, ST67]
G4 EUW B M Keeling, 41 Regent Rd, Brightlingsea, Colchester, CO7 0NN [JO01MT, TM01]
G4 EUY J McNulty, 28 Woodhead Ln, Haltwhistle, NE49 9DT [IO84SX, NY76]
G4 EUZ K Watson Gt Lumley AR &, Electronics Society, 34 Hermitage Park, Chester-le-Street, DH3 3JZ [IO94FU, NZ25]
G4 EVA C J Roberts, Byfield, 19 Burleigh Rd, Addlestone, KT15 1PN [IO91SI, TQ06]
G4 EVC K Chadwick, 14 Fell View, Crossens, Southport, PR9 8JX [IO83MQ, SD32]
G4 EVD E E Parry, 60 Hunters Forstal, Herne Bay, Kent, CT6 7DW [JO01NI, TR16]
G4 EVE P K Webster, The Old School, School Hill, Stratton, Cirencester Glos, GL7 2LS [IO91AR, SP00]
G4 EVI J W Howard, 127 Goldcroft, Yeovil, BA21 4DD [IO80QW, ST51]
GW4 EVJ G Watson, 19 Kelvin Rd, Clydach, Swansea, SA6 5JP [IO81BQ, SN60]
G4 EVK I E Shepherd, Grosvenor House, Watsons Ln, Harby, Melton Mowbray, LE14 4DD [IO92NU, SK73]
GW4 EVL T G Hopkins, 39 Glen Rd, Norton, West Cross, Swansea, SA3 5PR [IO71XN, SS68]
G4 EVN S S E Garrett, Winterfold, The St., Ashfield, Stowmarket, IP14 6LX [JO02OF, TM26]
G4 EVP C O N McPartland, 55 Elliotts Ln, Codsall, Wolverhampton, WV8 1PG [IO82VP, SJ80]
G4 EVR Dr A L M Davies, Fleet Cottage, 8 Half Acres, Bishops Stortford, CM23 2QP [JO01BU, TL42]
GM4 EVS D N Johnstone, Sycamore House, Kirkloan, Stormontfield, Perth, PH2 6BL [IO86GK, NO12]
G4 EVW G W Sutton, 2 Orchard Cl, Uttoxeter, ST14 7DZ [IO92BV, SK03]
GW4 EVX R Price, Glendale, New Brighton Rd, Sychdyn, Mold, Clwyd, CH7 6EF [IO83KE, SJ26]
G4 EVY P J Poole, 5 River Dr, Strood, Rochester, ME2 3JW [JO01FJ, TQ76]
G4 EVZ M I Powrie, 31 The Grove, Billericay, CM11 1AU [JO01FP, TQ69]
G4 EWB J L Duffus, Chapter Farm, Braunston, Oakham, Rutland Leics, LE15 8QZ [IO92OP, SK80]
G4 EWE O T Overton, 163 Woodford Rd, Woodford, Stockport, SK7 1QD [IO83WI, SJ88]
G4 EWI F E Warner, 48 Brookfield Rd, Walsall, WS9 8JE [IO92AO, SK00]
G4 EWJ B W Jordan, 42 Ben Nevis Rd, Birkenhead, L42 6QY [IO83LI, SJ38]
G4 EWK D R Mellor, 18 Briar Cl, Newhall, Swadlincote, DE11 0RX [IO92FS, SK22]
GM4 EWL R M Macleod, 22 Firthview, Dingwall, IV15 9PF [IO77SO, NH55]
GM4 EWM E W McLean, 21 Milnefield Ave, Elgin, IV30 3EJ [IO87IP, NJ26]
G4 EWS Details withheld at licensee's request by SSL.
G4 EWT S Mason, 15 Northfield Cl, Bishops Waltham, Southampton, SO32 1EW [IO90JX, SU51]
G4 EWW T R James, 2 The Green, Bottom St., Northend, Leamington Spa, CV33 0TL [IO92GD, SP35]
G4 EWX Details withheld at licensee's request by SSL.[Op: J Devaney.]
G4 EWY R F E Leigh, 22 Western Rd, Lancing, BN15 8RX [IO90TT, TQ10]
GM4 EWZ J R Halford, Greystone, Sanday, Orkney, Scotland, KW17 2AT [IO89QD, HY62]
G4 EXD Details withheld at licensee's request by SSL.
GW4 EXE B D Hope, Oriel, Moelfre, Isle of Anglesey, Gwynedd, LL72 8HN [IO73VI, SH58]
G4 EXF A Grindrod, Mullions, Church St., Kings Stanley, Stonehouse, GL10 3HX [IO81UR, SO80]
G4 EXG G F Reid Ex 'G' Radio Club, 65 Rowelfield, Luton, LU2 9HL [IO91TV, TL12]
G4 EXJ G H Cook, Maples, 10 Rosewall Terr, St. Ives, TR26 1QJ [IO70GF, SW54]
G4 EXK P Bradbury, 96 Woodhouse Rd, Davyhulme, Urmston, Manchester, M41 7WX [IO83TL, SJ79]
G4 EXM G P Cullum, 37 Coniston Cres, Humberston, Grimsby, DN36 4AY [IO93XM, TA30]
G4 EXN L J Dolman, 139 Rupert St., Norwich, NR2 2AX [JO02PO, TG20]
G4 EXP R E Pell, 381 Broad Gate, Weston Hills, Spalding, Lincs, PE12 6DB [IO92WS, TF22]
G4 EXQ Details withheld at licensee's request by SSL.
G4 EXT J B Corben, 65 Oatley Park Ave, Oatley, New South Wales 2223, Australia, X X
G4 EXU D A Fisher, 1 Francolin Cl, Woodhaven, Durban, Natal South Africa, ZZ1 9FR
G4 EXW B May, 25 Welford Gdns, Abingdon, OX14 2BN [IO91IQ, SU59]
G4 EXX D B Durance, Nansland, Dovecote Ln, Wainfleet All Saints, Lincs, PE24 4AD [JO03CC, TF55]
G4 EXZ R B Fidler, 55 Sunnyvale Dr, Longwell Green, Bristol, BS15 6YQ [IO81SK, ST67]
G4 EYA C D Evans, 64 Boyd Ave, Toftwood, Dereham, NR19 1ND [JO02LP, TF91]
G4 EYB D J Fernie, Shepherds Cl, Reigate Rd, Leatherhead, KT22 8RD [IO91UG, TQ15]
G4 EYE A J Free, Homeric, Harwich Rd, Little Oakley, Harwich, CO12 5JF [JO01OV, TM22]
G4 EYF G K Cundell, 4 Cranbrook Rd, Acomb, York, YO2 5JA [IO93KX, SE55]
G4 EYJ D A Davies, 124 Leigh Sinton Rd, Malvern, WR14 1LF [IO82UD, SO74]
G4 EYL A E Woollerton, 114 Grampian Way, Luton, LU3 3HE [IO91SW, TL02]
G4 EYM J Shardlow, 19 Portreath Dr, Darley Abbey, Allestree, Derby, DE22 2BJ [IO92GW, SK33]
G4 EYN K Wright, 61 Albert Rd, Chaddesden, Derby, DE21 6SH [IO92GW, SK33]
GW4 EYO C S Carver, 13 Plas Yn Rhos, Penyffordd, Chester, CH4 0JU [IO83LD, SJ26]
G4 EYP P Brown, 8 Websters Holt, Manorside Cl, Wirral, L49 4RG [IO83KJ, SJ28]
G4 EYQ E A Coward, 3 Bedford Rd, Eccles, Manchester, M30 9LA [IO83TL, SJ79]
G4 EYR S H Phillips, Millbank Cottage, Mill Half, Whitney on Wye, Herefordshire, HR3 6HY [IO82LD, SO24]
G4 EYT C R Williams, 35 Heath Cres, Norwich, NR6 6XF [JO02PP, TG21]
G4 EYU C W Roberts, 58 Oak Ave, Newport, TF10 7EF [IO82TS, SJ71]
G4 EYV Dr P J Skolar, 7 Greenhalgh Walk, London, N2 0DJ [IO91VO, TQ28]
G4 EYX P J Davies, 14 Saville Rd, Blackpool, FY1 6JP [IO83LT, SD33]
G4 EYY A W R Watts, 4 Harper Ave, Wednesfield, Wolverhampton, WV11 1HF [IO82WO, SJ90]
G4 EYZ J W Pearson, 30 Creighton Ave, Morphett Vale, South Australia, 5162
G4 EZE J C Hinton, 10 Moathouse Cl, Acton Trussell, Stafford, ST17 0QY [IO82WS, SJ91]
G4 EZF W D Logan, 27 Shaw St., Mottram, Hyde, SK14 6LE [IO83XK, SJ99]
G4 EZG Details withheld at licensee's request by SSL.
G4 EZH C R Bardfield, 15 Rowntree Way, Saffron Walden, CB11 4DF [JO02CA, TL53]
G4 EZI D Hughes, 3 Primley Park Cres, Leeds, LS17 7HY [IO93FU, SE33]
GM4 EZJ K B Glendinning, 14 Craiglockhart Ave, Edinburgh, EH14 1HW [IO85JW, NT27]
G4 EZK D J Grant, Orchard House, Canterbury Rd, Lyminge, Folkestone, CT18 8HU [JO01ND, TR14]
G4 EZL A E Macleod, 5 Cornflower, Harold Wood, Romford, Essex, RM3 0XY [JO01CO, TQ59]
G4 EZM E M R Green, 6 Downham Pl, Blackpool, FY4 1QS [IO83LS, SD33]
G4 EZN Dr J H Keeler, 8 Gwydir St., Cambridge, CB1 2LL [JO02BE, TL45]
G4 EZO A S Warne, Mereton, 113 Queens Rd, Vicars Cross, Chester, CH3 5HF [IO83NE, SJ46]
G4 EZP I T Melville, 3 Cres Rd, South Benfleet, Benfleet, SS7 1JL [JO01GN, TQ78]
G4 EZQ C G Doman, 6 Churnet Cl, Bedford, MK41 7ST [IO92SD, TL05]
G4 EZR M D Dubery, 5 Newstead Ave, Orpington, BR6 9RJ [JO01BI, TQ46]
G4 EZS H C Foster, 23 Ghyllroyd Dr, Birkenshaw, Bradford, BD11 2ET [IO93DR, SE22]
G4 EZT Details withheld at licensee's request by SSL.
G4 EZU W G Peterson, 124 Darnley Rd, Gravesend, DA11 0SN [JO01EK, TQ67]
G4 EZV Details withheld at licensee's request by SSL.
GW4 EZW M A Hill Newport Amateur Radio Soc, 13 Maesglas Gr, Newport, NP9 3DJ [IO81LN, ST28]
G4 EZX D E Titheridge, 2 The Oaks, Stapper Green, Wilsden, Bradford, BD15 0HH [IO93BT, SE03]
G4 EZZ D B Andrews, 16 Gedling Cl, Northampton, NN3 9UT [IO92OF, SP76]
G4 FAA L O Atkinson, 56 The Spinney, North Cray, Sidcup, DA14 5NF [JO01BK, TQ47]
G4 FAB S J Fox, 16 The Teasels, Bingham, Nottingham, NG13 8TY [IO92MW, SK63]
G4 FAD R J Langford, The Parsonage, Wellington Village, Hereford, HR4 8AZ [IO82PD, SO44]
G4 FAG G J Morris, 4 Southwest Rd, Leytonstone, London, E11 4AW [JO01AN, TQ38]
G4 FAH D Jones, 41 Sorrel Walk, Brierley Hill, DY5 2QG [IO82WL, SO98]
G4 FAI A Smith, 13 Morley Rd, Sheringham, NR26 8JE [JO02OW, TG14]
G4 FAJ R E Sadler, East Lynne, 202 Shire Oak, Walsall Wood, Walsall, WS9 9PD [IO92AP, SK00]
G4 FAK R H D Heed, 39 Deeds Gr, High Wycombe, HP12 3NT [IO91OP, SU89]
G4 FAL N H Totterdell, 35 Meadow Bank Ave, Sheffield, S7 1PB [IO93GI, SK38]
G4 FAM C A P Henderson, Bella Pais, Kelsey Ln, Beckenham, BR3 3NF [IO91XJ, TQ36]
G4 FAP R J Painting, 15 Surrey Walk, Aldridge, Walsall, WS9 8JG [IO92AO, SK00]
G4 FAQ D P Jones, 7 Camrose Gdns, Pendeford, Wolverhampton, West Midlands, WV9 5RN [IO82WP, SJ90]
G4 FAS G Royle, 56 Branksome Dr, Heald Green, Cheadle, SK8 3AJ [IO83VI, SJ88]
G4 FAT N S Trollope, 174 Newtown Rd, Malvern, WR14 1PJ [IO82UC, SO74]
GM4 FAU J Walker, 20 Cairneyhill Rd, Crossford, Dunfermline, KY12 8NZ [IO86FB, NT08]
G4 FAV A J Bevan, 8 Lawnsdown Rd, Quarry Bank, Brierley Hill, DY5 2EP [IO82WL, SO98]
G4 FAW D G Cutts, 4 Ipswich Rd, Newbourne, Woodbridge, IP12 4NS [JO02PA, TM24]
G4 FAX R J Macfie, 31 Fairgreen Rd, Caddington, Luton, LU1 4JG [IO91SU, TL01]
G4 FAZ G S J Brownett, 82 Mudford Rd, Yeovil, BA21 4AH [IO80QW, ST51]
G4 FB F Barnard, 8 Hillside Ave, Seaford, BN25 3JT [JO00BS, TQ50]
G4 FBA R E Edeson, 22 Ferrybridge Rd, Knottingley, WF11 8JF [IO93IR, SE42]
G4 FBB D J Ellis, 17 Victoria Ave, Yeadon, Leeds, LS19 7AS [IO93DU, SE24]
G4 FBC R M Heron, 66 Derwentwater Rd, Mirehouse, Whitehaven, CA28 9RH [IO84FM, NX91]
G4 FBE J D Gill, Spring Lodge, Ladywell, Wrington, Bristol, BS18 7LT [IO81OI, ST46]
G4 FBF M Giles, Summers End, 109 Marldon Rd, Paignton, TQ3 3NN [IO80FK, SX86]
G4 FBG D P Shone, 6 Windlehurst Rd, High Ln, Stockport, SK6 8AB [IO83XI, SJ98]
G4 FBH J Bytheway, 411 Himley Rd, Gornall Wood, Gornal Wood, Dudley, DY3 2RA [IO82WM, SO99]
G4 FBI E L Creasy, 16 Birchwood Cl, Horley, RH6 9TX [IO91WE, TQ24]
G4 FBJ Details withheld at licensee's request by SSL.
G4 FBK M A Kipp, 17 York Rd, Northwood, HA6 1JJ [IO91TO, TQ08]
G4 FBL Details withheld at licensee's request by SSL.
G4 FBN B M Neale, 30 Pine Cl, South Wonston, Winchester, SO21 3EB [IO91IC, SU43]
G4 FBO W R Fish, Chestnut Cottage, 2A Broom Ln, Whickham, Newcastle upon Tyne, NE16 4QP [IO94DW, NZ26]
GM4 FBP J C S Dean, 33 Woodstock Rd, Aberdeen, AB1 5EX [IO87TH, NJ72]
G4 FBQ J H Hobley, 31 The Pippins, Glemsford, Sudbury, CO10 7PQ [JO02IC, TL84]
GM4 FBR G M Brash, 19 Craigmount Gdns, Edinburgh, EH12 8EB [IO85IW, NT17]
G4 FBS S W Swain Horndean&Dis AR, 35 Mavis Cres, Havant, PO9 2AE [IO90MU, SU70]
GM4 FBU H Macdougall, 17 Prospecthill St., Greenock, PA15 4HH [IO75OW, NS27]
G4 FBV D P E Vaughan, 35A St. Georges Rd, Felixstowe, IP11 9PL [JO01QX, TM33]
G4 FBW Details withheld at licensee's request by SSL.
G4 FBZ W F Kitching, 3 Prince Charles Cres, Telford, TF3 2JX [IO82SQ, SJ60]
G4 FCA J A Haddon, Redbrook Gate Cott, Bosbury Rd, Cradley, Malvern Worcs, WR13 5JA [IO82TC, SO74]
G4 FCB N L Edwards, 40 Camden St., Walsall, WS1 4HF [IO92AN, SP09]
G4 FCC G C A Freeman, 12 The Haven, Beadnell, Chathill, NE67 5AW [IO95EN, NU22]
G4 FCD R A Girling, Heath Farmhouse, Cottisford, nr Brackley, Northants, NN13 5SN [IO91KX, SP53]
G4 FCE R L Jenson, 132 Barr Common, Walsall, West Midlands, WS9 0TD [IO92BO, SP09]
G4 FCF W H Wade, 11 St. Marys Rd, Bluntisham, Huntingdon, PE17 3XA [JO02AI, TL37]
G4 FCH E W P Stankiste, 23 High Mill Dr, Scarborough, YO12 6RN [IO94SH, TA09]
G4 FCI A N Cullup, 201 Elm Low Rd, Elm, Wisbech, PE14 0DF [JO02CP, TF40]
G4 FCL P Lawson, 14 Hazel Cl, Chandlers, Ledbury, Herefordshire, HR8 2XX [IO82SA, SO73]
G4 FCN C J Coker, 46 Clarendon Rd, Ipplepen, Newton Abbot, TQ12 5QS [IO80EL, SX86]
G4 FCO D K Stevens, 3 Boyleston Rd, Hall Green, Birmingham, B28 9JN [IO92CK, SP18]
GM4 FCP F S Sharp, 92 Hutchison Dr, Darvel, KA17 0BN [IO75UO, NS53]
G4 FCQ J F Benton, 10 Troydale Gdns, Pudsey, LS28 9JZ [IO93ES, SE23]
G4 FCR K H Williams, 106 Sparrowhawk Way, Hartford, Huntingdon, PE18 7XY [IO92WI, TL27]
G4 FCT J N Gunn, 8 College Gdns, Hornsea, HU18 1EF [IO93VV, TA14]
GW4 FCV R E Jones, 2 Pen y Cwarel Rd, Wyllie, Blackwood, NP2 2HP [IO81JP, ST19]
GI4 FCW J A McGuigan, 18 Tullybrannigan, Brae, Newcastle, Co Down N Ireland, N Ireland, BT33 0DG [IO74BF, J33]
G4 FCX B F Pearl, 66 Benfleet Rd, Benfleet, SS7 1QB [JO01HN, TQ88]
G4 FCY I R Smith, 228 Dover St, Los Gatos, California 95032, USA, X X
G4 FCZ M H Thomas, The Old School, 1 Church Ln, Lound, Lowestoft, NR32 5LL [JO02UM, TM59]
G4 FDA W B Chapman, Kiloran, 72 Norbury Hill, Norbury, London, SW16 3RT [IO91WK, TQ37]
G4 FDB M D Biart, Appletree Cottage, Mill Rd, Shiplake, Henley-on-Thames, RG9 3LP [IO91NM, SU77]
G4 FDC A Korda, 5 Windmill Ct, North St., Tunbridge Wells, TN2 4SU [JO01DD, TQ53]
G4 FDD J A Livingston, 26 Dikelands Ln, Upper Poppleton, York, YO2 6JB [IO93KX, SE55]
G4 FDF V D Cunningham, 9 Lacon Rd, Bramford, Ipswich, IP8 4HD [JO02NB, TM14]
G4 FDG R G W Taylor, Courtlands, Dulford, Cullompton, Devon, EX15 2EQ [IO80IU, ST00]
G4 FDI S C Giles, 11 Beeches Rd, Chelmsford, CM1 2RS [JO01FR, TL60]
G4 FDJ Details withheld at licensee's request by SSL.
G4 FDL M E France, 106 Harvey Ln, Golborne, Warrington, WA3 3QL [IO83QL, SJ69]
GM4 FDM Details withheld at licensee's request by SSL.[Station located near Johnstone.]
G4 FDN P G McGuinness, 9 Farmdale Rd, Carshalton, SM5 3NG [IO91VI, TQ26]
G4 FDO Details withheld at licensee's request by SSL.
G4 FDP R E Miller, 65 West Rd, Oakham, LE15 6LT [IO92PQ, SK80]
G4 FDR Details withheld at licensee's request by SSL.
G4 FDS J A C Ingram, 170 Churchill Rd, Parkstone, Poole, BH12 2JF [IO90AR, SZ09]
GM4 FDT R B Kerr, Rosskeen Bridge, Invergordon, Ross-shire, IV18 0PR [IO77VQ, NH66]
G4 FDX I C Offer, Southease, Balmer Lawn Rd, Brockenhurst, SO42 7TT [IO90FT, SU30]
G4 FEA C J Beezley, 19 Beech Ave, Claverton Down, Bath, BA2 7BA [IO81UI, ST76]
G4 FEB D L Emery, 424 Clement Ave, Charlotte, North Carolina, USA 28204
G4 FEF F E Finch, 175 Harold Rd, Hastings, TN35 5NQ [JO00HU, TQ81]
G4 FEH G L Sidgwick, 42 Parkland Dr, Darlington, DL3 9DU [IO94EM, NZ21]
G4 FEI A G Marsden, 19 Buddon Dr, Monifieth, Dundee, DD5 4TH [IO86OL, NO53]
G4 FEJ B Fawcett, 361 Main Rd, Bilton, Hull, HU11 4DS [IO93VS, TA13]
G4 FEM P M Greatorex, 2 Briar Briggs Rd, Bolsover, Chesterfield, S44 6SE [IO93IF, SK47]
G4 FEO J G Gaughan, 14 Willow Rd, Thame, OX9 3BE [IO91MS, SP70]
G4 FEQ H Stogdale, Meadowcroft, 14 Main St., Ledston, Castleford, WF10 2AA [IO93HS, SE42]
G4 FET T H Ransom, 9 Lyndhurst Ave, Hastings, TN34 2BD [JO00HU, TQ81]
G4 FEU T Southwell, 12 Chequer Ln, Upholland, Skelmersdale, WN8 0DE [IO83PM, SD50]
G4 FEV D K Whitty, 146 Ave Rd, Rushden, NN10 0SW [IO92RG, SP96]
G4 FFA R A Harris, 98 Evelyn Ave, Ruislip, HA4 8AJ [IO91TN, TQ08]
G4 FFC M C Packer, Ricmaes Cottage, Chadwell Ln, Pertenhall, Bedford, MK44 2AU [IO92TG, TL06]
G4 FFD Dr J Clarke, Rosebank, Canon Pyon, Hereford, Herefordshire, HR4 8NT [IO82OD, SO44]
G4 FFE L P Marriott, 94 Lyndhurst Rd, Worthing, BN11 2DW [IO90TT, TQ10]
G4 FFH Details withheld at licensee's request by SSL.
GI4 FFL J A Finnegan, 65 Barrack Hill, Armagh, BT60 1BL [IO64QI, H84]
G4 FFM D J C Bailey, 10 Manor Rd, Stutton, Tadcaster, LS24 9BR [IO93IU, SE44]
G4 FFN C A Baker, 78 Station Rd, Whittlesey, Peterborough, PE7 1UE [IO92WN, TL29]
GM4 FFP I D Campbell, 35 Radernie Pl, St. Andrews, KY16 8QR [IO86OH, NO41]
G4 FFS D E R Hodge, 15 Buckland Cl, Peterborough, PE3 9UH [IO92UN, TL19]
G4 FFU C Forster, 48 Woolsington Gdns, Woolsington, Newcastle upon Tyne, NE13 8AR [IO95DA, NZ16][Op: Colin Forster. Prestel mbx: 919995287. RSARS: 1295.]
G4 FFV I Forster, 48 Woolsington Gdns, Woolsington, Newcastle upon Tyne, NE13 8AR [IO95DA, NZ16][Op: Mrs Irene Forster, BYLARA 296.]
G4 FFW M E Betts, 56 Kingswood Rd, Fallowfield, Manchester, M14 6RX [IO83VK, SJ89]
G4 FFX R S Clear, 33 Cedars Rd, Beddington, Croydon, CR0 4PU [IO91WI, TQ36]
G4 FFY R T Howells, 9 Aultone Way, Sutton, SM1 3LD [IO91VJ, TQ26]
GW4 FGC D A Cutts, 62 Forest Dr, Broughton, Chester, CH4 0QJ [IO83MD, SJ36]
GM4 FGD A M Dalziel, 90 Mayburn Ave, Loanhead, EH20 9HE [IO85KV, NT26]
G4 FGF J E Draklely, 186 Conway Rd, Fordbridge, Birmingham, B37 5LD [IO92DL, SP18]
GI4 FGH W J Tweedy, 11 Beechgrove Rise, Belfast, BT6 0NH [IO74BN, J37]
GM4 FGI Rev M A McCarthy, St. Athanasius Pres, 21 Mount Stewart St., Carluke, ML8 5EB [IO85BR, NS85]
G4 FGJ G S McGowan, Flat 3A, 4 Clapham Rd, Bedford, Beds, MK41 7PP [IO92SD, TL05]
GW4 FGL G Williams, Maes y Fro, Colwinston Rd, Llysworney, nr Cowbridge S Wale, XX77 1AA
G4 FGO A Oliver, Beaver Lodge, Dale Rd, Elloughton Brough, North Humberside, HU15 1HY [IO93RR, SE92]
G4 FGP F G Preece, 44 Broadmeadow, Aldridge, Walsall, WS9 8JA [IO92BO, SK00]
G4 FGQ R J Edwards, 13 Belle Vue Rd, Rowley Regis, Warley, B65 9ND [IO82XL, SO98]
G4 FGR S H Porter, 138 Broad Ln, Essington, Wolverhampton, WV11 2RQ [IO82XP, SJ90]
GM4 FGS I Douglas, 47 Meadowpark, Ayr, KA7 2LW [IO75QK, NS32]
G4 FGW C S Hall, 14 Copenhagen Gdns, Chiswick, London, W4 5NN [IO91UL, TQ27]
G4 FGX L S Loewenthal, 79C Woodmancote, Dursley, Glos, GL11 4AG [IO81TQ, ST79]
G4 FGY J K Maltby, Ingle Nook, Lenton Rd, Ingoldsby, Grantham, NG33 4HA [IO92RU, TF03]
GM4 FH Dr E G Walsh, 64 Liberton Dr, Edinburgh, EH16 6NW [IO85JV, NT26]
G4 FHA M H Lister, 42 Cartmel Dr, Ulverston, LA12 9PF [IO84KE, SD27]
GI4 FHB W A Mc Faul, 9 Durham Park, Kilfennan, Londonderry, BT47 1YD [IO64IX, C41]
GI4 FHD E C Sass, 696 Doagh Rd, Newtownabbey, BT36 4TP [IO74AQ, J38]
G4 FHE E E Payne, 59 Hanworth Rd, Redhill, RH1 5HS [IO91VF, TQ24]

G4 FHF J H Walker, Rosebery, Ownby Rd, Searby, Barnetby, DN38 6BD [IO93TM, TA00]

G4 FHI Details withheld at licensee's request by SSL.

G4 FHK T A Knight, 3 Eaton Cl, Rainworth, Mansfield, NG21 0AR [IO93KC, SK55]

G4 FHN R J Lovell, 3 Point View, West End, Porlock, Minehead, TA24 8NP [IO81EF, SS84]

G4 FHO D Nicklin, 114 Main St., Barton U Needwood, Barton under Needwood, Burton on Trent, DE13 8AB [IO92DS, SK11]

G4 FHT I H Murray, 52 Wellesley Ave, Norwich, Norfolk, NR1 4NU [JO02PP, TG20]

G4 FHU R H Pearson, 13 Mill Drove, Bourne, PE10 9BX [IO92TS, TF02]

G4 FHV D E Bumstead, 34 Thanet Rd, Ipswich, IP4 5LB [JO02OB, TM14]

G4 FIA M Mucklow, 7 Burns Cl, Newport Pagnell, MK16 8PL [IO92PC, SP84]

GM4 FIB R B Wilson, Glenmayne, 17 Pendreich Terr, Bonnyrigg, EH19 2DT [IO85KV, NT36]

G4 FIC Details withheld at licensee's request by SSL.

G4 FIE P B Groom, 2A The Chestnuts, Countesthorpe, Leicester, LE8 5TL [IO92KN, SP59]

G4 FIF D A Cherrington, 4 Bloomfield Cl, Wombourne, Wolverhampton, WV5 8HQ [IO82VM, SO89]

G4 FIG B B Callaway, 44 Grover Ave, Lancing, BN15 9RQ [IO90UU, TQ10]

G4 FIH E N J Fernandes, 78 Queens Head St., Islington, London, N1 8NG [IO91WM, TQ38]

G4 FIJ A C Thompson, Little Pippin, 11 Lodge Rd, Sharnbrook, Bedford, MK44 1JP [IO92RF, SP95]

G4 FIK M G Smith, 131 Coppice Rd, Poynton, Stockport, SK12 1SN [IO83WI, SJ98]

G4 FIM Details withheld at licensee's request by SSL.

G4 FIP K Collins, 48 Westville Oval, Harrogate, HG1 3JW [IO94FA, SE25]

G4 FIQ G Clegg, 2 Crowfields, Deeping St. James, Peterborough, PE6 8NY [IO92UQ, TF11]

G4 FIS M J Barton, 52 Newtown Rd, Uppingham, Oakham, LE15 9TS [IO92PO, SP89]

G4 FIT J D Chapman, 7 Beechland Cttgs, Mogador Rd, Tadworth, KT20 7EW [IO91VG, TQ25]

G4 FIV P H Morley, Skeena, North Petherwin, Launceston, Cornwall, PL15 8LR [IO70SQ, SX28]

GM4 FIX Details withheld at licensee's request by SSL.

GM4 FIZ A J L Murray, The Gables, 12 West High St., Greenlaw, Duns, TD10 6XA [IO85SQ, NT74]

G4 FJA B J Prew, Quintrell, Stafford Rd, Haughton, Stafford, ST18 9EX [IO82VS, SJ82]

G4 FJB J F Dodd, 7 Hornbrook Gr, Olton, Solihull, B92 7HH [IO92CK, SP18]

G4 FJC R A Delarue, 80 Keable Rd, Marks Tey, Colchester, CO6 1XR [JO01JU, TL92]

G4 FJD A Bradshaw, 17 Gurnard Heights, Gurnard, Cowes, PO31 8EF [IO90IS, SZ49]

G4 FJF M N Thacker, Kings Walden, 25 London Rd, River, Dover, CT17 0SF [JO01PD, TR34]

G4 FJH D J Powell, 18 Exley Cl, North Common, Warmley, Bristol, BS15 5YD [IO81SK, ST67]

GD4 FJI R Allison, 20 Droghadfayle Park, Port Erin, IM9 6EP

G4 FJJ D Bayliss, 20 Midhill Dr, Rowley Regis, Warley, B65 9SD [IO82XL, SO98]

G4 FJK T J S Hugill, Swandham House, Sampford Peverell, Tiverton, Devon, EX16 7ED [IO80HV, ST01]

G4 FJO G S Gwilliam, 61 Carlton Rd, Boston, PE21 8PA [IO92XX, TF34]

G4 FJP J F Perry, 108 Elm Rd, New Malden, KT3 3HP [IO91UJ, TQ26]

G4 FJQ A C Weaver, The Hollies, Sedgeford, Whitchurch, SY13 1EX [IO82PX, SJ54]

G4 FJR Details withheld at licensee's request by SSL.

G4 FJT C R Cuthbert, 44 Towse Cl, Clacton on Sea, CO16 8US [JO01NT, TM11]

G4 FJU Details withheld at licensee's request by SSL.[Town: Walsall. Postcode: WS3 1PA. County: West Midlands. National Grid Ref: SK00. WAB Book No: 2891. All QSL via RSGB Bureau.]

G4 FJW C D Hook, 5 Cat Hill, East Barnet, Barnet, EN4 8HG [IO91WP, TQ29]

G4 FJX I M Perera, 1 Francis Rd, Perivale, Greenford, UB6 7AD [IO91UM, TQ18]

G4 FJY A L Ward, 488 Earlham Rd, Norwich, NR4 7HP [JO02PP, TG20]

G4 FKA G P J Plucknett, 32 West Rd, Malden Rushett, Chessington, KT9 2NR [IO91UI, TQ16]

G4 FKB R H Benstead, Outlanes Bungalow, Oulton, Stone, Staffs, ST15 8UU [IO82WW, SJ83]

G4 FKC L West, 5 Fairview Dr, Colkirk, Fakenham, NR21 7NT [JO02KT, TF92]

GM4 FKD D A Smillie, Muirhouse, Daviot Muir, Inverness, Highland, IV1 2ER [IO77WK, NH74]

G4 FKE C H White, 111 Waterbeach Rd, Slough, SL1 3JU [IO91QM, SU98]

G4 FKG G C Kirk, 124 Star Rd, Peterborough, PE1 5HF [IO92VN, TL29]

G4 FKH G Williams, 21 Borda Cl, Chelmsford, CM1 4JY [JO01FR, TL70]

G4 FKI D Thorpe, 70 Willow Way, Ampthill, Bedford, MK45 2SP [IO92SA, TL03]

GW4 FKJ S J H Cotton, Ystradaman House, 127 Pontamman Rd, Ammanford, Dyfed, SA18 2JD [IO81AT, SN61]

G4 FKK Details withheld at licensee's request by SSL.

G4 FKM Details withheld at licensee's request by SSL.

GM4 FKO A Roach, 144 Huron Ave, Howden, Livingston, EH54 6LQ [IO85FV, NT06]

G4 FKP B J Tarry, 6 Beech Gdns, Rainford, St. Helens, WA11 8DJ [IO83OM, SD40]

G4 FKQ M Barnwell, 77 Elmfield Rd, Peterborough, PE1 4HA [IO92VO, TF10]

G4 FKR R E Hammond, Hilden Way, Littleton, Winchester, Hants, SO22 6QH [IO91HC, SU43]

G4 FKS P Auty, 43 Carr Manor Rd, Leeds, LS17 5AW [IO93FU, SE23]

G4 FKU K G Salter, The Pines, 18 Grange Rd, Ellacombe, Torquay, TQ1 1JZ [IO80FL, SX96]

GW4 FKW F Wilde, Ffynonest, Llangenny Ln, Crickhowell, Powys, NP8 1AN [IO81KU, SO21]

G4 FKX R M Abel, 7 Kenilworth Gdns, Southall, UB1 2QL [IO91TM, TQ18]

G4 FKY R J Sharpe, 1 Park Copse, Horsforth, Leeds, LS18 5UN [IO93EU, SE23]

G4 FLA F L Aldridge, 22 Cedar Dr, Clifton, Swinton, Manchester, M27 6WF [IO83TM, SD70]

G4 FLF W Collins, 14 Moss Valley, Alwoodley, Leeds, LS17 7NS [IO93FU, SE24]

G4 FLJ S N Bradshaw, 20 Croftlands, Green Ln, Idle, Bradford, BD10 8RW [IO93DU, SE13]

G4 FLL J M Trotter, 25 Calderwood Cres, Low Fell, Gateshead, NE9 6PH [IO94EW, NZ25]

G4 FLM F E Crofts, 43 Broadlands Dr, East Ayton, Scarborough, YO13 9ET [IO94SG, SE98]

G4 FLN Details withheld at licensee's request by SSL.

GM4 FLP I S Strachan, 238 Coupar Angus Rd, Muirhead By, Muirhead, Dundee, DD2 5QN [IO86LL, NO33]

G4 FLQ J C B Rider, 20 Paradise Ln, Hall Green, Birmingham, B28 0DS [IO92BK, SP08]

G4 FLR D Tanner, 4 Duckpitts Bungalow, Bramling, Canterbury, Kent, CT3 1LY [JO01OG, TR25]

G4 FLS A W Snow, Springfield, Sunnyfield Ln, Up Hatherley, Cheltenham, GL51 6JE [IO81WV, SO92]

G4 FLU E M McQuade, 10 Ilminster Ct, Ouseburn Park, Newcastle upon Tyne, NE3 2QY [IO95DA, NZ26]

G4 FLW F W White, Delamain, Brakens Rise, Paignton, TQ4 6JU [IO80FJ, SX85]

GM4 FLX A T Lovegreen, 16 Grahams Ave, Lochwinnoch, PA12 4EG [IO75QT, NS35]

G4 FLY G R M Haynes, 39 Zinzan St., Reading, RG1 7UG [IO91MK, SU77]

GW4 FLZ H D Fennah, Bryn Mor, 14 Highfield, Hawarden, Deeside, CH5 3LR [IO83LE, SJ36]

G4 FM R H Kelsall, 25 Main St., Cockerham, Lancaster, LA2 0EF [IO83OX, SD45]

G4 FMA K J Fraser, 40 Whitmore Rd, Beckenham, BR3 3NS [IO91XJ, TQ36]

G4 FMB D R Taylor, 3 Howard Dr, Hale, Altrincham, WA15 0LT [IO83UI, SJ78]

G4 FMC M A Constable, 104 Coventry Rd, Coleshill, Birmingham, B46 3EE [IO92DL, SP28]

GI4 FME R J Beattie, 8 The Willows, Coolshinney Rd, Magherafelt, BT45 5HN [IO64QT, H89]

G4 FMG J L J Chapman, 61 Park Rd, Peterborough, PE1 2TH [IO92VN, TL19]

G4 FMH W H Beacham, 9 Fairdean Rd, Highbridge, TA9 3JR [IO81MF, ST34]

G4 FMI F Connor, 29 Parkdale Rd, Paddington, Warrington, WA1 3EN [IO83RJ, SJ68]

G4 FMJ L A Cooke, 23 Widecombe Rd, Birches Head, Hanley, Stoke on Trent, ST1 6SL [IO83WA, SJ84]

G4 FMK A E Smith, 84 St. Clements Hill, Norwich, NR3 4BY [JO02PP, TG21]

G4 FML P Readings, Kilrush, Philpot Ln, Chobham, Woking, GU24 8AP [IO91RI, SU96]

G4 FMM T Walsh, 106 Westgate, Elland, West Yorks, HX5 0BB [IO93BQ, SE12]

G4 FMO C M Palmer, Five Palms, 29 Paget Rise, Abbots Bromley, Rugeley, WS15 3EF [IO92BT, SK02]

G4 FMQ H S Charlesworth, 195 Fylde Rd, Southport, PR9 9XZ [IO83MQ, SD31]

G4 FMS Details withheld at licensee's request by SSL.

G4 FMU J C Clarke, 23 Woodstock Ave, Sutton, SM3 9EG [IO91VJ, TQ26]

G4 FMY D V Larsen, c/o Salbu (Ply) Ltd, Private Bag X 2352, Wingate Park, 0153, Republic of South Africa, X X

G4 FNC L Harper, Three Oaks, Braydon, Swindon, Wilts, SN5 0AD [IO91BO, SU08]

G4 FND D J Yeates, 58 Salisbury Rd, Burton, Christchurch, BH23 7JJ [IO90CR, SZ19]

GM4 FNE B T Spence, Easterhouse, Cullivoe, Yell Island, Shetland Isles, ZE2 9DD [IP90LR, HP50]

G4 FNG R W Walker, South Moor Farm, Langdale End, Scarborough, YO13 0LW [IO94SH, SE99]

G4 FNI K J Nichols, 11 Tregonwell Rd, Bournemouth, BH2 5NR [IO90BR, SZ09]

G4 FNJ P Fuller, 38 Tennyson Rd, Harold Hill, Romford, RM3 7AD [JO01CO, TQ59]

G4 FNK A M Jackson, Bezuidenhoutseweg, 213A 2594 Ak, The Hague, Netherlands

G4 FNL G Bubloz, 42 Hillcrest, Westdene, Brighton, BN1 5FN [IO90WU, TQ20]

GW4 FNO G Lloyd, 15 Budden Cres, Caldicot, Newport, NP6 4PP [IO81OO, ST48]

G4 FNP J R Guite, 15 Marlborough Ave, Falmouth, TR11 2RW [IO70LD, SW83]

G4 FNQ C P Wedgbury, 32 Cloverdale, Stoke Prior, Bromsgrove, B60 4NF [IO82XH, SO96]

G4 FNR D C Rabone, 6 Cranwell Gr, Kesgrave, Ipswich, IP5 7YN [JO02OB, TM24]

GI4 FNU M McDowell, 50 Dunraven Parade, Belfast, BT5 6BT [IO74BO, J37]

GM4 FNV J H Gillard, 5 Mortonhall Park, Ave, Edinburgh, EH17 8BP [IO85JV, NT26]

G4 FNZ D J Bannister, 7 Sudeley Cl, Malvern, WR14 1LP [IO82UD, SO74]

G4 FOB J M Samuels, 14 Edgebury, Woolavington, Bridgwater, TA7 8ES [IO81ME, ST34]

G4 FOC D J Andrews, 18 Downsview Cres, Uckfield, TN22 1UB [JO00BX, TQ42]

G4 FOD T C Yeomans, 15 Turner Rd, Woodfield, Cam, Dursley, GL11 6LT [IO81TQ, ST79]

G4 FOE Details withheld at licensee's request by SSL.

G4 FOH S B Foote, Willowbank, 14 High St., Chrishall, Royston, SG8 8RP [JO02BA, TL43]

GW4 FOI J J Doyle, The Pines, 54 Bryncatwg, Cadoxton, Neath, SA10 8BG [IO81CQ, SS79]

G4 FOJ H H Dyball, Los Altos, 2 Four Acre, Mellor, Blackburn, BB2 7ES [IO83RS, SD63]

GW4 FOK Details withheld at licensee's request by SSL.

G4 FOL J L Bell, 4 Sycamore House, Brookfield Cl, Chineham, Basingstoke, RG24 8RS [IO91LH, SU65]

GW4 FOM R Rowles, 51 Cowbridge Rd West, Cardiff, CF5 5BQ [IO81JL, ST17]

G4 FON R C Goff, 27 Harley Rd, Oxford, OX2 0HS [IO91IS, SP40]

G4 FOQ J W Golightly, 35 Front St., Pity Me, Durham, DH1 5DW [IO94ET, NZ24]

G4 FOR D W Hawkes, 7 Higham Dr, Luton, LU2 9SP [IO91TV, TL12]

G4 FOT H H Exley, 9 Granary Way, Horncastle, LN9 5SR [IO93WF, TF26]

G4 FOW R A Strangeway, 5 Jade Ct, Beaconsfield Rd, Waterlooville, Hants, PO7 7SP [IO90LV, SU60]

G4 FOX G A Griffiths Melton Mowby AR, 11 The Grove, Asfordby, Melton Mowbray, LE14 3UF [IO92MS, SK71]

G4 FOY K Scott, 20 Tower St., Alton, GU34 1NU [IO91MD, SU73]

GM4 FOZ D M Moodie, Taigh An Crannagh, Lageonan Rd, Grandtully, By Aberfeloy, Perthshire, PH15 2QY [IO86CP, NN95]

G4 FP E R Price, 42 Housman Park, School Dr, Bromsgrove, B60 1AZ [IO82XI, SO97]

G4 FPA J E Shorthouse, 20 Boxgrove Rd, Sale, M33 6QW [IO83TK, SJ79]

G4 FPB C W Roper, 3 Dorset Rd, Wallasey, L45 5DB [IO83LK, SJ39]

G4 FPC R A Stone, 48 Cromwell Rd, Winchester, SO22 4AF [IO91IB, SU42]

G4 FPD R A Saunderson, Wildwood, 73 Tinsley Ln, Three Bridges, Crawley, RH10 2AT [IO91WD, TQ23]

G4 FPE G A Butterfield, 137 Chequers Cl, Pontefract, WF8 2TF [IO93IQ, SE42]

G4 FPG R D Hawke, Basque Cl, Hastingleigh, Ashford, Kent, TN25 5JB [JO01LE, TR04]

G4 FPH Details withheld at licensee's request by SSL.

G4 FPI Details withheld at licensee's request by SSL.

G4 FPJ Details withheld at licensee's request by SSL.

G4 FPK Details withheld at licensee's request by SSL.

G4 FPM E G Keeler, 18 Clyde Rd, Worthing, BN13 3LG [IO90SU, TQ10]

G4 FPN K Knowles, 40 Hazelwood Rd, Birmingham, B27 7XP [IO92CK, SP18]

G4 FPO K M Wilson, 14 Stuart Gr, Hut Green, Eggborough, Goole, DN14 0JX [IO93KQ, SE52]

G4 FPP A J Ogg, Clearbury Cottage, Woodgreen, Fordingbridge, Hants, SP6 2QU [IO90DW, SU11]

G4 FPS Details withheld at licensee's request by SSL.

G4 FPV S J Perkins, 6 Delamere Rd, Malvern, WR14 2BQ [IO82UC, SO74]

GW4 FPX J Challenger, 4 Marshfield Rd, Crumlin, Pentwyn Crumlin, Newport, NP1 4JL [IO81KQ, SO20]

G4 FPY K T Jones, 20 Pinecroft Ct, Oakwood, Derby, DE21 2LL [IO92GW, SK33]

GM4 FPZ M E T Lisle, Cromalt, 50 Lade Braes, St. Andrews, KY16 9DA [IO86OI, NO51]

G4 FQA T E W Quinn, 30 Poplar Rd, Rayleigh, SS6 8SL [JO01HN, TQ88]

GM4 FQE E R Thirkell, 20 The Glebe, Crail, Anstruther, KY10 3UJ [IO86QG, NO60]

G4 FQF P R Herring, 34 Woodlands Rd, Romford, RM1 4HD [JO01CO, TQ58]

GM4 FQG R McLaren, Lethendry, North Rd, Dunbar, EH42 1AY [IO86RA, NT67]

G4 FQH B E A Nelmes, Birchgrove, 17 Woodfield Rd, Dursley, GL11 6HB [IO81TQ, ST79]

G4 FQI M J Smith, Wilson Hall Farm, Slade Ln, Melbourne, Derby, DE73 1AG [IO92HT, SK42]

G4 FQN G T Kelly, Holly Gr, 5 Rawlinson Rd, Hesketh Park, Southport, PR9 9LU [IO83MP, SD31]

G4 FQP C R B Bamford, 12 Lincoln Dr, Caistor, Waddington, Lincoln, LN5 9NH [IO93RE, SK96]

G4 FQQ R E Deane, 11 Park Hill, Carshalton, SM5 3RS [IO91WI, TQ26]

G4 FQR D M Jones, 56 Grebe Cres, Horsham, RH13 6ED [IO91UB, TQ13]

G4 FQS J T White, 66 Friars Walk, Southgate, London, N14 5LP [IO91WP, TQ29]

G4 FQT R D Gregory, 33 Southbourne Gr, Hockley, SS5 5EB [JO01IO, TQ89]

GW4 FQU I W Jones, Rhandirmyn, Llandegai, Bangor, Gwynedd, LL57 4LD [IO73WF, SH57]

G4 FQV D M Gray, Auldtownhill Cttgs, Carnousie, Turriff, Aberdeenshire, AB53 4LQ [IO87RN, NJ65]

G4 FQW B Dunn, 17 Duke St., Clayton-le-Moors, Accrington, BB5 5NQ [IO83TS, SD73]

G4 FQX A E Mead, 13 East Dr, Abbey Caravans, Blunsdon, Swindon, SN2 4DP [IO91CO, SU18]

G4 FQZ D H Simms, 1 Old Barn Cl, Morley Ln, Little Eaton, Derby, DE21 5AX [IO92GX, SK34]

G4 FRA V W Lane, 3 Lawnway, York, YO3 0JD [IO93LX, SE65]

G4 FRC R Codling, 17 Briarwood, Golden Bank, Liskeard, PL14 3QQ [IO70SK, SX26]

G4 FRD J E Walton, 2 Billy Mill Ave, North Shields, NE29 0QX [IO95GA, NZ36]

G4 FRE Details withheld at licensee's request by SSL.

G4 FRG B L Goddard, 2 Greenfield Park, Portishead, Bristol, BS20 8NQ [IO81OL, ST47]

GW4 FRH R G Dawkins, 22 Derwen Fawr, Crickhowell, NP8 1DQ [IO81KU, SO21]

G4 FRI G D Bird, Holmwood, 101 Brookfield Rd, Churchdown, Gloucester, GL3 2PN [!O81WV, SO82]

G4 FRJ M R Doyle, Trekelyn, Rezare, Launceston, PL17 9NX [IO70UM, SX37]

G4 FRK J E Rodway, 9 York Ave, Thornton Cleveleys, FY5 2UG [IO83LV, SD34]

G4 FRL N C Ambridge, 53 The Avenue, Chinnor, OX9 4PE [IO91NQ, SP70]

G4 FRM P V Hill, 8 Davenport Park Rd, Davenport Park, Stockport, SK2 6JS [IO83WJ, SJ98]

G4 FRN W T S Hall, Largo, 52 Mizen Way, Cobham, KT11 2RL [IO91TH, TQ16]

G4 FRO G Orford, 10 Laurie Cres, Henleaze, Bristol, BS9 4TA [IO81QL, ST57]

G4 FRP B Buxton, Leef, 320 Wimborne Rd Wes, Uddens Cross, Wimborne,Dorset, BH21 7NN [IO90AT, SU00]

G4 FRQ D M Thomas, 4 Catterick Rd, Colburn, Catterick Garrison, DL9 4QZ [IO94DJ, SE29]

G4 FRR K M Redford, 2 Cressington Cttgs, Westend, Stonehouse, GL10 3SN [IO81US, SO70]

G4 FRS L V Mayhead Farnbrgh&Dist Rd, Halycon, Lawday Link, Upper Hale Farnham, GU9 0BS [IO91OF, SU84]

GW4 FRU Details withheld at licensee's request by SSL.

G4 FRV R C Vincent, 71 Lagoon View, West Yelland, Barnstaple, EX31 3LE [IO71XA, SS52]

G4 FRW M Nicholson, 9 Shelley Ct, Coleport Cl, Cheadle Hulme, Cheadle, SK8 6JH [IO83VI, SJ88]

GW4 FRX J H Nelson, Bank Cottage, Bausley Hill, Crew Green, Shrewsbury, SY5 9BN [IO82MR, SJ31][Station located in Powys.]

G4 FRZ A E Jarrett, 14 Long Copse Ln, Emsworth, PO10 7UL [IO90MU, SU70]

GM4 FSA A Wilson, Allaleckie Farm, Dollar, Clackmannanshire, Scotland, FK14 7NF [IO86ED, NS99]

GM4 FSB G B Millar, 30 Albert Cres, Newport on Tay, DD6 8DT [IO86MK, NO42]

G4 FSC Details withheld at licensee's request by SSL.

G4 FSD J A Creasey, 14 Harlech Dr, Oswaldtwistle, Accrington, BB5 4NW [IO83TR, SD72]

G4 FSE P M Biner, 295 Daws Heath Rd, Rayleigh, SS6 7NS [JO01HN, TQ88]

GM4 FSF K R Horne, 28 Viewfield Terr, Dunfermline, KY12 7HZ [IO86GB, NT08]

G4 FSG P G Murchie, 42 Catherine Rd, Woodbridge, IP12 4JP [JO02PC, TM24]

G4 FSH J Bagnall, Rainow, Timbersbrook, Congleton, Ches, CW12 3PL [IO83WD, SJ86]

G4 FSI K L Taylor, 36 Brooklands Rd, Heath End, Farnham, GU9 9BS [IO91OF, SU84]

G4 FSJ D Whittaker, 152 Cop Ln, Penwortham, Preston, PR1 9AD [IO83PR, SD52]

G4 FSK P H Vallow, 17 Mottram Cl, Ipswich, IP2 9XQ [JO02NA, TM14]

G4 FSM Details withheld at licensee's request by SSL.[Op: J E Phillips, 152 Essex Way, Benfleet, Essex, SS7 1LN.]

G4 FSN E C M Walton, 44 St. Leonards Ave, Lostock, Bolton, BL6 4JE [IO83SN, SD60]

G4 FSQ J P Morley, 65 Longfield Ave, Golcar, Huddersfield, HD7 4BT [IO93BP, SE11]

G4 FSR S R Hall, 7 Howard Cl, Eastern Green, Coventry, CV5 7GN [IO92FJ, SP27]

G4 FSS D W Hocking, 10 Garfit Rd, Kirby Muxloe, Leicester, LE9 2DE [IO92JP, SK50]

G4 FSU I K Greenshields, Grebe House, North Poulner Rd, Ringwood, Hants, BH24 2SW [IO90BT, SU10]

GW4 FSY R Church, Awel y Mor, Dwyran, Llanfairpwll, Anglesey, LL61 6LQ [IO73UD, SH46]

G4 FTA D A Earle, Shenlo, Buckingham Rd, Bicester, OX6 7EX [IO91KV, SP52]

G4 FTC Details withheld at licensee's request by SSL.

GW4 FTF Details withheld at licensee's request by SSL.

G4 FTG C R Norton, 12 Sherbourne Dr, Cox Green, Maidenhead, SL6 3EP [IO91PM, SU87]

G4 FTI A J Bowhill, 9 West Park Dr, East, Leeds, LS8 2EE [IO93FU, SE33]

G4 FTJ R J Eckersley, 2 Berkeley Mews, Dedmere Rise, Marlow, SL7 1XT [IO91ON, SU88]

G4 FTK N S Cridland, 21 Saddleback Way, Fleet, GU13 8UR [IO91OH, SU85]

G4 FTL G C King, 24 Mount View, Barnet Rd, London Colney, St. Albans, AL2 1AT [IO91UR, TL10]

G4 FTN Details withheld at licensee's request by SSL.[Op: P Grainger.]

G4 FTO R W Weber, Hilbre, St. Johns Gdns, Flushing, Falmouth, TR11 5TU [IO70LD, SW83]

G4 FTP E Kraft, 6 The Nook, Wivenhoe, Colchester, CO7 9NH [JO01LU, TM02]

G4 FTQ P Clutterbuck, 1 Starmount Cottage, Horsham Rd, Beare Green, Dorking, Surrey, RH5 4QY [IO91UE, TQ14]

G4 FTW Dr M G M Rowland, The Conifers, 16 Hayter Cl, West Wratting, Cambridge, CB1 5LY [JO02ED, TL65]

G4 FTX G J Knock, 31 Northmead, Ledbury, HR8 1BE [IO82SB, SO73]

G4 FTY R A Page, Mercury House, 19 Green Ln, Coleshill, Birmingham, B46 3NE [IO92DL, SP18]

G4 FTZ B Togwell, 26 Garraways, Wootton Bassett, Swindon, SN4 8LL [IO91BM, SU08]

G4 FUB A D Frost, 5 Cunningham Cl, Bovington, Wareham, BH20 6NL [IO80VQ, SY88]

GI4 FUE C Morrison, 3 Sandringham Park, Carrickfergus, BT38 9EP [IO74CR, J48]

G4 FUF Details withheld at licensee's request by SSL.

G4 FUG P J Clark, 42 Shooters Hill Rd, Blackheath, London, SE3 7BG [JO01AL, TQ47]

G4 FUH J H Stace Scunthorpe ARC, 38 Skippingdale Rd, Scunthorpe, DN15 8NU [IO93QO, SE81]

G4 FUI M J Rigby, 16 Juniper Way, Heights, Penrith, CA11 8UF [IO84PP, NY52]

G4 FUJ G Wright, 9 Lypiatt St., Tivoli, Cheltenham, GL50 2UA [IO81WV, SO92]

GM4 FUL D S Milne, 19 Craigston Ave, Park, Ellon, AB41 9JW [IO87XI, NJ93]

GI4 FUM Dr W D Hutchinson, 40 Oldstone Hill, Muckamore, Antrim, BT41 4SB [IO64VQ, J18]

G4 FUN Details withheld at licensee's request by SSL.

G4 FUO J L Nowell, 50 Hudson Way, Wighill Ln, Tadcaster, LS24 8JF [IO93IV, SE44]

G4 FUP N M Braeman, 17 Michelgrove Rd, Boscombe, Bournemouth, BH5 1JH [IO90BR, SZ19]

G4 FUR Details withheld at licensee's request by SSL.[(Coulsdon Amateur Transmitting Society.)]

G4 FUS D W Fletcher, 10 Mildred Rd, Walton, St, BA16 9QP [IO81OC, ST43]

G4 FUT J J Dougherty, Firholme, Folly Top, Eggleston, Barnard Castle, DL12 0DH [IO94AO, NZ02]

G4 FUU M W Pothecary, 34 Hartscroft, Linton Glade, Addington, Croydon, CR0 9LB [IO91XI, TQ36]

G4 FUY P D Bonson, 162 Nine Mile Ride, Finchampstead, Wokingham, RG40 4JA [IO91NI, SU76]

G4 FUZ Details withheld at licensee's request by SSL.

G4 FVA P O Catling, 11 Trinity Cl, Fordham, Ely, CB7 5PB [JO02EH, TL67]

G4 FVC R S Terry, 85 Lake Rise, Romford, RM1 4EF [JO01CO, TQ59]

G4 FVD J A Newport, 10 Poolmans Rd, Windsor, SL4 4PA [IO91QL, SU97]

G4 FVF R J Earnshaw, 18 Fen Gr, Blackfen, Sidcup, DA15 8QN [JO01BK, TQ47]

G4 FVG M R T Hannan, Pentire, 29 Yarnhams Cl, Four Marks, Alton, GU34 5DH [IO91LC, SU63]

G4 FVI M L Willis, 188 Shorncliffe Rd, Folkestone, CT20 3PH [JO01NC, TR23]

G4 FVJ P Ellis Kodak ARS, Research Division, Kodak Ltd, Headstone Dr Harrow, HA1 4TY [IO91TO, TQ18]

G4 FVK D J Sewell, 11 Haddon Cl, Stanground, Peterborough, PE2 8LS [IO92VN, TL29]

G4 FVL G M Rankin, 25 The Chase, Coulsdon, CR5 2EJ [IO91WH, TQ26]

GI4 FVM J S Edgar, 13 Glendarragh, Belfast, Co. Antrim, BT4 2WB [IO74BO, J37]
G4 FVN Details withheld at licensee's request by SSL.
GM4 FVO C J Evans, East Cottage, Mount Melville, St Andrews, Fife, KY16 8NT [IO86NH, NO41]
G4 FVP C L Davies, 28 Neville Rd, Darlington, DL3 8HY [IO94FM, NZ21]
GM4 FVS G W Cusiter, 4 Elphin Hill, Ellon, AB41 8BH [IO87XI, NJ93]
G4 FVU A F Sweetapple, Bent Oak, Axminster Rd, Musbury, Axminster, EX13 6AQ [IO80LR, SY29]
G4 FVV B Vincent, 27 Naseby Walk, Leeds, LS9 7SY [IO93FT, SE33]
G4 FVW D A Hooper, 8 Barn Cl, Crewkerne, TA18 8BL [IO80OV, ST40]
G4 FVX R W Johns, 42 Lansdown Rd, Redland, Bristol, BS6 6NS [IO81QL, ST57]
G4 FVZ Dr P L Gould, Rowan House, Mill Ln, Mickleton Barnard, Castle Co Durham, DL12 0RG [IO84XO, NY92]
G4 FWC R E Deakin Tamworth A.R.S., 12 Henley Cl, Perrycrofts, Tamworth, B79 8TQ [IO92DP, SK20]
G4 FWF A P Gurney, 3 Morden Rd, Sandford, Wareham, BH20 7AA [IO80WQ, SY98]
G4 FWG Details withheld at licensee's request by SSL.[Op: M J Sheppard, 7 Beacon Gardens, Crowborough, E Sussex, TN6 1BD.]
G4 FWH G Barrett, 128 Seymour Dr, Ancaster, Ontario, Canada, L9G 4N6
G4 FWI D L Sumner, Corner House, Coulton, Hovingham, York, YO6 4NH [IO94LD, SE67]
GI4 FWK P Mooney, 57 Johnstown Rd, Dun Laoghaire, Co Dublin, Republic of Ireland
G4 FWM C A Webb, 5 Leyburn Ave, Fleetwood, FY7 8HG [IO83LV, SD34]
G4 FWN N A May, Sandock Nurseries, Middle Dimson, Gunnislake, Cornwall, PL18 9NG [IO70VM, SX47]
G4 FWP C L Longworth, 21 Westland Ave, Bolton, BL1 5NP [IO83SO, SD61]
GD4 FWQ C Matthewman, 26 King Orry Rd, Marown, Glen Vine, Douglas, IM4 4ES
G4 FWR A Johnson, 86 Meadow Cl, Thatcham, RG19 3RL [IO91IJ, SU56]
G4 FXA V Arnold, 435 Manchester Rd, Clifton, Swinton, Manchester, M27 6WH [IO83TM, SD70]
G4 FXB Details withheld at licensee's request by SSL.
G4 FXE Details withheld at licensee's request by SSL.
GW4 FXF G E Swan, Long Acre, New Rd, Aberdulais, Neath, SA10 8HT [IO81CQ, SN70]
G4 FXG R F Redhead, 7 Brocklewood Ave, Poulton-le-Fylde, FY6 8BZ [IO83MT, SD33]
G4 FXI P H Overell, 48 Bedgrove, Aylesbury, HP21 7BD [IO91OT, SP81]
G4 FXJ F D Pratt, Evescote, School Ln, Burghfield Common, Reading, RG7 3ES [IO91LJ, SU66]
GM4 FXL A W Docherty, 10 Dumyat Rd, Menstrie, FK11 7DG [IO86BD, NS89]
G4 FXM G H Farnie, 16 Stratford Rd, Sandy, SG19 2AB [IO92UC, TL14]
G4 FXQ C D L Williams, 92 Beechwood Gdns, Ilford, IG5 0AQ [JO01AN, TQ48]
G4 FXR W G R Wunderlich, Automotive Eng Serv, 31 College Rd, Bromley, BR1 3PU [JO01AJ, TQ46]
G4 FXT N D Burkitt, 31 Loxwood, Earley, Reading, RG6 5QZ [IO91MK, SU77]
G4 FXU R J Napper, 12 Brumell Dr, Lancaster Park, Morpeth, NE61 3RB [IO95DE, NZ18]
G4 FXW Details withheld at licensee's request by SSL.
GM4 FXX H G Brooks, 26 Alloa Rd, Causewayhead, Stirling, FK9 5LN [IO86AD, NS89]
G4 FXY P C Staton, 52 School Rd, Newborough, Peterborough, PE6 7RG [IO92VP, TF20]
G4 FYB Details withheld at licensee's request by SSL.
G4 FYD K S English, The Beeches, 40 Aigburth Hall Ave, Liverpool, L19 9EB [IO83NI, SJ38]
G4 FYE G F Coggon, 45 Ansten Cres, Cantley, Doncaster, DN4 6EZ [IO93LM, SE60]
G4 FYG M J Newlands, 44 Orchard Dr, Tonbridge, TN10 4LG [JO01DF, TQ64]
G4 FYI T M Fallick, 34 Fieldfare Rd, Newport, PO30 5FH [IO90IQ, SZ48]
G4 FYJ J J Lemon, 30 Iveagh Ct, Farm Hill, Exwick, Exeter, EX4 2LR [IO80FR, SX99]
G4 FYL D P Godfrey, 3 Honiton Way, Bedford, MK40 3AN [IO92SD, TL05]
G4 FYM D G Wiggs, 57 Alpine St., Reading, RG1 2PY [IO91MK, SU77]
G4 FYO T B Foley, 16 Buckingham Rd, Winslow, Buckingham, MK18 3DY [IO91NW, SP72]
G4 FYP J Roney, 76 Hilton Ln, Great Wyrley, Walsall, WS6 6DT [IO82XP, SJ90]
G4 FYQ M D Robins, 36 Wolverley Ave, Wollaston, Stourbridge, DY8 3PJ [IO82VL, SO88]
G4 FYR L Porter, 19 Harwood St., New Bradwell, Milton Keynes, MK13 0EH [IO92OB, SP84]
G4 FYS P Grimshaw, 55 Combe St. Ln, Yeovil, BA21 3PD [IO80QW, ST51]
G4 FYT D D Lawrence, 23 Parkmead Rd, Wyke Regis, Weymouth, DT4 9AL [IO80SO, SY67]
G4 FYW J V Coles-Macgregor, Tallyho Eyhurst Ave, Elm Park Hornchurch, Essex, RM12 4RA [JO01CN, TQ58]
G4 FYX P F Hartigan, Doonagore, Doolin, Co. Clare, Ireland, X X
G4 FYY P J Head, 97 Malthouse Rd, Southgate, Crawley, RH10 6BJ [IO91VC, TQ23]
G4 FZB J C Woodrow, 53 Burford Rd, Carterton, OX18 3AQ [IO91ES, SP20]
GI4 FZD P A Menown, 34 Cairnburn Rd, Belfast, BT4 2HS [IO74BO, J37]
G4 FZE Details withheld at licensee's request by SSL.
G4 FZG B Sirignano, Eversholt, 22 Cleevelands Dr, Cheltenham, GL50 4QB [IO81XV, SO92]
G4 FZH Dr C V Smith, Grey Gables, Humphrey Gate, Taddington, Buxton, SK17 9TS [IO93CF, SK17]
G4 FZJ C Williamson, 57 Stuart St., Thurnscoe East, Thurnscoe, Rotherham, S63 0ED [IO93IN, SE40]
G4 FZK G J Jackson, 68 Commerce St., Melbourne, Derby, DE73 1FT [IO92GT, SK32]
G4 FZL L B Povoas, 43 Hilders Rd, Western Park, Leicester, LE3 6HE [IO92JP, SK50]
G4 FZM M R Davey, Hyaway, Hill Green, Clavering, Saffron Walden, CB11 4QS [JO01BX, TL43]
G4 FZN J C Kirby, 2 Kneeton Park, Middleton Tyas, Richmond, DL10 6SB [IO94EK, NZ20]
G4 FZP A J Drury, 21 Kentmere Cl, Potters Green, Coventry, CV2 2GE [IO92GK, SP38]
G4 FZR K R Dally, Ealand Grange, Ealand, nr Scunthorpe, S Humberside, DN17 4DG [IO93OO, SE81]
G4 FZS H S Bulmer, Searchlight, Claremont Rd, Mount Pleasant, Newhavan, BN9 0NQ [JO00AT, TQ40]
GM4 FZT D Dempster, Belgrove, Culbokie, Dingwall, Ross-shire, IV7 8JY [IO77UO, NH65]
G4 FZV Dr P Redall, 106 Stowey Rd, Yatton, Bristol, BS19 4EB [IO81OJ, ST46]
G4 FZY J M Turner, 26 Clydesdale Gdns, Richmond, TW10 5EF [IO91UL, TQ17]
G4 FZZ D W J Holmes, 13 Park Ln, Norwich, NR2 3EE [JO02PP, TG20]
G4 GAB R M Padbury, 8 Osbourne Dr, Holton-le-Clay, Grimsby, DN36 5DS [IO93XM, TA20]
G4 GAC G E G Fairbrass obo Glenfield ATC 2070 Sqn, 58 Sonning Way, Glen Parva, Leicester, LE2 9RU [IO92KN, SP59]
G4 GAD Details withheld at licensee's request by SSL.
GW4 GAF A B McCann, Lower Fiddlers Gree, Felindre Knighton, Powys, Wales, LD7 1YT [IO82JK, SO18]
G4 GAI K M Taylor, 31 Stonehill Dr, Rochdale, OL12 7JN [IO83VP, SD81]
G4 GAJ M E Adams, 25 Allenfield Rd, Cheltenham, GL53 0LX [IO81WV, SO92]
G4 GAK M P Sykes, 1 Springhall Dr, Halifax, HX2 0BH [IO93BR, SE02]
G4 GAN F J Knight, 49 Middle Deal Rd, Deal, CT14 9RG [JO01QF, TR35]
G4 GAP H B Fitzherbert, 36 Westover Rd, Broadstairs, CT10 3ES [JO01RI, TR36]
G4 GAR J H Cook, 28 Hill Burn, Bristol, BS9 4RH [IO81QL, ST57]
G4 GAS Details withheld at licensee's request by SSL.[Op: B F Shaw.]
G4 GAT B F Denton, 2 Seacroft Rd, Broadstairs, CT10 1TL [JO01RI, TR36]
G4 GAU Details withheld at licensee's request by SSL.
G4 GAX N S B Martin, 17 Mynn Cres, Bearsted, Maidstone, ME14 4AR [JO01GG, TQ75]
G4 GBA C B Brookson, The St., Aspall, Stonham Aspal, Stowmarket, Suffolk, IP14 6AJ [JO02NE, TM15]
G4 GBB W F Mark, Woodlands, 6 Tormore Park, Deal, CT14 9UY [JO01QF, TR35]
G4 GBC F P Orchard, 39B Breach Rd, Marlpool, Heanor, DE75 7NJ [IO93HA, SK44]
G4 GBE R Blacker, Karingal, 6 Main Rd, West Keal, Spilsby, PE23 4BE [JO03AD, TF36]
G4 GBF K Watson, 34 Hermitage Park, Chester-le-Street, DH3 3JZ [IO94FU, NZ25]
G4 GBG Details withheld at licensee's request by SSL.
G4 GBH J Benson, 2 Saxon Walk, Lichfield, WS13 8AJ [IO92BQ, SK10]
G4 GBI A M Edwards, 96 Bathurst Rd, Winnersh, Wokingham, RG41 5JF [IO91NK, SU77]
G4 GBK C D Appleton, 249 Devonshire Rd, Atherton, Manchester, M46 9QB [IO83SM, SD60]
G4 GBM J M Wolfenden, Mallards, 14 The Island, Wraysbury, Staines, TW19 5AS [IO91RK, TQ07]
G4 GBN G W Canning, 10 Mill Cl, East Coker, Yeovil, BA22 9LF [IO80QV, ST51]
G4 GBR P B Parnaby, Icknield, Aylesbury Rd, Princes Risborough, HP27 0JW [IO91OR, SP80]
G4 GBS T R Calvert, 16 Baulk Ln, Harworth, Doncaster, DN11 8PE [IO93LK, SK69]
G4 GBT I T Coleman, 12 Headington Cl, Bradwell, Great Yarmouth, NR31 8DN [JO02UN, TG50]
G4 GBV E A Jeffries, 62 Cuckoo Dene, Hanwell, London, W7 3DR [IO91TM, TQ18]
G4 GBW Dr J J Wilcox, 533 Upper Brentwood Rd, Gidea Park, Romford, RM2 6LD [JO01CO, TQ58]
G4 GBX W Greed, 18 Nursteed Park, Devizes, SN10 3AN [IO91AI, SU06]
G4 GBZ C R H George, 22 Pinewood Rd, Hordle, Lymington, SO41 0GP [IO90ES, SZ29]
GW4 GCB Details withheld at licensee's request by SSL.[Station located near Colwyn Bay.]
G4 GCD Details withheld at licensee's request by SSL.
G4 GCE W R Godden, Wayfield Cottage, The Clump, Rickmansworth, WD3 4BG [IO91SP, TQ09]
G4 GCI N T Palmer, 17 Heathlands, Shedfield, Southampton, SO3 2JD [IO90HW, SU41]
G4 GCJ F Fuller, 7 Prestwick Cl, Bletchley, Milton Keynes, MK3 7RQ [IO91OX, SP83]
G4 GCL J Tyler, 32 Fountain Dr, Roberttown, Liversedge, West Yorks, WF15 7PX [IO93DQ, SE12]
G4 GCM Details withheld at licensee's request by SSL.[Op: R A Henshaw, c/o 18 Hogarth Close, Woodley, Romsey, Hants, SO51 7TF.]
G4 GCP N M Sheard, 2 Broadcroft Way, Tingley, Wakefield, WF3 1TT [IO93FR, SE22]
G4 GCQ R H Thomas, 7 Sandringham Cl, Rushden, NN10 9ER [IO92QG, SP96]
G4 GCT K J Ottrey Nth Bristol ARC, 51 Priory Ct Rd, Westbury on Trym, Bristol, BS9 4DB [IO81QL, ST57]
G4 GCU Z Kowalczyk, 6 St. Georges Cres, New Marske, Redcar, TS11 8BT [IO94LN, NZ62]
G4 GCV E M Spector, 4 Solsbury Way, Bath, BA1 6HH [IO81TJ, ST76]
G4 GCW D J Morrison, 5 Cowdry Cl, Thornhill, Dewsbury, WF12 0LW [IO93EP, SE21]
G4 GCX Details withheld at licensee's request by SSL.
G4 GD N G V Anslow, 56 Wolsey Rd, East Molesey, KT8 9EW [IO91TJ, TQ16]
G4 GDC S A Wiles, Conifers, Aisthorpe, Scampton, Lincoln, LN1 2SG [IO93RH, SK98]
GM4 GDF J M C G Cain, 24 Agnew Cres, Wigtown, Newton Stewart, DG8 9DT [IO74SU, NX45]
G4 GDK K J Taylor, 27 Hazlemere Rd, Thundersley, Benfleet, SS7 4AF [JO01GN, TQ78]
G4 GDL M I Ellis, 32 Pegholme Dr, Bradford Rd, Otley, LS21 3NZ [IO93DV, SE14]
GW4 GDM J S Owens, yr Hafan I Maes Gyn, An Llanarmon-Yn-lal, Mold, Clwyd N Wales, CH7 4PY [IO83JC, SJ15]
G4 GDP J OShea, 108 Orme Rd, Kingston upon Thames, KT1 3SB [IO91UJ, TQ26]
G4 GDQ R H E Found, 54 Lincoln Rd, North Hykeham, Lincoln, LN6 8HB [IO93RE, SK96]
G4 GDR Rev A Heath, 227 Windrush, Highworth, Swindon, SN6 7EB [IO91DP, SU29]
G4 GDS D A Jones, 3 Kingfisher Dr, Benfleet, SS7 5ES [JO01GN, TQ78]
G4 GDT D L Wood, 20 Varndean Gdns, Brighton, BN1 6WL [IO90WU, TQ30]
G4 GDU I S Hoskin, 14 Trevingey Parc, Redruth, TR15 3BZ [IO70JF, SW64]
G4 GDX I G Smith, 25 Windrush Ave, Brickhill, Bedford, MK41 7BS [IO92SD, TL05]
G4 GDY M R Edwards, 9 Earls Walk, Binley Woods, Coventry, CV3 2AJ [IO92GJ, SP37]
G4 GDZ Details withheld at licensee's request by SSL.
G4 GED D F Richardson, 92 Betham Rd, Greenford, UB6 8SA [IO91TM, TQ18]
G4 GEE Dr R J Nash, 135 Farren Rd, Wyken, Coventry, CV2 5EH [IO92GK, SP38]
G4 GEI P F Newall, 102 Park Ln, Castle Donnington, Castle Donington, Derby, DE74 2JG [IO92HU, SK42]
G4 GEK R F W Badcock, PO Box 15440, Farrarmere 1518, Gauteng, South Africa
G4 GEL R A Penn, 46 Ladbroke Dr, Potters Bar, EN6 1QR [IO91VQ, TL20]
G4 GEM Details withheld at licensee's request by SSL.
G4 GEN A R Morriss, Pippingford Prk Man, Nutley, Sussex, TN22 3HW [JO01AB, TQ43]
G4 GEO C G Tomkinson, 214 Bramhall Ln South, Bramhall, Stockport, SK7 3AA [IO83WI, SJ88]
G4 GEP V R T Peake, 25 Neales Cl, Harbury, Leamington Spa, CV33 9JQ [IO92GF, SP35]
G4 GER Details withheld at licensee's request by SSL.
G4 GES S M Jordan, 70 Hungerhill Rd, Kimberworth, Rotherham, S61 3NP [IO93GK, SK39]
G4 GET I A Jordan, 70 Hungerhill Rd, Kimberworth, Rotherham, S61 3NP [IO93GK, SK39]
G4 GEU J C Terry, 126 Dawberry Fields Rd, Birmingham, B14 6NZ [IO92BK, SP08]
G4 GEV E H Otten, 1 High St., Wellow, Bath, BA2 8QQ [IO81TH, ST75]
G4 GEW P W Lee, 190 Chaldon Way, Coulsdon, CR5 1DH [IO91WH, TQ35]
G4 GEY J S Carter, 30 Braemar Rd, Hazel Gr, Stockport, SK7 4QG [IO83WJ, SJ98]
G4 GEZ R W Evans, Redcote End, West Common, Harpenden, AL5 2JW [IO91TT, TL11]
G4 GFA F A H Dell, 2 Carterswood Dr, Nuthall, Nottingham, NG16 1AS [IO92JX, SK54]
G4 GFB W R Reason, Kellita, 14 Willow Dr, Camborne, TR14 7HR [IO70IF, SW64]
G4 GFC S C Wright, 163 Croham Valley Rd, South Croydon, CR2 7RE [IO91XI, TQ36]
G4 GFD A J Gilman, 10 Hanwell Cl, Pennington, Leigh, WN7 3NU [IO83RL, SJ69]
G4 GFE D H Foulds, Comms Branch, (Tels Divn), HQ Baor, BFP0 14O
G4 GFH G S Smith, 3 Carters Cl, Clacton on Sea, CO16 7AT [JO01NT, TM11]
G4 GFI M V L Broadway, 91 Tattenham Gr, Epsom Downs, Epsom, KT18 5QT [IO91VH, TQ25]
G4 GFJ L W J Frankham, 47 St. Marys Gdns, Hilperton Marsh, Trowbridge, BA14 7PH [IO81VI, ST86]
G4 GFK Details withheld at licensee's request by SSL.
GW4 GFL J B Davies, Abergavenny Rd Soc, 109 Croesonnen Parc, Abergavenny, NP7 6PF [IO81LT, SO31]
G4 GFM D R Hessom, 89 Pond Cl, Overton, Basingstoke, RG25 3LZ [IO91IF, SU54]
G4 GFN S N Dabbs, 9 Windover Cl, Bitterne, Southampton, SO19 5JS [IO90HV, SU41]
G4 GFQ R A W Sheppard, 48 Tennyson Ave, Thornton Cleveleys, FY5 2ET [IO83LV, SD34]
GM4 GFR S S Smith, Bank Pl, 67 Cash Feus, Strathmiglo, Cupar, KY14 7QP [IO86IG, NO20]
GW4 GFS H R Jones, Hafod y Grug, Mynytho, Pwllheli, Gwynedd, LL53 7RF [IO72RU, SH33]
G4 GFT Dr V Leach, Lakehead Cottage, Wellow Ln, Rufford, Newark Notts, NG22 9DG [IO93LE, SK66]
G4 GFU G H Costin, The Rosary, London Rd, Ryarsh, West Malling, ME19 5AW [JO01EH, TQ65]
G4 GFV J T Simpson, 19 Hollinside Cl, Whickham, Newcastle upon Tyne, NE16 5QZ [IO94DW, NZ26]
G4 GFX S Bevan, 41 St. Andrews Rd, Malvern, WR14 3PT [IO82UC, SO74]
G4 GFY P L King, 78 Gweal Wartha, Helston, TR13 0SN [IO70IC, SW62]
GW4 GFZ S R Dunkerley, PO Box Hm 2215, Hamilton Hmjx, Bermuda, ZZ9 9PO
GJ4 GG C D S Wintle, Ivy-Stone House, Rue de La Croix, St Clement, Jersey, Channel Islands, JE2 6LQ
G4 GGB D E E Owen, 48 Ridgemere Rd, Pensby, Wirral, L61 8RW [IO83KI, SJ28]
G4 GGC M J Marsh, 21 Stour Gdns, Great Cornard, Sudbury, CO10 0JN [JO02IA, TL83]
G4 GGE D T Nicholson, 41 Thurstons Barton, Whitehall, Bristol, BS5 7BQ [IO81RL, ST67]
GM4 GGF V Mason, 19 Sherwood Cres, Bonnyrigg, EH19 3LQ [IO85KU, NT36]
G4 GGI R Williamson, Acacia Cottage, 21 Onslow Rd, Burwood Park, Walton on Thames, KT12 5BB [IO91SI, TQ06]
G4 GGK K Lewis, 10 Woodland Rd, Rushden, NN10 6UT [IO92QH, SP96]
G4 GGL T Grainger, 34 Maple Ave, Ripley, DE5 3PY [IO93HB, SK34]
G4 GGO A C Beresford, 1258 Greenbriar Way, North Vancouver BC, Canada, V7R 1L9, ZZ1 2GR
G4 GGP J H Saynor, 28 Lune Rd, Norton, Stockton on Tees, TS20 1AZ [IO94IO, NZ42]
G4 GGR F S Gemmell, 89 Coach Rd, Guiseley, Leeds, LS20 8AY [IO93DU, SE14]
G4 GGS Details withheld at licensee's request by SSL.
G4 GGT M R Masterson, 44 Highstone Ave, London, E11 2PP [JO01AN, TQ48]
G4 GGV V R Moll, 1 Clifton Cl, Braywick Rd, Maidenhead, SL6 1DF [IO91PM, SU87]
G4 GGX S F Randall, 66 Park Ct, Harlow, CM20 2PZ [JO01BS, TL41]
G4 GGZ J R Birch, 13 Alison Way, Aldershot, GU11 3JX [IO91OF, SU85]
G4 GHA J Cleaton, 1 Avon Dr, Northmoor Park, Wareham, BH20 4EL [IO80WQ, SY98]
G4 GHB B Kitchen, 73 Birch St., Ashton under Lyne, OL7 0JD [IO83WL, SJ99]
G4 GHC G H Collis, 3 Summer Dr, Wirksworth, Matlock, DE4 4EL [IO93FB, SK25]
G4 GHD Details withheld at licensee's request by SSL.
G4 GHH Details withheld at licensee's request by SSL.
G4 GHJ Details withheld at licensee's request by SSL.
G4 GHK J Donovan, 6 Manor Pl, Church, Accrington, BB5 4DX [IO83TS, SD72]
G4 GHL Dr M C L Ward, 9 Woodshears Dr, Malvern, WR14 3EA [IO82UC, SO74]
G4 GHO S A Webb, 9 Fifth Ave, Chelmsford, CM1 4HB [JO01FR, TL70]
G4 GHP B J Gordon, 113 Pound Ln, Oakdale, Poole, BH15 3RS [IO90AR, SZ09]
G4 GHQ P R Fisher, 95 Slaithwaite Rd, Thornhill Lees, Dewsbury, WF12 9DN [IO93EQ, SE21]
G4 GHR D F R Humphreys, 64 Holne Chase, Plymouth, PL6 7UB [IO70WK, SX56]
G4 GHS H G Cohen, 41 South Station Rd, Gateacre, Liverpool, L25 3QE [IO83NJ, SJ48]
G4 GHT M R Skyner, Grassendale, Newcastle Rd, Smallwood, Sandbach, CW11 2UB [IO83UD, SJ76]
G4 GHU Details withheld at licensee's request by SSL.
G4 GHX J A Phipps, 133 Slad Rd, Stroud, GL5 1RD [IO81VR, SO80]
G4 GHZ P S Collins, 18 Linksway, Hendon, London, NW4 1JR [IO91VO, TQ29]
GI4 GID J Heasley, 1 Cairnmore Park, Lisburn, BT28 2DN [IO64XM, J26]
GM4 GIF Details withheld at licensee's request by SSL.
G4 GIG J E Mullany, 4 Delamere Ct, Highfield Rd, Hall Green, Birmingham, B28 0HN [IO92BK, SP18]
GM4 GIH W Goldie, 5 Laurel Gdns, Uddingston, Glasgow, G71 6SD [IO75XT, NS66]
G4 GIK D J Chambers, Orchard House, Woodstock Cres, Dorridge Solihull, West Midlands, B93 8DA [IO92CJ, SP17]
GM4 GIL A Marshall, 17 Ochil View, Kincardine, Alloa, FK10 4QG [IO86DB, NS98]
G4 GIM B T Waters, 60 Whitewood Way, Whittington, Worcester, WR5 2LN [IO82VE, SO85]
GM4 GIO R M Marshall, 15 Craigleith Hill, Edinburgh, EH4 2EF [IO85JW, NT27]
G4 GIQ R D Marshall, 5 Wilson Cres, Lostock Gralam, Northwich, CW9 7QH [IO83SG, SJ67]
G4 GIR I C Frith, 50 Rowallan Dr, Putnoe, Bedford, MK41 8AS [IO92SD, TL05]
G4 GIS J A Darbyshire, The Nook, 7 Sandle Rd, Bishops Stortford, CM23 5HY [JO01CU, TL42]
G4 GIT L Swaine, 47 Springdale Cres, Idle, Bradford, BD10 9QR [IO93DU, SE13]
G4 GIV R L Howarth, 91 Armadale Rd, Ladybridge, Bolton, BL3 4PB [IO83SN, SD60]
G4 GIX T M Kearns, East Lodge, Priorswood, Compton, Guildford, GU3 1DR [IO91QF, SU94]
G4 GIY R J Harris, 303 Northgate, Cottingham, HU16 5RL [IO93TS, TA03]
G4 GIZ E Waddington, 18 Barnwood Rd, Earby, Colne, BB8 6PB [IO83WV, SD94]
G4 GJA K P Austen, 101 Ufton Ln, Sittingbourne, ME10 1JA [JO01II, TQ96]
G4 GJB G J Bond, Wolvelay House, Woolley, Wakefield, W Yorks, WF4 2JJ [IO93FO, SE31]
G4 GJC N G Smith, 64 Ventress Farm Ct, Cambridge, CB1 4HD [JO02CE, TL45]
G4 GJE D Davis, 6 Regina Dr, Walsall, WS4 2HB [IO92AO, SP09]
GM4 GJG A N King, 47 Greenbank Cres, Edinburgh, EH10 5TD [IO85JW, NT27]
G4 GJH L Milner, West Grange Farm, Shincliffe, Co Durham, DH1 2TB [IO94FS, NZ23]
G4 GJL Details withheld at licensee's request by SSL.
G4 GJN Details withheld at licensee's request by SSL.
G4 GJO D C Blampied, 113 Green St., Enfield, EN3 7JF [IO91XP, TQ39]
G4 GJP Details withheld at licensee's request by SSL.
G4 GJR T J Aldridge, 6 St. Margarets Cl, Newport Pagnell, MK16 9EF [IO92PC, SP84]
G4 GJS W C Owens, Rothenbach Bp 42, Vlodrop, Prov Limburg, Netherlands 6063 Nm, ZZ4 2RO
GW4 GJT H C Taylor, 2 Pen y Dre, Highfields, Caerphilly, CF83 2NZ [IO81JO, ST18]
G4 GJU P G O Moxham, 233 Walsall Rd, Aldridge, Walsall, WS9 0QA [IO92AO, SP09]
G4 GJV A J Horne, 14 Penton Hook Rd, Staines, TW18 2PF [IO91RK, TQ07]
G4 GJW Details withheld at licensee's request by SSL.[Op: F M Wall. Station located in London E4.]
G4 GJX W Weir, 40 King Richard Dr, Bearwood, Bournemouth, BH11 9PE [IO90AS, SZ09]
G4 GJY S Simmonds, 14 Lindsey Cres, Kenilworth, CV8 1FL [IO92FH, SP27]
G4 GKA Details withheld at licensee's request by SSL.
G4 GKB A J Crofts, Woodside Cottage, Grenadier Rd, Ash Vale, Aldershot, GU12 5DT [IO91PG, SU85]
G4 GKC C G Willoughby, 79 Liskeard Rd, Walsall, WS5 3ES [IO92AN, SP09]
G4 GKD C F Lloyd, 9 Lime Kiln, Wootton Bassett, Swindon, SN4 7HF [IO91BN, SU08]
G4 GKE D J Archer, 121 Parliament Rd, Ipswich, IP4 5EP [JO02OB, TM14]
G4 GKG H C Chadwick, Trees Thurstaston Rd, Thurstaston, Wirral, Merseyside, L61 0HG [IO83KI, SJ28]
G4 GKH G R Duke, 3 Thurstable Rd, Tollesbury, Maldon, CM9 8SF [JO01KS, TL91]
G4 GKK A C Hawkins, 101 Tobyfield Rd, Bishops Cleeve, Cheltenham, GL52 4NZ [IO81XW, SO92]
G4 GKO R Roden, 27 Wilmington Cl, Hassocks, BN6 8QB [IO90WW, TQ31]
G4 GKQ E Trethewey, 60 Norwich Ave, Whitleigh, Plymouth, PL5 4JQ [IO70WK, SX46]
G4 GKR A L Briscoe, 63 Moor Ln, Crosby, Liverpool, L23 2SG [IO83LL, SD30]
G4 GKT F Delaney, 6 Stour Rd, Astley, Tyldesley, Manchester, M29 7HH [IO83SM, SD70]

G4 GKW F A Harwood, 1 Waverley Ct, St. Leonards on Sea, TN37 6QR [JO00GU, TQ80]
G4 GKX J L W Trevett, 6 Towers Way, Corfe Mullen, Wimborne, BH21 3UB [IO80XS, SY99]
G4 GKY C S Williams, 12A Parc An Dix Ln, Phillack, Hayle, TR27 5AB [IO70HE, SW53]
G4 GKZ R Revill, 2 Osborne Cres, Stafford, ST17 0AD [IO82WT, SJ92]
G4 GLB Details withheld at licensee's request by SSL.
G4 GLC D G Hamilton, Rome Lea, Four Ln Ends, Giggleswick, Settle, BD24 0AG [IO84UB, SD86]
GM4 GLD P M Coull, 66 Pitfour Ct, Peterhead, AB42 2YG [IO97CM, NK14]
GM4 GLE R J Cuthbert, 4 Peveril Ave, Burnside, Rutherglen, Glasgow, G73 4RD [IO75VT, NS66]
G4 GLF C N Davies, 2 St. Andrews Gr, Hatfield, Doncaster, DN7 6SW [IO93LN, SE60]
G4 GLG C D G Edwards, Cross House, The Cross, Bilsington, Ashford, Kent, TN25 7JX [JO01KB, TR03]
G4 GLH D G Bennett, 3 Beechacre, Ramsbottom, Bury, BL0 9LS [IO83UP, SD71]
G4 GLI M J King, 8 Lexington Cl, Hemsby, Great Yarmouth, NR29 4ES [JO02UQ, TG41]
G4 GLJ N Grundy, 4 Beechwood Cres, Astley, Tyldesley, Manchester, M29 7AH [IO83SM, SD60]
G4 GLL G D Rodwell, The Dadpad, Brooklands, Willitoft, Goole, DN14 7NY [IO93NT, SE73]
G4 GLM Dr G L Manning, 63 The Dr, Edgware, HA8 8PS [IO91UO, TQ19] [Special interest: collection of aircraft equipment and instruments. Visitors welcome by appointment. Tel: (0181) 958 5113.]
G4 GLN A R Bellfield, 41 Melody Rd, Biggin Hill, Westerham, TN16 3PH [JO01AH, TQ45]
G4 GLP D J Dale-Green, 31 Robins Bow, Forest Hills, Camberley, GU15 3NP [IO91OH, SU85]
G4 GLQ J P Tysiorowski, 52 Meadow Croft, Barco, Penrith, CA11 8EH [IO84PQ, NY53]
G4 GLV A Burgess, 12 Middleway, Grotton, Oldham, OL4 5SH [IO83XM, SD90]
G4 GLW C V Redmayne, 20 Kings Rd, Accrington, BB5 6BS [IO83TS, SD72]
G4 GLX L C H Glenister, Talgarth, 37 Ashcombe Gdns, Weston Super Mare, BS23 2XD [IO81MI, ST36]
G4 GLY E A Turner, 74 Beech Ln, Stretton, Burton on Trent, DE13 0DU [IO92ET, SK22]
GM4 GM D Smillie Clyde Valley DX Group, 20 Greenrig Rd, Hawksland, Lesmahagow, Lanark, ML11 9QA [IO85CP, NS84]
G4 GMA Details withheld at licensee's request by SSL.
G4 GMB D E J Hitchins, 1 Fyfe Cres, Baildon, Shipley, BD17 6DR [IO93CU, SE13]
G4 GMD A Riches, 54 Avondale Rd, Wigston, LE18 1ND [IO92KO, SP69]
G4 GME Details withheld at licensee's request by SSL.
G4 GMG J W Holmes, 7 Castle Dr, Ilford, IG4 5AE [JO01AN, TQ48]
GW4 GMI J P Seddon, 15 Pine Gr, Rhos on Sea, Colwyn Bay, LL28 4LW [IO83DH, SH88]
G4 GMJ Details withheld at licensee's request by SSL.[Op: J H Nelson, Bank Cottage, Bausley Hill, Crew Green, Shrewsbury, SY5 9BN. Station located in Powys.]
G4 GMK M R North, 50 Grange Park Ave, Wilmslow, SK9 4AL [IO83VH, SJ88]
G4 GML A J Watson, 93 St. Dunstans Dr, Gravesend, DA12 4BJ [JO01EK, TQ67]
G4 GMN R Caswell, 15 Murtwell Dr, Chigwell, IG7 5ED [JO01BO, TQ49]
G4 GMO Details withheld at licensee's request by SSL.
G4 GMQ Details withheld at licensee's request by SSL.
G4 GMS L J Hicks, Sleepy Hollow, Station Rd, Edenbridge, Kent, TN8 6HG [JO01AE, TQ44]
G4 GMT A E Aedy, 31 Tanyard Rd, Oakes, Huddersfield, HD3 4YW [IO93CP, SE11]
G4 GMU R F Berry, 99 Keighley Rd, Colne, BB8 0QG [IO83WU, SD84]
G4 GMV Details withheld at licensee's request by SSL.
G4 GMW M E Weaver, 22 Greenhill Rd, Alveston, Bristol, BS12 2LZ [IO81RO, ST68]
G4 GMY Details withheld at licensee's request by SSL.[Station located near Wantage.]
G4 GMZ Prof J F Alder, 104 Park Ln, Congleton, CW12 3DE [IO83VD, SJ86]
G4 GNA D Townend, 442 Blackmoorfoot Rd, Crosland Hill, Huddersfield, HD4 5NS [IO93CP, SE11]
G4 GNC A M Gibbins, 40 Loxley Dr, Mansfield, NG18 4FB [IO93JD, SK56]
G4 GND R M Culpan, 23 Aldreth Rd, Haddenham, Ely, CB6 3PP [JO02BI, TL47]
G4 GNG C Pemberton, 2 Henthorn St., Shaw, Oldham, OL2 7AY [IO83WN, SD90]
GD4 GNH R N Ferguson, Moaney Moar House, Corlea Rd, Ballasalla, IM9 3BA
G4 GNI A Walker, 77 The Bridle, Glen Parva, Leicester, LE2 9HR [IO92KO, SP59]
G4 GNK P R Jones, 27 Westerdale, Thatcham, RG19 3XA [IO91IJ, SU56]
G4 GNN G F Payne, 152 Limes Rd, Hardwick, Cambridge, CB3 7XX [JO02AF, TL35]
G4 GNP S J McGrory, The Paddock, 66 High St., Hook, Goole, DN14 5NY [IO93NR, SE72]
G4 GNQ G Sims, 85 Surrey St., Glossop, SK13 9AJ [IO93AK, SK09]
GM4 GNR W M Thow, 11 St. Marys Pl, Ellon, AB41 8QW [IO87XI, NJ93]
GU4 GNS S T Henry, The Hermitage, Lancresse, Vale, Guernsey, GY3 5AZ
GI4 GNT J Taggart, Windybrae, 5 Glasvey Dr, Ballykelly, Limavady, BT49 9HQ [IO65LA, C62]
G4 GNW T Hennigan, 128 Dimsdale View West, Newcastle, ST5 8EL [IO83VA, SJ84]
G4 GNX A K Baker, 15 Sunninghill Ave, Hove, BN3 8JB [IO90VU, TQ20]
GW4 GNY M R Davies, Laburnum House, Guilsfield, Welshpool, Powys, SY21 9PX [IO82KQ, SJ21]
GI4 GNZ B McCagherty, 38 Shelling Hill Rd, Cullybackey, Ballymena, BT42 1NR [IO64TV, D00]
G4 GOA J W Harris, 42 Cooks Cl, Bradley Stoke, Bristol, BS12 0BA [IO81RN, ST68]
G4 GOC Details withheld at licensee's request by SSL.
G4 GOG T F Densham, 37 Bovingdon Park, Roman Rd, Hereford, HR4 7SW [IO82OB, SO44]
G4 GOH Details withheld at licensee's request by SSL.
G4 GOJ G Porter, High St, North Thoresby, Grimsby, South Humberside, DN36 5PW [IO93XL, TF29]
G4 GOK A J Mellors, 11 The Paddocks, Litton Cheney, Dorchester, Dorset, DT2 9AF [IO80QR, SY59]
G4 GOM F Smith, 3 Moreton Ave, Whitefield, Manchester, M45 8GN [IO83UN, SD80]
G4 GON Dr J M Guest, 6 The Tyning, Bath, BA2 6AL [IO81TJ, ST76]
G4 GOO M H G Kimmitt, Stretton Mill, Tilston, Malpas, Ches, SY14 7HS [IO83OB, SJ45]
G4 GOP D Benn, 1 The Birches, Bramhope, Leeds, LS16 9DP [IO93EV, SE24]
GW4 GOQ Details withheld at licensee's request by SSL.
G4 GOR J B Cross, 15 Lune Gr, Blackpool, FY1 5PL [IO83LT, SD33]
GI4 GOS H A Sinclair, 43 Edgcumbe Gdns, Belfast, BT4 2EH [IO74BO, J37]
G4 GOT R G Bradbury-Harrison, 11 Derwent Dr, Goring By Sea, Worthing, BN12 6LA [IO90TT, TQ10]
G4 GOU M D Wilson, Thisuldo, Stixwould Rd, Horsington, Woodhall Spa, LN1 5EJ [IO93RG, SK97]
GI4 GOV P R Barr, 5 Rosewood Park, Belfast, BT6 9RX [IO74BN, J37]
GI4 GOW R L Armstrong, Knockraven, 49 Sea Rd, Castlerock, Coleraine, BT51 4TW [IO65OD, C73]
G4 GOX R Pearson, 33 Livedge Hall Ln, Liversedge, West Yorks, WF15 7DP [IO93DQ, SE22]
G4 GOY J Hague, Moorwood, Stone Ln, Wimborne, BH21 1HD [IO90AT, SU00]
G4 GOZ E Cockerill, 6 Richmond Ave, Barnoldswick, Colne, BB8 5JB [IO83VW, SD84]
GI4 GPA W J R Otterson, 8 Greenmount Cres, Coleraine, BT51 3QD [IO65PC, C83]
G4 GPB R A Cooper, 17 Cavendish Dr, Claygate, Esher, KT10 0QE [IO91TI, TQ16]
GI4 GPC J A Ferguson, 7 Lairds Rd, Katesbridge, Banbridge, BT32 5NN [IO64WH, J24]
G4 GPD W A Horn, 172 Gunnersbury Ave, London, W3 8LB [IO91UM, TQ17]
G4 GPI R H Hislop, Berry Knoll, Coach Hill Ln, Burley, Ringwood, BH24 4HN [IO90DU, SU20]
G4 GPJ N Bailey, Manchester, Helicopter Centre, Barton Airport, Eccles Manchester, M30 7SA [IO83TL, SJ79]
G4 GPK Details withheld at licensee's request by SSL.
G4 GPL A G Fish, 32 Deacons Hill Rd, Elstree, Borehamwood, WD6 3LH [IO91UP, TQ19]
GM4 GPP C Auty, Valsgarth, Haroldswick, Shetland, ZE2 9EF [IP90OT, HP61]
G4 GPR A K Mills, 116 Mays Ln, Barnet, EN5 2LS [IO91VP, TQ29]
G4 GPU C J Wilson, 19 Chace Ave, Potters Bar, EN6 5LX [IO91VQ, TL20]
G4 GPV A L Brown, 4 Old Farm Cl, Hankerton, Malmesbury, SN16 9LR [IO81XO, ST99]
G4 GPW B D Ainsworth, 23 Cokeham Rd, Sompting, Lancing, BN15 0AE [IO90TT, TQ10]
G4 GPX R E Bannister, 43 North Farm Rd, Lancing, BN15 9BT [IO90UT, TQ10]
G4 GPY S J R Edwards, 71 St. Leonards Rd, Molescroft, Beverley, HU17 7HP [IO93SU, TA04]
G4 GPZ S D Culpan, 10 South Ln West, New Malden, KT3 5AQ [IO91UJ, TQ26]
G4 GQA J J Chmielewski, 33 Dora Rd, Wimbledon, London, SW19 7EZ [IO91VK, TQ27]
G4 GQC C A Lingard, Smithy Ln Farm, Smithy Ln, Marple Bridge, Stockport, SK6 5NS [IO83XJ, SJ99]
G4 GQG F R Maxfield, 47 Crooks Barn Ln, Norton, Stockton on Tees, TS20 1LU [IO94IO, NZ42]
G4 GQJ G P Clay, Lynesmead, Tyrells Ln, Burley Ringwood, Hants, BH24 4DA [IO90DU, SU20]
G4 GQL Details withheld at licensee's request by SSL.
GM4 GQM G Y Firmin, Prestegaard, Uyeasound, Unst, Shetlandisle, ZE2 9DL [IP90NQ, HP60]
G4 GQO P R Chamberlain, 165 Revidge Rd, Blackburn, BB2 6EE [IO83RS, SD62]
G4 GQP R Foote, 9 Winchester Rd, Padiham, Burnley, BB12 7DN [IO83UT, SD83]
G4 GQR P F Thompson Brighton & Dist, Flat 3, 29 Cannon Pl, Brighton, BN1 2FB [IO90WT, TQ30]
G4 GQS B W Bentley, 25 Edinburgh Dr, North Anston, Anston, Sheffield, S31 7HD [IO93GJ, SK38]
G4 GQV J Barrett, 13 Church Bank, Church, Accrington, BB5 4JQ [IO83TS, SD72]
G4 GQW Can W Beswick, 23 Princess Ct, Princess Rd, Malton, YO17 0HL [IO94OD, SE77]
G4 GQY C W Lee, 74 Ilkeston Rd, Trowell, Nottingham, NG9 3PX [IO92IW, SK44]
G4 GQZ D P Tweedie, 14 Sandgate Ct, Sandgate, Penrith, CA11 7TG [IO84PP, NY53]
G4 GRA H A Dodd, 18 Copse View, East Preston, Littlehampton, BN16 1AY [IO90ST, TQ00]
GM4 GRC J Haliburton Glenrothes and District Arc., 21 Bolam Dr, Burntisland, KY3 9HP [IO86JB, NT28]
G4 GRI Details withheld at licensee's request by SSL.
G4 GRJ D O Gower, 2 Norview Rd, Whitstable, CT5 4DN [JO01MI, TR16]
G4 GRK P England, 2 Firs Cl, Cowes, PO31 7NF [IO90IS, SZ49]
G4 GRM L R Horton, 37 Chilton Rd, Richmond, TW9 4JD [IO91UL, TQ17]
G4 GRN T G Griffiths, 75 Central Ave, Waltham Cross, EN8 7JJ [IO91XQ, TL30]
G4 GRO P M D Cronin, Derhil, Church Rd, Killiney, Co. Dublin, Eire
G4 GRQ M A J Dufeu, 8/686 Mowbray Rd, Ln Cove, New South Wales, Australia 2066
G4 GRR Dr G W Searle, Magnolias, 26 Goldney Rd, Camberley, GU15 1DH [IO91PH, SU85]
G4 GRS M P Williams, 2 High Point, North Hill, Highgate, London, N6 4BA [IO91WN, TQ28]
G4 GRT D J Mounter, 36 Norwich Rd, Watton, Thetford, IP25 6DB [JO02JN, TF90]
G4 GRU D G Jones, 36 Moor Ln, Woodford, Stockport, SK7 1PP [IO83VI, SJ88]
G4 GRV R V Lee, Gothic Cottage, The St., Trowse, Norwich, NR14 8SS [JO02PO, TG20]
GW4 GRW Details withheld at licensee's request by SSL.
G4 GRY Dr J Groves, 25 Barkers Rd, Sheffield, S7 1SD [IO93GI, SK38]
G4 GRZ R T Marsh, 54 Waverton Rd, Bentilee, Stoke on Trent, ST2 0QY [IO83WA, SJ94]
G4 GSA P J Milsom, 214 Ormonds Cl, Bradley Stoke North, Bradley Stoke, Bristol, BS12 0DZ [IO81RM, ST68]
G4 GSB M C Hall, 63 Wantage, Telford, TF7 5PB [IO82SP, SJ60]
G4 GSC J H Osborne, 3 Temple Gdns, Chertsey Ln, Staines, TW18 3NQ [IO91RK, TQ06]
G4 GSE J Gray, Caravelle, 9 New Rd, Hextable, Swanley, BR8 7LS [JO01CJ, TQ57]
GW4 GSG E L Warner, 99 St. Peters Park, Northop, Mold, CH7 6YU [IO83KF, SJ26]
GW4 GSH M Beynon, 16 Hardy Cl, Woodfield Heights, Barry, CF62 9HJ [IO81IK, ST16]
G4 GSK P Barnett, Dunelm, Barley Hill, Dunbridge, Romsey Hants, SO51 0LF [IO91FA, SU32]
GW4 GSL R Prince, Ty-Capel Bungalow, Penmynydd Rd, Llangefni, Anglesey Gwynedd, LL77 7HR [IO73UG, SH47]
G4 GSM J OConnor, 11 Dalston Gr, Winstanley, Wigan, WN3 6EN [IO83PM, SD50]
G4 GSO H Elliott, 40 Dene House Rd, Seaham, SR7 7BQ [IO94HU, NZ44]
G4 GSP S R Lane, 51A Lindenthorpe Rd, Broadstairs, CT10 1BQ [JO01RI, TR36]
G4 GSQ Details withheld at licensee's request by SSL.
G4 GSR D Roberts, The Mead, Beaconsfield Rd, Woolton, Liverpool, L25 6EJ [IO83NJ, SJ48]
GW4 GSS R Bennett, Penrhiw Old Rd, Bwlchgwyn, Wrexham, Clwyd, LL11 5UH [IO83KB, SJ25]
G4 GSV Details withheld at licensee's request by SSL.
GW4 GSW Details withheld at licensee's request by SSL.
G4 GSY M J R Bainbridge, 21 Cockey Moor Rd, Bury, BL8 2HD [IO83TO, SD71]
G4 GSZ K L Court, 36 Elmsleigh Rd, Paignton, TQ4 5AU [IO80FK, SX86]
G4 GTA Details withheld at licensee's request by SSL.
GW4 GTC A Quayle Coleg Menai Bangor Site, Ty Ar y Gornel, Tafarn y Grisiau Rd, Port Dinorwic, Gwynedd, LL56 4NZ [IO73VE, SH56]
G4 GTD R G Ford, 2 Jersey Ave, St. Annes, Bristol, BS4 4RA [IO81RK, ST67]
GW4 GTE D E Evans, Glendale, Mount Pleasant Rd, Buckley, CH7 3ET [IO83LE, SJ26]
G4 GTG J K Reynolds, 47 Lulworth Dr, Roborough, Plymouth, PL6 7DT [IO70WK, SX46]
G4 GTH M J Linda, 16 Woodlinken Cl, Verwood, BH31 6BS [IO90BU, SU00]
GW4 GTI Details withheld at licensee's request by SSL.
G4 GTN P J Reeve, 2 Ct Rd, Tunbridge Wells, TN4 8ED [JO01DD, TQ53]
G4 GTO R McEwan Reid, 21 Byron Ave, Coulsdon, CR5 2JS [IO91WH, TQ35]
G4 GTP D M Austin, 20 Minerva Rd, Kingston upon Thames, KT1 2QA [IO91UJ, TQ16]
G4 GTR Details withheld at licensee's request by SSL.
G4 GTS D F Fairhurst, Stone Barn, Main St., Cleeve Prior, Evesham, WR11 5LG [IO92BD, SP04]
G4 GTT L Allwood London Airways ARC, London Atcc, Porters Way, West Drayton, Middx, UB7 9AX [IO91SM, TQ07]
G4 GTU S Pocock, 105 Arthurs Ave, Harrogate, HG2 0EB [IO93FX, SE25]
G4 GTW C S Kidd, 3 Forsythia Cl, Hedge End, Southampton, SO30 4TP [IO90IW, SU41]
G4 GTX W S Craigen, 19 Nilverton Ave, Sunderland, SR2 7TS [IO94HV, NZ35]
GI4 GTY J C Henry Lagan Valley AR, 3 Kirkwoods Park, Lisburn, BT28 3RR [IO64XM, J26]
G4 GTZ M L Phillips, 12 Reydon Ave, Wanstead, London, E11 2JD [JO01AN, TQ48]
G4 GUA Dr J G Overton, 7 Eden Pl, Eden Rd, Totland Bay, PO39 0HP [IO90FQ, SZ38]
G4 GUB Details withheld at licensee's request by SSL.[Op: M Catesby, 13 Colledge Close, Brinklow, nr Rugby, Warks, CV23 0NT.]
G4 GUC D Bailey, 3 Eastbanks, Sleaford, Lincs, NG34 7TL [IO92TX, TF04]
G4 GUD A E W Clarke, 29 Borrowdale Ave, Ramsgate, CT11 0PS [JO01QI, TR36]
G4 GUE I R Pope, P O Box 662, Durbanville 7551, Rep of South Africa
GM4 GUF Details withheld at licensee's request by SSL.
G4 GUG M J Meadows, Grass Roots Bodden, Shepton Mallet, Somerset, BA4 4PU [IO81RE, ST64]
GI4 GUH J M Clarke, 2 Areema Heights, Banbridge, BT32 3EX [IO64UI, J14]
G4 GUI M L Kershaw, 16 Pinderfields Rd, Wakefield, WF1 3NQ [IO93GQ, SE32]
G4 GUJ R P Merrell, 40 Fanton Walk, Wickford, SS11 8QT [JO01GO, TQ79]
G4 GUK K W Scott-Green, 1 Pickwick, Corsham, SN13 0JD [IO81VK, ST87]
GM4 GUL S F Macdonald, 5 Lower Glebe, Aberdour, Burntisland, KY3 0XJ [IO86IB, NT18]
G4 GUN G P Le Good, 45 Kingsfield Cres, Witney, OX8 6JB [IO91GS, SP31]
G4 GUO C H Brain, 2 Daisy Ct, Springfield, Chelmsford, CM1 6QU [JO01GR, TL70]
GM4 GUQ E C Crawford, 13 Meenans Cove Rd, Gondola Point, Rothesay, New Brunswick, Canada, E2E IM7
G4 GUS J L Firmin, 7 Orchard Way, Tasburgh, Norwich, NR15 1NJ [JO02PM, TM29]
G4 GUU D H Burrage, 11 Bannister Pl, Avondale, Auckland 7, New Zealand, ZZ1 1BA
G4 GUV J D Aindow, Autumn Cottage, 2 Cutlers Cl, Sydling St. Nicholas, Dorchester, DT2 9RG [IO80RT, SY69]
G4 GUW G G Baggott, 105 The Cres, Walsall, WS1 2DA [IO92AN, SP09]
G4 GUX J H Kuipers, 27 Shirley St., Hove, BN3 3WJ [IO90VT, TQ20]
G4 GUY T C Eaves, 3 Barons Rd, Dousland, Yelverton, PL20 6NG [IO70XL, SX56]
G4 GVB A Floyd, 95 Old Worthing Rd, East Preston, Littlehampton, BN16 1DU [IO90ST, TQ00]
G4 GVC Details withheld at licensee's request by SSL.[Op: J W Moore. Station located near Leicester.]
G4 GVD S C Jones, 11 Ainslie Rd, Fulwood, Preston, PR2 3DB [IO83PS, SD53]
G4 GVE J A Hawkings, 2 Balfour Gr, Biddulph, Stoke on Trent, ST8 7SZ [IO83WC, SJ85]
G4 GVF E Scott, 7 Wick Ln, Tuckton, Bournemouth, BH6 4JT [IO90CR, SZ19]
G4 GVG V Gormley, Applethwaite, Billinge End Rd, Pleasington, Blackburn, BB2 6QY [IO83RR, SD62]
G4 GVI B Spencer, 74 Micawber Rd, Poynton, Stockport, SK12 1UP [IO83WI, SJ98]
GM4 GVJ G N Marshall, Drummorlie, Wallyford Toll, Wallyford, Musselburgh, EH21 8JT [IO85LW, NT37]
GM4 GVK I Munro, 57 Craigiebuckler Ave, Aberdeen, AB1 8SF [IO87TH, NJ72]
G4 GVM D R R Alexander, 52 Brockington Rd, Bodenham, Hereford, HR1 3LP [IO82PD, SO55]
G4 GVN D J Barrott, 18 Church St., Somersham, Huntingdon, PE17 3EG [JO02AJ, TL37]
G4 GVR R E Mason, 3 Coronation Cl, Hellesdon, Norwich, NR6 5HF [JO02PP, TG21]
GI4 GVS P J Hallam, 95 Belfast Rd, Carrickfergus, BT38 8BX [IO74CR, J48]
G4 GVV S W Fox, Flat, 3 Woodford House, Cargate Terr, Aldershot, GU11 3EL [IO91OF, SU85]
G4 GVW P J Gillen, 86 Meadowlands, Kirton, Ipswich, IP10 0PP [JO02PA, TM23]
G4 GVY G M Hope, 34 Rowan Way, Exeter, EX4 2DS [IO80FR, SX99]
G4 GVZ D Morris, 40E Lansdown Cres, Cheltenham, GL50 2NG [IO81WV, SO92]
G4 GW G A Parris, The Hollies, Harley Ln, Heathfield, TN21 8AG [JO00DX, TQ52]
G4 GWB I Gibbs, 61 The Gables, Widdrington, Morpeth, NE61 5QZ [IO95EF, NZ29]
G4 GWC R Young, 22 Westerdale, Pickering, N Yorks, YO18 8DS [IO94OF, SE78]
G4 GWD Details withheld at licensee's request by SSL.
G4 GWF H Haden, 28 Welch Hill St., Gtr, Leigh, WN7 4DU [IO83RL, SD60]
G4 GWG D Snape, 30 Culcross Ave, Highfield, Wigan, WN3 6AA [IO83PM, SD50]
G4 GWH M K Steventon, Pippins, Bookers Ln, Earnley, Chichester, PO20 7JG [IO90NS, SZ89]
G4 GWI Details withheld at licensee's request by SSL.
G4 GWJ J M Butcher, 116 Cres Rd, Reading, RG1 5SN [IO91MK, SU77]
G4 GWK I T Budding, 48 Linketty Ln West, Hartley Vale, Plymouth, PL3 5RY [IO70WJ, SX45]
G4 GWO Details withheld at licensee's request by SSL.
G4 GWP B A Langford, Dulce Verano, 29M San Jaime, 03720 Benissa, Alicante, Spain, X X
GD4 GWQ A Matthewman, 26 King Orry Rd, Marown, Glen Vine, Douglas, IM4 4ES
G4 GWR A Scott-Green, 6 Aintree Dr, Chippenham, SN14 0FA [IO81WL, ST97]
GW4 GWS W J F Moss, 31 Alanbrooke Ave, Woodlands, Newport, NP9 6QH [IO81LO, ST39]
G4 GWT A D Kittle, 28 Clare Cres, Towcester, NN12 6QQ [IO92MD, SP64]
G4 GWU T Chapman, 11 Ash Ct, Brampton, Huntingdon, PE18 8FH [IO92VH, TL26]
G4 GWV R A Hookham, 50 Billy Mill Ave, North Shields, NE29 0QN [IO95GA, NZ36]
G4 GWW R W Broom, Sore Thumb Cottage, 13 Westgate, Oakham, LE15 6BH [IO92PQ, SK80]
G4 GWZ R Whitehead, 14 Southgate Cres, Rodborough, Stroud, GL5 3TS [IO81VR, SO80]
G4 GXA L P Bennett, 4 Ferndale Rd, Whiteshill, Stroud, GL6 6BA [IO81VS, SO80]
G4 GXB P L Butcher, 52 Chandos Rd, Rodbourogh, Stroud, GL5 3QZ [IO81VR, SO80]
G4 GXD D B Travis, 10 Greenfields Rise, Whitchurch, SY13 1EP [IO82PX, SJ54]
G4 GXE Details withheld at licensee's request by SSL.
G4 GXF V H Harris, 8 Toronto Rd, Tilbury, RM18 7RL [JO01EL, TQ67]
G4 GXI P H Pearson, 3 Lorimers Cl, Peterlee, SR8 2NH [IO94HS, NZ43]
G4 GXK K Hale Saltash Dist AR, 58 St. Stephens Rd, Saltash, PL12 4BJ [IO70VJ, SX45]
G4 GXM R C Corr, 15 Waterdell Ln, St. Ippolyts, Hitchin, SG4 7RA [IO91UW, TL12]
G4 GXN M L Wright, 5 Woodview Park, The Donahies, Raheny, Dublin 13
G4 GXO R W M Taylor, 89 Belthorn Rd, Belthorn, Blackburn, BB1 2PA [IO83SR, SD72]
G4 GXP B P C Hitchins Kidderminster &, 22 Granville Crest, Kidderminster, DY10 3QS [IO82VJ, SO87]
G4 GXQ P B W Swain, 39 Coniston Dr, Handforth, Wilmslow, SK9 3NN [IO83VI, SJ88]
GM4 GXR J B Higginbotham, Pollowick, Dunmore, Beauly, IV4 7AB [IO77SL, NH54]
G4 GXS E R Wilson, PO Box 141, Nuenen, The Netherlands, ZZ9 9PO
G4 GXT M J Pym, Woodbury House, Lincoln Cl, Exmouth, Devon, EX8 5QN [IO80HP, SY08]
G4 GXU G W Grieveson, 19 Hastings Ct, Church Ave, Stourport on Severn, DY13 9DB [IO82UI, SO87]
G4 GXW G Cahill, 21 Moresby Cl, Westlea, Swindon, SN5 7BX [IO91CN, SU18]
G4 GXZ A J Warrilow, Lanterns, The Row, Redlynch, Salisbury, SP5 2JT [IO90DX, SU22]
G4 GYA R Williscroft, Blackheath Farm, Cowhill Ln, Fradley, Lichfield, WS13 8NX [IO92CR, SK11]
G4 GYB Details withheld at licensee's request by SSL.
G4 GYD M G O'Hanlon, 61 Marshall Ln, Welwyn Garden City, AL7 3TG [IO91VT, TL21]
G4 GYF G A Hiscoe, 1 Greendale Cl, Fleetwood, FY7 8BQ [IO83LV, SD34]
G4 GYI P Ward, 23 Ropewalk, Alcester, B49 5DD [IO92BF, SP05]
G4 GYJ Mt R J Littlefield, 7 Carron Mead, Ferrers, South Woodham Ferrers, Chelmsford, CM3 5GH [JO01HP, TQ79]
G4 GYL M M Denby, 31 Scotland Way, Horsforth, Leeds, LS18 5SQ [IO93EU, SE23]
G4 GYN Dr R H Colson, 23 Grimsdells Ln, Amersham, HP6 6HF [IO91QQ, SU99]
G4 GYO Dr G Humpston, 10 Winwood Dr, Quainton, Aylesbury, HP22 4AZ [IO91MU, SP72]

G4 GYP L A C Ratcliff, 96 Crest Dr, Enfield, EN3 5QE [IO91XP, TQ39]
G4 GYQ A J Maguire, 42 Fairways, Dyke Rd, Brighton, BN1 5AD [IO90WU, TQ30]
GM4 GYR A Abel, 5A Orchard Pl, Aberdeen, AB2 3DH [IO87TH, NJ72]
G4 GYS J A Plested, 24 Farm Way, Bushey, Watford, WD2 3SS [IO91TP, TQ19]
G4 GYU J M Coates, 30 Abbott Rd, Mansfield, NG19 6DD [IO93JD, SK56]
G4 GYY H C Rumbelow, 79 Princes Way, London, SW19 6HY [IO91VK, TQ27]
G4 GZ J T Anglin, 23 Summerfields, Kings Rd, Cleethorpes, DN35 0AF [IO93XN, TA30]
G4 GZA D Ayris, 16 Chapel Ln, Northorpe, Gainsborough, DN21 4AF [IO93QL, SK89]
G4 GZB K W Turner, Clifton, High St., Epworth, Doncaster, DN9 1JS [IO93OM, SE70]
G4 GZC P M Teanby, 34 High St., Belton, Doncaster, DN9 1LR [IO93ON, SE70]
GM4 GZD G M Smith, Ardvourlie, Kiltarlity, By Beauly, Inverness-shire, IV4 7JQ [IO77SK, NH53]
G4 GZG L A Stringer, 2 Lion Cttgs, Toot Hill Rd, Ongar, CM5 9QL [JO01CQ, TL50]
G4 GZH D W M Andrew, 29 Great Hampden, Great Missenden, HP16 9RF [IO91OQ, SP80]
G4 GZJ Details withheld at licensee's request by SSL.
G4 GZK H G Dalton, 24 Church Ln, Coven, Wolverhampton, WV9 5DE [IO82WP, SJ90]
G4 GZL D E Barker, 79 South Parade, Boston, PE21 7PN [IO92XX, TF34]
G4 GZM A H McMillan, 70 Aller Park Rd, Newton Abbot, TQ12 4NQ [IO80EM, SX86]
G4 GZN K R Andreang, 62 Castleton Ave, Barnehurst, Bexleyheath, DA7 6QU [JO01CL, TQ57]
G4 GZO A D J Thurbon, 37 Lealand Rd, Drayton, Portsmouth, PO6 1LZ [IO90LU, SU60]
G4 GZQ J McGinty, 4 Barley Cl, Thatcham, RG19 4YJ [IO91JJ, SU56]
G4 GZS K A Wallace, 367 Dunchurch Rd, Rugby, CV22 6HU [IO92II, SP47]
G4 GZT P R Jensen, 7 Union St, Mosman, N.S.W, Australia 2088
G4 GZU R M Woodcock, 143 Berry Hill Rd, Mansfield, NG18 4RT [IO93JD, SK55]
G4 GZV S L Simon, 80 Hull Rd, Coniston, Hull, HU11 4LA [IO93VT, TA13]
GM4 GZW Dr E J Simon, 100 Findhorn Pl, Edinburgh, EH9 2NZ [IO85JW, NT27]
GW4 GZX J Hunter, 4 Fidlas Rd, Llanishen, Cardiff, CF4 5NB [IO81JM, ST18]
G4 GZZ A D Banks, 160 Ridgacre Rd, Quinton, Birmingham, B32 2SU [IO92AL, SP08]
GM4 HAA J K Barnes, 13 Marchhill Dr, Dumfries, DG1 1PP [IO85EB, NX97]
G4 HAB R F Rubins, 28 Dudley Rd, South Harrow, Harrow, HA2 0PR [IO91TN, TQ18]
G4 HAC C F Denscombe, High Holme, 4 Kendricks Bank, Bayston Hill, Shrewsbury, Salop, SY3 0EX [IO82OP, SJ40]
G4 HAD T P Moseley, 76 Station Rd, Hatton, Warwick, CV35 8XJ [IO92DH, SP26]
G4 HAE J E P Lewis, Harvard School of Education, 6 Appian Way 4th Floor, Cambridge, Ma 02138-3704, USA
G4 HAG J Long, 58 Allen Croft, Birkenshaw, Bradford, BD11 2AD [IO93DS, SE22]
G4 HAI P Levitt, The Bungalow, Middlegate Ln, Orby, Skegness, Lincs, PE24 5HZ [JO03CE, TF46]
G4 HAJ D J Magee, 2 Holt Park Vale, Holt Park Est, Leeds, LS16 7QX [IO93EU, SE24]
G4 HAK P Torrance, 1 Warburton Ln, Partington, Manchester, M31 4NR [IO83SK, SJ79]
GM4 HAO R M Mackean, Top Flat, 1 Kelvinside Terr West, Glasgow, G20 6DA [IO75UV, NS56]
G4 HAP H F Lavin, 30 Greenslate Rd, Billinge, Wigan, WN5 7BG [IO83PM, SD50]
G4 HAS D A Buck, 2605 Credit Valley Rd, Mississauga, Ontario, Canada, L5M 4K7
GW4 HAT P A Jones, 68 Pastoral Way, Tycoch, Sketty, Swansea, SA2 9LY [IO81AO, SS69]
G4 HAW T W Murgatroyd, Swallows, Gully Rd, Seaview, PO34 5BY [IO90KR, SZ69]
G4 HAY C J Baker, 12 Westland Dr, Brookmans Park, Hatfield, AL9 7UQ [IO91VR, TL20]
G4 HAZ K D Hope, 34 Rowan Way, Exeter, EX4 2DS [IO80FR, SX99]
G4 HBA R W Horne, Hayne Ln, Weston, Gittisham, Honiton, Devon, EX14 0PD [IO80JS, SY19]
G4 HBD P H Trepass, Barons Lodge, 43 Alton Rd, Poole, BH14 8SW [IO90AR, SZ09]
G4 HBH Details withheld at licensee's request by SSL.
G4 HBI F Cassidy, 55 High Bank Rd, Droylesden, Droylsden, Manchester, M43 6FS [IO83WL, SJ89]
G4 HBJ Details withheld at licensee's request by SSL.
GW4 HBK D D Lewis, 23 Gelligroes Rd, Pontllanfraith, Pontllanfraith, Blackwood, NP2 2JU [IO81JP, ST19]
G4 HBL Nr G Hardy, The Mill House, Thearne, Beverley, N H'side, HU17 0RU [IO93ST, TA03]
G4 HBP J H K Redman, Scolt, Ploughmans Piece, Thornham, Hunstanton, PE36 6NE [JO02GX, TF74]
GM4 HBQ A Taylor, 6 Bowling Green St., Methil, Leven, KY8 3DH [IO86LE, NT39]
G4 HBR J H McGee, 3 Hedgelea Rd, East Rainton, Houghton-le-Spring, DH5 9RR [IO94GT, NZ34]
GW4 HBS S A Illidge, 24 Maes Briallen, Off Deganwy Rd, Llanrhos, Llandudno, Gwynedd, LL30 1JJ [IO83CH, SH78]
G4 HBT M Foreman, 27 Winsbury Way, Bradley Stoke, Bristol, BS12 9BF [IO81RM, ST68]
G4 HBU L O Trembeth, 30 Fairview Rd, Kingswood, Bristol, BS15 2UT [IO81SL, ST67]
G4 HBV A J Martin, 21 Ashwood Way, Hucclecote, Gloucester, GL3 3JE [IO81VU, SO81]
G4 HBW Nr R Stevenson, 39 Croftway, Selby, YO8 9DD [IO93LS, SE63]
G4 HBY M W Cotton, 113 Belvedere Rd, Burton on Trent, DE13 0RF [IO92ET, SK22]
GW4 HBZ B C Clowes, 3 Lon Howell, Denbigh, LL16 4AN [IO83HE, SJ06]
G4 HCB Dr J E Harrison, 36 Elmlea Ave, Bristol, BS9 3UU [IO81QL, ST57]
G4 HCC M Hodgkinson, 68 Parker St., Colne, BB8 9QA [IO83VU, SD84]
G4 HCD A Reed, 28 Russell St., Sutton in Ashfield, NG17 4BE [IO93ID, SK45]
GM4 HCE K L Kirkland, 1 Villa Rd, South Queensferry, EH30 9RF [IO85HX, NT17]
G4 HCF G H Roe, 2 Repton Cl, Washingborough, Lincoln, LN4 1SB [IO93SF, TF07]
G4 HCG Dr R B Gordon, The Old School, 2 The Village, Orton Longueville, Peterborough, PE2 7DN [IO92UN, TL19]
G4 HCH D N Marsden, Windyridge, Red Ln, Colne, BB8 7JR [IO83VU, SD84]
G4 HCI M J Foreman, 213 Mountain Circle, Airdrie, Alberta, Canada, T4A 1X6
G4 HCJ S M Arkless, 80 Spring Ln, Whittington, Lichfield, WS14 9NA [IO92CQ, SK10]
G4 HCK N J Wilkinson, 12 Woodlands Cl, King Edwards Dr, Grays, RM16 2GB [JO01EL, TQ67]
G4 HCL Details withheld at licensee's request by SSL.[Station located near Eastleigh.]
GI4 HCN J P Clarke, 154 Galgorm Rd, Ballymena, BT42 1DE [IO64UU, D00]
GM4 HCO V J Kusin, East Overhill Farm, Stewarton, Ayrshire, KA3 5JP [IO75SQ, NS44]
G4 HCQ J W Westwood, Flat 1, 59 Nottingham Rd, Alfreton, DE55 7HL [IO93HC, SK45]
G4 HCT Details withheld at licensee's request by SSL.
GW4 HCV C G Snook, 3 Western Cres, Tredegar, NP2 3RQ [IO81JS, SO10]
GI4 HCX I A Magill, 205 Whitechurch Rd, Ballywalter, Newtownards, BT22 2LA [IO74GN, J67]
G4 HCY M D Stokes, 14 Shillitoe Ave, Potters Bar, EN6 3HG [IO91VQ, TL20]
G4 HCZ L Fellows, 19 Grosvenor Rd, Lower Gornal, Dudley, DY3 2PS [IO82WM, SO99]
GW4 HDB M Greatrex, 4 Lee St., St. Thomas, Swansea, SA1 8HQ [IO81AO, SS69]
G4 HDD S S Rose, 3 Grange Rd, Highgate, London, N6 4AR [IO91WN, TQ28]
GM4 HDE S R Green, 6 Poveys Mead, Kingsclere, Newbury, RG20 5ER [IO91JH, SU55]
GI4 HDJ B G McGarry, 74 Beechmount Park, Randalstown, Antrim, BT41 2AR [IO64UR, J09]
G4 HDK K C Hadley, 9 Compton Cl, Sandhurst, Camberley, Surrey, GU47 9RS [IO91OI, SU86]
G4 HDL N G Sedgwick, 4 Westbourne House, Farcroft Ave, Birmingham, B21 8AE [IO92AM, SP09]
G4 HDM T A Busby, 4 Hospital of Christ, Johnson Rd, Uppingham, Oakham, LE15 9RY [IO92PO, SP89]
G4 HDO A G Kirkland, 4 Laurelwood Rd, Droitwich, WR9 7SE [IO82WG, SO96]
G4 HDP K Griffin, 97 Woodlands Rd, Allestree, Derby, DE22 2HH [IO92GX, SK34]
G4 HDQ Details withheld at licensee's request by SSL.
GW4 HDR A R Evans, 4 Elm Gr, Rhyl, Denbighshire, LL18 3PE [IO83GH, SJ08]
G4 HDS P Unwin, 15 Gray Ave, Blackhall, Hesleden, Hartlepool, TS27 4PE [IO94IR, NZ43]
G4 HDU B C Keal, 46 Eastway, Maghull, Liverpool, L31 6BS [IO83MM, SD30]
GW4 HDW F S Hillier, 38 Purcell Rd, Penarth, CF64 3QN [IO81JK, ST17]
G4 HDY G M Burgess, 10 Malthouse Cl, North St., Wincanton, BA9 9TA [IO81TB, ST72]
GW4 HDZ D J Birch, 16 Llanharry Rd, Brynsadler, Pontyclun, CF7 9DB [IO81HN, ST08]
G4 HEB P Tuffs, 48 Mackie Dr, Guisborough, TS14 6DJ [IO94LM, NZ61]
G4 HEC P M Stracey, 14 Portfield Rd, Christchurch, BH23 2AG [IO90CR, SZ19]
G4 HEE W Dallas, 21 Jubilee Ave, Asfordby, Melton Mowbray, LE14 3RY [IO92MS, SK71]
G4 HEH A A H Smith, PO Box 15582, Westmead, 3608, South Africa, X X
G4 HEJ W J Reid, Comphurst, Comphurst Ln, Herstmonceux, Hailsham, BN27 4TX [JO00EV, TQ61]
GM4 HEL B P Spink Helensburgh ARC, 9 St. Andrews Cres, Mansewood Est, Dumbarton, G82 3ER [IO75RW, NS47]
G4 HEN A J Mountcastle, 268 Longfellow Rd, Wyken, Coventry, CV2 5HJ [IO92GJ, SP37]
G4 HEQ Details withheld at licensee's request by SSL.
GW4 HER S P Rogers, Green Tops, Megs Ln, Buckley, CH7 2AG [IO83LD, SJ26]
G4 HES W S Ray, 54 Gladstone Rd, Chesham, HP5 3AD [IO91QQ, SP90]
G4 HEV G H Cass, 18 Rawcliffe Dr, Clifton, York, YO3 6PE [IO93KX, SE55]
G4 HEW G Hancock, 12122-244th St, Maple Ridge BC, Canada, V4R 1L1
G4 HEZ W G Appleton, 453 Margate Rd, Ramsgate, CT12 6SN [JO01QI, TR36]
GI4 HFB Details withheld at licensee's request by SSL.
GI4 HFC Details withheld at licensee's request by SSL.
GM4 HFD G Clark, 12 Achaphubil, Fort William, Inverness-shire, PH33 7AL [IO76KU, NN07]
G4 HFF S Wraige, 137 Maidavale Cres, Coventry, CV3 6GE [IO92FJ, SP37]
G4 HFG G Eckersall, 65 Lowside Dr, Roundthorn, Oldham, OL4 1AS [IO83WM, SD90]
G4 HFH H L Wardill, 45 Driffield Way, Billingham, TS23 3RD [IO94IO, NZ42]
G4 HFJ S Robinson, Eastwood, The Green, Stowupland, Stowmarket, IP14 4AF [JO02ME, TM05]
G4 HFL D T Busby, 24 Wycombe Rd, Princes Risborough, HP27 0DH [IO91OR, SP80]
GM4 HFM L D Rollo, 5 Plewlands Ave, Edinburgh, EH10 5JY [IO85JW, NT27]
G4 HFN R A Fox, Ennerdale, Willow Dr, Twyford, Reading, RG10 9BA [IO91NL, SU77]
G4 HFO M S Blythe, Trethullan Farmhouse, Sticker, Saint Austell, Cornwall, PL26 7EH [IO70NH, SW95]
G4 HFQ G R Freeth, 9 South Ave, New Milton, BH25 6EY [IO90ES, SZ29]
G4 HFR D E Evans, 16 Honeybourne, Thorley Park, Bishops Stortford, CM23 4EF [JO01BU, TL42]
G4 HFS M K Davies, The Granary, Chequers Ln, Cadmore End, High Wycombe, HP14 3PH [IO91NP, SU79]
G4 HFT S J Richards, 2 Well Cross Rd, Gloucester, GL4 6SN [IO81VU, SO81]
G4 HFU P M Spooner, 196 The Fairway, South Ruislip, Ruislip, HA4 0SL [IO91TN, TQ18]
G4 HFV R S Beaumont, 54 Ballard Chase, Abingdon, OX14 1XQ [IO91IQ, SU59]
G4 HFX J L Mc Fall, Ballafayle, Copse Hill Ln, Homer, Much Wenlock, TF13 6NJ [IO82RO, SO69]
G4 HFZ S M McCann, 14 Queensway House, Queensway, Scunthorpe, DN16 2BU [IO93QN, SE81]
G4 HGB D W France, 28 Arbury Rd, Woodfield, Northampton, NN3 8QJ [IO92OG, SP86]
G4 HGD V P Cotton, Lower Meadow, Ayot St Lawrence, Welwyn, Herts, AL6 9BW [IO91UU, TL11]
G4 HGF P C Chadwick, 60 Gilderdale Cl, Gorse Covert, Birchwood, Warrington, WA3 6TH [IO83RK, SJ69]
G4 HGG P Riminton, 17 River Mount, Walton on Thames, KT12 2PP [IO91SJ, TQ06]
G4 HGI Details withheld at licensee's request by SSL.
GW4 HGJ G P Carruthers, Henllys, Tanygroes, Cardingham, Dyfed, SA43 2HR [IO72RC, SN24]
G4 HGK J D T Davis, Hurstbourne, Westdown Rd, Bexhill on Sea, TN39 4DY [JO00FU, TQ70]
G4 HGL Dr J M Buckley, Sandringham, Neston Rd, Ness, South Wirral, L64 4AT [IO83LG, SJ37]
G4 HGM M Gregory, Holly Cottage, Oakley Wood Rd, Bishops Tachbrook, Warks, CV33 9RW [IO92FF, SP36]
G4 HGN D W Hoyle, The Pharmacy, Fountain Sq, Tideswell Buxton, Derbyshire, SK17 8JT [IO93CG, SK17]
G4 HGO Details withheld at licensee's request by SSL.[Op: K G Cowman. Station located near Exmouth.]
GI4 HGQ Details withheld at licensee's request by SSL.
G4 HGR M A Baker, 39 The Cherry Orchard, Hadlow, Tonbridge, TN11 0HU [JO01EF, TQ65]
GW4 HGS G A Passmore, Victoria House, 127 High St., Neyland, Milford Haven, SA73 1TR [IO71MR, SM90]
G4 HGT J Wilkinson, 10 Tredgold Ave, Bramhope, Leeds, LS16 9BU [IO93EV, SE24]
G4 HGU S Fox, 9 Blackmore Rd, Tiverton, EX16 4AU [IO80GV, SS91]
G4 HGV M J Leach, 15 Beech Lea, Blunsdon, Swindon, SN2 4DE [IO91CO, SU19]
G4 HGZ W V Elliott, 8 St. Michaels Ave, Houghton Regis, Dunstable, LU5 5DN [IO91RV, TL02]
G4 HHA K D Stalley, Shottisham, Woodbridge, Suffolk, IP13 3EJ [JO02PE, TM26]
G4 HHB R A Kingstone, 8 Norris Cl, Ashley Heath, Ringwood, BH24 2HX [IO90BU, SU10]
GW4 HHD J D Hutchinson, 1 Upper Cwrt, Cwrt, Pennal, Machynlleth, SY20 9LA [IO82AO, SH60]
G4 HHH Maj P B Walker, East Rigg, Fylingdales, Robin Hoods Bay, Whitby North Yorks, YO22 4QG [IO94RJ, NZ90]
G4 HHI Details withheld at licensee's request by SSL.
G4 HHJ D Thomas, 35 Shelford Rd, Trumpington, Cambridge, CB2 2LZ [JO02BE, TL45]
G4 HHL V A Gorny, 22 Park Rd, Shirehampton, Bristol, BS11 0EF [IO81PL, ST57]
G4 HHM D J Ryder, Rock Bottom, Main St., Horsington, Woodhall Spa, LN1 5EX [IO93RG, SK97]
GW4 HHO Rev C F Buckley, Curraghmore, Cherry Gr, Model Farm Rd, Cork S Ireland
G4 HHP S Emery, 4 Dryden Rd, Walsall, WS3 1DR [IO92AO, SK00]
G4 HHS L May, 20 Cres Rd, Marland, Rochdale, OL11 3LF [IO83VO, SD81]
G4 HHT J Ward, 3 Shirley Rd, Walsgrave, Coventry, CV2 2EL [IO92GK, SP38]
G4 HHU C J Jones, Flat, 3 Teneriffe, Marine Parade, Lyme Regis, DT7 3JE [IO80MR, SY39]
G4 HHV Details withheld at licensee's request by SSL.
G4 HHX R J Edmonds, 15 Rose Gdns, Eythorne, Dover, CT15 4BS [JO01PE, TR24]
GM4 HHY C N S Goode, Piece, 55 Belwood Rd, Milton Bridge, Penicuik, EH26 0QN [IO85JU, NT26]
G4 HHZ A J Harwood, 55 Nichol Rd, Chandlers Ford, Eastleigh, SO53 5AX [IO91HA, SU42]
G4 HIA M J Nicholls, 12 Bents Dr, Ecclesall, Sheffield, S11 9RP [IO93FI, SK38]
G4 HIB L M Nicholls, 12 Bents Dr, Ecclesall, Sheffield, S11 9RP [IO93FI, SK38]
G4 HIC M H Maddison, 34 Maple Ave, Sandiacre, Nottingham, NG10 5EF [IO92IW, SK43]
G4 HIE M J Hammond, 53 Chiltern Rd, Baldock, SG7 6LT [IO91VX, TL23]
G4 HIF D L Mallet, Gwynant, Belmont Rd, Maidenhead, Berks, SL6 6JL [IO91PM, SU88]
G4 HIH R W Wilson, 4 Dinmont Pl, Hall Cl Grange, Cramlington, NE23 6DN [IO95FB, NZ27]
G4 HIJ R C Woolley, 29 Belle Vue Rd, Ashbourne, DE6 1AT [IO93DA, SK14]
G4 HIM M R Gershon, 10 Hogarth Ave, Brentwood, CM15 8BE [JO01DO, TQ69]
G4 HIN R C Twiggs, 31 Westlands Ave, Slough, SL1 6AH [IO91QM, SU98]
G4 HIP G W Rowlands, 26A Laburnum Rd, Hayes, UB3 4JX [IO91SL, TQ07]
GD4 HIT C A Buckler, Fairy Oak, Ardonan Ln, Regaby, Ramsey, IM7 3HN
G4 HIV B C Milne, C.S.O.S., Two Boats Village, Bfpo 677
G4 HIW C F Vernon, 2 Standing Butts Cl, Walton on Trent, Swadlincote, DE12 8NJ [IO92DS, SK21]
G4 HIX P Duncan, 89 Felstead Cres, Ford Est, Sunderland, SR4 0AE [IO94GV, NZ35]
G4 HIY B Burke, 43 Station Rd, Warboys, Huntingdon, PE17 2TH [IO92XJ, TL38]
G4 HIZ J N Easdown, 33 Abinger Dr, Chatham, ME5 8UL [JO01GI, TQ76]
G4 HJB C A Hall, 10 Porlock Ct, Northburn Chase, Cramlington, NE23 9TT [IO95EC, NZ27]
G4 HJD A M Goy, 352 Chanterlands Ave, Hull, HU5 4ED [IO93TS, TA03]
G4 HJE S G Small, 102 Crestway, Chatham, ME5 0BH [JO01GI, TQ76]
G4 HJF W C Dredge, 70 Elm Tree Rd, Locking, Weston Super Mare, BS24 8EH [IO81MI, ST35]
G4 HJG Details withheld at licensee's request by SSL.
G4 HJH M L Hardaker, Pacific View, Tower 3,Apartment 33B, 38 Tai Tam Rd, Hong Kong, X X
G4 HJI J G Bright, 22 Westfield Rd, Long Wittenham, Abingdon, OX14 4RF [IO91JP, SU59]
G4 HJJ G Mountain, 129 New Bank St., Morley, Leeds, LS27 8NT [IO93ER, SE22]
GM4 HJK R W Mitchell, 9 Pine Way, Perth, PH1 1DT [IO86JJ, NO02]
G4 HJL M Zarattini, The Pippens, Orchard St., Mickleover, Derby, DE3 5DF [IO92FV, SK33]
G4 HJM Details withheld at licensee's request by SSL.
G4 HJN L J Tucker, 110 Ringwood Rd, Poole, BH14 0RW [IO90AR, SZ09]
GM4 HJO M H Mozolowski, The Auld Manse, Sandport, Kinross, KY13 7DN [IO86HE, NO10]
GM4 HJQ D M Mackenzie, 58 High St., East Linton, EH40 3BH [IO85QX, NT57]
G4 HJS P Tempest, 15 Charles Ave, Leeds, LS9 0AE [IO93FS, SE33]
G4 HJT D K Lloyd, 39 High St., Twerton, Twerton on Avon, Bath, BA2 1DB [IO81TJ, ST76]
G4 HJW B A Wright, 39 High St., Little Wilbraham, Cambridge, CB1 5JY [JO02DE, TL55]
G4 HJX D E Reid, 25 Back Ln, Charlesworth, Broadbottom, Hyde, SK14 6HJ [IO93AK, SK09]
G4 HJY M P Black, 28 Cricketers Cl, Chessington, KT9 1NL [IO91UI, TQ16]
G4 HKA Dr H G C King, 9 Townsend Ln, Harpenden, AL5 2PY [IO91TT, TL11]
G4 HKB P J Turner, 1 Longridge, Briar Gr, Colchester, CO4 3FD [JO01LV, TM02]
G4 HKC I R Butson, 60 Churnwood Rd, Parsons Heath, Colchester, CO4 3EY [JO01LV, TM02]
G4 HKF Details withheld at licensee's request by SSL.
G4 HKG B J Butcher, 12 Caverswall Old Rd, Forsbrook, Stoke on Trent, ST11 9BL [IO82XX, SJ94]
GM4 HKH R W Hardie, 21 Clermiston Medway, Edinburgh, EH4 7EB [IO85IW, NT27]
G4 HKI M Pritchard, 18 Newey Rd, Wyken, Coventry, CV2 5HA [IO92GJ, SP37]
G4 HKM Details withheld at licensee's request by SSL.
G4 HKP C G G Turner, PO Box 4928, Halfway House, 1685, South Africa
G4 HKQ C E Marsh, 149 Mellow Purgess, Basildon, SS15 5XA [JO01FN, TQ68]
G4 HKR A Reed, 85 Ringway, Garforth, Leeds, LS25 1BZ [IO93HS, SE33]
G4 HKS Details withheld at licensee's request by SSL.
G4 HKT Details withheld at licensee's request by SSL.
G4 HKU D H Leonard, 8 Brownberrie Walk, Station Rd, Horsforth, Leeds, LS18 5PG [IO93EU, SE23]
GM4 HKV J Henderson, 45 Watts Gdns, Cupar, KY15 4UG [IO86LH, NO31]
GM4 HKW J W Henderson, 1 Rossiebank Cres, Westmuir, Kirriemuir, DD8 5LB [IO86LP, NO35]
GW4 HKX R E Rowlands, Berwyn, Hermon, Bodorgan, Anglesey, Gwynedd, LL62 5LH [IO73TE, SH36]
G4 HKY L R Bower, 1 Elmfield Dr, Skelmanthorpe, Huddersfield, HD8 9BT [IO93EO, SE21]
G4 HKZ J A Butcher, 116 Cres Rd, Reading, RG1 5SN [IO91MK, SU77]
G4 HL N D Whitehead, 97 Beauxfield, Whitfield, Dover, CT16 3JH [JO01PD, TR34]
G4 HLA J B Sullivan, 1 Godley Hill Rd, Godley, Hyde, SK14 3BW [IO83XK, SJ99]
G4 HLB R F Hallam, 9 Beaulieu Ct, Eye, Peterborough, PE6 7XT [IO92VO, TF20]
G4 HLE D J Turnell, 31 Greenbank Terr, Ringstead, Kettering, NN14 4DD [IO92RI, SP97]
G4 HLF P A Westwell, 11 Ches Park, Warfield, Bracknell, Berks, RG42 3XA [IO91PK, SU87]
G4 HLI J D Friend, 62 St. Catherines Hill, Bramley, Leeds, LS13 2LE [IO93ET, SE23]
G4 HLK J Kendall, Chittlebirch, Cripps Corner, Robertsbridge, E Sussex, TN32 5SA [JO00GX, TQ72]
G4 HLL C G Willoughby Walsall ARC, 79 Liskeard Rd, Park Hall, Walsall, WS5 3ES [IO92AN, SP09]
G4 HLM A O T Le-Mottee, 31 Marbury Rd, Vicars Cross, Chester, CH3 5PH [IO83NE, SJ46]
G4 HLN L C Bennett, 16 Axbridge Cl, Burnham on Sea, TA8 2FA [IO81MF, ST34]
GW4 HLO W E Davies, Erw Deg, 11 Madoc St., Porthmadog, LL49 9BU [IO72WW, SH53]
G4 HLP M W Haywood, Lacton Barn, The Street, Willesborough, Ashford, Kent, TN24 0JD [JO01KD, TR04]
G4 HLR Details withheld at licensee's request by SSL.
G4 HLS F C Bowen-Lock, 1 Blanchard Cl, Kirkby Cross, Kirby Cross, Frinton on Sea, CO13 0NE [JO01OU, TM22]
G4 HLT Details withheld at licensee's request by SSL.[Op: M E Eckhoff.]
G4 HLW K J Turnell, 31 Greenbank Terr, Ringstead, Kettering, NN14 4DD [IO92RI, SP97]
G4 HLX Dr N P Taylor, 46 Hunters Field, Stanford in The Vale, Faringdon, SN7 8LX [IO91FP, SU39]
G4 HLZ M Wood, Mark Wood, 2001 Jasmine Ln, Plano, Dallas, Texas 75074
G4 HMA M K Smith, The Brick House, 26 East St., Ashburton, Newton Abbot, TQ13 7AZ [IO80CM, SX76]
G4 HMC J D Oliver, Lilford Cottage, 24 Sixty Acres Rd, Prestwood, Great Missenden, HP16 0PE [IO91PQ, SP80]
G4 HMD H D Drury, 11 Batchworth Ln, Northwood, HA6 3AU [IO91TO, TQ19]
G4 HME L W Bailey, 47 Millers Park, Wellingborough, NN8 2NQ [IO92PG, SP86]
G4 HMG J Betts, Hardings, Clayhidon, nr Cullompton, Devon, EX15 3TJ [IO80JW, ST11]
G4 HMH A N Watts, Sancroft, Douglas Cl, Holton, Halesworth Suffol, IP19 8QE [JO02SI, TM37]
GI4 HMI Details withheld at licensee's request by SSL.
G4 HMJ Details withheld at licensee's request by SSL.
G4 HMK A M Gay, 12 Royal Field Cl, Hullavington, Chippenham, SN14 6DY [IO81WM, ST88]
GM4 HML S A McLuckie, 12 Croft Pl, Eliburn, Livingston, EH54 6RJ [IO85FV, NT06]
G4 HMM B M Dearing, 10 Woodlands Way, Southwater, Horsham, RH13 7HZ [IO91TA, TQ12]
GM4 HMN A G Cumming, 18 South Covesea Terr, Lossiemouth, IV31 6NA [IO87IR, NJ27]
G4 HMO D R Muir, 27 Springfield Cres, Horsham, RH12 2PP [IO91UB, TQ13]
G4 HMQ R S E Groves, 13 Mill View, Saham Toney, Thetford, Norfolk, IP25 7HG [JO02JN, TF90]
GW4 HMR D H Morris, Hafodty Cottage, Lon Hafodty, Tregarth, Bangor, Gwynedd, LL57 4NS [IO73WE, SH66]
G4 HMS P Balaam RN ARS, 57 Ruby Rd, Walthamstow, London, E17 4RE [IO91XO, TQ38]
G4 HMU H G Dabbs, 25 Thorpes Ave, Denby Dale, Huddersfield, HD8 8SP [IO93EN, SE20]
G4 HMW T H Hartshorn, 54 Walgrove Rd, Brampton, Chesterfield, S40 2DR [IO93GF, SK37]

G4

G4 HMX J R Halliday, 10 The Poppins, Beaumont Leys, Leicester, LE4 1DL [IO92KQ, SK50]
G4 HNB C B Hall, 3 Heswall Rd, Great Sutton, South Wirral, L66 4RZ [IO83MG, SJ37]
G4 HNC P R Snow, 117 Beresford Rd, Chingford, London, E4 6EF [JO01AP, TQ39]
G4 HND A P Course, 5 Conway Dr, Burton Latimer, Kettering, NN15 5TA [IO92PI, SP87]
G4 HNE Details withheld at licensee's request by SSL.
G4 HNF D Waterworth, 116 Reading Rd, Woodley, Reading, RG5 3AD [IO91NK, SU77]
G4 HNG G H Poulton, The Leas, Higher Sea Ln, Charmouth, Bridport, DT6 6BB [IO80NR, SY39]
G4 HNH Details withheld at licensee's request by SSL.
G4 HNJ G N Wheatley, Hillside, 67 Moorlands Rd, Verwood, BH31 7PD [IO90BV, SU00]
GM4 HNK C I Ferguson, Leckuary Farm, Kilichael-Glassary, By Lochgilphead, Argyll, PA31 8QL [IO76HC, NR99]
G4 HNO S J Wilson, 40 Ashburn Rd, Stockport, SK4 2PU [IO83VK, SJ89]
G4 HNQ J F Bryden, 32 Jerusalem Rd, Skellingthorpe, Lincoln, LN6 5TW [IO93QF, SK97]
G4 HNR K K Howard, 112 Fir Trees Ave, Lostock Hall, Preston, PR5 5SJ [IO83PR, SD52]
G4 HNU P B Vaughan, 26 Canterbury Rd, Worthing, BN13 1AE [IO90TT, TQ10]
G4 HNW S J Walls, 14 Copperfield Cl, Malton, YO17 0YN [IO94OD, SE77]
G4 HNX E A Beal, 49 Ambersham Cres, East Preston, Littlehampton, BN16 1AJ [IO90ST, TQ00]
G4 HNY Details withheld at licensee's request by SSL.
G4 HNZ S R Bannister, 14 Amery Cl, Worcester, WR5 2HL [IO82VE, SO85]
G4 HOC M V Oliver, Ashdell, Newlands Ln, Marston Green, Birmingham, B37 7EE [IO92DL, SP18]
G4 HOD M T I Gunby, 128 Heath Rd, Runcorn, WA7 4XL [IO83PH, SJ58]
G4 HOF P Warrener, 178 Farebrother St, Grimsby North East, Lincs, DN32 0JR [IO93XN, TA20]
G4 HOH W V Thursfield, 12 Kingswood Ave, Cannock, WS11 1QT [IO82XQ, SJ90]
G4 HOI W R Skeels, 141 Woodward Rd, Dagenham, RM9 4ST [JO01BM, TQ48]
G4 HOJ P Hobson, 7 Waterloo Cl, Caythorpe, Grantham, Lincs, NG32 3DL [IO93QA, SK94]
G4 HOK J A McKay, 2 Bransghyll Terr, Horton-in-Ribblesdl, Settle, Yorks, BD24 0HE [IO84UD, SD87]
G4 HOL M L Holden, 74 Trevillis Park, Liskeard, PL14 4EQ [IO70SK, SX26]
G4 HOM F Garratt, 5 Windmill Ct, North St., Tunbridge Wells, TN2 4SU [JO01DD, TQ53]
G4 HON C C Ward, 2 Arlington Dr, Stockport, SK2 7EB [IO83WJ, SJ98]
GD4 HOO N F Rimmer, Milner, Athol Ave, Port Erin, IM9 6EY
G4 HOP S E Fordham, 63 Baslow Rd, Totley, Sheffield, S17 4DL [IO93FH, SK38]
GW4 HOQ R R Jones, Bryn Ynys, 13 Strawberry Pl, Morriston, Swansea, SA6 7AG [IO81AP, SS69]
G4 HOS A A Munn, Tolenhof 5, 6443 BC Brunssum, Zuid Limberg, The Netherlands, ZZ2 8NE
G4 HOU L J Anstead, 21 Tickenor Dr, Finchampstead, Wokingham, RG40 4UD [IO91NJ, SU76]
G4 HOW N J Cleaver, 20 Acacia Park, Bishops Cleeve, Cheltenham, GL52 4WH [IO81XW, SO92]
GD4 HOX E H Brooks, Elmwood, Somerset Rd, Douglas, IM2 5AE
G4 HOY Details withheld at licensee's request by SSL.[Op: J R Fennell.]
GD4 HOZ D A Osborn, 30 Slieau Whallian Park, St. Johns, Douglas, IM4 3JJ
G4 HPE S P Richards, 6 Heathfield, Royston, SG8 5BW [IO92XB, TL34]
GM4 HPF Details withheld at licensee's request by SSL.[Op: J Ferguson. Station located in Inverness IV2.]
G4 HPG Details withheld at licensee's request by SSL.
G4 HPI Details withheld at licensee's request by SSL.[Op: R L Thake.]
G4 HPJ G E Greenslade, 22 Shepherds Rd, Winnall, Winchester, SO23 0NP [IO91IB, SU42]
GM4 HPK D A B Moore, Mardavhal, Strone, Argyll, PA23 8TB [IO75NX, NS18]
G4 HPL Details withheld at licensee's request by SSL.
GD4 HPN R S Baker, Thie Corniel, Bayr Grianagh, Castletown, Isle of Man, IM9 1HJ
G4 HPS P Barker, 11 Dipton Gdns, Tunstall Est, Sunderland, SR3 1AN [IO94HV, NZ35]
G4 HPT D B Oliver, Ashdell, Newlands Ln, Marston Green, Birmingham, B37 7EE [IO92DL, SP18]
G4 HPU A C Keeble, Heater Field, Colchester Rd, Ardleigh, Colchester Essex, CO7 7PA [JO01LW, TM02]
G4 HPV S Brown, 42 Londesborough Rd, Scarborough, YO12 5AF [IO94TG, TA08]
G4 HPW Details withheld at licensee's request by SSL.
G4 HPX J Trotter, 27 Maynards Park, Bere Alston, Yelverton, PL20 7AR [IO70VL, SX46]
G4 HPY R J Spragg, 6 St. Monicas Ave, Luton, LU3 1PJ [IO91SV, TL02]
GI4 HPZ N C Doherty, Limefield Bungalow, Ballynally, Moville, Co. Donegal, Ireland, ZZ4 6LI
G4 HQA R J Knowles, 22 Thornley Rd, Ribbleton, Preston, PR2 6EY [IO83QS, SD53]
G4 HQC C J Wilcox, 125 Alstone Ln, Cheltenham, GL51 8HW [IO81WV, SO92]
G4 HQD R G Bagley, 3 Queens Ct, Haverhill, CB9 9AU [JO02FB, TL64]
G4 HQE C W Morle, Brook House, Woodhill Ave, Gerrards Cross, SL9 8DJ [IO91RN, TQ08]
GM4 HQF D Lindsay, 39 Seamount Ct, Aberdeen, AB2 1DQ [IO87TH, NJ72]
G4 HQH S K Parker, 20 Swaddale Ave, Chesterfield, S41 0SU [IO93HF, SK37]
G4 HQI R Fewtrell, Westwood Cottage, Brocks Copse Rd, Wootton Bridge, Ryde, PO33 4NP [IO90JR, SZ59]
G4 HQJ W C J Matcham, 6 Joyes Cl, Folkestone, CT19 6HN [JO01OC, TR23]
G4 HQK E A Brock, 78 Capel Gdns, Seven Kings, Ilford, IG3 9DG [JO01BN, TQ48]
G4 HQM D L Waspe, 28 Wilman Way, Salisbury, SP2 8QS [IO91CB, SU12]
GI4 HQP H McNerlan, 100 Dunraven Ave, Bloomfield Rd, Belfast, BT5 5JS [IO74BO, J37]
G4 HQQ P J Nicoll, 14 Dayspring, Guildford, GU2 6QN [IO91RG, SU95]
GM4 HQU N W Gent, 1 Oaklea, Carrington, Midlothian, EH23 4LX [IO85KU, NT36]
GM4 HQZ A Morrison, 148 South St., Lochgelly, KY5 9BE [IO86IC, NT19]
G4 HRA G A Brownell, 15 Roe Ln, Newcastle, ST5 3PH [IO82VX, SJ84]
G4 HRB D B Taylor, 8 Fambridge Cl, Maldon, CM9 6DJ [JO01IR, TL80]
G4 HRC D L Nuttall Havering Rad CB, Fairkytes Art Centr, 51 Billet Ln, Hornchurch Essex, RM11 1XA [JO01CN, TQ58]
G4 HRE D S Hollow, 2 Mays Model Cottage, Colt Hill, Odiham, Hook, RG29 1AL [IO91MG, SU75]
G4 HRG R F Denley, 50 Cranmere Ave, The Wergs, Wolverhampton, WV6 8TS [IO82VO, SJ80]
G4 HRH A C Allen, The Hollies, Sedgeford, Whitchurch, SY13 1EX [IO82PX, SJ54]
G4 HRI R Greenhalgh, 18 Westville Dr, West Heath, Congleton, CW12 4LD [IO83VD, SJ86]
GM4 HRJ J McNiff, East Cove Cottage, Main Rd, Langbank, Port Glasgow, PA14 6XP [IO75QW, NS37]
G4 HRK Details withheld at licensee's request by SSL.
GM4 HRL A H Sergeant, Distant Hills, 24 Academy Rd, Boness, EH51 9QD [IO86EA, NT08]
G4 HRM Details withheld at licensee's request by SSL.
G4 HRN J T C Aslett, 72 Brookdale, Rochdale, OL12 0UY [IO83WP, SD81]
G4 HRP D C J Dimes, 111 Hargrave Ave, Needham Market, Ipswich, IP6 8ES [JO02MD, TM05]
G4 HRS J R Matthews Horsham ARC, 46 Park Ln, West Grinstead, Horsham, RH13 8LT [IO91UB, TQ13]
G4 HRT E D Cooper, 11 Gledhow Wood Ave, Leeds, LS8 1NY [IO93FT, SE33]
G4 HRU R F Profitt, 10 Taunton Vale, Hunters Vale, Guisborough, TS14 7NB [IO94LM, NZ61]
G4 HRV D A Ashton, 12 Juniper Cl, Swindon, SN3 4DZ [IO91DN, SU18]
GM4 HRW H Elder, 15 Frogston Gdns, Edinburgh, EH10 7AF [IO85JV, NT26]
G4 HRY D R Farn, 14 Corfe Cl, Clifford Park, Coventry, CV2 2JG [IO92GK, SP38]
G4 HSA V C Whitchurch, 5 Underhill Ln, Midsomer Norton, Bath, BA3 2RT [IO81SG, ST65]
G4 HSB P M R Rovardi, 8 Cambridge Rd, Linthorpe, Middlesbrough, TS5 5NQ [IO94JN, NZ41]
G4 HSC H Hughes, 16 Dalton Dr, Goose Green, Wigan, WN3 6TQ [IO83QM, SD50]
G4 HSD R Smithers, 16 Derby Rd, Cheam, Sutton, SM1 2BL [IO91VI, TQ26]
G4 HSG J H Simpson, 17 Parsons Dr, Ellington, Huntingdon, PE18 0AU [IO92UH, TL17]
GW4 HSH R Williams, 114 West Cross Ln, West Cross, Swansea, SA3 5NQ [IO71XO, SS68]
G4 HSI E G Hudson, 24 The Badgers, Station Rd, Netley Abbey, Southampton, SO31 5PT [IO90HU, SU40]
G4 HSK S F Glass, 36 Pickwick Ave, Chelmsford, CM1 4UN [JO01FS, TL60]
G4 HSL K Mould, 4 Legrice Cres, North Walsham, NR28 9AF [JO02QT, TG22]
G4 HSM R F Hurrell, 39 Kiveton Dr, Ashton in Makerfield, Wigan, WN4 9EX [IO83QL, SJ59]
G4 HSN A Chorley, 11 Staples Rd, Loughton, IG10 1HP [JO01AP, TQ49]
G4 HSO P A E Baker, 4 Milton Cres, Dudley, DY3 3DR [IO82WM, SO99]
GM4 HSR D W Gillies, 56 Forehill Rd, Ayr, KA7 3DT [IO75QK, NS32]
G4 HSS P J Forshaw, 54 The Park, Penketh, Warrington, WA5 2SG [IO83QJ, SJ58]
G4 HST O C English, 59 Trafford Rd, Norwich, NR1 2QR [JO02PO, TG20]
G4 HSU Dr C R Ayling, 30 Limes Rd, Folkestone, CT19 4AU [JO01NC, TR23]
GJ4 HSW F P Le Quesne, Brookhill House, Princes Tower Rd, St. Saviour, Jersey, JE2 7UD
G4 HSY Details withheld at licensee's request by SSL.
G4 HSZ P M Thacker, 23 Lulworth Ave, Whitkirk, Leeds, LS15 8LW [IO93GT, SE33]
G4 HTB T L Rance, 2 Glenavon Gdns, Slough, SL3 7HN [IO91RM, SU97]
G4 HTD L E Mason, Forest Farm, Folly Drove, Ashill, Ilminster, Somerset, TA19 9HX [IO80MW, ST31]
G4 HTE E Sergeant, 36 Ormesby Dr, Potters Bar, EN6 3DZ [IO91VQ, TL20]
G4 HTF S J Walker, 7 Norwich Way, Croxley Green, Rickmansworth, WD3 3SP [IO91SP, TQ09]
G4 HTH R E Herringshaw, 35 Oxley Cl, Shepshed, Loughborough, LE12 9LS [IO92IS, SK41]
G4 HTJ R A Earle, 10 Crosslands, Fringford, Bicester, OX6 9DF [IO91KW, SP62]
G4 HTL A A McCulloch, 109 Wildeamandel Ave, Roodekrans Ext 3, Roodepoort 1725, South Africa, ZZ1 0WI
G4 HTO I A Myford, 33 Station Rd, Edingley, Newark, NG22 8BX [IO93MC, SK65]
G4 HTP P O Mann, 21 Wilkinson Cl, Eaton Socon, St. Neots, Huntingdon, PE19 3HJ [IO92UF, TL15]
G4 HTR R G Harvey, 5 Bafford Approach, Charlton Kings, Cheltenham, GL53 9HH [IO81XV, SO91]
G4 HTS W H Etheridge, 78 Manor Rd, Stretford, Manchester, M32 9JB [IO83UK, SJ79]
GM4 HTU A Langton, 71 Gray St., Aberdeen, AB1 6JD [IO87TH, NJ72]
G4 HTV R W Thompson Htv Radio Club, 179 Newbridge Hill, Bath, BA1 3PY [IO81TJ, ST76]
G4 HTX Details withheld at licensee's request by SSL.
G4 HTY D R Stokes, 3 Tregenna Hill, St. Ives, TR26 1SE [IO70GF, SW54]
G4 HTZ Details withheld at licensee's request by SSL.
G4 HUA Details withheld at licensee's request by SSL.
G4 HUC J W H Chappell, 138 Woodcote Valley Rd, Purley, CR8 3BF [IO91WI, TQ36]
G4 HUD J R Bramall, 55 Wood Ln, Louth, LN11 8RY [JO03AI, TF38]
G4 HUE A Nehan, 42 Wordsworth Ave, South Woodford, London, E18 2HE [JO01AO, TQ38]
G4 HUF P E Baguley, 16 Churchill Rd, Broadheath, Altrincham, WA14 5LT [IO83TJ, SJ78]
G4 HUG W J S Daniels, Chy Vean, 48 Mellanear Rd, Hayle, TR27 4QT [IO70GE, SW53]
G4 HUH P R Chapman, 1291 Los Amigos Ave, Simi Valley, California 93065, USA
GM4 HUL W G Savory, 20 Broomfield, Carradale East, Campbeltown, PA28 6RZ [IO75GO, NR83]
G4 HUM D R Hazzard, 11 Hooks Farm Way, Havant, PO9 3DX [IO90MU, SU70]
G4 HUN N Whiteside, 27 Whitethorn Ln, Letchworth, SG6 2DN [IO91VX, TL23]
G4 HUO M J Bennett, 9 Lavender Ave, Blythe Bridge, Stoke on Trent, ST11 9RN [IO82XX, SJ94]
G4 HUP D S Powis, Fircroft, Pound Ln, Dallinghoo, Woodbridge, IP13 0LN [JO02PD, TM25]
G4 HUQ M J Crake, 12 Bosburn Dr, Mellor Brook, Blackburn, BB2 7PA [IO83RS, SD63]
G4 HUR E H Ward, 15 Leigh Cl, Row Town, Addlestone, KT15 1EL [IO91RI, TQ06]
G4 HUS D H Sullivan, 19 Headcorn Gdns, Cliftonville, Margate, CT9 3ES [JO01RJ, TR37]
G4 HUT Details withheld at licensee's request by SSL.
G4 HUV Details withheld at licensee's request by SSL.
G4 HUW S A Faulkner, Vaarveien 8, 1182 Oslo, Norway
G4 HUX R Lindsay, 6 Rothley Cl, Ponteland, Newcastle upon Tyne, NE20 9TD [IO95CB, NZ17]
GU4 HUY R D Sarre, -le-Clercs, Clos Du Murier, St. Sampson, Guernsey, GY2 4HJ
G4 HVA B Tindill, Hunters Moon, Station Rd, Newton-le-Willows, Bedale, DL8 1SX [IO94DH, SE28]
G4 HVB K H Rushall, 2 Morson Cres, Rugby, CV21 4AL [IO92JI, SP57]
G4 HVC A R Kiddle, 19 Old Lincoln Rd, Caythorpe, Grantham, NG32 3DF [IO93QA, SK94]
G4 HVD T S Barnett, East View, Hangerberry, Lydbrook, Glos, GL17 9QG [IO81RU, SO61]
G4 HVE M E W Simson, 489 Sipson Rd, Sipson, West Drayton, UB7 0JB [IO91SL, TQ07]
G4 HVF C J Bracewell, 14 Woodlane, Falmouth, TR11 4RF [IO70LD, SW83]
G4 HVG J H Phipps, 5 Akeman Cl, St. Stephens, St. Albans, AL3 4NJ [IO91TR, TL10]
G4 HVH R R Chapman, 114 Oxford St, Smithfield, Sydney, Nsw 2164 Australia
GI4 HVI A T Hamilton, 11 Norwell Park, Castlerock, Coleraine, BT51 4TS [IO65OD, C73]
G4 HVK M F Thomas, 18 North Parade, Penzance, TR18 4SN [IO70FC, SW43]
GM4 HVM A L Douglas, 24 Plane Gr, Dunfermline, KY11 5RA [IO86HB, NT18]
GW4 HVN M R Davies Powys ARC, Laburnum House, Guilsfield, Welshpool Powy, SY21 9PX [IO82KQ, SJ21]
G4 HVO J F Fitzwater, The Old Cottage, Babylon Ln, Lower Kingsworth, Tadworth, KT20 6XE [IO91VG, TQ25]
G4 HVP P E Baguley TS Talisman RC, 16 Churchill Rd, Broadheath, Altrincham, WA14 5LT [IO83TJ, SJ78]
GM4 HVR G V Southwell, Craigend House Anne, Kilmun, Dunoon, Argyll, PA23 8SE [IO76MA, NS18]
GM4 HVS Dr R J Teperek, 8 Forest Park, Stonehaven, AB3 2GF [IO87TH, NJ72]
G4 HVT N Wilkinson, Breidablikkbakken 15, 3911 Porsgrunn, Norway
GM4 HVU Dr J A H Brown, Delgany, Old Cambus, Cockburnspath, Berwickshire, TD13 5YS [IO85UW, NT87]
G4 HVV J Houlihan Haven Valley Contest Club, 2 Boundary Rd, Red Lodge, Bury St. Edmunds, IP28 8JQ [JO02FH, TL77]
G4 HVW F Moody, 87 Whitegate Walk, Rockingham Est, Rotherham, S61 4LP [IO93HL, SK49]
G4 HVX Details withheld at licensee's request by SSL.
G4 HVZ P Cooke, 31 Kent Cres, Palmerston North, New Zealand
G4 HWA B J Morton, Fielders, Horton Park, Horton, Northampton, NN7 2BJ [IO92OE, SP85]
G4 HWC E F King, 22 Coast Rd, Marske By The Sea, Redcar, TS11 6JU [IO94LO, NZ62]
G4 HWF R F Rudd, Flat 3, 12 Eaton Pl, Brighton, BN2 1EH [IO90WT, TQ30]
G4 HWH A P Jandrell, 64 Heath Ln, Old Swinford, Stourbridge, DY8 1RQ [IO82WK, SO98]
G4 HWI M D Allin, 23 Green Leys, Maidenhead, SL6 7EZ [IO91PM, SU88]
G4 HWJ M D Dawson, 23 Channels Farm Rd, Swaythling, Southampton, SO16 2PF [IO90HW, SU41]
G4 HWK F B Pilling, 10 Hanover Cres, Norbreck, Blackpool, FY2 9DL [IO83LU, SD34]
G4 HWM Dr D R N Jeffery, 13 Palmerston St., Romsey, SO51 8GF [IO90GX, SU32]
G4 HWN R A Heath, 71 Moat Rd, Oldbury, Warley, B68 8ED [IO82XL, SO98]
GM4 HWO C J Wright, 9 Corbiehill Ave, Edinburgh, EH4 5DP [IO85IX, NT27]
GW4 HWR E J Case, 2 Abbey Cl, Taffs Well, Cardiff, CF4 7RS [IO81IN, ST18]
GM4 HWS S Roberts, Thistlebrae, Lower Plaidy, Turriff, Grampian, AB53 7RJ [IO87TM, NJ74]
G4 HWU B A Osbourne, 3 Pershore Gdns, Normoss, Blackpool, FY3 7SW [IO83LT, SD33]
G4 HWW R C H Scott, Flat 57 Tatton Cour, 35 Derby Rd, Stockport, Ches, SK4 4NL [IO83VK, SJ89]
G4 HWY B N Sorger, 47 Rochester Rd, Gravesham, Kent, DA1 2JN [JO01CK, TQ57]
G4 HXC D J Edwards, 179 Pallett Dr, Nuneaton, CV11 6JA [IO92GM, SP39]
G4 HXD Details withheld at licensee's request by SSL.
G4 HXE A R Tilbee, 26 Kingston Cl, River, Dover, CT17 0NQ [JO01PD, TR24]
G4 HXG E E Seal, 10658 Jacatree Cour, Lehigh Acres, Florida 33936, USA
G4 HXH R L Pope, 95 Northolt Ave, Bishops Stortford, CM23 5DS [JO01CV, TL42]
G4 HXI L P Maguire, 95 Lye Ave, Wood Gate Valley, Birmingham, B32 3UG [IO82XK, SO98]
GJ4 HXJ G Brown Jersey Elect Cl, 1 Belmont Gdns, Belmont Pl, St. Helier, Jersey, JE2 4SD
G4 HXK F B Rendell, 64 Rivermead, Stalham, Norwich, NR12 9PJ [JO02SS, TG32]
G4 HXM R L Jennison, 112 Welwyn Park Rd, Hull, HU6 7EA [IO93TS, TA03]
G4 HXN D M Kelly, 20 Cannonside, Fetcham, Leatherhead, KT22 9LE [IO91TH, TQ15]
GW4 HXO M J Probert, 1 Ynys Dawel, Solva, Haverfordwest, Dyfed, SA62 6UA [IO71JV, SM72]
G4 HXQ G M Burlington, Pedlars, Catts Hill, Mark Cross, Crowborough, TN6 3NH [JO01CB, TQ53]
G4 HXT M J Lord, 44 Bramble Hill, Chandlers Ford, Eastleigh, SO53 4TP [IO90HX, SU42]
G4 HXU D B McDermott, 6 Chiltern Gr, Thame, OX9 3NH [IO91MR, SP70]
G4 HXV P C Gregory, 3 Spring St., Mossley, Ashton under Lyne, OL5 0BS [IO83XM, SD90]
G4 HXX Prof C T Dollery Hammersmith Hospital Radio Clu, Arc Dept of Medicin, Ducane Rd, London, W12 0NN [IO91VM, TQ28]
G4 HXY S G Simmons, 48 Copland Rd, Stanford-le-Hope, SS17 0DF [JO01FM, TQ68]
G4 HYB Details withheld at licensee's request by SSL.
G4 HYC Details withheld at licensee's request by SSL.
GM4 HYF G M Allan, 22 Tynwald Ave, High Burnside, Rutherglen, Glasgow, G73 4RN [IO75VT, NS65]
G4 HYG C J Moulding, 37 Edgefold, Plodder Ln, Bolton, BL4 0LW [IO83SN, SD70]
G4 HYH J H Wolfenden, 9 Willow St, Rfd3 Box 451, Pelham N H 03076, USA, ZZ9 9WI
G4 HYI E P Towers, Occidental, Whittonditch, Ramsbury, Marlborough, SN8 2QA [IO91EK, SU27]
G4 HYJ F D Shaw, Chantreys, Newton Reigny, Penrith, Cumbria, CA11 0AY [IO84OQ, NY43]
G4 HYQ Details withheld at licensee's request by SSL.
GM4 HYR M A Bond, 1 Saughtonhall Cres, Edinburgh, EH12 5RF [IO85JW, NT27]
G4 HYT Dr P P Kurian, Kelachandra Medical Centre, Chingavanam, Pin 686 531, Kerala State, Southern India, X X
G4 HYU C Russell, Ravenscar, Top Rd, Little Cawthorpe, Louth, LN11 8NB [JO03AH, TF38]
G4 HYV C H Hender, 2 Sharaman Cl, Menear Rd, St. Austell, PL25 3DH [IO70OI, SX05]
G4 HYW A D Wilkes, 5 Quay Hill, Lymington, Hants, SO41 3AR [IO90FS, SZ39]
G4 HYX Details withheld at licensee's request by SSL.
G4 HYY T D Jackson, 13 Hill Crest, Hebden Bridge, West Yorks, HX7 6BQ [IO83XR, SD92]
GW4 HYZ B D Green, 28 Sunnybank Rd, Griffithstown, Pontypool, NP4 5LT [IO81LQ, ST29]
G4 HZA R Taylor, Raeburn, Bush Ln, Send, Woking, GU23 7HP [IO91RG, TQ05]
G4 HZB I Shardlow, 18 Newhall Ave, Bradley Fold, Bolton, BL2 6RX [IO83TN, SD70]
G4 HZE E P Hill, 14 Station Rd, Saltash, PL12 4DY [IO70VJ, SX45]
G4 HZF R J W Scarlett, 1 St. Martins Cres, Grimsby, DN33 1BG [IO93WN, TA20]
G4 HZG M R White, 38 Tamworth Rd, Amington, Tamworth, B77 3BT [IO92DP, SK20]
GW4 HZH Dr D J D Doherty, 3 Llys Penpant, Llangyfelach, Morriston, Swansea, SA6 6DA [IO81AQ, SS69]
G4 HZI W A Backhouse, 10 Farrier Cl, Weavering, Maidstone, ME14 5SR [JO01GG, TQ75]
G4 HZJ L Jackson, 1 Belvedere Ave, Atherton, Manchester, M46 9LQ [IO83SM, SD60]
GW4 HZM J Styles, 36 Heol Ewenny, Pencoed, Bridgend, CF35 5QA [IO81GM, SS98]
G4 HZN T Lockwood, 8 St. Nicholas Rd, Thorne, Doncaster, DN8 5BS [IO93MO, SE61]
G4 HZO S Wardle, 42 Hearthcote Rd, Swadlincote, DE11 9DU [IO92FS, SK21]
G4 HZP Details withheld at licensee's request by SSL.
G4 HZR D M N Saunders, 32 Richmond Ct, Osmond Rd, Hove, BN3 1TD [IO90WT, TQ30]
G4 HZT T J Morton, 3 Grandstand Rd, Hereford, HR4 9NE [IO82PB, SO54]
G4 HZU D Hayter, 72 Broadsands Ave, Paignton, TQ4 6JW [IO80FJ, SX85]
G4 HZV R C Bagwell, 30 Christmas Pie Ave, Normanby, Normandy, Guildford, GU3 2EN [IO91PF, SU95]
G4 HZW A C Usher, 118 Mobberley Rd, Knutsford, WA16 8EP [IO83TH, SJ77]
G4 HZX N J Squibb, 127 Copers Cope Rd, Beckenham, BR3 1NY [IO91XK, TQ37]
G4 IAB A Bell, 10 Longacre, Weaverham, Northwich, CW8 3PT [IO83RG, SJ67]
G4 IAD D P Crompton, The Beeches, 6 St. Johns Wood, Lostock, Bolton, BL6 4FA [IO83RN, SD60]
G4 IAF W Heath, 30 Ellesboro Rd, Harborne, Birmingham, B17 8PT [IO92AL, SP08]
G4 IAG T J Court, Breach Oak Ln, Corley Ash, Corley, Coventry, West Midlands, CV7 8AU [IO92FL, SP38]
G4 IAJ T M Jefferson, 73 Southgate, Crossgates, Scarborough, YO12 4LZ [IO94SF, TA08]
G4 IAL J Heywood, 46 The Close, Wyre Vale Park, Garstang, Preston, PR3 1PL [IO83OV, SD44]
GM4 IAO A M Robertson, Lanton, Jedburgh, Roxburghshire, TD8 6SU [IO85QL, NT62]
G4 IAP C H Hender Poltair Sch RC, Trevarthian Rd, St Austell, Cornwall, PL25 4BZ [IO70OI, SX05]
G4 IAQ J Brooks, 28 Avon Vale Rd, Loughborough, LE11 2AA [IO92JS, SK51]
G4 IAR D R Brooks, 28 Avon Vale Rd, Loughborough, LE11 2AA [IO92JS, SK51]
G4 IAS J Martin, 19 Thorncliff Cl, Wellswood, Torquay, TQ1 2QW [IO80FL, SX96]
G4 IAT B S Smith, 157 Revidge Rd, Blackburn, BB2 6EE [IO83RS, SD62]
G4 IAU D Lilley, 65 Peel St., Horbury, Wakefield, WF4 5AN [IO93FP, SE21]
G4 IAV Details withheld at licensee's request by SSL.
G4 IAW B J Williams, 20 Hawthorn Cres, Stapenhill, Burton on Trent, DE15 9QP [IO92ES, SK22]
G4 IAY F B Whittaker, 91 Oakdale, Worsbrough Dale, Worsbrough, Barnsley, S70 5NR [IO93GM, SE30]
G4 IAZ Details withheld at licensee's request by SSL.
G4 IBC T C M Wigg RAIBC, Setters, Hyde, Fordingbridge, Hants, SP6 2QB [IO90CV, SU11]
G4 IBE C Higham, 95 Weston Rd, Olney, MK46 5AA [IO92PD, SP85]
G4 IBH D Dockery, 20 Saffron Way, Sittingbourne, ME10 2EY [JO01II, TQ96]
G4 IBI W A L Mitchell, Wychwood, The Ridgeway, Cranleigh, GU6 7HR [IO91SD, TQ03]

G4 IBL R Runkee, 21 Roslyn Rd, Anlaby Rd, Hull, HU3 6XQ [IO93TR, TA02]
G4 IBM C J Murphy, 93 Higher Blandford, Broadstone, Dorset, BH18 9AE [IO80XS, SY99]
G4 IBN K P Pointon, 82 Pontefract Rd, Knottingley, WF11 8RN [IO93IQ, SE42]
G4 IBO P J Thomas, 24 Silverberry Rd, Worle, Weston, Weston Super Mare, BS22 0RT [IO81NI, ST36]
G4 IBS G Baxendale, Sarno, Granville Rd, Darwen, BB3 2SS [IO83SQ, SD62]
GI4 IBV S G Johnston, 61 Ravenhill Park, Belfast, BT6 0DG [IO74BN, J37]
G4 IBZ P J Richardson, 10 Mosgrove Cl, Gateford, Worksop, S81 8TD [IO93KH, SK58]
G4 ICB B D Clarke, 18 Orchard Dr, Ackworth, Pontefract, WF7 7DS [IO93HP, SE41]
G4 ICC M J E Gater, 268 Main Rd, New Duston, Duston, Northampton, NN5 6PP [IO92MG, SP76]
GJ4 ICD Details withheld at licensee's request by SSL.
G4 ICE A J Mitchell, 11 Poplar Ln, Cannock, WS11 1NQ [IO82XQ, SJ91]
G4 ICF A J Denison, 40 Leysholme Dr, Leeds, LS12 4HQ [IO93ES, SE23]
G4 ICH C G Wickenden, 62 Layer Rd, Colchester, Essex, CO2 7JH [JO01KV, TL92]
G4 ICI R S Perks, 8 Laurel Cl, Lichfield, WS13 6TT [IO92CQ, SK10]
G4 ICM D J Stockley Icom (Uk) AR, 10 The Leas, Chestfield, Whitstable, CT5 3QQ [JO01MI, TR16]
G4 ICP R M Witney, 145 Broadway, Silver End, Witham, CM8 3XN [JO01HU, TL81]
G4 ICS C D Street, Yewdown House, 7 Sharpthorne Cl, Ifield, Crawley, RH11 0LU [IO91VC, TQ23]
G4 ICT Details withheld at licensee's request by SSL.
G4 ICU A B Jones, 15 High St., Sedgley, Dudley, DY3 1RL [IO82WN, SO99]
G4 ICV I J Stirling, 1 School Rd, Gretton, Corby, NN17 3BY [IO92PM, SP89]
G4 ICX W R Waddington, 1077 Clare Ave, Cambridge, Ontario Canada, N3H 2E2
G4 ICZ B J Greatrix, 12 Swainsfield Rd, Yoxall, Burton on Trent, DE13 8PT [IO92CS, SK11]
G4 IDB W K McCulloch, 6 Cannon St., Deal, CT14 6QA [JO01QF, TR35]
GW4 IDC M E Rudge, Eirianfa, 8 Penrallt Est, Llanystumdwy, Criccieth, LL52 0SR [IO72UW, SH43]
G4 IDD D R Dockar, 49 Dixon Ln, Wortley, Leeds, LS12 4RR [IO93ES, SE23]
G4 IDE R A Barker, 79 South Parade, Boston, PE21 7PN [IO92XX, TF34]
G4 IDF D G Hobro, 60 Linksview Cres, Newtown, Worcester, WR5 1JJ [IO82VE, SO85][Op: David G Hobro. QRV 7MHz to 70cms. (YM70g).]
G4 IDG G Tonge, The Old Forge, Lapley Hall Mews, Lapley, Stafford, ST19 9JN [IO82VR, SJ81]
G4 IDH I D Harris, 15 Diana House, Wittonwood Rd, Frinton on Sea, CO13 9JY [JO01OU, TM22]
G4 IDK Details withheld at licensee's request by SSL.
G4 IDL T P Wade, 47 Rig Dr, Swinton, Mexborough, S64 8UL [IO93IL, SK49]
G4 IDO Details withheld at licensee's request by SSL.
G4 IDR D A Redman, 13 Halifax Rd, Scapegoat Hill, Golcar, Huddersfield, HD7 4NS [IO93BP, SE01]
G4 IDT F Heywood, 62 Southleigh Rd, Leeds, LS11 5SG [IO93FS, SE22]
G4 IDU K V Kniveton, 155 Cot Ln, Kingswinford, DY6 9SA [IO82VL, SO88]
GM4 IDV P W Brown, Lightcost, Dykeside, Westray, Orkney, KW17 2DW [IO89LH, HY44]
G4 IDW A Compton, Aysgarth, Durley Brook Rd, Durley, Southampton, SO32 2AR [IO90IW, SU51]
G4 IDX D I O Turner, 119 Tally Ho Rd, Shadoxhurst, Ashford, TN26 1HW [JO01JC, TQ93]
G4 IDY Details withheld at licensee's request by SSL.
G4 IDZ M I Hollinshead, 14 Berrington Cl, Botcheston, Leicester, LE9 9FQ [IO92IP, SK40]
G4 IEB C S Williamson, 7 Hanbury Hill, Stourbridge, DY8 1BE [IO82WK, SO98]
G4 IEC A J Everard, 2 Oak Wood Rd, Wetherby, West Yorks, LS22 4QY [IO93GW, SE34]
GW4 IED R S Keyes, 4 Glanmor Cres, Newport, NP9 8AX [IO81MO, ST38]
GM4 IEF A Hancock, 25 Sandylands Rd, Cupar, KY15 5JS [IO86IH, NO31]
G4 IEH S G Lindell, 60 Lakenheath, Oakwood, London, N14 4RP [IO91WP, TQ29]
G4 IEI G A Ross, 53 Tolworth Gdns, Chadwell Heath, Romford, RM6 5TJ [JO01BO, TQ48]
G4 IEN G Cross, 49 Albion St., Otley, LS21 1BZ [IO93DV, SE24]
GM4 IEO F R Graham, 32 Bayview Cres, Ross & Cromarty, IV11 8YW [IO77XQ, NH76]
GW4 IEQ M McIntosh, Cartref Melys, Drury Ln, Buckley, CH7 3DY [IO83LE, SJ26]
G4 IER C D Colbeck, Chestergate, Frog Ln, Milton under Wychwood, Chipping Norton, Oxon, OX7 6JZ [IO91EU, SP21]
G4 IES W G Pitt, 1 Windy Ridge, James St, Kinver, Stourbridge, DY7 6ED [IO82VK, SO88]
G4 IET J R French, 10 Sunridge Ave, Luton, LU2 7JL [IO91TV, TL02]
GW4 IEU W A Griffiths, 3 Garreglwyd Park, Holyhead, LL65 1NW [IO73QH, SH28]
G4 IEV P D Gill, 48 Meeting House Ln, Balsall Common, Coventry, CV7 7FX [IO92EJ, SP27]
G4 IEY F J T Harris, 4 Merestones Dr, The Park, Cheltenham, GL50 2SS [IO81WV, SO92]
G4 IEZ R E Senior, 18 Piecewood Rd, Leeds, LS16 6EH [IO93EU, SE23]
G4 IFB Dr G T Hinson, Littlebrook Cottage, 379 Quemerford, Calne, SN11 8LF [IO91AK, SU06]
GM4 IFC Details withheld at licensee's request by SSL.
G4 IFD S G Lydiate, 11 Avondale Rd, Pitsea, Basildon, SS16 4TT [JO01FN, TQ78]
GW4 IFE Dr A J Strachan, 19 St. Aiden Dr, Killay, Swansea, SA2 7AX [IO71XO, SS59]
G4 IFH W I L L Coxon, 17 Leicester Rd, East Finchley, London, N2 9DY [IO91WO, TQ28]
G4 IFI C Loftus, 33 Carlisle Cres, Ashton under Lyne, OL6 8UJ [IO83WM, SD90]
G4 IFJ M Daniels, 8 Hathersage Dr, Glossop, SK13 8RG [IO93AK, SK09]
G4 IFK W J Ellis, 7 Meadowcroft, Hagley, Stourbridge, DY9 0LJ [IO82WJ, SO87]
G4 IFM Dr S J Petraitis, 69 Eastcroft Rd, Wolverhampton, WV4 4NL [IO82VN, SO89]
G4 IFO Details withheld at licensee's request by SSL.
G4 IFQ Details withheld at licensee's request by SSL.
G4 IFR P M Hanson, 42 Oak Ave, Newport, TF10 7EF [IO82TS, SJ71]
G4 IFT D W Howorth, 11A Norwood Dr, Torrisholme, Morecambe, LA4 6LT [IO84NB, SD46]
G4 IFU P Griffin, 47 Greenfield Rd, Dentons Green, St. Helens, WA10 6RB [IO83OL, SJ59]
GI4 IFV Details withheld at licensee's request by SSL.
G4 IFX C J Deacon, 12 Abbey Rd, Darlington, DL3 7RD [IO94FM, NZ21]
G4 IFY Details withheld at licensee's request by SSL.
G4 IGB Details withheld at licensee's request by SSL.
G4 IGC L D Hall, Richmond Lodge, 57 Station Hill, Swannington, Leics, LE6 4RJ [IO92JP, SK50]
G4 IGD Details withheld at licensee's request by SSL.[Op: B Walton. Station located in Leeds LS12.]
GW4 IGF P B Higgs, Parkside, Rossett, Wrexham, Clwyd, LL12 0BP [IO83NC, SJ35]
G4 IGG N E Bennett, 1 Burnham Ave, Oxley, Wolverhampton, WV10 6DX [IO82WO, SJ90]
G4 IGK M E Wickham, 43 Main St, Bishopstone, Aylesbury, Bucks, HP17 8SH [IO91NS, SP81]
G4 IGL R J Coombes, 9 Beechwood Cl, Evington, Leicester, LE5 6SY [IO92LP, SK60]
G4 IGN D I Fisher, Lyndale, Prince Cres, Staunton, Gloucester, GL19 3RF [IO81UX, SO72]
G4 IGO Details withheld at licensee's request by SSL.
GW4 IGQ G R Taylor, 32 Rhiw Melin, Upper Cwmbran, Cwmbran, NP44 5HZ [IO81LP, ST29]
GM4 IGS R M Chapman, 16 Fullarton Cres, Troon, KA10 6LL [IO75QM, NS33]
GW4 IGT R G Roberts, Clydfan, Llandegfan, Menai Bridge, Gwynedd, LL59 5TH [IO73WF, SH57]
G4 IGU K Blackett, 52 Plovers Mead, Doddinghurst, Wyatts Green, Brentwood, CM15 0PS [JO01DQ, TQ59]
G4 IGV D Baldwin, 14 Stainton Dr, Grimsby, DN33 1EG [IO93WN, TA20]
G4 IGW G D Burrows, 84 Fruitlands, Malvern Wells, Malvern, WR14 4XB [IO82UC, SO74]
G4 IGX Dr J K Higgins, 8 Delph Top, Greetby Hill, Ormskirk, L39 2DX [IO83NN, SD40]
G4 IGY G L Southwell, Mill House, Atwick Rd, Hornsea, HU18 1DZ [IO93VW, TA14]
G4 IGZ D Pellowe, 191 Preston New Rd, Blackpool, Lancs, FY3 9TN [IO83LT, SD33]
GD4 IHA E R Robson, I Fuchsia Ln, Governor Hill, Douglas, Isle of Man, IM2 7EB
GD4 IHB J Whitmore, Sound Rd, Glen Maye via, Glen Maye, Peel, Isle of Man, IM5 3BJ
GD4 IHC R H Furness, Breryk, Windsor Rd, Ramsey, IM8 3EB
G4 IHE D ARTS, 17 Ravenglass Cl, Wesham, Preston, PR4 3HZ [IO83NS, SD43]
G4 IHF E Fielding, 6 Thornton Ave, St. Annes on Sea, Lytham St. Annes, FY8 3RL [IO83LS, SD32]
G4 IHH R F Laws, Bp24, 11600 Villegailhenc, France, X X
G4 IHI P J Ferrari, Maggie, Back Rd, Wenhaston, Halesworth, IP19 9DY [JO02SH, TM47]
GM4 IHJ J Branegan, 8 Whitehills, Saline, Dunfermline, KY12 9UJ [IO86FC, NT09]
G4 IHL Details withheld at licensee's request by SSL.
GW4 IHM I C Wingfield, Keyhaven, 2 Belmont Cl, Abergavenny, NP7 5HW [IO81LT, SO31]
GW4 IHN D E Howells, 2 Belmont Cl, Abergavenny, NP7 5HW [IO81LT, SO31]
G4 IHO D Carson, 21 Harris Rd, Harpurhill, Buxton, SK17 9JS [IO93BF, SK07]
G4 IHP W Taylor, 5 Ashtree Bank, Brereton, Rugeley, WS15 1HN [IO92AR, SK01]
G4 IHR N J Allen, 8 Shoulbard, Fleckney, Leicester, LE8 8TX [IO92LM, SP69]
G4 IHS G Donn, 35 Lehmusstrasse, 90766 Fuerth, Germany
G4 IHT R Riddington, 22 Knighton Grange Rd, Leicester, LE2 2LE [IO92KO, SK60]
G4 IHV R G Cartledge, 68 Norfolk Gdns, Littlehampton, BN17 5PF [IO90RT, TQ00]
G4 IHX P Honour, 21 Castle St., Marsh Gibbon, Bicester, OX6 0HJ [IO91LV, SP62]
G4 IHY R W Clarkson, 39 Stamford Dr, Groby, Leicester, LE6 0YD [IO92JP, SK50]
G4 IHZ M J Hyde, 23 Northd. Way, Barnsley, S71 5DH [IO93GN, SE30]
GW4 IIA M A S Stamford, 13 Broad St., Knighton, LD7 1BL [IO82LI, SO27]
G4 IIB K Marshall, The Paddocks, Ordley, Hexhamshire, Northd., NE46 1SX [IO84WW, NY95]
G4 IIC C J Clifford, 11 Halfcot Ave, Pedmore, Stourbridge, DY9 0YB [IO82WK, SO98]
G4 IIG Details withheld at licensee's request by SSL.
G4 IIH P J Henson, 3 Shamrock Cl, Tollesbury, Maldon, CM9 8SZ [JO01KS, TL91]
G4 III P S Godwin, Selby Rd, Whitley Bridge, Eggborough, Goole, North Humberside, DN14 0LN [IO93KQ, SE52]
G4 IIK C F Lodge, 28 Anchor Rd, Tiptree, Colchester, CO5 0AP [JO01IT, TL81]
G4 IIN N Evans, 56 Homerton Rd, Middlesbrough, TS3 8LX [IO94JN, NZ51]
G4 IIO P Howe, 52 Watermeadow Dr, Northampton, NN3 8SS [IO92OG, SP76]
G4 IIP K J Chester, 41 Sparrey Dr, Bournville, Birmingham, B30 2LX [IO92AK, SP08]
G4 IIQ Details withheld at licensee's request by SSL.
GM4 IIR A R Nelson, 5 Scarletmuir, Lanark, ML11 7PS [IO85CQ, NS84]
G4 IIS V H Dutton, 13 Brampton Rd, St. Albans, AL1 4PN [IO91US, TL10]
G4 IIU Details withheld at licensee's request by SSL.
G4 IIX C J Wherrett, 91 Crossways, Badger Hill, York, YO1 5JE [IO93LW, SE65]
G4 IIY I N Fugler, 9 Westover Rd, Fleet, GU13 9DG [IO91OG, SU85]
G4 IJA B M Barnes, 28 Oaklands Park, Horncastle Rd, Woodhall Spa, Lincs, LN10 6UU [IO93VD, TF26]
G4 IJB R E Butterworth, 3 Derriman Glen, Ecclesall, Sheffield, S11 9LQ [IO93FI, SK38]
G4 IJC Details withheld at licensee's request by SSL.[Op: J D Cadman.]
G4 IJD J Seddon, Seedalls Farm, Easington Rd, Cow Ark, Clitheroe, BB7 3DH [IO83SV, SD64]
G4 IJE P W Turner, 61 Primley Ln, Sheering, Bishops Stortford, CM22 7NH [JO01CT, TL51]
G4 IJF Details withheld at licensee's request by SSL.[Op: Nigel Roberts (DJ0QD, ON8QD) PO Box 49, Manningtree, Essex, CO11 2SZ.]
G4 IJG J R Owen, 75 Mersey Bank Ave, Manchester, M21 7NT [IO83UK, SJ89]
G4 IJH F P Stevens, 60 Childsbridge Ln, Kemsing, Sevenoaks, TN15 6QR [JO01CH, TQ55]
G4 IJI M C Walker, 19 Highbury Pl, Headingley, Leeds, LS6 4HD [IO93FT, SE23]
G4 IJJ A J Spratt, 8 Pheasant Rise, Copdock, Ipswich, IP8 3LF [JO02NA, TM14]
G4 IJL Details withheld at licensee's request by SSL.[Op: P G Chapman.]
G4 IJM I W Arnold, 44 Elwick Ave, Acklam, Middlesbrough, TS5 8NT [IO94IN, NZ41]
G4 IJO G P Gaunt, Rose Cottage, Red Briars, Osmotherly, Northallerton, North Yorks, DL6 3AQ [IO94II, SE49]
G4 IJP D T James, 29 Cherrygarth Rd, Catisfield, Fareham, PO15 5NA [IO90JU, SU50]
G4 IJR B M Moyse, 1327 Falling Leaf L, Seabrook, Texas, USA 77586
G4 IJS J Siddon, 33 Nottingham Dr, Wingerworth, Chesterfield, Derbyshire, S42 6ND [IO93GE, SK36]
G4 IJU J D R Coles, 84 Mansfield Ln, Calverton, Nottingham, NG14 6HL [IO93LA, SK64]
G4 IJV B L Dowling, Box Cottage, Box, Stroud, Glos, GL6 9HB [IO81VQ, ST89]
G4 IKD C F Pinder, The Old School, Lower Sticker, St. Austell, PL26 7JN [IO70OH, SW94]
GI4 IKF T J Black, 147 Old Westland Rd, Cavehill Rd, Belfast, BT14 6TE [IO74AP, J37]
G4 IKH Details withheld at licensee's request by SSL.
G4 IKI Capt P R Gabriel, 6 Ray Rd, Romford, RM5 2HB [JO01BO, TQ49]
G4 IKJ P A Edwards, 34 Albion Rd, Malvern Link, Malvern, WR14 1PU [IO82UD, SO74]
G4 IKK D C Parker, 68 Chatsworth Rd, Fairfield, Buxton, SK17 7QN [IO93BG, SK07]
G4 IKL R C Hibbin, 8 Silver Ln, West Wickham, BR4 0SQ [IO91XI, TQ36]
G4 IKO L W Cain, 15 Walmsley House, Colson Way, Streatham, London, SW16 1RH [IO91WK, TQ27]
G4 IKQ R W M Kitchener, 43 Haven Cl, Swanley, BR8 7JY [JO01CJ, TQ56]
G4 IKR Details withheld at licensee's request by SSL.[Op: D Boase. Station located near Camborne.]
GM4 IKT R D Purves, 5 Forth Ct, Port Seton, Prestonpans, EH32 0TN [IO85MX, NT47]
GM4 IKU R Brown, 17 Links View, Port Seton, Prestonpans, EH32 0EY [IO85MX, NT47]
G4 IKW W Simpson, 5 Downs View Cl, Brading, Sandown, PO36 0JA [IO90KQ, SZ68]
G4 IKX D W K Thomas, 18A Stockwell Ln, Aylburton, Lydney, GL15 6DN [IO81RR, SO60]
G4 IKY D I Sillars, 34 Sandown Rd, Stevenage, SG1 5SF [IO91WW, TL22]
G4 ILA Rev W J McKae, The Rectory, St. Marys Dr, South Reddish, Stockport, SK5 7AX [IO83WK, SJ89]
GM4 ILE J V Smy, 2 Dungavel Gdns, Hamilton, ML3 7PE [IO75XS, NS75]
GW4 ILF A F Hyde, 3 Horeb Terr, Dyffryn Ardudwy, LL44 2DN [IO72WS, SH52]
G4 ILG L G Logan, 19 Fenton Ave, Barnoldswick, Colne, BB8 6HB [IO83VW, SD84]
G4 ILH J M Acott, 2 Park Hill Rd, Sidcup, DA15 7NL [JO01BK, TQ47]
G4 ILI G W Cratchley, Lambda, The Reddings, Cheltenham, GL51 6RT [IO81WV, SO92]
G4 ILK J R Leigh, 1 Silver Birch Ct, Roundswell, Barnstaple, EX31 3RJ [IO71WA, SS52]
G4 ILL J Beynon, 4 Leslie Cres, Gosforth, Newcastle upon Tyne, NE3 4AN [IO94EX, NZ26]
G4 ILM Dr M H Turnbull, 20 Victoria Rd, Gillingham, SP8 4HY [IO81UA, ST82]
G4 ILN G V Fitt, 15 Sidegate Ave, Ipswich, IP4 4JJ [JO02OB, TM14]
G4 ILP C A Borkowski, 25 Stroud Rd, Wimbledon Park, London, SW19 8DQ [IO91VK, TQ27]
G4 ILR C J Howett, Hilltop, Stalham Rd, Hoveton, Norwich, NR12 8DJ [JO02QR, TG31]
GM4 ILS Details withheld at licensee's request by SSL.[Op: R Adam, 1 Woodlands Crescent, Bishopmill, Elgin, Morayshire, IV30 2LY.]
G4 ILV D A S Burge, The Warren, Spring Ln, Marham, Kings Lynn, PE33 9HX [JO02GP, TF70]
G4 ILW J A Dingwall, 3 Baltic House, St. Matthews Rd, Brixton, London, SW2 1NQ [IO91WL, TQ37]
G4 ILX S M Sliwinski, 110 Brincliffe Edge, Rd, Brincliffe, Sheffield, S11 9BX [IO93GI, SK38]
G4 ILY J M Townsend, 7 The Furlongs, Steyning, BN44 3PE [IO90UV, TQ11]
GI4 ILZ W R Sharpe, 22 Tweskard Park, Belfast, BT4 2JZ [IO74BO, J37]
G4 IMB P G Gascoigne, 108 Blandford Ave, Castle Bromwich, Birmingham, B36 9JD [IO92CM, SP19]
GW4 IMC T R Waters, 34 Woodlands Park, Bettws, Ammanford, SA18 2HF [IO81AS, SN61]
GM4 IMD Details withheld at licensee's request by SSL.
G4 IME J Marrow, 12 Holcombe Rd, Tottington, Bury, BL8 4AR [IO83TO, SD71]
G4 IMH V K Tatman, 271 London Rd, Bedford, MK42 0PX [IO92SC, TL04]
G4 IMJ A J Turner, 2 Harlaxton St., Burton on Trent, DE13 0QZ [IO92ET, SK22]
G4 IMK Details withheld at licensee's request by SSL.
G4 IML M P Giles-Holmes, 27 Birchfield Ave, Beacon Park, Plymouth, PL2 3LA [IO70WJ, SX45]
G4 IMN Details withheld at licensee's request by SSL.
G4 IMO Details withheld at licensee's request by SSL.
G4 IMP A G Phillpott, Southways, Stombers Ln, Hawkinge, Folkestone, CT18 7AP [JO01OC, TR24]
G4 IMQ J Haworth, 7 Woodvale Rd, Ainsdale, Southport, PR8 3SU [IO83LO, SD31]
G4 IMT B H Litherland, The Old School Hous, North Wraxall, Chippenham, Wilts, SN14 7AB [IO81UL, ST87]
G4 IMU K Holley, 11 The Cres, Loughton, IG10 4PY [JO01AP, TQ49]
G4 IMV J W Mollart, 8 Harrison St., Newcastle, ST5 1NH [IO83VA, SJ84]
GM4 IMX W Marshall, 15 Craigleith Hill, Edinburgh, EH4 2EF [IO85JW, NT27]
G4 INA P G Grice, 51 Co. Dr, Tamworth, B78 3XE [IO92DO, SK20]
G4 INB Dr B C Dupree, 3 Hillary Rd, Cheltenham, GL53 9LB [IO81XV, SO91]
G4 INC C R Inch, 12 Byfield Cl, Woodmancote, Cheltenham, GL52 4PZ [IO81XW, SO92]
G4 IND M F Hubbard, 107 Pittmans Field, Harlow, CM20 3LD [JO01BS, TL41]
GM4 INE J McLeod, 17 Johnston Terr, Port Seton, Prestonpans, EH32 0BB [IO85MX, NT47]
G4 INF B J Walpole, Lyndhurst 2106, PO Box 89118, Johannesburg, South Africa, ZZ5 6CB
G4 ING J Hartley, 25 Abney Rd, Mossley, Ashton under Lyne, OL5 0AG [IO83XM, SD90]
G4 INI J C Church, 54 Blackberry Ln, Four Marks, Alton, GU34 5DF [IO91LC, SU63]
G4 INK I Marsh, 41 Rufford Rise, Sothall, Sheffield, S19 6DW [IO93GJ, SK38]
G4 INM T W Frankland, 131 Rutland Rd, Chelmsford, CM1 4BN [JO01FS, TL70]
G4 INO G E Morris, Copperfield, Masefield Dr, Cliffe Woods, Rochester,Kent, ME3 8JW [JO01FK, TQ77]
G4 INP P L Newman, 3 Red House Ln, Leiston, IP16 4JZ [JO02SE, TM46]
G4 INQ P M Taylor, 36 Brooklands Rd, Heathend, Farnham, GU9 9BS [IO91OF, SU84]
G4 INT T A Young, 33 Mill Park Ave, Hornchurch, RM12 6HD [JO01CN, TQ58]
GD4 INU F R E Haighton, 2028 Cheviot Ct, Burlington, Ontario, Canada L7P 1W8, ZZ5 6BS
G4 INV H Rimmer, 16 Mossville Rd, Liverpool, L18 7JW [IO83NJ, SJ38]
G4 INX A E Harada, 57 Lache Park Ave, Chester, CH4 8HS [IO83NE, SJ36]
G4 IOA P N Hill, The Hawthorns, Stoke Rd, Bishops Cleeve, Cheltenham, GL52 4RH [IO81XW, SO92]
GM4 IOB R A C Smith, Hestivald, Downies Ln, Stromness, KW16 3EP [IO88IX, HY20]
G4 IOD W F Marshall, Hedgeways, 63 High Moor Ln, Cleckheaton, BD19 6LW [IO93DR, SE12]
G4 IOE F R Stevenson, Odins Vei 10, N-3155 Asgardstrand, Norway
G4 IOF M East, 41 Ave Cl, Ave Rd, St. Johns Wood, London, NW8 6DA [IO91WM, TQ28]
G4 IOG J R Blackett, 70 Church Ln, Newington, Sittingbourne, ME9 7JU [JO01II, TQ86]
GW4 IOI J Davies, 59 Hazel Ct, Sketty, Swansea, SA2 8HJ [IO81AO, SS69]
G4 IOJ M Fielding, 106 Lutterworth Rd, Northampton, NN1 5JL [IO92NF, SP76]
G4 IOK C Marshall, 100 Hailey Rd, Witney, OX8 5HQ [IO91GT, SP31]
GD4 IOM W J Smith Iom ARS Con Grp, 1 High View Rd, Douglas, IM2 5BQ
G4 ION Dr E M Warrington Ionspheric P Gr, Dept of Engineering, Univ of Leicester, Univ Rd Leicester, LE1 7RH [IO92KO, SK50]
GI4 IOO R A Chambers, 32 Victoria Rd, Sydenham, Belfast, BT4 1QU [IO74BO, J37]
G4 IOP Details withheld at licensee's request by SSL.
G4 IOQ A W White, Harebells, Trefonen Rd, Trefonen, Oswestry Salop, SY10 9DZ [IO82KT, SJ22]
G4 IOR B A Rowell, 2 The Willows, Burton on The Wolds, Loughborough, LE12 5AP [IO92KS, SK52]
G4 IOT A T Hunt-Duke, 18 Hawkins Rd, Folkestone, CT19 4JA [JO01NC, TR13]
G4 IOV P G Emmerton, 8 Redwood Pl, Bognor Regis, PO21 3BS [IO90PS, SZ99]
G4 IOX A L Donnison, 23 Layston Park, Royston, SG8 9DS [IO92XB, TL34]
G4 IOY D L St.George, 8 Asmuns Hill, London, NW11 6ET [IO91VN, TQ28]
G4 IP S H Shacklock, 23 Grange Rd, Halesowen, B63 3EF [IO82XK, SO98]
G4 IPA A J Fowler IPARC, 78 Beckingham Rd, Guildford, GU2 6BU [IO91QF, SU95]
G4 IPB P J Hodgkinson, 13 Briardene, Lanchester, Durham, DH7 0QD [IO94DT, NZ14]
G4 IPE Details withheld at licensee's request by SSL.
G4 IPF L Horseman, 55 Sackville Ave, Hayes, Bromley, BR2 7JS [JO01AJ, TQ46]
G4 IPG E M Bass, 292 Thornhills Ln, Clifton, Brighouse, HD6 4JQ [IO93CR, SE12]
G4 IPH R Beans, 292 Thornhills Ln, Clifton, Brighouse, HD6 4JQ [IO93CR, SE12]
G4 IPI D H Foster, 1 Thorn Ct, Four Marks, Alton, GU34 5BY [IO91LC, SU63]
G4 IPJ C Jones, 20 Parkfield Rd, Ickenham, Uxbridge, UB10 8LN [IO91SN, TQ08]
GM4 IPK A J Steven, 27 Dalsetter Wynd, Dunrossness, Shetland, ZE2 9JJ [IO99IW, HU41]
G4 IPL R S Winters, 39 Larkhall Ln, Harpole, Northampton, NN7 4DP [IO92MF, SP66]
G4 IPM N S Terry, 15 Baldwins Cl, Bourn, Cambridge, CB3 7TH [IO92XE, TL35]
G4 IPN W A Flindall, 3 Meadow Dr, Gressenhall, Dereham, NR20 4LR [JO02KR, TF91]
G4 IPP D E Edmondson, 8 Carter Mount, Whitkirk, Leeds, LS15 7BJ [IO93GT, SE33]
G4 IPQ Details withheld at licensee's request by SSL.
G4 IPR T D Jones, 18 Golf Dr, Brighton, BN1 7HZ [IO90WU, TQ30]
G4 IPT R H Elphick, Roe Wen, Chalfont Ln, West Hyde, Rickmansworth, WD3 2XN [IO91RO, TQ09]

G4 IPV G F Mayne, 228 Tutbury Rd, Burton on Trent, DE13 0NY [IO92ET, SK22]
G4 IPW Details withheld at licensee's request by SSL.
G4 IPX F C Scott-Stapleton, 50 Woodberry Ave, North Harrow, Harrow, HA2 6AX [IO91TO, TQ18]
G4 IPY Dr A M White, 3 Guarlford Rd, Malvern, WR14 3QW [IO82UC, SO74]
G4 IPZ Details withheld at licensee's request by SSL.
GW4 IQA R C Lloyd, Llwyn Celyn, Pandy, Abergavenny, Gwent, NP7 8DN [IO81MV, SO32]
G4 IQC G R Marshall, 73 Langley, Bretton, Peterborough, PE3 8QB [IO92UO, TF10]
G4 IQD N Sivapragasam, 1 Treve Ave, Harrow, HA1 4AL [IO91TN, TQ18]
G4 IQE Details withheld at licensee's request by SSL.
G4 IQF S Wilkinson, 18 Tansey Cres, Stoney Stanton, Leicester, LE9 4BT [IO92IN, SP49]
G4 IQI Details withheld at licensee's request by SSL.[Op: A J McKinnon. Station located in Bracknell, Berks.]
G4 IQJ P M Brannon, 90 Jacksmere Ln, Scarisbrick, Ormskirk, L40 9RS [IO83MO, SD31]
G4 IQK G Evans, 14 Beach Priory Gdns, Southport, PR8 1RT [IO83LP, SD31]
G4 IQM D L Hill, 14 The Garrones, Worth, Crawley, RH10 7YT [IO91WC, TQ33]
GW4 IQP M P Lawton, Bryngarw Lodge, Brynmenyn, Bridgend, Mid Glam, CF32 8UU [IO81FN, SS98]
G4 IQQ R O Phillips, Moonraker, 2 The Close, Wilmington, Dartford, Kent, DA2 7ES [JO01CK, TQ57]
G4 IQR N M Troop, 10 Mellowdew Rd, Coventry, CV2 5GL [IO92GJ, SP37]
GM4 IQU D W C Bell, 81 Whitehouse Rd, Edinburgh, EH4 6PB [IO85IX, NT17]
G4 IQV G T Menzies, 40 Epsom Ln North, Epsom, KT18 5PY [IO91VH, TQ25]
G4 IQW A C Langford, 16 Haven Ln, Ealing, London, W5 2HN [IO91UM, TQ18]
G4 IQZ J H Long, 51 Bratton Rd, Westbury, BA13 3ES [IO81VG, ST85]
G4 IRB J B Heath, 19 Anson Rd, Swinton, Manchester, M27 5GZ [IO83TM, SD70]
G4 IRC J Gee Ipswich Rad Club, 11 Charlton Ave, Ipswich, IP1 6BH [JO02NB, TM14]
G4 IRD R T Richards, 39 North Holme Ct, Northampton, NN3 8UX [IO92NG, SP76]
G4 IRG E H Turner, 9 Wallingford Rd, Handforth, Wilmslow, SK9 3JT [IO83VI, SJ88]
G4 IRH T J Pendleton, 53 Ashby Rd, Kegworth, Derby, DE74 2DJ [IO92IU, SK42]
G4 IRJ L R Owles, 36 Starmead Dr, Wokingham, RG40 2HX [IO91OJ, SU86]
G4 IRM D M Warburton, 9 Stothard Rd, Stretford, Manchester, M32 9HA [IO83UK, SJ79]
G4 IRN Details withheld at licensee's request by SSL.
G4 IRP F A Boocock, 109 Northd., North Harrow, Middx, HA2 7RB [IO91TO, TQ18]
G4 IRQ Details withheld at licensee's request by SSL.
G4 IRR F J Beard, 23 Spring Gdns, West Molesey, KT8 2JA [IO91TJ, TQ16]
G4 IRS R O Ball, 1 Mount Hindrance C, Chard, Somerset, TA20 1BA [IO80MV, ST30]
G4 IRT Details withheld at licensee's request by SSL.
G4 IRU N R Ashcroft, 11 Greenway, Wilmslow, SK9 1LU [IO83VH, SJ88]
G4 IRV J A Hastie, Bonera Platt, The St., High Halstow, Rochester, ME3 8SG [JO01GK, TQ77]
GI4 IRW Details withheld at licensee's request by SSL.
G4 IRX N P Button, 10 Whittingham Rd, Mapperley, Nottingham, NG3 6BL [IO92KX, SK54]
G4 IRY R R Gladden, 2 Reepham Rd, Briston, Melton Constable, NR24 2LJ [JO02MU, TG03]
G4 IRZ L A Winnert, 135 Whitton Ave Eas, Greenford, Middx, UB6 0QE [IO91UN, TQ18]
G4 ISB S Brown, 7 Hawkstone Ave, Whitefield, Manchester, M45 7PG [IO83UN, SD70]
G4 ISC Details withheld at licensee's request by SSL.
GW4 ISF W L C Browning, Glan-Camlas, Talybont on Usk, Brecon, Powys, LD3 7YP [IO81IV, SO12]
G4 ISI Details withheld at licensee's request by SSL.
G4 ISK D Brighton, Datchets, 180 Medstead Rd, Beech, Alton, GU34 4AJ [IO91LD, SU63]
GM4 ISM M Hughes, 6 Hawthorn Gdns, Larkhall, ML9 2TD [IO85AR, NS75]
G4 ISN A J Holmes, 45 Windmill Way, Kegworth, Derby, DE74 2FA [IO92IU, SK42]
G4 ISO F E Wilson, 15 Byrd Walk, Baldock, SG7 6LN [IO91VX, TL23]
G4 ISP Details withheld at licensee's request by SSL.
G4 ISQ B S Jones, 7 Timbertree Rd, Cradley Heath, Warley, B64 7LE [IO82XL, SO98]
GI4 ISR C R McClurg, 104 Ave Rd, Lurgan, Craigavon, BT66 7BH [IO64UL, J05]
G4 ISS J J Proudfoot, 82 Linden Terr, Harraby, Carlisle, CA1 3PH [IO84NV, NY45]
G4 IST Details withheld at licensee's request by SSL.
G4 ISU N Whittingham, 7 Ridgedale Mount, Pontefract, WF8 1SB [IO93IQ, SE42]
G4 ISW P R Wild, 6 Ringstead Ave, Crosspool, Sheffield, S10 5SN [IO93FJ, SK38]
GM4 ISY J T Keir, 9 Fernleigh Rd, Glasgow, G43 2UD [IO75UT, NS56]
G4 ITB J W Stone, 4 Frisby Rd, Leicester, LE5 0DL [IO92KP, SK60]
G4 ITC C J Claydon, 69 Abingdon Rd, Brachla, Dorchester on Thames, Wallingford, OX10 7LB [IO91JP, SU59]
G4 ITF Details withheld at licensee's request by SSL.
G4 ITG B G Davey, 31 Somervell Dr, Fareham, PO16 7QL [IO90JU, SU50]
GM4 ITH T Henderson, 49 High St., North Berwick, EH39 4HH [IO86PB, NT58]
G4 ITI E Harrison, 35 Pembroke Rd, Bootle, L20 7BB [IO83MK, SJ39]
GW4 ITJ C J J Hard, 3 Longbridge, Ponthir, Newport, Gwent, NP6 1GT [IO81MP, ST39]
G4 ITL B J Salt, 135 Kingsland, Harlow, CM18 6XW [JO01BS, TL40]
G4 ITM E S Simpson, 49 Vicarage Rd, Hastings, TN34 3LZ [JO00HU, TQ81]
GW4 ITO C Jacobsen, 13 Conway Gr, Prestatyn, LL19 8TL [IO83HH, SJ08]
G4 ITP C W Owen, 334 Beaumont Leys Ln, Leicester, LE4 2BJ [IO92KP, SK50]
G4 ITQ B T Lindley, 3 Orchard Way, Fontwell, Arundel, BN18 0SH [IO90QU, SU90]
G4 ITR K J Fisher, 51 Edgehill, Ponteland, Newcastle upon Tyne, Tyne & Wear, NE20 9RR [IO95CA, NZ17]
G4 ITS J H Brear, Glen Vine, 162A Slad Rd, Stroud, GL5 1RH [IO81VS, SO80]
G4 ITV B Dingle, 74 Fenay Ln, Almondbury, Huddersfield, HD5 8UJ [IO93DP, SE11]
G4 ITX M J Payne, 5 Penhale Dr, Hucknall, Nottingham, NG15 6FH [IO93JA, SK54]
G4 ITY D W Hardie, 42 Lagoon Rd, Pagham, Bognor Regis, PO21 4TJ [IO90PS, SZ89]
G4 IUA J J Campbell, 61 Telegraph Ln, Claygate, Esher, KT10 0DT [IO91TI, TQ16]
G4 IUB Details withheld at licensee's request by SSL.
G4 IUE R A Exley, 4 Ouselea, York, YO3 6SA [IO93KX, SE55]
G4 IUF M H Parker, Greenacres, 23 Pannal Ave, Pannal, Harrogate, HG3 1JR [IO93FW, SE35]
G4 IUG Details withheld at licensee's request by SSL.
G4 IUH R J Pye, 7 Meadow View, Potterspury, Towcester, NN12 7PH [IO92NB, SP74]
G4 IUJ J G Wroe, 25 Yew Tree Ln, Poynton, Stockport, SK12 1PU [IO83WI, SJ98]
GW4 IUK H W Morley, 63 Lewis Rd, Neath, SA11 1DJ [IO81CP, SS79]
GW4 IUL D S Pullin, 32 Clinton Rd, Penarth, CF64 3JD [IO81JK, ST17]
G4 IUM G P Spink, 2 Gladesmere Ct, Carew Rd, Northwood, Middx, HA6 3NH [IO91SO, TQ09]
GW4 IUN R K Janes, 3 Greenway Ave, Rumney, Cardiff, CF3 8HQ [IO81KM, ST27]
G4 IUO W R Marsh, 40 Baysham St., Hereford, HR4 0EU [IO82PB, SO54]
G4 IUP R P Limbert, 33 Bartle Gill Dr, Baildon, Shipley, BD17 6UE [IO93CU, SE13]
G4 IUR Details withheld at licensee's request by SSL.
GM4 IUS N C Bethune, 9 Links Gdns, Leith, Edinburgh, EH6 7JH [IO85KX, NT27]
G4 IUT G Craig, Green Ridges, Tibberton, Newport, Salop, TF10 8NF [IO82SS, SJ62]
G4 IUV P L Clark, 329 Humber Doucy Ln, Ipswich, IP4 3PJ [JO02OB, TM14]
G4 IUX R E Williams, 1 Innage Rd, Northfield, Birmingham, B31 2DX [IO92AJ, SP07]
GW4 IUY W H Gray, 2 Cae Rhianfa, Neptune Rd, Tywyn, LL36 0TF [IO72WN, SH50]
G4 IUZ I D Cope, 30 Drovers Way, Hatfield, AL10 0PX [IO91VS, TL20]
G4 IVB R Wollaston, 35 Main Rd, Bilton, Hull, HU11 4AP [IO93US, TA13]
G4 IVC F J Wood, 20A Lynwood Ave, Felixstowe, IP11 9HS [JO01QX, TM33]
G4 IVD Rev A N James, Holy Trinity Vicarg, Oaklands Rd, Harrow Hill, Drybrook Glos, GL17 9JX [IO81RU, SO61]
G4 IVF P G Mann, 22 Cresswell Dr, Cottesmore, Oakham, LE15 7DY [IO92QR, SK91]
G4 IVG G Malone, Holmebrook, 6 Audmore Rd, Gnosall, Stafford, ST20 0HA [IO82US, SJ82]
G4 IVH S J Barrett, 26 Bassenhally Rd, Whittlesey, Peterborough, PE7 1RN [IO92WN, TL29]
GI4 IVI A G Kerr, 29 Rose Garden, Mullavilly, Tandragee, Craigavon, BT62 2NJ [IO64SI, J04]
G4 IVJ J K Harvey, 19 Loynells Rd, Rednal, Birmingham, B45 9NS [IO82XJ, SO97]
G4 IVK R T Preston, Seefield, Cooks Bank, Acton Trussell, Stafford, ST17 0RF [IO82WS, SJ91]
G4 IVM J T OShea, 31 Robson Rd, Goring By Sea, Worthing, BN12 4EE [IO90TT, TQ10]
G4 IVN G M Hewitt, 79 Beach Rd, Caister on Sea, Great Yarmouth, NR30 5HD [JO02UP, TG51]
G4 IVO R Hargreaves, 23 Bracken Rd, Long Eaton, Nottingham, NG10 4DA [IO92IV, SK43]
G4 IVQ H Lee, Brambletye Pannell, Old Frensham Rd, Lower Bourne, Farnham Surrey, GU10 3PB [IO91OE, SU84]
G4 IVR T C Clarke, 254 Birchfield Rd East, Abington, Northampton, NN3 2SY [IO92NG, SP76]
G4 IVS Details withheld at licensee's request by SSL.
G4 IVT G C Coleman, 111 Woodland Dr, Watford, WD1 3DA [IO91TP, TQ09]
G4 IVU A J Dixon, 349 Meadgate Ave, Chelmsford, CM2 7NL [JO01FR, TL70]
G4 IVV J Perkins, 26 Petersfield Rd, Duxford, Cambridge, CB2 4SF [JO02BC, TL44]
G4 IVZ G W Harper, Saentisstr 4, Ch8123 Ebmatingen, Switzerland, ZZ9 9SA
G4 IWA J R Arrowsmith, 16 Mancetter Rd, Mancetter, Atherstone, CV9 1NZ [IO92FN, SP39]
G4 IWB M P Brooks, 45 Northfield Rd, Townhill Park, Southampton, SO18 2QE [IO90HW, SU41]
G4 IWC B Catterall, Portsend, Eastleigh Terr, North Country, Redruth Cornwall, TR16 4AJ [IO70JF, SW64]
G4 IWD G S Craig, 41 Clifton Rd, Greenford, UB6 8SP [IO91TM, TQ18]
G4 IWE G H Miles, 6 Richmond House, 86 Harestone Valley, Rd, Caterham Surrey, CR3 6HF [IO91WG, TQ35]
G4 IWF G F Mason, 51 Egerton Rd, Streetly, Sutton Coldfield, B74 3PG [IO92BN, SP09]
G4 IWG D Midgaff, 118 Stanley Rd, Hinckley, LE10 0HT [IO92HN, SP49]
G4 IWI J S Stocking, Gr Rd, Melton Constable, Norfolk, NR24 2DE [JO02MU, TG03]
G4 IWJ R Towle, Elwood, Barrasford, Hexham, Northd., NE48 4AN [IO85WB, NY97]
GM4 IWK W F C Hind, 32 Hawthorn Bank, Duns, TD11 3HH [IO85US, NT75]
G4 IWN J R Andrews, 5 Chapman Ave, Maidstone, ME15 8EG [JO01GG, TQ75]
GI4 IWP E W Maclaine, 105 Bencran Rd, Sixmilecross, Omagh, BT79 9QA [IO64JO, H57]
G4 IWQ D J Cannon, 57 Halswell Rd, Clevedon, BS21 6LE [IO81NK, ST47]
G4 IWR S A Berry, 16 Queen Elizabeth Way, Eastfield Rd, Barton upon Humber, DN18 6AJ [IO93SQ, TA02]
G4 IWS C R Caine, 10 Goodwood Cl, Burghfield Common, Reading, RG7 3EZ [IO91LJ, SU66]
GW4 IWU J Scrivens, 5 Briardene, Lanchester, Durham, DH7 0QD [IO94DT, NZ14]
G4 IWV I Parker, 12 Woodside, Siddington, Macclesfield, SK11 9LG [IO83VF, SJ87]
G4 IWW B S Waite, Tigaiga, 6 Lark Hill, Swanwick, Alfreton, DE55 1DD [IO93HB, SK45]
G4 IWZ R G Talbot, 41A Mathews Way, Abingdon, OX13 6JX [IO91IQ, SP40]
G4 IXB C F Tuvey, 1 Dorset Way, Heston, Hounslow, Middx, TW5 0NF [IO91TL, TQ17]
GW4 IXC W R Uzzell, 30 Oak Ct, Myrtle Cl, Penarth, CF64 3NQ [IO81JK, ST17]
G4 IXD I Palgrave Brown, The Abbey House, The St., Marham, Kings Lynn, PE33 9HP [JO02GP, TF70]
G4 IXE G M Walmsley, Warwick Farm House, Cracknore Hard Ln, Marchwood, Southampton, SO40 4UT [IO90GV, SU31]
G4 IXF D J Toon, 26 Reddish Ave, Whaley Bridge, Stockport, Ches, SK12 7DP [IO83XI, SJ98]
GM4 IXH Dr J K Finlayson, 7 Abbotshall Rd, Cults, Aberdeen, AB1 9JX [IO87TH, NJ72]
G4 IXI T L Grant, School House, Church Rd, Wittering, Peterborough, PE8 6AF [IO92SO, TF00]
G4 IXK R Bennett, 59 Hinderheath Rd, Wheelock, Sandbach, Ches, CW11 9LY [IO83TD, SJ76]
G4 IXL R Swinney, 9 Paulhan, Great Field, London, NW9 5TN [IO91VO, TQ29]
G4 IXM M J Rail, 16 The Cres, Truro, TR1 3ES [IO70LG, SW84]
G4 IXN Details withheld at licensee's request by SSL.
G4 IXP J H Ayers, 39 Southlands Way, Congresbury, Bristol, BS19 5BW [IO81OI, ST46]
G4 IXT I P Jefferson, 1 Blackberry Cl, Boughton Vale, Rugby, CV23 0UJ [IO92JJ, SP57]
G4 IYA M P Adams, 23 Springvale, Iwade, Sittingbourne, ME9 8RY [JO01IJ, TQ86]
G4 IYB LCol W A Guest, 21 Hill Rise, Hinchley Wood, Esher, KT10 0AL [IO91UJ, TQ16]
G4 IYC B T Couchman, 88 Grange Rd, Gillingham, ME7 2PX [JO01GJ, TQ76]
G4 IYE R Smith, 72 Worthing Rd, Patchway, Bristol, BS12 5HX [IO81RM, ST58]
G4 IYI N R Wright, 16 Casterton, Euxton, Chorley, PR7 6HN [IO83PP, SD51]
G4 IYJ Details withheld at licensee's request by SSL.
G4 IYK S Dixon, 33 Medhurst Cres, Gravesend, DA12 4HJ [JO01EK, TQ67]
G4 IYL Details withheld at licensee's request by SSL.
GI4 IYO K J G Burnside, 4 Cuttles Rd, Comber, Newtownards, BT23 5YX [IO74CN, J46]
G4 IYP F Dearden, 22 Claremont Rd, Chorley, PR7 3NH [IO83QP, SD51]
G4 IYS D P Burgess, 13 St. Margarets Dr, Leire, Lutterworth, LE17 5HW [IO92JM, SP58]
G4 IYT D R Lett, 79 Dinas Ln, Huyton, Liverpool, L36 2NN [IO83NK, SJ49]
G4 IYU Details withheld at licensee's request by SSL.
G4 IYW W Orr, 14 Dorchester Rd, Lawns, Swindon, SN3 1LH [IO91CN, SU18]
GM4 IYZ J J Potts, PO Box 1294, Kato Paphos, 8132, Cyprus
G4 IZ J M Passmore, Inf Systems Divisio, Shape, Bfpo 26, ZZ2 6IN
G4 IZA D P Howard, 18 Chatfield Pl, Painted Post, New York 14870, USA
G4 IZB G G Robinson, 3 Wellington Cl, Sandhurst, Camberley, Surrey, GU17 8AJ [IO91OH, SU85]
G4 IZC Details withheld at licensee's request by SSL.
G4 IZE A H Taylor, 5 Farm Cl, Castle Park, Whitby, YO21 3LS [IO94QL, NZ81]
GI4 IZF M R Weller, 58 Manse Rd, Ballycarry, Carrickfergus, BT38 9LF [IO74CS, J49]
G4 IZH P Robinson, 24 Haveroid Way, Crigglestone, Wakefield, WF4 3PG [IO93FP, SE31]
G4 IZI C D Thomas, 25 Loverock Cres, Rugby, CV21 4AJ [IO92JI, SP57]
GW4 IZJ P J Rennick, 41 Church Rd, Pontnewydd, Cwmbran, NP44 1AT [IO81LP, ST29]
G4 IZK M J Arnold, Langdale, 26 Hill Side, Kingsbury, Tamworth, B78 2NH [IO92DN, SP29]
G4 IZL Details withheld at licensee's request by SSL.
G4 IZM J H Duncan, 21 Montgomery Dr, Bilton, Rugby, CV22 7LA [IO92II, SP47]
GM4 IZN D Scott, Beechview, Enzie Slackhead, Buckie, Moray, AB5 2BR [IO87TH, NJ72]
G4 IZO R Cochrane, 151 Fawe Park Rd, Putney, London, SW15 2EG [IO91VL, TQ27]
G4 IZQ A Scarth, 1 Beechwood Ave, Whitley Bay, NE25 8EP [IO95GB, NZ37]
G4 IZR J Wall, 35 Latimer Rd, London, E7 0LQ [JO01AN, TQ48]
G4 IZS R D Sexton, Banavie, 50 Manor Ave, Cam, Dursley, GL11 5JF [IO81TQ, SO70]
G4 IZU D J Byers, 16 Tealby Ct, Georges Rd, London, N7 8HY [IO91WN, TQ38]
G4 IZV N J Coombs, 23 Rue Du Sartau, B-1325 Dion-le-Val, Belgium, ZZ2 3RU
G4 IZW K J Hatton, Hamilton House, Boat Rd, Bellingham, Northd., NE48 2AP [IO85UD, NY88]
G4 IZX P H Beards, 3 Elm Dr, Brightlingsea, Colchester, CO7 0LA [JO01MT, TM01]
GM4 IZY A J Wills, 23B South Guildry S, Elgin, Moray, IV30 1QN [IO87IP, NJ26]
G4 JA Details withheld at licensee's request by SSL.
G4 JAA P J Hawkins, 38 Davidson Cl, Great Cornard, Sudbury, CO10 0YU [JO02JA, TL84]
G4 JAC R J Emeny, 28 Manor House Way, Brightlingsea, Colchester, CO7 0QR [JO01MT, TM01]
GW4 JAD D C Rowlands, 17 Avondale Ct, Porth, CF39 9NH [IO81HO, ST09]
GM4 JAE I M G Miller, Moorgate, 5 Heathcote Gdns, Inverness, IV2 4AZ [IO77VL, NH64]
G4 JAG Details withheld at licensee's request by SSL.[Op: C D W Marcroft. Station located near Rossendale.]
G4 JAH A K Fleming, 171A Bells Hill, Barnet, Herts, EN5 2TB [IO91VP, TQ29]
G4 JAI Details withheld at licensee's request by SSL.
G4 JAJ B A Noble, 19 Ayrton Ave, Blackpool, FY4 2BW [IO83LT, SD33]
G4 JAK J A Kelly, 20 Philip Ave, Sticklepath, Barnstaple, EX31 3AQ [IO71XB, SS53]
G4 JAL M R Vaslet, Heatherlea, Adbury Holt, Newtown, Newbury, RG20 9BN [IO91II, SU46]
G4 JAQ M A Crofts, 43 Broadlands Dr, East Ayton, Scarborough, YO13 9ET [IO94SG, SE98]
G4 JAR I T Melville Hadrabs Cont Gr, 3 Cres Rd, Benfleet, SS7 1JL [JO01GN, TQ78]
G4 JAV W C Bird, 65 Kingsdown Rd, Chase Terr, Walsall, West Midlands, WS7 8PZ [IO92AQ, SK01]
G4 JAX A J Lunn, 11 Dibden Lodge Cl, Hythe, Southampton, SO45 6AY [IO90HU, SU40]
G4 JAZ W E James, 155 Welbeck Rd, Long Eaton, Nottingham, NG10 4GR [IO92IV, SK43]
G4 JBA J B Alderman, 38 Greenacres, Shoreham By Sea, BN43 5WY [IO90UU, TQ20]
G4 JBE D J Lacey, 16 Abbots Way, Monks Risborough, Princes Risborough, HP27 9JZ [IO91OR, SP80]
G4 JBF Dr G A Lester, 4 Napier Ct, Heaton Moor Rd, Stockport, SK4 4LB [IO83VK, SJ89]
G4 JBG A S Retter, 2 Gillingham Ct, Chard, TA20 1DR [IO80MV, ST30]
G4 JBH A C Dening, 80A Preston Gr, Yeovil, Somerset, BA20 2DA [IO80QW, ST51]
G4 JBI Details withheld at licensee's request by SSL.
G4 JBK A Maude, Anthony Fold Farm, Bye Rd, Shuttleworth, Lancs, BL0 0RY [IO83UP, SD81]
G4 JBL C H White, Pegasus, Penny St., Sturminster Newton, DT10 1DF [IO80UW, ST71]
GW4 JBQ J P Cleak, 71 Pillmawr Rd, Malpas, Newport, NP9 6WG [IO81LO, ST39]
G4 JBR P Dixon, Hardwick House, New Rd, South Molton, EX36 4BH [IO81CA, SS72]
G4 JBU D B Probert, 50 Ferndale Rd, Oldbury, Warley, B68 8AP [IO82XL, SO98]
G4 JBW D R Barber, 3 Vestry Rd, St, BA16 0HY [IO81PD, ST43]
G4 JBX A J Booth, 22 Cleeve, Glascote, Tamworth, B77 2QD [IO92DP, SK20]
G4 JBY G Bowden, 78 Lynwood Ave, Darwen, BB3 0HZ [IO83SQ, SD62]
G4 JCA C J How, 1 Peartree Rd, Luton, LU2 8AZ [IO91TV, TL12]
G4 JCC S P Richardson, 52 Saltems Ln, Hayling Island, PO11 9PJ [IO90MS, SZ79]
GW4 JCD D R Evans, 1 School Rd, Gurnos, Lower Cwmtwrch, Swansea, SA9 1EQ [IO81CS, SN70]
GW4 JCE E Hawker, 31 Ystad Celyn, Maesteg, CF34 9LT [IO81EO, SS89]
G4 JCF G E Hoey, Freiherr Vom Steinster 14, 63303 Dreieich Sprendlingen, Germany
G4 JCG P J Chapman, 77 Leicester Rd, Measham, Swadlincote, DE12 7JG [IO92FQ, SK31]
G4 JCH B F Hercombe, 13 Dovecote, Shepshed, Loughborough, LE12 9RW [IO92IS, SK41]
G4 JCI C R Newport, 5 Honiton Rd, Clevedon, BS21 6LR [IO81NK, ST47]
G4 JCJ C J Newman, 1 Arnhill Rd, Gretton, Corby, NN17 3DN [IO92PM, SP89]
G4 JCK N D Warnock, 39 Tewkesbury Walk, Newport, NP9 5HP [IO81MO, ST38]
G4 JCL D Bryan, 3 New Ln, Skelmanthorpe, Huddersfield, HD8 9EH [IO93EO, SE21]
GM4 JCM A G Glashan, 35 Lochinver Cres, Gowrie Park, Dundee, DD2 4UA [IO86LL, NO33]
G4 JCN Details withheld at licensee's request by SSL.
GW4 JCO A J Jolly, Kilmallock, Highfield Rd, Osbaston, Monmouth, NP5 3HR [IO81PT, SO51]
G4 JCP Dr P J Cadman, 21 Scotts Green Cl, Scotts Green, Dudley, DY1 2DX [IO82WM, SO98]
GM4 JCR G M Donald, The Honeyholm, Balfron, Stirlingshire, G63 0QE
G4 JCS J C Stevenson, Highfields Farm, Loftus, Saltburn By The Sea, Cleveland, TS13 4UG [IO94NM, NZ71]
GM4 JCW R A Clark, 68 Porthill Ct, Gallowgate, Aberdeen, AB2 1BH [IO87TH, NJ72]
G4 JCX C I Gallacher, Fairview, 345 London Rd, Clanfield, Waterlooville, Hants, PO8 0PJ [IO90MW, SU71]
G4 JCY R Thornton, 8 The Meadows, Hassocks, BN6 8EH [IO90WW, TQ31]
G4 JCZ A C Clifton, 87 Aubrey Rd, Birmingham, B32 2BA [IO92AL, SP08]
G4 JDC L J Boddington, 33 Sorrel House, Tyburn Rd, Birmingham, B24 0TQ [IO92CM, SP19]
G4 JDD K J Maddock, Pleinmont, 52 Beech Rd, Branston, Lincoln, Lincs, LN4 1PR [IO93SE, TF06]
G4 JDE G Evans, Penchateau, Wesley Ct, Whitecroft, Lydney, GL15 4RF [IO81RS, SO60]
G4 JDG C Aitchison, Bolton House, Windmill Hill, Hampstead, London, NW3 6SJ [IO91VN, TQ28]
G4 JDI A J Barratt, 20 Cross Ln, Burton Lazars, Melton Mowbray, LE14 2UH [IO92NR, SK71]
G4 JDK M Hopkinson, 11A Red Ln, South Normanton, Alfreton, DE55 3HA [IO93HC, SK45]
G4 JDL Details withheld at licensee's request by SSL.
G4 JDO R Tew, 4 Chetwode Cl, Allesley Park, Coventry, CV5 9NA [IO92FJ, SP27]
G4 JDP S J Pallett, 6 Lancaster Cl, Agar Nook, Coalville, LE67 4TG [IO92IR, SK41]
G4 JDR P Pickhaver, 9 Barnway, Englefield Green, Egham, TW20 0QU [IO91RK, SU97]
G4 JDS L R Radley, 34 Queens Rd, Chelmsford, CM2 6HA [JO01FR, TL70]
GM4 JDU I E McGarvie, 54 High St. South, Crail, Anstruther, KY10 3RB [IO86QG, NO60]
G4 JDV Details withheld at licensee's request by SSL.
G4 JDW L Nelson-Jones, 15 Gainsborough Rd, Bournemouth, BH7 7BD [IO90BR, SZ19]
GI4 JDX C S Robinson, Glenfaba Lodge, 22 Massey Ave, Belfast, BT4 2JT [IO74BO, J37]
GW4 JDZ D H Samuel, 61 Bolgoed Rd, Pontarddulais, Swansea, SA4 1JF [IO71XR, SN50]
G4 JEC C G Cox, 5620 36th Ave So, Minneapolis, Minnesota 55417, USA, X X
G4 JED K L Bird, 2 Half Moon Ln, Hildenborough, Tonbridge, TN11 9HU [JO01CF, TQ54]
G4 JEE R A Barton, Apple Acre, Avisford Park Rd, Walberton, Arundel, BN18 0AP [IO90QU, SU90]

GM4 JEF D E Wood, Little Isegarth, Sanday Island, Orkney, KW17 2BL [IO89RF, HY64]
G4 JEG Details withheld at licensee's request by SSL.
G4 JEH C A Luke, 4 The Poplars, Cudworth Park, Newdigate, Dorking, Surrey, RH5 5BL [IO91UD, TQ24]
G4 JEI N R Osborne, 17 Rogate Cl, Sompting, Lancing, BN15 0DY [IO90TU, TQ10]
GM4 JEJ M J Thomson, 61 Millgate, Friockheim, Arbroath, DD11 4TW [IO86QP, NO54]
GM4 JEM W P Redpath, 69 Ulster Cres, Edinburgh, EH8 7JL [IO85KW, NT27]
G4 JEN Details withheld at licensee's request by SSL.
G4 JEO F W Kemp, 42 Baker Rd, Abingdon, OX14 5LW [IO91IP, SU49]
G4 JEP E V Towndrow, Sarn, Pankridge St., Crondall, Farnham, GU10 5RG [IO91NF, SU74]
G4 JEQ Details withheld at licensee's request by SSL.
GI4 JER S W Bell, 42 Clogher Rd, Lisburn, BT27 5PQ [IO64XL, J26]
G4 JES M J Wells, 16 Delph Rd, North Hykeham, Lincoln, LN6 9RF [IO93QE, SK96]
G4 JET Details withheld at licensee's request by SSL.
G4 JFB Details withheld at licensee's request by SSL.
G4 JFC G W Hainsworth, Apartment 9, 2/4 Leicester Rd, Northampton, Northants, NN2 6AQ [IO92NF, SP76]
G4 JFD D Featherstone, Olicana, 6 Claremont Gdns, Tunbridge Wells, TN2 5DD [JO01DD, TQ53]
G4 JFF C J Webb, 68 Higgs Field Cres, Cradley Heath, Warley, B64 6RB [IO82XL, SO98]
G4 JFG J R May, 20 Thornbridge Rd, Iver, SL0 0QD [IO91RN, TQ08]
G4 JFH S Draycott, 4 Corn Market Hill, Howden, Goole, DN14 7BU [IO93NR, SE72]
G4 JFI D Shipman, 4 Uplands Ave, East Ayton, Scarborough, YO13 9EU [IO94SG, SE98]
G4 JFN R Hudson, 15 Fellows Rd, Farnborough, GU14 6NU [IO91PG, SU85]
G4 JFO Details withheld at licensee's request by SSL.
GI4 JFP D A Goodman, 60 Castlewood Ave, Coleraine, BT52 1EW [IO65PD, C83]
G4 JFR B W Edwards, 12 Salcombe Circus, Arnold, Redhill, Nottingham, NG5 8JJ [IO93KA, SK54]
G4 JFS J Fitzsimons, 27 Brese Ave, Warwick, CV34 5TS [IO92FH, SP26]
G4 JFT Details withheld at licensee's request by SSL.
G4 JFV R D Oldroyd, Greenwayes, Links Ln, Pleasington, Blackburn, BB2 5JL [IO83RR, SD62]
G4 JFX B Mount, 4 Maplestone Rd, Whitchurch, Bristol, BS14 0HH [IO81RJ, ST66]
G4 JFY H J Crane, 186 Harepath Rd, Seaton, EX12 2HE [IO80LR, SY29]
G4 JFZ G A Lacey, 7 Ash Cl, Walters Ash, High Wycombe, HP14 4TR [IO91OQ, SU89]
G4 JG J H Gurr, 14 Southborough Cl, Surbiton, KT6 6PU [IO91UJ, TQ16]
G4 JGA M Stone, 151 Heanor Rd, Smalley, Ilkeston, DE7 8TA [IO92IX, SK44]
G4 JGB Details withheld at licensee's request by SSL.
G4 JGE M Green, 9 Oaklands, Hayes Ln, Kenley, CR8 5LB [IO91WH, TQ36]
G4 JGF J G Fitzgerald, 21 St. Aidans Ave, Darwen, BB3 2BS [IO83SQ, SD62]
G4 JGG J W H Pether, 7 Celina Cl, Bletchley, Milton Keynes, MK2 3LS [IO91OX, SP83]
G4 JGH A B Allchin, 9 Ashfield Rd, Kingsheath, Birmingham, B14 7AS [IO92BK, SP08]
G4 JGJ Details withheld at licensee's request by SSL.
G4 JGL Details withheld at licensee's request by SSL.
GM4 JGO J Baguant, 28 Baberton Mains, Bra, Edinburgh, EH14 3HH [IO85IV, NT16]
G4 JGP N Else, 12 The Ridgeway, Higher Bebington, Bebington, Wirral, L63 5NR [IO83LI, SJ38]
G4 JGQ J E Bevan, 10 Streamdale, Abbeywood, London, SE2 0PD [JO01BL, TQ47]
G4 JGS S W Harding, 9 Lightsfield, Oakley, Basingstoke, RG23 7BL [IO91JG, SU55]
G4 JGT C L Call, 35 Polmennor Rd, Golden Bank, Falmouth, TR11 5UX [IO70KD, SW73]
GW4 JGU A J Green, 9 Westbourne Gr, Sketty, Swansea, SA2 9DT [IO81AO, SS69]
G4 JGV S D Sharred, 69 Petersfield Rd, Hall Green, Birmingham, B28 0AU [IO92BK, SP18]
GW4 JGW K R Simpson, 59 Midland Pl, Llansamlet, Swansea, SA7 9QX [IO81BP, SS69]
G4 JGX Details withheld at licensee's request by SSL.
G4 JHA R H Thomas, 2 Woodlands Rd, Astley, Tyldesley, Manchester, M29 7BH [IO83SM, SD60]
G4 JHC P R Blunn, 6 Cranbury Rd, Eastleigh, SO50 5HA [IO90HX, SU41]
G4 JHD P J Case, 62 Deacon Rd, Bittern, Southampton, SO19 7PW [IO90HV, SU41]
G4 JHE M Green, 1 Morley Hill, Corringham, Stanford-le-Hope, SS17 8HP [JO01FM, TQ68]
G4 JHF Details withheld at licensee's request by SSL.
GM4 JHG R R Smith, 49 Springbank Gdns, Falkirk, FK2 7DF [IO86CA, NS88]
GU4 JHH R K Harvey, Courtil Masse, Les Landes, Vale, Guernsey, GY3 5JD
G4 JHI D N Miller, 10 Fairview, Horsham, RH12 2PY [IO91TB, TQ13]
G4 JHK Dr M D Bayliss, 94 Birchlands, Bridgnorth, WV15 5ED [IO82TM, SO79]
G4 JHL Details withheld at licensee's request by SSL.
G4 JHM N Dean, 51 Abbotsbury Ct, Horsham, RH13 5PT [IO91UA, TQ22]
G4 JHN J M Unwin, 28 Wallett Ave, Beeston, Nottingham, NG9 2QR [IO92JW, SK53]
G4 JHO D Lawson, 11 Cartledge Ave, Grimsby, DN32 8ES [IO93XN, TA20]
G4 JHP J A Hawes, 13 Broadmead Rd, Colchester, CO4 3HB [JO01LV, TM02]
G4 JHQ Dr C S J Kear, 73 Park St., Wombwell, Barnsley, S73 0HJ [IO93HM, SE40]
G4 JHS P Hey, 47 Hillcrest Rd, Thornton, Bradford, BD13 3PQ [IO93BT, SE03]
G4 JHU N R Fineman, Deansway, 2 The Dr, Chorleywood, Rickmansworth, WD3 4EB [IO91SP, TQ09]
G4 JHW D C Morrison, 55 Barnmead Rd, Beckenham, BR3 1JF [IO91XJ, TQ36]
G4 JHY K P Macmillan, 17 Plaistow Cres, Higher St. Budeaux, Plymouth, PL5 2EA [IO70VJ, SX45]
G4 JIA J T Marshall, 28 Edenwall Rd, Milkwall, Coleford, GL16 7LA [IO81QS, SO50]
GM4 JIB Details withheld at licensee's request by SSL.[Op: R A Chapman, 185 Carlisle Road, Blackwood, Lanark, ML11 9SB.]
GI4 JIC P G McAuley, 54 Upper Malone Par, Upper Malone Rd, Belfast, N Ireland, BT9 6PP [IO74AN, J36]
G4 JIE R L Newman, 38 Vine Dr, Wivenhoe, Colchester, CO7 9HB [JO01LU, TM02]
G4 JIG E R J White, 7 Leyfield, Markstey, Colchester, CO6 1LZ [JO01JV, TL92]
G4 JIH K Adams, 12 Kennedy Cl, Purbrook, Waterlooville, PO7 5NZ [IO90LU, SU60]
G4 JII R Green, Kingswood, Red House Ln, Adwick-le-St., Doncaster, DN6 7EA [IO93JN, SE50]
G4 JIJ I J Kraven, 55 Cranfield Cres, Cuffley, Potters Bar, EN6 4DZ [IO91WR, TL30]
G4 JIK D G Bird, Church Farm, Main St., Taddington, Buxton, SK17 9TU [IO93CF, SK17]
G4 JIN V M Gledhill, 38 Downlands, Waltham Abbey, EN9 1UH [JO01AQ, TL30]
G4 JIO K C Mason, 5 Davenport Ave, Hessle, HU13 0RL [IO93SR, TA02]
GI4 JIP Details withheld at licensee's request by SSL.
G4 JIQ W R Barker, 69 Britten Rd, Brighton Hill, Basingstoke, RG22 4HN [IO91KF, SU64]
G4 JIR A J Rixon, 12 Vancouver St., Darlington, DL3 6HN [IO94FM, NZ21]
G4 JIU I J McGarrigle, 177 Tollers Ln, Coulsdon, CR5 1BJ [IO91WH, TQ35]
GI4 JIW Details withheld at licensee's request by SSL.
GW4 JIY U J N Dixon, 77 Hendre Rd, Llangennech, Llanelli, SA14 8TH [IO71WQ, SN50]
G4 JIZ J H Rosling, 43 Holywell, Bakewell, DE45 1BA [IO93DF, SK26]
G4 JJ J A Ward, 44 Northgate, Barnsley, S75 2QH [IO93GN, SE30]
G4 JJB J J Broadhurst, 45 The Derby, Marton Manor Park, Marton in Cleveland, Middlesbrough, TS7 8RA [IO94JM, NZ51]
G4 JJC G J Lodge, 8 Thirlmere Ave, Wyke, Bradford, BD12 9DS [IO93CR, SE12]
GI4 JJD G E Curran, Sandycove, Rathmullan Lower, Minerstown, Downpatrick Co Down, N Ireland, BT30 8SU [IO74DG, J43]
GI4 JJF K A McIlroy, 69 Morston Park, Bangor, BT20 3ER [IO74DP, J48]
GM4 JJG R J Marshall, Hillcrest, Hillside Rd, Gourock, PA19 1NP [IO75OW, NS27]
G4 JJH J J Herbert, 8 Falmouth Rd, Springfield, Chelmsford, CM1 6HY [JO01FR, TL70]
GM4 JJJ D G L Anderson, Braeside, Urquhart, By Crossford, Fife, KY12 8QL [IO86GB, NT08]
G4 JJK M E Townley, P O Box 413, Larnaca, Cyprus
G4 JJM M Allison, 19 Ash Gr, Kirklevington, Yarm, TS15 9NQ [IO94HL, NZ40]
G4 JJN A Sambrook, 73 Hayes Dr, Barnton, Northwich, CW8 4JX [IO83RG, SJ67]
G4 JJP R P Thomas, 18 Eason Dr, Abingdon, OX14 3YD [IO91JQ, SU59]
G4 JJQ J H A Wheway, 25 Mount View Ave, Scarborough, YO12 4EW [IO94TG, TA08]
GW4 JJR L H James, 65 Fflorens Rd, Treowen, Newbridge, Newport, NP1 4DW [IO81KQ, ST29]
G4 JJS S J Harrison John Jamesons School ARC, St Ives Manor, St. Ives Est, Harden Rd, Bingley, West Yorks, BD16 1AT [IO93BU, SE03]
GW4 JJV M A Bell, 6 Owain Cl, Cyncoed, Cardiff, CF2 6HN [IO81JM, ST17]
GW4 JJW A J Bell, 6 Owain Cl, Cyncoed, Cardiff, CF2 6HN [IO81JM, ST17]
G4 JJX M A Grange, 59 Windermere Dr, Rainham, Gillingham, ME8 9DX [JO01HI, TQ86]
G4 JJY J M Carline, 9 Hamilton Rd, Scunthorpe, North Lincs, DN17 1BD [IO93QN, SE80]
G4 JKA J D Ewen-Smith, 57 Ludlow Rd, Church Stretton, SY6 6AD [IO82OM, SO49]
GM4 JKB J K Barnes, Capricorn, 13 Marchhill Dr, Dumfries, DG1 1PP [IO85EB, NX97]
G4 JKC P R Howard, 72 Marlowe Way, Lexden, Colchester, CO3 4JP [JO01KV, TL92]
G4 JKD Details withheld at licensee's request by SSL.
G4 JKE Details withheld at licensee's request by SSL.
G4 JKF B Hodges, Glebe House, Rock Ln, Standon, nr Eccleshall Staff, ST21 6QZ [IO82UV, SJ83]
G4 JKH J M Phillips, 57 New Sturton Ln, Garforth, Leeds, LS25 2NW [IO93HT, SE43]
G4 JKK A J Bexley, 8 Warren Cl, Hartley Wintney, Hook, RG27 8DS [IO91NH, SU75]
G4 JKM D E Twigg, 31 Parklands, Malmesbury, SN16 0QH [IO81WO, ST98]
G4 JKN M A Paull, 49 Orchard Rd, Barnstaple, EX32 9JJ [IO71XB, SS53]
G4 JKO Details withheld at licensee's request by SSL.
G4 JKQ T Bowen, 40 Grange Rd, Ibstock, LE67 6LF [IO92HQ, SK41]
G4 JKS M H Claytonsmith, 2 Falcon Dr, Hartford, Huntingdon, PE18 7LP [IO92WI, TL27]
GM4 JKT Dr O A Thores, 5 Havens Edge, Limekilns, Dunfermline, KY11 3LJ [IO86GA, NT08]
GW4 JKV M M C Rackham, 31 Severn Rd, Crown Est, Pontllanfraith, Blackwood, NP2 2GA [IO81JP, ST19]
G4 JKW A Whitehead, 3 Darley Yard, Worsbrough Dale, Worsbrough, Barnsley, S70 4SB [IO93GM, SE30]
G4 JKY E M T Lennox, Blazefield Hse Farm, Blazefield, Pateley Bridge, Harrogate, HG3 5DR [IO94DC, SE16]
G4 JKZ K N Leggett, 83 Treloweth Way, Pool, Redruth, TR15 3TS [IO70IF, SW64]
G4 JLB Details withheld at licensee's request by SSL.
GM4 JLD P Woods, 320 Stewarton St., Wishaw, ML2 8DT [IO85BS, NS85]
GI4 JLF R G Russell, 1 Belmont Dr, Belfast, BT4 2BL [IO74BO, J37]
G4 JLG Dr D A Yorke, 40 Edge Fold Rd, Worsley, Manchester, M28 7QF [IO83TM, SD70]
G4 JLI Details withheld at licensee's request by SSL.
G4 JLJ L B Bailey, 20 Linden Cl, Hutton Rudby, Yarm, TS15 0HX [IO94IK, NZ40]
GW4 JLK J Ellwood, Derwen-Las, Vale View, Pontneathvaughan, Neath West Glams, SA11 5UN [IO81ES, SN90]
G4 JLO H Dyson, 15 Swallow Gr, Netherton, Huddersfield, HD4 7SR [IO93CO, SE11]
G4 JLP D A Clarkson, 27 The Hollies, Shefford, SG17 5BX [IO92TA, TL13]
G4 JLQ S K W Williams, 634 Course 1223/22, RAF Cosford, Wolverhampton, West Midlands, WV7 3EX [IO82UP, SJ70]
G4 JLW W S Hearn, 35 Glenthorne Gdns, Ilford, IG6 1LA [JO01AO, TQ48]
GM4 JLZ E M Philip, 1 Pitstruan Terr, Aberdeen, AB10 6QW [IO87WD, NJ90]
G4 JMA Details withheld at licensee's request by SSL.
G4 JMB P J Weaver, 14 Clarefield Ct, North End Ln, Sunningdale, Ascot, SL5 0EA [IO91QJ, SU96]
G4 JMC J W Trickett, 86 School Rd, Thurcroft, Rotherham, S66 9DL [IO93IJ, SK48]
G4 JME G Marsden, South View, 18 Booths Ln, Lymm, WA13 0PE [IO83SI, SJ68]
G4 JMF D Ollerhead, 36 Park Dr, Ellesmere Port, Whitby, South Wirral, L65 6RA [IO83NG, SJ37]
G4 JMG J M Gorton, 12 Apsley Cl, Harrow, HA2 6AP [IO91TO, TQ18]
G4 JMH Details withheld at licensee's request by SSL.
G4 JMJ R D B Wilkie, 11 Crowntree Cl, Osterley, Isleworth, TW7 5PF [IO91UL, TQ17]
G4 JMM Details withheld at licensee's request by SSL.
GW4 JMN M Haighton, Trewaen, Penisarwaen, Gwynedd, LL55 3PP [IO73VD, SH56]
G4 JMO A R Oakley, 144 Revidge Rd, Blackburn, BB2 6EB [IO83RS, SD62]
G4 JMP J J Kelk, 10 Burton Fields, Herne Bay, CT6 6JU [JO01NI, TR16]
G4 JMR P L Rudwick Mears, Emc Group Mss Ltd, Anchorage Rd, Portsmouth Hants, PO3 5PU [IO90LT, SU60]
G4 JMT M R Firth, 6 Eastfield Dr, Woodlesford, Leeds, LS26 8SQ [IO93GS, SE32]
GM4 JMU K J Maxted, 18 Castleton Ave, Newton Mearns, Glasgow, G77 5NF [IO75US, NS55]
G4 JMW C Howard, 22 Greenside Cl, Dukinfield, SK16 5HS [IO83XL, SJ99]
G4 JMX Details withheld at licensee's request by SSL.
G4 JMY D J Liversidge, 6 Yardley Way, Grimsby, DN34 5UQ [IO93WN, TA20]
GM4 JMZ D L Morrison, 61 Pennyfern Rd, Greenock, PA16 9HE [IO75OW, NS27]
G4 JNA R I Macfadyen, 94 Old North Rd, Kneesworth, Bassingbourn, Royston, SG8 5JR [IO92XB, TL34]
GM4 JNB N D Baird, 23 Scorguie Ave, Inverness, IV3 6SD [IO77UL, NH64]
G4 JNC P M Mellor, The Willows, 14 Padgbury Ln, Congleton, CW12 4LP [IO83VD, SJ86]
G4 JND J W Maddrell, 7 Beech Cl, Willand, Cullompton, EX15 2SD [IO80HV, ST01]
G4 JNE Dr C J Houghton, 22 Rainow Rd, Macclesfield, SK10 2PF [IO83WG, SJ97]
GM4 JNF W C Watson, 5 Cardwell Rd, Gourock, Renfrewshire, PA19 1UB [IO75OW, NS27]
G4 JNG W S Fletcher, Sunnyside, Hawkbatch Farm, Bewdley, Worcs, DY12 3AH [IO82UJ, SO77]
G4 JNH R Barker, 171 Leicester Rd, New Packington, Ashby de La Zouch, LE65 1TR [IO92GR, SK31]
G4 JNI C J Davies, 46 Wentworth Gdns, Alton, GU34 2BJ [IO91MD, SU73]
G4 JNJ C Rogers, 25 Parkland Ave, New Mills, Stockport, Ches, SK12 4DT [IO83XI, SJ98]
G4 JNK N J Kendall M.Ed, 47 Jupiter Rd, Ipswich, Suffolk, IP4 4NT [JO02OB, TM14]
G4 JNL P J Senior, St Helena, Egremont, Cumbria, CA22 2EL [IO84FL, NY01]
G4 JNM C G Wright, 94 New Romney Cres, Nether Hall Est, Leicester, LE5 1NH [IO92LP, SK60]
G4 JNN P R Corrigan, 26 Manor Rd, Cottingley, Bingley, BD16 1QA [IO93CT, SE13]
G4 JNQ E J E Allison, 7 Abbey Rd, Flitcham, Kings Lynn, Norfolk, PE31 6BT [JO02GT, TF72]
GI4 JNS Dr D W Hughes, 381 Castlereagh Rd, Belfast, BT5 6AB [IO74BN, J37]
G4 JNT A C Talbot, 15 Noble Rd, Hedge End, Southampton, SO30 0PH [IO90IV, SU41]
G4 JNU P E Smith, 248A Kidmore Rd, Caversham, Reading, RG4 7NE [IO91ML, SU77]
G4 JNV J E Watson, 8 Yewtree Dr, Bromsgrove, B60 1AL [IO82XI, SO97]
G4 JNW L Norton, 5B Port of Ness, Isle of Lewis, Western Isles, HS2 0XA [IO68VL, NB56]
G4 JNX N D Whyborn, Kimberlin, Southwood Rd, Beighton, Norwich, NR13 3AB [JO02SO, TG30]
G4 JNY Details withheld at licensee's request by SSL.[Station located near Lichfield.]
G4 JNZ C J Barron, Kitwood Cottage, Kitwood Ln, Ropley, Alresford, SO24 0DB [IO91LC, SU63]
G4 JOA K J Wood, Tower House, Church End, West Walton, Wisbech, PE14 7ET [JO02CQ, TF41]
G4 JOB J P Barker, 6 Larkswood Cl, Tilehurst, Reading, RG3 6NP [IO91LJ, SU66]
G4 JOC B H Page, Wagtails, Stourton Caundle, Sturminster, Newton, DT10 2JW [IO80TW, ST71]
G4 JOD F Rawlings, 14 Haddon Way, Carlyon Bay, St. Austell, PL25 3QG [IO70PI, SX05]
GW4 JOG P A Truberg, 106 Johnston Rd, Llanishen, Cardiff, CF4 5HJ [IO81JM, ST18]
G4 JOI R C Tidnam, 21 Manor Ln, Lewisham, London, SE13 5QW [JO01AK, TQ37]
G4 JOJ Details withheld at licensee's request by SSL.[Station located near Leamington Spa.]
G4 JOK Details withheld at licensee's request by SSL.
G4 JON Details withheld at licensee's request by SSL.
G4 JOO Details withheld at licensee's request by SSL.
G4 JOP Details withheld at licensee's request by SSL.
GI4 JOR J J Farrell, 34 Muskett Rd, Carryduff, Belfast, BT8 8QS [IO74BM, J36]
G4 JOT Details withheld at licensee's request by SSL.
G4 JOU R J Bowden, 27 Rookswood, Alton, GU34 2LD [IO91MD, SU74]
G4 JOV J H Maclaganwedderburn, Aldwickbury School, Wheathampstead Rd, Harpenden, AL5 1AD [IO91UT, TL11]
G4 JOW Details withheld at licensee's request by SSL.
G4 JOX D Dunphy, 27 Hillingdon Ave, Sevenoaks, TN13 3RB [JO01CG, TQ55]
G4 JPA R F Jarvis, 17 St. Catherines Ct, Irvine Rd, Littlehampton, BN17 5HP [IO90RT, TQ00]
G4 JPB Can J P Beaumont, 9 Warren Bridge, Oundle, Peterborough, PE8 4DQ [IO92LSL, TL08]
GW4 JPC G O Woods, 87 Frampton Rd, Gorseinon, Swansea, SA4 4XZ [IO71XQ, SS59]
G4 JPE B Hatley, 9 Somerstown Ct, Tilehurst Rd, Reading, RG1 7TY [IO91MK, SU77]
GM4 JPG I S Wilson, 52 Seton Ct, Port Seton, Prestonpans, EH32 0TU [IO85MX, NT47]
G4 JPI Details withheld at licensee's request by SSL.
GM4 JPJ Mdm H P Genon, Croftgloy, Deskford, Banffshire, AB56 2UX [IO87MQ, NJ46]
G4 JPK T McVicker, 106 Town St., Middleton, Leeds, LS10 3QP [IO93FS, SE32]
GW4 JPN P K Kemp, 259 Delffordd, Rhos, Pontardawe, Swansea, SA8 3EP [IO81CR, SN70]
G4 JPO R A Norman, 24 Deane Gate Dr, Houghton on The Hill, Houghton on The Hill, Leicester, LE7 9HA [IO92LP, SK60]
GW4 JPP E C Jones, 36 Maes Hyfryd, Bryncrug, Tywyn, LL36 9PS [IO72XO, SH60]
G4 JPQ J Hopewell, 2 Pyes Meadow, Elmswell, Bury St. Edmunds, IP30 9UF [JO02LF, TL96]
G4 JPS A B Williams Bristol Raynet, obo Bristol Raynet Group, 38 Seneca St., St. George, Bristol, BS5 8DX [IO81RK, ST67]
G4 JPT G S Hirons, 25 Mayfield Dr, Hucclecote, Gloucester, GL3 3DS [IO81VU, SO81]
G4 JPU B W Hirons, 6 Ashlawn Cres, Solihull, B91 1PR [IO92CJ, SP17]
G4 JPX I Harrison, 61 Charles St., Golborne, Warrington, WA3 3DF [IO83QL, SJ69]
G4 JPY J E Hatter, 3 The Bower, Sergison Cl, Haywards Heath, RH16 1HD [IO91WA, TQ32]
GM4 JPZ C R Hall, Hopepark House, 15 Victoria Rd, Broughty Ferry, Dundee, DD5 1BL [IO86NL, NO43]
GM4 JQA S P Benzie, 2 Newburgh Path, Bridge of Don, Aberdeen, AB22 8SY [IO87WE, NJ91]
G4 JQB R J Wickham, 58 Gaping Ln, Hitchin, SG5 2JE [IO91UW, TL12]
G4 JQD Details withheld at licensee's request by SSL.
G4 JQE P J Ramsey, 43 The Heys, Coppull, Chorley, PR7 4NX [IO83QP, SD51]
G4 JQF M J Key, 14 Ascot Rd, Wigginton, York, YO3 3QE [IO94KA, SE55]
G4 JQJ R T Field, 37 Cotswold Ave, Rayleigh, SS6 8AW [JO01HO, TQ89]
G4 JQK S Casey, 18 Poundley Cl, Castle Bromwich, Birmingham, B36 9SZ [IO92CM, SP18]
G4 JQL S E Wayman, 22 Oakwood, Broadmayne, Dorchester, DT2 8UN [IO80TQ, SY78]
G4 JQM Details withheld at licensee's request by SSL.
G4 JQN R J Ward, 1 Dursley Rd, Heywood, Westbury, BA13 4LG [IO81VG, ST85]
G4 JQO L Hobson, Willow End, 1 Parkhill, Middleton, Kings Lynn, PE32 1RJ [JO02FR, TF61]
G4 JQP Details withheld at licensee's request by SSL.[Station located near Shepton Mallet.]
GW4 JQQ R J Henry, Ael-y-Bryn, Wernffrwd, Llanmorlais, Penclawdd Swansea, SA4 3TY [IO71WO, SS59]
G4 JQS C M Boulton, 12 Cormorant Pl, College Town, Camberley, Surrey, GU15 4XY [IO91PI, SU86]
G4 JQT I L Liston-Smith, 48 Swansea Rd, Reading, RG1 8HA [IO91ML, SU77]
G4 JQU Z P Pokusinski, 47 Clivesdale Dr, Hayes, UB3 3PX [IO91TM, TQ18]
G4 JQV C W Mee, 26 de Lisle Ct, Loughborough, Leicester, LE11 4PP [IO92JS, SK51]
G4 JQW F J Lobban, 20 Evering Ave, Parkstone, Poole, BH12 4JQ [IO90AR, SZ09]
G4 JQX C E Riley, Pandown Farmhouse, Coppershell, Gastard, Corsham, SN13 9PZ [IO81WK, ST86]
G4 JQY R J Connell, 9 Alma Rd, Stockport, SK4 4PU [IO83VK, SJ89]
GI4 JRA J Harrigan, 124 Drones Rd, Pharis, Ballymoney, BT53 8JT [IO65TB, D02]
G4 JRB M B Hahn, 21 Stanley Rd South, Rainham, RM13 8AJ [JO01CM, TQ58]
G4 JRC Details withheld at licensee's request by SSL.
G4 JRD R D de Muth, 66 Perkins Rd, Ilford, IG2 7NQ [JO01BN, TQ48]
G4 JRE J M O'Halloran, 4 Over Nidd, Harrogate, HG1 3DB [IO94FA, SE25]
GM4 JRF H S Hamilton, 8 Ardlui Gdns, Milngavie, Glasgow, G62 7RL [IO75TW, NS57]
GM4 JRG J L Gowans, 58 Car Rd, Cumnock, KA18 1HL [IO75UK, NS51]
G4 JRJ S J North, 2 Robey Dr, Eastwood, Nottingham, NG16 3DP [IO93IA, SK44]
GW4 JRK R Kentish, Eryl, Ponthirwaun, Cardigan, Dyfed, SA43 2RJ [IO72RB, SN24]
G4 JRL Details withheld at licensee's request by SSL.
G4 JRM Details withheld at licensee's request by SSL.
G4 JRN Details withheld at licensee's request by SSL.
G4 JRO Details withheld at licensee's request by SSL.
GM4 JRP F Knapp, 48 Millhill Ave, Kilmaurs, Kilmarnock, KA3 2TA [IO75RP, NS44]

G4	JRQ	S Knapp, 78 Grange Rd, Bracebridge Heath, Lincoln, LN4 2PW [IO93RE, SK96]
G4	JRW	K A Burton, 60 Grant St, Milford, Ct. 06460, U.S.A.
G4	JRY	T F S Wislocki, 30 Kingston Rd, Old Brumby, Scunthorpe, DN16 2BE [IO93QN, SE80]
G4	JS	W Lishman, 28 Lightbown St., Darwen, BB3 0DY [IO83SQ, SD62]
G4	JSB	B A Berry, 343 Tyldesley Rd, Hindsford, Atherton, Manchester, M46 9AP [IO83SM, SD60]
G4	JSC	W J Jagger, 9 Mulberry Gdns, Langdonhills, Basildon, SS16 6RB [JO01FN, TQ68]
G4	JSD	J A Hamilton, 89 The Paddocks, Norwich, NR6 7HE [JO02PQ, TG21]
G4	JSE	R L Salaman, 39 Arthur St, Unley 5061, South Australia, ZZ3 9AR
G4	JSG	D I Harrison, 95 Lindsay Dr, Kenton, Harrow, HA3 0TH [IO91UN, TQ18]
G4	JSL	Details withheld at licensee's request by SSL.[Op: H F Macgregor. Station located near Billericay.]
G4	JSM	P Hart, 112 Shelton Ave, Hucknall, Nottingham, NG15 7QA [IO93JA, SK54]
G4	JSN	A R Willis, 16 Oatfield Cl, Hereford, HR4 0RP [IO82PB, SO44]
G4	JSQ	D Piper, 203 Coalway Rd, Wolverhampton, WV3 7NG [IO82WN, SO89]
G4	JSR	D W Stocker, 79 Walton Rd, Oldbury, Warley, B68 9DB [IO92AL, SP08]
G4	JSV	N K Hingley, 29 Mayfield Rd, Hurst Green, Halesowen, B62 9QW [IO82XL, SO98]
G4	JSW	H C Butler, 8 Linthorn Cres, Greenmount, Perth, Western Australia, 6056, X X
G4	JSX	M C F Owen, Box Cottage, 64 Main Rd, Crick, Northampton, NN6 7TX [IO92KI, SP57]
G4	JSY	Details withheld at licensee's request by SSL.
G4	JSZ	D J Fry, Lyth Bank, Great Lyth, Bayston Hill, Shrewsbury, Salop, SY3 0BE [IO82OP, SJ40]
G4	JT	D A W Clark, 99 Ennerdale Rd, Richmond, TW9 2DN [IO91UL, TQ17]
GM4	JTA	B P Elliott, 14 Thornlea Dr, Giffnock, Glasgow, G46 6BZ [IO75UT, NS55]
G4	JTC	J J Bautista, 47 Valient House, Varyl Begg Est, Gibraltar
G4	JTE	P K Djali, 177 Mount Pleasant, Kingswinford, DY6 9SS [IO82VL, SO88]
GI4	JTF	Dr E H Squance, 11 Ballymenoch Rd, Holywood, BT18 0HH [IO74CP, J47]
G4	JTI	Details withheld at licensee's request by SSL.
G4	JTJ	J T Joyce, 16 Curlew Cres, Brickhill, Bedford, MK41 7HX [IO92SD, TL05]
G4	JTK	J R Lee, 115 Capenhurst Ln, Ellesmere Port, Whitby, South Wirral, L65 7AQ [IO83NG, SJ37]
G4	JTL	P Hodgetts, 4 Woodthorne Cl, Lower Gornal, Dudley, DY3 2PL [IO82WM, SO99]
G4	JTM	J N Llewellyn, 5 The Rowans, St. Marys Park, Portishead, Bristol, BS20 8QR [IO81OL, ST47]
G4	JTO	H F Young, 72 Perrinsfield, Lechlade, GL7 3SD [IO91DQ, SP20]
G4	JTP	G H Parker, 76 Kestrel Park, Ashurst, Skelmersdale, WN8 6TB [IO83ON, SD40]
G4	JTR	V K Robinson, 4 Hilltop Rd, Caversham, Reading, RG4 7HR [IO91ML, SU67]
GI4	JTS	R Macrory, 22 Whiteways, Mountain Rd, Newtownards, BT23 4UW [IO74DO, J47]
G4	JTT	R F G Austin, 57 Pipers Green, Hall Green, Birmingham, B28 0NU [IO92BK, SP18]
G4	JTV	J P Barras, 26 Wimborne Dr, Blackhill, Keighley, BD21 2TR [IO93AU, SE04]
G4	JTZ	R C Brown, 151 Brampton Rd, Bexleyheath, DA7 4SR [JO01BL, TQ47]
GI4	JUA	W M Barron, 42 Glenhugh Park, Saintfield Rd, Belfast, BT8 4PQ [IO74BN, J36]
G4	JUB	G Armstrong, 20 Cater St., Kempston, Bedford, MK42 8DR [IO92SC, TL04]
G4	JUC	H C Woodward, 11 Pant yr Odyn, Sketty, Swansea, SA2 9GR [IO81AO, SS69]
G4	JUD	F A Loach, 40 Park Rd West, Wolverhampton, WV1 4PL [IO82WO, SO99]
GM4	JUE	J G Cormack, 9 Mowat Ln, Wick, KW1 4NP [IO88KK, ND35]
GW4	JUF	P D Lyon, 73 Rhos Rd, Rhos on Sea, Colwyn Bay, LL28 4RY [IO83DH, SH88]
G4	JUH	R E Wilkinson, 3 Anglesey Rd, Hilltop, Dronfield, Sheffield, S18 6UZ [IO93GH, SK37]
GW4	JUI	D W Draper, Bryn Erin, Penmarian, Llangoed, Gwynedd, LL58 8SU [IO73XH, SH68]
G4	JUJ	R A Medcraft, 127 Beverley Rd, Ruislip Manor, Ruislip, HA4 9AP [IO91TN, TQ18]
G4	JUK	M L Neville, Bryher House, 103 Walsall Rd, Great Wyrley, Walsall, WS6 6LD [IO82XQ, SJ90]
G4	JUL	Details withheld at licensee's request by SSL.
G4	JUM	B J Buller, 36 Gr Rd, Ashtead, KT21 1BE [IO91UH, TQ15]
GW4	JUN	V Winton, Ty Cerrig, Halkyn, Holywell, Flintshire, CH8 8DL [IO83JF, SJ27]
G4	JUR	R Harrod, 70 Grasmere Cres, Darton, Barnsley, S75 5BE [IO93FO, SE31]
G4	JUT	Details withheld at licensee's request by SSL.
G4	JUV	C N Bauers, 33 Low Rd, Grimston, Kings Lynn, PE32 1AF [JO02GS, TF72]
GW4	JUW	W G Cole, Brook Cottage, Pentreheylin Wtrmil, Maesbrook, Oswestry Shropshire, SY10 8QH [IO82LS, SJ22]
G4	JUY	R M Moseley, 76 Station Rd, Hatton, Warwick, CV35 8XJ [IO92DH, SP26]
G4	JUZ	N A Gabriel, 156 Clarence Ave, New Malden, KT3 3DY [IO91UJ, TQ26]
G4	JVA	G C Butler, 6 The Glade, Welwyn Garden City, AL8 7LG [IO91VT, TL21]
G4	JVC	I R Jones, 4 Gr Cres South, Boston Spa, Wetherby, LS23 6AY [IO93HV, SE44]
G4	JVD	P G Hainsworth, 29 Ravensbourne Dr, Woodley, Reading, RG5 4LH [IO91NK, SU77]
GW4	JVE	N F C Phillips, 6 Whitcliffe Dr, Penarth, CF64 5RY [IO81JK, ST16]
G4	JVG	S Telenius-Lowe, Belvista, 27 Hertford Rd, Stevenage, SG2 8RZ [IO91VV, TL22]
G4	JVH	G R Onions, 95 Hazelmere Ct, Brades Rise, Oldbury, Warley, B69 2EE [IO82XM, SO98]
GJ4	JVI	R H Ford, Sanaldi House, Plat Douet Rd, St. Saviour, Jersey, JE2 7PN
G4	JVJ	T A Beighton, 43 Tryon Cl, Swindon, SN3 6HG [IO91DN, SU18]
G4	JVM	F T Pearson, Coach House, The Park, Mistley, Manningtree, CO11 2AL [JO01MW, TM13]
GJ4	JVP	J W Arthur, 13 Quennevais Park, St Brelade, Jersey, CI, Jersey, JE3 8GB
G4	JVQ	F A Record, 61 Over Rd, Wyke Regis, Weymouth, Dorset, DT4 9DB [IO80SO, SY67]
G4	JVT	G A Howell, 37 Copperkins Rd, Hednesford, Cannock, WS12 5NW [IO92AQ, SK01]
G4	JVV	R H S Greaves, 15 Beech Ride, Sanshurst, Sandhurst, Camberley, Surrey, GU17 8PR [IO91OH, SU85]
G4	JVW	R B Blatchford, Rosewood House, Tawstock, Barnstaple, Devon, EX31 3HZ [IO71XB, SS52]
G4	JVX	D K Powell, Lowenna, 15 Turnberry Way, Dinnington, Sheffield, S25 2TA [IO93JI, SK58]
G4	JVY	J R F Young, Squirrels Thatch, West Edge, Marsh Gibbon, Bicester, OX6 0HA [IO91LV, SP62]
G4	JVZ	M J Glennon, 5 Whiteley Croft Rise, Otley, LS21 3NR [IO93DV, SE24]
G4	JWA	D S B Naylor, 19 Bindbarrow, Burton Bradstock, Dorset, DT6 4RG [IO80PQ, SY48]
G4	JWF	J A Gregory, 10 Allington Rd, Halesworth, IP19 8TG [JO02SI, TM37]
G4	JWL	P S Woodward, 17 de Chardin Dr, Hastings, TN34 2UD [JO00HV, TQ81]
G4	JWT	Details withheld at licensee's request by SSL.
G4	JWU	T C L Richards, 2 Melrose Ave, Mitcham, CR4 2EG [IO91WJ, TQ27]
GW4	JWV	N T Lyons, The Rooks, Coombe Keynes, Wareham, Dorset, BH20 5PP [IO80VP, SY88]
G4	JWX	C J T Martin, Pb Box 74 Inglewood, Mountain Rd, Inglewood Taranaki, New Zealand, ZZ9 9MO
G4	JWY	W H F Jennings, White House, The St., Eyke, Woodbridge, IP12 2QW [JO02QC, TM35]
G4	JXC	R J Butler, 6 Woodland Ave, Dursley, GL11 4EW [IO81TQ, ST79]
G4	JXE	P M King, 21 Compton Way, Olivers Battery, Winchester, SO22 4HS [IO91HB, SU42]
G4	JXF	T P Ullathorne, 42 Sycamore Cl, Stretton, Alfreton, DE55 6GQ [IO93HD, SK35]
G4	JXG	M G Kendall Braintree & Dar Soc, Amateur Radio Soc, 88 Coldnailhurst Ave, Braintree, CM7 5PY [JO01GV, TL72]
G4	JXH	P McGivern, 83 Birdhill Ave, Reading, RG2 7JU [IO91MK, SU77]
G4	JXI	H A Collier, 12 Coronation Dr, Leigh, WN7 2UU [IO83SM, SD60]
G4	JXJ	C M J Blewitt, 33 Turton St., Greenhill, Kidderminster, DY10 2TH [IO82VJ, SO87]
G4	JXK	D M Bonfield, 88 Laburnum Rd, Fareham, PO16 0SW [IO90JU, SU50]
G4	JXL	Details withheld at licensee's request by SSL.
GI4	JXM	J A Welsh, 20 Bryantang Brae, Doagh, Ballyclare, BT39 0RJ [IO64XR, J29]
GW4	JXN	G W Roberts, 4 Frondeg Cres, Ffordd Penmynydd, Penmynydd, Llanfairpwllgwyngyll, LL61 5AX [IO73UF, SH47]
G4	JXO	M A Lawrence, 23 Sandyfield Cres, Cowplain, Waterlooville, PO8 8SQ [IO90LV, SU61]
GM4	JXP	S C H Green, Elf Enterprise Caledonia, 1 Claymore Dr, Bridge of Don, Aberdeen, AB23 8GD [IO87WE, NJ91]
G4	JXR	G T Wilde, 26 Fleetham Gr, Hartburn, Stockton on Tees, TS18 5LH [IO94HN, NZ41]
G4	JXS	G J P Nicholls, 16 Doulton Rd, Holmbush, St. Austell, PL25 3JA [IO70OI, SX05]
G4	JXX	Details withheld at licensee's request by SSL.
G4	JXY	Details withheld at licensee's request by SSL.
G4	JXZ	I M Terrell, 10 Red Lion Cl, Cranfield, Bedford, MK43 0JA [IO92QB, SP94]
G4	JY	A H Davies, Rest Haven, Garden Fields, Kinver, Stourbridge, DY7 6DT [IO82VK, SO88]
G4	JYA	R A King, 12 Berwick Cl, Mount Nod, Coventry, CV5 7JE [IO92FJ, SP27]
G4	JYB	B R Sparks, 17 Lodge Cres, Orpington, BR6 0QE [JO01BJ, TQ46]
G4	JYC	Details withheld at licensee's request by SSL.
G4	JYE	D C Sargent, 15 Wilton Rd, Balsall Common, Coventry, CV7 7QW [IO92EJ, SP27]
G4	JYF	C R L Golley, 10 New Molinnis, Bugle, St. Austell, PL26 8QL [IO70OJ, SX05]
G4	JYH	A S Curtis, One Blagdens Cl, Southgate, London, N14 6DE [IO91WP, TQ29]
G4	JYI	J B Jones, Crowgey Farm, Ruan Major, Helston, Cornwall, TR12 7NA [IO70JA, SW71]
GI4	JYJ	G W A McMaw, 26 Watch Hill Rd, Ballyclare, BT39 9QW [IO74BT, J39]
G4	JYK	R P K Leather, 35 Somerset Cl, Congleton, CW12 1SE [IO83VE, SJ86]
G4	JYL	G Thomas, 83 Salisbury Rd, Totton, Southampton, SO40 3HY [IO90GW, SU31]
G4	JYN	T G Williams Waterside Amateur Radio Societ, 31 Manor Rd, Holbury, Southampton, SO45 2NQ [IO90HT, SU40]
G4	JYP	N F Shelley, 25 Threeways, Cuddington, Northwich, CW8 2XJ [IO83QF, SJ57]
G4	JYQ	J D Tierney, Hillcrest, 26 Shore Rd, Ainsdale, Southport, PR8 2PX [IO83LO, SD31]
G4	JYT	R A Armstrong, Torch House, North End, Hallaton, Market Harborough, LE16 8UJ [IO92NN, SP79]
G4	JYU	N J Bourner, Woodpeckers, 11 Richborough Rd, Sandwich, CT13 9JE [JO01QG, TR35]
G4	JYW	D C Proctor, 161 Tribune Dr, Houghton, Carlisle, CA3 0LF [IO84NW, NY45]
G4	JYX	A G Gapper, 19 Castle Cres, St. Briavels, Lydney, GL15 6UA [IO81QR, SO50]
GM4	JYZ	A McGugan, 29 Langmuir Ave, Kirkintilloch, Glasgow, G66 2JQ [IO75WW, NS67]
G4	JZA	S L Geary, 11 Oakfield Ct, Kings Ave, London, SW4 8EQ [IO91WK, TQ37]
GM4	JZB	D J Gardner, 7 Croft Rd, Auchterarder, PH3 1EW [IO86DH, NN91]
G4	JZC	A B Leigh, 50 Colbourne Rd, Hove, BN3 1TB [IO90WT, TQ30]
G4	JZD	P J Hopwood, 27 Woodruff Ave, Guildford, GU1 1XT [IO91RG, TQ05]
G4	JZE	Details withheld at licensee's request by SSL.
G4	JZF	G S Taylor, 1 Threshers Dr, Manor Farm Est, Willenhall, WV12 4AN [IO82XO, SJ90]
G4	JZI	C A Hodges, 5 Coneygar Cl, Bridport, DT6 3AR [IO80OR, SY49]
GM4	JZJ	W D Lothian, 6 Burnhead Rd, Glasgow, G43 2SU [IO75UT, NS56]
G4	JZK	N G Roberts, 29 Grange Gr, Islington, London, N1 2NP [IO91WN, TQ38]
G4	JZL	J J Adams, 1 Powell Cl, Creech St. Michael, Taunton, TA3 5TE [IO81LA, ST22]
G4	JZM	C Makin, 61 Gleneagles Dr, Ainsdale, Southport, PR8 3TJ [IO83LO, SD31]
G4	JZO	M J D Watts, 376 Scalby Rd, Scarborough, YO12 6ED [IO94SG, TA08]
G4	JZP	G Cramp, Farringford, Linden Gdns, Tunbridge Wells, TN2 5QU [JO01DC, TQ53]
G4	JZQ	M Noakes, 61 High St., Tadlow, Royston, SG8 0EU [IO92WC, TL24]
G4	JZR	E E Williams, 7 Laurel Dr, Willaston, South Wirral, L64 1TN [IO83MH, SJ37]
G4	JZS	D L Dix, 1 Highfield Cres, Northwood, HA6 1EZ [IO91SO, TQ09]
G4	JZT	M R Cawley, 13 Keats Cl, Eccleston, Chorley, PR7 5PF [IO83PP, SD51]
G4	JZU	R J Lock, 5 South View, Bishops Tawton, Barnstaple, EX32 0AU [IO71XB, SS53]
G4	JZV	R P Bellamy, 16 Cedar Cl, Grafham, Huntingdon, PE18 0DZ [IO92UH, TL16]
G4	JZZ	C M Gadd, 40 Stanley Mount, Sale, M33 4AE [IO83UK, SJ79]
G4	KAB	L M Rose, 164 Regal Way, Harrow, HA3 0SQ [IO91UN, TQ18]
G4	KAD	Details withheld at licensee's request by SSL.
G4	KAE	D J Wood, 4 Windsor Way, Alderholt, Fordingbridge, SP6 3BN [IO90CV, SU11]
G4	KAG	G W Norris, 488 Manchester Rd, Paddington, Warrington, WA1 3HT [IO83RJ, SJ68]
G4	KAJ	Details withheld at licensee's request by SSL.
G4	KAK	R Parker, Red Leas, Bent Ln, Colne, BB8 7AA [IO83WU, SD94]
G4	KAL	B E Thompson, 45 Meadowbank, Great Coates, Grimsby, DN37 9PL [IO93WN, TA21]
G4	KAM	C Greenwood, Wilbachweg 8, 4515 Oberdorf, Solothurn, Switzerland
G4	KAO	Details withheld at licensee's request by SSL.
G4	KAQ	R Matthews, 6 Grange View, Eastwood, Nottingham, NG16 3DE [IO93IA, SK44]
G4	KAR	R P Jeffries, 22 Ingrams Way, Hailsham, BN27 3NP [JO00CU, TQ50]
GM4	KAT	M C Westwater, 14 Caldwell Rd, West Kilbride, KA23 9LE [IO75NQ, NS24]
G4	KAU	Dr T G R Mansfield, Belaugh Lodge, Coltishall Rd, Belaugh, Norwich, NR12 8UX [JO02QR, TG21]
GM4	KAV	F K Bowles, 40 Craigbarnet Rd, Milngavie, Glasgow, G62 7RA [IO75TW, NS57]
G4	KAW	Details withheld at licensee's request by SSL.[Op: A B Radford, 71 Lan Coed, Winch Wen, Swansea, SA1 7LR.]
G4	KAX	M J Haswell, 5 Westcombe Ave, St. Reet Ln, Leeds, LS8 2BS [IO93FU, SE33]
G4	KAY	Details withheld at licensee's request by SSL.
GW4	KAZ	B V Davies, Garth, 2Glanllyn, Bethel, Caernarfon Gwynedd, LL55 1YL [IO73VE, SH56]
G4	KBA	K Boucher, 22 Emery Cl, Walsall, WS1 3AL [IO92AN, SP09]
G4	KBB	B M Bristow, Camelot, Princes St, Piddington, High Wycombe Bucks, HP14 3BN [IO91OP, SU89]
G4	KBC	A R L Howe, 5 Ranmore Cl, Tollgate Hill, Crawley, RH11 9RB [IO91VC, TQ23]
G4	KBD	Details withheld at licensee's request by SSL.
GW4	KBG	T C Challoner, Bryn Eglur, Vicarage Rd, Penygraig, Tonypandy, CF40 1HP [IO81GO, ST09]
G4	KBH	R J Hodgson, 29456 Trailway Ln, Agoura Hills, Ca 91301, USA
G4	KBI	C Wainman, 9 Willson Dr, Riddings, Alfreton, DE55 4AF [IO93HB, SK45]
G4	KBJ	E C Aguilar, 191 Rue de Londres, 59420 Mouvaux, France, ZZ1 9RU
G4	KBK	R S Fisher, 24 Sugarloaf Rd, Beaconsfield Upper, 3808, Victoria Australia
GJ4	KBM	B B Nelson, Les Marais, Route Du Marais, St Ouen, Jersey, JE9 9LE
G4	KBO	L Bateman, 49 Blakemore Rd, Walsall Wood, Walsall, WS9 9JW [IO92AP, SK00]
G4	KBQ	J H Haslam, 102 West Garth, Cayton, Scarborough, YO11 3SJ [IO94TF, TA08]
G4	KBS	N Benton, 73 Stonehill Rise, Doncaster, DN5 9HD [IO93JN, SE50]
GW4	KBT	F B Phillips, 5 Nant y Gro, Llangennech, Llanelli, Dyfed, SA14 8YT [IO71WQ, SN50]
GI4	KBW	P J Henderson, Site 2, Clonaslea, Newtownabbey, Co.Antrim, N Ireland, BT37 0UL [IO74BQ, J38]
G4	KBX	Dr C L Chapple, Woodend, Hebron, Morpeth, Northd., NE61 3LA [IO95DE, NZ18]
G4	KBY	E A Homewood, 6 Stembridge Rd, Anerley, London, SE20 7UF [IO91XJ, TQ36]
G4	KCB	G H Read, 45 Yarmouth Rd, Great Sankey, Warrington, WA5 3EJ [IO83QJ, SJ58]
G4	KCC	H M Holmden, 29 Cambridge Rd Wes, Farnborough, Hants, GU14 6QA [IO91PG, SU85]
G4	KCD	B G Dean, 3 Marchant Ct, Gunthorpe Rd, Marlow, SL7 1UW [IO91ON, SU88]
GI4	KCE	S J Boyd, 10 Jacksons Cres, Station Rd, Saintfield, Ballynahinch, BT24 7EW [IO74CL, J45]
G4	KCF	K E A Sanderson, 39 Kirkland St., Pocklington, York, YO4 2BX [IO93OW, SE84]
G4	KCI	K G Aaron, 44 Holly Bank, Ackworth, Pontefract, WF7 7PE [IO93HP, SE41]
G4	KCM	C W Sanders, 16 Stirling Cl, Totton, Southampton, SO40 3GD [IO90GW, SU31]
G4	KCN	D C Salmon, The Pines, 5A Westfield Ave, Harpenden, AL5 4HN [IO91TT, TL11]
GI4	KCO	K Wright, 5 Woodview Park, The Donahies, Dublin 13, Eire, ZZ5 6SQ
G4	KCP	D A Appleton, 28 Edgewood, Shevington, Wigan, WN6 8HR [IO83PN, SD50]
GW4	KCQ	Details withheld at licensee's request by SSL.
G4	KCR	S C P Dunn, 4 St. Ronans Rd, Harrogate, HG2 8LE [IO93FX, SE35]
G4	KCS	Details withheld at licensee's request by SSL.
G4	KCT	B H Firth, 8 Lyndale Ave, Osbaldwick, York, YO1 3QB [IO93LW, SE65]
G4	KCU	D Greatbatch, 1 Hilltop Way, Southwood Meadows, Dronfield, Sheffield, S18 6YL [IO93GH, SK37]
GW4	KCV	Dr R Murray-Shelley, 25 Ffordd y Capel, Efail Isaf, Pontypridd, CF38 1AP [IO81IN, ST08]
G4	KCW	Details withheld at licensee's request by SSL.
G4	KCX	P A Hicks, 14 Oakwood, Flackwell Heath, High Wycombe, HP10 9DW [IO91PO, SU89]
GW4	KCY	P G Trimmer, 15 Cypress Ct, Landacre Park, Aberdare, CF44 8YB [IO81GR, SN90]
G4	KCZ	Dr C P Conduit, 1 Nutmead Cl, Bexley, DA5 2DT [JO01CK, TQ57]
G4	KDB	I L Wadman, 4 Lynton Ct, Pelican Ln, Newbury, RG14 1NN [IO91IJ, SU46]
GW4	KDD	D A Howls, 23 St. Marys Rd, Penllwyn, Pontllanfraith, Blackwood, NP2 2NR [IO81JP, ST19]
G4	KDE	Dr A Lamont, 249 Thorpe Hall Ave, Thorpe Bay, Southend on Sea, SS1 3SG [JO01JM, TQ98]
G4	KDG	Details withheld at licensee's request by SSL.
G4	KDH	K D Howe, Woodlands, St. Peters Rd, Hockley, SS5 6AA [JO01HO, TQ89]
G4	KDI	Details withheld at licensee's request by SSL.
G4	KDK	J F W Riggs, 1 Park Cl, North Bradley, Trowbridge, BA14 0ST [IO81VH, ST85]
G4	KDL	A G Seago, 50 Kimberley Rd, Lowestoft, NR33 0TZ [JO02UL, TM59]
G4	KDM	J M Pearson, 291 Cumberworth Ln, Denby Dale, Huddersfield, HD8 8RU [IO93EN, SE20]
G4	KDN	J J G Phaff, 14 North Park, Iver, SL0 9DJ [IO91RM, TQ07]
G4	KDO	O C Turner, 44 Norman Cres, Pinner, HA5 3QN [IO91TO, TQ19]
GW4	KDP	B E Viney, 7 Pentre Bach, Heol Idris, Barmouth, LL42 1HT [IO72XR, SH61]
G4	KDQ	J G Whiting, Threeways, Burcombe, Chalford Hill, Stroud, GL6 8BW [IO81WR, SO80]
G4	KDR	I J H Wassell, 21 Speedwell Way, Horsham, RH12 5WA [IO91UB, TQ13]
G4	KDS	C Lafferty, Cahertymore, Athenry, Co. Galway, Rep of Ireland
G4	KDU	G J Baldwin, Alderstones, 31 Kilnhurst Rd, Todmorden, OL14 6AX [IO83WR, SD92]
G4	KDW	I Davidson, 24 Queenswood Dr, Hitchin, SG4 0LG [IO91VW, TL23]
G4	KDX	C Arundel, 25 Howard St., Fishergate, York, YO1 4BQ [IO93LW, SE65]
G4	KDY	Details withheld at licensee's request by SSL.
G4	KDZ	A W S Clements, 4 Woodward Cl, Grays, RM17 5RP [JO01DL, TQ67]
G4	KEB	L A Bright, 49 Fellows Ave, Wall Heath, Kingswinford, DY6 9ET [IO82VM, SO88]
G4	KEE	V A Tomkins, 58 Chancellors Way, Beacon Heath, Exeter, EX4 9DY [IO80GR, SX99]
G4	KEI	C D T Gaston, Flat, 62 Furze Croft, Furze Hill, Hove, BN3 1PD [IO90WT, TQ20]
G4	KEL	S J Kell, Long Briar, Higher Durston, Somerset, TA3 5AG [IO81LB, ST22]
G4	KEM	R W Davey, Tanglewood, 1 Lambscroft Way, Chalfont St. Peter, Gerrards Cross, SL9 9AY [IO91RO, TQ09]
G4	KEN	K Smith, 32 St. Clements Rd, Harrogate, HG2 8LX [IO93FX, SE35]
G4	KEO	A R D Murray Cbe, 11 Ardross Ave, Northwood, HA6 3DS [IO91SO, TQ09]
G4	KEP	H J Haria, 34 Larkfield Ave, Kenton, Harrow, HA3 8NF [IO91UO, TQ18]
GI4	KEQ	B J McMahon, 26 Bally Craigy Rd, Newtown Abbey, Co Antrim, N Ireland, N Ireland, BT36 8ST [IO74AQ, J38]
G4	KES	B Bloomer, 2 Magor Hill Cttgs, Magor Hill, Magor Downs, Camborne, TR14 0JF [IO70IF, SW64]
GW4	KEV	W K Judge, Tyddyn Mawr, Arthog, Gwynedd, North Wales, LL39 1LJ [IO82AR, SH61]
G4	KEW	R D Marshall, 60 Drake Rd, Harrow, HA2 9EA [IO91TN, TQ18]
G4	KEX	B Hubbard, 16 Shelf Moor, Halifax, HX3 7PW [IO93CS, SE12]
G4	KEY	D R Turner, Holly Croft, Hamel Way, Widdington, Saffron Walden, CB11 3SJ [JO01CX, TL53]
G4	KF	N O Miller, Avon, Gardiners Ln North, Crays Hill, Billericay, CM11 2XA [JO01FO, TQ79]
G4	KFA	T Bearpark, 50 Holmpton Rd, Withernsea, HU19 2QD [JO03AR, TA32]
G4	KFB	M L Bird, 36 Derwent Rd, Southall, UB1 2DJ [IO91TM, TQ18]
G4	KFC	A E Scandrett, 72 Hesketh Rd, Yardley Gobion, Towcester, NN12 7TX [IO92NC, SP74]
GW4	KFD	B Wilson, Hillcrest, Llansteffan, Dyfed, SA33 5JZ [IO71TS, SN31]
G4	KFE	K Hircock, 77 Haxey Ln, Craiseloung Haxey, Haxey, Doncaster, DN9 2ND [IO93NL, SK79]
G4	KFF	R Hewson, 2 Ribchester Way, Brierfield, Nelson, BB9 0YH [IO83VT, SD83]
G4	KFH	Details withheld at licensee's request by SSL.
GW4	KFI	D A Bromfield, 3 Warwick Rd, Bryumawr, Gwent, South Wales, NP3 4AR [IO81JT, SO11]
G4	KFK	Details withheld at licensee's request by SSL.[Op: M Gathergood.]
G4	KFL	R J H Rowney, 2 Paseo de Las, Collalbas, Almijara Ii Sur, Nerja Malaga Spain
G4	KFN	D J Brown, Sandysike Cottage, Sandysike, Longtown, Cumbria, CA6 5SS [IO84MX, NY36]
G4	KFO	Details withheld at licensee's request by SSL.
G4	KFP	J R Marshall, Hedgeways, 63 High Moor Ln, Cleckheaton, BD19 6LW [IO93DR, SE12]
G4	KFR	A Lyle-Hodges, 10 Mayfield Park, Perrotts Brook, Bagendon, Cirencester, GL7 7BJ [IO91AS, SP00]
G4	KFS	T R Wood, 47 Marsh View, Beccles, NR34 9RT [JO02SK, TM49]
G4	KFT	M E Rothwell, 3 Chiltern Rd, Prestbury, Cheltenham, GL52 5JQ [IO81XV, SO92]
G4	KFU	J C Parkhurst, 28 Dorchester Gdns, Worthing, West Sussex, BN11 5AY [IO90TT, TQ10]
G4	KFV	E H Gawthorne, 11 Hamilton Ln, Great Brington, Northampton, NN7 4JJ [IO92LG, SP66]
GW4	KFY	Details withheld at licensee's request by SSL.
G4	KFZ	R K J Stanton, 6 Gassiott Way, Sutton, SM1 3BA [IO91VI, TQ26]
G4	KGA	M T Hattam, 16 Kilpatrick Way, Yeading, Hayes, UB4 9SX [IO91TM, TQ18]

G4 KGC P Suckling, 314A Newton Rd, Rushden, NN10 0SY [IO92RG, SP96]
GW4 KGD R Bradley, Erw Fwyn, Bwlchtocyn, Abersoch Pwllheli, Gwynedd, LL53 7BT [IO72ST, SH32]
G4 KGE J V R Baldwin, 30 Petters Rd, Ashtead, KT21 1NE [IO91UH, TQ15]
G4 KGF G T Brooks, 29A Hastings Rd, Pembury, Tunbridge Wells, TN2 4PB [JO01DD, TQ64]
G4 KGG P A Crooks, 116 Park Rd, Loughborough, LE11 2HH [IO92JS, SK51]
G4 KGK N Munro, 25 Brunswick, Hanworth, Bracknell, RG12 7YY [IO91OJ, SU86]
G4 KGL M Lees, 15 Blacklock, Chelmsford, CM2 6QL [JO01GR, TL70]
G4 KGM N J Beldon, 20 Romney Gdns, Bexleyheath, DA7 5HA [JO01BL, TQ47]
G4 KGN D I Mitchinson, 123 Fleminghouse Ln, Almondbury, Huddersfield, HD5 8UE [IO93DP, SE11]
G4 KGO R A M Matthews, 11 Bromeswell Rd, Ipswich, IP4 3AS [JO02OB, TM14]
G4 KGP K G Pollard, 27 Far End, Hatfield, AL10 8TG [IO91VR, TL20]
G4 KGQ N Martin, Birchlea, Durham Ln, Easington Village, Peterlee, SR8 3BA [IO94HS, NZ44]
GW4 KGR N R Spry, Tygwyn Cottage, Tan-y-Graig Rd, Llysfaen, Colwyn Bay, LL29 8UB [IO83EG, SH87]
G4 KGS J A Bowyer, 11 Thornhill Rd, Huddersfield, HD3 3DD [IO93CP, SE11]
G4 KGT J A Hughes, 74 Fairacres, Prestwood, Great Missenden, HP16 0LF [IO91PQ, SP80]
G4 KGU B Thomas, 12 Link Rd, Sale, M33 4HP [IO83TJ, SJ79]
G4 KGV G F Green, 70 Cumberland Rd, Middlesbrough, TS5 6PW [IO94JN, NZ41]
G4 KGW E C L Hill, 29 Forest Dr Wes, Leytonstone, London, E11 1JZ [JO01AN, TQ38]
G4 KGX W T Green, 3 Amos Rd, Leicester, LE3 6NA [IO92JP, SK50]
G4 KGY T W Lawford, 3 Alfred Cl, Thanington, Canterbury, CT1 3UL [JO01MG, TR15]
GM4 KGZ S A S Low, 5 Charteris Park, Longniddry, EH32 0NX [IO85NX, NT47]
GM4 KHA Details withheld at licensee's request by SSL.
G4 KHB M C Lowe, 14 St. Michaels Rd, Woodlands, Cheltenham, GL51 5RR [IO81WV, SO92]
G4 KHC H Irwin, Hazeldene, Coalborn, Morpeth, NE61 6LG [IO95ED, NZ28]
GM4 KHE G M Phanco, 1 Carleith Terr, Duntocher, Clydebank, G81 6HZ [IO75SW, NS47]
G4 KHF D J Wilkinson, Lutton Gowts, Long Sutton, Lutton, Spalding, Lincs, PE12 9LQ [JO02BT, TF42]
G4 KHG E Scholes, 19 Castle Hill, Newton-le-Willows, WA12 0DU [IO83QK, SJ59]
GM4 KHI T Ferrie, 17 Bargarron Dr, Paisley, PA3 4LL [IO75TU, NS46]
G4 KHJ B L Geeson, 5 Dodgeons Cl, Poulton-le-Fylde, FY6 7DX [IO83LU, SD33]
G4 KHK P Martin, 24 Heddington Cl, Trowbridge, BA14 0LH [IO81VH, ST85]
G4 KHL T A Kellaway, 4 Wentworth Ave, Fleetwood, FY7 8HY [IO83LV, SD34]
G4 KHM J Whitington, 18 Somerset Rd, Ferring, Worthing, BN12 5QA [IO90ST, TQ00]
G4 KHN P F Macgovern, 14 Cobbetts Ave, Ilford, IG4 5JR [JO01AN, TQ48]
G4 KHO N Aslaksen, 5 Apremyr Vegen 14, 1560 Larkollen, Norway
GW4 KHQ J M Woodland, 122 Parc Nant Celyn, Efail Isaf, Pontypridd, CF38 1AA [IO81IN, ST08]
G4 KHR R B A North, 21 St. Augustine Gr, Bridlington, YO16 5DB [IO94VC, TA16]
GM4 KHS W B Cavers Kelso ARC, Abbey Row Centre, Abbey Row, Kelso, TD5 7BJ [IO85SO, NT73]
G4 KHT A Lord, 5 Wasdale Green, Cottingham, HU16 4HN [IO93TS, TA03]
G4 KHU P M Hawkins, Temple View, High St., Templecombe, BA8 0JG [IO81TA, ST72]
G4 KHV Details withheld at licensee's request by SSL.
G4 KHX P R Winchester, 27A Lower Rd, Milton Malsor, Northampton, Northants, NN7 3AW [IO92ME, SP75]
G4 KHY M S Edwards, Nafferton, Killerton Rd, Bude, EX23 8EN [IO70RT, SS20]
GM4 KIA A G R Gregor, Lyn Cranna, 31 Aird St., Portsoy, Banff, AB45 2RD [IO87PQ, NJ56]
G4 KIB J L Hambleton, Siemens Limited, Cham Issara Tower Ii 31st Fl., 2922/283 New Petchburi Rd, Bangkapi Huaykwanga, Bangkok 10320 Thailand
G4 KIC H Kay, 24 Church Ave, Dacre Banks, Harrogate, HG3 4EB [IO94DB, SE16]
G4 KID SLdr B C Partridge, 95 Norwich Rd, Wymondham, NR18 0SJ [JO02NN, TG10]
G4 KIE D Hawley, Yew Tree Cottage, 7 Yew Tree, Slaithwaite, Huddersfield, HD7 5UD [IO93BO, SE01]
G4 KIF A J Sansom, 1881-9 Ave S E, Salmon Arm, British Columbia, Canada Vie 2J6
G4 KIH W E H Bartlett, 48 Barrymore Walk, Rayleigh, SS6 8YF [JO01HO, TQ89]
G4 KIJ Details withheld at licensee's request by SSL.[Op: A Rhodes.]
G4 KIK D M Whyborn, 33 Church Rd, Trull, Taunton, TA3 7LG [IO80KX, ST22]
G4 KIL P S Williams, Fair Winds, 27 Hove Park Way, Hove, BN3 6PT [IO90VU, TQ20]
G4 KIM P Newman, 23 Market Pl, Tetbury, GL8 8DD [IO81WP, ST89]
G4 KIN P Taylor, 22 Summerhill Dr, Maghull, Liverpool, L31 3DW [IO83MM, SD30]
G4 KIO Details withheld at licensee's request by SSL.
G4 KIP J E Ball, Moss Nook Farm, Moss Nook Rd, Rainford, St. Helens, WA11 8AG [IO83OL, SD40]
G4 KIQ A J Brooks, 10 St James Ave Eas, Staford-le-Hope, Essex, SS17 7BQ [JO01FM, TQ68]
G4 KIR K W Chattenton, 41 Sycamore Walk, Saltburn, Loftus, Saltburn By The Sea, TS13 4XJ [IO94NN, NZ71]
GI4 KIS B J Sheepwash, 204 Donore Cres, Antrim, BT41 1JB [IO64VR, J18][Op: Brian J Sheepwash. WAB bk no: 7408. Interests: 144 ssb, cw, dxing, ms. hscw, 10001pm, ssb. Scheds tel (018494) 67948, w/ends, evenings only. /P+ Dxpeditions a speciality.]
G4 KIT E C Sandaver, 33 North Farm Rd, Lancing, BN15 9BT [IO90UT, TQ10]
G4 KIU N D Peacock, 64 Cleveland, Tunbridge Wells, TN2 3NH [JO01DD, TQ54]
G4 KIV S B Bowden, 36 Aspin Dr, Knaresborough, HG5 8HQ [IO94GA, SE35]
GI4 KIX D L Gilmore, The Overlook, 29 Ballymaconaghy Rd, Knockbracken, N Ireland, BT8 4SB [IO74BN, J36]
G4 KIY K H Hancock, 5 St. Andrews Pl, Whittlesey, Peterborough, PE7 1BX [IO92WN, TL29]
G4 KIZ D A Holmes, Lancaster House, Magna Mile, Ludford, Market Rasen, LN8 6AD [IO93VJ, TF18]
G4 KJA B J Preston, 24 Nursery Cl, Hucknall, Nottingham, NG15 6DQ [IO93JA, SK54]
GI4 KJC N J Quinn, 54 Moyle Rd, Newton Stewart, N Ireland, BT78 4JT [IO64HR, H48]
G4 KJD I M Pitkin, Clover Cottage, Kenny Ashill, nr Ilminster, Somerset, TA19 9NH [IO80MW, ST31]
G4 KJF J A Turner, Birchwood, Church Ln, Hepworth, Diss, IP22 2QE [JO02LI, TL97]
G4 KJG Details withheld at licensee's request by SSL.
G4 KJI A M Taylor, 38 Farm Rd, Frimley, Camberley, GU16 5TE [IO91PH, SU85]
G4 KJJ J S Smith, 30 Rookery Cl, St. Ives, Huntingdon, PE17 4FX [IO92XI, TL37]
G4 KJK D C Oliver, 15 Brixham Ave, Cheadle Hulme, Cheadle, SK8 6JG [IO83VI, SJ88]
G4 KJN A H J Dent, 7 Hesleyside Rd, South Wellfield, Whitley Bay, NE25 9HB [IO95GB, NZ37]
G4 KJO S Heys, 81 Lincoln Ave, Peacehaven, BN10 7JU [IO90XT, TQ40]
GM4 KJQ Details withheld at licensee's request by SSL.[Op: R D Crawford, 9 Homefarm Road, Plymstock, Plymouth, PL9 7BY.]
G4 KJU R Fisher, 85 Larkway, Brickhill, Bedford, MK41 7JP [IO92SD, TL05]
G4 KJV J H Densem, Cotswold, Startley, Chippenham, Wilts, SN15 5HG [IO81XM, ST98]
GW4 KJW W G H Jones, 24 Underhill Cres, Abergavenny, NP7 6DF [IO81LT, SO21]
G4 KJX S K Williams, 133 Kings Ln, Bebington, Wirral, L63 5LZ [IO83LI, SJ38]
G4 KKB K M Blamey, 123 St. Johns Rd, Wootton Bridge, Ryde, PO33 4JJ [IO90JR, SZ59]
G4 KKD A G Chapple, 103 Cruick Ave, South Ockendon, RM15 6EJ [JO01DM, TQ58]
G4 KKE D J Froggatt, 74 Circular Rd, Denton, Manchester, M34 6EY [IO83WK, SJ99]
G4 KKF Details withheld at licensee's request by SSL.
G4 KKG J C Taylor, 49 Milford Rd, Yeovil, BA21 4QF [IO80QW, ST51]
G4 KKI Details withheld at licensee's request by SSL.
G4 KKJ H L Perryman, 15 Queen Mary Cres, Kirk Sandall, Doncaster, DN3 1JU [IO93LN, SE60]
GI4 KKK J K McCullagh East Antrim ARC, 2 Holestone Rd, Doagh, Ballyclare, BT39 0SB [IO64XS, J29]
G4 KKL G H Whitfield, 1 Rivan Gr, Scartho, Grimsby, DN33 3BL [IO93WM, TA20]
G4 KKM D T Peace, 39 Arlington Rd, Twickenham, TW1 2AZ [IO91UK, TQ17]
G4 KKN P W Roberts, 2 Samuals Fold, Pendlebury Ln, Haigh, Wigan, WN2 1KT
G4 KKO J Walton, 17 Wychperry Rd, Haywards Heath, RH16 1HJ [IO91WA, TQ32]
G4 KKP C McAllister, 18 Dorset Ave, Fulwell, Sunderland, SR6 8EX [IO94HW, NZ35]
G4 KKQ J Agar, 80 Holtdale Ave, Cockridge, Leeds, West Yorks, LS16 7SG [IO93EU, SE24]
G4 KKU A C Imianowski, 97 Bloomfield Rd, Brislington, Bristol, BS4 3QP [IO81RK, ST67]
GM4 KKV P A Rucklidge, Drumlins, 3 Dene Park, Biggar, ML12 6DD [IO85FO, NT03]
GM4 KKW C McKellar, 44 Salvesen Cres, Alness, IV17 0UJ [IO77VQ, NH66]
G4 KKZ K Robinson, Frys Wallpaper Shop, Race Hill, Launceston, Cornwall, PL15 9BB [IO70TP, SX38]
G4 KLA Dr J M Nelson, 67 Swarthmore Rd, Birmingham, B29 4NH [IO92AK, SP08]
G4 KLB C T Watts, 42 Truscott Ave, Winton, Bournemouth, BH9 1DB [IO90BR, SZ09]
G4 KLC Dr D V T Baldwin, Woodford Cottage, Tittleshall, Kings Lynn, Norfolk, PE32 2PF [JO02JS, TF82]
G4 KLD C J Dewhurst, 5 Ford Rd, Peasedown St. John, Bath, BA2 8DG [IO81SH, ST75]
G4 KLE M S Foster, 7 Church St., Fenstanton, Huntingdon, PE18 9JL [IO92XH, TL36]
G4 KLF A G F Selmes, Royal Omani Radio Society, PO Box 981, Muscat, Sultanate of Oman, Arabia
G4 KLG Details withheld at licensee's request by SSL.
G4 KLJ D C Wellings, 41 Wroxham Dr, Wollaton, Nottingham, NG8 2QR [IO92JW, SK53]
G4 KLL Details withheld at licensee's request by SSL.
G4 KLM Rm P N Raven, Wedgewood, Green Ln West, Rackheath, Norwich, NR13 6LT [JO02QQ, TG21]
GM4 KLN I Moore, 7 Greenside Ave, Rosemarkie, Ross Shire, IV10 8XA [IO77WO, NH75]
GM4 KLO M Mistofsky, 18 Troon Pl, Newton Mearns, Glasgow, G77 5TQ [IO75US, NS55]
G4 KLQ E A Balley, 103 Wyre Hill, Bewdley, DY12 2UG [IO82UI, SO77]
G4 KLS J A Humphries, 34 Ley Rd, Felpham, Bognor Regis, PO22 7HU [IO90QS, SZ99]
G4 KLT L H Jones, 52 The Dr, Bury, BL9 5DL [IO83UO, SD81]
G4 KLV A T Taylor, 14 Conway Cl, Haslingden, Rossendale, BB4 6TQ [IO83UQ, SD72]
G4 KLX J S Naylor, 24 Castle View Dr, Cromford, Matlock, DE4 3RL [IO93FC, SK25]
G4 KLY B W Herrington, 165 Brent Lea, Brentford, TW8 8HY [IO91UL, TQ17]
G4 KLZ L J Osborne, 77 Lloyds Way, Beckenham, BR3 3QT [IO91XJ, TQ36]
G4 KMA T J Haddon, 11 Lovetot Ave, Aston, Sheffield, S31 0BQ [IO93GJ, SK38]
G4 KMB A C Griggs, Barleycombe, Banwell Rd, Christon, Axbridge, BS26 2XZ [IO81NH, ST35]
G4 KMC M S Craven, 16 Goodstone Ave, Lostock Hall, Preston, PR5 5TY [IO83PR, SD52]
G4 KME J H R Horley, Willowbrook, 50 Hillswood Dr, Endon, Stoke on Trent, ST9 9BW [IO83WB, SJ95]
G4 KMF E A Colmer, 31 Mosyer Dr, Orpington, BR5 4PN [JO01BI, TQ46]
G4 KMH S M Cottis, 3 Hoylake Gdns, Eastcote, Ruislip, HA4 9SJ [IO91TN, TQ18]
G4 KMJ D R Edwards, 15 Grange Rd, Hastings, E Sussex, TN34 2RL [JO00GV, TQ81]
G4 KMM Details withheld at licensee's request by SSL.
G4 KMN H J Stoney, 76 Inham Rd, Chilwell, Beeston, Nottingham, NG9 4GU [IO92JW, SK53]
G4 KMP Details withheld at licensee's request by SSL.
G4 KMU L Fryer, 7 Old Farm Dr, Southampton, SO18 2PX [IO90HW, SU41]
G4 KMV Details withheld at licensee's request by SSL.
G4 KMW R Greenhough, 36 Churchbalk Ln, Pontefract, WF8 2QQ [IO93IQ, SE42]
G4 KMX R S Cope, 41 Hall Ln, Witherley, Atherstone, CV9 3LT [IO92FN, SP39]
GM4 KND Details withheld at licensee's request by SSL.
GM4 KNH E M Simpson, 10 Cairns St, Kirkcaldy Fife, Scotland, KY1 2JA [IO86KD, NT29]
G4 KNI R D K Rickard, 12 Dabryn Way, St. Stephen, St. Austell, PL26 7PF [IO70NI, SW95]
G4 KNJ R W Pinnell, 10 Starling Cl, Wokingham, RG41 3YY [IO91NJ, SU76]
G4 KNK Details withheld at licensee's request by SSL.
G4 KNL F Goatcher, 4 St. Helens Rd, Hayling Island, PO11 0BT [IO90MS, SZ79]
G4 KNM Details withheld at licensee's request by SSL.
G4 KNN A J C Leggett, 17 Armada Dr, Didben Purlieu, Hythe, Southampton, SO45 5BS [IO90HU, SU40]
G4 KNO A G Summers, 6 Hawthorne Rd, Stapleford, Cambridge, CB2 5DU [JO02BD, TL45]
G4 KNP Details withheld at licensee's request by SSL.
G4 KNQ H M Smith, Grey Gables, Humphrey Gate, Taddington, Buxton, SK17 9TS [IO93CF, SK17]
G4 KNR S Mason, 9 Bempton Cl, Bridlington, YO16 5HL [IO94VC, TA16]
G4 KNS J J Wallett, 46 Aldreth Rd, Haddenham, Ely, CB6 3PW [JO02BI, TL47]
G4 KNT I H Morton, 81 Suffolk Rd, North Harrow, Harrow, HA2 7QF [IO91TO, TQ18]
G4 KNV Dr D J Wilkinson, Walnut House, The Nookin, Husthwaite, York, YO6 3SY [IO94JD, SE57]
G4 KNX A R Bennett, 20 Upton Cl, Winyates East, Redditch, B98 0PD [IO92BH, SP06]
G4 KNZ S J Davies, 14 Herondale, Birch Hill, Bracknell, RG12 7ZT [IO91PJ, SU86]
G4 KOA E Nash, Pitomy Farm House, Collingham, nr Newark, Notts, NG23 7NL [IO93OD, SK86]
GW4 KOE R J Lines, 65 Forsythia Dr, Cyncoed, Cardiff, CF2 7HP [IO81KM, ST18]
G4 KOH Details withheld at licensee's request by SSL.
GM4 KOI S G Milne, Ashvale, Cammachmore, Stonehaven, Kincardine, AB3 2NY [IO87TH, NJ72]
G4 KOJ J B Wilson, 54 Devonshire Dr, Mickleover, Derby, DE3 5HB [IO92FV, SK33]
G4 KOK Details withheld at licensee's request by SSL.
G4 KON L G Butt, 7 Linnet Cl, Blackbird Leys, Oxford, OX4 5EL [IO91JR, SP50]
GM4 KOO S A Cawthorne, Garten, 8 Captains Brae, Twynholm, Kirkcudbright, DG6 4PE [IO74XU, NX65]
GI4 KOP Details withheld at licensee's request by SSL.
G4 KOQ G E Birkhead, 103 Roselawn Rd, Castleknock, Dublin 15, Eire, ZZ1 3RO
G4 KOR A J Hughes, 55 Welford Rd, Shirley, Solihull, B90 3HX [IO92CJ, SP17]
GW4 KOS A Papps, Gwystre Cottage, Cross Gates, Llandrindod Wells, Powys, LD1 6RN [IO82HG, SO06]
G4 KOT G Lindsay, 53 Northlands, Chester-le-Street, DH3 3UN [IO94FU, NZ25]
G4 KOU G D Martin, 5 Adelaide Cl, Durrington, Worthing, BN13 3HN [IO90ST, TQ10]
G4 KOV H B Wright, Sandpiper Cottage, Standard Rd, Wells Next The Sea, NR23 1JY [JO02KW, TF94]
G4 KOW D B McLachlan, 48 Nursery Ave, Bexleyheath, DA7 4JZ [JO01BL, TQ47]
G4 KOY R H Gill, 87 Penkett Rd, Wallasey, L45 7QQ [IO83LK, SJ39]
G4 KPB G E Glover, 8 Queensway, Kearsley, Bolton, BL4 8LP [IO83TM, SD70]
GW4 KPD Dr A I Grant, Chandlers, Crossway Green, Chepstow, NP6 5LU [IO81PP, ST59]
G4 KPF T P Hart, 15 Whitefriars Meadow, Sandwich, CT13 9AS [JO01QG, TR35]
G4 KPG W Lam, 2 Wistaria Rd, Flat 3A, Yau Yat Chuen, Kowloon Hong Kong, X X
G4 KPH D Lewis, 4 Raymond Ct, Pembroke Rd, Muswell Hill, London, N10 2HS [IO91WO, TQ29]
G4 KPJ B F Yallop, 32 de Hague Rd, Norwich, NR4 7EY [JO02PO, TG20]
G4 KPL M G Young, 8 Tweed Cl, Worcester, WR5 1SD [IO82VE, SO85]
G4 KPM Details withheld at licensee's request by SSL.
G4 KPN B F Stubbs, 12 Wycliffe Ct, Bewbush, Crawley, RH11 6AD [IO91VC, TQ23]
G4 KPO D F Ireland, Briggan Farm, Sinns Common, Radnor Reduth, Cornwall, TR16 4BJ [IO70JG, SW74]
G4 KPP C P Kelly, 30 Upper Longlands, Dawlish, EX7 9DB [IO80GO, SX97]
G4 KPT L S J Stenner, 35 Plain Pond, Wiveliscombe, Taunton, TA4 2UD [IO81IB, ST02]
G4 KPU G Taylor, 179 Bradway Rd, Bradway, Sheffield, S17 4PF [IO93FH, SK38]
G4 KPV F A Dunn, 12 Streete Ct, Westgate on Sea, CT8 8BT [JO01QJ, TR36]
G4 KPW Details withheld at licensee's request by SSL.
G4 KPX R W Burton, 28 Mulberry Way, Ely, CB7 4TH [JO02DJ, TL58]
G4 KPY C J Cawthorne, 40 Westbourne Rd, West Kirby, Wirral, L48 4DH [IO83JI, SJ28]
G4 KPZ V Cracknell, 106 High St., Upwood, Ramsey, Huntingdon, PE17 1QE [IO92WK, TL28]
GW4 KQD D H Phillips, 3 Wern Gifford Est, Pandy, Abergavenny, NP7 8RS [IO81MV, SO32]
GI4 KQA T J Moffitt, 36 Greenview, Parkgate, Ballyclare, BT39 0JP [IO64WR, J28]
G4 KQC G Leatherbarrow, 6 Queens Walk, Thornton Cleveleys, FY5 1JW [IO83LV, SD34]
G4 KQD A J Down, 95 High St., Henlow, SG16 6AB [IO92UA, TL13]
G4 KQE A C Mead, 9 Abraham Dr, Silver End, Witham, CM8 3SP [JO01HU, TL81]
G4 KQH D A Howes, 14 Manitoba Way, Eydon, Daventry, NN11 3PR [IO92JD, SP55]
G4 KQI M J Bulcock, 22 Melton Dr, Hollins, Bury, BL9 8BE [IO83UN, SD80]
G4 KQJ W C A Carpenter, 27 Long Ridge, Brighouse, HD6 3RZ [IO93CQ, SE12]
G4 KQK Dr C N Barnes, Glebe Farmhouse, Billington, Stafford, ST18 9DQ [IO82WS, SJ82]
G4 KQL A J Daulman, 2 Trentham Rd, Hartshill, Nuneaton, CV10 0SN [IO92FN, SP39]
G4 KQM R Dress, 4 Hill End Cttgs, Hill End Ln, Mottram, Hyde, SK14 6JP [IO83XK, SJ99]
G4 KQO R T Ferguson, 8 Rutland Gdns, Croydon, CR0 5ST [IO91XI, TQ36]
G4 KQP S K Jones, 114 Portland Rd, Toton, Beeston, Nottingham, NG9 6EW [IO92IV, SK53]
G4 KQQ R L Jones, 2 Bubwith Walk, Wells, BA5 2EN [IO81QE, ST54]
GM4 KQS A Smith, 9 McGavin Way, Kilwinning, KA13 6JP [IO75PP, NS24]
G4 KQT F T Ullman, 20 Deepdale Dr, Leasingham, Sleaford, NG34 8LR [IO92SX, TF04]
G4 KQU P J Homer, 25 The Causeway, Yardley, Birmingham, B25 8UL [IO92CL, SP18]
G4 KQV C D Hands, 41 Coverdale Rd, Solihull, B92 7NU [IO92CK, SP18]
G4 KQY M J Pearce, 51 Gr Ave, New Costessey, Norwich, NR5 0JB [JO02OP, TG11]
G4 KQZ T Thorne, 17 Pine St. South, Bury, BL9 7BU [IO83UO, SD81]
G4 KRB T R Jones, 63 Borrowdale Cl, Carcroft, Doncaster, DN6 8QT [IO93JO, SE51]
G4 KRD M S Khalaf, 508 London Rd, Thornton Heath, CR7 7HQ [IO91WJ, TQ36]
G4 KRF R S Moore, 2 Riverside Cl, Bootle, L20 4QG [IO83LL, SJ39]
G4 KRH R F Cane, 34 South Ridge, Billericay, CM11 2ER [JO01FO, TQ69]
G4 KRL W A Prince, 12 Coberley Rd, Cheltenham, GL51 6DG [IO81WV, SO92]
G4 KRM Details withheld at licensee's request by SSL.
G4 KRN A E Troy, 29 Longfellow St., Liverpool, L8 0QU [IO83MJ, SJ38]
G4 KRO B J Hay, 2 Lapworth Oaks, Lapworth, Solihull, B94 6LE [IO92DI, SP17]
G4 KRT M J Davis, 35 Mullion Croft, Kings Norton, Birmingham, B38 8PH [IO92AJ, SP07]
G4 KRV Details withheld at licensee's request by SSL.
G4 KRW R C Waterman, 52 Catterick Dr, Mickleover, Derby, DE3 5TX [IO92FV, SK33]
G4 KRX L C Maunder, 26 Kenilworth Gdns, Rayleigh, SS6 9HS [JO01HO, TQ89]
G4 KSA D R Mountain, 195 Wragby Rd, Lincoln, LN2 4PY [IO93RF, SK97]
G4 KSB Details withheld at licensee's request by SSL.
G4 KSC G A Galloway, 11 Lime Way, Burnham on Crouch, CM0 8RH [JO01JP, TQ99]
G4 KSD Details withheld at licensee's request by SSL.[Station located near Hornchurch.]
G4 KSE Details withheld at licensee's request by SSL.[Op: P J Hone. Station located in London E17.]
G4 KSG R P Ralph, 62 Northdown Rd, Solihull, B91 3ND [IO92CJ, SP17]
GI4 KSH H J Morrow, 2 Carnhill Gr, Carnmoney, Newtownabbey, BT36 6LS [IO74AQ, J38]
G4 KSJ S Dabrowski, 21 Coberley Cl, Downhead Park, Milton Keynes, MK15 9BJ [IO92PB, SP84]
G4 KSK R P Benyon, C/ Zaragoza 9 10 A, 28804 Alcala de, Henares, Spain
G4 KSL E Duncan, 20 Dorset Ave, South Shields, NE34 7JA [IO94HX, NZ36]
G4 KSN P McKay, 64 Wingate Dr, Ampthill, Bedford, MK45 2XF [IO92SA, TL03]
GI4 KSO D Mawhinney, 233 Ballynahinch Rd, Annahilt, Hillsborough, BT26 6BH [IO64XK, J25]
G4 KSP A K Enright, Crossfields, Firhouse Rd, Templeogue, Dublin, ZZ9 9CR
G4 KSQ B J Morris, 22 Burdell Ave, Sandhills Est, Headington, Oxford, OX3 8ED [IO91JS, SP50]
G4 KSR A J Norris, 17 Montroy Cl, Henleaze, Bristol, BS9 4RS [IO81QL, ST57]
G4 KSS I C Parker, 5 Lower Hagg, Thongsbridge, Holmfirth, Huddersfield, HD7 2UD [IO93CO, SE11]
G4 KST T S Hughes, 42 Western Dr, Hanslope, Milton Keynes, MK19 7LD [IO92OC, SP84]
G4 KSU Dr K N Prettyjohns, 99 Richmond St., Sheerness, ME12 2QS [JO01JK, TQ97]
G4 KSY A R Street, 43 Ridgedale Rd, Bolsover, Chesterfield, S44 6TX [IO93IF, SK47]
G4 KTB T W Cottham, 4 Talisman Cl, Tiptree, Colchester, CO5 0DT [JO01JT, TL81]
G4 KTF A E Ruffles, 108 Ouseburn Ave, Boroughbridge Rd, York, YO2 5NW [IO93KX, SE55]
G4 KTG H A Wilson, 133 Alexandra Rd, Grantham, NG31 7AW [IO92QV, SK93]
G4 KTI R K Taylor, 63 Peace Rd, Stanway, Colchester, CO3 5HL [JO01KV, TL92]
G4 KTL R G Honer, 19 Arndale Way, Filey, YO14 9EW [IO94UF, TA18]
G4 KTP Details withheld at licensee's request by SSL.
G4 KTR D Burrell, 19 Avondale Rd, Nelson, BB9 0DA [IO83VT, SD83]
G4 KTS P J Fletcher, Severn Bank House, Blackstone, Bewdley, Worcs, DY12 1QD [IO82UI, SO77]
GW4 KTT P H Valerio, The Brackens, Reynoldston, Swansea, West Glam, SA3 1AE [IO71VO, SS48]
G4 KTU K A White, 22 Ridyard St., Great, Wigan, WN5 9PA [IO83QM, SD50]
G4 KTW E Dale, The Woodlands, Cotheridge, Worcester, WR6 5LZ [IO82UE, SO75]
G4 KTX J O Goldsmith, Maltings, Flacks Green, Terling, Chelmsford, CM3 2QS [JO01GT, TL71]
G4 KTY R K Davie, 8 Leaholme Gdns, Slough, SL1 6LD [IO91QM, SU98]
G4 KUA Maj M A Forrester, Little Etchden, Bethersden, Kent, TN26 3DS [JO01JD, TQ94]

G4 KUC J Goodier, 20 Poleacre Ln, Woodley, Stockport, SK6 1PG [IO83WK, SJ99]
G4 KUD B Whittles, 2 Meadow Ln, Darton, Barnsley, S75 5PF [IO93FN, SE30]
G4 KUE C J Raspin, 35 Allesley Hall Dr, Coventry, CV5 9NS [IO92FJ, SP37]
G4 KUF A J Redman, 42 Gallows Hill Ln, Abbots Langley, WD5 0DA [IO91SQ, TL00]
G4 KUI Details withheld at licensee's request by SSL.
G4 KUJ T J Groves, 31 Tunnel Wood Cl, Watford, WD1 3SW [IO91TQ, TQ09]
G4 KUK Details withheld at licensee's request by SSL.[Op: B J Tayler. Station located near Billericay.]
G4 KUL D J Hepplestone, 3 Penshurst Way, Boyatt Wood, Eastleigh, SO50 4RH [IO90HX, SU42]
GI4 KUM W Glenn, 1 Meadowside, Dublin Rd, Antrim, BT41 4HD [IO64VQ, J18]
G4 KUN A S Hussey, Viola, Meads Ave, Little Common, Bexhill on Sea, TN39 4SZ [JO00FU, TQ70]
G4 KUQ Details withheld at licensee's request by SSL.
G4 KUR S V Hammonds, 28 Kingshurst Rd, Shirley, Solihull, B90 2QP [IO92BJ, SP17]
GW4 KUS H C Hemmens, 44 Cecil St., Manselton, Swansea, SA5 8QN [IO81AP, SS69]
G4 KUU Dr W H Young, 16 Buccleuch Cl, Guisborough, TS14 7LP [IO94LM, NZ61]
G4 KUV R W Woodward, 6 Acre Rd, Great Sutton, South Wirral, L66 3PW [IO83MG, SJ37]
G4 KUW C R R Jacobs, 6A Pine Cl, Wolverhampton, WV3 0UT [IO82WN, SO99]
G4 KUX N L Peckett, Four Winds, Woodland, Bishop Auckland, Co Durham, DL13 5RH [IO94BP, NZ02]
G4 KUY M L Hill, 203 Biggleswade Rd, Caldecote, Biggleswade, SG18 9BJ [IO92UC, TL14]
GI4 KUZ W J Hamill, 47 Gracefield, Gracehill, Ballymena, BT42 2RP [IO64UU, D00]
G4 KVA Details withheld at licensee's request by SSL.
G4 KVB J V Moorhouse, 2B Duncombe St., Grimsby, DN32 7EG [IO93XN, TA20]
G4 KVC R W Mitchell, 6 Green St., Smethwick, Warley, B67 7BX [IO92AL, SP08]
G4 KVD J P McMahon, PO Box 24294, Dubai, United Arab Emirates
G4 KVE P Ross, 82 Cradlebridge Dr, Willesborough, Ashford, TN24 0RF [JO01KD, TR04]
G4 KVG F Broadhurst, 33 The Cres, New Mills, High Peak, SK22 3DB [IO83XI, SJ98]
G4 KVI C L Dunn, 163 Farnham Rd, Slough, SL1 4XP [IO91QM, SU98]
GW4 KVJ R J Gardner, y Blanfa, Nelson Rd, Ystrad Mynach, Hengoed, CF8 7EG [IO81HN, ST08]
G4 KVK P E Park, 2 Leyburn Dr, High Heaton, Newcastle upon Tyne, NE7 7AP [IO94EX, NZ26]
G4 KVL B A Tharme, 4 Abacus Rd, Stoneycroft, Liverpool, L13 3DT [IO83NK, SJ39]
G4 KVP A Woodland, 32 Rudgrave Sq, Wallasey, L44 0EL [IO83LK, SJ39]
G4 KVQ R D Scott, 20 Forest Hill, Carlisle, CA1 3HF [IO84NU, NY45]
G4 KVR P J Bell, 24 Onslow Gdns, Ongar, CM5 9BG [JO01DR, TL50]
G4 KVS E A Buckley, Milford House, High St., South Milford, Leeds, LS25 5AQ [IO93IS, SE43]
G4 KVU M J C Shearer, Orchard House, Mill Rd, Kedington, Haverhill, Suffolk, CB9 7NN [JO02FC, TL74]
G4 KVV F G Wingfield, 122 Heathy Brow, Peacehaven, BN10 7SA [JO00AT, TQ40]
GM4 KVY R S Lamont, Kirkholm, Crossmichael, Castle Douglas, DG7 3AU [IO84AX, NX76]
G4 KWB Details withheld at licensee's request by SSL.
G4 KWC S R Packington, 25 Methley St., Kennington, London, SE11 4AL [IO91WL, TQ37]
G4 KWD R M Stanton, 25 St. Michaels Ave, Cleeve, Bishops Cleeve, Cheltenham, GL52 4NX [IO81XW, SO92]
G4 KWF E Pickup, 13 Mouselow Cl, Hadfield, Hyde, SK14 8BQ [IO93AK, SK09]
G4 KWH C P Meadows, 16 Dart Rd, Bedford, MK41 7BT [IO92SD, TL05]
G4 KWJ J C Hakes, Commonbank Cottage, Dolphinholme, Lancaster, Lancs, LA2 9AN [IO83PX, SD55]
G4 KWK K T Hakes, Commonbank Cottage, Dolphinholme, Lancaster, LA2 9AN [IO83PX, SD55]
G4 KWL T D Walter, 10 Wychwood Cres, Earley, Reading, RG6 5RA [IO91MK, SU77]
G4 KWM P S Deville, Bexton, Doncaster Rd, Mexborough, S64 0JD [IO93IL, SE40]
G4 KWN B W Stuart-Cole, 10 Barlow Fold Cl, Blackford Bridge, Bury, BL9 9SZ [IO83UN, SD80]
G4 KWO G Phillips, 20 Eastfield Dr, Solihull, B92 9ND [IO92CK, SP18]
G4 KWP K W Perfect, Littleton House, Pipe Ridware, Rugeley, Staffs, WS15 3QL [IO92BS, SK02]
G4 KWR E Soltysik, 15 Tower Rd, Pye Green, Cannock, WS12 4LJ [IO82XR, SJ91]
G4 KWT D B Pibworth, 20 Marathon Cl, Woodley, Reading, RG5 4UN [IO91NK, SU77]
GW4 KWV P W H Hubber, Glen Esk, Craig yr Eos Rd, Ogmore By Sea, Bridgend, CF32 0PH [IO81EL, SS87]
G4 KWW J E Ilott, 5 Chad Rd, St. Marks, Cheltenham, GL51 7BJ [IO81WV, SO92]
G4 KWX B N Cox, 66C Havelock Rd, Wimbledon, London, SW19 8HD [IO91VK, TQ27]
G4 KWY D M J Gasser, 49 Pennycress, Locksheath, Locks Heath, Southampton, SO31 6SY [IO90IU, SU50]
G4 KWZ G J Harris, Ripon Rd, Kirby Hill, Boroughbridge, York, North Yorks, YO5 9DP [IO94GC, SE36]
G4 KXD H S Spooner, 71 Templecombe Way, Morden, SM4 4JF [IO91VJ, TQ26]
G4 KXE S J Hobson, 63 Gorse Farm Rd, Great Barr, Birmingham, B43 5LS [IO92AM, SP09]
G4 KXF J R Farrar, New Cottage, Messing Park, Messing, Colchester, CO5 9TD [JO01IT, TL81]
G4 KXG K P Jackson, 30 Queen Eleanor Rd, Geddington, Kettering, NN14 1AY [IO92PK, SP88]
G4 KXH D R Brooks, 16 Churchill Way, Peverell, Plymouth, PL3 4PR [IO70WJ, SX45]
G4 KXK J E Ward, 38 Stonechat Ave, Abbeydale, Gloucester, GL4 4XD [IO81VU, SO81]
G4 KXL J H Redman, Dunster Churchfield, Horsell Park, Woking, Surrey, GU21 4LX [IO91RH, TQ05]
G4 KXO M T Reynolds, Ilex House, Redwick Rd, Pilning, Bristol, BS12 3LQ [IO81QN, ST58]
G4 KXP J W Brockett, Moorlands, Yeoford Meadows, Yeoford, Crediton, EX17 5PW [IO80DS, SX79]
G4 KXQ M Wogden, 1 Park View, St. Giles, Torrington, EX38 7JE [IO70WW, SS51]
G4 KXR A R Tipper, 10 Tithebarn Copse, Exeter, EX1 3XP [IO80GR, SX99]
G4 KXU G T Robinson, The Oak, 1050 Holderness High Rd, Hull, N H'side, HU9 4AH [IO93US, TA13]
G4 KXV D C Rigby, 145 Knightlow Rd, Harborne, Birmingham, B17 8PY [IO92AL, SP08]
G4 KXW G W Redhead, 18 Paddock Way, Dronfield, Sheffield, S18 6FF [IO93GH, SK37][Op: G Don Redhead. Station located in Derbyshire.]
G4 KYA Details withheld at licensee's request by SSL.
G4 KYC K L Hayler, 1 Roundstone Cres, East Preston, Littlehampton, BN16 1DG [IO90ST, TQ00]
G4 KYD Details withheld at licensee's request by SSL.
G4 KYF R J Clark, 10 Priory Gdns, Berkhamsted, HP4 2DR [IO91RS, SP90]
G4 KYG D F Clarke, 12 Avebury Gr, Stirchley, Birmingham, B30 2UL [IO92BK, SP08]
G4 KYH A L Waddilove, 17 Dracaena Cres, Copperhouse, Hayle, TR27 4EN [IO70HE, SW53]
G4 KYI R G Shipton, 3 Fiery Ln, Uley, Dursley, GL11 5DA [IO81UQ, ST79]
G4 KYJ B A J Wheeler, 35 Willowfield Rd, Eastbourne, BN22 8AP [JO00DS, TV69]
GW4 KYK J M A Jones, 36 Maes Hyfryd, Bryncrug, Tywyn, LL36 9PS [IO72XO, SH60]
GW4 KYN Details withheld at licensee's request by SSL.[Station located near Mumbles.]
G4 KYO G Barber, 25 Queens Way, Hayle, TR27 4NJ [IO70HE, SW53]
GW4 KYT D B Thomas, 3 New Rd, Trebanos, Pontardawe, Swansea, SA8 4DL [IO81BQ, SN70]
G4 KYU R J Ringrose, Melford House, George St., Hintlesham, Ipswich, IP8 3NH [JO02MB, TM04]
G4 KYV Details withheld at licensee's request by SSL.
G4 KYX D H Gee, 13 Dart Rd, Brickhill, Bedford, MK41 7BT [IO92SD, TL05]
G4 KYY P L Day, 46 Beatrice Ave, Saltash, PL12 4NG [IO70VJ, SX45]
GW4 KYZ D S Morgan, Penybont, Gellilydan, Blaenau Ffestiniog, Gwynedd, LL41 4EP [IO82AW, SH63]
G4 KZA Details withheld at licensee's request by SSL.[Op: R J Matten. Station located near Sandown.]
G4 KZB P G Hazelwood, 12 Ryecroft, Stourbridge, DY9 9EH [IO82WK, SO98]
GW4 KZC J H McLeod, Mendoza, Gorsedd Park, Gorsedd nr Holywell, Clwyd, CH8 8RP [IO83IH, SJ17]
G4 KZD J Young, 30 Crofton Way, Enfield, EN2 8HS [IO91WP, TQ39]
G4 KZE R P Ward, 18 Lorraine Rd, Timperley, Altrincham, WA15 7NA [IO83UJ, SJ78]
G4 KZG C F Adams, 32 Hillcrest Cl, Pound Hill, Crawley, RH10 7EQ [IO91WC, TQ23]
G4 KZH Details withheld at licensee's request by SSL.
G4 KZI B Clark, 270 Upton Rd, Haylands, Ryde, PO33 3HX [IO90JR, SZ59]
G4 KZJ W G Deeley, 5 Oak Tree Gdns, Wardsley, Stourbridge, DY8 5YF [IO82WL, SO98]
G4 KZK R A J Smith, 92 Stuart Ave, Harrow, HA2 9AZ [IO91TN, TQ18]
G4 KZO A Keir, Benenden School, Cranbrook, Kent, TN17 4AA [JO01GB, TQ83]
G4 KZP Details withheld at licensee's request by SSL.
G4 KZQ R Bennett, 16 Emily St., Nutgrove, St. Helens, WA9 5LZ [IO83OK, SJ49]
G4 KZS R M Inman, Fernglen, 5 Home Furlong, Wellesbourne, Warwick, CV35 9TW [IO92EE, SP25]
G4 KZT B Ashdown, Corner Cottage, Church St., Princes Risborough, HP27 9AA [IO91NR, SP80]
G4 KZU N A Rathbone, 7 Foreland Way, Keresley, Coventry, CV6 2NN [IO92FK, SP38]
G4 KZV Details withheld at licensee's request by SSL.[Op: J I Parkin.]
G4 KZW S Haydock, 60 Tong St., Dudley Hill, Bradford, BD4 9LX [IO93DS, SE13]
G4 KZX A G Still, 17 Arundel Rd, Newhaven, BN9 0ND [JO00AT, TQ40]
G4 KZY Details withheld at licensee's request by SSL.
G4 KZZ N P Roberts, 13 Rosemoor Cl, Hunmanby, Filey, YO14 0NB [IO94UE, TA07]
G4 LAA C A Goddard, Fairfield, Newtown, Irthington, Carlisle Cumbria, CA6 4PG [IO84OW, NY46]
G4 LAC Details withheld at licensee's request by SSL.
G4 LAD M Howes Leeds&Dist ARC, Yarnbury RUFC, Brownberrie Ln, Horsforth, Leeds, West Yorks, LS18 5HB [IO93EU, SE23]
G4 LAE C J Wordley, Whispering Winds, 7 Fulcher Ave, Springfield, Chelmsford, CM2 6QN [JO01FR, TL70]
G4 LAF R J Brodrick, Southbank, Hallatrow Rd, Paulton, Bristol, BS18 5LJ [IO81RH, ST65]
G4 LAG D G Eccles, Old Mill Lodge, Old Mill Ln, Roughton, North Norfolk, NR11 8PF [JO02PV, TG23]
G4 LAH M A Amos, 20 Truscott Ave, Winton, Bournemouth, BH9 1DB [IO90BR, SZ09]
G4 LAI C J Fone, 12 Chiltern Rise, Ashby de La Zouch, LE65 1EU [IO92GR, SK31]
G4 LAJ R A Hackett, 4 Ryton Gr, Shard End, Birmingham, B34 7RS [IO92CL, SP18]
G4 LAK R G Procter, 83 Twickenham Rd, Newton Abbot, TQ12 4JG [IO80FM, SX87]
G4 LAL M G Wheeler, 51 Ocean View Rd, Ventnor, PO38 1DH [IO90JO, SZ57]
G4 LAM R Lamberton, 28A Newtown Rd, Raunds, Wellingborough, NN9 6LX [IO92RI, SP97]
G4 LAN P Conway, 14 Leahall Ln, Rugeley, WS15 1JE [IO92BR, SK01]
GM4 LAO A Waddell, Dheallish, Auchewglen Rd, Carluke Larnakshire, Scotland, ML8 5LX [IO85BR, NS85]
G4 LAP Capt M Winter-Kaines, Turpins, Broadhurst Ln, West Chiltington, Pulborough, RH20 2RB [IO90SW, TQ01]
G4 LAV J A Walton, 2 Corner Croft, Bleasby Rd, Thurgarton, Nottingham, NG14 7FW [IO93MA, SK64]
G4 LAW F J Craven, 2 Barn Owl Way, Stoke Gifford, Bristol, BS12 6RZ [IO81RM, ST68]
G4 LAY G G Dobbs, Chaka, Grimsby Rd, Binbrook, Market Rasen, LN8 6DH [IO93VK, TF29]
G4 LAZ P J Henderson, 9 Redacre Rd, Sutton Coldfield, B73 5DX [IO92BN, SP19]
G4 LBB Details withheld at licensee's request by SSL.
G4 LBC Details withheld at licensee's request by SSL.[Op: Dr Paul A Rusling. Correspondence via Box 26744, Elkins Park, Pa 19117, USA.]
GM4 LBE A Tait, 45 Fogralea, Lerwick, ZE1 0SE [IP90JD, HU44]
G4 LBF H M Pearson, 284A London Rd, St. Albans, AL1 1HY [IO91UR, TL10]
G4 LBH R W Giles, 40 Sowerby Ave, Luton, LU2 8AF [IO91TV, TL12]
G4 LBI Details withheld at licensee's request by SSL.
G4 LBJ L W Gurney, 2 Eaton Rd, Maghull, Liverpool, L31 5JU [IO83ML, SD30]
G4 LBM P Hardiman SW London Raynet, 7 Osborne Rd, Thornton Heath, CR7 8PD [IO91WJ, TQ36]
GM4 LBN B T Armistead, 45 Swanston Gdns, Fairmilehead, Edinburgh, EH10 7DF [IO85JV, NT26]
G4 LBO Details withheld at licensee's request by SSL.
G4 LBQ J A M Philipson, Clifton Farm House, Pullover Rd, Kings Lynn, Norfolk, PE3 4LS [IO92UO, TF10]
G4 LBS J T Macrae Borden Gram Sch, Park House, 1 Highsted Rd, Sittingbourne, ME10 4PS [JO01II, TQ96]
G4 LBT R Harmer-Knight, 3 Grendon Dr, Sutton Coldfield, B73 6QA [IO92BN, SP09]
G4 LBU E V Kersey, 98 Campbell Rd, Ipswich, IP3 9RE [JO02OA, TM14]
GM4 LBV J R Eaton, 20 Cairnsmore Dr, Bearsden, Glasgow, G61 4RQ [IO75TW, NS57]
G4 LBX W Stopforth, 10 Cedar Cres, Ormskirk, L39 3NT [IO83NN, SD40]
G4 LBY S A Wright, 22 Crown St., Mansfield, NG18 3JL [IO93KD, SK56]
G4 LCB Dr M H Goldman, 19 Myddelton Park, Whetstone, London, N20 0HT [IO91WP, TQ29]
G4 LCD Details withheld at licensee's request by SSL.
G4 LCE N J Watson, 7 Goodings Green, Wokingham, RG40 1SA [IO91OJ, SU86]
GW4 LCF G Williams, 2 The Paddocks, Lodge Hill, Caerleon, Newport, NP6 1BZ [IO81MO, ST39]
GM4 LCJ C A Gove, 16 Aboyne Gdns, Kirkcaldy, KY2 6EL [IO86JD, NT29]
G4 LCL E R I C Beardmore, Kilaguni, The Avenue, Stockton Brook, Stoke on Trent, ST9 9LW [IO83WB, SJ95]
G4 LCM P J Allsopp, 13 Linden Ave, Prestbury, Cheltenham, GL52 3DW [IO81XV, SO92]
G4 LCO G A Hocking, Coniston, College Rd, Newton Abbot, TQ12 1EG [IO80EM, SX87]
GM4 LCP J P Staruszkiewicz, 1 Mafeking Terr, Neilston, Glasgow, G78 3LP [IO75SS, NS45]
G4 LCS F L Watt, 26 Marjorie Gr, Battersea, London, SW11 5SJ [IO91WL, TQ27]
G4 LCU M L Brownlow, The Croft, 1 Byne Cl, Storrington, Pulborough, RH20 4BS [IO90SW, TQ01]
G4 LCX E L Horner, 5 New Ln, Green Hammerton, York, YO5 8BL [IO94IA, SE45]
GM4 LCZ Details withheld at licensee's request by SSL.
G4 LDA R D Lawrence, 36 Egloshayle Rd, Wadebridge, PL27 6AE [IO70OM, SW97]
G4 LDB T R Kendall, 86 Rockford Cl, Oakenshaw South, Redditch, B98 7YL [IO92AG, SP06]
G4 LDC A P Wallis, 35 Luccombe Rd, Upper Shirley, Southampton, SO15 7RN [IO90GW, SU41]
G4 LDD P Harling, Spread Eagle House, Kirkgate, Settle, BD24 9DZ [IO84UB, SD86]
G4 LDE F Barnes, 22 St. Marks Pl, Blackburn, BB2 6TA [IO83RR, SD62]
G4 LDG M I Swan, 18 Paragon Pl, West Pottergate, Norwich, NR2 4BL [JO02PP, TG20]
G4 LDJ F C Gabell, 25 Woodland Way, Crowborough, TN6 3BQ [JO01CB, TQ52]
G4 LDL A D Bettley, 1 Dovetrees, Covingham, Swindon, SN3 5AX [IO91DN, SU18]
GI4 LDN S A Mc Quaid, Mullaghrodden, Dungannon, Co Tyrone, Ireland, N Ireland, BT70 3LU [IO64ON, H76]
GI4 LDO J J McQuaid, Mullaghrodden, Dungannon, Co Tyrone, BT70 3LU [IO64ON, H76]
GW4 LDP I R Dobby, 43 Chestnut Ave, West Cross, Swansea, SA3 5NL [IO71XO, SS68]
G4 LDR N J Underwood, Innisfail, Mill Ln, Winterslow, Salisbury, SP5 1PX [IO91EC, SU23]
G4 LDS C W Baker, Flat 5 Curlew House, Arnheim Rd, Burham on Crouch, Essex, CM0 8JH [JO01JP, TQ99]
G4 LDT D J Holland, Gareloch House, 1 Marine Approach, South Shields, NE33 2TG [IO94GX, NZ36]
GM4 LDU D McRae, 16 Birnie Well Rd, Slamannan, Falkirk, FK1 3HN [IO85CW, NS87]
G4 LDW M I Morris, 14 Batavia Rd, Sunbury on Thames, TW16 5NB [IO91TJ, TQ16]
GM4 LDX M McForsyth, Haltoun, Eddleston, Tweeddale, Scotland, EH45 8PW [IO85JQ, NT24]
G4 LDZ Details withheld at licensee's request by SSL.
G4 LEA F R Veale, 6 Grantson Cl, Brislington, Bristol, BS4 4NA [IO81RK, ST67]
G4 LEB Details withheld at licensee's request by SSL.
G4 LED A A Wood, New Mill Rd, Thongsbridge, Holmfirth, Huddersfield, West Yorks, HD7 2SQ [IO93CN, SE10]
G4 LEF Details withheld at licensee's request by SSL.
G4 LEG P J J Brent, 14 Stagelands, Crawley, RH11 7PE [IO91VC, TQ23]
G4 LEH D Wilkinson, 43 Castleford Ave, London, SE9 2AH [JO01AK, TQ47]
G4 LEI C Thorpe, 6 Bucks Rd, Belmont, Durham, DH1 2BD [IO94FS, NZ34]
G4 LEL Details withheld at licensee's request by SSL.
G4 LEM E S Goodman, 61 Park Rd, Torquay, TQ1 4QS [IO80FL, SX96]
G4 LEP C A Jacobs, 16 Woodyard Cl, Mulbarton, Norwich, NR14 8AS [JO02ON, TG10]
G4 LEQ G L Adams Leqtronics R C, 2 Ash Gr, Knutsford, WA16 8BB [IO83TH, SJ77]
GM4 LER T A Goodlad, 72 North Lochside, Lerwick, ZE1 0PJ [IP90KD, HU44]
G4 LES L P Macvean, 2 Broad Hapenny, Boundstone Rd, Wrecclesham, Farnham, GU10 4TF [IO91OE, SU84]
GW4 LEU R S Thomas, Bridge Cottage, Llangenny, Crickhowell, Powys, NP8 1HD [IO81KU, SO21]
G4 LEV C F Veitch, 81 Gloucester Rd, Brighton, BN1 4AP [IO90WT, TQ30]
G4 LEX G S Train, 113 Hansford Sq, Combe Down, Bath, BA2 5LL [IO81TI, ST76]
G4 LEZ Details withheld at licensee's request by SSL.[Op: I T Ambrose. Station located near Wickford.]
GM4 LFA G Cooper, 1 Cormorant Ave, Houston, Johnstone, PA6 7LG [IO75RU, NS46]
G4 LFB H E White, 184 Thistle Gr, Welwyn Garden City, AL7 4AJ [IO91VS, TL21]
G4 LFD E A Morton, 45 Leys Dr, Little Clacton, Clacton on Sea, CO16 9PD [JO01NT, TM11]
G4 LFE R M Broom, Barnstone, Cottesmore Rd, Ashwell, Oakham, LE15 7LJ [IO92PR, SK81]
GW4 LFF Dr J E Devonshire, 2 Parc Cttgs, St Donats, Llantwit Major, South Glam, CF61 1ZF [IO81FJ, SS96]
G4 LFJ M J Davis, 8 Kingsway, South Shields, NE33 3NN [IO94HX, NZ36]
GM4 LFK L N McLean, South Grange, Errol, Perthshire, PH2 7SY [IO86JK, NO22]
GM4 LFL J L Rennie, c/o Social Work Office, UK Dsu Medical Centre, HQ Afcent, Bfpo 28, X X
G4 LFM Details withheld at licensee's request by SSL.
GW4 LFO D Benson, Highfields Centre, 26 Allensbank Rd, Heath, Cardiff, CF4 3RB [IO81JM, ST17]
G4 LFP Details withheld at licensee's request by SSL.[Op: E Gwynne, 15 Caroline Street, Dudley, DY2 7DZ.]
G4 LFQ J W Holloway, 3 Downing Mews, Cutler Way, Norwich, NR5 9PE [JO02OP, TG10]
G4 LFU A W Bircher, Damson Cottage, Catterlen, Penrith, Cumbria, CA11 0BQ [IO84OQ, NY43]
GW4 LFV B Crow, Lindisfarne, Penywaun, Pentyrch, Cardiff, CF4 8SJ [IO81IM, ST08]
GW4 LFW T Cross, 50 Ty Newydd, Whitchurch, Cardiff, CF4 1NQ [IO81JM, ST17]
G4 LFX Dr D Fielden, Ashley House, 27 Rectory Rd, Wokingham, RG40 1DP [IO91OJ, SU86]
G4 LFY S D Pool, Mayfield Normanby Rd, Thealby, Scunthorpe, South Humberside, DN15 9AD [IO93QP, SE81]
GM4 LFZ W L T Nicoll, 124 Hilton Ave, Aberdeen, AB2 4LH [IO87TH, NJ72]
G4 LGA D Gowland, 78 Pleasant View, Bridgehill, Consett, DH8 8LF [IO94BU, NZ05]
G4 LGB B P Graham, 19 Cannerby Ln, Sprowston, Norwich, NR7 8NQ [JO02PP, TG21]
G4 LGD Details withheld at licensee's request by SSL.[Correspondence: c/o 99-101 High Street, Lee on Solent, Hants, PO13 9BU.]
G4 LGE R T Howe, 20 Greenwood Rd, Crowthorne, RG45 6QU [IO91OJ, SU86]
G4 LGF Details withheld at licensee's request by SSL.
G4 LGH K S Garside, 191 Kenton Rd, Newcastle upon Tyne, NE3 4NR [IO95EA, NZ26]
G4 LGI R J S Longson, 20 Vicarage Cl, Collingham, Newark, NG23 7PQ [IO93OD, SK86]
GM4 LGM J McGregor, 1 Northfield Dr, Bonhill, Alexandria, G83 9BQ [IO75RX, NS38]
G4 LGO N J Thomas, 24 Victoria Rd, Teddington, TW11 0BG [IO91UK, TQ17]
GM4 LGR J R Paton, 18 Penzance Way, Moodiesburn, Chryston, Glasgow, G69 0PB [IO75WV, NS67]
G4 LGU W F J Hills, Alperton, Rowhill Rd, Dartford, DA2 7QQ [JO01CK, TQ57]
G4 LGX J R A Hall, 30 Chatsworth Rd, Harrogate, HG1 5HS [IO94FA, SE35]
G4 LGY P W Harber, 28 Regent Rd, Epping, CM16 5DL [JO01BQ, TL40]
G4 LGZ Details withheld at licensee's request by SSL.
GM4 LHA G A Reoch, 9303 Charter Pine, Houston, Texas 77070, United States, X X
G4 LHE J C Lee, 41 Orchard Rd, Seer Green, Beaconsfield, HP9 2XH [IO91QO, SU99]
G4 LHF S K Moffat, 14 Churchill Rise, Burstwick, Hull, HU12 9HP [IO93WR, TA22]
G4 LHI P Rosamond, 13 Newnham Cl, Hartford, Huntingdon, PE18 7RP [IO92WI, TL27]
GM4 LHJ J P Campbell, 23 Napier Ave, Bathgate, EH48 1DF [IO85EV, NS96]
GW4 LHL M Edwards, 28 Portland St., Aberystwyth, Ceredigion, SY23 2DX [IO72XJ, SN58]
GM4 LHM F J Ewing, 8 Alder Gr, Wester Pitcorthie, Dunfermline, KY11 5RP [IO86GB, NT18]
G4 LHN Details withheld at licensee's request by SSL.
GM4 LHQ W H Herron, 21 Southfield Ave, Paisley, PA2 8BY [IO75ST, NS46]
G4 LHR E D Williams, 10 Eastbourne Cl, Ingol, Preston, PR2 3YR [IO83PS, SD53]
G4 LHU D Axford, 141 Nelson Rd, Gillingham, ME7 4LT [JO01GJ, TQ76]
GM4 LHW S J Burnett, 17 Crusader Dr, Roslin, EH25 9NP [IO85JU, NT26]
G4 LHX Details withheld at licensee's request by SSL.
G4 LHY E H Coventon, The Dolphins, 2 Trelawney Cl, Maenporth, Falmouth, TR11 5HS [IO70KC, SW72]
G4 LIA J A Gordon, 19 Augusta Cl, Darlington, DL1 3HT [IO94FN, NZ31]
G4 LIB Details withheld at licensee's request by SSL.[Op: S G Hunt, 33 Kimberley Road, Solihull, West Midlands, B92 8PU.]
G4 LIC J A Graham, 14 Ashbourne Gr, Shiswick, London, W4 2JH [IO91UL, TQ27]
GI4 LIF R T G G Goligher, Mountjoy East, Omagh, Co Tyrone, N Ireland, N Ireland, BT79 7JJ [IO64IP, H47]
G4 LIG P A J Hesketh, 36 Coombs View, Broadbottom, Hyde, Ches, SK14 6BJ [IO83XK, SJ99]
G4 LIH Dr D W K Jones, 10 Friarswood Cl, Yarm, TS15 9JG [IO94IM, NZ41]
G4 LIJ R S Nutt, 4 Mercers Dr, Bradville, Milton Keynes, MK13 7AY [IO92OB, SP84]
G4 LIK LtCr L R Borley, Oaklands, 116 Rowner Ln, Rowner, Gosport, PO13 0ES [IO90JT, SU50]
G4 LIL C Brown, Sandysike Cottage, Sandysike, Longtown, Cumbria, CA6 5SS [IO84MX, NY36]
G4 LIM G Moody, 37 Pine St., Norton, Stockton on Tees, TS20 2SP [IO94IO, NZ42]

G4 LIO J D Marshman, 12 Neelands Gr, Cosham, Portsmouth, PO6 4QL [IO90KU, SU60]
G4 LIP Details withheld at licensee's request by SSL.[Correspondence to Parallel Lines Contest Grp, c/o Mark Turner, G4PCS, 35 Culverhouse Road, Luton, Beds, LU3 1PY.]
G4 LIQ P H Williams, 54 High St., Yelling, St. Neots, Huntingdon, PE19 4SD [IO92WF, TL26]
GM4 LIS D Wilkes, 11 Trinity Cres, Beith, KA15 2HG [IO75QS, NS35]
G4 LIT C R Anderson, 8 Willow Cl, St. Leonards, Ringwood, BH24 2RQ [IO90BU, SU10]
G4 LIW L I Greenwood, Burrwood, Stainland Rd, Holywell Green, Halifax, HX4 9AJ [IO93BQ, SE01]
G4 LIX G L Greenwood, Burrwood, Stainland Rd, Holywell Green, Halifax, HX4 9AJ [IO93BQ, SE01]
G4 LIY C H Ware, 4 Highfield Terr, Low Bentham, Lower Bentham, Lancaster, LA2 7EP [IO84RC, SD66]
G4 LJ G D Brewer, 28 Hillcrest, Downham Market, PE38 9ND [JO02EO, TF60]
G4 LJA M H Franklin, 22 Clinton Cl, Budleigh Salterton, EX9 6QD [IO80IP, SY08]
G4 LJB J H Wild, 6 Chestnut End, Headley, Bordon, GU35 8NA [IO91OC, SU83]
GU4 LJC Cmr B F Le Lievre, Calabar Forest Rd, Forest Guernsey, Channel Islands, GY8 0AB
G4 LJE J B Senior, 17 Sedgley Rd, Bishops Cleeve, Cheltenham, GL52 4DD [IO81XW, SO92]
G4 LJF Capt I H Shepherd, Hutts Farm, Blagrove Ln, Wokingham, RG41 4AX [IO91NJ, SU76]
G4 LJG D J T Seabrook, 16 Blinco Rd, Rushden, NN10 0DT [IO92QH, SP96]
G4 LJI R A Pellatt, 66 Berrylands, Surbiton, KT5 8JY [IO91UJ, TQ16]
G4 LJJ Details withheld at licensee's request by SSL.
G4 LJK R T McKee, 39 Brandy Carr Rd, Kirkhamgate, Wakefield, WF2 0RS [IO93FQ, SE22]
G4 LJN R G Bartlett, 22 Mayfield Cl, Ferndown, BH22 9HS [IO90BT, SU00]
G4 LJO G J Bennett, Kelvadine, Yadley Way, Winscombe, BS25 1AX [IO81OH, ST45]
G4 LJP N A Cox, 89 School Ln, Quedgeley, Gloucester, GL2 4UH [IO81UT, SO81]
G4 LJQ E Glossop, 62 Wath Rd, Netheredge, Sheffield, S7 1HE [IO93GI, SK38]
G4 LJR G Garden, 227 Oak Cres, Burlington, Ontario, L7L 1H3
GW4 LJS P Harding, Harbour Lights, Horeb Rd, Five Roads Llanelli, Dyfed, SA15 5YY [IO71VR, SN40]
G4 LJT W M Hayward, 104 Snodhurst Ave, Walderslade, Chatham, ME5 0TB [JO01GI, TQ76]
G4 LJU C P Howell, 43 Copsleigh Cl, Salfords, Redhill, RH1 5BJ [IO91WE, TQ24]
G4 LJW J D Jenkins, 73 Greville Rd, Southville, Bristol, BS3 1LE [IO81QK, ST57]
G4 LJX R V Salmon, 112 Mewstone Ave, Wembury, Plymouth, PL9 0HT [IO70XH, SX54]
G4 LJY J L Warren, Clifden Farm, Quenchwell Rd, Carnon Downs, Truro Cornwall, TR3 6LN [IO70LF, SW84]
G4 LJZ Details withheld at licensee's request by SSL.
G4 LKB B J Stocker, Katkins, Oaklands Ave, Wistow, Huntingdon, PE17 2QF [IO92WJ, TL28]
G4 LKC R E Stocker, Katkins, Oaklands Ave, Wistow, Huntingdon, PE17 2QF [IO92WJ, TL28]
G4 LKD J L Spurgeon, 38 Old Forge Rd, Layer de La Haye, Colchester, CO2 0JT [JO01KU, TL92]
GW4 LKE R E R Sabido, 29 Taffs Vale, Edwardsville, Mid Glam, CF46 5NJ [IO81IP, ST09]
G4 LKF B G C Thompson, 21 Birling Pl, Corby, NN18 0LZ [IO92PL, SP88]
GI4 LKG V J Tait, 30 Corby Dr, Park, Lisburn, BT28 3HG [IO64XM, J26]
G4 LKH S J Alpine, 58 Drove Cres, Portslade, Brighton, BN41 2TA [IO90VU, TQ20]
G4 LKI A J Andrews, 5 Chippenham Rd, Moulton, Newmarket, CB8 8SN [JO02FG, TL66]
G4 LKM G R Clarke, 33 Mulberry Ave, Penwortham, Preston, PR1 0LL [IO83PR, SD52]
G4 LKP Dr K W E Craven, 8 Melander Cl, York, YO2 5RP [IO93KW, SE55]
GW4 LKS W J Evans, Dan y Craig, Craig Rd, Glais, Swansea Glam, SA7 9HS [IO81BQ, SN70]
G4 LKT P B Goodman, 34 Fullers Rd, South Woodford, London, E18 2QA [JO01AO, TQ39]
G4 LKU D P Hill, 8 Lingfield Walk, Corby, NN18 9JS [IO92PL, SP88]
G4 LKV E S Hocking, 32 Nursery Rd, Nether Poppleton, York, YO2 6NN [IO93KX, SE55]
G4 LKW P W D Head, 36A Ashacre Ln, Worthing, BN13 2DH [IO90TU, TQ10]
G4 LKX D C Hepworth, 2 Granby Cres, Bennetthorpe, Doncaster, DN2 6AN [IO93KM, SE50]
G4 LKZ C Shuttleworth, 17 Stirling Cl, Clitheroe, BB7 2QW [IO83TU, SD74]
GW4 LLC A K Pearce, The Swallows, Gelly Clynderwen, Dyfed, SA66 7HS [IO71OU, SN01]
G4 LLG P S West, 5 Stonehill Cl, Appleton, Warrington, WA4 5QD [IO83RI, SJ68]
G4 LLI G P Matthews, 101 Trafalgar Rd, Horsham, RH12 2QL [IO91UB, TQ13]
G4 LLJ A Reed, 14 Church Ln, Micklefield, Leeds, LS25 4AX [IO93IT, SE43]
G4 LLL Dr N Rudgewick-Brown, 8 The Avenue, Wheatley, Oxford, OX33 1YL [IO91KR, SP60]
G4 LLN R C Connolly, Newfane, Temple Way, Farnham Common, Slough, SL2 3HE [IO91QN, SU98]
G4 LLQ A J Leeming, The Firs, Enstone Rd, Charlbury, Chipping Norton, OX7 3QR [IO91GU, SP31]
G4 LLS D McIver, 65 Doralto Rd, New Plymouth, New Zealand, ZZ9 9DO
G4 LLT M J Nicol, 20 Larch Ave, Bricket Wood, St. Albans, AL2 3SN [IO91TQ, TL10]
G4 LLU G W P Piper, 104 Swan Bank, Penn, Wolverhampton, WV4 5PZ [IO82WN, SO89]
G4 LLV R J Rogers, 63 Ebrington Ave, Solihull, B92 8HX [IO92CK, SP18]
G4 LLW A R Saunders, 59 Fleming Rd, Quinton, Birmingham, B32 1NB [IO92AK, SP08]
GI4 LLX Details withheld at licensee's request by SSL.
GM4 LLY A Young, 10 Morrison Ave, Stevenston, KA20 4ET [IO75PP, NS24]
G4 LLZ A Barr, 28 Roundway, Honley, Huddersfield, HD7 2DD [IO93CO, SE11]
G4 LMA J F Baylis, 41 Ailesbury Way, Burbage, Marlborough, SN8 3TD [IO91DI, SU26]
G4 LMC L McLaren, 15 Harcourt Dr, Canterbury, CT2 8DP [JO01MG, TR15]
G4 LMD Details withheld at licensee's request by SSL.
G4 LMF R A Harrison, 38 Paddock Dr, Sheldon, Birmingham, B26 1QP [IO92CL, SP18]
G4 LMG D E Heasman, 1 Halfpenny Cl, Maidstone, ME16 9AJ [JO01FG, TQ75]
G4 LMH T J Heasman, 1 Halfpenny Cl, Maidstone, ME16 9AJ [JO01FG, TQ75]
G4 LMK J A Morris, 17 Overbrook Grange, Watling St, Nuneaton, Warks, CV11 6BQ [IO92GN, SP39]
G4 LML W G Turner, 11 Field View Cl, Exhall, Coventry, Warks, CV7 9BJ [IO92GL, SP38]
G4 LMM P B Stears, 127 Hughenden Ave, High Wycombe, HP13 5SS [IO91OP, SU89]
G4 LMQ I D Reid, 54 Gaynes Park Rd, Upminster, RM14 2HP [JO01CN, TQ58]
G4 LMR G Sims British Rail AR, 85 Surrey St., Glossop, SK13 9AJ [IO93AK, SK09][obo British Rail ARS.]
G4 LMS A R Geering, 43 Croftlands, Batley, WF17 6DG [IO93EQ, SE22]
G4 LMT E S Saunders, 16 Kirby Rd, Newthorpe, Nottingham, NG16 3PZ [IO93IA, SK44]
GW4 LMU F D Martin, 25 Monmouth Rd, Borras Park, Wrexham, LL12 7TP [IO83MB, SJ35]
G4 LMV W E Loxley, 92 Needlers End Ln, Balsall Common, Coventry, CV7 7AB [IO92EJ, SP27]
G4 LMX S A Crosson Smith, 5 Cassia Dr, Lower Earley, Earley, Reading, RG6 5YH [IO91MK, SU77]
G4 LMY J C Piggott, 18 Ledbury Rd, Netherton, Peterborough, PE3 9RH [IO92UN, TL19]
G4 LNA P E Balaam, 57 Ruby Rd, Walthamstow, London, E17 4RE [IO91XO, TQ38]
G4 LNC A S Friis, 53 Wistmans, Furzton, Milton Keynes, MK4 1LB [IO92OA, SP83]
G4 LND Details withheld at licensee's request by SSL.
G4 LNE W A Howorth Rossendale Rynt, 47 Rossendale Cres, Bacup, OL13 9LN [IO83VQ, SD82]
G4 LNF Details withheld at licensee's request by SSL.
G4 LNG F M Hollis, 97 Manor Rd, Chesterfield, S40 1HZ [IO93GF, SK37]
GM4 LNH W T Cooper, Tigh-An-T-Sruthan, Tarbert, Argyll, PA29 6TR [IO75HU, NR86]
GW4 LNK J F V Harris, Jacdaw, 56 Belgrave Rd, Fairbourne, LL38 2BQ [IO72XQ, SH61]
G4 LNL Details withheld at licensee's request by SSL.[Station located near Betchworth.]
G4 LNM D S Brown, 26 The Brucks, Wateringbury, Maidstone, ME18 5PX [JO01FG, TQ65]
G4 LNO G R Donington, 17 Churchgate, Hallaton, Market Harborough, LE16 8TY [IO92NN, SP79]
GW4 LNP C Trotman obo Bridgend & District ARC, 12 Blackmill Rd, Bryncethin, Bridgend, CF32 9YW [IO81FN, SS98]
G4 LNQ K Marshall, 44 Rosemary Dr, Alvaston, Derby, DE24 0TA [IO92GV, SK33]
G4 LNR L A Miles, 130 Well Ln, Willerby, Hull, HU10 6HS [IO93SS, TA03]
G4 LNT B J Thompson, 113 Gordon Rd, Corringham, Stanford-le-Hope, SS17 7QZ [JO01FM, TQ68]
GM4 LNU J S Thomson, North Lodge, Beaufort, Beauly, Invernessshire, IV4 7BE [IO77SL, NH54]
G4 LNY A H Thurgood, 6 Holben Cl, Barton, Cambridge, CB3 7AQ [JO02AE, TL45]
G4 LNZ G G Langford, 15 Ambleside Dr, Hereford, HR4 0LP [IO82PB, SO53]
G4 LOA M Rudd, Williford House, Hermitage, Dorchester, Dorset, DT2 7BB [IO80SU, ST60]
G4 LOB A M Major, 33 Borough Rd, Bridlington, YO16 4HN [IO94VC, TA16]
GW4 LOD D A Parrott, 39 Groves Rd, Newport, NP9 3SP [IO81LN, ST28]
G4 LOE G S Tuppeny, 5 Ashlawn Cres, Solihull, B91 1PR [IO92CJ, SP17]
G4 LOF M J Adams, 19 Maythorne Cl, West Bridgford, Nottingham, NG2 7TE [IO92KV, SK53]
G4 LOG R B Farley, 16 Union St., Camborne, TR14 8HG [IO70IF, SW63]
G4 LOH T J Fern, Springfield Farm, Sleights Ln, High Birstwith, nr Harrogate, North Yorks, HG3 2LH [IO94EA, SE25]
G4 LOI B Howell, 13 Westfield, Plympton, Plymouth, PL7 2DY [IO70XJ, SX55]
G4 LOJ C J B Black, Charisma, Church Rd, Yelverton, Norwich, NR14 7PB [JO02QN, TG20]
G4 LOM J Boult, 84 Princess Rd, Seaham, SR7 7TF [IO94HT, NZ44]
G4 LON J R Berg, 25 Larch Cl, Billinge, Wigan, WN5 7PX [IO83PL, SD50]
G4 LOO D B Ross, 3 Little Ln, Clophill, Bedford, MK45 4BG [IO92TA, TL03]
G4 LOP C E Hannah, 5 Main Rd, Orby, Skegness, Lincs, PE24 5HT [JO03CE, TF46]
G4 LOR W W Mooney, 21 Windsor Ct, Poulton-le-Fylde, FY6 7UX [IO83MU, SD33]
G4 LOS J R Lowe, 192 Green Hill Rd, Leeds, LS13 4AN [IO93ET, SE23]
G4 LOT S H Murphy, 7 Rushmere Ave, Northampton, NN1 5SD [IO92NF, SP76]
G4 LOU P L Ostwind, 4 Hadley Ct, Cazenove Rd, London, N16 6JU [IO91XN, TQ38]
G4 LOV J A Sutherland, 31 Kensington Rd, Sandiacre, Nottingham, NG10 5PD [IO92IV, SK43]
G4 LOW K J Whitehead Lowe Electronic, Lowe Electronics Limited, Chesterfield Rd, Matlock, Derbyshire, DE4 5LE [IO93FD, SK36]
G4 LOX D A Morton, 9 Metford Gr, Redland, Bristol, BS6 7LG [IO81QL, ST57]
G4 LOY B Carr, 359 Brereton Ave, Cleethorpes, DN35 7UP [IO93XN, TA20]
GW4 LPA D R Williams, 6 St. Annes Cl, Bexhill on Sea, TN40 2EL [JO00FU, TQ70]
GW4 LPB Details withheld at licensee's request by SSL.
GW4 LPC D H Phillips, 20 Lower Terr, Stanleytown, Stanley Town, Ferndale, CF43 3ES [IO81GP, ST09]
G4 LPD R B Mills, 3 Whitfield Cl, Wilford, Nottingham, NG11 7AU [IO92KW, SK53]
G4 LPF S Swain, 19 Rouse St., Pilsley, Chesterfield, S45 8BE [IO93HD, SK46]
GM4 LPG W A R Maslen, Broomie Knowe, Skye of Curr Rd, Dulnain Bridge, Grantown on Spey, PH26 3PA [IO87DG, NH92]
G4 LPH H Holbrook, 68 Chepstow Cl, Callands, Warrington, WA5 5SJ [IO83QJ, SJ59]
GM4 LPJ Dr G Kolbe, Riccarton Farm, Necastleton, Roxburghshire, Scotland, TD9 0SN [IO85PG, NY59]
G4 LPK F W Erskine Sunderland ARC, 27 Westbourne Rd, Sunderland, SR1 3SQ [IO94HV, NZ35]
G4 LPL I J Davis, 26 Foxfield Way, Oakham, LE15 6PR [IO92PQ, SK80]
G4 LPO Dr J R Hampson, 25 Holmeswood Park, Rawtenstall, Rossendale, BB4 6JA [IO83UQ, SD82]
G4 LPP Dr P M Holt, 32 Carleton Rise, Welwyn, AL6 9RF [IO91VU, TL21]
G4 LPS T Stansfield, 174 Perry St., Billericay, CM12 0NX [JO01FP, TQ69]
GM4 LPT J W Hopkins, 19 Cairnport Rd, Stranraer, DG9 8BQ [IO74LV, NX06]
GW4 LPU J D Jones, 9 Aelybryn, Ceinws, Machynlleth, SY20 9EZ [IO82CP, SH70]
G4 LPV H Bennett, 2 Meadow Cl, Reepham, Lincoln, LN3 4ED [IO93SG, TF07]
G4 LPW C E C L Clarke, 5 The Cttgs, Low Rd, North Tuddenham, Dereham, NR20 3DG [JO02MQ, TG01]
G4 LPX P J Linton, 12 Farnham Cl, Cheadle Hulme, Cheadle, SK8 6PD [IO83VI, SJ88]
G4 LPY J H Carter, 147 Maidenway Rd, Paignton, TQ3 2PT [IO80FK, SX86]
G4 LPZ R Doran, 1 Maple Dr, Chellaston, Derby, DE73 1RD [IO92GU, SK33]
G4 LQA Details withheld at licensee's request by SSL.
G4 LQC T H M Gibson, Eco Info Apartado No 76, Montemar 49B, Benissa 03720, Alicante, Spain, X X
G4 LQD A E Alderman, Avalon, 21 Mayfield Rd, Weybridge, KT13 8XB [IO91SI, TQ06]
G4 LQE N H Bishop, 8 South View Rd, Danbury, Chelmsford, CM3 4DX [JO01GR, TL70]
G4 LQF N J Field, 14 Regent Rd, Harborne, Birmingham, B17 9JU [IO92AL, SP08]
G4 LQG C E Richardson, 42 Oakdale, Harrogate, HG1 2LS [IO93FX, SE25]
G4 LQH R G Sharpe, Owl Cottage, Royal Oak Ln, Aubourn, Lincoln, LN5 9DT [IO93QD, SK96]
G4 LQI E David, 22 Island Wall, Whitstable, CT5 1EP [JO01MI, TR16]
G4 LQJ J H Spridgen, 4 Cheveril Ln, Bury, Ramsey, Huntingdon, PE17 1NH [IO92WK, TL28]
G4 LQL D H Lander, 1 Colby Cl, Forest Town, Mansfield, NG19 0LS [IO93KD, SK56]
G4 LQM LtCd T McCrimmon, 11 Winchester Cl, Chippenham, SN14 0QU [IO81WK, ST97]
G4 LQN K Packman, 27 Hunters Hill, High Wycombe, HP13 7EW [IO91PO, SU89]
G4 LQO Details withheld at licensee's request by SSL.
G4 LQP Details withheld at licensee's request by SSL.
GM4 LQR Nr J P Reid, Castlehill Orchard, Gill Rd, Overtown, Wishaw Strathclyde, ML2 0QB [IO85BS, NS85]
GM4 LQS J N Macdonald, Achlean, Dale Cres, Stranraer, Wigtownshire, DG9 0HQ [IO74LV, NX06]
G4 LQT Details withheld at licensee's request by SSL.[Op: D F Morton. Station located near Stafford.]
GI4 LQU R M McKinney, 12 Old Coach Gdns, Belfast, BT9 5PQ [IO74AN, J36]
G4 LQV D Morton Ariel Radio Grp, BBC Bristol Club, Whiteladies Rd, Bristol, BS8 2LR [IO81QL, ST57]
G4 LQW D A C McNiel, 17 Manning Rd, Malvern East, Victoria 3145, Australia
G4 LQX R Coleman, 35 Meadowside Rd, Upminster, RM14 3YT [JO01DN, TQ58]
G4 LQY D A Cooper, 6 Exeter Rd, Cannock, WS11 1QE [IO82XQ, SJ90]
G4 LQZ P G Dolling, Time House, 30 The Cres, Carterton, OX18 3SJ [IO91ER, SP20]
G4 LRB K R C Geen, 34 Kensington Rd, Ipswich, IP1 4LD [JO02NB, TM14]
G4 LRC R D Wilson Louth & District Radio Club, 112 Upgate, Louth, LN11 9HG [IO93XI, TF38]
G4 LRD D M Holt, 241 New Hey Rd, Oakes, Huddersfield, HD3 4GH [IO93CP, SE11]
G4 LRG J Wesby, 28 Ladysmith Rd, Ivinghoe, Leighton Buzzard, LU7 9EE [IO91QU, SP91]
G4 LRH G J E Obermaier, Bracken Rigg, Whitby Rd Fylingdal, Robin Hoods Bay, N Yorks, YO22 4QH [IO94RJ, NZ90]
G4 LRI A J P Sweetman, 1 Montrose Ave, Whitton, Twickenham, TW2 6HA [IO91TK, TQ17]
G4 LRK N A Vass, 4 Ambleside Rd, Willesden, London, NW10 3UJ [IO91VN, TQ28]
G4 LRL P Wilkins, 12 Chadcote Way, Catshill, Bromsgrove, B61 0JT [IO82XI, SO97]
G4 LRM R J Adams, 2 Audley Gdns, Seven Kings, Ilford, IG3 9LB [JO01BN, TQ48]
G4 LRN N Barker, 19 Beechcroft Rd, Castle Bromwich, Birmingham, B36 9SH [IO92CM, SP18]
G4 LRO R E Talbott, 33 Highfield St., Anstey, Leicester, LE7 7DU [IO92JQ, SK50]
G4 LRP A Boyd, 5 Walmer Cl, Southwater, Horsham, RH13 7XY [IO91TA, TQ12]
G4 LRQ R Collett, 5 Miles Dr, Gr, Wantage, OX12 7JA [IO91HO, SU48]
G4 LRR Details withheld at licensee's request by SSL.
G4 LRS Details withheld at licensee's request by SSL.[Station located near Barnsley, S Yorks.]
G4 LRT S L Berry, Hillview, Stanford Cl, Cold Ashby, Northampton, NN6 6EW [IO92LJ, SP67]
GM4 LRU T J Hood, 29 Thomson Cres, Port Seton, Prestonpans, EH32 0AN [IO85MX, NT47]
G4 LRV N J Bundle, Avonhill, Luddington, Stratford-upon-Avon, CV37 9SD [IO92CD, SP15]
G4 LRX Details withheld at licensee's request by SSL.
G4 LRY R A Ratcliffe, 12 Palmer Cl, Nine Mile Ride, Wokingham, RG40 3EB [IO91OJ, SU86]
G4 LRZ H J Dee, 36 Heath Croft Rd, Sutton Coldfield, B75 6RL [IO92CN, SP19]
G4 LSA J P Bell, 23 Barn Common, Woodseaves, Stafford, ST20 0LR [IO82UT, SJ72]
G4 LSB D Harris, 28 Oxford St., Lydney, GL15 5DJ [IO81RR, SO60]
GM4 LSD Dr G P Newton, Natural Philosophy, The University, Glasgow, G12 8QQ [IO75UU, NS56]
G4 LSE K S Darton, 18 Highfield Ave, Bishops Stortford, CM23 5LS [JO01CU, TL52]
G4 LSF D C Jarrett Fakenham RC, 15 Groveside Est, East Rudham, Kings Lynn, PE31 8RL [JO02IT, TF82]
G4 LSG S A Smith, 27 Hazelwood, Gossops Green, Crawley, RH11 8DY [IO91VC, TQ23]
G4 LSK A J Sate, 11 Tennyson Cl, Pound Hill, Crawley, RH10 3BJ [IO91WC, TQ23]
G4 LSL B R Lawrence, Tranby Meanee Rd, Scotton, Catterick Garrison, North Yorks, DL9 3NB [IO94DI, SE19]
G4 LSP G J Hinde, 12A Brick Kiln Park, Barretts Ln, Needham Market, Ipswich, Suffolk, IP6 8SA [JO02MD, TM05]
G4 LSQ P M Elmer, 6 Elmers Ln, Kesgrave, Ipswich, IP5 7GW [JO02OB, TM24]
G4 LSS A M Smith, Greenacres, Main St, Sawdon, Scarborough, N Yorks, YO13 9DY [IO94RG, SE98]
G4 LSU A H Burnett, 9 Christopher Cl, Blackfen, Sidcup, DA15 8PU [JO01BK, TQ47]
G4 LSV C Herrett, 61 Mansfield Rd, Alfreton, DE55 7JN [IO93HC, SK45]
G4 LSX G R Pearce, 19 Cedar Way, Nailsea, Bristol, BS19 1QZ [IO81PK, ST47]
G4 LSZ Details withheld at licensee's request by SSL.
G4 LTC J E Diment, 16 Riverside Walk, Isleworth, TW7 6HW [IO91TL, TQ17]
G4 LTE K G Parker, 41 Valley Rd, Chaddesden, Derby, DE21 6QU [IO92GW, SK33]
G4 LTF M F Wileman, 5 Barlow Cl, Rothwell, Kettering, NN14 6YD [IO92OK, SP88]
G4 LTG E Parkinson, 31 Middle St., Corringham, Gainsborough, DN21 5QT [IO93PJ, SK89]
G4 LTH J B Allan, 13 Vincent Cl, Stanford-le-Hope, Corringham, Stanford-le-Hope, SS17 7QL [JO01FM, TQ78]
G4 LTI M Coverdale, 1A Halton Cl, Westhead, Ormskirk, Lancs, L40 6JR [IO83NN, SD40]
G4 LTJ Details withheld at licensee's request by SSL.
G4 LTK P R Hinks, 1 Richard Joy Cl, Holbrooks, Coventry, CV6 4EY [IO92FK, SP38]
G4 LTL N C Hyde, 9 Gazzard Rd, Winterbourne, Bristol, BS17 1NR [IO81SM, ST68]
G4 LTM G M Hudsmith, 17 Greenside Cl, Dukinfield, SK16 5HS [IO83XL, SJ99]
G4 LTP W T Purdy, 34 The Cunnery, Kirk Langley, Ashbourne, DE6 4LP [IO92FW, SK23]
G4 LTR J Bryant, The Whistlers, Abbotswell Rd, Frogham, nr. Fordingbridge, Hants, SP6 2HT [IO90CV, SU11]
G4 LTS B J Packington, 83 Fitzroy Rd, Whitstable, CT5 2LE [JO01MI, TR16]
G4 LTT Details withheld at licensee's request by SSL.
G4 LTU H S Whitten, 15 Cedar Ave, Kirkby in Ashfield, Nottingham, NG17 8BD [IO93IC, SK45]
G4 LTV Details withheld at licensee's request by SSL.[Op: R T Webb, 9 Iversgate Close, Rainham, Kent, ME8 7PA.]
G4 LTX S K Stokes, 14 St. Peters Rd, Pedmore, Stourbridge, DY9 0TY [IO82WK, SO98]
G4 LTY S A Richards, 39 Trenowah Rd, Bethel, St. Austell, PL25 3EB [IO70OI, SX05]
G4 LTZ J A Peake, 8 Surrey Dr, Congleton, CW12 1NU [IO83VE, SJ86]
G4 LU S F Brown, Idlewild, Briggs Ln, Pant, Oswestry, SY10 8LD [IO82LT, SJ22]
G4 LUA R F S Gathergood, 37 Hawkley Dr, Tadley, RG26 3YH [IO91KI, SU66]
G4 LUB D P Waldron, 1 Galbraithe Cl, Bilston, WV14 8HX [IO82XM, SO99]
GM4 LUD R S Bannerman, 20 Post Box Rd, Burkhill By, Birkhill, Dundee, DD2 5PX [IO86LL, NO33]
G4 LUE E Bailey, 8 Hild Ave, Cudworth, Barnsley, S72 8RN [IO93HN, SE30]
G4 LUF R T Irish, Salamus, Diptford, nr Totnes, Devon, TQ9 7NY [IO80CJ, SX75]
G4 LUG Details withheld at licensee's request by SSL.
G4 LUH R Greaves, 4 Heron Cl, Thornton Dale, Pickering, YO18 7SN [IO94PF, SE88]
G4 LUM L A Ullman, 24 Tilburg Rd, Canvey Island, SS8 9EW [JO01GM, TQ78]
G4 LUN A C R Stickland, 1 Yew Ln, Ashley, New Milton, BH25 5BA [IO90ES, SZ29]
G4 LUO C R M Morgans, Merlewood, Maidstone Rd, Borden, Sittingbourne, ME9 7QA [JO01IH, TQ86]
G4 LUP Details withheld at licensee's request by SSL.
G4 LUQ M W Tust, 28 Osprey Cl, Higher Heys, Beechwood, Runcorn, WA7 3JH [IO83PH, SJ58]
GM4 LUS S M Smith, 80 Deanburn Park, Linlithgow, EH49 6HA [IO85EX, NS97]
G4 LUT W T S Terry, Morston, 121 Lodge Ln, Grays, RM17 5SF [JO01EL, TQ67]
G4 LUU B S Brodie, 9 Eddystone Cl, North Hykeham, Lincoln, LN6 8UH [IO93RE, SK96]
G4 LUW E J Johnson, 29 Watering Ln, Collingtree, Northampton, NN4 0NJ [IO92NE, SP75]
G4 LUX P E Biddle, 22 Crompton Rd, Rubery, Rednal, Birmingham, B45 0LH [IO82XJ, SO97]
G4 LUY D G Chubb, Walls End Cottage, Heathfield Cl, Bembridge, PO35 5UG [IO90KQ, SZ68]
G4 LUZ Details withheld at licensee's request by SSL.
G4 LVA A T Lucas, 4 Hewell Cl, Kingswinford, DY6 7RQ [IO82VM, SO88]
GI4 LVC J Chapman, 21 Coolshinney Cl, Magherafelt, BT45 5DR [IO64QR, H88]
G4 LVD B R Durrant, 140 Fletcher Rd, Ipswich, IP3 0LA [JO02OA, TM14]
G4 LVE F A Denney, 66 North Rd, Clacton on Sea, CO15 4DF [JO01NT, TM11]
G4 LVF D J Draycott, Woodend, Fen Rd, Digby, Lincoln, LN4 3NG [IO93TB, TF05]
G4 LVG P D Hills, 454 Nacton Rd, Ipswich, IP3 9NE [JO02OA, TM14]
G4 LVH A W Joines, 2904 Kristie Ct, Santa Cruz, Ca 95065, USA
G4 LVJ C J Fitch, 457 Studlands Park, Newmarket, Suffolk, CB8 7BD [JO02EG, TL66]
G4 LVK A Kelly, 8 Green Slade Cres, Marlbrook, Bromsgrove, B60 1DS [IO82XI, SO97]
G4 LVM P R Sharpe, Thimble Cottage, West Heath, Baughurst, Tadley, RG26 5LE [IO91KH, SU55]
G4 LVN K C Robinson, 51 Picksley Cres, Holton-le-Clay, Grimsby, DN36 5DR [IO93XM, TA20]
G4 LVO V G Stretch, 5 Ledwych Rd, Droitwich, WR9 9LA [IO82VG, SO86]

G4 LVQ R D Williams, 25 Upper Carr Ln, Calverley, Pudsey, LS28 5PL [IO93DT, SE23]
G4 LVR Dr B G F Roe, 7 Abbey Fields, Crewe, CW2 8HJ [IO83SB, SJ65]
G4 LVT Details withheld at licensee's request by SSL.
GM4 LVV J Hignett, 8 Glen Gdns, Callander, FK17 8ES [IO76VF, NN60]
GM4 LVW M D Bowman, 19 Mansefield Rd, Prestwick, KA9 2DL [IO75QL, NS32]
G4 LVY W Hughes, 7 Cambridge Ave, Marton, Marton in Cleveland, Middlesbrough, TS7 8EH [IO94JM, NZ51]
G4 LW L H B Huntley, 118 Bradford Rd, Trowbridge, BA14 9AR [IO81VH, ST85]
G4 LWA A E N Gard, 4 Iveldale Dr, Shefford, SG17 5AD [IO92UA, TL13]
G4 LWB P M Smith, Crossways Cottage, Top Rd, Croxton Kerrial, Grantham Lincs, NG32 1QB [IO92OU, SK82]
G4 LWC L W Collins, 44 Hollybush Ln, Penn, Wolverhampton, WV4 4JJ [IO82WN, SO89]
GW4 LWD Details withheld at licensee's request by SSL.
G4 LWF P R Green, 1 Haddon Croft, Hayley Green, Halesowen, B63 1JQ [IO82XK, SO98]
G4 LWG I R Lambert, 21 East View Terr, Barnoldswick, Colne, BB8 5NW [IO83VW, SD84]
GW4 LWL K W Edwards, 25 Gareth Cl, Thornhill, Cardiff, CF4 9AF [IO81JM, ST18]
G4 LWM F S Jackson, 20 Tudor Ave, Lea, Preston, PR2 1YP [IO83OS, SD43]
G4 LWN R W Nock, 83 Coles Ln, Hill Top, West Bromwich, B71 2QW [IO82XM, SO99]
GW4 LWO F S Cartlidge, Parkwall Lodge, Portskewett, Newport, Gwent, NP6 4UT [IO81PO, ST59]
G4 LWQ G C Simpson, Cutterne Mill Annexe, Evercreech, Shepton Mallet, Somerset, BA4 6LY [IO81RD, ST63]
GI4 LWR Details withheld at licensee's request by SSL.
G4 LWT J Yu The London Weekend TV Radio Cl, 21 Langley Ave, Surbiton, KT6 6QN [IO91UJ, TQ16]
G4 LWU R W Moore, The Saplings, Bryhampton Ln, Little Hereford, Salop, SY8 4LN [IO82QH, SO56]
G4 LWV E F Videan, 18 Veritys, Maryland, Hatfield, AL10 8HH [IO91VS, TL20]
G4 LWW E A Wood, Kiln Haw, Garsdale, Sedbergh, Cumbria, LA10 5NT [IO84SH, SD79]
G4 LWY J Bryce, 6A Cawley Ave, Culcheth, Warrington, WA3 4DF [IO83RK, SJ69]
G4 LX L G Spencer, 1 West Rig, Newcastle upon Tyne, NE3 4LR [IO95EA, NZ26]
G4 LXA D R Gibson, 110 Aylestone Ln, Wigston, LE18 1BA [IO92KO, SP69]
G4 LXC P A Johnson, 8 Chapel Cl, Ryarsh, West Malling, ME19 5LT [JO01EH, TQ66]
G4 LXD Dr F W J de Bass, Hawthorns, 10 Melville Rd, Croxton, Thetford, IP24 1NG [JO02JK, TL88]
G4 LXE Details withheld at licensee's request by SSL.
G4 LXH D E Jones, 34 Alpha Gr, Isle of Dogs, London, E14 8LH [IO91XM, TQ37]
G4 LXI G J Swann, 22 Upland Gr, Bromsgrove, B61 0EL [IO82XI, SO97]
G4 LXJ C J Phillips, 2 Woodville Rise, Chineham, Basingstoke, RG24 8GR [IO91LG, SU65]
G4 LXK Details withheld at licensee's request by SSL.
GI4 LXL E O'Reilly, 5 Lanntara, Ballymena, BT42 3BE [IO64UU, D10]
GM4 LXM G M Low, 23 Bellfield Rd, North Kessock, Kessock, Inverness, IV1 1XU [IO77UM, NH64]
GW4 LXO J D Eastment, 211 Pantbach Rd, Rhiwbina, Cardiff, CF4 6AE [IO81JM, ST18]
G4 LXQ Details withheld at licensee's request by SSL.
G4 LXR R A Hooper, 16 Ninehams Gdns, Caterham, CR3 5LP [IO91WH, TQ35]
G4 LXT G C Muggeridge, 48 Woodham Waye, Woodham, Woking, GU21 5SJ [IO91RI, TQ06]
G4 LXU C H Lennox, Blazefield Hse Farm, Blazefield, Pateley Bridge, Harrogate, HG3 5DR [IO94DC, SE16]
G4 LXV A D Rose, 40 Wilson Dr, Outwood, Wakefield, WF1 3DN [IO93FQ, SE32]
G4 LXW A R Trousdale, 65 Low Moor Side, New Farnley, Leeds, LS12 5EA [IO93ES, SE23]
G4 LXY D Millin, 28 Lakeside Cres, Barnet, EN4 8QJ [IO91WP, TQ29]
G4 LXZ Details withheld at licensee's request by SSL.[Op: H K Allsopp. Station located near Wimborne.]
G4 LYA R Dickinson, 19 Ledbury Cl, Sharp Ln, Middleton, Leeds, LS10 4RT [IO93FR, SE32]
G4 LYB C J Kitchener, 32 Old Rope Walk, Haverhill, CB9 9DF [JO02FC, TL64]
G4 LYC P Collett, 7 Saxon Rise, Earls Barton, Northampton, NN6 0NY [IO92PG, SP86]
G4 LYD D A Palmer, 123 Bucklesham Rd, Kirton, Ipswich, IP10 0PF [JO02PA, TM24]
G4 LYE N Pilling, 22 Templar Way, Selby, YO8 9XH [IO93LS, SE63]
G4 LYF Dr A F Webb, 10 The Glebe, Haverhill, CB9 0DL [JO02FC, TL64]
G4 LYG J J Lavis, Briar Cottage, Turleigh, Bradford on Avon, Wilts, BA15 2HG [IO81UI, ST86]
G4 LYH W D Mitchell, 5 Hill Park Rd, Gosport, PO12 3EB [IO90KT, SU50]
G4 LYJ Details withheld at licensee's request by SSL.
G4 LYL H R Bonnor, 8 Gordon Ave, Winchester, SO23 0QQ [IO91IB, SU42]
G4 LYM G Schiffeldrin, 68 The Fairway, Alwoodley, Leeds, LS17 7PD [IO93FU, SE24]
G4 LYO R Kenney, 197 Cranbrook Rd, Redland, Bristol, BS6 7QU [IO81QL, ST57]
G4 LYP D Wastnidge, 32 Fletcher Cl, Priorswood, Taunton, Somerset, TA2 8SQ [IO81KA, ST22]
GM4 LYQ R I Lamont, Joiners House, Main St, Marykirk, Kincardineshire, AB30 1UT [IO86RS, NO66]
G4 LYR D Ricketts, 14 Cambridge Way, Otley, LS21 1DB [IO93DV, SE24]
G4 LYU B W G Gauntlett, Heamoor, 4 Sandbanks Gdns, Hailsham, BN27 3TL [JO00DU, TQ50]
GM4 LYV W M Hattie, 71 Meiklehill Rd, Kirkintilloch, Glasgow, G66 2JY [IO75WW, NS67]
G4 LYW J F R Weston Prior Park Coll, Chenery Lodge, 44 Old Newbridge Hill, Bath, BA1 3LU [IO81TJ, ST76]
G4 LYX J G Wylie, 15 Semley Rd, Hassocks, BN6 8PD [IO90WW, TQ31]
G4 LYY J R Schoolar, 140 Slades Rd, Bolster Moor, Golcar, Huddersfield, HD7 4JR [IO93BP, SE01]
G4 LZA M J Prince, White Cottage, Chamberlain Ln, Cookhill, Alcester, B49 5LD [IO92AF, SP05]
G4 LZD G J S Reading, 73 Mayflower Cl, Townstal, Dartmouth, TQ6 9JN [IO80EI, SX85]
G4 LZE C R Lugard, 30 Upper Shirley Rd, Shirley, Croydon, CR0 5HA [IO91XI, TQ36]
G4 LZF C Chapman, 101 Stoneygate Rd, Luton, LU4 9TL [IO91SV, TL02]
G4 LZJ P A Garnett, Drewen Garth, Church St, Aldbrough, Hull, HU11 4RN [IO93WT, TA23]
G4 LZK R P Broughton, 18 Elim Ct Gdns, Crowborough, TN6 1BS [JO01BB, TQ53]
GW4 LZL T J Smith, 46 Marlborough Cl, Llantwit Fardre, Pontypridd, CF38 2NP [IO81IN, ST08]
G4 LZM Details withheld at licensee's request by SSL.
GM4 LZO G McDonald, Ellrigg, Ballencrieff Toll, Bathgate, EH48 4LD [IO85EW, NS97]
GW4 LZP T C Bryant Meirion ARS, Maes y Crynwyr, Llwyngwril, LL37 2JQ [IO72XQ, SH51]
G4 LZQ G M Williams, 1 Portland Pl, The Green, Frampton on Severn, Gloucester, GL2 7ET [IO81TS, SO70]
GI4 LZR W J Turner, 31 Thiepval Ave, Cregagh, Belfast, BT6 9JF [IO74BN, J37]
GI4 LZS J Smyth, 12 Cleland Park Central, Bangor, BT20 3EP [IO74DP, J48]
G4 LZT R Brown, 2 Layton Park Cl, Rawdon, Leeds, LS19 6PJ [IO93EU, SE23]
G4 LZU E A Hayden, The Old School, Fitzhead, nr Taunton, Somerset, TA4 3JP [IO81IB, ST12]
G4 LZV K J Brazington, 38 Tamworth Rd, Amington, Tamworth, B77 3BT [IO92DP, SK20]
G4 LZW J R Wood, 16 Belgravia Gdns, Hereford, HR1 1RB [IO82PB, SO54]
G4 LZY R Smith, 2 Ravensgate Rd, Charlton Kings, Cheltenham, GL53 8NN [IO81XV, SO91]
G4 LZZ A Siemieniago, 3 Skye Cl, Highworth, Swindon, SN6 7HR [IO91DP, SU29]
G4 MAB Details withheld at licensee's request by SSL.
GI4 MAC Dr M S McKinney, 29 Saintfield Rd, Ballygowan, Newtownards, BT23 6HB [IO74CM, J46]
G4 MAD I J OReilly, 62 Harold Rd, Sittingbourne, ME10 3AJ [JO01JI, TQ96]
G4 MAE M J Pinnell, 10 Starling Cl, Wokingham, RG41 3YY [IO91NJ, SU76]
G4 MAG D H Lucas, 23 Rectory Cl, Wistaston, Crewe, CW2 8HG [IO83SB, SJ65]
GI4 MAJ J T L McClintock, Redhall, Ballycarry, Carrickfergus, Co Antrim, N Ireland, BT38 9JL [IO74DS, J49]
G4 MAK J R W Gregg, 63 Low Common, Methley, Leeds, LS26 9AF [IO93HR, SE42]
G4 MAN M R Mansell, 251 Long Rd, Canvey Island, SS8 0JG [JO01GM, TQ78]
G4 MAP Details withheld at licensee's request by SSL.
G4 MAQ D E Andreoli, 45 Broad Oak, Headington, Oxford, OX3 8TS [IO91JR, SP50]
G4 MAR Details withheld at licensee's request by SSL.[Op: C Rowe, 121 Victoria St, Willenhall, W Midlands, WV13 1DW.]
G4 MAS C Day, 6 Lower Bourne, Gdns Ware, Herts, SG12 0BL [IO91XT, TL31]
G4 MAU Dr D R Birchall, 6 Hillmorton Rd, Knowle, Solihull, B93 9JL [IO92DJ, SP17]
G4 MAW M A W Marment, 3 Spruce Way, Paignton, TQ3 3SG [IO80FK, SX86]
GI4 MAY N Mayes, 187 Clarawood Park, Belfast, BT5 6FW [IO74BO, J37]
G4 MB J L Bowes, 20 Broomfield Rd, Bexleyheath, DA6 7PA [JO01BK, TQ47]
G4 MBA A Cowsill, 21 Manor Cl, Bromham, Bedford, MK43 8JA [IO92RD, TL05]
G4 MBB L A Jupp, Wood Burcote, Towcester, Northants, NN12 7JR [IO92MC, SP74]
G4 MBC F C Handscombe Mid-Beds Contest Assoc, Sandholm, Bridge End Rd, Red Lodge, Bury St. Edmunds, IP28 8LQ [JO02FH, TL67]
G4 MBD I H Moth, 145 Carisbrooke Rd, Winterdyne, Newport, PO30 1DG [IO90IQ, SZ48]
G4 MBE R Scargill, 17 Springfield Ln, Morley, Leeds, LS27 9PL [IO93ES, SE22]
GM4 MBG I Simpson, 1 Knockhall Rd, Newburgh, Ellon, Aberdeenshire, AB41 0BJ [IO87WI, NJ93]
G4 MBH A Embleton, c/o F.C.O.(Havana), King Charles St, London, SW1A 2AH [IO91WM, TQ37]
G4 MBJ R E Hyett, 18 Escley Dr, Hereford, HR2 7LU [IO82PB, SO53]
G4 MBK J L Broadbent, Buttercross Cottage, Low Rd, Grayingham, Gainsborough, DN21 4ER [IO93QK, SK99]
GW4 MBL S P Elmore, Eirianfron, Llangoed, Beaumaris, Gwynedd, LL58 8PG [IO73WH, SH68]
GI4 MBM S McConnell, 8 Carnesure Dr, Old Ballygowan Rd, Comber, Newtownards, BT23 5LP [IO74CN, J46]
G4 MBN J M Tonks, 72 Muskoka Dr, Bents Green, Sheffield, S11 7RJ [IO93FI, SK38]
GI4 MBQ C C Black, 5Woodbrook Park, Warrenpoint, Newry, BT34 3HL [IO64VC, J11]
G4 MBS C G Elliott, Fairlands Farm, Landford Wood, Salisbury, SP5 2ES [IO90EX, SU22]
G4 MBT A Henderson, 3 Helmsley Lawn, Redcar, TS10 2LL [IO94LO, NZ62]
G4 MBU Details withheld at licensee's request by SSL.
G4 MBV W L Blofield, Silver Woods, 3 The Elms, Warfield Park, Bracknell, RG42 3RP [IO91PK, SU87]
G4 MBW J C G E Dufrane, 18 Fairoak Dr, Hedgerows, Bromsgrove, B60 3PN [IO82XH, SO96]
G4 MBZ P D Taylor, 12 Dunbar Rd, Paddock Hill, Frimley, Camberley, GU16 5UZ [IO91PH, SU85]
G4 MCA J M Hanton, 5 St. Davids Dr, Thorpe End, Norwich, NR13 5HR [JO02RP, TG31]
G4 MCE A E Audcent, 9 Woodlands, Axbridge, BS26 2AX [IO81OG, ST45]
G4 MCF C R Begg, 11 Lilac Rd, Normanby, Middlesbrough, TS6 0BS [IO94KN, NZ51]
G4 MCG B Conway, 55 Rhodesville Ave, Harare, P.O Highlands, Zimbabwe
GI4 MCH N J Crymble, 60 Princes Dr, Newtownabbey, BT37 0AZ [IO74BQ, J38]
G4 MCI N R Clayton, 18 Bishop Rd, Bollington, Macclesfield, SK10 5NX [IO83WH, SJ97]
G4 MCK P Fox, 28 The Hollies, Shefford, SG17 5BX [IO92TA, TL13]
G4 MCL Details withheld at licensee's request by SSL.
G4 MCM D J Hadaway, 66 St. Annes Cl, Winchester, SO22 4LQ [IO91HB, SU42]
G4 MCN Details withheld at licensee's request by SSL.
G4 MCQ S J Bailey, 50 Quantock Cl, North Common, Warmley, Bristol, BS15 5UT [IO81SK, ST67]
GD4 MCR D P Cannon, Rose Cottage, Derby Rd, Peel, Isle of Man, IM5 1HP
G4 MCU G J Stow, 15 Hawthorne Gdns, Hockley, SS5 4SW [JO01HO, TQ89]
GM4 MCV A Patterson, 10A Hermitage Pl, Leith, Edinburgh, EH6 8AF [IO85KX, NT27]
GI4 MCW G Edgar, 51 Kempe Stones Rd, Newtownards, BT23 4SQ [IO74DO, J47]
G4 MD P R Howett, 37 Imperial Ave, Kidderminster, DY10 2RA [IO82VJ, SO87]
G4 MDB R F G Tokley, 9 Peel Rd, Springfield, Chelmsford, CM2 6AQ [JO01FR, TL70]
G4 MDC J R Divall, 2 Brockswood Ln, Welwyn Garden City, AL8 7BG [IO91VT, TL21]
GI4 MDD I H Gibson, 4 Ilford Ave, Belfast, BT6 9SF [IO74BN, J36]
G4 MDE K E Hodkinson, 13 Glovelly Rd, Edenthorpe, Doncaster, South Yorks, DN3 2PE [IO93LN, SE60]
G4 MDF B M Bristow Mid-Thames Rdf, Club Jays Lodge, Crays Pond, Reading, RG8 7QG [IO91KM, SU68]
G4 MDG J W Baily, 13 Longleigh Ln, Bexleyheath, DA7 5SL [JO01BL, TQ47]
G4 MDH G E Feary, 76 Parsons Way, Wootton Bassett, Swindon, SN4 8DJ [IO91BN, SU08]
G4 MDI Details withheld at licensee's request by SSL.[Op: Mrs D J Dagley.]
G4 MDJ A Smith, White Acre, Thornfield Ave, Leek, ST13 5BP [IO83XC, SJ95]
G4 MDM P E Brassington, 42 Dartmouth Ave, Newcastle, ST5 3NY [IO82VX, SJ84]
G4 MDN D A Fowler, The Dees, Cross Lanes, Chalfont St. Peter, Gerrards Cross, SL9 0LR [IO91RO, TQ09]
GI4 MDO S W Hewitt, 23 Drumard Rd, Portadown, Craigavon, BT62 4HP [IO64RJ, H95]
G4 MDP Details withheld at licensee's request by SSL.
G4 MDQ K M Fox, 14 Plantation Ave, Anston, Sheffield, S31 7DA [IO93GJ, SK38]
G4 MDR A M Farmer, 42 Sunridge Cl, Newport Pagnell, MK16 0LT [IO92PB, SP84]
G4 MDS R V Beard, 18 The Avenue, Kidsgrove, Stoke on Trent, ST7 1AG [IO83VC, SJ85]
G4 MDT G L Fitton, 29 Okus Gr, Upper Stratton, Swindon, SN2 6QA [IO91CO, SU18]
G4 MDU J R Gudgeon, Shillingsworth Cottage, Leckhampstead Rd, Wicken, Milton Keynes, Bucks, MK19 6BY [IO92NB, SP73]
G4 MDX Details withheld at licensee's request by SSL.
GD4 MDY S C Keenan, Fenella Villa, Peveril Rd, Peel, Isle of Man, IM5 1PJ
G4 MDZ S P Cline, Falklands, 69 The St., Hawkinge, Folkestone, CT18 7DE [JO01OC, TR24]
G4 MEA R J Hutchings, 16-le-Marchant Rd, Frimley, Camberley, GU16 5RW [IO91PH, SU85]
G4 MEB P R Green, 1 Haddon Croft, Halesowen, B63 1JQ [IO82XK, SO98]
G4 MEC Details withheld at licensee's request by SSL.
G4 MED J Norfolk, 44 West Down Rd, Plymouth, PL2 3HF [IO70WJ, SX45]
G4 MEE D P Mobbs, 39 Bramwell Rd, Freckleton, Preston, PR4 1SS [IO83NR, SD42]
G4 MEF B J Rudkin, 18 Beechfield Ave, Birstall, Leicester, LE4 4DA [IO92KQ, SK50]
G4 MEG M H Riches, 5 Prospect Cttgs, Haigh, Wigan, WN2 1LA [IO83QN, SD60]
G4 MEH M E Hughes, 8 Elm Beds Rd, Poynton, Stockport, SK12 1TG [IO83WI, SJ98]
GW4 MEI I Williams, 27 y Glyn, Caernarvon, Gwynedd, LL55 1HF [IO73UD, SH46]
G4 MEK C Chappell, 24 North Cross Rd, Cowcliffe, Huddersfield, HD2 2NL [IO93CQ, SE11]
G4 MEM M J David, 5 Glenfall St., Cheltenham, GL52 2JA [IO81XV, SO92]
G4 MEN P J Cooper S.I.R.S, Sports & Social Club, The Newlands Bishop, Cleeve Cheltenham, GL52 4NZ [IO81XW, SO92]
G4 MEO B T Elliott, 4 Ivel View, Sandy, SG19 1AU [IO92UD, TL14]
G4 MEP C J Hughes, 175 Arle Rd, Cheltenham, GL51 8LJ [IO81WV, SO92]
GI4 MEQ M I Kelly, 6 Beechdene Gdns, Lisburn, BT28 3JH [IO64XM, J26]
G4 MES J Willis, Kings Cottage, London Rd, Barkway, Royston, SG8 8EZ [JO01AX, TL33]
G4 MET E G Robinson, 29 Folly Ln, Hereford, HR1 1LX [IO82PB, SO54]
G4 MEU R A Russell, 22 Hopelands, Heighington Village, Newton Aycliffe, DL5 6PQ [IO94EO, NZ22]
G4 MEX M R Care, 12 Hallowell Rd, Northwood, HA6 1DW [IO91SO, TQ09]
G4 MEY R H Care, Crooksell Hall, Potash Rd, Wyverstone, Stowmarket, Suffolk, IP14 4SL [JO02LG, TM06]
G4 MEZ Details withheld at licensee's request by SSL.[Op: J A Eastment. Station located near Newport Pagnell.]
GM4 MF Capt J Gardner, 6 Haygate Ave, Brightons, Falkirk, FK2 0TL [IO85DX, NS97]
G4 MFB C E Bowden, 13 Shaw Dr, Sandford, Wareham, BH20 7BS [IO80WQ, SY98]
G4 MFD W C W Dean, Bowerdene, Staplehay Trull, Taunton, Somerset, TA3 7HH [IO80KX, ST22]
G4 MFE R A Kaiser, Blackcombe Farm, Henwood, Liskeard, Cornwall, PL14 5BW [IO70SM, SX27]
G4 MFF R A F Freer, 37 Wicklands Ave, Saltdean, Brighton, E Sussex, BN2 8AN [IO90XT, TQ30]
G4 MFH J C Seddon, 96 Yewdale, Westbank, Skelmersdale, WN8 6EP [IO83ON, SD40]
G4 MFI D E Roberts, 88 Woodhouse Rd, Davyhulme, Urmston, Manchester, M41 7WX [IO83TL, SJ79]
G4 MFJ P E Shaw, 4 Henderson St., Preston, PR1 7XP [IO83PS, SD53]
G4 MFK G M Dennick, 29 Vale Cottage, Old Post Office Ln, Badsey, Evesham Worcs, WR11 5UF [IO92BC, SP04]
GM4 MFL R B Kerr Easter Ross RC, Rosskeen Bridge, Invergordon, Ross-shire, IV18 0PL [IO77VQ, NH66]
GI4 MFM R Mc Ateer, 23 Main St., Garvagh, Coleraine, BT51 5AA [IO64PX, C81]
G4 MFN M R Jones, 67 Dosthill Rd, Two Gates, Tamworth, B77 1JD [IO92DO, SK20]
G4 MFO M Mackenzie, 22 Milton Cres, Ravenshead, Nottingham, NG15 9BA [IO93KC, SK55]
G4 MFP W B Woollen, Greensward, Townsend, Harwell, Didcot, OX11 0DX [IO91IO, SU48]
G4 MFQ R Dunstan, 100 Trevithick Rd, St. Austell, PL25 4RJ [IO70OI, SX05]
G4 MFR C S Shanks, 225 Freshfield Rd, Brighton, BN2 2YE [IO90WT, TQ30]
G4 MFS M F Smith, 10 Killicks Hill, Portland, Dorset, DT5 1JW [IO80SN, SY67]
G4 MFU T J Riggott, 94 Hallow Rd, Worcester, WR2 6BY [IO82VE, SO85]
G4 MFV J W Marshall, 278 Derby Rd, Bramcote, Nottingham, NG9 3JN [IO92IW, SK53]
G4 MFW B P M Fletcher, c/o Roebuck House, Brighton Rd, Godalming, Surrey, GU7 1NS [IO91QE, SU94]
G4 MFX G R Davis, 6 Summerfield Rd, Hemsby, Great Yarmouth, NR29 4LY [JO02UQ, TG41]
G4 MGB H C Mayor, Lock House, Canal Bank, Tarleton, Preston, PR4 6HD [IO83OQ, SD42]
G4 MGC D S Litton, 9 Fairway, Warner Rd, Ware, SG12 9JP [IO91XT, TL31]
G4 MGD Details withheld at licensee's request by SSL.
G4 MGF G E Fraser, 11 George V Ave, Margate, CT9 5QA [JO01QJ, TR37]
G4 MGG S J Esposito, 21 Spencefield Ln, Leicester, LE5 6PT [IO92LP, SK60]
GW4 MGH R A Hampson, 30 Witts Ln, Purton, Swindon, SN5 9EX [IO91BO, SU08]
G4 MGI P B Kemmis, 14 Rochester Rd, London, NW1 9JH [IO91WN, TQ28]
G4 MGK W Brown, 2 Ashworth Park, Knutsford, WA16 9DE [IO83TH, SJ77]
G4 MGL Details withheld at licensee's request by SSL.[Op: A R Hay, 16 Newfield Avenue, Cove, Farnborough, Hants, GU14 9PQ.]
G4 MGN M G Norman, 7 Kingsway, Seaford, BN25 2NE [JO00BS, TV49]
G4 MGO J B H Newman, 38 Vine Dr, Wivenhoe, Colchester, CO7 9HB [JO01LU, TM02]
G4 MGP H Boddy, Greyholme, West Ln, Snainton, Scarborough, YO13 9AR [IO94QF, SE98]
G4 MGQ R N Boddy, Greyholme, West Ln, Snainton, Scarborough, YO13 9AR [IO94QF, SE98]
G4 MGR D Jones Wirral & Dist Rd, 39 Pensby Rd, Heswall, Wirral, L60 7RA [IO83KI, SJ28]
G4 MGU R Smith, 81 Ingleside Cres, Lancing, BN15 8EW [IO90UT, TQ10]
G4 MGV R J Pass, 6 High St, Hanslope, Milton Keynes, Bucks, MK19 7LQ [IO92OC, SP84]
G4 MGW Details withheld at licensee's request by SSL.
G4 MGX J E Freeman, 5A Beech Ave, Briars Bank Park, Wilstead, Bedford, Beds, MK45 3WE [IO92TC, TL04]
G4 MGZ R Curtis, Grovely Cottage, Water St., Berwick St. John, Shaftesbury, SP7 0HS [IO80XX, ST92]
G4 MH J H Fish, 28 Banks Ave, Golcar, Huddersfield, HD7 4LZ [IO93BP, SE01]
G4 MHB H Balen, 2 Winchester Ave, Beeston, Nottingham, NG9 1AU [IO92JW, SK53]
G4 MHC D G Hobro Malvern Hills Radio Amateurs C, 60 Linksview Cres, Newtown, Worcester, WR5 1JJ [IO82VE, SO85]
GI4 MHD G Quaite, 21 Broomhill Rd, Ballynahinch, BT24 8QD [IO74BI, J34]
GM4 MHE R N M M Musto, Hartwood Mains Farmhouse, West Calder, West Lothian, EH55 8LE [IO85FU, NT06]
G4 MHF J Marshall, Chasborough House, Village Hall Ln, Three Legged Cross, Wimborne, BH21 6SG [IO90BU, SU00]
G4 MHJ R C Hewitt, 38 Eastry Rd, Erith, DA8 1NN [JO01BL, TQ47]
G4 MHK T J Fougere, 48 Longland Rd, Eastbourne, BN20 8HY [JO00DS, TV59]
G4 MHN Details withheld at licensee's request by SSL.
G4 MHO W Wainwright, 243 Long Ln, Aughton Park, Aughton, Ormskirk, L39 5BY [IO83NN, SD40]
G4 MHQ A Bell, 22 Ryde Pl, Lee on The Solent, PO13 9AU [IO90JT, SZ59]
G4 MHR C J Norman, 23 Snells Mead, Buntingford, SG9 9JF [IO91XW, TL32]
G4 MHS N H Naish, 85 Wear Bay Rd, Folkestone, CT19 6PR [JO01OC, TR23]
G4 MHU Details withheld at licensee's request by SSL.[Op: P J Maxted. Station located near Sandwich.]
G4 MHW R G A Pigg, 58 The Fairway, Newcastle upon Tyne, NE3 5AQ [IO95EA, NZ26]
G4 MHX B G Smith, 6 Howbeck Cres, Wybunbury, Nantwich, CW5 7NX [IO83SB, SJ64]
G4 MHY F D Monson, 37 Hillfoot Rd, Woolton, Liverpool, L25 7UJ [IO83NI, SJ48]
G4 MIA G Y O'Keeffe-Wilson, 20 South Dr, Upton, Wirral, L49 6LA [IO83KJ, SJ28]
G4 MIB D S Senior, 78 Palace Rd, London, SW2 3JX [IO91WK, TQ37]
G4 MID E I Pratt, The Mays, 65 Barton Rd, Thurston, Bury St. Edmunds, IP31 3PD [JO02JG, TL96]
G4 MIE A M Weeden, Great Gable, Metton Rd, Cromer, NR27 9JH [JO02PV, TG24]
G4 MIF G W Beck, 16 Bader Ct, Chapel Ln, Hawley, Farnborough, GU14 9BS [IO91OH, SU85]
GM4 MIG I Giffen, Glen-Roma, Reddingmuirhead, Falkirk, Scotland, FK2 0DU [IO85DX, NS97]
GW4 MII P A Jenkins, 2 Gwynfi St., Treboeth, Swansea, SA5 7DW [IO81AP, SS69]
G4 MIJ R M Hunt, Penlee Cottage, 13 High Row, Gainford, Darlington, DL2 3DN [IO94DN, NZ11]
G4 MIK Dr M A Bull, Toad in The Hole, 25 Prospect Park, Southborough, Tunbridge Wells, TN4 0EQ [JO01DD, TQ54]
GM4 MIM Rev I C Morrison, 29 Philip Ave, Linlithgow, EH49 7BH [IO85EX, NS97]
G4 MIN Details withheld at licensee's request by SSL.[Station located near St Pauls Cray. Correspondence to: M A Minns, 78 Dunvegan Road, London SE9 1SB.]
G4 MIO P J Davies, 675 Terhulpense St, 3090 Overijse, Belgium
G4 MIP P Phillips, Woodreefe, Amroth, Narberth, Dyfed, SA67 8NR [IO71QR, SN10]

G4 MIQ L J Rowley, 13 Ruddington Walk, Abbey Rise West, Leicester, LE4 2FH [IO92KP, SK50]
G4 MIS N B Allen, White House, Flintergill, Dent, Sedbergh, LA10 5QR [IO84SG, SD78]
G4 MIT G M Hurst, The Hollies, Etwall, Derby, DE6 6NB [IO92DX, SK24]
G4 MIU K R Rawlings, 58 Brunwin Rd, Rayne, Braintree, CM7 5BU [JO01GV, TL72]
G4 MIV K J Gibson, 179 Ratcliffe Rd, Sileby, Loughborough, LE12 7PX [IO92KR, SK61]
G4 MIX M J Howland, 1 Swanton Farm Cott, Swanton Ln, Lydden, Dover, CT15 7EY [JO01OD, TR24]
G4 MIY Details withheld at licensee's request by SSL.
G4 MIZ T Wood, St. Antonys Cottage, 143 Station Rd, Burgess Hill, RH15 9ED [IO90WW, TQ31]
G4 MJA Details withheld at licensee's request by SSL.
G4 MJB D Maddison, 18 Keswick Dr, Cullercoats, North Shields, NE30 3EW [IO95GA, NZ37]
G4 MJC F Jul-Christensen, Bushy Lodge Cottage, Ripe Ln, Firle, Lewes, Sussex, BN8 6LS [JO00BU, TQ40]
GI4 MJD M J Dunne, 26 Duncreggan Rd, Londonderry, BT48 0AD [IO65IA, C41]
G4 MJF M J Hill, 42 Oaklands Dr, Weston, Northampton, NN3 3JL [IO92NG, SP76]
G4 MJH Details withheld at licensee's request by SSL.
G4 MJI T A H Sturmey, 302 Roblar Ave, Millbrae, Ca 94030, USA
G4 MJL M C Bramhill, 16 Charnock View Rd, Gleadless, Sheffield, S12 3HJ [IO93GI, SK38]
G4 MJM Details withheld at licensee's request by SSL.[Station located in South Essex.]
G4 MJN D S Tompkins, 18 Carter Ave, Broughton, Kettering, NN14 1LZ [IO92OI, SP87]
G4 MJO Details withheld at licensee's request by SSL.
G4 MJS P C C Bowyer, 12A West End Rd, Mortimer Common, Reading, RG7 3SY [IO91LJ, SU66]
G4 MJT F R Harrison, Five Gables, Welton Low Rd, Elloughton, Brough, HU15 1HR [IO93RR, SE92]
G4 MJU E M Smith, 256 Stone Rd, Hanford, Stoke on Trent, ST4 8NJ [IO82VW, SJ83]
G4 MJW S R Carey, Skylites, 17 Francis Way, Silver End, Witham, CM8 3QX [JO01HU, TL81]
G4 MJX G Fisher, 87 Ethersall Rd, Nelson, BB9 0RP [IO83VT, SD83]
GW4 MJY Details withheld at licensee's request by SSL.
G4 MJZ Details withheld at licensee's request by SSL.
G4 MKD W M Kendal, 63 New St., Three Bridges, Crawley, RH10 1LP [IO91WC, TQ23]
G4 MKE A E Plaice, 10 Stockhill Rd, Chilcompton, Bath, BA3 4JL [IO81RG, ST65]
G4 MKF Dr M K Franks, 13 Fifth Rd, Newbury, RG14 6DN [IO91IJ, SU46]
G4 MKG P G Opie, Timbers, Stockers Hill Rd, Rodmersham, Sittingbourne, ME9 0PL [JO01IH, TQ96]
G4 MKI D S Bray, 180 Greenhill Rd, Herne Bay, CT6 7RS [JO01NI, TR16]
GI4 MKJ W V Waugh, Sea View Farm, I Ballyvester Rd, Donaghadee, Co Down, N Ireland, BT21 0LL [IO74FP, J57]
G4 MKM A J McMillan, 205 London Rd, Wokingham, RG40 1SP [IO91OJ, SU86]
G4 MKQ K E Barnes, 91 The Ridings, Gatcombe Park, Hilsea, Portsmouth, PO2 0UF [IO90LT, SU60]
G4 MKR R N Byford, 7 Sutton Mill Rd, Potton, Sandy, SG19 2QB [IO92VD, TL24]
G4 MKS P J Walsh, 8 Whitewood Cl, Ashton in Makerfield, Wigan, WN4 8ED [IO83QM, SD50]
G4 MKT B C Jackson, 16 Peterborough Cl, Ashton under Lyne, OL6 8XW [IO83WM, SD90]
GM4 MKU J Flett, 40 Commerce St., Lossiemouth, IV31 6QH [IO87IR, NJ27]
G4 MKW P J Bowden, 12 Honeywood, Roffey, Horsham, West Sussex, RH13 6AE [IO91UB, TQ13]
G4 MKX C S Gericke, Bondene, 25 Highmead Gdns, Bishop Sutton, Bristol, BS18 4XB [IO81QH, ST55]
GM4 MKY W Learmonth, Flat 3/2, 22 Farmeloan Rd, Rutherglen, Glasgow, G73 1DP [IO75VT, NS66]
G4 MLA Details withheld at licensee's request by SSL.
G4 MLB R Padmore, 3 Uldale Cl, Nelson, BB9 0ST [IO83VT, SD83]
G4 MLD A S Wortman, 1 Lincoln Gdns, Ilford, IG1 3NF [JO01AN, TQ48]
G4 MLE B A Farrelly, Sore Renen 59, 5044, Nattland, Norway
G4 MLF G Charlton, 9 Charter Dr, East Herrington, Sunderland, SR3 3PG [IO94GU, NZ35]
G4 MLG A K Denyer, 2 Malta Cttgs, Ashmore Green Rd, Ashmore Green, Thatcham, RG18 9EZ [IO91IK, SU56]
G4 MLI B J Mitchell, Trevescan, Tintagel, Cornwall, PL34 0DT [IO70PP, SX08]
G4 MLJ Details withheld at licensee's request by SSL.
G4 MLK J A Pennock, Wilton Cottage, Partney, Spilsby, Lincs, PE23 4PF [JO03BE, TF46]
G4 MLL F L Whitehead, 18 Bath Rd, Mickleover, Derby, DE3 5BW [IO92FV, SK33]
G4 MLN B A Payne, 4 Woodland Croft, Horsforth, Leeds, LS18 5NE [IO93EU, SE23]
G4 MLO G D Rae, 45 Mapperley Dr, Northampton, Northants, NN3 9UF [IO92NF, SP76]
G4 MLP Details withheld at licensee's request by SSL.
G4 MLQ J Lamont, 49 Castle St., Barnsley, S70 1NT [IO93GN, SE30]
G4 MLR J S Norris, 5 Julian Cl, Bristol, BS9 1JX [IO81QL, ST57]
G4 MLS R K Woodman, 5 Evergreen Cl, Leybourne, West Malling, ME19 5PY [JO01FH, TQ65]
G4 MLV L Gaunt, 31 Moat Hill, Birstall, Batley, WF17 0DX [IO93ER, SE22]
G4 MLW Dr I F Jones, 51 Springfield Park, Mirfield, WF14 9PE [IO93DQ, SE22]
G4 MLX C W Kempe, PO Box 463, Hamilton 5, Bermuda, ZZ8 7MS
G4 MLY I D Vincent, 1 Lullingstone Rd, Belvedere, DA17 5NJ [JO01BL, TQ47]
G4 MM J M Miller, Flat 24 Guardian, Ct Moorend Rd, Charlton Kings, Cheltenham Glos, GL53 9DP [IO81XV, SO92]
G4 MMA K F Barnard, 89 Kings Rd, Harrow, HA2 9LD [IO91TN, TQ18]
G4 MMB M M Barnes, 22 Lordswood, Silchester, Reading, RG7 2PZ [IO91KI, SU66]
G4 MMD H Waldron, 33 Forest Rd, Dudley, DY1 4BU [IO82XM, SO99]
G4 MMF R G Holtom, 51 Hertford Rd, Stratford upon Avon, CV37 9AN [IO92DE, SP15]
G4 MMG A F Beecher, 27 Normandale, Bexhill on Sea, TN39 3LU [JO00FU, TQ70]
G4 MMH M W Evans, Pastors, Dodds Bank, Nutley, Uckfield, E Sussex, TN22 3LR [JO01AA, TQ42]
G4 MMI R T Hodge, Corner House, Manor Gdns, Hurstpierpoint, West Sussex, BN6 9UG [IO90VW, TQ21]
GM4 MMM A S McLellan, 16 Jeffrey Bank, Boness, EH51 0EH [IO86EA, NS98]
G4 MMT A P Haley, F C O (Mexico City), King Charles St, London, SW1A 2AH [IO91WM, TQ37]
G4 MMV B R Hudson, 74 Telford St., Hull, HU9 3DX [IO93US, TA13]
G4 MMY A T Vernon, 33 New Rd, Horbury, Wakefield, WF4 5LS [IO93FP, SE21]
G4 MMZ A G Warner, 8 Castle View Park, Mawnan Smith, Falmouth, TR11 5HB [IO70KC, SW72]
G4 MNA L D Meale, 57 Chestnut Dr, Newton Abbot, TQ12 4JZ [IO80FM, SX87]
G4 MNB R W Sharp, 77 Cloche Way, Upper Stratton, Swindon, SN2 6JN [IO91CO, SU18]
G4 MNE J F A White, 25 Fulwith Dr, Harrogate, HG2 8HW [IO93FX, SE35]
GI4 MNF N Foote, 4 Bushfield Rd, Moira, Craigavon, BT67 0JB [IO64WM, J16]
G4 MNI W W Loucks, 155 Brentwood Rd N, Toronto, Ontario M8X 2C8, Canada
G4 MNL Details withheld at licensee's request by SSL.
G4 MNM Details withheld at licensee's request by SSL.
GI4 MNO Details withheld at licensee's request by SSL.
G4 MNP M E Ward, Sage Cottage, Farley, Wilts, SP5 1AA [IO91DB, SU22]
GW4 MNQ Details withheld at licensee's request by SSL.
G4 MNT R A Calkin, Sea View Parade, St. Lawrence Bay, St. Lawrence, Southminster, Essex, CM0 7PB [JO01JR, TL90]
G4 MNZ T R Littler, 21 Collop Dr, Heywood, OL10 2LS [IO83VN, SD80]
GM4 MOA J Y Merson, Lynnwood, 8 Newlands Ln, Buckie, AB56 1JX [IO87MQ, NJ46]
G4 MOC P D Fawkes, 56 Woolstencroft Ave, Kings Lynn, PE30 2PB [JO02ES, TF62]
G4 MOE S D J Noke, 48 Hoadley Green, Salisbury, SP1 3HS [IO91CC, SU13]
G4 MOF Details withheld at licensee's request by SSL.
GW4 MOG C D Tombs, 70 Heol y Frenhines, Southra Park, Dinas Powys, CF64 4UH [IO81JK, ST17]
G4 MOI D M Bone, 69 Pick Hill, Waltham Abbey, EN9 3LD [JO01AQ, TL30]
G4 MOJ A L Jeffries, 5 Darnick Rd, Sutton Coldfield, B73 6PE [IO92BN, SP09]
GW4 MOK V A Cashmore, 12 Bretton Dr, Broughton, Chester, CH4 0RS [IO83MD, SJ36]
GW4 MOL K W Highley, 3 West Hill Dr, Hythe, Southampton, SO45 6DL [IO90HU, SU40]
G4 MOM Details withheld at licensee's request by SSL.[Station located near Basildon.]
G4 MOP Details withheld at licensee's request by SSL.
G4 MOT K Watson, 11 Gotts Park Ave, Armley, Leeds, LS12 2RW [IO93ET, SE23]
G4 MOU J A Oates, 3 Canadian Ave, Chester, CH2 3HG [IO83NE, SJ46]
G4 MOV E R Durey, 71 Orchard Rd, Maldon, CM9 6EW [JO01IR, TL80]
GM4 MOX Details withheld at licensee's request by SSL.
GW4 MOZ J Upstone, 1 Rhymney St., Cathays, Cardiff, CF2 4DF [IO81JL, ST17]
G4 MPA G J Squibb, 36 Frognal Gdns, Teynham, Sittingbourne, ME9 9HU [JO01JH, TQ96]
GM4 MPC D W Smith, 13 Fernie Pl, Dunfermline, KY12 9BX [IO86GB, NT08]
G4 MPG P D Grace, 3 Warwick Grange, Warwick Rd, Solihull, B91 1DD [IO92CK, SP18]
G4 MPH N R Simmonds, 5 Wildfield Cl, Wood St. Village, Guildford, GU3 3EG [IO91QF, SU95]
G4 MPI W J Sharples, Snaefell Cottage, 135 Belthorn Rd, Belthorn, Blackburn, BB1 2PE [IO83SR, SD72]
G4 MPJ M R Whitfield, Waterside, 26 Kingsclere Dr, Bishops Cleeve, Cheltenham, GL52 4TG [IO81XW, SO92]
G4 MPK S Foster, 91 Copthorne Rd, Leatherhead, KT22 7EF [IO91UH, TQ15]
G4 MPL T C Grimbleby, 18 Downfield Ave, Hull, HU6 7XF [IO93TS, TA03]
G4 MPN J R Brisley, 8 Chillington Dr, Codsall, Wolverhampton, WV8 1AG [IO82VP, SJ80]
G4 MPO C K Duffy, 87 Allenby Dr, Beeston, Leeds, LS11 5RX [IO93FS, SE22]
G4 MPP S D Ellis, 46 Sedgefield Cl, Worth, Crawley, RH10 7XG [IO91WC, TQ33]
G4 MPQ K P Clark, Rosemary Cottage, Bylane End, Liskeard, Cornwall, PL14 3PZ [IO70SK, SX26]
GM4 MPR D Miller, The Old School, Ackergill, Wick, Caithness, KW1 4RG [IO88KL, ND35]
G4 MPS B S Spence, 212 Mill Rd, Kettering, NN16 0RN [IO92PJ, SP87]
G4 MPT D G Abbott, 42 Rosebery Ave, Blackpool, FY4 1LB [IO83LS, SD33]
GM4 MPU Details withheld at licensee's request by SSL.
G4 MPV A Finbow, 135 Doncaster Rd, Goldthorpe, Rotherham, S63 9JA [IO93IM, SE40]
G4 MPW M Corbett, Braemar, Heath Mill Ln, Worplesdon, Guildford, GU3 3PR [IO91QG, SU95]
GW4 MPX Details withheld at licensee's request by SSL.
GM4 MPY A M Bell, 48 Greenlaw Cres, Barshaw Wood, Paisley, PA1 3RT [IO75TU, NS46]
G4 MQB M A J Pill, 5 St. Leonards Cl, Upton St. Leonards, Gloucester, GL4 8AL [IO81VT, SO81]
G4 MQC S Laing, 16 Firs View Rd, Hazelmere, Hazlemere, High Wycombe, HP15 7TD [IO91PP, SU89]
G4 MQD W R N Davie, Rosemundy Villa, St Agnes, Cornwall, TR5 0UF [IO70JH, SW75]
G4 MQE Details withheld at licensee's request by SSL.[Op: R C Hewitson. Station located near Grimsby.]
G4 MQF A Ramsey, 51 Queens Rd, Cadbury Heath, Warmley, Bristol, BS15 5EJ [IO81SK, ST67]
G4 MQJ C R Winters, 45 Blackbush Spring, Harlow, CM20 3DY [JO01BS, TL41]
G4 MQK C C Cubitt, Sunrise, Coast Rd, Walcott, Norwich, Norfolk, NR12 0NG [JO02SU, TG33]
G4 MQL R C Cuddington, 35 Lowndes Gr, Shenley Church End, Milton Keynes, MK5 6EE [IO92OA, SP83]
G4 MQM D Cressey, 8 Parklands Dr, Harlaxton, Grantham, NG32 1HX [IO92PV, SK83]
G4 MQN W E E Eason, 36 Potters Field, Harlow, CM17 9BZ [JO01BS, TL40]
G4 MQP P A Crowe, 22 Ringsbury Cl, Purton, Swindon, SN5 9DE [IO91BO, SU08]
G4 MQQ S C Jones, 15 Heritage Park, Hatchwarren, Basingstoke, RG22 4XT [IO91KF, SU64]
G4 MQR Dr G D Blower, 30 The Glebe, Cumnor, Oxford, OX2 9QA [IO91IR, SP40]
G4 MQS P R Elliott, Flat 1, 16 Cavendish Rd, Leicester, LE2 7PG [IO92KO, SK50]
G4 MQT P P Wheatley, 67 Moorlands Rd, Verwood, BH31 7PD [IO90BV, SU00]
G4 MQU F C C Wyatt, The Old Nursery, The Drift, Swardeston Common, Norwich Norfolk, NR14 8LQ [JO02ON, TG10]
G4 MQV J W Sanderson, 5 Babbacombe Dr, Ruddhill Est, Ferryhill, DL17 8DA [IO94FQ, NZ33]
G4 MQW R C Richardson, Polkerris, Gr Park, Hampton on The Hill, Warwick, CV35 8QR [IO92EG, SP26]
G4 MQX S J Vincent, Meldon, Hillside, Axbridge, BS26 2AN [IO81OG, ST45]
G4 MRA R C Ainge, 2 Norway Cl, Corby, NN18 9EG [IO92PL, SP88]
G4 MRB J F Feeley, 177 Rock St., Sheffield, S3 9JF [IO93GJ, SK38]
G4 MRD Dr C D Scrase, 3 Martingale Cl, Cambridge, CB4 3TA [JO02BF, TL46]
G4 MRH Details withheld at licensee's request by SSL.[Station located near Hove.]
G4 MRJ J R Skett, 38 Woodchurch Ln, Ellesmere Port, South Wirral, L66 3NQ [IO83MG, SJ37]
G4 MRK J B Veitch, 14 Dunmore Ave, Sunderland, SR6 8ET [IO94HW, NZ45]
G4 MRL N M Thomas, 31A Gloucester St., London, SW1V 2DB [IO91WL, TQ27]
GW4 MRM D Stonehouse, 49 Heol y Gelynen, Upper Brynamman, Ammanford, SA18 1SB [IO81BT, SN71]
GI4 MRN J W McCrea, 14 Fairfield Park, Bangor, BT20 4TX [IO74EP, J58]
G4 MRO J P Turner, 15 Princess Rd, Hoverland Park, Taunton, TA1 4SY [IO81KA, ST22]
G4 MRP R C Sadler, 14 Briardale, Wimbledon Park Rd, Southfields, London, SW19 6PF [IO91VK, TQ27]
G4 MRQ R S Marchington, 78 Buxton Rd, Dove Holes, Buxton, SK17 8DW [IO93BH, SK07]
G4 MRR H V McEvoy, 5 Blackmore, Letchworth, SG6 2SX [IO91VX, TL23]
G4 MRS A R J Cook Martlesham RS, British Telecom, Matlesham Heath, Ipswich Suffolk, IP5 7RE [JO02OB, TM24]
G4 MRT E W F Malone, 65 Grosvenor Ave, Newcastle upon Tyne, NE2 2NP [IO94EX, NZ26]
G4 MRU P W Sables, 54 Harvey St., Deepcar, Sheffield, S30 5QB [IO93GJ, SK38]
G4 MRW R Westmeckett, 3 Alton Gr, Portchester, Fareham, PO16 9NJ [IO90KU, SU60]
G4 MRX J P Cooch, 6 Blackthorn Cl, Newton, Preston, PR4 3TU [IO83NS, SD43]
GI4 MRZ E Smith, 21 Clandeboye Way, Bangor, BT19 1AD [IO74DP, J48]
G4 MSA D G Price, c/o 17 Grange Cross, Ln, West Kirby, Wirral Merseyside, L48 8BJ [IO83KI, SJ28]
G4 MSB Details withheld at licensee's request by SSL.
G4 MSC J R Watts, 33 Knightsbridge Way, Hemel Hempstead, HP2 5ES [IO91SS, TL00]
G4 MSE J A C Q Sivapragasam, 1 Treve Ave, Harrow, HA1 4AL [IO91TN, TQ18]
G4 MSF K Watt, 7 Turfside, Leam Ln Est, Gateshead, NE10 8EX [IO94FW, NZ26]
G4 MSG W F Turton, 22 Meadow Rd, Ripley, DE5 3EP [IO93HB, SK45]
G4 MSH R L Penticost, 67 Kings Rd, Horsham, RH13 5PP [IO91UA, TQ22]
GW4 MSI P M N Needham, Frondeg, Copperhill St, Aberdyfi, Gwynedd, LL35 0HT [IO72XN, SN69]
G4 MSJ T Moan, 23 Laurel Gr, Sunderland, SR2 9EE [IO94HV, NZ35]
G4 MSK W E Wilkinson, 24 Greenway, Bromley, BR2 8EY [JO01AJ, TQ46]
GM4 MSL G Wallace, 13 Fraser Terr, Perth, PH1 1BX [IO86GJ, NO12]
G4 MSM Rev P McArdle, Mount St Mary's College, Spinkhill, Sheffield, S31 9EG [IO93GJ, SK38]
G4 MSN R Slator, 27 The Dr, Alwoodley, Leeds, LS17 7QB [IO93FU, SE24]
GW4 MSO Details withheld at licensee's request by SSL.
G4 MSQ F E Watson, Synehurste, Kimbolton Rd, Bolnhurst, Beds, MK44 2EW [IO92TF, TL06]
G4 MSR J E Russell, 56 Normanton Ln, Keyworth, Nottingham, NG12 5HA [IO92KV, SK63]
G4 MSS R Levesconte, 15 Lumsden Rd, Southsea, Hants, PO4 9LN [IO90LS, SZ69]
G4 MSV S J O,Donnell, Men A Vaur, Wall Rd Gwinear, Hayle, Cornwall, TR27 5HA [IO70HE, SW53]
G4 MSW L H Morgan, 22 Stonelea Rd, Hemel Hempstead, HP3 9JY [IO91SR, TL00]
G4 MSY R Naylor, 55 Caistor Rd, Laceby, Grimsby, DN37 7JA [IO93VM, TA20]
G4 MSZ S W Lowe, 4 Ashdene Dr, Crofton, Wakefield, WF4 1PQ [IO93GP, SE31]
G4 MTA H T Wood, 14 Orpwood Way, Abingdon, OX14 5PX [IO91IP, SU49]
GM4 MTC Details withheld at licensee's request by SSL.
GW4 MTD H N McMurray, 14 Hopkin St., Brynhyfryd, Swansea, SA5 9HN [IO81AP, SS69]
G4 MTF G Williams, 8 Blythe Cl, Newport Pagnell, MK16 9DN [IO92PB, SP84]
G4 MTG J A Sillitoe, 42 Marsham Rd, Kings Heath, Birmingham, B14 5HD [IO92BJ, SP07]
G4 MTH A E Smith, 25 Lindsay Cl, Stanwell, Staines, TW19 7LF [IO91SL, TQ07]
GM4 MTI D S M Spence, Glencairn House, Corran Esplanade, Oban, PA34 5AQ [IO76GK, NM83]
G4 MTL D J Epton Mtl Amateur Radio Club, 61 Cartmel Dr, Dunstable, LU6 3PT [IO91RV, TL02]
G4 MTM R E S Pye-Smith, 49 Peaks Hill, Purley, CR8 3JJ [IO91WI, TQ36]
G4 MTN C S E Page, 229 Kingston Rd, Staines, TW18 1PA [IO91SK, TQ07]
G4 MTQ A P Tapp, Penjuan, Parsons Green, Kelly Bray, Callington, Cornwall, PL17 8EY [IO70UM, SX37]
G4 MTS K R Brownjohn, 22 Hobson Rd, Elloughton, Brough, HU15 1JU [IO93RR, SE92]
G4 MTT Details withheld at licensee's request by SSL.
GW4 MTU B J Dennis, 6 Poppy Dr, Neyland, Milford Haven, SA73 1SF [IO71MR, SM90]
GM4 MTV J Beveridge, 86 Main St., Thornton, Kirkcaldy, KY1 4AG [IO86KD, NT29]
G4 MTW F Cook, 2 Burford Gdns, Sunderland, SR3 1LX [IO94HV, NZ35]
GI4 MTZ G K Downs, 19 Mullaghboy Rd, Islandmagee, Larne, BT40 3TT [IO74DU, D40]
G4 MUB H M Joy, 22 Bovingdon Mobil Homes Est, Roman Rd, Hereford, X X
G4 MUD L J Affleck, Creedy House, Nether Ave, Littlestone, New Romney, TN28 8NB [JO00LX, TR02]
GI4 MUE J Gwilt, 207 Clandeboye Rd, Bangor, BT19 1AA [IO74DP, J48]
G4 MUF Details withheld at licensee's request by SSL.[Station located at 51 32N, 01 53W, NGR: SU08.]
G4 MUI D A Brown, 114 Telford Way, High Wycombe, HP13 5TA [IO91OP, SU89]
G4 MUL Details withheld at licensee's request by SSL.[Op: D S Mayo.]
GM4 MUN Details withheld at licensee's request by SSL.
G4 MUO S B Rawlinson, Flat, 9 Badger House, Badger Rd, Macclesfield, SK10 2SS [IO83WG, SJ97]
G4 MUP G P Rouse, 48 Shrewsbury Rd, Birkenhead, L43 2HZ [IO83LJ, SJ38]
G4 MUQ C P Ashlin, 13 Brantfell Gr, Bolton, BL2 5LY [IO83TO, SD70]
G4 MUR A M Harvey, 3 High Ln, Norton Tower, Halifax, HX2 0NW [IO93BR, SE02]
G4 MUS D R Elwell, 18 Padgetts Way, Hullbridge, Hockley, SS5 6LR [JO01HP, TQ89]
G4 MUT T M Hackwill, 6 Ramsbury Dr, Earley, Reading, RG6 7RT [IO91MK, SU77]
G4 MUU F J Westall, 4 Francesca Lodge, Somerford Way, Christchurch, BH23 3QN [IO90CR, SZ19]
G4 MUV S Q Nicolle, 17 Allensmore Cl, Matchborough, Redditch, B98 0AS [IO92BH, SP06]
G4 MUW G L J Weaver, 60 Crispin Rd, Winchcombe, Cheltenham, GL54 5JX [IO91AX, SP02]
G4 MUY Details withheld at licensee's request by SSL.[Op: D A Lane, 1 Norman Crescent, Metheringham, Lincoln, LN4 3DN.]
GM4 MUZ H A Angus, 51 Mains Terr, Dundee, DD4 7DB [IO86ML, NO43]
G4 MVA G B Burhouse, The Cedars, Foulbridge Ln, Snainton, Scarborough, YO13 9AY [IO94QF, SE98]
G4 MVB A E Berrow, 19 Stotfold Rd, Kingsheath, Birmingham, B14 5JD [IO92BJ, SP07]
G4 MVC Details withheld at licensee's request by SSL.
G4 MVE D Casey, 18 Sandholme Dr, Ossett, WF5 8QP [IO93FQ, SE22]
G4 MVF G F M Engel, 12 Tredgold Garth, Bramhope, Leeds, LS16 9BP [IO93EV, SE24]
G4 MVH Details withheld at licensee's request by SSL.
G4 MVJ L T Hughes, 18 Whitehouse Way, Aldridge, Walsall, WS9 0BB [IO92AO, SP09]
G4 MVL S P Ward, St. Rosanna, 55 Birchley Heath Rd, Birchley Heath, Nuneaton, CV10 0QY [IO92FN, SP29]
G4 MVM C J Sturley, 17 Oxford St., Southam, Leamington Spa, CV33 0NS [IO92HG, SP46]
G4 MVN T P Rawlance, 18 Royal Sussex Cres, Eastbourne, BN20 8PB [JO00DS, TQ50]
G4 MVO J M Readings, 23 High St., Prestbury, Cheltenham, GL52 3AR [IO81XV, SO92]
G4 MVP J A Paskins, 190 Gore Rd, New Milton, BH25 5NQ [IO90DS, SZ29]
GI4 MVQ D McCluney, 49 Upper Cairncastle Rd, Larne, BT40 2EG [IO74CU, D30]
G4 MVR V Reynolds, 7 Fenton Cl, Chislehurst, BR7 6ED [JO01AK, TQ47]
G4 MVS G Mellett, 39 Ash Cl, Crawley Down, Crawley, RH10 4PG [IO91XC, TQ33]
G4 MVU Details withheld at licensee's request by SSL.
G4 MVV M D Betts, Hardings, Clayhidon, nr Cullompton, Devon, EX15 3TJ [IO80JW, ST11]
G4 MVX M J Gardiner, 206 Caulfield Rd, East Ham, London, E6 2DQ [JO01AM, TQ48]
GW4 MVY D Davies, 48 Bryn Eglur Rd, Morriston, Swansea, SA6 7PQ [IO81AP, SS69]
G4 MVZ R B Pepper, 4 Marine Ave, Skegness, PE25 3ER [JO03ED, TF56]
GI4 MWA Dr F H Ruddell, 16 Riverside Ct, Glenavy, Crumlin, BT29 4DS [IO64VO, J17]
G4 MWC J N Coupe West Manc'Ter Rd, 15 Belvedere Ave, Atherton, Manchester, M46 9LQ [IO83SM, SD60]
G4 MWD I R Shaw, 33 Park Farm Cl, Horsham, RH12 5EU [IO91UB, TQ13]
G4 MWF P R Wilkinson, 28 Ibberson Ave, Mapplewell, Barnsley, S75 6BJ [IO93FO, SE30]
G4 MWG Dr I R Grant, 41 Lynwood Dr, Collier Row, Romford, RM5 2QX [JO01BO, TQ49]
G4 MWH R Blythe, 4 Ashlea Cl, Selby, YO8 0NY [IO93LS, SE63]
G4 MWI C F Barlow, 20 Overton Way, Birkenhead, L43 2LF [IO83LJ, SJ28]
G4 MWJ R D Featherstone, Hilltop Cottage Tof, Hill Tumby, Boston, Lincs, PE22 7TB [IO93WD, TF26]
GW4 MWK Details withheld at licensee's request by SSL.[Licencee Dr H A Kennedy. Station located near Llanfairpwllgwyngyll.]
G4 MWL A Keyworth, 14 Robinson Rd, Park, Sheffield, S2 5QW [IO93GJ, SK38]
G4 MWM F Avenia, 99 Horace St., St. Helens, WA10 4NA [IO83PK, SJ59]
G4 MWN F J J Brown, 55 Baldocks Ln, Melton Mowbray, LE13 1EW [IO92NS, SK71]
G4 MWO P D Gaskell, 131 Greenfield Rd, Dentons Green, St. Helens, WA10 6SH [IO83OL, SJ59]

G4 MWP T R Underhill, 5 Lyndhurst Croft, Eastern Green, Coventry, CV5 7QE [IO92FK, SP28]
G4 MWQ C A Weir, 62 Cambridge Dr, Otley, LS21 1DD [IO93DV, SE24]
G4 MWR Details withheld at licensee's request by SSL.
G4 MWS L Parrott Macclesfield ARS, 55 Brown St., Macclesfield, SK11 6RY [IO83WG, SJ97]
G4 MWU W H Pearson, 42 Chesterfield Rd, Barlborough, Chesterfield, S43 4TT [IO93IG, SK47]
G4 MWW J G McLeod, 91 Gorselands Way, Rowner, Gosport, PO13 0DG [IO90KT, SU50]
G4 MWX L A Payne, 40 Westmorland Dr, Costhorpe, Worksop, S81 9JT [IO93KI, SK58]
G4 MX H B Davis, 11 Nuneaton Ln, Higham on The Hill, Nuneaton, CV13 6AD [IO92GN, SP39]
G4 MXA Details withheld at licensee's request by SSL.
G4 MXB C R Rogers, Firview, Treverbyn Rd, Stenalees, St Austell, PL26 8TJ [IO70OJ, SX05]
G4 MXE R Swinnerton, 8 Maple Cl, Brereton Green, Brereton, Sandbach, CW11 1SQ [IO83UE, SJ76]
G4 MXF R G Wallace, 161 Alma Ave, Hornchurch, RM12 6AT [JO01CN, TQ58]
G4 MXH A M Abercrombie, 33 Silton Gr, Hartburn, Stockton on Tees, TS18 5AT [IO94HN, NZ41]
G4 MXI K D Bruntlett, Cronk Ny Mona, King St., Yarburgh, Louth, LN11 0PN [JO03AK, TF39]
G4 MXL B J Guinnessy, 99 Newlands Rd, Tunbridge Wells, TN4 9AR [JO01DD, TQ54]
G4 MXM W Hawkridge, 7 Langdale Gdns, Leeds, LS6 3HB [IO93ET, SE23]
G4 MXO J W James, 14 Wilton Ave, Bletchley, Milton Keynes, MK3 6BN [IO91PX, SP83]
G4 MXQ B D Acres, 187 Malling Rd, Snodland, ME6 5EE [JO01FH, TQ76]
G4 MXR Details withheld at licensee's request by SSL.
G4 MXS E Carver, 29 Elm Dr, Cherry Burton, Beverley, HU17 7RJ [IO93SU, SE94]
G4 MXT D Humphreys, 3 Karen Cl, Broadlands, Bideford, EX39 4PQ [IO71VA, SS42]
G4 MXU Details withheld at licensee's request by SSL.[Op: J Juleff, 44 Old Hardenwaye, Totteridge, High Wycombe, Bucks, HP13 6TJ.]
GI4 MXW D J McKinney, 13 Lynden Gate, Ballyhannon Rd, Portadown, Craigavon, BT63 5YH [IO64TK, J05]
G4 MXX T H Fenwick, Lower Rill Farm, Chillerton, Isle of Wight, PO30 3HQ [IO90IP, SZ48]
G4 MXY B M Sowerby, 3 Goughs Ln, Wick Hill, Bracknell, RG12 2JR [IO91PK, SU87]
G4 MXZ B W G M Beeching, Homeland, Hurston Ln, Storrington, Pulborough, RH20 4HH [IO90SW, TQ01]
G4 MYA G Kearns, 34 Blackstone Ave, St. Helens, WA11 9BZ [IO83PL, SJ59]
G4 MYB C G Barham, 10 Little Brook Rd, Sale, M33 4WG [IO83TJ, SJ79]
G4 MYC Details withheld at licensee's request by SSL.
G4 MYE B F Chase, 10 Claremont Dr, Galmington, Taunton, TA1 4JF [IO81KA, ST22]
G4 MYG J Dooley, 6 The Cres, Great Holland, Frinton on Sea, CO13 0JG [JO01OT, TM21]
G4 MYH R Gartley, 4 Hurst Rd, Southam, Leamington Spa, CV33 0HY [IO92HF, SP46]
G4 MYI Details withheld at licensee's request by SSL.
GM4 MYL R D Knorr, St Marys Well, Aboyne, Aberdeenshire, AB3 5BS [IO87TH, NJ72]
G4 MYN J E Thomas, 42 Allington Dr, High Grange Est, Billingham, TS23 3UA [IO94IO, NZ42]
G4 MYQ G V Pettican, 73 Furrow Way, Maidenhead, SL6 3NY [IO91OM, SU87]
G4 MYR H D Read, 18 Mays Farm Dr, Stoney Stanton, Leicester, LE9 4HA [IO92IM, SP49]
G4 MYS A R Sillence, Nerissa House, 74 Atherley Rd, Southampton, SO15 5DS [IO90GV, SU41]
GI4 MYT W Stewart, 11 Fairway Gdns, Castlereagh, Belfast, BT5 7PS [IO74BN, J37]
G4 MYU A Summers, 6 Rothesay Rd, Brierfield, Nelson, BB9 5RS [IO83VT, SD83]
G4 MYW B Mallinson, Springburn, 23C Lyefield Rd Eas, Charlton Kings, Cheltenham Glos, GL53 8BA [IO81XV, SO92]
G4 MYY E G Ball, 6 Poltair Rd, Penryn, TR10 8PB [IO70KD, SW73]
G4 MYZ R W Roscoe, 2 Pulley Ln, Bayston Hill, Shrewsbury, SY3 0JH [IO82OQ, SJ40]
GW4 MZB R S Mills, Heather Lea, Oaklands, Leighton, Welshpool Powys, SY21 8HL [IO82KP, SJ20]
G4 MZC B J Horsman, Vale View, Green Ln, Churt, Farnham, GU10 2PA [IO91OD, SU83]
G4 MZD Details withheld at licensee's request by SSL.
G4 MZE E A Chant, 14 Brabazon Rd, Heston, Hounslow, TW5 9LS [IO91TL, TQ17]
G4 MZF D Earnshaw, 7425 Walton Ln, Annandale, Va 22003, U S A, X X
GW4 MZG I B Price, 85 Amanwy, Trallwm, Llanelli, SA14 9AG [IO71WQ, SN50]
G4 MZI G A N Dunn, 20 The Grange, Wombourne, Wolverhampton, WV5 9HX [IO82VM, SO89]
G4 MZJ P W Ansell, 46 Rochford Way, Croydon, CR0 3AD [IO91WJ, TQ36]
G4 MZK P J Ansell, 46 Rochford Way, Croydon, CR0 3AD [IO91WJ, TQ36]
G4 MZL E J Ailsby, 15 Norman Cl, King Charles Est, Bridport, DT6 4ET [IO80PR, SY49]
G4 MZM P Harper, 42 Saxville Rd, St. Pauls Cray, Orpington, BR5 3AW [JO01BJ, TQ46]
G4 MZN M P Huntsman, 36 Brackyn Rd, Cambridge, CB1 3PQ [JO02BE, TL45]
G4 MZQ R Bickley, Kingsley House, 12 Cemetery Rd, Market Drayton, TF9 3BD [IO82SV, SJ63]
G4 MZS Details withheld at licensee's request by SSL.[Op: T W Bates, 29 Martins Lane, Wallasey, Merseyside, L44 1BA.]
G4 MZT P Helm, 14 Milton Cl, Jump, Barnsley, S74 0HU [IO93GM, SE30]
G4 MZU C A Harrison, 21 Sycamore Ave, Chandlers Ford, Eastleigh, SO53 5RJ [IO91HA, SU42]
G4 MZV R A Privett, 2 Stevenson Ct, Eaton Ford, St. Neots, Huntingdon, PE19 3LF [IO92UF, TL16]
G4 MZX E C Young, 1 Ashton Rd, Roade, Northampton, NN7 2LF [IO92ND, SP75]
G4 MZY D J Plater, Obr (Masirah), Foreign & Commonwealth Office, King Charles St, London, SW1A 2AH [IO91WM, TQ37]
G4 MZZ J L Powell, 40 Kent Rd, Formby, Liverpool, L37 6BQ [IO83LN, SD30]
G4 NAA J Bain, 5 St. Andrews Rd, Marton, Marton in Cleveland, Middlesbrough, TS7 8EQ [IO94JM, NZ51]
G4 NAB Details withheld at licensee's request by SSL.
G4 NAC D J Bosworth, 144 Rothwell Rd, Kettering, NN16 8UP [IO92PJ, SP87]
G4 NAD Details withheld at licensee's request by SSL.[Station located near Matlock.]
GI4 NAE D D Jackson, 27 Manse Rd, Bangor, BT20 3DA [IO74DP, J58]
G4 NAG R M Brown, 9 Gosling Cl, Canford Heath, Poole, BH17 8QR [IO90AR, SZ09]
G4 NAJ N A J Ashdown, Cobwebs, Wilderness Ln, Hadlow Down, Uckfield, TN22 4HT [JO00CX, TQ52]
G4 NAK R P Morey, 1 Bradfield Cttgs, Queens Rd, Freshwater, PO40 9HB [IO90FQ, SZ38]
G4 NAO G Mullender, 3 Fernie Cl, Fareham, PO14 3SQ [IO90JT, SU50]
G4 NAP B Laverack, Cartref, 159 Weetshaw Ln, Cudworth, Barnsley, S72 8BL [IO93HO, SE31]
G4 NAQ C D Maby, 8 Cleeve Pl, Nailsea, Bristol, BS19 2UF [IO81PK, ST47]
G4 NAR Details withheld at licensee's request by SSL.
GI4 NAT W D Mitchell, 6 The Spires, Holywood, BT18 9DY [IO74CP, J47]
G4 NAV G D Quayle, 10 Abbotsford Gr, West Timperley, Timperley, Altrincham, WA14 5AZ [IO83TJ, SJ78]
GW4 NAW C A Simcock, Administration Off, Aztec Business Cent, The Queensway, Fforestfach Swansea, SA5 4DH [IO81AP, SS69]
G4 NAZ Details withheld at licensee's request by SSL.
G4 NBC M A Hoare, 81 Sheepfold Rd, Guildford, GU2 6TU [IO91QG, SU95]
G4 NBD Details withheld at licensee's request by SSL.
G4 NBF W T Bird, Sunny Corner, Nr.Ruan Minor, Cornwall, TR12 7LW [IN79JX, SW71]
G4 NBG Dr C J Budd, 24 Ash Rd, Horfield, Bristol, BS7 8RN [IO81QL, ST57]
G4 NBH W Cockshaw, 14 Shropshire Rd, Leicester, LE2 8HW [IO92KO, SK50]
G4 NBI L E Everton, 18 Markham Rd, Sutton Coldfield, B73 6QR [IO92BN, SP09]
G4 NBK D J Jones, 2 Walker Gr, Stapleford, Nottingham, NG9 7GY [IO92IW, SK43]
GW4 NBM M H Jones, Rhos Eithin, Brynsiencyn, Anglesey, Gwynedd, LL61 6TZ [IO73UE, SH46]
G4 NBN A Bergman, River House, Suggs Ln, Broadway, Ilminster, TA19 9RJ [IO80MW, ST31]
GI4 NBO G S Barr, 4 Sinclair Dell, Bangor, BT19 1ED [IO74DP, J48]
G4 NBR J R Hill, The Gatehouse, Gubboles Drove, Surfleet Cheal, Spalding, Lincs, PE11 4AX [IO92VU, TF22]
G4 NBS A J Collett, 10 Quince Rd, The Limes, Hardwick, Cambridge, CB3 7XJ [JO02AF, TL35]
G4 NBU N A H Barker, 6 Treble Cl, Olviers Battery, Winchester, SO22 4JN [IO91HB, SU42]
G4 NBW J L Alford, 86 Grindleford Rd, Great Barr, Birmingham, B42 2SQ [IO92BM, SP09]
G4 NBX F R Burt, 15 Albemarle Cl, Brough, HU15 1EH [IO93RR, SE92]
GW4 NBY K Barrett, 36 Priory Ave, Ewenny, Bridgend, CF31 3LR [IO81FL, SS97]
GM4 NBZ Details withheld at licensee's request by SSL.
G4 NCA P J Cook, 4 Russell Way, Higham Ferrers, Rushden, Northants, NN10 8EJ [IO92QH, SP96]
G4 NCB K Wooffindin, 16 Milford Rd, Sherburn-in-Elmet, Leeds, LS25 6AF [IO93JS, SE43]
G4 NCD K G Morey, Iona, Colwell Rd, Totland Bay, PO39 0AH [IO90FQ, SZ38]
G4 NCE Details withheld at licensee's request by SSL.
G4 NCF W C E C Prickett, 105 High St., Wootton Bassett, Swindon, SN4 7AU [IO91BN, SU08]
G4 NCI R E Smith, 33 Warwick Rd, Upton, Wirral, L49 6NF [IO83KJ, SJ28]
G4 NCJ J M Short, 21 Highmoors, Chineham, Basingstoke, RG24 8XR [IO91LG, SU65]
G4 NCK P Shapero, Flat, 3 Princess Ct, Harrogate Rd, Leeds, LS17 8BY [IO93FU, SE33]
G4 NCL Details withheld at licensee's request by SSL.
G4 NCM Details withheld at licensee's request by SSL.
G4 NCO R D Elsey, 5 Dacre Cl, Norwich, NR4 7NS [JO02PO, TG20]
G4 NCS C Angove, 61 Stewart Rd, Chelmsford, CM2 9BD [JO01FR, TL70]
G4 NCT Details withheld at licensee's request by SSL.[Op: B J Arrowsmith.]
G4 NCU M D Hewitt, The Granary, Delley, Yarnscombe, Devon, EX31 3LX [IO71XA, SS52]
G4 NCW L J Hollingworth, 111 Morwenna Park Rd, Northam, Bideford, EX39 1ET [IO71VB, SS42]
G4 NCX R Wilson, 11 Macbeth Rd, Fleetwood, FY7 7HR [IO83LW, SD34]
G4 NCY I R B Woomans, 223 Umberslade Rd, Selly Oak, Birmingham, B29 7SG [IO92AK, SP08]
G4 NCZ J Ramsay, 79 Humphrey Ln, Urmston, Manchester, M41 9PT [IO83UK, SJ79]
G4 NDC D L G Mason, The Old Post Office, 133 Bath Rd, Atworth, Melksham, SN12 8LA [IO81VJ, ST86]
G4 NDD J R Lloyd, 72 Thornyville Villas, Plymouth, PL9 7LD [IO70WI, SX55]
G4 NDE Details withheld at licensee's request by SSL.
G4 NDF E Short, Sunnyside, Knapp Rd, Thornbury, Bristol, BS12 2HF [IO81RO, ST69]
G4 NDH Details withheld at licensee's request by SSL.
G4 NDL R H Davies, Torre Hill Cttgs, Godwell Ln, Ivybridge, PL21 0LT [IO80BJ, SX65]
GM4 NDO T B Caldwell, 12 Mosshill Rd, Bellshill, ML4 1NQ [IO75XT, NS76]
G4 NDP J Burnett, 13 Stuart St., Thurnsco, Thurnscoe, Rotherham, S63 0EF [IO93IN, SE40]
G4 NDQ R F Brett, 3 St Christophers D, Addingham, Ilkley, West Yorks, LS29 0RJ [IO93BW, SE04]
G4 NDS MA R W Crewson, 46A Senhouse St., Maryport, CA15 6BL [IO84GR, NY03]
G4 NDT R G Bramley, 8 Ivy Bank Park, Bath, BA2 5NF [IO81TI, ST76]
G4 NDU Prof A N Bramley, 8 Ivy Bank Park, Bath, BA2 5NF [IO81TI, ST76]
GM4 NDV W G Rattray, 17 Brownside Rd, Cambuslang, Glasgow, G72 8NL [IO75VT, NS66]
G4 NDW Details withheld at licensee's request by SSL.
GI4 NDX Details withheld at licensee's request by SSL.
G4 NEA S C Rice, 13 Wigram Way, Stevenage, SG2 9TP [IO91VV, TL22]
G4 NEC R J Chance, 16 Clover Rd, Flitwick, Bedford, MK45 1PQ [IO92RA, TL03]
G4 NEE D M Foreman, 1 Stour Valley Cl, Upstreet, Canterbury, CT3 4DB [JO01OH, TR26]
G4 NEG W J Glew, Carinya, Beltoft Rd, Beltoft, Doncaster S York, DN9 1MB
G4 NEH J E G Hankin, The Croft, Portinscale, Keswick, Cumbria, CA12 5TX [IO84KO, NY22]
GW4 NEI K B Hodge, Bryn Hyfryd, 16 Mold Rd, Mynydd Isa, Mold, CH7 6TD [IO83KD, SJ26]
G4 NEJ K Jackson, 20 Southern Haye, Hartley Wintney, Hook, RG27 8TZ [IO91NH, SU75]
G4 NEL D J Bird Nel Rynt L/B Wt, 154 Cherrydown Ave, Chingford, London, E4 8DZ [IO91XO, TQ39]
G4 NEM K Vardy, 5 Rawcliffe Ave, Clifton, York, YO3 6QD [IO93KX, SE55]
G4 NEO C L Digby, 7 Dagnall Rd, Olney, MK46 5BJ [IO92PD, SP85]
G4 NEQ R Welsh, Holme View, Farleton, Lancaster, LA2 9LF [IO84QC, SD56]
G4 NER P E C Lidbetter, 1 Moor Ln, Westfield, Hastings, TN35 4QU [JO00GV, TQ81]
G4 NES D A Wright, 38 Manor Cl, Harpole, Northampton, NN7 4BY [IO92MF, SP66]
GM4 NEV Details withheld at licensee's request by SSL.
G4 NEW F Briggs, 24 Shield Ave, Worsbrough, Barnsley, S70 5BH [IO93GM, SE30]
G4 NEX A J H Hales, 55 Bakers Ln, Epping, CM16 5DQ [JO01BQ, TL40]
G4 NEY J W Jarvis, 116 Balland Field, Willingham By Stow, Willingham, Cambridge, CB4 5JU [JO02AH, TL47]
GI4 NEZ P J Harbinson, 98 Drumgullion Ave, Armagh Rd, Newry, BT35 6PF [IO64TE, J02]
G4 NFA F E Austin, 45 Southdown Cres, Cheadle Hulme, Cheadle, SK8 6EQ [IO83VI, SJ88]
G4 NFB M Bracey, 18 Mallard Gdns, Hedge End, Southampton, Hants, SO30 2XJ [IO90IW, SU41]
GM4 NFC Details withheld at licensee's request by SSL.
G4 NFD P A Broadhurst, 2 The Old Granary, High St., Ashwell, Baldock, SG7 5NQ [IO92WA, TL23]
G4 NFE J Edwards, 5 Windmill Rise, York, YO2 4TU [IO93KW, SE55]
GW4 NFF S A Frisby, 12 Chapel St., y Drenewydd, Powys, SY16 2BP [IO82IM, SO19]
GI4 NFH R Jennings, 117 Belsize Rd, Lisburn, BT27 4BS [IO64XM, J26]
GM4 NFI D J C Leckie, 6 Galloway Pl, Fort William, PH33 6UH [IO76KT, NN07]
GW4 NFJ P Mainwaring, 9 Blackthorn Pl, Sketty, Swansea, SA2 9JW [IO81AO, SS69]
G4 NFL C J G Peel, The Ferns, Parkwood Dr, Baldwins Gate, Newcastle, ST5 5EU [IO82UX, SJ74]
G4 NFO D Porter, 10 Blakes Cres, Highbridge, TA9 3LE [IO81MF, ST34]
G4 NFP M Saunby, Teachmore Farm, Inwardleigh, Okehampton, EX20 3AJ [IO70XS, SX59]
G4 NFR R Tyson, 49 Strathaird Ave, Walney, Barrow in Furness, LA14 3DE [IO84JC, SD16]
G4 NFT G C McAvoy, 5 Lytchett Way, Upton, Poole, BH16 5LS [IO80XR, SY99]
G4 NFU Details withheld at licensee's request by SSL.
GI4 NFW J A Hegarty, The Nook, 1 Cookstown Rd, Moneymore, Co Derry, N Ireland, BT45 7QF [IO64PQ, H88]
G4 NFX K Atkinson, 23 Devonshire Rd, Scunthorpe, DN17 1ER [IO93QN, SE80]
G4 NFY A Clarke, 51 Wedgewood Dr, Wisbech, PE13 2DD [JO02CP, TF40]
G4 NFZ J Dowell, 16 Sawyers Rd, Tolleshunt Major, Maldon, CM9 8NE [JO01IS, TL81]
G4 NGB W F Gliddon, 83 Fern Way, Ilfracombe, EX34 8JS [IO71WE, SS54]
G4 NGE Details withheld at licensee's request by SSL.[Op: G R Holland, The Haven, Highlands Road, Fareham, Hants, PO16 7XJ.]
G4 NGF M J Chapman, Millway, Dunton Ln, Ashby Parva, Lutterworth, LE17 5HX [IO92JL, SP58]
G4 NGH Details withheld at licensee's request by SSL.
GM4 NGJ C W Bridges, Highfield, Ballinluig, Perthshire, PH9 0LG [IO86DP, NN95]
G4 NGL J D Gass, 212 The Rowans, Milton, Cambridge, CB4 6ZL [JO02BF, TL46]
GI4 NGP D J Paul, 4 Draperstown Rd, Tobermore, Magherafelt, BT45 5QG [IO64PT, H89]
G4 NGR C L Webber, Applegarth, Brixham Rd, Paignton, TQ4 7BD [IO80FJ, SX85]
G4 NGS Dr G O Towler, 77 Worrin Rd, Shenfield, Brentwood, CM15 8JL [JO01DO, TQ69]
G4 NGT N G Townley, 22 Knowles Ave, Crowthorne, RG45 6DU [IO91OI, SU86]
GW4 NGU D R Thomas, Haul y Bryn, Wick Rd, St. Brides Major, Bridgend, CF32 0SF [IO81EK, SS87]
G4 NGW R S Perkins, 38 Grange Park Dr, Leigh on Sea, SS9 3JZ [JO01HN, TQ88]
GM4 NGY H S Munro, Beechwood, Sangomore, Durness, Lairg, Sutherland, IV27 4PZ [IO78PN, NC46]
G4 NHA P W Hastilow, Harrowsley Grdn Cot, 138 Smallfield Rd, Horley, Surrey, RH6 9LS [IO91WE, TQ34]
GW4 NHB P J Boyce, 30 Windway Rd, Victoria Park, Cardiff, CF5 1AF [IO81JL, ST17]
G4 NHC F Armstrong, 52 Leicester Way, Jarrow, NE32 4XL [IO94GW, NZ36]
G4 NHD M B Brightman, 34 Norris Cl, Chiseldon, Swindon, SN4 0LR [IO91DM, SU17]
G4 NHE C Evans, 4 Fryers Copse, Wimborne, BH21 2HR [IO90AT, SU00]
G4 NHF A S Jones, 51 Wiclif Way, Stockingford, Nuneaton, CV10 8NH [IO92FM, SP39]
GW4 NHH Dr M F Buck, 23 Velindre Rd, Whitchurch, Cardiff, CF4 7JE [IO81JM, ST18]
GM4 NHI J R Cramond, Robson's Croft, Brae of Letter, Dunecht, Skene, AB3 7EQ [IO87TH, NJ72]
G4 NHJ Details withheld at licensee's request by SSL.[Op: G R Dodd. Station located near Hertford.]
G4 NHL T C Dixon, 30 Green Ln, Stamford, PE9 1HF [IO92SP, TF00]
G4 NHN C N Rowsell, 46 Missenden Acres, Hedge End, Southampton, SO30 2RE [IO90IW, SU41]
G4 NHO J M Pattemore, 2 Edes Cttgs, Ottways Ln, Ashtead, KT21 2PG [IO91UH, TQ15]
G4 NHP S E G Porter, 12 Dawes Cl, Greenhithe, Kent, DA9 9RA [JO01DK, TQ57]
G4 NHQ A P McMackin, 8 Stella Hall Dr, Blaydon on Tyne, NE21 4LB [IO94DX, NZ16]
G4 NHR I J Turner, 74 Diban Ave, Elm Park, Hornchurch, RM12 4YF [JO01CN, TQ58]
G4 NHT C F Beesley Moorlands&Dis A, 15 Byron Cl, Cheadle, Stoke on Trent, ST10 1XB [IO82XX, SJ94]
G4 NHU J D Crowe, Arundel House, 11 High St., Arundel, BN18 9AD [IO90RU, TQ00]
G4 NHW D Collins, 22 Stalyhill Dr, Stalybridge, SK15 2TR [IO83XL, SJ99]
GM4 NHX G G Brooks, The Old Post Office, Scotscalder, Halkirk, Caithness, KW12 6XJ [IO88FL, ND05]
G4 NHY C W Brooke, 43 Williams Ave, Wyke Regis, Weymouth, DT4 9BW [IO80SO, SY67]
G4 NHZ K A Burdon, 245 Derby Rd, Chaddesden, Derby, DE21 6SY [IO92GW, SK33]
G4 NIA J W Houghton, 38 Bedford Pl, Bridport, Dorset, DT6 3LZ [IO80OR, SY49]
G4 NIC Details withheld at licensee's request by SSL.[Station located near Bridgwater.]
G4 NID G Bromley, O/B/O Ndbars, c/o Plumtree House, Walk Cl, Draycott,Derbyshire, DE7 3PN [IO92IX, SK44]
G4 NIE S A Looney, 18 Cherrington Rd, Nantwich, CW5 7AW [IO83RB, SJ65]
G4 NIF D G Lee, 22 Woodland Rise, Parkend, Lydney, GL15 4JX [IO81RS, SO60]
G4 NII A Wadsworth, 9 Linley Rd, Cheadle Hulme, Cheadle, SK8 7HP [IO83VI, SJ88]
G4 NIJ K Sheldon, Whitehaven, May Tree Rd, Lower Moor, Pershore, WR10 2NY [IO82XC, SO94]
G4 NIK G C Rich, 55 Wyndham Cres, Woodley, Reading, RG5 3AY [IO91NL, SU77]
G4 NIL R Henshall, St. Anns, Staplehay, Trull, Taunton, TA3 7HB [IO80KX, ST22]
G4 NIO W Towers-Perkins, Church Rd, Cookham Dean, Cookham, Maidenhead, Berks, SL6 9PJ [IO91PN, SU88]
G4 NIP D R Lewis, 76 Reading Rd, Finchampstead, Wokingham, RG40 4RA [IO91NI, SU76]
G4 NIQ R Warriner, 36 Eskdaleside, Sleights, Whitby, YO22 5EP [IO94QK, NZ80]
G4 NIR D Williams, 9 Poplar Ave, Lydgate, Oldham, OL4 4JX [IO83XM, SD90]
G4 NIS Details withheld at licensee's request by SSL.
G4 NIV N Rowcroft, 8 Chatsworth Rd, Hazel Gr, Stockport, SK7 6BH [IO83WI, SJ98]
G4 NIW Details withheld at licensee's request by SSL.
G4 NIX G F Cooke, 7 Lime Gr, Royston, SG8 7DJ [IO92XB, TL34]
G4 NIY S J Cooke, 28 Weston Ave, Royston, SG8 5DR [IO92XB, TL34]
G4 NIZ R H King, 20 Woodside East, Thurlby, Bourne, PE10 0HT [IO92TR, TF11]
G4 NJA R Hewson, 6 Talisman Dr, Bottesford, Scunthorpe, DN16 3SW [IO93QN, SE80]
G4 NJB K R Hepke, 386 Kingston Rd, Willerby, Hull, HU10 6NG [IO93SS, TA02]
G4 NJC Capt P F Henny, 97 Heathermount Dr, Edgcumbe Park, Crowthorne, RG45 6HJ [IO91OI, SU86]
GM4 NJD J W Rhynas, 18 Riccarton, Mains Rd, Currie, Mid Lothian, EH14 5NG [IO85IV, NT16]
G4 NJE Details withheld at licensee's request by SSL.
G4 NJH Details withheld at licensee's request by SSL.
G4 NJI A Corker, 59 Foljambe Rd, Eastwood, Rotherham, S65 2UA [IO93IK, SK49]
G4 NJJ P W Cousins, 28 Church Rd, Clenchwarton, Kings Lynn, PE34 4EA [JO02ES, TF52]
G4 NJK R L C Elliott, 42 Exeter Rd, Okehampton, Devon, EX20 1NH [IO80AR, SX59]
G4 NJL D E Gerrard, 11 Burnside, Hadfield, Hyde, SK14 8DX [IO93AK, SK09]
G4 NJN A K Bowman, 18 Essex Ave, Isleworth, TW7 6LF [IO91TL, TQ17]
GI4 NJQ P Igo, 84 Glebetown Dr, Downpatrick, BT30 6PZ [IO74DH, J44]
G4 NJU D W McQue, 6 Laburnam Gr, Bletchley, Milton Keynes, MK2 2JW [IO91PX, SP83][Op: Dave McQue. Tel: (01908) 378277.]
G4 NJW G R R Lowes, Trenute, Huttoft Rd, Sutton on Sea, Mablethorpe, LN12 2QY [JO03DH, TF58]
GI4 NKB F T Hunter, 20 Wandsworth Gdns, Belfast, BT4 3NL [IO74BO, J37]
G4 NKC M W Jones, Racecourse Farm, Church Ln, Tasley, Bridgnorth, WV16 4NW [IO82SM, SO79]
GI4 NKD D F Campbell, 109 Drumgor Park, Drumgor, Craigavon, BT65 4AH [IO64TK, J05]
GI4 NKE A R Walker, Rook Lodge, 42 Orpins Mill Rd, Ballyclare, Co Antrim, BT39 0SX [IO64XS, J29]
GM4 NKF N G Vogan, 27 South Main St., Wigtown, Newton Stewart, Wigtownshire, DG8 9HG [IO74SU, NX45]
G4 NKI J P Shaw, Stookes, Greenways, Walton, Chesterfield, S40 3HF [IO93GF, SK36]
GI4 NKJ C H Quigley, 21 Innisgarry Park, Antrim, BT41 4LA [IO64VR, J18]
GI4 NKK K Planck, 50 Well Rd, Ballywalter, Newtownards, BT22 2PU [IO74GN, J66]
G4 NKM Details withheld at licensee's request by SSL.
G4 NKP D J Mellin, 24 Fore St., Topsham, Exeter, EX3 0HB [IO80GQ, SX98]
GI4 NKQ E J McGookin, 30 Bryansburn Rd, Bangor, BT20 3SB [IO74DP, J48]
GW4 NKR G J Lewis, The Old, Post Office Cottage, Llyswen, Brecon Powys, LD3 0UR [IO82IA, SO13]
G4 NKT D J Bailey, 18 Heath Rise, Warmley, Bristol, BS15 5DD [IO81SK, ST67]
G4 NKU D C L Dunn, 37 Ridgemead, Calne, SN11 9EW [IO81XK, ST97]

G4 NKV Details withheld at licensee's request by SSL.
G4 NKW G Hilton, 8 Sandwich Cl, St. Ives, Huntingdon, PE17 6DQ [IO92XI, TL37]
G4 NKX P C Digby, 8 Littlepark Ave, Bedhampton, Havant, PO9 3QY [IO90LU, SU60]
GI4 NKZ T Fusco, 85 Downshire Rd, Holywood, BT18 9LY [IO74CP, J47]
G4 NLB B M Horne, 12 Merefield Gdns, Tadworth, KT20 5JY [IO91VH, TQ25]
G4 NLC P Hedison, 85 Moorhouse Ln, Whiston, Rotherham, S60 4NH [IO93IJ, SK49]
GW4 NLD P A Frost, 11 Trehearn Dr, Rhyl, LL18 3RR [IO83GH, SJ08]
GW4 NLE J Hawkins, 25 Ilfracombe Cres, Llanrumney, Cardiff, CF3 9TB [IO81KM, ST28]
G4 NLF A K Haward, 56 Orchard Piece, Blackmore, Ingatestone, CM4 0RZ [JO01DQ, TL60]
G4 NLG F Askew, 11 Skirlaw Cl, Howden, Goole, DN14 7BH [IO93NR, SE72]
G4 NLH D R Haydon, 50 Ward Cl, Stratton, Bude, EX23 9BB [IO70RT, SS20]
G4 NLI R K Scott, 56 The Toose, Yeovil, BA21 3SN [IO80QW, ST51]
GM4 NLJ J G Martin, Whitehillfoot, Kelso, Roxburghsire, TD5 8LB [IO85SN, NT72]
G4 NLK R A C Southall, 7 The Willows, Brereton, Rugeley, WS15 1EP [IO92AR, SK01]
G4 NLL D Bell, 5 Byron Ct, Dalton on Tees, Darlington, DL2 2PX [IO94FL, NZ20]
G4 NLM S G Blackaller, The Tower, Devon Rd, Salcombe, TQ8 8HQ [IO80CF, SX73]
G4 NLN E Alexander, 14 Pitcher Ln, Leek, ST13 5DB [IO83XC, SJ95]
G4 NLO K A Butcher, 33 Kempe Rd, Enfield, EN1 4QT [IO91XQ, TQ39]
G4 NLP D G Chapman, 7 Reynard St., Spilsby, PE23 5JB [JO03BE, TF36]
GI4 NLQ P H Burns, 41 Lambeg Rd, Lambeg, Lisburn, BT27 4QA [IO64XM, J26]
G4 NLR Details withheld at licensee's request by SSL.
G4 NLS M J Harriman 91st Leics Scts, 70 Station Rd, Thurnby, Leicester, LE7 9PU [IO92LP, SK60]
G4 NLU R Williams, 50 Hemerdon Heights, Plympton, Plymouth, PL7 2EY [IO70XJ, SX55]
G4 NLV K G Taylor, 32 Stanford Gdns, Aveley, South Ockendon, RM15 4BU [JO01DL, TQ58]
G4 NLW J W Wilding, 8 Millbrook Way, Brierley Hill, DY5 3YY [IO82WL, SO98]
G4 NLX Details withheld at licensee's request by SSL.
G4 NMA A Townsend, 12 Fieldfare, Stevenage, SG2 9NJ [IO91WV, TL22]
GI4 NMB P E Sullivan, 5 Beechgrove Dr, Aughagallon, Craigavon, BT67 0BH [IO64UM, J16]
G4 NMC D R Willis, 41 Chadbrook Crest, Richmond Hill Rd, Edgbaston, Birmingham, B15 3RL [IO92AL, SP08]
G4 NMD Revd G J Smith, 41 Furners Mead, Henfield, BN5 9JA [IO90UW, TQ21]
G4 NME R Smith, 47 Cinder Rd, Dudley, DY3 2RH [IO82WM, SO99]
G4 NMF W Squire, 230 The Avenue, Seaham, SR7 8BG [IO94HT, NZ44]
G4 NMG B Raybould, Newlands, Bromyard Rd, Tenbury Wells, WR15 8BY [IO82QH, SO56]
G4 NMI D A Lowe, 11 Plantagenet Ct, Nottingham, NG3 1HJ [IO92KX, SK54]
G4 NMJ D P Miller, 31 Hillborough Rd, Luton, LU1 5EY [IO91SU, TL02]
G4 NMK K Petre, 24 Harrogate Terr, Murton, Seaham, SR7 9PQ [IO94HT, NZ34]
GW4 NML B A Saunders, Cartref, 53 Clevedon Rd, Newport, NP9 8NA [IO81MO, ST38]
G4 NMP B Dudhill, 12 Eilam Rd, Kimberworth Park, Rotherham, S61 3PQ [IO93HK, SK49]
GW4 NMQ D K Hutchinson, Plas Dyfi, Pennal, Machynlleth, Powys, SY20 9LB [IO82AN, SN69]
G4 NMR S C Harvey, 2 Draycote Cl, Worcester, WR5 3SY [IO82VE, SO85]
G4 NMS S W Burgess, 55 Channels Farm Rd, Swaythling, Southampton, SO16 2PF [IO90HW, SU41]
G4 NMU R A Jones, 86 Liverpool Rd, Birkdale, Southport, PR8 4DF [IO83LP, SD31]
G4 NMV J H Baxter, 16 Avon Cl, Weston Super Mare, BS23 4QS [IO81MH, ST35]
G4 NMX Details withheld at licensee's request by SSL.
G4 NMY P Bennett, 45 Ravenbank Rd, Luton, LU2 8EJ [IO91TV, TL12]
GI4 NMZ J C McDermott, 29 Lawson Park, Ballymagorry, Strabane, BT82 0AZ [IO64GU, C30]
G4 NNA Details withheld at licensee's request by SSL.[Op: C J Seymour.]
G4 NNB G C Mackie, 8 The Avenue, Biggleswade, SG18 0PS [IO92UC, TL14]
GM4 NNC C D S Rodgers, 5 Elder Ave, Lincluden, Dumfries, DG2 0NL [IO85EC, NX97]
GM4 NNH J S F Barber, 156 Jamphlars, Cardenden, By Lochgelly, Fife, KY5 0ND [IO86ID, NT29]
G4 NNI E J Bailey, 213 Ashby Rd, Hinckley, LE10 1SJ [IO92HN, SP49]
G4 NNJ A G Forster, School House, Naas Ln, Lydney, GL15 5AT [IO81RR, SO60]
GM4 NNK D R J Harkness, 13 Marcus Dr, Kinellar, Aberdeen, AB2 0XT [IO87TH, NJ72]
GW4 NNL M Jones, 72 Princes Dr, Colwyn Bay, LL29 8PW [IO83DH, SH87]
GI4 NNM J J McGillian, 48 Millfield, Ballymena, BT43 6PB [IO64UV, D10]
G4 NNN C G W Winterflood, 12 Bourne Rd, Colchester, CO2 7LQ [JO01KV, TM02]
G4 NNO T Hadley, 34 Rydal Bank, Bebington, Wirral, L63 7LL [IO83LI, SJ38]
G4 NNQ S F Farley, 79 Falkland Rd, Hornsey, London, N8 0NS [IO91WO, TQ38]
G4 NNS B R Coleman, Woodlands, Redenham, Andover, Hants, SP11 9AN [IO91FF, SU24]
G4 NNT D H Canty, 46 Hawthorn Ave, Gainsborough, DN21 1HA [IO93OJ, SK89]
G4 NNW Details withheld at licensee's request by SSL.
G4 NNX D A P Ward, 18 Henders, Stony Stratford, Milton Keynes, MK11 1RB [IO92NB, SP74]
G4 NNY D A Ransford, 52 Loughborough Rd, Bunny, Nottingham, NG11 6QD [IO92KU, SK53]
G4 NNZ G Martorano, 81 Sapcote Dr, Melton Mowbray, LE13 1HG [IO92NS, SK71]
G4 NOA B F Nickells, Linden House, Ruckamore Rd, Chelston, Torquay, TQ2 6HF [IO80FL, SX96]
G4 NOB R D Burbeck, 20 St. Johns Rd, Smalley, Ilkeston, DE7 6EG [IO92HX, SK44]
G4 NOC N R H Black, 19 Pinfold Way, Weaverham, Northwich, CW8 3NL [IO83RG, SJ67]
G4 NOD A B Nickells, Linden House, Ruckamore Rd, Chelston, Torquay, TQ2 6HF [IO80FL, SX96]
G4 NOE M R C Hollinghurst, 63 Gloucester Rd, Cheltenham, GL51 8NE [IO81WV, SO92]
G4 NOK D Lilley Nth Wakefield Rd, 65 Peel St., Horbury, Wakefield, WF4 5AN [IO93FP, SE21]
G4 NON D A Watson, 1 Bracken Rd, Dringhouses, York, YO2 2JT [IO93KW, SE54]
G4 NOO M E Oliver, 159 Clockhouse Ln, Romford, RM5 2TJ [JO01CO, TQ59]
G4 NOP R Simpson, 55 Upleatham, Saltburn, Cleveland, TS12 1LR [IO94MN, NZ62]
G4 NOQ C Guy, 3 Broadley Lathe, Moor End Rd, Pellon, Halifax, HX2 0RP [IO93BR, SE02]
G4 NOR C H D Heaps, 12 Oak Tree Cl, Eakring Rd, Mansfield, NG18 3EN [IO93KD, SK56]
GW4 NOS R Hopkins, 8 Shady Rd, Gelli, Pentre, CF41 7UG [IO81GP, SS99]
G4 NOT C R Sturgeon, Windyridge, Linkside West, Hindhead, GU26 6PA [IO91OC, SU83]
G4 NOU R A Wigmore, Little Landguard, Whitecross Ln, Shanklin, PO37 7EJ [IO90JP, SZ58]
G4 NOV Details withheld at licensee's request by SSL.
G4 NOW Details withheld at licensee's request by SSL.
G4 NOX K A Campbell, 4 Orchard Cl, Rowlands Gill, NE39 1EQ [IO94CW, NZ15]
G4 NOY J Mills, 103 Irby Rd, Wirral, L61 6UZ [IO83KI, SJ28]
G4 NPA A M Abbot, 26 Grange Hill Rd, Birmingham, B38 8RG [IO92AJ, SP07]
G4 NPB S D Abbot, 26 Grange Hill Rd, Birmingham, B38 8RG [IO92AJ, SP07]
GW4 NPC R M Bayney, 10 Cae Ffynnon, Church Village, Pontypridd, CF38 1UB [IO81IN, ST08]
G4 NPD G Chamberlain, 13 Mayford Cl, Beckenham, BR3 4XS [IO91XJ, TQ36]
G4 NPE E Dench, 30 Gravel Walk, Emberton, Olney, MK46 5JA [IO92PD, SP85]
G4 NPF Details withheld at licensee's request by SSL.
G4 NPG P J Duffy, 15 The Glade, Sheldon, Birmingham, B26 3PW [IO92CK, SP18]
G4 NPH J P Arnold, 2 Duck Ln, Haddenham, Ely, CB6 3UE [JO02BI, TL47]
G4 NPI C E Darby, 7 Andrew Gr, Dukinfield, SK16 4AY [IO83XL, SJ99]
G4 NPJ Details withheld at licensee's request by SSL.
G4 NPM B A Hancock, 33 Beauxfield, Whitfield, Dover, CT16 3JW [JO01PD, TR34]
G4 NPN D L Westwood, 1 Elmfield Dr, Hartlebury, Worcs, DY11 7LA [IO82UI, SO87]
G4 NPQ G Pollitt, 18 The Mount, Selby, YO8 9BH [IO93LS, SE63]
G4 NPR C B Nicholls, 6 Welholme Ave, Grimsby, DN32 0HP [IO93WN, TA20]
G4 NPS J D Woolliss, Wharfdale, 245 Scartho Rd, Grimsby, DN33 2EA [IO93WM, TA20]
G4 NPT R A Williamson, 1 Pygall Ave, Gotham, Nottingham, NG11 0JW [IO92JU, SK53]
G4 NPU P M Williams, 122 Longhurst Ln, Mellor, Stockport, SK6 5PG [IO83XJ, SJ98]
G4 NPW J J Thomas, 18 Moss Ln, Churchtown, Southport, PR9 7QR [IO83MP, SD31]
G4 NPX Details withheld at licensee's request by SSL.
G4 NPY C K Read, 72 Springhill Rd, Chasetown, Walsall, West Midlands, WS7 8UJ [IO92AQ, SK00]
G4 NQ L A Hunt, 27 Harvey Brown House, Restawyle Ave, Hayling Island, PO11 0PQ [IO90MT, SU70]
G4 NQC J E Neal, 93 Hazelbank Rd, Catford, London, SE6 1LS [JO01AK, TQ37]
G4 NQE R J Pirie, 42 Bron yr Eglwys, Mynydd Isa, Mold, CH7 6YQ [IO83KE, SJ26]
G4 NQF Details withheld at licensee's request by SSL.
G4 NQI Capt R S Atterbury, 14 Holway Rd, Taunton, TA1 2EY [IO81KA, ST22]
GW4 NQJ C D Brown, Kingsdown Cottage, Fron, Montgomery, Powys, SY15 6SB [IO82JN, SO19]
G4 NQK S R Brough, 9 Valentine Ct, Crownhill, Milton Keynes, MK8 0HA [IO92OA, SP83]
G4 NQL M Churms, 123 Casablanca, Urb Buenavista, La Marina, Alicante, Spain 03194
G4 NQM M B Drohan, 23 Lindholme Dr, Rossington, Doncaster, DN11 0UP [IO93LL, SK69]
G4 NQN T D Green, Flat, 4 Bitham Hall, Avon Dassett, Leamington Spa, CV33 0AH [IO92HD, SP45]
G4 NQO Details withheld at licensee's request by SSL.
G4 NQQ N P Hemmings, 3 Church Cl, Shapwick, Bridgwater, TA7 9LS [IO81OD, ST43]
G4 NQR I C Rule, 166 Maney Hill Rd, Sutton Coldfield, B72 1JW [IO92CN, SP19]
G4 NQS R F Young, 79 Cradge Bank, Spalding, PE11 3AF [IO92WS, TF22]
GM4 NQT G G Miller, Sunnydale, Allanton, Berwickshire, TD11 3LA [IO85VS, NT85]
G4 NQW P S Perrins, 9 Merrick Cl, Halesowen, B63 1JY [IO82XK, SO98]
G4 NQX Details withheld at licensee's request by SSL.
G4 NQY J R Wilford, 6 Park Rd, Narborough, Leicester, LE9 5GQ [IO92JN, SP59]
G4 NQZ R Riley, 103 St. Nicolas Park Dr, Nuneaton, CV11 6DZ [IO92GM, SP39]
G4 NRA J F Tisdale, 12 Digby Rd, Kingswinford, DY6 7RP [IO82WM, SO88]
GI4 NRB W J Watson, 6 Cromwells Highway, Lisburn, BT27 5DH [IO64XM, J26]
G4 NRC G A Griffiths Radio Amateurs Emergency Netwo, 11 The Grove, Asfordby, Melton Mowbray, LE14 3UF [IO92MS, SK71]
G4 NRD A C Lindsay, 21 Willow Rd, Four Pools, Evesham, WR11 6YW [IO92AB, SP04]
G4 NRE Details withheld at licensee's request by SSL.
G4 NRF H A R O Westwood, 23 Went Gr, Featherstone, Pontefract, WF7 5NJ [IO93HQ, SE41]
G4 NRG R E Greengrass, 19 Wecover Ave, Connaught Chase, Attleborough, NR17 1QU [JO02MM, TM09]
G4 NRH D R Whitehead, 50 Southey Ln, Kingskerswell, Newton Abbot, TQ12 5JG [IO80FL, SX86]
G4 NRL Details withheld at licensee's request by SSL.
G4 NRM J K Neary, 1 Ashling Ct, Tyldesley, Manchester, M29 8QS [IO83SM, SD70]
G4 NRO J N Coupe, 15 Belvedere Ave, Atherton, Manchester, M46 9LQ [IO83SM, SD60]
G4 NRP J P Fish, Bramley Cottage, Mill Ln, Hatton, Warwick, CV35 7HN [IO92EH, SP26]
GI4 NRQ J Hamill Coleraine & Dis, 67 Windsor Ave, Coleraine, BT52 2DR [IO65QD, C83]
G4 NRR N G A Rollason, 76 Solihull Rd, Shirley, Solihull, B90 3HL [IO92CJ, SP17]
G4 NRT D N Bondy, 2 Woodside, Cheshunt, Waltham Cross, EN7 5DE [IO91XQ, TL30]
G4 NRU G P Bowser, Bowser & Co, 15 North Brink, Wisbech, PE13 1JR [JO02BP, TF40]
G4 NRV A J Diplock, 24 Billings Hillsha, Hartley, Dartford, Kent, DA3 8EU [JO01DJ, TQ66]
G4 NRW D Reeve, Trincoe, Mill Rd, Wells-Next-Sea, Norfolk, NR23 1BZ [JO02KW, TF94]
G4 NRX S Mantell, 24 Bourne Ave, Fazeley, Tamworth, B78 3TB [IO92DO, SK10]
G4 NRY I Mantell, 24 Bourne Ave, Fazeley, Tamworth, B78 3TB [IO92DO, SK10]
G4 NRZ K W Moody, 5 Moore Rd, Mapperley, Nottingham, NG3 6EF [IO92KX, SK54]
G4 NS J Hudson, 22 Essex Gdns, Marsden, South Shields, NE34 7JQ [IO94HX, NZ36]
G4 NSA D Morgan, 12 Rosalind Ave, Bebington, Wirral, L63 5JR [IO83LI, SJ38]
G4 NSC W Weatherspoon, 12 Greenacres Cl, Crawcrook, Ryton, NE40 4TD [IO94CX, NZ16]
G4 NSD I E M Mitchell, Greenway Cottage, Greenway, Tatsfield, Westerham, TN16 2BT [JO01AH, TQ45]
G4 NSE J H Rank, 18 Coldyhill Ln, Newby, Scarborough, YO12 6SF [IO94SH, TA08]
G4 NSG Details withheld at licensee's request by SSL.[Station located 5km north-west of the city of Birmingham. Op: Stuart.]
G4 NSH S Robinson, Upp Hambleton Farm, Wainstalls, Halifax, HX2 7TX [IO93AS, SE02]
G4 NSI Details withheld at licensee's request by SSL.
GM4 NSL G S Greenless, St Brydes Cottage, Lochwinnoch, Renfrewshire, PA12 4HN [IO75RT, NS36]
G4 NSM A Bruce, 77 High St., Epworth, Doncaster, DN9 1JS [IO93OM, SE70]
G4 NSN M R Craft, 21 Lindsey Rd, Denham, Uxbridge, UB9 5BW [IO91SN, TQ08]
G4 NSO S Auckland, 77 George St., Low Valley, Wombwell, Barnsley, S73 8AQ [IO93HM, SE40]
G4 NSP K S Harris, 1 Whitfield Cl, Wilford, Nottingham, NG11 7AU [IO92KW, SK53]
GI4 NSS L M H Robinson, 24 Strangford Rd, Lisburn, BT27 4BL [IO64XM, J26]
G4 NST S A Thorpe, 11 Gr Rd, Hethersett, Norwich, NR9 3JP [JO02OO, TG10]
G4 NSU J S Pearce, 71 Horseshoe Rd, Pangbourne, Reading, RG8 7JL [IO91LL, SU67]
GI4 NSV A A M Donnelly, Tirgarve, Allistragh, Co Armagh, N Ireland, BT61 8EZ [IO64PJ, H85]
G4 NSW F A Lacey, 52 Church St., Long Bennington, Newark, NG23 5ES [IO92OX, SK84]
G4 NSY Details withheld at licensee's request by SSL.[Also GX4NSY. Op: NSY Amateur Radio Society, Room 99, New Scotland Yard, Broadway, London, SW1H 0BG. Locator: IO91WL. WAB: TQ27, City of Westminster. QSL via bureau or direct: SAE/IRC required for direct reply.]
GM4 NSZ M F Stanway, 30 Durham Ave, Edinburgh, EH15 1PA [IO85KW, NT27]
G4 NTA P Allan, 2 Park View, Queensbury, Bradford, BD13 1PL [IO93BS, SE03]
G4 NTB G A Hodge, Flat 4, 43 Ave Gdns, Acton, London, W3 8HB [IO91UM, TQ17]
G4 NTC D H Henderson, 4 Vincent St., Bolton, BL1 4SA [IO83SN, SD70]
G4 NTE J Hodgetts, 42 Tyninghame Ave, Wolverhampton, WV6 9PW [IO82WO, SJ80]
GI4 NTF Details withheld at licensee's request by SSL.
G4 NTG J R Williamson, 196 Birchfield Rd, Headless Cross, Redditch, B97 4NA [IO92AH, SP06]
G4 NTJ A Rick, 9 Sheldon Cl, Loughborough, LE11 5EZ [IO92JS, SK52]
GM4 NTL J M J McDermott, Maple Lodge, Strome Ferry, Ross-shire, IV53 8UP [IO77FH, NG83]
G4 NTM Details withheld at licensee's request by SSL.
G4 NTN Details withheld at licensee's request by SSL.
GI4 NTO R McCaughey, 31 Greenburn Way, Lambeg, Lisburn, BT27 4LU [IO64XM, J26]
G4 NTP P J James, 100 Mount Pleasant, Kingswinford, DY6 9SH [IO82VL, SO88]
G4 NTQ M A D Bowden, Ramelton, Letterkenny, Donegal, Ireland
GD4 NTR G Kelly, Santos, 5 Tynwald Cl, Peel, IM5 1JJ
G4 NTS J Ashton, Higher Rydon Ball, Denbury Rd, Ogwell, Newton Abbot, Devon, TQ12 6BZ [IO80EM, SX86]
G4 NTT P J Dunthorne, 277 Westleigh Ln, Leigh, WN7 5PW [IO83RM, SD60]
G4 NTV A R Wilkes, 34 Tideswell Rd, Great Barr, Birmingham, B42 2DT [IO92BM, SP09]
G4 NTW W Douglas, 10 Parkshiel, South Shields, NE34 8BU [IO94HX, NZ36]
GM4 NTX K M Elliott, Northfield Cottage, Northfield By Denny, Stirlingshire, FK6 6RB [IO86AB, NS88]
G4 NTY J R Higson, 24 St. Marys Rd, Worsley, Manchester, M28 3RF [IO83TM, SD70]
G4 NUB M Tyler, 27 Shakespeare Dr, Dinnington, Sheffield, S31 7RP [IO93GJ, SK38]
G4 NUF Details withheld at licensee's request by SSL.
G4 NUG R A Needs, 13 Greenway, Great Horwood, Milton Keynes, MK17 0QR [IO91NX, SP73]
G4 NUH M J Shorey, Rose Cottage, Upper Spring Ln, Kenilworth, CV8 2JR [IO92FI, SP27]
G4 NUJ J O Kellaway, Flat, 25 Elizabeth House, Wilton Orchard, Taunton, TA1 3SA [IO81KA, ST22]
G4 NUK J A Brown, 33 Balmoral Dr, Leicester, LE3 3AD [IO92JO, SK50]
G4 NUL Details withheld at licensee's request by SSL.
G4 NUM E S Wain, 35 Kings Mount, Leeds, LS17 5NS [IO93FU, SE33]
GM4 NUN G W A Mackenzie, 7 Bonaly Wester, Colinton, Edinburgh, EH13 0RQ [IO85IV, NT26]
G4 NUO A N Mackenzie, Ivy House, 145 High St., Marske By The Sea, Redcar, Cleveland, TS11 6JX [IO94LO, NZ62]
G4 NUS A J Layland, 16 Park Rd, Quarry Bank, Brierley Hill, DY5 2DA [IO82WL, SO98]
GM4 NUU H P Park, Carn Dearg, Upper Steelend, Dunfermline, KY12 9LP [IO86FC, NT09]
G4 NUV R W Richmond, 7 Bishopdale Dr, Watnall, Nottingham, NG16 1LE [IO93JA, SK54]
G4 NUX C J Smith, Grassways The Dr, Ifold, Loxwood, Billingshurst W Sus, RH14 0TE [IO91RB, TQ03]
G4 NUY E Wharton, Vandling, Well, Bedale, North Yorks, DL8 2QF [IO94EF, SE28]
G4 NUZ A G Davies, 22 Meadow Rd, Cornmeadow, Worcester, WR3 7PP [IO82VF, SO85]
G4 NVA J R Dyke, 2 Brooklands Dr, Goostrey, Crewe, CW4 8JB [IO83UF, SJ77]
G4 NVB Details withheld at licensee's request by SSL.
G4 NVC I M C Hollingsbee, 89 Swift Rd, Abbeydale Est, Abbeydale, Gloucester, GL4 4XJ [IO81VU, SO81]
G4 NVD E G Knight, 6 Hamilton Cl, Grimsby, DN34 5QW [IO93WN, TA20]
G4 NVF Details withheld at licensee's request by SSL.
G4 NVH G P Boull, 80 Ascot Rd, Baswich, Stafford, ST17 0AQ [IO82WT, SJ92]
GM4 NVI D G Chapman, 9 Baillieswells Terr, Bieldside, Aberdeen, AB1 9AR [IO87TH, NJ72]
G4 NVJ L C Chapman, Millway, Dunton Ln, Ashby Parva, Lutterworth, LE17 5HX [IO92JL, SP58]
G4 NVK C J E Coomber, Old Paygate Cottage, Lower Stoneham, Uckfield Rd, Lewes, BN8 5RL [JO00AV, TQ41]
G4 NVM J E Duddridge, 19 Ridgeway, Hurst Green, Etchingham, TN19 7PJ [JO01FA, TQ72]
G4 NVN G W Harper, Pathways, 41 Somerset Cl, Congleton, CW12 1SE [IO83VE, SJ86]
GW4 NVO Details withheld at licensee's request by SSL.
G4 NVP K B Kett, 24 Deancourt Dr, New Duston, Northampton, NN5 6PY [IO92MG, SP76]
G4 NVQ D W Shirley, 93 Alfred Rd, Hastings, TN35 5HZ [JO00HU, TQ81]
G4 NVR Details withheld at licensee's request by SSL.
G4 NVS P G Nowland, 101 Surrey Rd, Huntingdon, PE18 7JU [IO92VI, TL27]
G4 NVT M J Musgrave, 49 Vowler Rd, Langdon Hills, Basildon, SS16 6AQ [JO01FN, TQ68]
G4 NVU G S Fox, 3 Chestnut Gr, Eynesbury, St. Neots, Huntingdon, PE19 2DW [IO92UF, TL15]
G4 NVV R S English, 124 Hillside Rd, Redcliffe Bay, Portishead, Bristol, BS20 8LG [IO81OL, ST47]
G4 NVW C M Cole, 16 York Ride, Weedon, Northampton, NN7 4PF [IO92LF, SP65]
G4 NVX C R Ellis, 26 Cheltenham Ave, Bobblestock, Hereford, HR4 9TQ [IO82PB, SO44]
G4 NVY M R Juffs, 3 Green Acre, Brockworth, Gloucester, GL3 4NG [IO81WU, SO81]
G4 NVZ Details withheld at licensee's request by SSL.
G4 NWA D M Adams, Rri 23522 Kennedy Rd, Sutton West, Ontario, Canada, X X
G4 NWC Details withheld at licensee's request by SSL.
G4 NWF C D Gledhill, 15 Warrels Gr, Bramley, Leeds, LS13 3NN [IO93ET, SE23]
G4 NWG R L Jones, Flat One Manor Lodge, 22 Old Town Ln, Formby, Liverpool, L37 3HP [IO83LN, SD20]
G4 NWH Revd R P Butcher, 4 Field St. Ave, Kettering, NN16 8EP [IO92PJ, SP87]
G4 NWI Dr B J Harrison, Bay House, 27 Thatcher Ave, Torquay, TQ1 2PD [IO80GL, SX96]
G4 NWJ J C Chick, Moonrakers, Allington, Salisbury, SP4 0BX [IO91DD, SU23]
GM4 NWK T S Hill, 3 Swift Cres, Knightswood Gate, Glasgow, G13 4QN [IO75TV, NS56]
G4 NWM M S Talbott, 44 Tamworth Rd, Amington, Tamworth, B77 3BT [IO92DP, SK20]
G4 NWN W J Talbott, 44 Tamworth Rd, Amington, Tamworth, B77 3BT [IO92DP, SK20]
G4 NWO K C Chaplin, 1 Beechwood Cres, Amington, Tamworth, B77 3JH [IO92EP, SK20]
G4 NWP Details withheld at licensee's request by SSL.
G4 NWR H N Woolrych Nth Wilts Rayne, 20 Meadow Dr, Devizes, SN10 3BJ [IO91AI, SU06]
G4 NWS A J C Wheeler, 1 Plain Gate, Rothley, Leicester, LE7 7SQ [IO92KR, SK51]
G4 NWT N W Try, The Limes, 38 Harefield Rd, Uxbridge, UB8 1PH [IO91SN, TQ08]
G4 NWU F T Morgan, 64 Northville Rd, Bristol, BS7 0RG [IO81RM, ST67]
G4 NWY C Venables, 115 Howden Cl, Bessacarr, Doncaster, DN4 7JN [IO93KM, SE60]
G4 NXA D A Kirk, Glebelands, Vicarage Ln, Priors Marston, Rugby, CV23 8RT [IO92IF, SP45]
G4 NXB J L Evans, 11 Columbia Cl, Rosemary Hill, Selston, Nottingham, NG16 6GP [IO93IB, SK45]
GW4 NXD R S Johns, 12 Woodfield Rd, New Inn, Pontypool, NP4 0PT [IO81LQ, ST39]
G4 NXE A D Tosler, 22 Homestead Rd, Thorngumbald, Hull, HU12 9PT [IO93WR, TA22]
G4 NXG A J Birch, 17 The Stakes, Castlemeadow, Moreton, Wirral, L46 3SW [IO83KJ, SJ29]
G4 NXH Details withheld at licensee's request by SSL.[Op: A W Elkington.]
G4 NXI R J Light, 81 Mendip Rd, Portishead, Bristol, BS20 8AF [IO81OL, ST47]
GI4 NXJ M E McFall, 60 Richmond Rd, Glengormley, BT36 8LD [IO74AQ, J38]
G4 NXM K J Reed, Nicker Hill, Stanton on The Wold, Keyworth, Nottingham, Notts, NG12 5EA [IO92LU, SK63]
G4 NXN T K Parker, 13 Stubby Ln, Draycott in The Clay, Ashbourne, DE6 5HA [IO92CU, SK12]

G4 NXO A J Collett Sheppey Western, Cont Gp. 10 Quince, The Limes, Hardwick, Cambridge, CB3 7XJ [JO02AF, TL35]
G4 NXP D Taylor, View Farm, Robin Hill, Powick, Worcs, WR7 4QA [IO82WE, SO95]
G4 NXQ C T W Millner, 4 Monkreed Villas, Longfield Rd, Longfield, DA3 7AR [JO01EJ, TQ66]
G4 NXR P L Rose, 5 Malmesbury Rd, St. Leonards, Ringwood, BH24 2QL [IO90BT, SU10]
G4 NXS M Davis, 446 Upper Wortley Rd, Scholes, Rotherham, S61 2SS [IO93HK, SK39]
GM4 NXT W M Davidson, 7 South St., Aberchirder, Huntly, AB54 7XR [IO87QN, NJ65]
G4 NXU B R Price, 66 Ashchurch Dr, Wollaton, Nottingham, NG8 2RA [IO92JW, SK53]
G4 NXV D M Gadsden, 37 Cambridge St., Wymington, Rushden, NN10 9LG [IO92QG, SP96]
G4 NXW K B Chesters, 69 Greenlands Ave, Ramsey, IM8 2PQ
G4 NXY D Cope, 58 Green Ln, Halton, Leeds, LS15 7EW [IO93GT, SE33]
G4 NYA R C Hyams, Holly Bank, 38 Parsonage Rd, Withington, Manchester, M20 4PT [IO83VK, SJ89]
G4 NYB R J Knighton, Merry Ways, The Green, Weston on Trent, Derby, DE72 2BJ [IO92HU, SK42]
G4 NYC Details withheld at licensee's request by SSL.
G4 NYD I R Watling, 5 Claylands Ct, Bishops Waltham, Southampton, SO32 1JS [IO90JW, SU51]
G4 NYE R Hartnell, 3 Town Mills, Wiveliscombe, Taunton, Somerset, TA4 2LY [IO81IA, ST02]
G4 NYG D F Routledge, 52 Elmbridge Dr, Shirley, Solihull, B90 4YP [IO92CJ, SP17]
G4 NYH N Westwood, 71 Finstall Rd, Finstall, Bromsgrove, B60 3DF [IO82XH, SO97]
G4 NYI G V Toase, 4 Gill Croft, Thirsk Rd, Easingwold, York, YO6 3HH [IO94JC, SE56]
G4 NYJ C Webb, 65 Littlebeck Dr, Darlington, DL1 2TU [IO94FM, NZ31]
G4 NYK Dr R J W Williams, 67 Sea Mills Ln, Stoke Bishop, Bristol, BS9 1DR [IO81QL, ST57]
G4 NYL S J Brown, 115 Southwood Dr, Baxenden, Accrington, BB5 2TU [IO83TR, SD72]
G4 NYM N D Porter, 12 Ct Meadow, Stone, Berkeley, GL13 9LR [IO81SP, ST69]
G4 NYN Details withheld at licensee's request by SSL.
G4 NYS Details withheld at licensee's request by SSL.
GU4 NYT N K Le Page, Les Martins, St. Martin, Guernsey, Channel Islands, GY4 6QJ
G4 NYU S O'Donnell, 2 Rose Ct, Oakerside Park, Peterlee, SR8 1BZ [IO94HR, NZ43]
G4 NYV M G Katzmann, 501 Defense Highway, Annapolis, Maryland 21401-6923, U.S.A
G4 NYW R Schoales, The White Hart, Towngate, Bradwell, Derbys, S30 2JX [IO93GJ, SK38]
G4 NYY P A Ernster, 36 Forest End, Fleet, GU13 9XE [IO91NG, SU85]
G4 NYZ J Battle-Welch, 325 Bromsgrove Rd, Webheath, Redditch, B97 4NH [IO92AH, SP06]
GW4 NZ S Roberts, 70 Cimla Rd, Neath, SA11 3TR [IO81CP, SS79]
G4 NZB P J Neville, 66 Oak Lodge Ave, Chigwell, IG7 5HZ [JO01BO, TQ49]
G4 NZC B R D Manchett, 21 Heathfield Rd, Chandlers Ford, Eastleigh, SO53 5RP [IO91HA, SU42]
G4 NZD J F Henderson, Dauntless, Ferry Rd, Canvey Island, Essex, SS8 0QT [JO01GM, TQ78]
G4 NZE E B Third, 14 Harrington Dr, Bedford, MK41 8DB [IO92SD, TL05]
G4 NZF A D Peterkin, Ashfield, Wendron, Helston, Cornwall, TR13 0JR [IO70JD, SW63]
G4 NZG S J Parsons, End Cottage, Church Ln, Covington, Huntingdon, Cambs, PE18 0RT [IO92SH, TL07]
G4 NZK B V Laniosh, 47 Barley Mow Ln, Catshill, Bromsgrove, B61 0LU [IO82XI, SO97]
G4 NZN K Lowe, Springwood Hall, Priesthorpe Rd, Farsley, Pudsey, LS28 5RE [IO93DT, SE23]
G4 NZO G A Frederick, 130 Maureen St, Winnipeg, Manitoba, Canada Rk3 1M2, ZZ1 3MA
G4 NZQ P Brooks, 7 Lindford Dr, Church Ln, Eaton, Norwich, NR4 6LT [JO02PO, TG20]
G4 NZU R G Wilson, 9 Greythorn Dr, West Bridgford, Nottingham, NG2 7GG [IO92KV, SK53]
G4 NZV Details withheld at licensee's request by SSL.
G4 NZX D G Cooper, 110 Brecks Ln, Kirk Sandall, Doncaster, DN3 1PZ [IO93LN, SE60]
G4 NZY D E G Coles, 91 Gr Ln, Harborne, Birmingham, B17 0QT [IO92AK, SP08]
G4 NZZ B A Coulson, 19 St. Lukes Cl, Kettering, NN15 5HD [IO92PJ, SP87]
G4 OAA R Frisby, 2 Westfield Rd, Hoddesdon, EN11 8QX [IO91XS, TL30]
G4 OAB M Clutton, 8 Ash Gr, Runcorn, WA7 5LR [IO83PH, SJ58]
G4 OAC R Hayter, 22 Shawclough Way, Rochdale, OL12 7HF [IO83VP, SD81]
G4 OAD P J Hale, 86 Bruce Gr, Shotgate, Wickford, SS11 8QJ [JO01GO, TQ79]
G4 OAE D J Crisp, 2 Flaxman Cl, Earley, Reading, RG6 5TH [IO91MK, SU77]
G4 OAG A R Dymott, Old Cottage, 31 Newnham Rd, Ryde, PO33 3TE [IO90JR, SZ59]
G4 OAI H A Richter, 84 Roehampton Dr, Wigston, LE18 1HU [IO92KO, SK50]
G4 OAJ S M Richards, 272 Forest Rd, Loughborough, LE11 3HX [IO92JS, SK51]
G4 OAK S D Richards, Little Piece, Stocks Mead, Washington, Pulborough, RH20 4AU [IO90TV, TQ11]
G4 OAN L J Wild, 18 Ormonde Rd, Hythe, CT21 6DN [JO01NB, TR13]
G4 OAR N B McLaren, 596 Woodchurch Rd, Oxton, Birkenhead, L43 0TT [IO83LI, SJ28]
G4 OAS G F Liddle, 17 Sun Hill, Royston, SG8 9AU [IO92XB, TL34]
G4 OAT D Payne, 4 Woodland Croft, Horsforth, Leeds, LS18 5NE [IO93EU, SE23]
G4 OAU G Austin, 38 Willow Cres, Hatfield Peverel, Chelmsford, CM3 2LJ [JO01HS, TL71]
G4 OAV S A Ames, 21 Common Ln, Harpenden, AL5 5BT [IO91UT, TL11]
G4 OAW M Adderley, 43 The Heys, Coppull, Chorley, PR7 4NX [IO83QP, SD51]
G4 OAX W M Joiner, 10 Richmond Dr, Jaywick, Clacton on Sea, CO15 2PH [JO01NS, TM11]
G4 OAY R Jellett, 6 Tamar Ave, Durrington, Worthing, BN13 3JY [IO90SU, TQ10]
G4 OAZ J F Burford, 26 Shrubbery Rd, Bromsgrove, B61 7BH [IO82XH, SO97]
G4 OBA R C Jarvis, Rose Cottage, Spring Gr Farm, Wribbenhall, Bewdley, Worcs, DY12 1LF [IO82UI, SO87]
G4 OBB Dr D J Kaylor, 49 Headley Way, Headington, Oxford, OX3 0LS [IO91JS, SP50]
G4 OBC M A Taylor, 5 Hawford Ave, Comberton Est, Kidderminster, DY10 3BH [IO82VJ, SO87]
GM4 OBD G P Sangster, 36 St. Marys Dr, Ellon, AB41 9LW [IO87XI, NJ93]
G4 OBE R F Snary, 12 Borden Ave, Enfield, EN1 2BZ [IO91XP, TQ39]
G4 OBF E Street, 36 Base Green Rd, Sheffield, S12 3FH [IO93HI, SK38]
GM4 OBG J Love, 6 Boydfield Ave, Prestwick, KA9 2JJ [IO75QM, NS32]
GM4 OBJ T Murphy, 14 Glencairn Ave, Wishaw, ML2 7RG [IO85AS, NS75]
G4 OBK P J Catterall, 54 Westlands, Pickering, YO18 7HJ [IO94OG, SE88]
G4 OBL Details withheld at licensee's request by SSL.
G4 OBN S A J Harding, Herriard, Basingstoke, Hants, RG25 2PN [IO91LE, SU64]
G4 OBR J P Consitt, 5 Logan Cl, Bransholme, Hull, HU7 4PG [IO93US, TA13]
G4 OBS J Waters, 3 Elm Cl, Pitton, Salisbury, SP5 1EU [IO91DB, SU23]
G4 OBT N P Day, 11 Arthur Rd, Horsham, RH13 5BG [IO91UA, TQ22]
G4 OBV E J Day, 42 Grosvenor Cl, Thorley Park, Bishops Stortford, CM23 4JP [JO01BU, TL41]
G4 OBW R H Flynn, 78 Langstone Cl, Torquay, TQ1 3TY [IO80FL, SX96]
G4 OBX R T Dobson, 40 Dipton Gdns, Essen Way, Tunstall, Sunderland, SR3 1AN [IO94HV, NZ35]
G4 OC C E W Copping, Block 9 Flat 54, St Thomas's Houses, The Alms Houses, Wrotham Rd Gravesen, DA11 7LA [JO01EK, TQ67]
GM4 OCA P A Windsor, Greenleys Croft, Gamrie, Banff, AB4 3JD [IO87TH, NJ72]
G4 OCG Details withheld at licensee's request by SSL.
G4 OCH K Dickens, 149 Quince, Amington, Tamworth, B77 4ET [IO92EP, SK20]
G4 OCJ Dr E B Mullock, 19 Chester St., Cirencester, GL7 1HF [IO91AR, SP00]
GI4 OCK J S Mackay, 12 Lynne Rd, Bangor, BT19 1NT [IO74DP, J48]
GI4 OCL R T McCurry, 82 Cumberland Rd, Dundonald, Belfast, BT16 0BB [IO74CO, J47]
GW4 OCN E W Gratton, 83 Open Hearth Cl, Griffithstown, Pontypool, NP4 5LU [IO81LQ, ST29]
G4 OCO Details withheld at licensee's request by SSL.
G4 OCP Details withheld at licensee's request by SSL.
G4 OCQ I Blackman, 69 Thorntons Cl, Pelton, Chester-le-Street, DH2 1QH [IO94EU, NZ25]
G4 OCS I P Bexon, 21 Featherstone Cl, Off Arnold Ln, Gedling, Nottingham, NG4 4JA [IO92KX, SK64]
G4 OCU D E Gipp, 9 Yew Tree Rd, Elkesley, Retford, DN22 8AY [IO93MG, SK67]
GI4 OCV J Goodall, 34 Glenariff Cres, Ballymena, BT43 6EU [IO64UU, D10]
G4 OCX L J Jewell, 42 Common Rd, Stotfold, Hitchin, SG5 4DB [IO92VA, TL23]
G4 OCZ C D Richardson, 149 Old Fort Rd, Shoreham By Sea, BN43 5HL [IO90UT, TQ20]
G4 ODA B K Tatnall, 73 Acacia Ave, Spalding, PE11 2LW [IO92WT, TF22]
G4 ODC G D Agness, 71 Links Ave, Mellesdon, Norwich, NR6 5PG [JO02PP, TG21]
G4 ODD M J Mathers, Rose Cottage, Kirton Rd, Egmanton, Newark, NG22 0HF [IO93NF, SK76]
G4 ODE Dr N G Dovaston, 53 Elmway, Cop, Chester-le-Street, DH2 2LX [IO94EU, NZ25]
G4 ODF D W Faulkner, Amber Croft, Dale Cl, Langwith, Mansfield, NG20 9EB [IO93JF, SK57]
G4 ODG V Cawthron, 8 Clay Hill Rd, Sleaford, NG34 7TF [IO92SX, TF04]
G4 ODH L N Fennelow, 39 Clarence Rd, Wisbech, PE13 2ED [JO02CQ, TF41]
G4 ODI A Dyer, Springfield, Aylesbury Rd, Askett, Princes Risborough, Bucks, HP27 9LY [IO91OR, SP80]
G4 ODK Details withheld at licensee's request by SSL.[Located in Cambridge postcode area.]
G4 ODM C Mott-Gotobed, Cherry Trees, 17 Reading Rd, Chineham, Basingstoke, RG24 8LN [IO91LG, SU65]
GW4 ODN A S Whitticombe, 160 Haven Dr, Hakin, Milford Haven, SA73 3HN [IO71LR, SM80]
G4 ODO D Tomlinson, 88 Linby Rd, Hucknall, Nottingham, NG15 7TW [IO93JB, SK54]
G4 ODQ Details withheld at licensee's request by SSL.
G4 ODS W L Allan, 50 Byland Rd, Harrogate, HG1 4ET [IO94FA, SE35]
G4 ODV J B Coyne, Sunnydene, Trevarth Rd, Carharrack Redruth, Cornwall, TR16 5SE [IO70JF, SW74]
GM4 ODW D R Maclean, Gramaiche, Donavourd Rd, Donavourd, Pitlochry, PH16 5JS [IO86DQ, NN95]
GJ4 ODX S C Langlois, L'Amarrage, Route Orange, St Brelade, Jersey, JE3 8GP
G4 OEC E McPheat, Old School, Dyche, Holford, nr Bridgwater, Somerset, TA5 1SF [IO81JD, ST14]
G4 OED G C Perry, 12 Boydell Cl, Shaw, Swindon, SN5 9QT [IO91CN, SU18]
G4 OEE R Bailey, 125A College St., Long Eaton, Nottingham, NG10 4GE [IO92IV, SK43]
G4 OEF D L Bayliss, 38 Yarborough Cres, Lincoln, LN1 3LU [IO93RF, SK97]
G4 OEH N P Rumble, 89 Cavalier Rd, Old Basing, Basingstoke, RG24 7ER [IO91LG, SU65]
GW4 OEJ A England, 9 Priory Rd, Milford Haven, SA73 2DS [IO71LR, SM90]
G4 OEK R Jones, Tarn Hows, 17 Plumpton Park Rd, Bessacarr, Doncaster, DN4 6SQ [IO93LM, SE60]
GD4 OEL Details withheld at licensee's request by SSL.
G4 OEM G A Hooker, 42A Nether Hall Rd, Doncaster, DN1 2PZ [IO93KM, SE50]
G4 OEP Dr A J Smith, 15 Dyrham Cl, Henleaze, Bristol, BS9 4TF [IO81QL, ST57]
G4 OEQ C W Thomas, 69 Quakers Rd, Downend, Bristol, BS16 6JG [IO81RL, ST67]
G4 OER D C Warner, 15 Douglas Cres, Houghton Regis, Dunstable, LU5 5AS [IO91RV, TL02]
GW4 OES A R Pickard, 89 Ael y Bryn, Llanedeyrn, Cardiff, CF3 7LL [IO81KM, ST17]
G4 OET M C Fretwell, 468 Nottingham Rd, Derby, DE21 6PE [IO92GW, SK33]
G4 OEU Q G Campbell, Wardens Flat, Ethel Williams Hall, Eastfield Rd, Benton, Newcastle on Tyne, NE12 9TY [IO95FA, NZ26]
GM4 OEW Details withheld at licensee's request by SSL.
G4 OEX G M Jones, 3 Kings Mews, Bedford St., Stockton Heath, Warrington, WA4 6GY [IO83QI, SJ68]
G4 OEY Dr T R Sanderson, Backershagenlaan 32, 2243 Ad Wassenaar, Netherlands 2241
GM4 OEZ W R Taylor, 2 Jubilee Terr, Findochty, Buckie, AB56 4QA [IO87NQ, NJ46]
G4 OFA B Millican, 24 Wellington, Terr, Bramley, Leeds, LS13 2LH [IO93ET, SE23]
GM4 OFC J A Snelgrove, 1 Lothian Pl, Fort William, PH33 6UA [IO76KT, NN07]
G4 OFD J R Wood, 4 Almond Cres, Mastin Moor, Chesterfield, S43 3AX [IO93IG, SK47]
G4 OFE J G Shea, 6 Parsonage Dr, Brierfield, Nelson, BB9 5DX [IO83VT, SD83]
G4 OFF M A Purves, Larchwood, Tower House Ln, Wraxall, Bristol, BS19 1JU [IO81PK, ST47]
G4 OFK N Harrison, 2 Watlington Rd, South Benfleet, Benfleet, SS7 5DR [JO01GN, TQ78]
G4 OFL K R Hawkins, 37 Darley Ave, Toton, Beeston, Nottingham, NG9 6JP [IO92IV, SK43]
G4 OFN P J Edmonds, 17 Rockmead Ave, Great Barr, Birmingham, B44 9DR [IO92BN, SP09]
G4 OFO N W E Baynes, 62 Cromford Way, New Malden, KT3 3BA [IO91UJ, TQ26]
G4 OFP J S Knowles, 26 Harrier Park, East Hunsbury, Northampton, NN4 0QG [IO92NF, SP75]
GW4 OFQ R C Hunter, 11 Pantllyn Terr, Llandybie, Ammanford, SA18 3JT [IO71XT, SN61]
G4 OFR Details withheld at licensee's request by SSL.
G4 OFU R Burdess, 23 Hemmings Cl, Sidcup, DA14 4JR [JO01BK, TQ47]
G4 OFY Details withheld at licensee's request by SSL.
GM4 OFZ R Cook, 49 Carron Ave, Bellfield, Kilmarnock, KA1 3NF [IO75SO, NS43]
G4 OG D C Gordon, 5 Seaview Terr, Wellington Pl, Sandgate, Folkestone, Kent, CT20 3DL [JO01NB, TR13]
G4 OGB L W Elliott, Elm Lodge, Brethergate, Westwoodside, Doncaster, DN9 2AD [IO93NL, SK79]
GW4 OGC A G Jones, 2 Isallt Park, Trearddur Bay, Holyhead, LL65 2US [IO73QG, SH27]
G4 OGD Details withheld at licensee's request by SSL.
G4 OGF Details withheld at licensee's request by SSL.
G4 OGG J West, 6 Tithe Rd, Kempston, Wootton, Bedford, MK43 9BE [IO92RC, TL04]
G4 OGH D P White, 19 Tilesford Cl, Shirley, Solihull, B90 4YF [IO92CJ, SP17]
GM4 OGI N D Shaxted, Viewbank Cottage, Shieldhill Rd, Redding Muirhead, Falkirk Stirlingsh, FK2 0DU [IO85DX, NS97]
G4 OGJ R C G Saxelby, 10 Shelley Ave, Balby, Doncaster, DN4 8LB [IO93KM, SE50]
G4 OGK G Tibbetts, The Brewers Droop, 44 Wolverhampton St., Willenhall, WV13 2PS [IO82XO, SO99]
GM4 OGM S C Mather, 41 Roosevelt Rd, Kirknewton, EH27 8AD [IO85GV, NT16]
GW4 OGO S J Williams, 33 Badminton Rd, Newport, NP9 7NH [IO81MO, ST38]
G4 OGP Details withheld at licensee's request by SSL.
GI4 OGQ W Kernohan, 3 Camphill Park, Ballymena, BT42 2DQ [IO64UU, D10]
G4 OGR R Blaikie, 16 Mapleton Gr, Birmingham, B28 9RG [IO92CK, SP18]
GI4 OGS Details withheld at licensee's request by SSL.
G4 OGT Details withheld at licensee's request by SSL.
G4 OGW D J Thomas, Handleycross Cottage, Harewood End, Hereford, HR2 8JT [IO81PW, SO52]
G4 OGY J A Naylor, 96 Broadclyst Gdns, Thorpe Bay, Southend on Sea, SS1 3QY [JO01JN, TQ98]
G4 OGZ M J Walker, 52 Derwent Rd, Harpenden, AL5 3NX [IO91TT, TL11]
G4 OHA L J Lux, Hyde Brae, Hyde Hill, Chalford, Stroud, GL6 8NY [IO81WR, SO80]
G4 OHB P N A Taylor, 11 Romsley Hill Grange, Farley Ln, Romsley, Halesowen, B62 0LN [IO82XJ, SO97]
G4 OHC R Poore, 30 Melbourne Ave, Goring By Sea, Worthing, BN12 4RT [IO90ST, TQ10]
GM4 OHE Details withheld at licensee's request by SSL.
GI4 OHI G A Irvine, 22 Carnreagh Bend, Newtownabbey, BT37 9EQ [IO74AP, J38]
G4 OHJ J Porter, 77 Westholme Rd, Bidford on Avon, Alcester, B50 4AN [IO92BD, SP05]
G4 OHK Details withheld at licensee's request by SSL.
G4 OHM Details withheld at licensee's request by SSL.[Correspondence to D Eccles, G4LAG, obo The South Birmingham RAYNET Group, c/o Hampstead House, Fairfax Road, West Heath, Birmingham, B31 3QY.]
G4 OHN Details withheld at licensee's request by SSL.
G4 OHP L Sharps, 37 Redgate, Ormskirk, L39 3NN [IO83NN, SD40]
G4 OHQ J L Reade, 7 Wilmar Cl, Hayes, UB4 8ET [IO91SM, TQ08]
G4 OHR T E Proctor, 27 Brookthorpe, Stansmawe Est, Yate, Bristol, BS17 4HX [IO81SM, ST78]
G4 OHT T A T Mitchell, 434 Buckfield Rd, Leominster, HR6 8SD [IO82OF, SO45]
G4 OHV C J Addison-Lees, 18 Langley Ave, Somercotes, Alfreton, DE55 4LT [IO93HB, SK45]
GI4 OHW N W Bell, Rocklyn, 16 Dromore Rd, Fintona, Omagh, BT78 1QZ [IO64IO, H47]
G4 OHX Details withheld at licensee's request by SSL.
GM4 OHY R J Cameron, 88 Little Vennel, Cromarty, IV11 8XF [IO77XQ, NH76]
G4 OIA S Hartgroves, Glen Grant, Week St Mary, Holsworthy, Devon, EX22 6LD [IO70SR, SX29]
G4 OIB C H Haydock, 54 The Green, Cheadle, Stoke on Trent, ST10 1PH [IO82XX, SJ94]
G4 OID J G Storry, 99 Swineshead Rd, Wyberton Fen, Boston, PE21 7JG [IO92XX, TF24]
G4 OIE R Neale, Field House, Recreation Rd, New Houghton, Mansfield, NG19 8TL [IO93IE, SK46]
G4 OIF P W Northover, 94 Risley Ln, Breaston, Derby, DE72 3AU [IO92IV, SK43]
G4 OIG G W Peck, Meridian House, 45 Bentley Cl, Rectory Farm, Northampton, NN3 5JS [IO92OG, SP86]
G4 OIH D S Want, 47 Percival Rd, Rugby, CV22 5JU [IO92JI, SP57]
G4 OII M J Morley, Padagi, Town Rd, Tetney, Grimsby, DN36 5JE [IO93XL, TA30]
GM4 OIJ B Robertson, Silverwells, Knockbain, Munlochy, Ross Shire, IV8 8PG [IO77UN, NH65]
G4 OIK J E Price, 4 Housman Walk, Kidderminster, DY10 3XL [IO82VJ, SO87]
G4 OIL J E Price, 26 Hales Park, Bewdley, DY12 2HT [IO82UI, SO77]
G4 OIM P S Marchant, 29 Hilldrop Rd, Bromley, BR1 4DB [JO01AK, TQ47]
G4 OIN A G Reeley, Gibraltar House, 289 Bristol Rd, Quedgeley, Gloucester, GL2 4QP [IO81UT, SO81]
G4 OIP Details withheld at licensee's request by SSL.
G4 OIQ P R Storey, 5 Bennett Hill Cl, Wooton Bassett, Wootton Bassett, Swindon, SN4 8LR [IO91BN, SU08]
G4 OIR N A Robinson, Commcen, Rsgs,B Sqn, Shape, Bfpo 26
G4 OIS G F Reid, 65 Rowelfield, Luton, LU2 9HL [IO91TV, TL12]
G4 OIT J Gardiner, 18 Howard Rd, Horsham, RH13 6AB [IO91UB, TQ13]
G4 OIV D Mavin, 52 Bywell Rd, Ashington, NE63 0LE [IO95FE, NZ28]
G4 OIW Dr C S F Pine, Chestnut House, Stokesley, Middlesbrough, Cleveland, TS9 5AE [IO94JL, NZ50]
G4 OIX Details withheld at licensee's request by SSL.
G4 OIY Details withheld at licensee's request by SSL.
G4 OIZ Details withheld at licensee's request by SSL.
G4 OJB W C Tatum, 18 Elmbank Ave, Engelfield Green, Egham, TW20 0TJ [IO91RK, SU97]
G4 OJC Details withheld at licensee's request by SSL.
G4 OJF G A Ball, 37 Greenwood, Tweedmouth, Berwick upon Tweed, TD15 2EB [IO85XS, NT95]
G4 OJG J V Glass, 70 Canterbury Rd, Lydden, Dover, CT15 7ES [JO01OD, TR24]
G4 OJH A W Giles, 209 New Bristol Rd, Weston Super Mare, BS22 0BJ [IO81NI, ST36]
G4 OJI D A A Schofield, 22 Paddock Garden, Whitchurch, Bristol, BS14 0TG [IO81RJ, ST56]
G4 OJJ V A Holyoake, 281 Causeway, Green Rd, Warley, West Midlands, B68 8LT [IO82XL, SO98]
G4 OJK H D Baxendale, 60 Stuart Rd, Waterloo, Liverpool, L22 4QT [IO83LL, SJ39]
G4 OJM M Edwards, 26 Carlton Cl, Forest Town, Mansfield, NG19 0LE [IO93KD, SK56]
G4 OJN A K Semark, 22 Morris Ave, Billericay, CM11 2LB [JO01FO, TQ69]
G4 OJO H M Mullenger, 31 Eggars Field, Bentley, Farnham, GU10 5LD [IO91NE, SU74]
G4 OJP R Prosser, 45 Harvey Rd, Hampton Dene, Hereford, HR1 1XB [IO82PB, SO53]
G4 OJR Details withheld at licensee's request by SSL.
G4 OJS J B Rowlands, 70 Braces Ln, Marlbrook, Bromsgrove, B60 1DY [IO82XI, SO97]
GW4 OJT Details withheld at licensee's request by SSL.[Op: C A Long. Station located near Pontypool, Gwent.]
G4 OJV A S Picton, 5 Tuttles Ln East, Wymondham, NR18 0EN [JO02NN, TG10]
G4 OJW Details withheld at licensee's request by SSL.
GW4 OJX A J Williams, 57 Park St., Pembroke Dock, SA72 6BL [IO71MQ, SM90]
G4 OJY A C Wright, 14 Thorne Gr, Rothwell, Leeds, LS26 0HP [IO93GS, SE32]
G4 OKA N A Crook, 39 Mason St., Reading, RG1 7PD [IO91MK, SU77]
G4 OKB Details withheld at licensee's request by SSL.
G4 OKC A R Gardner, 19 Lower Rea Rd, Brixham, TQ5 9UD [IO80FJ, SX95]
G4 OKD A M Forryan, 21 Blakesley Rd, Wigston, LE18 3WD [IO92KO, SP69]
G4 OKE M A Gould, 10 Canterbury Cl, Pelsall, Walsall, WS3 4PB [IO92AP, SK00]
GW4 OKF P W Granby, 104 Priory Rd, Milford Haven, Pembrokeshire, SA73 2ED [IO71LR, SM90]
GM4 OKG M F Dawson, Spindrift, 22 Woodlands Ave, Kirkcudbright, DG6 4BP [IO74XT, NX65]
G4 OKH M R Fisher, 16 Chestnut Cl, Watlington, Kings Lynn, PE33 0HX [JO02EQ, TF61]
GW4 OKI M Driscoll, 74 Ventnor Rd, Cwmbran, NP44 3JY [IO81LP, ST29]
G4 OKK Details withheld at licensee's request by SSL.
G4 OKM M J Smith, 7 Russley Green, Wokingham, RG40 3HT [IO91NJ, SU86]
G4 OKO D H Rycroft, 1 Littlefields Ave, Banwell, Weston Super Mare, BS24 6BE [IO81NH, ST35]
G4 OKP J Winstanley, 3 Farndon Dr, West Kirby, Wirral, L48 9YA [IO83KJ, SJ28]
G4 OKQ D Steiner, 316 Silvergrove Bay NW, Calgary, Alberta, Canada, T3B 4R5
G4 OKS M C G Porter, 127 St. Marys Rd, Par, PL24 2HA [IO70PI, SX05]
GW4 OKT Details withheld at licensee's request by SSL.
G4 OKW C Trayner, 32 Moor Park Villas, Leeds, LS6 4BZ [IO93FT, SE23]
G4 OKY R G Wilkinson, 10 Mildenhall Rd, Loughborough, LE11 4SN [IO92JS, SK51]
G4 OKZ D H Wilson, Fernlea, Melton Rd, Wranby, Brigg, DN39 6TG [IO93UO, TA11]

G4 OLA R W McCubbin, 20 Wellesley Park, Wellington, TA21 8PY [IO80JX, ST11]
G4 OLB W J Snoddy, 2 West Cliffe Corner, Harrogate, HG2 0PJ [IO93FX, SE25]
G4 OLC B Potts, 16 Gordon Terr, Stakeford, Choppington, NE62 5UE [IO95FD, NZ28]
G4 OLE P Langdon, 28 Irving Rd, Bournemouth, BH6 5BQ [IO90CR, SZ19]
G4 OLF A Thomson, 7 Bloomstiles, Salthouse, Holt, NR25 7XJ [JO02NW, TG04]
GM4 OLH I Walsh, Fco (Moscow), King Charles St, London, SW1A 2AH [IO91WM, TQ37]
G4 OLI J Spence, 163 Kingsley Ave, Kettering, NN16 9ES [IO92PJ, SP88]
G4 OLK A G Mackay, 2 Highcliffe Gr, New Marske, Redcar, TS11 8DU [IO94LN, NZ62]
G4 OLN Details withheld at licensee's request by SSL.
G4 OLO G E Spencer, 5 Pitchcroft Ln, Church Aston, Newport, TF10 9AQ [IO82TS, SJ71]
G4 OLP R E Parker, 2 Laurel Rd, Thorpe St. Andrew, Norwich, NR7 9LL [JO02PP, TG21]
G4 OLQ D Roberts, 97 St. Peters Rd, Netherton, Dudley, DY2 9HN [IO82XL, SO98]
G4 OLS J C Lloyd, 16 Gillbanks Rd, Stourbridge, West Midlands, DY8 4RN [IO82WL, SO88]
G4 OLT D W Luing, Red Ln House, Red Ln, Rosudgeon, Penzance, TR27 9PU [IO70HE, SW53]
G4 OLU D D Steward, 30 Riffhams Dr, Great Baddon, Great Baddow, Chelmsford, CM2 7DD [JO01GR, TL70]
G4 OLV P Nicholas, 7 Brentwood Rd, Blacon, Chester, CH1 5DT [IO83MF, SJ36]
G4 OLW A Lloyd, 243 Stand Ln, Radcliffe, Manchester, M26 1JA [IO83UN, SD70]
G4 OLY C G D Morgan, 316 Middle Rd, Sholing, Southampton, SO19 8NT [IO90HV, SU41]
G4 OLZ I Sirley, Ivy Cottage, Coventry Rd, Burbage, LE10 2HL [IO92HM, SP49]
G4 OMD D J Wilson, 10 Haroldsway, Stamford Bridge, York, YO4 1DW [IO93NX, SE75]
G4 OME Details withheld at licensee's request by SSL.
G4 OMG C Prescott, 15 Sarabeth Dr, Tunley, Bath, BA3 1EA [IO81SH, ST65]
G4 OMH D Nobles, 14 Martin Cl, Rushden, NN10 6YZ [IO92QG, SP96]
G4 OMI A J Proudler, 4 Highmead, 59 Meadrow, Godalming, Surrey, GU7 3HS [IO91QE, SU94]
G4 OMJ G Yarnall, 27 Ringwood Cres, Wollaton, Nottingham, NG8 1HL [IO92JW, SK54]
GI4 OMK P Murphy, 2 The Gables, Cairnshill Rd, Belfast, BT8 4UJ [IO74BN, J36]
G4 OMM R Reid, 39 Queensway, Irlam, Manchester, M44 6ND [IO83SK, SJ79]
G4 OMN D K Thorndike, 48 Cressingham Rd, Reading, RG2 7JR [IO91MK, SU77]
GI4 OMO D McGuckin, 6 Highfield Rd, Magerafelt, Co. Derry, N Ireland, BT45 5JD [IO64QS, H89]
G4 OMP M A Nyman, 26 Silverstone Ct, River Brook Dr, Birmingham, West Midlands, B30 2SH [IO92BK, SP08]
G4 OMS R W J Reynolds, 90 Manchester Rd, Blackpool, FY3 8DP [IO83LT, SD33]
GM4 OMT M D Taylor, 3 Abbotsgrange Rd, Grangemouth, FK3 9JD [IO86DA, NS98]
G4 OMV D J Marlow, 53 The Lawns, Corby, NN18 0TA [IO92PL, SP88]
GM4 OMW R Watt, St. Margarets Flats, 11 West Park St., Huntly, AB54 8DY [IO87OK, NJ54]
G4 OMX Details withheld at licensee's request by SSL.
G4 OMZ L Welding, 23 Haley Rd North, Burtonwood, Warrington, WA5 4JD [IO83QK, SJ59]
G4 ONA Details withheld at licensee's request by SSL.
G4 ONC E C Westcott, 8 Portal Pl, Ivybridge, PL21 9BT [IO80AJ, SX65]
G4 OND P Stevens, 28 Monteagle Dr, Charterfield Est, Kingswinford, DY6 7RY [IO82WM, SO89]
G4 ONE K A Lomas, 7 Pedmore Ln, Pedmore, Stourbridge, DY9 0SP [IO82WK, SO98]
G4 ONF P G Sergent, 6 Gurney Cl, Costessey, Norwich, NR5 0HB [JO02OP, TG10]
G4 ONG W B Lowe, 34 Ridgeway, Lowton, Warrington, WA3 2QL [IO83RL, SJ69]
G4 ONH C W Weller, Mole End, Brent Hall Rd, Finchingfield, Braintree, CM7 4JZ [JO01FX, TL63]
G4 ONI D L Mears, St Josephs, Dovenby, Cockermouth, Cumbria, CA13 0PN [IO84HQ, NY03]
G4 ONJ Mt A K Lightfoot, 13 Midhurst Cl, Ifield, Crawley, RH11 0BS [IO91VC, TQ23]
G4 ONK Details withheld at licensee's request by SSL.
G4 ONN J Martin, 7 Roundhaye Rd, Bournemouth, BH11 9JB [IO90BS, SZ09]
G4 ONO J H Wager, 21 Beacon View Dr, Sutton Coldfield, B74 2AW [IO92BN, SP09]
G4 ONP D Thorpe Loughtn &Dis AR, 9 Albion Hill, Loughton, IG10 4RA [JO01AP, TQ49]
G4 ONR M Shaw, 163 Rawthorpe Ln, Dalton, Huddersfield, HD5 9NX [IO93CP, SE11]
G4 ONS M F Slade, 5 Pedder Rd, Clevedon, BS21 5HB [IO81NK, ST47]
G4 ONV G E S Parker, 49 Newlands, Dawlish, EX7 0EA [IO80GO, SX97]
G4 ONX G Purser, 41 St. Martins Gr, Leeds, LS7 3LJ [IO93FT, SE33]
G4 ONY J Purser, 41 St. Martins Gr, Leeds, LS7 3LJ [IO93FT, SE33]
G4 ONZ P Tebbutt, 23 The Burrows, Batley, WF17 8BE [IO93ER, SE22]
G4 OO D Hoult, Chespool House, Gosberton Risegate, Spalding, Lincs, PE11 4EU [IO92VU, TF23]
G4 OOA C R Shepherd, 3 Blacksmith Row, Bassingham, Lincoln, LN5 9JN [IO93QD, SK96]
G4 OOB J Westerman, 7 Gascoigne Ct, Barwick in Elmet, Leeds, LS15 4NY [IO93HT, SE43]
G4 OOC J Muzyka obo Yorkshire Pudding Contest, 2 Engine Fold, Wrenthorpe, Wakefield, WF2 0PP [IO93FQ, SE32]
G4 OOE A R N Langmead, 41 Castle Park, Hemyock, Cullompton, Devon, EX15 3SB [IO80JV, ST11]
G4 OOH S C Parker, 64 Cardigan Ln, Leeds, LS4 2LD [IO93FT, SE23]
G4 OOI C Parker, 64 Cardigan Ln, Leeds, LS4 2LD [IO93FT, SE23]
G4 OOJ C M Rose, 3 Harley Dr, Leeds, LS13 4QY [IO93ET, SE23]
G4 OOK S Stobbs, 78 Hershall Dr, Town Farm, Middlesbrough, TS3 8NX [IO94JN, NZ51]
G4 OOL J D Yeandel, 50 Coniston Cres, Loughborough, LE11 3RH [IO92JS, SK51]
G4 OON B W Wilson, 9 Coniston Ave, West Auckland, Bishop Auckland, DL14 9LL [IO94DP, NZ12]
G4 OOP Details withheld at licensee's request by SSL.
G4 OOQ N G R Ward, 8 Meadowview Rd, Kempston, Bedford, MK42 7BE [IO92RC, TL04]
G4 OOR Details withheld at licensee's request by SSL.
G4 OOS J Ball, 21 Wellinger Way, Leicester, LE3 1RG [IO92JP, SK50]
GM4 OOU J A Forsyth, 20 Abington Rd, Dunfermline, KY12 7XU [IO86GB, NT08]
G4 OOW N J Gray, 53 Duport Rd, Burbage, Hinckley, LE10 2RN [IO92HM, SP49]
G4 OOX L G Harvey, 27 Guernsey Dr, Chelmsley Wood, Birmingham, B36 0PB [IO92DL, SP18]
G4 OOY D S Bird, 13 Kilvington Rd, Arnold, Nottingham, NG5 7HQ [IO93KA, SK54]
G4 OPA A C Hawkins, 29 Bedford Rd, Cranfield, Bedford, MK43 0EU [IO92QB, SP94]
G4 OPB N S Hydes, 2 Stable Ct, Martlesham Heath, Ipswich, IP5 7UQ [JO02OB, TM24]
G4 OPD A C Blissett, 26 Cherry Orchard, Holt Heath, Worcester, WR6 6ND [IO82UG, SO86]
G4 OPE M J Hodges, 40 Ennersdale Rd, Coleshill, Birmingham, B46 1EP [IO92DM, SP18]
GI4 OPH T J Crawford, 24 Broomhill Park, Bangor, BT20 5QZ [IO74EP, J58]
G4 OPI A Easom, 1 Station Cl, West Ayton, Scarborough, YO13 9JQ [IO94SF, SE98]
GM4 OPJ A R Cockrill, 26 Ercall Rd, Brightons, Falkirk, FK2 0TS [IO85DX, NS97]
G4 OPK D N Carrett, 34 Swansea Rd, Reading, RG1 8HA [IO91ML, SU77]
G4 OPL Details withheld at licensee's request by SSL.
G4 OPM H M Barnett, 27 Sandhill Oval, Alwoodley, Leeds, LS17 8EB [IO93FU, SE34]
G4 OPN W Broxup, 9 Kingsway, Hapton, Burnley, BB11 5RB [IO83US, SD73]
G4 OPO C J Haddrell, 9 Counterpool Rd, Kingswood, Bristol, BS15 2DQ [IO81RK, ST67]
G4 OPP I Horsefield, 61 Lewis Ct Dr, Boughton Monchelsea, Maidstone, ME17 4LG [JO01GF, TQ75]
G4 OPQ Details withheld at licensee's request by SSL.
G4 OPR R G Hayward, Sunnyfields, Lighthouse Rd, St Margarets Bay, nr Dover Kent, CT15 6EJ [JO01QD, TR34]
G4 OPT Details withheld at licensee's request by SSL.[Correspondence to: A R M Kemsley, 1 Riversleigh Avenue, Lytham, Lytham St. Annes, Lancs FY8 5QZ.]
GM4 OPU J B Houston, 26 Clerk Dr, Corpach, Fort William, PH33 7LE [IO76KU, NN07]
G4 OPV J Jackson, 15 Jackson Cres, Stourport on Severn, DY13 0EW [IO82UH, SO87]
GW4 OPW B Jones, 25 Parc Glas, Brynna Farm, Cwmdare, Aberdare, CF44 8RP [IO81GR, SN90]
G4 OPX Details withheld at licensee's request by SSL.
G4 OPY S Balmer, 101 Marsh Ln, Sheply, Shepley, Huddersfield, HD8 8AP [IO93DN, SE10]
G4 OPZ N J Bloomer, 17 Peeks Ave, Plystock, Plymouth, PL9 9BZ [IO70XI, SX55]
G4 OQ G C Lidstone, 76 Thames Dr, Leigh on Sea, SS9 2XD [JO01HN, TQ88]
GW4 OQB A Greatrex, Clydfan, Dinas Cross, Newport Pembs, Dyfed, SA42 0XS [IO72NA, SM93]
G4 OQC A E Forrest, 12 Verity Cl, Manchester, M20 4AU [IO83VK, SJ89]
G4 OQD D Buckley, 31 Mount Rd, Canterbury, CT1 1YD [JO01NG, TR15]
G4 OQE B A Curran, Brentwood, 51 Bankhall Ln, Hale, Altrincham, WA15 0LF [IO83TI, SJ78]
G4 OQG Prof M L Ayres, 3 Wicks Dr, Chippenham, SN15 3EL [IO81WK, ST97]
G4 OQH Rev H Callaghan, St. Johns Vicarage, 7 Alford Cl, Breightmet Dr, Bolton, BL2 6NR [IO83TN, SD70]
G4 OQI C Daniels, 1 Blacksmith Ln, Happisburgh, Norwich, NR12 0QT [JO02ST, TG33]
G4 OQJ I M James, 4 Lancaster Gdns, Earley, Reading, RG6 7PA [IO91MK, SU77]
G4 OQK R P Alderton, 2 Harefield Rd, Croxton, Thetford, IP24 1NE [JO02JK, TL88]
G4 OQL Dr C J Bowley, Plum Tree House, Walk Cl, Draycott, Derby, DE72 3PN [IO92HV, SK43]
G4 OQM V L A D Francl, 184 Dalton Ln, Rotherham, S65 3QJ [IO93IK, SK49]
G4 OQN M Barr, 83 High St., Whitton, Twickenham, TW2 7LD [IO91TK, TQ17]
G4 OQP J W Gerrity, 14 Lostock Ave, Hazel Gr, Stockport, SK7 5JN [IO83WI, SJ98]
G4 OQR M J Huddart, 7 The Old Forge Cl, Ryde, PO33 3PL [IO90JR, SZ59]
G4 OQU J T Davenport, 1 Lowfields, Staveley, Chesterfield, S43 3QB [IO93HG, SK47]
G4 OQV R Beecham, 7 Crummock Cl, Holbrooks, Coventry, CV6 6GY [IO92FK, SP38]
G4 OQX G Cooper, 61 Fallowfield Rd, Hasbury, Halesowen, B63 1BZ [IO82XK, SO98]
G4 OQZ B R Dawson, Isca, 12 Lestock Way, Fleet, GU13 9EB [IO91OG, SU85]
G4 ORB G C Busby, The Terrace, Terr Rd South, Binfield, Bracknell, RG42 4DS [IO91OK, SU87]
G4 ORC G Oliver Oldham Am Radio Club, 158 High Barn St., Royton, Oldham, OL2 6RW [IO83WN, SD90]
G4 ORD H E Baxter, 29 Hoylake Dr, Tividale, Warley, B69 1QA [IO82XM, SO98]
G4 ORE A Charles, Cramond, Chorleywood Bottom, Chorleywood, Rickmansworth, WD3 5JD [IO91RP, TQ09]
G4 ORF Details withheld at licensee's request by SSL.
GI4 ORG T M Groves, 12 Brooklands Dr, Dundonald, Belfast, BT16 0PH [IO74CO, J47]
GI4 ORI J Hamill, 67 Windsor Ave, Coleraine, BT52 2DR [IO65QD, C83]
G4 ORJ A B Jones, 44 Prospect Ave, Irchester, Wellingborough, NN29 7DZ [IO92QG, SP96]
GI4 ORK H Kane, 47 Prospect Rd, Portstewart, Co Derry, N Ireland, BT55 7NG [IO65PE, C83]
G4 ORM Details withheld at licensee's request by SSL.
G4 ORP M S Parsons, 15 Sherbourne Rd, Hangleton, Hove, BN3 8BA [IO90VU, TQ20]
G4 ORQ A G Walker, 4A Winston Dr, Eston, Middlesbrough, TS6 9LY [IO94KN, NZ51]
G4 ORR Details withheld at licensee's request by SSL.
G4 ORS W L Ragg, 45 The Tarters, Sherston, Malmesbury, SN16 0NT [IO81VN, ST88]
G4 ORT D R Waters, 17 Nab Wood Rd, Shipley, BD18 4AG [IO93CT, SE13]
G4 ORU G R Wadwell, 7 Barkhart Dr, Wokingham, RG40 1TW [IO91OJ, SU86]
G4 ORV D L Whatmough, Flat 3, 170 Buxton Rd, Stockport, SK2 6HA [IO83WJ, SJ98]
G4 ORW A N Atherley, 23 Gadwall Croft, Erdington, Birmingham, B23 7RN [IO92BM, SP09]
G4 ORX P M Baggett, 33 Foxglove Way, Thatcham, RG18 4DL [IO91IJ, SU56]
G4 ORY C J Bates, 335 Clarence Rd, Four Oaks, Sutton Coldfield, B74 4LU [IO92BO, SK10]
G4 OSA R C Hampton, 11 Greenlands, Hutton Rudby, Yarm, TS15 0JQ [IO94IK, NZ40]
G4 OSB T M Arris, 7 Rowan Rd, North Hykenham, North Hykeham, Lincoln, LN6 8LY [IO93RE, SK96]
G4 OSC C H B Burrill, 434 Lytham Rd, Blackpool, FY4 1EB [IO83LT, SD33]
GI4 OSG D J Robinson, 17 Dalton Glen, Comber, Newtownards, BT23 5RJ [IO74CN, J46]
G4 OSH A Nevison, 10 Birchway, Tunbridge Wells, TN2 3DA [JO01DD, TQ54]
G4 OSI D Whitehouse, 10 Felsted St, Baddeley Green, Stoke on Trent, Staffs, ST2 7HJ [IO83WB, SJ95]
G4 OSJ P G Brewer, 2 Mill Cl, Wing, Oakham, LE15 8RH [IO92PO, SK80]
G4 OSK K Hall, 21 Eardulph Ave, Chester-le-Street, Co. Durham, DH3 3PR [IO94FU, NZ25]
G4 OSN F Tully, 3 Criccieth Cl, Llandudno, LL30 1GZ [IO83CH, SH78]
G4 OSO E V Binns, Fieldview, 5 Moorside, Cleckheaton, BD19 6JH [IO93DR, SE12]
G4 OSP M R Binns, Fieldview, 5 Moorside, Cleckheaton, BD19 6JH [IO93DR, SE12]
GM4 OSQ A A Chilles, 5 Adamton Terr, Prestwick, KA9 2DW [IO75QL, NS32]
G4 OSR S Roberts, 1 Lakeside Cres, Long Eaton, Nottingham, NG10 3GH [IO92IV, SK43]
GM4 OSS S B Campbell, 26 Kinloch Ave, Stewarton, Kilmarnock, KA3 3HQ [IO75RQ, NS44]
G4 OST P G K Cabban, Ivydene, Upper Tockington Rd, Tockington, Bristol, BS12 4LQ [IO81RN, ST68]
G4 OSU M P Dixey, 50 Sandon Rd, Fordhouses, Wolverhampton, WV10 6EN [IO82WO, SJ90]
GM4 OSV C Dunn, 66 Glen Doll Rd, Neilston, Glasgow, G78 3QP [IO75SS, NS45]
G4 OSW Details withheld at licensee's request by SSL.
G4 OSX Details withheld at licensee's request by SSL.
G4 OTB Dr N S J Hailes, Yew Tree Cottage, The Hollies Common, Gnosall, Stafford, ST20 0JD [IO82UT, SJ82]
G4 OTC Mrf P R Gagen, 61 West End Rd, Golcar, Huddersfield, HD7 4JF [IO93BP, SE01]
G4 OTE M B Grayson, One Elm, 58 Kaye Ln, Almondbury, Huddersfield, HD5 8XU [IO93CP, SE11]
G4 OTF J H Pickering, 8 Whitefield Cres, Houghton-le-Spring, DH4 7QT [IO94GU, NZ35]
GI4 OTG A H McNiece, 148 Doagh Rd, Newtownabbey, BT36 6BA [IO74BQ, J38]
GM4 OTH M S Thomas, 21 Corbie Pl, Atholl Park, Milngavie, Glasgow, G62 7NB [IO75TW, NS57]
G4 OTI Dr P J Stockbridge, 11 Fairways, Frodsham, Warrington, WA6 7RU [IO83PG, SJ57]
G4 OTJ Details withheld at licensee's request by SSL.
G4 OTK Details withheld at licensee's request by SSL.
G4 OTL M J Baker, 17 Clumber Dr, Gomersal, Cleckheaton, BD19 4RP [IO93DR, SE22]
G4 OTM G Buckley, 43 Broad Oak Cres, Oldham, OL8 2PX [IO83WM, SD90]
G4 OTN M C Cook, 4A Long Barn, Hoghton, Preston, PR5 0SA [IO83RR, SD62]
G4 OTQ M Carson, 41 The Poynings, Richings Park, Iver, SL0 9DS [IO91RL, TQ07]
G4 OTS G Eccleston, 24 Orton Ln, Wombourne, Wolverhampton, WV5 9AW [IO82VN, SO89]
G4 OTV D W Green, St. Annes, Poundfield Rd, Crowborough, TN6 2BG [JO01CB, TQ53]
G4 OTX G W Hibberd, 2 Carr Bank, Oakamoor, Stoke on Trent, ST10 3EA [IO93AA, SK04]
G4 OTY M L Mason, 15 White Horse Rd, Winsley, Bradford on Avon, BA15 2JZ [IO81UI, ST86]
GI4 OTZ B McCann, 16 Downview Park, Belfast, BT15 5HY [IO74AP, J37]
G4 OUA P A Taylor, 26 Kelso Dr, The Priorys, North Shields, NE29 9NS [IO95GA, NZ36]
G4 OUB J F Whetstone, 70 Heanor Rd, Smalley, Ilkeston, DE7 6DX [IO93HA, SK44]
GI4 OUC J A Smylie, 81 Lansdowne Rd, Belfast, BT15 4AB [IO74AP, J37]
G4 OUG C F Beesley, 15 Byron Cl, Cheadle, Stoke on Trent, ST10 1XB [IO92XX, SJ94]
G4 OUH J Brockway, 9 Warren Cl, Princes End, Tipton, DY4 9PQ [IO82XM, SO99]
G4 OUI M J Brockway, 9 Warren Cl, Princes End, Tipton, DY4 9PQ [IO82XM, SO99]
G4 OUJ S A Carrigan, 1 Milford Cres, Littleborough, OL15 9EF [IO83WP, SD91]
G4 OUK S J Bradshaw, 14 Sheringham Dr, Coppenhall, Crewe, CW1 3XJ [IO83SC, SJ65]
G4 OUL Details withheld at licensee's request by SSL.
G4 OUM T W Bolton, 25 Woodfield Dr, Lichfield, WS14 9HH [IO92CQ, SK10]
GI4 OUN D I Fulton, 120 Dunnalong Rd, Bready, Strabane, Co. Tyrone, BT82 0DP [IO64GW, C30]
G4 OUO Details withheld at licensee's request by SSL.
GI4 OUP S F Henderson, 47 Donaghedy Rd, Bready, Strabane, BT82 0DB [IO64HV, C30]
G4 OUS D Bean, 10 Willerby Dr, Howden, Goole, DN14 7JA [IO93NR, SE72]
G4 OUT I L Cornes, 6 Haywood Heights, Little Haywood, Stafford, ST18 0UR [IO92AT, SK02][Op: Ian L Cornes.]
GW4 OUU D Grace, The Laurels, Gwbert Rd, Cardigan, SA43 1AF [IO72QC, SN14]
G4 OUV Details withheld at licensee's request by SSL.[Station located near Bognor Regis.]
G4 OUW Details withheld at licensee's request by SSL.
G4 OUY A B Jones, 19 Hopperstyle, Bickington, Barnstaple, EX31 2LA [IO71WB, SS53]
G4 OUZ L J Day, Wood Ct, Blachford Rd, Ivybridge, PL21 0AD [IO80AJ, SX65]
G4 OVB A R Poulston, 26 Hesters Way Ln, Cheltenham, GL51 0LN [IO81WV, SO92]
G4 OVD G A Rugen, 24 Highgate Rd, Lydiate, Liverpool, L31 0DA [IO83MM, SD30]
GI4 OVE J J McElvanna, 26 Lissummon Rd, Newry, BT35 6NA [IO64TF, J03]
G4 OVG J R Thompson, 61 The Sorrells, Stanford-le-Hope, SS17 7ER [JO01FM, TQ68]
GW4 OVH H M Owen, 5 Arilwyn Cefn Rd, Bwlchgwyn, N Wrexham, Clwyd, LL11 5YF [IO83KB, SJ25]
G4 OVI R A Rideout, 1 Breach Ln, Enmoor Green, Shaftesbury, SP7 8LE [IO81VA, ST82]
G4 OVJ R C Read, 1 St. Lawrence Cres, Shaftesbury, SP7 8EG [IO81VA, ST82]
G4 OVK Details withheld at licensee's request by SSL.[Op: D Collet, 48 Millington Rd, Castle Bromwich, Birmingham, B36 8BN.]
G4 OVM C Barnes, 13 Waterworks Rd, Farlington, Portsmouth, PO6 1NG [IO90LU, SU60]
GI4 OVN S H Dawson, 2 Glencraig Park, Craigavad, Holywood, BT18 0BZ [IO74CP, J48]
G4 OVO J G Featherstone, 19 Victoria Rd, Torquay, TQ1 1HU [IO80FL, SX96]
G4 OVP Details withheld at licensee's request by SSL.
G4 OVR D Fillingham, 25 Lynwood Cres, Woodlesford, Leeds, LS26 8LJ [IO93GS, SE32]
G4 OVS F H P B Goddard, 4 St. Peters Cl, Barnburgh, Doncaster, DN5 7EN [IO93IM, SE40]
G4 OVT J M Grant, 3 Cragg Wood Cl, Horsforth, Leeds, LS18 4RL [IO93ET, SE23]
G4 OVU Details withheld at licensee's request by SSL.
G4 OVV N J Howard, 25 Whitecroft Rd, Hawkley Hall, Wigan, WN3 5PS [IO83QM, SD50]
G4 OVW B Jempson, 38 Filey Rd, Scarborough, YO11 2TU [IO94TG, TA08]
G4 OVX M Kennett, Toms Cottage, Kendal Ln, Tockwith, York, YO5 8QN [IO93IX, SE45]
G4 OVY Dr G Barnish, 22 Park Way, Hoylake, Meols, Wirral, L47 7BT [IO83KJ, SJ29]
G4 OVZ M E Clarke, Newbury Farm, Ampthill Rd, Silsoe, Bedford, MK45 4HB [IO92SA, TL03]
GI4 OWA G A Elliott, 4 Fernbrae Gdns, Kilfennan, Londonderry, BT47 1XS [IO64IX, C41]
GI4 OWB J W Fallows, 52 Onslow Parade, Belfast, BT6 0AS [IO74BN, J37]
G4 OWC Details withheld at licensee's request by SSL.
G4 OWD J Kemp, 42 Hawksworth Ave, Guiseley, Leeds, LS20 8EJ [IO93DU, SE14]
G4 OWF A Cockroft, 12 Lickless Gdns, Horsforth, Leeds, LS18 5QU [IO93EU, SE23]
G4 OWG R S Leighton, 7 Greenacre Park Mews, Rawdon, Leeds, LS19 6RT [IO93DU, SE24]
G4 OWH G R Gregor, Lodge No 3, Ammerdown, Kilmersdon, Bath, BA3 5SW [IO81TG, ST75]
G4 OWI Details withheld at licensee's request by SSL.[Op: Dr S M Wood. Station located near Newark.]
G4 OWK T M Pearsall, 6 Vernon Cl, Martley, Worcester, Worcs, WR6 6QX [IO82TF, SO75]
G4 OWN A Turner, 1 Milton Rd, Flitwick, Bedford, MK45 1QA [IO92RA, TL03]
G4 OWO Details withheld at licensee's request by SSL.
G4 OWQ D G Scott, 31 Avondale Rd, South Benfleet, Benfleet, SS7 1EH [JO01GN, TQ78]
GM4 OWR J J W Brand, 18 Cluny Ct, Blairgowrie, PH10 6XY [IO86IN, NO14]
G4 OWS N P Cunliffe, 44 Shore Rd, Hesketh Bank, Preston, PR4 6RB [IO83NR, SD42]
G4 OWT S R Harwood, 24 Firle Cres, Lewes, BN7 1QG [IO90XV, TQ31]
G4 OWU R Hill, 30A Croft Dr, Bramham, Wetherby, LS23 6RJ [IO93HV, SE44]
G4 OWV G C Davies, 17 Kennedy Ave, Gorleston, Great Yarmouth, NR31 6TB [JO02UN, TG50]
GW4 OWX R A E Hillson, 2 Glaslyn Cttgs, Prenteg, Porthmadog, LL49 9SR [IO72WW, SH54]
G4 OWY R J Howes, 202 Abbotsbury Rd, Weymouth, DT4 0NA [IO80SO, SY67]
GW4 OWZ L Lee, Flat 14, Ravenscourt Flats, Richmond Rd, Cardiff, CF2 3BW [IO81JL, ST17]
GW4 OXB T W Morgan, 1 Jersey St., Hafod, Swansea, SA1 2HF [IO81AP, SS69]
G4 OXD T M Rose, 41 Keats Way, Hitchin, SG4 0DP [IO91UW, TL22]
G4 OXF J R Topliss, 12 Rose Tree Ln, Newhall, Swadlincote, DE11 0LN [IO92FS, SK22]
GW4 OXG N J Wood, 43 Holmesdale Rd, London, N6 5TH [IO91WN, TQ28]
G4 OXH B Mulloch, 737 Chester Rd, Erdington, Birmingham, B24 0BY [IO92CM, SP19]
G4 OXK R E Waygood, 2 Brookside Cl, Bransgore, Christchurch, BH23 8BT [IO90DS, SZ19]
GW4 OXL W E Smith, Perhiw, Freshwater East, Pembroke, Dyfed, SA71 5LG [IO71NP, SS09]
G4 OXM C J W Mason, 35 Mutley Rd, Mannamead, Plymouth, PL3 4SB [IO70WJ, SX45]
GI4 OXO W T Fitzsimons, 83 Boghill Rd, Glengormley, Newtownabbey, BT36 4QT [IO74AQ, J38]
G4 OXR C R Mortimer, Meadow Bank, Dowlish Wake, Ilminster, Somerset, TA19 0NZ [IO80NV, ST31]
G4 OXS Details withheld at licensee's request by SSL.
G4 OXU Dr D A Whan, 1 Hillclose Ave, Darlington, DL3 8BH [IO94EM, NZ21]
G4 OXV M J Smith, 72 Park Rd, Lower Gornal, Dudley, DY3 2JL [IO82WM, SO99]
GW4 OXW Details withheld at licensee's request by SSL.
G4 OXX Details withheld at licensee's request by SSL.
G4 OXY Details withheld at licensee's request by SSL.[Station located near Biggleswade.]
G4 OXZ J Blenkin, 28 Tees St., Loftus, Saltburn By The Sea, TS13 4LW [IO94NN, NZ71]

G4 OY G Beaumont, 2 Hawke Garth, Hunmanby, Filey, YO14 0NH [IO94UE, TA17]
G4 OYA H E Davis, 28 Epping Rd, Stoke on Trent, ST4 6LL [IO82VX, SJ84]
G4 OYC J A Cook, The Bungalow Stores, 4 Vicarage Hill, Marldon, Paignton, TQ3 1NH [IO80EL, SX86]
GW4 OYD Details withheld at licensee's request by SSL.[Op: T Davies.]
GI4 OYE W G Heyburn, Sea Breeze, 30 Browns Bay Rd, Islandmagee, Larne, BT40 3RX [IO74CU, D40]
G4 OYF Details withheld at licensee's request by SSL.
GI4 OYG M Black, Site 37 Riverdale, Woodburn Ave, Carrickfergus, Co. Antrim
G4 OYH A P Bowyer, 37 St. Mark Dr, Colchester, CO4 4LP [JO01LV, TM02]
GI4 OYI D A Chambers, 238 Donaghadee Rd, Newtownards, BT23 7QP [IO74DN, J46]
G4 OYJ G A Cartwright, 17 Saxon Park, High St., Albrighton, Wolverhampton, WV7 3LZ [IO82UP, SJ80]
GM4 OYK J Cooper, 48 Hillfoot Ave, Wishaw, ML2 8TR [IO85BT, NS85]
GI4 OYL J Cuthbert, 79 Harmin Park, Newtownabbey, BT36 7UT [IO74AQ, J38]
GI4 OYM W Elliott, 27 Hopefield Ave, Portrush, BT56 8HB [IO65QE, C83]
G4 OYN A J Fuller, 17 Brington Dr, Barton Seagrave, Kettering, NN15 6UW [IO92PI, SP87]
G4 OYQ Details withheld at licensee's request by SSL.
G4 OYT H J T Moss, 101 Barnford Cres, Aoldbury, Oldbury, Warley, B68 8PR [IO82XL, SO98]
G4 OYU J A Philpot, Far Field, Mill Ln, Cranham, Gloucester, GL4 8HU [IO81WT, SO81]
G4 OYW E Parkinson, 14 Keyes Gdns, Tonbridge, TN9 2QD [JO01DE, TQ54]
G4 OYX D Porter, 8 Stanton Dr, Ludlow, SY8 2PH [IO82PJ, SO57]
G4 OYY J D Flegg, 'Oaklea', Ham, Axminster, Devon, EX13 7HL [IO80LT, ST20]
G4 OYZ M Spencer, 13 Marlborough Ave, Halifax, HX3 0DS [IO93BR, SE02]
GW4 OZB D D Sedgebeer, 5 Gwynfi Terr, Llanharan, Pontyclun, CF7 9PL [IO81HN, ST08]
G4 OZC W Smith, 15 Henbury Dr, Woodley, Stockport, SK6 1PY [IO83WK, SJ99]
G4 OZD R G Woolley, 29 Belle Vue Ave, Leicester, LE4 0DE [IO92KP, SK50]
G4 OZG E Haskett, 23 Gloucester Rd, Gaywood, Kings Lynn, PE30 4AB [JO02FS, TF62]
GW4 OZH D M James, 10 Pond Row, Abercanaid, Merthyr Tydfil, CF48 1YT [IO81HR, SO00]
GI4 OZI A J Kinghan, 14 Sunningdale Park North, Belfast, BT14 6RZ [IO74AP, J37]
GI4 OZJ G A Allen, 6 Lougherne Rd, Annahilt, Hillsborough, Co.Down, BT26 6BX [IO74AK, J35]
G4 OZL P M J Ingram, Rosehill Cottage, Palestine, Andover, Hants, SP11 7ER [IO91ED, SU24]
G4 OZM D J Bradberry, 6 The Close, Easton on The Hill, Stamford, PE9 3NA [IO92RO, TF00]
G4 OZN P D Badger, The Badgers, 3 High St., Sturton By Stow, Lincoln, Lincs, LN1 2AE [IO93QH, SK88]
G4 OZO J S Christmas, 12 Badlake Hill, Dawlish, EX7 9AY [IO80GN, SX97]
G4 OZP Details withheld at licensee's request by SSL.
G4 OZQ G Fisher, 9 Shrubbery Rd, Drakes Broughton, Pershore, WR10 2AX [IO82WD, SO94]
GW4 OZU P J Hyams, Tricklewood, Chapel Hill, St Twynnells, Pembroke Dyfed, SA71 5HY [IO71MP, SR99]
G4 OZW D Hulme, 10 Roslin Gdns, Halliwell, Bolton, BL1 8BX [IO83SO, SD71]
G4 OZX C F Goble, 12 Longfield Rd, Emsworth, PO10 7TR [IO90MU, SU70]
GM4 OZZ R H Kerr, Sirius, Avernish, Kyle, IV40 8EQ [IO77FG, NG82]
G4 PAA P Booth, 3 Main Rd, Gransmoor, Driffield, Yorks, YO25 8HU [IO94UA, TA15]
G4 PAC P A Caldwell, 31 Cedar Dr, Kingsclere, Newbury, RG20 5TD [IO91JH, SU55]
G4 PAD K J Thompson, 113 Gordon Rd, Corringham, Stanford-le-Hope, SS17 7QZ [JO01FM, TQ68]
GW4 PAF J Thomas, 2 Tudor Way, Crown Hill Est, Llantwit Fardre, Pontypridd, CF38 2NH [IO81HN, ST08]
G4 PAH Dr M L J Rollason, Ash Ridge, Clint, Harrogate, N Yorks, HG3 3DS [IO94EA, SE25]
G4 PAI P M Wells, 15 Apple Tree Gr, Ferndown, BH22 9LA [IO90BT, SU00]
G4 PAJ C E Davies, 112 High Ash Ave, Alwoodley, Leeds, LS17 8TQ [IO93FU, SE34]
G4 PAK J H Tew, 95 Kincaple Rd, Rushey Mead, Leicester, LE4 7YD [IO92KP, SK60]
GW4 PAL Details withheld at licensee's request by SSL.[Op: A G Tucker, Wrexham, Clwyd.]
G4 PAN Details withheld at licensee's request by SSL.
G4 PAO Details withheld at licensee's request by SSL.
G4 PAP Details withheld at licensee's request by SSL.
G4 PAQ N J McLeod, 67 Preston Rd, Brighton, E Sussex, BN1 4QE [IO90WU, TQ30]
G4 PAR Details withheld at licensee's request by SSL.
G4 PAS P A Searles, 63 Whalley Rd, Shuttleworth, Ramsbottom, Bury, BL0 0DP [IO83UP, SD81]
G4 PAT J P Thirsk, 16 Widford Walk, Blackrod, Bolton, BL6 5TD [IO83RO, SD61]
G4 PAV G E Brutnall, 66 Arkwright Rd, Irchester, Wellingborough, NN29 7EF [IO92QG, SP96]
G4 PAY B A Caines, 10 Shears Cres, West Mersea, Colchester, CO5 8AR [JO01LS, TM01]
G4 PAZ P de Carte, 40 Loxwood Rd, Lovedean, Waterlooville, PO8 9TU [IO90LV, SU61]
G4 PBA J R Abbott, 9 Hurstleigh Terr, Harrogate, HG1 4TF [IO93FX, SE35]
G4 PBC J A Kilroy, 119 Station Rd, Brimington, Chesterfield, S43 1LJ [IO93HG, SK37]
G4 PBD R G Hughes, 40 Collier Row Ln, Romford, RM5 3BE [JO01CO, TQ59]
G4 PBE L R Hewett, 19 Parkers Pl, Martlesham Heath, Ipswich, IP5 7UX [JO02OB, TM24]
G4 PBF P M Baker, 17 Mallard Cl, Basingstoke, RG22 5JP [IO91KF, SU54]
G4 PBI S Nash, 77 Endowood Rd, Sheffield, S7 2LY [IO93FI, SK38]
G4 PBJ B M Oakley, 6 Windmill Way, Haxby, York, YO3 3NL [IO94LA, SE65]
G4 PBK D H Meech, 25 Sydney Cl, Plympton, Plymouth, PL7 1PY [IO70XJ, SX55]
G4 PBL Details withheld at licensee's request by SSL.[Op: D Lomas.]
G4 PBM Details withheld at licensee's request by SSL.
G4 PBN J E Vivian, 7 Kemble Gr, The Reddings, Cheltenham, Gloucester, GL51 6TX [IO81WV, SO92]
G4 PBO D J Smith, 21 Sydney Rd, Benfleet, SS7 5RD [JO01GN, TQ78]
G4 PBP R A Stewart, 424 Wood End Rd, Wolverhampton, WV11 1YD [IO82XO, SJ90]
G4 PBR D F Suttenwood, Wheatfields, Lucy Ln North, Stanway, Colchester, CO3 5JQ [JO01KV, TL92]
GI4 PBS T D Wilson, 39 Woburn Rd, Ballyrolly, Millisle, Newtownards, BT22 2HY [IO74FO, J57]
GI4 PBT T M Wilson, Brambly Hedge, 39 Woburn Rd, Millisle, Newtownards, BT22 2HY [IO74FO, J57]
G4 PBV Details withheld at licensee's request by SSL.
G4 PBX Details withheld at licensee's request by SSL.
G4 PBY B A Jones, 13 Albert St., Cheltenham, GL50 4HS [IO81XV, SO92]
G4 PBZ T A Ashton, 90 Secker Ave, Warrington, WA4 2RE [IO83RJ, SJ68]
G4 PC Details withheld at licensee's request by SSL.[Op: C S Burnham.]
G4 PCA Details withheld at licensee's request by SSL.
G4 PCB A Cox, 6 Canterbury Rd, Exeter, EX4 2EQ [IO80FR, SX99]
G4 PCC Details withheld at licensee's request by SSL.
G4 PCD M A Dally, 11 Wrightson Terr, Bentley, Doncaster, DN5 9ST [IO93KM, SE50]
G4 PCE R Collins, 389 Lode Ln, Solihull, B92 8NN [IO92CK, SP18]
G4 PCF P W Goodson, 46 Southwold, Bracknell, RG12 8XY [IO91OJ, SU86]
G4 PCH W J Akines, 164 Sutcliffe Ave, Grimsby, DN33 1AR [IO93WN, TA20]
G4 PCI Details withheld at licensee's request by SSL.
GW4 PCJ R V Belcher, 12 Connaught Cl, Nottage, Porthcawl, CF36 3SL [IO81DL, SS87]
G4 PCK Details withheld at licensee's request by SSL.
G4 PCL B W Walker, 149 Yew Ln, Ecclesfield, Sheffield, S5 9AP [IO93GK, SK39]
G4 PCM P C See, 15 Manor Rd, Middle Littleton, Evesham, WR11 5LL [IO92BC, SP04]
G4 PCN C J Lambert, 23 Palmars Cross Hill, Rough Common, Canterbury, CT2 9BL [JO01MG, TR15]
GW4 PCO P L A Mogford, 27 Ynysymaerdy Rd, Neath, SA11 2TE [IO81CP, SS79]
G4 PCP C J Shelton, Hawthorn House, The Close, Averham, Newark, NG23 5RP [IO93NB, SK75]
GI4 PCQ J A Quinn, 86 Knocknacarry Rd, Cushendun, Ballymena, BT44 0NS [IO65XC, D23]
G4 PCR J B O'Hara, 12 Ray Ave, Nantwich, CW5 6HJ [IO83RB, SJ65]
G4 PCS M H Turner, 15 Witley Green, Luton, LU2 8TR [IO91TV, TL12]
GM4 PCT A Gordon, 2 Duchray St., Riddrie, Glasgow, G33 2DD [IO75VU, NS66]
G4 PCV Details withheld at licensee's request by SSL.[Op: J T Macdonald. Station located near Market Harborough.]
G4 PCW A Tucker, 5 Castle Cl, Falmouth, TR11 4PE [IO70LD, SW83]
GW4 PCX R Price, 2 Grassholm Pl, Broadway, Broad Haven, Haverfordwest, SA62 3HX [IO71LS, SM81]
GI4 PCY A L Sammon, 11 Drumclay Rd, Enniskillen, Co Fermanagh, BT74 6NG [IO64EI, H24]
G4 PCZ D St Quintin, 16 Cromwell Rd, Sprowston, Norwich, NR7 8XH [JO02PP, TG21]
G4 PDA P M Caines, 10 Shears Cres, West Mersea, Colchester, CO5 8AR [JO01LS, TM01]
GM4 PDB F J D Jackson, 5 Pigeni Akrita, 7530, Ormidhia, Cyprus, X X
G4 PDD F P Bibby, 14 St. Clare Terr, Chorley New Rd, Lostock, Bolton, BL6 4AZ [IO83RN, SD60]
G4 PDE R H Bradshaw, 44 Hawthorn St., Derby, DE24 8BD [IO92GV, SK33]
G4 PDF Details withheld at licensee's request by SSL.
G4 PDG B F Hillard, Farmlea, Hele Ln, South Petherton, TA13 5AP [IO80OW, ST41]
G4 PDH Details withheld at licensee's request by SSL.
G4 PDI B W Kenzie, 11 Trinity Cl, Balsham, Cambridge, CB1 6DW [JO02DD, TL55]
G4 PDK R R Davies, 42 The Ridings, Desborough, Kettering, NN14 2LP [IO92OK, SP78]
G4 PDL Details withheld at licensee's request by SSL.
G4 PDM D Mills, Strathnaver, 31 Claremont Dr, Hartlepool, TS26 9PD [IO94JQ, NZ43]
G4 PDO D M Oakley, 4 Cross Keys Ln, Low Fell, Gateshead, NE9 6DA [IO94EW, NZ26]
G4 PDP J Allen, 12 Homefield Rd, Chawston, Bedford, MK44 3BN [IO92UE, TL15]
G4 PDQ J Clayton, 217 Prestbury Rd, Cheltenham, GL52 3ES [IO81XV, SO92]
G4 PDR D J Hughes, 19 Burnsall Cl, Farnborough, SSL GU14 8NN [IO91OH, SU85]
G4 PDT Details withheld at licensee's request by SSL.
G4 PDU Details withheld at licensee's request by SSL.
G4 PDV Details withheld at licensee's request by SSL.
G4 PDW J G Allerton, 25 Belleisle Rd, Laceby Acres, Grimsby, DN34 5QY [IO93WN, TA20]
G4 PDX A Blears, 1 Cedarwood, Legh Cl, Poynton, Stockport, SK12 1JW [IO83WI, SJ98]
G4 PDY J A Brandhuber, 3 Brigham Pl, Felpham, Bognor Regis, PO22 7NW [IO90QS, SZ99]
G4 PDZ F R A Elliott, 40 Treasure Cl, Glenfield, Leicester, LE3 8LT [IO92JP, SK50]
G4 PEA R Flanders, 29 Western Way, Alverstoke, Gosport, PO12 2NE [IO90KS, SZ59]
G4 PEB Details withheld at licensee's request by SSL.
G4 PEC A R Fraser, 43 Edith St., Tynemouth, North Shields, NE30 2PN [IO95GA, NZ36]
G4 PED J Hinde, 12A Station Parade, Ockham Rd South, East Horsley, Leatherhead, KT24 6QN [IO91SG, TQ05]
G4 PEF Prof W Ingram, 141 Churchill Rd, Willesden Green, London, NW2 5EH [IO91VN, TQ28]
G4 PEK L J Dymond, 20 Ayres Cl, Bideford, EX39 4DY [IO71VA, SS42]
G4 PEL W R Threapleton, Cobbs Nook Farm, Newstead Ln, Belmesthorpe, Stamford, PE9 4JJ [IO92SQ, TF00]
G4 PEM S W A Rodda, Cliff Hotel, Chyandour Cliff, Penzance, TR18 2HH [IO70FC, SW43]
G4 PEN R A Potts, 6 Woodland Gr, Wath upon Dearne, Rotherham, S63 7TG [IO93HL, SK49]
G4 PEO J R Pitty, Little Orchard, 12 St. Leonards Rd, Horsham, RH13 6EJ [IO91UB, TQ12]
G4 PEQ A M Stewart, Glydfan, Main St, Thurlaston, Rugby, CV23 9JS [IO92II, SP47]
GI4 PES N Robinson, 3 Moorland Dr, Lisburn, Co. Antrim, BT28 2XU [IO64WM, J26]
G4 PET J H Smith, Pasturefields House, Pasturefields Ln, Great Haywood, Staffs, ST18 0RD [IO82XT, SJ92]
G4 PEU K N Smith, Pasturefields House, Pasturefields Ln, Great Haywood, Staffs, ST18 0RD [IO82XT, SJ92]
GW4 PEX W D Williams, 168 Mumbles Rd, West Cross, Swansea, SA3 5AN [IO71XO, SS69]
G4 PEY R W Wilmot, 1 Retreat Cttgs, Church Ln, Broadbridge Heath, Horsham, RH12 3ND [IO91TB, TQ13]
G4 PFA P A C Wheeler, Hadleigh, 5 Widmere Field, Prestwood, Great Missenden, HP16 0SP [IO91PQ, SP80]
GW4 PFC J McCann, 9 Kings Ave, Rhyl, LL18 1LT [IO83GH, SJ08]
G4 PFD Details withheld at licensee's request by SSL.
G4 PFE J W Laverick, 5 York Cres, Newton Hall Est, Durham, DH1 5PU [IO94FT, NZ24]
G4 PFF J M Potter, 38 West St., Great Gransden, Sandy, SG19 3AU [IO92WE, TL25]
G4 PFG M J Spooner, 6 Cross Rd, Starston, Harleston, IP20 9NQ [JO02PJ, TM28]
G4 PFH H Packington, 41 The Cres, Abbots Langley, WD5 0DR [IO91TQ, TL00]
G4 PFJ J H Backus, 2 Southview Villas, Dunmow Rd, Takeley, Bishops Stortford, CM22 6SW [JO01DU, TL52]
G4 PFK G V Gifford, 184 Chantrey Cres, Great Barr, Birmingham, B43 7PG [IO92BN, SP09]
GD4 PFL P J Freestone, 16 Brunswick Rd, Douglas, IM2 3LQ
G4 PFM A E Bishop, 64 Kestrel Way, Cheslyn Hay, Walsall, WS6 7LB [IO82XP, SJ90]
G4 PFO J Gregory, 22 Tower View Rd, Parklands, Great Wyrley, Walsall, WS6 6HE [IO82XP, SJ90]
G4 PFQ D Gowland Derwentside ARC, 78 Pleasant View, Bridgehill, Consett, DH8 8LF [IO94BU, NZ05]
G4 PFR J F Harding, 19 Carrington Cres, Wendover, Aylesbury, HP22 6AW [IO91PS, SP80]
G4 PFS Details withheld at licensee's request by SSL.
G4 PFT J Harris, Loughrea, Redehall Rd, Smallfield, Horley, RH6 9QA [IO91WD, TQ34]
G4 PFU D I Blunt, 12 Mallard Pl, East Grinstead, RH19 4TF [IO91XC, TQ33]
G4 PFV J Stevens, 68 Hoblands, Haywards Heath, RH16 3NB [IO90XX, TQ32]
G4 PFX D J Palmer, 14 Garibaldi Rd, Redhill, RH1 6PB [IO91VF, TQ24]
G4 PFY J O'Hagan, 13 Chapel Rd, Stanford in The Vale, Faringdon, SN7 8LE [IO91FP, SU39]
G4 PFZ J A Aspland, 6 Trilithon Cl, Hellesdon, Norwich, NR6 5EP [JO02PP, TG21]
G4 PGA B R Gage, 117 Kingshill Ave, Worcester Park, KT4 8BZ [IO91VJ, TQ26]
G4 PGB P C Hayward, 22 Falconers Park, Sawbridgeworth, CM21 0AU [JO01BT, TL41]
G4 PGC F Hopewell, 48 Gladstone St., Loughborough, LE11 1NS [IO92JS, SK52]
G4 PGE A Ironside, 67 Boscombe Rd, Worcester Park, KT4 8PJ [IO91VJ, TQ26]
G4 PGF D A Jarvis, 74 Halford Rd, Ickenham, Uxbridge, UB10 8QA [IO91SN, TQ08]
G4 PGG P C Beesley, 15 Byron Cl, Cheadle, Stoke on Trent, ST10 1XB [IO82XX, SJ94]
GI4 PGH J Crawford, 2 Holywood Rd, Ballybarnes, Newtownards, BT23 4TQ [IO74CO, J47]
G4 PGI Details withheld at licensee's request by SSL.
G4 PGJ D Ward, 48 Moat Bank, Bretby Ln, Bretby, Burton on Trent, DE15 0QJ [IO92ET, SK22]
G4 PGK J Fisher, 65 Cres Towers, Leeds, LS11 5UR [IO93FS, SE33]
GM4 PGL Details withheld at licensee's request by SSL.
GM4 PGM P M Brash, 4 Union St., Lossiemouth, IV31 6BA [IO87IR, NJ27]
GI4 PGN J B Bailie, 4 Quarry Rd, Greyabbey, Newtownards, BT22 2QF [IO74FN, J56]
G4 PGP Details withheld at licensee's request by SSL.
G4 PGQ D Harrison, 77 Leigh Rd, Hindley Green, Hindley, Wigan, WN2 4SZ [IO83RM, SD60]
G4 PGR S A Henwood, 29 Fishpond Way, Abbey Hulton, Stoke on Trent, ST2 8DE [IO83WA, SJ94]
G4 PGS P F Clark, 23 Nova Mews, Stonecot Hill, Sutton, SM3 9HY [IO91VJ, TQ26]
G4 PGT T J Lee, 15 Wantage Gdns, Little Paxton, St. Neots, Huntingdon, PE19 4EZ [IO92UF, TL16]
GM4 PGV P J Lawless, 37 Oaklands Ave, Irvine, KA12 0SE [IO75QO, NS33]
G4 PGW N M Puttick, 14 Rivercourt, Gosport St., Lymington, Hants, SO41 9BB [IO90FS, SZ39]
G4 PGX M J Williams, 137 Mill Hill Ln, Burton on Trent, DE15 0AX [IO92ET, SK22]
G4 PGY Dr R B White, Beech Hill, Flore, Northampton, NN7 4LL [IO92LF, SP66]
G4 PGZ G W S Hussey, 12 Waterside, Wooburn Green, High Wycombe, HP10 0HW [IO91PO, SU98]
GW4 PHB W J Vickers, Creigiau, Penrhyndeudraeth, Gwynedd, North Wales, LL48 6LS [IO72XW, SH63]
G4 PHC G Steam, Sunrise, Barton Rd, Alcombe, Minehead, TA24 6BZ [IO81GE, SS94]
G4 PHD P Lloyd, 82 Albert Rd South, Great, Malvern, WR14 3DX [IO82UC, SO74]
G4 PHE J D Brown, 18 Springfield Ave, Ashbourne, DE6 1BJ [IO93DA, SK14]
G4 PHF N J Baker, Gamage Ct, Lower Ley Ln, Minsterworth, Gloucester, GL2 8JT [IO81TU, SO71]
G4 PHH R Fraser, 19 Church Ln, Altham West, Clayton-le-Moors, Accrington, BB5 4DE [IO83TS, SD73]
G4 PHJ Details withheld at licensee's request by SSL.
G4 PHK P Colbeck, 1 Linden Cl, Winterbourne, Bristol, BS17 1LG [IO81RM, ST68]
G4 PHL P T Green, 6 Yews Cl, Worrall, Sheffield, S30 3BB [IO93GJ, SK38]
G4 PHM D Ferguson, Glenelg, 4 Coniston Cres, Workington, CA14 3NL [IO84FP, NY02]
G4 PHP D Foster, 147 Moseley Wood Gdns, Leeds, LS16 7JF [IO93EU, SE24]
G4 PHQ T Burrows, 81 Kingsway, Huyton, Liverpool, L36 2PL [IO83NJ, SJ49]
G4 PHR T J Clough, 37 Park Ave, Mirfield, WF14 9PB [IO93DQ, SE21]
GW4 PHT D W Dalling, 308 Townhill Rd, Mayhill, Swansea, SA1 6PD [IO81AP, SS69]
G4 PHV G R Bennison, 35 Ermine St., Thundridge, Ware, SG12 0SY [IO91XU, TL31]
G4 PHW Details withheld at licensee's request by SSL.
G4 PHY D H Buxton, 135 Chiltern Ave, Putnoe, Bedford, MK41 9EL [IO92SD, TL05]
G4 PHZ B F Hill, 73 Kinsale Rd, Knowle, Bristol, BS14 9EY [IO81RK, ST66]
G4 PIA L J Roberts, 102 Edgeley Rd, Clapham, London, SW4 6HB [IO91WL, TQ27]
G4 PIB F E Reed, 70 West St., Minehead, TA24 5HR [IO81GE, SS94]
GI4 PID B J Little, 8 Ballynoe Rd, Antrim, N Ireland, N Ireland, BT41 2QT [IO64WR, J18]
G4 PIF G Tynemouth, 45 Broadmeadows, East Herrington, Sunderland, SR3 3RF [IO94GU, NZ35]
G4 PII J B Bulteel, 15 Elmdale Rd, Clifton, Tyndalls Park, Bristol, BS8 1SF [IO81QK, ST57]
G4 PIJ J C Goodman, 8 Burley Cl, Verwood, BH31 6TQ [IO90BU, SU00]
G4 PIK H Green, Filnarry, Aughton Ln, Aston, Sheffield, S31 0AN [IO93GJ, SK38]
G4 PIM J M R Greatorex, 7 Causeway, Darley Abbey, Derby, DE22 2BW [IO92GW, SK33]
G4 PIO B J Chandler, 30 Windle Ave, Hull, HU6 7EE [IO93TS, TA03]
G4 PIP C H Bottoms, Treboro House, Ullenhall, Solihull, West Midlands, B95 5NN [IO92CH, SP16]
G4 PIQ A Cook, Fishers Farm, Colchester Rd, Tendring, Clacton on Sea, CO16 9AA [JO01NU, TM12]
G4 PIR R Beckett, 22 Saville Ct, Hoyland, Barnsley, S74 0NY [IO93GL, SK39]
G4 PIT M J Clapham, 3 Hamelsham Ct, Hailsham, BN27 3EL [JO00DU, TQ50]
G4 PIW J M Costello, 73 Rosslyn Cres, Luton, LU3 2AT [IO91SV, TL02]
G4 PIX M Connor, 12 Kirkham Cl, Eden Park, Chilton, Ferryhill, DL17 0RL [IO94FP, NZ22]
G4 PIY W G Cooper, Totara, Birch Platt, West End, Woking, GU24 9NR [IO91QI, SU96]
G4 PJA Details withheld at licensee's request by SSL.
G4 PJC Details withheld at licensee's request by SSL.
G4 PJD H A Hoare, Farvardale, The St., Sheering, Bishops Stortford, CM22 7LT [JO01CT, TL51]
G4 PJE R Kershaw, 13 Silver Hill, Milnrow, Rochdale, OL16 3UJ [IO83WO, SD91]
G4 PJF Details withheld at licensee's request by SSL.
G4 PJG Details withheld at licensee's request by SSL.
G4 PJJ Dr N I B Garbutt, Tudor Cottage, Main Rd, Minsterworth, Gloucester, GL2 8JP [IO81TU, SO71]
G4 PJK R T Mosedale, 21 Druids Ave, Aldridge, Walsall, WS9 8LA [IO92BO, SK00]
G4 PJL R M Bailey, 5 Braemar Rd, Doncaster, DN2 5HN [IO93KM, SE50]
G4 PJN D G Brown, 7 Colne Ave, West Drayton, UB7 7AJ [IO91SM, TQ07]
G4 PJO V J Baker, 16 Grass Royal, Yeovil, BA21 4JN [IO80QW, ST51]
G4 PJP M F Clay, 1 Welbeck Ave, Burbage, Hinckley, LE10 2JH [IO92HM, SP49]
GM4 PJR N Yarrow, 10 Coxburn Brae, Bridge of Allan, Stirling, FK9 4PS [IO86AD, NS79]
G4 PJS P J Shields, 41 Mill Ln, Upholland, Appley Bridge, Wigan, WN6 9DD [IO83PN, SD50]
G4 PJT S Schofield, 6 Co Operative St., Wath upon Dearne, Rotherham, South Yorks, S63 6QJ [IO93HM, SE40]
G4 PJV E F Smethurst, Hartfield House, Bescaby Ln, Waltham-on-the-Wold, Melton Mowbray Leic, LE14 4AB [IO92OT,
SK82]
G4 PJW H Southern, 57 Park Est, Shavington, Crewe, CW2 5AW [IO83SB, SJ75]
G4 PJX A M Smith, 35 Drew Ave, Grimsby, DN32 0AY [IO93XN, TA20]
G4 PJY G J Taylor, 23 Welland Way, Oakham, LE15 6SL [IO92PP, SK80]
G4 PJZ J A Towle, 63 Digby Ave, Mapperley, Nottingham, NG3 6DS [IO92KX, SK64]
G4 PKB A D Trevethick, La Casa Del Jardin, La Molineta, 29788 Frigiliana, Spain
G4 PKD J Addison, 29 Whitewell Rd, Accrington, BB5 6DA [IO83TS, SD72]
G4 PKE R H Badham, 71 Yeovil Rd, Owlsmoor, Sandhurst, GU47 0TD [IO91OI, SU86]
G4 PKF E M Wood, 68 Baswich Crest, Stafford, ST17 0HJ [IO82XT, SJ92]
G4 PKG Details withheld at licensee's request by SSL.[Licence holder abroad.]
G4 PKH A J P Stanborough, 38 Hempstead Ln, Potten End, Berkhamsted, HP4 2SD [IO91RS, TL00]
GM4 PKJ D M Smith, Haremuir Bungalow, Benholm, Inverbervie, Montrose Angus, DD10 0HX [IO86TT, NO77]
G4 PKK Dr S Juden, 17 Astonville St., Southfield, London, SW18 5AN [IO91VK, TQ27]
G4 PKM J A Derrick, 16 Rufus Isaacs Rd, Caversham, Reading, Berks, RG4 6DD [IO91ML, SU77]
G4 PKO D R French, 8 Hawthorne Rd, Tranmere, Birkenhead, L42 7LA [IO83LJ, SJ38]
G4 PKP J Jones, Jason Photographic, 122 Bold St., Liverpool, L1 4JA [IO83MJ, SJ38]
G4 PKT D S Lewin, 14A Warwick New Rd, Leamington Spa, CV32 5JG [IO92FG, SP36]
G4 PKU P F Beever, 34 Welby Ln, Melton Mowbray, LE13 0TB [IO92NS, SK72]

G4 PKV D G Griffiths, 61 The Dr, North Harrow, Harrow, HA2 7EJ [IO91TN, TQ18]
G4 PKW C J A Gerard, 7 Parkwood Rd, Sidemoor, Bromsgrove, B61 8UA [IO82XI, SO97]
G4 PKX F Gallimore, 3 Wilson Cres, Lostock Gralam, Northwich, CW9 7QH [IO83SG, SJ67]
G4 PKZ R Richardson, Manor Farm, Manor Ln, Langham, Oakham, LE15 7JL [IO92OQ, SK81]
G4 PLA M W Pickering, The Bungalow, Far Baulker Farm, Oxton, Southwell Notts, NG25 0RQ [IO93MB, SK65]
G4 PLC Details withheld at licensee's request by SSL.
G4 PLD L O Tostevin, 26 Jermyn Rd, Kings Lynn, PE30 4AE [JO02FS, TF62]
G4 PLE D R Chatterton, Foxboro, Dark Ln, Rodborough, Stroud, GL5 3UF [IO81VR, SO80]
G4 PLF A B Nelmes, 1 Peghouse Rise, Slad Rd, Stroud, GL5 1RU [IO81VS, SO80]
G4 PLH R G Hughes, 17 South Rise, Carshalton Beeches, Carshalton, SM5 4PD [IO91VI, TQ26]
GM4 PLI J T Nellis, 64 Kirkwood Ave, Linnvale, Clydebank, G81 2ST [IO75TV, NS57]
G4 PLK S K Lewis, 187 Ashburton Rd, Hugglescote, Coalville, LE67 2HE [IO92HR, SK41]
G4 PLL I G Thomas, 15 Wakefield Rd, Fitzwilliam, Pontefract, WF9 5AJ [IO93HP, SE41]
G4 PLM S Lewis, 181 Kent Dr, Helensburgh, G84 9RX [IO76PA, NS38]
G4 PLS A J Haigh, Jack O Walls, Leighton Rd, Northall, Dunstable, LU6 2EZ [IO91QU, SP91]
G4 PLT P Teather, 91 Chesterton Park, Cirencester, GL7 1XS [IO91AR, SP00]
G4 PLU J Bates, 63 Sunny Blunts, Peterlee, SR8 1LP [IO94HR, NZ43]
G4 PLV M Seton, 12 Chatsworth St., Roundthorn, Oldham, OL4 5LF [IO83WM, SD90]
G4 PLW Dr P J Walker, Willow Corner, Bendish, Hitchin, SG4 8JH [IO91UV, TL12]
G4 PLX A G Salata, 64 Wildwood Rd, London, NW11 6UP [IO91VN, TQ28]
G4 PLY V M Morris, 21 Cranhill Rd, Street, BA16 0BY [IO81PC, ST43]
G4 PLZ P F Connors, 6 Robins Cl, Bramhall, Stockport, SK7 2PF [IO83WI, SJ88]
G4 PM H Axon, 33 Norman Cl, Felixstowe, IP11 9NQ [JO01QX, TM33]
G4 PMA D J Pearson, 42 Church St., Stapleford, Nottingham, NG9 8DJ [IO92IW, SK43]
G4 PMB F Thompson, 38 Hanson Park, Orchard Leigh, Northam, Bideford, EX39 3SB [IO71VA, SS42]
GI4 PMF W Calvin, The Maltings, 62 Glenstall Rd, Co Antrim, Ballymoney Ni, N Ireland, BT53 7QN [IO65QB, C92]
G4 PMG M P Green, Huntley, Chesham Rd, Wigginton, Tring, HP23 6HH [IO91QS, SP90]
GM4 PMH S P Dunn, 4 Mid St., Rosehearty, Fraserburgh, AB43 7JS [IO87WQ, NJ96]
G4 PMJ A Santos, Baytree, 17 Elm Garth, Roos, Hull, HU12 0HH [IO93XR, TA22]
G4 PMK R P Blackwell, 5 Tollgate Rd, Culham, Abingdon, OX14 4NL [IO91IP, SU59]
G4 PMM R Williams, 43 Beeston Cl, Bestwood Village, Nottingham, NG6 8XG [IO93JA, SK54]
G4 PMN S Prince, 49 Hallowes Rise, Dronfield, Sheffield, S18 6YA [IO93GH, SK37]
GI4 PMP H B McLaughlin, 1A Taylor Park, Limavady, Co Derry, N Ireland, BT49 0NT [IO65MB, C62]
G4 PMQ Details withheld at licensee's request by SSL.
G4 PMR R F Skinner, 10 Stretton Ave, Cresswell Manor Far, Stafford, ST16 1UJ [IO82WT, SJ92]
G4 PMS P M Steele, 107 Lower Shelton Rd, Marston Moretaine, Marston Moreteyne, Bedford, MK43 0LW [IO92RB, SP94]
GM4 PMT A W Ross, 6 Burnside St., Findochty, Buckie, AB56 4QW [IO87NQ, NJ46]
G4 PMV K W Grime, 124 Queens Ave, Bromley Cross, Bolton, BL7 9BP [IO83TO, SD71]
G4 PMW R L Dunn, Ray Hall House, 1 Ray Hall Ln, Great Barr, Birmingham, B43 6JE [IO92AN, SP09]
G4 PMX Details withheld at licensee's request by SSL.
G4 PMY G J Bell, Linden Lea, Crewe Rd, Wheelock Heath, Crewe, Ches, CW11 4RE [IO83TC, SJ75]
G4 PMZ P Butcher, 9 Little Platt, Park Barn, Guildford, GU2 6JU [IO91QF, SU95]
G4 PNB A J E Bathurst, 64 Oakfields, Broadacres, Guildford, GU3 3AU [IO91QG, SU95]
G4 PNC D F Hood, 2 Greenfinch Cl, Crossgates, Scarborough, YO12 4TX [IO94SF, TA08]
G4 PND A J Daniel, 10 Tamarisk Cl, Hatch Warren, Basingstoke, RG22 4UX [IO91KF, SU64]
G4 PNE R T Finch, 18 Bratch Ln, Wombourne, Wolverhampton, WV5 9AD [IO82VM, SO89]
G4 PNF Details withheld at licensee's request by SSL.[Op: Peter Fullman.]
G4 PNH G Aungiers, 17 Broadwood Dr, Fulwood, Preston, PR2 9SS [IO83PT, SD53]
G4 PNI R A Bishop, 40 Auburn Gr, Blackpool, FY1 5NJ [IO83LT, SD33]
G4 PNK T D Crosland, Park Farm House, Stevington, Bedford, MK43 7QF [IO92RD, SP95]
G4 PNL A J Coe, 112 Harborough Rd, Desborough, Kettering, NN14 2QY [IO92OK, SP88]
G4 PNM A J W Wixon, 152 Brunton Gdns, Montgomery St., Edinburgh, EH7 5ET [IO85JX, NT27]
G4 PNO Details withheld at licensee's request by SSL.
G4 PNP B Deak, 57 Arundel Rd, Pearcehaven, Peacehaven, BN10 8RP [JO00AS, TQ40]
G4 PNT A Hellewell, 41 Woodlea Gr, Armthorpe, Doncaster, DN3 2HN [IO93LM, SE60]
GW4 PNV G H Powrie, 2 Glyn Garth Ct, Menai Bridge, LL59 5PB [IO73WF, SH57]
G4 PNW Details withheld at licensee's request by SSL.
G4 PNX D G Painter, 93 Oxclose Ln, Arnold, Nottingham, NG5 6FN [IO92KX, SK54]
GD4 PNY A P Cooper, Balladuke, 11 Cronk Drean, Douglas, IM2 6AX
GW4 PNZ W G K Evans, 16 Carnglas Rd, Sketty, Swansea, SA2 9BP [IO81AO, SS69]
GW4 POA Details withheld at licensee's request by SSL.
G4 POB T S Hutchings, 9 Little Dell, Welwyn Garden City, AL8 7HZ [IO91VT, TL21]
GI4 POC R J Drain, 29 Bryansglen Park, Bangor, BT20 3RS [IO74DP, J48]
G4 POD J W Gould, 30 Weymouth Ave, Marston, Middlesbrough, TS8 9AB [IO94JM, NZ51]
G4 POE R Fergusson, 49 Selborne Rd, Leek, ST13 5PJ [IO83XC, SJ95]
G4 POF J P Hart, 1 Meadow Ct, Fordingbridge, SP6 1LW [IO90CW, SU11]
G4 POG Details withheld at licensee's request by SSL.
G4 POH J F S Titherington, 140 Porlock Ave, Stafford, ST17 0XY [IO82XS, SJ92]
G4 POI D R Lambert, 27 Northaw Rd West, Northaw, Potters Bar, EN6 4NP [IO91WQ, TL20]
G4 POK D F Topham, 38 Balfour Ct, Balfour Cres, Wolverhampton, WV6 0BH [IO82WO, SO89]
G4 POL B S Robertson, 12 Green Ln, Woodstock, OX20 1JY [IO91HU, SP41]
G4 POP D Beschizza, 20 Trelispen Park, Gorran Haven, St. Austell, PL26 6HT [IO70OF, SX04]
G4 POQ Details withheld at licensee's request by SSL.
G4 POR B J Banks, 30 Hospital Rd, Burntwood, WS7 0ED [IO92BQ, SK00]
G4 POT D P Girling, Llawnroc, 20 Fore St, Praze-An-Beeble, Camborne, TR14 0JX [IO70IE, SW63]
G4 POU P E Dyer, 36 Margate Rd, Ipswich, IP3 9DE [JO02OB, TM14]
GI4 POV I H V Stewart, 164 Ballymoney Rd, Ballymena, BT43 5BZ [IO64UU, D10]
G4 POW A Owen, 60 Brighton Ave, Elson, Gosport, PO12 4BX [IO90KT, SU50]
G4 POY R S Kent, 243 Carr Ln, Tarleton, Preston, PR4 6YB [IO83NQ, SD42]
G4 PPB E T Marshall, 75 Acacia Cres, Beech Hill, Wigan, WN6 8NJ [IO83QN, SD50]
G4 PPC B H L Lowe, 19 Wolverhampton Rd, Bloxwich, Walsall, WS3 2EZ [IO82XO, SJ90]
G4 PPD V H Dann, 13 Cedar Gr, Southall, UB1 2XD [IO91TM, TQ18]
G4 PPE M C C Bell, 55 Park Rd, Hampton Hill, Hampton, TW12 1HX [IO91TK, TQ17]
G4 PPG J O,Sullivan, 40 Sheldon Ave, Standish, Wigan, WN6 0LW [IO83QO, SD51]
G4 PPH A Betts, 41 Long Ln, Shirebrook, Mansfield, NG20 8AZ [IO93JE, SK56]
G4 PPJ S W Bone, The Cottage, Crowfield, Brackley, Northants, NN13 5TW [IO92KB, SP64]
G4 PPK C M Everley, 5 Firs Cl, Hazlemere, High Wycombe, HP15 7TF [IO91PP, SU89]
G4 PPL P H Fisher, 26R Listers Residen, Caravan Park, North Sea Ln, Humberstow Grimsby, DN36 4HH [IO93XM, TA30]
G4 PPM Details withheld at licensee's request by SSL.
G4 PPN D Chapman, 37 Yeovilton Pl, Kingston upon Thames, KT2 5GP [IO91UK, TQ17]
G4 PPP R Bailey, 7 Parkhill Rd, Chase Terr, Walsall, West Midlands, WS7 8ER [IO92AQ, SK00]
G4 PPR M W Spencer, 13 Low Edges, Chesterfield Rd South, Sheffield, S8 8LW [IO93GI, SK38]
G4 PPS Dr D W Herbert, 44 The Grove, Marton, Marton in Cleveland, Middlesbrough, TS7 8AG [IO94JM, NZ51]
GM4 PPT R M Hodge, 34 Craig View, Coylton, Ayr, KA6 6LB [IO75SK, NS41]
G4 PPU M S Roy, 17 Elgar Ave, Tolworth, Surbiton, KT5 9JH [IO91UJ, TQ16]
G4 PPV G V Jarrett, 3 Carisbrooke Rd, Strood, Rochester, ME2 3SN [JO01FJ, TQ77]
G4 PPW A Keech, 2 Mountfield Rd, Irthlingborough, Wellingborough, NN9 5SY [IO92QH, SP97]
G4 PPX R J Martin, 23 Sandford Cl, Wivenhoe, Colchester, CO7 9NP [JO01LU, TM02]
G4 PPZ J C Young, Thatch Cottage, Hines Mead Ln, Litton Cheney, Dorchester, Dorset, DT6 9AD [IO80OR, SY49]
G4 PQA Details withheld at licensee's request by SSL.
G4 PQB D J Mathers, Jordans Farm House, Main St., Willersey, Broadway, WR12 7PJ [IO92BB, SP13]
GW4 PQE E R Clayson, Bryngwyn Farm, Cwrtneydd, Llanybydder, Dyfed, SA40 9YR [IO72VC, SN44]
G4 PQG R Britton, 7 Edward Cl, Hucknall, Nottingham, NG15 6SP [IO93JA, SK54]
G4 PQI J H Raybould, 3 Lawns Down Rd, Quarry Bank, Brierley Hill, West Midlands, DY5 2EP [IO82WL, SO98]
G4 PQJ R A W Crosby, 15 Churchill Way, Heckington, Sleaford, NG34 9RQ [IO92UX, TF14]
G4 PQL L N Fennelow Wisbech Radio & Electronics Cl, 39 Clarence Rd, Wisbech, PE13 2ED [JO02CQ, TF41]
G4 PQM E R James, 59 Queensway, Euxton, Chorley, PR7 6PN [IO83QP, SD51]
G4 PQN C N H Hodges, 55 Filton Gr, Bristol, BS7 0AW [IO81RL, ST57]
G4 PQP P S Malme, 24 Clevedon House, Prince of Wales Rd, Cromer, NR27 9HR [JO02PW, TG24]
G4 PQQ J M Malme, Newhaven, Mill Ln, East Runton, Cromer, NR27 9PH [JO02PW, TG24]
G4 PQS W H Tedbury, Tyting House, Exeter Rd, Honiton, EX14 8AX [IO80JT, ST10]
G4 PQT Details withheld at licensee's request by SSL.
G4 PQU A Harwood, 125 Parsloes Ave, Dagenham, RM9 5PT [JO01BN, TQ48]
GM4 PQV T Pollock, 3 Middlefield, Whitehills, East Kilbride, Glasgow, G75 0HJ [IO75VR, NS65]
G4 PQW M P Davis, 478 Eastern Ave, Gants Hill, Ilford, IG2 6EQ [JO01AN, TQ48]
G4 PQX Details withheld at licensee's request by SSL.
G4 PQY A Williams, 7 Bower Hall Dr, Steeple Bumpstead, Haverhill, CB9 7ED [JO02FA, TL64]
G4 PQZ W E Cotton, Cottons Corner, 17 Ombersley St Wes, Droitwich Spa, Worcs, WR9 8HZ [IO82WG, SO86]
G4 PRD P R Dakin, 12 Spinney Cl, Kidderminster, DY11 6DQ [IO82UJ, SO87]
G4 PRF S I Brown, 17 Wood End Rd, Heanor, DE75 7PP [IO93HA, SK44]
G4 PRG M J W Smith, 8 Holly Dr, Waterlooville, PO7 8HT [IO90LU, SU60]
GI4 PRH D C Simpson, 31 Beech Green, Doagh, Ballyclare, BT39 0QB [IO64XR, J28]
G4 PRI R T W Walker, 161 Long Ln, Hillingdon, Uxbridge, UB10 9JN [IO91SM, TQ08]
G4 PRJ M Worsfold, 5 Turner Cl, Langney, Eastbourne, BN23 7PF [JO00DT, TQ60]
G4 PRK B D Roost, 52 Fairfield Rd, Eastwood, Leigh on Sea, SS9 5SB [JO01HN, TQ88]
G4 PRL Details withheld at licensee's request by SSL.
GM4 PRO T E ONeil, 187 Main St., Chapelhall, Airdrie, ML6 8SF [IO85AU, NS76]
GW4 PRP S F Lane, 12 Carlos St., Port Talbot, SA13 1YD [IO81CO, SS79]
G4 PRQ I A Hooper, 5 Vinings Gdns, Sandown, PO36 8DX [IO90KP, SZ58]
G4 PRR N D Nelson, 23 Tetbury Gdns, Nailsea, Bristol, BS19 2TJ [IO81PK, ST47]
G4 PRS V F Cotton Poole Ras, 45 Branksome Hill Rd, Bournemouth, BH4 9LF [IO90BR, SZ09]
G4 PRU G H F Cousins, Grayways, Barnes Green, Brinkworth, Chippenham, SN15 5AQ [IO91AN, SU08]
GW4 PRV Details withheld at licensee's request by SSL.
G4 PRX Details withheld at licensee's request by SSL.
G4 PRZ R S Cardy, 35 Eland Rd, Aston Manor, Kempton Park, Rep Sth Africa 1620
G4 PSA Details withheld at licensee's request by SSL.[Op: Miss S S Tomlinson.]
G4 PSB Details withheld at licensee's request by SSL.
G4 PSC Lord A Newton of Brauncenell, 24 Nether Croft Rd, Brimington, Chesterfield, S43 1QD [IO93HG, SK47]
G4 PSE M Grime, 10 East Park Ave, Darwen, BB3 2SQ [IO83SQ, SD62]
GM4 PSF C W Martindale, 16 Murray Ave, Saltcoats, KA21 6DA [IO75OP, NS24]
G4 PSG Details withheld at licensee's request by SSL.[Prestwood Scout Group, c/o G4KGT QTHR.]
G4 PSH T H Owen, Touchwood, The St., Sutton, Norwich, NR12 9RF [JO02SS, TG32]
G4 PSI G Smith, 8 Byron Ct, Thistleflat, Crook, DL15 9TS [IO94DR, NZ13]
G4 PSJ R Stroud, 53 Adames Rd, Fratton, Portsmouth, PO1 5QE [IO90LT, SU60]
G4 PSL T Grice, 11 Durham St., Wallsend, NE28 7RZ [IO94FX, NZ36]
G4 PSN Details withheld at licensee's request by SSL.
G4 PSO A R Little, 20 Vicarage Cl, Shillington, Hitchin, SG5 3LS [IO91TX, TL13]
G4 PSP S H Gardner, 191 Charlton Park, Midsomer Norton, Bath, BA3 4BR [IO81SG, ST65]
G4 PSR C A Sartorius, 339 Fullwell Ave, Ilford, IG5 0RR [JO01AO, TQ49]
G4 PSS Details withheld at licensee's request by SSL.
G4 PST D S Turner, 16 Tiptree Gr, Wick Meadows, Wickford, SS12 9AL [JO01GO, TQ79]
GM4 PSW Capt P S Burn, Wilan Cottage, Newton Rd, Innellan Dunoon, Argyll, PA23 7SY [IO75MV, NS17]
GM4 PSX A Dorsett, West Howar, Isle of Sanday, Orkney, KW17 2BJ [IO89QF, HY63]
G4 PTA A Deighton, Ivy Cottage, Sandhills, Thorner, Leeds, LS14 3DF [IO93GU, SE33]
G4 PTB Details withheld at licensee's request by SSL.
G4 PTC V C Sievey, 3 Stratton Rd, Castlemead, Bournemouth, BH9 3PG [IO90BS, SZ19]
G4 PTE Dr K R Lown, 14 The Ridings, Cliftonville, Margate, CT9 3EJ [JO01RJ, TR37]
G4 PTF C Keeping, 12 St. Francis Ave, Bitterne, Southampton, SO18 5QJ [IO90HW, SU41]
G4 PTJ Details withheld at licensee's request by SSL.
G4 PTK G C Mason, 120 Scalford Rd, Melton Mowbray, LE13 1JZ [IO92NS, SK71]
G4 PTN Details withheld at licensee's request by SSL.
GM4 PTQ Revd S J G Bennie, St. Peters House, 10 Springfield Rd, Stornoway, HS1 2PT [IO68TF, NB43]
GW4 PTS Details withheld at licensee's request by SSL.
G4 PTT R G Davis, 9 Old Rectory Cl, Instow, Bideford, EX39 4LY [IO71WB, SS43]
G4 PTU C Charlesworth, Bere Rd, Blandford, Winterborne Kingston, Blandford Forum, Dorset, DT11 9BA [IO80VS, SY89]
GD4 PTV B W Brough, Kimmeragh View, Ballacorey Rd, Bride, Ramsey, IM7 4AW
G4 PTW A N Brunning, 6 Newstead Rd, Barnwood, Gloucester, GL4 3TQ [IO81VU, SO81]
G4 PTX J R Andrews, 12 Corville Rd, Lapal, Halesowen, B62 9TJ [IO82XL, SO98]
G4 PUA P A Williams, 42 Pynes Ln, Bideford, EX39 3EE [IO71VA, SS42]
G4 PUB S F Wensley Basildon Dist Rd, 15 Lamden Way, Burghfield Common, Reading, RG7 3LZ [IO91LJ, SU66]
GW4 PUC R Rees, 16 Brynheulog, Penygoer, Llanelli, SA14 8AE [IO71WQ, SN50]
G4 PUD B T Langdon, 80 Glen Rise, Kings Heath, Birmingham, B13 0EJ [IO92BK, SP08]
G4 PUE W Dixon, 24 Alder Hill Gr, Leeds, LS7 2PT [IO93FT, SE23]
G4 PUF Details withheld at licensee's request by SSL.
G4 PUG R Hopkins, 2 Church Rd West, Farnborough, GU14 6RT [IO91OG, SU85]
G4 PUH C P Jones, 39 Pensby Rd, Heswall, Wirral, L60 7RA [IO83KI, SJ28]
G4 PUI J T Browell, 9 Langholm Ave, Chirton Park, North Shields, NE29 8DH [IO95GA, NZ36]
GM4 PUJ Details withheld at licensee's request by SSL.
G4 PUK W F Dunbar, Flat 2, 6 Cornwall Rd, Harrogate, HG1 2PL [IO93FX, SE25]
G4 PUM R L W Brookes, 11 Rydal Way, Alsager, Stoke on Trent, ST7 2EH [IO83UC, SJ75]
G4 PUN E G Pollard, New Inn, 77 Bath St., Ilkeston, DE7 8AJ [IO92IX, SK44]
G4 PUO W W Stumpf, 18 Saxhorn Rd, Ln End, High Wycombe, HP14 3JN [IO91OO, SU89]
G4 PUP B H Philipp, The New Haven, 2 Red Lion Park, Hooe near Battle, E Sussex, TN33 9EW [JO00EU, TQ61]
G4 PUQ P J McEwen, 7 Pound Cttgs, Pound Corner, Easton, Woodbridge, IP13 0EH [JO02QE, TM25]
GM4 PUS Dr J A Murray, Cnoc Gorm House, Half 2 Upper Breakish, Isle of Skye, IV42 8PY [IO77CF, NG62]
GW4 PUX R Cardwell, 23 Russell Ave., Colwyn Bay, Clwyd, LL29 7TR [IO83DG, SH87]
G4 PUY Details withheld at licensee's request by SSL.
G4 PUZ N H D Barrington, 78 Little Meadow, Acorn Way Great Oakle, Great Oakley, Corby, NN18 8JP [IO92PL, SP88]
G4 PVA Details withheld at licensee's request by SSL.[Station located near Wembley.]
G4 PVB R W Houlston, 12 Longacres, St. Albans, AL4 0DR [IO91US, TL10]
GM4 PVC A C F Smith, PO Box 687, Morley 6943, Western Australia
GM4 PVF F R A Rankin, 4 Everard Quadrant, Colston, Glasgow, G21 1XP [IO75VV, NS66]
GM4 PVH G H Brown, 20 City Rd, Brechin, DD9 6DW [IO86QR, NO56]
G4 PVM P J Tittensor, 47 St. Johns Rd, Chelmsford, CM2 0TY [JO01FR, TL70]
G4 PVN A Taylor, 3 Mond Cres, Billingham, TS23 1DL [IO94IO, NZ42]
G4 PVO E Cotton Droitwich ARS, Cottons Corner, 17 Ombersley St. West, Droitwich, WR9 8HZ [IO82WG, SO86]
G4 PVP P V Painter, 80 Willowsbrook Rd, Hurst Green, Halesowen, B62 9RF [IO82XL, SO98]
GM4 PVQ D A Ross, 11 Edinview Gdns, Stonehaven, AB3 2EG [IO87TH, NJ72]
G4 PVR M Nichols, 255 Lichfield Rd, Rushall, Walsall, WS4 1EB [IO92AO, SK00]
G4 PVS J Maude, Anthony Fold Farm, Bury Old Rd, Shuttleworth, Ramsbottom Lancs, BL0 0RY [IO83UP, SD81]
GW4 PVU A J Wilkinson, 1 Langley Cl, Penrhyn Bay, Llandudno, LL30 3LN [IO83CH, SH88]
G4 PVV L Watkins-Field, Sharlion, 27 Bosvigo Rd, Truro, TR1 3DG [IO70LG, SW84]
G4 PVX A Daws, 19 Ashley Ave, Lower Weston, Bath, BA1 3DS [IO81TJ, ST76]
G4 PVY R C Limb, 3 Canford Heights, Western Rd, Poole, BH13 7BE [IO90BQ, SZ08]
G4 PVZ G A Loach, 40 Park Rd West, Wolverhampton, WV1 4PL [IO82WO, SO99]
G4 PWA P G Dane, Bank House, 9 High St., Ditchling, Hassocks, BN6 8SY [IO90WW, TQ31]
G4 PWB G W Smith, 9A Lansdowne Dr, Rayleigh, SS6 9AL [JO01HO, TQ89]
G4 PWD M J McHale, 41 Sheringham Dr, Etching Hill, Rugeley, WS15 2YG [IO92AS, SK01]
G4 PWE J A Veness, 45 Berlin Rd, Hastings, TN35 5JD [JO00HU, TQ81]
G4 PWG J V Hubbard, 2 Carlton Rd, Portchester, Fareham, PO16 8JW [IO90KU, SU60]
GM4 PWH Details withheld at licensee's request by SSL.[Op: J A Fraser, 26 Clunes Ave, Caol, Fort William, Inverness-shire, PH33 7BJ.]
G4 PWI P C Heredge, 118 Oxford Cres, Didcot, OX11 7AX [IO91IO, SU58]
G4 PWJ A Donnelly, Highlands, Durton Ln, Broughton, Preston, PR3 5LE [IO83PT, SD53]
G4 PWK M Henry, 2 Trentham Walk, Liverpool, L32 4UD [IO83NL, SJ49]
G4 PWM D A East, 39 Chapel Ln, Navenby, Lincoln, LN5 0ER [IO93RC, SK95]
G4 PWP D J Blackwell, 58 Bleadon Hill, Weston Super Mare, BS24 9JW [IO81MH, ST35]
GM4 PWQ J T Foster, 185 Sea Rd, Methil, Leven, KY8 2EQ [IO86LE, NT39]
G4 PWS S M Keen, 34 Unwin Rd, Isleworth, TW7 6HX [IO91TL, TQ17]
G4 PWT D Branham, 41 Aylesbury Ct, Wilbraham Rd, Manchester, M21 0US [IO83UK, SJ89]
G4 PWV D A Topping, 4 Stonepine Cl, Wildwood, Stafford, ST17 4QS [IO82WS, SJ92]
G4 PWX J A Hopkins, 8 Princess Cl, Chase Terr, Walsall, West Midlands, WS7 8BP [IO92AQ, SK00]
GW4 PWZ W T Evans, Windy Ridge Bungalo, Mount View, Mountain Hare, Merthyr Tydfil, CF47 0UX [IO81HR, SO00]
G4 PXA Details withheld at licensee's request by SSL.[Op: D C Simmons.]
GM4 PXB M G Gale, 48 Fairview Circle, Danestone, Aberdeen, AB22 8ZQ [IO87WE, NJ91]
G4 PXC A J Lord, 16 Lark Valley Dr, Fornham St. Martin, Bury St. Edmunds, IP28 6UG [JO02IG, TL86]
G4 PXD Details withheld at licensee's request by SSL.[Station located near Great Barr, Birmingham.]
G4 PXE A G Brend, 42 West Garth Rd, Exeter, EX4 5AJ [IO80FR, SX99]
G4 PXF M Harries, 14 Princess Rd, Dronfield, Sheffield, S18 6LY [IO93GH, SK37]
GM4 PXG T Worthington, 1 Sandy Loch Dr, Lerwick, ZE1 0SR [IP90JD, HU44]
G4 PXH E A P Southwell, Flat 2 Solent Breezes, Hock Ln, Walsash, Southampton, Hants, SO31 9HG [IO90IU, SU50]
GI4 PXI Details withheld at licensee's request by SSL.
G4 PXJ J Peet, 30 Sulgrave Rd, Northampton, NN5 7BL [IO92MF, SP76]
G4 PXL Details withheld at licensee's request by SSL.
GI4 PXM W D Nelson, 111 Rutherglen St., Belfast, BT13 3LR [IO74AO, J37]
G4 PXN C Sissons, 8 Penarth Rd, Falmouth, TR11 2NY [IO70LD, SW83]
G4 PXO Details withheld at licensee's request by SSL.
GW4 PXQ G R Griffiths, Rosalea, Meredith Terr, Dolwyddelan, Gwynedd, LL25 0NQ [IO83BB, SH75]
G4 PXR T G Geldart, Langdale, Coast Rd, Bardsea, Ulverston, LA12 9QZ [IO84LD, SD37]
G4 PXS P M Freeman, 16 York Ave, Sandiacre, Nottingham, NG10 5HB [IO92IW, SK43]
G4 PXU T P S Batchelor, 34 Beech Cl, Willand, Cullompton, EX15 2SD [IO80HV, ST01]
G4 PXV J V E Brown, 67 Water St., Chase Terr, Walsall, West Midlands, WS7 8AW [IO92AQ, SK00]
G4 PXW Details withheld at licensee's request by SSL.
G4 PXX P L Styles, PO Box 270, Frankston, Victoria 3199, Australia
G4 PXY G A Bloomfield, 15 Beaulieu Dr, Pinner, HA5 1NB [IO91TN, TQ18]
G4 PXZ Details withheld at licensee's request by SSL.
G4 PYA A D Ledger, 32 St. Augustines Cres, Whitstable, CT5 2NW [JO01MI, TR16]
G4 PYC H Massey, Fintry, 7 Lees Rd, Anderton, Chorley, PR6 9PP [IO83QO, SD61]
G4 PYD C Johnson, 51 Newstead Ave, Holton-le-Clay, Grimsby, DN36 5BQ [IO93XM, TA20]

G4 PYG I G Bennett, Collins Green, School Rd, Messing, Colchester, CO5 9TH [JO01JU, TL81]
G4 PYH D P Jackson, 9 Stour Cl, Altrincham, Ches, WA14 4UE [IO83TJ, SJ78]
GM4 PYJ J W Balfour, Cressington, 34 Causewayhead Rd, Stirling, FK9 5EU [IO86AD, NS89]
GW4 PYK D C Delfosse, 25 Mill Pl, Ely, Cardiff, CF5 4AJ [IO81JL, ST17]
G4 PYM G P Allsop, 39 Gateford Rise, Worksop, S81 7DU [IO93KH, SK58]
G4 PYQ A Hill, 37 Rock St., Gee Cross, Hyde, SK14 5JX [IO83XK, SJ99]
G4 PYS E M Devereux, 15 Severn Cl, Paulsgrove, Portsmouth, PO6 4BB [IO90KU, SU60]
G4 PYU S Harding, Fosse House, Burlescombe, nr Tiverton, Devon, EX16 7JH [IO80IW, ST01]
G4 PZB J A Wilson, 7 Sutherland Way, Stamford, PE9 2TA [IO92SP, TF00]
G4 PZD W H Agnew, 3 Clifton Dr, Morecambe, LA4 6SR [IO84NB, SD46]
G4 PZE Details withheld at licensee's request by SSL.[Op: A Henry, 1 Southfield Road, Tuffley, Glos, GL4 9UG.]
G4 PZF A Jennings, 22 Tennyson Rd, Wellingborough, NN8 3NH [IO92PH, SP86]
G4 PZH T D Hibberd, 19 Palatine Gr, Heatherton, Littleover, Derby, DE23 7RR [IO92FV, SK33]
G4 PZJ C H Christopher, 15 Inman Rd, Earlsfield, London, SW18 3BB [IO91VK, TQ27]
G4 PZK Details withheld at licensee's request by SSL.
G4 PZM G E Lane, 2 School Ln, Horbury, Wakefield, WF4 5LN [IO93FP, SE21]
G4 PZQ D A Coe, 1 Hoton Rd, Wymeswold, Loughborough, LE12 6UA [IO92KT, SK62]
G4 PZR Sir K A Cradock-Hartopp, Keepers, Yeovilton, Yeovil, Somerset, BA22 8EX [IO81QA, ST52]
G4 PZS R Connell, 10 William Plows Ave, Heslington Rd, York, YO1 5BU [IO93LW, SE65]
G4 PZT H Jow, 41 Powis Rd, Ashton on Ribble, Preston, PR2 1AD [IO83PS, SD53]
G4 PZU F A Day, 27 Prince Charles Rd, Lewes, BN7 2HY [JO00AV, TQ41]
G4 PZV A G Terry, 6 Seaton Cl, Stubbington, Fareham, PO14 2PX [IO90JT, SU50]
G4 PZW R J Proctor, 4 Ganwick Cl, Haverhill, CB9 9JX [JO02FC, TL64]
G4 PZX Nr A F Tracey, Well 'N' Garden, Abberton Rd, Fingringhoe, Colchester Essex, CO5 7AS [JO01LU, TM01]
G4 PZY J M Robson, 31 Melton Ave, Littleover, Derby, DE23 7FY [IO92FV, SK33]
G4 QK J B Roscoe, 6 Park Ave, Bridgwater, TA6 7EE [IO81LD, ST23]
G4 RAA M P B Brown, 12 Mead Way, Burnham, Slough, SL1 6HD [IO91QM, SU98]
G4 RAB D V N Ellis, 98 Ct Rd, Kingswood, Bristol, BS15 2QP [IO81RK, ST67]
G4 RAC J F Cooper, 134 Jordan Ave, Stretton, Burton on Trent, DE13 0JD [IO92EU, SK22]
G4 RAD Details withheld at licensee's request by SSL.
G4 RAE R E R Laney, 54 Durban Rd, Patchway, Bristol, BS12 5HQ [IO81RM, ST68]
GW4 RAF A Ward RAF Sealand ARC, 158 Mold Rd, Alltami, Mynydd Isa, Mold, CH7 6TF [IO83KD, SJ26]
GD4 RAG J E Martin, Tradewinds, Mount Gawne Rd, Port St. Mary, IM9 5LX
GM4 RAH P Robertson, 32 Crosswood Cres, Balerno, EH14 7HS [IO85HV, NT16]
G4 RAJ J S Shaw, 31 Dartmouth Ave, Almondbury, Huddersfield, HD5 8UP [IO93DP, SE11]
G4 RAK J C Hornby, 21 West Wools, Portland, DT5 2EA [IO80SM, SY67]
G4 RAL R V Fleetwood, 13 Far Dene, Kirkburton, Huddersfield, HD8 0QZ [IO93DO, SE11]
G4 RAM R A Mackney, Noyon, East Burton Rd, Wool, Wareham, BH20 6HF [IO80VQ, SY88]
GM4 RAO J R Sangster, 6 Cranford Terr, Aberdeen, AB1 7NQ [IO87TH, NJ72]
G4 RAP V J Seaman, Bambara, 156 Stonecliff Park, Welton, Lincs, LN2 3LL [IO93SH, TF08]
G4 RAR P A Clemens, 18 Ladylea Rd, Horsley, Derby, DE21 5BN [IO92GX, SK34]
G4 RAV P R Evans, 7 Pound Cl, Harleston, IP20 9HF [JO02PJ, TM28]
G4 RAW S P Ortmayer, 14 The Cres, Hipperholme, Halifax, HX3 8NQ [IO93CR, SE12]
G4 RAY G Pickering, 22 Northfield Ave, Wells Next The Sea, NR23 1LL [JO02KW, TF94]
GM4 RAZ B A H Smith, 8 Moss Side Dr, Portlethen, Aberdeen, Aberdeenshire, AB12 4NY [IO87WB, NO99]
GW4 RBA R Beedles, 50 Chester Cl, Shotton, Deeside, CH5 1AU [IO83LF, SJ36]
G4 RBB Details withheld at licensee's request by SSL.[Op: J W Turner, 21 Croft Way, Horsham, W Sussex, RH12 2AS.]
G4 RBC C J D Hawkridge, 2 Windward Cl, Littlehampton, BN17 6QX [IO90RT, TQ00]
G4 RBD D W Batchelor, 14 Oakleigh Heath, Hallow, Worcester, WR2 6NQ [IO82UF, SO85]
G4 RBE J W Harvey, 41 Churchdown Rd, Liverpool, L14 7PE [IO83NK, SJ49]
G4 RBH T A Farmer, 12 Rose Ave, Mitcham, CR4 3JS [IO91WJ, TQ26]
G4 RBI J E Hutson, 34 York Rd, Hitchin, SG5 1XB [IO91UW, TL13]
G4 RBO Details withheld at licensee's request by SSL.
G4 RBP R B Purdy, San Gain Industrial Co. Ltd, Rm 612 Blk B Hoplite Ind Cent, 3-5 Wang Tai Rd, Kowloon Bay, Hong Kong,
G4 RBQ D R Love, Evander, Green Rd, Wivelsfield Green, Haywards Heath, RH17 7QL [IO90XX, TQ32]
G4 RBR C J Randall, 38 Kilmorey Gdns, St. Margarets, Twickenham, TW1 1PY [IO91UL, TQ17]
G4 RBS M D Adams, The Old Coach House, Salthouse Rd, Millom, Cumbria, LA18 5AD [IO84IF, SD18]
G4 RBU T F Crookes, 167 Willow Dr, Handsworth, Sheffield, S9 4AU [IO93HJ, SK48]
G4 RBV A M M Marriott Blackmore Vale A R S, 8 Melway Gdns, Child Okeford, Blandford Forum, DT11 8EP [IO80VV, ST81]
G4 RBW A D J Horder, 24 Elizabeth Cl, Thornbury, Bristol, BS12 2YN [IO81RO, ST68]
G4 RBZ C Dervin, 20 Stocker Ave, Alvaston, Derby, DE24 0QS [IO92HV, SK33]
G4 RCA Details withheld at licensee's request by SSL.
G4 RCB D M Thorp, 50 Victoria Ave, Droitwich, WR9 7DF [IO82WG, SO86]
G4 RCC A J Wright Caravan & Camping ARC, 34 Webbs Way, Stoney Stanton, Leicester, LE9 4BW [IO92IN, SP49]
G4 RCD M T Capstick, 38 Grotto Rd, Weybridge, KT13 8PN [IO91SJ, TQ06]
GW4 RCE M H Capstick, 180 Hull Rd, York, YO1 3LF [IO93LW, SE65]
G4 RCF J F ODell, 5 Further Ends Rd, Freckleton, Preston, PR4 1RL [IO83NS, SD42]
G4 RCG J Muzyka, 2 Engine Fold, Wrenthorpe, Wakefield, WF2 0PP [IO93FQ, SE32]
G4 RCH S Thompson, 44 Blakeney Gr, Leeds, LS10 3BL [IO93FS, SE33]
G4 RCI D E Protheroe, 29 Burges Rd, Norwich, NR3 2LP [JO02PP, TG21]
G4 RCJ D A Underwood, Nicklety, Inchfield Rd, Walsden, Todmorden, OL14 7QP [IO83WQ, SD92]
GI4 RCK W McMillen, 26 Maymount St., Belfast, BT6 8BH [IO74BO, J37]
G4 RCL B Walters, The Highlands, Beccles Rd, Barnby, Beccles, NR34 7QW [JO02TK, TM48]
GW4 RCM W D O Williams, 35 Mostyn Ave, Craig y Don, Llandudno, LL30 1YY [IO83CH, SH78]
GM4 RCN J W Young, 13 Craig Cres, Causewayhead, Stirling, FK9 5LR [IO86AD, NS89]
G4 RCP C J Marriott, 19 Beechey Cl, Denver, Downham Market, PE38 0DH [JO02EO, TF60]
G4 RCR P J Starley, Oak House, Birmingham Rd, Budbrooke, Warwick, CV35 7DX [IO92EG, SP26]
G4 RCS M J Robbins, 1 Wyld Ct, Allesley Park, Coventry, CV5 9LQ [IO92FK, SP28]
G4 RCT R E Williams, 8 Highgate Ave, Urmston, Manchester, M41 8GG [IO83TL, SJ79]
G4 RCU J V Shirley, 10 Peat Cl, Highland, Rugby, CV22 6SA [IO92II, SP47]
G4 RCV H A C Cooke, West View, Higher Ln, Ashton, Helston, TR13 9SB [IO70HC, SW62]
G4 RCY Details withheld at licensee's request by SSL.
G4 RCZ I B Dempster, 54 Fashoda Rd, Selly Park, Birmingham, B29 7QJ [IO92BK, SP08]
G4 RD F Briggs, 4 Mount Boone Way, Dartmouth, TQ6 9PL [IO80FI, SX85]
G4 RDA U M Harris, Preston Lodge, Kentisbury, Barnstaple, Devon, EX31 4NH [IO81AE, SS64]
GM4 RDB M P Cheasley, 39 Douglas Terr, Stirling, FK7 9LW [IO86AC, NS79]
G4 RDC J A G Gumb, 19 Castle Ln, Bolsover, Chesterfield, S44 6PS [IO93IF, SK47]
G4 RDD P Knight, 51 Fordwater Rd, Streetly, Sutton Coldfield, B74 2BG [IO92BN, SP09]
G4 RDF M J R Bradbury, 3 Salisbury Ave, East Leake, Loughborough, LE12 6NJ [IO92JT, SK52]
G4 RDG Dr G E Murray, 176 Golfwood Dr, Hamilton, Ontario, Canada, L9C 7B8
G4 RDH M J Wirthner, 51 College Rd, Beeding, Upper Beeding, Steyning, BN44 3TB [IO90UV, TQ21]
GM4 RDI J White, 1 Banknowe Rd, Tayport, DD6 9LG [IO86NK, NO42]
G4 RDJ Details withheld at licensee's request by SSL.[Op: A J Morgan.]
G4 RDL P Auty, 43 Carr Manor Rd, Leeds, LS17 5AW [IO93FU, SE23]
G4 RDM P R Rouget, 7 Palmer Cl, Wellingborough, NN8 5NX [IO92PH, SP86]
G4 RDS B W Wood, 100 Lower Rd, Coventry Hill, Hullbridge, Hockley, SS5 6DD [JO01HO, TQ89]
GW4 RDW I D Jones, 6 Norton Terr, Glyncorrwg, Port Talbot, SA13 3AN [IO81EQ, SS89]
G4 RDY L Harrison, 48 Bleasdale Ave, Anchorsholme, Staining, Blackpool, FY3 0DW [IO83MT, SD33]
G4 REB W J Thorpe, 38 Woolston Rd, Butlocks Heath, Netley Abbey, Southampton, SO31 5FQ [IO90HV, SU40]
G4 REC A C Marrows, 7 Victoria Cl, Yeadon, Leeds, LS19 7AU [IO93DU, SE24]
G4 REE P M Miller, 2 The Pavilions End, Camberley, GU15 2LD [IO91PH, SU85]
GM4 REF W McLean, 159 Castlemilk Rd, Glasgow, G44 4NA [IO75VT, NS66]
G4 REG A Boocock, 2 Vine Garth, Clifton, Brighouse, HD6 4JZ [IO93CQ, SE12]
G4 REH R H England, 5 Weir Rd, Congresbury, Bristol, BS19 5HL [IO81OI, ST46]
GW4 REI D O May, 19 Sycamore St., Pembroke Dock, SA72 6QN [IO71MQ, SM90]
G4 REK J E Tylee, 40 Luna Rd, Thornton Heath, CR7 8NY [IO91WJ, TQ36]
GM4 REN B Strathdee, 85 Weavers Knowe Cres, Currie, EH14 5PP [IO85IV, NT16]
G4 REQ D J Aveling, Ridgeway, Plough Garth, Kellington nr Goole, North Humberside, DN14 0PD [IO93KR, SE52]
G4 RES D M Walters, 25A Milton Rise, Weston Super Mare, BS22 8AS [IO81MI, ST36]
G4 RET L M Boxwell, 83 Lockington Cres, Stowmarket, IP14 1DA [JO02LE, TM05]
G4 REU J R Taylor, 33 Brick Kiln Ln, Mansfield, NG18 5LA [IO93JD, SK56]
G4 REV Details withheld at licensee's request by SSL.
GW4 REX P R Hassmann, 51 Heol Coed Cae, Whitchurch, Cardiff, CF4 1HJ [IO81JM, ST17]
G4 REY Details withheld at licensee's request by SSL.
G4 REZ T L Watkins-Field, 62 Kingsbridge Rd, Whitley, Reading, RG2 7RQ [IO91MK, SU77]
G4 RFA M I Tew, 25 Broad Oak Ln, Penwortham, Preston, PR1 0UX [IO83PR, SD52]
G4 RFC S C Fletcher, 90 Westcombe Park Rd, Blackheath, London, SE3 7QS [JO01AL, TQ37]
GI4 RFH T R Robinson, 21 Carnhill Rd, Carnmoney, Newtownabbey, BT36 6LA [IO74AQ, J38]
G4 RFI R Linden, 24 Hartland Dr, Edgware, HA8 8RH [IO91UP, TQ19]
G4 RFJ I G McCann, Rosslyn Ave, Preesall, Blackpool Lancs, FY6 0HE [IO83MW, SD34]
GD4 RFK M E Dodd, Ellan Geay, Ballayockey Ln, Regaby, Ramsey, IM7 3HP
G4 RFL F Tennant, 3 The Orchard Gr, Shurdington, Cheltenham, GL51 5TN [IO81WU, SO91]
G4 RFM T J McLoughlin, 229 Barking Rd, London, E6 1LB [JO01AM, TQ48]
G4 RFN A L Robey, 54 Jarrett Ave, Wainscott, Rochester, ME2 4NL [JO01GJ, TQ77]
G4 RFO B D Wood, 11 Oakdale Ave, Wibsey, Bradford, BD6 1RP [IO93CS, SE13]
G4 RFP A L Goodall, 10 Beacon Cl, Everton, Lymington, SO41 0LQ [IO90ER, SZ29]
G4 RFQ Details withheld at licensee's request by SSL.[Station located near Northolt.]
G4 RFR M J Owen Flt Refueling AR, 3 Canford View Dr, Wimborne, BH21 2UW [IO90AT, SU00]
G4 RFS S A Eade, The Annexe, 2A St. Johns Rd, Farnham, Surrey, GU9 8NT [IO91OE, SU84]
G4 RFU D L Abbott, 21 Leckhampton Rd, Cheltenham, GL53 0AZ [IO81XV, SO92]
G4 RFV B N C Adams, 38 Waterloo Rd, Darbys Corner, Poole, BH17 7LF [IO90AR, SZ09]
GD4 RFW E I Kelly, Shenvalla, Victoria Rd, Castletown, Isle of Man, IM9 1EE
G4 RFX Details withheld at licensee's request by SSL.
G4 RGA J M Dunnett, 43 Oakfield Park, Wellington, TA21 8EX [IO80JX, ST11][Op: Jim Dunnett. Tel: (01823) 664911.]
G4 RGB M A R K Rogers, The Old Water Mill, Kingsland, Herefordshire, HR6 9SW [IO82OG, SO46]
GM4 RGC G P Sangster Robert Gordons, College Arc, Physics Department, Schoolhill Aberdee, AB9 1FR [IO87TH, NJ72]
G4 RGE N P Lovely, Dolphin Cottage, Upper Green Rd, St. Helens, Ryde, PO33 1XE [IO90KQ, SZ68]
G4 RGF P McCall, 11 Elworthy Dr, Wellington, TA21 9AT [IO80JX, ST11]
G4 RGH D J McLaughlin, 150 Moor Ln South, Ravenfield, Rotherham, South Yorks, S65 4QR [IO93IK, SK49]
GW4 RGI W R Baker, 51 Angle Village, Angle, Pembroke, SA71 5AX [IO71KQ, SM80]
G4 RGJ J M Gillard, 23 Sapphire Cres, St. Johns, Worcester, WR2 5PT [IO82VE, SO85]
G4 RGK Details withheld at licensee's request by SSL.[Op: D A Dibley, 6 Oaktree Road, Marlow, Bucks, SL7 3EE.]
GW4 RGL I Purnell, 26 The Cres, Cwmbran, NP44 7JG [IO81LP, ST29]
G4 RGM Details withheld at licensee's request by SSL.
G4 RGN D M Gibson, Marlow Westwell Ln, Ashford, Kent, TN26 1JA [JO01KE, TQ94]
G4 RGO J C Crocker, 8 Oakwood Ave, Littlepark Est, Bedhampton, Havant, PO9 3RA [IO90LU, SU60]
G4 RGP E P Hall, 93 Sthbourne Coast, Bournemouth, Dorset, BH6 4DX [IO90CR, SZ19]
GD4 RGR K R Grattan, 41 Carrick Park, Sulby, Ramsey, IM7 2EY
GM4 RGS R G Smith, 8 Moss Side Dr, Portlethen, Aberdeen, Aberdeenshire, AB12 4NY [IO87WB, NO99]
GW4 RGT G J Padfield, 32 Forest Hill, The Bryn, Pontllanfraith, Blackwood, NP2 2PN [IO81JP, ST19]
GM4 RGU D B Nicolson, South Cottage, Feddinch, St. Andrews, Fife, KY16 8NR [IO86OH, NO41]
GI4 RGV J O Ruff, P O Box 159, Horley, Surrey, RH6 0FW [IO91VD, TQ24]
G4 RHA R H Anderson, 4 Elmwood Cl, Stokesley, Middlesbrough, TS9 5HX [IO94JL, NZ50]
G4 RHB M J Bailey, 97 New Mills Rd, Birch Vale, High Peak, SK22 1BX [IO93AJ, SK08]
G4 RHC M Kellett, 16A The Lyons, Hetton-le-Hole, Easington Ln, Houghton-le-Spring, DH5 0HT [IO94GT, NZ34]
G4 RHD Details withheld at licensee's request by SSL.
G4 RHF J M Coggan, 31 Belcourt Rd, Brecks, Rotherham, S65 3JF [IO93IK, SK49]
G4 RHI H J Spanswick, 14 Millbrook Dale, Axminster, EX13 5EF [IO80MS, SY39]
G4 RHJ P J Vickers, 5 Cropmark Way, Hatchwarren, Basingstoke, RG22 4TA [IO91KF, SU64]
G4 RHK L S Woodcock, Poolhay Cl, Linkend Rd, Corse Lawn, Gloucester, Glos, GL19 4NY [IO81VX, SO83]
G4 RHL R H Langdon, Stonesdale, Pittington Rd, Rainton Gate, Houghton-le-Spring, DH5 9RG [IO94FT, NZ34]
G4 RHM R D Kingston, 32 Ford End, Woodford Green, IG8 0EG [JO01AO, TQ49]
G4 RHN Details withheld at licensee's request by SSL.
G4 RHP L G Page, 45 Poole Ln, Kinson, Bournemouth, BH11 9DY [IO90BS, SZ09]
G4 RHQ H G Tait, 16 Pentland Cl, Spring Cottage Est, Hull, HU8 9LN [IO93US, TA13]
G4 RHR K J Backhouse, 113 Bucklesham Rd, Kirton, Ipswich, IP10 0PF [JO02PA, TM24]
G4 RHS S L Bell, Redstone Mensions, 20 Carlisle Rd, Eastbourne, BN20 7EN [JO00DS, TV69]
G4 RHT G D Brooks, 29 Queen St., Cubbington, Leamington Spa, CV32 7NB [IO92FH, SP36]
G4 RHX A J Moore, 1 St. Andrews Rd, New Marske, Redcar, TS11 8AU [IO94LN, NZ62]
G4 RHY M A Pratt, 3 Nene Gr, Auckley, Doncaster, DN9 3JJ [IO93LM, SE60]
G4 RHZ B Coupe, 9 School Ln, Auckley, Doncaster, DN9 3JR [IO93LL, SE60]
G4 RIA K Lydall, 28 Netherwood Cl, Fixby, Huddersfield, HD2 2LR [IO93CQ, SE11]
GW4 RIB D Stoole, Fairview Cottage, Upper House Farm, Henllys, Cwmbran, Gwent, NP44 6HZ [IO81LP, ST29]
G4 RIC Details withheld at licensee's request by SSL.
G4 RIE D A Littler, 16 Lee Bank, Westhoughton, Bolton, BL5 3HQ [IO83RN, SD60]
G4 RIG W H Cunliffe, 63 Rupert St., Radcliffe, Manchester, M26 1BE [IO83UN, SD70]
GW4 RII G Williams, 12 Oliver Jones Cres, Tredegar, NP2 3BJ [IO81JS, SO10]
G4 RIK R I Kirkwood, 43 Clover Rd, Flitwick, Bedford, MK45 1PH [IO92RA, TL03]
G4 RIM A C Day, Wood Ct, Blachford Rd, Ivybridge, PL21 0AD [IO80AJ, SX65]
GM4 RIN N Cram, 3 Erskine Rd, Giffnock, Glasgow, G46 6TQ [IO75US, NS55]
G4 RIO S F Williams, 89 Marlborough Rd, Castle Bromwich, Birmingham, B36 0EL [IO92CM, SP18]
G4 RIP J Creasey, 4 Low Farm Dr, Folkingham, Sleaford, NG34 0SP [IO92TV, TF03]
G4 RIQ L F Rushforth, 90 Brearley Ave, New Whittington, Chesterfield, S43 2DZ [IO93HG, SK37]
G4 RIR Details withheld at licensee's request by SSL.
G4 RIS B F Didmom, 45 Millstrood Rd, Whitstable, CT5 1QF [JO01MI, TR16]
G4 RIT J Tansley, 79A Edmund Rd, Hastings, TN35 5LE [JO00HU, TQ81]
G4 RIU R W Jones, 67 Plover Rd, Larkfield, Aylesford, ME20 6LA [JO01FH, TQ65]
GM4 RIV A E Gaston Wigtownshire AR, Ellena, Lochans Mill, Lochans, Stranraer, Dumfries and Galloway, DG9 9BA [IO74LU, NX05]
GM4 RIW R B Donaldson, 33 Broomwell Gdns, Monikie, Broughty Ferry, Dundee, DD5 3QP [IO86OM, NO43]
GW4 RIX D G Jenkins, School House, Meifod, Powys, SY22 6DE [IO82IR, SJ11]
G4 RIZ P O Robertson, Kloiberweg 12, 82541 Ammerland, West Germany
G4 RJ B Farleigh, 22 Hillhead Park, Brixham, TQ5 0HG [IO80FI, SX95]
G4 RJA I Wilkinson, 8 Weir Bank, Stapenhill Bank, Burton on Trent, DE15 9RB [IO92ES, SK22]
G4 RJC R J Coleman, 31 Kingfisher Rd, Upminster, RM14 1ER [JO01DN, TQ58]
G4 RJD K T Ward, 3 Levetts Hollow, Cannock, WS12 5AW [IO92AR, SK01]
GM4 RJE J C Harsant, 24 Balfour St., Edinburgh, EH6 5EP [IO85JX, NT27]
GM4 RJF J M Weatherer, 20 Gilloch Cres, Dumfries, DG1 4DW [IO85FB, NX97]
G4 RJG I P Toon, 18 Barrowfield Rd, Farmhill, Stroud, GL5 4DF [IO81VS, SO80]
G4 RJH C W Harratt, 5 Coniston Ave, Congleton, CW12 4LY [IO83VD, SJ86]
G4 RJK Capt J A Kay, Treetops, Gentles Ln, Headley, Hants, GU35 8NH [IO91OC, SU83]
G4 RJM G J Heward, 54 Foley Rd East, Streetly, Sutton Coldfield, B74 3JD [IO92BO, SP09]
G4 RJO B C Robertson, 28 Heath Ln, Blackfordby, Swadlincote, DE11 8AA [IO92FS, SK31]
G4 RJT C J Chown, 5 Aldwell Cl, Wootton, Northampton, NN4 6AX [IO92NF, SP75]
G4 RJU A K Gregory, 2 White Horse Cl, Worcester, WR2 4EB [IO82VE, SO85]
GW4 RJW L Willis, 36 Tyn y Celyn, Glan Conwy, Colwyn Bay, LL28 5NN [IO83CG, SH87]
GM4 RJX J W Hatton, 64 Abercromby Cres, Helensburgh, G84 9DN [IO76PA, NS38]
G4 RJY C J Sidney, 10 Colville Cl, Bampton, OX18 2NN [IO91FR, SP30]
G4 RKB J N Conlon, 24 Goldcrest Cl, Longridge Park, Colchester, CO4 3FN [JO01LV, TM02]
GI4 RKC S E Jennings, 34 Palmer Ave, Belsize Rd, Lisburn, BT28 3QB [IO64XM, J26]
G4 RKD T E Clarke, 19 Ratby Ln, Markfield, LE67 9RJ [IO92IQ, SK40]
G4 RKE C A Toomer, 11 Ottervale Cl, Upottery, Rawridge, Honiton, EX14 9TA [IO80KU, ST20]
G4 RKF B A C Peart, 44 St. Annes Rd, Headington, Oxford, OX3 8NL [IO91JS, SP50]
G4 RKG J Whiting, 19 Watermore Cl, Cotterel, Frampton Cotterell, Bristol, BS17 2NQ [IO81SM, ST68]
GM4 RKH Details withheld at licensee's request by SSL.
GW4 RKI K R Perryman, 93 St. Pauls Rd, Aberavon, Port Talbot, SA12 6PH [IO81CO, SS78]
G4 RKJ Details withheld at licensee's request by SSL.
G4 RKK I Welford, The Cottage, Hall Ln, Morley St. Botolph, Wymondham, NR18 9TB [JO02MN, TM09]
G4 RKL W V Welford, The Cottage, Hall Ln, Morley St. Botolph, Wymondham, NR18 9TB [JO02MN, TM09]
GM4 RKM T G Cassidy, Ariadne, 36 Dunrobin Rd, Raithview, Kirkcaldy, KY2 5YT [IO86JC, NT29]
G4 RKN P J Wells, 24 Common Cl, West Winch, Kings Lynn, PE33 0LB [JO02ER, TF61]
G4 RKO B D Cooper, 1 Strouds Meadow, Cold Ash, Thatcham, RG18 9PQ [IO91IK, SU56]
G4 RKP R J Groom, Tryst, Rackhams Corner, Corton, Lowestoft, NR32 5LB [JO02UM, TM59]
G4 RKQ B R Mason, 10 Junction Rd, Kingsley, Northampton, NN2 7JQ [IO92NG, SP76]
G4 RKR D R Geddes, 61 Great Holme Cour, Thorplands, Northampton, NN3 1YE [IO92NG, SP76]
G4 RKU B J Bamber, 27 Cowley Ln, Gnosall, Stafford, ST20 0DS [IO82US, SJ82]
G4 RKV L E Adams, 2 Reculver Ln, Herne Bay, CT6 6SP [JO01OI, TR26]
G4 RKX G C Cook, 22 Northlands Park, Bishopston, Swansea, SA3 3JW [IO71XO, SS58]
GW4 RKZ R A Cleverley, 33 Tylchawen Cres, Tonyrefail, Porth, Mid Glam, CF39 8AL [IO81GN, ST08]
G4 RL Dr L D Philp, The Rookery, 15 Scotby Village, Scotby, Carlisle, CA4 8BS [IO84NV, NY45]
G4 RLA C Butcher, 7 Lascelles Hall Rd, Kirkheaton, Huddersfield, HD5 0AT [IO93DP, SE11]
G4 RLC T O Isom, 64 Cuffling Dr, Braunstone Frith, Leicester, LE3 6NF [IO92JP, SK50]
G4 RLD H W Hayes, 48 The Turnpike, Marple, Stockport, SK6 6HG [IO83XJ, SJ98]
G4 RLF M E Wright, 24 Wessex Rd, Wilton, Salisbury, SP2 0LW [IO91BB, SU03]
G4 RLI P M Edis, 339 Bentley Ln, Walsall, WS2 8TT [IO82XO, SO99]
G4 RLJ W E Edis, 2 Lynmouth Cl, Aldridge, Walsall, WS9 0JR [IO92AO, SK00]
G4 RLK R G King, 3 Cope Park, Almondbury, Almondsbury, Bristol, BS12 4EZ [IO81RN, ST68]
G4 RLL J Woods, 21 Doddington Rd, Wilby, Wellingborough, NN8 2UA [IO92PH, SP86]
G4 RLM J P K Hiscock, 62 East Borough, Wimborne, BH21 1PL [IO90AT, SU00]
G4 RLN D C Rosevear, 20 Bay View Park, St. Austell, PL25 3TR [IO70OI, SX05]
G4 RLO G J L Seal, 4 Salcombe Dr, London Rd, Shrewsbury, SY2 6SH [IO82PQ, SJ51]
GW4 RLP T H Varney, 29 Ffordd Eryri, Hendre, Caernarvon, LL55 2UR [IO73UD, SH46]
G4 RLQ J W Crocker, 8 School Ln, Fulford, York, YO1 4LS [IO93LW, SE64]
G4 RLR J R Hogan, 17 High St., Croxton, St. Neots, Huntingdon, PE19 4SX [IO92VF, TL25]
G4 RLS J Elsdon, 15 Union Rd, Lowestoft, NR32 2BZ [JO02UL, TM59]
G4 RLT R P Langford, 43 Oldminster Rd, Sharpness, Berkeley, GL13 9US [IO81SR, SO60]
G4 RLU P B Quickfall, Quickfall, 14 Lade Fort Cres, Lydd on Sea, Romney Marsh, TN29 9YF [JO00LW, TR02]
GM4 RLV W K R Duguid, Villach, 7 Hawthorn Pl, Ballater, AB35 5QH [IO87LB, NO39]
G4 RLX H G Cave, 3 Grace Meadow, Whitfield, Dover, CT16 3HA [JO01PD, TR34]

G4	RLZ	D J Sleep, 9 Hanson Rd, Liskeard, PL14 3NT [IO70SK, SX26]
GI4	RMA	L McCullough, Down Lodge, Strangford, Co Down, N Ireland, N Ireland, BT30 7LY [IO74FI, J54]
G4	RMC	D R Marsden, 67 Fourth Ave, Garston, Watford, WD2 6QH [IO91TQ, TQ19]
G4	RMD	J A Cobley, 4 Briars Cl, Hatfield, AL10 8DQ [IO91VS, TL20]
GW4	RMI	Details withheld at licensee's request by SSL.
G4	RMJ	A L Cattani, Rozel, Marsh Rd, Gedney Drove End, Spalding, PE12 9PJ [JO02CU, TF42]
G4	RMK	Details withheld at licensee's request by SSL.[Station located near Newport Pagnell.]
GW4	RML	D G Davies, 101 Westlands, Port Talbot, SA12 7DE [IO81CO, SS79]
G4	RMM	Details withheld at licensee's request by SSL.[Op: D Ashley, 43a Fore St, Eastcote, Pinner, Middx, HA4 2JF.]
G4	RMN	M F Hogan, 16 Freshfield Cl, West Earlham, Norwich, NR5 8RA [JO02OP, TG10]
G4	RMS	G W Dover Rushey Mead School AR Club, Rushey Mead School Ar Club, Melton Rd, Leicester, Leics, LE4 7PA [IO92KP, SK60]
G4	RMT	P R Johnson, 8 Culzean Gdns, Lowestoft, NR32 4UE [JO02UL, TM59]
G4	RMV	M E J Buckle, No 3.Tilesford Pk Rd, Throckmorton, nr Pershore, Worcs, WR10 2LA [IO82XD, SO95]
G4	RMX	J F Phelps, 33 Thirlmere Dr, North Anston, Anston, Sheffield, S31 7JP [IO93GJ, SK38]
G4	RNA	P J Dronfield, White Lodge Farm, High Bradfield, Sheffield, S6 6LG [IO93EK, SK29]
G4	RNB	Dr B H W Hill, 49 Woodstock Rd, London, E17 4BH [JO01AO, TQ39]
GI4	RNC	C R Blezard, Newlands, Monkton Deverill, Warminster, Wilts, BA12 7EX [IO81VD, ST83]
G4	RND	C Hawkins, 5 Springfield Ave, Accrington, BB5 0EZ [IO83TR, SD72]
G4	RNE	J C E Bailey, 53 Moor Ln, Ainsdale, Southport, PR8 3NY [IO83LO, SD31]
G4	RNF	J A Handley, Marquis The Bungalo, Kirkham Rd, Freckleton, Preston Lancs, PR4 1HY [IO83NS, SD43]
G4	RNI	Details withheld at licensee's request by SSL.
G4	RNJ	K J Whiffin, 42 Canute Rd, Birchington, CT7 9QH [JO01PJ, TR26]
G4	RNK	R J Dodson, Trevilwood, Five Lanes, Dobwalls, Liskeard, Cornwall, PL14 6JB [IO70RK, SX26]
G4	RNL	Details withheld at licensee's request by SSL.[Correspondence via PO Box 141, Warrington, Cheshire.]
G4	RNN	J H Wilcox, 51 Marlow Bottom, Marlow, SL7 3LZ [IO91OO, SU88]
G4	RNO	P A F Callow, Glaziers Forge Cottage, Dallington, Heathfield, E Sussex, TN21 9JJ [JO00EX, TQ62]
GI4	RNP	V S McFarland, 10 Railway St., Derriaghy, Belfast, BT17 9EU [IO64XN, J26]
G4	RNR	J L Maunder, 56 Conery Ln, Enderby, Leicester, LE9 5AB [IO92JO, SP59]
G4	RNS	C J G Peel obo North Staffs Raynet Group, The Ferns, Parkwood Dr, Baldwins Gate, Newcastle, ST5 5EU [IO82UX, SJ74]
G4	RNT	D Thorpe, 10 Stoke Rd, Taunton, TA1 3EJ [IO81KA, ST22]
G4	RNU	G H Harris, 1 Balfour Cres, Bracknell, RG12 7JA [IO91PJ, SU86]
G4	RNV	V G Long, 177 Fulford Rd, York, YO1 4HH [IO93LW, SE64]
G4	RNW	M S Stewart, 29 Elstree Rd, Bushey, Watford, WD2 3QU [IO91TP, TQ19]
GM4	RNX	A W Walker, Old Smithy Cottage, Boreland, Lockerbie, Dumfriesshire, DG11 2LL [IO85IE, NY19]
G4	RNZ	K R Page, 51 Bournville Rd, Weston Super Mare, BS23 3RR [IO81MH, ST35]
G4	ROA	A J Chamberlain, 16 Okehampton Rd, Styvechale, Coventry, CV3 5AU [IO92GJ, SP37]
G4	ROB	R J Taylor, 18 Spruce Ave, Selston, Nottingham, NG16 6DX [IO93IB, SK45]
G4	ROC	L R ODell, 110 Salisbury Rd, Grays, RM17 6DQ [JO01EL, TQ67]
G4	ROE	Details withheld at licensee's request by SSL.
G4	ROG	R C Trelease, Gwelo, 15 Springfield Park, Barripper, Camborne, TR14 0QZ [IO70IE, SW63]
G4	ROH	W H Smith, 10 Playford Rd, Rushmere, Ipswich, IP4 5RH [JO02OB, TM24]
G4	ROI	S Kiernan, 60 Riverview Rd, Ewell, Epsom, KT19 0LB [IO91UI, TQ26]
G4	ROJ	R W Stafford, Brackens, Market Ln, Blundeston, Lowestoft, NR32 5AN [JO02UM, TM59]
G4	ROK	G Sambrook, 73 Hayes Dr, Barnton, Northwich, CW8 4JX [IO83RG, SJ67]
G4	ROM	M M Ellis, The Squirrels, Tower Rd, Hindhead, GU26 6SN [IO91PC, SU83]
G4	RON	Details withheld at licensee's request by SSL.
G4	ROO	Details withheld at licensee's request by SSL.
G4	ROP	C A White, 18 Ashton Gdns, Old Tupton, Chesterfield, S42 6JF [IO93GE, SK36]
G4	ROR	D J Harrison, 1 Winnipeg Cl, Lower Wick, Worcester, WR2 4XT [IO82VE, SO85]
G4	ROS	F E Sweetingham, 38 Pippin Green Ave, Kirkhamgate, Wakefield, West Yorks, WF2 0RU [IO93FQ, SE22]
G4	ROU	W A Maudsley, 42 Crawford St., Clock Face, St. Helens, WA9 4XH [IO83PK, SJ59]
G4	ROX	A W J Capel, 33 Romney Ave, Lockleaze, Bristol, BS7 9ST [IO81RL, ST67]
G4	ROY	Details withheld at licensee's request by SSL.
G4	ROZ	Details withheld at licensee's request by SSL.[Op: P Burgin. Station located near Littlehampton.]
G4	RPA	D C Court, 4 Rucrofts Cl, Aldwick Felds, Bognor Regis, PO21 3SL [IO90PS, SZ99]
G4	RPB	G H Beacall, Pendelyn, Station Rd, Brompton on Swale, Richmond, DL10 7HN [IO94EJ, SE29]
G4	RPC	R N L Cassling, 14 Canada Way, Lower Wick, Worcester, WR2 4DJ [IO82VE, SO85]
G4	RPD	A Else, Fen View, Tattershall Bridge Rd, Tattershall Bridge, Lincoln, LN4 4JW [IO93VC, TF16]
GM4	RPE	J McCabe, 109 Weirwood Ave, Baillieston, Glasgow, G69 6LQ [IO75WU, NS66]
G4	RPF	P J Osborne, 78 Mountbatten Rd, Braintree, CM7 9TP [JO01GV, TL72]
G4	RPG	C P Bourne, 8 Glebe Ave, Broughton, Kettering, NN14 1NE [IO92OI, SP87]
G4	RPI	C Parrish, 92 West End Rd, Wyberton, Boston, PE21 7LS [IO92XW, TF34]
G4	RPJ	I J Flaherty, 10 Highfield Park, Heaton Mersey, Stockport, SK4 3HD [IO83VJ, SJ89]
G4	RPK	J R Kaine, 74 Camden Mews, London, NW1 9BX [IO91WN, TQ28]
G4	RPL	D I Ingham, Auchengray, 51 Helena St., Mexborough, S64 9PF [IO93IL, SE40]
G4	RPM	J Andrew, 39 Straker Ave, Ellesmere Port, South Wirral, L65 3BD [IO83MG, SJ37]
GM4	RPO	T K Gemmell, 30 Goldie Cres, Nithside, Dumfries, DG2 0AJ [IO85EB, NX97]
G4	RPP	Details withheld at licensee's request by SSL.
G4	RPR	T Taylor, 35 Ramillies Rd, Sidcup, DA15 9JA [JO01BK, TQ47]
G4	RPT	M J E Edis, 28 High St., Broughton, Kettering, NN14 1NG [IO92OI, SP87]
G4	RPV	K Baker, 48 Heycott Gr, Walkers Heath, Birmingham, B38 0BQ [IO92BJ, SP07]
G4	RPW	F N Charnley, 30 Dunkirk Ave, Fulwood, Preston, PR2 3RY [IO83PS, SD53]
G4	RPX	P Conway, 77 Ennerdale Dr, Congleton, CW12 4FJ [IO83VD, SJ86]
G4	RQA	T L Mills, 14 Oxford Cl, Hapton, Padiham, Burnley, BB12 7DB [IO83UT, SD73]
G4	RQC	C E Lewis, 93 Woodgrove Rd, Burnley, BB11 3EJ [IO83VS, SD83]
G4	RQD	Details withheld at licensee's request by SSL.
G4	RQF	R F Langer, 6 Oak Green Dr, Wales, Sheffield, S Yorks, S31 8NA [IO93GJ, SK38]
G4	RQG	S Baggaley, 35 Hayner Gr, Weston Coyney, Stoke on Trent, ST3 6PQ [IO82WX, SJ94]
G4	RQH	Dr T J Cole, 100 Barton Rd, Cambridge, CB3 9LH [JO02BE, TL45]
G4	RQI	D P Warr, 5 Ashton Rd, Castleford, WF10 5AU [IO93HR, SE42]
G4	RQJ	R J Hannan, 87 Plymouth St., Walney Island, Walney, Barrow in Furness, LA14 3AN [IO84IC, SD16]
G4	RQK	A T Johnson, South Lodge, 38 South St., Bourne, PE10 9LY [IO92TS, TF01]
G4	RQL	M A I Wilson, The Old Chapel, Poulshot Rd, Poulshot, Devizes, SN10 1RW [IO81XI, ST96]
G4	RQN	Details withheld at licensee's request by SSL.
G4	RQO	J P E C Pulford, 68 York Ave, Droitwich, WR9 7DQ [IO82WG, SO86]
G4	RQP	G R Hallett, 9 Dolcroft Rd, Rookley, Ventnor, PO38 3NT [IO90IP, SZ58]
GW4	RQQ	T Jones, 19 Penlon, Cae Tros Lon, Menai Bridge, LL59 5LR [IO73VF, SH57]
G4	RQR	B J Radford, 9 New Rd, Bolehill, Matlock, DE4 4GL [IO93FC, SK25]
GW4	RQS	J Leighton, 12 Morley Ave, Connahs Quay, Deeside, CH5 4RE [IO83LF, SJ26]
G4	RQU	D M Young, 43 Bates Ln, Weston Turville, Aylesbury, HP22 5SN [IO91OS, SP81]
GM4	RQW	A W McEwen, 3 Spencer House, St. Pauls Sq, Carlisle, CA1 1AE [IO84MV, NY45]
G4	RQX	C F Riley, 80 Fylde Rd, Southport, PR9 9XL [IO83MQ, SD31]
G4	RQY	Details withheld at licensee's request by SSL.
G4	RQZ	A J West, 12 Alicia Ave, Garlinge, Margate, CT9 5JY [JO01QJ, TR36]
G4	RRA	P Pasquet, 64 Bricksbury Hill, Farnham, GU9 0LY [IO91OF, SU84]
G4	RRC	Details withheld at licensee's request by SSL.
G4	RRD	I H Reynolds, The Heathers, 4 Chappel Hill, Fakenham, NR21 9HW [JO02KU, TF93]
G4	RRI	S C Jarvis, Larcombe House, Post Office Ln, South Chard, Chard, TA20 2RR [IO80MU, ST30]
G4	RRJ	C I Pemberton, 28 Ashgate Ln, Wincham, Northwich, CW9 6PN [IO83SG, SJ67]
G4	RRK	F I Timmins, 63 Park Ave, Longlevens, Gloucester, GL2 0EA [IO81VV, SO82]
G4	RRM	Dr B G F Roe Rolls Royce Mtr, 7 Abbey Fields, Crewe, CW2 8HJ [IO83SB, SJ65]
G4	RRN	C H Harrold, Boundary Farm, Cromer Rd, Felbrigg, Norwich, NR11 8PD [JO02PV, TG23]
GM4	RRP	R E Morris, Wyvis View, Strathpeffer, Ross-shire, Scotland, IV14 9DJ [IO77RO, NH45]
G4	RRQ	I H Wright, Gwel Pennow, 5 Parc An Gate, Mousehole, Penzance, TR19 6TT [IO70FC, SW42]
GW4	RRR	T J L Bunce, 2 The Gdns, Birch Rd, Ellesmere, SY12 9AA [IO82NV, SJ43]
G4	RRT	A J Wood, Glentworth, Cawdor Hill, Ross on Wye, HR9 7DL [IO81RW, SO62]
G4	RRU	R Crooks, 6 Whylands Ave, Worthing, BN13 3HG [IO90TU, TQ10]
G4	RRW	K J Taylor, 7 Bentinck Rd, Fairfield, Stockton on Tees, TS19 7PU [IO94HN, NZ41]
G4	RRX	R J S Saxton, 7 Huxley Rd, Lakenham, Norwich, NR1 2JR [JO02PO, TG20]
G4	RRZ	Details withheld at licensee's request by SSL.
G4	RS	K G King Royal Signals Amateur Radio So, 18 Alexandria Ct, Glenmoor Rd, West Parley, Ferndown, BH22 8PW [IO90BS, SZ09][Station located near Catterick Garrison, N Yorks.]
G4	RSB	R R Evans Bolsover A.R.S, Mansfield, Mooracre Ln, Bolsover, Chesterfield, Derbyshire, S44 6ER [IO93IF, SK47]
G4	RSD	D P Bones, Flint Cottage, Ipswich Rd, Charsfield, Woodbridge, IP13 7PP [JO02PD, TM25]
G4	RSE	K J Hendry South Essex Amateur Radio Soci, 48 Downer Rd North, Benfleet, SS7 3EG [JO01GN, TQ78]
G4	RSF	M M Booth, 45 Park Ave, Thackley, Bradford, BD10 0RJ [IO93DU, SE13]
G4	RSG	R Booth, 45 Park Ave, Thackley, Bradford, BD10 0RJ [IO93DU, SE13]
G4	RSH	Details withheld at licensee's request by SSL.
GI4	RSI	K G A Allen, 25 Knockgreenan Ave, Omagh, BT79 0EB [IO64IO, H47]
GM4	RSJ	J Dixon, 45 Berelands Rd, Prestwick, KA9 2JT [IO75QM, NS32]
G4	RSK	G A E Ford, 25 Margaret Rd, Penwortham, Preston, PR1 9QT [IO83PR, SD52]
G4	RSM	R Sanders, 126 Hertford Rd, Edmonton, London, N9 7HL [IO91XP, TQ39]
G4	RSN	R A H Burman, Kinnagh, Hancocks Mount, Sunningdale, Ascot, SL5 9PQ [IO91QJ, SU96]
G4	RSO	D N Smith, Poyle Cottage, 12 Poyle Rd, Guildford, GU1 3SJ [IO91RF, TQ04]
G4	RSP	D H Sandy, The Chestnuts, Dumbs Ln, Hainford, Norwich, NR10 3BH [JO02PR, TG21]
G4	RSQ	K J Stevenson, Marchfield House, Taynton, Gloucester, GL19 3AN [IO81TV, SO72]
G4	RSR	Details withheld at licensee's request by SSL.
G4	RSS	Details withheld at licensee's request by SSL.
G4	RST	D R Martin, 111 Arkwright Rd, Irchester, Wellingborough, NN29 7EE [IO92QG, SP96]
G4	RSU	P T Winnett, 54 Littleton Rd, Ashford, Middx, TW15 1UQ [IO91SK, TQ07]
G4	RSW	A C Bairstow, 63 Barnes Rd, Stafford, ST17 9RL [IO82WS, SJ92]
G4	RSX	M C Dean, 117 Waltham Way, Chingford, London, E4 8HD [IO91XO, TQ39]
G4	RSY	Details withheld at licensee's request by SSL.
G4	RSZ	K G Smith, 17 Lovelace Ave, Southend on Sea, SS1 2QU [JO01IM, TQ88]
G4	RTA	K W Mellor, 2 Clune St., Clowne, Chesterfield, S43 4NJ [IO93JG, SK57]
G4	RTB	Details withheld at licensee's request by SSL.
G4	RTC	X Iona, 13 Vicars Cl, Enfield, EN1 3DW [IO91XP, TQ39]
G4	RTD	Y Taylor, Silver Birches, Effingham Rd, Burstow, Horley, RH6 9RP [IO91WD, TQ34]
G4	RTF	D W R French, 3 Morgan Ct, Rolle Rd, Exmouth, EX8 2AD [IO80HO, SY08]
G4	RTG	G Campbell, Clementine, 49 Firs Rd, Hellesdon, Norwich, NR6 6UP [JO02PP, TG21]
G4	RTH	R T Hamstead, Tall Oaks, 6 New Rd, North Walsham, NR28 9DF [JO02QT, TG23]
G4	RTI	E J Handy, 80 Watwood Rd, Shirley, Solihull, B90 2HY [IO92BJ, SP17]
G4	RTJ	C Howe, Fatfield, 106 Fatfield Park, Washington, NE38 8BP [IO94FV, NZ35]
G4	RTK	Details withheld at licensee's request by SSL.
GM4	RTN	T J Morton, 15 Craig Cres, Causewayhead, Stirling, FK9 5LR [IO86AD, NS89]
G4	RTO	G A Calkin, Av Des Rhododendrons 4, B-1950 Kraainem, Belgium, X X
G4	RTP	Dr A G Shattock, Scurlocks Leap, Manor Kilbridge, Blessington, Co Wicklow Ireland
G4	RTQ	I R Whitehead, 3 Botany Cl, Thatcham, RG19 4GJ [IO91JJ, SU56]
G4	RTR	I D Lusty, Roselands, Selsley Hill, Stroud, Glos, GL5 5LL [IO81VR, SO80]
G4	RTS	W M Bateson, 10 Priestfield Ave, Colne, BB8 9QJ [IO83VU, SD84]
G4	RTU	P Davidson, 108 Greville Rd, Warwick, CV34 5PL [IO92FH, SP26]
G4	RTV	C E Tucker, 4 Kelsey Park Rd, Beckenham, BR3 6LJ [IO91XJ, TQ36]
G4	RTW	G L Rolf, 35 Hunnyhill, Newport, PO30 5HJ [IO90IQ, SZ48]
G4	RTX	G K Kingdon, 9 Buckingham Mews, Shoreham By Sea, BN43 6AJ [IO90UU, TQ20]
G4	RTY	R M Hayward, Chester Lodge Hotel, 7 Beachfield Rd, Sandown, PO36 8NA [IO90KP, SZ58]
G4	RUA	R C F Medcalf, 21 Greenbank, Falmouth, Cornwall, TR11 2SW [IO70LD, SW83]
G4	RUB	F C Mountain, 11 Cliff Way, Sandown, PO36 8PR [IO90KP, SZ58]
G4	RUC	A Parker, 41 Greenways, Northwood, Cowes, PO31 8AN [IO90IR, SZ49]
G4	RUE	I Worsdale, 10 Manton Rd, Stores Park, Lincoln, LN2 2JL [IO93RF, SK97]
G4	RUF	Details withheld at licensee's request by SSL.
G4	RUI	A B Keeble, 9 Horsley Ave, Shiremoor, Newcastle upon Tyne, NE27 0UF [IO95FA, NZ37]
G4	RUJ	P Evans, 706 St. Johns Rd, Clacton on Sea, CO16 8BN [JO01NT, TM11]
GU4	RUK	A B Jefferys, 2 Les Douze Maisons, Collings Rd, St. Peter Port, Guernsey, GY1 1FQ
G4	RUL	A P Turner, 42 Brassey Ave, Hampden Park, Eastbourne, BN22 9QG [JO00DT, TQ60]
G4	RUN	M A Beesley, 60 Ainsbury Rd, Canley Gdns, Coventry, CV5 6BB [IO92FJ, SP37]
GM4	RUP	Rev J Campbell, 3 Herries Rd, Glasgow, G41 4DE [IO75UU, NS56]
G4	RUR	Details withheld at licensee's request by SSL.
G4	RUS	J L Childs, 38 Carlton Rd, Wilbarston, Market Harborough, LE16 8QD [IO92OL, SP88]
G4	RUV	Details withheld at licensee's request by SSL.
G4	RUW	R F Daniel, 4 Gloucester Rd, Newbury, RG14 5JP [IO91IJ, SU46]
GW4	RUX	A Jones, 13 Glynhafod St., Cwmaman, Aberdare, CF44 6LD [IO81GP, SS99]
GW4	RUY	G P Jones, 71 Cleviston Park, Llangennech, Llanelli, SA14 9UP [IO71WQ, SN50]
G4	RUZ	A P Athawes, Holly Croft, Great Coxwell, Faringdon, SN7 7NG [IO91EP, SU29]
GW4	RVA	T A Nicholas, 15 Maes Llewelyn, Carmarthen, SA31 1JJ [IO71UU, SN41]
G4	RVC	I Henderson, Edgefield, High St., Medstead, Alton, GU34 5LN [IO91LD, SU63]
G4	RVE	C R Andrews, 29 Dell Dr, Angmering, Littlehampton, BN16 4HE [IO90ST, TQ00]
GI4	RVF	Details withheld at licensee's request by SSL.
G4	RVG	I T S Binding, 40 Parklands, South Molton, EX36 4EW [IO81BA, SS72]
G4	RVH	D M M Hird, 27 Red Beck Park, Wath Brow, Cleator Moor, CA25 5EU [IO84FM, NY01]
G4	RVJ	D Jones, 6 Priory Cl, Pilton, Barnstaple, EX31 1QX [IO71XC, SS53]
G4	RVK	D J Bentley, 9 Tinkers Castle Rd, Seisdon, Seisdon, Wolverhampton, WV5 7HF [IO82UN, SO89]
G4	RVL	D A Gentle, 1 Sunny Hill, Milford, Belper, DE56 0QR [IO93GA, SK34]
G4	RVM	Details withheld at licensee's request by SSL.
G4	RVN	A J Evans, Wyngates, 1 Calder Dr, Mossley Hill, Liverpool, L18 3HX [IO83NJ, SJ48]
G4	RVO	T G Collinson, 8 Brownberrie Dr, Horsforth, Leeds, LS18 5PP [IO93EU, SE23]
G4	RVP	S M O'Donnell, Men A Vaur, Wall Rd Gwinear, Hayle, Cornwall, TR27 5HA [IO70HE, SW53]
GD4	RVQ	J Wornham, 64 Seafield Cl, Onchan, Douglas, IM3 3BU
G4	RVR	J E S Marshall, 16 Gr Rd, Whittington Moor, Chesterfield, S41 8LN [IO93GG, SK37]
G4	RVS	A W Rodgers, 278 Norton Ln, Norton, Sheffield, S8 8HE [IO93GH, SK38]
GI4	RVT	R N Jenkins, 11 Willowvale Cres, Ballystruder Rd, Islandmagee, Larne, BT40 3SQ [IO74DS, J49]
G4	RVU	P J Wigley, 27 Garth Cres, Alvaston, Derby, DE24 0GX [IO92GV, SK33]
G4	RVV	M W Stoneham, Hafnia, 139 Hever Ave, West Kingsdown, Sevenoaks, TN15 6DT [JO01DI, TQ56]
G4	RVW	P D Southwart, 2 Alan Moss Rd, Loughborough, LE11 5LX [IO92JS, SK52]
G4	RVX	E Benson, 43 Oak Rd, Hale, Altrincham, WA15 9JA [IO83TJ, SJ78]
G4	RVY	G Richardson, 14 Falmouth Cl, Dalton-le-Dale, Seaham, SR7 8HE [IO94HT, NZ44]
G4	RVZ	G Smith, Dormie House, 61 Cable Rd, Hoylake, Wirral, L47 2AZ [IO83JJ, SJ28]
G4	RW	R A Wilson, 1 Newry Ave, Felixstowe, IP11 7SA [JO01QX, TM23]
G4	RWA	N D Van Stigt, 93 Park Rd, Teddington, TW11 0AW [IO91UK, TQ17]
G4	RWB	G R Gurden, Shrublands, Greystones Park, Sundridge, Sevenoaks, TN14 6EB [JO01BG, TQ45]
G4	RWC	Details withheld at licensee's request by SSL.
G4	RWD	K J Cheetham, Callingwood Hall, Tatenhill, Burton-on-Trent, Staffs, DE13 9SH [IO92DT, SK12]
GM4	RWE	D W Brown, Willow Crook, Turin, Aberlemno By, Forfar, DD8 2UZ [IO86OP, NO55]
G4	RWF	Details withheld at licensee's request by SSL.
G4	RWG	R W Guest, 67 Hanbury Rd, Dorridge, Solihull, B93 8DN [IO92CJ, SP17]
G4	RWJ	G F Pollard, 69 Fountains Ave, Harrogate, HG1 4ER [IO94FA, SE35]
G4	RWK	W J M Tolman, Pulland Cottage, West Down, Ilfracombe, Devon, EX34 8NH [IO71WD, SS54]
G4	RWL	J S Walters, 94 Goldcroft Rd, Weymouth, DT4 0EB [IO80SO, SY67]
G4	RWM	P J Titherington, 5 Hayland Green, Hailsham, BN27 1SR [JO00DU, TQ51]
G4	RWN	F J Rowan, 1 Massey Walk, Peel Hall, Wythenshawe, Manchester, M22 5JY [IO83VI, SJ88]
G4	RWO	Details withheld at licensee's request by SSL.[Op: D J York. Station located near Barrow-in-Furness.]
G4	RWP	E R Perry, 29 Pick Hill, Upshire, Waltham Abbey, EN9 3LB [JO01AQ, TL30]
G4	RWQ	B E Wilkes, 3 Alsop Crest, Acton Trussell, Stafford, ST17 0SJ [IO82XS, SJ91]
GW4	RWR	J R Thomas, Ystrad Isa, Denbigh, Clwyd, LL16 4RL [IO83HE, SJ06]
G4	RWS	S F Valentine, 65 Holland St., Astley Bridge, Bolton, BL1 8PA [IO83SO, SD71]
G4	RWU	Details withheld at licensee's request by SSL.
G4	RWV	P R Paling, 15 Longfellow Rd, Banbury, OX16 9LB [IO92HB, SP43]
G4	RWW	P E Glaisher, The Firs, 279 Addiscombe Rd, Croydon, CR0 7HY [IO91XI, TQ36]
G4	RWY	A S Jones, 81 Barston Rd, Oldbury, Warley, B68 0PU [IO82XL, SO98]
G4	RXB	K F Hawkings, 2 Balfour Gr, Biddulph, Stoke on Trent, ST8 7SZ [IO83WC, SJ85]
G4	RXD	R Gasken, 3 Hamaraverin, Glendale, Isle of Skye, Scotland, IV55 8WL [IO77TL, NH54]
G4	RXE	R S Squire, 10 Baddesley Rd, Chandlers Ford, Eastleigh, SO53 5NG [IO90HX, SU42]
G4	RXF	G Bence, 10 Valley Rd, Mangotsfield, Bristol, BS17 3HN [IO81SL, ST67]
G4	RXG	A M Mumford, Jacaranda, 22 Island Rd, Upstreet, Canterbury, CT3 4DA [JO01OH, TR26]
G4	RXH	H H Fowler, 63 Grain Rd, Wigmore, Gillingham, ME8 0ND [JO01HI, TQ86]
G4	RXI	J P Harris, 20 Thornwick Ave, Kingston Rd, Willerby, Hull, HU10 6LP [IO93SS, TA02]
G4	RXK	P McMullan, 6 Deepdale Rd, Blackpool, FY4 4UD [IO83MT, SD33]
GI4	RXM	T R Stitt, 51 Lakeland Rd, Hillsborough, BT26 6PW [IO64XK, J25]
G4	RXN	Details withheld at licensee's request by SSL.
GW4	RXO	P R G Alexander, 21 Church Rd, Llansamlet, Swansea, SA7 9RH [IO81BP, SS69]
G4	RXQ	R L Buck, 2 Talbot Cttgs, Birtley, Chester-le-Street, DH3 1AR [IO94FV, NZ25]
G4	RXR	R W Raine, 47 Buckingham Rd, Peterlee, SR8 2DT [IO94HS, NZ44]
GI4	RXS	R J G Burnside, 67 Earlswood Rd, Belfast, BT4 3EB [IO74BO, J37]
G4	RXT	Details withheld at licensee's request by SSL.
G4	RXU	Details withheld at licensee's request by SSL.
GM4	RXW	N Webster, Meric, 7 Woodmuir Cres, Newport on Tay, DD6 8HL [IO86MK, NO42]
GI4	RXX	S Tweedie, 38 Carwood Dr, Newtownabbey, BT36 5LP [IO74AQ, J38]
G4	RXZ	J Rigby, Lacock, Perran Downs, Goldsithney, Penzance, TR20 9HJ [IO70GC, SW53]
G4	RYB	J H Baldwin, 31 Beech Rd, Branston, Lincoln, LN4 1PG [IO93SE, TF06]
GI4	RYD	P A Bird, 44 Churchill Rd, Larne, BT40 2EN [IO74BU, D30]
G4	RYE	D J Cocker, 34 Beechfield, New Farnley, Leeds, LS12 5QS [IO93ES, SE23]
G4	RYG	J K Pendleton, 12 Hempshaw Ave, Woodmansterne, Banstead, SM7 3PG [IO91WH, TQ25]
G4	RYH	F B Appleby, 10 Buckingham Orchard, Chudleigh Knighton, Newton Abbot, Devon, TQ13 0EW [IO80EO, SX87]
G4	RYI	Details withheld at licensee's request by SSL.
GW4	RYJ	A E Salisbury, 133 Bollingbroke Heights, Flint, Clwyd, CH6 5AW [IO83KF, SJ27]
GW4	RYK	A J Richards, Castell Forwyn, Abermule, Montgomery, Powys, SY15 6JH [IO82JN, SO19]
GI4	RYL	M M McCallan, 16 Abbey Cres, Newtownabbey, BT37 9PD [IO74BP, J38]
G4	RYM	I N Spalding, Marraway Barn, Briery Lands, Snitterfield, Stratford upon Avon, CV37 0PP [IO92DG, SP26]
GI4	RYN	Details withheld at licensee's request by SSL.

G4 RYO P D Allan, 7 Homelands Pl, Trebble Park, Kingsbridge, TQ7 1QU [IO80CG, SX74]
GI4 RYP J A Ferguson, Drumbee-More, Armagh, N Ireland, BT60 1HP [IO64QI, H94]
G4 RYQ K A Edwards, 10 Bala Dr, Rogerstone, Newport, NP1 9HN [IO81LO, ST28]
G4 RYS N Black, 23 Moor Allerton, Ave, Leeds, L17 6SG [IO83MI, SJ38]
G4 RYT J W Pickup, 274 Mauldeth Rd Wes, Chorlton Cum Hardy, Manchester, M21 2RJ [IO83UK, SJ89]
G4 RYU Details withheld at licensee's request by SSL.
G4 RYV D A Rumbold, 15 Lodge Gr, Yateley, Camberley, Surrey, GU17 7AD [IO91OH, SU85]
GW4 RYW Dr A H Jones, Ty Gwyn, New Dixton Rd, Monmouth, NP5 3PL [IO81PT, SO51]
G4 RYY C Leyden, 55 Cross Green, Otley, LS21 1HE [IO93DV, SE24]
G4 RZC L C F Ingerslev, 7 Worchester Ln, Princeton Junction, Nj 08550-1509, USA, X X
G4 RZD L A Bradley, 138 Templeton Rd, Birmingham, B44 9BY [IO92BN, SP09]
G4 RZE A W J Taylor, 51 Caird St., Chepstow, NP6 5DX [IO81PP, ST59]
G4 RZF G May, 4 The Mulberrys, Wootton Bassett, Swindon, SN4 8BB [IO91BM, SU08]
G4 RZG Details withheld at licensee's request by SSL.
G4 RZH L M Winters, 43 Manor Cl, Harpole, Northampton, NN7 4BX [IO92MF, SP66]
G4 RZI M J Nagle, 21 The Brambles, Crowthorne, RG45 6EF [IO91OI, SU86]
GM4 RZJ J A Peddie, 70 Duncan St., Thurso, KW14 7HS [IO88FO, ND16]
G4 RZK Details withheld at licensee's request by SSL.
G4 RZN R J Leeds, Sunholme, The Green, Aldborough, Norwich, NR11 7AA [JO02OU, TG13]
G4 RZQ K Russell, Courtiles, Main Rd, Rookley, Ventnor, PO38 3NH [IO90IP, SZ58]
G4 RZR R W F Tooth, 5 Laurelside Walk, Dunstable, LU5 4PJ [IO91RV, TL02]
G4 RZS Details withheld at licensee's request by SSL.
GW4 RZU J W Redfern, 24 St. Brides View, Roch, Haverfordwest, SA62 6AZ [IO71KU, SM82]
GM4 RZW D F Taylor, 27 St. Clair Terr, Edinburgh, EH10 5PS [IO85JW, NT27]
G4 RZY L F Baker, 62 Ct Farm Rd, Whitchurch, Bristol, BS14 0EG [IO81RJ, ST66]
G4 RZZ I Griffin, 15 Hesselyn Dr, Rainham, RM13 7EJ [JO01CM, TQ58]
G4 SAA J Edson, Casita, Limes Ave, Nether Langwith, Mansfield, NG20 9EU [IO93JF, SK57]
G4 SAB Details withheld at licensee's request by SSL.
G4 SAC S A Collings, 16 Gleneagles Dr, Waterlooville, PO7 8RX [IO90LV, SU61]
GW4 SAE D Williams, Beechwood, 66 Rowans Ln, Bryncethin, Bridgend, CF32 9LQ [IO81FN, SS98]
G4 SAF Details withheld at licensee's request by SSL.
GW4 SAG Details withheld at licensee's request by SSL.
G4 SAI A T Harrison, 41 Kingswood Ave, Carlton Colville, Lowestoft, NR33 8BZ [JO02UK, TM59]
G4 SAJ C A Green, 76 Dibleys, Blewbury, Didcot, OX11 9PU [IO91JN, SU58]
G4 SAL G Gordon, 1 Surrey St., Hetton-le-Hole, Houghton-le-Spring, DH5 9LX [IO94GT, NZ34]
GI4 SAM S M C Noble, 19 New Line, Dundonald, Belfast, BT16 0UU [IO74CN, J47]
G4 SAN R H Ashton, 6 Lincoln Cl, Eastbourne, BN20 7TZ [JO00DS, TV59]
G4 SAP H J Tribe, 3 Old Garden Cl, Locks Heath, Southampton, SO31 6RN [IO90IU, SU50]
G4 SAQ R E Tribe, 3 Old Garden Cl, Locks Heath, Southampton, SO31 6RN [IO90IU, SU50]
G4 SAR E J Banks, Burnaby Withers Ln, High Legh, Knutsford, Ches, WA16 0SF [IO83SI, SJ68]
G4 SAS R C Jones, 37 Keer Ct, Bordesley Village, Birmingham, B9 4PQ [IO92BL, SP08]
G4 SAU B L Cuff, 51 Valley Rd, Dover, CT17 0QW [JO01PD, TR24]
G4 SAV F A Hepworth, 5 Snydale Ave, Normanton, WF6 1SS [IO93HQ, SE32]
G4 SAW M Arbon, 106 The Tideway, Rochester, ME1 2NN [JO01GI, TQ76]
G4 SAX D Corns, 24 Blue Hill Cres, Wortley, Leeds, LS12 4PB [IO93ES, SE23]
G4 SAZ Details withheld at licensee's request by SSL.
GI4 SBA K R Branagh, 17 Rathmoyle Park West, Carrickfergus, BT38 7NG [IO74CR, J48]
GW4 SBB Dr C W Fay, Mayfield, Sway Rd, Brockenhurst, SO42 7RX [IO90FT, SU20]
G4 SBD G J Bax, 8 Hockeredge Gdns, Westgate on Sea, CT8 8AN [JO01QJ, TR36]
G4 SBE K D Bowden, Warren Cottage, 14 Pool Hey Ln, Scarisbrick, Southport, PR8 5HS [IO83MP, SD31]
G4 SBF P W Fry, 54 Studley Ave, Holbury, Southampton, SO45 2PP [IO90HT, SU40]
G4 SBG H N Crossland, 9 Newlaithes Cres, Normanton, WF6 1SX [IO93HQ, SE32]
G4 SBI L Wilkie, 46 Main St., Cranswick, Driffield, YO25 9QY [IO93SW, TA05]
G4 SBM R H Harding, 12 Keswick Ave, Loughborough, LE11 3RL [IO92JS, SK51]
G4 SBN J R Ayers, 3 Sovereign Way, Ryde, PO33 3DL [IO90JR, SZ59]
GM4 SBP G T Allan, 342 Aitken Rd, Glenrothes, KY7 6SQ [IO86KE, NO20]
G4 SBQ M Rushton, 14 Acorn Cl, Mayfield, Leyland, Preston, PR5 2AF [IO83PQ, SD52]
G4 SBS R W Phillips, 4 Cumberland Dr, Off Bitterscote Ln, Fazeley, Tamworth, B78 3YA [IO92DO, SK20]
G4 SBT Details withheld at licensee's request by SSL.
G4 SBU B H C Gundry, 40 Haig Rd, Aldershot, GU12 4PS [IO91PF, SU85]
G4 SBV L R Portnoy, 95 Cavendish Rd, Salford, M7 4NB [IO83UM, SD80]
G4 SBW T M Carberry, 10 Honeymeade Cl, Stanton upon Hine Heath, Stanton, Bury St. Edmunds, IP31 2EF [JO02KH, TL97]
G4 SBX P L Shuffell, 15 Silver St., Bridgwater, TA6 3EG [IO81LD, ST23]
G4 SBZ S Darwood, Dalewood, Wembley Rd, North Somercotes, Louth Lincs, LN11 7NP [JO03BK, TF49]
G4 SCA Details withheld at licensee's request by SSL.
G4 SCB M J Sargent, 8 Sevenoaks Dr, Littledown Park, Bournemouth, BH7 7JG [IO90CR, SZ19]
G4 SCD C C Webber, 3 Orchard Rd, St, BA16 0BT [IO81PD, ST43]
G4 SCE P T Whitehead, c/o Church End, Weston Colville, Cambridge, CB1 5PE [JO02ED, TL65]
G4 SCG C W Curson, 3 Cranmer Rd, Edgware, HA8 8UA [IO91UO, TQ19]
G4 SCH K C V OConnor, 7 Hillcrest Cl, Wembury, Plymouth, PL9 0HA [IO70XH, SX54]
G4 SCI G F Carter, 40 Havisham Cl, Brichwood, Birchwood, Warrington, WA3 7NB [IO83RK, SJ69]
G4 SCJ D Meakins, 19 Booth Ln North, Boothville, Northampton, NN3 6JQ [IO92NG, SP76]
GW4 SCK R Hancock, 6 Alexandra Terr, Abernant, Aberdare, CF44 0RG [IO81GR, SO00]
G4 SCL M W Starkey, Cutlers Forth Farm, Radley Rd, Halam, Newark, NG22 8AP [IO93LB, SK65]
G4 SCM J F Claxton, 109 Downs Rd, South Wonston, Winchester, SO21 3EH [IO91IC, SU43]
G4 SCO N E Drury, 3 Northam Cl, Marshside, Southport, PR9 9GA [IO83MQ, SD32]
G4 SCQ P G Haworth, 85 Shorrock Ln, Blackburn, BB2 4PS [IO83RR, SD62]
G4 SCR A R Berg, 14 Grenville Ct, Kent Ave, London, W13 8BQ [IO91UM, TQ18]
G4 SCU I B Greenwood, 6 St. James Rd, Long Sutton, Spalding, PE12 9AZ [JO02BS, TF42]
G4 SCV I G Gammon, The Haven, Craddock Farm Cott, Craddock, Devon, EX15 3LH [IO80IV, ST01]
G4 SCX Details withheld at licensee's request by SSL.
G4 SCZ J R E Aylmer-Kelly, Willow Cottage, 137A Whitley, Melksham, Wilts, SN12 8QZ [IO81VJ, ST86]
G4 SDE J M Stuart, Churchgates, 62 Church St., Mexborough, S64 0ER [IO93IL, SK49]
G4 SDF B Thomson, 7 Bloomstiles, Salthouse, Holt, NR25 7XJ [JO02NW, TG04]
G4 SDG Details withheld at licensee's request by SSL.
G4 SDH R G Catterall-Annal, 33 Hill Mount, Dukinfield, SK16 5HT [IO83XL, SJ99]
G4 SDI L M Footring, 26 Ernest Rd, Wivenhoe, Colchester, CO7 9LG [JO01LU, TM02]
G4 SDJ R T G Freeman, Highfield House, Osmington Mills, Weymouth, Dorset, DT3 6HA [IO80TP, SY78]
G4 SDL B D Dorricott, 6 Knowsley Ave, Urmston, Manchester, M41 7BT [IO83TK, SJ79]
G4 SDM Details withheld at licensee's request by SSL.
G4 SDN Details withheld at licensee's request by SSL.
GW4 SDO D M Phillips, Tri Thy, Coed Talon, Mr Mold, Clwyd, CH7 4TU [IO83LC, SJ25]
G4 SDP Details withheld at licensee's request by SSL.
GM4 SDQ J Barclay, 23 Main St., Dalmellington, Ayr, KA6 7QL [IO75TH, NS40]
G4 SDR M J Jackson, 21 Walsh Ave, Hengrove, Bristol, BS14 9DF [IO81RK, ST66]
GW4 SDT S G Lansown, 177 Pilton Vale, Malpas, Newport, NP9 6LL [IO81MO, ST39]
G4 SDU P K Smart, 6 Nobold Cl, Newtown, Baschurch, Shrewsbury, SY4 2EH [IO82NS, SJ42]
G4 SDV J Blackadder, 6 Denmark Rd, Exeter, EX1 1SL [IO80FR, SX99]
G4 SDW G A Evans, 13 Hardens Cl, Chippenham, SN15 3AA [IO81WK, ST97]
G4 SDX G M Townend, 9 Warren Park Cl, Crow Nest Park, Hove Edge, Brighouse, HD6 2RU [IO93CR, SE12]
G4 SDY N J Green, 143 Netherhampton Rd, Salisbury, SP2 8NB [IO91CB, SU12]
G4 SDZ M J Gayler, 39 Holmefield Av Ws, Leicester Forest Es, Leicester, LE3 3FF [IO92JO, SK50]
G4 SEA Details withheld at licensee's request by SSL.[Station located near Solihull.]
G4 SEB R C Beaumont, 1 New Rd, Sutton, Norwich, NR12 9RB [JO02SS, TG32]
G4 SEC G B Hayes, 9 Malpas Rd, Rudheath, Northwich, CW9 7AY [IO83SG, SJ67]
G4 SEE R V Evison, 29 Princess St., Immingham, Grimsby, DN40 1LH [IO93VO, TA11]
G4 SEF R J Jenkinson, 1 Winslow Dr, Immingham, Grimsby, DN40 2AY [IO93VO, TA11]
G4 SEG A Clayton, 448 Gisburn Rd, Blacko, Nelson, BB9 6LZ [IO83VU, SD84]
GM4 SEH B Sherriff, 18 North Loch Rd, Forfar, DD8 3LS [IO86NP, NO45]
G4 SEJ B E Vane, Little Croft, 5 Green Leas, Chestfield, Whitstable, CT5 3JY [JO01MI, TR16]
G4 SEK K G Aylwin, 9 Hockeredge Gdns, Westgate on Sea, CT8 8AN [JO01QJ, TR36]
G4 SEL G J Wilkes, 49 Charlemont Rd, Walsall, WS5 3NQ [IO92AN, SP09]
G4 SEM W G Winteridge, 16 Bullar Rd, Bitterne Park, Southampton, SO18 1GS [IO90HW, SU41]
G4 SEN N D Whitham, 29 Grosvenor Rd, Congleton, CW12 4PG [IO83VE, SJ86]
G4 SEO B Weathered, 11 Churchbank, Stalybridge, SK15 2QJ [IO83XL, SJ99]
G4 SEP C B Turner, Saxavord, Humberston Rd, Tetney, Grimsby, DN36 5NJ [IO93XL, TA30]
G4 SEQ D Vickers, 48 Bromley Rd, Hanging Heaton, Batley, WF17 6EH [IO93EQ, SE22]
G4 SET C N Hall, Irene Rd, Stoke D Abernon, Stoke Dabernon, Cobham, Surrey, KT11 2SR [IO91TH, TQ16]
G4 SEU J Russell, 60 Berrington Rd, Nuneaton, CV10 0LB [IO92FM, SP39]
G4 SEV N D Le Gresley, 9 Belmont Cl, Wickford, SS12 0HR [JO01GO, TQ79]
GM4 SEW K F Weir, 10 Mannan Dr, Clackmannan, FK10 4ST [IO86DC, NS99]
G4 SEZ C Greenland, 20 Barn Cl, Corsham, SN13 9XB [IO81VK, ST87]
GM4 SFA A Keenan, Darwin, Coalhall By Ayr, Scotland, KA6 6ND [IO75SK, NS41]
G4 SFB D J Knowler, Little Fawsley, Fawsley Park, Daventry, Northants, NN11 6BU [IO92KF, SP55]
G4 SFC R Farrington, 59 Olive Gr, Forest Town, Mansfield, NG19 0AR [IO93KD, SK56]
G4 SFD D L Birks, Flat, 1 Pewsham House, Pewsham, Chippenham, SN15 3RX [IO81XK, ST97]
GI4 SFE J T McCullough, 12 Bramble Grange, Newtownabbey, BT37 0XH [IO74BQ, J38]
G4 SFG P R OConnor, 12 Culmore Rd, Halesowen, B62 9HP [IO82XL, SO98]
G4 SFH N M R Richardson, 22 Bramshott Dr, Hook, RG27 9EY [IO91MG, SU75]
G4 SFJ S Stott, 20 Lingfield Cres, Beech Hill, Wigan, WN6 8QA [IO83QN, SD50]
G4 SFK Details withheld at licensee's request by SSL.
G4 SFL L Stockdale, Ashton Cove, 15 Hilperton Rd, Trowbridge, BA14 7JL [IO81VH, ST85]
G4 SFM Details withheld at licensee's request by SSL.
G4 SFN J H Tetlow, 14 Fountains Cres, Hebburn, NE31 2HT [IO94FX, NZ36]
G4 SFO N W Chiverton, Suttons View, Castle Ln, Woolscott, Rugby, CV23 8DE [IO92IH, SP46]
G4 SFP J F Nash, 259 Weald Dr, Furnace Green, Crawley, RH10 6PN [IO91VC, TQ23]
G4 SFS P Grosjean, Garden House, West Horrington, Wells, Somerset, BA5 3ED [IO81QF, ST54]
GM4 SFT D McAlonan, Glenlonan House, Cromlech Rd, Sandbank, Dunoon, Argyllshire, X X
G4 SFU D R Mirams, 5 Shaftesbury Ave, Cheadle Hulme, Cheadle, SK8 7DB [IO83VI, SJ88]
GI4 SFV A Miller, 21 Marmont Dr, Belfast, BT4 2GT [IO74BO, J37]
GM4 SFW J C Stuart, Balavil, Conon Bridge, Dingwall E, Ross-shire, IV7 8AJ [IO77SN, NH55]
GI4 SFX Details withheld at licensee's request by SSL.
G4 SFY R E Baker, 25 Princes St., North Walsham, NR28 0HX [JO02QT, TG23]
GI4 SFZ C G Hought, 28 Ballymoney Rd, Ballymena, BT43 5BY [IO64UU, D10]
G4 SGA G M Barnes, 3 Blandford Ave, Castle Bromwich, Birmingham, B36 9HX [IO92CM, SP18]
GM4 SGB Details withheld at licensee's request by SSL.[Station located near Wishaw.]
G4 SGC E Buck, Dobroyd Castle, Pexwood Rd, Todmorden, West Yorks, OL14 7JJ [IO83WR, SD92]
G4 SGD S Simpson, 4 Balmoral Rd, Mountsorrel, Loughborough, LE12 7EN [IO92KR, SK51]
G4 SGE B C Hughes, 86 Lewis Ave, Willenhall Rd North, Wolverhampton, WV1 2AR [IO82WO, SO99]
G4 SGF Details withheld at licensee's request by SSL.[Ken, West Sheffield, SK28. Can be contacted through ZB2GR. See the International Call Book, any year since 1981]
G4 SGG D Earp, 92 Somersby Rd, Woodthorpe, Nottingham, NG5 4LT [IO92KX, SK54]
G4 SGH P Frampton, 1 Warwick Pl, Penrith, CA11 7DT [IO84OP, NY52]
G4 SGI S Collings, 46 St. Michaels Rd, Cheltenham, GL51 5RR [IO81WV, SO92]
G4 SGJ J P C Campbell, 303 Frost Ln, Hythe, Southampton, SO45 3NB [IO90HU, SU40]
G4 SGL A J Horne Feltham Sea Cadets ARC, 14 Penton Hook Rd, Staines, TW18 2PF [IO91RK, TQ07]
G4 SGN P P Playle, 6 Walnut Tree Cl, Cheshunt, Waltham Cross, EN8 8NH [IO91XQ, TL30]
G4 SGO Details withheld at licensee's request by SSL.
GI4 SGP Details withheld at licensee's request by SSL.
G4 SGQ P Hruza, Limes Flat, Tettenhall Green, Wolverhampton, WV6 8NU [IO82WO, SO89]
GW4 SGR S D Richards S.Glam Rynt Grp, Crydon, St. Andrews Rd, Wenvoe, Vale of Glam, CF5 6AF [IO81IK, ST17]
G4 SGS K Hanton, Keramet, Toprow, Wreningham, Norwich, NR16 1AR [JO02OM, TM19]
G4 SGU G R Gilbertson, 6 The Stray, South Cave, Brough, HU15 2AL [IO93QS, SE93]
G4 SGV K B Jones, 228 Evesham Rd, Headless Cross, Redditch, B97 5EP [IO92AG, SP06]
G4 SGW W F Prouse, 1 Springfield Cttgs, Bishops Tawton, Barnstaple, EX32 0DF [IO71XB, SS53]
G4 SGY J F Winters, 94 Wharncliffe Rd, Loughborough, LE11 1SN [IO92JS, SK51]
G4 SGZ S H Garner, Flat Warren House, Warren Rd, Crowborough, TN6 1TX [JO01BB, TQ43]
G4 SHA M P Webb, 9 Steele Cl, The Paddock, Devizes, SN10 3SL [IO91AI, SU06]
G4 SHB P Sheridan, 17 Boakes Dr, The Mallards, Stonehouse, GL10 3QW [IO81UR, SO80]
G4 SHC R S Bentham, 108 Stanycliffe Ln, Middleton, Manchester, M24 2QL [IO83VN, SD80]
G4 SHD Details withheld at licensee's request by SSL.
G4 SHF S J Purser, 80 John Bold Ave, Stoney Stanton, Leicester, LE9 4DN [IO92IN, SP49]
G4 SHH P Brooking, 110 Binstead Lodge, Rd, Binstead Ryde, Isle of Wight, PO33 3UD [IO90JR, SZ59]
G4 SHI M E Beasley, 10 Manor Rd, Chippenham, SN14 0LH [IO81WL, ST97]
G4 SHJ N A Douglas, 87 Hutton Ave, Hartlepool, TS26 9PR [IO94JQ, NZ43]
G4 SHK G Clifford, The Forge, Hyde Farm, Turnpike Rd, Blunsdon Swindon, SN2 4EA [IO91CO, SU19]
G4 SHM M R Baker, 92 Moy Ave, Eastbourne, BN22 8UQ [JO00DS, TQ60]
G4 SHN J W Pitts, 32 Winding Way, Dagenham, RM8 2TB [JO01BN, TQ48]
G4 SHO F J Dibden, 127 Mayola Rd, Clapton, London, E5 0RG [IO91XN, TQ38]
G4 SHP Details withheld at licensee's request by SSL.
G4 SHQ D J West, 12 Church Ln, Farnborough, GU14 9RJ [IO91OH, SU85]
G4 SHU J B Gibbs, 1 Berry Hill Rd, Mansfield, NG18 4RU [IO93JD, SK55]
G4 SHV G H Tilsed, 6 Nansen Ave, Oakdale, Poole, BH15 3DA [IO90AR, SZ09]
G4 SHX J Brindle, 41 Heversham Ave, Fullwood, Fulwood, Preston, PR2 9TD [IO83PT, SD53]
G4 SHY L Afford, 54 Mancetter Rd, Nuneaton, CV10 0HN [IO92GM, SP39]
G4 SIA J H Spershott, Longview, Barn Cl, Yorkletts, Whitstable, CT5 3AF [JO01MH, TR06]
G4 SIB A A Butcher, The Glade, Broad Ln, Newdigate, Dorking, RH5 5AT [IO91TD, TQ13]
GM4 SID S Will, 53 Bishop Forbes Cres, Kinellar, Aberdeen, AB2 0TW [IO87TH, NJ72]
G4 SIE R A Mason, 8 Tweedy St., Cowpen Village, Blyth, NE24 5NB [IO95FD, NZ28]
G4 SIF R Rowsell, North Lodge, 61 Barrack Rd, Bexhill on Sea, TN40 2AZ [JO00FU, TQ70]
G4 SIG Dr S P Vahl, Downlands, Petersham Cl, Petersham, Richmond, TW10 7DZ [IO91UK, TQ17]
GW4 SII P Garston, 85 Wood Ln, Hawarden, Deeside, CH5 3JG [IO83LE, SJ26]
G4 SIJ B Hammond, 2 McDowell Way, Narborough, Leicester, LE9 5RA [IO92JO, SP59]
GM4 SIK S D Ryall, Sunhoney, East Camaloun, Fyvie, Turriff, Aberdeenshire, AB53 8JY [IO87TK, NJ73]
G4 SIL E J Tubman, 54 Summerfield Ave, Whitstable, CT5 1NS [JO01MI, TR16]
G4 SIN M A Wulwick, 88 Princes Park Ave, London, NW11 0JX [IO91VO, TQ28]
GI4 SIP J F Mc Kavanagh, 28 Thompsons Grange, Carryduff, Belfast, BT8 8TG [IO74BM, J36]
G4 SIQ W Wilkie, 14 Horseshoe Cl, Northwood, Cowes, PO31 8PZ [IO90IR, SZ49]
G4 SIS R C Keefe, 28 Burstead Dr, South Green, Billericay, CM11 2QN [JO01FO, TQ69]
G4 SIV Details withheld at licensee's request by SSL.[Correspondence to Five Bells Group, c/o B K Tatnell, G4ODA, 73 Acacia Avenue, Spalding, Lincs, PE11 2LW.]
GI4 SIW Dr W D Hutchinson Antrim & Dis AR, 40 Oldstone Hill, Muckamore, Antrim, BT41 4SB [IO64VQ, J18]
GI4 SIZ T M Thomson, 28 Josephine Ave, Limavady, BT49 9BA [IO65MB, C62]
GM4 SJA Details withheld at licensee's request by SSL.
GI4 SJB Details withheld at licensee's request by SSL.
GM4 SJC H Wilson, Esperance, 21 Woodcot Park, Stonehaven, AB39 2HG [IO86VX, NO88]
G4 SJD S J Davis, 33 Pollard Cl, Hooe, Plymstock, Plymouth, PL9 9RR [IO70WI, SX45]
G4 SJG G K Upton, 18 Cranthorne Dr, Bakersfield, Nottingham, NG3 7HD [IO92KX, SK64]
G4 SJH Details withheld at licensee's request by SSL.[Station located near Lightwater.]
G4 SJJ Details withheld at licensee's request by SSL.[Station located near Burnham-on-Sea.]
GM4 SJL S J Thomson, 158 Goldhurst Terr, South Hampstead, London, NW6 3HP [IO91VM, TQ28]
G4 SJM J D Reeves, 5 Arrows Cres, Boroughbridge, York, YO5 9LP [IO94HC, SE36]
G4 SJN B T Hunt, Tralee, Oakbridge Lynch, Stroud, Glos, GL6 7NY [IO81WR, SO90]
GW4 SJO M J Edwards, Aelwyd-y-Don, Tresaith, Cardigan, Dyfed, SA43 2JH [IO72RD, SN25]
G4 SJP S J Prior, 47 Yelland Rd, Yelland, Fremington, Barnstaple, EX31 3DS [IO71WB, SS53]
GI4 SJQ G E Frazer, 20 Old Rectory Park, Portadown, Craigavon, BT62 3QH [IO64SJ, J05]
G4 SJR A Thornes, 32 Cornmill Dr, Liversedge, WF15 7EE [IO93DQ, SE22]
G4 SJS Details withheld at licensee's request by SSL.[Op: J R Lill. Station located in London NW3.]
G4 SJT W J Young, Kiplings, Graynfylde Dr, Bideford, EX39 4AP [IO71VA, SS42]
G4 SJU S R Browne, 38 Aldrin Rd, Pennsylvania, Exeter, EX4 5DN [IO80FR, SX99]
G4 SJV D J Chapman, 24 Broad Ln, Moulton, Spalding, PE12 6PN [IO92XT, TF32]
G4 SJX Details withheld at licensee's request by SSL.
G4 SJZ R J J G Cole, 22 Pilley Rd, Tupsley, Hereford, HR1 1NB [IO82PB, SO54]
GM4 SKB M J Whyatt, Backburn Cottage, Castleton, Auchterarder, Pertshire, PH3 1JS [IO86DH, NN91]
GW4 SKC F J Lynch, 48 Llanllienwen Rd, Morriston, Cwmrhydyceirw, Swansea, SA6 6NA [IO81AQ, SS69]
G4 SKE D C Rigby Aes Soc, 145 Knightlow Rd, Harborne, Birmingham, B17 8PY [IO92AL, SP08]
G4 SKH J B H Trever, 6 Ash Cl, Westlands Rd, Sproatley, Hull, HU11 4XE [IO93VT, TA13]
G4 SKJ Details withheld at licensee's request by SSL.
G4 SKM P Goben Maltby A R S, 1 Petal Cl, Maltby, Rotherham, S66 7HJ [IO93JK, SK59]
G4 SKN K J Romang, 27 Elley Green, Neston, Corsham, SN13 9TX [IO81VK, ST86]
G4 SKO M Brooke, 1 Kenmore Way, Cleckheaton, BD19 3EL [IO93DR, SE12]
GW4 SKP A T D Clark, 27 Heol St Bridget, St Bridges Major, Bridgend Mid Glam, CF32 0SL [IO81EL, SS87]
G4 SKQ G D Horsfield, 2 Linden Rd, Ecclesfield, Sheffield, S30 3XL [IO93GJ, SK38]
G4 SKR Details withheld at licensee's request by SSL.
G4 SKS W H Bradshaw, 11 Mayfair Ave, Romford, RM6 6UB [JO01BN, TQ48]
G4 SKT E A Ayres, 219 Ashingdon Rd, Rochford, SS4 1RS [JO01IO, TQ89]
G4 SKU Details withheld at licensee's request by SSL.
G4 SKW D A Redfern, 57A Queen St., Waingroves, Ripley, DE5 9TJ [IO93HA, SK44]
G4 SKX D Fields, 43 Barrington Ave, Elm Tree Park, Stockton on Tees, TS19 0UE [IO94HN, NZ41]
G4 SKY M Dyson, 44 Dawlish Cl, Hucknall, Nottingham, NG15 6NY [IO93JA, SK54]
G4 SLD E Thomas, 24 Goldsmid Rd, Tonbridge, TN9 2BX [JO01DE, TQ54]
G4 SLE K A Lambert, 38 Beechers Rd, Portslade, Brighton, BN41 2RG [IO90VU, TQ20]
G4 SLF G Mottram, 71 Sycamore House Rd, Sh1Regreen, Sheffield, S5 0UD [IO93GK, SK39]
G4 SLG K Oliver, 42 Minster Dr, Cherry Willingham, Lincoln, LN3 4NA [IO93SF, TF07]
G4 SLH K Thompson Stanford-le-Hop, Amateur Radio Club, 113 Gordon Rd, Corringham, Stanford-le-Hope, SS17 7QZ [JO01FM, TQ68]
GW4 SLI J C Plumley, 34 Graigwen Cres, Abertridwr, Caerphilly, CF8 4BN [IO81HN, ST08]
GW4 SLK R E Craddock, Garden Cottage, Mostyn Hall, Mostyn Holywell, Clwyd, CH8 9DX [IO83JH, SJ17]
G4 SLL J E Buckland, Linden, Mackham Ln, Dunkeswell, Devon, EX14 0RH [IO80JU, ST10]
G4 SLN E Jobling, 15 Warnhead Rd, Bedlington, NE22 5RE [IO95FD, NZ28]
GI4 SLQ K J Boyd, 40 Killyman St., Moy, Dungannon, BT71 7SJ [IO64PK, H85]

GW4 SLS F Jones, 45 Linkside Dr, Pennard, Southgate, Swansea, SA3 2BR [IO71WN, SS58][Op: F Jones, CWAO, c/o 9th Sigs Regt, BFPO 58. Cyprus call: ZC4FJ.]
G4 SLU Details withheld at licensee's request by SSL.[Op: C A Hardy. Station located near Poole, Dorset. IARU locator: IO90AR. Op: C A Hardy.]
G4 SLW Details withheld at licensee's request by SSL.
G4 SLX Details withheld at licensee's request by SSL.
GM4 SLY J F Bell, 54 Lochgreen Ave, Barassie, Troon, KA10 6UN [IO75QN, NS33]
GW4 SLZ G F Holtum, Grassmere, Abergele Rd, Rhuddlan, Rhyl, LL18 5UE [IO83GG, SJ07]
G4 SMA M G Goode, Meadowgreen, Batch Valley, All Stretton, Church Stretton, SY6 6JW [IO82ON, SO49]
G4 SMB Capt M R Briggs, 6 Rosedale, High Stile, Leven, Beverley, HU17 5NE [IO93UV, TA14]
G4 SMC Details withheld at licensee's request by SSL.
G4 SMD Details withheld at licensee's request by SSL.
G4 SME J Rogers Skmrsdle & D AR, 186 Beavers Ln, Birleywood, Skelmersdale, WN8 9BP [IO83OM, SD40]
GW4 SMG S M Green, 9 Westbourne Gr, Sketty, Swansea, SA2 9DT [IO81AO, SS69]
GM4 SMH E Metcalfe, 14 Harperland Dr, Kilmarnock, KA1 1UH [IO75RO, NS43]
GM4 SMI E A Shields, 34 Moorfield Ave, Kilmarnock, KA1 1TT [IO75RO, NS43]
G4 SMK K Wilson, 15 Woodside Ave, Cottingley, Bingley, BD16 1RB [IO93BT, SE13]
G4 SMM M R Manley, Rolleston, Parkgate Rd, Saughall, Chester, CH1 6JS [IO83MF, SJ37]
G4 SMP M H Cooper, Gorse Farm, Lazy Hill Rd, Stonnall, Walsall, WS9 9DS [IO92BO, SK00]
G4 SMQ Details withheld at licensee's request by SSL.
G4 SMS R Stubbs, 8 Gwylan Ave, Deeside, CH5 4AT [IO83LF, SJ26]
G4 SMT J T Short, 27 Hawksworth Dr, Formby, Liverpool, L37 7EY [IO83LN, SD30]
G4 SMU Details withheld at licensee's request by SSL.[Station located near Evesham.]
G4 SMV J M Markwell, 29 Lyndale Dr, Wrenthorpe, Wakefield, WF2 0JZ [IO93FQ, SE32]
G4 SMX J K Wilson, 168 Elms Vale Rd, Dover, CT17 9PN [JO01PC, TR34]
G4 SMZ W A Grant, 2 Newlyn Way, Parkstone, Poole, BH12 4EA [IO90AR, SZ09]
GI4 SNA D Ross, 127 Pond Park Rd, Lisburn, Co Antrim, N Ireland, BT28 3RE [IO64XM, J26]
GI4 SNC K Moore, 57 Meadowvale Park, Limavady, BT49 0RD [IO65MA, C62]
G4 SND M C Newey, 2 Caldy Walk, Stourport on Severn, DY13 8QX [IO82UI, SO87]
G4 SNG J Molyneux, 3 Bollin Walk, Reddish, Stockport, SK5 7JW [IO83WK, SJ89]
G4 SNI H C Farley, 11 College Ln, Hatfield, AL10 9PB [IO91VS, TL20]
G4 SNJ C D Frenzel, 26 Foxglove Cl, Bishops Stortford, CM23 4PU [JO01BU, TL42]
G4 SNK J T Frenzel, 29 Tremlett Gr, 1st Floor Flat, London, N19 5JY [IO91WN, TQ28]
G4 SNL I Dunworth, 42 Gloucester Rd, Waterlooville, PO7 7BJ [IO90LU, SU60]
G4 SNN N Booth-Isherwood, 65 Burnaby St., Alvaston, Derby, DE24 8RN [IO92GV, SK33]
G4 SNO A F Duggins, Dowles Bungalow, Dowles Rd, Bewdley, DY12 3AA [IO82TJ, SO77]
GM4 SNP H H Christie, 100 Dunlop Terr, Ayr, KA8 0SP [IO75QL, NS32]
G4 SNQ T F Wadsworth, 20 Rook Wood Way, Little Kingshill, Great Missenden, HP16 0DF [IO91PQ, SP80]
G4 SNR E Meekers, 5 Frobisher Cl, Mudeford, Christchurch, BH23 3SN [IO90DR, SZ19]
G4 SNS Details withheld at licensee's request by SSL.
G4 SNT Details withheld at licensee's request by SSL.
G4 SNU K Hardie, Vessey House, Rodden Rd, Langton Herring, Dorset, DT3 4JA [IO80RP, SY68]
G4 SNV G M Eastgate, 103 Western Rd, Leigh on Sea, SS9 2PB [JO01HN, TQ88]
G4 SNY J A Marsdon, 39 Hill Common, Hemel Hempstead, HP3 8JH [IO91SR, TL00]
GM4 SNZ D J McCrandles, 21 Wall St., Camelon, Falkirk, FK1 4QB [IO86CA, NS88]
G4 SOA P E W Instone, 32 Melfort Cl, Sparcells, Swindon, SN5 9FG [IO91CN, SU18]
G4 SOB W A E Hammond, 29 Wordsworth Rd, Lexden, Colchester, CO3 4HR [JO01KV, TL92]
GW4 SOC V T Shaw, 130 Aberthaw Rd, Ringland Est, Newport, NP9 9QS [IO81MO, ST38]
G4 SOF J J Blight, Lowbell, Handy Cross, Bideford, N Devon, EX39 3ET [IO71VA, SS42]
G4 SOH M G Spence, 5 St. Helens Ave, Benson, Wallingford, OX10 6RY [IO91KO, SU69]
G4 SOI M Wray, Ivy Nook Cottage, Laneham St., Rampton, Retford, DN22 0JX [IO93OH, SK87]
G4 SOK R K Hollow, The Beeches, Gr Ln, Goldsithney, Penzance, TR20 9HN [IO70GC, SW53]
G4 SOL I J Bale, 11 Maplewell, Coalville, LE67 4RE [IO92HR, SK41]
G4 SOM J Spiteri, 27 Hillcote Cl, Fulwood, Sheffield, S10 3PT [IO93FI, SK38]
G4 SON T Williams-Berry, 15 Bredon Ave, Wrose, Shipley, BD18 1LU [IO93DT, SE13]
G4 SOO H Pilling, 183 Wigan Rd, New Springs, Aspull, Wigan, WN2 1DU [IO83QN, SD50]
G4 SOP K L Lillingstone, 6 Taylors Ln, Old Catton, Norwich, NR6 7BE [JO02PP, TG21]
G4 SOQ B W Lyons, 9 Brendon Way, Westcliff on Sea, SS0 0JD [JO01IN, TQ88]
G4 SOR A Collins, 19 Cavendish Rd, Skegness, PE25 2QZ [JO03ED, TF56]
G4 SOT D G Goodwin, Dawn View, Bogshole Ln, Broomfield, Herne Bay, CT6 7BZ [JO01NI, TR16]
G4 SOX Details withheld at licensee's request by SSL.
GI4 SOY M P Dornan, 6 Grange Park, Saintfield, Ballynahinch, BT24 7NT [IO74BK, J45]
G4 SOZ D J P Herd, Southacres, Low Common, Hellington, Norwich, NR14 7BU [JO02RN, TG30]
G4 SPA R S Marchington Buxton RA, 78 Buxton Rd, Dove Holes, Buxton, SK17 8DW [IO93BH, SK07]
G4 SPC B A Escreet, 198 Front St., Sowerby, Thirsk, YO7 1JN [IO94HF, SE48]
G4 SPD N J Jarvis, 25 St. Augustines Cl, Aldershot, GU12 4SF [IO91PF, SU85]
G4 SPE G W Callaghan, 1 Wessex Cl, Semington, Trowbridge, BA14 6SA [IO81WI, ST86]
G4 SPI R E Endersby, Delph Cottage, Victoria Ct, Southwell Ln Horbur, Wakefield W Yorks, WF4 5BN [IO93FP, SE21]
G4 SPK R S Milligan, 6 Haydon Rd, Aylesbury, HP19 3NN [IO91OT, SP81]
GW4 SPL P H Ace, 116 Gellionen Rd, Clydach, Swansea, SA6 5HF [IO81BQ, SN60]
G4 SPM T R Barrett, 15 Hill Top Est, South Kirkby, Pontefract, WF9 3EW [IO93HO, SE41]
G4 SPN R Heyes, 11 The Links, Cambridge Rd, Newmarket, CB8 0TG [JO02EF, TL66]
G4 SPQ B H Edwards, 2 Springfields, Dursley, GL11 6PF [IO81TQ, ST79]
G4 SPR F Rattray, 4 Winton Manor Ct, Winton, Kirkby Stephen, CA17 4HR [IO84UL, NY71]
G4 SPS Details withheld at licensee's request by SSL.
GI4 SPT H J Golding, 12 St. Marys Terr, Stream St., Newry, BT34 1HL [IO64UE, J02]
G4 SPU N Alcock, Vivenda Montihno, Caixa 105Z, Picota, 8100 Loule, Portugal
G4 SPV A T Adamson, 520 York Rd, Stevenage, SG1 4EP [IO91VW, TL22]
G4 SPW T B Devlin, The Bungalow, 17 Fair View, Dalton in Furness, LA15 8RZ [IO84JD, SD27]
G4 SPX Details withheld at licensee's request by SSL.[Op: F Butterworth, 26 Torwood Road, Chadderton, Oldham, OL9 0RA.]
G4 SPY A Kay, 36 Yorks Wood Dr, Birmingham, B37 6DL [IO92CL, SP18]
G4 SPZ P N Harris, 22 Bramley Way, Hill, Bewdley, DY12 2PU [IO82UJ, SO77]
G4 SQA D Yeoman, 29 Fenside Dr, Newborough, Peterborough, PE6 7SF [IO92VP, TF10]
G4 SQF B Harrington, 237 Norton East Rd, Norton Canes, Cannock, WS11 3RW [IO92AQ, SK00]
G4 SQG R Nicholson, 7 Half Mile Gdns, Bramley, Leeds, LS13 1BL [IO93ET, SE23]
G4 SQI G Gulliford, 18 Purslane, Abingdon, OX14 3TR [IO91IQ, SU59]
G4 SQJ G Turner, 51 Caernarvon Rd, Chichester, PO19 2YH [IO90OT, SU80]
G4 SQK G Middleton, 803 Park Ridge Rd, Apartment A7, Durham, North Carolina, USA, S19 5SF [IO93GJ, SK38]
GI4 SQL S J Adrain, 10 Highgate Dr, Mallusk, Newtownabbey, BT36 4WQ [IO74AQ, J38]
GM4 SQM D M Anderson, 34 Culzean Cres, Kilmarnock, KA3 7DT [IO75SO, NS43]
GM4 SQO R J Riddiough, 1 Cedar Rd, Ayr, KA7 3PE [IO75QK, NS31]
G4 SQP C Wilding, 92 Ravenhill Dr, Codsall, Wolverhampton, WV8 1BW [IO82VP, SJ80]
G4 SQQ C R Hollister, Rosemead, 326 Passage Rd, Cribbs Causeway, Bristol, BS10 7TE [IO81QM, ST58]
G4 SQR D F Townsend, 66 Norman Rd, Swindon, SN2 1AX [IO91CN, SU18]
GM4 SQS J S Holdsworth, 3 Balrymonth Ct, St. Andrews, KY16 8XT [IO86OH, NO51]
GM4 SQT Rev K Holdsworth, 3 Balrymonth Ct, St. Andrews, KY16 8XT [IO86OH, NO51]
G4 SQU Details withheld at licensee's request by SSL.
G4 SQV J C Hart, Buckhurst, Morley Ln, Little Eaton, Derby, DE21 5AH [IO92GX, SK34]
G4 SQW K Watson, 18 Rufford Ave, Athersley North, Barnsley, S71 3EF [IO93GN, SE30]
G4 SQY R Brewitt, 19 Woods Gr, Burniston, Scarborough, YO13 0JD [IO94SH, TA09]
G4 SRA Details withheld at licensee's request by SSL.
G4 SRB E A Turner Staffs Rg(Burto, 74 Beech Ln, Stretton, Burton on Trent, DE13 0DU [IO92ET, SK22]
G4 SRC M J Elliott Swale Am Rad Cl, 20 Haysel, Sittingbourne, ME10 4QE [JO01IH, TQ96]
G4 SRD R K Sealy, 10 Mallard Cl, Bowerhill, Melksham, SN12 6TQ [IO81WI, ST96]
G4 SRE J Reeves, 5 Lon y Bryn, Glynneath, Neath, SA11 5BG [IO81ER, SN80]
G4 SRF C M Radcliffe, 85 Brian Ave, Cleethorpes, DN35 9DE [IO93XN, TA20]
G4 SRG Details withheld at licensee's request by SSL.
G4 SRH Prof M Harrison Staffs Raynet, 1 Church Fields, Keele, Newcastle, ST5 5HP [IO82UX, SJ84]
G4 SRI A C D Gray, 5 Wesley Way, Tamworth, B77 3JQ [IO92DP, SK20]
GI4 SRK Details withheld at licensee's request by SSL.
GM4 SRL R M Cowan, 85 Eastwoodmains Rd, Clarkston, Glasgow, G76 7HG [IO75UT, NS55]
GW4 SRO R F Cashmore, 65 Michaelston Rd, Culverhouse Cross, Cardiff, CF5 4SX [IO81IL, ST17]
G4 SRP G T Brierley, 6 Yeo Dr, Appledore, Bideford, EX39 1RD [IO71VB, SS43]
GI4 SRQ W G McHugh, 47 Main St., Hamiltons Bawn, Hamiltonsbawn, Armagh, BT60 1LP [IO64RI, H94]
G4 SRS S W Smith Stroud ARS, 60 Grange Rd, Tuffley, Gloucester, GL4 0PG [IO81UU, SO81]
GM4 SRU J Watt, 15 Murrayston, Lerwick, ZE1 0RE [IP90JD, HU44]
G4 SRV D M Darby, 28 Coleshill Cl, Hunt End, Redditch, B97 5UN [IO92AG, SP06]
G4 SRX G P Sampson, 47 Netherthorpe Way, North Anston, Anston, Sheffield, S31 7FE [IO93GJ, SK38]
GM4 SSA H Hassel, Sumra, Eshaness, Shetland, ZE2 9RS [IP90FM, HU28]
G4 SSB S S Bosley Ssbrc, 67 Holly Ave, New Haw, Addlestone, KT15 3UD [IO91SI, TQ06]
G4 SSC Dr A W Taylor, 38 Summershades Ln, Grasscroft, Oldham, OL4 4ED [IO83XM, SD90]
G4 SSD J S May Sth Devon RC, 6 Hodson Cl, Paignton, TQ3 3NU [IO80FK, SX86]
G4 SSE J C Sartin, The Gantry House, 9 Graces Maltings, Tring, Herts, HP23 6DL [IO91QT, SP91]
GI4 SSF S F Craig, 10 Grange Ave, Killfennan, Londonderry, BT47 1YN [IO64IX, C41]
G4 SSG Details withheld at licensee's request by SSL.
G4 SSH R Clayton, 9 Green Island, Irton, Scarborough, YO12 4RN [IO94SF, TA08][Op: Roy Clayton. Tel: (01723) 862924. RSGB Chief Morse Examiner. RNARS: 2770. WAB: 7666 TA08 Scarborough.]
G4 SSJ R W Bolton, 83 Sandicroft Cl, Locking Stumps, Birchwood, Warrington, WA3 7LY [IO83RK, SJ69]
G4 SSL W E Lancashire, 64 Ashpole Furlong, Loughton, Milton Keynes, MK5 8DX [IO92OA, SP83]
G4 SSN J E Tottle, 20 Abbey Rd, Cheadle, SK8 2JW [IO83VJ, SJ88]
G4 SSO A K McMillan, Marlbank, St. Marys Rd, Ramsey, Huntingdon, PE17 1SN [IO92WL, TL28]
G4 SSP K B Waghorne, 23 Bramley Hill, Mere, Warminster, BA12 6JX [IO81UC, ST83]
G4 SSQ D Lawton, 197 Lowerhouses Ln, Longley, Huddersfield, HD5 8LA [IO93CP, SE11]
G4 SSR B Taylor, 188 Walstead Rd, Walsall, WS5 4DN [IO92AN, SP09]
G4 SST Details withheld at licensee's request by SSL.
G4 SSV S Smith, 1 Buckfast Rd, Lincoln, LN1 3JS [IO93RF, SK97]
G4 SSW J R Walker, 15 Hillfield Rd, Bilton, Rugby, CV22 7EW [IO92II, SP47]
G4 SSX Details withheld at licensee's request by SSL.
G4 SSZ D J W Fox, Cornbury, Cliff Rd, Hythe, CT21 5XA [JO01NB, TR13]
G4 STA Details withheld at licensee's request by SSL.[Op: E J Mills, Royal Greenwich Observatory, Apartado De Correos 368, Santa Cruz de la Palma, Tenerife, Islas Canarias.]
G4 STB P Lock, 36 Parkengear Vean, Probus, Truro, TR2 4JT [IO70MG, SW94]
G4 STD B H W West, 13 Tanglewood Cl, Shirley, Croydon, CR0 5HX [IO91XI, TQ36]
G4 STE S N Farr, 7 Fenwick Cl, Hill, Alcester, B49 6JZ [IO92BF, SP05]
G4 STF L P Straus, 12 Vicarage Rd, Maidenhead, SL6 7DS [IO91PM, SU88]
G4 STH G G Timbrell, Crossing Cottage, Lamyatt, Shepton Mallet, Somerset, BA4 6NG [IO81RC, ST63]
G4 STI M B Pinder, 36 West Ridge, Allesley Park, Coventry, CV5 9LN [IO92FJ, SP27]
GI4 STJ H J Bell, 109 Mullanahoe Rd, Dungannon, BT71 5AX [IO64RO, H97]
G4 STK E Brown, 16 Springfield, Ovington, Prudhoe, NE42 6EH [IO94BX, NZ06]
G4 STL M J Acton, 51 Tintern Ave, Whitefield, Manchester, M45 8WY [IO83UN, SD80]
G4 STM Details withheld at licensee's request by SSL.
G4 STN Details withheld at licensee's request by SSL.
G4 STO P G Rose, Pinchbeck Farmhouse, Mill Ln, Sturton By Stow, Lincoln, LN1 2AS [IO93PH, SK87]
G4 STP T Mangles, 46 Cedar Cres, Wear Valley View, Willington, Crook, DL15 0DA [IO94DR, NZ23]
G4 STR Details withheld at licensee's request by SSL.
G4 STT A J M Cameron, 5 Westminster Gate, Burn Bridge, Harrogate, HG3 1LU [IO93FX, SE35]
G4 STV S Blayer Hadley Wood Contest Group, 452 Goffs Ln, Cheshunt, Goffs Oak, Waltham Cross, EN7 5EN [IO91XQ, TL30]
G4 STW A Buckley, 8 Greengarth, Bottesford, Scunthorpe, DN17 2UH [IO93QN, SE80]
G4 STX H W Adcock, 70 Park Ln, Norwich, NR2 3EF [JO02PP, TG20]
GW4 STY C E Bates, 3 Shearwater Cl, Porthcawl, CF36 3TU [IO81DL, SS87]
G4 STZ T H O M Bromsgrove, 53 Dublin Croft, Great Sutton, South Wirral, L66 2TD [IO83MG, SJ37]
G4 SUA B K Common, 59 Thornbera Gdns, Bishops Stortford, CM23 3NP [JO01BU, TL41]
GM4 SUC M W Dalrymple, 11 Shawfield Ave, Ayr, KA7 4RE [IO75QK, NS31]
GW4 SUD K M Jones, 111 Ewenny Rd, Bridgend, CF31 3LN [IO81FL, SS97]
GW4 SUE M A Hill, 13 Maesglas Gr, Newport, NP9 3DJ [IO81LN, ST28]
GM4 SUF P M Gane, The Studios, Ardmore Lodge, Edderton By Tain, Ross-shire Scotland, IV19 1LB [IO77VU, NH78]
G4 SUG S G B Heuser, 34 Long Gr, Seer Green, Beaconsfield, HP9 2YW [IO91QO, SU99]
G4 SUH Details withheld at licensee's request by SSL.
G4 SUI S M Bolt, 112 Leeds Rd, Mirfield, WF14 0JE [IO93DQ, SE12]
G4 SUJ L G Greville-Smith, Hilton Lea, Woodhayes Rd, Wolverhampton, WV10 8QH [IO82WO, SJ90]
G4 SUK M J Kent, 304 Reculver Rd, Herne Bay, CT6 6SR [JO01OI, TR26]
G4 SUL H Little, 10 Bamburgh Rd, Ferryhill, DL17 8QH [IO94FQ, NZ23]
G4 SUM K G Summerhill, 23 Ellis Cl, Cottenham, Cambridge, CB4 4UN [JO02BG, TL46]
GW4 SUN B A Le Carpentier, Cartref, 43 Abernant Rd, Aberdare, CF44 0PY [IO81GR, SO00]
G4 SUO P J Barwick, Brook Cottage, 94 Ambleside Rd, Lightwater, GU18 5UJ [IO91PI, SU96]
G4 SUP A M Morgan, 32 Bury Ln, Codicote, Hitchin, SG4 8XX [IO91VU, TL21]
GM4 SUR R Aitken, 2 Eskdale Dr, Bonnyrigg, EH19 2LD [IO85KV, NT36]
G4 SUS S F Morgan, Hollaway, Northbourne Rd, Great Mongeham, Deal, CT14 0LA [JO01QF, TR35]
G4 SUT R E Moxon, 19 Edenbridge Rd, Hall Green, Birmingham, B28 8QB [IO92CK, SP18]
G4 SUX R E H Payne, 9 Shelburne Way, Derry Hill, Calne, SN11 9PA [IO81XK, ST97]
G4 SUY P A Shaw, 12 Coombe Way, Hartburn, Stockton on Tees, TS18 5PX [IO94HN, NZ41]
G4 SUZ W A Wells, 9 Clumber Dr, Radcliffe on Trent, Nottingham, NG12 1DA [IO92LW, SK64]
G4 SVA J K Appleton, 53 Chapel St., Leigh, WN7 2PB [IO83RL, SD60]
G4 SVB A E Gatrell, Sunnyside, Muddles Green, Chiddingly, Lewes, BN8 6HW [JO00CV, TQ51]
G4 SVC T S Axtell, 146 Olivers Battery Rd South, Winchester, SO22 4LF [IO91HB, SU42]
G4 SVD A M Ames, Kismet, 10 Milner Dr, Cobham, KT11 2EZ [IO91TI, TQ16]
G4 SVE J Hewett, 84 Dunsgreen, Ponteland, Newcastle upon Tyne, NE20 9EJ [IO95DB, NZ17]
G4 SVF Details withheld at licensee's request by SSL.
G4 SVG S F Wensley, 15 Lamden Way, Burghfield Common, Reading, RG7 3LZ [IO91LJ, SU66]
G4 SVI C J Ames, 33 Keswick Cl, Cringleford, Norwich, NR4 6UW [JO02OO, TG10]
G4 SVK J E Philipp, 2 Red Lion Park, Hooe, Battle, E Sussex, TN33 9EW [JO00EU, TQ61]
G4 SVL I Hewitt, 51 Three Oaks Rd, Wythall, Birmingham, B47 6HG [IO92BJ, SP07]
GM4 SVM G J Hudson, 56 Vancouver Ave, Howden, Livingston, EH54 6BS [, NT06][Member: BATC, RSGB, ICFM. Mostly operational on HF CW.]
G4 SVN K Johnson, 74 Deepdale Rd, Kimberworth, Rotherham, S61 2NT [IO93HK, SK49]
GI4 SVO Rev H L F Bolster, 17 Fernmore Rd, Bangor, BT19 6DY [IO74EP, J58]
G4 SVP Details withheld at licensee's request by SSL.
G4 SVQ P J Haffenden, Badgers Mount, 12 Brookside Ave, Wraysbury, Staines, Middx, TW19 5HB [IO91RL, TQ07]
G4 SVR W J Baddeley, 12 Stockport Rd, Altrincham, WA15 8ET [IO83TJ, SJ78]
G4 SVS D B Bush, 12 Bridle Way, Orpington, BR6 7TJ [JO01AI, TQ46]
G4 SVT R E Halliwell, 30 Sherborne Rd, Kitt Green, Orrell, Wigan, WN5 0JA [IO83PN, SD50]
G4 SVU G L Holdom, 229 Southwell Rd West, Mansfield, NG18 4HF [IO93JD, SK56]
GM4 SVW W McLaren, 4 Firth Cres, Gourock, PA19 1EW [IO75NW, NS27]
G4 SVY J A Perez, Kenbury Private Htl, Clarence Rd, Shanklin, Isle of Wight, PO37 7BH [IO90JP, SZ58]
G4 SWA S W Authers, 9 Conway Ave, Birmingham, B32 1DR [IO82XL, SO96]
G4 SWB G Cox, 39 Midland Rd, Bramhall, Stockport, SK7 3DY [IO83WJ, SJ88]
G4 SWD Details withheld at licensee's request by SSL.[Station located near Crowborough.]
G4 SWE D A Screen, 16 Hollydale Cl, Reading, RG2 8LL [IO91MK, SU77]
G4 SWF D H Redmond, 210 Gershwin Rd, Basingstoke, RG22 4HL [IO91KF, SU64]
G4 SWH R Jones, Tangmere, 40 Lordship Ln, Letchworth, SG6 2BL [IO91VX, TL23]
G4 SWK E Chinn, 4 Kempton Cres, Lillington, Leamington Spa, CV32 7TS [IO92FH, SP36]
G4 SWL Details withheld at licensee's request by SSL.
G4 SWM D W Crosby-Clarke, 55 Robin Ln, Sandhurst, GU47 9AU [IO91OI, SU86]
G4 SWN S R Griffiths, New House, Thompsons Hill, Sherston, Malmesbury, SN16 0PZ [IO81VN, ST88]
G4 SWO S R Griffiths, 25 Hanks Cl, Reeds Farm Est, Malmesbury, SN16 9UA [IO81WO, ST98]
G4 SWQ R Torence-Smith, 29 Horsley Rd, Chingford, London, E4 7HX [IO91XP, TQ39]
G4 SWR I M Hill, 7 Cosford Cl, Matchborough East, Redditch, B98 0BH [IO92BH, SP06]
GI4 SWS A Larmour, 15 Rhanbuoy Park, Craigavad, Holywood, BT18 0DX [IO74CP, J48]
G4 SWT J Moore, 33 Belford Terr, North Shields, NE30 2DA [IO95GA, NZ36]
GM4 SWU F S Sinclair, Parkhall, Gott, Shetland, ZE2 9SF [IP90JE, HU44]
G4 SWV Details withheld at licensee's request by SSL.
G4 SWW Details withheld at licensee's request by SSL.[Station located near Borehamwood.]
G4 SWX Details withheld at licensee's request by SSL.[Op: J Regnault. Station located near Woodbridge.]
G4 SWY D Turner, 24 Titian Ave, Bushey Heath, Bushey, Watford, WD2 1LU [IO91TP, TQ19]
G4 SWZ D Cross, 9 Ardmere Rd, Hither Green, London, SE13 6EL [IO91XK, TQ37]
GW4 SXA R Williams, 44 Duffryn St., Mountain Ash, CF45 3NL [IO81HQ, ST09]
GW4 SXB J Williams, 84 Phillip St., Mountain Ash, CF45 4BG [IO81HQ, ST09]
G4 SXC A Wilson, 14 Vale Cl, Lower Bourne, Farnham, GU10 3HR [IO91OE, SU84]
G4 SXD D E Jones, Edorene, Tincleton, Dorchester, Dorset, DT2 8QR [IO80UR, SY79]
G4 SXE B Holden, 76 The Lawns, Rolleston, Rolleston on Dove, Burton on Trent, DE13 9DE [IO92EU, SK22]
G4 SXF R E Chastell, 4 Fairley Way, Cheshunt, Waltham Cross, EN7 6LG [IO91XR, TL30]
G4 SXH L A Fletcher, 4 Manor Rd, Barton-le-Clay, Bedford, MK45 4NP [IO91SX, TL03]
G4 SXI B M Hartley, 62 Whiteholme Dr, Charleton, Poulton-le-Fylde, FY6 7PP [IO83LU, SD34]
GM4 SXJ R Malcolm, 43 Kinghorne St., Hospitalfield, Arbroath, DD11 2LZ [IO86QN, NO64]
G4 SXK A R Leighs, 16 Spode Cl, Cheadle, Stoke on Trent, ST10 1DT [IO92AX, SK04]
G4 SXL B J V Gardiner, The Homestead Farm, Pond Hill, Wanborough, Guildford, Surrey, GU3 2JW [IO91QF, SU94]
GU4 SXM P J Bannier, 10-le-Bouet, Longstore, St Peter Port, Guernsey, GY1 2BA
GW4 SXN W A C Roe, 5 Gwylfa Est, Penybonc, Amlwch, LL68 9DU [IO73TJ, SH49]
G4 SXO P W Newcombe, 17 South View, Chaddiford Ln, Barnstaple, EX31 1RD [IO71XC, SS53]
G4 SXQ D Lempriere, Harewarren Lodge, Wilton, Salisbury, Wilts, SP2 0NF [IO91BB, SU02]
G4 SXR C M Bracher, 29 Bungalow Park, Holders Rd, Amesbury, Salisbury, SP4 7PJ [IO91CE, SU14]
G4 SXS Details withheld at licensee's request by SSL.
G4 SXT D Ayers, 18 Kinglake Ct, Woking, GU21 1AL [IO91QH, SU95]
GI4 SXV W S E Barker, 18 Main St., Seskanore, Omagh, BT78 1UG [IO64IM, H46]
G4 SXX J P Key, Pendennis, 7 Milton Park, Brixham, TQ5 0AT [IO80FJ, SX95]
G4 SXY G E Haines, 26 High View Cl, Upper Norwood, London, SE19 2DS [IO91XJ, TQ36]
G4 SXZ R Hiles, 19 Station Rd, Kirton Lindsey, Gainsborough, DN21 4BB [IO93QL, SK99]
G4 SYA C J Allen, 11 Chandos Ct, Martlesham, Woodbridge, IP12 4SU [JO02PB, TM24]
G4 SYB P J Loveland, 25 White Acres Rd, Mytchett, Camberley, GU16 6JJ [IO91PH, SU85]
G4 SYC G Lomas, 2 Linney Rd, Bramhall, Stockport, SK7 3JW [IO83WJ, SJ88]

G4 SYD H S Cook, 24 Front St., Sherburn Hill, Durham, DH6 1PA [IO94GS, NZ34]
G4 SYE M D Wilson, 31 Eastfield Rd, Burnham, Slough, SL1 7EH [IO91PM, SU98]
GM4 SYF N F Wallace, 2 Mansefield, Leitholm, Coldstream, TD12 4JQ [IO85UQ, NT74]
G4 SYG P R Tattersall, Anchor Cottage, Nacton, Ipswich, Suffolk, IP10 0EU [JO02OA, TM24]
G4 SYJ W A Bibby, 5 Wellington Rd, Oxton, Birkenhead, L43 2JG [IO83LJ, SJ38]
G4 SYK Details withheld at licensee's request by SSL.
G4 SYL J S Frost, 68 Wessex Rd, Didcot, OX11 8BP [IO91JO, SU58]
GI4 SYM W M J Donaldson, 44 Drumman Hill, Armagh, BT61 8RW [IO64QJ, H94]
GW4 SYO H Thomas, 1 Cambrian Terr, Llwynypia, Tonypandy, CF40 2HN [IO81GP, SS99]
G4 SYP T G Morgan, 26 Rowan Ave, Newton Rd, Lowton, Warrington, WA3 2DD [IO83RL, SJ69]
GU4 SYQ L M E Le Page, Les Martins, St. Martin, Guernsey, Channel Islands, GY4 6QJ
G4 SYR S G Field, 4 Lyndale, Kelvedon Hatch, Brentwood, CM15 0BQ [JO01DP, TQ59]
G4 SYT D S Chambers, 26 Drummond Gdns, Epsom, KT19 8RP [IO91UI, TQ26]
G4 SYV R Ackroyd, 16 Pipits Croft, Bicester, OX6 0XW [IO91KV, SP52]
G4 SYW A A Hargreaves, 50 Deysbrook Ln, West Derby, Liverpool, L12 8RG [IO83NK, SJ49]
G4 SYY L W Colwell, 28 Oakdene, Cheshunt, Waltham Cross, EN8 9JA [IO91XQ, TL30]
GM4 SZA I A Donaldson, 25 Alwyn Rd, Maidenhead, SL6 5EG [IO91PM, SU88]
G4 SZB E P Flannigan, Beachcomber Hotel, 7 Barton Ave, Blackpool, FY1 6AP [IO83LT, SD33]
G4 SZC Details withheld at licensee's request by SSL.
G4 SZD B Hamilton, 13 Moorside, Middlestone Moor, Spennymoor, DL16 7DY [IO94EQ, NZ23]
GM4 SZG J J Freeland, 48 Elgin Pl, Shawhead, Coatbridge, ML5 4JQ [IO75XU, NS76]
G4 SZI A M T Parry, 18 Spinney Ln, Rabley Heath, Welwyn, AL6 9TF [IO91VU, TL21]
GM4 SZJ I Walker, Old Smithy Cottage, Boreland, Lockerbie, Dumfriesshire, DG11 2LL [IO85IE, NY19]
G4 SZO K H Painter, 38 Meadowridge, Hatch Warren, Basingstoke, RG22 4QH [IO91KF, SU64]
GI4 SZP N M C K Hughes, 32 Kinedale Park, Ballynahinch, BT24 8YS [IO74BJ, J35]
GI4 SZQ K A Murphy, 8 Rosscolban Meadows, Kesh, Enniskillen, BT93 1UH [IO64DM, H16]
G4 SZR Prof W Hirst, 19 Lynch Hill Park, Whitchurch, RG28 7NF [IO91IF, SU44]
G4 SZS T T Harber, Oak Tree Farm, Northay, Whitestaunton, nr Chard Somerset, TA20 3DN [IO80LV, ST21]
G4 SZT D J Parry, 36 Park Rd, Duffield, Belper, DE56 4GR [IO92GX, SK34]
GI4 SZU J A McCurry, 30 Carrowadoon Rd, Dunloy, Ballymena, BT44 9DL [IO65TA, D01]
GW4 SZV D P Kirby Aberpporth YMCA ARC, Arosfa, 7 Heol Enlli, Tanygroes, Cardigan, SA43 2JE [IO72RC, SN24]
GI4 SZW M P J Keenan, 30 Ballynabee Rd, Camlough, Newry, BT35 7HD [IO64TE, J02]
G4 SZX M Stockton, 7 The Croft, Thorne, Doncaster, DN8 5TL [IO93MO, SE61]
GI4 SZY R E Wilson, 6 Oaklands Dr, Newtownabbey, BT37 0XE [IO74BQ, J38]
GI4 SZZ Nr L W Wilson, 31 Woodford Rd, Doagh Rd, Newtownabbey, BT36 6TS [IO74AQ, J38]
G4 TAA Details withheld at licensee's request by SSL.
G4 TAD M R Wooltorton, Chailly Hall Ln, Blundeston, Suffolk, NR32 5BL [JO02UM, TM59]
G4 TAE B G Robbins, 35 Berkeley House, Berkeley Rd, Staple Hill, Bristol, BS16 5HS [IO81RL, ST67]
G4 TAG T A Gammage, 31 Kennet Dr, Congleton, CW12 3RH [IO83VD, SJ86]
G4 TAH I P Conibear, 165 Goldcrest Rd, Chipping Sodbury, Bristol, BS17 6XL [IO81TM, ST78]
GI4 TAJ J T Bingham, 35 Rathmena Dr, Ballyclare, BT39 9HZ [IO64XS, J29]
G4 TAK J Hancock, 3548 Sanctuary Way South, Jacksonville Beach, Fl 32250, USA
GM4 TAL A Blyth, 73 Glassel Park Rd, Longniddry, EH32 0TA [IO85NX, NT47]
G4 TAM B L Chambers, 93 Main Rd, Hoo, Rochester, ME3 9EU [JO01GK, TQ77]
GI4 TAN F N Roberts, 38 Silverbirch Rd, Bangor, Co Down, N Ireland, N Ireland, BT19 2EU [IO74EP, J58]
G4 TAO S G N Rogers, Gaythorpe, Blacketts Wood Dr, Chorleywood, Rickmansworth, WD3 5QQ [IO91RP, TQ09]
GI4 TAP S F McCabe, 27 Baronscourt Rd, Carryduff, Belfast, BT8 8BQ [IO74BM, J36]
G4 TAS Details withheld at licensee's request by SSL.
G4 TAT D K Ruth, 1 Derwent Dr, Appley, Ryde, PO33 1NT [IO90KR, SZ69]
GW4 TAU E G Davies, Esgair y Gwynt, Caergelach, Llandegfan, Ynys Mon, LL59 5UF [IO73WF, SH57]
GI4 TAV M Doherty, 20 Drumcairn Cl, Belfast, BT8 8HQ [IO74AN, J36]
GJ4 TAW N R Perrott, 87 Winona Rd, Mount Eliza, Victoria 3930, Australia, X X
G4 TAX Details withheld at licensee's request by SSL.
G4 TAY D Sommerfield, 51 Spindletree Dr, Oakwood, Derby, DE21 2DG [IO92GW, SK33]
G4 TAZ F T Rewaj, 4 Dunham Cl, Home Farm Grange, Alsager, Stoke on Trent, ST7 2XR [IO83UC, SJ75]
G4 TBA Details withheld at licensee's request by SSL.
G4 TBD S J Pain, 12 Naunton Way, Ashbury Park, Weston Super Mare, BS22 9QW [IO81MI, ST36]
G4 TBG D F Smith, 49 Suffield Rd, High Wycombe, HP11 2JN [IO91OP, SU89]
G4 TBI P J Cornell, 22 Ravine Rd, Bournemouth, BH5 2DU [IO90CR, SZ19]
G4 TBJ Details withheld at licensee's request by SSL.[Op: R J Smith. Station located near Solihull.]
G4 TBK D Nix, 75 Mayfield Rd, Chaddesden, Derby, DE21 6FX [IO92GW, SK33]
G4 TBL Details withheld at licensee's request by SSL.
G4 TBM M T Tester, 6 Harvard Cl, Lewes, BN7 2EJ [JO00AV, TQ41]
G4 TBN C M Anderson, Wyke Cottage, Stubbs Rd, Everdon, Daventry, NN11 3BN [IO92KF, SP65]
G4 TBO C A Harris, 28 Castle Hill, Barnwell, Banwell, Weston Super Mare, BS24 6NY [IO81NH, ST45]
GI4 TBP N D Greer, Ticino, 82 Purdysburn Hill, Ballylesson, Belfast, BT8 8JZ [IO74AM, J36]
G4 TBR A A E Carter, Nordlys, Brays Ln, Hyde Heath, Amersham, HP6 5RT [IO91QQ, SP90]
G4 TBU J Sutherland, 3 Quantock Rd, Scartho, Grimsby, DN33 3AU [IO93WM, TA20]
GI4 TBV J Dilworth, Oakdene, 58 Corr Rd, Dungannon, Co. Tyrone, N Ireland, BT71 6HH [IO64PM, H86]
G4 TBY E Bockhoefer, 56 The Portway, Kingswinford, DY6 8HL [IO82WL, SO88]
G4 TBZ C J G Knowles, 67 Commercial Rd, Skelmanthorpe, Huddersfield, HD8 9DX [IO93EO, SE21]
G4 TCA R W Hawthorn, 23 Quail Green, Wightwick, Wolverhampton, WV6 8DF [IO82VO, SO89]
G4 TCB P Holland, 9 Garmont Rd, Leeds, LS7 3LY [IO93FT, SE33]
G4 TCC A J Hawkins, 23 Scotland Ln, Bartley Green, Birmingham, B32 4BP [IO82XK, SO98]
G4 TCE P T Wade, 356 Shirehall Rd, Sheffield, S5 0JP [IO93GK, SK39]
G4 TCF W H C Nichols, 90 Ravenhill Dr, Codsall, Wolverhampton, WV8 1BL [IO82VP, SJ80]
G4 TCG R Tams, 4 Langdale Cl, Fryston, Castleford, WF10 2RB [IO93IR, SE42]
G4 TCI M J Soars, 84 Ridge Rd, Kingswinford, DY6 9RG [IO82VL, SO88]
G4 TCJ Details withheld at licensee's request by SSL.
G4 TCK N L Nicholls, 4 Bridgwater Cl, Telford, TF4 3TP [IO82SP, SJ60]
G4 TCL Details withheld at licensee's request by SSL.
G4 TCM W T Smith, 65 Larchwood Rd, Yew Tree Est, Walsall, WS5 4HE [IO92AN, SP09]
G4 TCN Details withheld at licensee's request by SSL.
G4 TCO D L Preece, 12 South Dene, Stoke Bishop, Bristol, BS9 2BW [IO81QL, ST57]
G4 TCP J M Caddick, 1 Newmarket Cl, Chippenham, SN14 0FE [IO81WL, ST97]
G4 TCQ P J Vaughan, 1 Pound Piece Cottage, Main Rd, Hallow, Worcester, WR2 6PW [IO82UF, SO85]
GI4 TCR A C J Jackson, Shantara, 21 Carnreagh, Hillsborough, BT26 6LJ [IO64XL, J25]
GI4 TCS W J Jackson, Shantara, 21 Carnreagh, Hillsborough, BT26 6LJ [IO64XL, J25]
G4 TCT P W Johnson, 5 Moorside Dr, Drighlington, Bradford, BD11 1HE [IO93ES, SE22]
G4 TCU Details withheld at licensee's request by SSL.
G4 TCX D H Shingler, 39 Oaklands, Bridgnorth, WV15 5DU [IO82TM, SO79]
G4 TCZ Details withheld at licensee's request by SSL.
G4 TDA K Rimmer, 13 Crowthorns, Brownsover, Rugby, Warks, CV21 1PP [IO92JJ, SP57]
G4 TDB D J Waterhouse, 19 Finsbury Dr, Amblecote Bank, Brierley Hill, DY5 3NY [IO82WL, SO98]
G4 TDC L Wilson, Eastwood, Common Rd, Barkston Ash, Tadcaster, LS24 9PQ [IO93JT, SE43]
G4 TDE T B Lawrence, 21 Henderson Dr, Dartford, DA1 5LE [JO01CK, TQ57]
G4 TDF R A Copsey, 19 Windsor Dr, Solihull, B92 8HS [IO92CK, SP18]
G4 TDG R J Dowson, 12 Northd. Ave, Bishop Auckland, DL14 6AW [IO94DP, NZ22]
G4 TDI D J Day, 40 Hirstlands Dr, Off Kingsway, Ossett, WF5 8EJ [IO93EQ, SE22]
G4 TDL D Lemin, Beehive Manor, Cox Green Ln, Maidenhead, SL6 3ET [IO91PM, SU87]
G4 TDN C J A Campbell, 30 Temple Gr, Burgess Hill, RH15 9XN [IO90WW, TQ21]
G4 TDO B A Fereday, 16 Glentworth Gdns, Dunstall Hill, Wolverhampton, WV6 0SF [IO82WO, SJ90]
G4 TDP S L Bowden, 62 Manor House Rd, Wednesbury, WS10 9PH [IO82XN, SO99]
G4 TDQ L R Ashton, Bacton Wood Mill, Spa Common, North Walsham, Norfolk, NR28 9SH [JO02QT, TG23]
G4 TDR C E Jay, Hill House, Badgers Orchard, Bricklehampton, Pershore, WR10 3HJ [IO82XB, SO94]
G4 TDS Details withheld at licensee's request by SSL.
G4 TDU B J Brandon, 8 Moor Park Ave, Castleton, Rochdale, OL11 3JG [IO83VO, SD81]
G4 TDV R Stokes, Webbers Farm, Woodbury, Exeter, Devon, EX5 1EA [IO80HQ, SY08]
G4 TDW R Maskew, 23 Daventry Rd, Rochdale, OL11 2LN [IO83WO, SD81]
G4 TDY C C Eaton, 10 Chatteris Dr, Breadsall, Derby, DE21 4SF [IO92GW, SK33]
G4 TDZ A Jones, Ordsall Post Office, Ollerton Rd, Retford, Notts, DN22 7TH [IO93MH, SK77]
GI4 TEA D Callen, 35 Hillside Park, Ransevyn, Whitehead, Carrickfergus, BT38 9LJ [IO74DS, J49]
G4 TEB P R Larbalestier, 54 Churchill Ave, Halstead, CO9 2BE [JO01HW, TL83]
G4 TEC W Hartshorne, 11 Pentland Gdns, Wolverhampton, WV3 9JY [IO82WO, SO89]
GI4 TED K J Doherty, 77 Drumflugh Rd, Benburb, Dungannon, BT71 7QF [IO64OK, H85]
GW4 TEE V G Jones, 2 Grasmere Cres, Buckley, CH7 3LB [IO83LE, SJ26]
G4 TEG G G E Ellis, 12 St. Veronica Rd, Deepcar, Sheffield, S30 5TP [IO93GJ, SK38]
G4 TEH J B Andrews, 30 Northill Rd, Ickwell, Biggleswade, SG18 9ED [IO92UC, TL14]
G4 TEI A T Johnson, Flat 1, 38 South St., Bourne, PE10 9LY [IO92TS, TF01]
GW4 TEJ D J Honeysett, 23 Williamson St., Orange Gdns, Pembroke, SA71 4ER [IO71MQ, SM90]
G4 TEL N S Balmforth, 10 Weatherhill Cres, Birchencliffe, Huddersfield, HD3 3QZ [IO93CP, SE11]
G4 TEM J S Freestone, c/o Kololi Holiday Services, Pigeon Cottage, Ilsington, Newton Abbot, TQ13 9RE [IO80DN, SX77]
G4 TEP L A Kennedy, 69 Drayton Rd, Borehamwood, WD6 2DA [IO91UP, TQ19]
GW4 TEQ J R Mattocks, The Cottage, High St, Bangor Is y Wed, Wrexham, LL13 0AU [IO83NA, SJ34]
G4 TER R S J Hall, 238 Whitmore Way, Basildon, SS14 2NN [JO01FN, TQ78]
GW4 TES K T I Hyatt, 55 Llanerch Cres, Brynteg, Gorseinon, Swansea, SA4 4FP [IO71XQ, SS59]
G4 TET Details withheld at licensee's request by SSL.
G4 TEU J W Burr, Lorne Croft, Wellpond Green, Standon, Ware, SG11 1NJ [JO01AV, TL42]
G4 TEX M L Crookall, 35 Barn Ln, Budleigh Salterton, EX9 6QG [IO80IP, SY08]
G4 TEY Fr J F Morris, Our Ladys Convent, Park Rd, Loughborough, LE11 2EF [IO92JS, SK51]
G4 TEZ J J E Lever, 2 Chilham Cl, Over, Winsford, CW7 1LS [IO83RE, SJ66]
G4 TFA B Neilson, 14 Argyle Rd, Ferrybridge, Knottingley, WF11 8LZ [IO93IR, SE42]
G4 TFB B Gray, 29 Verity Walk, Wordsley, Stourbridge, DY8 4XS [IO82WL, SO88]
G4 TFC D Gray, 29 Verity Walk, Wordsley, Stourbridge, DY8 4XS [IO82WL, SO88]
G4 TFD B Pickard, 407 Highfield Rd, Idle, Bradford, BD10 8RS [IO93DU, SE13]
G4 TFE Details withheld at licensee's request by SSL.[Station located near Hastings.]
G4 TFF P Ayre, 95 Quantock Rd, Bridgwater, TA6 7EJ [IO81LD, ST23]
G4 TFH P J Mackinven, 43 Stringers Ln, Aston, Stevenage, SG2 7EF [IO91WV, TL22]
G4 TFI W E White, Twin Trees, White Elm Rd, Danbury, Chelmsford, Essex, CM3 4LR [JO01HQ, TL70]
GM4 TFJ E R I C Wallace, Lochiel Villa, Achintore Rd, Fort William, PH33 6RQ [IO76KT, NN07]
GU4 TFM Rev A D H Davis, The Flat, Rodney House, Newtown Rd, Alderney, Guernsey, GY9 3XP
G4 TFN J F Stadon, 47 Stonesby Ave, Leicester, LE2 6TX [IO92KO, SK50]
G4 TFP P A Cranmer, 38 Barbrook Ln, Tiptree, Colchester, CO5 0EF [JO01JT, TL81]
GW4 TFS A G Jones, 6 Gower View, Llanelli, SA15 3SN [IO71WQ, SN50]
G4 TFT C Greenwood, 3 Moorfield Dr, Oakworth, Keighley, BD22 7EX [IO93AU, SE03]
G4 TFU Dr A F Gerrard, 13 Wentworth Ave, Timperley, Altrincham, WA15 6NG [IO83UJ, SJ78]
G4 TFV F Kelly, 2 Victoria Terr, Leeds, LS3 1BX [IO93FT, SE23]
G4 TFW H W Guy, 24 The Mall, Binstead, Ryde, PO33 3SF [IO90JR, SZ59]
GW4 TFX B G James, 9 Brangwyn Cl, Clasemont, Morriston, Swansea, SA6 6AS [IO81AQ, SS69]
G4 TFZ D R Jarvis, 21 Ashurst Pl, Stannington, Sheffield, S6 5LN [IO93FJ, SK38]
GW4 TGA S E Marvelley, 12 Sunnybank Cl, West Cross, Swansea, SA3 5HB [IO71XO, SS68]
G4 TGB D Meadows, Alston, 39 Sylvester St., Mansfield, NG18 5QS [IO93JD, SK56]
GM4 TGC E R J Orbell, 1 Macrae Cres, Kincraig, Kingussie, PH21 1NN [IO87AD, NH80]
G4 TGE J W Pullen, 71 Barrow Rd, Barton upon Humber, DN18 6AE [IO93SQ, TA02]
GW4 TGF C Romano, The Glen, Glen Rd, West Cross, Swansea, SA3 5QJ [IO71XN, SS68]
G4 TGG Details withheld at licensee's request by SSL.[Op: G M Sifford, Hotel Corniche, 21 Mount Wise, Newquay, Cornwall, TR7 2BQ.]
G4 TGJ R P Tomlinson, 6 Badger Gate, Threshfield, Skipton, BD23 5EN [IO84XB, SD96]
G4 TGK W J Wimble, 87 Rolfe Ln, New Romney, TN28 8JL [JO00LX, TR02]
GW4 TGL W J Protheroe-Thomas, 51 Penygroes Rd, Caerbryn, Ammanford, SA18 3DQ [IO71XT, SN51]
G4 TGM A F Sherratt, Anlyn, Norbury Dr, Brierley Hill, DY5 3DP [IO82WL, SO98]
G4 TGN B J Newman, 405 Sipson Rd, Sipson Village, West Drayton, UB7 0HY [IO91SL, TQ07]
GW4 TGO G M Davies, 25 Bridgend Rd, Maesteg, CF34 0NN [IO81EO, SS89]
G4 TGP R Barling, Grovehill Farm, Turweston, Brackley, Northants, NN13 5JH [IO92KA, SP63]
G4 TGQ P J Creighton, 6 Kirkwall Dr, Penketh, Warrington, WA5 2HX [IO83QJ, SJ58]
GI4 TGR T Greer, Ticino, 82 Purdysburn Hill, Ballylesson, Belfast, BT8 8JZ [IO74AM, J36]
G4 TGS D Swarbrook, 6 Westview Cl, Leek, ST13 8ES [IO83XC, SJ95]
GW4 TGT T R Threlfall, 14 Clarence St., Pembroke Dock, SA72 6JP [IO71MQ, SM90]
G4 TGU A Mason, 5 Birch Rd, Kippax, Leeds, LS25 7DY [IO93HS, SE43]
G4 TGV I K Soaft, 55 The Close, Thurleigh, Bedford, MK44 2DT [IO92SF, TL05]
G4 TGW D C Starmer, 20 Garners Way, Harpole, Northampton, NN7 4DN [IO92MF, SP66]
G4 TGX C M Bigg-Wither, 49 Delamere Rd, Colchester, CO4 4NH [JO01LV, TM02]
G4 TGZ J K Chaudhry, 3 Norfolk Gdns, Littlehampton, BN17 5PE [IO90RT, TQ00]
G4 THA M G Crook, 140 Deepdale Rd, Preston, PR1 6PY [IO83PS, SD53]
G4 THC M J Arnison, 57 Heywood Rd, Cinderford, GL14 2QU [IO81ST, SO61]
G4 THD Details withheld at licensee's request by SSL.
G4 THE M J Round, 4 James Dee Cl, Quarry Bank, Brierley Hill, DY5 1DH [IO82WL, SO98]
G4 THF B Smith, 63 Hitchin Rd, Stotfold, Hitchin, SG5 4HT [IO92VA, TL23]
G4 THG J E Thompson, 1 Berkley Cl, Chippenham, SN14 0PS [IO81WK, ST97]
G4 THH D J Patterson, 4 Cliffe View, Prune Park Ln, Allerton, Bradford, BD15 9JQ [IO93CT, SE13]
G4 THI A J Robson, 27 Gray Fallow, The Chine, South Normanton, Alfreton, DE55 3BQ [IO93IC, SK45]
G4 THJ Details withheld at licensee's request by SSL.
GW4 THK W A Moore, 2 Heol Cae Glas, Bryncethin, Bridgend, CF32 9UG [IO81FM, SS98]
G4 THL Details withheld at licensee's request by SSL.
G4 THN M G Anthony, Middlewood House, Blacksmiths Ln, Forward Green, Stowmarket, IP14 5ET [JO02ME, TM06]
GM4 THP D S Last, 96 Dunlin Rd, Cove Bay, Aberdeen, AB1 3WD [IO87TH, NJ72]
G4 THR Details withheld at licensee's request by SSL.
G4 THT M N J Veasey, Linden Cottage, Upton, Snodsbury, Worcester, WR7 4NH [IO82WE, SO95]
G4 THU J R Read, 35 Maytree Hill, Droitwich, WR9 7QU [IO82WG, SO96]
G4 THV R Biddlecombe, Mystique Too, Maldon Rd, Steeple, Essex, CM0 7RT [JO01JQ, TL90]
G4 THW J A C Ingram, Highwood House, London Rd, Shrewton, Salisbury, Wilts, SP3 4DN [IO91BE, SU04]
G4 THX G Donoughue, 1 Kings Mount, Leeds, LS17 5NS [IO93FU, SE33]
G4 THY K J Cornes, 6 Haywood Heights, Little Haywood, Stafford, ST18 0UR [IO92AT, SK02]
G4 TIA S J Jarvis, 1 Wakenslade Cttgs, School House, Thorncombe, Chard, TA20 4PJ [IO80NT, ST30]
G4 TIC B Helman, 26 Fitzroy Dr, Leeds, LS8 1RW [IO93FT, SE33]
G4 TID D Hall, 6 St. Augustine Dr, Droitwich, WR9 8QR [IO82WG, SO96]
GW4 TIE R T Jago, 30 St. Margarets Cl, Haverfordwest, SA61 1LD [IO71MS, SM91]
G4 TIF M L Jones, 5 Congreve Cl, Warwick, CV34 5RQ [IO92FH, SP26]
G4 TIG W J Pope, Maldon House, 69 Sidford Rd, Sidmouth, EX10 9LR [IO80JQ, SY18]
G4 TIH C R Kay, 26 Clare Cl, Elstree, Borehamwood, WD6 3NJ [IO91UP, TQ19]
G4 TIJ Details withheld at licensee's request by SSL.
G4 TIK J F A Doyle, Avondale, 4 Mavis Gr, Cookridge, Leeds, LS16 7LN [IO93EU, SE24]
G4 TIL C A Finnis, 28 Lime Rd, Southam, Leamington Spa, CV33 0EQ [IO92HG, SP46]
G4 TIM G T Clementson, 74 Pentley Park, Welwyn Garden City, AL8 7SG [IO91VT, TL21]
G4 TIN Details withheld at licensee's request by SSL.
G4 TIO S P Blackmore, Les Camelias, 15 Ave Saint Donatien, 06600 Antibes, France
G4 TIS L F de Faubert Maunder, 39 Fermor Way, Crowborough, TN6 3BE [JO01CB, TQ52]
G4 TIU J Owens, 23 Oaktree Way, Little Sandhurst, Camberley, Surrey, GU17 8QS [IO91OH, SU85]
G4 TIV C C Roper, 12 Canford Dr, Allerton, Bradford, BD15 7AU [IO93CT, SE13]
G4 TIW D A Wood, 13 Brunswick Rd, Pudsey, LS28 7NA [IO93ET, SE23]
G4 TIX H N Woolrych, 20 Meadow Dr, Devizes, SN10 3BJ [IO91AI, SU06]
GW4 TIZ P W Wyles, The Lawns, Halkyn Rd, Holywell, CH8 7SJ [IO83JG, SJ17]
G4 TJA P E Prosser, 47 Devereux Dr, Watford, WD1 3DD [IO91TQ, TQ09]
G4 TJB Details withheld at licensee's request by SSL.
G4 TJC Dr S J Melhuish, 18 Thistleton Cl, Macclesfield, SK11 8BE [IO83WG, SJ97]
GM4 TJD M J Maclennan, 10 Ruilick, Beauly, Inverness Shire, IV4 7EY [IO77SL, NH54]
G4 TJE Details withheld at licensee's request by SSL.[Op: K Lewis. Station located near Sidcup.]
G4 TJH T G Sumner, 13 Yew Tree Gdns, Harrogate, HG2 9JU [IO93FX, SE25]
G4 TJI R A Slim, 55 Fairways Dr, Harrogate, HG2 7ER [IO93GX, SE35]
G4 TJJ A Shelbourne, 10 Church St. West, Brampton, Chesterfield, S40 3AG [IO93GF, SK37]
G4 TJK M C Porter, 6287 Tweedholm Ct, San Jose, California 95120, U S A
GM4 TJL J Hebborn, Elysian Fields, Achindaul, Spean Bridge, Inverness-shire, PH34 4EX [IO76MV, NN18]
G4 TJM E J Mordas, 18 Ellsworth Rd, High Wycombe, HP11 2TX [IO91OO, SU89]
GW4 TJN G E Smallwood, Rushbrit, Old Rd, Bwlchgwyn, Wrexham, LL11 5UF [IO83KB, SJ25]
G4 TJO J W Orson, 24 Stanhope Ave, Horsforth, Leeds, LS18 5AR [IO93EU, SE23]
G4 TJP Details withheld at licensee's request by SSL.
GW4 TJQ J H Wallis, 27 Wingfield Rd, Whitcurch, Cardiff, CF4 1NJ [IO81JM, ST17]
G4 TJT J L E Finnis, 28 Lime Rd, Southam, Leamington Spa, CV33 0EQ [IO92HG, SP46]
G4 TJU F S Richards, 9 Dales Gr, Worsley, Manchester, M28 7JW [IO83TM, SD70]
G4 TJX J R Blythe, Trethullan Farmhouse, Sticker, St Austell, Cornwall, PL26 7EH [IO70NH, SW95]
G4 TJY L M Barker, 98 Quinton Rd, Needham Market, Ipswich, IP6 8DA [JO02MD, TM05]
G4 TJZ W L Stacey, Ifield, Brady Rd, Lyminge, Folkestone, CT18 8EY [JO01ND, TR14]
G4 TKA Details withheld at licensee's request by SSL.
G4 TKF P J Tuck, 178 St. Ediths Marsh, Bromham, Chippenham, SN15 2DJ [IO81XJ, ST96]
G4 TKI T A Wilson, Woodlands Bungalow, Dog Kennel Hill, Kiveton Park Statio, Sheffield, S31 8NG [IO93GJ, SK38]
G4 TKJ Details withheld at licensee's request by SSL.[Station located in London SW4.]
G4 TKM I Jinks, 5 Pineapple Gr, Kings Heath, Birmingham, B30 2TJ [IO92BK, SP08]
G4 TKN Details withheld at licensee's request by SSL.[Op: A R Unsworth. Station located near Northwich.]
G4 TKO J Sharman, Forge Cottage, 102 Commercial Rd, Skelmanthorpe, Huddersfield, HD8 9DS [IO93EO, SE21]
G4 TKP R W Peel, 3 Martins Hill Ln, Burton, Christchurch, BH23 7NJ [IO90CS, SZ19]
G4 TKQ P A Steel, 154 Station Rd, Ratby, Leicester, LE6 0JP [IO92JP, SK50]
G4 TKR Details withheld at licensee's request by SSL.
G4 TKS J Clancy, 60 Brunels Way, Highbridge, TA9 3LG [IO81MF, ST34]
G4 TKT W L Tregonning, 41 Porthpean Rd, St. Austell, PL25 4PN [IO70OI, SX05]
G4 TKV A R Hooppell, 41 Moorland Way, Exwick, Exeter, EX4 2ER [IO80FR, SX99]
G4 TKW D E Hamilton, 38 Gosport Rd, Lee on The Solent, PO13 9EN [IO90JT, SZ59]
G4 TKX A H Freeman, 6 Holly Cl, Threemilestone, Truro, TR3 6TX [IO70KG, SW74]
G4 TKY E L Beamer, 79 Viking Rd, Bridlington, YO16 5TW [IO94VC, TA16]
G4 TKZ A C Bartram, 36 Park Ln, Norwich, NR2 3EF [JO02PP, TG20]
G4 TLE H E G Kennard, Chestnut Cottage, Main St., Peasmarsh, Rye, TN31 6UL [JO00IX, TQ82]
G4 TLJ A M Hall, 200 Nottingham Rd, Ripley, DE5 3AZ [IO93HB, SK45]

G4 TLK O W Kemp, 45 Merlin Mews, Sprowston, Norwich, NR7 8BZ [JO02PP, TG21]
G4 TLL J H T Jones, Amusement Depot, Station Rd, Cullompton, EX15 1BQ [IO80HU, ST00]
G4 TLM B Jennings, 28 Dickinson St., Wakefield, WF1 3PR [IO93GQ, SE32]
G4 TLN T S Appleby, 22 Kirklinton Rd, Marden Est, North Shields, NE30 3AX [IO95GA, NZ37]
G4 TLO P A Johnson, 79 Wheatlands, Fareham, PO14 4SU [IO90IU, SU50]
G4 TLQ H Diffey, 3 Slinn Rd, Christchurch, BH23 3AN [IO90CR, SZ19]
G4 TLR B Richards, 60 Twycross Gr, Castle Bromwich, Birmingham, B36 8LD [IO92CM, SP18]
G4 TLS J A Norton, 47 Farhalls Cres, Horsham, RH12 4BT [IO91UB, TQ13]
G4 TLV R Vincent, 68 Shelley Dr, Broadbridge Heath, Horsham, RH12 3NT [IO91TB, TQ13]
G4 TLW H Allen, 425 Broadway, Chadderton, Oldham, OL9 8AP [IO83WM, SD80]
G4 TLY E C Holmes, 36 Corn Gastons, Malmesbury, SN16 0DR [IO81WO, ST98]
G4 TMA P R Fielding, 14 Caldicot Way, Carleton, Poulton-le-Fylde, FY6 7LP [IO83LU, SD34]
GI4 TMB M T Beggs, 46 Donard Ave, Bangor, BT20 3QD [IO74DP, J48]
G4 TMC P N Barnett, 8 Parsonage Rd, Horsham, RH12 4AR [IO91UB, TQ13]
G4 TMD L P Downes, 357 Stone Rd, Stafford, ST16 1LD [IO82WT, SJ92]
G4 TME W J Black, 5 River View, Dalgety Bay, Dunfermline, KY11 5YE [IO86HA, NT18]
G4 TMF P G Aisthorpe-Buckley, Bewick, 17 Pine View Rd, Verwood, Dorset, BH31 6LQ [IO90BV, SU00]
G4 TMG C J Sherwood, 14 Amberley Rd, Rustington, Littlehampton, BN16 2EF [IO90RT, TQ00]
G4 TMK P F Morrall, 22 Chudleigh Rd, Erdington, Birmingham, B23 6HB [IO92BM, SP19]
G4 TML B S Parr, 5 Ashes Ln, Almondbury, Huddersfield, HD4 6TE [IO93CP, SE11]
G4 TMO Details withheld at licensee's request by SSL.[Licence holder abroad.]
G4 TMQ J Martin, 22 Wansbeck Ct, Front St. East, Bedlington, NE22 5BU [IO95ED, NZ28]
GM4 TMS H D W Martin Stirling&Dis AR, 11 Ewing Ct, Broomridge, Stirling, FK7 0QP [IO86AC, NS89]
G4 TMT Details withheld at licensee's request by SSL.
G4 TMU T Eglin, 27 Axbridge Ave, Sutton Leach, St. Helens, WA9 4NZ [IO83PK, SJ59]
G4 TMV E A Gale, 4 Waingap Cres, Whitworth, Rochdale, OL12 8PX [IO83VP, SD81]
G4 TMW Details withheld at licensee's request by SSL.
G4 TMX W T Armstrong, 121 Bede St., Roker, Sunderland, SR6 0NT [IO94HW, NZ45]
G4 TMY N I Hounslow, 18 Crompton Pl, Blackburn, BB2 6LW [IO83SR, SD62]
G4 TMZ D A Gillott, 132 Racecommon Rd, Barnsley, S70 6JY [IO93GN, SE30]
G4 TNA K J Pope, 305 Hulton Ln, Bolton, BL3 4LF [IO83SN, SD60]
G4 TNB P N Dollery, Greenways, 22 Barley Mead, Danbury, Chelmsford, CM3 4RP [JO01HR, TL70]
G4 TNE D J Horseman, 33 Chanters Hill, Barnstaple, EX32 8DN [IO71XB, SS53]
GW4 TNF T W Jones, 9 Hinsley Dr, Goulbourne Park, Wrexham, LL13 9QH [IO83MB, SJ35]
G4 TNG A I Henson, 5 Jackson Cl, Oadby, Leicester, Leics, LE2 4US [IO92LO, SP69]
G4 TNI Details withheld at licensee's request by SSL.
GM4 TNJ R A T O Milenkovic, 10 Loganbarns Rd, Dumfries, DG1 4BU [IO85FB, NX97]
G4 TNL A J Lee, 41 North Rd, Earls Barton, Northampton, NN6 0LP [IO92PG, SP86]
G4 TNM P A Smith, 116 Cromer Rd, Mundesley, Norwich, NR11 8DF [JO02RV, TG33]
G4 TNN Details withheld at licensee's request by SSL.
GM4 TNP J Burke, 25 Duncan Rd, Auchmuty, Glenrothes, KY7 4HS [IO86JE, NO20]
G4 TNQ E W Harvey, 8 Holly Cl, West Chiltington, Pulborough, RH20 2JR [IO90SW, TQ01]
G4 TNT Details withheld at licensee's request by SSL.
G4 TNU A Scott, 79 Westwood Dr, Amersham, HP6 6RR [IO91RQ, SU99]
GM4 TNV Details withheld at licensee's request by SSL.
GM4 TNW Details withheld at licensee's request by SSL.
G4 TNX L West, Cherry Tree Cottage, Church Ln, Bradley, Grimsby, South Humberside, DN37 0AE [IO93WM, TA20]
G4 TNY Details withheld at licensee's request by SSL.
G4 TNZ S I Cowley, 19 Melbourn Cl, Duffield, Belper, DE56 4FX [IO92GX, SK34]
G4 TOA G D Carlton, Sandyridge, 1 Horton Rd, Kinver, Stourbridge, West Midlands, DY7 6AL [IO82VK, SO88]
G4 TOC Details withheld at licensee's request by SSL.
GW4 TOD T W Hughes, Madryn, Llanbedrgoch, Ynys Mon, Gwynedd, LL76 8TZ [IO73VH, SH58]
GM4 TOE B J Horning, Kirkmichael School, Tomachlaggan, Tomintoul, Ballindalloch, AB37 9AR [IO87HG, NJ12]
G4 TOF C Key, 5 Griffiths Way, Aston Lodge Park, Little Stoke, Stone, ST15 8SB [IO82WV, SJ93]
G4 TOG B A Grainger, 23 Heath Rd, Hordle, Lymington, SO41 0GG [IO90ES, SZ29]
G4 TOH R T Russell, 23 Milfoil Ave, Conniburrow, Milton Keynes, MK14 7DY [IO92OB, SP83]
G4 TOI P C Andrews, 12 Cedarwood Ave, Tunstall, Sunderland, Tyne & Wear, SR2 9EJ [IO94HV, NZ35]
G4 TOJ Details withheld at licensee's request by SSL.
G4 TOK Details withheld at licensee's request by SSL.
G4 TOL Details withheld at licensee's request by SSL.
G4 TOM T H Turbert, 200 Salisbury Terr, Leeman Rd, York, YO2 4XP [IO93KX, SE55]
G4 TON P Smith, 15 Seaview Cttgs, Grimsby Rd, Louth, Lincs, LN11 0DY [IO93XI, TF38]
G4 TOO M R Final, 3 Borda Cl, Chelmsford, CM1 4JY [JO01FR, TL70]
GM4 TOQ A H Stewart, Three Acres, Cochno Rd, Hardgate, Clydebank Scotland, G81 6PU [IO75TW, NS47]
GI4 TOR A A Kincaid, 63 Carolhill Park, Toome Rd, Ballymena, BT42 2DG [IO64UU, D10]
G4 TOT R F James, Brantholme, Hasty Brow Rd, Slyne, Lancaster, LA2 6AG [IO84OB, SD46]
GW4 TOU W D Johns, 129 Cecil Rd, Gowerton, Swansea, SA4 3DN [IO71XP, SS59]
GD4 TOW K Bearpark, 2 Holmes Ct, Rowany Dr, Port Erin, IM9 6LW
G4 TOX J S Glover, Toot Hill, Ongar, Essex, CM5 9QW [JO01CQ, TL50]
G4 TOY R Handstock, 38 Watson Cl, Upavon, Pewsey, SN9 6AE [IO91CH, SU15]
G4 TOZ Details withheld at licensee's request by SSL.[Op: S Marchini.]
G4 TPB A H Greenfield, 72 Grecian Cres, London, SE19 3HH [IO91WK, TQ37]
G4 TPC C A Harrington, 8 Norton Springs, Norton Canes, Cannock, WS11 3TX [IO92AQ, SK00]
G4 TPD A C Hopkins, 16 Ashfields Rd, Shrewsbury, SY1 3SB [IO82PR, SJ41]
GM4 TPE G Allan Glasgow Batt., The Boys Brigade Ar, 22 Tynwald Ave, Rutherglen, Glasgow, G73 4RN [IO75VT, NS65]
GM4 TPF G L Davidson Inverness ARC, Balthangie, 37 Ballifeary Ln, Inverness, IV3 5PH [IO77VL, NH64]
GW4 TPG M Evans, 14 Heol Dewi, Hengoed, CF8 7NP [IO81HN, ST08]
G4 TPH T H Brockman, Spring Cottage, 6 Bagnor, Newbury, Berks, RG20 8AQ [IO91HK, SU46]
GI4 TPI Dr G H I Anderson, 13 Ashley Park, Bangor, BT20 5RQ [IO74EP, J58]
G4 TPJ R W Mepham, 36 Bramble Cl, Hildenborough, Tonbridge, TN11 9HQ [JO01DE, TQ54]
G4 TPK P F Phillips, 83 Arundel Rd, Benfleet, SS7 4EE [JO01GN, TQ78]
G4 TPL C D Prust, 32 Pengelly, Delabole, PL33 9AR [IO70PO, SX08]
G4 TPM A G Malcher, 68 Maryatt Ave, South Harrow, Harrow, HA2 0SX [IO91TN, TQ18]
GI4 TPN S A Nicholl, 40 Killane Rd, Limavady, BT49 0DN [IO65MB, C62]
G4 TPO S J McCulloch, 63 Comptons Ln, Horsham, RH13 5NL [IO91UA, TQ22]
G4 TPP A C Prichard, 22 The Firs, Coventry, CV5 6QD [IO92FJ, SP37]
GM4 TPQ W Milligan, 45 Hillview Rd, Darvel, KA17 0DQ [IO75UO, NS53]
GM4 TPR J Mitchell, 65 Robb Pl, Castle Douglas, DG7 1LW [IO84AW, NX76]
G4 TPS P J Stinton, 2 Rodinghead, Springwood, Kings Lynn, PE30 4TQ [JO02FS, TF62]
G4 TPV G Jones, Merlin Villa, 397 Fishponds Rd, Eastville, Bristol, BS5 6RJ [IO81RL, ST67]
G4 TPW H J Igglesden, Treeways, Littleworth Ln, Partridge Green, Horsham, RH13 8ER [IO91UB, TQ12]
GM4 TPX K J Gerard, 9 Overdale Cres, Prestwick, KA9 2DB [IO75QL, NS32]
GI4 TPY K B Boag, 12 Plantation Rd, Bangor, BT19 6AF [IO74EP, J58]
G4 TQA I Brookes, 20 School Ave, Guidepost, Choppington, NE62 5DN [IO95ED, NZ28]
G4 TQB Dr P K Grannell, 6 Fermain Cl, Seabridge, Newcastle, ST5 3EF [IO82VX, SJ84]
G4 TQC M J Anson, 15 Clover Ridge, Cheslyn Hay, Walsall, WS6 7DP [IO82XP, SJ90]
GW4 TQD J V P Gulley, 42 Burnt Barn Rd, Bulwark, Chepstow, NP6 5NG [IO81PP, ST59]
G4 TQE J G Burton, 43 Perry Rd, Rhewl, Gobowen, Oswestry, SY10 7BX [IO82LV, SJ33]
G4 TQF S K Jones, Ivyhouse, Pitt Ln, Lower Withington, Macclesfield, SK11 9ED [IO83UF, SJ86]
G4 TQG R Gray, 16 Goddards Cl, Little Berkhamsted, Hertford, SG13 8NA [IO91WS, TL20]
G4 TQH J T Heesom, 11 Bower Cres, Stretton, Warrington, WA4 4NF [IO83RI, SJ68]
G4 TQJ Details withheld at licensee's request by SSL.
G4 TQK H V Wright, 6 Great Croft, Dronfield Woodhouse, Dronfield, S18 8XR [IO93GH, SK37]
G4 TQL K G Prince, 75 Queens Ave, Bromley Cross, Bolton, BL7 9BJ [IO83TO, SD71]
GM4 TQN B Q Deans, 3 Beechwood Cres, Southmuir, Kirriemuir, DD8 5EE [IO86MQ, NO35]
G4 TQO P Fowler, 7 Ormonde Ave, Orpington, BR6 8JP [JO01AI, TQ46]
G4 TQP A E Hall, 11 Hornsea Cl, Brunswick Green, Wideopen, Newcastle upon Tyne, NE13 7HG [IO95EB, NZ27]
G4 TQR R Wilkes, 25 Lutterworth Rd, Aylestone, Leicester, LE2 8PH [IO92KO, SK50]
G4 TQS A Wallis, 27 Church Farm Rd, Upchurch, Sittingbourne, ME9 7AG [JO01HJ, TQ86]
G4 TQT I T Waller, 25 Livingstone Rd, Burncross, Chapeltown, Sheffield, S30 4UG [IO93GJ, SK38]
GW4 TQU A D Mason, 25 Hawarden Way, Mancot, Deeside, CH5 2EL [IO83LE, SJ36]
G4 TQV W G Ingleby, 18 Ross, Wentworth Park Est, Ouston, Chester-le-Street, DH2 1LB [IO94EV, NZ25]
G4 TQW J P Chambers, 216 High Barn Rd, Royton, Oldham, OL2 6RR [IO83WN, SD90]
G4 TQY M R B Addison, 6 Hanley Orchard, Hanley Swan, Worcester, WR8 0DS [IO82UB, SO84]
G4 TRA S M Redway, The Brambles, Synwell Ln, Edge, Wotton under Edge, GL12 7HQ [IO81TP, ST79]
G4 TRB R C Redway, The Pear Tree Inn, 6 Wotton Rd, Charfield, Wotton under Edge, GL12 8TP [IO81TP, ST79]
GM4 TRC L R Alexander Turriff Academy Radio Club, 97 Land St., Keith, AB55 5AP [IO87MN, NJ45]
G4 TRD R J E Dafter, 49 Balmoral Rd, Salisbury, SP1 3PZ [IO91CC, SU13]
G4 TRE B S Boon, 73 Avon Rd, Chelmsford, CM1 2JX [JO01FR, TL60]
G4 TRF I E Boon, 327 Broomfield Rd, Chelmsford, CM1 4DU [JO01FS, TL70]
G4 TRG I P Willmer, 30 Portland Rd, East Grinstead, RH19 4EA [IO91XC, TQ33]
GM4 TRH A R Macdonald, Greenacres, Stein, Waternish, Dunvegan, Isle of Skye, IV55 8GA [IO67RM, NG25]
G4 TRI S M March, 36 Handel Cl, Brighton Hill, Basingstoke, RG22 4DL [IO91KF, SU65]
G4 TRM S F P Burgess, Muston Farm, Winterborne Muston, Blandford, Dorset, DT11 9BV
G4 TRN J N Everingham, 17 Collingwood Rd, Redland, Bristol, BS6 6PD [IO81QL, ST57]
G4 TRP M J Fell, 77 Norfolk Gdns, Littlehampton, BN17 5PF [IO90RT, TQ00]
GM4 TRS A R Pierce, Mains of Auchreddie, New Deer, Turriff, Aberdeenshire, AB53 6SL [IO87VM, NJ84]
G4 TRT D J Dye, 15 Kingston Vale, Royston, SG8 9UG [IO92XB, TL34]
G4 TRU A C Thompson, Four Windows & A Door, 19 Wellswood Gdns, Redhills, Exeter, EX4 1RH [IO80FR, SX99]
G4 TRV R A Pears, 24 Westbourne Dr, St. Austell, PL25 5EA [IO70OI, SX05]
G4 TRW K E Prior, 14 Bincombe Rise, Weymouth, DT3 6AS [IO80SP, SY68]
GI4 TRX J O Nelson, 29 Carmavy Rd, Nutts Corner, Crumlin, BT29 4TG [IO64WP, J28]
GM4 TRY A G Moriarty, 1 Maclean Ct, East Kilbride, Glasgow, G74 4TH [IO75VS, NS65]
GM4 TRZ T G McLeod, 1 Lochside Cttgs, Otterston, Aberdour, Burntisland Fife, KY3 0RZ [IO86HB, NT18]
G4 TSA Details withheld at licensee's request by SSL.
G4 TSB Dr R F Cooper, 8 Hollyfield Dr, Barnt Green, Birmingham, B45 8HP [IO82XI, SO97]
G4 TSD R O Edwards, 8 Boney Hay Rd, Burntwood, Walsall, West Midlands, WS7 9AB [IO92BQ, SK00]
G4 TSF P R Scholefield, 10 Gainsborough Ave, Leeds, LS16 7PG [IO93EU, SE24]
GW4 TSG J R Williams, Cartref, Capel Garmon, Llanrwst, Gwynedd, LL26 0RG [IO83CB, SH85]
G4 TSH J M S Snow, 104 Redfern Ave, Hounslow, TW4 5LZ [IO91TK, TQ17]
GI4 TSK J M Skillen, Shalimar, 3 Copeland Dr, Comber, Newtownards, BT23 5JJ [IO74DN, J46]
G4 TSM Details withheld at licensee's request by SSL.
G4 TSN J W Lee, 46 Little Ln, Huthwaite, Sutton in Ashfield, NG17 2RA [IO93ID, SK45]
G4 TSO P J Oliver, 39 Long Wools, Paignton, TQ4 6HU [IO80FJ, SX85]
G4 TSQ M J Levett, 5 Park Rd, Yapton, Arundel, BN18 0JE [IO90QT, SU90]
G4 TSR Details withheld at licensee's request by SSL.[Op: M J Lee.]
G4 TSS A A Lock, 32 Chessington Ave, Hengrove, Bristol, BS14 9NN [IO81RJ, ST66]
G4 TST D W E Richardson, Holmside, Hambledon Rd, Denmead, Waterlooville, PO7 6PS [IO90LV, SU61]
G4 TSV J Robinson, 2 Bridge Mill Rd, Nelson, BB9 7BD [IO83VU, SD83]
G4 TSW S White Tiverton (Sw) Radio Club, 3 PO Box, Tiverton, EX16 6RS [IO80GV, SS91]
GW4 TTA J E Parry Dragon ARC, 2 Bryn Poeth, Tregarth, Bangor, LL57 4PG [IO73WE, SH66]
G4 TTB A Gordon, Valkyriegata 9, Oslo 0366, Norway, X X
G4 TTC P Howes, 43 Tanzieknowe Rd, Cambuslang, Glasgow, G72 8RD [IO75WT, NS65]
GM4 TTD N P Loughrey, 47 Obsdale Rd, Alness, IV17 0TU [IO77VQ, NH66]
G4 TTF T O Bevan Bishop Auckland AR Club, 6 Buttermere Gr, West Auckland, Bishop Auckland, DL14 9LG [IO94DP, NZ12]
G4 TTG H Bryant, 141 Shakespeare Rd, Fleetwood, FY7 7HH [IO83LW, SD34]
G4 TTI Details withheld at licensee's request by SSL.[Station located near Camberley.]
G4 TTJ J D Lee, 25 Coralin Gr, Cowplain, Waterlooville, PO7 8QY [IO90LV, SU61]
GI4 TTL M G Corcoran, 9 Demesne Gr, Moira, Craigavon, BT67 0DS [IO64VL, J16]
G4 TTM J W Betts, 65 Poplar Cl, Stone, ST15 0JB [IO82WV, SJ83]
G4 TTN G A Redgewell, 121 Gubbins Ln, Harold Wood, Romford, RM3 0DL [JO01CO, TQ59]
G4 TTO M Ellis Oswestry&Dis AR, Eagle Communications, Unit E3 Bank Top Ind. Est., St. Martins, Oswestry, Salop, SY10 7BB [IO82LW, SJ33]
G4 TTP J Tinsley, 141 Ruxley Rd, Bucknall, Stoke on Trent, ST2 9BT [IO83WA, SJ94]
G4 TTQ R F Philpot, 68 Brocksparkwood, Hutton, Brentwood, CM13 2TJ [JO01EO, TQ69]
G4 TTR R F Philpot, Stock Am Radio Group, 68 Brocksparkwood, Brentwood, CM13 2TJ [JO01EO, TQ69]
G4 TTS C H Harrison, 5 Windsor Ct, Wychwood Rd, Bingham, Nottingham, Notts, NG13 8TL [IO92MW, SK63]
GW4 TTU Details withheld at licensee's request by SSL.
G4 TTW Details withheld at licensee's request by SSL.
G4 TTX R F Smith, 405 Windmill Ave, Kettering, NN15 6PS [IO92PJ, SP87]
G4 TTY E N Macdonald, 7 Alder Cl, Crawley Down, Crawley, RH10 4UL [IO91XC, TQ33]
G4 TTZ R J Margolis, 12 Wyndham Ct, Yateley, Camberley, Surrey, GU17 7TT [IO91OH, SU85]
G4 TUA T Higgs, Town Head Farmhouse, Ravenstonedale, Kirkby Stephen, Cumbria, CA17 4NQ [IO84SK, NY70]
GW4 TUC D L Williams, 20 Clarach Rd, Borth, Dyfed, SY24 5NP [IO72XL, SN68]
GW4 TUD I W Williams, 4 Queen St., Aberystwyth, SY23 1PU [IO72XJ, SN58]
GI4 TUE Details withheld at licensee's request by SSL.
G4 TUH S T Elsdon, 22 The Swallows, Welwyn Garden City, AL7 1BY [IO91VT, TL21]
G4 TUI G Duffin Bromsgrove ARS, 20 Byron Rd, Headless Cross, Redditch, B97 5EB [IO92AG, SP06]
GI4 TUJ W G Konos, 27 Hillhead Rd, Ballynahinch, BT24 8LB [IO74AJ, J35]
G4 TUK R S Scarfe, Freshfields, Great Melton Rd, Little Melton, Norwich, NR9 3NR [JO02OO, TG10]
GW4 TUL F C Wybrew, 35 Turberville Rd, Northville, Cwmbran, NP44 1QQ [IO81LP, ST29]
G4 TUM J H Speakman, 33 Leyburn Ave, Bispham, Blackpool, FY2 9AQ [IO83LU, SD34]
G4 TUO E S Whitworth, 129A Broomhill, Downham Market, PE38 9QU [JO02EO, TF60]
G4 TUP D W Norris, 148 Sefton St., Southport, PR8 5DA [IO83MP, SD31]
G4 TUR S P Turner, 2 Painswick Rd, Hall Green, Birmingham, B28 0HH [IO92BK, SP18]
G4 TUS Details withheld at licensee's request by SSL.
G4 TUT R L Brunton, 181 Kestrel House, Alma Rd, Ponders End, Enfield, EN3 4QF [IO91XP, TQ39]
GI4 TUV R W Bailie, 26 Moatview Park, Dundonald, Belfast, BT16 0BE [IO74CO, J47]
G4 TUW F T Basden, Barnaway Farm, Needham, Harleston, Norfolk, IP20 9LW [JO02PJ, TM28]
G4 TUX J G Baines, 163 Windmill Ln, Sneinton, Nottingham, NG3 2BH [IO92KW, SK54]
G4 TUY Details withheld at licensee's request by SSL.
G4 TUZ P Boyd, 13 Stackbraes Rd, Longtown, Carlisle, CA6 5UR [IO85MA, NY36]
G4 TVA J E Wiles, 38 Northwood Ln, Clayton, Newcastle, ST5 4BN [IO82VX, SJ84]
G4 TVC J C Darby, 58 Cloverlands, Northgate, Crawley, RH10 2EH [IO91VC, TQ23]
G4 TVD G Hector, 54 Hudson Cl, Chipping Sodbury, Yate, Bristol, BS17 4NP [IO81TM, ST78]
GW4 TVE S M Edwards, Aelwyd-y-Don, Tresaith, Cardigan, Dyfed, SA43 2JH [IO72RD, SN25]
GD4 TVG T V Gill, 4 Orestal, Colby, Castletown, IM9 4AS
G4 TVJ A R Johns, 5 Oakfields, Greenways, Loddon, Norwich, NR14 6UT [JO02RM, TM39]
G4 TVK Details withheld at licensee's request by SSL.
G4 TVL D A W Hammond, 43 Barnes Cl, Blandford Forum, DT11 7NG [IO80WU, ST80]
G4 TVN B Yates, Bowland View, 39 Moss Ln, Garstang, Preston, PR3 1PD [IO83OV, SD44]
G4 TVP M C Taylor, 23 Wilding Rd, Belstead, Ipswich, IP8 3SG [JO02NA, TM14]
GW4 TVQ R J Thomas, 3 Tor y Mynydd, Baglan, Port Talbot, SA12 8LE [IO81CP, SS79]
G4 TVR S M Hawkins, 23 Micklehill Dr, Shirley, Solihull, B90 2PU [IO92CJ, SP17]
G4 TVT G Spencer, 83 Tuddenham Ave, Ipswich, IP4 2HG [JO02OB, TM14]
GW4 TVU V R Sedgebeer, 13 Ty Heddwch, Pen y Bryn, Coesern Cymmer, Port Talbot W.Glam, SA13 3SD [IO81EP, SS89]
G4 TVW R Stone, 51 Elaine Ave, Strood, Rochester, ME2 2YW [JO01FJ, TQ76]
G4 TVX R Lamb, 56 Hanbury, Orton Goldhay, Peterborough, PE2 5QU [IO92UN, TL19]
G4 TVZ P C Prosser, Piglets, King Ln, Clutton Hill, Clutton, Somerset, BS18 4QQ [IO81RI, ST65]
GW4 TWB F A Openshaw, 9 The Dale, Abergele, LL22 7DS [IO83EG, SH97]
G4 TWC D P Powell, 8 Cranbrook Dr, Sittingbourne, ME10 1RF [JO01IH, TQ86]
GW4 TWE M W Reid, 11 Hawthorne Ave, Buckley, Clwyd, CH7 2NQ [IO83KE, SJ26]
G4 TWG S A Greenwood, 39 Rydal Rd, Haslingden, Rossendale, BB4 4EF [IO83UQ, SD72]
G4 TWH G Wood-Hill, 26 Bramerton Rd, Hockley, SS5 4PJ [JO01HO, TQ89]
G4 TWI Details withheld at licensee's request by SSL.[Op: T R Harris.]
GW4 TWJ T W Jones, Egryn Hotel, Main St, Abersoch, Pwllheli, Gwynedd, LL53 7EE [IO72RT, SH32]
G4 TWK H J Hart, 20 Cowdray Dr, Goring By Sea, Worthing, BN12 4LH [IO90TT, TQ10]
G4 TWL T W Lee, 19A Imperial Ave, Mayland, Chelmsford, CM3 6AQ [JO01JQ, TL90]
G4 TWN A J Keates, 16 Hall Rd, Uttoxeter, ST14 7PN [IO92BV, SK03]
G4 TWP L G Miles, 5 Beechey Cl, Copthorne, Crawley, RH10 3LS [IO91WD, TQ33]
G4 TWR T T Kettlewell, 33 Westlode St., Spalding, PE11 2AF [IO92WS, TF22]
G4 TWS S R Holmes, 7 Parkland Cres, Old Catton, Norwich, NR6 7RQ [JO02PP, TG21]
G4 TWT H W M Holmes, 7 Parkland Cres, Old Catton, Norwich, NR6 7RQ [JO02PP, TG21]
G4 TWU I M Blake, 39 College Rd, Fishponds, Bristol, BS16 2HP [IO81RL, ST67]
G4 TWV M Allman, 52 Nairn Ave, Derby, DE21 6BU [IO92GW, SK33]
G4 TWW T M Bevan, 98 Heage Rd, Ripley, DE5 3GH [IO93GB, SK35]
G4 TXA D K McCartney, 3 Cwmcarn, Caversham, Reading, RG4 8LE [IO91ML, SU77]
GJ4 TXB Details withheld at licensee's request by SSL.
G4 TXC J Brennan, Drimla, Racecourse Rd, Dormansland, Lingfield, RH7 6PP [JO01AE, TQ44]
G4 TXD M J Robbins, 2 Tolview Terr, Hayle, TR27 4AG [IO70GE, SW53]
G4 TXE A M Goode, Tudor House, Chenhalls Rd, St Erth, Cornwall, TR27 6HJ [IO70GE, SW53]
G4 TXF C E White, 8 Hudson Cl, Pershore, WR10 1QL [IO82XC, SO94]
G4 TXG N C Hamilton, North View, Chawston, Beds, MK44 3BH [IO92UE, TL15]
G4 TXI B A Leader Chew, 26 Buckmans Rd, West Green, Crawley, RH11 7DR [IO91VC, TQ23]
G4 TXK T J Stanley, 35 Moorgate Rd, Kippax, Leeds, LS25 7ET [IO93HS, SE43]
G4 TXL A H Stevenson, Szabadsag U. 32, 2112 Veresegyhaz, Hungary, X X
G4 TXM G R Porter, 15 Cottesmore Ave, Barton Seagrave, Kettering, NN15 6QU [IO92PJ, SP87]
GM4 TXN A J Newlands, 5 Dyers Ct, Kelso, TD5 7NQ [IO85SO, NT73]
G4 TXO J Middleton, 29 Clementhorpe, York, YO2 1AN [IO93LW, SE65]
GM4 TXP Details withheld at licensee's request by SSL.
G4 TXR Details withheld at licensee's request by SSL.
G4 TXS Details withheld at licensee's request by SSL.
G4 TXT D C Wales, 91 Tower Dr, Neath Hill, Milton Keynes, MK14 6JX [IO92PB, SP84]
G4 TXV A C Turner, 13 Shakespeare Dr, Offmore Farm, Kidderminster, DY10 3QW [IO82VJ, SO87]
GM4 TXX Details withheld at licensee's request by SSL.
G4 TXY Details withheld at licensee's request by SSL.
G4 TYA C Carter, 26 Hopton Crofts, Leamington Spa, CV32 6NT [IO92FH, SP36]
GW4 TYB Details withheld at licensee's request by SSL.[Op: D S Hughes. Station located near Pwllheli.]
G4 TYD A N Kelly, Crowhurst Ln, Plaxtol, Borough Green, Sevenoaks, Kent, TN15 8PE [JO01DG, TQ65]

G4 TYF E Aston, 64 Gurney Valley, Cl House, Bishop Auckland, DL14 8RW [IO94EP, NZ22]
G4 TYG C A R M Armstrong, 32 Gloucester St., New Hartley, Whitley Bay, NE25 0RH [IO95FB, NZ37]
GW4 TYH R J Roberts, 31 Lon Hedyn, Rhyl, LL18 4JR [IO83GH, SJ08]
G4 TYI O C Wendels, 50 King St., Southwell, NG25 0EN [IO93MB, SK75]
G4 TYK Details withheld at licensee's request by SSL.[Op: M H Cartwright, 75 Whites Road, Bitterne, Southampton, SO2 7NR.]
G4 TYN D Bell, 22 Bosworth Way, Long Eaton, Nottingham, NG10 1EA [IO92IV, SK43]
G4 TYO G K Lilley, 100 Trentham Dr, Aspley, Nottingham, NG8 3NE [IO92JX, SK54]
G4 TYP K Ward, 9 Porlock Cl, Longeaton, Long Eaton, Nottingham, NG10 4NZ [IO92IV, SK43]
GM4 TYQ F A Rae, 18 Hawthorn Ave, Lenzie, Kirkintilloch, Glasgow, G66 4RA [IO75WW, NS67]
G4 TYR C R Miles, 23 Redacre Rd, Boldmere, Sutton Coldfield, B73 5EA [IO92BN, SP19]
G4 TYS B Willoughby, Willowbeck, Churchlaneham, Retford, Notts, DN22 0NQ [IO93OG, SK87]
G4 TYT A G Hunt, 141 Pickhurst Ln, Bromley, BR2 7HU [JO01AJ, TQ36]
GM4 TYU D R Clanachan, Alva Cottage, Berstane Rd, St. Ola, Kirkwall, KW15 1SZ [IO88MX, HY41]
G4 TYW R Wilson, 95 Longfield Rd, Todmorden, OL14 6ND [IO83WR, SD92]
G4 TYY J P Worley, 37 Fall Rd, Heanor, DE75 7PQ [IO93HA, SK44]
G4 TZA C L Read, 58 Somerset Rd, Chiswick, London, W4 5DN [IO91UL, TQ27]
G4 TZB T Brown, 20 Hillcrest, Middleton, Manchester, M24 5JA [IO83VN, SD80]
G4 TZE J Wilkinson, 34 Park Ln, Great Houghton, Barnsley, S72 0AX [IO93HN, SE40]
G4 TZF C Toby, 32 Swallow Rd, Langley Green, Crawley, RH11 7RF [IO91VC, TQ23]
G4 TZH R W Noble, 19 Foxglove Rd, Eastbourne, BN23 8BU [JO00DT, TQ60]
G4 TZI C R Perry, 58 Hadrian Ave, Dunstable, LU5 4SP [IO91SV, TL02]
G4 TZJ Details withheld at licensee's request by SSL.
G4 TZK G Prater, 297 Highfield Rd Nt, Chorley, Lancs, PR7 1PH [IO83QP, SD51]
G4 TZL K Rogers, 7 Buckleigh Rd, Wath upon Dearne, Rotherham, S63 7JB [IO93HL, SK49]
G4 TZM I A Paterson, 37 Remercie Rd, Mistley, Manningtree, CO11 1NF [JO01NW, TM13]
G4 TZN M W Randall, 29 Victoria St, Dinnington, Sheffield, S25 2SF [IO93JI, SK58]
G4 TZO P V Pledger, Mas Trabuch, Brunyola, Gerona 17441, Spain[Spanish callsign EA3FWZ]
G4 TZP Details withheld at licensee's request by SSL.
G4 TZQ D P Rouse, 14 Kestrel Cl, Downley, High Wycombe, HP13 5JN [IO91OP, SU89]
G4 TZR R Stringfellow, 48 Richmond Rd, Eccleston, Chorley, PR7 5SR [IO83PP, SD51]
G4 TZS A Atherton, 84 Cambourne Dr, Hindley Green, Hindley, Wigan, WN2 4TU [IO83RM, SD60]
G4 TZT K F Horton, 1 Mulberry Cl, Woolston, Warrington, WA1 4ED [IO83RJ, SJ68]
G4 TZV E P Cronin, 10 Wellington Gr, Waterloo Rd, Pudsey, LS28 8DG [IO93DT, SE23]
G4 TZW P Davidson, 86 Overchurch Rd, Upton, Wirral, L49 4NN [IO83KJ, SJ28]
G4 TZX G C Everest, 20 Seaway Rd, St. Marys Bay, Romney Marsh, TN29 0RU [JO01LA, TR02]
G4 TZZ A S Buick, 16 Berkeley Vale Park, Berkeley, GL13 9TG [IO81SQ, ST69]
G4 UAA J P Gaffney, 16 High Ridge, Cuffley, Potters Bar, EN6 4JH [IO91WR, TL30]
G4 UAC R Grant, 21 Eastgate, Deeping St. James, Peterborough, PE6 8HH [IO92UQ, TF10]
G4 UAE J R Hesford, 93 Old Ln, Rainford, St. Helens, WA11 8JJ [IO83OM, SD40]
G4 UAF J Higgins, 124 Cromwell Rd, South Kensington, London, SW7 4ET [IO91VL, TQ27]
G4 UAH P G Hutton, 40 Aston Mead, Windsor, SL4 5PP [IO91QL, SU97]
G4 UAI P M Cockman, 29 Kensington Rd, Southend on Sea, SS1 2SX [JO01IM, TQ88]
G4 UAL J P V Guffogg, 12 Vantage Walk, St. Leonards on Sea, TN38 0YP [JO00GU, TQ71]
G4 UAM A T Gould, 3 Clarkson Rd, Lingwood, Norwich, NR13 4BA [JO02RO, TG30]
G4 UAN G F Thompson, Tor View, Trevanson Rd, Wadebridge, PL27 7HD [IO70NM, SW97]
G4 UAQ I K Weston, 53 Dickens Rd, Maidstone, ME14 2QR [JO01GG, TQ75]
G4 UAS P C Rolfe, 4 Windgate, Waterside Park, Silsden, Keighley, BD20 0LG [IO93AV, SE04]
G4 UAT A Thomas, 10 Brisco Ave, Loughborough, LE11 5HB [IO92JS, SK52]
G4 UAU J T P Parish, 50 Far Hey Cl, Radcliffe, Manchester, M26 3GL [IO83TN, SD70]
G4 UAW M H Spillett, 56 North Ln, Rustington, Littlehampton, BN16 3PW [IO90RT, TQ00]
G4 UAX P S York, 1 Keswick Ave, Bromborough, Wirral, L63 0NP [IO83MH, SJ38]
G4 UAY D Grant, 115 Clayton Rd, Newcastle, ST5 3EW [IO82VX, SJ84]
G4 UAZ J G Hawes, 193 Leckhampton Rd, Cheltenham, GL53 0AD [IO81XV, SO91]
G4 UBB Eur Ing J R D Brown, 17 St. Ursula Gr, Pinner, HA5 1LN [IO91TO, TQ18]
G4 UBC K F Durrant, 26 Dozule Cl, Leonard Stanley, Stonehouse, GL10 3NL [IO81UR, SO80]
GM4 UBF A P Pontiero, 1 Dalmeny Rd, Hamilton, ML3 6PP [IO75XS, NS75]
G4 UBH H R White, Croft Cottage, Bere Ct Rd, Pangbourne, Reading, RG8 8JY [IO91KL, SU67]
G4 UBI A Priddy, 44 Frys Hill, Kingswood, Bristol, BS15 4QJ [IO81SL, ST67]
GM4 UBJ W J Tracey, 4 Finnie Wynd, Motherwell, ML1 2JJ [IO85AS, NS75]
G4 UBK K Martin, Nanscawen House, Luxulyan Valley, St Blazey Park, Cornwall, PL24 2SR [IO70PI, SX05]
G4 UBL L Bean, 2 Scotton Gr, Knaresborough, HG5 9HQ [IO94FA, SE35]
G4 UBM R D Bryant, 40 Hall Dr, Harefield, Uxbridge, UB9 6LA [IO91SO, TQ09]
GW4 UBQ C Powles, 14 Willow Cl, Four Crosses, Llanymynech, Powys, SY22 6NF [IO82LT, SJ32]
G4 UBR P J Richardson, Old Mill, The Dimple, Fritchley, Belper, DE56 2HP [IO93GB, SK35]
G4 UBT K J Stone, 63 Banks Rd, Pound Hill, Crawley, RH10 7BS [IO91WC, TQ23]
G4 UBV C P Murray, 175 Wolsey Dr, Kingston upon Thames, KT2 5DR [IO91UK, TQ17]
G4 UCC W Trinder, 354 Livesey Branch, Rd, Blackburn, BB2 4QJ [IO83RR, SD62]
G4 UCE B Davies, 9 Paisley Ave, Eastham, Wirral, L62 8DL [IO83MH, SJ37]
G4 UCH W D Hughes, 8 Coape Rd, Stockwood, Bristol, BS14 8TN [IO81RJ, ST66]
G4 UCI D A Luckhurst, 6 Dartmouth Ave, Stourbridge, DY8 5QE [IO82VL, SO88]
G4 UCJ Details withheld at licensee's request by SSL.
GW4 UCK G R Jones, Frondeg Chapel Rd, Three Crosses, Swansea, SA4 3PU [IO71XP, SS59]
G4 UCL A G Fallows, 72 Soutergate, Ulverston, LA12 7ES [IO84KE, SD27]
G4 UCM M M Clarkson, 55 Nares Rd, Witton, Blackburn, BB2 2TH [IO83RR, SD62]
G4 UCN E C Streatfield, 2 Gardiners Cl, Churchdown, Gloucester, GL3 2DX [IO81VV, SO82]
G4 UCO R J Seabrook, 1 Neal Rd, West Kingsdown, Sevenoaks, TN15 6DD [JO01DI, TQ56]
G4 UCR G C Watt, 5 Singleton, Chichester, PO18 0HA [IO90OV, SU81]
G4 UCT A Cooke, 1 Spring Gdns, Crewe, CW1 4AP [IO83SC, SJ75]
G4 UCU S A Hebel, 35 Rushton St., Barrowford, Nelson, BB9 6EA [IO83VU, SD83]
G4 UCW A J Hamilton, 8 Icepits Cl, Great Barton, Bury St. Edmunds, IP31 2PB [JO02JG, TL96]
G4 UCX P F Johnson, 5 Wellesley Rd, Ipswich, IP4 1PP [JO02OB, TM14]
G4 UCY C J Laird, 31 Fox Lease, Bedford, MK41 8AP [IO92SD, TL05]
G4 UCZ M E Kirk, 2 Denton Gdns, East Cowes, PO32 6EJ [IO90IS, SZ59]
G4 UDA T M Burton, 23 Foldyard Cl, Oakwood, Walmley, Sutton Coldfield, B76 1QZ [IO92CM, SP19]
G4 UDB C J Fay, 36 Shooters Hill Cl, Sholing, Southampton, SO19 1FW [IO90HV, SU41]
GM4 UDC Details withheld at licensee's request by SSL.
G4 UDD S Chapman, 2 Birds Croft, Great Livermere, Bury St. Edmunds, IP31 1JJ [JO02JH, TL87]
G4 UDE M W Ellis, Leighton, Bronygarth Rd, Weston Rhyn, Oswestry, SY10 7RQ [IO82LV, SJ23]
G4 UDF I A Fox, 40 Calow Dr, Leigh, Lancs, WN7 3DA [IO83SL, SJ69]
G4 UDG C S Fawkes, 31 Burland Rd, Waterhayes Village, Red St., Newcastle, ST5 7ST [IO83VA, SJ84]
G4 UDH P A Harley, 6 Hunts Back Dr, Waterhayes Village, Newcastle, Staffs, ST5 7TB [IO83UB, SJ85]
GI4 UDI J B F McCullagh, 53 Fernagh Rd, Omagh, BT79 0PL [IO64KP, H57]
G4 UDJ Details withheld at licensee's request by SSL.
G4 UDK R F Wood, 102 Ombersley Cl, Woodrow South, Redditch, B98 7UT [IO92BG, SP06]
G4 UDN C G Peake, 54 Farnham Rd, Poole, BH12 1PP [IO90BR, SZ09]
G4 UDO R Heron, 66 Derwentwater Rd, Mirehouse, Whitehaven, CA28 9RH [IO84FM, NX91]
G4 UDQ Details withheld at licensee's request by SSL.
G4 UDR Details withheld at licensee's request by SSL.[Op: T V Allen. Station located near the Wirral.]
G4 UDT Y A G Remedios, 44 Kingsway, Wembley, HA9 7QR [IO91UN, TQ18]
G4 UDU P L Godbold, 13 Dawn Cres, Beeding, Upper Beeding, Steyning, BN44 3WH [IO90UV, TQ11]
G4 UDV R A Green, 8 Briarbeck, Shelfield, Walsall, WS4 1XA [IO92AO, SK00]
G4 UDW P J Hersey, St Farm Cottage, Newchurch, Romney Marsh, Kent, TN29 0DZ [JO01LB, TR03]
GM4 UDX J Bell, 27 Grange Terr, Kilmarnock, KA1 2JR [IO75RO, NS43]
G4 UDY B Moorecroft, 4 St. Davids Rd, Locksheath, Locks Heath, Southampton, SO31 6EP [IO90IU, SU50]
G4 UDZ S R Tyler, 2 John Ct, Hoddesdon, EN11 9LZ [IO91XS, TL31]
G4 UEA P Robinson, 27 Warwick Dr, Brierfield, Nelson, BB9 0PP [IO83VT, SD83]
G4 UED G N Henstridge, 21 John Gay Rd, Amesbury, Salisbury, SP4 7NN [IO91CE, SU14]
G4 UEE D Jennison, 8 New Rd, Crich, Matlock, DE4 5BX [IO93GB, SK35]
G4 UEF K J Dalton, 79 Longcroft Rd, Kingsclere, Newbury, RG20 5TL [IO91JH, SU55]
GM4 UEH Rev A A Ford, The Manse, 7 Raebog Rd, Glenmavis, Airdrie, ML6 0NW [IO85AV, NS76]
GW4 UEJ M S Charman, Illimani, Stop and Call Hill, Goodwick, Dyfed, SA64 0ES [IO72LA, SM93]
G4 UEK M E Biddle, 8 Sunnindale Dr, Tollerton, Nottingham, NG12 4ES [IO92KV, SK63]
G4 UEL G Hollebon, 5 Evelyn Ave, Aldershot, GU11 3QB [IO91OF, SU84]
G4 UEN K E R Foskett, 2 Ambleside Gdns, Sholing, Southampton, SO19 8EY [IO90HV, SU41]
G4 UEO D G H Stewart, The Cottage, Chesterwood, Haydon Bridge, Northd., NE47 6HW [IO84UX, NY86]
GW4 UEP J K Morgan, 34 Maescurig, Newport, SA42 0RQ [IO72NA, SN03]
G4 UEQ T G Raybould, 9 Upper Albert Rd, Sheffield, S8 9HR [IO93GI, SK38]
G4 UER Details withheld at licensee's request by SSL.[Op: I Rundle.]
GM4 UEU Details withheld at licensee's request by SSL.
G4 UEV A C Gray, 59 Littlemead, Hatfield, AL10 0UQ [IO91VS, TL20]
G4 UEW D Hoose, 41 Parker Cres, Ormskirk, L39 1PJ [IO83NN, SD40]
G4 UEY B R Curtis, 10 Coppern Way, Stalbridge, Sturminster Newton, DT10 2NH [IO80TW, ST71]
G4 UFC P H Fretwell, 53 Stratford St., Cotmanhay, Ilkeston, DE7 8QZ [IO92IX, SK44]
GM4 UFD R B Gall, Ingleneuk, 49 Ugie St., Peterhead, AB42 1NX [IO97CM, NK14]
G4 UFF K Q Hickey, 12 Lewes House, Castle Dr, Reigate, RH2 8DF [IO91VF, TQ24]
G4 UFG A J Johnson, 27 Walden Ave, Moorside, Oldham, OL4 2PW [IO83XN, SD90]
G4 UFH Details withheld at licensee's request by SSL.[Op: D G Phillips. Station located near Christchurch.]
G4 UFJ N Taylor, The Olde Barn, 369A Leymoor Rd, Golcar, Huddersfield, West Yorks, HD7 4QQ [IO93BP, SE01]
G4 UFK A R Watts, 23 St. Marys Cl, Taddiport, Torrington, EX38 8AS [IO70WW, SS41]
G4 UFL N Wood, 244 Leymoor Rd, Golcar, Huddersfield, HD7 4QP [IO93BP, SE11]
G4 UFM T S Ellison, 116 Avondale Rd, Sparkhill, Birmingham, B11 3JY [IO92BK, SP08]
G4 UFN D E Knox, 50 Butt Ln, Manuden, Bishops Stortford, CM23 1BZ [JO01BW, TL42]
G4 UFO R Preston, 13 Partridge Mead, Banstead, SM7 1LN [IO91VH, TQ25]
GM4 UFP C D Ross, The Old Cttgs, Middlestead, Selkirk, TD7 5EY [IO85NM, NT42]
GW4 UFQ B Jackson, Bryn Tirion, Maes y Waen, Bala, Gwynedd, LL23 7SF [IO82EW, SH83]
G4 UFR E Horsfield, 13 St. Leonards Way, Ardsley, Barnsley, S71 5BS [IO93GN, SE30]
G4 UFS D F Pearson, 9 Nuneham Gr, Westcroft, Milton Keynes, Bucks, MK4 [IO92OA, SP83]
G4 UFU B D W Steen, 30 Shady Gr, Alsager, Stoke on Trent, ST7 2NH [IO83UC, SJ75]
G4 UFV B M Peacey, The Old Wagon Works, Pennsylvania, Chippenham, Wilts, SN14 8LD [IO81TK, ST77]
G4 UFW N Polson, 7 Whitefield Cl, Lymm, WA13 9QG [IO83SJ, SJ68]
G4 UFX D Blackwell, Rosegarth, 31 Main St., Horsley Woodhouse, Ilkeston, DE7 6AU [IO93HA, SK34]
G4 UFY W A Eccles, 22 Mount Ave, Worksop, S81 7JL [IO93KH, SK58]
G4 UFZ R C Greenwood, 15 Denham Dr, Netherthong, Holmfirth, Huddersfield, HD7 2FA [IO93CN, SE10]
G4 UGA A Jones, 215 Fernside Ave, Almondbury, Huddersfield, HD5 8PH [IO93DP, SE11]
GM4 UGB R C Bracegirdle, 69 High Parksail, Park Mains, Erskine, PA8 7HY [IO75SV, NS46]
G4 UGD I Clover, West Lodge, Oulton Park, Little Budworth, nr Tarporley, Ches, CW6 9BN [IO83QE, SJ56]
G4 UGE Details withheld at licensee's request by SSL.
GM4 UGF D W Duff, Felcanty, Craigton Monikie, Broughty Ferry, Dundee, DD5 3QN [IO86OM, NO53]
GM4 UGG D Fantom, Rhian Cottage, Borgue, Berriedale, Caithness, Scotland, KW7 6HA [IO88GE, ND12]
GW4 UGH S M Morgan, Sion Mari, Roman Way, Caerleon, Newport, South Wales, NP6 1DY [IO81MO, ST39]
GW4 UGI R C Crowley, 15 Rudry St., Penarth, CF64 2TZ [IO81JK, ST17]
G4 UGK C Cattrall, 57 Stonebridge, Orton Malborne, Peterborough, PE2 5NT [IO92UN, TL19]
G4 UGM D Wade, 28 Hazel Rd, Altrincham, WA14 1JL [IO83TJ, SJ78]
GM4 UGN D K Duckworth, Damar 9Blaich, Byfort William, Invernessshire, Scotland, PH33 7AN [IO76JU, NN07]
G4 UGO T A Farmer, York House, Old Gloucester Rd, Alveston, Bristol, BS12 2LQ [IO81RO, ST68]
GW4 UGP C H Emery, 16 Maes y Coed Rd, Cardiff, CF4 4HF [IO81JM, ST18]
G4 UGQ Dr J E Davies, Essex House, 42 Boxworth Rd, Elsworth, Cambridge, CB3 8JQ [IO92XG, TL36]
G4 UGR T Burke, 32 Sunnywood Dr, Tottington, Bury, BL8 3EN [IO83UO, SD71]
G4 UGT D E Hockin, Highlands, 18 Lower Down Rd, Portishead, Bristol, BS20 9PF [IO81OL, ST47]
G4 UGU R W Hall, 19 Buckingham Pl, Downend, Bristol, BS16 5TN [IO81SL, ST67]
G4 UGV M J Hurrell, 52 Marine Parade, Gorleston, Great Yarmouth, NR31 6EY [JO02UN, TG50]
G4 UGW P A Jones, 6 Rivermead Ave, Halebarns, Altrincham, WA15 0AN [IO83UI, SJ78]
G4 UGX J J J Meadowcroft, 17 Chantry Rd, Thornbury, Bristol, BS12 1ER [IO81RO, ST69]
GI4 UHA J J Maguire, 4 Lawnakilla Park, Enniskillen, BT74 7JN [IO64EI, H24]
GD4 UHB J B Parslow, Traie Vane, Lhergy Dhoo, German, Peel, Isle of Man, IM5 2AE
G4 UHC Details withheld at licensee's request by SSL.
G4 UHD Details withheld at licensee's request by SSL.[Op: N N Salmon. Station located near Market Drayton.]
G4 UHE N E Stevens, 23 Downleaze, Fishponds, Downend, Bristol, BS16 6JR [IO81RL, ST67]
G4 UHI D P Westby, 55 Tarn Rd, Thornton Cleveleys, FY5 5AY [IO83MU, SD34]
G4 UHJ D J Lee, West View Cottage, The Level, Pillowell Lydney, Glos, GL15 4QU [IO81RS, SO60]
GW4 UHK J F Newell, 35 St. Andrews Cres, Abergavenny, NP7 6HL [IO81LT, SO31]
G4 UHM S G Parsons, Shangri-La, Maldon Rd, Margaretting, Ingatestone, CM4 9JW [JO01FQ, TL60]
G4 UHN Details withheld at licensee's request by SSL.
G4 UHO G Hayes, 47 Oak Gr, Poynton, Stockport, SK12 1AD [IO83WI, SJ98]
GI4 UHP M Hutchinson, 16 Dernavogy Rd, Fivemiletown, Co. Tyrone, N Ireland, BT75 0SL [IO64HI, H44]
G4 UHQ B E Carr, 23 Belford Dr, Bramley, Rotherham, S66 3YW [IO93IK, SK49]
G4 UHR D C Reekie, 37 Harvey Way, Saffron Walden, CB10 2AP [JO02DA, TL53]
G4 UHS R W Rowlands, 18 Green Cres, Rowner, Gosport, PO13 0DP [IO90JT, SU50]
G4 UHT W R O'Reilly, Troy Nelson Park Rd, St Margarets At Cli, Dover, Kent, CT15 6HL [JO01QD, TR34]
G4 UHU M D Rumens, 90 West Bay Rd, Bridport, Dorset, DT6 4AX [IO80OR, SY49]
G4 UHW D J Allsopp, Karveden, 7 Dowthorpe Hill, Earls Barton, Northampton, NN6 0PB [IO92PG, SP86]
G4 UHX S P Bradshaw, 10 Dawlish Rd, Irby, Wirral, L61 2XP [IO83KI, SJ28]
G4 UHZ D C Goulsbra, Delfour, 3 Chapel St., Market Rasen, LN8 3AG [IO93TJ, TF18]
G4 UIA D J Johnson, 9 Iburndale Ln, Sleights, Whitby, YO22 5EL [IO94QK, NZ80]
GM4 UIB W B Cavers, Edenside Rd, Kelso, Roxburgh, Kelso, Roxburghshire, TD5 7BS [IO85SO, NT73]
G4 UIC M C Hamon, 5 Eastbank Rd, Newcastle, Ontario L1B 1B7, Canada
G4 UIG H L Smith, 29 Hill View, Henleaze, Bristol, BS9 4QD [IO81QL, ST57]
G4 UIH S Woolgar, 73 Cedar Ave, Hazelmere, Hazlemere, High Wycombe, HP15 7EE [IO91PP, SU89]
G4 UII D C Woolgar, 73 Cedar Ave, Hazelmere, Hazlemere, High Wycombe, HP15 7EE [IO91PP, SU89]
GM4 UIJ J A L Spiers, Nether Craigow, By Milnathort, Kinross, KY13 7RR [IO86GF, NO00]
GW4 UIL R Lewis, Siop Penygraig, Llangwnnadl, Pwllheli, Gwynedd, LL53 8NT [IO72PU, SH13]
G4 UIM A D Brown, 3 West Way, Kettering, NN15 7LE [IO92PJ, SP87]
G4 UIO A J Cochrane, 136 Osward, Ct Wood Ln, Croydon, CR0 9HE [IO91XI, TQ36]
G4 UIQ J E Greenhough, 58 Gorsey Bank, Wirksworth, Matlock, DE4 4AD [IO93FB, SK25]
GW4 UIR J D Patterson, Fairhaven, Caerwys Rd, Dyserth, Rhyl, LL18 6HT [IO83HH, SJ07]
G4 UIT D J Geraghty, 38 Clematis Dr, Pendeford, Wolverhampton, West Midlands, WV9 5SD [IO82WP, SJ90]
G4 UIU B H Cullen, Chestnut View, 17 Knowle Ln, Wookey, Wells, BA5 1LB [IO81PF, ST54]
GI4 UIV A N M McBride, 102 Kilcoole Park, Belfast, BT14 8LD [IO74AP, J37]
G4 UIW P Wood, 61 Stoke Rd, Aston Fields, Bromsgrove, B60 3EP [IO82XH, SO96]
G4 UIX Details withheld at licensee's request by SSL.[Op: J P J Birch. Station located near Weymouth.]
G4 UIY S A Hamilton, 89 The Paddocks, Old Catton, Norwich, NR6 7HE [JO02PQ, TG21]
G4 UJA J E Adshead, 91 Juers St, Kingston, Queensland 4114, Australia
G4 UJB A Rogers, 7 Ruskin Gr, Hapton, Burnley, BB11 5RE [IO83US, SD73]
G4 UJC F Williams, 180 Leicester Rd, Wigston, LE18 1DS [IO92KO, SP69]
G4 UJD Details withheld at licensee's request by SSL.[Op: D J Wood.]
G4 UJE A Dowdell, 76 Hartburn Ave, Stockton on Tees, TS18 4HF [IO94HN, NZ41]
GW4 UJF M J Finnigan, 3 Frances Ave, Rhyl, LL18 2LW [IO83GH, SJ08]
G4 UJG Details withheld at licensee's request by SSL.
G4 UJI E Cowperthwaite, Woodlands, Garstang Rd, Cockerham, Lancaster, LA2 0EG [IO83OX, SD45]
G4 UJJ D R G Bodman, 56 Martins Rd, Keevil, Trowbridge, BA14 6NA [IO81WH, ST95]
G4 UJK O G Winship, Hillside, Main Rd, Brigsley, Grimsby, DN37 0RF [IO93WL, TA20]
G4 UJL B J F Poole, 1 Hungerford Piece, Studley, Calne, SN11 9LR [IO81XK, ST97]
G4 UJM S M Ogden, 42 Pynes Ln, Bideford, EX39 3EE [IO71VA, SS42]
G4 UJN J W Newstead, 72 Copse Hill, Harlow, CM19 4PW [JO01BS, TL40]
G4 UJO M Greer, The Pines, 5A Leek Rd, Mossley, Congleton, CW12 3HS [IO83VD, SJ86]
G4 UJR C G Harper, 34 Neva Rd, Bitterne Park, Southampton, SO18 4FJ [IO90HW, SU41]
G4 UJS R P Harrison, Green Ln House, Whixall, Shropshire, SY13 2PT [IO82PV, SJ53]
GW4 UJT R S Johnson, (Mur Madog), Trawsfynydd, Gwynedd, LL41 4SE [IO82AV, SH73]
G4 UJV J C Fenn, 40 Mildenhall Rd, Fordham, Ely, CB7 5NR [JO02EH, TL67]
G4 UJW C Elliott, 52 Wellfield Rd, Alrewas, Burton on Trent, DE13 7EZ [IO92DR, SK11]
G4 UJX A F Ackroyd, 54 Santon Way, Seascale, CA20 1NG [IO84GJ, NY00]
G4 UJY G F B Sparkes, 10B Eskdale Way, Grimsby, DN37 9EA [IO93WN, TA20]
GM4 UJZ D F Gorrill, 18 Hillside Ave, Dalgety Bay, Dunfermline, KY11 5XF [IO86HA, NT18]
G4 UKA C J Hawkridge, 57 Wilkes Wood, Creswell, Stafford, ST18 9QR [IO82WT, SJ82]
G4 UKD B Gibson, 161 Torbay Rd, South Harrow, Harrow, HA2 9QF [IO91TN, TQ18]
G4 UKE K A Hunt, 22 Hitchin Rd, Stotford, Stotfold, Hitchin, SG5 4HP [IO92VA, TL23]
GM4 UKG M J E Manekshaw, 32 Inchcolm Dr, North Queensferry, Inverkeithing, KY11 1LD [IO86HA, NT18]
GI4 UKH C Duignan, 362 Castlereagh Rd, Belfast, BT5 6AE [IO74BN, J37]
G4 UKI W A Hill, 3 Reddings Cl, Wendover, Aylesbury, HP22 6LG [IO91PS, SP80]
G4 UKK Details withheld at licensee's request by SSL.
G4 UKL R W J Humphries, Wayside, Treverva, nr Falmouth, Cornwall, TR10 9BN [IO70KD, SW73]
G4 UKM E K Page, 26 Colne Rd, High Wycombe, HP13 7XN [IO91PP, SU89]
G4 UKO N J Hill, 4 The Wickets, Bentley Rd, Willesborough, Ashford, TN24 0HU [JO01KD, TR04]
G4 UKP D A Ford, Longshoot, Woodside Cttgs, Mow Cop, Stoke on Trent, Staffs, ST7 4NB [IO83VC, SJ85]
G4 UKQ J S Cresswell, Kenley Cottage, 1 Bramley Ave, Coulsdon, CR5 2DR [IO91WH, TQ25]
G4 UKR Details withheld at licensee's request by SSL.[Station located near Waltham Cross.]
G4 UKS A Alexandrou, 62 Hartsdown Rd, Margate, CT9 5RD [JO01QJ, TR36]
GW4 UKU P A Jones, 22 Glyn y Mor, Llanbedrog, Pwllheli, LL53 7NW [IO72SU, SH33]
G4 UKV I E J Leonard, Cross Heyes, 231 Hale Rd, Hale, Altrincham, WA15 8DN [IO83UJ, SJ78]
G4 UKW K Wevill, 51 Danehill, Ratby, Leicester, LE6 0NG [IO92JP, SK50]
G4 UKX R H Miller, 6 Mill Ln, Corton, Lowestoft, NR32 5HZ [JO02UM, TM59]
G4 UKZ R H Rounce, Field House Farm, Blakeney Rd, Hindringham, Fakenham, NR21 0BU [JO02LV, TF93]
G4 ULA Details withheld at licensee's request by SSL.
G4 ULB A J Watkins, 5 Culmore Rd, Halesowen, B62 9HP [IO82XL, SO98]
G4 ULD R K Todd, 11 Alexandra Dr, Surbiton, KT5 9AA [IO91UJ, TQ16]
GI4 ULE T J Wilson, Lisbanoe House, 45 Ballyards Rd, Milford, Armagh, BT60 3NS [IO64PH, H84]
G4 ULF Details withheld at licensee's request by SSL.
GW4 ULG J M Rawlings, 14 The Paddock, Chepstow, NP6 5BW [IO81PP, ST59]
G4 ULH J A Organ, 181 Manor Rd, Fishponds, Bristol, BS16 2EL [IO81RL, ST67]
G4 ULI B W Long, 2 The Limes, Castor, Peterborough, PE5 7BH [IO92TO, TF10]
G4 ULK P Greener, 158 Grosvenor Dr, Loughton, IG10 2LE [JO01AP, TQ49]

G4 ULM J A Martin, 27 Beeson Cl, Little Paxton, St. Neots, Huntingdon, PE19 4NE [IO92UF, TL16]
G4 ULP D J Pritchard, Sandstone Cottage, Walton in Gordano, Clevedon, Avon, BS21 7AJ [IO81OK, ST47]
G4 ULQ G P Judd, 1 Mayfield Way, Ferndown, BH22 9HP [IO90BT, SU00]
G4 ULR R W Skinner, 220 Bluebell Rd, Norwich, NR4 7LW [JO02PP, TG10]
GM4 ULS V M Thompson, The Larches, Harper Way, Scone, Perth, PH2 6PW [IO86HK, NO12]
G4 ULT L R Walker, Oaklea, 13 Manor Rd, Lake, Sandown, PO36 9JA [IO90JP, SZ58]
G4 ULU R M Tucker, 1 Nightingale Cl, Winchester, SO22 5QA [IO91HB, SU42]
G4 ULV D Woodman, 66 Southfield Ave, Kingswood, Bristol, BS15 4BQ [IO81RL, ST67]
G4 ULW M J Clarke, 49 Layland Rd, Saltburn, Skelton in Cleveland, Saltburn By The Sea, TS12 2AQ [IO94MN, NZ61]
G4 ULZ R W Ottway, 10 Stirling Ct Rd, Burgess Hill, RH15 0PT [IO90WX, TQ31]
GM4 UMA S J Maclennan, 10 Ruilick, Beauly, Inverness Shire, IV4 7EY [IO77SL, NH54]
G4 UMB P A Howard, 188 Dashwood Ave, High Wycombe, HP12 3DD [IO91OP, SU89]
G4 UME H U G H Park, 11A Morecambe Rd, Morecambe, LA3 3AA [IO84NB, SD46]
G4 UMF F S Martin, Stirchley Lodge, Stirchley Rd, Stirchley, Telford, TF3 1DY [IO82SP, SJ60]
G4 UMG M Brassington, 42 Dartmouth Ave, Newcastle, ST5 3NY [IO82VX, SJ84]
G4 UMH R Ashworth, 90 Gawthorpe Edge, Park Padiham Rd, Burnley, Lancs, BB12 6PA [IO83UT, SD83]
G4 UMI P Brooks, 11 Graylands Cl, Horsell, Woking, GU21 4LR [IO91RH, TQ05]
G4 UMJ R E R Carslake, 38 Loppets Rd, Tilgate, Crawley, RH10 5DW [IO91VC, TQ23]
G4 UMM A M Curran, 9 Swallowfield Cl, Priorslee, Telford, TF2 9TG [IO82SQ, SJ71]
G4 UMN C D R Ashley, 3 Over Innox, Frome, BA11 2LB [IO81UF, ST74]
G4 UMO Details withheld at licensee's request by SSL.
G4 UMQ R E Luff, 7 Hill Cl, Stannington, Sheffield, S6 6BH [IO93FJ, SK38]
G4 UMS M E Kinger, 5 Fore St., Gunnislake, PL18 9BN [IO70VM, SX47]
G4 UMV P B Johnson, 52 Evesham Rd, Cookhill Alcester, Cookhill, Alcester, B49 5LJ [IO92AF, SP05]
G4 UMW R E W Browning, 28 Mowbray Cl, Bromham, Bedford, MK43 8LF [IO92RD, TL05]
GW4 UMX A R Davis, 23 Dock St., Cogan, Penarth, CF64 2LA [IO81JK, ST17]
G4 UMY M R Strong, 92 Cobham Rd, Halesowen, B63 3JX [IO82XK, SO98]
G4 UNB D Williams, 2 Phillipstown, Whitewell Bottom, Waterfoot, Rossendale, BB4 9NZ [IO83UR, SD82]
G4 UNC J Quash, 20 Silvergarth, Weelsby Meadows, Grimsby, DN32 8QR [IO93XN, TA20]
GM4 UND D G Pople, 22 Stratherrick Gdns, Inverness, IV2 4LZ [IO77VK, NH64]
G4 UNE S J Sharples, 1 Garners End, Chalfont St. Peter, Gerrards Cross, SL9 0HE [IO91RO, TQ09]
G4 UNF K East, 39 Chapel Ln, Navenby, Lincoln, LN5 0ER [IO93RC, SK95]
G4 UNG S J Patterson, 44 Stoke Farthing, Broadchalke, Broad Chalke, Salisbury, SP5 5ED [IO91AA, SU02]
G4 UNH A F Pyne, 414 Beacon Rd, Bank Top, Bradford, BD6 3DJ [IO93CS, SE13]
G4 UNI T J Hepple, 18 King Charles Walk, Princes Way, London, SW19 6JA [IO91VK, TQ27]
G4 UNJ B J Walters, 36 Croft House Gdns, Morley, Leeds, LS27 8NY [IO93ES, SE22]
G4 UNL R A Charlesworth, 6 Curzon Ave, Enfield, EN3 4UD [IO91XP, TQ39]
G4 UNM R J Bushell, 12 Sandham Cl, Sandown, PO36 9DS [IO90KP, SZ58]
G4 UNP A A Eames, 28 Pinewood, Feniscowles, Blackburn, BB2 5AD [IO83RR, SD62]
G4 UNR D K A Stephenson, 57 Kingsbury Rd, Coundon, Coventry, CV6 1PT [IO92FK, SP38]
G4 UNS D S Brown, 8 Gaynes Ct, Little Gaynes Ln, Upminster, RM14 2JH [JO01CN, TQ58]
G4 UNT Details withheld at licensee's request by SSL.
G4 UNU Details withheld at licensee's request by SSL.
GW4 UNV A J Green Swansea R.A.C.C, 9 Westbourne Gr, Sketty, Swansea, SA2 9DT [IO81AO, SS69]
G4 UNW P Everard, The Bungalow, Toynton Fenside, Spilsby, Lincs, PE23 5DB [JO03BD, TF36]
GW4 UNY C C Gibson, 1184 Carmarthen Rd, Fforestfach, Swansea, SA5 4BL [IO81AP, SS69]
G4 UOA G T Howes, 14 Manor Walk, Thornbury, Bristol, BS12 1SW [IO81RO, ST69]
G4 UOB Details withheld at licensee's request by SSL.
GM4 UOD L M Drake-Brockman, 59 Sunnyside, Culloden Moor, Inverness, IV1 2ES [IO77XL, NH74]
G4 UOF D B Lowe, 49 Town Ln, Southport, PR8 6NJ [IO83MP, SD31]
G4 UOG P H J Lord, Route de Colomars, 06790, Aspremont France
G4 UOI R A Butterfield, 33 Orchard Sq, Wormley, Broxbourne, EN10 6JA [IO91XR, TL30]
G4 UOJ R M Williams, The Corner House, 11 Old Bridge St., Truro, TR1 2AQ [IO70LG, SW84]
G4 UOL S Muster, Flat 4, 60 Genesta Rd, Westcliff on Sea, SS0 8DB [JO01IM, TQ88]
G4 UON P V Prowse, 9 Fairway, Carlyon Bay, St. Austell, PL25 3QE [IO70PI, SX05]
G4 UOO J C B Bleaney, 58 Jeans Way, Dunstable, LU5 4PW [IO91SV, TL02]
G4 UOQ Details withheld at licensee's request by SSL.
G4 UOR C J Bourke, 36 The Dr, Fareham, PO16 7NL [IO90JU, SU50]
G4 UOS G Newton, 5 Southend Gdns, Highbridge, TA9 3LD [IO81MF, ST34]
G4 UOT P Sweeny, 170 Worsley Rd, Winton, Eccles, Manchester, M30 8LT [IO83TL, SJ79][Station located at Sheffield S10.]
G4 UOV D G Scofield, 30 Challock Cl, Biggin Hill, Westerham, TN16 3XP [JO01AH, TQ45]
G4 UOW J Cosgrove, 8 Wandsworth Rd, Newcastle upon Tyne, NE6 5AD [IO94EX, NZ26]
G4 UOX R W Buddle, 2 Aberdour Rd, Goodmayes, Ilford, IG3 9SB [JO01BN, TQ48]
G4 UOY J P Botfield, Oaklodge, Love Ln, Stourbridge, DY8 2DH [IO82WK, SO98]
G4 UOZ E H Ball, 8 Highgate Ct, Highgate, Beverley, HU17 0DW [IO93SU, TA03]
G4 UPA J W Poxon, 22 Sandhills Rd, Bolsover, Chesterfield, S44 6EY [IO93IF, SK47]
GI4 UPC W V G Millar, 121 Ballypollard Rd, Magheramorne, Larne, BT40 3JG [IO74CT, J49]
G4 UPD M J Parks, 240 Stainbeck Rd, Leeds, LS7 2NN [IO93FT, SE23]
G4 UPE Details withheld at licensee's request by SSL.
G4 UPF Details withheld at licensee's request by SSL.
G4 UPG V T Jackson, 40 Salisbury Rd, Gloucester, GL1 4JQ [IO81VU, SO81]
G4 UPH T G Palmer, 1 Tavistock Cl, Tamworth, B79 8TJ [IO92DP, SK20]
G4 UPI J Green, St. Annes, Poundfield Rd, Crowborough, TN6 2BG [JO01CB, TQ53]
G4 UPJ A G Stone, 86A Joy Ln, Whitstable, CT5 4DD [JO01MI, TR06]
G4 UPK D J Thompson, 112 Lexton Dr, Churchtown, Southport, PR9 8QW [IO83MP, SD31]
GM4 UPL R M Drake Brockman, 59 Sunnyside, Culloden Moor, Inverness, IV1 2ES [IO77XL, NH74]
G4 UPM R J Taggart, 38 Morecambe Rd, Scalehall, Lancaster, LA1 5JA [IO84OB, SD46]
G4 UPO D P Ince, 64 Newton Rd, Lowton St. Marys, Lowton, Warrington, WA3 1EB [IO83RL, SJ69]
G4 UPP E J Ince, 72 Newton Rd, Lowton St. Marys, Lowton, Warrington, WA3 1EB [IO83RL, SJ69]
G4 UPQ C D Carter, Lynden, 1 Jermyn Rd, Kings Lynn, PE30 4AD [JO02FS, TF62]
G4 UPR J M Dickson, Woodside Lodge, 33 Ringwood Gr, Weston Super Mare, BS23 2UA [IO81MI, ST36]
G4 UPS E H Collins, 27 Parklands, Hemyock, Cullompton, EX15 3RY [IO80JV, ST11]
G4 UPT D Corney, 24 Springfield, Rotherham, South Yorks, S60 3AW [IO93HJ, SK49]
G4 UPU R G Ainsworth, 14 Edge Fold Cres, Worsley, Manchester, M28 7EX [IO83TM, SD70]
G4 UPW G W Trevelyan, 53A Blackboy Rd, Exeter, EX4 6TB [IO80FR, SX99]
GM4 UPX I H Wilson, 30 Howdenburn Ct, Jedburgh, TD8 6NP [IO85RL, NT62]
G4 UPY A N Hodge, 116 Broad Rd, Lower Willingdon, Eastbourne, BN20 9RD [JO00CT, TQ50]
G4 UPZ Details withheld at licensee's request by SSL.
G4 UQA M J Goodman, 54 Church St., North Borough, Northborough, Peterborough, PE6 9BN [IO92UP, TF10]
G4 UQB Details withheld at licensee's request by SSL.
G4 UQC Details withheld at licensee's request by SSL.
GM4 UQD Details withheld at licensee's request by SSL.
G4 UQF M Sole, Whitecourt, 80 Linkside, Bretton, Peterborough, PE3 8PA [IO92UO, TF10]
GM4 UQG R J Aitkenhead, 7 Waterside Gdns, Hamilton, ML3 7PY [IO75XS, NS75]
G4 UQH G A D Everley, 21 Cleveland Rd, Uxbridge, UB8 2DR [IO91SM, TQ08]
G4 UQI A M Lether, 16 The Dingle, Fulwood, Preston, PR2 3EX [IO83PS, SD53]
G4 UQJ K D Monk, 19 Sunnyfield Rd, Hardwicke, Gloucester, GL2 4QF [IO81UT, SO71]
G4 UQK J D Roberts, 11 Bar Ln, Astley Bridge, Bolton, BL1 7JD [IO83SO, SD71]
G4 UQL P Woodhead, 2 Hilltop Cttgs, Knott Hill Ln, Delph, Oldham, OL3 5RJ [IO83XN, SD90]
G4 UQM D A Grainger, 25 Westwood Heath Rd, Leek, ST13 8LN [IO83XC, SJ95]
G4 UQN K J Stockley, 22 The Lawns, Leverington Rd, Wisbech, PE13 1SW [JO02BQ, TF41]
GD4 UQO P R Parker, 46 Ballaquane Park, Peel, IM5 1PX
G4 UQP I G Hunter, 46 Station Rd, Scalby, Scarborough, YO13 0QA [IO94SH, TA09]
G4 UQR J R Gibbs, 3 Holts Green, Great Brickhill, Milton Keynes, MK17 9AJ [IO91PX, SP92]
G4 UQS J W Blundell, 68 Alton Rd, Leicester, LE2 8QA [IO92KO, SK50]
G4 UQU D W Smith, 2 Niton Rd, Weddington, Nuneaton, CV10 0BX [IO92GM, SP39]
GD4 UQV M J Charleton, Coreen, Peveril Rd, Peel, IM5 1PJ
G4 UQW D A Beckett, 433 New St., Biddulph Moor, Stoke on Trent, ST8 7NG [IO83WC, SJ95]
G4 UQX S R Follows, 9 The Beeches, First Ave, Porthill, Newcastle, ST5 8RX [IO83VA, SJ84]
G4 UQY Details withheld at licensee's request by SSL.
G4 URA A J Haynes, 5 Wallis Ct, Pilborough Way, Colchester, CO3 5XU [JO01KU, TL92]
GW4 URB R J Teesdale, 22 Cwmgelli Dr, Treboeth, Swansea, SA5 9BS [IO81AP, SS69]
G4 URD R A Caira, 12 West Hill Rd, Herne Bay, CT6 8HG [JO01NI, TR16]
G4 URE Details withheld at licensee's request by SSL.
G4 URG S R Richardson, The Bungalow, 2 Norbury Ln, Oldham, OL8 2EW [IO83XM, SD90]
G4 URH Details withheld at licensee's request by SSL.
G4 URI A M Hyde, 2 Ardsley Rd, Worbrough, Worsbrough, Barnsley, S70 4RN [IO93GM, SE30]
GW4 URJ G B B Chavasse, Woodlawn, Penmaenpool, Dolgellau, Gwynedd, LL40 1YE [IO82AR, SH61]
G4 URL L M Haynes, 62 Mountbatten Dr, Ashford Lodge, Colchester, CO2 8BH [JO01LU, TM02]
G4 URM P W Butler, Tanglewood, Elms Ln, Shareshill, Wolverhampton, WV10 7JS [IO82WP, SJ90]
G4 URN M F Turvey, 106 Foxwell St., Worcester, WR5 2ET [IO82VE, SO85]
G4 URO J Macfarlane, 26 The Oval, Farncombe, Godalming, GU7 3JW [IO91QE, SU94]
G4 URP R Powell, 57 Bartons Dr, Yateley, Camberley, Surrey, GU17 7DW [IO91OH, SU85]
G4 URS J D Osborne, 2 Sedge Rd, Scarning, Dereham, NR19 2UA [JO02LP, TF91]
G4 URT Details withheld at licensee's request by SSL.[Station located near Hailsham.]
G4 URU M J Hancock, 22 The Hodges, Rynal St., Evesham, WR11 4QL [IO92AC, SP04]
G4 URV Dr W E Peel, 34 Carlyn Ave, Sale, M33 2EA [IO83UK, SJ79]
G4 URW J B Allison, 17 Gordon Terr, Stakeford, Choppington, NE62 5UE [IO95FD, NZ28]
G4 URX T E Robinson, 26 Keeble Dr, Washingborough, Lincoln, LN4 1DZ [IO93SF, TF07]
GW4 URY D N F Whitehouse, Pendyffryn, Pentraeth, Anglesey, Gwynedd, LL75 8UN [IO73VG, SH57]
G4 USA B A Skipworth, 4 Short Rd, Hill Head, Fareham, PO14 3HP [IO90JT, SU50]
G4 USB N E Pascoe, Westwynds, Loscombe Ln, Four Lanes, Redruth, TR16 6LP [IO70JE, SW63]
G4 USD D H Brill Bomchil Castro Goodrich Claro, 25 Boulevard Barbes, 75018 Paris, France, X X
G4 USE Details withheld at licensee's request by SSL.
G4 USG C G Barker, 40 Lowick Ct, Moulton, Northampton, NN3 7TZ [IO92NG, SP76]
G4 USH O Heathershaw Skirlaugh Rayne, Ch Em/Y Pl/G Office, Em/Y Plan/G Service, 39 Meaux Road, Wawn, HU7 5XD [IO93TT, TA03]
G4 USI D P Tilley, Clovelly, 20 The Woodcroft, Diseworth, Derby, DE74 2QT [IO92IT, SK42]
G4 USK B J Finlay, 101 Tithe Barn Dr, Bray, Maidenhead, SL6 2DD [IO91PL, SU97]
G4 USL P S Fischer, 3 Hollybush Cl, London, E11 1PZ [JO01AN, TQ48]
G4 USM S A Kelly, 88 Kings Hedges Rd, Cambridge, CB4 2PA [JO02BF, TL46]
G4 USN S E Havard, Altonswood, 1 Merricks Ln, Bewdley, DY12 2PA [IO82UI, SO77]
G4 USO G J Gustar, 68 Selworthy Rd, Weston Super Mare, Somerset, BS23 3SX [IO81MH, ST35]
G4 USQ T I D Hodgetts, 14 St. Peters Rd, Portishead, Bristol, BS20 9QY [IO81OL, ST47]
G4 USS Details withheld at licensee's request by SSL.
G4 UST Details withheld at licensee's request by SSL.
G4 USV Details withheld at licensee's request by SSL.[Station located in London SE9.]
G4 USW W R Jenkins, 5 Seatoller Pl, Barrow in Furness, LA14 4NH [IO84JD, SD27]
G4 USX E Pritchard, 18 New Ridd Rise, Hyde, SK14 5DD [IO83XK, SJ99]
GM4 USY H H Templeton, 3 Kirk Rd, Newport on Tay, DD6 8JD [IO86MK, NO42]
G4 UTA R Walker, 216 Milnrow Rd, Rochdale, OL16 5BB [IO83WO, SD91]
GM4 UTC D A Abraham, 42 Lower Greenfield, Ingol, Preston, PR2 3ZT [IO83PS, SD53]
G4 UTE M Y Chaudhry, No 668, near Wahid Public, School G 10/4, Islamabad Pakistan, ZZ7 9PO
G4 UTF A P Cockman, 31 Kensington Rd, Southend on Sea, SS1 2SX [JO01IM, TQ88]
G4 UTG F C Collins, 31 Mount Pleasant Rd, Poole, BH15 1TU [IO90AR, SZ09]
G4 UTJ J R Gorton, 43B Mill Rd, Mile End, Colchester, CO4 5LE [JO01KV, TL92]
GM4 UTK D R James, Baltic House, Baltic St, Montros, Angus, DD10 8EX [IO86SR, NO75]
G4 UTM B J Dennis, Thistledown, Yallands Hill, Monkton Heathfield, Taunton, TA2 8NA [IO81LA, ST22]
G4 UTN G Bromley, 46 Independent Hill, Alfreton, DE55 7DG [IO93HC, SK45]
GM4 UTP S Cameron, 5 Frankfield Pl, Dalgety Bay, Dunfermline, KY11 5LR [IO86HA, NT18]
G4 UTQ M Adamson, 13 Towers Cl, Bedlington, NE22 5ER [IO95ED, NZ28]
G4 UTR M J Alder, 342 Church St., Edmonton, London, N9 9HP [IO91XP, TQ39]
GW4 UTS E M Bracey, 3 Dyffryn Rd, Waunlwyd, Ebbw Vale, NP3 6UA [IO81JS, SO10]
G4 UTT C M Broadbent, 1 Walnut Cl, Winchcombe, Cheltenham, GL52 3AF [IO81XV, SO92]
G4 UTV A S Cockerill, 90 Stockton Rd, Middlesbrough, TS5 4AJ [IO94IN, NZ41]
G4 UTW Details withheld at licensee's request by SSL.
G4 UTX G G Eagle, Dewerstone Cottage, Goodameavy, Roborough, Plymouth, PL6 7AP [IO70XL, SX56]
G4 UTY G P Fooks, 10 Bincombe Dr, Crewkerne, TA18 7BE [IO80OV, ST41]
G4 UUA M Robinson, 2 Bridge Mill Rd, Nelson, BB9 7BD [IO83VU, SD83]
G4 UUB M H Lemin, Mill House, Lingwood Rd, Blofield, Norwich, NR13 4AH [JO02RP, TG30]
GI4 UUC W Thompson, Maranne, 21 Watch Hill Rd, Straid, Ballyclare, BT39 9QW [IO74BT, J39]
G4 UUE Dr L McGrogan, Holly House, Haslingden Rd, Rawtenstall, Lancs., BB4 6RX [IO83UQ, SD72]
G4 UUF N A Kelly, 4 Southdowns, Plumpton Green, Lewes, BN7 3EB [IO90XW, TQ31]
G4 UUG D K Payne, 20 Huntingdon Rd, Leicester, LE4 9GF [IO92KP, SK60]
G4 UUH S M Rogers, 4 Brookside Cl, Yelvertoft, Northampton, NN6 6LP [IO92KJ, SP57]
G4 UUI S A Rooker, 211 Faversham Rd, Kennington, Ashford, TN24 9AF [JO01KE, TR04]
G4 UUJ E E M Russell, 11 Fyfe Rd, Shanklin, PO37 7PA [IO90JP, SZ58]
G4 UUK N Sanderson, 48 Mill Cres, Kingsbury, Tamworth, B78 2NN [IO92DN, SP29]
G4 UUM P P Skivington, 25 Churchgate, Cheshunt, Waltham Cross, EN8 9NB [IO91XQ, TL30]
G4 UUP J J Taylor West London ARS, 20 Aylett Rd, Isleworth, TW7 6NP [IO91TL, TQ17]
G4 UUQ T H Tallis, The Croft, Creamery Ln, Parwich Ashbourne, DE6 1QB [IO93DC, SK15]
G4 UUR T Thompson, No3 St Helier Ave, Broadwey, Weymouth, Dorset, DT3 5DU [IO80SP, SY68]
G4 UUS G G P Thorburn, 27 Coniston Rd, Fulwood, Preston, PR2 8AX [IO83PS, SD53]
G4 UUT A J Turner, 30 Wheatlands Rd, Paignton, TQ4 5HU [IO80FK, SX85]
G4 UUU C P Clayton, 9 Green Island, Irton, Scarborough, YO12 4RN [IO94SF, TA08]
G4 UUV P D Watson, 10 Ross Ave, Wirral, L46 2SB [IO83LK, SJ29]
G4 UUW D F Williams, Holly Tree Cottage, Upton Pyne, Exeter, Devon, EX5 5JA [IO80FS, SX99]
G4 UUX L W Williams, 12 Low Ln, Torrisholme, Morecambe, LA4 6PN [IO84OB, SD46]
G4 UUZ A K Kaye, Flat C2 Woodland Grange, 31 Dean Park Rd, Bournemouth, Dorset, BH1 1HY [IO90BR, SZ09]
G4 UVA P R Money, Meadow View, Podmore Ln, Scarning, Dereham, NR19 2NS [JO02KQ, TF91]
G4 UVB P D Gibson, Rivendell, 9 Mallard Cl, Aughton, Ormskirk, L39 5QJ [IO83NN, SD40]
GW4 UVC J D Stephens, 105 Wern St., Tonypandy, CF40 2DH [IO81GP, SS99]
G4 UVD D S Asquith, 516 Old Bedford Rd, Luton, LU2 7BY [IO91SV, TL02]
G4 UVE Details withheld at licensee's request by SSL.
G4 UVF J E Taylor, 6 The Stray, Little Wold Ln, South Cave, Brough, HU15 2AL [IO93QS, SE93]
G4 UVG D P Stewart, 4 Towles Pastures, Castle Donington, Derby, DE74 2RX [IO92HU, SK42]
G4 UVJ D S Speechley, 19 Ambleside Walk, Canvey Island, SS8 9TD [JO01GM, TQ78]
G4 UVL J Brown, 15 Silton Gr, Hartburn, Stockton on Tees, TS18 5AT [IO94HN, NZ41]
GW4 UVN J Travers, 54 Crymlyn Rd, Skewen, Neath, SA10 6EA [IO81BP, SS79]
G4 UVO F E V Green, 72 Junction Rd, Andover, SP10 3QX [IO91GF, SU34]
G4 UVP Details withheld at licensee's request by SSL.
GW4 UVT P Wensley, 8 Chepstow Pl, Winch Wen, Bonymaen, Swansea, SA1 7HP [IO81BP, SS69]
G4 UVV D J Pike, 29 Galmington Dr, Taunton, TA1 5AQ [IO81KA, ST22]
G4 UVW E D Underhill, 5 Lyndhurst Croft, Eastern Green, Coventry, CV5 7QE [IO92FK, SP28]
G4 UVX P G Lee, Cdf Pointe Noir, C.For Sec, Shell Centre, London, SE1 7NA [IO91WM, TQ38][Op: P G Lee, 37 Stonebridge Way, Faversham, Kent, ME13 7SF.]
G4 UVZ A K Whatmore, Hollybank, Sellicks Green, Taunton, Somerset, TA3 7SD [IO80KX, ST21]
G4 UWA M A Styne, 268 Burton Rd, Midway, Swadlincote, DE11 7LY [IO92FS, SK32]
G4 UWB C I D Dodge, 28, Yanworth, Cheltenham, Glos, GL54 3LQ [IO91BT, SP01]
GW4 UWD O P C Roberts, Tan y Bryn, 3 Frankwell St., Tywyn, LL36 9EP [IO72WO, SH50]
G4 UWE P Sheppard Wawne Raynet Gr, Council Em/Y Pl/G Office, 39 Meaux Rd, Wawne, Hull, North Humberside, HU7 5XD [IO93TT, TA03]
G4 UWF M E Kebbell, 56 King Edward Ave, Hastings, TN34 2NQ [JO00GU, TQ81]
G4 UWG N R Dunn, Lower Woodside, Whinfell, Penrith, Cumbria, CA10 2AP [IO84QP, NY52]
GW4 UWI J B Thorogood, 75 Victoria Park, Colwyn Bay, LL29 7YY [IO83DH, SH87]
G4 UWJ A H Hayter, Appletrees, Fernden Ln, Haslemere, GU27 3LA [IO91PB, SU83]
G4 UWK I Birkenshaw, 7 Wesley Rd, Ambergate, Belper, DE56 2GT [IO93GB, SK35]
G4 UWL B W D Lewis, Homelea, Longdowns, Penryn, Cornwall, TR10 9DR [IO70KD, SW73]
G4 UWM Details withheld at licensee's request by SSL.
GM4 UWN R W Kane, 39 Tollohill Dr, Kincorth, Aberdeen, AB1 5DQ [IO87TH, NJ72]
GM4 UWO J G Campbell, 17 Cotter Dr, Wellpark, Kilmarnock, KA3 7EA [IO75SO, NS43]
G4 UWP L J Flynn, 4 Courtenay Gdns, St. Annes, Nottingham, NG3 4QG [IO92KX, SK54]
G4 UWQ R Wilkins, 38 Winchester Dr, Linton, Swadlincote, DE12 6PP [IO92ER, SK21]
GW4 UWR V T Thomas, 8 Dinas Path, Fairwater, Cwmbran, NP44 4QQ [IO81LP, ST29]
G4 UWS A Walker, 53 Parkstone Ave, Parkstone, Poole, BH14 9LW [IO90AR, SZ09]
GI4 UWT P R Bolton, 17 Harbour Rd, Kilkeel, Newry, BT34 4AR [IO64XB, J31]
G4 UWW A G Prior, 27 Seafield Ave, Mistley, Manningtree, CO11 1UE [JO01NW, TM13][Station located near Wotton-under-Edge.]
GM4 UWX J Rennie, 19 Harbour Pl, Portknockie, Buckie, AB56 4NR [IO87NQ, NJ46]
G4 UWY S L Ince, 6 Marie Cl, Cantley, Norwich, NR13 3RN [JO02SN, TG30]
G4 UWZ V Harper, 12 Bradford Park, Come Down, Bath, BA2 5PR [IO81TI, ST76]
G4 UXA J D Alldridge, 95 Rose Dr, Chesham, HP5 1RT [IO91QQ, SP90]
G4 UXB R Ball, 144 Broad Ln, Hampton, TW12 3BW [IO91TK, TQ17]
G4 UXC M J Butler, 16 Clevedon Green, South Littleton, Evesham, WR11 5TY [IO92BC, SP04]
G4 UXD D Brandon, 1 Woodlands Rd, Saltney, Chester, CH4 8LB [IO83NE, SJ36]
G4 UXE C D Humphries, 254 Sprotborough Rd, Sprotbrough, Doncaster, DN5 8BY [IO93KM, SE50]
G4 UXF S Humphries, 254 Sprotborough Rd, Sprotbrough, Doncaster, DN5 8BY [IO93KM, SE50]
G4 UXG J E Dew, 12 Tanglewood, Finchampstead, Wokingham, RG40 3PR [IO91NJ, SU86]
G4 UXH C Wilkinson, 14 Ryleyfield Rd, Milnthorpe, LA7 7PT [IO84OF, SD48]
G4 UXJ T R A Ager, 5 Matthews Cl, Bedhampton, Havant, PO9 3NJ [IO90LU, SU70]
G4 UXL K M Jones, 10 Whetstone Hey, Great Sutton, South Wirral, L66 3PH [IO83MG, SJ37]
G4 UXM H V Jenkins, Annecy, Middle Hill, Chalford Hill, Stroud, GL6 8BU [IO81WR, SO80]
G4 UXN S W Skinner, Avalon, 34 Lansdown, Stroud, GL5 1BW [IO81VR, SO80]
G4 UXO N J Emson, 9 Sands Cl, Pattishall, Towcester, NN12 8LU [IO92LE, SP65]
G4 UXP M J Huxham, 34 The Close, Furzeham, Brixham, TQ5 8RF [IO80FJ, SX95]
G4 UXR P J Gilligan, 7 Eton Ct, Old Trafford, Manchester, M16 7WR [IO83UL, SJ89]
G4 UXS L M Moore, Magwitch, Pips View, Cooling, Rochester, ME3 8DH [JO01GK, TQ77]
G4 UXT D T Skinner, 62 Bridgewater Rd, Harold Hill, Romford, RM3 7UB [JO01CO, TQ59]
G4 UXU A M Clifton, 26A Thomas Rd, Whitwick, Coalville, LE67 5FY [IO92HR, SK41]
G4 UXV C J Osborn, 19 Maple Dr, Oxmoor, Huntingdon, PE18 7JE [IO92VI, TL27]
GI4 UXW D K Hanna, 19 Roeview Park, Limavady, BT49 9BQ [IO65MB, C62]

G4

GM4 UXX A Hood, 4 Murray Rd, Law, Carluke, ML8 5HR [IO85BR, NS85]
G4 UXY C Boulter, 18 Red Lion St., Chesham, HP5 1EZ [IO91QQ, SP90]
G4 UYA P B Henson, 11 Potters Croft, Horsham, RH13 5LR [IO91UA, TQ22]
G4 UYB J T Overington, 96 Sherwood Rd, South Harrow, Harrow, HA2 8AT [IO91TN, TQ18]
GM4 UYE H D W Martin, 11 Ewing Ct, Broombridge, Stirling, FK7 0QP [IO86AC, NS89]
G4 UYF L R G Aldhous, 5 Banks Ln, Heckington, Sleaford, NG34 9QY [IO92UX, TF14]
G4 UYI R J Heselwood, 31 Berwick St., Workington, CA14 3EN [IO84FP, NY02]
G4 UYJ B Crow, 298 Walmersley Rd, Bury, BL9 6NH [IO83UO, SD81]
G4 UYL Details withheld at licensee's request by SSL.
G4 UYM Dr F D Roberts, The Spinney, Welford, Northampton, NN6 7HG [IO92KH, SP66]
G4 UYN D Robson, 36 Orford Rd, Newton Heath, Manchester, M40 1JY [IO83VL, SD80]
GM4 UYP J Smith, 10 Witchknowe Ave, Riccarton, Kilmarnock, KA1 4LQ [IO75SO, NS43]
G4 UYQ E J Ford, 14 Brooke Rd, Cirencester, GL7 1SX [IO91AQ, SP00]
G4 UYR R J Noble, Chandler Rd, Holy Cross, Stoke Holy Cross, Norwich, Norfolk, NR14 8RG [JO02PN, TG20]
GW4 UYT R B Jenkins, 1 Lon y Bryn, Glynneath, Neath, SA11 5BG [IO81ER, SN80]
GW4 UYU J Jones, Oakdene, Upper Denbigh Rd, St. Asaph, LL17 0RR [IO83GG, SJ07]
GM4 UYZ R E Glasgow, 7 Castle Terr, Port Seton, Prestonpans, EH32 0EE [IO85MX, NT47]
GW4 UZC D V Ralph, Tal-y-Maes, Llanbedr, Crickhowell, Powys, NP8 1SY [IO81KV, SO22]
G4 UZE C N Mason, 145 Park Ave, Ruislip, HA4 7UN [IO91SO, TQ08]
G4 UZF B J Matthews, 12 School Rd, Thurston, Bury St. Edmunds, IP31 3SP [JO02JG, TL96]
G4 UZG G E Price, 6 Windmill Rd, Walkden, Worsley, Manchester, M28 3RP [IO83TM, SD70]
GW4 UZH G J Mills, 83 Usk Rd, Pontypool, NP4 8AF [IO81LQ, SO20]
GW4 UZL P N Oneill, Mount Pleasant, Ambleston, Haverfordwest, Dyfed, SA62 5DP [IO71NV, SM92]
G4 UZN A M Quest, 445 St. Ln, Leeds, LS17 6HQ [IO93FU, SE33]
G4 UZO K C Richards, Lyndhurst, 27 Trewetha Ln, Port Isaac, PL29 3RW [IO70OO, SX08]
GM4 UZP A S S Low, 21 Earn Cres, Menzieshill, Dundee, DD2 4BS [IO86LL, NO33]
GM4 UZR J E Low, 4 Smith Ave, Inverness, IV3 5ES [IO77VL, NH64]
G4 UZS C Dawson, 44 Cranborne Rd, Cosham, Portsmouth, PO6 2BQ [IO90LU, SU60]
G4 UZT T J Cooper, 17 Fraser Cl, Chelmsford, CM2 0TD [JO01FR, TL70]
G4 UZU G Cobb, 25 Friendly Ave, Burnley Rd, Sowerby Bridge, HX6 2TY [IO93AR, SE02]
GW4 UZW R H Walsh, 10 Hazel Gr, Longmeadow, Dinas Powys, CF64 4TE [IO81JK, ST17]
GM4 UZY C W M Wilson, Goldenacre, 1 Borrowfield Cres, Montrose, DD10 9BR [IO86SR, NO75]
G4 UZZ A D Wragg, New Orchard, Mansfield Rd, Creswell, Worksop, S80 4AB [IO93JG, SK57]
G4 VAA R C Powell, 11 North Park, Fakenham, NR21 9RG [JO02KU, TF93]
G4 VAB Details withheld at licensee's request by SSL.
GM4 VAC J L Plunkett, Blar Samhraidh, Bunchrew, Inverness, Invernessshire, IV3 6TA [IO77UL, NH64]
G4 VAD Details withheld at licensee's request by SSL.[Op: I Rush. Station located near Brentwood.]
G4 VAF N P Dorrington, Bahnhof Strasse 19, 64347, Griesheim, Germany
GW4 VAG H G D C Green, 2 Whitchurch Rd, Bangor on Dee, Bangor Isycoed, Wrexham, LL13 0AY [IO83NA, SJ34]
G4 VAH P A Hudson, 15 Fellows Rd, Farnborough, GU14 6NU [IO91PG, SU85]
G4 VAJ P W Handley, 12 The Bungalows, Kirkham Rd, Freckleton, Preston, PR4 1HY [IO83NS, SD43]
G4 VAL V Pellowe, 191 Preston New Rd, Blackpool, Lancs, FY3 9TN [IO83LT, SD33]
G4 VAM P Harrison, 39 Priory Rd, Peterborough, PE3 9ED [IO92UN, TL19]
G4 VAO M E Jordan, Edgefield, Easthaugh Rd, Lyng, Norwich, NR9 5LN [JO02NR, TG01]
G4 VAP I Kenyon, 12 Leamington Rd, Morecambe, LA4 4RL [IO84NB, SD46]
G4 VAS E T Cooper, 39 Violet Rd, Bassett, Southampton, SO16 3GZ [IO90HW, SU41]
G4 VAV A A Brooks, 17 Grosvenor Ave, Carshalton, SM5 3EJ [IO91WI, TQ26]
G4 VAX S T Cope, 24 Metcalf Rd, Newthorpe, Nottingham, NG16 3NL [IO93IA, SK44]
GM4 VAY A Newlands, 7 Muir Cl, Stewarton, Kilmarnock, KA3 3HG [IO75RQ, NS44]
GD4 VBA R Harrison, 64 Friary Park Rd, Ballabeg, Castletown, Isle of Man, IM9 4EP
G4 VBB Details withheld at licensee's request by SSL.
G4 VBC R V Kay, 8 Fairthorne Way, Shrivenham, Swindon, SN6 8EB [IO91EO, SU28]
G4 VBD S L McAdam, 39 The Mallards, St. Ives, Huntingdon, PE17 4HT [IO92WI, TL37]
GM4 VBE R Fairholm, 28 Queensberry Ave, Clarkston, Glasgow, G76 7DU [IO75US, NS55]
G4 VBH A G Fisher, 108 Heston Grange, North Hyde Ln, Heston, Hounslow, TW5 0HD [IO91TL, TQ17]
G4 VBI R J Harte, 32 Kingsgate Ave, Kingsgate, Broadstairs, CT10 3QP [JO01RJ, TR37][QSL via G4VAA.]
G4 VBJ B Kay, 19 Langham Gr, Timperley, Altrincham, WA15 6DY [IO83UJ, SJ78]
G4 VBK R J Deeprose, 70 Hollington Old Ln, St. Leonards on Sea, TN38 9DP [JO00GU, TQ71]
GW4 VBM C Leighton, 12 Morley Ave, Connahs Quay, Deeside, CH5 4RE [IO83LF, SJ26]
G4 VBO J J Mattock, The Coach House, South Rd, Taunton, Somerset, TA1 3DT [IO81KA, ST22]
G4 VBP B Patchett, 107 Handsworth Ave, Sheffield, S9 4BU [IO93HJ, SK38]
G4 VBQ S Largent, 1 Church Cl, Bucklesham, Ipswich, IP10 0DU [JO02PA, TM24]
G4 VBR G E Keggen, 2 Lansdowne Cl, Worthing, BN11 5HF [IO90TT, TQ10]
G4 VBS P J Chapman, 10 School Cttgs, Hargrave, Bury St. Edmunds, IP29 5HR [JO02HE, TL75]
G4 VBT Details withheld at licensee's request by SSL.
GW4 VBV R J Beckers, 13 Taplow Terr, Pentrechwyth, Swansea, SA1 7AD [IO81AP, SS69]
G4 VBW M Calvert, 163 Milner Rd, Heswall, Wirral, L60 5RY [IO83KH, SJ28]
G4 VBX A J Currell, Chandler Rd, Holy Cross, Stoke Holy Cross, Norwich, Norfolk, NR14 8RG [JO02PN, TG20]
G4 VBY A W Dickson-Smith, 13 Charmile Ct, Spa Rd, Weymouth, DT3 5EP [IO80SP, SY68]
GI4 VBZ G W Esler, 8 Cushendall Rd, Ballymena, BT43 6HE [IO64UU, D10]
G4 VCA G Mason, 25 Farthingloe Rd, Dover, CT17 9LD [JO01PC, TR34]
G4 VCB C E Melvin, 2 Burvill Ct, Langham Rd, London, SW20 8TP [IO91VJ, TQ26]
G4 VCD Details withheld at licensee's request by SSL.
G4 VCE S L Sewell, Medway, The Rosery, Mulbarton, Norwich, NR14 8AL [JO02ON, TG10]
G4 VCF F E Carr, Kelpie Marine, Great North Rd, Roxton, Bedford, MK44 3DS [IO92UE, TL15]
G4 VCG J M Travers, 76 Boundstone Rd, Wrecclesham, Farnham, GU10 4TR [IO91OE, SU84]
G4 VCJ R Percival, 6 Bulmer Pl, Hartlepool, TS24 9BQ [IO94JQ, NZ43]
G4 VCK Details withheld at licensee's request by SSL.
G4 VCL R P I Parry, 6 Rosehill Terr, Coltham Fields, Cheltenham, GL52 6SW [IO81XV, SO92]
G4 VCN A J Soars, 118 Braddon Rd, Loughborough, LE11 5YZ [IO92JS, SK52]
G4 VCO D M Seddon, 15 Fairfield Dr, Greenford, UB6 7AX [IO91UM, TQ18]
G4 VCP C T B Smith, 2 The Poplars, Hall Ln, Huyton Duarny, Liverpool, L36 6AT [IO83OJ, SJ49]
G4 VCQ R J Hogan, 6 Tadden Walk, Broadstone, BH18 9NU [IO80XS, SY99]
G4 VCT R Aitken, 31 Hawkshead Dr, Westgate, Morecambe, LA4 4SP [IO84NB, SD46]
G4 VCV A J Crompton, 36 Priory Ave, London, N8 7RN [IO91WO, TQ28]
GM4 VCW P H Drysdale, 121 Chapelle Cres, Tillicoultry, FK13 6NL [IO86DD, NS99]
G4 VCX M J Beaumont, 75 Middlecotes, Tile Hill, Coventry, CV4 9AX [IO92FJ, SP27]
G4 VCY J Eley, 94 Raphael Cl, Whoberley, Coventry, CV5 8LS [IO92FJ, SP37]
GI4 VCZ P J Donnelly, 9 Hollybrook Park, Newtownabbey, BT36 4ZN [IO74AQ, J38]
G4 VDB D Brocklehurst, 73 Ridgeway, Clowne, Chesterfield, S43 4BD [IO93IG, SK47]
G4 VDC Details withheld at licensee's request by SSL.
G4 VDD G B Sutton, 25 Brick Kiln Ln, Rufford, Ormskirk, L40 1SY [IO83OP, SD41]
G4 VDE Details withheld at licensee's request by SSL.
G4 VDF A M Palmer, 11 Cedar Ave, Blackwater, Camberley, GU17 0JE [IO91OI, SU86]
GM4 VDG J R Rankin, 64 Forrest Walk, Stankards, Uphall, Broxburn, EH52 5PN [IO85FW, NT07]
G4 VDH W E Peak, 10 The Oval, Scarborough, YO11 3AP [IO94TG, TA08]
G4 VDJ B Lee, 61 Pendleway, Pendlebury, Swinton, Manchester, M27 8QS [IO83UM, SD70]
G4 VDK H T Schmitz-Goertz, Park Residenz 712, Am Spitzenbach 2, 53604 Bad Honnef, Germany, X X
GM4 VDL C G Van Arman, Eilean Dubh, Post Office, West Argyle St, Ullapool, Ross Shire, IV26 2TY [IO77KV, NH19]
GW4 VDP D G James, Sea View, Llanfawr Rd, Holyhead, LL65 2PP [IO73QH, SH28]
G4 VDQ G C Butterworth, 233 Rockingham Rd, Corby, NN17 2AB [IO92PM, SP89]
G4 VDT R I Burns, 16 Park Gr, Frizinghall, Bradford, BD9 4JY [IO93CT, SE13]
G4 VDU Details withheld at licensee's request by SSL.
GW4 VDW J D Taylor, Gofer Glas, Maesybont, Llanelli, Dyfed, SA14 7HH [IO71WT, SN51]
G4 VDX J Menguy, 6 Laurel Gr, Lowton St. Lukes, Lowton, Warrington, WA3 2EE [IO83RL, SJ69]
G4 VDY Details withheld at licensee's request by SSL.
G4 VDZ F Williamson, Spring View, 113 Marlborough Ave, Ince, Wigan, WN3 4PR [IO83QM, SD50]
G4 VEA M F Piercy, 87 Morshead Mansion, Morshead Rd, London, W9 1LG [IO91VM, TQ28]
GW4 VEB D T Lintern, 108 Pontygwindy Rd, Caerphilly, CF8 3HF [IO81HN, ST08]
G4 VEC M J Elliott, 20 Haysel, Sittingbourne, ME10 4QE [JO01IH, TQ96]
GM4 VEE J B Moir, 31 Skerry Dr, Peterhead, AB42 2YJ [IO97CM, NK14]
G4 VEF A G Sanders, 35 Burlea Dr, Shavington, Crewe, CW2 5BZ [IO83SB, SJ65]
G4 VEG S H Hall, 15 Hurst Cl, Staplehurst, Tonbridge, TN12 0BX [JO01GD, TQ74]
G4 VEH L K Skinner, 15 Ridge Cl, Portishead, Bristol, BS20 8RQ [IO81OL, ST47]
GW4 VEI H Bew, 38 Limeslade Cl, Hirwaun, Aberdare, CF44 9RN [IO81FR, SN90]
GM4 VEJ J K McDonald, 18 Winston Rd, Galashiels, TD1 2EJ [IO85OO, NT53]
GW4 VEK Details withheld at licensee's request by SSL.
G4 VEL J A W Smith, 43 Ash Cl, Thetford, IP24 3HQ [JO02IJ, TL88]
G4 VEO G A Barratt, Charnwood, North Rd, Ranskill, Retford, DN22 8NL [IO93LJ, SK68]
G4 VEP R Barratt, The Lodge, North Rd, Torworth, Retford, DN22 8NW [IO93LI, SK68]
GW4 VEQ T W Jones, Penrhiw Bach, Brungwran, Anglesey, Gwynedd, LL65 3RD [IO73SG, SH37]
GW4 VES D E Williams, 3 Shetland Walk, St. Julians, Newport, NP9 7TJ [IO81MO, ST38]
G4 VET N Greaves, Flat, 10 Cramhurst House, Sutton Gr, Sutton, SM1 4TH [IO91VI, TQ26]
G4 VEU J R McTomney, 54 Oulston Rd, Fairfield, Stockton on Tees, TS18 4HU [IO94HN, NZ41]
G4 VEW G Davies, 8 Browning Gr, Talke, Stoke on Trent, ST7 1PD [IO83UB, SJ85]
G4 VEY F J Havard, Whitehalgh Farm, Whitehalgh Ln, Langho, Blackburn, BB6 8ET [IO83ST, SD73]
G4 VEZ E D P Pether, 35 Penponds Rd, Porthleven, Helston, TR13 9LP [IO70IC, SW62]
G4 VFA Details withheld at licensee's request by SSL.
G4 VFC D C Monnery, 8 Reeds Ln, Southwater, Horsham, RH13 7DQ [IO91UA, TQ12]
G4 VFD A D Janaway, Haydon Bungalow, Ashbrittle, Wellington, Somerset, TA21 0LG [IO80HX, ST02]
GW4 VFE C J Davies, 200 Sealand Rd, Chester, CH1 4LH [IO83ME, SJ36]
G4 VFF C B Waldron, 31 Dawlish Dr, Pinner, HA5 5LL [IO91TO, TQ18]
G4 VFG P F A Lewis, 18 Bittaford Wood, Bittaford, Ivybridge, PL21 0ET [IO80BJ, SX65]
G4 VFH G Fuller, 41 Burrage Rd, Woolwich, London, SE18 7LN [JO01AL, TQ47]
G4 VFJ S J Shenton, 36 Walleys Dr, Basford, Newcastle, ST5 0NG [IO83VA, SJ84]
G4 VFK C J Archer, Sports Office, Technology Dr, Beeston, Nottingham, Notts, NG9 1LA [IO92JW, SK53]
G4 VFL A J Holland, 17 Hillview Dr, Redhill, RH1 4DQ [IO91WF, TQ24]
G4 VFM Details withheld at licensee's request by SSL.
G4 VFO C J Pearson, Tangle Trees, 6 Enderley Cl, Bloxwich, Walsall, WS3 3PF [IO82XP, SJ90]
G4 VFQ D R Nuttall, The Mount, Church Ln, Bearsted, Maidstone, ME14 4EF [JO01GG, TQ75]
G4 VFR K L Hackwell, 15 Standish Ave, Billinge, Wigan, WN5 7TF [IO83PL, SJ59]
GW4 VFS Details withheld at licensee's request by SSL.
G4 VFT Details withheld at licensee's request by SSL.
G4 VFU Details withheld at licensee's request by SSL.
G4 VFV Details withheld at licensee's request by SSL.[Op: G J King, AMIERE TEng, 1 Chiseldon Farm, Follafield Park, Brixham, Devon, TQ5 0AD.]
G4 VFW K C Millar, 103 Roman Way, Farnham, GU9 9RQ [IO91OF, SU84]
G4 VFX C J Perkins, Signal Row, St Martins, Isles of Scilly, Cornwall, TR25 0QL [IN69UX, SV91]
GW4 VGB R I C Harper, 114 Pantbach Rd, Cardiff, CF4 1UE [IO81JM, ST18]
G4 VGE Details withheld at licensee's request by SSL.
G4 VGG D Packwood, 65 Mary Fell, Sedbergh, LA10 5AW [IO84RH, SD69]
G4 VGH Details withheld at licensee's request by SSL.
GD4 VGL S Luckhaus, Ludwigstrabe 42, 8753 Obernburg, West Germany
GD4 VGM E Grindel, Bischofsheimer, Platz 24, D 6000 Frankfurt, Germany, X X
GD4 VGN V Havran, Kurt-Schumacher, Ring 31, D-6072 Dreieich, West Germany
G4 VGQ B A Tooke, Deansfield, 17 Broad Ln, Moulton, Spalding, PE12 6PN [IO92XT, TF32]
GM4 VGR J Buchanan, 114 Glasgow Rd, Whins of Milton, Stirling, FK7 0LJ [IO86AC, NS79]
G4 VGT C B Butcher, Country Ways, Willow Rd, Great Mongeham, Deal, CT14 0HN [JO01QF, TR35]
GM4 VGU A M Lyttle, 23 Heathfield Dr, Kirkmuirhill, Lanark, ML11 9SR [IO85AQ, NS74]
G4 VGY D G Cline, 68 Frenchgate, Richmond, DL10 7AG [IO94DJ, NZ10]
G4 VHB C E J Averill-Elias, 12 Bubwith Cl, Chard, TA20 2BL [IO80MU, ST30]
G4 VHD A E B Jehle, Jolliffe, 13 Crediton Hill, London, NW6 1HT [IO91VN, TQ28]
G4 VHE R Haase, 674 Valley View Ln, Strafford, Pennsylvania 19087, USA, ZZ6 7VA
G4 VHG J A Fowler, 6 Cridlake, Axminster, EX13 5BS [IO80MS, SY39]
G4 VHH F Atkin, 18 East Ln, Corringham, Gainsborough, DN21 5QU [IO93PJ, SK89]
G4 VHI M R Sawyers, 36 Frome Rd, Bath, BA2 2QB [IO81TI, ST76]
G4 VHJ J S D Taylor, 29 Meadow Walk, Ewell, Epsom, KT17 2EF [IO91VI, TQ26]
G4 VHK Prof R A Leslie, Tranquil, Rectory Ln, Kingston, Cambridge, Cambs, CB3 7NL [IO92XE, TL35]
G4 VHL T B Langford, 11 The Grove, Blackawton, Totnes, TQ9 7BA [IO80DI, SX85]
G4 VHM M R H Hindley, 203 Gillshill Rd, Sutton Rd, Hull, HU8 0JP [IO93US, TA13]
GW4 VHO D W Calderwood, 23 Ravensbrook, Radyr, Morganstown, Cardiff, CF4 8LT [IO81IM, ST18]
GW4 VHP S A Murdoch, 55 Pendre Ave, Prestatyn, LL19 9SH [IO83HH, SJ08]
G4 VHQ E A W Smith, Borodino, Manor Rd, Edington, Bridgwater, TA7 9HB [IO81ND, ST33]
G4 VHR Details withheld at licensee's request by SSL.
GW4 VHS E J Williams, Newhaven, Kinmel Way, Towyn, Abergele, LL22 9NE [IO83FH, SH97]
G4 VHT R Kimberley, 18 Edinburgh Dr, Rushall, Walsall, WS4 1HR [IO92AO, SK00]
GM4 VHU B A J Meldrum, 7 Main St., New Deer, Turriff, AB53 6TA [IO87VM, NJ84]
G4 VHV A F Sims, 38 Giffard Dr, Bishopswood, Welland, Malvern, WR13 6SE [IO82UB, SO74]
G4 VHX J A Allen, 18 Horsley Cl, Chesterfield, S40 4XD [IO93GF, SK37]
GM4 VHZ N Brown, 7 Mid Rd, Beith, KA15 2AJ [IO75QS, NS35]
G4 VIA J McSherry, 1 Station Houses, Corkickle, Whitehaven, Cumbria, CA28 7XG [IO84EN, NX91]
GW4 VIB J H Endersby, Minycoed, 2 Kinmel Ave, Abergele, LL22 7LW [IO83FG, SH97]
G4 VIC Details withheld at licensee's request by SSL.
G4 VID D P Rogers, 5 Braemar Cl, Kettering, NN15 5DD [IO92PJ, SP87]
G4 VIE T J Wilkie, 33 Shepherds Hill, Haslemere, GU27 2NB [IO91PC, SU93]
G4 VIF R Watts, 116 Hassall Rd, Sandbach, CW11 4HL [IO83TD, SJ75]
G4 VII J L Lawrence, 25 Sylvia Cres, Totton, Southampton, SO40 3LP [IO90GW, SU31]
G4 VIJ Details withheld at licensee's request by SSL.
GM4 VIK T F R Irwin, Balmore Cottage, Balvraid, Tomatin, Inverness, IV13 7XY [IO87AI, NH83]
G4 VIL J Fielding, 35 Amos Ave, Litherland, Liverpool, L21 7QH [IO83ML, SJ39]
G4 VIM B Pulfrey, 18 Lavenham Rd, Grimsby, DN33 3EX [IO93WM, TA20]
G4 VIN Details withheld at licensee's request by SSL.
GI4 VIP P Murphy The Sth Belfast, Vhf Cont Grp, 2 The Gables, Cairnshill Rd Belfs, N Ireland, BT8 4UJ [IO74BN, J36]
G4 VIQ P D Brushwood, 2 High Trees, Waterlooville, PO7 7XP [IO90LV, SU60]
GM4 VIR A Anderson, Kirkbride, Skyreburn, Gatehouse of Fleet, Castle Douglas, DG7 2HE [IO74VU, NX55]
GM4 VIS H Cameron, 14 Queen St., Castledouglad, Castle Douglas, DG7 1HX [IO84AW, NX76]
G4 VIT M A Wood, 42 Buckingham Dr, Short Heath, Willenhall, WV12 5TD [IO82XO, SJ90]
GI4 VIV P G Mercer, 4 Cranley Park, Bangor, BT19 7HF [IO74EP, J57]
G4 VIW W A Watts, 11 Meadow Way, Upminster, RM14 3AA [JO01DN, TQ58]
G4 VIX D J Bartlett The East Coast VHF Group, 80 Burnway, Hornchurch, RM11 3SG [JO01CN, TQ58]
G4 VIY A G Pevy, Brambletye, Ashtead Ln, Godalming, GU7 1SY [IO91QE, SU94]
GI4 VIZ M T Jamieson, 61 Belraugh Rd, Ringsend, Garvagh, Co.Londonderry, BT51 5HB [IO65OA, C71]
G4 VJB V J Bloor, 22 Regency Cl, Talke Pits, Stoke on Trent, ST7 1RH [IO83UB, SJ85]
GI4 VJC J S McIlroy, 165 Roguery Rd, Toomebridge, Antrim, BT41 3RR [IO64TS, J09]
G4 VJE G A Leadbetter, 24 Harebell Dr, Witham, CM8 2XB [JO01HT, TL81]
G4 VJF D F Barnes, 7A Partlands Ave, Swanmore, Ryde
G4 VJG Details withheld at licensee's request by SSL.
G4 VJI C W Lindsay, 31 Barnes Cl, Blandford, Blandford Forum, DT11 7NG [IO80WU, ST80]
G4 VJJ L W Lawrence, Cotswold, Bashley Rd, New Milton, BH25 5RX [IO90ES, SZ29]
G4 VJL J Oldfield, 62 Chipperfield Dr, Bristol, BS15 4DR [IO81SL, ST67]
G4 VJM Details withheld at licensee's request by SSL.[Op: Nigel J Ludlow, c/o 5 Laburnum Avenue, Laffak, St Helens, Merseyside, WA11 9DZ. Please QSL via Bureau.]
G4 VJO G S Hague, 137 Waddington Ave, Great Barr, Birmingham, B43 5JD [IO92AN, SP09]
G4 VJQ K A Partridge, Downalong, 58 Old Penkridge Rd, Cannock, WS11 1HX [IO82XQ, SJ91]
G4 VJS J Sayner, 3 Bulls Copse Ln, Waterlooville, PO8 9QX [IO90LV, SU61]
G4 VJT K W Farmer, 61 Queens, Beckenham, Kent, BR3 4JJ [IO91XJ, TQ36]
G4 VJU Details withheld at licensee's request by SSL.[Op: H J Bloomer. Station located near Halesowen.]
GM4 VJV I Fairbairn, 1 Callander Pl, Cockburnspath, TD13 5XY [IO85TW, NT77]
G4 VJW Details withheld at licensee's request by SSL.
G4 VJX Details withheld at licensee's request by SSL.
GM4 VJY Details withheld at licensee's request by SSL.
GI4 VJZ T J Wilson, Wilden, 18 Ahoghill Rd, Randalstown, Antrim, BT41 3BJ [IO64US, J09]
G4 VKA K P Kozma, 11 Pyenot Dr, Pyenot Hall Ln, Cleckheaton, BD19 5AX [IO93DR, SE12]
G4 VKB Details withheld at licensee's request by SSL.
G4 VKC D F Lawrence, 6 Hollycombe Cl, Liphook, GU30 7HR [IO91OB, SU83]
G4 VKE R Pearce, 1A Green Ln, Dalton in Furness, LA15 8LZ [IO84JD, SD27]
GW4 VKG W J Weston, 3 Factory Terr, Aberkenfig, Bridgend, CF32 9AF [IO81EM, SS88]
GM4 VKI M R Kavanagh, 4 Old Auchans View, Dundonald, Kilmarnock, KA2 9EX [IO75QN, NS33]
G4 VKJ T P Grant, 81 Hillworth Rd, Devizes, SN10 5HD [IO81XI, ST96]
G4 VKK K H Kirby, 36 Hornbeam Dr, Cottingham, HU16 4RU [IO93TS, TA03]
G4 VKM R H Williams, 72 Botham Hall, Longwood, Huddersfield, West Yorks, HD3 4RJ [IO93BP, SE11]
G4 VKO J G W Whittock, Orchard Cottage, East Coker, nr Yeovil, Somerset, X X
GI4 VKS A P McCallion, 3 Lisky Rd, Strabane, BT82 8NW [IO64GT, H39]
G4 VKV T Linacre, 69 Elizabeth Rd, Walton, Liverpool, L10 4XL [IO83NL, SJ39]
G4 VKX I L Wade, 59 St. Annes Rd, Kettering, NN15 5EQ [IO92PJ, SP87]
G4 VKY R I Brown Blythe ARC, 17 Rowley St., Blyth, NE24 2HQ [IO95FC, NZ38]
G4 VLA R C Trudgill, 24 Farm Cl, Ampthill, Bedford, MK45 2UE [IO92SA, TL03]
G4 VLB K Dunne, 19 Fox St., Edgeley, Stockport, SK3 9EL [IO83WJ, SJ88]
G4 VLC Details withheld at licensee's request by SSL.
G4 VLE Details withheld at licensee's request by SSL.
G4 VLF J A Wingfield, 14 Brook Cl, Staines, TW19 7AW [IO91SK, TQ07]
G4 VLH B W Pash, Dales Stores, Station Rd, Andoversford, Cheltenham, GL54 6HP [IO91BV, SP12]
G4 VLI R Allen, 39 Deerpark, Ashbourne, Co Meath, Ireland
G4 VLK D Haslehurst, 23 Yew Tree Dr, Shirebrook, Mansfield, NG20 8QH [IO93JE, SK56]
G4 VLL C G Denham, 58 Oak Tree Rd, Marlow, SL7 3EG [IO91ON, SU88]
G4 VLM J J McTaggart, 56 Lower House Ln, Liverpool, L11 2SQ [IO83MK, SJ39]
G4 VLN M C E Evans, 2A Moreton Rd, Worcester Park, Surrey, KT4 8EZ [IO91VJ, TQ26]
G4 VLP H N Knatchbull, 19 Riverside Rd, West Moors, Ferndown, BH22 0LG [IO90BT, SU00]
GW4 VLQ V E Hodges, Greenfields, Eagleswell Rd, Boverton, Llantwit Major, CF61 1UF [IO81GJ, SS96]
G4 VLR Details withheld at licensee's request by SSL.[Station located near Wimborne.]
G4 VLS P R Turnham, 71 Theobald Rd, Norwich, NR1 2NX [JO02PO, TG20]
G4 VLT C D Tunna, 52 Shaftoe Rd, Springwell, Sunderland, SR3 4EZ [IO94GV, NZ35]

GW4 VLU M T Hatwood, Calgary, Denbigh Circle, Kinmel Bay, Rhyl, LL18 5HW [IO83FH, SH97]
G4 VLV A Flint, The Gables, Friday St., Painswick, Stroud, GL6 6QJ [IO81VS, SO80]
G4 VLW R Davey, 35 The Pines, Faringdon, SN7 8AT [IO91EP, SU29]
GM4 VLX J S Brown, 33 Gartmore Rd, Paisley, PA1 3NG [IO75TU, NS56]
G4 VLY J Smith, 252 Firs Ln, Leigh, WN7 4TT [IO83RL, SD60]
G4 VLZ M D Nettleship, 141 Hollybank Dr, Sheffield, S12 2BU [IO93HI, SK38]
G4 VMA M Anderson, 26 Skipton Rd, Earby, Colne, BB8 6PX [IO83WV, SD94]
G4 VMB D R Stoddart, 81 Fir Tree Cl, Flitwick, Bedford, MK45 1NY [IO91SX, TL03]
G4 VMC P R Coates, 10 Glenthorne Cl, Stafford, ST17 4RW [IO82WT, SJ92]
G4 VMD C E Hackney, 20 St. Matthews Way, All Hallows, Allhallows, Rochester, ME3 9SH [JO01HL, TQ87]
GW4 VMF S Schofield, 155 Withycombe Village Rd, Exmouth, EX8 3AN [IO80HP, SY08]
G4 VMG J Holmes, 52 Earlspark Dr, Bieldside, Aberdeen, AB1 9AH [IO87TH, NJ72]
G4 VMH T Everall, 1 Thorburn St., Edge Hill, Liverpool, L7 1QS [IO83MJ, SJ39]
G4 VMI M C Pickworth, 71 Queensway, Grantham, NG31 9RW [IO92QW, SK93]
G4 VMJ R Jolly, 17 Mylen Rd, Andover, SP10 3HD [IO91GF, SU34]
G4 VML R H Knight, 81 Edinburgh Ave, Bentley, Walsall, WS2 0HT [IO82XO, SO99]
G4 VMM S J Tidmarsh, 23 Johnson Rd, Birstall, Leicester, LE4 3AT [IO92KQ, SK50]
G4 VMN M J Alexander, 3 Rockingham Cl, Dorridge, Solihull, B93 8EH [IO92CI, SP17]
G4 VMO J F Harris, 23 Brookvale Gr, Olton, Solihull, B92 7JH [IO92CK, SP18]
G4 VMP N E Parker, 43 Coombe Meadows, Chillington, Kingsbridge, TQ7 2JL [IO80DG, SX74]
G4 VMR J Watkins, One Ash, Frogshall Ln, Haultwick, Ware, SG11 1JH [IO91XV, TL32]
GW4 VMT G G Williams, 12 Heol Johnson, Talbot Green, Pontyclun, CF7 8HR [IO81HN, ST08]
G4 VMU E Gordon, 11 Apperley, West Denton, Newcastle upon Tyne, NE5 2JS [IO94DX, NZ16]
G4 VMV R Hopkins, 28 Connolly Dr, Carterton, OX18 1BH [IO91ES, SP20]
G4 VMW J G Curtis, 18 Carlidnack Cl, Mawnan Smith, Falmouth, TR11 5HF [IO70KC, SW72]
G4 VMX A C Ritchie, 24 Swift Cl, Newport Pagnell, MK16 8PP [IO92PC, SP84]
G4 VMY A Cooper, 3 Marina Way, Ripon, HG4 2LJ [IO94FC, SE36]
G4 VMZ A I Jones, 16 Hawe Farm Way, Broomfield, Herne Bay, CT6 7UD [JO01NI, TR16]
G4 VNA R A Bell, Haddockstones Farm, Bishop Thornton, Harrogate, North Yorks, HG3 3LA [IO94EC, SE26]
G4 VNC M J Leak, 9 The Vineyards, Leven, Beverley, HU17 5LD [IO93UV, TA14]
G4 VND Details withheld at licensee's request by SSL.
G4 VNE D Hunt, 233 Kingsley Rd, Kingswinford, DY6 9RP [IO82VL, SO88]
G4 VNG R J McCallum, 103 Walgrave, Orton Malborne, Peterborough, PE2 5NS [IO92UN, TL19]
G4 VNI A L J J Sleeman, 44 Burnside, Exmouth, EX8 3AH [IO80HP, SY08]
G4 VNK L Smith, 1 Jarvis Fields, Bursledon, Southampton, SO31 8AF [IO90IV, SU40]
G4 VNM S L Frost, 15 Bartlett Cl, Fareham, PO15 6BQ [IO90JU, SU50]
GW4 VNO Details withheld at licensee's request by SSL.[Op: T B Tilley, Pen-y-Bryn, Mount Road, Llanfairfechan, Gwynedd, LL33 0HA.]
GM4 VNQ D McLean, Carolina, Main Rd, Aberuthven, Auchterarder, PH3 1HB [IO86EH, NN91]
G4 VNR R E J Sharp, Limesgasse 330, Petronell A-2404, Austria, X X
GW4 VNS R W Sims, 61 Constable Dr, Newport, NP9 7QB [IO81MO, ST38]
G4 VNU I Stacey, The Lodge, 1 May Park, Calcot, Reading, RG31 7RU [IO91LK, SU67]
G4 VNV W C Brock, 22 Fountain Head Bank, Seaton Sluice, Whitley Bay, NE26 4HU [IO95GC, NZ37]
G4 VNX W J Wood, 11 Walbert Ave, Thurnscoe, Rotherham, S63 0TN [IO93IN, SE40]
G4 VOB G E M Sunter, 16 Waindale Cl, Mount Tabor, Halifax, HX2 0UL [IO93AR, SE02]
G4 VOC R W Ludgate, The Haywain, North Mills Rd, Bridport, Dorset, DT6 3AH [IO80OR, SY49]
G4 VOG A G Hepworth, 9 Linden Gr, Kirkby in Ashfield, Nottingham, NG17 8JJ [IO93IC, SK45]
G4 VOJ A Tennant, 43 Oak Ave, Morecambe, LA4 6HY [IO84OB, SD46]
G4 VOK P Dresser, 6 Acacia Ave, Fence House, Houghton-le-Spring, DH4 6JG [IO94GU, NZ35]
GW4 VOL Details withheld at licensee's request by SSL.[Op: L B Tilley, Pen-y-Bryn, Mount Road, Llanfairfechan, Gwynedd, LL33 0HA.]
G4 VOM Details withheld at licensee's request by SSL.
G4 VON Details withheld at licensee's request by SSL.
G4 VOQ D S Gibson, 7 Rowlatt Cl, Wilmington, Dartford, DA2 7BT [JO01CK, TQ57]
G4 VOT R E Bullimore, 103 Podsbrook House, Guithavon St., Witham, CM8 1DR [JO01HT, TL81]
G4 VOU A Pinkney, 1 Hester Gdns, New Hartley, Whitley Bay, NE25 0SH [IO95FC, NZ37]
G4 VOW B G Pluckrose, 104 Edward Rd, West Bridgford, Nottingham, NG2 5GB [IO92KW, SK53]
G4 VOX Details withheld at licensee's request by SSL.
G4 VOY R W Powell, Old School House, Broxwood, nr Leominster, Herefordshire, HR6 9JQ [IO82ME, SO35]
G4 VOZ J R Jennings, Mill Side, Mill Rd, Ullesthorpe, Lutterworth, LE17 5DE [IO92IL, SP58]
GM4 VPA J H Martindale, 6 Springfield Ave, Uddingston, Glasgow, G71 7LY [IO75XT, NS66]
G4 VPC E S Ikin, 30 Kelsborrow Way, Kelsall, Tarporley, CW6 0NL [IO83PE, SJ56]
G4 VPD M A Pugh, 44 Simms Ln, Hollywood, Birmingham, B47 5HY [IO92BJ, SP07]
G4 VPE Dr R T Derricott, 1 Glenelg Dr, Pedmore, Stourbridge, DY8 2PF [IO82WK, SO98]
G4 VPF O Davies, 16 Central Way, Horninglow, Burton on Trent, DE13 0UU [IO92ET, SK22]
G4 VPG W A Harrop, Apartdos Correos 373, 07300 Inca, Mallorca, Spain
G4 VPI R W N Riley, 161 Botany Rd, Kingsgate, Broadstairs, CT10 3SD [JO01RJ, TR37]
G4 VPJ D K Bridgnell, Penvale, 1 Tretherras Rd, Newquay, TR7 2RB [IO70LJ, SW86]
GW4 VPK P J M Roberts, Derwendeg, Peniel, Carmarthen, Dyfed, SA32 7AB [IO71UV, SN42]
G4 VPL J Villena Bota, Santa Ana 74, Estartit, Gerona, Spain 17258, X X
G4 VPM A D Stafford, 13 Hospital Ln, South Petherton, TA13 5AR [IO80OW, ST41]
G4 VPN F Matthews, 37 Buckingham Ave, Horwich, Bolton, BL6 6NS [IO83RO, SD61]
G4 VPO P Ashman, 52 Parkway, Gildersome, Morley, Leeds, LS27 7DY [IO93ES, SE22]
G4 VPP B G Beardmore, 30 Madrona, Amington, Tamworth, B77 4EJ [IO92EP, SK20]
G4 VPS B W Lewis, Juno, Richmond Rd, Ingleton, Carnforth Lancs, LA6 3AN [IO84TD, SD77]
G4 VPU J F Armstrong, 3 Priory Ave, Whitley Bay, NE25 8RU [IO95GA, NZ37]
G4 VPW P Wilcock, 12 Napier Rd, Monton, Eccles, Manchester, M30 8AG [IO83TL, SJ79]
GW4 VPX A J Jones, Ffynnon-Wen, Llanllwni, Pencader, Dyfed, SA39 9DH [IO72VA, SN43]
G4 VPZ A A Hill, 36 Narrow Ln, Halesowen, B62 9NQ [IO82XL, SO98]
G4 VQB Details withheld at licensee's request by SSL.
G4 VQF P R White, 4 Barnett Ln, Wonersh, Guildford, GU5 0SA [IO91RE, TQ04]
G4 VQH M G Clutton, Cumberwell, Cumberland Ln, Whixall, Whitchurch, SY13 2NJ [IO82PW, SJ53]
G4 VQI P H Hughes, 145 Blatchcombe Rd, Paignton, TQ3 2JP [IO80FK, SX86]
G4 VQJ D J Northwood, 5 Beech Grange, Landford, Salisbury, SP5 2AL [IO90EX, SU21]
GI4 VQK A G Ward, 50 Derry Rd, Strabane, BT82 8LD [IO64GU, H39]
G4 VQL G R Dymond, 74 Rangoon Rd, Solihull, B92 9DD [IO92CK, SP18]
G4 VQP C Smith, 7 Church Ln, Sutton Waldron, Blandford Forum, DT11 8PB [IO80VW, ST81]
G4 VQQ Details withheld at licensee's request by SSL.
G4 VQR S Reed, 20 Vicarage Cl, Outwood, Wakefield, WF1 2LX [IO93GR, SE32]
G4 VQS H J Jones, 47 Penkett Rd, Wallasey, L45 7QG [IO83LK, SJ39]
G4 VQT M V G Pinnell, Greenfields, Westfield Ln, Wrecclesham, Farnham, GU10 4QP [IO91OE, SU84]
G4 VQU A Lynam, 118 The Patchills, Mansfield, NG18 3BS [IO93JD, SK56]
G4 VQX D M Noble, 6 Oak Rd, Morley, Leeds, LS27 0PU [IO93ER, SE22]
GM4 VQY G R Leiper, 76 Martin Dr, Stonehaven, AB3 2LU [IO87TH, NJ72]
G4 VQZ J W Oakley, 152 Little Breach, Chichester, PO19 4UA [IO90OU, SU80]
G4 VRA A J Ridgway, Broadview, Peasmarsh, Rye, E Sussex, TN31 6YJ [JO00HX, TQ82]
G4 VRC R J Doran, 28 Buckingham Rd, Petersfield, GU32 3AZ [IO91MA, SU72]
GM4 VRE P I Henderson, 134 Gray St., Aberdeen, AB1 6JU [IO87TH, NJ72]
GI4 VRF F G Macdonald, 5 Glenview Cres, Castlereagh, Belfast, BT5 7LX [IO74BN, J37]
G4 VRG F W Margrave, 23 Steventon, Sandymoor, Runcorn, WA7 1UB [IO83QI, SJ58]
GW4 VRH R Hall, 4 Highmead, Pontllanfraith, Blackwood, NP2 2PE [IO81JP, ST19]
G4 VRI L F Lockyer, 91 Shakespeare Rd, Gillingham, ME7 5QJ [JO01GJ, TQ76]
G4 VRJ R Clifft, 11 Hambleton St., Wakefield, WF1 3NW [IO93GQ, SE32]
G4 VRL D G Cobbledick, Treburyse Rd, Launceston, Cornwall, PL15 7EL [IO70TP, SX38]
G4 VRM A G Berry, 22 The Venn, Hill Farm Est, Shaftesbury, SP7 8EB [IO81VA, ST82]
G4 VRN M J Blewett, 32 Miltons Cres, Godalming, GU7 2NT [IO91QE, SU94]
GW4 VRO P O Parsons, S3 St Teilos Rd, Pembroke Dock, Dyfed, SA72 6LH [IO71MQ, SM90]
G4 VRP R E Porter, 47 Milford Ave, Wick, Bristol, BS15 5PP [IO81SK, ST77]
G4 VRQ Details withheld at licensee's request by SSL.
G4 VRR Details withheld at licensee's request by SSL.[Op: S J Gray. Station located near Ashford.]
G4 VRS I J Eamus Aylesbury VI RS, 10A Vale Rd, Aylesbury, HP20 1JA [IO91OT, SP81]
G4 VRT B Barker, 28 Alder Gr, Darfield, Barnsley, South Yorks, S73 9JL [IO93HM, SE40]
G4 VRU B J Stephenson, 12 Claremont Terr, Gillygate, York, YO3 7EJ [IO93LX, SE65]
G4 VRW K Newbould, 18 Swarcliffe Rd, Leeds, LS14 5LE [IO93GT, SE33]
G4 VRX Dr G D Brown, 1 Dog Kennel Ln, Oldbury, Warley, B68 9LU [IO92AL, SP08]
G4 VSB A R Brown, 67 Willow Way, Flitwick, Bedford, MK45 1LN [IO91SX, TL03]
GI4 VSC P Jones, 43 Strathmore Park, South Belfast, N Ireland, BT15 5HJ [IO74AP, J37]
G4 VSD P J Samuels, 3 Station Rd, Stallingborough, Grimsby, DN37 8AF [IO93UN, TA10]
GW4 VSE M R Carey, 47 Heol Ty Gwyn, Maesteg, CF34 0BD [IO81EO, SS89]
G4 VSF J F Burn, 7 Atkinson St., Haverigg, Millom, LA18 4HA [IO84IE, SD17]
G4 VSI A M Stone, 23 Cecil St., Derby, DE22 3GQ [IO92GW, SK33]
G4 VSJ K Drakeford, Sunnyside, Frolesworth Ln, Claybrooke Magna, Lutterworth, Leics, LE17 5AS [IO92IL, SP48]
G4 VSK B Skelton, 15 Birch Gr, The Elms, Torksey, Lincoln, LN1 2EZ [IO93PH, SK87]
G4 VSL T A Watkins, One Ash, Frogshall Ln, Haultwick, Ware, SG11 1JH [IO91XV, TL32]
G4 VSN D L Bickerton, 2 Elaine Ct, Elaine Ave, Strood, Rochester, ME2 2YR [JO01FJ, TQ76]
G4 VSO R N Carter, 336 Peppard Rd, Caversham, Reading, RG4 8UY [IO91ML, SU77]
G4 VSP P Tomlinson, 41 Burlington Rd, Gillhill Rd, Hull, HU8 0HN [IO93US, TA13]
G4 VSQ A Bolton, 16 Cardinal Cl, Caversham, Reading, RG4 8BZ [IO91ML, SU77]
G4 VSR S R Alston, 21 Hilltop Rd, Wingerworth, Chesterfield, S42 6RX [IO93GE, SK36]
G4 VSS M E Isherwood, 52 St. Bridgets Cl, Cinnamon Brow, Fearnhead, Warrington, WA2 0EW [IO83RJ, SJ69]
GM4 VST K U Macleod, 84 Drumossie Ave, Inverness, IV2 3SX [IO77VL, NH64]
G4 VSU H G Jones, 62 Highgate Rd, Sileby, Loughborough, LE12 7PR [IO92KR, SK61]
G4 VSV G I Ingham, Courthaven, South Duffield Rd, Osgodby, Selby, YO8 7HP [IO93LT, SE63]
G4 VSW M S Taylor, 2 Bickerton Dr, Hazel Gr, Stockport, SK7 5QY [IO83WI, SJ98]
G4 VSX P T Reilly, 40 Bollin Dr, Lymm, WA13 9QA [IO83SJ, SJ68]
G4 VSZ Details withheld at licensee's request by SSL.[Op: C Darley. Station located near Gillingham.]
G4 VTA J W Taylor, 219 Mandarin Way, Cheltenham, GL50 4SB [IO81WW, SO92]
GM4 VTB M Budas, 20 Oak Ave, Bearsden, Glasgow, G61 3HD [IO75UW, NS57]
G4 VTC A J Croydon, Harvesters, Couldharbour Ln, Dorking, Surrey, RH4 3JH [IO91TF, TQ14]
G4 VTD I B Daniels, 71 Firsby Ave, Shirley, Croydon, CR0 8TP [IO91XJ, TQ36]
GW4 VTG E D W Smith, 21 St. Davids Rd, Pembroke, SA71 5JH [IO71NQ, SM90]
G4 VTH Details withheld at licensee's request by SSL.
G4 VTK J E Rollason, Ash Ridge, Clint, Harrogate, North Yorks, HG3 3DS [IO94EA, SE25]
G4 VTL T R Lawry, Ballacraine, 3 Carnarth Cove, Caharrack Redruth, Cornwall, TR16 5SA [IO70JF, SW74]
G4 VTM J L Hicks, Cory House, Kilworth Rd, Husbands Bosworth, Lutterworth, LE17 6JW [IO92LK, SP68]
G4 VTN J H Rodda, 22 Balmoral Dr, Felling, Gateshead, NE10 9TZ [IO94FW, NZ26]
G4 VTO P M Tanner, 4 Maddacombe Rd, Kingskerswell, Newton Abbot, TQ12 5LF [IO80EL, SX86]
G4 VTP G W Greenwood, 10 Triangle Dr, North Ferriby, HU14 3AU [IO93RR, SE92]
G4 VTQ D J R Rainer, 64 Mill Rd, Millcroft Cottage, Burgess Hill, RH15 8DZ [IO90WW, TQ31]
G4 VTR F Baker, 2 Carlidnack Cl, Mawnan Smith, Falmouth, TR11 5HF [IO70KC, SW72]
G4 VTS C J Cowling, 119 Agar Rd, Illogan Highway, Redruth, TR15 3EF [IO70IF, SW64]
G4 VTT C K Lloyd, 35 St. Lukes Rd, Newton Abbot, TQ12 4NE [IO80EM, SX87]
G4 VTU R P George, 67 The Oaks, Milton, Cambridge, CB4 6ZG [JO02BF, TL46]
G4 VTX Details withheld at licensee's request by SSL.[Op: D Richards, 23 Jubilee Avenue, W Bromwich, W Midlands, B71 2QT.]
G4 VTY B R Southwell, 7 Briggate, Silsden, Keighley, BD20 9JS [IO93AV, SE04]
G4 VUA A K Burton, 26 Woffindin Cl, Great Gonerby, Grantham, NG31 8LP [IO92PW, SK83]
GW4 VUC M Smith, 18 Tenby Cl, Llanyrafon, Llanyravon, Cwmbran, NP44 8TA [IO81LP, ST39]
G4 VUD J D Head, 21 Reynell Ave, Buckland, Newton Abbot, TQ12 4HE [IO80EM, SX87]
G4 VUE W Kostryca, 9 Cherry Tree Rd, Gainsborough, DN21 1RG [IO93OJ, SK89]
G4 VUF C C Tugman, 28 Seagate Rd, Hunstanton, PE36 5BD [JO02FW, TF64]
G4 VUG C Green, 14 Cranmore Cl, Broadmead, Trowbridge, BA14 9BU [IO81VH, ST85]
GW4 VUH J W Washington, 15 Sundawn Ave, Holywell, CH8 7BH [IO83JG, SJ17]
G4 VUI P B Sweeney, 9 Bury Rd, Shefford, SG17 5AP [IO92UA, TL13]
G4 VUK L Wolfson, 7 Gilmore Dr, Prestwich, Manchester, M25 1NB [IO83UM, SD80]
G4 VUM D R S Stocks, 78 Moor St., Mansfield, NG18 5SQ [IO93JD, SK56]
G4 VUN P F Norris, Thirn Farm, Thirn, Ripon, N Yorks, HG4 4AU [IO94DG, SE28]
G4 VUO Details withheld at licensee's request by SSL.
G4 VUP G D Newton, 107 Armstrong St., Grimsby, DN31 2QQ [IO93WN, TA21]
G4 VUR I Daniels, 18 Stalham Rd, Hoveton, Norwich, NR12 8DG [JO02QR, TG31]
GW4 VUT Details withheld at licensee's request by SSL.
G4 VUU W E Dick, Carpenters Cottage, Main Rd, Scamblesby, Louth Lincs, LN11 9XH [IO93WG, TF27]
G4 VUW R A Kemp, 35 Rushett Dr, Dorking, RH4 2NR [IO91UF, TQ14]
G4 VUX G M Stannett, 314 Watford Rd, Croxley Green, Rickmansworth, WD3 3DE [IO91SP, TQ09]
G4 VUY W G Duschek, Devereux, 78 Wray Park Rd, Reigate, RH2 0EH [IO91VF, TQ25]
G4 VVC P M Barry, 39 Constable Rd, Lockleaze, Bristol, BS7 9YF [IO81RL, ST67]
G4 VVD P D Taylor, 8 High St., Clive, Shrewsbury, SY4 3JL [IO82PT, SJ52]
G4 VVE E Macmanus, 41 Oldfield Cres, Stainforth, Doncaster, DN7 5PE [IO93LO, SE61]
G4 VVF N T Allen, Hill Barn Bungalow, Aymestrey, Leominster, Hereford, HR6 9SR [IO82NG, SO46]
G4 VVG T H Bower, 13 Holtspur Cl, Holtspur, Beaconsfield, HP9 1DP [IO91PO, SU98]
G4 VVH G Hird, 29 Milton Keynes, Village, Milton Keynes, MK10 9AH [IO92PB, SP83]
G4 VVI J D Hughes, 63 Queens Cres, Stubbington, Fareham, PO14 2QQ [IO90JT, SU50]
G4 VVJ Details withheld at licensee's request by SSL.
G4 VVK B Murray, La Casa, 30 Middlegate Green, Rossendale, BB4 8PY [IO83UR, SD82]
GW4 VVL K G Arrowsmith, 35 Meadow Rise, Guidfa Meadows, Crossgates, Llandrindod Wells, LD1 6TA [IO82II, SO17]
G4 VVM A Bennett, The Oaks, 43 School Ln, Exhall, Coventry, CV7 9GE [IO92GL, SP38]
G4 VVO R H Parker, Hazelhurst, Stamfordham Rd, Ponteland, Newcastle upon Tyne, NE20 9TN [IO95DB, NZ17]
G4 VVP B N Gillard, Charmaine, Broadway, Chilcompton, Bath, BA3 4JW [IO81RG, ST65]
G4 VVQ F H Shead, 7 White Cttgs, Fuller St., Fairstead, Chelmsford, CM3 2AY [JO01GT, TL71]
G4 VVR D Bowden, 41 Low Seaton, Seaton, Workington, Cumbria, CA14 1PX [IO84FP, NY03]
G4 VVS J S Blanchard, 41 Deane Dr, Galmington, Taunton, TA1 5PQ [IO81KA, ST22]
G4 VVT A S Moss, 9 Summerfield Dr, Boarshaw, Middleton, Manchester, M24 2TQ [IO83VN, SD80]
G4 VVU Details withheld at licensee's request by SSL.
GW4 VVX Details withheld at licensee's request by SSL.[Op: Clive O'Hennessy, c/o Newsagents, Commercial Road, Pontllanfraith, Blackwood, Gwent, NP2 2PG. Station located in Gwent at IO81JP. Active as GB2XS in IO78WA from time to time, and as GB0LCS in August each year.]
G4 VVY D H Davis, 7 The Courtyard, Fisherwick Wood Ln, Fisherwick Wood, Lichfield, WS13 8QQ [IO92DQ, SK10]
G4 VVZ C N Wilson, 2 Bainton Cl, Bradford on Avon, BA15 1SE [IO81VI, ST86]
G4 VWA S E Ward, 88 Little Barn Ln, Mansfield, NG18 3JJ [IO93JD, SK56]
G4 VWB R W Osgathorpe, 9 Woodside Dr, Allestree, Derby, DE22 2UN [IO92GW, SK33]
GI4 VWC G Christie, The Brambles, 9 Burnet Park, Old Carrick Rd, Newtownabbey, BT37 0XY [IO74BQ, J38]
G4 VWE M N Courteney, 36 Nursery Cl, Hellesdon, Norwich, NR6 5SJ [JO02PP, TG21]
G4 VWG S G R Van Kassel, 50 Rodney Rd, Mitcham, CR4 3DG [IO91VJ, TQ26]
G4 VWH Details withheld at licensee's request by SSL.[Vale of White Horse ARS, c/o G3SEK.]
G4 VWI D F Hatton, 9 Gregory Cl, Colby Lodge, Thurmaston, Leicester, LE4 8BP [IO92KQ, SK61]
G4 VWK N Sinclair, Robins Nest, Rockbeare Hill, Rockbeare, Exeter, EX5 2EZ [IO80HR, SY09]
G4 VWL J Owen, 27 Sutherland Dve, Eastham, Wirral, Merseyside, L62 8DY [IO83MH, SJ37]
G4 VWN C Drobnica, Falcons, St. Georges Ave, Weybridge, KT13 0BS [IO91SI, TQ06]
G4 VWO J Bloodworth, 88 Clare Gdns, Petersfield, GU31 4EU [IO91NA, SU72]
G4 VWP R P Bloodworth, 88 Clare Gdns, Petersfield, GU31 4EU [IO91NA, SU72]
G4 VWQ J S Noble, 67 Broadlee, Henley Grange, Wilnecote, Tamworth, B77 4PG [IO92EO, SK20]
G4 VWS C A Davies, Essex House, 42 Boxworth Rd, Elsworth, Cambridge, CB3 8JQ [IO92XG, TL36]
G4 VWU P J Shaw, 39 Bestwood Park, Clay Cross, Chesterfield, S45 9LD [IO93HD, SK36]
GM4 VWV R McEwan, Oakwood, 12 Valleyfield Dr, Cumbernauld, Glasgow, G68 9NW [IO75XW, NS77]
G4 VWX A Shone, 14 Weslea Cl, Tithe Meadow, Bromyard Rd, St Johns Worcs, WR2 5UH [IO82UE, SO85]
GW4 VWY G Whiteway, The Wardens Flat, 27 Crown Ave, Ynyswen, Treorchy, CF42 6DY [IO81FP, SS99]
GM4 VWZ P S Williams, Sandgate, Kilmichael Glassary, By Lochgilphead, Argyll, PA31 8QL [IO76HC, NR99]
G4 VXB M W T Ellis, 16 Fielding St., Faversham, ME13 7JZ [JO01KH, TR06]
G4 VXC Details withheld at licensee's request by SSL.[Op: K A Christensen.]
G4 VXD R L King, 1 Emmas Cres, Stanstead Abbotts, Ware, SG12 8AZ [IO91XS, TL31]
G4 VXE T H Kirby, 2 Penoweth, Mylor Bridge, Falmouth, TR11 5NQ [IO70LE, SW83]
G4 VXG Details withheld at licensee's request by SSL.[Station located at Catterick Garrison. Correspondence to: G A Buxton, 52 Coronation Road, Loftus, Saltburn, Cleveland, TS13 4SL.]
G4 VXH A J Rean, 17 Mount Pleasant Rd, Dawlish Warren, Dawlish, EX7 0NA [IO80GO, SX97]
G4 VXI S Hitchens, 16 Skipton Rd, Chandlers Ford, Eastleigh, SO53 3BN [IO90HX, SU41]
G4 VXL Details withheld at licensee's request by SSL.
GM4 VXM I R Munro, 7 Canisp Cres, Gowrie Park, Dundee, DD2 4TP [IO86LL, NO33]
G4 VXN C Bennett, 37 Carisbrooke Ave, Sands, High Wycombe, HP12 4NL [IO91OP, SU89]
G4 VXP R F Want, 19 Canterbury Rd, Leyton, London, E10 6EE [IO91XN, TQ38]
G4 VXQ Details withheld at licensee's request by SSL.
G4 VXR D G Manning, 3 Palgrave Cl, Taverham, Norwich, NR8 6LP [JO02OQ, TG11]
G4 VXT B J Phillimore, 19 Russell Dr, Ampthill, Bedford, MK45 2UA [IO92SA, TL03]
G4 VXU J E Haig, 3 Hartland Ct, Gaping Ln, Hitchin, SG5 2JH [IO91UW, TL12]
G4 VXV R F Boulton, 23 York Rd, Stamford Bridge West, Riccall, York, YO4 6QG [IO93LU, SE63]
G4 VXW R N Seddon, 255 Westleigh Ln, Leigh, WN7 5PN [IO83RM, SD60]
G4 VXY Details withheld at licensee's request by SSL.
G4 VYA J Jacobs, 17 Cotswold Dr, Albrighton, Wolverhampton, WV7 3DQ [IO82UP, SJ80]
G4 VYB G Williams, 9 Poplar Ave, Lydgate, Oldham, OL4 4JX [IO83XM, SD90]
G4 VYC V D Packman, 90 Wentworth Way, Sanderstead, South Croydon, CR2 9EW [IO91XH, TQ36]
G4 VYE Details withheld at licensee's request by SSL.
G4 VYG B F Roberts, 52 School Ln, Toft, Cambridge, CB3 7RE [IO92XE, TL35]
G4 VYH N J Baker, Roffensis, 16 Boulderside Cl, Thorpe St. Andrew, Norwich, NR7 0JJ [JO02QP, TG20]
G4 VYI M A M Dalley, 195 Marcliffe Rd, Sheffield, S6 4AH [IO93FJ, SK39]
G4 VYJ J O'Sullivan, 13 Bickerton Ave, Frodsham, Warrington, WA6 7RE [IO83PH, SJ57]
G4 VYK R Vaughan, 73 Westward Dr, Pill, Bristol, BS20 0JR [IO81PL, ST57]
G4 VYL R Reilly, 4 Moreton Dr, Blackpool, FY3 0DR [IO83MT, SD33]
G4 VYM Details withheld at licensee's request by SSL.[Op: K R Tovey. Station located near Selby.]
G4 VYN Dr J B Lawrence, 7 The Woodlands, Corton, Lowestoft, NR32 5EZ [JO02UM, TM59]
G4 VYP D Rimmer, Church View, Mere Ln, Halsall, Ormskirk, L39 8RT [IO83MN, SD30]
GM4 VYQ W A Harvey, 32 Upper Glenfyne Park, Ardrishaig, Lochgilphead, PA30 8HF [IO76GA, NR88]
G4 VYR G G McCartney, 12 Timway Dr, West Derby, Liverpool, L12 4YR [IO83NK, SJ49]
G4 VYS Details withheld at licensee's request by SSL.[Op: Mrs S Meehan, PO Box 73, London, N10 2JW.]

GM4 VYU O Jackson, Cossarshill, Ettrick, Selkirk, Scotland, TD7 5JB [IO85JK, NT21]
G4 VYV D D Steward Anglia Coll ARS, Victoria Rd South, Chelmsford, Essex, CM1 1LL [JO01FR, TL70]
G4 VYW Details withheld at licensee's request by SSL.
G4 VYX W J Halliwell, 130 Chestnut St., Ashington, NE63 0BS [IO95FE, NZ28]
G4 VYZ T Smith, 21 Rufford St., Alverthorpe Rd, Wakefield, WF2 9PB [IO93FQ, SE32]
G4 VZA K J Parsons, Apartment 7, Crellen House, Priory Rd, Malvern, Worcs, WR14 3DN [IO82UC, SO74]
G4 VZB I C Ray, 20 Westbury Ln, Newport Pagnell, MK16 8JA [IO92PC, SP84]
G4 VZC P A Stokes-Herbst, 72 Devonshire Rd, Linthorpe, Middlesbrough, TS5 6DP [IO94JN, NZ41]
G4 VZF P Womack, 20 Church Cl, Mountnessing, Brentwood, CM15 0TJ [JO01EP, TQ69]
G4 VZH A G P Hobkirk, 216 Northwick Rd, Worcester, WR3 7EH [IO82VF, SO85]
GM4 VZI W W Lawrie, Oldwoodhouse Lee, Roslin, Midlothian, Scotland, EH25 9QJ [IO85JU, NT26]
GW4 VZJ R E Knott, 35 Tyr y Sarn Rd, Rumney, Cardiff, CF3 8BD [IO81KM, ST27]
G4 VZK D K Perkins, 10 The Foxes, Sutton Hill, Telford, TF7 4NH [IO82SP, SJ70]
G4 VZL J C Caddick, 11 Charles Ave, Essington, Wolverhampton, WV11 2TE [IO82XP, SJ90]
G4 VZN Details withheld at licensee's request by SSL.
G4 VZO M Deeley, 14 Carnforth Cl, Kingswinford, DY6 9BL [IO82VM, SO88]
G4 VZR D S Cormack, Lukes Orchard, Far Green Coaley, Dursley, Glos, GL11 5EL [IO81UQ, SO70]
G4 VZS L H Andrew, 20 Hookstone Grange Ct, Harrogate, HG2 7BP [IO93GX, SE35]
G4 VZT P J Green, 61 Gravel Hill, Wimborne, BH21 3BJ [IO90AS, SZ09]
G4 VZU A N Toogood, 18 Cranthorne Dr, Bakersfield, Nottingham, NG3 7HD [IO92KX, SK64]
G4 VZV Details withheld at licensee's request by SSL.
GM4 VZW K J R Rankine, 8 Craigrigg Cttgs, Westfield, Bathgate, EH48 3DH [IO85DW, NS97]
G4 VZX S Hatton, 207 Newton Rd, Lowton, Warrington, WA3 2BG [IO83RL, SJ69]
GM4 VZY D S Deans, 17 Montrose Way, Dunblane, FK15 9JL [IO86AE, NN70]
G4 WAB K M Wragg The Worked All, 11A Fall Rd, Heanor, Derbyshire, DE75 7PQ [IO93HA, SK44]
G4 WAC D G Dawkes Wythall Rad Club, 83 Alcester Rd, Hollywood, Birmingham, B47 5NR [IO92BJ, SP07]
G4 WAF A F Fewkes, 19 St. Pauls Cres, Pelsall, Walsall, WS3 4EP [IO92AP, SK00]
GI4 WAH J H Keenan, 24 Leode Rd, Hilltown, Newry, Co Down, N Ireland, BT34 5TJ [IO64VE, J12]
G4 WAI P J Bramley, 31 Evering Ave, Parkstone, Poole, BH12 4JF [IO90AR, SZ09]
G4 WAK N C Rumbol, 66 The Avenue, Hadleigh, Benfleet, SS7 2DL [JO01HN, TQ88]
G4 WAL P Walton, 6 Gorse Gr, Longton, Preston, PR4 5NP [IO83OR, SD42]
G4 WAM M Lockley, 37 Farmside Ln, Biddulph Moor, Stoke on Trent, ST8 7LY [IO83WC, SJ95]
GW4 WAN E G Williams, 12 Trefelin St., Port Talbot, SA13 1DQ [IO81CO, SS79]
G4 WAO J N Kimpton, 28 Clifton St., Stourbridge, DY8 3XT [IO82WK, SO88]
G4 WAP R Southern, 31 Burnsall Rd, Brighouse, HD6 3JS [IO93CQ, SE12]
G4 WAQ D E J Skipper, 18 Shipfield, Norwich, NR3 4DX [JO02PP, TG21]
G4 WAR H A Richter Wigston ARC, 84 Roehampton Dr, Wigston, LE18 1HU [IO92KO, SK50]
G4 WAS K M Atack, 29 High Hill, Essington, Wolverhampton, WV11 2DW [IO82XO, SJ90]
G4 WAU Details withheld at licensee's request by SSL.[Op: R Whittaker, 23 Station Rd, Marple, Stockport, Cheshire, SK6 6JY.]
G4 WAW L F Baker South Bristol, Arcade, 62 Ct Farm Rd, Whitchurch, Bristol, BS14 0EG [IO81RJ, ST66]
G4 WAX J O Moon, 25 Shotley Gdns, Low Fell, Gateshead, NE9 5DP [IO94EW, NZ26]
G4 WAY R M Holyoake, 105 Onslow Rd, Croydon, CR0 3NZ [IO91WJ, TQ36]
G4 WAZ N Mackinnon, 49 Balmoral Way, Worle, Weston Super Mare, BS22 9AL [IO81MI, ST36]
G4 WBA B A Westbrook, 14 Pickering St., Loose, Maidstone, ME15 9RS [JO01GF, TQ75]
G4 WBB Details withheld at licensee's request by SSL.
G4 WBC D C Watson West Bromwich Rd, 72 Dawes Ave, West Bromwich, B70 7LS [IO92AM, SP09]
G4 WBE J S Ball, Mellomasvn 128, 1414, Trollasen, Norway, X X
G4 WBG R J Dunn, 13 Horton Gate, Giffard Park, Milton Keynes, MK14 5JQ [IO92PB, SP84]
G4 WBH P Jackson, 15 Bankside, Retford, DN22 7UW [IO93MH, SK77]
G4 WBI S Haydon, 58 Deanfield Rd, Henley on Thames, RG9 1UU [IO91NM, SU78]
G4 WBK M G Moss, 65 Wolds Rise, Cavendish Park, Matlock, DE4 3HJ [IO93FD, SK36]
G4 WBN T N Nettleship, 36 Westfield Ave, Aughton, Sheffield, S31 0XR [IO93GJ, SK38]
GM4 WBO K R Johnson, 102 Balmacaan Rd, Drumnadrochit, Inverness, IV3 6UR [IO77SH, NH52]
G4 WBP D C Hatfield, 29 Awbridge Rd, Netherton, Dudley, DY2 0HZ [IO82WL, SO98]
G4 WBR Details withheld at licensee's request by SSL.
GW4 WBT S P Clifton, 15 Cae Clyd, Craig y Don, Llandudno, LL30 1BL [IO83CH, SH78]
GM4 WBU W Swinburne, 29 Murray Pl, Dollar, FK14 7HP [IO86DD, NS99]
G4 WBV A R Fry, 128 Sylvan Way, Sea Mills, Bristol, BS9 2LU [IO81QL, ST57]
G4 WBW K Odlum, 17 Glebe St., Talke, Stoke on Trent, ST7 1NP [IO83UC, SJ85]
G4 WBX M N D Mustard, Redcroft, Venlake End, Uplyme, Lyme Regis, DT7 3SF [IO80MR, SY39]
GD4 WBY M J Jerrome-Jones, Fairfield, Jurby Rd, Lezayre, Ramsey, IM7 2EB
G4 WBZ A F de Maio, 145 Hughenden Rd, Hastings, TN34 3TA [JO00HU, TQ81]
G4 WCB M D Bradley, 4 Castle Ct, Lower Burraton, Saltash, PL12 4SE [IO70VJ, SX45]
G4 WCC J R Horton, Penbracken, Penruan Ln, St. Mawes, Truro, TR2 5UH [IO70LD, SW83]
G4 WCD D Longstaff, 100 Hawthorn Ave, Anlaby Rd, Hull, HU3 5QR [IO93TR, TA02]
G4 WCE P J Kirsop, 8 Warburton Cl, Lymm, WA13 9QE [IO83SJ, SJ68]
G4 WCG N B McLaren Wirral Cont Grp, 596 Woodchurch Rd, Oxton, Birkenhead, L43 0TT [IO83LI, SJ28]
G4 WCH W C Houghton, 23 Poplar Rd, Haydock, St. Helens, WA11 0SW [IO83RL, SJ59]
G4 WCI O E Searle, Gore Ct, North Rd, Goudhurst, Cranbrook, TN17 1JR [JO01FD, TQ73]
G4 WCJ B H Baverstock, 28 Kingston Rd, Poole, BH15 2LP [IO90AR, SZ09]
G4 WCK C A Baverstock, 43 Tatnam Rd, Longfleet, Poole, BH15 2DW [IO90AR, SZ09]
GW4 WCM M Meredith, 19 Heol Cefnydd, Cefn Mawr, Wrexham, LL14 3ND [IO82LX, SJ24]
G4 WCO D J Foy, 37 Gorsey Croft, Eccleston Park, Prescot, L34 2RS [IO83OK, SJ49]
G4 WCP S D Richardson, 56 St. Marys Rd, Hastings, TN34 3LW [JO00HU, TQ81]
G4 WCQ T Jordan, The Cres, Havenview Rd, Seaton, Devon, EX12 2PF [IO80LQ, SY29]
G4 WCS S A Collier, 4 Manston Gr, Chorley, PR7 2QH [IO83QP, SD51]
G4 WCY H Bottomley, 8 Leyburn Pl, Filey, YO14 0DQ [IO94UF, TA18]
G4 WDC G J Cooke, 106 Wirral Dr, Winstanley, Wigan, WN3 6LD [IO83PM, SD50]
G4 WDD G H Clark, The Spinney, 1D Merton Rd, Southsea, PO5 2AE [IO90KS, SZ69]
G4 WDG Details withheld at licensee's request by SSL.[Station located near Warwick.]
G4 WDH B J Cowley, 1 Coracle Cl, Warsash, Southampton, SO31 9AT [IO90IU, SU50]
G4 WDI Details withheld at licensee's request by SSL.
G4 WDJ B L Frisby, 30 Heaton Rd, Canterbury, CT1 3PY [JO01MG, TR15]
G4 WDL Details withheld at licensee's request by SSL.
G4 WDM P J Binks, 18 Bibby Cl, Corringham, Stanford-le-Hope, SS17 7QB [JO01FM, TQ78]
G4 WDN J F Dickson, 85 Western Rd, Wolverton, Milton Keynes, MK12 5AY [IO92OB, SP84]
GM4 WDO A Douglas, 5 Walterstead, Ladykirk, Norham, Berwick upon Tweed, TD15 1XW [IO85VR, NT84]
G4 WDP D J Preston, 77 Wensley Rd, Woodthorpe, Nottingham, NG5 4JX [IO92KX, SK54]
G4 WDQ J Preston, 77 Wensley Rd, Woodthorpe, Nottingham, NG5 4JX [IO92KX, SK54]
G4 WDR J Allen W Devon Rynt Gr, Redferns, Brentor, Tavistock, Devon, PL19 0LR [IO70WO, SX48]
G4 WDS R B W Silvera, 24 White Hill, Kinver, Stourbridge, DY7 6AD [IO82VK, SO88]
G4 WDT Details withheld at licensee's request by SSL.
G4 WDU M W Fisher, 12 Boultham Park Rd, Lincoln, LN6 7AY [IO93RF, SK97]
G4 WDW W H Spencer, 38 Kingjohn Ave, Poole, Dorset, BH11 9RW [IO90AS, SZ09]
G4 WDZ K M Bennett, 5 Montagu Sq, Eynesbury, St. Neots, Huntingdon, PE19 2TL [IO92UF, TL15]
G4 WEA Details withheld at licensee's request by SSL.[Op: J H Wells.]
G4 WEC G Leesley, 6 Grosvenor Cl, Retford, DN22 7HP [IO93MH, SK77]
G4 WED N J Tipping, Milcote, Kings Ride, Ascot, SL5 8AB [IO91PJ, SU96]
G4 WEE S F Leech, 9 Parkside Dr, Old Catton, Norwich, NR6 7DP [JO02PP, TG21]
G4 WEG K G Jackson, Hilltop House, 145 The Hill, Glapwell, Chesterfield, S44 5LU [IO93IE, SK46]
G4 WEH M S Pepper, 56 Meadow Ln, Burgess Hill, RH15 9JA [IO90WW, TQ31]
G4 WEL J E Bolton, 110 Vale Rd, Ash Vale, Aldershot, GU12 5HS [IO91PG, SU85]
G4 WEM A P Penney, 110 Vale Rd, Ash Vale, Aldershot, GU12 5HS [IO91PG, SU85]
G4 WEN K M Porter, 53 Hunters Cl, Oakley, Basingstoke, RG23 7BG [IO91JG, SU55]
G4 WEO C A Gardner, 65 Beverley Rd, Whyteleafe, CR3 0DU [IO91XH, TQ35]
G4 WEP W Hewitt, 101 Sunnyside Ave, Tunstall, Stoke on Trent, ST6 6DZ [IO83VB, SJ85]
G4 WER S W Green, 56 Carlson Gdns, Lutterworth, LE17 4DR [IO92JK, SP58]
G4 WES C S Weston, Shadow Lawns, 65 Station Rd, Lutterworth, LE17 4AP [IO92JK, SP58]
G4 WET M J Butler Triple B.C.G, 16 Clevedon Green, South Littleton, Evesham, WR11 5TY [IO92BC, SP04]
G4 WEV Dr A M Russell, 6 Bartlemy Rd, Newbury, RG14 6JX [IO91IJ, SU46]
GM4 WEW C M Brown, Clengraig, Ballentrae, Girvan, Ayrshire, KA26 0PA [IO75MC, NX08]
GM4 WEX K Brown, Clengraig, Ballentrae, Girvan, Ayrshire, KA26 0PA [IO75MC, NX08]
G4 WEY B E Bush, 45 Mimosa Ave, Wimborne, BH21 1TU [IO90AS, SZ09]
G4 WEZ K Westley, 29 The Limes, Sawston, Cambridge, CB2 4DH [JO02CD, TL44]
G4 WFC M Morris, Field Head Farm, Keighley Rd, Denholme, Bradford W Yorks, BD13 4LZ [IO93BT, SE03]
G4 WFE J G Longton, Moorlands, Moor Rd, Croston, Lancs, PR5 7HP [IO83OQ, SD41]
G4 WFF C McGuire, 6 Montgomery Cl, Catshill, Bromsgrove, B61 0PG [IO82XI, SO97]
G4 WFI J Constable, 15 Waverley Cres, Ettingshall Park Es, Lanesfield, Wolverhampton, WV4 6PS [IO82WN, SO99]
G4 WFK C F Seddon, 20 Pinfold Ln, Bottesford, Nottingham, NG13 0AR [IO92OW, SK83]
G4 WFL P E Ford, 24 Tonstall Rd, Epsom, KT19 9DP [IO91UI, TQ26]
GW4 WFM G Moller, 31 Wyngarth, Winchwen, Winch Wen, Swansea, SA1 7EF [IO81BP, SS69]
GM4 WFN W E Clayworth, Low Barr Cottage, Sanquhar, Dumfriesshire, Scotland, DG4 6LQ [IO85AI, NS70]
G4 WFO S Speake, 89 The Majestic, Clifton Dr North, St Annes on Sea, Lancs, FY8 2PH [IO83LS, SD32]
GW4 WFP R Gibson, 1184 Carmarthen Rd, Fforestfach, Swansea, SA5 4BL [IO81AP, SS69]
G4 WFR R I Cooper, Post Office Stores, Little Oakley, Harwich, Essex, CO12 5JF [JO01OV, TM22]
G4 WFS A F Wisher, 13 Ennerdale Cl, Little Lever, Bolton, BL3 1UQ [IO83TN, SD70]
G4 WFT T D Kearsley, 13 Newman St., Higham Ferrers, Rushden, NN10 8JP [IO92QH, SP96]
GM4 WFV S Duguid, Oakwood, 4 Corsehill Pl, Ayr, KA7 2SU [IO75QK, NS32]
G4 WFW P Edwards, 1 Radley Ave, Wickersley, Rotherham, S66 0HZ [IO93IK, SK49]
G4 WFZ P Marsh, Columbia, 28 Orcheston Rd, Charminster, Bournemouth, BH8 8SR [IO90BR, SZ19]
G4 WGA R G Hall, Hillside, Potten End Hill, Water End, Hemel Hempstead, HP1 3BN [IO91RS, TL00]
G4 WGB R T Vaughan, 6 Dellside Gr, St. Helens, WA9 5AR [IO83OK, SJ59]
GM4 WGC P Naughton, 16 Holton Cres, Sauchie, Alloa, FK10 3DZ [IO86CD, NS89]
G4 WGE A D Cross, 31 Mountcombe Cl, Surbiton, KT6 6LJ [IO91UJ, TQ16]
G4 WGF G K Fairhurst, 42 Chorley Rd, Standish, Wigan, WN1 2SS [IO83QN, SD50]
G4 WGH A K Dicken, 73 Brocklehurst Ave, Macclesfield, SK10 2RF [IO83WG, SJ97]
G4 WGJ M J Collins, 185 Church Rd, Haydock, St. Helens, WA11 0NB [IO83QL, SJ59]
G4 WGK G C Kemp, 38 Merlin Way, Leckhampton, Cheltenham, GL53 0LU [IO81WV, SO92]
G4 WGL P V Whitehouse, Bell View, Harleston Rd, Fressingfield, Eye, IP21 5TE [JO02PI, TM27]
G4 WGN K Wilson, 102 Waddicar Ln, Melling, Liverpool, L31 1DY [IO83NL, SJ39]
G4 WGO D R Currie, Dancing Beggars, Overseas Est, Stoke Fleming, Dartmouth, TQ6 0PJ [IO80EH, SX84]
G4 WGP V J Dowse, 12 St. Jamess Ave, Brighton, BN2 1QD [IO90WT, TQ30]
G4 WGR R S Gibson, 52 Broomfields, Denton, Manchester, M34 3TH [IO83WL, SJ99]
G4 WGT W G Taylor, 27 Netherley, Coppull, Chorley, Lancs, PR7 5EH [IO83QO, SD51]
G4 WGU G E Tarry, 8 Wareham Rd, Blaby, Leicester, LE8 4BE [IO92KN, SP59]
G4 WGW W G Winter, Yellowstones, 4 Sea Mills Ln, Stoke Bishop, Bristol, BS9 1DW [IO81QL, ST57]
G4 WGX B A Rivers, 8 Bateman Gr, Ash, Aldershot, GU12 6QG [IO91PF, SU85]
G4 WGY S M Hughesdon, 3 Lyndhurst Rd, Gosport, PO12 3QY [IO90KT, SZ69]
G4 WGZ A D Brooker, 36 Pope Rd, Bromley, BR2 9QB [JO01AJ, TQ46]
G4 WHA G Harper, 5 Maple Dr, Penrith, CA11 8TU [IO84PP, NY52]
GM4 WHD D Calder, 53 Coach Rd, Wick, KW1 4HA [IO88KK, ND35]
G4 WHF K Wilson, 32 Lowside Dr, Roundthorn, Oldham, OL4 1AS [IO83WM, SD90]
G4 WHK R P Cann, 9 Beacon Hill Ave, Dovercourt, Harwich, CO12 3NR [JO01PW, TM23]
G4 WHL P W Callaghan, 8 Abbey Rd, Edwinstowe, Mansfield, NG21 9LQ [IO93LE, SK66]
G4 WHM P Callaghan, 8 Abbey Rd, Edwinstowe, Mansfield, NG21 9LQ [IO93LE, SK66]
G4 WHN C Walker, 14 Collingwood Rd, Long Eaton, Nottingham, NG10 1DR [IO92IV, SK43]
G4 WHO N E Foot, Oakfield Farm, Horton Rd, Verwood, Dorset, BH31 6JJ [IO90BU, SU00]
GW4 WHP W Pryce, 24 Alma St., Dowlais, Merthyr Tydfil, CF48 3RP [IO81HS, SO00]
G4 WHQ R Howell, Tunstal, Station Rd, North Thoresby, Grimsby, DN36 5QP [IO93XL, TF29]
G4 WHT W H Tattersall, 45 Russell Ave, Alsager, Stoke on Trent, ST7 2BN [IO83UC, SJ75]
G4 WHV M M Langdon, 58 Upper Marsh Rd, Warminster, BA12 9PN [IO81VE, ST84]
G4 WHY M R Foot, Oakfield Farm, Horton Way, Verwood, BH31 6JJ [IO90BU, SU00]
G4 WHZ D B Cater, 104 St. Johns Rd, Great Clacton, Clacton on Sea, CO16 8DB [JO01NT, TM11]
G4 WIA I A Whitmore, 2 de Vere Cl, Hemingford Grey, Huntingdon, PE18 9BH [IO92WH, TL27]
G4 WIE C G Holloway, 7 Cavell Cres, Harold Wood, Romford, RM3 0WL [JO01CO, TQ59]
G4 WIF A F Fishpool, 38 James Rd, Dartford, DA1 3NF [JO01CK, TQ57]
G4 WIG P A Lees, 107 Balmoral Rd, Worosley, Stourbridge, DY8 5JJ [IO82VL, SO88]
G4 WIH R R Fenton, 22 Gwyns Piece, Lambourn, Hungerford, RG17 8YZ [IO91FM, SU37]
G4 WII M J Sesemann, 10 Kemerton Rd, Beckenham, BR3 6NJ [IO91XJ, TQ36]
G4 WIK H N Woolrych Devizes Dist AR, 20 Meadow Dr, Devizes, SN10 3BJ [IO91AI, SU06]
G4 WIL J Wilkinson, 147 Alder Ln, Hindley Green, Hindley, Wigan, WN2 4ET [IO83RM, SD60]
G4 WIN C J Mabbutt, 17 Lakeside, Werrington, Peterborough, PE4 6QZ [IO92UO, TF10]
G4 WIP A J Crickett, 11 Musgrave Cl, Cheshunt, Waltham Cross, EN7 6TZ [IO91XR, TL30]
G4 WIQ D T S Whitbread, Lismore, Fernleigh Rd, Mannameaz, Plymouth, PL3 5AN [IO70WJ, SX45]
G4 WIR I M Page, 127 Whyke Ln, Chichester, PO19 2AU [IO90OT, SU80]
G4 WIS V Gleek, 5 Highview Ave, Edgware, HA8 9TX [IO91UO, TQ29]
G4 WIV D Bevin, 7 Greenoak Dr, Worsley, Manchester, M28 3QR [IO83TM, SD70]
G4 WIX Details withheld at licensee's request by SSL.
G4 WIY A L Clark, Hunts End, Waterworks Ln, Martin, Dover, CT15 5JW [JO01QE, TR34]
G4 WIZ Details withheld at licensee's request by SSL.
GM4 WJA J C Fraser, Cherrybrae Croft, Aultmore, Keith, AB55 6QU [IO87MN, NJ45]
G4 WJB R E Barratt, 172 Coneygree Rd, Stanground, Peterborough, PE2 8LQ [IO92VN, TL29]
G4 WJE M Fenelon, 72 Fieldside, Epworth, Doncaster, DN9 1DP [IO93OM, SE70]
G4 WJF J E Farish, 6 Woodhall Park Ave, Stanningley, Pudsey, LS28 7HF [IO93DT, SE23]
G4 WJG D Nicolson, 2 Woodley Ave, Accrington, BB5 2LF [IO83TR, SD72]
G4 WJH P L Mathews, 1 Erith Rd, Belvedere, DA17 6HB [JO01BL, TQ47]
G4 WJI Details withheld at licensee's request by SSL.
G4 WJJ Details withheld at licensee's request by SSL.
GM4 WJL D D Archibald, 3 Larkfield Dr, Eskbank, Dalkeith, EH22 3HJ [IO85KV, NT36]
G4 WJM W J M Cooper, 32 High St., Thurlby, Bourne, PE10 0EE [IO92TR, TF01]
G4 WJN R S Alexander, 113 Casterton Rd, Stamford, PE9 2UF [IO92SP, TF00]
GW4 WJO R S Thomas, Braemar, 6 Ty Mawr Est, Holyhead, LL65 2DN [IO73QH, SH28]
G4 WJR J Singleton, 3 Willow Dr, Skelmersdale, WN8 8PR [IO83ON, SD40]
G4 WJS W J Somerville, Glendella, Wycombe Rd, Stokenchurch, High Wycombe, HP14 3RP [IO91NP, SU79]
G4 WJT D S Phelps, 32 Galley Ln, Barnet, EN5 4AJ [IO91VP, TQ29]
GW4 WJU E F Goodall, Plough, Blaencillech, Newcastle Emlyn, Dyfed, SA38 9EP [IO72SB, SN34]
G4 WJV J E Forrest, 3 Martindale Park, Hall Ln, Houghton-le-Spring, DH5 8EX [IO94GU, NZ34]
G4 WJW T M Murphy, 7 The Knapp, Ct Rise, Templecombe, BA8 0JP [IO81TA, ST72]
G4 WJX M J Kessel, 4 Harington Dr, Stoke on Trent, ST3 5ST [IO82WX, SJ94]
G4 WJY A B Durnford, 32 The Meadows, Hanham, Bristol, BS15 3PA [IO81RK, ST67]
G4 WJZ A J M Kerr, Braemoray, Dalditch Ln, Budleigh Salterton, EX9 7AS [IO80HP, SY08]
G4 WKB H J Poulton, 1 Marnhull Cl, Walsgrave, Coventry, CV2 2JS [IO92GJ, SP37]
G4 WKC K Clowes, 42 Sides Rd, Pontefract, WF8 3PN [IO93IQ, SE42]
G4 WKD K Dunstan, 41 Gravel Ln, Wilmslow, SK9 6LS [IO83VH, SJ88]
G4 WKG G G Rowley, Glassonby Lodge, Penrith, Cumbria, CA10 1DT [IO84QR, NY53]
G4 WKH Details withheld at licensee's request by SSL.
G4 WKJ T R Hughes, 260 Farthing Gr, Netherfield, Milton Keynes, MK6 4JF [IO92PA, SP83]
G4 WKN R J Gardner, 35 Hamsterly Park, Northampton, NN3 5DA [IO92OG, SP76]
GM4 WKO J G Harris, 30 Springfield Rd, New Elgin, Elgin, IV30 3BZ [IO87IP, NJ26]
G4 WKP T Taylor, 25 Suffolk St., Barrow in Furness, LA13 9QH [IO84JC, SD26]
GW4 WKQ L A Jones, 22 Glyn y Mor, Llanbedrog, Pwllheli, LL53 7NW [IO72SU, SH33]
G4 WKS E D Walker Warwick School, Amateur Radio Soc, 8 Primrose Hill, Warwick, CV34 5HW [IO92FH, SP26]
G4 WKT N Bleek, 35 George Ave, Easington Colliery, Peterlee, SR8 3NG [IO94IT, NZ44]
G4 WKU N A Reedman, Kennels Cottage, Bury Ln, Marchwood, Southampton, Hants, SO4 4UD [IO90HW, SU41]
G4 WKV Details withheld at licensee's request by SSL.[Op: Mrs D S Elliston. Station located near Ingatestone.]
G4 WKY N D Lee, Bourne End, 68 Andlers Ash Rd, Liss, GU33 7LR [IO91NA, SU72]
G4 WKZ G L Fitch, 1 Woodgreen Cl, Callow Hill, Redditch, B97 5YR [IO92AG, SP06]
G4 WLA D Dell, Bushmead, 12 Penfield Gdns, Dawlish, EX7 9NQ [IO80GN, SX97]
G4 WLC N G Pearce, 4 Dunster Gr, Cheltenham, GL51 0PE [IO81WV, SO92]
G4 WLE N T Slater, 17 Hall Park Dr, Lytham St. Annes, FY8 4QR [IO83MR, SD32]
G4 WLF A F Suffolk, 2 Holdens Way, Curry Rivel, Langport, TA10 0JL [IO81NA, ST32]
G4 WLG K Dunwell, 8 Cochrane Cl, Thatcham, RG19 4QX [IO91IJ, SU56]
G4 WLI G K Nutt, 1 Ashtree Farm Ct, Willaston, South Wirral, L64 2XL [IO83LG, SJ37]
G4 WLJ N S Bell, 16 Amersham Cl, Davyhulme, Urmston, Manchester, M41 7WH [IO83TL, SJ79]
G4 WLK M L Morgan, 6 Blakeley Heath Dr, Wombourne, Wolverhampton, WV5 0HW [IO82VM, SO89]
GM4 WLL D A Dodds, 60 Stevenson Rd, Penicuik, EH26 0RH [IO85JU, NT26]
GM4 WLN R Cassells, Caorunn, Naemoor Rd, Crook of Devon, Kinross, KY13 7UH [IO86FE, NO00]
G4 WLP S McCombe, Willow Dene, Lower Broad Oak Rd, West Hill, Ottery St Mary, Devon, EX11 1XH [IO80IR, SY09]
G4 WLS C G Smith, 8 Priory Cl, Horley, RH6 8AX [IO91VE, TQ24]
GW4 WLT J Williams, 7 Tynewydd, Nantybwch, Tredegar, NP2 3SG [IO81IT, SO11]
G4 WLV D Gladwin, Dorset House, St. Annes Rd, Eastbourne, BN21 2HR [JO00DS, TV69]
GW4 WLZ J D Williams, Glas-y-Dorlan, Holyhead, Ynys Mon, North Wales, LL65 1DS [IO73QH, SH28]
G4 WMA P J Haslam, 24 Ashleigh Gr, Jesmond, West Jesmond, Newcastle upon Tyne, NE2 3DL [IO94EX, NZ26]
G4 WMB W M Bell, 33 King Harry Ln, St. Albans, AL3 4AS [IO91TR, TL10]
GW4 WMD W M David, Sirmione, Freestone Cross, Cresselly, Kilgetty Dyfed, SA68 0SX [IO71OR, SN00]
GI4 WME F A Hull, 44 Killynether Walk, Park, Belfast, BT8 4DB [IO74AN, J36]
G4 WMF G Blake, Flat 5, 46 Marlborough Rd, Ipswich, IP4 5AX [JO02OB, TM14]
GU4 WMG J F Gallienne, Westward, Rue Des Marettes, St. Martin, Guernsey, GY4 6JW
G4 WMH W M Hall, 45 Dorchester Rd, Solihull, B91 1LN [IO92CJ, SP17]
G4 WMI Details withheld at licensee's request by SSL.[Op: F Crease, PO Box 29, Hayes, Middx, UB4 QTS.]
GW4 WMK D G Turner, 20 Long Mains, Monktown, Monkton, Pembroke, SA71 4NB [IO71MQ, SM90]
GM4 WMM W S McMillan, 1 Churchill Wood, Inverneil, Ardrishaig, Lochgilphead, PA30 8ES [IO75GX, NR88]
G4 WMN J Robb, 19 Woodfoot Quadrant, Glasgow, G53 7JP [IO75TT, NS55]
G4 WMO P Stainton, Willow Lodge, Ferry Rd, Fiskerton, Lincoln, LN3 4HU [IO93TF, TF07]
G4 WMP M Bangle, 21 Oakhill Rd, Addlestone, KT15 1DH [IO91RI, TQ06]
G4 WMQ A W Richardson, 1 Silverton Terr, Rothbury, Morpeth, NE65 7QS [IO95BH, NU00]
G4 WMS T J Clough Wyke Manor Sc Rd, 37 Park Ave, Mirfield, WF14 9PB [IO93DQ, SE21]
G4 WMT Details withheld at licensee's request by SSL.
G4 WMV R J Bridge, 20 Hamp Green Rise, Bridgwater, TA6 6AZ [IO81MC, ST33]
G4 WMW J A Budd, 45 Sevenoaks Dr, Hastings Hill, Sunderland, SR4 9LS [IO94GV, NZ35]
G4 WMX C Francks, 16 Oakpool Gdns, Leicester, LE2 9FL [IO92KO, SP59]

G4 WMY G Kay, High Trees, Stockland Bristol, Bridgwater, Somerset, TA5 2PZ [IO81LE, ST24]
G4 WMZ K H Law, The Sidings, 2 The Bank, Somersham, Huntingdon, PE17 3DJ [JO02AJ, TL37]
G4 WNA H A Williams, 37 Mickledales Dr, Marske By The Sea, Redcar, TS11 6DF [IO94LO, NZ62]
G4 WNB W N Brown, 48 Sunnybank Rd, Sutton Coldfield, B73 5RJ [IO92CN, SP19]
G4 WNC Details withheld at licensee's request by SSL.
G4 WND R M Banks, 33 New Rd, Shuttington, Tamworth, B79 0DT [IO92EP, SK20]
G4 WNE Details withheld at licensee's request by SSL.
G4 WNF F R Rhodes, 248 Woolwich Rd, London, SE2 0DW [JO01BL, TQ47]
G4 WNG T Furness, 129 North Ridge, Bedlington, NE22 6DF [IO95ED, NZ28]
GI4 WNH E G Loughran, 6 Oaklea Rd, Ballymulderg, Magherafelt, BT45 6NH [IO64RR, H98]
G4 WNI J A Howarth, 80 John F Kennedy Est, Washington, NE38 7AL [IO94FV, NZ35]
G4 WNJ A R Bleach, Lych Gate, 1 St Bartholomews, . Gdns, Southsea, PO5 1RD [IO90LS, SZ69]
G4 WNM P D Stokes, 8 Aldred Rd, London, NW6 1AN [IO91VN, TQ28]
G4 WNO W J Cullen, 9 Brushford Cl, Furzton, Milton Keynes, MK4 1EG [IO92OA, SP83]
G4 WNP R G Tant, 34 Manor Rd, Wheathampstead, St. Albans, AL4 8JD [IO91UT, TL11]
GM4 WNQ P Ramsey, 25 Caledonian Rd, Stevenston, KA20 3LW [IO75PP, NS24]
G4 WNR D J Daws, 75 Buckland Rd, Tadworth, KT20 7EF [IO91VG, TQ25]
G4 WNT Details withheld at licensee's request by SSL.
G4 WNU J Smith, 9A Roseberry Rd, Langley Vale, Epsom Downs, Epsom, Surrey, KT18 6AF [IO91UH, TQ25]
G4 WNV Dr S R Robinson, 18 Headley Gr, Tadworth, KT20 5JF [IO91VH, TQ25]
G4 WNW T V Almond, Maranatha, Lumber Ln, Burtonwood, Warrington, WA5 4AX [IO83QK, SJ59]
G4 WNY D D Sullivan, Yvenlia 66, N - 1712 Gralum, Norway, X X
G4 WOB J Bazyk, 123 Wolf Ln, Windsor, SL4 4YY [IO91QL, SU97]
G4 WOC R F Hyatt, 2 Hine Rd, Newbarn Park, Galmington, Taunton, TA1 4NE [IO81KA, ST22]
G4 WOD J M A Sheppard, 37 Oakfield Rd, Kingswood, Bristol, BS15 2NT [IO81RK, ST67]
G4 WOE Details withheld at licensee's request by SSL.[Op: D Hudson. Station located near Bexhill-on-Sea.]
G4 WOF Details withheld at licensee's request by SSL.
G4 WOH P L Thwaytes, 1 Sunningdale, Waltham, Grimsby, DN37 0UA [IO93WM, TA20]
G4 WOI R Allen, 23 Tredour Rd, Newquay, TR7 2EY [IO70LJ, SW86]
GW4 WOJ W O John, 5 Ffordd Ganol, Rhuddlan, Rhyl, Denbighshire, LL18 2ST [IO83GH, SJ07]
G4 WOL R Tenwolde, 121 Whirley Rd, Whirley, Macclesfield, SK10 3JL [IO83VG, SJ87]
G4 WOP M R Harding, 307 Woodlands Rd, Gillingham, ME7 2TA [JO01GJ, TQ76]
G4 WOQ A L Leach, Wendover, Park Rd, Scotby, Carlisle, CA4 8AT [IO84NV, NY45]
G4 WOS D C Flello, 1 St. Andrews Way, Tilmanstone, Deal, CT14 0JH [JO01PF, TR35]
G4 WOT Details withheld at licensee's request by SSL.[Station located near Dorking.]
GW4 WOV Details withheld at licensee's request by SSL.
GD4 WOW J A Jones, The Spinney, Glen Auldyn, Lezaure, Ramsey, Isle of Man, IM7 2AQ
G4 WOX G H Hedley, 3 Glendale Gdns, Stakeford, Choppington, NE62 5AW [IO95ED, NZ28]
GW4 WPA T D Leary, 21 Gelli Glas Rd, Morriston, Swansea, SA6 7PS [IO81AP, SS69]
G4 WPB P S Bruce, 2 Constance Rd, Croydon, CR0 2RS [IO91WJ, TQ36]
G4 WPC P F W Cardy, 20 Station Rd, Benfleet, SS7 1NG [JO01GN, TQ78]
G4 WPD A B K Kenward, 64 Squires Bridge Rd, Shepperton, TW17 0QA [IO91SJ, TQ06]
G4 WPE M N Bland, 63 Bernard St., Woodville, Swadlincote, DE11 8BY [IO92FS, SK31]
G4 WPG L Hatton, 51 Castner Ave, Weston Point, Runcorn, WA7 4EH [IO83PH, SJ58]
GW4 WPH S J Valentine, Unit, 15 Industrial Est, Bala, LL23 7NL [IO82EV, SH93]
G4 WPI J T Fuller, 42 Kitchener Rd, Amesbury, Salisbury, SP4 7AD [IO91CE, SU14]
GW4 WPJ D D G Jones, Hafandeg, Wellfield Rd, Abergwili, Carmarthen, SA31 2JQ [IO71UU, SN42]
G4 WPL L J Connolly, 38 Mayfield Gr, South Reddish, Stockport, SK5 7JB [IO83WK, SJ89]
G4 WPN R P Bennellick, Cherry Cottage, Fidges Ln, Eastcombe, Stroud, GL6 7DW [IO81WR, SO80]
G4 WPO D D N Bevan, 46 Park Ln, Hartford, Northwich, CW8 1PY [IO83RF, SJ67]
G4 WPP L Dobson, 25 Wadham Rd, Liskeard, PL14 3BD [IO70SK, SX26]
G4 WPR D E Trotman, 9 Keeps Mead, Kingsclere, Newbury, RG20 5EZ [IO91JH, SU55]
G4 WPS Details withheld at licensee's request by SSL.
G4 WPT D R Jackman, 48 Crane Dr, Verwood, BH31 6QB [IO90BV, SU00]
GM4 WPU F E Wright, 9 Scott Cres, Tayport, DD6 9PN [IO86NK, NO42]
G4 WPW R Frettsome, 7 Wheatfield Cres, Mansfield Woodhouse, Mansfield, NG19 9HH [IO93JE, SK56]
G4 WPZ K J Edwards, 20 Westbourne Ave, Clevedon, BS21 7UA [IO81NK, ST37]
G4 WQB K Hamlyn, Ty Dedwydd, 33 Chalfont Ln, Chorleywood, Rickmansworth, WD3 5PR [IO91RP, TQ09]
GW4 WQC D E A Williams, 149 Rhiwr Ddar, Taffs Well, Cardiff, CF4 7PD [IO81IN, ST18]
G4 WQD J B Jocys, 28 Vaudrey Dr, Timperley, Altrincham, WA15 6HQ [IO83UJ, SJ78]
G4 WQG S L S Harris, 20 New Rd, Chiseldon, Swindon, SN4 0LU [IO91DM, SU17]
GM4 WQH J Naughton, 124 Churchill St., Alloa, FK10 2JU [IO86CC, NS89]
G4 WQI D M Saxton, 1 Roothings, Heybridge, Maldon, CM9 4NA [JO01IR, TL80]
G4 WQJ J C Carr, 95 The Wye, Grange Est, Daventry, NN11 4PX [IO92JG, SP56]
G4 WQL M L Bender, Ivy Chimney Villa, Skinners Bottom, Scorrier, Redruth Cornwall, TR16 5DT [IO70JG, SW74]
G4 WQO P B Truitt, 11 Kelso Pl, London, W8 5QD [IO91VL, TQ27]
GU4 WQP R A Attwater, 3 Stanley Villas, Stanley Rd, St. Peter Port, Guernsey, GY1 1QW
GM4 WQQ J Campbell, 85 Drumcavel Rd, Muirhead, Glasgow, G69 9EP [IO75WV, NS66]
G4 WQR Details withheld at licensee's request by SSL.
G4 WQS N E Reading, 30 Clifton Rise, Windsor, SL4 5TD [IO91QL, SU97]
G4 WQT R J Watson, 3 Layton Rd, Larches, Ashton on Ribble, Preston, PR2 1PB [IO83OS, SD53]
G4 WQU P D Barrett, 9 Froghall Dr, Wokingham, RG40 2LE [IO91OJ, SU86]
G4 WQW P S Newberry, 3 Willow Cl, Kingsbury, Tamworth, B78 2JP [IO92DN, SP29]
G4 WQX K Nicolaides, 63 Rowstock Gdns, Camden Rd, London, N7 0BH [IO91WN, TQ28]
G4 WQZ J N R Wiles, 12 Ashling Gdns, Denmead, Waterlooville, PO7 6PR [IO90LV, SU61]
G4 WRA S E Sands Wordsley ARC, Gabledown, Bridgnorth Rd, Stourton, Stourbridge, DY7 6RW [IO82VL, SO88]
G4 WRB K J Beech, 42 Star St., Bradmore, Wolverhampton, WV3 9BL [IO82WN, SO89]
G4 WRC A J C Wheeler Wanlip Rad Club, Field House, 1 Plain Gate, Rothley, Leicester, LE7 7SQ [IO92KR, SK51]
GW4 WRD N Underwood, 12 Rockwood House, Gravel Hill Rd, Yate, Bristol, BS17 5BW [IO81TN, ST78]
GI4 WRJ R B Jennings, 12 Garnerville Gdns, Belfast, BT4 2PA [IO74BO, J37]
G4 WRK I R T Edwards, 9 Long Ln, nr Wellington, Telford, Shop, TF6 6HH [IO82RR, SJ61]
G4 WRL T E Myatt, 23 Whitegate Cl, Minehead, TA24 5ST [IO81GE, SS94]
G4 WRM M K Colebourne, 6 Lea Dr, Mickleover, Derby, DE3 5HJ [IO92FV, SK33]
G4 WRN K Eyre, 1 Ford Ln, Willington, Derby, DE65 6DQ [IO92FU, SK32]
GU4 WRO E J Cobb, Mont St Michel, Portinfer Ln, Vale, Guernsey, Channel Islands, GY6 8LH
GU4 WRP D R Fletcher, Celicia, La Neuve Rue, St Peter Port, Guernsey, GY1 1SF
G4 WRQ D P Wring, 42 Ham Green, Pill, Bristol, BS20 0HA [IO81PL, ST57]
GJ4 WRR F Leighton, 4 Victoria Village, Est Trinity, Jersey, Chanel Islands, Jersey, JE4 9VI
G4 WRS A R Kiddle William Robrtso, School A.R.C., Welbourn, Lincoln, LN5 0PA [IO93RB, SK95]
G4 WRT D S Hunter, 16 Shottsford Rd, Oakdale, Poole, BH15 3DU [IO90AR, SZ09]
G4 WRU M Davidson, 52 Overton Dr, Wanstead, London, E11 2NJ [JO01AN, TQ48]
G4 WRW Details withheld at licensee's request by SSL.
G4 WRX D K Cherrington, 4 Bloomfield Cl, Wombourne, Wolverhampton, WV5 8HQ [IO82VM, SO89]
G4 WRY Details withheld at licensee's request by SSL.
G4 WRZ A T Bennett, Three Ways, St. End Ln, Broad Oak, Heathfield, TN21 8TT [JO00DX, TQ62]
G4 WSB A W M Bowditch, 7 The Peak, Purton, Swindon, SN5 9AT [IO91BO, SU08]
G4 WSD Details withheld at licensee's request by SSL.
G4 WSE T B Saggerson, 18 Ploughmans Way, Haymakers Park, Great Sutton, South Wirral, L66 2YJ [IO83MG, SJ37]
G4 WSF R L Smith, 4 Cherry Tree Way, Doniford, Watchet, TA23 0UB [IO81IE, ST04]
G4 WSH F J Blaxland, Wenman The Sq, The Lizard, Helston, Cornwall, TR12 7NZ [IN79JX, SW71]
G4 WSI A A J Brown, 81 Ipswich Cres, Great Barr, Birmingham, B42 1LY [IO92BM, SP09]
G4 WSK J W Poole, 16 Laburnum Gr, Beech Hill, Wigan, WN6 8QY [IO83QN, SD50]
G4 WSL R E Cable, 165 Cranborne Cres, Potters Bar, EN6 3AF [IO91VQ, TL20]
G4 WSM Details withheld at licensee's request by SSL.[Weston-Super-Mare RS - details c/o G4ZUX, 36 Tormynton Road, Worle, Weston-Super-Mare, BS22 9HT.]
G4 WSN L G Brown, 17 Regents Dr, Severn Meadows, Shrewsbury, SY1 2TN [IO82PR, SJ51]
G4 WSO J D Green, 31 Canny Croft, Penrith, CA11 9HA [IO84OQ, NY53]
G4 WSP M Green, 31 Canny Croft, Penrith, CA11 9HA [IO84OQ, NY53]
G4 WSR S J Peck, 15 Heather Ave, Ipswich, IP3 9EP [JO02OA, TM14]
GW4 WSU P I C Roberts, 7 Llys Erw, Off Maes Hafod Rd, Ruthin, LL15 1LZ [IO83IC, SJ15]
G4 WSX J R A Fogden, 38 Green Ln, Chichester, PO19 4NP [IO90OU, SU80]
GM4 WSY J Kincaid, 12 Norlands, Errol, Perth, PH2 7QU [IO86JJ, NO22]
G4 WSZ S V Platt, 8 Cambridge Rd, Fulbourn, Cambridge, CB1 5HQ [JO02CE, TL55]
G4 WTA M V Jones, 57 Mountway Rd, Bishops Hull, Taunton, TA1 5DS [IO81KA, ST22]
G4 WTB D M Lauder Univ Herts RC, Division of Elec Engineering, University of Herts, College Ln, Hatfield, Herts, AL10 9SB [IO91US, TL20]
G4 WTD C J Flatman, 36 Skoner Rd, Bowthorpe, Norwich, NR5 9AX [JO02OP, TG10]
G4 WTE M D Rye, 40 Victoria Dr, Southdowns, South Darenth, Dartford, DA4 9NA [JO01DJ, TQ56]
G4 WTF Details withheld at licensee's request by SSL.
G4 WTH W T Holden, 81 Rookery Ln, Rainford, St. Helens, WA11 8BL [IO83OL, SD40]
GM4 WTJ T D M Merritt, 13 Toner Gdns, Priory Gate Overtow, Wishaw, ML2 0RP [IO85BS, NS75]
GM4 WTK R G Fortune, Stewarton Lodge, Eddleston, Peebles, Scotland, EH45 8PP [IO85JQ, NT24]
G4 WTL L H H Robertson, 51 Parker Ct, Foredown Rd, Portslade, Brighton, BN41 2FT [IO90VU, TQ20]
GU4 WTN A P Hamon, Goose Hollow, Rue A La Terre, St. Saviour, Guernsey, GY7 9YX
G4 WTO Details withheld at licensee's request by SSL.
G4 WTP W C Kirkham, 8 Long Valley Rd, Biddulph, Gillow Heath, Stoke on Trent, ST8 6RA [IO83VD, SJ85]
G4 WTQ N A Harvey, 5 Harvey Gdns, Loughton, IG10 2AD [JO01AP, TQ49]
GM4 WTS W T Stevenson, 11 West Dr, Petersburn, Airdrie, ML6 8BL [IO85AU, NS76]
GI4 WTT T Mc'Donnell, 52 Moira Rd, Glenavy, Crumlin, BT29 4JL [IO64VO, J17]
G4 WTU D J Pay, Longmeadow House, Reedy Dunsford, Exeter, Devon, EX6 7AD [IO80EQ, SX88]
G4 WTV Details withheld at licensee's request by SSL.[Please QSL via G6WOR. Station located near Worthing.]
G4 WTW P N A Stickland, 28 Fairway, Market Harborough, LE16 9QL [IO92ML, SP78]
G4 WTX V G Hansford, Whitehouse Farm, Hardway, South Brewham, Bruton, Somerset, BA10 0RJ [IO81SC, ST63]
G4 WTZ J M Hall, 23 St. James Ave, Congleton, CW12 4DY [IO83VD, SJ86]
G4 WUA G S Brown, 13 Francis Ave, Moreton, Wirral, L46 6DH [IO83KJ, SJ28]
G4 WUB D S Farr, 94 Ridgeway Ln, Whitchurch, Bristol, BS14 9PH [IO81RJ, ST66]
G4 WUF G Beech, 4 Archers Way, Great Sutton, South Wirral, L66 2RY [IO83MG, SJ37]
G4 WUH Details withheld at licensee's request by SSL.
G4 WUI J Marr, 11 Morley Cres, Kelloe, Durham, DH6 4NN [IO94GR, NZ33]
G4 WUJ N D Plant, 73 Robert Burns Ave, Cheltenham, GL51 6NX [IO81WV, SO92]
G4 WUK D Dyer, 64 Churchill Cl, Sturminster Marshall, Wimborne, BH21 4BH [IO80XT, SY99]
G4 WUL A W Blayney, 6 The Chase, Parkside, West Moor, Newcastle upon Tyne, NE12 0EW [IO95EA, NZ27]
G4 WUM F Amos, 53 Valley View, Jarrow, NE32 5QT [IO94GX, NZ36]
G4 WUN K Boot, 7 Balmoral Rd, Parkstone, Poole, BH14 8TJ [IO90AR, SZ09]
G4 WUO P L Bullock, Gull Ln, Framingham Pigot, Framingham Earl, Norwich, Norfolk, NR14 7PN [JO02QN, TG20]
G4 WUP M W Schofield, 15 Holm Oak Gdns, Broadstairs, CT10 2JF [JO01RI, TR36]
GW4 WUR C V Phillips, Aelwyd, Hill St., Rhosllanerchrugog, Wrexham, LL14 1LW [IO83LA, SJ24]
G4 WUS W A Bingham, 67 Coronation St., Saltburn, Carlin How, Saltburn By The Sea, TS13 4DW [IO94NN, NZ71]
G4 WUT A W Sampson, 5 Lilian Rd, Spixworth, Norwich, NR10 3PZ [JO02PQ, TG21]
G4 WUU P S Williamson, The Laurels, Norwich Rd, Horstead, Norwich, NR12 7EE [JO02QR, TG21]
G4 WUV C J Baker, 48 Hazell Rd, Farnham, GU9 7BP [IO91OF, SU84]
G4 WUW M W Baker, 48 Hazell Rd, Farnham, GU9 7BP [IO91OF, SU84]
G4 WUY L W Thomas, Ye Olde Bakehouse, Main Rd, Saltfleetby, Louth, LN11 7TL [JO03CJ, TF49]
G4 WUZ B Hargreaves, 27 Stanley St., Colne, BB8 9DD [IO83VU, SD84]
GW4 WVB J T Williams, Windsor Rd, Rhos, Rhosllanerchrugog, Wrexham, Clwyd, LL14 1ST [IO83LA, SJ24]
G4 WVC Details withheld at licensee's request by SSL.
G4 WVD M W Bundy, 5 Dawe Cres, Bodmin, PL31 1PY [IO70PL, SX06]
G4 WVF C Farley, The White House, Mellor, Stockport, Ches, SK6 5ND [IO83XJ, SJ98]
G4 WVG Details withheld at licensee's request by SSL.
G4 WVH T W Hathway, 16 Bonnington Way, Newcastle upon Tyne, NE5 3RF [IO94DX, NZ26]
G4 WVJ J A Hedley, 30 Rosewood Cres, Seaton Sluice, Whitley Bay, NE26 4BL [IO95GB, NZ37]
G4 WVK D M Davies, 20 Dale Park Rd, Upper Norwood, London, SE19 3TY [IO91WJ, TQ36]
G4 WVL D J Golledge, Blue House, Witham Rd, Fairstead, Chelmsford, CM3 2BS [JO01HT, TL71]
G4 WVM M R Keating, 45 Gollands, Brixham, TQ5 8JY [IO80FJ, SX95]
GI4 WVN Prof H B Gilbody, 5 The Plateau, Piney Hills, Belfast, BT9 5QP [IO74AN, J37]
GW4 WVO K J Winnard Wenvoe Transmitter Club ARS, Wenvoe Transmitting Station, St.Lythans Down, Wenvoe, Cardiff, CF5 6BQ [IO81IK, ST17]
G4 WVR W Hodgson Welland Valley ARS, Welland Valley Ars, 82 Arden Way, Market Harborough, LE16 7DD [IO92NL, SP78]
G4 WVS J White, East Wood, 30 Stockhill Rd, Chilcompton, Bath, BA3 4JL [IO81RG, ST65]
G4 WVT J C W Stageman, Arcade, 50 High St., Dymchurch, Romney Marsh, TN29 0NL [JO01LA, TR12]
G4 WVU Details withheld at licensee's request by SSL.
G4 WVW J F Lane, 41 Ravenswood Cres, Harrow, HA2 9JL [IO91TN, TQ18]
G4 WVX Details withheld at licensee's request by SSL.[Station located in south Bucks. Correspondence to: B StJ C Gilson, P O Box 434, Ascot, Berks, SL5 0QY.]
G4 WVY J F Joynt, Masonbrook, Loughrea, Co. Galway, Ireland, ZZ9 9MA
G4 WWA Dr A H E Williams, Cranesbie, 6 Dore Rd, Dore, Sheffield, S17 3NB [IO93FH, SK38]
G4 WWB W Bennett, 49 Wellbrow Rd, Liverpool, L4 6TX [IO83MK, SJ39]
GW4 WWE P S Carver, 37 Squirrel Walk, Pontarddulais, Swansea, SA4 1UH [IO71XR, SN50]
GI4 WWF V M Fails, 31 Lismurphy Ave, Coleraine, BT51 3QN [IO65PC, C83]
G4 WWG A S Brown, 92 Enstone, Tanhouse, Skelmersdale, WN8 6AS [IO83ON, SD40]
G4 WWH P E Pavelin, Flat 20, 30 Mariner Ave, Edgbaston, Birmingham, B16 9DL [IO92AL, SP08]
GM4 WWJ J C Williamson, 2 Laburnum Gr, Burntisland, KY3 9EU [IO86JB, NT28]
G4 WWL I M Rowe, 19 Poplar Ave, Wetherby, LS22 7RA [IO93HW, SE44]
GW4 WWN M Rowles, 3 Gower Villas, Cadoxton, Neath, SA10 8BN [IO81CQ, SS79]
G4 WWO C Jay, 4 Bramall Mount, Bramhall Ln, Stockport, Ches, SK2 6JQ [IO83WJ, SJ88]
G4 WWP D G Barry, 8 Dell Rd, Little Hallingbury, Bishops Stortford, CM22 7SJ [JO01CU, TL41]
G4 WWQ D J Kay, 5 Newlands Rd, Ruishton, Taunton, TA3 5JZ [IO81LA, ST22]
G4 WWR R S Coombes Three Counties, Amateur Radio Club, Pinewood Lodge, Sandhills, Wormley, Surrey, GU8 5TD [IO91QD, SU93]
GM4 WWT J Christie, 48 Forman Dr, Peterhead, AB42 2XL [IO97CM, NK14]
GM4 WWU R J Steel, 16 Lower Burnmouth, Burnmouth, Eyemouth, TD14 5SP [IO85XU, NT96]
G4 WWX I G Mant, 28 Welbourne Rd, Childwall, Liverpool, L16 6AJ [IO83NJ, SJ49]
G4 WWY P Brown, White Cottage, Woodside, Thornwood, Epping, CM16 6LF [JO01BR, TL40]
GI4 WWZ Details withheld at licensee's request by SSL.
GI4 WXA F B Shaw, 63 Wheatfield Gdns, Belfast, BT14 7HW [IO74AO, J37]
G4 WXB G Lees, 14 Rathan Rd, Urmston, Manchester, M41 7BA [IO83TK, SJ79]
G4 WXC S A Vaughan, 37 Mons Way, Abingdon, OX14 1NJ [IO91IQ, SU49]
G4 WXD J H Day, 35 West View, Creech St. Michael, Taunton, TA3 5QP [IO81LA, ST22]
G4 WXF R P Logan, 13 Mulberry Cl, Hereford, HR2 7UT [IO82OA, SO43]
G4 WXG C M Lees, 152 Birmingham Rd, Enfield, Redditch, B97 6EN [IO92AH, SP06]
G4 WXH G T Cook, 8 Rodney Rd, Hartford, Huntingdon, PE18 7RZ [IO92WI, TL27]
G4 WXI F W Mc Keown, 1 Thirlmere Rd, Preston, PR1 5TR [IO83QS, SD53]
G4 WXJ R Harnett, 6 Crab Ln, Scarborough, YO12 4JY [IO94SF, TA08]
G4 WXK G Sears, 36 Cedars Rd, Exhall, Coventry, CV7 9NJ [IO92GL, SP38]
G4 WXL J C Brandon, 1 Woodlands Rd, Chester, CH4 8LB [IO83NE, SJ36]
GW4 WXM J F Roberts Wrexham ARC, Tan-y-Cefn, Llanarmon y N Ial, Mold, Clwyd, CH7 4QU [IO83JB, SJ15]
G4 WXO J S Pemberton, Dunkirk Cottage, Dunkirk Ln, Dunkirk, Chester, CH1 6LU [IO83MG, SJ37]
G4 WXP Details withheld at licensee's request by SSL.
GM4 WXQ W Goudie, 5 North Lochside, Lerwick, ZE1 0PA [IP90KD, HU44]
G4 WXR B Hayes, 363 Watnall Rd, Hucknall, Nottingham, NG15 6EP [IO93JA, SK54]
G4 WXS D I Jackson, 3 Pedmore Cl, Woodrow South, Redditch, B98 7XB [IO92BG, SP06]
G4 WXT G H S Shead, 37 Shalford Rd, Rayne, Braintree, CM7 5BY [JO01GV, TL72]
G4 WXX J L Charnock, 20 Clifton Rd, Ashton in Makerfield, Wigan, WN4 0AZ [IO83QL, SD50]
G4 WXY R K A Martin, The Willows, 14B Warren Park Rd, Bengeo, Hertford, SG14 3JA [IO91XT, TL31]
G4 WXZ Details withheld at licensee's request by SSL.[Station located near St. Helens.]
G4 WYA H G McCall, 37 Shaw Head Dr, Failsworth, Manchester, M35 0SA [IO83WM, SD80]
G4 WYC S D Powell, 10 Foresters Sq, Bullbrook, Bracknell, RG12 9ES [IO91PJ, SU86]
G4 WYD H K Gibson, 14 Ribbleton Gr, Pollard Park, Bradford, BD3 0RH [IO93DT, SE13]
GI4 WYE P Doran, 143 Gransha Rd, Bangor, BT19 7RB [IO74EP, J57]
G4 WYF C R Ellison, 31 Dudley Ave, Blackpool, FY2 0TU [IO83LU, SD33]
G4 WYG B Rowe, 28 Malyons Rd, Hextable, Swanley, BR8 7RE [JO01CJ, TQ57]
G4 WYH A McPhail, 300 Fletcher Rd, Preston, PR1 5HJ [IO83PS, SD53]
G4 WYI A D Huff, 4 Greding Walk, Hutton, Brentwood, CM13 2UF [JO01EO, TQ69]
G4 WYJ Details withheld at licensee's request by SSL.[Op: J T Gale. Station located near Tadworth.]
G4 WYL C Levett, 5 Park Rd, Yapton, Arundel, BN18 0JE [IO90QT, SU90]
G4 WYM R J Chamberlain, 15 Cheviot Ct, Cheviot Cl, Harlington, Middx, UB3 5LR [IO91SL, TQ07]
G4 WYN D G Harries, 1 St. Michaels Cl, Ashby de La Zouch, LE65 1ES [IO92GR, SK31]
G4 WYO K Brewer, 14 Poplar Rd, Kensworth, Dunstable, LU6 3RS [IO91RU, TL01]
G4 WYP P T Ruse, 19 Purston Park Ct, Purston, Featherstone, Pontefract, WF7 5LR [IO93HQ, SE41]
G4 WYQ P S Murray, 19 Hafod Bryan Ct, Llwynhendy, Llanelli, SA14 9DT [IO71WQ, SS59]
G4 WYW Details withheld at licensee's request by SSL.
GW4 WYX W R Thomas, 24 Pontnth Vaughn Rd, Glynneath, Neath, West Glam, SA11 5NT [IO81ES, SN80]
G4 WYZ M J Prescott, Rathgael, 44 Glamis Dr, Chorley, PR7 1LX [IO83QP, SD51]
G4 WZA A R Nokes, 24 Braces Ln, Marlbrook, Bromsgrove, B60 1DY [IO82XI, SO97]
G4 WZB H E R Worley, 22 Cross Rd, Wellingborough, NN8 4AT [IO92PH, SP86]
G4 WZD J H Nicholl, 17 Pendle Cres, Billingham, TS23 2RA [IO94IO, NZ42]
G4 WZE Details withheld at licensee's request by SSL.
G4 WZF D A Grundy, Brent Lodge, 119 Broadway, Chilton Polden, Bridgwater, TA7 9EW [IO81ND, ST33]
G4 WZG B McIntosh, 8 Driftwell Dr, Stockton on Tees, TS19 7LA [IO94HN, NZ41]
G4 WZH A Le Couteur Bisson, 36 Gibson Way, Porthleven, Helston, TR13 9AW [IO70IC, SW62]
G4 WZI R Hilton, 14 Roberts Rd, Balby, Doncaster, DN4 0JW [IO93KM, SE50]
G4 WZJ M V Wiblin, 60 Shepherds Ln, Bracknell, RG42 2BT [IO91OK, SU87]
G4 WZK J P Cox, 20 Renals Way, Calverton, Nottingham, NG14 6PH [IO93LA, SK64]
GM4 WZL J R Scott, 5 Barrwood Gate, Galston, KA4 8NA [IO75TO, NS43]
G4 WZM S Johnston, Burn Moor End Farm, Wheathead Ln, Blacko, Nelson, BB9 6LD [IO83VV, SD84]
G4 WZN B H Turner, 15 Smardon Ave, Brixham, TQ5 8JN [IO80FJ, SX95]
GM4 WZP J D Gentles, Culra, 19 Clufflat Brae, South Queensferry, EH30 9YQ [IO85HX, NT17]
G4 WZQ I E Smith, 24 Sea View Rd, Herne Bay, CT6 6JA [JO01NI, TR16]
G4 WZS A J Glynn, 36 Paris Ave, Winstanley, Wigan, WN3 6FA [IO83PM, SD50]
G4 WZT Details withheld at licensee's request by SSL.
G4 WZU L F Thompson, 12 Long St., Great Gonerby, Grantham, NG31 8LN [IO92QW, SK83]

G4

G4 WZV J D Aizlewood, 36 King St., Winterton, Scunthorpe, DN15 9TP [IO93QP, SE91]
GM4 WZY W Pennycook, 21 Kinghorn Pl, Brechin, DD9 6BT [IO86QR, NO56]
G4 WZZ B K Housden, 5 Osprey Cl, Whitstable, CT5 4DT [JO01MI, TR16]
GI4 XAA C D McCann, Drumbally Hue House, Rock, Dungannon, Tyrone, N Ireland, BT70 3JY [IO64ON, H77]
G4 XAB R D Hunt, 3 Osprey Cl, Whitstable, CT5 4DT [JO01MI, TR16]
G4 XAD Details withheld at licensee's request by SSL.
G4 XAE A D Cole, 328 Raglan St., Lowestoft, NR32 2LB [JO02VL, TM59]
G4 XAF R G Dawe, Pendaross, Langore, Launceston, Cornwall, PL15 0EH [IO70TP, SX38]
G4 XAG B F Mahany, 3 Portland Rd, Frome, BA11 4JA [IO81UF, ST74]
G4 XAH D J Mahany, 3 Portland Rd, Frome, BA11 4JA [IO81UF, ST74]
G4 XAK Details withheld at licensee's request by SSL.
G4 XAL P G Lawrence, 39 Sarum Cres, Wokingham, Berks, RG40 1XF [IO91OK, SU86]
G4 XAM J Edwards, 3 Bank Sq, St. Just, Penzance, TR19 7HH [IO70DD, SW33]
G4 XAN C Goddard, 9 Clovelly Ave, Ickenham, Uxbridge, UB10 8PR [IO91SN, TQ08]
GI4 XAP J W Seaman, 109 Belvoir Dr, Belfast, BT8 4DN [IO74AN, J36]
G4 XAQ K H Smith, 12 Clarks Ln, Aston on Trent, Derby, DE72 2AB [IO92HU, SK42]
G4 XAR M A Kearns, 16 Fieldton Rd, Norris Green, Liverpool, L11 9AG [IO83NK, SJ39]
G4 XAT R G Evans, 7 Westland Dr, Hayes, Bromley, BR2 7HE [JO01AI, TQ36]
GW4 XAU J M A Rutkowski, 7 Beach Rd, Holyhead, LL65 1ES [IO73QH, SH28]
GM4 XAV J Stevens, 10 Tiel Path, Glenrothes, KY7 5AX [IO86KE, NO20]
GM4 XAW P C Nelson, Croc Ard, Botany St., Wigtown, Newton Stewart, DG8 9JG [IO74SU, NX45]
G4 XAX Details withheld at licensee's request by SSL.
G4 XAY E R Urda, 40 Roman Way, Daventry, NN11 5RW [IO92JG, SP56]
GW4 XAZ I M Mitchell, Llys Alaw, Ton-Breigam, Llanharry, Mid-Glamorgan, CF7 9JX [IO81HN, ST08]
G4 XBA M J Taylor, 4 Rosier Cl, Thatcham, RG19 4FN [IO91JJ, SU56]
G4 XBC A J Turner, 19 Trelawney Rd, St. Austell, PL25 4JA [IO70OI, SX05]
G4 XBD G A H Nash, 36 Lynton Ave, Arlesey, SG15 6TS [IO92UA, TL13]
G4 XBE J D Easey, Bojangles, Hackmans Ln, Cock Clarks, Chelmsford, CM3 6RE [JO01HQ, TL80]
G4 XBF M A Ray, 12 Sunnyhill, Witley, Godalming, GU8 5RN [IO91QD, SU94]
G4 XBG K B Murphy, 34 Hawkenbury Way, Lewes, BN7 1LT [IO90XU, TQ41]
G4 XBI D E Parslow, 1 Willington Cl, Little Harlescott L, Shrewsbury, SY1 3RH [IO82PR, SJ51]
G4 XBJ P T Kemp, 9 Moorfield Way, Wilberfoss, York, YO4 5PL [IO93NW, SE75]
G4 XBK A D Somerfield, 142 Hatherton Rd, Cannock, WS11 1HH [IO82XQ, SJ91]
G4 XBL J A Bell, 22 Brayton Rd, Aspatria, Carlisle, CA5 3DN [IO84IS, NY14]
G4 XBQ L Stevenson, 274 Alfreton Rd, Jubilee Hill, Pye Bridge, Alfreton, DE55 4PB [IO93HB, SK45]
GM4 XBR R B Wilson, 17 Pendreich Terr, Bonnyrigg, EH19 2DT [IO85KV, NT36]
G4 XBS C J Smith, 1 Langley Ct, Ramsey Rd, St. Ives, Huntingdon, PE17 4WX [IO92XH, TL37]
G4 XBT Details withheld at licensee's request by SSL.
G4 XBU J B Atkinson, 8 Woodcock Rd, Flamborough, Bridlington, YO15 1LJ [IO94WC, TA27]
G4 XBV J M Quin, 9A High St., Windsor, SL4 1LD [IO91QL, SU97]
G4 XBW R Head, 4 Uplands Cl, Totteridge, High Wycombe, HP13 6JX [IO91PP, SU89]
G4 XBY A G Martin, 103 Walker St., Eastwood, Nottingham, NG16 3FP [IO93IA, SK44]
G4 XBZ Dr A Roberts, 90 Hiltingbury Rd, Chandlers Ford, Eastleigh, SO53 5NZ [IO90HX, SU42]
G4 XCA R J Valder, 41 Tollgate, Peacehaven, BN10 8ED [JO00AT, TQ40]
G4 XCB Details withheld at licensee's request by SSL.
G4 XCE A F K Tamplin, Browtop, Old Ln, Crowborough Hill, Crowborough, TN6 2AD [JO01CB, TQ53]
G4 XCK S A Boden, 14 Potters Way, Ilkeston, DE7 5EX [IO92IX, SK44]
G4 XCM J Tavener, The Cube, North Dr, Heswall, Wirral, L60 0BD [IO83KH, SJ28]
GI4 XCO H Dunn, 18 Richill Park, Kilfennan, Londonderry, BT47 1QY [IO64IX, C41]
G4 XCR G T F Wood, The Old Corner Smithy, New Rd, Whissonsett, Dereham, Norfolk, NR20 5TA [JO02KS, TF92]
G4 XCS N Moorcraft, Oakview, 40 Never Wood Dr, Fownhope, Hereford, HR1 4PN [IO82QA, SO53]
G4 XCT J F Bennett, 10 Birch Gr, Martlesham Heath, Ipswich, IP5 7TD [JO02OB, TM24]
G4 XCV R M Barnett, Trewellard Arms Hotel, Trewellard, Pendeen, Penzance, Cornwall, TR19 7TA [IO70DD, SW33]
G4 XCW M J Page, 51 Lodge Rd, Rugeley, WS15 1HT [IO92AR, SK01]
G4 XCX C J Clarke, 33 James Rd, Kidderminster, DY10 2TP [IO82VJ, SO87]
G4 XCY F J Mills, 14 Seagram Cl, Aintree, Liverpool, L9 0NA [IO83MK, SJ39]
G4 XCZ D A C Medhurst, 3 Western Rd, Tunbridge Wells, TN1 2JJ [JO01DD, TQ54]
GM4 XDA Details withheld at licensee's request by SSL.
G4 XDB A G Parry, 189 Kimbolton Rd, Bedford, MK41 8DR [IO92SD, TL05]
G4 XDC M J Taylor, 4 Langstone Way, Westlea, Swindon, SN5 7BU [IO91CN, SU18]
G4 XDD Details withheld at licensee's request by SSL.[Station located in Kensington, London W8.]
G4 XDE S S Crosskey, 25 Meadow Gdns, Baddesley Ensor, Atherstone, CV9 2DA [IO92EN, SP29]
G4 XDG D G Humphreys, 129 Chester Rd, Greenbank, Northwich, CW8 4AA [IO83RG, SJ67]
G4 XDH D G Humphreys, 15 Burton Rd, Orford, Warrington, WA2 9AJ [IO83RJ, SJ69]
G4 XDK L M Neasman, 15 Brooklands Rd, Brantham, Manningtree, CO11 1RN [JO01MX, TM13]
G4 XDL M N Norman, 83 Caernarvon Rd, Hatherley, Cheltenham, GL51 5LH [IO81WV, SO92]
G4 XDM S D Comis, 178 Lordswood Rd, Birmingham, B17 8QH [IO92AL, SP08]
G4 XDN A R Allcock, 6 Marchesi Cl, Hucknall, Nottingham, NG15 6JY [IO93JA, SK54]
G4 XDO L Holden, 11 Highgate Ave, Warstones, Penn, Wolverhampton, WV4 4QY [IO82WN, SO89]
G4 XDQ M G Lavell, No 8 Bungalow, Zona Benidorm, Callas de Mallorca, Mallorca, Spain
G4 XDR A V Walker, 14 Matlock Ave, Flixton, Urmston, Manchester, M41 9FW [IO83TK, SJ79]
G4 XDU D J Chislett, Hilltops, 2A St Marks Rd, Maidenhead, Berks, SL6 6DA [IO91PM, SU88]
G4 XDV R N Hall, 53 Stoney Ln, Hall Green, Wakefield, WF4 3JR [IO93FP, SE31]
G4 XDW A E J Chidwick, 4 Burgess Cl, Whitfield, Dover, CT16 3NP [JO01PD, TR34]
G4 XDX S C T Garbett, 33 Mere Rd, Wigston, LE18 3RJ [IO92KO, SP69]
GU4 XEA P R Carre, La Petite Miellette, La Miellette Ln, Vale, Guernsey, GY3 5EN
G4 XEB B Dobson, 5 Stirling Cl, Leyland, Preston, PR5 2UU [IO83PQ, SD52]
G4 XEC Details withheld at licensee's request by SSL.
G4 XED S V Cowdell, 6 Pearl St., Chessels, Bedminster, Bristol, BS3 3EA [IO81QK, ST57]
G4 XEE D Bate, 24 Finchdean Cl, Stoke on Trent, ST3 7UT [IO82WX, SJ94]
GW4 XEF B Passmore, 16 Epworth Rd, Rhyl, LL18 2NU [IO83GH, SJ07]
G4 XEG H P Remmert, 37 Faygate Way, Lower Earley, Reading, RG6 4DA [IO91MK, SU77]
G4 XEH Details withheld at licensee's request by SSL.
G4 XEI D A McLoughlin, 23 St. Marys Ct, Clayton-le-Moors, Accrington, BB5 5LA [IO83TS, SD73]
G4 XEJ A Allen, 86 Grayswood Park Rd, Quinton, Birmingham, B32 1HE [IO92AL, SP08]
G4 XEL S A Evans, 72 Sandown Rd, Toton, Beeston, Nottingham, NG9 6JW [IO92IV, SK43]
G4 XEM J S Palfrey, 44 Grange Rd, Wellingborough, NN9 5YQ [IO92PH, SP87]
G4 XEN J L Palfrey, 44 Grange Rd, Wellingborough, NN9 5YQ [IO92PH, SP87]
G4 XEO D L M Holmes, 32 Eastcheap, Rayleigh, SS6 9JZ [JO01HO, TQ89]
GM4 XEP C A Grant, 21 Nether Currie Cres, Currie, EH14 5JJ [IO85IV, NT16]
GM4 XEQ G McCarlie, 20 Peggieshill Rd, Ayr, KA7 3RJ [IO75QK, NS31]
G4 XER E A Bonnett, 5 Royce Rd, Alwalton, Peterborough, PE7 3UR [IO92UN, TL19]
GW4 XES D A Johns, 129 Cecil Rd, Gowerton, Swansea, SA4 3DN [IO71XP, SS59]
G4 XET J G Rawlinson, Hollydene, Newbiggin, Temple Sowerby, Cumbria, CA10 1TA [IO84RP, NY62]
G4 XEV R Jenkins, 7 Claire Ct, Lymington Rd, Highcliffe on Sea, Christchurch, BH23 5DZ [IO90DR, SZ29]
G4 XEW I Rosenberg, 11 Parkside Dr, Edgware, HA8 8JU [IO91UO, TQ19]
G4 XEX P D Rivers, 20 Pendell Ave, Harlington, Hayes, UB3 5HH [IO91SL, TQ07]
G4 XEZ A P Smith, 4 Betony Cl, Tamebridge Est, Walsall, WS5 4RY [IO92AN, SP09]
G4 XFB P D McAuliffe, 53 Hunt Rd, Christchurch, BH23 3BW [IO90DR, SZ19]
G4 XFD H M Field, 9 Shepherds Fold Dr, Winsford, CW7 2UE [IO83RE, SJ66]
GI4 XFE A G Calvin, 20 Orangefield Cres, Armagh, BT60 1DS [IO64QI, H84]
G4 XFF J B Holdsworth, 37 Harewood Cres, Old Tupton, Chesterfield, S42 6HS [IO93GE, SK36]
G4 XFG N K Goodman, 24 Greenacres, Westfield, Hastings, TN35 4QT [JO00GV, TQ81]
G4 XFH G H Mullin, 13 Burnmoor Ave, Mirehouse, Whitehaven, CA28 9JN [IO84FM, NX91]
G4 XFL C Allen, 40 Abbots Cl, Knowle, Solihull, B93 9PP [IO92DJ, SP17]
G4 XFM D J Steer, 24 Manor Dr, Ivybridge, PL21 9BD [IO80AJ, SX65]
G4 XFO S Martindale, 34 Redcar Ln, Redcar, TS10 2HN [IO94LO, NZ62]
GI4 XFR D W Beattie, 14 Joanmount Gdns, Belfast, BT14 6NX [IO74AO, J37]
GI4 XFS C M Gilbody, 5 The Plateau, Piney Hills, Belfast, BT9 5QP [IO74AN, J37]
G4 XFT J S Tranter, 275 Bosty Ln, Aldridge, Walsall, WS9 0QE [IO92AO, SP09]
G4 XFV A M McKechnie, 46 St. Magdalenes, Linlithgow, EH49 6AQ [IO85EX, NT07]
G4 XFW A R West, 11 Oak Hill, Burpham, Guildford, GU4 7JF [IO91RG, TQ05]
GI4 XFX R J Reid, 6 Sperrin Heights, Townhill Rd, Portglenone, Ballymena, Co. Antrim, BT44 8AD [IO64SV, C90]
GI4 XFY E E M Townley, 27 Windmill Rd, Cranfield, Kilkeel, Newry, BT34 4LP [IO64XA, J21]
G4 XFZ R Griffin, 53 St. Johns Ave, Warley, Brentwood, CM14 5DG [JO01DO, TQ59]
GM4 XGA D J Iles, 37 Stratherrick Gdns, Inverness, IV2 4LX [IO77VK, NH64]
GU4 XGB A L Bichard, 17 Clos Du Murier, Rue de Bas, St. Sampson, Guernsey, Channel Islands, GY2 4HJ
G4 XGC L Ward, 16 Fishers Cl, Blandford Forum, Blandford, Blandford Forum, DT11 7EL [IO80WU, ST80]
G4 XGD P H J Harman, 25 Pitts Rd, Slough, SL1 3XG [IO91QM, SU98]
G4 XGE H Ling, Flat, 5 Yantlet, London Rd, Leigh on Sea, SS9 3JD [JO01HN, TQ88]
G4 XGF H B Roberts, White Sands, 20 Riverside Cres, Newquay, TR7 1PJ [IO70KJ, SW76]
GU4 XGG R G Grove, Villa D Arc-En-Ciel, Rue Du Passeur Vale, Guernsey, Channel Islands, Guernsey, GY3 5JP
G4 XGI R G Hales, 239 Charlton Rd, Shepperton, TW17 0SH [IO91SJ, TQ06]
G4 XGL G C Stringfellow, Lovelace Cl, Effingham, Effingham Junction, Leatherhead, Surrey, KT24 5HJ [IO91SG, TQ15]
G4 XGM J H Stacey, 31 Gorsehill Rd, Poole, BH15 3QH [IO90AR, SZ09]
G4 XGN P M Riggott, 1 Mill Ln, Queensbury, Bradford, BD13 1LP [IO93BS, SE03]
GI4 XGO G Armstrong, 45 Rathmena Dr, Ballyclare, BT39 9HZ [IO64XS, J29]
G4 XGP M M Kelly, 28 Skelwith Dr, Barrow in Furness, LA14 4PF [IO84JD, SD27]
GI4 XGQ T J Devine, 26 Carmoney Park, Campsie, Londonderry, BT47 3JN [IO65JA, C42]
G4 XGR S D Clark, 1 Holcroft, Orton Malborne, Peterborough, PE2 5SL [IO92UN, TL19]
G4 XGT J D Dawson, 21 Church St., Needingworth, St. Ives, Huntingdon, PE17 3TB [IO92XH, TL37]
GU4 XGU Details withheld at licensee's request by SSL.[Station located in the parish of St. Saviour. Locator: IN89QK. Op: Neil P Martin, The Flat, Charroterie, St Peter Port, Guernsey, C.I. Telephone (01481) 22327.]
G4 XGV W E Setterfield, 54 Hallam Rd, Nelson, BB9 8AB [IO83VU, SD83]
GM4 XGW T C McCabe, 8 East Park Cres, Kilmaurs, Kilmarnock, KA3 2QT [IO75RP, NS44]
GM4 XGY G C Smith, 37 Dalrymple Dr, East Mains, East Kilbride, Glasgow, G74 4LG [IO75VS, NS65]
G4 XHC F Jackson, Dun Roamin, 34 High St., Blyton, Gainsborough, DN21 3JY [IO93PK, SK89]
G4 XHE R L Cook, 3 Pyecombe Ct, Cuckfield Cl, Bewbush, Crawley, RH11 8UF [IO91VC, TQ23]
G4 XHF P C Chamberlain, 114 Grattons Dr, Crawley, RH10 3JP [IO91WD, TQ23]
G4 XHG P L Thornton, Low Ghyll, Hebers Ghyll Dr, Ilkley, LS29 9QH [IO93BW, SE14]
G4 XHH R Franklin, 10 Baltimore Ct, Belvedere Rise, Newhall, Swadlincote, DE11 0JW [IO92FS, SK22]
GM4 XHJ Details withheld at licensee's request by SSL.
G4 XHK L D L Soutter, 2 Hyde Barton, Churchill Way, Northam, Bideford, EX39 1NX [IO71VA, SS42]
GI4 XHO F E Orr, 29A McCraes Brae, Whitehead, Carrickfergus, Co. Antrim, BT38 9NZ [IO74DS, J49]
G4 XHP D Daniels, 21 Newton St., Dean Bank, Ferryhill, DL17 8PW [IO94FQ, NZ23]
GM4 XHQ G A McInnes, 14 East Croft, Ratho, Newbridge, EH28 8PD [IO85HW, NT17]
G4 XHS Details withheld at licensee's request by SSL.
G4 XHT E J Wilkinson, 33 Redesmere Park, Flixton, Urmston, Manchester, M41 9ER [IO83TK, SJ79]
G4 XHU J D Theobald, 78 West Ave, Maylandsea, Mayland, Chelmsford, CM3 6AF [JO01JQ, TL90]
GM4 XHV G Horsburgh, 3 Dumyat Rd, Alva, FK12 5NN [IO86CD, NS89]
G4 XHX M Powers, 16 Roman Ave North, Stamford Bridge, York, YO4 1DP [IO93NX, SE75]
G4 XHY P B Powers, 16 Roman Ave North, Stamford Bridge, York, YO4 1DP [IO93NX, SE75]
G4 XHZ F Jolley, 30 Oban Dr, Shadsworth, Blackburn, BB1 2HY [IO83SR, SD72]
G4 XIB D D Horton, 33 Trajan Rd, Stratton St. Margaret, Swindon, SN3 4BW [IO91DN, SU18]
G4 XIE R P Shard, 76 Clipsey Ln, Haydock, St Helens, Merseyside, WA11 0UD [IO83PL, SJ59]
G4 XIF Details withheld at licensee's request by SSL.[Op: J E Machin. Station located near Chatham.]
G4 XIG A D Figg, 5 Notgrove Cl, Benhall, Cheltenham, GL51 6BB [IO81WV, SO92]
G4 XIL B P Hurst, 25 Hoadly Rd, Cambridge, CB3 0HX [JO02BF, TL46]
G4 XIM R G Bradfield, 118 East Rd, Langford, Biggleswade, SG18 9QP [IO92UB, TL14]
G4 XIN A Henstock, 16 The Coppice, Enfield, EN2 7BY [IO91WP, TQ39]
G4 XIP J L Baker, 11 London Rd, Old Basing, Basingstoke, RG24 7JE [IO91LG, SU65]
G4 XIQ J A Lainchbury, 33 Ennersdale Rd, Coleshill, Birmingham, B46 1EP [IO92DM, SP18]
GI4 XIR W D Bird, 198 Ashmount Gdns, Lisburn, BT27 5DB [IO64XM, J26]
GU4 XIT R A Bird, Mill View Barn, Rue Des Petits Houg, Es Bordeaux Vale, Guernsey, Guernsey, GY99 1AA
G4 XIU M F Kelleway, Greenhills, Newport Rd, Whitwell, Ventnor, PO38 2QW [IO90IO, SZ57]
G4 XIV Details withheld at licensee's request by SSL.
G4 XIW S Mackenzie, 6 Bridge Farm Cl, Gr, Wantage, OX12 7QF [IO91GO, SU48]
G4 XIX M R Purnell, The Olde Cottage, Lewdown, Okehampton, Devon, EX20 4DQ [IO70VP, SX48]
G4 XIY Details withheld at licensee's request by SSL.[Op: B D Wright. Station located near Matlock.]
G4 XIZ R G Heath, Springways, 9 Woodside Ln, Ladderidge, Leek, ST13 7AN [IO93AC, SK05]
GI4 XJA Details withheld at licensee's request by SSL.
G4 XJC J A Colley, Flat 1, Block 3, Matson Ave, Matson, Gloucester, GL4 9LP [IO81VU, SO81]
GI4 XJD J Doherty, 75 Drumflugh Rd, Benburb, Dungannon, BT71 7QF [IO64OK, H85]
G4 XJE D J Brawn, 16 Mansel Cl, Cosgrove, Milton Keynes, MK19 7JQ [IO92NB, SP74]
G4 XJG R G Hirst, 47A Rowley Ln, Fenay Bridge, Huddersfield, HD8 0JG [IO93DP, SE11]
GI4 XJJ W J McCaughey, 60 Ballynure Rd, Newtownabbey, BT36 5SJ [IO74AQ, J38]
G4 XJK R King, 8 Kipling Rd, Kettering, NN16 9JZ [IO92PJ, SP88]
G4 XJL D H Rogers, Gabwell House, Lower Gabwell, Stoke in Teignhead, Newton Abbot S Devo, TQ12 4QS [IO80FM, SX97]
G4 XJM G B Stanley, Gabwell House, Stoke in Teignhead, Newton Abbot, Devon, TQ12 4QS [IO80FM, SX97]
G4 XJN J Williamson, 13 Furnival Ave, Slough, SL2 1DH [IO91QM, SU98]
G4 XJO M W Smith, 19 Prestwich Dr, Fixby Park, Huddersfield, HD2 2NU [IO93CQ, SE11]
G4 XJP Details withheld at licensee's request by SSL.
GD4 XJR R W Moore, 16 Tynwald Gr, Castletown, IM9 1BU
G4 XJS J G Smith, 84 Oakwood Dr, St. Albans, AL4 0XA [IO91US, TL10]
G4 XJU L M Harvey, 30 Wingate Way, Cambridge, CB2 2HD [JO02BE, TL45]
G4 XJV S J Bassam, 3 Glaisdale Gr, Seaton Carew, Hartlepool, TS25 1DU [IO94JP, NZ52]
G4 XJW Details withheld at licensee's request by SSL.
GM4 XJX R S Longley, Toll Cottage, Ardersier, Inverness Shire, IV1 2SX [IO77XN, NH75]
GM4 XJY D McMinn, Crestholme, East Bay, Mallaig, PH41 4QF [IO77CA, NM69]
G4 XKA P J Adams, 52 Blewbury Dr, Tilehurst, Reading, RG3 5HL [IO91LJ, SU66]
G4 XKC A J S Sieroslawski, 26 Raikes Ln, Birstall, Batley, WF17 9QU [IO93ER, SE22]
G4 XKD K Dixon, 23 Dorking Walk, Corby, NN18 9JL [IO92PL, SP88]
GW4 XKE D K Egan, 19 Sycamore Cl, Longmeadow, Dinas Powys, The Vale of Glam, CF64 4TG [IO81JK, ST17]
G4 XKF D F Browne, 67 Benfield Way, Portslade, Brighton, BN41 2DN [IO90VU, TQ20]
GM4 XKG G L Davidson, Balthangie, 37 Ballifeary Ln, Inverness, IV3 5PH [IO77VL, NH64]
GI4 XKI J J O Neill, 225 Dungannon Rd, Killeshill, Dungannon, BT70 1TH [IO64ML, H66]
G4 XKK K D Burston, Can Singala, Hope Corner Ln, Taunton, TA2 7PB [IO81KA, ST22]
G4 XKL R A Whetton, 117 Tutbury Rd, Burton on Trent, DE13 0NU [IO92ET, SK22]
G4 XKM P L Andrews, 88 Connegar Leys, Blisworth, Northampton, NN7 3DF [IO92ME, SP75]
GM4 XKP K G Macgillivray, 87 Castle St., Forfar, DD8 3AG [IO86NP, NO45]
G4 XKQ R J Janes, 8 Bell Meadow Rd, Hook, RG27 9HJ [IO91MG, SU75]
G4 XKR R M Coward, 10 Market St., Hambleton, Poulton-le-Fylde, FY6 9AP [IO83MV, SD34]
G4 XKS A A J Banks, Long Acre Cottage, 25 Wolverhampton Rd, Stourton, Staffs, DY7 5AF [IO82VL, SO88]
G4 XKV G J Pigott, 67 Mayplace Rd West, Bexleyheath, DA7 4JL [JO01BK, TQ47]
G4 XKW D W Furness, 38 Chesterhill, Collingwood Grange, Cramlington, NE23 6JN [IO95FB, NZ27]
G4 XKX N McGuigan, The Vinnery, Ad, Abberley, Worcester, Worcs, WR6 6BX [IO82TH, SO76]
G4 XKY Details withheld at licensee's request by SSL.
G4 XLA T Carruthers, Flat 43, House 119, Nevsily Pospect, 193024 St Petersburg, Russia
GI4 XLB G H Curry Sunspots Rac, 4 Rocklands, Annahilt, Hillsborough, BT26 6NU [IO74AK, J25]
G4 XLC E Metcalfe, 18 Kirkstone Dr, Morecambe, LA4 5XP [IO84NB, SD46]
G4 XLF B J Sandford, 30 Fishponds Rd, Kenilworth, CV8 1EZ [IO92EI, SP27]
G4 XLG M J Rollings, 39 Summerleys, Eaton Bray, Dunstable, LU6 2HR [IO91QU, SP91]
GM4 XLI G Bell, 52 Manor Gdns, Blairgowrie, PH10 6JS [IO86IO, NO14]
GW4 XLK E W S Meredith, 5 Woodfield Rd, Llandybie, Ammanford, SA18 3UR [IO71XT, SN61]
G4 XLM J W Todd, 60 Worcester Rd, Cowley, Uxbridge, UB8 3TH [IO91SM, TQ08]
GM4 XLN J Durrand, 9 Breadalbane Cres, Wick, KW1 5AS [IO88KK, ND35]
GW4 XLP J E Williams, 12 Maes Clyd, Craig y Don, Llandudno, LL30 1YA [IO83CH, SH78]
GW4 XLS J B K Pauwels, Ferndale, Afoneitha Rd, Penycae, Wrexham, LL14 2DH [IO83LA, SJ24]
GM4 XLU E Wallace, 10 Gean Ct, Abronhill, Cumbernauld, Glasgow, G67 3LU [IO85AX, NS77]
G4 XLY E S Grint, 15 Ivythorn Rd, St, BA16 0TE [IO81PC, ST43]
GM4 XLZ J D Wilson, 7 Loudoun Ave, Galston, KA4 8DB [IO75TO, NS43]
G4 XMA B W M Easton, 8 Church View Rd, Camborne, TR14 8RQ [IO70IF, SW64]
GM4 XMD W S McDicken, 4 Baillie Dr, Logan, Cumnock, KA18 3HS [IO75VL, NS52]
G4 XME L P Shone, 3 Ascot Dr, Dudley, DY1 2SN [IO82WM, SO99]
GM4 XMF R F E Beames, 2 Coranbae Pl, Alloway, Ayr, KA7 4JB [IO75QK, NS31]
G4 XMG Details withheld at licensee's request by SSL.
G4 XMH Details withheld at licensee's request by SSL.
GW4 XMI D C Edwards, 3 Hamilton St., Mountain Ash, CF45 3RH [IO81HQ, ST09]
G4 XMK Details withheld at licensee's request by SSL.
G4 XMM S M Walker, 4A Winston Dr, Eston, Middlesbrough, TS6 9LY [IO94KN, NZ51]
G4 XMO S Ashfield, 28 Long Gr, Baughurst, Tadley, RG26 5NY [IO91JI, SU56]
G4 XMP C R Balderston, 14 Achilles Cl, Chineham, Basingstoke, RG24 8XB [IO91LH, SU65]
G4 XMQ T R Cooling, 32 Larne Rd, Off Brant Rd, Lincoln, LN5 9TY [IO93RE, SK96]
G4 XMR M W Richardson, 131 Recreation Rd, Burghfield Common, Reading, RG7 3EN [IO91LJ, SU66]
G4 XMS C G F Munton, 86 Amsbury Rd, Hunton, Maidstone, ME15 0QH [JO01FF, TQ75]
G4 XMT J B Clarke, 21 Gorge Rd, Sedgley, Dudley, DY3 1LF [IO82WN, SO99]
GW4 XMU D H Jones, 5 Clos Llyswen, Bryn Siriol, Penpedairheol, Hengoed, CF8 7TQ [IO81HN, ST08]
GW4 XMV D J Palmer, Hazelgrove, Mwtswr Ln, St Dogmaels, Cardigan Dyfed, SA43 3HZ [IO72PB, SN14]
G4 XMX Details withheld at licensee's request by SSL.
G4 XMY J D Colson, 94 Herongate Rd, Wanstead Park, London, E12 5EQ [JO01AN, TQ48]
G4 XMZ P J Toms, The Pines, 42 Wareham Rd, Lytchett Matravers, Poole, BH16 6DR [IO80XS, SY99]
G4 XNA A K Willis, 20 Oakenbrow, Sway, Lymington, SO41 6DY [IO90ES, SZ29]
GM4 XND W F Clark, 173 Dunnikier Rd, Kirkcaldy, KY2 5AD [IO86KD, NT29]
G4 XNE F C Handy, 429 Penn Rd, Penn, Wolverhampton, WV4 5LN [IO82WN, SO89]
G4 XNF J R Cameron, 62 Erringden Rd, Mytholmroyd, Hebden Bridge, HX7 5AR [IO93AR, SE02]
G4 XNG Details withheld at licensee's request by SSL.
G4 XNH Details withheld at licensee's request by SSL.
G4 XNI A R Cross, 11 Brindles Cl, Linford, Stanford-le-Hope, SS17 0RS [JO01EL, TQ67]
G4 XNJ Details withheld at licensee's request by SSL.
G4 XNK H Johnson, 52 Kirkham Gdns, Bromyard, HR7 4EA [IO82RE, SO65]

G4 XNL J G J Alblas, 36 Mill Rd, Eastbourne, BN21 2PG [JO00DS, TQ50]
G4 XNO M J Goodearl, Woodlands, 49 Wycombe Rd, Princes Risborough, HP27 0EE [IO91NR, SP80]
G4 XNP D J Rayner, 69 Saracen Rd, Hellesdon, Norwich, NR6 6PB [JO02PQ, TG21]
GM4 XNQ D Muir, 28 Grange Cres, Edinburgh, EH9 2EH [IO85JW, NT27]
G4 XNR P Morris, 14 Storrs Hill Rd, Ossett, WF5 0DL [IO93FQ, SE21]
G4 XNS N E Speak, 17 Stiles Ave, Hutton, Preston, PR4 5FL [IO83OR, SD42]
GW4 XNT S J Morrison, 5 The Beeches, Holywell, CH8 7SW [IO83JG, SJ17]
G4 XNV D W Owen, 58 Stonecross Ln, Lowton, Warrington, WA3 2SE [IO83QL, SJ69]
G4 XNW J K Simmonds, 15 Meriden Ave, Wollaston, Stourbridge, DY8 4QN [IO82WL, SO88]
G4 XNY A D Turner, 10 Jervis Cres, Sutton Coldfield, B74 4PW [IO92BO, SP09]
G4 XOC P N Dunn, 4 Poulton Ave, Sutton, Surrey, SM1 3PY [IO91VI, TQ26]
GD4 XOD W Jones, Ballanard Rd, Onchan, Abbeylands, Douglas, Isle of Man, IM4 5EA
G4 XOF Details withheld at licensee's request by SSL.
G4 XOH D C Blackwell, 5 Blenheim Gr, Offord Darcy, St. Neots, Huntingdon, PE18 9RD [IO92VG, TL26]
GM4 XOI A McGill, 37 Barlae Ave, Eaglesham, Glasgow, G76 0DA [IO75US, NS55]
G4 XOJ N C Wade, 19 Bradfield Cres, Hadleigh, Ipswich, IP7 5EU [JO02LB, TM04]
G4 XOL M Osborne, 27 Silverdale Rd, Newton-le-Willows, WA12 0JT [IO83QL, SJ59]
G4 XOM R Egan, 56 Walker Ave, Stourbridge, DY9 9EL [IO82WK, SO98]
G4 XOO Details withheld at licensee's request by SSL.
G4 XOP T J Cooper, 55 Meadway, Parkway, St. Austell, PL25 4HT [IO70OI, SX05]
G4 XOR N I Gibson, 52 Parkfield Cres, Feltham, TW13 7LD [IO91TK, TQ17]
G4 XOS W L Beresford, 4 Valley View, Bredenbury, Bromyard, HR7 4UJ [IO82RE, SO65]
G4 XOT Details withheld at licensee's request by SSL.
G4 XOU R Hague, 37 Vernon Dr, Horsendale Est, Nuthall, Nottingham, NG16 1AR [IO92JX, SK54]
G4 XOW D A Lomas, Galmpton, Cannon Ln, Maidenhead, SL6 3NR [IO91OM, SU87]
G4 XOX Details withheld at licensee's request by SSL.
G4 XPA V Edwards, Tanglewood, Higher Rydons, Brixham, Devon, TQ5 8QD [IO80FJ, SX95]
G4 XPB W A Keen, 31 Borrowdale Dr, Norwich, NR1 4LX [JO02PP, TG20]
G4 XPC Details withheld at licensee's request by SSL.
G4 XPE Details withheld at licensee's request by SSL.
G4 XPF Details withheld at licensee's request by SSL.
G4 XPH C E Farr, 1 Webb Cl, Pewsham, Chippenham, SN15 3XF [IO81WK, ST97]
G4 XPI P J ODea, 41 Oakwood Cl, Blackpool, South Shore, Blackpool, FY4 5FD [IO83LS, SD33]
G4 XPJ A B Gridley, 13 Brockwell, Oakley, Bedford, MK43 7TD [IO92RE, TL05]
G4 XPK N G Powell, 42 Beach Rd, Carlyon Bay, St. Austell, PL25 3PH [IO70PI, SX05]
G4 XPL Details withheld at licensee's request by SSL.
G4 XPM Details withheld at licensee's request by SSL.
GW4 XPN W G Davies, Hendref, Red Wharf Bay, Pentraeth Anglesey, North Wales, LL75 8YG [IO73VG, SH57]
G4 XPO A J Telford, 16 Springfields, Wigton, CA7 9JS [IO84KT, NY24]
G4 XPP J D Bolton, 20 Appleton Cres, Willington, Crook, DL15 0DX [IO94DR, NZ13]
G4 XPR G A Wall, 143 Vine Rd, Stoke Poges, Slough, SL2 4DH [IO91QN, SU98]
G4 XPT A J T M Fernandez, 2 Siverston Ave, Bognor Regis, West Sussex, PO21 2RB [IO90PS, SZ99]
G4 XPU M P Bennett, 7 Woburn Ave, Bolton, BL2 3AY [IO83TO, SD71]
G4 XPV P J Maisey, 155 Parkfield Dr, Birmingham, B36 9TY [IO92CM, SP19]
G4 XPY D P Fuller, 51 Evenlode, Banbury, OX16 7PQ [IO92HB, SP44]
G4 XQA K James, 59 Pelham Rd, Thelwall, Warrington, WA4 2HA [IO83RJ, SJ68]
G4 XQD D W A Cast, 24 Station Rd, Tiptree, Colchester, CO5 0AJ [JO01JT, TL81]
G4 XQE A E Turner, 4 Taylor Rd, Ashtead, KT21 2HY [IO91UH, TQ15]
G4 XQF N Clacher, 3 Annan Cres, Marston, Blackpool, FY4 4RQ [IO83MT, SD33]
G4 XQG K P Icke, The Hops, 6 Brewers Ln, Badsey, Evesham, WR11 5EU [IO92BC, SP04]
GW4 XQH J B Davies, 109 Croesonnen Parc, Abergavenny, NP7 6PF [IO81LT, SO31]
G4 XQI J Burton, 8 Peel Moat Rd, Heaton Moor, Stockport, SK4 4PL [IO83VK, SJ89]
GM4 XQJ B J Waddell, Carsemount, 3A Polmont Rd, Laurieston, Falkirk, FK2 9QQ [IO85DX, NS97]
GW4 XQK T J Perry, 70 Lawrenny St., Neyland, Milford Haven, SA73 1TB [IO71MR, SM90]
G4 XQN A E Aubury, 110 Rayleigh Dr, Wideopen, Newcastle upon Tyne, NE13 6AJ [IO95EB, NZ27]
G4 XQQ J Freeman, Nb Ecat Canal Wharf, Leighton Rd Bridge, Leighton Buzzard, Beds, LU7 7LA [IO91PV, SP92]
G4 XQR W Young, Post Office, 18 Front St., Sherburn, Durham, DH6 1HA [IO94FS, NZ34]
G4 XQS Details withheld at licensee's request by SSL.
G4 XQT G R Haydon, 138 Lincoln Rd, Skegness, PE25 2DN [JO03DD, TF56]
G4 XQV T G Sismey, 12 Layton Ln, Rawdon, Leeds, LS19 6RG [IO93EU, SE23]
G4 XQW R J Stacey, 8 Rookery Hill, Rockland St. Mary, Norwich, NR14 7EW [JO02QO, TG30]
G4 XQX D A E Oliver, 6 Kensington Rd, Gosport, PO12 1QY [IO90KS, SZ69]
G4 XQY Details withheld at licensee's request by SSL.[Op: A R Clack.]
G4 XQZ J Fisher, 6 Castle Way, Havant, PO9 2RZ [IO90MU, SU70]
G4 XRA R V G Avery, 137 Kingsley Ave, Kettering, NN16 9ES [IO92PJ, SP88]
G4 XRB J A Gagg, 20 Stanstead Ave, Tollerton, Nottingham, NG12 4EA [IO92KV, SK63]
G4 XRD G V Pope, 4A Main St, Rockingham, Market Harborough, Leics, LE16 8TG [IO92PM, SP89]
GM4 XRF N Lowson, 39 Lordburn Pl, Forfar, DD8 2DE [IO86NP, NO45]
G4 XRG E R Godlieb, 4 Tytherington Park Rd, Macclesfield, SK10 2EL [IO83WG, SJ97]
G4 XRI Details withheld at licensee's request by SSL.
G4 XRJ J Mills, Aquila, 4 Westhill Rd South, South Wonston, Winchester, SO21 3HP [IO91IC, SU43]
G4 XRK J A Lord, 16 Lark Valley Dr, Fornham St. Martin, Bury St. Edmunds, IP28 6UG [JO02IG, TL86]
G4 XRL E J Otty, 103 Fifers Ln, Hellesdon, Norwich, NR6 6EF [JO02PP, TG21]
G4 XRN J C Lucas, 10 Laton Rd, Hastings, TN34 2ET [JO00GU, TQ81]
G4 XRO S G Hill, 25 Southfield Dr, Epworth, Doncaster, DN9 1DG [IO93OM, SE70]
GM4 XRP J T Porter, 1 Loney Cres, Denny, FK6 5EG [IO86BA, NS88]
G4 XRQ E J McKeand, 1 John Rushout Ct, Northwick Park, Blockley, Moreton in Marsh, GL56 9RJ [IO92CA, SP13]
G4 XRS A R Fry, Vande, Moreton on Lugg, Herefordshire, HR4 8DG [IO82PC, SO54]
G4 XRT M Taylor, 71 Dartford Ave, Edmonton, London, N9 8HE [IO91XP, TQ39]
G4 XRU J W Hicks, 3 Lockitt Way, Kingston, Lewes, BN7 3LG [IO90XU, TQ30]
G4 XRV R Bullock, 9 The Braid, Chesham, HP5 3LU [IO91QR, SP90]
GW4 XRW R W Wood, Bwlcyn, Eifl Rd, Trefor, Caernarvon, LL54 5HG [IO72SX, SH34]
G4 XRX R B Headland, 18 Blucher St., Liverpool, L22 8QB [IO83LL, SJ39]
GM4 XRY A A Rimmer, 16 Johnston Dr, Barassie, Troon, KA10 6SD [IO75QN, NS33]
G4 XSA A C Boniface, 33 Caraway Pl, Wallington, SM6 7AG [IO91WJ, TQ26]
G4 XSB W Bridgen, The Sea Holme, 13 Alleyne Way, Elmer on Sands, Bognor Regis, PO22 6JZ [IO90QT, SU90]
G4 XSC G E Trim, 29 Casterbridge Rd, Dorchester, DT1 2AH [IO80SQ, SY78]
G4 XSD R F Wood, 66 Elm Gr South, Barnham, Bognor Regis, PO22 0EL [IO90QU, SU90]
GI4 XSF M W Stevenson, 69 Portaferry Rd, Cloughey, Newtownards, BT22 1HP [IO74GK, J65]
G4 XSG S Stuart, Oakdene, 102 Mitton Rd, Whalley, Clitheroe, BB7 9JN [IO83TT, SD73]
G4 XSH S R Senior, 13 Aintree Cl, Kippax, Leeds, LS25 7HY [IO93HS, SE43]
G4 XSI Details withheld at licensee's request by SSL.
G4 XSJ S Johnston, 6 Lytham Rd, Wrottesley Fields, Perton, Wolverhampton, WV6 7YY [IO82VO, SO89]
G4 XSK G H Bragg, 22 Cemetery Rd, Earby, Colne, BB8 6QX [IO83WW, SD94]
G4 XSL L Wagstaff, 4 Plompton Cl, Harrogate, HG2 7DT [IO93GX, SE35]
G4 XSM G A Davey, 49 Maltward Ave, Bury St. Edmunds, IP33 3XQ [JO02IF, TL86]
G4 XSN A Bailey, 36 Redcar St., Tuebrook, Liverpool, L6 0AJ [IO83MK, SJ39]
GW4 XSP A B Gillies, 10 Cysgod y Graig, Denbigh, LL16 3TD [IO83GE, SJ06]
G4 XSS J P Brown, 16 Springfield, Ovington, Prudhoe, NE42 6EH [IO94BX, NZ06]
G4 XST P I Cheeseman, Delmar, 10 Lymden Cl, Stonegate, Wadhurst, TN5 7EG [JO01EA, TQ62]
G4 XSV S H Forshaw, Pound Lodge, Crabtree Gdns, Headley Bordon, Hants, GU35 8LN [IO91OC, SU83]
G4 XSW E B Ward, 7 Blenheim Cl, Ruddington, Nottingham, NG11 6DL [IO92KV, SK53]
G4 XSX M J Tovey, Fron Farm, Bontnewydd, Aberystwyth, Dyfed, SY23 4G [IO72XI, SN67]
G4 XSZ I S Moreton, Plough House, Wisbech Rd, Walpole St Andrew, Wisbech, Cambs, PE14 7LH [JO02CR, TF51]
G4 XTA P D Godolphin, 3 Knipe View, Bampton, Penrith, CA10 2RF [IO84ON, NY51]
GI4 XTC W M Armstrong, 8 Killowen Cres, Lisburn, BT28 3DS [IO64XM, J26]
G4 XTE J Johnson, Winterwood, West Lodge Cres, Fixby Huddersfield, West Yorks, HD2 2EH [IO93CQ, SE11]
G4 XTF N S J Hancocks, Wesley House, Allensmore, Hereford, HR2 9BE [IO82OA, SO43]
G4 XTG I Brown, The Old Smithy, 451 Blackburn Rd, Turton, Bolton, BL7 0PW [IO83TP, SD71]
GM4 XTI I R Paterson, Paterson, 9 Elm Gr, Newton Stewart, DG8 6JT [IO74SW, NX46]
G4 XTK A R G Kurnatowski, 24 Eversleigh Rise, South Darley, Darley Bridge, Matlock, DE4 2JW [IO93ED, SK26]
G4 XTL D M Hall obo The Dukeries A.R.S., 72 Mansfield Rd, Edwinstowe, Mansfield, NG21 9NH [IO93LE, SK66]
G4 XTM Dr T A Morris, 16 Balbec Ave, Headingley, Leeds, LS6 2BB [IO93FT, SE23]
G4 XTN Details withheld at licensee's request by SSL.[Op: P M Harvey.]
G4 XTO W Reade, 106 Wellington Rd, Bollington, Macclesfield, SK10 5HT [IO83WH, SJ97]
G4 XTR N J A Hearn, Horsebrook Farm, Avonwick, South Brent, Devon, TQ10 9EU [IO80CJ, SX75]
G4 XTS J M Strutt, Woodland, Gardiners Ln North, Crays Hill, Billericay Essex, CM11 2XE [JO01FO, TQ79]
GD4 XTT W D G Brown, Cleckheaton, Ballaragh, Lonan, Isle of Man, IM4 7PW
G4 XTU J A Jones, 3 Blackstope Ln, Retford, DN22 6NW [IO93MH, SK78]
G4 XTV G Spink, 34 Kelfield Rd, Riccall, York, YO4 6PH [IO93LT, SE63]
G4 XTW A E Bowes, 19 Queens Dr, Mildenhall, Bury St. Edmunds, IP28 7JW [JO02GI, TL77]
G4 XTX C C Cooper, 26 Burlington Rd, Skegness, PE25 2EW [JO03DD, TF56]
GW4 XTY J Hammond, 10 Davids Way, Waterloo Rd, Penygroes, Llanelli Dyfed, SA14 7NP [IO71XT, SN51]
G4 XTZ A H Taylor, 36 Bodmin Ave, Northborough Est, Slough, SL2 1SL [IO91QM, SU98]
G4 XUA R S Walton, 275 Ridgacre Rd, Quinton, Birmingham, B32 1EG [IO92AL, SP08]
GM4 XUC G F H Gottschlich, 26 Millbank Rd, Stranraer, DG9 0EJ [IO74LV, NX06]
GW4 XUE Details withheld at licensee's request by SSL.
G4 XUG G Patterson, 4 Ruislip Gdns, Nye Timber, Bognor Regis, PO21 4LB [IO90PS, SZ99]
G4 XUH S J Bates, 83 Ferndale Rd, Thurmaston, Leicester, LE4 8JE [IO92KQ, SK60]
G4 XUI J C D Gordon, 54 Guibal Rd, London, SE12 9LX [JO01AK, TQ47]
GM4 XUJ K Traill, 57 Ashfield Dr, Summerville, Dumfries, DG2 9BP [IO85EB, NX97]
G4 XUK A V Cornwell, Woodlands, Bridle Dene, Shelf, Halifax, HX3 7NR [IO93CR, SE12]
G4 XUL K Haworth, 51 Yew Tree Dr, Oswaldtwistle, Accrington, BB5 3AX [IO83TR, SD72]
G4 XUM M J Platt, 451 Newcastle Rd, Shavington, Crewe, CW2 5JU [IO83SB, SJ75]
G4 XUO J S Ball, 64 Castlehey, Skelmersdale, WN8 9DS [IO83PM, SD50]
G4 XUQ S R Winters, 90 Hookfield, Harlow, CM18 6QJ [JO01BS, TL40]
G4 XUR D L Smith, 47 Laburnum St., Taunton, TA1 1LB [IO81KA, ST22]
GM4 XUS G J A Smith, 80 Deanburn Park, Linlithgow, EH49 6HA [IO85EX, NS97]
G4 XUT R J Loss, 84 Thorne Rd, Eldene, Swindon, SN3 6DU [IO91DN, SU18]
G4 XUU D M Halberg, Hempstead Gr, Hempstead Ln, Hailsham, BN27 3AA [JO00DU, TQ51]
G4 XUV D A Bevan, 46 Park Ln, Hartford, Northwich, CW8 1PY [IO83RF, SJ67]
G4 XUW D R Hudson, 18 Durham Cl, St Margarets, Ware, SG12 8DZ [JO01AS, TL31]
G4 XUX M Scott, 11 Barnstone Vale, Wakefield, WF1 4TJ [IO93GQ, SE32]
G4 XUY E J Gridley, 13 Brockwell, Oakley, Bedford, MK43 7TD [IO92RE, TL05]
G4 XUZ R W Chandler, 43 The Dr, Shoreham By Sea, BN43 5GD [IO90UU, TQ20]
G4 XVE J G Francis, 9 Glendower Rd, East Sheen, London, SW14 8NY [IO91UL, TQ27]
G4 XVF A J Henk, 10 Aston Way, Epsom, Surrey, KT18 5LZ [IO91UH, TQ26]
G4 XVG P R E Finch, 10 Clandon Cl, Stoneleigh, Epsom, KT17 2NQ [IO91VI, TQ26]
G4 XVH D M Van Haaren, Tillers, Hoe Ln, Abridge, Romford, RM4 1AU [JO01BP, TQ49]
G4 XVI J E Ames, 33 Keswick Cl, Cringleford, Norwich, NR4 6UW [JO02OO, TG10]
G4 XVJ G P Brooker, 66 Butler Cl, Woodstock Rd, Oxford, OX2 6JQ [IO91IS, SP50]
G4 XVK F W Wootten, 83 Park Ave, Orpington, BR6 9EG [JO01BI, TQ46]
G4 XVL A Geary, 6 Windles Row, Lyppard Woodgreen, Worcester, WR4 0RS [IO82VE, SO85]
G4 XVM M J Brett, 4 Third Ave, Galley Hill, Waltham Abbey, EN9 2AP [JO01AQ, TL40]
G4 XVN R Siddens, 45 Brooklands Rd, Hall Green, Birmingham, B28 8LA [IO92BK, SP18]
G4 XVO Dr S D Bate, 73 Golf Ln, Whitnash, Leamington Spa, CV31 2QB [IO92FG, SP36]
G4 XVP P L Hart, 4 Kings Ride, Tylers Green, Penn, High Wycombe, HP10 8BL [IO91PP, SU89]
G4 XVQ L W Caine, 10 Firs House, Drylea Gr, Castle Bromwich, Birmingham, B36 8DD [IO92CM, SP18]
G4 XVR B A Hughes, Glen Cottage, Blue Cap Ln, Hampton Malpas, Ches, SY14 8JQ [IO83PA, SJ44]
G4 XVS K A Hughes, Glen Cottage, Blue Cap Ln, Hampton Malpas, Ches, SY14 8JQ [IO83PA, SJ44]
G4 XVV E Davies, 9 Rosary Ct, Potters Bar, EN6 1HA [IO91VQ, TL20]
G4 XVW I A Dobson, Pine View, Forest Dale Rd, Marlborough, SN8 2AS [IO91DK, SU16]
G4 XVY D E Bastin, 94 Clyfton Cl, Broxbourne, EN10 6NY [IO91XR, TL30]
GW4 XVZ Details withheld at licensee's request by SSL.
G4 XWA M H Cohen, Quatrems, 7 Northdale Park, Swanland, North Ferriby, HU14 3RH [IO93SR, SE92]
GD4 XWB A J Wills Browne, 1 Holmes Ct, Rowany Dr, Port Erin, IM9 6LW
GW4 XWC W G Crooks, 52 St. Catherines Rd, Baglan, Port Talbot, SA12 8AS [IO81CO, SS79]
G4 XWD J E Cookson, Fen Hill, Hall Rd, Ludham, Great Yarmouth, NR29 5NU [JO02SQ, TG31]
G4 XWE L G R J Perrett, 1 Churchill Cl, Wells, BA5 3HY [IO81QF, ST54]
GD4 XWF J E Harrison, 11 The Bretney, Jurby, Ramsey, IM7 3BL
G4 XWI B F Parrish, 61 Cirrus Cres, River View Park Est, Gravesend, DA12 4QR [JO01EK, TQ67]
G4 XWJ Details withheld at licensee's request by SSL.[Station located near Christchurch.]
GM4 XWL S J Gaw, 10 Scotstoun Park, South Queensferry, EH30 9PQ [IO85HX, NT17]
G4 XWM F R Walton, 13 Ranelagh Gdns, Newport Pagnell, MK16 0JP [IO92PC, SP84]
GW4 XWN H Walker, 5 Shaun Dr, Rhyl, LL18 4LF [IO83GH, SJ08]
G4 XWO P Twells, 422 Foxhill Rd, Carlton, Nottingham, NG4 1JY [IO92KX, SK54]
G4 XWP C S Boyce, 19 The Glebe, Prestwood, Great Missenden, HP16 9DN [IO91PQ, SP80]
G4 XWQ D C Cottle, The Brambles, Landkey Rd, Newport, Barnstaple, EX32 9BW [IO71XB, SS53]
G4 XWR P W Grainger, 26 Beattie St., South Shields, NE34 0NJ [IO94GX, NZ36]
GM4 XWS D G Munro, Duil Alvinn, Green Ln, Kingussie, Highland, PH21 1JU [IO77XB, NH70]
G4 XWT F Donachie, 11 Crabtree Cl, Burton, Christchurch, BH23 7HG [IO90CS, SZ19]
G4 XWV M G E Walden, Serendipity Kennels, Brightstone Ln, Farringdon, Alton, GU34 3EU [IO91LC, SU63]
G4 XWW G Winyard, 76 West Elloe Ave, Spalding, PE11 2BJ [IO92WT, TF22]
G4 XWY C H Curryer, 5 Pines Rd, Devizes, SN10 3DJ [IO91AI, SU06]
G4 XWZ D J Lerner, 6 Willow Rd, Kings Stanley, Stonehouse, Glos, GL10 3HS [IO81UR, SO80]
G4 XXA F W Mills, 66 Beeches Rd, Charlton Kings, Cheltenham, GL53 8NQ [IO81XV, SO91]
G4 XXB E Mills, 66 Beeches Rd, Charlton Kings, Cheltenham, GL53 8NQ [IO81XV, SO91]
G4 XXD S J White, 15 Spurway Rd, Canal Hill, Tiverton, EX16 4ER [IO80GV, SS91]
GD4 XXE D R Stoutt, Ellanbane, Lezayre, Ramsey, Isle of Man, IM7 2AU
GW4 XXF B A E Morris, 62 Gerllan, Tywyn, LL36 9DE [IO72XO, SH50]
G4 XXG D Burton Stockton&Dis AR, 49 Aske Rd, Redcar, TS10 2BP [IO94LO, NZ62]
G4 XXH R Miles, Lone Oak, Clappers Farm Rd, Silchester, Reading, RG7 2LH [IO91LI, SU66]
G4 XXI G S Lee, 5 Morton, Tadworth Park, Tadworth, KT20 5UA [IO91VH, TQ25]
GW4 XXJ N R V Jones, 2 Pentrosfa Rd, Llandrindod Wells, LD1 5NL [IO82HF, SO05]
G4 XXK D G Hart, Kingswood, 52 Scalwell Ln, Seaton, EX12 2DJ [IO80LR, SY29]
G4 XXM D J Frederick, 64 Sevenoaks Rd, Langney, Eastbourne, BN23 7LW [JO00DT, TQ60]
GM4 XXO I Carby, 24 Craigenhill Rd, Kilncadzow, Carluke, ML8 4QT [IO85CR, NS84]
GW4 XXP J R Bancroft, 101 Meliden Rd, Prestatyn, LL19 8LU [IO83HH, SJ08]
G4 XXS G W Cooper, 44 Nursery Cl, Hucknall, Nottingham, NG15 6DQ [IO93JA, SK54]
G4 XXT J T Cassidy, 137 Heath Park Rd, Gidea Park, Romford, RM2 5XJ [JO01CN, TQ58]
G4 XXU B I D Lovelock, Priest Croft Farm, Wootton Rivers, Marlborough, Wilts, SN8 4NQ [IO91DI, SU16]
G4 XXW J R Groeger, Deninex UK Oil and Gas Ltd, Bowater House, 114 Knightsbridge, London, SW1X 7LD [IO91WM, TQ27]
G4 XXX F J Pullen, 35 Berrycroft, Willingham By Stow, Willingham, Cambridge, CB4 5JX [JO02AH, TL46]
G4 XXY M B Masterman, 3 Glenwood, Ashington, NE63 8EL [IO95FE, NZ28]
G4 XXZ D J Palfreman, 464 Uppingham Rd, Leicester, LE5 2GG [IO92LP, SK60]
G4 XYA R E Penny, 2 Chapel Cttgs, West St., Aldbourne, Marlborough, SN8 2BS [IO91EL, SU27]
G4 XYB M G Kingdon, Corner Cottage, Church Ln, Wymondham, Melton Mowbray, LE14 2AB [IO92PS, SK81]
G4 XYC A J Reynolds, 90 Windfield, Leatherhead, KT22 8UJ [IO91UH, TQ15]
G4 XYD R Young, 5 Edge Hill, Chellaston, Derby, DE73 1RP [IO92GV, SK33]
G4 XYE J R Elliott, 25 Inglewood Ave, Mickleover, Derby, DE3 5RT [IO92FV, SK33]
GM4 XYF W R Swan, Rowan Cottage, 23 Seagate, Kingsbarns, St. Andrews, KY16 8SR [IO86QH, NO51]
G4 XYH W O Stock, The Cottage, Hollow Rd, Shipham, Winscombe, BS25 1TG [IO81OH, ST45]
GW4 XYI P A Coombs, 28 Cae Braenar, Holyhead, LL65 2PN [IO73QH, SH28]
G4 XYK P Mitchell, 19 Ashbourne Ave, Whetstone, London, N20 0AL [IO91WP, TQ29]
GW4 XYL B Sweeting, Beuno Cottage, Pen y Ball Top, Holywell, CH8 8SU [IO83JG, SJ17]
G4 XYM J Hoskins, 37 Green Cl, Didcot, OX11 8TE [IO91JO, SU58]
G4 XYN R A J Savin, 7 Bannard Rd, Maidenhead, SL6 4NG [IO91OM, SU88]
G4 XYO B R Baker, 2 School Way, Back Ln, Horsmonden, Tonbridge, TN12 8NJ [JO01FD, TQ74]
G4 XYP R Jobes, The Wessington Public House, Donwell, Washington, Tyne & Wear, NE37 1EE [IO94FW, NZ25]
G4 XYR W F Clarkson, 30 South View Terr, Yeadon, Leeds, LS19 7QL [IO93DU, SE24]
G4 XYS J F Mundy, 14 Blackwood Gr, Pellon, Halifax, HX1 4RH [IO93BR, SE02]
G4 XYT Details withheld at licensee's request by SSL.[Op: A I Morrison.]
G4 XYU P H Davis, 20 The Croft, Harwell, Didcot, OX11 0ED [IO91IO, SU48]
G4 XYW A P Pevy, 6 Hormer Cl, Sandhurst, Owlsmoor, Camberley, Surrey, GU15 4QW [IO91PI, SU86]
G4 XYY J E England, 2 Clifford Rd, Bramham, Wetherby, LS23 6RN [IO93HV, SE44]
G4 XZA E A Wardle, 57 Brook View Dr, Keyworth, Nottingham, NG12 5RA [IO92KU, SK63]
G4 XZC P R Gass, 4 Una Rd, Bowers Gifford, Basildon, SS13 2HU [JO01GN, TQ78]
G4 XZD R W Goodall, 24 Silk Mill Dr, Leeds, LS16 6DX [IO93EU, SE23]
G4 XZE N Sheard, 156 Elmfield Dr, Odsal, Bradford, BD6 1PS [IO93CS, SE12]
G4 XZF B A W Irwin, Smythen Farm, Sterridge Valley, Berrynarbor, Ilfracombe, EX34 9TB [IO71XE, SS54]
G4 XZG P E Lawton, 5 Belvedere Gdns, Leeds, LS17 8BS [IO93FU, SE33]
G4 XZI G J Hall, 22 Templenewsam View, Leeds, LS15 0LW [IO93GT, SE33]
GW4 XZJ H M Jones, Hafan, 7 Tan y Bryn St., Abergynolwyn, Tywyn, Gwynedd, LL36 9UY [IO82AP, SH60]
G4 XZL A D Vare, 48 Dore Ave, Fareham, PO16 8BX [IO90KU, SU60]
G4 XZM K J Pickles, 79 Mill Ln, Hanging Heaton, Batley, WF17 6DZ [IO93EQ, SE22]
GM4 XZN J L Macdonald, 15 Muir Wood Dr, Currie, EH14 5EZ [IO85IV, NT16]
GW4 XZP E Wood, Bwlcyn, Eifl Rd, Trefor, Caernarvon, LL54 5HG [IO72SX, SH34]
G4 XZS R G Weston, 2 Gill Park, Elford, Plymouth, PL3 6LX [IO70WJ, SX45]
G4 XZT R H Toft, 63 Dawson Rd, Kingston upon Thames, KT1 3AU [IO91UJ, TQ16]
G4 XZU Details withheld at licensee's request by SSL.
G4 XZV Details withheld at licensee's request by SSL.
GM4 XZZ L Fairbairn, 34 Crofts Acres, Cockburnspath, TD13 5YD [IO85TW, NT77]
GM4 YAA J W Sinclair, 3 Ben More Dr, Paisley, PA2 7NU [IO75TT, NS56]
G4 YAB J A Livesley, 79 Mellor Rd, New Mills, High Peak, SK22 4DP [IO93AI, SK08]
G4 YAC N J Howarth, 2 Sudeley Gr, Hardwick, Cambridge, CB3 7XS [JO02AF, TL35]
GJ4 YAD P G M Boyden, Peterborough House, High St, St Aubin, Jersey, Channel Islands, JE3 8BR
G4 YAF A Trudgen, 14 Park An Pyth, Pendeen, Penzance, TR19 7ET [IO70DD, SW33]
G4 YAG F A G Belfield, 9 Pelham Rd, Bexleyheath, DA7 4LT [JO01BL, TQ47]
G4 YAH G Warnes, 20 Clivedon Way, Halesowen, B62 8TB [IO82XL, SO98]
G4 YAJ S L Woodhead, 804 Huddersfield Rd, Dewsbury, WF13 3LZ [IO93DQ, SE22]
G4 YAK C J Dobinson, 37 Ladram Rd, Thorpe Bay, Southend on Sea, SS1 3PX [JO01JM, TQ98]

G4 YAL S J Tuffin, 21 Garraways, Wootton Bassett, Swindon, SN4 8NQ [IO91BN, SU08]
G4 YAM C C Atkin, 23 Brewster Ave, Immingham, Grimsby, DN40 1DW [IO93VO, TA11]
G4 YAN R B Page, 26 Colne Rd, High Wycombe, HP13 7XN [IO91PP, SU89]
G4 YAO H H Loffler, 9 Lower Hillside, East Taphouse, Liskeard, PL14 4LL [IO70SK, SX26]
G4 YAP G South, 76 Lilac Cress, Hoyland, Barnsley, S.Yorks, S74 9PW [IO93GM, SE30]
G4 YAQ B W Setter, Briarwood, Alexandra Rd, Crediton, EX17 2DH [IO80ET, SS80]
G4 YAR P E Read, 30 Talbot Rd, Paddington, London, W2 5LJ [IO91VM, TQ28]
G4 YAS E D Lucas-Davis, 27 Cadbury Rd, Sunbury on Thames, TW16 7NA [IO91SK, TQ07]
GM4 YAT T H Turner, 11 Henderson Row, Fort William, PH33 6HT [IO76KT, NN17]
GM4 YAU W K Scott, 44 Patrick Allan, Fraser St, Arbroath, Angus, DD11 1DR [IO86RN, NO64]
G4 YAV R Dixon, 30 Albatross Way, South Beach Est, Blyth, NE24 3QH [IO95FC, NZ37]
GW4 YAW J Lawton, Elford, Cathedral View, Llangawsai, Aberystwyth, SY23 1HH [IO72XJ, SN58]
G4 YAX D S Diss, 130 Beridge Rd, Halstead, CO9 1JU [JO01HW, TL83]
G4 YAZ H A Sheer, Sea Echo, 53 Leonard Rd, Greatstone, New Romney, TN28 8RX [JO00LW, TR02]
G4 YBA G Collis, 13 Westbrook Cl, Horsforth, Leeds, LS18 5RQ [IO93EU, SE23]
G4 YBB M G Coombs, Rose Cottage, Hoops, Horns Cross, Bideford, Devon, EX39 5DJ [IO70UX, SS32]
G4 YBC Details withheld at licensee's request by SSL.
G4 YBD G E Reed, The Ridings, Fullerton Terr, Collingham, Wetherby, N Yorks, LS22 5AT [IO93HV, SE34]
GW4 YBE A C Evans, 110 Thorney Rd, Port Talbot, SA12 8LS [IO81CO, SS79]
G4 YBG A E White, Northdale, Goughs Ln, Bracknell, RG12 2RA [IO91PJ, SU86]
G4 YBH B Hawkins, Andorra, Haw Ln, Bledlow Ridge, High Wycombe, HP14 4JG [IO91NQ, SU89]
G4 YBI S H Aust, 7 Amberley Cl, Wivenhoe, Colchester, CO7 9RB [JO01LU, TM02]
GM4 YBJ J Davies, 2 South Moa, Rendall, Mainland, Orkney, KW17 2PB [IO89LB, HY41]
G4 YBL J W Rees, 17 East Moor Cres, Leeds, LS8 1AD [IO93FU, SE33]
GJ4 YBM A C Alexandre, Merryvale Cottage, La Vallee de St. Pierre, St. Lawrence, Jersey, JE3 1EZ
G4 YBN I T Ansell, 9 Sewell Harris Cl, Harlow, CM20 3HB [JO01BS, TL41]
G4 YBO Details withheld at licensee's request by SSL.[Station located near Bishops Waltham.]
G4 YBP Details withheld at licensee's request by SSL.
G4 YBQ D C Jones, 7 Barnfield Cl, Coulsdon, CR5 1QR [IO91WH, TQ35]
G4 YBS B A Watson Morecambe By AR, 7 Branksome Dr, Morecambe, LA4 5UJ [IO84NB, SD46]
G4 YBT E J Tracey, 100 Booth Cl, Crestwood Park, Kingswinford, DY6 8SP [IO82WL, SO98]
G4 YBU Details withheld at licensee's request by SSL.
GW4 YBV B H Simpson, Eryl y Don, Fford Uchaf, Harlech, Gwynedd, LL46 2SS [IO72WU, SH53]
GU4 YBW P M Wadley, Gironde, Lorier Ln, Vale, Guernsey, GY3 5JG
G4 YBX E Scleparis, 17 Gloster Gdns, Wellesbourne, Warwick, CV35 9TQ [IO92EE, SP25]
G4 YCA P L Massey, 171 Newhall Rd, Chester, CH2 1TB [IO83NF, SJ46]
G4 YCB Details withheld at licensee's request by SSL.[Op: S Riddell.]
G4 YCC Details withheld at licensee's request by SSL.
G4 YCD M S Lowe, Crossley Farm, Winterbourne, Bristol, BS17 1RH [IO81RM, ST68]
G4 YCE L M Ball, 16 Kelston View, Whiteway, Bath, BA2 1NW [IO81TI, ST76]
G4 YCG C S Beeston, 74 Liss Rd, Southsea, PO4 8AS [IO90LT, SZ69]
G4 YCH M P Sharp, Anchor House, Main Rd, Woodham Ferrers, Chelmsford, CM3 8RN [JO01HQ, TL70]
G4 YCJ A R Clift-Jones, Filkins End, Filkins Rd, Langford, Lechlade Glos, GL7 3LW [IO91ER, SP20]
G4 YCK Details withheld at licensee's request by SSL.
G4 YCL J F Sanders, 10 Crawley Ln, Pound Hill, Crawley, RH10 7EE [IO91WC, TQ23]
G4 YCN R Sharman, 12 Church Mead, Keymer, Hassocks, BN6 8BN [IO90WW, TQ31]
GW4 YCO D Gill, 19 Rowling St., Williamstown, Tonypandy, CF40 1QY [IO81GO, ST09]
G4 YCP G A Newman, 2 Grange Rd, Eldwick, Bingley, BD16 3DH [IO93CU, SE13]
G4 YCQ D A Moss, 183A Eastbourne Rd, Darlington, DL1 4ES [IO94FM, NZ31]
GJ4 YCR D H Le Brocq, Autumn Shade, Poplar Cl, St. Saviour, Jersey, JE2 7JP
G4 YCS I C D Carby, 6 Winscar Gr, Clifton Moor, York, YO3 4SZ [IO93KX, SE55]
GW4 YCT I Hughes Carmarthen ARS, 31 Ystrad Dr, Johnstown, Carmarthen, SA31 3PQ [IO71UU, SN31]
GW4 YCU P A H Beckett, 24 Glas y Pant, Whitchurch, Cardiff, CF4 7DB [IO81JM, ST18]
G4 YCV C S Vickery, Santis, 7 Higher Redgate, Tiverton, EX16 6RJ [IO80GV, SS91]
G4 YCW C F Croxford, Bodley Cottage, Parracombe, North Devon, EX31 4PR [IO81BE, SS64]
GM4 YCY E S Munro, Mulben, 56 St. Ternans Rd, Newtonhill, Stonehaven, AB3 3PF [IO87TH, NJ72]
GI4 YCZ J D Rainey, 40 Cranny Ln, Portadown, Craigavon, BT63 5SW [IO64TK, J05]
G4 YDA D A W Akerman, Hillside, Tufthorn Rd, Milkwall, nr Coleford, Glos, GL16 8PY [IO81QS, SO50]
G4 YDB F R Heald, Harewood, 229 Holt Rd, Horsford, Norwich, NR10 3EB [JO02OQ, TG11]
GM4 YDC S K Hunt, 5 Highland Rd, Crieff, PH7 4LE [IO86BJ, NN82]
G4 YDD W P Davidson, 171 Ramsey Rd, St. Ives, Huntingdon, PE17 4TZ [IO92XI, TL37]
G4 YDE E Metcalf, Beech Lee, Vicarage Ln, Ropley, Alresford, SO24 0DU [IO91KB, SU63]
GW4 YDF M E P Thomas, 6 Heathfield, Llanharan, Pontyclun, Mid Glam, CF72 9RU [IO81GM, SS98]
G4 YDG H W D Maude, 85 Watkinson Rd, Halifax, HX2 9DA [IO93BR, SE02]
G4 YDH G A Innes, 4 Althorp Cl, Tuffley, Gloucester, GL4 0XP [IO81UT, SO81]
G4 YDI R D Benbow, 54 Park Lea, Bradley Grange, Huddersfield, HD2 1QH [IO93DQ, SE12]
G4 YDJ K G Landrebe, Rosedene, High St, Hildersham, Cambridge, CB1 6BU [JO02DC, TL54]
G4 YDK Details withheld at licensee's request by SSL.[Op: R W Goghan, 27a Harlington Road, Hillingdon, Middx, UB8 3HX.]
G4 YDM J T Allsopp, 30 Manor Park, Concord, Washington, NE37 2BT [IO94FV, NZ35]
G4 YDN V L Crow, 2 Village Way, Ashford, TW15 2LB [IO91SK, TQ07]
G4 YDQ D W Hannant, 75 Woodcock Cl, Norwich, NR3 3TB [JO02PP, TG21]
G4 YDR D A Reed, Glen Lee, 14 Glenholt Rd, Glenholt, Plymouth, PL6 7JA [IO70WK, SX56]
G4 YDW P Grant, 3 Cragg Wood Cl, Horsforth, Leeds, LS18 4RL [IO93ET, SE23]
GW4 YDX K T Gill, 16 Hafodarthen Rd, Llanhilleth, Abertillery, NP3 2RY [IO81KQ, SO20]
G4 YDY R H Harbord, 349 Reepham Rd, Norwich, NR6 5QJ [JO02PQ, TG21]
G4 YDZ M G Massen, The Old School, Horning Rd, Hoveton St John, Norfolk, NR12 8JH [JO02RQ, TG31]
G4 YEA Details withheld at licensee's request by SSL.
G4 YEB D J Whitton, 61 Greenacre Park South, Gilberdyke, Brough, HU15 2TY [IO93PS, SE82]
GM4 YED R A Alexander, 15 Gardenrose Path, Maybole, KA19 8AG [IO75PI, NS21]
G4 YEE P J Hall, Barnlea, Knapp Ln, Ampfield, Romsey, Hants, SO51 9BT [IO91GA, SU42]
G4 YEF B T Renner M.I.S.M., 95 Reids Piece, Purton, Swindon, SN5 9BA [IO91BO, SU08]
GW4 YEG R Paganuzzi, 32 Avon Cl, Cove, Farnborough, GU14 9LN [IO91OH, SU85]
G4 YEH E J Hill, 13 St. Marys Cres, Yeovil, BA21 5RP [IO80QW, ST51]
G4 YEI S F Masterman, 11 Castellain Rd, Maida Vale, London, W9 1EY [IO91VM, TQ28]
G4 YEJ A A Ayton, 3 Links Ave, Norwich, NR6 5PE [JO02PP, TG21]
G4 YEK S P Clack, 23 Cameron Gr, York, YO2 1LE [IO93KW, SE65]
G4 YEN Details withheld at licensee's request by SSL.[Op: Clive. Station located at Leigh-on-Sea, near Southend-on-Sea, Essex - 65km East of London on North coast of Thames Estuary.]
G4 YEO L J Gillain, 8 Kingston Rd, Portsmouth, PO1 5RZ [IO90LT, SU60]
G4 YEP E G Herwig, 4 Tadburn Rd, Romsey, SO51 5AU [IO90GX, SU32]
GM4 YEQ J G Campbell Gala & Dist ARS, 9 Brunton Park, Bowden, Melrose, TD6 0SZ [IO85PN, NT53]
G4 YES J P Thompson, 3 Newport Mount, Hadingley, Leeds, LS6 3DB [IO93FT, SE23]
G4 YET R E Littlewood, Manor House Farm, Scotton, Gainsborough, Lincs, DN21 3QZ [IO93QL, SK89]
G4 YEV Details withheld at licensee's request by SSL.[Station located in London E13.]
G4 YEW J Gill, 1 Towers Way, Stonegate Rd, Meanwood, Leeds, LS6 4PJ [IO93FU, SE23]
G4 YEX J E Kennedy, 60 Burnway, Albany, Washington, NE37 1QQ [IO94FV, NZ35]
G4 YEY Details withheld at licensee's request by SSL.
G4 YEZ Details withheld at licensee's request by SSL.
G4 YFB S A Coleman, 263 Wykeham Rd, Reading, RG6 1PL [IO91MK, SU77]
G4 YFC T J Hill, 11 Paget Cttgs, Munden Rd, Dane End, Ware, SG12 0NL [IO91XV, TL32]
G4 YFD F B Fish, 36 Limetrees Cl, Sandpiper Ct Two, Port Clarence, Middlesbrough, TS2 1SL [IO94JO, NZ42]
G4 YFE P M Hounslow, 53 Leighton Ct, Copperdale Cl, Earley, Reading, RG6 5SG [IO91MK, SU77]
G4 YFF E G Reynolds, 4 Underwood Cl, Stafford, ST16 1TB [IO82WT, SJ92]
G4 YFH G E Overend, 1 Tatton Dr, Sandbach, CW11 1DR [IO83TD, SJ76]
G4 YFI R S Larter, 12 Ashby Rd, Hinckley, Leics, LE10 1SL [IO92HN, SP49]
G4 YFJ N M P Von Fircks, 48 Appledown Cl, Alresford, SO24 9ND [IO91KB, SU53]
G4 YFK M R Aitchison, 21 St. Pauls Rd West, Dorking, RH4 2HT [IO91UF, TQ14]
G4 YFN C L Crane, West Barn, The Old Farmhouse, Winnersh, Wokingham, RG41 5JJ [IO91NK, SU77]
G4 YFO G Wardy, Marbury, 7 Yew Tree Rd, Crewe, Ches, CW2 8BN [IO83SB, SJ65]
G4 YFP Details withheld at licensee's request by SSL.[Operator: A Reed. Station located Hythe, Kent.]
G4 YFQ Details withheld at licensee's request by SSL.
G4 YFS S S Van Praag, 106 Leyburn Rd, Darlington, DL1 2ES [IO94FN, NZ21]
G4 YFT B Page, 7 Muirfield Cres, Oakham Green, Tividale, Warley, B69 1PW [IO82XL, SO98]
G4 YFU M C Parker, 85 Elston Rd, Aldershot, GU12 4HZ [IO91PF, SU84]
G4 YFV N B Banham, Hawthorns, Glebe Rd, Gissing, Diss, IP22 3UY [JO02OK, TM18]
G4 YFX P Johnson, 2A Lister Rd, Wellingborough, NN8 4EN [IO92PH, SP86]
G4 YFY P G Drawmer, Post Office Cottage, Clifton, Deddington, Oxon, OX15 0PD [IO91IX, SP43]
G4 YFZ B A Jones, Jetza, Rose Ave, Stretton, Burton on Trent, Staffs, DE13 0DQ [IO92EU, SK22]
G4 YGA S J Howarth, 163 Moore Ave, Sprowston, Norwich, NR6 7LQ [JO02PP, TG21]
G4 YGD G W Aldred, 212 Reepham Rd, Norwich, NR6 5SW [JO02PQ, TG21]
G4 YGE C G O Oakley, 9 Fitzroy Rd, Landport Est, Lewes, BN7 2UB [IO90XV, TQ41]
G4 YGF M Errington, 123 Essex Dr, Washington, NE37 2NU [IO94FW, NZ35]
G4 YGH D B Hart, 30 Dartford Ave, Edmonton, London, N9 8HD [IO91XP, TQ39]
G4 YGI Details withheld at licensee's request by SSL.
G4 YGJ L M Rozentals, 21 Mallard Cl, Eastbourne, BN22 9NA [JO00DT, TQ60]
G4 YGL G H Reece, 34 Priestley Gdns, Romford, RM6 4SL [JO01BN, TQ48]
G4 YGM P C Reynolds, 34 Thurlstone Rd, Ruislip, HA4 0BT [IO91TN, TQ18]
G4 YGN F R Albers, 14 Hartwith Green, Summerbridge, Harrogate, HG3 4HX [IO94DB, SE26]
G4 YGO W Young, Mickleton, David Terr, Bowburn, Durham, DH6 5EF [IO94FR, NZ33]
G4 YGP J M Vinson, Akhurst Cottage, Shepherds Hill, Selling, Faversham, ME13 9RS [JO01KG, TR05]
G4 YGQ M J Tate, Stone Cottage, The St., Halvergate, Norwich, NR13 3PL [JO02SO, TG40]
G4 YGS G A Wells, 351 Holyhead Rd, Coventry, CV5 8LD [IO92FJ, SP37]
G4 YGT W T Waldron, 16 Barke St., Highley, Bridgnorth, WV16 6LQ [IO82TK, SO78]
G4 YGU J G Hetherington, 44 Brookside Rd, Istead Rise, Gravesend, DA13 9JJ [JO01EJ, TQ67]
G4 YGV F/Lt R K Barnes, 7 Maycroft Cl, Ipswich, IP1 6RG [JO02NC, TM14]
G4 YGX M J Holmes, 50 Harvington Rd, Weoley Castle, Birmingham, B29 5EL [IO92AK, SP08]
G4 YGY R J Fawke, 25 Derwent Dr, Mitton, Tewkesbury, GL20 8BA [IO82WA, SO93]
G4 YGZ K C Ghillyer, 69 Hillside Ave, Mutley, Plymouth, PL4 6PS [IO70WJ, SX45]
GM4 YHA Col E P Grier, Top Croft, Banff, Scotland, AB4 3RB [IO87TH, NJ72]
G4 YHB G D Sanders, 10 Spring Cttgs, Horsham Rd, Holmwood, Dorking, RH5 4LU [IO91UE, TQ14]
G4 YHC R Sato, 3-16-7 Minamidai, Sagamihara - Shi, Kanagawa, 228 Japan, X X
G4 YHE Details withheld at licensee's request by SSL.
G4 YHF Details withheld at licensee's request by SSL.
G4 YHG J Hubner, 30 Orchard Rise, Olveston, Bristol, BS12 3DZ [IO81RN, ST68]
G4 YHN A U Gehammar, The Lodge, 119 Ashdon Rd, Saffron Walden, CB10 2AJ [JO02DA, TL53]
GM4 YHO J W Smith, 1110 Fourth St, Radford, Va, 24141 - 1314, United States, X X
G4 YHP C H Jobling, Joycliff, 20A Poplar Rd, Healing, Grimsby, DN37 7RD [IO93VN, TA20]
G4 YHQ D A Webb, The Laurels, Drove Rd, Chilbolton, Stockbridge, SO20 6AD [IO91GD, SU33]
G4 YHR Details withheld at licensee's request by SSL.
GM4 YHS S W Grant, 16 Netherton Pl, Westmuir, Kirriemuir, DD8 5LD [IO86LP, NO35]
G4 YHV Details withheld at licensee's request by SSL.
G4 YHW Details withheld at licensee's request by SSL.
G4 YHY Details withheld at licensee's request by SSL.
G4 YHZ B W Mitchell, 66 Hungerford Dr, Maidenhead, SL6 7UU [IO91PM, SU88]
G4 YIA B Robinson, 196 Bristol Ave, Leyland, Farington, Preston, PR5 2QZ [IO83PQ, SD52]
G4 YIB Details withheld at licensee's request by SSL.
G4 YIC A J Pitt, 16 Highlands Dr, North Nibley, Dursley, GL11 6DX [IO81TP, ST79]
GW4 YID M James, 14 Carmel Rd, Winch Wen, Swansea, SA1 7JY [IO81BP, SS69]
G4 YIE C J Kelley, 5 Russell Way, Wootton, Bedford, MK43 9EX [IO92RC, TL04]
G4 YIF R V Lumbard, Old Forge, Rusper Rd, Ifield, Crawley, RH11 0LQ [IO91VC, TQ23]
G4 YIG J C Cluley, 24 Avon Green, Wyre Piddle, Pershore, WR10 2JE [IO82XC, SO94]
G4 YIH W D Maycey, 21 Brook Dr, Wickford, SS12 9EQ [JO01GO, TQ79]
G4 YIJ F W Robinson, 97 Chestnut Dr, Castle Bromwich, Birmingham, B36 9BH [IO92CM, SP18]
G4 YIK G F Whetstone, 8 Thorne Ave, Wardley, Gateshead, NE10 8TE [IO94FW, NZ26]
G4 YIM J G Cameron, 20 Fellmead, East Peckham, Tonbridge, TN12 5EQ [JO01EF, TQ64]
G4 YIQ C Allison, 13 Newick Ave, Pallister Park Est, Middlesbrough, TS3 8QJ [IO94JN, NZ51]
G4 YIR J Charles, 40 Gurdon Rd, Colchester, CO2 7PP [JO01KU, TL92]
G4 YIS J C OFarrell, Shannon, Gunville Rd, Winterslow, Salisbury, SP5 1PP [IO91EC, SU23]
G4 YIT I P Toon, 56 Cockbank, Turves, Whittlesey, Cambs, PE7 2HN [IO92XM, TL39]
G4 YIU H V Gammon, Howards Cottage, 33 Howards Wood Dr, Gerrards Cross, SL9 7HR [IO91RN, TQ08]
G4 YIV B R Whyle, Highways, High St, Pembridge, Leominster, Herefordshire, HR6 9DT [IO82NF, SO35]
G4 YIX D R Godwin, 65 Lilliesfield Ave, Barnwood, Gloucester, GL3 3AH [IO81VU, SO81]
G4 YIZ A F Mansfield, 42 Cavendish Way, Mickleover, Derby, DE3 5BL [IO92FV, SK33]
G4 YJA B M Lambert, Firbank, East St., Rusper, Horsham, RH12 4RE [IO91UC, TQ23]
G4 YJB R J Briggs, 32 Waterside, Evesham, WR11 6BU [IO92AC, SP04]
G4 YJC G A Thornton, 115 High St., Studley, B80 7HN [IO92BG, SP06]
G4 YJD J A Donin, 3 High St., Oving, Chichester, PO20 6DD [IO90PU, SU90]
G4 YJF L H Ferris, 29 Compton Ct, Chidham Cl, Havant, PO9 1DT [IO90MU, SU70]
G4 YJH J Hockey, Fieldview, The Pound, Ashwick, Oakhill, Somerset, BA3 5BB [IO81RF, ST64]
GW4 YJI Details withheld at licensee's request by SSL.
G4 YJJ C Berry, 12 Moorside Ave, Ainsworth, Bolton, BL2 5RP [IO83TO, SD71]
G4 YJK P Duke, 4 Doggets Ln, Fulbourn, Cambridge, CB1 5BT [JO02CE, TL55]
GM4 YJL R J Turner, Mondraiks, Upper Yetts, Dollar, Clackmannanshire, FK14 7JU [IO86EE, NO00]
G4 YJM M S Leonard, 7 Moorside Parade, Drighlington, Bradford, BD11 1HR [IO93ER, SE22]
G4 YJN M J Griggs, Tudor Rose Cottage, Malting Green, Layer de La Haye, Colchester, CO2 0JE [JO01KU, TL92]
G4 YJO Details withheld at licensee's request by SSL.
G4 YJP S K Stanton, 7 Wentworth Way, Links View, Northampton, NN2 7LW [IO92NG, SP76]
G4 YJQ D C Cutts, 36 Lodge Rd, Little Oakley, Harwich, CO12 5EE [JO01OW, TM22]
G4 YJS B Parsons, 33 Baguley Ave, Halebank, Widnes, WA8 8UY [IO83OI, SJ48]
GW4 YJT C Roberts, 20 Bridge St., Shotton, Deeside, CH5 1DU [IO83LF, SJ36]
G4 YJU C A Mount, 6 Almond Cl, Countesthorpe, Leicester, LE8 5TG [IO92KN, SP59]
G4 YJV C W Abbott, 21 Hillview Rd, Irby, Wirral, L61 4XH [IO83KI, SJ28]
G4 YJW G Grundy, 47 Northiam Rd, Eastbourne, BN20 8LP [JO00DS, TV59]
G4 YJX J Boyes, 11 Heath Rd, Hammer, Haslemere, GU27 3QN [IO91PB, SU83]
G4 YJY W F Marsden, 33 Kilton Cres, Worksop, S81 0AX [IO93KH, SK57]
G4 YJZ R V Atkinson, 20 Cothelstone Cl, Durleigh, Bridgwater, TA6 7JH [IO81LC, ST23]
G4 YK B M Morrissey, 50 Fingringhoe Rd, Langenhoe, Colchester, CO5 7LB [JO01LU, TM01]
G4 YKA B Shaw, 15 Bayswood Ave, Boston, PE21 7RT [IO92XX, TF34]
G4 YKB H W Ramsden, 23 Nandywell, Little Lever, Bolton, BL3 1JU [IO83TN, SD70]
G4 YKC A R Barker, 90 Pilsdon Dr, Canford Heath, Poole, BH17 9HS [IO90AS, SZ09]
G4 YKF T S Clarke, 16 Brende Gdns, West Molesey, KT8 2PW [IO91TJ, TQ16]
G4 YKG J P Pether, 2 Belle Ct, Bellevue Rd, Totterdown, Bristol, BS4 2BG [IO81RK, ST57]
G4 YKH C B Littler, 11 Richards Rd, Stoke Dabernon, Cobham, KT11 2SX [IO91TH, TQ16]
G4 YKJ Details withheld at licensee's request by SSL.
G4 YKK A K Babbage, 248 Molesey Ave, West Molesey, KT8 2ET [IO91TJ, TQ16]
GW4 YKM K Martich, 25 Pentwyn Isaf, Energlyn, Caerphilly, CF8 2NR [IO81HN, ST08]
G4 YKN J Crowhurst, 1 Rugby Rd, Poole, BH17 7HJ [IO90AR, SZ09]
G4 YKO G R Wilkinson, 8 Laughton Ave, Scarborough, YO12 5DB [IO94SG, TA08]
G4 YKQ P B Welford, 11 Ridgeside, Bledlow Ridge, High Wycombe, HP14 4JN [IO91NQ, SU89]
G4 YKR K Ramsdale, 770 Warrington Rd, Birchwood, Risley, Warrington, WA3 6AQ [IO83RK, SJ69]
G4 YKS Details withheld at licensee's request by SSL.
G4 YKT D H Davies, 1 Tees Rd, Guisborough, TS14 8AP [IO94LM, NZ61]
G4 YKU Details withheld at licensee's request by SSL.
G4 YKV K Wood, 29 Coronation Dr, Haydock, St. Helens, WA11 0RB [IO83QL, SJ59]
GW4 YKW A T Hopkins, 30 Wavell Dr, Newport, NP9 6QN [IO81LO, ST39]
G4 YKX M E Blakeley, Richmond, Old Wood, Shrewsbury, Shropshire, SY4 3AX [IO82OS, SJ42]
G4 YKZ R G Harvey, Richlyn House, Cedar Rd, Hethersett, Norwich, NR9 3JY [JO02OO, TG10]
G4 YLA Details withheld at licensee's request by SSL.
G4 YLB Details withheld at licensee's request by SSL.[Op: J A Welsh. Station located near Darwen.]
G4 YLC M P Godfrey, 23 York Ave, Hayes, UB3 2TN [IO91SM, TQ08]
G4 YLD D Watson, 6 Cross Henley Rd, Bramley, Leeds, LS13 3AN [IO93ET, SE23]
GW4 YLF J T Dixon, 73 Brynhyfryd, Ferndale, CF43 4HT [IO81GP, SS99]
G4 YLI R W Barnes, Cottage, Hutton Row, Skelton, Penrith, CA11 9TR [IO84NR, NY43]
G4 YLJ F H Normington, 116 The Cray, Milnrow, Rochdale, OL16 4EA [IO83WO, SD91]
G4 YLK D A Adams, 21 Lowton Rd, Sale, M33 4LD [IO83TJ, SJ79]
G4 YLM M W Cockroft, 12 Lickless Gdns, Woodside, Horsforth, Leeds, LS18 5QU [IO93EU, SE23]
GM4 YLN C Grierson, 174 Baberton Mains, Dr, Edinburgh, EH14 3DZ [IO85IV, NT16]
G4 YLO H J Timbrell, Crossing Cottage, Lamyatt, Shepton Mallet, Somerset, BA4 6NG [IO81RC, ST63]
GJ4 YLP C P Landor, Lauge, Rue Des Raisies, St Martin, Jersey Cl, Jersey, JE9 9LA
G4 YLQ R May, 1 Green Ln, Hadfield, Hyde, SK14 8DT [IO93AK, SK09]
G4 YLS J E Spittle, 9 Woden Rd West, Wednesbury, WS10 7SF [IO82XN, SO99]
G4 YLT J F Smith, 31 Macaulay Cl, Lunsford Park, Larkfield, Aylesford, ME20 6TZ [JO01FH, TQ75]
GM4 YLU W J Wilson, 34 Oxgangs Green, Edinburgh, EH13 9JS [IO85JV, NT26]
G4 YLV Details withheld at licensee's request by SSL.[Station located near Liversedge.]
G4 YLW P G L Yaxley, 103 Portland Rd, Hove, BN3 5DP [IO90VT, TQ20]
G4 YLX P V Crotty, 1 New Bungalows, Crewe Green, Crew Green, Shrewsbury, SY5 9AT [IO82MR, SJ31]
GM4 YLY D B Reid, 8 Wellesley Rd, Buckhaven, Leven, KY8 1HS [IO86LE, NT39]
G4 YLZ V K Wyles, Brockhall, 51 Badgers Cl, Bugbrooke, Northampton, NN7 3BA [IO92ME, SP75]
GM4 YMA T Dunlop, 38 Leslie Ave, Newton Mearns, Glasgow, G77 6JE [IO75US, NS55]
G4 YMB M Brass, 11 Lealholm Way, Guisborough, TS14 8LN [IO94LM, NZ61]
G4 YMC J R McCallum, 31 Meadow Rd, Lemington, Newcastle upon Tyne, NE15 7LP [IO94DX, NZ16]
GM4 YMD W S Macdiarmid, Willis, 167 Glasgow Rd, Whins of Milton, Stirling, FK7 0LH [IO86AC, NS79]
G4 YME M Elsey, 27 North Fawley, Nr.Wantage, Oxon, OX12 9NJ [IO91GM, SU38]
G4 YMG P J Chorley, 9 Conference Cl, Imberwood, Warminster, BA12 8TF [IO81VE, ST84]
G4 YMH M W Harney, 14 Cob Cl, Crawley Down, Crawley, RH10 4EX [IO91XC, TQ33]
GW4 YMJ W A Fitzgerald, 23 Clos yr Eos, South Cornelly, Bridgend, CF33 4RJ [IO81DM, SS88]
G4 YMK Details withheld at licensee's request by SSL.[Op: J H Orford. Station located near Kings Lynn.]
G4 YML E I Jones, 8 Wellfield Gr, Penistone, Sheffield, S30 6GP [IO93GJ, SK38]
GM4 YMM E M C Dons, 37 Ashley Dr, Edinburgh, EH11 1RP [IO85JW, NT27]
G4 YMN D Hobson, 13 Elizabeth Ave, Stalybridge, SK15 1DJ [IO83XL, SJ99]
G4 YMO R Cowgill, 27 Park Rd, Barnoldswick, Colne, BB8 5BQ [IO83VV, SD84]

G4 YMP T D Boreham, 22 Fairlight Ave, Hastings, TN35 5HS [JO00HU, TQ81]
G4 YMQ A J Kimm, 9 Tennis St., Burnley, BB10 3AG [IO83VT, SD83]
G4 YMS J R Russell, Penhowe, 7 Towthorpe Rd, Haxby, York, YO3 3LY [IO94LA, SE65]
G4 YMT M E Taylor, Walnut Cottage, One Manor Ct, Wagstaff Cl, Harbury, Warks, CV33 9ND [IO92GF, SP35]
G4 YMU A G Wallis, 10 Middlewood Rd, Lanchester, Durham, DH7 0HL [IO94DT, NZ14]
GU4 YMV Details withheld at licensee's request by SSL.
G4 YMW J A Brookes, 2 Havana Cl, Beswick, Manchester, M11 3JE [IO83VL, SJ89]
GJ4 YMX D J Warncken, 10 St. Lukes Cres, Green Rd, St. Clement, Jersey, JE2 6QH
GD4 YMY R I Nicol, 9 Hillberry Lakes, Governors Hill, Douglas, IM2 7BD
G4 YMZ J P May, 58 Blackthorn Croft, Clayton-le-Woods, Chorley, PR6 7TZ [IO83QQ, SD52]
GM4 YNA Details withheld at licensee's request by SSL.
G4 YNC C R Philpott, 4 Footways, Wootton Bridge, Ryde, PO33 4NQ [IO90JR, SZ59]
G4 YND Revd D J Evans, The Rectory, Rectory Ln, Somersham, Huntingdon, Cambs, PE17 3EL [IO92XJ, TL37]
G4 YNE D R Evans, 25 Offerton Ave, Derby, DE23 8DU [IO92GV, SK33]
G4 YNG M P Garlick, 37 Weekley Glebe Rd, Kettering, NN16 9NR [IO92PJ, SP88]
G4 YNH S G Payas, 36 Tintern Cl, Popley, Basingstoke, RG24 9HE [IO91KG, SU65]
G4 YNI H Beckman, 16 Wilton Rd, Crumpsall, Manchester, M8 4WQ [IO83VM, SD80]
G4 YNJ D P Cook, 211 Brodie Ave, Allerton, Liverpool, L19 7NB [IO83NI, SJ48]
G4 YNK N Fletcher, 11 Parkgate Dr, Astley Bridge, Bolton, BL1 8SD [IO83SO, SD71]
G4 YNL R M Banks The A Team Contest Group, 33 New Rd, Shuttington, Tamworth, B79 0DT [IO92EP, SK20]
G4 YNM B C Spencer, Enterprise House, 33 New King St., Bath, BA1 2BL [IO81TJ, ST76]
G4 YNN D H Smith, 90 Normanton Rd, Derby, DE1 2GP [IO92GV, SK33]
G4 YNO P W Barnes, 69 Southborne, Overcliff Dr, Southborne, Dorset, BH6 3NN [IO90CR, SZ19]
GW4 YNP D J A Collins, 54 Darwin Dr, Newport, NP9 6FR [IO81LO, ST39]
GW4 YNR D McIntyre, 11 Severn Ave, Tutshill, Chepstow, NP6 7EF [IO81QP, ST59]
G4 YNS S Shakeshaft, Willow Bank, Apsley Gr, Bowdon, Altrincham, WA14 3AH [IO83TI, SJ78]
G4 YNT M E Yallop, 4 Cottle Mead, Corsham, SN13 9UP [IO81VK, ST86]
G4 YNU J Scriven, 1 Holgate Rd, Pontefract, WF8 4ND [IO93IQ, SE42]
G4 YNV H A Snaden, 92 Avon Way, Portishead, Bristol, BS20 9LU [IO81OL, ST47]
G4 YNW A P W St Aubyn, Kingswood Cottage, Vinehall Rd, Mountfield, Robertsbridge, TN32 5JN [JO00FW, TQ72]
G4 YNX B W Bassford, 12 Little Brum, Grendon, Atherstone, CV9 2ET [IO92EO, SP29]
G4 YNY J Millward, 27 Ash Gr, Walsall, West Midlands, WS7 8TG [IO92AQ, SK00]
G4 YNZ J W Redfearn, 64 Park House, Bridge Rd, St. Austell, PL25 5HD [IO70OI, SX05]
G4 YOA F Harvey, 137 Epping New Rd, Buckhurst Hill, IG9 5TZ [JO01AP, TQ49]
G4 YOB Details withheld at licensee's request by SSL.[Station located in London SE13.]
G4 YOC D J Gully, 46 Shellards Rd, Longwell Green, Bristol, BS15 6DU [IO81SK, ST67]
G4 YOD A K Steggles, 22 Catherine Cl, Chafford Hundred, Grays, Essex, RM16 6QH [JO01DL, TQ67]
G4 YOF G K Coomber, 13 Sinnott Rd, Walthamstow, London, E17 5PL [IO91XO, TQ39]
G4 YOH W R Paternoster, 18 Winslow Gr, Chingford, London, E4 6EU [JO01AO, TQ39]
G4 YOI M Bulbeck, Kalena Cottage, Higher Tremarcoombe, Liskeard, Cornwall, PL14 5HP [IO70SL, SX26]
GD4 YON Dr T M Adair, Ballavartyn House, Ballavartyn Rd, Santon, Douglas, IM4 1HT
G4 YOP E Popple, 14 Rudbeck Cres, Harrogate, HG2 7AQ [IO93GX, SE35]
G4 YOR R Greenwood, 26 Littlefield Walk, Bradford, BD6 1UU [IO93CS, SE12]
G4 YOS D E Corallini, 35 The Green, Kings Hill, Ware, SG12 0QW [IO91XT, TL31]
G4 YOT L R Zalicks, 3 Retford Path, Harold Hill, Romford, RM3 9NL [JO01CO, TQ59]
G4 YOU Details withheld at licensee's request by SSL.
G4 YOV J Metcalfe, 14 Ludham Gr, The Park, Stockton on Tees, TS19 0XH [IO94HN, NZ42]
G4 YOW Details withheld at licensee's request by SSL.
GU4 YOX R W Beebe, San Grato, La Houguette, Castel, Guernsey, GY5 7DZ
G4 YOY R P Sumner, 64 Lock Ln, Long Eaton, Nottingham, NG10 3DD [IO92IV, SK43]
G4 YOZ B J Starkey, 87 Northd. Ave, Nuneaton, CV10 8EP [IO92GM, SP39]
G4 YPA J H Aisher, 44 Cranleigh Rd, Portchester, Fareham, PO16 9DN [IO90KU, SU60]
G4 YPC P Croucher, 66 Loop Rd, Westfield, Woking, GU22 9BQ [IO91RH, TQ05]
G4 YPD F J Thorne, 1 Burwen Dr, Orrell Park, Liverpool, L9 8DE [IO83ML, SJ39]
G4 YPE N F Hanney, 62 Avonfield Ave, Bradford on Avon, BA15 1JF [IO81VI, ST86]
G4 YPF W F Taylor, 3 Westcroft, Leominster, HR6 8HE [IO82PF, SO45]
G4 YPG R G M Mason, 5 Countryside Farm, Church Ln, Upper Beeding, Steyning, West Sussex, BN44 3HP [IO90UV, TQ11]
G4 YPH D J Rothwell, 37 Eamont Ave, Crossens, Southport, PR9 9YX [IO83MQ, SD32]
G4 YPI A I Maires, 26 Dunmow Rd, Thelwall, Warrington, WA4 2HQ [IO83RJ, SJ68]
G4 YPK P G Knowles, 6 Dorchester Cl, Basingstoke, RG23 8EX [IO91KG, SU65]
GM4 YPL R M Thompson, Sharphil, Edinburgh Rd, Linlithgow, EH49 6QT [IO85FX, NT07]
G4 YPN D T E Howroyd, 78 New Zealand Way, Rainham, RM13 8JD [JO01CM, TQ58]
GW4 YPO G H Watkins, 137 Powys Ave, Townhill, Swansea, SA1 6PJ [IO81AP, SS69]
G4 YPP Details withheld at licensee's request by SSL.
G4 YPQ K J Simpson, 62 Myott Ave, Newcastle, ST5 2ER [IO83VA, SJ84]
GI4 YPR W D Swail, 41 Church St., Portaferry, Newtownards, BT22 1LT [IO74FJ, J55]
G4 YPS A H Bradley, 11 James St., Failsworth, Manchester, M35 9PY [IO83WM, SD90]
G4 YPT M E Frear Hazelrigg ARC, 18 Boulsworth Rd, North Shields, NE29 9EN [IO95GA, NZ37]
G4 YPU P Farley, Exeter House, 11 East St., Ashburton, Newton Abbot, TQ13 7AD [IO80CM, SX76]
G4 YPV D Ramsden, 76 Brigg Ln, Camblesforth, Selby, YO8 8HD [IO93LR, SE62]
G4 YPW C R J Wells, Flat, 5 Churchgate, London Rd, Westcliff on Sea, SS0 9HS [JO01IN, TQ88]
G4 YPX Details withheld at licensee's request by SSL.
G4 YPY Details withheld at licensee's request by SSL.
G4 YQA M W Lawson, Hill Top Cottage, Pasture Rd, Embsay, Skipton, BD23 6PN [IO93AX, SE05]
G4 YQC P J Whiting, 77 Melford Way, Felixstowe, IP11 8UH [JO01PX, TM23]
G4 YQD T M Mayfield, 14 Wheatley Grange, Coleshill, Birmingham, B46 3LZ [IO92DL, SP18]
G4 YQE S Graham, 59 Belvedere Rd, Birmingham, B24 9RW [IO92CM, SP19]
G4 YQF A T E Pacewicz, Fir Tree Cottage, Herberts Way, Oldcroft, Lydney, GL15 4NS [IO81RS, SO60]
G4 YQG B A Hodgetts, 15 Wiltons, Wrington, Bristol, BS18 7LS [IO81OI, ST46]
G4 YQH J A Frampton, 161 Longmead Ave, Bristol, BS7 8QG [IO81QL, ST57]
G4 YQI Details withheld at licensee's request by SSL.
G4 YQJ F J H Collie, Cerf, Island, Seychelles, Indian Ocean
G4 YQK K R Taylor, 22 Anderson, Dunholme, Lincoln, LN2 3SR [IO93SI, TF08]
G4 YQL R N K Silvey, 20 Wixfield Park, Great Bricett, Ipswich, IP7 7DW [JO02LC, TM05]
G4 YQM A Dickens, The Poplars, Branxton, Cornhill on Tweed, Northd., TD12 4SL [IO85WP, NT83]
G4 YQN C N Clark-Booth, 26 Pillar Ave, Brixham, TQ5 8LB [IO80FJ, SX95]
G4 YQO N W Hutchings, 2 Burn Cl, Verwood, BH31 6DN [IO90BV, SU00]
G4 YQP M J Simmens, 1 Meaver Cottage, Mullion, Helston, Cornwall, TR12 7DN [IO70JA, SW61]
G4 YQQ A V Booth, 656 Southmead Rd, Filton Park, Filton, Bristol, BS12 7RD [IO81RM, ST57]
G4 YQR B J Collins, 144 Longmead Ave, Horfield, Bristol, BS7 8QQ [IO81QL, ST57]
G4 YQS T B K White, Rosewall Bungalow, Towednack Rd, St. Ives, TR26 3AL [IO70FE, SW43]
G4 YQT E F Morgan, 26 Ok Mobile Home Park, Lyndhurst Rd, Christchurch, BH23 4SE [IO90DR, SZ19]
G4 YQU Details withheld at licensee's request by SSL.
G4 YQV P J Rose, 15 Lynchets Rd, Amesbury, Salisbury, SP4 7HZ [IO91CE, SU14]
G4 YQW K Lawton, 52 Gamble Ln, Leeds, LS12 5LP [IO93ET, SE23]
G4 YQY Details withheld at licensee's request by SSL.
G4 YQZ F L Neufville, 25 Pullar Ct, Pullar Cl, Bishops Cleeve, Cheltenham, GL52 4RW [IO81XW, SO92]
G4 YRA J W C Hills, 27 Wellington Rd, Denton, Newhaven, BN9 0RD [JO00AT, TQ40]
G4 YRB Details withheld at licensee's request by SSL.[Op: J M Learoyd, Rose Cottage, Bradford, West Riding, Yorks]
G4 YRC V Long York Radio Club, 19 Chapel Walk, Riccall, York, YO4 6NU [IO93LT, SE63]
G4 YRD D Critchlow Doncaster & District Raynet So, 7 Willow Rd, Armthorpe, Doncaster, DN3 3HD [IO93LM, SE60]
GM4 YRE E C Marcus, 11 Parkview, Fettercairn, Fettercairn, Laurencekirk, AB30 1XZ [IO86RU, NO67]
G4 YRF K Amos, 1 Byron Cl, Caldecote, Biggleswade, SG18 9DF [IO92UC, TL14]
G4 YRH H Turner, Riding Hill House, 3 Riding Hill, Shelf, Halifax, HX3 7TS [IO93CS, SE12]
GM4 YRI J Hedtmann, Gartenstr 16, D-15831 Mahlow, Germany
G4 YRJ Details withheld at licensee's request by SSL.
G4 YRL A C Jeffery, Trevounder, Tresowas, Ashton, Helston, TR13 9SY [IO70HC, SW52]
G4 YRM R F Maynard, Clamard, 7 Phillipps Ave, Exmouth, EX8 3HY [IO80HP, SY08]
G4 YRN L J Carpenter Rickmansworth, Corps, 87 Kings Ave, Watford, WD1 7SB [IO91TP, TQ19]
GM4 YRO W P Patterson, 28 Maxtone Terr, Gilmerton, Crieff, PH7 3ND [IO86CJ, NN82]
GI4 YRP T Hutchinson, 47 Ballylough Rd, Donaghcloney, Craigavon, BT66 7PQ [IO64UJ, J15]
G4 YRR R T Rushton, 53 Crossfield Ave, Blythe Bridge, Stoke on Trent, ST11 9PL [IO82XX, SJ94]
G4 YRS Details withheld at licensee's request by SSL.[WAB: NZ10. Op: G A Vallely. QSL via RSGB Bureau.]
G4 YRT G I Cogger, 88 Gardner Rd, Portslade, Brighton, BN41 1PL [IO90UU, TQ20]
G4 YRU Details withheld at licensee's request by SSL.
G4 YRV M C White, 34 Pains Way, Amesbury, Salisbury, SP4 7RG [IO91CE, SU14]
G4 YRX J G Hilliard, 68 Lodge Rd, Portswood, Southampton, SO14 6RG [IO90HW, SU41]
G4 YRY M A Holloway, Barn Cottage, Hengistbury Head, Bournemouth, BH6 4EW [IO90CR, SZ19]
G4 YRZ R Denton, 399 Gateford Rd, Worksop, S81 7BN [IO93KH, SK58]
G4 YSB J R Leary, 18 Chestnut Ave, Andover, SP10 2HE [IO91GE, SU34]
G4 YSE G E Ring, 31 Studland Park, Westbury, BA13 3HQ [IO81VG, ST85]
G4 YSF J E W Stimpson, 12 Fairhaven Ct, Pittville Circus Roa, Cheltenham, GL52 2QR [IO81XV, SO92]
G4 YSG A Cooper, 85 Mansfield Rd, Aston, Sheffield, S31 0BR [IO93GJ, SK38]
G4 YSH C O J Bowers, Cornbury, Seymour Plain, Marlow, SL7 3BZ [IO91OO, SU88]
G4 YSI J R Rayner, Kalaw, Faris Ln, Woodham, Addlestone, KT15 3DN [IO91RI, TQ06]
G4 YSJ P J Rowland, 32 Norfolk Rd, Barnet, EN5 5LU [IO91VP, TQ29]
G4 YSL A J Boot, 2 Trent Villas, Farndon Rd, Farndon, Newark, NG24 4SL [IO93OB, SK75]
G4 YSM K A Wilkinson, 9 Kingsmead Dr, Branton, Doncaster, DN3 3LY [IO93LM, SE60]
G4 YSN I A Brown, 1 Yew Tree Villas, Preston Rd, Charnock Richard, Chorley, PR7 5LF [IO83PP, SD51]
G4 YSO M Nixon, 103 Patrick St., Grimsby, DN32 9PQ [IO93XN, TA20]
G4 YSP K J Metcalf, 34 Framland Dr, Melton Mowbray, LE13 1HY [IO92NS, SK72]
G4 YSQ T J Rogers, 5 Severalls Farm, Shillingford, Oxon, OX10 8LH [IO91KO, SU59]
G4 YSR Dr W R Sadler, Bowman's Lodge, Longashton Rd, Longashton, Bristol, BS18 9LD [IO81QK, ST57]
G4 YSS J D Earnshaw, Dunelm, Ayton Rd, Irton, Scarborough, YO12 4RQ [IO94SF, TA08][Op: John Earnshaw. Tel: (01723) 863137. WAB: 7664. N Yorks. Member of RSGB and Scarborough Special Events Group G0OOO. Interests outdoor HF/P and /M.]
G4 YSU D Hodkinson, Ferndale, Liverpool Rd, Much Hoole, Preston, PR4 4RJ [IO83OQ, SD42]
GW4 YSV A Copestake, 19 Glan-Pwll, Nefyn, Pwllheli, Gynedd, LL53 6EH [IO72RW, SH34]
G4 YSW M McLoughlin, 41 Croft Holm, Moreton in Marsh, GL56 0JH [IO91DX, SP23]
G4 YSZ R E Painton, 17 Brookside, Pill, Bristol, BS20 0JX [IO81PL, ST57]
G4 YTA M J Owen, Hamden, 3 Canford View Dr, Wimborne, BH21 2UW [IO90AT, SU00]
G4 YTB D H Slade, 67 Meadow Ln, Moulton, Northwich, CW9 8QQ [IO83RF, SJ66]
G4 YTD T Booth, Langdale, 7 Cherry Dr, Nafferton, Driffield, YO25 0LY [IO94TA, TA05]
G4 YTF P V Godber, 3 Chalvington Cl, Evington, Leicester, LE5 6XT [IO92LO, SK60]
G4 YTG A W Gilbey, 83 Chignal Rd, Chelmsford, CM1 2JA [JO01FR, TL60]
G4 YTH T J Handford, 20 Minehead Rd, Knowle, Bristol, BS4 1BN [IO81RK, ST57]
G4 YTJ J M Pagett, 26 Rednal Hill Ln, Rubery, Rednal, Birmingham, B45 9LR [IO82XJ, SO97]
G4 YTK S W Hopley, 35 Norton Grange, Norton Canes, Cannock, WS11 3QZ [IO92AQ, SK00]
G4 YTL D Hilton-Jones, Home Farm, Lillingstone Lovell, Buckingham, Bucks, MK18 5BJ [IO92MB, SP74]
G4 YTM I Pettinger, 266 West St., Hoyland, Barnsley, S74 9EQ [IO93GL, SE30]
G4 YTN L G Thomas, 31 Claude Ave, Oldfield Park, Bath, BA2 1AE [IO81TJ, ST76]
G4 YTO M J Yeomans, 6 Badsey Cl, Northfield, Birmingham, B31 2EJ [IO92AJ, SP07]
G4 YTP W D Montagu, Garlands Cottage, Beards Yard, Bow St., Langport, TA10 9PS [IO81OA, ST42]
G4 YTT M W B Curran, 15 Willow Ln, Goostrey, Crewe, CW4 8PP [IO83TF, SJ77]
G4 YTU K J Maskell, 19 Arethusa Rd, Rochester, ME1 2TZ [JO01GI, TQ76]
G4 YTV R J W Guttridge, Ivy House, Rise Rd, Skirlaugh, Hull, HU11 5BH [IO93UU, TA14]
G4 YTW J F Vowles, 49 Strode Rd, St, BA16 0DJ [IO81PC, ST43]
G4 YTX L A R Holloway, 27 Fairfield Rd, Tasley Park, Bridgnorth, Salop, WV16 4RY [IO82SM, SO79]
G4 YTY A J Dodd, 39 Beech Gr, Ellesmere Port, Whitby, South Wirral, L66 2PA [IO83NG, SJ37]
G4 YUA M Rowland, 27 Wilmot Cl, Witney, OX8 7NL [IO91GS, SP30]
G4 YUB A J Briscoe, 33 Keats Dr, Bilston, WV14 8SQ [IO82XN, SO99]
G4 YUC L J Fraser, 62 King Alfred Ave, Catford, London, SE6 3HE [IO91XK, TQ37]
G4 YUD L N Harris, 11 Ash Gr, Wollescote, Stourbridge, DY9 7JL [IO82WK, SO98]
G4 YUE T R Parkin, 32 Walker Ave, Woolescote, Stourbridge, DY9 9EL [IO82WK, SO98]
G4 YUF C C Sharon, 7 Waverley Gdns, Barkingside, Ilford, IG6 1PJ [JO01BO, TQ49]
G4 YUI R E Smith, 27 Laburnum Rd, Bournville, Birmingham, B30 2BA [IO92AK, SP08]
G4 YUJ W R Williams, 22 Beechfield Rd, Doncaster, DN1 2AH [IO93KM, SE50]
G4 YUK A Ince, 9 Wensleydale Rd, Scawsby, Doncaster, South Yorks, DN5 8SR [IO93JM, SE50]
G4 YUL R D Hudson, 5 Common Ln, Hemingford Abbots, Huntingdon, PE18 9AN [IO92WH, TL27]
G4 YUM Details withheld at licensee's request by SSL.
G4 YUO E Hodges, 2 Joeys Field, Bishops Nympton, South Molton, EX36 4PX [IO80CX, SS72]
G4 YUP Details withheld at licensee's request by SSL.
G4 YUQ Details withheld at licensee's request by SSL.
G4 YUT A N Fuller, 53 Sambrook Rd, Fallings Park, Wolverhampton, WV10 0ST [IO82WO, SJ90]
G4 YUU J D Leonard, 1 Freeman Dr, Walmley, Sutton Coldfield, B76 1NT [IO92CN, SP19]
G4 YUV H Boehner, Hans Boeckler Str 5, D8510 Fuerth, West Germany, ZZ9 9HA
G4 YUW A J Perry, 45 Windsor Rd, Kingshurst, Castle Bromwich, Birmingham, B36 0JR [IO92CL, SP18]
GW4 YUX D J Jones, Maesyglan, Heol Glantawe, Ystradgynlais, Swansea, SA9 1ES [IO81CS, SN70]
G4 YVA J S Guest, 57 Park Rd, Quarry Bank, Brierley Hill, DY5 2HT [IO82WL, SO98]
G4 YVB J E Finch, Hillside House, Chapel St, Camelford, Cornwall, PL32 9PJ [IO70PO, SX18]
G4 YVD P W Challinor, 1 Abbeyside, Ranton, Stafford, ST18 9JF [IO82VT, SJ82]
G4 YVE C T Herwig, 29 New Rd, Cupernham, Romsey, SO51 7LL [IO90GX, SU32]
G4 YVF F D C James, 70 Broadway West, Walsall, WS1 4DZ [IO92AN, SP09]
G4 YVG J Beech, 127 Lower Lodge, Mobile Home Park, Armitage, Rugeley Staffs, WS15 4BG [IO92BR, SK01]
G4 YVH R L Van Haaren, 24 Lake Rise, Romford, RM1 4DY [JO01CO, TQ58]
G4 YVI Dr P G Rimmer, 1 Pear Tree Cl, Weaverham, Northwich, CW8 3HD [IO83RG, SJ67]
G4 YVJ L J Beal, The Post Office, South Somercotes, Louth, Lincs, LN11 7BH [JO03BK, TF49]
G4 YVK J E Felgate, 31 Melbourne Rd, Ipswich, IP4 5PP [JO02OB, TM14]
G4 YVL Details withheld at licensee's request by SSL.[Station located near Sidcup.]
G4 YVM Details withheld at licensee's request by SSL.
GW4 YVN G G Bertos, Beach House, Beach Rd, Llanreath, Pembroke Dock, SA72 6TP [IO71MQ, SM90]
G4 YVO G W H Howarth, 79 Eden Ave, Edenfield, Ramsbottom, Bury, BL0 0LD [IO83UP, SD71]
G4 YVP J F Foster, 5 James St., Sheerness, ME12 2QE [JO01JK, TQ97]
G4 YVQ R W Hargreaves, Lawnswood, Lee Rd, Blackpool, Lancs, FY4 4QS [IO83LT, SD33]
GM4 YVR I M J Nicol, 11 Duff St., Keith, AB55 5EA [IO87MN, NJ45]
G4 YVU A Wilkinson, 48 Hopwood Rd, Hollins, Middleton, Manchester, M24 6HY [IO83VN, SD80]
G4 YVV P I Leetham, 26 Petersham Dr, Alvaston, Derby, DE24 0JU [IO92GV, SK33]
G4 YVW P R Speed, 52 Hunter Ave, Shenfield, Brentwood, CM15 8PF [JO01DP, TQ69]
G4 YVX J L Rafferty, 130 Hurst Rd, Coseley, Bilston, WV14 9EU [IO82WN, SO99]
G4 YVY T G Williams, 31 Manor Rd, Holbury, Southampton, SO45 2NQ [IO90HT, SU40]
G4 YWA P J Cartwright, 3 Blythwood Gdns, Stansted, CM24 8HG [JO01CV, TL52]
G4 YWB J W White, Olive House, Rock Rd, Rock, Wadebridge, Cornwall, PL27 6NW [IO70NN, SW97]
G4 YWC T A F Stuttard, 22 Taranis Cl, Wavendon Gate, Milton Keynes, MK7 7SJ [IO92PA, SP93]
G4 YWD W J Davies, 104 Bromborough Village Rd, Wirral, L62 7EX [IO83MH, SJ38]
G4 YWE J F Leonard, 42 Hermitage Rd, Hale, Altrincham, WA15 8BW [IO83UJ, SJ78]
G4 YWG D W Fowler, 22 Larchwood Cres, Leyland, Preston, PR5 1RJ [IO83PQ, SD52]
GM4 YWI T Ross, 15 Parkgrove Rd, Edinburgh, EH4 7NG [IO85IX, NT17]
G4 YWJ P D Ganley, 95 Park Gr, Barnsley, S70 1QE [IO93GN, SE30]
G4 YWK F Lord, 8 Langdale Ave, Clitheroe, BB7 2PG [IO83TU, SD74]
G4 YWL D R Allardyce, Vohnenstr 76, 28201 Bremen, Germany
GW4 YWM W Matthews, Cornerways, William St., Ystradgynlais, Swansea, SA9 1AT [IO81CS, SN70]
G4 YWN G T Morris, 43 Chester Rd, Chigwell, IG7 6AH [JO01AO, TQ49]
G4 YWO J Kenworthy, 99 High St., Wombwell, Barnsley, S73 8HS [IO93HM, SE30]
G4 YWP D W Howells, Boswergy, Trebarvah Ln, Perranuthnoe, Penzance, Cornwall, TR18 3AF [IO70FD, SW43]
GM4 YWQ A J Bowie, 8 Maxwell St., Fochabers, IV32 7DE [IO87KO, NJ35]
G4 YWR N Lill, 15 Gloucester Rd, Bingley, BD16 4RW [IO93CU, SE13]
GM4 YWS G McKay, Reay House, St Vigeans Brae, St Vigeans, Arbroath, DD11 4RD [IO86QN, NO64]
GI4 YWT J D Crichton, 10 Bann Dr, Lisnagelvin, Londonderry, BT47 2HW [IO64IX, C41]
GM4 YWU J A Bledowski, 13 Riggend Rd, Arbroath, DD11 2DR [IO86QN, NO64]
GM4 YWV W A Watson, 21 Cameron Way, Bridge of Don, Aberdeen, AB23 8QD [IO87WE, NJ91]
G4 YWX A F Bell, 43 Bigsby Rd, Retford, DN22 6SF [IO93MH, SK78]
G4 YWY W H Slater, 44 Cheviot Way, Oakwood Park, Verwood, BH31 6UG [IO90BU, SU00]
G4 YWZ C D T Winning, 31 Astral Gdns, Hamble, Southampton, SO31 4RQ [IO90IU, SU40]
G4 YXA G J Cooper, 8 Hugh St., Glossop, SK13 8HF [IO93AK, SK09]
G4 YXB A W Utley, 3 Dene Gr, Silsden, Keighley, BD20 9NR [IO93AW, SE04]
G4 YXC Maj W H Edwards, c/o V H Edwards, Long Orchard, Godshill Wood, Godshill, Fording Bridge Hants, SP6 2LR [IO90CW, SU11]
G4 YXE E Brown, 23 Langshaw Dr, Clitheroe, BB7 1EY [IO83TU, SD74]
G4 YXF B A Tappin, 102 Sandcross Ln, Reigate, RH2 8EY [IO91VF, TQ24]
G4 YXG Details withheld at licensee's request by SSL.
GM4 YXI Dr K M Kerr, East Loanhead, Auchnagatt, Ellon, Aberdeenshire, AB41 8YH [IO87WK, NJ93]
G4 YXJ T D Trethewey, 8 Sunningdale Rd, Saltash, PL12 4BN [IO70VJ, SX45]
GM4 YXK I C Ferguson, 52 Bute Dr, Perth, PH1 3BL [IO86GJ, NO12]
G4 YXL Details withheld at licensee's request by SSL.[Station located near Edgware. Op: R B Ford.]
G4 YXO C T Marley, Hill Toft, Kirby Park, West Kirby, Wirral, L48 2HA [IO83JI, SJ28]
G4 YXP K J White, 4 Barnett Ln, Wonersh, Guildford, GU5 0SA [IO91RE, TQ04]
G4 YXQ T J Tilston, 43 Sanderson Cl, Bridge Farm, Whetstone, Leicester, LE8 6ER [IO92JN, SP59]
G4 YXR P W Waygood, 18A Howard Rd, Wellington, TA21 8RU [IO80JX, ST12]
G4 YXS F I Lake, 77 Wood Ln, Chapmanslade, Westbury, BA13 4AT [IO81VF, ST84]
G4 YXT S A Miller, 47 Scotswood Cres, Leicester, LE2 9QE [IO92KO, SP59]
G4 YXU S W R Rawcliffe, 9 Oak Dr, Colwall, Malvern, WR13 6RA [IO82TB, SO74]
G4 YXV G Pogoda, 7 Browning Cl, South Shields, NE34 9JR [IO94GX, NZ36]
G4 YXX N J Varnes, Kelneath, West Hill, Wincanton, BA9 9BZ [IO81SB, ST72]
G4 YXZ C D Whittle, Stoker Farm, Off Mill Ln, Ryther, Tadcaster, LS24 9EQ [IO93KU, SE53]
G4 YYC D G Craig, 48 Fairholme, Putnoe, Bedford, MK41 9DD [IO92SD, TL05]
G4 YYD A Birtwistle, 6 Solness St., Bury, BL9 6PP [IO83UO, SD81]
G4 YYE K N Smith, 27 Fairleas, Branston, Lincoln, LN4 1NW [IO93SE, TF06]
GM4 YYF S Collings, Glengyre Farm, Kirkcolm, Stranraer, DG9 0RG [IO74KW, NW96]
G4 YYH R E Blemings, 1 Trethem Cl, Troon, Camborne, TR14 9ER [IO70IE, SW63]
G4 YYI S J T Tear, 18 The Chase, Sinfin, Derby, DE24 9PD [IO92GV, SK33]
G4 YYL T D Anderton, Half Acre, 8 Malthouse Ln, Cantley, Norwich, NR13 3AD [JO02SO, TG30]

G4 YYM M J Remnant, The Count House, 2 Carn Entral, Beacon, Camborne, TR14 9AJ [IO70IF, SW63]
G4 YYO P C K Sutcliffe, Rosemead, Cheadle Rd, Alton, Stoke on Trent, ST10 4BH [IO92BX, SK04]
G4 YYR S R Gibbs, 17 Kites Nest Ln, Lightpill, Stroud, GL5 3PQ [IO81VR, SO80]
G4 YYS S M Jones, 2 Rutland Way, Southampton, SO18 5PG [IO90HW, SU41]
G4 YYV W M Cliffe, 53 Easedale Dr, Southport, PR8 3TT [IO83LO, SD31]
GW4 YYY Details withheld at licensee's request by SSL.
G4 YZC E F Smith, Willow Cottage, Mill Rd, Donington on Bain, Louth, LN11 9TF [IO93WH, TF28]
G4 YZD F J Benstead, 24 Homefield, Thornbury, Bristol, BS12 2EW [IO81RO, ST69]
G4 YZE W G Learmonth, 36 Oakwood Dr, Bolton, BL1 5EH [IO83SO, SD60]
G4 YZF B R Alston-Pottinger, 43 Wadhurst Cl, St. Leonards on Sea, TN37 7AZ [JO00GV, TQ71]
G4 YZG A R Morrison, 12 Coleman Ave, New Balderton, Newark, NG24 3DR [IO93OB, SK85]
G4 YZH B M Calvert-Toulmin, Brandesby House, 31 West End, Winteringham, Scunthorpe, DN15 9NR [IO93QQ, SE92]
GM4 YZI G Pirie, 112 Newmill Rd, Elgin, IV30 2BP [IO87IP, NJ26]
G4 YZK R Marsh, 59 Tixall Rd, Hall Green, Birmingham, B28 0RS [IO92BK, SP18]
G4 YZL G J Woollams, 82 Guy Rd, Wallington, SM6 7LY [IO91WI, TQ26]
G4 YZM S B Green, 4 Countess Dr, Walsall, WS4 1HT [IO92AO, SK00]
G4 YZN K J Chapman, 10 Beck Ln, Collingham, Wetherby, West Yorks, LS22 5BW [IO93HV, SE34]
G4 YZO J F Badger, 87 Blackberry Ln, Four Oaks, Sutton Coldfield, B74 4JF [IO92BO, SP19]
G4 YZP J McMahon, 5 Sandown Cl, Runcorn, WA7 4YU [IO83PH, SJ58]
G4 YZQ B Tucker, 18 Clements Rise, Norton, Stockton on Tees, TS20 1HD [IO94IO, NZ42]
G4 YZR M G Baker, 62 Ct Farm Rd, Whitchurch, Bristol, BS14 0EG [IO81RJ, ST66]
GM4 YZT J Simpson, 55 Strathtay Rd, Perth, PH1 2NA [IO86GJ, NO02]
GM4 YZU J Borland, 64 Ailsa Rd, Gourock, PA19 1DY [IO75OW, NS27]
GU4 YZV G H Guilbert, Araucaria House, Les Varendes, St Andrews, Guernsey, GY9 9LE
G4 YZW J T Onions, 18 St. Annes Cl, Coggeshall, Colchester, CO6 1ST [JO01IU, TL82]
G4 YZX G J Maccourt, Longfield, 39 Dover Rd, Walmer, Deal, CT14 7JJ [JO01QE, TR35]
G4 YZZ E C Hastings, The Battle, Pound Ln, Gillingham, SP8 4NP [IO81UA, ST72]
GM4 ZAA Details withheld at licensee's request by SSL.[Op: Mrs M V Swankie. Station located near South Queensferry.]
G4 ZAB L B H R Taylor, 3 Howard Dr, Hale, Altrincham, WA15 0LT [IO83UI, SJ78]
G4 ZAC M S Lewis, 10 Belgrave Rd, Slough, SL1 3RE [IO91QM, SU98]
G4 ZAD G L Kitson, 33 Pace Cres, Bradley, Bilston, WV14 8BJ [IO82XN, SO99]
G4 ZAF H Thompson, 4 Appleton Rd, Skelmersdale, WN8 8RP [IO83ON, SD40]
GW4 ZAG G S Woodworth, 136 Wepre Park, Connahs Quay, Deeside, CH5 4HW [IO83LF, SJ26]
GI4 ZAH F E Anderson, 6 Perrymount, 25 Main St, Castlerock, Co Londonderry, N Ireland, BT51 4RA [IO65OE, C73]
G4 ZAI A W Stevenson, 11 Alexandra Rd, Malvern, WR14 1HA [IO82UC, SO74]
G4 ZAK B Newton, 67 Greenways, Delves Ln, Consett, DH8 7DG [IO94CU, NZ14]
G4 ZAL N P Head, 21 Besley Cl, Tiverton, EX16 4JF [IO80GV, SS91]
G4 ZAM R A Lowe, The Chimes, 4 Broadway, Swindon, SN2 3BT [IO91CO, SU18]
G4 ZAO D J Holmes, 17 Green Ln, Newby, Scarborough, YO12 6HL [IO94SG, TA08]
G4 ZAP C N Wilson, 2 Bainton Cl, Bradford on Avon, BA15 1SE [IO81VI, ST86]
GW4 ZAR D Flanagan, 4 Henblas, Flint Mountain, Flint, CH6 5PW [IO83KF, SJ27]
G4 ZAS J H Searle, 21 Chetwynd Dr, Southampton, SO16 3HY [IO90HW, SU41]
G4 ZAT Details withheld at licensee's request by SSL.
G4 ZAU Details withheld at licensee's request by SSL.[Station located near Oswestry.]
G4 ZAV Dr P Pay, Longmeadow House, Dunsford, Exeter, EX6 7AD [IO80EQ, SX88]
G4 ZAW J B D B Aspinall, 66 Lake Rd East, Cardiff, South Glam, CF2 5NN [IO81JM, ST17][Tel: (01227) 373511. RNARS 2462.]
G4 ZAX S E Jones, 1 Frogmore Cttgs, Norley Wood, Lymington, SO41 5RX [IO90GS, SZ39]
G4 ZAY J D Tench, 12 Rye Cl, North Walsham, NR28 9EY [JO02OT, TG23]
G4 ZAZ D B Ineson, 13 Churchward Ave, Swindon, SN2 1NJ [IO91CN, SU18]
G4 ZBA Details withheld at licensee's request by SSL.
G4 ZBC L P Brazier, 65 Cadman Cres, Fallings Park, Wolverhampton, WV10 0SH [IO82WO, SJ90]
G4 ZBE T J Anderson, 38 Redwood Dr, Chase Terr, Walsall, West Midlands, WS7 8AS [IO92AQ, SK00]
G4 ZBF G C Starkey, 45 Chalcot Dr, Hednesford, Cannock, WS12 4SF [IO82XR, SJ91]
G4 ZBH A F Holder, 47 Church Rd, Gurnard, Cowes, Isle of Wight, PO31 8JP [IO90IS, SZ49]
G4 ZBI J A Shores, 15 Woodlea Gdns, Cantley Manor, Doncaster, DN4 6TE [IO93LM, SE60]
G4 ZBK J M Olsen, c/o Mwsc Ltd, P.O. Box 20148, Male, Republic of Maldives
G4 ZBM D Bowker, 62 Michaelson Ave, Torrisholme, Morecambe, LA4 6SE [IO84OB, SD46]
GW4 ZBN L D Connery, 37 Thomas St., Abertridwr, Caerphilly, CF8 4AU [IO81HN, ST08]
G4 ZBO R Parker, 34 Sandgate, Kendal, LA9 6HT [IO84PH, SD59]
G4 ZBP H W Westbrook, 102 Garden Rd, Walton on The Naze, CO14 8SJ [JO01PU, TM22]
G4 ZBQ J L Thomas, 113 Southwood Dr, Coombe Dingle, Bristol, BS9 2QR [IO81QL, ST57]
G4 ZBS A Appleyard, 27 Ash Rd, Horfield, Bristol, BS7 8RN [IO81QL, ST57]
G4 ZBT D Treasure, 228 Lodge Causeway, Fishponds, Bristol, BS16 3QJ [IO81RL, ST67]
GW4 ZBU J A Johns, 129 Cecil Rd, Gowerton, Swansea, SA4 3DN [IO71XP, SS59]
G4 ZBV Details withheld at licensee's request by SSL.
G4 ZBW J S C Sceal, 655 Lightwood Rd, Longton, Stoke on Trent, ST3 7HD [IO82WW, SJ94]
G4 ZBY Details withheld at licensee's request by SSL.
G4 ZBZ Details withheld at licensee's request by SSL.
G4 ZCA R E McCormick, 22 Eric Rd, Liscard, Wallasey, L44 5RQ [IO83LK, SJ39]
G4 ZCD I C Wilson, New House, 65 Cupernham Ln, Romsey, SO51 7LE [IO91GA, SU32]
G4 ZCF W Leung, 74 Hollyfield, Harlow, CM19 4NB [JO01BS, TL40]
G4 ZCG A Ashworth, 296 Shelley Rd, Ashton Ribble, Ashton on Ribble, Preston, PR2 2EJ [IO83PS, SD53]
G4 ZCJ H E Cresswell, 34 Kingsgate Ave, Birstall, Leicester, LE4 3HB [IO92KQ, SK50]
G4 ZCK R B Miller, 140 New Fosseway Rd, Bristol, BS14 9LJ [IO81RK, ST66]
GW4 ZCL P Jones, 14 Fonmon Rd, Rhoose, Barry, CF62 3DZ [IO81HJ, ST06]
GW4 ZCM K W Frowd, 290 Pilton Vale, Newport, NP9 6LS [IO81MO, ST39]
G4 ZCN B Grylls, 76 Wharton Terr, Hartlepool, TS24 8NX [IO94JQ, NZ53]
G4 ZCP B D Roberts, 231 Clifton Rd, Rugby, CV21 3QU [IO92JI, SP57]
G4 ZCR C S Glenn, 7 Rover Dr, Castle Bromwich, Birmingham, B36 9LA [IO92CM, SP19]
G4 ZCS C D Saunders, Trinco, Gloucester Rd, Burgess Hill, RH15 8QD [IO90WW, TQ31]
G4 ZCT C J Thomas, 13 Tyndale Ave, Fishponds, Bristol, BS16 3SJ [IO81RL, ST67]
G4 ZCV E W Prietzel, 34 Manse Way, Swanley, BR8 8DD [JO01CJ, TQ56]
G4 ZCW R D Reed, 1 Curlew Cl, Bakersfield, Nottingham, NG3 7GA [IO92KW, SK64]
G4 ZCX P K Ranner, 2 Rockfield Cttgs, Crouchfield, Chapmore End, Ware, Herts, SG12 0HA [IO91WT, TL31]
G4 ZCZ J D Whitfield, 12 Low Mill Ct, Shaw Mills, Harrogate, HG3 3HJ [IO94EB, SE26]
G4 ZDA B C S Llewellyn Mudhoppers Ct G, 110 South Ave, Southend on Sea, SS2 4HU [JO01IN, TQ88]
G4 ZDB A D Robinson, 1 Nightingale Cl, University Park, Nottingham, NG7 2QG [IO92JW, SK53]
G4 ZDD B Brookfield, 17 St.Stephens Dr, Aston, Sheffield, S26 2EP [IO93II, SK48]
G4 ZDE R M Boss, 64 Chiltern Rd, Church Gresley, Swadlincote, DE11 9SJ [IO92FS, SK21]
G4 ZDF T W Langham, 1 Chatsworth Ave, Radcliffe on Trent, Nottingham, NG12 1DG [IO92LW, SK63]
G4 ZDG R W Mather, Dudley, Main Rd, Ansty, Coventry, CV7 9JA [IO92HK, SP38]
G4 ZDJ Details withheld at licensee's request by SSL.
G4 ZDK Details withheld at licensee's request by SSL.
G4 ZDL Details withheld at licensee's request by SSL.
G4 ZDM P M Shaw-Brookman, 157 St. Pancras Way, London, NW1 0SY [IO91WN, TQ28]
G4 ZDO L G Newman, 55 Kingsleigh Rd, Heaton Mersey, Stockport, SK4 3PP [IO83VK, SJ89]
G4 ZDP S D Williams, 18 Croft Rd, Newbury, RG14 7AL [IO91IJ, SU46]
G4 ZDQ A C R Siddons, 18 Earlswood Rd, Evingotn, Leicester, LE5 6JB [IO92LP, SK60]
G4 ZDR A H R J Perrett, 99 Welsford Ave, Wells, BA5 2HZ [IO81QF, ST54]
G4 ZDS J B Stringer, Flat 5, 29 Langham St, London, W1N 5RE [IO91WM, TQ28]
G4 ZDT D W Brodie, Top of The Hill, Hospital Rd, Norfolk, NR18 9PR [JO02MN, TG00]
G4 ZDX A J Staniforth, 2 Park View, Mapperley, Nottingham, NG3 5FD [IO92KX, SK54]
G4 ZDY Details withheld at licensee's request by SSL.[Op: R C Haining. Station located near Frinton-on-Sea.]
GW4 ZEA E J Hawkins, 12 Marine Dr, Ogmore By Sea, Bridgend, CF32 0PJ [IO81EL, SS87]
G4 ZEB A T Richardson, 117 Polgrean Pl, St. Blasey, St. Blazey, Par, PL24 2LH [IO70PI, SX05]
G4 ZEC Details withheld at licensee's request by SSL.[Op: R F Colwell. Station located near Buckingham.]
G4 ZED Details withheld at licensee's request by SSL.
G4 ZEF A Rostron, 8 Barberry Bank, Egerton, Bolton, BL7 9UJ [IO83SP, SD71]
G4 ZEG E J Cross, 15 Carisbrooke Cres, Barrow in Furness, LA13 0HU [IO84JC, SD26]
G4 ZEJ R S Coombes, Pinewood Lodge, Sandhills, Wormley, Surrey, GU8 5TD [IO91QD, SU93]
G4 ZEK Details withheld at licensee's request by SSL.
G4 ZEL D Rampton, Chalemar, Eddeys Ln, Headley Down, Hants, GU35 8HU [IO91OC, SU83]
G4 ZEM P J Hargreaves, Broomfield, New Rd, Gomshall, Guildford, GU5 9LZ [IO91SF, TQ04]
G4 ZEN G W Gardner, Flat 1, 15 Claremont Gr, Hale, Altrincham, WA15 9HH [IO83TJ, SJ78]
G4 ZEO R L Greaves, Bannut Tree Farm, Bannut Tree Ln, Kentchurch, Herefordshire, HR2 0DB [IO81NW, SO42]
G4 ZEP Details withheld at licensee's request by SSL.
G4 ZER D Whittaker, 89 Sherwood Ave, Spring Park, Northampton, NN2 8TA [IO92NG, SP76]
G4 ZES R M Mills, 5 Summerlands Rd, Marshalswick, St. Albans, AL4 9XB [IO91US, TL10]
GM4 ZET W N Connolly, 4 Lomond Bank, Glenfarg, Perth, PH2 9PF [IO86HG, NO11]
G4 ZEU W H Dix, Colne House, Wells Rd, Radstock, Bath, BA3 3RL [IO81SG, ST65]
G4 ZEV R J Kemp, 10 Prescot St., Wallasey, L45 9JW [IO83LK, SJ39]
G4 ZEW D A Adams, 48 Baddow Hall Cres, Great Baddow, Chelmsford, CM2 7BY [JO01GR, TL70]
GM4 ZEX G G H Duncan, Mansewood, Woodhead, Turriff, Aberdeenshire, AB53 8LT [IO87TK, NJ73]
G4 ZEY E W Gough, 41 Matlock Green, Matlock, DE4 3BT [IO93FD, SK35]
G4 ZEZ J M Curwen, The Post Office, Main St, Wray, Lancaster, LA2 8QD [IO84QC, SD66]
GI4 ZFA J Neely, 16 Bridgewater, Caw, Londonderry, BT47 1YA [IO65IA, C41]
G4 ZFB G E Parmiter, 186 St. Michaels Ave, Yeovil, BA21 4LX [IO80QW, ST51]
G4 ZFC R H Marks, 41 Bridge Meadows, Mercury Way, New Cross, London, SE14 5SU [IO91XL, TQ37]
G4 ZFD D E Roper, Lunesdale, 87 Halifax Rd, Brierfield, Nelson, BB9 5BY [IO83VT, SD83]
G4 ZFE R H O Everitt, 8 Oasthouse Way, Ramsey, Huntingdon, PE17 1SB [IO92WK, TL28]
G4 ZFF S Archer, 96 Rocky Ln, Monton, Eccles, Manchester, M30 9LY [IO83TL, SJ79]
GM4 ZFG F C Abberley, Cairnsmore Cottage, Pinwherry, Ayrshire, Scotland, KA26 0RT [IO75ND, NX18]
G4 ZFJ C M Roberts, 26 Mercer Ave, Great Wakering, Southend on Sea, SS3 0ER [JO01JN, TQ98]
G4 ZFK A E Lalley, 8 Daleview Ave, Wix, Manningtree, CO11 2SB [JO01NV, TM12]
G4 ZFL Details withheld at licensee's request by SSL.
GJ4 ZFM M L Jouault, Chant Du Vent, Maufant, St Saviour, Jersey Cl, Jersey, JE9 9CH
G4 ZFN G S C Crabbe, 39 Templeway West, Lydney, GL15 5JD [IO81RR, SO60]
G4 ZFO Details withheld at licensee's request by SSL.
G4 ZFP P J Lewis, 12 St. James Park, Tunbridge Wells, TN1 2LH [JO01DD, TQ54]
G4 ZFQ A E Reeves, Wayside, 41 Nodes Rd, Northwood, Cowes, PO31 8AD [IO90IR, SZ49]
G4 ZFR P J Whiting Felixstowe DARS, 77 Melford Way, Felixstowe, IP11 8UH [JO01PX, TM23]
GM4 ZFS S Graham, 8 Kirkton Cres, Dundee, DD3 0BN [IO86ML, NO33]
G4 ZFT M E Nurse, 67 Grasleigh Way, Allerton, Bradford, BD15 9BD [IO93BT, SE13]
G4 ZFU C H Wright, 55 Eton Rd, Oxbridge, Stockton on Tees, TS18 4DL [IO94IN, NZ41]
G4 ZFV D R Green, 56 Southfields Rd, Littlehampton, BN17 6PA [IO90RT, TQ00]
G4 ZFX J C Blades, 3 Briery Croft, Stainburn, Workington, CA14 1XJ [IO84FP, NY02]
G4 ZFY J S Bowes, 8 Coxford Drove, Coxford, Southampton, SO16 5FD [IO90GW, SU31]
GM4 ZGB J G Piekarski, 17 Dumyat Rd, Menstrie, FK11 7DH [IO86BD, NS89]
G4 ZGC P D A Crouch, 85 Stomp Rd, Burnham, Slough, SL1 7NA [IO91QM, SU98]
G4 ZGE T E Stocks, 91 Dewsbury Ave, Scunthorpe, DN15 8BT [IO93PO, SE81]
G4 ZGG B L Storey, 9 Chadleigh Ln, Godmanchester, Huntingdon, PE18 8AL [IO92VH, TL27]
G4 ZGH W Bradford, 11 Eston Ct, Cowpen, Blyth, NE24 5JE [IO95FD, NZ28]
G4 ZGJ I Y Aizlewood, 36 King St., Winterton, Scunthorpe, DN15 9TP [IO93QP, SE91]
G4 ZGL J G Crane, 3 Cross Rd, Clarendon Park, Leicester, LE2 3AA [IO92KO, SK60]
G4 ZGM C S Macdonald, 3 Shaftesbury Ave, Intake, Doncaster, DN2 6DT [IO93KM, SE50]
G4 ZGO F W Clements, 32 Chestnut Gr, Southend on Sea, SS2 5HQ [JO01IN, TQ88]
G4 ZGP G P Pritchard, 45 Fairfield Cres, Scarborough, YO12 6TL [IO94SH, TA08][Op: Geoffrey P Pritchard. Tel: (01723) 372275. RSGB Deputy Chief Morse Examiner and Senior Morse Examiner for North Yorkshire. RNARS: 3652.]
G4 ZGQ D W Richardson, 40 Berlin Rd, Stockport, Ches, SK3 9QD [IO83VJ, SJ88]
G4 ZGR K Thompson, C/Ripolles No 5 Apt U, S Agro, Gerona, Spain, 172X 48X
G4 ZGS E F Morgan, 2 Hopleys Cl, Tamworth, B77 3JU [IO92DP, SK20]
GM4 ZGV D G Davidson, 12 The Paddock, Postcliffe, Peterculter, AB1 0UE [IO87TH, NJ72]
G4 ZGX R J Blatchford, Church View, North Rd, Ranskill, Retford, DN22 8NL [IO93LJ, SK68]
G4 ZHA D B Raine, 160 Aragon Rd, Morden, SM4 4QN [IO91VJ, TQ26]
G4 ZHD P J Crofts, 8 Sandown Ave, Mickleover, Derby, DE3 5QQ [IO92FV, SK33]
G4 ZHE D M Cannon, 69 Hayfield Rd, Oxford, OX2 6TX [IO91IS, SP50]
G4 ZHG J Nevin, 26 Beech Ave, Newark, NG24 4DY [IO93OB, SK85]
G4 ZHH A Wilkinson, 15 Homestead Dr, Four Oaks, Sutton Coldfield, B75 5LN [IO92CO, SP19]
G4 ZHI J B Howell-Pryce, Robins Mead, South Stoke, nr Wallingford, Oxon, RG8 0JH [IO91KN, SU68]
G4 ZHK D Lennard, 24 Southdown Rd, Shoreham By Sea, BN43 5AN [IO90UU, TQ20]
GM4 ZHL G Graham, 18 Hamarsgarth, Mossbank, Shetland, ZE2 9TH [IP90JL, HU47]
G4 ZHN D E Young, The Gables, Aerodrome Rd, Bekesbourne, Canterbury, CT4 5EX [JO01NG, TR25]
G4 ZHR A A Lock, Chase Cottage, Bures Rd, Nayland, Colchester, CO6 4LZ [JO01KX, TL93]
G4 ZHS J Diaz Aguilar, 92 Langdale Gate, Witney, OX8 6EY [IO91GS, SP30]
G4 ZHT C J R Strevens, 11 Kenley Rd, London, SW19 3JJ [IO91VJ, TQ26]
G4 ZHU D R Stanley, Mosman, Vicar St, Oakengates, Telford, TF2 6BJ [IO82SQ, SJ61]
G4 ZHW D Nicholson, 25B Glenholme Rd, Farsley, Pudsey, LS28 5BY [IO93DT, SE23]
G4 ZHX J L Burton, 23 Dorchester Cl, Dartford, DA1 1ND [JO01CK, TQ57]
G4 ZHY R J T Nicholls, 20 Herrick Rd, Coventry, CV2 5JL [IO92GJ, SP37]
G4 ZHZ A P Nash, 83 Frimley Rd, Camberley, GU15 3EQ [IO91OH, SU85]
G4 ZIA Details withheld at licensee's request by SSL.
G4 ZIB A F Roberts, 25 Sebright Rd, Wolverley, Kidderminster, DY11 5TZ [IO82UJ, SO87]
G4 ZID L E Chapman, 6 Barholm Ave, Lutton, Spalding, PE12 9HS [JO02BT, TF42]
G4 ZIF M P Taylor, Holly House, Faussett Hill, St. End, Canterbury, CT4 7AH [JO01MF, TR15]
G4 ZIG R Reed Norfolk Vhf/Uhf, Contest Group, Oak Cott Dereham R, Bawdeswell Norfol, NR20 4AA [JO02MR, TG02]
G4 ZIH R West, 101 College Park Cs, Lewisham, London, SE13 5EZ [IO91XL, TQ37]
G4 ZII A Taylor, 50 Long Hill Rise, Hucknall, Nottingham, NG15 6GN [IO93JA, SK54]
GM4 ZIL A Brown, Skellies Knowes Eas, Cairnbrock, Ervie, Stranraer, DG9 ORY
G4 ZIO Details withheld at licensee's request by SSL.
G4 ZIP L Piper, 21 Longford Ave, Bedfont, Feltham, TW14 9TQ [IO91SK, TQ07]
GI4 ZIQ E M F Morrison, 26 Malone Park Ln, Belfast, BT9 6NQ [IO74AN, J37]
G4 ZIS R J Beech, 131 Bounces Rd, Lower Edmonton, London, N9 8LJ [IO91XP, TQ39]
GM4 ZIT J L Brown, 19 Tor View, Contin, Strathpeffer, IV14 9EF [IO77RN, NH45]
G4 ZIU M R O'Connell, Braesyde, Blackmoor Ln, Bardsey, Leeds, LS17 9DY [IO93GV, SE34]
G4 ZIW M E Hutchings, 31 Newtown Rd, Little Irchester, Wellingborough, NN8 2DX [IO92PG, SP96]
G4 ZIX R C R Hooper, 5 Northcote House, Larch Cres, Hayes, Middx, UB4 9DS [IO91TM, TQ18]
G4 ZIY M J Haddon, 1 Victoria Pl, Easton, Portland, DT5 2AA [IO80SN, SY67]
G4 ZIZ R F Barrett, Willow Lodge, Links Rd, Kirby Muxloe, Leicester, LE9 2BP [IO92JP, SK50]
G4 ZJA J A Ward, 18 Lorraine Rd, Timperley, Altrincham, WA15 7NA [IO83UJ, SJ78]
G4 ZJB H Mace, 75 Crawshaw Gr, Sheffield, S8 7EA [IO93GH, SK38]
G4 ZJC P Berry, Willow Cottage, Dalton Ln, Halsham, North Humberside, HU12 0DG [IO93XR, TA22]
G4 ZJD Dr L J Taylor, 14 Spring Gr, Chiswick, London, W4 3NH [IO91UL, TQ17]
G4 ZJE K D Faichney, 57 Moorside Rd, Brook House, Brookhouse, Lancaster, LA2 9PJ [IO84PB, SD56]
G4 ZJF Details withheld at licensee's request by SSL.
G4 ZJH I Tickle, 21 St. James, Beaminster, Dorset, DT8 3PW [IO80PT, SY49]
GM4 ZJI C J C Claydon, 28 Inveraray Ave, Glenrothes, KY7 4QN [IO86KE, NT29]
G4 ZJJ B L Powell, 18 Chestnut Gr, Maltby, Rotherham, S66 8DX [IO93JK, SK59]
G4 ZJK R J White, Peartree Cottage, Chapel Rd, Beaumont, Clacton on Sea, CO16 0AR [JO01NV, TM12]
G4 ZJL D H Wood, 29 Oakville Rd, Heysham, Morecambe, LA3 2TB [IO84NA, SD46]
G4 ZJN M P Sables, 54 Harvey St., Deepcar, Sheffield, S30 5QB [IO93GJ, SK38]
G4 ZJO H A Docherty, 17 Yorks St., Morecambe, LA3 1QE [IO84NB, SD46]
G4 ZJP M A Ward, 4 Glenway, Bognor Regis, PO22 8BU [IO90PS, SZ99]
G4 ZJQ J A Thornton, 7 Glebe Cres, Stanley, Ilkeston, DE7 6FL [IO92HX, SK44]
G4 ZJR E C Knibb, The Cottage, Main St, Cold Newton, Leicester, LE7 9DA [IO92MP, SK70]
G4 ZJT J R Lowther, 17 Keats Way, Hitchin, SG4 0DP [IO91UW, TL22]
G4 ZJU M Batchelor, 31 Youell Ave, Gorleston, Great Yarmouth, NR31 6HT [JO02UN, TG50]
GM4 ZJV A Clark, Mount Pleasant Schl, Hendry St, Portsoy, Banff Scotland, AB45 2RS [IO87PQ, NJ56]
G4 ZJW P J Brown, 60 Walton Cres, Ashbourne, DE6 1FZ [IO93DA, SK14]
G4 ZJX K G Thompson, 123 Hainault Rd, Leytonstone, London, E11 1DT [JO01AN, TQ38]
G4 ZJY J H Rawson, Tollymore, Limekiln Bank, St. Georges, Telford, TF2 9NU [IO82SQ, SJ71]
G4 ZKA J G Watson, 3 Layton Rd, Ashton, Ashton on Ribble, Preston, PR2 1PB [IO83OS, SD53]
G4 ZKB I W Thompson, 2 Sandpit Cttgs, Garton on The Wolds, Driffield, YO25 9LP [IO93RX, SE95]
G4 ZKC P J Williams, 3 Burnham Gr, Scawthorpe, Doncaster, DN5 9JY [IO93KN, SE50]
G4 ZKD J P Moule, Silver Dale, Callow Hill Rock, Kidderminster, Worcs, DY14 9DB [IO82TI, SO77]
G4 ZKE M B Chapman, 1 Henley Ave, North Cheam, Sutton, SM3 9SQ [IO91VJ, TQ26]
G4 ZKF K S Ford, 12 Buttermere Gr, Beechwood, Runcorn, WA7 2RF [IO83PH, SJ58]
G4 ZKG J A Corfield, 5 Beasley Cl, Great Sutton, South Wirral, L66 2SX [IO83MG, SJ37]
G4 ZKH A M Curtis, 11 Pentreath Terr, Lanner, Redruth, TR16 6HP [IO70JF, SW73]
G4 ZKI M H Day, 76 Freeman Rd, Didcot, OX11 7DB [IO91IO, SU59]
G4 ZKJ W F J Applebee, 9 The Glade, Bucks Horn Oak, Bucks Horn Oak, Farnham, GU10 4LU [IO91NE, SU84]
G4 ZKK Details withheld at licensee's request by SSL.
G4 ZKM W E Ingram, 39 Ainsdale Dr, Peterborough, PE4 6RL [IO92UO, TF10]
G4 ZKN Details withheld at licensee's request by SSL.
G4 ZKQ J F Garner, Cobwebs, Lewes Rd, Scayne Hill, Sussex, RH17 7PG [IO90XX, TQ32]
G4 ZKR J Olbrien, 14 Ryecroft Cl, Middlewich, CW10 0PJ [IO83SE, SJ66]
G4 ZKS A M Howland, Hollydene, Station Rd, Great Bentley, Colchester, CO7 8LJ [JO01MU, TM12]
G4 ZKT A R Hale, 32 Russell Rd, Northolt, UB5 4QS [IO91TN, TQ18]
G4 ZKU Details withheld at licensee's request by SSL.
G4 ZKW T E Davies, 7 Medway, Sturton By Stow, Lincoln, LN1 2DY [IO93QH, SK88]
G4 ZKX A P O'Flanagan, 3 Egret Gr, Birchwood Est, Lincoln, LN6 0JL [IO93QF, SK96]
G4 ZKY J C D Mackay, Cluniac House, Abbey Ct, Faversham, Kent, ME13 7BG [JO01KH, TR06]
G4 ZLB L J Zilberberg, 3 Queenswood Park, London, N3 1UN [IO91VO, TQ29]
GI4 ZLD G M A Breslin, 85 Whitehouse Park, Derry City, Londonderry, BT48 0QA [IO65HA, C41]
G4 ZLF L S J Forde, 23 Laburnum Rd, Wellington, TA21 8EL [IO80JX, ST12]
G4 ZLH T J Winters, 26 Elizabeth Cl, Grudy St., Poplar, London, E14 6DW [IO91XM, TQ38]
G4 ZLI T F Schofield, 34 Oaks Dr, St. Leonards, Ringwood, BH24 2QT [IO90BT, SU10]
G4 ZLJ P Asinall, 20 Carr Ln, Rawtenstall, New Hall Hey, Rossendale, BB4 6BE [IO83UQ, SD82]
G4 ZLK S C Pike, 42 Stanley Rd, Chingford, London, E4 7DB [JO01AP, TQ39]

G4 ZLL M J Holgate, 10 Zinnia Cl, Ensbury Park, Bournemouth, Dorset, BH10 4HR [IO90BS, SZ09]
G4 ZLN B N E Phillips, 2 Oriole Gr, Kidderminster, DY10 4HG [IO82VI, SO87]
G4 ZLP N Crook, 44 King Georges Rd, Rossington, New Rossington, Doncaster, DN11 0PW [IO93LL, SK69]
G4 ZLQ D C Brunsden, Tadley Lodge, 91 Blewbury Rd, East Hagbourne, Didcot, Oxon, OX11 9LE [IO91JO, SU58]
G4 ZLT R I Winkup, 92 Barnes Cres, Ensbury Park, Bournemouth, BH10 5AW [IO90BS, SZ09]
G4 ZLU T Clark, Thaw House, Brunswick St., Nelson, BB9 0HZ [IO83VT, SD83]
G4 ZLW G W Howard, 80 Falkland Rd, Greatfield Est, Hull, HU9 5EZ [IO93US, TA12]
G4 ZLX A K Whillock, 74 Chettell Way, Blandford St. Mary, Blandford Forum, DT11 9PH [IO80WU, ST80]
G4 ZLY Details withheld at licensee's request by SSL.
G4 ZLZ Details withheld at licensee's request by SSL.
G4 ZMA J H Smith, 7A The Green, East Leake, Loughborough, LE12 6LD [IO92JT, SK52]
G4 ZMB D Sharples, 11 Lina St., Accrington, BB5 1SL [IO83TS, SD72]
G4 ZMC G Allsop, 82 Charter Ave, Canley, Coventry, CV4 8ED [IO92FJ, SP37]
G4 ZME E Collins, 24 Durham Cl, Canterbury, CT1 3QL [JO01MG, TR15]
GM4 ZMG Details withheld at licensee's request by SSL.
G4 ZMH G E Robinson, Hamsden Garth, Cadney Rd, Howsham, Lincoln, LN7 6LA [IO93SM, TA00]
G4 ZMI Details withheld at licensee's request by SSL.
G4 ZMJ R G Browell, 42 Thorncliffe Rd, Norwood Green, Southall, UB2 5RQ [IO91TL, TQ17]
GM4 ZMK R Coyle, 216 Faifley Rd, Clydebank, G81 5EG [IO75TW, NS57]
G4 ZML D P Coupe, 14 Maltby Rd, Thornton, Middlesbrough, TS8 9BU [IO94IM, NZ41]
G4 ZMM J C Roberts, Vernann House, Gayton, Staffs, ST18 0HJ [IO82XU, SJ92]
G4 ZMN P P Shepherd, 315 Daws Heath Rd, Benfleet, SS7 2TY [JO01HN, TQ88]
G4 ZMO P Rogers, 68 Southfield Rd, Princes Risborough, HP27 0JB [IO91OR, SP80]
G4 ZMP D K Butler, 42 Coombe Farm Ave, Fareham, PO16 0TR [IO90JU, SU50]
G4 ZMQ M E Pemberton, 38 Welson Rd, Folkestone, CT20 2NP [JO01NB, TR23]
G4 ZMR M J Reynolds, 4 St. Albans Dr, Nantwich, CW5 7DW [IO83RB, SJ65]
G4 ZMS E R W Rennie, 26 Kingshill Ave, Collier Row, Romford, RM5 2SD [JO01CO, TQ59]
G4 ZMT P R White, 5A Bingham Rd, Croydon, Surrey, CR0 7EA [IO91XJ, TQ36]
G4 ZMU A M Vernon, 13 Vicarage Ln, Great Baddow, Chelmsford, Essex, CM2 8HY [JO01GR, TL70]
G4 ZMW M P Williams, 23 The Park, Kingswood, Bristol, BS15 4BL [IO81RL, ST67]
G4 ZMX A E Collar, 40 Princes Ave, South Croydon, CR2 9BA [IO91XH, TQ35]
G4 ZMY A J Wakely, 3 Kings Rd, Hayling Island, PO11 0PD [IO90MT, SU70]
GM4 ZNA Dr J N McCormick, 27 Ferryhills Rd, North Queensferry, Inverkeithing, KY11 1HE [IO86HA, NT18]
GM4 ZNC W Findlay, 9 Scott Pl, Troon, KA10 6XD [IO75QN, NS33]
G4 ZND G D Dodds, 15 Acton Cres, Coquet Park, Felton, Morpeth, NE65 9NF [IO95DH, NU10]
G4 ZNE J P Houghton, 32 Delamere Dr, Macclesfield, SK10 2PW [IO83WF, SJ97]
GM4 ZNG J G Dunn, 54 Park St., Crosshill, Lochgelly, KY5 8BH [IO86ID, NT19]
G4 ZNI A Wragg, 11A Fall Rd, Heanor, Derbyshire, DE75 7PQ [IO93HA, SK44]
G4 ZNK R V Walsh, 16 Pinewood Gr, Midsomer Norton, Bath, BA3 2RH [IO81SG, ST65]
G4 ZNL K Cork, 1 Goddard Cl, Kilby, Wigston, LE18 3TN [IO92LN, SP69]
G4 ZNM N T Muggeridge, 8 Rowan Cl, Seaford, BN25 4NW [JO00BS, TV49]
G4 ZNN P K Bee, 5 Gibraltar Ln, Laceby, Grimsby, DN37 7AU [IO93VN, TA20]
GM4 ZNS J Callaghan, 25 Clearmount Ave, Newmilns, KA16 9ER [IO75UO, NS53]
G4 ZNT K E Barrow, 18 Somerfield Walk, Leicester, LE4 0QQ [IO92KP, SK50]
GW4 ZNU M Jones, 62 Duffryn St., Mountain Ash, CF45 3HR [IO81HQ, ST09]
GM4 ZNX D Stockton, 13 Dunvegan Ct, Crossford, Dunfermline, KY12 8YL [IO86GB, NT08]
G4 ZNY K Brown, 54 Ousebank Way, Stony Stratford, Milton Keynes, MK11 1LB [IO92NB, SP74]
G4 ZNZ D R Neilson, 4 North St., Nafferton, Driffield, YO25 0JW [IO94TA, TA05]
GM4 ZOA S J D ' McGregor Mudhoppers Co G, 35 Pentland Gdns, Edinburgh, EH10 6NN [IO85JV, NT26]
G4 ZOB P G Harris, 47 North Park Gr, Roundhay, Leeds, LS8 1EW [IO93FU, SE33]
G4 ZOC J M Lawton, Grenehurst, Pinewood Rd, High Wycombe, HP12 4DD [IO91OP, SU89]
G4 ZOD Details withheld at licensee's request by SSL.[Station located near Edgware.]
G4 ZOE Details withheld at licensee's request by SSL.
G4 ZOF A Hughes, c/o Kemble Motors, Coundon, Bishop Auckland, DL14 8PD [IO94EP, NZ22]
G4 ZOG D N Andrews, 3 St. Davids Rd, Thornbury, Bristol, BS12 1AE [IO81RO, ST69]
G4 ZOH C C Hetherington, Hafan, 1 Fron Farm, Menai Bridge, LL59 5QY [IO73WF, SH57]
G4 ZOI D Hunsdale, 57 Newbarns Rd, Barrow in Furness, LA13 9PG [IO84JC, SD26]
G4 ZOJ J Fulton, 7 Tulacorr Gdns, Victoria Rd, Strabane, Co. Tyrone, N Ireland, BT82 8RB [IO64GT, H39]
G4 ZOK K M Mills, 75 Caistor Ln, Caistor St. Edmunds, Caistor St. Edmund, Norwich, NR14 8RB [JO02PN, TG20]
GW4 ZOM Details withheld at licensee's request by SSL.
G4 ZON A D Edwards, 68 Whittington Rd, Hutton, Brentwood, CM13 1JX [JO01EP, TQ69]
G4 ZOO T Lowe, 24 Severn Dr, Hindley Green, Hindley, Wigan, WN2 4TW [IO83RM, SD60]
G4 ZOQ J R Dennis, 44 The Dr, Uckfield, TN22 1BZ [JO00BX, TQ42]
G4 ZOR E A Wand, 76A Brunswick Park Rd, London, N11 1JJ [IO91WO, TQ29]
GI4 ZOS W Boyd, 51 South Sperrin, Knock, Belfast, BT5 7HW [IO74CO, J47]
G4 ZOU D G Nuthall, Clarendon School Lodge, Hanworth Rd, Hampton, TW12 3DH [IO91TK, TQ17]
G4 ZOW P de Cadenet, 22 Beeching Cl, Harpenden, AL5 4LZ [IO91TT, TL11]
G4 ZOX C S Moore, Spion Cop, Blacksmiths Ln, Harmston, Lincoln, LN5 9SW [IO93RD, SK96]
G4 ZOY D Elliott, 6 Linden Cl, Stakeford, Choppington, NE62 5LD [IO95ED, NZ28]
G4 ZOZ J F Burford Bromsgrove QRP Club, 26 Shrubbery Rd, Bromsgrove, B61 7BH [IO82XH, SO97]
G4 ZPA B N Watling, 10 Nutbourne Rd, Farlington, Portsmouth, PO6 1NR [IO90LU, SU60]
G4 ZPB R L Alexander, 1 Locarno Rd, Swanage, BH19 1HY [IO90AO, SZ07]
GM4 ZPC P P Collier, 5 Oxford St., Dundee, DD2 1TJ [IO86LL, NO33]
G4 ZPE M W Wheaton, 2 Grouch View, Rettendon, nr Chelmsford, Essex, CM3 5DS [JO01HP, TQ89]
G4 ZPH F Machniak, 18 Wyatt Rd, Kempston, Bedford, MK42 7EN [IO92RC, TL04]
G4 ZPI D Madden, 29 Raven Hays Rd, Birmingham, B31 5JP [IO92AJ, SP07]
G4 ZPJ C Marks, 63 Alvis Walk, Chelmsley Wood, Birmingham, B36 9JZ [IO92CM, SP19]
G4 ZPL Details withheld at licensee's request by SSL.
G4 ZPN M L B Brown, 78 York Rd, Paignton, TQ4 5NS [IO80FK, SX85]
G4 ZPO G Belt, The Flat, Co. Hall, Colliton Park, Dorchester, Dorset, DT1 1XJ [IO80SR, SY69]
G4 ZPP B C Hope, 60 West Green Dr, West Green, Crawley, RH11 7DL [IO91VC, TQ23]
G4 ZPQ S N Drury, 24 Mollison Rd, Hull, HU4 7HB [IO93SR, TA02]
G4 ZPR R E Wilson, 1 Larkfield Ave, Kenton, Harrow, HA3 8NQ [IO91UO, TQ18]
G4 ZPS W J Upson, 45 Lampits Ln, Corringham, Stanford-le-Hope, SS17 9AE [JO01FM, TQ78]
G4 ZPU Details withheld at licensee's request by SSL.
G4 ZPV W H Cook, 54 Sandringham Rd, Petersfield, GU32 2AA [IO91MA, SU72]
G4 ZPW D J McKie, 11 Guys Cl, Ringwood, BH24 1PQ [IO90CU, SU10]
G4 ZPX Details withheld at licensee's request by SSL.
G4 ZPY G M Crowhurst, 41 Mill Dam Ln, Burscough, Ormskirk, L40 7TG [IO83NO, SD41]
G4 ZQA L Standen, 134 Wakeham, Portland, DT5 1HP [IO80SM, SY67]
G4 ZQB G B Tapp, 23 Blacksmith Dr, Weavering, Maidstone, ME14 5SZ [JO01GG, TQ75]
G4 ZQC R H Alderson, Old School House, Tattersett, King's Lynn, Norfolk, PE31 8RS [JO02IT, TF82]
G4 ZQD J Hamer, 24 Horsepool Rd, Connor Downs, Hayle, TR27 5DZ [IO70HE, SW53]
G4 ZQF R Worth, 9 Lingfield Park, Windsor Rd, Bray, Maidenhead, SL6 2DS [IO91PL, SU97]
G4 ZQG B Jones, 9 Barmby Cl, Ossett, WF5 0DS [IO93FQ, SE21]
GM4 ZQH J G Howell, 26 Bonaly Cres, Colinton, Edinburgh, EH13 0EW [IO85IV, NT26]
G4 ZQI A H G Farmer, 29 Ashmead Rd, Maybush, Southampton, SO16 5DJ [IO90GW, SU31]
G4 ZQJ A J Mayes, 103 Lionel Rd, Canvey Island, SS8 9DJ [JO01HM, TQ78]
G4 ZQK Details withheld at licensee's request by SSL.
G4 ZQL N R Higgins, 68 Broomfield Cres, Langley Est, Middleton, Manchester, M24 4FW [IO83VN, SD80]
G4 ZQM J M J Neary, 283 Old Rd East, Gravesend, DA12 1PW [JO01EK, TQ67]
G4 ZQO T A Lewis, 23 Northiam Cl, Hemlington, Middlesbrough, TS8 9PT [IO94JM, NZ41]
G4 ZQP Details withheld at licensee's request by SSL.
GM4 ZQQ I R Pattullo, 659 South Rd, Dundee, DD2 4SG [IO86LL, NO33]
G4 ZQS L Brown, 17 Chaucer Walk, Poets Est, Langney, Eastbourne, BN23 7QT [JO00DT, TQ60]
G4 ZQT J W Wright, 85 Kingfisher Dr, Beacon Park Home Vil, Skegness, PE25 1TQ [JO03DD, TF56]
GW4 ZQV I J Bradford, The Meadows, Penyrheol, Pontypool, Gwent, NP4 5XJ [IO81LQ, ST29]
GW4 ZQY M G Couch, 37 Heol Rhosyn, Morriston, Swansea, SA6 6ER [IO81AQ, SS69]
GM4 ZQZ Details withheld at licensee's request by SSL.
G4 ZRA G W Moffatt, 30 Rose Walk, St. Albans, AL4 9AF [IO91US, TL10]
G4 ZRB Dr W A Gerrard, 12 Brighton Rd, Addlestone, KT15 1PJ [IO91SI, TQ06]
G4 ZRC M Cole, 25 Holly Gdns, West Drayton, UB7 9PE [IO91SM, TQ07]
G4 ZRD Details withheld at licensee's request by SSL.
G4 ZRE Details withheld at licensee's request by SSL.
G4 ZRF T W F Emery, 23 Richmondfield Way, Barwick in Elmet, Leeds, LS15 4HJ [IO93HT, SE43]
GM4 ZRH A R Hutton, 4 The Pheasantry, Linn Rd, Stanley, Perth, PH1 4QS [IO86GL, NO13]
G4 ZRI U D Shields, 41 Mill Ln, Upholland, Appley Bridge, Wigan, WN6 9DD [IO83PN, SD50]
GW4 ZRK H V Stephens, Tycoch, Rhyddwen Pl, Craig Cefn Parc, Swansea, SA6 5RN [IO81BQ, SN60]
G4 ZRL Details withheld at licensee's request by SSL.
G4 ZRM D J Line, 28 Wykeham Rd, Higham Ferrars, Wellingborough, Northants, NN10 8HU [IO92QH, SP96]
G4 ZRP Details withheld at licensee's request by SSL.
GM4 ZRQ I F McClure, 21 Milton Park, Aviemore, PH22 1RR [IO87CE, NH81]
GM4 ZRR I G Watt, 12 Lastingham Ct, Laleham Rd, Staines, TW18 2NW [IO91RK, TQ07]
G4 ZRT M T Johnson, Dellmar House, Donigers Dell, Swanmore, Southampton, Hants, SO32 2TL [IO90JW, SU51]
G4 ZRU C H Field, 54 Kingston Dr, Urmston, Manchester, M41 9FG [IO83TK, SJ79]
G4 ZRV T W Russell, 57 Windmill Ln, Worksop, S80 2SQ [IO93KG, SK57]
GW4 ZRW T S George, 80 Yew St., Troedyrhiw, Merthyr Tydfil, CF48 4EE [IO81HR, SO00]
GM4 ZRX J G Lindsay, 6 Netherhouse Ave, Lenzie, Kirkintilloch, Glasgow, G66 5NG [IO75WW, NS67]
G4 ZRY A E Clemons, 2 Cherry Tree Rd, Rainham, Gillingham, ME8 8JU [JO01HI, TQ86]
G4 ZRZ G R Baker, 13 Thorn Ln, Four Marks, Alton, GU34 5BX [IO91LC, SU63]
G4 ZSA A R Smith, 16 Burley Cl, South Milford, Leeds, LS25 5BT [IO93JS, SE43]
G4 ZSB L W Wood, 61 Denison St., Beeston, Nottingham, NG9 1AX [IO92JW, SK53]
G4 ZSC A M Kent, 9 Tolmers Gdns, Cuffley, Potters Bar, EN6 4JE [IO91WQ, TL30]
G4 ZSD W Guy, 102 Bonington Rd, Mansfield, NG19 6QQ [IO93JD, SK56]
G4 ZSF Details withheld at licensee's request by SSL.
G4 ZSG J D Savegar, 39 Roundfields, Little Ln, Upper Bucklebury, Reading, RG7 6RA [IO91JJ, SU56]
G4 ZSH Details withheld at licensee's request by SSL.
G4 ZSI H Hargreaves, 7 Harwood Walk, Tottington, Bury, BL8 3NT [IO83TO, SD71]
G4 ZSJ G Bray, 65 High St., Barkway, Royston, SG8 8EB [JO02AA, TL33]
G4 ZSK G Dougherty, 77 Ravensworth, Ryhope, Sunderland, SR2 0BH [IO94HU, NZ35]
G4 ZSL D P Speed, 105 Thelwall New Rd, Thelwall, Warrington, WA4 2HR [IO83RJ, SJ68]
G4 ZSM G P Driver, 216 Ct Ln, Birmingham, B23 5RH [IO92BM, SP19]
G4 ZSN C J Troop, 10 Mellowdew Rd, Coventry, CV2 5GL [IO92GJ, SP37]
G4 ZSO N E Dakin, The Grange, Chapel Ln, Granby, Nottingham, NG13 9PW [IO92NW, SK73]
G4 ZSP G Marriott, 9 Seward St., Loughborough, LE11 3BU [IO92JS, SK51]
G4 ZSQ Details withheld at licensee's request by SSL.
G4 ZSR D Woollams, 82 Guy Rd, Wallington, SM6 7LY [IO91WI, TQ26]
G4 ZSS S L Simpson, 8 Hallfield Ave, Micklefield, Leeds, LS25 4AU [IO93IT, SE43]
G4 ZST D L Nuttall, 34 Douglas Rd, Hornchurch, RM11 1AR [JO01CN, TQ58]
G4 ZSV J H Broughton, 12 Warnborough Rd, Oxford, OX2 6HZ [IO91IS, SP50]
G4 ZSW P Withall, 19 Highfield Dr, Ewell, Epsom, KT19 0AU [IO91UI, TQ26]
G4 ZSZ B C Watts, 74 Westfield Rd, Caversham, Reading, RG4 8HJ [IO91ML, SU77]
G4 ZTA J W Reed, Easton Villa, Grangemoor Rd, Widdrington, Morpeth, NE61 5PU [IO95EF, NZ29]
G4 ZTC T Cleghorn, 12 Rennington Cl, Stobhill Farm, Morpeth, NE61 2TQ [IO95DD, NZ28]
G4 ZTF J L Scott, Kemsley St Cottage, Bredhurst, Kent, ME7 3LS [JO01HH, TQ86]
G4 ZTG K A Ford, 51 Druids Ln, Maypole, Birmingham, B14 5SR [IO92BJ, SP07]
G4 ZTH R I Brown, 17 Rowley St., Blyth, NE24 2HQ [IO95FC, NZ38]
G4 ZTI Details withheld at licensee's request by SSL.
G4 ZTK T S Kay, 11 Tangmere Cl, Bowerhill, Melksham, SN12 6XW [IO81WI, ST96]
G4 ZTL R A Barlow, 2 The Bassett, Langwith Junction, Mansfield, NG20 9AR [IO93JF, SK56]
G4 ZTM N B Rohsler, 107 Quinton Ln, Quinton, Birmingham, B32 2TT [IO92AL, SP08]
GM4 ZTO J McIlwraith, 22 Foreland, Ballantrae, Girvan, KA26 0NQ [IO75LC, NX08]
G4 ZTP Details withheld at licensee's request by SSL.[Station located in the Bournemouth area.]
G4 ZTQ S J Chappell, 4 Blanford Rd, Chichester, PO19 4TW [IO90OU, SU80]
G4 ZTR J K Lemay, Carlton House, White Hart Ln, West Bergholt, Colchester, CO6 3DB [JO01KW, TL92]
G4 ZTS C J Thorne, Woodland House, Westleigh, nr Tiverton, Devon, EX16 7EP [IO80HW, ST01]
G4 ZTT Dr M W Dixon Mid Cheshire AR, Woodstock, Gad Bank, Norley, Warrington, WA6 8LL [IO83QF, SJ57]
GI4 ZTU H A Morgan, 42 Ardmore Rd, Holywood, BT18 0PJ [IO74CP, J47]
GW4 ZTW C D Gallagher, 20 Bridget Dr, Sedbury, Chepstow, NP6 7AR [IO81QP, ST59]
G4 ZTX F Hall, 7 Furlong Ct, Goldthorpe, Rotherham, S63 9PZ [IO93IM, SE40]
G4 ZTY D Dalton, 22 Fernleigh Ave, Mapperley, Nottingham, NG3 6FL [IO92KX, SK54]
GW4 ZUA W P Webb, 5 Shop Houses, Llwydcoed, Aberdare, Mid Glam, CF44 0TH [IO81GR, SN90]
G4 ZUC G E Allin, 5 Brookhill Ct, Sutton in Ashfield, NG17 1EP [IO93IC, SK45]
G4 ZUD B G Perry, Wynstone, Cunnery Rd, Church Stretton, SY6 6AF [IO82OM, SO49]
G4 ZUE R Hopkins, 259 Croft Rd, Nuneaton, CV10 7EE [IO92GM, SP39]
G4 ZUF Details withheld at licensee's request by SSL.
G4 ZUH J F Rowles, 38 Swan Cl, St. Ives, Huntingdon, PE17 4HX [IO92XI, TL37]
G4 ZUI P A Bevington, 40 Carnarthen St., Camborne, TR14 8UP [IO70IF, SW64]
GW4 ZUJ B L Willis, 5 Park St., Penrhiwceiber, Mountain Ash, CF45 3YW [IO81HQ, ST09]
GM4 ZUK Details withheld at licensee's request by SSL.[Aberdeen VHF Group, Please QSL via Allan Duncan, GM4ZUK, 69 Abbotshall Drive, Cults, Aberdeen, AB1 9JJ.]
G4 ZUL S J Cocks, 9 Church Rd, Noak Bridge, Basildon, SS15 5SJ [JO01FN, TQ68]
G4 ZUM T King, 46 Old Barn Way, Southwick, Brighton, BN42 4NT [IO90VU, TQ20]
G4 ZUN C S Gee, 100 Plantation Hill, Kilton, Worksop, S81 0QN [IO93KH, SK57]
G4 ZUP R Angel, The Olde Cheese Hse, Upton Lovell, Warminster, Wilts, BA12 0JW [IO81XD, ST94]
G4 ZUQ B C Hunter, 47 Ashdene Rd, Ashurst, Southampton, SO40 7BW [IO90FV, SU31]
G4 ZUR G P Albrighton, 3 Martins Dr, Atherstone, CV9 3AU [IO92FO, SP39]
GW4 ZUS G Smith, 40 Llys Gwenllian, Kidwelly, SA17 5JT [IO71UR, SN40]
G4 ZUU P D Chambers, Holly House, Bleasby Rd, Fiskerton, Southwell, NG25 0XL [IO93NB, SK75]
G4 ZUV J E Barrington, 50 Greenway Ln, Chippenham, SN15 1AE [IO81WL, ST97]
GW4 ZUW A S Hockley, 44 Brookfields, Crickhowell, NP8 1DJ [IO81KU, SO21]
G4 ZUX N J Sparks, 36 Tormynton Rd, Worle, Weston Super Mare, BS22 9HT [IO81MI, ST36]
G4 ZUY M C Goodwill, 9 Kemerton Rd, Beckenham, BR3 6NJ [IO91XJ, TQ36]
G4 ZVA T I Webster, 16 Long Shoot, Bates Caravan Park, Goostrey, Crewe, CW4 8JX [IO83TF, SJ77]
G4 ZVB G Mantovani, 74 Barnsley Rd, South Kirkby, Pontefract, WF9 3QE [IO93IO, SE41]
G4 ZVC A R Brackenborough, 41 Poets Corner, Margate, CT9 1TR [JO01QJ, TR37]
G4 ZVD J C Birse, 178 Long Lee Ln, Keighley, BD21 4TT [IO93BU, SE04]
G4 ZVE K D Galloway, 6 Welbeck Ct, Kettering, NN15 5NP [IO92PI, SP87]
G4 ZVF M Sheriff, Defence Animal Centre, Welby Ln, Elmhurst Ave, Melton Mowbray, LE13 0SL [IO92NS, SK72]
G4 ZVG Details withheld at licensee's request by SSL.
G4 ZVH I W Gardner Darlington & District Amateur, 30 Pierremont Cres, Darlington, Co. Durham, DL3 9PB [IO94FM, NZ21]
G4 ZVJ A T Chadwick, 5 Thorpe Chase, Ripon, HG4 1UA [IO94FD, SE37]
G4 ZVK J Taylor, 23 High St., Atherton, Manchester, M46 9DW [IO83SM, SD60]
G4 ZVN P Baxter, 16 Mortomley Cl, High Green, Sheffield, S30 4HZ [IO93GJ, SK38]
GW4 ZVO Details withheld at licensee's request by SSL.
G4 ZVP B Rhodes, 13 Amanda Rd, Harworth, Doncaster, DN11 8HP [IO93LK, SK69]
G4 ZVR T Case, 4 Blackthorn Way, Leavenheath, Colchester, CO6 4UR [JO01KX, TL93]
G4 ZVS C F Ford, 19 Listowel Rd, Kings Heath, Birmingham, B14 6HH [IO92BK, SP08]
G4 ZVT F S Peacock, 41 Weir Rd, Hemingford Grey, Huntingdon, PE18 9EH [IO92WH, TL27]
G4 ZVU T W Chadwick, 515 Staines Rd, West, Ashford, Middx, TW15 2AB [IO91SK, TQ07]
GW4 ZVV B Raby, 4 Tyfica Rd, Pontypridd, CF37 2DA [IO81HO, ST09]
G4 ZVW D A Ilsley, 3 Foxcote Cl, Shirley, Solihull, B90 4PR [IO92CJ, SP17]
G4 ZVX M Russell, 67 Dugard Rd, Cleethorpes, DN35 7SD [IO93XN, TA20]
GW4 ZVY Details withheld at licensee's request by SSL.
G4 ZVZ S K Josko, 69 Newborough Rd, Shirley, Solihull, B90 2HB [IO92BJ, SP17]
G4 ZWA G M Johnson, The Cottage, Horncastle Rd, Mareham on The Hill, Horncastle Lincs, LN9 6PQ [IO93XE, TF26]
G4 ZWB R F Ayers, 6 Cherry Walk, Wythall, Hollywood, Birmingham, B47 5RL [IO92BJ, SP07]
GW4 ZWC R L Huckfield, Stockton Smithy, Stockton Marton, Welshpool, Powys, SY21 8JL [IO82KO, SJ20]
G4 ZWD C A Barber, 27 Carden Cres, Patcham, Brighton, BN1 8TQ [IO90WU, TQ30]
G4 ZWE P D Burtenshaw, 21 Frenchgate Cl, Hampden Park, Eastbourne, BN22 9EX [JO00DT, TQ60]
G4 ZWI Details withheld at licensee's request by SSL.
GM4 ZWJ S D Macfarlane, 8 Balgray Rd, Newton Mearns, Glasgow, G77 6PB [IO75TS, NS55]
G4 ZWM H L Hoy, The Meadows Cottage, Stow Heath Rd, Felmingham, North Walsham, NR28 0LR [JO02PT, TG22]
GW4 ZWN T P Anziani, Ty-Coch, Penrhyd, Amlwch, Anglesey, Gwynedd, LL68 1AA [IO73TJ, SH49]
GW4 ZWO A W Green, 1 Kelston Ct, Gwespyr, Holywell, CH8 9LN [IO83HI, SJ18]
G4 ZWP Details withheld at licensee's request by SSL.[Station located near Ware.]
G4 ZWQ P M Smith, 16 Church St., Owston Ferry, Doncaster, DN9 1RG [IO93OL, SE80]
G4 ZWR D J Edwards, 2 Mason Cl, Headless Cross, Redditch, B97 5DF [IO92AG, SP06]
GW4 ZWS E G Woodruff, Chantier, Pentre Rd, Halkyn, Holywell, CH8 8BS [IO83JF, SJ27]
G4 ZWU W H Waddle, 12 Mallory Rd, Norton, Stockton on Tees, TS20 1TJ [IO94IO, NZ42]
G4 ZWV B M Taylor, 9 Penland Cl, Haywards Heath, RH16 1PR [IO91WA, TQ32]
G4 ZWX Dr P R Harrison, 3 Vaughan Copse, Willow Brook, Eton, Windsor, SL4 6HL [IO91QL, SU97]
G4 ZWY S W Icke, 11 Church Ln, Bromyard, HR7 4DZ [IO82RE, SO65]
G4 ZWZ E J Murphy, 4 Spinney Rd, Long Eaton, Nottingham, NG10 4HW [IO92IV, SK43]
G4 ZXA R C Smith, 1 Hall Ln, Wolvey, Hinckley, LE10 3LF [IO92HL, SP48]
G4 ZXC T D Walker, 11 Sullivan Cl, Shefford, SG17 5SG [IO92UA, TL13]
GW4 ZXD S M Hodson, Ivy Cottage, Bryn Sannon, Brynford, Holywell, CH8 8AX [IO83JG, SJ17]
G4 ZXE M G Bowden, 31 Shepton Walk, Bedminster Rd, Bristol, BS3 5NU [IO81QK, ST57]
G4 ZXF K G Kimber, 15 Belinus Dr, Billingshurst, RH14 9BX [IO91SA, TQ02]
GW4 ZXG L F G Thomas, 'Roughton', Corntown Rd, Ewenny nr Bridgend, Mid Glam, CF35 5BH [IO81FL, SS97]
GW4 ZXH A L Taylor, Hafan, Esguryn Rd, Llandudno Junction, Gwynedd, LL31 9QE [IO83CH, SH87]
G4 ZXI N Parnell, Shangri Ln, 4 Forge Ln, Headcorn, Ashford, TN27 9QQ [JO01HE, TQ84]
GM4 ZXJ J A Burns, 7 Johns Rd, Eyemouth, TD14 5DX [IO85XU, NT96]
G4 ZXK G T Knight, 60 Wellington Rd, Sandhurst, Camberley, Surrey, GU17 8AY [IO91OH, SU85]
GW4 ZXL W D I Hughes, 31 Ystrad Dr, Johnstown, Carmarthen, SA31 3PQ [IO71UU, SN31]
G4 ZXM S M Marshall, 1 Graham Cl, Stanford-le-Hope, SS17 8EB [JO01FM, TQ68]
G4 ZXN M Ward, 25 Margeson Cl, Hill, Coventry, CV2 5NU [IO92GJ, SP37]
G4 ZXO P S Horbaczewskyj, 11 Tadworth Ave, New Malden, KT3 6DJ [IO91VJ, TQ26]
G4 ZXP V M Legge, 26 Goldcroft Ave, Weymouth, DT4 0ET [IO80SO, SY67]
G4 ZXQ L T Mainwaring, Conway, Newlands Dr, Leominster, HR6 8PR [IO82OF, SO45]
G4 ZXR J E Douglas, 44 Westernmoor, Blackfell2, Washington, NE37 1LP [IO94FV, NZ25]

G4	ZXS	E R Loach, 16 Cheriton Gr, Wolverhampton, WV6 7SP [IO82VO, SO89]
G4	ZXT	M V Twigg, 30 Valley Dr, Yarm, TS15 9JQ [IO94IM, NZ41]
G4	ZXU	D R Angel, 67 Ashey Rd, Ryde, PO33 2UZ [IO90KR, SZ59]
G4	ZXV	W Bailey, 225 Holburne Rd, Kidbrooke, London, SE3 8HF [JO01AL, TQ47]
G4	ZXX	Details withheld at licensee's request by SSL.[Station located near Hatfield.]
G4	ZXZ	W A Johnson, 7 Monks Meadow, Crowland, Peterborough, PE6 0LJ [IO92WQ, TF20]
G4	ZYC	N W Hubbard, 16 Smithville Cl, St. Briavels, Lydney, GL15 6TN [IO81QR, SO50]
G4	ZYE	L V Bird, 159 Longford Ln, Gloucester, GL2 9HD [IO81VV, SO82]
G4	ZYF	D H Collins, 63 Church Rd, Hanham, Bristol, BS15 3AF [IO81RK, ST67]
G4	ZYH	M G Timms, 5 Lytchett Way, Nythe, Swindon, SN3 3PJ [IO91DN, SU18]
G4	ZYI	H S Wilson, 71 Worksop Rd, Woodsetts, Worksop, S81 8RW [IO93KI, SK58]
G4	ZYJ	Details withheld at licensee's request by SSL.
G4	ZYL	J L Anderson, 38 Redwood Dr, Chase Terr, Walsall, West Midlands, WS7 8AS [IO92AQ, SK00]
GW4	ZYM	W Williams, Plum Tree Farm, Commonwood Rd, Holt, Wrexham, LL13 9TA [IO83NB, SJ35]
G4	ZYN	M Sherlock, 9 Conway Rd, Eccleston, Chorley, PR7 5SW [IO83PP, SD51]
G4	ZYO	B M Lawrance, Penhale Jakes, Ashton, Helston, Cornwall, TR13 9SD [IO70HC, SW62]
G4	ZYP	M A Chandler, 5 Glendower Cl, Southdale, Hereford, HR2 7QG [IO82PA, SO53]
G4	ZYQ	A G Boswell, 57 Low Moorlands, Dalston, Carlisle, CA5 7PA [IO84MU, NY35]
G4	ZYR	H T Webber, 6 Barn Ground, Highnam, Gloucester, GL2 8LJ [IO81UV, SO72]
G4	ZYT	Details withheld at licensee's request by SSL.
GW4	ZYV	J T Raymond, 23 Castle Pill Cres, Steynton, Milford Haven, SA73 1HD [IO71LR, SM90]
G4	ZYX	G Astington, 2B Forest Hill, Yeovil, BA20 2PE [IO80QW, ST51]
G4	ZYY	G T Fildes, 67 Cranbrook Dr, Esher, KT10 8DN [IO91TJ, TQ16]
G4	ZYZ	T R Seymour, 77 Bucklesham Rd, Ipswich, IP3 8TR [JO02OA, TM24]
G4	ZZB	Details withheld at licensee's request by SSL.
G4	ZZD	A R Hellier, 64 Penlee Park, Torpoint, PL11 2PZ [IO70VJ, SX45]
G4	ZZF	Details withheld at licensee's request by SSL.[Station located near Fareham.]
G4	ZZG	C Wells, 1 Greenfield Cl, Forest Town, Mansfield, NG19 0DX [IO93KD, SK56]
GM4	ZZH	H M Firth, Edan, Berstane Rd, Kirkwall, KW15 1NA [IO88MX, HY41]
G4	ZZI	R Smith, 17 Belmont View, Harwood, Bolton, BL2 3QJ [IO83TO, SD71]
G4	ZZJ	J J Purdy, 66A Rustat Rd, Cambridge, CB1 3QN [JO02BE, TL45]
G4	ZZK	B L Tutt, 76 Reculver Rd, Beltinge, Herne Bay, Kent, CT6 6ND [JO01NI, TR16]
G4	ZZL	R A Lyons, 15 Winston Ave, Tiptree, Colchester, CO5 0JU [JO01JT, TL91]
G4	ZZM	J Griffiths, 32 Ashley Rd, Newmarket, CB8 8DA [JO02FF, TL66]
G4	ZZN	A T Rickward, 167 Earlshall Rd, Eltham Park, London, SE9 1PJ [JO01AL, TQ47]
GW4	ZZO	C Stowe, 40 Sycamore Way, Carmarthen, SA31 3QE [IO71UU, SN42]
G4	ZZP	K C Lock, 5 Copthorne Crest, Shrewsbury, SY3 8RU [IO82OR, SJ41]
G4	ZZR	Details withheld at licensee's request by SSL.
G4	ZZS	D J Bamber, 8 Brunel Rd, Stevenage, SG2 0AA [IO91VV, TL22]
G4	ZZU	Details withheld at licensee's request by SSL.
GM4	ZZW	R G Watts, 8 Oxford Ave, Gourock, PA19 1XU [IO75OW, NS27]
G4	ZZY	T M Watts, Carne Grey Cottage, Trethurgy, St Austell, Cornwall, PL25 3TB [IO70OI, SX05]
G4	ZZZ	T A Smith, Lower Carneggy Farm, Greenbottom, Chacewater, Truro, Cornwall, TR4 8QL [IO70KG, SW74]

G5

GW5	AF	Dr W F Floyd, Rhiw Wen, Pennant, Aberystwyth, Dyfed, SY23 5PD [IO72VF, SN56]
G5	AL	A B May, 14 High St., Sandridge, St. Albans, AL4 9DJ [IO91US, TL11]
G5	BH	M H Coleman, Flat A, 53 de Parys Ave, Bedford, MK40 2TR [IO92SD, TL05]
GW5	BI	J Evans, 2 Clos Coedydafarn, Lisvane, Cardiff, CF4 5ER [IO81JM, ST18]
G5	BK	J J Clayton C.A.R.A., 217 Prestbury Rd, Cheltenham, Glos, GL52 3ES [IO81XV, SO92]
G5	BM	F H Watts, Woodland View, Birches Ln, Newent, GL18 1DN [IO81TW, SO72]
G5	BR	G F Mason, 8 Highbury Rd, Streetly, Sutton Coldfield, B74 4TF [IO92BO, SP09]
GU5	BVQ	C M J Hurel, Pennyfarthing, 4-le-Bourgage, Alderney, Guernsey
G5	BW	W J Waugh, 67 Cragside, Whitley Lodge, Whitley Bay, NE26 3EF [IO95GB, NZ37]
G5	BZ	G G E Bennett, Lostwithiel, Northdown Rd, Woldingham, Caterham, CR3 7BB [IO91XG, TQ35]
G5	CO	Details withheld at licensee's request by SSL.
G5	CW	E S Wilson, 66 Horseshoe Dr, Romsey, SO51 7TP [IO91GA, SU32]
G5	DQ	P J Broom, 7 Flamsteed Rd, Cambridge, CB1 3QU [JO02BE, TL45]
G5	DS	J L Danks, 57 Queens Dr, Surbiton, KT5 8PW [IO91UJ, TQ16]
G5	FD	F D Clough, Trees, Firs Ln, Off Old Rd, Bromyard, HR7 4BA [IO82RE, SO65]
G5	FH	L H Lee, 18 Kilmington Way, Highcliffe, Christchurch, BH23 5BL [IO90DR, SZ29]
G5	FS	A Melia Brunel Amateur Radio Club, 17 Grange Rd, Saltford, Bristol, BS18 3AH [IO81SJ, ST66]
G5	FZ	P G Rose Lincoln Short Wave Club, Pinchbeck Farmhouse, Mill Ln, Sturton By Stow, Lincoln, LN1 2AS [IO93PH, SK87]
G5	GC	G A H Eckles, 15 Cooper Ct, Salisbury Rd, Farnborough, GU14 7AZ [IO91PH, SU85]
G5	GW	S Bowden, 20 Richmond Ct, Oldway Rd, Paignton, TQ3 2TX [IO80FK, SX86]
G5	GZ	Dr G L Grisdale, Ramsau, The Avenue, Langport, Somerset, TA10 9SA [IO81OA, ST42]
G5	HD	L N G Hawkyard, The Eyry, Newton St Petrock, Torrington, N Devon, EX38 8LU [IO70VV, SS41]
G5	HF	H R Heap, Meadowside, Rectory Ln, Chelmsford, CM1 1RQ [JO01FR, TL70]
G5	HN	E M Handcocks, 1 Conisboro Way, Caversham, Reading, RG4 7HT [IO91ML, SU77]
G5	HY	D M Wilkins, 802 Kenton Ln, Harrow Weald, Harrow, HA3 6AG [IO91UO, TQ19]
G5	IJ	I J P James, 20 St. Ursula Gr, Pinner, HA5 1LN [IO91TO, TQ18]
G5	IX	W A Dix, Upgate, Carding Mill Valley, Church Stretton, SY6 6JE [IO82ON, SO49]
G5	JJ	D C Hall, Heather Cottage, Churchstanton, Taunton, Somerset, TA3 7PU [IO80KV, ST11]
G5	JP	Details withheld at licensee's request by SSL.
G5	JR	R J C Buckstone, 20 Greenleas Ave, Emmer Green, Reading, RG4 8TA [IO91ML, SU77]
G5	KA	Details withheld at licensee's request by SSL.
G5	KC	G W Kelley, 57 St. Georges Pl, York, YO2 2DT [IO93KW, SE55]
G5	KM	H H Eyre, Robin Hill, Ln Head Rd, Cawthorne, Barnsley, S75 4AA [IO93FN, SE20]
G5	KN	R F Smith obo Kettering and District, 405 Windmill Ave, Kettering, NN15 6PS [IO92PJ, SP87]
G5	KS	A C Bevington, 53 Knottsall Ln, Oldbury, Warley, B68 9LG [IO82XL, SO98]
G5	KW	Maj K E Ellis, 18 Joyes Rd, Folkestone, CT19 6NX [JO01OC, TR23]
G5	LK	R T Collins Reigate ARS, 8 Sylvan Way, Redhill, RH1 4DE [IO91WF, TQ24]
G5	LL	A H Lunn, 3 Tennyson Ave, Mablethorpe, LN12 1HF [JO03DI, TF58]
G5	LO	B M Crook Oxford & District Amateur Radi, 41 Radley Rd, Abingdon, OX14 3PL [IO91IQ, SU59][obo Oxford DARS.]
G5	LP	L Parker, 128 Northampton Rd, Wellingborough, NN8 3PJ [IO92PH, SP86]
G5	LW	A Littlewood, 29 Besthorpe Rd, Attleborough, NR17 2AN [JO02MM, TM09]
G5	LY	K C Lay, 53 Riders Bolt, Bexhill on Sea, TN39 4JY [JO00FU, TQ70]
G5	MS	Details withheld at licensee's request by SSL.
GI5	MV	M McVeigh, 46 North St., Lurgan, Craigavon, BT67 9AH [IO64UL, J05]
G5	MW	J Hale Medway A.R.T., obo Medway A R T, 136 Bush Rd, Cuxton, Rochester, ME2 1HB [JO01FJ, TQ76]
G5	MY	H Mee, 268 Victoria Rd Eas, Leicester, LE5 0LF [IO92KP, SK60]
G5	NB	N I Brown, 14 Gordon Terr, Middle Market Rd, Great Yarmouth, NR30 2EF [JO02UO, TG50]
GW5	NF	R J Ward, Lower Tonyfeln Farm, Croespenmane, Crumlim, Newport Gwent, NP1 4BE [IO81KQ, ST19]
GJ5	NO	A G Chambers, -le-Petit Chateau, La Route Des Landes, St. Ouen, Jersey, JE3 2AA
GM5	NU	W B H Lord, 2 Orchard Brae, Edingburgh, Edinburgh, EH4 1NY [IO85JW, NT27]
G5	NV	H S C Crownshaw, 46 Butler Rd, Sheffield, S6 5HS [IO93FJ, SK38]
G5	NZ	R Stokes, 102 Bury St., Ruislip, HA4 7TG [IO91SN, TQ08]
G5	OD	A Ogden, 21 Amberley Cl, Send, Woking, GU23 7BX [IO91RG, TQ05]
G5	OG	Lord C I Orr-Ewing, 6 Caroline Terr, London, SW1W 8JS [IO91WL, TQ27]
G5	OW	W O Wigg, 7 Brendon Way, Long Eaton, Nottingham, NG10 4JS [IO92IV, SK43]
GW5	PH	Details withheld at licensee's request by SSL.
G5	PI	J F Wilson Philips Tele Lt, 24 PO Box, St. Andrews Rd, Cambridge, CB4 1DP [JO02BF, TL46]
G5	PQ	D H Allerston, 343 Main Rd, Bilton, Hull, HU11 4DS [IO93VS, TA13]
G5	PW	Dr B Crisp, 19 Westroyd Ave, Hunsworth, Cleckheaton, BD19 4DS [IO93DR, SE12]
G5	QK	A P Radley Southend&Dis AR, 16 Kingsley Ln, Thundersley, Benfleet, SS7 3TU [JO01HN, TQ78]
G5	RI	F J U Ritson, Red Lion House, Corbridge Rd, Hexham, NE46 1UL [IO84XX, NY96]
G5	RL	B K Rowell, 14 Market Hill, St Ives, Hunteingdon, Cambs, PE17 4AL [IO92XH, TL37]
G5	RP	Dr I F White Vowhars, 52 Abingdon Rd, Drayton, Abingdon, OX14 4HP [IO91IP, SU49]
G5	RQ	L Cd G P Tonkin, Old Westmoreland, Farmhouse, Crow Ln, Henbury, BS10 7AG [IO81QM, ST57]
G5	RR	P Hart Hucknall Rolls Royce Amateur R, 112 Shelton Ave, Hucknall, Nottingham, NG15 7QA [IO93JA, SK54]
G5	RS	P Croucher Guildford Co Gr, Model Engineers HQ, London Rd, Guildford, GU1 1TU [IO91RF, TQ05]
G5	RV	R L Varney, 82 Folders Ln, Burgess Hill, RH15 0DX [IO90WW, TQ31]
G5	SD	D E Campbell, 96 Climping Park, Bognor Rd, Climping, Littlehampton, BN17 5DW [IO90RT, TQ00]
G5	SG	N A F Edwards, 9 Foxwood Ave, Mudeford, Christchurch, BH23 3JZ [IO90DR, SZ19]
GI5	SJ	S N Johnson, Apartment No 7, Rockfield, Greengraves Rd, Dundonald Belfast, N Ireland, BT16 0UZ [IO74CO, J47]
G5	TA	Details withheld at licensee's request by SSL.
GI5	TK	A R Irwin, Braeside, 250 Ballygowan Rd, Crossnacreevy, Belfast, BT5 7UB [IO74BN, J36]
G5	TU	J C H Tucker, Woodlands, Mawnan Smith, Falmouth, Cornwall, TR11 5HT [IO70KC, SW72]
GM5	UI	R H Perkis, Knockoudie, Kirkcolm, Stranraer, DG9 0PA [IO74KX, NX07]
G5	UM	D G Wills Leicester Radio Soc Contes, 70 Hidcote Rd, Oadby, Leicester, LE2 5PF [IO92LO, SP69]
GM5	VG	W B Miller obo Windy Yett Contest Group, Whiteleys Farm, Alloway, Ayr, KA7 4EG [IO75QJ, NS31]
G5	VH	P R Chapman, 112 Sharpland, Leicester, LE2 8UP [IO92KO, SK50]
G5	VO	J H Hargreaves, 87 High St., Bempton, Bridlington, YO15 1HP [IO94VD, TA17]
G5	VQ	E W Taylor, 168 Westbourne Gr, Westcliff on Sea, SS0 9TY [JO01IN, TQ88]
GM5	VS	Details withheld at licensee's request by SSL.
G5	VU	Details withheld at licensee's request by SSL.
G5	WW	P M Carment, Alberta House, Cess Rd, Martham, Gt Yarmouth Norfolk, NR29 4RQ [JO02TQ, TG41]
G5	XG	E J B Butcher, 8 Kelso Cl, The Ridings, Worth, Crawley, RH10 7XH [IO91WC, TQ33]
G5	XJ	J W Moorhouse, 15 Broadway, Royton, Oldham, OL2 5DD [IO83WN, SD90]
G5	XV	R Y Parry, 14 Old Bath Rd, Newbury, RG14 1QL [IO91IJ, SU46]
G5	XX	Details withheld at licensee's request by SSL.[Station located near Daventry.]
G5	YC	C J Isham Imperial Coll Rd, Prof. of Physics Dp, Blackett Lab Prince, Consort Rd London, SW7 2BZ [IO91VL, TQ27]
G5	YM	Details withheld at licensee's request by SSL.
G5	YN	Sir E Y Nepean, Allwyns, 43 Queens Rd, Devizes, Wilts, SN10 5HR [IO91AI, SU06]
G5	ZG	A J Judge Bishops Stortford AR Soc, 44 Thorley Ln, Bishops Stortford, CM23 4AD [JO01BU, TL41]
G5	ZK	R N Lawson, 2 Edenhurst Dr, Timperley, Altrincham, WA15 7AU [IO83UJ, SJ78]
G5	ZN	P Nicoll, 8 Egroms Ln, Withernsea, East Riding, Yorks, HU19 2LZ [JO03AR, TA32]

G6

G6	AAB	T H A Sloane, 42 Ashbury Dr, Hawley, Blackwater, Camberley, GU17 9HH [IO91OH, SU85]
G6	AAD	Details withheld at licensee's request by SSL.
GW6	AAG	F A Steadman, 10 Oaktree Ave, Sketty Park, Sketty, Swansea, SA2 8LL [IO81AO, SS69]
GM6	AAJ	G Scattergood, 10 Taranty Rd, Forfar, DD8 1JY [IO86NP, NO45]
G6	AAK	J H Smith, Red Lodge, 16 Cross Keys Osset, W Yorks, WF5 9SJ [IO93FQ, SE22]
G6	AAL	Details withheld at licensee's request by SSL.
G6	AAU	T J Booth, 12 Jerrymoor Hill, Finchampstead, Wokingham, RG40 4UG [IO91NJ, SU76]
G6	AAW	Details withheld at licensee's request by SSL.
G6	AAZ	K Woodward, 19 Hazel Gr, Winchester, SO22 4PQ [IO91IB, SU42]
G6	ABA	P H C Dobson, 16 Glenair Ave, Parkstone, Poole, BH14 8AD [IO90AR, SZ09]
G6	ABF	Details withheld at licensee's request by SSL.
G6	ABG	D Coldbeck, 101 Westlands Rd, Hull, HU5 5NX [IO93TS, TA02]
G6	ABL	Details withheld at licensee's request by SSL.
G6	ABM	Details withheld at licensee's request by SSL.
G6	ABO	R C Campbell, 2 Marlborough Rd, Wallasey, L45 1JE [IO83LK, SJ39]
G6	ABP	C J R Cave, 31 Mill Rd, Rearsby, Leicester, LE7 4YN [IO92LR, SK61]
G6	ABQ	Details withheld at licensee's request by SSL.
G6	ABR	Details withheld at licensee's request by SSL.[Station located near Willenhall.]
G6	ABU	M G Dale, 2 Ward Ave, Mapperley, Nottingham, NG3 6EQ [IO92KX, SK54]
G6	ABW	R G Daniels, 55 Dart Cl, Strood, Rochester, ME2 2HE [JO01FJ, TQ76]
G6	ABZ	Details withheld at licensee's request by SSL.
G6	ACA	Details withheld at licensee's request by SSL.
G6	ACC	Details withheld at licensee's request by SSL.
G6	ACD	C J Frost, 23 Stirling Way, Christchurch, BH23 4JJ [IO90DR, SZ19]
G6	ACE	Details withheld at licensee's request by SSL.
G6	ACJ	D J Frampton, 28 Horsham Rd, Owlsmoor, Sandhurst, GU47 0YY [IO91OI, SU86]
G6	ACL	A C Lewis, 2 Crosswood Cl, Loughborough, LE11 4BP [IO92JS, SK51]
G6	ACT	S C Harvey, 3 High Ln, Norton Tower, Halifax, HX2 0NW [IO93BR, SE02]
G6	ACY	Details withheld at licensee's request by SSL.
G6	ADD	T Hallam, 98 Keppel Rd, Sheffield, S5 0TY [IO93GK, SK39]
G6	ADG	M W Kennedy, 96 Kingsway, Boston, PE21 0AU [JO02AX, TF34]
G6	ADK	A C Leech, 8 Filey Ave, London, N16 6NT [IO91XN, TQ38]
GW6	ADM	Details withheld at licensee's request by SSL.
G6	ADO	S E Nicholas, Greenbank, Chester High Rd, Neston, South Wirral, L64 7TR [IO83LH, SJ37]
G6	ADQ	Details withheld at licensee's request by SSL.
G6	ADR	E Williams, 26 Andreas Ave, Walney Isle, Walney, Barrow in Furness, LA14 3JN [IO84IC, SD16]
G6	ADW	Details withheld at licensee's request by SSL.
G6	ADX	M J Waterman, 92 Spring Rd, Letchworth, SG6 3SJ [IO91VX, TL23]
G6	ADY	W Yates, 32 Lark Hill St., Preston, PR1 4JJ [IO83PS, SD52]
G6	AEB	S P Neil, 55 Colne Rd, Brightlingsea, Colchester, CO7 0DU [JO01MT, TM01]
G6	AEC	D C J Nicholls, 22 Yeo Way, Clevedon, BS21 7UP [IO81NK, ST37]
G6	AED	A R Neve, 5 Grasmere Ave, Sompting, Lancing, BN15 9UQ [IO90TT, TQ10]
G6	AEN	T J Haddon obo Sheffield (Sth Yorks) Rayn, 11 Lovetot Ave, Aston, Sheffield, S31 0BQ [IO93GJ, SK38]
GW6	AEO	J N McHardy, 4 Bryn Arnold, High St., Connahs Quay, Deeside, CH5 4ED [IO83LF, SJ26]
G6	AER	P H Forge, 18 Conference Way, Colkirk, Fakenham, NR21 7JJ [JO02KT, TF92]
GM6	AES	M Clark, 12 Achaphubil, Fort William, Inverness-shire, PH33 7AL [IO76KU, NN07]
G6	AFA	P Paskin, Palm Springs, 36 Lewarne Rd, Newquay, TR7 3JT [IO70LK, SW86]
G6	AFB	N I Bazley, 30 Sunningdale Way, Neston, South Wirral, L64 0UY [IO83LG, SJ27]
G6	AFE	R C Plested, 33 Hartbury Cl, Cheltenham, GL51 0NZ [IO81WV, SO92]
G6	AFG	A N Afford, 2 Holly Ct, Sandiway, Northwich, CW8 2PP [IO83PF, SJ57]
G6	AFK	J M Adams, 6 Austen Rd, Guildford, GU1 3NP [IO91RF, TQ04]
G6	AFL	P J Blay, Treetops, Mount Pleasant, Crewkerne, TA18 7AH [IO80OV, ST40]
G6	AFS	Details withheld at licensee's request by SSL.
G6	AFT	Details withheld at licensee's request by SSL.
G6	AFU	Details withheld at licensee's request by SSL.
G6	AFY	Details withheld at licensee's request by SSL.
G6	AFZ	M Tipper, 10 Lowdale Ave, Scarborough, YO12 6JW [IO94TG, TA08]
G6	AG	C McClelland, Four Elements, 12 Garners Rd, Chalfont St. Peter, Gerrards Cross, SL9 0EZ [IO91RO, TQ09]
G6	AGA	G S Clark, 2 Whitton Manor Rd, Isleworth, TW7 7NL [IO91TK, TQ17]
G6	AGF	R H C Chipperfield, 3 Dukes Dr, Halesworth, IP19 8DS [JO02RI, TM37]
G6	AGI	M Dale, 11 Cherry Gdns, Littlestone, New Romney, TN28 8QR [JO00LX, TR02]
G6	AGN	D Darby, 24 Bishops Hall Rd, Pilgrims Hatch, Brentwood, CM15 9NX [JO01DP, TQ59]
G6	AGO	B A M Bean, 81 Park Rd, Sutton Coldfield, B73 6BT [IO92CN, SP19]
G6	AGP	A G Patterson, 10 Pear Tree Cl, Barnston, Heswall, Wirral, L60 1YD [IO83LH, SJ28]
G6	AGQ	Details withheld at licensee's request by SSL.
G6	AGR	M E L V Taylor, 8 Clifford Cl, Long Eaton, Nottingham, NG10 3BT [IO92IV, SK43]
GW6	AGS	R G Thomas, 4 Duffryn Ave, Lakeside, Cardiff, CF2 6LF [IO81JM, ST18]
G6	AGT	R P Thomas, The Cottage, Village Farm, Runcorn Rd, Warrington Cheshir, WA4 6XH [IO83QI, SJ58]
G6	AGY	A Smith, 103 Station Rd, Seaham, SR7 0BD [IO94HU, NZ44]
G6	AGZ	G Smith, 103 Station Rd, Seaham, SR7 0BD [IO94HU, NZ44]
G6	AHA	M L Surplice, 43A Cremorne Rd, Four Oaks, Sutton Coldfield, B75 5AQ [IO92CO, SP19]
G6	AHC	T J Snook, 22 Garnault Rd, Enfield, EN1 4TS [IO91XP, TQ39]
G6	AHD	M E Sumner, Jaggen, Maldon Rd, Latchingdon, Chelmsford, CM3 6LF [JO01IQ, TL80]
G6	AHE	P J Young, 41 The Broadway, Gustard Wood, Wheathampstead, St. Albans, AL4 8LW [IO91UU, TL11]
G6	AHF	C Waterworth, 16 Fountains Walk, Lowton, Warrington, WA3 1EU [IO83RL, SJ69]
G6	AHH	C M Walden, The Briers, Scures Hill, Nately Scures, Hook, Hants, RG27 9JS [IO91MG, SU75]
G6	AHI	R A C Watkins, 15 Tweed Cl, Worcester, WR5 1SD [IO82VE, SO85]
G6	AHK	C J Wallwork, Baileys Farm Cttgs, 40-44 Henwood Green Rd, Pembury, Tunbridge Wells, TN2 4LF [JO01DD, TQ64]
G6	AHN	S J Reynolds, 12 Lowlands Cres, Great Kingshill, High Wycombe, HP15 6EG [IO91PQ, SU89]
G6	AHO	A H Oakes, 1 Culbeck Ln, Euxton, Chorley, PR7 6EP [IO83PP, SD51]
G6	AHQ	F Rhodes, 37 Croftway, Selby, YO8 9DD [IO93LS, SE63]
G6	AHR	R E Redpath, 18 Warwick Rd, Coulsdon, CR5 2EE [IO91WH, TQ26]
G6	AHV	J W Spriggs, 8 Kensington Dr, Spalding, PE11 2UU [IO92WS, TF22]
G6	AHX	S P Evans, 18 Hillview Ln, Twyning, Tewkesbury, GL20 6JW [IO82WA, SO83]
GW6	AHY	M S Freeta, 46 Grantchester Rise, Burwell, Cambridge, CB5 0BE [JO02DG, TL56]
G6	AIB	G K Farline, Willow Cottage, Manor View Rd, Lebberston, Scarborough, YO11 3PB [IO94TF, TA08]
G6	AID	M J Thomson N London Ryt As, 2 Bencroft Rd, Hemel Hempstead, HP2 5UY [IO91SS, TL00]
G6	AIF	Details withheld at licensee's request by SSL.
G6	AIG	H J E Gibson, 10 Trafalgar St., Cambridge, CB4 1ET [JO02BF, TL45]
G6	AIH	E J George, Orestans, Charlton Rd, Singleton, Chichester, PO18 0HT [IO90OV, SU81]
G6	AII	R P George, Timbers, Lake Ln, Barnham, Bognor Regis, PO22 0AD [IO90QT, SU90]
G6	AIK	J V L Gill, 36 Friar Rd, Brighton, BN1 6NH [IO90WU, TQ30]
G6	AIQ	M J Homer, 69 Ashdown Way, Saxon Meadows, Romsey, SO51 5QR [IO90GX, SU32]
G6	AIU	L T Harland, 332 Bonham Rd, Dagenham, RM8 3BP [JO01BN, TQ48]
G6	AIZ	M D Holmes, 15 Anderton Way, Garstang, Preston, PR3 1RF [IO83OV, SD44]
G6	AJ	E Bailey Barnsley & Dist Radio Soc, 8 Hild Ave, Cudworth, Barnsley, S72 8RN [IO93HN, SE30]
G6	AJA	M R Hunt, 10 Clayton Ct, Longridge, Preston, PR3 3UD [IO83QT, SD63]
G6	AJC	I Hodgkins, 2 Seagrave Rd, Coventry, CV1 2AA [IO92GJ, SP37]
GU6	AJE	M P Johnson, 6 Clos Des Jardinieres, Pointues Rocques, St. Sampson, Guernsey, GY2 4HW
G6	AJF	S B Jebb, 30 Runnymede, Nunthorpe, Middlesbrough, TS7 0QL [IO94JM, NZ51]
G6	AJG	T F Jenkins, 134 Frankland Rd, Croxley Green, Rickmansworth, WD3 3AU [IO91SP, TQ09]
G6	AJS	A J Sharp, 17 Beechwood Ave, Flanshaw Park, Wakefield, WF2 9JZ [IO93FQ, SE32]
G6	AJT	B G J Kenneally, 5 Havengore, Pitsea, Basildon, SS13 1JU [JO01GN, TQ78]
G6	AJU	J D Lees, Mini Manor, 37 High St., Amblecote, Stourbridge, DY8 4DG [IO82WL, SO88]

G6 AJW D L Lucas, The Maltings, 98 Crib St., Ware, SG12 9HG [IO91XT, TL31]
G6 AJX S Lampard, 111 Whitworth Way, Wilstead, Bedford, MK45 3EF [IO92SB, TL04]
G6 AJY N J Lawlor, 22 Old Lodge Ln, Purley, CR8 4DF [IO91WH, TQ36]
G6 AKC G Lewczenko, 51 Main Rd, Brookville, Thetford, IP26 4RG [JO02GM, TL79]
G6 AKE E R Arnold, 5 Nanjivey Terr, Higher Stannack, St. Ives, TR26 1BQ [IO70GF, SW54]
G6 AKF B H Afford, 419 Oundle Rd, Orton Longue Ville, Orton Longueville, Peterborough, PE2 7DA [IO92UN, TL19]
G6 AKG R W Ayley, 1 Ballam Cl, Upton, Poole, BH16 5QT [IO80XR, SY99]
G6 AKK P S Archer, 3 Wheatfield Cl, Macclesfield, SK10 2TT [IO83WG, SJ97]
G6 AKL R Boddy, 3 Forefield Green, Springfield, Chelmsford, CM1 6YU [JO01GS, TL70]
G6 AKN M J Bentley, 9 Tinkers Castle Rd, Selsdon, Seisdon, Wolverhampton, WV5 7HF [IO82UN, SO89]
G6 AKP A J Bruce, 30 Longleat Cres, Chilwell, Beeston, Nottingham, NG9 5EU [IO92JW, SK53]
G6 AKS F Barwell, 104 Stamford Rd, Kettering, NN16 8LW [IO92PJ, SP87]
G6 ALA Details withheld at licensee's request by SSL.
G6 ALB Details withheld at licensee's request by SSL.
G6 ALG N A F Cutmore, 3 Linden Cl, Tadworth, KT20 5UT [IO91VH, TQ25]
G6 ALI Details withheld at licensee's request by SSL.
G6 ALJ T Collins, 11 Sutton Rd, Maidstone, ME15 9AE [JO01GF, TQ75]
G6 ALK W E Cobbett, 57 Brantwood Way, St. Pauls Cray, Orpington, BR5 3WA [JO01BJ, TQ46]
G6 ALM J Cheetham, 43 Green Ln, Garden Suburb, Oldham, OL8 3BA [IO83WM, SD90]
G6 ALN G Colclough, 20 Pembroke Dr, Eccesmere Port, Whitby, South Wirral, L65 6TD [IO83NG, SJ37]
G6 ALR R G Delamare, 8 Hollycroft Rd, Compton, Plymouth, PL3 6PP [IO70WJ, SX45]
G6 ALS C M Dix, 15 Hackleton Rise, Stratton, Swindon, SN3 4EF [IO91DN, SU18]
G6 ALU S A Drury, 25 Crosslands, Stantonbury, Milton Keynes, MK14 6AY [IO92OB, SP84]
G6 ALW B D Darby, 96 Bassnage Rd, Halesowen, B63 4HG [IO82XK, SO98]
G6 ALZ A D Davis, 2 Wolverhampton Rd, Essington, Wolverhampton, WV11 2DB [IO82XO, SJ90]
G6 AMF B R Elliott, Frogmoor Cottage, 41 Henwick Ln, Thatcham, RG18 3BN [IO91IJ, SU56]
GM6 AMI K J Ogston, 39 Dale Ave, East Kilbride, Glasgow, G75 9AW [IO75VS, NS65]
GW6 AMK W H E Needham, 16 Heol Tredeg, Upper Cwmtwrch, Swansea, SA9 2XD [IO81CS, SN71]
G6 AMM P L Needham, 217 Newbold Rd, Chesterfield, S41 7AB [IO93GF, SK37]
GM6 AMP D J McInerney, 7 Clent Rd, Reading, RG2 0ES [IO91MK, SU77]
G6 AMW C D Williams, 133 Devon Dr, Chandlers Ford, Eastleigh, SO53 3GJ [IO90HX, SU41]
G6 AMX P Helm, 90 Horne St., Bury, BL9 9HS [IO83UO, SD80]
G6 AMY R H Mills, 35 Nethan Dr, Aveley, South Ockendon, RM15 4RT [JO01DM, TQ58]
GI6 ANC A Murphy, 4 Brackenridge Gdns, Carrickfergus, BT38 8FN [IO74CR, J48]
G6 ANF Details withheld at licensee's request by SSL.[Op: Dr T E A Palmer.]
G6 ANI J E Baverstock, Meadow View, Newbridge, Cadnam, Southampton, SO40 2NW [IO90FW, SU21]
G6 ANJ C G Perrott, 15 Chestnut Dr, Claverham, Bristol, BS19 4LN [IO81OJ, ST46]
G6 ANM Details withheld at licensee's request by SSL.
G6 ANN Details withheld at licensee's request by SSL.
G6 ANO L P Goodwin, 16 Drummond Pl, Cranfield Park Rd, Wickford, SS12 9LL [JO01GO, TQ79]
G6 ANR S Garfirth, 19 Ingleside Dr, Stevenage, SG1 4RN [IO91VW, TL22]
G6 ANV B O Gulliford, 12 Hawthorn Rd, Eynsham, Witney, OX8 1NT [IO91HS, SP40]
G6 ANW J P Gatland, 36 Chatsworth Ave, Winnwesh, Winnersh, Wokingham, RG41 5EU [IO91NK, SU77]
G6 AOB A OBrien, 25 Baslow Dr, Heald Green, Cheadle, SK8 3HW [IO83VI, SJ88]
G6 AOF J P T Henshaw, Honeywood, 7A High St, Piddlehinton, Dorchester, Dorset, DT2 7DT [IO80SU, ST60]
G6 AOH R J P Hoblin, 4 Portiswood Cl, Pamber Heath, Tadley, RG26 3UQ [IO91KI, SU66]
G6 AOI D E Hayes, 19 Marsden Rd, Welwyn Garden City, AL8 6YQ [IO91VT, TL21]
GM6 AOJ W W Hay, 11 Lovat Rd, Glenrothes, KY7 4RU [IO86KE, NT29]
G6 AON D A Rhymes, Woodacre, 190 Woodstock Rd, Yarnton, Kidlington, OX5 1PW [IO91IT, SP41]
G6 AOS S V Pilbeam, 74 Southbank Ave, Marston, Blackpool, FY4 5BX [IO83LT, SD33]
G6 AOV C White, Hamdon, 20 The Beacon, Ilminster, TA19 9AH [IO80NW, ST31]
G6 AOW E Palmer, 6 Tonford Ln, Canterbury, CT1 3XU [JO01MG, TR15]
G6 APB G G Taylor, 26 Moor Ln, Kirkburton, Huddersfield, HD8 0QS [IO93DO, SE11]
G6 APD L H Sawford, 6 Prospect Sq, Westbury, BA13 3ET [IO81VG, ST85]
G6 APF C R Sweeting, 47 Barnett Cl, Wonersh, Guildford, GU5 0SD [IO91RE, TQ04]
G6 APH C R Shiradski, 69 Masefield Ave, Borehamwood, WD6 2HG [IO91UP, TQ19]
G6 APJ Revd G J Smith, 41 Furners Mead, Henfield, BN5 9JA [IO90UW, TQ21]
G6 APN B Dixon, 97 Sunny Blunts, Peterlee, SR8 1LN [IO94HR, NZ43]
G6 APO Details withheld at licensee's request by SSL.[Op: Jon Stanley, ACGI.]
G6 APQ F W Hill, 12 Woodbine Walk, Chelmsley Wood, Birmingham, B37 6SB [IO92DL, SP18]
GW6 APR D S Hughes, 28 Meadows View, Marford, Wrexham, LL12 8LS [IO83MC, SJ35]
G6 APT Details withheld at licensee's request by SSL.
G6 APU Details withheld at licensee's request by SSL.
G6 APV B J Hood, 14 Droveway Gdns, St. Margarets Bay, Dover, CT15 6BS [JO01QD, TR34]
G6 APZ D Hardy Derbyshire Hill, Con Gp Thorntree Hs, Wensley Matlock, Derbyshire, DE4 2LL [IO93ED, SK26]
GM6 AQB A Riddell, 16 Lewis Dr, Old Kilpatrick, Glasgow, G60 5LE [IO75SW, NS47]
G6 AQC S Wyatt, 55 Ridgefield Rd, Oxford, OX4 3BX [IO91JR, SP50]
G6 AQF Details withheld at licensee's request by SSL.
G6 AQG M Scott, 69 Farmanby Cl, Thornton Dale, Pickering, YO18 7TE [IO94PF, SE88]
G6 AQH I M Scott, 69 Farmanby Cl, Thornton Dale, Pickering, YO18 7TE [IO94PF, SE88]
G6 AQI T N Smith, 1 St. Jude Gdns, St. Johns, Colchester, CO4 4QJ [JO01LV, TM02]
G6 AQL A Ryan, 16 Bradgate Ave, Heald Green, Cheadle, SK8 3AQ [IO83VI, SJ88]
GW6 AQN Details withheld at licensee's request by SSL.
G6 AQP Details withheld at licensee's request by SSL.[Op: P A Williams.]
G6 AQR Details withheld at licensee's request by SSL.
G6 AQV A C Wright, 16 Kennedy Cl, Faversham, ME13 7DW [JO01KH, TR06]
G6 AQW N B Wiltshire, 66 Neville Rd, Shirley, Solihull, B90 2QW [IO92BJ, SP17]
G6 AQX Details withheld at licensee's request by SSL.
G6 AQY C D Wilson, Essington Hall, Cottage Bognop Rd, Essington, Wolverhampton, WV11 2AZ [IO82XP, SJ90]
G6 AQZ B G Whitcombe, 20 Wordsworth Cl, Lichfield, WS14 9BY [IO92CQ, SK10]
GM6 ARB A N Hutchinson, 30 Fettercairn Gdns, Bishopbriggs, Glasgow, G64 1AY [IO75VV, NS67]
G6 ARM N S Kett, 40 The Fields, Tacolneston, Norwich, NR16 1DG [JO02NM, TM19]
G6 ARN Details withheld at licensee's request by SSL.
G6 ARO I J Kendall, 24 Deben Cres, Swindon, SN2 3QB [IO91CO, SU18]
G6 ART S Langton, Corner Cottage, Harlow Rd, Sheering, Bishops Stortford, CM22 7NB [JO01CT, TL41]
G6 ARU Dr D B Lowrie, 28 West End Ln, Pinner, HA5 1AQ [IO91TO, TQ18]
G6 ASA Prof N H Lipman, Meadowcroft, Cotswold Rd, Cumnor Hill, Oxford, OX2 9JG [IO91IR, SP40]
G6 ASH N D Ash, 30 Granta Rd, Sawston, Cambridge, CB2 4HT [JO02CC, TL44]
G6 ASK J Matthews, Highcroft Cottage, Rose Ash, South Molton, Devon, EX36 4RA [IO80DX, SS72]
G6 ASO Details withheld at licensee's request by SSL.[Op: R F Moore.]
G6 ASP Details withheld at licensee's request by SSL.
G6 ASW Details withheld at licensee's request by SSL.
GI6 ATD G Rodgers, 23 Rathmore Park, Bangor, BT19 1DQ [IO74DP, J48]
G6 ATG J C Raynor, Barley Cl, Church Ln, Great Longstone, Bakewell, DE45 1TB [IO93DF, SK27]
G6 ATJ Details withheld at licensee's request by SSL.
G6 ATK K R Austin, 13 North End Gr, North End, Portsmouth, PO2 8NF [IO90LT, SU60]
G6 ATL A C Andrew, Yonder Marsh, Northgate Ln, Grimoldby, Louth, LN11 8TG [JO03BJ, TF38]
G6 ATM Details withheld at licensee's request by SSL.
G6 ATS D J Bowen, 41 Paddock Caravan Park, Bristol Rd, Weston Super Mare, BS22 0BW [IO81NI, ST36]
GW6 ATT M Bryan, 10 Woodlands Rd, Barry, CF63 4EF [IO81IJ, ST16]
G6 ATW Details withheld at licensee's request by SSL.
GI6 ATZ G H Curry, 91 Burren Rd, Ballynahinch, Co. Down, BT24 8LF [IO74AJ, J35]
G6 AUC Prof H Whitfield, Chapel House, Whittonstall, Consett, DH8 9JP [IO94BV, NZ05]
G6 AUD S Challis, 22 Allens Rd, Ramsden Heath, Billericay, CM11 1JF [JO01FP, TQ79]
G6 AUE G R Cosham, Westgate, 7 Crakell Rd, Reigate, RH2 7DT [IO91VF, TQ24]
G6 AUK G Evans, 36 Walton Rd, Folkestone, CT19 5QS [JO01OC, TR23]
G6 AUO M F Graffham, 25 Southfield Ave, Edgbaston, Birmingham, B16 0JN [IO92AL, SP08]
G6 AUP B Goodyear, 13 Moorland Ave, Barnsley, S70 6PQ [IO93FN, SE30]
G6 AUQ J M Greenbank, Close Ln, Crouch Ln, Borough Green, Sevenoaks, Kent, TN15 8LU [JO01DG, TQ65]
G6 AUR B G Golding, 174 Forest Rd, Hanham, Kingswood, Bristol, BS15 2EN [IO81RK, ST67]
G6 AUS P A Humby, 126 Middleton Rd, Oswestry, SY11 2XA [IO82LU, SJ22]
G6 AUX P Henson, 154 Mercer Cres, Helmshore, Haslingden, Rossendale, BB4 4DQ [IO83UQ, SD72]
G6 AUY D Hawley, 3 Horse Croft Ln, Wharncliffe Side, Sheffield, S30 3EB [IO93GJ, SK38]
G6 AVF M Leeder, 15 Hazlemere View, Hazlemere, High Wycombe, HP15 7BY [IO91PP, SU89]
G6 AVI R E Tucker, Foxhall Cottage, Caston, Attleborough, Norfolk, NR17 1BL [JO02KN, TL99]
G6 AVK C J Thomson, 160 Downhall Rd, Rayleigh, SS6 9PD [JO01HO, TQ89]
G6 AVL H B Thompson, 21 Windsor Cres, Whitley Bay, NE26 2NT [IO95GA, NZ37]
G6 AVP A Rowe, 68 Cedarcroft Rd, Chessington, KT9 1RP [IO91UI, TQ16]
G6 AVT G T Stanhope, 39 Denham Cl, Stubbington, Fareham, PO14 2BQ [IO90JT, SU50]
G6 AVY D Lane, 230 Raeburn Ave, Eastham, Wirral, L62 8BB [IO83MH, SJ38]
GM6 AWA A W A Anderson, Mynwhirr, Greystone Ave, Kirkconnel, Dumfriesshire, DG4 6JU [IO85AJ, NS71]
G6 AWM C P Montgomery, 70 Campbell Rd, Twickenham, TW2 5BY [IO91TK, TQ17]
G6 AWO R R Mansel, Ashcroft House, Ashfield Rd, Elmswell, Bury St Edmunds, Suffolk, IP30 9HJ [JO02LF, TL96]
G6 AWP A R McHardy, The Haven, Hull Rd, Easington, Hull, HU12 0TE [JO03BP, TA31]
G6 AWT Details withheld at licensee's request by SSL.[Op: N Negus, 41 Oxstalls Lane, Longlevens, Gloucester GL2 9HP]
G6 AWV M W Needham, 21 Willows Cres, Birmingham, B12 9NS [IO92BK, SP08]
G6 AWY N Armstrong, 16 Clay Hill Rd, Sleaford, NG34 7TF [IO92SX, TF04]
G6 AXB J B Ashworth, 19 Montague Rd, Burnley, BB11 4JQ [IO83US, SD83]
G6 AXC R E G Beaumont, 11 Chaytor Cl, Southfield Park, Hedon, Hull, HU12 8PU [IO93VR, TA12]
G6 AXE G R Broad, 14 Albion Rd, Westcliff on Sea, SS0 7DR [JO01IN, TQ88]
G6 AXK P S Butler, 25 Orrishmere Rd, Cheadle Hulme, Cheadle, SK8 5HP [IO83VJ, SJ88]
G6 AXM Details withheld at licensee's request by SSL.
G6 AXO A P Bell, 24 Onslow Gdns, Ongar, CM5 9BG [JO01DR, TL50]
G6 AXR F J Bayliss, 33 Buckingham Rd, Castle Bromwich, Birmingham, B36 0JP [IO92CL, SP18]
G6 AXW P S Chapman, 82 East St., Leven, Beverley, HU17 5NG [IO93UV, TA14]
G6 AXY P J Coombes, Two Farthing, Soldiers Rise, Finchampstead, Wokingham, RG40 3NF [IO91OJ, SU86]
GM6 AXZ K V Cocks, 60 Palmerston Pl, Edinburgh, EH12 5AY [IO85JW, NT27]
G6 AYD D J Chorley, Sunnylands, Sandpitts Hill, Curry Rivel, Langport, TA10 0NG [IO81NA, ST42]
G6 AYG A Chell, 65 High St., Newchapel, Stoke on Trent, ST7 4PU [IO83VC, SJ85]
G6 AYH K D Cooke, 28 Curland Pl, Longston, Stoke on Trent, ST3 5JL [IO82WX, SJ94]
G6 AYK Details withheld at licensee's request by SSL.
GW6 AYM P D A Roberts, 21 Copley Lodge, Bishopston, Swansea, SA3 3JJ [IO71XO, SS58]
G6 AYN L Smith, 20 Kirkcroft Dr, Killamarsh, Sheffield, S31 8GY [IO93GJ, SK38]
GW6 AYR R J Shearing, Woodstock 6 Fairvie, Clwydyfagwr, Merthyr Tydfil, Mid Glam, CF48 1HW [IO81HS, SO00]
G6 AYS T Ramsden, 9 Clevedon Ct, Somerset Est, Battersea, London, SW11 3NN [IO91VL, TQ27]
G6 AYU P H Rice, 4 Council St., Peterborough, PE4 6AQ [IO92UO, TF10]
GU6 AYV J E Ranklin, Blanches Pierres Ln, St. Martins, St. Martin, Guernsey, Channel Islands, GY4 6SA
GM6 AYW B W Robson, Brambleside, Dunmore St, Balfron, Glasgow, G63 0PZ [IO76UB, NS58]
G6 AYX D A Robinson, 60 Langer Rd, Felixstowe, IP11 8HS [JO01QW, TM23]
G6 AYY T G Rumbold, 23 Montague Rd, Saltford, Bristol, BS18 3LA [IO81SJ, ST66]
GM6 AZA Details withheld at licensee's request by SSL.
G6 AZE A P Roberts, 9 Littlemoor Ln, Newton, Alfreton, DE55 5TY [IO93HD, SK45]
G6 AZL P H Tarmey, 17 Drews Ln, Ward End, Birmingham, B8 2QE [IO92BL, SP18]
G6 AZN P J Galer, 320 Broadway, Gillingham, ME8 6DU [JO01GJ, TQ76]
G6 AZP D J Glover, 16 Cardigan Gr, New Park, Trentham, Stoke on Trent, ST4 8XY [IO82VX, SJ84]
G6 AZQ Details withheld at licensee's request by SSL.
G6 AZR A Granshaw, 38 Tudor Gdns, Stony Stratford, Milton Keynes, MK11 1HX [IO92NB, SP73]
GI6 AZS Details withheld at licensee's request by SSL.
G6 AZU Details withheld at licensee's request by SSL.[Op: T J Gallon.]
G6 AZV J Goodier, 16 Firgrove Cl, North Baddesley, Southampton, SO52 9JP [IO90GX, SU31]
GW6 AZX R W Hughes, 4 Brittania Terr, Porthmadog, Gwynedd, LL49 9NB [IO72WW, SH53]
G6 BAA M R Dyde, 61 Eastfield Rd, Southsea, PO4 9EJ [IO90LS, SZ69]
G6 BAF W Furby, 25 Brecklands, Mundford, Thetford, IP26 5EF [JO02HM, TL89]
GW6 BAH G Davis, 2 New House, Ponthir Rd, Caerleon, Gwent, NP6 1PE [IO81MP, ST39]
G6 BAI Details withheld at licensee's request by SSL.
GI6 BAJ Details withheld at licensee's request by SSL.
G6 BAL D C de La Haye, 18 Crossley Moor Rd, Kingsteignton, Newton Abbot, TQ12 3LQ [IO80EN, SX87]
G6 BAM J G Draper, 42 Pitt St, Broadwaters, Kidderminster, Worcs, DY10 2UN [IO82VJ, SO87]
GM6 BAO A M Devine, 12 Auchengate, Barassie, Troon, KA10 6UG [IO75QN, NS33]
G6 BAT D D Falstein, 37 The Copse, Fareham, PO15 6EG [IO90JU, SU50]
G6 BAZ Details withheld at licensee's request by SSL.
G6 BBB Details withheld at licensee's request by SSL.
G6 BBC D A Pick Ariel Rad Group, BBC Birmingham Club, 168 Pebble Mill Roa, Edgbaston Birmingha, B5 7QQ [IO92BK, SP08]
G6 BBD R Hancock, 16 Buttermere, Off Windermere Dr, Wellingborough, NN8 3ZA [IO92PH, SP86]
G6 BBE A R Hudson, 16 Bridge Bungalows, Burstwick, Hull, HU12 9JS [IO93WR, TA22]
G6 BBG A M Harland, 3 Woodbury Rise, Malvern Link, Malvern, WR14 1QZ [IO82UD, SO74]
G6 BBH N J Burton, 63 Salcombe Dr, Glenfield, Leicester, LE3 8AG [IO92JP, SK50]
G6 BBI P T S Ward, 63 Salcombe Dr, Glenfield, Leicester, LE3 8AG [IO92JP, SK50]
G6 BBK S A Nelson, 10 Wragg Dr, Newmarket, CB8 7SD [JO02EG, TL66]
G6 BBM G D Tremain, 20 Grafton Way, West Molesey, KT8 2NW [IO91TJ, TQ16]
G6 BBN J N Temple-Heald, 12 Joscelynes, Stapleford, Cambridge, CB2 5EA [JO02BD, TL45]
G6 BBR M R Thomas, 17 Rectory Park Ave, Sutton Coldfield, B75 7BL [IO92CN, SP19]
G6 BBW J A Witts, 35 Warton Rd, Basingstoke, RG21 5HL [IO91LG, SU65]
G6 BCE S Wilders, 24 St. Annes Cl, Claines, Worcester, WR3 7PS [IO82VF, SO85]
G6 BCG R C Whitehouse, 2 Rivermead Ave, Darlington, DL1 3SG [IO94FN, NZ31]
G6 BCL N O Miller, 178 Warley Hill, Warley, Brentwood, CM14 5HF [JO01DO, TQ59]
G6 BCM S J Ward, 33 All Saints Way, Aston, Sheffield, S31 0FJ [IO93GJ, SK38]
G6 BCQ V A Mitchell, Wychwood, The Ridgeway, Cranleigh, GU6 7HR [IO91SD, TQ03]
G6 BCS A Harries Bromham Cmpsl, Arg, 38 Carterweys, Dunstable, LU5 4RB [IO91SV, TL02]
G6 BCV D E Horton, Millstone Shay Ln, Forton, Newport, Shropshire, TF10 8DA [IO82TS, SJ72]
G6 BCZ F M Heathcote, 91 Bishopscote Rd, Luton, LU3 1PA [IO91SV, TL02]
G6 BDA Details withheld at licensee's request by SSL.[Op: D P Harvey.]
G6 BDH J I Kennard, 52 Lavender Ln, Norton, Stourbridge, DY8 3EF [IO82VK, SO88]
GI6 BDI A R J King, 43 Orby Gdns, Belfast, BT5 5HS [IO74BO, J37]
GW6 BDM C M Parker, Charlotte Villa, Cleeve Hill, Cheltenham, Glos, GL52 3QE [IO81XW, SO92]
GI6 BDN R W Larke, 11 Ballymaconnell Rd South, Bangor, BT19 6DG [IO74EP, J58]
G6 BDS H H Roberts, 40 Derwent Rd, Meols, Wirral, L47 8XZ [IO83KJ, SJ28]
G6 BDU Details withheld at licensee's request by SSL.
G6 BDV Details withheld at licensee's request by SSL.[Station located near Harpenden.]
G6 BDW A Sibley, 25 Vesta Ave, St. Albans, AL1 2PG [IO91TR, TL10]
G6 BDY R A Southern, 208 Puxton Dr, Kidderminster, DY11 5HJ [IO82UJ, SO87]
G6 BEA Details withheld at licensee's request by SSL.[Op: A R Parker, 100 Truro Road, St Austell, Cornwall, PL25 5HH.]
G6 BEB J Lines, 2 Meadowcroft, Hagley, Stourbridge, DY9 0LJ [IO82WJ, SO87]
G6 BEH K S Penaluna, 5 Holkham Cl, Rushmere, Ipswich, IP4 5DW [JO02OB, TM14]
G6 BEI G D Pendrick, 23 Hazel Dr, Spondon, Derby, DE21 7DS [IO92HW, SK43]
G6 BEL S J Fairweather, Welcome Home, 65 Ambleside Ave, Elm Park, Hornchurch, RM12 5EU [JO01CN, TQ58]
G6 BEN A J Burke, Collingwood, 121C Upper Hale Rd, Farnham, GU9 0JG [IO91OF, SU84]
G6 BER S R Boote, 21 Cobham Way, Merley, Wimborne, BH21 1SJ [IO90AS, SZ09]
G6 BEZ Details withheld at licensee's request by SSL.
G6 BFD Details withheld at licensee's request by SSL.
G6 BFE G F Edwards, Hillcrest, Grange Rd, nr Ellesmere, Shropshire, SY12 9DL [IO82NV, SJ33]
G6 BFH Details withheld at licensee's request by SSL.[Op: R L Freail, 5 Dahmahoy Close, Nuneaton, Warks, CV11 6UB.]
G6 BFM A Green, 117 Acanthus Rd, Stoneycroft, Liverpool, L13 3DY [IO83NK, SJ39]
G6 BFN Dr C D Hill, 6 Roden Ave, Kidderminster, DY10 2RF [IO82VJ, SO87]
G6 BFP L S Humphrey, Four Gables, Gilletts Ln, High Wycombe, HP12 4BB [IO91OP, SU89]
G6 BFW C T Smith, First Impressions Ltd, 37-42 Compton St., London, EC1V 0AP [IO91WM, TQ38]
G6 BGA K J Turvey, Treforest, 12 Oak Lodge Ave, Chigwell, IG7 5HZ [JO01BO, TQ49]
G6 BGG L B Walker, 7 Chardstock Ave, Bristol, BS9 2RY [IO81QL, ST57]
G6 BGH I P Macdiarmid, 73 Stadium Ave, Blackpool, FY4 3QA [IO83LS, SD33]
GM6 BGQ D Small, Cartref, Barnyards, Kilconquhar, Leven, KY9 1LB [IO86OF, NO40]
G6 BGW Details withheld at licensee's request by SSL.
G6 BGY J M Meek, Flat, 1 St. Clements Ct, Hallam Rd, Clevedon, BS21 7SQ [IO81NK, ST47]
G6 BHA R M Smart, 67 Corkland Rd, Chorlton Cum Hardy, Manchester, M21 8XT [IO83UK, SJ89]
G6 BHB J D Seager, 14 The Guelders, Purbrook, Waterlooville, PO7 5QT [IO90LU, SU60]
G6 BHC A V Rawls, 2 Claude St., Crawcrook, Ryton, NE40 4DW [IO94CX, NZ16]
G6 BHE N V Rogers, 31 Feeches Rd, Prittlewell, Southend on Sea, SS2 6TE [JO01IN, TQ88]
G6 BHH D Palmer, Firdene, Abbey Rd, Medstead, Alton, GU34 5PB [IO91LD, SU63]
G6 BHI A F Palmer, Firdene, Abbey Rd, Medstead, Alton, GU34 5PB [IO91LD, SU63]
G6 BHK Details withheld at licensee's request by SSL.
G6 BHL V H Thomas, Stiperstones, King St., West Deeping, Peterborough, PE6 9HP [IO92TP, TF10]
G6 BHM Details withheld at licensee's request by SSL.
GW6 BHQ K Williams, 12 Llanyravon Way, Llanyravon, Cwmbran, NP44 8HN [IO81LP, ST39]
GM6 BHR R T Warbrick, 8 Bathurst Dr, Alloway, Ayr, KA7 4QN [IO75QK, NS31]
G6 BHS J M Watson, 88 Bath Rd, Cheltenham, GL53 7JT [IO81XV, SO92]
G6 BHW Details withheld at licensee's request by SSL.
G6 BHY R G V Vicarage, 10 Fleming Ave, Sidford, Sidmouth, EX10 9NY [IO80JQ, SY18]
G6 BIA R Thompson, 39 Grotto Rd, South Shields, NE34 7AQ [IO94HX, NZ36]
G6 BIF P N Amey, 163 Bath Rd, Bradford on Avon, BA15 1SW [IO81UI, ST86]
GM6 BIG D S Anderson, 20 Greenrig Rd, Hawksland, Lesmahagow, Lanark, ML11 9QA [IO85CP, NS84]
G6 BIL Details withheld at licensee's request by SSL.
G6 BIM J M Bowers, Mizpah, 20 Martin Ct, Ayton Village, Washington, NE38 0EP [IO94FV, NZ25]
G6 BIT Dr D N Crossley, 25 Newhaven Cl, Brandlesholme, Bury, BL8 1XX [IO83UO, SD71]
G6 BIU D J Carter, 23 First St., Low Moor, Bradford, BD12 0JQ [IO93CS, SE12]
G6 BIW Details withheld at licensee's request by SSL.
G6 BIX E J Donbavand, 6 Springmeadow, Charlesworth, Broadbottom, Hyde, SK14 6HP [IO93AK, SK09]
G6 BJB Details withheld at licensee's request by SSL.
G6 BJC P B Flatman, 44 Dryden Rd, Ipswich, IP1 6QP [JO02NB, TM14]
G6 BJG I R Hancock, 64 Swanswell Rd, Solihull, B92 7EY [IO92CK, SP18]
G6 BJJ I D Harley, 292 Tavy House, Duke St, Devonport, Plymouth, Devon, PL1 4HL [IO70VI, SX45]
G6 BJK P N Humphries, 7 Coniston Cl, Verwood, BH31 6HW [IO90BU, SU00]
G6 BJL R P H Harding, Daisy Mount Farm, Exeter Rd, Ottery St. Mary, EX11 1LE [IO80IS, SY09]

G6 BJO P N Mc Taggart, 33 Manor Farm Cl, Barton-le-Clay, Bedford, MK45 4TB [IO91SX, TL03]
G6 BJP J C R Middleton, 4 Bronyon Cl, Bury St. Edmunds, IP33 3XB [JO02IF, TL86]
G6 BJQ R J Hanrahan, 53 Main St., Walton, St, BA16 9QQ [IO81OC, ST43]
G6 BJR K H Hulbert, 15 St. Germans Rd, Forest Hill, London, SE23 1RH [IO91XK, TQ37]
G6 BJS J J W Howells, Tanglewood, Belle Vue Rd, Duryard, Exeter, EX4 5BP [IO80FR, SX99]
G6 BJY D S Vivash, 16 Whitchurch Cl, Cliveden View, Maidenhead, SL6 7TZ [IO91PM, SU88]
G6 BKD J P Scotney, 30 Trinity Rd, Rothwell, Kettering, NN14 6HY [IO92OK, SP88]
GM6 BKE F N Ryan, Cargenriggs, Islesteps, Dumfries, DG2 8ES [IO85EA, NX97]
GM6 BKK E R Meldrum, 53 Burnhead Rd, Hawick, TD9 8HB [IO85OK, NT51]
G6 BKL P Metcalfe, 65 Saville Rd, Whiston, Rotherham, S60 4DZ [IO93IJ, SK49]
G6 BKQ E Langford, Ponderosa Caravan Site, Eastmoor, Sutton-on-The Forest, Yorks, YO6 1ET [IO94KB, SE56]
G6 BKR M J Lee, 26 Cotton Dr, Ormskirk, L39 3AZ [IO83NN, SD40]
G6 BKT J L M Male, 4 Watford Cl, Witherwack, Sunderland, SR5 5SS [IO94HW, NZ35]
G6 BKY N B Arkwright, Greystone, Penrith Ave, Heysham, Lancs, LA3 2DJ [IO84NB, SD46]
G6 BLA S R Woodford, The Red Lion, Tedburn St. Mary, Exeter, Devon, EX6 6EQ [IO80DR, SX89]
G6 BLB Dr W C A Carey, Albion Cottage, 24 Bekesbourne Ln, Littlebourne, Canterbury, CT3 1UY [JO01OG, TR25]
G6 BLC B M Conway, 29 Mandeville Rd, Southgate, London, N14 7NJ [IO91WP, TQ29]
G6 BLH Details withheld at licensee's request by SSL.
G6 BLK A J J Johnson, Edelweiss, Boxley Rd, Walderslade, Chatham, ME5 9JG [JO01GH, TQ76]
GM6 BLL Details withheld at licensee's request by SSL.
G6 BLS R Spittle, 6 Honeybourne Way, Willenhall, WV13 1HN [IO82XO, SO99]
G6 BLU B J Nicholls, 29 Wittmead Rd, Mytchett, Camberley, GU16 6ER [IO91PH, SU85]
G6 BME D J Gibb, 4 Rosemount, Clarendon Rd, Wallington, Surrey, SM6 8RW [IO91WI, TQ26]
G6 BMG J E Hind, 14 The Slade, Silverstone, Towcester, NN12 8UH [IO92LC, SP64]
GM6 BML A J Ramsay, Tighnduin, 2 Queen St., Monifieth, Dundee, DD5 4HG [IO86OL, NO43]
GW6 BMP A Roberts, 14 Maeshyfryd Rd, Llangefni, LL77 7PY [IO73UG, SH47]
G6 BMQ E J Pinsent, 5 Crossways, Uplowman, Tiverton, EX16 7DL [IO80HW, ST01]
GW6 BMR S D Roberts, 3 West Gr, Merthyr Tydfil, CF47 8HJ [IO81HS, SO00]
G6 BMV J S Stokes, 13 Mill Hill, Braintree, CM7 3QR [JO01GU, TL72]
G6 BMY R Satterthwaite, 47 Aberford Rd, Baguley, Manchester, M23 1JY [IO83UJ, SJ88]
G6 BMZ M G Williams, 8 Scalpcliffe Rd, Stapenhill, Burton on Trent, DE15 9AA [IO92ET, SK22]
GI6 BNI D J M Mawhinney, 271 Old Belfast Rd, Bangor, BT19 1LU [IO74DP, J48]
G6 BNJ J A Bonnett, 87 Well Rd, Otford, Sevenoaks, TN14 5PT [JO01CH, TQ55]
G6 BNM F C Colvin, 98 Billericay Rd, Botney Hill, Billericay, CM12 9SL [JO01EO, TQ69]
G6 BNW Dr J Garcia-Rodriguez, 20 Broomfield, Martlesham Heath, Ipswich, IP5 7TP [JO02OB, TM24]
G6 BOB D M Lauder Univ Herts RC, Div of Electronic Engineering, University of Herts, College Ln, Hatfield, Herts, AL10 9AB [IO91VR, TL20]
G6 BOF G T Hollidge, Clifton Cl, Boundstone, Wrecclesham, Farnham, Surrey, GU10 4TP [IO91OE, SU84]
G6 BOI M Kendrick-Finn, 1 Orchard Cl, Yardley Gobion, Towcester, NN12 7UG [IO92NC, SP74]
G6 BOK P B King, 10 Heath Hey, Liverpool, Merseyside, L25 4TJ [IO83NJ, SJ48]
GW6 BOQ E A Parker, Charlotte Villa, Cleeve Hill, Cheltenham, Glousestershire, GL52 3QE [IO81XW, SO92]
G6 BOS J E Wilson, Queensgate, Ingoldsby Ave, Ingoldisthorpe, Kings Lynn, PE31 6NH [JO02GU, TF63]
G6 BOX S J Wilson, 13 Burma Rd, Duxford, Cambridge, CB2 4QP [JO02BC, TL44]
G6 BPB R W Panting, 27 Clayton Walk, Reading, RG2 7TT [IO91MK, SU77]
GI6 BPF Dr K Adamson, 29 Croft Rd, Crof Manor, Ballygally, Larne, BT40 2QP [IO74BV, D30]
G6 BPG R M Bennewitz, 83A Dalecroft Rise, Allerton, Bradford, BD15 9AT [IO93BT, SE13]
G6 BPH F O M Bennewitz, 1 Millfield Ave, Saxilby, Lincoln, LN1 2QN [IO93QG, SK87]
G6 BPJ P N Britten, 49 Rock Hill, Chipping Norton, OX7 5BA [IO91FW, SP32]
G6 BPK Dr S Cook, 19 Gloster Gdns, Wellesbourne, Warwick, CV35 9TQ [IO92EE, SP25]
G6 BPN R J Edmondson, 91 Lewin Rd, London, SW16 6JX [IO91WK, TQ27]
G6 BPY W R Roe, 4 Verdon Pl, Barford, Warwick, CV35 8BT [IO92EF, SP26]
G6 BQC M Stuart, 207 Saunders Ln, Mayford, Woking, GU22 0NT [IO91QH, SU95]
G6 BQH B C Ward, 1 Elizabeth Cres, Stoke Gifford, Bristol, BS12 6NY [IO81RM, ST67]
G6 BQJ C F Lee, 32B Station Rd, Harpenden, AL5 4SE [IO91TT, TL11]
G6 BQM P G T Bentley, Sandy Ridge, Church St., Rookery, Stoke on Trent, ST7 4RS [IO83VC, SJ85]
G6 BQQ M C Barnes, 23 Fullbrooks Ave, Worcester Park, KT4 7PE [IO91VJ, TQ26]
G6 BQW Details withheld at licensee's request by SSL.
G6 BRA I W N Pawson, 3 Orion, Roman Hill, Bracknell, RG12 7YX [IO91OJ, SU86]
G6 BRB N E Head, 13 Frobisher Rd, Stivichall, Coventry, CV3 6LW [IO92FJ, SP37]
GW6 BRC C S Beynon, 16 Hardy Cl, Woodfield Heights, Barry, CF62 9HJ [IO81IK, ST16]
G6 BRD W Hammond, 245 Broadoak Rd, Ashton under Lyne, OL6 8RP [IO83XM, SD90]
G6 BRE Details withheld at licensee's request by SSL.
G6 BRH Details withheld at licensee's request by SSL.
G6 BRJ R H Oakley, 21 Forton Cl, Compton, Wolverhampton, WV6 8AY [IO82VO, SO89]
G6 BRL C J K Timmins, 15 Blackthorn Cl, St. Albans, AL4 9RP [IO91US, TL10]
G6 BRM A M McMath, 96 Malvern Rd, North Shields, NE29 9ES [IO95GA, NZ36]
G6 BRO Details withheld at licensee's request by SSL.
G6 BRP P G Walter, 27 Longridge Rd, Woodthorpe, Nottingham, NG5 4LX [IO92KX, SK54]
G6 BRS P E Smith Bury Radio Soc, 52 Grantham Dr, Bury, BL8 1XW [IO83UO, SD71]
G6 BRU J P M Steele, 2 Cardigan Rd, Reading, RG1 5QL [IO91MK, SU77]
G6 BRV Details withheld at licensee's request by SSL.
G6 BRW S K Sumner, 7 St. Marys Cl, Pirton, Hitchin, SG5 3RG [IO91UX, TL13]
G6 BRY C L Thomas, 47 Johnson Rd, Erdington, Birmingham, B23 6PX [IO92BM, SP19]
G6 BSE P Brindley Bury St.Eds ARC, 2 Beech Park, School Rd, Great Barton, Bury St. Edmunds, IP31 2JL [JO02JH, TL97]
G6 BSF A M Booth, 71 Oversetts Rd, Newhall, Swadlincote, DE11 0SL [IO92FS, SK22]
G6 BSK D Crye, 68 Barco Ave, Penrith, CA11 8LY [IO84PP, NY53]
G6 BSO I P Findlay, 25 Musgrove Rd, New Cross, London, SE14 5PP [IO91XL, TQ37]
G6 BSU Details withheld at licensee's request by SSL.[Op: P I Harris.]
G6 BSX G D Johnson, 5 Ardmillan Cl, Oswestry, SY11 2JZ [IO82LU, SJ22]
G6 BTB C W Pringle, 38 Priory Rd, Littlemore, Oxford, OX4 4NE [IO91JR, SP50]
G6 BTC A I Layton, 17 Maplehurst, Leatherhead, KT22 9NB [IO91TG, TQ15]
G6 BTH B Blount, 14 Denver Ct, Stapleford, Nottingham, NG9 8LN [IO92IW, SK43]
GI6 BTN Details withheld at licensee's request by SSL.
G6 BTO A W Lloyd, Little Plawhatch, Plawhatch Ln, Sharpthorne, East Grinstead, RH19 4JL [IO91XB, TQ33]
G6 BTP E T Beswarick, 34 Bury Ln, Codicote, Hitchin, SG4 8XX [IO91VU, TL21]
G6 BTR M S Challis, 24 Petersfield Rd, Swindon, SN3 2AH [IO91DN, SU18]
G6 BTX K D Holmes, 313 Havering Rd, Romford, RM1 4BZ [JO01CO, TQ59]
G6 BUC Details withheld at licensee's request by SSL.
GJ6 BUK R P Taylor, 21 Samares Ave, La Grande Route de South, St. Clement, Jersey, JE2 6NY
G6 BUS S A Heather, 216 High Rd, Woodford Green, IG8 9HH [JO01AO, TQ39]
G6 BUT K Mott Harlow&Dist ARC, Mark Hall Barn, First Ave, Harlow Essex, CM20 2LE [JO01BS, TL41]
G6 BUU R J Costello, 6 Qua Fen Common, Soham, Ely, CB7 5DH [JO02EI, TL57]
G6 BUV A B Cutts, Highthorns Cottage, North Frodingham, Driffield, North Yorks, YO25 8LS [IO93UW, TA15]
GW6 BUW I C Davies, Llys yr Haul, Rhosmeirch, Llangefni, Mon, Gwynedd, LL77 7SX [IO73UG, SH47]
G6 BUY R M Gingell, 23 Woodfarm Rd, Malvern Wells, Malvern, WR14 4PL [IO82UB, SO74]
G6 BVN S P Daniels, 9 Moss Fold, Astley, Tyldesley, Manchester, M29 7FP [IO83SM, SD70]
GI6 BVQ T R Finlay, 4 Station Rd, Eglinton, Londonderry, BT47 3PR [IO65JA, C52]
G6 BVR R W S Gammage, Best's Cottage, Farley, Salisbury, Wilts, SP5 1AY [IO91DB, SU22]
G6 BWA C G Clarke, 11 Eastmoor Villas, Epworth Rd, Haxey, Doncaster, DN9 2LH [IO93OL, SK79]
G6 BWE K A V Edwards, 289 Monks Walk, Buntingford, SG9 9DY [IO91XW, TL32]
G6 BWK T G Wallis, 34 Belmont Way, Rochdale, OL12 6HR [IO83WP, SD81]
G6 BWN J G Stewart, 101 West Way, Lancing, BN15 8LZ [IO90UT, TQ10]
G6 BWP D W Weaver, 72 Harefield Ave, Worthing, BN13 1DR [IO90TT, TQ10]
G6 BWR Details withheld at licensee's request by SSL.
G6 BWT A S Bajjon, 35 Blackford Rd, Shirey, Shirley, Solihull, B90 4BU [IO92CJ, SP17]
G6 BWW P M Dudek, 104 Cranmere Rd, Melton Mowbray, LE13 1TB [IO92NS, SK72]
GW6 BWX Details withheld at licensee's request by SSL.
G6 BXA Details withheld at licensee's request by SSL.
G6 BXK S D G Crowther, 7 Slyne Rd, Torrisholme, Morecambe, LA4 6PA [IO84OB, SD46]
G6 BXO C H Blackwell, 20 Southworth Ave, Marston, Blackpool, FY4 3LH [IO83LS, SD33]
G6 BXR R Calvert, 97 Salisbury Rd, Great Yarmouth, NR30 4LS [JO02UO, TG50]
G6 BXS D E Ellison, 45 Somerville Dr, Bicester, OX6 7TU [IO91KV, SP52]
G6 BXV D W Willis, Rivendell, Upper Farringdon, Alton, Hants, GU34 3EJ [IO91MC, SU73]
GI6 BXY P D Stevens, 15 Beechcote Ave, Portadown, Craigavon, BT63 5DG [IO64SK, J05]
G6 BYF C D A Gomez, Keepers Cottage, Nightingale Ln, Ide Hill, Sevenoaks, TN14 6JA [JO01BF, TQ45]
G6 BYK J R Parkes, 65 Ferrier Rd, Chells, Stevenage, SG2 0NZ [IO91WV, TL22]
G6 BYL D C Lycett, 1 Saredon Cl, Pelsall, Walsall, WS3 4DH [IO92AO, SK00]
G6 BYT M N Bates, 11 Mannington Pl, South Wootton, South Wootton, Kings Lynn, PE30 3UD [JO02FS, TF62]
G6 BZ GpCt M C Bunting, Estcourt Mile Path, Hook Heath, Woking, Surrey, GU22 0JX [IO91QH, SU95]
G6 BZB Details withheld at licensee's request by SSL.
G6 BZE M James, Southwinds, Cadgwith, Ruan Minor, Helston, Cornwall, TR12 7JZ [IN79JX, SW71]
G6 BZG L A Green, 37 Park Rd, Northville, Bristol, BS7 0RH [IO81RM, ST67]
G6 BZH R E Harper, 31 Somerford Rd, Broughton, Chester, CH4 0SY [IO83MD, SJ36]
G6 BZP G W Keene, 10 Pembroke Cl, Horwich, Bolton, BL6 7TB [IO83RO, SD61]
G6 BZQ G J E Doubleday, 1 St. Johns Ave, Chelmsford, CM2 0UA [JO01FR, TL70]
G6 BZV C Barker, 1 Balfour Rd, Southport, PR8 6LE [IO83MP, SD31]
GW6 BZX Details withheld at licensee's request by SSL.
G6 CAA P J West, Alpine House, Quarry Rd, Winchester, Hants, SO23 0JG [IO91IB, SU42]
G6 CAC J E Hallett, 16 Streche Rd, Swanage, BH19 1NF [IO90AO, SZ08]
G6 CAF T A Morgan, 24 Kinloss Rd, Carshalton, SM5 1BH [IO91VJ, TQ26]
G6 CAO Details withheld at licensee's request by SSL.
G6 CAR A Baldwin, Rathlin, Dromnea, Kilcrohane, Co Cork, Ireland
G6 CAT L Hine, 9 Well St., Ulverston, LA12 7EG [IO84KE, SD27]
G6 CBB D A Beddow, 34 Loweswater Rd, Stourport on Severn, DY13 8LP [IO82UI, SO87]
G6 CBL D S Leslie, 8 The Avenue, Swarland, Morpeth, NE65 9JL [IO95DH, NU10]
G6 CBP A B Pidgeon, 106 Winchester Ave, St. Johns, Worcester, WR2 4JQ [IO82VE, SO85]
G6 CBU A A Clarke, 25 Bagshot Green, Bagshot, GU19 5JR [IO91PI, SU96]
G6 CBY M L Jeeves, 52 Castlefields, Stead Rise, Gravesend, DA13 9EJ [JO01EJ, TQ66]
G6 CCA Details withheld at licensee's request by SSL.
G6 CCE M W Neyman, Elysium, 1A Daybrook Rd, London, SW19 3DJ [IO91VJ, TQ26]
G6 CCQ R K Powell, Manuela, Jack Haye Ln, Light Oaks, Stoke on Trent, ST2 7NG [IO83WB, SJ94]
G6 CCV N J Boid Bcd Electr Serv, Somerset House, Somerset St, Hull, HU3 3QH [IO93TR, TA02]
G6 CDK E W Prime, 81 Rope Ln, Shavington, Crewe, CW2 5DA [IO83SB, SJ65]
G6 CDT G Henshaw, 18 Queens Ave, Ilkeston, DE7 4DL [IO92IW, SK44]
G6 CDU G G Keeble, 25 Columbine Gdns, Walton on The Naze, CO14 8NL [JO01PU, TM22]
G6 CDV A G Morling, 33 Russell Ct, Chesham, HP5 3JH [IO91QR, SP90]
G6 CDW N L Miller, 3 Upwood Gorse, Tupwood Ln, Caterham, CR3 6DQ [IO91XG, TQ35]
G6 CEK I A Chivers, 3 Watts Cl, Hook, RG27 9NT [IO91MG, SU75]
G6 CEM E M Weir, 10 St. Georges Cres, Whitley Bay, NE25 8BJ [IO95GA, NZ37]
G6 CEP A F Kneebone, 34 Henver Rd, Newquay, Cornwall, TR7 3BN [IO70LK, SW86]
G6 CEQ Details withheld at licensee's request by SSL.[Op: R A Haynes, 17 Ellon Avenue, Rainhill, Merseyside, L35 0NZ.]
G6 CEZ R H Brand, 17 Park Rd, Fordingbridge, SP6 1EQ [IO90CW, SU11]
G6 CFA J P Carrick Smith, 15 The Vale, Oakley, Basingstoke, RG23 7LB [IO91JF, SU55]
G6 CFC G Purchon, 33 Lancaster Ave, Hitchin, SG5 1PA [IO91UW, TL12]
G6 CGB Details withheld at licensee's request by SSL.
G6 CGC R K Sheppard, 51 Marks Rd, Wokingham, RG41 1NR [IO91NK, SU86]
G6 CGD M W Ferriday Polytechnic of, West London, Wellington St, Slough Berks, SL1 1YD [IO91QM, SU98]
G6 CGF P L Denton, 42 Trafalgar Rd, Wallasey, L44 0EB [IO83LK, SJ39]
G6 CGN C Morris, 51 Rose Cres, Scawthorpe, Doncaster, DN5 9EW [IO93KM, SE50]
G6 CGO E T Parr, 18 Arundel Cl, Macclesfield, SK10 2NS [IO83WG, SJ97]
G6 CGQ R J B Hatch, 4 Springfield Cres, Parkstone, Poole, BH14 0LL [IO90AR, SZ09]
G6 CHA E V Povey, Schoolfields, Shiplake, Shiplake Cross, Henley on Thames, Oxon, RG9 4DH [IO91NL, SU77]
G6 CHC V Appleton, 17 Youd St., Leigh, WN7 4BY [IO83RL, SD60]
G6 CHD P G Bridle, 118 Ludlow Ave, Crewe, CW1 6DY [IO83SC, SJ75]
G6 CHH Details withheld at licensee's request by SSL.
G6 CHI A J Bowley, Plum Tree House, Walk Cl, Draycott, Derby, DE72 3PN [IO92HV, SK43]
G6 CHJ M P Carter, 29 Neale Rd, Halstead, Essex, CO9 1DL [JO01HW, TL83]
G6 CHO J F Duell, 7 Somerset House, The Farmlands, Northolt, UB5 5EP [IO91TN, TQ18]
G6 CHX P T Holland, 123 New Rd, Aston Fields, Bromsgrove, B60 2LJ [IO82XH, SO97]
G6 CIE R N Townsend, 3 Cranfield View, Darwen, BB3 2HP [IO83SQ, SD72]
G6 CIF D J Taylor, 6 Garrett Gr, Clifton Village, Nottingham, NG11 8PU [IO92JV, SK53]
G6 CII K J Sutton, 9 Babbacombe Dr, Ferryhill, DL17 8DA [IO94FQ, NZ33]
G6 CIO J L Robinson, 31 Church Rd, Banks, Southport, PR9 8ET [IO83MQ, SD32]
G6 CIP P Ralston, Laund House, 9 College Ave, Formby, Liverpool, L37 3JL [IO83LN, SD20]
G6 CIT R J F Young, 143 Rodmell Ave, Saltdean, Brighton, BN2 8PH [IO90XT, TQ30]
G6 CIZ T I Williams, 49 High St., Bodicote, Banbury, OX15 4BP [IO92IA, SP43]
G6 CJB P A P White, 8 Kingswood Ct, Maidenhead, SL6 1DD [IO91PM, SU88]
G6 CJF Details withheld at licensee's request by SSL.
GW6 CJJ Dr J L Alexander, Awel-Ingli, Cilgwyn, Newport, Dyfed, SA42 0QS [IO71OX, SN03]
G6 CJK K G Allen, 14 Beechwood Ave, Melton Mowbray, LE13 1RT [IO92NS, SK72]
G6 CJN M J Bowyer, 173 Brettenham Rd, London, E17 5AX [IO91XO, TQ39]
G6 CJR Details withheld at licensee's request by SSL.
G6 CJT B I Bradshaw, 28 Park House Walk, Lowmoor, Low Moor, Bradford, BD12 0PL [IO93CS, SE12]
G6 CJW R S Linney, Sunny Bank, Oak Ln, Bicton Heath, Shrewsbury, SY3 5BW [IO82OR, SJ41]
G6 CKD L Newbury, 37 Johns Ave, Hendon, London, NW4 4EN [IO91VO, TQ28]
G6 CKE C J Evans, 21 Snowdrop Cl, Crawley, RH11 9EG [IO91VC, TQ23]
G6 CKG H M Neary, 19 St. Pauls St., Burslem, Stoke on Trent, ST6 4BZ [IO83VB, SJ84]
G6 CKH J Muir, 150 Thorntree Rd, Thornaby, Stockton on Tees, TS17 8LX [IO94IN, NZ41]
G6 CKJ D J Morris, 10 Addison Pl, Bilston, WV14 7BD [IO82XN, SO99]
G6 CKK Details withheld at licensee's request by SSL.
G6 CKL I A M Martin, 24 Heddington Cl, Trowbridge, BA14 0LH [IO81VH, ST85]
G6 CKM D S Langdon, 2 Ennerdale Rd, Wistaston, Crewe, CW2 8RT [IO83SC, SJ65]
G6 CKN R G Morrison, 37 Grosvenor Cres, Hyde, SK14 5AN [IO83XK, SJ99]
G6 CKR R N Beever, 48 Granville Ave, Northborough, Peterborough, PE6 9DE [IO92UP, TF10]
G6 CKW R J Beattie, 11 Pine Gr, Bricket Wood, St. Albans, AL2 3ST [IO91TQ, TL10]
G6 CKY M Gray, 20 Ravenstone St., London, SW12 9SS [IO91WK, TQ27]
G6 CLA G Blacksell, 152 Hawthorn Ave, Colchester, CO4 3YA [JO01KW, TL92]
G6 CLD G C Coker, 46 Clarendon Rd, Ipplepen, Newton Abbot, TQ12 5QS [IO80EL, SX86]
G6 CLE Details withheld at licensee's request by SSL.[Op: J Curran. Station located near Hayling Is.]
G6 CLK P J Carter, 107 Cranford Ln, Heston, Hounslow, TW5 9HQ [IO91TL, TQ17]
G6 CLL J Chadwick, 47 Red Hall Ln, Leeds, LS14 1NT [IO93GU, SE33]
G6 CLO Details withheld at licensee's request by SSL.
G6 CLP J J Miller, 7 Malvern Cres, Ashby de La Zouch, LE65 2JZ [IO92GS, SK31]
G6 CLU D J P Lawes, 87 Glebelands, Crayford, Dartford, DA1 4RY [JO01CK, TQ57]
G6 CLW B Lloyd, 243 Stand Ln, Radcliffe, Manchester, M26 1JA [IO83UN, SD70]
G6 CLX D C Lloyd, 252 Swan Ln, Hindley Green, Hindley, Wigan, WN2 4EY [IO83RM, SD60]
GI6 CMA R H Dawson, 11 Deramore Ave, Backwood Rd, Moira, Craigavon, BT67 0PY [IO64VL, J16]
G6 CMD C M Driver, 14 Worcester Way, Daventry, NN11 4TY [IO92KF, SP56]
G6 CMF A F Daborn, 39 Juniper Cl, Guildford, GU1 1NX [IO91RG, SU95]
G6 CMG Details withheld at licensee's request by SSL.
G6 CMK Details withheld at licensee's request by SSL.
G6 CML J C Sykes, 20 Woodend Rd, Winton, Bournemouth, BH9 2JQ [IO90BR, SZ09]
G6 CMN A R Shaw, 14 Delph Cres, Clayton, Bradford, BD14 6RY [IO93CS, SE13]
G6 CMP C G Sheldon, Whitehaven, May Tree Rd, Lower Moor, Pershore, WR10 2NY [IO82XC, SO94]
GM6 CMQ D J F Robson, 4F2, 316 Morningside Rd, Edinburgh, Midlothian, EH10 4QH [IO85JW, NT27]
G6 CMS M A Robertson, 13 Orchard Cttgs, Main Rd, Boreham, Chelmsford, CM3 3AD [JO01GS, TL71]
G6 CMV D A Palmer, 15 Albion Walk, Malvern Link, Malvern, WR14 1PX [IO82UD, SO74]
G6 CMX J E Pell, 33 Low St., Winterton, Scunthorpe, DN15 9RT [IO93QP, SE91]
G6 CMZ A Paterson, 70 Ernest Rd, Wivenhoe, Colchester, CO7 9LQ [JO01LU, TM02]
G6 CNB K G Holley, Greenacres, The Hill, Sandbach, CW11 1FG [IO83TD, SJ76]
G6 CND J R Oliver, 67 High St., Great Houghton, Barnsley, S72 0AU [IO93HN, SE40]
G6 CNF J A Payne, 71 Waarden Rd, Canvey Island, SS8 9AB [JO01HM, TQ78]
G6 CNL P H Farnell, 40 Thorney Ln, Midgley, Luddendenfoot, Halifax, HX2 6UX [IO93AR, SE02]
G6 CNO S Fernie, 40 Muswell Ave, Muswell Hill, London, N10 2EG [IO91WO, TQ29]
G6 CNQ T M Genes, 2 Coltishall Cl, Shotgate, Wickford, SS11 8XN [JO01GO, TQ79]
G6 CNR F A S Gomm, 8 Heathmoors, Bracknell, RG12 7NR [IO91PJ, SU86]
GW6 CNS J Graham, 23 Somerset Rd, Barry, CF62 8BL [IO81IJ, ST16]
G6 CNX J M Goodwin, 22 Willingham Gdns, Sothall, Sheffield, S19 6PE [IO93GJ, SK38]
G6 CNZ C Gower, The Shrubbery, 3 Fortis Way, Salendine Park, Huddersfield, HD3 3WW [IO93BP, SE11]
G6 COB J W Hodkinson, 3 Wolvesey Pl, Winsford, CW7 1HE [IO83RE, SJ66]
G6 COE C Hill, 16 Rombalds Gr, Leeds, LS12 2BB [IO93ET, SE23]
G6 COG D W Holdsworth, Middle Pasture, Heath Ln, Halifax, HX3 0AG [IO93BR, SE02]
G6 COH Details withheld at licensee's request by SSL.
G6 COK L W Hill, The Hollies, 21 Old Croft Ln, Castle Bromwich, Birmingham, B36 0AR [IO92CM, SP18]
G6 COL P G Rose Lincoln Shortwave, Pinchbeck Farmhouse, Mill Ln, Sturton By Stow, Lincoln, LN1 2AS [IO93PH, SK87]
G6 CPE K R Stanley, 35 St. Blaize Rd, Romsey, SO51 7JY [IO90GX, SU32]
G6 CPF J M Stephenson, 16 Greenways, Driffield, YO25 7HX [IO94SA, TA05]
G6 CPI Details withheld at licensee's request by SSL.
G6 CPO N Wysocki, Fistral, Summerfield Rd, Stourport on Severn, DY13 9BE [IO82UI, SO87]
G6 CPS A R Yates, Trumeau, Graham Dr, Middleton, Kingslynn, Norfolk, PE32 1RL [JO02FR, TF61]
G6 CPX M I Waples, 54 Oakley Dr, Wellingborough, NN8 3JZ [IO92PH, SP86]
G6 CPY E G Whitham, 72 Bole Hill, Treeton, Rotherham, S60 5RE [IO93HJ, SK48]
G6 CPZ Details withheld at licensee's request by SSL.
G6 CQA Details withheld at licensee's request by SSL.
G6 CQB M P Wilson, 23 Claydown Way, Slip End, Luton, LU1 4DU [IO91SU, TL01]
G6 CQC A A Varty, Wisteria, Hillcrest, Burnhope, Durham, DH7 0BQ [IO94DT, NZ14]
G6 CQD Details withheld at licensee's request by SSL.
G6 CQF R Tomalski, 18 Fitch Ct, Laburnum Rd, Mitcham, CR4 2ND [IO91WJ, TQ26]
G6 CQG L K Constantine, The Old Exchange, Burnley Rd, Mytholmroyd, Hebden Bridge, HX7 5PD [IO93AR, SE02]
G6 CQH J B Abbishaw, Hastings House Farm, Littletown, Sherburn Hill, Co Durham, DH1 2SQ [IO94FS, NZ34]

G6 CQM Details withheld at licensee's request by SSL.
G6 CQO J Britton, 12 Bulkeley Ave, Windsor, SL4 3LP [IO91QL, SU97]
G6 CQR C A Bailey, 32 Ryland Rd, Moulton, Northampton, NN3 7RE [IO92NG, SP76]
G6 CQT P Baugh, 87 Hazel Gr, Wombourne, Wolverhampton, WV5 9EH [IO82VM, SO89]
G6 CQZ Details withheld at licensee's request by SSL.
GW6 CRB Details withheld at licensee's request by SSL.
G6 CRC J Crabbe Cheshunt&Dis AR, 37 Warner Rd, Ware, SG12 9JN [IO91XT, TL31]
G6 CRD S Brown, 11 The Hedges, Poolhouse Meadow, Wombourne, Wolverhampton, WV5 8LD [IO82VM, SO89]
G6 CRF T Bailey, 65 Edge Ln, Chorlton Cum Hardy, Manchester, M21 9JU [IO83UK, SJ89]
G6 CRG B J Bowes, 1 Rockall Cl, Lordshill, Southampton, SO16 8EH [IO90GW, SU31]
G6 CRL Details withheld at licensee's request by SSL.
G6 CRM G L Smith, 43 Spinney Hill, Melbourne, Derby, DE73 1GT [IO92GT, SK32]
G6 CRR R Solomons, 32 Church Rd, Pembury, Tunbridge Wells, TN2 4BT [JO01DD, TQ64]
G6 CRV D J Staniforth, 3 Ferncliffe Dr, Heysham, Morecambe, LA3 1NZ [IO84NB, SD46]
G6 CRX F McLeod-Stangroom, 35 Parker Rd, Grays, RM17 5YW [JO01DL, TQ67]
G6 CSC W N Skidmore, Mires Ln, Rowland, nr Bakewell, DE4 1NP [IO93FC, SK25]
G6 CSK A S Beal, 115 Maldon Rd, Witham, CM8 1HR [JO01HT, TL81]
G6 CSN G Chadwick, 15 Winifred St., Passmonds, Rochdale, OL12 7ND [IO83VO, SD81]
G6 CSR H R Calloway, 6 Franchise Gdns, Wednesbury, WS10 9RQ [IO82XN, SO99]
G6 CSW I W Carpenter, 21 Jays Mead, Wotton under Edge, GL12 7JF [IO81TP, ST79]
G6 CSX E R Clark, 14 Bognor Rd, Chichester, PO19 2NF [IO90OU, SU80]
G6 CSY Details withheld at licensee's request by SSL.[Op: G Caselton, 19 Cowden Road, Orpington, Kent, BR6 0TP.]
G6 CTA J D Davidson, 12 Hanbury Cl, Dronfield, Sheffield, S18 6RF [IO93GH, SK37]
G6 CTC J Witt Covtry Tech ARC, 67 Dillotford Ave, Coventry, CV3 5DS [IO92FJ, SP37]
G6 CTH E J Dunne, 16 Ulleswater Cl, Little Lever, Bolton, BL3 1UD [IO83TN, SD70]
G6 CTK J L Davis, 29 Bytham Heights, Castle Bytham, Grantham, NG33 4ST [IO92RR, SK91]
G6 CTP H C Wakefield, 32 Mandene Gdns, Great Gransden, Sandy, SG19 3AP [IO92WE, TL25]
G6 CTR R M Williams, 4 Larkfield Cl, Farnham, GU9 7DA [IO91OF, SU84]
G6 CTV E J Eggs, 21 Nightingale Rd, Carshalton, SM5 2DN [IO91WI, TQ26]
G6 CTY C J Edwards, 54 Thoroughgood Rd, Clacton on Sea, CO15 6DP [JO01NT, TM11]
G6 CUA H C Erridge, 1 Oaklands Gdns, Titchfield-Common, Fareham, Hants, PO14 4LG [IO90IU, SU50]
G6 CUE J R Frampton, 54 Hudson Rd, Bexleyheath, DA7 4PG [JO01BL, TQ47]
G6 CUI R A Fallowfield, 73 Campsall Field Rd, Wath upon Dearne, Rotherham, S63 7SR [IO93HL, SE40]
G6 CUK A Fisher, 11 Rosedale Way, Forest Town, Mansfield, NG19 0QR [IO93KD, SK56]
G6 CUQ N E Wedgbury, 12 The Ridgeway, Astwood Bank, Redditch, B96 6LT [IO92AG, SP06]
GW6 CUR Details withheld at licensee's request by SSL.
G6 CUT J Whitehurst, 45 Carisbrooke Rd, Newport, PO30 1BU [IO90IQ, SZ48]
G6 CUV K R Wyeth, 3 West Palace Gdns, Weybridge, KT13 8PU [IO91SJ, TQ06]
G6 CUY J H Wildsmith, Lingmoor, 7 Lambert Rd, Uttoxeter, ST14 7QG [IO92BV, SK03]
G6 CVB J R H Taylor, 12 Fairview Dr, Westcliff on Sea, SS0 0NY [JO01IN, TQ88]
G6 CVD C Thornley, Sylvastone House, Herne St, Herne, Kent, CT6 7HG [JO01NI, TR16]
G6 CVE R M Tanfield, 8 Rede Cl, Bedford, MK41 7UH [IO92SD, TL05]
G6 CVH G A Stafford, 8 Flatholme Rd, Leicester, LE5 1LR [IO92LP, SK60]
G6 CVI Details withheld at licensee's request by SSL.
G6 CVP D C T Wilkins, 74 Wood Lodge Ln, West Wickham, BR4 9NA [IO91XI, TQ36]
G6 CVV M A Gumbrell, 47 Rycroft Ave, Deeping St. James, Peterborough, PE6 8NT [IO92UQ, TF10]
G6 CVW W E Griffiths, 6 Stanway Cl, Middleton, Manchester, M24 1HE [IO83VM, SD80]
G6 CVX Details withheld at licensee's request by SSL.
G6 CVY H C Gibbons, 15 Kilbride Ave, Bolton, BL2 6UQ [IO83TN, SD70]
G6 CW M C Shaw Amateur Radio Club of Nottingh, 50 White Rd, Nottingham, NG5 1JR [IO92JX, SK54]
G6 CWF C H H Hazell, 18 Cleeve Hill, Downend, Bristol, BS16 6HN [IO81RL, ST67]
G6 CWL Details withheld at licensee's request by SSL.
G6 CWR Details withheld at licensee's request by SSL.
G6 CWS Details withheld at licensee's request by SSL.
G6 CWU R G Hathaway, 105 Killigrew St., Falmouth, TR11 3PU [IO70LD, SW83]
G6 CWW V E B Holbrook, 84 Haddon St., Derby, DE23 6NQ [IO92GV, SK33]
GW6 CWZ D N McCallum, Trosgol, Deiniolen, Caernarfon, Gwynedd, LL55 3LU [IO73WD, SH56]
G6 CXI A P Long, 43 Heath Ct, Grampian Way, Sinfin, Derby, DE24 9NG [IO92GV, SK33]
G6 CXM G Lees, 35 Meadow Cl, Hockley Heath, Solihull, B94 6PF [IO92CI, SP17]
G6 CXO D J Lloyd, 16 Kingsley Rd, Brighton, BN1 5NH [IO90WU, TQ20]
G6 CXQ G A Goldman, 39 Beechwood Rise, Watford, WD2 5SE [IO91TQ, TQ19]
G6 CXY R L Revan, 8 Thelusson Ct, Woodfield Rd, Radlett, Herts, WD7 8JF [IO91UQ, TQ19]
G6 CYA R King, 6 Holland Villas, Main Rd, Great Holland, Frinton on Sea, CO13 0JJ [JO01OT, TM21]
G6 CYE A Read, 36 West St., Tollesbury, Maldon, CM9 8RJ [JO01JS, TL91]
G6 CYH I Roberts, 15 Manor Rd, Plymouth, PL9 7DP [IO70WI, SX55]
G6 CYL C J Bridger, 22A Barons Rd, Bury St. Edmunds, Suffolk, IP33 2LW [JO02IF, TL86]
G6 CYN J A Jones, 8 Ingham Ln, Watlington, OX9 5EJ [IO91LP, SU69]
G6 CYO I M Jarvis, Vale View, Knapp Ln, Besbury, Minchinhampton, Stroud, GL6 9EP [IO81VR, SO80]
G6 CYR R M Jenkins, 29 Ebnal Cl, Baronscross, Leominster, HR6 8SL [IO82OF, SO45]
G6 CYT R C Kempton, 14 Bloxam Gdns, Rugby, CV22 7AP [IO92II, SP47]
G6 CYU M S Kendrick, 79 Chanctonbury Rd, Burgess Hill, RH15 9EZ [IO90WW, TQ31]
G6 CYV P W Kirkham, 8 Long Valley Rd, Biddulph, Gillow Heath, Stoke on Trent, ST8 6RA [IO83VD, SJ85]
G6 CYX E R Acton, 27 Fenton Cl, Congleton, CW12 3TH [IO83VD, SJ86]
G6 CYZ A Jenkins, 92 Graham Ave, Patcham, Brighton, BN1 8HD [IO90WU, TQ30]
G6 CZA Details withheld at licensee's request by SSL.
G6 CZB R M Poffley, 8 Bowerhill Rd, Salisbury, SP1 3DN [IO91CB, SU13]
G6 CZC Details withheld at licensee's request by SSL.
G6 CZE C D Peacock, 22 Chaucer Rd, Wellingborough, NN8 3NL [IO92PH, SP86]
G6 CZF Details withheld at licensee's request by SSL.
G6 CZL Details withheld at licensee's request by SSL.[Station located near Warlingham.]
GM6 CZM I C McAulay, 9 Randolph Cliff, Edinburgh, EH3 7TZ [IO85JW, NT27]
G6 CZO D I McGhie, 54 Schoolrd, Newborough, Peterborough, PE6 7RG [IO92VP, TF20]
G6 CZQ T F Minns, 64 Springdale Ave, Broadstone, BH18 9EX [IO80XS, SY99]
G6 CZS C J Moore, 57 Park Rd, Bury St. Edmunds, IP33 3QW [JO02IF, TL86]
G6 CZT M J Prince, 4 Ladymead, Woolbrook, Sidmouth, EX10 9XN [IO80JQ, SY18]
G6 CZV Details withheld at licensee's request by SSL.
G6 CZX W W Aitchison, 18 Kerensa Green, Falmouth, Cornwall, TR11 2HE [IO70KD, SW73]
G6 CZZ J N Abram, 3 Frenchies View, Denmead, Waterlooville, PO7 6SH [IO90KW, SU61]
G6 DAA Details withheld at licensee's request by SSL.
G6 DAD D J Blagburn, 10 Tottington Ave, Springhead, Oldham, OL4 4RY [IO83XN, SD90]
G6 DAH D J Budd, 81 Bohemia Chase, Leigh on Sea, SS9 4PW [JO01HN, TQ88]
G6 DAI N M Brickwood, 4 Vale Cttgs, Shillingstone, Blandford Forum, DT11 0SS [IO80VV, ST81]
G6 DAK P Bloomer, 86 Trafalgar Rd, Barclay Ct, Cirencester, GL7 2EN [IO91AR, SP00]
G6 DAN B J Daniels, 113 Orchard Way, Wymondham, NR18 0NZ [JO02NN, TG10]
G6 DAO G Bradbury, Bradlea, 46 Clifton Ave, Barlborough, Chesterfield, S43 4HF [IO93IG, SK47]
G6 DAP J J W Balmford, 5 Limes Way, Shabbington, Aylesbury, HP18 9HB [IO91LS, SP60]
G6 DAQ A Y Boonham, 1 Oakleigh Dr, Sedgley, Dudley, DY3 3LH [IO82WM, SO99]
G6 DAU Details withheld at licensee's request by SSL.
G6 DAW G J Barton, 32 Drummond Cl, Erith, DA8 3QS [JO01CL, TQ57]
G6 DAY M J Pemberton, 37 Bardsley Cl, Park Hill, Croydon, CR0 5PS [IO91XI, TQ36]
G6 DAZ Details withheld at licensee's request by SSL.[Op: M J Phillips.]
G6 DB D N Biltcliffe obo Aylesbury Vale Repeater Cl, Spindles, 24 Hambleside, Bicester, OX6 8GA [IO91JV, SP52]
G6 DBC N J Parkinson, 42 West St., Winterton, Scunthorpe, DN15 9QF [IO93QP, SE91]
G6 DBE Details withheld at licensee's request by SSL.
G6 DBI Details withheld at licensee's request by SSL.
G6 DBJ J T Fairhurst, 8 Galley Field, Abingdon, OX14 3RS [IO91IQ, SU59]
G6 DBL J Fedyk, 56 St. Ervans Rd, North Kensington, London, W10 5QT [IO91VM, TQ28]
GW6 DBP J W Firmstone, 26 Melwood Cl, Penyffordd, nr Chester, Flintshire, CH4 0NB [IO83LD, SJ36]
G6 DBQ D W Fryer, Norwood, 105 Chester Rd, Hazel Gr, Stockport, SK7 6HG [IO83WI, SJ98]
G6 DBT Details withheld at licensee's request by SSL.
G6 DBU R J Gambles, 28 Beta Boulevard, Kings Copse Park, Garsington, Oxford, OX44 9BJ [IO91JR, SP50]
G6 DBX A M Grover, 44 Stirling Ct Rd, Burgess Hill, RH15 0PT [IO90WX, TQ31]
G6 DBY P A Gould, Derna, Surrey Ln, Tiptree, Colchester, CO5 0QT [JO01IT, TL81]
G6 DBZ S M Griffin, 50 Cherrybrook Dr, Broseley, TF12 5SH [IO82SO, SJ60]
G6 DCH J A Molyneux, 8 Bay Cl, Horley, RH6 8LF [IO91VE, TQ24]
G6 DCJ I Miles, 65 Horsey Rd, Kirby-le-Soken, Frinton on Sea, CO13 0EQ [JO01OU, TM22]
G6 DCM S T O'Leary, 76 Harrogate Rd, Reddish, Stockport, SK5 6EX [IO83WK, SJ89]
G6 DCS G Norris, 5 Bakewell Green, Newhall, Swadlincote, DE11 0TE [IO92FS, SK22]
G6 DCT D Littlewood, 572 Herries Rd, Galsworthy, Sheffield, S5 8TR [IO93GJ, SK39]
GM6 DCU T S Lawson, 6 Broomieknowe Park, Bonnyrigg, EH19 2JA [IO85KV, NT36]
G6 DCV R J Lindsey, 87 Station Rd, Whittlesey, Peterborough, PE7 1UE [IO92WN, TL29]
GI6 DCX E Lyons, Creevy Tennant Lodg, 17 Brae Rd, Ballynahinch, BT24 8UN [IO74BK, J35]
G6 DDA A D Moss, 23 Short St., Nuneaton, CV10 8JF [IO92FM, SP39]
G6 DDE Details withheld at licensee's request by SSL.
GW6 DDF J A Morris, 45 Branksome Est, Llanaber Rd, Barmouth, Gwynedd Wales, LL42 1YP [IO72XR, SH61]
G6 DDH G Morgan, c/o 16 Gresley Cour, Beckfield Pl, Acomb, York, YO2 5FF [IO93KX, SE55]
G6 DDJ S Pillinger, 28 Reddenhill Rd, Babbacombe, Torquay, TQ1 3RQ [IO80FL, SX96]
G6 DDK G A H Owen, Olivedene, 53 Hervey St., Ipswich, IP4 2ET [JO02NB, TM14]
G6 DDO R A Owen, 12 Bromsgrove Rd, Hagley, Stourbridge, DY9 9LX [IO82WK, SO98]
G6 DDP R Oakden, 38 Brookfield Ave, Hucknall, Nottingham, NG15 6FF [IO93JA, SK54]
G6 DDQ Details withheld at licensee's request by SSL.
G6 DDS Details withheld at licensee's request by SSL.
G6 DDU G W Goddard, 10 Stukeley Rd, Holbeach, Spalding, PE12 7LQ [JO02AT, TF32]
G6 DEA G F Goss, Orchard Bungalow, 66 Cucumber Ln, Brundall, Norwich, NR13 5QR [JO02RP, TG30]
G6 DEG T Hampson, 6 Rushmere Dr, Bury, BL8 1DW [IO83UO, SD71]
G6 DEK M R Hunt, 39 Ridge St., Wollaston, Stourbridge, DY8 4QF [IO82VL, SO88]
G6 DEN D Batten, 7 Bayham Rd, Bristol, BS4 2DY [IO81RK, ST67]
GI6 DEO H E Hunniford, 23 Portadown Rd, Tandragee, Craigavon, BT62 2BE [IO64SI, J04]
GW6 DEP M P Harris, 11 Lower Rawlinson Terr, Tredegar, NP2 4JD [IO81JS, SO10]
G6 DEQ Details withheld at licensee's request by SSL.[Op: I A Harrison. Station located in London E11.]
G6 DER K Hewitt, 6 Church Gr, Monk Bretton, Barnsley, S71 2EY [IO93GN, SE30]
G6 DEV D A Harris, 15 Millwood Rd, Orpington, BR5 3LG [JO01BJ, TQ46]
G6 DFA C J Willies, 17 Campion Way, Sheringham, NR26 8UN [JO02OW, TG14]
G6 DFB C S T Smith, 83 Sledmore Rd, Dudley, DY2 8DY [IO82XM, SO98]
G6 DFC P Johnson, 3 Lance Dr, Burntwood, WS7 8FA [IO92AQ, SK01]
G6 DFF Details withheld at licensee's request by SSL.
G6 DFH J A Roberts, 155 Langley Hall Rd, Olton, Solihull, B92 7HB [IO92CK, SP18]
G6 DFL P J Plested, 11 Dulwich Way, Croxley Green, Rickmansworth, WD3 3PX [IO91SP, TQ09]
G6 DFM J V Phelps, Windy Dene, Green Ln, Chessington, KT9 2DT [IO91UI, TQ16]
G6 DFR T S Parfitt, 4 Back St., Lakenheath, Brandon, IP27 9HF [JO02GJ, TL78]
GI6 DFU S Patrick, 11 Dunboyne Ave, Larne, BT40 1PS [IO74CU, D30]
G6 DFV A S Parker, 13 Hartley St., Colne, BB8 9DF [IO83VU, SD84]
GW6 DFX D W James, 27 Alfreda Rd, Whitchurch, Cardiff, CF4 2EH [IO81JM, ST17]
G6 DFY Dr G C Joly, 14 Flora Cl, London, E14 6DX [IO91XM, TQ38]
G6 DFZ M A T Jones, Wester Green Villa, 83 London Rd, Braintree, CM7 2LF [JO01GU, TL72]
G6 DGA A R E Jones, 8 New Rd, Cam, Dursley, GL11 6PN [IO81TQ, ST79]
G6 DGC Details withheld at licensee's request by SSL.
GI6 DGJ Details withheld at licensee's request by SSL.
G6 DGK Capt G J Keegan, Hurstfields, Allingtonrd, Newick, Lewes E Sussex, BN8 4NA [JO00AX, TQ42]
G6 DGR N K Bean, 81 Park Rd, Sutton Coldfield, B73 6BT [IO92CN, SP19]
G6 DGT Details withheld at licensee's request by SSL.
GW6 DGU J R Britton, 22 Hardy Cl, Woodfield Heights, Barry, CF62 9HJ [IO81IK, ST16]
G6 DGV C H Brock, 37 Ashington Dr, Lowercroft, Bury, BL8 2TS [IO83TO, SD71]
G6 DGX J E Raby, Cedar House, Coppenhall, Stafford, Staffs, ST18 9DA [IO82WS, SJ91]
G6 DHB M L Robinson, 1050 Holderness Rd, Hull, HU9 4AH [IO93US, TA13]
GW6 DHC O C Roberts, 15 Ffordd Alban, Tywyn, LL36 9EA [IO72WO, SH50]
G6 DHD A J A Rollason, 76 Solihull Rd, Shirley, Solihull, B90 3HL [IO92CJ, SP17]
G6 DHI D Kennedy, 33 Spring Bridge Rd, Manchester, M16 8PW [IO83VK, SJ89]
G6 DHN R T Cunningham, 47 Westfield Ave, Skelmanthorp, Skelmanthorpe, Huddersfield, HD8 9AH [IO93EO, SE21]
G6 DHS Details withheld at licensee's request by SSL.
G6 DHT P H Chace, 16 Broyle Cl, Chichester, PO19 4BG [IO90OU, SU80]
G6 DHU M H Chace, 1 Chedworth Cl, Claverton Down, Bath, Avon, BA2 7AF [IO81UI, ST76]
G6 DHW I D Clayton, 15 Ashbourne Dr, Desborough, Kettering, NN14 2XG [IO92NK, SP78]
G6 DHY C G Deane, 5 Queensberry Rd, Amesbury, Salisbury, SP4 7PU [IO91CE, SU14]
G6 DIA W T Donoghue, 58 Stannington Cres, Totton, Southampton, SO40 3QB [IO90GW, SU31]
G6 DID J R K Davis, 38 Dover Cl, Southwater, Horsham, RH13 7XX [IO91TA, TQ12]
G6 DIE G Drohan, 23 Lindholme Dr, Rossington, Doncaster, DN11 0UP [IO93LL, SK69]
G6 DIK Details withheld at licensee's request by SSL.
G6 DIM T R E Eves, Banks Farm, Manor Rd, Abridge, Romford, RM4 1NH [JO01BO, TQ49]
G6 DIO R A Everson, High Trees, The Sycamores, Thorn Gr, Bishops Stortford, CM23 5JR [JO01CU, TL42]
G6 DIR M P Wray, 6 Prince Ave, Lancing, BN15 8NH [IO90UT, TQ10]
G6 DIS M G Welch, 23 Longdown Rd, West Heath, Congleton, CW12 4QH [IO83VE, SJ86]
G6 DIV Details withheld at licensee's request by SSL.
G6 DIW Details withheld at licensee's request by SSL.
G6 DIZ D Feeley, 177 Rock St., Sheffield, S3 9JF [IO93GJ, SK38]
G6 DJA Details withheld at licensee's request by SSL.[Station located near Farnborough.]
G6 DJC Details withheld at licensee's request by SSL.
G6 DJE W H Smart, 33 Parkfield Rd, Willesden, London, NW10 2BG [IO91VN, TQ28]
G6 DJH D J Harvey, 23 Sprules Rd, Brockley, London, SE4 2NL [IO91XL, TQ37]
G6 DJI Details withheld at licensee's request by SSL.
G6 DJJ Details withheld at licensee's request by SSL.
G6 DJO E Ellis-Brown, 14 Staindale Gdns, Stockton on Tees, TS19 0RW [IO94IN, NZ41]
G6 DJQ G P Tomlinson, 10 Ashbourne Rd, Underwood, Nottingham, NG16 5EH [IO93IB, SK45]
G6 DJS D J Sojkowski, 7 Spenlow Dr, Chelmsford, CM1 4UQ [JO01FS, TL60]
G6 DJT Details withheld at licensee's request by SSL.
G6 DJV R D A Sivyer, 22 Boxley Rd, Walderslade, Chatham, ME5 9LF [JO01GI, TQ76]
G6 DKA K G Twist, 125 Needlers End Ln, Balsall Common, Coventry, CV7 7AA [IO92EJ, SP27]
G6 DKB D E Twist, 235 Little Ridge Ave, St. Leonards on Sea, TN37 7HN [JO00GV, TQ81]
G6 DKC Details withheld at licensee's request by SSL.
G6 DKE E A Reynolds, 11 New St., Sudbury, CO10 6JB [JO02IA, TL84]
G6 DKF L E Marsh, 18 Northgate, Hornsea, East Ridings, HU18 1ES [IO93VW, TA24]
G6 DKI R N Tew, 57 Papist Way, Cholsey, Wallingford, OX10 9QH [IO91KN, SU58]
G6 DKK S F Simes, 53 Waterford Ln, Cherry Willingham, Lincoln, LN3 4AN [IO93SF, TF07]
G6 DKM L Sandford, 150 Tipton Rd, Woodsetton, Dudley, West Midlands, DY3 1AL [IO82WM, SO99]
G6 DKS R P Saverton, Flat 7, 6 Surbiton Hill Park, Berrylands, Surbiton, Surrey, KT5 8EX [IO91UJ, TQ16]
G6 DKT Details withheld at licensee's request by SSL.
G6 DKW D K Walton, 149 Randall Ave, Cricklewood, London, NW2 7TA [IO91VN, TQ28]
G6 DKX Details withheld at licensee's request by SSL.
G6 DKY R C Alcock, 12 Fairway, Shelfield, Walsall, WS4 1RP [IO92AO, SK00]
G6 DKZ P C Astfalck, Gables Lodge, 66 Westhill Rd, Coundon, Coventry, CV6 2AA [IO92FK, SP38]
G6 DLA T W Anderson, 1 Storths Rd, Huddersfield, HD2 2XN [IO93CP, SE11]
G6 DLF E J Boone, 16 Shawhurst Croft, Hollywood, Birmingham, B47 5PB [IO92BJ, SP07]
G6 DLJ P C Bridges, Sm House, School Cl, Chandlers Ford Ind Est, Eastleigh, Hants, SO53 4BY [IO90HX, SU42]
G6 DLO C S Benson, 98 Victoria Ave, Grays, RM16 2RW [JO01EL, TQ67]
G6 DLT J Bartlett, 9 Beaufort Rd, Doncaster, South Yorks, DN2 6EP [IO93KM, SE50]
G6 DLX E Buxton, 63 Dock Rd, Tilbury, RM18 7DB [JO01EL, TQ67]
G6 DLZ P W Bosanquet-Bryant, 41 Skelmersdale Rd, Clacton on Sea, CO15 6DA [JO01NT, TM11]
G6 DMA Details withheld at licensee's request by SSL.
G6 DMG S L Wellon, 71 Toftdale Green, Lyppard Bourne, Worcester, WR4 0PE [IO82VE, SO85]
G6 DMJ M Wright, 20 Lincoln Cl, Woolston, Warrington, WA1 4LU [IO83RJ, SJ68]
G6 DMM K C Webster, 27 Glendale Cl, Horsham, RH12 4GR [IO91UB, TQ13]
G6 DMN I A Walton, 1 Press Ln, Aylsham Rd, Norwich, NR3 2JY [JO02PP, TG21]
G6 DMR H M Saagi, 1 Woodbridge Walk, Hollesley, Woodbridge, IP12 3LA [JO02RB, TM34]
G6 DMV F A Vine, Park Lodge, Burtons Ln, Chorleywood, Rickmansworth, WD3 5PJ [IO91RP, TQ09]
G6 DMW D J Prince, 24 Buchanan Rd, Walsall, WS4 2EN [IO92AO, SP09]
G6 DMY A Bosansko, 26 The Hambros, Thurston, Bury St. Edmunds, IP31 3PS [JO02JG, TL96]
G6 DNA T J Cattermole, 24 Cromwell Rd, Colchester, CO2 7EN [JO01KV, TL92]
G6 DNH M J Carvell, 12 Liskeard Dr, Allestree, Derby, DE22 2GW [IO92GW, SK33]
GI6 DNI D W M Chapman, 13 Andraid Cl, Stiles East, Antrim, BT41 1RF [IO64VR, J18]
G6 DNJ D J Cook, 7 Heath Ct, Trimley St. Martin, Trimley, Felixstowe, IP11 0YQ [JO01PX, TM23]
G6 DNK Details withheld at licensee's request by SSL.
G6 DNL K M Snellin, 3 Turnberry, Home Farm, Bracknell, RG12 8ZJ [IO91OJ, SU86]
G6 DNN Details withheld at licensee's request by SSL.
G6 DNO R J Laver, 34 Providence Way, Waterbeach, Cambridge, CB5 9QJ [JO02CG, TL46]
G6 DNV G D Taylor, 10 Scott Cl, Beaumont Park, Hexham, NE46 2QB [IO84WX, NY96]
G6 DNX P E Trickett, 25 Spring St., Halesowen, B63 2SY [IO82XL, SO98]
GW6 DNZ C M Treadwell, 2 Wynn Cres, Old Colwyn, Colwyn Bay, LL29 9DF [IO83DG, SH87]
G6 DOD M Wheeler, 24 Goodwood Way, Chippenham, SN14 0SY [IO81WK, ST97]
G6 DOE M Parnell, 31 Hughenden Ave, Kenton, Harrow, HA3 8HA [IO91UO, TQ18]
G6 DOF C G Wankling, 6 Marthorne Cres, Wealdstone, Harrow, HA3 5PL [IO91TO, TQ19]
G6 DOI C J Wigginton, 4 Copes Haven, Shenley Brook End, Milton Keynes, MK5 7HA [IO92OA, SP83]
GW6 DOK C Williams, Caermai, 134 Gaerwen Uchaf, Gaerwen, Ynys Mon Gwynedd, LL60 6HN [IO73UF, SH47]
G6 DOM P J Wieland, 3 Troutbeck Rd, Timperley, Altrincham, WA15 7JD [IO83UJ, SJ78]
G6 DON Details withheld at licensee's request by SSL.
G6 DOP Details withheld at licensee's request by SSL.
G6 DOQ H Davies, 76 Brook Ln, Timperley, Altrincham, WA15 6RS [IO83TJ, SJ78]
G6 DOR D A J Durrant, 22 St. Martinsfield, Martinstown, Dorchester, DT2 9JU [IO80RQ, SY68]
G6 DOT S R Bullock, Kunance, 10 Woodville Rd, Canvey Island, SS8 8JU [JO01HM, TQ88]
G6 DOV L E Dunn, 24 Mynchen Rd, Beaconsfield, HP9 2BA [IO91QO, SU99]
G6 DOW A Deacon, 1 Connaught Gdns, West Green, Crawley, RH10 2NB [IO91VC, TQ23]
G6 DOX D M Dodd, 5 Orchard Garth, Low Hurst, Wreay, Carlisle, Cumbria, CA4 0RN [IO84NT, NY44]
G6 DOY Details withheld at licensee's request by SSL.

G6 DP Dr D E Palin, Ashfield, Tarvin Rd, Manley, Warrington, WA6 9EW [IO83PG, SJ57]
G6 DPA B J English, 24 James Ave, Paignton, Devon, TQ3 3LZ [IO80EK, SX86]
G6 DPC Details withheld at licensee's request by SSL.
G6 DPH B J Flinn, 10 Porlock Cl, Penketh, Warrington, WA5 2QE [IO83QJ, SJ58]
G6 DPL L P Green, 76 Dibleys, Blewbury, Didcot, OX11 9PU [IO91JN, SU58]
G6 DPP Details withheld at licensee's request by SSL.
G6 DPS M L Harrison, 33 Campion Park, Up Hatherley, Cheltenham, GL51 5WA [IO81WV, SO92]
G6 DPV J Nicholson, The Old Rectory, Nether Denton, Brampton, Cumbria, CA8 2LY [IO84QX, NY56]
G6 DPW D P Waghorne, 5 Freelands Dr, Church Crookham, Fleet, GU13 0TE [IO91NG, SU85]
GI6 DPZ Details withheld at licensee's request by SSL.
G6 DQD D W Orme, 22 Seniors Dr, Thornton Cleveleys, FY5 2RD [IO83LV, SD34]
GW6 DQH D W Moore, 71 Woodlands Ave, Talgarth, Brecon, LD3 0AT [IO81JX, SO13]
G6 DQM Details withheld at licensee's request by SSL.
G6 DQO I Martin, 3 The Stiles, Delamere Park, Cuddington, Northwich, CW8 2UR [IO83QF, SJ57]
G6 DQT W E Lasbury, Sonserra Flats, Flat 4 Fekruna St, Bugibba, St Pauls Bay Malta
G6 DQU Details withheld at licensee's request by SSL.
G6 DQW Details withheld at licensee's request by SSL.
G6 DQY J T Orrells, Perry Willows, Yeaton, Baschurch, Shrewsbury, SY4 2HY [IO82NS, SJ41]
G6 DQZ N A Perry, 5 Catherine Cttgs, Droitwich Rd, Hartlebury, Kidderminster, DY10 4EL [IO82VI, SO87]
G6 DRC D R Cooper, Linden House, Greenhill Park Rd, Evesham, WR11 4NL [IO92AC, SP04]
G6 DRF R A Platt, 72 Churchbury Rd, Enfield, EN1 3HP [IO91XP, TQ39]
G6 DRG T Place, 73 Williams St., Langold, Worksop, S81 9NX [IO93KJ, SK58]
G6 DRH D R Hickton, 27 Vanguard Rd, Longeaton, Long Eaton, Nottingham, NG10 1DX [IO92IV, SK43]
G6 DRM K Harris, 86 Swan Gdns, Erdington, Birmingham, B23 6QG [IO92BM, SP19]
G6 DRN P Haylor, 76 Beauchamp Rd, Billesley, Birmingham, B13 0NR [IO92BK, SP08]
G6 DRO D R Hodgkisson, 32 Faraday Ave, Stretton, Burton on Trent, DE13 0FX [IO92ET, SK22]
G6 DRP D N Hemmins, 18 Burn Walk, Burnham, Slough, SL1 7EW [IO91QM, SU98]
G6 DRW G P Hudson, 6 Crestmount Dr, Salisbury, SP2 9LH [IO91CB, SU13]
G6 DRX R A W Jarvis, 5 Ariel Way, Bilton, Rugby, CV22 6LR [IO92II, SP47]
G6 DRZ Details withheld at licensee's request by SSL.
G6 DSB S D Jarvis, 74 Halford Rd, Ickenham, Uxbridge, UB10 8QA [IO91SN, TQ08]
G6 DSD R A Jones, 20 Bibsworth Ave, Moseley, Birmingham, B13 0BA [IO92BK, SP08]
G6 DSG N Austin, 10 Ridge Rd, Sandyford, Stoke on Trent, ST6 5LG [IO83VB, SJ85]
GI6 DSH W Armstrong, 9 Barry St., Londonderry, BT48 7PJ [IO65IA, C41]
G6 DSJ R A Alcock, Silver How, Bay Hill, Ilminster, TA19 0AT [IO80NW, ST31]
G6 DSP C J Addis, 1 Newchurch Ln, Culcheth, Warrington, WA3 5RW [IO83RK, SJ69]
G6 DSQ N Abbott, 50 Poplar Gr, Forest Town, Mansfield, NG19 0HN [IO93KD, SK56]
G6 DSW T A Kellett, 10 Alexandra St., Warrington, WA1 3SE [IO83RJ, SJ68]
G6 DTB Details withheld at licensee's request by SSL.
G6 DTH A P Allnutt, 24 Harrowlands Park, Dorking, RH4 2RA [IO91UF, TQ14]
G6 DTN Details withheld at licensee's request by SSL.
G6 DTO P I Clarke, 746 Alexandria Dr, Naperville, Illinois 60565, USA, X X
G6 DTR Details withheld at licensee's request by SSL.
G6 DTT A J Campbell, 53 Oxford Rd, London, W5 3SR [IO91UM, TQ18]
G6 DTW A S Challen, Links Corner Cottage, Links Rd, Ashtead, KT21 2EG [IO91UH, TQ15][Op: Alvin. QRV 6, 4. 2m, 70, 23cms - most modes. Station located in Epsom, Surrey.]
G6 DTX P Chambers, 11 Addison Sq, Ringwood, BH24 1NY [IO90CU, SU10]
G6 DUB S D Richards, 23 Sycamore Rise, Bracknell, RG12 9BU [IO91PJ, SU86]
G6 DUG Details withheld at licensee's request by SSL.
G6 DUI I R Castle, 26 Lonsdale Dr, Sittingbourne, ME10 1TS [JO01II, TQ86]
G6 DUN R A Burrows, 40 Fairmile Rd, Christchurch, BH23 2LL [IO90CR, SZ19]
G6 DUQ A P Bridgeland, 17 Oldfield Ln, Wisbech, PE13 2RJ [JO02BP, TF40]
G6 DUS Details withheld at licensee's request by SSL.
G6 DVE A L Redshaw, 417 Marston Rd, Marston, Oxford, OX3 0JG [IO91JS, SP50]
G6 DVH G Roberts, 64 Lynwood Gr, Audenshaw, Manchester, M34 5TE [IO83WL, SJ99]
G6 DVO H B Warehand, 79 Woodlands Rd, Hertford, SG13 7JF [IO91XT, TL31]
G6 DVQ C E Thornton, 51 Glebe Rd, Wickford, SS11 8ET [JO01GO, TQ79]
G6 DVR J H Thompson, 291 Beechings Way, Rainham, Gillingham, ME8 7BP [JO01HI, TQ86]
GM6 DVZ Details withheld at licensee's request by SSL.
G6 DWB B J Thompson, 4 Wellington Gdns, Newton-le-Willows, WA12 9LT [IO83QK, SJ59]
G6 DWI Details withheld at licensee's request by SSL.
G6 DWM G S Sohal, 15 Icknield Rd, Luton, LU3 2NY [IO91SV, TL02]
G6 DWS N S L Shearer, 64 Balsall Heath Rd, Edgbaston, Birmingham, B5 7NE [IO92BL, SP08]
G6 DWW K I Deacon, 18 Waterloo Cres, Countesthorpe, Leicester, LE8 5SU [IO92KN, SP59]
G6 DWX Details withheld at licensee's request by SSL.
G6 DXD A J Edwards, 35 Eldon Rd, Cheltenham, GL52 6TX [IO81XV, SO92]
G6 DXH P R Etchells, 1 Cromley Rd, Stockport, SK2 7DT [IO83WJ, SJ98]
G6 DXM Details withheld at licensee's request by SSL.
G6 DXN T J Fosbrook, 13 Brook St., Stotfold, Hitchin, SG5 4LA [IO92VA, TL23]
G6 DXP M A Gentry, 12 Albert Cl, Grays, RM16 2RB [JO01EL, TQ67]
G6 DXU W K Gardner, 11 Spring Ln, Whittington, Lichfield, WS14 9LX [IO92CQ, SK10]
G6 DYA P G Marrison, 43 Park Rd, Alrewas, Burton on Trent, DE13 7AG [IO92DR, SK11]
G6 DYD Details withheld at licensee's request by SSL.[Station located in Lodon SE6.]
G6 DYI G A Hughes, 113 Reservoir Rd, Erdington, Birmingham, B23 6DL [IO92BM, SP19]
G6 DYK S W Hicks, 78 Highover Way, Hitchin, Herts, SG4 0RQ [IO91UX, TL13]
G6 DYM G Hudgell, 18 Fellowes Ln, Colney Heath, St. Albans, AL4 0QA [IO91VR, TL20]
G6 DYU L Horn, 149 Finedon Rd, Irthlingborough, Wellingborough, NN9 5TY [IO92QH, SP97]
G6 DYX B Hines, 41 Crosslands Dr, Abingdon, OX14 1JU [IO91IQ, SU49]
GI6 DZC Details withheld at licensee's request by SSL.
G6 DZD Details withheld at licensee's request by SSL.
G6 DZH K S Killigrew, 26 Gorsey Cl, Astwood Bank, Redditch, B96 6AG [IO92AG, SP06]
G6 DZJ S M Kitchener, 101 Highfield Rd, Tring, HP23 4DS [IO91PT, SP91]
G6 DZM Details withheld at licensee's request by SSL.[Op: R C Knight.]
G6 DZR Details withheld at licensee's request by SSL.
G6 DZT D A Anstock, 12 Raymoor Ave, St. Marys Bay, Romney Marsh, TN29 0RD [JO01LA, TR02]
G6 DZX J D Beardmore, 6 Essex Cl, Congleton, CW12 1SH [IO83VE, SJ86]
G6 DZY W J Bottrell, 17 Eastleigh Dr, Romsley, Halesowen, B62 0PA [IO82XK, SO97]
G6 EAL B E Johnson, 387 Hither Green Ln, Lewisham, London, SE13 6TR [JO01AK, TQ37]
G6 EAR P Dowler, 84 Highfield Rd, Chelmsford, CM1 2NQ [JO01FR, TL60]
G6 EAS Details withheld at licensee's request by SSL.
G6 EAV Details withheld at licensee's request by SSL.
G6 EAY M R Hughes, The Limes, 425 Evesham Rd, Crabs Cross, Redditch, B97 5JA [IO92AG, SP06]
G6 EAZ R J Hildebrand, Meadow View, Cunningham Pl, Bakewell, DE45 1DD [IO93DF, SK26]
G6 EBC G J Jack, 86 Southend Rd, Gateshead, NE9 6XU [IO94EW, NZ25]
G6 EBG J P Allen, 99 Maybury Rd, Holderness Rd, Hull, HU9 3LB [IO93US, TA13]
G6 EBI P A Aust, 10 Bradwell Cl, Hornchurch, RM12 5PR [JO01CM, TQ58]
G6 EBJ Details withheld at licensee's request by SSL.
G6 EBK Details withheld at licensee's request by SSL.
G6 EBL M J Brundle, 36 Campion St., Derby, DE22 3EF [IO92GW, SK33]
G6 EBO B H Beckers, 6 Patmore Way, Collier Row, Romford, RM5 2HF [JO01BO, TQ49]
G6 EBQ M E Brooks, Leafield, Killerby, Cayton, Scarborough, YO11 3TW [IO94TF, TA08]
GI6 EBX S J Bird, 20 Irvington Park, Knockchree Ave, Kilkeel, Newry, BT34 4LX [IO64XB, J31]
GI6 EBY B G Barron, 55 Henderson Ave, Belfast, BT15 5FN [IO74AP, J37]
G6 EBZ A J Brake, 157 Avondale Rd, Kettering, NN16 8PN [IO92PJ, SP87]
GI6 ECD M G E Camley, 1 Gardenville Ave, Omagh, BT79 7DB [IO64IO, H47]
G6 ECF Details withheld at licensee's request by SSL.
G6 ECJ J T Clements, 26 Glebe Rd, Norwich, NR2 3JG [JO02PO, TG20]
G6 ECK I F Cox, 20 The Hill, Old Harlow, Harlow, CM17 0BJ [JO01BS, TL41]
G6 ECN B E Clay, 64 Coppice Rd, Poynton, Stockport, SK12 1SN [IO83WI, SJ98]
G6 ECO Details withheld at licensee's request by SSL.
G6 ECR R F Buckingham, 1C Biddenham Turn, Bedford, MK40 4AT [IO92SD, TL05]
G6 ECS P R Buckingham, 1C Biddenham Turn, Bedford, MK40 4AT [IO92SD, TL05]
GI6 ECV N G Bingham, 87 Hawthorn Ave, Carrickfergus, BT38 8EQ [IO74CR, J48]
G6 EDB D J T Davies, 30 Curlew Cres, Bedford, MK41 7HX [IO92SD, TL05]
G6 EDD S Donald, 5 Windsor Rd, Royston, SG8 9JF [IO92XB, TL34]
G6 EDE M A Elliott, 224 Oakwood Ln, Leeds, LS8 2PE [IO93GT, SE33]
G6 EDF D E J Evans, 107 Bradbury Rd, Olton, Solihull, B92 8AL [IO92CK, SP18]
G6 EDJ S C Edwards, 10 Ermin Cl, Baydon, Marlborough, SN8 2JQ [IO91EM, SU27]
G6 EDL Details withheld at licensee's request by SSL.
G6 EDM D Evans, Caithness, Greenlands Rd, Kemsing, Sevenoaks, TN15 6PG [JO01CH, TQ55]
G6 EDR R L Fletcher, 31 Snowdrop Cl, Broadfield, Crawley, RH11 9EG [IO91VC, TQ23]
G6 EDT M Fletcher, Chusan, 32A Mill Rd, Sharnbrook, Beds, MK44 1NX [IO92RF, TL05]
G6 EDU M L Firth, Kasamily, 73 Lions Ln, Ashley Heath, Ringwood, BH24 2HH [IO90BT, SU10]
G6 EEA R O Challis, 2 Grange Park Ave, Wilmslow, SK9 4AH [IO83VH, SJ88]
G6 EEB W T Moodie, 141 Wood Ln, Handsworth Wood, Handsworth, Birmingham, B20 2AQ [IO92AM, SP09]
G6 EED N Mockridge, 6 Dunkerton Rise, Norton Fitzwarren, Taunton, TA2 6TF [IO81KA, ST22]
G6 EEE A J Mead, 17 Beadle Way, Great Leighs, Chelmsford, CM3 1RT [JO01GT, TL71]
G6 EEF D Malekout, 21 Raglan Ave, Wolverhampton, WV6 7RZ [IO82VO, SO89]
GI6 EEH S J McCullagh, 18 Village Walk, Portadown, Craigavon, BT63 5TL [IO64TK, J05]
G6 EER G H Middleton, 37 Hamdon Cl, Stoke Sub Hamdon, TA14 6QN [IO80PW, ST41]
G6 EES P A Morris, 8 Millfield, Lambourn, Hungerford, RG17 8YQ [IO91FM, SU37]
G6 EET D L Monk, 311 Birmingham Rd, Lickey End, Bromsgrove, B61 0ER [IO82XI, SO97]
G6 EEU M J Meredith, Lyndhurst, 43 Evesham Rd, Bishops Cleeve, Cheltenham, GL52 4SA [IO81XX, SO92]
G6 EEV W G Littlewood, 27 Williams Orchard, Highnam, Gloucester, GL2 8EL [IO81UV, SO71]
G6 EEY Details withheld at licensee's request by SSL.
GU6 EFB K D Le Boutillier, Bailiffs Cross Rd, St. Andrews, Guernsey, Channel Islands, GY6 8RT
G6 EFC J A Lyons, 187 Broad Oak Way, Hatherley, Cheltenham, GL51 5LN [IO81WV, SO92]
G6 EFE S W Weiss, 7 Tennyson Ave, Grays, RM17 5RG [JO01EL, TQ67]
GW6 EFK J A Walters, 2 Caerleon Ct, Caerphilly, Mid Glam, CF83 2UF [IO81JN, ST18]
G6 EFO K T Goodchild, 2 Westfield Cl, Norden, Rochdale, OL11 5XB [IO83VP, SD81]
G6 EFR N J Gough, 20 Earlsfield, Branston, Lincoln, LN4 1NP [IO93SE, TF06]
G6 EFY C A Hill, 15 The Moat, Quedgeley, Gloucester, GL2 4TB [IO81UT, SO81]
G6 EGG Details withheld at licensee's request by SSL.
G6 EGI Details withheld at licensee's request by SSL.
G6 EGL Details withheld at licensee's request by SSL.
G6 EGM D I C Patrick, Birkthwaite Farm, Wreay, Carlisle, Cumbria, CA4 0RZ [IO84NT, NY44]
G6 EGO D C Pink, 15 Primrose Hill, Daventry, NN11 5BX [IO92KG, SP56]
G6 EGU B W Nixon, 87 Field Ave, Canterbury, CT1 1TS [JO01NG, TR15]
G6 EGV Details withheld at licensee's request by SSL.
G6 EGY I H Niven, Keepers Cottage, Naseby Rd, Sulby, Northampton, Northants, NN6 6EZ [IO92LK, SP68]
G6 EHB T A Webb, 10 Woodview, Grays, Essex, RM17 5UW [JO01EL, TQ67]
G6 EHJ Details withheld at licensee's request by SSL.
G6 EHL D A Partington, 6 Celandine Ave, Cowplain, Waterlooville, PO8 9BE [IO90LV, SU61]
G6 EHN D Bateman, 6 Oxford Dr, Whitehills, Kippax, Leeds, LS25 7JG [IO93HS, SE43]
G6 EHO D Harrison, 69 Norton Rd, Southwick, Sunderland, SR5 2PA [IO94HW, NZ35]
G6 EHP J Hawkes-Bayliss, 36 Bagley Cl, Kennington, Oxford, OX1 5LT [IO91JR, SP50]
G6 EHU Details withheld at licensee's request by SSL.[Op: D Harper, 5 Poplar Place, Armthorpe, Doncaster, S Yorks, DN3 2EA.]
G6 EHV L Harper, Ivy Cottage, 28 Wroot Rd, Finningley, Doncaster, DN9 3DN [IO93ML, SK69]
G6 EIA Details withheld at licensee's request by SSL.
G6 EID N B Johnson, Edensor Post Office, Edensor, Bakewell, Derbyshire, DE4 1PH [IO93FC, SK25]
G6 EIG M B Kelly, 158 Roman Way, Bicester, OX6 7FL [IO91KV, SP52]
G6 EIH R L McCracken, 16 Station Rd, Rolleston on Dove, Burton on Trent, DE13 9AA [IO92EU, SK22]
G6 EII A S Langer, 1 Moss Side Ln, Moore, Warrington, WA4 6XA [IO83QI, SJ58]
G6 EIL Details withheld at licensee's request by SSL.
G6 EIO A R Mitchell, 33 Bramble Park, Taunton, TA1 2QT [IO81LA, ST22]
GI6 EIR D J Mullan, 17 Coleraine Rd, Garvagh, Coleraine, BT51 5HP [IO64PX, C81]
G6 EIY H Welchman, Greenhill Hse Ches Home, South Rd, Timsbury, Wilts, BA3 1ES [IO81SH, ST65]
G6 EIZ J Austin, 17 New Rd, Ascot, SL5 8QB [IO91PK, SU97]
G6 EJA Details withheld at licensee's request by SSL.
G6 EJD D L Bird, 59 Speedwell Cl, Melksham, SN12 7TE [IO81WJ, ST96]
G6 EJF R C W Amos, 141 Lister Rd, Braintree, CM7 1XW [JO01GU, TL72]
G6 EJG Details withheld at licensee's request by SSL.
G6 EJH J F Bradley, 66 Belmont Rd, Parkstone, Poole, BH14 0DB [IO90AR, SZ09]
G6 EJI P Barrett, 91 Victoria St., Shaw, Chadderton, Oldham, OL9 0HJ [IO83WN, SD90]
G6 EJJ Details withheld at licensee's request by SSL.
G6 EJM T R Burrows, 11 Louis Cl, Old Catton, Norwich, NR6 7BG [JO02PQ, TG21]
G6 EJP Details withheld at licensee's request by SSL.
G6 EJR D J Ball, 55 Sherriffs Dr, Tyldesley, Manchester, M29 8PQ [IO83SM, SD70]
G6 EJU C O Biddles, 129 Hallam Cres, Braunstone, Leicester, LE3 1FG [IO92KO, SK50]
G6 EJV Details withheld at licensee's request by SSL.
GW6 EJY Details withheld at licensee's request by SSL.
G6 EKA J V Bers, 4 Royal View, Links Gate, Lytham St. Annes, FY8 3LF [IO83LS, SD32]
G6 EKB Details withheld at licensee's request by SSL.
G6 EKG Details withheld at licensee's request by SSL.
G6 EKJ Details withheld at licensee's request by SSL.
G6 EKM R S C Perks, 51 Badgers Croft, Eccleshall, Stafford, ST21 6DS [IO82VU, SJ82]
G6 EKQ Details withheld at licensee's request by SSL.
G6 EKS A N Stelfox, 29 Waterside Way, Middlewich, CW10 9HP [IO83SE, SJ66]
G6 EKT R Guttridge Hornsea ARC, Ivy House, Rise Rd, Skirlaugh, Hull, HU11 5BH [IO93UU, TA14]
G6 EKW R A Smith, 190 Lynmouth Ave, Morden, SM4 4RP [IO91VJ, TQ26]
GW6 ELA Details withheld at licensee's request by SSL.
G6 ELD Details withheld at licensee's request by SSL.[Station located near South Manchester.]
G6 ELE D E Wells, 4 Stanley Cres, Bracebridge Heath, Lincoln, LN4 2JP [IO93RE, SK96]
G6 ELG M D Wright, 6 Tregalister Gdns, St. Germans, Saltash, PL12 5NQ [IO70UJ, SX35]
G6 ELH C M Slater-Walker, 152 Woodland Dr, Watford, WD1 3DB [IO91TP, TQ09]
G6 ELO R A Binns, 37 Bloomsbury Ct, Donnington, Telford, TF2 8DL [IO82SR, SJ71]
G6 ELQ Details withheld at licensee's request by SSL.
G6 ELR E L Richards, Sea View, Penwithick, St Austell, Cornwall, PL26 8UR [IO70OI, SX05]
G6 ELW M J Culkin, White Lodge, Verwood Rd, Three Legged Cross, Wimborne, BH21 6RR [IO90BU, SU00]
G6 ELZ C P Coad, Cockleshells, Polzeath, Wadebridge, Cornwall, PL27 6SX [IO70NN, SW97]
G6 EMB G S Collins, 175 Windmill Cottage, Kemble, Cirencester, Glos, GL7 6AN [IO81XQ, ST99]
G6 EME I R Dorian, 8 Southcourt Cl, Rustington, Littlehampton, BN16 3JD [IO90ST, TQ00]
G6 EMH I Davidson, 40 Dove Croft, New Ollerton, Newark, NG22 9RG [IO93LE, SK66]
G6 EMJ E M Jarrett, Egremont, Newport Rd, Stafford, ST16 1DH [IO82WT, SJ92]
G6 EMK Details withheld at licensee's request by SSL.
G6 EMP A T Filmer, 84 Barton Tors, Bideford, EX39 4HA [IO71VA, SS42]
G6 ENA M R Pyrah, 53 St. Georges Rd, Ramsgate, CT11 7EF [JO01RI, TR36]
G6 ENC R C Lish, 10 Rushfield, Sawbridgeworth, CM21 9NF [JO01BT, TL41]
G6 ENH A M Patrick-Gleed, 39 Ludlow Cl, Basingstoke, RG23 8QW [IO91KG, SU65]
G6 ENK K Fisher, 49 Brixham Dr, Wyken, Coventry, CV2 3LA [IO92GK, SP38]
G6 ENM R Gurowich, 1 St Cuthbert Villa, Haybridge, Wells, Somerset, BA5 1AH [IO81PF, ST54]
G6 ENN D T Gordon, 38 Deerpark Rd, Langtoft, Peterborough, PE6 9RB [IO92UQ, TF11]
G6 ENO B M Garrett, 226 Rydal Dr, Bexleyheath, DA7 5DG [JO01BL, TQ47]
GJ6 ENP J L E Q Gready, Flat 2 Sunnydene, La Route de Maufant, St. Saviour, Jersey, Channel Islands, JE2 7HX
G6 ENR S P Grant, Rectory Farm, Station Rd, Lockington, Driffield, YO25 9SQ [IO93SV, TA04]
G6 ENS S J Gordon, 2 Flamborough Cl, Lower Earley, Reading, RG6 3XB [IO91NK, SU77]
G6 ENU I P Gordon, 9 Park Rd, Camberley, GU15 2SP [IO91OH, SU85]
GM6 ENX Dr J L Grieve, Garramore, Glassel Rd, Banchory, Kincardineshires, AB31 4AE [IO87RB, NO69]
G6 ENY N T Graham, 2 Cravens Ln, Habrough, DN40 3AW [IO93UO, TA11]
G6 ENZ G M Holmes, 3 New Row, Deeping St. James, Peterborough, PE6 8NA [IO92UQ, TF10]
G6 EOA G F Hill, 14 Oak Dr, Colwall, Malvern, WR13 6RA [IO82TB, SO74]
G6 EOB Details withheld at licensee's request by SSL.
GW6 EOL R A Morgan, 76 Harvey Cres, Victoria Mews, Aberavon, Port Talbot, SA12 6DF [IO81CO, SS78]
G6 EON P A Martin, Unit 7, Scorrier Rural Workshops, Scorrier, Redruth, Cornwall, XX XX
G6 EOO R Machin, 236 Tamworth Rd, Kettlebrook, Tamworth, B77 1BY [IO92DO, SK20]
GM6 EOP J W Mitchell, 99 Elgin Dr, Glenrothes, KY6 2JS [IO86JE, NO20]
G6 EOR W Power, 31 Darbys Hill Rd, Tividale, Warley, B69 1SE [IO82XM, SO98]
G6 EOS K Partington, 14 Napier Rd, Eccles, Manchester, M30 8AG [IO83TL, SJ79]
G6 EOX B A Morrall, 61 Archer Rd, Walsall, WS3 1AW [IO92AO, SK00]
G6 EOY Details withheld at licensee's request by SSL.
G6 EPD Details withheld at licensee's request by SSL.
G6 EPJ D Jones, 24 Meadow Rise, Tiffield, Towcester, NN12 8AP [IO92MD, SP65]
G6 EPL R T Jonas, 4 Cripps Cl, Aylsham, Aylesham, Canterbury, CT3 3BX [JO01OF, TR25]
G6 EPN P A Knight, Hawkwind, Elcot Ln, Marlborough, SN8 2AZ [IO91DK, SU16]
G6 EPX P E Shuttleworth, 12 Oak Ave, Penwortham, Preston, PR1 0XQ [IO83PR, SD52]
G6 EQB J Singleton, 48 Pennine Way, Ashby de La Zouch, LE65 1EW [IO92GR, SK31]
G6 EQD B R Stirk, 25 Edinburgh Dr, Rushall, Walsall, WS4 1HW [IO92AO, SK00]
G6 EQF R W Skinner, 23 Woodstock Rd, Worcester, WR2 5ND [IO82VE, SO85]
G6 EQI R Smith, Roseville, Pennymoor, Tiverton, Devon, EX16 8LF [IO80EV, SS81]
G6 EQJ Details withheld at licensee's request by SSL.
G6 EQL S G Thomas, 64 Victoria Rd, Aigburth, Liverpool, L17 0DP [IO83MJ, SJ38]
G6 EQN C R Taylor, 1 Valley View, Park Rd, Elland, HX5 9HU [IO93BQ, SE12]
G6 EQS J E H Hastings, 17 Forrest Cres, Luton, LU2 9AR [IO91TV, TL12]
G6 EQT J J F Aston, 2 The Clock Tower, New York Industrial, Summerbridge, Harrogate, HG3 4LA [IO94DB, SE26]
G6 EQV Details withheld at licensee's request by SSL.
G6 EQY T V Baldwin, 51 Queens Rd, Broadstairs, CT10 1PG [JO01RI, TR36]
G6 EQZ R L Bracken, 72 Brampton Way, Portishead, Bristol, BS20 9YT [IO81OL, ST47]
G6 ERB H N Brooke, 33 St. Wilfrids Rd, Bessacarr, Doncaster, DN4 6AA [IO93KM, SE60]
G6 ERC Details withheld at licensee's request by SSL.
G6 ERG E R Godfrey, 30 Chapman House, Hazelwood Cl, Hitchin, SG5 1PS [IO91UW, TL12]
G6 ERI R A Couch, 72 Jervis Rd, Portsmouth, PO2 8PS [IO90KT, SU60]

G6 ERJ A R Croucher, 73 Loxley Cl, The Firs, Church Hill, Redditch, B98 9JH [IO92BH, SP06]
G6 ERK A W Cunliffe, 28 Rosebank Cl, Ainsworth, Bolton, BL2 5QU [IO83TO, SD71]
G6 ERN R Taylor South Staffs Raynet, 89 St. Johns Rd, Pelsall, Walsall, WS3 4EZ [IO92AP, SK00]
G6 ERQ G P Goddard, 49 Legbourne Rd, Louth, LN11 8ES [JO03AI, TF38]
G6 ERV Details withheld at licensee's request by SSL.
G6 ERZ G B J Jones, 46 Ryedale Way, West Ardsley, Tingley, Wakefield, WF3 1AJ [IO93FR, SE22]
GW6 ESI Details withheld at licensee's request by SSL.
G6 ESJ P E Wookey, 59 Cobden Ave, Copnor, Portsmouth, PO3 6NB [IO90LT, SU60]
G6 ESK D S Wh1Ttle, 15 Risdale Cl, Beverley Hills, Leamington Spa, CV32 6NN [IO92FH, SP36]
G6 ESM D Tankaria, 113 Burns Ave, Southall, UB1 2LT [IO91TM, TQ18]
G6 ESN J H Vardon, 50 Collingwood Ave, March, PE15 9EG [JO02BN, TL49]
G6 ESQ P H Baker, 12 College Cl, Coltishall, Norwich, NR12 7DT [JO02QR, TG21]
G6 ESR N P Bourke, 12 Yarm Ln, Great Ayton, Middlesbrough, TS9 6PJ [IO94KL, NZ51]
G6 ESZ Details withheld at licensee's request by SSL.
G6 ETA M D Barson, 61 Plantation Rd, Chestfield, Whitstable, CT5 3LQ [JO01MI, TR16]
G6 ETC J Brown, 44 Perowne Way, Sandown, PO36 9BX [IO90KP, SZ58]
GI6 ETD M T Bell, 10 Seapark Ave, Holywood, BT18 0LL [IO74CP, J47]
G6 ETE Details withheld at licensee's request by SSL.
G6 ETI W T Birch, 120 Biddulph Rd, Chell Green, Stoke on Trent, ST6 6TB [IO83VB, SJ85]
G6 ETP J H Cookson, Barker Fold, Tockholes, Darwen, Lancs, BB3 0LU [IO83RQ, SD62]
GI6 ETQ A A Campbell, 16 Parkwood, Harmony Heights, Lisburn, BT27 4EF [IO64XM, J26]
G6 ETX M A Carter, 22 John Morgan Cl, Hook, RG27 9RP [IO91MG, SU75]
G6 ETZ C Chalmers, 11 Maidenhead Ct Park, Maidenhead, SL6 8HN [IO91PM, SU88]
GM6 EUC D Cruickshank, 61 Woodside Rd, Banchory, Kincardineshire, AB31 4EN [IO87SB, NO69]
G6 EUF P N Raynor, 29 Kilvin Dr, Beverley, HU17 9PG [IO93TU, TA04]
G6 EUG C Slater, 70 Windsor Ave, Ashton, Ashton on Ribble, Preston, PR2 1JD [IO83PS, SD53]
G6 EUI C B Shaw, 19 Church Rd, Teversham, Cambridge, CB1 5AW [JO02CE, TL45]
G6 EUO J W Slater, 47 Broom Rd, Lakenheath, Brandon, IP27 9EZ [JO02GJ, TL78]
G6 EUP T H W Wray, 6 Laburnum Pl, Englefield Green, Egham, TW20 0SX [IO91RK, SU97]
GW6 EUT A L N Williams, Erw Deg Forden, Forden, Welshpool, Powys, SY21 8TS [IO82KO, SJ20]
G6 EUW A R Sheridan, 6 Mill Rd, Burnham on Crouch, CM0 8PZ [JO01JP, TQ99]
G6 EVA F J Roberts, Byfield, 19 Burleigh Rd, Addlestone, KT15 1PN [IO91SI, TQ06]
G6 EVC C F Sleight, Orchard House, School Hill, Napton, Rugby, CV23 8NN [IO92IF, SP46]
G6 EVD Details withheld at licensee's request by SSL.
G6 EVI J G Williams, 61 Longfield Rd, South Woodham Ferrers, Chelmsford, CM3 5JJ [JO01HP, TQ89]
G6 EVK P Wright, 46 Rosehill Rd, Burnley, BB11 2JL [IO83VS, SD83]
G6 EVT Details withheld at licensee's request by SSL.
G6 EVV R J Whittaker, 190 Water St., Accrington, BB5 6QU [IO83TS, SD72]
G6 EVX A E Wood, 27 Newbury Gdns, Upminster, RM14 2PS [JO01CN, TQ58]
G6 EVY H N Woolrych, 20 Meadow Dr, Devizes, SN10 3BJ [IO91AI, SU06]
G6 EWC P G Turner, 69 Windmill Ave, Bicester, OX6 7DZ [IO91KV, SP52]
G6 EWJ D R Taylor, Holmehurst, Church St., Rudgwick, Horsham, RH12 3ET [IO91SC, TQ03]
GI6 EWM Rev S Reid, 30 Ashwood, Lurgan, Craigavon, BT66 8PF [IO64TL, J05]
GI6 EWO B C Davis, 49 The Roddens, Larne, BT40 1QL [IO74CU, D30]
G6 EWP D W Davy, 22 Scott Gdns, St. Giles, Lincoln, LN2 4LX [IO93RF, SK97]
GW6 EWQ C Dormer, 33 Brynderwen Rd, Newport, NP9 8LQ [IO81MO, ST38]
G6 EWV J M Edwards, 4 Martha Cres, Mount Pleasant, New South Wales, Australia, 2749 X
G6 EWX N J Evans, 183 Church St., Wolverton, Milton Keynes, MK12 5JY [IO92OB, SP84]
G6 EWZ V Fairhurst, 44 Harold Rd, Stoke, Coventry, CV2 5LG [IO92GJ, SP37]
G6 EXE M J Graham, 11 Robert Moffat, High Legh, Knutsford, WA16 6PS [IO83SI, SJ78]
G6 EXG M S M S Gee, 100 Plantation Hill, Kilton, Worksop, S81 0QN [IO93KH, SK57]
G6 EXN E P Hall, 9 Valance Ave, Chingford, London, E4 6DR [JO01AP, TQ39]
G6 EXO Details withheld at licensee's request by SSL.
G6 EXP M J Hargreaves, 11 Hind St., Burnley, BB10 1EQ [IO83VT, SD83]
G6 EXQ D J Holt, 2 Bovington Ave, Thornton, Thornton Cleveleys, FY5 3DW [IO83LU, SD34]
G6 EXU A R Jobber, Church Hill, Kings North, Kingsnorth, Ashford, Kent, TN23 3EG [JO01KC, TR03]
G6 EXY Details withheld at licensee's request by SSL.
G6 EXZ A N Kent, 166 Louth Rd, Scartho, Grimsby, DN33 2LG [IO93WM, TA20]
G6 EYA P S Kershaw, Southview Cottage, Grittleton Rd, Yatton Keynell, Chippenham, Wilts, SN14 7JZ [IO81VL, ST87]
G6 EYD A G Mott, 1 Glenfield Cres, Newbold, Chesterfield, S41 8SF [IO93GG, SK37]
G6 EYH R McDonough, 12 Fulmere Ct, Swinton, Manchester, M27 0FD [IO83TM, SD70]
G6 EYJ D P Morton, 27 Beechfield Way, Hazlemere, High Wycombe, HP15 7TP [IO91PP, SU89]
G6 EYS A C Morne, 16 Warmden Ave, Baxenden, Accrington, BB5 2PR [IO83TR, SD72]
G6 EYT J W K Melhuish, 7 Cuthburga Rd, Wimborne, BH21 1LH [IO90AT, SZ09]
G6 EYX Details withheld at licensee's request by SSL.
G6 EYY M C J Posen, 74 Edwin Panks Rd, Hadleigh, Ipswich, IP7 5JL [JO02LA, TM04]
G6 EZA J M Douglass, 28 High St., Methwold, Thetford, IP26 4NT [JO02GM, TL79]
G6 EZG I R Prince, 31 Gillshill Rd, Hull, HU8 0JG [IO93US, TA13]
G6 EZJ Details withheld at licensee's request by SSL.
G6 EZK Details withheld at licensee's request by SSL.
G6 EZM D Winters, 14 Sunnyhill Rd, Southbourne, Bournemouth, BH6 5HP [IO90CR, SZ19]
G6 EZW J A Lewer, 3 Fold View, Egerton, Bolton, BL7 9TG [IO83SP, SD71]
G6 EZY D J Powell, 82 Belmont St., Southport, PR8 1JH [IO83LP, SD31]
G6 FAB Details withheld at licensee's request by SSL.
G6 FAF C Narroway, 26 Fern Way, Garston, Watford, WD2 6HG [IO91TQ, TQ19]
G6 FAH K R Lawrence, 54 Sheldrake Rd, Mudeford, Christchurch, BH23 4BP [IO90DR, SZ19]
G6 FAL R P Stoneman, 9 Winchester Rd, Northampton, NN4 8AZ [IO92NF, SP75]
G6 FAZ D Breckell, 82 Park Farm Rd, Feniscowles, Blackburn, BB2 5HP [IO83RR, SD62]
G6 FBA J R Butters, 21 Erleigh Rd, Reading, RG1 5LR [IO91MK, SU77]
G6 FBB R A K Chidgey, 14 Drury Rd, Colchester, CO2 7UX [JO01KV, TL92]
G6 FBC J H Evans, 103 Shorncliffe Rd, Coventry, CV6 1GQ [IO92FK, SP38]
G6 FBF T J Drew, 10 Sparkey Cl, Witham, CM8 1QR [JO01HS, TL81]
G6 FBH G Davis, Westbury House, 3 Windermere, Wilnecote, Tamworth, B77 5TD [IO92EO, SK20]
G6 FBI G P Dowsett, 7 Dorset Ave, Great Baddow, Chelmsford, CM2 9TZ [JO01FR, TL70]
G6 FBJ J Q Endicott, 37 Shetland Cl, Pound Hill, Crawley, RH10 7YZ [IO91WC, TQ33]
GW6 FBM N G Garside, 31 Lilac Way, Wrexham, LL11 2BB [IO83LB, SJ35]
G6 FBR R Garwood, 4 Charnwood Ave, Castle Ln West, Bournemouth, BH9 3LU [IO90BS, SZ09]
G6 FBS M R J Hardiman, Valetta House, Doles Ln, Wokingham, Berks, RG11 4EB [IO91LJ, SU66]
G6 FBT R J Hawkesworth, Hawken House, Greenwood Rd, Pateley Bridge, Harrogate, HG3 5LR [IO94CC, SE16]
G6 FBV T D Howell, 14 Winton Ave, Audenshaw, Manchester, M34 5NS [IO83WL, SJ99]
GM6 FBZ T Hoggan, 23 Wallace Pl, Hamilton, ML3 7DE [IO75XS, NS75]
G6 FCI C J McMahon, 1 Riversway, Layton, Blackpool, FY3 8PD [IO83LT, SD33]
G6 FCJ P A Magnus-Watson, 30 Barchester Rd, Langley, Slough, SL3 7HA [IO91RM, TQ07]
G6 FCL J C Mahoney, 89 Tyefields, Pitsea, Basildon, SS13 1JA [JO01GN, TQ78]
G6 FCN A Marshall, 16 Gr Rd, Whittington Moor, Chesterfield, S41 8LN [IO93GG, SK37]
GM6 FCW I M Macarthur, 22 Royal Park Terr, Edinburgh, EH8 8JB [IO85KW, NT27]
G6 FCX A P Nilsson, Melrose House, 18 Park St., Grimsby, DN32 7QU [IO93XN, TA21]
G6 FDB J Newby, 3 Greenshields Rd, Grindon, Sunderland, SR4 9RQ [IO94GV, NZ35]
G6 FDC A Owen, 2 Randle Bennett Cl, Elworth, Sandbach, CW11 3GA [IO83TD, SJ76]
G6 FDD R Pinchin, 10 Epping Dr, Melton Mowbray, LE13 1UH [IO92NS, SK72]
G6 FDG I A Rivers, Oakdene, 4 Glenview Ave, Heaton, Bradford, BD9 5PA [IO93CT, SE13]
G6 FDI B Raymer, 19 Caithness Dr, Crosby, Liverpool, L23 0RG [IO83LL, SJ39]
G6 FDK S Maskrey, The Hayloft, Stamford Ln, Cotton Edmunds, Chester, CH3 7QD [IO83OE, SJ46]
G6 FDO I D A Moody, 54 Lansdowne Rd, Studley, B80 7RD [IO92BG, SP06]
GM6 FDQ G Allan, Corse Farm, Kininmouth, Peterhead, Aberdeenshire, AB4 8JU [IO87TH, NJ72]
G6 FDS D A Allen, 21 Kelvin Rd, Thornton Cleveleys, FY5 3AF [IO83LU, SD34]
G6 FDU R C Butterworth, 49 Swandene, Pagham, Bognor Regis, PO21 4UR [IO90PS, SZ89]
G6 FDX C N Bicknell, 42 Palace Rd, New Southgate, London, N11 2PR [IO91WO, TQ39]
GW6 FED D Corsi, 4 Horsley Dr, Wrexham, LL12 8BE [IO83MB, SJ35]
G6 FEI D L Harris, 53 Welwyn Dr, Salford, M6 7PQ [IO83UM, SD70]
G6 FEJ R A Hawkes, 1 The Fairway, Wellingborough, NN9 5YS [IO92PH, SP87]
G6 FEO D G G Jones, 8 Kenilworth Rd, Lighthorne Heath, Leamington Spa, CV33 9TH [IO92GE, SP35]
G6 FEP M S Jackson, 403 Wheatley Ln Rd, Fence, Burnley, BB12 9ED [IO83UU, SD83]
G6 FES S Jones, 12 Meadow Croft, Nant Clwyd Park, Cross Lanes, Wrexham, LL13 0UJ [IO83NA, SJ34]
G6 FFH T J Sallis, 3 Stanley St., Brighton, BN2 2GP [IO90WT, TQ30]
G6 FFL T E Short, The Warren, Woodlands Rd, Tytherington, Wotton under Edge, GL12 8UU [IO81SO, ST68]
G6 FFM W D Scott, 29 Hazlewood Cres, Asfordby, Melton Mowbray, LE14 3UB [IO92MS, SK71]
G6 FFQ F Bilton, 50 Coldwell Rd, Crossgates, Leeds, LS15 7HA [IO93GT, SE33]
G6 FFR B M Berry, 7 Barlow Cl, Randlay, Telford, TF3 2NQ [IO82SP, SJ70]
G6 FFU P J Coogan, 24 High St., Kingsley, Stoke on Trent, ST10 2AE [IO93AA, SK04]
G6 FFX T F Chung, 1 Poole St., Allenton, Derby, DE24 9DA [IO92GV, SK33]
G6 FGE W A Kirby, 2 St. Peters Pl, Haslingden, Rossendale, BB4 4BT [IO83UQ, SD72]
G6 FGJ C G Tandy, Caradon House, Bassetsbury Ln, High Wycombe, HP11 1RB [IO91PO, SU89]
G6 FGK Details withheld at licensee's request by SSL.
G6 FGO A J Luckman, 23 Sandpiper Cl, Quedgeley, Gloucester, GL2 4LZ [IO81UU, SO81]
G6 FGP Details withheld at licensee's request by SSL.
G6 FGV J Webber, 25 Rossall Rd, Lancaster, LA1 5HQ [IO84OB, SD46]
G6 FGW R J Weekes, 84 Vera Rd, Yardley, Birmingham, B26 1TT [IO92CL, SP18]
G6 FGY E B Westbrook, 66 Nelson Cl, Croydon, CR0 3SW [IO91WJ, TQ36]
GI6 FHD A M McPartland, 4 Clanbrassil Gdns, Portadown, Craigavon, BT63 5YD [IO64SK, J05]
G6 FHE C D Martin, 4 Brendon Gdns, Wollaton, Nottingham, NG8 1HY [IO92JX, SK54]
G6 FHJ C V Lascelles, 26 Tudor Way, Dersingham, Kings Lynn, PE31 6LX [JO02GU, TF63]
G6 FHK C C Leonard, 138 Sundridge Dr, Chatham, ME5 8JD [JO01GI, TQ76]
G6 FHM D Sunderland, 1 Allfield Cttgs, Condover, Shrewsbury, SY5 7AP [IO82PP, SJ50]
G6 FHO G A Onion, 4 Mill Rd, Woodhouse Eaves, Loughborough, LE12 8RD [IO92JR, SK51]
G6 FHQ Details withheld at licensee's request by SSL.
G6 FHR R C Plant, Plant Engineering, Park Ln, Langport, Somerset, TA10 0NF [IO81NA, ST42]
G6 FHW Details withheld at licensee's request by SSL.
G6 FIB T E Wicks, 12 Kensington Ave, Watford, WD1 7RY [IO91TP, TQ19]
G6 FID Details withheld at licensee's request by SSL.
GM6 FIK D R Stevenson, 2 Melbourne Ct, Braidpark Dr, Giffnock, Glasgow, G46 6LA [IO75UT, NS55]
G6 FIL D H Smith, 323 Colchester Rd, Ipswich, IP4 4SF [JO02OB, TM14]
G6 FIN A R Stevens, 16 Tremlett Gr, Ipplepen, Newton Abbot, TQ12 5BZ [IO80EL, SX86]
G6 FIO J A Slater, 154 Ralph Rd, Shirley, Solihull, B90 3JZ [IO92CJ, SP17]
G6 FIP I E Tebboth, 20 Glebe Rd, Stratford upon Avon, CV37 9JU [IO92DE, SP15]
G6 FIQ M L Wright, 6 North St, Swinford, Leics, LE17 6BE [IO92KJ, SP57]
G6 FIT A Lewis, 81 Ashton Ave, Rainhill, Prescot, L35 0QR [IO83OJ, SJ49]
G6 FIU D M Leitch, 31 Heybrook Ave, Preston Grange, North Shields, NE29 9HG [IO95GA, NZ36]
G6 FIV Details withheld at licensee's request by SSL.
G6 FJA F J Aunger, 20 Hillside Gr, Penketh, Warrington, WA5 2DP [IO83QJ, SJ58]
G6 FJE L A Plewa, 174 Dorset Ave, St. Baddow, Chelmsford, CM2 8YY [JO01FR, TL70]
G6 FJF D R Parker, 10 Priory Rd, Wollaston, Wellingborough, NN29 7PW [IO92PG, SP96]
G6 FJG N A B Pinkney, Mdems, RAF (H), Wegberg, Bfpo 40, X X
G6 FJH R G Robinson, 50 Kenilworth Rd, Whitley Bay, NE25 8BD [IO95GA, NZ37]
G6 FJI M C Richards, 27 Burlington Rd, Enfield, EN2 0LL [IO91XP, TQ39]
G6 FJM Details withheld at licensee's request by SSL.
G6 FJO S Turner, 13 Gaydon Walk, Bicester, OX6 7YY [IO91KV, SP52]
G6 FJQ Details withheld at licensee's request by SSL.
G6 FJT R K Mannix, 26 Caroline Pl, London, W2 4AN [IO91VM, TQ28]
G6 FKA G H Valler, 35 Howard Rd, Woodside, South Norwood, London, SE25 5BU [IO91XJ, TQ36]
GW6 FKB E A Taylor, Waysdide, 19 Chester Rd, Saltney Ferry, Chester, CH4 0AQ [IO83ME, SJ36]
G6 FKE K P Redmond, Flat 1, 53 Church Rd, Moseley, Birmingham, B13 9EB [IO92BK, SP08]
G6 FKN M B R Lee, 55 Wodeland Ave, Guildford, GU2 5LA [IO91RF, SU94]
GW6 FKP S J Moore, 25 Overdale Ave, Mold Rd, Mynydd Isa, Mold, CH7 6US [IO83KD, SJ26]
G6 FKQ Details withheld at licensee's request by SSL.
G6 FKR D R Roberts, 24 Rosebery Ave, Lancaster, LA1 4DJ [IO84OA, SD46]
G6 FKS S G Robinson, 4 Grayling Cl, Cambridge, CB4 1NP [JO02BF, TL45]
G6 FKY T S Norris, 7 Johnson Way, Ford, Arundel, BN18 0TD [IO90QT, SU90]
G6 FLE N L J Scott, 1 Lakeside, Fareham, PO17 5EP [IO90JU, SU50]
G6 FLH J P Smith, 33 Chyngton Way, Seaford, BN25 4JB [JO00BS, TV49]
G6 FLJ J T Wade, Tresco, 63 Yarmouth Rd, Blofield, Norwich, Norfolk, NR13 4LG [JO02RP, TG30]
GM6 FLL A Simpson-Fraser, 430 Millcroft Rd, South Carbrain, Cumbernauld, Glasgow, G67 2QW [IO85AW, NS77]
GM6 FLM D D Simpson-Fraser, 430 Millcroft Rd, South Carbrain, Cumbernauld, Glasgow, G67 2QW [IO85AW, NS77]
G6 FLQ G C Smith, Hockerton Rd, Kirklington, Notts, NG22 8PB [IO93MC, SK65]
GW6 FLU C H Mock, Colden, Newport Rd, St. Mellons, Cardiff, CF3 9UA [IO81KM, ST28]
G6 FLV Details withheld at licensee's request by SSL.
G6 FLW C B Thompson, 27 Queensland Dr, Colchester, CO2 8UD [JO01KU, TM02]
G6 FLX Details withheld at licensee's request by SSL.
G6 FLY H M Lee, 43 Heath Rd, Stapenhill, Burton on Trent, DE15 9LG [IO92ES, SK22]
G6 FMC Details withheld at licensee's request by SSL.
G6 FMF D Timson, 40 Rockwood Rd, Woodhall Park, Calverley, Pudsey, LS28 5AA [IO93DT, SE23]
G6 FMN P J Rogers, 12 St. Peters Rise, Headley Park, Bristol, BS13 7LY [IO81QK, ST56]
G6 FMQ A P Polding, 17 Josephine Rd, Cowlersley, Huddersfield, HD4 5UD [IO93CP, SE11]
G6 FMS M Peers, 46 Lowndes Park, Driffield, North Humberside, YO25 7BG [IO94SA, TA05]
G6 FMU I H Muir, 62 Lennard Rd, Dunton Green, Sevenoaks, TN13 2UU [JO01CH, TQ55]
G6 FMW R A Mantle, 33 Purleigh Cl, Basildon, SS13 1RJ [JO01GN, TQ79]
G6 FMZ Details withheld at licensee's request by SSL.
G6 FNA B W Lambert, 43 Mount Rd, Prestwich, Manchester, M25 2GP [IO83UM, SD80]
G6 FNB D S Morris, 21 Crosfield Ct, Cambridge, CB4 2SA [JO02BF, TL46]
G6 FNC Details withheld at licensee's request by SSL.
GM6 FND I W Reid, 38 Muirside Gr, Cairney Hill, Cairneyhill, Dunfermline, KY12 8RB [IO86FB, NT08]
G6 FNE S R Saunders, 5 Morvale Cl, Belvedere, DA17 5HS [JO01BL, TQ47]
G6 FNG Dr P J Wilson, 2 Homestead Rd, Thorngumbald, Hull, HU12 9PT [IO93WR, TA22]
G6 FNI S Woodroffe, 12 Fulwood Cl, Chilwell, Beeston, Nottingham, NG9 5LG [IO92JV, SK53]
G6 FNJ R A Oglesby, Littleton Mill, Semington, Trowbridge, Wilts, BA14 6LQ [IO81WI, ST96]
G6 FNQ A E Smith, 16 Hazel Way, Barwell, Leicester, LE9 8GP [IO92HN, SP49]
G6 FNY T D Strand, 129 Malmesbury Rd, Chippenham, SN15 1PZ [IO81WL, ST97]
GW6 FOF C J Morris, Glantivy, Penglais Terr, Aberystwyth, Dyfed, SY23 2ET [IO72XJ, SN58]
G6 FOH C M McLean, 31 Basingbourne Rd, Fleet, GU13 9TG [IO91NG, SU85]
G6 FOI Dr A C Regan, 153 Acre Ln, Cheadle Hulme, Cheadle, SK8 7PB [IO83VI, SJ88]
G6 FOL Dr S R Williams, 32 Hill Dr, Hove, BN3 6QL [IO90VU, TQ20]
G6 FOM Details withheld at licensee's request by SSL.
G6 FOO D E Morgan, 82 Hoddern Ave, Peacehaven, BN10 7QY [IO90XT, TQ40]
GI6 FOR N Lane, Turnpike House, 117A Hillhall Rd, Lisburn, BT27 5BT [IO64XM, J26]
GM6 FOT T D J Armour, 86 Hillend Rd, Clarkston, Glasgow, G76 7XT [IO75US, NS55]
G6 FOV D J Aldridge, 17 Priory Cl, Tavistock, PL19 9DJ [IO70WM, SX47]
G6 FPC J E Body, 201 Sig Sqn, 1 (Uk) Adsr, Herford, Bfpo 15, X X
GM6 FPD J Bray, 12 Millbank Rd, Stranraer, DG9 0EJ [IO74LV, NX06]
G6 FPF Dr G H Barnes, Rockleigh, 17 Savile Park, Halifax, HX1 3EA [IO93BR, SE02]
G6 FPH M J M Cole, 45 Gainsborough Rd, Tilgate, Crawley, RH10 5LD [IO91VC, TQ23]
G6 FPK N A B Cooper, 53 Stanway Rd, Benhall, Cheltenham, GL51 6BU [IO81WV, SO92]
G6 FPP A J M Ford, 143 Avocet Way, Langford Village, Bicester, OX6 0YW [IO91KV, SP52]
G6 FPQ C Groves, 223 Markfield, Ct Wood Ln, Croydon, CR0 9HU [IO91XI, TQ36]
G6 FPX S F Hill, 1 Flavian Ct, Halton Brow, Runcorn, WA7 2JW [IO83PI, SJ58]
G6 FQL R T Heath, Friarcroft, Gayton Rd, Ashwicken, Kings Lynn, PE32 1LW [JO02GR, TF71]
G6 FQM Details withheld at licensee's request by SSL.
G6 FQN S W Jarmyn, 2 Strathmore, Fenlands Park, East Tilbury, Grays, Essex, RM18 8RW [JO01FL, TQ67]
G6 FQP B L Kneebone, 1 Chapel Terr, Carnkie, Helston, TR13 0DT [IO70JD, SW73]
GI6 FQT D M C McConville, 28 Derrycor Ln, Derryadd, Craigavon, BT66 6QW [IO64SL, J06]
G6 FQZ C G Potter, 12 Beech Rd, Headington, Oxford, OX3 7RR [IO91JS, SP50]
G6 FRA P J Phippin, 69 Chesterton Ave, Harpenden, AL5 5SU [IO91TT, TL11]
G6 FRB J Pearce, 25 Boughton St., St. Johns, Worcester, WR2 4HE [IO82VE, SO85]
G6 FRL Maj J A H West, 6 Lynwood Ct, Priestland Pl, Lymington, SO41 9GA [IO90FS, SZ39]
G6 FRS M Hearsey Farnborough&Dis, Rs. Halycon, Lawday Link Upper-Hale Farnham Surrey, GU9 0BS [IO91OF, SU84]
G6 FRT S E Wedge, 21 Holmesland Walk, Botley, Southampton, SO30 2DZ [IO90IV, SU51]
G6 FRU Details withheld at licensee's request by SSL.
G6 FS D M Ferguson, 3 Aldeburgh Rd, Leiston, IP16 4JY [JO02SE, TM46]
G6 FSA J N Coupe, 15 Belvedere Ave, Atherton, Manchester, M46 9LQ [IO83SM, SD60]
G6 FSD A R Bird, 3 Cudham Gdns, Cliftonville, Margate, CT9 3HG [JO01RJ, TR37]
G6 FSF A T Carter, 64 Marina Dr, May Bank, Newcastle, ST5 0RS [IO83VA, SJ84]
GM6 FSG W B Chamberlain, Cwmmelyn Kings Rd, Whithorn, Newton Stewart, Wigtownshire, DG8 8PP [IO74TR, NX44]
G6 FSP Details withheld at licensee's request by SSL.[Op: D Helliwell. Station located near Newton Abbot.]
G6 FSU M B Apperly, Church Cottage, Church Way, Little Stukeley, Huntingdon Cambs, PE17 5BQ [IO92VI, TL27]
G6 FSW S C Bakin, 11 Highgrove Dr, Chellaston, Derby, DE73 1XA [IO92GV, SK33]
GM6 FT R T Frost, St. Felix, 93 Vardar Ave, Clarkston, Glasgow, G76 7QW [IO75US, NS55]
G6 FTA M R Everall, 37 Tudor Ave, Cheshunt, Waltham Cross, EN7 5AU [IO91XQ, TL30]
G6 FTB F S Jackson, 27 Prairie Cres, Burnley, BB10 1EU [IO83VT, SD83]
G6 FTC M J King, 3 Craythorne Cl, Hythe, CT21 5SP [JO01NB, TR13]
G6 FTE P Bondar, 5 Mayes Meadow, Moulton, Newmarket, CB8 8SZ [JO02FG, TL66]
G6 FTH D Clark, 43 Glenfield Cres, Chesterfield, S41 8SF [IO93GG, SK37]
G6 FTI S G Coughlan, 123 Ravenstone Dr, Greetland, Halifax, HX4 8DY [IO93BQ, SE02]
G6 FTJ P Carter, 145 Wakefield Rd, Dewsbury, WF12 8AJ [IO93EQ, SE22]
GI6 FTM J B Dynes, 30 Breagh Rd, Portadown, Craigavon, BT63 5LT [IO64TK, J05]
G6 FTN Details withheld at licensee's request by SSL.
G6 FTP W J Ginty, 9 Belmont Rd, Bolton, BL1 7AF [IO83SO, SD71]
GJ6 FTU B Johnson, Cyprus Cottage, Rue Horman, Gorey, Jersey, JE3 9EJ
G6 FTY R A Miles, 60 Aylesham Way, Yateley, Camberley, Surrey, GU17 7NT [IO91OH, SU85]
G6 FUB Details withheld at licensee's request by SSL.
G6 FUT I Donn, The Old Post Office, Boxford, Newbury, Berks, RG20 8DH [IO91HK, SU47]
G6 FUW Details withheld at licensee's request by SSL.

G6 FUY R G Fishwick, 7 Howard Ave, Monton, Eccles, Manchester, M30 9GF [IO83TL, SJ79]
G6 FVB R J Baker, Ashmeadfield Rd, Apple Gr, Emsworth, Hants, PO10 8EU [IO90MU, SU70]
G6 FVF P M Fenn, 399B Kenton Ln, Belmont Circle, Harrow, HA3 8RZ [IO91UO, TQ19]
G6 FVH A R Holland, 36 Crown Lea Ave, Malvern, WR14 2DP [IO82UC, SO74]
G6 FVJ A L James, 82 Sandringham Dr, Spondon, Derby, DE21 7QA [IO92HW, SK43]
GW6 FVK Details withheld at licensee's request by SSL.
G6 FVL R A Young, 134 Harport Rd, Redditch, B98 7PD [IO92AH, SP06]
G6 FVM A B Williamson, 69 Hall St., Southport, PR9 0RF [IO83MP, SD31]
GM6 FVP W S Wilkins, 7 Hermitage Ave, Aberdeen, AB2 3LU [IO87TH, NJ72]
G6 FVT N V Smith, 33 Constitution Hill, Benfleet, SS7 1EB [JO01GN, TQ78]
G6 FVZ R V Munns, 37 Goldfinch Way, South Wonston, Winchester, SO21 3SG [IO91IC, SU43]
G6 FWK D Booth, 54 Shaw Dr, Knutsford, WA16 8JR [IO83TH, SJ77]
G6 FWT C Knowles, 77 Sandstone Rd, Sheffield, S9 1AF [IO93GJ, SK39]
GW6 FXB C K O'Brien, 33 Pleasant St., Morriston, Swansea, SA6 6HH [IO81AQ, SS69]
G6 FXE C M Walton, 49 Blandford Dr, Walsgrave, Coventry, CV2 2JD [IO92GK, SP38]
G6 FXG K A Brabon, 128 Fornham Rd, Bury St. Edmunds, IP32 6AP [JO02IG, TL86]
G6 FXH K D Brailey, 62 Widgery Rd, Whipton, Exeter, EX4 8BB [IO80GR, SX99]
G6 FXL J C Worthington, Field House, Thorpe Langton, Market Harborough, Leics, LE16 7UN [IO92NM, SP79]
G6 FXM F E Wright, 21 St. Fabians Dr, Chelmsford, CM1 2PR [JO01FR, TL60]
G6 FXN M J Aynge, 20 St. Mildreds Rd, Lee, London, SE12 0RA [JO01AK, TQ37]
G6 FXO R J Bennison, The Old Smithy, Wilsill, Harrogate, HG3 5EB [IO94DB, SE16]
G6 FXR D W Ainslie, 16 Blount Cres, Binfield, Bracknell, RG42 4UH [IO91OK, SU87]
GI6 FXY E L Connolly, 21 Clanrye Ave, Newry, BT35 6EH [IO64TE, J02]
GM6 FXZ A J Carnall, 3 Main St., Glenluce, Newton Stewart, DG8 0PN [IO74OV, NX15]
G6 FYC J W Cowee, 26 Arundel Rd, Old Heatherside, Camberley, GU15 1DL [IO91PH, SU95]
G6 FYE C Das Neves Pedro, 37 Madresfield Rd, Great Malvern, Worcs, WR14 2AS [IO82UC, SO74]
G6 FYL N J Harris, 104 Blandford Dr, Walsgrave, Coventry, CV2 2NE [IO92GK, SP38]
G6 FYT G D Whiting, Gdn Flat, 5 South View, Addington, Liskeard, PL14 3EX [IO70SL, SX26]
G6 FYU J D Walker, 4 Manor Fields, Horsham, RH13 6SB [IO91UB, TQ13]
G6 FYW R L Gordon, 16 Penerley Rd, Catford, London, SE6 2LQ [IO91XK, TQ37]
GM6 FYY S S Bremner, 21 Braal Terr, Halkirk, KW12 6YN [IO88GM, ND15]
G6 FYZ C Cromar, 201 Salisbury Rd, Testwood, Totton, Southampton, SO40 3LL [IO90GW, SU31]
G6 FZC D Hall, 75 Hunter Terr, Grangetown, Sunderland, SR2 8SA [IO94HV, NZ45]
G6 FZV W H Day, 4 Queenswood Dr, Norton Park, Worcester, WR5 3SZ [IO82VE, SO85]
G6 FZW A Eaves, 3 Station Cttgs, Station Rd, Cheddington, Leighton Buzzard, LU7 0SQ [IO91QU, SP91]
G6 GA R W Ainge, Happy Cottage, Majors Barn, Cheadle, Stoke on Trent, ST10 1PY [IO92AX, SK04]
G6 GAB W A J Honey, 20 Pennor Dr, St. Austell, PL25 4UW [IO70OI, SX05]
G6 GAD R P Kelly, 118 Brooksfield, Panshanger, Welwyn Garden City, AL7 2AN [IO91VT, TL21]
G6 GAF A J Franklin, 56 Kersland St., Hillhead, Glasgow, G12 8BT [IO75UV, NS56]
GI6 GAG N Orr, 405 Enniskeen, Drumgor, Craigavon, BT65 4AB [IO64TK, J05]
G6 GAK M J Tyrrell, 189 Runcorn Rd, Barnton, Northwich, CW8 4HR [IO83RG, SJ67]
G6 GAN J W Sykes, 14 Heddon Cl, Isleworth, TW7 7DP [IO91UL, TQ17]
GI6 GAQ H Warke, 5 Meadow View, Ballymoney, BT53 7AH [IO65RB, C92]
G6 GAV P A Sharp, 54 Burton Fields Rd, Stamford Bridge, York, YO4 1JJ [IO93NX, SE75]
G6 GAW D J L Peters, 123 Grinstead Ln, Lancing, BN15 9DR [IO90UT, TQ10]
G6 GBC D W I Ayers, 13 Battlemead Cl, Maidenhead, SL6 8LB [IO91PM, SU98]
G6 GBH Details withheld at licensee's request by SSL.
G6 GBI P S Tehara, 64 Oakway, Crawley, RH10 2HS [IO91VC, TQ23]
G6 GBL A D Abbott, 164 Bath Rd, Reading, RG30 2HA [IO91LK, SU67]
G6 GBO Details withheld at licensee's request by SSL.
G6 GBT I D Cole, 21 Quincey Dr, Erdington, Birmingham, B24 9LX [IO92CM, SP19]
G6 GBU Details withheld at licensee's request by SSL.
G6 GCE M J Carter, 4 Herrings Way, Fordham, Colchester, CO6 3NB [JO01JW, TL92]
G6 GCI C G Burnett, 36 Mill Ln, Romsey, SO51 8EQ [IO91FA, SU32]
G6 GCJ J W Burnett, 8 Kintbury Mill, Kintbury, Hungerford, RG17 9UN [IO91GJ, SU36]
GW6 GCK J I Cook, The Hollies, 52 Kelston Rd, Whitchurch, Cardiff, CF4 2AH [IO81JM, ST18]
G6 GCO D J Davies, 79A Spenser Rd, Bedford, MK40 2BE [IO92SD, TL05]
G6 GCU R W Golledge, Blue House, Witham Rd, Fairstead, Chelmsford, CM3 2BS [JO01HT, TL71]
G6 GCY J Robinson, 84 Hereford Way, Scowcroft Farm, Middleton, Manchester, M24 2NN [IO83VN, SD80]
G6 GDI V I Gerhardi, 24 Putnams Dr, Aston Clinton, Aylesbury, HP22 5HH [IO91PT, SP81]
GI6 GDM K R Galbraith, 14 Millars Forge, Dundonald, Belfast, BT16 0UT [IO74CN, J47]
GU6 GDO M J Skillett-Habin, Somerset House, La Rue Des Prevosts, St. Saviour, Guernsey, GY7 9UH
GW6 GDR C D Price-Gore, Brynsadler PO, Cowbridge Rd, Brynsadler, Pontyclun Mid Gla, CF7 9BS [IO81HN, ST08]
G6 GDT Details withheld at licensee's request by SSL.
G6 GEA M J Walker, The Saddlery, The Manor House, Manor Park, Ruddington, Notts, NG11 6DS [IO92KV, SK53]
G6 GEC P J E Carey GEC Avionics, 99 Bells Ln, Hoo St. Werburgh, Hoo, Rochester, ME3 9HX [JO01GK, TQ77]
G6 GEG L M Abrahams, 1 The Grove, Edgware, HA8 9QA [IO91UO, TQ19]
G6 GEJ Details withheld at licensee's request by SSL.[Op: Mike. Station located near Long Eaton.]
G6 GEK A P Elliott, 4 Quakers Way, Fairlands, Guildford, GU3 3NF [IO91QG, SU95]
G6 GEL K Inman, 15 Waterbridge Ct, Appleton, Warrington, WA4 3BJ [IO83RI, SJ68]
G6 GEN R Ainsworth, 29 Broom House, Reddington Dr, Langley, Slough, SL3 7QY [IO91RL, TQ07]
G6 GEP Details withheld at licensee's request by SSL.
G6 GES R Haywood, 16 The St., Kingston, Canterbury, CT4 6JB [JO01NF, TR15]
G6 GEV D R Ashton, Stonewalls, Sturt Rd, Charlbury, Chipping Norton, OX7 3EP [IO91GU, SP31]
G6 GEX C Farley, 1 Wesley Cttgs, Peverell, Mutley, Plymouth, PL3 4RB [IO70WJ, SX45]
G6 GEY D J Gorman, 18 Tamarisk Cl, Eastney, Southsea, PO4 9TS [IO90LS, SZ69]
G6 GFA P J Arscott, 122 Woodlands Rd, Ashurst, Southampton, SO40 7AL [IO90FV, SU31]
G6 GFC J E Burrows, 4 Cavendish Cres, Alsager, Stoke on Trent, ST7 2EF [IO83UC, SJ75]
G6 GFG P R Cook, 20 Brookside Rd, Loughborough, LE11 3PQ [IO92JS, SK51]
GM6 GFH D R Craig, Lettershuna House, Appin, Argyll, Scotland, PA38 4BN [IO76HO, NM94]
G6 GFI R B Gamble, Hardcastle House, Thorpe, Skipton, Nrth Yorks, BD23 6BJ [IO94AB, SE06]
G6 GFJ H B Goozee, 45 Brighton Rd, Purley, Surrey, CR8 2LR [IO91WI, TQ36]
GM6 GFL Dr D M Begg, 12 Broomhill Rd, Penicuik, EH26 9EE [IO85JT, NT25]
G6 GFO N J Atrill, 17 Wye Ct, River Way, Andover, SP10 5HD [IO91GF, SU34]
G6 GFQ C J Barnard, 47 Revidge Rd, Blackburn, BB2 6JH [IO83RS, SD62]
G6 GFR T D E Crook, 21 Cleveland Cl, Maidenhead, SL6 1XE [IO91PM, SU88]
G6 GGD R Forster-Pearson, 202A Far Laund, Belper, DE56 1FP [IO93GA, SK34]
G6 GGN M E Hoskin, 7 Worrall Mews, Worrall Rd, Clifton, Bristol, BS8 2TX [IO81QL, ST57]
G6 GGU Details withheld at licensee's request by SSL.
G6 GGV B Hollingworth, 62 Illingworth Ave, Bradshaw, Halifax, HX2 9JD [IO93BS, SE02]
G6 GGW N Gautrey, Bradgate House, Hunston, Bury St. Edmunds, Suffolk, IP31 3EL [JO02KG, TL96]
G6 GGY L J Fitzwater, The Old Cottage, Babylon Ln, Lower Kingswood, Tadworth, KT20 6XE [IO91VG, TQ25]
G6 GGZ M A L Ferne, 24 Essex Gdns, Leigh on Sea, SS9 4HG [JO01HN, TQ88]
G6 GHD A S Burleton, 27 Doncaster Rd, Bristol, BS10 5PN [IO81QL, ST57]
G6 GHE A J Rawdon, 44 Southgate, Hornsea, HU18 1AL [IO93VV, TA24]
G6 GHP R Vansittart, 4486 Winnetka Ave, Woodland Hills, California 91364, USA
G6 GHU R Wood, 12 Roundhead Dr, Thame, OX9 3DG [IO91MS, SP70]
GI6 GIE J W D Pinkerton, Clonkeen, 218 Seacon Rd, Ballymoney, BT53 6PZ [IO65RB, C92]
G6 GIF M A Oram, 25 Jerome Cl, Marlow, SL7 1TX [IO91ON, SU88]
G6 GIG R C Ralton, 62 Hall Farm Rd, Duffield, Belper, DE56 4FS [IO92GX, SK34]
G6 GIH P R Sewell, 6 Hawthorne Ave, Bedford, MK40 4HJ [IO92SD, TL04]
G6 GIU A S Stephens, 4 Falcon House, Gurnell Gr, West Ealing, London, W13 0AE [IO91UM, TQ18]
G6 GIY Details withheld at licensee's request by SSL.
G6 GJD C Harper, 6 Bela Gr, Blackpool, FY1 5JZ [IO83LT, SD33]
G6 GJN T K Biggs, 3 Pentathlon Way, Cheltenham, GL50 4SE [IO81WW, SO92]
G6 GJQ Details withheld at licensee's request by SSL.
G6 GJV W D Willis, 22 Kenilworth Gdns, Ilford, IG3 8DU [JO01BN, TQ48]
GM6 GJW J L Leith, 87 Mandarin Way, Wymans Brook, Cheltenham, GL50 4RS [IO81WW, SO92]
G6 GJX B Revill, 9 Margaret Ave, Ilkeston, DE7 5DD [IO92IX, SK44]
G6 GJY S R Smith, 36 Greenfields, Earith, Huntingdon, PE17 3QH [JO02AI, TL37]
G6 GKG W L Hodson, 2 Beaulieu House, Raymond Rd, Wimbledon, London, SW19 4AW [IO91VK, TQ27]
G6 GKI Details withheld at licensee's request by SSL.
G6 GKK A W Barton, 81 Trent St., Retford, DN22 6NG [IO93MH, SK78]
G6 GKL M J Borrow, 235 Crofton Rd, Orpington, BR6 8JE [JO01AI, TQ46]
G6 GKO B L Clarke, 4 Croyde Cl, Ethel Rd, Leicester, LE5 4WG [IO92KP, SK60]
GW6 GKP J J Coyne, 44 Brompton Ave, Rhos on Sea, Colwyn Bay, LL28 4TF [IO83DH, SH87]
G6 GKT I M Houldridge, 115 Summergangs Rd, Hull, HU8 8JX [IO93US, TA13]
G6 GLD P Trippear, 721 Rochdale Rd, Royton, Oldham, OL2 5UT [IO83WN, SD90]
G6 GLH Details withheld at licensee's request by SSL.
G6 GLO J W Pallister Gloucestershire County Raynet, 1 Third Ave, Highfields, Dursley, GL11 4NT [IO81TQ, ST79]
G6 GLP A P Rider, 52 Clarendon Rd, Ipplepen, Newton Abbot, TQ12 5QS [IO80EL, SX86]
G6 GLR B Corker Gt Lumley ARS, 46 Danelaw, Great Lumley, Chester-le-Street, DH3 4LU [IO94FU, NZ24]
G6 GLT Dr R J M Bennett, 59 Blacburn Rd, Edenfield, Bury, BL0 0JD [IO83UP, SD71]
G6 GLV B Dent, 1 The Green, Ribble Village, Ribbleton, Preston, PR2 6QF [IO83QS, SD53]
G6 GLZ D J Clews, 11 Roping Rd, Yeovil, Somerset, BA21 4BD [IO80QW, ST51]
G6 GMC Details withheld at licensee's request by SSL.
GW6 GMF M J Inness, 6 Denning Rd, Borras Park, Wrexham, LL12 7UG [IO83MB, SJ35]
G6 GMH G M Russell, Eastgate, Overend, Baslow, Derbyshire, DE45 1SG [IO93FF, SK27]
G6 GML W J J L Devenish, 13 Riverside, Hendon, London, NW4 3TU [IO91VN, TQ28]
G6 GMR E C M Walton, 44 St. Leonards Ave, Lostock, Bolton, BL6 4JE [IO83SN, SD60]
G6 GMW D J Ward T.C.A.R.S., 8 The Grove, Thornton Cleveleys, FY5 2JD [IO83LV, SD34]
G6 GN H J Gratton, 250 Wordsworth Rd, Bristol, BS7 0ED [IO81RL, ST67]
GI6 GNA H A Wright, 2 Duncans Rd, Lisburn, BT28 3LP [IO64XM, J26]
G6 GNC J C P Thornber, 7 Buckland Cl, Peterborough, PE3 9UH [IO92UN, TL19]
G6 GND R S Lambert, 10 Ambleside, Rugby, CV21 1JB [IO92JJ, SP57]
G6 GNE J E Sugden, Brooklyn, 12 Cliffe Ln, Cleckheaton, BD19 4ET [IO93DR, SE12]
G6 GNG E J T Pearce, 2 Forest Rd, Fishponds, Bristol, BS16 3XJ [IO81RL, ST67]
G6 GNH A F Mason, 18 Merrydale Rd, Stapenhill, Burton on Trent, DE15 9DQ [IO92ET, SK22]
G6 GNN M R Millington, 8 Ryles Cres, Macclesfield, SK11 8DD [IO83WG, SJ97]
G6 GNO B Cooper, 8 Stanley Rd, Doncaster, DN5 8RR [IO93JM, SE50]
G6 GNS N Shergold, 15 Marne Rd, Bitterne, Southampton, SO18 6AJ [IO90HW, SU41]
G6 GNW A P Chaddock, 8 Maple Rd, Faringdon, SN7 8BE [IO91EP, SU29]
G6 GNY D W L Clamp, 9 Tidkin Ln, Guisborough, TS14 8BX [IO94LM, NZ61]
G6 GO J H M G Goodacre, Ct Farm House, Frolesworth Rd, Ullesthorpe, Lutterworth, LE17 5BZ [IO92IL, SP58]
G6 GOG A G Kerr, 10 Hillcrest Rd, Crosby, Liverpool, L23 9XS [IO83LL, SJ39]
G6 GOL S J Lawrence, Franklands, Walden Rd, Sewards End, Saffron Walden, Essex, CB10 2LF [JO02DA, TL53]
G6 GOR L Carrick-Smith, Highfield House, Sheffield Rd, Clowne, Chesterfield, S43 4AP [IO93JG, SK57]
G6 GOV B Wood, 8 Chichester Dr, Chelmsford, CM1 7RY [JO01FR, TL70]
G6 GOW R A Wheeler, 2 Heather Cl, Brereton, Rugeley, WS15 1BB [IO92AR, SK01]
G6 GOY Capt A L Taylor, 7 Berkeley Cres, Lydney, GL15 5SH [IO81RR, SO60]
G6 GPF J K Woodhouse, 10 Foxes Meadow, Walmley, Sutton Coldfield, B76 1AW [IO92CN, SP19]
GM6 GPH J C Robertson, 41 Balgarvie Cres, Cupar, KY15 4EF [IO86LH, NO31]
G6 GPL G W Massingham, 4 The Garlings, Aldbourne, Marlborough, SN8 2DT [IO91EL, SU27]
G6 GPM P D Tambini, Flat 4, 11 Streatfield Rd, Heathfield, E Sussex, TN21 8LA [JO00CX, TQ52]
G6 GPS Details withheld at licensee's request by SSL.
G6 GQF Dr G R Martin, 9 Clarkes Ave, Kenilworth, CV8 1HX [IO92FI, SP27]
G6 GQG I D Moston, 24 Upton Rd, Southville, Bristol, BS3 1LP [IO81QK, ST57]
G6 GQI R Swann, 3 Elizabeth Ave, Newmarket, CB8 0DJ [JO02EG, TL66]
G6 GQJ J Davies, 1 Woodland Rd, Halesowen, B62 8JS [IO82XL, SO98]
G6 GQL J A Carter, 6 Green Ct, Wilton, Ross on Wye, HR9 6BH [IO81QV, SO52]
G6 GQP R J E Moreton, 86 Tyrrells Way, Sutton Courtenay, Abingdon, OX14 4DH [IO91IP, SU49]
GM6 GQT W S Gray, 11 Laghall Ct, Kingholm Quay, Dumfries, DG1 4SY [IO85EB, NX97]
G6 GRG R White Gordano ARC, 3 Robin Ln, Clevedon, BS21 7EX [IO81NK, ST47]
G6 GRU J S Brayshaw, 12 Weatherdon Dr, Ivybridge, PL21 0DD [IO80BJ, SX65]
GI6 GRV J C Barnett, 2 Donegall Park, Whitehead, Carrickfergus, BT38 9ND [IO74DS, J49]
G6 GRX Details withheld at licensee's request by SSL.
G6 GS A Pevy Guildford & D Rd, 6 Hormer Cl, Sandhurst, Owlsmoor, Camberley, Surrey, GU15 4QW [IO91PI, SU86]
G6 GSC H I Thomas, 28 Munday Cl, Manor Village, Bussage, Stroud, GL6 8DG [IO81WR, SO80]
G6 GSF K R Edwards, 16 Garden House Ln, East Grinstead, RH19 4JT [IO91XB, TQ33]
G6 GSI D J Millington, 15 Collett Way, Priorslee, Telford, TF2 9SL [IO82SQ, SJ71]
G6 GSP Details withheld at licensee's request by SSL.
G6 GSV J A Williams, 22 New Ave, Kirkheaton, Huddersfield, HD5 0JD [IO93DP, SE11]
G6 GSW Details withheld at licensee's request by SSL.
G6 GTA C E Lawley, 20 Planks Ln, Wombourne, Wolverhampton, WV5 9HG [IO82VM, SO89]
G6 GTB J C C Tracey, 100 Booth Cl, Kingswinford, DY6 8SP [IO82WL, SO98]
G6 GTC P J Willson, 37 The Grove, North Cray, Sidcup, DA14 5NG [JO01BK, TQ47]
G6 GTD Details withheld at licensee's request by SSL.
G6 GTH S E C Leonard, 231 Hale Rd, Hale, Altrincham, WA15 8DN [IO83UJ, SJ78]
G6 GTJ L V Baldwin, Pinewood Rd, Ashley Heath, Hookgate, Market Drayton, Salop, TF9 4QE [IO82TV, SJ73]
G6 GTM Details withheld at licensee's request by SSL.
G6 GTN Details withheld at licensee's request by SSL.
GW6 GTS P Dudman, 52 Gardden Rd, Rhosllanerchrugog, Wrexham, LL14 2EP [IO83LA, SJ24]
G6 GTU L Hemming, 2 River Ln, Blandford, Charlton Marshall, Blandford Forum, DT11 9NZ [IO80WT, ST90]
G6 GTZ P S Wilson, 162 Bowerdean Rd, High Wycombe, HP13 6XW [IO91PP, SU89]
G6 GUC D T Ellis, 94 Caernarvon Rd, Hatherley, Cheltenham, GL51 5JR [IO81WV, SO92]
G6 GUD M Everley, 5 Firs Cl, Hazlemere, High Wycombe, HP15 7TF [IO91PP, SU89]
G6 GUH P E Bonds, The Gables, Astley Abbotts, Bridgnorth, Shropshire, WV16 4SH [IO82SN, SO79]
G6 GUN R P Kimpton, 15A Buckden Rd, Brampton, Huntingdon, PE18 8PR [IO92VH, TL27]
G6 GUP Details withheld at licensee's request by SSL.
G6 GUT B J Turley, 12 Legh Dr, Woodley, Stockport, SK6 1PT [IO83WK, SJ99]
G6 GUU T B Monaghan, 48 Aspin Oval, Knaresborough, HG5 8EL [IO93GX, SE35]
G6 GUW M O Newsome, Wits End, Main St., Stillington, York, YO6 1LF [IO94KC, SE56]
GM6 GVE J M Moulds, Arranlea, 6 Douglas St., Carluke, ML8 5BH [IO85BR, NS85]
G6 GVF K G Waters, 68 Howe St., Derby, DE22 3ER [IO92GW, SK33]
G6 GVG Details withheld at licensee's request by SSL.
G6 GVH J A Marks, 124 Stowey Rd, Yatton, Bristol, BS19 4EB [IO81OJ, ST46]
G6 GVJ P H Charlesworth, Ribby Bank, Scronkey, Pilling, Preston Lancs, PR3 6SQ [IO83NW, SD44]
G6 GVK D A Hulme, 25 Beach Rd, Preesall, Lancs, FY6 0HQ [IO83MW, SD34]
G6 GVL M J Longley, 78 Priory Rd, Eastbourne, BN23 7BE [JO00DT, TQ60]
G6 GVM P T Martin, 21 Baldwin Ave, Eastbourne, BN21 1UJ [JO00DS, TQ50]
G6 GVO M J Pearce, 19 Neo Pee Tech Ln, Pasir Panjang, Singapore, 0511 X
G6 GVR G Whittle, 5 Chantry Cl, Houghton, Westhoughton, Bolton, BL5 2LY [IO83RM, SD60]
G6 GVS R J Wood, 6 Timberlaine Rd, Pevensey Bay, Pevensey, BN24 6DE [JO00ET, TQ60]
G6 GVU S J Wood, 141 Latimer Rd, Eastbourne, BN22 7JB [JO00DS, TQ60]
G6 GVZ E Rigby, 12 Sorrel Ave, Tean, Stoke on Trent, ST10 4LY [IO92AW, SK03]
GW6 GW N Davies Blackwood & District Amateur R, 2 The Alders, Wuan Vale Est, Oakdale, Blackwood, NP2 0LQ [IO81JQ, ST19]
GM6 GWB C D Prentice, 27 Lanark Rd, Ravenstruther, Lanark, ML11 7SS [IO85DQ, NS94]
G6 GWE M J Ranger, 13 Springfield Cl, Old Ln, Crowborough, TN6 2BN [JO01CB, TQ53]
G6 GWI A H Faulkner, 19 Long Ln Cl, Holbury, Southampton, SO45 2LE [IO90HT, SU40]
GW6 GWK H T John, 11 Penylan, Litchard, Bridgend, CF31 1QW [IO81FM, SS98]
G6 GWP J D Briggs, Wood-Lea, Bawtry Rd, Everton, Doncaster, DN10 5BS [IO93MJ, SK69]
G6 GWV J Hopkinson, 24 Arundel Rd, Mitton, Tewkesbury, GL20 8AU [IO82WA, SO93]
G6 GWX C F Hore, 20 Well St., Delabole, PL33 9BJ [IO70PO, SX08]
G6 GWY K Dodd, 1 Nansen St., Bulwell, Nottingham, NG6 9JE [IO92JX, SK54]
G6 GXE L L Jordan, 20 Coniston Rd, Folkestone, CT19 5JF [JO01NC, TR23]
G6 GXG G D Ridden, 9 Woodlands Gdns, Romsey, SO51 7TE [IO90GX, SU32]
G6 GXK D Wrigley, 45 Norford Way, Rochdale, OL11 5QS [IO83VO, SD81]
G6 GXL Details withheld at licensee's request by SSL.
G6 GXO C E Parks, 29 Heighams, Harlow, CM19 5NU [JO01AS, TL40]
G6 GXS D McLean, Chestnut House, Kewstoke Rd, Kewstoke, Weston Super Mare, BS22 9YH [IO81MI, ST36]
G6 GXZ B M E Vaslet, Heatherlea, Adbury Holt, Newtown, Newbury, RG20 9BN [IO91II, SU46]
G6 GYC D A Oultram, 61 Bolton Rd, Westhoughton, Bolton, BL5 3DN [IO83RN, SD60]
G6 GYF M J Marshman, 12 Neelands Gr, Cosham, Portsmouth, PO6 4QL [IO90KU, SU60]
G6 GYJ D J Langridge, The Glades, 26 Whitecross Dr, Weymouth, DT4 9PA [IO80SO, SY67]
G6 GYK M E Wright, 12 Celandine Rise, Swinton, Mexborough, S64 8PL [IO93IL, SK49]
G6 GYM D C Popely, 4 West End, Whittlesey, Peterborough, PE7 1HR [IO92WN, TL29]
G6 GYN P J Price, 67 Bennetts Rd, Keresley End, Coventry, CV7 8HY [IO92FL, SP38]
G6 GYZ Details withheld at licensee's request by SSL.
G6 GZB S P Henderson, Dauntless, Ferry Rd, Canvey Island, Essex, SS8 0QT [JO01GM, TQ78]
G6 GZH D A Brooke, 15A Fen End, Over, Cambridge, CB4 5NE [JO02AH, TL37]
G6 GZJ R L Bailey, 55 Windermere Ave, Scathoe, Grimsby, DN33 3DB [IO93WM, TA20]
GW6 GZM Details withheld at licensee's request by SSL.
GW6 GZN Details withheld at licensee's request by SSL.
G6 GZR P Dexter, 6 Gerrards Rd, Shipston on Stour, CV36 4HH [IO92EB, SP24]
G6 GZS P Empringham, Paulzanne, Bank End, North Somercotes, Louth, LN11 7LN [JO03BK, TF49]
GM6 GZX Details withheld at licensee's request by SSL.
G6 HAA A S Johns, Crossleigh, The Cross, Offenham, Evesham, WR11 5RB [IO92BC, SP04]
G6 HAC Details withheld at licensee's request by SSL.
G6 HAE A M Shone, 50 Whitefield Ave, Norden, Rochdale, OL11 5YG [IO83VO, SD81]
G6 HAG Details withheld at licensee's request by SSL.
G6 HAI B N Thompson, Springfield Gdns, Peasmarsh Rd, Playden, Rye, E Sussex, TN31 7UL [JO00IX, TQ92]
G6 HAJ R D Tabberer, Coronation Cttgs, 86 Kilby Rd, Fleckney, Leicester, LE8 8BN [IO92LM, SP69]
G6 HAL Details withheld at licensee's request by SSL.
G6 HAT P C Calpin, 82 Meadow Croft, Harrogate, HG1 3LH [IO94FA, SE35]
G6 HBJ T J Charman, 1 Bowler Lea, Downley, High Wycombe, HP13 5UD [IO91OP, SU89]
G6 HBK P J Delamere, 49 Storthes Hall Ln, Kirkburton, Huddersfield, HD8 0PT [IO93DO, SE11]
G6 HBN G M Ereaut, 31 Twyford Ave, Shirley, Southampton, SO15 5NZ [IO90GW, SU41]
G6 HBQ A J Ford, 1 Hem Heath Cottage, Longton Rd, Trentham, Stoke on Trent, ST4 8HP [IO82WX, SJ84]
G6 HBS G R Guest, Cedar Wood, Field Rd, Chedworth, Cheltenham, GL54 4NQ [IO91BT, SP01]
G6 HBY G Isles, 9 Sycamore Chase, Kent Rd, Pudsey, LS28 9BP [IO93ET, SE23]
G6 HBZ S R A Jenkinson, Woodlands, West Ln, Sutton in Craven, Keighley, BD20 7AS [IO93AV, SE04]

G6 HC Details withheld at licensee's request by SSL.
GD6 HCB A M Kennaugh, 34 Seafield Cl, Onchan, Douglas, IM3 3BU
G6 HCF L J Carter, Hattersbrick Farm, Lancaster Rd, Out Rawcliffe, nr Preston Lancs, PR3 6BN [IO83NV, SD44]
G6 HCH C D Back, The Fells, The Launches, West Lulworth, Wareham, BH20 5SF [IO80VP, SY88]
G6 HCI B M Byrne, 13 Tittensor Rd, Newcastle, ST5 3BS [IO82VX, SJ84]
G6 HCL P R Blay, 7 Rowson Dr, Cadishead, Manchester, M44 5YW [IO83SK, SJ79]
G6 HCQ S K C Crawford, 86 Harrowden Rd, Bedford, MK42 0SP [IO92SC, TL04]
G6 HCT Details withheld at licensee's request by SSL.
G6 HCV B G Fletcher, 2 Slade Gdns, Codsall, Wolverhampton, WV8 1BJ [IO82VP, SJ80]
G6 HCW D L Fieldsend, 20 Downsell Rd, Webheath, Redditch, B97 5RT [IO92AH, SP06]
G6 HD T L Herdman, Ave Lodge, 37 North Cray Rd, Bexley, DA5 3ND [JO01BK, TQ47]
G6 HDD P Ingham, 14 Baker St., Kearsley, Bolton, BL4 8QU [IO83TM, SD70]
G6 HDF S P Kelly, 4 Franklyn Cl, Green Fields, Perton, Wolverhampton, WV6 7SB [IO82VO, SJ80]
G6 HDG M Mace, 99 Broadsands Dr, Alverstoke, Gosport, PO12 2TJ [IO90KS, SZ59]
G6 HDI A E Parkinson, 239 Heyhouses Ln, Lytham St. Annes, FY8 3RQ [IO83MS, SD32]
G6 HDJ J C Oswin, 10 Highfields Dr, Loughborough, LE11 3JT [IO92JS, SK51]
G6 HDP Details withheld at licensee's request by SSL.
G6 HEA Details withheld at licensee's request by SSL.
G6 HEB P W Ballance, 6 Coronation Terr, Thistle Hill, Knaresborough, HG5 8JN [IO93GX, SE35]
G6 HEE A Grimmett, 17B Market Pl, Market Deeping, Lincs, PE6 8EA [IO92UQ, TF10]
G6 HER C A Wells, Darwells, Talskiddy, St Columb, Cornwall, TR9 6EB [IO70MK, SW96]
G6 HEW R W Tribe, Kentish Town Section House, 10A Holmes Rd, London, NW5 3AD [IO91WN, TQ28]
G6 HFD H F Wood, 5 Naseby Cl, Hatfield, Doncaster, DN7 6AN [IO93LN, SE60]
G6 HFF G Bates, 3 Braemar Gdns, Ladybridge, Bolton, BL3 4TU [IO83SN, SD60]
GM6 HFH I M Baker, 31 Strathaven Rd, Stonehouse, Larkhall, ML9 3EN [IO85AQ, NS74]
G6 HFI Details withheld at licensee's request by SSL.
G6 HFK L A Dutton, 12 Exmouth Gr, Burslem, Stoke on Trent, ST6 2JX [IO83VA, SJ84]
G6 HFS B R Shaw, 43 Egremont Rd, Hardwick, Cambridge, CB3 7XR [JO02AF, TL35]
G6 HFW J Graham, 142 Shakerley Ln, Tydesley, Tyldesley, Manchester, M29 8LZ [IO83SM, SD60]
G6 HFZ S G Homer, 31 Shaftmoor Ln, Acocks Green, Birmingham, B27 7RU [IO92BK, SP18]
G6 HGD R H Waller, Mauray, 22 Nightingale Ln, Feltwell, Thetford, IP26 4AR [JO02GL, TL79]
G6 HGE D C Heale, 54 Alameda Way, Purbrook, Waterlooville, PO7 5HB [IO90LU, SU60]
G6 HGG R A Ireson, 6 Walker Sq, The Links, Wellingborough, NN8 5PQ [IO92PH, SP86]
G6 HGK E G D Kesterton, 6 Box Rd, Bathford, Bath, BA1 7RN [IO81UJ, ST76]
G6 HGM R K Buckle, 22 Tylers Hill Rd, Chesham, HP5 1XH [IO91QR, SP90]
G6 HGN M W Maidens, No 3 Main Rd, Spilsby, Lincs, PE23 4BU [IO93XD, TF36]
G6 HGQ J A Perry, 30 Hillside Ave, Atherton, Manchester, M46 9LX [IO83RM, SD60]
G6 HGR K H Potts, 10 South Park Cl, Redruth, TR15 3AR [IO70JF, SW64]
G6 HGT P W Robinson, 135 Wensley Dr, Leeds, LS7 2LT [IO93FT, SE23]
G6 HGU P J Saunders, 17 Northfield Ave, Monkwood, Rawmarsh, Rotherham, S62 7JZ [IO93HL, SK49]
GM6 HGW C H Topping, 32 Maryknowe, Gauldry, Newport on Tay, DD6 8SL [IO86LJ, NO32]
G6 HGX B G Waterloo, 55 Solent Rd, Hillhead, Fareham, PO14 3LB [IO90JT, SU50]
G6 HH T H Ransom Hastings Electronics & Radio C, 9 Lyndhurst Ave, Hastings, TN34 2BD [JO00HU, TQ81]
G6 HHB Details withheld at licensee's request by SSL.
G6 HHD W E Downes, 40 Southwood Dr, Baxenden, Accrington, BB5 2PZ [IO83TR, SD72]
G6 HHE J H W Avern, 8 Napier Cres, Fareham, PO15 5BL [IO90JU, SU50]
G6 HHF C P Harding, 2 Rose Cttgs, Duncan Rd, Park Gate, Southampton, Hants, SO31 1BD [IO90IU, SU50]
G6 HHH G J Dowse, 60 Lower Mortimer Rd, Southampton, SO19 2HF [IO90HV, SU41]
G6 HHP Details withheld at licensee's request by SSL.
G6 HHS J W Broadwater, Tarn Hows, 9 Molinnis Rd, Bugle, St. Austell, PL26 8QJ [IO70OJ, SX05]
G6 HIA A M Cook, 3 Hollyhock Cl, Hemel Hempstead, HP1 2HW [IO91RR, TL00]
G6 HIC Details withheld at licensee's request by SSL.[Correspondence to: GGHI Contest Group, c/o The Three Tuns, Knayton, Thirsk, North Yorkshire, YO7 4AN.]
G6 HIE B P Edwards, 9 Coleridge Cl, Goring By Sea, Worthing, BN12 6LD [IO90ST, TQ10]
G6 HIG G B Edmonds, 48 Springcroft, Hartley, Longfield, DA3 8AS [JO01DJ, TQ66]
G6 HIJ E J Foy, 37 Gorsey Croft, Eccleston Park, Prescot, L34 2RS [IO83OK, SJ49]
G6 HIL Details withheld at licensee's request by SSL.
G6 HIO Details withheld at licensee's request by SSL.
G6 HIQ R C J Lavis, Glen Orchard, Church Ln, East Lydford, Somerton, Somerset, TA11 7HD [IO81QB, ST53]
G6 HIU N Lasher, 29 Sefton Ave, London, NW7 3QB [IO91UO, TQ29]
G6 HIV Details withheld at licensee's request by SSL.
G6 HIX J A O Hagan, 13 Chapel Rd, Stanford in The Valley, Stanford in The Vale, Faringdon, SN7 8LE [IO91FP, SU39]
G6 HIZ Details withheld at licensee's request by SSL.
G6 HJD M D Morrish, 10 Hayle Rd, Moorside, Oldham, OL1 4NW [IO83WN, SD90]
GW6 HJO C Eynon, 14 York Terr, Porth Rhondda, Mid Glam, CF39 9UP [IO81HO, ST09]
G6 HJP Details withheld at licensee's request by SSL.
G6 HJR Details withheld at licensee's request by SSL.
G6 HJU J C Binns, 2 Gawsworth Cl, Poynton, Stockport, SK12 1XB [IO83WI, SJ98]
G6 HJV J M Evill, 54 Copsey Gr, Farlington, Portsmouth, PO6 1NB [IO90LU, SU60]
G6 HKE W D Leitch, 34 White House Croft, Long Newton, Stockton on Tees, TS21 1PJ [IO94HN, NZ31]
G6 HKF R P Mew, Tehig, Back St., Garboldisham, Diss, IP22 2SD [JO02JJ, TM08]
G6 HKH P J Randell, 12 Chaseley Rd, Salford, M6 7DZ [IO83UL, SJ89]
G6 HKK S C Mammatt, 9 Manor Park, Kingsbridge, TQ7 1BB [IO80CG, SX74]
G6 HKL D J Martin, 9 Twinberrow Ln, Woodmancote, Dursley, GL11 4AP [IO81TQ, ST79]
G6 HKM E B Martyr, 1 High Houses, Mashbury Rd, Great Waltham, Chelmsford, CM3 1EL [JO01FT, TL61]
G6 HKN W J McCue, 2 Downham Ave, Culcheth, Warrington, WA3 5RU [IO83RK, SJ69]
G6 HKP D T Merrington, 31 North Rd, Wellington, Telford, TF1 3ED [IO82RQ, SJ61]
G6 HKQ Details withheld at licensee's request by SSL.
G6 HKS R A Mason, 19 Salts Rd, West Walton, Wisbech, PE14 7EJ [JO02CQ, TF41]
GW6 HKT D R Newton-Goverd, 2 Blaen y Morfa, Morfa, Llanelli, SA15 2BG [IO71WQ, SS59]
G6 HKY W H G Metcalfe, 81 Westminster Dr, Bromborough, Wirral, L62 6AN [IO83MH, SJ38]
G6 HKZ R Moses, 80 Edgeworth, Yate, Bristol, BS17 4YW [IO81SM, ST78]
G6 HL W/Cdl E Hill, Marlows, 7 Thame Rd, Piddington, Bicester, OX6 0PT [IO91LU, SP61]
G6 HLH D A Arnold, Harbour View, 4 Battery Terr, Mevagissey, St. Austell, PL26 6QS [IO70OG, SX04]
G6 HLL B J Allman, 38 Whinchat Dr, Birchwood, Warrington, WA3 6PB [IO83RK, SJ69]
G6 HLP L C Loosley, 17 Orchard Way, Stanbridge, Leighton Buzzard, LU7 9JE [IO91QV, SP92]
G6 HLR G B Marshall, 118 Heather Rd, Small Heath, Birmingham, B10 9TB [IO92BL, SP18]
GM6 HLT J Melville, Shirva, 6 Dixon Ave, Kirn, Dunoon, PA23 8NA [IO75NX, NS17]
G6 HLU T J Miller, 23 Manchester Rd, Altrincham, WA14 4RQ [IO83TJ, SJ78]
G6 HLV Details withheld at licensee's request by SSL.
G6 HLW P Morrell, 1318A Stratford Rd, Hall Green, Birmingham, B28 9EE [IO92BK, SP18]
G6 HM E R A Henman, 8 Doubledays Ln, Burgh-le-Marsh, Skegness, PE24 5EN [JO03CD, TF46]
G6 HMA P J Matthews, The Maples, 2 Rogersfield, Langho, Blackburn, BB6 8HB [IO83ST, SD73]
G6 HMF R S Venison, Brookland, Sharnbrook Rd, Souldrop, Bedford, MK44 1EX [IO92RF, SP96]
G6 HMG R M Trowsdale, 422 Leatherhead Rd, Chessington, KT9 2NN [IO91UI, TQ16]
G6 HMJ A E Upcott, 66 Chandlers Way, Hertford, SG14 2EF [IO91WT, TL31]
G6 HMN R E Sunter, 15 Wellhead, Winewall, Trawden, Colne, BB8 8BW [IO83WU, SD93]
G6 HMO P J Tournant, 164 Parson St., Bedminster, Bristol, BS3 5QT [IO81QK, ST57]
G6 HMS E Veall, 24 Meadow Dr, Tickhill, Doncaster, DN11 9ET [IO93KK, SK69]
G6 HMU S M Thomas, 19 Norfolk Rd, Maldon, CM9 6AZ [JO01IR, TL80]
G6 HMV R E Tilley, 41 Rookery Rd, Knowle, Bristol, BS4 2DX [IO81RK, ST67]
G6 HMX D Tucker, 121 Lockerbie, Thornton-Cleveleys, Blackpool, FY5 3EP [IO83LU, SD34]
G6 HNF M J Butterworth, 26 Torwood Rd, Chadderton, Oldham, OL9 0RA [IO83WN, SD80]
G6 HNI D F Baker, 16 Warners Bridge Chase, Rochford, SS4 1JE [JO01IN, TQ88]
G6 HNJ I A Bennett, Ravenswood, The Shires, Hedge End, Hants, SO30 4BA [IO90IV, SU41]
G6 HNN M Bugg, 39 Glencoe Rd, Ipswich, IP4 3PP [JO02OB, TM14]
G6 HNP P R Beever, 33 Masterton Rd, Stamford, PE9 1SN [IO92SP, TF00]
G6 HNQ K W Blackburn, 57 Hope St., Leigh, WN7 1NB [IO83RL, SD60]
G6 HNR J H Ball, 94 Marshall Lake Rd, Shirley, Solihull, B90 4PN [IO92CJ, SP17]
G6 HNS R J Ball, 2 Honeybone Rd, Sutton Coldfield, B75 6DA [IO92CN, SP19]
G6 HNY F A Bradley, 18 Hollygate Cl, Melton Mowbray, LE13 1HD [IO92NS, SK71]
G6 HOB D Brebner, 30 Old Town Mews, Old Town, Stratford upon Avon, CV37 6GP [IO92DE, SP25]
G6 HOC A J Bird, 95 Hundred Acre Rd, Streetly, Sutton Coldfield, B74 2BS [IO92BN, SP09]
G6 HOH Details withheld at licensee's request by SSL.[Op: C J Vousden, living in Netherlands.]
GW6 HOJ Details withheld at licensee's request by SSL.
G6 HOK Details withheld at licensee's request by SSL.
G6 HOL Details withheld at licensee's request by SSL.
G6 HOM Details withheld at licensee's request by SSL.
G6 HOR D N Padfield, Shalom, 27 Fownes Rd, Alcombe, Minehead, TA24 6AF [IO81GE, SS94]
G6 HOS C D Playford, 6 Nutberry Cl, Teynham, Sittingbourne, ME9 9SP [JO01JH, TQ96]
G6 HOT M Palmer, 21 Ibbett Cl, Kempston, Bedford, MK43 9BT [IO92RC, TL04]
GW6 HOU Details withheld at licensee's request by SSL.
G6 HOV A F Whitehouse, 285 West Boulevard, Birmingham, B32 2PE [IO92AK, SP08]
G6 HOW G D Pease, 29 Edinburgh Dr, Darlington, DL3 8BD [IO94EM, NZ21]
G6 HPC G R Stephens, 25 Lullington Rd, Knowle, Bristol, BS4 2LH [IO81RK, ST67]
G6 HPE P R Simms, 61 Ipswich Cres, Great Barr, Birmingham, B42 1LY [IO92BM, SP09]
G6 HPK D Scott, 8 Lynton Rd, Chesham, HP5 2BU [IO91QR, SP90]
G6 HPO A E Smith, 69 Victoria Rd, Finsbury Park, London, N4 3SN [IO91WN, TQ38]
G6 HPQ Details withheld at licensee's request by SSL.
G6 HPS N T Smith, 10 Peacock Hill, Alveley, Bridgnorth, WV15 6JX [IO82TL, SO78]
G6 HPU N R Scales, 2 Hudsons Yard, Flowergate, Whitby, YO21 3BG [IO94QL, NZ81]
G6 HPY D T Seabrook, Lyndene, 5 Mill View, Gazeley, Newmarket, CB8 8RN [JO02GG, TL76]
GW6 HQA E Jones, Braemar, Acton Gdns, Wrexham, LL12 8DE [IO83MB, SJ35]
G6 HQD N P E Jones, Ty Newydd, 3 Adaston Ave, Eastham, Wirral, L62 8BT [IO83MH, SJ37]
G6 HQE D J Inskip, 47 Charnwood Dr, Thurnby, Leicester, LE7 9PD [IO92LP, SK60]
G6 HQI E T Jacobs, 26 Pondfield Rd, Colchester, CO4 3EG [JO01LV, TM02]
G6 HQQ Details withheld at licensee's request by SSL.
G6 HQR Details withheld at licensee's request by SSL.
G6 HQS D Cawthorne, 174 Greenbank Rd, Wirral, L48 6DF [IO83JJ, SJ28]
G6 HQX A J Cook, 90 Ramsbury Walk, Holbrook Park, Trowbridge, BA14 0UX [IO81VH, ST85]
G6 HRA J A Chesterman, Danby, 69 Heath Ln, Bladon, Woodstock, OX20 1RZ [IO91HT, SP41]
G6 HRF P L Carter, 1 Bentham Way, Staincross, Mapplewell, Barnsley, S75 5QA [IO93FO, SE31]
GW6 HRG R Cretney, 2 Hunters Meadow, Cross Lanes, Wrexham, LL13 0TQ [IO83MA, SJ34]
G6 HRH Details withheld at licensee's request by SSL.
G6 HRK S J Wilson, 59 Greenway Ave, Upper Walthamstow, London, E17 3QJ [JO01AO, TQ38]
G6 HRL H G Winter, High House Marina, Heyford Ln, Weedon, Northampton, NN7 4SF [IO92LF, SP65]
G6 HRM J C Wood, 13 Brunswick Rd, Pudsey, LS28 7NA [IO93ET, SE23]
G6 HRN Details withheld at licensee's request by SSL.[Station located near Coterave, Nottingham.]
G6 HRX T J Whitehead, 14 Somerset Rd, Co. Bridge, Willenhall, WV13 2RY [IO82XO, SO99]
G6 HRZ F T Wilkinson, 25 Austral Ave, Woolston, Warrington, WA1 4ND [IO83RJ, SJ68]
G6 HSC V R Williamson, Oak Lodge, 25 Springwood Dr, Oakwood, Derby, DE21 2HE [IO92GW, SK33]
G6 HSD R P Willmott, 85 Malthouse Ln, Ealswood, Earlswood, Solihull, B94 5RZ [IO92CI, SP17]
G6 HSG A T Walsh, 46 Greenway, Braunston, Daventry, NN11 7JT [IO92JH, SP56]
G6 HSI S A Wallace, 26 Parsons Dr, Glen Parva, Leicester, LE2 9NS [IO92JO, SP59]
G6 HSL Details withheld at licensee's request by SSL.
G6 HSM Details withheld at licensee's request by SSL.
G6 HSS P J Hardiman, 12 Brempsons, Basildon, SS14 2AZ [JO01FN, TQ78]
G6 HST W G Hayball, Lennox Wood, Petham Green, Gillingham, ME8 6SZ [JO01HI, TQ86]
G6 HSW L R Hagger, 48 Little Meadow, Bar Hill, Cambridge, CB3 8TD [JO02AF, TL36]
G6 HTA P J Hartas, 6 Newton St., Whitby, YO21 1QX [IO94QL, NZ81]
G6 HTB E A D Brodie, 116 Pagham Rd, Pagham, Bognor Regis, PO21 4NN [IO90PS, SZ89]
G6 HTH C T Hall, Oakenhill, North Pole Rd, Barming, Maidstone, ME16 9HH [JO01FG, TQ75]
G6 HTL G W Hall, 60 Mount Pleasant Rd, Collier Row, Romford, RM5 3YL [JO01CO, TQ59]
G6 HTM B D Hedgecock, 15 Kennet Way, Oakley, Basingstoke, RG23 7AA [IO91JF, SU55]
G6 HTS R S G Hooper, 125 Woodlands Rd, Ditton, Aylesford, ME20 6HF [JO01FH, TQ75]
G6 HTT G P Reece, 69 Ranelagh Rd, Ipswich, IP2 0AD [JO02NB, TM14]
G6 HTY A M Rollitt, St. Peters, 29 High St., Navenby, Lincoln, LN5 0EE [IO93RC, SK95]
G6 HTZ A B Rogers, 11 Avebury Cl, Curzon Park, Calne, SN11 0EP [IO81XK, ST97]
G6 HU J E Hunter, 1 Cotman Cl, Lowestoft, NR32 4NW [JO02UM, TM59]
G6 HUA Details withheld at licensee's request by SSL.
GW6 HUD R Rees, 5 Rhydyffynnon, Pontyates, Llanelli, SA15 5UG [IO71VS, SN40]
G6 HUH S R Ross, 55 Lincoln Park, Amersham, HP7 9HD [IO91QQ, SU99]
G6 HUI B C Tanner, 116 Ascot Cardens, Sothall, Middx, UB1 2SB [IO91TM, TQ18]
GJ6 HUL S Taylor, The Cottage, Sunshine Ave, St. Saviour, Jersey, JE2 7TS
G6 HUM S J Thomas, 17 Rectory Park Ave, Sutton Coldfield, B75 7BL [IO92CN, SP19]
G6 HUN A R Thompson, Canal Lodge, Bath Rd, Padworth, Reading, RG7 5HR [IO91KJ, SU66]
G6 HUO J S Thompson, Newstead, 38 Matford Ave, Exeter, EX2 4PL [IO80FR, SX99]
G6 HUP M Thompson, 2 Cotman Rd, Lincoln, LN6 7PA [IO93RE, SK96]
G6 HUR N J Thursfield, Ash Gr, Ifton Heath, St. Martins, Oswestry, SY11 3DG [IO82LW, SJ33]
GW6 HUV R H Tyson, Tan y Cedr, 12 Trillo Ave, Rhos on Sea, Colwyn Bay, LL28 4NS [IO83DH, SH88]
G6 HV D K Bradley, Thelbridge Hall, Wirtheridge, Tiverton, Devon, EX16 8NZ [IO80DV, SS71]
GW6 HVA M Vernon, Flat 2, 11 Lloyd St., Llandudno, Gwynedd, LL30 2UU [IO83CH, SH78]
G6 HVD N C Dunford, 8 Fair Mead, Mountsorrel, Loughborough, LE12 7BN [IO92KR, SK51]
G6 HVE N Martin, Pooh Corner, 1 Edendale Rd, Cheltenham, GL51 0TX [IO81WV, SO92]
G6 HVI Details withheld at licensee's request by SSL.
G6 HVL J A Dawes, 30 Trapstyle Rd, Parkside, Ware, SG12 0BB [IO91XT, TL31]
G6 HVQ L S Dodson, 16 Cannon Way, West Molesey, KT8 2NB [IO91TJ, TQ16]
G6 HVX D J Gladwish, 36 Oakfield Rd, Ore, Hastings, TN35 5AX [JO00HU, TQ81]
G6 HVZ C A Gilbert, 4 Cliffords, Cricklade, Swindon, SN6 6BU [IO91BP, SU09]
G6 HWA F W Glover, 19 Gerrard St., Spennymoor, DL16 6DY [IO94EQ, NZ23]
G6 HWB E A Clarke, 4 Church Ln, Milton, Cambridge, CB4 6AB [JO02CF, TL46]
G6 HWD E Gethin, 2 Hazel Gr, Heswall, Irby, Wirral, L61 4UZ [IO83KI, SJ28]
G6 HWF W Green, 58 Fecitt Brow, Blackburn, BB1 2AZ [IO83SR, SD72]
G6 HWH L J Gregory, 390 Ings Rd, Hull, HU8 0NP [IO93US, TA13]
G6 HWI B Wilson, 102 Woodlands Rd, Woodlands, Doncaster, DN6 7JZ [IO93JN, SE50]
G6 HWM M Evans, Iona, 111 Church Rd, Tiptree, Colchester, CO5 0AB [JO01JT, TL81]
G6 HWO M B Frampton, 57 Sedgemoor Rd, Bridgwater, TA6 5NP [IO81MC, ST33]
G6 HWR M Fern, Handsworth, Hackney Rd, Matlock, Derbyshire, DE4 2PW [IO93FD, SK26]
G6 HWT C J Freeman, The Lees, 66 London Rd, Newington, Sittingbourne, ME9 7NR [JO01HI, TQ86]
GM6 HWZ T F L Akers, 143 St. Michael St., Dumfries, DG1 2PP [IO85EB, NX97]
G6 HXB M Aston, 51 Acton House, Horn Ln, Acton, London, W3 9EJ [IO91UM, TQ28]
G6 HXK Details withheld at licensee's request by SSL.
G6 HXL D Latham, 89 Kestrel Park, Skelmersdale, WN8 6TA [IO83ON, SD40]
G6 HXR D W J Lawrence, 31 Taylor Rd, Snodland, ME6 5HJ [JO01FH, TQ66]
G6 HXU Dr E J Loader, 13 Vale Rd, Hartford, Northwich, CW8 1PL [IO83RF, SJ67]
G6 HXV J Lanham, 123 Bradbury Rd, Olton, Solihull, B92 8AL [IO92CK, SP18]
G6 HXW L Leighton, Hunters Moon, 177 Terringes Ave, Worthing, BN13 1JS [IO90TT, TQ10]
G6 HXX D J Lister, 25 Roberts Ave, Huthwaite, Sutton in Ashfield, NG17 2JP [IO93IC, SK45]
G6 HXZ P J Lovett, St Farmhouse, Ashford Rd, High Halden, Kent, TN26 3LY [JO01IC, TQ83]
G6 HY R Healey, 9 Mary Rd, Eastwood, Nottingham, NG16 2AH [IO93IA, SK44]
G6 HYD Details withheld at licensee's request by SSL.
G6 HYF C S Ironmonger, 77 Boston Rd, Spilsby, PE23 5HH [JO03BE, TF36]
G6 HYI P M Ingle, Old Farm Cottage, Olmstead Green, Castle Camps, Cambridge, Cambs, CB1 6TW [JO02EB, TL64]
G6 HYJ E R Ibrahim, 5 Farm Cl, High Wycombe, HP13 7YA [IO91PP, SU89]
G6 HYN P R James, 32 Beacon Rd, Herne Bay, CT6 6DJ [JO01NI, TR16]
G6 HYP N G Jones, 7 Church Terr, Church Rd, East Harling, Norwich, NR16 2NA [JO02LK, TL98]
G6 HYU Details withheld at licensee's request by SSL.
G6 HZG K J Purser, Burnt Oak, 6 Parkway, Binstead, Ryde, PO33 3UX [IO90JR, SZ59]
G6 HZH S J Prosser, 53 Broadlands Rise, Off Pentire Rd, Lichfield, WS14 9SF [IO92CQ, SK10]
G6 HZK G S Partridge, 53 Acres Rd, Brierley Hill, DY5 2XY [IO82WL, SO98]
G6 HZV R J Plant, 10 The Cres, Bracebridge Heath, Lincoln, LN4 2NP [IO93RE, SK96]
G6 HZW E R Payne, Stonyric, 1 Bakers Ln, Tolleshunt Major, Maldon, CM9 8JS [JO01IS, TL81]
G6 HZX R Purdy, 49 Mansfield Rd, Eastwood, Nottingham, NG16 3DY [IO93IA, SK44]
G6 IAB D R Pink, Marlbern, Upper London Rd, Black Notley, Braintree Essex, CM7 8QH [JO01GU, TL72]
G6 IAE F G Watts, 4 Sydney Taylor Ct, Peaks Ln, New Waltham, Grimsby, DN36 4NJ [IO93XM, TA20]
G6 IAJ W E J Saunders, 61 Nether Ct, Halstead, CO9 2HN [JO01HW, TL83]
G6 IAK G Wyles, 61 Platts Cres, Stourbridge, DY8 4YU [IO82WL, SO88]
G6 IAN I D Brooks, 10 Windermere Cl, Dunstable, LU6 3DD [IO91RV, TL02]
G6 IAT T N Bruce, 17 Blaydon Rd, Luton, LU2 0RP [IO91TV, TL12]
G6 IAW D P Birt, 10 Ashwin Ave, Copford, Colchester, CO6 1BS [JO01JV, TL92]
G6 IBD D A Bowles, 23 Broughton Way, Rickmansworth, WD3 2GW [IO91RO, TQ09]
G6 IBE Details withheld at licensee's request by SSL.
G6 IBH G Butterworth, 8 Denmark Cl, Corby, NN18 9EH [IO92PL, SP88]
G6 IBI R Beck, 19 Blakeney Fields, Great Shefford, Hungerford, RG17 7BX [IO91GL, SU37]
G6 IBK Prof M A Bartle, Four Winds, Hugus Rd, Threemilestone, Truro, TR3 6DD [IO70KG, SW74]
GI6 IBL M S Barr, 4 Sandelwood Ave, Coleraine, BT52 1JW [IO65QC, C83]
G6 IBN M T Bodill, 24 Dalbeattie Cl, Arnold, Nottingham, NG5 8QX [IO92KX, SK54]
G6 IBO Details withheld at licensee's request by SSL.
G6 IBP Details withheld at licensee's request by SSL.
G6 IBQ Details withheld at licensee's request by SSL.[Op: S Brown. Station located near Chesterfield.]
G6 IBW R J Savigar, 8 Hancock Cl, Chippenham, SN15 3UZ [IO81WK, ST97]
G6 IBZ P J Simpson, 3 Stour Cl, Brook Farm, Saxmundham, IP17 1XX [JO02SE, TM46]
G6 ICE Details withheld at licensee's request by SSL.
G6 ICF Details withheld at licensee's request by SSL.
G6 ICH R D Brothwood, Amberley, Coombe Cross, Bovey Tracey, Newton Abbot, TQ13 9EP [IO80EO, SX87]
G6 ICJ Details withheld at licensee's request by SSL.
G6 ICL R N Taylor Amcom ARC, Personal Common Club, 25 Daven Rd, Congleton, CW12 3RA [IO83VD, SJ86]
G6 ICP Details withheld at licensee's request by SSL.
GD6 ICR M J Webb, 1 Mount Morrison, Peel, IM5 1PN
G6 ICS K R Whittaker, Hillhead Rd, Budock, Kergilliack, Falmouth, Cornwall, TR11 5PA [IO70KD, SW73]
G6 ICX K Wyatt, 3 Prescott, Baschurch, Shropshire, SY4 2DP [IO82NS, SJ42]
G6 ICZ R S Waller, 12 Kelton Rd, Brotton, Saltburn By Sea, Cleveland, TS12 2TJ [IO94MN, NZ61]

G6 IDA P Russell, 6 Bartlemy Rd, Newbury, RG14 6JX [IO91IJ, SU46]
G6 IDD R B Stevens, 6 Goosander Way, Thamesmead, London, SE28 0ER [JO01BL, TQ47]
G6 IDF Dr H C Stinton, 118 Offington Ave, Worthing, BN14 9PR [IO90TT, TQ10]
G6 IDG C H Stringer, Meadowbank, Back Ln, Newton Poppleford, Sidmouth, EX10 0EY [IO80IQ, SY08]
G6 IDI Details withheld at licensee's request by SSL.
G6 IDJ D C Smith, 93 Digby Rd, Coleshill, Birmingham, B46 3NL [IO92DL, SP18]
GW6 IDK J S P Woodward, 29 Maes yr Awel, Ponterwyd, Aberystwyth, SY23 3JT [IO82BJ, SN78]
G6 IDL M A Waud, 2 Wrights Ln, Friskney, Friskney, Boston, PE22 8RW [JO03CB, TF45]
G6 IDQ P J Wright, Willow House, Landford Wood, Salisbury, Wilts, SP5 2ES [IO90EX, SU22]
G6 IDU I D Rose, 56 Sunbury Ct, Shoeburyness, Southend on Sea, SS3 8TR [JO01JN, TQ98]
G6 IDW T M Roberts, The Firs, 15 Castle End Rd, Maxey, Peterborough, PE6 9EP [IO92UP, TF10]
G6 IDZ R E Rigby, 64 Toms Ln, Kings Langley, WD4 8NB [IO91SR, TL00]
G6 IEA Details withheld at licensee's request by SSL.
G6 IEE M L Elsley, 25 Elmsdale Rd, Wootton, Bedford, MK43 9JW [IO92RC, TL04]
G6 IEI P R M Williams, Peanjays, 4 Cutbush Cl, Lower Earley, Reading, RG6 4XA [IO91MK, SU76]
G6 IEQ P G Hawkridge, 211 Goring Way, Goring By Sea, Worthing, BN12 5BU [IO90ST, TQ10]
GI6 IES C W Hagan, 16 Shore Rd, Portaferry, Newtownards, BT22 1JY [IO74FJ, J55]
G6 IEZ Details withheld at licensee's request by SSL.
G6 IFA D G C Hicks, Beggars Roost, 12 Toll Bar Rd, Cristleton, Chester, CH3 5QX [IO83NE, SJ46]
G6 IFC M Hill, 62 St. Catherines Dr, Bramley, Leeds, LS13 2JZ [IO93ET, SE23]
G6 IFD Details withheld at licensee's request by SSL.
G6 IFE P F Holland, 3 Manor Villas, Chilton Rd, Chearsley, Aylesbury, HP18 0DN [IO91MS, SP71]
G6 IFF Details withheld at licensee's request by SSL.
G6 IFH J P Rimington, 35 Long Meadow Dr, Wickford, SS11 8AY [JO01GO, TQ79]
G6 IFI G M Roberts, 42 Avoca Cl, Leicester, LE5 4RA [IO92LP, SK60]
G6 IFK H E Rand, 198 Oakfield Rd, Benfleet, SS7 1DU [JO01GN, TQ78]
G6 IFN L A Rouse, 69 Shackerdale Rd, Wigston, LE18 1BR [IO92KO, SK50]
G6 IFO Details withheld at licensee's request by SSL.
G6 IFQ S Howcroft, Warwick Cottage, 5 Ecclesgate Rd, Blackpool, FY4 5DW [IO83LS, SD33]
G6 IFR G M Horwood, 25 Briar Rd, Shepperton, TW17 0JB [IO91SJ, TQ06]
G6 IFS N Hollinshead, 35 Parkside Dr, May Bank, Newcastle, ST5 0NL [IO83VA, SJ84]
G6 IFT M A Homer, 18 Weatheroaks, Walsall Wood, Walsall, West Midlands, WS9 9RN [IO92AP, SK00]
G6 IFV J A Hunt, 77 Scott St., Burnley, BB12 6NJ [IO83UT, SD83]
G6 IFX Details withheld at licensee's request by SSL.
G6 IFZ Details withheld at licensee's request by SSL.
G6 IGA Details withheld at licensee's request by SSL.
G6 IGB Details withheld at licensee's request by SSL.
G6 IGK L J Glasscock, 37 Huntingfield Rd, Bury St. Edmunds, IP33 2JA [JO02IF, TL86]
G6 IGU A C Greenleaf, The Lindons, Frating Rd, Ardleigh, Colchester, Essex, CO7 7SY [JO01MV, TM02]
G6 IGV D W Gregson, 8 Lennox Gate, Blackpool, FY4 3JQ [IO83LS, SD33]
G6 IGW N J Gutten, 5 Clare Dr, Wistaston, Crewe, CW2 8ED [IO83SC, SJ65]
GW6 IGY J C Mead, 1 Tudor Ct, Hope, Wrexham, LL12 9PJ [IO83LC, SJ35]
G6 IHB F A Norton, 13 Regents Way, Minehead, TA24 5HW [IO81GE, SS94]
G6 IHC Details withheld at licensee's request by SSL.
G6 IHD S L Maxwell, 11 Gerard Rd, Wallasey, L45 6UQ [IO83LJ, SJ29]
G6 IHF A C Murphy, Captains Ct, 38 Captain French Ln, Kendal, LA9 4HP [IO84PH, SD59]
G6 IHG H J Mitchell, 17 Burners Cl, Burgess Hill, RH15 0QA [IO90WW, TQ31]
G6 IHH D Harding Arborfield ARC, c/o OIC Pri, School of Elec. Eng. REME, Aborfield, Reading, Berks, RG2 9NH [IO91MJ, SU76]
G6 IHK Details withheld at licensee's request by SSL.
G6 IHL Details withheld at licensee's request by SSL.
GI6 IHM R J McDowell, 3 Lord Wardens Park, Rathgael Rd, Bangor, BT19 1YG [IO74DP, J47]
G6 IHN Details withheld at licensee's request by SSL.
G6 IHU J M Measom, 108 Carlton Dr, Wigston, LE18 1DH [IO92KO, SP69]
G6 IHV D Moore, 47 Stretton Rd, Willenhall, WV12 5EJ [IO82XO, SJ90]
G6 IHW A J Mackinlay, 26 Anderson Rd, Erdington, Birmingham, B23 6NN [IO92BM, SP19]
GI6 IHX R D Mc Cullough, 21 Rostrevor Way, Bangor, BT19 1AE [IO74DP, J48]
G6 IHY J D Mills, Fan Cottage, Lyne Ln, Lyne, Chertsey, KT16 0AJ [IO91RJ, TQ06]
G6 IIA A D Stansfield, 22 Low Stobhill, Morpeth, NE61 2SG [IO95DD, NZ28]
G6 IIF R H Sharpe, 14 Dansie Ct, Compton Rd, Colchester, CO4 4EA [JO01LV, TM02]
G6 IIG S C Stearn, 86 Summerland Ave, Minehead, TA24 5BW [IO81GE, SS94]
G6 III J Stirrup, 105 Crow Ln East, Newton-le-Willows, WA12 9UG [IO83QK, SJ59]
G6 IIK D Gill, 79 Heather Walk, Bolton on Dearne, Bolton upon Dearne, Rotherham, S63 8BZ [IO93IM, SE40]
G6 IIM P Jones, 12 Coronation Dr, Bromborough, Wirral, L62 3LF [IO83MI, SJ38]
G6 IIN P F Currigan, 5 Gayton Ave, New Brighton, Wallasey, L45 9LJ [IO83LK, SJ39]
G6 IIO V L Chandler, 62 Compass Rd, Hull, HU6 7AW [IO93TS, TA03]
G6 IIP J K Clarke, 19 Kensington Rd, Gaywood, Kings Lynn, PE30 4AT [JO02FS, TF62]
G6 IIU R I Cooper, 69 Vicarage Ln, Elworth, Sandbach, CW11 3BU [IO83TD, SJ76]
G6 IIZ J N Clark, Brooklyn Cottage, Milton Combe, Yelverton, Devon, PL20 6HP [IO70WL, SX46]
G6 IJG W Carter, 43 Marsh Ln, Misterton, Doncaster, DN10 4DL [IO93OK, SK79]
G6 IJJ Details withheld at licensee's request by SSL.
G6 IJK A E Clayphon, 71 Blagrove Dr, Wokingham, RG41 4BD [IO91NJ, SU76]
G6 IJL G Craighead, 1 Kew Dr, Oadby, Leicester, LE2 5TS [IO92LO, SP69]
G6 IJN R J Clarke, 4 Highfield Rd, Stowupland, Stowmarket, IP14 4DA [JO02ME, TM05]
G6 IJQ W H Cartwright, 3 Masefield Rise, Abbeyfield, Halesowen, B62 8SH [IO82XK, SO98]
G6 IJW G J Stoelwinder, Anneth Lowen, St Ive Cross, Liskeard, Cornwall, PL14 3LZ [IO70TL, SX36]
G6 IJX H Sullivan, 3 King John St., Sleaford, NG34 7QH [IO92TX, TF04]
G6 IKC S W Saunders, 16 Hill Cl, Pensylvania, Exeter, EX4 6HG [IO80FR, SX99]
G6 IKE S Lynch, 34 Scott Ave, Baxenden, Accrington, BB5 2XA [IO83TR, SD72]
G6 IKK R J Tomlinson, 403 Red Lees Rd, Cliviger, Burnley, BB10 4TF [IO83VS, SD83]
G6 IKM D Tarbuck, 53 Bradlegh Rd, Newton-le-Willows, WA12 8RA [IO83QK, SJ59]
G6 IKN K W Towns, 75 Lancaster Gate, London, W2 3NN [IO91VM, TQ28]
G6 IKQ B C Trigger, 2 Stocking Ln, Shenington, Banbury, OX15 6NF [IO92GB, SP34]
G6 IKS J Sumner, Jaggen, Maldon Rd, Latchingdon, Chelmsford, CM3 6LF [JO01IQ, TL80]
G6 IKT Details withheld at licensee's request by SSL.
G6 IKW K Smith, 14 Upper St. Helens Rd, Hedge End, Southampton, SO30 0LH [IO90IV, SU41]
G6 ILC G C Sword, 28 Saxonhurst, Moot Ln, Downton, Salisbury, SP5 3JN [IO90DX, SU12]
G6 ILD P J Southgate, 6 Lippits Hill, Langdon Hills, Basildon, SS16 6LN [JO01FN, TQ68]
G6 ILE D A Liquorice, 14 Oakfield Mansion, Oakfield Gr, Clifton, Bristol, BS8 2BN [IO81QL, ST57]
G6 ILH A W R Davies, Dunster, 68 Branch Rd, Mellor Brook, Blackburn, Lancs, BB2 7NY [IO83RS, SD63]
G6 ILM P D Donovan, Oakfield, High St., Buxted, Uckfield, TN22 4JZ [JO00BX, TQ42]
G6 ILN J W Dodge, 5 Moat Way, Queenborough, ME11 5BU [JO01JJ, TQ97]
G6 ILT E E Elliston, 117 Willbye Ave, Diss, IP22 3NP [JO02NJ, TM18]
G6 ILX B B Edward, 27 Barford Cl, Ainsdale, Southport, PR8 2RS [IO83LO, SD31]
G6 ILY W Evans, Tree Tops, Holly Bush, Bangor-On Dee, Clwyd, LL13 0BH [IO82NX, SJ44]
G6 ILZ I R Fullerton, 27 Crest Rd, Checkley Grange, St. Georges, Telford, TF2 9NG [IO82SQ, SJ71]
G6 IM W B Smith, Beechfield, Croft Rd, Cosby, Leicester, LE9 1RE [IO92JN, SP59]
G6 IMG J H Finlay, 9 Ravenswood Gdns, Clarendon Rd, Southsea, PO5 2LU [IO90LS, SZ69]
G6 IMH R L Firth, Kasamily, 73 Lions Ln, Ashley Heath, Ringwood, BH24 2HH [IO90BT, SU10]
G6 IMJ K Walker, 37 Willingdon Rd, Childwall, Liverpool, L16 3NE [IO83NJ, SJ49]
G6 IML I T Walsh, 7 Winchester Ave, Hartshead Est, Ashton under Lyne, OL6 8BU [IO83XM, SD90]
G6 IMM C P Woolley, 177 Thetford Rd, Brandon, IP27 0DF [JO02HK, TL78]
G6 IMN K A Wetherell, Treviskey Cottage, Lanner Moor, Lanner, Redruth Cornwall, TR16 6JF [IO70JF, SW73]
G6 IMQ J P Wild, 20 Sandy Ln, Cholsey, Wallingford, OX10 9PY [IO91KN, SU58]
GW6 IMS T P Vernalls, 5 Min-y-Treath, Minfordd, Penryndeudraeth, Gwynedd, LL48 6EG [IO72WW, SH53]
G6 IMT Details withheld at licensee's request by SSL.
G6 IMV M W Richardson, 18 Mossvale Gr, Washwood Heath, Birmingham, B8 3QJ [IO92BL, SP18]
G6 IMY D N Ridyard, 47 The Oaks, Walton-le-Dale, Preston, PR5 4LT [IO83PR, SD52]
G6 INA P M Reidy, 17 Parker Ave, Calow, Chesterfield, S44 5AX [IO93HF, SK47]
GW6 INF J H F Markham, 4 Ty Afon, Ffordd Gwynedd, Tywyn, LL36 0TA [IO72WN, SH50]
G6 ING S J Meigh, 75 Botteslow St., Hanley, Stoke on Trent, ST1 3NE [IO83WA, SJ84]
G6 INI C D Mahony, 25 Wheelers Ln, Bradville, Milton Keynes, MK13 7HN [IO92OB, SP84]
G6 INK E G McGlen, 22 Stratford Ave, City of, Sunderland, SR2 8RX [IO94HV, NZ45]
G6 INM K J OReilly, 1 Evesham Way, Weston Park, Longton, Stoke on Trent, ST3 5TP [IO82WX, SJ94]
G6 INO R Ottolini, 154 Barwick Rd, Stanks, Leeds, LS15 8SW [IO93GT, SE33]
G6 INR Details withheld at licensee's request by SSL.
G6 INU D G Port, 8 Betterton Dr, Sidcup, DA14 4PS [JO01BK, TQ47]
G6 INV D C Pratley, 2 Haseldine Meadows, Hatfield, AL10 8HE [IO91VS, TL20]
G6 INX G A T Pryke, 38 Colne Dr, Walton on Thames, KT12 3SQ [IO91TJ, TQ16]
GW6 IOA C M Crow, Lindisfarne, Penywaun, Pentyrch, Cardiff, CF4 8SJ [IO81IM, ST08]
G6 IOB P G Coghlan, The End House, 34 Quarry Ln, Swaffham Bulbeck, Cambridge, CB5 0LU [JO02DF, TL56]
G6 IOE G I Crawford, 4 Beverley Gdns, Gedling, Nottingham, NG4 3LF [IO92LX, SK64]
G6 IOM M P Cunningham, 16 Cherry Waye, Eythorne, Dover, CT15 4BT [JO01PE, TR24]
G6 ION R A Civil, 7 Sunnybanks, Hatt, Saltash, PL12 6SA [IO70UK, SX36]
GM6 IOU M E Pollock, 3 Middlefield, Whitehills, East Kilbride, Glasgow, G75 0HJ [IO75VR, NS65]
G6 IOV P I Phelps, 152 Cherry Tree Ave, Waterlooville, PO8 8AX [IO90LV, SU61]
G6 IOW D C Peachey, 4 Windermere Dr, Braintree, CM7 8UA [JO01GU, TL72]
G6 IOX A G Pearce, 49 Bishopswood Rd, Tadley, RG26 4HF [IO91KI, SU56]
G6 IPB B Doyle, 52 Appleton Rd, Catisfield, Fareham, PO15 5QH [IO90JU, SU50]
G6 IPC I A Downes, 21 Caldbeck Ct, Chilwell, Beeston, Nottingham, NG9 5NH [IO92JW, SK53]
G6 IPH R V Distin, 16 The Hawthornes, Broad Oak, Rye, TN31 6EN [JO00HW, TQ81]
G6 IPN M D Davies, 54 Helmside Rd, Oxenholme, Kendal, LA9 7HA [IO84PH, SD58]
G6 IPQ R P Deakin, 55 Pendeen Cres, Southway, Plymouth, PL6 6RE [IO70WK, SX46]
GW6 IPR P Drew, 10 Merlin Cl, Malory Park, Thornhill, Cardiff, CF4 9AW [IO81JM, ST18]
G6 IPU G A Edwards, 34 Haden Way, Willingham By Stow, Willingham, Cambridge, CB4 5HB [JO02AH, TL37]
G6 IPW S A Featherstone, 36 Denton Ave, Grantham, NG31 7JL [IO92QV, SK93]
G6 IQA C H Leon, 43 Milton Lawns, Chesham Bois, Amersham, HP6 6BH [IO91QQ, SU99]
G6 IQE D Firmager, 26 Brownleaf Rd, Brighton, BN2 6LB [IO90XT, TQ30]
G6 IQF R P Harris, 11 Closemead, Clevedon, BS21 5EG [IO81NK, ST47]
G6 IQH R J Wickenden, 24 Buttermere Ave, Dunstable, LU6 3PD [IO91RU, TL02]
G6 IQL V R While, 11 Innage Rd, Northfield, Birmingham, B31 2DX [IO92AJ, SP07]
G6 IQM M J Wooding, 5 Ware Orchard, Barby, Rugby, CV23 8UF [IO92JH, SP57]
G6 IQP D Marriott, 10 Springfield Dr, Meadow Rise Est, Bulwell, Nottingham, NG6 8WD [IO92JX, SK54]
G6 IQQ N S Rossiter, 12 Holyhead Dr, Oakwood, Derby, DE21 2TD [IO92GW, SK33]
G6 IQT R McConnell, 27 St. Marks Ct, Pool Cl, Bilton, Rugby, CV22 7RW [IO92II, SP47]
G6 IQU P D Mulvany, 28 Church Ln, Chalgrove, Oxford, OX44 7TA [IO91LP, SU69]
G6 IQY J A Price, 31 Wattis Rd, Bearwood, Smethwick, Warley, B67 5BB [IO92AL, SP08]
G6 IRB Details withheld at licensee's request by SSL.
G6 IRE R G Aynge, 9 Sedgebrook Rd, Blackheath, London, SE3 8LR [JO01AL, TQ47]
G6 IRF S R Atwell, 10 Belding Ave, Manchester, M40 3SE [IO83WM, SD80]
G6 IRG M Andrews, 23 Brelades Cl, Dudley, DY1 2UZ [IO82WM, SO99]
G6 IRH G Andrews, 2 Tudor Gdns, Stourbridge, DY8 3RX [IO82WK, SO88]
G6 IRI Details withheld at licensee's request by SSL.
G6 IRJ G Andronov, 90 Overbury Cl, Northfield, Birmingham, B31 2HD [IO92AJ, SP07]
GI6 IRL J T Agnew, 40 Downhill Ave, Park, Belfast, BT8 4EF [IO74AN, J36]
G6 IRR B Gooch, 21 Lowndes Ln, Offerton, Stockport, SK2 6DP [IO83WJ, SJ98]
G6 IRW K F Holmes, Gable Cottage, Low Hill, Dunham Hill, Helsby Warrington, WA6 0NW [IO83OF, SJ47]
G6 IRX C S Holt, 1 Vale View, Wincanton, BA9 9RB [IO81TB, ST72]
G6 IRY R L Hobbs, 120 Misbourne Rd, Hillingdon, Uxbridge, UB10 0HP [IO91SM, TQ08]
G6 IRZ Details withheld at licensee's request by SSL.
G6 ISB A J Hunt, 10 Sturton St., Forest Fields, Nottingham, NG7 6HU [IO92KX, SK54]
G6 ISC Details withheld at licensee's request by SSL.
G6 ISD Details withheld at licensee's request by SSL.
G6 ISG P J Hancock, 2 Gulistan Rd, Leamington Spa, CV32 5LU [IO92FH, SP36]
G6 ISM J Hancock, 7 Hollies Cl, Houghton on The Hill, Leicester, LE7 9GW [IO92MP, SK60]
G6 ISN M Hancock, 9 Tansey Cres, Stoney Stanton, Leicester, LE9 4BT [IO92IN, SP49]
GI6 ISQ R Harron, 45 Upper Ballyboley Rd, Ballyclare, BT39 9ST [IO74AT, J29]
GI6 ISW Details withheld at licensee's request by SSL.
G6 ISX R Higgins, 6 Links Ave, Little Sutton, South Wirral, L66 1QT [IO83MG, SJ37]
G6 ISY L P Hill, 21 Liddiards Way, Purbrook, Waterlooville, PO7 5QW [IO90LU, SU60]
GW6 ITB J F Imperato, 118 Heol Uchaf, Rhiwbina, Cardiff, CF4 6SS [IO81JM, ST18]
G6 ITD Details withheld at licensee's request by SSL.
G6 ITF Details withheld at licensee's request by SSL.
G6 ITH Dr C D Jones, 98 Avon Dr, Alderbury Park, Alderbury, Salisbury, SP5 3TH [IO91DA, SU12]
GW6 ITJ Details withheld at licensee's request by SSL.
G6 ITK Details withheld at licensee's request by SSL.
G6 ITM D V Jupp, 20 Plumtree Gr, Hempstead, Gillingham, ME7 3RW [JO01GI, TQ76]
G6 ITO P A Kelly, 39 Copeland Ave, Tittensor, Stoke on Trent, ST12 9JA [IO82VW, SJ83]
G6 ITU M C Bunn, 86 Liverpool Rd South, Burscough, Ormskirk, L40 7TA [IO83NO, SD41]
G6 ITV L Parker, 15 Savile Pl, Mirfield, WF14 0AJ [IO93DQ, SE22]
G6 ITW M W Blundell, 68 Alton Rd, Leicester, LE2 8QA [IO92KO, SK50]
G6 ITY J H Beardall, 5 Meadow Walk, Great Abington, Abington, Cambridge, CB1 6AZ [JO02CC, TL54]
G6 IUF R E Brittain, Sarenchel, St Cross, South Elham, Harleston Norfolk, IP20 0NY [JO02QJ, TM28]
G6 IUH P L Bailey, 15 Essex Pl, Newcastle, ST5 3PS [IO82VX, SJ84]
GW6 IUK R R Bastable, Gwynfryn, Holyhead Rd, Llanfairpwllgwngyll, Anglesey, LL61 5SZ [IO73VF, SH57]
G6 IUQ M R Gaylard, 66 Runnymede Rd, Yeovil, BA21 5SU [IO80QW, ST51]
G6 IUS B J D Gilbert, 22 Oaklands Way, Hildenborough, Tonbridge, TN11 9DA [JO01DF, TQ54]
G6 IVB A Gornall, 28 Woodward Cl, Winnersh, Wokingham, RG41 5NW [IO91NK, SU77]
G6 IVC M Griffiths, 25 Lethbridge Rd, Southport, PR8 6JA [IO83MP, SD31]
G6 IVD Details withheld at licensee's request by SSL.
G6 IVE D L Barnes, 30 Royds Ave, Accrington, BB5 2LE [IO83TR, SD72]
GI6 IVJ R Brown, 157 Newtownards Rd, Bangor, BT20 4HS [IO74DP, J58]
G6 IVN Details withheld at licensee's request by SSL.
G6 IVP J F Burton, 22 Pear Tree Ln, Hempstead, Gillingham, ME7 3PT [JO01GI, TQ76]
G6 IVQ J R N L Beveridge, 36 Nursery Hill, Shamley Green, Guildford, GU5 0UN [IO91RE, TQ04]
G6 IVR P J Baxter Itchen Valley Amateur Radio Cl, 8 Birch Cl, Romsey, SO51 5SQ [IO90GX, SU32]
G6 IVT A M Burnett, 6 Cedar Cres, North Baddesley, Southampton, SO52 9FT [IO90GX, SU32]
G6 IVU V W Baldry, 4 Gr Pl, Padstow, PL28 8AX [IO70MN, SW97]
G6 IVW R E Balderson, 15 Woodrush Way, Evergreens, Moulton, Northampton, NN3 7HU [IO92NG, SP76]
G6 IWB J Skelly, 26 Princes Ave, Maylandsea, Mayland, Chelmsford, CM3 6BA [JO01JQ, TL90]
G6 IWC M A Saunders, 32 Chinchilla Rd, Southend on Sea, Essex, SS1 2QL [JO01IM, TQ88]
G6 IWD L M Sherratt, Alwyn, Norbury Dr, Brierley Hill, West Mids, DY5 3DP [IO82WL, SO98]
G6 IWK Details withheld at licensee's request by SSL.
G6 IWT M Rea, Osmary, Station Rd, Elsenham, Bishops Stortford, CM22 6LG [JO01CW, TL52]
G6 IWZ D J Jefferys, 22 Cleveland Gdns, Cricklewood, London, NW2 1DY [IO91VN, TQ28]
G6 IX S P Mason, 71 Melrose Ave, Sutton Coldfield, B73 6NS [IO92BN, SP19]
GW6 IXA C H P Jones, Pen-y-Berth, Pen-y-Garth, Caernarfon, Gwynedd, LL55 1EY [IO73UD, SH46]
GI6 IXC A M Stewart, 12 Donegall Dr, Whitehead, Carrickfergus, BT38 9LT [IO74DS, J49]
G6 IXE M D James, 9 Meadow Ln, Hamble, Southampton, SO31 4RB [IO90IU, SU40]
G6 IXH D G Hodges, 5 Greenlands, Leighton Buzzard, Beds, LU7 8UJ [IO91QV, SP92]
G6 IXK W Hall, 80 Thirlmere Rd, Ridge Est, Lancaster, LA1 3LL [IO84OB, SD46]
G6 IXN G S Hewitt, 66 Portland Dr, Forsbrook, Stoke on Trent, ST11 9AU [IO82XX, SJ94]
G6 IXQ B N Howell, 7 Hampton Rd, Southport, PR8 6SX [IO83MP, SD31]
G6 IXS Dr S N Henson, 4 Monaco Pl, Westlands, Newcastle, ST5 2QT [IO83VA, SJ84]
G6 IXT P Hearl, 49 Milford Hill, Harpenden, AL5 5BL [IO91UT, TL11]
G6 IYA H G Woodnutt, 155 Llanrwst Rd, Colwyn Bay, LL28 5YS [IO83DG, SH87]
G6 IYD Details withheld at licensee's request by SSL.
G6 IYE V J O'Herlihy, 130 Hitchin Rd, Upper Caldecote, Biggleswade, SG18 9BU [IO92UC, TL14]
G6 IYJ M J Plested, 26 Pudding Ln, Gadebridge, Hemel Hempstead, HP1 3JS [IO91SS, TL00]
G6 IYM N Perkins, Heathcote, Kents Rd, Torquay, TQ1 2NL [IO80FL, SX96]
GW6 IYP R Parry, Glen Garriff, Rhyl Rd, Rhuddlan, Rhyl, LL18 2TP [IO83GH, SJ07]
G6 IYR C A Porter, 2 Osborne, Coton Farm, Tamworth, B79 7SZ [IO92DP, SK10]
G6 IYS I R Porter, 2 Osborne, Coton Farm, Tamworth, B79 7SZ [IO92DP, SK10]
G6 IZA I R J Alderton, 6 Hurford Dr, Thatcham, RG19 4WA [IO91IJ, SU56]
G6 IZE R G Curry, 21 Sophia Gdns, Weston Super Mare, BS22 0DS [IO81NI, ST36]
G6 IZF K M Cahill, 4 Keynes Cl, Newport Pagnell, MK16 9AT [IO92PC, SP84]
G6 IZG W Campion, 58 Okehampton Cres, Mapperley, Nottingham, NG3 5SE [IO92KX, SK54]
G6 IZK A J Collier, 19 Lullington Garth, Woodside Park, North Finchley, London, N12 7LT [IO91VO, TQ29]
G6 IZN Details withheld at licensee's request by SSL.[Station located near Fareham.]
G6 IZO P C Cook, 87 Brodie Ave, Liverpool, L18 4RF [IO83NI, SJ38]
G6 IZP P M Clements, 22 Lynton Ave, Hateley Heath, West Bromwich, B71 2QZ [IO92AM, SP09]
G6 IZQ J R Coad, Bybrook, 17 Dilly Ln, Barton on Sea, New Milton, BH25 7DQ [IO90ER, SZ29]
G6 IZS Details withheld at licensee's request by SSL.
GM6 IZU K R Frame, 3 Scotts Pl, Melrose, TD6 9QZ [IO85PO, NT53]
G6 IZZ C T E Evans, 20 Cabot Cl, Yate, Bristol, BS17 4NN [IO81TM, ST78]
G6 JAC P A Dalley, 32 Albert Rd, Erdington, Birmingham, B23 7LT [IO92BM, SP09]
G6 JAD Details withheld at licensee's request by SSL.
G6 JAF I P Evans, 19 Grange Rd, Stone, ST15 8PR [IO82WV, SJ93]
GM6 JAG E E Evans, 56 Southhouse Rd, Edinburgh, EH17 8EU [IO85KV, NT26]
G6 JAK S C Deacon, 3 Blenheim Gr, Offord Darcy, St. Neots, Huntingdon, PE18 9RD [IO92VG, TL26]
G6 JAM M J Dainty, 56 Church Hill, Wednesbury, WS10 9DJ [IO82XN, SO99]
G6 JAR M J Drake, 6 Coniston Rd, Ringwood, BH24 1PF [IO90CU, SU10]
G6 JAS A J R Balding, 8 Winston Way, Farcet, Peterborough, PE7 3BU [IO92VM, TL29]
G6 JAY R J Luckett, 20 Leicester Villas, Hove, BN3 5SQ [IO90VT, TQ20]
G6 JAZ R H Leburn, 40 Brunswick, Hanworth, Bracknell, RG12 7YY [IO91OJ, SU86]
GM6 JBF D P Mardlin, 35 Uist Rd, West Sheddocksley, Aberdeen, AB1 6FN [IO87TH, NJ72]
G6 JBG J Mathers, 8 Providence Terr, Chippenham, SN15 1HD [IO81WL, ST97]
G6 JBL G C Moore, 26 Southwick Rd, Canvey Island, SS8 0EP [JO01GM, TQ78]
GW6 JBN R F Thomas, Post Office, Llanbedr, Gwynedd, LL45 2HH [IO72WT, SH52]
G6 JBQ G B Taylor, 16 Roundwood Cl, Hitchin, SG4 0RD [IO91VX, TL23]
G6 JBY J W Bibby, 24 Assarts Ln, Malvern, WR14 4JR [IO82UB, SO74]
G6 JCE Details withheld at licensee's request by SSL.

G6 JCI W T Henson, 1 Bonser Cl, Carlton, Nottingham, NG4 1DP [IO92KX, SK64]
GI6 JCL H B Hawthorne, 110 Morgans Hill Rd, Cookstown, BT80 8BW [IO64OP, H87]
G6 JCM J H Hatfield, Tenter Cl, Husthwaite, York, YO6 3SF [IO94JD, SE57]
G6 JCT H E Haslehurst, Westlands, Stinting Ln, Shirebrook, Mansfield, NG20 8EQ [IO93JE, SK56]
G6 JCV A Haslehurst, Westlands, Stinting Ln, Shirebrook, Mansfield, NG20 8EQ [IO93JE, SK56]
G6 JCX F W Hewitt, Whitegables, 1 Northmead Dr, North Walsham, NR28 0AU [JO02QT, TG23]
G6 JCY B Hedge, 11 Robert Smith Ct, Stalham, Norwich, NR12 9EH [JO02SS, TG32]
G6 JDC Mt T Kemp, 85 Rosehill Rd, Rawmarsh, Rotherham, S62 7BX [IO93HL, SK49]
GW6 JDF G R Walker, Lluesty, Bryn Hyfryd Rd, Tywyn, LL36 9HG [IO72WN, SH50]
G6 JDH G C Webster, 37 Coleford Bdg Rd, Mytchett, Camberley, Surrey, GU16 6DH [IO91PG, SU85]
GW6 JDJ P T Weaver, 24 Montclaire Ave, Blackwood, NP2 1EE [IO81JP, ST19]
G6 JDO N K Wright, 25 Penny Park Ln, Coventry, CV6 2GU [IO92FK, SP38]
G6 JDP J A M Mott-Gotobed, Cherry Trees, 17 Reading Rd, Chineham, Basingstoke, RG24 8LN [IO91LG, SU65]
G6 JDY A B Smith, 149 Cottingham Gr, Bletchley, Milton Keynes, MK3 5AJ [IO91PX, SP83]
GM6 JDZ G Taylor, 31 Commissioner St., Crieff, PH7 3AY [IO86BI, NN82]
G6 JEB J E Bailey, 46 Greswolde Rd, Solihull, B91 1DY [IO92CK, SP18]
G6 JEF S H Wardley, 5 Swindon St., Bridlington, YO16 4JD [IO94VC, TA16]
G6 JEG Details withheld at licensee's request by SSL.
G6 JEJ Details withheld at licensee's request by SSL.
G6 JEK A V Williams, Wolstanton, 95 Downton Rd, Salisbury, SP2 8AT [IO91CB, SU12]
G6 JEL M W Williams, Wolstanton, 95 Downton Rd, Salisbury, SP2 8AT [IO91CB, SU12]
G6 JEM P J Stoneman, 111 Fletemoor Rd, St. Budeaux, Plymouth, PL5 1UL [IO70VJ, SX45]
G6 JEN J Askin, 54 York Rd, Greenwood Ave, Hull, HU6 9RA [IO93TS, TA03]
GM6 JEP F J Cassidy, 9 Spey Rd, Troon, KA10 7DY [IO75QN, NS33]
G6 JEU P Chrysostomou, 45 Leyborne Ave, Ealing, London, W13 9RA [IO91UM, TQ17]
G6 JEV Details withheld at licensee's request by SSL.
G6 JEY M E Cooper, Huntley, Littlehampton Rd, Ferring, Worthing, BN12 6PN [IO90ST, TQ00]
G6 JFJ L C Bacon, 100 Etherington Dr, Hull, HU6 7JT [IO93TS, TA03]
G6 JFK D R Binnington, 5 Colsons Way, Olney, MK46 5EQ [IO92PD, SP85]
GW6 JFM S C Brooks, 79 Ffordd Dryden, Killay, Swansea, SA2 7PD [IO71XP, SS69]
G6 JFN P C Brazier, The Stud House, Mentmore, Leighton Buzzard, Beds, LU7 0QE [IO91PU, SP92]
GM6 JFP D A Brown, 7 Young St., Peebles, EH45 8JX [IO85JP, NT24]
G6 JFS Details withheld at licensee's request by SSL.
G6 JFU A S Mayman, Lingmell, Cedar Gr, Aldbrough, E Yorks, HU11 4QH [IO93WT, TA23]
G6 JFV T C Morris, 558 Bromford Rd, Hodge Hill, Birmingham, B36 8AL [IO92CL, SP18]
GW6 JFX K A Oliver, 34 Tolcarne St., Camborne, TR14 8JH [IO70IF, SW63]
G6 JFZ Details withheld at licensee's request by SSL.
GI6 JGB M S McNinch, 23 Hazeldene Gdns, Bangor, BT20 4RD [IO74EP, J58]
GW6 JGE N T P Lewis, 1 Clyne Dr, Mayals, Blackpill, Swansea, SA3 5BU [IO81AO, SS69]
G6 JGF A J Morgan, 8 Shaftesbury Rd, Watford, WD1 2RQ [IO91TP, TQ19]
G6 JGL A McKechnie, 12 Fraser Ct, Handbridge, Chester, CH4 7DL [IO83NE, SJ46]
G6 JGM F E B Lewington, 24 Torcross Cl, Glenfield, Leicester, LE3 8AP [IO92JP, SK50]
G6 JGP A W J Lawrence, Columbine Cottage, Ford Ln, Ford, Salisbury, SP4 6DJ [IO91CC, SU13]
G6 JGR J Richardson, 65 Campbell Rd, Eastleigh, SO50 5AA [IO90HX, SU41]
G6 JGT A D Davis, 22 Tamar Cl, Fareham, PO16 8QF [IO90KU, SU60]
G6 JHD A J Fox, 19 Pemberton Gdns, Chadwell Heath, Romford, RM6 6SH [JO01BN, TQ48]
G6 JHE Details withheld at licensee's request by SSL.
G6 JHG J G Grieve, 65 Royal Ln, Hillingdon, Uxbridge, UB8 3QU [IO91SM, TQ08]
GM6 JHH W D Gunn, 16 Aytoun Gr, Baldridgeburn, Dunfermline, KY12 9YA [IO86GB, NT08]
GW6 JHM Details withheld at licensee's request by SSL.
G6 JHR Details withheld at licensee's request by SSL.
G6 JID C D Powell, 3 The Willows, Bradley Stoke, Bristol, BS12 9BJ [IO81RM, ST68]
G6 JIE Details withheld at licensee's request by SSL.
G6 JIF J R M Purdy, 267 St. Helens Rd, Hastings, TN34 2NF [JO00GU, TQ81]
G6 JIM J W King, 4 Glenhurst Ave, Ruislip, HA4 7LZ [IO91SN, TQ08]
G6 JIY R L Bamber, Hermes, 84 Paynesfield Rd, Tatsfield, Westerham, TN16 2BQ [JO01AH, TQ45]
G6 JJ W N Craig, 16 The Mount, Rickmansworth, WD3 4DW [IO91SP, TQ09]
G6 JJA N Billingham, 35 Scalwell Park, Seaton, EX12 2DB [IO80LR, SY29]
G6 JJB B J Banks, 16 Park Rd, Burntwood, WS7 0EE [IO92BQ, SK00]
G6 JJE C S Barker, 22 Melton Rd, Wymondham, NR18 0DB [JO02NN, TG10]
G6 JJF C Byrne, 22 Cromer Rd, Bury, BL8 1ES [IO83UO, SD71]
G6 JJG K J Breakwell, 91 Lynton Ave, Cloregate, Wolverhampton, WV6 9NQ [IO82WO, SJ80]
G6 JJH Details withheld at licensee's request by SSL.[Op: D B Bowell. Station located near Mitcheldean.]
G6 JJI A J Bromfield, 11 Blackthorn Croft, Woodend, Clayton-le-Woods, Chorley, PR6 7TZ [IO83QQ, SD52]
G6 JJK J G Bourne, 91 Burwell Rd, Exning, Newmarket, CB8 7DU [JO02EG, TL66]
GM6 JJN R I Berry, Sylvan House, Glenmoriston, Inverness, IV3 6YJ [IO77NE, NH21]
G6 JJO P G Beck, 269 Birmingham Rd, Walsall, WS5 3AA [IO92AN, SP09]
G6 JJP J J Pinson, 10 Kenelm Cl, Clifton on Teme, Worcester, WR6 6EB [IO82SG, SO76]
GI6 JJR N Loughrey, 8 Oak Vale Ave, Newry, BT34 2BQ [IO64UE, J02]
G6 JJT E J Ferguson, 33 High St., Cranfield, Bedford, MK43 0DP [IO92QB, SP94]
GW6 JJV G D Lacy, 2 Godrer Gaer, Llwyngwril, LL37 2JZ [IO72XP, SH50]
GW6 JJX R F Price, Far Cottage, Penoyre, Brecon, Powys, LD3 9LP [IO81GX, SO03]
G6 JKG Details withheld at licensee's request by SSL.
G6 JKK G D Orchard, 34 Crusader Rd, Bearwood, Bournemouth, BH11 9TZ [IO90AS, SZ09]
G6 JKP M Howden, 11 Marsh Ln Garde, Kellinton, Goole, Humberside, DN14 0PG [IO93KR, SE52]
GM6 JKU M R Henderson, 34 Soutar Cres, Perth, PH1 1QB [IO86GJ, NO02]
G6 JKV R J Henneman, 14 Savernake Cl, Tilehurst, Reading, RG3 4LY [IO91LJ, SU66]
G6 JKX A M Hembery, 15 Wivenhoe Ct, Frome, BA11 2DF [IO81UF, ST74]
GW6 JLH A E Davies, Noddfa, Lower Rd, Harlech, Gwynedd, LL46 2UB [IO72WU, SH53]
G6 JLI P Dixey, Rose Cottage, 197 Raikes Ln, Birstall, Batley, WF17 9QF [IO93ER, SE22]
G6 JLL S E Douglas, 1030 Shields Rd, Walkerville, Newcastle upon Tyne, NE6 4SR [IO94FX, NZ26]
GM6 JLM L J T Dairon, 6 Kingsley Ave, Wootton Bassett, Swindon, SN4 8LF [IO91BN, SU08]
G6 JLP I OToole, Our House, 270 Victoria St., Newton, Hyde, SK14 4DT [IO83XL, SJ99]
G6 JLU A G Millar, 24 Springwood Cres, Edgware, HA8 8SD [IO91UP, TQ19]
G6 JMA G Smith, 1 Edinburgh Dr, North Anston, Anston, Sheffield, S31 7HD [IO93GJ, SK38]
G6 JMB J D M Mountain, 44 Townmead Rd, Fulham, London, SW6 2RR [IO91VL, TQ27]
GW6 JMC D L Miller, Bryn Awel, Llangwm, Corwen, Clwyd, LL21 0RB [IO82FX, SH94]
GI6 JMD J G Moller, 29 Carlton Heights, Bangor, BT19 6ZB [IO74EP, J58]
G6 JME R A Powell, Sunningdale, 39 Compton Dr, Eastbourne, BN20 8DA [JO00DS, TV59]
G6 JMF Details withheld at licensee's request by SSL.
G6 JMG P N Parton, 1 Clee Rise, Highley, Bridgnorth, WV16 6EL [IO82TK, SO78]
G6 JMJ K R Renton, 87 Shirley Gdns, Barking, IG11 9XB [JO01BM, TQ48]
G6 JMX B J Wendon, 89 Palewell Park, East Sheen, London, SW14 8JJ [IO91UL, TQ27]
G6 JMZ Details withheld at licensee's request by SSL.
G6 JNB R T Shergold, 28 Knightstons, Heights, Frome, Somerset, BA11 1NR [IO81UF, ST74]
GW6 JNE R D Sartin, 7 Penrhos Cres, Rumney, Cardiff, CF3 8PB [IO81KM, ST27]
GM6 JNQ I C Cox, 8 Traill St., Castletown, Thurso, KW14 8UG [IO88HO, ND16]
G6 JNS P L Crosland, Sprackets Orchard, Curry Rivel, Langport, Somerset, TA10 0PP [IO81NA, ST32][Op: Peter L Crosland, QSL manager for C30AKA, ON9CP, G1EME, G7EME. E-mail: g6jns@amsat.org.]
G6 JNV M E B Carter, 17 McWilliam Rd, Woodingdean, Brighton, BN2 6BE [IO90XU, TQ30]
G6 JNW S R Carter, 84 Barnett Rd, Brighton, BN1 7GH [IO90WU, TQ30]
G6 JNZ W T Caine, 116B Hill St., Cannock, WS12 5DR [IO92AQ, SK01]
GM6 JOA A White, 6 Suffolk St., Helensburgh, G84 8EH [IO76PA, NS28]
GM6 JOD T Lawless, 2 Lawers Pl, Middleton Park, Bourtreehill North, Irvine, KA11 1LR [IO75QO, NS33]
G6 JOI Details withheld at licensee's request by SSL.
G6 JOL R E Young, 55 Kirby Rd, Leicester, LE3 6BD [IO92KP, SK50]
G6 JON Details withheld at licensee's request by SSL.
GI6 JOP A H Wallace, 61 Locksley Park, Belfast, BT10 0AS [IO74AN, J36]
G6 JOR D W Webb, 1 Corelli Rd, Basingstoke, RG22 4NB [IO91KF, SU64]
G6 JOS K M Arrowsmith, 27 Clifton Ave, Eaglescliffe, Stockton on Tees, TS16 9AZ [IO94HM, NZ41]
G6 JOX C Inman, 12 Ditchfield Rd, Penketh, Warrington, WA5 2NN [IO83QJ, SJ58]
G6 JOY M J Timbrell, 92 Honeyborne Rd, Sutton Coldfield, B75 6BN [IO92CN, SP19]
G6 JOZ C J Timbrell, 92 Honeyborne Rd, Sutton Coldfield, B75 6BN [IO92CN, SP19]
G6 JP G R Jessop, 32 North View, Eastcote, Pinner, HA5 1PE [IO91TN, TQ18]
G6 JPA Details withheld at licensee's request by SSL.
GU6 JPE A Jephcott, 12 Clos Du Beauvoir, Rue Cohu, Castel, Guernsey, CI, X X
G6 JPG J P Gilliver, 44 Templeton Park, Bakers Ln, West Hanningfield, Essex, CM2 8LF [JO01FQ, TL70]
G6 JPI I S Kirby, 27 The Orchards, Sutton, Ely, CB6 2PX [JO02BJ, TL47]
G6 JPJ R H Knight, 48 Edinburgh Dr, Uxbridge, UB10 8QY [IO91SN, TQ08]
G6 JPM S N Green, The Bears Nest, 5 Bridge Terr, Tuckenhay, Totnes, TQ9 7EH [IO80EJ, SX85]
G6 JPN D Grimshaw, 66 St. Marys Gdns, Mellor, Blackburn, BB2 7JP [IO83RS, SD63]
G6 JPQ J Gould, Weir House, 108 Newton Rd, Burton on Trent, DE15 0TT [IO92ET, SK22]
G6 JPS Details withheld at licensee's request by SSL.
G6 JPT D Gleave, 1 Fearnley Way, Newton-le-Willows, WA12 8SQ [IO83QK, SJ59]
G6 JQB R P Swancutt, 27 Kingstanding Rd, Perry Barr, Birmingham, B44 8BA [IO92BM, SP09]
G6 JQD R H Skinner, 6 Springfield, Ashford, Middx, TW15 2LR [IO91SK, TQ07]
G6 JQE G Tannahill, 6 Ryal Walk, Newcastle upon Tyne, NE3 3YE [IO95EA, NZ26]
GU6 JQF M Trenchard, Mont Gibel, 3 Clifton Stairs, St. Peter Port, Guernsey, GY1 2PL
G6 JQH J Williams, 89 Marlborough Rd, Castle Bromwich, Birmingham, B36 0EL [IO92CM, SP18]
GW6 JQS Details withheld at licensee's request by SSL.
GW6 JQT R B Brown, 34 Fishguard Cl, Llanishen, Cardiff, CF4 5QG [IO81JM, ST18]
G6 JQV Details withheld at licensee's request by SSL.[Op: Archie Roberts. Station located near Long Eaton.]
G6 JQW Details withheld at licensee's request by SSL.
G6 JQX Details withheld at licensee's request by SSL.[Station located 3/4 mile from threshold of Runway 27 Left, Heathrow Airport. Telephone: John (0181) 890 0900 (working hours).]
G6 JR Details withheld at licensee's request by SSL.[Op: J W Rhind. Station located in London SE1.]
G6 JRE S J Stanton, 6 Trevor Rd, Beeston, Nottingham, NG9 1GR [IO92JW, SK53]
G6 JRH T R Steele, 6 Leigh Dr, Elsenham, Bishops Stortford, CM22 6BY [JO01DW, TL52]
G6 JRI I Wright, 25 Stray Rd, Burnholme, York, YO3 0NE [IO93LX, SE65]
G6 JRL M Bernard, 36 Garth Dr, Hambleton, Selby, YO8 9QD [IO93JS, SE53]
G6 JRM H M Bottomley, Nerefield, Aylesbury Rd, Chearsley, Aylesbury, HP18 0BL [IO91MT, SP71]
GW6 JRP Details withheld at licensee's request by SSL.
GM6 JRX D J Fraser, 18 Harland Rd, Castletown, Thurso, KW14 8UB [IO88HO, ND16]
GI6 JRY R J Getty, 6 Rocheville, Cookstown, BT80 8QE [IO64PP, H87]
G6 JRZ S J Gunn, 55 Station Rd, West Byfleet, KT14 6DT [IO91SI, TQ06]
G6 JS A A Jones, 37 Green Acres Dr, London, Ontario, Canada N6G 2S4
GU6 JSC T P Hodkinson, Arama Les Baissiere, St Peter Port, Guernsey, Guernsey, GY1 2UD
G6 JSF M R Hayward, 1 Station Rd, Grateley, Andover, SP11 8LG [IO91EE, SU24]
G6 JSI A W Haswell, 66 White Hart Ln, Fareham, PO16 9BQ [IO90KU, SU60]
G6 JSN A G Sym, 11 Linden Ave, Ruislip Manor, Ruislip, HA4 8TW [IO91TN, TQ18]
G6 JSR A Mason, 5 Birch Rd, Kippax, Leeds, LS25 7DY [IO93HS, SE43]
G6 JSV Details withheld at licensee's request by SSL.
G6 JSZ D B Wilkinson, 19 Highfield Gr, Exley Ln, Elland, HX5 0SP [IO93BQ, SE12]
G6 JTC M J Whiteley, 29 Stavanger Cl, Corby, NN18 9HT [IO92PL, SP88]
G6 JTI H Martin, 80 Topcliffe Rd, Sowerby, Thirsk, YO7 1RT [IO94HF, SE48]
G6 JTJ Details withheld at licensee's request by SSL.
G6 JTK R W Nokes, 20 Hayes Ln, Slinfold, Horsham, RH13 7SQ [IO91TB, TQ13]
G6 JTL M V Parks, 34 Lingfield Ave, Fordhouses, Wolverhampton, WV10 6PD [IO82WP, SJ90]
G6 JTN Details withheld at licensee's request by SSL.
G6 JTT J T Trett, 1 Moorland Way, Bridgwater, TA6 4JL [IO81MD, ST33]
G6 JTV R C Allen, 65 Atherstone Rd, Measham, Swadlincote, DE12 7EG [IO92FQ, SK31]
G6 JTW H D W Marshall, Fen Farm, Side Bar Ln, Heckington, Sleaford Lincs, NG34 9LY [IO92VX, TF14]
GW6 JTX E R Bielawski, 1 Lakeside, Lake Farm, Gresford, Wrexham, Clwyd, LL12 8PU [IO83MC, SJ35]
G6 JTZ Details withheld at licensee's request by SSL.
GM6 JUA D A T Brown, 40 Abbots Rd, Grangemouth, FK3 8JE [IO86DA, NS98]
G6 JUG E Cheneler, 173 Church St., Witham, CM8 2JW [JO01HT, TL81]
G6 JUI K Dare, One Bee, 1 Gloucester Rd, Reading, RG3 2TH [IO91LJ, SU66]
G6 JUJ R G Bredge, 8 Hoadley Green, Bishopdown Est, Salisbury, SP1 3HS [IO91CC, SU13]
GW6 JUL P A Eglington, 1 Dan y Coed, Cwmavon, Port Talbot, SA12 9NH [IO81CO, SS79]
G6 JUO V J Fernyhough, Hill Farm House, School Ln, Moss Pit, Stafford, ST17 9JB [IO82WS, SJ92]
G6 JUP J Sutton, 252 Rawling Rd, Gateshead, NE8 4UH [IO94EW, NZ26]
G6 JUQ G M Williams, 32 Brook St., Crossens, Southport, PR9 8HY [IO83MQ, SD31]
G6 JUT J C Whiting, 97 Barrowby Gate, Grantham, NG31 8RB [IO92QV, SK83]
GM6 JUU Details withheld at licensee's request by SSL.
G6 JUZ Details withheld at licensee's request by SSL.
G6 JVA J B Greevy, 24 Truro Rd, Park Hall, Walsall, West Midlands, WS5 3EH [IO92AN, SP09]
GW6 JVB R S Griffiths, 26 Brynglas, Gilwern, Abergavenny, NP7 0BP [IO81KT, SO21]
G6 JVG Details withheld at licensee's request by SSL.
G6 JVK M W Jeffery, 9 Cassia Dr, Earley, Reading, RG6 5YH [IO91MK, SU77]
G6 JVO M N Kidd, 99 Ferry Rd West, Scunthorpe, DN15 8UG [IO93PO, SE81]
G6 JVP P J Cole, 190 Regents Park Rd, Southampton, SO15 8NY [IO90GW, SU31]
G6 JVS Details withheld at licensee's request by SSL.
G6 JVT C S Santer, 2 The Haven, Beaumont Park, Littlehampton, BN17 6NS [IO90RT, TQ00]
G6 JVX H G Schofield, 15 Deerfield Rd, March, PE15 9AH [JO02BN, TL49]
G6 JVY Details withheld at licensee's request by SSL.
GW6 JWD J W Davies, Welfare House, Borth, Dyfed, SY24 5JD [IO72XL, SN68]
GM6 JWH D R Taylor, 3 Abbotsgrange Rd, Grangemouth, FK3 9JD [IO86DA, NS98]
GW6 JWL H M Roberts, Pen-yr-Erw, Graigfechan, Ruthin, Clwyd N Wales, LL15 2EY [IO83IB, SJ15]
G6 JWM D W Le Grove, 91 Kings Rd, Ilkley, LS29 9BZ [IO93BW, SE14]
G6 JWO A W Legg, Riverside, Westborough, Long Bennington, Nr.Newark, Notts, NG23 5HN [IO92PX, SK84]
G6 JWU E Ormerod, 86 Chapel Hill, Longridge, Preston, PR3 2YB [IO83QT, SD63]
G6 JWV M R Ratcliffe, Middle Gates House, Wick Ln, Englefield Green, Egham, Surrey, TW20 0HT [IO91QK, SU97]
G6 JWX C D Benton, 48 Langley Rd, Chedgrave, Norwich, NR14 6HD [JO02RM, TM39]
G6 JWZ D Bagnall, 20 Clifton Cl, Swadlincote, DE11 9SQ [IO92FS, SK21]
G6 JXA Details withheld at licensee's request by SSL.[Op: K M Brown. Station located near Morden.]
GI6 JXG W T Collins, 33 New Row, Kilrea, Coleraine, BT51 5TA [IO64RW, C91]
G6 JXN Details withheld at licensee's request by SSL.
G6 JXP N Hakes, Thornlea, 25 Carr Ln, Rawdon, Leeds, LS19 6PD [IO93DU, SE23]
GW6 JXR Details withheld at licensee's request by SSL.
G6 JXS W Hughes, 27 Winchester Cl, Ashington, NE63 9QJ [IO95FE, NZ28]
G6 JXW E M Jolliffe, Cromer Rd, Overstand, Overstrand, Cromer, Norfolk, NR27 0JJ [JO02PW, TG24]
G6 JY Dr F T Farmer, 41 Goldspink Ln, Newcastle upon Tyne, NE2 1NQ [IO94EX, NZ26]
G6 JYB M J Niman, 55 Harrow Way, Chelmsford, CM2 7AU [JO01GR, TL70]
GM6 JYC A I Mutch, 58 West Ferryfield, Edinburgh, EH5 2PU [IO85JX, NT27]
G6 JYG Details withheld at licensee's request by SSL.[Op: C M J Page. Station located near Newton Abbot.]
G6 JYO C W Allen F.B.S, 20 Hollywood Ln, Hollywood, Birmingham, B47 5PX [IO92BJ, SP07]
G6 JYX R H Drew, Derwent House, Landing Ln, Hemingbrough, Selby, YO8 7RA [IO93MS, SE63]
G6 JZE P J Graham, 14 Carlaw Rd, Prenton, Birkenhead, L42 8QA [IO83LI, SJ38]
G6 JZN A Ogden, 12 Flying Fields Rd, Southam, Leamington Spa, CV33 0GA [IO92GF, SP35]
G6 JZS Details withheld at licensee's request by SSL.
G6 JZV K R Lummis, 10 Church Rd, Old Newton, Stowmarket, IP14 4ED [JO02MF, TM06]
G6 JZW C M Muller, 5 Ash Cl, Flitwick, Bedford, MK45 1JY [IO92SA, TL03]
G6 KAA F G Wright, 30 Field Cl, Hilton, Derby, DE65 5GL [IO92EU, SK23]
G6 KAC A M Trim, 29 Casterbridge Rd, Dorchester, DT1 2AH [IO80SQ, SY78]
G6 KAE J A Bailey, 54 Dimsdale Rd, Northfield, Birmingham, B31 5RD [IO92AJ, SP07]
G6 KAI M D Brighton, 11 West Cl, Norwich, NR5 0NH [JO02OP, TG11]
G6 KAM A Drummond, 18 Peacock Pl, Ecclefechan, Lockerbie, DG11 3EQ [IO85IB, NY17]
G6 KAR Details withheld at licensee's request by SSL.
GW6 KAV Dr H G A Hughes, Talwrn Glas, Denbigh Rd, Afonwen, Mold, CH7 5UB [IO83IF, SJ17]
G6 KAW G C Instone, 32 Melfort Cl, Sparcells, Swindon, SN5 9FG [IO91CN, SU18]
G6 KAX Details withheld at licensee's request by SSL.
GM6 KAY C K Bates, 9 Winton Terr, Edinburgh, EH10 7AP [IO85JV, NT26]
G6 KBC C J Philpot, 17 Jervis Ct, Ilkeston, DE7 8PX [IO92IX, SK44]
GW6 KBD D M J Potts, 11 Walmer Rd, Newport, NP9 8NU [IO81MO, ST38]
GM6 KBG G Reynolds, 22 Duncan Rd, Helensburgh, G84 9DQ [IO76PA, NS38]
G6 KBI J M Maunder, 23 Englehurst, Egham, TW20 0EE [IO91RK, SU97]
G6 KBJ R A Newell, 59 Western Rd, Burgess Hill, RH15 8QW [IO90WW, TQ31]
G6 KBQ T J P Williams, 2 Hazelwood, Greasby, Wirral, L49 2RQ [IO83KJ, SJ28]
G6 KBR Details withheld at licensee's request by SSL.
G6 KBS J . M . Musgrave, 57 Chiltern Rd, Baldock, SG7 6LT [IO91VX, TL23]
GI6 KBX Rev J Turner, 45 Gloonan Hill, Ahoghill, Ballymena, BT42 1PU [IO64TU, D00]
G6 KCE R Thornton, 50 Springfield Ave, Brough, HU15 1BU [IO93RR, SE92]
G6 KCG P R Sharpe, 46 Beaumont Rd, New Costessey, Norwich, NR5 0HG [JO02OP, TG11]
G6 KCJ D A J Wynters, 11 Heritage Ln, Ascott under Wychwood, Chipping Norton, OX7 6AD [IO91FU, SP31]
G6 KCV P C L Willmott, c/o Oil Management Serv. Ltd., P.O. Box Hm 1751, Hamilton Hm Gx, Bermuda, X X
GI6 KCX J H P Madden, 17 Avondale, Springfarm Rd, Antrim, BT41 2AT [IO64VR, J18]
GM6 KDB K J Lee, 3 Braikley Ave, Tarves, Ellon, AB41 7PU [IO87VI, NJ83]
GM6 KDD D Scobbie, 17 Roselea Dr, Brightons, Falkirk, FK2 0TJ [IO85DX, NS97]
G6 KDI Details withheld at licensee's request by SSL.
G6 KDJ J A Smallwood, 4 Leece Dr, Dalton in Furness, LA15 8NP [IO84JD, SD27]
GM6 KDN K E N N McInnes, Flat 3/2, Holmlea Rd, Langside, Glasgow, Lanarkshire, G44 4BL [IO75UT, NS56]
G6 KDP D M Power, 118 High Rd, Islington St. Germa, Islington, Kings Lynn, PE34 3BJ [JO02DQ, TF51]
G6 KDQ Details withheld at licensee's request by SSL.
G6 KDY A Perkins, 44 Holly Gr Ln, Burntwood, WS7 8QA [IO92AQ, SK01]
GM6 KEC I M Sinclair, 3 Ben More Dr, Paisley, PA2 7NU [IO75TT, NS56]
G6 KEJ Details withheld at licensee's request by SSL.
G6 KEN K G C Dasilva-Hill, 12 St. Stephens Cres, Thornton Heath, CR7 7NP [IO91WJ, TQ36]
G6 KEQ T J Heavingham, 39 Harvey St., Halstead, CO9 2LH [JO01HW, TL83]
G6 KES J Hopper, 11 Orchard Ave, Barrow in Furness, LA13 9JA [IO84JD, SD27]
GM6 KEV D A Smith, 38 Queens Pl, Dunbar, EH42 1YA [IO85RX, NT67]
G6 KEZ P Pattison, 18 Broadgate Ln, Deeping St. James, Peterborough, PE6 8NW [IO92UQ, TF10]
G6 KFD P F G Stockwell, 62 Golden Cross Rd, Ashingdon, Rochford, SS4 3DQ [JO01IO, TQ89]

GW6 KFH B Gayther, Coed y Parc Waun, Pensiarwaun, Gwynedd, LL55 3PW [IO73WD, SH56]
GM6 KFO G Gordon, 31 Stoneyhill Ave, Musselburgh, EH21 6SB [IO85LW, NT37]
G6 KFR Details withheld at licensee's request by SSL.
G6 KFW Details withheld at licensee's request by SSL.
G6 KFY P J Bennett, 10 The Croft, Didcot, OX11 8HR [IO91JO, SU58]
G6 KGA L J Coleman, Lilac Cottage, Coley Ln, Little Haywood, Stafford, ST18 0XB [IO92AT, SK02]
G6 KGB J R Coulson, Holystone House, 2 Whitley Rd, Holystone, Newcastle upon Tyne, NE27 0DB [IO95FA, NZ37]
G6 KGG W G Duffner, 23 Grange Cl, Misterton, Doncaster, DN10 4EN [IO93NK, SK79]
G6 KGI Details withheld at licensee's request by SSL.
GW6 KGR M R Buck, Glaisfor Uchaff Far, Llangynidr, Crickhowell, Powys, NP8 1LN [IO81JU, SO11]
G6 KGU Dr D C Craig, Pear Tree Cottage, Cripps Corner, Staplecross, E Sussex, TN32 5QS [JO00GX, TQ72]
G6 KGW K Driver, 5 Hill St., Colne, BB8 0DH [IO83VU, SD83]
GM6 KGZ Details withheld at licensee's request by SSL.[Op: D Henderson.]
G6 KHA T J Hyde, 14 Wyley Rd, Radford, Coventry, CV6 1NW [IO92FK, SP38]
G6 KHC P S Boden, 1 Hitherfield Ln, Harpenden, AL5 4JD [IO91TT, TL11]
G6 KHD K A Bierton, 42 Thrift Rd, Heath An Roach, Heath & Reach, Leighton Buzzard, LU7 0AX [IO91QW, SP92]
G6 KHG R S Champion, 25 Congreve Rd, Worthing, BN14 8EL [IO90TT, TQ10]
GW6 KHH D W Davies, Pemberley, 59 Queensway, Haverfordwest, SA61 2NU [IO71MT, SM91]
G6 KHM L M Edwards, 71 Gleneagles Rd, Yardley, Birmingham, B26 2HT [IO92CL, SP18]
G6 KHN S Harvey, 53 Winleigh Rd, Handsworth Wood, Birmingham, B20 2HN [IO92AM, SP09]
G6 KHP D Hamm, 29 Brow Hey, Bamber Bridge, Preston, PR5 8DS [IO83QR, SD52]
G6 KHW I A Bultitude, 2 The Quest, Ampthill Rd, Houghton Conquest, Bedford, MK45 3JP [IO92SB, TL04]
G6 KIA C K Duckles, 8 Railway Cttgs, Skillings Ln, Brough, HU15 1EN [IO93RR, SE92]
G6 KIB P G Duesbury, The Bungalow, Robins Ln, Lolworth, Cambridge, CB3 8HH [JO02AG, TL36]
G6 KIE D R Banks, 145 Compton Cres, Chessington, KT9 2HG [IO91UI, TQ16]
GM6 KIW M C Dennis, 199 Maxwell Ave, Bearsden, Glasgow, G61 1HS [IO75UV, NS57]
G6 KIZ M E Griffiths, 8 Lapwing Cl, East Hunsbury, Northampton, NN4 0RT [IO92NE, SP75]
GI6 KJC Dr W P Abram, 11 Glebe Manor, Hillsborough, BT26 6NS [IO74AK, J25]
GM6 KJD J M Cowie, 122 Cornhill Rd, Aberdeen, AB2 2EH [IO87TH, NJ72]
G6 KJE Details withheld at licensee's request by SSL.
G6 KJF J J Else, 96 Ben Nevis Rd, Tranmere, Birkenhead, L42 6QZ [IO83LI, SJ38]
G6 KJH P J Horobin, 12 Laurel Rd, Blaby, Leicester, LE8 4DL [IO92KN, SP59]
G6 KJK J Chappell, 15 Edmund Ave, Castle House Gdns, Stafford, ST17 9FT [IO82WT, SJ92]
G6 KJM J A Mirams, Hawkstone House, 33 The Moors, Kidlington, OX5 2AH [IO91IT, SP41]
G6 KJO M Edwards, 24 Kelham Green, Gordon Rd, Nottingham, NG3 2LP [IO92KX, SK54]
GM6 KJQ J G G Farquhar, 91 Park View, Fauldhouse, Bathgate, EH47 9JZ [IO85DU, NS96]
G6 KJR Details withheld at licensee's request by SSL.
G6 KJT S W Brabbins, Bramleigh, 8 Park Dr, Eldwick, Bingley, BD16 3DF [IO93CU, SE14]
G6 KJU Details withheld at licensee's request by SSL.
G6 KJY L C Cartwright, 18 High Causeway, Much Wenlock, TF13 6BZ [IO82RO, SO69]
G6 KKD N S Bancil, 41 St. Thomas Rd, Derby, DE23 8RF [IO92GV, SK33]
GI6 KKG R J M Baxter, 29 Largy Rd, Portglenone, Ballymena, BT44 8BX [IO64SU, C90]
GM6 KKL J Anderson, 16 Castledykes Rd, Kirkcudbright, DG6 4AN [IO74XT, NX65]
G6 KKM T B Bailey, 80 Homefield Rise, Orpington, BR6 0RW [JO01BI, TQ46]
G6 KKN P Clowes, 14 Derek Dr, Sneyd Green, Stoke on Trent, ST1 6BY [IO83WA, SJ84]
G6 KKO M F Dolby, 66 Winslow Dr, Immingham, Grimsby, DN40 2BZ [IO93VO, TA11]
G6 KLA J D Woodward, 13 Martins Cl, Wells, BA5 2ES [IO81QE, ST54]
GW6 KLC A G D Morris, Bodvel Hall, Pwllheli, Gwynedd, LL53 6DW [IO72SV, SH33]
G6 KLE B L Malone, 39 Goldsmiths Ln, Wallingford, OX10 0DJ [IO91KO, SU68]
G6 KLF A G Lythaby, 25 Greenhill Rd, Otford, Sevenoaks, TN14 5RR [JO01CH, TQ55]
G6 KLH R L Taylor, 57 Walnut Tree Rd, Shepperton, TW17 0RP [IO91SJ, TQ06]
G6 KLK R J Townshend, 128 Heath Dr, Chelmsford, CM2 9HQ [JO01FR, TL70]
GM6 KLL J H Paterson, 29 Newfield Cres, Hamilton, ML3 9DS [IO75XS, NS75]
G6 KLM Details withheld at licensee's request by SSL.[Op: Michael. Station located in central Wandsworth.]
G6 KLO C A Rae, 65 Westbourne Park, Bourne, PE10 9QS [IO92TS, TF02]
G6 KLQ J Laing, 80 Sandy Ln, Stretford, Manchester, M32 9BX [IO83UK, SJ79]
G6 KLR E W Shaw, 8 Riley Dr, Runcorn, WA7 4NZ [IO83PH, SJ58]
G6 KLS Details withheld at licensee's request by SSL.
G6 KLW Details withheld at licensee's request by SSL.
G6 KMA Details withheld at licensee's request by SSL.
G6 KMD R C Smith, 15 Chanston Ave, Kings Heath, Birmingham, B14 5BD [IO92BJ, SP07]
G6 KMG I Turnbull, 47 Norfolk Cres, Ormesby, Middlesbrough, TS3 0LZ [IO94JN, NZ51]
GM6 KMK S Windsor, Greenleys Croft, Gamrie, Banff, AB4 3JD [IO87TH, NJ72]
G6 KML S L V Vickers, 1 Green Hill Chase, Leeds, LS12 4HF [IO93ES, SE23]
G6 KMM Details withheld at licensee's request by SSL.
G6 KMQ C E Meadows, 47 Widney Ln, Solihull, B91 3LL [IO92CJ, SP17]
G6 KMT Details withheld at licensee's request by SSL.
G6 KMV Details withheld at licensee's request by SSL.
G6 KMY D A Nicolaou, 17 Carlyle Rd, Gosport, PO12 3NH [IO90KT, SU60]
G6 KNE J S Wright, Rhumbles, Coles Ln, Capel, Dorking, Surrey, RH5 5HT [IO91TD, TQ14]
G6 KNF Details withheld at licensee's request by SSL.
G6 KNI D M Williams, 22 Goldcrest Ct, Netherton, Huddersfield, HD4 7LN [IO93CO, SE11]
G6 KNK J E Solomon, 11 Angle Cl, Hillingdon, Uxbridge, UB10 0BS [IO91SM, TQ08]
G6 KNM R S Suttenwood, 9 Poppy Gdns, Abbots Heath, Colchester, CO2 8AE [JO01LU, TM02]
G6 KNU Details withheld at licensee's request by SSL.
GW6 KNX Details withheld at licensee's request by SSL.
G6 KOA Details withheld at licensee's request by SSL.
G6 KOE A W Reilly, 14 Carleton Gdns, Carleton, Poulton-le-Fylde, FY6 7PB [IO83LU, SD34]
GI6 KOI Details withheld at licensee's request by SSL.
GM6 KON T A Wilkins, Midtown, Freswick, Caithness, KW1 4XX [IO88LO, ND36]
GM6 KOR K H Osborne, 42 India St., Edinburgh, EH3 6HB [IO85JW, NT27]
G6 KOY Details withheld at licensee's request by SSL.
G6 KPD J P R J Perrett, 29 Bath View, Strattton on Fosse, Stratton on The Fosse, Bath, BA3 4RE [IO81SG, ST65]
G6 KPJ P G Vaughan, The Views, Bedford Rd West, Yardley Hastings, Northampton, NN7 1HB [IO92PE, SP85]
GM6 KPL A Y Wilson, 1 Union St., Newmilns, KA16 9BJ [IO75UO, NS53]
GW6 KPN Details withheld at licensee's request by SSL.
G6 KPR F G Trussler, 126 The Causeway, Petersfield, GU31 4LL [IO90MX, SU72]
G6 KPV P A Stapleton, 16 Falkland Cl, Boreham, Chelmsford, CM3 3DD [JO01GS, TL70]
G6 KPX A C Thorne, 4 Heddon Walk, Farnborough, Hants, GU14 8UG [IO91OH, SU85]
G6 KQ K Spicer, Gr Cottage, Dallinghoo, Woodbridge, Suffolk, IP13 0LR [JO02PD, TM25]
GW6 KQC L J McCarthy, Cliff Cottage, Hillgrove, Caswell, Swansea, SA3 4RE [IO71XN, SS68]
G6 KQD G D Morris, 20 Victoria Way, Stafford, ST17 0NU [IO82XS, SJ92]
G6 KQF J Laycock, 12 Catherine Slack, Brighouse, HD6 2LL [IO93CR, SE12]
G6 KQI Details withheld at licensee's request by SSL.
G6 KQJ H L Moon, 25 Shotley Gdns, Low Fell, Gateshead, NE9 5DP [IO94EW, NZ26]
G6 KQN G J Robertson, 24 Begonia Ave, Farnworth, Bolton, BL4 0DS [IO83TN, SD70]
G6 KQP Details withheld at licensee's request by SSL.
G6 KQS J A Newton, 47 Markfield, Ct Wood Ln, Selsdon, Croydon, CR0 9HL [IO91XI, TQ36]
G6 KQZ B M Wiseman, 307 Kempshott Ln, Kempshott, Basingstoke, RG22 5LY [IO91KF, SU64]
G6 KRC A F Hartland Kidderminster Rd, 22 Granville Crest, Kidderminster, DY10 3QS [IO82VJ, SO87]
GM6 KRD A C Dun, 21 Henry Bell St., Helensburgh, G84 7RF [IO76PA, NS38]
G6 KRF Details withheld at licensee's request by SSL.
GW6 KRK E C Karklins, Lonlas House, Lonlas, Skewen, Neath, SA10 6SD [IO81BP, SS79]
G6 KRN M C Everitt, Lark Rise, 5 Pine Cl, South Wonston, Winchester, SO21 3EB [IO91IC, SU43]
GM6 KRO D L Gray, 21 Queens Ave, Methilhill, Leven, KY8 2DD [IO86LE, NT39]
GW6 KRQ D H James, 20 Cae Nant Terr, Neath, SA10 6UP [IO81BP, SS79]
G6 KRS N P Ashall, 21 Buxton Ln, Droylsden, Manchester, M43 6HL [IO83WL, SJ89]
G6 KRY C Pieters, 32 Olde Farm Dr, Darby Green, Blackwater, Camberley, GU17 0DU [IO91OI, SU86]
G6 KSE H Dawson, 68 Whalley Rd, Great Harwood, Blackburn, BB6 7TF [IO83TT, SD73]
G6 KSF R I H Grosvenor, 8 Berrington Gdns, Tenbury Wells, WR15 8ET [IO82QH, SO56]
G6 KSH A Henshaw, 5 Bosworth Dr, Newthorpe, Nottingham, NG16 3RF [IO93IA, SK44]
G6 KSK A J Hodgson, 33 Higham Rd, Frindsbury, Wainscott, Rochester, ME3 8BE [JO01GJ, TQ77]
G6 KSM Details withheld at licensee's request by SSL.
G6 KSN A R Knott, 65 Bolton Ct, Ocker Hill, Tipton, DY4 0UU [IO82XN, SO99]
G6 KSO P W Lash, 63 Sextant Cl, Murdishaw, Runcorn, Ches, WA7 6DR [IO83QH, SJ58]
G6 KSQ R N Olding, Woodland View, Poole Rd, Lytchett Matravers, Poole, BH16 6AF [IO80XS, SY99]
G6 KSR F G Patman, Northcote, 31 Church Rd, Skellingthorpe, Lincoln, LN6 5UW [IO93QF, SK97]
G6 KSU P R Rogers, 36 Eastling Cl, Gillingham, ME8 6XT [JO01HI, TQ86]
G6 KSV A J Sayers, 145 Campkin Rd, Cambridge, CB4 2NP [JO02BF, TL46]
G6 KTB W P Curtis, Innsbruck, Trevingey Cres, Redruth, TR15 3DF [IO70JF, SW64]
G6 KTE D J Brunt, 31 The Green, Kingsley, Stoke on Trent, ST10 2AG [IO93AA, SK04]
G6 KTF Revd J A Carr, 7 Mickleby Way, Meir Heath, Stoke on Trent, ST3 7RU [IO82WX, SJ94]
G6 KTG D R Clements, Orchard House, 89 Oyster Ln, Byfleet, West Byfleet, KT14 7JF [IO91SI, TQ06]
G6 KTK C Handley, 21 Marton Dr, Wellington, Telford, TF1 3HL [IO82RQ, SJ61]
G6 KTO J E Martyn Clark, 12 St. Thomas Rd, Lythan, St. Annes, Lytham St. Annes, FY8 1JL [IO83LR, SD32]
GM6 KTP K F M Morrison, 8 St. Helena Cres, Hardgate, Clydebank, G81 5PD [IO75TW, NS57]
G6 KTW Details withheld at licensee's request by SSL.
G6 KTX A D King, 2 Longstaff Gdns, Fareham, PO16 7RR [IO90JU, SU50]
G6 KUA R P Matthews, 7 West Ave, Aldwick, Bognor Regis, PO21 3QN [IO90PS, SZ99]
GM6 KUC Details withheld at licensee's request by SSL.
G6 KUG Details withheld at licensee's request by SSL.
G6 KUH Details withheld at licensee's request by SSL.
G6 KUI P J Walker, 23 Denstone Dr, Alvaston, Derby, DE24 0HZ [IO92GV, SK33]
G6 KUJ F Moulding, 28 Woodbine Rd, Bolton, BL3 3JH [IO83SN, SD70]
GM6 KUL R Shields, 1 The Barn, Olaf Rd, Kyleakin, Isle of Skye, IV41 8PJ [IO77DG, NG72]
G6 KUN D J Richardson, 5 Gleneagles, Orton Waterville, Peterborough, PE2 5UZ [IO92UN, TL19]
G6 KUS P D Quilter, Wheatsheaf Cottage, Shop Rd, Little Bromley, Manningtree, Essex, CO11 2PY [JO01MV, TM02]
G6 KVE C D Payne, 57 Compton Ave, Leagrave, Luton, LU4 9AY [IO91SV, TL02]
G6 KVG S C R Reid, 1 Caenwood Rd, Ashtead, KT21 2JA [IO91UH, TQ15]
G6 KVI B Gosling, 48 Barbrook Ln, Tiptree, Colchester, CO5 0EF [JO01JT, TL81]
G6 KVK G H Howell, 29 Blackmore, Letchworth, SG6 2SX [IO91VX, TL23]
G6 KVR P N Roberts, 30 Baldwins Ln, Hall Green, Birmingham, B28 0QX [IO92BK, SP18]
GI6 KVS H H Porter, 30 Twinburn Rd, Monkstown, Newtownabbey, BT37 0EL [IO74AQ, J38]
G6 KVX G E Snaith, 14 Linburn Dr, Greenfields Dene, Bishop Auckland, DL14 0RG [IO94DP, NZ12]
G6 KVY S D Trotter, 61 Trinity Rd, Billericay, CM11 2RY [JO01FO, TQ69]
G6 KWA D M King, 20 Trinity Cl, Haslingfield, Cambridge, CB3 7LS [JO02AD, TL45]
G6 KWC K J Beesley, 30 Kentwell, Riverside, Tamworth, B79 7UB [IO92DP, SK10]
G6 KWJ Details withheld at licensee's request by SSL.
G6 KWO Details withheld at licensee's request by SSL.
GW6 KWU R M Adamson, 55 Sandy Ln, Garden City, Deeside, CH5 2JF [IO83LF, SJ36]
G6 KWZ J G Manning, 280 Ledbury Rd, Hereford, HR1 1QL [IO82PB, SO54]
G6 KXB R N Linzey, 29 Arkle Ct, Alnwick, NE66 1BS [IO95DJ, NU11]
G6 KXD M G Parker, Hazel House, Talkin, Brampton, Cumbria, CA8 1LE [IO84PV, NY55]
G6 KXJ K Turner, 16 Orford St., Wavertree, Liverpool, L15 8HX [IO83MJ, SJ38]
G6 KXK G G Smith, 26 Send Cl, Send, Woking, GU23 7EL [IO91RG, TQ05]
G6 KXN R P Perry, 6 Morgan Cl, Arley, Coventry, CV7 8PR [IO92FM, SP28]
GM6 KXP D M C Flanagan, Ryan Mar, Stair Dr, Stranraer, DG9 8EY [IO74LV, NX06]
G6 KXW A J Blair, Gotherment House, Wigmore, Leominster, Herefordshire, HR6 9UF [IO82NH, SO46]
G6 KYD Details withheld at licensee's request by SSL.
G6 KYE M M E Davis, 86 Upper Shaftesbur, Ave, Highfield, Southampton, SO2 3RT [IO90HW, SU41]
G6 KYK E M Blackburn, 15 Chapel Cl, Warwick Bridge, Carlisle, CA4 8RH [IO84OV, NY45]
G6 KYR A R G Peto, 16 The Lugger, Portscatho, Truro, TR2 5HE [IO70MD, SW83]
G6 KYW G Murray, 33 Greenlaw, West Denton, Newcastle upon Tyne, NE5 5DD [IO94DX, NZ16]
G6 KYX Details withheld at licensee's request by SSL.
G6 KZA T E Deakin, 10 Begonia Cl, Hinckley, LE10 2SS [IO92HM, SP49]
G6 KZF Details withheld at licensee's request by SSL.
G6 KZI R W Gregory, 75 Station Rd South, Belton, Great Yarmouth, NR31 9LZ [JO02TN, TG40]
G6 KZJ Details withheld at licensee's request by SSL.
G6 KZR K Knutton, 51 Derwent Rd, Honley, Huddersfield, HD7 2EL [IO93CO, SE11]
G6 KZU S J Cain, 12 Lockswell Cl, Pewsham, Chippenham, SN15 3UR [IO81WK, ST97]
GW6 KZZ Details withheld at licensee's request by SSL.[Op: C J Dixon. Station located near Ash.]
G6 LAE J E Clifton, 6 Chester Cl, Greenham, Newbury, RG14 7RR [IO91IJ, SU46]
G6 LAH Details withheld at licensee's request by SSL.
G6 LAJ A R Gunbie, 1 Thorney Cl, Millers Green, Lower Earley, Reading, RG6 3AF [IO91NK, SU77]
G6 LAM Details withheld at licensee's request by SSL.
G6 LAU D R Tanswell, Highstead Farmhouse, Highstead Cross, Bradford - Holemoor, Nr.Holsworthy, Devon, EX22 7AA [IO70VT, SS40]
G6 LAW Details withheld at licensee's request by SSL.
G6 LAX E N D Wellington, 25 The Spinney, Swanley, BR8 7YA [JO01CJ, TQ56]
G6 LBA Details withheld at licensee's request by SSL.
G6 LBC Details withheld at licensee's request by SSL.
G6 LBE J Massey, 10 Rapley Ave, Storrington, Pulborough, RH20 4QL [IO90SW, TQ01]
G6 LBG N J Orgill, 32 Upland Ave, Chesham, HP5 2EB [IO91QR, SP90]
G6 LBJ P E Shadbolt, 54 Madeira Rd, Palmers Green, London, N13 5SS [IO91WO, TQ39]
G6 LBL J D Scarr, Kawana, 40 Northway, Thatcham, RG18 3FG [IO91IJ, SU56]
G6 LBO K Batty, 19 Breckland Cl, St. Pauls Garden, Stalybridge, SK15 2QQ [IO83XL, SJ99]
G6 LBQ A S Hunter, 16 Farmside Ave, Irlam, Manchester, M44 6WX [IO83TK, SJ79]
G6 LBR A A Ledger, 92 Crawford Cl, Freshbrook, Swindon, SN5 8PU [IO91BN, SU18]
G6 LBV W T Large, 3 Hall Dr, Willaston, Nantwich, CW5 6NA [IO83SB, SJ65]
G6 LCC Details withheld at licensee's request by SSL.[Op: Mrs S R Metcalfe. Station located near Morden.]
G6 LCH R Myers, 1 Spitfire Cl, Marske By The Sea, Redcar, TS11 6NH [IO94LO, NZ62]
G6 LCL T G Mallett, 26 Ullswater Rd, Chester-le-Street, DH2 3HG [IO94EU, NZ25]
G6 LCM D Mangles, 46 Cedar Cres, Wear Valley View, Willington, Crook, DL15 0DA [IO94DR, NZ23]
G6 LCS J A McNeill, 9 Stonyford Rd, Sale, M33 2FJ [IO83UK, SJ79]
G6 LCU J W C P Retter, 12 Palmerstone Rd, Grays, Essex, RM16 1YR [JO01EL, TQ67]
G6 LD I C I Lamb, The Hirsel, Jumble Wood, Fenay Bridge, Huddersfield, HD8 0AN [IO93DP, SE11]
G6 LDA J Round, 53 Furlong Ln, Halesowen, B63 2TB [IO82XL, SO98]
G6 LDH J Cliff, 36 Heronscroft, Putnoe, Bedford, MK41 9LP [IO92SD, TL05]
G6 LDI Details withheld at licensee's request by SSL.
G6 LDJ R Wilkinson, 2 Conway Ave, Billingham, TS23 2HX [IO94IO, NZ42]
G6 LDL Details withheld at licensee's request by SSL.
G6 LDM D J Shippen, 5 Shelley Ave, Wincham, Northwich, CW9 6PH [IO83SG, SJ67]
G6 LDN Details withheld at licensee's request by SSL.
G6 LDO C R Seeney, 91 Dovehouse Cl, Eynsham, Witney, Oxon, OX8 1EW [IO91HS, SP40]
G6 LDP D I Scott, 7 Greenfield Mount, Wrenthorpe, Wakefield, WF2 0TJ [IO93FQ, SE32]
G6 LDV R E Unsworth, 6 Teesdale Ave, Davyhulme, Urmston, Manchester, M41 8BY [IO83TK, SJ79]
G6 LDW J Tottle, 20 Abbey Rd, Cheadle, SK8 2JW [IO83VJ, SJ88]
G6 LDY J M Seddon, 11 Hilda St., Leigh, WN7 5DG [IO83RM, SD60]
G6 LEB T Leader-Chew, 5 The Dingle, West Green, Crawley, RH11 7JD [IO91VC, TQ23]
G6 LED M K Little, 11 Courtway, Rodborough, Stroud, GL5 3TR [IO81VR, SO80]
G6 LEI S G Meadwell, 42 Braunston Rd, Oakham, LE15 6LD [IO92PQ, SK80]
G6 LEL J H D Morris, 36 Elizabeth Rd, Moseley, Birmingham, B13 8QJ [IO92BK, SP08]
G6 LES I Lloyd, Woodvale, 2 Lime Gr, Elton, Chester, CH2 4PX [IO83OG, SJ47]
G6 LEU D W Last, The Millers Cottage, Ruan Highlanes, Truro, Cornwall, TR2 5LE [IO70MF, SW93]
G6 LEY D C R Miller, 44 Long Ln, Ickenham, Uxbridge, UB10 8TA [IO91SN, TQ08]
GM6 LEZ J McDermott, 12 Margaret St., Greenock, PA16 8AS [IO75OW, NS27]
G6 LFA Details withheld at licensee's request by SSL.
G6 LFD J H Corderoy, 1 Alandale Dr, Pinner, HA5 3UP [IO91TO, TQ19]
G6 LFG J G Bradbury, 281 Peter St., Macclesfield, SK11 8EX [IO83WG, SJ97]
G6 LFJ J J Aslan, 4 Denmark St., Oxford, OX4 1QS [IO91JR, SP50]
G6 LFQ Details withheld at licensee's request by SSL.
G6 LFR L A Carr, 29 Hill Dr, Whaley Bridge, Stockport, Ches, SK12 7BH [IO83XI, SJ98]
G6 LFT Dr G A Cooke, 35 Coniston Rd, Earlsdon, Coventry, CV5 6GU [IO92FJ, SP37]
G6 LFW J P Ford, 24 Tonstall Rd, Epsom, KT19 9DP [IO91UI, TQ26]
G6 LGC E W Ingarfill, 44 Pitts Ln, Earley, Reading, RG6 1BU [IO91MK, SU77]
G6 LGH A Rushton, 698 Walsall Rd, Great Barr, Birmingham, B42 1EY [IO92AM, SP09]
G6 LGM I Rogers, 47 Bitteswell Rd, Lutterworth, LE17 4EN [IO92JL, SP58]
G6 LGN J M Redhead, 22 Woodgrove Park, Polgoth, Polgooth, St. Austell, PL26 7BN [IO70OH, SW95]
G6 LGO N O Connor, 50 Hannell Rd, Fulham, London, SW6 7RB [IO91VL, TQ27]
G6 LGR A J Picot, 14 Ringshall Rd, St. Pauls Cray, Orpington, BR5 2LZ [JO01BJ, TQ46]
G6 LGW A S Pollard, 75 Ridgeway Ave, Dunstable, LU5 4QL [IO91RV, TL02]
G6 LGZ Details withheld at licensee's request by SSL.
G6 LHA F G Priestnall, 56 Badger Gate, Threshfield, Skipton, BD23 5EN [IO84XB, SD96]
G6 LHB E G Pollard, New Inn, 77 Bath St., Ilkeston, DE7 8AJ [IO92IX, SK44]
GW6 LHD Details withheld at licensee's request by SSL.
G6 LHG B K O'Shea, 51 Beridge Rd, Halstead, CO9 1JZ [JO01HW, TL83]
G6 LHI P F Nicholson, 25B Glenholme Rd, Farsley, Pudsey, LS28 5BY [IO93DT, SE23]
G6 LHJ Details withheld at licensee's request by SSL.
G6 LHQ R P Harber, 7 Hamilton Ave, Cobham, KT11 1AU [IO91SH, TQ16]
G6 LIB J M Baker, 5 Parsonage Cl, Bishops Stortford, CM23 4LL [JO01BU, TL42]
G6 LIC B Keedy, 14 Lingwell Gate Cres, Wakefield, WF1 2PA [IO93FR, SE32]
G6 LIJ D Chilton, 12 Limpton Gate, Yarm, TS15 9JA [IO94HL, NZ41]
G6 LIK L Clark, 346 Alice St., South Shields, NE33 5PH [IO94GX, NZ36]
G6 LIQ Details withheld at licensee's request by SSL.
G6 LIY R N Deakin, 7 Reservoir Rd, Stockport, SK3 9BZ [IO83WJ, SJ88]
G6 LJ S K Lewer, 54 Chaldon Common Rd, Chaldon, Caterham, CR3 5DD [IO91WG, TQ35]
G6 LJA J Webster, 13 The Bank, Eccleshill, Bradford, BD10 8BL [IO93DT, SE13]
G6 LJB H G Williams, 5 Watford Rd, New Mills, High Peak, SK22 4EJ [IO93AJ, SK08]
G6 LJC Details withheld at licensee's request by SSL.
GM6 LJE R J Waitt, Orchard Cottage, Canonbie, Dumfriesshire, DG14 0RZ [IO85MC, NY47]
G6 LJF J L White, Taeping, Manse Chase, Maldon, CM9 5EA [JO01IR, TL80]
G6 LJH M S W Wilson, Alcrest, Chapel Ln, Denford, Kettering, NN14 4EA [IO92RJ, SP97]

G6 LJJ Details withheld at licensee's request by SSL.
G6 LJR D P Twyman, Dunlin House, 77 Essex Rd, Maldon, CM9 6JH [JO01IR, TL80]
G6 LJU J G J Whitehouse, 1 Eagles Nest, Sandhurst, Camberley, Surrey, GU17 8RR [IO91OH, SU85]
G6 LKA B E Woolnough, 57 Cranborne Rd, Potters Bar, EN6 3AB [IO91VQ, TL20]
G6 LKB D C Warburton, 36 Bigland Dr, Ulverston, LA12 9PD [IO84KE, SD27]
G6 LKG R Milne, 9 Brunstath Cl, Barnston, Wirral, L60 1UH [IO83LH, SJ28]
G6 LKH C J Dunlop, 32 Ct Way, Twickenham, TW2 7SN [IO91UK, TQ17]
G6 LKJ J C Depledge, 96 Compstall Rd, Romiley, Stockport, SK6 4DE [IO83XJ, SJ99]
G6 LKM Details withheld at licensee's request by SSL.
G6 LKQ E W Fry, Gr House, Elm Gr, Maidenhead, SL6 6AD [IO91PM, SU88]
G6 LKS D A Alldridge, 31 William Sim Wood, Winkfield Row, Bracknell, RG42 6PW [IO91PK, SU87]
G6 LKT Details withheld at licensee's request by SSL.
G6 LKV G D Ashbee, 6 The Green, Wimbledon Common, London, SW19 5AZ [IO91VK, TQ27]
G6 LKW T G Ashbee, 6 The Green, Wimbledon Common, London, SW19 5AZ [IO91VK, TQ27]
G6 LKY S J Bennetts, Bendel, Carthew Way, St. Ives, TR26 1RJ [IO70GF, SW54]
G6 LKZ D J Bentley, 4 Highway, Edgcumbe Park, Crowthorne, RG45 6HE [IO91OI, SU86]
G6 LLB Details withheld at licensee's request by SSL.
G6 LLD G S Bell, 4 Dallymore Dr, Bowburn, Durham, DH6 5ES [IO94FR, NZ33]
G6 LLF P J Bennett, 1 The Briars, St. Michaels Rd, Newcastle, ST5 9PU [IO83VA, SJ84]
G6 LLG M Broadway, Birchwood, 69 The Brambles, Crowthorne, RG45 6EF [IO91OI, SU86]
G6 LLL D Burrows, 32 Whitfield Cross, Glossop, SK13 8NW [IO93AK, SK09]
G6 LLP R E Farey, 45 Malvern Cl, Woodley, Reading, Berks, RG5 4HL [IO91NK, SU77]
G6 LLR S A Raistrick, 135 Lister Ave, Bradford, BD4 7QU [IO93DS, SE13]
G6 LLT T O Raper, Hawthorne Cottage, 5 Main St, Drax, Selby, North Yorks, YO8 8PB [IO93MR, SE62]
G6 LMB P Steadman, 31 Walter May House, Whitehawk Rd, Brighton, BN2 5GF [IO90WT, TQ30]
G6 LMC P W Webb, 63 Trinity Rd, Halstead, CO9 1ED [JO01HW, TL83]
GW6 LMF P I Edwards, 58 Park Ave, Whitchurch, Cardiff, CF4 7AN [IO81JM, ST18]
G6 LMG Details withheld at licensee's request by SSL.
GW6 LMI J Evans, 91 Queens Ave, Flint, CH6 5JP [IO83KF, SJ27]
G6 LMJ G Eardley, Dawn Hill, Little Moss, Scholar Green, Stoke on Trent, ST7 3BL [IO83VC, SJ85]
G6 LMK Details withheld at licensee's request by SSL.
G6 LMR K Fisher, 26 Manila St., Sunderland, SR2 8RS [IO94HV, NZ45]
G6 LMU A C Franklin, 51 Banks Rd, Pound Hill, Crawley, RH10 7BS [IO91WC, TQ23]
G6 LMW J P Coogan, Fairlands, 2 Holbeck Park Ave, Barrow in Furness, LA13 0RE [IO84JC, SD27]
G6 LMZ Details withheld at licensee's request by SSL.
G6 LNF N J Clare, 9 Southfield Ave, Weymouth, DT4 7QN [IO80SO, SY68]
G6 LNL I R Dobson, 1 Tempest Rd, Seaham, SR7 7BA [IO94IU, NZ44]
G6 LNP J E B Davage, 31 Springfield Ave, Holbury, Southampton, SO45 2LN [IO90HT, SU40]
G6 LNS J M Duxbury, Pinewood Garden Centre, Wallace Ln, Forton, Preston, PR3 0BB [IO83OX, SD45]
G6 LNU J Durban, 62 Westfield Way, Charlton Heights, Wantage, OX12 7EP [IO91HO, SU48]
G6 LNV J W D Cunliffe, 142 Hall Rd, Hull, HU6 8SB [IO93TS, TA03]
G6 LNW J S Wheelhouse, 105 Leads Rd, Hull, HU7 0BX [IO93US, TA13]
G6 LOB R W F Stanley, 22 Creighton Rd, Millbrook, Southampton, SO15 4JF [IO90GV, SU31]
G6 LOC T D Stirrup, 23 Round Wood, Penwortham, Preston, PR1 0BN [IO83PS, SD52]
G6 LOG F G J Reed, 21 Goffenton Dr, Oldbury Ct, Fishponds, Bristol, BS16 2QB [IO81RL, ST67]
G6 LOH Details withheld at licensee's request by SSL.[Op: Julian Tether. Station located 5km S Towcester.]
G6 LOJ N D Pettit, 10 Broom Rd, Lakenheath, Brandon, IP27 9ES [JO02GJ, TL78]
G6 LOY R E Payne, 4 Featherdell, Hatfield, AL10 8DD [IO91VS, TL20]
G6 LPC A J F Samways, 41 Bodmin Cl, Eastbourne, BN20 8HZ [JO00DS, TV59]
G6 LPD A J W Tucker, 63 Oakes Rd, Bury St. Edmunds, IP32 6PU [JO02IG, TL86]
G6 LPF Details withheld at licensee's request by SSL.[Op: S G Turner. Station located in London SE6.]
G6 LPG S P Taylor, 305B Wells Rd, Bristol, BS4 2PT [IO81RK, ST67]
G6 LPS T F Biddle, 10 Erwood Cl, Headless Cross, Redditch, B97 5XD [IO92AH, SP06]
G6 LPX P Brown, 5 Fairview Cl, Amington, Tamworth, B77 3LA [IO92EP, SK20]
G6 LPZ Details withheld at licensee's request by SSL.[Station located in mid-Sussex.]
G6 LQD Details withheld at licensee's request by SSL.
G6 LQE D T Byrom, 206 Didsbury Rd, Heaton Mersey, Stockport, SK4 2AA [IO83VJ, SJ89]
G6 LQG D Bate, 24 Finchdean Cl, Meir Heath, Stoke on Trent, ST3 7UT [IO82WX, SJ94]
G6 LQI N S Bird, 25 Sarisbury Gate, Dove Gdns, Park Gate, Southampton, Hants, SO31 7FP [IO90IU, SU50]
G6 LQM G R A H Barker, 99 Sheffield Rd, Wymondham, NR18 0HS [JO02NN, TG10]
G6 LQP D A Brown, Keepers Nook, Holmes Chapel Rd, Somerford, Congleton, CW12 4SN [IO83UE, SJ86]
G6 LQR B Walker, 5 Cochrane Terr, Willington, Crook, DL15 0HN [IO94DR, NZ13]
G6 LQV Details withheld at licensee's request by SSL.[Op: F J V Martin.]
G6 LQY L E Walters, 25 Lime Gr, Chaddesden, Derby, DE21 6WL [IO92GW, SK33]
G6 LRA R J Martin Colchester Borg, 2282 PO Box, Town Hall, Colchester, CO1 1AP [JO01KV, TL92]
G6 LRD R Taylor, 89 St. Johns Rd, Pelsall, Walsall, WS3 4EZ [IO92AP, SK00]
G6 LRG Details withheld at licensee's request by SSL.
G6 LRL S G Jones, 6 Leacroft Cl, Aldridge, Walsall, WS9 8RX [IO92BO, SK00]
G6 LRQ Details withheld at licensee's request by SSL.
G6 LRT C O L A Johnson, 52 Evesham Rd, Cookhill, Alcester, B49 5LJ [IO92AF, SP05]
G6 LRU R B Jones, 53 Wavertree Rd, Blacon, Chester, CH1 5AF [IO83ME, SJ36]
G6 LSB N G Key, Thimble Cottage, Front St., Denford, Kettering, NN14 4EG [IO92RJ, SP97]
G6 LSC M G Kitchener, 5 Whinbush Gr, Hitchin, SG5 1PT [IO91UW, TL12]
G6 LSD P Kerry, 63 Holcombe St., Derby, Derbyshire, DE23 8JA [IO92GV, SK33]
GM6 LSG A S Kay, 29 Lanark Rd, Ravenstruther, Lanark, ML11 7SS [IO85DQ, NS94]
GW6 LSL S M J Wood, 2 Radyr Rd, Llandaff North, Cardiff, CF4 2FU [IO81JM, ST17]
G6 LST D Rhodes, 1 Tanpit Cttgs, Tanpit Ln, Winstanley, Wigan, WN3 6JY [IO83QM, SD50]
G6 LTB P A Townrow, 64 Millham Rd, Bishops Cleeve, Cheltenham, GL52 4BG [IO81XW, SO92]
G6 LTD P C Sutton-Atkins, 3 Little Mill Cttgs, Ulting, Maldon, CM9 6PZ [JO01HR, TL80]
G6 LTK K D Wilson, 44 Campbell Rd, Caterham, CR3 5JN [IO91WH, TQ35]
G6 LTL H R Wilders, 105 Greenstead Rd, Colchester, CO1 2ST [JO01LV, TM02]
G6 LTM A M Wanford, 4 Willows Cl, Tydd St. Mary, Wisbech, PE13 5QR [JO02BR, TF41]
G6 LTR J L Warner, 64 Stanfell Rd, Clarendon Park, Leicester, LE2 3GA [IO92KO, SK50]
G6 LTZ Details withheld at licensee's request by SSL.[Op: A D Smith, c/o 59 Willow Grove, Old Stratford, Milton Keynes, MK19 6AY. Station dismantled and removed.]
G6 LUD Details withheld at licensee's request by SSL.
G6 LUE T O Yates, 5 Manor Garth, Kellington, Goole, DN14 0NW [IO93KR, SE52]
G6 LUF A C Yates, 153 Fox Ln, Leyland, Preston, PR5 1HE [IO83PQ, SD52]
G6 LUH Details withheld at licensee's request by SSL.
G6 LUJ R G Perry, Thornaby, Queen St., Colyton, EX13 6JU [IO80LR, SY29]
G6 LUK J Russell, 60 Berrington Rd, Nuneaton, CV10 0LB [IO92FM, SP39]
G6 LUM J S Papworth, 339 Gayfield Ave, Withymoor Village, Brierley Hill, DY5 3JE [IO82WL, SO98]
G6 LUO B M Maynard, 11 Denham Rd, Canvey Island, SS8 9HB [JO01GM, TQ78]
G6 LUP V F Morhall, 52 Church Rd, Boreham, Chelmsford, CM3 3ER [JO01GS, TL70]
G6 LUZ S Morgan, Bank House, Bunsley Bank, Audlem, Crewe, CW3 0HS [IO82SX, SJ64]
G6 LVC J M Shergold, 35 Orchard Gr, New Milton, Hants, BH25 6NZ [IO90ER, SZ29]
G6 LVE A Nicholson, Buena Vista, New Rd, Old Snydale, Pontefract, WF7 6HD [IO93HQ, SE42]
G6 LVG S J Normandale, 5 The Beacon, Ilminster, TA19 9AH [IO80NW, ST31]
G6 LVJ R G P Hickey, Mallorin, Blackfield Rd, Fawley, Southampton, SO45 1EG [IO90HT, SU40]
G6 LVM C J Holderness, 7 Oakfield Ave, Clayton-le-Moors, Accrington, BB5 5XG [IO83TS, SD73]
G6 LVN R W Hope, 15 Birchfield Gdns, Mulbarton, Norwich, NR14 8BT [JO02ON, TG10]
G6 LVO G A Haydock, 21 Willian Way, Letchworth, SG6 2HQ [IO91VX, TL23]
G6 LVS D M Howard, 21 Nelson Cres, Ryde, PO33 3QN [IO90JR, SZ59]
G6 LVT C Harvey, Primrose Hill Cottage, 619 West St., Crewe, CW2 8SH [IO83SC, SJ65]
G6 LVW Details withheld at licensee's request by SSL.
G6 LWA C A Hall, 147 Gordon Ave, Camberley, GU15 2NR [IO91PH, SU85]
G6 LWC R W Hardman, 4 Alverston Rd, Wallasey, Merseyside, L44 9AA [IO83LJ, SJ39]
G6 LWE G Hall, 26 Chapel Walk, Lowton St. Marys, Lowton, Warrington, WA3 1EE [IO83RL, SJ69]
G6 LWH Details withheld at licensee's request by SSL.[Station located near Newcastle-under-Lyme.]
G6 LWK M Horsfield, 13 St. Leonards Way, Ardsley, Barnsley, S71 5BS [IO93GN, SE30]
G6 LWO A P Lee, Newlands, Upfold Ln, Cranleigh, GU6 8PD [IO91SD, TQ04]
G6 LWY M J McKay, 72 Rupert Rd, Chaddesden, Derby, DE21 4ND [IO92GW, SK33]
G6 LWZ L J Miles, 23 Redacre Rd, Boldmere, Sutton Coldfield, B73 5EA [IO92BN, SP19]
G6 LX R L Glaisher, 279 Addiscombe Rd, Croydon, CR0 7HY [IO91XI, TQ36]
G6 LXF P D Duley, 4 Brean Rd, Stafford, ST17 0PA [IO82XS, SJ92]
G6 LXI J H Dyson, 2 Grange Rd, Barnton, Northwich, CW8 4PE [IO83RG, SJ67]
G6 LXJ R J Dawson, 20 St. Johns Rd, Wembley, HA9 7JD [IO91UN, TQ18]
G6 LXL D J Ellis, Goosters Green, Hope Bagot, nr Ludlow, Shropshire, SY8 3AE [IO82QI, SO57]
G6 LXP D J English, 6 Station Ave, Rayleigh, SS6 9AD [JO01HO, TQ89]
G6 LXU S G Westall, 4 South View, Great Harwood, Blackburn, BB6 7NL [IO83TS, SD73]
G6 LXW J Weigh, 167 Farm View Rd, Kimberworth, Rotherham, S61 2BL [IO93HK, SK39]
G6 LXX Details withheld at licensee's request by SSL.
G6 LYA P J Whysall, 12 Neville Dr, Fishlake Meadows, Romsey, SO51 7RP [IO90GX, SU32]
G6 LYE G Whiles, 18 Ferndale Rd, Off Brownshore Ln, Essington, Wolverhampton, WV11 2JG [IO82XP, SJ90]
G6 LYK D Latham, 2 Purchase Ave, Loscoe, Heanor, DE75 7GB [IO93HA, SK44]
G6 LYM D J Miller, 21 Sunningdale Ave, Leigh on Sea, SS9 1JY [JO01IN, TQ88]
G6 LYY Details withheld at licensee's request by SSL.
G6 LYZ R Anderson, 10 Clifton Ave, Horbury, Wakefield, WF4 6JW [IO93FP, SE21]
G6 LZB P D Adams, 464 Whippendell Rd, Watford, WD1 7PT [IO91TP, TQ09]
GW6 LZH J L Adams, 35 Braunton Ave, Llanrumney, Cardiff, CF3 9HW [IO81KM, ST28]
G6 LZJ Details withheld at licensee's request by SSL.
G6 LZM G Beddington, 11 Happisburgh Rd, North Walsham, NR28 9HA [JO02QT, TG23]
GM6 LZO Details withheld at licensee's request by SSL.
G6 LZU Details withheld at licensee's request by SSL.
G6 LZV G W Barnett, 9 Marlow Cl, Daventry, NN11 4HN [IO92JG, SP56]
G6 LZX B J Broad, 27 Mellow Purgess, Laindon, Basildon, SS15 5UY [JO01FN, TQ68]
G6 LZZ J Bolton, Huds House, Cowgill, Dent, Cumbria, LA10 5TQ [IO84TG, SD78]
G6 MAA G L Bishop, Oyston Lodge, Lynstone Rd, Bude, Cornwall, EX23 8LR [IO70RT, SS20]
G6 MAC B McDonnell, 68 Chaigley Rd, Longridge, Preston, PR3 3TQ [IO83QU, SD63]
G6 MAE M J Baxter, 69 Lincoln Rd, Skegness, PE25 2ED [JO03ED, TF56]
G6 MAF Details withheld at licensee's request by SSL.
G6 MAJ A J Mulvaney, 38 Ramwells Brow, Bromley Cross, Bolton, BL7 9LL [IO83TO, SD71]
G6 MAO L H West, 70 Werneth Rd, Simmondley, Glossop, SK13 9NJ [IO93AK, SK09]
G6 MAP Details withheld at licensee's request by SSL.
G6 MAR G E Wratten, 8 Hood Cl, Eastbourne, BN23 6BS [JO00DS, TQ60]
G6 MAW R D Underwood, 41 Dovecote Rd, Eastwood, Nottingham, NG16 3EY [IO93IA, SK44]
G6 MAY D W Pool, 6 Rivett Cl, Clothall Common, Baldock, SG7 6TW [IO91VX, TL23]
G6 MB F Hicks-Arnold, Ingleside, Junction Rd, Alderbury, Salisbury, SP5 3AZ [IO91DB, SU12]
G6 MBD J R Durston-Wyatt, 62 St. Johns Rd, Epping, CM16 5DP [JO01BQ, TL40]
G6 MBG W T Sheppard, 4 St. Marys Cl, Thrapston, Kettering, NN14 4PR [IO92RJ, TL07]
G6 MBH W G Stiling, 11 Carrol Gr, Cheltenham, GL51 0PP [IO81WV, SO92]
G6 MBI D E Stainton, Tilton House, 39 Redland Gr, Carlton, Nottingham, NG4 3ET [IO92KX, SK64]
G6 MBL M B S Snow, 32 Orchard Ave, Worthing, BN14 7PY [IO90TT, TQ10]
G6 MBR I McIver Mid Beds R/Net, 31 Harts Hill, Bedford, MK41 9AL [IO92SD, TL05]
G6 MBW P L Santillo, 173 Holcombe Ln, Bathampton, Bath, BA2 6UU [IO81UJ, ST76]
G6 MC J C Martin, 151 Park Rd, Bingley, BD16 4EJ [IO93CU, SE13]
G6 MCB M C Baldry, 10 Kingfisher Ct, Lowestoft, NR33 8PJ [JO02UL, TM59]
G6 MCC L C Crompton, 12 Hartington Rd, Preston, PR1 8PP [IO83PS, SD52]
G6 MCD Details withheld at licensee's request by SSL.
G6 MCE P G F Garde, 21 Leicester Ave, Timperley, Altrincham, WA15 6HR [IO83UJ, SJ78]
G6 MCG C J Garnham, 1 Ennerdale Cl, Felixstowe, IP11 9SS [JO01QX, TM33]
G6 MCI A J Gledhill, 14 Butt Ln, Milton, Cambridge, CB4 6DG [JO02BF, TL46]
G6 MCJ Details withheld at licensee's request by SSL.[Op: A S V Grainger-Allen. Station located near St Austell.]
G6 MCN A W Gillespie, Elmtree Cottage, Chilbolton, Stockbridge, Hants, SO20 6BA [IO91GD, SU33]
G6 MCO J M Gardner, Rivington, 2 Dibden Lodge Cl, Hythe, Southampton, SO45 6AY [IO90HU, SU40]
G6 MCQ J A Grane, 15 Pinelands Way, Osbaldwick, York, YO1 3QJ [IO93LW, SE65]
G6 MCS P Gleave, 22 Vose Cl, Great Sankey, Warrington, WA5 1EW [IO83QJ, SJ58]
G6 MCT H C Glanville, 9 Austin Ave, North Prospect, Plymouth, PL2 2LB [IO70WJ, SX45]
GM6 MCV J C McVicar, 2 Lilliardsedge Par, Mr Ancrum, Roxburghshire, TD8 6TZ [IO85QM, NT62]
G6 MCX P N J Garland, 4 Bronte Cl, Testbourne Farm, Totton, Southampton, SO40 8SR [IO90FV, SU31]
G6 MCY M J Goddard, 65 Langley Hall Rd, Solihull, B92 7HE [IO92CK, SP18]
GM6 MD A L Dunn Clyde Coast Contest Club, Shankston, Patna, Ayr, KA6 7LD [IO75RJ, NS41]
G6 MDB J M Gamble, 283/285 Halifax Rd, Todmorden, OL14 5SQ [IO83XR, SD92]
G6 MDC M Green, 9 Greencroft Ave, Northowram, Halifax, HX3 7EP [IO93CR, SE12]
G6 MDF M S Pethen, 1059 Garratt Ln, Tooting, London, SW17 0LN [IO91VK, TQ27]
G6 MDG J E Tyson, 1102 Rochdale Rd, Blackley, Manchester, M9 7EQ [IO83VM, SD80]
G6 MDM W J Smyth, 4 Dereham Rd, Pudding Norton, Fakenham, NR21 7NA [JO02KT, TF92]
G6 MDN G J Tillett, 43 Chippenham Rd, Harold Hill, Romford, RM3 8HJ [JO01CO, TQ59]
G6 MDR I H Stanley, 6 Kennedy Ave, Long Eaton, Nottingham, NG10 3GF [IO92IV, SK43]
G6 MDS A P Scott, 3 Majestic Rd, Hatch Warren, Basingstoke, RG22 4XD [IO91KF, SU64]
G6 MDT Details withheld at licensee's request by SSL.
G6 MDU Details withheld at licensee's request by SSL.[Op: G H Smith.]
G6 MDW Details withheld at licensee's request by SSL.
G6 MDY Details withheld at licensee's request by SSL.
G6 MEB P R Green, 1 Haddon Croft, Halesowen, B63 1JQ [IO82XK, SO98]
G6 MED Details withheld at licensee's request by SSL.
G6 MEE A P Copper, 12 Mopley Cl, Blackfield, Southampton, Hants, SO45 1YL [IO90HT, SU40]
G6 MEH J R A Turner, 44B Foxgrove Rd, Beckenham, BR3 5DB [IO91XJ, TQ36]
G6 MEI C Thacker, 62 Old Clough Ln, Walkden, Worsley, Manchester, M28 3HG [IO83TM, SD70]
G6 MEJ E M Tait, Birch Glen, 71 Twemlows Ave, Higher Heath, Whitchurch, SY13 2HD [IO82QW, SJ53]
G6 MEK Details withheld at licensee's request by SSL.
GM6 MEN Details withheld at licensee's request by SSL.[Op: P Thompson.]
G6 MER M E Roberts, 15 Henley Cl, Perrycrofts, Tamworth, B79 8TQ [IO92DP, SK20]
G6 MFD Details withheld at licensee's request by SSL.
G6 MFH D A Hurr, 20 Andrew Ave, Cosby, Leicester, LE9 1SB [IO92JN, SP59]
G6 MFM D R Hills, 14 Curzon Rd, Thornton Heath, CR7 6BR [IO91WJ, TQ36]
G6 MFR D R M Hunt, 67 Dorchester Ave, North Harrow, Harrow, HA2 7AX [IO91TN, TQ18]
G6 MFT L S Hartman, 178 East Barnet Rd, New Barnet, Barnet, EN4 8RD [IO91WP, TQ29]
G6 MFU N A Cowley, 126 Racecourse Rd, Swinton, Mexborough, S64 8DS [IO93IL, SK49]
G6 MGA S J Cooper, 27 Huntsmans Gate, Bretton, Peterborough, PE3 9AU [IO92UO, TL19]
G6 MGB Details withheld at licensee's request by SSL.
G6 MGE S M Cuthbert, 44 Towse Cl, Clacton on Sea, CO16 8US [JO01NT, TM11]
G6 MGH M Cooper, 3 Marina Way, Ripon, HG4 2LJ [IO94FC, SE36]
G6 MGN D R Richardson, 29 Mill Rd, Lakenheath, Brandon, IP27 9DU [JO02GJ, TL78]
G6 MGP E A Robinson, 16 Shaw Green, Storth, Milnthorpe, LA7 7JB [IO84OF, SD48]
G6 MGQ A Reddish, 33 Station Rd, Smallford, St. Albans, AL4 0HB [IO91US, TL10]
G6 MGR K Richards, 56 Seafield Ave, Holderness Rd, Hull, HU9 3JG [IO93US, TA13]
GM6 MGS Details withheld at licensee's request by SSL.[Op: Bill Robertson, 6 Pinewood Terr, Aberdeen AB1 8LS. Member Aberdeen VHF Grp GM0FRT. IO87WB. Mainly /P tropo/RSGB contests from ideal sites IO86RW 87XC 97BJ prefer 70 23 13cm long Yagi arrays 1.5m dish]
G6 MGX B J Pearce, 6 Buckbury Cl, Pedmore, Stourbridge, DY9 0TF [IO82WK, SO98]
G6 MGY D Moffat, 5 Humber Cl, Didcot, OX11 7RU [IO91JO, SU59]
G6 MGZ J W Middleton, 9 Thorndyke Cl, Miltons Brook, Maidenbower, Crawley, RH10 7WL [IO91WC, TQ23]
G6 MHC R E McNaught, 11 Glenshee Dr, Ladybridge, Bolton, BL3 4QG [IO83SN, SD60]
G6 MHG Details withheld at licensee's request by SSL.
G6 MHK J M Rickatson, 27 Sandford Rd, Mapperley, Nottingham, NG3 6AL [IO92KX, SK54]
G6 MHM M H Meadows, 17 Riverdene Dr, Winnersh, Wokingham, RG41 5TF [IO91NK, SU77]
G6 MHO I Pomfret, 20 Sandown Rd, Bury, BL9 8HN [IO83UN, SD80]
G6 MHR Details withheld at licensee's request by SSL.
G6 MHU M F Chester, 72 Wellington Way, Salisbury, SP2 9BX [IO91CB, SU13]
GW6 MHV B K Cooke, 51 Celyn Ave, Lakeside, Cardiff, CF2 6EJ [IO81JM, ST18]
G6 MIC M J Clayden, 121 North Ln, East Preston, Littlehampton, BN16 1HB [IO90ST, TQ00]
G6 MID P D Croft, Exchange Building, Exchange St, Normanton, West Yorks, WF6 2AA [IO93GQ, SE32]
G6 MIF Details withheld at licensee's request by SSL.
GW6 MIH M Cleverley, 33 Tylchawen Terr, Tonyrefail, Porth, CF39 8AL [IO81GN, ST08]
G6 MII Details withheld at licensee's request by SSL.
G6 MIJ C C Cater, 42 Moors Bank, St. Martins, Oswestry, SY10 7BG [IO82LW, SJ33]
G6 MIS S D Ransom, 1 Bilberry Rd, Clifton, Shefford, SG17 5HB [IO92UA, TL13]
G6 MIT Details withheld at licensee's request by SSL.
G6 MIU Mrr W Livesey, 73 Hertford Rd, Tyldesley, Manchester, M29 8LU [IO83SM, SD60]
G6 MIW Details withheld at licensee's request by SSL.
G6 MJA M J Addison, Berrymead, Oxford St., Lee Common, Great Missenden, HP16 9JH [IO91PR, SP90]
G6 MJB D Lloyd, Rangelands, Old Guildford Rd, Frimley Green, Camberley, GU16 6PH [IO91PH, SU95]
G6 MJG Details withheld at licensee's request by SSL.
G6 MJM S M Parker, 19 Sundour Cres, Wednesfield, Wolverhampton, WV11 1AP [IO82WO, SJ90]
G6 MJN Details withheld at licensee's request by SSL.
G6 MJR P Patterson, 4 Cliffe View, Prune Park Ln, Allerton, Bradford, BD15 9JQ [IO93CT, SE13]
G6 MJV Details withheld at licensee's request by SSL.
G6 MJW G F Davis, 30 Bonny Wood Rd, Hassocks, BN6 8HR [IO90WW, TQ31]
GM6 MJY C Donald, 9 Perwinnes Path, Bridge of Don, Aberdeen, AB22 8PH [IO87WE, NJ91]
G6 MKD M J Douglass, 31 Mullion Gr, Padgate, Warrington, WA2 0QW [IO83RJ, SJ69]
GW6 MKI R R Evans, 54 Oakfield Rd, Newport, NP9 4LP [IO81LO, ST28]
G6 MKJ N P D Ellis, 140 Wollaston Rd, Irchester, Wellingborough, NN29 7DH [IO92QG, SP96]
G6 MKL R D Ellis, 4 Elmdale Rd, Earl Shilton, Leicester, LE9 7HQ [IO92IN, SP49]
G6 MKO S P Everett, 10 Rivers St., Ipswich, IP4 4BG [JO02OB, TM14]
G6 MKQ G K Evans, 31 Queen Elizabeth, Cres, Accrington, Lancs, BB5 2AS [IO83TR, SD72]
GW6 MKR C A Foster, Pentyn House, Penyard, Llwycoed, Aberdare Mid Glam, CF44 0PX [IO81HR, SO00]
GW6 MKV S A Fletcher, Llanwenarth, 10 Tyr Common, Gilwern, Abergavenny, NP7 0BB [IO81KT, SO21]
G6 MKZ P M Fisher, 43 Goring Rd, Ipswich, IP4 5LR [JO02OB, TM14]
GW6 MLF Details withheld at licensee's request by SSL.
G6 MLJ Details withheld at licensee's request by SSL.
GW6 MLL B M Murphy, 22 Deepglade Cl, Grenfell Park, St. Thomas, Swansea, SA1 8EJ [IO81AO, SS69]

G6	MLM	R Mullee, 12 Burfield Dr, Appleton, Warrington, WA4 5DB [IO83RI, SJ68]
G6	MLS	T Abson, 177 Meadowhall Rd, Kimberworth, Rotherham, S61 2JW [IO93HK, SK39]
G6	MLV	K E Barker, 8 Shelley Gdns, Wembley, HA0 3QG [IO91UN, TQ18]
G6	MLX	Details withheld at licensee's request by SSL.
G6	MMB	J S Bulbrook, 25 Fairfax Way, March, PE15 9HP [JO02BM, TL49]
G6	MMC	Details withheld at licensee's request by SSL.
G6	MMD	S J Burrows, 2 Luscombe Farm Cot, Snitterfield, Stratford-on-Avon, CV37 0PP [IO92DG, SP26]
G6	MMG	D Brown, 28 Bishop Dr, Whiston, Prescot, L35 3JL [IO83OJ, SJ49]
G6	MMJ	P M J Bromley, 3 Georgia Ave, Broadwater, Worthing, BN14 8AZ [IO90TT, TQ10]
G6	MML	V A Bates, 3 Braemar Gdns, Great, Bolton, BL3 4TU [IO83SN, SD60]
GW6	MMM	J L Bowen, 24 Parklands View, Sketty, Swansea, SA2 8LX [IO81AO, SS69]
G6	MMS	J Young, 45 Summercroft, Chadderton, Oldham, OL9 7JW [IO83WM, SD90]
G6	MMT	J M Ward, 64 Gladstone Rd, Ipswich, IP3 8AT [JO02OB, TM14]
G6	MNA	C C Underwood, Bobtail Lodge, Finedon Sidings Indu, Finedon, Wellingborough, Northants, NN9 5NY [IO92PI, SP87]
G6	MNB	M N Bulmer, Highfield, 7 Fountain Ave, Hale, Altrincham, WA15 8LY [IO83UJ, SJ78]
GW6	MNC	W G Turner, 37 Dan y Bryn Ave, Radyr, Cardiff, CF4 8DD [IO81IM, ST18]
G6	MNI	R L Andrews, Heneva, Bedmond Rd, Pimlico, Hemel Hempstead, HP3 8SH [IO91TR, TL00]
G6	MNJ	P W Andrews, Heneva, Bedmond Rd, Pimlico, Hemel Hempstead, HP3 8SH [IO91TR, TL00]
G6	MNK	R A Burfield, 12 Horbury Cl, Scunthorpe, DN15 8DD [IO93PO, SE81]
G6	MNL	A R Butler, 45 Roewood Cl, Holbury, Southampton, SO45 2JT [IO90HT, SU40]
G6	MNN	M R Broad, 18 Ramptons Meadow, Tadley, RG26 3UR [IO91KI, SU66]
GW6	MNQ	Details withheld at licensee's request by SSL.
G6	MNZ	S J Bates, 3 Shearwater Cl, Porthcawl, CF36 3TU [IO81DL, SS87]
G6	MOD	P M Boden, 54 Avill, Hockley, Tamworth, B77 5QF [IO92EO, SK20]
G6	MOG	Details withheld at licensee's request by SSL.
G6	MOI	A R Bates, 38 Finsbury Ave, Sileby, Loughborough, LE12 7PJ [IO92KR, SK61]
G6	MON	A W H Bruce, 2 The Mullions, Mountnessing Rd, Billericay, CM12 9XG [JO01EP, TQ69]
G6	MOT	S R Kilmister, Parc Brause House, Penmenner Rd, The Lizard, Helston, TR12 7NR [IN79JX, SW71]
G6	MOY	Details withheld at licensee's request by SSL.
G6	MPB	S J Stromqvist, 6 Frederick Rd, Malvern, WR14 1RS [IO82UD, SO74]
G6	MPE	J A R Simmons, 92A Montgomery St., Hove, BN3 5BD [IO90VU, TQ20]
G6	MPH	Details withheld at licensee's request by SSL.
G6	MPK	T A Smith, 222 Boothferry Rd, Hessle, HU13 9AU [IO93SR, TA02]
G6	MPN	A P Shalders, 1 South Parade, Weston Point, Runcorn, WA7 4HZ [IO83OH, SJ48]
G6	MPT	P H Pritchard, 5 Charlemont Rd, Stone Cross, West Bromwich, B71 3HX [IO92AN, SP09]
GW6	MPW	R L Nelson, Speculation Inn, Hundleton, Pembroke, Dyfed, SA71 5RM
GW6	MPX	D W Prince, 29 Vernon St., Wrexham, LL11 2LW [IO83MB, SJ35]
G6	MQF	M P Morris, 29 Norwich Cl, Lichfield, WS13 7SJ [IO92CQ, SK11]
G6	MQG	S H Northrop, Buena Vista, Garstang Rd East, Singleton, Poulton-le-Fylde, FY6 7SX [IO83MU, SD33]
G6	MQI	G Pointon, 10 Woodlands Dr, Thelwall, Warrington, WA4 2EU [IO83RJ, SJ68]
G6	MQJ	P A Racher, 2 Heron Way, Horsham, RH13 6DG [IO91UB, TQ13]
G6	MQN	R E Wyatt, Ivy Villa, 1 Kings Hill, Kedington, Haverhill, CB9 7NA [JO02FC, TL74]
G6	MQR	W F Wratten, 43 Crabble Ln, River, Dover, CT17 0NY [JO01PD, TR24]
G6	MQU	B Plumtree, 65 Abbey Rd, Kirkby in Ashfield, Nottingham, NG17 7PA [IO93JC, SK55]
G6	MQY	P J Wilson, Laurel Cottage, 43 Newnham Rd, Binstead, Ryde, PO33 3TE [IO90JR, SZ59]
G6	MQZ	T A Wilson, Orchard House, Whitmoor Ln, Guildford, GU4 7QB [IO91RG, TQ05]
G6	MR	Details withheld at licensee's request by SSL.
G6	MRK	D K Richardson, Home Farm, Liverpool Rd, Ashton in Makerfield, Wigan, WN4 9LX [IO83QL, SJ59]
G6	MRN	N Parr, 24 Park Ave, Awsworth, Nottingham, NG16 2RA [IO92IX, SK44]
G6	MRP	K Playford, 1 Cherwell Cl, Abingdon, OX14 3TD [IO91IQ, SU59]
G6	MRT	P A Gibbons, 68 Hill Rd, Pinner, HA5 1LE [IO91TN, TQ18]
G6	MRU	J S T Gilmore, 439 Greenford Rd, Greenford, UB6 8RQ [IO91TM, TQ18]
G6	MRW	P A Grant, 117 Hazel Ave, Cove, Farnborough, GU14 0DW [IO91OG, SU85]
G6	MRY	C Guy, 78 Park Rd, Bolton, BL1 4RQ [IO83SN, SD70]
G6	MRZ	P Gregory, 25 Wye Rd, Clayton, Newcastle, ST5 4AZ [IO82VX, SJ84]
G6	MSC	T W Glover, 70 Sandown Rd, Toton, Beeston, Nottingham, NG9 6JW [IO92IV, SK43]
G6	MSD	F B Glaze, Trees, Stourbridge Rd, Wombourne, Wolverhampton, WV5 9BN [IO82VM, SO89]
G6	MSH	A L J Harper, y Worry, 37 Wheatfields, Hagbourne Down, Didcot, OX11 0BQ [IO91IO, SU58]
G6	MSN	I D Halson, 5 Manor Rd, Moulton, Northampton, NN3 7QU [IO92NG, SP76]
G6	MSQ	C P Hunt, 28 Argyle St., Oxford, OX4 1SS [IO91JR, SP50]
GM6	MSS	Details withheld at licensee's request by SSL.
G6	MSW	R Hirst, Wishington, Ln End Cl, Bembridge, PO35 5UF [IO90LQ, SZ68]
G6	MTB	D A Hesketh, 30 Pembridge Cl, Winyates West, Redditch, B98 0JP [IO92BH, SP06]
G6	MTF	T A T Horn, 9 Gipton Wood Ave, Oakwood, Leeds, LS8 2TA [IO93FT, SE33]
G6	MTG	D Knight, 61 Bracebridge St., Nuneaton, CV11 5PB [IO92GM, SP39]
G6	MTH	J T Hackett, 148 College St., Nuneaton, CV10 7BJ [IO92GM, SP39]
GI6	MTL	M McCutcheon, 1 Prospect Terr, Gilford, Craigavon, BT63 6JL [IO64TJ, J04]
G6	MTY	M A Matthews, Lake View, Spion Cop Farm, Langoft, Peterborough, PE6 9QB [IO92UR, TF11]
G6	MTZ	Details withheld at licensee's request by SSL.
G6	MUJ	C B James, 72 Chester Ave, Luton, LU4 9SQ [IO91SV, TL02]
GW6	MUP	W L Jones, Pen-y-Berth, Pen-y-Garth, Caerarfon, Gwynedd, LL55 1EY [IO73UD, SH46]
G6	MUQ	T Jones, 38 Netherwood Rd, West Kensington, London, W14 0BJ [IO91VL, TQ27]
G6	MUS	C A E Johnson, 110 The Greenway, The Hyde, London, NW9 5AP [IO91UO, TQ28]
G6	MUV	R C Kent, 20 St. Thomas, West Parade, Bexhill on Sea, TN39 3YA [JO00FU, TQ70]
G6	MUW	B G Kent, 105 Church Rd, Dordon, Tamworth, B78 1RN [IO92EO, SK20]
GM6	MUZ	Dr C G Duncan, 12 Juniper Park Rd, Edinburgh, EH14 5DX [IO85IV, NT16]
G6	MVF	C R Stokes, 12 High St. Ave, Arnold, Nottingham, NG5 7DF [IO93KA, SK54]
G6	MVO	P G Shoosmith, Four Winds, Risborough Rd, Little Kimble, Aylesbury, HP17 0UE [IO91OS, SP80]
G6	MVP	J R Storey, 65 Priory Green, Highworth, Swindon, SN6 7NU [IO91DP, SU29]
G6	MVQ	J D Simmonds, Stanghow, Saltburn, Lingdale, Saltburn By The Sea, Cleveland, TS12 3JU [IO94MM, NZ61]
G6	MVR	D J S Scott, 20 Belmont View, Harwood, Bolton, BL2 3QN [IO83TO, SD71]
G6	MVS	M C Sandler, 7 Hill Gr, Romford, RM1 4JP [JO01CO, TQ58]
G6	MVW	E W Sayer, 27 Glenmere Park Ave, Benfleet, SS7 1SS [JO01HN, TQ78]
G6	MWB	T J Gordon, Pippins, 1 Blenheim Way, Flimwell, Wadhurst, TN5 7PQ [JO01FB, TQ73]
G6	MWD	Dr C J Goodhand, 25 Columbia Way, Lammack, Blackburn, BB2 7DT [IO83RS, SD62]
GW6	MWG	J T Harston, 14 Little Castle Gr, Herbrandston, Milford Haven, SA73 3SP [IO71KR, SM80]
G6	MWJ	Details withheld at licensee's request by SSL.
GW6	MWN	C G Hopkins, 25 Normandy Way, Crossways Green, Chepstow, NP6 5NB [IO81PP, ST59]
G6	MWP	Details withheld at licensee's request by SSL.
G6	MWS	A S Hueck, 9 Corden Ave, Mickleover, Derby, DE3 5AQ [IO92FV, SK33]
G6	MWY	V H Harris, 72 Elmore Ave, Lee on The Solent, PO13 9ES [IO90JT, SU50]
G6	MX	M C Crowley-Milling, 142 The Green, Worsley, Manchester, M28 2PA [IO83TL, SD70]
GW6	MXB	R C Hancock, 6 Alexandra Terr, Abernant, Aberdare, CF44 0RG [IO81GR, SO00]
G6	MXE	R W Peeling, 1 Station View, Station Hill, Wadhurst, TN5 6RY [JO01DB, TQ63]
GW6	MXG	A A Paxton, 33 Bryn Castell, Conwy, LL32 8LF [IO83CG, SH77]
G6	MXJ	Details withheld at licensee's request by SSL.
G6	MXL	C Redwood, 45 Lulworth Ave, Hamworthy, Poole, BH15 4DH [IO80XR, SY99]
G6	MXO	G K M Russell, Bellmere, Brightling Rd, Robertsbridge, TN32 5EJ [JO00FX, TQ72]
G6	MXR	Details withheld at licensee's request by SSL.
G6	MXT	A Riley, 64 Dig Ln, Wybunbury, Nantwich, CW5 7EY [IO83SB, SJ65]
G6	MXV	A D Poupard, Ryefield, Four Elms Rd, Eoenbridge, Edenbridge, TN8 6AF [JO01AE, TQ44]
G6	MXW	T J Parkinson, 239 Heyhouses Ln, Lytham St. Annes, FY8 3RQ [IO83MS, SD32]
GM6	MYA	M W McVicar, 2 Lilliardsedge Par, nr Ancrum, Roxburghshire, TD8 6TZ [IO85QM, NT62]
G6	MYE	Details withheld at licensee's request by SSL.
G6	MYG	M Innes, 8 Cricketers Cl, Ackworth, Pontefract, WF7 7PW [IO93IP, SE41]
G6	MYH	L D Jackson, 11 Elderbush Ln, Catfield, Great Yarmouth, NR29 5BZ [JO02SR, TG32]
G6	MYI	P Jackson, 22 Cliff Gdns, Oswald Rd, Scunthorpe, DN15 7PJ [IO93QO, SE81]
G6	MYO	B W Johnson, 2 Plumtree Cttgs, Hill St., Donisthorpe, Swadlincote, DE12 7PW [IO92FR, SK31]
GI6	MYQ	J A Kelly, 9 Eden Terr, Northland Rd, Londenderry, N Ireland, BT48 0DH [IO65IA, C41]
G6	MYT	C J King, 12 Rolls Dr, Bournemouth, BH6 4NA [IO90CR, SZ19]
G6	MYW	Details withheld at licensee's request by SSL.
GW6	MYY	D A Davies, Cil y Graig, Hen Dumpike, Tregarth, Bangor, Gwynedd, LL57 4NN [IO73XE, SH66]
G6	MZJ	R Eaton, 13 Henry St., Wakefield, WF2 9NX [IO93FQ, SE32]
GI6	MZL	D Eames, 29 Pond Park Rd, Lisburn, BT28 3LA [IO64XM, J26]
G6	MZN	M E Esser, 10 Van Diemans Rd, Wombourne, Wolverhampton, WV5 0BQ [IO82VM, SO89]
G6	MZS	Details withheld at licensee's request by SSL.
G6	MZW	D M Speak, 42 Penn Lea Rd, Weston, Bath, BA1 3RB [IO81TJ, ST76]
G6	MZX	Details withheld at licensee's request by SSL.
G6	NA	H C Spencer, Tilshead Toms Field, Langton Matravers, Swanage, Dorset, BH19 3HN [IO80XO, SY97]
G6	NAD	M McDermott, 91 Hargwyne St., London, SW9 9RH [IO91WL, TQ37]
G6	NAG	D H A Lang, 8 Church Hill, Cheddington, Leighton Buzzard, LU7 0SY [IO91QU, SP91]
G6	NAH	P J Proudlove, 14 Heath Ave, Rode Heath, Stoke on Trent, ST7 3RY [IO83UC, SJ85]
G6	NAJ	Revd T J B Leyland, 129 Essington Rd, New Invention, Willenhall, WV12 5DT [IO82XO, SJ90]
G6	NAK	R B Luker, 17 Heaton Rd, Brockurst, Gosport, PO12 4PL [IO90KT, SU50]
G6	NAL	R F Pain, The Hornbeams, Boxley Rd, Walderslade, Kent, ME5 9JG [JO01GH, TQ76]
G6	NAN	J A Lowden, 3 Boscobel Rd, Great Barr, Birmingham, B43 6BB [IO92AN, SP09]
G6	NAP	E M Lester, 50 Mayfield Rd, Worcester, WR3 8NT [IO82VE, SO85]
GI6	NAQ	S J McCullagh, 62 Oakdene Park, Bleary, Co. Down, N Ireland, BT63 5SB [IO64TK, J05]
G6	NAV	R E Martin, Bywoner, Binscombe, Godalming, Surrey, GU7 3QJ [IO91QE, SU94]
G6	NAX	S P C Moring, Whites Cottage, Tawney Common, Epping, Essex, CM16 7PU [JO01BQ, TL40]
G6	NB	D N Biltcliffe, Spindles, 24 Hambleside, Bicester, OX6 8GA [IO91JV, SP52]
G6	NBD	C E Ansell, 5 Kelso Dr, Gravesend, DA12 4NR [JO01EK, TQ67]
G6	NBE	P R Bethell, 6 Givendale Dr, Higher Crumpsall, Manchester, M8 4PY [IO83VM, SD80]
G6	NBF	J C Baron, 18 Spotborough Rd, Doncaster, DN5 8AU [IO93KM, SE50]
G6	NBI	J L Brett, 37 Lyneham Gdns, Pinkneys Green, Maidenhead, SL6 6SJ [IO91PM, SU88]
G6	NBK	Capt R M J Berry, 41 Elliotts Ln, Codsall, Wolverhampton, WV8 1PG [IO82VP, SJ80]
G6	NBL	R J Barnett, 5 Overbrook, Evesham, WR11 6DE [IO92AC, SP04]
G6	NBP	P Blease, 23 Beech Rd, Stockton Heath, Warrington, WA4 6LT [IO83RI, SJ68]
G6	NCF	Details withheld at licensee's request by SSL.
G6	NCL	R Pickstone, 33 Shore Mount, Littleborough, OL15 8EN [IO83WP, SD91]
G6	NCM	T D Pickstone, 33 Shore Mount, Littleborough, OL15 8EN [IO83WP, SD91]
G6	NCP	P D Wright, 4 Clivedon Rd, Connahs Quay, Deeside, CH5 4LN [IO83LF, SJ26]
G6	NCR	Details withheld at licensee's request by SSL.
G6	NCU	P H White, 13 Lin Brook Dr, Poulner, Ringwood, BH24 3LJ [IO90CU, SU10]
GU6	NCZ	P J Wild, Timbertops, Les Hurettes Ln, St. Martin, Guernsey, GY4 6QS
G6	NDA	C J Venn, Stantor High Rd, Horsington, Templecombe, Somerset, BA8 0DN [IO81SA, ST62]
GD6	NDE	J Winn, 18 Hilltop Rise, Farmhill, Braddan, Douglas, IM2 2LE
G6	NDG	D J Wall, 19 Dukes Dr, Newbold, Chesterfield, S41 8QB [IO93GG, SK37]
G6	NDH	Details withheld at licensee's request by SSL.
G6	NDI	Details withheld at licensee's request by SSL.
G6	NDJ	A C Wilson, 23 Claydown Way, Slip End, Luton, LU1 4DU [IO91SU, TL01]
GI6	NDM	G D O'Boyle, 27 Drapersfield Rd, Cookstown, BT80 8RS [IO64PP, H87]
G6	NDS	Details withheld at licensee's request by SSL.[Correspondence to: Northampton Scout ARG, c/o I Rivett, 25 Masefield Way, Kinsley Park, Northampton, NN2 7JT.]
GI6	NDZ	G S Devenney, 6 Gormley Cres, Strabane, BT82 9HZ [IO64GT, H39]
G6	NEA	J T Dean, 15 Park Cl, Sonning Common, Reading, RG4 9RY [IO91MM, SU78]
G6	NEB	Details withheld at licensee's request by SSL.
G6	NEK	P E Diss, 130 Beridge Rd, Halstead, CO9 1JU [JO01HW, TL83]
G6	NEM	Details withheld at licensee's request by SSL.
G6	NEP	J W Dean, 42 New Park Rd, Newgate St., Hertford, SG13 8RF [IO91WR, TL20]
G6	NER	Details withheld at licensee's request by SSL.
G6	NET	J H Dale, 39 Daleview Rd, Carlton Hill, Nottingham, NG3 7AJ [IO92KX, SK54]
G6	NEX	Details withheld at licensee's request by SSL.[Op: D L Evans, Flat 23, 28 St Johns Road, Buxton, Derbys, SK17 6XQ.]
G6	NEZ	R Emerson, 4 Freeford Gdns, Boley Park, Lichfield, WS14 9RJ [IO92CQ, SK10]
G6	NFB	T B Wright, 27 Grosvenor Pl, Oxton, Birkenhead, L43 1UA [IO83LJ, SJ38]
G6	NFC	A W G Young, Trelawney House, Attwood Ln, Pensilva, Liskeard, PL14 5QU [IO70SL, SX26]
G6	NFE	R J White, 1 Grasslands, Langley, Maidstone, ME17 3JJ [JO01HF, TQ85]
G6	NFF	Details withheld at licensee's request by SSL.
G6	NFJ	J P F Eden, 23 Elm Green Cl, Worcester, WR5 3HD [IO82VE, SO85]
G6	NFU	I Foster, The Rookery, Raisbeck, Penrith, Cumbria, CA10 3SG [IO84RL, NY60]
G6	NFZ	P R Longstaff, 77 Finedon Rd, Irthlingborough, Wellingborough, NN9 5TY [IO92QH, SP97]
G6	NGA	J R Lamble, 18 Rochford Rd, Bishops Stortford, CM23 5EX [JO01CV, TL42]
G6	NGE	C E Thaiss, 3 Drax Ct, Middle Rasen, Market Rasen, LN8 3UE [IO93TJ, TF08]
G6	NGF	M A Tatlow, 19 Arbour Cl, Mickleton, Chipping Campden, GL55 6RR [IO92CC, SP14]
G6	NGK	V C Mackney, Noyon, East Burton Rd, Wool, Wareham, BH20 6HF [IO80VQ, SY88]
G6	NGN	D R Simpson, The Hawthorne, Slacken Ln, Talke, Stoke on Trent, ST7 1NQ [IO83UC, SJ85]
G6	NGO	Details withheld at licensee's request by SSL.
G6	NHA	M Malyon, 16 Tintern Rd, Gosport, PO12 3QN [IO90KT, SZ69]
GW6	NHB	G Mahoney, 684 Beechley Dr, Pentrebane, Cardiff, CF5 3SS [IO81JL, ST17]
GM6	NHF	J C Miller, 2 Pundeavon Ave, Kilbirnie, KA25 7BH [IO75PS, NS35]
G6	NHG	S E Marshall, 25 Carlcroft, Wilnecote, Tamworth, B77 4DL [IO92EO, SK20]
G6	NHK	N J Martin, Stonea House, Middle Rd, March, PE15 0AJ [JO02AN, TL39]
GW6	NHL	A McCallum, Trosgol, Deiniolen, Caernarfon, Gwynedd, LL55 3LU [IO73WD, SH56]
G6	NHO	R E Smith, 5 Wykeham Rd, Higham Ferrers, Rushden, NN10 8HU [IO92QH, SP96]
G6	NHQ	T B Silk, 9 Spode Cl, Tilehurst, Reading, RG3 6DW [IO91LJ, SU66]
G6	NHU	K L Maton, 41 Bemerton Gdns, Kirkby Cross, Kirby Cross, Frinton on Sea, CO13 0LQ [JO01OU, TM22]
G6	NHW	P D Minchin, 122 Mildenhall Rd, Great Barr, Birmingham, B42 2PQ [IO92AN, SP09]
G6	NHX	K Millichamp, 46 Teddington Gr, Perry Barr, Birmingham, B42 1RG [IO92BM, SP09]
G6	NHY	K Marriott, 1 Holbeck Rd, Hucknall, Nottingham, NG15 7SR [IO93JB, SK54]
GM6	NIA	D A McCall, 11 Craiglockhart Dell Rd, Edinburgh, EH14 1JW [IO85JW, NT27]
GM6	NIC	J I McAulay, 9 Randolph Cliff, Edinburgh, EH3 7TZ [IO85JW, NT27]
G6	NID	J Matthew, 4 Woodland Rd, Heywood, OL10 4DB [IO83VO, SD81]
G6	NII	N H Lancaster, 5 Hoscote Park, West Kirby, Wirral, L48 0QN [IO83JI, SJ28]
G6	NIO	M W Smith, 39 Seliot Cl, Poole, BH15 2HQ [IO90AR, SZ09]
G6	NIS	J G Stephenson, 57 Kingsbury Rd, Coundon, Coventry, CV6 1PT [IO92FK, SP38]
G6	NIW	F T Shaw, 43 Egremont Rd, Hardwick, Cambridge, CB3 7XR [JO02AF, TL35]
G6	NIX	E G Samuels, 156 Fulmer Cl, Hampton, TW12 3YN [IO91TK, TQ17]
GI6	NIY	E F Stanford, 33 Glenview Gdns, Belfast, BT5 7LY [IO74BN, J37]
G6	NIZ	A J Scott, The Conifers, Back Ln, Newton on Ouse, York, YO6 2DF [IO94JA, SE56]
G6	NJA	Details withheld at licensee's request by SSL.
G6	NJF	T Shilton, 30 Nabbs Ln, Hucknall, Nottingham, NG15 6NS [IO93JA, SK54]
G6	NJH	A J Saunders, 25 Gorsey Cl, Astwood Bank, Redditch, B96 6AG [IO92AG, SP06]
G6	NJI	Details withheld at licensee's request by SSL.
G6	NJJ	M Swift, Spa Cottage, Spa Ln, Lathom, Ormskirk, L40 6JQ [IO83ON, SD40]
G6	NJL	M R Spittle, 1A Manor Dr, Kirkham, Preston, PR4 2ZN [IO83NS, SD43]
G6	NJO	P H Askham, 1 Park House Cottage, Carr Ln, Asenby, Thirsk, YO7 3PF [IO94HE, SE37]
G6	NJR	Details withheld at licensee's request by SSL.
G6	NJT	F C Neill, 7 Bellevue Terr, Southampton, SO14 0LB [IO90HV, SU41]
G6	NJW	D W Oatley, 9 Ashby Rd, Woodville, Swadlincote, DE11 7BZ [IO92FS, SK31]
G6	NKD	A G Amies, Farm View, 1 Flatts Cl, Treeton, Rotherham, South Yorks, S60 5RR [IO93HJ, SK48]
GW6	NKG	B T N Brock, 33 Heol Glyndwr, Fishguard, SA65 9LN [IO71MX, SM93]
G6	NKI	K E Brown, 73 Church Ave, Preston, PR1 4UD [IO83QS, SD53]
G6	NKJ	J H Bowden, 112 Jewell Rd, Bournemouth, BH8 0JS [IO90CR, SZ19]
G6	NKL	D W Baldock, Deeside, Platts Ln, Bucknall, Woodhall Spa, LN1 5DY [IO93RG, SK97]
G6	NKM	G S D Brown, 16 Stott Gdns, Cambridge, CB4 1FJ [JO02BF, TL46]
G6	NKS	G W Pearn, 7 Wensleydale, Carlton Colville, Lowestoft, NR33 8TL [JO02UK, TM59]
G6	NLC	C M Rabe, 40 Felton Rd, Parkstone, Poole, BH14 0QS [IO90AR, SZ09]
G6	NLD	A Reed, 6 Brancaster Cl, Cinderhill, Nottingham, NG6 8SL [IO92JX, SK54]
G6	NLE	H W O Roberts, Rowen House, Weston Underwood, Derbyshire, DE6 4PA [IO92FX, SK24]
G6	NLG	M P Ritchie, Blythe Ct, 7 Norfolk Rd, Edgbaston, Birmingham, B15 3QD [IO92AL, SP08]
G6	NLH	T Roberts, 46 Woodrow, Pennylands, Skelmersdale, WN8 8AH [IO83ON, SD40]
G6	NLI	J T Roberts, 10 Glynswood, Chard, TA20 1AH [IO80MV, ST30]
G6	NLL	Details withheld at licensee's request by SSL.
G6	NLM	P A Boorman, Readers Barn, Readers Ln, Iden, Rye, TN31 7UU [JO00IX, TQ92]
GW6	NLP	M J S Bryant, The Nook, Llanarmon Rd, Bwlchgwyn, Wrexham, LL11 5YP [IO83KB, SJ25]
G6	NLQ	T A Fradley, 34 The Ridings, Saughall, Chester, CH1 6AX [IO83MF, SJ36]
G6	NLR	J P Bugg, 96 Horringer Rd, Bury St. Edmunds, IP33 2DP [JO02IF, TL86]
G6	NLS	D Budd, 81 Bohemia Chase, Leigh on Sea, SS9 4PW [JO01HN, TQ88]
G6	NLU	S Buxton, 54 Southview Rd, Carlton, Nottingham, NG4 3QL [IO92KX, SK64]
G6	NLW	I D Rylett, 212 Hague Ave, Rawmarsh, Rotherham, S62 7PR [IO93HL, SK49]
G6	NLZ	M C Reynolds, 24 Mill Rd, Lydd, Romney Marsh, TN29 9EJ [JO00KW, TR02]
G6	NMA	P R Ayers-Hunt, 7 The Model Village, Long Itchington, Rugby, CV23 8RB [IO92HG, SP46]
G6	NMH	J A Gwillam, Park House, 4 Tudor Ln, Southam, Leamington Spa, CV33 0HS [IO92HF, SP46]
G6	NMI	R M Garment, 24 Newlands Rd, Westoning, Bedford, MK45 5LD [IO91SX, TL03]
G6	NMK	M R Grimes, Orchard End, 73 Ryston Rd, Denver, Downham Market, PE38 0DP [JO02EO, TF60]
GW6	NMO	R E Goff, Marle Farm, Boncath, Dyfed, Wales, SA37 0JS [IO71QX, SN23]
G6	NMQ	G H Goodyer, Flat, 54 Wyndham Rd, Petworth, GU28 0EQ [IO90QX, SU92]
G6	NMR	C J Goodhead, Flat 6, 6 Uxbridge Rd, Kingston upon Thames, KT1 2LL [IO91UJ, TQ16]
G6	NNK	P W Frampton, 118 Ramnoth Rd, Wisbech, PE13 2JD [JO02CP, TF40]
G6	NNO	J E Evans, 74 Trejon Rd, Cradley Heath, Warley, B64 7HJ [IO82XL, SO98]
GI6	NNP	V Hagan, 7 Emania Terr, Armagh, BT60 4AS [IO64QI, H84]
GW6	NNR	C Hatch, 530 Chepstow Rd, Newport, NP9 9DA [IO81MO, ST38]
G6	NNV	Details withheld at licensee's request by SSL.
G6	NOI	J F Whelan, 8 Welland Rd, Higher Bebington, Wirral, L63 2JU [IO83LI, SJ38]
G6	NOL	J Weir, 17 Pasteur Dr, Mytton Priory, Leegomery, Telford, TF1 4PQ [IO82RR, SJ61]
G6	NOQ	C N Wood, Llwyn Onn, Ruthin Rd, Bwlchgwyn, Wrexham, Clwyd, LL11 5UR [IO83KB, SJ25]
G6	NOS	Details withheld at licensee's request by SSL.
G6	NOT	Details withheld at licensee's request by SSL.
G6	NOW	R A Beck, 50 Bushbury Rd, Kittsgreen, Birmingham, B33 9NG [IO92CL, SP18]
G6	NOY	R L Blake, 1 Carnforth Cl, Hadrian Park, Wallsend, NE28 9TG [IO95FA, NZ36]

G6 NOZ C P Bootles, Highland Farm, Leicester Rd, Broughton Astley, Leicester, LE9 6RB [IO92JN, SP59]
G6 NPC Details withheld at licensee's request by SSL.
GW6 NPD J R Charles, 43 Penrhyn Isaf Rd, Penrhyn Bay, Llandudno, LL30 3LT [IO83CH, SH88]
G6 NPE A P Coates, 33 Hunter Ave, Burntwood, Walsall, West Midlands, WS7 9AF [IO92BQ, SK00]
GI6 NPF M G Conlon, 79A Kinrush Rd, Cookstown, BT80 0HP [IO64RP, H97]
G6 NPJ J M Copeland, Little Cophall, Dowlands Ln, Copthorne, Crawley, RH10 3HX [IO91WD, TQ34]
G6 NPK A F Foxall, Darinian, Panwell Rd, Bitterne, Southampton, SO18 6BJ [IO90HW, SU41]
G6 NPL M R Vickers, 42 Green Ln, Great Barr, Birmingham, B43 5LE [IO92AM, SP09]
G6 NPP R C Willis, 24 Elizabeth Ave, Abingdon, OX14 2NS [IO91IQ, SU59]
G6 NPQ Details withheld at licensee's request by SSL.
G6 NPR Details withheld at licensee's request by SSL.
G6 NPW S P Caine, 19 Turner Dr, West Ardsley, Tingley, Wakefield, WF3 1UD [IO93FR, SE22]
G6 NPZ P E W Chard, 29 Nettle Gap Cl, Wootton, Northampton, NN4 6AH [IO92NF, SP75]
G6 NQB S Clements, 39 Redland Cl, Marlbrook, Bromsgrove, B60 1DZ [IO82XI, SO97]
G6 NQF P M Crayden, 19 Lammas Cl, Wick, Littlehampton, BN17 6HU [IO90RT, TQ00]
G6 NQL J Wilkinson, The Old Joinery, Garsdale, nr Sedbergh, Cumbria, LA10 5PJ [IO84TH, SD79]
G6 NQM K A Whitchurch, 65 Honey Hill Rd, Kingswood, Bristol, BS15 4HN [IO81SL, ST67]
G6 NQO D J Hawken, 421 Chiswick High Rd, London, W4 4AR [IO91UL, TQ17]
GW6 NQU J L Hoy, 39 Blackbird Rd, Caldicot, Newport, NP6 4RE [IO81PO, ST48]
G6 NRA B L Goodall, 45 Ennerdale Terr, Seacliffe, Whitehaven, CA28 9PF [IO84EM, NX91]
G6 NRF A C Hawkins, 6 Chiltern Way, Huntington, York, YO3 9RS [IO93LX, SE65]
G6 NRK A S Hunt, 39 Circular Rd West, Norris Green, Liverpool, L11 1AY [IO83MK, SJ39]
G6 NRL C E Hargreaves, Viridis, Retford Rd, South Leverton, Retford, DN22 0BY [IO93OH, SK78]
G6 NRQ Details withheld at licensee's request by SSL.
G6 NRU I G Watson, The Observatory Rtmt Home, 79 Sea Rd, Westgate on Sea, CT8 8QG [JO01QJ, TR37]
G6 NRY J A Walker, 90 Surbiton Rd, Stockton on Tees, TS18 5QE [IO94HN, NZ41]
G6 NRZ B E Jordan, 6 Cherry Orchard, Holt Heath, Worcester, WR6 6ND [IO82UG, SO86]
G6 NSB Details withheld at licensee's request by SSL.
GW6 NSG J D Jones, 26 Spring Rd, Rhosddu, Wrexham, LL11 2LU [IO83MB, SJ35]
GW6 NSK A V Jones, Maen Derwydd, Llanddew, Brecon, Powys, LD3 9SY [IO81HX, SO03]
G6 NSM R Jones, 155 Victoria Rd, Garswood, Ashton in Makerfield, Wigan, WN4 0RG [IO83QL, SD50]
G6 NSQ P A James, Fairfields, 9 Small Holding, Tutbury Rd, Burton on Trent, DE13 0AL [IO92EU, SK22]
G6 NSY Details withheld at licensee's request by SSL.[Also GX6NSY. Op: NSY Amateur Radio Society, Room 99, New Scotland Yard, Broadway, London, SW1H 0BG. Locator: IO91WL. WAB: TQ27, City of Westminster. QSL via bureau or direct; SAE/IRC required for direct reply.]
G6 NSZ D Lawton, 28 Sheardown St., Hexthorpe, Doncaster, DN4 0BH [IO93KM, SE50]
G6 NTE J D Lyons, 3 Cromwell Rd, Devizes, SN10 3EJ [IO91AI, SU06]
G6 NTI Details withheld at licensee's request by SSL.
G6 NTK Details withheld at licensee's request by SSL.
G6 NTM E A Murphy, 25 Warrington Rd, Ashton in Makerfiel, Ashton in Makerfield, Wigan, WN4 9PJ [IO83QL, SJ59]
G6 NTQ R S Morgan, 1 Hillmeads Dr, Oakham, Dudley, DY2 7TS [IO82XM, SO98]
G6 NTW K H Gosbee, 64 Connaught Gdns, Palmers Green, London, N13 5BS [IO91WO, TQ39]
G6 NTY B E Griffiths, 18 Julius Dr, Coleshill, Birmingham, B46 1HL [IO92DM, SP19]
G6 NUA Details withheld at licensee's request by SSL.
G6 NUD A B Goddard, 58 Outer Circle, Southampton, SO16 5GY [IO90GW, SU31]
G6 NUI D J Chamberlain, 44 Parsonage Chase, Minster on Sea, Sheerness, ME12 3JX [JO01JJ, TQ97]
G6 NUK H W Critchley, 36 Priestley Way, Shaw, Oldham, OL2 8HU [IO83XN, SD90]
G6 NUL R C Crawford, Moreton Grange, Moreton St., Prees, Whitchurch, SY13 2EF [IO82PV, SJ53]
G6 NUO S H Clark, 90 Hamstead Rd, Great Barr, Birmingham, B43 5BN [IO92AM, SP09]
G6 NUQ S Coward, 100 Lytham Rd, Freckleton, Preston, PR4 1XB [IO83NS, SD42]
G6 NUR S G Coe, 37 Alexandra Rd, Yeovil, BA21 5AL [IO80QW, ST51]
G6 NUS A V Croft, 22 Wollaston Rd, Bozeat, Wellingborough, NN29 7LT [IO92PF, SP95]
G6 NUT Details withheld at licensee's request by SSL.
G6 NUX S L Clack, 7 Carlingford Ct, Victoria Dr, Bognor Regis, PO21 2EN [IO90PS, SZ99]
G6 NUZ A B Charlton, 26 Saundergate Ln, Wyberton, Boston, PE21 7BZ [IO92XW, TF34]
G6 NVA Details withheld at licensee's request by SSL.
G6 NVE K Moore, 10 Alms Hill Cres, Sheffield, S11 9QZ [IO93FI, SK38]
G6 NVF D J Mc Glasson, 19 Kennedy St., Ulverston, LA12 9EA [IO84LE, SD37]
G6 NVI S Mindel, 33 Snaresbrook Dr, Stanmore, HA7 4QN [IO91UO, TQ19]
G6 NVJ C G Marlow, 27 Sandy Way, Connahsquay, Connahs Quay, Deeside, CH5 4SH [IO83LF, SJ26]
G6 NVL B C Mumford, Cedarwood, Manston Rd, Manston, Ramsgate, CT12 5BE [JO01QI, TR36]
G6 NVP D A Gittins, 43 Maybank Cl, Boley Park, Lichfield, WS14 9UJ [IO92CQ, SK10]
G6 NVS P J Harrison, 41 Chestnut Cl, Hands Acre, Handsacre, Rugeley, WS15 4TH [IO92BR, SK01]
G6 NVT Details withheld at licensee's request by SSL.
G6 NVU M P Haynes, 10 Cypress Gr, Denton, Manchester, M34 6EA [IO83WK, SJ99]
G6 NVV J Hebden, 23 Moss Cl, Wickersley, Rotherham, S66 0ET [IO93IK, SK49]
G6 NVY A C Hilbourne, Chimneys, 50 Laburnham Rd, Maidenhead, SL6 4DE [IO91PM, SU88]
G6 NW F James, 18 The Banks, Long Buckby, Northampton, NN6 7QQ [IO92LH, SP66]
G6 NWF R O Parry, 57 Chessel St., Bristol, BS3 3DN [IO81QK, ST57]
G6 NWK M P Parker, 31 Sandholme Dr, Burley in Wharfedale, Ilkley, LS29 7RG [IO93DV, SE14]
G6 NWM H R Smith, 26 Cherry Tree Cres, Wickersley, Rotherham, S66 0LS [IO93IK, SK49]
G6 NWN I Poyser, 24 Overstone Cl, Sutton in Ashfield, NG17 4NL [IO93ID, SK46]
G6 NWO P Rawlins, 135 High Ln, Stoke on Trent, ST6 7BS [IO83VB, SJ85]
G6 NWR J R Arrowsmith N Warks Raynet, Co, 16 Mancetter Rd, Mancetter, Atherstone, CV9 1NZ [IO92FN, SP39]
G6 NWT W J Taylor, 33 Lancaster Ave, Dawley, Telford, TF4 2HS [IO82SP, SJ60]
G6 NWV Details withheld at licensee's request by SSL.
G6 NWW M Hawkey, Hawley, Tuesley Ln, Busbridge, Godalming, GU7 1SG [IO91QE, SU94]
GM6 NX D M K Harrower, 26 Millar Pl, Stirling, FK8 1XD [IO86AD, NS89]
GW6 NXH W L Rees, 1 St. Marys Cl, Briton Ferry, Neath, SA11 2JU [IO81CP, SS79]
GW6 NXL T A Rees, Tygoleu, Llwyngwril, Gwynedd, North Wales, LL37 2UZ [IO72XP, SH50]
G6 NXM R N E Rixon, 11 The Ridings, Waltham Chase, Southampton, SO32 2TR [IO90JW, SU51]
G6 NXP S Rafferty, 11 Gilbert Rd, Peterlee, SR8 2AN [IO94HS, NZ44]
G6 NXR A H Barfield, 43 James Dawson Dr, Allesley, Coventry, CV5 9QJ [IO92EK, SP28]
G6 NXV M Shannon, Woodhayes, 129 Hampton Ln, Blackfield, Southampton, SO45 1WF [IO90HT, SU40]
G6 NXW K Sykes, 68 Newtown Ave, Cudworth, Barnsley, S72 8DY [IO93HN, SE30]
G6 NYC R P Ashman, 44 Conan Doyle Walk, Swindon, SN3 6JB [IO91DN, SU18]
G6 NYF K Aylward, Fairfield, 3 Agbrigg Rd, Wakefield, WF2 6AA [IO93GP, SE31]
G6 NYG R A Adams, 28 Greenside, Stoke Prior, Bromsgrove, B60 4EB [IO82XG, SO96]
G6 NYH G J Austin, 21 St. Georges Pl, Northampton, NN2 6EP [IO92NF, SP76]
G6 NYL P J Baylis, 118 Eastgate, Deeping St. James, Peterborough, PE6 8RD [IO92UP, TF10]
GW6 NYR A B Davis, 70 St. Andrews Rd, Upper Colwyn Bay, Colwyn Bay, LL29 6DL [IO83DG, SH87]
GM6 NYT J E Danton, 12 Laburnum Rd, Methil, Leven, KY8 2HA [IO86LE, NO30]
G6 NZ M A Newnham, 75 Liss Rd, Southsea, PO4 8AS [IO90LT, SZ69]
G6 NZA M A Davenport, 213 Old Hall Rd, Brampton, Chesterfield, S40 1HQ [IO93GF, SK37]
G6 NZB A R Dickey, Homestead Cottage, Godshill, Fordingbridge, Hants, SP6 2LG [IO90CW, SU11]
G6 NZC B H Dean, 2 Whitehouse Rd, Sawtry, Huntingdon, PE17 5UA [IO92UK, TL18]
G6 NZG S K Edson, 58 Blake Rd, Stapleford, Nottingham, NG9 7HR [IO92IW, SK43]
G6 NZL P A Fletcher, 43 Merlin Way, Woodville, Swadlincote, DE11 7QU [IO92FS, SK31]
G6 NZN G C E Fowler, Gull Cottage, 10 Ullswater Rd, Merley, Wimborne, BH21 1QT [IO90AS, SZ09]
G6 NZO M S Finney, 49 Ashcroft Dr, Old Whittington, Chesterfield, S41 9PA [IO93GG, SK37]
G6 NZS D M Freeman, Rear Flat, 24 London Rd, Southborough, Tunbridge Wells, TN4 0QB [JO01DD, TQ54]
G6 NZT E C Fitzgerald, 154 Chapelhill Rd, Wirral, L46 9RP [IO83KJ, SJ28]
G6 NZW D E Smith, 90 Endhill Rd, Kingstanding, Birmingham, B44 9RP [IO92BN, SP09]
G6 NZY B J Sparke, 101 Hedingham Rd, Halstead, CO9 2DW [JO01HW, TL83]
G6 NZZ R J Smith, 2 Sutton Rd, Kegworth, Derby, DE74 2DX [IO92IT, SK42]
GM6 OA G H Miller, 7 Windsor Gdns, Largs, KA30 9DN [IO75NT, NS26]
G6 OAI S C Baverstock, 43 Tatnam Rd, Poole, BH15 2DW [IO90AR, SZ09]
G6 OAJ Details withheld at licensee's request by SSL.
G6 OAN C W Bryan, 113 Hoe View Rd, Cropwell Bishop, Nottingham, NG12 3DJ [IO92MV, SK63]
G6 OAS A W Inglis, 15 Morris Dr, Kingston Hill, Stafford, ST16 3YE [IO82WT, SJ92]
G6 OAU M R Jones, 3 Burghclere Dr, Maidstone, ME16 8UQ [JO01GG, TQ75]
G6 OAV C J Jones, 1 Stonehill Cl, Leigh on Sea, SS9 4AZ [JO01HN, TQ88]
GW6 OAW T O V Jones, Lifeboat House, Moelfre, Anglesey, Gwynedd, LL72 8LG [IO73VI, SH58]
G6 OAY Details withheld at licensee's request by SSL.
G6 OBA M Kaznowski, 85 St. Albans Rd, Kingston upon Thames, KT2 5HH [IO91UK, TQ17]
G6 OBD T G Keeling, 1 New Ln, Brown Edge, Stoke on Trent, ST6 8TQ [IO83WC, SJ95]
G6 OBE S Kimblin, 22 Lorton Gr, Breightmet, Bolton, BL2 6PR [IO83TO, SD70]
G6 OBG J J Kay, 12 Williams Ave, Newton-le-Willows, WA12 0NN [IO83QL, SJ59]
G6 OBJ G M Webster, 21 Markham Rd, Capel, Dorking, RH5 5JT [IO91UD, TQ14]
G6 OBO L F Weiss, 7 Tennyson Ave, Grays, RM17 5RG [JO01EL, TQ67]
G6 OBP R Wilson, 47 Arbrook Ave, Bradwell Common, Milton Keynes, MK13 8BW [IO92OB, SP83]
G6 OBT H Wilson, 11 Palmerston Ct, Haslington, Crewe, CW1 5QE [IO83TC, SJ75]
G6 OBU S C Wright, 21 Poplars Cl, Woodside, Watford, WD2 7EW [IO91TQ, TL10]
G6 OBW R B Burton, 262 Carlton Hill, Carlton, Nottingham, NG4 1FY [IO92KX, SK64]
G6 OBY R J Bacon, 29 Nettlefold Cres, Melbourne, Derby, DE73 1DA [IO92GT, SK32]
G6 OCA M W Barnes, 451 Hough Fold Way, Harwood, Bolton, BL2 3PU [IO83TO, SD71]
G6 OCB D M Byers, 2 Beechwood Ave, Ashton in Makerfiel, Ashton in Makerfield, Wigan, WN4 9LZ [IO83QL, SJ59]
GI6 OCC K C J Brennan, 1 Ballyscullion Ln, Bellaghy, Co Derry, N Ireland, N Ireland, BT45 8NQ [IO64ST, H99]
G6 OCD Details withheld at licensee's request by SSL.
G6 OCE C Stapleton, 18 The Woodlands, Melbourne, Derby, DE73 1DP [IO92GT, SK32]
G6 OCF R Wallis, 15 Montagu Dr, Eaglestone, Milton Keynes, MK6 5ER [IO92PA, SP83]
G6 OCL Details withheld at licensee's request by SSL.
G6 OCO A J Bruce, 26 Regency Cl, Wigmore, Gillingham, ME8 0LA [JO01HI, TQ86]
G6 OCP P J Brownlow, 22 Church Field View, Balby, Doncaster, DN4 0XD [IO93JM, SE50]
G6 OCQ Details withheld at licensee's request by SSL.
G6 OCT Details withheld at licensee's request by SSL.
G6 OCW Details withheld at licensee's request by SSL.
G6 ODA A E Bardy, 28 Gladsmuir Rd, London, N19 3JX [IO91WN, TQ28]
G6 ODD K M Balcombe, 40 Shaw Dr, Scartho, Grimsby, DN33 2JB [IO93XM, TA20]
G6 ODE B R Aylward, 49 The Ruffetts, South Croydon, CR2 7LT [IO91XI, TQ36]
G6 ODF Details withheld at licensee's request by SSL.
G6 ODI P Archer, 16 Park Dingle, Bewdley, DY12 2JY [IO82UI, SO77]
G6 ODO J Wilson, 10 Charnell Ave, Maltby, Rotherham, S66 7DB [IO93JK, SK59]
G6 ODR Details withheld at licensee's request by SSL.
G6 ODT K J Lamford, 41 Drayton Rd, Irthlingborough, Wellingborough, NN9 5TA [IO92QH, SP97]
G6 ODU R G W Leong, 55 Liverpool Rd, Aughton, Ormskirk, L39 5AP [IO83NN, SD40]
G6 ODW L B Liffchak, 6 Ashmore Gr, Welling, DA16 2RU [JO01BL, TQ47]
G6 OEI T Blest, 2 Gayton Ave, Littleover, Derby, DE23 7GA [IO92FV, SK33]
G6 OEJ A R Barnard, 36 St. Pauls Rd, Walton Highway, Wisbech, PE14 7DN [JO02CQ, TF51]
G6 OEM J G Bolland, 18 Ward Ave, Formby, Liverpool, L37 2JD [IO83LN, SD20]
G6 OEN Details withheld at licensee's request by SSL.
G6 OER Details withheld at licensee's request by SSL.
G6 OES M G Smith, 16 Digby Rd, Kingswinford, DY6 7RP [IO82WM, SO88]
G6 OET M H Telford, 11 Twyford Cl, Swinton, Mexborough, S64 8UH [IO93HL, SK49]
G6 OEW C S Thorn, 20 Kiln Rd, Shaw, Newbury, RG14 2HA [IO91IJ, SU46]
G6 OFA J T Lesurf, 110 Hunters Sq, Dagenham, RM10 8BG [JO01CM, TQ58]
GM6 OFB J S Mc Ardle, 40 Rodney Dr, Girvan, KA26 9DZ [IO75NF, NX19]
G6 OFE Details withheld at licensee's request by SSL.
GM6 OFO M C Clark, Lynemore, Madderty By Crieff, Perthshire, PH7 3NY [IO86DI, NN92]
G6 OFT S R Cruise, 26A The Street, Bapchild, Sittingbourne, Kent, ME9 9AH [JO01JI, TQ96]
G6 OFU F O Cowell, Oakwood, 16 Hollin Cl, Rossington, Doncaster, DN11 0XX [IO93LL, SK69]
G6 OFV S Crossland, 16 Holland Rd, High Green, Sheffield, S30 4HF [IO93GJ, SK38]
G6 OFZ J H Day, 16 Blenheim Way, Market Harborough, LE16 7LQ [IO92ML, SP78]
G6 OGD S Dawber, 7 Ashby Dr, Sandbach, CW11 3NY [IO83TC, SJ75]
G6 OGF J Darby, 272 Spring Rd, Ipswich, IP4 5NN [JO02OB, TM14]
GM6 OGN A W Sives, 4 Fir Gr, Craigshill East, Livingston, EH54 5JP [IO85GV, NT06]
G6 OGT A A Scholes, 45 Howden Rd, Higher Blackley, Manchester, M9 0RQ [IO83VM, SD80]
G6 OGY Details withheld at licensee's request by SSL.
GI6 OHA A McGurgan, 40 Meenagh Park, Coalisland, Dungannon, BT71 4NG [IO64PM, H86]
G6 OHB Details withheld at licensee's request by SSL.[Op: A G Marlow. Station located in London N8.]
G6 OHD A P J Medway, 34 Irvine Dr, Farnborough, GU14 9HG [IO91OH, SU85]
GM6 OHG Details withheld at licensee's request by SSL.
G6 OHM A L Dunham, 28 Kingfisher Cl, Chatteris, PE16 6TP [JO02AL, TL38]
G6 OHR R L Edwards, 11 Litlington Ct, Surrey Rd, Seaford, E Sussex, BN25 2NZ [JO00BS, TV49]
G6 OHT J T Ellner, 21 Cranmer Rd, Hampton Hill, Hampton, TW12 1DW [IO91TK, TQ17]
GW6 OHX G C Ellis, 30 Bryn Hyfryd, Coedpoeth, Wrexham, LL11 3YB [IO83LB, SJ25]
G6 OHY C L Edmunds, 51 Whiston Rd, Kingsthorpe, Northampton, NN2 7RR [IO92NG, SP76]
G6 OI Details withheld at licensee's request by SSL.
G6 OIA P E Rattenbury, 1 Holmewood, Bletchley, Furzton, Milton Keynes, MK4 1AR [IO92OA, SP83]
G6 OIB J A Riley, 8 Lydcott Cres, Widegates, Looe, PL13 1QG [IO70TJ, SX25]
G6 OIH R Phillips, 23 Princes St., Kettering, NN16 8RW [IO92PJ, SP87]
G6 OIL Details withheld at licensee's request by SSL.
GW6 OIO M . R . Thomas, Delaville, 42 Wyndham Rd, Abergavenny, NP7 6AF [IO81LT, SO31]
GI6 OIR C T Robinson, 21 Carnhill Rd, Carnhoney, Newtownabbey, BT36 6LA [IO74AQ, J38]
G6 OIX J M J Roberts, 9 Tower Cl, North Weald, Epping, CM16 6HA [JO01CR, TL50]
G6 OIY J A Roberts, 85 London Rd, Braintree, CM7 2LF [JO01GU, TL72]
G6 OJA B A I Robinson, 102 Highways Ave, Euxton, Chorley, PR7 6QD [IO83PP, SD51]
G6 OJB S Nimmo, Woodham Halt Vety C, Woodham Halt, South Woodham Ferre, Essex, CM3 5JB [JO01HP, TQ89]
GI6 OJC O OKane, 39 Harberton Park, Ballymena, BT43 6NF [IO64UV, D10]
G6 OJI F O'Grady, 87 Grimthorpe Ave, Whitstable, CT5 4PY [JO01MI, TR16]
GW6 OJK A G Mayall, 17 Bear St., Hay on Wye, Hereford, Herefordshire, HR3 5AN [IO82KB, SO24]
G6 OJN Details withheld at licensee's request by SSL.
G6 OJO C J Handy, 9 Kent Rd, Whiteshill, Whitehill, Bordon, GU35 9PZ [IO91NC, SU73]
G6 OJU Details withheld at licensee's request by SSL.
G6 OJV J V Greenley, 22 Langley Dr, Langton Rd, Norton, Malton, YO17 9AR [IO94OC, SE77]
G6 OJX E A Grayson, 3 Spurrier Ave, Knottingley, WF11 0ER [IO93IQ, SE42]
G6 OJZ P A Anstock, 12 Raymoor Ave, St. Marys Bay, Romney Marsh, TN29 0RD [JO01LA, TR02]
G6 OKA C B Glover, 16 Woodfield Rd, Radlett, WD7 8JD [IO91UQ, TQ19]
G6 OKB R Gilham, Wren Cottage, Wayborough Hill, Minster, Ramsgate, CT12 4HR [JO01QI, TR36]
GW6 OKC M Gerrard, 29 Forest Dr, Woodlands, Broughton, Chester, CH4 0QT [IO83MD, SJ36]
GM6 OKJ H H Tilbrook, 17 Castle St., St. Monans, Anstruther, KY10 2AP [IO86OE, NO50]
G6 OKN R A Turner, 53 Queens Dr, Sandbach, CW11 1BN [IO83TD, SJ76]
G6 OKU M R Law, 23 Yeldersley Cl, Holme Hall Est, Chesterfield, S40 4LG [IO93GF, SK37]
G6 OLJ Dr D B Hill, 33 Cleveland Cl, Chiltern Park, Thornbury, Bristol, BS12 2YD [IO81RO, ST68]
G6 OLM R J Hendry, Genazzano, 224 London Rd, Wickford, SS12 0JX [JO01FO, TQ79]
G6 OLR D S Hannam, 24 Mercia Way, Barwick Rd, Leeds, LS15 8UA [IO93GT, SE33]
G6 OLS Details withheld at licensee's request by SSL.
G6 OLU P R Hobson, 220 Station Rd, Burton Latimer, Kettering, NN15 5NT [IO92PI, SP87]
G6 OLV A J White, 20 Wyles St., Gillingham, ME7 1ND [JO01GJ, TQ76]
G6 OLY A M Williams, 10 Goose Cttgs, Chelmsford Rd, Rawreth, Wickford, SS11 8TB [JO01GO, TQ79]
G6 OM K W Niebuhr, 156 Wadards Meadow, Witney, OX8 6TZ [IO91GS, SP30]
G6 OMA R C S Williams, 274 Studfall Ave, Corby, NN17 1LH [IO92PL, SP88]
G6 OMH B G Staddon, 311 Cheney Manor Rd, Swindon, SN2 2PE [IO91CN, SU18]
GM6 OML J Porter, 147 Glasgow Rd, Dumbarton, G82 1RQ [IO75RW, NS47]
G6 OMN G H Rogers, 7 Flordon, Birch Green, Skelmersdale, WN8 6PA [IO83ON, SD40]
G6 OMT Details withheld at licensee's request by SSL.
G6 OMX K Ratcliffe, 28 Carr Ln, Middleston, Middleton, Morecambe, LA3 3LA [IO84NA, SD45]
G6 OMY G W Russell, 5 Park Ln, Harbury, Leamington Spa, CV33 9HX [IO92GF, SP35]
G6 ONC Details withheld at licensee's request by SSL.
G6 ONE D J Williams, 16 Church St., Owston Ferry, Doncaster, DN9 1RG [IO93OL, SE80]
G6 ONM M W Stonton, 26 Queen Eleanor Rd, Northampton, NN4 8NS [IO92NF, SP75]
G6 ONN Details withheld at licensee's request by SSL.
G6 ONW S L Jackson, 256 Perry Rd, Sherwood, Nottingham, NG5 1GP [IO92KX, SK54]
G6 ONZ P Kinsey, Glyn Elwy, Allt Goch, Trefnant, Denbighshire, LL17 0BP [IO83GF, SJ07]
GM6 OOA D M King, Marionville, nr Donisbristle, Cowdenbeath, Fife, KY4 8EU [IO86HC, NT18]
G6 OOC G L Scott, 128 Pack Ln, Kempshott, Basingstoke, RG22 5HP [IO91KF, SU65]
G6 OOH J R Stone, 13 Winchester Cl, Newport, PO30 1DR [IO90IQ, SZ48]
G6 OOK M E Stewart, Tollbar, Gate House, Lower Bentham, near Lancaster, LA2 7DD [IO84RC, SD66]
G6 OOO Details withheld at licensee's request by SSL.
G6 OOT C P Atkins, 112 Redmayne Dr, Chelmsford, CM2 9XE [JO01FR, TL70]
G6 OOY A Barnes, 18 Volunteer St., St. Helens, WA10 2AY [IO83PK, SJ59]
G6 OPB Details withheld at licensee's request by SSL.[Op: J Nicholls, 1 Allotment Cottages, Fridaybridge Road, Elm, Wisbech, Cambs, PE14 0AT.]
G6 OPD L Middleton, 24 Townshend Rd, Worle, Weston Super Mare, BS22 0FW [IO81NI, ST36]
GM6 OPK M Lee, The Anchorage, St Marys, Holm, Orkney, KW17 2RT [IO88NV, HY40]
G6 OPL F G Little, 122 Little Pynchons, Harlow, CM18 7DF [JO01BS, TL40]
G6 OPM Details withheld at licensee's request by SSL.
G6 OPO N T Martin, 14 Canterbury Cl, Cambridge, CB4 3QQ [JO02BF, TL45]
G6 OPP P C Martin, 47 Bryant Rd, Strood, Rochester, ME2 3EP [JO01FJ, TQ76]
G6 OPU C Melville, 36 Louise Dr, Blurton, Stoke on Trent, ST3 2DT [IO82WX, SJ84]
G6 OQC S H Cook, 3 Devon Rd, Felixstowe, IP11 9AF [JO01QX, TM23]
G6 OQJ W D Castle, 42 Saddle Rise, Springfield, Chelmsford, CM1 6SX [JO01FS, TL70]
G6 OQK Details withheld at licensee's request by SSL.
GI6 OQL J Craig, 8 Muckamore View, Muckamore, Antrim, BT41 2EU [IO64WQ, J18]
GM6 OQN R M Campbell, 32 Harvie Ave, Newton Mearns, Glasgow, G77 6LQ [IO75US, NS55]
G6 OQO J D Douthwaite, 38 Burnside Rd, Newcastle upon Tyne, NE3 2DU [IO95EA, NZ26]
G6 OQV B N Everitt, The Hermitage, The Rookery, Galley Common, Nuneaton, CV10 9PB [IO92FM, SP39]
G6 ORA Details withheld at licensee's request by SSL.
G6 ORC Details withheld at licensee's request by SSL.
G6 ORG G R Travis, 24 Orchard St., Fearnhead, Warrington, WA2 0QF [IO83RJ, SJ69]
GW6 ORE R H Trangmar, 5 Middle St., Bethesda, Bangor, LL57 3BU [IO73XE, SH66]
G6 ORH D Wright, 23 Oakenhall Ave, Hucknall, Nottingham, NG15 7TF [IO93JA, SK54]

G6 ORJ A W Weller, 104 Medina Ave, Newport, PO30 1HG [IO90IQ, SZ58]
G6 ORM S J Whiley, 34 Oulton Cl, Marlpool Gdns, Kidderminster, DY11 5DY [IO82UJ, SO87]
G6 ORN G M Wicks, 78 Kynaston Rd, Didcot, OX11 8HA [IO91JO, SU58]
G6 ORO G L J Walsh, 36 Westminster St., Newtown, Wigan, WN5 9BH [IO83QM, SD50]
G6 ORQ E H S T Beech, 86 Morris Ave, Bently, Walsall, WS2 0EE [IO82XO, SO99]
G6 ORS A Bennett, 39 Westview, Parbold, Wigan, WN8 7NT [IO83OO, SD41]
GW6 OSE G R Perry, 25 Moira Terr, Adamsdown, Cardiff, CF2 1EJ [IO81KL, ST17]
G6 OSG J K Rattey, 44 Whitedown Ln, Alton, GU34 1PS [IO91MD, SU73]
G6 OSH D T Ridley, 37 Harewood Cl, Whickham, Newcastle upon Tyne, NE16 5SZ [IO94DW, NZ25]
G6 OSJ J P Roberts, 1 Ollerdale Cl, Allerton, Bradford, BD15 9BT [IO93CT, SE13]
G6 OSK E J Robinson, 32 Ardeen Rd, Intake, Doncaster, DN2 5EU [IO93KM, SE50]
G6 OSO D E Parr, Lordings, Station Rd, Pulborough, RH20 1AH [IO90RW, TQ01]
G6 OSR Details withheld at licensee's request by SSL.
GW6 OSS J H Owen, Hafod y Rhos, Gaernarvon, Llanrug, Gwynedd, LL55 4AN [IO73VD, SH56]
G6 OSV I Woodward, 20 Boyle Ave, Orford, Warrington, WA2 0EZ [IO83RJ, SJ69]
G6 OSW G Walker, 28 Manor Cl, Thanington, Canterbury, CT1 3XA [JO01MG, TR15]
G6 OSZ B Williams, 29 St. Ternans Rd, Newtonhill, Stonehaven, AB3 3PF [IO87TH, NJ72]
G6 OTA Details withheld at licensee's request by SSL.
G6 OTB N S Young, 42 Central Way, Oxted, RH8 0LZ [IO91XG, TQ35]
GW6 OTD P R Sizer, Gambo End, Reynoldston, Gower, Swansea, SA3 1BR [IO71VO, SS49]
G6 OTE D Shaw, 19 Upper Moors, Great Waltham, Chelmsford, CM3 1RB [JO01FS, TL61]
G6 OTG G H A Spicer, 186 St. Williams Way, Rochester, ME1 2PE [JO01GI, TQ76]
G6 OTJ Details withheld at licensee's request by SSL.[Op: Dr D R Homer.]
G6 OTL G D Blades, 11 Willard Gr, Stanhope, Bishop Auckland, DL13 2XY [IO94AR, NZ03]
G6 OTM Details withheld at licensee's request by SSL.
G6 OTP M A J Rainbow, Hillview, 6 Shurdington Rd, Brockworth, Gloucester, GL3 4PS [IO81WU, SO81]
G6 OTQ T Roddy, North View, 26 Chapeltown Rd, Radcliffe, Manchester, M26 1YF [IO83UN, SD70]
G6 OTS W P Peck, 5 Stirling Cres, Scotland Ln, Horsforth, Leeds, LS18 5SJ [IO93EU, SE23]
G6 OTU Details withheld at licensee's request by SSL.[Op: R W Raft.]
G6 OTV A Ricalton, 166 Downs Barn Boulevard, Downs Barn, Milton Keynes, MK14 7QQ [IO92PB, SP84]
G6 OTW A Ricalton, 4 South Rd, Longhorsley, Morpeth, NE65 8UW [IO95CG, NZ19]
G6 OTX R S Sobey-Smith, 51 Leysters Cl, Redditch, B98 0NJ [IO92BH, SP06]
G6 OTZ A J Shaw, 92 Woodhouses Rd, Burntwood, WS7 9EJ [IO92BQ, SK00]
G6 OUA Details withheld at licensee's request by SSL.
G6 OUF Details withheld at licensee's request by SSL.
G6 OUG L F Blackaller, 60 Hathersage Ct, Newington Green, London, N1 4RF [IO91WN, TQ38]
G6 OUJ B J Bozman, 33 Maple Rd, Loughborough, LE11 2JL [IO92JS, SK51]
GM6 OUL N J Bowry, 18 Mortonhall, Park Gdns, Edinburgh, EH17 8SR [IO85JV, NT26]
G6 OUM B G Breaden, 10 Partridge Cl, Lewsey Farm, Luton, LU4 0YD [IO91SV, TL02]
G6 OUO P M Burgess, 232 Hightown Rd, Luton, LU2 0DN [IO91TV, TL02]
G6 OUT D Andrew, 30 Woodhill Ave, Morecambe, LA4 4PF [IO84NB, SD46]
G6 OUX Details withheld at licensee's request by SSL.
G6 OVA N J Styne, 2 Greenway, Off Mill Hill Ln, Burton-upon-Trent, Staffs, DE15 0AR [IO92ET, SK22]
G6 OVC B R Thurlow, 1 Sheffield Way, Earls Barton, Northampton, NN6 0PF [IO92PG, SP86]
G6 OVN Details withheld at licensee's request by SSL.[Op: M G Horton, 10 Hawthorne Rd, Delves, Walsall, W Midlands, WS5 4NA.]
G6 OVO D J Hudson, 62 Derron Ave, Yardley, Birmingham, B26 1LA [IO92CK, SP18]
G6 OVU Details withheld at licensee's request by SSL.
G6 OVY Details withheld at licensee's request by SSL.
G6 OWD J A Harris, Flat 64, 29 Cressingham Gr, Sutton, Surrey, SM1 4DR [IO91VI, TQ26]
G6 OWG R E Holborn, Greytiles, Nounsley Rd, Hatfield Peverel, Chelmsford, CM3 2NQ [JO01HS, TL71]
G6 OWI P Haworth, 2 Heys Ct, Blackburn, BB2 4PQ [IO83SR, SD62]
G6 OWO N C Johnson, The Willows, Hough Side Rd, Pudsey, LS28 9JW [IO93ET, SE23]
GW6 OWQ C J Jones, Frondeg Hall, Aberoer, Wrexham, Clwyd, LL14 4LG [IO83LA, SJ24]
G6 OWS Details withheld at licensee's request by SSL.
G6 OWT G A R Kelly, Crowhurst Ln, Plaxtol, Borough Green, Sevenoaks, Kent, TN15 8PE [JO01DG, TQ65]
G6 OWX M A King, 24 Carteret Way, Deptford, London, SE8 3QA [IO91XL, TQ37]
GD6 OXG J F Williams, Brookfield, Douglas Rd, Ballabeg, Castletown, IM9 4EF
G6 OXH G A Watkins, 1 Nursery Cl, Charnock Richard, Chorley, PR7 5UA [IO83QP, SD51]
G6 OXI C R Webb, 50 Ridgeway, Eynesbury, St. Neots, Huntingdon, PE19 2QY [IO92UF, TL15]
G6 OXL I K Wilkins, 20 Hellier St., Dudley, DY2 8RE [IO82WM, SO98]
G6 OXN I Walker, 66 Wood St., Kettering, NN16 9SB [IO92PJ, SP87]
G6 OXP D C Doody, 10 Chapel Ln, Knighton, Market Drayton, TF9 4HW [IO82TW, SJ74]
G6 OXQ A C Day, 7 Seagers, Hall Rd, Great Totham, Maldon, CM9 8PB [JO01IS, TL81]
G6 OXW W P Dibden, 33 Lyndhurst Ave, Bredbury, Stockport, SK6 2AJ [IO83WK, SJ99]
G6 OXY E W Chinn, 10 Ironstone Ln, Northampton, NN4 8TR [IO92MF, SP75]
G6 OXZ M Charlton, 20 Bailey Cres, South Elmsall, Pontefract, WF9 2TL [IO93IO, SE41]
G6 OYB Details withheld at licensee's request by SSL.
G6 OYE Details withheld at licensee's request by SSL.[Op: C D Coleman. Station located near Tonbridge.]
G6 OYF M Matthews, 39 Buddleia Cl, Preston Downs, Weymouth, DT3 6SG [IO80SP, SY68]
G6 OYU R E Spinner, 12 Ballechroisk Terr, Killin, FK21 8TH [IO76UL, NN53]
G6 OYV T J Silvers, 15 Stanford Way, Walton, Chesterfield, S42 7NH [IO93GG, SK37]
GW6 OZA D E Turner, Grangemoor, Mountain West, Newport, Dyfed, SA42 0QY [IO71NX, SN03]
G6 OZH D Martin, 2 Farm View Rd, Kirkby in Ashfield, Nottingham, NG17 7HF [IO93JC, SK55]
GI6 OZI R S McVea, 2 Abbeyview Rd, Crossgar, Downpatrick, BT30 9JD [IO74CK, J45]
G6 OZJ Y E Moore, 33 Belford Terr, North Shields, NE30 2DA [IO95GA, NZ36]
G6 OZN Details withheld at licensee's request by SSL.
G6 OZR B Walker, 35 Slipper Ln, Mirfield, WF14 0HE [IO93DQ, SE12]
G6 OZT P White, 3 South View, Whitwell, Worksop, S80 4NP [IO93JG, SK57]
G6 OZU A K Wilson, 67 Sandpits, Leominster, HR6 8HT [IO82PF, SO45]
GW6 OZV S R Williams, 52 Tymawr St., St. Thomas, Port Tennant, Swansea, SA1 8NE [IO81BO, SS69]
G6 OZW Details withheld at licensee's request by SSL.
G6 PAA J F C Brownsett, 24 Lancaster Rd, Shortdtown, Shortstown, Bedford, MK42 0UB [IO92SC, TL04]
G6 PAE R A Hillum, 48 Lydiard Way, Trowbridge, BA14 0UJ [IO81VH, ST85]
G6 PAJ R J Green, 1 Knightsbridge Rd, Messingham, Scunthorpe, DN17 3RA [IO93QM, SE80]
G6 PAN Details withheld at licensee's request by SSL.
G6 PAP S M Hale, 437 Littleworth Rd, Hednesford, Cannock, WS12 5HZ [IO92AQ, SK01]
G6 PAR P A Rhodes, 1 Killinghall Ave, Bradford, BD2 4SA [IO93DT, SE13]
G6 PAS Details withheld at licensee's request by SSL.
G6 PAY E Moore, 93 Mornington Rd, Sneyd Green, Stoke on Trent, ST1 6EN [IO83WB, SJ84]
GI6 PAZ W H McConnell, 17 Beech Green, Doagh, Ballyclare, BT39 0QB [IO64XR, J28]
G6 PBI K J Partington, 38 Queensgate Dr, Royton, Oldham, OL2 5SD [IO83WN, SD90]
G6 PBJ W F Pugh, 30 Warbreck Rd, Orrell Park, Liverpool, L9 8EG [IO83ML, SJ39]
G6 PBL E M Whitlam, Oak Farm, Barningham Rd, Edgefield, Melton Constable, Norfolk, NR24 2AW [JO02NU, TG13]
G6 PBO J S Tobin, 17 Holmwood Garth, Hightown, Ringwood, BH24 3DT [IO90CU, SU10]
G6 PBQ F Watson, 3 Hargreaves St., Nelson, BB9 7DB [IO83VT, SD83]
G6 PBS C J McCarthy-Stewart, 71 Beverley Cl, Gillingham, ME8 9HQ [JO01HI, TQ86]
GI6 PBV Rev. J Shea, Christian Brothers, Kelvin Rd, Omagh, Co. Tyrone, N Ireland, BT78 1LD [IO64IO, H47]
G6 PBW J Wainwright, 8 Common Ln, Cutthorpe, Chesterfield, S42 7AN [IO93GG, SK37]
G6 PBZ P R Wright, Preston Rd, Abingdon, Oxon, OX14 5NG [IO91IP, SU49]
G6 PCA J Spinks, 5 Spencer Cl, The Prinnels, Swindon, SN5 6NE [IO91BN, SU18]
G6 PCC R J Slade, 6 Bullen Cl, Bury St. Edmunds, IP33 3JP [JO02IF, TL86]
G6 PCD Details withheld at licensee's request by SSL.
G6 PCE R J Stamford, 30 Craft Way, Steeple Morden, Royston, SG8 0PF [IO92WB, TL24]
G6 PCJ D E Allen, 162 Wood Ln, Newhall, Swadlincote, Derbyshire, DE11 0LY [IO92FS, SK22]
G6 PCN J H Bailey, 3 Maple Dr, Hucknall, Nottingham, NG15 6GG [IO93JA, SK54]
G6 PCP J C Brown, 19 Church Rd, Rayleigh, SS6 8PL [JO01HO, TQ89]
GM6 PCW P A Boyd, 144 Brown St., Paisley, PA1 2JE [IO75SU, NS46]
G6 PCX J M Beresford, 1 Russell Pl, Maltby, Rotherham, S66 7HB [IO93JK, SK59]
G6 PDE C J Irish, 20 Ransome Rd, Ipswich, IP3 9BD [JO02OB, TM14]
G6 PDH Details withheld at licensee's request by SSL.
G6 PDK Details withheld at licensee's request by SSL.
G6 PDM S H Procter, 1B York Villas, York St, Colne, Lancs, BB8 0ND [IO83WU, SD84]
G6 PDR S P Riggs, 10 Chaplin Rd, Easton, Bristol, BS5 0JU [IO81RL, ST67]
G6 PDV Details withheld at licensee's request by SSL.
G6 PEA B W Terry, 389 Sutton Rd, Maidstone, ME15 9BU [JO01GF, TQ75]
G6 PEG C T Price, 42 Kipling Rd, Kettering, NN16 9JZ [IO92PJ, SP88]
G6 PEH A H Rands, 20 Riby Rd, Keelby, Grimsby, DN37 8ER [IO93WH, TA10]
G6 PEI R Newby, 10 Pennine Cres, Brierfield, Nelson, BB9 5EU [IO83VT, SD83]
G6 PEJ J R Moreton, 16 Oakwood Rd, Rode Heath, Stoke on Trent, ST7 3TG [IO83UC, SJ85]
G6 PEP J A G Morris, 22 St. Amand Dr, Abingdon, OX14 5RQ [IO91IP, SU49]
GM6 PER I J McMillan, 188 Abbeygreen Rd, Lesmahagow, Lanark, ML11 0AL [IO85BP, NS83]
G6 PEZ R H Whalley, 94 Wellington Rd, Bollington, Macclesfield, SK10 5HT [IO83WH, SJ97]
G6 PFB D Wain, Hillsborough, 148 Congleton Rd, Biddulph, Stoke on Trent, ST8 6QN [IO83WC, SJ85]
G6 PFD J A Wilkinson, The Bungalow, Winchester Way, Brinsworth, Rotherham, S60 5NS [IO93HJ, SK49]
G6 PFE M K Walker, 12 Atkinson Rd, Sale, M33 6FY [IO83UK, SJ79]
G6 PFF A M Willis, Kilncroft, Broad Layings, Woolton Hill, Newbury, Berks, RG15 9TS [IO91LJ, SU66]
GM6 PFJ G W J Gott, 21 Hamilton Ave, Broomlands, Dumfries, DG2 7LW [IO85EB, NX97]
GW6 PFK L Griffiths, 5 Heol Sarn, Llantrisant, Mid Glam, CF7 8DA [IO81HN, ST08]
G6 PFN A Hewitt, 29 Brabazon Rd, Oadby, Leicester, LE2 5HF [IO92KO, SK60]
G6 PFP S G Hill, 5 Tavistock Ave, St. Albans, AL1 2NQ [IO91TR, TL10]
G6 PFU J B Halfpenny, 82 Mellington Ave, East Didsbury, Manchester, M20 5NH [IO83VJ, SJ88]
G6 PFX I Harris, 4 Hopton Cl, Bartestree, Hereford, HR1 4DQ [IO82QB, SO54]
G6 PGB Details withheld at licensee's request by SSL.
G6 PGC Details withheld at licensee's request by SSL.
G6 PGD R J Healy, 9 Homer Cl, Rowner, Gosport, PO13 9TJ [IO90JT, SU50]
G6 PGG Dr D M Jones, Westwood, 8 Chapel Cl, Comberbach, Northwich, CW9 6BA [IO83RH, SJ67]
G6 PGJ J D Kyle, 9 Cromwell Rd, Alperton, Wembley, HA0 1JS [IO91UM, TQ18]
G6 PGM D R Kaye, 3 Anderson Cl, Needham Market, Ipswich, IP6 8UA [JO02MD, TM05]
G6 PGN C J King, 18812 Thornwood Circle, Huntington Beach, California, USA 92646, X X
G6 PGO B J Key, 65 Ravenhurst Rd, Harborne, Birmingham, B17 9TB [IO92AL, SP08]
G6 PGP S N Kinton, 7 Ferndale Dr, Ratby, Leicester, LE6 0LH [IO92IP, SK50]
G6 PGQ M B Karazy-Kulin, The Old Cottage Farm, Nuthurst Rd, Maplehurst, West Sussex, RH13 6RE [IO91UA, TQ12]
G6 PGT J G H Chapman, 19 Church Farm Rd, Upchurch, Sittingbourne, ME9 7AG [JO01HJ, TQ86]
G6 PGV N P Coppack, 15 Wyndham Rd, Salisbury, SP1 3AA [IO91CB, SU13]
G6 PHA Details withheld at licensee's request by SSL.
G6 PHC P W Dewick, Glen Coe, 6 Station Rd, Kirton Lindsey, Gainsborough, DN21 4BB [IO93QL, SK99]
G6 PHF M C Dent, 2 Wythop Croft, Westgate, Morecambe, LA4 4UT [IO84NB, SD46]
G6 PHH P H Dickens, 2 Millfield Ave, Marsh Gibbon, Bicester, OX6 0HP [IO91LV, SP62]
G6 PHJ P R Daines, 4 Hornbeam Cl, Narborough, Leicester, LE9 5YQ [IO92JN, SP59]
G6 PHQ R J Elmer, 44 Cavalier Rd, Old Basing, Basingstoke, RG24 7ER [IO91LG, SU65]
G6 PHT S C Fitzhugh, 12 The Piece, Cogenhoe, Northampton, NN7 1LX [IO92OF, SP86]
G6 PHU D A Ford, 49 Devon Rd, Luton, LU2 0RJ [IO91TV, TL12]
G6 PHV D R G Ford, 17 Crowland Rd, Stopley, Luton, LU2 8EH [IO91TV, TL12]
G6 PHY Lady B H Plowden, Martells Manor, High Easter Rd, Barnston, Dunmow, Essex, CM6 1NB [JO01EU, TL61]
G6 PHZ P F Maddox, 7 Keats Rd, Flitwick, Bedford, MK45 1QD [IO92RA, TL03]
G6 PIB M J Livingston, The Electricity Directorat, Training Department, PO Box 2, Bahrain
G6 PIF F W Thompson, The Rookery, Village Green, Allington, Grantham, Lincs, NG32 2EA [IO92PW, SK84]
G6 PII D Simpson, 37 Moore Cl, Claypole, Newark, NG23 5AU [IO93PA, SK84]
G6 PIM P T Lawford, 44 Clarendon Rd, Broadstone, BH18 9HY [IO80XS, SY99]
G6 PIR H Simmonds, 118 Paris Ave, Westlands, Newcastle, ST5 2QX [IO83VA, SJ84]
G6 PJC P W Brown, 13 Hillside Cl, Biddulph Moor, Stoke on Trent, ST8 7PF [IO83WC, SJ85]
GI6 PJD M R Belshaw, 155 Dunlady Manor, Dundonald, Co. Down, BT16 0YS [IO74CO, J47]
G6 PJE Details withheld at licensee's request by SSL.
G6 PJK Details withheld at licensee's request by SSL.
G6 PJL Details withheld at licensee's request by SSL.
G6 PJM B A Bucknall, 38 Hillside, Brownhills, Walsall, WS8 7AF [IO92AP, SK00]
G6 PJP L L Bealing, 18 Avon Rd, Oakley, Basingstoke, RG23 7DJ [IO91JG, SU55]
G6 PJS W R Chenoweth, 30 Whitland Rd, Hartcliffe, Bristol, BS13 9QG [IO81QJ, ST56]
G6 PJW D J Coles, 34 Riverview Gdns, Twickenham, TW1 4RT [IO91UK, TQ17]
G6 PKA J S Colls, 44 Willow Ln, Appleton, Warrington, WA4 5EA [IO83RI, SJ68]
G6 PKM J Allen, 27 Grafton Rd, Whitley Bay, NE26 2NR [IO95GA, NZ37]
GM6 PKP J W Allardyce, 17 Hallglen Terr, Glen Village, Falkirk, FK1 2AP [IO85CX, NS87]
G6 PKR Details withheld at licensee's request by SSL.
G6 PKS B J Bean, 9 Groombridge Cl, Welling, DA16 2BP [JO01BK, TQ47]
G6 PKT J R Brenchley, 14 Linton Dr, Artists Way, Andover, SP10 3TT [IO91GF, SU34]
G6 PKV G Branagan, 434 Manchester Rd West, Little Hulton, Manchester, M38 9XU [IO83SM, SD70]
G6 PKY R B Bush, 1 Woodview, Cotgrave, Nottingham, NG12 3LA [IO92LV, SK63]
G6 PKZ B Woodley, 1 Melton Dr, Didcot, OX11 7JP [IO91JO, SU59]
G6 PLB P S Ridley, 8 Locksley Dr, Ferndown, BH22 8JY [IO90BT, SZ09]
G6 PLF J L Smoker, 9 Anson Way, Bicester, OX6 7UH [IO91KV, SP52]
GM6 PLG P Sloan, 36 Trenchard Cres, Kinloss, Forres, IV36 0UP [IO87FP, NJ06]
G6 PLH C C Smith, 86 Woodland Rd, Handsworth, Birmingham, B21 0EP [IO92AM, SP08]
G6 PLN Details withheld at licensee's request by SSL.
GI6 PLO I M Bell, 3 Stratford Dr, Bangor, BT19 6ZW [IO74EP, J57]
G6 PLP Details withheld at licensee's request by SSL.
G6 PLR L C Chandless, 16 Crest Gdns, South Ruislip, Ruislip, HA4 9HD [IO91TN, TQ18]
G6 PLT E M Cheetham, 172A Hesketh Ln, Tarlton, Preston, PR4 6UD [IO83NQ, SD42]
G6 PLU A B Chenery, 43 Wessex Est, Ringwood, BH24 1XD [IO90CU, SU10]
G6 PLV P Cole, 42 Beechwood Park, Highlands Rd, Leatherhead, KT22 8NL [IO91UG, TQ15]
GW6 PMC R W Evans, 16 Monmouth Gr, Prestatyn, LL19 8TS [IO83GH, SJ08]
G6 PMD C J Eagling, 96 Regent Rd, Brightlingsea, Colchester, CO7 0NZ [JO01MT, TM01]
GI6 PME C Foley, 5 Woodland Dr, Cookstown, BT80 8PL [IO64PP, H87]
GM6 PMG R C McGregor, 3 Hillcrest View, Larkhall, ML9 2HX [IO85AR, NS75]
G6 PMJ S P C Murphy, 1 Orchard Cottage, Golden Valley, Upleadon, Newent, GL18 1HN [IO81TW, SO72]
G6 PMO I T Parker, 27 St. Audries Rd, Worcester, WR5 2AL [IO82VE, SO85]
G6 PMP P L Rawson, 23 Little Hey, Nooking Ln, Maltby, Rotherham, S66 8AR [IO93JK, SK59]
G6 PMR P Shaw, 52 Belvedere Parade, Bramley, Rotherham, S66 0WA [IO93IK, SK49]
G6 PMS I K Gregg, 7 Partlands Ave, Ryde, PO33 3DS [IO90JR, SZ59]
G6 PMT P M Townshend, 48 Cabrera Ave, Virginia Water, GU25 4HA [IO91RJ, SU96]
G6 PMW G C Goodier, 12 The Ridings, Whittle-le-Woods, Chorley, PR6 7QH [IO83QQ, SD52]
G6 PMX T G Green, Draidean, Rosslyn Ave, Presall, Poulton-le-Fylde, FY6 0HE [IO83MW, SD34]
G6 PNB E J Harris Nrth Bristol AR, 4 Maisemore Ave, Patchway, Bristol, BS12 6BT [IO81RM, ST68]
G6 PNG P S Hill, 95 Walton Way, Newbury, RG14 2LL [IO91IJ, SU46]
G6 PNH Details withheld at licensee's request by SSL.
G6 PNI C Hawkes, 6 Simpson Cl, Maidenhead, SL6 8RZ [IO91PM, SU88]
G6 PNJ S C Hammond, 9 North Rd, Belvedere, DA17 6JX [JO01BL, TQ47]
G6 PNM J P Haywood, 18 Barnard Cl, Eynesbury, St. Neots, Huntingdon, PE19 2UP [IO92UF, TL15]
G6 PNO P Hill, 5 Eider Cl, South Beach Est, Blyth, NE24 3QD [IO95FC, NZ37]
G6 PNT Details withheld at licensee's request by SSL.
G6 POC R E Kinrade, 23 Crofthill Rd, Slough, SL2 1HG [IO91QM, SU98]
G6 POD K V T Knott, 85 Tollgate Rd, Colney Heath, St. Albans, AL4 0PX [IO91VR, TL20]
G6 POE J K Knott, 3 Lords Wood, Welwyn Garden City, AL7 2HF [IO91VT, TL21]
G6 POF G W Greenfield, 279 Lords Wood Ln, Chatham, ME5 8JU [JO01GI, TQ76]
G6 POI J Wright, Chez Mon, Burton Rd, Holme, Carnforth, LA6 1QN [IO84PE, SD57]
G6 POJ I Worthy, 7 The Paddocks, Pilsley, Chesterfield, S45 8ET [IO93HD, SK46]
GW6 POO R Smallwood, 8 Queensway, Shotton, Deeside, CH5 1HT [IO83LF, SJ36]
G6 POQ A W Stone, Kilpton, Banbury Rd, Bicester, OX6 7NH [IO91KV, SP52]
G6 POV M G Walker, 232 Bideford Green, Leighton Buzzard, LU7 7TS [IO91PW, SP92]
G6 POW D J Pow, 16 Ancaster Cl, Trowbridge, BA14 9DA [IO81VH, ST85]
G6 POY Details withheld at licensee's request by SSL.
G6 PPA T M Farmer, 35 Ascot Dr, Dudley, DY1 2SN [IO82WM, SO99]
G6 PPD A R Morgan, 14 Hawthorn Gdns, Stargate, Ryton, NE40 3ED [IO94DX, NZ16]
G6 PPJ Details withheld at licensee's request by SSL.[Op: P L Nicholls. Station located in London N16.]
G6 PPT P Thompson, Highfield, Park Ln, Pickmere, Knutsford, Ches, WA16 0JX [IO83SG, SJ67]
G6 PPU H G Chappell, Oanley, East Lyng, Taunton, Somerset, TA3 5AU [IO81MB, ST32]
G6 PPV P Caswell, 94 Dewsbury Rd, Luton, LU3 2HJ [IO91SV, TL02]
G6 PPY S Carter, 3 Mary St, Burnley, Lancs, BB10 4AJ [IO83VS, SD83]
G6 PQB Details withheld at licensee's request by SSL.
G6 PQI J T Finch, Ash Cottage, Chapel Ln, Old Dalby, Melton Mowbray, LE14 3LA [IO92LT, SK62]
GW6 PQT J A Barlow, 19 Tanymarian, Llanddulas, Clwyd, Nth Wales, LL22 8ER [IO83EG, SH97]
G6 PQW N J Redmayne, Garden Cottage, Middle Row, Cambo, Morpeth, Northd., NE61 4AZ [IO95AD, NZ08]
G6 PRA J Whittaker, 6 Bradley Gdns, Rosegrove, Burnley, BB12 6JT [IO83US, SD83]
G6 PRE E W Snell, 156 Brookdale Ave South, Greasby, Wirral, L49 1SS [IO83KJ, SJ28]
G6 PRI C R Brown, 1 College Cl, Coltishall, Norwich, NR12 7DT [JO02QR, TG21]
G6 PRK K Barrett, 114 William St., Long Eaton, Nottingham, NG10 4GD [IO92IV, SK43]
G6 PRL D C Brown, 63A Great Northern St., Huntingdon, PE18 6HJ [IO92VH, TL27]
G6 PRO R Brown, 36 Hazelbury Rd, Bristol, BS14 9ER [IO81RK, ST66]
G6 PRP W S Barker, 297 Williamthorpe Rd, North Wingfield, Chesterfield, S42 5NT [IO93HE, SK46]
G6 PRV Details withheld at licensee's request by SSL.
G6 PS A F M Parsons, Wyvern, Eridge Rd, Crowborough, TN6 2SY [JO01CB, TQ53]
G6 PSA N R Turnham, 153 Canterbury Rd, Davyhulme, Urmston, Manchester, M41 0PY [IO83TK, SJ79]
G6 PSC M G Horn, The Grove, Rectory Rd, Coltishall, Norwich, Norfolk, NR12 7HG [JO02QR, TG21]
GM6 PSK C J Pollard, 87 Netherwood Park, Deans, Livingston, EH54 8RW [IO85FV, NT06]
G6 PSO I D Russell, 24 Standard Ave, Tile Hill, Coventry, CV4 9BW [IO92FJ, SP27]
G6 PSZ T K Shackleton, 27 Ct Cres, Kingswinford, DY6 9RJ [IO82VL, SO88]
G6 PTE B Underwood, The Hollies, 30 Heanor Rd, Codnor, Ripley, DE5 9SH [IO93HA, SK44]
G6 PTF J Wilson, Belle Vue House, Common Side, Distington, Workington, CA14 4PU [IO84FO, NY02]
G6 PTI L J Gooch, 27 Stanstead Way, Frinton on Sea, CO13 0BG [JO01PU, TM22]
G6 PTJ T C Houghton, 93 Redfern Cl, Solihull, B92 8SJ [IO92CK, SP18]
G6 PTL Details withheld at licensee's request by SSL.
GM6 PTX G A T Gane, 28 Queens Croft, Kelso, TD5 7NN [IO85SO, NT73]

G6 PTY M G Greenow, 7 Elm Cl, Little Stoke, Bristol, BS12 6RG [IO81RM, ST68]
G6 PTZ W R Hinds, 22 Manta Rd, Dosthill, Tamworth, B77 1NQ [IO92DO, SK20]
G6 PUA E G Hodder, 7 Chard Rd, Axminster, EX13 5HN [IO80MS, SY39]
G6 PUC Details withheld at licensee's request by SSL.
G6 PUE M J Mackmin, 89 Wellingborough Rd, Rushden, NN10 9YJ [IO92QG, SP96]
G6 PUH E A Patterson, 174 Lowe Ave, Wednesbury, WS10 8NT [IO82XN, SO99]
G6 PUV W Holding, 20 Lingfield Cres, Beech Hill, Wigan, WN6 8QA [IO83QN, SD50]
G6 PVA M A J Green, 97 Langley Hall Rd, Solihull, B92 7HD [IO92CK, SP18]
G6 PVB Details withheld at licensee's request by SSL.
G6 PVC P V Coates, Jacaranda, Cotswold Cl, Staines, Middx, TW18 2DD [IO91RK, TQ07]
G6 PVG Details withheld at licensee's request by SSL.
GW6 PVK G L Jones, 18 Mountain Cl, Hope, Wrexham, LL12 9SE [IO83LC, SJ35]
G6 PVQ Details withheld at licensee's request by SSL.
G6 PVS R N M Alston, 3 Nursery Cl, Biggleswade, SG18 0HR [IO92UC, TL14]
G6 PVU A C Appleyard, 18 Earl St., Nelson, BB9 9JA [IO83VU, SD83]
G6 PVV C Burt, 2 Regatta House, Thames Side, Staines, TW18 2HA [IO91RK, TQ07]
G6 PWF S E Choules, 43 Ashbrook Rd, Old Windsor, Windsor, SL4 2LT [IO91RK, SU97]
G6 PWJ J D Chiddick, Itamerenkatu 8A6, 00180 Helsinki, Finland, X X
G6 PWL R W G Cloutman, 44 Carlisle Ave, St. Albans, AL3 5LX [IO91TS, TL10]
G6 PWO Details withheld at licensee's request by SSL.
G6 PWQ D K D Dick, 140 Chatham St., Edgeley, Stockport, SK3 9JU [IO83VJ, SJ88]
G6 PWS R J Fuller, 18 St. Leonards Cres, Sandridge, St. Albans, AL4 9EH [IO91US, TL11]
G6 PWT D H Francis, High Point, Long Park, Crawley, Winchester Hants, SO21 2QE [IO91HC, SU43]
G6 PWY C Baron, 9 Beech St., Great Harwood, Blackburn, BB6 7RB [IO83TS, SD73]
G6 PWZ B Boothby, 16 Wandales Ct, Burniston, Scarborough, YO13 0HE [IO94SH, TA09]
G6 PXA A L Abbott, 12 Moorside Gdns, Drighlington, Bradford, BD11 1HZ [IO93ER, SE22]
GW6 PXF G R Hughes, 35 Penrhos Rd, Bangor, LL57 2AX [IO73WF, SH57]
G6 PXJ A Harrison, Nirvana Cottage, 42 Bell Ln, Rawcliffe, Goole, DN14 8RP [IO93MQ, SE62]
G6 PXK J I Henderson, 10 Exeter Rd, Felixstowe, IP11 9AS [JO01QX, TM23]
G6 PXN R Bee, 80 Hospital Rd, Burntwood, WS7 0EQ [IO92BQ, SK00]
G6 PXQ R Boyce, 3 Castleton Cttgs, Westhide, Hereford, HR1 3RF [IO82QC, SO54]
G6 PXV N D Chapman, Holyfield Farm, Waltham Abbey, Essex, EN9 2ED [IO91XQ, TL30]
G6 PXX L Dempsey, 24 James St., Great Harwood, Blackburn, BB6 7JE [IO83TS, SD73]
G6 PXZ M D Foote, 31 New Rd, Wonersh, Guildford, GU5 0SF [IO91RE, TQ04]
GM6 PYD A J S Dunnett, 11 Silverknowes View, Edinburgh, EH4 5PY [IO85IX, NT27]
G6 PYF D Hills, 4 Pinfield Ln, Chippenham, SN15 3SU [IO81WK, ST97]
G6 PYH D S James, 33 Wolverley House, Gardiners Ln, Ashwell, Baldock, SG7 5LZ [IO92WA, TL23]
G6 PYL P Hatter, 14 Morland Ave, Bromborough, Wirral, L62 6BE [IO83MH, SJ38]
G6 PYM A K Hedges, 4 Haselette Way, Upper Hatherley, Up Hatherley, Cheltenham, GL51 5RQ [IO81WV, SO92]
GI6 PYP A R Gault, 8 Spring Vale, Coles Hill, Enniskillen, BT74 7FA [IO64EI, H24]
G6 PYR H W Adams, Hill Sixty, Happisburgh, Norwich, Norfolk, NR12 0RP [JO02ST, TG32]
G6 PZ P R Beecham, 56 Moorland Rd, Weston Super Mare, BS23 4HR [IO81MH, ST35]
G6 PZE D M Jefferson, 48 Neston Rd, Walshaw, Bury, BL8 3DB [IO83TO, SD71]
G6 PZF P W James, 33 Headley Chase, Warley, Brentwood, CM14 5BN [JO01DO, TQ59]
G6 PZN M A McLoughlin, Greystone Barn, Banbury Rd, Finmere, Buckingham, MK18 4AJ [IO91KX, SP63]
G6 PZS D H Carr, 5 Church Meadow, Hyde, SK14 4RT [IO83WK, SJ99]
G6 PZT K Davy, 4 Malvern Cl, Chelmsford, CM1 2HL [JO01FS, TL60]
G6 PZW Details withheld at licensee's request by SSL.
G6 QA L V Jopson, 68 Greenmount Park, Kearsley, Bolton, BL4 8NT [IO83TN, SD70]
G6 QI R Walker, Tregellas, Cury Cross Lanes, Helston, Cornwall, TR12 7AZ [IO70JA, SW62]
G6 QM F J T Harris Qn Mary Cont Gr, 4 Merestones Dr, The Park, Cheltenham, GL50 2SS [IO81WV, SO92]
G6 QN T J Blakeman, 194 Seaforth Ave, New Malden, KT3 6JW [IO91VJ, TQ26]
G6 QQ R D L Dutton, 55 Stalham Rd, Hoveton, Norwich, NR12 8DU [JO02RR, TG31]
G6 QY W B Brown, 34 Hulbert Rd, Bedhampton, Havant, PO9 3TF [IO90LU, SU70]
G6 RAD J W Crook, 19 Rylands Rd, Kennington, Ashford, TN24 9LH [JO01KD, TR04]
G6 RAE R Davis, 43 Oldfields Cl, Leominster, HR6 8TL [IO82PF, SO55]
G6 RAF R Hyde RAF North Luffenham ARC, 25 The Pastures, Cottesmore, Oakham, LE15 7DZ [IO92QR, SK91]
G6 RAJ I P Bower, 4 Havelock St., Ilkeston, DE7 5RJ [IO92IX, SK44]
GM6 RAK D T Brown, 14 Nevin Cres, Carnoustie, DD7 6HW [IO86PM, NO53]
G6 RAQ S W Hayter, 2 Shelsley Dr, Basildon, SS16 6NA [JO01FN, TQ68]
G6 RAR Details withheld at licensee's request by SSL.
G6 RAU P Johnstone, 22 Red Hall Dr, Barwell, Leicester, LE9 8BY [IO92HN, SP49]
G6 RAV W J Keeley, 7 Parr Cl, Myton Rd, Warwick, CV34 6NE [IO92FG, SP36]
G6 RAZ J S Paton, 16 Homefield, Thornbury, Bristol, BS12 2EW [IO81RO, ST69]
G6 RBH Details withheld at licensee's request by SSL.
G6 RBJ R B Jones, 39 Ashton Cl, Needingworth, St. Ives, Huntingdon, PE17 3UA [IO92XH, TL37]
G6 RBM R A Jeffery, 15 Greenway, Hulland Ward, Ashbourne, DE6 3FE [IO93EA, SK24]
G6 RBO W J Bennett, 44 Wood Ln, Streetly, Sutton Coldfield, B74 3LR [IO92BO, SP09]
G6 RBP R B Pearsey, 21 Ashwood Dr, Newbury, RG14 2PN [IO91IJ, SU46]
G6 RBR M H Allen, Oakwood House, 15 East Park St., Chatteris, PE16 6LQ [JO02AK, TL38]
G6 RC R E R Carslake Crawley Amateur Radio Club, 38 Loppets Rd, Tilgate, Crawley, RH10 5DW [IO91VC, TQ23]
G6 RCD P J Clark, 166 Attenborough Ln, Attenborough, Beeston, Nottingham, NG9 6AB [IO92JV, SK53]
G6 RCH W M Cutler, 2 Rudgard Ave, Cherry Willingham, Lincoln, LN3 4JG [IO93SF, TF07]
G6 RCJ J E Fagg, 5 Singledge Ave, Whitfield, Dover, CT16 3LQ [JO01PD, TR34]
GW6 RCK H A Fray, 6 Meirwen Dr, Culverhouse Cross, Cardiff, CF5 4ND [IO81IL, ST17]
GW6 RCP Details withheld at licensee's request by SSL.
G6 RCT T A Stellar, 27 Blackmore Chase, Wincanton, BA9 9SB [IO81TB, ST72]
GW6 RCV Details withheld at licensee's request by SSL.
G6 RCY D A Reed, 11 Grenville Cl, Corby, NN17 2RP [IO92PL, SP88]
G6 RDD I F Senter, 21 King Coel Rd, Colchester, CO3 5AG [JO01KV, TL92]
G6 RDG Details withheld at licensee's request by SSL.
G6 RDL D J Leach, 8 Park St., Cheslyn Hay, Walsall, WS6 7EF [IO82XP, SJ90]
G6 RDO A C B Shaw, 45 Aire Rd, Wetherby, LS22 7UE [IO93HW, SE44]
GW6 RDV B J Clarke, 47 Oakway, South Pentrebane, Cardiff, CF5 3EH [IO81JL, ST17]
G6 RDZ D P Jolley, 101 Bishops Rd, Kings Lynn, PE30 4NU [JO02FS, TF62]
G6 REA J A Gilpin, 149 Elm Rd, New Malden, KT3 3HX [IO91UJ, TQ26]
G6 REC M L Hart, 7 Ullswater Ave, South Wootton, Kings Lynn, PE30 3NJ [JO02FS, TF62]
G6 REG A P Joyce, 16 Princethorpe Dr, Falcon Rise, Banbury, OX16 8FS [IO92IB, SP44]
G6 REH J C S Staplehurst, 27 Brumfield Rd, West Ewell, Epsom, KT19 9PA [IO91UI, TQ26]
G6 REM I Atkinson, 45 Fd Sp Sqn, 21 Engr Regt, Bfpo 48, X X
G6 REN R G Lomas, 9 Nettleton Ln, Harpurhill, Buxton, SK17 9JX [IO93BF, SK07]
GW6 REQ W H Vize, Cefn Rhos, Bethel, Caernarvon, Gwynedd, LL55 1YB [IO73VD, SH56]
G6 RES Details withheld at licensee's request by SSL.
G6 REV J Yam, 58 Rosemary Ave, West Molesey, KT8 1QE [IO91TJ, TQ16]
G6 REW G C Seymour-Smith, Trelawney, Old Post Office Hill, Stratton, Bude, EX23 9DB [IO70RT, SS20]
G6 REY R Q McMinn, 41 Overton Dr, Water Orton, Birmingham, B46 1QL [IO92DM, SP19]
GJ6 RFE R P Thomson, 20 Rosemount Est, James Rd, St. Saviour, Jersey, JE2 7RR
GM6 RFG K Russell, 36 Hawkhill Rd, Alloa, FK10 1SA [IO86CC, NS89]
G6 RFH D E J Ruck, 38 Kincardine Dr, Bletchley, Milton Keynes, MK3 7PG [IO92OA, SP83]
G6 RFM A J Warren, 20 Wolverhampton Rd, Stafford, ST17 4BP [IO82WT, SJ92]
GM6 RFQ Details withheld at licensee's request by SSL.
G6 RFR A Brammer, Rosskeen, The Green, Reepham, Lincoln, LN3 4DH [IO93SG, TF07]
G6 RFU P Csapo, 87 Latchmere Rd, Kingston upon Thames, KT2 5TU [IO91UK, TQ17]
G6 RFV P D Green, 82 Suffolk Ave, Derby, DE21 6ER [IO92GW, SK33]
G6 RGA J A Ewen, 7 Nutbourne Cl, Eastbourne, BN23 7EN [JO00DT, TQ60]
GM6 RGD T A H Murray, 2 The Glebe, Edzell, Brechin, DD9 7SZ [IO86QT, NO56]
G6 RGN W C Stockley, 10 Swan Rd, Timperley, Altrincham, WA15 6BX [IO83UJ, SJ79]
G6 RGO L C Tootill, 27 Rossett St., Liverpool, L6 4AN [IO83MK, SJ39]
GW6 RGT M L Morgan, 11 Marlborough Rd, Meas y Rliw, Greenmeadow, Cwmbran, NP44 5EJ [IO81LP, ST29]
GM6 RGU R J Nicolson, Daisybank, Braehead Rd, Letham, Forfar, DD8 2PG [IO86OO, NO54]
G6 RGV R J Paxton, Mount Pleasant Cottage, Hambrook Hill North, Hambrook, nr Chichester, West Sussex, PO18 8UQ [IO90NU, SU70]
GM6 RGY W R Hardie, 50 Braemar Dr, Falkirk, FK2 9HA [IO86CA, NS88]
G6 RGZ Details withheld at licensee's request by SSL.
G6 RHJ M S Swain, 17 Sponnes Rd, Towcester, NN12 6ED [IO92MD, SP64]
G6 RHK C G Spencer, 83 Oulton Rd, Ipswich, IP3 0QE [JO02OB, TM14]
G6 RHL P J West, 6 Iveldale Dr, Shefford, SG17 5AD [IO92UA, TL13]
G6 RHN Dr D L Morris, Rowley Farm, Rowley Ln, Borehamwood, Herts, WD6 5PE [IO91VQ, TQ29]
G6 RHO Dr N Y L Yue, 20 Broad Walk, Winchmore Hill, London, N21 3DB [IO91WP, TQ39]
G6 RHP F A Law, 47 Springcroft, Hartley, Longfield, DA3 8AR [JO01DJ, TQ66]
G6 RHQ F Martland, 6 Eskdale Ave, Fleetwood, FY7 8LU [IO83LV, SD34]
G6 RHV I C R Smith, 8 Tudor Ct, Castle Way, Feltham, TW13 7QQ [IO91TK, TQ17]
G6 RHX Details withheld at licensee's request by SSL.
G6 RIB A J Evans, 26 Deanwater Cl, Locking Stumps, Birchwood, Warrington, WA3 6ER [IO83RK, SJ69]
G6 RIC M J Ellis, 28 High Meadows, Romiley, Stockport, SK6 4PT [IO83XK, SJ99]
G6 RIG N A Golding, Coppice View, 16 Littlewood Rd, Walsall, WS6 7EU [IO82XP, SJ90]
G6 RII M B Dodson, 56 Campion Gr, Harrogate, HG3 2UG [IO94FA, SE25]
G6 RIJ R C Fletcher, 33 Littlewood Ln, Cheslyn Hay, Walsall, WS6 7EJ [IO82XQ, SJ90]
G6 RIM C Berry, 258 Lowerhouse Ln, Padiham, Burnley, BB12 6NG [IO83US, SD83]
G6 RIO Details withheld at licensee's request by SSL.
G6 RIP Details withheld at licensee's request by SSL.
G6 RIQ G S Dunn, 11 Ellesmere Rise, Grimsby, DN34 5PE [IO93WN, TA20]
G6 RIY A C Wilkinson, 4 Royd St., Cowling, Keighley, BD22 0BN [IO83XV, SD94]
G6 RJ A Robinson, High Garth, Well Bank, Well, Bedale, DL8 2QQ [IO94EF, SE28]
G6 RJH J M Proffitt, 38 Hockley Rd, Poynton, Stockport, SK12 1RW [IO83WI, SJ98]
G6 RJS A C Miles, 44 Clarendon Gdns, Ilford, IG1 3JW [JO01AN, TQ48]
G6 RJT Details withheld at licensee's request by SSL.
G6 RJW R J Woodgate, c/o Counter Nz Post, Princess St, Centrall, Palmerston North, New Zealand
G6 RJZ P R Mead, The Buntings, 83 Boniface Rd, Ickenham, Uxbridge, UB10 8BY [IO91SN, TQ08]
G6 RKE Details withheld at licensee's request by SSL.
G6 RKF R G Taylor, 20 Scraley Rd, Heybridge, Maldon, CM9 4BL [JO01IR, TL80]
G6 RKG J C M Walters, The Gables, Lavenham Rd, Great Waldingfield, Sudbury, CO10 0RN [JO02JB, TL94]
G6 RKJ R C Butland, 7 Heathfield Ave, Binfield Heath, Henley on Thames, RG9 4ED [IO91MM, SU77]
G6 RKS R Domville, 3 Eden Terr, Low Willington, Willington, Crook, DL15 0BA [IO94DR, NZ23]
G6 RKV M S Gillett, Iona, 113 Pk Ln, Sanbach, Ches, CW11 9EE [IO83TD, SJ76]
G6 RLG J A Kerr, 82 Avon Dr, Alderbury, Salisbury, SP5 3TH [IO91DA, SU12]
G6 RLH R L Lintorn, 220 Mayplace Rd Eas, Barnehurst, Kent, DA7 6EW [JO01CL, TQ57]
G6 RLS J A Husk Looe School RC, Sunrising, East Looe, Cornwall, PL13 1NL [IO70SI, SX25]
G6 RMA D Glover, 56 Selbourne Rd, Gillingham, ME7 1QP [JO01GJ, TQ76]
G6 RMC T R Cross, 2 Green Ln, Wincham, Northwich, CW9 6EF [IO83SG, SJ67]
G6 RMJ W A Clements, 3 May St., Durham, DH1 4EN [IO94ES, NZ24]
G6 RML M H Girdham, 19 Charles Ave, Laceby, Grimsby, DN37 7EZ [IO93VM, TA20]
GI6 RMO D H Johnston, Olanda, Lisreagh, Lisbellaw, Co Fermanagh, N Ireland, BT94 5BX [IO64FH, H34]
G6 RMR Details withheld at licensee's request by SSL.[Station located near Haslemere.]
G6 RMS Details withheld at licensee's request by SSL.
G6 RMV R E Sharp, School, Bridge St., Netherbury, Bridport, DT6 5LS [IO80PS, SY49]
G6 RMW H R Thompson, 176 Collingwood Rd, Chorley, PR7 2QF [IO83QP, SD51]
G6 RMX K Wilson, 40 Bournesfield, Hoghton, Preston, PR5 0EH [IO83QR, SD52]
G6 RMY C Wilson, 44 Callow Rd, Wavertree, Liverpool, L15 0HP [IO83MJ, SJ38]
GJ6 RND S W Sole, Heatherdene, 15 Longueville Rd, St. Saviour, Jersey, JE2 7SA
G6 RNE Details withheld at licensee's request by SSL.
G6 RNF R W Smith, 40 Burnside Rd, West Bridgford, Nottingham, NG2 7HW [IO92KW, SK53]
G6 RNN Details withheld at licensee's request by SSL.
G6 RNP Details withheld at licensee's request by SSL.[Op: R E Hale, OBE, FCIS, Veryan, 11 Meriden Road, Hampton-in-Arden, Solihull, West Midlands, B92 0BS.]
G6 RNT G R Kingdon, Wymering, Copley Dr, Sticklepath, Barnstaple, EX31 2BH [IO71XB, SS53]
GW6 RNV C E Brewster, 35 Ffordd Las, Sychdyn, Mold, CH7 6DU [IO83KE, SJ26]
G6 RNW A J Collins, 22 Butterfield Rd, Southampton, SO16 7EE [IO90HW, SU41]
G6 RO Dr R C Kaye, 6 Belmont Ave, Baildon, Shipley, BD17 5AJ [IO93CU, SE13]
GI6 ROP Details withheld at licensee's request by SSL.
G6 ROQ R J Skobelski, 2 Hemsley Ct, Stoughton Rd, Guildford, GU2 6PW [IO91QG, SU95]
G6 RPD R L Montford, 390 Selbourne Rd, Luton, LU4 8NU [IO91SV, TL02]
G6 RPF L J Pamment, 5 New Captains Rd, West Mersea, Colchester, CO5 8QP [JO01KS, TM01]
G6 RPV B M Davies, 62 Lincoln Rd, Leasingham, Sleaford, NG34 8JT [IO93SA, TF04]
G6 RPW Maj S I Andrews, 5 Radcliffe Rd, Salisbury, SP2 8EH [IO91CB, SU12]
G6 RQA D E Nicholls, 44 Hall Orchards, Middleton, Kings Lynn, PE32 1RY [JO02FR, TF61]
GW6 RQG F R Wightman, Bod Hyfryd, Pen y Cefn, Caerwys, Mold, CH7 5BN [IO83HG, SJ17]
G6 RQJ R Sherlock, 34 St. Cecilias Rd, Belle Vue, Doncaster, DN4 5EG [IO93KM, SE50]
G6 RQL W J McCormack, 39 Sackville Rd, Windle, St. Helens, WA10 6JD [IO83OL, SJ49]
G6 RQP S I Bateman, 31 Coberley, Footshill, Hanham, Bristol, BS15 2ET [IO81RK, ST67]
G6 RQT Details withheld at licensee's request by SSL.
GM6 RQU B A Hamilton, 51 Grange Rd, Grange, Edinburgh, EH9 1UF [IO85JW, NT27]
GM6 RQW T N Christie, 26 Alexandra Dr, Boghall, Bathgate, EH48 1ST [IO85EV, NS96]
G6 RQY C M Haddad, 8 Charmouth Ct, Kings Rd, Richmond, TW10 6EW [IO91UL, TQ17]
G6 RQZ B J Cripps, 3 Sabre Ct, Beaumont Park, Aldershot, GU11 1YY [IO91OF, SU85]
G6 RRB L G Logan Rolls Royce ARC, 19 Fenton Ave, Barnoldswick, Colne, BB8 6HB [IO83VW, SD84]
G6 RRM Dr B G F Roe Rolls Royce Mtr, 7 Abbey Fields, Crewe, CW2 8HJ [IO83SB, SJ65]
G6 RRP P Metson, 159 Lonsdale Dr, Enfield, EN2 7NB [IO91WP, TQ39]
G6 RRU B E Smith, Stables Cottage, 136 Main St., Ticknall, Derby, DE73 1JZ [IO92GT, SK32]
G6 RRV R A Weston, The Old Dairy, Slate Cross, Chedzoy Ln, Bridgwater, Somerset, TA7 8QR [IO81MD, ST33]
G6 RRY M R Dunbar, 42 Wickham Way, Shepton Mallet, BA4 5YG [IO81RE, ST64]
G6 RRZ E J Doyle, 33 Bodenham Rd, Northfield, Birmingham, B31 5DP [IO92AJ, SP07]
G6 RSB S J Burston, 14 Rheola Gdns, Thornbury, Plymouth, PL6 8UB [IO70WJ, SX55]
G6 RSC Details withheld at licensee's request by SSL.
G6 RSE K J Hendry South Essex Amateur Radio Soci, 48 Downer Rd North, Benfleet, SS7 3EG [JO01GN, TQ78]
G6 RSI L A L Hart, 25 Murcroft Rd, Stourbridge, DY9 9HT [IO82WK, SO98]
GW6 RSP I G Howson, 8 Craig-y-Dorth, Field House Farm, Monmouth, Gwent, NP5 4FH [IO81PT, SO41]
G6 RSQ R H Anderton, 38 Acacia Rd, Hampton, TW12 3DS [IO91TK, TQ17]
G6 RST F G Shepherd Horndean and District Amateur, 20 Frances Rd, Purbrook, Waterlooville, PO7 5HH [IO90LU, SU60]
G6 RSV P J Baxter, 49 Saltems Ave, Milton, Southsea, PO4 8QJ [IO90LT, SU60]
GW6 RSY Details withheld at licensee's request by SSL.
GI6 RTB J Jackson, 21 Carnreagh, Hillsborough, BT26 6LJ [IO64XL, J25]
G6 RTD G T Small, 6 Mary St., Longridge, Preston, PR3 3WN [IO83QU, SD63]
G6 RTG J Sutton, 29 Victory Ave, Darlaston, Wednesbury, WS10 7RR [IO82XN, SO99]
G6 RTJ Details withheld at licensee's request by SSL.
G6 RTM R J Ashberry, 44 Thames Rd, Langley, Slough, SL3 8DZ [IO91RL, TQ07]
G6 RTN K Bone, 1 Castle Cttgs, Kielder, Hexham, NE48 1EP [IO85QF, NY69]
G6 RTV Details withheld at licensee's request by SSL.
G6 RTY D M Johnson, 166 Whitmore Rd, Harrow, HA1 4AQ [IO91TN, TQ18]
GW6 RUE J E R Newey, 8 Llandaff Row, Penpentre, Brecon, LD3 8DH [IO81HW, SO02]
G6 RUK Details withheld at licensee's request by SSL.
G6 RUM I Nice, Stoneville, 6 Malden Rd, Sidmouth, EX10 9LS [IO80JQ, SY18]
GW6 RUO Details withheld at licensee's request by SSL.
G6 RUP J D Horner, 43 Birch Cl, Patchway, Bristol, BS12 5SA [IO81QM, ST58]
G6 RUS Details withheld at licensee's request by SSL.
G6 RUY J L Heaney, 6 Prince Charles Rd, Monkwick, Colchester, CO2 8NS [JO01KU, TM02]
G6 RVB G R K Cording, 10 Cover Green, Home Meadow, Worcester, WR4 0JF [IO82VE, SO85]
G6 RVH R G Jamieson, 3 Waterpark Rd, Prenton, Birkenhead, L42 9NZ [IO83LI, SJ38]
GW6 RVI D L Williams, 15 Mundy Pl, Cardiff, CF2 4BZ [IO81JL, ST17]
GM6 RVL A Latta, 12 Wolsey Ave, Bonnyrigg, EH19 3LU [IO85KU, NT36]
G6 RVP C W Pung, 6 Prince Charles Rd, Monkwick, Colchester, CO2 8NS [JO01KU, TM02]
G6 RVQ R F Pearce, 70 Mundesley Rd, North Walsham, NR28 0DB [JO02QT, TG23]
G6 RVS R V Sohst, 2 Shaftesbury Dr, Maidstone, ME16 0JS [JO01GG, TQ75]
G6 RVZ D E Carruthers, 47 Heigham Rd, London, E6 2JL [JO01AM, TQ48]
GU6 RWD S D Hancock, L'Hirondelle, Les Hubits de Bas, St Martins, Guernsey, Channel Islands, GY4 6NB
G6 RWI Details withheld at licensee's request by SSL.
G6 RWJ D B Silcox, Troedyrhiw, Penparc, Cardigan, Dyfed, SA43 2AE [IO72QC, SN24]
G6 RWL Details withheld at licensee's request by SSL.
G6 RWT Details withheld at licensee's request by SSL.
GM6 RWW L A Christie, Tigh Ban, Kilchrenan, Taynuilt, Argyll, PA35 1HD [IO76JJ, NN02]
GW6 RXA R D Railton, Glas Coed, Rhydargaeau, Carmarthen, Dyfed, SA32 7JT [IO71VW, SN52]
G6 RXD M J Yirrell, 40 St. Albans Hill, Hemel Hempstead, HP3 9NG [IO91SR, TL00]
G6 RXF G R Priestley, 7 Affleck Ave, Stoneclough, Radcliffe, Manchester, M26 1HN [IO83TN, SD70]
GM6 RXJ D F Mearns, Beechwood, East Main St., Harthill, Shotts, ML7 5QW [IO85DU, NS96]
G6 RXK A C Orchard, 6 Great Severals, Kintbury, Hungerford, RG17 9SN [IO91GJ, SU36]
G6 RXP S Dwyer, 10 Swan St., Darwen, BB3 2LW [IO83SQ, SD62]
GM6 RXQ A R B Gordon, Martello Ct, 87/20 Pennywell Gdn, Edinburgh, EH4 4TE [IO85IX, NT27]
G6 RXZ A J Humphreys, 25 Childscroft Rd, Gillingham, ME8 7SW [JO01HI, TQ86]
G6 RYD S T Hubbard, 8 Birch Gr, Gosmoor Ln, Elm, Wisbech, PE14 0AP [JO02CP, TF40]
G6 RYF P E King, 42 Beaconsfield Rd, Stoke, Coventry, CV2 4AR [IO92GJ, SP37]
G6 RYM Details withheld at licensee's request by SSL.
G6 RYW V J Smith, 9 Pinewood Dr, Mansfield, NG18 4PG [IO93JC, SK55]
G6 RZA Details withheld at licensee's request by SSL.[Station located near Newhaven.]
G6 RZG W Dewhurst, Twin Oaks, Whins Ln, Simonstone, Burnley, BB12 7QR [IO83TT, SD73]
G6 RZJ O R Somers, Houghwood House, Red Barn Rd, Billinge, Wigan, WN5 7UA [IO83PM, SD50]
G6 RZQ W S Lawrence, 41 Wheat Cl, Sandridge, St. Albans, AL4 9NN [IO91US, TL10]
G6 RZS R H Wood, 115 Anchorway Rd, Green Ln, Coventry, CV3 6JH [IO92FJ, SP37]
G6 RZY N Harper, 88 Kineton Green Rd, Solihull, B92 7EE [IO92CK, SP18]
G6 SAF Details withheld at licensee's request by SSL.
GM6 SAG W B L Miller, 4 Ross Rd, Edinburgh, EH16 5QN [IO85JW, NT27]
G6 SAL D N Stapleford, 182 Ashburton Rd, Hugglescote, Coalville, LE67 2HD [IO92HR, SK41]

G6 SAP Details withheld at licensee's request by SSL.
G6 SAQ G A Hines, 11 Montagu Gdns, Wallington, SM6 8EP [IO91WI, TQ26]
GW6 SBD G Davies, 2 Ffordd Aled, Park View, Wrexham, LL12 7PP [IO83MB, SJ35]
G6 SBF Details withheld at licensee's request by SSL.
G6 SBG A Lubrani, Bell Ln, Bell Bar, Brookmans Park, Hatfield, Herts, AL9 7AY [IO91VR, TL20]
G6 SBI D A Smith, 11 Churchill Ave, Horsham, RH12 2JP [IO91TB, TQ13]
G6 SBN M J Searl, 130 Chatham St., Reading, RG1 7HT [IO91MK, SU77]
G6 SBR Details withheld at licensee's request by SSL.
G6 SBU A A F Gookey, 115 Farndale Ave, Palmers Green, London, N13 5AJ [IO91WO, TQ39]
G6 SBW A D Alcock, Vivenda Montinho, Caixa 105Z, Picota, 8100 Loule, Portugal
G6 SC S R Chapple, 12 Hayes Hill Rd, Bromley, BR2 7HT [JO01AJ, TQ36]
G6 SCD Details withheld at licensee's request by SSL.
G6 SCG M J Lockwood, 33 Elmtree Rd, Calverton, Nottingham, NG14 6QA [IO93KA, SK64]
G6 SCM J D Webb, Oakdene, 22 Meeting House Ln, Balsall Common, Coventry, CV7 7FX [IO92EJ, SP27]
G6 SCP E A G Bennett, 20 Chichester Cl, Exmouth, EX8 2JU [IO80HO, SY08]
G6 SCZ G Allen, 45 Cookson St., Kirkby in Ashfield, Nottingham, NG17 8DZ [IO93IC, SK45]
G6 SDC B G Simmons, Wootton Leas, 35 Benenden Green, Alresford, SO24 9PE [IO91KC, SU53]
G6 SDE A R Curley, 21 Trinity Rise, Penton Mewsey, Andover, SP11 0RE [IO91FF, SU34]
G6 SDG N J Knowles, 53 Fox St., Swarthmoor, Ulverston, LA12 0HT [IO84KE, SD27]
G6 SDI P S Hall, 28 Grangeway, Rushden, NN10 9EZ [IO92QG, SP96]
G6 SDJ Details withheld at licensee's request by SSL.
G6 SDO J C C Stevens, 187 Bexley Ln, Sidcup, DA14 4JQ [JO01BK, TQ47]
G6 SDP Details withheld at licensee's request by SSL.
G6 SDQ J H Stacey, 16 Crane Dr, Verwood, BH31 6QB [IO90BV, SU00]
GM6 SDV J A More, 51 Hilton Dr, Aberdeen, AB24 4NJ [IO87WD, NJ90]
G6 SDY R Beecroft, 28 Hall Garth Ln, West Ayton, Scarborough, YO13 9JA [IO94SF, SE98]
G6 SEE Z I D Feast, 3 Granville Rd, Westerham, TN16 1RU [JO01AG, TQ45]
G6 SEF J K Feast, 3 Granville Rd, Westerham, TN16 1RU [JO01AG, TQ45]
G6 SEG Details withheld at licensee's request by SSL.
G6 SEJ N R Porter, 111 Station Rd, Brimington, Chesterfield, S43 1LJ [IO93HG, SK37]
G6 SEK P Ovey, 35 Lower Fairfield, St. Germans, Saltash, PL12 5NH [IO70UJ, SX35]
G6 SEM M B T O'Gara, 73 Ascot Cres, Martins Wood, Stevenage, SG1 5SU [IO91VW, TL22]
G6 SEN Details withheld at licensee's request by SSL.
G6 SET Details withheld at licensee's request by SSL.
GM6 SEV I C Carr, 14 Durham Pl, Bonnyrigg, EH19 3EX [IO85KU, NT36]
G6 SFB J W Stone, 44 Astor Cl, Brockworth, Gloucester, GL3 4AS [IO81WU, SO81]
G6 SFC T K Foulds, Elgar Cottage, Long Mill Ln, Platt, Sevenoaks, TN15 8NB [JO01DG, TQ65]
G6 SFE Sa'Ad A W Al-Katan, 12 Nimrod Cl, Loddon Fields, Woodley, Reading, RG5 4UW [IO91NK, SU77]
G6 SFF C Hewes, 15 Heathfield Rd, Nottingham, NG5 1NL [IO92KX, SK54]
G6 SFH P Barber, 17 Wheelwright Ave, Leeds, LS12 4UW [IO93ES, SE23]
G6 SFN Details withheld at licensee's request by SSL.
G6 SFR J C Goodman Flt Refueling Amateur Radio Soc, 8 Burley Cl, Verwood, BH31 6TQ [IO90BU, SU00]
G6 SFW P D Dodd, 9 Rudge Croft, Kitts Green, Birmingham, B33 9NZ [IO92CL, SP18]
G6 SFX M S James, 3 Southend Cl, Hursley, Winchester, SO21 2LJ [IO91HA, SU42]
G6 SFY G Barker, 8 Padstow Cl, Nuneaton, CV11 6FN [IO92GM, SP39]
G6 SGA S K Thornber, 31 Balmoral Dr, Braunstone, Leicester, LE3 3AD [IO92JO, SK50]
G6 SGD D J Carding, The Mill House, Walcot, Telford, Shropshire, TF6 5ER [IO82QQ, SJ51]
G6 SGJ S G Johnstone, 22 Red Hall Dr, Barwell, Leicester, LE9 8BY [IO92HN, SP49]
G6 SGM C P Macey, 29 Burleigh Rd, Sutton, SM3 9NE [IO91VJ, TQ26]
G6 SGQ M Abbott, 35 Denholme, Upholland, Skelmersdale, WN8 0AX [IO83PM, SD50]
G6 SGR S D Richards S.Glam Rynt Grp, Croydon, St. Andrews Rd, Wenvoe, Vale of Glam, CF5 6AF [IO81IK, ST17]
G6 SGW J A Miller, 21 Foxley Dr, Portsmouth, PO3 5TG [IO90LT, SU60]
G6 SGX Details withheld at licensee's request by SSL.[Station located near Robertsbridge.]
G6 SGY B Sales, 2 Highview, Hurley, Atherstone, CV9 2RP [IO92EN, SP29]
GM6 SHB Details withheld at licensee's request by SSL.
G6 SHD G McBrien, 26 Lumb Carr Ave, Ramsbottom, Bury, BL0 9QG [IO83UP, SD71]
G6 SHF M J Trolan, 12 Broadwell Rd, Solihull, B92 8QH [IO92CK, SP18]
G6 SHS K H Eldridge, 44 Merley Gdns, Merley, Wimborne, BH21 1TB [IO90AS, SZ09]
G6 SHZ S R Congrave, 67 Shores Green Dr, Wincham, Northwich, CW9 6EJ [IO83SG, SJ67]
G6 SIC Details withheld at licensee's request by SSL.
G6 SIM L J W Simarpi, 6 Holmlea, Wookey, Wells, BA5 1LG [IO81PE, ST54]
G6 SIQ W J Whitcombe, 2 New Cttgs, Buckland, Sittingbourne, ME9 9LF [JO01KH, TQ96]
G6 SIR N Rank, 6 Lee Dale, Harpur Hill, Buxton, SK17 9LQ [IO93BF, SK07]
GW6 SIX P Macmillen, 8 Watkin St., Conwy, LL32 8RL [IO83CG, SH77]
G6 SJA W J Barnes, 17 Greenhill Rd, Long Buckby, Northampton, NN6 7PU [IO92LH, SP66]
G6 SJE Details withheld at licensee's request by SSL.
G6 SJG T W Hurton, 4 Athlone Cl, Enham Alamein, Andover, SP11 6JY [IO91GF, SU34]
G6 SJH Details withheld at licensee's request by SSL.
G6 SJQ A D Briggs, 8 High Greeve, Wootton, Northampton, NN4 6BA [IO92NE, SP75]
G6 SJX D J Edwards, 30 Third Ave, Wickford, SS11 8RF [JO01GO, TQ79]
G6 SJY J G Elkins, Park Farm Bungalow, 45 Witham Rd, Black Notley, Braintree, CM7 8LQ [JO01GU, TL72]
G6 SKF L S Hopson, 10 Cascade Cl, Buckhurst Hill, IG9 6DY [JO01AO, TQ49]
G6 SKK T Parkin, 8 Horsley Cres, Holbrook, Belper, DE56 0UB [IO93GA, SK34]
G6 SKL Details withheld at licensee's request by SSL.
G6 SKM W A Taylor, 5 Gadbury Ave, Atherton, Manchester, M46 0LQ [IO83RM, SD60]
G6 SKN P D Wilson, 59 Oakfield Rd, Bishops Cleeve, Cheltenham, Glos, GL52 4LA [IO81XW, SO92]
G6 SKO Details withheld at licensee's request by SSL.
G6 SKP A Whitgreave, 2 Oaklea Ave, Hoole, Chester, CH2 3RE [IO83NE, SJ46]
G6 SKR G A Walker, 81 Normanshire Dr, Chingford, London, E4 9HE [IO91XO, TQ39]
G6 SKS C J Learoyd, Leofric House, 31 Leofric Ave, Bourne, PE10 9QT [IO92TS, TF02]
G6 SKT W A J Learoyd, Leofric House, 31 Leofric Ave, Bourne, PE10 9QT [IO92TS, TF02]
G6 SKU R J Leddington, 314 Stourbridge Rd, Catshill, Bromsgrove, B61 9LH [IO82XI, SO97]
G6 SKX J F Mackey, 16 Albion Terr, Sewardstone Rd, London, E4 7SB [IO91XP, TQ39]
G6 SKZ C R Lowe, 1A Main Rd, Willows Riverside P, Maidenhead Rd, Windsor Berks, SL4 5TT [IO91QL, SU97]
G6 SL C A Pettitt, Eddystone Radio Ltd, Unit 8/9 Birkdale Ave, Selly Oak, Birmingham, B29 6UB [IO92AK, SP08]
G6 SLD J K Duncan, 54 Ennerdale Dr, Astbury, Mere, Congleton, CW12 4FL [IO83VD, SJ86]
G6 SLE R F Bateman, Waterloo, Whixall, Whitchurch, Shropshire, SY13 2PX [IO82PV, SJ43]
G6 SLG A H Berry, Emberley Leys, Chapel Ln, Ratley, Banbury Oxon, OX15 6DS [IO92GC, SP34]
G6 SLH A M Bland, 63 Bernard St., Woodville, Swadlincote, DE11 8BY [IO92FS, SK31]
G6 SLK Details withheld at licensee's request by SSL.
G6 SLM Details withheld at licensee's request by SSL.
G6 SLN C Gleave, 10 Henley Rd, Neston, South Wirral, L64 0SG [IO83LG, SJ27]
G6 SLU Details withheld at licensee's request by SSL.
G6 SLY D M Lewis, 30 Printers Park, Hollingworth, Hyde, SK14 8QH [IO93AL, SK09]
G6 SLZ J A V Lee, Bourne End, 68 Andlers Ash Rd, Liss, GU33 7LR [IO91NA, SU72]
G6 SMC Details withheld at licensee's request by SSL.
G6 SMG H Parker, 17 Oaklea, Pinnex Moor, Tiverton, EX16 6NS [IO80GV, SS91]
G6 SMJ M A Pitts, 30 Sandhurst Ave, Berrylands, Surbiton, KT5 9BS [IO91UJ, TQ16]
G6 SMK H G G Powell, Versalles, Red Ln, Colne, Lancs, BB8 7JT [IO83VU, SD84]
G6 SMX Details withheld at licensee's request by SSL.
G6 SMZ B M Hatt, 1 Fir Tree Cl, Hemel Hempstead, HP3 8NG [IO91SR, TL00]
G6 SNA D Heather, 65 Ashurst Cottage, Fairview Rd, Headley Down, Hants, GU35 8HQ [IO91OC, SU83]
G6 SND W A Jarvis, 74 Halford Rd, Ickenham, Uxbridge, UB10 8QA [IO91SN, TQ08]
G6 SNH Details withheld at licensee's request by SSL.[Op: A C Radford. Station located near Acton.]
G6 SNI G J Platt, 15 Mount Cl, Nantwich, CW5 6JJ [IO83RB, SJ65]
G6 SNJ M R Parker, 14 Orchard Cl, Ockbrook, Derby, DE72 3RQ [IO92HV, SK43]
G6 SNL Details withheld at licensee's request by SSL.
G6 SNN D H Ramsey, 11 Pendle Cl, Basildon, SS14 3NA [JO01FN, TQ79]
G6 SNO Details withheld at licensee's request by SSL.
GJ6 SNQ A J Leighton, 2-le-Petit Menage, Fountain Ln, St Saviour, Jersey, Channel Islands, JE2 7RL
G6 SNX D O Smith, 38 Prospect Rd, Childs Hill, Cricklewood, London, NW2 2JU [IO91VN, TQ28]
G6 SNY R M H Tedbury, Flat 1, The Wintergardens, Fore St, Sidmouth, Devon, EX10 8AG [IO80JQ, SY18]
G6 SOE H M Shaw, 19 Wren Cl, Poulton-le-Fylde, FY6 7QL [IO83LU, SD33]
G6 SOY P B Berry, 134A South Rd, Haywards Heath, West Sussex, RH16 4LP [IO90WX, TQ32]
G6 SOZ A J Byrne, Holly Cottage, Deacons Ln, Hermitage, Thatcham, RG18 9RJ [IO91IK, SU57]
G6 SPB D J Corder, 140 Edward Rd, Somerford, Christchurch, BH23 3EW [IO90DR, SZ19]
G6 SPG P A Cesnavicius, 13 Holt St., Eccles, Manchester, M30 7HQ [IO83TL, SJ79]
G6 SPH J C C Crookbain, 11 Champlain Ave, Canvey Island, SS8 9QL [JO01GM, TQ78]
G6 SPI S Carwood, 34 Flemming Ave, Ruislip, HA4 9LF [IO91TN, TQ18]
G6 SPN R J C Barton, 82 Buckingham Rd, South Woodford, London, E18 2NJ [JO01AO, TQ39]
G6 SPR K R Hudders, 9 Railway Cttgs, Old Station Way, Bordon, GU35 9HH [IO91NC, SU73]
G6 SPS A J Hutley, 120 Constance Cl, Witham, CM8 1XZ [JO01HS, TL81][Station location details as for G0VDP.]
G6 SQS F J Sivyer, 22 Boxley Rd, Walderslade, Chatham, ME5 9LF [JO01GI, TQ76]
G6 SQT C A Wall, 151 Bisley Rd, Stroud, GL5 1HS [IO81VR, SO80]
G6 SRC M J Elliott Swale ARC, 20 Haysel, Sittingbourne, ME10 4QE [JO01IH, TQ96]
G6 SRE B J Stone, Reindene, Faversham Rd, Boughton Aluph, Ashford, TN25 4PQ [JO01KE, TR04]
G6 SRJ A E Waring, 2 Wroxton Cl, Thornton Cleveleys, FY5 3EY [IO83LU, SD34]
G6 SRN J Pearson, Tangle Trees, 6 Enderley Cl, Bloxwich, Walsall, WS3 3PF [IO82XP, SJ90]
G6 SRQ V M L Alexandrou, 62 Hartsdown Rd, Margate, CT9 5RD [JO01QJ, TR36]
G6 SRR Details withheld at licensee's request by SSL.
G6 SRS Details withheld at licensee's request by SSL.[Correspodence to Stourbridge & DARS, Robin Woods Centre, Stourbridge]
G6 SRT D P Armstrong, 18 Leyburne Gdns, Chinnor, OX9 4EL [IO91NQ, SP70]
G6 SRU C G J Alford, 6 Meadow Rise, Bournville, Birmingham, B30 1UZ [IO92AK, SP08]
G6 SRV R D Andrews, 11 Holly Gr, Verwood, BH31 6XA [IO90BU, SU00]
G6 SRW Capt B Armstrong, Wensley, Rusper Rd, Horsham, RH12 5QW [IO91UC, TQ13]
G6 SRX Details withheld at licensee's request by SSL.
G6 SRY S J Aram, 5 The Croft, West Hanney, Wantage, OX12 0LD [IO91HP, SU49]
G6 SRZ W A Baxter, 19 Westbury Rd, Basford, Nottingham, NG5 1EP [IO92KX, SK54]
G6 SSH V W Bates, 167 St. Georges Ave, Westhoughton, Bolton, BL5 2ER [IO83RM, SD60]
G6 SSM O W Burgess, 14 Leys Cl, Harefield, Uxbridge, UB9 6QB [IO91SO, TQ09]
G6 SSN K M Burton, 84 Aire Rd, Grantham, NG31 7QS [IO92QV, SK93]
G6 SSQ D Clark, Huds House, Congill Dent, Sedbergh, Cumbria, LA10 5TQ [IO84TG, SD78]
G6 SSV J Cott, 50 Parkside Way, North Harrow, Harrow, HA2 6DG [IO91TO, TQ18]
G6 SSX K M Coleman, 4 Bridge Cl, Thurmaston, Leicester, LE4 8GY [IO92KQ, SK60]
G6 STA Details withheld at licensee's request by SSL.
G6 STD D Macey, 12 Carwinard Cl, Angarrack, Hayle, TR27 5JA [IO70HE, SW53]
G6 STE B J Stevens, 7 Pembroke Dr, Darras Hall, Ponteland, Newcastle upon Tyne, NE20 9HS [IO95CA, NZ17]
G6 STI H F Staddon, 45 Saxony Parade, Hayes, UB3 2TQ [IO91SM, TQ08]
G6 STJ S M Smith, Bachedonna, Bagton, nr Kingsbridge, Devon, TQ7 3EE [IO80CG, SX74]
GW6 STK R E Sweet, 9 Seafield Rd, Colwyn Bay, LL29 7HB [IO83DG, SH87]
G6 STL M Smith, 49 St. Johns Rd, Walsall, WS2 9TJ [IO82XN, SO99]
GW6 STS G J Jones, The Bungalow, Castle St., Blaenavon, Gwent, NP4 9QL [IO81KS, SO20]
G6 STW D A Jappy, 6 Fieldhouse Rd, Huddersfield, HD1 6NX [IO93CP, SE11]
G6 SUD Details withheld at licensee's request by SSL.
G6 SUK B A Trevor, 39 Clayton Cres, Brentford, TW8 9PT [IO91UL, TQ17]
G6 SUN C O Tate, 23 Hawthorn Villas, Holmes Chapel, Crewe, CW4 7AR [IO83TE, SJ76]
G6 SUP A P Niven, The Bunk Barn, Orlham Farm, Leddington, Ledbury, Herefordshire, HR8 2LN [IO82SA, SO63]
G6 SUR P Thornsby, 20 Stowupland Rd, Stowmarket, IP14 5AG [JO02ME, TM05]
G6 SUV J C Barlow, 3 Shaw Brook Cl, Rishton, Blackburn, BB1 4ES [IO83SS, SD72]
G6 SVB Details withheld at licensee's request by SSL.
G6 SVF L Herbert, 17 Richmond Rd, Scawsby, Doncaster, DN5 8SX [IO93JM, SE50]
G6 SVH K Henderson, 36 Frith View, Chapel En-le-Frith, High Peak, SK23 9TT [IO93AH, SK08]
G6 SVJ S J Harvey, 148 Smithfield Rd, Uttoxeter, ST14 7LB [IO92BV, SK03]
G6 SVV R D Gray, Willett, 28 Hoe Ln, Abridge, Romford, RM4 1AX [JO01BP, TQ49]
G6 SVX R Goodge, 36 Henty Cl, Sullivan Dr, Crawley, RH11 6AL [IO91VC, TQ23]
G6 SVY J F Gold, 5 The Cres, White Cross Rd, Swaffham, PE37 7QZ [JO02IP, TF80]
G6 SVZ A E Godwin, 27 Melbourne Ave, Dronfield Woodhouse, Sheffield, S18 5YW [IO93GH, SK37]
G6 SW K T Ward Cannock Chase Rd, 3 Levetts Hollow, Cannock, WS12 5AW [IO92AR, SK01]
G6 SWD A A J Gibbings, 3 Fairview Cres, Tidcombe Park, Tiverton, EX16 4BN [IO80GV, SS91]
G6 SWJ J D Askey, 1 Acorn Mews, Horslow St., Potton, Sandy, SG19 2NS [IO92VD, TL24]
G6 SWT S J French, 47 Horn Ln, Woodford Green, IG8 9AA [JO01AO, TQ49]
G6 SWW S J Ellin, 7 Crawshaw Ave, Beauchief, Sheffield, S8 7DZ [IO93GH, SK38]
G6 SWZ M D Davy, 22 Scott Gdns, St. Giles, Lincoln, LN2 4LX [IO93RF, SK97]
G6 SXB L E S Dunham, 5 King St., Wimblington, March, PE15 0QF [JO02BM, TL49]
G6 SXD E J Drinkwater, 57 Ludlow Rd, Bridgnorth, WV16 5AH [IO82SM, SO79]
G6 SXJ S J Collins, 44 Hollybush Ln, Wolverhampton, WV4 4JJ [IO82WN, SO89]
GM6 SXK J N Conn, 33 Alford Dr, Glenrothes, KY6 2HH [IO86JE, NO20]
G6 SXM C P Collier, 18 Palm Tree Way, Lyminge, Folkestone, CT18 8JL [JO01ND, TR14]
G6 SXN G W Dixon, 4 Yarborough Rd, Keelby, Grimsby, DN37 8HG [IO93UN, TA10]
G6 SXO S J Phillippo, 11 Heacham Cl, Lower Earley, Reading, RG6 4AG [IO91MK, SU76]
G6 SYA M F Mills, 6 Bower Rd, Hextable, Swanley, Kent, BR8 7SE [JO01CJ, TQ57]
G6 SYB J Malcolm, 62 Linden Ave, Ruislip Manor, Ruislip, HA4 8UA [IO91TN, TQ18]
GM6 SYC E G Tennant, 21 Strathbeg Dr, Dalgety Bay, Dunfermline, KY11 5XQ [IO86HA, NT18]
G6 SYI P J Somerfield, 27 Ormerod St., Worsthorne, Burnley, BB10 3NU [IO83VS, SD83]
G6 SYN B R Nye, 160 Church Rd, Swanscombe, DA10 0HP [JO01DK, TQ67]
G6 SYT Details withheld at licensee's request by SSL.
G6 SYV F Ackroyd, 1 Bedford Green, Tinshill, Leeds, LS16 6DR [IO93EU, SE23]
G6 SYW B J Bauly, Poplar Farm, Mendelsham, Stowmarket, Suffolk, IP14 5SN [JO02MG, TM06]
G6 SYX G J Brookes, 6 Abbeydale Oval, Leeds, LS5 3RF [IO93ET, SE23]
G6 SZB J H Barton, 76 Elvaston Rd, North Wingfield, Chesterfield, S42 5HH [IO93HE, SK46]
GW6 SZF Details withheld at licensee's request by SSL.
G6 SZG H Barnes, 38 Gayhurst Dr, Sittingbourne, ME10 1UD [JO01II, TQ86]
G6 SZP N Carter, 23 Sandhall Dr, Highroad Well, Halifax, HX2 0DL [IO93BR, SE02]
G6 SZQ Details withheld at licensee's request by SSL.
G6 SZS R P Crook, 26 Chapel St., Rishton, Blackburn, BB1 4NP [IO83TS, SD73]
G6 TAF P A Penny, 13 Newnham Cl, Braintree, CM7 2PR [JO01GV, TL72]
G6 TAH D J Palmer, 17 Atyeo Cl, Burnham on Sea, TA8 2EJ [IO81MF, ST34]
G6 TAI J S Peel, 9 Hillspring Rd, Springhead, Oldham, OL4 4SJ [IO83XN, SD90]
G6 TAK P Reay, 26 Clifton Ct, Workington, CA14 3HR [IO84FP, NY02]
G6 TAL G E Ring, 9 Milburn Rd, Weston Super Mare, BS23 3BE [IO81MI, ST36]
GW6 TAM R G Meal, 4 Sparrow Cl, Brampton, Huntingdon, PE18 8PY [IO92VH, TL26]
GM6 TAN M G Shread, 15 Hardie Ct, Aberchirder, Huntly, AB54 7TG [IO87QN, NJ65]
G6 TAP D J Squire, Green Valley, Raleigh Rd, Barnstaple, EX31 4HY [IO71XC, SS53]
G6 TAS R Wroe, 318 Warwick Rd, Banbury, OX16 7AZ [IO92HB, SP44]
G6 TAU Details withheld at licensee's request by SSL.
G6 TAV R A Vigar, 33 Gloucester Rd, Old Fletton, Peterborough, PE2 8BH [IO92VN, TL19]
GI6 TBC V P Loughran, 10 Oakwood, Ballynahonemore, Armagh, BT60 1QR [IO64QI, H84]
GM6 TBE P S Lowrie, 4 Eschiehaugh, Kelso, TD5 7SJ [IO85SO, NT73]
G6 TBJ J . O Larssen, 228 Barnsole Rd, Gillingham, ME7 4JB [JO01GI, TQ76]
G6 TBN J L Rozentals, Talsu Apr, Lubes Pagasta, Bilavas Farme, Latvia, Lv 3262
G6 TBT G Davey, 18 Grantham Rd, Chiswick, London, W4 2RS [IO91UL, TQ27]
G6 TC T E Rowley, 136 Blackhalve Ln, Wolverhampton, WV11 1AA [IO82WO, SJ90]
G6 TCD R Gleeson, 47 Shore Ave, Shaw, Oldham, OL2 8DA [IO83WO, SD91]
G6 TCJ R J Gillingham, 184 High St., Roydon, Harlow, CM19 5EQ [JO01AS, TL41]
G6 TCK P D Green, 35 Trenchard Cl, Hersham, Walton on Thames, KT12 5QT [IO91TI, TQ16]
G6 TCP S H Hook, 20 Buttermere Dr, Millom, LA18 4PL [IO84IE, SD17]
G6 TCQ B R Cheese, 12 Carbonels, Great Waldingfield, Sudbury, CO10 0RQ [JO02JB, TL94]
G6 TCV J W Halliday, 19 Cathrine St. East, Denton, Manchester, M34 3RQ [IO83WK, SJ99]
G6 TCW D Howarth, Lancastria, Maidstone Rd, Borden, Sittingbourne, ME9 7QA [JO01IH, TQ86]
G6 TDB B W Hayman, 19 Manor Rd, Plymstock, Plymouth, PL9 7DP [IO70WI, SX55]
G6 TDG K L Hodges, 18 Leycester Cl, Birmingham, B31 4SS [IO92AJ, SP07]
G6 TDL A T Worsley, 8 Winston Dr, Hensingham, Whitehaven, CA28 8RB [IO84FM, NX91]
G6 TDM G C Robinson, 3 Highlands, Stone, ST15 0LA [IO82WV, SJ83]
G6 TDR R S McDermott, 3 Coombe Ln, Whiteley Village, Walton on Thames, KT12 4EL [IO91SI, TQ06]
G6 TDW J C Mead, Castlemead, 40 Eastbourne Rd, Pevensey Bay, Pevensey, BN24 6HJ [JO00ET, TQ60]
G6 TDX D A Yarrow, 193 Ladygate Ln, Ruislip, HA4 7RD [IO91SO, TQ08]
G6 TEB A P Varga, 2 Yew Tree Ln, Malvern, WR14 4LJ [IO82UB, SO74]
G6 TEC Details withheld at licensee's request by SSL.
G6 TED E W Chamberlain, 69 Kingsland, Harlow, CM18 6XL [JO01BS, TL40]
GW6 TEO G W Smith, 11 Sandy Leys, Castlemartin, Pembroke, SA71 5HJ [IO71LP, SR99]
G6 TEQ I C Stuckey, 4 Ingsdon Burn Farm, Ingsdon, Newton Abbot, TQ12 6NW [IO80DN, SX87]
G6 TER R J Monksummers, 29 Cloverfields, Peacemarsh, Gillingham, SP8 4UP [IO81UB, ST82]
G6 TET B Smith, 8 Devon St., Leigh, WN7 2NG [IO83SL, SJ69]
G6 TEX T B Speak, Ashley House, 18A Elson Rd, Formby, Liverpool, L37 2EG [IO83LN, SD20]
G6 TEZ N P Stevenson, 27 Green St., Stevenage, SG1 3DS [IO91VV, TL22]
G6 TFC T W Seddon, 1 Armadale Rd, Ladybridge, Bolton, BL3 4QE [IO83SN, SD60]
G6 TFE D C Standen, 54 Park Way, Hastings, TN34 2PJ [JO00GU, TQ81]
GI6 TFF R Symington, 8 Thorndene Park, Carrickfergus, BT38 9EA [IO74CR, J48]
G6 TFJ R S Smith, 68 Queens Cres, Eastbourne, BN23 6JR [JO00DS, TQ60]
G6 TFM B Shuttleworth, 25 Chelford Cres, High Acres, Kingswinford, DY6 8PB [IO82WL, SO98]
G6 TFV D Owen, 18 Prescott Ave, Atherton, Manchester, M46 9LN [IO83RM, SD60]
G6 TGB A K Pennells, 58 Chelveston Cres, Southampton, SO16 5SB [IO90GW, SU31]
G6 TGD S N Harvey, 3 High Ln, Norton Tower, Halifax, HX2 0NW [IO93BR, SE02]
G6 TGE G Holman, 5 Ingleton Rd, Newsome, Huddersfield, HD4 6QX [IO93CP, SE11]
G6 TGI G R Harley, 4 Grosvenor Ave, Crosby, Liverpool, L23 0SB [IO83LL, SJ39]
G6 TGJ J E Hirons, Furlong House, Racecourse Ln, Bicton Heath, Shrewsbury, SY3 5BJ [IO82OR, SJ41]
G6 TGM W J Howard, 2 Heather Dr, Rise Park, Romford, RM1 4SP [JO01CO, TQ59]
G6 TGP Details withheld at licensee's request by SSL.
G6 TGQ S A Houghton, 259B St.Faith Rd, Old Catton, Norwich, Norfolk, NR6 7BB [JO02PP, TG21]
GW6 TGR G J Jones, Glyn Dwr, Bethel, Caernarfon, Gwynedd, LL15 1UN [IO83IC, SJ15]
G6 TGW R D Jones, 22 The Pound, St. Ives, Huntingdon, PE17 6XQ [IO92XH, TL37]

G6 TGY M J Inch, 18 Glencastle Way, Stoke on Trent, ST4 8QE [IO82VX, SJ84]
G6 THB M Jackson, 23 Sedgebrook Cl, Oakwood, Derby, DE21 2DX [IO92GW, SK33]
G6 THC M Johnson, 91 Harrow Rd, Linthorpe, Middlesbrough, TS5 5LJ [IO94IN, NZ41]
G6 THG Details withheld at licensee's request by SSL.
G6 THH A E Kiddy, 39 Garden Dr, Brampton, Barnsley, S73 0TN [IO93HM, SE40]
G6 THM C T Smith, 8 Terry Cl, Weston Coyney, Stoke on Trent, ST3 6NS [IO82WX, SJ94]
G6 THN A K Smith, 8 Terry Cl, Weston Coyney, Stoke on Trent, ST3 6NS [IO82WX, SJ94]
G6 THP L N Curwen, 12 Garden Cl, St. Ives, Huntingdon, PE17 6XZ [IO92XI, TL37]
G6 THR Details withheld at licensee's request by SSL.
GM6 TIB I A Campbell, 35 Thornwood Ave, Lenzie, Kirkintilloch, Glasgow, G66 4EL [IO75VW, NS67]
G6 TID M Coleman, 1 Burdon Dr, Bartestree, Hereford, HR1 4DL [IO82QB, SO54]
G6 TIF Details withheld at licensee's request by SSL.[Op: Steve Pratt, PO Box 217, Portslade, Brighton, BN4 1QE.]
G6 TIH Details withheld at licensee's request by SSL.
G6 TIM Details withheld at licensee's request by SSL.
G6 TIQ A J Price, Flat, 2 South Elms, Silverdale Rd, Eastbourne, BN20 7EU [JO00DS, TV69]
G6 TIW D Reeve, 12 Lambourne Rd, Birstall, Leicester, LE4 4FU [IO92KQ, SK50]
G6 TIX Details withheld at licensee's request by SSL.
G6 TIY Details withheld at licensee's request by SSL.[Op: M W Rider. Station located in north Warwickshire.]
G6 TIZ Details withheld at licensee's request by SSL.
G6 TJC S Deville, 39 Acre Cl, Maltby, Rotherham, S66 8BL [IO93JK, SK59]
GM6 TJD J A Doull, 52 Howburn Rd, Thurso, KW14 7ND [IO88FO, ND16]
G6 TJE R G Dawson, 3 Silver St., Walgrave, Northampton, NN6 9QB [IO92OI, SP87]
G6 TJH I A A Dorrell, Unsers, Culford Rd, Ingham, Bury St. Edmunds, IP31 1NP [JO02IH, TL87]
G6 TJI D I Day, 13 Edgecombe Ave, Weston Super Mare, BS22 9AY [IO81MI, ST36]
G6 TJJ R G Downham, 133 Carlyon Ave, South Harrow, Harrow, HA2 8SN [IO91TN, TQ18]
G6 TJK A M Dowell, 54 Station St., Castle Gresley, Burton on Trent, DE14 1BS [IO92ET, SK22]
G6 TJP I J Record, West View, 12 Hauxton Rd, Little Shelford, Cambridge, CB2 5HJ [JO02BD, TL45]
G6 TJR Details withheld at licensee's request by SSL.
G6 TJY I K Randle, 12 Cuckoo Ave, Hanwell, London, W7 1BT [IO91TM, TQ18]
G6 TJZ P J Rendell, Hillside, Winterbourne Hill, Winterbourne, Bristol, BS17 1JW [IO81RM, ST68]
G6 TKA P M Reed, Lily Farm, Castle Cary, Somerset, BA7 7NF [IO81SB, ST63]
G6 TKB K M Sutcliffe, 652 Newchurch Rd, Newchurch, Rossendale, BB4 9HG [IO83UQ, SD82]
G6 TKC M J Walker, Smirthwaite House, Mickletown Rd, Methley, Leeds, LS26 9HY [IO93HR, SE32]
GU6 TKE C J Wild, Honfleur-le-Hurel, Vale, Guernsey, Channel Islands, GY3 5AF
G6 TKH J R Torring, 41 The Avenue, Moordown, Bournemouth, BH9 2UW [IO90BS, SZ09]
GM6 TKL Details withheld at licensee's request by SSL.[Op: W S Wood.]
G6 TKM A K Thompson, 3 Rufford Rd, Long Eaton, Nottingham, NG10 3FP [IO92IV, SK43]
G6 TKR E H J Tratt, 5 Oaklands, Cheddar, BS27 3BS [IO81OG, ST45]
G6 TKV T N Tyrer, 85 Swann Ln, Cheadle Hulme, Cheadle, SK8 7HU [IO83VI, SJ88]
G6 TKW P Tomlinson, 158 Seamore Ave, South Benfleet, Benfleet, SS7 4LA [JO01GN, TQ78]
GD6 TKX T G Sayle, Archallagan Park, Markown, Isle of Man, IM4 2HJ
G6 TKY E Caligari, 209 Ormskirk Rd, Upholland, Skelmersdale, WN8 0AA [IO83PM, SD50]
G6 TLA P A Curran, 29 Wingfield Ave, Worksop, S81 0SY [IO93KH, SK58]
G6 TLB P A Curran, 29 Wingfield Ave, Worksop, S81 0SY [IO93KH, SK58]
G6 TLN D J Allen, 35 Fortescue Chase, Thorpe Bay, Southend on Sea, SS1 3SS [JO01JM, TQ98]
G6 TLR Details withheld at licensee's request by SSL.
G6 TLX C J Bull, 35 Manor Rd, Wokingham, RG41 4AR [IO91NJ, SU86]
GW6 TM R Jones Conway Vall ARC, Woodcote, 37 Coed Pella Rd, Colwyn Bay, LL29 7BB [IO83DH, SH87]
G6 TME A M Blackman, 10 Cromwell St., Hounslow, TW3 3LQ [IO91TL, TQ17]
G6 TMF A H Bird, Dunnings Ln, West Horndon, Brentwood, Essex, CM13 3HE [JO01EN, TQ68]
GM6 TMH D E Bell, 11 Shebster Ct, Thurso, KW14 7ES [IO88FO, ND16]
G6 TMK H A Burnham, 13 The Close, Chequers Park, Wye, Ashford, TN25 5BD [JO01LE, TR04]
GJ6 TMM Details withheld at licensee's request by SSL.[Station located at St Helier, Jersey. Op: K S Boleat, PO Box 437, Jersey.]
G6 TMN H Bryan, 2 Rivington Cl, Tarleton, Preston, PR4 6DS [IO83NQ, SD42]
G6 TMQ L Saagi, 13 Braunstone Ave, Leicester, LE3 0JF [IO92KP, SK50]
GW6 TMW N C Weedon, 10 Lark Rise, Coity, Brackla, Bridgend, CF31 2NU [IO81FM, SS98]
G6 TNA C B Walton, 6 Gorse Gr, Longton, Preston, PR4 5NP [IO83OR, SD42]
G6 TNE G Z Skulski, 83 Long Riding, Basildon, SS14 1QU [JO01FN, TQ78]
G6 TNH G S Taylor, 11 Philip Gr, Skegness, PE25 2JH [JO03DD, TF56]
G6 TNI B Telford, 18 Kirkstall Cl, South Anston, Anston, Sheffield, S31 7BA [IO93GJ, SK38]
G6 TNJ D D Thomas, 241 Ashburton Ave, Kings, Ilford, IG3 9EJ [JO01BN, TQ48]
G6 TNK J C Turnbull, 34 Bridge Ave, Hanwell, London, W7 3DJ [IO91TM, TQ18]
G6 TNO Details withheld at licensee's request by SSL.
G6 TNQ N F Bosanquet-Bryant, 44 Purley Way, Hartley Gdns, Clacton on Sea, CO16 8YX [JO01NT, TM11]
G6 TNR D I Blackman, 115 Ringwood, Bracknell, RG12 8XU [IO91OJ, SU86]
G6 TNW I T Webb, Corner Ways, Orchard Rd Eaton Ford, St. Neots, Huntingdon, Cambs, PE19 3AN [IO92UF, TL15]
G6 TNZ R D Taylor, 55 Park Ave, Driffield, YO25 7EN [IO94SA, TA05]
G6 TOB Details withheld at licensee's request by SSL.
G6 TOC W J Forster, 19 Ravens Way, Burton on Trent, DE14 2JS [IO92ET, SK22]
G6 TOI A J Edgcombe, 18 Elmwood Park, Loddiswell, Kingsbridge, Devon, TQ7 4SA [IO80CH, SX74]
G6 TOT A B White, 10 Slade Cl, Walderslade, Chatham, ME5 8RD [JO01GI, TQ76]
G6 TOW Details withheld at licensee's request by SSL.
GW6 TOX B Taylor, Swn-y-Don, Beaumaris, Anglesey, North Wales, LL58 8RW [IO73XH, SH67]
G6 TOY D J Williams, Hollybank, Royston Rd, Churchinford, Taunton, Somerset, TA3 7RE [IO80KV, ST21]
G6 TOZ Details withheld at licensee's request by SSL.
G6 TPB S C Lines, 7 Sharon Cl, Felmingham, North Walsham, NR28 0LJ [JO02QT, TG22]
G6 TPC P Randall, 75 Brookmead Dr, Wallingford, OX10 9BH [IO91KO, SU68]
G6 TPE Details withheld at licensee's request by SSL.
G6 TPH A P Lawrence, 23 Brocks Dr, North Cheam, Sutton, SM3 9UJ [IO91VJ, TQ26]
G6 TPI C J Bryan, 3 Hales Pl, Longton, Stoke on Trent, ST3 4NF [IO82WX, SJ94]
G6 TPK Details withheld at licensee's request by SSL.[Op: W Booth, 2 Windermere Avenue, Dane Bank, Denton, Manchester, M34 2EN.]
G6 TPO D E Bathe, Moel Tryfan, Grange Ln, Roydon, Harlow, CM19 5HG [JO01AS, TL40]
G6 TPQ L R Bartlett, 15 Haggate Cres, Royton, Oldham, OL2 5NF [IO83WN, SD90]
G6 TPT D J T Burchell, 21 Stokes Croft, Calne, SN11 9AQ [IO81XK, ST97]
G6 TPV Details withheld at licensee's request by SSL.
G6 TQ R R Smith, 58 Deakin Leas, Tonbridge, TN9 2JX [JO01DE, TQ54]
G6 TQC W J S George, 28 Melbourn Cl, Duffield, Belper, DE56 4FX [IO92GX, SK34]
G6 TQD G Green, The Knowles, Drakelow Ln, Wolverley, Kidderminster, DY11 5RY [IO82UK, SO88]
G6 TQF P B Game, 15 Nightingale Cl, Gosport, PO12 3EU [IO90KT, SU50]
GW6 TQH G Giudice, 31 Woodfield Cross, Tredegar, NP2 4JG [IO81JS, SO10]
G6 TQL T S W Law, 96 Hertford Rd, East Finchley, London, N2 9BU [IO91WO, TQ28]
G6 TQM A G Lawrence, 48 Water Ln, Oakington, Cambridge, CB4 5AL [JO02AG, TL46]
G6 TQS J Maxwell, 10 Elwyn Rd, Bradford, BD5 7HN [IO93DS, SE13]
G6 TQZ R G Andrews, 33 Highmoor, Amersham, HP7 9BU [IO91QQ, SU99]
G6 TRA D H Andrew, Neatmoor Hall Farm, Nordelph, Downham Market, Norfolk, PE38 0BY [JO02DO, TF50]
G6 TRD H G H Annis, The Limes, Mattersey Rd, Lound, Retford, DN22 8RP [IO93MJ, SK68]
G6 TRG P F Rigg Todmorden Ryt G, Birchwood House, Walsden Todmorden, West Yorks, OL14 6QX [IO83WQ, SD92]
G6 TRN D J Best, 64 New Hey Rd, Cheadle, SK8 2AQ [IO83VJ, SJ88]
G6 TRO B M Wilcox, 135 Windsor Rd, Wellingborough, NN8 2NB [IO92PG, SP86]
G6 TRP I R Pearn, 24 Wychwood Rise, Great Missenden, HP16 0HB [IO91PQ, SU89]
G6 TRQ J Wright, 12 Ayscough Ave, Nuthall, Nottingham, NG16 1BY [IO92JX, SK54]
G6 TRS M C Barton, 35 Shrubbery Ave, Worcester, WR1 1QN [IO82VE, SO85]
GW6 TRV I J Taylor, 6 Maeslan, Mynyddcerrig, Llanelli, SA15 5BE [IO71WT, SN51]
G6 TRW A W Toas, 116 Rownhams Rd, North Baddesley, Southampton, SO52 9EU [IO90GX, SU31]
G6 TRX J P Sugrue, 124 Hall Ln, Upminster, RM14 1AL [JO01DN, TQ58]
G6 TRY G S Smillie, 5 Fleckers Dr, Hatherley, Cheltenham, GL51 5BB [IO81WV, SO92]
G6 TSC V J Simmons, 88 Wellcome Ave, Dartford, DA1 5JW [JO01CK, TQ57]
G6 TSF P R J Shayler, 38 Maryside, Langley, Slough, SL3 7ET [IO91RM, TQ07]
G6 TSH Details withheld at licensee's request by SSL.
G6 TSJ P J Hannam, 7 Bodenham Cl, Buckingham, MK18 7HR [IO91MX, SP73]
G6 TSM W P Hirst, 4 Warwick Sq, Carlisle, CA1 1LB [IO84MV, NY45]
G6 TSQ E Hagger, 3 Minsmere Way, Great Cornard, Sudbury, CO10 0LB [JO02JA, TL84]
G6 TSS M Hird, 7 Forster St., Consett, DH8 7JU [IO94CU, NZ15]
G6 TSX C K Heater, 18 Shawfield Rd, Ash, Aldershot, GU12 6PF [IO91PF, SU85]
G6 TSZ D H Hall, 282 Dereham Rd, Norwich, NR2 3TL [JO02PP, TG20]
G6 TTE Details withheld at licensee's request by SSL.[Station located near Hayes.]
GW6 TTK B E Insole, 42 St. Annes Dr, Crownhill Est, Llantwit Fardre, Pontypridd, CF38 2PD [IO81IN, ST08]
G6 TTL A Jarvis, 31 The Downings, Herne Bay, CT6 7EJ [JO01NI, TR16]
G6 TTX W R Kenyon, 13 Baskerfield Gr, Green, Woughton on The Green, Milton Keynes, MK6 3ES [IO92PA, SP83]
GW6 TUD M E Prosser, 18 Thornhill Way, Rogerstone, Newport, NP1 9FT [IO81LO, ST28]
GM6 TUE P A B McLaren, Dalriada, Fogwatt, nr Elgin, Moray, IV30 3SL [IO87IO, NJ25]
G6 TUG I R Metcalfe, 15 Farnborough Cres, Bromley, BR2 7DL [JO01AI, TQ46]
G6 TUH Details withheld at licensee's request by SSL.
G6 TUI J P Machin, 11 East Butts Rd, Rugeley, WS15 2LU [IO92AS, SK01]
G6 TUO Details withheld at licensee's request by SSL.
G6 TUS R M Page, 124 The Spinney, Bar Hill, Cambridge, CB3 8TW [JO02AF, TL36]
G6 TUT Details withheld at licensee's request by SSL.
G6 TUX B S Searby, 18 Manor House Rd, Kimberworth, Rotherham, S61 1NT [IO93HK, SK49]
G6 TUZ Details withheld at licensee's request by SSL.
G6 TVA Details withheld at licensee's request by SSL.
G6 TVB R W Steele, 98 Obelisk Rise, Boughton Green, Northampton, NN2 8QU [IO92NG, SP76]
G6 TVC D J Spooner, Bank Cottage, Abbots Morton, Worcester, WR7 4NA [IO92AE, SP05]
G6 TVD E W T Sims, Hafan, Engedi, Bryngwran, Holyhead, LL65 3RR [IO73SG, SH37]
G6 TVE J L Tyreman, 39 Heather Ave, Shaw, Oldham, OL2 8HL [IO83XN, SD90]
G6 TVI G Sculthorpe, 50 Station Rd, Dersingham, Kings Lynn, PE31 6PR [JO02GU, TF63]
G6 TVJ I F Bennett, 6 Dighton Gate, Stoke Gifford, Bristol, BS12 6XA [IO81RM, ST68]
G6 TVK A Baker, Heathfield, 48 Harnham Rd, Salisbury, SP2 8JJ [IO91CB, SU12]
G6 TVW A Pattinson, 7 Heysham Hall Dr, Heysham, Morecambe, LA3 2QX [IO84NA, SD46]
G6 TWA A N Woollard, 1 The Rowans, Victoria St., Wellington, TA21 8HR [IO80JX, ST12]
G6 TWB Details withheld at licensee's request by SSL.[Correspondence to S Cheshire ARC, c/o T Lester, Cloverdale, Ravenshall, Betley, Crewe, CW2 6AP.]
G6 TWD W J West, Lectric, Alexandra Rd, Crediton, EX17 2DH [IO80ET, SS80]
GD6 TWF C P Wood, 2 Lyndale Ave, Peel, IM5 1JY
G6 TWJ J L Willis, 24 Handbury Rd, Malvern Link, Malvern, WR14 1NN [IO82UD, SO74]
G6 TWL A Tickle, 290 Common Rd, Newton-le-Willows, WA12 9JN [IO83QK, SJ59]
G6 TWR K R Simmonds, 1 Oak Tree Rd, Knaphill, Woking, GU21 2RN [IO91QH, SU95]
G6 TWW P A Spyrakis, 38 Tormead Rd, Guildford, GU1 2JB [IO91RF, TQ05]
G6 TWX A P Tatterton, 15 Ulleswater Cl, Little Lever, Bolton, BL3 1UD [IO83TN, SD70]
G6 TXB B J Thompson, 258 Walderslade Rd, Chatham, ME5 0PA [JO01GI, TQ76]
G6 TXL G E Castle, 26 Grantley Cres, Kingswinford, DY6 9EH [IO82VL, SO88]
G6 TXR P D A Clewes, 19 Church Mews, Denton, Manchester, M34 3GL [IO83WK, SJ99]
G6 TXV S P Calver, 132 Imperial Ave, Maryland Sea, Mayland, Chelmsford, CM3 6AJ [JO01JQ, TL90]
G6 TXW C R Campbell, 35 Bentley Dr, Kiln Ln, Harlow, CM17 9PA [JO01BS, TL40]
G6 TXY B J Coulstock, 51 Windward Cl, Littlehampton, BN17 6QX [IO90RT, TQ00]
G6 TXZ A J Coulstock, 28 Stanley Rd, Northampton, NN5 5DT [IO92NF, SP76]
G6 TYB J E Cooke, 106 Wirral Dr, Winstanley, Wigan, WN3 6LD [IO83PM, SD50]
G6 TYF R J Duley, Denova, Hornsby Ln, Orsett Heath, Orsett, Grays, Essex, RM16 3AU [JO01EL, TQ67]
G6 TYG Details withheld at licensee's request by SSL.
GW6 TYJ J Downer, 7 y Gwernydd, Glais, Swansea, SA7 9HF [IO81BQ, SN60]
GM6 TYL G I Davidson, 11 Dunedin St., Edinburgh, EH7 4JD [IO85JX, NT27]
G6 TYS Details withheld at licensee's request by SSL.
G6 TYT S A White, 38 Mayfield Rd, Rainbow Hill, Worcester, WR3 8NT [IO82VE, SO85]
G6 TYU C M Wigham, 18 Marystow Cl, Allesley, Coventry, CV5 9EA [IO92FK, SP28]
GM6 TYX C Macleod, Morven, Marybank, Stornoway, Isle of Lewis, PA86 0DD [IO76JB, NS08]
G6 TYZ K E Miller, 13 Victoria Mill, Belmont Bridge, Skipton, BD23 1RL [IO83XX, SD95]
G6 TZD T A Mellor, 17 Millbrook Ave, Atherton, Manchester, Lancs, M46 9LL [IO83SM, SD60]
G6 TZE J K Richards, 23 Pitwood Green, Tadworth, KT20 5HZ [IO91VH, TQ25]
G6 TZG C E A Pickett, Fourwinds, Hay Grn, Terrington St Clem, Kings Lynn, PE34 4PU [JO02DR, TF51]
G6 TZI C J Robins, 167 Park Ln, Macclesfield, SK11 6UB [IO83WG, SJ97]
G6 TZJ R L Rudkin, 28 Dalkeith Ave, Alvaston, Derby, DE24 0BG [IO92GV, SK33]
G6 TZO N M Roberts, 6 Parsons Dr, Sileby, Loughborough, LE12 7QH [IO92KR, SK61]
G6 TZT T P Rogers, 36 Goodacre Rd, Ullesthorpe, Lutterworth, LE17 5DL [IO92JL, SP58]
G6 UAC Details withheld at licensee's request by SSL.
G6 UAD Details withheld at licensee's request by SSL.
GW6 UAK P E Lubbock, 34 Birchwood Gdns, Whitchurch, Cardiff, CF4 1HY [IO81JM, ST17]
G6 UAM R J Murphy, 27 Monson Way, Oundle, Peterborough, PE8 4QG [IO92SL, TL08]
G6 UAP E R Magnuszewski, 49 Elvaston Rd, Wollaton, Nottingham, NG8 1JU [IO92JW, SK54]
G6 UAW M J Egerton, 13 Acacia Ave, Sandhurst, Owlsmoor, Camberley, Surrey, GU15 4RU [IO91PI, SU86]
G6 UAX Details withheld at licensee's request by SSL.
G6 UAY Details withheld at licensee's request by SSL.
G6 UBD Details withheld at licensee's request by SSL.
G6 UBH M J Faithfull, 99 Bramble Rd, Hatfield, AL10 9SB [IO91US, TL20]
G6 UBL Details withheld at licensee's request by SSL.[Locator: JO01CE, WAB: TQ54, Sevenoaks district.]
G6 UBM Details withheld at licensee's request by SSL.[Locator: JO01CE, WAB: TQ54, Sevenoaks district.]
G6 UBU S L McCann, 7 Brown St., Radcliffe, Manchester, M26 4HW [IO83TN, SD70]
G6 UBW D C Monckton, 68 Beechfield, Hoddesdon, EN11 9QJ [IO91XS, TL31]
G6 UCI G J Miller, 11 Friars Ave, Great Sankey, Warrington, WA5 2AR [IO83QJ, SJ58]
G6 UCJ Details withheld at licensee's request by SSL.
G6 UCO G S Bee, 80 Hospital Rd, Burntwood, WS7 0EQ [IO92BQ, SK00]
G6 UCQ M J Burgess, 20 Norfolk Rd, Luton, LU2 0RE [IO91TV, TL12]
G6 UCT P J Bowron, 52 Eastcotes, Tile Hill, Coventry, CV4 9AU [IO92FJ, SP27]
G6 UCW Details withheld at licensee's request by SSL.
G6 UCY K R Porter, 180 Durand Cl, Carshalton, SM5 2BY [IO91WJ, TQ26]
G6 UDA R W Pyrah, 20 Eddington Rd, Fairhaven, Lytham, Lytham St. Annes, FY8 1BS [IO83LR, SD32]
G6 UDB F R Scott, 13 Cromwell St, Mansfield, Notts, NG18 2SF [IO93JD, SK56]
G6 UDE B S Powell, 5 Briarwood Ave, Clacton on Sea, CO15 5QX [JO01OT, TM21]
G6 UDF P H Phelps, 14 The Warren, Hazlemere, High Wycombe, HP15 7ED [IO91PP, SU89]
G6 UDG A Price, Myrtle House, Main Rd, Pontesbury, Shrewsbury, SY5 0PY [IO82NP, SJ30]
G6 UDI S P Phillips, 79 Selwyn St., Stoke, Stoke on Trent, ST4 1ED [IO82VX, SJ84]
G6 UDN Details withheld at licensee's request by SSL.
G6 UDO Details withheld at licensee's request by SSL.
G6 UEB Details withheld at licensee's request by SSL.
G6 UEC J A Robinson, Leewood, Upton, Huntingdon, Cambs, PE17 5YQ [IO92UJ, TL17]
G6 UED T S Lloyd, 18 Coleville Rd, Minworth, Sutton Coldfield, B76 1XR [IO92CM, SP19]
G6 UEG I A Norman, Crown & Falcon, High St., Puckeridge, Ware, SG11 1RN [JO01AV, TL32]
G6 UEI P Norman, 20 Meadow Cl, Budleigh Salterton, EX9 6JN [IO80IP, SY08]
G6 UEJ Details withheld at licensee's request by SSL.[Op: F J Newton.]
G6 UEO Details withheld at licensee's request by SSL.
G6 UEQ R E Rix, Patterdale, Roe Downs Rd, Medstead, Alton, GU34 5LG [IO91LC, SU63]
G6 UER V D Rice, 24 Harewood Cl, Tuffley, Gloucester, GL4 0SR [IO81UT, SO81]
G6 UEU G S Reid, 18 Richard Rd, Walsall, WS5 3QW [IO92AN, SP09]
G6 UEV M T Piper, 26 Hare Law Gdns, Stanley, DH9 8DG [IO94CU, NZ15]
G6 UEX R F Bain, 17 Fallowfield Ave, Newcastle upon Tyne, Newcastle upon Tyne, NE3 3NN [IO95EA, NZ26]
GW6 UFH S M Fry, 10 Heaseland Pl, Killay, Swansea, SA2 7EQ [IO71XO, SS59]
GM6 UFJ Details withheld at licensee's request by SSL.
G6 UFL S K Farrell, 35 Fowlmere Rd, Foxton, Cambridge, CB2 6RT [JO02AC, TL44]
G6 UFM Dr E Foxley, 31 Greenfield St., Dunkirk, Nottingham, NG7 2JN [IO92JW, SK53]
GW6 UFO J Fitzgerald, 1 Ellesmere Ln, Penley, Wrexham, LL13 0LP [IO82NW, SJ43]
G6 UFP R W Caine, 22 Freeth Rd, Brownhills, Walsall, WS8 6JG [IO92AP, SK00]
GI6 UFS J Corey, 19 Orritor Cres, Cookstown, BT80 8BQ [IO64PP, H87]
G6 UFT D Carr, 17 Woodland Way, Theydon Bois, Epping, CM16 7DY [JO01BQ, TQ49]
GI6 UFU J R Campbell, 2 Downshire Park, Canreagh, Hillsborough, BT26 6HB [IO64XL, J25]
G6 UGA M J Pinkney, 169 Sandringham Rd, Perry Barr, Birmingham, B42 1PZ [IO92BM, SP09]
GW6 UGC G C Phillips, 83 Heol y Llwynau, Trebanos, Pontardawe, Swansea, SA8 4DB [IO81BQ, SN70]
GW6 UGD Details withheld at licensee's request by SSL.
G6 UGE C H Power, 296 Alderley, Little Digmore, Skelmersdale, WN8 9NB [IO83OM, SD40]
G6 UGG A J Panton, 35 Long Water Dr, Gosport, PO12 2UP [IO90KS, SZ69]
G6 UGH A R Parsons, 12 Hall Green, Upton upon Severn, Worcester, WR8 0NQ [IO82VB, SO83]
G6 UGO R Alexander, 8 Kent House Rd, Sydenham, London, SE26 5LB [IO91XK, TQ37]
G6 UGP D M J Arnold, 50 Freeman Rd, Didcot, OX11 7DD [IO91IO, SU59]
G6 UGS M A Allison, 6 Eden Rd, Beverley, HU17 7HD [IO93SU, TA04]
G6 UGT T Aherne, 21 Burbage Pl, Alvaston, Derby, DE24 8NP [IO92GV, SK33]
G6 UGW M J Bell, 61 Oldbury Orchard, Churchdown, Gloucester, GL3 2PU [IO81WV, SO81]
G6 UGX T M Arnold, 72 Chepstow Dr, Bletchley, Milton Keynes, MK3 5NB [IO91OX, SP83]
GM6 UHC A C Stewart, 25 Henderson Dr, Skene, Westhill, AB32 6RA [IO87UD, NJ80]
G6 UHD B R Scott, 58 Ridgemead, Calne, SN11 9EN [IO81XK, ST97]
GM6 UHE A C Wilson, Lochend, Beith, Ayrshire, KA15 2LN [IO75RS, NS45]
G6 UHL B C Ritchie, Shenandoah, 25 Abbeyfields Dr, Studley, B80 7BF [IO92BG, SP06]
G6 UHQ J S Crawley, 11 Robinswood, Luton, LU2 7YU [IO91TV, TL02]
G6 UHS A K Coe, 22 St. Annes Way, Spalding, PE11 3PN [IO92WT, TF22]
GW6 UHY L D E Crompton, 6 Morgans Terr, Pontrhydyfen, Port Talbot, SA12 9TP [IO81DP, SS79]
G6 UIA Details withheld at licensee's request by SSL.
G6 UIC M W Cooper, 14 Dale Rd, Redditch, B98 8HJ [IO92AH, SP06]
G6 UIL D Durell, 27 Old Pasture Rd, Frimley, Camberley, GU16 5SA [IO91PH, SU85]
G6 UIM S B Daniels, 270 Dartmouth Rd, Paignton, TQ4 6LH [IO80FK, SX85]
G6 UIP G B Dodd, 28 Leen Mills Ln, Hucknall, Nottingham, NG15 8BZ [IO93JB, SK55]
G6 UJB R G Delaforce, Trewillen House, Trecrogo Ln End, South Petherwin, Launceston Cornwall, PL15 7LF [IO70TO, SX38]
G6 UJC C R Dukes, 56 Uppergate Rd, Stannington, Sheffield, S6 6BY [IO93FJ, SK38]

GM6 UJG V Simpson, 43 Fortingall Pl, Perth, PH1 2NF [IO86GJ, NO02]
G6 UJH Details withheld at licensee's request by SSL.
G6 UJI B W Staton, 99 Linden Ave, Prestbury, Cheltenham, GL52 3DT [IO81XV, SO92]
GM6 UJO S Sinagoca, North Cassingray Hs, Largoward, Leven, Fife, KY9 1JD [IO86OG, NO40]
G6 UJR M A Severs, 18 Wincanton Rd, Redcar, TS10 2HR [IO94LO, NZ62]
G6 UJU P Saville, 573 Manchester Rd, Linthwaite, Huddersfield, HD7 5QX [IO93BP, SE11]
G6 UKB S A Beldon, 18 Meller Cl, Beddington, Croydon, CR0 4UB [IO91WI, TQ36]
G6 UKK Details withheld at licensee's request by SSL.
G6 UKN W Bailey, 35 Elton Ln, Winterley, Sandbach, CW11 4TN [IO83TC, SJ75]
G6 UKQ J Riley, 56 Church St., Bignall End, Stoke on Trent, ST7 8PE [IO83UB, SJ85]
G6 UKV P Hallard, 127 Witton Ln, West Bromwich, B71 2AE [IO82XM, SO99]
G6 UKZ J M Heward, 25 Frankburn Rd, Streetly, Sutton Coldfield, B74 3QH [IO92BN, SP09]
G6 ULD R C Humphrys, 19 Flamingo Cl, Walderslade, Chatham, ME5 7RF [JO01GI, TQ76]
GW6 ULE Details withheld at licensee's request by SSL.
G6 ULI J C Gardner, 14 Totley Grange Rd, Totley, Sheffield, S17 4AF [IO93FH, SK37]
G6 ULP P D Gronbech, 22 Abelwood Rd, Long Hanborough, Witney, OX8 8DD [IO91HT, SP41]
G6 ULS P G Kent-Woolsey, Kyakamena, The Green, Flempton, Bury St Edmunds, Suffolk, IP28 6EL [JO02HH, TL87]
G6 ULX A C Sanders, 99 Wyken Ave, Wyken, Coventry, CV2 3BZ [IO92GK, SP38]
G6 UMA B Smith, 13 Moss Park Ave, Werrington, Stoke on Trent, ST9 0LS [IO83WA, SJ94]
G6 UMF P Tainton, 31 Abbeyfields, Randley, Telford, Shropshire, TF3 2AH [IO82SP, SJ70]
G6 UMK R G Jones, 473 Staines Rd West, Ashford, TW15 2AB [IO91SK, TQ07]
G6 UML T Reader, 76 West View Rd, Dartford, DA1 1TR [JO01CK, TQ57]
G6 UMN L Gibson, 44 Latona St., Walney, Barrow in Furness, LA14 3QS [IO84JC, SD16]
G6 UMP G P Gould, 32 Archer Rd, Kenilworth, CV8 1DJ [IO92EI, SP27]
G6 UMX J Hibbert, 125 Chase Hill Rd, Arlesey, SG15 6UF [IO92UA, TL13]
G6 UNC J E Mather, 18 Regent Rd, Kirkheaton, Huddersfield, HD5 0LW [IO93DP, SE11]
GM6 UNL J S Leslie, 7 Charters St., Stirling, FK7 0QE [IO86AC, NS79]
G6 UNN R K Lewis, 5 Popham Cl, Bridgwater, TA6 4LD [IO81MD, ST33]
GM6 UNQ E Leask, Lochpark, Inverinan, Taynuilt, Argyll, PA35 1HH [IO76JH, NN01]
G6 UNR Details withheld at licensee's request by SSL.
G6 UNW Details withheld at licensee's request by SSL.
G6 UNX B L Lockhart, 15 Tiverton Ave, Chapel House, Skelmersdale, WN8 8PA [IO83ON, SD40]
G6 UNY Details withheld at licensee's request by SSL.
G6 UOD Details withheld at licensee's request by SSL.
G6 UOH I A Thacker, 85 Winchester Rd, Grantham, NG31 8RN [IO92QV, SK93]
G6 UOJ I H Underwood, Finedon Sidings, Finedon Sidings Indu, Finedon, Wellingborough, Northants, NN9 5NY [IO92PI, SP87]
G6 UOO D S Wilde, Canal House, Buxworth, Stockport, Ches, SK12 7NF [IO83XI, SJ98]
G6 UOU L G Goudge, 12 Elizabeth Dr, Haslingden, Rossendale, BB4 4JB [IO83UQ, SD72]
G6 UOX M J Walker, 94 Lambert Rd, Uttoxeter, ST14 7QY [IO92BV, SK03]
G6 UOZ M G Wilshaw, 19 Milton Cres, The Coppice, Talke, Stoke on Trent, ST7 1PF [IO83UB, SJ85]
G6 UPA D S Wiseman, 22 Queens Cres, Clapham, Bedford, MK41 6DA [IO92SD, TL05]
G6 UPH M J Hackney, 20 St. Matthews Way, Allhallows, Rochester, ME3 9SH [JO01HL, TQ87]
G6 UPI B B Hurrell, 33 Meadow Way, Hellesdon, Norwich, NR6 5NN [JO02PP, TG21]
G6 UPQ D W L Holloway, 48 Wenrisc Dr, Minster Lovell, Witney, OX8 5RQ [IO91FT, SP31]
G6 UPR B E Hingston, Hazelwood Farm, Marldon, Paignton, Devon, TQ3 1SQ [IO80EL, SX86]
G6 UPX Details withheld at licensee's request by SSL.
G6 UQ B Naylor Stockport RS, 47 Chester Rd, Poynton, Stockport, SK12 1HA [IO83WI, SJ98]
G6 UQB Details withheld at licensee's request by SSL.
G6 UQC A C Wildsmith, 11 Rushford St., Manchester, M12 4WZ [IO83VK, SJ89]
G6 UQI E J Payne, 12 Cowley Dr, Worthy Down, Winchester, SO21 2QW [IO91IC, SU43]
G6 UQN E J R May, 9 Post House Ln, Great Bookham, Bookham, Leatherhead, KT23 3EA [IO91TG, TQ15]
G6 UQO D Oliver, 4 Meadowfield Dr, Hoyland, Barnsley, S74 0QE [IO93GL, SE30]
GI6 UQU A E Rattray, 20 Charlemont Gdns, Armagh, BT61 9BB [IO64QI, H84]
G6 UQW N D Ruffle, 21 Gorse Way, Burntwood, WS7 8TB [IO92AQ, SK00]
G6 UQX H Purves, 2 Bourtree Cl, Wallsend, NE28 9AA [IO95FA, NZ36]
G6 UQZ Details withheld at licensee's request by SSL.
G6 URA Details withheld at licensee's request by SSL.
G6 URF A Hartley, 18 Smithy Cl, Cronton, Widnes, WA8 5BT [IO83OJ, SJ48]
G6 URJ S J Jelly, 6 Jasmine Terr, West Drayton, UB7 9AN [IO91SM, TQ07]
G6 URK J W E Jennings, 354 Williamthorpe Rd, North Wingfield, Chesterfield, S42 5NS [IO93HE, SK46]
G6 URM B Johnson, 12 Hessary Terr, Princetown, Yelverton, PL20 6RB [IO80AN, SX57]
G6 URP Details withheld at licensee's request by SSL.
G6 URR I Kirk, 12 Edinbane Cl, Rise Park, Nottingham, NG5 5DU [IO93JA, SK54]
G6 URT C Kapoutsis, 78 South Worple Way, Mortlake, London, SW14 8NG [IO91UL, TQ27]
G6 URX M Gallon, 44 Broomridge Ave, Fenham, Newcastle upon Tyne, NE15 6QP [IO94EX, NZ26]
G6 URY Details withheld at licensee's request by SSL.
G6 USD M J Matthews, 213 Hucclecote Rd, Brockworth, Gloucester, GL3 3TZ [IO81VU, SO81]
G6 USG G M Comer, 56 Woodhall Rd, Kidsgrove, Stoke on Trent, ST7 4QY [IO83VC, SJ85]
G6 USH Details withheld at licensee's request by SSL.
G6 USL B J Cowell, Oakwood, 16 Hollin Cl, Rossington, Doncaster, DN11 0XX [IO93LL, SK69]
G6 USN R B Chester, 15 Davids Ln, Springhead, Oldham, OL4 4RZ [IO83XM, SD90]
G6 USO P R Charnings, 52 Crown St., Redbourn, St. Albans, AL3 7PF [IO91TT, TL11]
G6 USU T S Derbyshire, 214 Greasby Rd, Greasby, Wirral, L49 2PN [IO83KJ, SJ28]
G6 USX W P Dennison, 41 Tarbert Walk, Stepney, London, E1 0EE [IO91XM, TQ38]
G6 UT I T Ansell Harlow and District A.R.S., 9 Sewell Harris Cl, Harlow, CM20 3HB [JO01BS, TL41]
G6 UTC R G Harrison, Cotswold Villa, All Saints Rd, Uplands, Stroud, Glos, GL5 1TT [IO81VR, SO80]
G6 UTD Details withheld at licensee's request by SSL.
G6 UTK G J Fisher, 85 Audley Park Rd, Bath, BA1 2XN [IO81TJ, ST76]
G6 UTL S R Foulser, 6032 PO Box, Basingstoke, RG22 5YU [IO91KF, SU65]
G6 UTN B J Frost, Coppins, 8 Robins Wood Dr, Ferndown, BH22 9RZ [IO90BT, SU00]
G6 UTP Details withheld at licensee's request by SSL.
G6 UTT P Sheppard, Round Corners, 7 First Ave, Middleton on Sea, Bognor Regis, PO22 6ED [IO90QT, SU90]
GI6 UUC J Thompson, 21 Watchill Rd, Ardboley, Ballyclare, Co. Antrim, BT39 9QW [IO74BT, J39]
GW6 UUF Details withheld at licensee's request by SSL.
G6 UUO M G Smith, Nova Cinq Windsor Rd, Jarvis Brook, Crowborough, E Sussex, TN6 2JB [JO01CB, TQ52]
G6 UUR S A Whitehead, 94 Cranmore Boulevard, Shirley, Solihull, B90 4RU [IO92CJ, SP17]
G6 UUS T J Robinson, 291 Preston New Rd, Blackburn, BB2 6PL [IO83SS, SD62]
G6 UUY J J Maxwell, Hillcrest, Castle View, Egremont, Cumbria, CA22 2NA [IO84FL, NY01]
G6 UUZ S A T Gascoigne, The Bungalow, Newstead Abbey Park, Nottingham, Notts, NG15 8GD [IO93JC, SK55]
G6 UV J G Brown, Uplands, 2 Larks Field, Hartley, Longfield, DA3 7EJ [JO01DJ, TQ66]
G6 UVL G Hall, 23 Monarch Cl, Chatham, ME5 7PD [JO01GI, TQ76]
G6 UVN M Henman, 4 Lyne Walk, Hackleton, Northampton, NN7 2BW [IO92OE, SP85]
G6 UVO C F L Heritage, 20 Comet Rd, Hatfield, AL10 0SX [IO91VS, TL20]
G6 UVQ C M J Hall, Linby, 4 Wideatts Rd, Cheddar, BS27 3AP [IO81OG, ST45]
G6 UVT Details withheld at licensee's request by SSL.
G6 UVU J P Handy, 77 Abbeyfield Rd, Old Hall Park, Moseley, Wolverhampton, WV10 8TH [IO82WP, SJ90]
G6 UW J H Keeler Cambridge Uws, 41 Enniskillen Rd, Cambridge, CB4 1SQ [JO02BF, TL46]
GM6 UWF J L Allan, Roseville, 87 Needless Rd, Perth, PH2 0LD [IO86GJ, NO12]
G6 UWH Details withheld at licensee's request by SSL.
G6 UWI N Bradshaw, 22 Madeira Terr, South Shields, NE33 3AQ [IO94GX, NZ36]
G6 UWK J P Barden, 2 Pond Hall Cttgs, Bradfield Rd, Wix, Manningtree, CO11 2SP [JO01NW, TM12]
G6 UWM G J Boskett, 7 Lapwing Cl, Northway, Tewkesbury, GL20 8TN [IO82WA, SO93]
G6 UWO D C Bullock, 1 Selby Cl, Toton, Beeston, Nottingham, NG9 6HS [IO92IV, SK43]
G6 UWS M P Byles, 108 Kingsway, Wellingborough, NN8 2EN [IO92PG, SP86]
GW6 UWV H F Postel, Ty-Isaf, Clynnog Fawr, Caernarvon, Gwynedd, LL54 5NH [IO73TA, SH44]
G6 UWW M P Williams-Davies, 16 Hoyle Rd, Wirral, L47 3AQ [IO83JJ, SJ28]
G6 UX A Simmons, 161 Upper Bond St., Hinckley, LE10 1RT [IO92HN, SP49]
G6 UXD Details withheld at licensee's request by SSL.
G6 UXF K D Young, 8 Magnolia Cl, Worcester, WR5 3SJ [IO82VE, SO85]
G6 UXG A D Webb, 35 Hill House Dr, Minster, Ramsgate, CT12 4BE [JO01PI, TR36]
G6 UXL G Rigg, 100 Collingwood St., South Shields, NE33 4JY [IO94GX, NZ36]
G6 UXM S W J Vinnicombe, 8A Cross Rd, Cholsey, Wallingford, Oxon, OX10 9PE [IO91KN, SU58]
G6 UXN Details withheld at licensee's request by SSL.
G6 UXR Details withheld at licensee's request by SSL.
G6 UXW P Mundy, 25 Lonsdale Ave, Cosham, Portsmouth, PO6 2PU [IO90LU, SU60]
G6 UXX P T V Leese, 4 Harefield, Harlow, CM20 3EF [JO01BS, TL41]
GW6 UXY A L Lightly, 8 Smithville Cl, St. Briavels, Lydney, GL15 6TN [IO81QR, SO50]
GM6 UYC R D Matheson, 1 Margaret St., Avoch, IV9 8PX [IO77VN, NH75]
G6 UYI W E Rutter, 136 Trispen Cl, Helwood Village, Liverpool, L26 7YS [IO83OI, SJ48]
G6 UYJ A S Page, 127 Whyke Ln, Chichester, PO19 2AU [IO90OT, SU80]
G6 UYK P G Russell, High Park Barn, Oxenholme, nr Kendal, Cumbria, LA9 7RE [IO84PH, SD59]
G6 UYM D J Richards, 25 Burnivale, Malmesbury, SN16 0BL [IO81WO, ST98]
G6 UYN A E Rumney, 21 Beverley Gdns, Woodmancote, Cheltenham, GL52 4QD [IO81XW, SO92]
G6 UYT M Havard, 6 The Jetties, North Luffenham, Oakham, LE15 8JX [IO92QO, SK90]
G6 UYU D Iveson, 30 Thornlaw North, Thornley, Durham, DH6 3EY [IO94GR, NZ33]
G6 UYY P G Jupp, 28 Shenley Rd, Dartford, DA1 1YE [JO01CK, TQ57]
G6 UZA A J Kotowicz, 47 Portree Dr, Rise Park, Nottingham, NG5 5DT [IO93JA, SK54]
G6 UZB Details withheld at licensee's request by SSL.
G6 UZF Details withheld at licensee's request by SSL.
G6 UZG P S Ashby, 26 Van Diemens Ln, Bath, BA1 5TW [IO81TJ, ST76]
G6 UZH Details withheld at licensee's request by SSL.
G6 UZJ J M Austen, Station House, Breedon Rd, Worthington, Ashby de La Zouch, LE65 1RA [IO92HS, SK42]
G6 UZL P J Bunn, 29 Griffiths Rd, West Bromwich, B71 2EH [IO92AN, SP09]
G6 UZM S C Byford, 21 Clarke Dr, Shaw, Swindon, SN5 9SH [IO91CN, SU18]
G6 UZO M K Brunsdon, 7 Oldberg Gdns, Brighton Hill, Basingstoke, RG22 4NP [IO91KF, SU64]
G6 UZP Details withheld at licensee's request by SSL.
G6 UZR A M Brown, 24 Salway Dr, Salway Ash, Bridport, DT6 5LD [IO80OR, SY49]
G6 UZT D P Brown, 5 Old St., Upton upon Severn, Worcester, WR8 0HN [IO82VB, SO84]
G6 UZY M J Owen, 9 Valeside Gdns, Colwick, Nottingham, NG4 2EL [IO92KW, SK64]
G6 UZZ A J Pickles, 103 Norristhorpe Ln, Liversedge, WF15 7AL [IO93DQ, SE22]
G6 VAA G E T Perks, 55 Andrew Rd, Tipton, DY4 0AJ [IO82XN, SO99]
G6 VAD P W Purdy, 4 Hethersett Rd, East Carleton, Norwich, NR14 8HX [JO02ON, TG10]
G6 VAH J O N Le Bon, 7 Langham Gr, Timperley, Altrincham, WA15 6DY [IO83UJ, SJ78]
G6 VAL A M Oughton, 176 South Lodge Dr, Southgate, London, N14 4XN [IO91WP, TQ39]
G6 VAT T J Sheils, 35 Quarry Ln, Halesowen, B63 4PB [IO82XK, SO98]
G6 VAW C E Soars, 118 Braddon Rd, Loughborough, Leics, LE11 5YZ [IO92JS, SK52]
G6 VAX R M Saunders, 93 Oaks Ave, Worcester Park, KT4 8XG [IO91VI, TQ26]
G6 VAZ D J Thomas, 2 Riverside Cl, Mildenhall, Bury St. Edmunds, IP28 7LH [JO02GI, TL77]
G6 VBB A E V Thomas, 4 Hayes Ave, Prescot, L35 5BJ [IO83OK, SJ49]
GW6 VBC K J Tomlinson, Jubliee Cottage, The Coffee Tavern, Cwmavon Rd, Blaemavon Gwent, NP4 9ED [IO81LS, SO20]
G6 VBD J H Savage, 2 Alvecote Cttgs, Alvecote Ln, Alvecote, Tamworth, B79 0DJ [IO92EP, SK20]
G6 VBE R J Ransom, 1 Bilberry Rd, Clifton, Shefford, SG17 5HB [IO92UA, TL13]
G6 VBJ P L Tasker, Chenar House, Mile Path, Hook Heath, Woking, GU22 0JL [IO91RH, SU95]
G6 VBK D Hatton, Cotton Arms, Cholmondeley Rd, Wrenbury, Nantwich, CW5 8HG [IO83QA, SJ54]
GW6 VBN M R Hunt, 23 Swansea Rd, Pontardawe, Swansea, SA8 4AL [IO81BR, SN70]
GW6 VBO J J Harrison, 31 Erasmus St., Penmaenmawr, LL34 6LH [IO83AG, SH77]
G6 VBQ A E Haddock, 1 Heron Way, St. Ives, Huntingdon, PE17 4SS [IO92XI, TL37]
G6 VBR L G Cowley, 4 Coronation Rd, Wath upon Dearne, Rotherham, S63 7AP [IO93HM, SE40]
GI6 VCG J L Brownlees, 8 Cairnbeg Park, Larne, BT40 1UB [IO74CU, D40]
G6 VCH N J Carter, 101 Broadhurst, Farnborough, GU14 9XA [IO91OH, SU85]
GI6 VCL K Cunningham, 4 Garvaghy Rd, Portglenone, Ballymena, BT44 8EF [IO64SU, C90]
G6 VCM D J Cann, 33 Cleveland Cl, Highwoods, Colchester, CO4 4RD [JO01LV, TM02]
G6 VCN H Dunn, 3 Roper St., Workington, CA14 3BZ [IO84FP, NY02]
G6 VCR H Eden, 24 Hillcrest Rd, Langho, Blackburn, BB6 8EP [IO83ST, SD73]
GM6 VCV W Ferguson, 1 Gartcows Ave, Falkirk, FK1 5QJ [IO85CX, NS87]
G6 VCY P A Eley, 94 Raphael Cl, Whoberley, Coventry, CV5 8LS [IO92FJ, SP37]
G6 VDA A C Sutton, 151 Sherwood Rd, Hall Green, Birmingham, B28 0EY [IO92BK, SP18]
G6 VDH Details withheld at licensee's request by SSL.
G6 VDK P W Lutas, 616 Queens Dr, Swindon, SN3 1AZ [IO91CN, SU18]
G6 VDL W McMullen, 25 Stetchworth Rd, Walton, Warrington, WA4 6JE [IO83QI, SJ68]
GW6 VDR J Miller, Tirionfa, 41 Bodelwydden Ave, Old Colwyn, Colwyn Bay, LL29 9NP [IO83DG, SH87]
G6 VDT S M D Picco, 52 Athelstan Rd, Bitterne, Southampton, SO19 4DD [IO90HV, SU41]
G6 VDU B L Miles, Carwithen Cottage, Fore St., Goldsithney, Penzance, TR20 9LQ [IO70GD, SW53]
GW6 VDY H J Wright, Glan Menai, Conwy Old Rd, Penmaen Mawr, Gwynedd, LL34 6YE [IO83BG, SH77]
GW6 VED R M Straughan, 1 Crossroads, Gilwern, Abergavenny, NP7 0DX [IO81LT, SO21]
G6 VEG T W Gray, 56 Knightside Walk, Chapel Park Est, Newcastle upon Tyne, NE5 1TP [IO95DA, NZ16]
GW6 VEN A Rose, 4 Llys Clwyd, Foryd Rd, Kinmel Bay, Rhyl, LL18 5EW [IO83FH, SH98]
G6 VEP G Reid, Woodlands, 51 Doncaster Rd, Costhorpe, Worksop, S81 9QW [IO93KJ, SK58]
G6 VEQ C P Rybak, Caywood, Newlands Dr, Maidenhead, Berks, SL6 4LL [IO91OM, SU88]
G6 VET J G Goodson, 29 Bilberry Dr, Marchwood, Southampton, SO40 4YR [IO90GV, SU31]
G6 VEX M L Harris, 105 Walmley Rd, Sutton Coldfield, B76 1QL [IO92CN, SP19]
G6 VEY I M Haver, 115 Harrington St., Bourne, PE10 9HB [IO92TS, TF12]
G6 VEZ G Helm, 31 Faringdon Ave, Blackpool, FY4 3QQ [IO83LS, SD33]
G6 VF S M Illman, 25 The Heythrop, Ingatestone, CM4 9HG [JO01EP, TQ69]
G6 VFA M D Hine, Tall Trees, Lime Ln, Oakwood, Derby, DE21 4RF [IO92GW, SK33]
G6 VFB W Hogan, 279 Halliwell Rd, Bolton, BL1 3PE [IO83SO, SD71]
G6 VFC D Hooton, 80 Portland Rd, Rushden, NN10 0DJ [IO92QG, SP96]
GW6 VFH R H Jenkins, 29 Pemberton St., Llanelli, SA15 2RB [IO71WQ, SS59]
G6 VFN T W Scott, 82 Northfield Cres, Wells Next The Sea, NR23 1LR [JO02KW, TF94]
G6 VFO J Stokes, 109 Hollyhedge Rd, West Bromwich, B71 3BT [IO92AM, SP09]
G6 VGA C S G Mc Call, Three Oaks, Curtisden Green, Goudhurst, Cranbrook, TN17 1LE [JO01FD, TQ74]
G6 VGC R E Woolley, 12 Princess Rd, Uttoxeter, ST14 7DN [IO92BV, SK03]
G6 VGH I Allen, The Bungalow, Kingsley Green, Kingsley Rd, Frodsham, Ches, WA6 6YA [IO83PG, SJ57]
GM6 VGL M A Brunton, 2 Easter Pl, Portlethen, Aberdeen, AB1 4XL [IO87TH, NJ72]
G6 VGN S F Buckmaster, 173 Whinney Ln, New Ollerton, Newark, NG22 9TJ [IO93ME, SK66]
G6 VGO M H Barrett, The Inch Cottage, Pallinsburn, Cornhill on Tweed, Northd., TD12 4SJ [IO85WP, NT93]
G6 VGT D R Bowlas, 38 Senneleys Park Rd, Northfield, Birmingham, B31 1AL [IO92AK, SP08]
GI6 VGW Details withheld at licensee's request by SSL.
G6 VGY J A Curtis, 189 Whitchurch Rd, Harlescott, Shrewsbury, SY1 4EY [IO82PR, SJ51]
G6 VGZ D G Cheriton, 9 Garlick Dr, Kenilworth, CV8 2TT [IO92FI, SP37]
G6 VHE Dr M D Entwistle, 3 Barton Cl, Bradenstoke, Chippenham, SN15 4EZ [IO91AM, SU07]
G6 VHG A J Foster, 35 Gloucester Pl, Peterlee, SR8 2HB [IO94HS, NZ44]
G6 VHL J E Guppy, 16 Barnfield Cl, Hastings, TN34 1TS [JO00GU, TQ80]
G6 VHO L Indri, 39 Turnpike House, Goswell Rd, London, EC1V 7PD [IO91WM, TQ38]
G6 VHW H Newell, 1047A Bradford Rd, Birstall, Batley, West Yorks, WF17 9HX [IO93DR, SE22]
G6 VIF B T Morris, 21 Loxley Gdns, Southdown Rd, Bath, BA2 1HS [IO81TI, ST76]
G6 VIK I A King, 11 Cockhall Cl, Litlington, Royston, SG8 0RB [IO92WB, TL34]
G6 VIN J G Walker, 44 Albany Rd, Kilnhurst, Rotherham, S62 5UG [IO93HL, SK49]
G6 VIO M Wingrove, 46 Clifford Rd, Wembley, HA0 1AE [IO91UM, TQ18]
G6 VIQ M R Watson, 2 Hill Top View, Dacre Banks, Harrogate, HG3 4BH [IO94DB, SE16]
GM6 VIU Details withheld at licensee's request by SSL.
G6 VIX Details withheld at licensee's request by SSL.
G6 VIY A S Wood, 55 Warrington Dr, New Oscott, Birmingham, B23 5YP [IO92BM, SP09]
G6 VJA I Taylor, 97 George St., Cleethorpes, DN35 8PL [IO93XN, TA30]
G6 VJB L Levoir, 51 Crow Green Rd, Pilgrims Hatch, Brentwood, CM15 9RB [JO01DP, TQ59]
G6 VJC J R Taylor, 17 Aintree Way, Barons Keep, Dudley, DY1 2SL [IO82WM, SO99]
G6 VJE Details withheld at licensee's request by SSL.
GI6 VJI J P McGuckin, 30 Hightown Rise, Glengormley, Newtownabbey, BT36 7XA [IO74AP, J38]
G6 VJM D E Lynch, 30 Whitecroft View, Baxenden, Accrington, BB5 2QP [IO83TR, SD72]
G6 VJN J T Pinson, 170 Worcester Rd, Malvern Link, Malvern, WR14 1AA [IO82UD, SO74]
G6 VJR D W Reading, 51 Selworthy Rd, Castle Bromwich, Birmingham, B36 0HR [IO92CM, SP18]
G6 VKA C Thompson, Fourwinds, Walton Hill, Deerhurst, Gloucester, GL19 4BT [IO81WW, SO82]
GW6 VKD C V S Smith, 13 Barham Rd, Trecwn, Haverfordwest, SA62 5XX [IO71MW, SM93]
G6 VKG Details withheld at licensee's request by SSL.
G6 VKI C J Richardson, Laburnum Cottage, Railswood Dr, Pelsall, Walsall, WS3 4BD [IO92AP, SK00]
G6 VKJ Details withheld at licensee's request by SSL.
G6 VKL D R Mayers, 21 Moreton Dr, Poynton, Stockport, SK12 1FA [IO83WI, SJ98]
G6 VKS Dr I D P Morgan, Leigh House, 64 Widney Rd, Bentley Heath, Solihull, B93 9AW [IO92CJ, SP17]
G6 VKX M A Webber, 23 Ramsey Cl, Horley, RH6 8RE [IO91VE, TQ24]
GW6 VKY A E J White, 89 Mill Gdns, Blackpill, Swansea, SA3 5AZ [IO81AO, SS69]
GW6 VLA G Williams, 11 Haulfryn, Clydach, Abergavenny, NP7 0LZ [IO81KT, SO21]
G6 VLB Details withheld at licensee's request by SSL.
G6 VLC S D Paxton, 43 Curtis Ave, Abingdon, OX14 3UL [IO91IQ, SU59]
G6 VLE R I Lawrence, 468 Bexhill Rd, St. Leonards on Sea, TN38 8AU [JO00GU, TQ70]
G6 VLP D E Barnes, No3 The Chennells, High Halden, Ashford, Kent, TN26 3NB [JO01IC, TQ83]
G6 VLT J A Higgins, 190 Little Glen Rd, Glen Parva, Leicester, LE2 9TT [IO92KN, SP59]
G6 VLV D Colman, 4 South Rd, Bisley, Woking, GU24 9ES [IO91QH, SU95]
GI6 VLY Dr J A P Earle, 25 Carnesure Park, Comber, Newtownards, BT23 5LT [IO74DN, J46]
GW6 VMB C S Gibson, 1184 Carmarthen Rd, Forestfach, Swansea, SA5 4BL [IO81AP, SS69]
GM6 VME Details withheld at licensee's request by SSL.
G6 VMF R Hope, 30 Greendale Gdns, Hetton-le-Hole, Houghton-le-Spring, DH5 0EF [IO94GT, NZ34]
G6 VMI D S Milne, 22 Eastnor Rd, Reigate, RH2 8NE [IO91VF, TQ24]
G6 VMR Details withheld at licensee's request by SSL.
G6 VMV S Brocklehurst, 1 Overton Cl, Congleton, CW12 1JZ [IO83VD, SJ86]
G6 VNI G A Duggan, Fern Cottage, 28 Higher Rads End, Eversholt, Milton Keynes, MK17 9ED [IO91RX, SP93]
G6 VNO N Hanson, 31 Nicholls Ct, Thorplands, Northampton, NN3 8AP [IO92NG, SP76]
G6 VNT Details withheld at licensee's request by SSL.
G6 VNW B Major, 18 Cuckoo Ln, Gateacre Park, Liverpool, L25 4UQ [IO83NJ, SJ48]
G6 VNZ M Owen, 21 Dartmouth Hill, Greenwich, London, SE10 8AJ [IO91XL, TQ37]
G6 VOC C R Raine, Heathric, Front St., Ingleton, Darlington, DL2 3HS [IO94DN, NZ12]

G6 VOE D R Simpkins, 18 Oundle Rd, Weldon, Corby, NN17 3JT [IO92QL, SP98]
G6 VOG W C Seaney, 9 Vernon Villas, St Blazeygate, Par, Cornwall, PL24 2EE [IO70PI, SX05]
G6 VOH Details withheld at licensee's request by SSL.
G6 VOV R L Leavold, 8 Wilkinson Way, North Walsham, NR28 9BB [JO02QT, TG22]
G6 VOX Details withheld at licensee's request by SSL.
G6 VPF B M Cawthorne, 20 Huntsmans Gate, Burntwood, WS7 9LL [IO92BQ, SK00]
G6 VPH R J Gorton, 3 Pickford Ave, Little Lever, Bolton, BL3 1PN [IO83TN, SD70]
G6 VPJ G C Hall, 54 Townfields, Sandbach, CW11 4PQ [IO83TD, SJ76]
G6 VPK K M Higbee, 5 Davoren Walk, Bury St. Edmunds, IP32 6QA [JO02IG, TL86]
G6 VPL J E Hopkinson, 4 Marwood Croft, Streetly, Sutton Coldfield, B74 3JU [IO92BO, SP09]
G6 VPN B C Jameson, 16 Osborne Ct, Park View Rd, Ealing, London, W5 2JE [IO91UM, TQ18]
G6 VPV J A Wake, 15 Deepdale Way, Red Hall Est, Darlington, DL1 2TA [IO94FM, NZ31]
G6 VPW R B Stoate, 19 Jean Rd, Brislington, Bristol, BS4 4JT [IO81RK, ST67]
G6 VPZ R A Phillips, 422 Birchgrove Rd, Birchgrove, Swansea, SA7 9NR [IO81BQ, SS79]
G6 VQC A G Read, 6 Oast House Rd, Icklesham, Winchelsea, TN36 4BN [JO00IW, TQ81]
G6 VQN A D Morris, 67 Broad Oak Way, Cheltenham, GL51 5LL [IO81WV, SO92]
G6 VQV S D Shenfield, 15 Buckeridge Way, Bradwell on Sea, Southminster, CM0 7QQ [JO01KR, TM00]
G6 VQW R K Seaton, Welsh Rd, Off Church, Offchurch, Leamington Spa, Warks, CV33 9AQ [IO92GG, SP36]
G6 VQX Details withheld at licensee's request by SSL.
G6 VQY Details withheld at licensee's request by SSL.
GM6 VQZ S Windwick, 31 Craigie Cres, Kirkwall, KW15 1ER [IO88MX, HY41]
G6 VRA S G Tamlin, 18 Cater Rd, Barnstaple, Devon, EX32 9JU [IO71XB, SS53]
G6 VRF B Crowther, 11 Houseman Pl, Marston, Blackpool, FY4 5AE [IO83LT, SD33]
G6 VRH Details withheld at licensee's request by SSL.
G6 VRM S K Hall, 77 Glenton St., Peterborough, PE1 5HH [IO92VN, TL29]
GW6 VRN M Jones, 30 Bryntaf, Cefn Coed, Merthyr Tydfil, CF48 2PU [IO81HS, SO00]
G6 VRT Details withheld at licensee's request by SSL.
G6 VRU G W Giles, 42 Owls Rd, Verwood, BH31 6HJ [IO90BU, SU00]
G6 VS W H G Metcalfe, 7 Cottage Cl, Bromborough, Wirral, L63 0PW [IO83LH, SJ38]
G6 VSC S M Adams, 39 Wirehill Dr, Lodge Park, Redditch, B98 7JU [IO92AH, SP06]
G6 VSD Details withheld at licensee's request by SSL.
GI6 VSK J B Savage, 61 Ardcarn Park, Newry, BT35 8PD [IO64TE, J02]
G6 VSQ F A Whitehurst, Roselands, Clarke Ln, Bollington, Macclesfield, SK10 5AH [IO83WG, SJ97]
G6 VTA M S Fisher, 41 Granary Rd, Stoke Heath, Bromsgrove, B60 3QH [IO82XH, SO96]
G6 VTH Details withheld at licensee's request by SSL.
G6 VTN P A Green, 79 The Spinney, Bar Hill, Cambridge, CB3 8SU [JO02AF, TL36]
G6 VTQ J George, 2 Hollybank Ct, Highfield Rd, Widnes, WA8 7DP [IO83PI, SJ58]
G6 VTU Details withheld at licensee's request by SSL.
G6 VTX M D Brindley, 1 Shillingford Dr, Hunters Oak Trentha, Stoke on Trent, ST4 8YG [IO82VX, SJ84]
G6 VTY D W Chardin, 28 Brook Rd, Brentwood, CM14 4PT [JO01DO, TQ59]
G6 VUE S Butler, 231 Newman Rd, Wincobank, Sheffield, S9 1LU [IO93GK, SK39]
G6 VUF Details withheld at licensee's request by SSL.
G6 VUG R A Collins, 113 Alexandra Rd, Farnborough, GU14 6RR [IO91PG, SU85]
G6 VUI Details withheld at licensee's request by SSL.
G6 VUJ J M Davis, 69 Bryanston Rd, Solihull, B91 1BS [IO92CK, SP18]
G6 VUK R R Gray, 19 Giller Dr, Penwortham, Preston, PR1 9LT [IO83PR, SD52]
GM6 VUL Details withheld at licensee's request by SSL.
G6 VUN M C Hodson, 17 Marshfield Cl, Church Hill North, Redditch, B98 8RW [IO92BH, SP06]
G6 VUP Details withheld at licensee's request by SSL.
G6 VUX S P Challoner, Grosvenor Farm, Holme St., Tarvin, Chester, CH3 8EQ [IO83OE, SJ46]
G6 VUY Details withheld at licensee's request by SSL.
G6 VVB P Banwatt, 219 St. Pauls Rd, Peterborough, PE1 3EH [IO92VO, TF10]
G6 VVC M R Brickley, Willows, Poundpool, Somerton, Somerset, TA11 6LZ [IO81PB, ST42]
G6 VVE S F Banks, 29 Froxmere Cl, Crowle, Worcester, WR7 4AP [IO82WE, SO95]
G6 VVL K C Hotchen, 6 Nourse Cl, Leckhampton, Cheltenham, GL53 0NQ [IO81WU, SO91]
G6 VVS R Jackson, 32 Broadway Ave, Halesowen, B63 3DD [IO82XK, SO98]
G6 VVU Details withheld at licensee's request by SSL.
G6 VVV Details withheld at licensee's request by SSL.[Op: L D Dunn, 4 Linburn Drive, Bishop Auckland, Co Durham, DL14 0RG.]
GM6 VVX N Dunnachie, 12 Blackhill View, Law, Carluke, ML8 5JZ [IO85BR, NS85]
G6 VVZ T O Butler, 103 Spring Gdns, Anlaby Common, Hull, HU4 7QH [IO93TR, TA02]
G6 VWC Details withheld at licensee's request by SSL.
G6 VWF J W Holbrook, 1 Segrave Gr, Willerby Rd, Hull, HU5 5DJ [IO93TS, TA02]
G6 VWH Details withheld at licensee's request by SSL.[Vale of White Horse ARS, c/o G3SEK.]
G6 VWP G Skacel, 32 Weelsby Way, Hessle, HU13 0JW [IO93SR, TA02]
GI6 VWS J A Quigg, 9 Springhill Terr, Limavady, BT49 9BS [IO65MB, C62]
G6 VWV S H Cresswell, 7 Japonica Dr, Nottingham, NG6 8PU [IO92JX, SK54]
G6 VWZ P J Kelly, 74 Buckingham Rd, Bletchley, Milton Keynes, MK3 5HL [IO91PX, SP83]
G6 VX Details withheld at licensee's request by SSL.[Correspondence to: Gt Western Contest Group, Cheltenham, Glos. QSL via G3NKS.]
GM6 VXB Details withheld at licensee's request by SSL.[Locator: IO97AQ. WAB: NK06.]
G6 VXC R J Callaghan, 6 Barn Meadow, Birmingham, B25 8YT [IO92CL, SP18]
G6 VXD J Devereux, 440 Birchfield Ln, Oldbury, Warley, B69 1AF [IO82XL, SO98]
G6 VXE D B Hilton, 35 Sandringham Rd, Northolt, UB5 5HN [IO91TN, TQ18]
G6 VXL T G Buck, 178 Rover Dr, Castle Bromwich, Birmingham, B36 9LL [IO92CM, SP19]
G6 VXM M D Barratt, 28 Penrhyn Cres, Beeston, Nottingham, NG9 5PA [IO92JV, SK53]
G6 VXO Details withheld at licensee's request by SSL.
G6 VXR H J Metcalf, Beech Lee, Vicarage Ln, Ropley, Alresford, SO24 0DU [IO91KB, SU63]
G6 VXS N E Morris, 45 High Mount St., Hednesford, Cannock, WS12 4BL [IO82XR, SJ91]
G6 VXZ A Sorab, Woodgaston Cottage, Woodgaston Ln, Northney, Hayling Island, Hants, PO11 0RL [IO90MT, SU70]
GW6 VYD M W Shortman, Wenmor Lodge, Benllech, Anglesey, LL74 8UH [IO73VH, SH58]
GW6 VYE J A Williams, 2 Ty Nant, Caerphilly, CF83 2RA [IO81JO, ST18]
G6 VYK E A Williams, 12 St. Andrews, Grantham, NG31 9PE [IO92QW, SK93]
G6 VYM R J Dove, 9 Vaughton Dr, Sutton Coldfield, B75 6AQ [IO92CN, SP19]
G6 VYR R A Balderson, 4 Cottesmore Rd, Stamford, PE9 2SQ [IO92RP, TF00]
G6 VYS S A D'Arcy, The Gatehouse, Red Cap Ln, Boston, PE21 9LZ [IO92XX, TF34]
G6 VYT R Goring, 54 The Glade, Coulsdon, CR5 1SL [IO91WH, TQ35]
G6 VYV V Head, 50 Coronation Gdns, Hurst Green, Etchingham, TN19 7PH [JO01FA, TQ72]
GM6 VYY A McMinn, Siar-Ard, Mallaig, Inverness-shire, Scotland, PH41 4QY [IO77CA, NM69]
GM6 VYZ W S McMinn, Glengyle, East Bay, Mallaig, PH41 4QF [IO77CA, NM69]
G6 VZB M N Lennox, Vaila, 83 Cheyne Walk, Hornsea, HU18 1BX [IO93VV, TA14]
G6 VZF A Dawes, 8 Peel Cl, Romsey, SO51 7UQ [IO90GX, SU32]
G6 VZM J C Johnson, 62 Julien Rd, Ealing, London, W5 4XA [IO91UL, TQ17]
G6 VZS D J Goodall, 94 Camp Mount, Pontefract, WF8 4BX [IO93IQ, SE42]
G6 VZU C P Hunt, 41 Blickling Rd, Norwich, NR6 6DQ [JO02PP, TG21]
GW6 VZW P Baker, 5 Moseley Terr, Pontrhydyrun, Cwmbran, NP44 1SA [IO81LQ, ST29]
G6 VZZ J J Hackett, 66 Staghills Rd, Waterfoot, Rossendale, BB4 7TS [IO83UQ, SD82]
GW6 WAG D E Jones, Bradford House, The Sq, Corwen, LL21 0DL [IO82HX, SJ04]
G6 WAI C W Parkinson, 25 Moorview Gr, Long Lee, Keighley, BD21 4RR [IO93BU, SE04]
G6 WAO N A Austin, 310 Hatfield Rd, St. Albans, AL1 4UN [IO91US, TL10]
G6 WAR B D Clulee Mid Warwickshire A.R.S., 11 Ascot Ride, Lillington, Leamington Spa, CV32 7TT [IO92FH, SP36]
G6 WAY J M Randall, 3 Steins Ln, Humberstone, Leicester, LE5 1ED [IO92LP, SK60]
G6 WBG P R Smith, 9 Oldenburg, Whiteley, Fareham, PO15 7EJ [IO90IV, SU50]
G6 WBR D R Whitters, 9 Morning St., Spring Bank, Keighley, BD21 5BW [IO93BU, SE03]
G6 WBS S P Siggins, 94 Woodplumpton Rd, Fulwood, Preston, Lancs, PR2 2LR [IO83PS, SD53]
G6 WBT I R Thorp, Pinelodge, Carleton Green, Pontefract, WF8 3NJ [IO93IQ, SE42]
GM6 WBV G J Welch, 43 Lady Nairne Rd, Dunfermline, KY12 9YD [IO86GB, NT08]
G6 WBX Details withheld at licensee's request by SSL.
G6 WCI M J Richards, 72 Carlton Ave, Westcliff on Sea, SS0 0QL [JO01IN, TQ88]
G6 WCM Details withheld at licensee's request by SSL.
G6 WCT D C Taylor, 13 Doughty St., Stamford, PE9 1UT [IO92SP, TF00]
G6 WCX D P Mardle, 138 Sibthorpe Rd, Lee, London, SE12 9DP [JO01AK, TQ47]
G6 WDC L R Baldwin, 26 Cheney Hill, Heacham, Kings Lynn, PE31 7BS [JO02FV, TF63]
G6 WDF W F Head, 1 Hemmingway Dr, Bicester, OX6 8FY [IO91KV, SP52]
G6 WDJ C J Haji-Michael, 175 Okehampton Rd, St. Thomas, Exeter, EX4 1ES [IO80FR, SX99]
GJ6 WDK M J Monteil, Kalimera, 1, Six Rues Villas, St Lawrence, Jersey, JE3 1GL
GJ6 WDN D L Speight, Sandown, 13 Jardin A Pommiers, St Saviour, Jersey, JE2 7LT
G6 WDR A F Tett, 137 Poplar Ave, Hove, BN3 8PL [IO90VU, TQ20]
GW6 WDS F Deravi, Dept Elec Eng, University College, Swansea, SA2 8PP [IO81AO, SS69]
G6 WDZ R L Welsh, 45 Galpin St., Modbury, Ivybridge, PL21 0QB [IO80BI, SX65]
G6 WEH R W Burrows, 102 Roundmoor Dr, Cheshunt, Waltham Cross, EN8 9HH [IO91XQ, TL30]
G6 WEI G M Cockcroft, 31 Holt Rd, Kintbury, Hungerford, RG17 9UY [IO91GJ, SU36]
G6 WEL J R Reynolds, 14 Alumbrook Ave, Holmes Chapel, Crewe, CW4 7BX [IO83TE, SJ76]
G6 WEM Details withheld at licensee's request by SSL.
GW6 WEU K O Turner, 115 Newton Rd, Newton, Swansea, SA3 4SW [IO71XN, SS68]
G6 WEW J Fitzsimons, 63 School Ln, Chapel House, Skelmersdale, WN8 8EN [IO83ON, SD40]
G6 WEX A J Binnington, 47 Sundew Rd, Creekmoor, Poole, Dorset, BH17 7NX [IO90AR, SZ09]
G6 WFF G A Solkow, 12A Manor Ct, Penkhull, Stoke on Trent, Staffs, ST4 5DW [IO82VX, SJ84]
GI6 WFI J K Creighton, 77 Ballyrashane Rd, Coleraine, BT52 2LJ [IO65QD, C83]
G6 WFK S Fox, 33 Mellor Rd, Leyland, Preston, PR5 3JL [IO83PQ, SD52]
G6 WFM K G Farr, 25 Sheppard Dr, Chelmer Village, Chelmsford, CM2 6QE [JO01GR, TL70]
G6 WFS G Quantrill, Innisfree, 1 Ironwell Ln, Hawkwell, Hockley, SS5 4JY [JO01IO, TQ89]
GW6 WFW A Humphreys, 45 Cwm Pl, Llandudno, Conwy, LL30 1LP [IO83CH, SH78]
GI6 WFX R Johnston, 51 Kennedy Dr, Lisburn, BT27 4JA [IO64XM, J26]
G6 WGA A Swift, 56 Birch Hall Ave, Darwen, BB3 0JB [IO83SQ, SD62]
G6 WGD Details withheld at licensee's request by SSL.
G6 WGE N J Riding, 15 Church Ln, Dewsbury Moor, Dewsbury, WF13 4EN [IO93EQ, SE22]
G6 WGK Details withheld at licensee's request by SSL.
GW6 WGP Details withheld at licensee's request by SSL.
GW6 WGY R Clague, 11 Trebor Ave, Bryntirion Park, Bagillt, CH6 6DP [IO83KG, SJ27]
G6 WHH S J D Martin, 14 Barham Rd, Tatlers Farm, Stevenage, SG2 9HX [IO91WV, TL22]
G6 WHL C C Chadburn, 58 Drewry Rd, Keighley, BD21 2HB [IO93BU, SE04]
G6 WHS N A Read, 296 Westdale Ln, Mapperley, Nottingham, NG3 6EU [IO92KX, SK54]
G6 WHY K A W Daniels, 71 Little Yeldham Rd, Little Yeldham, Halstead, CO9 4LN [JO02HA, TL73]
GI6 WHZ R A Freeburn, 60 Tullagh Rd, Cookstown, BT80 9RJ [IO64OP, H77]
G6 WI R J C R Crutchley, 24A Baskerville Rd, Kidderminster, DY10 2YF [IO82VJ, SO87]
G6 WIA A D Mercer, 13 Ollerton St., Eagley Bank, Bolton, BL1 7JU [IO83SO, SD71]
G6 WIC C Newman, Stacombe Barn, Doccombe, Moretonhampstead, Newton Abbot, Devon, TQ13 8SS [IO80DQ, SX78]
G6 WID M R Haselup, Bevenden Oast, Great Chart, Ashford, Kent, TN26 1JP [JO01JE, TQ94]
G6 WIE B H Tompkins, 15 Newick Dr, Newick, Lewes, BN8 4NY [JO00AX, TQ42]
G6 WIG G J Crowton, 64 Atlantic Rd, Old Oscott, Birmingham, B44 8LQ [IO92BN, SP09]
G6 WIM G A King, Forest Edge, Hythe Rd, Marchwood, Southampton, SO40 4WT [IO90GV, SU30]
G6 WIO B A McHugh, 63 Three Butt Ln, Liverpool, L12 7HE [IO83MK, SJ39]
G6 WIR P Harman Burnham Beeches, Radio Club, 25 Pitts Rd, Slough, SL1 3XG [IO91QM, SU98]
G6 WIT G Anderton, Half Acre, 8 Malthouse Ln, Cantley, Norwich, NR13 3AD [JO02SO, TG30]
G6 WJC J Pemberton, Mill Ln, Little Saredon, Shareshill Wolv'To, WV10 7LJ [IO82WP, SJ90]
G6 WJD J Dobson, 15 Gordon Terr, Stakefield, Choppington, NE62 5UE [IO95FD, NZ28]
G6 WJH M P Noake, 127 Cres Rd, Warley, Brentwood, CM14 5JB [JO01DO, TQ59]
G6 WJJ Details withheld at licensee's request by SSL.
GW6 WJM I W F Smith, Freshwinds, 4 Lakeside, Rhosneigr, Anglesey Gwynedd, LL64 5JW [IO73RF, SH37]
G6 WJP Details withheld at licensee's request by SSL.
G6 WJR G Woodward, 34 Osborne Rd, Southport, PR8 2RJ [IO83LO, SD31]
G6 WJS B D Worland, 1 Boscombe Ct, Pixmore Ave, Letchworth, SG6 1RN [IO91VX, TL23]
G6 WJT A M Miles, 23 Redacre Rd, Boldmere, Sutton Coldfield, B73 5EA [IO92BN, SP19]
G6 WJW H C Hutton, 174 Stradbroke Rd, Pakefield, Lowestoft, NR33 7HY [JO02UK, TM59]
G6 WJX E E Jackson, Melford, 36 Ickleton Rd, Duxford, Cambridge, CB2 4RT [JO02BC, TL44]
G6 WKI R F Lewis, 42 Launceston Cl, Romford, RM3 8HQ [JO01CO, TQ59]
G6 WKL G R Pryke, 23 Gillsmans Park, St. Leonards on Sea, TN38 0SN [JO00GU, TQ71]
G6 WKN C A R Reed, 63 Somerset Gdns, Hornchurch, RM11 3QT [JO01CN, TQ58]
G6 WKO J A Richards, 8 Swan Rd, Harrogate, HG1 2SS [IO93FX, SE25]
G6 WKQ P J Rowe, 31 Thorpe Way, Cambridge, CB5 8UJ [JO02CF, TL45]
G6 WKW Details withheld at licensee's request by SSL.
G6 WKX Details withheld at licensee's request by SSL.
G6 WLA K T Jarratt, Belvoir, Rose, Goonhavern, Truro, Cornwall, TR4 9PF [IO70KI, SW75]
G6 WLE R H Bailey, The Malt House, Great Shefford, Hungerford, RG17 7ED [IO91GL, SU37]
G6 WLI F Medlock, 27 Silverdale, Hesketh Bank, Preston, PR4 6RZ [IO83NQ, SD42]
GM6 WLJ D O Milne, 30 Bruceland Rd, Elgin, IV30 1SF [IO87IP, NJ26]
GI6 WLL Details withheld at licensee's request by SSL.
G6 WLM S Simmonds, 3 Robert Cramb Ave, Tile Hill South, Coventry, CV4 9LA [IO92FJ, SP27]
GM6 WLN Details withheld at licensee's request by SSL.
G6 WLP G Smith, Brantfell, Rowrah, Frizington, Cumbria, CA26 3XJ [IO84GN, NY01]
G6 WLX A R Davey, Ty Bryn, 4 Ilminster Cl, Nailsea, Bristol, BS19 2YU [IO81OK, ST46]
G6 WLZ A L Evans, 12 Minson Rd, Hackney, London, E9 7HG [IO91XM, TQ38]
GW6 WMB I V Fearn, 12 Tan y Bwlch, Mynydd Llandegai, Bangor, LL57 4DX [IO73WE, SH56]
G6 WME B Gray, Cruachan, The St., Lessingham, Norwich, NR12 0DE [JO02ST, TG32]
G6 WMG D J Hastings, Westering, Salhouse, Norwich, Norfolk, NR13 6RQ [JO02RQ, TG31]
G6 WML J T A Barrasford, 34 Barnard Ave, Ludworth, Durham, DH6 1LS [IO94GS, NZ34]
G6 WMR M Nyman, 26 Silverstone Ct, River Brook Dr, Stirchley, Birmingham, B30 2SH [IO92BK, SP08]
G6 WMT B G Roper, Glyndowns, Eastcliff, Porthtowan, Truro, TR4 8AP [IO70JG, SW64]
G6 WMU D I Pearce, 247 Wigston Ln, Leicester, LE2 8DJ [IO92KO, SK50]
G6 WMV R W Nunn, The Common, Dunston, Norwich, Norfolk, NR14 8PF [JO02PN, TG20]
GJ6 WMZ Details withheld at licensee's request by SSL.
G6 WNA D G Alliker, 44 Gordon Rd, Grays, RM16 2GW [JO01EL, TQ67]
G6 WNB G S Bennett, 6 Danescroft, Bridlington, YO16 5PZ [IO94VC, TA16]
G6 WNG J R Haines, 18 Pale Meadow Rd, Castle Meadows, Bridgnorth, WV15 6BE [IO82TM, SO79]
G6 WNO Details withheld at licensee's request by SSL.
G6 WNW Details withheld at licensee's request by SSL.
GM6 WNX D J S Mitchell, 65 Robb Pl, Castle Douglas, DG7 1LW [IO84AW, NX76]
G6 WOB H A Stevens, Berkeley Cottage, Westgate, Thornton Dale, Pickering, YO18 7SG [IO94PF, SE88]
G6 WOC J C Bibby, 19 Lampits Ln, Corringham, Stanford-le-Hope, SS17 9AD [JO01FM, TQ78]
GM6 WOE Details withheld at licensee's request by SSL.
GM6 WOF A J Firth, Edan, Berstane Rd, Kirkwall, KW15 1NA [IO88MX, HY41]
G6 WOI G N Flint, 782 College Rd, Erdington, Birmingham, B44 0AL [IO92BN, SP09]
G6 WOL Details withheld at licensee's request by SSL.
G6 WOR R Stephens Worth & Dist. Vrg, 21 St. James Ave, Lancing, BN15 0NN [IO90UU, TQ10]
G6 WOV K Harding, Lavender Leas, 12 Station Rd, Shapwick, Bridgwater, TA7 9NJ [IO81OD, ST43]
G6 WOZ D Wozencroft, 12 Tenby Tower, Willetts Rd, Northfield, Birmingham, B31 4BE [IO92AJ, SP07]
G6 WPE S Mason, 11 East Raynham, Fakenham, Norfolk, NR21 7EL [JO02JT, TF82]
G6 WPJ M Phillips, 80 Colne Rd, Halstead, CO9 2HP [JO01HW, TL83]
G6 WPK J C Puttock, Ambleside, Millfield, St. Margarets At Cliffe, Dover, Kent, CT15 6JL [JO01QD, TR34]
GW6 WPL S J Lawson, Ceinwen, Wrexham Rd, Mold, CH7 1HT [IO83KD, SJ26]
G6 WPO A R Brislin, 14 Banbrook Cl, Solihull, B92 9NE [IO92CK, SP18]
G6 WPR D M Fleetwood, 31 Upper Highway, Hunton Bridge, Kings Langley, WD4 8PP [IO91SQ, TL00]
G6 WQB Details withheld at licensee's request by SSL.
GW6 WQF R I Mason, 25 Hawarden Way, Mancot, Deeside, CH5 2EL [IO83LE, SJ36]
GM6 WQH J O Wilson, Jocksbughts, Avonbridge, Falkirk, FK1 2LD [IO85DW, NS97]
G6 WQI S V E Newman, 2 Winchelsea Ct, Folkestone Rd, Dover, CT17 9TB [JO01PC, TR34]
GW6 WQJ A J V Tidswell, 9 Dewi Ave, Holywell, CH8 7UG [IO83JG, SJ17]
G6 WQK Details withheld at licensee's request by SSL.[Op: P E J Watts. Station located near Egham.]
G6 WQN W Convery, Pyghtle House, Norwich Rd, Strumpshaw, Norwich, Norfolk, NR13 4NT [JO02RO, TG30]
G6 WQS G C Longman, 6 Colbourne Cl, Bransgore, Christchurch, BH23 8BW [IO90DS, SZ19]
G6 WQY B Higgott, 26 Link Rd, Anstey, Leicester, LE7 7BW [IO92JQ, SK50]
G6 WRA Details withheld at licensee's request by SSL.
G6 WRB Details withheld at licensee's request by SSL.[Station located at Farley Hill, Luton.]
G6 WRC G Wood Warrington RC, 3 Cleveleys Rd, Great Sankey, Warrington, WA5 2SR [IO83QJ, SJ58]
G6 WRG Details withheld at licensee's request by SSL.
GW6 WRP R C Brown, The Manse, West End, Penclawdd, Swansea, SA4 3YX [IO71WP, SS59]
GM6 WRY G Smith, 41 Glebe Pl, Galashiels, TD1 3JW [IO85OO, NT43]
G6 WSF M R M Strickland, 25 Coniston Dr, Aylsham, Aylesham, Canterbury, CT3 3HZ [JO01OF, TR25]
G6 WSI Details withheld at licensee's request by SSL.
G6 WSN D R Westgate, 72 Bosworth St., Leicester, LE3 5RA [IO92KP, SK50]
G6 WSX W Carter, 49 The Oval, Netherfield, Holmfirth, Huddersfield, HD7 2YR [IO93CN, SE10]
GW6 WSY K Watkins, 16 Glenside, Pontnewydd, Cwmbran, NP44 1BN [IO81LP, ST29]
G6 WSZ J OHara, 4 Lower Mill Cl, Goldthorpe, Rotherham, S63 9BY [IO93IM, SE40]
G6 WT Details withheld at licensee's request by SSL.
G6 WTA S K Hurst, 14 Boars Head Ave, Standish, Wigan, WN6 0BH [IO83QN, SD50]
G6 WTD R A Kenward, The Bungalow, 20 Church Rd, Ryton on Dunsmore, Coventry, CV8 3ET [IO92GI, SP37]
GM6 WTH T E Callaghan, 18 Kingswell Ave, Onthank, Kilmarnock, KA3 2EZ [IO75SP, NS44]
G6 WTM M A Higlett, 89 Calmore Rd, Totton, Southampton, SO40 8GR [IO90FW, SU31]
GM6 WTP N R Sanders, 8 Danube St., Edinburgh, EH4 1NT [IO85JW, NT27]
G6 WTS A Rankin, 68 Dinting Rd, Glossop, SK13 9EB [IO93AK, SK09]
G6 WUR T D Price, 54 Medeway, Lake, Sandown, PO36 9HQ [IO90JP, SZ58]
G6 WUX T Patterson, 64 Whitemere Gdns, Wardley, Gateshead, NE10 8BE [IO94FW, NZ36]
G6 WVC P Harrison, Santis, 1 Shelley Dr, Wistaston, Crewe, CW2 8DW [IO83SC, SJ65]
GW6 WVD J Williams, 48 Belvedere Dr, Plas Coch, Wrexham, LL11 2BG [IO83LB, SJ35]
G6 WVK Details withheld at licensee's request by SSL.
G6 WVL J Parr, 114 Ashton Rd, Golborne, Warrington, WA3 3UX [IO83QL, SJ69]
G6 WVM D G Harrison, 22 Oswin Gr, Wyken, Coventry, CV2 5GJ [IO92GJ, SP37]
G6 WVR Details withheld at licensee's request by SSL.
G6 WVS P I Child, 36 Higham Rd, Barton-le-Clay, Beds, MK45 4LT [IO91SX, TL03]
G6 WVX Details withheld at licensee's request by SSL.

G6 WWA T E Banham, 28 Norwood Ave, High Ln, Stockport, SK6 8BJ [IO83XI, SJ98]
G6 WWM Details withheld at licensee's request by SSL.
G6 WWR R Coombes Three Counties, Amateur Radio Club, Pinewood Lodge, Sandhills, Wormley, Surrey, GU8 5TD [IO91QD, SU93]
G6 WWT C A Steward, 31 Sycamore Ave, Hayes, UB3 2NT [IO91SM, TQ08]
G6 WWV P J Mann, 345 Devonshire Rd, North Shore, Blackpool, FY2 0RA [IO83LU, SD33]
G6 WWW W R Monk, 6 Hexham Walk, Billingham, TS23 2EA [IO94IO, NZ42]
G6 WWY G B Miller, Silvermine, Cooks Ln, Raymonds Hill, Axminster, EX13 5SQ [IO80MS, SY39]
G6 WWZ R Milne, 12 Warrington Gdns, Hornchurch, RM11 2AG [JO01CN, TQ58]
G6 WXC Details withheld at licensee's request by SSL.
G6 WXI G J Ball, Ciss Green Farm, Watery Ln, Astbury, Congleton, CW12 4RS [IO83VD, SJ86]
G6 WXK I E Buckie, 156 Greenfield Cres, Horndean, Waterlooville, PO8 9EW [IO90LV, SU71]
G6 WXM A P Burt, 17 Western Rd, Wolverton, Milton Keynes, MK12 5AY [IO92OB, SP84]
G6 WXN C L Bennett, 129 Lovedean Ln, Lovedean, Waterlooville, PO8 9RW [IO90LV, SU61]
G6 WXS R C Archer, 37 Caroline St., Preston, PR1 5UY [IO83PS, SD53]
G6 WXZ A J Collier, Hollow Tree, Mill Ln, Horndon on The Hill, Stanford-le-Hope, SS17 8LY [JO01EM, TQ68]
G6 WYC Details withheld at licensee's request by SSL.
G6 WYD S A Chambers, 52 Chapel Ln, Spondon, Derby, DE21 7JW [IO92HW, SK43]
G6 WYF R R Cook, The Beeches, 8 School Ln, Watton At Stone, Hertford, SG14 3SF [IO91WU, TL31]
G6 WYL J R Williamson, 5 Frensham Cl, Stanway, Colchester, CO3 5HP [JO01KV, TL92]
G6 WYQ S H Quade, 8 Highview Cl, Sudbury, CO10 6LY [JO02IB, TL84]
G6 WYS W R Patching, 7 Bursledon Rd, Hedge End, Southampton, SO30 0BP [IO90IV, SU41]
G6 WZA D H Wickens, Auchensail, Bews Ln, Chard, TA20 1JU [IO80MV, ST30]
G6 WZC D W Slatter, 5 Opendale Rd, Burnham, Slough, SL1 7LY [IO91QM, SU98]
G6 WZD C Sillence, 85 Eaton Ave, Bletchley, Milton Keynes, MK2 2HN [IO91PX, SP83]
G6 WZE P S Robinson, 108 Station Rd, Mickleover, Derby, DE3 5FP [IO92FV, SK33]
G6 WZG F A Shelmerdine, 14 Elton Dr, Hazel Gr, Stockport, SK7 6EP [IO83WI, SJ98]
G6 WZK D Wallace, 29 Jerrard Cl, Honiton, EX14 8EA [IO80JT, ST10]
G6 WZL B Walker, 81 Stacey Ave, Wolverton, Milton Keynes, MK12 5DN [IO92OB, SP84]
G6 WZM H I Collinson, 28 Tadcaster Ave, Eyres Monsell, Leicester, LE2 9GA [IO92KO, SP59]
G6 WZN M G Hodges, 5 Ridgeway Ct, Northam, Westward Ho, Bideford, EX39 1TP [IO71VB, SS42]
G6 WZO G R Ratcliffe, 7 Lilac Ave, Ainsdale, Southport, PR8 3RY [IO83LO, SD31]
G6 WZP D B Rogers, 39 Fore St., Seaton, EX12 2AD [IO80LQ, SY29]
G6 WZQ J F Rowley-Guyon, 3 Upper Park Rd, Clacton on Sea, CO15 1HU [JO01NS, TM11]
G6 WZR M H G Ruddock, 36 Brockhurst Rd, Gosport, PO12 3DE [IO90KT, SU50]
G6 XAC N P Harrison, 35 Appletree Gr, The Farriers, Burwell, Cambridge, CB5 0BF [JO02DG, TL56]
G6 XAG A J F Higgs, Neatsfold, Hilton, Blandsford, Dorset, DT11 0DQ [IO80UT, ST70]
G6 XAI Details withheld at licensee's request by SSL.
G6 XAJ Details withheld at licensee's request by SSL.[Op: W Houlton. Station located near Amersham.]
G6 XAK C A Harding, 15 The Stampers, Tovil, Maidstone, ME15 6FF [JO01GG, TQ75]
G6 XAN S J Harding, 29 Wegbarton, Byfleet, Surrey, KT14 7EF [IO91SI, TQ06]
G6 XAR L Hall, 170 Macers Ln, Wormley, Broxbourne, EN10 6EE [IO91XR, TL30]
G6 XAT D L Armstrong, 69 Station Cres, Rayleigh, SS6 8AR [JO01HO, TQ89]
G6 XAU S Kemp, 23 Highclere, Sunninghill, Ascot, SL5 0AA [IO91QJ, SU96]
G6 XAV G J Lawrence, 5 Longwood View, Furnace Green, Crawley, RH10 6PB [IO91VC, TQ23]
G6 XAW D J L Lawrence, 23 Crow Hill Ln, Freehold, New Jersey 07728, USA
G6 XBC Details withheld at licensee's request by SSL.
G6 XBG J H Lines, 6 Hawthorn Rd, Denmead, Waterlooville, PO7 6LJ [IO90LV, SU61]
G6 XBH S M Owen, 2 Plantation Rd, Wollaton, Nottingham, NG8 2ER [IO92JW, SK53]
G6 XBS J K Newman, 36 Gipson Park Cl, Eastwood, Leigh on Sea, SS9 5PW [JO01HN, TQ88]
G6 XBV K Simpson, 35 Dargle Rd, Sale, M33 7FN [IO83UK, SJ79]
G6 XCC J P Sayer, 19 Arras Boulevard, Hampton Magna, Warwick, CV35 8TY [IO92EG, SP26]
G6 XCD E V Ashworth, 11 Como Ave, Burnley, BB11 5LU [IO83US, SD83]
G6 XCE S G Ashby, 24 Princess Dr, Sawston, Cambridge, CB2 4DL [JO02CD, TL45]
G6 XCK S J Bishop, 1 Walsh Cl, Hitchin, SG5 2HP [IO91UW, TL12]
GU6 XCM M G Paul, La Jaoniere, Les Ruettes, St Andrew, Guernsey, Channel Islands, GY6 8UG
G6 XCO R Piper, 13 The Cres, Tanfield Lea, Stanley, DH9 9NQ [IO94DV, NZ15]
G6 XCR F G Yeo, 130 Wanstead, Ln, Ilford, Essex, IG1 3SF [JO01AN, TQ48]
G6 XCU R C Willis, 10 Nayling Rd, Braintree, CM7 2RZ [JO01GV, TL72]
G6 XCV R K Williams, 23 Finchdean Rd, Rowlands Castle, PO9 6DA [IO90MV, SU71]
G6 XCX E G Walker, 15 Hogarth Cl, Sheerwater Rd, London, E16 3SR [JO01AM, TQ48]
G6 XCY R A White, 40 Deanery Gdns, Bocking, Braintree, CM7 5SU [JO01GV, TL72]
G6 XCZ A G Walker, 43 Bourn Ave, Hillingdon, Uxbridge, UB8 3AR [IO91SM, TQ08]
G6 XD J Taylor, Greensleeves, 14 Woodway Cl, Teignmouth, TQ14 8QG [IO80GN, SX97]
G6 XDB J R Woodnutt, 17 Hill Farm Rd, Chalfont St. Peter, Gerrards Cross, SL9 0DD [IO91RO, TQ09]
G6 XDC I Wilkinson, 58 Rossendale Rd, Heald Green, Cheadle, SK8 3HF [IO83VI, SJ88]
G6 XDG J F Page, 18 Winifred Rd, Dagenham, RM8 1PP [JO01BN, TQ48]
G6 XDH D Preston, 97 Littlehaven Ln, Horsham, RH12 4JE [IO91UB, TQ13]
G6 XDI C J Packman, 4 Angel Ln, Hayes, UB3 2QX [IO91SM, TQ08]
G6 XDN D A A Lindop, 44 Young Rd, Galleons Reach, London, E16 3RR [JO01AM, TQ48]
G6 XDP M F Memory, The Bungalow (Woodville), Westfield Ln, Cumberworth, Lincs, LN13 9LQ [JO03CF, TF47]
G6 XDQ Details withheld at licensee's request by SSL.
G6 XDS A G Smith, 2 Sutton Rd, Kegworth, Derby, DE74 2DX [IO92IT, SK42]
G6 XDY K R Gibson-Ford, 181 Newcome Rd, Fratton, Portsmouth, PO1 5DS [IO90LT, SU60]
G6 XEB D L Green, 48 Walnut Cres, Kingswood, Bristol, BS15 4HU [IO81SL, ST67]
G6 XEE P B Hackett, 148 College St., Nuneaton, CV10 7BJ [IO92GM, SP39]
G6 XEF P V Hammond, 31 Honey Way, Royston, SG8 7ES [IO92XB, TL34]
G6 XEG Details withheld at licensee's request by SSL.
G6 XEN R A Hill, 114 Moorside Cres, Sinfin, Derby, DE24 9PT [IO92GV, SK33]
G6 XEV R J Jones, Gable House, Pond Hall Rd, Hadleigh, Ipswich, IP7 5PQ [JO02LA, TM04]
G6 XEX A B Croft, Exchange Building, Exchange St, Normanton, West Yorks, WF6 2AA [IO93GQ, SE32]
G6 XFB B J Roe, Two Hoots, 6 Colworth Cl, Hadleigh, Benfleet, SS7 2SP [JO01HN, TQ88]
G6 XFC P J Richards, 34 Norton Rd, Bristol, BS4 2HA [IO81RK, ST67]
G6 XFD Details withheld at licensee's request by SSL.
G6 XFI Details withheld at licensee's request by SSL.[Op: A J Smith, 194 Court Road, Orpington, Kent, BR6 9DF.]
G6 XFN V L G Stoneman, -le-Mar, 17 Manor Park, Woolsery, Bideford, EX39 5RH [IO70TX, SS32]
G6 XFP Details withheld at licensee's request by SSL.
G6 XFQ P Fairhurst, 9 Berkeley Cl, Leigh, WN7 3QJ [IO83RL, SJ69]
G6 XFR F A Fielder, 97 Pennine Ave, Sundon Park, Luton, LU3 3EJ [IO91SW, TL02]
G6 XGF D Cadman, 32 Breedon Hill Rd, Derby, DE23 6TG [IO92GV, SK33]
G6 XGG S A Clayton, 48 Hankins Ln, Mill Hill, London, NW7 3AJ [IO91VP, TQ29]
G6 XGJ J A Davis, 446 Upper Wortley Rd, Scholes, Rotherham, S61 2SS [IO93HK, SK39]
G6 XGK M L Drinkall, 215 Broad Ln, Bramley, Leeds, LS13 2NJ [IO93ET, SE23]
G6 XGL Details withheld at licensee's request by SSL.[Op: Glynis. Station located at Leigh-on-Sea, near Southend-on-Sea, Essex - 65km East of London on, on the North coast of Thames Estuary).]
G6 XGQ Details withheld at licensee's request by SSL.
G6 XGT M Thornsby, 25 Highfield Rd, Stowupland, Stowmarket, IP14 4DA [JO02ME, TM05]
G6 XGV M R Valenti, 545 Gander Green Ln, North Cheam, Sutton, SM3 9RF [IO91VJ, TQ26]
G6 XHF S M Richards, 65 Shrewsbury Rd, Oxton, Birkenhead, L43 6TE [IO83LJ, SJ38]
G6 XHG E J Rixon, 119 Pensby Rd, Heswall, Wirral, L60 7RD [IO83KI, SJ28]
G6 XHI K E D Ridgwell, 9 Laburnum Dr, Chelmsford, CM2 9NR [JO01FR, TL70]
G6 XHK I M Roper, Flat 109, Birstall Park Ct, Birstall, W Yorks, WF17 9DL [IO93ER, SE22]
G6 XHM C P Seearam, 209 Glyn Rd, Clapton, London, E5 0JR [IO91XN, TQ38]
G6 XHQ Details withheld at licensee's request by SSL.[Op: E J Saunders, tel: (0181) 876 1108. Station located near Richmond.]
G6 XHZ Details withheld at licensee's request by SSL.[Station located near Sale.]
G6 XID S J K Mann, Mannleigh, 32 Gawer Park, Liverpool Rd, Chester, CH1 4DA [IO83NE, SJ46]
G6 XIF C D Milton, 31 Morley Rd, Tiptree, Colchester, CO5 0AA [JO01JT, TL81]
G6 XIH R Morton, 10 Rythergate, Cawood, Selby, YO8 0TP [IO93KU, SE53]
GW6 XII J H H Miller, 29 The Halfpennys, Lower Common, Gilwern, Abergavenny, NP7 0EA [IO81LT, SO21]
G6 XIK L A Mobbs, 39 Keighley Ave, Broadstone, BH18 8HS [IO80XR, SY99]
G6 XIO G Smith, 40 Highfield Cres, Rock Ferry, Birkenhead, L42 2DR [IO83LI, SJ38]
G6 XIP Details withheld at licensee's request by SSL.
G6 XIR M Bennett, Ravenswood, The Shires, Hedge End, Hants, SO3 4BA [IO90HW, SU41]
G6 XJB D R Wratten, 47 Telephone Rd, Southsea, PO4 0AU [IO90LT, SZ69]
G6 XJC L Whitehead, 30 Kempe Rd, Finching Field, Finchingfield, Braintree, CM7 4LE [JO01FX, TL63]
G6 XJD D Whysall, 55 Trowell Park Dr, Trowell, Nottingham, NG9 3RA [IO92IW, SK43]
G6 XJE J Whysall, 55 Trowell Park Dr, Trowell, Nottingham, NG9 3RA [IO92IW, SK43]
G6 XJI D Wiblin, 98 Pemberton Rd, Slough, SL2 2JY [IO91QM, SU98]
G6 XJJ S J McKay, 13 Common Ln, Sawston, Cambridge, CB2 4HW [JO02CC, TL44]
GW6 XJM P T White, Glyst Cottage, Llangynhafal, near Ruthin, Clwyd, LL15 1RT [IO83ID, SJ16]
G6 XJN G R Valenti, 545 Gander Green Ln, North Cheam, Sutton, SM3 9RF [IO91VJ, TQ26]
G6 XJP Details withheld at licensee's request by SSL.
G6 XJU S J Reed, 23 Donnington Dr, Chandlers Ford, Eastleigh, SO53 3PL [IO90HX, SU41]
G6 XJZ D J Rowe, 5 Kelburn Cl, South Millers Dale, Chandlers Ford, Eastleigh, SO53 2PU [IO90HX, SU42]
G6 XKC D W Parrott, 59 Birch Rd, Farncombe, Godalming, GU7 3NU [IO91QE, SU94]
G6 XKE H A Papworth, 339 Gayfield Ave, Withymoor Village, Brierley Hill, DY5 3JE [IO82WL, SO98]
G6 XKF A R Parfitt, 8 Upper High St., Epsom, KT17 4QJ [IO91UI, TQ26]
G6 XKJ I M Pinkard, Casa Verand, 1A Millstone Park, Haward Rd, Penyffordd, Chester, CH4 0JF [IO83LD, SJ36]
G6 XKK H Parrott, 55 Brown St., Macclesfield, SK11 6RY [IO83WG, SJ97]
G6 XKO R A McLellan, Wynwood House, 74 Mount Ambrose, Redruthe, Cornwall, TR15 1QR [IO70JF, SW74]
G6 XKV D C Bodenham, 75 Rosedale Ave, Stonehouse, GL10 2QH [IO81UR, SO80]
G6 XKX R Newell, 57 Evendene Rd, Evesham, WR11 6QA [IO92AC, SP04]
G6 XKY G J Ogden, 10 Hartington Dr, Standish, Wigan, Lancs, WN6 0UA [IO83QN, SD50]
G6 XKZ F Maras, 10 Chepstow Gdns, Grangewood, Chesterfield, S40 2UQ [IO93GF, SK36]
G6 XLB R J Morris, 10 Danetre Dr, Daventry, NN11 5HT [IO92KG, SP56]
G6 XLC J Mills, 6 Borrowdale Rd, Halfway, Sheffield, S19 5HL [IO93GJ, SK38]
G6 XLG P J Pulley, 60 Redesdale Pl, Moreton in Marsh, GL56 0EF [IO91DX, SP23]
G6 XLI Details withheld at licensee's request by SSL.
G6 XLL L Segal, 1 Masons House, 1/3 Valley Dr, Kingsbury, London, NW9 9NQ [IO91UN, TQ18]
G6 XLQ M S Messenger, Nevach, 48 High St., Fenstanton, Huntingdon, PE18 9LA [IO92XH, TL36]
G6 XLU Details withheld at licensee's request by SSL.
G6 XLV Details withheld at licensee's request by SSL.
G6 XM Details withheld at licensee's request by SSL.[Op: W James. Station located near Okehampton.]
G6 XMA S L Butler, 45 Roewood Cl, Holbury, Southampton, SO45 2JT [IO90HT, SU40]
G6 XMG J G Borrington, 52 Hollowood Ave, Littleover, Derby, DE23 6JD [IO92FV, SK33]
G6 XMH D Brown, 12 Milton Dr, Liversedge, WF15 7AX [IO93DQ, SE22]
G6 XMI K L Blayney, 33 Fraser Cl, Cowes, PO31 7QB [IO90IS, SZ49]
G6 XML W T R Barnes, 363 Kenton Ln, Harrow, HA3 8RY [IO91UO, TQ19]
G6 XMM T D Bugg, Gravel Hill, Nayland, Colchester, Essex, CO6 4BJ [JO01KW, TL93]
G6 XMO Details withheld at licensee's request by SSL.
G6 XMP S M Smith, 19 Chatsworth Ave, Nottingham, NG7 7EW [IO92JX, SK54]
G6 XMQ D A Swift, 8 Gr Ln, Buxton, SK17 9HG [IO93BF, SK07]
G6 XMR D M Smith, 21 Millbrook, Leybourne, West Malling, ME19 5QJ [JO01FH, TQ65]
G6 XMT M J Samson, 19 Far Gosford St., Coventry, CV1 5DT [IO92GJ, SP37]
G6 XMU G Smith, 71 Mount Pleasant Rd, Wisbech, PE13 3NQ [JO02BQ, TF41]
G6 XMY J W Stokes, 64 Holmesfield Rd, Marlpool, Heanor, Derbyshire, DE75 7BT [IO93HA, SK44]
G6 XMZ G Slater, 3 Oldcott Cres, Kidsgrove, Stoke on Trent, Staffs, ST7 4HF [IO83VB, SJ85]
G6 XN L A Moxon, Gorsehill, Tilford Rd, Hindhead, GU26 6SJ [IO91PC, SU83]
G6 XNC Details withheld at licensee's request by SSL.
G6 XND P J Smith, 6 Nuthatch, Longfield, DA3 7NS [JO01DJ, TQ66]
G6 XNI A Taylor, 20 Mythop Rd, Marston, Blackpool, FY4 4UZ [IO83MT, SD33]
G6 XNJ J Taylor, 23 Greystone Ave, Elland, HX5 0QH [IO93BQ, SE12]
G6 XNK J S Theedom, 83 Caulfield Rd, Shoeburyness, Southend on Sea, SS3 9LP [JO01JM, TQ98]
G6 XNN E A Townsend, 10 Little Oak Ave, Kirkby in Ashfield, Nottingham, NG17 9BG [IO93JB, SK55]
G6 XNP A C Trett, 81 The Chantrys, Farnham, GU9 7AQ [IO91OF, SU84]
G6 XNQ R E Taylor, 3 Dundridge Gdns, St. George, Bristol, BS5 8SZ [IO81RK, ST67]
G6 XNR S R Taylor, 24 Marston Rd, Tockwith, York, YO5 8PR [IO93IX, SE45]
G6 XNU V Williams, 24 Sunny Bank Ave, Blackpool, FY2 9EQ [IO83LU, SD33]
G6 XO R J Newman, 117 Nalders Rd, Chesham, HP5 3DA [IO91QR, SP90]
G6 XOB F Box, 32 Lancaster Gdns, Herne Bay, CT6 6PU [JO01NI, TR16]
G6 XOD E P Whitby, 1 Gloucester Ave, Chilwell, Beeston, Nottingham, NG9 1HE [IO92JW, SK53]
G6 XOE F A Whitby, 1 Gloucester Ave, Chilwell, Beeston, Nottingham, NG9 1HE [IO92JW, SK53]
G6 XOG C Wells, Troutbeck, Arthington Ln, Pool in Wharfedale, Otley, LS21 1JZ [IO93EV, SE24]
G6 XOI Details withheld at licensee's request by SSL.
G6 XOU H J Yeldham, Belle Fleurs, Wade Reach, Walton on The Naze, Essex, CO14 8RG [JO01PU, TM22]
GI6 XOV J D Clarke, 16 Lochinver Ave, Holywood, BT18 0NQ [IO74CP, J47]
G6 XOW Details withheld at licensee's request by SSL.
G6 XOX A J B P Patrick, 22 Falcon Way, Dinnington, Sheffield, S31 7NY [IO93GJ, SK38]
G6 XPB R Partner, 22 Moordale Ave, Priestwood, Bracknell, RG42 1RT [IO91OK, SU86]
G6 XPL Details withheld at licensee's request by SSL.
G6 XPM C P Perry, 7 Exmoor Cl, Irby, Wirral, L61 9QN [IO83KI, SJ28]
G6 XPP P S Rice, 159 Clapgate Ln, Ipswich, IP3 0RF [JO02OA, TM14]
G6 XPS L T Richards, 30 Wilbert Gr, Beverley, HU17 0AN [IO93SU, TA03]
G6 XPT K W Maxwell, 9 Heaton Cl, Baildon, Shipley, BD17 5PL [IO93CU, SE13]
G6 XPU J S Maxwell, 9 Heaton Cl, Baildon, Shipley, BD17 5PL [IO93CU, SE13]
G6 XPY R Chappell, 17 Redcar Ave, Hereford, HR4 9TJ [IO82PB, SO44]
G6 XPZ A N Carter, 28 Smithwell Ln, Heptonstall, Hebden Bridge, HX7 7NX [IO83XR, SD92]
G6 XQB R J Carter, Redroof, 56 Main Rd, Naphill, High Wycombe, HP14 4QB [IO91OP, SU89]
G6 XQE P Bradley, 75 New Rd, Sawston, Cambridge, CB2 4BN [JO02BD, TL44]
G6 XQG N B Coombs, Chapel Cottage, Chapel Ln, Hempsted, Holt Norfolk, NR25 6LA [JO02NV, TG13]
G6 XQH Details withheld at licensee's request by SSL.
G6 XQO P J Gait, 6 Martindale Rd, Churchdown, Gloucester, GL3 2DW [IO81VV, SO82]
G6 XQP G D J Garner, Tredore, Haugh Rd, Banham, Norwich, NR16 2DE [JO02ML, TM08]
G6 XQR F Gizzi, 19 Kings Field, Bursledon, Southampton, SO31 8EN [IO90IV, SU41]
G6 XQT N S Godwin, 199 Dodworth Rd, Barnsley, S70 6HR [IO93GN, SE30]
G6 XQW W Goulden, 110 The Pastures, Downley, High Wycombe, HP13 5RU [IO91OP, SU89]
GM6 XQX S T Gray, 38 Nelson St., Rosyth, Dunfermline, KY11 2JU [IO86GB, NT18]
G6 XQY J A Griffin, 6 Heathfield, Royston, SG8 5BW [IO92XB, TL34]
G6 XRE B A Helsdon, 23 Kintore Dr, Great Sankey, Warrington, WA5 3NW [IO83QJ, SJ58]
G6 XRF J A Hicks, 26 Studley Ave, Highams Park, London, E4 9PS [JO01AO, TQ39]
G6 XRH J E Hoare, 8 Sunnyheath, Havant, PO9 3BW [IO90MU, SU70]
G6 XRI S A Hobbs, 19 Ashfield Rd, Kenilworth, CV8 2BE [IO92FI, SP27]
G6 XRK M C Huggins, 67 Talbot Rd, Isleworth, TW7 7HG [IO91UL, TQ17]
G6 XRL J Hunt, 11 Vicarage Ln, Poynton, Stockport, SK12 1BG [IO83WI, SJ98]
G6 XRQ Details withheld at licensee's request by SSL.
G6 XRS R E Talbott obo Leicester Radio Soc, Thornfield, 33 Highfield St., Anstey, Leicester, LE7 7DU [IO92JQ, SK50]
G6 XRT L E Jones, 71 Dartford Ave, Edmonton, London, N9 8HE [IO91XP, TQ39]
G6 XRX Details withheld at licensee's request by SSL.
G6 XRY G D Kobiela, 61 Earith Rd, Willingham By Stow, Willingham, Cambridge, CB4 5LS [JO02AH, TL47]
G6 XRZ Details withheld at licensee's request by SSL.
G6 XSA J G Darby, 3A Milton Rd, Wimborne, BH21 1NY [IO90AT, SU00]
G6 XSB M J Dower, 19 Fullwell Ct, Fullwell Ave, Clayhall, Ilford, IG5 0RZ [JO01AO, TQ49]
G6 XSC I J Denison, 152 Broad Oak Way, Hatherley, Cheltenham, GL51 5LL [IO81WV, SO92]
G6 XSF C J Enticknap, 15D Grenfell Rd, Mitcham, Surrey, CR4 2BZ [IO91WK, TQ27]
G6 XSI T Farrell, 12 Salmon Walk, Bury St. Edmunds, IP32 6PS [JO02IG, TL86]
G6 XSK E A Firth, 2 Gladstone Cl, Little Moor, Weymouth, DT3 6RH [IO80SP, SY68]
G6 XSP Details withheld at licensee's request by SSL.
G6 XSR Details withheld at licensee's request by SSL.
G6 XSS B P Gell, 27 Park Rd, Barnstone, Nottingham, NG13 9JF [IO92NV, SK73]
G6 XSY J A Goodey, 62 Rose Hill, Binfield, Bracknell, RG42 5LG [IO91OK, SU87]
G6 XSZ D Graham, 127 Shephall View, Stevenage, SG1 1RP [IO91VV, TL22]
G6 XTC A Tripp, 3 Ash Cl, Chester Rd, Oathills, Malpas, SY14 8JB [IO83OA, SJ44]
G6 XTD R I Hallsworth, 27 Westfield Ave, Heanor, DE75 7BN [IO93HA, SK44]
G6 XTJ R S Harris, 88 Earles Meadow, Horsham, RH12 4HR [IO91UB, TQ13]
G6 XTK D G Harris, 72 Canons Ln, Burgh Heath, Tadworth, KT20 6DP [IO91VH, TQ25]
G6 XTL Details withheld at licensee's request by SSL.
G6 XTM R S Harris, 10 Lewis Cl, Ashill, Thetford, IP25 7BH [JO02JO, TF80]
G6 XTN Details withheld at licensee's request by SSL.[Op: Mrs P M Harvey.]
G6 XTO Details withheld at licensee's request by SSL.
G6 XTR Details withheld at licensee's request by SSL.
G6 XTT R F Holgate, 228 Hanson Ln, Halifax, HX1 4QW [IO93BR, SE02]
G6 XTV D Hough, 18 Jeffreys Dr, Greasby, Wirral, L49 2NJ [IO83KJ, SJ28]
G6 XTW Details withheld at licensee's request by SSL.
G6 XTY C W Jarvis, 20 Sheepbridge Caravan Park, Snettisham, Kings Lynn, Norfolk, PE31 7QR [JO02FU, TF63]
G6 XTZ T Jarvis, 1 Whitehall Ave, Mirfield, WF14 0AQ [IO93DQ, SE22]
G6 XUB Details withheld at licensee's request by SSL.
G6 XUD S Justin, 57 Birchmead Ave, Pinner, HA5 2BQ [IO91TO, TQ18]
G6 XUJ G M Collins, Little Orchard, Hemp Ln, Wigginton, Tring, HP23 6HF [IO91QS, SP91]
G6 XUL F B Honnor, 46 York Rd, Mosscroft Est, Huyton, Liverpool, L36 1XB [IO83OJ, SJ49]
G6 XUU G D V Mancini, Templars Lodge, 7 Crown Row, Bracknell, Berks, RG12 0TH [IO91PJ, SU86]
G6 XUV D R Lee, 188 Manstone Ave, Sidmouth, EX10 9TJ [IO80JQ, SY18]
G6 XUX R W Mettam, 12 School Ln, Marsh Ln, Eckington, Sheffield, S31 9RS [IO93GJ, SK38]
G6 XVM N Bruce, 82 Western Ave, Bentley, Walsall, WS2 0AQ [IO82XO, SO99]
G6 XVN C R Blayney, 74 Clarence Rd, East Cowes, PO32 6HA [IO90IS, SZ59]
G6 XVO M K Brown, 10 Epping Walk, Furnace Green, Crawley, RH10 6LX [IO91VC, TQ23]
G6 XVQ P A Braybrooke, 6 Tubbenden Ln, Orpington, BR6 9PN [JO01BI, TQ46]
G6 XVT Details withheld at licensee's request by SSL.
G6 XVW Details withheld at licensee's request by SSL.[Station located near Tadworth.]
G6 XVY T G Barker, 55 Acre Ave, Eccleshill, Bradford, BD2 2LL [IO93DT, SE13]
G6 XVZ A J Barker, 25 Chestnut Ave, Euxton, Chorley, Lancs, PR7 6BT [IO83PQ, SD51]
G6 XWD C M W Breckons, Low Wood Farm, Lamonby, Nr.Penrith, Cumbria, CA11 9SS [IO84NQ, NY43]

G6 XWE Details withheld at licensee's request by SSL.
G6 XWF M Hallam, 13 Silk St., Glossop, SK13 8QQ [IO93AK, SK09]
G6 XWK S J Branton, 28 Warren Ln, Martlesham Heath, Ipswich, IP5 7SH [JO02OB, TM24]
G6 XWM R L King, 52 Ford Rd, Tiverton, EX16 4BE [IO80GV, SS91]
G6 XWN Details withheld at licensee's request by SSL.
G6 XWO Details withheld at licensee's request by SSL.
G6 XWT Details withheld at licensee's request by SSL.
G6 XWU J Chandler, 1 East Cres, Windsor, SL4 5LD [IO91QL, SU97]
G6 XWY D B Clarke, 90 The Willows, Colchester, CO2 8PX [JO01KU, TM02]
G6 XWZ T M Cooper, Tamarisk, Exbury Rd, Blackfield, Southampton, SO45 1XD [IO90HT, SU40]
G6 XXB D A Cook, The Bungalow Stores, 4 Vicarage Hill, Marldon, Paignton, TQ3 1NH [IO80EL, SX86]
G6 XXC W J Cartledge, 46 Davian Way, Chesterfield, S40 3HX [IO93GF, SK36]
G6 XXE S Crowther, 17 Carr Gate Cres, Wrenthorpe, Carr Gate, Wakefield, WF2.0QR [IO93FR, SE32]
G6 XXJ D A Clubley, The Cattery, 181 Main St., Newthorpe, Nottingham, NG16 2DL [IO93IA, SK44]
G6 XXN A O Clarke, 138 High St., Barwell, Leicester, LE9 8DR [IO92HN, SP49]
G6 XXO D S Day, 32 Somerset Rd, Basildon, SS15 6PE [JO01FN, TQ68]
G6 XXQ B Dodds, 21 Lynton Dr, Lordswood, Chatham, ME5 8QA [JO01GI, TQ76]
G6 XXU Details withheld at licensee's request by SSL.
G6 XXX Details withheld at licensee's request by SSL.
G6 XYA Dr D E Clark, 21 St. Brannocks Rd, Chorleton Cum Hardy, Manchester, M21 0UP [IO83UK, SJ89]
G6 XYD K G Elsworth, 88 Mungo Park Way, Orpington, BR5 4EQ [JO01BJ, TQ46]
GW6 XYE M C Ellett, 14 Canon Dr, Fairfield Park, Bagillt, CH6 6LS [IO83KG, SJ27]
G6 XYF R C Ediss, 5 Stirling Cres, Totton, Southampton, SO40 3BN [IO90GW, SU31]
G6 XYH B J Seddon, Rockfield, 74 Pellhurst Rd, Ryde, PO33 3BS [IO90KR, SZ59]
G6 XYO J N Fazey, 90 Beecher Rd, Halesowen, B63 2DW [IO82XK, SO98]
G6 XYP E E Fuller, 26 Waltham Rise, Melton Mowbray, LE13 1EJ [IO92NS, SK71]
G6 XYR J A Scothern, 24 Cavendish Cres, Annesley Woodhouse, Kirkby in Ashfield, Nottingham, NG17 9BN [IO93JB, SK55]
G6 XYS A D Searle, 22 Crowther Cl, Sholing, Southampton, SO19 1BX [IO90HV, SU41]
G6 XYT M O Gale, 17 Cheltenham Rd, Parkstone, Poole, BH12 2ND [IO90AR, SZ09]
G6 XYU J J Stanton, 4 Fambridge Rd, South Fambridge, Rochford, Essex, SS4 3LY [JO01IP, TQ89]
G6 XYV E Strode, 26 Churchill Cl, Congleton, CW12 4QU [IO83VE, SJ86]
G6 XYW A D Sole, 6 Blackthorne Rd, Gee Cross, Hyde, SK14 5EG [IO83XK, SJ99]
G6 XYX B D Slater, 47 Broom Rd, Lakenheath, Brandon, IP27 9EZ [JO02GJ, TL78]
G6 XZA M N Scott, 28 Penwarden Way, Bosham, Chichester, PO18 8LF [IO90NU, SU80]
G6 XZC C G Shaw, 13 Chapel St., Stonebroom, Alfreton, DE55 6JX [IO93HD, SK45]
G6 XZF B Spink, Intake Lodge, Yearsley, Brandsby, York, YO6 4SW [IO94KD, SE57]
GW6 XZI M S Smith, 42 Westgate, Barnoldswick, Colne, BB8 5QF [IO83VV, SD84]
G6 XZK M G Sanders, 5 Staddon Park Rd, Plymstock, Plymouth, PL9 9HL [IO70WI, SX55]
G6 XZL N E Smith, 14 Pirbright Rd, South Shields, London, SW18 5LZ [IO91VK, TQ27]
G6 XZM C T SmiTh, 104 Warren Rd, Banstead, SM7 1LB [IO91VH, TQ26]
G6 XZP R H Sammons, Peacehaven, 42 Woodcote Ave, Wallington, SM6 0QY [IO91WI, TQ26]
G6 XZS J M J Thorn, 20 Kiln Rd, Shaw, Newbury, RG14 2HA [IO91IJ, SU46]
GW6 XZU G H Thomas, 103 Llangorse Rd, Cwmbach, Aberdare, CF44 0LD [IO81HR, SO00]
G6 XZV Details withheld at licensee's request by SSL.
G6 XZW D Tyers, 31 High St., Clapham, Bedford, MK41 6AG [IO92SD, TL05]
G6 YAC W H Watkins, 134 Vicarage Rd, Wollaston, Stourbridge, West Midlands, DY8 4QY [IO82WL, SO88][Correspondence to Mrs Debra Barton, G1PEP, Secretary Wordsley RC, Rose & Crown, High St, Wordsley, Stourbridge, W Midlands.]
G6 YAF C M Ward, 16 Daleside Rd, West Ewell, Epsom, KT19 9SR [IO91UI, TQ26]
G6 YAH C Wheeler, 11 Brooklands Way, Redhill, RH1 2BN [IO91VF, TQ25]
G6 YAI I M Wilson, 2 Kingswood Cl, Owlthorpe, Sheffield, S19 6SD [IO93GJ, SK38]
G6 YAK P J Willetts, 49 Summervale Rd, Hagley, Stourbridge, DY9 0LX [IO82WK, SO98]
G6 YAL P J Walters, 6 Victoria St., Alfreton, DE55 7GS [IO93HC, SK45]
G6 YAN J C Wappett, 18 Greenacres Dr, Halifax, West Yorks, HX3 7QS [IO93CS, SE12]
G6 YAQ F Barker, 13 Ashbourne Rd, Eccles, Manchester, M30 0HW [IO83TL, SJ79]
G6 YAR R Porteus, 22 North View, Meadowfield, Durham, DH7 8SQ [IO94ER, NZ23]
G6 YAT G Baker, 4 Waltham Rd, Scartho, Grimsby, DN33 2LX [IO93WL, TA20]
G6 YAY D Allen, 19 Stoneyhill, Abbotskerswell, Newton Abbot, TQ12 5LH [IO80EL, SX86]
G6 YB D J Bailey City Bristol Gr, 18 Heath Rise, Warmley, Bristol, BS15 5DD [IO81SK, ST67]
G6 YBC D P Anderson, 97 Leigh Rd, Atherton, Manchester, M46 0LX [IO83RM, SD60]
G6 YBF Details withheld at licensee's request by SSL.
G6 YBH A T White, Amara, Brewers End, Takeley, Bishops Stortford, CM22 6QJ [JO01DU, TL52]
G6 YBM G Wright, Hazeldene, Market Pl, Kessingland, Lowestoft, NR33 7TE [JO02UK, TM58]
G6 YBN A W West, 82 Mount Pleasant Rd, Alton, GU34 2RS [IO91MD, SU73]
G6 YBO C W J Wyatt, 14 Church Rd, Coxley, Wells, BA5 1RJ [IO81PE, ST54]
G6 YBY Details withheld at licensee's request by SSL.
G6 YCE A S Brooke, 33 St. Wilfrids Rd, Doncaster, DN4 6AA [IO93KM, SE60]
G6 YCG A W Bennett, 5 Fifth Ave, Northville, Bristol, BS7 0LP [IO81RM, ST67]
G6 YCI M J Buck, 178 Rover Dr, Castle Bromwich, Birmingham, B36 9LL [IO92CM, SP19]
G6 YCK Details withheld at licensee's request by SSL.
G6 YCL M J Banner, 37 Chesterfield Ave, Gedling, Nottingham, NG4 4GE [IO92KX, SK64]
G6 YCM R Brookes, 52 Larch Gr, Kendal, LA9 6AU [IO84PH, SD59]
G6 YCN R Brassington, Above Park Farm, Leek Rd, Dilhorne, Stoke on Trent, ST10 2PT [IO83XA, SJ94]
G6 YCO J E Baddeley, 52 Stephens Way, Bignall End, Stoke on Trent, ST7 8PL [IO83UB, SJ85]
G6 YCT M T Le Ves Conte, 28 Woodlands, Paddock Wood, Tonbridge, TN12 6AR [JO01EE, TQ64]
G6 YCU R Lord, 7 Harewood Rd, Norden, Rochdale, OL11 5TG [IO83VP, SD81]
G6 YCV T Leach, 19 Brook Vale, Charlton Kings, Cheltenham, GL52 6JD [IO81XV, SO92]
G6 YCW B R Lancaster, 1 Belgrave Cl, Boydell Park, Dodleston, Chester, CH4 9NU [IO83MD, SJ36]
G6 YCX Details withheld at licensee's request by SSL.
G6 YCZ J M Massey, 10 Rapley Ave, Storrington, Pulborough, RH20 4QL [IO90SW, TQ01]
GW6 YDA P A Mitchell, Tyn-y-Giat, Llandyfrydog, Llanerchymedd, Anglesey, LL17 8AL [IO83GG, SJ07]
G6 YDC Details withheld at licensee's request by SSL.
G6 YDD R Boyce Hereford ARS, 3 Castleton Cttgs, Westhide, Hereford, HR1 3RF [IO82QC, SO54]
G6 YDK Details withheld at licensee's request by SSL.
G6 YDN J R Mountain, 15 Eldon Cl, Chapel En-le-Frith, Stockport, Ches, SK12 6PX [IO83XI, SJ98]
G6 YDO F N Mirams, 10 Ravenoak Park Rd, Cheadle Hulme, Cheadle, SK8 7EH [IO83VI, SJ88]
G6 YDQ K Gardiner, 4 Crossways South, Wheatley Hills, Doncaster, DN2 5SJ [IO93KM, SE50]
G6 YDS K G Giddings, 8 Latchmere Cl, Richmond, TW10 5HQ [IO91UK, TQ17]
GW6 YDT E W Gittins, Hedge Rows, 59 Hillock Ln, Marford, Wrexham, LL12 8YG [IO83MC, SJ35]
G6 YEA N W Guy, 78 Park Rd, Bolton, BL1 4RQ [IO83SN, SD70]
G6 YEC G F Hancock, 115 Berkeley Ave, Chesham, HP5 2RS [IO91QR, SP90]
G6 YEK D W Heard, 17 Abbey Rd, Mount Pleasant, Exeter, EX4 7BG [IO80FR, SX99]
G6 YES E Kennedy, 33 Spring Bridge Rd, Manchester, M16 8PW [IO83VK, SJ89]
G6 YET M M I Knight, 81 Edinburgh Ave, Bentley, Walsall, WS2 0HT [IO82XO, SO99]
G6 YEX R G Hooper, 12 Sleaford Rd, Bracebridge Heath, Lincoln, LN4 2NA [IO93RE, SK96]
G6 YEY R M Hope, 26 Chaucer Ave, Andover, SP10 3DS [IO91FF, SU34]
G6 YFB G J Hughes, 2 Benlaw Gr, Felton, Morpeth, NE65 9NG [IO95DH, NU10]
G6 YFF G A Hunter, 7 Torver Way, Orpington, BR6 8DX [JO01AI, TQ46]
G6 YFG S E F Iles, 3 Petersway Gdns, St. George, Bristol, BS5 8TA [IO81RK, ST67]
G6 YFH R S Ingle, 48 Barlborough Rd, Clowne, Chesterfield, S43 4RF [IO93IG, SK47]
G6 YFJ A K Jakusz-Gostomski, 15 Goodliffe Gdns, Tilehurst, Reading, RG31 6FZ [IO91LL, SU67]
G6 YFL H N Jones, 15 Bonchurch Walk, Manchester, M18 8BP [IO83VL, SJ89]
G6 YFY I P Pitfield, 27 Winchester Cres, Fulwood, Sheffield, S10 4ED [IO93FI, SK28]
G6 YFZ D J Paul, Enfield, Gunton Rd, Wymondham, NR18 0QP [JO02NN, TG10]
G6 YGB R L Preston, 60 Heywood Rd, Prestwich, Manchester, M25 1FN [IO83UM, SD80]
G6 YGD Details withheld at licensee's request by SSL.
G6 YGH Details withheld at licensee's request by SSL.
G6 YGI B G Rogers, Fronucha Rhewl, Gobowen, Oswestry, Shropshire, SY10 7AS [IO82LV, SJ33]
G6 YGJ R Robinson, 128 Norman Ave, Bradford, BD2 2NE [IO93DT, SE13]
G6 YGP D R Lee, 4 Blythe Ln, Lathom, Ormskirk, L40 5TY [IO83NO, SD41]
G6 YGQ C P Lilley, 662 Leek New Rd, Milton, Stoke on Trent, ST2 7EF [IO83WB, SJ85]
G6 YGR R M Lamble, 14 Portway Mews, Portway, Wantage, OX12 9BT [IO91GO, SU38]
G6 YGV M T Lane, Harewood Villa, Harewood Pl, Warley Rd, Halifax, West Yorks, HX2 7PN [IO93BR, SE02]
G6 YGW B R Finch, 352 Worsley Rd, Swinton, Manchester, M27 0FH [IO83TM, SD70]
G6 YHD J H Martin, 21 Victoria Rd, Wisbech, PE13 2QL [JO02BP, TF40]
G6 YHF R F Marchant, 16 Chestnut Cl, Huntingdon, PE18 7BS [IO92VI, TL27]
G6 YHK T S Miller, 6 Captains Walk, Swanpool, Falmouth, TR11 4HR [IO70KD, SW73]
G6 YHL C Miller, 5 Lodge Ln, Bewsey, Warrington, WA5 5AG [IO83QJ, SJ58]
G6 YHS A Matthews, 20 Hendon Rd, Nelson, BB9 9JL [IO83VU, SD83]
G6 YHT Details withheld at licensee's request by SSL.
G6 YHU Details withheld at licensee's request by SSL.
G6 YHV Mrt R W Moore, 59 Broadway, South Shore, Blackpool, FY4 2HF [IO83LS, SD33]
G6 YHW G D Murly, 15 Winrose Approach, Leeds, LS10 3PZ [IO93FS, SE32]
G6 YIE S Forbes, 8 Nutmeg Cl, Earley, Reading, RG6 5GX [IO91MK, SU77]
G6 YIG D R J Enticknap, 52 Granes End, Great Linford, Milton Keynes, MK14 5DX [IO92OB, SP84]
G6 YII K Everington, 1 Norfolk Rd, Wigston, LE18 4WH [IO92KO, SP59]
G6 YIJ J Elford, 7 Cunliffe Rd, Stoneleigh, Epsom, KT19 0RJ [IO91VI, TQ26]
G6 YIN J V Clark, 71 Leysholme Cres, Leeds, LS12 4HH [IO93ES, SE23]
G6 YIP A D Cohen, 32 Grafton Cl, Whitehill, Whitehill, Bordon, GU35 9QY [IO91NC, SU83]
G6 YIQ J A Dixon, 24 Glenwood, Welwyn Garden City, AL7 2JS [IO91WT, TL21]
G6 YIU P W Dawson, Ivy Dene, Middle Ln, Oaken, Wolverhampton, WV8 2BE [IO82VO, SJ80]
G6 YIW W K S Gilroy, 10 Hawk Cl, Stoke on Trent, ST3 7GB [IO82WX, SJ94]
G6 YJA D P Gorvett, 30 St. Pauls Ct, Westminster Rd, Worsley, Manchester, M28 3BL [IO83TM, SD70]
G6 YJD J Govier, 14 Witham Ct, Higham, Barnsley, S75 1PX [IO93FN, SE30]
G6 YJI S J Halleron, 5 Hameldon Ave, Baxenden, Accrington, BB5 2QD [IO83TR, SD72]
G6 YJJ P C Hambly, 80 North Dr, Hounslow, TW3 1PU [IO91TL, TQ17]
G6 YJK E F Mack, 15 Keynes Way, Lowfields, Dovercourt, Harwich, CO12 3UA [JO01OW, TM23]
G6 YJN Details withheld at licensee's request by SSL.
G6 YJO D C Arscott, 68 Sandygate Mill, Kingsteignton, Newton Abbot, TQ12 3PE [IO80EN, SX87]
G6 YJR J M Angel, 33 Grovewood Cl, Chorleywood, Rickmansworth, WD3 5PX [IO91RP, TQ09]
G6 YKP Details withheld at licensee's request by SSL.
G6 YLA J Howard, 6 Wansford Green, Goldsworth Park, Woking, GU21 3QH [IO91QH, SU95]
G6 YLB G A Howse, 1 Sutherland Cl, Woodloes Park, Warwick, CV34 5UJ [IO92FH, SP26]
G6 YLD G P Hope, 17 Church Rd, Sutton At Hone, Dartford, DA4 9EX [JO01CJ, TQ57]
G6 YLJ Details withheld at licensee's request by SSL.
G6 YLN M B Hobbs, 3 Beighton Cl, Lower Earley, Reading, RG6 4HZ [IO91MK, SU76]
G6 YLO P A Hizzey, 5 Rue Du Mont Vallier, 31650 St Orens de Gameville, France, X X
G6 YLP R W Harwood, 229 Radley Rd, Abingdon, OX14 3SQ [IO91IQ, SU59]
G6 YLQ D G Harrop, Apartdos Correos 373, 07300 Inca, Mallorca, Spain
G6 YLR K W Harris, 20 Rose Walk, Wicken Green Village, Sculthorpe, Fakenham, Norfolk, NR21 7QE [JO02IU, TF83]
G6 YLS E Chambers, 3 Whitegate Ct, Rainham, Gillingham, ME8 9LG [JO01HI, TQ86]
G6 YLT R Cotton, 60 Stanley St., Bloxwich, Walsall, WS3 3EQ [IO92AO, SK00]
G6 YLV J T Cromack, 13 Hyde Ln, Nash Mills, Hemel Hempstead, HP3 8RY [IO91SR, TL00]
G6 YLW T C Cannon, 36 St. Margarets Dr, Wigmore, Gillingham, ME8 0NR [JO01HI, TQ86]
G6 YLZ P Cornes, 46 Newland Ave, Stafford, ST16 1NL [IO82WT, SJ92]
GI6 YM J T Barnes City of Belfast, YMCA Radio Club, 95 Crawfordsburn Rd, Bangor, BT19 1BJ [IO74DP, J48]
G6 YMA N R Clark, 2 Barley Croft, Stevenage, SG2 9NP [IO91WV, TL22]
GI6 YMC J T Barnes Belfast YMCA AR, White Gables, 95 Crawfordsburn Rd, Bangor, BT19 1BJ [IO74DP, J48]
G6 YME Dr N R Carlson, 92 Castle Dr, Pevensey Bay, Pevensey, BN24 6LA [JO00ET, TQ60]
G6 YMH C Hughes, 85 Benson Gdns, Wortley, Leeds, LS12 4LA [IO93ES, SE23]
G6 YMP Details withheld at licensee's request by SSL.
G6 YMQ E A Hudson, 63 Priory Gdns, The Fieldings, Burnham on Sea, TA8 1QW [IO81MF, ST34]
GW6 YMS P R Humphreys, Ty'N Llan, Heneglwys Bodffordd, Llangefni, Gwynedd, LL77 7SL [IO73UG, SH47]
G6 YMU D Hutchings, 65 Thorley Park Rd, Bishops Stortford, CM23 3NG [JO01BU, TL41]
G6 YMV S P K Hutchings, 65 Thorley Park Rd, Bishops Stortford, CM23 3NG [JO01BU, TL41]
G6 YMW K B Jackson, 13 Thornley Cl, Radford Semele, Leamington Spa, CV31 1UL [IO92GG, SP36]
G6 YMY P J Jacques, Caprius, The Parks, Aldington, Evesham Worcs, WR11 5JP [IO92BC, SP04]
G6 YMZ E Johnson, 32 Moor Ave, Lee Moor, Stanley, Wakefield, WF3 4EJ [IO93GR, SE32]
G6 YNA A R Johnston, 70 Queensbury Ave, Gillingham, ME8 9NZ [JO01HI, TQ86]
G6 YNL R T K Perry, Straight Mile Cott, Gloucester Rd, Rudgeway, Bristol, BS12 2SB [IO81RN, ST68]
G6 YNT S Pentecost, 16 Elmshurst Cres, London, N2 0LP [IO91VO, TQ28]
GW6 YNV S Raddy, 15 Ty To Maen Cl, St. Mellons, Cardiff, CF3 9EY [IO81KM, ST28]
G6 YNW M G Reeves, 17 Newark Ave, Putnoe, Bedford, MK41 8NX [IO92SD, TL05]
G6 YNX L J Rae, 24 Glynne St., Bootle, L20 6DF [IO83ML, SJ39]
G6 YNZ Details withheld at licensee's request by SSL.
G6 YOG M P J Rutt, 31 Succombs Pl, Southview Rd, Warlingham, CR6 9JQ [IO91XH, TQ35]
G6 YON Details withheld at licensee's request by SSL.
G6 YOP P R Harding, 54 Manor Rd, Stretford, Manchester, M32 9JB [IO83UK, SJ79]
G6 YOR J Gillott, 132 Racecommon Rd, Barnsley, S70 6JY [IO93GN, SE30]
G6 YOS Details withheld at licensee's request by SSL.
GW6 YOV Details withheld at licensee's request by SSL.
G6 YOY C A Foulsham, 173 Henley Ave, North Cheam, Sutton, SM3 9SD [IO91VI, TQ26]
G6 YOZ J L Addison, 20 Wychwood Rise, Great Missenden, HP16 0HB [IO91PQ, SU89]
G6 YPE J R Allen, Old Hall Dr, Sulby, nr Welford, Northampton, NN6 7EZ [IO92LI, SP67]
G6 YPF J V Armstrong, 17 Caenwood Rd, Ashtead, KT21 2JA [IO91UH, TQ15]
G6 YPK Details withheld at licensee's request by SSL.
G6 YPM J A Willats, 16 Eden Rd, Gossops Green, Crawley, RH11 8LZ [IO91VC, TQ23]
GM6 YPQ F R Capocci, Duncraggan, Duncraggen Rd, Oban, Argyll, PA35 5DU [IO76JJ, NN02]
G6 YPU Details withheld at licensee's request by SSL.
G6 YPV S H Davies, 14 Haileybury Ave, Liverpool, L10 6LP [IO83ML, SJ39]
G6 YPY S M Davis, 30 Bonny Wood Rd, Hassocks, BN6 8HR [IO90WW, TQ31]
GM6 YQA C M Davies, 57D Menzies Rd, Torry, Aberdeen, AB11 9AS [IO87WD, NJ90]
G6 YQI E Fletcher, 18 Buckingham Ave, Horwich, Bolton, BL6 6NR [IO83RO, SD61]
G6 YQJ D B Fisher, 86 Parsons Ln, Littleport, Ely, CB6 1JS [JO02DK, TL58]
G6 YQN N Fox, 32 Westmorland Ave, Clough Hall, Kidsgrove, Stoke on Trent, ST7 1AT [IO83VB, SJ85]
G6 YQT S K Forbes, 11 Henfield View, Warborough, Wallingford, OX10 7DB [IO91KP, SU59]
G6 YQU R W Fuller, The New House, Main St, Mowsley, Lutterworth, Leics, LE17 6NT [IO92LL, SP68]
G6 YQY Details withheld at licensee's request by SSL.
G6 YRB J F Stewart, 107 Turnberry, Prospect Gdns, Skelmersdale, WN8 8EG [IO83ON, SD40]
G6 YRC A Smith, 4 Wesley Gr, Clifton Farm Est, Burnley, BB12 0JJ [IO83UT, SD83]
G6 YRE S A Sweet, 69 Norton Ave, Herne Bay, CT6 7TA [JO01NI, TR16]
G6 YRG N J M Spiteri, 49 Greenlea Rd, Yeadon, Leeds, LS19 7SN [IO93DU, SE14]
GM6 YRH A C Smith, Robsland, Strathaven Rd, Lesmahagow, Lanark, ML11 0HY [IO85BP, NS84]
G6 YRI S Sizmur, 38 Longbourne Way, Chertsey, KT16 9ED [IO91RJ, TQ06]
G6 YRJ T P Simmons, 3 West Hill Pl, Brighton, BN1 3RU [IO90WT, TQ30]
G6 YRK S P Wright, 6 The Grove, Hadfield, Hyde, SK14 8BA [IO93AK, SK09]
G6 YRM Details withheld at licensee's request by SSL.
GM6 YRN A D Stewart, Ropeway Cottage, Dunkeld Rd, Aberfeldy, Perthshire, PH15 2EJ [IO86BO, NN84]
G6 YRQ H S Stevens, 4 Leonards Ave, Easton, Bristol, BS5 6BG [IO81RL, ST67]
G6 YRV D F Bedford, 28 Durfold Dr, Reigate, RH2 0QA [IO91VF, TQ25]
G6 YRX J E Bignall, 5 Hillway, Amersham, HP7 0JL [IO91QP, SU99]
G6 YRY R J Bearchell, 81 Leaves Green Rd, Keston, BR2 6DG [JO01AI, TQ46]
GM6 YRZ R N Blake, Middle Cottage, Whitehill Foot, Heiton, Kelso, Roxburghshire, TD5 8LB [IO85SN, NT72]
G6 YSB J P Bates, 16 Harewood Ave, Great Barr, Birmingham, B43 6QE [IO92AN, SP09]
G6 YSK K H Bilton, 6 Park Vale Dr, Thrybergh, Rotherham, S65 4HZ [IO93IK, SK49]
G6 YSL S A Watts, 15 Churchill Way, Northam, Bideford, EX39 1DF [IO71VA, SS42]
G6 YSN K W Ward, 8 Hinckley Rd, St. Helens, WA11 9HU [IO83PL, SJ59]
G6 YSO Details withheld at licensee's request by SSL.
G6 YSZ P Tonge, 1 Hill View, Stalybridge, SK15 2TH [IO83XL, SJ99]
G6 YTB R J Watts, 41 Watford Rd, Crick, Northampton, NN6 7TT [IO92KI, SP57]
G6 YTC K Taylor, 7 Merryfields, Rochester, ME2 3ND [JO01FJ, TQ77]
G6 YTH J Thornett, 19 Gill Sike, Bungalows, Wakefield, West Yorks, WF2 8BP [IO93FQ, SE32]
G6 YTN Details withheld at licensee's request by SSL.
G6 YTP R G Cracknell, 35 Tennyson Ave, Houghton Regis, Dunstable, LU5 5UQ [IO91RV, TL02]
G6 YTR R E Broughton, Brookside, Blagdon Terr, Seaton Burn, Newcastle upon Tyne, NE13 6EY [IO95EB, NZ27]
G6 YTT Details withheld at licensee's request by SSL.
G6 YTV A R Black, Redholme, The St., Great Cressingham, Thetford, IP25 6NL [JO02IN, TF80]
G6 YTW A J Bennett, 29 Kennington Rd, Kennington, Oxford, OX1 5NZ [IO91JR, SP50]
G6 YTX R C Burnett, 46 Dorset Waye, Heston, Hounslow, TW5 0ND [IO91TL, TQ17]
GW6 YTY A K Bournes, 39 Windmill Rd, Halstead, CO9 1JL [JO01HW, TL83]
G6 YTZ W L Van Den Bergh, 44 Northfields Ln, Brixham, TQ5 8RS [IO80FJ, SX95]
G6 YUB P J Blanking, 21 Beechwood Gdns, Caterham, CR3 6NH [IO91XG, TQ35]
GW6 YUC E Brooksbank, Ffynnon Ddu, O Heol Ddu, Ammanford, Carmarthenshire, SA18 3SP [IO71XS, SN61]
G6 YUD D P Brailsford, 6 Norwith Rd, Bessacarr, Doncaster, DN4 7HW [IO93LL, SE60]
G6 YUH Details withheld at licensee's request by SSL.
G6 YUI Details withheld at licensee's request by SSL.
GW6 YUN Details withheld at licensee's request by SSL.
G6 YUV Details withheld at licensee's request by SSL.
G6 YUX B Clough, 31 Countess Rd, Amesbury, Salisbury, SP4 7AS [IO91CE, SU14]
G6 YVD G Wood, Tethers End, Angarrack Ln, Connor Downs, Hayle, TR27 5JF [IO70HE, SW53]
G6 YVE Details withheld at licensee's request by SSL.
G6 YVJ Details withheld at licensee's request by SSL.
G6 YVN J J Ward, 22 Holm Rd, Westwoodside, Doncaster, DN9 2EZ [IO93NL, SK79]
G6 YVS J R Wilson, 36 North Warren Rd, Gainsborough, DN21 2TU [IO93OJ, SK89]
G6 YVX Details withheld at licensee's request by SSL.
G6 YWB Details withheld at licensee's request by SSL.
G6 YWC G M Moorhouse, Baywatch, 36 Trerieve, Downderry, Torpoint, Cornwall, PL11 3LY [IO70TI, SX35]
G6 YWI S Godwin, 9 The Brindles, Banstead, SM7 1AE [IO91VH, TQ25]
G6 YWK F A Goodman, 28 Amal An Avon, Phillack, Hayle, TR27 4QD [IO70HE, SW53]
G6 YWN P B Groom, Midulun, Newton Cttgs, Scropton, Derby, DE65 5PS [IO92DU, SK23]
G6 YWT G A Hankin, 45 Durnsford Ave, Fleet, GU13 9TB [IO91NG, SU85]

G6 YWU D C Harding, 20 Darcy Rd, Tolleshunt Knights, Tiptree, Colchester, CO5 0RP [JO01JT, TL91]
G6 YWV M G Harrison, 41 Belper Rd, Ashbourne, DE6 1BB [IO93DA, SK14]
G6 YWW J H Harwood, South Haven, Tennyson Cl, Yarmouth, PO41 0PT [IO90GQ, SZ38]
G6 YWX J H P Hartnell, 3 Fairmead, Sidmouth, EX10 9SU [IO80JQ, SY19]
G6 YXB D A Hewson, 2 Valley Way, Norwich Rd, Fakenham, NR21 8PH [JO02KT, TF92]
G6 YXG A Davey, Oakcottage, Wennington, Lancaster, LA2 8NU [IO84QD, SD67]
G6 YXH P L Frost, 7Arterial Rd, Rayleigh, Essex, SS6 7TR [JO01HN, TQ78]
G6 YXI Details withheld at licensee's request by SSL.[Op: Mrs M E Grant.]
G6 YXO S W Fisher, 37 Elmlands Gr, Stockton Ln, York, YO3 0ED [IO93LX, SE65]
G6 YXT B J Evans, 153 Preston Down Rd, Preston, Paignton, TQ3 1DW [IO80FK, SX86]
G6 YXU D J Eccles, 28 Seven Acres, Clayton Brook, Bamber Bridge, Preston, PR5 8EY [IO83QR, SD52]
G6 YXV K M Faulkner, 5 Tregarrick, West Looe, Looe, PL13 2SD [IO70SI, SX25]
G6 YXW Details withheld at licensee's request by SSL.
G6 YXX N Frederick, 77 Greengate St., Barrow in Furness, LA14 1EZ [IO84JC, SD26]
G6 YXY C J Edwards, Elmhurst, Bledlow Rd, Saunderton, Princes Risborough, HP27 9NG [IO91NR, SP70]
G6 YXZ Details withheld at licensee's request by SSL.
G6 YYA M A Doe, 2 Summerfields, Yarnfield, Stone, ST15 0RH [IO82VV, SJ83]
G6 YYN K W McCann, Treverven, Back Ln, Hemingbrough, Selby, YO8 7QP [IO93MS, SE63]
G6 YYQ N A S Munro, 40 The Oaks, New Barn Park, Swanley, BR8 7YR [JO01CJ, TQ56]
G6 YYU A H Mutimer, 52 Sycamore Ave, Wymondham, NR18 0HX [JO02NN, TG10]
G6 YYX Details withheld at licensee's request by SSL.
G6 YZB L Nunn, 103 Bladindon Dr, Bexley, DA5 3BT [JO01BK, TQ47]
G6 YZC P Newcombe, 6 Church Ln, Bessacarr, Doncaster, DN4 6QB [IO93LM, SE60]
G6 YZF S A Alston-Pottinger, 43 Wadhurst Cl, St. Leonards on Sea, TN37 7AZ [JO00GV, TQ71]
G6 YZH Details withheld at licensee's request by SSL.
G6 YZK B J Palfrey, 9 Marsh Ln, Penkridge, Stafford, ST19 5BY [IO82WR, SJ91]
G6 YZR M Smith, 5 Derwent Cl, North Anston, Anston, Sheffield, S31 7GD [IO93GJ, SK38]
G6 YZU L Nixon, 87 Field Ave, Canterbury, CT1 1TS [JO01NG, TR15]
G6 YZZ D L Stanley, 3 Warwick Cl, Market Bosworth, Nuneaton, CV13 0JX [IO92HO, SK40]
G6 ZAA J A Wellard, 19 South Motto, Park Farm, Kingsnorth, Ashford, TN23 3NJ [JO01KD, TQ94]
G6 ZAB Details withheld at licensee's request by SSL.
G6 ZAC A R C Wilson, 19 Bedford House, Onslow St., Guildford, GU1 4TL [IO91RF, SU94]
G6 ZAF D J Walker, 27 Daltons Cl, Langley Mill, Nottingham, NG16 4GP [IO93HA, SK44]
G6 ZAG Details withheld at licensee's request by SSL.
G6 ZAM P I Waldron, 5 Dalston Cl, Camberley, GU15 1BT [IO91PH, SU95]
GW6 ZAN J D Williams, Caeabergam Cottage, Llanbedr, Gwynedd, LL45 2HT [IO72WU, SH52]
G6 ZAO G Wright, 55 Homefield Rd, Hemel Hempstead, HP2 4BZ [IO91SS, TL00]
G6 ZAX R C Hollick, 7 Grenfell Rd, Moordown, Bournemouth, BH9 2UD [IO90BS, SZ09]
G6 ZAY R E Hope, 129 Lunedale Rd, Fleet Esate, Dartford, DA2 6JX [JO01CK, TQ57]
G6 ZB H Benford, 10 Edgebrook, Sheringham, NR26 8HA [JO02OW, TG14]
G6 ZBI Details withheld at licensee's request by SSL.
G6 ZBO M G Julians, 29 Trentdale Rd, Carlton, Nottingham, NG4 1BU [IO92KW, SK64]
G6 ZBT D Green, 6 Garth Villas, Rimswell, Withernsea, HU19 2DB [IO93XR, TA32]
G6 ZBV A M Higham, 12 Lakenheath Dr, Sharples, Bolton, BL1 7RJ [IO83SO, SD71]
G6 ZBY J W Delaney, 59 Coleford Bridge, Rd, Mytchett, Camberley Surrey, GU16 6DN [IO91PG, SU85]
G6 ZC J Watt, Oak Tree Cottage, Writtle Rd, Margaretting, Ingatestone, CM4 0EJ [JO01EQ, TL60]
G6 ZCH J M Agass, 5 The Orchard, Chiseldon, Swindon, SN4 0PH [IO91DM, SU17]
G6 ZCI J A Anderson, 72 Saffron, Amington, Tamworth, B77 4EP [IO92EP, SK20]
GW6 ZCR J A Phillips, Gernant, 39 Bryn Glas, Rhosllanerchrugog, Wrexham, LL14 2EA [IO83LA, SJ24]
GW6 ZCS V Priamo, 58 Ffordd Glyn, Wrexham, LL13 7QW [IO83MA, SJ34]
G6 ZCX M Rochester, 3 Forth Cl, Oakham, LE15 6JW [IO92PP, SK80]
G6 ZCY M B Rochester, 3 Forth Cl, Oakham, LE15 6JW [IO92PP, SK80]
G6 ZDB G G Reddington, 2 South St., Newton, Alfreton, DE55 5TT [IO93HC, SK45]
G6 ZDE D F Ellingworth, 59 Hawkeridge Park, Westbury, BA13 4HJ [IO81VG, ST85]
GW6 ZDM P Benson, 16 Penrhos, Radyr, Cardiff, CF4 8RJ [IO81IM, ST18]
G6 ZDP K Baum, 10 Edgefield Cl, Whitebushes Est, Redhill, RH1 5LD [IO91WF, TQ24]
G6 ZDQ C L Baker, 16 Mountford Rd, Solihull Lodge, Shirley, Solihull, B90 1JA [IO92BJ, SP07]
G6 ZDS R D Baldwin, 1 Queens Ave, Kidlington, OX5 2JQ [IO91IT, SP51]
G6 ZDT T E Burgess, 210 Lumb Ln, Audenshaw, Manchester, M34 5RX [IO83WL, SJ99]
G6 ZDV A J Beales, 78 Ainley Rd, Birchencliffe, Huddersfield, HD3 3QX [IO93CQ, SE11]
GM6 ZDW J A Bakewell, 17 Coronation St., Monkton, Prestwick, KA9 2QW [IO75QM, NS32]
G6 ZDY R J Bethell, 5 Dorset Way, Maidstone, ME15 7EL [JO01GG, TQ75]
G6 ZEA F Gambles, 479 Hucknall Rd, Sherwood, Nottingham, NG5 1FW [IO92KX, SK54]
G6 ZEH M J Griffin, 2 Witherslack Cl, Westgate, Morecambe, LA4 4UN [IO84NB, SD46]
G6 ZEL M P Holdaway, 106 Mill Ln, Chadwell Heath, Romford, RM6 6UU [JO01BN, TQ48]
G6 ZEN J A Homan, 55 Ark Royal, Bilton, Hull, HU11 4BN [IO93US, TA13]
G6 ZEQ R D Hubert, 11 Norwood Cttgs, Cooling Rd, Cliffe, Rochester Kent, ME3 7RY [JO01FK, TQ77]
G6 ZER R W Hulands, 14 Nene Rise, Cogenhoe, Northampton, NN7 1NT [IO92OF, SP86]
G6 ZES S Hutton, 4 Oatfield, Quedgeley, Gloucester, GL2 4GY [IO81UT, SO81]
G6 ZET D W Jackson, 19 Shelley Cl, Bolton-le-Sands, Carnforth, LA5 8HQ [IO84OC, SD46]
G6 ZEW M Jennings, 16 High St., Worsbro Dale, Worsbrough, Barnsley, S70 4AE [IO93GM, SE30]
G6 ZEZ C A Jones, 709 Bath Rd, Taplow, Maidenhead, SL6 0PB [IO91PM, SU98]
G6 ZFA J C Justice, 6 Stanley Terr, Pans Ln, Devizes, SN10 5AJ [IO91AI, SU06]
G6 ZFK E W Toohey, No6 Block E, Peabody Ave, Pimlico, London, SW1V 4AS [IO91WL, TQ27]
G6 ZFO D Tate, 73 Sparth Ave, Clayton-le-Moors, Accrington, BB5 5QH [IO83TS, SD73]
G6 ZFT Details withheld at licensee's request by SSL.
G6 ZFU C J Stephen, 12 Beaufort Cl, Leegomery, Telford, TF1 4XU [IO82SR, SJ61]
G6 ZFV T A Wynne-Jones, 37 Oakleigh Dr, Croxley Green, Rickmansworth, WD3 3EE [IO91SP, TQ09]
G6 ZFW D P Smith, 11 Birch Cl, Birkenhead, L43 5XE [IO83LJ, SJ38]
G6 ZFZ M Turner, 14 Lauderdale Gdns, Bushbury, Wolverhampton, WV10 8AY [IO82WP, SJ90]
G6 ZGA M F Smith, 6 Norton Cres, Towcester, NN12 6DN [IO92MD, SP64]
G6 ZGB C F Salmon, 20 Lime Cl, Sandbach, CW11 1BZ [IO83TD, SJ76]
G6 ZGC M Tann, 29 Arthur St., Redcar, TS10 1BW [IO94LO, NZ52]
G6 ZGD S T Waller, 72 Birchwood Ave, Hatfield, AL10 0PS [IO91VS, TL20]
G6 ZGF D Simpson, 8 Rutland Rd, Ellesmere Park, Eccles, Manchester, M30 9FA [IO83TL, SJ79]
G6 ZGH R K E York, 44 Denmark St., Lancaster, LA1 5LY [IO84OB, SD46]
G6 ZGI C Butler, 15 Pearce Manor, Chelmsford, CM2 9XH [JO01FR, TL60]
G6 ZGK G Weston, 2 Gill Park, Plymouth, PL3 6LX [IO70WJ, SX45]
G6 ZGU Details withheld at licensee's request by SSL.
GW6 ZH T Winchcombe, Croes y Llan, Llangoedmor, Cardigan, Dyfed, SA43 2LG [IO72QB, SN14]
G6 ZHB J G Booth, 6 Withington Ct, Abingdon, OX14 3QB [IO91IQ, SU49]
G6 ZHC B V Blyth, Kleefeld, Wallingford Rd, Cholsey, Oxon, OX10 9LB [IO91KN, SU58]
G6 ZHJ Details withheld at licensee's request by SSL.
G6 ZHL M J W Leack, Arnold House, 68 Dale St., Lancaster, LA1 3AW [IO84OB, SD46]
GW6 ZHM R W Lannon, 16 Heol Mabon, Cardiff, CF4 6RL [IO81JM, ST18]
G6 ZHO G P Lattin, 5 Seymour Rd, Broadfield, Crawley, RH11 9ES [IO91VC, TQ23]
G6 ZHS K J Lupton, 162 Harpur Hill Rd, Buxton, SK17 9LJ [IO93BF, SK07]
G6 ZHU P B Lightfoot, 9 Conway Rd, Knypersley, Stoke on Trent, ST8 7AL [IO83VC, SJ85]
G6 ZIC J E McComb, 7 Pikestone Cl, Lambton, Washington, NE38 0QE [IO94FV, NZ25]
G6 ZIM Details withheld at licensee's request by SSL.[Op: A C Day.]
G6 ZIO N Dessau, 30 Thumwood, Chineham, Basingstoke, RG24 8TE [IO91LG, SU65]
G6 ZIP E G Hodby, 17 Holmewood Cl, Wokingham, RG41 4AS [IO91NJ, SU86]
G6 ZIY A P Fairhurst, 44 Leyburn Cl, Wigan, WN1 3NF [IO83QN, SD50]
G6 ZJD E J Fensome, 4 Penina Cl, Bletchley, Milton Keynes, MK3 7TL [IO91OX, SP83]
GW6 ZJG B A Meredith, 27 Hyde Pl, Llanhilleth, Abertillery, NP3 2RT [IO81KQ, SO20]
G6 ZJI A Washby, Greenwood Lock Hous, Low Mills Ln, Ravensthorpe, Dewsbury W Yorks, WF13 3LX [IO93DQ, SE22]
G6 ZJJ J Freeman, 19 Elwyn Cl, Stretton, Burton on Trent, DE13 0BG [IO92ET, SK22]
G6 ZJK F A D Webster, 1 Fir Tree Cttgs, Lower Ansford, Ansford, Castle Cary, BA7 7JY [IO81RC, ST63]
G6 ZJM D J Leese, 31 Millhaven Ave, Stirchley, Birmingham, B30 2QH [IO92BK, SP08]
GW6 ZJO K P Watkins, 38 Maerdy Park, Pencoed, Bridgend, CF35 5HX [IO81FM, SS98]
G6 ZJS L Wright, 17 Drayton St., Alumwell Est, Walsall, WS2 9QB [IO82XO, SO99]
G6 ZJV A Winterbottom, 38 Heaton Ave, Wakefield Rd, Dewsbury, WF12 8AQ [IO93EQ, SE22]
G6 ZJW J Wealleans, 8 Hawkeys Ln, North Shields, NE29 0JF [IO95GA, NZ36]
G6 ZKC D J Usher, Bryce, Bloomfield Cl, Timsbury, Bath, Avon, BA3 1LP [IO81SH, ST65]
G6 ZKH Details withheld at licensee's request by SSL.
G6 ZKL W G Seedhouse, 1 Charlmont Cl, Cannock, WS12 5NH [IO92AQ, SK01]
G6 ZKM J A S Cornell, 36 Parkstone Ave, Southsea, PO4 0QZ [IO90LS, SZ69]
G6 ZKS M T Staniland, 2 Epsom Rd, Cantley, Doncaster, DN4 6HX [IO93KM, SE60]
G6 ZKT P A Sawyers, 36 Frome Rd, Bath, BA2 2QB [IO81TI, ST76]
G6 ZKU B Sawyers, 36 Frome Rd, Bath, BA2 2QB [IO81TI, ST76]
G6 ZKY E J Stebbings, 1 Coupland Rd, Abingdon, OX13 6DU [IO91IQ, SP40]
G6 ZLD P R W Bent, 7 Bandon Rise, Wallington, SM6 8PT [IO91WI, TQ26]
G6 ZLE M A Harris, 21 Broadmead Ave, Worcester Park, KT4 7SN [IO91VJ, TQ26]
G6 ZLG M J Anderton, Rodenna, Mullion, Helston, Cornwall, TR12 7HW [IO70JA, SW61]
G6 ZLI Details withheld at licensee's request by SSL.
G6 ZLJ M P Adams, 62 Woodlands Rd, Holmcroft, Stafford, ST16 1QP [IO82WT, SJ92]
G6 ZLM G D Bloomfield, 50 Girdlestone Rd, Headington, Oxford, OX3 7NA [IO91JS, SP50]
G6 ZLP N Burbidge, 21 Quernstone Ln, Danesfield, Northampton, NN4 8UN [IO92MF, SP75]
G6 ZLQ R C Bishop, 33 Greenfields Rd, Reading, RG2 8SG [IO91MK, SU76]
G6 ZLS G D Ashbee, 6 The Green, Wimbledon Common, London, SW19 5AZ [IO91VK, TQ27]
G6 ZLT H P Ashbee, 12 Wellington Rd, London, SW19 8EQ [IO91VK, TQ27]
G6 ZLU A P Barclay, 21 Priors Wood, Crowthorne, RG45 6BZ [IO91OI, SU86]
G6 ZLV S A Boden, 54 Avill, Hockley, Tamworth, B77 5QF [IO92EO, SK20]
G6 ZLY D H Brasenell, 17 Bradlaugh Terr, Wibsey, Bradford, BD6 1JY [IO93CS, SE13]
G6 ZMC Details withheld at licensee's request by SSL.
G6 ZMD S G Roberts, 36 Hill Cres, Dudleston Heath, Ellesmere, SY12 9NA [IO82MW, SJ33]
G6 ZME J R Wakenell Telford Dist AR, 15 Cuckoo Oak Green, Madeley, Telford, TF7 4HT [IO82SP, SJ70]
G6 ZMF Details withheld at licensee's request by SSL.
G6 ZMG G E Mills, 57 Holborough Rd, Snodland, Kent, ME6 5PA [JO01FH, TQ76]
G6 ZMK Details withheld at licensee's request by SSL.
GW6 ZMN W J Mc Dowall, 36 Adenfield Way, Rhoose, Barry, CF62 3EA [IO81HJ, ST06]
G6 ZMU G K Randall, 23 Chaffinch Cl, Broadwey, Weymouth, DT3 5QU [IO80SP, SY68]
G6 ZMY Details withheld at licensee's request by SSL.
G6 ZMZ Details withheld at licensee's request by SSL.
G6 ZNF J W Riley, 15 Ashby Rd, Moira, Burton on Trent, DE15 0LA [IO92ET, SK22]
G6 ZNJ A B Reeve, 188 Dorset Ave, Great Baddon, Chelmsford, CM2 8YY [JO01FR, TL70]
GD6 ZNL N G Povall, 12 Ballasteen Rd, Andreas, Ramsey, IM7 4HG
G6 ZNO I D Martin, 21 Baldwin Ave, Eastbourne, BN21 1UJ [JO00DS, TQ50]
G6 ZNR P E Chapman, 48 Oakleigh Gr, Wirral, L63 7QT [IO83LI, SJ38]
G6 ZNW Details withheld at licensee's request by SSL.[Station located near Folkestone.]
G6 ZOB A Crowther, Ivy Cottage, Great North Rd, Tuxford, Newark, Notts, NG22 0JB [IO93NF, SK77]
G6 ZOD Details withheld at licensee's request by SSL.
G6 ZOE C L English, 124 Hillside Rd, Redcliffe Bay, Portishead, Bristol, BS20 8LG [IO81OL, ST47]
G6 ZOI A Bond, 14 Broadmanor, North Duffield, Selby, YO8 7RZ [IO93MT, SE63]
G6 ZOJ A V Buchan, 5 Copythorne Cl, Brixham, TQ5 8QG [IO80FJ, SX95]
G6 ZOL Details withheld at licensee's request by SSL.
G6 ZOO Details withheld at licensee's request by SSL.
G6 ZOS P H Lovelock, 82 Chaworth Rd, West Bridgford, Nottingham, NG2 7AD [IO92KW, SK53]
G6 ZOT J Leary, 24 Howard Dr, Old Whittington, Chesterfield, S41 9JU [IO93GG, SK37]
G6 ZOV Details withheld at licensee's request by SSL.[Station located near Dorking.]
GM6 ZPD Details withheld at licensee's request by SSL.[Station located near Arbroath.]
G6 ZPJ A R Mudie, The Station Inn, Southwaite, Carlisle, Cumbria, CA4 0LB [IO84NT, NY44]
G6 ZPL P J Manning, 21 Whitethorn Way, Blackbird Leys, Oxford, OX4 5ER [IO91JR, SP50]
G6 ZPR A C Morris, Follywood, 32 New Rd, Wonersh, Guildford, GU5 0SE [IO91RE, TQ04]
G6 ZPV N D Mansfield, 2 Little Halt, Portishead, Bristol, BS20 8JQ [IO81OL, ST47]
G6 ZQA G Nolan, 94 St. Andrews Rd, Burgess Hill, RH15 0PH [IO90WX, TQ31]
G6 ZQC J Naylor, 17 Green Ave, Canvey Island, SS8 0LH [JO01GM, TQ78]
G6 ZQE Details withheld at licensee's request by SSL.[Op: J A Newton. Station located near Wymondham.]
G6 ZQH A V Darley, 141 Copperfield, Limes Farm Est, Chigwell, IG7 5NJ [JO01BO, TQ49]
G6 ZQJ A M Doughty, 42 Thornton Rd, Ilford, IG1 2ER [JO01AN, TQ48]
G6 ZQL V S Dorrance, 18 Ruskin Ave, Rowley Regis, Warley, B65 9QW [IO82XL, SO98]
GM6 ZQQ Details withheld at licensee's request by SSL.
G6 ZQS M P Charlton, 26 Saundergate Ln, Wyberton, Boston, PE21 7BZ [IO92XW, TF34]
G6 ZQU D A Crook, Sherington Nurseries, Bedford Rd, Sherington, Newport Pagnell, MK16 9NQ [IO92PC, SP84]
G6 ZQX Details withheld at licensee's request by SSL.
G6 ZQY Details withheld at licensee's request by SSL.
G6 ZRC M R P Smith, Sheerland, Blackness Rd, Crowborough, TN6 2NB [JO01CB, TQ52]
G6 ZRF J C Sharratt, 222 Old London Rd, Hastings, E Sussex, TN34 3NS [JO00HU, TQ81]
G6 ZRL K L Sellens, 14 Adhara Rd, Northwood, Middx, HA6 3LR [IO91TO, TQ09]
G6 ZRS B Starr, 121 Pretoria Rd, Patchway, Bristol, BS12 5PY [IO81RM, ST58]
G6 ZRT Details withheld at licensee's request by SSL.
G6 ZRU F S Southwell, 40 Downsview, Smalldole, Small Dole, Henfield, BN5 9YB [IO90UV, TQ21]
G6 ZRV P Stainton, 42 Stoke Ln, Gedling, Nottingham, NG4 2QP [IO92LX, SK64]
GW6 ZRX J H L Thomas, 6 Dunraven St., Treherbert, Treorchy, CF42 5BH [IO81FQ, SS99]
G6 ZSF Dr D Neely, 3 Sidestrand Rd, Newbury, RG14 6HP [IO91HJ, SU46]
G6 ZSJ Details withheld at licensee's request by SSL.
GW6 ZSK C B D Owen, 3 Hermitage Gr, Haverfordwest, SA61 2PS [IO71MT, SM91]
G6 ZSN Details withheld at licensee's request by SSL.
G6 ZSQ J C Pepper, 36 Westbourne Heights, Redruth, TR15 2TQ [IO70JF, SW64]
G6 ZSR K Poole, 1 The Craven, Heelands, Milton Keynes, MK13 7QR [IO92OB, SP83]
G6 ZSU P P Payton, 11 Hexham Way, Milking Bank, Dudley, DY1 2UN [IO82WM, SO99]
G6 ZTC S A Roberts, 4 Langworthy Ave, Little Hulton, Manchester, M38 9GQ [IO83TM, SD70]
G6 ZTD B L Robinson, 23 Croft Dr, Millhouse Green, Sheffield, S30 6NE [IO93GJ, SK38]
G6 ZTF A Reynolds, 169 Bell Green Rd, Coventry, CV6 7GW [IO92GK, SP38]
GW6 ZTG B J Robson, 3 Cendl Cres, Rassau, Ebbw Vale, NP3 5PR [IO81JT, SO11]
G6 ZTI A D Russell, 126 Boothferry Rd, Hessle, HU13 9AX [IO93SR, TA02]
G6 ZTM D J Redmill, 38 Whitland Rd, Carshalton, SM5 1QT [IO91VJ, TQ26]
G6 ZTP G M Down, 9 Broad View, Trispen, Truro, TR4 9RQ [IO70LH, SW85]
G6 ZTR S H Davies, 111 South End, Garsington, Oxford, Oxon, OX44 9DL [IO91KR, SP50]
G6 ZTT M V Baguley obo Mid-Cheshire Contest Gr, 21 Sovereign Cl, Rudheath, Northwich, CW9 7XN [IO83SG, SJ67]
G6 ZTV C R M Fairall, 32 Cherry Tree Rd, Chinnor, OX9 4QZ [IO91NQ, SP70]
G6 ZTZ S P French, 22 Amity St., Newtown, Reading, RG1 3LP [IO91MK, SU77]
G6 ZUC Details withheld at licensee's request by SSL.
G6 ZUE W G Edwards, 31 Cumberland Ave, Benfleet, SS7 5NU [JO01GN, TQ78]
G6 ZUO D I Gibson, 14 Lowfield Rd, Dewsburymoor, Dewsbury, WF13 3SR [IO93EQ, SE22]
GW6 ZUQ D J F Gordon, 6 Oak Cl, Bulwark, Chepstow, NP6 5RL [IO81PP, ST59]
GW6 ZUS J E Gray, 36 Heol Pentre Felen, Morriston, Swansea, SA6 6BY [IO81AQ, SS69]
G6 ZUT M J Gray, 32 Abbots Cl, Cambridge, CB4 2SY [JO02BF, TL46]
G6 ZUV J S Griffin, 35 Cottage St., Kingswinford, DY6 7QE [IO82WL, SO88]
G6 ZUW C D Gurnhill, 32 Harrington Ave, Borrowash, Derby, DE72 3JB [IO92HV, SK43]
G6 ZUZ J A Hampshire, 14 Fellows Rd, Cowes, PO31 7JN [IO90IS, SZ49]
G6 ZVE I R Hill, 1 Bretton Ave, Bolsover, Chesterfield, S44 6XN [IO93IF, SK47]
G6 ZVJ E M Hogan, 19 South Rd, Corfe Mullen, Wimborne, BH21 3HY [IO80XS, SY99]
G6 ZVK Details withheld at licensee's request by SSL.
G6 ZVO K Howarth, 79 Eden Ave, Edenfield, Ramsbottom, Bury, BL0 0LD [IO83UP, SD71]
G6 ZVR D J Inskip, 42 Jubilee Ct, Belper, DE56 1NN [IO93GA, SK34]
G6 ZVU S Hughes, 50 Albany Rd, Dalton, Huddersfield, HD5 9UW [IO93DP, SE11]
G6 ZVV N J Hull, 126 Coval Ln, Chelmsford, CM1 1TG [JO01FR, TL70]
G6 ZWA Details withheld at licensee's request by SSL.
G6 ZWC C T Brown, 16 Old Croft Cl, Good Easter, Chelmsford, CM1 4SJ [JO01ES, TL61]
G6 ZWF G I Banham, 42 Lower Outwoods Rd, Burton on Trent, DE13 0QX [IO92ET, SK22]
G6 ZWG L A Alldridge, 95 Rose Dr, Chesham, HP5 1RT [IO91QQ, SP90]
GW6 ZWH T P A Armstrong, 7 Fairy Gr, Killay, Swansea, SA2 7BY [IO71XO, SS69]
G6 ZWI A W Angove, 22 Bramble Cl, Newquay, TR7 2SU [IO70LJ, SW86]
G6 ZWL C S Wright, 19 Redwood Glen, Chapeltown, Sheffield, S30 4EA [IO93GJ, SK38]
G6 ZWM R W Wade, 104 Brookehowse Rd, Bellingham, London, SE6 3TW [IO91XK, TQ37]
G6 ZWY P Vickers, 13 Prospect Rd, Hartshead, Liversedge, WF15 8BA [IO93DQ, SE12]
G6 ZWZ J L Sexton, 31 Hurst Green, Mawdesley, Ormskirk, L40 2QS [IO83OP, SD41]
G6 ZXF T H Fawbert, 34 Brangwyn Ave, Brighton, BN1 8XG [IO90WU, TQ30]
G6 ZXO J T Crowe, 15 Lambert Rd, Kendray, Barnsley, S70 3AA [IO93GN, SE30]
G6 ZXV P J Baizley, 12 St. Georges Park, Broadwater Down, Tunbridge Wells, TN2 5NT [JO01DC, TQ53]
G6 ZY S C Ingram, 6 Swift St., London, SW6 5AG [IO91VL, TQ27][Please QSL to Box 89, Santa Eulalia, Ibiza, Balaerics, Spain.]
G6 ZYG Details withheld at licensee's request by SSL.
GW6 ZYI B G Jones, 10 Hughes St., Penygraig, Tonypandy, CF40 1LX [IO81GO, SS99]
G6 ZYM K J Keeble, Hall Cottage, Starston, Harleston, IP20 9PU [JO02PK, TM28]
G6 ZYP A Keye, The Studio, 45 Mansfield Rd, Swallownest, Sheffield, S31 0UA [IO93GJ, SK38]
G6 ZYQ D G Kirby, 3 The Glebelands, Great Glen, Leicester, LE8 9FR [IO92LN, SP69]
G6 ZYS J Spence, 15 Maltings Rd, Gretton, Corby, NN17 3BZ [IO92PM, SP89]
G6 ZYX G Spruce, 158 Wolverhampton St., Wednesbury, WS10 8UB [IO82XN, SO99]
G6 ZYZ P Skerritt, 5 Oxford Rd, West Bromwich, B70 8PE [IO82XM, SO99]
G6 ZZA A Stansfield, 4 Lime Ave, Todmorden, OL14 5NW [IO83WR, SD92]
GW6 ZZF P Thomas, Delaville, 42 Wyndham Rd, Abergavenny, NP7 6AF [IO81LT, SO31]
G6 ZZN M M H Wright, 116 Wynchgate, Winchmore Hill, London, N21 1QU [IO91WP, TQ39]
GW6 ZZP R P Wood, Bwlcyn, Eifl Rd, Trefor, Caernarvon, LL54 5HG [IO72SX, SH34]
G6 ZZR D G Wiltshire, 19 Heron Way, Kempshott, Basingstoke, RG22 5QF [IO91KF, SU65]
G6 ZZS D G A Watts, 62 Petersfield Rd, Hall Green, Birmingham, B28 0AT [IO92BK, SP18]
G6 ZZX Details withheld at licensee's request by SSL.
G6 ZZZ T J Cleaver Daventry Amateur Radio Club, 2 Ruskin Way, Daventry, NN11 4TT [IO92KF, SP56]

G7

G7 AAC Details withheld at licensee's request by SSL.
G7 AAF G D Dunford, 62 Paddock Ln, Oaken Shaw, Redditch, B98 7XP [IO92AG, SP06]
GI7 AAH P J Smiley, 100 Lislaban Rd, Cloughmills, Ballymena, BT44 9HZ [IO65UA, D02]
G7 AAP P Kiff, 1 Trent Ave, Milnrow, Rochdale, OL16 3EX [IO83WO, SD91]
G7 AAQ K D Miller, 94 Deerhurst Cres, Paulsgrove, Portsmouth, PO6 4EJ [IO90KU, SU60]
G7 AAR P E Comben, 2 Putnams Dr, Aston Clinton, Aylesbury, HP22 5HH [IO91PT, SP81]
G7 AAS D M Hawkins, 8 Braybrook St., East Acton, London, W12 0AP [IO91VM, TQ28]
GW7 AAU H P Studdart, 33 Linden Ave, Connahs Quay, Deeside, CH5 4SN [IO83LF, SJ26]
GW7 AAV S K R Studdart, 33 Linden Ave, Connahs Quay, Deeside, CH5 4SN [IO83LF, SJ26]
G7 AAW A Smedley, 5 Rosemary Ln, Haddenham, Aylesbury, HP17 8JS [IO91MS, SP70]
G7 AAY K A Richardson, 180 Beech Ave, Abington, Northampton, NN3 2JW [IO92NG, SP76]
G7 ABB Details withheld at licensee's request by SSL.[Station located near Basildon. Op: M R Smith, via P O Box 761, Basildon, Essex, SS15 5YL, or via the bureau.]
G7 ABF K Austin, 6 Boothey Cl, Biggleswade, SG18 0DG [IO92UC, TL14]
G7 ABL A J Goodwin, 23 Priory Rd, Bicknacre, Chelmsford, CM3 4EY [JO01HQ, TL70]
G7 ABP D B Ferns, 2 Lancaster Dr, Martlesham, Martlesham Heath, Ipswich, IP5 7TH [JO02OB, TM24]
G7 ABQ D A Ferns, 2 Lancaster Dr, Martlesham Heath, Ipswich, IP5 7TH [JO02OB, TM24]
G7 ABR A P Clark, Brookside, Milford, Bakewell, DE45 1DX [IO93DF, SK26]
G7 ABV S G Bowler, 87 Birchover Way, Allestree, Derby, DE22 2QH [IO92GW, SK33]
G7 ABZ M Bromage, 14 Rhuddlan Way, Kidderminster, DY10 1YH [IO82VI, SO87]
G7 ACA B Pearce, 90 Park Hall Rd, Mansfield Woodhouse, Mansfield, NG19 8PY [IO93JE, SK56]
G7 ACC J E Evans, Maenir, Sandy Beach, Llanfwrog, nr Hollyhead, Isle of Anglesey, LL65 4YH [IO73RH, SH28]
G7 ACD R J Cariss, Greenways, 4 Granville Ave, Newport, TF10 7DX [IO82TS, SJ71]
G7 ACE Details withheld at licensee's request by SSL.
G7 ACG J Baker, Green Ln Farmhous, Rugeley, Staffs, WS15 2AR [IO92AS, SK01]
G7 ACJ G G Mantle, 4 Brown St., Wolverhampton, WV2 1HR [IO82WN, SO99]
G7 ACM S M Pinkney, 169 Sandringham Rd, Perry Barr, Birmingham, B42 1PZ [IO92BM, SP09]
G7 ACN C J G Hayes, Gladstone House, 15 The Beck, Feltwell, Thetford, IP26 4DB [JO02GL, TL79]
G7 ACQ S C J Bunting, The Old Vicarage, 161 High St., Tibshelf, Alfreton, DE55 5NE [IO93HD, SK46]
G7 ACR P Blakemore, 110 Fife St., Wincobank, Sheffield, S9 1NQ [IO93GK, SK39]
G7 ACU Details withheld at licensee's request by SSL.
G7 ACW Details withheld at licensee's request by SSL.
G7 ADE P R Taylor, 18 Park Meadow, Princes Risborough, HP27 0EB [IO91NR, SP80]
G7 ADP C F J Baker, 17 Coronation Rd, Illogan, Redruth, TR16 4SG [IO70IF, SW64]
G7 ADQ Details withheld at licensee's request by SSL.
G7 ADS A R Cresswell, Studland, Melford Rd, Cavendish, Sudbury, CO10 8AA [JO02HC, TL84]
GM7 ADU Dr L M Morrison, 22 Lodge Park, Kilmacolm, PA13 4PY [IO75QV, NS37]
G7 ADW G R D Laycock, 18 Montague Cres, Garforth, Leeds, LS25 2EP [IO93HT, SE43]
GM7 ADY Dr M Morrison, 22 Lodge Park, Kilmacolm, PA13 4PY [IO75QV, NS37]
G7 ADZ Details withheld at licensee's request by SSL.
G7 AEA P A J Swinbank Ullenwood Rayne, Emer Planning Dept, Shire Hall, Gloucester, GL1 2TG [IO81VU, SO81]
G7 AEB D W Long Glos Raynet HQ, Emer Planning Dept, Shire Hall, Gloucester, GL1 2TG [IO81VU, SO81]
G7 AEC R P I Parry Cheltenham Rynt, Emer Planning Dept, Shire Hall, Gloucester, GL1 2TG [IO81VU, SO81]
G7 AED I W Carpenter Stroud Raynet, Emer Planning Dept, Shire Hall, Gloucester, GL1 2TG [IO81VU, SO81]
G7 AEE W G Littlewood Tewkesbury Rynt, Emer Planning Dept, Shire Hall, Gloucester, GL1 2TG [IO81VU, SO81]
G7 AEF G D Crawshaw Frst of Dean Ry, Emer Planning Dept, Shire Halle Rd, Gloucester, GL1 2TG [IO81VU, SO81]
G7 AEG N R Negus Glos City Rayne, Emer Planning Dept, Shire Hall, Gloucester, GL1 2TG [IO81VU, SO81]
G7 AEH V S Smith Cotswold Raynet, Emer Planning Dept, Shire Hall, Gloucester, GL1 2TG [IO81VU, SO81]
G7 AEJ Details withheld at licensee's request by SSL.
GW7 AEL W Crompton, 9 Ronald Ave, Llandudno Junction, LL31 9HA [IO83CG, SH77]
G7 AEO Details withheld at licensee's request by SSL.
G7 AEP Details withheld at licensee's request by SSL.
G7 AEQ R J Murphy, 17 Valley View Rd, Paulton, Bristol, BS18 5QB [IO81SH, ST65]
G7 AET S A W Saunders, 17 Bure Rd, Friars Cliffe, Christchurch, BH23 4ED [IO90DR, SZ19]
G7 AEU Details withheld at licensee's request by SSL.
GM7 AEX Details withheld at licensee's request by SSL.
G7 AEY D P Martin, 38 Yarrow Rd, Weedswood, Walderslade, Chatham, ME5 0RY [JO01GI, TQ76]
G7 AFE A F Erwood, 19 Fir Tree Way, Fleet, Hants, GU13 9NB [IO91OG, SU85]
G7 AFL A A Fountaine, 19 Metcalfe Gr, Blakelands, Milton Keynes, MK14 5JY [IO92PB, SP84]
G7 AFP K Hollingsworth, 59 Newark Ave, Peterborough, PE1 4NH [IO92VO, TF20]
G7 AFQ K A Marlow, Computer Science, PO Box 363, University of B'ham, Edgbaston Birmingha, B15 2TT [IO92AK, SP08]
G7 AFS Details withheld at licensee's request by SSL.
G7 AFW M Towers, 44 Ravenscroft Dr, Ehaddesden, Chaddesden, Derby, DE21 6NX [IO92GW, SK33]
G7 AFZ D Payea, 29 Orchid Cl, Langney, Eastbourne, BN23 8DE [JO00DT, TQ60]
G7 AGA K Askew, 3 Craven Dr, Broadheath, Altrincham, WA14 5JF [IO83TJ, SJ78]
G7 AGB Details withheld at licensee's request by SSL.
G7 AGC M A Collis, 2 Westwood Ave, Urmston, Manchester, M41 9NG [IO83TK, SJ79]
GW7 AGG R Ricketts, 2 Brynystwyth, Penparcau, Aberystwyth, SY23 1SS [IO72XJ, SN58]
G7 AGO J M Lee, 188 Manstone Ave, Sidmouth, EX10 9TJ [IO80JQ, SY18]
G7 AGR R P Clark Aylesbury Ryn Group, 9 Conigree, Chinnor, OX9 4JY [IO91MQ, SP70]
GW7 AGW M Weale, Top Flat, 8 Richards Terr, Cardiff, CF2 1RU [IO81KL, ST17]
G7 AGY J M Barton, Prenton, 27 Francis Way, Salisbury, SP2 8EF [IO91CB, SU12]
G7 AGZ D J Spratt, 66 Glenthorne Rd, Threemilestone, Truro, TR3 6UA [IO70KG, SW74]
GM7 AHA V M I Turnbull, 18 Easterfield Ct, Livingston Village, Livingston, EH54 7BZ [IO85FV, NT06]
G7 AHB T G G Green, 13 Pelican Pl, Eynsham, Witney, OX8 1NW [IO91HS, SP40]
G7 AHE M S Painter, 26 Hamp Brook Way, Bridgwater, TA6 6JZ [IO81LC, ST23]
G7 AHJ Details withheld at licensee's request by SSL.
G7 AHO P G Russell, 23 Everette House, East St, London, SE17 2DY [IO91WL, TQ37]
G7 AHP S A Crask, 107 Highland Rd, Chelston, Torquay, TQ2 6NH [IO80FL, SX86]
G7 AHQ Details withheld at licensee's request by SSL.
G7 AHR M D Brett, 3 Rectory Cl, Chingford, London, E4 8BG [IO91XO, TQ39]
G7 AHS Details withheld at licensee's request by SSL.
G7 AHT D T Smith, Orchardside, 14, Ashmead Green, Dursley, Glos, GL11 5EW [IO81TQ, ST79]
G7 AHV Details withheld at licensee's request by SSL.
G7 AHZ E W K Ford, 58 West Coker Rd, Yeovil, BA20 2JA [IO80QW, ST51]
G7 AIC V H Newman, 35 Netherton Rd, Yeovil, BA21 5NY [IO80QW, ST51]
G7 AIE Details withheld at licensee's request by SSL.
G7 AIF D M Grevatt, 7 Shelley Rd, East Grinstead, RH19 1SX [IO91XD, TQ33]
G7 AIH R L Whitenstall, 4 Monksmead, Borehamwood, WD6 2LQ [IO91UP, TQ29]
G7 AII J M Sedgley, 51 Gordon Cl, Haywards Heath, RH16 1ER [IO91WA, TQ32]
G7 AIK C B Edmunds, Galadhone, Rockalls Rd, Polstead, Colchester, CO6 5AR [JO02KA, TL93]
G7 AIL M J Harris, 8 Hollinsmoor, Swindon, SN3 6NJ [IO91DN, SU18]
GW7 AIN J M Taylor, 2 Bryn Tirion Park, Gyffin, Conwy, LL32 8ND [IO83BG, SH77]
G7 AIS Details withheld at licensee's request by SSL.
G7 AIV Details withheld at licensee's request by SSL.
GW7 AIY V Lamb, 19 Pemba Dr, Buckley, CH7 2HQ [IO83LE, SJ26]
G7 AJB A J Pratt, 2 Clarence Pl, Didcot, OX11 8NT [IO91JO, SU58]
G7 AJC A J Clark, 167 Meadow Way, Bradley Stoke, Bristol, BS12 8BP [IO81RM, ST68]
G7 AJE F Salt, 6 Bodycoats Rd, Chandlers Ford, Eastleigh, SO53 2GX [IO90HX, SU42]
G7 AJG T R Ellis, 29 St. Annes Rd, Clacton on Sea, CO15 3NF [JO01NT, TM11]
G7 AJJ A D Hammond, 5 Durness Cl, Kettering, NN15 5BN [IO92PJ, SP87]
G7 AJM D Smith, 1 Edinburgh Dr, Anston, Sheffield, S31 7HD [IO93GJ, SK38]
G7 AJN H Lister, 68 Spring Ave, Gildersome, Morley, Leeds, LS27 7BT [IO93ES, SE22]
G7 AJO B J Staniforth, 5 Windsmoor Rd, Brookenby, Binbrook, Market Rasen, LN8 6EF [IO93VK, TF29]
G7 AJP S Godfrey, 7 Laburnam Cl, North Baddesley, Southampton, SO52 9JT [IO90VX, SU32]
G7 AJS T B Green, 34 Thorn Cl, Kettering, NN16 9BU [IO92PJ, SP88]
G7 AJT M J Carter, 17 Ash Cres, Higham, Rochester, ME3 7BA [JO01FJ, TQ77]
G7 AJU N Owens, 20 Glebe Way, Burnham on Crouch, CM0 8QJ [JO01JP, TQ99]
G7 AJW C E Deacon, 121 Ladywell, Prospect, Sawbridgeworth, Herts, CM21 9PS [JO01BT, TL41]
G7 AJX R H King, 31 Lambert Rd, Sprowston, Norwich, NR7 8AA [JO02PP, TG21]
G7 AKD A J P Ripley, 70 Druids Meadow, Boroughbridge, York, YO5 9NF [IO94GB, SE36]
G7 AKF L Surguy, 1 Orchard Cl, Woolhampton, Reading, RG7 5SD [IO91JJ, SU56]
G7 AKJ W R H Wrench, 23 Ridgewayk, Battishorne Park, Honiton, Devon, EX13 5NG [IO80MS, SY29]
G7 AKL Details withheld at licensee's request by SSL.
G7 AKM D J Pearson, 29 Aspen Cl, Swanley, BR8 7UA [JO01CJ, TQ56]
G7 AKO L G Talmage, 18 The Causeway, East Hanney, Wantage, OX12 0JN [IO91HP, SU49]
G7 AKP Y N Branch, 38 Kynaston Rd, Didcot, OX11 8HD [IO91JO, SU58]
G7 AKQ R E S Cotterill, Greycot, Golf Ln, Whitehill, Bordon, GU35 9EH [IO91NC, SU73]
G7 AKV E J Grantham, Dunroamin, Badby Rd West, Daventry, NN11 4HJ [IO92JG, SP56]
G7 AKX Details withheld at licensee's request by SSL.
G7 AKZ W H Bundy, 30 Ashley Dr, Tylers Green, Penn, High Wycombe, HP10 8BQ [IO91PP, SU99]
G7 ALB Details withheld at licensee's request by SSL.
G7 ALD Details withheld at licensee's request by SSL.
G7 ALG Details withheld at licensee's request by SSL.
GI7 ALH Details withheld at licensee's request by SSL.
G7 ALK G Leach, 115 Churchill Ave, Lakeview, Northampton, NN3 6PF [IO92NG, SP76]
G7 ALN A P Senior, 7 Kelston Rd, Keynsham, Bristol, BS18 2JH [IO81RJ, ST66]
GI7 ALP B M Kennedy, 1 Legaterriff Rd, Ballinderry Upper, Lisburn, BT28 2EY [IO64VN, J16]
GI7 ALQ I T Kennedy, 1 Legaterriff Rd, Ballinderry Upper, Lisburn, BT28 2EY [IO64VN, J16]
G7 ALR P L Goodman, 85 Rantree Fold, Lee Chapel South, Basildon, SS16 5TW [JO01FN, TQ68]
G7 ALT Details withheld at licensee's request by SSL.
G7 ALW P A Shackleton, 37 Merlin Walk, Eaglestone, Milton Keynes, MK6 5EP [IO92PA, SP83]
G7 ALZ Details withheld at licensee's request by SSL.
G7 AMD G J Blakemore, 37 Mount Side St., Cannock, WS12 4DD [IO82XR, SJ91]
G7 AME Details withheld at licensee's request by SSL.
G7 AMF Details withheld at licensee's request by SSL.
G7 AMG Details withheld at licensee's request by SSL.
G7 AMH Details withheld at licensee's request by SSL.
GM7 AMJ Details withheld at licensee's request by SSL.
GI7 AMK Details withheld at licensee's request by SSL.
G7 AMP S M Boreham, 22 Fairlight Ave, Hastings, TN35 5HS [JO00HU, TQ81]
G7 AMQ J G Morstatt, 32 Elwy Circle, Ash Green, Coventry, CV7 9AU [IO92FL, SP38]
G7 AMU B M Smith, 8 Berrows Mead, Rangeworthy, Bristol, BS17 5QQ [IO81SN, ST68]
G7 AMW P C Jackson, 4 Abbotsbury, Orton Malborne, Peterborough, PE2 5PS [IO92UN, TL19]
G7 AMY Details withheld at licensee's request by SSL.
G7 ANB D G Ball, 16 Kelston View, Whiteway, Bath, BA2 1NW [IO81TI, ST76]
GM7 ANE W Jamieson, 90 Highpark Ave, New Cumnock, Cumnock, KA18 4HH [IO75VJ, NS61]
G7 ANF R Hathway, 5 Percy Terr, Leamington Spa, CV32 5PG [IO92FG, SP36]
G7 ANG A R Santagata, 69 Chisbury Cl, Bracknell, RG12 0TX [IO91PJ, SU86]
G7 ANH F M Pattinson, 10 High Hall Cl, Trimley St. Martin, Trimley, Ipswich, IP10 0TJ [JO02PA, TM24]
G7 ANJ K P J King, 11 Somerton Rd, Martham, Great Yarmouth, NR29 4QF [JO02TR, TG41]
G7 ANK C Bryan, 175 Lichfield Rd, Shire Oak, Walsall Wood, Walsall, WS9 9NX [IO92AP, SK00]
G7 ANO T A Hyde, 10 Castleton Ave, Riddings, Alfreton, DE55 4AG [IO93HB, SK45]
G7 ANP G Blessin, 68 Churchill Way, Stafford, ST17 9PB [IO82WS, SJ92]
G7 ANQ J E Hedges, 31 Meadow Rd, Hartshill, Nuneaton, CV10 0NL [IO92FM, SP39]
G7 ANR Details withheld at licensee's request by SSL.
G7 ANV S A O'Malley, Eaver Cottage, Rothbury Rd, Long Framlington, Northd., NE65 8AE [IO95CH, NU10]
G7 ANX A J Smith, Rose Ln Cottage, Higher Huxham, Stoke Canon, Exeter, EX5 4EP [IO80GS, SX99]
G7 ANY A B King, 31 Springhill, Pennycross, Plymouth, PL2 3QZ [IO70WJ, SX45]
G7 ANZ I W Spray, 27 Laburnum House, The Beeches, Cambridge, CB4 1FY [JO02BF, TL46]
G7 AOA P T Gash, 2 Betjeman Walk, Yateley, Camberley, Surrey, GU46 6YP [IO91NH, SU85]
GJ7 AOG C R Eve, 2 The Elms, La Rue Des Cosnets, St. Ouen, Jersey, JE3 2BJ
G7 AOK R Swynford-Lain, 19 Kenton Rd, Earley, Reading, RG6 7LQ [IO91MK, SU77]
GM7 AOM J Curr, 56 Drygate St., Larkhall, ML9 2DA [IO85AR, NS75]
GM7 AON R Henry, Woodyard House, Woodyard Rd, Dumbarton, G82 4BG [IO75RW, NS37]
G7 AOO Details withheld at licensee's request by SSL.
G7 AOP Details withheld at licensee's request by SSL.
G7 AOQ J C Johnson, 8 Cheviot Cl, Knutton, Newcastle, ST5 6HU [IO83UA, SJ84]
G7 AOR Details withheld at licensee's request by SSL.
G7 AOW U Siebert, 36 Narbonne Ave, Eccles, Manchester, M30 9DL [IO83UL, SJ79]
G7 AOX W L Furze, 2 Lynwood Rd, Cromer, NR27 0EE [JO02PW, TG24]
GW7 AOZ Details withheld at licensee's request by SSL.
G7 APA P Ash, 33 The Avenue, Leighton Bromswold, Ramsey, Huntingdon, PE17 1AS [IO92WK, TL28]
G7 APC Details withheld at licensee's request by SSL.
G7 APD S A Tompsett Rugby Am Tra So, 9 Ashlawn Rd, Rugby, CV22 5ET [IO92JI, SP57]
G7 APH M J Johnson, 17 Cedar Rd, Kettering, NN16 9PU [IO92PJ, SP87]
G7 API S E Garlick, 37 Edith Rd, Kettering, NN16 0QB [IO92PJ, SP87]
GM7 APK K J Martin, 2 Martnaham Dr, Coylton, Ayr, KA6 6JE [IO75RK, NS41]
G7 APL S N Bonham, 4 St. Martins Ave, Studley, B80 7JJ [IO92BG, SP06]
G7 APM C E Mockford, 146 Fulwell Park Ave, Twickenham, TW2 5HB [IO91TK, TQ17]
G7 APO C W Monckton, The Hawthorns, Eastbourne Rd, Blindley Heath, Lingfield, RH7 6JR [IO91XE, TQ34]
G7 APQ A N Jones, 258 Windmill Ave, Kettering, NN15 6PF [IO92PJ, SP87]
G7 APS S Ellison, 16 Beechtree Rd, Walsall, WS9 9LS [IO92AP, SK00]
G7 APU L V Young, 124 Wolvey Rd, Burbage, Hinckley, LE10 2JJ [IO92HM, SP49]
G7 APV A Fryer, 17D Cedar Ct, Alsager, Stoke on Trent, ST7 2DZ [IO83UC, SJ85]
G7 APY A C Wilson, 38 Lancaster Rd, Basingstoke, RG21 5UE [IO91KG, SU65]
G7 AQA M R Hawkshaw, 9 Hawthorn Dr, Rodley, Leeds, LS13 1NJ [IO93DT, SE23]
G7 AQD W C Williams, 2 Lightfoot Ln, Fulwood, Preston, PR2 3LP [IO83PT, SD53]
G7 AQF A Gregory, 13 Combe Ave, Portishead, Bristol, BS20 9JR [IO81OL, ST47]
G7 AQI R W Dimon, 26 Old Banwell Rd, Locking, Weston Super Mare, BS24 8BS [IO81NH, ST35]
G7 AQK N McGrath, 48 Willersley Ave, Orpington, BR6 9RS [JO01BI, TQ46]
G7 AQL B Burbage, 54 Malvern Cl, Woodley, Reading, Berks, RG5 4HL [IO91NK, SU77]
G7 AQN I Cooper, 58 Fairford Cres, Downhead Park, Milton Keynes, MK15 9AE [IO92PB, SP84]
GI7 AQO R M Todd, 14 Glencroft Rd, Newtownabbey, BT36 5GD [IO74AQ, J38]
G7 AQU Details withheld at licensee's request by SSL.
G7 AQV A R Russell, 73 Seymour Rd, Newton Abbot, TQ12 2PX [IO80EM, SX87]
G7 ARB R G Barker, 50 Baldocks Ln, Melton Mowbray, LE13 1EN [IO92NS, SK71]
G7 ARC Details withheld at licensee's request by SSL.
GW7 ARD R Lee, 283 PO Box, Cardiff, CF2 3YF [IO81JL, ST17]
G7 ARF S P Wright, 54 Clarence Rd, Ponders End, Enfield, EN3 4BW [IO91XP, TQ39]
G7 ARG R Wasteney, 8 Rivermead Rd, Rosehill, Oxford, OX4 4UD [IO91JR, SP50]
G7 ARI W G Holder, 24 Long Gr Rd, Epsom, KT19 8TE [IO91UI, TQ26]
G7 ARJ J L Baber, 8 Orion Cl, Stubbington, Fareham, PO14 2SQ [IO90JT, SU50]
G7 ARM M R Aylott, 40 Rue Du Champ Meloin, Thorigne-Fouillard, 35235, France
G7 ARP R Orchard, 51 Lea House, Woodview Dr, Edgbaston, Birmingham, B15 2HE [IO92BL, SP08]
G7 ARQ R F Shearman, 29 Dryleaze Ct, Wotton under Edge, GL12 7BL [IO81TP, ST79]
GD7 ARS W H A Wrigley, 20 Fairy Hill Cl, Ballafesson, Port Erin, IM9 6TJ
G7 ART M J Blake, 7 Greenbank Ave, St. Judes, Plymouth, PL4 9BT [IO70WI, SX45]
G7 ARU Details withheld at licensee's request by SSL.
G7 ARV E J S Bourne, Wayside, St. Dunstans Rd, Salcombe, TQ8 8AR [IO80CF, SX73]
G7 ARW Details withheld at licensee's request by SSL.
G7 ASF A Bennett Coventry ARS, The Oaks, 43 School Ln, Exhall, Coventry, CV7 9GE [IO92GL, SP38]
G7 ASH Details withheld at licensee's request by SSL.
G7 ASJ Details withheld at licensee's request by SSL.
G7 ASK Capt P Cope, 5 Pembroke Rd, Clifton, Bristol, BS8 3AU [IO81QK, ST57]
GW7 ASL P W Jones, 27 Hawthorn Rd East, Llandaff North, Cardiff, CF4 2LR [IO81JM, ST17]
G7 ASM A S Marcer, 57 Ct St., Woodville, Swadlincote, DE11 7JJ [IO92FS, SK31]
GM7 ASN C L Baird, Fair View, Saval, Lairg, IV27 4ED [IO78TA, NC50]
G7 AST Details withheld at licensee's request by SSL.
G7 ASX Details withheld at licensee's request by SSL.
G7 ASY P L Matkin, 31 Southgate, Cannock, WS11 1PS [IO82XQ, SJ90]
GW7 ASZ N Blair, Pendre, Llanrhystud, Dyfed, SY23 5DL [IO72WH, SN56]
G7 ATD J D R Rowland, 7 Pangbourne St., Reading, RG3 1HS [IO91LJ, SU66]
G7 ATP J R Smith, 25 Meadway St., Burntwood, WS7 8TW [IO92AQ, SK00]
GM7 ATQ Details withheld at licensee's request by SSL.
G7 ATV Details withheld at licensee's request by SSL.
G7 ATW G P Johnson, 63 Laurel Dr, Bradwell, Great Yarmouth, NR31 8PB [JO02UN, TG50]
G7 ATX M D Walsh, 27 Sinclair Ct, Copers Cope Rd, Beckenham, BR3 1PA [IO91XJ, TQ37]
G7 ATY B Purdy, 7 Hawthorn Ave, Higher Walton, Preston, PR5 4EE [IO83QR, SD52]
G7 ATZ F McKie, 1 Clover Cl, Vange, Basildon, SS16 4SS [JO01FN, TQ78]
G7 AUB A Spencer, 8 Brooklands Ave, Chapel En-le-Frith, High Peak, SK23 0PR [IO93BH, SK08]
G7 AUD Details withheld at licensee's request by SSL.
G7 AUE B L Oubridge, 12 Wilmin Gr, Loughton, Milton Keynes, MK5 8EU [IO92OA, SP83]
G7 AUF K D Gebhardt, 15 Jubilee Rd, Corfe Mullen, Wimborne, BH21 3NH [IO80XS, SY99]
G7 AUG Details withheld at licensee's request by SSL.
G7 AUL J Hope, 70 Belmont Ave, Wickford, SS12 0HG [JO01GO, TQ79]
G7 AUP A D White, 19 Chapel Walk, Riccall, York, YO4 6NU [IO93LT, SE63]
G7 AUQ N M Smith, 51 Wentwood Gnds, Thornbury, Plymouth, Devon, PL6 8TD [IO70WJ, SX55]
G7 AUR S M Davis, 22 Tamar Cl, Fareham, PO16 8QF [IO90KU, SU60]
G7 AUS N J Cheesman, 2 South Cres, Coxheath, Maidstone, ME17 4QB [JO01FF, TQ75]
GM7 AUW J M Milne, 24 Lorne St., Edinburgh, EH6 8QP [IO85JX, NT27]
GM7 AUX E M Ramsay, Tighnduin, 2 Queen St., Monifieth, Dundee, DD5 4HG [IO86OL, NO43]
GI7 AUY I W A Potts, 46 Richmond Park, Omagh, Co. Tyrone, BT79 7SJ [IO64IO, H47]
GW7 AVB D S Daymond, The Foxgloves, 23 Lavender Way, Rogerstone, Newport, Gwent, NP1 9BF [IO81LO, ST28]
G7 AVD L N Duggan, 5 Clydesdale Cl, Whitchurch, Bristol, BS14 0RL [IO81RK, ST66]
G7 AVF P W Honeybone, 30 The Hordens, Barns Green, Horsham, RH13 7PJ [IO91TA, TQ12]
G7 AVT C J Walker, 173 Walmley Rd, Sutton Coldfield, B76 1PX [IO92CN, SP19]
G7 AVU R D Fisk, 25 Cromwell St., Gainsborough, DN21 1DH [IO93OJ, SK88]
G7 AVZ L Collings, 37 Armstrong Rd, Mansfield, NG19 6HZ [IO93JD, SK56]

G7 AWC B Park, 30 Heugh Rd, Craster, Alnwick, NE66 3TJ [IO95EL, NU21]
G7 AWG V C Watson, 3 Anderton Rise, Millbrook, Torpoint, PL10 1DA [IO70VI, SX45]
GM7 AWK D F Easton, 86 Dryburn Ave, Kelloholm, Sanquhar, Dumfriesshire, DG4 6SN [IO75XJ, NS71]
G7 AWL L J Steele, 36 Fairoaks, Aldershot Rd, Worplesdon, Guildford, GU3 3HG [IO91QG, SU95]
GW7 AWO T C Jones, 10 Waunscil Ave, Bridgend, CF31 1TX [IO81FM, SS97]
G7 AWP T Noyes, 59 Abbots Leys Rd, Winchcombe, Cheltenham, GL54 5QX [IO91AW, SP02]
G7 AWR S R Povey, 24 Beech Ave, Chelmsley Wood, Birmingham, B37 7PS [IO92DL, SP18]
G7 AWU D W Hodgetts, 97 Sandon Rd, Wolverhampton, WV10 6EL [IO82WO, SJ90]
G7 AWV M S Pilkington, 1 Norton Cl, Matchborough East, Redditch, B98 0BL [IO92BH, SP06]
G7 AWW M Gynane, 164 Stockbridge Ln, Huyton, Liverpool, L36 8EH [IO83NK, SJ49]
GM7 AWY J H R Bodle, 14A Hanover Sq, Stranraer, DG9 7AF [IO74LV, NX06]
G7 AWZ Prof H Whitfield Newcastle Com, Laboratory Ars, Computing Laboratory, Newcastle upon Tyne, NE1 7RU [IO94EX, NZ26]
G7 AXC Details withheld at licensee's request by SSL.
G7 AXE P G Cross Axe Vale ARC, Balls Farm Cottage, Musbury Rd, Axminster, EX13 5TT [IO80LS, SY29]
G7 AXF Details withheld at licensee's request by SSL.
G7 AXG Details withheld at licensee's request by SSL.
G7 AXH Details withheld at licensee's request by SSL.
G7 AXI L Connor, 29 Kings Rd, Evesham, WR11 5BP [IO92AC, SP04]
G7 AXJ Details withheld at licensee's request by SSL.
G7 AXM J Smith, Thames Ditton Marina, M.V. Marpessa, Portsmouth Rd, Surbiton, KT6 5QD [IO91UJ, TQ16]
G7 AXN R C Moxham, 37 Sherwood Dr, Marske By The Sea, Redcar, TS11 6DB [IO94LO, NZ62]
GM7 AXQ W C Thomson, 25 Beechlands Dr, Clarkston, Glasgow, G76 7XA [IO75US, NS55]
G7 AXU Details withheld at licensee's request by SSL.
G7 AXV Details withheld at licensee's request by SSL.
G7 AXW Dr A W Parkes, 96 Oakham Rd, Dudley, DY2 7TQ [IO82XM, SO98]
G7 AYA G P Jessup, 14 Langley Lodge Gdns, Langley, Blackfield, Southampton, SO45 1FZ [IO90HT, SU40]
G7 AYC Details withheld at licensee's request by SSL.
G7 AYD A W Grange, 7 West St., Isleham, Ely, CB7 5SD [JO02EI, TL67]
G7 AYE R J Phipps, 39 Perrinsfield, Lechlade, GL7 3SD [IO91DQ, SP20]
G7 AYF H R Richardson, 20 Hillhead, Parkway, Chapel House, Newcastle upon Tyne, NE5 1ER [IO94DX, NZ16]
G7 AYI L J Faragher, 4 Kirloe Ave, Leicester Forest East, Leicester, LE3 3LA [IO92JO, SK50]
GW7 AYK Details withheld at licensee's request by SSL.
G7 AYO L A Hutt, The Crossing House, Fenwick Ln, Doncaster, South Yorks, DN6 0EZ [IO93KP, SE51]
G7 AYP A Gregory, 3 Irwin Ct, 470 London Rd, Ashford, TW15 3AD [IO91SK, TQ07]
G7 AYQ A T Tregay, The Castings, High St., Foulsham, Dereham, NR20 5AD [JO02MS, TG02]
G7 AYS D J Webster, 5 Eastfield Rd, Princes Risborough, HP27 0JA [IO91OR, SP80]
GM7 AYW J Hunter, 1 Mitchell Dr, Rutherglen, Glasgow, G73 3QP [IO75VT, NS66]
G7 AYX W T Scott, 2 Castweazle, Rolvenden Rd, Tenterden, TN30 6UA [JO01IB, TQ83]
G7 AZA J N Cash, 66 The Hides, Harlow, CM20 3QN [JO01BS, TL41]
G7 AZD Details withheld at licensee's request by SSL.
G7 AZH Details withheld at licensee's request by SSL.
G7 AZI H F Stone, 14 Marion Cl, Wymondham, NR18 0ND [JO02NN, TG10]
G7 AZJ B J Sayers, 158 Pheasant Rise, Bar Hill, Cambridge, Cambs, CB3 8SD [JO02AF, TL36]
G7 AZK P Hodges, 15 Middlewich St., Crewe, CW1 4BS [IO83SC, SJ75]
G7 AZM W A Knowler, 33 Cherry Tree Rd, Rainham, Gillingham, ME8 8JY [JO01HI, TQ86]
G7 AZP Details withheld at licensee's request by SSL.[Op Peter. Correspondence via PO Box 703, West Moors, Ferndown, Dorset, BH22 0YB.]
G7 AZU M Macdonnell, Swanston, Bicknells Bridge, Huish Episcopi, Langport Somerset, TA10 9HH [IO81OB, ST42]
G7 AZV R J Quick, 3 Harvesters Dr, St, BA16 0UB [IO81PC, ST43]
G7 AZW M J Ney, 4 Rathen Rd, Withington, Manchester, M20 4GH [IO83VK, SJ89]
G7 BAB W R Foden, 209 Lord Ln, Failsworth, Manchester, M35 0PX [IO83WL, SD80]
G7 BAC M J G Gohl, 142 Well Ln, Willerby, Hull, HU10 6HS [IO93SS, TA03]
G7 BAD T M Chapman, 21 Orton Dr, Witchford, Ely, Cambs, CB6 2JG [JO02CJ, TL47]
G7 BAE T J Searle, Haven Orchard, Exwick Ln, Exeter, EX4 2AP [IO80FR, SX99]
G7 BAI F J Bell Ryl Grammar Scl, Am Rad Satellite Cl, c/o Royal Gmr Sch, High St Guildford, GU1 3BB [IO91RF, TQ04]
G7 BAL Details withheld at licensee's request by SSL.
G7 BAP B A Pilkington, 79 Millfield Rd, Morton, Bourne, PE10 0NU [IO92TT, TF02]
G7 BAR A D Goodhew Brickfields Amateur Radio Soci, Strathwell Manor, Strathwell Park, Whitwell, Isle of Wight, PO38 2QU [IO90IO, SZ57]
GM7 BAS W M Hunter, 9 Mill Park, Southend, Campbeltown, Argyll, PA26 6RH [IO76MG, NN11]
G7 BAU A J Comber, Lowfield House, Church Sq, Blakeney, GL15 4DS [IO81SS, SO60]
G7 BAV A J Roberts, 17 Welbeck Ave, North Rd East, Plymouth, PL4 6BG [IO70WJ, SX45]
G7 BAX A P Newman, 140 Brambleside, Kettering, NN16 9BT [IO92PJ, SP88]
G7 BAY E A Miles, 18 Pearson Garth, West Ayton, Scarborough, YO13 9LH [IO94SF, SE98]
G7 BBC G W Rowlands BBC Clb Ariel Rd, 26A Laburnum Rd, Hayes, UB3 4JX [IO91SL, TQ07]
G7 BBD C S Hartley, 102A Bedford Rd, Cranfield, Bedford, MK43 0HA [IO92QB, SP94]
G7 BBG Details withheld at licensee's request by SSL.
G7 BBI Details withheld at licensee's request by SSL.
G7 BBJ B Jenkinson, 79 Beech Rd, Harrogate, HG2 8DZ [IO93FX, SE35]
G7 BBL Details withheld at licensee's request by SSL.
G7 BBM D J Perry, 20 Barmouth Ct, Cromer Pl, Ingol, Preston, PR2 3XR [IO83PS, SD53]
G7 BBN E J Crookall, 17 Dundee St., Moorlands, Lancaster, LA1 3DS [IO84OB, SD46]
G7 BBP Details withheld at licensee's request by SSL.
G7 BBQ Details withheld at licensee's request by SSL.
GM7 BBU Dr C M Sharp, Dept. of Astronomy, University of Arizona, Tucson Az 85721, USA
G7 BBY M J Jones, 9 Nailers Cl, Stoke Heath, Bromsgrove, B60 3PL [IO82XH, SO96]
GM7 BCC R G Sutherland, Tigh - Na - Coille, Mill Rd, Nairn, Scotland, IV12 5EW [IO87BN, NH85]
G7 BCI V R White, 6 Laburnum Cl, Anston, Sheffield, S31 7GL [IO93GJ, SK38]
G7 BCK N Gillies, 5 Pickmere Terr, Dukinfield, SK16 4JJ [IO83WL, SJ99]
G7 BCL Details withheld at licensee's request by SSL.
G7 BCN Details withheld at licensee's request by SSL.
G7 BCO A G Adey, 8 Spinners Ct, Telford, TF5 0PG [IO82RR, SJ61]
G7 BCP J S Parkinson, 45 Livingstone Rd, Blackburn, BB2 6NE [IO83RR, SD62]
G7 BCR Details withheld at licensee's request by SSL.
G7 BCS K J Wood, 15 Robin Hill Dr, Camberley, GU15 1EG [IO91PH, SU85]
G7 BCW H Fitches, 29 Harrington Way, Oakham, LE15 6SE [IO92PP, SK80]
G7 BDB Details withheld at licensee's request by SSL.
GW7 BDG L D W Jones, Rio Balsas 1009, Col. Jardine de La Salle, Monclova, Coah. 25770, Mexico, X X
GI7 BDJ D Reade, 30 Mullaghdrin Rd East, Dromara, Dromore, BT25 2AQ [IO74AJ, J35]
G7 BDK P M Blackett, 32 Woodstock Rd, Carshalton, SM5 3DZ [IO91WI, TQ26]
G7 BDO Details withheld at licensee's request by SSL.
G7 BDQ Details withheld at licensee's request by SSL.
G7 BDR A Davis, 5 Ludlow Cl, Loughborough, LE11 3TB [IO92JS, SK51]
G7 BDS J S Mills, 42 Maple Gr, Welwyn Garden City, AL7 1NL [IO91VT, TL21]
G7 BDV Details withheld at licensee's request by SSL.
G7 BDW A J Taylor, 33 North Rd, East Dene, Rotherham, S65 2RR [IO93IK, SK49]
G7 BEA R D D Dickson, 5 Sedge Rise, Tadcaster, LS24 9LQ [IO93IV, SE44]
G7 BEB H C T Penman, Wayma, 16 Chapmans Cl, Frome, BA11 2SH [IO81UF, ST74]
G7 BED Details withheld at licensee's request by SSL.
G7 BEE M Hood, 59 Ross Lea, Shiney Row, Houghton-le-Spring, DH4 4PQ [IO94GU, NZ35]
GW7 BEF Details withheld at licensee's request by SSL.
G7 BEP M Van Der Steeg, Horsebrook Farm, Avonwick, South Brent, Devon, TQ10 9EU [IO80CJ, SX75]
GI7 BET R Griffin, 1 Dobbin Hill Rd, Armagh, BT60 1AU [IO64QI, H84]
G7 BEV Details withheld at licensee's request by SSL.
G7 BEX Details withheld at licensee's request by SSL.[Op: R Trezise, 95 Osprey Park, Thornbury, Bristol, Avon, BS12 1LZ.]
GW7 BEY J R Knott, 35 Tyr y Sarn Rd, Rumney, Cardiff, CF3 8BD [IO81KM, ST27]
GM7 BFD Details withheld at licensee's request by SSL.
G7 BFE C J Broom, 24 The Weal, Weston, Bath, BA1 4EX [IO81TJ, ST76]
G7 BFG H A Smith, 2A Durham Rd, Wollaston, Stourbridge, DY8 4SX [IO82WL, SO88]
G7 BFH N M Lambert, 5 Foxwell Sq, Southfields, Northampton, NN3 5AT [IO92OG, SP76]
G7 BFO Details withheld at licensee's request by SSL.
G7 BFS Details withheld at licensee's request by SSL.
G7 BFT G E V Gomes, 19 Shadwell Cl, Weeting, Brandon, IP27 0RH [JO02HL, TL78]
G7 BFX T R Melton, 63 Beresford Rd, Chandlers Ford, Eastleigh, SO53 2JZ [IO90HX, SU42]
G7 BGL J H Russell, 26 Cecil Rd, Seaforth, Liverpool, L21 1DD [IO83LL, SJ39]
G7 BGM D N Allison, 3 Church Ln, Chalgrove, Oxford, OX44 7TB [IO91LP, SU69]
G7 BGO P M Toll, New Lodge, Water Millock, Penrith, Cumbria, CA11 0JH [IO84NO, NY42]
G7 BGP C J Lord, 154 Plants Brook Rd, Sutton Coldfield, B76 1HL [IO92CM, SP19]
G7 BGT R K Pykett, 20 Bolton St., Swanwick, Alfreton, DE55 1BU [IO93HB, SK35]
G7 BGY M J Bellaby, 77 Southchurch Dr, Clifton, Nottingham, NG11 8AS [IO92JV, SK53]
G7 BGZ M W Hickman, 235 Amersall Rd, Doncaster, DN5 9PN [IO93KM, SE50]
G7 BHB J R M Stark, 4 Fort Royal Mews, London Rd, Worcester, WR5 2DL [IO82VE, SO85]
G7 BHG R E Gill, 24 Larkfield Cres, Rawdon, Leeds, LS19 6EH [IO93DU, SE23]
G7 BHH L R Kemp, 11 Florence Terr, Kingston Vale, London, SW15 3RU [IO91UK, TQ27]
G7 BHN P C Chapman, 6 Pickhurst Green, Hayes, Bromley, BR2 7QT [JO01AJ, TQ36]
G7 BHR D S Coombes, 2 Ormesby Dr, Potters Bar, EN6 3DZ [IO91VQ, TL20]
G7 BHU T P Mayfield, 184 Wharf Rd, Pinxton, Nottingham, NG16 6LQ [IO93IC, SK45]
G7 BHW J D Wilson, 61 New Ln, Hilcote, Alfreton, DE55 5HT [IO93IC, SK45]
G7 BHX S Mellor, 124 Ryknield Rd, Kilburn, Belper, DE56 0PF [IO93GA, SK34]
G7 BHY P J Bailey, 21 Westhall Rd, Mickleover, Derby, DE3 5PA [IO92FW, SK33]
G7 BIA Details withheld at licensee's request by SSL.
G7 BIK D Clark, Rosemary Cottage, Bylane End, Liskeard, Cornwall, PL14 3PZ [IO70SK, SX26]
G7 BIL W J Knox, 5 Farne Cl, Bristol, BS9 4HU [IO81QL, ST57]
G7 BIM S J Beazley, 26 Church Rd, Wretton, Kings Lynn, PE33 9QR [JO02GN, TF70]
G7 BIP G J Roffey, 16 Killewarren Way, Orpington, BR5 4DJ [JO01BJ, TQ46]
G7 BIQ K F Lloyd, 9 Hornbeam Walk, Witham, CM8 2SZ [JO01HT, TL81]
G7 BIV R C Hudson, 27 Riverside Way, Kelvedon, Colchester, CO5 9LX [JO01IU, TL81]
G7 BIX T S Houlihane, 60 Lucerne Cl, Cambridge, CB1 4SA [JO02CE, TL45]
G7 BJB C Foster, 10 Handel St., Derby, DE24 8AZ [IO92GV, SK33]
G7 BJC M J Wiggins, 158 Prince Charles, Ave, Mackworth, Derby, DE3 4LQ [IO92FV, SK33]
G7 BJD R C Ward, 12 Meadow Lea, Knighton Fields, Worksop, S80 3QJ [IO93KH, SK57]
G7 BJH Details withheld at licensee's request by SSL.
G7 BJL J Collett, 58 St. Ives Rd, Wyken, Coventry, CV2 5FX [IO92GJ, SP37]
G7 BJN J Barlow, 26 York St., Newcastle, ST5 1DE [IO83VA, SJ84]
G7 BJQ Details withheld at licensee's request by SSL.
G7 BJR G J Mitchell, 8 Addison Rd, Mexborough, S64 0DJ [IO93IL, SE40]
G7 BJU Details withheld at licensee's request by SSL.
G7 BJY Details withheld at licensee's request by SSL.
G7 BKC Details withheld at licensee's request by SSL.
G7 BKF Details withheld at licensee's request by SSL.
G7 BKH Details withheld at licensee's request by SSL.
G7 BKI Details withheld at licensee's request by SSL.
G7 BKL B Moulton, 70 St. Georges Ave, Houghton, Westhoughton, Bolton, BL5 2EU [IO83RM, SD60]
G7 BKM S G Griggs, 262 Seabrook Rd, Hythe, CT21 5RL [JO01NB, TR13]
G7 BKN Details withheld at licensee's request by SSL.
G7 BKR Details withheld at licensee's request by SSL.
G7 BKU Details withheld at licensee's request by SSL.
G7 BKV Details withheld at licensee's request by SSL.
G7 BKX Details withheld at licensee's request by SSL.
G7 BLB T Pearce, 11 Clover Laid, Sawmills Est, Brandon, Durham, DH7 8BB [IO94ER, NZ23]
G7 BLD C C Holden, 6 Payton House, Kings Way, Burgess Hill, RH15 0UB [IO90WW, TQ31]
G7 BLJ R H Maytum, 38 The Ryde, Hatfield, AL9 5DL [IO91VS, TL20]
G7 BLK T H Dicks, 4 Nicholas Dr, Reydon, Southwold, IP18 6RE [JO02UI, TM57]
G7 BLM G W Yates, 65 Baysham St., Whitecross, Hereford, HR4 0ET [IO82PB, SO54]
G7 BLQ E Thornton, 60 Towngate House, Westgate, Elland, HX5 0DN [IO93BQ, SE12]
G7 BLR Details withheld at licensee's request by SSL.
G7 BLT Details withheld at licensee's request by SSL.
G7 BLX M J Rowe, 31 Thornhill Ave, Thornhill, Southampton, SO19 6PS [IO90HV, SU41]
G7 BMA Details withheld at licensee's request by SSL.
G7 BMC P J Hollis, 47 Spring Ln, Whittington, Lichfield, WS14 9NA [IO92CQ, SK10]
G7 BMD Details withheld at licensee's request by SSL.
G7 BME D Watts, 9 Filwood Dr, Kingswood, Bristol, BS15 4HT [IO81SL, ST67]
GD7 BMG M J H Parnell, 15 Cronk y Berry Ave, Cronk y Berry Gdns, Douglas, IM2 6HE
G7 BMM S T Hallam, 125 Charnwood Rd, Shepshed, Loughborough, LE12 9NL [IO92IS, SK41]
G7 BMN Details withheld at licensee's request by SSL.
G7 BMO W B Crossley, 313 South Rd, Walkley, Sheffield, S6 3TD [IO93GJ, SK38]
G7 BMS Details withheld at licensee's request by SSL.[Bedford Modern School ARS, Manton Lane, Bedford. Details c/o G0BKN QTHR.]
G7 BMT S Hodges, 15 Middlewich St., Crewe, CW1 4BS [IO83SC, SJ75]
G7 BMW A J Grant, 3 Cragg Wood Cl, Horsforth, Leeds, LS18 4RL [IO93ET, SE23]
G7 BMY J A Moore, Fairview, 28 Bulmer Ln, Winterton on Sea, Great Yarmouth, NR29 4AF [JO02UR, TG41]
G7 BMZ D A C Wormald, 15 Sabrina Dr, Bewdley, DY12 2RJ [IO82UJ, SO77]
G7 BNB S E Reynolds, 48 Westmorland Rd, Felixstowe, IP11 9TE [JO01QX, TM33]
G7 BND B A Welthy, 8 Du Cane Pl, Witham, CM8 2UQ [JO01HT, TL81]
G7 BNE J R Moore, 2 Elm Rd, Shoeburyness, Southend on Sea, SS3 9PB [JO01JM, TQ98]
G7 BNF M Harrington, 9 High House Est, Sheering Rd, Harlow, CM17 0LL [JO01BS, TL41]
G7 BNI N S Pope, 6 Elsee Rd, Rugby, CV21 3BA [IO92II, SP57]
G7 BNK D H Wood, 16 Church Rd, Pelsall, Walsall, WS3 4QN [IO92AP, SK00]
G7 BNL A J Creek, 7 Green Park, Brinkley, Newmarket, CB8 0SQ [JO02EE, TL65]
G7 BNM A T Ison, 32 Station Rd, Lode, Cambridge, CB5 9HB [JO02CF, TL56]
G7 BNN K Sheppard, Woodlands, Orby, Skegness, PE24 5HT [JO03CE, TF46]
G7 BNO C S Sheppard, Woodlands, Orby, Skegness, PE24 5HT [JO03CE, TF46]
G7 BNS T L Healey, 5 St. Johns Cres, Huddersfield, HD1 5DY [IO93CP, SE11]
G7 BNT J R Squires, 44 St. Marys Rd, Doncaster, DN1 2NP [IO93KM, SE50]
G7 BNV Details withheld at licensee's request by SSL.
G7 BNW G L Ingmire, 72 Farley Farm Rd, Farley Hill, Luton, LU1 5PB [IO91SU, TL02]
G7 BOB R Kin, The Poplars, Kingsmead Rd, Loudwater, High Wycombe, HP11 1JL [IO91PO, SU89]
G7 BOD A F Davy, 7 Trevia, Camelford, Cornwall, PL32 9UX [IO70PO, SX18]
G7 BOH P M Craig, 219 Oakley Rd, Shirley, Southampton, SO16 4NN [IO90GW, SU31]
G7 BOJ M I A Rodgers, 33 Dewhurst Rd, Harwood, Bolton, BL2 3NF [IO83TO, SD71]
GI7 BON R J Johnston, 31 Glenallen St., Belfast, BT5 4HT [IO74BO, J37]
G7 BOR K H Kirkland, Cymbeline, 51 The Borough, Downton, Salisbury, SP5 3LX [IO90DX, SU12]
GM7 BOU Details withheld at licensee's request by SSL.
GM7 BOW R M King, 25 Swinton Ave, Rowanbank, Baillieston, Glasgow, G69 6JW [IO75WU, NS66]
GW7 BOY B Hodgkinson, 16 Swain Ave, Buckley, CH7 3BR [IO83LD, SJ26]
GM7 BOZ A L Mackenzie, 34 Blair Dr, Milton of Campsie, Glasgow, G65 8DS [IO75WX, NS67]
G7 BPF D G Rose, 8 Ambrose Ave, Hatfield, Doncaster, DN7 6QQ [IO93LN, SE60]
G7 BPG D P Dixon, 21 Silverdale St., Knutton, Newcastle, ST5 6BY [IO83UA, SJ84]
G7 BPM N S Hemingway, 30 Links View, Half Acre, Rochdale, OL11 4DD [IO83VO, SD81]
G7 BPN K G Pentecost, 46 Austen Way, High Farm, Crook, DL15 9UT [IO94CR, NZ13]
G7 BPO C J Hoare PO Research Cnt, Ars Room Fw01, Wheatstone Rd, Dorcan Swindon, SN3 4RD [IO91DN, SU18]
G7 BPR S R Winlove-Smith, 4 Clay Dyke Bank, Amber Hill, Boston, Lincs, PE20 3RL [IO93VA, TF24]
G7 BPU Details withheld at licensee's request by SSL.
G7 BPV C J Wright, 25 Pembroke Rd, Weston Est, Macclesfield, SK11 8RT [IO83WG, SJ97]
G7 BPX Details withheld at licensee's request by SSL.[Op: A J]
G7 BQD B Ward, 44 Symons St., Salford, M7 4AP [IO83UM, SD80]
G7 BQI L Lee, 36 Sycamore Lodge, Paynes Rd, Southampton, SO15 3SE [IO90GV, SU41]
G7 BQM Dr E P Turk, Sunny Meadow Farm, Long Buckby Wharf, Long Buckby, Northampton, Northants, NN6 7PP [IO92KG, SP66]
G7 BQS M Rodgers, 14 North St., Rawmarsh, Rotherham, S62 5NH [IO93IL, SK49]
G7 BQT B E Miller, 11 Neptune Rd, Fareham, PO15 6SW [IO90JU, SU50]
G7 BQX B Thompson, 2 Sandpit Cttgs, Garton on The Wolds, Driffield, YO25 9LP [IO93RX, SE95]
G7 BQY A W Brighton, 22 Langport Dr, Vicars Cross, Chester, CH3 5LY [IO83NE, SJ46]
G7 BRA Dr J P M Riley, 132 Barrs Rd, Cradley Heath, Warley, B64 7EZ [IO82XL, SO98]
G7 BRB J B Marsh, 11 Howard Rd, Kings Heath, Birmingham, B14 7PB [IO92BK, SP08]
G7 BRC M J Pearson Bredhurst Receiving & Transmit, 56 Parkwood Green, Parkwood, Gillingham, ME8 9PP [JO01HI, TQ86]
G7 BRF D K Oliver, 36 Baker Ave, Stratford upon Avon, CV37 9PN [IO92DE, SP15]
G7 BRG A Merrix Birmingham Raynet, 56 Morris St., West Bromwich, B70 7SP [IO92AM, SP09]
G7 BRJ J C Mills, 70 Cres Rd, Rochdale, OL11 3LG [IO83VO, SD81]
GM7 BRL Details withheld at licensee's request by SSL.
G7 BRM G T Jeffery, 48 Minnis Ln, River, Dover, CT17 0PR [JO01PD, TR24]
G7 BRN A C Sparkes Breckland Raynet, 13 Swathing, Cranworth, Thetford, IP25 7SJ [JO02LO, TF90]
G7 BRP J R Biggs, 26 Laceys Ln, Exning, Newmarket, CB8 7HL [JO02EG, TL66]
G7 BRR R A Wood, 96 West Dr, Tintwistle, Hadfield, Hyde, SK14 7NB [IO93AL, SK09]
G7 BRS J Rushton, 391 Rossendale Rd, Burnley, BB11 5HP [IO83US, SD83]
G7 BRU J B Rhodes, 30 Preston Rd, Overcombe, Weymouth, DT3 6PZ [IO80SO, SY68]
G7 BRV M R Hackett, 12 Hawkins Cl, Rose Green, Bognor Regis, PO21 3LW [IO90PS, SZ89]
G7 BRX R J Bell, 10 Old Mill Way, Weston Super Mare, BS24 7AS [IO81NI, ST36]
G7 BRZ C J Gaunt, 39 Sonja Crest, Immingham, Grimsby, DN40 2EQ [IO93VO, TA11]
G7 BSA K F Corser, 85 Coleshill Rd, Marston Green, Birmingham, B37 7HT [IO92DL, SP18]
G7 BSB Details withheld at licensee's request by SSL.
GW7 BSC R K Snelling, 91 Oakfield Rd, Newport, South Wales, NP9 4LP [IO81LO, ST28]
G7 BSE M Ashmore, 4 Marl Croft, Chester, CH3 5SH [IO83NE, SJ46]
G7 BSF A M Lewis, Sky Waves, 20 Annes Walk, Caterham, CR3 5EL [IO91WH, TQ35]
G7 BSG A N Carter Bemerton Sc Grp, 75 Main St., Brampton, Cumbria, CA8 1SH [IO84PW, NY56]
G7 BSK J A Ingram, 16 Lime Gr, Long Eaton, Nottingham, NG10 4LD [IO92IV, SK43]
G7 BSL P R Bedford, 9 Woodland Ave, Kirkthorpe, Wakefield, WF1 5TD [IO93GQ, SE32]
G7 BSM Details withheld at licensee's request by SSL.

G7 BSO A J Noble, 19 Foxglove Rd, Eastbourne, BN23 8BU [JO00DT, TQ60]
G7 BSP S E Farmer, Horton Brook Cottage, Horton, Wem, Shrewsbury, Salop, SY4 5NB [IO82OU, SJ42]
G7 BSS Details withheld at licensee's request by SSL.
G7 BSW A Rushby, 19 Holtdale Pl, Cookridge, Leeds, LS16 7RH [IO93EU, SE24]
G7 BTA Details withheld at licensee's request by SSL.
G7 BTB Details withheld at licensee's request by SSL.
GW7 BTC S D Richards B.T.(Wales)A.R.S., Crydon, St. Andrews Rd, Wenvoe, Vale of Glam b, CF5 6AF [IO81IK, ST17]
G7 BTD Details withheld at licensee's request by SSL.
G7 BTI N R Prosser Madley Amateur Radio Group, 35 Holmfirth Cl, Belmont, Hereford, HR2 7UG [IO82PA, SO43]
G7 BTK J E Scott, 154 Ingram Rd, Bulwell, Nottingham, NG6 9GQ [IO92JX, SK54]
GM7 BTL J C Q Hunter, 31 Hamilton Dr, New Cumnock, Cumnock, KA18 4JP [IO75VJ, NS61]
G7 BTP P Jensen, 16 Hawthorne Ave, Immingham, Grimsby, DN40 1AR [IO93VO, TA11]
G7 BTW W D Peel, 17 Arthur St., Whitburn, Sunderland, SR6 7NE [IO94HX, NZ46]
G7 BTX D R P Drew, 34 Church St., Eastwood, Nottingham, NG16 3HS [IO93IA, SK44]
G7 BUF P R Parkin, 2 The Knoll, Dronfield, Sheffield, S18 6EH [IO93GH, SK37]
G7 BUK D N Brinnen, 134 Victoria Rd, Mablethorpe, LN12 2AJ [JO03DI, TF58]
G7 BUN H V Ellis, Home Cottage, Drury Sq, Beeston, Kings Lynn, PE32 2NA [JO02JQ, TF91]
G7 BUQ Details withheld at licensee's request by SSL.
G7 BUR Details withheld at licensee's request by SSL.
G7 BUS G B Payne, 155 Camping Hill, Stiffkey, Wells Next The Sea, NR23 1QL [JO02LW, TF94]
G7 BUT Details withheld at licensee's request by SSL.
G7 BUZ Details withheld at licensee's request by SSL.
G7 BVG Details withheld at licensee's request by SSL.
G7 BVH M P Gurr, Elan, Sandown Rd, Sandwich, CT13 9NY [JO01QG, TR35]
G7 BVK Details withheld at licensee's request by SSL.
G7 BVL K Lambert, 1 Langton Rd, Chichester, PO19 3LY [IO90OU, SU80]
G7 BVN S H Sheppard, Silverlea, Scotland St., Stoke By Nayland, Colchester, CO6 4QF [JO01KX, TL93]
G7 BVS Details withheld at licensee's request by SSL.
G7 BVV J Gardiner, 20 Hillside Ave, Oakworth, Keighley, BD22 7QQ [IO93AU, SE03]
G7 BVX Capt W Miller, The Breakers, 27 Lusty Glaze Rd, Newquay, TR7 3AE [IO70LK, SW86]
GW7 BVY Details withheld at licensee's request by SSL.
G7 BVZ B R Allen, 84 Holland Rd, Little Clacton, Clacton on Sea, CO16 9RS [JO01NT, TM11]
G7 BWD J H Knowles, 4 Mardale Cres, Worden Park, Leyland, Preston, PR5 2BT [IO83PQ, SD52]
G7 BWE E T Gibbons, 19 Queens Park Rd, Caterham, CR3 5RB [IO91WG, TQ35]
G7 BWF A L Gray, 147 Kirby Rd, Walton on The Naze, CO14 8RL [JO01PU, TM22]
G7 BWH W W Hattersley, 1 Huntingdon Rd, Kempston, Bedford, MK42 7EX [IO92SC, TL04]
G7 BWI J C White, 14 The Old Station Yard, Lambourn, Hungerford, RG17 8QA [IO91FM, SU37]
G7 BWO D R Morgan, 26 Lyndhurst Rd, Exmouth, EX8 3DT [IO80HP, SY08]
G7 BWP Details withheld at licensee's request by SSL.[Op: R G Beaumont. Station located near Cirencester, Glos.]
G7 BWQ R J Somerville Roberts, 16 Walton, High Ercall, Telfordcall, Shropshire, TF6 6AR [IO82QS, SJ51]
G7 BWS Details withheld at licensee's request by SSL.
G7 BWV R A Fosbraey, 122 East St., Sittingbourne, ME10 4RX [JO01II, TQ96]
G7 BWW M Widdows, 27 Market Cl, Barnham, Bognor Regis, PO22 0LH [IO90QT, SU90]
G7 BWY J A Clarke, 83 Grange Rd, Erdington, Birmingham, B24 0ET [IO92CM, SP19]
G7 BXA P G Austin, 24 Fairfield Terr, Bramley, Leeds, LS13 3DH [IO93ET, SE23]
G7 BXB S M Burrows, 249A High St, Waltham Cross, Herts, EN8 7BE [IO91XQ, TL30]
G7 BXE A Whitfield, Flat 3, 137A Forster St., Warrington, WA2 7AX [IO83QJ, SJ68]
G7 BXG F K Clarke, 95 Kirklington Rd, Rainworth, Mansfield, NG21 0JZ [IO93KC, SK55]
G7 BXJ P Turner, 260 New Ln, Huntington, York, YO3 9LY [IO93LX, SE65]
G7 BXN T W Jones, 17 Beacon St., Wibsey, Bradford, BD6 3BE [IO93CS, SE13]
G7 BXO R J P Pollard, 85 Greystones Rd, Whiston, Rotherham, S60 4DB [IO93IJ, SK49]
G7 BXS R A Wake, 55 Bearsdown Rd, Eggbuckland, Plymouth, PL6 5TR [IO70WJ, SX55]
G7 BYA I H Herbert, 53 New Beacon Rd, Grantham, NG31 9JS [IO92QV, SK93]
GM7 BYB A I McIntyre, 71 Carman View, Bellsmyre, Dumbarton, G82 3AU [IO75RW, NS47]
G7 BYD E C Goddard, 25 Falcon Rd, Hampton, TW12 2RA [IO91TK, TQ17]
G7 BYE J Hersom, 10 Young St., Gilesgate, Durham, DH1 2JU [IO94FS, NZ24]
G7 BYF P D May, Stable Cottage, Quarrendon House Farm, Bicester Rd, Aylesbury, HP18 0PS [IO91NT, SP71]
G7 BYG A Found, 4 Tilehurst Cl, Kersal Way, Salford, M7 3ST [IO83UM, SD80]
G7 BYH S J J Crump, 14 Brunswick St., Bungalows, Wakefield, WF1 4NY [IO93GQ, SE32]
G7 BYI M J E Baldry, 10 Kingfisher Ct, Lowestoft, NR33 8PJ [JO02UL, TM59]
G7 BYJ Dr S M Murray, 36 Foxwood Dr, St. Georges Park, Kirkham, Preston, PR4 2DS [IO83NS, SD43]
G7 BYN D S Bendrey, 73 Kestrel Cl, Chipping Sodbury, Bristol, BS17 6XB [IO81TM, ST78]
G7 BYS J P R Pollard, 25 Bridgemere Cl, Radcliffe, Manchester, M26 4FS [IO83TN, SD70]
G7 BYU W G McGuffie, 1 Norbury Dr, Marple, Stockport, SK6 6LL [IO83XJ, SJ98]
G7 BYV T M Connolly, 20 Belmore Park, Ashford, TN24 8UW [JO01KD, TR04]
G7 BYW C G Stone, 35 Sycamore Dr, Torpoint, PL11 2NA [IO70VI, SX45]
G7 BZB K N Waller, 29 Rosslyn Rd, Bath, BA1 3LQ [IO81TJ, ST76]
G7 BZC R A Pickett, Cottage 1 Hospitalr, Wingland, Sutton Bridge, Spalding, PE12 9YR [JO02CS, TF42]
G7 BZD P M Yates, 74 Olton Rd, Shirley, Solihull, B90 3NN [IO92CK, SP18]
G7 BZE W Gillott, 20 Stretton Rd, Monk Bretton, Barnsley, S71 1XQ [IO93GN, SE30]
G7 BZH I R Graham, 36 Lime Tree Rd, Ulverston, LA12 9EY [IO84KE, SD27]
G7 BZI P J Wallington, Walk Farm, Chipping Norton, Oxon, OX7 5TG [IO91FX, SP32]
G7 BZM R S Brown, 2 Hay Green Cl, Bournville, Birmingham, B30 1RQ [IO92AK, SP08]
G7 BZN J A Richards, 1 Foxglade, Oulton Broad, Lowestoft, NR32 3JJ [JO02UL, TM59]
G7 BZP A F P St Aubyn, Son-Bou, Meres Valley, Laflounder Ln, Mullion Cornwall, TR12 7HX [IO70IA, SW61]
G7 BZQ G R Reynolds, 187 Steelhouse Ln, Wolverhampton, WV2 2AU [IO82WN, SO99]
G7 BZU A McColl, 49 Botley Rd, Southampton, SO19 0NQ [IO90HV, SU41]
G7 BZV Details withheld at licensee's request by SSL.
G7 BZX Details withheld at licensee's request by SSL.
G7 BZZ Details withheld at licensee's request by SSL.
G7 CAB Details withheld at licensee's request by SSL.
G7 CAF S F Bond, 38 Hampsfell Dr, Morecambe, LA4 4TU [IO84NB, SD46]
G7 CAG A Malpass, 48 Geoffrey Barbour Rd, Abingdon, OX14 2ES [IO91IQ, SU49]
G7 CAH M J Astley, 14 Basemoors, Bullbrooke, Bracknell, RG12 2RG [IO91PK, SU86]
G7 CAK Details withheld at licensee's request by SSL.
GM7 CAN Details withheld at licensee's request by SSL.
G7 CAQ Dr C A Gerrard, 13 Wentworth Ave, Timperley, Altrincham, WA15 6NG [IO83UJ, SJ78]
G7 CAR Details withheld at licensee's request by SSL.
G7 CAS W E Welburn, P 0 Box 2114, Wellington, New Zealand
G7 CAU Details withheld at licensee's request by SSL.
G7 CAV A V Lewis, 36 Trent Cl, Stevenage, SG1 3RT [IO91VW, TL22]
G7 CAW Details withheld at licensee's request by SSL.
G7 CAY Details withheld at licensee's request by SSL.
GI7 CBD T H Hopkins, The Cottage, 142 Ballymoney Rd, Banbridge, BT32 4HN [IO64VI, J14]
G7 CBE F M Clarke, Marchmont, Horsecombe Gr, Combe Down, Bath, BA2 5QP [IO81TI, ST76]
G7 CBG P A Marshall, 31 Clement Mews, Kimberworth, Rotherham, S61 2JU [IO93HK, SK49]
G7 CBL F W Gillard, 140 Putnoe St., Bedford, MK41 8HJ [IO92SD, TL05]
GW7 CBU J Griffiths, 143 Brynglas, Hollybush, Cwmbran, NP44 7LL [IO81LP, ST29]
G7 CBW S T Duffield, 64 Jays Ave, Tipton, West Midlands, DY4 8UZ [IO82XM, SO99]
G7 CBY L J Cotton, 34 Whittaker Ln, Norden, Rochdale, OL11 5PL [IO83VO, SD81]
G7 CBZ S Cotton, 34 Whittaker Ln, Norden, Rochdale, OL11 5PL [IO83VO, SD81]
GW7 CCC Details withheld at licensee's request by SSL.
GW7 CCD Details withheld at licensee's request by SSL.
GW7 CCG Details withheld at licensee's request by SSL.
G7 CCH D Woods, 26 Compton Rd, Southport, PR8 4HA [IO83LP, SD31]
G7 CCI Details withheld at licensee's request by SSL.
G7 CCL S A Cullingworth, 66 Meadow Rd, Garforth, Leeds, LS25 2EN [IO93HT, SE43]
GW7 CCM Details withheld at licensee's request by SSL.
G7 CCN G C Batiste, 36 Old Mead Rd, Wick, Littlehampton, BN17 7PU [IO90RT, TQ00]
GW7 CCR Details withheld at licensee's request by SSL.
G7 CCS G Oliver, 24 Pendennis Rd, Penzance, TR18 2BA [IO70FD, SW43]
G7 CCV P J Upton, 73 Allington Cl, Taunton, TA1 2NA [IO81LA, ST22]
GW7 CCW Details withheld at licensee's request by SSL.
G7 CCX D J Spooner, 42 Colebrook Rd, Wick, Littlehampton, BN17 7NU [IO90RT, TQ00]
G7 CCZ Details withheld at licensee's request by SSL.
G7 CDA Details withheld at licensee's request by SSL.
G7 CDH G M Finch, 36 Admers Wood, Vigo Village, Meopham, Gravesend, DA13 0SP [JO01EH, TQ66]
GM7 CDI P R Matthews, Finningley House, 71 The Avenue, Girvan, KA26 9DT [IO75NF, NX19]
G7 CDK A J Florence, Apple Tree Cottage, 23 North End, Meldreth, Royston, SG8 6NR [JO02AC, TL34]
G7 CDO A A Corps, 27 Watlington Rd, South Benfleet, Benfleet, SS7 5DS [JO01GN, TQ78]
G7 CDP R Keene, Sweetnap, 4 Blackberry Ln, Four Marks, Alton, GU34 5BN [IO91LC, SU63]
G7 CDU G E Prince, 75 Queens Ave, Bromley Cross, Bolton, BL7 9BJ [IO83TO, SD71]
G7 CDX Details withheld at licensee's request by SSL.
G7 CDZ Details withheld at licensee's request by SSL.
G7 CEB Details withheld at licensee's request by SSL.
G7 CEC J L Evans, 55 Aylestone Dr, Leicester, LE2 8QE [IO92KO, SK50]
G7 CED K V Barnett, 126 Oldham St., Latchford, Warrington, WA4 1EX [IO83RJ, SJ68]
G7 CEH H G Bampton, 50 Onslow Gdns, Sanderstead, South Croydon, CR2 9AT [IO91XI, TQ36]
GW7 CEI D G Davies, 36 Millfield Dr, Cowbridge, CF7 7BR [IO81HN, ST08]
G7 CEK D Grevatt, 7 Shelley Rd, East Grinstead, RH19 1SX [IO91XD, TQ33]
G7 CEL G Packham, 9 Locksley Cl, Walderslade, Chatham, ME5 9BT [JO01GI, TQ76]
G7 CEM G M Lewis, 108 Tottington Rd, Bury, BL8 1LR [IO83UO, SD71]
G7 CEN T B Martin, 248 Camp Hill Rd, Nuneaton, CV10 0JN [IO92FM, SP39]
GW7 CEP D Lewis, Woolloomooloo, 19 Gr Park Ave, Rhyl, LL18 3RG [IO83GH, SJ08]
GW7 CEQ W J Jones, 26 Cwm Silyn, Hendre Park, Caernarvon, LL55 2AG [IO73UD, SH46]
G7 CER C E Riley-Moxon, 51 Nuttall Ln, Ramsbottom, Bury, BL0 9JX [IO83UP, SD71]
G7 CEU P N Bushell, Well Ln, Neston, Ness, South Wirral, Merseyside, L64 4AN [IO83LG, SJ27]
G7 CEV S M Chambers, 9La Tourne Gdns, Orpington, Kent, BR6 8EJ [JO01AI, TQ46]
G7 CEW D Everard, 6 Leith Hill Green, St. Pauls Cray, Orpington, BR5 2SB [JO01BJ, TQ46]
G7 CFC A J Chamberlain, Acl (G), Bfpo 39, Germany
G7 CFF R F C Wells, 44 Basing Rd, Mill End, Rickmansworth, WD3 2QJ [IO91SP, TQ09]
G7 CFG R J T Coakes, 50 Main St., Kirkby Lonsdale, Carnforth, LA6 2AJ [IO84QE, SD67]
G7 CFJ Details withheld at licensee's request by SSL.
G7 CFM Details withheld at licensee's request by SSL.
G7 CFO Details withheld at licensee's request by SSL.
G7 CFP L W Reilly, 20 Tewit Well Gdns, Tewit Well Rd, Harrogate, HG2 8JG [IO93FX, SE35]
G7 CFR Details withheld at licensee's request by SSL.
G7 CFT G W Taylor, 48 Westwood Heath Rd, Leek, ST13 8LL [IO83XC, SJ95]
G7 CFU Details withheld at licensee's request by SSL.
G7 CFV Details withheld at licensee's request by SSL.
G7 CFW K E Morrison, 7 Turnstone End, Yateley, GU46 6PE [IO91NI, SU86]
G7 CFX Details withheld at licensee's request by SSL.[Op: A Newell, 87 Tintagel Close, Andover, Hants, SP10 4DB.]
G7 CFY Details withheld at licensee's request by SSL.
G7 CGB Details withheld at licensee's request by SSL.
G7 CGC P D Oliver, 52 Epping Way, Witham, CM8 1NQ [JO01HT, TL81]
G7 CGG Details withheld at licensee's request by SSL.
G7 CGI Details withheld at licensee's request by SSL.
G7 CGK Details withheld at licensee's request by SSL.
G7 CGO S Horton, 64 Hawthorne Rd, Little Sutton, South Wirral, L66 1PU [IO83MG, SJ37]
G7 CGS S R Webb, Astrid House, The Green, Swinton, Malton, North Yorks, YO17 0SY [IO94ND, SE77]
G7 CGT A M Day, 13 Moore Cl, Tongham, Farnham, GU10 1YZ [IO91PF, SU84]
G7 CGV C D S Horsfield, 1 Vale View Pl, Claremont Rd, Bath, BA1 6QW [IO81TJ, ST76]
G7 CGW K Ireson, 5 Mannings Rise, Rushden, NN10 0LY [IO92QG, SP96]
G7 CHB R H P Montague, 142 Davenport Dr, Cleethorpes, DN35 9NJ [IO93XN, TA20]
G7 CHH P N Finegan, 116 Parkside Rd, Bebington, Wirral, L63 7NS [IO83LI, SJ38]
G7 CHJ P K Adams, 25 Abingdon Garth, Bransholme, Hull, HU7 4LA [IO93UT, TA13]
G7 CHN L P Evans, 184 West St., Dunstable, LU6 1NX [IO91RV, TL02]
G7 CHO Details withheld at licensee's request by SSL.
G7 CHT Details withheld at licensee's request by SSL.
G7 CHW A F Hartland Mid Severn Valley Raynet, 22 Granville Crest, Kidderminster, DY10 3QS [IO82VJ, SO87]
GM7 CHX L Hall, 5 Thistle Ct, Virkie, Shetland, ZE3 9JZ [IO99IV, HU31]
G7 CIB Details withheld at licensee's request by SSL.
G7 CIC Details withheld at licensee's request by SSL.
G7 CIE Details withheld at licensee's request by SSL.
G7 CII J N Burgess, Zarcrest, 5 Church Meadow, Lilleshall, Newport, TF10 9HD [IO82TR, SJ71]
G7 CIK M J Kielthy, 82 Dellfield Cres, Cowley, Uxbridge, UB8 2EU [IO91SM, TQ08]
G7 CIQ A Wiseman, 61 Hilton Ave, Horwich, Bolton, BL6 5RH [IO83RO, SD61]
G7 CIT T J McGuigan, 16 Earlswood Ave, Lowfell, Gateshead, NE9 6AH [IO94EW, NZ25]
G7 CIU P R Burbury, Holly House, 15 West Rd, Melsonby, Richmond, DL10 5ND [IO94DL, NZ10]
G7 CIV B W W Perrin, Glebe Cottage, 8 Station Rd, Gretton, Corby, NN17 3BU [IO92PM, SP89]
G7 CIY K I Gaunt, 79 Cobbold Rd, Woodbridge, IP12 1HA [JO02PC, TM25]
G7 CJC J D Hughes, The Hollows, Prescott, Cleobury Mortimer, Salop, DY14 8RR [IO82SK, SO68]
G7 CJD C J Dyer, 6 Witcombe, Yate, Bristol, BS17 4SA [IO81SM, ST78]
G7 CJG G M Stringer, 13 Garfield St., Kettering, NN15 7HX [IO92PJ, SP87]
G7 CJJ G J Doyle, 15 Whitby Cl, Biggin Hill, Westerham, TN16 3NX [JO01AH, TQ45]
G7 CJO J A Graves, 26 Ratton Dr, Eastbourne, BN20 9BS [JO00DS, TQ50]
G7 CJS D C Evans, 16 Cruden Rd, Gravesend, DA12 4HD [JO01EK, TQ67]
G7 CJW J V Wardle, 9 Leefield Rd, Chapel En-le-Frith, Stockport, Ches, SK12 6LF [IO83XI, SJ98]
G7 CKK Details withheld at licensee's request by SSL.
G7 CKL B J Taylor, 32 Marples Ave, Mansfield Woodhouse, Mansfield, NG19 9HA [IO93JE, SK56]
G7 CKP J Surman, 74 Colwell Dr, Burwell Farm Est, Witney, OX8 7NQ [IO91GS, SP30]
GW7 CKR R S Abramczyk, 27 Gr Pl, Griffithstown, Pontypool, NP4 5DH [IO81LQ, ST29]
G7 CKS D P Davies, 2 London Rd, Battle, TN33 0EU [JO00FW, TQ71]
G7 CKV J D Smith, 17 Marshall St., Leicester, LE3 5FA [IO92KP, SK50]
G7 CKX M N Steele, 4 Royville Pl, Stoke on Trent, Staffs, ST6 1RP [IO83WB, SJ84]
G7 CLG P J Ord, 52 Hillside Rd, Norton, Stockton on Tees, TS20 1JQ [IO94IO, NZ42]
G7 CLH D L Smith, 7 Dunlin Cl, Norton, Stockton on Tees, TS20 1SJ [IO94IO, NZ42]
G7 CLO C A Gaukroger, Meadow View, Lansallos, Looe, Cornwall, PL13 2PU [IO70RI, SX15]
G7 CLQ N Hall, 10 Moulton Gr, Westwood, Peterborough, PE3 7JG [IO92UO, TF10]
G7 CLR F N Charnley, 30 Dunkirk Ave, Fulwood, Preston, PR2 3RY [IO83PS, SD53]
G7 CLX K Marsden, 3 Ln Head, Heptonstall, Hebden Bridge, HX7 7PB [IO83XR, SD92]
G7 CLY J Hill, 55 The Oval, Welton Rd, Brough, HU15 1DA [IO93RR, SE92]
G7 CLZ Details withheld at licensee's request by SSL.
G7 CMB D A Kennett, 18 Bramble Rd, Hatfield, AL10 9SA [IO91US, TL20]
GI7 CMC N E Moore, 164 Ardenlee Ave, Belfast, BT6 0AE [IO74BN, J37]
GW7 CMF R P Jones, 3 Bryn Pandy, Llangefni, LL77 7NT [IO73UG, SH47]
GU7 CMH G Simon, 3 Mahaut Villas, Collings Rd, St. Peter Port, Guernsey, GY1 1FP
G7 CMI Prof B J Birch, 14 Duchy Cl, Chelveston, Wellingborough, NN9 6AW [IO92RH, SP96]
G7 CMJ Details withheld at licensee's request by SSL.
GW7 CMM G P Jones, 13 Palace Cl, Flint, CH6 5YE [IO83KF, SJ27]
G7 CMN B M Williams Cheshire County Raynet, 3 Welton Cl, Wilmslow, SK9 6HD [IO83VH, SJ87]
G7 CMP B A Fielding, The Copse, Charmouth Rd, Raymonds Hill, Axminster, EX13 5SZ [IO80MS, SY39]
G7 CMQ D R Stevens, 43 Spring Ln, Bottisham, Cambridge, CB5 9BL [JO02DF, TL56]
GM7 CMR M Murphy, 21 Earlston Way, Macedonia, Glenrothes, KY6 1JJ [IO86JE, NO20]
G7 CMS Details withheld at licensee's request by SSL.
G7 CMU B M Jones, Delana, Lutton Cornwood, Ivybridge, Devon, PL21 9SL [IO80AJ, SX55]
G7 CMX A E Smith, 34 Briery Rd, Halesowen, B63 1AT [IO82XK, SO98]
G7 CNC D Gray, 3 Lodge Gdns, Plymouth, Devon, PL6 5DP [IO70WJ, SX45]
G7 CND F J Bell, Coturnix House, Rake Ln, Milford, Godalming, GU8 5AB [IO91QD, SU94]
G7 CNH Details withheld at licensee's request by SSL.
GU7 CNI M W Elliston, La Guillard Ln, St. Andrews, St. Andrew, Guernsey, Channel Islands, GY6 8YJ
G7 CNP H S Weatherhead, 5 Hanks Cl, Malmesbury, SN16 9UA [IO81WO, ST98]
G7 CNQ L Williams, 4 Ridge Cres, Hawk Green Marple, Marple, Stockport, SK6 7JA [IO83XJ, SJ98]
GI7 CNS E M Clarke, 18 Drumsnade Rd, Ballynahinch, BT24 8NG [IO74BI, J34]
GI7 CNT H J Clarke, 18 Drumsnade Rd, Ballynahinch, BT24 8NG [IO74BI, J34]
GM7 CNW G Dryburgh, 86 Normand Rd, Dysart, Kirkcaldy, KY1 2XP [IO86KD, NT39]
G7 CNX P F Hammond, 26 Hawthorn Rd, Hoddesdon, EN11 0PJ [IO91XS, TL30]
G7 CNZ K Ford, 8 Blakedon Rd, Wednesbury, WS10 7HY [IO82XN, SO99]
G7 COA R K Johnson, 7 West Parade, Warminster, BA12 8LY [IO81VE, ST84]
GW7 COB S J Coburn, 54 Queensway, Hope, Wrexham, LL12 9PE [IO83LC, SJ35]
G7 COC N R Whelan, 47 Romanby Rd, Northallerton, DL7 8NG [IO94GI, SE39]
G7 COD A Kitchen, 213 Melton High St., Wath on Dearne, Wath upon Dearne, Rotherham, S63 6RQ [IO93HM, SE40]
G7 CON D E Gibbs, 83 Wookey Hole Rd, Wells, BA5 2NH [IO81QF, ST54]
G7 COP P J Payton, 3 Astor Cl, Winnersh, Wokingham, RG41 5JZ [IO91NK, SU77]
G7 COQ K J Raxworthy, 9 Harrow Dr, Edmonton, London, N9 9EQ [IO91XP, TQ39]
G7 COT K H Sparks, 14 Felden Cl, Garston, Watford, WD2 6QW [IO91TQ, TL10]
G7 COU R C Stormes, 31 Farndale Rd, Knaresborough, HG5 0NY [IO94GA, SE35]
G7 COV Details withheld at licensee's request by SSL.
G7 COW Details withheld at licensee's request by SSL.
G7 COY Details withheld at licensee's request by SSL.
G7 COZ Details withheld at licensee's request by SSL.
G7 CPB Details withheld at licensee's request by SSL.
G7 CPC Details withheld at licensee's request by SSL.
GM7 CPJ G A L Currie, The Old Post Office, Garmond, Cuminestown, Aberdeenshire, AB53 7TQ [IO87TM, NJ74]
GM7 CPL C B Scott, 115 Tarvit Terr, Springfield, Cupar, KY15 5SE [IO86LH, NO31]
GM7 CPM D A Inglis, 6 Durham St., Monifieth, Dundee, DD5 4PG [IO86OL, NO43]
G7 CPN S J Burgess, 27 Freemantle St., Edgeley, Stockport, SK3 9LF [IO83VJ, SJ88]
G7 CPQ C R Ambrose, 47 Whitton Cl, Swavesey, Cambridge, CB4 5RT [JO02AH, TL36]
GM7 CPR J A Wright, 9 Meadowhead Rd, Plains By, Plains, Airdrie, ML6 7JF [IO85AV, NS76]
GM7 CPY S Leggat, Ailach, St. Aethans Rd, Burghead, Elgin, IV30 2YR [IO87GQ, NJ16]
G7 CQA R Ginn, 91 High St., Shoeburyness, Southend on Sea, SS3 9AR [JO01JM, TQ98]
G7 CQB D G Locock, 1 Rose Cttgs, Station Rd, Styal, Wilmslow, Ches, SK9 4JW [IO83VI, SJ88]

G7 CQD R C Perry, Westwinds, Bredenbury, Bromyard, Herefordshire, HR7 4TF [IO82RE, SO65]
G7 CQI P G Fieldhouse, 60 South Coast Rd, Peacehaven, E Sussex, BN10 8SH [JO00AS, TQ40]
G7 CQK Capt I R Phillips, Goldsworthy Farm, Calstock Rd, Gunnislake, PL18 9BX [IO70VM, SX47]
GU7 CQN J P Gardner, Wykeham, Richmond Ave, St. Peter Port, Guernsey, GY1 1QQ
GM7 CQQ A Donaldson, 36 Rothes Park, Leslie, Glenrothes, KY6 3LH [IO86JE, NO20]
G7 CQT C J Finnegan, 25 Westcliff Gdns, Margate, CT9 5DT [JO01QJ, TR37]
G7 CQW A B Riley, 4 Birtle Dr, Astley, Tyldesley, Manchester, M29 7RE [IO83SM, SD70]
G7 CQX D W Read, 31 Grace Gdns, Bishops Stortford, CM23 3EU [JO01BU, TL41]
G7 CQZ E G Courtnell, 12 Bray St., Birkenhead, L41 8BX [IO83LJ, SJ38]
G7 CRA I C Brelsford, 38 Woodside Ave, Burnholme, York, YO3 0QS [IO93LX, SE65]
G7 CRG P P Fox Central Cheshire Raynet Group, 5 Llandovery Cl, Winsford, CW7 1NA [IO83RE, SJ66]
G7 CRK S P Halbertsma, 65 Gareth Gr, Bromley, BR1 5EG [JO01AK, TQ47]
G7 CRM D R Stump, 9 Shipton Gr, Walcot West, Swindon, SN3 1BZ [IO91CN, SU18]
G7 CRN R F Phin, 35 Parkland Cl, Newquay, Cornwall, TR7 3EB [IO70LK, SW86]
G7 CRO Details withheld at licensee's request by SSL.
G7 CRQ A Haywood, Flat 21, 178 Woodcote Rd, Wallington, SM6 0PB [IO91WI, TQ26]
G7 CRR J T Wheeler, 8 Slimbridge Cl, Greenhills, Worcester, WR5 3SH [IO82VE, SO85]
G7 CRS R E S Evans Maxpac Ru-Group, 6 Park End, Lichfield, WS14 9US [IO92CQ, SK10]
G7 CRY J S Bain, 28 Christchurch Rd, Southend on Sea, SS2 4JS [JO01IM, TQ88]
G7 CSE Details withheld at licensee's request by SSL.
G7 CSI M Morris, 20 Bracken Way, Chobham, Woking, GU24 8PR [IO91QI, SU96]
G7 CSJ K D Chapman, 19 St. Johns Rise, Woking, GU21 1PN [IO91RH, SU95]
GW7 CSK D J Winter, 25 Pembroke St., Thomastown, Tonyrefail, Porth, CF39 8DU [IO81GN, ST08]
G7 CSL R J Dent, 70 Dart Cl, St. Ives, Huntingdon, PE17 6JB [IO92XI, TL37]
G7 CSP G D T Brean, 83 Meadow Ln, Burgess Hill, RH15 9JD [IO90WW, TQ31]
G7 CSS M A Budd, 20 The Grove, St. Margarets Rd, Twickenham, TW1 1RB [IO91UK, TQ17]
G7 CSV T J Clough Spen Valley ARS, 37 Park Ave, Mirfield, WF14 9PB [IO93DQ, SE21]
G7 CSX A M Keen, 20 Horam Park Cl, Horam, Heathfield, TN21 0HW [JO00CW, TQ51]
G7 CSY B J Smallshaw, 58 Loch Ave, Nairn, IV12 4TF [IO87BN, NH85]
G7 CSZ P A Snart, 22 Ringwood, Bretton, Peterborough, PE3 9SH [IO92UN, TL19]
G7 CTA Details withheld at licensee's request by SSL.
G7 CTE N C B Farmer, 8 Mill Hill Ln, Sandbach, CW11 4PN [IO83TD, SJ75]
G7 CTG E Ives, 9 Northlands, Lutterworth Dr, Adwick-le-St., Doncaster, DN6 7AX [IO93JN, SE50]
G7 CTH R J Benson, 11 Redruth Dr, Carnforth, LA5 9TT [IO84OC, SD46]
GM7 CTI M L Durrant, 65 Kepplehills Rd, Bucksburn, Aberdeen, AB2 9DN [IO87TH, NJ72]
G7 CTN M W Puddephatt, 9 Pine Cl, Irchester, Wellingborough, NN29 7BY [IO92QG, SP96]
G7 CTP P J Eavis, 33 Welldon Cres, Harrow, HA1 1QP [IO91TO, TQ18]
G7 CTT S Lock, 82 Picton Rd, Rhoose, Barry, CF62 3HU [IO81HJ, ST06]
GM7 CTV A Smith, 13 Park Terr, Markinch, Glenrothes, KY7 6BN [IO86KE, NO20]
GI7 CTW E Regan, 4 Lecumpher Rd, Derry, Desertmartin, Magherafelt, BT45 5LY [IO64PS, H89]
G7 CTX T M Wears, 7 Sullington Way, Shoreham By Sea, BN43 6PJ [IO90UU, TQ20]
G7 CUA R I Cookson, 7 Birkdale Ave, Longton, Preston, PR4 5ZH [IO83OR, SD42]
G7 CUB D C Price, Mount House School, Mount Tavy Rd, Tavistock, Devon, PL19 9JL [IO70WN, SX47]
G7 CUD Details withheld at licensee's request by SSL.
G7 CUE Details withheld at licensee's request by SSL.
G7 CUF Details withheld at licensee's request by SSL.
G7 CUG Details withheld at licensee's request by SSL.
G7 CUH Details withheld at licensee's request by SSL.
G7 CUI Details withheld at licensee's request by SSL.
G7 CUL R J Bourn, 7 Clitheroes Ln, Freckleton, Preston, PR4 1SD [IO83NS, SD42]
G7 CUP P D I Ingle, 12 East Elloe Ave, Holbeach, Spalding, PE12 7NB [JO02AT, TF32]
G7 CUT B L Myatt, 18 Church Ln, Arlesey, SG15 6UL [IO92UA, TL13]
G7 CVA E Curnow, 9 Moreton Bay, Fleet Est, Bilton, Hull, HU11 4ER [IO93VS, TA13]
G7 CVC M S Porter, 1 Larchdale Gr, Liverpool, L9 2BB [IO83MK, SJ39]
GW7 CVF S D Evans, 65 Renwick Rd, Liverpool, L9 2DE [IO83ML, SJ39]
G7 CVH Details withheld at licensee's request by SSL.
G7 CVJ Details withheld at licensee's request by SSL.
G7 CVK S E Hack, 25 Relko Gdns, Sutton, SM1 4TJ [IO91VI, TQ26]
G7 CVM D E Illman, 27 Blackborough Rd, Reigate, RH2 7BS [IO91VF, TQ24]
G7 CVS W D Lindseth, 273 Upper Elmers En, Beckenham, Kent, BR3 3QR [IO91XJ, TQ36]
G7 CVX H Nortcliffe, 96 North Rd, Withernsea, HU19 2AY [JO03AR, TA32]
G7 CVY K H Helgesen, 172 New Kent Rd, London, SE1 4YT [IO91WL, TQ37]
G7 CVZ A B Bevins, 43 Victoria St., Burscough, Ormskirk, L40 0SN [IO83NO, SD41]
G7 CWA Details withheld at licensee's request by SSL.
GM7 CWC Details withheld at licensee's request by SSL.
G7 CWE D Ishmael, 38 Greenford Cl, Orrell, Wigan, WN5 8RH [IO83PM, SD50]
G7 CWH Details withheld at licensee's request by SSL.
G7 CWI I J Green, 25 Riley Ave, Lytham St. Annes, FY8 1HZ [IO83LR, SD32]
G7 CWK G S Noble, 2 Hawthorn Ave, Stopsley, Luton, LU2 8AW [IO91TV, TL12]
G7 CWM J W V Denton, The Rowans, 48 Seas End Rd, Surfleet, Spalding, PE11 4DQ [IO92WU, TF22]
G7 CWN F V Merchant, 3 Cottington Ct, Hanham, Bristol, BS15 3SJ [IO81RK, ST67]
G7 CWQ Details withheld at licensee's request by SSL.
G7 CWR D Cordwell, 6 Carr Hill Gr, Calverley, Pudsey, LS28 5QB [IO93DT, SE23]
G7 CWV Details withheld at licensee's request by SSL.
G7 CWX Details withheld at licensee's request by SSL.
G7 CWY Details withheld at licensee's request by SSL.
G7 CWZ N J Kimber, The Walled Garden, Back Ln, Malham, West Yorks, BD23 6LL [IO84XA, SD95]
G7 CXA A L Miles, 1 Pond Cross Cttgs, London Rd, Newport, Saffron Walden, CB11 3PT [JO01CX, TL53]
G7 CXB T A McInnes, 7 Hilary Dr, Merry Hill, Wolverhampton, WV3 7NJ [IO82WN, SO89]
G7 CXK K Brown, 8 Station Rd, Arlesey, SG15 6RG [IO92UA, TL13]
G7 CXO V Cassar, 51 Aylesford Ave, Beckenham, BR3 3SB [IO91XJ, TQ36]
GM7 CXP Details withheld at licensee's request by SSL.
G7 CXT S G Haywood, 30 Bull Rd, Stratford, London, E15 3HQ [JO01AM, TQ38]
G7 CXU S M Power, 8 Green Ln, Chislehurst, BR7 6AG [JO01AK, TQ47]
G7 CXV R J Elliot, 8 Maple Cres, Newbury, RG14 1LL [IO91IJ, SU46]
G7 CXZ M J Dawkins, 16 Westwood Rd, Bridgwater, TA6 4HJ [IO81MD, ST33]
G7 CYA Details withheld at licensee's request by SSL.
G7 CYC M J Bulcock, 8 Coles Hill Ave, Burnley, Lancs, BB10 4LH [IO83VS, SD83]
G7 CYD A T Jenkins, 15 Tilstone Ave, Eton Wick, Windsor, SL4 6NF [IO91QL, SU97]
G7 CYF C J Wardill, 3 The Cres, Alvaston, Derby, DE24 0AD [IO92GV, SK33]
G7 CYM P A Cowburn, 125 Slater Ln, Leyland, Preston, PR5 3SE [IO83PQ, SD52]
G7 CYO Details withheld at licensee's request by SSL.
G7 CYP Details withheld at licensee's request by SSL.
G7 CYQ E J Hornby, 14 Essex Rd, Stevenage, SG1 3EZ [IO91VV, TL22]
G7 CYZ A Colclough, 10 The Holdings, Oxford St., Church Gresley, Swadlincote, DE11 9NS [IO92FS, SK21]
G7 CZA T W Wilson, 5 Rosefield Ave, Bebington, Wirral, L63 5JN [IO83LI, SJ38]
GM7 CZC R W Johnson, 3 Hamilton Gdns, Edinburgh, EH15 1NH [IO85KW, NT27]
G7 CZL G P Miles, 7 Dobbin Cl, Rawtenstall, Rossendale, BB4 7TH [IO83UQ, SD82]
G7 CZP Details withheld at licensee's request by SSL.
G7 CZT Details withheld at licensee's request by SSL.
GM7 CZU J K McLaughlan, 2 Donaldson Dr, Irvine, KA12 0QG [IO75QO, NS33]
G7 CZV S J Metcalfe, 69 Bourn Lea, Houghton-le-Spring, DH4 4PF [IO94GU, NZ35]
G7 CZW W G Jones, Sylvan, 34 Fairfield Way, Totland Bay, PO39 0EF [IO90FQ, SZ38]
G7 CZZ Details withheld at licensee's request by SSL.
G7 DAA Details withheld at licensee's request by SSL.
G7 DAC A H A D Surooprajally, 26 Walton Ave, North Cheam, Sutton, SM3 9UB [IO91VI, TQ26]
GW7 DAF Details withheld at licensee's request by SSL.
G7 DAH C R Moore, 168 Church End Ln, Runwell, Wickford, SS11 7DN [JO01GO, TQ79]
G7 DAL J S Ratigan, 81 Cunningham Dr, Unsworth, Bury, BL9 8PD [IO83UN, SD80]
GM7 DAP A Lord, 5 Windsor Terr, Fern By, Brechin, DD9 6SD [IO86OR, NO56]
G7 DAR R L Charlton, 13 Hollywood Ave, Walkerville, Newcastle upon Tyne, NE6 4TN [IO94FX, NZ26]
G7 DAY R S Billups, 5 Woodville Pl, Caterham, CR3 5NY [IO91WG, TQ35]
G7 DBA Details withheld at licensee's request by SSL.
G7 DBC G W Yoxall, 38 Jubilee Cl, Pamber Heath, Tadley, RG26 3HP [IO91KI, SU66]
G7 DBD D G Cooper, 88 Thealby Gdns, Bessacarr, Doncaster, DN4 7EG [IO93KM, SE60]
G7 DBE Details withheld at licensee's request by SSL.
G7 DBH Details withheld at licensee's request by SSL.
G7 DBJ Details withheld at licensee's request by SSL.
GM7 DBK Details withheld at licensee's request by SSL.
G7 DBL Details withheld at licensee's request by SSL.
G7 DBM Details withheld at licensee's request by SSL.
GI7 DBN E N Turner, 6 Park Ln, Hillsborough, BT26 6AQ [IO64XL, J25]
G7 DBO I C Leaver, 2 Marnhull Cl, Coventry, CV2 2JS [IO92GJ, SP37]
G7 DBQ Details withheld at licensee's request by SSL.
G7 DBR C R Ewing, 12 Linford Cl, Handsacre, Rugeley, WS15 4EF [IO92BR, SK01]
G7 DBS Details withheld at licensee's request by SSL.
G7 DBT R D Claridge, 3 Wentworth Ave, Leagrave, Luton, LU4 9EN [IO91SV, TL02]
G7 DBV D Ritson, 8 Maritime Cres, Horden, Peterlee, SR8 3SU [IO94IS, NZ44]
GI7 DBZ W R Hollinger, 51 Collin Rd, Ballyclare, BT39 9JS [IO64XS, J29]
G7 DCJ S P Warner, 96 Walter Nash Rd East, Birchencoppice, Kidderminster, DY11 7BY [IO82UI, SO87]
G7 DCK S M Kirk, 4 Melville Gdns, Woodhouse, Leeds, LS6 2TR [IO93FT, SE23]
G7 DCL Details withheld at licensee's request by SSL.
G7 DCM L F Bant, 41 Newburn Croft, Quinton, Birmingham, B32 1QU [IO82XK, SO98]
G7 DCO Details withheld at licensee's request by SSL.
G7 DCQ P J B Aykroyd, 32 Coach Rd, Astley, Tyldesley, Manchester, M29 7ER [IO83SM, SD70]
G7 DCT A Horsfall, 2 Temple Walk, Halton, Leeds, LS15 7SQ [IO93GT, SE33]
G7 DDD P J Scott, Brixmis, 4 Prospect Ln, Solihull, B91 1HJ [IO92CK, SP18]
G7 DDF P E Johnson, 139 St.Nicolas Pk D, Nuneaton, Warks, CV11 6EF [IO92GM, SP39]
G7 DDI M J Ferry, 29 Mill View Ave, Fulwell, Sunderland, SR6 9HU [IO94HW, NZ35]
G7 DDK Details withheld at licensee's request by SSL.
G7 DDQ W G Roberts, 12 Camberley Dr, Penn, Wolverhampton, WV4 5RP [IO82WN, SO99]
G7 DDR R H Murray, Woodcroft, 8 Church Ln, Kirk Langley, Ashbourne, DE6 4NG [IO92FW, SK23]
G7 DDY Details withheld at licensee's request by SSL.
G7 DEC A R Grundy, 647 Preston Old Rd, Feniscowles, Blackburn, BB2 5ER [IO83RR, SD62]
G7 DEE K J Denniss, 125 Newstead St., Hull, HU5 3NF [IO93TS, TA02]
G7 DEH H F Carpenter, 44 Bowbridge Rd, Newark, NG24 4BZ [IO93OB, SK85]
G7 DEJ Details withheld at licensee's request by SSL.
G7 DEL Details withheld at licensee's request by SSL.
G7 DEM Details withheld at licensee's request by SSL.
G7 DEN Details withheld at licensee's request by SSL.
G7 DEQ Details withheld at licensee's request by SSL.
G7 DER Details withheld at licensee's request by SSL.
G7 DES Details withheld at licensee's request by SSL.
GW7 DET Details withheld at licensee's request by SSL.
G7 DEU G W Renton, 58 St.Christophersr, Humberston, DN36 4EA [IO93XM, TA30]
G7 DEX C Foster, 13 Breck Bank, New Ollerton, Newark, NG22 9XQ [IO93LE, SK66]
G7 DEY P J Knowles, 35 Raby Park Rd, Neston, South Wirral, L64 9SW [IO83LH, SJ27]
G7 DFC J Worsnop, 35 Westwood Ave, Eccleshill, Bradford, BD2 2NJ [IO93DT, SE13]
G7 DFD A V Scovell, 30 Station St., Ryde, PO33 2QH [IO90KR, SZ59]
G7 DFE E T Hill, 8 Rockdown Ct, Swindon, SN2 5BU [IO91CO, SU18]
G7 DFF A A Waring, 29 Ballogie Ave, Neasden, London, NW10 1SU [IO91UN, TQ28]
GM7 DFI K R Trinder, 29 Woodside Rd, Brookfield, Johnstone, PA5 8UB [IO75RU, NS46]
G7 DFM Details withheld at licensee's request by SSL.
G7 DFO Details withheld at licensee's request by SSL.
G7 DFP J T Fitz Patrick, 22 Angel Rd, Norwich, NR3 3HP [JO02PP, TG20]
G7 DFQ B P Doyle, 23 Carlton Ave, Tunstall, Stoke on Trent, ST6 7HR [IO83VB, SJ85]
G7 DFS R Doran DFS ARS, Cont Rm Fire Serv H, Burton Rd Littleove, Derby, DE3 6EH [IO92FV, SK33]
G7 DFV G M Jelley, 28 Blanches Rd, Partridge Green, Horsham, RH13 8HZ [IO91UB, TQ12]
G7 DFW A D Crisp, 17 Gaitskell House, Howard Dr, Borehamwood, WD6 2PB [IO91UP, TQ29]
G7 DFX G S Allan, Kent House, 106 Kent Rd, Sheffield, S8 9RL [IO93GI, SK38]
G7 DGC M J Lewis, 21 Woodlands Rd, Ashton under Lyne, OL6 9DU [IO83XM, SD90]
G7 DGE A Biggin, 14 Coultas Ave, Deepcar, Sheffield, S30 5PT [IO93GJ, SK38]
G7 DGF D W Coupe, 22 West St., South Normanton, Alfreton, DE55 2AJ [IO93HC, SK45]
G7 DGH R C Jones, 51 Tennyson Dr, Ormskirk, L39 3PJ [IO83NN, SD40]
GJ7 DGJ J G Poole C.I. Arcg, Cheriton, Manor Park Rd, St. Helier, Jersey, JE2 3GJ
G7 DGK M F Myles, 17 Ashlyn Cl, Bushey, Watford, WD2 2EJ [IO91TP, TQ19]
G7 DGP B Barrass, 7 The Cres, Eastow on The Hill, Easton on The Hill, Stamford, PE9 3LZ [IO92RP, TF00]
G7 DGR Details withheld at licensee's request by SSL.
G7 DGU Details withheld at licensee's request by SSL.
G7 DGW T Fell, 24 Ardmay Gdns, Surbiton, KT6 4SW [IO91UJ, TQ16][Please contact at G7KCR, QTHR.]
G7 DGX Details withheld at licensee's request by SSL.
G7 DGZ Details withheld at licensee's request by SSL.
GM7 DHA K Pugh, 28 Pladda Rd, Saltcoats, KA21 6AQ [IO75OP, NS24]
G7 DHD D H Dyson, 5 Warwick St., Church, Accrington, BB5 4AL [IO83TS, SD72]
GU7 DHI A G Harvey, 2 Old Farm, Petit Bouet, St. Peter Port, Guernsey, GY1 2AH
G7 DHJ M E Harding, 43 Duxbury Rd, Leicester, LE5 3LR [IO92KP, SK60]
G7 DHL Details withheld at licensee's request by SSL.
G7 DHM W H Holt, 20 Lingfield Mount, Moortown, Leeds, West Yorks, LS17 7EP [IO93FU, SE33]
G7 DHP K J Hoon, 14 The Villas, West End, Stoke on Trent, ST4 5AH [IO82VX, SJ84]
G7 DHW D A Neale, Greyhills Barn, Diptford, South Devon, TQ9 7NQ [IO80DJ, SX75]
GW7 DHX B J Gulliver, y Berllan Fach, Port Rd, Wenvoe, Cardiff, CF5 6AB [IO81IK, ST17]
G7 DIB A Finon, Radford House, Hall Ln, Morley St. Botolph, Wymondham, NR18 9TB [JO02MN, TM09]
G7 DIE S G Salmon, 7 Swindon Ave, Marston, Blackpool, FY4 3DX [IO83LT, SD33]
G7 DIG R W Dee, 10 Sanderson St., Coxhoe, Durham, DH6 4DG [IO94FR, NZ33]
GW7 DIL P Davis, The Willows, Park View, Pontnewydd, Cwmbran, NP44 1RB [IO81LP, ST29]
G7 DIO C J Harper, 5660 Oak Gr Dr, Acworth, Ga 30102-1752, USA, X X
G7 DIR A W Brinton, 136 Efford Rd, Plymouth, PL3 6NQ [IO70WJ, SX45]
G7 DIS C M Baker, Moffat House, Church Rd, Broughton Moor, Maryport, CA15 7SS [IO84GQ, NY03]
GI7 DIT D A W Roberts, 6 Plantation Rd, Bangor, BT19 6AF [IO74EP, J58]
G7 DIU D C Leech, 4 Rydal Cl, Huntingdon, PE18 6UF [IO92VI, TL27]
G7 DIW A Saul, 18 Elm Bank Cl, Lillington, Leamington Spa, CV32 6LR [IO92FH, SP36]
G7 DIZ M A Beatrup, 34 Springfield Dr, Halesowen, B62 8EU [IO82XL, SO98]
GW7 DJL T T Davis, 127 Lon Glanyrafon, Vaynor, Newtown, SY16 1QT [IO82HM, SO09]
G7 DJN A M Coates, 19 Bretton Ave, Bolsover, Chesterfield, S44 6XN [IO93IF, SK47]
G7 DJR Details withheld at licensee's request by SSL.
G7 DJT D J Tinley, 2 Rosemount Cl, Loose, Maidstone, ME15 0AJ [JO01GF, TQ75]
G7 DJW Details withheld at licensee's request by SSL.
G7 DJX R Plume, Barracks Ln, Gorefield, Wisbech, Cambs, PE13 4PQ [JO02AQ, TF31]
G7 DJY N Monkman, 71 Church Ln, Normanton, WF6 1HB [IO93GQ, SE32]
G7 DKB D A Simons, 65 Dolphin Ct Rd, Paignton, TQ3 1AB [IO80FK, SX86]
G7 DKE Details withheld at licensee's request by SSL.
GW7 DKI I P Jones, 24 Beech St., Summerhill, Wrexham, LL11 4UF [IO83LB, SJ35]
G7 DKJ Details withheld at licensee's request by SSL.
G7 DKK Details withheld at licensee's request by SSL.
GW7 DKL Details withheld at licensee's request by SSL.
G7 DKM Details withheld at licensee's request by SSL.
G7 DKN Details withheld at licensee's request by SSL.
G7 DKQ Details withheld at licensee's request by SSL.
G7 DKR B O Williams, 16 Mandale Rd, West Howe, Bournemouth, BH11 8ET [IO90BS, SZ09]
G7 DKS T M Smith, 11 Broadway Gdns, Stevensons Cl, Wimborne, BH21 1LS [IO90AT, SZ09]
G7 DKW Details withheld at licensee's request by SSL.
G7 DKX D F Green, 6 Brick Rd, Patrington, Sunk Island, Hull, HU12 0QN [IO93XP, TA22]
G7 DKZ J Stearn, Half Acre, Hatch Green, Hatch Beauchamp, Taunton, TA3 6TN [IO80MX, ST21]
G7 DLC P P McCluskey, 216 Shelbourne Rd, Charminster, Bournemouth, BH8 8RB [IO90BR, SZ09]
G7 DLD R M Hilton, 2 Clarendon St., Bloxwich, Walsall, WS3 2HT [IO82XO, SJ90]
G7 DLE J S Hodges, 26 Wisley Rd, Andover, SP10 3UQ [IO91GE, SU34]
G7 DLF G L Daines, 2 Sperrin Cl, Buckskin, Basingstoke, RG22 5BT [IO91KG, SU65]
G7 DLI G Abbott, 15 Beaconsfield Ct, Haverhill, CB9 8JW [JO02FB, TL64]
G7 DLJ Details withheld at licensee's request by SSL.
G7 DLK Details withheld at licensee's request by SSL.
G7 DLM Details withheld at licensee's request by SSL.
G7 DLO Details withheld at licensee's request by SSL.
G7 DLP Details withheld at licensee's request by SSL.
G7 DLS Details withheld at licensee's request by SSL.
G7 DLU Details withheld at licensee's request by SSL.
G7 DLV Details withheld at licensee's request by SSL.
GM7 DLW Details withheld at licensee's request by SSL.
G7 DLY J M Whitcomb, The Pines, Glenmore Park, Tunbridge Wells, Kent, TN2 5NZ [JO01DC, TQ53]
G7 DMD Details withheld at licensee's request by SSL.
G7 DME R C Gornall, 50 Blackman Ave, Hollington, St. Leonards on Sea, TN38 9EE [JO00GU, TQ71]
G7 DMF Details withheld at licensee's request by SSL.
G7 DMG W A Hetherington, 27 Princess Ave, Knaresborough, HG5 0AW [IO94GA, SE35]
G7 DMH S A Hetherington, 1 Acacia Ave, Lutterworth, LE17 4UR [IO92JL, SP58]
G7 DMJ Details withheld at licensee's request by SSL.
G7 DMK P M Drage, 53 Spencer St., Burton Latimer, Kettering, NN15 5SQ [IO92PI, SP87]
G7 DMM J R Hudson, 1 Johnson Cl, North Luffenham, Oakham, LE15 8LL [IO92OP, SK80]
G7 DMO Details withheld at licensee's request by SSL.
G7 DMP J P Barnes, 26 Fairthorn Rd, Firth Park, Sheffield, S5 6LX [IO93GK, SK39]
G7 DMS M Horsfall, 8 Greenbrook Rd, Burnley, BB12 6NZ [IO83US, SD83]
G7 DMU R W G Page, 1A Laity Rd, Troon, Camborne, TR14 9EL [IO70IE, SW63]
G7 DMX N J Taylor, 5 Miranda Rd, Preston, Paignton, TQ3 1LE [IO80FK, SX86]
G7 DMZ B Knight, 81 Abbotswood Rd, Brockworth, Gloucester, GL3 4PD [IO81WU, SO81]

G7 DNB C D Draycott, Grantchester Cottage, 3 Sycamore Gdns, Dymchurch, Romney Marsh, TN29 0LA [JO01MA, TR12]
G7 DND N Read, Flat 3, The Croft, 32 Church Rd, Great Bookham, Surrey, KT23 3PW [IO91TG, TQ15]
G7 DNE D E Walker, Coombe Gables, Ongar Hill, Addlestone, KT15 1DF [IO91RI, TQ06]
G7 DNF T Haye, Gillswood House, Rotherfield Park, East Tisted/Alton, Hants, GU34 3QL [IO91MC, SU73]
G7 DNG I J Casey, 38 Wordsworth Rd, Salisbury, SP1 3BH [IO91CB, SU13]
GJ7 DNI S C J McAdams, 3 Tyneville Apt's, Tyneville Ln, First Tower, St Helier, JE2 3WD
GJ7 DNJ I G Meade, Lucton Cottage, La Grand Rte Des Sa, Fauvic Grouville, Jersey, C I, JE3 9BA
G7 DNM E J Ashworth, 6 Queens Terr, St. Acksteads, Bacup, OL13 0EQ [IO83VQ, SD82]
G7 DNP A S Patel, 93 Christchurch Ave, Harrow, HA3 8LZ [IO91UO, TQ18]
G7 DNQ S J Howard, Hazeldene, 96 Worlds End Ln, Chelsfield, Orpington, BR6 6AP [JO01BI, TQ46]
G7 DNS T J Lovelock, 18 Litchfield Cl, Clacton on Sea, CO15 3SZ [JO01NT, TM11]
G7 DNV L C Carter, 3 Cleviscroft, Stevenage, SG1 1UJ [IO91VV, TL22]
G7 DNW P C Greenfield, Brooklyn, Lynn Rd, West Rudham, Kings Lynn, Norfolk, PE31 8RW [JO02IT, TF82]
G7 DNX M W Morris, 4 Meadow Brook Rd, Northfield, Birmingham, B31 1NE [IO92AK, SP08]
G7 DNY D A F Light, 9 Grange Gdns, Banstead, SM7 3RF [IO91VH, TQ26]
G7 DOA D S Morris, 8 Coplow Cres, Syston, Leicester, LE7 2JE [IO92LQ, SK61]
G7 DOB R Whittaker, 60 Kings Rd, Rushall, Walsall, WS4 1JB [IO92AO, SK00]
G7 DOE R D Mount, 4 Hermitage Rd, Abingdon, OX14 5RN [IO91IQ, SU49]
G7 DOF K E Mount, 4 Hermitage Rd, Abingdon, OX14 5RN [IO91IQ, SU49]
G7 DOH Details withheld at licensee's request by SSL.
G7 DOI Details withheld at licensee's request by SSL.
G7 DOL B T Osborne Submarine A.R.C, 12 Arminers Cl, Gosport, PO12 2HB [IO90KS, SZ69]
G7 DOO Details withheld at licensee's request by SSL.
G7 DOR R A Bye Dorking&Dist.Rs, 154 Woodland Way, Surrey Hills Park, Boxhill Rd, Tadworth, KT20 7NB [IO91UG, TQ25]
G7 DOS B Smith, 7 School Walk, Chase Terr, Walsall, West Midlands, WS7 8NQ [IO92AQ, SK00]
G7 DOT Details withheld at licensee's request by SSL.
G7 DOW G A Smith, 59 Radipole Ln, Weymouth, DT4 9RR [IO80SO, SY68]
G7 DOY J Baddeley, 22 Scott Rd, Denton, Manchester, M34 6FT [IO83WK, SJ99]
G7 DPB P Bailey, Panache, Bingley Ct, Littlethorpe, Leicester, LE9 5XA [IO92JN, SP59]
G7 DPC M Howorth, 11A Norwood Dr, Torrisholme, Morecambe, LA4 6LT [IO84NB, SD46]
G7 DPE J E Glass, 13 Elora Rd, High Wycombe, HP13 7LL [IO91PP, SU89]
G7 DPF G M Brightman, 5 Meadow Rise, Lacey Green, Princes Risborough, HP27 0QY [IO91OQ, SP80]
GD7 DPG J Wrigley, 20 Fairy Hill Cl, Ballafesson, Port Erin, IM9 6TJ
GJ7 DPH Maj. J A Campbell, Westbourne House, Westbourne Ave, St. Saviour, Jersey, JE2 7TJ
GM7 DPI P F Blacklaw, 1A Dalgleish Rd, Dundee, DD4 7JN [IO86ML, NO43]
G7 DPK P A Jarrett, 254 Church Path, Upper Deal, Deal, CT14 9DD [JO01QF, TR35]
G7 DPN Details withheld at licensee's request by SSL.
GI7 DPP Details withheld at licensee's request by SSL.
G7 DPR F G Overbury, 47 The Maltings, Great Dunmow, Dunmow, CM6 1BY [JO01EU, TL62]
G7 DPU D L Reynolds, 19 Clipstone Cl, Wigston, LE18 3QS [IO92KN, SP69]
G7 DPV A Marks, 7 Saunby Cl, Arnold, Nottingham, Notts, NG5 7LA [IO93KA, SK54]
G7 DPW A Butterworth, 3 Fir Tree Ave, Boothstown, Worsley, Manchester, M28 1LP [IO83TM, SD70]
G7 DPY Details withheld at licensee's request by SSL.
G7 DPZ E Curd, 11 Ashkirk Cl, Waldridge Park, Chester-le-Street, DH2 3HY [IO94EU, NZ25]
G7 DQA J A Hallin, 2A Corporation Rd, Plymouth, PL2 3NT [IO70WJ, SX45]
G7 DQC P R Anderson, 11 Kelly Cl, St. Budeaux, Plymouth, PL5 1DS [IO70VJ, SX45]
G7 DQD Details withheld at licensee's request by SSL.
G7 DQE M J M Meerman, University of Surre, Dept Elec. Eng., Guildford, Surrey, GU2 5XH [IO91QF, SU95]
G7 DQF J G R Munn, Little Downs, Sandgate Ln, Storrington, Pulborough, RH20 3HJ [IO90SW, TQ11]
G7 DQG M A Hind, 43 Bushy Hill Dr, Merrow, Guildford, GU1 2UH [IO91RF, TQ05]
GI7 DQI K H McConnell, 336 Glebe Rd, Carnmoney, Newtownabbey, BT36 6RL [IO74AQ, J38]
G7 DQL P J Perkins, 87 Whittington Ave, Hayes, UB4 0AE [IO91TM, TQ08]
G7 DQM Details withheld at licensee's request by SSL.
G7 DQX C D Lee, 3 Keith Ave, Great Sankey, Warrington, WA5 3NZ [IO83QJ, SJ58]
G7 DQZ D J Keen, Hollytrees, Gr Hill, Hellingly, Hailsham, BN27 4HG [JO00DV, TQ61]
G7 DRC G A Merritt, 2 Uplands Rd, Rowlands Castle, PO9 6BU [IO90MV, SU71]
G7 DRF M J Drabble, 52 Potters Rd, New Barnet, Barnet, EN5 5HN [IO91VP, TQ29]
G7 DRG R A Moseley, 307 Archer Rd, Stevenage, SG1 5HF [IO91VV, TL22]
G7 DRI D C Jones, 54 Hardwick Rd, Streetly, Sutton Coldfield, B74 3DL [IO92BO, SP09]
G7 DRK C M Byrne, 50 Tiller Rd, London, E14 8NN [IO91XL, TQ37]
G7 DRL M L Walsh, 84 Burford, Brookside, Telford, TF3 1LJ [IO82SP, SJ70]
G7 DRM J M Pilling, 22 Templar Way, Selby, YO8 9XH [IO93LS, SE63]
GW7 DRN M S Cull, Englefield, 10 Salisbury Rd, Wrexham, LL13 7AS [IO83MA, SJ34]
G7 DRO W Webber, Springfield Lodge, Broadway, Shipham, Winscombe, BS25 1UE [IO81OH, ST45]
G7 DRQ M Perkins, 1 Tavis Rd, Paignton, TQ3 2PU [IO80FK, SX86]
G7 DRR D J Bellinger, Holly Cottage, Deacons Ln, Hermitage, Thatcham, RG18 9RJ [IO91IK, SU57]
GI7 DRS Details withheld at licensee's request by SSL.
G7 DRT M H Dickinson, 1 Tregaron Ave, Cosham, Portsmouth, PO6 2JU [IO90LU, SU60]
G7 DRU A D Tink, 13 The Wicketts, Filton, Bristol, BS7 0SR [IO81RM, ST57]
GW7 DRX J Stelmasiak, Rhych-Gwyn, Dolforgan View, Kerry, Newtown Powys, SY16 4PZ [IO82IL, SO18]
GM7 DRY S R Graham, 1 Beveridge Cl, Mansfield Park, Mayfield, Dalkeith, EH22 5TP [IO85LU, NT36]
G7 DSA S J Wright, 167 Delamere St., Over, Winsford, CW7 2LY [IO83RE, SJ66]
GU7 DSB P L Blampied, 33 Courtil Bris, St Peter Port, Guernsey, Channel Islands, GY1 1SA
GM7 DSC J D Simpson, 6 Roman Ct, Cleghorn, Lanark, ML11 7RU [IO85DQ, NS94]
G7 DSD J W Blake, The Coach House, Tapton House Rd, Sheffield, South Yorks, S10 5BY [IO93FJ, SK38]
G7 DSG Details withheld at licensee's request by SSL.
G7 DSI Details withheld at licensee's request by SSL.
G7 DSJ Details withheld at licensee's request by SSL.
G7 DSK Details withheld at licensee's request by SSL.
G7 DSN M A Squire, 24 Peter St., Deal, CT14 6DG [JO01QF, TR35]
G7 DSP F A Bell, 77 Westwood St., Accrington, BB5 4BL [IO83TS, SD72]
G7 DSQ R J Roberts, c/o Portavon Marina, Bitton Rd, Keynsham, Bristol, BS18 2DD [IO81SK, ST66]
GM7 DST D S Thomson, 13 Westwood Dr, Westhill, Skene, Westhill, AB32 6WW [IO87UD, NJ80]
G7 DSU C D Tong, 150 Ordnance St., Chatham, ME4 6SE [JO01GJ, TQ76]
G7 DSW J R Dymock, 33 Grosvenor Rd, East Grinstead, RH19 1HS [IO91XD, TQ33]
G7 DSZ P L Gregory, 8 Chestnut Cl, Burgess Hill, RH15 8HN [IO90WX, TQ32]
GJ7 DTA A Lange, -le-Tournesol, Sunnycrest Cl, La Route de Maufant, St Saviour, Jersey, JE2 7HX
GW7 DTB R W Dixon, 19 The Burrows, Porthcawl, CF36 5AJ [IO81DL, SS87]
GM7 DTC J F Arthur, 15 St. Andrews Pl, Alder Gr, Beith, KA15 1JE [IO75QR, NS35]
G7 DTD Details withheld at licensee's request by SSL.
G7 DTG Details withheld at licensee's request by SSL.
G7 DTH Details withheld at licensee's request by SSL.
G7 DTK T M Timms, 16 Claverdon Rd, Mount Nod, Coventry, CV5 7HP [IO92FJ, SP27]
G7 DTL M E Timms, 16 Claverdon Rd, Mount Nod, Coventry, CV5 7HP [IO92FJ, SP27]
G7 DTN Details withheld at licensee's request by SSL.
G7 DTO Details withheld at licensee's request by SSL.
G7 DTR M D Penny, Thirtynine Steps, 13 Newnham Cl, Braintree, CM7 2PR [JO01GV, TL72]
G7 DTS A J Lowe, 33 Dandies Chase, Eastwood, Leigh on Sea, SS9 5RF [JO01HN, TQ88]
G7 DTT A V K Reeves, Pendeen, 13 Maple Way, Leavenheath, Colchester, CO6 4PQ [JO01KX, TL93]
G7 DTV H Partridge, 11 Elm Cl, Norton, Stourbridge, DY8 3JH [IO82VK, SO88]
G7 DTX M J Bellamy, 19 Wickenby Cres, Ermine Est West, Lincoln, LN1 3TJ [IO93RF, SK97]
G7 DTY C W Hartley, 105 Gold St., Wellingborough, NN8 4EQ [IO92PH, SP86]
G7 DUA M A Barnes, 3 Woodcroft Cl, St. Annes, Bristol, BS4 4QP [IO81RK, ST67]
G7 DUB P S Spencer, Kinson, Trevanion Rd, Wadebridge, PL27 7PA [IO70NM, SW97]
G7 DUC B Tonkin, 9 Penhallick Rd, Carn Brea, Redruth, TR15 3YJ [IO70IF, SW64]
GM7 DUG S W Mattravers, 91 Sundrum Pl, Pennyburn, Kilwinning, KA13 6SU [IO75PP, NS24]
G7 DUI C P Teague, 40 Kestrel Park, Ashurst, Skelmersdale, WN8 6TB [IO83ON, SD40]
G7 DUK M R Clarke, 1 Clarence House, Queens Rd, Hersham, Walton on Thames, KT12 5JT [IO91TI, TQ16]
G7 DUL Details withheld at licensee's request by SSL.
G7 DUO Details withheld at licensee's request by SSL.
G7 DUP Details withheld at licensee's request by SSL.
G7 DUR Details withheld at licensee's request by SSL.
G7 DUW Details withheld at licensee's request by SSL.
GJ7 DUX Details withheld at licensee's request by SSL.
GD7 DUZ S P Kelly, 8 PO Box, Castletown, IM99 5TH
GW7 DVB M M Lewis, 26 Tremynoddfa, Carno, Caersws, SY17 5LJ [IO82FN, SN99]
G7 DVC M M Perry, Cavroche, Histons Hill, Codsall, Wolverhampton, WV8 2EY [IO82VO, SJ80]
G7 DVD J W Collier, 37 Galmington Rd, Taunton, TA1 5NL [IO81KA, ST22]
G7 DVG S J White, Whitkirk, Winston, Darlington, Co.Durham, DL2 3RN [IO94CN, NZ11]
G7 DVI D J N Kirkby, Rocksands, East Camps Bay, Downderry, Torpoint, Cornwall, PL11 3LQ [IO70TI, SX35]
G7 DVJ D Maxted, 22 Station Ln, Paignton, TQ4 5AR [IO80FK, SX86]
G7 DVO T S Spearing, 139 Holt Rd, Hellesdon, Norwich, NR6 6UA [JO02PQ, TG21]
G7 DVP M E Stockwell, 92 North Park, Fakenham, NR21 9RH [JO02KU, TF93]
G7 DVY Details withheld at licensee's request by SSL.
G7 DWB Details withheld at licensee's request by SSL.
G7 DWC D W Cotterill, 96 Coleman Rd, Fleckney, Leicester, LE8 8BH [IO92LM, SP69]
G7 DWD Details withheld at licensee's request by SSL.
GW7 DWE W L Brown, 33 Brookfield Ave, Barry, CF63 1EP [IO81JK, ST16]
GI7 DWF J A Murphy, 18 Ogle St., Armagh, BT61 7EN [IO64QI, H84]
G7 DWH E Shaw, 5 Charlock Gr, Heath Hayes, Cannock, WS11 2FR [IO92AQ, SK01]
G7 DWK Details withheld at licensee's request by SSL.
G7 DWM C D Hadjigeorgiou, 26 Priory Gdns, Hampton, TW12 2PZ [IO91TK, TQ17]
G7 DWN A Keen, 79 Cobbold Rd, Woodbridge, IP12 1HA [JO02PC, TM25]
G7 DWO S J Prisk, 86 Wycliffe Gr, Werrington, Peterborough, PE4 5DF [IO92UO, TF10]
G7 DWP R J Tibbit, Ilv Apartdo Aereo 100602, Santafe de Bogota D.C., Colombia
G7 DWQ Details withheld at licensee's request by SSL.
GW7 DWR J Butterfield, 59 Harding Cl, Boverton, Llantwit Major, CF61 1GX [IO81GJ, SS96]
G7 DWV Dr K A Webster, 6 Myrtle Gr, Baffins, Portsmouth, Hants, PO3 6HB [IO90LT, SU60]
G7 DWX M Twells, 40 The Schools, 13 Aston Rd, Shrewsbury, Salop, SY3 7AP [IO82OQ, SJ41]
G7 DXB L Martin, Bayshore, Gilcrux, Carlisle, Cumbria, CA5 2QD [IO84HR, NY13]
G7 DXC J J W Maxwell, Breckland House, 27 Warfield Rd, Bracknell, RG42 2JY [IO91PK, SU87]
G7 DXE A D Todd, Rose Cottage, 22 High St., Westcott, Aylesbury, HP18 0PH [IO91MU, SP71]
G7 DXG A J Squire, 11 Trinity Pl, Deal, CT14 9HH [JO01QF, TR35]
G7 DXK L H Clements, 143 Cherry Garden Ln, Newport, Saffron Walden, CB11 3QW [JO01CX, TL53]
G7 DXL P J Bratt, 4 Compton Dr, Dudley, DY2 7ES [IO82XM, SO98]
G7 DXM I J Bratt, 27 Ryebank Rd, Ketley Grange, Ketley Bank, Telford, TF2 0EF [IO82SQ, SJ61]
G7 DXP Details withheld at licensee's request by SSL.
G7 DXQ K E Salt, 4 Burghwood Rd, Ormesby St. Michael, Ormesby, Great Yarmouth, NR29 3LT [JO02TQ, TG41]
G7 DXS Details withheld at licensee's request by SSL.
GM7 DXT J Macleod, Top Flat Right, 406 Dumbarton Rd, Partick, Glasgow, G11 6SB [IO75UU, NS56]
G7 DXV P J Shepherd, 25 Tomkins Cl, Stanford-le-Hope, SS17 8QU [JO01FM, TQ68]
G7 DXX J A Walker, 121 Park Dr, Upminster, RM14 3AU [JO01DN, TQ58]
G7 DYA I K Finney, 77 Earle Gdns, Kingston upon Thames, KT2 5TB [IO91UK, TQ17]
G7 DYB A T Rudling, 1 St. Anthonys Cl, Ottery St. Mary, EX11 1EN [IO80IR, SY19]
G7 DYD M Green, 59 Brand End Rd, Butterwick, Boston, PE22 0JD [JO02AX, TF34]
G7 DYG Details withheld at licensee's request by SSL.
G7 DYI Details withheld at licensee's request by SSL.
G7 DYJ Details withheld at licensee's request by SSL.
G7 DYL E Gallagher, 9 Brookhouse Dr, Crewe, CW2 6NQ [IO83SB, SJ75]
G7 DYO C Greaves, 71 March St., Burnley, BB12 0BU [IO83VT, SD83]
G7 DYU C N Gilbert, 214 Uxbridge Rd, Harrow Weald, Harrow, HA3 6SW [IO91TO, TQ19]
G7 DYV J Springett, 1 Sycamore Lodge, Paynes Rd, Freemantle, Southampton, SO15 3SE [IO90GV, SU41]
GW7 DZA A A Griffiths, 20 Cross Roads, Holywell, CH8 7UN [IO83JG, SJ17]
GW7 DZB K P Johnson, 34 Eagle Rd, St. Athan, Barry, South Glam, CF62 4NR [IO81GJ, SS96]
GW7 DZC A C Griffiths, 20 Cross Roads, Holywell, CH8 7UN [IO83JG, SJ17]
G7 DZD Dr K M Quinlan, 12 Firs Cl, Hazlemere, High Wycombe, HP15 7TF [IO91PP, SU89]
GI7 DZE S J Thompson, 19 Windsor Heights, Larne, BT40 1UL [IO74CU, D40]
G7 DZG Details withheld at licensee's request by SSL.
G7 DZH J Williamson, 28 Oakwood Rd, Romiley, Stockport, SK6 4DX [IO83WJ, SJ99]
G7 DZI M J Adlam, 7 Bloomfield Rd, Wellsway, Bath, BA2 2AD [IO81TI, ST76]
GM7 DZK J Malone, 8 St. Margarets Cres, Polmont, Falkirk, FK2 0UP [IO85DX, NS97]
G7 DZP Details withheld at licensee's request by SSL.
G7 DZR G C Shand, 5 Bromyard Dr, Chellaston, Derby, DE73 1PF [IO92GV, SK33]
G7 DZX Details withheld at licensee's request by SSL.
G7 DZY B Daniel, Tamar Bay Rd, Freshwater Bay, Freshwater, Isle of Wight, PO40 9QS [IO90FQ, SZ38]
G7 EAG Details withheld at licensee's request by SSL.
G7 EAH Dr P Stewart, 1 Hannah Lodge, Palatine Rd, Manchester, M20 2QH [IO83VK, SJ89]
G7 EAK G D Mulholland, 17 Cleasby Cl, Westlea, Swindon, SN5 7AE [IO91CN, SU18]
G7 EAM J Plumtree, 86 Heatherfield, Astley Bridge, Bolton, BL1 7QF [IO83SO, SD71]
G7 EAN Details withheld at licensee's request by SSL.
G7 EAR Details withheld at licensee's request by SSL.[Correspondence to G A Lamb, obo Echelford Amateur Radio Society, c/o 5 Georgian Close, Staines, Middx, TW18 4NR. Tel: (01784) 456555. Club address: St Martins Court, Kingston Crescent, Ashford, Middx. QSL via RSGB.]
G7 EAT Dr J P Hatfield, 22 Blackhorse Cres, Amersham, HP6 6HP [IO91QQ, SU99]
G7 EBA H C Owen, Woodcote North Dr, Heswall, Merseyside, L60 0BB [IO83KH, SJ28]
G7 EBF P A Langfield, 153 Acre Ln, Derker, Oldham, OL1 4DN [IO83WN, SD90]
G7 EBI M D Evans, 1 Northfield Cottage, Wildhern, Andover, SP11 0JD [IO91GG, SU35]
G7 EBL Details withheld at licensee's request by SSL.
GI7 EBM A J Stewart, 19 Dunleath Ave, Cookstown, BT80 8JA [IO64PP, H87]
G7 EBR R W McLachlan M'Hd & E.Berk Rd, Heathersett, Lightlands Ln, Cookham, Maidenhead, SL6 9DH [IO91PN, SU88]
G7 EBX W Parkin, 40 Cliffe Ave, Saltburn, Carlin How, Saltburn By The Sea, TS13 4DT [IO94NN, NZ71]
G7 EBY Details withheld at licensee's request by SSL.
G7 ECE S T Holmes, 22 Wilton Green, Lazenby, Middlesbrough, TS6 8DP [IO94KN, NZ51]
G7 ECG K Thompson, 17 King St., Carnforth, LA5 9DU [IO84OD, SD47]
G7 ECI M C T Hill, Pilgrims Rest, Church Rd, Offham, West Malling, ME19 5NX [JO01EH, TQ65]
G7 ECK P C Richmond, Glemmtal, 10 Church Fields, Headley, Bordon, Hants, GU35 8PE [IO91OC, SU83]
G7 ECL Details withheld at licensee's request by SSL.
G7 ECN Details withheld at licensee's request by SSL.
G7 ECO R J Carter, The Little House, Main St., Chackmore, Buckingham, MK18 5JF [IO92LA, SP63]
G7 ECQ N E Murray, East End House, Oak Ln, Minster on Sea, Sheerness, ME12 3QR [JO01JJ, TQ97]
G7 ECR J E Belfield N.Ferriby U ARS, 100 Etherington Dr, Hull, HU6 7JT [IO93TS, TA03]
G7 ECV Details withheld at licensee's request by SSL.
G7 ECY E I Bentley, 28 Frost Hole Cres, Fareham, PO15 6AQ [IO90JU, SU50]
G7 EDA R D Woodcock, 5 Walton Rd, Sidcup, DA14 4LJ [JO01BK, TQ47]
G7 EDC Details withheld at licensee's request by SSL.
G7 EDF D R Hall, 68 Common Ln, Tickhill, Doncaster, DN11 9UF [IO93KK, SK59]
G7 EDK I H Barkley, 39 Fulbeck Ave, Goose Green, Wigan, WN3 5QN [IO83QM, SD50]
G7 EDM Details withheld at licensee's request by SSL.
G7 EDQ Details withheld at licensee's request by SSL.
G7 EDT Details withheld at licensee's request by SSL.
G7 EDU K R Hailey, 21 Cedar Gr, Amersham, HP7 9BG [IO91QQ, SU99]
G7 EDX C F Baker, 1 Astley Green, Darley Heights, Luton, LU2 8TS [IO91TV, TL12]
G7 EEA Details withheld at licensee's request by SSL.
G7 EED Details withheld at licensee's request by SSL.
G7 EEE Cpt. A P Mothew, Forest View, 7 Ashfields, Loughton, IG10 1SB [JO01AP, TQ49]
G7 EEG T A Cole, Kinnellan, Longage Hill, Rhodes Minnis, Canterbury, CT4 6XT [JO01MD, TR14]
G7 EEI S H Chell, 56 Gordon St., Agbrigg, Wakefield, WF1 5AT [IO93GP, SE31]
G7 EET Details withheld at licensee's request by SSL.
G7 EEU H W Skinner, 97 Gr Rd, Harpenden, AL5 1ER [IO91TT, TL11]
G7 EEW Details withheld at licensee's request by SSL.
GM7 EEY P A Webster, 16 Abbey Pl, Aberdeen, AB1 9QH [IO87TH, NJ72]
G7 EFC R A England, 24 Glenway, Penwortham, Preston, PR1 9AJ [IO83PR, SD52]
G7 EFD R A Farman, 81 Bushy Cl, Bletchley, Milton Keynes, MK3 6PX [IO92PA, SP83]
G7 EFG M E Marston, 7 Powell St., Harrogate, HG1 4BY [IO94FA, SE35]
G7 EFL A P Crooks, 7 The Cleave, Harpenden, AL5 5SJ [IO91TT, TL11]
G7 EFM R D Nicholson, 75 Portway, Wells, BA5 2BJ [IO81QF, ST54]
G7 EFO Details withheld at licensee's request by SSL.
G7 EFQ Details withheld at licensee's request by SSL.
G7 EFS Details withheld at licensee's request by SSL.
G7 EFV W Waterton, 23 Mill Dr, Leven, Beverley, HU17 5NR [IO93UV, TA14]
G7 EGK Details withheld at licensee's request by SSL.
G7 EGN Details withheld at licensee's request by SSL.
G7 EGQ I J Harrop, 35 Langdale Cres, Dalton in Furness, LA15 8NR [IO84JD, SD27]
G7 EGU P D Adams, 12 The Birches, South Benfleet, Benfleet, SS7 4NT [JO01GN, TQ78]
G7 EGX M J Miller, 20 Tamar Dr, Aveley, South Ockendon, RM15 4NB [JO01DM, TQ58]
G7 EHD D Cotton, 22 Dales Cl, Biddulph Moor, Stoke on Trent, ST8 7LZ [IO83WC, SJ95]
G7 EHN J Baird, 23 Crewton Way, Alvaston, Derby, DE24 8XH [IO92GV, SK33]
G7 EHR R E Watson, 7 Glatton Rd, Sawtry, Huntingdon, PE17 5SY [IO92UK, TL18]
G7 EHS I C Tooley, 18 Hermes Dr, Burnham on Crouch, CM0 8SW [JO01JP, TQ99]
G7 EHT G E Maplestone, 14 Angela Cl, Martlesham, Woodbridge, IP12 4TG [JO02PB, TM24]
G7 EHU B M Walley, 52 Main St., Rosliston, Swadlincote, DE12 8JW [IO92ER, SK21]
G7 EHY D Robinson, 130 Magnolia Dr, Colchester, CO4 3LX [JO01LV, TM02]
G7 EIA S F Ralph, 12 Gladstone Dr, Sittingbourne, ME10 3BH [JO01JI, TQ96]
G7 EIC J R Newby S.A.D.A.R.S., Council Offices, Argyle Rd, Sevenoaks Kent, TN13 1HG [JO01CG, TQ55]
G7 EID M J Jacobs, 43 Winfields, Pitsea, Basildon, SS13 1HA [JO01GN, TQ78]
G7 EIE T J Jacobs, 43 Winfields, Pitsea, Basildon, SS13 1HA [JO01GN, TQ78]
G7 EIF N A Barnes, 10 Parsons Hill, Lexden, Colchester, CO3 4DT [JO01KV, TL92]
G7 EIK G E Edlin, 2 Ashby Rd, Donisthorpe, Burton on Trent, DE15 0LA [IO92ET, SK22]
G7 EIM Details withheld at licensee's request by SSL.

G7 EIP Details withheld at licensee's request by SSL.
G7 EIR Details withheld at licensee's request by SSL.
G7 EIS E I Shaddick, 6 Haylands, Portland, DT5 2JZ [IO80SN, SY67]
G7 EIT C J Fisher, 2 Gammage St., Dudley, West Midlands, DY2 8XL [IO82WM, SO98]
G7 EIX R A Clafton, 18 Burwell Cl, Lower Earley, Reading, RG6 4BB [IO91MK, SU76]
G7 EJF Details withheld at licensee's request by SSL.
G7 EJG Details withheld at licensee's request by SSL.
G7 EJH J W Tyerman, 7 Veronica Cl, Branston, Lincoln, LN4 1PU [IO93SE, TF06]
G7 EJK J W Turner, Moorland House, 2 Byrds Ln, Uttoxeter, ST14 7NU [IO92BV, SK03]
G7 EJN G Prosser, 23 Guiting Rd, Selly Oak, Birmingham, B29 4RD [IO92AK, SP08]
G7 EJO I M Oxley, 25 Meadow View Rd, Newhall, Swadlincote, DE11 0UL [IO92FS, SK22]
G7 EJS Details withheld at licensee's request by SSL.
G7 EKA D K Banks, 10 Berrylands Ct, Blackbush Cl, Sutton, SM2 6BA [IO91VI, TQ26]
G7 EKC A R J Taylor, 24 Pilgrim Cl, Ranby, Retford, DN22 8JU [IO93LH, SK68]
G7 EKD P G Kneebone, 16 Greenheart, Amington, Tamworth, B77 4NG [IO92EP, SK20]
G7 EKI Details withheld at licensee's request by SSL.
G7 EKJ S D Cox, 118 Chipstead Rd, Erdington, Birmingham, B23 5EZ [IO92BM, SP19]
G7 EKM K C McGeough, 57 Stonehouse Park, Thursby, Carlisle, CA5 6NS [IO84LU, NY35]
G7 EKT G L Aucott, 21 Riverway, Wednesbury, WS10 0DN [IO82XN, SO99]
G7 EKU M Barber, 471 Kings Rd, Stretford, Manchester, M32 8QN [IO83UK, SJ89]
G7 EKX R Howard, 13 Ches Cl, Newton-le-Willows, WA12 8PY [IO83QK, SJ59]
G7 EKY N A Bettney, 9 Meadow Cl, Stoney Middleton, Sheffield, S30 1TQ [IO93GJ, SK38]
G7 ELA M Dunn, 45 Chaddock Ln, Boothstown, Worsley, Manchester, M28 1DE [IO83SM, SD70]
G7 ELB P A Legan, 11 Hanley Rd, Sneyd Green, Stoke on Trent, ST1 6BG [IO83VA, SJ84]
G7 ELC N A Fountain, The Venture, Green Ln Upton, Huntingdon, Cambs, PE17 5YE [IO92UJ, TL17]
G7 ELD J M Tresadern, The Steps 18, Cheswardine, Market Drayton, Shropshire, TF9 2RS [IO82SU, SJ72]
G7 ELF C E Dennis, 6 River View, Sturry, Canterbury, CT2 0PD [JO01NH, TR16]
G7 ELG A L Scarisbrick, 51 Mayfield Rd, Eastrea, Whittlesey, Peterborough, PE7 2AY [IO92WN, TL29]
G7 ELI S R Dean, 3 Easton Rd, New Ferry, Wirral, L62 1DR [IO83MI, SJ38]
G7 ELK R N Garth, 44 Lulworth Ave, Whitkirk, Leeds, LS15 8LN [IO93GT, SE33]
G7 ELM Details withheld at licensee's request by SSL.
G7 ELR M Higgin East Lanc Rnt G, 24 Tiverton Dr, Briercliffe, Burnley, BB10 2JT [IO83VT, SD83]
G7 ELS A A Bartram, 47 Temple Gate Cres, Leeds, LS15 0EZ [IO93GT, SE33]
G7 ELV P Lockwood, 25 Percy Ave, Whitley Bay, NE26 3PR [IO95GB, NZ37]
G7 ELX J E Northfield, 23 Hill Ct Dr, Leeds, LS13 2AN [IO93ET, SE23]
G7 ELZ Details withheld at licensee's request by SSL.
G7 EMA Details withheld at licensee's request by SSL.
G7 EMC Details withheld at licensee's request by SSL.
G7 EMD A E Roberts, 248 Valley Rd, Flixton, Urmston, Manchester, M41 8RQ [IO83TK, SJ79]
G7 EME D A Palmer Worcs Mnbounce, 15 Albion Walk, Malvern Link, Malvern, WR14 1PX [IO82UD, SO74]
G7 EMF Details withheld at licensee's request by SSL.
G7 EMH Details withheld at licensee's request by SSL.
G7 EMJ Details withheld at licensee's request by SSL.
G7 EMK Details withheld at licensee's request by SSL.
G7 EML Details withheld at licensee's request by SSL.
GW7 EMV M O'Reilly, 40 St. Anthony Rd, Heath, Cardiff, CF4 4DJ [IO81JM, ST17]
G7 EMZ I Mellor, 124 Ryknield Rd, Kilburn, Belper, DE56 0PF [IO93GA, SK34]
G7 ENA D N Clifton, 4 Breedon Dr, Lincoln, LN1 3XA [IO93RF, SK97]
G7 ENC D M Flatters, 7 Cornwall Cres, Diggle, Oldham, OL3 5PW [IO93AN, SE00]
G7 END A J Court, 22 Parkside Rd, Handsworth Wood, Birmingham, B20 1EL [IO92AM, SP09]
G7 ENI J A Davies, 75 New Rd, Great Wakering, Southend on Sea, SS3 0AR [JO01JN, TQ98]
G7 ENM S C Yates, 37 Flaxpiece Rd, Clay Cross, Chesterfield, S45 9HB [IO93HD, SK36]
G7 ENQ W K Donald, 53 Millfield, Sittingbourne, ME10 4TP [JO01II, TQ96]
G7 ENS N R Swift, 3 Donkin Hill, Caversham, Reading, RG4 5DG [IO91ML, SU77]
G7 ENT S M Alexander, 18 Southbourne Gr, Hockley, SS5 5EE [JO01IO, TQ89]
G7 ENY W A Baldock, 66 Port Rd, New Duston, Northampton, NN5 6NL [IO92MG, SP76]
G7 EOA G D Harrold, Ln Morley, St Boltoph Wymondha, Norfolk, NR18 9DD [JO02MN, TM09]
G7 EOB C G Deal, 52 Sweechgate, Broadoak, Broad Oak, Canterbury, CT2 0QX [JO01NH, TR16]
G7 EOC C J Hopkins, 6 The Paddock, Meopham, Gravesend, Kent, DA13 0TE [JO01EH, TQ66]
G7 EOE E L David, 22 Island Wall, Whitstable, CT5 1EP [JO01MI, TR16]
G7 EOG C D Flynn, 2 Trafalgar Ave, Laceby Acres, Grimsby, DN34 5RE [IO93WN, TA20]
G7 EOH G L Newnham, 22 Warren Pl, Calmore, Southampton, SO40 2SD [IO90FW, SU31]
G7 EOK Details withheld at licensee's request by SSL.
G7 EOL Details withheld at licensee's request by SSL.
G7 EOM Details withheld at licensee's request by SSL.
GW7 EOO Details withheld at licensee's request by SSL.
G7 EOR Details withheld at licensee's request by SSL.
G7 EOT Details withheld at licensee's request by SSL.
G7 EOV Details withheld at licensee's request by SSL.
G7 EOX J Wardle, Fletchers Thorns, The St., Bramley, Tadley, RG26 5BP [IO91LH, SU65]
G7 EOY Details withheld at licensee's request by SSL.
G7 EOZ Details withheld at licensee's request by SSL.
G7 EPC Details withheld at licensee's request by SSL.
G7 EPG Details withheld at licensee's request by SSL.
G7 EPJ Details withheld at licensee's request by SSL.
G7 EPL Details withheld at licensee's request by SSL.
G7 EPM J H Beddoes, 1 Glenridding Cl, Warndon Green, Worcester, WR4 9EX [IO82VE, SO85]
G7 EPP Revd R P Tickle, 5 Bramley Ct, Orchard Ln, Harrold, Bedford, MK43 7BG [IO92QE, SP95]
G7 EPR R S Jeeves, 78 Willowhale Green, Bognor Regis, PO21 4LW [IO90PS, SZ99]
G7 EPS L M Marchant, Staple Farm House, Durlock Rd, Staple, Canterbury, CT3 1JX [JO01PG, TR25]
G7 EPU A C Myers Tyne&Wear ARC, Tyne & Wear Epu Floor 2, Portman House, Portland Rd, Newcastle upon Tyne, NE2 1AQ [IO94EX, NZ26]
G7 EPW H W Lichfield, 12 Tudor Rd, Sutton Coldfield, B73 6BA [IO92CN, SP19]
G7 EPX B V Coffin, 27 Barnes Wallis, Bowerhill, Melksham, Wilts, SN12 6UH [IO81WI, ST96]
G7 EPY W E Wright, 53 Forshaw Ave, Grange Park, Blackpool, FY3 7PW [IO83LT, SD33]
G7 EQB J V Harper, 34 Neva Rd, Bitterne Park, Southampton, SO18 4FJ [IO90HW, SU41]
GW7 EQC K I Dwyer, 1 Dolydd Terr, Capel Curig, Betws y Coed, LL24 0EH [IO83BC, SH75]
G7 EQG J W Sturman, 41 Jay Cl, Haverhill, CB9 0JR [JO02FB, TL64]
G7 EQK A Grime, 23 Claremont Rd, Milnrow, Rochdale, OL16 4EZ [IO83WO, SD91]
G7 EQM N M Buckley, 25 Rainsbrook Cl, Southam, Leamington Spa, CV33 0GL [IO92GF, SP35]
G7 EQO Details withheld at licensee's request by SSL.
G7 EQP M J Smith, St Columb, 28 Pin Hill, Exeter, Devon, EX1 3TQ [IO80GR, SX99]
G7 EQQ L A Dawson, 169 Downs Rd, Istead Rise, Northfleet, Gravesend, DA13 9HF [JO01EJ, TQ66]
G7 EQR D C R Gammans, 35 Chute Ave, High Salvington, Worthing, BN13 3DS [IO90TU, TQ10]
G7 EQU G T James, 58 Cromwell Ct, Cromwell Rd, Rushden, NN10 0DS [IO92QG, SP96]
G7 EQX P F Wickers, 11 Paddock Dr, Springfield, Chelmsford, CM1 6SS [JO01FS, TL70]
G7 EQY A M Ford, 46 Ruspidge Rd, Cinderford, GL14 3AD [IO81RT, SO61]
G7 EQZ C J N Bradford, 18 Morton Gdns, Rugby, CV21 3TG [IO92JI, SP57]
G7 ERB G L Daft, 6 Prospect Ave, Irchester, Wellingborough, NN29 7DZ [IO92QG, SP96]
G7 ERC D R Edwards E Sussex/Hastin, & Rother Radio Gp, 2 London Rd, Battle, E Sussex, TN33 0EU [JO00FW, TQ71]
G7 ERH V Houghton, 8 Wheatfield Rd, Cronton, Widnes, WA8 5BU [IO83OJ, SJ48]
GW7 ERI A Brown, 3 Wyebank View, Tutshill, Chepstow, NP6 7DR [IO81QP, ST59]
G7 ERK Details withheld at licensee's request by SSL.
G7 ERL Details withheld at licensee's request by SSL.
G7 ERM Details withheld at licensee's request by SSL.
G7 ERN J T C Sladden, 5 Knavewood Rd, Kemsing, Sevenoaks, TN15 6RH [JO01CH, TQ55]
G7 ERO R C Willis Raynet/Oxford-, Shire, 24 Elizabeth Ave, Abingdon, OX14 2NS [IO91IQ, SU59]
G7 ERP Details withheld at licensee's request by SSL.
G7 ERS Details withheld at licensee's request by SSL.
G7 ERZ Details withheld at licensee's request by SSL.
G7 ESB Details withheld at licensee's request by SSL.
G7 ESC Details withheld at licensee's request by SSL.
G7 ESD Details withheld at licensee's request by SSL.
G7 ESE C V C Tully, 15 Bingham Rd, Verwood, BH31 6TU [IO90BU, SU00]
GW7 ESF T W Ford, 14 Hillsnook Rd, Ely, Cardiff, CF5 5DD [IO81JL, ST17]
G7 ESG C F Fogden, 157 Ringwood Rd, Totton, Southampton, SO40 8DX [IO90GW, SU31]
G7 ESI S Gregory, 73 Princess Way, Walshes Est, Stourport on Severn, DY13 0EL [IO82UH, SO87]
G7 ESJ Details withheld at licensee's request by SSL.
G7 ESK P M Leybourne, 25 Paddocks Rd, Rushden, NN10 6RY [IO92QH, SP96]
GD7 ESM J H Grundey, Edeville, Falcon Cliff Terr, Douglas, IM2 4AU
G7 ESO K L Reynolds, 20 Wentwood Gdns, Thornbury, Plymouth, PL6 8TD [IO70WJ, SX55]
GD7 ESR R Wernham, Fiar Isle, Lhoores Rd, Foxdale, Isle of Man, IM4 3AT
GD7 ESU C A Gill, Ballahams, 43 Governors Hill, Douglas, IM2 7AT
G7 ESV J Gregory, 17 Hazelton Rd, Marlbrook, Bromsgrove, B61 0JG [IO82XI, SO97]
G7 ESX S J Bowman, 39 Pearson St., Spennymoor, DL16 6HP [IO94FQ, NZ23]
G7 ESY I P Bowman, 22 Bryan St., Spennymoor, DL16 6DW [IO94EQ, NZ23]
G7 ESZ D Foster, 145 Tailyour Rd, Plymouth, PL6 5DH [IO70WJ, SX45]
G7 ETA Details withheld at licensee's request by SSL.
G7 ETC S J Abel, 121 Angela Rd, Horsford, Norwich, NR10 3HF [JO02OQ, TG11]
GI7 ETI Details withheld at licensee's request by SSL.
G7 ETK S Yates, 66 Bramley Rd, Sharples, Bolton, BL1 7RW [IO83SO, SD71]
G7 ETN Details withheld at licensee's request by SSL.
G7 ETR Details withheld at licensee's request by SSL.
G7 ETS S E Hambleton, Park Cottage, Birmingham Rd, Shenstone, Lichfield, Staffs, WS14 0JY [IO92BP, SK10]
G7 ETT Details withheld at licensee's request by SSL.
G7 ETU Details withheld at licensee's request by SSL.
G7 ETW Details withheld at licensee's request by SSL.
G7 ETX Details withheld at licensee's request by SSL.
G7 ETY M Tymon, 26 Nelson Rd, Caterham, CR3 5PP [IO91WG, TQ35]
G7 ETZ Details withheld at licensee's request by SSL.
G7 EUA Details withheld at licensee's request by SSL.
G7 EUB Details withheld at licensee's request by SSL.
G7 EUE Details withheld at licensee's request by SSL.
G7 EUF G D Rhodes, 54 Chell Green Ave, Stoke on Trent, ST6 7JY [IO83VB, SJ85]
G7 EUG M T Graham, 53 Goirle Ave, Canvey Island, SS8 8AW [JO01HM, TQ88]
G7 EUH Details withheld at licensee's request by SSL.
G7 EUI Details withheld at licensee's request by SSL.
G7 EUK M G Phillips, 1 Whitehouse Cttgs, Preston Cross, Ledbury, Herefordshire, HR8 2LH [IO82SA, SO63]
G7 EUL M Prince, 25 Chiltern Rd, Burnham, Slough, SL1 7NF [IO91QM, SU98]
G7 EUM S A J Parrett, 30 Heacham Cl, Lewsey Farm, Luton, LU4 0YJ [IO91SV, TL02]
G7 EUO Details withheld at licensee's request by SSL.
G7 EUR D E Stevenson, 22 Jubilee Ave, Crewe, CW2 7PR [IO83SC, SJ65]
G7 EUT D Richards, Orchard Cottage, Ashbourne Rd, Kirk Langley, Ashbourne, DE6 4NJ [IO92FW, SK23]
G7 EUV G Y Koszegi, 356 Nottingham Rd, Ripley, DE5 3JX [IO93HB, SK45]
G7 EUW S A Hall, 92 Summer St., Stroud, GL5 1PE [IO81VR, SO80]
G7 EVA Details withheld at licensee's request by SSL.
G7 EVC P H Stone, 2 The Russetts, Lees Cl, Barbourne, Ashford Kent, TN25 6RW [JO01LD, TR04]
G7 EVE M C Martyn, Aspiration, Queens Rd, Crowborough, TN6 1QQ [JO01BB, TQ52]
G7 EVF G E Keene, 28 Little Hoddington, Upton Grey, Basingstoke, RG25 2RN [IO91MF, SU74]
GW7 EVG G A Nicholas, y Rofft, Mount Rd, St. Asaph, LL17 0DF [IO83GG, SJ07]
G7 EVK A G Wade, 3 Ashendene Gr, Hanford, Stoke on Trent, ST4 8NW [IO82VX, SJ84]
G7 EVN A Jackson, Flat, 1 Hope Ct, Hope St., Rochdale, OL12 0NX [IO83WO, SD81]
G7 EVP M Pritchard, 155 Elliott Rd, March, PE15 8HF [JO02AN, TL49]
G7 EVQ M J R Jordan, 139 Camping Hill, Stiffkey, Wells Next The Sea, NR23 1QL [JO02LW, TF94]
G7 EVR S E Sprint, 35 Bradley Ln, Pudsey, LS28 8LH [IO93DT, SE23]
G7 EVT C Mills, 16 Broom Cl, Wath upon Dearne, Rotherham, S63 7JU [IO93HL, SK49]
G7 EVW N Kaberry, 45 Regents Dr, Tynemouth, North Shields, NE30 2NR [IO95GA, NZ37]
G7 EVY G Lawton, 4 Holmeswood Rd, Holmeswood, Rufford, Ormskirk, L40 1TX [IO83NP, SD41]
G7 EWA A D Hampton, 115 Overdale, Ashtead, KT21 1PZ [IO91UH, TQ15]
GW7 EWD D J Siviter, Cilgeraint Farm, St. Anns, Bethesda, Gwynedd, LL57 4AX [IO73XE, SH66]
G7 EWF J T Campbell, 8 Stenton Cl, Abingdon, OX14 5LN [IO91IP, SU49]
G7 EWH D R J Hughes, Dunedin, Chapel Rd, Alderley Edge, Ches, SK9 7DX [IO83VH, SJ87]
G7 EWK V T Thomas, 5 Bayfield Ave, Dereham, NR19 1PH [JO02LQ, TF91]
G7 EWL S Hogarth, 34 High St., Irthlingborough, Wellingborough, NN9 5TN [IO92QH, SP97]
G7 EWR J C Martin East Warwickshire Raynet, 2 Bishopton Lodge, 346 Birmingham Rd, Stratford upon Avon, Warks, CV37 0RE [IO92DF, SP15]
G7 EWS P A Breck, 79 Wordsworth Rd, Penge, London, SE20 7JF [IO91XJ, TQ37]
G7 EWU P Atkinson, 27 Ranworth Rd, Bramley, Rotherham, S66 0SP [IO93IK, SK49]
G7 EWX N Price, 245 Anchor Rd, Longton, Stoke on Trent, ST3 5DX [IO82WX, SJ94]
G7 EWZ M Hinton, 7 Cardinal Dr, Kidderminster, DY10 4RZ [IO82VI, SO87]
G7 EXC Dr P A M Holland, The Poplars, Hardy Barn, Shipley, Heanor, DE75 7LY [IO93HA, SK44]
G7 EXD R M Fletcher, 160 Barnsley Rd, Denby Dale, Huddersfield, HD8 8QW [IO93EN, SE20]
G7 EXG D R Goodacre, 2 Heath Cttgs, Sandford, Wareham, BH20 7DF [IO80WQ, SY98]
GW7 EXH N B Mee, Anncott, Hylas Ln, Rhuddlan, Rhyl, LL18 5AG [IO83GG, SJ07]
G7 EXM Dr R Artym, 22 Welbeck Rise, Harpenden, AL5 1SN [IO91TT, TL11]
GI7 EXN P R Hutchinson, 40 Oldstone Hill, Muckamore, Antrim, BT41 4SB [IO64VQ, J18]
G7 EXO K T Brown, 15 Gloucester Rd, Aldershot, GU11 3SL [IO91OF, SU84]
G7 EXP I D Matthews, 9 Shelley Cl, Risinghurst, Headington, Oxford, OX3 8HB [IO91JS, SP50]
G7 EXQ R J Morris, 20 Cornwallis Ave, Beltinge, Herne Bay, CT6 6UQ [JO01NI, TR16]
G7 EXT H A Jarvis, 31 The Downings, Herne Bay, CT6 7EJ [JO01NI, TR16]
GW7 EXW T G Simmons, 8 Tennyson Rd, Penarth, CF64 2RY [IO81JK, ST17]
G7 EXX E M Gwilliam, 15 Sheppard Way, Minchinhampton, Stroud, GL6 9BZ [IO81VQ, SO80]
G7 EYA R G Hatcher, Somerset, 39 Moor Ln, Hutton, Weston Super Mare, BS24 9QL [IO81MH, ST35]
GW7 EYB F H Terry, 12 Kings Ave, Rhyl, Denbighshire, LL18 1LT [IO83GH, SJ08]
G7 EYD R Attwood, 2 Highland Cttgs, Acton, Langton Matravers, Swanage, BH19 3LA [IO80XO, SY97]
G7 EYE S E Finnegan, 25 Westcliff Gdns, Margate, CT9 5DT [JO01QJ, TR37]
G7 EYG P Baxter, 49 Salterns Ave, Milton, Southsea, PO4 8QJ [IO90LT, SU60]
G7 EYL M J Dixon, 70 Shelley Dr, Broadbridge Heath, Horsham, RH12 3NT [IO91TB, TQ13]
G7 EYM M K Parkyn, Brookfield, Clee St Margaret, Craven Arms, Shropshire, SY7 9DX [IO82QK, SO58]
G7 EYO W G W Griggs, Flat 2, 86 South Eastern Rd, Ramsgate, CT11 9QE [JO01QI, TR36]
GW7 EYP M G Jenkins, 23 Guenever Cl, Thornhill, Cardiff, CF4 9AH [IO81JM, ST18]
GW7 EYQ C Lock, 4 Charles St., Caerphilly, CF8 3AQ [IO81HN, ST08]
G7 EYR P Wiles, 16 Churchill Rd, Altrincham, WA14 5LT [IO83TJ, SJ78]
G7 EYS C M Chance, 34 Main Rd, Emsworth, Hants, PO10 8AU [IO90MU, SU70]
G7 EYT P J Marsh, 17 Ramley Rd, Pennington, Lymington, SO41 8HF [IO90FS, SZ39]
G7 EYV A B Little, 444 Dunsbury Way, Leigh Park, Havant, PO9 5BJ [IO90MU, SU70]
G7 EYX Details withheld at licensee's request by SSL.
G7 EZD J A Jones, 61 The Sackville, de La Warr Parade, Bexhill on Sea, TN40 1LS [JO00FU, TQ70]
GI7 EZF F J Bowman, 127 Gortnagola Rd, Dungannon, BT70 3BH [IO64NM, H76]
G7 EZG Details withheld at licensee's request by SSL.
G7 EZH Details withheld at licensee's request by SSL.
G7 EZM Details withheld at licensee's request by SSL.
G7 EZP Details withheld at licensee's request by SSL.
GI7 EZR Details withheld at licensee's request by SSL.
G7 EZT I D Pucknell, 7 St. Peters Ct, Addingham, Ilkley, LS29 0RL [IO93BW, SE04]
G7 EZU Details withheld at licensee's request by SSL.
G7 EZV Details withheld at licensee's request by SSL.
G7 EZW Details withheld at licensee's request by SSL.
G7 EZY Details withheld at licensee's request by SSL.
G7 FAD V C Ritson, 15 The Chantry, Rooksbridge, Axbridge, BS26 2TR [IO81NG, ST35]
GW7 FAE W Brown, 15 Maes Heulog, Caernarvon, LL55 1PS [IO73UD, SH46]
G7 FAG Details withheld at licensee's request by SSL.
G7 FAI R E Quaintance, 18 Queens Ave, Ilfracombe, EX34 9LN [IO71WE, SS54]
G7 FAL Details withheld at licensee's request by SSL.
G7 FAN Details withheld at licensee's request by SSL.
G7 FAQ Details withheld at licensee's request by SSL.
G7 FAR R Pickles obo RAF Waddington ARC, The Bell Inn, Far Ln, Coleby, Lincoln, LN5 0AH [IO93RD, SK96]
G7 FAS Details withheld at licensee's request by SSL.
G7 FAX Details withheld at licensee's request by SSL.
G7 FAZ A B Mould, 12 Windmill Ave, Rubery, Rednal, Birmingham, B45 9TA [IO82XJ, SO97]
G7 FBD M P Adlard, 22 Ash Gr, Bristol, BS16 4JT [IO81RL, ST67]
G7 FBE E G Dare, 17 Montgomery Dr, Spencers Wood, Reading, RG7 1BQ [IO91MJ, SU76]
G7 FBF A J Cockburn, Pound Cottage, West St., Odiham, Hook, RG29 1NR [IO91MG, SU75]
G7 FBO D A Fasham, 29 Granville Ave, Ramsgate, CT12 6DX [JO01QI, TR36]
G7 FBT E M Woodhouse, 51A Avernon Rd, London, E11 4QT [JO01AN, TQ38]
G7 FBU G W Sandilands, 35 Clare Rd, Greenford, UB6 0DF [IO91TN, TQ18]
G7 FBX M E Pugh, Treetops, 6 Prospect Rd, Gornal Wood, Dudley, DY3 2TJ [IO82WM, SO99]
G7 FBY R Furniss, 387 Ct Ln, Erdington, Birmingham, B23 5JX [IO92BM, SP19]
G7 FCE M J Warren, 23 Shakespear Ct, Chaucer Way, Hoddesdon, Herts, EN11 9QS [IO91XS, TL31]
G7 FCJ P J Honeywell, 230 Lodge Ln, Grays, RM16 2TH [JO01DL, TQ67]
G7 FCK R J Clark, 23 The Close, Cheltenham, GL53 0PG [IO81XU, SO91]
G7 FCL D Hames, 5 Sidney Cooper Cl, Rough Common, Canterbury, CT2 9BQ [JO01MG, TR15]
GI7 FCM S T Fleming, 21 Sandyknowes Park, Newtownabbey, BT36 5DE [IO74AQ, J38]
G7 FCO G D Howat, 44 Castle St., Thornbury, Bristol, BS12 1HB [IO81RO, ST69]
GI7 FCP J B McCormick, 14 Ballyoran Park, Portadown, Craigavon, BT62 1JN [IO64SK, J05]
G7 FCR Details withheld at licensee's request by SSL.
G7 FCT Details withheld at licensee's request by SSL.
G7 FCU Details withheld at licensee's request by SSL.
G7 FCV Details withheld at licensee's request by SSL.
GI7 FCW P J J Quinn, 53 Dernanaught Rd, Crosscavanagh, Dungannon, Co. Tyrone, N Ireland, BT70 2NR [IO64NM, H76]

G7 FCY J H Kemp, 13 Springwood Cres, Grimsby, DN33 3HG [IO93WM, TA20]
G7 FDC J S May South Devon RC, 6 Hodson Cl, Paignton, TQ3 3NU [IO80FK, SX86]
G7 FDD W L Cooper, 15 Ashby Ave, Hartsholme, Lincoln, LN6 0ED [IO93RF, SK96]
G7 FDF D M M Jallands, 314 Porchester Rd, Mapperley, Nottingham, NG3 6GR [IO92KX, SK54]
G7 FDG R S Eley, Bridge House, Purton, Berkeley, Glos, GL13 9HS [IO81SR, SO60]
G7 FDL M J Clift, 4 Sheridan Cl, Hawkslade Farm, Aylesbury, HP21 9HL [IO91OT, SP81]
G7 FDN C A J Day, 80 Bidwell Hill, Houghton Regis, Dunstable, LU5 5EP [IO91RV, TL02]
G7 FDP Details withheld at licensee's request by SSL.
G7 FDR Details withheld at licensee's request by SSL.
GI7 FDT Details withheld at licensee's request by SSL.
G7 FDU Details withheld at licensee's request by SSL.
GI7 FDZ Details withheld at licensee's request by SSL.
G7 FEA M R Codling, 18 Ash Gr, Pinehurst, Swindon, SN2 1RX [IO91CN, SU18]
G7 FED E L Wells, 1A Brocklewood Ave, Poulton-le-Fylde, FY6 8BZ [IO83MT, SD33]
G7 FEE F W Harvey, 39 Simonside Terr, Heaton, Newcastle upon Tyne, NE6 5JY [IO94FX, NZ26]
G7 FEF J D Pipkin, 46 Charles Ave, Albrighton, Wolverhampton, WV7 3LF [IO82UP, SJ80]
G7 FEG I Kilkenny, 23 Hazelhurst Rd, Stalybridge, SK15 1HD [IO83XL, SJ99]
G7 FEK M J Dennis, 103 Willis Rd, Haddenham, Aylesbury, HP17 8HG [IO91MS, SP70]
G7 FEL J P Slater, 33 Delius Way, Stanford-le-Hope, SS17 8RG [JO01FM, TQ68]
G7 FEO A Greenhalgh, Colby, Appleby-in-Westmorland, Cumbria, CA16 6BD [IO84RN, NY62]
G7 FEP D P Birt, 6 Buckland Green, Worle, Weston Super Mare, BS22 0HL [IO81NI, ST36]
G7 FEQ B J Shipton, 4 School Cl, Hillesley, Wotton under Edge, GL12 7RH [IO81UO, ST78]
G7 FER P H Baird, 87 Inglenook, Clacton on Sea, CO15 4SG [JO01OT, TM11]
G7 FEV Details withheld at licensee's request by SSL.
GI7 FEW Details withheld at licensee's request by SSL.
G7 FEX Details withheld at licensee's request by SSL.
GW7 FEZ Details withheld at licensee's request by SSL.
G7 FFA Details withheld at licensee's request by SSL.
G7 FFB D M Egleton, Popley, 20 Marlowe Cl, Basingstoke, RG24 9DD [IO91LG, SU65]
G7 FFC V H Prall, 20 Marlowe Cl, Basingstoke, RG24 9DD [IO91LG, SU65]
G7 FFE Details withheld at licensee's request by SSL.
GI7 FFF E Barr, Ed Mar, 1 Onslow Dr, Bangor, BT19 7HQ [IO74EP, J57]
G7 FFG D T Seeds, 8 Westwood Cl, Swanpool, Lincoln, LN6 0HG [IO93RF, SK96]
G7 FFI K E Harding, 21 Doulton Way, Ashingdon, Rochford, SS4 3BX [JO01IO, TQ89]
G7 FFK A E James, 66 Rydal Cres, Worsley, Manchester, M28 7JD [IO83TM, SD70]
GI7 FFL Details withheld at licensee's request by SSL.
G7 FFM P M Malpass, 30 Countisbury Rd, Norton, Stockton on Tees, TS20 1PZ [IO94IO, NZ42]
G7 FFR J Rutherford, 270 Milburn Rd, Ashington, NE63 0PL [IO95FE, NZ28]
G7 FFS A R Pike, 24 Birchfield Cl, Wood End, Atherstone, CV9 2QT [IO92EN, SP29]
G7 FFT D K Chatterton, 51 Lovely Ln, Warrington, WA5 1NB [IO83QJ, SJ58]
GW7 FFU B J Shelley, Sunray, Pendine, Carmarthen, Dyfed, SA33 4PD [IO71RR, SN20]
G7 FFW S A Lonsdale, 16 Hinkler St., Cleethorpes, DN35 8PR [IO93XN, TA30]
G7 FFZ J Humphries, 23 Sycamore Dr, Lutterworth, LE17 4TR [IO92JL, SP58]
G7 FGA D M Moreland, Mayfield, 179 Carr Ln, Acomb, York, YO2 5HQ [IO93KX, SE55]
G7 FGD J C Brown, St Winnolls House, St Winnolls, Torpoint, Cornwall, PL11 3DX [IO70UI, SX35]
GM7 FGF A A Calderwood, Cnoc An Fhraoich, Glenlomond, Kinross, KY13 7HF [IO86HF, NO10]
GM7 FGH J J Bartolo, 84 Calderbraes Ave, Uddingston, Glasgow, G71 6ED [IO75WT, NS66]
G7 FGI C W Pond, 115 Sandwich Rd, Cliffsend, Ramsgate, CT12 5JA [JO01QH, TR36]
G7 FGK R L Dobbs, 11 Plantation Cres, Bredon, Tewkesbury, GL20 7QG [IO82WA, SO93]
GW7 FGL D G Griffiths, Mount Villa, 188 St Teilo, Pontardulais, SA4 1LQ [IO71XR, SN50]
G7 FGO J Atherton, 31 Hillside Ave, St. Helens, WA10 6LX [IO83PL, SJ59]
GI7 FGQ P J Faulkner, 40 Glenariff Dr, Comber, Newtownards, BT23 5HA [IO74CN, J46]
G7 FGR G S Cluley, 36 Great Ln, Greetham, Oakham, LE15 7NG [IO92QR, SK91]
G7 FGY I D R Hirst, 30 Lincoln Rd, Skellingthorpe, Lincoln, LN6 5UU [IO93QF, SK97]
G7 FGZ D J Mitchell, 55 Halewick Ln, Sompting, Lancing, BN15 0ND [IO90UU, TQ10]
G7 FHA C D Brailsford, 65 Cherry Orchard, Codford St. Mary, Codford, Warminster, BA12 0PW [IO81XD, ST93]
G7 FHB L J Boorman, 2 Bull Ln Cttgs, Bull Ln, Bethersden, Ashford, Kent, TN26 3HA [JO01ID, TQ93]
GI7 FHC Details withheld at licensee's request by SSL.
G7 FHD Details withheld at licensee's request by SSL.
G7 FHE Details withheld at licensee's request by SSL.
G7 FHI Details withheld at licensee's request by SSL.
G7 FHJ Details withheld at licensee's request by SSL.
G7 FHK Details withheld at licensee's request by SSL.
G7 FHN Details withheld at licensee's request by SSL.
G7 FHQ Details withheld at licensee's request by SSL.
G7 FHR Details withheld at licensee's request by SSL.
GI7 FHT Details withheld at licensee's request by SSL.
G7 FHV T A Beeching, 11 Kents Rd, Haywards Heath, RH16 4HL [IO90WX, TQ32]
G7 FHW J A Short, North Trew Farm, Highampton, Beaworthy, Devon, EX21 5JG [IO70VT, SS40]
G7 FHY Details withheld at licensee's request by SSL.
GI7 FHZ E I McCrystal, 33 Richmond Park, Omagh, BT79 7SJ [IO64IO, H47]
G7 FIA P M Page, 67 Teesdale Rd, Dartford, DA2 6LB [JO01CK, TQ57]
GM7 FIE I M Stevenson, 5/1 Gilmours Entry, Edinburgh, Midlothian, EH8 9XL [IO85JW, NT27]
G7 FIF G C H Bell, 16 Buckland Cl, Farnborough, GU14 8DH [IO91PH, SU85]
G7 FIJ B W Parsons, 20 High Park Rd, Halesowen, B63 2JA [IO82WK, SO98]
G7 FIM K J Keen, 124 Southbury Rd, Enfield, EN1 1YE [IO91XP, TQ39]
G7 FIQ Details withheld at licensee's request by SSL.
G7 FIR Details withheld at licensee's request by SSL.
GM7 FIS J E Russell, 15 Glen View, Kildrum, Cumbernauld, Glasgow, G67 2DA [IO85AW, NS77]
G7 FIU T R Harris, 5 Bickington Lodge, Bickington, Barnstaple, Devon, EX31 2LH [IO71WB, SS53]
GI7 FIY Details withheld at licensee's request by SSL.
G7 FJA Details withheld at licensee's request by SSL.
G7 FJC P J Fellingham, 26 Fitch Dr, Bevendean, Brighton, BN2 4HX [IO90WU, TQ30]
G7 FJK T R Brown, 22 Moss Rd, Congleton, CW12 3BN [IO83VD, SJ86]
G7 FJN B M Woods, 275 Scotter Rd, Scunthorpe, DN15 7EH [IO93PO, SE81]
G7 FJO Details withheld at licensee's request by SSL.
G7 FJU Details withheld at licensee's request by SSL.
GI7 FJY N Gamble, 174 Lisnafin Park, Strabane, BT82 9DJ [IO64GT, H39]
G7 FJZ P Selley, 32 Kala Fair, Golf Links Rd, Westward Ho!, Bideford, North Devon, EX39 1TX [IO71VA, SS42]
G7 FKB Details withheld at licensee's request by SSL.
G7 FKE Details withheld at licensee's request by SSL.
G7 FKF Details withheld at licensee's request by SSL.
G7 FKG A E Evans, Boskerrow, Castle Gate, Ludgvan, Penzance, TR20 8BG [IO70FD, SW43]
GM7 FKH Details withheld at licensee's request by SSL.
G7 FKI Details withheld at licensee's request by SSL.
G7 FKJ C Holloway, 23 Ryecroft Rd, Stretford, Manchester, M32 9BS [IO83UK, SJ79]
G7 FKP D Henderson, 14 Sandfield Cl, Market Weighton, York, YO4 3ET [IO93PU, SE84]
G7 FKS H E A Ellis, 49 Kingsway, Banbury, Oxon, OX16 9EN [IO92HB, SP43]
G7 FKX C N E Wood, 2 Plain Cttgs, Plain Rd, Marden, Tonbridge, TN12 9LS [JO01FD, TQ74]
G7 FKY P B Pettman, 1 Staddon Cttgs, Staddon Terr Ln, Plymouth, Devon, PL1 5DN [IO70WI, SX45]
G7 FKZ R Bowden, 35 Glebelands, Biddenden, Ashford, TN27 8EA [JO01HC, TQ83]
GW7 FLA R J Swan, Hafod, 7 Anchor Down, Solva, Haverfordwest, SA62 6TQ [IO71JU, SM82]
G7 FLE S J Hemmings, 27 Heath Hill Rd, Mount Tabor, Halifax, HX2 0UT [IO93AR, SE02]
GM7 FLG D D Pegg, 11 Glenward Ave, Lennoxtown, Glasgow, G65 7EP [IO75VX, NS67]
G7 FLI J A J Moyse, Spire View, Kestle Dr, Truro, Cornwall, TR1 3PT [IO70LG, SW84]
G7 FLK G T D Davies, 161 Meadowcroft, Upper Stratton, Swindon, SN2 6LW [IO91CO, SU18]
GM7 FLM R F Smith, 80 High St., Leslie, Glenrothes, KY6 3DB [IO86JE, NO20]
G7 FLO Details withheld at licensee's request by SSL.
G7 FLQ G Parsons, 106 Princess Cres, Halesowen, B63 3QG [IO82XL, SO98]
G7 FLR C Capewell, School House, Bridge Rd, Stoke Bruerne, Towcester, NN12 7SD [IO92MD, SP74]
G7 FLT Details withheld at licensee's request by SSL.
G7 FLX B G Timms, 74 Park Gwyn, St. Stephen, St. Austell, PL26 7PN [IO70NI, SW95]
GM7 FLZ E J Chesters, Dunessa, Fintray, Dyce, Aberdeen, AB2 0HY [IO87TH, NJ72]
G7 FMB R W Burns, 43 Gibson St., Bickershaw, Wigan, WN2 5TF [IO83RM, SD60]
G7 FMI M J Kensall, 40 Eskdale Ave, Ramsgate, CT11 0PB [JO01QI, TR36]
GM7 FMJ G C Mitchell, 16A Prospect Terr, Newport on Tay, Fife, DD6 8AW [IO86MK, NO42]
G7 FML P J Clarke, Chris Wyn, Race Ground, Little London, Spalding, PE11 3AP [IO92VS, TF22]
G7 FMM Details withheld at licensee's request by SSL.
GI7 FMN Details withheld at licensee's request by SSL.
G7 FMO Details withheld at licensee's request by SSL.
G7 FMP Details withheld at licensee's request by SSL.
G7 FMQ J M Sutton, 10 Cathcart Rd, Stourbridge, DY8 3UZ [IO82WK, SO88]
G7 FMT A J Wesley, 12 Woodford Ave, Ramsgate, CT12 6RD [JO01QI, TR36]
G7 FMU D T Appleby, Flat 2, 18 High St., Holt, NR25 6BH [JO02NV, TG03]
G7 FMV D W Sweet, 50 Mereside, Soham, Ely, CB7 5XE [JO02DI, TL57]
G7 FMW P D Olson, 10 Euston St., Liverpool, L4 5PR [IO83MK, SJ39]
G7 FMX Details withheld at licensee's request by SSL.
G7 FNA Details withheld at licensee's request by SSL.
G7 FND E T Millership, 16 Bramble Way, Moreton, Wirral, L46 7UP [IO83KJ, SJ29]
G7 FNE Details withheld at licensee's request by SSL.
GI7 FNJ P F O'Toole, 51 Annsborough Park, Castlewellan, BT31 9NH [IO74AG, J33]
G7 FNM J F Walmsley, Bank Field, 5 Dimples Ln, Garstang, Preston, PR3 1RD [IO83OV, SD44]
G7 FNN A N Morley, 28 Dryden Rd, Ipswich, IP1 6QN [JO02NB, TM14]
GI7 FNP H W I Massey, 156 Killaughey Rd, Donaghadee, BT21 0BQ [IO74FP, J57]
GW7 FNQ D A Jones, Noddfa, High St., Malltraeth, Bodorgan, LL62 5AS [IO73TE, SH46]
GW7 FNR R M Rees, 12 Colemere St., Wrexham, LL13 7PD [IO83MA, SJ34]
G7 FNU R C Beaumont, 49 Vincent Cl, Broadstairs, CT10 2ND [JO01QI, TR36]
G7 FNV C E Birks, 1 Daffodil Cl, Blackhall, Blackhall Colliery, Hartlepool, TS27 4NU [IO94IS, NZ44]
G7 FOE Details withheld at licensee's request by SSL.
G7 FOF Details withheld at licensee's request by SSL.
GW7 FOJ Details withheld at licensee's request by SSL.
G7 FOP C D D Duncan, 8 Kennedy Cres, Gosport, PO12 2NN [IO90KS, SZ59]
G7 FOV M R Newton, 51 Moss Ln, Timperley, Altrincham, WA15 6LQ [IO83UJ, SJ78]
G7 FOX R G Barker Melton Mowbray ARS, 50 Baldocks Ln, Melton Mowbray, LE13 1EN [IO92NS, SK71]
G7 FPB Details withheld at licensee's request by SSL.
G7 FPE Details withheld at licensee's request by SSL.
G7 FPG K E Carter, The Poplars, Gwealfolds Rd, Helston, TR13 8UB [IO70IC, SW62]
G7 FPM Details withheld at licensee's request by SSL.
GM7 FPN A J C McPherson, 91 Robertson Rd, Lhanbryde, Elgin, IV30 3PQ [IO87JP, NJ26]
G7 FPO P S Edwards, 63 Kilvert Rd, Wednesbury, WS10 0QP [IO82XN, SO99]
G7 FPR G W Flanagan, 9 Beau Ct, Portarlington Rd, Bournemouth, Dorset, BH4 8BX [IO90BR, SZ09]
G7 FPS S D Harrison, 11 Waverley Rd, Worsley, Manchester, M28 7UW [IO83TM, SD70]
G7 FPU C Collett, 111 Berkeley Vale Park, Hook St, Berkeley, Glos, GL13 9TQ [IO81SQ, ST69]
G7 FPW P M Ives-Whitaker, 17 Bedford Rd, Wells, BA5 3NH [IO81QE, ST54]
G7 FPY P M Chapman, Fraser, The Avenue, Hullbridge, Hockley, SS5 6LP [JO01HP, TQ89]
G7 FPZ Details withheld at licensee's request by SSL.
G7 FQB C A Wilson, 25 Highfield Ave, Scunthorpe, DN15 7DZ [IO93QO, SE81]
G7 FQE J R Whiffen, 2 Arundel Cl, Alresford, SO24 9PJ [IO91KB, SU53]
G7 FQF Details withheld at licensee's request by SSL.
G7 FQG Details withheld at licensee's request by SSL.
G7 FQI Details withheld at licensee's request by SSL.
G7 FQL L W Rawlinson, 106 Whitby Ave, Ingol, Preston, PR2 3ZP [IO83PS, SD53]
G7 FQP S D Earle, Bayleigh, Croft Ln, Chipperfield, Kings Langley, WD4 9DX [IO91SQ, TL00]
G7 FQY G E Spark, 5 Emsworth Dr, Brooklands, Sale, M33 3PR [IO83UJ, SJ79]
G7 FRA Details withheld at licensee's request by SSL.
G7 FRB Details withheld at licensee's request by SSL.
GM7 FRC J Burke Fife Raynet, 25 Duncan Rd, Glenrothes, KY7 4HS [IO86JE, NO20]
GW7 FRD H McKee, 15 Pen y Fan, Trallwn, Llansamlet, Swansea, SA7 9XB [IO81BP, SS69]
G7 FRE Details withheld at licensee's request by SSL.
G7 FRH A M Russ, 21 Francis Rd, St. Pauls Cray, Orpington, BR5 3LY [JO01BJ, TQ46]
G7 FRL P M Roberts, Dunlins, 18 Yannon Dr, Teignmouth, TQ14 9JP [IO80FN, SX97]
G7 FRW K R McCaffery, 2 Finham Green Rd, Finham, Coventry, CV3 6EP [IO92FI, SP37]
G7 FSA R Colclough, 8 Parker Jervis Rd, Parhall Est, Longton, Stoke on Trent, ST3 5RP [IO82WX, SJ94]
GW7 FSF P M Thomas, 49 Cwmaman Rd, Godreaman, Aberdare, CF44 6DT [IO81GQ, SO00]
G7 FSH F N Pearson, 15 York Rd, Driffield, YO25 7AT [IO94SA, TA05]
G7 FSI B M H Williams, 3 Beach Rd, Sheringham, NR26 8BH [JO02OW, TG14]
G7 FSL Details withheld at licensee's request by SSL.
G7 FSR A R Wyard, 11 Roebuck Est, Binfield, Bracknell, RG42 4DG [IO91OK, SU87]
G7 FTD K W Taber, 27 Woodlands Walk, Harrogate, HG2 7BB [IO93GX, SE35]
G7 FTF B T L Lee, 43 Longview Rd, Saltash, PL12 6EF [IO70VJ, SX45]
G7 FTH A D Duff, Woodlea, Highfield Rd, Croston, Preston Lancs, PR5 7HH [IO83OP, SD41]
G7 FTJ W A Graham-Kerr, The Thatched Cottage, Whitchurch Hill, Oxon, RG8 7NY [IO91LM, SU67]
G7 FTM S J Clayton, 22 Orchard Ave, North Anston, Anston, Sheffield, S31 7BW [IO93GJ, SK38]
G7 FTN Details withheld at licensee's request by SSL.
GI7 FTO I T McAlpine, 17 Tandragee Rd, Portadown, Craigavon, BT62 3BQ [IO64SJ, J05]
G7 FTP Details withheld at licensee's request by SSL.
G7 FTQ Details withheld at licensee's request by SSL.
G7 FTR Details withheld at licensee's request by SSL.
G7 FTS O J H Whiteside, Bedford House, 82 Cornwall Rd, Harrogate, HG1 2NE [IO93FX, SE25]
G7 FTW S E Howes, The Pantiles, Palace Rd, Ripon, North Yorks, HG4 1HA [IO94FD, SE37]
G7 FTY Details withheld at licensee's request by SSL.
G7 FUM N P Seath, 27 Summerfield Ave, Whitstable, CT5 1NR [JO01MI, TR16]
G7 FUQ A E W Vincent, 12 Spelman Rd, Norwich, NR2 3NJ [JO02PO, TG20]
G7 FUR W Lee, 8 Thorns Rd, Quarry Bank, Brierley Hill, DY5 2JT [IO82WL, SO98]
G7 FUV S Gray, Allington, Old Dartford Rd, Farningham, Dartford, DA4 0EB [JO01CJ, TQ56]
G7 FUW J F Birch, 49 Crowhurst Rd, Birmingham, B31 4PB [IO92AJ, SP07]
G7 FVB Details withheld at licensee's request by SSL.
G7 FVD Details withheld at licensee's request by SSL.
G7 FVF Details withheld at licensee's request by SSL.
G7 FVG Details withheld at licensee's request by SSL.
G7 FVH R Barrick, Orchard Bungalow, Pasture Ln, Middlesbrough, TS6 8EH [IO94KN, NZ51]
G7 FVN M D Peskett, 24 Tweed Cl, Thornbury, Bristol, BS12 2HA [IO81RO, ST68]
G7 FVP S L Hildreth, 1 Ruxton Ct, Ruxton Cl, Swanley, BR8 7DA [JO01CJ, TQ56]
G7 FVR C Heywood, 7 Old Quay Cl, Neston, Parkgate, South Wirral, L64 6UA [IO83LG, SJ27]
G7 FVV Details withheld at licensee's request by SSL.
G7 FWD B G Nicholls, 18 Somerfield Cl, Shelfield, Walsall, WS4 1PP [IO92AO, SK00]
G7 FWE A P Smith, 57 Woodhouse Ln, Sale, M33 4JZ [IO83TJ, SJ79]
G7 FWF A G J Beaumont, 18 Richmond Gdns, Canterbury, CT2 8ES [JO01MG, TR15]
G7 FWG A F Bareham, Pond Cl Cottage, Warlands Ln, Shalfleet, Newport IoW, PO30 4NG [IO90HQ, SZ48]
G7 FWL M J Knight, 28 Cumberland Cl, Aylesbury, HP21 7HH [IO91OT, SP81]
G7 FWM T G Henson, 8 Dartmouth Walk, Basingstoke, RG22 6QU [IO91KG, SU65]
GU7 FWO D C Williams, Les Venelles, Alderney, Channel Islands, X X
G7 FWV S Metson, 19 Neville Ct, Dropmore Rd, Burnham, Slough, SL1 8BQ [IO91QM, SU98]
G7 FWW G . W . Ross, 48 Penhurst Rd, Bedhampton, Havant, PO9 3NX [IO90LU, SU60]
G7 FXH R J Crane, 21 Coombe Dr, Sittingbourne, ME10 3DA [JO01JI, TQ96]
G7 FXM Details withheld at licensee's request by SSL.
G7 FXO P R Werba, 47 Ulwell Rd, Swanage, BH19 1LG [IO90AO, SZ07]
G7 FXU R Wishart, 92 Firs Dr, Rugby, CV22 7AQ [IO92II, SP47]
G7 FXV Details withheld at licensee's request by SSL.
GW7 FXX B J Harries, 12 Panteg, Llanelli, SA15 3TF [IO71WQ, SN50]
G7 FXY P H Hallett, 30 Summerdown Walk, Trowbridge, BA14 0LJ [IO81VH, ST85]
G7 FXZ G E Hodgetts, 2 Friars Gorse, Stourton, Stourbridge, DY7 6SP [IO82VL, SO88]
GM7 FYB D J Wemyss, 24 Brucklay Ct, Peterhead, AB42 2UF [IO97CM, NK14]
G7 FYC Details withheld at licensee's request by SSL.
G7 FYE Details withheld at licensee's request by SSL.
GW7 FYG C S Wright, 66 Llys Celyn, Newtown, SY16 1PT [IO82IM, SO09]
G7 FYM R Shirt, 4 Trent Ave, Milnrow, Rochdale, OL16 3EX [IO83WO, SD91]
G7 FYN Details withheld at licensee's request by SSL.
G7 FYR Details withheld at licensee's request by SSL.
G7 FYX Details withheld at licensee's request by SSL.
G7 FYZ R C Whitehouse, Kia Ka Mina, Polkirt Hill, Mevagissey, St. Austell, PL26 6UR [IO70OG, SX04]
G7 FZB Dr G M Ridgeway, 15 Blenheim Ct, Alsager, Stoke on Trent, ST7 2BY [IO83UC, SJ75]
GM7 FZC R P Cripps, The School House, Knock Point, Isle of Lewis, PA86 0BW [IO76JB, NS08]
G7 FZD S G Vanstone, 28 Williams Ave, Weymouth, DT4 9BP [IO80SO, SY67]
G7 FZE Details withheld at licensee's request by SSL.
G7 FZJ M J H Whatley, Woodside West, Wood Ln, Hipperholme, Halifax, HX3 8HB [IO93CR, SE12]
G7 FZL Details withheld at licensee's request by SSL.
G7 FZN P G Du Plessis, 51 La Providence, Rochester, ME1 1NB [JO01GJ, TQ76]
G7 FZV Details withheld at licensee's request by SSL.
GW7 FZW Details withheld at licensee's request by SSL.
G7 FZX Details withheld at licensee's request by SSL.
G7 GAB R M Hagues, 10 Garrett Cl, Long Lawford, Rugby, Warks, CV23 9DL [IO92IJ, SP47]
GM7 GAE I Mackenzie, Napier University, 219 Colinton Rd, Edinburgh, EH14 1BJ [IO85JW, NT27]
GM7 GAF Details withheld at licensee's request by SSL.
GW7 GAH R C Dore, Maespoeth Cottage, Corris, Nr.Machynlleth, Powys, SY20 9RD [IO82BP, SH70]
G7 GAM Details withheld at licensee's request by SSL.
G7 GAN Details withheld at licensee's request by SSL.
G7 GAP J J Cartwright, 24 Tudor Ct, Castle Way, Feltham, TW13 7QQ [IO91TK, TQ17]
GI7 GAQ A C McElhinney, 39 Knockmoyle Rd, Omagh, BT79 7TB [IO64IP, H47]
G7 GAR A S Fairweather, Rooks Pawn, 46 Lower Wyche Rd, Malvern Wells, Malvern, WR14 4ET [IO82TC, SO74]
G7 GAZ B J Kerrison, Mariners, Imperial Ave, Minster on Sea, Sheerness, ME12 2HG [JO01JK, TQ97]
G7 GBA Details withheld at licensee's request by SSL.

G7 GBC R T Arnold, Stonehouse, Hartfield Rd, Forest Row, RH18 5DA [JO01AC, TQ43]
GM7 GBD G L Macgregor, 6 Kincaidfield, Milton of Campsie, Glasgow, G65 8ER [IO75WX, NS67]
G7 GBF G C W Hattley, 36 Gaskell St., Union Rd, London, SW4 6NS [IO91WL, TQ37]
G7 GBJ J J Kaczmarek, 2 Westgate Terr, London, SW10 9BJ [IO91VL, TQ27]
G7 GBK G M N Cater, 53 Poundfield Rd, Loughton, IG10 3JN [JO01AP, TQ49]
G7 GBN P K Baird, 168 Plumberow Ave, Hockley, SS5 5AT [JO01HO, TQ89]
G7 GBQ Details withheld at licensee's request by SSL.
G7 GBS Details withheld at licensee's request by SSL.
G7 GBY L F Wickers, 11 Paddock Dr, Springfield, Chelmsford, CM1 6SS [JO01FS, TL70]
G7 GBZ P R Leach, 127 Robin Way, Chelmsford, CM2 8AU [JO01FR, TL70]
G7 GCB R A Bishop, Ye Olde Mitre Inne, 58 High St., Barnet, EN5 5SJ [IO91VP, TQ29]
GW7 GCD S G Lee, 52 Heol Mabon, Cwmafan, Cwmavon, Port Talbot, SA12 9PD [IO81CO, SS79]
G7 GCF P E Kell, 56A Central Parade, New Addington, Surrey, CR0 0JL [IO91XI, TQ36]
G7 GCI Details withheld at licensee's request by SSL.
G7 GCJ Details withheld at licensee's request by SSL.
G7 GCK Details withheld at licensee's request by SSL.
G7 GCO Details withheld at licensee's request by SSL.
G7 GCP Details withheld at licensee's request by SSL.
G7 GCQ Details withheld at licensee's request by SSL.
G7 GCR Details withheld at licensee's request by SSL.
G7 GCS Details withheld at licensee's request by SSL.
G7 GCU M G Edge, 10 Selby Way, Mossley Est, Walsall, WS3 2RS [IO82XO, SJ90]
G7 GCW P Andrew, 17 St. James Cl, Kettering, NN15 5HB [IO92PJ, SP87]
G7 GCX A R Thomas, 2 Elm Ave, Sandiacre, Long Eaton, Nottingham, NG10 4LR [IO92IV, SK43]
G7 GDA Details withheld at licensee's request by SSL.
G7 GDC A J Gosden, 10 Radcliffe Way, Northolt, UB5 6HP [IO91TM, TQ18]
GM7 GDE A J Hood, 39 Broomhill Cres, Erskine, PA8 7AN [IO75SV, NS46]
GW7 GDH G M Jones, 12 Lynmouth Cres, Rumney, Cardiff, CF3 9AT [IO81KM, ST27]
G7 GDI J A Jenkinson, 28 Northville Rd, Northville, Bristol, BS7 0RG [IO81RM, ST67]
G7 GDJ A M B Knight, 8 Clifton Rd, Sidcup, DA14 6PY [JO01BK, TQ47]
G7 GDQ J H Guy, 40 Parkfield Rd, Cheadle Hulme, Cheadle, SK8 6EX [IO83VI, SJ88]
G7 GDT J T G Collins, 35 Wedmore Park, Southdown, Bath, BA2 1JZ [IO81TI, ST76]
G7 GDV A M Porteous, 73 Dowgate Cl, Tonbridge, TN9 2EJ [JO01DE, TQ54]
G7 GEB L M Dare, 17 Montgomery Dr, Spencers Wood, Reading, RG7 1BQ [IO91MJ, SU76]
G7 GED Details withheld at licensee's request by SSL.
G7 GEE J M Gee, 51 Hattons Ln, Childwall, Liverpool, L16 7QR [IO83NJ, SJ48]
GM7 GEF G Duthie, 16 St. Swithins Ct, Polehampton Cl, Twyford, Reading, RG10 9RP [IO91NL, SU77]
G7 GEI M R Arliss, 55 Hartland Cres, Edenthorpe, Doncaster, DN3 2PQ [IO93LN, SE60]
G7 GEL J D Mansell, 8 Himley Gdns, The Straits, Lower Gornal, Dudley, DY3 3AS [IO82WM, SO89]
G7 GEQ Details withheld at licensee's request by SSL.
G7 GER Details withheld at licensee's request by SSL.
G7 GES B A Norcott, 5 The Shrubbery, Upminster, RM14 3AH [JO01DN, TQ58]
G7 GEU V D Bruntnell, 4 Cypress Ave, Sedgley, Dudley, DY3 2JF [IO82WM, SO99]
G7 GEX N J Potter, 4 Eastleigh Dr, Mickleover, Derby, DE3 5HZ [IO92FV, SK33]
G7 GFC D M A Mullock, 83 Bluebell Cl, Huntington, Chester, CH3 6RS [IO83NE, SJ46]
G7 GFD A J Cook, The Cottage, Powder Mill Ln, Leigh, Tonbridge, TN11 8PZ [JO01CE, TQ54]
G7 GFH W T Baker, 41 Kenwood Park Rd, Sheffield, S7 1NE [IO93GI, SK38]
G7 GFK K H Percival, 1608 Scant Row, Chorley Old Rd, Horwich, Bolton, BL6 6PZ [IO83RO, SD61]
G7 GFM J H Hunt, 15 Badger Cl, Winyates West, Redditch, B98 0JE [IO92BH, SP06]
G7 GFP I G R Bishop, 115 Burman Rd, Shirley, Solihull, B90 2BQ [IO92BJ, SP17]
G7 GFR B Clifford, 8 Caldbeck Pl, North Anston, Anston, Sheffield, S31 7JY [IO93GJ, SK38]
G7 GFS Details withheld at licensee's request by SSL.
G7 GFU Details withheld at licensee's request by SSL.
G7 GFX P Everard, 7 Broomfield Cres, Mile Oak, Portslade, Brighton, BN41 2YU [IO90VU, TQ20]
G7 GFY W Weedon, Chapel Lodge, Oxburgh Hall, Oxborough, Kings Lynn Norfolk, PE33 9PS [JO02GN, TF70]
G7 GGA N P Dawson, 8 Manor Park, Norton Fitzwarren, Taunton, TA2 6SG [IO81KA, ST12]
G7 GGF Details withheld at licensee's request by SSL.
G7 GGG G E Richardson, 11 Queensway, Forest Town, Mansfield, NG19 0BX [IO93KD, SK56]
G7 GGH G Hurrell, The Otters, 15 Church Rd, Wootton Bridge, Ryde, PO33 4PT [IO90JR, SZ59]
G7 GGJ A W Edwards, 45 Chilton Gr, Yeovil, BA21 4AW [IO80QW, ST51]
G7 GGM D Thomalla, 14 Walkers Ln, Penketh, Warrington, WA5 2PA [IO83QJ, SJ58]
G7 GGN J H Williams, 41 Cote Green Ln, Marple Bridge, Stockport, SK6 5EB [IO83XJ, SJ99]
G7 GGP Details withheld at licensee's request by SSL.
G7 GGS Details withheld at licensee's request by SSL.
G7 GGT S J Mullins, 549 Bromford Ln, Washwood Heath, Birmingham, B8 2EA [IO92CL, SP18]
G7 GGU Details withheld at licensee's request by SSL.
G7 GGV A C Fielding, 4 Fen Ln, Sawtry, Huntingdon, PE17 5TG [IO92UK, TL18]
G7 GGX Details withheld at licensee's request by SSL.
G7 GGY H Goddard, 15 Allendale Rd, Wingerworth, Chesterfield, S42 6PX [IO93GE, SK36]
G7 GGZ Details withheld at licensee's request by SSL.
G7 GHA Details withheld at licensee's request by SSL.
G7 GHB B S A Beavan, 22B West Hill, Portishead, Bristol, BS20 9LQ [IO81OL, ST47]
GW7 GHE D R Pearson, Warren Cottage, Pontfadog, Llangollen, Clwyd, LL20 7AT [IO82KW, SJ23]
G7 GHH I Wraith, 25 Gleadless Ave, Sheffield, S12 2QG [IO93GI, SK38]
G7 GHI W E Stainforth, 2 Grangefield Terr, New Rossington, Doncaster, DN11 0LT [IO93LL, SK69]
GM7 GHL P E Meikle, 37 Turfholm, Lesmahagow, Lanark, ML11 0ED [IO85BP, NS83]
GI7 GHM T E Jensen, 12 Glenview Cres, Belfast, BT5 7LX [IO74BN, J37]
G7 GHO Details withheld at licensee's request by SSL.
G7 GHP G F Fellows, 34 The Ridings, Bexhill on Sea, TN39 5HU [JO00FU, TQ70]
G7 GHU Details withheld at licensee's request by SSL.
G7 GHW Details withheld at licensee's request by SSL.
GM7 GIB Details withheld at licensee's request by SSL.
G7 GIC Details withheld at licensee's request by SSL.
GM7 GIE Details withheld at licensee's request by SSL.
GM7 GIF Details withheld at licensee's request by SSL.
G7 GIG R R Vincent, 26 Tremayne Park, Pengegon, Camborne, TR14 7UT [IO70IF, SW63]
G7 GIJ J Barnett, 22 Highclere Rd, Bassett, Southampton, SO16 7AW [IO90GW, SU41]
GM7 GIL A J Munro, Flat, 36 Old Distillery, Dingwall, IV15 9XE [IO77SO, NH55]
G7 GIN P J Morris, 35 Milner St., Birkenhead, L41 8HE [IO83LJ, SJ38]
GM7 GIO W H B Mackinnon, 31 Kirk Bauk, Symington, Biggar, ML12 6LB [IO85EO, NS93]
GM7 GIS M Glendinning, 148 Gala Park, Galashiels, TD1 1HD [IO85OO, NT43]
G7 GIV C Guttridge, 103 Commercial Rd, Hazel Gr, Stockport, Ches, SK7 4BP [IO83WJ, SJ98]
G7 GJA P L Cockayne, 47 Queen St., Bilston, WV14 7ER [IO82XN, SO99]
G7 GJC S C Johnson, 54 Beechwood Cl, Chandlers Ford, Eastleigh, SO53 5PB [IO90HX, SU42]
G7 GJD K R Johnson, 54 Beechwood Cl, Chandlers Ford, Eastleigh, SO53 5PB [IO90HX, SU42]
G7 GJE D Johnson, 54 Beechwood Cl, Chandlers Ford, Eastleigh, SO53 5PB [IO90HX, SU42]
G7 GJF Details withheld at licensee's request by SSL.
G7 GJG Details withheld at licensee's request by SSL.
G7 GJH Details withheld at licensee's request by SSL.
G7 GJI D P Sager, 29 Station Rd, Mickleover, Derby, DE3 5GH [IO92FV, SK33]
G7 GJM C Unsworth, 3 Waterworks House, Hurleston, Nantwich, Ches, CW5 6BU [IO83RC, SJ65]
G7 GJN A Khachaturian, 377 Watford Rd, St. Albans, AL2 3DD [IO91TR, TL10]
G7 GJO P R Morris, 117 Lonsdale Ave, Doncaster, DN2 6HF [IO93KM, SE60]
G7 GJP K S Cassell, 74B Tomkinson Dr, Kidderminster, DY11 6NP [IO82UJ, SO87]
G7 GJQ Details withheld at licensee's request by SSL.
G7 GJS N L Cheesewright, 5 Duberly Cl, Perry, Huntingdon, Cambs, PE18 0BY [IO92UG, TL16]
G7 GJT W R Everton, Fencott, Fen Rd, Washingborough, Lincoln, LN4 1AE [IO93SF, TF07]
G7 GJU G M Darby, 60 Pine St., Grange Villa, Chester-le-Street, DH2 3LX [IO94EU, NZ25]
G7 GJV A Gordon, 1 Surrey St., Hetton-le-Hole, Houghton-le-Spring, DH5 9LX [IO94GT, NZ34]
GI7 GJX N C Simmons, 116 Killyglen Rd, Larne, BT40 2HX [IO74BU, D30]
G7 GJY J Chapman, 4 Chichester Rd, Bognor Regis, PO21 2XE [IO90PS, SZ99]
G7 GJZ C R Brown, 73 Ringstone, West Huntspill, Highbridge, TA9 3RF [IO81ME, ST34]
G7 GKA B Allen, The Rectory, Church Rd, Leonard Stanley, Stonehouse, Glos, GL10 3WP [IO81UR, SO80]
GI7 GKC I F Boyd, Site 17, Old Ridge Park, Derriaghy Rd, Lisburn, BT28 X [IO74AO, J37]
G7 GKD L A Tryhorn, 46 Mill Green Rd, Amesbury, Salisbury, SP4 7RE [IO91CE, SU14]
G7 GKH V I Bartoloni, 80 Long Valley Rd, Gillow Heath, Stoke on Trent, ST8 6QZ [IO83VD, SJ85]
G7 GKN S J Gordon, 6 Hartford Rise, Camberley, GU15 4HT [IO91PI, SU86]
G7 GKP C F Bates, 51 Curlew Dr, Tilehurst, Reading, RG3 4TA [IO91LJ, SU66]
G7 GKQ L Measures, 163 Huddersfield Rd, Meltham, Huddersfield, HD7 3AJ [IO93BO, SE11]
GM7 GKT R M Smith, 27 Elm Ln, Foresters Lodge, Glenrothes, KY7 5TD [IO86JE, NO20]
GW7 GKX J M L Rough, 10 Beaconsfield Rd, Shotton, Deeside, CH5 1EZ [IO83LF, SJ36]
G7 GLA J A Mitchinson, 93 Hinckley Rd, Leicester Forest East, Leicester, LE3 3GN [IO92JO, SK50]
G7 GLF Details withheld at licensee's request by SSL.
GI7 GLI Details withheld at licensee's request by SSL.
GM7 GLJ A T Potter, 12 Morlich Ct, Dalgety Bay, Dunfermline, KY11 5XU [IO86HA, NT18]
G7 GLL Details withheld at licensee's request by SSL.
G7 GLM C W R Benham, 11 Sandford Cl, Bransholme, Hull, HU7 4HJ [IO93UT, TA03]
G7 GLN G C H Garrett, 37 Pollard Ln, Bradford, BD2 4RN [IO93DT, SE13]
G7 GLO C Davey, 2 Gloucester Rd, Lupset, Wakefield, WF2 8NF [IO93FQ, SE32]
G7 GLQ D G Cottrell, 4 Gloucester Rd Nrt, Bristol, BS7 0SF [IO81RM, ST57]
G7 GLR I R Jackson Grtr Lndn Rayne, 5 Vivien Cl, Chessington, KT9 2DE [IO91UI, TQ16]
G7 GLS J Pinna, 19 Oxford Rd, Little Lever, Bolton, BL3 1DY [IO83TN, SD70]
G7 GLT A Marland, 71 Church Rd, Kearsley, Bolton, BL4 8AW [IO83TN, SD70]
G7 GLW R J Cains, 58 Sunnydale Rd, Lee, London, SE12 8JN [JO01AK, TQ47]
G7 GLY R A Da Silva Curiel, 1 Dartnell Cres, West Byfleet, KT14 6QG [IO91SI, TQ06]
G7 GLZ R C Hourston, 3 Aylward Gdns, Chesham, HP5 2QX [IO91QR, SP90]
GM7 GMC G M Christie, Burnbank, Hillside Rd, Stromness, KW16 3HR [IO88IX, HY21]
G7 GMD M B Ollerton, 7 Lakeside, Brighton Rd, Lancing, BN15 8LN [IO90UT, TQ10]
G7 GME I D Torr, 17 Fairways Dr, Kirkby in Ashfield, Nottingham, NG17 8NY [IO93IC, SK45]
G7 GMF Details withheld at licensee's request by SSL.
G7 GMG Details withheld at licensee's request by SSL.
GI7 GMJ Details withheld at licensee's request by SSL.
G7 GMM Details withheld at licensee's request by SSL.
G7 GMN Details withheld at licensee's request by SSL.
G7 GMQ D G Smith, 65 St. Anthonys Rd, Kettering, NN15 5JB [IO92PJ, SP87]
G7 GMU G J Lamb, Parisfield, Headcorn Rd, Staplehurst, Tonbridge, TN12 0BT [JO01GD, TQ74]
G7 GMV G Jenkins, 11 Scott Cl, Lichfield, WS14 9DB [IO92CQ, SK10]
G7 GMY J P Brown, 63 Lancaster Terr, Chester-le-Street, DH3 3NP [IO94FU, NZ25]
G7 GMZ W F Newton, 7 Moss Cl, Bridgwater, TA6 4NA [IO81MD, ST33]
G7 GNA L W Smith, 79 Laburnum Rd, Waterlooville, PO7 7EW [IO90LV, SU60]
G7 GNB Details withheld at licensee's request by SSL.
G7 GNE R J Strawson, 11 Stevenson Dr, Abingdon, OX14 1SN [IO91IQ, SU49]
GW7 GNF Details withheld at licensee's request by SSL.
G7 GNG Details withheld at licensee's request by SSL.
G7 GNH Details withheld at licensee's request by SSL.
G7 GNI Details withheld at licensee's request by SSL.
GM7 GNK Details withheld at licensee's request by SSL.
G7 GNL Details withheld at licensee's request by SSL.
G7 GNM A S Ramsdale, 2 Chestnut Ave, Kingston Rd, Willerby, Hull, HU10 6PA [IO93SS, TA02]
GM7 GNO N J D Goodall, Flat 61, 124 Lothian Rd, Edinburgh, Midlothian, EH3 9DD [IO85JW, NT27]
G7 GNU P Brayshaw, 38 Chilfrome Cl, Canford Heath, Poole, BH17 9WE [IO90AR, SZ09]
G7 GNY N Preedy, Holmlea Cottage, Church Ln, Paulton, BS18 5LF [IO81RH, ST65]
G7 GOA S A Constable, 20 Guildford Rd, Worthing, BN14 7LL [IO90TT, TQ10]
G7 GOC Details withheld at licensee's request by SSL.
GM7 GOD R H Williams, 5 Faulds Cres, Aberdeen, AB12 5QR [IO87WC, NJ90][Locator: IO92UN, WAB: TL19. All direct QSL's to this address please. Also from Leverton, nr Boston, Lincs (JO03BA TF44), and from RAF Waddington, Lincs (IO93RE SK96) Op: Riley H Williams, Interests: Raynet, WAB, packet.]
GM7 GOE Details withheld at licensee's request by SSL.
G7 GOH D A Osborne, 9 Glebe Cl, Osmington, Weymouth, DT3 6EY [IO80TP, SY78]
G7 GOK N A Breckell, 32 Greenbank Ave, St. Judes, Plymouth, PL4 8PS [IO70WI, SX45]
G7 GOP D W Searle, 40 Truro Ln, Penryn, TR10 8BW [IO70KE, SW73]
G7 GOQ S P Bracewell, 17 Wood St., Hapton, Burnley, BB12 7JU [IO83US, SD73]
G7 GOR N A Leonard, 329 Abbey Rd, Basingstoke, RG24 9EH [IO91KG, SU65]
G7 GOV M C Hill, 9 St. Annes Cl, Brackley, NN13 6DT [IO92JB, SP53]
G7 GOX J Brown, 24 Helford Rd, Peterlee, SR8 1ER [IO94HS, NZ43]
G7 GOZ E M Weiner, Brookside Farm, 1 Brookside Cl, Cheadle, SK8 1HP [IO83VJ, SJ88]
G7 GPA M P Knell, 13 Northd. Rd, Leamington Spa, CV32 6HE [IO92FH, SP36]
G7 GPB M W Knowles, 44 Park Est, Shavington, Crewe, CW2 5AP [IO83SB, SJ75]
GW7 GPD C A Hall, Lower Sea View, Greenfield Rd, Holywell, CH8 7PZ [IO83JG, SJ17]
GM7 GPG A Jakowuik, 78 Higholmside, Galashiels, Selkirkshire, Scotland, TD1 2HW [IO85OO, NT43]
G7 GPI A C Baily, 13 Longleigh Ln, Bexleyheath, DA7 5SL [JO01BL, TQ47]
G7 GPJ R C Banks, 23 North Park, Eltham, London, SE9 5AW [JO01AK, TQ47]
G7 GPT A Eisenberg, 19 Trevelyan Cres, Kenton, Harrow, HA3 0RN [IO91UN, TQ18]
GW7 GPU A M Sharman, 15 Airdale Spinney, Oulton Cross, Stone, ST15 8AZ [IO82WV, SJ93]
G7 GPY N A Barfoot, 11 Boulnois Ave, Parkstone, Poole, BH14 9NX [IO90AR, SZ09]
G7 GPZ S W Hall, 17 Staplefield Cl, Streatham Hill, London, SW2 4AE [IO91WK, TQ37]
G7 GQB M H Woodhouse, 18 Soame Cl, Aylsham, Norwich, NR11 6JF [JO02PS, TG12]
G7 GQC G D Beckingham, 20 Baptist Cl, Abbeymead, Gloucester, GL4 5GD [IO81VU, SO81]
G7 GQD D Pearce, 8 New St., Ardsley, Barnsley, South Yorks, S71 5AJ [IO93GN, SE30]
G7 GQF Details withheld at licensee's request by SSL.
G7 GQH R B Hannemann, 33 Malvern Cl, High Wycombe, HP13 5SA [IO91OP, SU89]
G7 GQJ F Stewart, 116 Covert Rd, Holts Est, Oldham, OL4 5PH [IO83XM, SD90]
G7 GQK O Bowden, 2 Commercial Rd, Coxhoe, Durham, DH6 4HJ [IO94FR, NZ33]
G7 GQM E B Sutton, 15 Lowther St., Penrith, CA11 7UW [IO84OQ, NY53]
G7 GQN Details withheld at licensee's request by SSL.
G7 GQO R Harman, 58 Pickford Ln, Bexleyheath, DA7 4QT [JO01BL, TQ47]
G7 GQP Details withheld at licensee's request by SSL.
G7 GQQ Details withheld at licensee's request by SSL.
G7 GQS I N Hunnisett, 69 Cornwall Rd, Ruislip, HA4 6AJ [IO91TN, TQ08]
G7 GQW D A Williams, 28 Mill Ln, Great Sutton, South Wirral, L66 3PF [IO83MG, SJ37]
G7 GQX D W Howard, 6 Draycote Cl, Damsonwood, Solihull, B92 9PT [IO92CK, SP18]
G7 GRA D G Cooke, 98 Penhill Dr, Swindon, SN2 5LL [IO91CO, SU18]
G7 GRB H D F Ewing, 17 Walkley Rd, Dartford, DA1 3BH [JO01CK, TQ57]
G7 GRC J C Whiting Grantham Radio Club, 97 Barrowby Gate, Grantham, NG31 8RB [IO92QV, SK83]
GM7 GRH N J Hardie, 38 Sentry Knowe, Selkirk, TD7 4BG [IO85ON, NT42]
G7 GRN R C Selby, 63 Globe Farm Ln, Darby Green, Blackwater, Camberley, GU17 0DZ [IO91OH, SU86]
G7 GRQ A T Gray, 79 Brougham Terr, Hartlepool, TS24 8EU [IO94JQ, NZ53]
G7 GRR S R Everett, 118 Queens Park Gdns, Crewe, CW2 7SW [IO83SC, SJ65]
GI7 GRY S J Gordon, 138 Mullalelish Rd, Richhill, Armagh, BT61 9LT [IO64RI, H94]
GI7 GSB Details withheld at licensee's request by SSL.
G7 GSC N Godden, 23 Rapsons Rd, Willingdon, Eastbourne, BN20 9RJ [JO00CT, TQ50]
G7 GSD K F Osborn, 207 Lyndhurst Rd, Worthing, BN11 2DN [IO90TT, TQ10]
G7 GSF S J Blandford, 19 Amwell Rd, Kings Hedges, Cambridge, CB4 2UH [JO02BF, TL46]
GI7 GSH C Henderson, 7 Legaloy Rd, Ballyclare, BT39 9PS [IO74AS, J39]
G7 GSL Details withheld at licensee's request by SSL.
G7 GSN Details withheld at licensee's request by SSL.
G7 GSO Details withheld at licensee's request by SSL.
G7 GSP Details withheld at licensee's request by SSL.
G7 GSQ Details withheld at licensee's request by SSL.
G7 GSU Details withheld at licensee's request by SSL.
G7 GSW Details withheld at licensee's request by SSL.
G7 GSX C F Penfold, Moray, 149 Shuttlewood Rd, Bolsover, Chesterfield, S44 6NX [IO93IF, SK47]
G7 GTA Details withheld at licensee's request by SSL.
G7 GTB Details withheld at licensee's request by SSL.
G7 GTD R Broadbent, 69 Albany Dr, Herne Bay, CT6 8PU [JO01NI, TR16]
G7 GTG A J Hyndman, Norman House, Railway Terr, Kings Langley, Herts, WD4 8JE [IO91SR, TL00]
G7 GTH A V Marriott, Norman House, Railway Terr, Kings Langley, Herts, WD4 8JE [IO91SR, TL00]
GM7 GTK K G Small, 167 Alloway Dr, Kirkintilloch, Glasgow, G66 2SB [IO75WW, NS67]
G7 GTN M J Stevens, 13 Downs Rd, Westbury on Trym, Bristol, BS9 3TX [IO81QL, ST57]
G7 GTO Details withheld at licensee's request by SSL.
G7 GTP J G Hale, 19 Cardigan Rd, Winton, Bournemouth, BH9 1BD [IO90BR, SZ09]
G7 GTQ S M Laddiman, Kingsleigh, 15 The St., South Walsham, Norwich, NR13 6AH [JO02RP, TG31]
GM7 GTS C Richman, 18 Nigel Rise, Livingston, EH54 6LT [IO85FV, NT06]
G7 GTV Details withheld at licensee's request by SSL.
GW7 GTW Details withheld at licensee's request by SSL.
G7 GUA A P Dennis, Fishermans House, Goonlaze, Stithians, Cornwall, TR3 7AR [IO70JE, SW73]
G7 GUB A F Alderton, 7 Bigland Dr, Ulverston, LA12 9NU [IO84KE, SD27]
G7 GUC Details withheld at licensee's request by SSL.
G7 GUD Details withheld at licensee's request by SSL.
G7 GUF Details withheld at licensee's request by SSL.
G7 GUG G L Wales, 7 Montgomery Ave, Hampton on The Hill, Budbrooke, Warwick, CV35 8QP [IO92EG, SP26]
G7 GUJ Details withheld at licensee's request by SSL.
G7 GUK Dr D Nelson, 101 Gledhow Ln, Roundhay, Leeds, LS8 1NE [IO93FT, SE33]
GI7 GUM Details withheld at licensee's request by SSL.
G7 GUN Details withheld at licensee's request by SSL.
G7 GUO S D Falconer, 6 Ogilvie Rd, High Wycombe, HP12 3DS [IO91OP, SU89]
G7 GUR Details withheld at licensee's request by SSL.
G7 GUS B S Wilkins, 14 Cromwell Rd, High Wycombe, HP13 7AN [IO91PO, SU89]
GI7 GUT D Watt, 51 Rashee Rd, Ballyclare, BT39 9HT [IO64XS, J29]
G7 GUW J C Owen, 26 Redfern Ave, Sale, M33 2TJ [IO83UJ, SJ89]
G7 GVB J H Dawber, 26 Whitecroft Rd, Hawkley Hall, Wigan, WN3 5PS [IO83QM, SD50]
G7 GVC R Townley, 9 Cooks Cl, Fugglestone Red, Salisbury, SP2 9PS [IO91CC, SU13]

GM7 GVD D G Innes, 6 Mamore Terr, Kinmylies, Inverness, IV3 6PF [IO77UL, NH64]
G7 GVH S I Parry, 1 Lopes Dr, Roborough, Plymouth, PL6 7PH [IO70WK, SX55]
GI7 GVI T M Henderson, 7 Legaloy Rd, Ballyclare, BT39 9PS [IO74AS, J39]
G7 GVJ S R Fletcher, Fernleigh, Ash Ln, Down Hatherley, Gloucester, GL2 9PS [IO81VV, SO82]
G7 GVK Details withheld at licensee's request by SSL.
G7 GVP C Price, 10 School Rd, Dursley, GL11 4PB [IO81TQ, ST79]
G7 GVV Details withheld at licensee's request by SSL.
G7 GVZ K Spencer, 92 Melbury Rd, Bilborough, Nottingham, NG8 4AU [IO92JX, SK54]
G7 GWA A J Jakins, 11 Abbott Way, Yaxley, Peterborough, PE7 3YF [IO92UM, TL19]
GI7 GWB Details withheld at licensee's request by SSL.
G7 GWF D A Kennedy, 11 Silverwood Dr, Laverstock, Salisbury, SP1 1SH [IO91CB, SU13]
G7 GWH E W Leaver, 2 Beltana Dr, Gravesend, DA12 4BS [JO01EK, TQ67]
G7 GWI K D Lovegrove, 96 Woodrow Cres, Knowle, Solihull, B93 9EQ [IO92DJ, SP17]
GW7 GWM T W Jones Gwynedd County Council Ry H, Emergency Planning Unit, Isle of Anglesey Co. Count., Council Offices, Llangefni, Anglesey, LL77 7TW [IO73UG, SH47]
GW7 GWO S J Evans, Hazelbrook, Melin y Coed, Cardigan, Dyfed, SA43 1PE [IO72QC, SN14]
G7 GWP Details withheld at licensee's request by SSL.
G7 GWR Details withheld at licensee's request by SSL.
G7 GWZ K J Dixon, 50 Greenfield Cres, Cowplain, Waterlooville, PO8 9EJ [IO90LV, SU61]
G7 GXE P D Kitson, 15 Louvain Rd, Derby, DE23 6DA [IO92GV, SK33]
G7 GXF M A Kent, 8 Chelmorton Pl, Chaddesden, Derby, DE21 4QL [IO92GW, SK33]
G7 GXG Details withheld at licensee's request by SSL.
GM7 GXI G M Cowan, 85 Eastwoodmains Rd, Clarkston, Glasgow, G76 7HG [IO75UT, NS55]
G7 GXK Details withheld at licensee's request by SSL.
G7 GXM M Wynn, Howards Croft, 187 Trysull Rd, Merryhill, Wolverhampton, WV3 7JP [IO82WN, SO89]
G7 GXO G Angelou, 16 Brookside, Watlington, OX9 5AQ [IO91LP, SU69]
G7 GXQ Details withheld at licensee's request by SSL.
G7 GXR B Clewes, 19 Church Mews, Denton, Manchester, M34 3GL [IO83WK, SJ99]
G7 GXX J A Gwillam Daventry Xx C G, Park House, 4 Tudor Ln, Southam, Leamington Spa, CV33 0HS [IO92HF, SP46]
GI7 GXZ S Dornan, 108 Glen Rd, Ballygowan Rd, Castlereagh, Belfast, BT5 7LU [IO74BN, J37]
G7 GYB Details withheld at licensee's request by SSL.
G7 GYF Details withheld at licensee's request by SSL.
G7 GYH G G Clarke, 19 Park Ave, Hildenborough, Tonbridge, TN11 9DE [JO01DF, TQ54]
G7 GYN C Barlow, 7 Strokins Rd, Kingsclere, Newbury, RG20 5RH [IO91JH, SU55]
G7 GYR K R Wade, Eccleston Hall, Lydiate Ln, Eccleston, Chorley, PR7 6LY [IO83PP, SD51]
G7 GYY J W Bewley 2309 Sqn ATC, Plymth & Cornwall West, 21 Duloe Gdns, Pennycross, Plymouth, PL2 3RS [IO70WJ, SX45]
G7 GZB C J Davies, 84 Hob Hey Ln, Culcheth, Ches, WA3 4NW [IO83RK, SJ69]
G7 GZC D H E Coles, 8A Park Ave, Hounslow, TW3 2LZ [IO91TK, TQ17]
G7 GZE Details withheld at licensee's request by SSL.
G7 GZG Details withheld at licensee's request by SSL.
G7 GZK I P Croft, 34 Laburnum Dr, Armthorpe, Doncaster, DN3 3HE [IO93LM, SE60]
G7 GZL Details withheld at licensee's request by SSL.
G7 GZS M G Daniels, 21 Hornes End Rd, Flitwick, Bedford, MK45 1JH [IO91SX, TL03]
G7 GZU S M Selwyn, 7 Elizabeth Cl, Thornbury, Bristol, BS12 2YN [IO81RO, ST68]
G7 GZV H G Houldershaw, The (First)Bungalow, Fen Rd, Stickford, Boston Lincs, PE22 8EX [JO03AC, TF35]
G7 GZW G E Bennett, Tanglewood, 134 The Shields, Ilfracombe, EX34 8JX [IO71WE, SS54]
G7 GZZ E M Gaffney, 1 White Hart Ln, Wistaston, Crewe, CW2 8EX [IO83SB, SJ65]
G7 HAA Details withheld at licensee's request by SSL.
GW7 HAE C M Davies, Afallon, 3 Penygraig, Aberystwyth, Dyfed, SY23 2JA [IO72XK, SN58]
G7 HAF W Hunton, Hatch Green, 60 Bondgate, Helmsley, York, YO6 5EZ [IO94LF, SE68]
G7 HAH G Boothroyd Finningley Amateur Radio Soc., Finningley Ar Society, 38 Ascot Ave, Cantley, Doncaster, DN4 6HE [IO93KM, SE60]
G7 HAJ K Barnard, 56 George St., Keadby, Scunthorpe, DN17 3DB [IO93PO, SE81]
G7 HAR B H Ferris, 70 Braithwell Rd, Maltby, Rotherham, S66 8JU [IO93JK, SK59]
G7 HAT Details withheld at licensee's request by SSL.
G7 HAW Details withheld at licensee's request by SSL.
G7 HBB C C Dodshon, 62 Moor Rd, Melsonby, Richmond, DL10 5PE [IO94DL, NZ10]
G7 HBC D R Waldren, Ippleden, Weavering St., Weavering, Maidstone, ME14 5JN [JO01GG, TQ75]
G7 HBF J C Sheehan, 2 Link Way, Thatcham, RG18 3DY [IO91IJ, SU56]
G7 HBH R Leach, Kirklands The Avenue, Eaglescliffe, Stockton on Tees, Cleveland, TS16 9AS [IO94HM, NZ41]
G7 HBI R P Isaac, Northleigh House, Northleigh Hill, Goodleigh, Barnstaple, EX32 7NR [IO81AC, SS63]
G7 HBJ D R Stevens, 63 Byron Ave, Coulsdon, CR5 2JS [IO91WH, TQ35]
G7 HBN P R Osborne, 11 Galston Rd, Luton, LU3 3JZ [IO91SW, TL02]
G7 HBO R E Cornell, 18 Holland Park Ave, Newbury Park, Ilford, IG3 8JR [JO01BN, TQ48]
G7 HBR A E Edgar, 27 Earlsfield, Branston, Lincoln, LN4 1NP [IO93SE, TF06]
G7 HBS J R McCormack, Albamor, 25 Kingfisher Cl, Hayling Island, PO11 9NS [IO90MS, SZ79]
GM7 HBT L L Crompton, 23 Haugh Rd, Dalbeattie, DG5 4AR [IO84CW, NX86]
G7 HBU T E Hickling, 6 Harrold Rd, Bozeat, Wellingborough, NN29 7LP [IO92PF, SP95]
G7 HBV C F Heard, 42 Hallowell Down, South Woodham Ferrers, Chelmsford, CM3 5FS [JO01HP, TQ79]
G7 HBW V G M Yates, 2 Meadow Cl, Mundesley, Norwich, NR11 8LW [JO02RV, TG33]
G7 HCB B A Atterbury, 90 Ely Cl, Stevenage, SG1 4NR [IO91VW, TL22]
G7 HCC D N Jones, Mill Cottage, 120 Heathfield Rd, Keston, BR2 6BA [JO01AI, TQ46]
G7 HCD S N Wilton, 32 Deerhurst Chase, Bicknacre, Chelmsford, CM3 4XG [JO01HQ, TL70]
G7 HCE D Boult, 2 Headingley Cl, Exeter, EX2 5UH [IO80GR, SX99]
G7 HCF M Ashfield, 24 in The Ray, Maidenhead, SL6 8DH [IO91PM, SU88]
G7 HCI G N Burton, 33 Hollins Park, Moor Row, CA24 3LQ [IO84FM, NY01]
G7 HCJ A D Houlton, 9 Manvers Rd, West Bridgford, West Bridgford, Nottingham, NG2 6DJ [IO92KW, SK53]
G7 HCL P Good, 80 Meredith Rd, Stevenage, SG1 5QS [IO91VV, TL22]
G7 HCO N J Lambert, Bradfields Farm, Burntmills Rd, North Benfleet, Wickford, SS12 9JX [JO01GO, TQ79]
G7 HCQ D F Browne, 293 St. Albans Rd, Hemel Hempstead, HP2 4RP [IO91SR, TL00]
G7 HCR G J Richardson, The Homestead, Washway Rd, Holbeach, Spalding, PE12 7PP [JO02AT, TF32]
G7 HCT K E Moore, 8 Lilac Cl, Toftwood, Dereham, NR19 1JY [JO02LP, TF91]
G7 HCU R M Boardman, 76 Gravel Ln, Hemel Hempstead, HP1 1SA [IO91SS, TL00]
G7 HCV A S Boyes, 7 Thornwood Covert, Foxwood, Acomb, York, YO2 3LF [IO93KW, SE55]
G7 HDA Details withheld at licensee's request by SSL.
G7 HDB Details withheld at licensee's request by SSL.
G7 HDC Details withheld at licensee's request by SSL.
G7 HDI Details withheld at licensee's request by SSL.
G7 HDK Details withheld at licensee's request by SSL.
G7 HDL Details withheld at licensee's request by SSL.
G7 HDN Details withheld at licensee's request by SSL.
G7 HDO Details withheld at licensee's request by SSL.
GW7 HDP Details withheld at licensee's request by SSL.
G7 HDQ Details withheld at licensee's request by SSL.
G7 HDR D A Horder, 77 Gr Ave, Harpenden, AL5 1EZ [IO91UT, TL11]
G7 HDT G E Conn, 80 St. Cuthberts Dr, Heworth, Gateshead, NE10 9AD [IO94FW, NZ26]
G7 HDW J M Bigger, 18 Wallace Rd, Loughborough, Leics, LE11 3NX [IO92JS, SK51]
GW7 HDX P G Jones, Great House, Church St., Llangadog, Dyfed, SA19 9AA [IO81BW, SN72]
G7 HDZ A Rowell, 105 Hedgehope Rd, Newbiggin Hall Est, Newcastle upon Tyne, NE5 4LB [IO95DA, NZ26]
GW7 HEC P A Redman, 87 Wepre Park, Connahs Quay, Deeside, CH5 4HL [IO83LF, SJ26]
GW7 HEE R Roberts, 14 Aled Ave, Rhyl, LL18 2HN [IO83GH, SJ08]
G7 HEF K Howard, 13 Haningwood Terr, Felling, Gateshead, NE10 0UE [IO94FW, NZ26]
G7 HEL Details withheld at licensee's request by SSL.
G7 HEN M B Priestley, 29 Birchlands Ave, Wisden, Wilsden, Bradford, BD15 0HB [IO93BT, SE03]
G7 HEO Details withheld at licensee's request by SSL.
G7 HEP A S Ellis, 29 Corfe Cl, Hillhead, Stubbington, Fareham, PO14 3NN [IO90JT, SU50]
G7 HER Y M Campbell, 5 Farne Rd, Shiremoor, Newcastle upon Tyne, NE27 0PQ [IO95FA, NZ37]
G7 HES W J Robe, Whitewood, Sandy Bank, Riding Mill, NE44 6HU [IO94AW, NZ06]
G7 HET Details withheld at licensee's request by SSL.
G7 HEV B L Phillips, 49 Langley Gr, Nyetimber, Bognor Regis, PO21 4LJ [IO90PS, SZ99]
GI7 HEW B Bennett, 31 Fernisky Park, Kells, Ballymena, BT42 3LL [IO64VT, J19]
G7 HEY L A Morrell-Cross, Delta Lodge, 14 Rushton Cres, Bournemouth, BH3 7AF [IO90BR, SZ09]
G7 HEZ T J French, 2 Pepper Hill, Stourbridge, DY8 1BJ [IO82WK, SO98]
G7 HFC D W Bradley, 1 Traddles Ct, Chelmsford, CM1 4XZ [JO01FS, TL60]
G7 HFE S L Jallands, 314 Porchester Rd, Mapperley, Nottingham, NG3 6GR [IO92KX, SK54]
G7 HFH D G Widdus, 4 Jennifer Gdns, Margate, CT9 3XX [JO01QI, TR36]
G7 HFL C D Elphick, 12 Nairn House, Cameron Cl, Warley, Brentwood, CM14 5BP [JO01DO, TQ59]
G7 HFP C E Castle, 42 Saddle Rise, Springfield, Chelmsford, CM1 6SX [JO01FS, TL70]
G7 HFQ W E Betts, 4 Victoria Rd, Addlestone, KT15 2PL [IO91SI, TQ06]
G7 HFS I K Harling, 224 Sevenoaks Rd, Eastbourne, BN23 7SA [JO00DT, TQ60]
G7 HFU G T Huxtable, Cherry Bean, St. Margarets, Great Gaddesden, Hemel Hempstead, HP1 3BZ [IO91RT, TL01]
G7 HFX B L Nelhams, 2 St James's Villas, Hampton Ln, Hanworth, Middx, TW13 6NP [IO91TK, TQ17]
G7 HGB J Dunn, 10 Endsleigh Cl, Upton, Chester, CH2 1LX [IO83NF, SJ46]
G7 HGF I Simpson, 1 Ryton Ct, The Meadows, Nottingham, NG2 2JH [IO92KW, SK53]
G7 HGI R A Roberts, 13 Tudor Way, Wickford, SS12 0HS [JO01GO, TQ79]
GW7 HGJ M Parry, 2 Jubilee Dr, Halkyn, Holywell, CH8 8DT [IO83JF, SJ27]
G7 HGQ D J Horwood, 12 Curtis Cl, Mill End, Rickmansworth, WD3 2QA [IO91SP, TQ09]
G7 HGR J Collett, 58 St. Ives Rd, Wyken, Coventry, CV2 5FX [IO92GJ, SP37]
G7 HGS G S Squires, 7 Daleside Dr, Potters Bar, EN6 2LL [IO91VQ, TL20]
GW7 HGU M J Howard, 64 Lawrenny St., Neyland, Milford Haven, SA73 1TB [IO71MR, SM90]
G7 HGX C R Shelley, Mallards, Church Rd, Peldon, Colchester, CO5 7PT [JO01KT, TL91]
G7 HGY P S Bryant, 14 Bugby Way, Raunds, Wellingborough, NN9 6SX [IO92RI, SP97]
GM7 HHB J P Brown, 22 Meadow Ln, Renfrew, PA4 8TD [IO75TV, NS56]
G7 HHC J M Sedgley Haywards Hth AR, 51 Gordon Cl, Haywards Heath, RH16 1ER [IO91WA, TQ32]
G7 HHG P S Baldwin, 13 Maple Hatch Cl, Godalming, GU7 1TQ [IO91QE, SU94]
G7 HHI S F Curry, Barnabus Cottage, Egley Rd, Mayford, Woking, Surrey, GU22 0NQ [IO91RH, SU95]
G7 HHK R G Johnson, Honeysuckle Cottage, Appleton Wiske, Northallerton, North Yorks, DL6 2AA [IO94HK, NZ30]
G7 HHL R P Garner, 3 Cozens Hardy Rd, Sprowston, Norwich, NR7 8QE [JO02PP, TG21]
G7 HHM L S Dring, 22 Castle St., Eastwood, Nottingham, NG16 3GW [IO93IA, SK44]
G7 HHN K R Glover, Oaklea, Swan St., Chappel, Colchester, Essex, CO6 2ED [JO01JV, TL82]
G7 HHO J K Halliwell, 17 Birtley Ave, Tynemouth, North Shields, NE30 2RR [IO95GA, NZ36]
G7 HHQ R A Saunders, The Grange, High Rd, Wisbech St. Mary, Wisbech, PE13 4RG [JO02BP, TF40]
G7 HHT M S Gotts, 65 Willow St., Romford, RM7 7LB [JO01CN, TQ58]
G7 HHU A J Edwards, 3 Simonside Cl, Morpeth, NE61 2XY [IO95DD, NZ18]
G7 HHW G D Phillips, 87 Hemerdon Heights, Plymouth, PL7 2EZ [IO70XJ, SX55]
G7 HHZ J F Whelan, 23 Drury Ave, Spondon, Derby, DE21 7FZ [IO92HW, SK33]
G7 HIA J Heath, Chestnuts, Desford Ln, Kirkby Mallory, Leics, LE9 7QF [IO92IO, SK40]
G7 HIC Details withheld at licensee's request by SSL.
G7 HID M Burgess, 63 Chalvey Park, Slough, SL1 2HX [IO91QM, SU97][Op: Mike Burgess, Tel: (01753) 76362.]
G7 HIF S M O Walker, 8 Browning Ave, Kettering, NN16 8NP [IO92PJ, SP88]
G7 HIH R Pedro, 65 Glebe Cres, Kenton, Harrow, HA3 9LB [IO91UO, TQ18]
G7 HII D J Lloyd, 47 Burton Wood, Weobley, Hereford, HR4 8SZ [IO82ND, SO45]
G7 HIJ J R Gunia, 21 Campbell Ave, Leek, ST13 5RR [IO83XC, SJ95]
G7 HIK J Doherty, 4 St. Austins Cl, Ivybridge, PL21 9BZ [IO80AJ, SX65]
G7 HIM K P W Leeder, 99 Vale Green, Norwich, NR3 2EL [JO02PP, TG21]
G7 HIN P Riddell, 13 Pear Tree Rd, Addlestone, KT15 1SR [IO91SI, TQ06]
G7 HIO W G Austin, 53 Giantswood Ln, Congleton, CW12 2HQ [IO83VE, SJ86]
GM7 HIR A J Pert, 56 Lochiel Dr, Milton of Campsie, Glasgow, G65 8ET [IO75WX, NS67]
G7 HIT P A Chambers, 7 Redland Cl, Chilwell, Beeston, Nottingham, NG9 5LA [IO92JV, SK53]
G7 HIW R A Hull, 56 Lead Ln, Ripon, HG4 2LN [IO94FC, SE36]
G7 HIX R H Gray, 12 St. Francis Cl, Deal, CT14 9LS [JO01QF, TR35]
G7 HIY T J Jefford, 7 Bellevue St., Folkestone, CT20 1HY [JO01OB, TR23]
G7 HJB Details withheld at licensee's request by SSL.
G7 HJD G J Holland, 11 Swanton Dr, Dereham, NR20 4DW [JO02LQ, TF91]
G7 HJH N V Pickering, The Hawthorns, The Village, Westbury on Severn, GL14 1LN [IO81TT, SO71]
G7 HJJ H M Holman, 62 The Ridge, Kennington, Ashford, TN24 9EU [JO01KD, TR04]
G7 HJK R J Kearnes, 25 Epsom Cl, Clacton on Sea, CO16 8FE [JO01NT, TM11]
G7 HJN S L Tweed, 42 Ophir Rd, Worthing, BN11 2SS [IO90TT, TQ10]
G7 HJP J C Whittaker, Riverside, 48 Baunton, Cirencester, Glos, GL7 7BB [IO91AR, SP00]
G7 HJQ M E Erber, 75 St Andrews Rd Nt, Lytham St Annes, Lancs, FY8 2JF [IO83LS, SD32]
G7 HJR T S P Rudderham, 25 Mount Pleasant, Holton-le-Clay, Grimsby, South Humberside, DN36 5ED [IO93XM, TA20]
G7 HJT T Reynard, 19 Oakbank Dr, Keighley, BD22 7DX [IO93AU, SE04]
G7 HJU N M D Smith, The Flat, 95 High St., Brackley, NN13 7BW [IO92KA, SP53]
G7 HJV C R Johnson, 6 Broadway Ave, Wallasey, L45 6TA [IO83LK, SJ29]
G7 HJX D R Raybould, 63 Rochester Ave, Chase Terr, Walsall, West Midlands, WS7 8DL [IO92AQ, SK00]
G7 HJY Details withheld at licensee's request by SSL.
G7 HJZ R W Bateman, 24 Shakespeare Rd, Cheltenham, GL51 7HA [IO81WV, SO92]
G7 HKA Details withheld at licensee's request by SSL.
G7 HKD Details withheld at licensee's request by SSL.
G7 HKK Details withheld at licensee's request by SSL.
G7 HKM Details withheld at licensee's request by SSL.
G7 HKN P J Walsh, 2 Elm Rd, Winwick, Warrington, WA2 9TW [IO83QK, SJ69]
G7 HKP Details withheld at licensee's request by SSL.
G7 HKQ I S Tideswell, 2 Pangbourne Ave, Davyhulme, Urmston, Manchester, M41 0GF [IO83TK, SJ79]
GW7 HKR Details withheld at licensee's request by SSL.
G7 HKT C A Fowle, 69 Essetford Rd, Ashford, TN23 5BP [JO01KD, TR04]
G7 HKU J W Turner, 7 Highfield Cres, Baildon, Shipley, BD17 5NR [IO93CU, SE13]
G7 HKW R I Penn, 11 Feneley Cl, Deeping St. James, Peterborough, PE6 8HN [IO92UQ, TF10]
G7 HKY Details withheld at licensee's request by SSL.
G7 HKZ T K Allen, 1 Tetherdown, Prestwood, Great Missenden, HP16 0RY [IO91PQ, SP80]
G7 HLC Details withheld at licensee's request by SSL.
G7 HLD C P Woolley, 12 Heathfield Rd, Farmhill, Stroud, GL5 4DQ [IO81VS, SO80]
G7 HLE Details withheld at licensee's request by SSL.
G7 HLG B E Morrell, 10 Batchelor Cres, West Howe, Bournemouth, BH11 8HE [IO90BS, SZ09]
G7 HLH Details withheld at licensee's request by SSL.
GM7 HLI W J Finlay, 34 Balnafettack Rd, Inverness, IV3 6TF [IO77UL, NH64]
G7 HLJ Details withheld at licensee's request by SSL.
G7 HLL Details withheld at licensee's request by SSL.
G7 HLP K D Baldock, 66 Port Rd, New Duston, Northampton, NN5 6NL [IO92MG, SP76]
G7 HLT J C Ricketts, 82 Main St., Weston on Trent, Derby, DE72 2BL [IO92HU, SK42]
G7 HLU V R Meads, 18 Chatsworth Dr, Wellingborough, NN8 5FB [IO92PH, SP86]
G7 HLV J E Jordan, 4 Cliffe Cres, Riddlesden, Keighley, BD20 5LB [IO93BU, SE04]
G7 HLW G P W Burn, 4 Goston Gdns, Thornton Heath, CR7 7NQ [IO91WJ, TQ36]
GW7 HLZ R J Davies, 27 Belmont Rd, Abergavenny, NP7 5HN [IO81LT, SO31]
G7 HMA I H Smith, 75 Hesketh Rd, Yardley Gobion, Towcester, NN12 7TS [IO92NC, SP74]
G7 HMB G Bull, 48 Spragghouse Ln, Norton, Stoke on Trent, Staffs, ST6 8DX [IO83WB, SJ85]
G7 HMC P E Glasscock, 37 Huntingfield Rd, Bury St. Edmunds, IP33 2JA [JO02IF, TL86]
G7 HMF E E Last, 134 New Queens Rd, Sudbury, CO10 6PJ [JO02IB, TL84]
GD7 HMG W J Smith, 1 High View Rd, Douglas, IM2 5BQ
G7 HMI R C Shelford, 2 Edwards Walk, Earith, Huntingdon, PE17 3QX [JO02AI, TL37]
G7 HMK A L Baldwin, B.M. Box 6902, London, WC1N 3XX [IO91WM, TQ38]
G7 HMN C Boutell, 48 Cambridge Rd, Clacton on Sea, CO15 3QL [JO01NT, TM11]
G7 HMQ B G Boult, Prince St Post Office, Prince St, Bristol, BS1 4PJ [IO81QK, ST57]
G7 HMR P G Lee, 18 Fleckers Dr, Hatherley, Cheltenham, GL51 5BD [IO81WV, SO92]
G7 HMS P Balaam, 57 Ruby Rd, Walthamstow, London, E17 4RE [IO91XO, TQ38]
G7 HMT J B Stratton, 45 Paddenswick Rd, Hammersmith, London, W6 0UA [IO91VL, TQ27]
G7 HMV M J Wood, 126 Valley Rd, Codicote, Hitchin, SG4 8YN [IO91VU, TL21]
G7 HMW W Knight, 30 Stretford Rd, Urmston, Manchester, M41 9JZ [IO83TK, SJ79]
G7 HMZ A Murrin, 31 Kings Rd, St. Neots, Huntingdon, PE19 1LD [IO92UF, TL16]
G7 HNC A D Tambini, 12 Bramdene Ave, Weddington, Nuneaton, CV10 0DH [IO92GM, SP39]
G7 HNF J M Baldwin, B.M. Box 6902, London, WC1N 3XX [IO91WM, TQ38]
G7 HNL M J Carter, Meadow View, Bransford Rd, Rushwick, Worcester, WR2 5SJ [IO82UE, SO85]
G7 HNM G P Greatrix, 80 Liquorpond St., Boston, PE21 8UJ [IO92XX, TF34]
G7 HNO D B Pickerill, 61 Worthington Rd, Balderton, Newark, NG24 3RE [IO93OB, SK85]
G7 HNR A T Ball, 94 Marshall Lake Rd, Shirley, Solihull, B90 4PN [IO92CJ, SP17]
GM7 HNU G V Banks, 9 Powis Cres, Aberdeen, AB2 3YS [IO87TH, NJ72]
G7 HNW W A Eyre, Conholt, 11 Parkhurst Rd, Guildford, GU2 6AP [IO91QF, SU95]
G7 HNX H A Campbell, 2 Roebuck Green, Cippenham, Slough, SL1 5QY [IO91QM, SU98]
G7 HNY M J Lenzi, 12 Putton Ln, Chickerell, Weymouth, DT3 4AG [IO80SO, SY68]
G7 HOA D Wilson Widnes & Runcorn ARC, 12 New St., Elworth, Sandbach, CW11 3JF [IO83TD, SJ76]
G7 HOB Details withheld at licensee's request by SSL.
G7 HOC D J Warburton, 19 Walkers Heath Rd, Kings Norton, Birmingham, B38 0AB [IO92BJ, SP07]
G7 HOD Details withheld at licensee's request by SSL.
G7 HOE P Goode, 23 Byworth Rd, Farnham, GU9 7BT [IO91OF, SU84]
G7 HOG B N Munro, 56 Purbrook Gdns, Purbrook, Waterlooville, PO7 5LD [IO90LU, SU60]
G7 HOI Y V Munro, 56 Purbrook Gdns, Purbrook, Waterlooville, PO7 5LD [IO90LU, SU60]
G7 HOJ B R Bennett, 7 Maple Ave, Fishponds, Bristol, BS16 4HJ [IO81RL, ST67]
G7 HOK P C Kellingley, 4 Gage Cl, Old Basing, Lychpit, Basingstoke, RG24 8SE [IO91LG, SU65]
G7 HOL D R Martin, 2 Watermill Cl, Maidstone, ME16 0NE [JO01GG, TQ75]
GW7 HOM V B Cole, 5 Manobier Cl, Tonteg, Pontypridd, CF38 1HL [IO81IN, ST08]
G7 HON S S Martin, Broad Oak, Pheasant Ln, Loose, Maidstone Kent, ME15 9QR [JO01GG, TQ75]
G7 HOP K P Mullett, 16 Trent Cres, Melksham, SN12 8BG [IO81WJ, ST96]
G7 HOS J H Bowles, 23 Stirtingale Rd, Bath, BA2 2NF [IO81TI, ST76]
G7 HOT J M Scott, 81 Churchill Dr, Newark, NG24 4LU [IO93OB, SK75]
G7 HOV J C R Bertram, 68 Belmont Ave, New Malden, KT3 6QD [IO91VJ, TQ26]
G7 HOW Details withheld at licensee's request by SSL.
G7 HOX Details withheld at licensee's request by SSL.
G7 HOY Details withheld at licensee's request by SSL.
G7 HPB Details withheld at licensee's request by SSL.
G7 HPD Details withheld at licensee's request by SSL.

G7

G7 HPE Details withheld at licensee's request by SSL.
G7 HPF Details withheld at licensee's request by SSL.
G7 HPG Details withheld at licensee's request by SSL.
G7 HPI C P Vance, 64 Caulfield Rd, Gorse Hill, Swindon, SN2 6BT [IO91CN, SU18]
G7 HPK Details withheld at licensee's request by SSL.
G7 HPL Details withheld at licensee's request by SSL.
G7 HPM Details withheld at licensee's request by SSL.
G7 HPO Details withheld at licensee's request by SSL.
G7 HPP Details withheld at licensee's request by SSL.
G7 HPS Details withheld at licensee's request by SSL.
G7 HPT Details withheld at licensee's request by SSL.
GW7 HPW Details withheld at licensee's request by SSL.
G7 HPX Details withheld at licensee's request by SSL.
G7 HPY Details withheld at licensee's request by SSL.
G7 HPZ D W Sillence, 4 St. Margarets Dr, Sibsey, Boston, PE22 0ST [JO03AA, TF35]
G7 HQA J C Burns, 37 Gypsy Ln, Marton in Cleveland, Middlesbrough, TS7 8NF [IO94JM, NZ51]
G7 HQB G B Slater, 54 The Dunterns, Alnwick, NE66 1AW [IO95DJ, NU11]
G7 HQC I B Sorrell, 67 Northfield Dr, Pontefract, WF8 2DJ [IO93IQ, SE42]
G7 HQE M A Holmes, 193 Greasby Rd, Greasby, Wirral, L49 2PE [IO83KJ, SJ28]
G7 HQF P G Smith, 17 Beverley Ave, Canvey Island, SS8 0DN [JO01GM, TQ78]
G7 HQH M W Croxford, 34 Brington Rd, Long Buckby, Northampton, NN6 7RW [IO92LH, SP66]
G7 HQM C D Ruffle, 6 Briar Cl, Church Rd, Yapton, Arundel, BN18 0ES [IO90QT, SU90]
G7 HQP J J L Woods, 1 Dean Rd, Cosham, Portsmouth, PO6 3DG [IO90LU, SU60]
G7 HQQ J M Keal, New Lodge, Holnicote, Nr.Minehead, Somerset, TA24 8TQ [IO81FE, SS94]
G7 HQR K A White, 12 Hamilton Cl, Mudesford, Christchurch, BH23 3LS [IO90CR, SZ19]
G7 HQU T W Richter, 2 Hillside Cottage, Cooks Ln, Walderton, Chichester, West Sussex, PO18 9EF [IO90NV, SU71]
G7 HQW B Currie, 21 Mercury Pl, Waterlooville, Hants, PO7 8BA [IO90LU, SU60]
G7 HQY K G Walton, Lansdown, Marshlands Ln, Heathfield, TN21 8EX [JO00DX, TQ52]
G7 HRB Details withheld at licensee's request by SSL.
G7 HRD N Smith, 5 Knowles Walk, Staplehurst, Tonbridge, TN12 0SG [JO01GD, TQ74]
G7 HRF S J Crutchley, 40 Ufton Cres, Shirley, Solihull, B90 3SA [IO92CJ, SP17]
G7 HRH R D Conway, 8 Turnberry Cl, Alwoodley Park, Leeds, LS17 7TE [IO93FU, SE23]
G7 HRI H H Dixon, 84 Hardwick Rd, Eynesbury, St. Neots, Huntingdon, PE19 2SD [IO92UF, TL15]
G7 HRJ T D Jackson, 6 The Ridge, Letchworth, SG6 1PP [IO91VX, TL23]
G7 HRK P B Hastelow, 63 Whitechapel Rd, Moorend, Cleckheaton, BD19 6HU [IO93DR, SE12]
G7 HRL T J Turner, 21 Spurgate, Hutton, Brentwood, CM13 2LA [JO01DO, TQ69]
G7 HRM S D Baker, 30 Thames Rd, Huntingdon, PE18 7QW [IO92VI, TL27]
G7 HRN M J Jeffery, 28 Thelwall New Rd, Thelwall, Warrington, WA4 2JF [IO83RJ, SJ68]
G7 HRP I N Booth, 164 Henconner Ln, Bramley, Leeds, LS13 4JH [IO93ET, SE23]
G7 HRQ D G Gorse, 4 St. Michaels Cl, Southport, PR9 9QY [IO83MQ, SD31]
G7 HRU E A Falconer, 95 Green Farm Cl, Chesterfield, S40 4UR [IO93GF, SK37]
G7 HRW P M Oldham, 110 Green Ln, Dronfield, Sheffield, S18 6FH [IO93GH, SK37]
G7 HRY A D J Gray, 14 Tower Rd, Mere Green, Sutton Coldfield, B75 5EW [IO92CO, SP19]
G7 HRZ S Haynes, 10 Cypress Gr, Denton, Manchester, M34 6EA [IO83WK, SJ99]
G7 HSA A W Cramp, 51 Julian Rd, Ludlow, SY8 1HD [IO82PI, SO57]
G7 HSB A P Green, Moss View, Southport Rd, Ormskirk, Lancs, L39 7JU [IO83MN, SD30]
G7 HSE Details withheld at licensee's request by SSL.
G7 HSG Details withheld at licensee's request by SSL.
G7 HSN J A Calder, Grassington, Station Rd, Newton-le-Willows, Bedale, DL8 1SX [IO94DH, SE28]
G7 HSO D S Hardinges, 4 The Close, North View, Eastcote, Pinner, HA5 1PH [IO91TN, TQ18]
G7 HSP M T D Cruz, 1 Pennine Cl, Tibshelf, Alfreton, DE55 5PR [IO93HD, SK46]
G7 HSQ R V Thompson, 326 Westmount Rd, Eltham, London, SE9 1NL [JO01AL, TQ47]
G7 HSR R L Halstead, 8 Kingswear Cl, Crossgates, Leeds, LS15 8RX [IO93GT, SE33]
G7 HSS J A East, 102 Westfield Ln, Wyke, Bradford, BD12 9LS [IO93CR, SE12]
G7 HST K Eley, 71 Max Rd, Chaddesden, Derby, DE21 4GY [IO92GW, SK33]
G7 HSV B M Gill, 24 Larkfield Cres, Rawdon, Leeds, LS19 6EH [IO93DU, SE23]
GD7 HSX J Ireland, 16 Kermode Cl, Ballastowell Garden, Ramsey, IM8 2AT
G7 HSY B Stanton, 107 Beaconside, Cleadon Manor, South Shields, NE34 7PT [IO94HX, NZ36]
G7 HTB C R Endersby, 3 Woodside, Vigo Village, Meopham, Gravesend, DA13 0SU [JO01EH, TQ66]
G7 HTS G J Davies, Walmore House, Walmore Hill, Minsterworth, Gloucester, GL2 8LA [IO81TT, SO71]
GW7 HTU P H Jenkins, 3 Gwalia Building, Nantyglo, Brynmawr, NP3 4LE [IO81JS, SO11]
GJ7 HTV A J Mourant, Little Mead, Claremont Rd, St. Saviour, Jersey, JE2 7RT
G7 HUC M A M Florentini, 20 Pytchley Cres, Upper Norwood, London, SE19 3QT [IO91WK, TQ37]
G7 HUD A D Sinclair, 67 Saughall Massie, Rd, Upton, Wirral, L49 6LZ [IO83KJ, SJ28]
G7 HUJ S A Telford, 44 Northcote Cres, Beeston, Leeds, LS11 6NN [IO93FS, SE33]
G7 HUK P B Hart, 104 St. Austell Dr, Wilford, Nottingham, NG11 7BQ [IO92KW, SK53]
G7 HUO C G Terry, 38 Salt Hill Way, Slough, SL1 3TR [IO91QM, SU98]
G7 HUP M Terry, 38 Salt Hill Way, Slough, SL1 3TR [IO91QM, SU98]
G7 HUU J Bonning, 5 Hillcross Walk, Bromford, Castle Bromwich, Birmingham, B36 8NN [IO92CM, SP18]
G7 HUV C F Bailey, 38 Augustus Cl, Cambridge, CB4 2JD [JO02BF, TL46]
G7 HUW W Gelder, 27 Martin Cl, Coseley, Bilston, WV14 8JG [IO82XM, SO99]
G7 HUZ A W Kempster, Spinners, High St., South Woodchester, Stroud, GL5 5EL [IO81VR, SO80]
GI7 HVC T Kennedy, 19 Orchard Ave, Newtownards, BT23 7AF [IO74DN, J46]
G7 HVD M H Wood, 13 Tregwary Rd, St. Ives, TR26 1BL [IO70GF, SW54]
G7 HVG E Pond, 16 Headcorn Gdns, Cliftonville, Margate, CT9 3ES [JO01RJ, TR37]
G7 HVK D Welsh, 3 Cow Ln, Wareham, BH20 4RE [IO80WQ, SY98]
G7 HVL C M Spires, 5 Springhead, Sutton Veny, Warminster, BA12 7AG [IO81WE, ST94]
G7 HVN M Templeman, Aratingas, 45 Nightingale Rise, Portishead, Bristol, BS20 8LN [IO81OL, ST47]
G7 HVO R T Gerrard, 12 Goldrill Gdns, Breightmet, Bolton, BL2 5NL [IO83TO, SD70]
G7 HVR J Sumner, 3 Wesley Cl, Westhoughton, Bolton, BL5 3SY [IO83RN, SD60]
G7 HVT Details withheld at licensee's request by SSL.
G7 HVU Details withheld at licensee's request by SSL.
G7 HVW Details withheld at licensee's request by SSL.
G7 HVZ Details withheld at licensee's request by SSL.
G7 HWB Details withheld at licensee's request by SSL.
G7 HWC Details withheld at licensee's request by SSL.
G7 HWD Details withheld at licensee's request by SSL.
G7 HWF Details withheld at licensee's request by SSL.
G7 HWG Details withheld at licensee's request by SSL.
G7 HWJ Details withheld at licensee's request by SSL.
G7 HWK Details withheld at licensee's request by SSL.
G7 HWL Details withheld at licensee's request by SSL.
G7 HWN Details withheld at licensee's request by SSL.
G7 HWS Details withheld at licensee's request by SSL.
G7 HWT Details withheld at licensee's request by SSL.
G7 HWU Details withheld at licensee's request by SSL.
G7 HWW Details withheld at licensee's request by SSL.
G7 HWX Details withheld at licensee's request by SSL.
G7 HWZ Details withheld at licensee's request by SSL.
G7 HXB Details withheld at licensee's request by SSL.
GW7 HXF Details withheld at licensee's request by SSL.
G7 HXI Details withheld at licensee's request by SSL.
G7 HXJ Details withheld at licensee's request by SSL.
G7 HXM Details withheld at licensee's request by SSL.
G7 HXN P W Wagg, 23 Courland Rd, Addlestone, KT15 2UQ [IO91SI, TQ06]
G7 HXQ Details withheld at licensee's request by SSL.
G7 HXS Details withheld at licensee's request by SSL.
G7 HXT Details withheld at licensee's request by SSL.
G7 HXU C M Loader, 13 Vale Rd, Hartford, Northwich, CW8 1PL [IO83RF, SJ67]
G7 HXV Details withheld at licensee's request by SSL.
G7 HXW Details withheld at licensee's request by SSL.
G7 HXX Details withheld at licensee's request by SSL.
G7 HYA R J Chuter, 1 Murrell Rd, Ash, Aldershot, GU12 6ST [IO91PG, SU85]
G7 HYG R W Barber, 180 Beechfield, Hoddesdon, EN11 9QN [IO91XS, TL31]
G7 HYH J Kemley, 56 Lynton Gr, Corby, NN18 8BP [IO92PL, SP88]
G7 HYK K W Mak, 25 Calderdale, Wallsend, NE28 8SN [IO95FA, NZ26]
G7 HYM N A Singer, 11 Langley Rd, Beckenham, BR3 4AE [IO91XJ, TQ36]
G7 HYO J G Brennan, Hollies Farm, Stone St, Petham, Kent, CT4 5PU [JO01ME, TR14]
G7 HYS D J Germaney, 13 Lower Chapel Ln, Frampton Cotterell, Bristol, BS17 2RL [IO81SM, ST68]
GI7 HYU J A Adams, 2 Dorset Cl, Old Galgorm Rd, Ballymena, BT42 1QP [IO64UU, D00]
G7 HYZ M A Thompson, 19 Cambridge Cres, Crofton, Wakefield, WF4 1RZ [IO93GP, SE31]
G7 HZH Details withheld at licensee's request by SSL.
G7 HZI Details withheld at licensee's request by SSL.
G7 HZP G A R Major, 17 Jubilee Cttgs, Station Rd, Marston Moreteyne, Bedford, MK43 0PN [IO92RB, SP94]
G7 HZQ S R Breen, 29 Ashdale Ave, Kempston, Bedford, MK42 8NT [IO92SC, TL04]
G7 HZR V Willerton, 80 Bellhouse Way, Foxwood Ln, York, YO2 3LN [IO93KW, SE54]
G7 HZS G S West, Sunnydale, 27 Horbling Ln, Stickney, Boston, PE22 8DG [JO03AC, TF35]
G7 HZU A R Bateman, 4 Fair Meadows, High St., Colton, Rugeley, WS15 3LD [IO92AS, SK02]
G7 HZV R G Baal, 26 Weston Ave, Mt Albert, Auckland, New Zealand
G7 HZX Details withheld at licensee's request by SSL.
G7 HZZ A R Clayton, 6 Albert Rd, Bunny, Nottingham, NG11 6QE [IO92KU, SK53]
G7 IAE R J Lindley, 23 Quadrant Cl, Muroishaw, Murdishaw, Runcorn, WA7 6DW [IO83QH, SJ58]
G7 IAG R J Wragg, 6 Arlington Ave, Aston, Sheffield, S31 0AA [IO93GJ, SK38]
G7 IAK C M Hughes, 4 Fallowfield Cl, Hillside, Hereford, HR2 7NZ [IO82PB, SO53]
G7 IAM M K Chrzanowski, 8 Pooles Rd, Biddulph Moor, Stoke on Trent, ST8 7LS [IO83WC, SJ95]
G7 IAN I M J Berry, 18 Highfield Rise, Stannington, Sheffield, S6 6BT [IO93FJ, SK38]
G7 IAS R W Bell, 58 Swallowtail Rd, Horsham, RH12 5YG [IO91UB, TQ13]
GW7 IAT M J Howells, 34 Cobden St., Cross Keys, Newport, NP1 7PF [IO81KO, ST29]
G7 IAU C D Penney, Coppers, Oak Ln, Minster on Sea, Sheerness, ME12 3QW [JO01JK, TQ97]
G7 IAW C M Walsh, 5 Carr Mill St., Haslingden, Rossendale, BB4 5BU [IO83UR, SD72]
G7 IAY R C Smith, 114 Teversham Drift, Cambridge, CB1 3JY [JO02CE, TL45]
G7 IBA H J Newell, 90 Arthur St., Kenilworth, CV8 2HG [IO92FI, SP27]
G7 IBB B J Barnes, 434 Hall Rd, Tuckswood, Norwich, NR4 6NF [JO02PO, TG20]
G7 IBD S P Beaumont, 6 Coral Way, Aughton, Sheffield, S31 0RE [IO93GJ, SK38]
G7 IBF R W Waller, 6 Pitchcombe, Yate, Bristol, BS17 4JX [IO81SM, ST78]
G7 IBH K R Ashton, 13 Laceys Ave, Leverton, Boston, PE22 0BG [JO03AA, TF34]
G7 IBI Details withheld at licensee's request by SSL.
G7 IBL M G Whatley, The Bungalow, Holcombe Hill, Holcombe, Bath, BA3 5DG [IO81SF, ST64]
GM7 IBM M M Robertson, Silverwells, Knockbain, Munlochy, Ross Shire, IV8 8PG [IO77UN, NH65]
G7 IBN F M Goodes, 17 Ashmead Cl, Lordswood, Chatham, ME5 8NY [JO01GI, TQ76]
G7 IBP P J Sims, 76 Mount Rd, Canterbury, CT1 1YF [JO01NG, TR15]
G7 IBR M J Watson Ipswich Borough Raynet, The Tubbery, Henley, Ipswich, Suffolk, IP6 0BR [JO02NC, TM15]
GW7 IBT A T Earp, 42 Tudor Gdns, Neath, SA10 7RX [IO81CQ, SS79]
G7 IBU D A Nicholls, 19 Kimmeridge, Crown Wood, Bracknell, RG12 0UD [IO91PJ, SU86]
G7 IBX V Finlayson, 92 Herlington, Orton Malborne, Peterborough, PE2 5PR [IO92UN, TL19]
G7 IBY Details withheld at licensee's request by SSL.
GW7 IBZ A M J Howell-Walmsley, 16 Wepre Ln, Deeside, CH5 4JS [IO83LF, SJ26]
G7 ICC E O Proctor, 185 Boldmere Rd, Sutton Coldfield, B73 5UL [IO92BN, SP19]
G7 ICH Details withheld at licensee's request by SSL.
G7 ICI Details withheld at licensee's request by SSL.
GW7 ICL Details withheld at licensee's request by SSL.
G7 ICO Details withheld at licensee's request by SSL.
G7 ICS Details withheld at licensee's request by SSL.
G7 ICV S C R Hardes, 28 Weybridge Cl, Lordswood, Chatham, Kent, ME5 8RW [JO01GI, TQ76]
G7 ICW Details withheld at licensee's request by SSL.
G7 ICX Details withheld at licensee's request by SSL.
G7 ICY Details withheld at licensee's request by SSL.
G7 IDB Details withheld at licensee's request by SSL.
G7 IDD Details withheld at licensee's request by SSL.
G7 IDE I D Evans, 18 Plemstall Way, Mickle Trafford, Chester, CH2 4QJ [IO83OF, SJ47]
G7 IDJ M Schonborn, 112 Hough Ln, Wombwell, Barnsley, S73 0EF [IO93HM, SE30]
G7 IDL Details withheld at licensee's request by SSL.
G7 IDP T M Woods, 275 Scotter Rd, Scunthorpe, DN15 7EH [IO93PO, SE81]
G7 IDS Details withheld at licensee's request by SSL.
G7 IDW Details withheld at licensee's request by SSL.
G7 IEA D C Kerr-Munslow, Orbital Mobile Communications, Keytech Centre, Ashwood Way, Basingstoke, RG23 8BG [IO91KG, SU65]
G7 IED R Martin, 45 Quailholme Rd, Knott End on Sea, Poulton-le-Fylde, FY6 0BT [IO83MW, SD34]
G7 IEF D M Roadnight, 14 Newquay Cres, Harrow, HA2 9LJ [IO91TN, TQ18]
G7 IEG J D Mitchell, 53 Radcliffe Rd, Hitchin, SG5 1QH [IO91UW, TL12]
GD7 IEH M Blackburn, 63 Westbourne Dr, Douglas, IM1 4BB
G7 IEI S Pasquill, 374 Manchester Rd, Westhoughton, Bolton, Lancs, BL5 3JT [IO83RN, SD60]
G7 IEO A J Cook, 26 Worcester Rd, Stourport on Severn, DY13 9PB [IO82UI, SO87]
G7 IEQ H N J Scott, 12 Lyndale Rd, Braunstone, Leicester, LE3 2QD [IO92JO, SK50]
G7 IER A Bent, 3 Overton St., Leigh, WN7 4HZ [IO83RL, SJ69]
G7 IET Dr A J Dunlop, High View, Milton Ave, Badgers Mount, Sevenoaks Kent, TN14 7AU [JO01BI, TQ46]
GM7 IEU A D Steele, 58 Stockethill Ct, Aberdeen, AB1 5UQ [IO87TH, NJ72]
G7 IEW I V C Teer, 28 Pear Tree Ln, Hempstead, Gillingham, ME7 3PT [JO01GI, TQ76]
G7 IEX A R B Allen, 136 Markland Rd, Dover, CT17 9NJ [JO01PC, TR24]
GI7 IEZ T J McGeown, 52 Alexander Cres, Armagh, BT61 7HZ [IO64QI, H84]
G7 IFA P J Williamson, 29 Alexandra Rd, Southport, PR9 9HA [IO83MP, SD31]
G7 IFB S R Thompson, 31 Bentleigh Ct, Greenstead Rd, Colchester, CO1 2TL [JO01LV, TM02]
G7 IFD R Hodder, 42 St. Marys Cl, Marston Moreteyne, Bedford, MK43 0QZ [IO92RB, SP94]
G7 IFE S R Evans, 16 Kynaston Dr, Wem, Shrewsbury, SY4 5DE [IO82PU, SJ52]
G7 IFF Details withheld at licensee's request by SSL.
G7 IFH Details withheld at licensee's request by SSL.
G7 IFI B G Jones, 25 Milton Dr, Wistaston, Crewe, CW2 8BT [IO83SB, SJ65]
G7 IFK J E Thornton, 65 Johnson St., Mirfield, WF14 8PQ [IO93DQ, SE21]
G7 IFL P J A King, 21 Greystones, Great Sutton, South Wirral, L66 3PD [IO83MG, SJ37]
G7 IFM J A Hewitt, 9 Alford Fold, Fulwood, Preston, PR2 3UU [IO83PT, SD53]
G7 IFO N Rigby, 2 Mill Brow Cl, Sutton, St. Helens, WA9 4JR [IO83PK, SJ59]
G7 IFR T I N Chibnell-Smith, Bank House, Bollingham, Kington, Herefordshire, HR5 3JG [IO82MF, SO35]
GI7 IFW S W Boskett, 314 Shore Cres, Belfast, BT15 4JU [IO74AP, J37]
GM7 IFX J W Barnett, 72 Cameron Toll Gdns, Edinburgh, EH16 4TG [IO85KW, NT27]
G7 IGC Details withheld at licensee's request by SSL.
G7 IGF I G Fields, Boarzell Cottage, London Rd, Hurst Green, Etchingham, TN19 7QY [JO01FA, TQ72]
G7 IGG A T Wells, 14 Gresham Rd, Blackpool, Lancs, FY5 3GE [IO83LU, SD34]
G7 IGJ A J Foster, 42 Ripon St., Parkinson Ln, Halifax, HX1 3UG [IO93BR, SE02]
G7 IGK J Wadsworth, 6 Pether Hill, South Parade, Stainland, Halifax, HX4 9HW [IO93BQ, SE01]
G7 IGQ D A Fielder, 4 Harefield Rise, Linton, Cambridge, CB1 6LS [JO02DC, TL54]
GM7 IGS A S Orr, 4 Bennochy Ct, Bennochy Rd, Kirkcaldy, KY2 5YU [IO86JC, NT29]
G7 IGU L Evans, 58 Westminster Dr, Bromborough, Wirral, L62 6AW [IO83MH, SJ38]
G7 IGV W H Ballard, 9 Fife Cl, Stamford, PE9 2YX [IO92RP, TF00]
G7 IGZ Details withheld at licensee's request by SSL.
G7 IHA B R Steadman, Glendalough, Kilcot, Newent, Glos, GL18 1NN [IO81SW, SO62]
G7 IHC S J Hatton, 7 Grasmere House, Lovett Ave, Oldbury, Warley, B69 1DQ [IO82XL, SO98]
G7 IHD J E Bolsover, 12 Mill Hill Gr, Middleton, Morecambe, LA3 3JZ [IO84NA, SD45]
G7 IHE H Robinson, 16 Coniston Ave, Ashton in Makerfiel, Ashton in Makerfield, Wigan, WN4 8AY [IO83QL, SJ59]
G7 IHF R A Cochrane, 69 Tranquil Vale, Blackheath, London, SE3 0BP [JO01AL, TQ37]
GM7 IHJ M J Alexander, 15 Gardenrose Path, Maybole, KA19 8AG [IO75PI, NS21]
G7 IHL W J F Humphries, 47 Fouracre Cres, Downend, Bristol, BS16 6PT [IO81SL, ST67]
G7 IHN S R Harvey, 5 Bafford Approach, Charlton Kings, Cheltenham, GL53 9HH [IO81XV, SO91]
G7 IHP K M Weston, 2 Beech Gr, Pinewood, Somerton, TA11 6LG [IO81PB, ST42]
GM7 IHR R A Brodie, Midgeloch Cottage, Arbuthnott, Kincardineshire, Scotland, AB3 1NX [IO87TH, NJ72]
G7 IHS Details withheld at licensee's request by SSL.
G7 IHV G R Havell, 13 Waldron House, Brixton Water Ln, London, SW2 1PA [IO91WK, TQ37]
G7 IHW M C Southall, 7 The Willows, Rugeley, WS15 1EP [IO92AR, SK01]
GM7 IHZ Dr G M Hayes, Flat 6, 87 London Rd, Edinburgh, EH7 5TT [IO85KW, NT27]
G7 IIB P J Shields, 3 Hawthorn Rd, Bishopsmead, Tavistock, PL19 9DL [IO70WM, SX47]
G7 IIC S C Oliphant, 8 Boveridge, Cranborne, Wimborne, BH21 5RX [IO90AW, SU01]
G7 IIF A Barrett, 7 Acacia Gr, Wallasey, L44 7BH [IO83LJ, SJ39]
G7 IIH J S Bond, The Old House, 2 Kent Rd, Fleet, GU13 9AH [IO91OG, SU85]
G7 III I R P Young, 10 Bowleymead, Swindon, SN3 3TD [IO91DN, SU18]
GI7 IIJ H W Best, 31 Jervis St., Portadown, Craigavon, BT62 3HA [IO64SK, J05]
GM7 IIL A Ferguson, 13 Isla Pl, Tayport, DD6 9AS [IO86NK, NO42]
G7 IIM Details withheld at licensee's request by SSL.
G7 IIN M D Hewitt, 43 Manor Rd, North Walsham, NR28 9LH [JO02QT, TG23]
G7 IIO B F Bellamy, 71 High Rd, Benfleet, SS7 5LH [JO01GN, TQ78]
G7 IIP T R Lennon, Hamnavoe Uplands Rd, Spital Park, Bromborough Wirral, Merseyside, L62 2BZ [IO83MI, SJ38]
G7 IIQ J Davis, 95 Buxton Rd, Stratford, London, E15 1QX [JO01AN, TQ38]
G7 IIS C Beatrup, Bon Air, 34 Springfield Dr, Halesowen, B62 8EU [IO82XL, SO98]
GI7 IIU G McHugh, 47 Main St., Hamiltonsbawn, Armagh, Co. Armagh, BT60 1LP [IO64RI, H94]
G7 IIV A G Pascoe, Rosetop, Old Hill, Bickington, Newton Abbot Devo, TQ12 6JU [IO80DM, SX77]
G7 IJC Cllr D R Wells, Peace Villa, 21 Kings Rd, Barnet, EN5 4EF [IO91VP, TQ29]
G7 IJD A Carter, Harmony Cottage, 26 The Green, Hempton, Fakenham, NR21 7LG [JO02KT, TF92]
G7 IJE S Down, 43 Deacon Ave, Kempston, Bedford, MK42 7DU [IO92RC, TL04]
G7 IJK Details withheld at licensee's request by SSL.
G7 IJL D G McClew, 135 Parkview Terr, Hermitage St, Rishton, Lancs, BB1 4ND [IO83SS, SD73]
G7 IJM Details withheld at licensee's request by SSL.
G7 IJN Details withheld at licensee's request by SSL.
GM7 IJO Details withheld at licensee's request by SSL.
G7 IJQ M S Gough, Applegate, Newtown, Langport, TA10 9SE [IO81OB, ST42]

G7 IJR Details withheld at licensee's request by SSL.
G7 IJS Details withheld at licensee's request by SSL.
G7 IJT Details withheld at licensee's request by SSL.
G7 IJU Details withheld at licensee's request by SSL.
G7 IJW Dr B Rushton, 4 Winters Ln, Walkern, Stevenage, SG2 7NZ [IO91WW, TL22]
G7 IJY B Evans, 51 Katrina Gr, Featherstone, Pontefract, WF7 5LW [IO93HQ, SE41]
GM7 IKA C Gray, 8 Bruce St., Dundee, DD3 6RG [IO86ML, NO33]
G7 IKB G T Riddell, 28 Lovat St., Newport Pagnell, MK16 0EF [IO92PC, SP84]
G7 IKD J L Edwards, 19 Churchill Dr, Wem, Shrewsbury, SY4 5HX [IO82PU, SJ52]
G7 IKG A Thynne, 1 Earlston Way, Birmingham, B43 5JR [IO92AN, SP09]
G7 IKK H G T Morrison, Arncliffe, Marston Rd, Sherborne, DT9 4BL [IO80RW, ST61]
G7 IKL S Hales, 55 Bakers Ln, Epping, CM16 5DQ [JO01BQ, TL40]
G7 IKM W K Willan, 31 St. Oswalds Ln, Bootle, L30 5QD [IO83ML, SJ39]
G7 IKO A G Collins, The Close, Prinsted Ln, Prinsted, Emsworth, PO10 8HT [IO90NU, SU70]
GI7 IKP A B B Smith, 5 Pypers Hill, Portavogie, Newtownards, BT22 1EJ [IO74GL, J65]
G7 IKR Details withheld at licensee's request by SSL.
G7 IKS A S Raistrick, 10 Orchard Way, Chinnor, OX9 4UD [IO91NQ, SP70]
G7 IKU J A Holden, 48 Marlborough, Accrington, Lancs., BB5 6AY [IO83TS, SD72]
G7 IKY Details withheld at licensee's request by SSL.
G7 ILA Details withheld at licensee's request by SSL. [JO01QX, TM23]
G7 ILD P W Brown, 75 Barton Ct Ave, Barton on Sea, New Milton, BH25 7ET [IO90ER, SZ29]
G7 ILE T E Mills, 45 Langdon Ave, Aylesbury, HP21 9UW [IO91OT, SP81]
G7 ILI A M Page, 28 Chaucer Cl, Fareham, PO16 7PD [IO90JU, SU50]
G7 ILJ B Thornton Banbury Ray Grp, 21 Valley Rd, Greenhills Est, Banbury, OX16 9BQ [IO92IB, SP43]
G7 ILK Details withheld at licensee's request by SSL.
G7 ILL Details withheld at licensee's request by SSL.
G7 ILN M D Young, Hillsview, 9 Binkham Hill, Yelverton, PL20 6BD [IO70XL, SX56]
GM7 ILO C Milne, 24 Church Pl, Coupar Angus, Blairgowrie, PH13 9BP [IO86IN, NO24]
G7 ILP K J Naylor, 3 Windrush Cl, Bicester, OX6 8AR [IO91KV, SP52]
G7 ILQ M L Jolliff, 34 College Rd, Newport, PO30 1HB [IO90IQ, SZ58]
G7 ILR K J Williams, Caerwys, Bartlow Rd, Castle Camps, Cambridge, Cambs, CB1 6SX [JO02EB, TL64]
G7 ILS I C Warrilow, 84 Marple Rd, Stockport, SK2 5RN [IO83WJ, SJ98]
GI7 ILU R J Ferguson, 38 Castleward Park, Belfast, BT8 4DG [IO74AN, J36]
G7 ILV Details withheld at licensee's request by SSL.
G7 ILX R P Voges, 9 Douglas Rd, Fulwood, Preston, PR2 3AP [IO83PS, SD53]
G7 IMB S E Jeffcoate, 38 Oliffe Cl, Aylesbury, HP20 2BJ [IO91OT, SP81]
G7 IMD A B Spittlehouse, Pentlands, High St., Wroot, Doncaster, DN9 2BT [IO93MM, SE70]
G7 IME R Padgett, 60 Campion Gr, Marton in Cleveland, Middlesbrough, TS7 8SL [IO94JM, NZ51]
G7 IMF Details withheld at licensee's request by SSL.
G7 IMH M T Fortescue, 98 Campbell Rd, Florence Park, Oxford, OX4 3NU [IO91JR, SP50]
G7 IML Details withheld at licensee's request by SSL.
G7 IMO L T Handley, 8 Croxden Cl, Cheadle, Stoke on Trent, ST10 1NL [IO92AX, SK04]
G7 IMT D M Gerard, 15 Nyetimber Ln, Pagham, Bognor Regis, PO21 3HQ [IO90PS, SZ99]
GI7 IMU A D Reid, 30 Cooks Cove, Kircubbin, Newtownards, BT22 2ST [IO74FL, J56]
G7 IMV R S King, Old Orchard, South Milton, Kingsbridge, Devon, TQ7 3JZ [IO80CG, SX64]
G7 IMY S P Kemp, 24 Stocklake, Aylesbury, HP20 1DN [IO91OT, SP81]
G7 IMZ G A Smith, 19 Parker Rd, Humberston, Grimsby, DN36 4TT [IO93XM, TA30]
G7 INA G J Mitchell, 38 Highmoor Rd, Corfe Mullen, Wimborne, BH21 3PT [IO80XS, SY99]
G7 INB Details withheld at licensee's request by SSL.
G7 INC G R Bacon, 36 Warnadene Rd, Sutton in Ashfield, NG17 5BD [IO93IC, SK45]
G7 IND R M Bowden, 38 Gorse Ln, West Kirby, Wirral, L48 8BH [IO83KI, SJ28]
G7 ING M S Darbyshire, 50 Gaythorne Ave, Preston, PR1 5TA [IO83QS, SD53]
G7 INJ C R Hughes, 175 Percival Rd, Eastbourne, BN22 9LB [JO00DT, TQ60]
G7 INK J L Clarke, 16 Sandringham Ct, Cavendish Mews, Wilmslow, SK9 1PW [IO83VH, SJ88]
G7 INM Details withheld at licensee's request by SSL.
G7 INO Details withheld at licensee's request by SSL.
G7 INP Details withheld at licensee's request by SSL.
GI7 INR Details withheld at licensee's request by SSL.
G7 INV Details withheld at licensee's request by SSL.
G7 INX Details withheld at licensee's request by SSL.
G7 INY Details withheld at licensee's request by SSL.
G7 INZ Details withheld at licensee's request by SSL.
G7 IOC Details withheld at licensee's request by SSL.
G7 IOD Details withheld at licensee's request by SSL.
G7 IOF K Roebuck, 53 Hollingthorpe Rd, Hall Green, Wakefield, WF4 3NL [IO93FP, SE31]
G7 IOH R Turton, 77 Sparks Ln, Thingwall, Heswall, Wirral, L61 7XF [IO83KI, SJ28]
G7 IOI N Telford, 18 Kirkstall Cl, Anston, Sheffield, S31 7BA [IO93GJ, SK38]
GU7 IOJ P R Plant, Figtree Cottage, Landes Du Marche, Vale, Guernsey, GY6 8DG
G7 ION M J Tennant, 64 Aldenham Rd, Kemplah Park, Guisborough, TS14 8LD [IO94LM, NZ61]
G7 IOO P G Horton, 48 Siberts Cl, Mill Ln, Shepherdswell, Dover, CT15 7LW [JO01OE, TR24]
G7 IOS M T J Hunter, Hazel Down, Hazel Ln, Tockington, Bristol, BS12 4PL [IO81RO, ST68]
G7 IOT Dr C S Hunter, Hazel Down, Hazel Ln, Tockington, Bristol, BS12 4PL [IO81RO, ST68]
G7 IOW Details withheld at licensee's request by SSL.
GI7 IOX Details withheld at licensee's request by SSL.
G7 IOY Details withheld at licensee's request by SSL.
G7 IPA M G Clements, 23 Pudding Ln, Gadebridge, Hemel Hempstead, HP1 3JU [IO91SS, TL00]
G7 IPB R S Harvey, 5 Bafford Approach, Charlton Kings, Cheltenham, GL53 9HH [IO81XV, SO91]
G7 IPG Details withheld at licensee's request by SSL.
G7 IPH P G Baker, 6 Firework Cl, Kingswood, Bristol, BS15 4LT [IO81SL, ST67]
G7 IPI P L Crane, 64 Bridge Ave, Cheslyn Hay, Walsall, WS6 7EP [IO82XQ, SJ90]
G7 IPK A OBrien, 32 Conway Ave, Bolton, BL1 6AZ [IO83SO, SD61]
G7 IPN Cllr M D Green, Eyot House, Church St., Shoreham, Sevenoaks, TN14 7RY [JO01CH, TQ56]
GI7 IPO H Stokes, 32 Islay St., Springfarm, Antrim, BT41 2TS [IO64VR, J18]
G7 IPP G Hudson, 36 Windsor Rd, Millfields Est, West Bromwich, B71 2NT [IO82XN, SO99]
GW7 IPS S E Hamlyn, 6 New Rd, Newcastle Emlyn, SA38 9BA [IO72SA, SN34]
G7 IPT Details withheld at licensee's request by SSL.
G7 IPX C L Bowden, 36 Aspin Dr, Knaresborough, HG5 8HQ [IO94GA, SE35]
G7 IQA R W Fisher, 83 Hurley Rd, Little Corby, Carlisle, CA4 8QY [IO84OV, NY45]
G7 IQC A M Southwart, 2 Alan Moss Rd, Loughborough, LE11 5LX [IO92JS, SK52]
G7 IQD R V Cook, 147 The Grove, Hartlebury Park, Stourport on Severn, DY13 9NE [IO82UI, SO87]
GM7 IQG I B Dunn, 38 Barnton Park Gdns, Edinburgh, EH4 6HN [IO85IX, NT17]
G7 IQH L H G Wakefield, 36 Hamilton Rd, Dover, CT17 0DA [JO01PD, TR34]
G7 IQI Details withheld at licensee's request by SSL.
G7 IQJ Details withheld at licensee's request by SSL.
G7 IQM P J Jaggs, 74 Brindwood Rd, Chingford, London, E4 8BA [IO91XO, TQ39]
G7 IQO D R Flatters, 87 Albert Promenade, Loughborough, LE11 1RD [IO92JS, SK51]
GI7 IQR Details withheld at licensee's request by SSL.
G7 IQU L R Andrews, West Libbear, Shebbear, Devon, EX21 5SZ [IO70VU, SS40]
G7 IQV R M Roulstone S N A R C, 5 Havenwood Rise, Clifton Est, Nottingham, NG11 9HD [IO92JV, SK53]
G7 IQW Details withheld at licensee's request by SSL.
G7 IQZ R P Norman, 25 Hazlemere Gdns, Worcester Park, KT4 8AH [IO91VJ, TQ26]
G7 IRC N J Hull Chelmsford Ry G, 126 Coval Ln, Chelmsford, CM1 1TG [JO01FR, TL70]
G7 IRD T A B Jones, 3 Woodlands Cl, St. Arvans, Chepstow, NP6 6EF [IO81PQ, ST59]
G7 IRE D G Walker, 162 North Sea Ln, Humberston, Grimsby, DN36 4XB [IO93XM, TA30]
G7 IRF A W Saunders, 61 Southlands Dr, Timsbury, Bath, BA3 1HB [IO81SH, ST65]
G7 IRH W D Moth, 145 Carisbrooke Rd, Winterdyne, Newport, PO30 1DG [IO90IQ, SZ48]
G7 IRI K M Baxter, 144 Boughton Green Rd, Northampton, NN2 7AA [IO92NG, SP76]
GI7 IRJ P J McAteer, 68 Oriel Rd, Antrim, BT41 4HR [IO64VR, J18]
G7 IRK Details withheld at licensee's request by SSL.
G7 IRN M R Scott, 3 Summerhill, Ticehurst, Wadhurst, TN5 7JA [JO01EB, TQ63]
G7 IRP D J Williams, 53 Jeffries Hill, Bottom, Hanham, Bristol Avon, BS15 3BE [IO81RK, ST67]
G7 IRS Dr D J Mellor, 31 High St., Swinderby, Lincoln, LN6 9LW [IO93PD, SK86]
G7 IRT J M Dolby, 36 Gregory St., Ilkeston, DE7 8AE [IO92IX, SK44]
G7 IRU Details withheld at licensee's request by SSL.
GW7 IRV Details withheld at licensee's request by SSL.
G7 IRW T A W Reynolds, Hilbre, Kingsway Ln, Ruardean, GL17 9XT [IO81RU, SO61]
G7 IRY Details withheld at licensee's request by SSL.
G7 ISB J M Crabtree, 15 Ewood Hall Ave, Hebden Bridge, HX7 5PH [IO93AR, SE02]
G7 ISD C A E Rizzo, 65 Tasker Rd, Sheffield, S10 1UY [IO93FJ, SK38]
G7 ISE G N Walters, 54 Church Way, Weston Favell, Northampton, NN3 3BX [IO92NF, SP76]
G7 ISF Details withheld at licensee's request by SSL.
G7 ISG G J Phillips, 7 Millfield Rd, Carisbrooke, Newport, PO30 5RH [IO90IQ, SZ48]
G7 ISO Details withheld at licensee's request by SSL.
G7 ISR G I Lines, 11 Gloucester Rd, Filton, Almondsbury, Bristol, BS12 4HD [IO81RN, ST68]
G7 ISV G R Smith, 12 Andrew Cl, Short Heath, Willenhall, WV12 5PL [IO82XO, SJ90]
GI7 ISX S E Butler, 25 Chippendale Ave, Bangor, BT20 4PX [IO74EP, J58]
G7 ISY Details withheld at licensee's request by SSL.
G7 ISZ Details withheld at licensee's request by SSL.
G7 ITD S R Cole, 10 Stanshawe Cres, Yate, Bristol, BS17 4EB [IO81TM, ST78]
G7 ITF P Kane, The Hollies, Hornby Rd, Appleton Wiske, Northallerton, DL6 2AF [IO94HK, NZ30]
GM7 ITG R L Young, 3 Gordon Rd, Alford, AB33 8AL [IO87PF, NJ51]
G7 ITM G F Clarkson, 32 Osborne Rd, Ryde, PO33 2TH [IO90KR, SZ59]
G7 ITN Details withheld at licensee's request by SSL.
G7 ITO M J de Banks, 56 Blackwater Dr, Walton Ct, Aylesbury, HP21 9RX [IO91OT, SP81]
G7 ITQ G R W Reader, 12 South View Gdns, Ravenshead, Nottingham, NG15 9GB [IO93KC, SK55]
G7 ITS M D Fasham, 29 Granville Ave, Ramsgate, CT12 6DX [JO01QI, TR36]
G7 ITT S Import, The Old Rectory, Dufton, Appleby in Westmorland, Cumbria, CA16 6DA [IO84SP, NY62]
G7 ITU S D Marlow, 329 Norcot Rd, Tilehurst, Reading, RG3 6AG [IO91LJ, SU66]
G7 ITW D L Fennelly, 23 Trent View Gdns, Radcliffe on Trent, Nottingham, NG12 1AY [IO92LW, SK64]
G7 ITX T P Emblem English, 36 Horsley Rd, Chingford, London, E4 7HX [IO91XP, TQ39]
G7 ITZ B Cohen, Wall To Wall Comms, 13 Manor Rd, Wallington, SM6 0BW [IO91WI, TQ26]
G7 IUE G F Clem, 25 Alexander Cl, Waterlooville, PO7 5TB [IO90LU, SU60]
GI7 IUH A Cardwell, 1 Glendale Park, Saintfield Rd, Newtownbreda, Belfast, BT8 4HT [IO74BN, J36]
G7 IUI W F Fagan, 56 Western Ave, New Milton, BH25 7PZ [IO90DR, SZ29]
GI7 IUJ Details withheld at licensee's request by SSL.
GW7 IUK Details withheld at licensee's request by SSL.
G7 IUL G E Woolmore, 40 Dalkeith Rd, Ilford, IG1 1JE [JO01AN, TQ48]
G7 IUO Details withheld at licensee's request by SSL.
GI7 IUR J M Moore, 1 Brett Gr, Lurgan, Craigavon, BT66 6QY [IO64SM, H96]
G7 IUT B A Johnson, 10 Saffron Rd, Tickhill, Doncaster, DN11 9PW [IO93KK, SK59]
G7 IUU Details withheld at licensee's request by SSL.
G7 IUV A E Collins, 37 Royal Rd, Ramsgate, CT11 9LF [JO01QH, TR36]
G7 IUW J P Woof, 64 Leaway, Greasby, Wirral, L49 2PZ [IO83KJ, SJ28]
G7 IUY Details withheld at licensee's request by SSL.
G7 IVC J M S Scott, 31 Bracken Park, Scarcroft, Leeds, LS14 3HZ [IO93GU, SE34]
G7 IVE A J Siddon, 11 Floret Cl, Ravenstone, Coalville, LE67 2NY [IO92HR, SK41]
G7 IVF C J Flux, 13 Main Ave, Sandford on Thames, Oxford, OX4 4YT [IO91JR, SP50]
G7 IVG K Graham, 7 Beaufort Dr, Barton Seagrave, Kettering, NN15 6SF [IO92PJ, SP87]
G7 IVJ Details withheld at licensee's request by SSL.
G7 IVM C A Vaslet, 16 Golden Rd, Oxford, OX4 3AR [IO91JR, SP50]
G7 IVN P S Jagdev, 28 St. Dunstans Rd, Hanwell, London, W7 2HB [IO91TM, TQ17]
G7 IVP Details withheld at licensee's request by SSL.
G7 IVQ Details withheld at licensee's request by SSL.
G7 IVR C N Chatwin, 18 Oakridge Cl, Churchill North, Redditch, B98 9JU [IO92BH, SP06]
G7 IVW P A A Hester, Soughley Bungalow, Soughley Ln, Wortley, Sheffield, S30 7DJ [IO93GJ, SK38]
GI7 IVX R A Connolly, 21 Eleastan Park, Kilkeel, Newry, BT34 4DA [IO64XB, J31]
G7 IVY Details withheld at licensee's request by SSL.
G7 IWA A G Maunder, 25 Mallard Cl, Bishops Waltham, Southampton, SO32 1LW [IO90JW, SU51]
G7 IWF Details withheld at licensee's request by SSL.
G7 IWG Details withheld at licensee's request by SSL.
G7 IWK G R Blackburn, 10 Lodge Cl, Redhill, Nottingham, NG5 8NZ [IO93KA, SK54]
G7 IWS Details withheld at licensee's request by SSL.
G7 IWU H F Judge, 8 Fontenoy Rd, Balham, London, SW12 9LU [IO91WK, TQ27]
G7 IWW R P Gibbs, 32 Beswick Ave, Ensbury Park, Bournemouth, BH10 4EY [IO90BR, SZ09]
G7 IWZ R Murray, 92 North Ln, East Preston, Littlehampton, BN16 1HE [IO90ST, TQ00]
G7 IXA Details withheld at licensee's request by SSL.
G7 IXC R D Barkley, 39 Fulbeck Ave, Goose Green, Wigan, WN3 5QN [IO83QM, SD50]
G7 IXD R D A Rees, 23 Twiss Green Dr, Culcheth, Warrington, WA3 4HY [IO83RK, SJ69]
G7 IXG D Dormann, 19 Jackman Cl, Fradley, Lichfield, WS13 8PW [IO92CR, SK11]
G7 IXH P J Lawton, 126 Woodlands Farm Rd, Erdington, Birmingham, B24 0PQ [IO92CM, SP19]
G7 IXI H Davies Altrincham D Sr, 76 Brook Ln, Timperley, Altrincham, WA15 6RS [IO83TJ, SJ78]
G7 IXK G A Smith, 129 Chiltern Way, Duston, Northampton, NN5 6BW [IO92MF, SP76]
G7 IXL Details withheld at licensee's request by SSL.
G7 IXM M J Hooks, 299 Cotton End Rd, Wilstead, Bedford, MK45 3DT [IO92SC, TL04]
G7 IXO Details withheld at licensee's request by SSL.
G7 IXP P Hammersley, 30 Bonner Gr, Aldridge, Walsall, WS9 0DU [IO92AO, SK00]
G7 IXQ T Middlemass, 9 Bent House Ln, Durham, DH1 2EA [IO94FS, NZ24]
G7 IXS E H R Middel, 15 Wakefield Cl, Grantham, NG31 8RT [IO92PV, SK83]
G7 IXT J E Downing, 27 Saxon Way, Old Windsor, Windsor, SL4 2PT [IO91RK, SU97]
G7 IYA B G Whittock, 12 Hillside Cres, Midsomer Norton, Bath, BA3 2NB [IO81SG, ST65]
G7 IYB J Dunn, 12 Lawrence Hill Rd, Dartford, DA1 3AG [JO01CK, TQ57]
G7 IYD Details withheld at licensee's request by SSL.
G7 IYF P D Van Klinkenberg, 59 Watlington St., Reading, RG1 4RF [IO91MK, SU77]
G7 IYG N A J Hobbs, 7 Maygoods Ln, Cowley, Uxbridge, UB8 3TE [IO91SM, TQ08]
G7 IYH L A Hobbs, 7 Maygoods Ln, Cowley, Uxbridge, UB8 3TE [IO91SM, TQ08]
G7 IYI B A Goddard, 3 Spring Gdns, Quenington, Cirencester, GL7 5BG [IO91CR, SP10]
G7 IYM T J Mann, 1 Charlton Ave, Northavon, Bristol, BS12 7QX [IO81RM, ST57]
G7 IYN A C Attack, 8 Lewis Rd, Hornchurch, RM11 2AJ [JO01CN, TQ58]
G7 IYO J F Gale, 303 Dover Rd, Folkestone, Kent, CT19 6NY [JO01OC, TR23]
G7 IYQ K G Fulcher, Derventio, Studley Cres, New Barn, Longfield Kent, DA3 7JL [JO01DJ, TQ66]
G7 IYR Details withheld at licensee's request by SSL.
G7 IYS Details withheld at licensee's request by SSL.
G7 IYT Details withheld at licensee's request by SSL.
G7 IYW Details withheld at licensee's request by SSL.
G7 IYX R T Dodds, 33 Westgate, Lion Farm Est, Oldbury, Warley, B69 1BA [IO82XL, SO98]
G7 IZA G Griffiths, The Old White Hart, Oxford Rd, Stokenchurch, High Wycombe, HP14 3SX [IO91NP, SU79]
G7 IZC R R D Phillips, 30 Sandringham Ave, Burton on Trent, DE15 9BJ [IO92ET, SK22]
G7 IZD G M Hourston, 3 Aylward Gdns, Chesham, HP5 2QX [IO91QR, SP90]
G7 IZF K B Allen, 55 Brandish Cres, Clifton Est, Nottingham, NG11 9JZ [IO92JV, SK53]
G7 IZH Details withheld at licensee's request by SSL.
G7 IZI Details withheld at licensee's request by SSL.
G7 IZJ Details withheld at licensee's request by SSL.
G7 IZM F J Lucas, Ivella, Recreation St., Netherton, Dudley, DY2 9EU [IO82XL, SO98]
G7 IZN M J Dunn, 39 Gainsbrook Cres, Norton Canes, Cannock, WS11 3TN [IO92AQ, SK00]
GM7 IZO D D Meek, Athole Cottage, 31 High St., Edzell, Brechin, DD9 7TE [IO86QT, NO66]
G7 IZQ Details withheld at licensee's request by SSL.
G7 IZR Details withheld at licensee's request by SSL.
G7 IZS D J McLean, 13 St. Margarets Dr, Brandon, IP27 0JW [JO02HK, TL78]
G7 IZT K R J Townsend, 36 Dark St. Ln, Plympton, Plymouth, PL7 1PN [IO70XJ, SX55]
G7 IZU A P Smith, 44 Parsons Cl, Plymouth, PL9 9UY [IO70XI, SX55]
G7 IZV C Funnell, 25 Broadyates Rd, Yardley, Birmingham, B25 8JF [IO92CL, SP18]
G7 IZW F E Chilton, 127 Nicholls Field, Harlow, CM18 6EB [JO01BS, TL40]
G7 IZX Details withheld at licensee's request by SSL.
G7 IZY F B Rothera, Threeways, Church Rd, Mersham, Ashford, TN25 6NS [JO01LC, TR03]
G7 JAB Details withheld at licensee's request by SSL.
G7 JAC Details withheld at licensee's request by SSL.
G7 JAD Details withheld at licensee's request by SSL.
G7 JAE C F Pritchard, 11 Willow Green, Needingworth, St. Ives, Huntingdon, PE17 3SW [IO92XH, TL37]
G7 JAF A D Lambert, 24 Mill Ln, Amersham, HP7 0EH [IO91QQ, SU99]
G7 JAH J S Denmead, 47 Holland Rd, Clevedon, BS21 7YJ [IO81NK, ST37]
G7 JAI E J House, 4 Elizabeth Way, Kenilworth, CV8 1QP [IO92EI, SP27]
G7 JAK D A Webb, 28 Holland Copse, Pathfinder Village, Exeter, EX6 6DB [IO80ER, SX89]
GI7 JAM Dr K F Gibson, 4 Ilford Ave, Belfast, BT6 9SF [IO74BN, J36]
G7 JAN J C Martyn, Aspiration, Queens Rd, Crowborough, TN6 1QQ [JO01BB, TQ52]
G7 JAO C R King, 39 West St., Huntingdon, PE18 6RT [IO92VH, TL27]
G7 JAQ R A Adam, 8 Lexington Ct, Purley, CR8 1JA [IO91WI, TQ36]
G7 JAU D R Slee, Station House, Kielder, Hexham, Northd., NE48 1EG [IO85QF, NY69]
G7 JAV D A Wilkins, 6 Crown Ln, Rothwell, Kettering, NN14 6LR [IO92OK, SP88]
G7 JAW A C Bond, 45 Cherry Cl, Knebworth, SG3 6DS [IO91VU, TL21]
G7 JBB Details withheld at licensee's request by SSL.
G7 JBD Details withheld at licensee's request by SSL.
G7 JBE Details withheld at licensee's request by SSL.
G7 JBN Details withheld at licensee's request by SSL.
G7 JBQ Details withheld at licensee's request by SSL.
G7 JBT Details withheld at licensee's request by SSL.
G7 JBW P J Hoath, 1 Red Lodge Dr, Bilton, Rugby, CV22 7TT [IO92II, SP47]
G7 JBZ R A Cone, 5 Wren Ct, Gateford, Worksop, S81 8TU [IO93KH, SK58]
G7 JCC M J Barnett, 55 Westbrook Rd, Weston Super Mare, BS22 8JY [IO81MI, ST36]
G7 JCD M J Jones, 4 Bell St., Tipton, DY4 8HZ [IO82XM, SO99]
G7 JCE M P Crimes, 4 Courteenhall Cl, Northampton, NN2 8PQ [IO92NG, SP76]
G7 JCF S J Beamish, The Old Vicarage, Vicarage Rd, Laxfield, Woodbridge, IP13 8DT [JO02QH, TM27]
G7 JCI G A Smerdon, 47 Greenway Rd, Weymouth, DT3 5BD [IO80SP, SY68]

G7	JCP	P C Nott, 87 Powney Rd, Maidenhead, SL6 6EG [IO91PM, SU88]
G7	JCR	F R Lugg, 4 Newbury Cl, Walsall, WS6 6DF [IO82XP, SJ90]
G7	JCU	H R Hawkins, 16 Primrose Hill, Bexhill on Sea, TN39 4LP [JO00FU, TQ70]
G7	JCX	J R Price, 28 Lindsey Cl, Chaddesden, Derby, DE21 6DG [IO92GW, SK33]
G7	JDA	A J Roberts, 67 Falmouth Rd, Springfield, Chelmsford, CM1 6JA [JO01FR, TL70]
G7	JDE	G K Dickie, 17 Ash Gr, Southall, UB1 2UN [IO91TM, TQ18]
G7	JDF	J J Hope, 48 Holbeck, Bracknell, RG12 8XE [IO91OJ, SU86]
G7	JDH	A J Nevill, 43 Allenby Cres, New Rossington, Doncaster, DN11 0JX [IO93LL, SK69]
G7	JDK	R A Rothwell, Lambes House, High St., Sutton Valence, Maidstone, ME17 3AG [JO01HF, TQ84]
G7	JDN	M J Collins, 8 Newfield Rd, Marlow, SL7 1JW [IO91ON, SU88]
G7	JDQ	M W Softley, 15 Castle Cttgs, Thornham, Hunstanton, Norfolk, PE36 6NF [JO02GX, TF74]
G7	JDR	C A Bennett, The Old Cottage, Waterside, Bradwell on Sea, Southminster, CM0 7QT [JO01KR, TL90]
GM7	JDS	B W Reid, Flat 11, 300 Springburn Rd, Glasgow, Scotland, G21 1RX [IO75VV, NS66]
G7	JDU	Details withheld at licensee's request by SSL.
G7	JDV	K M Perkins, 1 Tavis Rd, Paignton, TQ3 2PU [IO80FK, SX86]
G7	JDW	A L Fell, 66 Wayman Rd, Corfe Mullen, Wimborne, BH21 3PN [IO80XS, SY99]
GW7	JDX	Dr M Ghassempoory, 102 Colchester Ave, Penylan, Cardiff, CF3 7AZ [IO81KL, ST17]
G7	JDY	E L L Williams, 27 Maria Rd, Walton, Liverpool, L9 1EG [IO83MK, SJ39]
G7	JDZ	C D Keeling, Orchard Cottage, Woodland, Newton Abbot, Devon, TQ13 7JS [IO80DM, SX76]
GI7	JEB	M Gibson, 1 Downshire Park, Bangor, BT20 3TP [IO74DP, J48]
GM7	JED	I M Macdonald, 3 Anderson Rd, Stornoway, HS1 2PG [IO68TF, NB43]
G7	JEF	S Lauwers, 170 Coolgardie Ave, Highams Park, London, E4 9HX [JO01AO, TQ39]
G7	JEJ	T K Edwards, 10 Bury Ct Cttgs, Bentley, Farnham, GU10 5LZ [IO91NE, SU74]
GI7	JEM	D W Branagh, 17 Rathmoyle Park West, Carrickfergus, BT38 7NG [IO74CR, J48]
G7	JEQ	A I McMullin Poly Tech SW AR, Plymouth Business S, Polytechnics S West, Plymouth, PL4 8AA [IO70WJ, SX45]
G7	JER	R J Heaton, 34 Dale View Rd, Keighley, BD21 4YF [IO93BU, SE04]
GW7	JES	Details withheld at licensee's request by SSL.
G7	JET	R A White Braintree Air, Training Corps Ars, 40 Deanery Gdns, Bocking, Braintree, CM7 5SU [JO01GV, TL72]
G7	JEW	Details withheld at licensee's request by SSL.
G7	JEX	Details withheld at licensee's request by SSL.
G7	JEZ	Details withheld at licensee's request by SSL.
G7	JFA	Details withheld at licensee's request by SSL.[Op: B L Davies. Station located in Westcombe Park, Blackheath, London SE3. RAFARS 3289. RSGB. Please QSL via RAFARS or RSGB Bureaux.]
G7	JFB	Details withheld at licensee's request by SSL.
G7	JFF	Details withheld at licensee's request by SSL.
G7	JFH	Details withheld at licensee's request by SSL.
G7	JFI	Details withheld at licensee's request by SSL.
G7	JFJ	Details withheld at licensee's request by SSL.
G7	JFM	S D Smith, 81 Haytor Ave, Roselands, Paignton, TQ4 7TB [IO80FK, SX85]
GM7	JFN	K D Maclean, 10B Knockaird, Port of Ness, Isle of Lewis, Scotland, PA86 0XF [IO76JB, NS08]
GM7	JFR	P Glanville, 35 Boclair Rd, Bearsden, Glasgow, G61 2AF [IO75UW, NS57]
G7	JFS	R J Brown, 27 The Heathers, Boughton, Newark, NG22 9HE [IO93ME, SK66]
G7	JFU	R D Evison, 15 Chewter Ln, Windlesham, GU20 6JP [IO91PJ, SU96]
G7	JGD	D M Clarke, 15 Grig Pl, Alsager, Stoke on Trent, ST7 2SU [IO83UC, SJ75]
GM7	JGE	C A Hobson, 17 Well Brae, Pitlochry, PH16 5HH [IO86DQ, NN95]
G7	JGF	T J Froggatt, 23 Water Ln, Hollingworth, Hyde, SK14 8HT [IO93AL, SK09]
GM7	JGH	A Bruce, 20 Weir Cres, Milton, Wick, KW1 5SS [IO88KK, ND35]
G7	JGI	J M Dilks, Handley Farm Bungalow, Brant Broughton Clays, Beckingham, Lincs, LN5 0RN [IO93QB, SK95]
G7	JGL	R P Davies, 33 Sandown Rd, Ocean Heights, Paignton, TQ4 7RL [IO80FJ, SX85]
GM7	JGN	Details withheld at licensee's request by SSL.
GM7	JGP	C Macleod, 9 South Shawbost, Isle of Lewis, PA86 9BJ [IO76JB, NS08]
GM7	JGR	J Howie, 198 Craigleith Rd, Edinburgh, EH4 2EE [IO85JW, NT27]
GI7	JGT	M T McNamee, 22 St. Patricks Park, Roslea, Rosslea, Enniskillen, BT92 7QY [IO64IF, H43]
G7	JGU	Details withheld at licensee's request by SSL.
G7	JGW	W H Holroyd, 8 Carr Dene Ct, Preston St., Kirkham, Preston, PR4 2XA [IO83NS, SD43]
G7	JGY	P Smith, 174 Willerby Rd, Hull, HU5 5JW [IO93TS, TA02]
G7	JGZ	R A Brooks, 8 Chichester Pl, Tiverton, EX16 4BW [IO80GV, SS91]
GI7	JHA	W J Reed, 39 Cogry Hill, Doagh, Ballyclare, BT39 0RY [IO64XS, J29]
G7	JHC	T A Christie, 29 High Fawr Ave, Oswestry, SY11 1TB [IO82LU, SJ22]
G7	JHE	G Beckett, Royston, 2A Cadewell Ln, Shiphay, Torquay, TQ2 7AG [IO80FL, SX86]
GJ7	JHF	Details withheld at licensee's request by SSL.
G7	JHG	Details withheld at licensee's request by SSL.
GI7	JHJ	Details withheld at licensee's request by SSL.
GW7	JHK	P D Brettle, 27 Neath Rd, Resolven, Neath, SA11 4AA [IO81DR, SN80]
G7	JHL	Details withheld at licensee's request by SSL.
G7	JHM	J H McCollin, 5 Saltergate Hill Cottage, Skipton Rd, Killinghall, Harrogate, North Yorks, HG3 2BU [IO94EA, SE25]
G7	JHN	Details withheld at licensee's request by SSL.
G7	JHO	Details withheld at licensee's request by SSL.
G7	JHP	Details withheld at licensee's request by SSL.
G7	JHQ	Details withheld at licensee's request by SSL.
GI7	JHR	Details withheld at licensee's request by SSL.
G7	JHS	J C Wood Central Yorks, 13 Brunswick Rd, Pudsey, LS28 7NA [IO93ET, SE23]
GI7	JHV	D W Gervais, 12 Omagh Rd, Drumquin, Omagh, BT78 4QY [IO64GO, H37]
G7	JHX	Dr J A R Williams, 15 Deanbrook Cl, Shirley, Solihull, B90 4XS [IO92CJ, SP17]
G7	JHZ	D J Randles, 20 Felix Rd, Ealing, London, W13 0NT [IO91UM, TQ18]
G7	JIB	Details withheld at licensee's request by SSL.
G7	JIC	Details withheld at licensee's request by SSL.
G7	JIF	S L Ruffell, 2 Beulah Cottage, Church St., West Stour, Gillingham, Dorset, SP8 5RL [IO81UA, ST72]
G7	JIJ	D J Child, 17 Shannon Cl, Gr, Wantage, OX12 7PT [IO91HO, SU49]
G7	JIM	W J G Barton, 13 Paynes Cl, Piddlehinton, Dorchester, DT2 7TF [IO80TS, SY79]
G7	JIN	C S Willis, 9 Avington Cl, Sedgley, Dudley, DY3 3LN [IO82WM, SO99]
G7	JIP	J Wright, 13 Hall Hills, Diss, IP22 3LP [JO02NJ, TM18]
G7	JIQ	D P Marvin, 11 Sapcote Rd, Burbage, Hinckley, LE10 2AS [IO92HM, SP49]
G7	JIW	Details withheld at licensee's request by SSL.
G7	JIY	Details withheld at licensee's request by SSL.
G7	JJC	P Gerrard, 6 Ellabank Rd, Heanor, DE75 7HF [IO93HA, SK44]
G7	JJD	A V Harrington, 38 Pilgrims Rd, Halling, Rochester, ME2 1HW [JO01FI, TQ76]
G7	JJF	J Welch, 50 Quarrydale Rd, Sutton in Ashfield, NG17 4DR [IO93ID, SK45]
G7	JJG	K J Watts, 74 Westfield Rd, Caversham, Reading, RG4 8HJ [IO91ML, SU77]
G7	JJJ	C A Marshall, 11 Park Cres, Cuddington, Northwich, CW8 2TY [IO83QF, SJ57]
G7	JJP	L P Towler, 2 Willoughby Ct, Peterborough, PE1 4SZ [IO92VO, TF20]
G7	JJQ	Details withheld at licensee's request by SSL.
G7	JJT	Details withheld at licensee's request by SSL.
G7	JJW	S J Coffin, 5 Colt Cl, Streetly, Sutton Coldfield, B74 2EA [IO92BN, SP09]
G7	JJX	R I Wallace, 25 Walnut Tree Way, Meopham, Gravesend, DA13 0EH [JO01EJ, TQ66]
G7	JJZ	P R Johnson, 28 Hillside Rd, Blidworth, Mansfield, NG21 0TR [IO93KC, SK55]
GI7	JKA	J M McCullagh, 2 Holestone Rd, Doagh, Ballyclare, BT39 0SB [IO64XS, J29]
G7	JKC	D Humphreys, 7 Howard Park, Greystoke, Penrith, CA11 0TU [IO84NQ, NY43]
G7	JKD	M J Coward, Stonethwaite, Hayton, Carlisle, Cumbria, CA4 9JE [IO84OV, NY55]
G7	JKH	C G Hyde, 42 Fern Rd, Ellesmere Port, Whitby, South Wirral, L65 6PB [IO83NG, SJ37]
G7	JKI	B R W Hillier, Little Bourne Farm, Upton, Andover, Hants, SP11 0JN [IO91GG, SU35]
G7	JKK	J Mossman, 31 Marlborough Rd, Stretford, Manchester, M32 0AW [IO83UK, SJ79]
G7	JKL	R Beachill, Yew Tree Cottage, Ashford Carbonell, nr Ludlow, Shropshire, SY8 4DG [IO82PH, SO57]
GI7	JKM	S A Glendinning, 2 Scotts Rd, Ballyrogully, Moneymore, Magherafelt, BT45 7TW [IO64QQ, H98]
G7	JKN	J H Owen, 12 Dellcott Cl, Welwyn Garden City, AL8 7BD [IO91VT, TL21]
G7	JKO	Details withheld at licensee's request by SSL.
G7	JKU	Details withheld at licensee's request by SSL.
G7	JKV	Details withheld at licensee's request by SSL.
G7	JKW	S J Avery, Wilding Farm Cottage, Cinder Hill, North Chailey, E Sussex, BN8 4HP [IO90XW, TQ41]
G7	JKX	Details withheld at licensee's request by SSL.
G7	JKY	Dr S D Smith, Bighton Ln, Gundleton, Bishops Sutton, Alresford, Hants, SO24 9SW [IO91KC, SU63]
G7	JKZ	Details withheld at licensee's request by SSL.
G7	JLA	Details withheld at licensee's request by SSL.
G7	JLC	A S Edwards, 34 Albion Rd, Malvern Link, Malvern, WR14 1PU [IO82UD, SO74]
GI7	JLD	J T Hunter, 6 Glenholm Dr, Newtownbreda, Belfast, BT8 4LW [IO74BN, J36]
G7	JLE	R Clappison, 83 Bernadette Ave, Hull, HU4 7QB [IO93TR, TA02]
GW7	JLG	A Williams, 2 Nant y Berllan, Llanfairfechan, LL33 0SN [IO83AF, SH67]
G7	JLH	R H Wyman, 6 St. Marys View, St. Leonards Ave, Kenton, Harrow, HA3 8ED [IO91UN, TQ18]
G7	JLI	M L Wyman, 6 St. Marys View, St. Leonards Ave, Kenton, Harrow, HA3 8ED [IO91UN, TQ18]
G7	JLK	R W Elliott, 11 Meadland, Corsham, SN13 9DU [IO81VK, ST87]
G7	JLL	E R Jarvis, 1 Prospect Terr, Station Rd, Heathfield, TN21 8DD [JO00DX, TQ52]
G7	JLM	Details withheld at licensee's request by SSL.
G7	JLQ	Details withheld at licensee's request by SSL.
G7	JLR	Details withheld at licensee's request by SSL.
G7	JLS	D Bryant, Jessamine Cottage, Chapel Ln, Ashford Hill, Newbury Berks, RG15 8BE [IO91LJ, SU66]
G7	JLT	K D Bryant, Jessamine Cottage, Chapel Ln, Ashford Hill, Newbury Berks, RG15 8BE [IO91LJ, SU66]
G7	JLX	D W Wilson, 65 Laverick Rd, Jacksdale, Nottingham, NG16 5LQ [IO93HB, SK45]
G7	JLZ	R Davies, 11 Hillside Dr, Yealmpton, Plymouth, PL8 2NT [IO70XI, SX55]
G7	JMB	J M Baker, Green Ln Farmhous, Rugeley, Staffs, WS15 2AR [IO92AS, SK01]
G7	JME	Details withheld at licensee's request by SSL.
G7	JMI	Details withheld at licensee's request by SSL.[Station located near Stowmarket.]
G7	JMK	Details withheld at licensee's request by SSL.
GM7	JMO	C F Allan, 4 Moor of Balvack, Monymusk, Inverurie, Aberdeenshire, AB51 7SQ [IO87RF, NJ61]
G7	JMQ	Details withheld at licensee's request by SSL.
G7	JMU	D B Butterworth, 420 Blackmoorfoot Rd, Huddersfield, HD4 5NS [IO93CP, SE11]
G7	JMV	M A Newton, 18 Wellington Cl, Heckington, Sleaford, NG34 9GZ [IO92TX, TF04]
G7	JMW	A C W Weaver, 116 Maldon Rd, Tiptree, Colchester, CO5 0BN [JO01IT, TL81]
G7	JMX	B Bowles, 23 St. Leodegars Cl, Wyberton, Boston, PE21 7DU [IO92XW, TF34]
G7	JMZ	J P Bache, 65 Winchester Rd, Whitchurch, RG28 7HW [IO91HF, SU44]
G7	JNF	Details withheld at licensee's request by SSL.
G7	JNM	Details withheld at licensee's request by SSL.
G7	JNN	H A Goddard-Watson, 21 Lynn Cl, Leigh Sinton, Malvern, WR13 5DU [IO82UD, SO75]
G7	JNS	S J McLennan, 179 King John Ave, Bearwood, Bournemouth, BH11 9SJ [IO90AS, SZ09]
G7	JNT	Details withheld at licensee's request by SSL.
G7	JNX	S D Wood, 9 Charmouth Cl, Croxteth Park, West Derby, Liverpool, L12 0PG [IO83NK, SJ49]
G7	JNY	Details withheld at licensee's request by SSL.
G7	JOA	C Rickerby Rsc of Cheshire, 113 Cliftonville Rd, Woolston, Warrington, WA1 4BJ [IO83RJ, SJ68]
G7	JOB	Details withheld at licensee's request by SSL.
G7	JOC	L J Timbrell, 149 Darbys Hill Rd, Tividale, Warley, B69 1SG [IO82XM, SO98]
G7	JOD	P H Pegram, 90 Lime Ave, Leamington Spa, CV32 7DQ [IO92FH, SP36]
G7	JOE	K R Francks, 63 Parc Godrevy, Pentire, Newquay, TR7 1TY [IO70KJ, SW86]
G7	JOF	J Covel, 25 Epsom Mews, Bury New Rd, Salford, M7 2BZ [IO83UM, SD80]
G7	JOG	A J Allen, 58 Wolds Rise, Matlock, DE4 3HJ [IO93FD, SK36]
GI7	JOJ	P W Little, 8 Ballynoe Rd, Antrim, Co Antrim, N Ireland, BT41 2QT [IO64WR, J18]
GM7	JOM	Details withheld at licensee's request by SSL.
G7	JON	Details withheld at licensee's request by SSL.
G7	JOO	J Holroyd, 48 Blounts Ct Rd, Sonning Common, Reading, Berks, RG4 9RS [IO91MM, SU78]
G7	JOQ	Details withheld at licensee's request by SSL.
G7	JOS	M Bedford-White, 16 Westfield Rd, Acocks Green, Birmingham, B27 7TL [IO92CK, SP18]
GM7	JOV	Details withheld at licensee's request by SSL.
G7	JOW	J Q Ashbee, 2 Singledge Ave, Whitfield, Dover, CT16 3LQ [JO01PD, TR34]
G7	JOX	Details withheld at licensee's request by SSL.
G7	JPB	S Wood, Brownhills Cott Far, Market Drayton, Shropshire, TF9 4BE [IO82SW, SJ63]
GW7	JPC	T J Cole, 154 Lon-Dol-Afon, Newtown, Powys, SY16 1QL [IO82IM, SO19]
G7	JPD	Details withheld at licensee's request by SSL.[Op: P J Dearman. Station located at Moulton Chapel, Lincs.]
G7	JPE	Details withheld at licensee's request by SSL.
G7	JPJ	M Wilkinson, 124 Doncaster Rd, Darfield, Barnsley, S73 9JA [IO93HM, SE40]
G7	JPL	I Macfarlane, 30 Victoria St., Burscough Bridge, Burscough, Ormskirk, L40 0SN [IO83NO, SD41]
G7	JPN	M J Bateman, 22 Bowling Green Ln, Albrighton, Wolverhampton, WV7 3HL [IO82UP, SJ80]
G7	JPP	R D Lane, 130 Alumhurst Rd, Alum Chine, Bournemouth, BH4 8HU [IO90BR, SZ09]
G7	JPS	D C Leech Huntingdons ARS, 4 Rydal Cl, Huntingdon, PE18 6UF [IO92VI, TL27]
G7	JPU	S W Hoy, 5 Beech Rd, Hullbridge, Hockley, SS5 6JF [JO01HO, TQ89]
GI7	JPW	G M M Quinn, 69 Killeeshill Rd, Cabragh, Dungannon, BT70 1TJ [IO64ML, H65]
G7	JQE	Details withheld at licensee's request by SSL.
G7	JQF	W Booth, 8 Park Cres, New Line, Bacup, OL13 9RL [IO83VQ, SD82]
GD7	JQI	Details withheld at licensee's request by SSL.
G7	JQJ	J H Preston, 31 Richmondfield Ln, Barwick in Elmet, Leeds, LS15 4HB [IO93HT, SE43]
GM7	JQK	Details withheld at licensee's request by SSL.
G7	JQT	E G J Barry, 26 Friars Rd, Scunthorpe, DN17 1HQ [IO93QN, SE80]
G7	JQW	H C F Derrick, 28 Great Parks, Holt, Trowbridge, BA14 6QP [IO81VI, ST86]
G7	JQY	P E K Donaldson, Highclere, Greenhill Rd, Otford, Sevenoaks Kent, TN14 5RR [JO01CH, TQ55]
G7	JQZ	D Beadle, 4 Harlaxton Dr, Doddington Park, Lincoln, LN6 3NR [IO93QE, SK96]
G7	JRC	D C Smith, 10 Kibroyd Dr, Kexborough, Darton, Barnsley, S75 5DF [IO93FN, SE30]
G7	JRD	T J Alwyn-Clark, 106 Oxford Rd, Cambridge, Cambs, CB4 3PL [JO02BF, TL46]
G7	JRI	N Smith, 11 Devas Gdns, Spondon, Derby, DE21 7AD [IO92HW, SK33]
G7	JRJ	C A Wainwright, 31 Queens Rd, Leytonstone, London, E11 1BA [JO01AN, TQ38]
G7	JRK	P A Dixon, 7 Pincey Nead, Pitsea, Basildon, Essex, SS13 3EW [JO01FN, TQ78]
G7	JRL	Dr M R Price, 13 Charles Ct, Wake Green Park, Birmingham, B13 9YW [IO92BK, SP08]
G7	JRM	C D Hinton, Terringe, 65/67 South St., Tarring, Worthing, West Sussex, BN14 7NE [IO90TT, TQ10]
G7	JRN	G L Gooding, 195 Mount Rd, Penn, Wolverhampton, WV4 5RT [IO82WN, SO99]
G7	JRP	T C Pratley, 28 Charles Ave, Watton, Thetford, IP25 6BZ [JO02JN, TF90]
GW7	JRT	J R Tonge, Elstone, Llanfaelog, Ty-Cross, Anglesey, Gwynedd, LL63 5SR [IO73SF, SH37]
G7	JRV	P L Hills, 5 Washington Cttgs, Storrington Rd, Washington, Pulborough, RH20 4AQ [IO90TV, TQ11]
G7	JRW	Details withheld at licensee's request by SSL.
G7	JRX	P J C Snow, 1 Sanvino Ave, Ainsdale, Southport, PR8 3NB [IO83LO, SD31]
G7	JSA	E J Toome, 38 Sandling Ln, Maidstone, ME14 2DY [JO01GG, TQ75]
G7	JSB	J S Buxton, 38 Maulden Rd, Flitwick, Bedford, MK45 5BW [IO92SA, TL03]
G7	JSC	R P Brotherton, 167 Pershore Rd, Evesham, WR11 6NB [IO92AC, SP04]
G7	JSD	D A Spreadbury, Little Reeds, Ford Ln, Trottiscliffe, West Malling, ME19 5DP [JO01EH, TQ65]
G7	JSE	S C Almond, 26 Moorland Cres, Castleside, Consett, DH8 9RG [IO94BT, NZ04]
G7	JSF	A E Gerrard, 13 Wentworth Ave, Timperley, Altrincham, WA15 6NG [IO83UJ, SJ78]
G7	JSH	J C Field, 35 Mill Rd, Twickenham, TW2 5HA [IO91TK, TQ17]
G7	JSK	A D Grieve, 5 Wesley Way, Elms Park, Ruddington, Nottingham, NG11 6GZ [IO92KV, SK53]
G7	JSO	C Solly, 29 Hopes Ln, Ramsgate, CT12 6RN [JO01QI, TR36]
G7	JSQ	P S Domachowski, 24 Dunrose Cl, Coventry, CV2 5PF [IO92GJ, SP37]
G7	JSS	C F Watson, 242 Fir Tree Rd, Epsom Downs, Epsom, KT17 3NL [IO91VH, TQ26]
G7	JST	J H Wheatley Jubilee Sailing Trust(Ars), 13 Elstead Gdns, Purbrook, Waterlooville, PO7 5EX [IO90LU, SU60]
G7	JSU	E F J Bartlett, 65 Ryan Ct, Bryanston St., Blandford, Blandford Forum, DT11 7XE [IO80WU, ST80]
G7	JSV	W J McAreavey, 28 Alverstone Rd, Wembley, HA9 9SB [IO91UN, TQ18]
G7	JSW	R J Steward, 2 Glenister House, Avondale Dr, Hayes, UB3 3PP [IO91TM, TQ18]
G7	JSX	R E Bishop, 99 St. Pauls Ave, Kenton, Harrow, HA3 9PR [IO91UO, TQ18]
G7	JTB	R G Pluck, 31 Coalway Rd, Penn, Wolverhampton, WV3 7LU [IO82WN, SO99]
G7	JTD	D J Lockett, 10 Cornwall Dr, Baystonhill, Bayston Hill, Shrewsbury, SY3 0ER [IO82OQ, SJ40]
G7	JTF	A G Harvey, 31 Ivanhoe Rd, Edenthorpe, Doncaster, DN3 2JG [IO93LN, SE60]
G7	JTH	J D Carter, 30 Swift Way, Pledwick Ln, Sandal, Wakefield, WF2 6SR [IO93GP, SE31]
G7	JTJ	Details withheld at licensee's request by SSL.
G7	JTK	S K Bell, 10 Sycamore Gr, Prudhoe, NE42 6QA [IO94BX, NZ06]
G7	JTL	S Livesey, 7 Gartside St., Ashton under Lyne, OL7 0DY [IO83WL, SJ99]
G7	JTN	C J Miles, 231 Howlands, Welwyn Garden City, AL7 4HG [IO91VS, TL21]
G7	JTO	E C Bird, Sunny Corner, nr Ruan Minor, Cornwall, TR12 7LW [IN79JX, SW71]
G7	JTQ	P Underhill, 37 James Ave, Herstmonceux, Hailsham, BN27 4PD [JO00DV, TQ61]
G7	JTR	D C Lock, Fountain Ct, Bramshaw, Hants, SO43 7JB [IO90EW, SU21]
G7	JTS	L E C Nutbrown, Merlin House, Bagshot Rd, Sunninghill, Ascot, SL5 9JL [IO91QJ, SU96]
G7	JTT	J C P McCarthy, 29 Harefield Rd, Swaythling, Southampton, SO17 3TG [IO90HW, SU41]
G7	JTV	J J E Caswell, 3 Birch Rd, Finchampstead, Wokingham, RG40 3LB [IO91NJ, SU86]
G7	JTY	R D Tiller, 5 Tylney Rd, Halstead, CO9 2BG [JO01HW, TL83]
G7	JTZ	R Smith, 17 Julian Rd, Spixworth, Norwich, NR10 3QA [JO02PQ, TG21]
G7	JUA	M L Hate, 1 Green Fall, Poringland, Norwich, NR14 7SP [JO02QN, TG20]
G7	JUB	T L Jones, 37 Pickford Hill, Batford, Harpenden, AL5 5HE [IO91TT, TL11]
G7	JUC	K J Marsh, 21 Edward Rd, Eynesbury, St. Neots, Huntingdon, PE19 2QF [IO92UF, TL15]
G7	JUD	H W Aviss, Newlands, Lewes Rd, Wilmington, Polegate, E Sussex, BN26 5SG [JO00CT, TQ50]
G7	JUE	Details withheld at licensee's request by SSL.
G7	JUH	T J E Cox, 5 The Acre, Pillerton Priors, Warwick, CV35 0PT [IO92FD, SP35]
GW7	JUJ	P A Moss, 17 Clarendon St., Dukinfield, SK16 4LP [IO83WL, SJ99]
G7	JUL	A E Whitcher, 12 Battersby St., Fairfield, Bury, BL9 7SG [IO83UO, SD81]
G7	JUN	M R Steadman, Glendalough, Kilcot, Newent, Glos, GL18 1NN [IO81SW, SO62]
G7	JUO	B Dunn, 61 Pretoria Rd, Patchway, Bristol, BS12 5PU [IO81RM, ST58]
G7	JUP	J A Beckingham, 20 Baptist Cl, Abbeymead, Gloucester, GL4 5GD [IO81VU, SO81]
G7	JUR	P C Lock, 1 Carters Walk, Farnham, GU9 9AY [IO91OF, SU84]
GM7	JUX	W N Dyer, 24 Southfield Rd, Balloch, Cumbernauld, Glasgow, G68 9DZ [IO75XW, NS77]
G7	JUY	J A Denham, 6 Old Forge Cl, Digswell, Welwyn Garden City, AL7 1SR [IO91VT, TL21]
G7	JUZ	R Shams-Nia, 1090 Eastern Ave, Ilford, IG2 7SF [JO01BN, TQ48]
G7	JVB	P F Wade, 41 Prospect Ave, Stanford-le-Hope, SS17 0NH [JO01FM, TQ68]
G7	JVC	M A Hewitt, 1 Harpswell Hill Park, Hemswell, Gainsborough, Lincs, DN21 5UT [IO93QJ, SK99]
G7	JVD	M A Rigby, 41 Browning Dr, Hitchin, SG4 0QR [IO91UW, TL12]
G7	JVE	N M Cook, 35 Glanville Rd, Hadleigh, Ipswich, IP7 5SQ [JO02LA, TM04]
G7	JVF	S R Mobley, 2 Lingham Cl, Solihull, B92 9NW [IO92CK, SP18]
G7	JVG	A E White, 19 Haswell Cl, Wardley, Gateshead, NE10 8UE [IO94FW, NZ36]
G7	JVK	R H Hardie, 12 Hopland Cl, Longwell Green, Bristol, BS15 6XB [IO81SK, ST67]
G7	JVL	Details withheld at licensee's request by SSL.

G7 JVN D M Blake, 3 Denham Cl, St. Leonards on Sea, TN38 9RS [JO00GV, TQ71]
G7 JVO K S Saxby, 184 Brodrick Rd, Eastbourne, BN22 9RH [JO00DT, TQ60]
G7 JVP M P Booth, The Hollies, St James Rd, Kingsdown, Deal, CT14 8BQ [JO01QE, TR34]
G7 JVQ F M Sparks, 36 High View Rd, Guildford, GU2 5RT [IO91QF, SU94]
GW7 JVS K G Dyer, 34 Lundy Dr, West Cross, Swansea, SA3 5QL [IO71XN, SS68]
G7 JWB J M Merriman, Chantry, Edlington, Horncastle, Lincs, LN9 5RJ [IO93VF, TF27]
G7 JWD T F Place, 34 Holcroft, Orton Malborne, Peterborough, PE2 5SL [IO92UN, TL19]
G7 JWE A J Liddell, 4 Russet Ct, Kingswood, Wotton under Edge, GL12 8SG [IO81TO, ST79]
G7 JWH A N Butler, 371 Grangemouth Rd, Radford, Coventry, CV6 3FH [IO92FK, SP38]
G7 JWJ E Hickman, Eriska, 33 Romany Way, Norton, Stourbridge, DY8 3JR [IO82VK, SO88]
G7 JWL E J Oakes, 30 Linden Ave, Stourport on Severn, DY13 0EQ [IO82UH, SO87]
G7 JWN A G Smith, 2 Windsor Cl, Weedon, Northampton, NN7 4PE [IO92LF, SP65]
G7 JWO R Allcock, 44 Newmount Rd, Fenton, Stoke on Trent, ST4 3HQ [IO82WX, SJ94]
G7 JWQ B A Priestley, Ct Lodge, Walkley Hill, Rodborough, Stroud, GL5 3TU [IO81VR, SO80]
G7 JWU S W Corris, 17 Tyne Cl, Nutgrove Hall Est, Thatto Heath, St. Helens, WA9 5NP [IO83OK, SJ49]
G7 JWV R F Ebbetts, Markway House, Blackbush Rd, Milford on Sea, Hants, SO41 0PB [IO90ER, SZ29]
G7 JWX B P Maley, 203 Tankerton Rd, Whitstable, CT5 2AT [JO01MI, TR16]
G7 JWY C M Ainley, 87 Ryedale Way, Tingley, Wakefield, WF3 1AL [IO93FR, SE22]
G7 JXB K R Cox, 14 The Meadows, Walberton, Arundel, BN18 0PB [IO90QU, SU90]
G7 JXC D Tetlow, 2 Maple Rd, Woolston, Warrington, WA1 4DP [IO83RJ, SJ68]
G7 JXD J W Pritchard, 22 Osborne Way, Haslingden, Rossendale, BB4 4DZ [IO83UQ, SD72]
G7 JXF M L Forknell, 24 Sherbourne Ave, Nuneaton, CV10 9JH [IO92FM, SP39]
G7 JXJ C M Smith, 30 Rookery Cl, St. Ives, Huntingdon, PE17 4FX [IO92XI, TL37]
G7 JXK M A Statt, 19 Heathercroft Rd, Wickford, SS11 8YA [JO01GO, TQ79]
G7 JXL D Kerridge, 17 Venmore Dr, Dunmow, CM6 1HN [JO01EU, TL62]
G7 JXN S J Yates, 38 Kestrel Dr, Bury, BL9 6JE [IO83UO, SD81]
G7 JXQ D A Greenwood, 36 Cubbington Rd, Lillington, Leamington Spa, CV32 7AB [IO92FH, SP36]
G7 JXR G M Wiseman, 7 South Cl, Melton, Woodbridge, IP12 1QR [JO02QC, TM25]
G7 JXT I Ballantyne, 21 Loddon Rd, Cove, Farnborough, GU14 9NR [IO91OH, SU85]
G7 JXU M Barker, 103 Friarswood Rd, Newcastle, ST5 2EF [IO83VA, SJ84]
G7 JXV J L Bourner, 14 Fulmar Rd, Worle, Weston Super Mare, BS22 8UU [IO81MI, ST36]
G7 JXX I M Thaiss, 16 Frederick St., Lincoln, LN2 5NL [IO93RF, SK97]
G7 JXY I Guffick, 13 Alderwood Cl, Hartlepool, TS27 3QR [IO94IR, NZ43]
G7 JXZ O J Chivers, 27 Farley Dell, Coleford, Bath, BA3 5PN [IO81SF, ST64]
G7 JYD G D Leggett, Burghfield, Junction Rd, Cold Norton, Chelmsford, CM3 6HU [JO01IQ, TL80]
G7 JYE Details withheld at licensee's request by SSL.
G7 JYF S R Barron, 74 Salisbury Rd, Gravesend, DA11 7DE [JO01EK, TQ67]
G7 JYG H G Odd, Verona, Harrow Rd, Knockholt, Sevenoaks, TN14 7JU [JO01BH, TQ45]
GW7 JYJ T S Gittoes, 1 Oaklands Cres, Builth Wells, LD2 3EP [IO82HD, SO05]
GI7 JYK P S Lowrie, 13 Carwood Park, Glengormley, Newtownabbey, BT36 5JU [IO74AQ, J38]
G7 JYL J P Sage, 40 Wentworth Dr, Cliffe, Rochester, ME3 8UL [JO01GK, TQ77]
G7 JYM Details withheld at licensee's request by SSL.
G7 JYN Details withheld at licensee's request by SSL.
G7 JYQ Details withheld at licensee's request by SSL.
G7 JYR Details withheld at licensee's request by SSL.
G7 JYS Details withheld at licensee's request by SSL.
G7 JYU P R Thrower, Clevelands, Tamworth Rd, Keresley End, Coventry, CV7 8JJ [IO92FK, SP38]
GM7 JYW P D Lawrence, Gateside Smithy, Munlochy, Black Isle, Ross-shire, IV8 8PA [IO77VN, NH65]
G7 JYY M J Penn, 53 Manfield Ave, Walsgrave, Coventry, CV2 2QF [IO92GK, SP38]
G7 JYZ S J Turley, 22 Powlers Cl, Pedmore, Stourbridge, DY9 9HH [IO82WK, SO98]
G7 JZD M Downes, 49 Cherry Garth, Beverley, HU17 0EP [IO93TU, TA03]
G7 JZE Details withheld at licensee's request by SSL.[Op: M Eccles. Station located near Lytham St Annes.]
G7 JZI W Hilton, 8 Ashfield Ave, Hindley Green, Hindley, Wigan, WN2 4RG [IO83RM, SD60]
G7 JZJ M J Doyle, 133A Pope Ln, Penwortham, Preston, Lancs, PR1 9DD [IO83PR, SD52]
G7 JZK W Hancox, 30 Barlows Cl, Aintree, Liverpool, L9 9HH [IO83ML, SJ39]
G7 JZL Details withheld at licensee's request by SSL.
G7 JZM G D Priestley, 24 Saxton Ave, Bradford, BD6 3SW [IO93CS, SE13]
G7 JZO Details withheld at licensee's request by SSL.
G7 JZP Details withheld at licensee's request by SSL.
G7 JZR Details withheld at licensee's request by SSL.
G7 JZT Details withheld at licensee's request by SSL.
G7 JZU Details withheld at licensee's request by SSL.
G7 JZW Details withheld at licensee's request by SSL.
G7 JZY K Long, Grimston, Garton, Aldbrough, Hull, HU11 4QE [IO93XS, TA23]
G7 KAC J W Derrick, Essex House, 39 Princes Rd, Clevedon, BS21 7NQ [IO81NK, ST47]
G7 KAE Details withheld at licensee's request by SSL.
G7 KAG W J Preece, 41 Parkfield Rd, Ruskington, Sleaford, NG34 9HT [IO93TB, TF05]
G7 KAH Details withheld at licensee's request by SSL.
G7 KAJ Details withheld at licensee's request by SSL.
G7 KAK I D Clewley, 31 Kenilworth Rd, Basingstoke, RG23 8JF [IO91KG, SU65]
G7 KAO D P Clarke, 2 Wilmot Rd, Dartford, DA1 3BA [JO01CK, TQ57]
G7 KAT P J Hulse, 56 The Platters, Rainham, Gillingham, ME8 0DJ [JO01HI, TQ86]
G7 KAV N Stemp, 3 Loxwood, East Preston, Littlehampton, BN16 1DT [IO90ST, TQ00]
GI7 KAW Details withheld at licensee's request by SSL.
GW7 KAX R E Jones, 13 Tir Estyn, Deganwy, Conwy, LL31 9PY [IO83CG, SH77]
G7 KBB Details withheld at licensee's request by SSL.
G7 KBD A P Carlton, 21 Glover Rd, Totley Rise, Sheffield, S17 4HN [IO93FH, SK38]
G7 KBE B S McIntyre, 39 Hillbrook Ct, Acreman St, Sherborne, Dorset, DT9 3NZ [IO80RW, ST61]
G7 KBF Details withheld at licensee's request by SSL.
G7 KBH Details withheld at licensee's request by SSL.
GW7 KBI G Dreiling, Picton Farm, Picton, Holywell, Clwyd, CH8 9JQ [IO83IH, SJ18]
G7 KBJ Details withheld at licensee's request by SSL.
G7 KBL Details withheld at licensee's request by SSL.
G7 KBN Details withheld at licensee's request by SSL.
GW7 KBP Details withheld at licensee's request by SSL.
G7 KBR P S Phillips, 45 Maple Dr, East Grinstead, RH19 3UR [JO01AD, TQ43]
G7 KBV Details withheld at licensee's request by SSL.
G7 KBY Details withheld at licensee's request by SSL.
G7 KCA Details withheld at licensee's request by SSL.
G7 KCB C Bradley, 30 Peabody St., Great Lever, Bolton, BL3 6SW [IO83SN, SD70]
G7 KCE J L Hannaford, 22 Barn Park, Stoke Gabriel, Totnes, TQ9 6SR [IO80EJ, SX85]
G7 KCF Details withheld at licensee's request by SSL.
G7 KCG A P Deane, 5 Osbourne Terr, London Rd, Thrupp, Stroud, GL5 2BJ [IO81VR, SO80]
G7 KCK P M Langley, 58 Binnacle Rd, Rochester, ME1 2XP [JO01GI, TQ76]
G7 KCN Details withheld at licensee's request by SSL.
G7 KCO Details withheld at licensee's request by SSL.
G7 KCR T Fell Kingston Coll Rd, Kingston College, Kingston Hall Rd, Kingston upon Thames, KT1 2AQ [IO91UJ, TQ16]
G7 KCS Details withheld at licensee's request by SSL.
G7 KCT G J Rayner, 8 Caernarvon Ct, Caernarvon Cl, Hemel Hempstead, HP2 4DH [IO91SS, TL00]
G7 KCU Details withheld at licensee's request by SSL.
G7 KCV Details withheld at licensee's request by SSL.
G7 KCY Details withheld at licensee's request by SSL.
G7 KCZ Details withheld at licensee's request by SSL.
G7 KDA Details withheld at licensee's request by SSL.
G7 KDC Details withheld at licensee's request by SSL.
G7 KDF L E Whiddington, 29 Fortescue Cl, Tattershall, Lincoln, LN4 4LN [IO93VC, TF25]
G7 KDG P Edmondson, Springfield Lodge, Sutton Ln, Eastburn, Keighley, West Yorks, BD20 7AH [IO93AV, SE04]
G7 KDH D Edmondson, Springfield Lodge, Sutton Ln, Eastburn, Keighley, West Yorks, BD20 7AH [IO93AV, SE04]
G7 KDJ A Chadwick, 2 Auden Pl, Longton, Stoke on Trent, ST3 1SJ [IO82WX, SJ94]
G7 KDM C G Campbell, 136 York Rd, Carterton, OX18 1DP [IO91ES, SP20]
G7 KDP D J Brown, Poplar House Farm, Main Rd, Saltfleetby, Louth, Lincolnshire, LN11 7SS [JO03BI, TF48]
G7 KDR B L Hopkins, 14 Falkenham Rise, Basildon, SS14 2JQ [JO01FN, TQ78]
G7 KDS A J Drage Kettering Sc AR, Wyndthorpe, 51 Greenbank Ave, Kettering, NN15 7EF [IO92PJ, SP87]
GW7 KDU M J Lewis, 14 Hornbeam Cl, St. Mellons, Cardiff, CF3 0JA [IO81KM, ST28]
G7 KDX R T Bell, 5 Byron Ave, Blyth, NE24 5RN [IO95FC, NZ38]
GW7 KDZ J Condron, 5 Fairway, Sandycroft, Deeside, CH5 2PJ [IO83LE, SJ36]
G7 KEA R Chapman, 1 Henley Ave, North Cheam, Sutton, SM3 9SQ [IO91VJ, TQ26]
GI7 KEC J C Stafford, 1 Shimna Cl, Cregagh Est, Belfast, BT6 0DZ [IO74BN, J37]
G7 KEE B Daw, 19 Rowan Cl, Cherryfields, Stone, ST15 0EP [IO82WV, SJ93]
G7 KEI B W Edgley, Fernlea, 30 Oak Ave, Romiley, Stockport, SK6 4DN [IO83WJ, SJ99]
G7 KEJ P Leather, 219 Speakman Rd, Dentons Green, St. Helens, WA10 6TH [IO83OL, SJ59]
G7 KEK R Horsfall, 7 Lytham Cl, Doncaster, DN4 6UT [IO93LM, SE60]
G7 KEL T H Bradley, 28 Bracken Cl, Birchwood, Warrington, WA3 7NS [IO83RK, SJ69]
G7 KEM M A Bradley, 28 Bracken Cl, Birchwood, Warrington, WA3 7NS [IO83RK, SJ69]
G7 KEP A V Reeve, 97 Mendip Vale, Coleford, Bath, BA3 5PP [IO81SF, ST64]
G7 KES Details withheld at licensee's request by SSL.
G7 KET Details withheld at licensee's request by SSL.
G7 KEX Details withheld at licensee's request by SSL.
G7 KEZ Details withheld at licensee's request by SSL.
G7 KFA Details withheld at licensee's request by SSL.
G7 KFC Details withheld at licensee's request by SSL.
G7 KFD Details withheld at licensee's request by SSL.
G7 KFE Details withheld at licensee's request by SSL.
G7 KFG Details withheld at licensee's request by SSL.
G7 KFH Details withheld at licensee's request by SSL.
G7 KFI Details withheld at licensee's request by SSL.
G7 KFJ Details withheld at licensee's request by SSL.
G7 KFL Details withheld at licensee's request by SSL.
G7 KFM Details withheld at licensee's request by SSL.
G7 KFN Details withheld at licensee's request by SSL.
G7 KFO Details withheld at licensee's request by SSL.
G7 KFP Details withheld at licensee's request by SSL.
G7 KFQ N J Camp, 48 Northfield Rd, Harpenden, AL5 5HZ [IO91TT, TL11]
G7 KFT Details withheld at licensee's request by SSL.
G7 KFU Details withheld at licensee's request by SSL.
G7 KFW Details withheld at licensee's request by SSL.
G7 KFX Details withheld at licensee's request by SSL.
G7 KFY Details withheld at licensee's request by SSL.
G7 KFZ Details withheld at licensee's request by SSL.
GW7 KGD H T Wrighton, 43 Bryn Celyn, Colwyn Heights, Colwyn Bay, LL29 6DH [IO83DG, SH87]
G7 KGH M J Forder, 157 Kennington Rd, Kennington, Oxford, OX1 5PE [IO91JR, SP50]
G7 KGI E J Gould, 53 Green Rd, Kidlington, OX5 2EU [IO91IT, SP41]
G7 KGK Details withheld at licensee's request by SSL.
G7 KGL J A Norman, 103 Hardman Ave, Rawtenstall, Rossendale, BB4 6BW [IO83UQ, SD82]
G7 KGM T D J Smith, 60 Station Rd, Chiseldon, Swindon, SN4 0PW [IO91DM, SU17]
G7 KGN B N Hill, 410 Watnall Rd, Hucknall, Nottingham, NG15 6FQ [IO93JA, SK54]
G7 KGP J Chisholm, 162 Ardington Rd, Abington, Northampton, NN1 5LT [IO92NF, SP76]
G7 KGR C Saunders, 11 Abbey Dr, Luton, LU2 0LG [IO91TV, TL12]
G7 KGS G D Hipwell, 340 Hillmorton Rd, Rugby, CV22 5EY [IO92JI, SP57]
G7 KGT Details withheld at licensee's request by SSL.
G7 KGV I C Lewis, 64 Craven Park Rd, Stamford Hill, London, N15 6AB [IO91XN, TQ38]
GM7 KHA S Grant, 16 Corse Ave, Springside, Irvine, KA11 3AF [IO75QO, NS33]
G7 KHB D J Birks, 334 Sarehole Rd, Birmingham, B28 0AQ [IO92BK, SP18]
G7 KHC T A G Wyatt, 11 Claremont Ave, Hersham, Walton on Thames, KT12 4NS [IO91TI, TQ16]
G7 KHE M B Knowlson, 23 Hawthorne Ave, Shipley, BD18 2JB [IO93CT, SE13]
GD7 KHG D W Walton, Ballaghennie, Lezayre Rd, Ramsey, IM8 2TA
G7 KHL S A Smith, 287 Campkin Rd, Cambridge, CB4 2LD [JO02BF, TL46]
G7 KHN N J Downs, 10 Cale Cl, Kettlebrook, Tamworth, B77 1DB [IO92DO, SK20]
GI7 KHR W Smyth, 35 Davarr Ave, Dundonald, Belfast, BT16 0NT [IO74CO, J47]
G7 KHT A J Haw, 2 Windmill Ln, Yeadon, Leeds, West Yorks, LS19 7TQ [IO93DU, SE24]
G7 KHV R J L Irvine, Chequers, 1 Nutana Ave, Hornsea, HU18 1JU [IO93VW, TA24]
G7 KHW D Nock, 2 Lessingham Rd, Widnes, WA8 9FU [IO83PJ, SJ58]
G7 KHX H Andrews, 72 Cornfield Dr, Boley Park, Lichfield, WS14 9UG [IO92CQ, SK10]
G7 KHY L Miles, Astwell House, Cotefields Ave, Farsley Pudsey, Leeds, W Yorks, LS28 5EJ [IO93DT, SE23]
G7 KHZ R C Hobbs, 3 Duncombe Cl, Bridgwater, TA6 4UT [IO81MD, ST33]
G7 KIA J Miles, 31 Poplar Rise, Bramley, Leeds, LS13 4SQ [IO93ET, SE23]
G7 KID C J Baily, 13 Longleigh Ln, Bexleyheath, DA7 5SL [JO01BL, TQ47]
G7 KIE N A Kirkman, 4 Woodhall Cres, Saxilby, Lincoln, LN1 2HZ [IO93QG, SK87]
G7 KIF C L Davis, Glebe Farm, Haunton, Tamworth, Staffs, B79 9HN [IO92EQ, SK21]
G7 KII M J Chilcott, 16 Mount Gould Ave, St. Judes, Plymouth, PL4 9EZ [IO70WJ, SX45]
GW7 KIL C A Hunt, 39 Withdean Cres, Brighton, BN1 6WG [IO90WU, TQ30]
GM7 KIM R A Woods, 136 Birch Rd, Abronhill, Cumbernauld, Glasgow, G67 3PB [IO85AX, NS77]
G7 KIN B S Kinsella, 8 Sherwood Park Rd, Sutton, SM1 2SQ [IO91VI, TQ26]
GW7 KIO G M Hawthorn-Slater, Pant y Moeliaid, Pencaenewydd, Pwllheli, Gwynedd, LL53 6RD [IO72TW, SH44]
GW7 KIP S M Insole, 42 St. Annes Dr, Fardre, Llantwit Fardre, Pontypridd, CF38 2PD [IO81IN, ST08]
G7 KIQ P P Hyde, 10 Highfield Cres, Comeytrowe Ln, Taunton, TA1 5JH [IO81KA, ST22]
GW7 KIS M J Brookman, 6 Antrim St., Liverpool, L13 8DF [IO83MK, SJ39]
G7 KIT D Hogg, 26 Grenville Dr, Church Crookham, Fleet, GU13 0NR [IO91NG, SU85]
GW7 KIU G E Morris, 7 Park Gr, Connahs Quay, Deeside, CH5 4HU [IO83LF, SJ26]
GW7 KIV R N Gadney, 6 Dan yr Eppynt, Tirabad, Llangammarch Wells, LD4 4DR [IO82EB, SN84]
G7 KIW R F Henery, 117 Marlborough Rd, Swindon, SN3 1NJ [IO91CN, SU18]
G7 KJA R Early, 11 Wenlock Dr, Newport, TF10 7HH [IO82TS, SJ71]
G7 KJD J Smallwood, 6 Thatchers Croft, Copmanthorpe, York, YO2 3YD [IO93KW, SE54]
G7 KJE A Wilkes, 51 Shrewsbury Dr, Newcastle, ST5 7RQ [IO83VB, SJ85]
GM7 KJL D I Turner, Balgowan, Colpy, Insch, Aberdeenshire, AB52 6TR [IO87QJ, NJ63]
G7 KJO M Wray, Ashdene, 8 The Avenue, Moulton, Northampton, NN3 7TL [IO92NG, SP76]
G7 KJP R McMahon, 8 Meadow Cl, Holburn Dene Est, Ryton, NE40 3RU [IO94CX, NZ16]
G7 KJR C J Baxter, 6 Merrington Cl, Kirk Merrington, Spennymoor, DL16 7HU [IO94EQ, NZ23]
G7 KJT S Mills, 49 Temple Gate Cres, Leeds, LS15 0EZ [IO93GT, SE33]
G7 KJW P N Haylock, 25 Whitehouse Rd, Sawtry, Huntingdon, PE17 5UA [IO92UK, TL18]
G7 KJX R T Tebbutt, 3 Trinity Cl, Daventry, NN11 4RN [IO92JF, SP56]
G7 KJY P H Walker, 20 Arbury Banks, Chipping Warden, Banbury, OX17 1LU [IO92ID, SP44]
G7 KKC S Newsham, Elm House, Bridgnorth Rd, Enville, Stourbridge, DY7 5HA [IO82UL, SO88]
G7 KKE B M Mowle, 5 Chesham Dr, Steepleview, Laindon, Basildon, SS15 4AH [JO01FO, TQ69]
G7 KKH Details withheld at licensee's request by SSL.
G7 KKJ D R Medley, 32 Poppleton Rise, Tingley, Wakefield, WF3 1UT [IO93FR, SE22]
G7 KKL Details withheld at licensee's request by SSL.
GW7 KKN Details withheld at licensee's request by SSL.
GW7 KKP Details withheld at licensee's request by SSL.
G7 KKR Details withheld at licensee's request by SSL.
G7 KKS Details withheld at licensee's request by SSL.
G7 KKT Details withheld at licensee's request by SSL.
G7 KKV Details withheld at licensee's request by SSL.
G7 KKW Details withheld at licensee's request by SSL.
G7 KKY Details withheld at licensee's request by SSL.
G7 KLA Details withheld at licensee's request by SSL.
GW7 KLC Details withheld at licensee's request by SSL.
G7 KLD Details withheld at licensee's request by SSL.
G7 KLJ Details withheld at licensee's request by SSL.
G7 KLN J R Abbey, 4 Northway, Curzon Park, Chester, CH4 8BB [IO83NE, SJ36]
G7 KLP C Hartigan, Doonagore, Doolin, Co. Clare, Republic of Ireland, X X
G7 KLQ K Parnham, 91 Marlborough Rd, Romford, RM7 8AP [JO01BN, TQ48]
G7 KLR L D C Pooley, 134 Aylsham Dr, Ickenham, Uxbridge, UB10 8UE [IO91SN, TQ08]
G7 KLS A Macaulay, 27 Hewson Pl, Sheriff Hill, Gateshead, NE9 6QS [IO94EW, NZ26]
G7 KLT T A Hassall, 5 Ashworth St., Bacup, OL13 9LS [IO83VQ, SD82]
G7 KLU R M Ruttle, 63 Biddulph Rd, Chell, Stoke on Trent, ST6 6SW [IO83VB, SJ85]
G7 KLV G N Lovegrove, 64 Vicarage Ln, Great Baddow, Chelmsford, CM2 8HY [JO01GR, TL70]
G7 KLZ J V P Fowler, Burton Mill Cottage, Gr Rd, Burton Bradstock, Bridport, DT6 4QU [IO80PQ, SY48]
G7 KMA S J Balkham, 28 Plynlimmon Rd, Hastings, TN34 3LT [JO00HU, TQ81]
GI7 KMC J Magee, 13 Woodcroft Heights, Lower Braniel Rd, Belfast, BT5 7NX [IO74BN, J37]
G7 KMD N J Hilton, 2 Clarendon St., Bloxwich, Walsall, WS3 2HT [IO82XO, SJ90]
G7 KME D M Silverton, Heron Lodge, Cornsland, Brentwood, CM14 4JN [JO01DO, TQ59]
G7 KMH S T Smith, Thurne House, Middle St, Swinton, Malton N Yorks, YO17 0SR [IO94ND, SE77]
GI7 KMJ J G Mc Guinness, 11 Rosemount Gdns, Belfast, BT15 5AG [IO74AO, J37]
G7 KMK N H Deacon, 30 Harcourt Rd, Wigston, LE18 3SB [IO92KN, SP69]
G7 KMO P B Butler, 15 Roxby Cl, Bessacarr, Doncaster, DN4 7JH [IO93KL, SE60]
G7 KMR A C Williams, 18 Sturdee Cl, Thetford, IP24 2LF [JO02JK, TL88]
G7 KMW Details withheld at licensee's request by SSL.
G7 KMX Details withheld at licensee's request by SSL.
G7 KMY Details withheld at licensee's request by SSL.
G7 KMZ R J Paul, 1 Celestine Rd, Yate, Bristol, BS17 5DZ [IO81SN, ST78]
G7 KNA Details withheld at licensee's request by SSL.
G7 KNB Details withheld at licensee's request by SSL.
G7 KNE Details withheld at licensee's request by SSL.
G7 KNF Details withheld at licensee's request by SSL.
G7 KNI Details withheld at licensee's request by SSL.
G7 KNK H Arrowsmith, 15 Hermitage Cl, Frimley, Camberley, GU16 5LP [IO91PH, SU85]
G7 KNM M Giles, 9 Bower Green, Lordswood, Chatham, ME5 8TN [JO01GH, TQ76]
GW7 KNN B Jones, Rivendell, Llewelyn Rd, Coedpoeth, Wrexham, Clwyd, LL11 3PB [IO83LB, SJ25]
G7 KNP P J Barnes, 24 Harrowfield Rd, Stechford, Birmingham, B33 9BU [IO92CL, SP18]
G7 KNQ C S Martin, 17 St. Marys Cres, East Leake, Loughborough, LE12 6QR [IO92JU, SK52]

G7 KNS G M Bubb, Clearways, Hadlow Stair, Tonbridge, TN10 4HD [JO01DE, TQ64]
G7 KNT P D Schofield, 98 Mountfield Rd, Waterloo, Huddersfield, HD5 8RB [IO93DP, SE11]
G7 KNU P R M Davis, 2 Mead Park, Atworth, Melksham, SN12 8JS [IO81VJ, ST86]
G7 KNV Details withheld at licensee's request by SSL.
GM7 KNW Details withheld at licensee's request by SSL.
G7 KNZ Details withheld at licensee's request by SSL.
G7 KOA Details withheld at licensee's request by SSL.
G7 KOF L Barr, 7 Southwold Gdns, Silksworth, Sunderland, SR3 1LG [IO94HV, NZ35]
G7 KOG P M Black, 43 Malvern Ave, Hillmorton, Rugby, CV22 5JN [IO92JI, SP57]
G7 KOI G Russ, 12 Marconi Rd, Chelmsford, CM1 1QB [JO01FR, TL70]
G7 KOL M A Newton, 35 West St., Blaby, Leicester, LE8 4GY [IO92JN, SP59]
G7 KOO Details withheld at licensee's request by SSL.
G7 KOQ Details withheld at licensee's request by SSL.
G7 KOS S J McCormick, 22 Eric Rd, Liscard, Wallasey, L44 5RQ [IO83LK, SJ39]
G7 KOT H Gray, 65 Parkview Dr, Liverpool, L27 6WQ [IO83NJ, SJ48]
G7 KOY M P Robinson, 2 Bell Cres, Burham, Rochester, ME1 3SZ [JO01FH, TQ76]
G7 KPB A C E Ottaway, 3 Swan Cl, Worle, Weston Super Mare, BS22 8XR [IO81MI, ST36]
G7 KPC J R Page, 2 Fee Terr, Sunderland, SR2 0JY [IO94HU, NZ45]
G7 KPD A Pluck, 31 Coalway Rd, Penn, Wolverhampton, WV3 7LU [IO82WN, SO99]
GM7 KPE J I Reid, 8 Balmoral Pl, Leven, Fife, KY8 4RQ [IO86ME, NO30]
G7 KPF A R Gayne, 119 Lower Lickhill Rd, Stourport on Severn, DY13 8UQ [IO82UI, SO87]
G7 KPH M J Wood, New Mill Rd, Thongsbridge, Holmfirth, Huddersfield, West Yorks, HD7 2SQ [IO93CN, SE10]
GI7 KPJ R D Irwin, 23 Rhone Rd, Drummond, Dungannon, BT71 7EN [IO64PL, H85]
G7 KPL A J Stringer, 264 Jardine Cres, Tile Hill, Coventry, CV4 9QS [IO92FJ, SP27]
G7 KPM J R Haywood, 31 Oak Way, Cleethorpes, DN35 0RA [IO93XM, TA20]
G7 KPN A A Mee, 37B Cuckfield Rd, Hurstpierpoint, Hassocks, West Sussex, BN6 9RW [IO90VW, TQ21]
G7 KPP Details withheld at licensee's request by SSL.
G7 KPQ Details withheld at licensee's request by SSL.
G7 KPR Details withheld at licensee's request by SSL.
G7 KPS C G R Pickles, Ovington House, Far Ln, Coleby, Lincoln, LN5 0AH [IO93RD, SK96]
G7 KPT Details withheld at licensee's request by SSL.
G7 KPV Details withheld at licensee's request by SSL.
G7 KPX Details withheld at licensee's request by SSL.
G7 KPZ Details withheld at licensee's request by SSL.
G7 KQA Details withheld at licensee's request by SSL.
G7 KQC Details withheld at licensee's request by SSL.
G7 KQE Details withheld at licensee's request by SSL.
G7 KQF Details withheld at licensee's request by SSL.
GM7 KQG Details withheld at licensee's request by SSL.
GW7 KQI Details withheld at licensee's request by SSL.
G7 KQL Details withheld at licensee's request by SSL.
G7 KQM Details withheld at licensee's request by SSL.
G7 KQT S J Schrier, 163 West Ln, Hayling Island, PO11 0JW [IO90MT, SU70]
G7 KQW G Beresford, 17 Frobisher Gr, Maltby, Rotherham, S66 8QU [IO93JK, SK59]
G7 KQY B R Strawson, 11 Stevenson Dr, Abingdon, OX14 1SN [IO91IQ, SU49]
G7 KRB S J Wells, 55 Staverton Rd, Daventry, NN11 4EY [IO92JG, SP56]
G7 KRC K A Conlon Keighley ARS, 76 Deanwood Cres, Allerton, Bradford, BD15 9BL [IO93CT, SE13]
G7 KRE T P Benjamin, 24 Moat Farm Dr, Rugby, CV21 4HG [IO92JI, SP57]
G7 KRG T Binns Keighley Ray Gr, Crossfarm Cottage, Oxenhope, Keighley, West Yorks, BD22 9LE [IO93AT, SE03]
G7 KRH T D Hurley, 22 Honeyhill, Wootton Bassett, Swindon, SN4 7DX [IO91BM, SU08]
G7 KRI J G Tilley, 40 Marlborough Rd, Stretford, Manchester, M32 0AN [IO83UK, SJ79]
G7 KRM B M Walker, 3 Moorlands Dr, Mayfield, Ashbourne, DE6 2LP [IO93CA, SK14]
G7 KRO D S J Willis, 27 Marennes Cres, Brightlingsea, Colchester, CO7 0RU [JO01MT, TM01]
G7 KRP A B Morrow, 31 Rutherford Way, Bushey, Watford, WD2 1NJ [IO91TP, TQ19]
GM7 KRQ H M E Gordon, Woodlands, Polesburn, Methlick, Grampian, AB41 0DU [IO87WI, NJ93]
G7 KRS L J L Davies Kettering ARS, 2 Kettonby Gdns, Headlands, Kettering, NN15 6BT [IO92PJ, SP87]
G7 KRU H Tarrant, Springbank, 46 Melrose Ave, Fulwood, Preston, PR2 8DE [IO83PS, SD53]
GW7 KRY A C Ryall, Tanglewood, Mountain Rd, Pentyrch, Cardiff, CF4 8QP [IO81IM, ST18]
G7 KRZ S M Pountain, 33 Milldale Ave, Buxton, SK17 9BE [IO93AG, SK07]
GM7 KSA R R S Vennard, 4 Braehead, Girdle Toll, Irvine, KA11 1BD [IO75QP, NS34]
GM7 KSC C C Ritchie, Craven, 31 Cross Rd, Paisley, PA2 9QJ [IO75SU, NS46]
G7 KSG M R Bailey, 47 Willowbed Dr, Chichester, PO19 2HX [IO90OT, SU80]
G7 KSI Details withheld at licensee's request by SSL.
G7 KSL D A C Tanner, 55 Arundel Dr, Bramcote Hills, Beeston, Nottingham, NG9 3FN [IO92JW, SK53]
G7 KSM Details withheld at licensee's request by SSL.
G7 KSN Details withheld at licensee's request by SSL.
G7 KSP G R Hampson, 11 Gladstone Gr, Heaton Moor, Stockport, SK4 4BX [IO83VK, SJ89]
G7 KSQ S D Little, 84 Brockenhurst Way, Bicknacre, Chelmsford, CM3 4XW [JO01HQ, TL70]
G7 KSS M A Watts, 74 Westfield Rd, Caversham, Reading, RG4 8HJ [IO91ML, SU77]
G7 KSV Details withheld at licensee's request by SSL.
G7 KSW Details withheld at licensee's request by SSL.
G7 KTC E T Baxter, 62 Cowlersley Ln, Huddersfield, HD4 5UB [IO93CP, SE11]
G7 KTD W D Walker, 6 Romford St., Burnley, BB12 8AF [IO83UT, SD83]
G7 KTE Details withheld at licensee's request by SSL.
GI7 KTF Details withheld at licensee's request by SSL.
G7 KTH Details withheld at licensee's request by SSL.
G7 KTK Details withheld at licensee's request by SSL.
G7 KTL Details withheld at licensee's request by SSL.
G7 KTM Details withheld at licensee's request by SSL.
G7 KTN Details withheld at licensee's request by SSL.
GW7 KTP T J Daniels, Foresters House, Tair Onen, Cowbridge, S Glam, CF7 7UA [IO81HN, ST08]
G7 KTQ J R Klunder, 58 Windsor Dr, Brinscall, Chorley, PR6 8PX [IO83RQ, SD62]
G7 KTR A J Slinn, Santon, Pound Ln, Knockholt, Sevenoaks, TN14 7NA [JO01BH, TQ45]
GM7 KTY P R May, 6 Hillpark Way, Edinburgh, EH4 7BJ [IO85IX, NT27]
G7 KUA E T G Hewett, 65 Ferry St., London, E14 3DT [IO91XL, TQ37]
G7 KUB R G Warrell, Rose Cottage, Brookbottom, New Mills, High Peak, Derbyshire, SK22 3AY [IO83XI, SJ98]
G7 KUD Details withheld at licensee's request by SSL.
G7 KUF Details withheld at licensee's request by SSL.
G7 KUG Dr D A Rutherford, 9 College Dr, Ruislip, HA4 8SD [IO91TN, TQ18]
G7 KUJ N G Hull, 77 Beeches Cres, Southgate, Crawley, RH10 6BU [IO91VC, TQ23]
GW7 KUK A W Pritchard, 10 St. Christophers Rd, Porthcawl, CF36 5RY [IO81DL, SS87]
G7 KUM A H Yorke, 45 Ling Rd, Chesterfield, S40 3HT [IO93GF, SK36]
GM7 KUN C Schofield, Airidh Ghrianach, Knock, Carloway, Isle of Lewis, HS2 9AU [IO68OG, NB24]
G7 KUU K C W Bates, Newhaven Cottage, Star Green, Whiteshill, Stroud, GL6 6AD [IO81VS, SO80]
G7 KUW Details withheld at licensee's request by SSL.
G7 KUX A W G Tavinor, 78 Conant House, St. Agnes Pl, London, SE11 4AY [IO91WL, TQ37]
G7 KUZ Details withheld at licensee's request by SSL.
GM7 KVB A J Whyte, 3 Glenfield Rd, Cowdenbeath, KY4 9EP [IO86HC, NT19]
G7 KVH T R Burn, 40 Colegrave St., Lincoln, LN5 8DR [IO93RF, SK96]
G7 KVK Details withheld at licensee's request by SSL.
G7 KVN Details withheld at licensee's request by SSL.
GI7 KVR P P McDonald, 13 Heathfield, Culmore Rd, Londonderry, BT48 8JD [IO65IA, C42]
G7 KVT B A Moorey, 132 Queensway, Hereford, HR1 1HQ [IO82PB, SO54]
GM7 KVU G A Kilgour, 6 Thornwood Pl, Glasgow, G11 7PP [IO75UU, NS56]
GW7 KVW K Scott, 48 Gourley Rd, Liverpool, Merseyside, L13 4AY [IO83NJ, SJ39]
G7 KVX Details withheld at licensee's request by SSL.
G7 KVY E W D Whittaker, 17 Booth Rise, Northampton, NN3 6HP [IO92NG, SP76]
G7 KVZ J Ashmore, 46 Mease Cl, Measham, Swadlincote, DE12 7NA [IO92FQ, SK31]
G7 KWA J E Billam, 46 Rugby Rd, Rainworth, Mansfield, NG21 0AU [IO93KC, SK55]
GW7 KWB D C Cornes, 86 Stairhaven Rd, Liverpool, L19 7NW [IO83NI, SJ38]
G7 KWC B J Broxup, 39 Woodgrove Rd, Burnley, BB11 3EL [IO83VS, SD83]
G7 KWD M C Savin, 8 Waterloo Rise, Reading, RG2 0LN [IO91MK, SU77]
G7 KWF A B Richards, 18 Orchard Way, Lower Kingswood, Tadworth, KT20 7AD [IO91VG, TQ25]
GW7 KWG Details withheld at licensee's request by SSL.
G7 KWH Details withheld at licensee's request by SSL.
G7 KWL Details withheld at licensee's request by SSL.
G7 KWN A J Dance, 8 Eversley Rd, Arborfield, Arborfield Cross, Reading, RG2 9PU [IO91NJ, SU76]
G7 KWO J A Lewis, 103 Lennard Rd, Beckenham, BR3 1QS [IO91XK, TQ37]
G7 KWP G W Lewis, 103 Lennard Rd, Beckenham, BR3 1QS [IO91XK, TQ37]
G7 KWQ T F Holliday, 131 Skinburness Rd, Silloth, Carlisle, CA5 4QH [IO84HV, NY15]
G7 KWS S E Riches, 5 Norfolk St., Forest Gate, London, E7 0HN [JO01AN, TQ48]
G7 KWT J M Ruddock, 164 Studland Rd, Hanwell, London, W7 3QZ [IO91TM, TQ18]
G7 KWZ Details withheld at licensee's request by SSL.
G7 KXA P C Chester-Brown, 23 Eden Ave, Culcheth, Warrington, WA3 5HX [IO83SK, SJ69]
GM7 KXB Details withheld at licensee's request by SSL.
G7 KXD Details withheld at licensee's request by SSL.
G7 KXG P Fleming, 94 Pigott St., London, E14 7DW [IO91XM, TQ38]
GM7 KXJ R Donnet, 13 Coranbae Pl, Doonfoot, Ayr, KA7 4JB [IO75QK, NS31]
G7 KXM Details withheld at licensee's request by SSL.
G7 KXN M J Bonser, 24 Meend Garden Terr, Cinderford, GL14 2EB [IO81ST, SO61]
G7 KXP A J T Nicholls, 1 Brayshaw Cl, Heywood, OL10 3EE [IO83VO, SD81]
G7 KXR A R Green Nottingham Raynet ARS, 37 Bramcote Ln, Wollaton, Nottingham, NG8 2NA [IO92JW, SK53]
G7 KXS P G Adam, 50 Lower Edge Rd, Rastrick, Brighouse, HD6 3LD [IO93CQ, SE12]
G7 KXU A G Edwards, The Hazels, 13 Barby Ln, Rugby, CV22 5QJ [IO92JI, SP57]
G7 KXV Details withheld at licensee's request by SSL.
G7 KXW Details withheld at licensee's request by SSL.
G7 KXX Details withheld at licensee's request by SSL.
G7 KXY S C Veitch, 31 St. Johns Ct, Lagham Rd, South Godstone, Godstone, RH9 8HD [IO91XF, TQ34]
G7 KYB Details withheld at licensee's request by SSL.
G7 KYC Details withheld at licensee's request by SSL.
G7 KYD S Walker-Kier, 45 Anstey Rd, Peckham, London, SE15 4JX [IO91XL, TQ37]
G7 KYF T J Fellows, 38 Bedser Dr, Greenford, UB6 0SE [IO91TN, TQ18]
G7 KYG J W Hope, 29 Horner Rd, Taunton, TA2 8DZ [IO81KA, ST22]
G7 KYH J N Mann, 44 Hardwick Park, Banbury, OX16 7YF [IO92HB, SP44]
G7 KYJ K R Clark, 1 Honington Cres, Ermine West, Lincoln, LN1 3UT [IO93RF, SK97]
G7 KYL M S Lack, 8 Westwater Way, Ladygrove, Didcot, OX11 7SN [IO91JO, SU59]
G7 KYM Details withheld at licensee's request by SSL.
G7 KYN Details withheld at licensee's request by SSL.
G7 KYO Details withheld at licensee's request by SSL.
G7 KYR Details withheld at licensee's request by SSL.
GW7 KYT T A Hulmes, Ashley House, 12 Blackmill Rd, Bryncethin, Bridgend, CF32 9YW [IO81FN, SS98]
G7 KYV J W Sharp, 82 Delaval Cres, Newsham, Blyth, NE24 4BD [IO95FC, NZ27]
G7 KYW C I Mellings, 7 Todds Cl, Horley, RH6 8LB [IO91VE, TQ24]
GM7 KYX G M Stones, Hillhead Cottage, Cross Rd Grange, Keith, Scotland, AB55 3LU [IO87LM, NJ34]
G7 KYY R N Scott, 117 Horley Rd, Redhill, RH1 5AS [IO91VF, TQ24]
G7 KYZ D A Shaw, 159 Saltwells Rd, Dudley, DY2 0BN [IO82WL, SO98]
G7 KZA Details withheld at licensee's request by SSL.
G7 KZB Details withheld at licensee's request by SSL.
G7 KZG C C Cain, 62 Lancaster Cres, St. Eval, Wadebridge, PL27 7TP [IO70ML, SW86]
G7 KZH N K Swain, 234 Chesford Rd, Luton, LU2 8DT [IO91TV, TL12]
G7 KZI M J Hillary, 45 Frances Rd, Purbrook, Waterlooville, PO7 5HH [IO90LU, SU60]
G7 KZJ Details withheld at licensee's request by SSL.
GM7 KZL J S Mawson, 5 Forth View, Kirknewton, EH27 8AN [IO85HV, NT16]
G7 KZN H T Davies, 47 Vincent Rd, Rainhill, Prescot, L35 8PE [IO83OK, SJ49]
GW7 KZS J Goodwin, Woodfield House, Chalfont Ln, Chorleywood, Rickmansworth, WD3 5PP [IO91RP, TQ09]
G7 KZT B C Chessell, Popes Oak Farm, West Grinstead Aprk, Cowfold Rd, West Grinstead, RH13 8LU [IO91UB, TQ13]
G7 KZV C Dodson, 64 Stoneleigh Rd, Solihull, B91 1DQ [IO92CK, SP18]
G7 KZW S V Powell, 19 Moor Farm Ln, Three Elms Rd, Hereford, HR4 0NT [IO82PB, SO44]
G7 KZX Details withheld at licensee's request by SSL.
G7 KZZ Details withheld at licensee's request by SSL.
GI7 LAA J R D Martin, 29 Clonroot Rd, Portadown, Craigavon, BT62 4HG [IO64RJ, H95]
GM7 LAC P J Green, Clochcan School Cot, Auchnagatt, Ellon, Aberdeenshire, AB41 8UJ [IO87WL, NJ94]
G7 LAE Details withheld at licensee's request by SSL.
G7 LAF M R Kidman, 465 Gr Green Rd, Leytonstone, London, E11 4AA [JO01AN, TQ38]
GI7 LAI D Dempster, 4 Muskett Ave, Carryduff, Belfast, BT8 8QH [IO74BM, J36]
G7 LAK C T Wilkinson, 15 Woodland Cl, Farnsfield, Newark, NG22 8DN [IO93LC, SK65]
G7 LAL I M Mazura, 45 Bolingbroke Rd, Scunthorpe, DN17 2NQ [IO93QN, SE80]
G7 LAN D T Halsey, 9 Kings Rd, Luton, Chatham, ME5 7JY [JO01GI, TQ76]
GW7 LAU Details withheld at licensee's request by SSL.
GD7 LAV A J Gawne, Keristal House, Marine Dr, Port Soderick, Douglas, IM4 1BJ
G7 LAW J W Danner, 16 Batemans Acre South, Coventry, CV6 1BE [IO92FJ, SP37]
G7 LAX W Keeys, 5 Admirals Ct, Swaffham, PE37 7TE [JO02IP, TF80]
G7 LBA R D Nixon, 52 Brookfield Ave, Bredbury, Stockport, SK6 1DF [IO83WK, SJ99]
G7 LBC Details withheld at licensee's request by SSL.
G7 LBD L A Lewis, 29 Sefton Ave, Hove Edge, Brighouse, HD6 2NA [IO93CR, SE12]
G7 LBG Details withheld at licensee's request by SSL.
G7 LBH A Champion, 5 Airedale Cliff, Pollard Ln, Leeds, LS13 1EA [IO93ET, SE23]
G7 LBJ M D Danfer, 72 Warwick Ave, Woodbridge, IP12 1JY [JO02PC, TM24]
G7 LBL Dr A D Batey, Quarter House, Haddows Cl, Longstanton, Cambridge, CB4 5DJ [JO02AG, TL36]
G7 LBM T L Howard, 21 Church Ln, Thornhill, Dewsbury, WF12 0JZ [IO93EP, SE21]
G7 LBO D A Wilson, 4 The St., Sparham, Lenwade, Norwich, NR9 5SD [JO02NR, TG01]
G7 LBP M E Akiki, 16 Reed Dr, Marchwood, Southampton, SO40 4YG [IO90GV, SU31]
G7 LBS P A Smith, 47 Wildern Ln, East Hunsbury, Northampton, NN4 0SN [IO92NE, SP75]
G7 LBT P A J Bundock, 115 Apple Gr, Enfield, EN1 3DB [IO91XP, TQ39]
G7 LBV G Middleton Stanchester, Stanchester Community School, Stoke Sub Hamdon, Somerset, TA14 6UG [IO80PW, ST41]
GW7 LCB Details withheld at licensee's request by SSL.
GI7 LCC M S Sheils, 56 Jocelyn Ave, Belfast, BT6 9AX [IO74BO, J37]
G7 LCD A J Sermons, 17 Wellside, Marks Tey, Colchester, CO6 1XG [JO01JV, TL92]
G7 LCG Details withheld at licensee's request by SSL.
G7 LCI Details withheld at licensee's request by SSL.
G7 LCK J H Berry, Roseneath, Walcote Rd, South Kilworth, Lutterworth, LE17 6EQ [IO92KK, SP58]
GW7 LCL Details withheld at licensee's request by SSL.
G7 LCM Details withheld at licensee's request by SSL.
GW7 LCP A L Mackay, 23 Dock St., Cogan, Penarth, CF64 2LA [IO81JK, ST17]
GI7 LCQ J C Serplus, 14 Claggan Park, Aghadowey, Coleraine, N Ireland, N Ireland, BT51 4BD [IO65QA, C81]
G7 LCS A J Daniels, Wiscombe, Cleveland Rd, Worcester Park, KT4 7JQ [IO91UJ, TQ26]
G7 LCU Details withheld at licensee's request by SSL.
G7 LCW C J Simpson, 32 Mill Rd, Stilton, Peterborough, PE7 3XY [IO92UL, TL18]
G7 LCY Details withheld at licensee's request by SSL.
G7 LDD R H Newton, 114 Kingston Rd, Taunton, TA2 7SP [IO81KA, ST22]
GI7 LDE Details withheld at licensee's request by SSL.
GW7 LDF Details withheld at licensee's request by SSL.
G7 LDJ S W Harding Sony Brdcast AR, Sony Broadcast Ltd, Jays Cl, Basingstoke Hants, RG22 4SB [IO91KF, SU65]
GW7 LDP P S Martin, 1 Well Cttgs, Coychurch, Bridgend, CF35 5HD [IO81FM, SS97]
G7 LDR E A Woolfenden, 6 Cliff Grange, Bury New Rd, Salford, M7 4EZ [IO83UM, SD80]
GM7 LDU W L Adie, 9 Shore Rd, Ballantrae, Girvan, KA26 0NG [IO75LC, NX08]
G7 LEB F L Stevens, 4 Pennine Rd, Putnoe, Bedford, MK41 9AS [IO92SD, TL05]
G7 LED D L Miles, 2 Barrington Rd, Olton, Solihull, B92 8DP [IO92CK, SP18]
G7 LEE Details withheld at licensee's request by SSL.
G7 LEG R D Martin Solpak Group, 2 Elgin Gdns, Strood, Rochester, ME2 2PU [JO01FJ, TQ76]
G7 LEL D J Hawkins, 93 Buxton Dr, Bexhill on Sea, TN39 4AS [JO00FU, TQ70]
G7 LEN A S Clark West Lincs Rynt, Emergency Planning Department, Fire Brigade Headquarters, South Park Ave, Lincoln, LN5 8EL [IO93RF, SK96]
G7 LEQ Details withheld at licensee's request by SSL.
G7 LER W Marshall, 43 Pilton Rd, Newcastle upon Tyne, NE5 4PP [IO94DX, NZ16]
G7 LEV G Wilson, 6 The Chase, Calcot, Reading, RG3 7DN [IO91LJ, SU66]
G7 LEX S Wilkes, 44 Banklands Rd, Dudley, DY2 8BT [IO82XL, SO98]
G7 LEY G B Lloyd Essex Packet Gr, 9 Hornbeam Walk, Witham, CM8 2SZ [JO01HT, TL81]
G7 LFB D English, 1 Howrigg Bank, Wigton, CA7 9JF [IO84KT, NY24]
G7 LFC D J Hughes, 86 Colinmander Gdns, Ormskirk, L39 4TF [IO83NN, SD40]
G7 LFM A J Cocker, 30 Shaw Rd, Rochdale, OL16 4SH [IO83WO, SD90]
G7 LFP Details withheld at licensee's request by SSL.
G7 LFQ J A White, 62 Cornwallis Ave, Aylsham, Aylesham, Canterbury, CT3 3HQ [JO01OF, TR25]
G7 LFR Details withheld at licensee's request by SSL.
G7 LFT A C Monk, 23 Trinity Rd, Hurstpierpoint, Hassocks, BN6 9UY [IO90VW, TQ21]
G7 LFU J Morley, 72 Princess Anne Rd, Boston, PE21 9AP [IO92XX, TF34]
G7 LFV S Young, 24 East Carr, Cayton, Scarborough, YO11 3TS [IO94TF, TA08]
G7 LFY Details withheld at licensee's request by SSL.
G7 LFZ W H Mumford, Agden Green Farm, Perry Rd, Grt Staughton, Huntingdon, Cambs, PE19 4DH [IO92UG, TL16]
G7 LGD Details withheld at licensee's request by SSL.
G7 LGG M W Smith, 24 Priors Cl, Beeding, Upper Beeding, Steyning, BN44 3HT [IO90UV, TQ11]
G7 LGH A E Moore, 4 The Parks, Sundorne Gr, Shrewsbury, SY1 4TJ [IO82PR, SJ51]
G7 LGL Details withheld at licensee's request by SSL.
G7 LGO Details withheld at licensee's request by SSL.
G7 LGQ J McNulty, 59 Briar Cl, Evesham, WR11 4JQ [IO92AC, SP04]
G7 LGS N M Green, Walnut View, Kingstone, Uttoxeter, Staffs, ST14 8QH [IO92BU, SK02]
G7 LGV M C Smith, 8B Aldsworth Ct, Witney, OX8 5FP [IO91FS, SP30]
G7 LGW Details withheld at licensee's request by SSL.
G7 LGX Details withheld at licensee's request by SSL.
G7 LGY H N E Abbott, 1 St. Lawrence Cl, Heanor, DE75 7AN [IO93HA, SK44]

G7 LHB Details withheld at licensee's request by SSL.
GI7 LHF Details withheld at licensee's request by SSL.
G7 LHH Details withheld at licensee's request by SSL.
GW7 LHI P D Tomlinson, 17 Heol yr Orsedd, Margam, Port Talbot, West Glam, SA13 2HL [IO81CN, SS78]
G7 LHL Details withheld at licensee's request by SSL.
G7 LHO D Oubridge, 120A North View Rd, Hornsey, London, N8 7LP [IO91WO, TQ28]
G7 LHS S Gray, 23 Craig Cl, Heaton Mersey, Stockport, SK4 2BH [IO83VJ, SJ89]
G7 LHT F R Wilson, 3A Vernon Rd, Kirkby in Ashfield, Nottingham, NG17 8EJ [IO93IC, SK55]
G7 LHU J E Hickingbottom, 50 Tensing Rd, Ashby, Scunthorpe, DN16 3DS [IO93QN, SE90]
G7 LHV G Beaumont, 1 Manor House Farm, Ln Bottom, Milnrow, Rochdale, OL16 3TD [IO83XO, SD91]
G7 LIC D A P Rickerby, Green Park, Dalston, Carlisle, Cumbria, CA5 7AF [IO84MT, NY34]
G7 LIE B W Lovatt, Daleholme, Masterman Pl, Middleton in Teesdale, Barnard Castle, DL12 0ST [IO84XO, NY92]
G7 LIH S J Warren, 41 Barton Rd, Rugby, CV22 7PT [IO92II, SP47]
G7 LIK C A Tunbridge, 12 Burnham Rd, Latchingdon, Chelmsford, CM3 6EU [JO01IQ, TL80]
G7 LIL M Argyle, 8 Bardon Rd, Coalville, LE67 4BH [IO92HR, SK41]
G7 LIN P M Wilkes, 20 Redgates Pl, Chelmsford, CM2 6BG [JO01FR, TL70]
G7 LIO J Hillman, 105 High St., Wootton Bridge, Ryde, PO33 4LU [IO90JR, SZ59]
G7 LIP W F Overett, 61 Crown Rd, Clacton on Sea, CO15 1AU [JO01NS, TM11]
G7 LIR P Otto, 27 Gosfield Rd, Dagenham, RM8 1JY [JO01BN, TQ48]
G7 LIT G Blaxall, 27 St. Davids Rd, Hextable, Swanley, BR8 7RJ [JO01CJ, TQ57]
G7 LIW D A Kent, 10 Goldgarth, Weelsby Meadows, Grimsby, DN32 8QS [IO93XN, TA20]
GW7 LIY G J Price, 56 Bron y Crug, Brecon, LD3 9LF [IO81HW, SO02]
G7 LIZ Details withheld at licensee's request by SSL.
G7 LJA P A Gibson, 62 Glen Park, Pensilva, Liskeard, PL14 5PW [IO70TM, SX26]
G7 LJB C E Mott-Gotobed, 5 Cotswold Cl, Basingstoke, RG22 5BA [IO91KG, SU65]
GM7 LJE J A Freer, 489 Castlemilk Rd, Croftfoot, Glasgow, G44 5PQ [IO75VT, NS66]
GW7 LJG B E Jones, 41 Penrhys Rd, Ystrad Rhondda, Ystrad, Pentre, CF41 7SJ [IO81GP, SS99]
G7 LJH Details withheld at licensee's request by SSL.
GJ7 LJJ N V T Utting, Oberon, Bagatelle Rd, St. Saviour, Jersey, JE2 7TX
G7 LJL S E Murton, 44 Crofton Cl, Kennington, Ashford, Kent, TN24 9BU [JO01KD, TR04]
G7 LJM R F J Keal, New Lodge, Holnicote, nr Minehead, TA24 8TQ [IO81FE, SS94]
G7 LJN M R Jeffs, 18 The Avenue, Tiverton, EX16 4HR [IO80GV, SS91]
G7 LJP P A Edwards, 2 Newham Ln, Steyning, BN44 3LR [IO90TV, TQ11]
G7 LJQ G A Roser, 26 Willow Rd, Larkfield, Aylesford, ME20 6QZ [JO01FH, TQ65]
G7 LJR W A Harding, 11 Mallard Cl, Kempshott, Basingstoke, RG22 5JP [IO91KF, SU54]
G7 LJU W G Deadman, Willine, Eden Vale, East Grinstead, West Sussex, RH19 2JH [IO91XD, TQ33]
G7 LJV Details withheld at licensee's request by SSL.
G7 LJW Details withheld at licensee's request.by SSL.
G7 LJY Details withheld at licensee's request by SSL.
G7 LJZ K E J Seeley, 16 Riley Cl, Bracebridge Heath, Lincoln, LN4 2QS [IO93RE, SK96]
G7 LKB Details withheld at licensee's request by SSL.
G7 LKC J M Radtke, 22 Spinney Dr, Banbury, OX16 9TA [IO92IB, SP43]
G7 LKG L K Groves, 17 Little Linford Ln, Newport Pagnell, MK16 8DE [IO92PB, SP84]
G7 LKI D J Connor, 101 Millington Rd, Castle Bromwich, Birmingham, B36 8BW [IO92CM, SP18]
G7 LKL S Titterington, 33 Victoria Rd, Urmston, Manchester, M41 5BZ [IO83TK, SJ79]
G7 LKN Details withheld at licensee's request by SSL.
G7 LKP J T Banks, 14 Zamenhof Gr, Smallthorne, Stoke on Trent, ST6 1RX [IO83VA, SJ84]
G7 LKR A E Maciver, 55 Nordale Park, Norden, Rochdale, OL12 7RT [IO83VP, SD81]
G7 LKV R Spray, 132 Mansfield St., Sherwood, Nottingham, NG5 4BD [IO92KX, SK54]
G7 LKY D W Parkinson, 36 Henley Rd, Ipswich, IP1 3SA [JO02NB, TM14]
G7 LLC K G Wanstall, 73 Rosemary Ave, Minster, Minster on Sea, Sheerness, ME12 3HU [JO01JK, TQ97]
G7 LLD M A Bewley, 21 Duloe Gdns, Pennycross, Plymouth, PL2 3RS [IO70WJ, SX45]
G7 LLE V N Mitchell, 7 Rollis Park Cl, Oreston, Plymouth, PL9 7NW [IO70WI, SX55]
GW7 LLF G Waters, Poulton House, Station Rd, Dinas Rhondda, Sth Wales, Mid Glam, CF4 2PL [IO81JM, ST17]
G7 LLG R A Middleton, Fairwinds, Southella Rd, Yelverton, PL20 6AT [IO70XL, SX56]
G7 LLK N H Taylor, 46 Ralph Rd, Netherthorpe, Staveley, Chesterfield, S43 3PY [IO93HG, SK47]
G7 LLP D H Brown, 48 Trinity Ave, Northampton, NN2 6JN [IO92NG, SP76]
G7 LLQ Details withheld at licensee's request by SSL.
G7 LLS Details withheld at licensee's request by SSL.
G7 LLT Details withheld at licensee's request by SSL.
G7 LLY J Wharton, 74 Brompton Park, Brompton on Swale, Richmond, DL10 7JP [IO94EJ, SE29]
G7 LMA Details withheld at licensee's request by SSL.
G7 LMC Details withheld at licensee's request by SSL.
G7 LME Details withheld at licensee's request by SSL.
G7 LMF Details withheld at licensee's request by SSL.
G7 LMH Details withheld at licensee's request by SSL.
G7 LMI Details withheld at licensee's request by SSL.
G7 LMJ Details withheld at licensee's request by SSL.
G7 LMN Details withheld at licensee's request by SSL.
G7 LMP Details withheld at licensee's request by SSL.
G7 LMR K C Lavin, 19 Kingsway Ave, Paignton, TQ4 7AA [IO80FJ, SX85]
G7 LMT D Pantrey, 10 Columbine Cl, East Malling, West Malling, ME19 6ES [JO01FH, TQ65]
GW7 LMW D J Jenkins, 36 Brynawel Rd, Gorseinon, Swansea, SA4 4UX [IO71XQ, SS59]
G7 LMX L V Pearse, Hill Farm, East Pennard, Shepton Mallet, Somerset, BA4 6TS [IO81QD, ST53]
G7 LMY R R Loxton-Gear, 4 Pitmore Ln, Pennington, Lymington, SO41 8LL [IO90FS, SZ29]
G7 LNB A E West, 142 The St., Kingston, Canterbury, CT4 6JQ [JO01NF, TR15]
G7 LNG J L Tucker, 2 Ivydene Rd, Ivybridge, PL21 9BH [IO80AJ, SX65]
G7 LNI S P Czarnota, 154 Buriton Rd, Harestock, Winchester, SO22 6JG [IO91IB, SU43]
G7 LNJ R E J Woolridge, 8 Alastair Dr, Yeovil, BA21 3BT [IO80QW, ST51]
G7 LNL B Nicholson, 25B Glenholme Rd, Farsley, Pudsey, LS28 5BY [IO93DT, SE23]
G7 LNM D E P Gilham, 53 The Close, Bradwell, Great Yarmouth, NR31 8DR [JO02UN, TG50]
G7 LNO G Cash, 14 Crossfield Park, Felling, Gateshead, NE10 9SA [IO94FW, NZ26]
G7 LNP A M Jones, 1 Abbey Way, Rushden, NN10 9HF [IO92QG, SP96]
G7 LNT P Cundall, Red Roofs, 15 West Busk Ln, Otley, LS21 3LW [IO93DV, SE14]
G7 LNU N A Sparrow, 16 Hadley Cl, Bocking, Braintree, CM7 5LP [JO01GV, TL72]
G7 LNV N A Turland, 2 Ludlow Cl, Beeston, Nottingham, NG9 3BY [IO92JW, SK53]
G7 LNW S D Thirlaway, 10 Sea View, Blackhall Rocks, Blackhall Colliery, Hartlepool, TS27 4AX [IO94IR, NZ43]
G7 LNX A E Hemmings, 2 Tierney St., Northwood, Stoke on Trent, ST1 2EE [IO83WA, SJ84]
G7 LNY H J Mascall, 4 Rushleydale, Chelmsford, CM1 6JX [JO01FR, TL70]
G7 LNZ A Banks, 14 Zamenhof Gr, Smallthorne, Stoke on Trent, ST6 1RX [IO83VA, SJ84]
G7 LOA L J Fisher, 195 Malvern Rd, Billingham, TS23 2PJ [IO94IO, NZ42]
G7 LOB G F Poulson, 113 Bank Hall Rd, Burslem, Stoke on Trent, Staffs, ST6 7DR [IO83VB, SJ85]
G7 LOE J S Bhogal, 36 Titford Rd, Oldbury, Warley, B69 4QA [IO82XL, SO98]
G7 LOF Details withheld at licensee's request by SSL.
G7 LOG T J Smallwood, 51 Barlow Rd, Barlow Village, Barlow, Blaydon on Tyne, NE21 6JU [IO94CW, NZ16]
G7 LOH Details withheld at licensee's request by SSL.
G7 LOI Details withheld at licensee's request by SSL.
GM7 LOK Details withheld at licensee's request by SSL.
G7 LOL Details withheld at licensee's request by SSL.
G7 LOM Details withheld at licensee's request by SSL.
GI7 LOQ Details withheld at licensee's request by SSL.
G7 LOR Details withheld at licensee's request by SSL.
G7 LOS Details withheld at licensee's request by SSL.
G7 LOV E A Farrar, 23 Grovehill Rd, Filey, YO14 9NL [IO94UF, TA18]
G7 LOX Details withheld at licensee's request by SSL.
G7 LPB A E Sellick, 12 Sutherland Farm, Tibberton, Newport, Shropshire, TF10 8NN [IO82SS, SJ62]
G7 LPE S Wragg, The Cavalier Public House, Eastleigh Rd, Devizes, Wilts, SN10 3EG [IO91AI, SU06]
G7 LPF C Hewitt, 9 Alford Fold, Fulwood, Preston, PR2 3UU [IO83PT, SD53]
G7 LPG Details withheld at licensee's request by SSL.[Operator: Peter. Station located near Eastbourne, East Sussex.]
G7 LPK R J Hilliard, 103 Sleaford Rd, Newark-on-Trent, Notts, NG24 1NG [IO93OB, SK85]
GW7 LPM L J La Traille, 33 Festival Cres, New Inn, Pontypool, NP4 0NB [IO81LQ, ST39]
G7 LPN T R Snape, 4 Back St., Abbotsbury, Weymouth, DT3 4JP [IO80QP, SY58]
G7 LPO A Perry, 63A Brookland Rd, Huish Episcopi, Langport, TA10 9TH [IO81OB, ST42]
G7 LPP F Rice, 42 Donegal Rd, Knowle, Bristol, BS4 1PL [IO81QK, ST57]
G7 LPT A V Page, Homeleigh House, East Taphouse, Liskeard, Cornwall, PL14 4NQ [IO70RK, SX16]
G7 LPV Details withheld at licensee's request by SSL.
G7 LPY N R Wootton, 13A Grange Rd, Tettenhall, Wolverhampton, WV6 8RQ [IO82VO, SJ80]
G7 LQD M V Baguley, 21 Sovereign Cl, Rudheath, Northwich, CW9 7XN [IO83SG, SJ67]
G7 LQG Details withheld at licensee's request by SSL.
GI7 LQI G S Duffy, Carrageen Cottage, Tullycreevy, Monea, Enniskillen, Co Fermanagh, BT14 8ET [IO74AP, J27]
G7 LQK R Dunn, 12 Roseberry St., Beamish, Stanley, DH9 0QR [IO94EU, NZ25]
G7 LQN C A King, 33 Alexandra Rd, Swallownest, Sheffield, S31 0TA [IO93GJ, SK38]
G7 LQO L B Brown, 49 Barholm Rd, Sheffield, S10 5RR [IO93FJ, SK38]
G7 LQP E T F Livermore, Lanmore, The St., Takeley, Bishops Stortford, CM22 6QW [JO01CU, TL52]
GM7 LQR C Francey, 2 Woodlands Cttgs, Newbridge, Dumfries, DG2 0HZ [IO85EC, NX97]
G7 LQV A H C Cossens, 55 Nicholas Cres, Fareham, PO15 5AJ [IO90JU, SU50]
G7 LRB P G P Stevens, 80 Portland Cres, Shrewsbury, SY2 5NW [IO82PQ, SJ51]
G7 LRD Details withheld at licensee's request by SSL.
G7 LRE Details withheld at licensee's request by SSL.
G7 LRF Details withheld at licensee's request by SSL.
G7 LRG E Bailey Leeds Raynet Gr, 49 Brander St., Leeds, LS9 6QH [IO93GT, SE33]
G7 LRK Details withheld at licensee's request by SSL.
G7 LRM Details withheld at licensee's request by SSL.
G7 LRN Details withheld at licensee's request by SSL.
G7 LRO T S Cotter, P O Box 478, 9 Jeremy Moore Ave, Port Stanley, Falkland Islands, South Atlantic
G7 LRQ Details withheld at licensee's request by SSL.
G7 LRR Details withheld at licensee's request by SSL.
G7 LRS Details withheld at licensee's request by SSL.
G7 LRU S R Allen, 100 Highfields Rd, Witham, CM8 2HH [JO01HT, TL81]
G7 LSA J J O Williams, 92 Cantley Ln, Bessacarr, Doncaster, DN4 6NH [IO93LM, SE60]
G7 LSB L S Brown, 4 Loraine Gdns, Ashtead, KT21 1PD [IO91UH, TQ15]
G7 LSD P Wainwright, 30 Leven Way, Walsgrave, Coventry, CV2 2RA [IO92GK, SP38]
G7 LSF J M Blain, Churchfield Cottage, Fawley, Henley-on-Thames, Oxon, RG9 6HZ [IO91MN, SU78]
G7 LSG Details withheld at licensee's request by SSL.
GM7 LSI J Stuart, 11 North St., Bishopmill, Elgin, IV30 2EG [IO87IP, NJ26]
G7 LSL Details withheld at licensee's request by SSL.
G7 LSM Details withheld at licensee's request by SSL.
G7 LSO Details withheld at licensee's request by SSL.
G7 LSP P Harness, 74 Rosebery Ave, Boston, PE21 7QR [IO92XX, TF34]
G7 LSR Details withheld at licensee's request by SSL.
G7 LSS Details withheld at licensee's request by SSL.
G7 LSV Details withheld at licensee's request by SSL.
G7 LSW J Simons, 51A Ave Gdns, Acton, London, W3 8HB [IO91UM, TQ17]
G7 LSY L J Hughes, 3 Warland Rd, Plumstead, London, SE18 2EX [JO01BL, TQ47]
G7 LSZ M Foreman, 12 Groveland Rd, Beckenham, BR3 3QA [IO91XJ, TQ36]
G7 LTD Details withheld at licensee's request by SSL.[Op: B C Thomas. Correspondence via PO Box 703, West Moors, Ferndown, Dorset, BH22 0YB.]
G7 LTE K Pluck, 31 Coalway Rd, Penn, Wolverhampton, WV3 7LU [IO82WN, SO99]
G7 LTG P J Savage, 60 Colonial Rd, Bordesely Green, Birmingham, B9 5NG [IO92BL, SP18]
G7 LTH A J Everitt, 6 Belmont Rd, Rednal, Birmingham, B45 9LW [IO82XJ, SO97]
G7 LTL C Croasdale, 38 Aysgarth Ave, Fulwood, Preston, PR2 9TG [IO83PT, SD53]
G7 LTM G R Skate, 34 Millfield Rd, Deeping St. James, Peterborough, PE6 8QX [IO92UQ, TF11]
G7 LTO M D Milns, 3 Merlin Ct, Birstall, Batley, WF17 0RG [IO93ER, SE22]
G7 LTP P E Sawyer, 45 Charmouth Rd, Welling, DA16 1RH [JO01BL, TQ47]
G7 LTQ J R Newgas, 32 Merton Ln, Highgate Village, London, N6 6NB [IO91WN, TQ28]
G7 LTR D J Ingham, 81 Belvedere Rd, Ashton in Makerfield, Wigan, WN4 8RX [IO83QL, SJ59]
G7 LTT Details withheld at licensee's request by SSL.
G7 LTU G R Smith, 36 Sandalwood Rd, Loughborough, LE11 3PS [IO92JS, SK51]
G7 LTW T R Metcalfe, 92 Sheridan Rd, Frimley, Camberley, GU16 5EH [IO91PH, SU85]
GM7 LTX A Warner, 41 Gaynor Ave, Loanhead, EH20 9LU [IO85KV, NT26]
G7 LTZ A C Jenno, c/o 259 Signal Squadron, Episkopi Garrison, Bfpo 53
GW7 LUB J D Broome, Rheidol View, Cwm Rheidol, Aberwystwyth, Dyfed, SY23 3NB [IO82BJ, SN77]
G7 LUC Details withheld at licensee's request by SSL.
G7 LUD Details withheld at licensee's request by SSL.
G7 LUF G R Whitehouse, 8 Shaftesbury Cl, Parklands, Bromsgrove, B60 2RB [IO82XH, SO96]
G7 LUK P J Preston, 45 Saxons Heath, Long Wittenham, Abingdon, OX14 4PU [IO91JP, SU59]
G7 LUL F P Russell, 63 Fleet St., Keyham, Plymouth, PL2 2BU [IO70VJ, SX45]
GM7 LUN J W Keddie, Garrion, Bowland Rd, Clovenfords, Galashiels, TD1 3ND [IO85NO, NT43]
G7 LUO N M Head, 12 Heston Walk, Redhill, RH1 5JB [IO91VF, TQ24]
GM7 LUP A Young, Newhaven, Forsyth St., Hopeman, Elgin, IV30 2SY [IO87GQ, NJ16]
G7 LUQ L J Ladd, 4 Church End, Potterspury, Towcester, NN12 7PX [IO92NB, SP74]
G7 LUR J M Nolan, 40 The Mixies, Stotfold, Hitchin, SG5 4LF [IO92VA, TL23]
G7 LUS J R Arey, 622 Scott Hall Rd, Chapel Allerton, Leeds, LS7 3QJ [IO93FT, SE33]
G7 LUU R W Milroy, 1 Avenswood Ln, Scunthorpe, DN15 8TB [IO93PO, SE81]
G7 LUV K A Seeley, 66 Matlock Dr, North Hykeham, Lincoln, LN6 8PU [IO93QE, SK96]
G7 LUW R A Stanley, 15 Harradon Rd, Walton, Liverpool, L9 0HE [IO83ML, SJ39]
G7 LVE D A Wright, 21 Pickett Croft, Stanmore, HA7 1HY [IO91UO, TQ19]
G7 LVF Details withheld at licensee's request by SSL.
G7 LVG J Ashton-Jones, Kiddleys Copse, Mordiford, Hereford, HR1 4LR [IO82QA, SO53]
G7 LVM J D Maule, 12 Edith Cavell Way, Steeple Bumpstead, Haverhill, CB9 7EE [JO02FA, TL64]
G7 LVN M E B Odam, 14 Norbins Rd, Glastonbury, BA6 9JE [IO81PD, ST43]
G7 LVS M Unsworth, 41 Aylesbury Cres, Hindyey Green, Hindley, Wigan, WN2 4TY [IO83RM, SD60]
G7 LVU G M Garratt, 126 Dorset Ave, Chelmsford, CM2 8YY [JO01FR, TL70]
GM7 LVV S Thomas, 265 Deans South, Deans, Livingston, EH54 8ED [IO85FV, NT06]
G7 LVW Details withheld at licensee's request by SSL.
G7 LVX Details withheld at licensee's request by SSL.
G7 LVY Details withheld at licensee's request by SSL.
GM7 LWC M Wallace, 146 Georgetown Rd, Dumfries, DG1 4DT [IO85FB, NX97]
GD7 LWE J P Want, 43 Tromode Park, Douglas, Isle of Man, IM2 5LG
G7 LWF J A Totten, 1 Bell Ln Cl, Fetcham, Leatherhead, KT22 9NE [IO91TG, TQ15]
G7 LWJ C R Forrester, Scottwood, Noads Way, Dibden Purlieu, Southampton, SO45 4PB [IO90HU, SU40]
G7 LWK Details withheld at licensee's request by SSL.
G7 LWM Details withheld at licensee's request by SSL.
G7 LWN Details withheld at licensee's request by SSL.
G7 LWT Details withheld at licensee's request by SSL.
G7 LWU Details withheld at licensee's request by SSL.
G7 LWW Details withheld at licensee's request by SSL.
G7 LWY D J Northeast, 11 Repton Rd, Earley, Reading, RG6 7LJ [IO91MK, SU77]
G7 LXA Details withheld at licensee's request by SSL.
G7 LXB Details withheld at licensee's request by SSL.
G7 LXC Details withheld at licensee's request by SSL.
G7 LXH D I Hayzen, 3 Greenwood Cl, Romsey, Hants, SO51 7QT [IO90GX, SU32]
GW7 LXI J H Baines, Pentre Clawdd, Cottage, Gobowen Oswestry, Shrops, SY10 7AE [IO82LV, SJ23]
G7 LXK G K Hyde, 16 Burnholme Gr, York, YO3 0LN [IO93LX, SE65]
G7 LXN Dr M A Burdon, 15 Daybrook Rd, London, SW19 3DJ [IO91VJ, TQ26]
G7 LXP D Remnant, 3 Berkeley Ct, Mayfare, Croxley Green, Rickmansworth, WD3 3DF [IO91SP, TQ09]
G7 LXQ S D Houlton, 50 Church Walk, Brinsley, Nottingham, NG16 5AT [IO93IA, SK44]
G7 LXT J B Whistlecraft, 19 Wellingborough Rd, Olney, MK46 4BJ [IO92PD, SP85]
G7 LXU T Knott, 14 Hound Cl, Abingdon, OX14 2LU [IO91IQ, SU59]
G7 LXV N D Hobbs, 224 Belchers Ln, Bordesley Green, Birmingham, B9 5RY [IO92CL, SP18]
G7 LYB R Brown, 61 Paddockhurst Rd, Gossops Green, Crawley, RH11 8EU [IO91VC, TQ23]
GW7 LYD P E Jenkins, 5 Briar Cl, Undy, Magor, Newport, NP6 3LQ [IO81ON, ST48]
G7 LYG R O N I Gail, 8 Primley Park East, Paignton, TQ3 3JN [IO80FK, SX86]
G7 LYH J L S Briggs, 16 Belmont Pl, Colchester, CO1 2HU [JO01LV, TM02]
G7 LYL C R Nixon, 6 Mallard Cl, Knutsford, WA16 8ES [IO83TH, SJ77]
G7 LYN S J Laugher, Jasmine Cottage, Healey, Masham, North Yorks, HG4 4LH [IO94DF, SE18]
G7 LYP Details withheld at licensee's request by SSL.
G7 LYQ P J Clayton, 41 Scammerton, Wilnecote, Tamworth, B77 4LA [IO92EO, SK20]
G7 LYR Details withheld at licensee's request by SSL.
G7 LYS C F Plummer, Barley House Farm, Biddulph Park, Stoke on Trent, Staffs, ST8 7SW [IO83WD, SJ96]
G7 LYX C M Burnett, 11 Elizabeth Cl, Bridlington, YO15 2TQ [IO94VB, TA16]
G7 LZB A R Howat, 12 Dover Rd, Birkdale, Southport, PR8 4SY [IO83LP, SD31]
G7 LZC R J Varrow, 3 Park Rd, St. Dominick, Saltash, PL12 6TL [IO70UL, SX46]
G7 LZE A Page, 17 Barlborough Rd, Clowne, Chesterfield, S43 4RA [IO93IG, SK47]
G7 LZF D E Mockett, 3 Adrian Dr, Barwell, Leicester, LE9 8GB [IO92HN, SP49]
G7 LZI Details withheld at licensee's request by SSL.
GW7 LZK Details withheld at licensee's request by SSL.
G7 LZM L Mountain, 45 Westway Gdns, Redhill, RH1 2JB [IO91WG, TQ25]
GW7 LZO D M Allen, 32 Hillfield Pl, Parcllyn, Cardigan, SA43 2DJ [IO72RD, SN25]
G7 LZX Details withheld at licensee's request by SSL.
G7 LZY W A Eatwell, 45 Admirals Walk, Minster on Sea, Sheerness, ME12 3BB [JO01JJ, TQ97]
G7 LZZ T M Moore, 9 Moores Green, Wokingham, RG40 1QG [IO91OK, SU86]
G7 MAB M G Dodson, 64 Stoneleigh Rd, Solihull, B91 1DQ [IO92CK, SP18]
GI7 MAC W P Mc Menamin, 91 Cashelmore Park, Londonderry, Co. Londonderry, BT48 0RU [IO65HA, C41]
G7 MAD Details withheld at licensee's request by SSL.
G7 MAG P Budgen, 6 Barnmead, Haywards Heath, RH16 1UZ [IO91WA, TQ32]
G7 MAI Details withheld at licensee's request by SSL.
G7 MAJ S M Hoyle, 20 Spring Hall Pl, Halifax, HX1 4TF [IO93BR, SE02]
GD7 MAN A C Kissack Three Legs VHF Contest Group, 30 High View Rd, Douglas, IM2 5BH
G7 MAO G Byron, 7 Cambridge Rd, Hessle, HU13 9DB [IO93SR, TA02]
G7 MAP M J Bellamy Lincoln & District Amateur Rad, 19 Wickenby Cres, Ermine Est West, Lincoln, LN1 3TJ [IO93RF, SK97]

G7 MAQ K Bellamy, 33 Torksey St., Kirton Lindsey, Gainsborough, DN21 4PW [IO93QL, SK99]
G7 MAR J Rivers, 1 Hazelwood Cl, Ryde, PO33 2UP [IO90KR, SZ59]
G7 MAT K Hinton, Tyn y Coed, 69 Grainger Cl, Basingstoke, RG22 4EA [IO91KF, SU65]
G7 MAY D A Heath, 14 Meadow Park, Bannerdown, Bathford, Bath, BA1 7PX [IO81UJ, ST76]
G7 MAZ B J Isaacson, Denmede, Eddington Rd, Nettlestone, Seaview, PO34 5EE [IO90KR, SZ69]
GM7 MBB L Millar, 34 Brora Dr, Deanpark, Renfrew, PA4 0XA [IO75TU, NS56]
G7 MBD A G Buck, 21 Southview Rd, Biscovey, Par, PL24 2HJ [IO70PI, SX05]
G7 MBH M E Davis, 3 Thornley Cl, Ushaw Moor, Durham, DH7 7NN [IO94ES, NZ24]
G7 MBI N Davies, 66 Hope St., Dukinfield, SK16 4EB [IO83WL, SJ99]
GI7 MBP W Kane, 21 Mount Coole Gdns, Belfast, BT14 8JY [IO74AP, J37]
G7 MBS Details withheld at licensee's request by SSL.
G7 MBU Details withheld at licensee's request by SSL.
G7 MBX A V Profitt, 199 Markfield, Ct Wood Ln, Croydon, CR0 9HR [IO91XI, TQ36]
G7 MBY D F Richards, 6 Kingley Cl, Wickford, SS12 0EN [JO01GO, TQ79]
G7 MCD A D Utting 51 Squadron Air Training Corps, 20 Davenport Rd, Goodwood, Leicester, LE5 6SA [IO92LP, SK60]
G7 MCE D A Wilkinson, 56 Cobden St., Dalton in Furness, LA15 8SE [IO84JD, SD27]
G7 MCK K Singleton, Spring Cottage, Barcombe Ln, Paignton, TQ3 2QS [IO80FK, SX86]
G7 MCP R A Davies, 3 Fair View Cl, Barnton, Northwich, CW8 4HH [IO83RG, SJ67]
G7 MCS B J McShea, 5 Frensham Ave, Fleet, GU13 9EL [IO91OG, SU85]
G7 MCT C M Taylor, 36 Harewood Rd, Shaw, Oldham, OL2 8EA [IO83WN, SD90]
GI7 MCY Dr G Logan, 5 The Lawns, Waringstown, Craigavon, BT66 7GD [IO64UK, J15]
GW7 MDH A J Myatt, 14 Daisy Bank Cl, Bersham Rd, Wrexham, LL14 4JB [IO83LA, SJ34]
GI7 MDK D Robinson, 4 Ballylesson Rd, Magheramorne, Larne, BT40 3HL [IO74CT, J49]
G7 MDM Details withheld at licensee's request by SSL.
GI7 MDP S D J McIlvenna, 13 Trossachs Gdns, Belfast, BT10 0HX [IO74AN, J36]
G7 MDR D J Tunbridge Essex Raynet Maldon District, 12 Burnham Rd, Latchingdon, Chelmsford, CM3 6EU [JO01IQ, TL80]
G7 MDT Details withheld at licensee's request by SSL.
G7 MDV C J Prowse, Tewin Cottage, The Mount, Warlingham, CR6 9JF [IO91XH, TQ35]
GI7 MDX Details withheld at licensee's request by SSL.
G7 MDY W A Collins, 99 Barkers Ln, Bedford, MK41 9TB [IO92SD, TL04]
G7 MEA R Thomas, 3 Blackberry Way, Kingsteignton, Newton Abbot, TQ12 3QX [IO80EN, SX87]
G7 MED Details withheld at licensee's request by SSL.
G7 MEE A Wood, 33 Woodberry Dr, Sittingbourne, ME10 3AT [JO01JI, TQ96]
G7 MEG D P Cash, 46 Woden Rd, Park Village, Wolverhampton, WV10 0BB [IO82WO, SO99]
G7 MEI P Craigen, 90 Dunstanburgh Rd, Newcastle upon Tyne, NE6 2PX [IO94FX, NZ26]
G7 MEJ Details withheld at licensee's request by SSL.
G7 MEQ H C Mowthorpe, 31 Bramblys Cl, Basingstoke, RG21 8UP [IO91KG, SU65]
G7 MER C L Hurst, 12 Burgess Cl, Woodley, Reading, RG5 3LJ [IO91NK, SU77]
G7 MES M E Stevens, 22 Lytham Dr, Waltham, Grimsby, DN37 0DG [IO93WM, TA20]
G7 MET D H Hoyle, 20 Vernon Ct, London Rd, Portsmouth, PO2 9JR [IO90LT, SU60]
G7 MEU D J Hughes, 1 Kilford Cl, Callands, Warrington, WA5 5SR [IO83QJ, SJ59]
G7 MEX T Sheppard Mexborough & District ARC, 4 Lindrick Ave, Swinton, Mexborough, S64 8TE [IO93IL, SK49]
G7 MEY D S W Collins, 120 Plumberow Ave, Hockley, SS5 5AT [JO01HO, TQ89]
G7 MEZ J Arter, 18 Essex Rd, Westgate on Sea, CT8 8AP [JO01QJ, TR36]
G7 MFA A Sejwacz, 20 Wellington Gdns, Newton-le-Willows, WA12 9LT [IO83QK, SJ59]
G7 MFD Details withheld at licensee's request by SSL.
G7 MFE S Morris, 5 Green Ln, Walton on Thames, KT12 5HD [IO91TI, TQ16]
G7 MFH B E Fifield, 165 Warwick Rd, Rayleigh, SS6 8SG [JO01HN, TQ89]
G7 MFK Details withheld at licensee's request by SSL.
G7 MFL L Darkes, 68A Drift Rd, Waterlooville, Hants, PO8 0NX [IO90LW, SU71]
G7 MFN Details withheld at licensee's request by SSL.
G7 MFO R J Parkes, 7 Main St., Preston, Hull, HU12 8UB [IO93VS, TA13]
G7 MFP A M Dresser, 66 Starlings Dr, Tilehurst, Reading, RG3 4SS [IO91LJ, SU66]
G7 MFQ T J Boylan, 3 Evergreen Cl, Higham, Rochester, ME3 7EL [JO01FJ, TQ77]
G7 MFR Details withheld at licensee's request by SSL.
G7 MFW D J Burdett, 17 Brambledown, Chatham, ME5 0DY [JO01GI, TQ76]
G7 MFX Details withheld at licensee's request by SSL.
G7 MFY A J Wakeling, 86 Laurel Cres, Rush Green, Romford, RM7 0RT [JO01CN, TQ58]
G7 MFZ M L Sherratt, 21 Tweedale Cl, Station Rd, Mursley, Milton Keynes, MK17 0SB [IO91OW, SP82]
G7 MGA R J Thorley, The Post Office and Stores, Leigh, Stoke on Trent, Staffs, ST10 4PT [IO92AW, SK03]
G7 MGC T Gerrard, 139 Paulhan St., Bolton, BL3 3DT [IO83SN, SD70]
G7 MGE Details withheld at licensee's request by SSL.
G7 MGG R S Shirley, 78 Godwin Rd, Hove, BN3 7FR [IO90VU, TQ20]
G7 MGH A L Treacher, 7 High Gr, Welwyn Garden City, AL8 7DW [IO91VT, TL21]
G7 MGK Details withheld at licensee's request by SSL.
G7 MGL Details withheld at licensee's request by SSL.
G7 MGM D Barnes, 36 Westbrook Cres, New Barnet, Barnet, EN4 9AS [IO91VP, TQ29]
G7 MGN Details withheld at licensee's request by SSL.
G7 MGP W R Gray, 5 Kathleen Dr, Kettering, NN16 0QE [IO92PJ, SP87]
G7 MGQ M R Ball, 11 Plantation Rd, Thorne, Doncaster, DN8 5EA [IO93MO, SE61]
GW7 MGR H T John Glamorgan Raynet, 11 Penylan, Litchard, Bridgend, CF31 1QW [IO81FM, SS98]
G7 MGS Details withheld at licensee's request by SSL.
GW7 MGW E M Palmer, 102 Park Ave, Bryn y Baal, Mold, CH7 6TP [IO83KE, SJ26]
G7 MGX Details withheld at licensee's request by SSL.
G7 MGY S A Welger, 55 Burford Ave, Walcot, Swindon, SN3 1BX [IO91CN, SU18]
GW7 MHB M H Burt, 7 Bod Offa Dr, Buckley, CH7 2PB [IO83KD, SJ26]
G7 MHD A Thorp, 34 Third Ave, Liversedge, WF15 8JU [IO93DR, SE12]
G7 MHE Details withheld at licensee's request by SSL.
G7 MHF Details withheld at licensee's request by SSL.
G7 MHG Details withheld at licensee's request by SSL.
G7 MHK B R Hall, 11 Chelmer Way, Burnham on Crouch, CM0 8TN [JO01JP, TQ99]
G7 MHL Details withheld at licensee's request by SSL.
G7 MHM Details withheld at licensee's request by SSL.
G7 MHO S Fell, 14 Rectory Ave, Corfe Mullen, Wimborne, BH21 3EZ [IO80XS, SY99]
G7 MHQ G Taylor, 21 New Rd, Kirkheaton, Huddersfield, HD5 0JB [IO93DP, SE11]
G7 MHS M R Vaslet Mary Hare Grammar School, Arlington Manor, Smelsmore Common, Newbury, RG16 9BQ [IO91LJ, SU66]
G7 MHT J C Fowler, Old Bakery, 4 Langley St., Derby, DE22 3GL [IO92GW, SK33]
G7 MHV S J Stillwell, 36 Capel Rd, East Barnet, Barnet, EN4 8JE [IO91WP, TQ29]
G7 MHX P G Brobin, 64 Langton Ct Rd, St. Annes, Bristol, BS4 4EQ [IO81RK, ST67]
G7 MHZ J C Bunting, 4 Springfield Rd, Sittingbourne, ME10 2NB [JO01II, TQ86]
G7 MID A J Haydon, Basement Flat, 118 Carisbrooke High St., Newport, PO30 1NL [IO90IQ, SZ48]
G7 MIE S W Hudson, 50 Swanley Ln, Swanley, BR8 7JG [JO01CJ, TQ56]
G7 MIF B W Dickenson, 47 Albatross Ave, Rochester, ME2 2XW [JO01FJ, TQ76]
G7 MII D F Burgin, 7 Bramble Cl, Halliford, Shepperton, TW17 8RR [IO91SJ, TQ06]
G7 MIJ Details withheld at licensee's request by SSL.
G7 MIL Details withheld at licensee's request by SSL.
G7 MIM T M Wheeler, 259 Lonsdale Dr, Rainham, Gillingham, ME8 9JP [JO01HI, TQ86]
G7 MIN Details withheld at licensee's request by SSL.
G7 MIP H O Kneale, 5 Keightley Way, Tuddenham, Ipswich, IP6 9BJ [JO02OC, TM14]
G7 MIR Details withheld at licensee's request by SSL.
G7 MIS A J Trott, 8A Wyatt Rd, Kempston, Bedford, MK42 7EH [IO92RC, TL04]
G7 MIT Details withheld at licensee's request by SSL.
G7 MIW K D Williams, 61 Horsham Rd, Pease Pottage, Crawley, RH11 9AW [IO91VB, TQ23]
G7 MIX Details withheld at licensee's request by SSL.
G7 MIZ J W Locker, Delamere, 8 Concordia Ave, Wirral, Merseyside, L49 6JD [IO83KJ, SJ28]
G7 MJA Details withheld at licensee's request by SSL.
G7 MJB Dr W T Ramsey, 6 Chesterfield Rd, Copnor, Portsmouth, PO3 6LZ [IO90LT, SU60]
G7 MJC A J Fell, 28 Hedgeway, East Hunsbury, Northampton, NN4 0SP [IO92NE, SP75]
G7 MJD A Bruring, Ouse Cottage, 5 Church Ln, Hartford, Huntingdon, PE18 7XP [IO92WI, TL27]
G7 MJF J Buckland, 110 Peerless Dr, Harefield, Uxbridge, UB9 6JQ [IO91SO, TQ08]
G7 MJG Details withheld at licensee's request by SSL.
G7 MJH P S Rae, 11 Ruxley Cl, Sidcup, DA14 5LS [JO01BJ, TQ47]
G7 MJI P E Sayers, 23 Roseveare Rd, Eastbourne, BN22 8RS [JO00DS, TQ60]
G7 MJJ R W J Delves, 66 Palmeira Rd, Bexleyheath, DA7 4UX [JO01BL, TQ47]
G7 MJK Details withheld at licensee's request by SSL.
G7 MJL Details withheld at licensee's request by SSL.
G7 MJM W J Fear, Nance Cottage, Nance, Carbis Bay, St. Ives, TR26 3JF [IO70GE, SW53]
G7 MJN P R Hunter, 40 Hadrian Ave, Chester-le-Street, DH3 3RS [IO94FU, NZ25]
G7 MJP C Edwards, The Bungalow, Tweed Hill Farm, Park Ln, Swanley, Kent, BR8 8DT [JO01CJ, TQ56]
G7 MJR Details withheld at licensee's request by SSL.
G7 MJS G A Davies, 32 Oakwood Dr, Southport, PR8 3HZ [IO83LO, SD31]
G7 MJV Details withheld at licensee's request by SSL.
G7 MJW Details withheld at licensee's request by SSL.
G7 MJX D Hanson, 36 The Avenue, Stotfold, Hitchin, SG5 4LY [IO92VA, TL23]
G7 MJZ Details withheld at licensee's request by SSL.
G7 MKB J Humphries, 25 Wrekenton Row, Wrekenton, Gateshead, NE9 7JD [IO94FW, NZ25]
G7 MKD Details withheld at licensee's request by SSL.
G7 MKF B M Stracey, 31 Westfield Rd, Westbrook, Margate, CT9 5PA [JO01QJ, TR36]
G7 MKG P J Bradbury, 37 Brandish Cres, Clifton, Nottingham, NG11 9JZ [IO92JV, SK53]
G7 MKO Details withheld at licensee's request by SSL.
G7 MKP A N Brooks, 7 Lindford Dr, Eaton, Norwich, NR4 6LT [JO02PO, TG20]
G7 MKQ Details withheld at licensee's request by SSL.
G7 MKR Details withheld at licensee's request by SSL.
G7 MKS Details withheld at licensee's request by SSL.
G7 MKT G Wincott, 12 Walmer Rd, Lytham St. Annes, FY8 3HL [IO83LS, SD32]
G7 MKW K A Woodall, 21 Napier Ave, Jaywick, Clacton on Sea, CO15 2LH [JO01NS, TM11]
G7 MLC G D Bunn, 19 Southwell Rd, Norwich, NR1 3HU [JO02PO, TG20]
G7 MLJ P Skinner, 84 Beresford Ave, Tolworth, Surbiton, KT5 9LW [IO91UJ, TQ26]
G7 MLK D M Rose, 87 Second Ave, Sudbury, CO10 6QX [JO02IB, TL84]
G7 MLL Details withheld at licensee's request by SSL.
G7 MLO J D J Large, 5 Raynsford Rise, Stanningfield Rd, Great Whelnetham, Bury St. Edmunds, IP30 0TS [JO02JE, TL86]
G7 MLQ P Edwards, Petdore, 41 Chaddiford Ln, Barnstaple, EX31 1RE [IO71XC, SS53]
G7 MLR D A Buxton, 14 Allens Ave, Sprowston, Norwich, NR7 8EP [JO02PP, TG21]
G7 MLT A M Armstrong-Bednall, 24 Wellington St., Heanor, DE75 7FW [IO93HA, SK44]
G7 MLU S Y Murray, 10 Calvert Cl, Upper Belvedere, Belvedere, DA17 6EU [JO01BL, TQ47]
G7 MLX G P Crisp, Hoppers Farm, Gt Kings Hill, High Wycombe, Bucks, HP15 6EY [IO91PQ, SU89]
G7 MLY Details withheld at licensee's request by SSL.
G7 MLZ Details withheld at licensee's request by SSL.
G7 MMA R N Cook, 104 Bleak Hill Rd, Windle, St. Helens, WA10 6DR [IO83OL, SJ49]
G7 MMC S P Reed, 36 Plantation Rd, Amersham, HP6 6HL [IO91QQ, SU99]
G7 MME Details withheld at licensee's request by SSL.
G7 MMF Details withheld at licensee's request by SSL.
GW7 MMG P H Pike, 19 Hillrise Park, Clydach, Swansea, SA6 5DX [IO81BQ, SN60]
GW7 MMH E J Cooke, Oldacres, Three Crosses, Swansea, SA4 3PU [IO71XP, SS59]
GM7 MMI J Wilson, 32 Silverburn Rd, Bridge of Don, Aberdeen, AB22 8RW [IO87WE, NJ91]
G7 MMJ Details withheld at licensee's request by SSL.
G7 MMK D M Baggaley, 6 Bylands Pl, Westlands, Newcastle under Lyme, Staffs, ST5 3PQ [IO82VX, SJ84]
G7 MML L J S Evans, 3 Thorndale Ct, Thorndale Mews, Clifton, Bristol, BS8 2JA [IO81QL, ST57]
G7 MMM T Kershaw, 47 Balls Rd, Birkenhead, L43 1UT [IO83LJ, SJ38]
G7 MMQ D J Hill, 86 The Downs, Silverdale, Wilford, Nottingham, NG11 7EB [IO92KV, SK53]
G7 MMR G E Seaton, 15 Forest Cl, Selston, Nottingham, NG16 6QZ [IO93IB, SK45]
G7 MMS Details withheld at licensee's request by SSL.
G7 MMT Details withheld at licensee's request by SSL.
G7 MMW P J Francis, 56 Hillside Ave, Gravesend, DA12 5QR [JO01EK, TQ67]
G7 MMY S Wright, 33 Eastdale Cl, Kempston, Bedford, MK42 8LY [IO92SC, TL04]
G7 MND W N South, Dufonis, Dorchester Rd, Wool, Wareham, BH20 6EQ [IO80VQ, SY88]
G7 MNE B Altman, 1 The Woodlands, Southgate, London, N14 5RL [IO91WP, TQ29]
G7 MNG M P Whale, 499 Maidstone Rd, Wigmore, Gillingham, ME8 0JX [JO01HI, TQ86]
G7 MNH Details withheld at licensee's request by SSL.
G7 MNK J V Lambe, 33 Bronhill Terr, Lansdowne Rd, Tottenham, London, N17 0LN [IO91XO, TQ39]
G7 MNL C J Coker South Devon Raynet Group, 46 Clarendon Rd, Ipplepen, Newton Abbot, TQ12 5QS [IO80EL, SX86]
G7 MNM J P M Webberley, 4 Gr Rd, Turvey, Bedford, MK43 8EA [IO92QD, SP95]
G7 MNN Details withheld at licensee's request by SSL.
G7 MNO R P Nightingale, 58 Nutfield Gr, Filton, Bristol, BS12 7LJ [IO81RM, ST67]
G7 MNP Details withheld at licensee's request by SSL.
G7 MNQ P J Bland, 19 Sookholme Dr, Warsop, Mansfield, NG20 0DN [IO93KE, SK56]
G7 MNR Details withheld at licensee's request by SSL.
G7 MNS A M Cartwright, 23 Kentford Rd, Felixstowe, IP11 8XZ [JO01PX, TM23]
G7 MNT B K Woods, 64 Yarningale Rd, Weeford Est, Coventry, CV3 3EQ [IO92GJ, SP37]
G7 MNY J J McGillivray, 39B Church Rd, St. Leonards on Sea, TN37 6HB [JO00GU, TQ80]
G7 MOA L J Everitt, 6 Belmont Rd, Rednal, Birmingham, B45 9LW [IO82XJ, SO97]
G7 MOB P Thain, 89 Cemetery Rd, Darwen, BB3 2LZ [IO83SQ, SD62]
G7 MOD D J Rust, Fairfield House, 26 Mill Rd, St. Germans, Kings Lynn, PE34 3HL [JO02EQ, TF51]
G7 MOF P A Barcroft, 10 Sowclough Rd, Stacksteads, Bacup, OL13 8LB [IO83VQ, SD82]
G7 MOH E A D Middleton, Fairwinds, Southella Rd, Yelverton, PL20 6AT [IO70XL, SX56]
GM7 MOI Details withheld at licensee's request by SSL.
G7 MOJ K B Zealand, Gardeners Cottage, Sheringham Park, Upper Sheringham, Sheringham, Norfolk, NR26 8TB [JO02OW, TG14]
G7 MOK N P Ridd, 3 Ormesby Dr, Swaffham, PE37 7SL [JO02IP, TF80]
G7 MOM R Hodson, 6 Eastbourne Gdns, Trowbridge, BA14 7HR [IO81VH, ST85]
G7 MON R W Powell, 15 Chelsea Way, Kingswinford, DY6 9EG [IO82VL, SO88]
G7 MOO K Parker, 11 Ringer Way, Clowne, Chesterfield, S43 4DW [IO93IG, SK47]
G7 MOR D Vickers, 69 Hallowes Ln, Dronfield, Sheffield, S18 6ST [IO93GH, SK37]
G7 MOT Details withheld at licensee's request by SSL.
G7 MOW K J Starnes, 19 Stoneham Cl, South Malling, Lewes, BN7 2ET [JO00AV, TQ41]
G7 MOX W D Jones, 62 Mallings Dr, Bearsted, Maidstone, ME14 4HG [JO01HG, TQ85]
G7 MOY Details withheld at licensee's request by SSL.
G7 MPB K Haddaway, 11 Aviemore Rd, Hemlington, Middlesbrough, TS8 9HW [IO94JM, NZ41]
G7 MPF R C Ransome, High Winds, High Town Green, Rattlesden, Bury St. Edmunds, IP30 0SZ [JO02KD, TL95]
G7 MPH R Cole, 64 Neville Shaw, Basildon, SS14 2AJ [JO01FN, TQ78]
G7 MPJ J Tweedy, 85 St. Pauls Rd, Jarrow, NE32 3AS [IO94GX, NZ36]
G7 MPL A E Stebbing, 4 Ripley Cres, Davyhulme, Urmston, Manchester, M41 8PH [IO83TL, SJ79]
G7 MPM Details withheld at licensee's request by SSL.
G7 MPS Details withheld at licensee's request by SSL.
GW7 MPW D T R Simmons, 15 Abbey Cl, Taffs Well, Cardiff, CF4 7RS [IO81IN, ST18]
G7 MPZ C H Atkins, 278 Walderslade Rd, Chatham, ME5 9AA [JO01GI, TQ76]
G7 MQB I P Sparkes, 58 Broadwater Cres, Welwyn Garden City, AL7 3TU [IO91VT, TL21]
G7 MQC C E Thomas, 1 George Gent Cl, Steeple Bumpstead, Haverhill, CB9 7EW [JO02FA, TL64]
G7 MQD Details withheld at licensee's request by SSL.
GW7 MQE D K Smith, 19 St. Marks Ave, Connahs Quay, Deeside, CH5 4XN [IO83LF, SJ26]
G7 MQF A H Kirkham, 24 Winghay Rd, Kidsgrove, Stoke on Trent, ST7 4XJ [IO83VC, SJ85]
G7 MQL J M Robson, 9 Acorn Way, Hurst Green, Etchingham, TN19 7QG [JO01FA, TQ72]
G7 MQM Details withheld at licensee's request by SSL.
G7 MQN N J Wall, 11 Coombe View, Teignmouth, TQ14 9UY [IO80FN, SX97]
GW7 MQO Details withheld at licensee's request by SSL.
G7 MQQ H D Griffiths, 2 Buckhurst Pl, Bexhill on Sea, E Sussex, TN39 3PA [JO00FU, TQ70]
G7 MQS R H King, Balcarres, Milton Ave, Badgers Mount, Sevenoaks, TN14 7AU [JO01BI, TQ46]
G7 MQT Details withheld at licensee's request by SSL.
G7 MQU D J Sandever, 57 Hayes Ln, Wimborne, BH21 2JB [IO90AT, SU00]
G7 MQV Details withheld at licensee's request by SSL.
G7 MQW R V Carroll, 71 Pelham St., Manton, Worksop, S80 2TT [IO93KH, SK57]
G7 MQY Details withheld at licensee's request by SSL.
G7 MQZ D L W Lappage, 25 The Chase, Wallington, SM6 8NA [IO91WI, TQ36]
G7 MRC Details withheld at licensee's request by SSL.
G7 MRD Details withheld at licensee's request by SSL.
G7 MRF M Farmer, 3 Brackenberry, St. Michaels Meadow, Cross Heath, Newcastle, ST5 9PS [IO83VA, SJ84]
G7 MRG R G Barker Melton Mowbray Raynet Group, 50 Baldocks Ln, Melton Mowbray, LE13 1EN [IO92NS, SK71]
G7 MRH E R C Cole, 9 West Hill Pl, Brighton, BN1 3RU [IO90WT, TQ30]
G7 MRJ Details withheld at licensee's request by SSL.
G7 MRK Details withheld at licensee's request by SSL.
G7 MRL N Williams, 1 Dorset Cl, Whitehaven, CA28 8JP [IO84FM, NX91]
G7 MRN Details withheld at licensee's request by SSL.
G7 MRO B Bowker, 205 Smallshaw Ln, Ashton under Lyne, OL6 8RJ [IO83WL, SD90]
G7 MRP Details withheld at licensee's request by SSL.
G7 MRQ R L Wilkinson, 70 Anchorsholme Ln East, Thornton Cleveleys, FY5 3QL [IO83LU, SD34]
GI7 MRT Details withheld at licensee's request by SSL.
G7 MRU Details withheld at licensee's request by SSL.
G7 MRV M Barfield, 33 Eccleston Rd, Kirk Sandall, Doncaster, DN3 1NX [IO93LN, SE60]
G7 MRY M D L Pratt, Bell Cottage, Ringwell Ln, Norton St. Philip, Bath, BA3 6LZ [IO81UH, ST75]
G7 MRZ R J R Thompson, 4 Hill Top Rd, Birdwell, Barnsley, S70 5QZ [IO93GM, SE30]
G7 MSC B L Knight, Milton Cottage, 35 Upper Strand St., Sandwich, CT13 9EL [JO01QG, TR35]
G7 MSD Details withheld at licensee's request by SSL.
G7 MSF K C Sanderson, 45 Bygrove, Fieldway, New Addington, Croydon, CR0 9DG [IO91XI, TQ36]
G7 MSH H M Samwells, 2 Dudley Walk, Macclesfield, SK11 8SD [IO83WG, SJ87]
GI7 MSJ P Hannaway, 4 Ballyhenry Gr, Glengormley, Newtownabbey, BT36 5BQ [IO74AQ, J38]
G7 MSK T Mann, 21 Glastonbury Ct, Yeovil, BA21 3TW [IO80PW, ST51]
G7 MSL P C Martin, 2 Bowles Rd, Falmouth, TR11 2PW [IO70LD, SW73]
G7 MSQ G Shelley, 41 Thornley Rd, Stanfields, Burslem, Stoke on Trent, ST6 7AL [IO83VB, SJ85]
G7 MSR M J Watson Mid Suffolk Raynet, The Tubbery, Henley, Ipswich, Suffolk, IP6 0BR [JO02NC, TM15]

G7 MSS D N Forward, Flat C, 16 King St., Deal, Kent, CT14 6HX [JO01QF, TR35]
G7 MST T B Bennett, Rose Cottage, High St., Rawmarsh, Rotherham, S62 6LN [IO93HL, SK49]
GI7 MSY G P Woods, 181 Ballyclare Rd, Newtownabbey, BT36 5JP [IO74AQ, J38]
G7 MTA C Parr M.B.E., 13 Peartree Ave, Bitterne, Southampton, SO19 7JN [IO90HV, SU41]
G7 MTE M Coote, 22 Tennyson Cl, Boston, Lincs, PE21 8DL [IO92XX, TF34]
G7 MTF T E J Foley, 143 Okus Rd, Swindon, SM1 4JY [IO91VI, TQ26]
G7 MTG A M Blakeston, 8 Andersen Ct, Townville, Castleford, WF10 3HY [IO93IR, SE42]
G7 MTI F T Paley, 68 Dennill Rd, Manston, Leeds, West Yorks, LS15 8SD [IO93GT, SE33]
G7 MTJ C W C Chase, Asholt, Ermine St., Appleby, Scunthorpe, DN15 0AD [IO93RO, SE91]
GM7 MTK R J M Gibson, 26 Woodlands Ct, 22/24 Woodlands Terr, Thornliebank, Glasgow, G4 7SA [IO75UU, NS56]
G7 MTQ Details withheld at licensee's request by SSL.
G7 MTS Details withheld at licensee's request by SSL.
G7 MTT W M Johns, 6 Wakefield Ave, Northbourne, Bournemouth, BH10 6DS [IO90BS, SZ09]
G7 MTU R Donson, 5 Overend Cl, Sheffield, S14 1JG [IO93GI, SK38]
G7 MTV M Bourne, 100 Dimsdale View West, Newcastle, ST5 8EL [IO83VA, SJ84]
G7 MTW R Powell, 2 The Limes, Bramley, Tadley, RG26 5UW [IO91LH, SU65]
G7 MTX P A Mullis, 39 Buckingham Rd, Lawn, Swindon, SN3 1HZ [IO91CN, SU18]
G7 MTZ Details withheld at licensee's request by SSL.
G7 MUB R A Harcourt, 7 Lightfoot Cl, Newark, NG24 2HT [IO93OB, SK85]
G7 MUD D I Layne Siemens Plessey, Siemens Plessey, Christchurch Ars, Grange Rd, Christchurch, Dorset, BH23 4JE [IO90DR, SZ19]
G7 MUE S J Roper, 1 Holywell Rd, Kilnhurst, Mexborough, S64 5UQ [IO93IL, SK49]
G7 MUF E A C Fowler, Rydal, Bedford Cres, Frimley Green, Camberley, GU16 6HH [IO91PH, SU85]
G7 MUG J O Lam, 381 Mountnessing Rd, Billericay, CM12 0EU [JO01EP, TQ69]
G7 MUH Details withheld at licensee's request by SSL.
G7 MUI Details withheld at licensee's request by SSL.
G7 MUL Details withheld at licensee's request by SSL.
G7 MUM Details withheld at licensee's request by SSL.
G7 MUN J S Smith, 77 Trafalgar Rd, Downham Market, PE38 9RT [JO02EO, TF60]
G7 MUS Details withheld at licensee's request by SSL.
G7 MUT T W Cannon, 71 Northport Dr, Wareham, BH20 4DN [IO80WQ, SY98]
G7 MUV D C Spencer, 217 Marlborough Ave, Princes Ave, Hull, HU5 3LF [IO93TS, TA02]
G7 MUY A J Sadler, 15 Scarborough Rise, Breadsall Est, Derby, DE21 4DG [IO92GW, SK33]
G7 MVB Details withheld at licensee's request by SSL.
G7 MVC Details withheld at licensee's request by SSL.
G7 MVE M S Cotton, 2 Redhill View, Castleford, WF10 4QL [IO93IR, SE42]
GW7 MVG H S Millington, Arran, Clayton Rd, Mold, CH7 1SU [IO83KE, SJ26]
G7 MVJ S A Steed, 37 Greg St., South Reddish, Stockport, SK5 7LB [IO83WK, SJ89]
G7 MVM Details withheld at licensee's request by SSL.
G7 MVN Details withheld at licensee's request by SSL.
G7 MVO M Armfield, 21 Lees Rd, Mossley, Ashton under Lyne, OL5 0PG [IO83XM, SD90]
G7 MVP Details withheld at licensee's request by SSL.
GW7 MVQ P M Pearl, Camsigns, 34 Jeffrey St., Newport, NP9 0DA [IO81MO, ST38]
G7 MVU N J Brown, 6 Hundon Pl, Haverhill, CB9 0AP [JO02FC, TL64]
G7 MVY R Stockley, 10 Swan Rd, Timperley, Altrincham, WA15 6BX [IO83UJ, SJ79]
GI7 MWA S M Stewart, 3 Killyfaddy Rd, Magherafelt, BT45 6EX [IO64QR, H88]
G7 MWB Details withheld at licensee's request by SSL.
G7 MWC R Moss, 6 Adelaide Gdns, Stonehouse, GL10 2PZ [IO81US, SO80]
G7 MWD P G Turner, 21 Glendale Tower, Beechmount Dr, Erdington, Birmingham, B23 5UE [IO92CM, SP19]
G7 MWF Details withheld at licensee's request by SSL.
G7 MWG Details withheld at licensee's request by SSL.
G7 MWH P S Cross, 11 Daniell House, Cranston Est, London, N1 5EH [IO91WM, TQ38]
G7 MWI L C Hansen, 19 Market St., Appledore, Bideford, EX39 1PW [IO71VB, SS43]
G7 MWJ A J Holloway, 31 Gays Rd, Hanham, Bristol, BS15 3JR [IO81RK, ST67]
GM7 MWL H Murray, 23 Denmore Gdns, Bridge of Don, Aberdeen, AB22 8LJ [IO87WE, NJ90]
G7 MWM A J Scott, 22 Planters Gr, Oulton Broad, Lowestoft, NR33 9QL [JO02UL, TM59]
G7 MWN Details withheld at licensee's request by SSL.
G7 MWP G W Veary, 5 Milton Fields, Chalfont St. Giles, HP8 4ER [IO91RP, SU99]
G7 MWQ Details withheld at licensee's request by SSL.
G7 MWR Details withheld at licensee's request by SSL.
G7 MWS P Parrish, 5 Kestrel Ln, Cheadle, Stoke on Trent, ST10 1RU [IO92AX, SK04]
G7 MWT Details withheld at licensee's request by SSL.
G7 MWU G Haswell, 16 Hither Green, Jarrow, NE32 4LP [IO94GW, NZ36]
G7 MWW J R Smith, 19 The Cres, Carisbrook Gdns, Mitcheldean, GL17 0SB [IO81SU, SO61]
GM7 MWX R H Raynor, La Pergola, Kilmuir, North Kessock, Inverness, Scotland, IV1 1XG [IO77VM, NH65]
G7 MWY Details withheld at licensee's request by SSL.
G7 MWZ R Watson, Plantation, Church Hill, Ravensden, Bedford, MK44 2RL [IO92SE, TL05]
G7 MXA Details withheld at licensee's request by SSL.
G7 MXB Details withheld at licensee's request by SSL.
G7 MXD A K Williams, 106 Sparrowhawk Way, Hartford, Huntingdon, PE18 7XY [IO92WI, TL27]
GM7 MXE Details withheld at licensee's request by SSL.
G7 MXG S J Cox, 62 Bridle Rd, Eastcote, Pinner, HA5 2SH [IO91TO, TQ18]
G7 MXJ Details withheld at licensee's request by SSL.
G7 MXL T S R Grange, 14 Eastbourne Rd, Stratford, London, E15 3LJ [JO01AM, TQ38]
G7 MXM C J Turner, 55 Fordfield Rd, Sunderland, SR4 6XG [IO94GV, NZ35]
G7 MXP R A Bebbington, 64 Stafford Rd, Toll Bar, St. Helens, WA10 3JH [IO83OK, SJ49]
G7 MXQ B S Gilbraith, 19 Bullcote Green, Royton, Oldham, OL2 6NJ [IO83WN, SD90]
G7 MXS A C Harrison, 22 Sherburn Gate, Chapeltown, Sheffield, S30 4EU [IO93GJ, SK38]
G7 MXT D F Harris, Woodacre, Rock Rd, Storrington, Pulborough, RH20 3AG [IO90TW, TQ11]
G7 MXX S J Greenall, 57 Weydon Ln, Farnham, GU9 8UW [IO91OE, SU84]
G7 MXY Details withheld at licensee's request by SSL.
G7 MXZ Details withheld at licensee's request by SSL.
G7 MYB B A Moye, 5 Ixworth Rd, Troston, Bury St. Edmunds, IP31 1EZ [JO02JH, TL87]
GW7 MYD P Williams, 5 Bright St., Cross Keys, Newport, NP1 7PB [IO81KO, ST29]
GM7 MYF C Dennett, 35 Hill St., Alexandria, G83 0DU [IO75RX, NS38]
G7 MYG A J O Maguire, 55 Mendip Vale, Coleford, Bath, BA3 5PR [IO81SF, ST64]
G7 MYI A E Stride, Poppy Cottage, 3 Barnfield, Edmondsham, Wimborne, Dorset, BH21 5RD [IO90AV, SU01]
G7 MYJ R Ball, 5 Miller Fold Ave, Accrington, BB5 0NT [IO83TR, SD72]
G7 MYK Capt M Corner, 18 Yannon Dr, Teignmouth, TQ14 9JP [IO80FN, SX97]
G7 MYL Details withheld at licensee's request by SSL.
G7 MYM D K Roberts, 22 Norfolk Rd, Turvey, Bedford, MK43 8DU [IO92QD, SP95]
G7 MYN C M George, Tolmers Scout Camp, Tolmers Rd, Cuffley, Potters Bar, EN6 4JS [IO91WR, TL30]
G7 MYO C E McIver, 31 Harts Hill, Putnoe, Bedford, Beds, MK41 9AL [IO92SD, TL05]
G7 MYQ J Mead, Ebdale, Moat Ln, Sedlescombe, Battle, E Sussex, TN33 0RZ [JO00GV, TQ71]
G7 MYR D Hicks, 13 Gregory Rd, Glasshoughton, Castleford, WF10 4PH [IO93HR, SE42]
G7 MYS Details withheld at licensee's request by SSL.
G7 MYT P J Hilton, 40 Megstone Ave, Whitelea Chase, Cramlington, NE23 6TU [IO95EB, NZ27]
G7 MYW Details withheld at licensee's request by SSL.
G7 MYY L Fuller, 1 Nightingale Rd, Kemsing, Sevenoaks, TN15 6RU [JO01CH, TQ55]
G7 MYZ Details withheld at licensee's request by SSL.
G7 MZA R K Loukes, White Cottage, Bushmead Rd, Colmworth, Bedford, MK44 2LH [IO92TF, TL16]
G7 MZB S G S Rudd-Clarke, Padmore Orchard, Leebotwood, Church Stretton, Shropshire, SY6 6NH [IO82OO, SO49]
G7 MZE T R Ingle, 68 Wooldale Dr, Filey, YO14 9ER [IO94UF, TA18]
G7 MZF Details withheld at licensee's request by SSL.
G7 MZH Details withheld at licensee's request by SSL.
G7 MZI J P Head, 36A Ashacre Ln, Worthing, BN13 2DH [IO90TU, TQ10]
G7 MZJ I S Mitchell, 20 Ward St., Penistone, Sheffield, S30 6EP [IO93GJ, SK38]
G7 MZK D Mitchell, 28 Southgate, Penistone, Sheffield, S30 6EA [IO93GJ, SK38]
G7 MZL N C Baker, 17 Heathview, Holden Rd, Southborough, Tunbridge Wells, Kent, TN4 0QE [JO01DD, TQ54]
G7 MZN K J Hartley, 72 Seymour Rd, Trowbridge, BA14 8LU [IO81VH, ST85]
G7 MZO Details withheld at licensee's request by SSL.
G7 MZR Details withheld at licensee's request by SSL.
G7 MZS L G Terry, 10 Cypress Rd, Bellfields, Guildford, GU1 1NB [IO91RG, SU95]
G7 MZT P Cobley, 83 Stonebury Ave, Eastern Green, Coventry, CV5 7NY [IO92EJ, SP27]
G7 MZU Details withheld at licensee's request by SSL.
G7 MZW A Calvert, 7 Half Mile, Bramley, Leeds, LS13 1BN [IO93ET, SE23]
G7 MZY I P Sharp, 6 Ullswater Dr, Fairfield Park, Bath, BA1 6NP [IO81TJ, ST76]
GM7 MZZ G Whiting, 21 Leckethill Ct, Westfield, Cumbernauld, Glasgow, G68 9EG [IO75XW, NS77]
GM7 NAA I T Skeoch, 20 Strath Elgin, Law, Carluke, ML8 5LF [IO85BS, NS85]
G7 NAD Details withheld at licensee's request by SSL.
G7 NAH Details withheld at licensee's request by SSL.
G7 NAI J T Stock, 31 Grange Rd, Wickham Bishops, Witham, CM8 3LT [JO01IS, TL81]
G7 NAL H E Wood, 31 Goring Ave, Gorton, Manchester, M18 8WW [IO83VL, SJ89]
G7 NAM Details withheld at licensee's request by SSL.
G7 NAN A Griffin, 32 Mill Cres, Hebburn, NE31 1UQ [IO94FX, NZ36]
G7 NAO I Langmuir, 2 Nelson Rd, Newport, PO30 1QT [IO90IQ, SZ48]
G7 NAP Details withheld at licensee's request by SSL.
G7 NAR S Goodwin South Gloucestershire Raynet, 14 Greenhill, Alveston, Bristol, BS12 2QX [IO81RO, ST68]
G7 NAS Details withheld at licensee's request by SSL.
G7 NAT A A Robinson, 23 Pine Gr, Sale, M33 5WN [IO83TK, SJ79]
GW7 NAU Details withheld at licensee's request by SSL.
GI7 NBB J M A McMaster, 71 Lismurn Park, Ahoghill, Ballymena, BT42 1JW [IO64TU, D00]
G7 NBC Details withheld at licensee's request by SSL.
G7 NBE M J Goodwin, 23 Saxon Way, Ashby de La Zouch, LE65 2JR [IO92GR, SK31]
G7 NBF A Sadler, 23 Wolsey Rd, Moor Park, Northwood, HA6 2HN [IO91SP, TQ09]
G7 NBG L J McGuire, 68 Kingsmead Park, Bedford Rd, Rushden, NN10 0NF [IO92QG, SP96]
G7 NBI P J Webb, 42 Holland Rd, Ampthill, Bedford, MK45 2RS [IO92SA, TL03]
G7 NBJ D Corfield, 177 Hurst Rise, Matlock, DE4 3EU [IO93FD, SK36]
G7 NBK Details withheld at licensee's request by SSL.
G7 NBL C M King, 70 Russell Dr, Ampthill, Bedford, MK45 2TU [IO92SA, TL03]
G7 NBM P R Smith, 67 Harestone Hill, Caterham, CR3 6DX [IO91XG, TQ35]
G7 NBN Details withheld at licensee's request by SSL.
G7 NBP C J Williams, 1 Home Farm Cttgs, Leebotwood, Church Stretton, SY6 6LX [IO82ON, SO49]
G7 NBR W R Hayward, 15 Whitehouse Rd, South Woodham Ferrers, Chelmsford, CM3 5PF [JO01HP, TQ89]
G7 NBU K B Bassett, 52 Tamworth Rd, Sutton Coldfield, B75 6DH [IO92CN, SP19]
G7 NBV F Young, 23 Llewellyn Walk, Corby, NN18 0RY [IO92PL, SP88]
G7 NBW J M Vincent, 12 Palmdale Cl, Longwell Green, Bristol, BS15 6UH [IO81SK, ST67]
G7 NBX M F Slaney, 23 Creswell Gr, Creswell, Stafford, ST18 9QP [IO82WT, SJ82]
G7 NBY Details withheld at licensee's request by SSL.
G7 NBZ P J Millerchip, 6 Washbrook View, Ottery St. Mary, EX11 1EP [IO80IS, SY19]
G7 NCB D Geldart, 5 Crabtree Dr, Great Houghton, Barnsley, S72 0AF [IO93HN, SE40]
G7 NCC J A Hother, 10 Woodlands, Barrowfield Dr, Hove, BN3 6TJ [IO90WU, TQ20]
G7 NCD T L Willis, 15 Cedar Ct, Congleton, CW12 3JP [IO83VD, SJ86]
G7 NCE Details withheld at licensee's request by SSL.
G7 NCF J E Foster, 2 Kent Cl, Little Hulton, Worsley, Manchester, M28 0TT [IO83TM, SD70]
G7 NCG E M Turner, Cranes Farm, Manea Fifties, Tipps End Welney, Wisbech, Cambs, PE14 9SP [JO02CM, TL59]
G7 NCH Details withheld at licensee's request by SSL.
G7 NCJ Details withheld at licensee's request by SSL.
G7 NCK Details withheld at licensee's request by SSL.
G7 NCM Details withheld at licensee's request by SSL.
G7 NCN Details withheld at licensee's request by SSL.
G7 NCO Details withheld at licensee's request by SSL.
G7 NCP Details withheld at licensee's request by SSL.
G7 NCR G Foote, 64 Stable Yard, Tyntesfield, Wraxall, Bristol, BS19 1NS [IO81PK, ST57]
G7 NCT D I Lees obo Nationwide Scout Communica, 80 Delrene Rd, Shirley, Solihull, West Midlands, B90 2HL [IO92BJ, SP17]
G7 NCU W J Henney, 74 Moughland Ln, Runcorn, WA7 4SQ [IO83PH, SJ58]
G7 NCV K G Hobbs, 20 Holding St., Rainham, Gillingham, ME8 7JP [JO01HI, TQ86]
G7 NCW G D Hinton, 25 Linden Cl, Prestbury, Cheltenham, GL52 3DX [IO81XV, SO92]
G7 NCY M I Baines, 17 Fir Tree Ln, Littleton, Chester, CH3 7DN [IO83NE, SJ46]
GU7 NCZ Dr N R Turner, Camellia Lodge, L'Aumone, Castel, Guernsey, Channel Islands, GY5 7RT
G7 NDB K C Marshall, Doveysmead, Chapel St., North Waltham, Basingstoke, RG25 2BZ [IO91JF, SU54]
G7 NDC P Hirst, 47A Rowley Ln, Lepton, Fenay Bridge, Huddersfield, HD8 0JG [IO93DP, SE11]
G7 NDD Details withheld at licensee's request by SSL.
G7 NDF Details withheld at licensee's request by SSL.
G7 NDG Details withheld at licensee's request by SSL.
G7 NDJ M C Burns, 338 Carlton Hill, Carlton, Nottingham, NG4 1JD [IO92KX, SK54]
G7 NDL S A Yeomans, 17 Talbot Rd, Farnham, GU9 8RP [IO91OE, SU84]
G7 NDN S D C Ward, 16 St. Peters Way, Chorleywood, Rickmansworth, WD3 5QE [IO91RP, TQ09]
G7 NDO H S Blackburn, 4 Hawkridge, Furzton, Milton Keynes, MK4 1BQ [IO92OA, SP83]
G7 NDP Details withheld at licensee's request by SSL.
G7 NDQ G R Bettyes, 8 Orchard Cres, Woodnewton, Peterborough, PE8 5EN [IO92SM, TL09]
G7 NDR Details withheld at licensee's request by SSL.
G7 NDS M R Fry, Holly Cottage, Farnsfield Rd, Bilsthorpe, Newark, NG22 8SJ [IO93LD, SK66]
G7 NDT R Walker, 46 Lodge Rd, Little Houghton, Northampton, NN7 1AE [IO92OF, SP85]
G7 NDW Details withheld at licensee's request by SSL.
G7 NDY Details withheld at licensee's request by SSL.
G7 NEA K Walters, 5 Manor Cl, Bengeo, Hertford, SG14 3JZ [IO91XT, TL31]
GI7 NEB J Conlon, 3 Meadowbank, Dungannon, BT71 7DA [IO64OM, H76]
G7 NEC C D Watson, Holly Cottage, 11 Turf Ln, Cullingworth, Bradford, BD13 5EJ [IO93BT, SE03]
G7 NEE E W Maloney, 56 Westonfields Dr, Longton, Stoke on Trent, ST3 5JA [IO82WX, SJ94]
G7 NEG R O Smith, 32 Water Ln, Wootton, Northampton, NN4 6HE [IO92NE, SP75]
G7 NEH G A Pemberton, 2 Hockenhull Ave, Tarvin, Chester, CH3 8LP [IO83OE, SJ46]
G7 NEI Details withheld at licensee's request by SSL.
G7 NEK I M Thomas, 69 Magnaville Rd, Thorley Park, Bishops Stortford, CM23 4DW [JO01BU, TL41]
G7 NEL Details withheld at licensee's request by SSL.
G7 NEM J Barber, 1 Crown Cttgs, The Street, Haddiscoe, Norwich, Norfolk, NR14 6AA [JO02TM, TM49]
G7 NEN J W Foster, 2 Cowley Ave, Chertsey, KT16 9JH [IO91RJ, TQ06]
G7 NEP Details withheld at licensee's request by SSL.
G7 NEV S Cheetham, 107 Cemetery Rd, Southport, PR8 5EQ [IO83MP, SD31]
GM7 NEW S J Smallwood, Beulah, 140 Main St, Neilston, Glasgow, Scotland, G78 3JX [IO75SS, NS45]
G7 NEX M S Bland, 17 Sandy Ln, Warrington, WA2 9BX [IO83RJ, SJ69]
G7 NFA Details withheld at licensee's request by SSL.
GI7 NFB M J Robinson, 92A Dromore Rd, Hillsborough, Co. Down, BT26 6HU [IO64WK, J25]
G7 NFD J D Pilgrim, 1 Geraldine Terr, Forder, Saltash, Cornwall, PL12 4QR [IO70VJ, SX45]
G7 NFE R D McKay, Sun Cottage, Tarrandean Ln, Perranwell Station, Truro, TR3 7NW [IO70KE, SW73]
G7 NFF P Salmon, Lindeth, 95 Craig Walk, Windermere, LA23 2JS [IO84NI, SD49]
G7 NFG M D Holdsworth, 1 Killinghall Ave, Bradford, BD2 4SA [IO93DT, SE13]
GW7 NFI Details withheld at licensee's request by SSL.
G7 NFJ Details withheld at licensee's request by SSL.
G7 NFK J A Watmough, 78 Victoria Park Rd, Fairfield, Buxton, SK17 7PE [IO93BG, SK07]
GW7 NFM E Jones, Maes y Coed, Vownog Rd, Sychdyn, Mold, CH7 6ED [IO83KE, SJ26]
G7 NFN R E K Bailey, 29 Priory Cl, Bath, BA2 5AL [IO81TI, ST76]
G7 NFO M F Hall, 30 Kingsley Ave, Rugby, CV21 4JY [IO92JI, SP57]
G7 NFP J Walsh, 8 Swinburne St., Derby, DE1 2HJ [IO92GV, SK33]
G7 NFQ Me E C Bright, 59 Vancouver Ave, Kings Lynn, PE30 5RD [JO02ER, TF61]
G7 NFR J P Thomas, 2 Alexandra Rd, Uxbridge, UB8 2PQ [IO91SM, TQ08]
G7 NFS Details withheld at licensee's request by SSL.
GW7 NFT J A Parry, Charlbury, Usk Rd, Caerleon, Newport, NP6 1LP [IO81MO, ST39]
G7 NFU Details withheld at licensee's request by SSL.
G7 NFW E Skoyles, 29 Gordon Ave, Thorpe St. Andrew, Norwich, NR7 0DW [JO02QP, TG20]
G7 NFX R P Maher, 2 Grange View, Abbotskerswell, Newton Abbot, TQ12 5PB [IO80EM, SX86]
GW7 NFY M Mee, Ann Cott, Hylas Ln, Rhuddlan, Clwyd, LL18 5AG [IO83GG, SJ07]
G7 NFZ Details withheld at licensee's request by SSL.
G7 NGB J K S Alger, 15 Tollers Ln, Coulsdon, CR5 1BE [IO91WH, TQ35]
G7 NGC M Wooldridge, 29 Conifer Gr, Gosport, PO13 0TP [IO90JT, SU50]
G7 NGD Details withheld at licensee's request by SSL.
G7 NGF D J Hatcher, 8 Churchfield, Monks Eleigh, Ipswich, IP7 7JH [JO02KC, TL94]
G7 NGG Details withheld at licensee's request by SSL.
G7 NGH D A W Collins, 44 Underlane, Plymstock, Plymouth, PL9 9JZ [IO70WI, SX55]
G7 NGI J Price, 32 Wilts Dr, Trowbridge, BA14 0RE [IO81VH, ST85]
G7 NGM N G J Mothew, 7 Ashfields, Loughton, IG10 1SB [JO01AP, TQ49]
G7 NGN R D Williams, 73 Quedgeley Park, Greenhill Dr, Gloucester, GL2 5NZ [IO81UU, SO81]
G7 NGO H E Scott, 112 Manor St., Accrington, BB5 6EA [IO83TS, SD72]
G7 NGQ A D Bauer, 33 Tudor Rd, Eastham, London, E6 1DP [JO01AM, TQ48]
G7 NGS Details withheld at licensee's request by SSL.
GW7 NGU J B Vaughan, Montrose, 5 Trewarren Dr, St. Ishmaels, Haverfordwest, SA62 3TR [IO71KS, SM81]
G7 NGV Details withheld at licensee's request by SSL.
G7 NGZ Details withheld at licensee's request by SSL.
G7 NHB R W Griffiths, 4 Wolrige Way, Plympton, Plymouth, PL7 2RU [IO70XJ, SX55]
G7 NHC D W Scholes, 71 Pelham St., Ashton under Lyne, OL7 0DU [IO83WL, SJ99]
G7 NHD T J Carroll, 14 Glenpark Dr, Southport, PR9 9FA [IO83MQ, SD31]
G7 NHE J I Turnbull, 32 Haydon, Fatfield, Washington, NE38 8PF [IO94FV, NZ35]
G7 NHF K C Williams, 1 St. Ives Way, Halewood, Liverpool, L26 7YW [IO83OI, SJ48]
G7 NHK F A Cartwright, 30 Pear Trees, Ingrave, Brentwood, CM13 3RP [JO01EO, TQ69]
G7 NHL K M Mitchell, 2 Ripon Gdns, Foxlow Park, Buxton, SK17 9PL [IO93BF, SK07]
G7 NHP L V Carter, 4 Croft House Mount, Morley, Leeds, LS27 8NN [IO93ES, SE22]
G7 NHQ K J Cotterill, 1 Molineux Ave, Broadgreen, Liverpool, L14 3LT [IO83NJ, SJ49]
G7 NHR P R Dunlop, 4 Birket Ave, Moreton, Wirral, L46 1QZ [IO83KJ, SJ29]
GM7 NHS Dr R P C Johnson, 3 Hopetoun Green, Bucksburn, Aberdeen, AB2 9QX [IO87TH, NJ72]
GM7 NHU G P Devereux, 44 Greenhead Rd, Dumbarton, G82 2PN [IO75RW, NS47]
G7 NHV K Johnson, 43 Glencoe Rd, Great Sutton, South Wirral, L66 4NA [IO83MG, SJ37]

G7 NHW K P McKane, Four Winds, Liskeard Rd, Callington, PL17 7EZ [IO70UL, SX36]
GU7 NHX A J Dorrian, -le-Petit Jardin, Clos de Emrais, Castel, Guernsey, GY5 7YB
G7 NHY V J Brooker, Flat 6, 46 Foxglove Way, Wallington, Surrey, SM6 7JU [IO91WI, TQ26]
G7 NHZ Details withheld at licensee's request by SSL.
G7 NIA H C Seldon, 83 Dorchester Ave, Whitleigh, Plymouth, PL5 4AZ [IO70WJ, SX45]
G7 NIC J Wallis, 129 Braithwaite Ave, Keighley, BD22 6QE [IO93AU, SE04]
GU7 NIE Details withheld at licensee's request by SSL.
G7 NIG Details withheld at licensee's request by SSL.
G7 NIH A E Davies, 1 Fire Station Yard, Rochdale, OL11 1DT [IO83WO, SD81]
G7 NII Prof R S Kalawsky, Holt Lodge, The Holt of Loughborough Ui., Holt Dr, Loughborough, LE11 3JB [IO92JS, SK51]
GI7 NIL J Chin, Site 29, Meadowbrook, Islandmagee, Larne, Co. Antrim, BT40 3UG [IO74CT, J49]
G7 NIM Details withheld at licensee's request by SSL.
G7 NIN D J Townsend, 40 Popes Ln, Sturry, Canterbury, CT2 0JZ [JO01NH, TR16]
G7 NIO H Russell, 212 Argyle St., Heywood, OL10 3LS [IO83VO, SD81]
GW7 NIQ J Gould, y Llest House, Penyfai, Bridgend, Mid Glam, CF31 4NA [IO81EM, SS88]
G7 NIR L B Jones, 53 Ennisdale Dr, West Kirby, Wirral, L48 9UF [IO83KI, SJ28]
GW7 NIS M A Smith, 35 Heol Llansantffraid, Sarn, Bridgend, CF32 9NH [IO81EM, SS98]
G7 NIT Details withheld at licensee's request by SSL.
G7 NIU S Tanner, 53 Waynflete Ave, Brackley, NN13 6AG [IO92KA, SP53]
G7 NIV Details withheld at licensee's request by SSL.
GW7 NIW G S Durno, Lothlorien, Upper Denbigh Rd, St Asaph, Clwyd, LL17 0BH [IO83GF, SJ07]
G7 NIX C R Shurety, O Fran Villa, Camp Rd, Taverham, Norwich, NR8 6LD [JO02OQ, TG11]
G7 NIY M J Stanley, Low Moor, Fenton, Keswick, Cumbria, CA12 4AZ [IO84KO, NY22]
G7 NIZ Details withheld at licensee's request by SSL.
G7 NJB G P Harris, 58 The Leas, Minster on Sea, Sheerness, ME12 2NL [JO01JK, TQ97]
G7 NJD J Stewart, 22 Garden Rd, Kendal, LA9 7ED [IO84PH, SD59]
G7 NJE C White, 13 Peel St., Heywood, OL10 4QD [IO83VO, SD81]
G7 NJF D G Roberts, 1 Kelsey Ave, Scunthorpe, DN15 8BW [IO93PO, SE81]
G7 NJG D J Godwin, Chestnut Glade, 2 Barncroft Dr, Hempstead, Gillingham, ME7 3TJ [JO01GI, TQ76]
G7 NJH B Thomas, 45 Manor Dr, Ivybridge, PL21 9BE [IO80AJ, SX65]
G7 NJI M L Tribe, Wix Hill, West Morsley, West Horsley, Leatherhead, Surrey, KT24 6ED [IO91SG, TQ05]
G7 NJK Details withheld at licensee's request by SSL.
GW7 NJO Details withheld at licensee's request by SSL.
G7 NJP M R Neal, 3 Nursery Way, Grimston, Kings Lynn, PE32 1DQ [JO02GS, TF72]
GW7 NJQ S J Richardson, Holmleigh, Chapel Rd, Broughton, Cowbridge, South Glam, CF7 7QR [IO81HN, ST08]
GW7 NJT J C Jones, 21 Pen yr Heol, Penyfai, Pen y Fai, Bridgend, CF31 4ND [IO81EM, SS88]
G7 NJX S F Mullen, 86 Homeleaze Rd, Bristol, BS10 6BZ [IO81QM, ST57]
G7 NJZ J O Rourke, 39 Rutherglen Rd, Corby, NN17 1ER [IO92PL, SP88]
G7 NKB C E Goodwin, 2 Crawley Ave, Eccles, Manchester, M30 0DW [IO83UL, SJ79]
G7 NKC D A Field, The Porcupine, Rushmore Hill, Knockholt, Sevenoaks, TN14 7NS [JO01BH, TQ46]
G7 NKE M Jordan, 17 Gilbert Ave, Tuxford, Newark, NG22 0JB [IO93NF, SK77]
G7 NKF Details withheld at licensee's request by SSL.
G7 NKH D C Smith, 7 Salisbury Rd, Carshalton, SM5 3HA [IO91WI, TQ26]
G7 NKI A K Davin, 10 Beech Ln, Eye, Peterborough, PE6 7YG [IO92VO, TF20]
G7 NKJ T H Westbrook, 5 Newlands, Northallerton, DL6 1SJ [IO94GI, SE39]
G7 NKM D R Byham, 63 Minsmere Way, Great Cornard, Sudbury, CO10 0LD [JO02JA, TL84]
GM7 NKP Details withheld at licensee's request by SSL.
G7 NKS Details withheld at licensee's request by SSL.
G7 NKT Details withheld at licensee's request by SSL.
G7 NKU C E Prout, Cromwell House, 14 Lenham Ave, Saltdean, Brighton, BN2 8AE [IO90XT, TQ30]
G7 NKW H Edwards, 312 Stourbridge Rd, Catshill, Bromsgrove, B61 9LH [IO82XI, SO97]
G7 NLA C Milburn, 11 Freer Gdns, Louth, LN11 8AW [JO03AI, TF38]
G7 NLC Details withheld at licensee's request by SSL.
G7 NLE W G Newman, 18 Reeds Ave East, Moreton, Wirral, L46 1RQ [IO83KJ, SJ29]
G7 NLF G Bandara Mieee, 26 Undine St., London, SW17 8PR [IO91WK, TQ27]
G7 NLI P J Oliver, 45 Charlotte Cl, Mount Hawke, Truro, TR4 8TS [IO70JG, SW74]
G7 NLJ R Gardner, 94 Salisbury Ave, Newbold, Chesterfield, S41 8PN [IO93GG, SK37]
G7 NLL D H Meakin, 47 Sheridan St., Pleck, Walsall, WS2 9QX [IO82XN, SO99]
G7 NLP J Greenacre, 21 Pilling St., Elton, Bury, BL8 1NE [IO83UO, SD71]
G7 NLQ Details withheld at licensee's request by SSL.
G7 NLR D Andrew North Lancashire Raynet Group, 30 Woodhill Ave, Morecambe, LA4 4PF [IO84NB, SD46]
G7 NLS D M C Chamberlin, 6 Yew Tree Ct, Hockering, Dereham, NR20 3JR [JO02MQ, TG01]
GM7 NLT Details withheld at licensee's request by SSL.
GM7 NLU Details withheld at licensee's request by SSL.
G7 NLW G R Williams, 65 Tudor Ave, North Watford, Watford, WD2 4NU [IO91TQ, TQ19]
GW7 NLX Details withheld at licensee's request by SSL.
G7 NLY J S James, 14 Fairview Dr, Bayston Hill, Shrewsbury, SY3 0LE [IO82OQ, SJ40]
G7 NMA D C Perryman, 50 Bellingham Cres, Plympton, Plymouth, PL7 2QP [IO70XJ, SX55]
G7 NMB K Bennett Cumbria Emergency Planning Uni, Arroyo Block, The Castle, Carlisle, Cumbria, CA3 8UR [IO84MV, NY35]
GM7 NMC Details withheld at licensee's request by SSL.
G7 NME B C Caldicott, Woodbury, Parkenhead, Trevone, Padstow, PL28 8QH [IO70MN, SW87]
G7 NMG Details withheld at licensee's request by SSL.
G7 NMH Details withheld at licensee's request by SSL.
G7 NMI K Osborne, 42 Barbrook Ln, Tiptree, Colchester, CO5 0EF [JO01JT, TL81]
G7 NMJ P F Jordan, 27 Wembley Hill Rd, Wembley, HA9 8AS [IO91UN, TQ18]
GI7 NMK L M Breadon, 42 Copeland Ave, Millisle, Newtownards, BT22 2DP [IO74FO, J57]
G7 NMP J E Bailey, Acorn Cottage, Langham Rd, Blaxhall, Woodbridge, IP12 2EF [JO02RD, TM35]
G7 NMR K H Clapton, 37 Blackwell Rd, East Grinstead, RH19 3HP [IO91XD, TQ33]
G7 NMT M J Beach, 25 Poplar Walk, Farnham, GU9 0QL [IO91OF, SU84]
G7 NMU H B Steele, 79 Trinity Rd, Narborough, Leicester, LE9 5BW [IO92JO, SP59]
G7 NMZ A J Gibson, 28 Lacey Ave, Wilmslow, SK9 4BB [IO83VI, SJ88]
G7 NNA A L Watkin BA Hnd, Aarburg, Windsor Cl, Oswestry, Shropshire, X X
G7 NNC J Andrews, 15 Edgehill Cl, Fulwood, Preston, PR2 3JQ [IO83PS, SD53]
G7 NND J R Ranson, 18 Beaufort Ave, Brooklands, Sale, M33 3WL [IO83UJ, SJ79]
G7 NNE Details withheld at licensee's request by SSL.
G7 NNF Details withheld at licensee's request by SSL.
GM7 NNG Details withheld at licensee's request by SSL.
GM7 NNH S A Fortune, 26 Newton Gr, Newton Mearns, Glasgow, G77 5QJ [IO75US, NS55]
G7 NNI A J Vickers, 16 Waveney Dr, Springfield, Chelmsford, CM1 7PY [JO01FR, TL70]
G7 NNL Details withheld at licensee's request by SSL.
G7 NNN R Bell, 36 Leith Towers, Grange Vale, Sutton, SM2 5BY [IO91VI, TQ26]
G7 NNR B C Hughes, Obspringener Str 4, 52525 Waldfeucht-Haaren, Germany, X X
GM7 NNS A C Strachan, Rose Lea, Mormond View, New Leeds, Peterhead, AB42 8HX [IO97BL, NK04]
G7 NNT M D Evans, 3 Dalkeith Cl, Bransholme, Hull, HU7 5AS [IO93UT, TA13]
G7 NNU J F Wood, 19 Arbour Cres, Macclesfield, SK10 2JB [IO83WG, SJ97]
GW7 NNW A S Pritchard, 10 St. Christophers Rd, Newtown, Porthcawl, CF36 5RY [IO81DL, SS87]
G7 NNX Details withheld at licensee's request by SSL.
GM7 NOA G J M Cowan, 85 Eastwoodmains Rd, Clarkston, Glasgow, G76 7HG [IO75UT, NS55]
G7 NOB Details withheld at licensee's request by SSL.
G7 NOC P N Paterson Division of Science and Techno, Division Science & Technology, North Oxon College, Broughton Rd, Banbury, OX16 9QA [IO92HB, SP44]
G7 NOD Details withheld at licensee's request by SSL.
G7 NOF D D Hall, 214 Ashby Rd, Loughborough, LE11 3AG [IO92JS, SK51]
G7 NOI G P Evans, 13A Regent St., Longridge, Preston, PR3 3FH [IO83QT, SD63]
G7 NOK Details withheld at licensee's request by SSL.
G7 NOP Details withheld at licensee's request by SSL.
G7 NOQ J Howarth, 270 Bolton Rd, Radcliffe, Manchester, M26 3GW [IO83TN, SD70]
G7 NOT Details withheld at licensee's request by SSL.
GI7 NOW R G McMaster, 1 Willowvale Dr, Islandmagee, Larne, BT40 3SF [IO74DS, J49]
G7 NOX Details withheld at licensee's request by SSL.
G7 NOY J L Macfarlane, 10 Sydney Building, Bathwick, Bath, BA2 6BZ [IO81TJ, ST76]
GM7 NOZ Details withheld at licensee's request by SSL.
G7 NPG D Moon, Lower House Farm, Towngate, Eccleston, Chorley, PR7 5QS [IO83PP, SD51]
G7 NPJ S Wall, 578 Manchester Rd, Hollinwood, Oldham, OL9 7LS [IO83WM, SD90]
G7 NPK Details withheld at licensee's request by SSL.
G7 NPL C Daniel, 24 Canterbury Rd, Dewsbury, WF12 7LA [IO93EQ, SE22]
GM7 NPO Details withheld at licensee's request by SSL.
G7 NPW D Caulton, 21 Main St., Horsley Woodhouse, Ilkeston, Derbyshire, DE7 6AU [IO93HA, SK34]
G7 NPY P Martell, 40 Tonfield Rd, Sutton, SM3 9JP [IO91VJ, TQ26]
G7 NQB J J Dignan, 4 Forest Dr, Standish, Wigan, WN6 0SG [IO83PO, SD51]
G7 NQE Details withheld at licensee's request by SSL.
G7 NQG Details withheld at licensee's request by SSL.
GW7 NQI Details withheld at licensee's request by SSL.
G7 NQN V J Criswick, 34 The Grove, Meadowside, Jarrow, NE32 4RZ [IO94GW, NZ36]
G7 NQO Details withheld at licensee's request by SSL.
GM7 NQP G Kinnell, 22 Greenbrae Cres, Bridge of Don, Aberdeen, AB23 8NJ [IO87WE, NJ91]
G7 NQQ K G R Hopkins, 1 Hayley Rd, Lancing, BN15 9EL [IO90UT, TQ10]
G7 NQS Details withheld at licensee's request by SSL.
G7 NQU J A Varnham, 1 Burgin Rd, Anstey, Leicester, LE7 7FA [IO92JQ, SK50]
G7 NQV R S Roberts, 58 Botany Rd, Walsall, WS5 4NE [IO92AN, SP09]
G7 NQX J H Bailes, 48 Harlech Cl, Eston, Middlesbrough, TS6 9SZ [IO94KN, NZ51]
GM7 NQY Details withheld at licensee's request by SSL.
G7 NQZ D Harrison, 18 The Cres, Eaglescliffe, Stockton on Tees, TS16 0JB [IO94HM, NZ41]
G7 NRE Details withheld at licensee's request by SSL.
G7 NRG P Atkinson, 8 Thanet Terr, Appleby in Westmorland, CA16 6TU [IO84SN, NY62]
G7 NRH Details withheld at licensee's request by SSL.
G7 NRI Details withheld at licensee's request by SSL.
G7 NRK Details withheld at licensee's request by SSL.
G7 NRM H W Tombs, 3 The Sadlers, Tilehurst, Reading, RG31 6QZ [IO91LL, SU67]
G7 NRN P J Rose, 25 Lunedale Ave, Tollesby, Middlesbrough, TS5 7LA [IO94JN, NZ41]
G7 NRO C Flanagan, 21 Pentland Ave, Billingham, TS23 2PG [IO94IO, NZ42]
G7 NRR P P Morris, Antler Cottage, High St., Scaldwell, Northampton, NN6 9JS [IO92NI, SP77]
G7 NRS A W Saunders, 14 Heaton Ct, Ostrich Ln, Prestwich, Manchester, M25 1JJ [IO83UM, SD80]
G7 NRT P A Walden, 2 The Oaks, Ashill, Thetford, IP25 7AN [JO02JO, TF80]
GM7 NRV R A Wheeldon, Sunny Acres, Oldhill Wood, Studham, Dunstable, LU6 2NE [IO91RU, TL01]
G7 NRX B M Bentham, 89 Westborough Way, Hull, HU4 7SW [IO93TR, TA02]
G7 NSA C Schofield, 45 Church Hill, Royston, Barnsley, S71 4NQ [IO93GO, SE31]
GI7 NSC Details withheld at licensee's request by SSL.
GM7 NSG D J Stewart, Leffnoll Cottage, Cairnryan, DG9 8QU [IO74LW, NX06]
G7 NSK P P Blunden, Green Croft, Fiskerton Rd, Reepham, Lincoln, LN3 4EB [IO93SG, TF07]
G7 NSM A J Bull, 18 Meon Rd, Bournemouth, Dorset, BH7 6PN [IO90CR, SZ19]
G7 NSP F C S Kok, 31 Oakdale Glen, Harrogate, HG1 2JY [IO93FX, SE25]
G7 NSQ R A Crallan, Dawn Cottage, Sutton Grange, Ripon, North Yorks, HG4 3JZ [IO94FD, SE27]
G7 NSR K Hollinshead obo North Staffs Raynet Group, 35 Parkside Dr, May Bank, Newcastle, ST5 0NL [IO83VA, SJ84]
GM7 NSS T S Goody Strathallan School Radio Club, Strathallan School, Forgandenny, Perth, PH2 9EG [IO86GI, NO01]
G7 NSU G Elliott, 70 Stradbroke Rd, Richmond, Sheffield, S13 8SQ [IO93HI, SK48]
G7 NSY Details withheld at licensee's request by SSL.
G7 NSZ N G A Powell, 42 Parkfield Rd, Pucklechurch, Bristol, BS17 3PS [IO81SL, ST67]
GW7 NTA T Blunsdon, 3 Railway Terr, Aberbeeg, Abertillery, NP3 2AD [IO81KQ, SO20]
G7 NTB Details withheld at licensee's request by SSL.
G7 NTI A Wood, 23 Cross Ryecroft St., Ossett, WF5 9EW [IO93FQ, SE22]
GM7 NTJ D Gray, 3 Castle Gdns, Dingwall, IV15 9HY [IO77SO, NH55]
G7 NTO Details withheld at licensee's request by SSL.
GW7 NTP P A Banks, Ysgol Emrys Ap Iwan, Faenol Ave, Abergele, Clwyd, LL22 7HE [IO83FG, SH97]
GU7 NTR Details withheld at licensee's request by SSL.
G7 NTS Details withheld at licensee's request by SSL.
GW7 NTU Details withheld at licensee's request by SSL.
G7 NTV Details withheld at licensee's request by SSL.
G7 NTY M J Sables, 1 Langley Hill Cl, Hurst, Tilehurst, Reading, RG3 4EJ [IO91LJ, SU66]
GW7 NTZ D H Green, 116 Crogen, Chirk, Wrexham, LL14 5BJ [IO82LW, SJ23]
G7 NUB B A Pope, 12 Wilton Cl, Christchurch, BH23 2PL [IO90CR, SZ19]
G7 NUC D G A Poulet, 19 Elmside, Exeter, EX4 6LW [IO80FR, SX99]
G7 NUE Details withheld at licensee's request by SSL.
G7 NUG T M Brown, 125 Godinton Rd, Ashford, TN23 1LN [JO01KD, TR04]
G7 NUK M J Bethell, Victoria Mews, 50 Avern Cl, Tipton, DY4 7ND [IO82XM, SO99]
G7 NUM M J Exton, Apollo, Tarry Hill, Swineshead, Boston, Lincs, PE20 3LL [IO92WW, TF24]
G7 NUN S R Latham, 4 Shaston Rd, Stourpaine, Blandford Forum, DT11 8TA [IO80VV, ST80]
G7 NUO J Popplewell, 25 The Poplars, Bramhope, Leeds, LS16 9DL [IO93EV, SE24]
GM7 NUQ C J Mair, 47 Durness St., Thurso, KW14 8BQ [IO88FO, ND16]
G7 NUR A S Gale, 115 Bruce St., Swindon, SN2 2EN [IO91CN, SU18]
G7 NUT Details withheld at licensee's request by SSL.
G7 NUU Details withheld at licensee's request by SSL.
G7 NUV K D Browning, 9 Fowbridge Gdns, Walford, Coughton, Ross on Wye, HR9 5RZ [IO81QV, SO52]
G7 NUX J Allen, 18 Clover Braes, Donaghcloney, Craigavon, BT66 7LQ [IO64UJ, J15]
G7 NVA Details withheld at licensee's request by SSL.
G7 NVD M Neale, 57 Willis Rd, Haddenham, Aylesbury, HP17 8HG [IO91MS, SP70]
G7 NVE Details withheld at licensee's request by SSL.
G7 NVG C J C Park, Flat 4, Corrow Farm, Lochgoilhead, Cairndow, Argyll, PA24 8AD [IO76NE, NN10]
G7 NVH R A Rigley, 91 Melford Rd, Bilborough, Nottingham, NG8 4AN [IO92JX, SK54]
G7 NVI P Lancaster, 33 Dinsdale Gdns, Rustington, Littlehampton, BN16 3NH [IO90RT, TQ00]
G7 NVJ Details withheld at licensee's request by SSL.
G7 NVK Details withheld at licensee's request by SSL.
G7 NVL M A Brook, The Gatehouse, Claypit Ln, Carlton, Poole, DN14 9PR [IO93LR, SE62]
G7 NVM Rd M R Skelton, 122 Woodcote, Killay, Swansea, SA2 7AU [IO71XO, SS59]
G7 NVN D G Priston, 5 Kings Way, Manor Lodge Rd, Rowlands Castle, PO9 6AZ [IO90MV, SU71]
G7 NVO Details withheld at licensee's request by SSL.
G7 NVS D M Poulton, 33 Ark Royal Cl, Plymouth, PL5 1EZ [IO70VJ, SX45]
G7 NVT J R Andrews, 36 Haighs Cl, Chatteris, PE16 6HU [JO02AK, TL38]
G7 NVZ Details withheld at licensee's request by SSL.
G7 NWB J W Clark, 34 Stationfields, Halwill Junction, Beaworthy, EX21 5XX [IO70VS, SS40]
G7 NWC R Booth, 12 Meerbrook Pl, Ilkeston, DE7 4NY [IO92IW, SK44]
G7 NWG M Rosevear, Bosanath Mill, Bosanath Valley, Mawnan Smith, Falmouth, TR11 5LL [IO70KC, SW72]
G7 NWH A H Hill, Hillcrest School & Comm Col., Simms Ln, Netherton, Dudley, West Midlands, DY3 0PB [IO82WM, SO99]
GI7 NWI Details withheld at licensee's request by SSL.
G7 NWJ Details withheld at licensee's request by SSL.
G7 NWM W A Riley, 25 Oak Tree Rd, Branton, Doncaster, DN3 3QB [IO93LM, SE60]
G7 NWR B J Watson North Wiltshire Raynet Group, 20 St. Marys Gdns, Hilperton Marsh, Trowbridge, BA14 7PG [IO81VI, ST85]
G7 NWS J T Down, 2 Huntingdon Way, Clacton on Sea, CO15 4EZ [JO01NT, TM11]
G7 NWU K Franklin, 19 Castle Mead, Kings Stanley, Stonehouse, GL10 3LB [IO81UR, SO80]
G7 NWX Details withheld at licensee's request by SSL.
G7 NWY Details withheld at licensee's request by SSL.
G7 NWZ Details withheld at licensee's request by SSL.
G7 NXA K Keenan, 25 Harris Rd, Harpur Hill, Buxton, SK17 9JS [IO93BF, SK07]
G7 NXB Details withheld at licensee's request by SSL.
G7 NXC Details withheld at licensee's request by SSL.
GM7 NXI M W Hepburn, Toll House, Corse, Lumphanan, Banchory, Kincardineshire, AB31 4RY [IO87PD, NJ50]
G7 NXN Details withheld at licensee's request by SSL.
G7 NXQ R P Horton Christchurch Amateur Radio Clu, 31 Furze Ln, Purley, CR8 3EJ [IO91WI, TQ36]
G7 NXV D Ross, 37 Cartmell Dr, Leeds, LS15 0NQ [IO93GT, SE33]
G7 NXW Details withheld at licensee's request by SSL.
GM7 NYB N Macfarlane, 3 Kilmore Terr, Dervaig, Tobermory, Isle of Mull, PA75 6QN [IO66VO, NM45]
G7 NYC Details withheld at licensee's request by SSL.
G7 NYD B P Collinge, 6 Wyresdale Caravan Park, Kiln Ln, Hambleton, Poulton-le-Dylde, Lancs, FY6 9DZ [IO83MV, SD34]
G7 NYF P D Ridley, 11 Thorney Cl, Fareham, PO14 3AF [IO90JU, SU50]
G7 NYK Details withheld at licensee's request by SSL.
GW7 NYN L Bradley, 25 Fairfield Cl, Penrhyn Bay, Llandudno, LL30 3HU [IO83CH, SH88]
G7 NYO N A Hamer, 75 Somers Rd, Stockport, SK5 6SL [IO83WK, SJ89]
G7 NYP N R Negus Gloucester Repeater Group, 41 Oxstalls Ln, Longlevens, Gloucester, GL2 9HP [IO81VV, SO81]
GW7 NYR R L Williams Gogledd Cymru Packet Group, The Basement, 9 Charlton St., Llandudno, LL30 2AA [IO83CH, SH78]
G7 NYU Details withheld at licensee's request by SSL.
G7 NYV N A Bush, Llamedos, Occupation Ln, New Bollingbroke, Boston, Lincs, PE22 7NQ [IO93XC, TF35]
G7 NYW D Evans, Edith Neville School, 174 Ossulston St., London, NW1 1DN [IO91WM, TQ28]
G7 NYY E Steed, 1 Falcon Green, Farlington, Portsmouth, PO6 1LW [IO90LU, SU60]
G7 NZC S J Butt, 30 Jessopp Ave, Bridport, DT6 4AN [IO80PR, SY49]
G7 NZE Details withheld at licensee's request by SSL.
G7 NZF A J Thornton, 2 Thorngate, Penwortham, Preston, PR1 0XL [IO83PR, SD52]
G7 NZH A J Tietjen, 75 Rushleydale, Springfield, Chelmsford, CM1 6JX [JO01FR, TL70]
GM7 NZI R Simpson, 53 Jedworth Ave, Drumchapel, Glasgow, G15 7QE [IO75TV, NS57]
G7 NZJ J S Hayward, 32 Brendon Gr, Bingham, Nottingham, NG13 8TN [IO92MW, SK63]
G7 NZK M Aylett, 22 Pleasant Hill, Tadley, RG26 4LS [IO91KI, SU56]
G7 NZM G M Clifton, 21 Park Rd, Hilton Est, Featherstone, Wolverhampton, WV10 7HS [IO82XP, SJ90]
G7 NZO S I Bate, 26 Ilfracombe Way, Lower Earley, Reading, RG6 3AQ [IO91NK, SU77]
G7 NZQ Details withheld at licensee's request by SSL.
G7 NZR A Haslam, The Goldings, Hayton Village, Nr.Brampton, Carlisle, Cumbria, CA4 9JA [IO84OV, NY55]
G7 NZS Details withheld at licensee's request by SSL.

G7 NZU L Elliott, 14 Carlton Ave, Hull Rd, York, YO1 3JZ [IO93LW, SE65]
G7 NZV R Easting, 48 Church Rd, Fordham, Colchester, CO6 3NE [JO01JW, TL92]
G7 NZY C R Wells, 6 Craister Ct, Cambridge, CB4 2SH [JO02BF, TL46]
G7 OAA J C Davies, 32 Oakwood Dr, Ainsdale, Southport, PR8 3HZ [IO83LO, SD31]
G7 OAC Details withheld at licensee's request by SSL.
G7 OAE J Bate, 20 Lilydale Rd, Stoke on Trent, ST2 9HH [IO83WA, SJ94]
GM7 OAF E G Capstick, 24 Dalmore Cres, Helensburgh, G84 8JP [IO76OA, NS28]
G7 OAH K J Adams, 28B Padnell Ave, Cowplain, Waterlooville, PO8 8DY [IO90LV, SU61]
G7 OAI J E Cannell, 53 Thimble Cl, Hurstead, Rochdale, OL12 9QP [IO83WP, SD91]
G7 OAK Details withheld at licensee's request by SSL.
G7 OAL P M Knapper, 19 Montgomery St., Hollinwood, Oldham, OL8 3PR [IO83WM, SD90]
G7 OAM G H Tollefson, 41 Palmer Rd, Salisbury, SP2 7LX [IO91CB, SU13]
G7 OAP G V Kirkham, 11 Penn Rd, Richmond, DL10 4BE [IO94DJ, NZ10]
G7 OAS A R J White, 6 Lansdown Gr, Chippenham, SN15 1TE [IO81WL, ST97]
G7 OAT Details withheld at licensee's request by SSL.
G7 OAU Details withheld at licensee's request by SSL.
G7 OAV Dr A L Holohan, 8 School House Terr, Kirk Deighton, Wetherby, LS22 4EH [IO93HW, SE35]
GM7 OAW A H Irvine, 41 Craighead Rd, Bishopton, PA7 5DT [IO75SV, NS47]
G7 OAZ Details withheld at licensee's request by SSL.
G7 OBA Details withheld at licensee's request by SSL.
G7 OBC R G Hudman, 27 Egerton Rd, Streetly, Sutton Coldfield, B74 3PQ [IO92BN, SP09]
G7 OBD M J Peach, 48 Melrose Ave, Portslade, Brighton, BN41 2LS [IO90VU, TQ20]
G7 OBE D R Peach, 48 Melrose Ave, Portslade, Brighton, BN41 2LS [IO90VU, TQ20]
G7 OBF J Aston, 26 Sandhills Rd, Reigate, RH2 7RJ [IO91VF, TQ24]
G7 OBG Details withheld at licensee's request by SSL.
G7 OBH Details withheld at licensee's request by SSL.
GM7 OBI R Webster, 1 Hallyards Farm Cottage, Kirkliston, EH29 9DZ [IO85HW, NT17]
G7 OBK Details withheld at licensee's request by SSL.
GM7 OBM D A Macpherson, 40 Murroch Cres, Alexandria, G83 9QA [IO75RX, NS47]
G7 OBN C L Sermons, 17 Wellside, Marks Tey, Colchester, CO6 1XG [JO01JV, TL92]
G7 OBO Details withheld at licensee's request by SSL.
G7 OBP G Turner, 11 Royds Cres, Rhodesia, Worksop, S80 3HF [IO93KH, SK58]
G7 OBR M D Biddles, 12 Manor Gdns, Glenfield, Leicester, LE3 8FN [IO92JP, SK50]
G7 OBS M P Simkins, 32 Hertford Rd, Enfield, EN3 5AN [IO91XP, TQ39]
G7 OBU E L Frazer, 42 Oliver Dr, Calcot, Reading, RG3 4XN [IO91LJ, SU66]
G7 OBW D Wilson, 12 New St., Elworth, Sandbach, CW11 3JF [IO83TD, SJ76]
G7 OBX N D Finbow, 10 Old Heath Rd, Colchester, CO1 2ES [JO01KV, TM02]
G7 OBY D R Burrows, Newborough End, Newborough, Burton on Trent, Staffs, DE13 8SR [IO92CT, SK12]
G7 OBZ Details withheld at licensee's request by SSL.
G7 OCA Details withheld at licensee's request by SSL.
G7 OCB Details withheld at licensee's request by SSL.
G7 OCC A H Graham, 19 Talbot Rd, Rushden, NN10 9NS [IO92QG, SP96]
G7 OCD Details withheld at licensee's request by SSL.
G7 OCG C S Ghiotti, 15 The Grove, Upminster, RM14 2ER [JO01CN, TQ58]
G7 OCH J Doy, 11A Shrubland Ave, Ipswich, Suffolk, IP1 5EA [JO02NB, TM14]
G7 OCI J A Rudd, 23 Grange Gdns, Ware, SG12 9NE [IO91XT, TL31]
G7 OCJ J Roberts, Llwyn-Celyn, Bedwellty, Blackwood, Gwent, NP2 0BD [IO81JQ, SO10]
G7 OCK B Phillipson, 27 Victoria Ave, Crook, DL15 9DB [IO94DR, NZ13]
G7 OCL G C Lovett, 20 Mapledurham Walk, Maidenhead, SL6 7UL [IO91PM, SU88]
G7 OCM Details withheld at licensee's request by SSL.
G7 OCO Details withheld at licensee's request by SSL.
G7 OCP P Adshead, 5 Hazel Ave, Killamarsh, Sheffield, S31 8GD [IO93GJ, SK38]
G7 OCQ W G Horwood, Willow Lodge, Ipswich Rd, Yaxley, Eye, IP23 8BX [JO02NH, TM17]
G7 OCS Details withheld at licensee's request by SSL.
G7 OCT Details withheld at licensee's request by SSL.
GM7 OCU G N Rule, 105 Causewayside, Edinburgh, EH9 1QG [IO85JW, NT27][Op: Graham Rule.]
G7 OCV T M Pirrie, Walnut Thatch, Radway, Warks, CV35 0UE [IO92GD, SP34]
G7 OCW Details withheld at licensee's request by SSL.
G7 OCX J M Turton, 60 Shafton Ln, Leeds, LS11 9RE [IO93FS, SE23]
G7 OCY D K Norton, 52 Letchworth Rd, Leicester, LE3 6FG [IO92KP, SK50]
GM7 ODA M G W Gibbons, 4 Elm Gdns, Bearsden, Glasgow, G61 3BH [IO75UW, NS57]
G7 ODB R J Evans, 18 Lilac Cl, Keyworth, Nottingham, NG12 5DN [IO92LU, SK63]
G7 ODD Details withheld at licensee's request by SSL.
G7 ODE Details withheld at licensee's request by SSL.
G7 ODG R L Beadle, 8 Erica Gdns, Croydon, CR0 8LG [IO91XI, TQ36]
G7 ODL P I Dodd, 18 Lyndon Mead, Sandridge, St. Albans, AL4 9EX [IO91US, TL11]
G7 ODM G G Wane, 2 Plantation Ave, Arnside, Carnforth, LA5 0HT [IO84OE, SD47]
G7 ODO Details withheld at licensee's request by SSL.
GW7 ODP M A Jones, 50 Upper Kings Head Rd, Gendros, Swansea, SA5 8BR [IO81AP, SS69]
G7 ODQ Details withheld at licensee's request by SSL.
G7 ODR D J Cartwright, Halum, The Cross, Bonsall, Matlock, Derbyshire, DE4 2AA [IO93FC, SK25]
G7 ODT C J Wright, 6 Lee Ct, Eynesbury, St. Neots, Huntingdon, PE19 2SZ [IO92UF, TL16]
G7 ODW T M Green, 1 Forest Rd, Blidworth, Mansfield, NG21 0SJ [IO93KC, SK55]
G7 ODX R J Kirby, 46 Hedley Dr, Brimington, Chesterfield, S43 1BF [IO93HG, SK37]
G7 ODY N C Peck, 46 Hedley Dr, Brimington, Chesterfield, S43 1BF [IO93HG, SK37]
G7 OEA P M Foulkes, 26 Moorfield Rd, Crosby, Liverpool, L23 9UD [IO83LL, SD30]
G7 OEB G H Brookes, 38 Mersey St., St. Helens, WA9 2JX [IO83PK, SJ59]
G7 OEC Details withheld at licensee's request by SSL.
G7 OED R C Cowles, 58 Wells Gdns, Basildon, SS14 3QS [JO01FN, TQ78]
GW7 OEE D A J Rees, 1 Bessant Cl, Brookfield Park, Cowbridge, CF7 7HP [IO81HN, ST08]
G7 OEF D W Hills, 156 Fronks Rd, Dovercourt, Harwich, CO12 4EF [JO01PW, TM23]
G7 OEI M Kiddy, 39 Garden Dr, Brampton, Barnsley, S73 0TN [IO93HM, SE40]
G7 OEK Details withheld at licensee's request by SSL.
G7 OEL Details withheld at licensee's request by SSL.
G7 OEM A Hulme, 142 Adelaide St, Blackpool, Lancs, FY1 4LU [IO83LT, SD33]
G7 OEP Details withheld at licensee's request by SSL.
G7 OEQ Details withheld at licensee's request by SSL.
G7 OES Details withheld at licensee's request by SSL.
G7 OET L M Hitchin, 40 Methuen Ave, Fulwood, Preston, PR2 9QX [IO83PS, SD53]
G7 OEV R W Ledger, 34 Little Norton Ln, Norton, Sheffield, S8 8GD [IO93GH, SK38]
G7 OEW S T Yohn, 16 Barehill St., Littleborough, OL15 9BL [IO83WP, SD91]
G7 OEY G J Hodges, 12 Linwal Ave, Houghton on The Hill, Leicester, LE7 9HD [IO92LP, SK60]
G7 OFB J W Selley, 5 Pine Gr Mews, Pine Gr, Weybridge, KT13 9BD [IO91SI, TQ06]
G7 OFC Details withheld at licensee's request by SSL.
G7 OFF Details withheld at licensee's request by SSL.
G7 OFI P Smith, 11 Springwell Cl, Maltby, Rotherham, South Yorks, S66 7HG [IO93JK, SK59]
G7 OFJ D S McFarlane, 4405 Twin Oaks Ct, Monmouth Junction, New Jersey 08852-2257, USA, KT22 8QJ [IO91UG, TQ15]
G7 OFM R A Squires, 84 Queens Ave, Meols, Wirral, L47 0NA [IO83KJ, SJ28]
G7 OFT Details withheld at licensee's request by SSL.
G7 OFU N M Patterson, 4 Lytham Cl, Fulwood, Preston, PR2 2HB [IO83PS, SD53]
G7 OFV T W Wordsworth, 61 Crane Rd, Kimberworth Park, Rotherham, S61 3HN [IO93HK, SK49]
G7 OGA Details withheld at licensee's request by SSL.
G7 OGB R J Hutton, 57 Sandy Ln, Upton, Poole, BH16 5EJ [IO80XR, SY99]
G7 OGG Details withheld at licensee's request by SSL.
G7 OGJ A P Craft, 16 Marsham Cl, Aylesbury, HP21 9XB [IO91OT, SP81]
G7 OGK A R Ellis, 10 Jarvis Brook Cl, Bexhill on Sea, TN39 3UQ [JO00FU, TQ70]
G7 OGL D J Parker, 10 Priory Rd, Wollaston, Wellingborough, NN29 7PW [IO92PG, SP96]
G7 OGM F V Parker, 5 Wyndham St. East, North Rd, Plymouth, PL1 5HE [IO70WI, SX45]
G7 OGN D J Arnold, 18 Pheasant Way, Spring Park, Kingsthorpe, Northampton, NN2 8BJ [IO92NG, SP76]
G7 OGO K P Mann, 89 Wootton Village, Boars Hill, Oxford, OX1 5HW [IO91IR, SP40]
G7 OGP M J Griffiths, 1 Heol Nest, Whitchurch, Cardiff, CF4 1SY [IO81JM, ST18]
G7 OGR D J Arthurs, 32 Lowfield Ave, Greasbrough, Rotherham, S61 4PD [IO93HK, SK49]
G7 OGS D J Rushmer, 21 Upper End Rd, Peak Dale, Buxton, SK17 8AU [IO93BG, SK07]
G7 OGT T S McDonald, 64 Hollowood Ave, Littleover, Derby, DE23 6JD [IO92FV, SK33]
G7 OGX Details withheld at licensee's request by SSL.
GM7 OHB J T Deas, 81 Speirs Rd, Bearsden, Glasgow, G61 2LT [IO75UV, NS57]
G7 OHD P T Martindale, 4 The Crayke, Marton Fields, Bridlington, YO16 5YP [IO94VC, TA16]
G7 OHE B Overend, Flat 14 Edentown Ct, Eden St, Carlisle, Cumbria, CA3 9LQ [IO84MV, NY35]
G7 OHI Dr J R Richardson, 47B Gore Rd, London, E9 7HN [IO91XM, TQ38]
G7 OHK D J Hooker, 137 Cross St., Arnold, Nottingham, NG5 7AT [IO93KA, SK54]
G7 OHM R A Jarvis, 8 Walker St., Cockermouth, CA13 0AB [IO84HP, NY13]
G7 OHO J D Hislop, 10 Park Wood Cl, Broadstairs, CT10 2XN [JO01RI, TR36]
GM7 OHQ S R Houston, 2 Brownlee Rd, Law, Carluke, ML8 5JD [IO85BR, NS85]
G7 OHS J Wilkinson, 17 Oaks Cl, Leatherhead, KT22 7SH [IO91UH, TQ15]
G7 OHT Details withheld at licensee's request by SSL.
G7 OHU Details withheld at licensee's request by SSL.
G7 OHW Details withheld at licensee's request by SSL.
G7 OHX Details withheld at licensee's request by SSL.
G7 OIA M A Larcombe, 52 Orchard Rd, Burgess Hill, RH15 9PL [IO90WW, TQ31]
G7 OIC H B Smith, Flat 1, 29 Wensley Cres, Cantley, Doncaster, DN4 6QL [IO93LM, SE60]
G7 OII Details withheld at licensee's request by SSL.
G7 OIJ Details withheld at licensee's request by SSL.
GW7 OIK D A Todd, Ty'R Cae, Gwernogle Bach, Brechfa, Carmarthenshire, SA32 7SA [IO71WX, SN53]
G7 OIL H Blakely, 21 Water Ln, Lowestoft, NR32 2NH [JO02UL, TM59]
GM7 OIN J H Cowan, 1 Treebank Cres, Ayr, KA7 3NF [IO75QK, NS32]
G7 OIR Details withheld at licensee's request by SSL.
G7 OIT I P Guest, 1 Conifer Rise, Banbury, OX16 7FP [IO92HB, SP44]
GW7 OIV J R Hughes, Watsmead, Kiln Ln, Hope, Wrexham, LL12 9PH [IO83LC, SJ35]
G7 OIW R Shaw, 35 Barrington Ave, Hull, HU5 4AZ [IO93TS, TA03]
GW7 OIX Details withheld at licensee's request by SSL.
GM7 OIZ Details withheld at licensee's request by SSL.
G7 OJA P Mann, Jasmine Cottage, 11 New Mills Rd, Hayfield, High Peak, Derbyshire, SK22 2JG [IO93AJ, SK08]
G7 OJB J E Butterworth, Woodcombe, Gr Rd, Blue Anchor, Minehead, TA24 6JX [IO81HE, ST04]
G7 OJC R Marston, 41 Drake Rd, Willesborough, Ashford, TN24 0UZ [JO01KD, TR04]
G7 OJD P E Fisher, 5 Friars Cl, Cheadle, Stoke on Trent, ST10 1AT [IO92AX, SK04]
G7 OJE J Dyson, 9 Green View, Lymm, WA13 9RB [IO83SJ, SJ68]
G7 OJH Details withheld at licensee's request by SSL.
G7 OJI C A N McAlister, 24 Lodge Gr, Yateley, Camberley, Surrey, GU17 7AD [IO91OH, SU85]
GM7 OJJ J A Alexander, Newton of Kinmundy Cottage, Kinmundy, By Mintlaw, Grampian, AB42 8AY [IO97BL, NK04]
G7 OJM Dr G J Leese, 15 Ewden Way, Pogmoor, Barnsley, S75 2JW [IO93FN, SE30]
G7 OJN Details withheld at licensee's request by SSL.
G7 OJO Details withheld at licensee's request by SSL.
G7 OJP Details withheld at licensee's request by SSL.
G7 OJQ A E Foster, 18 Talbot St., Hazel Gr, Stockport, SK7 4BH [IO83WJ, SJ98]
G7 OJR Details withheld at licensee's request by SSL.
G7 OJT E J Wolfenden, 6 Ward Cl, Wokingham, RG40 1XE [IO91OK, SU86]
G7 OJU F G Dixon, 9 Lincoln Rd, Fenton, Lincoln, LN1 2EP [IO93PH, SK87]
G7 OJW Details withheld at licensee's request by SSL.
G7 OJX K J Trigg, 41 Veasey Rd, Hartford, Huntingdon, PE18 7TA [IO92WI, TL27]
G7 OJY V A Holyoake, 14 Maudlin Ct, de Cham Rd, St. Leonards on Sea, TN37 6JY [JO00GU, TQ80]
G7 OJZ J A Neale, 25 Ismere Way, Hill, Kidderminster, DY10 2YG [IO82VJ, SO87]
GM7 OKA M Sutcliffe, 65 Larchfield Ave, Newton Mearns, Glasgow, G77 5QN [IO75US, NS55]
G7 OKB J D C Taylor, 8 Kenmore Rd, Swarland, Morpeth, NE65 9JS [IO95DH, NU10]
G7 OKC K J Robertson, 1359 Melton Rd, Syston, Leicester, LE7 2EP [IO92LQ, SK61]
GM7 OKD Details withheld at licensee's request by SSL.
G7 OKE A D Smith, 17A Mulroy Rd, Sutton Coldfield, B74 2QA [IO92CN, SP19]
G7 OKF M Hawkins, 23 Micklehill Dr, Shirley, Solihull, B90 2PU [IO92CJ, SP17]
G7 OKI W M Cornish, 33 Munster Rd, North End, Portsmouth, PO2 9BS [IO90LT, SU60]
GI7 OKJ B McKittrick, 13 Fern Gr, Bangor, BT19 1FG [IO74DP, J48]
GW7 OKM Details withheld at licensee's request by SSL.
G7 OKO F J Webb, 50 Hassam Ave, Newcastle, ST5 9ET [IO83VA, SJ84]
G7 OKR J P Campbell, 3 The Leas, Thingwall, Wirral, L61 1BA [IO83LI, SJ28]
G7 OKT S P Smith, 17 Thackers Way, Deeping St. James, Peterborough, PE6 8HP [IO92UQ, TF11]
G7 OKU Details withheld at licensee's request by SSL.
G7 OKV K Porter, 14 Meadow Cross, Waltham Abbey, EN9 3DJ [JO01AQ, TL30]
GM7 OKX G H Chesworth, Auchinway, Skares, Cumnock, Ayrshire, KA18 2RE [IO75UK, NS51]
G7 OKY D Schofield, 54 Brighton Rd, Purley, Surrey, CR8 2LJ [IO91WI, TQ36]
G7 OKZ A Burnside, 30 Lime Cres, Sandal, Wakefield, WF2 6RY [IO93GP, SE31]
G7 OLC P A Broadhead, 17 Waterside Way, Pendeford, Wolverhampton, WV9 5LL [IO82WO, SJ90]
G7 OLD Details withheld at licensee's request by SSL.
G7 OLE K A Silkstone, 46 Royston Dr, Belper, DE56 0EL [IO93GA, SK34]
G7 OLF W G Levick, 1 Sycamore Ave, The Elms, Torksey, Lincoln, LN1 2NJ [IO93PH, SK87]
G7 OLG M P Hodge, 71 Rawcliffe Rd, Liverpool, Merseyside, L9 1AN [IO83MK, SJ39]
G7 OLH A B Barnett, 71 Winsham Gr, London, SW11 6NB [IO91WK, TQ27]
G7 OLI M J Evans, 19 Nightingale House, Connaught Mews, London, SE18 6SU [JO01AL, TQ47]
GW7 OLL M I Jones, 8 Heol Dderwen, The Coppice, Tonteg, Pontypridd, CF38 1TD [IO81IN, ST08]
GM7 OLN G G Di Ponio, 5 Milton Cres, Edinburgh, EH15 3PF [IO85KW, NT37]
G7 OLO C R Dicks, 280 Addington Rd, Irthlingborough, Wellingborough, NN9 5UT [IO92QI, SP97]
GM7 OLP Details withheld at licensee's request by SSL.
GM7 OLQ J Innes, 142 Easter Rd, Edinburgh, EH7 5RJ [IO85JX, NT27]
G7 OLT Details withheld at licensee's request by SSL.
G7 OLU W M Wood, 8 Home Farm Rd, Stanion, Kettering, NN14 1DQ [IO92QL, SP98]
G7 OLX J D Chambers, 26 Ringwood Rd, Bingham, Nottingham, NG13 8SZ [IO92MW, SK63]
G7 OLZ B Zarucki, 9 Dads Ln, Moseley, Birmingham, B13 8PG [IO92BK, SP08]
G7 OME G J Albone, 79 Drove Rd, Biggleswade, SG18 0HN [IO92UC, TL14]
G7 OMF K Gill, 358 Moor End Rd, Halifax, HX2 0RH [IO93BR, SE02]
G7 OMG J L Everitt, 55 Risborough Rd, Bedford, MK41 9QR [IO92SD, TL05]
G7 OMI J Patel, 1 The Glade, Furnace Green, Crawley, RH10 6JS [IO91VC, TQ23]
GU7 OMJ M Adams, 64 Maple Dr, Port Soif, Vale, Guernsey, GY6 8HX
G7 OMK Details withheld at licensee's request by SSL.
G7 OML K Maynard, Flat 1, 24 Madrid Rd, Guildford, GU2 5NU [IO91QF, SU94]
GM7 OMM R S Stroud, 376 Auchmill Rd, Bucksburn, Aberdeen, AB2 9NL [IO87TH, NJ72]
G7 OMN J D Eyes, 96 Haig Ave, Southport, PR8 6JX [IO83MP, SD31]
G7 OMO M R Legg, 8 Horse Fields, Peacmarsh, Gillingham, Dorset, SP8 4UQ [IO81UA, ST82]
G7 OMP Details withheld at licensee's request by SSL.
G7 OMT Details withheld at licensee's request by SSL.
GM7 OMU Details withheld at licensee's request by SSL.
GM7 OMV Details withheld at licensee's request by SSL.
G7 ONB Details withheld at licensee's request by SSL.
GI7 ONC C W Archer, 3 Pinehill Ct, Bangor, BT19 6SG [IO74EP, J58]
G7 OND I T Hilton, 129 Upton Rd, Moreton, Wirral, L46 0SQ [IO83KJ, SJ28]
G7 ONE R M Jacobs, Rose Mount, Gr Rd, Ventnor, PO38 1TH [IO90JO, SZ57]
G7 ONF M P Whitley, Malton Gate, Langtoft, Driffield, East Yorks, YO25 0TN [IO94SC, TA06]
G7 ONG S C W Guscin, 11 Hatton Cl, Bradford, BD6 1JS [IO93CS, SE12]
G7 ONH Details withheld at licensee's request by SSL.
G7 ONI J M Churchill, 68 Anthony Rd, Woodside, London, SE25 5HB [IO91XJ, TQ36]
GM7 ONJ A Martin, Abercorn, 11 Langlee Pl, Broughty Ferry, Dundee, DD5 3RP [IO86NL, NO43]
G7 ONK Details withheld at licensee's request by SSL.
G7 ONL Details withheld at licensee's request by SSL.
G7 ONM T R Eyre, St. Michael Mead, The Common, Barton Turf, Norwich, NR12 8BA [JO02RR, TG32]
G7 ONO A J Hobbis, 45 Deene Cl, Adderbury, Banbury, OX17 3LD [IO92IA, SP43]
G7 ONR Details withheld at licensee's request by SSL.
G7 ONV D N Das, 4 Farcliff, Sprotbrough, Doncaster, DN5 7RE [IO93JM, SE50]
G7 ONZ S W Brierley, 18 Aspen Gr, Paddington, Warrington, WA1 3ET [IO83RJ, SJ68]
G7 OOA Details withheld at licensee's request by SSL.
G7 OOD Details withheld at licensee's request by SSL.
G7 OOE J S Bone, 3 Claremont Pl, Gateshead, NE8 1TL [IO94EW, NZ26]
G7 OOF E Cottle, 29 Chalk Hill, West End, Southampton, SO18 3BY [IO90HW, SU41]
G7 OOG K Whitney, Gr View, Upthorpe Rd, Stanton, Bury St. Edmunds, IP31 2AP [JO02KH, TL97]
G7 OOH K Mc Allister, 109A Kaye Ln, Almondbury, Huddersfield, HD5 8XT [IO93CP, SE11]
G7 OOI R Phillipson, 22 Bagmere Cl, Brereton Green, Brereton, Sandbach, CW11 1SG [IO83UE, SJ76]
G7 OOJ Details withheld at licensee's request by SSL.
G7 OOK C J Roberson, 111 Norwood Rd, London, SE24 9AE [IO91WK, TQ37]
G7 OOL Details withheld at licensee's request by SSL.
GI7 OOM D J Magowan, 35 Princeton Ave, Lurgan, Craigavon, BT66 8LW [IO64TK, J05]
G7 OOO R Clayton Scarborough SEG, 9 Green Island, Irton, Scarborough, YO12 4RN [IO94SF, TA08]
G7 OOP A J Constantine, 1 The Spinney, Bexhill on Sea, TN39 3SW [JO00FU, TQ70]
GM7 OOR Details withheld at licensee's request by SSL.
G7 OOS J R Smalley, 117 Marine Dr, Rottingdean, Brighton, BN2 7GE [IO90XT, TQ30]
G7 OOU D J Witts, 4A Wilton Rd, Redhill, RH1 6QR [IO91VF, TQ24]
G7 OOV R I Nunn, 49 Lulworth Dr, Roborough, Plymouth, PL6 7DT [IO70WK, SX46]
G7 OOW R D Keen, 13 Mill View Cl, Woodbridge, IP12 4HR [JO02PC, TM24]
G7 OOX Details withheld at licensee's request by SSL.
GM7 OOY Details withheld at licensee's request by SSL.
G7 OOZ Details withheld at licensee's request by SSL.
G7 OPB A L Adams, 13 Daubigny Mead, Brize Norton, Carterton, OX18 3QE [IO91FS, SP30]
G7 OPD P E J Hundy, 101 Goodway Rd, Great Barr, Birmingham, B44 8RS [IO92BM, SP09]
G7 OPJ J A Buttery, 38 Wigmore Gdns, Worle, Weston Super Mare, BS22 9AQ [IO81MI, ST36]
G7 OPM N B Couch, 245 Wishing Tree Rd, St. Leonards on Sea, TN38 9LA [JO00GU, TQ71]
G7 OPN W E Cairns, Leadgate Cottage, Wark, Hexham, Northd., NE48 3DS [IO85UC, NY87]
G7 OPO N E Keen, 6 Naunton Rd, Woodbridge, IP12 4HP [JO02PC, TM24]

G7 OPS P J Whiting, 77 Melford Way, Felixstowe, IP11 8UH [JO01PX, TM23]
G7 OPW C T Wilkinson 1st Farnsfield Scout Group, 15 Woodland Cl, Farnsfield, Newark, NG22 8DN [IO93LC, SK65]
G7 OPY L A Kelly, Clamerkin Park, Newtown, Newport, Isle of Wight, PO30 4PD [IO90HR, SZ49]
G7 OQA A J Bahlke, 52 Adisham Green, Kemsley, Sittingbourne, ME10 2SR [JO01II, TQ96]
G7 OQB Details withheld at licensee's request by SSL.
G7 OQC D C McKenzie, 5 Druim Shiadair, Point, Shulishader, Isle of Lewis, HS2 0PP [IO68VF, NB53]
GM7 OQE G M Murray, 3 Glebelands, Tongue, Lairg, Sutherland, IV27 4XL [IO78TL, NC55]
G7 OQG S T Williams, Chestnut Rise, Jacksons Ln, Hazel Gr, Stockport, SK7 5JS [IO83WI, SJ98]
G7 OQH D J Hone, 5 Chesterfield Ave, New Whittington, Chesterfield, S43 2BX [IO93HG, SK47]
G7 OQJ M W Wrighton, 23A Cotham Park, Cotham, Bristol, BS6 6BZ [IO81QL, ST57]
G7 OQK Details withheld at licensee's request by SSL.
G7 OQL C D Broad, 96 Kingsley Ct, Penhale, Fraddon, St. Columb, TR9 6PD [IO70MJ, SW95]
G7 OQM Details withheld at licensee's request by SSL.
G7 OQO A C N Thomas, Catherine House, Millgate Rd, Whaplode Saint Catherine, Spalding, Lincs, PE12 6SF [IO92XS, TF31]
G7 OQQ S E Ayers, 20 Wytham View, Eynsham, Witney, OX8 1LU [IO91HS, SP40]
G7 OQS C E H Everitt, 61 Arnfield Rd, Withington, Manchester, M20 4AG [IO83VK, SJ89]
G7 OQT K Peacock, Gkn Defence, 106 PO Box, Hadley Castle Works, Telford, TF1 4QW [IO82SQ, SJ61]
G7 OQU Details withheld at licensee's request by SSL.
G7 OQV Details withheld at licensee's request by SSL.
G7 OQZ Details withheld at licensee's request by SSL.
G7 ORB D P Cullen, Hollytree Farmhouse, Town Drove, Quadring, Spalding, Lincs, PE11 4PU [IO92VV, TF23]
G7 ORE J Binning, 293 Perry St., Billericay, CM12 0RB [JO01FP, TQ69]
GW7 ORF Details withheld at licensee's request by SSL.
G7 ORG Details withheld at licensee's request by SSL.
G7 ORH Details withheld at licensee's request by SSL.
G7 ORI P H Bernard, 4 Boston Castle Gr, Rotherham, S60 2BA [IO93HK, SK49]
G7 ORJ A Ross, 16 Croft Rd, Kiltarlity By Kiltarlity, Beauly, IV4 7HZ [IO77SK, NH54]
G7 ORK D J Love, 9 Harefield Pl, St. Albans, AL4 9JQ [IO91US, TL10]
G7 ORL S V Platt O.R.L. ARC, Olivetti Research Lab, 24A Trumpington St, Cambridge, CB2 1QA [JO02BE, TL45]
G7 ORM Details withheld at licensee's request by SSL.
G7 ORR M A Jeffery, Monks House, Westerleigh Hill, Westerleigh, Bristol, BS17 4RD [IO81TM, ST78]
GM7 ORS J P Lorenzen, 40 Boundary Rd, Ramsgate, CT11 7NW [JO01RI, TR36]
G7 ORT D Buckley, 22B Anerley Gr, Kingstanding, Birmingham, B44 9QH [IO92BN, SP09]
G7 ORU Dr C W Leese, 15 Ewden Way, Pogmoor, Barnsley, S75 2JW [IO93FN, SE30]
G7 ORV S E Edmonds, 17 Rockmead Ave, Great Barr, Birmingham, B44 9DR [IO92BN, SP09]
G7 ORW R E Garnett-Frizelle, 17 Bridport Ave, New Moston, Manchester, M40 3WP [IO83VM, SD80]
GM7 ORX J J Lee, 2/5 Heriot Bridge, Edinburgh, EH1 2HR [IO85JW, NT27]
G7 OSB C S Baker, 36 Bracken Dr, Chigwell, IG7 5RF [JO01AO, TQ49]
G7 OSC P S Duxbury, The Paddock, Bovingdon Green, Marlow, SL7 2JL [IO91ON, SU88]
G7 OSE Details withheld at licensee's request by SSL.
GE7 OSH H Davies Nortel(Paigton)Amateur Radio, 33 Sandown Rd, Paignton, TQ4 7RL [IO80FJ, SX85]
G7 OSJ A E M Ramm, 17 Sharrington Rd, Bale, Fakenham, NR21 0QX [JO02LV, TG03]
G7 OSK D J Ramm, 17 Sharrington Rd, Bale, Fakenham, NR21 0QX [JO02LV, TG03]
G7 OSL B D Morrissey, 91 Hillview Cl, Rowhedge, Colchester, CO5 7HS [JO01LU, TM02]
G7 OSM S M Lovelass, 9 Claxton Ct, Newton Aycliffe, DL5 7LA [IO94EP, NZ22]
G7 OSN Details withheld at licensee's request by SSL.
G7 OSO A Kinnersley, Weathertop, Barthomley Rd, Audley, Stoke on Trent, ST7 8HU [IO83UB, SJ75]
GM7 OSQ D M Clark, Benmhor, Baluachrach, Tarbert, PA29 6TF [IO75HU, NR86]
G7 OST S E Timms, 7 Portway Dr, High Wycombe, HP12 4AU [IO91OP, SU89]
G7 OSW A Moss, 1A Southfield Cres, Norton on Tees, Cleveland, TS20 2ET [IO94IN, NZ42]
G7 OTG J C Mahoney obo Kepac Packet Radio Grou, 89 Tyefields, Pitsea, Basildon, SS13 1JA [JO01GN, TQ78]
GW7 OTI L Lewis, 5 Maes y Parc, Cwmafan, Cwmavon, Port Talbot, SA12 9PU [IO81CO, SS79]
G7 OTM Details withheld at licensee's request by SSL.
G7 OTQ A J Dibbins, 16 Peverel Dr, Whittington, Oswestry, SY11 4PN [IO82LU, SJ33]
GM7 OTT I M Alexander, Newton of Kinmundy Cottage, Kinmundy, By Mintlaw, Aberdeenshire, AB42 8AY [IO97BL, NK04]
GW7 OTU S G Hibbert, Pantgwyn, Llanddewibrefi, Tregaron, Dyfed, SY25 6PE [IO82AE, SN65]
G7 OTV Details withheld at licensee's request by SSL.
G7 OTY Details withheld at licensee's request by SSL.
G7 OUC Details withheld at licensee's request by SSL.
G7 OUE Details withheld at licensee's request by SSL.
G7 OUF T R Farr, 97 Templeton Rd, Birmingham, B44 9DA [IO92BN, SP09]
G7 OUL Details withheld at licensee's request by SSL.
G7 OUQ J E Lewis, 24 Forest Rd, Southport, PR8 6ST [IO83MP, SD31]
GM7 OUR G D Morrison, East Murkle, Thurso, Caithness, KW14 8SR [IO88GO, ND16]
G7 OUT M J Reynolds, 15 Foxfield Dr, Stanford-le-Hope, SS17 8HH [JO01FM, TQ68]
GI7 OUW Details withheld at licensee's request by SSL.
G7 OUZ B R Donkin, 13 Saddlebow Rd, Kings Lynn, PE30 5BQ [JO02ER, TF61]
G7 OVC Details withheld at licensee's request by SSL.
G7 OVE D M Brown, 9 Lancaster Way, East Winch, Kings Lynn, PE32 1NY [JO02GR, TF61]
G7 OVJ Details withheld at licensee's request by SSL.
G7 OVK C L Carson, 15 Sudbury Way, Beacon Hill Green, Cramlington, NE23 8HG [IO95EC, NZ27]
G7 OVM N P Ward, Seale Cottage, Worth Matravers, Swanage, Dorset, BH19 3LQ [IO80XO, SY97]
G7 OVO D A Roffey, 32 Hertford Rd, Digswell, Welwyn, AL6 0DB [IO91VT, TL21]
G7 OVS J P Dilks, Handley Farm Bungalow, Beckingham, Lincs, LN5 0RN [IO93QB, SK95]
G7 OVT Details withheld at licensee's request by SSL.
GW7 OVV D Gibbs, Gwynfryn, Pentre Llyn, Llanilar, Aberystwyth, SY23 4NS [IO72XI, SN67]
G7 OWB Details withheld at licensee's request by SSL.
G7 OWE S W Mumby, 44 Theaker Ave, Gainsborough, DN21 1RH [IO93OJ, SK89]
G7 OWF L A Miles obo East Yorkshire Hospital, Works Department, Castle Hill Hospital, Castle Rd, Cottingham, North Humberside, HU16 5JQ [IO93SS, TA03]
G7 OWH Details withheld at licensee's request by SSL.
G7 OWM R E Shortland, Aboyne Cottage, 22 Kendal Rd, Kemnay, Aberdeenshire, AB51 9ND [IO87SG, NJ72]
G7 OWN S A J Mawson, 29 Westfields, Stanley, DH9 7BP [IO94DU, NZ15]
G7 OWP J Oliphant, 27 Byron St., Amble, Morpeth, NE65 0ER [IO95FH, NU20]
G7 OWQ B D Clifton, 3 Kirton Rd, Drayton, Cosham, Portsmouth, PO6 2ES [IO90LU, SU60]
G7 OWR P V Blake, 46 Marlborough Rd, Ipswich, IP4 5AX [JO02OB, TM14]
GM7 OWU D Brander, 3 Spartleton Pl, Dundee, DD4 0UJ [IO86NL, NO43]
G7 OWX D Allen, 130 Seamer Rd, Scarborough, YO12 4EY [IO94TG, TA08]
G7 OWZ D L Payne, 147 Upper Marehay, Ripley, DE5 8JG [IO93HA, SK34]
G7 OXA D W Giles, 73 Barsby Dr, Loughborough, LE11 5UJ [IO92JS, SK52]
G7 OXB S Richardson, 6 Dane Ghyll, Barrow in Furness, LA14 4PZ [IO84JD, SD27]
G7 OXF B Preston, 34 Meadows Ave, Haslingden, Rossendale, BB4 5NR [IO83UQ, SD72]
G7 OXG B V Newman, 83 Cranleigh Rd, Bournemouth, BH6 5JX [IO90CR, SZ19]
G7 OXH R Wilkins, 85 St. Richards Rd, Otley, LS21 2AL [IO93DV, SE14]
G7 OXJ Details withheld at licensee's request by SSL.
G7 OXK J F Leach, 2 Andover Cl, Feltham, TW14 9XG [IO91SK, TQ07]
G7 OXN B Burdis, St Annes Head Lighthouse, Dale, Haverfordwest, Dyfed, SA62 3RS [IO71JQ, SM80]
G7 OXP Details withheld at licensee's request by SSL.
G7 OXS Details withheld at licensee's request by SSL.
G7 OXV R F Blott, 24 Campion Rd, London, SW15 6NW [IO91VL, TQ27]
G7 OXX Details withheld at licensee's request by SSL.
G7 OXY J P Brandram, The Limes, Titley Bawk Ave, Earls Barton, Northampton, NN6 0LA [IO92PG, SP86]
G7 OYB Details withheld at licensee's request by SSL.
G7 OYC Details withheld at licensee's request by SSL.
G7 OYD P Thompson, 87 Selby Rd, Garforth, Leeds, LS25 1LN [IO93HS, SE43]
G7 OYF B J Gilbert, 2 Beeches Cl, West Row, Bury St. Edmunds, IP28 8QE [JO02FI, TL67]
G7 OYN C W Rodgers, 37 Heatherset Gdns, Norbury, London, SW16 3LS [IO91WJ, TQ37]
G7 OYP S Hollis, 34 Bowood Rd, Enfield, EN3 7LH [IO91XP, TQ39]
GU7 OYU A R Stoaling, Carando, La Petite Mare de Lis Clos, Castel, Guernsey, GY5 7BN
G7 OYW R A Newman, 11 Pine View Cl, Woodfalls, Salisbury, SP5 2LR [IO90DX, SU11]
G7 OYX B P Dorey, 8 Richmond Rd, Swanage, BH19 2PZ [IO90AO, SZ07]
G7 OYZ J M Spashett, Llys-y-Coed, Trefriw, Gwynedd, LL27 0QA [IO83BD, SH76]
G7 OZA G W E Johnston, 94 Abercorn Cres, Harrow, HA2 0PU [IO91TN, TQ18]
G7 OZC Details withheld at licensee's request by SSL.
G7 OZE K G Baldry, 160 Rover Dr, Castle Bromwich, Birmingham, B36 9LL [IO92CM, SP19]
G7 OZF Details withheld at licensee's request by SSL.
G7 OZH D J Albury, 29 The St., Rockland St. Mary, Norwich, NR14 7ER [JO02QO, TG30]
G7 OZI A S C Morley, 6 Millway, Chudleigh, Newton Abbot, TQ13 0JN [IO80EO, SX87]
G7 OZJ S A Morley, 6 Millway, Chudleigh, Newton Abbot, TQ13 0JN [IO80EO, SX87]
G7 OZK Details withheld at licensee's request by SSL.
G7 OZL A C Page, 22 Magnolia Cl, Kempston, Bedford, MK42 7RY [IO92SC, TL04]
G7 OZM G J Rolle, 30 Hillview Gdns, Cheshunt, Waltham Cross, EN8 0PE [IO91XR, TL30]
G7 OZO Details withheld at licensee's request by SSL.
GW7 OZP J M Barrett, Tree Tops, Comins Coch, Aberystwyth, Dyfed, SY23 3BL [IO72XK, SN68]
G7 OZQ T J Pluck, 16A Field End Rd, Leeds, LS15 0PZ [IO93GT, SE33]
G7 OZU M A R Knight, 2 Ash Gr, Biggleswade, SG18 0HG [IO92UC, TL14]
G7 OZV Details withheld at licensee's request by SSL.
G7 OZW Details withheld at licensee's request by SSL.
G7 OZX Details withheld at licensee's request by SSL.
G7 OZY Details withheld at licensee's request by SSL.
G7 OZZ H Longley, Coach House, Towers Ln, Cockermouth, Cumbria, CA13 9ED [IO84HP, NY12]
G7 PAA W C Hobson, 50 Newland Ave, Scunthorpe, DN15 7HQ [IO93PO, SE81]
G7 PAB Details withheld at licensee's request by SSL.
G7 PAC C J Shadlock, 29 Malvern Ave, Washingborough, Lincoln, LN4 1EB [IO93SF, TF07]
G7 PAD Details withheld at licensee's request by SSL.
G7 PAE C J Bailes, 48 Harlech Cl, Eston, Middlesbrough, TS6 9SZ [IO94KN, NZ51]
G7 PAF Details withheld at licensee's request by SSL.
G7 PAG P A Gould, 31 Schofield St., Mexborough, S64 9NJ [IO93IL, SE40]
G7 PAJ Details withheld at licensee's request by SSL.
G7 PAK A D Smith, 117 Stratton Rd, Sunbury on Thames, TW16 6PG [IO91SJ, TQ06]
G7 PAL Details withheld at licensee's request by SSL.
G7 PAM W Turner, 16 Ozanan Cl, Dudley, Cramlington, NE23 7BW [IO95FB, NZ27]
G7 PAN Details withheld at licensee's request by SSL.
G7 PAO G Plant, 43 Forest Rd, Almondbury, Huddersfield, HD5 8EU [IO93CP, SE11]
G7 PAV Details withheld at licensee's request by SSL.
G7 PAZ Details withheld at licensee's request by SSL.
G7 PBA B E Fry, Beggars Roost, 12 Thistleton Ln, South Witham, Grantham, NG33 5QE [IO92QS, SK91]
G7 PBB Details withheld at licensee's request by SSL.
G7 PBC P D Sherburn, 70 Briarwood Rd, Stoneleigh, Epsom, KT17 2NG [IO91VI, TQ26]
GI7 PBE W R Nicholl, 114 Garvaghy Rd, Cullybackey, Ballymena, BT42 1EN [IO64TV, D00]
G7 PBH H B Parrish, 5 Kestrel Ln, Cheadle, Stoke on Trent, ST10 1RU [IO92AX, SK04]
G7 PBK J H Oliver, 7 The Quayside, Apperley Bridge, Bradford, West Yorks, BD10 0UL [IO93DU, SE13]
G7 PBL H Taylor, 6 Broadway Cl, Swinton, Mexborough, S64 8HG [IO93IL, SK49]
G7 PBN V A Langton, 4 The Limes, Beckingham, Doncaster, DN10 4QN [IO93OJ, SK79]
G7 PBO M A Sewell, 95 Austcliffe Park, Austcliffe Ln, Cookley, Kidderminster, DY10 3UR [IO82VK, SO88]
GW7 PBP V D Roberts, Llwyn-Celyn, Bedwellty, Blackwood, Gwent, NP2 0BD [IO81JQ, SO10]
GI7 PBQ R J Young, 8 Glenside Ave, Drumbo, Lisburn, BT27 5LQ [IO74AM, J36]
GW7 PBR C A Jones, 75 Bowleaze, Greenmeadow, Cwmbran, NP44 4LF [IO81LP, ST29]
G7 PBT R A Spirrell, 16 Balsam Park, Wincanton, BA9 9HB [IO81TB, ST72]
G7 PBU J W K Rowe, The Old Rectory, Wickenby, Lincs, LN3 5AB [IO93TH, TF08]
G7 PBV J L Mason, 56 Skegby Rd, Sutton in Ashfield, NG17 4EZ [IO93JD, SK55]
G7 PBX I J Dovey, 11 Chapel Walk, Dudley, DY3 2NJ [IO82WM, SO99]
G7 PCA S P Ashton, 6 Hillock Ln, Woolston, Warrington, WA1 4NF [IO83RJ, SJ68]
G7 PCC R A Fletcher, 8 Broad Oaks Cl, Dewsbury, WF12 8RH [IO93EQ, SE22]
G7 PCE K E Toop, 10 Hunt Rd, Blandford Forum, DT11 7LZ [IO80WU, ST80]
G7 PCF J D Banfield, Highbury, 81 Clophill Rd, Maulden, Bedford, MK45 2AD [IO92SA, TL03]
G7 PCG Details withheld at licensee's request by SSL.
G7 PCH A D Goodman, 23 Evelyn Dr, Pinner, HA5 4RL [IO91TO, TQ19]
G7 PCI D E Simmonds, 5 Westminster Cl, Melton Mowbray, Leics, LE13 0PF [IO92NS, SK72]
G7 PCJ Details withheld at licensee's request by SSL.
G7 PCL Details withheld at licensee's request by SSL.
G7 PCM Details withheld at licensee's request by SSL.
G7 PCN Details withheld at licensee's request by SSL.
G7 PCQ Details withheld at licensee's request by SSL.
G7 PCT P C Treadwell, 22 Meynell Cl, Melton Mowbray, LE13 0RA [IO92NS, SK71]
GW7 PCU Details withheld at licensee's request by SSL.
GW7 PCX G Bellis, 70 Osborne St., Rhos, Rhosllanerchrugog, Wrexham, LL14 2HT [IO83LA, SJ24]
G7 PCY E E S Ruiz, Electronics Lab, University of Kent, Canterbury, CT2 7NT [JO01MH, TR15]
G7 PCZ S Dickinson, 16 Heathfield Gr, Beeston, Nottingham, NG9 5EB [IO92JV, SK53]
G7 PDA M A Binns, 22 Rydings Dr, Brighouse, HD6 2DA [IO93CQ, SE12]
G7 PDH J Swallow, 19 Sunningdale Walk, Herne Bay, CT6 7TR [JO01NI, TR16]
G7 PDL Details withheld at licensee's request by SSL.
G7 PDO M E Galea, 17 Waterloo Rd, Horsham St. Faith, Norwich, NR10 3HS [JO02PQ, TG21]
G7 PDP A P Jablonski, 32 Downleaze, South Woodham Ferrers, Chelmsford, CM3 5SN [JO01HP, TQ89]
G7 PDR J E Martin, 45 Quailholme Rd, Knott End on Sea, Poulton-le-Fylde, FY6 0BT [IO83MW, SD34]
G7 PDS T Simpkins, 16 Hesketh Ave, Blackpool, Lancs, FY2 9JX [IO83LU, SD33]
G7 PDU S J Willis, 180 Thisselt Rd, Canvey Island, SS8 9BL [JO01HM, TQ78]
G7 PDY Details withheld at licensee's request by SSL.
G7 PDZ P L Balm, 140 Coppice Rd, Arnold, Nottingham, NG5 7GT [IO93KA, SK54]
G7 PEA Details withheld at licensee's request by SSL.
G7 PEB J Edwards, The Bungalow, Tweedhill Farm, Park Ln, Swanley Kent, BR8 8DT [JO01CJ, TQ56]
G7 PEC C A Tunbridge Essex Raynet Maldon District, 12 Burnham Rd, Latchingdon, Chelmsford, CM3 6EU [JO01IQ, TL80]
G7 PEE T J Griffiths, 49 Lombardy Cl, Hempstead, Gillingham, ME7 3SQ [JO01GI, TQ76]
G7 PEF Details withheld at licensee's request by SSL.
G7 PEG O Gray, 18 Arrow View, Lower Hergest, Kington, HR5 3ER [IO82LE, SO25]
G7 PEH P D Gater, 166 Rolls Ave, Crewe, CW1 3QD [IO83SC, SJ65]
G7 PEN E M Penn, 53 Manfield Ave, Walgrave, Coventry, CV2 2QF [IO92GK, SP38]
G7 PER D C Boughton, 59 Redland Dr, Kirk Ella, Hull, HU10 7UX [IO93SS, TA03]
G7 PES C G A Walters, 48 Ramsons Way, Abingdon, OX14 3TJ [IO91IQ, SU59]
G7 PET Details withheld at licensee's request by SSL.
G7 PEU R A Chapman, 12 Lynton Rd, Chesham, HP5 2BU [IO91QR, SP90]
G7 PEW Details withheld at licensee's request by SSL.
G7 PEX K Barker, 72 Keel Dr, Slough, SL1 2XY [IO91QM, SU97]
GW7 PEY Details withheld at licensee's request by SSL.
G7 PEZ G P J Thompson, 81 Runnymede Ave, Bournemouth, BH11 9SG [IO90AS, SZ09]
G7 PFB Details withheld at licensee's request by SSL.
G7 PFD A M Thompson, 73 Wickersley Rd, Rotherham, S60 3PX [IO93IK, SK49]
G7 PFE Details withheld at licensee's request by SSL.
G7 PFG J F Smith, 23 Feering Dr, Basildon, SS14 1TG [JO01FN, TQ78]
G7 PFH Details withheld at licensee's request by SSL.
G7 PFI M Green, 9 Greencroft Ave, Northowram, Halifax, HX3 7EP [IO93CR, SE12]
G7 PFJ Details withheld at licensee's request by SSL.
GW7 PFK E Gillet, 26 Gower Ct, Claude Rd West, Barry, South Glam, CF62 7JG [IO81IJ, ST16]
G7 PFM P F Cook, 55 Bushey Croft, Harlow, CM18 6RH [JO01BS, TL40]
G7 PFN Details withheld at licensee's request by SSL.
G7 PFO Details withheld at licensee's request by SSL.
G7 PFQ Details withheld at licensee's request by SSL.
G7 PFR E Beardmore, 9 Sutherland Ct, Longton Rd, Stoke on Trent, ST4 8BT [IO82VX, SJ84]
G7 PFT J J Richardson, 107 Pinehurst Rd, West Moors, Ferndown, BH22 0AL [IO90BT, SU00]
G7 PFU Details withheld at licensee's request by SSL.
G7 PFX Details withheld at licensee's request by SSL.
G7 PFY J R Ellis, 11 Moorland Cres, Guiseley, Leeds, LS20 9EF [IO93DV, SE14]
G7 PGA Details withheld at licensee's request by SSL.
G7 PGB Details withheld at licensee's request by SSL.
G7 PGC J G McNab, 36 Wellington St., Matlock, DE4 3GS [IO93FD, SK36]
G7 PGG A R Huntley, 47 Wellfield Cl, Throckley, Newcastle upon Tyne, NE15 9JL [IO94CX, NZ16]
G7 PGH M Card, 11 Manifold Rd, Eastbourne, BN22 8EH [JO00DS, TV69]
G7 PGK Details withheld at licensee's request by SSL.
GM7 PGM R Darroch, 57 Barrisdale Rd, Maryhill, Glasgow, G20 0JL [IO75UV, NS56]
G7 PGN Details withheld at licensee's request by SSL.
G7 PGP Details withheld at licensee's request by SSL.
G7 PGQ G J Parker, 44 Evergreen Rd, Lowestoft, NR32 2SA [JO02UL, TM59]
G7 PGS Details withheld at licensee's request by SSL.
GM7 PGT A R Foulis, 12 Richmond Gdns, Chryston, Glasgow, G69 9PA [IO75WV, NS67]
G7 PGY G R Sleeman, Whitebeams, St. Johns Cl, Penn, High Wycombe, HP10 8HX [IO91PP, SU89]
G7 PGZ N R Allchin, 62 Saltwood Rd, Seaford, BN25 3SR [JO00BS, TV49]
G7 PHB S T C Beesley, 15 Byron Cl, Cheadle, Stoke on Trent, ST10 1XB [IO82XX, SJ94]
G7 PHC M A Porter, 94 Oaken Gr, Haxby, York, YO3 3QZ [IO94LA, SE65]
G7 PHD I L Connor, 130 Chandlers Dr, Ocean Park, Erith, Kent, DA8 1LW [JO01BL, TQ47]
G7 PHE A R Lassman, 69 St. Ladoc Rd, Keynsham, Bristol, BS18 2EQ [IO81RJ, ST66]
G7 PHF G D Lloyd, Little Haven, Crickheath Hill, Llynclys, Oswestry, SY10 8LG [IO82LT, SJ22]
G7 PHG D B Harbron, 48 Sheridan Rd, Biddick Hall, South Shields, NE34 9JJ [IO94GX, NZ36]
G7 PHH W J Alderman, Outspan, Quethiock, Liskeard, Cornwall, PL14 3SQ [IO70TK, SX36]
G7 PHI T R Larsen, 92 Gaskell Ave, South Shields, NE34 9TA [IO94GX, NZ36]
G7 PHK J L Wilkes, 229 Merland Rise, Tadworth, KT20 5JQ [IO91VH, TQ25]
G7 PHL R J Marshall, 113 Carsic Rd, Sutton in Ashfield, NG17 2BQ [IO93ID, SK45]
G7 PHM S H Cocker, 21 Mayfield Rd, Chorley, PR6 0DG [IO83QP, SD51]
G7 PHN D L Partridge, 50 Edison Rd, Walsall, WS2 7EZ [IO82XO, SJ90]
G7 PHP R G Shepherd, 38 Ridge Way, Hixon, Stafford, ST18 0NZ [IO92AT, SK02]

G7 PHQ E J Roberts, 3 Haywood Grange, Little Haywood, Stafford, ST18 0UB [IO92AS, SK02]
G7 PHR W Stewart, 1 Laing Cl, Bardney, Lincoln, LN3 5XS [IO93UE, TF16]
G7 PHT K J Dennis, 4 Ash Gr, Sheringham, NR26 8PT [JO02OW, TG14]
G7 PHU Details withheld at licensee's request by SSL.
G7 PHV K P Lawless, 33 River View Rd, Southampton, SO18 1NW [IO90HW, SU41]
G7 PHW S Dobson, 166 Lynfield Dr, Bradford, BD9 6EZ [IO93CT, SE13]
G7 PHY D S Dobson, 166 Lynfield Dr, Bradford, BD9 6EZ [IO93CT, SE13]
G7 PHZ Details withheld at licensee's request by SSL.
G7 PIA Details withheld at licensee's request by SSL.
G7 PIB J Challenger, Claverley, Chepstow Rd, Langstone, Newport, NP6 2JP [IO81NO, ST39]
G7 PIC J Harwood, 9 Holland Park Dr, Hedworth Ln, Jarrow, NE32 4LL [IO94GW, NZ36]
G7 PIG C F Wood, 4 Mount Scar View, Scholes, Holmfirth, Huddersfield, HD7 1XH [IO93CN, SE10]
G7 PIJ M H Beeson, 1 Tamar Gr, Highfield Park, Cheadle, Stoke on Trent, ST10 1QQ [IO92AX, SK04]
G7 PIK B G Bashford, 51 Broadwater Rd, Worthing, BN14 8AH [IO90TT, TQ10]
G7 PIM Details withheld at licensee's request by SSL.
G7 PIO J H Thorpe, Highcroft, Butts Ln, Harbury, Leamington Spa, CV33 9JL [IO92GF, SP35]
G7 PIP R S Oswald, 17 Dunclutha Rd, Hastings, TN34 2JA [JO00HV, TQ81]
G7 PIQ Details withheld at licensee's request by SSL.
G7 PIR J Briggs, 20 Druids Ln, Maypole, Birmingham, B14 5SN [IO92BJ, SP07]
G7 PIT Details withheld at licensee's request by SSL.
G7 PIV G R Ottaway, Garden Cottage, Abermarlais, Llangadog, Dyfed, Wales, SA19 9NG [IO81BW, SN62]
G7 PIX D E Atkins, 97 South St., Tillingham, Southminster, CM0 7TH [JO01KQ, TL90]
GI7 PIZ S W Hewitt, 24 Ballymacruise Dr, Millisle, Newtownards, BT22 2NN [IO74FO, J57]
G7 PJC C A Harris, 10 Caister Ave, Whitefield, Manchester, Lancs, M45 6EL [IO83UN, SD80]
G7 PJD R M Howes, 58 Mayfield Ave, Orpington, BR6 0AQ [JO01BJ, TQ46]
G7 PJE Details withheld at licensee's request by SSL.
GI7 PJF R A Stewart, 12 Donegall Dr, Whitehead, Carrickfergus, BT38 9LT [IO74DS, J49]
G7 PJG Details withheld at licensee's request by SSL.
G7 PJH D W Kent, 41 Summerfield Rd, Malvern Link, Malvern, WR14 1DZ [IO82UD, SO74]
G7 PJI P Wood, 2 Central Cres, Hethersett, Norwich, NR9 3EP [JO02OO, TG10]
G7 PJJ W F Smith, 135 Allestree Cl, Derby, DE24 8SX [IO92GV, SK33]
G7 PJL S A Hammond, 10 Russet Glade, Burghfield Common, Reading, Berks, RG7 3DZ [IO91LJ, SU66]
G7 PJN S C Phillips, 48 Woodbridge Ave, Audenshaw, Manchester, M34 5LL [IO83WL, SJ99]
G7 PJQ Details withheld at licensee's request by SSL.
G7 PJR Details withheld at licensee's request by SSL.
G7 PJT P E S Lovejoy, 144 Westbourne Ave, Walkergate, Newcastle upon Tyne, NE6 4HQ [IO94FX, NZ26]
G7 PJV Details withheld at licensee's request by SSL.
G7 PJW P J Dann, 151 West End, March, PE15 8DD [JO02AN, TL49]
G7 PJZ H M A M Robinson, 5 Coppice Cl, Haxby, York, YO3 3RR [IO94LA, SE65]
G7 PKD R J Brown, 1 Octavian Cl, Hatch Warren, Basingstoke, RG22 4TY [IO91KF, SU64]
G7 PKF Details withheld at licensee's request by SSL.
G7 PKG B J Jenkins, 1 Wyken Cl, Luton, LU3 3XL [IO91SW, TL02]
G7 PKH P M Precious, 99 Sherwood Ave, St. Albans, AL4 9PW [IO91US, TL10]
G7 PKJ Details withheld at licensee's request by SSL.
G7 PKM Details withheld at licensee's request by SSL.
G7 PKO Details withheld at licensee's request by SSL.
G7 PKQ P J Troll, 18 Bowness Rd, Millom, LA18 4LS [IO84IE, SD17]
G7 PKS Details withheld at licensee's request by SSL.
G7 PKT R M Morrison, Corran Garden, Onich, Fort William, PH33 6SE [IO76JR, NN06]
G7 PKU P R Finch, 164 Godstow Rd, Abbeywood, London, SE2 9AY [JO01BL, TQ47]
G7 PKV N A Foley, 164 Godstow Rd, Abbeywood, London, SE2 9AY [JO01BL, TQ47]
G7 PKW Details withheld at licensee's request by SSL.
G7 PKX Details withheld at licensee's request by SSL.
G7 PKY E D A Plant, Plant Engineering, Park Ln, Langport, Somerset, TA10 0NF [IO81NA, ST42]
G7 PKZ K Pickford, 23 Green Farm End, Kineton, Warwick, CV35 0LD [IO92FD, SP35]
G7 PLA M D Armstrong, 41 North Parade, Whitley Bay, NE26 1NX [IO95GB, NZ37]
G7 PLE J Goodliffe, 56 Robert Jennings Cl, Cambridge, CB4 1YU [JO02BF, TL46]
G7 PLF Details withheld at licensee's request by SSL.
GM7 PLG Details withheld at licensee's request by SSL.
G7 PLH T Wesley, 12 Elm Rd, Hemsworth, Pontefract, WF9 4JD [IO93HO, SE41]
G7 PLI Details withheld at licensee's request by SSL.
G7 PLL N A Smith, 18 Halstead Rd, Earls Colne, Colchester, CO6 2NG [JO01IW, TL82]
G7 PLM J E Williamson, 39 Marriott Ave, Mansfield, NG18 5NQ [IO93JD, SK56]
G7 PLN Details withheld at licensee's request by SSL.
G7 PLP I J Brown, 5 Gower Ave, Kingswinford, DY6 8NE [IO82WL, SO88]
G7 PLQ Prof D G Graham Cbe, Gilmour Mews, Battlebarrow, Appleby in Westmorland, CA16 6XT [IO84SN, NY62]
G7 PLR G B Done, Lodge Cottage, The Green, Ainderby Steeple, Northallerton, DL7 9QA [IO94GH, SE39]
G7 PLS N L Partridge, 13 Alderney Way, Immingham, Grimsby, DN40 1RB [IO93VO, TA11]
G7 PLV A C Miller, 10 Limerick Cl, Ipswich, IP1 5LR [JO02NB, TM14]
G7 PLX J P Milton, 46 Coles Ln, Oakington, Cambridge, CB4 5AF [JO02AG, TL46]
G7 PLY Details withheld at licensee's request by SSL.
G7 PLZ Details withheld at licensee's request by SSL.
G7 PMA J F Beach, 18 Tintagel Way, Cosham, Portsmouth, PO6 4SS [IO90KU, SU60]
G7 PMB R Cooke, 305 Haigh Rd, Aspull, Wigan, WN2 1RR [IO83QN, SD60]
GU7 PMC K J Barlow, Rue Maze, St. Martins, St. Martin, Guernsey, Channel Islands, GY4 6LJ
GU7 PMD J G Fraser, Rue Maze Dental Surgery, Lindfield, Rue Maze, St Martins, Guernsey, GY4 6LJ
G7 PME P M Eaton, 81 Alcuin Ave, York, YO1 3TN [IO93LW, SE65]
G7 PMF V J Lennox, 9 Old Newark Rd, Mansfield, NG18 4SZ [IO93JC, SK55]
G7 PMG S G Rose, 22 Ramsden Rd, Orpington, BR5 4LT [JO01BJ, TQ46]
G7 PMI D W Williams, 6 Raven Cres, Billericay, CM12 0JF [JO01EP, TQ69]
G7 PMJ Details withheld at licensee's request by SSL.
G7 PMK W A Mitchell, 2 Mariners Ct, Great Wakering, Southend on Sea, SS3 0DR [JO01JN, TQ98]
G7 PMO K E Walton, 38 Honister Heights, Purley, CR8 1EU [IO91WH, TQ36]
G7 PMQ P J Slight, 2 St. Matthews Cl, Cherry Willingham, Lincoln, LN3 4LS [IO93SF, TF07]
G7 PMS Details withheld at licensee's request by SSL.
G7 PMT G Slater, 12 Storrington Cl, Hove, BN3 8JE [IO90VU, TQ20]
G7 PMU S M Lawrence, 85 West Ave, Clacton on Sea, Essex, CO15 1HB [JO01NS, TM11]
G7 PMV G L Gimber, 10 Harrowdene Gdns, Teddington, TW11 0DH [IO91UK, TQ17]
G7 PMW G B Wall, Westerland, Sandhill Rd, East Claydon, Buckingham, MK18 2LZ [IO91MW, SP72]
G7 PMX M S Firth, 5 Courtenays, Seacroft, Leeds, LS14 6JZ [IO93GT, SE33]
G7 PMY B F Whittington, Flat 3, 45 Susans Rd, Eastbourne, BN21 3TJ [JO00DS, TV69]
G7 PMZ P F Rushton, 59 Eastwood Rd, New Moston, Manchester, M40 3TE [IO83WM, SD80]
G7 PNA D B Stewart, 26 Speak Cl, Wakefield, WF1 4TG [IO93GQ, SE32]
G7 PNB S P Phillips, 2 Glenfield Cres, Mickleover, Derby, DE3 5RF [IO92FV, SK33]
G7 PND K W Marriott, 99 Stapleford Ln, Toton, Beeston, Nottingham, NG9 6FZ [IO92IV, SK53]
G7 PNE D J Head, 76 Dryden Cres, Stevenage, SG2 0JH [IO91WV, TL22]
G7 PNF M Cleverley, 43 Friesian Gdns, Red St, Chesterton, Newcastle, Staffs, ST5 6BB [IO83UA, SJ84]
G7 PNG R J Owen, 53 Huntingdon Dr, Castle Donington, Derby, DE74 2SR [IO92HU, SK42]
G7 PNI Details withheld at licensee's request by SSL.
G7 PNK S G Stephen, 15 Winton Ave, Westcliff on Sea, SS0 7QU [JO01IM, TQ88]
G7 PNM P R Smith, 8 Heather Cl, Outwood, Wakefield, WF1 3HE [IO93GR, SE32]
G7 PNP Details withheld at licensee's request by SSL.
G7 PNQ Details withheld at licensee's request by SSL.
G7 PNR Details withheld at licensee's request by SSL.
G7 PNS A Spendiff, 36 Sunholme Dr, Hadrian Lodge Est, Wallsend, NE28 9YT [IO95FA, NZ26]
G7 PNU Details withheld at licensee's request by SSL.
G7 PNW B G Bond, 5 Stourbank Rd, Christchurch, BH23 1LH [IO90CR, SZ19]
G7 PNX N Armstrong, 99 Cedarland Cres, Nuthall, Nottingham, NG16 1AG [IO92JX, SK54]
GW7 PNY Details withheld at licensee's request by SSL.
G7 POA Dr F H Greaves, 10 Brocks Cl, Dibden Purlieu, Southampton, SO45 5ST [IO90GU, SU40]
G7 POB M Slade, 7 Glebe Field, Chaddleworth, Newbury, RG20 7EZ [IO91HL, SU47]
G7 POC Details withheld at licensee's request by SSL.
G7 POF R A Munt, Tyn y Coed, Kilnwood Ln, South Chailey, Lewes, BN8 4AU [IO90XW, TQ31]
G7 POG M H Mountford, 1 Bowker House, Moor St., West Bromwich, B70 7AZ [IO92AM, SP09]
G7 POI L Moss, 13 Perth, Stonehouse, GL10 2PT [IO81UR, SO80]
G7 POJ Details withheld at licensee's request by SSL.
GM7 POK J J Cook, 2 Springwells, St. Anns, Lockerbie, DG11 1HG [IO85GF, NY09]
G7 POL P E Duggan, 90 Marton Dr, Marston, Blackpool, FY4 3EU [IO83LT, SD33]
GI7 POP S W Miller, 2 Ransevyn Dr, Whitehead, Carrickfergus, BT38 9NQ [IO74DS, J49]
G7 POR J Turner, 18 Birks Rd, Mansfield, NG19 6JU [IO93JD, SK56]
G7 POS Details withheld at licensee's request by SSL.
G7 POT S A Eastwood, 12 Cross St., Savile Rd, Castleford, WF10 1PF [IO93HR, SE42]
G7 POU M E Willard, 44 Wilmington Way, Haywards Heath, West Sussex, RH16 3HZ [IO91WA, TQ32]
G7 POV A Lickley, 18 Byron St., Macclesfield, SK11 7PL [IO83WF, SJ97]
G7 POW J H Sanderson, 40 Sheldon Cl, Honiton Rd, Bransholme, Hull, HU7 4RU [IO93US, TA13]
G7 POZ S Quantrill, 7 Clare Rd, Kessingland, Lowestoft, NR33 7PS [JO02UK, TM58]
G7 PPB Details withheld at licensee's request by SSL.
G7 PPC B H L Lowe, 19 Wolverhampton Rd, Bloxwich, Walsall, WS3 2EZ [IO82XO, SJ90]
G7 PPD Details withheld at licensee's request by SSL.
G7 PPE Details withheld at licensee's request by SSL.
G7 PPF A P Wells, 106 Tor O Moor Rd, Woodhall Spa, LN10 6SB [IO93VD, TF26]
G7 PPK M A O Pendleton, 17 Layard Rd, Forty Hill, Enfield, EN1 4BA [IO91XP, TQ39]
G7 PPL A P Pardivalla, 27 Wilberforce Rd, Leicester, LE3 0GT [IO92KP, SK50]
G7 PPM Details withheld at licensee's request by SSL.
GM7 PPN H G B Waugh, 87 Hercus Loan, Musselburgh, EH21 6BA [IO85LW, NT37]
G7 PPP Details withheld at licensee's request by SSL.
G7 PPQ T J Davies, 34 Treforthlan, Paynters Ln End, Redruth, Cornwall, TR16 4RN [IO70IF, SW64]
G7 PPS H I Lodge, 69 Helena Rd, Rayleigh, SS6 8LQ [JO01HO, TQ89]
G7 PPT Details withheld at licensee's request by SSL.
G7 PPU Details withheld at licensee's request by SSL.
G7 PPV Details withheld at licensee's request by SSL.
G7 PPW J R Jordon, 18 Priory Cres, Binham, Fakenham, NR21 0DB [JO02LV, TF93]
G7 PPX K B Gibbs, 10 Bala Ct, Gunthorpe, Peterborough, PE4 7UQ [IO92UO, TF10]
G7 PPY J K Richards, 13 Creak A Vose, St. Stephen, St. Austell, Cornwall, PL26 7NZ [IO70NI, SW95]
G7 PQB K H Hooker, 340 Long Rd, Lowestoft, NR33 9DL [JO02UK, TM59]
G7 PQD B A Harrison, 145 St. Leonard St., Hendon, Sunderland, SR2 8QB [IO94HV, NZ45]
G7 PQE Details withheld at licensee's request by SSL.
G7 PQF Details withheld at licensee's request by SSL.
GW7 PQG T Roberts, Highfield, 126 Llysfaen Rd, Old Colwyn, Colwyn Bay, LL29 9HL [IO83DG, SH87]
G7 PQI P D Clark, 2 Barley Croft, Stevenage, SG2 9NP [IO91WV, TL22]
G7 PQL M Robinson, 1 Selby Cl, Baxenden, Accrington, BB5 2TQ [IO83TR, SD72]
G7 PQM R Buchan, 16 Lomond Dr, Kettering, NN15 5DE [IO92PJ, SP87]
G7 PQN Details withheld at licensee's request by SSL.
G7 PQO S G Fuller, 150 Plum Ln, Shooters Hill, London, SE18 3HF [JO01AL, TQ47]
G7 PQP D F Little, 20 Vicarage Cl, Shillington, Hitchin, SG5 3LS [IO91TX, TL13]
G7 PQS M Waite, 186 Frampton Rd, Gorseinon, Swansea, SA4 4YG [IO71XQ, SS59]
G7 PQT M R Grubb, 7 Falstone Ave, Newark, NG24 1SH [IO93OB, SK85]
G7 PQU Details withheld at licensee's request by SSL.
G7 PQV Details withheld at licensee's request by SSL.
G7 PQW N Wills, Lhotse, 18 Hollis Way, Southwick, Trowbridge, BA14 9PH [IO81VH, ST85]
G7 PQX P W Smith, 6 Greate House Farm Rd, Layer de La Haye, Colchester, CO2 0LP [JO01KU, TL92]
G7 PQY G R Wilson, 33 Church Rd, Illogan, Redruth, TR16 4SR [IO70IF, SW64]
G7 PRA Details withheld at licensee's request by SSL.
G7 PRB W Ross, 12 Manvers View, Boughton, Newark, NG22 9HJ [IO93MF, SK66]
G7 PRC S W Newstead, Rectory Cottage, Church Rd, Ashmanhaugh, Norwich, NR12 8YL [JO02RR, TG32]
G7 PRD R F Ware, 72 Alan Rd, Heaton Moor, Stockport, SK4 4DF [IO83VK, SJ89]
G7 PRE J C Huke, 137 Church Parade, Canvey Island, SS8 9RD [JO01GM, TQ78]
G7 PRH D G W Browne, 92 Keir Hardie Way, Barking, IG11 9NX [JO01BM, TQ48]
G7 PRI P K Newton, 60 The Lynch, Winscombe, BS25 1AR [IO81OH, ST45]
G7 PRJ R P Clark, 201 South Park Dr, Ilford, IG3 9AH [JO01BN, TQ48]
GW7 PRK R W S Zeal, Brynteg, Danygraig, Croesyceiliog, Cwmbran, Gwent, NP44 2PR [IO81LQ, ST39]
G7 PRM P R Munn, 34 Bannock Rd, Whitwell, Ventnor, PO38 2RD [IO90IO, SZ57]
G7 PRN Details withheld at licensee's request by SSL.
G7 PRO Details withheld at licensee's request by SSL.
G7 PRP Details withheld at licensee's request by SSL.
G7 PRQ Details withheld at licensee's request by SSL.
G7 PRR Details withheld at licensee's request by SSL.
GW7 PRT W S Townsend, 133 Hazeldene Ave, Brackla, Bridgend, CF31 2JR [IO81FM, SS97]
G7 PRW M A Price, 3 Regency Mews, Regency Rd, Malvern Link, Malvern, WR14 1EB [IO82UD, SO74]
G7 PRY Details withheld at licensee's request by SSL.
G7 PSA Details withheld at licensee's request by SSL.
G7 PSC P Rivers, 39 Ashton Rd, Birmingham, West Midlands, B25 8NZ [IO92CL, SP18]
G7 PSF J D Block, 6 Greystoke Ct, Cherry Hinton Rd, Cambridge, CB1 4DG [JO02CE, TL45]
GM7 PSH A Stevens, 9 Rope Walk, Prestonpans, EH32 9BN [IO85MW, NT37]
G7 PSJ Details withheld at licensee's request by SSL.
G7 PSK Details withheld at licensee's request by SSL.
G7 PSL M A Lamb, 11 Vanburgh Gdns, Kirkhill, Morpeth, NE61 2YF [IO95DD, NZ18]
G7 PSQ A J Hammond, 13 Syon Rd, Minchinhampton, Stroud, GL6 9LD [IO81VQ, SO80]
G7 PSR M H Curtis, 324 Halifax Rd, Sheffield, S6 1AB [IO93GK, SK39]
G7 PST T Ellis, 19 Cavendish Ave, Colchester, CO2 8BP [JO01LU, TM02]
G7 PSU A J Drummond, 10 Bull's Head Cttgs, Turton, Bolton, BL7 0HS [IO83TP, SD71]
G7 PSV M D Bennett, 83 Middlethorpe Rd, Cleethorpes, DN35 9PP [IO93XN, TA20]
G7 PSW D K Wieloch, 43 Northampton Gr, Langdon Hills, Basildon, SS16 6ED [JO01EN, TQ68]
G7 PSX D G Harmer, 19 Horsmonden Rd, Crofton Park, London, SE4 1RG [IO91XK, TQ37]
G7 PSZ A S Laurence, Brookvale, Nooklands, Fulwood, Preston, PR2 4XN [IO83PS, SD53]
G7 PTA M J Masterman, 7 Pond Bank, Blisworth, Northampton, NN7 3EL [IO92ME, SP75]
G7 PTB R K Player, 49 St. Johns Rd, Tilney St. Lawrence, Kings Lynn, PE34 4QJ [JO02DQ, TF51]
G7 PTC M Watson, 56 Woodrush, Coulby Newham, Middlesbrough, TS8 0XB [IO94JM, NZ51]
G7 PTD J M R Midwood, The Vicarage, Reeth, Richmond, North Yorks, DL11 6TR [IO94AJ, SE09]
G7 PTE J E Fox, 41 Plymouth Gr, Radcliffe, Manchester, M26 3WU [IO83TN, SD70]
G7 PTH R Graham, 48 Gudge Heath Ln, Fareham, PO15 5AU [IO90JU, SU50]
G7 PTL Details withheld at licensee's request by SSL.
G7 PTM J M Taylor, 100 Kilgrimol Gdns, St. Annes, Lytham St. Annes, FY8 2RB [IO83LS, SD33]
G7 PTN Details withheld at licensee's request by SSL.
G7 PTO Details withheld at licensee's request by SSL.
G7 PTQ Details withheld at licensee's request by SSL.
G7 PTT A C Myers, 25 Bletchley Ave, Townend Farm, Sunderland, SR5 4LX [IO94GW, NZ35]
G7 PTV M J Howse, 28 Courtiers Dr, Bishops Cleeve, Cheltenham, GL52 4NU [IO81XW, SO92]
G7 PTW R P Elliott, 8 Bridge Pl, Amersham, HP6 6JF [IO91QQ, SU99]
G7 PTX P R Castle, 26 Chestnut Walk, Pulborough, RH20 1AW [IO90RX, TQ01]
G7 PTZ R A Mold, 179 Braeside Ave, Patcham, Brighton, BN1 8SP [IO90WU, TQ30]
G7 PUA J C Campbell, 10 Howarth Ave, Church, Accrington, BB5 4AF [IO83TS, SD72]
G7 PUB A J Rowe, 12 The Knapps, Semington, Trowbridge, BA14 6JG [IO81WI, ST86]
G7 PUC D Mitton, 225 Henhurst Hill, Burton on Trent, DE13 9SX [IO92DT, SK22]
G7 PUD Details withheld at licensee's request by SSL.
G7 PUF Details withheld at licensee's request by SSL.
G7 PUI D A Dearing, 40 Christchurch Rd, Southend on Sea, SS2 4JN [JO01IN, TQ88]
G7 PUJ W F Jones, 19 Jarman House, Silwood Est, Rotherhithe, London, SE16 2PW [IO91XL, TQ37]
G7 PUL R A Hoggard, 1 Whiphill Cl, Bessacarr, Doncaster, DN4 6DX [IO93LM, SE60]
G7 PUM Details withheld at licensee's request by SSL.
G7 PUN D I Russon, 123 Queens Dr, Newton-le-Willows, WA12 0LN [IO83QL, SJ59]
G7 PUP A C Hurd, 34 Sandrock Dr, Bessacarr, Doncaster, DN4 6DT [IO93LM, SE60]
G7 PUQ Details withheld at licensee's request by SSL.
G7 PUR C P Miles, 61 Ashe Cres, Chippenham, Wilts, SN15 1RN [IO81WL, ST97]
G7 PUU Details withheld at licensee's request by SSL.
G7 PUV Details withheld at licensee's request by SSL.
G7 PUW S Frizzell, 85 Gibbon Rd, Newhaven, BN9 9ER [JO00AS, TQ40]
GW7 PUX A J Rollin, Flat 20, St. Marys Mews, Church Ln, Mold, Clwyd, CH7 1NT [IO83KD, SJ26]
G7 PUZ L N Martyn, 1 Canewdon Hall Cl, Canewdon, Rochford, SS4 3PY [JO01IO, TQ89]
G7 PVA J C Davidson, 14 Petworth Gr, Yardley, Birmingham, B26 1JF [IO92CL, SP18]
GW7 PVD G A M Palmer, 102 Park Ave, Bryn y Baal, Mold, CH7 6TP [IO83KE, SJ26]
G7 PVE G C Eddy, 102 Springfield Cl, Andover, SP10 2QT [IO91GF, SU34]
G7 PVF M E Folland, 14 High St., Shoreham, Sevenoaks, TN14 7TD [JO01CI, TQ56]
G7 PVG B J H Fox, 10 Materman Rd, Stockwood, Bristol, BS14 8SS [IO81RJ, ST66]
GI7 PVI M Major, Mon Reve, Ave Beauvais, Ville Au Roi, St Peter Port, Guernsey, GY1 1PQ
G7 PVJ C J Collins, 19 Fletcher Way, Bognor Regis, PO21 2NU [IO90PS, SZ99]
G7 PVK Details withheld at licensee's request by SSL.
G7 PVL C J Watts, 81 Tetbury Gdns, Nailsea, Bristol, BS19 2TN [IO81PK, ST47]
G7 PVM P K Bown, 12 Hawkins St., Rodbourne, Swindon, SN2 2AQ [IO91CN, SU18]
G7 PVQ Details withheld at licensee's request by SSL.
G7 PVR Details withheld at licensee's request by SSL.
G7 PVS Details withheld at licensee's request by SSL.
GM7 PVT W McArthur, 16 Arden Rd, Greenock, PA15 3AB [IO75PW, NS27]
GM7 PVV S Fairfull, 10 Canmore Gdns, Kirkcaldy, KY2 6XR [IO86JD, NT29]
G7 PVW J C Raybould, 38 Church St., West Hagley, Hagley, Stourbridge, DY9 0NA [IO82WK, SO98]
G7 PVX J C Russell, 10 Loyd Rd, Didcot, OX11 8JT [IO91IO, SU58]
G7 PVZ Details withheld at licensee's request by SSL.
G7 PWA N D Padley, 49 Willow Way, Farnham, GU9 0NU [IO91OF, SU84]
G7 PWB Details withheld at licensee's request by SSL.
G7 PWD R A Baker, 5 Bakers Rd, Halvergate, Norwich, NR13 3PY [JO02SO, TG40]
G7 PWE Details withheld at licensee's request by SSL.
G7 PWH S Smith, Flat, 8 Maunsell Ct, Western Rd, Haywards Heath, RH16 3LJ [IO90WX, TQ32]

GM7 PWI C E Thornton, 14 Kevock Rd, Lasswade, EH18 1HT [IO85KV, NT26]
G7 PWJ J E Churchill, 68 Anthony Rd, Woodside, London, SE25 5HB [IO91XJ, TQ36]
G7 PWK M Bibb, 28 Netherfield Rd, Kettering, NN15 6DY [IO92PJ, SP87]
G7 PWL M D Turnbull, 11 Waverley Ave, Whitley Bay, NE25 8AU [IO95GA, NZ37]
G7 PWO Details withheld at licensee's request by SSL.
G7 PWP D Blackman, 20 Catherine St., Winton, Eccles, Manchester, M30 8JB [IO83TL, SJ79]
GI7 PWQ G J McCormick, 8 Kinallen Cttgs, Kinallen, Dromara, Dromore, BT25 2PW [IO64XJ, J24]
G7 PWS C J Collins, 15 Holly Rd, St. Marys Bay, Romney Marsh, TN29 0XB [JO01LA, TR02]
G7 PWU H Tomlinson, 132 Main Rd, Wybunbury, Nantwich, CW5 7LR [IO83SB, SJ64]
G7 PWV P J Woolhouse, 21 Coombe Wood Hill, Purley, CR8 1JQ [IO91WH, TQ36]
G7 PWW Details withheld at licensee's request by SSL.
G7 PWX Details withheld at licensee's request by SSL.
G7 PWY N A Spencer, 28 Hazelwood Rd, Corby, NN17 1HS [IO92PL, SP88]
G7 PXA Details withheld at licensee's request by SSL.
G7 PXB Details withheld at licensee's request by SSL.
G7 PXC M K J Howard, 145 Pump Ln, Springfield, Chelmsford, CM1 6TA [JO01GS, TL70]
G7 PXE Details withheld at licensee's request by SSL.
GM7 PXJ L I Michie, 15 Upper Wellheads, Limekilns, Dunfermline, KY11 3JQ [IO86GA, NT08]
GM7 PXL D W Warner, Achnacon, Glencoe, Argyll, PA39 4LA [IO76KP, NN15]
G7 PXP Details withheld at licensee's request by SSL.
G7 PXR P J Gilbert, 39 Kenn Moor Dr, Clevedon, BS21 5AR [IO81NK, ST47]
G7 PXS G A Mape, 34 Amberwood Dr, Baguley, Manchester, M23 9NZ [IO83UJ, SJ78]
G7 PXT Details withheld at licensee's request by SSL.
G7 PXV S Vivian, 26 Kilpeck Ave, Newton Farm, Hereford, HR2 7DS [IO82PA, SO43]
G7 PXX E W Spires, 336 Hainton Ave, Grimsby, DN32 9LX [IO93XN, TA20]
G7 PXY R N Astle, 52 Bevan St. West, Lowestoft, NR32 2AD [JO02UL, TM59]
G7 PYA Details withheld at licensee's request by SSL.
G7 PYG B L Stockley, 18 Repton Ave, Perton, Wolverhampton, WV6 7TD [IO82VO, SO89]
GW7 PYH D H Williams, 45 Llys Nercwys, Mold, CH7 1HR [IO83KD, SJ26]
G7 PYI Details withheld at licensee's request by SSL.
G7 PYK Details withheld at licensee's request by SSL.
G7 PYL R B Rush, Solucki, Old Bridge, Gressenhall, Dereham, NR19 2QE [JO02KQ, TF91]
G7 PYN A Wentworth, 5 York Ave, Off Kings Rd, Prestwich, Manchester, M25 0FZ [IO83UM, SD80]
G7 PYQ N Gell, 1 Lawton Rd, Rushden, NN10 0DX [IO92QH, SP96]
G7 PYR V G Tuff, 49 Solingen Est, Blyth, NE24 3ER [IO95FC, NZ38]
G7 PYT G A Coleman, 23 Graham Ave, Patcham, Brighton, BN1 8HA [IO90WU, TQ30]
G7 PYV A R Turner, 20 Kipling Gdns, Upper Stratton, Swindon, SN2 6LJ [IO91CO, SU18]
G7 PYW K J Houghton, 42 Pear Tree Ave, Coppull, Chorley, PR7 4NL [IO83QP, SD51]
G7 PZA Details withheld at licensee's request by SSL.
G7 PZB R P Dewsbery, 8 Westfield Cl, Market Harborough, LE16 9DX [IO92ML, SP78]
G7 PZE F J Eastham, 19 Southern Parade, Preston, PR1 4NH [IO83PS, SD52]
G7 PZF W D Bailey, 15 Norfolk Rd, Congleton, CW12 1NY [IO83VE, SJ86]
G7 PZG J Warham, 17 Pexhill Dr, Macclesfield, SK10 3LP [IO83WG, SJ87]
GM7 PZH M Drennan, 6 Hillpark Way, Edinburgh, EH4 7BJ [IO85IX, NT27]
G7 PZI Details withheld at licensee's request by SSL.
G7 PZK Details withheld at licensee's request by SSL.
G7 PZM Details withheld at licensee's request by SSL.
G7 PZP S J Porritt, 15 Fountains Cres, Eston, Middlesbrough, TS6 9DF [IO94KN, NZ51]
G7 PZQ P A Breese, 15 Willey Gr, Erdington, Birmingham, B24 9RQ [IO92CM, SP19]
G7 PZS L E Taylor, 27 Wharfdale Dr, Wirral, L62 8EZ [IO83MH, SJ38]
G7 PZU A F Haworth, Coach Cottage, Trewince Manor, Portscatho, Truro, TR23 5ET [IN69TW, SV81]
G7 PZX Details withheld at licensee's request by SSL.
G7 PZY A E G McCabe, 36 West Park Rd, South Shields, NE33 4LB [IO94GX, NZ36]
G7 PZZ Details withheld at licensee's request by SSL.
G7 RAB D A Evans, 31 Kinsbourne Way, Thornhill, Southampton, SO19 6HB [IO90IV, SU41]
G7 RAC Details withheld at licensee's request by SSL.
G7 RAD D A Reed, 27 Acacia Dr, Sutton, SM3 9NJ [IO91VJ, TQ26]
G7 RAF Details withheld at licensee's request by SSL.
G7 RAG J F Dennis, The Old Chapel House, Farlesthorpe, Alford, Lincs, LN13 9PH [JO03CF, TF47]
GI7 RAH T D Tweedie, 17 Schomberg Park, Belmont Rd, Belfast, BT4 2HH [IO74BO, J37]
G7 RAI M J Moorhouse, 11 Hazel Gr, Huddersfield, HD2 2JP [IO93CQ, SE12]
G7 RAJ D G R Eggett, Copper Beeches, 68 Forest Ln, Kirklevington, Yarm, TS15 9ND [IO94HL, NZ40]
GM7 RAK J Boyd, 102 Provost Milne Gr, South Queensferry, EH30 9PL [IO85HX, NT17]
G7 RAL I A Hewitt, 26 Outwoods Dr, Loughborough, LE11 3LT [IO92JS, SK51]
GI7 RAM J T Christie, 3 Victoria Dr, Sydenham, Belfast, BT4 1QT [IO74BO, J37]
G7 RAO M E Hall, Morar, Altwood Cl, Maidenhead, Berks, SL6 4PP [IO91PM, SU88]
G7 RAP Details withheld at licensee's request by SSL.
G7 RAR Details withheld at licensee's request by SSL.
G7 RAT R T Collins Reigate ATS, 8 Sylvan Way, Redhill, RH1 4DE [IO91WF, TQ24]
G7 RAU D A Edwards, 37 Barton Cl, Whippingham, East Cowes, PO32 6LS [IO90IR, SZ59]
G7 RAW D J Linsley, 4 Sanders Gdns, Birtley, Chester-le-Street, DH3 1NA [IO94FV, NZ25]
G7 RAY R S Biddle, 6 Hornby Croft, Moss Side, Leyland, Preston, PR5 3UT [IO83PQ, SD52]
G7 RAZ M J Wager, 36 Pickards Way, Wisbech, PE13 1SD [JO02BQ, TF41]
G7 RBA M C Sims, 23 Winding Way, Alwoodley, Leeds, LS17 7RB [IO93FU, SE24]
G7 RBB A K Perkins, 9 St. Martins Cl, Canterbury, CT1 1QG [JO01NG, TR15]
G7 RBC I G Rodgers, 89 Braemar Rd, Worcester Park, KT4 8SN [IO91VI, TQ26]
G7 RBF R P Day, Resting Oak Cottage, Resting Oak Hill, Cooksbridge, Lewes, BN8 4PS [IO90XV, TQ31]
G7 RBG Details withheld at licensee's request by SSL.
G7 RBH E A J Dell, Church Farm, Church Ln, Mattishall, Dereham, NR20 3QZ [JO02MP, TG01]
G7 RBI Details withheld at licensee's request by SSL.
G7 RBK Details withheld at licensee's request by SSL.
G7 RBL C G Johnson, 11 Maple Ave, Newcastle, ST5 7BN [IO83VB, SJ85]
G7 RBN F Benson, 55 Lowes Rd, Bury, BL9 6PJ [IO83UO, SD81]
GM7 RBP S B Wilkins, Firlee, Hayhillock, Ellon, AB41 8DH [IO87WJ, NJ93]
G7 RBQ P C Dodman, 26 Beaumont Lawns, Marlbrook, Bromsgrove, B60 1HZ [IO82XI, SO97]
G7 RBR J Pavia, 85 Bodmin Rd, Chelmsford, CM1 6LL [JO01FR, TL70]
G7 RBS A J Sercombe, 28 Strumpshaw Rd, Brundall, Norwich, NR13 5PA [JO02RO, TG30]
G7 RBT S G Worrall, 41 The Grove, Stourport on Severn, DY13 9ND [IO82UI, SO87]
G7 RBV Details withheld at licensee's request by SSL.
GM7 RBW D A Saunders, 172 Colinton Mains Rd, Edinburgh, EH13 9DB [IO85JV, NT26]
G7 RBX B A Waddington, 12 Princes End, Lawley Bank, Dawley Bank, Telford, TF4 2JN [IO82SQ, SJ60]
G7 RBZ J J Hudson, 3 Park Ln, Harvington, Kidderminster, DY10 4LW [IO82VI, SO87]
G7 RCC R W Wendes, 108 Osborne Rd, East Cowes, PO32 6RZ [IO90IS, SZ59]
G7 RCD Details withheld at licensee's request by SSL.
G7 RCE P D Beeson, 52 Milford Cl, Abbey Wood, London, SE2 0DT [JO01BL, TQ47]
G7 RCG Details withheld at licensee's request by SSL.
G7 RCK Details withheld at licensee's request by SSL.
G7 RCL R J Abbott, 2 Leybourne Dr, Springfield, Chelmsford, CM1 6TX [JO01FS, TL70]
G7 RCN R A Barker, 6 The Oval, Otley, LS21 2ED [IO93DV, SE14]
G7 RCO J Cooper, 72 Chelmsford Rd, Edgeley, Stockport, SK3 9LS [IO83VJ, SJ88]
G7 RCP D Baines, 39 Hydes Rd, Wednesbury, WS10 9SX [IO82XN, SO99]
G7 RCQ Details withheld at licensee's request by SSL.
GW7 RCR C R Evans, y Rhosfa, High St., Llanfyllin, SY22 5AF [IO82IS, SJ11]
G7 RCU A Walker, 2 Shredicote Ln, Bradley, Staffs, ST18 9EB [IO82VR, SJ81]
G7 RCW J A Matthews, The Forge, Norton Heath, Ingatestone, Essex, CM4 0LJ [JO01DR, TL60]
G7 RCX Details withheld at licensee's request by SSL.
G7 RDA P J Brownsett, 24 Lancaster Rd, Shortstown, Bedford, MK42 0UB [IO92SC, TL04]
G7 RDE S A Newton, 7 Little Pynchons, Harlow, CM18 7DB [JO01BS, TL40]
G7 RDG N C Wainman, 9 Ascot Dr, Arnold, Redhill, Nottingham, NG5 8LR [IO93KA, SK54]
GM7 RDH R W R Spence, Leylan, Harray, Orkney, Scotland, KW17 2LQ [IO89JB, HY31]
G7 RDI Details withheld at licensee's request by SSL.
G7 RDJ R C Middleton, 10 Crossbeck Rd, Ilkley, LS29 9JN [IO93CW, SE14]
G7 RDL J M Davison, obo Leeds Raynet Group, 29 Glenfield Ave, Wetherby, LS22 6RN [IO93HW, SE44]
G7 RDM Details withheld at licensee's request by SSL.
G7 RDN Details withheld at licensee's request by SSL.
GM7 RDO Details withheld at licensee's request by SSL.
G7 RDQ T L Rochford, 41 Lynwood Dr, Blakedown, Kidderminster, DY10 3JZ [IO82VJ, SO87]
GM7 RDR J A Ryan, 182 Dunearn Dr, Kirkcaldy, KY2 6LD [IO86JD, NT29]
G7 RDT C G Hampson, 7 Merryfield Cl, Bransgore, Christchurch, BH23 8BS [IO90DS, SZ19]
GW7 RDV Details withheld at licensee's request by SSL.
G7 RDX Details withheld at licensee's request by SSL.
GM7 RDY J W Mowat, Nether Bigging, Shapinsay, Orkney, KW17 2EB [IO89OB, HY52]
G7 REA T G Hayton, Longroyd, Little Humby, Grantham, Lincs, NG33 4HW [IO92RV, TF03]
G7 REB P J Rimmer, Heather Lea Cottage, Well Ln, Brinscall, Chorley, Lancs, PR6 8QX [IO83RQ, SD62]
G7 REE S B Meager, 7 Kinsley St., Wymondham, NR18 0NU [JO02NN, TG10]
G7 REF M Harrington Epping Forest Raynet Group, 9 High House Est, Sheering Rd, Old Harlow, Harlow, CM17 0LL [JO01BS, TL41]
GM7 REG J A Robertson, 13 Swanston View, Edinburgh, EH10 7DG [IO85JV, NT26]
G7 REH P R Evans, 9 Cressingham Rd, Reading, RG2 7RT [IO91MK, SU77]
G7 REJ S J Hutchinson, 5 Rokeby Sq, Merryoaks, Durham, DH1 3QE [IO94ES, NZ24]
G7 REK D Bradley, 159 Liverpool Rd, Irlam, Manchester, M44 6DA [IO83SK, SJ79]
G7 REL F L Libby, Gweltek, Point Rd, Carnon Downs, Truro, TR3 6JN [IO70LF, SW84]
G7 REM M E O'Rourke, Valley Farm, Laundry Ln, Huntingfield, Halesworth, IP19 0PY [JO02RH, TM37]
G7 REO A J White, 4 Haine Farm Cottage, Haine Rd, Ramsgate, CT12 5AF [JO01QI, TR36]
GI7 REP H W R J Browne, 56 Ballylinney Rd, Ballyclare, BT39 9RH [IO64XR, J28]
G7 RER R B L Clegg, Cleggs Castle, Monks Ln, Freshwater, PO40 9ST [IO90FQ, SZ38]
G7 RES G P O'Neill, 4D Solent Gdns, Freshwater, PO40 9PN [IO90FQ, SZ38]
G7 REV S R Hurst, 25 Florence Rd, Abington, Northampton, NN1 4NA [IO92NF, SP76]
G7 REX Details withheld at licensee's request by SSL.
GM7 REY J Macdonald, 58 Carn Dearg Rd, Claggan, Fort William, PH33 6QD [IO76LT, NN17]
GW7 RFA A Lord, The Mount, Trefecca, Brecon, Powys, LD3 0PW [IO81JX, SO13]
G7 RFB A T Clarke, Highfield, Dunsford, near Exeter, EX6 7DE [IO80DQ, SX88]
G7 RFC S D Little, 89 Hunts Dr, Writtle, Chelmsford, CM1 3HQ [JO01FR, TL60]
G7 RFD P Johnson, Sixpenny Cottage, Farthings Fold, Hanthorpe, Bourne, Lincs, PE10 0RN [IO92TT, TF02]
G7 RFE R E D Johnson, Sixpenny Cottage, Farthings Fold, Hanthorpe, Bourne, Lincs, PE10 0RN [IO92TT, TF02]
G7 RFF A L Lloyd, 52 Church Rd, Oxley, Wolverhampton, WV10 6AF [IO82WO, SJ90]
G7 RFG Details withheld at licensee's request by SSL.
G7 RFH J Fearns, 14 Grange Dr, Heswall, Wirral, L60 7RU [IO83KI, SJ28]
G7 RFI Details withheld at licensee's request by SSL.
G7 RFK J Taylor, 47 Delamere Dr, Macclesfield, SK10 2PW [IO83WF, SJ97]
G7 RFM G L Hunt, 7 Kevington Dr, St. Pauls Cray, Orpington, BR5 2NT [JO01BJ, TQ46]
GM7 RFN D W Mitchell, 10 Braehead, Cupar, KY15 4BE [IO86LH, NO31]
GW7 RFP P M N Barry, 42 Derwen Way, Abergavenny, NP7 6BP [IO81LT, SO31]
G7 RFR Details withheld at licensee's request by SSL.
G7 RFS K Abeynayake, 28 Mallow Park, Maidenhead, SL6 6SQ [IO91PM, SU88]
G7 RFT K L A Whittle, 26 Beachs Dr, Chelmsford, CM1 2NJ [JO01FR, TL60]
G7 RFV D P Kamm, 134 Haslemere Rd, Liphook, GU30 7BX [IO91OB, SU83]
G7 RFW R K Bridges, 31 Barnes Green, Bebington, Wirral, L63 9LU [IO83MI, SJ38]
G7 RFX R N Oxlade, 3 Thyme Ct, Lumbertubs, Northampton, NN3 8HY [IO92NG, SP76]
G7 RFY G T Annett, 2 Hay Cl, Balsham, Cambridge, CB1 6EJ [JO02DD, TL55]
G7 RFZ J G Bilmen, 145 The Maples, Harlow, CM19 4RD [JO01BR, TL40]
G7 RGA P H Cattanach, 21 Darlington Cl, Amersham, HP6 5AD [IO91QQ, SU99]
GW7 RGB Details withheld at licensee's request by SSL.
GM7 RGC R G Brown, 12 Warriston Terr, Edinburgh, EH3 5LZ [IO85JX, NT27]
G7 RGF G L Malynicz, 22 Coolgardie Ave, Chigwell, Essex, IG7 5AY [JO01AO, TQ49]
G7 RGG S R Emmett, 141 Wood St., Chelmsford, CM2 8BH [JO01FR, TL70]
G7 RGI H M Yates Jones, Chestnut Cottage, Chard Rd, Drimpton, Beaminster, DT8 3RF [IO80OU, ST40]
G7 RGJ Details withheld at licensee's request by SSL.
G7 RGK K R Bland, 75 Moseley Wood Gdns, Leeds, LS16 7HX [IO93EU, SE24]
G7 RGN F H Duggins, 28 Hill Dr, Exmouth, EX8 4QQ [IO80HP, SY08]
G7 RGO E J Allan, 55B Kingston Rd, South Wimbledon, London, SW19 1JW [IO91VJ, TQ27]
G7 RGR R Jones, 18 Ashgrove, Burnham on Crouch, CM0 8DP [JO01JP, TQ99]
G7 RGS Details withheld at licensee's request by SSL.
G7 RGV H K Jump, 85 Liverpool Rd, Bickerstaffe, Ormskirk, L39 0EG [IO83NM, SD40]
GM7 RGW Details withheld at licensee's request by SSL.
G7 RGX G Henderson, 20 Weston Park Ave, Shelton Lock, Derby, DE24 9EQ [IO92GV, SK33]
G7 RGY R N Hannigan, 4 Westlands Ave, Tetney, Grimsby, DN36 5LP [IO93XL, TA30]
G7 RGZ A Drain, 34 Helsby Rd, Sale, M33 2XD [IO83UJ, SJ89]
G7 RHC Details withheld at licensee's request by SSL.
G7 RHD M R Clarke, 45 Eyebrook Rd, Bowdon, Altrincham, WA14 3LQ [IO83TI, SJ78]
G7 RHF Details withheld at licensee's request by SSL.
G7 RHG T D Moore, High Ferry, Tattershall Rd, Woodhall Spa, LN10 6TP [IO93VD, TF16]
G7 RHI A S McLocklin, 43 Forbes Ave, Potters Bar, EN6 5NB [IO91WQ, TL20]
G7 RHL Details withheld at licensee's request by SSL.
G7 RHM Details withheld at licensee's request by SSL.
G7 RHO Details withheld at licensee's request by SSL.
GM7 RHP D H Brattle, 5 Downshire Rd, Bangor, BT20 3TW [IO74DP, J48]
G7 RHT P G Bennett, 6 Dighton Gate, Stoke Gifford, Bristol, BS12 6XA [IO81RM, ST68]
GW7 RHV S J Bailey, 37 Richards Terr, Roath, Cardiff, CF2 1RW [IO81KL, ST17]
G7 RHX C B E Hopkins, 13 Brantfell Dr, Burnley, BB12 8AW [IO83UT, SD83]
G7 RHY Details withheld at licensee's request by SSL.
G7 RIA M J Cook, 70 Listowel Rd, Dagenham, RM10 7QP [JO01BN, TQ48]
GW7 RIB P G Nicholls, 11 Ifor Hael Rd, Rogerstone, Newport, NP1 9FB [IO81LO, ST28]
G7 RIE J D Glenn, School House, Norwich Rd, Horstead, Norwich, NR12 7EG [JO02QR, TG21]
G7 RIF Details withheld at licensee's request by SSL.
G7 RIH N J Camp, 48 Northfield Rd, Harpenden, AL5 5HZ [IO91TT, TL11]
G7 RIJ E C Devine, 23 Radley Ave, Wickersley, Rotherham, S66 0HZ [IO93IK, SK49]
G7 RIL Details withheld at licensee's request by SSL.
G7 RIM M C Atkinson, 54 Wellington Rd, Enfield, EN1 2PH [IO91XP, TQ39]
G7 RIN G E Brough, 30 Forest Rd, Ford Est, Sunderland, SR4 0DS [IO94GV, NZ35]
G7 RIO W Care, Grenville, 36 Rosedale Rd, Truro, TR1 3RZ [IO70LG, SW84]
G7 RIR E J Winthorpe, 11 Gallimore Cl, Glenfield, Leicester, LE3 8HA [IO92JP, SK50]
G7 RIS Details withheld at licensee's request by SSL.
G7 RIU D F Kirk, 19 The Meads, Hildersley, Ross on Wye, HR9 7NF [IO81RV, SO62]
G7 RIY G Webster, Wallage Lodge, Wallage Ln, Rowfant, Crawley, RH10 4NL [IO91WC, TQ33]
GW7 RJC D J Morris, 62 Gerllan, Tywyn, LL36 9DE [IO72XO, SH50]
G7 RJD P P K Ho, Gonville & Caius College, Trinity St, Cambridge, CB2 1TA [JO02BE, TL45]
G7 RJF Details withheld at licensee's request by SSL.
GM7 RJG A Forbes, 28 Innes St, Inverness, IV1 1NS [IO77VL, NH64]
G7 RJI R W Lupton, 18 Florey Walk, Clifton Est, Nottingham, NG11 8RE [IO92JV, SK53]
G7 RJK Details withheld at licensee's request by SSL.
G7 RJL Details withheld at licensee's request by SSL.
G7 RJO C W Compton, 55 Lulot Gdns, Highgate New Town, London, N19 5TR [IO91WN, TQ28]
G7 RJP Details withheld at licensee's request by SSL.
G7 RJR Details withheld at licensee's request by SSL.
G7 RJV Details withheld at licensee's request by SSL.
G7 RJW D P Lamden, 25 Ashwood Dr, Turnpike Est, Newbury, RG14 2PN [IO91IJ, SU46]
G7 RJX D J Nicholas, 41 High Ridge, Hazlerigg, Newcastle upon Tyne, NE13 7ND [IO95EA, NZ27]
G7 RKB Details withheld at licensee's request by SSL.
GM7 RKD I C Millar, Seklie Stanes, Scatness, Virkie, Shetland, ZE3 9JW [IO99IV, HU31]
G7 RKE J P Bottomley, 10 Gyllyng St, Falmouth, TR11 3EL [IO70LD, SW83]
G7 RKH J M Kenny, 3 Priory View, Minting, Horncastle, LN9 5RU [IO93VF, TF17]
G7 RKI Details withheld at licensee's request by SSL.
G7 RKJ C N Hindmarsh, 3 Newlay Wood Dr, Horsforth, Leeds, LS18 4LL [IO93ET, SE23]
G7 RKK Details withheld at licensee's request by SSL.
G7 RKM Details withheld at licensee's request by SSL.
G7 RKO R Kennedy, 6 Oaky Balks, Alnwick, NE66 2QE [IO95DJ, NU11]
GW7 RKQ S C Rudge, Glanafon Nant, Fairy Glen Rd, Dwygyfylchi, Penmaenmawr, LL34 6YU [IO83BG, SH77]
G7 RKT P Jones, 14 Westerleigh Rd, Clevedon, BS21 7US [IO81NK, ST37]
G7 RKU P C Dickinson, 152 Rostrevor Rd, Stockport, Ches, SK3 8UT [IO83WJ, SJ88]
G7 RKV D R Wilson, 5 Hamilton Terr, Whitehaven, CA28 7TT [IO84FN, NX91]
G7 RKX C E R Hill-Smith, Seale Stoke, Holne, Newton Abbot, Devon, TQ13 7SS [IO80CM, SX76]
G7 RKY Details withheld at licensee's request by SSL.
GM7 RLA A G Sheret, 57 Glenview Ave, Banknock, Bonnybridge, FK4 1JX [IO85AX, NS77]
G7 RLB R J Thomson, 20 Cadzaw Cres, Boness, EH51 7AZ [IO86EA, NS98]
G7 RLE Details withheld at licensee's request by SSL.
GE7 RLF M J Watson Raynet Lowestoft Fire, The Tubbery, Henley, Ipswich, Suffolk, IP6 0BR [JO02NC, TM15]
G7 RLG S G Hale, 4 Prescott Rd, Wilmslow, SK9 4DL [IO83VI, SJ88]
G7 RLH D E Brocklesby, 34 Sinnington End, Highwoods, Colchester, CO4 4RE [JO01LV, TM02]
G7 RLI Details withheld at licensee's request by SSL.
G7 RLK S J Drury, 37 Harrington St., Cleethorpes, DN35 7AY [IO93XN, TA20]
G7 RLN J N Palmer, 13 Chandos Rd, Chorlton Cum Hardy, Manchester, M21 0SS [IO83UK, SJ89]
G7 RLO L J Van Beers, 4 Cwmbeth Cl, Crickhowell, NP8 1DX [IO81KU, SO21]
G7 RLP Details withheld at licensee's request by SSL.
G7 RLQ T H G Winton, Corriehead, 41 Oxford Rd, Benson, Wallingford, OX10 6LX [IO91KO, SU69]
GW7 RLS J E Gray, City & Co. of Swansea, Emergency Planning Unit, The Guildhall, Swansea, Glam, SA1 4PH [IO81AO, SS69]
G7 RLV C L Pitchford, 84 New Rd, Rubery, Rednal, Birmingham, B45 9HY [IO82XJ, SO97]
G7 RLX G V Coleman, 120 Kidderminster Rd South, Hagley, Stourbridge, DY9 0JH [IO82WJ, SO87]
GW7 RLZ G L Roberts, 4 Fawnog Wen, Penrhyndeudraeth, LL48 6PS [IO72XW, SH63]
G7 RMA R Paddock, 9 Woodstock Rd, Old Trafford, Manchester, M16 0HR [IO83UK, SJ89]

G7 RMC Details withheld at licensee's request by SSL.
G7 RMD D S Devlin, The Bungalow, 17 Fair View, Dalton in Furness, LA15 8RZ [IO84JD, SD27]
G7 RME M W Buckley, 8 Highthorne St., Armley, Leeds, West Yorks, LS12 3LB [IO93ET, SE23]
GM7 RMF E P Walker, 38 Greenbank Gdns, Edinburgh, EH10 5SN [IO85JW, NT27]
G7 RMG G K J Chapman, Crockers Farm, Stoke Wake, Blandford Forum, Dorset, DT11 0HF [IO80TU, ST70]
G7 RMH Details withheld at licensee's request by SSL.
G7 RMI Details withheld at licensee's request by SSL.
G7 RML D Woodall, 28 Glenluce, Birtley, Chester-le-Street, DH3 2HY [IO94FV, NZ25]
G7 RMO N J Smith, 7 Quarry Ln, Sheffield, S11 9EA [IO93FI, SK38]
G7 RMP M J Watson Raynet Mildenhall Police, The Tubbery, Henley, Ipswich, Suffolk, IP6 0BR [JO02NC, TM15]
G7 RMQ R B Scarce, East Side, The Common, Tunstall, Woodbridge, IP12 2JR [JO02RD, TM35]
G7 RMR T Matsuoka, 102 Holland Rd, London, W14 8BD [IO91VM, TQ27]
G7 RMS Details withheld at licensee's request by SSL.
GM7 RMV T Dennis, 12 Sunnyside Gdns, Aberdeen, AB2 3LZ [IO87TH, NJ72]
G7 RMW R Medcalf Mid Warks Raynet Group, 19 All Saints Rd, Warwick, CV34 5NL [IO92FG, SP26]
G7 RMX N R Taylor, 16 Rosemount Ln, Honiton, EX14 8RJ [IO80JT, ST10]
G7 RMZ B M Williams East Chehire Raynet Group, 3 Welton Cl, Wilmslow, SK9 6HD [IO83VH, SJ87]
G7 RNA K M Kent obo North Anglia Raynet, 5 Jubilee Rd, Heacham, Kings Lynn, PE31 7AR [JO02FV, TF63]
G7 RNB S W Bieber, Tonkins Quay House, Mixtow, Lanteglos-By-Fowey, Cornwall, PL23 1NB [IO70QI, SX15]
GW7 RNC T G Heywood-Bell, 21 Tregarn Cl, Langstone, Newport, NP6 2JL [IO81NO, ST39]
G7 RNG K Sim, 3 Thorngate Cl, Penwortham, Preston, PR1 0XN [IO83PR, SD52]
GM7 RNJ M Dennis, 12 Sunnyside Gdns, Aberdeen, AB2 3LZ [IO87TH, NJ72]
G7 RNL N J Hull, 126 Coval Ln, Chelmsford, CM1 1TG [JO01FR, TL70]
G7 RNN C H Harrold, Boundary Farm, Cromer Rd, Felbrigg, Norwich, NR11 8PD [JO02PV, TG23]
G7 RNO E Birtwell, 29 Westgate, Flixton, Urmston, Manchester, M41 9EL [IO83TK, SJ79]
G7 RNQ R C Young, 12 Elmwood Cl, Stokesley, Middlesbrough, TS9 5HX [IO94JL, NZ50]
G7 RNT A M Blackburn, 5 Mill View, Eppleby, Richmond, DL11 7BJ [IO94DM, NZ11]
G7 RNX A Linney, 5 Elliscales Ave, Dalton in Furness, LA15 8BW [IO84JD, SD27]
G7 ROC J Armstrong, 111 Moreton Way, Cippenham, Slough, SL1 5LR [IO91QM, SU98]
G7 ROF Details withheld at licensee's request by SSL.
G7 ROI J Naylor, 46 Loxley Dr, Mansfield, NG18 4FB [IO93JD, SK56]
G7 ROJ Details withheld at licensee's request by SSL.
G7 ROL G F S Heaney, 194 Beckenham Rd, Beckenham, BR3 4RJ [IO91XJ, TQ36]
G7 ROM A S Boardman, 51 Tonge Moor Rd, Bolton, BL2 2DL [IO83TO, SD71]
G7 RON Details withheld at licensee's request by SSL.
G7 ROQ R Collins, c/o Room 4.4 Swindon College, North Star Ave, Swindon, SN2 1DY [IO91CN, SU18]
G7 ROS Details withheld at licensee's request by SSL.
G7 ROY R Clayton, 9 Green Island, Irton, Scarborough, YO12 4RN [IO94SF, TA08]
G7 RPC J Herold, 18 Asquith St., Stockport, SK5 6UR [IO83WK, SJ89]
G7 RPG Details withheld at licensee's request by SSL.
G7 RPJ J B Barnard, 21 Burleigh Cl, Great Yarmouth, NR30 2RU [JO02UO, TG50]
G7 RPK L C Goffin, The Hollies, Belaugh Green, Coltishall, Norwich, Norfolk, NR12 7AJ [JO02QR, TG21]
G7 RPM C W Sedgwick, 28 Woodall Ave, Scarborough, YO12 7TH [IO94TG, TA08]
G7 RPO Details withheld at licensee's request by SSL.
G7 RPP I K Gurney, 14 Temple Towers, Richmond Rd, Basingstoke, RG21 5PA [IO91KG, SU65]
G7 RPR Details withheld at licensee's request by SSL.
G7 RPS D M Gaston, 8 Linden Cl, Chelsfield, Orpington, BR6 6JJ [JO01BI, TQ46]
GM7 RPT D I Hutchison, 30 Seright Sq, Crookedholm, Kilmarnock, KA3 6LE [IO75SO, NS43]
G7 RPV S Fisher, 8 Tubbenden Ln, Orpington, BR6 9PN [JO01BI, TQ46]
G7 RPW S D Pike, 63 Otterfield Rd, West Drayton, UB7 8PE [IO91SM, TQ08]
G7 RPX Details withheld at licensee's request by SSL.
G7 RPZ Details withheld at licensee's request by SSL.
G7 RQA Details withheld at licensee's request by SSL.
G7 RQB K J Greenfield, 36 Barttelot Rd, Horsham, RH12 1DQ [IO91UB, TQ13]
G7 RQD Details withheld at licensee's request by SSL.
G7 RQE A Laver, 34 Providence Way, Waterbeach, Cambridge, CB5 9QJ [JO02CG, TL46]
G7 RQG Details withheld at licensee's request by SSL.
GW7 RQI D R Pearson, 142 Heol Bryngwili, Crosshands, Cross Hands, Llanelli, SA14 6LY [IO71WS, SN51]
GM7 RQK S R Skidmore, 6 Blairlinn View, Luggiebank, Cumbernauld, Glasgow, G67 4AD [IO85AW, NS77]
G7 RQL C Jenkins Oulder Hill Radio Soc, 31 Ashbrook Cres, Smallbridge, Rochdale, OL12 9AJ [IO83WP, SD91]
GW7 RQM Details withheld at licensee's request by SSL.
GW7 RQN Details withheld at licensee's request by SSL.
G7 RQO Details withheld at licensee's request by SSL.
G7 RQS Details withheld at licensee's request by SSL.
G7 RQT Details withheld at licensee's request by SSL.
G7 RQU D J W Earnshaw, 7 Conisborough Ln, Garforth, Leeds, LS25 2LR [IO93HT, SE43]
GW7 RQV N Jenkins, 37 Manorbier Dr, Llanyravon, Cwmbran, NP44 8TQ [IO81LP, ST39]
GW7 RQW P W Lewis, 7 Bryn Tirion, Clydach, Swansea, SA6 5LB [IO81BQ, SN60]
G7 RQX Details withheld at licensee's request by SSL.
G7 RQZ J E Plant, 4 Wennington Rd, Southport, PR9 7EU [IO83MP, SD31]
G7 RRA L D Dalton, 45 Falcon Dr, Patchway, Bristol, BS12 5RB [IO81QM, ST58]
G7 RRC A M Harvey Calderdale Raynet ARC, 3 High Ln, Norton Tower, Halifax, HX2 0NW [IO93BR, SE02]
G7 RRD G Smith, Signpost, High Rd, Whaplode, Spalding, PE12 6TJ [IO92XT, TF32]
G7 RRH Details withheld at licensee's request by SSL.
G7 RRI Details withheld at licensee's request by SSL.
G7 RRJ A McConnachie, 38 Farmdale Gr, Rubery, Rednal, Birmingham, B45 9NA [IO82XJ, SO97]
G7 RRK Details withheld at licensee's request by SSL.
G7 RRL M J Whitaker, Meadow View, Mill Ln, Charlton Mackrell, Somerton, Somerset, TA11 7BQ [IO81PB, ST52]
G7 RRM S J Whitehouse, 7 New Yatt Rd, Witney, OX8 6NU [IO91GT, SP31]
G7 RRN W J R Rosser, The Little House, Lower Dunton Rd, Dunton, Brentwood, CM13 3SJ [JO01EN, TQ68]
G7 RRO Details withheld at licensee's request by SSL.
G7 RRP M L Cleary, 18 Mayswood Gr, Quinton, Birmingham, B32 2RQ [IO92AK, SP08]
G7 RRQ Details withheld at licensee's request by SSL.
G7 RRR P N Griffiths, 190 Tythebarn Ln, Shirley, Solihull, B90 1PF [IO92BJ, SP17]
G7 RRW G C S Ingram, 7 Cheriswood Ave, River View, Exmouth, EX8 4HG [IO80HP, SY08]
G7 RRY M Saltmer, 35 The Green, Newby, Scarborough, YO12 5JA [IO94SH, TA08]
G7 RRZ H Evans, 14 Castle Cl, Roch, Haverfordwest, SA62 6AG [IO71KU, SM82]
G7 RSA C Hawkes, 84 Gordon Ave, Hateley Heath, West Bromwich, B71 2HU [IO92AM, SP09]
G7 RSD Details withheld at licensee's request by SSL.
GW7 RSE L J Clarke, 83 Lancaster St., Blaina, NP3 3EQ [IO81KS, SO20]
G7 RSF M J Watson obo Raynet Fire, The Tubbery, Henley, Ipswich, Suffolk, IP6 0BR [JO02NC, TM15]
G7 RSG Details withheld at licensee's request by SSL.
G7 RSI Details withheld at licensee's request by SSL.
G7 RSJ Details withheld at licensee's request by SSL.
G7 RSK A M Scott, 62 Berry Meade, Ashtead, KT21 1SG [IO91UH, TQ15]
G7 RSL Details withheld at licensee's request by SSL.
G7 RSM Dr R N Bloor, Pinewood House, Pinewood Dr, Ashley Heath, Market Drayton, TF9 4PA [IO82TW, SJ73]
G7 RSO Details withheld at licensee's request by SSL.
G7 RSP M J Watson Raynet Suffolk Police, The Tubbery, Henley, Ipswich, Suffolk, IP6 0BR [JO02NC, TM15]
G7 RSR J Chandler, 33 Kingfisher Dr, Chelmsley Wood, Birmingham, B36 0RD [IO92DM, SP18]
G7 RSS M J Watson Raynet South Suffolk, The Tubbery, Henley, Ipswich, Suffolk, IP6 0BR [JO02NC, TM15]
G7 RST A Taylor, 5 High Ln, Norton Tower, Halifax, HX2 0NW [IO93BR, SE02]
G7 RSV Details withheld at licensee's request by SSL.
G7 RSW C W Watson, 92 Cromwell Rd, Saffron Walden, CB11 4BE [JO02CA, TL53]
G7 RSY Details withheld at licensee's request by SSL.
G7 RSZ Details withheld at licensee's request by SSL.
G7 RTA S D Harding, 39 Victoria Ave, Cliftron, Brighouse, HD6 1QT [IO93CQ, SE12]
GI7 RTB P F McCrory, 13 Knocknamoe Bungalows, Omagh, BT79 7LA [IO64IO, H47]
G7 RTC G R Darby, 69 Churchill Rd, Earls Barton, Northampton, NN6 0PQ [IO92PG, SP86]
G7 RTD R M Rich, 13 Campfield Rd, Whippingham, East Cowes, PO32 6LT [IO90IR, SZ59]
G7 RTI K P Werner, 85 Brecon Way, Downley, High Wycombe, HP13 5NW [IO91OP, SU89]
G7 RTJ D A P Bransby, 7 West Cliff Ave, Whitby, YO21 3JB [IO94QL, NZ81]
GW7 RTK A W Jones, 85 Rowan Way, Malpas Park, Newport, NP9 6JT [IO81LO, ST39]
G7 RTL M A Pell Radio-Tele Lincolnshire Group, 7 Churchfleet Ln, Gosberton, Spalding, PE11 4NE [IO92WU, TF23]
G7 RTM A Bushell, 121 Rickmansworth Rd, Watford, WD1 7JD [IO91TP, TQ09]
G7 RTN Details withheld at licensee's request by SSL.
G7 RTO B A Theaker, 99 Truro Dr, Badgers Wood, Plymouth, PL5 4TR [IO70WK, SX46]
G7 RTP Details withheld at licensee's request by SSL.
G7 RTQ M K Cowley, 72 Warley Rd, Oldbury, Warley, B68 9TB [IO92AL, SP08]
G7 RTR D J Freeman, 65 Verulam Way, Cambridge, CB4 2HJ [JO02BF, TL46]
G7 RTU B Fisher, 4 Field View, Shepton Mallet, BA4 5RW [IO81RE, ST64]
G7 RTY N L Hall, 31 Tyndale Walk, Batley, WF17 8PX [IO93ER, SE22]
G7 RTZ Details withheld at licensee's request by SSL.
G7 RUC D R Millen, 38 Hythe Rd, Sittingbourne, ME10 2LS [JO01II, TQ86]
G7 RUE J C Hind, 16 Dawney Dr, Four Oaks, Sutton Coldfield, B75 5JA [IO92CO, SP19]
G7 RUF A J Hind, 16 Dawney Dr, Four Oaks, Sutton Coldfield, B75 5JA [IO92CO, SP19]
G7 RUG Details withheld at licensee's request by SSL.
G7 RUH R A Peggram, Cherry Trees, Broad Ln, Bracknell, RG12 9BY [IO91PJ, SU86]
GM7 RUI Details withheld at licensee's request by SSL.
G7 RUJ D F Brain, 15 Swaffham Rd, Burwell, Cambridge, CB5 0AN [JO02DG, TL56]
G7 RUK H W Martin, 23 St. Marks Rd, Gorefield, Wisbech, PE13 4QQ [JO02BQ, TF41]
G7 RUL Details withheld at licensee's request by SSL.
G7 RUN M R Graves, 20 Stace Way, Pound Hill, Worth, Crawley, RH10 7YW [IO91WC, TQ33]
G7 RUQ L J Murphy, Flat 3, 187 South Coast Rd, Peacehaven, E Sussex, BN10 8NS [JO00AS, TQ40]
G7 RUR S Todorovic, 13 Camborne St., Yeovil, BA21 5DF [IO80QW, ST51]
G7 RUS R W Parkin, 25 Kent House Ln, Beckenham, BR3 1LE [IO91XK, TQ37]
G7 RUT J M Dunster, 11 Barrie Cl, Whiteley, Fareham, PO15 7HN [IO90IV, SU50]
G7 RUV S B Dunster, 11 Barrie Cl, Whiteley, Fareham, PO15 7HN [IO90IV, SU50]
G7 RUW Details withheld at licensee's request by SSL.
G7 RUX J A Gardner, 67 Woodside Rd, Tunbridge Wells, TN4 8PY [JO01CD, TQ53]
G7 RUY D Ager, 11 Tilbury Cl, St. Pauls Cray, Orpington, BR5 2JR [JO01BJ, TQ46]
G7 RVC P Sutherland, 9 Lely Cl, Bedford, MK41 7LS [IO92SD, TL05]
G7 RVD H K Goodman, 23 Evelyn Dr, Pinner, HA5 4RL [IO91TO, TQ19]
G7 RVF K Grimshaw, 38 Pilling Ave, Lytham St. Annes, FY8 3QG [IO83LS, SD32]
G7 RVG P J Doble, 23 West St., Stoke Sub Hamdon, TA14 6PZ [IO80OW, ST41]
G7 RVH R A Bush, 19 Fanns Rise, Purfleet, RM19 1GP [JO01CL, TQ57]
G7 RVI T R F Hankins, Cawdor House, Cawdor, Ross 0N Wye, Hereford, HR9 7DN [IO81RW, SO62]
G7 RVJ Details withheld at licensee's request by SSL.
G7 RVL Details withheld at licensee's request by SSL.
G7 RVM R I Jinks, 52 Frederick Rd, Malvern, WR14 1RS [IO82UD, SO74]
GD7 RVP S W Rand, Tides Reach, Queens Dr, Ramsey, Isle of Man, IM8 2JD
GM7 RVR N G Moir, 34 Souter Dr, Inverness, IV2 4XJ [IO77VK, NH64]
G7 RVS J Leader, 76 St. Philips Rd, Newmarket, CB8 0EN [JO02EF, TL66]
G7 RVT T T Smith, 9 Crofters Way, Westlands, Droitwich, WR9 9HU [IO82VG, SO86]
G7 RVW R W L Crofts, Little Isle, Knighton on Teme, nr Tenbury Wells, Worcs, WR15 8LX [IO82RH, SO67]
G7 RVX G Fisk, 18 Millcroft Rd, Cliffe, Rochester, ME3 7QN [JO01FK, TQ77]
G7 RVY H E Branch, 326 Springfield Rd, Chelmsford, CM2 6BA [JO01FR, TL70]
G7 RVZ D E F Phillips, 14 Seymour Rd, Newton Abbot, TQ12 2PU [IO80EM, SX87]
G7 RWA Details withheld at licensee's request by SSL.
G7 RWC C M Halbert, 21 Rookswood, Morpeth, NE61 2UB [IO95DD, NZ28]
G7 RWF J L H Buck, 14 Crosstree Walk, Colchester, CO2 8QF [JO01KU, TM02]
G7 RWG A J Blount, 22 Barton St., West Bromwich, B70 8AJ [IO92AM, SP09]
G7 RWH J E Kessel, Penagar, New Rd, Boscastle, PL35 0AB [IO70PQ, SX09]
G7 RWI D C Reeve, 41 Priors Way, Maidenhead, SL6 2EL [IO91PM, SU87]
G7 RWK E M Sully, 10 The Paddock, Pound Hill, Crawley, RH10 7RQ [IO91WC, TQ23]
GM7 RWM C C Ward, 9 Rose St., Kirkintilloch, Cumbernauld, Glasgow, G67 4DE [IO75XW, NS77]
G7 RWN D Taylor, 41 Melville Rd, Gosport, PO12 4QU [IO90KT, SU60]
G7 RWO G J Ellis, Batson, 1 Highfield Park, Marlow, SL7 2DE [IO91ON, SU88]
G7 RWP K J Moseley, 11 St. Peters Dr, Whetstone, Leicester, LE8 6JT [IO92JN, SP59]
G7 RWQ N C Jones, 198 Leicester Rd, Narborough, Leicester, LE9 5BF [IO92JN, SP59]
GJ7 RWT A P Cutland, Little Gables, Route Orange, St. Brelade, Jersey, Channel Islands, JE3 8GQ
G7 RWU Details withheld at licensee's request by SSL.
G7 RWV M J Watson Raynet Waveney, The Tubbery, Henley, Ipswich, Suffolk, IP6 0BR [JO02NC, TM15]
G7 RWX Details withheld at licensee's request by SSL.
G7 RWY B K Sankey, 121 Green Ln South, Green Ln, Coventry, West Midlands, CV3 6EB [IO92FJ, SP37]
G7 RWZ J G C McLusky, 11 Ripon Rd, Killinghall, Harrogate, HG3 2DG [IO94FA, SE25]
G7 RXB N J Larson, 90 Lingfield Ash, Coulby Newham, Middlesbrough, TS8 0SU [IO94JM, NZ51]
G7 RXD Details withheld at licensee's request by SSL.
G7 RXE V A Donald, 85 Stafford St., Long Eaton, Nottingham, NG10 2EA [IO92IV, SK43]
G7 RXG Details withheld at licensee's request by SSL.
G7 RXH Details withheld at licensee's request by SSL.
G7 RXI V J Ball, 30 Park Dr, Worlingham, Beccles, NR34 7DJ [JO02TK, TM48]
G7 RXJ J A Dyson, 21 Highmoor, Kirkhill, Morpeth, NE61 2AS [IO95DD, NZ18]
G7 RXK R Thompson, 7 Rufford Cl, Skegby, Sutton in Ashfield, NG17 4BX [IO93ID, SK46]
GM7 RXL D Winton, 273 Hilton Dr, Aberdeen, AB2 4NT [IO87TH, NJ72]
G7 RXO Details withheld at licensee's request by SSL.
G7 RXP Details withheld at licensee's request by SSL.
G7 RXQ T G Martindale, The Beeches, Raby Dr, Bromborough, Wirral, L63 0NL [IO83MH, SJ38]
G7 RXR Details withheld at licensee's request by SSL.
G7 RXS J A Senior, 34 Shelley Rd, Enderby, Leicester, LE9 5QX [IO92JO, SP59]
G7 RXT Details withheld at licensee's request by SSL.
GI7 RXV Details withheld at licensee's request by SSL.
G7 RXW M F Lockitt, 19 Roundway Down, Perton, Wolverhampton, WV6 7SX [IO82VO, SO89]
G7 RXX S S Cooper, 169 Deansway, Friarscroft, Bromsgrove, B61 7PJ [IO82XH, SO97]
G7 RXY Details withheld at licensee's request by SSL.
G7 RYA D Tomlin, 154 Ct Ln, Erdington, Birmingham, B23 5RG [IO92BM, SP19]
G7 RYC R S Hampson, 11 Gladstone Gr, Heaton Moor, Stockport, SK4 4BX [IO83VK, SJ89]
G7 RYE Details withheld at licensee's request by SSL.
G7 RYG N Evans, 5 Mitchells Cl, Woodfalls, Salisbury, SP5 2LG [IO90DX, SU12]
GM7 RYK G J Pollard, 2 Auchinleck Terr, Port Glasgow, PA14 5SP [IO75QW, NS37]
G7 RYL D F Sheridan, 78 Oaklands Park, Buckfastleigh, TQ11 0BP [IO80CL, SX76]
G7 RYM R H G Pugh, 63 Globe Rd, Hornchurch, RM11 1BN [JO01CN, TQ58]
G7 RYN D Proctor, 11 Bedford Rise, Winsford, CW7 1NE [IO83RE, SJ66]
G7 RYO K J Turner, 34 Amherst Rd, Kenilworth, CV8 1AH [IO92EI, SP27]
GM7 RYR A B T Watson, 38 Brunstane Rd, Joppa, Edinburgh, EH15 2QR [IO85KW, NT37]
GM7 RYT D A Weller, Mither Tap, Bridge Rd, Kemnay, Inverurie, AB51 5QT [IO87SF, NJ71]
G7 RYV Details withheld at licensee's request by SSL.
G7 RYW F Trainer, 31 Hurst Park Dr, Liverpool, L36 1TE [IO83OK, SJ49]
G7 RYZ K Lucas, 143 Mason Lathe Rd, Shiregreen, Sheffield, S5 0TQ [IO93GK, SK39]
G7 RZA J I Roberts, 12 High St., Horsell, Woking, GU21 4UR [IO91RH, SU95]
G7 RZB K J Davison, 24 Suffolk Ave, Chaddesden, Derby, DE21 6ER [IO92GW, SK33]
GW7 RZC G H Bodley, 34 Claremont Rd, Pant, Newbridge, Newport, NP1 5DL [IO81KQ, ST29]
G7 RZD Details withheld at licensee's request by SSL.
G7 RZF Details withheld at licensee's request by SSL.
G7 RZK Details withheld at licensee's request by SSL.
GW7 RZN E G Taylor, 8 First Ave, Prestatyn, LL19 7LP [IO83HI, SJ08]
G7 RZQ N M Waterman, 71 Elizabeth Ave, Staines, TW18 1JW [IO91SK, TQ07]
G7 RZT J R Sinclair, 5 Cambridge Rd, Waterloo, Liverpool, L22 1RR [IO83LL, SJ39]
G7 RZU S N Adlem, 18 Alinora Ave, Goring By Sea, Worthing, BN12 4ND [IO90TT, TQ10]
G7 RZV A D Earl, 35 St. Marks Ave, Harrogate, HG2 8AF [IO93FX, SE35]
G7 RZW A Davies, 16 Sutton Rd, Deane, Bolton, BL3 4QR [IO83SN, SD60]
G7 RZZ H J S Middleton, Lilac Cottage, Marshwood, Bridport, Dorset, DT6 5QD [IO80NT, ST30]
GW7 SAB Details withheld at licensee's request by SSL.
G7 SAC J L Puttock, 53 Alexandra Ave, Sutton, SM1 2PA [IO91VI, TQ26]
GI7 SAH S P C Milliken, 4 Lisnabreen Cres, Bangor, BT20 4XF [IO74EP, J58]
G7 SAI E J Birt, 52 Powderham Ave, Berkeley Hunderton, Worcester, WR4 0DN [IO82VE, SO85]
GM7 SAK A J Jardine, 15 Louisa Dr, Girvan, KA26 9AH [IO75NF, NX19]
G7 SAL D J Newman, 89 Sea Pl, Goring By Sea, Worthing, BN12 4BH [IO90TT, TQ10]
G7 SAM R Goddard, 6 Upper Ley Dell, Greenside Ct, Chapeltown, Sheffield, S30 4AL [IO93GJ, SK38]
G7 SAN Details withheld at licensee's request by SSL.
G7 SAO Details withheld at licensee's request by SSL.
G7 SAP S J Taylor, 9 Purdy Rd, Carisbrooke Park, Carisbrooke, Newport, PO30 5SU [IO90IQ, SZ48]
GM7 SAQ R P Turner, 8 Leighton Ct, Dunblane, Perthshire, FK15 0ED [IO86AE, NN70]
GM7 SAR D A Turnbull, 13 Almond Cres, Paisley, PA2 0NQ [IO75ST, NS46]
G7 SAT D M Marritt, 17 Crossfield Rd, Hessle, HU13 9DA [IO93SR, TA02]
G7 SAX R H Newman, 31 Oval Gdns, Alverstoke, Gosport, PO12 2RA [IO90KT, SZ59]
GM7 SBB G A Cowie, 44 Buchanan Dr, Bearsden, Glasgow, G61 2EP [IO75UW, NS57]
G7 SBC S G Solomon, 26 Elm Dr, Cherry Burton, Beverley, HU17 7RJ [IO93SU, SE94]
G7 SBD I R C Davies, 56 Roselyn, Harlescott, Shrewsbury, SY1 4LP [IO82PR, SJ51]
G7 SBG R Tomkinson, 24 Beech Dr, Wistaston, Crewe, CW2 8RE [IO83SC, SJ65]
G7 SBH Details withheld at licensee's request by SSL.
G7 SBK D J Hunt, 26 Barwell Dr, Strelly Est, Nottingham, NG8 6LU [IO92JX, SK54]
G7 SBL P Trembath, 4 Penrose Terr, Penzance, TR18 2HQ [IO70FC, SW43]
G7 SBN M R Royal, 3 Bethany Pl, St. Just, Penzance, TR19 7HB [IO70DC, SW33]
GW7 SBO R G Thomas, 3 Gilfach Goch, Menai Bridge, LL59 5QL [IO73WF, SH57]
G7 SBP N H Hancock, 52 Heabrook Parc, Heamoor, Penzance, TR18 3QR [IO70FD, SW43]
GM7 SBR J Mitchell, 31 Bailies Dr, New Elgin, Elgin, IV30 3JW [IO87IP, NJ26]
G7 SBV D Moore, East View, The Common, East Stour, Gillingham, SP8 5NB [IO81UA, ST82]
G7 SBW Details withheld at licensee's request by SSL.
G7 SBX Details withheld at licensee's request by SSL.
G7 SBY Details withheld at licensee's request by SSL.
G7 SBZ M E Newton, 24 Chestnut Ave, Stockton Ln, York, YO3 0BR [IO93LX, SE65]

G7 SCC T J Jacobs Thurrock Sea Cadets, 43 Winfields, Pitsea, Basildon, SS13 1HA [JO01GN, TQ78]
G7 SCE P A Farman, Yew Tree Cottage, West St., Odiham, Hook, Hants, RG29 1NT [IO91MG, SU75]
G7 SCF Details withheld at licensee's request by SSL.
GM7 SCJ G L Deas, 81 Speirs Rd, Bearsden, Glasgow, G61 2LT [IO75UV, NS57]
G7 SCK D Stocks, 63 Welland Ct, Higham, Barnsley, S75 1PZ [IO93FN, SE30]
G7 SCL J R Robinson, 11 Wordsworth Rd, Loughborough, LE11 4LG [IO92JS, SK51]
G7 SCM C J Laws, 5 Fieldgate Ln, Whitnash, Leamington Spa, CV31 2QJ [IO92FG, SP36]
G7 SCO D L Brooke, 179 Pinza Cl, Newmarket, CB8 7AR [JO02EG, TL66]
G7 SCP D A Wain, 45 Park Grange Cl, Norfolk Park, Sheffield, S2 3SG [IO93GI, SK38]
G7 SCQ Details withheld at licensee's request by SSL.
G7 SCR M J Watson Suffolk Coastal Raynet, The Tubbery, Henley, Ipswich, Suffolk, IP6 0BR [JO02NC, TM15]
G7 SCT G P Rutherford, 24 Chestnut Ave, Hedon, Hull, HU12 8NH [IO93VR, TA12]
G7 SCV J Straughan, 16 Garner Cl, Chapel Park, Westerhope, Newcastle upon Tyne, NE5 1SQ [IO94DX, NZ16]
G7 SCX P O'Rourke, 186 Cottingham Rd, Corby, NN17 1SY [IO92PL, SP88]
G7 SCY Details withheld at licensee's request by SSL.
G7 SCZ D F Kiteley, 13 Chiltern Cl, Astley Cross, Stourport on Severn, DY13 0NU [IO82UH, SO86]
GW7 SDB N J Jones, Persondy, Llandefaelog Fach, Brecon, Powys, LD3 9TT [IO81HX, SO03]
G7 SDC D L Coe, 6 Hodson Cl, Bury St. Edmunds, IP32 6RS [JO02IG, TL86]
G7 SDD M R Smith, The Knap, Milborne Port, Sherborne, Dorset, DT9 5AR [IO80SX, ST61]
GW7 SDE I G Jones, 14 Clare Ct, Glan y Mor Park, Loughor, Swansea, SA4 6UH [IO71XP, SS59]
G7 SDF H E Peacock, 8 Mulberry Pl, Chesterton, Newcastle, ST5 7AZ [IO83VB, SJ84]
G7 SDG R J Martin, 40 St. Lawrence Way, Bricket Wood, St. Albans, AL2 3XN [IO91TQ, TL10]
G7 SDH R E Daly Dengie Hundred Amateur Radio S, 51 Queen St., Southminster, CM0 7BB [JO01JP, TQ99]
G7 SDL A R Payne, 155 Camping Hill, Stiffkey, Wells Next The Sea, NR23 1QL [JO02LW, TF94]
G7 SDM G L Davies, 11 Ninfield Cl, Carlton Colville, Lowestoft, NR33 8SD [JO02UK, TM59]
G7 SDP D M Ryan, 7A Beech Rd, Erdington, Birmingham, B23 5QJ [IO92BM, SP19]
G7 SDQ M J Smith, 7 North Rd, Combe Down, Bath, BA2 5DE [IO81TI, ST76]
G7 SDR C R England, 24 Glenway, Penwortham, Preston, PR1 9AJ [IO83PR, SD52]
G7 SDT P A Hill, 35 Ave Rd, Stratford upon Avon, CV37 6UW [IO92DE, SP25]
G7 SDV Details withheld at licensee's request by SSL.
GB7 SDX R P James DX Cluster Support Group, 4 Pentland Pl, Bearsden, Glasgow, G61 4JU [IO75TW, NS57]
GM7 SDX Details withheld at licensee's request by SSL.
G7 SEA Details withheld at licensee's request by SSL.
G7 SEC Details withheld at licensee's request by SSL.
G7 SED M A Cook, 64 Claremont Rd, Coton Green, Tamworth, B79 8EW [IO92DP, SK20]
G7 SEG A P Harrison, 44 Rosslyn Rd, Whitwick, Coalville, LE67 5PT [IO92HR, SK41]
G7 SEH A D Turner, Holydene, The Hill, Acle, Norwich, NR13 3DW [JO02SP, TG40]
GI7 SEI Details withheld at licensee's request by SSL.
G7 SEJ M C Baskeyfield, 14 Eaton Rd, Camberley, GU15 3EF [IO91OH, SU85]
G7 SEK R H Newham, 14 Ashdown Cl, Loughborough, LE11 4TQ [IO92JS, SK52]
G7 SEO R J Plant, 22 The Woodlands, Wokingham, RG41 4UY [IO91NJ, SU76]
G7 SEP E Richardson, Military Cottage, Irthington, Carlisle, Cumbria, CA6 4NE [IO84OW, NY45]
G7 SEQ R L Eddy, 85 Wembdon Hill, Wembdon, Bridgwater, TA6 7QA [IO81LD, ST23]
G7 SER J V Trickey Sutton Coldfield & Dist Raynet, 59 Shelley Dr, Four Oaks, Sutton Coldfield, B74 4YD [IO92BO, SK10]
G7 SEU E F Everett, 61 Hurrell Down, Boreham, Chelmsford, CM3 3JP [JO01GS, TL71]
G7 SEV P Vukasinovic, 3 Stanhope St., Ashton under Lyne, OL6 9QY [IO83XL, SJ99]
G7 SEY P Simpson, 10 Abelia Way, Priorslee, Telford, TF2 9TJ [IO82SQ, SJ71]
G7 SEZ I Bennett Scart, Ravenswood, The Shires, Hedge End, Hants, SO30 4BA [IO90IV, SU41]
G7 SFA M P Stevens, 127 St. Marys Dr, Pound Hill, Crawley, RH10 3BG [IO91WC, TQ23]
G7 SFB Details withheld at licensee's request by SSL.
G7 SFC M C Sleeman, Whitebeams, St. Johns Cl, Penn, High Wycombe, HP10 8HX [IO91PP, SU89]
G7 SFD M King, 4 Keith Ave, Ramsgate, CT12 6JQ [JO01QI, TR36]
GM7 SFE R Lawrie, 84 Redlawood Rd, Newton, Cambuslang, Glasgow, G72 7TP [IO75WT, NS66]
G7 SFF D Hartshorn, 60 Baden Powell Rd, Birdholme, Chesterfield, S40 2SL [IO93GF, SK37]
G7 SFG A A Anderson, 14 Pinehurst Park, Aldwick, Bognor Regis, PO21 3DL [IO90PS, SZ99]
G7 SFI S L Merrifield, Penang, St. Martins Rd, Gobowen, Oswestry, SY11 3NP [IO82LV, SJ33]
G7 SFK Details withheld at licensee's request by SSL.
G7 SFL Details withheld at licensee's request by SSL.
G7 SFM R G Wiltshire, 30 Cearns Rd, Oxton, Birkenhead, L43 2JP [IO83LJ, SJ38]
G7 SFQ J M Miller, 12 Leafield Rd, Darlington, DL1 5DE [IO94FM, NZ21]
G7 SFR Details withheld at licensee's request by SSL.
G7 SFS L J Banner, 37 Chesterfield Ave, Gedling, Nottingham, NG4 4GE [IO92KX, SK64]
G7 SFV Details withheld at licensee's request by SSL.
G7 SFW Details withheld at licensee's request by SSL.
G7 SFY B Purkiss, 99 Westland Rd, Yeovil, BA20 2AZ [IO80QW, ST51]
G7 SGA A J Lole, 5 Clayton Ave, Didsbury, Manchester, M20 6BL [IO83VK, SJ89]
G7 SGB D B Barker, 26 Poplar Cl, Haverhill, CB9 9EJ [JO02FC, TL64]
G7 SGC Details withheld at licensee's request by SSL.
G7 SGD J M Clark, 1 Neville St., Hazel Gr, Stockport, SK7 4EB [IO83WJ, SJ98]
G7 SGE Details withheld at licensee's request by SSL.
G7 SGG O B Schou, 2 Perry Cl, Woodhouse Eaves, Loughborough, LE12 8SB [IO92JR, SK51]
G7 SGI M Huke, 77 Raydons Rd, Dagenham, RM9 5JL [JO01BN, TQ48]
G7 SGJ Details withheld at licensee's request by SSL.
G7 SGK R H Ward, 9 Shelton Ave, East Ayton, Scarborough, YO13 9HB [IO94SG, SE98]
G7 SGL B Checkley, 12 Church Cl, Hartwell, Northampton, NN7 2HU [IO92ND, SP75]
G7 SGM R M Gifford, 100 Gadebridge Rd, Hemel Hempstead, HP1 3EW [IO91SS, TL00]
G7 SGO Details withheld at licensee's request by SSL.
G7 SGS Details withheld at licensee's request by SSL.
G7 SHC D F Holman, 28 The Ridings, Saughall, Chester, CH1 6AX [IO83MF, SJ36]
G7 SHG Details withheld at licensee's request by SSL.
G7 SHI C V Conce, 196 Canterbury Rd, Davyhulme, Urmston, Manchester, M41 0QS [IO83TK, SJ79]
G7 SHM A J Parnell, Meadow Cottage, Moor Ln, Irton, Scarborough, YO12 4RW [IO94SG, TA08]
G7 SHQ M D Brewer, 5 Park St., Ampthill, Bedford, MK45 2LR [IO92SA, TL03]
G7 SHS Details withheld at licensee's request by SSL.
G7 SHT J F Taverner, 31 Broadmeadow, Droitwich, WR9 8SZ [IO82WG, SO86]
G7 SHU G J Hughes, 77 Parkfield Rd, Taunton, TA1 4SD [IO81KA, ST22]
G7 SHW G N Stephens, 46 Newall Dr, Beeston, Nottingham, NG9 6NX [IO92IV, SK43]
G7 SIB N J Chilton, 30 Bleakhouse Rd, Oldbury, Warley, B68 9DR [IO92AL, SP08]
GM7 SIC Details withheld at licensee's request by SSL.
G7 SIQ D A Banks, 9 Woodbank, Egremont, Cumbria, CA22 2RL [IO84FL, NY01]
G7 SIR R B Leete, 10 Troughton Rd, Charlton, London, SE7 7QH [JO01AL, TQ47]
G7 SIS C M Hill, 35 Ave Rd, Stratford upon Avon, CV37 6UW [IO92DE, SP25]
GW7 SIT R Chatwin obo Torfaen Scouts A.R.C., New St Post Office, Pontnewydd, Cwmbran, Gwent, NP44 1EE [IO81LP, ST29]
G7 SIU G D W Parfitt, Stresemannstrasse 375, Haus 11, D 22761, Hamburg, Germany, X X
G7 SIV A W Mitchell, 27 Hanmer Rd, Simpson, Milton Keynes, MK6 3AY [IO92PA, SP83]
G7 SIZ Details withheld at licensee's request by SSL.
GM7 SJC G S Crowley, Rinnes View, Keith, AB55 6RJ [IO87LN, NJ35]
G7 SJF G R Cavie, Dawn, Maypole Rd, Tiptree, Colchester, CO5 0EN [JO01IT, TL81]
G7 SJK T D Masson, Addle Tree Cottage, Neath Gdns, Tilehurst, Reading, RG3 4UL [IO91LJ, SU66]
G7 SJP Details withheld at licensee's request by SSL.
G7 SJS P B Roberts, 5 Snelston Cres, Littleover, Derby, DE23 6BL [IO92FV, SK33]
G7 SJV E J Chapman, 2 Friars Bank, Friars Hill, Guestling, Hastings, TN35 4ET [JO00HV, TQ81]
G7 SJW A J Ashley, 40 Mill Hill, Brancaster, Kings Lynn, PE31 8AQ [JO02HX, TF74]
G7 SJX B A Shields, 20 Gresley Ct, Grantham, NG31 7RH [IO92QV, SK93]
G7 SKA P D Burnett, 4 Lavendon Ct, Barton Seagrave, Kettering, NN15 6QH [IO92PJ, SP87]
GM7 SKB Dr D Fortune, 26 Newton Gr, Newton Mearns, Glasgow, G77 5QJ [IO75US, NS55]
GW7 SKC J E Gray obo West Glamorgan CC, City and Co. of Swansea, Emergency Planning Unit, The Guildhall, Swansea, SA1 4PH [IO81AO, SS69]
G7 SKE Details withheld at licensee's request by SSL.
G7 SKF P Morgan, 15 Ardmore Cl, Tuffley, Gloucester, GL4 0BJ [IO81VU, SO81]
G7 SKH G O Murray, Brookside, Thirlby, Thirsk, North Yorks, YO7 2DJ [IO94IF, SE48]
G7 SKI Details withheld at licensee's request by SSL.
G7 SKJ Details withheld at licensee's request by SSL.
G7 SKM K E Johnson Maltby and District ARS, 20 Rolling Dales Cl, Maltby, Rotherham, S66 8EJ [IO93JK, SK59]
G7 SKO Details withheld at licensee's request by SSL.
G7 SKQ A N P Furness, 30 Radfield Dr, Odsal, Bradford, BD6 1BY [IO93CS, SE13]
G7 SKR D A Tarbatt, 9 Walkers Dr, Victoria Park, Leigh, WN7 2JP [IO83SL, SJ69]
G7 SKS Details withheld at licensee's request by SSL.
G7 SKU Details withheld at licensee's request by SSL.
G7 SKW B G R McInnes, 4 Lindrick Rd, Hatfield Woodhouse, Doncaster, DN7 6PF [IO93MN, SE60]
G7 SKX A J Wilkinson, 21 Solbys Rd, Basingstoke, RG21 7TG [IO91KG, SU65]
G7 SKZ Details withheld at licensee's request by SSL.
G7 SLB Details withheld at licensee's request by SSL.
G7 SLD Details withheld at licensee's request by SSL.
G7 SLE Details withheld at licensee's request by SSL.
G7 SLF Details withheld at licensee's request by SSL.
G7 SLH H C S Pipe, Field House, Green Ln, High Hatton, Shrewsbury, Shropshire, SY4 4HA [IO82RT, SJ62]
G7 SLK Details withheld at licensee's request by SSL.
G7 SLL G L J Peach, 120 Craven Rd, Newbury, RG14 5NR [IO91HJ, SU46]
GI7 SLN G A Mc Afee, 12 Skerryview, Craigahullier, Portrush, BT56 8NJ [IO65QE, C83]
G7 SLO Rev P S Midwood, The Vicarage, Reeth, Richmond, North Yorks, DL11 6TR [IO94AJ, SE09]
G7 SLP P R Hardcastle, 19 Dunkirk Terr, Halifax, HX1 3RB [IO93BR, SE02]
G7 SLQ R Cornthwaite, 18 Slaidburn Dr, Accrington, BB5 0JJ [IO83TR, SD72]
G7 SLR T S Cooper, 292 Chartridge Ln, Chesham, HP5 2SG [IO91QR, SP90]
G7 SLS Details withheld at licensee's request by SSL.
G7 SLT Details withheld at licensee's request by SSL.
GJ7 SLU C Whittaker, Coeur Joyeux, La Rue Des Sapins, St. Peter, Jersey, JE3 7AD
G7 SLV R J Walker, Ambergate, London Rd, Hook, RG27 9EQ [IO91MG, SU75]
G7 SLY P Taylor, 46 Ralph Rd, Staveley, Chesterfield, S43 3PY [IO93HG, SK47]
GM7 SLZ T C Gill, 25 Eastleigh Rd, Taunton, Somerset, TA1 2YA [IO81KA, ST22]
G7 SMA Details withheld at licensee's request by SSL.
G7 SMB R M Glover, 89 Cambridge Rd, Linthorpe, Middlesbrough, TS5 5LD [IO94IN, NZ41]
G7 SMC G Jameson, 17 Lansbury Ave, Mastin Moor, Chesterfield, S43 3AG [IO93IG, SK47]
G7 SMD R Buttery, 96 Rupert St., Lower Pilsley, Chesterfield, S45 8DE [IO93HD, SK46]
G7 SME Details withheld at licensee's request by SSL.
G7 SMF R F Fry, 23 Nine Acres, Hadleigh Rd, Ipswich, IP2 0DL [JO02NB, TM14]
G7 SMH G A Newton, 8 Lynch Mead, Winscombe, BS25 1AT [IO81NH, ST45]
G7 SMJ R Jones, 15 Derby Rd, Sandiacre, Nottingham, NG10 5HW [IO92IW, SK43]
G7 SMK Details withheld at licensee's request by SSL.
G7 SML B Farrington, 650 Blackburn Rd, Rising Bridge, Accrington, BB5 2RY [IO83TR, SD72]
G7 SMN A P Holden, 1 Rose Cottage, Bramford Ln, Ipswich, IP1 2PH [JO02NB, TM14]
G7 SMO S J Morley, 21 Landseer Cl, Sheffield, S14 1BT [IO93GI, SK38]
G7 SMQ B S Cottee, 41 Colesbourne Rd, Clifton, Nottingham, NG11 8JG [IO92JV, SK53]
G7 SMT F J Claydon, 5 Mill Gdns, Ringmer, Lewes, BN8 5JD [JO00AV, TQ41]
G7 SMU Details withheld at licensee's request by SSL.
GW7 SMV Details withheld at licensee's request by SSL.
G7 SMW Details withheld at licensee's request by SSL.
G7 SMX D E Johnson, 12 Mill Ln, Felixstowe, IP11 7RN [JO01QX, TM23]
G7 SMY Details withheld at licensee's request by SSL.
G7 SMZ R D Walker, 24 Colin St., Alfreton, DE55 7HT [IO93HC, SK45]
G7 SNA P D Luckcuck, 36A Rushmere Walk, Leicester Forest East, Leicester, LE3 3PD [IO92JO, SK50]
G7 SNB O E Newland, 22A Cromwell Rd, Basingstoke, RG21 5NR [IO91KG, SU65]
G7 SNC I R Palmer, 19 Stonehill, Green Ln West, Rackheaths, Norfolk, NR1X 6LX [JO02QP, TG21]
G7 SNE Details withheld at licensee's request by SSL.
GW7 SNF H D Joynes, The Shrubbery, Straits Ln, Nash, Newport, NP6 2BY [IO81MN, ST38]
G7 SNJ R G Chaytor, 16 Eastfields, Kitty Frisk, Hexham, NE46 1LE [IO84XX, NY96]
G7 SNK S J Murphy, 26 Westwood Cl, Shortstown, Bedford, MK42 0JS [IO92SC, TL04]
G7 SNN R J Heathcote, The Lane use, Top Rd, Acton Trussell, Stafford, ST17 0RQ [IO82WS, SJ91]
G7 SNO Dr G Mitchener, 10 Pine Cl, Brantham, Manningtree, CO11 1TP [JO01MX, TM13]
G7 SNP K Jordan, 7 Park Ave, Grange Park Est, Bedlington, NE22 7EH [IO95FD, NZ28]
G7 SNQ S C Taylforth, 1 Clough Terr, Barnoldswick, Colne, BB8 5PD [IO83VV, SD84]
G7 SNR S J Brodie, Top of The Hill, Hospital Rd, Wicklewood, Wymondham, Norfolk, NR18 9PR [JO02MN, TG00]
G7 SNT B J Jordan, Gr House, Coltishall, Norwich, NR12 7HD [JO02QR, TG21]
G7 SNU Details withheld at licensee's request by SSL.
G7 SNW J G Ward, 68 Moreton Rd North, Luton, LU2 9DP [IO91TV, TL12]
G7 SNX M A Pearce, 42 Pine Cl, Rudloe, Corsham, SN13 0LB [IO81VK, ST87]
GW7 SOA D E Walters, 17 Heol Islwyn, Llanrhystud, SY23 5BW [IO72WH, SN56]
GI7 SOB J K Elgin, 806 Farranseer Park, Macosquin, Coleraine, BT51 4NB [IO65PC, C82]
G7 SOC Details withheld at licensee's request by SSL.
G7 SOE M J Howard, East Dean House, East End, Langtoft, Peterborough, PE6 9LP [IO92TQ, TF11]
G7 SOF G J Batho, 44 The Spinney, Bridle Ln, Ripley, DE5 3LX [IO93HB, SK45]
G7 SOH C G Brown, 9 Marjorie St., Rhodesia, Worksop, S80 3HR [IO93KH, SK58]
G7 SOI Details withheld at licensee's request by SSL.
G7 SOJ B J Clifford, 49 Mearns, High Littleton, Bristol, BS18 5JR [IO81SH, ST65]
GW7 SOK R A Williams, 43 St. Margarets Rd, Whitchurch, Cardiff, CF4 7AB [IO81JM, ST18]
G7 SOM M R Gibson, Treguddick Cottage, Tideford Rd, Landrake, Saltash, PL12 5DP [IO70UK, SX36]
G7 SON P Brunnschweiler, 11 Glen Eyre Rd, Southampton, SO16 3GA [IO90HW, SU41]
G7 SOO J W Arnold, 2 Newbourne Gdns, Felixstowe, IP11 8PW [JO01PW, TM23]
G7 SOP S G Frank, 36 Melksham Rd, Bestwood Park, Nottingham, NG5 5RX [IO93KA, SK54]
G7 SOQ R N Benson, 19 Laburnum Cl, Horsford, Norwich, NR10 3EP [JO02OQ, TG11]
G7 SOR L Maughan, 151 Locks Rd, Locks Heath, Southampton, SO31 6LF [IO90IU, SU50]
G7 SOU A R Sillence, 74 Atherley Rd, Shirley, Southampton, SO15 5DS [IO90GV, SU41]
G7 SOV C E Howarth, 5 West Mount, Orrell, Wigan, WN5 8LX [IO83PM, SD50]
G7 SOX G E L Galliver, 29 Archery Fields, Odiham, Hook, RG29 1AE [IO91MG, SU75]
G7 SOY P E Pearce, 17B Chard Rd, St. Budeaux, Plymouth, PL5 2EG [IO70VJ, SX45]
G7 SOZ S R Jude, 7 Caister Way, Winsford, CW7 1LT [IO83RE, SJ66]
GM7 SPA J A Brown, 13 Boghead Rd, Kirkintilloch, Glasgow, G66 4EG [IO75WW, NS67]
GM7 SPB M J Garrington, 3 Sutherland Ave, Fort William, PH33 6JS [IO76KT, NN07]
G7 SPE R E Keep, 14 Foster Rd, Kempston, Bedford, MK42 8BU [IO92SC, TL04]
G7 SPG Details withheld at licensee's request by SSL.
G7 SPH D Tunmer, 47 Anson Rd, Denton, Manchester, M34 2HG [IO83WK, SJ99]
G7 SPJ Details withheld at licensee's request by SSL.
G7 SPK Details withheld at licensee's request by SSL.
G7 SPL D N S Pomfret, 19 Chadderton Dr, Unsworth, Bury, BL9 8NL [IO83UN, SD80]
G7 SPM C A Jones, 1 Omega Terr, Marsh Ln, Rowde, Devizes, SN10 2NS [IO81XI, ST96]
G7 SPN Details withheld at licensee's request by SSL.
G7 SPP H A Stansfield, 22 Low Stobhill, Morpeth, NE61 2SG [IO95DD, NZ28]
G7 SPR M S Bradley, 20 Allport Terr, Barrow Hill, Chesterfield, S43 2NQ [IO93HG, SK47]
G7 SPS A Bradley, 8 Church St., Brimington, Chesterfield, S43 1JG [IO93HG, SK47]
G7 SPT Details withheld at licensee's request by SSL.
G7 SPU Details withheld at licensee's request by SSL.
G7 SPV P R Mooney, 41 High St., Moorsholm, Saltburn By The Sea, TS12 3JH [IO94MM, NZ61]
G7 SPW Details withheld at licensee's request by SSL.
G7 SPZ R J Brown, 19 Comberton Rd, Toft, Cambridge, CB3 7RY [JO02AE, TL35]
G7 SQC P Young, 31 Cygnet Walk, North Bersted, Bognor Regis, PO22 9LY [IO90PT, SU90]
G7 SQD Details withheld at licensee's request by SSL.
G7 SQE M B Brown, Highbury, Carrs Hill, Badingham, Woodbridge, IP13 8NE [JO02QG, TM36]
G7 SQH G G Chew, Ove Arup & Partners, 13 Fitzroy St., London, W1P 5AA [IO91WM, TQ28]
G7 SQI M Colgan, 37 The Signals, Feniton, Honiton, EX14 0UP [IO80IT, ST10]
G7 SQK Details withheld at licensee's request by SSL.
G7 SQM N P Crawford, 20 Fearnley Cres, Kempston, Bedford, MK42 8NL [IO92SC, TL04]
GW7 SQP H A M Dublon, Tyn-y-Waun Farm, Dare Rd, Cwmdare, Aberdare, Mid Glam, CF44 8UB [IO81GR, SN90]
G7 SQR Details withheld at licensee's request by SSL.
G7 SQU M K Meldrum, 52 Briardale, Stevenage, SG1 1TR [IO91VV, TL22]
G7 SQV R C Stanley, Bryerry, Rectory Rd, Tivetshall St. Mary, Norwich, NR15 2AL [JO02OK, TM18]
G7 SQW A J Woods, 10 Radcliffe Rd, Thorpe Marriott, Drayton, Norwich, NR8 6XZ [JO02OQ, TG11]
G7 SQX C A Agass, 25 Mays Ave, Balsham, Cambridge, CB1 6ER [JO02DD, TL55]
G7 SQY D P Colton, 5 Candidus Ct, Werrington, Peterborough, PE4 5DB [IO92UO, TF10]
G7 SQZ Details withheld at licensee's request by SSL.
G7 SRA J D J Large, 5 Raynsford Rise, Stanningfield Rd, Great Whelnetham, Bury St. Edmunds, IP30 0TS [JO02JE, TL86]
G7 SRB D Shorten, 32 Stoneleigh Dr, Carterton, Oxon, OX18 1ED [IO91ES, SP20]
G7 SRC H T Brock, 56 Chapel Ln, Hadleigh, Benfleet, SS7 2PP [JO01HN, TQ88]
G7 SRE D Brown, 15 Monkseaton Terr, Ashington, NE63 0UB [IO95FE, NZ28]
G7 SRG G R Reynolds obo Sandwell Raynet Group, 187 Steelhouse Ln, Wolverhampton, WV2 2AU [IO82WN, SO99]
G7 SRH M S T J Harper, 31 Lorland Rd, Cheadle Heath, Stockport, SK3 0JJ [IO83VJ, SJ88]
G7 SRI Details withheld at licensee's request by SSL.
G7 SRJ S R Jones, 38 Albury Gr Rd, Cheshunt, Waltham Cross, EN8 8NS [IO91XQ, TL30]
G7 SRK R O Carder, 21 Mill Ln, Sawston, Cambridge, CB2 4HY [JO02BC, TL44]
G7 SRM J Bellis, 1 Sherbourne Cl, Newstead Est, Blurton, Stoke on Trent, ST3 3NR [IO82WX, SJ84]
G7 SRN M Rose, 32 Goddard St., Longton, Stoke on Trent, ST3 1JH [IO82WX, SJ94]
G7 SRU M J Higgins, 82 Balliol Rd, Kempston, Bedford, MK42 7HX [IO92RC, TL04]
G7 SRV Details withheld at licensee's request by SSL.
G7 SRX Details withheld at licensee's request by SSL.
G7 SRZ J Trybulski, 78 Ditchling Rd, Brighton, E Sussex, BN1 4SG [IO90WU, TQ30]
G7 SSA M J Addicott, Orchardleigh, The St., Chilcompton, Bath, BA3 4HG [IO81RG, ST65]
G7 SSB D E Jones, 69 Woodland Rd, Hellesdon, Norwich, NR6 5RW [JO02PP, TG21]
G7 SSC C E Parker, 10 Linden Ave, Sheffield, S8 0GA [IO93GI, SK38]
G7 SSD J A Edwards, 17 Marlowe Cl, Galley Common, Nuneaton, CV10 9QP [IO92FM, SP39]
G7 SSF B Bartley, 5 Cookes Wood, Deerness Vale, Broompark, Durham, DH7 7RL [IO94ES, NZ24]

G7 SSG J R Smye, 24 Eastfield Rd, Wincanton, BA9 9LT [IO81TB, ST72]
G7 SSH B H Giles, 171 St. Stephens Rd, Saltash, PL12 4NJ [IO70VJ, SX45]
GW7 SSI W S Monk, Wyndcliff, 7 Craig yr Eos Ave, Ogmore By Sea, Bridgend, CF32 0PF [IO81EL, SS87]
G7 SSK R S Walton, 4 Cornwall Cl, Upton Priory, Macclesfield, SK10 3HE [IO83WG, SJ87]
GW7 SSN Details withheld at licensee's request by SSL.
G7 SSO Details withheld at licensee's request by SSL.
GW7 SSQ P W Cole, 9 Perry Ct, Thornhill, Cwmbran, NP44 5UD [IO81LP, ST29]
G7 SSR D C Norman Southend on Sea, 19 Eastcote Gr, Southend on Sea, SS2 4QA [JO01IN, TQ88]
G7 SSS C W Kemp, 1 The Yeolands, Stoke Gabriel, Totnes, TQ9 6SY [IO80EJ, SX85]
G7 SST D Rands, 25 Downsland Park, Woodrow Ln, Great Moulton, Norwich, NR15 2DR [JO02OL, TM18]
G7 SSU J R Stewart, 15 Beverley Way, Trumpington, Cambridge, CB2 2JS [JO02BE, TL45]
G7 SSW J T Haywood, 7 Anna Walk, Burslem, Stoke on Trent, ST6 3BX [IO83VB, SJ84]
G7 SSZ Details withheld at licensee's request by SSL.
G7 STA Details withheld at licensee's request by SSL.
G7 STC K T Gater, 110 Byrds Ln, Uttoxeter, ST14 7NB [IO92BV, SK03]
G7 STF Details withheld at licensee's request by SSL.
G7 STG B V Spavins, 3 Lancaster Rd, Shortstown, Bedford, MK42 0UA [IO92SC, TL04]
GM7 STI I Pearce, Craigkniwe, The Mount, Cupar, Fife, KY15 4NA [IO86LI, NO31]
G7 STL M C Anderson, 12 Reedswood Rd, St. Leonards on Sea, TN38 8DN [JO00GU, TQ70]
G7 STM M F Wyatt, Chestnut House, 239 Broadgate, Sutton St. Edmund, Spalding, PE12 0LT [JO02AQ, TF31]
G7 STN E C Ball, 4 Ferndown Cl, Leicester, LE3 6UT [IO92JP, SK50]
G7 STO I Johnston, Foster House, Butterknowle, Bishop Auckland, Co. Durham, DL13 5JY [IO94BP, NZ12]
G7 STQ M J Oura, Quoins, Gloucester Rd, Upper Swainswick, Bath, BA1 8AD [IO81TK, ST76]
G7 STR M Tallarigo, via Flaminia 322, 00196 Roma, Italy
G7 STS A Wiseman, 28 Inchfield, Worsthorne, Burnley, BB10 3PS [IO83VS, SD83]
G7 STT J E Baker, 13 The Croftlands, Bredon, Tewkesbury, Glos, GL20 7NL [IO82WA, SO93]
G7 STU S Capstick, 160 Keighley Rd, Bingley, BD16 2DZ [IO93BU, SE04]
GW7 STV R G Smith, 21 Gaer Park Hill, Newport, NP9 3NP [IO81LN, ST28]
G7 STX J R Yeats, 27 Torcross Way, Parkside Grange, Cramlington, NE23 9PQ [IO95FC, NZ27]
G7 STZ C B Almey, Flying Field Farm, Wheatley Bank, Walsoken, Wisbech, PE14 7AZ [JO02CQ, TF41]
G7 SUA D W Wiseman, 12 Hamilton Way, Acomb, York, YO2 4LE [IO93KW, SE55]
GW7 SUC E D C Hughes, 13 Glan y Mor, Aberaeron, SA46 0BH [IO72UF, SN46]
G7 SUF G Reilly-Cooper, 6 Booth Rd, Hartford, Northwich, CW8 1RB [IO83RF, SJ67]
G7 SUI Details withheld at licensee's request by SSL.
G7 SUM G Fewings, 22 Watcombe Rd, Southbourne, Bournemouth, BH6 3LU [IO90CR, SZ19]
G7 SUP Details withheld at licensee's request by SSL.
G7 SUQ A K Jobson, Middleton House Farm, Elwick, Hartlepool, Cleveland, TS27 3EN [IO94IQ, NZ43]
G7 SUR G A Howard, 489 Liverpool Rd, Southport, PR8 3BP [IO83LO, SD31]
G7 SUS R C Biss, 1 Fairey Cres, Gillingham, SP8 4PE [IO81UB, ST82]
G7 SUT C E R James, Lower Keneggy Farm, Rosudgeon, Penzance, Cornwall, TR20 9AR [IO70HC, SW52]
G7 SUU R C Wolk, 11 Dashmonden Cl, Wainscott, Rochester, ME2 4PB [JO01GJ, TQ77]
G7 SUV J Patterson, 161 Ringwood Rd, Eastbourne, BN22 8UW [JO00DS, TQ60]
GM7 SUW A T Cowlin, 3 Alness Gr, Dunfermline, KY12 7XH [IO86GB, NT08]
G7 SUZ Details withheld at licensee's request by SSL.
G7 SVB Details withheld at licensee's request by SSL.
G7 SVD S Kubiesa, 103 Fifers Ln, Hellesdon, Norwich, NR6 6EF [JO02PP, TG21]
G7 SVE A W Jackson, 14 West Field Gdns, Sandy, SG19 1HF [IO92UD, TL14]
G7 SVF K C Ingram, 15 Kent Ave, East Cowes, PO32 6QN [IO90IS, SZ59]
GW7 SVG Details withheld at licensee's request by SSL.
G7 SVH J C Churchill, 43 Berry Head Rd, Brixham, TQ5 9AA [IO80FJ, SX95]
G7 SVI C N R Lambert-Hutchinson, 63 Chalbury Cl, Canford Heath, Poole, BH17 8BP [IO90AR, SZ09]
G7 SVJ D E H Gregory, Meadowview, 3 The Green, Thriplow, Royston, SG8 7QX [JO02AC, TL44]
GM7 SVK Details withheld at licensee's request by SSL.
G7 SVL R Briggs, 25 Barley Croft, Cheadle, Stoke on Trent, ST10 1NA [IO92AX, SK04]
G7 SVM D W Bradley, 22 Grosvenor Rd, Ettingshall Park, Wolverhampton, WV4 6QY [IO82WN, SO99]
G7 SVP W R Burns, 40 Wainbody Ave North, Green Ln, Coventry, CV3 6DB [IO92FJ, SP37]
G7 SVQ R A Holmes, 18 Dresden Cl, Mickleover, Derby, DE3 5RD [IO92FV, SK33]
G7 SVT D A Bultitude, 2 The Quest, Ampthill Rd, Houghton Conquest, Bedford, MK45 3JP [IO92SB, TL04]
G7 SVU N Hinchcliffe, 19 Grange Rd, Blidworth, Mansfield, NG21 0RN [IO93KC, SK55]
G7 SVV A G Hartnell, 61 The Avenue, Lewes, BN7 1QU [JO00AU, TQ41]
G7 SVX G Hare, Flat 4, 26 Hove Park Villas, Hove, E Sussex, BN3 6HG [IO90VU, TQ20]
G7 SVZ M P Addison, 68 Pillar Ave, Brixham, TQ5 8LB [IO80FJ, SX95]
G7 SWA C R Coe, 10 Kings Cl, Hatfield, Doncaster, DN7 6QY [IO93LN, SE60]
GW7 SWB R R Bambrey, Glangwili, Ystrad Aeron, Lampeter, Dyfed, SA48 7PG [IO72WE, SN55]
G7 SWC Details withheld at licensee's request by SSL.
G7 SWD B Marshland, 2 Tunstall Hill Cl, Sunderland, SR2 9DU [IO94HV, NZ35]
G7 SWE F W Rowbotham, 56 Farnborough Rd, Clifton Est, Nottingham, NG11 8GF [IO92JV, SK53]
G7 SWH A Howell, 35 Melton Rd, Wakefield, WF2 7PR [IO93FP, SE31]
G7 SWK D M Horton, 3 May Rd, Turvey, Bedford, MK43 8DY [IO92QD, SP95]
GW7 SWN J M Carey, Mount Pleasant, New Mills, Whitebrook, Monmouth, Gwent, NP5 4TY [IO81PS, SO50]
G7 SWQ I D Wild, 153 Alexandra Rd, Heeley, Sheffield, S2 3EH [IO93GI, SK38]
G7 SWR M A Prentice, 11 Wheatley, Great Hollands, Bracknell, RG12 8UF [IO91OJ, SU86]
G7 SWS Details withheld at licensee's request by SSL.
G7 SWU R M B Ledger, 33 Third Ave, York, YO3 0TY [IO93LX, SE65]
G7 SWV C Smith, 11 Woods Cl, Downholland, Haskayne, Ormskirk, L39 7JL [IO83MN, SD30]
G7 SWW R M G Jones, 138 Rutland St., Derby, DE23 8PS [IO92GV, SK33]
GM7 SWX D Curran, 104 McPherson Cres, Chapelhall, Airdrie, ML6 8XL [IO85AU, NS76]
G7 SWZ J Halliday, 14 Heath Gdns, Halifax, HX3 0BD [IO93BR, SE02]
G7 SXA Details withheld at licensee's request by SSL.
G7 SXB D S Phillips, 6 Lea Side Gdns, Longwood, Huddersfield, HD3 4XP [IO93BP, SE11]
G7 SXC C R Barlow, 14 Everest Rd, High Wycombe, HP13 7RD [IO91PP, SU89]
G7 SXE Details withheld at licensee's request by SSL.
G7 SXG D G Dean, 24 Hathaway Rd, Gateacre, Liverpool, L25 4ST [IO83NJ, SJ48]
G7 SXH Details withheld at licensee's request by SSL.
GM7 SXI A D Williams, 43 Church Dr, Mossblown, Ayr, KA6 5AX [IO75RL, NS42]
G7 SXJ J Farrow, 11 Dale Rd, Swanley, BR8 7HP [JO01BJ, TQ56]
G7 SXK F J Willis, 99 Kenilworth Ct, Asthill Gr, Coventry, CV3 6JB [IO92FJ, SP37]
G7 SXL M D Hawthorne, 29 Richards Cres, Monkton Heathfield, Taunton, TA2 8NR [IO81LA, ST22]
G7 SXM Details withheld at licensee's request by SSL.
GW7 SXN D W Davies, Rock House, Heol Goch, Pentyrch, Cardiff, CF4 8PN [IO81IM, ST18]
G7 SXP G G Cornes, 13 Liddle St., Stoke on Trent, ST4 5RL [IO82VX, SJ84]
G7 SXQ Details withheld at licensee's request by SSL.
G7 SXR Details withheld at licensee's request by SSL.
GW7 SXU I W Harries, Gwastad, Maen-y-Groes, New-Quay, Dyfed, SA45 9RJ [IO72TE, SN35]
G7 SXW R G Payne, 58 Sheepcote Ln, Amington, Tamworth, B77 3JW [IO92EP, SK20]
G7 SXX J A F Vining, 7 Silchester Cl, Andover, SP10 3RL [IO91GF, SU34]
G7 SXZ B J Johnson, 38 Hele Rd, Torquay, TQ2 7PR [IO80FL, SX96]
GW7 SYB Details withheld at licensee's request by SSL.
G7 SYC W M Jarvill, 16 The Cres, Guildford Rd, Horsham, RH12 1NB [IO91TB, TQ13]
G7 SYE Details withheld at licensee's request by SSL.
G7 SYI Details withheld at licensee's request by SSL.
G7 SYJ Details withheld at licensee's request by SSL.
G7 SYK Details withheld at licensee's request by SSL.
G7 SYL Details withheld at licensee's request by SSL.
G7 SYN Details withheld at licensee's request by SSL.
G7 SYO Details withheld at licensee's request by SSL.
G7 SYP Details withheld at licensee's request by SSL.
G7 SYQ A R Orchiston, 52 Marsden Hall Rd, Nelson, BB9 9PA [IO83VU, SD83]
G7 SYS R N Baxter, 107 Kendale Rd, Bridgwater, Somerset, TA6 3QE [IO81LD, ST23]
G7 SYT C B Denman, 3 Lombardy Ct, The Arbours, Northampton, NN3 3RX [IO92NG, SP76]
G7 SYU D S Bowers, 101 St Petersgdns, Wrecklesham, Farnham, Surrey, GU10 4QX [IO91OE, SU84]
G7 SYY S J Howarth, 14 Eaves Ln, Chorley, PR6 0PY [IO83QP, SD51]
G7 SYZ Details withheld at licensee's request by SSL.
G7 SZA S M Mussell, 13 Tugwell Rd, Eastbourne, BN22 9LH [JO00DT, TQ60]
G7 SZC J R Foster, 4 Limes Rd, Linthorpe, Middlesbrough, TS5 6RQ [IO94JN, NZ41]
G7 SZD M D Harrison, 2 Hopton Cl, Chaddesden, Chaddesden, Derby, DE21 4PR [IO92GW, SK33]
G7 SZF N J Hartley, 66 Broad Ln, Norris Green, Liverpool, L11 1AN [IO83MK, SJ39]
G7 SZG K Gardner, 27 Lindon Dr, Alvaston, Derby, DE24 0LP [IO92GV, SK33]
G7 SZH Details withheld at licensee's request by SSL.
G7 SZI E A Pulleyblank, 26 Oxford St., Mexborough, S64 9RL [IO93IL, SE40]
G7 SZJ Details withheld at licensee's request by SSL.
G7 SZL F P Wilson, 5 Harford House, Tavistock Cres, London, W11 1AY [IO91VM, TQ28]
G7 SZM Dr S H Brock, 40 Bramley Ln, Lightcliffe, Halifax, HX3 8NS [IO93CR, SE12]
G7 SZN Details withheld at licensee's request by SSL.
G7 SZO R C Collinson, 56 Orchard Valley, Hythe, CT21 4EA [JO01MB, TR13]
G7 SZT Details withheld at licensee's request by SSL.
G7 SZU Details withheld at licensee's request by SSL.
GI7 SZV Details withheld at licensee's request by SSL.
G7 SZZ R J Roberts, Priory Pl, Grange Ln, Little Dunmow, Dunmow, CM6 3HY [JO01EU, TL62]
GW7 TAA M L Hodges, 47 Faenol Isaf, Tywyn, LL36 0DW [IO72WN, SH50]
G7 TAC Details withheld at licensee's request by SSL.
G7 TAE S J Wersby, Perrandale, Merstone, Newport, Isle of Wight, PO30 3DF [IO90IP, SZ58]
G7 TAF M J Hawes, 2 Carlisle Ave, Swindon, SN3 1PY [IO91CN, SU18]
G7 TAG Details withheld at licensee's request by SSL.
G7 TAH Details withheld at licensee's request by SSL.
G7 TAI Details withheld at licensee's request by SSL.
G7 TAJ S C Duckling, Many Oaks, Collington Ln West, Bexhill on Sea, E Sussex, TN39 3TD [JO00FU, TQ70]
GM7 TAN C C Rogers, 51 Reed Dr, Newtongrange, Dalkeith, EH22 4SW [IO85LU, NT36]
G7 TAO S K Sheriff, 70 Clayfield View, Mexborough, S64 0HT [IO93IM, SE40]
G7 TAP Details withheld at licensee's request by SSL.
G7 TAR Details withheld at licensee's request by SSL.
G7 TAS Details withheld at licensee's request by SSL.
G7 TAT Details withheld at licensee's request by SSL.
G7 TAU Details withheld at licensee's request by SSL.
G7 TAV S D Houghton, 61 Manors Way, Silver End, Witham, CM8 3QP [JO01HU, TL81]
G7 TAX F T Roullier, 19 Terling Rd, Dagenham, Essex, RM8 1DS [JO01BN, TQ48]
G7 TBB Details withheld at licensee's request by SSL.
G7 TBC Details withheld at licensee's request by SSL.
G7 TBE E G Pease, 75 Welling Way, Welling, DA16 2RN [JO01BL, TQ47]
G7 TBF N P Smith, 47 Kiveton Ln, Todwick, Sheffield, S31 0HJ [IO93GJ, SK38]
G7 TBH Details withheld at licensee's request by SSL.
G7 TBJ J A Kewn, 18 Hillside Parc, Madron, Penzance, TR20 8RU [IO70FC, SW43]
G7 TBM T M Goodwin, 41 Mount Rd, Prestwich, Manchester, M25 2GP [IO83UM, SD80]
G7 TBN Details withheld at licensee's request by SSL.
G7 TBO Details withheld at licensee's request by SSL.
G7 TBQ M H Levens, 1 Horrocksford Way, Westbourne Park, Lancaster, LA1 5UU [IO84OB, SD46]
G7 TBR R Masson, 16 Catharine Pl, Bath, BA1 2PS [IO81TJ, ST76]
G7 TBS S H Collins, 8 Upper St. Marys Rd, Smethwick, Warley, B67 5JR [IO92AL, SP08]
G7 TBT L Fitzjohn, 43 Sydney St., Brightlingsea, Colchester, CO7 0BE [JO01MT, TM01]
G7 TBU S T Fitzjohn, 43 Sydney St., Brightlingsea, Colchester, CO7 0BE [JO01MT, TM01]
G7 TBV P W Brooke, 43 Willow Rd, Downham Market, PE38 9PG [JO02EO, TF60]
G7 TBW T A Polain, 22 Hilltop Ave, Hullbridge, Hockley, SS5 6BN [JO01HO, TQ89]
G7 TBX A G Siddle, 5 Neneside, Benwick, March, PE15 0YF [IO92XL, TL39]
G7 TCC G C Eddy obo Emergency Planning Hcc, 102 Springfield Cl, Andover, SP10 2QT [IO91GF, SU34]
G7 TCD G D Ward, 162 Greenbank Rd, Darlington, DL3 6ES [IO94FM, NZ21]
G7 TCH D E Grandfield obo Tech Coll Hastings Radio C, Hastings College, Engineering Dept, Archery Rd, St Leonards on Sea, Sussex, TN38 0HX [JO00GU, TQ70]
G7 TCK D K Vallins, 40 Curteys Walk, Crawley, RH11 8NP [IO91VC, TQ23]
G7 TCL Details withheld at licensee's request by SSL.
G7 TCM Details withheld at licensee's request by SSL.
G7 TCR Details withheld at licensee's request by SSL.
G7 TCU P O Close, Carosa, Hervines Rd, Amersham, HP6 5HS [IO91QQ, SU99]
G7 TCV J Wright, 17 Claremont View, Woodlesford, Leeds, LS26 8TA [IO93GS, SE32]
G7 TCW C S Haslewood, 66 Hunter Rd, Cannock, WS11 3AF [IO82XQ, SJ90]
G7 TCY P Marwood, 44 Rutland Rd, Goole, DN14 6LX [IO93NR, SE72]
GI7 TDA J McKeever, 19 Corrycroar Rd, Pomeroy, Dungannon, BT70 3DY [IO64NN, H77]
G7 TDD P J Rose, 27 Zealand Rd, Canterbury, CT1 3QW [JO01MG, TR15]
GI7 TDF D A S Ardill, At, 2 Lonsdale Ct, Shore Rd, Newtownabbey, BT37 0FA [IO74BQ, J38]
G7 TDJ L A Coalston, 2 Meadow End, Bembridge, PO35 5YB [IO90LQ, SZ68]
GW7 TDK D Lewis, Leyshon Rd, Ynysmeudwy, Pontardawe, Swansea, West Glam, SA8 4LP [IO81BR, SN70]
G7 TDL C H Lucas, 9 Cornford Cl, Crowborough, TN6 1EZ [JO01BB, TQ53]
G7 TDM D P Mann, 10 Hawkins Way, Wootton, Abingdon, OX13 6LB [IO91IQ, SP40]
G7 TDN A D Baines, 60 Norton Dr, Halifax, HX2 7RB [IO93BR, SE02]
G7 TDO Details withheld at licensee's request by SSL.
G7 TDP M Langford, 18 High St., Princes End, Tipton, DY4 9HW [IO82XM, SO99]
GW7 TDQ D F Banister, 41 Tynycoed Rd, Great Orme, Llandudno, LL30 2QA [IO83BH, SH78]
G7 TDR R J Smith, 47 Kiveton Ln, Todwick, Sheffield, S31 0HJ [IO93GJ, SK38]
G7 TDW J F Williams, 233 Church Plantation, Keele, Newcastle, ST5 5AX [IO83UA, SJ84]
G7 TDX Details withheld at licensee's request by SSL.
G7 TDZ Details withheld at licensee's request by SSL.
G7 TEA A Goddard, 52 Tramore Walk, Peel Est, Wythenshawe, Manchester, M22 5QW [IO83UJ, SJ88]
GI7 TEB J M M Mathers, Rosslea, 14 Castlewood Ave, Coleraine, BT52 1JR [IO65QC, C83]
G7 TEC D J Redman, 15 Price Rd, Cubbington, Leamington Spa, CV32 7LG [IO92GH, SP36]
GW7 TED G J Rowlands, 20 Oxwich Rd, Mochdre, Colwyn Bay, LL28 5AG [IO83CH, SH87]
G7 TEE L R Hughes, 23 Ebrington Ave, Solihull, B92 8HU [IO92CK, SP18]
G7 TEF Details withheld at licensee's request by SSL.
G7 TEG G K Fletcher, 171 Obelisk Rise, Northampton, NN2 8TX [IO92NG, SP76]
G7 TEI G Mills, 14 Hedgeway, East Hunsbury, Northampton, NN4 0SP [IO92NE, SP75]
G7 TEJ R Wallbank, 32 Truro Pl, Heath Hayes, Cannock, WS12 5YJ [IO92AQ, SK01]
G7 TEN J Richards, 11 Mount Rd, Etching Hill, Rugeley, WS15 2TL [IO92AS, SK01]
GW7 TEO Details withheld at licensee's request by SSL.
G7 TEP Details withheld at licensee's request by SSL.
G7 TER R V Herr, 6 Palace Green, Addington, Croydon, CR0 9AG [IO91XI, TQ36]
G7 TES G O Fletcher, 3 Pennine Terr, Winskill, Penrith, CA10 1PJ [IO84QR, NY53]
G7 TET I Mowbray, 64 Hawkwood Cres, Chingford, London, E4 7PJ [IO91XP, TQ39]
GI7 TEU Dr C J Clotworthy, 34 Glencregagh Park, Belfast, BT6 0NT [IO74BN, J37]
G7 TEW P K Harrison, Flat 24, Shore Acre, Marine Dr East, Barton on Sea, Hants, BH25 7DT [IO90ER, SZ29]
G7 TEX D W Paddock, 2 Longford Cl, Bidford on Avon, Alcester, B50 4EB [IO92BE, SP15]
G7 TEY Details withheld at licensee's request by SSL.
G7 TEZ Details withheld at licensee's request by SSL.
G7 TFA Details withheld at licensee's request by SSL.
G7 TFB J M Carter, 14 Lillywhite Cres, Andover, SP10 5NA [IO91GF, SU34]
G7 TFD Details withheld at licensee's request by SSL.
GW7 TFE Details withheld at licensee's request by SSL.
G7 TFG H M Orchel, Gildertofts, Ingleby Greenhow, Great Ayton, Middlesborough, Cleveland, TS9 6JF [IO94KL, NZ50]
G7 TFH Details withheld at licensee's request by SSL.
G7 TFJ A J Postill, 2 Selby Rd, New Lodge Est, Barnsley, S71 1TA [IO93GN, SE30]
GI7 TFK S McCormick, 74 Belsize Rd, Lisburn, BT27 4BH [IO64XM, J26]
G7 TFL S D Dodds, 4 Claremont Rd, Wisbech, PE13 2JR [JO02CP, TF40]
GM7 TFN C J Paton, Millgate, 4 Gateside Gdns, Greenock, PA16 7DA [IO75OW, NS27]
G7 TFR G Armstrong, 142 Main St., Shadwell, Leeds, LS17 8JB [IO93GU, SE33]
G7 TFU B P George, 43 Claverton Rd West, Saltford, Bristol, BS18 3DU [IO81SJ, ST66]
G7 TFV M Hunter, 57 Australia Gr, South Shields, NE34 9DF [IO94GX, NZ36]
G7 TFW Details withheld at licensee's request by SSL.
G7 TFX Details withheld at licensee's request by SSL.
G7 TFY J B Pollard, 8 Dombey Rd, Ipswich, IP2 0JS [JO02NB, TM14]
G7 TFZ D S Thomas, 14 Whitworth Jones Ave, Henlow, SG16 6HR [IO92UA, TL13]
G7 TGB C J Bristow, 25 Westfield Way, Charlton, Wantage, OX12 7EW [IO91HO, SU48]
G7 TGE J Gibson, 100 Top Row, Woolley Colliery, Darton, Barnsley, S75 5JQ [IO93FO, SE31]
G7 TGF A K Gibson, 100 Top Row, Woolley Colliery, Darton, Barnsley, S75 5JQ [IO93FO, SE31]
G7 TGG C G Preston, 29 Darby House, Caledon St., Walsall, WS2 9HZ [IO92AN, SP09]
GI7 TGJ Details withheld at licensee's request by SSL.
G7 TGK C M Coombe, 123 Farleigh Rd, Pershore, WR10 1JY [IO82XC, SO94]
G7 TGL M T Else, 8 Allen Rd, Coningsby, Lincoln, LN4 4RW [IO93VC, TF25]
G7 TGM R J Thornhill, 206 Marrowbrook Ln, Cove, Farnborough, GU14 0AD [IO91OG, SU85]
G7 TGN P M J Dawson, 1 Eastfield Rd, Bridlington, YO16 5DZ [IO94VC, TA16]
G7 TGO Details withheld at licensee's request by SSL.
G7 TGP J McHale, 150 Newton Rd, Burton on Trent, DE15 0TR [IO92ET, SK22]
G7 TGR Details withheld at licensee's request by SSL.
G7 TGV Details withheld at licensee's request by SSL.
GM7 TGX Details withheld at licensee's request by SSL.
G7 TGY S D Gander, 39 Hamsey Cres, Lewes, BN7 1NP [IO90XV, TQ31]
GW7 TGZ T G Jones, Pen-y-Bryn, Pen-y-Lan, Penclawdd, Swansea, SA4 3LJ [IO71WP, SS59]
G7 THA T E Pettitt, 2 Chassen Ave, Flixton, Urmston, Manchester, M41 5DS [IO83TK, SJ79]
G7 THB E Oord, 5 Rectory Farm Cl, West Hanney, Wantage, OX12 0LR [IO91HP, SU49]
GI7 THC R Boyd, 7 Rockmore, Banbridge, BT32 4QP [IO64UI, J14]
GI7 THD Details withheld at licensee's request by SSL.
GW7 THE R D Salusbury, 34 St. Margarets Dr, Rhyl, LL18 2HU [IO83GH, SJ08]
G7 THF M D Thompson, 95 Buxton Ave, Carlton, Nottingham, NG4 3RR [IO92KX, SK64]
G7 THG Cpt R J I Bell, Gough Rd, Sandgate, Folkestone, Kent, CT20 3BE [JO01NB, TR23]
GI7 THH T White, Shallamar, 3 Park Rd, Strabane, BT82 8EL [IO64GT, H39]

G7 THI F Gillespie, 453 Halifax Rd, Springside, Todmorden, OL14 8SS [IO83XR, SD92]
G7 THJ B W Mills, 37 Ashley Rd, Hildenborough, Tonbridge, TN11 9ED [JO01DF, TQ54]
G7 THK K W Grover, 3 Ludlow Cl, Southfields, Northampton, NN3 5LJ [IO92OG, SP86]
G7 THL Details withheld at licensee's request by SSL.
GW7 THM E F Lucocq, 96 Carisbrooke Way, Cyncoed, Cardiff, CF3 7HX [IO81KM, ST17]
G7 THO A D Whittaker, 1 Nursery Dr, Ecclesfield, Sheffield, S30 3XU [IO93GJ, SK38]
GW7 THP A J Richards, 229 Elm Dr, Ty Sign, Risca, Newport, NP1 6PP [IO81KO, ST29]
G7 THT Details withheld at licensee's request by SSL.
G7 THU D H Eaton-Watts, 129 Blake Rd, West Bridgford, Nottingham, NG2 5LA [IO92KW, SK53]
G7 THV P A Clay, Ranville, Billington Ln, Derrington, Stafford, ST18 9LR [IO82WS, SJ82]
GI7 THY R F Larimer, 85 Orritor Rd, Cookstown, BT80 8BN [IO64OP, H87]
G7 THZ A F Shipp, 36 Pell Ct, Lumbertubs, Northampton, NN3 8HL [IO92NG, SP76]
G7 TIA P Jones, 74 Bullfinch Ln, Riverhead, Sevenoaks, TN13 2EB [JO01CG, TQ55]
G7 TIB D L Cross, 91 Ilges Ln, Cholsey, Wallingford, OX10 9PA [IO91KN, SU58]
G7 TIC Details withheld at licensee's request by SSL.
G7 TID Details withheld at licensee's request by SSL.
G7 TIE I F Chamberlain, 17 Sturmy Cl, Long Stratton, Norwich, NR15 2XU [JO02OL, TM19]
GW7 TIH R E Fry, Frys Croft, Parc Aelas, Llangernyw, Clwyd, N Wales, LL22 8PJ [IO83DE, SH86]
G7 TII L J Shufflebotham, 64 Monks Dyke Rd, Louth, LN11 8DX [JO03AI, TF38]
G7 TIJ F Boele, 301 Cell Barnes Ln, St. Albans, AL1 5QB [IO91UR, TL10]
G7 TIK C S McQueen, 1 Westminster Walk, Corby, NN18 9JA [IO92PL, SP88]
G7 TIL Details withheld at licensee's request by SSL.
G7 TIM G R Jones, 42 Everard Rd, Southport, PR8 6NA [IO83MP, SD31]
G7 TIN R L Martin, 2 Eastview, North Walsham Rd, Trunch, NR28 0JP [JO02QT, TG22]
G7 TIO S C Robertson, 5 Sear Hill Cl, Balsall Common, West Midlands, CV7 7QL [IO92EJ, SP27]
G7 TIP Details withheld at licensee's request by SSL.
G7 TIR D Thomas, 9 Bonville Chase, Altrincham, WA14 4QA [IO83TJ, SJ78]
G7 TIS Details withheld at licensee's request by SSL.
G7 TIU J Gossage, 2 Barnes Wallis Way, Churchdown, Gloucester, GL3 2TR [IO81VV, SO82]
G7 TIV J P Askew, 55 Melsome Rd, Lyneham, Chippenham, SN15 4QP [IO91AL, SU07]
G7 TIW A P Morris, 9 Otter Way, Wootton Bassett, Swindon, SN4 7SH [IO91BM, SU08]
GW7 TIX D T Price, Cartref, 9 Traddydan Terr, Fron Ln, Newtown, SY16 2ER [IO82IM, SO19]
G7 TIY D J Miller, 139 Town Ln, Bebington, Wirral, L63 8LB [IO83LI, SJ38]
G7 TIZ A E Grudzinski, 126 Main St., Willoughby on The Wolds, Loughborough, LE12 6SZ [IO92LT, SK62]
G7 TJD M L Crosfill, Polmennor Farmhouse, Heamoor, Penzance, TR20 8UL [IO70FD, SW43]
G7 TJE Details withheld at licensee's request by SSL.
G7 TJH D T Pywell, Austr. 22, 71034 Boebuneen, Germany, X X
GW7 TJI Details withheld at licensee's request by SSL.
GW7 TJK Details withheld at licensee's request by SSL.
G7 TJO C A Dudbridge, 42B Duncansby Rd, Stanmore Bay, Whangaparaoa 1463, Auckland, New Zealand
G7 TJR Details withheld at licensee's request by SSL.
GW7 TJS Details withheld at licensee's request by SSL.
G7 TJT Details withheld at licensee's request by SSL.
GM7 TJV C M Ho, Flat D, 9/F, Fu Wah Ct, 44 Hiu Kwong St, Kowloon, Hong Kong, X X
G7 TJW Details withheld at licensee's request by SSL.
G7 TJX K L Davis, 109 Ermine St., Ancaster, Grantham, NG32 3QL [IO92RX, SK94]
G7 TJY Details withheld at licensee's request by SSL.
G7 TJZ I G Smith, 39 Hollingsworth Rd, Lowestoft, NR32 4AU [JO02UL, TM59]
GM7 TKA J Maclean, 36 Morrison Ave, Tranent, EH33 2AR [IO85MW, NT47]
G7 TKB F E Coles, 50 Grange Rd, Darlington, DL1 5NP [IO94FM, NZ21]
G7 TKD D G Hedges, 8 New Rd, Fleet End, Warsash, Southampton, SO31 9SB [IO90IU, SU50]
G7 TKE Details withheld at licensee's request by SSL.
G7 TKF B Nichols, Long Clandon, The St., West Clandon, Guildford, GU4 7TF [IO91RG, TQ05]
G7 TKG B Mersi, 32 The Close, Sutton, SM3 9EQ [IO91VJ, TQ26]
G7 TKH T A Nichols, Ivy Cottage, 15 Ivy Gr, Moorhead, Shipley, BD18 4JZ [IO93CT, SE13]
G7 TKI R M Pettett, 59 Edinburgh Ave, Sawston, Cambridge, CB2 4DW [JO02CD, TL45]
G7 TKJ D Dunn, 19 The Oval, Shildon, DL4 1EU [IO94EP, NZ22]
G7 TKK Details withheld at licensee's request by SSL.
G7 TKL Details withheld at licensee's request by SSL.
G7 TKM M J Hewitt, 102 Valley Green, Hemel Hempstead, HP2 7RG [IO91SS, TL01]
G7 TKN K Oliver, 94 Hollyhill Gdns West, South Stanley, Stanley, DH9 6NP [IO94DU, NZ15]
G7 TKO M D Smith, 9 Rowan Ct, Melksham, SN12 6HS [IO81WI, ST96]
G7 TKP M J Hewitt, 3 Orchard Rise, Beenham, Reading, RG7 5NS [IO91KJ, SU56]
G7 TKS Details withheld at licensee's request by SSL.
G7 TKT Dr P Ashford, 3 Valley Rd, Cheadle, Ches, SK8 1HY [IO83VJ, SJ88]
G7 TKW M J Peppiatt, 31E Llverton St, Kentish Town, London, NW5 2PE [IO91WN, TQ28]
G7 TKY O R Peters, 1 The Dr, Station Rd, Wilburton, Ely, CB6 3RP [JO02CI, TL47]
GW7 TKZ Details withheld at licensee's request by SSL.
G7 TLA Details withheld at licensee's request by SSL.
G7 TLC D J Benton, Trelowen, Penrose, Wadebridge, Cornwall, PL27 7TB [IO70ML, SW87]
G7 TLD M J Clare, 17 Birchfield Cl, Blackbird Leys, Oxford, OX4 5DL [IO91JR, SP50]
G7 TLG A R Embling, 12 Valley Rd, Cinderford, GL14 2PD [IO81RT, SO61]
G7 TLH A P Mayfield, 23 Sandringham Ave, Whiston, Rotherham, S60 4DS [IO93IJ, SK49]
G7 TLK K I Hemsil, 40 Rogate Dr, Thornbury, Plymouth, PL6 8SY [IO70WK, SX55]
G7 TLL H L B Hodson, 1 Chevin Ave, Borrowash, Derby, DE72 3HR [IO92HV, SK43]
G7 TLM D C Prior, 3 The Conifers, Cleeve Rd, Gotherington, Cheltenham, GL52 4EW [IO81XX, SO92]
G7 TLN Details withheld at licensee's request by SSL.
G7 TLP R D Scutchings, 45 Overhill Way, Beckenham, BR3 6SN [IO91XJ, TQ36]
G7 TLR K R Marshall, 16 Burnside, Esh Winning, Durham, DH7 9NA [IO94DS, NZ14]
G7 TLS J D Down, 12 Coach Rd, Henlow, SG16 6BT [IO92UA, TL13]
GW7 TLU D M Faraday, Hillcrest, Glascoed, nr Pontypool, Gwent, NP4 0UB [IO81MQ, SO30]
G7 TLX Details withheld at licensee's request by SSL.
G7 TLY J I Bennett, 18 Raleigh Cres, Goring By Sea, Worthing, BN12 6EF [IO90TT, TQ10]
G7 TMA N S Turner, 36 Yardley Cl, Woodloes Park, Warwick, CV34 5EX [IO92FH, SP26]
G7 TMD E W Allison, 10 Spinney Rd, Sundon Park, Luton, LU3 3DG [IO91SW, TL02]
G7 TMF T Foster, 98 Station Rd, Carlton, Nottingham, NG4 3DA [IO92LX, SK64]
G7 TMG Details withheld at licensee's request by SSL.
G7 TMH A F K Hunt, 63A Toms Ln, Kings Langley, WD4 8NJ [IO91SR, TL00]
G7 TMJ Details withheld at licensee's request by SSL.
G7 TMM A Kirkham, 32 Heath Park Rd, Buxton, SK17 6NX [IO93BG, SK07]
G7 TMO P A Foster, 41 Howden Cl, Bessacarr, Doncaster, DN4 7JN [IO93KM, SE60]
GI7 TMQ J M Bell, 72 Coleraine Rd, Portrush, BT56 8HN [IO65QE, C83]
G7 TMR R M Nelson, 115 Royal George Rd, Burgess Hill, RH15 9SJ [IO90WW, TQ31]
G7 TMS I G Dussold, 1 Sturt Rd, Charlbury, Chipping Norton, OX7 3EP [IO91GU, SP31]
GM7 TMT S H P Spence, Alton House, Berstane Rd, Kirkwall, Orkney, KW15 1NA [IO88MX, HY41]
G7 TMU V H Swanwick, 43 Hormare Cres, Storrington, Pulborough, RH20 4QX [IO90SW, TQ01]
G7 TMV Details withheld at licensee's request by SSL.
G7 TMX R S K Argall, 7 Dozmere, Feock, Truro, TR3 6RJ [IO70LE, SW83]
G7 TMY Details withheld at licensee's request by SSL.
G7 TMZ B M Shepherd, 178 Rhodesway, Bradford, BD8 0DB [IO93CT, SE13]
G7 TNA Details withheld at licensee's request by SSL.
G7 TNC Details withheld at licensee's request by SSL.
G7 TND Details withheld at licensee's request by SSL.
GW7 TNF Details withheld at licensee's request by SSL.
G7 TNG Details withheld at licensee's request by SSL.
G7 TNH M J Herbert, 6 School Rd, Joys Green, Lydbrook, GL17 9QY [IO81RU, SO61]
G7 TNK R M Stanton, 39 Brooklands Dr, Kidderminster, DY11 5EB [IO82UJ, SO87]
G7 TNL Details withheld at licensee's request by SSL.
G7 TNO D I Lunn, 23 Moynton Cl, Crossways, Dorchester, DT2 8TX [IO80UQ, SY78]
G7 TNP R F Kent, 2 Hicks Cl, Probus, Truro, TR2 4NE [IO70MH, SW85]
G7 TNQ M Mrzyglod, 152 Saxton Rd, Abingdon, OX14 5JA [IO91IP, SU49]
G7 TNS M Davies, 3 Park St., Pontycymer, Bridgend, CF32 8HD [IO81FO, SS99]
G7 TNT B L Scarsbrook, Salix, 96 Moss Ln, Alderley Edge, SK9 7HW [IO83VH, SJ87]
G7 TNU P G D W Sparke, 21 Charlesworth Park, Haywards Heath, RH16 3JG [IO90XX, TQ32]
G7 TNV L G Petts, 3 Buckingham Rd, Mitcham, CR4 1QR [IO91WJ, TQ36]
G7 TNX A T George, 45 Slater Ave, Derby, DE1 1GT [IO92GW, SK33]
G7 TNY P M Gater, 10 Wyre Cl, Roselands, Paignton, TQ4 7RU [IO80FK, SX85]
G7 TNZ M J Wells, 37 Water Meadows, Worksop, S80 3DF [IO93KH, SK57]
G7 TOB R Wardell, 1 Enfield Cl, Norden, Rochdale, OL11 5RT [IO83VO, SD81]
G7 TOC Details withheld at licensee's request by SSL.
G7 TOD D R Wardell, 1 Enfield Cl, Norden, Rochdale, OL11 5RT [IO83VO, SD81]
G7 TOF I Pardington, 36 Rivermeads Ave, Twickenham, TW2 5JJ [IO91TK, TQ17]
G7 TOG Details withheld at licensee's request by SSL.
G7 TOI P J Goodayle, Brookside, Jackies Ln, Newick, Lewes, BN8 4QX [JO00AX, TQ42]
G7 TOJ M P McDermott, 9 St. Alban Gr, Leeds, LS9 6LD [IO93GT, SE33]
G7 TOL B L Cox, 13 The Graylands, Coventry, CV3 6EW [IO92FJ, SP37]
G7 TON Details withheld at licensee's request by SSL.
G7 TOO P J Crabtree, 106 Sagecroft Rd, Thatcham, RG18 3BF [IO91IJ, SU56]
G7 TOP Details withheld at licensee's request by SSL.
G7 TOS M M O'Brien, 35 Kennington Ave, Benfleet, SS7 4BS [JO01GN, TQ78]
G7 TOT Details withheld at licensee's request by SSL.
G7 TOU M J Mussard, 35 Oakfield Gdns, Beckenham, BR3 3AY [IO91XJ, TQ36]
GI7 TOW Details withheld at licensee's request by SSL.
GI7 TOX Rev S Kelly, 23 Thornhill Park, Londonderry, BT48 8PB [IO65IA, C42]
G7 TOY A Peet, 62 Sandown Rd, Rugby, CV21 3LG [IO92JJ, SP57]
G7 TOZ J N Whytock, 48 Lythe Fell Ave, Halton, Lancaster, LA2 6NL [IO84OB, SD56]
G7 TPA R J Powell, 3 Powis Dr, Wellington, Telford, TF1 3HJ [IO82RQ, SJ61]
G7 TPB J H Kilminster, 499 Hagley Rd West, Quinton, Birmingham, B32 2AA [IO92AL, SP08]
G7 TPD T Morton, Brayside, Brays Ln, Hyde Heath, Amersham, HP6 5RU [IO91QQ, SP90]
G7 TPF Details withheld at licensee's request by SSL.
G7 TPG B M Barber, 114 Scrogg Rd, Walker, Newcastle upon Tyne, NE6 4HA [IO94FX, NZ26]
G7 TPH R B Hand, 2 Sturges Rd, Bognor Regis, PO21 2AH [IO90PS, SZ99]
G7 TPI Details withheld at licensee's request by SSL.
G7 TPJ Details withheld at licensee's request by SSL.
G7 TPK A Quinn, 31 Glebe Rd, Bayston Hill, Shrewsbury, SY3 0PN [IO82OQ, SJ40]
G7 TPL S Parker, 159 Belvedere Rd, Newton-le-Willows, WA12 0LQ [IO83QL, SJ59]
G7 TPM D W Allen, 5 West Heath, Pirbright, Woking, GU24 0JQ [IO91QH, SU95]
G7 TPN Details withheld at licensee's request by SSL.
GI7 TPO Details withheld at licensee's request by SSL.
GI7 TPP Details withheld at licensee's request by SSL.
G7 TPV J D Nunn, 13 Bakers Ln, Ingatestone, CM4 0BZ [JO01EQ, TQ69]
G7 TPW A B Grigor, 48 Valebridge Dr, Burgess Hill, RH15 0RW [IO90WX, TQ32]
GI7 TPX E P Brannigan, 8 Edward St., Downpatrick, BT30 6JD [IO74DH, J44]
G7 TPZ J E Luck, 20 McDermott Rd, Borough Green, Sevenoaks, TN15 8SA [JO01DG, TQ65]
G7 TQA D Legge, 28 Dresser Rd, Prestwood, Great Missenden, HP16 0NA [IO91PQ, SP80]
G7 TQB Details withheld at licensee's request by SSL.
G7 TQC C J Banister, 64 Mayfield Dr, Newport, PO30 2DR [IO90IQ, SZ58]
G7 TQE T J Brown, 138 Holmesdale Rd, South Norwood, London, SE25 6HY [IO91WJ, TQ36]
G7 TQJ N J Morrison, 51 Scarborough Walk, Corby, NN18 0NR [IO92PL, SP88]
G7 TQK Details withheld at licensee's request by SSL.
G7 TQR A B Williams obo Bristol ARC, 38 Seneca St., St. George, Bristol, BS5 8DX [IO81RK, ST67]
G7 TQS Details withheld at licensee's request by SSL.
G7 TQT R E Denton, 37 Tenby Rd, Cheadle Heath, Stockport, SK3 0UN [IO83VJ, SJ88]
G7 TQU Details withheld at licensee's request by SSL.
G7 TQV Details withheld at licensee's request by SSL.
G7 TQW D N Gardner, 11 Wheatsheaf Parade, Over 120 High St, West Wickham, Kent, BR4 0LZ [IO91XJ, TQ36]
G7 TQX E J Grisley, Spinney, Setley, Brockenhurst, SO42 7UH [IO90FT, SU30]
G7 TQY Details withheld at licensee's request by SSL.
G7 TQZ Details withheld at licensee's request by SSL.
G7 TRA K Furness, 55 Upper Belmont Rd, Chesham, HP5 2DE [IO91QR, SP90]
G7 TRB P R Stevenson, 361 Queens Rd West, Chilwell, Beeston, Nottingham, NG9 1GX [IO92JW, SK53]
G7 TRD Details withheld at licensee's request by SSL.
G7 TRE Details withheld at licensee's request by SSL.
G7 TRF M D Bennett, 3 Highview Gdns, Exmouth, EX8 2JR [IO80HO, SY08]
G7 TRG K Liddle, 36 Vicarage Ln, Grasby, Barnetby, DN38 6AU [IO93TM, TA00]
GM7 TRH R Innocenti, 18 Bentinck St., Greenock, PA16 7RN [IO75OX, NS27]
G7 TRJ Details withheld at licensee's request by SSL.
G7 TRL D B Wright, 24809035 L/Cpl Wright Db, Lad Mt (Gene Bay), 200 Sig Sqn, Bfpo 22, X X
G7 TRM K J White, 20 Agnes Cl, Bude, EX23 8SB [IO70RT, SS20]
G7 TRN Details withheld at licensee's request by SSL.
GM7 TRP Details withheld at licensee's request by SSL.
G7 TRT R M Powell, 39B London Rd, Wembley, HA9 7ET [IO91UN, TQ18]
G7 TRU D I Collier, 1 Epsom Pl, Cranleigh, GU6 7ET [IO91SD, TQ03]
G7 TRV D Glover, 13 Clevelands, Abingdon, OX14 2EQ [IO91IQ, SU59]
G7 TRW Details withheld at licensee's request by SSL.
G7 TRZ Details withheld at licensee's request by SSL.
GI7 TSA N F Kearney, 30 Bedeque House, Annesley St., Belfast, BT14 6AJ [IO74AO, J37]
G7 TSB E A Jones, 91 Way Ln, Waterbeach, Cambridge, CB5 9NQ [JO02CG, TL46]
GU7 TSI P L Lanoe, Les Adoubez, Ville Amphrey, St. Martin, Guernsey, GY4 6DP
G7 TSJ Details withheld at licensee's request by SSL.
G7 TSL Details withheld at licensee's request by SSL.
G7 TSM P I Lingard, 122 Plodder Ln, Farnworth, Bolton, BL4 0BU [IO83SN, SD70]
G7 TSN J D Soakell, 47 Lentune Way, Lymington, SO41 3PE [IO90FS, SZ39]
GW7 TSO K G Jones, 2 Bryn Ffynon, Penrhyndeudraeth, LL48 6RE [IO72XW, SH63]
G7 TSQ J B Stafford, 8 Lower Broadacre, Stalybridge, SK15 2UE [IO83XL, SJ99]
G7 TSR W Ditchburn, 42 de Mowbray Way, Morpeth, NE61 3RF [IO95DE, NZ18]
G7 TSS Details withheld at licensee's request by SSL.
G7 TSW S J White Tiverton (Sw)Rc, 3 PO Box, Tiverton, EX16 6RS [IO80GV, SS91]
G7 TSX Details withheld at licensee's request by SSL.
G7 TSY L A Thorne, 41 Ingleby Rd, Long Eaton, Nottingham, NG10 3DG [IO92IV, SK43]
G7 TSZ G D Steabler, 1 Westhill Rd, Grimsby, DN34 4SG [IO93WN, TA20]
G7 TTB Details withheld at licensee's request by SSL.
G7 TTD J W G Allen, 149 Penistone Rd, Waterloo, Huddersfield, HD5 8RP [IO93DP, SE11]
G7 TTE J D Wilkinson, 12 Orwell Rd, Walsall, WS1 2PJ [IO92AN, SP09]
G7 TTF P Cooper, 200 Kingsnorth Rd, Ashford, TN23 6LS [JO01KD, TR04]
G7 TTH G D Quint, 8 Alexandra Mews, Alexandra Ln, Malvern, WR14 1JF [IO82UC, SO74]
GI7 TTJ N A Baird, 110 Cedar Gr, Holywood, BT18 9QB [IO74BO, J37]
G7 TTK C J Walker, 25 Hill Park Cres, North Hill, Plymouth, PL4 8JP [IO70WJ, SX45]
G7 TTL D C N To, St Hughs College, Oxford, OX2 6LE [IO91IS, SP50]
G7 TTM E W Y Leung, 7 Palace Mews, Fulham, London, SW6 7TQ [IO91VL, TQ27]
GI7 TTO D Dunlop, 63 Cloyfin Rd, Coleraine, BT52 2NY [IO65QD, C83]
G7 TTP R O K Hughes, Top Flat, 24 Grosvenor Rd, Norwich, NR2 2PY [JO02PP, TG20]
G7 TTR Details withheld at licensee's request by SSL.
G7 TTT Details withheld at licensee's request by SSL.
G7 TTV T E Atkinson, 37 Burniston Gdns, Burniston, Scarborough, YO13 0HW [IO94SH, TA09]
G7 TTW T F Bootyman obo Welbeck College ARS, Welbeck College, Welbeck, Worksop, Notts, S80 3LN [IO93KG, SK57]
GW7 TTX M Tahla, Maes y Cambren, Bryn Coedifor, nr Dolgellau, LL40 2BN [IO82CS, SH82]
G7 TTY A D Hubbard, 12C Park Aveue, Mansfield, Notts, NG18 2AU [IO93JD, SK56]
G7 TTZ I C Hamilton obo Dacorum Packet Radio Group, 139 Elstree Rd, Woodhall Farm, Hemel Hempstead, HP2 7QW [IO91SS, TL01]
G7 TUA Details withheld at licensee's request by SSL.
GM7 TUD Details withheld at licensee's request by SSL.
G7 TUF T E W Johnson, 23 Greenock Rd, Hartlepool, TS25 4ES [IO94JP, NZ42]
G7 TUG N Mitchell, 49 Kersey Rd, Felixstowe, IP11 8UL [JO01PX, TM23]
G7 TUH P H Ferguson, 152 Chestnut Dr, Sale, M33 4HR [IO83TJ, SJ79]
G7 TUI K Patel, 647 Honeypot Ln, Stanmore, HA7 1JE [IO91UO, TQ19]
G7 TUJ Details withheld at licensee's request by SSL.
G7 TUK J Steel, 25 Denhead Cres, Potterton, Aberdeen, AB23 8UA [IO87WF, NJ91]
G7 TUL D Wilson, 8 Mulberry Cl, Hempstead, Gillingham, ME7 3SJ [JO01GI, TQ76]
G7 TUM J R Moore, Abbotts Way, Bush Est, Eccles on Sea, Norwich, Norfolk, NR12 0TA [JO02ST, TG42]
G7 TUP R H Irwin, 36 Heath Gr, Barming Heath, Maidstone, ME16 9AS [JO01FG, TQ75]
G7 TUQ B R Forhead, 67 Gale Moor Ave, Gosport, PO12 2SZ [IO90KT, SZ59]
G7 TUS R G Munden, 2 Hain Villa, Forest Rd, Ruardean Woodside, Ruardean, GL17 9XR [IO81RU, SO61]
GW7 TUU M W Griffiths, 1 Heol Nest, Cardiff, CF4 1SY [IO81JM, ST18]
G7 TUV J R Hewitt, 6 Crawley Walk, Cradley Heath, Warley, B64 5EX [IO82XL, SO98]
G7 TUY Details withheld at licensee's request by SSL.
G7 TUZ P R McLaughlin, 68 Calthorpe St., Liverpool, L19 1RF [IO83NI, SJ48]
G7 TVC L McCann, 34 Seacole Cl, Acton, London, W3 6TE [IO91UM, TQ28]
G7 TVD Details withheld at licensee's request by SSL.
G7 TVE Details withheld at licensee's request by SSL.
G7 TVH A Wallbanks, 36 Blyth Ave, Melton Mowbray, LE13 0HF [IO92NS, SK71]
G7 TVL E H Roberts, 800 Walsall Rd, Great Barr, Birmingham, B42 1EU [IO92AM, SP09]
GW7 TVM J G Mayo, Greig View House, Grosmont, Abergavenny, Gwent, NP7 8HN [IO81NV, SO32]
G7 TVP Details withheld at licensee's request by SSL.
G7 TVR Details withheld at licensee's request by SSL.
G7 TVS J Brittain, 42 Recreation Rd, Guildford, GU1 1HP [IO91RF, SU95]
G7 TVT L K Whiteside, 34 Ennerdale Ave, Dunstable, LU6 3AR [IO91RV, TL02]
G7 TVX Details withheld at licensee's request by SSL.
G7 TVY D B Battell, 6 Linden Ave, Sheffield, S8 0GA [IO93GI, SK38]
G7 TWA D P Bullard, 20 Chaney Rd, Wivenhoe, Colchester, CO7 9QZ [JO01LU, TM02]
G7 TWB Details withheld at licensee's request by SSL.

G7 TWC M J Ruttenberg, 5 Hampstead Heights, Heath View, London, N2 0PX [IO91VO, TQ28]
G7 TWD Details withheld at licensee's request by SSL.
G7 TWE D J Thomas, 204 Watchouse Rd, Galleywood, Chelmsford, CM2 8NF [JO01FQ, TL70]
G7 TWF Details withheld at licensee's request by SSL.
G7 TWI Details withheld at licensee's request by SSL.
G7 TWJ P F Edwards, Cleveland, Blackberry Rd, Lingfield, RH7 6NQ [IO91XD, TQ34]
GM7 TWM I H L Hipkin, 28 Tocher Terr, Drummuir, Keith, AB55 5JD [IO87LL, NJ34]
G7 TWO C Van Zuilen, Oksholm 185, 2133 Kt, Hoofddorp, The Netherlands
G7 TWP Details withheld at licensee's request by SSL.
G7 TWR Details withheld at licensee's request by SSL.
G7 TWU F G Clarkson, 313 Normanby Rd, Middlesbrough, TS6 0BQ [IO94KN, NZ51]
G7 TWV K Liddle obo Caistor School ARC, Caistor Grammar School, 2 Church St., Caistor, Market Rasen, LN7 6UG [IO93UL, TA10]
G7 TWW C Papaioannou, Pond House, Temple, Marlow, SL7 1SA [IO91ON, SU88]
G7 TWX Details withheld at licensee's request by SSL.
G7 TWY Details withheld at licensee's request by SSL.
G7 TXA Details withheld at licensee's request by SSL.
G7 TXD R E Snelling Portsmouth ARC, Highbury College, Dovercourt Rd, Portsmouth, PO6 2SA [IO90LU, SU60]
G7 TXE K S Goodman, Nicholas Farm, Winforton, Hereford, HR3 6EB [IO82LC, SO24]
G7 TXF A J Scott, 26 The Mount, Driffield, YO25 7JW [IO94SA, TA05]
G7 TXH Details withheld at licensee's request by SSL.
G7 TXI Details withheld at licensee's request by SSL.
GM7 TXJ Details withheld at licensee's request by SSL.
G7 TXL T H Ruth, 19 Blanchard Cl, Leominster, HR6 8SH [IO82OF, SO45]
G7 TXN J M Woolnough, Purnells House, 2 Mill Ln, Bulkington, Devizes, SN10 1SW [IO81XH, ST95]
G7 TXO Capt W A Langworthy, 10 Meadow Green, Welwyn Garden City, AL8 6SS [IO91VT, TL21]
G7 TXP J W Langworthy, 10 Meadow Green, Welwyn Garden City, AL8 6SS [IO91VT, TL21]
G7 TXQ D W Brodie obo Assoc Norfolk Raynet Group, Top of The Hill, Hospital Rd, Wicklewood, Wymondham, Norfolk, NR18 9PR [JO02MN, TG00]
G7 TXR S A Whitley, 5 Montford Cl, Burwell, Cambridge
G7 TXS Details withheld at licensee's request by SSL.
G7 TXT Dr D R Lewis, 14 Central Way, Oxted, RH8 0LS [IO91XG, TQ35]
G7 TXV C W Wood, 20 Tedworth Cl, Guisborough, TS14 7PR [IO94LM, NZ61]
G7 TXW M W J Oliver, 14 Harwood Rd, Bridgemary, Gosport, PO13 0TT [IO90JT, SU50]
G7 TXX D Williams, 57 Hillside Ave, Kidsgrove, Stoke on Trent, ST7 4LW [IO83VB, SJ85]
G7 TXY Details withheld at licensee's request by SSL.
G7 TXZ M C Stribblehill, 149 Hagley Rd, Halesowen, B63 4JN [IO82XK, SO98]
G7 TYB J Hawley, 89 Mansfield Ave, Denton, Manchester, M34 3NS [IO83WL, SJ99]
GW7 TYC Details withheld at licensee's request by SSL.
G7 TYH S J Furminger, 23A Sinodun Rd, Wallingford, OX10 8AD [IO91KO, SU69]
G7 TYI B G Isted, 170 Westfields Rd, Corby, NN17 1HQ [IO92PL, SP88]
G7 TYJ J J Pennington, 94 Rutland Ave, Stockingford, Nuneaton, CV10 8EG [IO92GM, SP39]
G7 TYK J Shephard, 19 Duffryn, Hollinswood, Telford, TF3 2BU [IO82SQ, SJ70]
G7 TYN Details withheld at licensee's request by SSL.
G7 TYO S Birtwhistle, 14 Woodley St., Fishpool, Bury, BL9 9HZ [IO83UN, SD80]
G7 TYP B A Cook, 40 Preston Ave, Alfreton, DE55 7JY [IO93HC, SK45]
G7 TYQ R Pugsley, 156 Thistledown Rd, Clifton Est, Nottingham, NG11 9ED [IO92JV, SK53]
G7 TYR D J Collins obo West Kent Raynet, 71 Trench Rd, Tonbridge, TN10 3HG [JO01DF, TQ54]
G7 TYS T Worsey, 6 Wenlock Cl, Sedgley, Dudley, DY3 3NJ [IO82WM, SO99]
G7 TYT M J Claxton, 9 Thompson Ave, Beverley, HU17 0BG [IO93SU, TA03]
G7 TYU W H Cooper, 16 Beaumont Hill, Great Dunmow, Dunmow, CM6 2AP [JO01EV, TL62]
G7 TYV Details withheld at licensee's request by SSL.
G7 TYW W D J Hibberd, 169 Highbury Gr, Cosham, Portsmouth, PO6 2RL [IO90LU, SU60]
G7 TYX J J English, 62 Old Rd, Failsworth, Manchester, M35 0AW [IO83WM, SD80]
G7 TYY Details withheld at licensee's request by SSL.
G7 TZA Details withheld at licensee's request by SSL.
G7 TZB D L Vincent, 18 Oakfield Rd, Poole, BH15 3BG [IO90AR, SZ09]
G7 TZC K J Nice, 48 Longfleet Rd, Poole, Dorset, BH15 2JA [IO90AR, SZ09]
G7 TZD P R Jones, 361 Wellingborough Rd, Rushden, NN10 6BA [IO92QH, SP96]
GW7 TZG P G Kelly, 33 Yeo St., Resolven, Neath, SA11 4HS [IO81DR, SN80]
GW7 TZI M C Tonkin, 48 Iorwerth St., Manselton, Swansea, SA5 9NP [IO81AP, SS69]
G7 TZM . P E Smith, 36 Mornington Ave, Ipswich, IP1 4LA [JO02NB, TM14]
G7 TZN S M Buckingham, 8 Tedder Ave, Harpur Hill, Buxton, SK17 9JU [IO93BF, SK07]
G7 TZO C W Turner, Two Oaks, Manor Rd, Rangeworthy, Bristol, BS17 5LR [IO81SN, ST68]
G7 TZQ R Darby, 25 Bramley Rd, Marsh Ln, Sheffield, S31 9RD [IO93GJ, SK38]
G7 TZS K Hopps, 100 Etherley Ln, Bishop Auckland, DL14 6TU [IO94DP, NZ22]
GM7 TZU Details withheld at licensee's request by SSL.
G7 TZV G Broughton, 20 Ravensbury St., Clayton, Manchester, M11 4GF [IO83VL, SJ89]
G7 TZW R J Wheatley, 288 Bennett St., Long Eaton, Nottingham, NG10 4JA [IO92IV, SK43]
G7 TZX D Johnson, 6 Denford Cl, Broughton, Chester, CH4 0SP [IO83MD, SJ36]
G7 TZY Details withheld at licensee's request by SSL.
G7 TZZ J M Eyre, 20 Jean Rd, Kettering, NN16 0PS [IO92PJ, SP87]
G7 UAA Details withheld at licensee's request by SSL.
G7 UAC E J Edwards, 12 Highfield Pl, Girdle Toll, Irvine, KA11 1BW [IO75QP, NS34]
G7 UAD A S K Murphy, 10 Byron St., Ilkeston, DE7 5JG [IO92IX, SK44]
GW7 UAE G L Jones, 48 Neath Rd, Resolven, Neath, SA11 4AH [IO81DQ, SN80]
G7 UAF B D Greenway, Green Furlong, Henley Ln, Butleigh, Glastonbury, BA6 8TJ [IO81PC, ST53]
G7 UAG D A Greenway, Green Furlong, Henley Ln, Butleigh, Glastonbury, BA6 8TJ [IO81PC, ST53]
G7 UAH B J Titmarsh, 61 Tudor Way, Sele Farm Est, Hertford, SG14 2DY [IO91WT, TL31]
G7 UAJ A G Taylor, 4 Lower St., Stutton, Ipswich, IP9 2SQ [JO01NX, TM13]
G7 UAK S V Hunter, 30 Adelaide St., Barrow in Furness, LA14 5TX [IO84JC, SD16]
G7 UAL Details withheld at licensee's request by SSL.
G7 UAM J B Otter, Corner Cottage, 97 Hilltop Rd, Dronfield, Sheffield, S18 6UN [IO93GH, SK37]
G7 UAN Details withheld at licensee's request by SSL.
G7 UAO Details withheld at licensee's request by SSL.
G7 UAP Details withheld at licensee's request by SSL.
G7 UAQ J W Pears, 8 The Hawthorns, Ellesmere, SY12 9ER [IO82MW, SJ33]
G7 UAR G W King, 267 Wensley Rd, Blackburn, BB2 6SU [IO83RR, SD62]
G7 UAT J G Johnson, 55 Magdalen Rd, Stoke on Trent, ST3 3HT [IO82WX, SJ84]
GM7 UAU Details withheld at licensee's request by SSL.
G7 UAV I Morris, 103 Moorland Ave, Lincoln, LN6 7NH [IO93RE, SK96]
G7 UAW D B A Welsh, 18 Waterloo Cl, Brampton, Huntingdon, PE18 8UN [IO92VH, TL27]
G7 UAX R J Le Piez, 279 Oakley Rd, Millbrook, Southampton, SO16 4NR [IO90GW, SU31]
G7 UAY D K Pickering, 10 Lighthurst Ave, Chorley, PR7 3HY [IO83QP, SD51]
G7 UBA C E R Banner, 42 High House Dr, Lickey, Rednal, Birmingham, B45 8ET [IO82XI, SO97]
G7 UBB E J Knight, 34 Balcombe Rd, Peacehaven, BN10 7RP [IO90XT, TQ40]
G7 UBC D W Brodie, Top of The Hill, Hospital Rd, Wicklewood, Wymondham, Norfolk, NR18 9PR [JO02MN, TG00]
G7 UBD T A W Thomas, 7 Noble Rd, Hedge End, Southampton, SO30 0PH [IO90IV, SU41]
G7 UBE E W Thomas, 5 Ripon Gdns, Foxlow Park, Buxton, SK17 9PL [IO93BF, SK07]
G7 UBF Details withheld at licensee's request by SSL.
G7 UBJ Details withheld at licensee's request by SSL.
G7 UBK G Reddecliffe, 5 Stanley Cl, Dymchurch, Romney Marsh, TN29 0TY [JO01MB, TR13]
GW7 UBL Details withheld at licensee's request by SSL.
G7 UBO J L Pearson, Blakeney, Ashburnham Cl, Norton, Doncaster, Sth Yorks, DN6 9HJ [IO93KP, SE51]
G7 UBP Details withheld at licensee's request by SSL.
G7 UBQ T M Bray, 2 Cambourne Dr, Fixby, Huddersfield, HD2 2NF [IO93CQ, SE11]
G7 UBT Details withheld at licensee's request by SSL.
G7 UBU Details withheld at licensee's request by SSL.
GI7 UBV S A Higgs, 5 Ardenlee Dr, Belfast, BT6 8QL [IO74BN, J37]
G7 UBW C D Wilson, 18 Guildown Ave, Guildford, GU2 5HB [IO91RF, SU94]
G7 UBX P Pleydell, 42 West Ave, Castle Bromwich, Birmingham, B36 0DY [IO92CM, SP18]
GI7 UBY C Lunnon, 9 Slievecoole Park, Belfast, BT14 8JN [IO74AP, J37]
GW7 UCA J Lightfoot, Chapel Cottage, Bleddfa, Knighton, Powys, LD7 1PA [IO82JH, SO26]
G7 UCB P M Hudson, 47 Hall Farm Rd, Duffield, Belper, DE56 4FJ [IO92GX, SK34]
G7 UCC J Bagley obo Raynet Ncc, 109 Norwich Rd, Wymondham, NR18 0SJ [JO02NN, TG10]
G7 UCE S Collins, 85 Malvern Rd, Swindon, SN2 1AU [IO91CN, SU18]
G7 UCG J G Woodward, 108 Tamworth Rd, Sutton Coldfield, B75 6DH [IO92CN, SP19]
G7 UCH Details withheld at licensee's request by SSL.
G7 UCJ R W J Crabtree, 46 Hill Village Rd, Four Oaks, Sutton Coldfield, B75 5BD [IO92CO, SP19]
G7 UCL S J Dixon, 24 Bodle Cres, Bexhill on Sea, TN39 4BG [JO00FU, TQ70]
G7 UCM K A Dixon, 19 Goldhurst Green, Icklesham, Winchelsea, TN36 4BU [JO00IW, TQ81]
G7 UCN A R Allport, 55 Byrds Ln, Uttoxeter, ST14 7NF [IO92BV, SK03]
G7 UCO D T Reed, 8 Wolverstone Dr, Hollingdean, Brighton, BN1 7FB [IO90WU, TQ30]
G7 UCP D G Hornby, 2 Broad Ing, Rochdale, OL12 7AR [IO83VO, SD81]
G7 UCQ P A Staniforth, 3 Springwood Cl, Plympton, Plymouth, PL7 1PE [IO70XJ, SX55]
G7 UCR K A Yeo, 46 Dairymans Walk, Guildford, GU4 7FE [IO91RG, TQ05]
G7 UCT B Lord, 7 Harewood Rd, Norden, Rochdale, OL11 5TG [IO83VP, SD81]
G7 UCU B Rowland, 16 Queens Gr, Ashley, New Milton, BH25 5DA [IO90ES, SZ29]
G7 UCV D W Lambert, 8 Primrose Ln, Winnersh, Wokingham, RG41 5UR [IO91NK, SU77]
G7 UCW Details withheld at licensee's request by SSL.
G7 UCX Details withheld at licensee's request by SSL.
G7 UCY Details withheld at licensee's request by SSL.
G7 UCZ D M Evans, 3 Dalkeith Cl, Bransholme North, Bransholme, Hull, HU7 5AS [IO93UT, TA13]
G7 UDB O McClea, Cartref, Pool Rd, Pool in Wharfedale, Otley, LS21 1EG [IO93EV, SE24]
G7 UDD D J Veale, 5 Heathfield Cl, Dronfield, Sheffield, S18 6RJ [IO93GH, SK37]
G7 UDE D J Clark, 6 Lower Rock Gdns, Brighton, BN2 1PG [IO90WT, TQ30]
G7 UDG Details withheld at licensee's request by SSL.
G7 UDH Details withheld at licensee's request by SSL.
GM7 UDI R A Duncan, South Backieley, Turriff, Aberdeenshire, AB53 7GS [IO87TM, NJ74]
G7 UDJ C J Edwards, 6 Blacksmiths Cl, Nether Broughton, Melton Mowbray, LE14 3EW [IO92MT, SK62]
G7 UDK H W Williams, Laurels, 9 Edward Rd, Biggin Hill, Westerham, TN16 3HN [JO01AH, TQ45]
G7 UDL Details withheld at licensee's request by SSL.
G7 UDM D G Bonfield, 49 Linden Gr, Chandlers Ford, Eastleigh, SO53 1LE [IO90HX, SU42]
G7 UDN M J Buckland, The Homestead, Bollow, Westbury on Severn, Glos, GL14 1QX [IO81TT, SO71]
G7 UDO Details withheld at licensee's request by SSL.
G7 UDP Details withheld at licensee's request by SSL.
G7 UDQ Details withheld at licensee's request by SSL.
G7 UDS Details withheld at licensee's request by SSL.
GI7 UDT Details withheld at licensee's request by SSL.
G7 UDU J Selwyn, 13 Crown Meadow, Braintree, CM7 9EX [JO01GV, TL72]
GI7 UDV W H Weir, 9 Ripley Terr, Portadown, Craigavon, BT62 3ED [IO64SJ, J05]
G7 UDX C E Harris, Stoke Damerel Comm. College, Somerset Pl, Stoke, Plymouth, Devon, PL3 4BD [IO70WJ, SX45]
G7 UEC D E Goodall, 181 Farhalls Cres, Horsham, RH12 4BX [IO91UB, TQ13]
G7 UED Details withheld at licensee's request by SSL.
G7 UEG P G Austin obo Northern VHF Activity Grou, 24 Fairfield Terr, Bramley, Leeds, LS13 3DH [IO93ET, SE23]
G7 UEH A J Murrell, 16 Sorrell Rd, Horsham, RH12 5FL [IO91UB, TQ13]
G7 UEI D W Longhurst, Burston, Wood Rd, Hindhead, GU26 6PZ [IO91PC, SU83]
G7 UEJ S P Kitchen, 344 Windward Way, Castle Bromwich, Birmingham, B36 0UH [IO92CM, SP18]
G7 UEK A J Jones, Welbeck Cottage, St Efrides Rd, Torquay, Devon, TQ2 5SG [IO80FL, SX96]
G7 UEL R J Dean, 127 Kenpas Highway, Styvechale, Coventry, CV3 6PF [IO92FJ, SP37]
G7 UEM S M Freegard, 5 Coppins Ct, Flansham Park, Felpham, Bognor Regis, PO22 6RJ [IO90QT, SU90]
G7 UEN M D Deary, 35 Milwain Dr, Heaton Chapel, Stockport, SK4 5AT [IO83VK, SJ89]
G7 UEO T Mills, 13 Morella Walk, Lenham, Maidstone, ME17 2JX [JO01IF, TQ85]
G7 UEP Details withheld at licensee's request by SSL.
G7 UEQ Details withheld at licensee's request by SSL.
G7 UES H E E Giles, 19 Hawkendon Rd, Grange Park, Clacton on Sea, CO16 7LE [JO01NT, TM11]
G7 UET A Levy, 47 Dudley Cl, Tilehurst, Reading, RG31 6JJ [IO91LL, SU67]
G7 UEU S Roberts, 62 Barford Rd, Hunts Cross, Liverpool, L25 0PP [IO83NI, SJ48]
G7 UEV C J Fox, 41 Alton Rd, Luton, LU1 3NS [IO91TU, TL02]
G7 UEX P A Cardwell, 2 Hayfield Pl, Frecheville, Sheffield, S12 4XH [IO93HI, SK38]
G7 UEY Details withheld at licensee's request by SSL.
G7 UFA Details withheld at licensee's request by SSL.
G7 UFD D V Hall, 320 Bexhill Rd, Woodingdean, Brighton, BN2 6QH [IO90XU, TQ30]
G7 UFF L H Brackstone, 276 Ladyshot, Harlow, CM20 3EY [JO01BS, TL41]
G7 UFG Details withheld at licensee's request by SSL.
GW7 UFH C R Barrett, 30 Brecon Rd, Hirwaun, Aberdare, CF44 9ND [IO81FR, SN90]
G7 UFI B Courtenay, 137 Vine Rd, Stoke Poges, Slough, SL2 4DH [IO91QN, SU98]
GM7 UFN T Graham, 265 Gilmartin Rd, Linwood, Paisley, PA3 3SU [IO75RU, NS46]
G7 UFO N A Bartley, 5 Cookes Wood, Broompark, Durham, DH7 7RL [IO94ES, NZ24]
G7 UFP Details withheld at licensee's request by SSL.
G7 UFQ P R Child, 1 The Mead, West Wickham, BR4 0BA [IO91XJ, TQ36]
G7 UFR Details withheld at licensee's request by SSL.
G7 UFT R D Elliott, 16 Prince Philip Rd, Monkwick Est, Colchester, CO2 8PA [JO01KU, TM02]
G7 UFV D B Riseborough, 2 The Barn, Rawley Rd, East Harling, Norwich, NR5 9AE [JO02OP, TG10]
GI7 UFX W J Belshaw, 24 Harmin Dr, Glengormley, Newtownabbey, BT36 7UJ [IO74AQ, J38]
G7 UFY Details withheld at licensee's request by SSL.
G7 UFZ P S Haselgrove, 28 Wickfield Ave, Christchurch, BH23 1JA [IO90CR, SZ19]
G7 UGA M K Turner, 41 Cleveland Ave, Draycott, Derby, DE72 3NR [IO92HV, SK43]
G7 UGB M E Howard, Netherwood, Short Thorn Rd, Stratton Strawless, Norwich, NR10 5NT [JO02OR, TG11]
G7 UGC A E J Fellows, Richeldis, 343 Wake Green Rd, Moseley, Birmingham, B13 0BH [IO92BK, SP08]
G7 UGE Details withheld at licensee's request by SSL.
G7 UGF Details withheld at licensee's request by SSL.
G7 UGG Details withheld at licensee's request by SSL.
G7 UGH A C Bawden, 11 Clegg Ave, Torpoint, PL11 2DR [IO70VJ, SX45]
GI7 UGJ P M Heaney, 38 Derryvore Ln, Portadown, Craigavon, BT63 5RS [IO64SK, J05]
G7 UGK Details withheld at licensee's request by SSL.
G7 UGN Details withheld at licensee's request by SSL.
GI7 UGP D J Robertson, 27 Ann St., Newtownards, BT23 7AD [IO74DN, J46]
G7 UGR D W Barnett, 102 El Alamein Way, Bradwell, Great Yarmouth, NR31 8SY [JO02UN, TG50]
G7 UGS Details withheld at licensee's request by SSL.
G7 UGT Details withheld at licensee's request by SSL.
GM7 UGV J M L Grant, 17 Cooper St., Hopeman, Elgin, IV30 2SB [IO87GQ, NJ16]
G7 UGW J C Smith, 32 Aberdeen St., Hull, HU9 3JU [IO93US, TA13]
G7 UGY N A Thornley, The Vicarage, 9 Langton Dr, Horncastle, Lincs, LN9 5AJ [IO93WE, TF26]
GM7 UGZ W J Seatter, 8 Earl Thorfinn St., Kirkwall, KW15 1QF [IO88MX, HY41]
G7 UHB M S Gutkowski, 14 Wheatfield Dr, Cranbrook, TN17 3LU [JO01GC, TQ73]
G7 UHE G D Tiller, 2 Fernlea, Sandleheath, Fordingbridge, SP6 1PN [IO90CW, SU11]
G7 UHF Details withheld at licensee's request by SSL.
G7 UHG D J Tropman, 91 Reindeer Rd, Deer Park, Fazeley, Tamworth, B78 3SW [IO92DO, SK10]
G7 UHH P K Bicknell, 15 Netherleigh Rd, Torquay, TQ1 3HB [IO80FL, SX96]
G7 UHI H W Parker, 21 Mayfield St., Spring Bank, Hull, HU3 1NS [IO93TS, TA02]
GI7 UHJ S W McNally, Edenbrook, Eastbourne Rd, Blindley Heath, Lingfield, Surrey, RH7 6JX [IO91XE, TQ34]
G7 UHK D C Lovatt, The Bay Tree, 5 North Hill, Highgate, London, N6 4AB [IO91WN, TQ28]
G7 UHL S J Yuill, 79 Firs Cl, Mitcham, CR4 1AX [IO91WJ, TQ26]
GW7 UHM P C Dunstan, 66 Allt yr Yn Rd, Newport, NP9 5EF [IO81LO, ST28]
G7 UHN A J Webster, 52 Montague Rd, West Harnham, Salisbury, SP2 8NL [IO91CB, SU12]
G7 UHO Details withheld at licensee's request by SSL.
G7 UHP G E R Smith, 256 Jockey Rd, Sutton Coldfield, B73 5XP [IO92BN, SP19]
G7 UHQ D J Bower, Chesterfield Rd, Sutton Cum Duckmanton, Duckmanton, Chesterfield, Derbyshire, S44 5JE [IO93HF, SK47]
G7 UHS C J F Merriman, Chantry, Edlington, Horncastle, Lincs, LN9 5RJ [IO93VF, TF27]
G7 UHT C Griffiths, 29 West Hill Rd, Ryde, PO33 1LG [IO90KR, SZ69]
G7 UHU D K Thorne, 11 Bower Hall Dr, Steeple Bumpstead, Haverhill, CB9 7ED [JO02FA, TL64]
G7 UHV Details withheld at licensee's request by SSL.
G7 UHW M J McDermott, 4 Tolcairn Ct, Lessness Park, Belvedere, DA17 5BT [JO01BL, TQ47]
G7 UHX B T Anderson, 57 Selby Rd, Plaistow, London, E13 8NB [JO01AM, TQ48]
G7 UHY Details withheld at licensee's request by SSL.
G7 UHZ K F Whittaker, 32 Ashleigh Mount Rd, Exeter, EX4 1SW [IO80FR, SX99]
G7 UIA J Thomas, 18 Jermyn Rd, Gaywood, Kings Lynn, PE30 4AE [JO02FS, TF62]
G7 UIB Details withheld at licensee's request by SSL.
G7 UID S R Clarke, 47 Peartree Rd, Enfield, EN1 3DE [IO91XP, TQ39]
G7 UIE A J Phillips, 24 Scott Rd, Chell, Stoke on Trent, ST6 6NG [IO83VB, SJ85]
G7 UIG Details withheld at licensee's request by SSL.
G7 UII C J Savage, 20 Croft Cres, Awsworth, Nottingham, NG16 2QY [IO92IX, SK44]
G7 UIJ J D Johnson, 34 Gr Ave, Gosport, PO12 1JX [IO90KT, SU60]
G7 UIL B W Allen, 4 City Rd, Tividale, Warley, B69 1RB [IO82XL, SO98]
GI7 UIM S Hand, 12 Church St., Rosslea, Enniskillen, BT92 7DD [IO64IF, H43]
G7 UIO N Johnson, 42 Brownlow Cres, Pinchbeck, Spalding, PE11 3XB [IO92WT, TF22]
GI7 UIP K O'Reilly, 165 Crom Rd, Mullaghboy, Newtown Butler, Fermanagh, BT92 8AX [IO64HE, H42]
G7 UIQ A W Wingrove, 11 Hillcrest, Mayland, Chelmsford, CM3 6AZ [JO01JQ, TL90]
G7 UIR L E Eyre, 18 Kylesku Cres, Kettering, NN15 5BH [IO92PJ, SP87]
GJ7 UIT C Totty, 34 Clos Paumelle, St Saviour, Jersey, Channel Islands, JE2 7TW
G7 UIU S M Palmer, 54 Hawthorn Rd, Exeter, EX2 6EA [IO80FR, SX99]
G7 UIV D P Jones, 26 Springhill Cres, Madeley, Telford, TF7 4DF [IO82SP, SJ70]
G7 UIX D Corden, Whisper Corner, Filance Ln, Penkridge, Stafford, ST19 5HQ [IO82WR, SJ91]
G7 UIY J G Drew, 64 Haselworth Dr, Alverstoke, Gosport, PO12 2UH [IO90KS, SZ69]
GW7 UIZ J D Hughes, Maes y Ffynnon, 7 Meadow Gdns, Llandudno, LL30 1UW [IO83CH, SH78]
G7 UJC G Taylor, 34 Hockley Rd, Poynton, Stockport, SK12 1RW [IO83WI, SJ98]
G7 UJE W G Dalzell, 9 Pyms Ln, Crewe, CW1 3PJ [IO83SC, SJ65]
GI7 UJH Details withheld at licensee's request by SSL.
GI7 UJI S F McBrien, Cushwash, Lisnaskea, Co Fermanagh, BT92 0DW [IO64GG, H33]
GM7 UJJ J D Scott, 1 Carrick Knowe Dr, Edinburgh, EH12 7EB [IO85IW, NT27]

G7 UJK T A Medlicott, 17 Kestrel Rd, Oldbury, Warley, B68 8AS [IO82XL, SO98]
G7 UJL M R D Haines, Park Cottage, Beech Ln, Grasscroft, Oldham, OL4 4EP [IO83XM, SD90]
G7 UJM Details withheld at licensee's request by SSL.
G7 UJN Details withheld at licensee's request by SSL.
G7 UJO S Maxwell, 44 Cairngorm Cl, Buckskin, Basingstoke, RG22 5DS [IO91KG, SU65]
G7 UJR F E Winter, 3 Parkfield Cl, Lower Weston, Totnes, TQ9 5YJ [IO80EK, SX86]
G7 UJY M P Poole, 184 Woodgates Ln, Swanland, North Ferriby, HU14 3PR [IO93SR, SE92]
G7 UJZ Details withheld at licensee's request by SSL.
G7 UKA T Collier, 23 The Riggs, Brandon, Durham, DH7 8PQ [IO94ES, NZ24]
G7 UKF M W Ellis, 40 Wigford Rd, Dosthill, Tamworth, B77 1LY [IO92DO, SP29]
G7 UKG Details withheld at licensee's request by SSL.
G7 UKH A E Beckers, 96 Rush Green Gdns, Romford, RM7 0NP [JO01CN, TQ58]
GW7 UKJ R A Cullis, 18 Church Cl, Croesyceiliog, Cwmbran, NP44 2EJ [IO81LP, ST39]
G7 UKK A G Firth, Highfield, 41 near Bank, Shelley, Huddersfield, HD8 8LT [IO93DO, SE21]
GM7 UKL K Dickinson, 3 Langton Gdns, East Calder, Livingston, EH53 0DZ [IO85GV, NT06]
G7 UKM E J Fisher, 35 Moorland Rd, Witney, OX8 5LS [IO91GS, SP31]
G7 UKN K Riley, 27 Limewood Cl, Blythe Bridge, Stoke on Trent, ST11 9NZ [IO82XX, SJ94]
G7 UKP P D Wood, Kiln Haw, Garsdale, Sedbergh, Cumbria, LA10 5NT [IO84SH, SD79]
G7 UKQ Details withheld at licensee's request by SSL.
G7 UKR M Blackburn, 36 Mardale Gr, Barrow in Furness, LA13 9QG [IO84JC, SD26]
GM7 UKS H Lyttle, 85 Brown Ave, Whitecrook, Clydebank, G81 1AW [IO75TV, NS56]
G7 UKT Details withheld at licensee's request by SSL.
G7 UKW Details withheld at licensee's request by SSL.
G7 UKX Details withheld at licensee's request by SSL.
G7 ULC Details withheld at licensee's request by SSL.
G7 ULD G R Rundle, 8 Belfield Cl, Marldon, Paignton, TQ3 1NZ [IO80EK, SX86]
G7 ULE Details withheld at licensee's request by SSL.
GI7 ULF D G Elliott, 15 Hanwood Park, Dundonald, Belfast, BT16 0XW [IO74CO, J47]
GI7 ULG S Murdoch, 22 Society St., Coleraine, Co. Londonderry, BT52 1LA [IO65PD, C83]
G7 ULH J D Collinson, 17 West Busk Ln, Otley, LS21 3LW [IO93DV, SE14]
G7 ULJ P White, 79 Chiltern Way, Bestwood Park, Nottingham, NG5 5NP [IO92KX, SK54]
G7 ULL P M Craig, 6 Marsham Cl, Chislehurst, BR7 6JD [JO01AK, TQ47]
G7 ULM P D Howarth, 5 Lushington Rd, London, NW10 5UX [IO91VM, TQ28]
G7 ULN J A Grundy, 47 Northiam Rd, Eastbourne, BN20 8LP [JO00DS, TV59]
G7 ULS K W Hurst, 94 East Park, Harlow, CM17 0SB [JO01BS, TL41]
G7 ULT M R Albon, 8 Byron Rd, Hutton, Brentwood, CM13 2RU [JO01EP, TQ69]
G7 ULU M C Evans, 2 Aller Brake Rd, Newton Abbot, TQ12 4NJ [IO80EM, SX86]
GI7 ULV Details withheld at licensee's request by SSL.
G7 ULZ I R Williamson, 47 Ochre Dike Walk, Rotherham, S61 4DL [IO93HK, SK49]
G7 UMA N M Tindall, 15 Maizebrook, Dewsbury, WF13 3TG [IO93EQ, SE21]
G7 UMB P Palmer, 6 Balcombe Cl, Newcastle, ST5 2DX [IO83VA, SJ84]
G7 UMC C J Price, 32 Cornwall Way, Ruskington, Sleaford, NG34 9HW [IO93TB, TF05]
G7 UMD Details withheld at licensee's request by SSL.
G7 UME J R Smith, Havigal House, Main St., Corbridge, NE45 5LE [IO84XX, NY96]
G7 UMF D J Griffiths, Home Farmhouse Cottage, Cardington Rd, Leebotwood, Shropshire, SY6 6LX [IO82ON, SO49]
GM7 UMG G Black, 9 McCulloch Rd, Girvan, KA26 0EF [IO75NF, NX19]
G7 UMH D L J Spalding Medway Raynet, 171 Minster Rd, Minster on Sea, Sheerness, ME12 3LH [JO01JK, TQ97]
G7 UMK G Moberley, 22 Larkfield Rd, Redditch, B98 7PL [IO92AH, SP06]
G7 UML E Owen, 55 Ruskin Ave, Feltham, TW14 9HY [IO91TK, TQ17]
G7 UMN Details withheld at licensee's request by SSL.
G7 UMP Details withheld at licensee's request by SSL.
G7 UMR D J Owen, 23 Beech Gr, Netherton, Bootle, L30 1SA [IO83ML, SJ39]
GW7 UMS K J Keepin, 14 Cwm Ln, High Cross, Rogerstone, Newport, NP1 9AF [IO81LO, ST28]
G7 UMT Details withheld at licensee's request by SSL.
G7 UMU J L Renshaw, 15 Mare Bay Cl, St. Leonards on Sea, E Sussex, TN38 8EQ [JO00GU, TQ71]
G7 UMW A D Banner, 32 Westbourne Rd, West Bromwich, B70 8LD [IO82XM, SO99]
G7 UMX Details withheld at licensee's request by SSL.
G7 UMY D J Rockliffe, 3 Hewell Ln, Barnt Green, Birmingham, B45 8NZ [IO92AI, SP07].
G7 UNB A J Bevington, 11 Raby Cl, Tividale, Warley, B69 1US [IO82XM, SO98]
G7 UNE Details withheld at licensee's request by SSL.
G7 UNF Details withheld at licensee's request by SSL.
G7 UNH Details withheld at licensee's request by SSL.
GW7 UNI Details withheld at licensee's request by SSL.
GW7 UNJ Details withheld at licensee's request by SSL.
G7 UNK Dr B Jordan, 57 Moorlands Cres, Huddersfield, HD3 3UF [IO93BP, SE01]
G7 UNL P A Irwin, Hatchell House, 15 Goldington Rd, Bedford, MK40 3JY [IO92SD, TL05]
G7 UNN Details withheld at licensee's request by SSL.
GI7 UNQ Details withheld at licensee's request by SSL.
G7 UNS S Manthri, 97 Abbey Rd, Barrow in Furness, Cumbria, LA14 5ES [IO84JC, SD26]
G7 UNU N D Davies, 16 Chapel Cl, Fressingfield, Eye, IP21 5QQ [JO02PI, TM27]
GW7 UNV E Jones, Crungoed, Llanbister Rd, Llandrindod Wells, Powys, LD1 5UR [IO82JI, SO17]
G7 UNW N P Othen, 205 Pine Rd, Bournemouth, BH9 1LY [IO90BR, SZ09]
G7 UNY J G Gabbatiss, 2 Linden Gr, Bramhall, Stockport, SK7 1HT [IO83VI, SJ88]
G7 UNZ W G Scott, Rose Brae, Lazonby, Penrith, Cumbria, CA10 1AJ [IO84PR, NY53]
G7 UOC Details withheld at licensee's request by SSL.
G7 UOD Details withheld at licensee's request by SSL.
G7 UOE J A Sanderton, Pond Farm House, Bentley Rd, Chapel St Mary, Ipswich, IP9 2JN [JO02MA, TM13]
GW7 UOH S N Lupton, 12 Bryn Tyddyn, Pentre Felin, Pentrefelin, Criccieth, LL52 0PE [IO72VW, SH53]
G7 UOI R P Hughes, 52 Hermes Cres, Henley Green, Coventry, CV2 1HZ [IO92GK, SP38]
GM7 UOJ J B Marshall, 8 Wilson St., Thurso, KW14 8AQ [IO88FO, ND16]
G7 UOL R C Bennion, 8 Manor Dr, Sawtry, Huntingdon, PE17 5UU [IO92UK, TL18]
G7 UOM B W Milford, 82 Oakley Ln, Oakley, Basingstoke, RG23 7JX [IO91JG, SU55]
G7 UON A R Price, 5A Ridge Way, Shaftesbury, SP7 9HB [IO81VA, ST82]
G7 UOP Details withheld at licensee's request by SSL.
G7 UOQ N J Birt, 60 Church Rd, Woodley, Reading, RG5 4QB [IO91NK, SU77]
G7 UOS B Yates, 14 Hathaway Ave, Braunstone, Leicester, LE3 2SN [IO92JO, SK50]
G7 UOT Details withheld at licensee's request by SSL.
G7 UOU A Colville, 10 The Oaks, Woodside Ave, North Finchley, London, N12 8AR [IO91VO, TQ29]
G7 UOV Details withheld at licensee's request by SSL.
G7 UOW Details withheld at licensee's request by SSL.
G7 UOY A Strong obo Dept of Electronics, Dept of Electronics, University of York, Heslington, YO1 5DD [IO93LW, SE65]
G7 UOZ M T Lock, 45 Blewbury Dr, Tilehurst, Reading, RG3 5HJ [IO91LJ, SU66]
G7 UPA Details withheld at licensee's request by SSL.
G7 UPC Details withheld at licensee's request by SSL.
GM7 UPD C W Edwards, Hillhead Croft, Chapel of Garioch, Inverurie, Aberdeenshire, AB51 5HE [IO87SH, NJ72]
G7 UPE Details withheld at licensee's request by SSL.
G7 UPF P S J England, 115 Trinity Rd, Cleethorpes, DN35 8UN [IO93XN, TA30]
G7 UPG J D Hislop obo Ursuline Convent School, 10 Park Wood Cl, Broadstairs, CT10 2XN [JO01RI, TR36]
G7 UPI H J Felstead, 69 Sullington Gdns, Worthing, BN14 0HS [IO90TU, TQ10]
G7 UPL Details withheld at licensee's request by SSL.
G7 UPN C D Jackson, 39 Fitzjohn Cl, Guildford, GU4 7HB [IO91RG, TQ05]
G7 UPP R O Hall, Greenviews, Lower Kingsbury, Milborne Port, Sherborne, DT9 5ED [IO80SX, ST61]
GI7 UPQ M Cunningham, 15 Fortwilliam Cres, Belfast, BT15 3RB [IO74AP, J37]
G7 UPS Dr P Bosworth, Little Wilverley, 9 Shaston Rd, Stourpaine, Blandford Forum, DT11 8TA [IO80VV, ST80]
G7 UPT M A Savage, Flat 2, 41 Norfolk Rd, Brighton, BN1 3AB [IO90WT, TQ30]
GI7 UPU F W Gillespie, 157 Lone Moor Rd, Londonderry, BT48 9LA [IO64IX, C41]
G7 UPV C D Bruce, 24 Shakespeare Rd, Thatcham, RG18 3FQ [IO91IJ, SU56]
G7 UPW L S Palmer, 77 Cambridge Rd, West Wimbledon, London, SW20 0PU [IO91VJ, TQ26]
G7 UPX S Graham, 4 Oakland Ave, Ellenborough, Maryport, CA15 7BU [IO84GR, NY03]
GM7 UPY J G H Craig, Bloomfield House, Bloomfield Rd, Lesmahagow, Lanark, ML11 0DF [IO85BP, NS84]
G7 UPZ I L Sansom, 86 Alexandra Rd, Wellingborough, NN8 1EE [IO92PH, SP86]
G7 UQC C Giuliano, Casa Giuliano, St Monica St, G/Maga, Malta, Msd 07, X X
G7 UQE D G Place, 61 Knowle Ave, Southport, PR8 2PB [IO83LO, SD31]
G7 UQF P Salt, 8 Grange Bottom, Royston, SG8 9UQ [IO92XB, TL34]
G7 UQG Dr R N Bloor Newcastle District Scouts Radi, Pinewood House, Pinewood Dr, Ashley Heath, Market Drayton, TF9 4PA [IO82TW, SJ73]
G7 UQH D Brown, 2 Old Tanyard, Tank BottOM, Ingleton, Carnforth, Lancs, LA6 3HB [IO84SD, SD67]
G7 UQJ M J Mee, 2A Thelda St, Keyworth, Nottingham, NG12 5HU [IO92KU, SK63]
GW7 UQL S C Utting, 31 Park View, Crossway Green, Chepstow, Gwent, NP6 5NA [IO81PP, ST59]
GM7 UQM M A Horne, 10 Blair Pl, Kirkcaldy, KY2 5SQ [IO86JC, NT29]
GW7 UQN M G Jones, 70 Ffordd Tudur, Holyhead, LL65 2DU [IO73QH, SH28]
G7 UQP Details withheld at licensee's request by SSL.
G7 UQR Details withheld at licensee's request by SSL.
G7 UQV M A Willoughby, 30 Kipling Rd, Ipswich, IP1 6EW [JO02NB, TM14]
GI7 UQW B A Neill, 81 Orangefield Rd, Belfast, BT5 6DD [IO74BO, J37]
G7 UQX Details withheld at licensee's request by SSL.
GI7 UQY J D Smith, 144 Bangor Rd, Holywood, BT18 0EY [IO74CP, J48]
G7 URA Details withheld at licensee's request by SSL.
GI7 URC A Brown, 3 Clara Rd, Belfast, BT5 6FN [IO74BO, J37]
G7 URG Details withheld at licensee's request by SSL.
G7 URH I M Johns, 1 Cambridge Terr, Redruth Highway, Redruth, Cornwall, TR15 1RS [IO70JF, SW74]
G7 URJ Details withheld at licensee's request by SSL.
G7 URL D E Foster, Pentlow, Crowle Bank Rd, Althorpe, Scunthorpe, South Humberside, DN17 3HZ [IO93ON, SE80]
G7 URM T A Heartfield, 69 Great Thrift, Orpington, BR5 1NF [JO01AJ, TQ46]
G7 URN Dr J P A Williamson, Claycombe House, Burleigh, Stroud, Glos, GL5 2PS [IO81VR, SO80]
G7 URP Details withheld at licensee's request by SSL.
G7 URR S J Easter, 89 Buckingham Dr, Luton, LU2 9RB [IO91TV, TL12]
G7 URS R J Bird, 9 Orchard Ln, Wembdon, Bridgwater, TA6 7QY [IO81LD, ST23]
G7 URT C E Langham, 9 Lawrence Cl, Shurdington, Cheltenham, GL51 5SZ [IO81WU, SO91]
G7 URU Details withheld at licensee's request by SSL.
G7 URV C E Harris Stoke Damerel Community Colleg, Stoke Damerel Comm. College, Somerset Pl, Stoke, Plymouth, Devon, PL3 4BD [IO70WJ, SX45]
G7 URW N R Tucker, 7/357 Montgomery Dr, Prospect, Waterfalls, Harare, Zimbabwe
G7 URX M P Skinner Barton Scout Radio Group, 1 Fullbrook Ave, Barton under Needwood, Burton on Trent, DE13 8HD [IO92DS, SK11]
G7 URZ B Fisher Whitstone Community School AR, 4 Field View, Shepton Mallet, BA4 5RW [IO81RE, ST64]
GI7 USA A J Niblock, 45 Greenland Cres, Larne, BT40 1HE [IO74CU, D40]
G7 USB J Ainsworth, 42 Buttfield Rd, Hessle, HU13 0AS [IO93SR, TA02]
GM7 USC Details withheld at licensee's request by SSL.
G7 USE C H T R Smith 2339 (Plymstock Mt Batten) Sqn, 162 Tamar House, James St, Plymouth, Devon, PL1 3HJ [IO70WI, SX45]
G7 USG J R Sutherland, 4 Cherbury Cl, Bracknell, RG12 9HT [IO91PJ, SU86]
GM7 USH M J Beavis, 52 Arduthie Rd, Stonehaven, Kincardineshire, AB39 2EH [IO86VX, NO88]
G7 USI R J Hayselden, 18 Swan Cl, Talke, Stoke on Trent, ST7 1TA [IO83UB, SJ85]
G7 USJ T W Cogan, 6 Birch Cl, Locking, Weston Super Mare, BS24 8BP [IO81NH, ST35]
G7 USL Details withheld at licensee's request by SSL.
G7 USM K Reavill, 11 Clarence Rd, Beeston, Nottingham, NG9 5HY [IO92JV, SK53]
G7 USP S J Garwood, 25 South Hill Cres, Horndon on The Hill, Stanford-le-Hope, SS17 8PH [JO01EM, TQ68]
G7 USQ B P Siddall, 6 Delside Ave, Manchester, M40 9LF [IO83VM, SD80]
G7 USR C J Pitchford, 9 Pickhurst Ln, Bromley, BR2 7JE [JO01AJ, TQ46]
G7 USS Details withheld at licensee's request by SSL.
G7 UST C Shackleton, 159 Halifax Rd, Brighouse, HD6 2EQ [IO93CQ, SE12]
G7 USV Dr D J Atkins, 14 Ryde Pl, Lee on The Solent, PO13 9AU [IO90JT, SZ59]
G7 USX M E Woollard, Barnside, Colchester Rd, Elmstead, Colchester, CO7 7EG [JO01LV, TM02]
G7 USZ Details withheld at licensee's request by SSL.
G7 UTB H J S Scott-Telford, 9 Squires Cl, Rochester, ME2 2TZ [JO01FJ, TQ76]
G7 UTC M R Bean, Ashmore, Belle Vue Rd, Sudbury, CO10 6PP [JO02IA, TL84]
GM7 UTD D A Forrest, 1 Laggan Rd, Newton Mearns, Glasgow, G77 6LP [IO75US, NS55]
G7 UTE B K Spencer, 31 Bell Ln, Fosdyke, Boston, PE20 2BS [IO92XV, TF33]
G7 UTF E A Gallon, 44 Broomridge Ave, Newcastle upon Tyne, NE15 6QP [IO94EX, NZ26]
G7 UTH R S Banks, 70 Bates Rd, Brighton, BN1 6PG [IO90WU, TQ30]
G7 UTI G M Du Cros, 24 Boringdon Cl, Plymouth, PL7 4HR [IO70XJ, SX55]
G7 UTJ Details withheld at licensee's request by SSL.
G7 UTK Details withheld at licensee's request by SSL.
G7 UTL S B Thompson, 33 Church Ln, Barton under Needwood, Burton on Trent, DE13 8HU [IO92DS, SK11]
G7 UTN R J Bailey, 3 Anchor Rd, Clacton on Sea, CO15 1HP [JO01NS, TM11]
G7 UTO Details withheld at licensee's request by SSL.[Op: Robert Adlington, 33 Columbine Way, Harold Wood, Romford, Essex. County: Greater London. Council area: Havering. Loc: JO01CN, WAB Sq: TQ59. QRV on 6m, 2m and 70cms. Member UK 6m Group. Pse QSL via bureau. E-mail: 106461.2440@compuserve.com.]
G7 UTQ Details withheld at licensee's request by SSL.
G7 UTR G Kelsall, 105 Brunswick House, Crosley Wood Rd, Bingley, BD16 4QG [IO93CU, SE13]
G7 UTS M J James, 7 Greenfield Park, Portishead, Bristol, BS20 8NG [IO81OL, ST47]
G7 UTT R Pearce, 15 St. Andrews Rd, Backwell, Bristol, BS19 3NR [IO81PJ, ST46]
G7 UTU Details withheld at licensee's request by SSL.
G7 UTV R R Homer, 45 Pinhaw Rd, Skipton, BD23 2SJ [IO83XW, SD95]
G7 UTW Details withheld at licensee's request by SSL.
G7 UTY C Lewis, 3 Jacobs Cl, Stantonbury, Milton Keynes, MK14 6EJ [IO92OB, SP84]
G7 UTZ L Smith, 45 Station Rd, Harpenden, AL5 4XE [IO91TT, TL11]
G7 UUA P Matthews, 6 West Rd, Halstead, CO9 1EH [JO01HW, TL83]
G7 UUB F W Gibbs, 62 Wenvoe Ave, Bexleyheath, DA7 5BT [JO01BL, TQ47]
G7 UUC M West, 69 Frampton Cres, Bristol, BS16 4JD [IO81RL, ST67]
G7 UUD K R Matthews, 6 West Rd, Halstead, CO9 1EH [JO01HW, TL83]
G7 UUF L J Preston, 17 Rosedale Gdns, Thatcham, RG19 3LE [IO91IJ, SU56]
G7 UUG N Griffiths, 125 Coleridge Way, Crewe, CW1 5LF [IO83SC, SJ75]
GW7 UUH A Hughes, 30 Liddell Dr, Llandudno, LL30 1UH [IO83CH, SH78]
G7 UUJ Details withheld at licensee's request by SSL.
G7 UUK M W Hooper, 12 Meare, Dunster Cres, Weston Super Mare, BS24 9DY [IO81MH, ST35]
G7 UUL Details withheld at licensee's request by SSL.
G7 UUN R H Wood, 7 Lilac Gr, Luston, Leominster, HR6 0EF [IO82PF, SO46]
G7 UUP J A Chapman, 11 Holmfield Rd, Beeston, Nottingham, NG9 5GJ [IO92JV, SK53]
G7 UUQ R J Moorhouse, 232 Wollaton Rd, Beeston, Nottingham, NG9 2PL [IO92JW, SK53]
G7 UUS N J Tolcher, 209 Dunraven Dr, Plymouth, PL6 6BA [IO70WK, SX46]
G7 UUT A M Wilson, Westcroft Cottage, 20B High Green, Great Shelford, Cambridge, Cambs, CB2 5EG [JO02BD, TL45]
G7 UUV Details withheld at licensee's request by SSL.
G7 UUW S W Hearn, 28 Neithrop Ave, Banbury, OX16 7NF [IO92HB, SP44]
G7 UUX Details withheld at licensee's request by SSL.
G7 UVB D A Anstie, 20 Keyes Rd, Norwich, NR1 2JX [JO02PO, TG20]
G7 UVC A R Cope, 5 First Ave, Wickford, SS11 8RA [JO01GO, TQ79]
G7 UVD P Franklin, 8 Swallowfields Ct, Skegness, PE25 2JR [JO03DD, TF56]
G7 UVE R N Slater, 37 Grosvenor St., Radcliffe, Manchester, M26 4BQ [IO83UN, SD70]
G7 UVF G M Cheetham, 35 South Park Gr, New Malden, KT3 5BZ [IO91UJ, TQ26]
G7 UVG Details withheld at licensee's request by SSL.
G7 UVH Details withheld at licensee's request by SSL.
G7 UVK Details withheld at licensee's request by SSL.
G7 UVL D M Croot, 15 Barker Ave, Jacksdale, Nottingham, NG16 5JH [IO93IB, SK45]
G7 UVN I A Cross, 9 Garthfield Cres, Newcastle upon Tyne, NE5 2LY [IO94DX, NZ26]
G7 UVO R Moss, 2 Hill End Lodge, Hatfield Park, Hatfield, AL9 5PH [IO91VS, TL20]
G7 UVP J P Swanwick, Ramblers, Clarks Farm Rd, Danbury, Chelmsford, CM3 4PH [JO01HR, TL70]
GM7 UVS J W Graham, 265 Gilmartin Rd, Linwood, Paisley, PA3 3SU [IO75RU, NS46]
G7 UVT Details withheld at licensee's request by SSL.
G7 UVU D J Heath, 22 Freeby Ave, Mansfield Woodhouse, Mansfield, NG19 9HS [IO93JE, SK56]
G7 UVV I M Perry, Meadow Cottage, Mill Ln, Pebmarsh, Halstead, CO9 2NW [JO01IX, TL83]
G7 UVW D Mills, 11 Northfield Rd, Dagenham, RM9 5XH [JO01BN, TQ48]
G7 UVY C D Carr, 10 Bonds Rd, Hemblington, Norwich, NR13 4QF [JO02RP, TG31]
G7 UVZ P J M Addison, 68 Pillar Ave, Brixham, TQ5 8LB [IO80FJ, SX95]
G7 UWA Details withheld at licensee's request by SSL.
G7 UWB B A Wright, 22 Melville Rd, Ipswich, IP4 1PN [JO02OB, TM14]
G7 UWC C A Wright, 22 Melville Rd, Ipswich, IP4 1PN [JO02OB, TM14]
G7 UWE P A Smith, 12A Sandicroft Pl, Preesall, Poulton-le-Fylde, FY6 0PB [IO83MW, SD34]
G7 UWF Details withheld at licensee's request by SSL.
G7 UWG R J Hancox, 13 Regnum Cl, Eastbourne, BN22 0XH [JO00DU, TQ60]
G7 UWH Details withheld at licensee's request by SSL.
G7 UWI M J Jones, 21 Crediton, Tamar Rd, Weston Super Mare, Somerset, BS22 0LG [IO81MI, ST36]
G7 UWJ G G Dyson-Bawley, Grahil, 33 Ridgetor Rd, Liverpool, L25 6DG [IO83NJ, SJ48]
G7 UWK B J Duffy, 8 Mirfield Cl, Liverpool, L26 9XP [IO83OI, SJ48]
G7 UWL D J Cottage, 14 Wellington Ln, Farnham, GU9 9BA [IO91OF, SU84]
GW7 UWM M T J Beynon, Sunrise, Kilgetty-Lane, Stepaside, Narberth, Pembrokeshire, SA67 8JL [IO71PR, SN10]
GM7 UWN A Stewart, 22 Seggielea Rd, Glasgow, G13 1XJ [IO75TV, NS56]
G7 UWO G W Holland, 15 Rollis Park Rd, Oreston, Plymouth, PL9 7LU [IO70WI, SX55]
G7 UWP P J Groves, Flat 4, 147 St. Peters Rise, Bristol, Avon, BS13 7ND [IO81QK, ST56]
G7 UWR C M Hillman, 119 Porter Rd, Basingstoke, RG22 4JT [IO91KF, SU65]
G7 UWS D L Brunt, 91 Shaftesbury Ave, Feltham, TW14 9LW [IO91TK, TQ17]
G7 UWT Details withheld at licensee's request by SSL.
G7 UWU N D Goddard, 113 Linden Walk, Louth, LN11 9HT [IO93XI, TF38]
G7 UWV I K Brown, 58 Highfields Rd, Bilston, WV14 0SF [IO82XN, SO99]
G7 UWW C S Harding, 1 Saddleton Gr, Saddleton Rd, Whitstable, CT5 4LY [JO01MI, TR16]
G7 UWZ D J Pooley, 25 Wharncliffe Rd, The Lanterns, Highcliffe, Christchurch, BH23 5DB [IO90DR, SZ29]
G7 UXA M Thompson, 17 Conduit Rd, Bolsover, Chesterfield, S44 6TN [IO93IF, SK46]
G7 UXC Details withheld at licensee's request by SSL.
G7 UXD R B Wade, The Limes, The Street, Hunston, Bury St. Edmunds, Suffolk, IP31 3EL [JO02KG, TL96]

G7 UXE D Wilson, 14 Jerome Way, Shipton on Cherwell, Kidlington, OX5 1JT [IO91IU, SP41]
G7 UXF A Jackson, 16 Melrose Park, Beverley, HU17 8JL [IO93SU, TA03]
G7 UXG Details withheld at licensee's request by SSL.
G7 UXH E J Gaunt, 94 Bullamoor Rd, Northallerton, DL6 1JS [IO94GI, SE39]
G7 UXI Details withheld at licensee's request by SSL.
G7 UXJ Details withheld at licensee's request by SSL.
G7 UXK S A Hedges, 25 Rudland Cl, Thatcham, RG19 3XW [IO91IJ, SU56]
G7 UXL S Ramsden, 24 Reddington Rd, Plymouth, PL3 6PS [IO70WJ, SX45]
G7 UXM M P Walsh, 305 Ripple Rd, Barking, IG11 7RJ [JO01BM, TQ48]
G7 UXN A E Barrow, 13 Manston Rd, Penketh, Warrington, WA5 2HS [IO83QJ, SJ58]
G7 UXO P J Stokes, 1A Moonsfield, Callington, PL17 7BQ [IO70UM, SX36]
G7 UXP M E Franks, 57 Thistle Downs, Northway, Tewkesbury, GL20 8RE [IO82WA, SO93]
G7 UXQ J W Manwaring, 24 Priestman Ave, Consett, DH8 8AT [IO94BU, NZ05]
G7 UXR M F Taylor, 27 Lincoln Rd, Newark, NG24 2BU [IO93OC, SK85]
G7 UXS R P Lapham, 14 Rookery Cl, St. Ives, Huntingdon, PE17 4FX [IO92XI, TL37]
G7 UXT M J Thompson, 19 Cavalier Cl, Norwich, NR7 0TE [JO02QP, TG20]
G7 UXU H Andrews, 24 Belvoir Rd, Widnes, WA8 6HR [IO83PI, SJ58]
G7 UXV Details withheld at licensee's request by SSL.
G7 UXW K Smith, 3 Cranley Lodge, Cranley Rd, Guildford, GU1 2EH [IO91RF, TQ05]
G7 UXX Details withheld at licensee's request by SSL.
G7 UXY L Williams, 3 Sawel Ct, Hendy, Pontarddulais, Swansea, SA4 1TF [IO71XR, SN50]
G7 UYA P R Williams, 3 Upper Holway Rd, Taunton, TA1 2HF [IO81LA, ST22]
G7 UYB C L Major, 17 Jubilee Cttgs, Station Rd, Marston Moreteyne, Bedford, MK43 0PN [IO92RB, SP94]
G7 UYC Details withheld at licensee's request by SSL.
G7 UYD Details withheld at licensee's request by SSL.
G7 UYE D M Whitehead, 5 Manorstead, Skelmanthorpe, Huddersfield, HD8 9DW [IO93EO, SE21]
GW7 UYF E Thomas, Graig-y-Don, Trefriw, Gwynedd, LL27 0JJ [IO83CD, SH76]
G7 UYG Details withheld at licensee's request by SSL.
G7 UYH Details withheld at licensee's request by SSL.
G7 UYI B C Williamson, 12 Middleton Cl, Southampton, SO18 2FP [IO90HW, SU41]
G7 UYJ J C Jardine, 3 Beaumont Lodge Rd, Leicester, LE4 1BT [IO92KQ, SK50]
G7 UYK A E Collis, 2 Sidcliffe House, Sidcliffe, Sidmouth, EX10 9QA [IO80JQ, SY18]
G7 UYM L Edmondson, 160 Torquay Rd, Paignton, TQ3 2AH [IO80FK, SX86]
G7 UYN S L Gargett, 25 Prospero Rd, London, N19 3QX [IO91WN, TQ28]
G7 UYO S Passmore, 332A Torquay Rd, Paignton, TQ3 2DZ [IO80FK, SX86]
G7 UYP J A Balls, 7 Rowan Cl, Holbeach, Spalding, PE12 7BT [JO02AT, TF32]
G7 UYR R D Houghton, 2 Beanhill Cres, Alveston, Bristol, BS12 2JG [IO81RO, ST68]
G7 UYT J J T O'Toole, 16 Paul St., London, E15 4QD [JO01AM, TQ38]
G7 UYU C Kells, 117 Briar Cres, Exeter, EX2 6DR [IO80FR, SX99]
G7 UYV Details withheld at licensee's request by SSL.
G7 UYW J H Kitchener, 101 Highfield Rd, Tring, HP23 4DS [IO91PT, SP91]
G7 UYX M C Dibdin, 6 St. Johns Cl, Newport, PO30 1LU [IO90IQ, SZ48]
G7 UYY Details withheld at licensee's request by SSL.
G7 UYZ K N Read, 15 Harlaxton Cl, Eastleigh, SO50 4QX [IO90HX, SU42]
G7 UZA J Rodinson, 21 Graylands Rd, Liverpool, L4 9UG [IO83MK, SJ39]
G7 UZB Details withheld at licensee's request by SSL.
G7 UZE B G W Lilley, 7 Pankhurst Ave, Brighton, BN2 2YP [IO90WT, TQ30]
G7 UZF J A Ford, 8 Blakes Green, West Wickham, BR4 0RA [IO91XJ, TQ36]
G7 UZG A J McWilliam, 26 Rosehill, Great Ayton, Middlesbrough, TS9 6BH [IO94KL, NZ51]
G7 UZI P G Pullen, 12 Kimpton Rd, Cheam, Sutton, SM3 9QJ [IO91VI, TQ26]
G7 UZM Details withheld at licensee's request by SSL.
G7 UZN D C Dawson, Flat 8, 238 Royal College St., London, NW1 9QR [IO91WN, TQ28]
G7 UZO D Lock, 34 Southgate, Honley, Huddersfield, HD7 2NT [IO93CO, SE11]
G7 UZP Details withheld at licensee's request by SSL.
G7 UZQ R S Marsh, 58 Mill Ln, Marston, Oxford, OX3 0QA [IO91JS, SP50]
G7 UZS N Thompson, 30 Dene View, Ashington, NE63 8JF [IO95EE, NZ28]
G7 UZW Details withheld at licensee's request by SSL.
G7 UZX G G A Saunders, 50 Mellow Purgess, Basildon, SS15 5UY [JO01FN, TQ68]
G7 UZY A P Musther, 2 Fakenham Cl, Lower Earley, Reading, RG6 4AB [IO91MK, SU76]
G7 UZZ Details withheld at licensee's request by SSL.
G7 VAB M R Richards, 4 Marlborough Rd, London, E4 9AL [IO91XO, TQ39]
G7 VAC Details withheld at licensee's request by SSL.
G7 VAD M J Beeney, Fairways, Ghyll Rd, Heathfield, TN21 0XL [JO00CX, TQ52]
G7 VAE J M Beeney, Fairways, Ghyll Rd, Heathfield, TN21 0XL [JO00CX, TQ52]
GI7 VAF C White, Shallamar, 3A Park Rd, Strabane, BT82 8EL [IO64GT, H39]
G7 VAG G D J Podmore, 10 Synge St., Warrington, WA2 7QB [IO83RJ, SJ68]
G7 VAH S M Rutter, Old Stores, The St., Sutton, Norwich, NR12 9RF [JO02SS, TG32]
G7 VAJ Details withheld at licensee's request by SSL.
G7 VAK Details withheld at licensee's request by SSL.
G7 VAM Details withheld at licensee's request by SSL.
G7 VAP T M Richards, 59 Abshot Rd, Titchfield Common, Fareham, PO14 4NB [IO90IU, SU50]
G7 VAR H E Wright, 46 Holmpton Rd, Withernsea, HU19 2QD [JO03AR, TA32]
G7 VAS Details withheld at licensee's request by SSL.
G7 VAU Details withheld at licensee's request by SSL.
G7 VAV Details withheld at licensee's request by SSL.
G7 VAX Details withheld at licensee's request by SSL.
G7 VAY R Dingle, 87 Eighth Ave, Bridlington, YO15 2NA [IO94VC, TA16]
G7 VBA P Fonseca, Swinton Castle, Masham, North Yorks, HG4 4JH [IO94DF, SE27]
G7 VBB Details withheld at licensee's request by SSL.
G7 VBC Details withheld at licensee's request by SSL.
G7 VBD M J Ennis, 19 Mowbray Terr Front, Choppington, NE62 5QP [IO95ED, NZ28]
GW7 VBE D Harris, 29 Queen St., Blaengarw, Pontycymer, Bridgend, CF32 8AH [IO81FP, SS99]
G7 VBF J H Barwell, 7 Alloa Rd, Ilford, IG3 9SW [JO01BN, TQ48]
G7 VBI F W Russell, 8 Boscombe Rd, Worcester Park, KT4 8PL [IO91VJ, TQ26]
G7 VBJ D J Wager, 14 The Paddock, Harston, Cambridge, CB2 5PR [JO02BD, TL45]
G7 VBL J G Munday, 56 Osea Way, Springfield, Chelmsford, CM1 6JT [JO01FR, TL70]
G7 VBN B J D Richards, 2 Craddock Row, Sandhutton, Thirsk, YO7 4RT [IO94HF, SE38]
G7 VBP Details withheld at licensee's request by SSL.
G7 VBQ Details withheld at licensee's request by SSL.
G7 VBR R Ferrand, 38 Hills Rd, Steyning, BN44 3QG [IO90UV, TQ11]
GI7 VBS D Aughey, 239 Bridge St., Portadown, Craigavon, BT63 5AR [IO64SK, J05]
G7 VBT I D Main, Croft Cottage, 93 Tewit Ln, Illingworth, Halifax, HX2 9SD [IO93BS, SE02]
G7 VBU D G Firth, Highfield, 41 near Bank, Shelley, Huddersfield, HD8 8LT [IO93DO, SE21]
G7 VBV A C Watson, 19 Walnut Tree Ln, Longwick, Princes Risborough, HP27 9SJ [IO91NR, SP70]
G7 VBW S F Smith, 33 Babbacombe Gdns, Ilford, IG4 5LZ [JO01AN, TQ48]
G7 VBX Details withheld at licensee's request by SSL.
G7 VBY Dr D J Morrison-Smith, 4 Edgefield Cl, Cranleigh, GU6 8PX [IO91SD, TQ03]
G7 VBZ P A Bunce, 47 Hobbs Cres, Saltash, PL12 4JJ [IO70VJ, SX45]
G7 VCA W Carrington, 4 Newthorn, Oswaldtwistle, Accrington, BB5 3SB [IO83TR, SD72]
G7 VCB L C Tooze, Flat 1, 91 Harbour Rd, Seaton, EX12 2NJ [IO80LQ, SY28]
G7 VCC A Roberts, 67 Parnel Rd, Ware, SG12 7LE [IO91XT, TL31]
G7 VCD Details withheld at licensee's request by SSL.
G7 VCE M D Flack, 31 Harebell Cl, Cambridge, CB1 4YL [JO02CE, TL45]
G7 VCF J O'Donnell, 3 Linden Ave, Altrincham, WA15 8HA [IO83TJ, SJ78]
G7 VCG I A Firby, 19 St. Georges Dr, Manchester, M40 5HL [IO83VM, SD80]
GW7 VCH Rev S L C Marsh, The Presbytery, Meyrick St., Dolgellau, LL40 1LR [IO82BR, SH71]
G7 VCI S J Nicholls, 29 Coanwood Dr, Cramlington, NE23 6TL [IO95FB, NZ27]
G7 VCJ C J Hansford, 14 Parsonage Cres, Castle Cary, BA7 7LT [IO81RC, ST63]
G7 VCK M F Robertson, 67 Oatland Gdns, Leeds, LS7 1SL [IO93FT, SE33]
GW7 VCL Details withheld at licensee's request by SSL.
G7 VCM C M Kemp, 10 Laurel Cl, Dartford, DA1 2QL [JO01CK, TQ57]
G7 VCN R Bawley, 52 Firth Ave, Liverpool, L18 7JG [IO83NJ, SJ38]
G7 VCO A C Callender, 2 Masham Gr, Stockton on Tees, TS19 7QX [IO94HN, NZ41]
G7 VCP P Stubbs, 6 Sydney St., Weston Point, Runcorn, WA7 4JG [IO83OH, SJ48]
G7 VCQ Details withheld at licensee's request by SSL.
GI7 VCR S J Robertson, 5 East Mount, Newtownards, BT23 8SE [IO74DN, J46]
GI7 VCS B J Russell, 10 Ferngrove Ave, Aghagallon, Craigavon, BT67 0HA [IO64UM, J16]
G7 VCT G Mould, 25 Kingsley Rd, Talke Pits, Stoke on Trent, ST7 1RB [IO83UB, SJ85]
G7 VCU R K Perrin, 11 Firs View Rd, Hazlemere, High Wycombe, HP15 7TD [IO91PP, SU89]
GM7 VCV A Brown, 41 Main Rd, Gateside, Beith, KA15 2LF [IO75QR, NS35]
G7 VCW Details withheld at licensee's request by SSL.
G7 VCX A A Spencer, 80 Woodland Rd, Whitby, South Wirral, L65 6PS [IO83NG, SJ37]
G7 VCY D C Seymour, 27 Magdalene Rd, Radstock, Bath, BA3 3LB [IO81SG, ST75]
G7 VCZ P A Weaver, 1 The Willows, Wootton, Boars Hill, Oxford, OX1 5LD [IO91IQ, SP40]
G7 VDA I G Singer, 197 Rosalind St., Ashington, NE63 9BB [IO95FE, NZ28]
G7 VDC P S Betts, 29 Waldegrave Rd, Brighton, BN1 6GR [IO90WU, TQ30]
G7 VDD P McGowan, 11 Pankhurst Gdns, Gateshead, NE10 8EN [IO94FW, NZ26]
G7 VDE W D Pitt, 194 Nuthurst Rd, Birmingham, B31 4TD [IO92AJ, SP07]
G7 VDF Details withheld at licensee's request by SSL.
G7 VDG R E Turner, 58 Parkway, London, SW20 9HF [IO91VJ, TQ26]
G7 VDH J R Brook, Windsor Cottage, New Laithe Ln, Holmfirth, West Yorks, HD7 2HL [IO93CO, SE11]
G7 VDJ S A Henry, 11 Chancery Ln, Alsager, Stoke on Trent, ST7 2HE [IO83UC, SJ75]
G7 VDK G D Taylor, 22 Crathie, Birtley, Chester-le-Street, DH3 1QJ [IO94FV, NZ25]
G7 VDN H W May, 18 Pennant Hills, Bedhampton, Havant, PO9 3JZ [IO90LU, SU60]
G7 VDO W Brown, 126 Hawthorn Rd, Ashington, NE63 9BG [IO95FE, NZ28]
G7 VDP M I Reid, 41 Allhallowgate, Ripon, HG4 1LF [IO94FD, SE37]
G7 VDQ D Butterworth, 9 Meadow Rd, Weaverham, Northwich, CW8 3BS [IO83RG, SJ67]
G7 VDR Details withheld at licensee's request by SSL.
G7 VDS M S Hudson, 14 Claremont Mews, King St., Wellington, Telford, TF1 3PF [IO82RQ, SJ61]
G7 VDT G Baines, 59 Thornleigh Ave, Wakefield, WF2 7SF [IO93GQ, SE31]
G7 VDU A C Marston, 17 Bramble Cl, Hamilton, Leicester, Leics, LE5 1UB [IO92LP, SK60]
G7 VDV N Keech, 14 Simpson Ct, Ashington, NE63 9SD [IO95FE, NZ28]
G7 VDW J Renshaw, 74 Lanhydrock Rd, Plymouth, PL4 9HH [IO70WJ, SX45]
G7 VDX S J Taverner, 8 The Rye Lea, Droitwich, WR9 8SS [IO82VG, SO86]
G7 VDY C M Williamson, 5 Orchard Dr, South Hiendley, Barnsley, S72 9BQ [IO93HO, SE31]
G7 VDZ D M Williamson, 5 Orchard Dr, South Hiendley, Barnsley, S72 9BQ [IO93HO, SE31]
G7 VEB D Wilson, 210 Stanks Ln South, Swarcliffe, Leeds, LS14 5PD [IO93GT, SE33]
G7 VEC Details withheld at licensee's request by SSL.
G7 VED M I Brown, 61 Valley Rd, Pudsey, LS28 9EU [IO93ES, SE23]
G7 VEE A R Saunders, 25 Southern Dr, South Woodham Ferrers, Chelmsford, CM3 5NY [JO01HP, TQ89]
G7 VEF R J Parkin, 17 Roberts Rd, Watford, WD1 8AY [IO91TP, TQ19]
G7 VEG P T Nicholls, 12 Magpie Cl, Burnham on Sea, TA8 2QH [IO81LG, ST25]
GW7 VEH J H Humphrey, 1 Pentwyn Terr, Abersychan, Pontypool, NP4 7TH [IO81LR, SO20]
G7 VEI A D Stripp, 83 Kingswood Rd, Shortlands, Bromley, BR2 0NG [IO91XJ, TQ36]
G7 VEJ B L Harrison, 7 Romsey Cl, Mickleover, Derby, DE3 5SD [IO92FV, SK33]
G7 VEK M P Dixon, 6 Paddock Rise, Verney Fields, Stonehouse, GL10 2RD [IO81UR, SO80]
GW7 VEL C J T Lee, 23 Forest Cl, Coed Eva, Cwmbran, NP44 4TE [IO81LP, ST29]
G7 VEM K G White, 3 Hutton Ave, Worsley, Manchester, M28 1JP [IO83SM, SD70]
G7 VEN B E E Sweet, End Cottage, Holton, Wincanton, Somerset, BA9 8AN [IO81SA, ST62]
G7 VEO Details withheld at licensee's request by SSL.
G7 VEP C D Chantler, 45 Longhurst Ln, Marple Bridge, Stockport, SK6 5AE [IO83XJ, SJ98]
G7 VEQ Details withheld at licensee's request by SSL.
G7 VER Revd J G Slee, The Rectory, St Mawgan, Newquay, Cornwall, TR8 4EZ [IO70MK, SW86]
G7 VES C M J Slee, The Rectory, St Mawgan, Newquay, Cornwall, TR8 4EZ [IO70MK, SW86]
GM7 VET J J Keenan, 36 Eriskay Cres, Newton Mearns, Glasgow, G77 6XE [IO75TS, NS55]
GW7 VEU K G Dancer, 118 Fairwater Gr West, Cardiff, CF5 2JR [IO81JL, ST17]
G7 VEV C M Toseland, 4 Windmill Ln, Belper, DE56 1GN [IO93GA, SK34]
G7 VEX N J Hindle, 15 Willow Rd, Stamford, PE9 2FF [IO92RP, TF00]
G7 VEY J Martin, 3 The Rise, Calne, SN11 0LQ [IO91AK, SU06]
G7 VEZ Details withheld at licensee's request by SSL.
G7 VFA J W Juggins, 52 Gweal Darras Est, Mabe Burnthouse, Penryn, TR10 9HQ [IO70KD, SW73]
G7 VFD P J Westwood, 62 Blackbrook Ln, Bickley, Bromley, BR2 8AY [JO01AJ, TQ46]
G7 VFE R Wills, 14 Penwood Heights, Penwood, Burghclere, Newbury, RG20 9EY [IO91HI, SU46]
G7 VFF B Konemann, 36 Main St., Lower Bentham, Bentham, Lancaster, LA2 7HN [IO84RC, SD66]
G7 VFH B S Ford, 8 Lee Cl, Honiton, EX14 8NS [IO80JT, ST10]
G7 VFI A A Cussans, 66 Henley St., Lincoln, LN5 8BA [IO93RF, SK96]
G7 VFJ Details withheld at licensee's request by SSL.
G7 VFL K Sherman, 12 Portland Dr, Stourbridge, DY9 0SD [IO82WK, SO98]
G7 VFM Details withheld at licensee's request by SSL.
G7 VFN D J W Watts-Read, 43 Whyteleafe Hill, Whyteleafe, CR3 0AJ [IO91XH, TQ35]
G7 VFO R D Rawlinson, 17 Walmer Pl, Winsford, CW7 1HA [IO83RE, SJ66]
GW7 VFP H G Morris, 1 Bryn Ffynnon, Beddgelert, Nantmor, Caernarvon, LL55 4YG [IO72WX, SH54]
G7 VFQ A Latham, 273 Adswood Rd, Stockport, SK3 8PA [IO83WJ, SJ88]
GM7 VFR J R Smith, 28 Tollerton Dr, Irvine, KA12 0QE [IO75QO, NS33]
G7 VFS P McCormack, 3 Greenway Cl, Torquay, TQ2 8EF [IO80FL, SX96]
G7 VFU A J Mullord, 296 City Way, Rochester, ME1 2BL [JO01GI, TQ76]
G7 VFV G M Somers, 76 Fowler Rd, Aylesbury, HP21 8QG [IO91OT, SP81]
G7 VFW A Clark, 9 Jackson St., Burtonwood, Warrington, WA5 4HN [IO83QK, SJ59]
G7 VFX R J Watts-Read, 43 Whyteleafe Hill, Whyteleafe, CR3 0AJ [IO91XH, TQ35]
G7 VFY S I Walters, 42 Achilles Rd, London, NW6 1EA [IO91VN, TQ28]
G7 VFZ P J Westerby, 2 Faldo Cl, Abbeymead, Gloucester, GL4 5BN [IO81VU, SO81]
G7 VGA S A Bonney, 22 Gordon Dr, Abingdon, OX14 3SW [IO91IQ, SU59]
GW7 VGB D Beynon, 129 Eureka Pl, Ebbw Vale, NP3 6LN [IO81JS, SO10]
G7 VGC B D Goody, 121 Tilkey Rd, Coggeshall, Colchester, CO6 1QN [JO01IV, TL82]
G7 VGD G H Hutchings, 21 Gaviots Cl, Gerrards Cross, SL9 7EJ [IO91RN, TQ08]
G7 VGE R H Teague, 70 Leggatt Dr, Bramford, Ipswich, IP8 4EY [JO02NB, TM14]
G7 VGF W J Appleyard, 45 Shakespeare Rd, Lexden, Colchester, CO3 4HZ [JO01KV, TL92]
G7 VGG Details withheld at licensee's request by SSL.
G7 VGH M A Smith, 57 Lynn Rd, Ely, CB6 1DD [JO02DJ, TL58]
G7 VGJ A B Cole, 21 Tudor Cl, Chigwell, IG7 5BG [JO01AO, TQ49]
G7 VGK W R Parrett, 35 Fazakerley Rd, Liverpool, L9 2AH [IO83MK, SJ39]
G7 VGL D J Pemberton, 12 Victor Rd, Thatcham, Berks, RG19 4LX [IO91JJ, SU56]
G7 VGM S J Taylor, 28 Parton St., Hartlepool, TS24 8NN [IO94JQ, NZ53]
G7 VGN A W E Chamberlain, 7 McCalmont Way, Newmarket, CB8 8HU [JO02FF, TL66]
G7 VGO A M Hurst, 12 Spilsby Cl, Hartlepool, TS25 2RD [IO94JP, NZ42]
G7 VGP Details withheld at licensee's request by SSL.
G7 VGQ J Jeffels, 22 Impala Cl, Norwich, NR6 7PN [JO02PP, TG21]
GI7 VGR C J Dorrian, 36 Windmill Ave, Carrickfergus, BT38 8DH [IO74CR, J48]
G7 VGU A J Gibson, Oakwood, The Hill, Millom, Cumbria, LA18 5HG [IO84IF, SD18]
G7 VGV H Quigg, Flat 8, 2A Strawberry Dale, Harrogate, HG1 5EF [IO93FX, SE35]
G7 VGW Details withheld at licensee's request by SSL.
G7 VGX Details withheld at licensee's request by SSL.
G7 VGY D Childs, 26 Granby Gr, Highfield, Southampton, SO17 3RZ [IO90HW, SU41]
G7 VGZ M R Jones, 1 Poplar Walk, Stradishall, Newmarket, CB8 9YB [JO02GD, TL75]
G7 VHA T Knight, 6 The Close, Great Horwood, Milton Keynes, MK17 0QG [IO91NX, SP73]
G7 VHC Maj C E Spires, 5 Springhead, Sutton Veny, Warminster, BA12 7AG [IO81WE, ST94]
GW7 VHD A P Jones, Westwinds, 57 Dinerth Rd, Colwyn Bay, Clwyd, LL28 4YG [IO83CH, SH87]
G7 VHE M R Bowman, 1 Birkbeck Cl, South Wootton, Kings Lynn, Norfolk, PE30 3LH [JO02FS, TF62]
G7 VHF A M Durrant East Anglian Six Meter Group, 2 Ramsey Hall Cttgs, Wix Rd, Ramsey, Harwich, CO12 5LS [JO01OW, TM12]
G7 VHG S A Bryan, 14 Cowleaze Cres, Wroughton, Swindon, SN4 9EN [IO91CM, SU18]
G7 VHH N D Purchon, 54 Gondar Gdns, London, NW6 1HG [IO91VN, TQ28]
G7 VHJ P K J Gow, 11 Rodley Sq, Lydney, GL15 5AZ [IO81RR, SO60]
G7 VHK M B Kelleher, Flat C, 44 Cambridge Ave, Kilburn, London, NW6 5BA [IO91VM, TQ28]
G7 VHL P F J Eames, 6 Sirius Cl, Seaview, Isle of Wight, PO34 5LH [IO90KR, SZ69]
G7 VHM P D Bailey, Top Flat, 278 Heneage Rd, Grimsby, DN32 9NP [IO93XN, TA20]
G7 VHN J L Hart, 20 Church Ave, Pinner, HA5 5JQ [IO91TN, TQ18]
G7 VHO B D Hart, 20 Church Ave, Pinner, HA5 5JQ [IO91TN, TQ18]
G7 VHP Details withheld at licensee's request by SSL.
GM7 VHQ I Helie, 25 Ard Rd, Renfrew, PA4 9DD [IO75TV, NS46]
G7 VHR M Burke, 236 Pooltown Rd, Ellesmere Port, South Wirral, L65 8HZ [IO83NG, SJ37]
G7 VHT Details withheld at licensee's request by SSL.
G7 VHU D Beastall, 11 Hopwood Bank, Horsforth, Leeds, LS18 5AW [IO93EU, SE23]
G7 VHV D D Bott, 11 Catcliffe Way, Lower Earley, Reading, RG6 4HX [IO91MK, SU76]
GW7 VHW C S McLaughlin, Garn-Wen, Llansadwrn, Llanwrda, Dyfed, SA19 8NS [IO81AX, SN63]
G7 VHX J L Golding, 65 Longworth Ave, Tilehurst, Reading, RG3 5JU [IO91LJ, SU66]
G7 VHY F Ball, 3 Merepark Dr, Southport, PR9 9FB [IO83MQ, SD31]
G7 VHZ E A Gilowski, 652 Huddersfield Rd, Lees, Oldham, OL4 3PZ [IO83XN, SD90]
G7 VIA M D Airs, 3 East St., Didcot, OX11 8EJ [IO91JO, SU58]
G7 VIB D P Airs, 3 East St., Didcot, OX11 8EJ [IO91JO, SU58]
G7 VIC J R H Lythgoe, 12 Oak Tree Cl, Bedale, DL8 1UG [IO94EG, SE28]
G7 VID S J Gunn, 51 Spruce Hill, Harlow, CM18 7SS [JO01BR, TL40]
G7 VIE J Cooper, 9 Highfield Cres, Halesowen, B63 2BD [IO82XL, SO98]
G7 VIF Details withheld at licensee's request by SSL.
G7 VIG A B Smith, 19 Gibsons Gdns, North Somercotes, Louth, LN11 7QH [JO03BK, TF49]
G7 VIH P Wilson, 115 Avondale Rd, Kettering, NN16 8PN [IO92PJ, SP87]
G7 VIJ N Lawton, 10 Highfield Cres, Halesowen, B63 2BD [IO82XL, SO98]
G7 VIK N C Higgins, 6 Larksfield Ave, Bournemouth, BH9 3LP [IO90BS, SZ19]
G7 VIL G D Mason, 142 Bankhead Rd, Northallerton, DL6 1JD [IO94GI, SE39]
G7 VIM W W Brock, 7 Galtres Rd, Northallerton, DL6 1QN [IO94GI, SE39]
G7 VIN I J Prater, 470 Bishport Ave, Bristol, BS13 0HS [IO81QJ, ST56]
G7 VIO Details withheld at licensee's request by SSL.

G7 VIP F W G Marston, 1 Weaver Rd, Leicester, LE5 2RL [IO92LP, SK60]
GI7 VIQ R G Cummings, 19 Castle Rd, Carrickfergus, BT38 7JY [IO74CR, J48]
G7 VIR A James, 19 Coach Ln, Redruth, Cornwall, TR15 2TP [IO70JF, SW64]
G7 VIT E J Robinson, 18 Broadway, Wellingborough, NN8 2DA [IO92PH, SP86]
G7 VIU D Sinclair, 17 Mount Pleasant Rd, Wath upon Dearne, Rotherham, S63 7JE [IO93HL, SK49]
G7 VIV A Nussey, 9 Brent St., Brent Knoll, Highbridge, Somerset, TA9 4DU [IO81MF, ST35]
GI7 VIW A F Harvey, Creg-Ny-Baa, Kilmaine Rd, Bangor, Co. Down, BT19 6DT [IO74EP, J58]
G7 VIX I A Rogers, 21 Hunnington Cres, Halesowen, B63 3DJ [IO82XK, SO98]
G7 VIY A J Harper, 3 Eskdale Cres, Blackburn, BB2 5DT [IO83RR, SD62]
G7 VIZ I G Shields, 38A Westlands Way, Leven, Beverley, HU17 5LG [IO93UV, TA14]
G7 VJA K W J Sharman, 82 Bedfont Cl, East Bedfont, Feltham, TW14 8LF [IO91SK, TQ07]
G7 VJB Details withheld at licensee's request by SSL.
G7 VJC Details withheld at licensee's request by SSL.
G7 VJD Details withheld at licensee's request by SSL.
G7 VJE C C Rohrer, Alpenrose, Bedlars Green, Great Hallingbury, Bishops Stortford, CM22 7TP [JO01CU, TL52]
G7 VJF M J Connolly, 34 Brown Ln East, Leeds, LS11 9LA [IO93FS, SE23]
G7 VJG P P Veitch, 78 Hughes St., Swindon, SN2 2HG [IO91CN, SU18]
G7 VJH T A D Scanlon, 46 Fairfax Ave, Drighlington, Bradford, BD11 1DN [IO93ES, SE22]
G7 VJI Details withheld at licensee's request by SSL.
G7 VJK Details withheld at licensee's request by SSL.
G7 VJL Details withheld at licensee's request by SSL.
G7 VJM C R Margetts, 16 Lahn Dr, Droitwich, WR9 8TQ [IO82WG, SO86]
GI7 VJN Details withheld at licensee's request by SSL.
G7 VJP D M Almond, 23 Frithwald Rd, Chertsey, KT16 9EZ [IO91RJ, TQ06]
G7 VJQ Details withheld at licensee's request by SSL.
G7 VJR M J Wells, 52 Clotherholme Rd, Ripon, HG4 2DL [IO94FD, SE27]
G7 VJS Details withheld at licensee's request by SSL.
G7 VJT S P Tibbetts, 24 Crimmond Rise, Halesowen, B63 3RA [IO82XK, SO98]
G7 VJU R J Jones, 31 Virginia Ave, Liverpool, L31 2NN [IO83MM, SD30]
G7 VJV N P Massey, 105 Blakedown Rd, Halesowen, B63 4NR [IO82XK, SO98]
G7 VJW C R Gardiner, 9 Mill Ln, Upholland, Skelmersdale, WN8 0HH [IO83PN, SD50]
G7 VJX A Lowe, 18 Cob Moor Rd, Billinge, Wigan, WN5 7EH [IO83PM, SD50]
G7 VJY T C Hill, 15 Catkin Walk, Rugeley, WS15 2NS [IO92AS, SK01]
G7 VJZ Details withheld at licensee's request by SSL.
G7 VKA K Wandless, 61 Harle Rd, Backworth, Newcastle upon Tyne, NE27 0RZ [IO95FA, NZ37]
G7 VKB A P Cunnington, 131 Colson Rd, Loughton, IG10 3QY [JO01AP, TQ49]
G7 VKC A J Butcher, 4 Northern Ave, Benfleet, SS7 5SN [JO01GN, TQ78]
G7 VKF C R White, 5 St. James Ln, Coventry, CV3 3GR [IO92GJ, SP37]
G7 VKG M A Gibson, 70 Ruskin Rd, Mansfield, NG19 7LX [IO93JD, SK56]
G7 VKH Details withheld at licensee's request by SSL.
GW7 VKI S J Brewer, 12 Oakfield St., Llanbradach, Caerphilly, CF83 3NX [IO81JO, ST19]
G7 VKJ C Buckley, 14 Sunny Dr, Prestwich, Manchester, M25 3JJ [IO83UM, SD80]
G7 VKK Details withheld at licensee's request by SSL.
G7 VKL Details withheld at licensee's request by SSL.
G7 VKM R O G Booth, 3 Golden Oak Cl, Farnham Common, Slough, SL2 3SY [IO91QN, SU98]
G7 VKN R A Beharie, 51 Winslow Way, Feltham, TW13 6QF [IO91TK, TQ17]
G7 VKO J E Carden, 7 Rowland Cl, Gillingham, ME7 3DJ [JO01GI, TQ76]
G7 VKP Details withheld at licensee's request by SSL.
G7 VKQ Details withheld at licensee's request by SSL.
G7 VKR D L Eccles, 34 Westwood Rd, Stoke on Trent, ST3 6BH [IO82WX, SJ94]
G7 VKS M J Hamilton, 58 Kingswood Way, Selsdon, South Croydon, CR2 8QQ [IO91XI, TQ36]
G7 VKT H Hamilton, Windrush, 58 Kingswood Way, Selsdon, South Croydon, CR2 8QQ [IO91XI, TQ36]
G7 VKU Details withheld at licensee's request by SSL.
G7 VKW S Yates, 5 Camellia Dr, Warminster, BA12 7RN [IO81VE, ST84]
GW7 VKX G P Judge, 1 West View Dr, Mynydd Isa, Mold, CH7 6YF [IO83KE, SJ26]
G7 VKY B J Shrimpling, 3 Oliver St., Cleethorpes, DN35 7QQ [IO93XN, TA20]
G7 VKZ P Carrington, 10 Bluebell Cres, Sanderson Park, Wednesfield, Wolverhampton, WV11 3XB [IO82XO, SJ90]
G7 VLA M E Sandham, 7 Mill Cl, Caverswall, Stoke on Trent, ST11 9HA [IO82XX, SJ94]
GM7 VLC A Haines, 164 North High St., Musselburgh, EH21 6AR [IO85WW, NT37]
G7 VLD K J Howard, 43 Hazeldell, Watton At Stone, Hertford, SG14 3SN [IO91WU, TL21]
G7 VLF N A Faiz, 48 Cox House, Field Rd, London, W6 8HN [IO91VL, TQ27]
G7 VLG Details withheld at licensee's request by SSL.
G7 VLH Details withheld at licensee's request by SSL.
G7 VLI K Nickolls, Wold View, Main Rd, Covenham St. Bartholomew, Louth, LN11 0PF [JO03AK, TF39]
G7 VLJ E Baker, 10 Newbury Rd, Ipswich, IP4 5EX [JO02OB, TM14]
G7 VLK D M Woodhouse, 14 Turbinia Gdns, Newcastle upon Tyne, NE7 7LP [IO94FX, NZ26]
G7 VLL J Woodhouse, 14 Turbinia Gdns, Newcastle upon Tyne, NE7 7LP [IO94FX, NZ26]
G7 VLM D J P Brooks, 14 Kidderminster Rd, Bridgnorth, WV15 6BW [IO82TM, SO79]
G7 VLN J McLellan, 31 Lindisfarne Rd, Corby, NN17 2EL [IO92PM, SP89]
G7 VLO Details withheld at licensee's request by SSL.
G7 VLP P F Edwards, 10 Shefford Rd, Meppershall, Shefford, SG17 5LJ [IO92TA, TL13]
G7 VLQ K Garrett, 19 Stoneleigh Gr, Ossett, WF5 8QN [IO93FQ, SE22]
G7 VLR I M Johnson Leicester Raynet Group, 78 Croft Rd, Cosby, Leicester, Leics, LE9 1SE [IO92JN, SP59]
GW7 VLT Details withheld at licensee's request by SSL.
G7 VLU S J Elliott, 52 Gerrard St., Lancaster, LA1 5LZ [IO84OB, SD46]
G7 VLV D R Hunt, 60 Woolslope Rd, West Moors, Ferndown, BH22 0PE [IO90BT, SU00]
G7 VLX Details withheld at licensee's request by SSL.
GM7 VLZ A Pearce, Craigknowe, The Mount, Cupar, Fife, KY15 4NA [IO86LI, NO31]
G7 VMB A J Bostock, 9 Sunningdale Cl, Jupps Ln, Goring By Sea, Worthing, BN12 4TU [IO90ST, TQ10]
G7 VMC D P Way, 56 Fernside Rd, Poole, BH15 2JJ [IO90AR, SZ09]
G7 VMD A P Griffin, 10 Vespasian Cl, Stratton St. Margaret, Swindon, SN3 4BX [IO91DN, SU18]
G7 VME P Schofield, Fives Ct Flat, Eton College, Windsor, Berks, SL4 6DU [IO91QL, SU97]
G7 VMF Details withheld at licensee's request by SSL.
G7 VMG Details withheld at licensee's request by SSL.
G7 VMH D Spittlehouse, Pentlands, High St., Wroot, Doncaster, DN9 2BT [IO93MM, SE70]
G7 VMI K M Greenhalgh, 3 Tavis Rd, Paignton, TQ3 2PU [IO80FK, SX86]
G7 VMJ J W Golton, 688 New North Rd, Ilford, IG6 3XP [JO01BO, TQ49]
G7 VML L H Alden, 189 Upland Rd, London, SE22 0DG [IO91XK, TQ37]
G7 VMN Details withheld at licensee's request by SSL.
G7 VMO S G Fawcett, Ison Cottage, West End Rd, Norton, Doncaster, DN6 9EF [IO93JP, SE51]
G7 VMP M L Howes, Rendalls Row, 7A High St., Puddletown, Dorchester, DT2 8RT [IO80TR, SY79]
G7 VMQ T Jones, Ockton House, 24 Station Rd, Okehampton, EX20 1EA [IO70XR, SX59]
G7 VMS G J Ellis, Hurles Cottage, Oxford St, Ramsbury, Marlborough, Wilts, SN8 2PQ [IO91EK, SU27]
GW7 VMT E C Graham-Brown, 44 Priory Ville, Milford Haven, SA73 2JR [IO71LR, SM90]
G7 VMU Details withheld at licensee's request by SSL.
G7 VMV Details withheld at licensee's request by SSL.
G7 VMX Details withheld at licensee's request by SSL.
G7 VMY M Prior, 114 Livingstone Walk, Hemel Hempstead, HP2 6AL [IO91SS, TL00]
G7 VNB Details withheld at licensee's request by SSL.
G7 VNC C T Cave, Little Meadow, Brewham Rd, Bruton, BA10 0JD [IO81SC, ST63]
G7 VND Dr G W Morris, 17 Bradshaw Rd, Inkersall, Chesterfield, S43 3HJ [IO93HF, SK47]
G7 VNE P Oates, 22 Sunny Bank Walk, Mirfield, WF14 0NH [IO93DQ, SE22]
G7 VNG A J Elmes, Pookeezows, 10 Farnham Ave, Keymer, Hassocks, BN6 8NS [IO90WW, TQ31]
G7 VNH Details withheld at licensee's request by SSL.
G7 VNI Details withheld at licensee's request by SSL.
G7 VNJ M P B Swain, 13 Gatley Dr, Guildford, Surrey, GU4 7JJ [IO91RG, TQ05]
G7 VNK R Cronshaw, Flat 2, 28 Adelaide Terr, Blackburn, BB2 6ET [IO83SS, SD62]
G7 VNL C E Lambert, Flat 5, 5 Semaphore Rd, Guildford, GU1 3PS [IO91RF, TQ04]
G7 VNM A Melham, 201 Bridgewood Rd, Worcester Park, KT4 8XU [IO91VI, TQ26]
G7 VNN C I Backhouse, 2 Medusa Villas, Denmark Ln, Roydon, Diss, IP22 3BL [JO02NI, TM17]
G7 VNO C A Brown, 6 Ellesmere Ave, Wilmorton, Derby, DE24 8WD [IO92GV, SK33]
G7 VNP K D P Knights, 26 Kirtling Rd, Saxon St., Newmarket, CB8 9RR [JO02FE, TL65]
G7 VNQ K Staddon, 1 Allergrove Cttgs, Whimple, Exeter, Devon, EX5 2TJ [IO80HS, SY09]
G7 VNR D H Jones, 2 Exeter Rd, Urmston, Manchester, M41 0RE [IO83TK, SJ79]
GD7 VNS W H Heaps, 87 Friary Park, Ballabeg, Arbory, Isle of Man, IM9 4ES
G7 VNU Details withheld at licensee's request by SSL.
G7 VNV E Watson, 49 Beechfield Rd, Bolton, BL1 6HZ [IO83SO, SD61]
G7 VNW D Hooper, 21 High St., Great Linford, Milton Keynes, MK14 5AX [IO92OB, SP84]
G7 VNX G Birch, 58 Baldwins Ln, Birmingham, B28 0QE [IO92BK, SP18]
G7 VNY Details withheld at licensee's request by SSL.
G7 VNZ J Buck, 31 Newgate Cl, St. Albans, AL4 9JF [IO91US, TL10]
G7 VOA D Hughes, 1 Lytham Dr, Bramhall, Stockport, SK7 2LD [IO83WI, SJ98]
G7 VOB Details withheld at licensee's request by SSL.
G7 VOC M E Fishburn, Brent Hill, Bellevue Rd, Ventnor, Isle of Wight, PO38 1DB [IO90JO, SZ57]
G7 VOE R Gregory, Highfield, Dunham Hill, Warrington, Ches, WA6 0LT [IO83OG, SJ47]
G7 VOF Details withheld at licensee's request by SSL.
GI7 VOG Details withheld at licensee's request by SSL.
G7 VOH S Holland, Woodlands, Pink Moors, St. Day, Redruth, TR16 5NL [IO70JF, SW74]
G7 VOI T J Nicholas, Talmont, Chester Rd, Kelsall, Tarporley, CW6 0SD [IO83PF, SJ56]
G7 VOK A Bracey, 42 Lampton Gr, Bristol, BS13 0QA [IO81QJ, ST56]
G7 VOL C D Brayshaw, 50 Victoria Terr, Mill Bank, Sowerby Bridge, HX6 3EF [IO93AQ, SE02]
G7 VOM J Snelgrove, 22 Plains Ave, Maidstone, ME15 7AU [JO01GG, TQ75]
G7 VON A J Snelgrove, 22 Plains Ave, Maidstone, ME15 7AU [JO01GG, TQ75]
G7 VOO P Hockey, 96 Merthyr Rd, Pontypridd, CF37 4DD [IO81IO, ST09]
GM7 VOP A Pentland, 110 Craigmount, Kirkcaldy, KY2 6NZ [IO86JD, NT29]
G7 VOQ S Casey, 102 Dunkirk Ln, Leyland, Preston, PR5 3SQ [IO83PQ, SD52]
G7 VOS G W Greaves, 11 Colonels Ln, Chertsey, KT16 8RH [IO91RJ, TQ06]
G7 VOT A D C Moseley, 46 Harford St., Middlesbrough, TS1 4PR [IO94JN, NZ41]
G7 VOV H W Anderson, 1 Holly Bank, Ayr, KA7 3PW [IO75QK, NS32]
G7 VOX M C Ingram, Foxhill, Lower Daggons, nr Fordingbridge, Hants, SP6 3EE [IO90BW, SU01]
GM7 VOZ S Johnston, 43 Woodlands Bank, Dalgety Bay, Dunfermline, KY11 5SX [IO86HA, NT18]
G7 VPA N Muncey, 2 Ladysmith Ave, Whittlesey, Peterborough, PE7 1XX [IO92WN, TL29]
G7 VPB B Jewell, 44 Clovelly Rd, Bideford, EX39 3DF [IO71VA, SS42]
G7 VPC G Rowe, 28 Lawn Dr, Chudleigh, Newton Abbot, TQ13 0LT [IO80EO, SX87]
G7 VPD R R J Friend, High Hedges, Church Rd, Ashmanhaugh, Norwich, NR12 8YL [JO02RR, TG32]
G7 VPE K Riley, 33 Thwaites Rd, Oswaldtwistle, Accrington, BB5 4QT [IO83TR, SD72]
GI7 VPF C Tunnah, 76 Abbey Park, Belfast, BT5 7HR [IO74CO, J47]
G7 VPG G J Phillips, 13 Oak Dene Cl, Claverdon, Warwick, CV35 8PZ [IO92DG, SP16]
G7 VPI Details withheld at licensee's request by SSL.
GW7 VPJ R T Rees, 888 Heol y Ffynon, Penrhys, Ferndale, CF43 3RN [IO81GP, ST09]
G7 VPK Details withheld at licensee's request by SSL.
G7 VPL W Jackson, 4 Beaumaris Ave, Blackburn, BB2 4TW [IO83RR, SD62]
G7 VPM T Worrall, 40 Clement Mews, Kimberworth, Rotherham, South Yorks, S61 2JU [IO93HK, SK49]
G7 VPN A J Berry, 24 Collimer Cl, Chelmondiston, Ipswich, IP9 1HX [JO01OX, TM23]
G7 VPO Details withheld at licensee's request by SSL.
G7 VPP G Moffat, 4 Crosswood Ave, Balerno, EH14 7HT [IO85HV, NT16]
G7 VPQ J Bishop, 27 Southway, Blacon, Chester, CH1 5NW [IO83ME, SJ36]
G7 VPS D A W Barwood, 49 The Oaklands, Swaffham, PE37 7EW [JO02IP, TF80]
GM7 VPT D Leask, Avonmuir, The Loan, Muiravonside, Linlithgow, EH49 6LW [IO85DX, NS97]
G7 VPV M E Kirkman, 98A Cotmanhay Rd, Ilkeston, DE7 8NG [IO92IX, SK44]
G7 VPW C Kirkman, 98A Cotmanhay Rd, Ilkeston, DE7 8NG [IO92IX, SK44]
G7 VPX G P J Parish, 24 Dunveth Rd, West Hill, Wadebridge, PL27 7XD [IO70ML, SW86]
G7 VPY J Pearson, 82 Devoke Ave, Worsley, Manchester, M28 7EN [IO83TM, SD70]
G7 VPZ Details withheld at licensee's request by SSL.
G7 VQA N A Parker, 112 Stockport Rd, Marple, Stockport, SK6 6AH [IO83XJ, SJ98]
G7 VQB T Roy, 1 Rose Terr, Leven, KY8 4DF [IO86ME, NO30]
G7 VQC D P Driver, 13 Wenny Rd, Chatteris, PE16 6UT [JO02AK, TL38]
G7 VQD Details withheld at licensee's request by SSL.
G7 VQE D A Frost, 14 Banbury Cl, Northampton, NN4 9UA [IO92MF, SP75]
G7 VQF D L Courts, Hamel Down, Harrowbeer Ln, Yelverton, PL20 6DZ [IO70WL, SX56]
G7 VQG C I Perrin, 6 The Acreage, Goostrey, Crewe, CW4 8JY [IO83TF, SJ77]
G7 VQH Details withheld at licensee's request by SSL.
G7 VQI G D Burch, 6 The Barracks, Parkend, Lydney, GL15 4HR [IO81RS, SO60]
G7 VQJ W J Willmott, 5 Reedham Cres, Cliffe, Rochester, ME3 8HT [JO01GK, TQ77]
G7 VQK C M G Endean, Hazeldown, 12 Quentin Ave, Brixham, TQ5 0AG [IO80FJ, SX95]
G7 VQL M M Endean, 12 Quentin Ave, Brixham, TQ5 0AG [IO80FJ, SX95]
G7 VQM C J Davis, 100 Gorsey Ln, Wallasey, L44 4AG [IO83LJ, SJ39]
G7 VQN Details withheld at licensee's request by SSL.
G7 VQO W R Good, 433 The Ridge, Hastings, E Sussex, TN34 2RT [JO00GV, TQ81]
G7 VQP D K Houlden Portland Amateur Radio Club, 29 Ct Barton, Portland, DT5 2HJ [IO80SN, SY67]
G7 VQR P M Mawdsley, 56 Merriefield Ave, Broadstone, BH18 8DE [IO90AS, SZ09]
G7 VQS J T Williams, The Old Vicarage, Brook Ln, Moreton Morrell, Warwick, CV35 9AT [IO92FE, SP35]
G7 VQT T J Capon, 5 Spinney Ave, Goostrey, Crewe, CW4 8JE [IO83TF, SJ77]
G7 VQU Details withheld at licensee's request by SSL.
G7 VQV A C Cook, 10 Suffolk Cl, Chandlers Ford, Eastleigh, SO53 3GZ [IO90HX, SU41]
G7 VQW P Jessup, 20 Lawrence Dr, Cobham, Gravesend, DA12 3BU [JO01EJ, TQ66]
G7 VQX J J Hunter, 58 Ecos Ct, Frome, BA11 1HZ [IO81UF, ST74]
G7 VQY Details withheld at licensee's request by SSL.
GJ7 VQZ S M Powell, La Petella, La Rue Des Vignes, St. Peter, Jersey, JE3 7BE
G7 VRB R P Burgess, 22 Lee Cl, Kidlington, OX5 2XZ [IO91IT, SP41]
G7 VRE J M Tyrrell, The Old Mill, Nortonbury Ln, nr Baldock, Herts, SG6 1AN [IO91VX, TL23]
G7 VRH J R Hughes, 33 Beech Rd, Skellow, Doncaster, DN6 8HT [IO93JO, SE51]
G7 VRI Details withheld at licensee's request by SSL.
G7 VRJ M R Holland, 31 The Avenue, Andover, SP10 3EP [IO91GF, SU34]
G7 VRK S G Balding, Beechedge, Church Cl, Banningham, Norwich, NR11 7DY [JO02PT, TG22]
G7 VRL N Greenwood, 35 Station Rd, North Ferriby, HU14 3DG [IO93RR, SE92]
G7 VRM R E George, 85 Devonshire Way, Croydon, CR0 8BW [IO91XI, TQ36]
G7 VRN Details withheld at licensee's request by SSL.
G7 VRO Details withheld at licensee's request by SSL.
GM7 VRP Details withheld at licensee's request by SSL.
G7 VRR S P Wilshaw, 9 Delaney Dr, Stoke on Trent, ST3 5RL [IO82WX, SJ94]
G7 VRS R H Segall, 28 Acacia Cl, Stanmore, HA7 3JR [IO91TO, TQ19]
G7 VRT Details withheld at licensee's request by SSL.
G7 VRU E P Pedretti, Lovaine Flats, University of Northumbria, Newcastle upon Tyne, Tyne & Wear, NE1 8SU [IO94EX, NZ26]
G7 VRV M Taylor, 38 Manor Rd, Slyne, Lancaster, Lancs, LA2 6LB [IO84OC, SD46]
G7 VRW R V C Hill, Conquermoor Cottage, 74, Tibberton, Newport, Salop, TF10 8PF [IO82SS, SJ62]
G7 VRX R J Croft, 6 Clover Ave, Bishops Stortford, CM23 4BW [JO01BU, TL42]
G7 VRY P J Bambridge, 8 Temple Ln, Tonwell, Ware, SG12 0HP [IO91XU, TL31]
GW7 VRZ C Dolphin, 32 Dyas Ave, Birmingham, B42 1HE [IO92AM, SP09]
G7 VSC S Shackleton, 19 Greenwood Cl, Thames Ditton, KT7 0BG [IO91UJ, TQ16]
G7 VSD Details withheld at licensee's request by SSL.
GW7 VSF W G Thomas, 2 Ffordd Trecastell, Llanharry, Pontyclun, CF72 9ND [IO81HM, ST08]
G7 VSG D J Webster, 1 Woodbine Cl, Potter Heigham, Great Yarmouth, Norfolk, NR29 5NF [JO02TR, TG41]
G7 VSH Details withheld at licensee's request by SSL.
G7 VSI A P Desoer, 53 Highfield Rd South, Chorley, PR7 1RH [IO83QP, SD51]
G7 VSJ D M Jones, 6 Eastville, Bath, BA1 6QN [IO81TJ, ST76]
G7 VSL R G Taylor, 11 Ranscombe Cl, Brixham, TQ5 9UR [IO80FJ, SX95]
G7 VSM J Skinner, 1 Fullbrook Ave, Barton under Needwood, Burton on Trent, DE13 8HD [IO92DS, SK11]
G7 VSN L Franklin, 4 Carres Sq, Billinghay, Lincoln, LN4 4EL [IO93UB, TF15]
GW7 VSO D A Lewis, 47 Glebelands, Johnston, Haverfordwest, SA62 3PW [IO71MS, SM91]
G7 VSP S Wooster, 44 King Johns Rd, North Warnborough, Hook, RG29 1EJ [IO91MG, SU75]
G7 VSQ Details withheld at licensee's request by SSL.
G7 VSR S Clink Scottish Digital Communication, Southsyde, Woodhead Ave, Bothwell, Lanarkshire, G71 8AR [IO75XT, NS75]
G7 VSS P J Bartlett, 43 Chamberlain Way, Pinner, HA5 2AU [IO91TO, TQ18]
GW7 VST G H Davies, 41 Woodlands Rd, Barry, CF63 4EF [IO81IJ, ST16]
G7 VSW S L H Weston, 11 Friars Rd, Stoke on Trent, ST2 8DQ [IO83WA, SJ94]
G7 VTA V M Casey, 239 Littleton Rd, Salford, M7 3TJ [IO83UL, SD80]
G7 VTB R Sharp, 6 Acklington Dr, Acklington, Morpeth, NE65 9BX [IO95EH, NU20]
G7 VTC L A Dodd, 10 Nelson Cl, Bracknell, RG12 2RN [IO91PK, SU86]
G7 VTD I E Schofield, 9 Ashdene Rd, Stockport, SK4 3AD [IO83VJ, SJ89]
G7 VTE G B Forster, Treloyhan, Deer Valley Rd, Holsworthy, Devon, EX22 6DA [IO70TT, SS30]
G7 VTF A Sweeney, 27 Southlands Rd, Goostrey, Crewe, CW4 8JF [IO83TF, SJ77]
G7 VTH D T Reacher, 33 Cator Cres, New Addington, Croydon, CR0 0BL [IO91XI, TQ36]
GW7 VTI H K Morgan, 6 Gwaunmiskin Rd, Beddau, Pontypridd, CF38 2AU [IO81HN, ST08]
G7 VTJ T Scott, 50 Davison Ave, Whitley Bay, NE26 1SH [IO95GB, NZ37]
G7 VTK J Barnes, 60 James St., Great Harwood, Blackburn, BB6 7JH [IO83TS, SD73]
G7 VTL Details withheld at licensee's request by SSL.
GM7 VTM Details withheld at licensee's request by SSL.
G7 VTN P McCaulay, 33 Millmoor Way, North Hykeham, Lincoln, LN6 9PJ [IO93QE, SK96]
G7 VTP Details withheld at licensee's request by SSL.
G7 VTQ D J Parsons, 1 Kent Dr, Congleton, CW12 1SD [IO83VE, SJ86]
G7 VTR Details withheld at licensee's request by SSL.
G7 VTS P M Green, 10 Alexandra Rd, Farnborough, GU14 6DA [IO91PG, SU85]
G7 VTT J A King, 7 Valerian Ave, Heddon on The Wall, Newcastle upon Tyne, NE15 0EA [IO94CX, NZ16]
G7 VTU Details withheld at licensee's request by SSL.
G7 VTV Details withheld at licensee's request by SSL.
G7 VTW D M Jacques, 33 North St., Otley, LS21 1AH [IO93DV, SE24]
G7 VTY A T Loukes, 24 Shaftesbury Ave, Southampton, SO17 1SD [IO90HW, SU41]
G7 VTZ A G Wilkinson, 1 Wayneflete Cl, Bishops Waltham, Southampton, SO32 1HY [IO90JW, SU51]
G7 VUB R L Walker, 3 Rachael Gdns, Wednesbury, WS10 0RW [IO92AN, SP09]

G7 VUD J J Hansell, 70 Brian Ave, Norwich, NR1 2PD [JO02PO, TG20]
G7 VUE D W Hook, 64 Acre Way, Northwood, HA6 1SX [IO91TO, TQ19]
GI7 VUF Details withheld at licensee's request by SSL.
G7 VUH C J Squire, 19 Southfield Rd, Burley in Wharfedale, Ilkley, LS29 7PA [IO93CV, SE14]
GM7 VUJ A J Gough, House 105, 30 Kennishead Ave, Thornliebank, Glasgow, Lanarkshire, G46 8RD [IO75UT, NS56]
G7 VUK A E Stirland, 98 Aldreds Ln, Heanor, DE75 7HG [IO93HA, SK44]
G7 VUL J R Cook, 7 Ash Bank Rd, Stoke on Trent, ST2 9DS [IO83WA, SJ94]
G7 VUM D M Riches, 38 Half Mile Cl, Norwich, NR3 2LW [JO02PP, TG21]
G7 VUN C Brodrick, 6 Carlton St., Hartlepool, TS26 9ES [IO94JQ, NZ53]
G7 VUO R T Suggitt, 10 Rectory Ave, Guisborough, TS14 6QL [IO94LM, NZ61]
G7 VUP J M Milner, 1 Didworthy Cottage, Moretonhampstead, Newton Abbot, Devon, TQ13 8SJ [IO80DP, SX78]
G7 VUQ Details withheld at licensee's request by SSL.
G7 VUR Details withheld at licensee's request by SSL.
G7 VUT Details withheld at licensee's request by SSL.
G7 VUU C L Coe, 19 Seton Rd, Taverham, Norwich, NR8 6QE [JO02OQ, TG11]
G7 VUV Details withheld at licensee's request by SSL.
G7 VUW Details withheld at licensee's request by SSL.
G7 VUX P D Edwards, 28 Sutcliffe Dr, Harbury, Leamington Spa, CV33 9LT [IO92GF, SP35]
G7 VUY J E Shepherd, 29 Stechford Rd, Birmingham, B34 6AA [IO92CL, SP18]
G7 VUZ Details withheld at licensee's request by SSL.
G7 VVA C K Dables, 66 Town St., Pinxton, Nottingham, NG16 6HN [IO93IC, SK45]
G7 VVB P C R Andrew, Thrift, Madles Ln, Stock, Ingatestone, CM4 9QA [JO01FP, TQ69]
G7 VVC C K Sharp, 2 Caters Pl, Dorchester, DT1 1YB [IO80SR, SY69]
G7 VVE Details withheld at licensee's request by SSL.
G7 VVF D J Rossiter, 37 Meadway, Enfield, EN3 6NT [IO91XQ, TQ39]
G7 VVG Details withheld at licensee's request by SSL.
GM7 VVJ B C Sim, 23 Willow Dr, Girvan, KA26 0DE [IO75NF, NX19]
G7 VVK P J Bradley, 5 Crabtree Ave, Chadwell Heath, Romford, RM6 5EX [JO01BO, TQ48]
G7 VVL N V Quest, 43 Stanley Rd, Hornchurch, RM12 4JS [JO01CN, TQ58]
G7 VVM T L Aldred, 31 Cock Rd, Bristol, BS15 2SH [IO81SK, ST67]
G7 VVO I R Anderson, 1 The Paddock, Bath Rd, Wick, Bristol, BS15 5RL [IO81TK, ST77]
G7 VVP R D Barbour, 93 Sidney St., Cleethorpes, DN35 7NQ [IO93XN, TA20]
G7 VVQ M J Norton, Flat 1, 8 Zetland Rd, Manchester, M21 8TH [IO83UK, SJ89]
G7 VVR A Diacakis, 141 Peters Ct, Porchester Rd, London, W2 5DT [IO91VM, TQ28]
G7 VVT M R Fazal, 5 Lucas Ave, Harrow, HA2 9UH [IO91TN, TQ18]
G7 VVU Details withheld at licensee's request by SSL.
G7 VVV Details withheld at licensee's request by SSL.
GM7 VVW Details withheld at licensee's request by SSL.
G7 VVX A Renton, 20 Hollin Head, Baildon, Shipley, BD17 7LJ [IO93DU, SE13]
G7 VVY Details withheld at licensee's request by SSL.
G7 VVZ M J Wilson, 26 Highgate Rd, Woodley, Reading, RG5 3QR [IO91NK, SU77]
G7 VWA D R Lever, 35 Carteret Rd, Luton, LU2 9JZ [IO91TV, TL12]
G7 VWC R Bleach, 33 Regency Ct, Withdean Rise, Brighton, BN1 6YG [IO90WU, TQ30]
G7 VWE L Green, Tullamarine, Wignall St., Lawford, Manningtree, CO11 2HX [JO01MW, TM03]
G7 VWG R Evans, 12 Great Common, Inkpen, Hungerford, RG17 9QR [IO91GJ, SU36]
G7 VWH S Wallace, 45 Stanley St., Southsea, PO5 2DS [IO90KS, SZ69]
G7 VWK G Fisher, 6 Totternhoe Rd, Dunstable, LU6 2AG [IO91RV, TL02]
G7 VWL F Selvestian, 453 Claremont Rd, Manchester, M14 5WP [IO83VK, SJ89]
G7 VWM J R Hazell, 7 Higher Rd, Woolavington, Bridgwater, TA7 8EA [IO81ME, ST34]
G7 VWN J D Blackwell, Apple Trees, 7 Church Rd, Darley Dale, Matlock, DE4 2GG [IO93ED, SK26]
G7 VWO D Lisle, Kent Ii, Broadmoor Hospital, Crowthorne, Berks, RG45 7EG [IO91OI, SU86]
G7 VWP Details withheld at licensee's request by SSL.
G7 VWQ Details withheld at licensee's request by SSL.
G7 VWS Details withheld at licensee's request by SSL.
G7 VWT Details withheld at licensee's request by SSL.
G7 VWW Details withheld at licensee's request by SSL.
GW7 VWY R R Morrison, 1 Bro Dawel, Menai Bridge, LL59 5LT [IO73VF, SH57]
GI7 VXC A D Crozier, 5 Meadowvale, Dromore, BT25 1BF [IO64WJ, J25]
G7 VXD M A May, PO Box 60229, Addis Ababa, Ethiopia
G7 VXE T Alemayehu, PO Box 31089, Addis Ababa, Ethiopia
G7 VXF D Firew, PO Box 6057, Addis Ababa, Ethiopia
G7 VXG B Aleemayehu, PO Box 3699, Adis Ababa, Ethiopia
G7 VXH Details withheld at licensee's request by SSL.
G7 VXJ B A Abebe, PO Box 62545, Addis Ababa, Ethiopia
G7 VXK B Amare, PO Box 30464, Addis Ababa, Ethiopia
G7 VXL Details withheld at licensee's request by SSL.
G7 VXN Details withheld at licensee's request by SSL.
G7 VXP R D High, Marquis House, Woodlands, Gulworthy, Devon, PL19 8JE [IO70VM, SX47]
G7 VXQ C M Selby, The Caravan, New Barn Farm, East Holme, Wareham, Dorset, BH20 6AG [IO80WQ, SY88]
GM7 VXR P F Crankshaw, 3 North Neuk, Troon, KA10 6TT [IO75QN, NS33]
G7 VXS G P Burchell, 23B Luff Meadow, Stowmarket Rd, Needham Market, Ipswich, IP6 8DP [JO02MD, TM05]
G7 VXU Details withheld at licensee's request by SSL.
G7 VXV Details withheld at licensee's request by SSL.
G7 VXW Details withheld at licensee's request by SSL.
G7 VXX Details withheld at licensee's request by SSL.
G7 VXY Details withheld at licensee's request by SSL.
G7 VXZ C B M Guest, Deveron House, Turriff, Aberdeenshire, AB53 7JB [IO87TM, NJ74]
G7 VYA M Richardson, South View, South Rise, Binbrook, Market Rasen, LN8 6DP [IO93VK, TF29]
G7 VYB R M Patel, 30 Buckingham Dr, Luton, LU2 9RA [IO91TV, TL12]
G7 VYC Details withheld at licensee's request by SSL.
G7 VYD L Wami, PO Box 11088, Addis Ababa, Ethiopia
G7 VYE D L Wami, PO Box 11088, Addis Ababa, Ethiopia
G7 VYF K Tadesse, PO Box 60258, Addis Ababa, Ethiopia
GM7 VYG S A McQuillian, 20E Lomond Cres, Cornton, Stirling, Stirlingshire, FK9 5DN [IO86AD, NS79]
G7 VYH M Richardson, 4 Crosslands, Stantonbury, Milton Keynes, MK14 6AX [IO92OB, SP84]
G7 VYI R T Smith, 17 Shackleton Way, Shrewsbury, Shropshire, SY3 8SW [IO82OQ, SJ41]
G7 VYJ N A Gorman, 14 Keston Park Cl, Keston, BR2 6DX [JO01AI, TQ46]
G7 VYK Details withheld at licensee's request by SSL.
G7 VYM R Hewison, 1 Queensholm Ave, Bristol, BS16 6LF [IO81RL, ST67]
G7 VYN M A Jarman, 26 Hanbury Cl, Monk Bretton, Barnsley, S71 2LA [IO93GN, SE30]
G7 VYP A R Stapley, 1 St. Botolphs Rd, Northfleet, Gravesend, DA11 8ES [JO01EK, TQ67]
G7 VYQ I R Holman, 41 Leat Walk, Roborough, Plymouth, PL6 7AT [IO70WK, SX56]
GM7 VYR I G Findlay, 2 Bothwell Rd, Uddingston, Glasgow, G71 7ET [IO75XT, NS66]
G7 VYT J F Graver, 15 Cartwright Rd, Charlton, Banbury, OX17 3DG [IO92JA, SP53]
G7 VYY S P Middleton, 22 Hall Villa Ln, Toll Bar, Doncaster, DN5 0LH [IO93KN, SE50]
G7 VYZ M R Lancastle, 23 Clarendon Rd, Bolton, BL2 6BT [IO83TN, SD70]
G7 VZA W D Barr, 14 Martin Cl, Heighington, Lincoln, LN4 1RL [IO93SF, TF06]
G7 VZB A Peirson, 133 Runley Rd, Luton, LU1 1TX [IO91SV, TL02]
G7 VZC Details withheld at licensee's request by SSL.
G7 VZD J S Payne, 62 Roundhill, Tiverton, EX16 5BH [IO80GV, SS91]
G7 VZE Details withheld at licensee's request by SSL.
GM7 VZF A P Thornton, The Stables Flat, Raddery, Fortrose, IV10 8SN [IO77WO, NH75]
G7 VZG T C Roedel, 15 Trafalgar Rd, Shoeburyness, Southend on Sea, SS3 9EJ [JO01JM, TQ98]
G7 VZI H J Charles, 6 Bridewell St., Wymondham, NR18 0AR [JO02NN, TG10]
G7 VZJ Details withheld at licensee's request by SSL.
G7 VZK S Briggs, 88 Beaconsfield Rd, Enfield, EN3 6AP [IO91XQ, TQ39]
G7 VZL D J Forster, 15 Bracondale, Norwich, NR1 2AL [JO02PO, TG20]
G7 VZM M Blacklock, 39 Birtwistle Ave, Colne, BB8 9RS [IO83VU, SD84]
G7 VZN Details withheld at licensee's request by SSL.
G7 VZP A I Abrahams, 15 Westmorland Terr, Holmes Chapel, Crewe, CW4 7EE [IO83TE, SJ76]
G7 VZQ D J Polley, 4 Windrum Cl, Horsham, RH12 1XR [IO91TB, TQ12]
G7 VZR Details withheld at licensee's request by SSL.
G7 VZS A Walsh, 21 Rydal Ave, Darwen, BB3 2SA [IO83SQ, SD62]
G7 VZT Details withheld at licensee's request by SSL.
G7 VZU T L Evans, Ty Newydd, Tregynon, Newton, Powys, SY16 3ER [IO82IN, SO19]
GI7 VZW Details withheld at licensee's request by SSL.
G7 VZX Details withheld at licensee's request by SSL.
G7 VZY M A Page-Jones, 34 Edwards Way, Hutton, Brentwood, CM13 1BT [JO01EP, TQ69]
G7 WAA E J Donaghy, Mendips, 36 Attwood Rd, Salisbury, SP1 3PR [IO91CB, SU13]
G7 WAB J R Thornley Worked All Britain Awards Grou, 270 Hurdsfield Rd, Macclesfield, SK10 2PN [IO83WG, SJ97]
G7 WAC L J Volante Wythall Contest Group, 200 Longmore Rd, Shirley, Solihull, B90 3EX [IO92CJ, SP17]
G7 WAD Details withheld at licensee's request by SSL.
G7 WAE T P Ward, 127 Lower Lime Rd, Oldham, OL8 3NP [IO83WM, SD90]
G7 WAF D Keeble, 4 Davey Ct, Bolsover, Chesterfield, S44 6RS [IO93IF, SK47]
G7 WAG Details withheld at licensee's request by SSL.
G7 WAH R C Purser, 12 Magnolia Cl, Kempston, Bedford, MK42 7RY [IO92SC, TL04]
G7 WAJ Details withheld at licensee's request by SSL.
G7 WAK S Aspland, 6 Trilithon Cl, Norwich, NR6 5EP [JO02PP, TG21]
G7 WAL M F Sadler, Hillview, Forest Mill Ln, Horton, Ilminster, Somerset, TA19 9QU [IO80MW, ST31]
G7 WAM Details withheld at licensee's request by SSL.
G7 WAO Details withheld at licensee's request by SSL.
G7 WAP Details withheld at licensee's request by SSL.
G7 WAQ A P Moss, Caretakers House, 2 Brickfields Rd, South Woodham Ferrers, Chelmsford, Essex, CM3 5JX [JO01HP, TQ89]
G7 WAR G Lowsley, 19 Barons Cres, Copmanthorpe, York, YO2 3YR [IO93KV, SE54]
G7 WAS S E Staines, 19 Campion Rd, Westoning, Bedford, MK45 5LB [IO91SX, TL03]
G7 WAT C I Platten, 31 Wetherby Cl, Stevenage, SG1 5RX [IO91WV, TL22]
G7 WAV Details withheld at licensee's request by SSL.
G7 WAW D Thompson, 12 Dam Head Rd, Barnoldswick, Colne, BB8 5NH [IO83VW, SD84]
G7 WAY Details withheld at licensee's request by SSL.
G7 WAZ J P Gilmartin, 24 Eversley Cres, London, N21 1EJ [IO91WP, TQ39]
G7 WBA R H Grandshaw, Treehaven, South Ln, Nomansland, Salisbury, SP5 2BZ [IO90EW, SU21]
G7 WBB Details withheld at licensee's request by SSL.
G7 WBC J Egglestone, 4 Castle View, Ushaw Moor, Durham, DH7 7NT [IO94ES, NZ24]
G7 WBE D Welch, 38 Little Sammons, Chilthorne Domer, Yeovil, BA22 8RB [IO80PX, ST51]
G7 WBF Details withheld at licensee's request by SSL.
G7 WBG Details withheld at licensee's request by SSL.
G7 WBH D L Small, 17 Claygate Rd, Wimblebury, Cannock, WS12 5RN [IO92AQ, SK01]
G7 WBI Details withheld at licensee's request by SSL.
G7 WBJ W Naylor, 8 Brentwood Cres, Altrincham, WA14 1NW [IO83TJ, SJ78]
G7 WBK Details withheld at licensee's request by SSL.
G7 WBL J Wheeler, 92 Holford Rd, Bridgwater, TA6 7NZ [IO81LD, ST23]
G7 WBM P W Longhurst, Burston, Wood Rd, Hindhead, GU26 6PZ [IO91PC, SU83]
G7 WBO M J Stenning, 3A Mafeking Rd, Brighton, BN2 4EL [IO90WU, TQ30]
G7 WBP Details withheld at licensee's request by SSL.
G7 WBQ A H Johnson, 131 Rylands Rd, Kennington, Ashford, TN24 9LU [JO01KD, TR04]
G7 WBR E C Davis, 33 Truggers, Handcross, Haywards Heath, RH17 6DQ [IO91VB, TQ22]
G7 WBT Details withheld at licensee's request by SSL.
G7 WBU A E Hopley, 41 Old Pound Cl, Lytchett Matravers, Poole, BH16 6BW [IO80XS, SY99]
G7 WBV B Johnson, 158 Skinnerthorpe Rd, Firvale, Sheffield, S4 8GH [IO93GJ, SK39]
G7 WBW A Millward, 6 Trafalgar Sq, Winterbourne Gunner, Salisbury, SP4 6HU [IO91DC, SU13]
G7 WBX Details withheld at licensee's request by SSL.
G7 WBY B M Nixon, 33 Longmead Rd, Ryde, PO33 2TN [IO90KR, SZ59]
G7 WBZ M Malone, 10 Cattle St., Great Harwood, Blackburn, BB6 7NG [IO83TS, SD73]
G7 WCA J P McMullen, Burlington Lodge, Burlington Gr, Barnstaple, EX32 9BU [IO71XB, SS53]
G7 WCB A J Bennett, 12 Barns Cl, Walsall, WS9 9BD [IO92AP, SK00]
G7 WCC A R Neville, 1 Bethel Ave, Bispham, Blackpool, FY2 9NA [IO83LU, SD33]
G7 WCD N Charlton, 94 Cromarty, Ouston, Chester-le-Street, DH2 1JU [IO94EV, NZ25]
GW7 WCE Details withheld at licensee's request by SSL.
G7 WCF C J Rose, 7 Washbrook Ave, Worsley, Manchester, M28 7UQ [IO83TM, SD70]
G7 WCG Details withheld at licensee's request by SSL.
G7 WCH Details withheld at licensee's request by SSL.
G7 WCI Z Randeria, 39 Norroy Rd, London, SW15 1PQ [IO91VL, TQ27]
G7 WCJ C Jones, 44 Dewar Ln, Kesgrave, Ipswich, IP5 7GJ [JO02OB, TM24]
G7 WCK P A Byng, 4 Commonside, Peal Ln, Stourport on Severn, DY13 0RB [IO82UH, SO86]
G7 WCM K E Jermey, 1 Beldowes, Basildon, SS16 4DS [JO01FN, TQ78]
G7 WCN K Packer, 47 Sheppard Rd, Basingstoke, RG21 3JH [IO91KG, SU65]
GM7 WCO Details withheld at licensee's request by SSL.
G7 WCP C Sharpe, 10 Windsor Rd, Quorn, Loughborough, LE11 4LL [IO92JS, SK52]
G7 WCQ P A Richards, 315 Deans Rd, Wolverhampton, WV1 2AD [IO82WO, SO99]
G7 WCR J A Pitkin, 29 Dolwerdd Est, Pen y Parc, Cardigan, Dyfed, SA43 1RF [IO72QC, SN24]
GI7 WCS J Stitt, 52 Ransevyn Park, Whitehead, Carrickfergus, BT38 9LY [IO74DS, J49]
G7 WCU Details withheld at licensee's request by SSL.
G7 WCW L J Dobson, Harrogate Ladies College, Clarance Dr, Harrogate, North Yorks, HG1 2QG [IO93FX, SE25]
G7 WCX C K F Tse, 17 York Rd, Tower House, Harrogate, HG1 2QN [IO93FX, SE25]
G7 WCZ L E Macculloch, York House, Harrogate Ladies College, Clarence Dr, Harrogate, North Yorks, HG1 2QG [IO93FX, SE25]
G7 WDA F K Y Yuen, Lancaster House, Harrogate Ladies College, Clarence Dr, Harrogate, North Yorks, HG1 2QG [IO93FX, SE25]
G7 WDC M P Kiteley, 13 Chiltern Cl, Astley Cross, Stourport on Severn, DY13 0NU [IO82UH, SO86]
G7 WDD B J Bird, 2 Tomlins Ave, Frimley, Camberley, GU16 5LJ [IO91PH, SU85]
G7 WDE D Whattingham, Wildwood, 62 Dartnell Ave, West Byfleet, KT14 6PJ [IO91SI, TQ06]
G7 WDF Details withheld at licensee's request by SSL.
G7 WDG P A Wyatt, 45 Bridge Rd, Coalville, LE67 3PW [IO92HR, SK41]
G7 WDH Details withheld at licensee's request by SSL.
G7 WDK Details withheld at licensee's request by SSL.
G7 WDL D J S Taylor, 45 Ardsley Rd, Chesterfield, S40 4DG [IO93GF, SK37]
G7 WDM N A Feetham, 154 Magdalen Ln, Hedon, Hull, HU12 8LB [IO93VR, TA12]
G7 WDN Details withheld at licensee's request by SSL.
G7 WDO C M Barker, 15 Epping Green, Hemel Hempstead, HP2 7JP [IO91SS, TL00]
G7 WDQ Details withheld at licensee's request by SSL.
G7 WDS A R James, 36 Lemon Hill, Mylor Bridge, Falmouth, TR11 5NA [IO70LE, SW83]
G7 WDT Details withheld at licensee's request by SSL.
G7 WDX Details withheld at licensee's request by SSL.
G7 WDY W E Harland, 192 Ock St., Abingdon, OX14 5DR [IO91IQ, SU49]
G7 WEA A P Snell, 44 Dennis Way, Liss, GU33 7HL [IO91NA, SU72]
G7 WEB D A G Raxter, Chamaecy, Webbs Garden Centre, Wychbold, Droitwich, Worcs, WR9 0DG [IO82WG, SO96]
G7 WEC S Birbeck, 9 Shelley Dr, Accrington, BB5 2QS [IO83TR, SD72]
GM7 WED R D Feilen, 28 Belvidere Cres, Aberdeen, AB2 2NH [IO87TH, NJ72]
GW7 WEE A N Brown, Oakridge, 6 Bro Hafan, Cross Inn, Llandysul, SA44 6NQ [IO72TE, SN35]
GM7 WEF I Owens, 1-95 Slateford Rd, Edinburgh, EH11 1NX [IO85JW, NT27]
G7 WEG E D Smith, 24 Carnforth Cl, Stapleford, Nottingham, NG9 7EZ [IO92IW, SK43]
G7 WEJ Details withheld at licensee's request by SSL.
G7 WEK C A Taylforth, 1 Clough Terr, Barnoldswick, Colne, BB8 5PD [IO83VV, SD84]
G7 WEL T I Hewitt, 39 Hollins Gr, Fulwood, Preston, PR2 3TT [IO83PS, SD53]
G7 WEM T D Hewitt, 5 Hollins Gr, Fulwood, Preston, PR2 3TT [IO83PS, SD53]
G7 WEN J H Marron, 30 York Rd, Nunthorpe, Middlesbrough, TS7 0EZ [IO94JM, NZ51]
G7 WEP M Williams, 59 Thistledene, Thames Ditton, KT7 0YH [IO91TJ, TQ16]
G7 WER P A Fisher, 21 Charlotte Cl, Mount Hawke, Truro, TR4 8TS [IO70JG, SW74]
GI7 WET T Millar, 8 Ballyhenry Ave, Newtownabbey, BT36 5AZ [IO74AQ, J38]
G7 WEU J R Selwood, 16 Priestman Ave, Consett, DH8 8AT [IO94BU, NZ05]
G7 WEV J C Wedge, 38 Upper Vauxhall, Wolverhampton, WV1 4SY [IO82WO, SO99]
G7 WEW A D Cossey, 11 Halden Ave, Norwich, NR6 6UX [JO02PQ, TG21]
G7 WEY Details withheld at licensee's request by SSL.
G7 WEZ D A Tuffin, 132 Fleming Ave, North Baddesley, Southampton, SO52 9FB [IO90GX, SU31]
G7 WFA Details withheld at licensee's request by SSL.
G7 WFB Details withheld at licensee's request by SSL.
G7 WFC Details withheld at licensee's request by SSL.
G7 WFD M Dockerty, 2 The Park, Grasscroft, Oldham, OL4 4ES [IO83XM, SD90]
G7 WFE Details withheld at licensee's request by SSL.
G7 WFF C J Hope, 18 Dunlop Ave, Rochdale, OL11 2NG [IO83WO, SD81]
G7 WFG G Beresford, 94 Burnside Ave, Skipton, BD23 2DA [IO83XW, SD95]
G7 WFH T A Ford, 7 Sunnyside, London, NW2 2QP [IO91VN, TQ28]
GW7 WFI J F Piggott, 32 East View, Bargoed, CF8 8LU [IO81HN, ST08]
G7 WFJ Details withheld at licensee's request by SSL.
G7 WFK G K Tew, 5 Hill Top Ave, Tamworth, B79 8QB [IO92DP, SK20]
G7 WFL A J Edwards, The Old House, North St., Haselbury Plucknett, Crewkerne, TA18 7RJ [IO80PV, ST41]
G7 WFM A A Robinson, 3 Chesil Ave, Nottingham, NG8 1QA [IO92JW, SK54]
GW7 WFO E J Howells, 72 Holywell Cres, Abergavenny, Gwent, NP7 5LG [IO81LT, SO31]
G7 WFP J O L Smith, 54 Albert Rd, Hythe, CT21 6BT [JO01MB, TR13]
G7 WFR C Bennett, 16 Denver Rd, Norton, Doncaster, DN6 9HN [IO93KP, SE51]
G7 WFS C S Backshell, 26 The Tithings, Halton Brook, Runcorn, WA7 2DT [IO83PI, SJ58]
G7 WFT G D Edwards, 15 Watson Cres, Edinburgh, EH11 1HA [IO85JW, NT27]
G7 WFU N C Toombes, 10 Roding Cl, Riverdene, Basingstoke, RG21 4DU [IO91LG, SU65]
G7 WFV J A Kendall, 44 Davenport Rd, London, SE6 2AZ [IO91XK, TQ37]
G7 WFW Details withheld at licensee's request by SSL.
G7 WFY Details withheld at licensee's request by SSL.
G7 WFZ R J Stone, 38 Cator Rd, Ln End, High Wycombe, Bucks, HP14 3JD [IO91OO, SU89]
G7 WGA J W Potter, 23 Marline Ave, St. Leonards on Sea, TN38 9HS [JO00GV, TQ71]
G7 WGB P Graham, 22 Tennyson Rd, Colne, BB8 9SD [IO83VU, SD84]
G7 WGC D R Isaacs, 4 Woodham Gr, Little Neston, South Wirral, L64 0UH [IO83LG, SJ27]
G7 WGD D Price, 26 Teviot Gdns, Brierley Hill, DY5 4QL [IO82WL, SO98]

G7 WGE M P A Forster, 6 Charsley Cl, Amersham, HP6 6QQ [IO91RQ, SU99]
G7 WGF Details withheld at licensee's request by SSL.
G7 WGH C A Robinson, 98 Carlisle Rd, Worcester, WR5 1HZ [IO82VE, SO85]
G7 WGI J L Gordon, 12 Harcourt Cl, Cowplain, Waterlooville, PO8 8JL [IO90LV, SU61]
G7 WGJ Details withheld at licensee's request by SSL.
G7 WGK D J Rowley, 105 Corndon Cres, Harlescott, Shrewsbury, SY1 4LG [IO82PR, SJ51]
G7 WGL K L R Firth, Chimneys, 30 Kingscroft, Dersingham, Kings Lynn, PE31 6QN [JO02GU, TF63]
GM7 WGM S F Andrew, Fereneze View, Gateside Rd, Barrhead, Glasgow, G78 1TT [IO75TT, NS45]
G7 WGN Details withheld at licensee's request by SSL.
G7 WGO G Bradshaw, 3 Falmouth Ave, Haslingden, Rossendale, BB4 6QN [IO83UQ, SD72]
G7 WGP A J Brook, 34 Westerley Ln, Shelley, Huddersfield, HD8 8HP [IO93DO, SE21]
G7 WGQ D Morris, 86 Richardson St., Carlisle, CA2 6AG [IO84MV, NY35]
G7 WGT Details withheld at licensee's request by SSL.
G7 WGU A W Maddams, 1 Pykestone Cl, Bransholme, Hull, HU7 5AT [IO93UT, TA13]
G7 WGV V W Wills, 19A Coombe Rd, Saltash, PL12 4ER [IO70VJ, SX45]
G7 WGW J J Osborne, 28 Lytton Ave, Enfield, EN3 6EN [IO91XP, TQ39]
G7 WGX D A Sayles, 82 Molineaux Rd, Shiregreen, Sheffield, S5 0JY [IO93GK, SK39]
G7 WGY D Wilson, 31 Beechfield Ave, Skelmanthorpe, Huddersfield, HD8 9BZ [IO93EO, SE21]
G7 WHA O M Morley, 36 Hazeldene Rd, Stoke on Trent, ST4 8DN [IO82VX, SJ84]
G7 WHB G J Targonski, 4 Woodville Gdns, Dudley, DY3 1LB [IO82WN, SO99]
G7 WHC Details withheld at licensee's request by SSL.
G7 WHD D H Clayfield, 17 Foley Rd, Claygate, Esher, KT10 0LU [IO91TI, TQ16]
G7 WHE Details withheld at licensee's request by SSL.
G7 WHF Details withheld at licensee's request by SSL.
G7 WHG E R Braithwaite, 33 Somers Park Ave, Malvern Link, Malvern, WR14 1SE [IO82UD, SO74]
G7 WHI Details withheld at licensee's request by SSL.
G7 WHJ Details withheld at licensee's request by SSL.
G7 WHK Details withheld at licensee's request by SSL.
G7 WHL Details withheld at licensee's request by SSL.
G7 WHM A I Howgate, 7 Caledonian Way, Belton, Great Yarmouth, NR31 9PQ [JO02TN, TG40]
G7 WHN P R S Clark, Lower Sandhill House, Bodle St Green, Hailsham, E Sussex, BN27 4QU [JO00EW, TQ61]
G7 WHO B L Clarkson, 231 Overdown Rd, Tilehurst, Reading, RG3 6NX [IO91LJ, SU66]
G7 WHP W G Jones, 7 Hampstead Gdns, Hockley, SS5 5HN [JO01IO, TQ89]
GM7 WHQ Details withheld at licensee's request by SSL.
G7 WHS J F Beckett, 23 Newton Ave, Wakefield, WF1 2PX [IO93FQ, SE32]
G7 WHT B Coulthard, 62 Rochester Cres, Crewe, CW1 5YF [IO83SC, SJ75]
G7 WHU M P Nock, Mesquida, Lorraine Rd, Newhaven, E Sussex, BN9 9QB [JO00AT, TQ40]
G7 WHV A M Langford, 31 Rosehall Cl, Oakenshaw, Redditch, B98 7YD [IO92AG, SP06]
G7 WHX J Bodle, 48 Bolsover Rd, Hove, BN3 5HP [IO90VU, TQ20]
G7 WHY Details withheld at licensee's request by SSL.
G7 WHZ T P Crane, 15 Belchamps Way, Hawkwell, Hockley, SS5 4NT [JO01HO, TQ89]
G7 WIC G S Probyn, 65 York Rd, Swindon, SN1 2JU [IO91CN, SU18]
G7 WID G L White, Coulters Farm, Station Rd, Old Leake, Boston, PE22 9QQ [JO03AB, TF35]
G7 WIE Details withheld at licensee's request by SSL.
G7 WIG R C Bilsland, 92 Fruitlands, Malvern Wells, Malvern, WR14 4XB [IO82UC, SO74]
G7 WIH Details withheld at licensee's request by SSL.
G7 WII Details withheld at licensee's request by SSL.
G7 WIJ Details withheld at licensee's request by SSL.
G7 WIK Dr S K Hole, 29 Cotswold Cl, Loughborough, LE11 3AN [IO92JS, SK51]
G7 WIL G G Yunnie, St Marys Lodge, Battenhall Rd, Worcester, Worcs, WR5 2HP [IO82VE, SO85]
G7 WIM S J Gleadall, 59 Old Chapel Rd, Smethwick, Warley, B67 6HU [IO92AL, SP08]
G7 WIN Details withheld at licensee's request by SSL.
G7 WIO Details withheld at licensee's request by SSL.
G7 WIP Details withheld at licensee's request by SSL.
G7 WIQ S E Pack, Flat 3, 77 Dovecote Ln, Beeston, Nottingham, NG9 1JG [IO92JW, SK53]
G7 WIR M G Abbott, 9 Chancel Cl, West Kingsdown, Sevenoaks, TN15 6UD [JO01DI, TQ56]
G7 WIS Details withheld at licensee's request by SSL.
G7 WIT Details withheld at licensee's request by SSL.
GI7 WIU Details withheld at licensee's request by SSL.
G7 WIV Details withheld at licensee's request by SSL.
G7 WIW Details withheld at licensee's request by SSL.
GW7 WIX K E Dancer, 63 Romilly Cres, Cardiff, CF1 9NQ [IO81JL, ST17]
G7 WIY M K Downing, 12 Martindale Rd, Woking, GU21 3PJ [IO91QH, SU95]
G7 WIZ R D Ainley, 136 Scartho Rd, Grimsby, DN33 2AX [IO93WN, TA20]
G7 WJC B J Webster, 50 Blackburn Rd, Rishton, Blackburn, BB1 4BH [IO83SS, SD72]
G7 WJD B Moreau, Bowdene, 24 The Avenue, Totland Bay, PO39 0DL [IO90FQ, SZ38]
G7 WJE H Coots, 40 Essex Cl, Romford, RM7 8BD [JO01BO, TQ48]
G7 WJF Details withheld at licensee's request by SSL.
G7 WJG D T Moore, 13 Flanderwell Ln, Bramley, Rotherham, S66 0QJ [IO93IK, SK49]
G7 WJH Details withheld at licensee's request by SSL.
GM7 WJI Details withheld at licensee's request by SSL.
G7 WJJ R F Morton, 29 Lanmor Est, Lanner, Redruth, TR16 6HN [IO70JF, SW73]
G7 WJK J C Stephens, 19 Aspen Fold, Oswaldtwistle, Accrington, BB5 4PH [IO83TS, SD72]
G7 WJL K F Scott, Kirklands, Craigend Rd, Stow, Galashiels, TD1 2RJ [IO85NQ, NT44]
G7 WJN Details withheld at licensee's request by SSL.
G7 WJP A F Anderson, 232 Annan Rd, Dumfries, DG1 3HE [IO85FB, NX97]
G7 WJQ J J Brennan, 51 Oak Dr, Larkfield, Aylesford, ME20 6NL [JO01FH, TQ75]
G7 WJT J Taylor, 3 Inhams Cl, Murrow, Parson Drove, Wisbech, PE13 4HS [JO02AP, TF30]
G7 WJU Details withheld at licensee's request by SSL.
G7 WJV R W Stroud, 55 Haymeads Ln, Bishops Stortford, CM23 5JJ [JO01CU, TL52]
G7 WJW G J Cripps, 2 Castlefields, Gravesend, DA13 9EJ [JO01EJ, TQ66]
G7 WJX J G Davies, 243A Bradford Rd, Winsley, Bradford on Avon, BA15 2HL [IO81UI, ST86]
GI7 WJY R Donnan, 71 Victoria Ave, Newtownards, BT23 7ED [IO74DN, J46]
G7 WJZ P Clarke, 21 Long Furlong Rd, Sunningwell, Abingdon, OX13 6BL [IO91IQ, SU49]
G7 WKA Details withheld at licensee's request by SSL.
G7 WKB D Riddle, 71 Newlands Park, Dearham, Maryport, CA15 7ED [IO84GQ, NY03]
G7 WKC T W Hasted, Springfield House, Birds End, Hargrave, Bury St. Edmunds, IP29 5HE [JO02HF, TL76]
G7 WKD D M Watts, 52 Victoria St., Wolverton, Milton Keynes, MK12 5HJ [IO92OB, SP84]
GM7 WKF Details withheld at licensee's request by SSL.
G7 WKG A P Roche, 22 Trident Ct, Butlers Rd, Birmingham, B20 2NX [IO92AM, SP09]
G7 WKH P R Clark, 21 Sandfield Rd, Arnold, Nottingham, NG5 6QA [IO92KX, SK54]
G7 WKI Details withheld at licensee's request by SSL.
G7 WKJ M J Ballard, 3 Sage Cl, Portishead, Bristol, BS20 8ET [IO81OL, ST47]
G7 WKL Dr W A Johnstone, 67 Station Ln, Birkenshaw, Bradford, BD11 2JE [IO93DS, SE22]
G7 WKM E Bluer, 7 Pinewood Sq, St. Athan, Barry, CF62 4JR [IO81HJ, ST06]
G7 WKN D A Santillo, 34 Wearde Rd, Saltash, PL12 4PP [IO70VJ, SX45]
G7 WKO Details withheld at licensee's request by SSL.
G7 WKP A D Andrew, Thrift, Madles Ln, Stock, Ingatestone, CM4 9QA [JO01FP, TQ69]
GM7 WKQ T G Costford, 14 Redburn Ct, Whitelees, Cumbernauld, Glasgow, G67 3NL [IO85AX, NS77]
G7 WKS G N Frykman Warwick School ARS, 8 Orchard Cl, Bishops Itchington, Leamington Spa, CV33 0QS [IO92GF, SP35]
G7 WKU Details withheld at licensee's request by SSL.
G7 WKV B E Jewell, 22 St. Anthonys Rd, Kettering, NN15 5HT [IO92PJ, SP87]
G7 WKX S N Davis, 120 Muxton Ln, Muxton, Telford, TF2 8PF [IO82SR, SJ71]
G7 WKY Details withheld at licensee's request by SSL.
G7 WKZ I S Walsh, 101 Charter St., Accrington, BB5 0SA [IO83TS, SD72]
GI7 WLA Details withheld at licensee's request by SSL.
GI7 WLB Details withheld at licensee's request by SSL.
G7 WLC D J Evans, 1 Hill Cttgs, Layer Breton Hill, Layer Breton, Colchester, CO2 0PR [JO01JT, TL91]
GW7 WLD T R Jones, 6 Maes y Felin, Llanrhystud, SY23 5AT [IO72WH, SN56]
GM7 WLE D McKay, 11 Mowat Ln, Wick, KW1 4NP [IO88KK, ND35]
GW7 WLF Details withheld at licensee's request by SSL.
G7 WLG S W Balderson, 116 Woodside Rd, Amersham, HP6 6NL [IO91QQ, SU99]
G7 WLI D L Page, Quarry Cottage, Maidenhill, Penrith, Cumbria, CA11 8SQ [IO84PQ, NY53]
G7 WLJ B Ringrose, 1 Thirkleby Row, West Lutton, Malton, YO17 8TB [IO94RC, SE96]
G7 WLL I J Irlam, 8 Hawley Terr, Hawley Rd, Dartford, DA2 7RN [JO01CK, TQ57]
GW7 WLM J Tamlyn, Miglenfa, Llangoedmor, Cardigan, Dyfed, SA43 2LP [IO72QB, SN24]
G7 WLN J H H Buckley, The Viking, Main St, Helperthorpe, Malton, North Yorks, YO17 8TQ [IO94RC, SE97]
GM7 WLO J G Burt, Olivet, Lanton Rd, Jedburgh, TD8 6SD [IO85RL, NT62]
G7 WLQ Details withheld at licensee's request by SSL.
G7 WLR J F Stringer, 53 Castle Gr, Fareham, PO16 9NY [IO90KU, SU60]
GI7 WLS P W Greer, 6 Mountain View Terr, Banbridge, BT32 3HL [IO64UI, J14]
G7 WLT Details withheld at licensee's request by SSL.
G7 WLV G R Southall, 6 Dudley Wood Ave, Dudley, DY2 0DG [IO82XL, SO98]
G7 WLW Details withheld at licensee's request by SSL.
GW7 WLX D C Williams, 43 St. Margarets Rd, Whitchurch, Cardiff, CF4 7AB [IO81JM, ST18]
G7 WLY R B Bagwell, 9 Grampian Gr, West Boldon, East Boldon, NE36 0NJ [IO94GW, NZ36]
G7 WLZ Details withheld at licensee's request by SSL.
G7 WMB M A Brown, 94 Bridge St., Barnsley, S71 1PW [IO93GN, SE30]
G7 WNN R A Jenkins North Notts Data Group ARC, 13 Baulk Ln, Worksop, S81 7DF [IO93KH, SK58]
G7 WOS A H Irvine West of Scotland Amateur Radio, 41 Craighead Rd, Bishopton, PA7 5DT [IO75SV, NS47]
G7 WOT Details withheld at licensee's request by SSL.
G7 WRA J Hatch Wincanton Amateur Radio Club, Peveril, 42 Bowden Rd, Templecombe, BA8 0LF [IO80TX, ST72]
G7 WRG B H L Lowe Walsall Raynet Group, 19 Wolverhampton Rd, Bloxwich, Walsall, WS3 2EZ [IO82XO, SJ90]
G7 WRO Details withheld at licensee's request by SSL.
G7 WRS Details withheld at licensee's request by SSL.
G7 WRW Details withheld at licensee's request by SSL.
G7 WSH Details withheld at licensee's request by SSL.
G7 WSN K P G Harris, 28 Pauntley Rd, Christchurch, BH23 3JN [IO90DR, SZ19]
G7 WST Details withheld at licensee's request by SSL.
G7 WWW B S Cole, 6 Parkstone Parade, Hastings, TN34 2PS [JO00GV, TQ81]
GW7 WXM R D Hall Marford & District Amateur Rad, 7 Oak Dr, Marford, Wrexham, LL12 8XT [IO83MC, SJ35]
G7 XAB Details withheld at licensee's request by SSL.
G7 XAL Details withheld at licensee's request by SSL.
G7 XJS Details withheld at licensee's request by SSL.
G7 XLR Details withheld at licensee's request by SSL.
G7 XOZ C Harding Bridgwater College Amateur Rad, 24 Bryer Cl, Bridgwater, TA6 6UR [IO81LC, ST23]
G7 XPC P A Chorley, Boone Hill House, Mount Boone Hill, Dartmouth, TQ6 9NZ [IO80FI, SX85]
G7 XWH R Horton Harrogate Ladies College ARS, 7 Carlton Rd, Harrogate, HG2 8DD [IO93FX, SE35]
G7 XYL R Savage, 148 Gravesend Rd, Rochester, ME2 3QT [JO01FJ, TQ77]
G7 XYZ Details withheld at licensee's request by SSL.
G7 YAK Details withheld at licensee's request by SSL.
G7 YEW D J Marsh, 39 Fir St., Cadishead, Manchester, M44 5AR [IO83SK, SJ79]
G7 ZDX J L Walmsley South Coast Hilltoppers, PO Box 1637, Yeovil, Somerset, BA21 4YF [IO80QW, ST51]
G7 ZMS M J Larcombe, 52 Orchard Rd, Burgess Hill, RH15 9PL [IO90WW, TQ31]
G7 ZPE Details withheld at licensee's request by SSL.
G7 ZRT R W Thayne, 7 McDonald Pl, Hartlepool, TS24 0PZ [IO94JQ, NZ53]
G7 ZZY P F Pile, 65 Mill Ln, Teignmouth, TQ14 9BB [IO80FN, SX97]
G7 ZZZ Details withheld at licensee's request by SSL.

G8

G8 AAD B M Blight, Hill View, High St., Kingston Blount, Chinnor, OX9 4SJ [IO91MQ, SU79]
G8 AAE D G Phillips, 2 Walkers Cl, Chelmsford, CM1 6UW [JO01FS, TL70]
G8 AAF F B Blake, 24 Highlands Rd, Seer Green, Beaconsfield, HP9 2XN [IO91QO, SU99]
GW8 AAG Details withheld at licensee's request by SSL.[Op: Bruce Carter. Station located near Brecon.]
G8 AAI M J Bues, 23 Chapel Way, Epsom, KT18 5TE [IO91VH, TQ25]
GM8 AAN Dr N R Webster, Wicker Inn Farm House, Raimoir, nr Banchory, Aberdeenshire, AB31 5QX [IO87SC, NJ70]
G8 AAP G Toulalan, Cornerways, Burleigh Gdns, Boston, Lincs, PE21 9DE [IO92XX, TF34]
G8 AAR F A May, Church Rd, Newton Green, Newton, Sudbury, Suffolk, CO10 0QP [JO02JA, TL94]
G8 AAU N D Stanners, 22 Brands Hill Ave, High Wycombe, HP13 5QA [IO91PP, SU89]
G8 ABB G Rogers, 10 The Laurels, Bletchley, Milton Keynes, MK1 1BL [IO92PA, SP83]
G8 ABI Details withheld at licensee's request by SSL.
G8 ABQ Details withheld at licensee's request by SSL.
G8 ABU M C Davidson, 21 Holm Gr, Oakwood Park, Hillingdon, Uxbridge, UB10 9LZ [IO91SN, TQ08]
G8 ABX G E J Catling, 27 Clothall Rd, Baldock, SG7 6PE [IO91VX, TL23]
G8 ACA H B Crockett, 28 Church Ln, Middleton, Tamworth, B78 2AW [IO92DN, SP19]
G8 ACC S F F Weber, 28 Appletree Dr, Hala, Lancaster, LA1 4QY [IO84OA, SD45]
G8 ACE Details withheld at licensee's request by SSL.[Op: J E Hazell. Station located near Winchester.]
G8 ACJ F Mathews, Easedale, Woodway, Merrow, Guildford, GU1 2TF [IO91RF, TQ05]
G8 ACK M L Latimer-Sufit, The Belsize, 40 Belsize Park Gdn, London, NW3 4NA [IO91WN, TQ28]
G8 ACL H Cosford, 3 Applewood, Park Gate, Southampton, SO31 7HQ [IO90IV, SU40]
G8 ACM R E Snelling, 12 Heathfield Ave, Catisfield, Fareham, PO15 5QA [IO90JU, SU50]
G8 ACO K A Smith, 7 Ogilvie Homes, Leiston Rd, Aldringham, Leiston, IP16 4PS [JO02SE, TM46]
G8 ACQ R C Whattam, The Aviary No1, Arkwright Rd, Milton Ernest, Beds, MK44 1SE [IO92RE, TL05]
G8 ACR R Yates, 28 Daimler Rd, Yardley Wood, Birmingham, B14 4JJ [IO92BJ, SP07]
G8 ACT G T Gunn, 18 Barnfield, Hatfield Broad Oak, Bishops Stortford, CM22 7JR [JO01DT, TL51]
G8 ADC J O Haile, 145 Dunstable Rd, Caddington, Luton, LU1 4AN [IO91SU, TL01]
G8 ADG R V Perkins, 104 Oakham Rd, Dudley, DY2 7TQ [IO82XM, SO98]
G8 ADH C G Slingsby, 25 The Woodlands, Ryall, Upton upon Severn, Worcester, WR8 0PQ [IO82VB, SO84]
GM8 ADK M D Ritchie, 99 Forest Ave, Aberdeen, AB1 4TN [IO87TH, NJ72]
G8 ADM D Mann, Norton House, Gr Ln, Chalfont St. Peter, Gerrards Cross, SL9 9LB [IO91RO, SU99]
G8 ADS A W Morgan, 64 Halfmoon Ln, Dunstable, LU5 4AB [IO91RV, TL02]
G8 ADV R Day, 37 Druids Ln, Hollywood, Birmingham, B14 5SL [IO92BJ, SP07]
G8 ADX E W Lawley, Buckland, 21 Burnt House Ln, Ingatestone, CM4 9AN [JO01EP, TQ69]
G8 ADY P Harrison, 2 The Barns, Bridge End, Carlton, Bedford, MK43 7LP [IO92QE, SP95]
G8 ADZ N W Shepherd, 7 High St., Kelvedon, Colchester, CO5 9AG [JO01IT, TL81]
G8 AED D J Emmett, 898 Wakefield Rd, Bradford, BD4 7RT [IO93DS, SE13]
G8 AEE T Sanders, 48 Rock Mount, Altham West, Accrington, Lancs, BB5 5EF [IO83TS, SD73]
G8 AEN P Helm, 74 Neston Rd, Walshaw, Bury, BL8 3DB [IO83TO, SD71]
G8 AER J E Tanner, Merlins Mill, Toadsmoor Rd, Brimscombe, Stroud, Glos, GL5 2UG [IO81VR, SO80]
G8 AEU J G A Nightingale, 6 Aubrey Cl, Broomfield Rd, Chelmsford, CM1 4EJ [JO01FS, TL70]
G8 AFC P L Entwistle, 3 Grange Ave, Flixton, Manchester, M31 3PF [IO83SK, SJ79]
G8 AFG D H Roe, 39 Station Rd, Cosham, Drayton, Portsmouth, PO6 1PJ [IO90LU, SU60]
G8 AFI P C Funnell, 25 Broadyates Rd, Yardley, Birmingham, B25 8JF [IO92CL, SP18]
G8 AFN P E Cleall, 139 Preston Gr, Yeovil, BA20 2DB [IO80QW, ST51]
G8 AFU P B Gilby, 191 Send Rd, Send, Woking, GU23 7ET [IO91RH, TQ05]
G8 AFV A D Benn, 9 Daisy Bank, Clover Hill, Halifax, HX1 2YJ [IO93BR, SE02]
G8 AFZ A R Martin, 6 Roselands, Waterlooville, PO8 9QZ [IO90LV, SU61]
GW8 AGI R J Robson, 47 Thornhill Way, Rodgerstone, Rogerstone, Newport, NP1 9FS [IO81LO, ST28]
G8 AGJ J H Evans, 2 Sheraton Ave, Hatch Warren, Basingstoke, RG22 4TS [IO91KF, SU64]
GM8 AGM M D Collar, Shoemakers Croft, Hatton, Peterhead, Aberdeenshire, AB4 7TB [IO87TH, NJ72]
G8 AGN Dr B Chambers, 31 Hallam Grange Cl, Sheffield, S10 4BN [IO93FI, SK38]
G8 AGQ Details withheld at licensee's request by SSL.
G8 AGR Dr S C Craddock, 38 Briardene, Lanchester, Durham, DH7 0QD [IO94DT, NZ14]
G8 AGT G K Otway, Northern Lights, Ham Ln, Dundry, Bristol, BS18 8JA [IO81QJ, ST56]
G8 AGY Dr S K Erents, 50 Blandy Ave, Southmoor, Abingdon, OX13 5DB [IO91GQ, SU49]
G8 AHA S G Taylor, 9 Holme Way, Barby, Rugby, CV23 8UQ [IO92JH, SP57]
G8 AHB P A J Swinbank, 15 Old Bath Rd, Cheltenham, GL53 7QE [IO81XV, SO92]
G8 AHE L Arnold, 227 Rednal Rd, Kings Norton, Birmingham, B38 8EA [IO92AJ, SP07]
G8 AHF F W G Parkman, 35 Beatrice Ave, East Cowes, PO32 6HR [IO90IR, SZ59]
G8 AHK M J Blewett Univ. Surrey AR, Electronic Eng Dept, University/Surrey, Guildford Surrey, GU2 5XH [IO91QF, SU95]
G8 AHN J P Barnes, 2 Mappins Rd, Catcliffe, Rotherham, S60 5TH [IO93HJ, SK48]
G8 AHR P W Rushworth, 2 Aberdeen Cl, Coventry, CV5 7NE [IO92FK, SP28]
G8 AHU Details withheld at licensee's request by SSL.
G8 AHW F Edmondson, 59 Grange Rd, Dorridge, Solihull, B93 8QS [IO92CI, SP17]
G8 AIE P W Willcocks, 27 Manor Rd, Barnet, EN5 2LE [IO91VP, TQ29]
G8 AIF R G Baker, 12 Westland Dr, Brookmans Park, Hatfield, AL9 7UQ [IO91VR, TL20]
G8 AIJ E A Darben, Rose Dene, Fieldside, Mareham-le-Fen, Boston, PE22 7QU [IO93WD, TF26]
G8 AIM F G Tarver, 14 Heralds Ct, Humphris St., Warwick, CV34 5RB [IO92FG, SP26]
G8 AIQ G J Willis, Mill Cottage, Mill Cl, Tredington, Shipston on Stour, CV36 4HQ [IO92EB, SP24]
GI8 AIR W K Parkes, 15 Bushfoot Park, Portballintrae, Bushmills, BT57 8YX [IO65RF, C94]
G8 AIV W E M Symes, 135 Moreton Rd, Upton, Wirral, L49 4NT [IO83KJ, SJ28]
GW8 AJA D H Hardy, 7 Coed y Go Cttgs, Coed y Go, Oswestry, SY10 9AU [IO82LU, SJ22]
G8 AJF R B Vieira, 6 Lechmere Ave, Chigwell, IG7 5ET [JO01BO, TQ49]
G8 AJL L W Wendon, 51 Beckford Rd, Cowes, PO31 7SJ [IO90IS, SZ49]
G8 AJM C J Payne, Rotherfield, 128 Windsor Rd, Bray, Maidenhead, SL6 2DW [IO91PL, SU97]
G8 AJN Details withheld at licensee's request by SSL.
G8 AJZ R S Boardall, 3 Parkway, Queensbury, Bradford, BD13 2HJ [IO93BS, SE03]
G8 AKA T R Wiltshire, Bramblings, Pelican Rd, Pamber Heath, Tadley, RG26 3EL [IO91KI, SU66]
G8 AKB G W Rolland, 7 Ashdale Park, Lower Wokingham Rd, Finchampstead, Wokingham, RG40 3QS [IO91OJ, SU86]
G8 AKC C F Bell, Croftner, Mary Tavy, Devon, PL19 9QD [IO70WO, SX58]
G8 AKE J D Warrington, 26 Lynton Rd, Melton Mowbray, LE13 0NN [IO92NS, SK72]
G8 AKL G A Ashcroft, 86 Avondale Ave, North Finchley, London, N12 8EN [IO91VO, TQ29]
G8 AKM G B Roper, 19 Normay Rise, Newbury, RG14 6RY [IO91HI, SU46]
G8 AKN G A Taylor, 5 Stanhope Ave, Audenshaw, Manchester, M34 5AZ [IO83WL, SJ99]
G8 AKP P J McQuade, Old Swan, Holt Rd, Sharrington, Melton Constable, NR24 2PH [JO02MV, TG03]
G8 AKQ S J Birkill, Dale House Farm, Chapel St., Monyash, Bakewell, DE45 1JJ [IO93CE, SK16]
G8 AKU B F Willson, Hilltop, Cryers Hill Rd, Cryers Hill, High Wycombe, HP15 6LJ [IO91PP, SU89]

G8 AKX M R Perry, 216 Marlpool Ln, Kidderminster, DY11 5DL [IO82UJ, SO87]
G8 ALB Details withheld at licensee's request by SSL.
G8 ALE M Brereton, Gleaston Watermill, Gleaston, Ulverston, Cumbria, LA12 0QH [IO84KD, SD27]
G8 ALG R Garry, 51 Renacres Ln, Shirdley Hill, Halsall, Ormskirk, L39 8SG [IO83MO, SD31]
G8 ALO Details withheld at licensee's request by SSL.[Op: R J Burrows. Station located near Sutton Coldfield.]
G8 ALQ A J Whitlock, 23 Daly Way, Aylesbury, HP20 1JW [IO91OT, SP81]
G8 ALR J T Cull, Drybrook Cottage, Amesbury Rd, Cholderton, Wilts, SP4 0ER [IO91DE, SU24]
G8 ALS M Stevenson, 15 Wall Hill Rd, Brownshill Green, Allesley, Coventry, CV5 9EN [IO92FK, SP38]
G8 AMD H L Bate, 88 Darnick Rd, Sutton Coldfield, B73 6PG [IO92BN, SP09]
G8 AMG M G Foster, 9 Norman Way, Irchester, Wellingborough, NN29 7AT [IO92QG, SP96]
G8 AMJ D L Woolley, Oaktree Cottage, Westbourne Rd, Coltishall, Norwich, NR12 7HT [JO02QR, TG22]
G8 AMK L J Parry, 13 Cannon Hill, Bracknell, RG12 7QA [IO91PJ, SU86]
G8 AMP Details withheld at licensee's request by SSL.
G8 AMU C E Saveker, 23 Southlands Ave, Horley, RH6 8BS [IO91WE, TQ24]
G8 AMV Details withheld at licensee's request by SSL.
G8 AMW Details withheld at licensee's request by SSL.
G8 AMY B A Yates, 17 Elswick Pl, St. Annes, Lytham St. Annes, FY8 3JY [IO83LS, SD32]
G8 AMZ Details withheld at licensee's request by SSL.
G8 ANA G W Cummings, 99 Dinsdale Cres, Darlington, DL1 1EZ [IO94FM, NZ31]
G8 ANK J Tranmer, 4 Grocot Rd, Evington, Leicester, LE5 6AB [IO92LO, SK60]
G8 ANN G H Townsend, 61 Richmond Park Rd, East Sheen, London, SW14 8JU [IO91UL, TQ27]
G8 ANR D W Perkins, Strawberry Hill House, Orchard Dr, Braintree, CM7 1EQ [JO01GU, TL72]
GM8 AOB J M Briscoe, 2 Peebles Pl, Fort William, PH33 6UG [IO76KT, NN07]
G8 AOG M W Browne, 143 Thatch Leach Ln, Whitefield, Manchester, M45 6EP [IO83UN, SD80]
G8 AOK A Porch, 17 Purcell Cl, Brighton Hill, Basingstoke, RG22 4EL [IO91KF, SU65]
G8 AOL B W Godwin, 20 Pembury Rd, Bexleyheath, DA7 5NB [JO01BL, TQ47]
G8 AOO B C Hills, 3 Frithmead Cl, Basingstoke, RG21 3JW [IO91KG, SU65]
G8 AOZ P Hughes, 247 High Greave, Ecclesfield, Sheffield, S5 9GS [IO93GK, SK39]
G8 APB C D Plummer, Newtown Rd, Biddulph Park, Stoke on Trent, Staffs, ST8 7SW [IO83WD, SJ96]
G8 APK A J D Cooke, 7 Ravenscourt Rd, Rough Common, Canterbury, CT2 9DH [JO01MH, TR15]
G8 APM G M White, 1 Drakes Cl, Dibden Purlieu, Hythe, Southampton, SO45 5BP [IO90HU, SU40]
G8 APW D J Taylor, 87 Grasmere Rd, Garden Farm Est, Chester-le-Street, DH2 3EU [IO94EU, NZ25]
GM8 APX W H Jarvis, 6 Peggysmill Rd, Edinburgh, EH4 6JY [IO85IX, NT17]
G8 APY J P Bond, Folly House, The Reddings, Cheltenham, GL51 6RL [IO81WV, SO92]
G8 APZ S R Lucas, 84 Woodman Rd, Warley, Brentwood, CM14 5AZ [JO01DO, TQ59]
G8 AQB M N Ballance, 24 Western Rd, Wolverton, Milton Keynes, MK12 5BE [IO92OB, SP84]
G8 AQN A J Hibbard, 20 Barby Ln, Hillmorton, Rugby, CV22 5QJ [IO92JI, SP57]
G8 AQO A E Copperwaite, 11A Byculiah Rd, Enfield, EN2 8EG [IO91WP, TQ39]
G8 AQP S J Warner, 86 Ellwood Ave, Stanground, Peterborough, PE2 8LY [IO92VN, TL29]
G8 AQT F W J Neale, Pines Edge, Forest Dale Rd, Marlborough, SN8 2AS [IO91DK, SU16]
G8 ARA B E King, 15 Newstead Rd, Southbourne, Bournemouth, BH6 3HJ [IO90CR, SZ19]
GW8 ARC Dr A Craggs, 15 Pen y Groes Ave, Rhiwbina, Cardiff, CF4 4SP [IO81JM, ST18]
G8 ARF L W Thompson, 44 Tillmouth Ave, Holywell, Whitley Bay, NE25 0NP [IO95FB, NZ37]
G8 ARG Details withheld at licensee's request by SSL.
G8 ARM B G Pickrell, Perrans, Carvossa Pl, Ludgvan, Penzance, Cornwall, TR20 8AJ [IO70GD, SW53]
GW8 ARR P J Edwards, Trevland, Felindre, Knighton, Powys, LD7 1YL [IO82JK, SO18]
GM8 ARV D J Taylor, 19 Buckstone Ct, Edinburgh, EH10 6UL [IO85JV, NT26]
GW8 ASA G J Wyatt, 52 Parcau Ave, Bridgend, CF31 4SY [IO81EM, SS87]
G8 ASC P J Richards, 134 Downhills Park Rd, Tottenham, London, N17 6BP [IO91WO, TQ39]
GW8 ASD A H Pugh, Willcroft, Mold Rd, Gwersyllt, Wrexham, LL11 4AF [IO83LB, SJ35]
G8 ASG M D Farrell, Hobberley House, Hobberley Ln, Shadwell, Leeds, LS17 8LX [IO93GU, SE33]
G8 ASI M C Hastings, 43 Delmar Ave, Leverstock Green, Hemel Hempstead, HP2 4LZ [IO91SR, TL00]
G8 ASJ G Swan, Morogar, Post Office Ln, Kempsey, Worcester, WR5 3NX [IO82VD, SO84]
G8 ASO B A Jones, 12 Woodside Rd, Larkhill, Worcester, WR5 2EG [IO82VE, SO85]
G8 ASW R Warrender, 102 Turnberry Rd, Great Barr, Birmingham, B42 2HT [IO92BM, SP09]
G8 ASX A Hoggan, 25 Clingan Rd, Boscombe East, Bournemouth, BH6 5PY [IO90CR, SZ19]
G8 ASZ Details withheld at licensee's request by SSL.
GM8 AT W M Beattie, 43 Invercauld Rd, Mastrick, Aberdeen, AB1 5RP [IO87TH, NJ72]
G8 ATA J P Chettle, Bank Cottage, Cart Ln, Grange Over Sands, LA11 7AB [IO84ME, SD47]
G8 ATB S Chettle, 16 Moors Ln, Winsford, CW7 1JX [IO83RE, SJ66]
G8 ATC Dr R J Gayton, 194 Booker Ave, Allerton, Liverpool, L18 9TB [IO83NI, SJ48]
G8 ATD A Barter, 63 Ringwood Rd, Luton, LU2 7BG [IO91SV, TL02]
G8 ATE R T Turlington, 44 Glenfield Frith Dr, Glenfield, Leicester, LE3 8PQ [IO92JP, SK50]
G8 ATG M Williamson, 120 Warbreck Hill Rd, North Shore, Blackpool, FY2 0TR [IO83LU, SD33]
G8 ATK M Hearsey, Halycon, Lawday Link, Upper Hale, Farnham, GU9 0BS [IO91OF, SU84]
G8 ATL M R Lankester, 154 Gorse Ln, Clacton on Sea, CO15 4RJ [JO01OT, TM11]
G8 ATO Details withheld at licensee's request by SSL.
G8 ATP K E Mintern, 43A Lower Addiscombe Rd, Croydon, CR0 6PQ [IO91WJ, TQ36]
G8 ATS J Reeve, 16 Junction Rd, Mildenhall, Bury St. Edmunds, IP28 7BZ [JO02GI, TL77]
G8 ATY D W Smith, 1 Rhymers Cl, Hanslope, Milton Keynes, MK19 7DA [IO92NC, SP74]
G8 AU D B Stanley-Jackson, 5 Marline Ave, Hollington, St. Leonards on Sea, TN38 9HP [JO00GV, TQ71]
G8 AUB A J Mulley, 10 Deards End Ln, Knebworth, SG3 6NL [IO91VU, TL22]
G8 AUC Details withheld at licensee's request by SSL.[Op: L J Hall. Station located near Hampton.]
G8 AUJ G G Papworth, 12 Brook Way, Cupernham, Romsey, SO51 7JZ [IO91GA, SU32]
G8 AUL P Buck, 41 Marion St., Brighouse, HD6 2BJ [IO93CQ, SE12]
G8 AUM D Davies, 10 Egerton Rd, Berkhamsted, HP4 1DT [IO91RS, SP90]
G8 AUN R H A Chiddick, 87 Aylsham Rd, Norwich, NR3 2HW [JO02PP, TG21]
G8 AUU C G Partridge, 6 Blagdon Walk, Teddington, TW11 9LN [IO91UK, TQ17]
G8 AUX A J Strike, 72 Laburnum Ave, Taverham, Norwich, NR8 6JZ [JO02OQ, TG11]
G8 AVA M J Good, 19 Bramble Rd, Daws Heath, Thundersley, Benfleet, SS7 2UN [JO01HN, TQ88]
G8 AVH S D Ferneyhough, 30 Bedford Dr, Sutton Coldfield, B75 6AU [IO92CN, SP19]
GM8 AVM I Macdonald, 11 Low Rd, Castlehead, Paisley, PA2 6AQ [IO75SU, NS46]
G8 AVO Details withheld at licensee's request by SSL.
G8 AVV Prof K H Bennett, 11 Dunelm Ct, South St., Durham, DH1 4QX [IO94FS, NZ24]
G8 AVX L L Williams, 25 Streetsbrook Rd, Shirley, Solihull, B90 3PB [IO92CK, SP18]
G8 AVZ M W Keeping, 8 Calderdale Cl, Southgate West, Crawley, RH11 8SQ [IO91VC, TQ23]
G8 AWB R M Lawrence, 24 The Sheeplands, Sherborne, DT9 4BS [IO80RW, ST61]
G8 AWE M J Wellspring, 2 Harold Rd, Maiden Bower, Worth, Crawley, RH10 7RD [IO91WC, TQ23]
G8 AWL I L Teager, 67 Bradgate Rd, Markfield, LE67 9SN [IO92IQ, SK41]
GW8 AWM F W Evans, Llwynbedw, Gwehelog, Usk, Gwent, NP5 1RB [IO81NR, SO30]
G8 AWN J B Procter, 28 Holme Gr, Burley in Wharfedale, Ilkley, LS29 7QB [IO93DV, SE14]
G8 AWO M Gray, 79 Stockbreach Rd, Hatfield, AL10 0AU [IO91VS, TL20]
G8 AWS A K Russell, 51 Brooklyn Dr, Ellesmere Port, Great Sutton, South Wirral, L65 7EF [IO83MG, SJ37]
GW8 AWT W J Mainwaring, Tyle Bach, Manordeilo, Llandeilo, Dyfed, SA19 7BA [IO81AW, SN62]
G8 AWV M A Pawley, 38 Tesimond Dr, Monteagle Park, Yateley, Camberley, Surrey, GU17 7FE [IO91OH, SU85]
G8 AWY J A Ward, 71 Rothschild Ave, Aston Clinton, Aylesbury, HP22 5LY [IO91PT, SP81]
G8 AWZ P E Le Fevre, Firbank, Green Ln, Horsford, Norwich, NR10 3ED [JO02OR, TG11]
G8 AXA Hon M G Wallace, 17 Leamington Ave, Orpington, BR6 9QA [JO01BI, TQ46]
G8 AXH G Smith, Mahon House, Main St., Cropwell Butler, Nottingham, NG12 3AB [IO92MW, SK63]
G8 AXK R E Jones, 27 Coneydale, Welwyn Garden City, AL8 7RX [IO91VT, TL21]
G8 AXN C B Amery, 122 Main St., Willoughby on Wolds, Willoughby on The Wolds, Loughborough, LE12 6SZ [IO92LT, SK62]
G8 AXO A J Nunn, 1 Andrew Cl, Leiston, IP16 4LE [JO02SE, TM46]
G8 AXV K V Shail, 28 Hornyold Rd, Malvern, WR14 1QH [IO82UC, SO74]
G8 AXW Details withheld at licensee's request by SSL.
G8 AXX S Barker, 402 Fonthill Dr, Doylestown, Pa 18901, USA, X X
G8 AYC N A Walker, 36 Meyrick Dr, Wash Common, Newbury, RG14 6SX [IO91HJ, SU46]
G8 AYG D H Myers, 9 Romway Cl, Shepshed, Loughborough, LE12 9DT [IO92IS, SK41]
G8 AYK D W Johnson, 2 Church House Rd, Berrow, Burnham on Sea, TA8 2NG [IO81LG, ST25]
G8 AYM N B Pritchard, 108 Kynaston Ave, Aylesbury, HP21 9DS [IO91OT, SP81]
G8 AYN R C Whitbread, 3 Mackaness Cl, Gilmorton, Lutterworth, LE17 5PP [IO92KL, SP58]
G8 AYV J Lewis, Newnham House, Shurton, Stogursey, Bridgwater, TA5 1QG [IO81KE, ST24]
G8 AYY P T Gaskin, 58 Elmcroft Rd, Yardley, Birmingham, B26 1PL [IO92CL, SP18]
GI8 AYZ I J Kyle, 1 Portulla Dr, Pond Park Rd, Lisburn, BT28 3JS [IO64XM, J26]
G8 AZA J E Agar, 291 Overdale, Southwold Est, Eastfield, Scarborough, YO11 3RE [IO94TF, TA08]
G8 AZB S A Smith, 121 St. Pauls Ave, Slough, SL2 5EN [IO91QM, SU98]
G8 AZC E F Higgins, 8 Beta Rd, Woking, GU22 8EF [IO91RH, TQ05]
G8 AZF W McInarlin, 16 Arncliffe Ct, Croft House Ln, Huddersfield, HD1 4PT [IO93CP, SE11]
G8 AZI Details withheld at licensee's request by SSL.
G8 AZM D A Johnson, Pippin Hill, 195 Staplers Rd, Newport, PO30 2DP [IO90IQ, SZ58]
G8 AZN R G S Barnes, 18 Battle Rd, Tewkesbury Park, Tewkesbury, GL20 5TZ [IO81WX, SO83]
G8 AZR J P Dimmock, 93 Barton Rd, Harlington, Dunstable, LU5 6LG [IO91SX, TL03]
G8 AZT J A Jones, 9 Queens Walk, Thornbury, Bristol, BS12 1SR [IO81RO, ST69]
G8 AZV D W Withey, Cambridge Cottage, 49 The Hill, Wheathampstead, St. Albans, AL4 8PR [IO91UT, TL11]
G8 BAA D A Earnshaw, The Boundary, Birch Rd, Bignall End, Stoke on Trent, ST7 8LB [IO83UB, SJ85]
G8 BAD D Donati, 53 Smithbarn, Horsham, RH13 6DT [IO91UB, TQ13]
G8 BAG G Rowley, School House, Aldbrough St John, Richmond, North Yorks, DL11 7SU [IO94DL, NZ21]
G8 BAS D Gardiner, 31 Alexander Dr, Cirencester, GL7 1UG [IO91AQ, SP00]
G8 BAT Details withheld at licensee's request by SSL.
G8 BAV J D Bosworth, 57 Livingstone Rd, Derby, DE23 6PS [IO92GV, SK33]
G8 BAZ P B Talbot, Spring Gr, Upper Ninety, Malmesbury, Wilts, SN16 9PR [IO91AO, SU09]
GM8 BBA E Bailey, 23 McCallum Gdns, Strath View, Bellshill, ML4 2SR [IO75XT, NS75]
G8 BBB A R Taylor, 12 The Rampart, Haddenham, Ely, CB6 3ST [JO02BI, TL47]
G8 BBC V Reynolds Ariel Radio Grp, White House, Motspur Park, New Maldon, KT3 6PQ [IO91VJ, TQ26][Correspondence to Ariel Radio Group, Broadcasting House, London W1]
G8 BBD R C Stevens, 4 Colbeck Rd, Harrow, HA1 4BS [IO91TN, TQ18]
G8 BBK R L Nelson, 10 Wragg Dr, Newmarket, CB8 7SD [JO02EG, TL66]
G8 BBN G A Moore, 2 Spinacre, Becton Ln, Barton on Sea, New Milton, BH25 7DF [IO90ER, SZ29]
G8 BBV J A Goulty, 1 Larksway, Felixstowe, IP11 8PN [JO01PW, TM23]
G8 BBW M C D Mann Audlem ARS, 18 Potton Rd, Eynesbury, St. Neots, Huntingdon, PE19 2NP [IO92UF, TL15]
G8 BBZ P Barker, 15 Epping Green, Hemel Hempstead, HP2 7JP [IO91SS, TL00]
G8 BCA R H Chambers, 11 Thetford Rd, Mildenhall, Bury St. Edmunds, IP28 7HX [JO02GI, TL77]
GM8 BCB J R M Campbell, 18 Riccarton Ave, Currie, EH14 5PQ [IO85IV, NT16]
G8 BCF G S Podmore, Granville, Langley Rd, Lower Penn, Wolverhampton, WV4 4XX [IO82VN, SO89]
G8 BCG P Taylor, Brewlyf, Herodsfoot, Liskeard, PL14 4QX [IO70RK, SX26]
G8 BCH G R Watts, 46 Links Rd, Weymouth, DT4 0PE [IO80SO, SY67]
G8 BCJ A E Unsworth, Meadow View, Clockhouse Ln, North Stifford, Grays, RM16 5UR [JO01DL, TQ67]
G8 BCL H F Bottomley, Nerefield, Aylesbury Rd, Chearsley, Aylesbury, HP18 0BL [IO91MT, SP71]
G8 BCO C A E Boys, 34 Firacre Rd, Ash Vale, Aldershot, GU12 5JT [IO91PG, SU85]
G8 BCQ W D Green, 2 Irkdale Ave, Forty Hill, Enfield, EN1 4BD [IO91XP, TQ39]
G8 BCT F H Townsend, 6 The Hatches, Farnham, GU9 8UE [IO91OE, SU84]
G8 BCU S Brown, 81 Brown Ln, Heald Green, Cheadle, SK8 3RG [IO83VI, SJ88]
G8 BDF J F Hanney, 16 Parsonage Barn Ln, Ringwood, BH24 1PX [IO90CU, SU10]
G8 BDO S L Norman, 6 Castle Cl, Bottesford, Nottingham, NG13 0EH [IO92OW, SK83]
G8 BDQ G Hedley, 1 Lisle Ln, Ely, CB7 4AS [JO02DJ, TL58]
GM8 BDX A J Scott, 9 Treaty Park, Birgham, Coldstream, TD12 4NG [IO85TP, NT73]
G8 BDZ K R Cowdell, 6 Pearl St., Bedminster, Bristol, BS3 3EA [IO81QK, ST57]
G8 BEA B Harcourt, 52 Westcott Way, Honiton, EX14 8JE [IO80JT, ST10]
G8 BEB P R Jackson, 5 Lower Rd, Eastbourne, BN21 1QE [JO00DS, TV59]
G8 BEG M G Carrington, 63 Elgar Dr, Shefford, SG17 5RZ [IO92UA, TL13]
G8 BEH D W Hill, Las Colinas, West Fen Drainside, Frithville, Boston, PE22 7EP [IO93XA, TF34]
G8 BEK C T Dunn, 11 Thames Ave, Burnley, BB10 2PZ [IO83VT, SD83]
G8 BEN K C Smith, 36 New Rd, Whittlesey, Peterborough, PE7 1SU [IO92WN, TL29]
G8 BEQ K S Greenough, 2 Bexley Cl, Heath, Glossop, SK13 9BG [IO93AK, SK09]
G8 BFA S Davis, 21 Cordville Cl, Chaddesden, Derby, DE21 6WX [IO92GW, SK33]
G8 BFC P C Johnson, 6 Wesley Rd, Alvaston, Derby, DE24 0LG [IO92GV, SK33]
GM8 BFE L P Dawkes, 29 Newton Cres, Rosyth, Dunfermline, KY11 2QW [IO86HA, NT18]
G8 BFF Details withheld at licensee's request by SSL.
G8 BFH J F Marriott, 104 Whinbush Rd, Hitchin, SG5 1PN [IO91UW, TL12]
G8 BFK S A C Ballard, 26 Crafts End, Chilton, Didcot, OX11 0SA [IO91IN, SU48]
G8 BFL Details withheld at licensee's request by SSL.[Op: B J Jayne, 38 Townfields, Lichfield, Staffs, WS13 8AA.]
G8 BFM A J Whittaker, 6 Kingsbridge Way, Beeston, Bramcote, Nottingham, NG9 3LW [IO92JW, SK53]
G8 BFO R C Hayter, 87 Clare St., Stoke on Trent, ST4 6EE [IO83VA, SJ84]
G8 BFS J F Dudeney, The Wheelhouse, Middle St., Clavering, Saffron Walden, CB11 4QL [JO01BX, TL43]
G8 BFT Details withheld at licensee's request by SSL.
G8 BFV D R Edwards, 34 Campkin Rd, Wells, BA5 2DG [IO81QE, ST54]
G8 BFW T A Wildman, 58 Carlton Mews, Castle Bromwich, Birmingham, B36 0AD [IO92CM, SP18]
G8 BGG H G Cane, 17 Combeland Rd, Alcombe, Minehead, TA24 6BS [IO81GE, SS94]
G8 BGL R H Gilliatt, 21 Main St., Thorpe on The Hill, Lincoln, LN6 9BG [IO93QE, SK96]
G8 BGM M J Lee, 3 Canal Ct, Barwell Ln, Wantage, Oxon, OX12 9YH [IO91GO, SU38]
G8 BGR W D Curtis, Saint Fillans, Orchard Rd, Farnborough, Orpington Kent, BR6 7BP [JO01AI, TQ46]
G8 BGT A A Dermont, 4 Fox Cl, Burghfield Common, Reading, RG7 3NA [IO91LJ, SU66]
G8 BGW D C Gaskell, Orchard House, Church St., Tansley, Matlock, Derbyshire, DE4 5FE [IO93FD, SK36]
G8 BHD P R Hudson, 50 Swanley Ln, Swanley, BR8 7JG [JO01CJ, TQ56]
G8 BHE N L Gutteridge, 68 Max Rd, Quinton, Birmingham, B32 1LB [IO92AK, SP08]
G8 BHG A Richards, 11 George St., Kings Lynn, PE30 2AQ [JO02ES, TF62]
G8 BHH R A Stewart, 424 Wood End Rd, Wolverhampton, WV11 1YD [IO82XO, SJ90]
G8 BHP Details withheld at licensee's request by SSL.
GM8 BHR G S Pearson, 2 Hamilton Terr, Edinburgh, EH15 1NB [IO85KW, NT37]
G8 BHX M Berry, 4 Wadsworth Cl, Handforth, Wilmslow, SK9 3AY [IO83VI, SJ88]
G8 BHY A J Heath, 7 Coral Cl, Coventry, CV5 7AD [IO92FJ, SP37]
GW8 BIA F R Hopwood, 1 Trem y Mynydd, Abergele, LL22 9YY [IO83FH, SH97]
G8 BIG M Stebbings, 15 St. Helena Way, Horsford, Norwich, NR10 3EA [JO02OQ, TG11]
G8 BIH Details withheld at licensee's request by SSL.
G8 BII B Hunt, 53 The Sands, Milton under Wychwood, Chipping Norton, OX7 6ER [IO91EU, SP21]
G8 BIJ J F Batten, 21 Squires Bridge Rd, Shepperton, TW17 0JZ [IO91SJ, TQ06]
G8 BIO A Cummings, 151 Heywood Old Rd, Middleton, Manchester, M24 4QR [IO83VN, SD80]
G8 BIQ R W Franks, 65 Kings Rd, Surbiton, KT6 5JE [IO91UJ, TQ16]
G8 BIR R J Harris, 35 Freemantle Rd, Eastville, Bristol, BS5 6SY [IO81RL, ST67]
G8 BIS P Lyon, Tyle House, Stocks Rd, Wittersham, Kent, TN30 7EH [JO01IA, TQ82]
G8 BIT A W Butler, 12 Lapwing Dell, Letchworth, SG6 2TE [IO91VX, TL23]
G8 BIW R Booth, 16 Darwynn Ave, Swinton, Mexborough, S64 8DU [IO93IL, SK49]
G8 BIX A J Parcell, 23 Castle St., Portchester, Fareham, PO16 9PY [IO90KU, SU60]
G8 BIZ P W Barnett, 71 Raymonds Plain, Welwyn Garden City, AL7 4TE [IO91VS, TL21]
G8 BJA D T Couchy, 8 Chapel St., Wincham, Northwich, CW9 6DA [IO83SG, SJ67]
G8 BJB G M King, 62 Heathfield Rd, Sholing, Southampton, SO19 1DP [IO90HV, SU41]
GM8 BJF Dr B W Flynn, 15 Riselaw Cres, Edinburgh, EH10 6HN [IO85JV, NT26]
G8 BJG Details withheld at licensee's request by SSL.
GM8 BJJ A T Morton, 4 Mountstuart St., Millport, KA28 0DP [IO75MS, NS15]
G8 BJK B D Cockell, 13 Clifton Rd, Watford, WD1 8DH [IO91TP, TQ19]
G8 BJL T N Tisdall, 28 Agden Brow Park, Agden Brow, Lymm, Ches, WA13 0UB [IO83SI, SJ78]
G8 BJN S J Taplin, 37 North Park, Fakenham, NR21 9RG [JO02KU, TF93]
G8 BJP A R Bond, 9 Ethelred Rd, Westgate on Sea, CT8 8SJ [JO01QJ, TR37]
G8 BJQ L Case, 58 Brookdale, Widnes, WA8 4TB [IO83OJ, SJ48]
G8 BJV D Jacklin, 15 Long Perry, Capel St. Mary, Capel, Ipswich, IP9 2XD [JO02MA, TM03]
G8 BJW G L Marshall, 9 Eastbourne Rd, Southport, PR8 4EN [IO83LP, SD31]
G8 BKD P Scotney, 30 Trinity Rd, Rothwell, Kettering, NN14 6HY [IO92OK, SP88]
G8 BKE C H Towns, 21 Seafield Cl, Barton on Sea, New Milton, BH25 7HR [IO90ER, SZ29]
G8 BKF W F Duckett, 57 Newchurch Rd, Wellington, Telford, TF1 1JX [IO82RQ, SJ61]
G8 BKG D F Wright, 61 Potton Rd, Eynesbury, St. Neots, Huntingdon, PE19 2NN [IO92UF, TL15]
G8 BKH G J P Shepherd, 64 Dawley Rd, Wellington, Arleston, Telford, TF1 2JF [IO82RQ, SJ61]
G8 BKL E R Danks, 18 Lichfield St., Stourport on Severn, DY13 9EU [IO82UI, SO87]
G8 BKQ C G Clark OBE, 21A Headland Park Rd, Paignton, Devon, TQ3 2EN [IO80FK, SX86]
G8 BKX O W Pearce, 23 Bush House, Elm Gr, Southsea, PO5 1JH [IO90KS, SZ69]
G8 BKY R A Nicholls, 88A Chestnut Dr, Broadway Park, Petersfield, GU31 4ND [IO90MX, SU72]
G8 BLB P R Blakeney, 45 Hampden Ave, Chesham, HP5 2HL [IO91QR, SP90]
G8 BLC D J Hebden, Herlins, Holling Hill Ln, Wickersley, Rotherham, S66 0BG [IO93IK, SK49]
G8 BLF D McCrum, 75 Greaves Rd, Sheffield, S5 9DE [IO93GK, SK39]
G8 BLP C J Bond, 6 Brincliffe Edge C, Sheffield, South Yorks, S11 9GG [IO93GI, SK38]
G8 BLS P W Best, 41 Dryden Rd, Ipswich, IP1 6QP [JO02NB, TM14]
G8 BLV L J Hakes, The Vicarage, Dolphingholme, Lancaster, LA2 9AH [IO83PX, SD55]
G8 BLW A J Stepney, 39 Southcote Rd, Bournemouth, BH1 3SH [IO90BR, SZ19]
G8 BMD D Stanley, 3 Silvington Cl, Stirchley, Telford, TF3 1RW [IO82SP, SJ70]
GW8 BME F H Burrow, Oxwich Green, Gower Swansea, West Glam, SA3 1LU [IO71VN, SS48]
G8 BMG D W Platt, 10 Colwyn Dr, Knypersley, Stoke on Trent, ST8 7BL [IO83WC, SJ85]
G8 BMH J R Parry, 29 Heath Rd, Uptin By Chester, Chester, CH2 1HT [IO83NF, SJ46]
G8 BMI G T Theasby, 31 Middleton, Cowling, Keighley, BD22 0DQ [IO83XV, SD94]
G8 BMJ P Wild, 9 Red Lion Cl, Talke, Stoke on Trent, ST7 1SZ [IO83UB, SJ85]
G8 BMP M A Taylor, 96 Woodhouses Rd, Burntwood, Walsall, West Midlands, WS7 9EJ [IO92BQ, SK00]
G8 BMQ B L Cedar, 29 Velsheda Ct, Hythe Marina Village, Hythe, Southampton, SO45 6DW [IO90HU, SU40]
G8 BMR Details withheld at licensee's request by SSL.
G8 BMZ P Cowling, 94 Welholme Rd, Grimsby, DN32 0NG [IO93XN, TA20]
G8 BNB R S Gibbs, 15 Gosford Hill Ct, Bicester Rd, Kidlington, OX5 2XP [IO91IT, SP41]
GI8 BNC J H McCann, 61 Glenpark Rd, Omagh, BT79 7SS [IO64IP, H47]
G8 BNE R E G Kendall, 2 Ash Hill Dr, Leeds, LS17 8JT [IO93GU, SE33]
G8 BNG A R Green, 37 Bramcote Ln, Nottingham, NG8 2NA [IO92JW, SK53]
GM8 BNH I J Gall, Cluaran, Murcar, Bridge of Don, Aberdeen, AB23 8BD [IO87WE, NJ91]
GW8 BNI G A Donovan, Briarfield, 15 Old Hill Cres, Christchurch, Newport, NP6 1JL [IO81MO, ST38]
G8 BNR Mt R J Wells, 279 Hatfield Rd, St. Albans, AL4 0DH [IO91US, TL10]
G8 BOB A E Robinson, 29 Thomas Manning Rd, Diss, IP22 3HL [JO02NI, TM17]
G8 BOE F G Davenport, 27 Buce Hayes Cl, Highcliffe, Christchurch, BH23 5HJ [IO90DR, SZ29]
G8 BOI M G Simpson, 2 Hanover Rd, Laira, Plymouth, PL3 6BY [IO70WJ, SX55]

G8 BOJ K C Agombar, 54 Julien Rd, London, W5 4XA [IO91UL, TQ17]
GM8 BOM E Somerville, 2 Hillcrest Pl, Kilwinning, KA13 7NH [IO75PP, NS34]
G8 BOP M J Palmer, 109 Longfellow Rd, Dudley, DY3 3EF [IO82WM, SO99]
GW8 BOQ K Phillips, 636 Owens Rd, Clover, S.C. 29710, USA, X X
GM8 BOW L P Farrell, 12 Harefield Ave, Dundee, DD3 6AW [IO86ML, NO33]
G8 BPE Details withheld at licensee's request by SSL.
G8 BPF Details withheld at licensee's request by SSL.
G8 BPH J E Rome, 1 Bridge Cttgs, Matching Rd, Hatfield Heath, Bishops Stortford, CM22 7AS [JO01CT, TL51]
G8 BPN G S Wilkerson, Hill House, Newton, Leominster, Herefordshire, HR6 0PF [IO82PE, SO55]
G8 BPQ J N Wiseman, 8 Eaton St., Nottingham, NG3 5QL [IO92KX, SK54]
G8 BPU H R Skelhorn, School House, Lower House, Bollington, Macclesfield, SK10 5HN [IO83WH, SJ97]
G8 BPW A C Stoker, 35A Church End Ln, Runwell, Wickford, SS11 7JE [JO01GO, TQ79]
G8 BPY P R Hollis, 5 Salisbury Rd, New Malden, KT3 3HZ [IO91UJ, TQ26]
G8 BQF A Dixon, 62 The Wolds, Cottingham, HU16 5LQ [IO93SS, TA03]
G8 BQH M J Marsden, Hunters Moon, Buckingham Rd, Hardwick, Aylesbury, HP22 4EF [IO91OU, SP81]
GW8 BQK G Oatway, 21 Victoria Park, Colwyn Bay, LL29 7AX [IO83DH, SH87]
G8 BQS R L Ames, Church Farm, Town Rd, Ingham, Norwich, NR12 9TA [JO02SS, TG32]
GM8 BQV Dr W N Thomson, 42 Spylaw Bank Rd, Edinburgh, EH13 0JG [IO85IV, NT26]
G8 BQX J Ridd, 8 Beaufort Rd, St. Leonards on Sea, TN37 6QA [JO00GU, TQ81]
GM8 BQY R McLennan, 69 Stewart Terr, Aberdeen, AB1 5SX [IO87TH, NJ72]
G8 BQZ P C Plunkett, 31 Rooksmead Rd, Sunbury on Thames, TW16 6PD [IO91TJ, TQ16]
G8 BRD Dr C Dawson, 33 Rough Common Rd, Rough Common, Canterbury, CT2 9DL [JO01MH, TR15]
G8 BRF A Hirst, 44 Newlands Ave, Cheadle Hulme, Cheadle, SK8 6ND [IO83VI, SJ88]
G8 BRG Details withheld at licensee's request by SSL.
G8 BRK D Geldart, 92 Rockingham St., Barnsley, S71 1JR [IO93GN, SE30]
G8 BRL B E Ward, 10 Upper Moorfield Rd, Woodbridge, IP12 4JW [JO02PC, TM24]
G8 BRU G Gallamore, 30 Orchard Ave, Partington, Manchester, M31 4DL [IO83SK, SJ79]
G8 BSD J P Ceresole, 73 Western Way, Alverstoke, Gosport, PO12 2NF [IO90KS, SZ59]
GM8 BSE I G Swan, 1 Crosslaw Ave, Lanark, ML11 9AY [IO85CQ, NS84]
G8 BSF P A Seldon, 22 Seaford Rd, Wokingham, RG40 2EL [IO91OJ, SU86]
G8 BSK P G Robins, 290 Priory Rd, St. Denys, Southampton, SO17 2LS [IO90HW, SU41]
G8 BSP A J Wicks, Sarahs Cottage, Oldborough, Morchard Bishop, Crediton Devon, EX17 6SQ [IO80DU, SS70]
GM8 BSQ A Shepherd, 2 Westwood Pl, Skene, Westhill, AB32 6WS [IO87UD, NJ80]
G8 BSW R J Ware, 1 Reddons Rd, Beckenham, BR3 1LY [IO91XJ, TQ37]
G8 BTC B Fenwick, 16 Pine Walk, Uckfield, TN22 1TU [JO00BX, TQ42]
G8 BTD P A Sladen, 2 Burlea Cl, Crewe, CW2 8SZ [IO83SC, SJ65]
G8 BTK C W Harlow, Flat, 12 Penhurst Ct, Gr Rd, Worthing, BN14 9DG [IO90TT, TQ10]
G8 BTL H G Futcher, Sarum, 12 Thursby Rd, Woking, GU21 3NZ [IO91QH, SU95]
G8 BTR Details withheld at licensee's request by SSL.[Op: J K Law. Station located near Bingley.]
G8 BTS Details withheld at licensee's request by SSL.
G8 BTU J R Dowson, The Granary, St. Peters Rd, Arnesby, Leicester, LE8 5WJ [IO92KM, SP69]
G8 BTV P J C Marlow, 1 Vineries Cl, Leckhampton, Cheltenham, GL53 0NU [IO81WU, SO91]
G8 BTX T J Storeton-West, 7 Wash Ln, Kessingland, Lowestoft, NR33 7QZ [JO02UK, TM58]
G8 BTY M C Dennis, Thistledown, Yallands Hill, Monkton Heathfield, Taunton, TA2 8NA [IO81LA, ST22]
G8 BUB B S Goodall, 10 Westoby Cl, Shepshed, Loughborough, LE12 9SS [IO92IS, SK41]
G8 BUE I D Rae, 72 Lynch Hill Ln, Slough, SL2 2QD [IO91QM, SU98]
G8 BUI Dr C V Nowikow, 19 Park Hill, Old Harlow, Harlow, CM17 0AE [JO01BS, TL41]
G8 BUJ J T Nightingale, 14 Lismore Rd, Eastbourne, BN21 3AU [JO00DS, TV69]
G8 BUR Details withheld at licensee's request by SSL.
G8 BUV C A Chapman, 6 Pickhurst Green, Hayes, Bromley, BR2 7QT [JO01AJ, TQ36]
G8 BUX K Buxcey, Cold Wall Farm, Cobden Edge, Mellor Stockport, Ches, SK6 5NH [IO83XJ, SJ98]
G8 BUZ J F Paine, 1 Elm Cl, London, SW20 9HX [IO91VJ, TQ26]
G8 BVB P Power, 8 The Fairway, Camberley, GU15 1EF [IO91PH, SU85]
G8 BVF J J A Wearing, 122 Dixon Dr, Chelford, Macclesfield, SK11 9BX [IO83UG, SJ87]
G8 BVJ A J Gilchrist, 37 Greenway, Buckden, St. Neots, Huntingdon, PE18 9TU [IO92VH, TL26]
G8 BVL M J Porter, 94 Oaken Gr, Haxby, York, YO3 3QZ [IO94LA, SE65]
G8 BVQ R G Straker, 26 Constance Cres, Hayes, Bromley, BR2 7QJ [JO01AJ, TQ36]
G8 BVR G Oddy, Rectory Cottage, Widworthy, Wilmington, Honiton, Devon, EX14 9JS [IO80KT, SY29]
G8 BVT K Alexander, 17 Ridding Ln, Greenford, UB6 0JX [IO91UN, TQ18]
G8 BVY G B Spinks, 40 Ferndale Ave, Walthamstow, London, E17 9EH [JO01AN, TQ38]
G8 BWA M D Pollard, 3 Highfield Rd, Chertsey, KT16 8BU [IO91RJ, TQ06]
G8 BWB R A Wysome, Moordene, Axtown Ln, Yelverton, PL20 6BU [IO70WL, SX56]
G8 BWC R Smith, 9 Lawrence Dr, Brinsley, Nottingham, NG16 5AU [IO93IA, SK44]
G8 BWG N J Toovey, Apartment 3R1, Scarborough Manor, Scarborough, New York 10510, USA, ZZ9 9SC
G8 BWH R E Robinson, 1 John Dixon Ln, Darlington, DL1 1HG [IO94FM, NZ31]
G8 BXA A T Nicol, Woodhouse, Lower End, Swaffham Prior, Cambridge, Cambs, CB5 0HT [JO02DG, TL56]
G8 BXD Dr R H Edgecombe, 48 Birchwood Rd, Netherend, Woolaston, Lydney, GL15 6PE [IO81QQ, SO50]
G8 BXH J J Pryke, 52 Oaklands Ave, Watford, WD1 4LW [IO91TP, TQ19]
G8 BXJ A J Pullen, 22700 Gault St, West Hills, California, USA Ca91 3O7
G8 BXM P Shield, 56 Station Rd, Tempsford, Sandy, SG19 2AX [IO92UE, TL15]
G8 BXO J H Stacey, 3 West Park, South Molton, EX36 4HJ [IO81BA, SS72]
G8 BXQ T N Hordley, Imble Laze, 9 Newtown, Charlton Marshall, Blandford Forum, DT11 9NN [IO80WT, ST90]
G8 BXU Details withheld at licensee's request by SSL.
G8 BXV G H Rackham, 49 St. Lukes Rd, Southend on Sea, SS2 4AA [JO01IN, TQ88]
G8 BYB A G Hebden, 7 Crabapple Green, Orton Wistow, Peterborough, PE2 6YR [IO92UN, TL19]
G8 BYC C F D Keen, 24 Kings Ave, Mount Pleasant, Newhaven, BN9 0NA [JO00AT, TQ40]
G8 BYH N Berry, 12 Blossom St., Bootle, L20 5BB [IO83MK, SJ39]
G8 BYI R W Burrows, 76 Southfield, Southwick, Trowbridge, BA14 9PW [IO81VH, ST85]
G8 BYP R B Adams, 13 Johnson Rd, Great Baddow, Chelmsford, CM2 7JL [JO01GR, TL70]
G8 BZJ Details withheld at licensee's request by SSL.
G8 BZL G Lindsay, 71 Woodland Ave, Hove, BN3 6BJ [IO90VU, TQ20]
G8 BZN D V Goadby, Ty Mawr, Bryncroes, Pwllheli, Gwynedd, LL53 8EH [IO72UU, SH23]
G8 BZR P J Clark, Hamlet, Saunders Ln, Awbridge, Romsey, SO51 0GP [IO91FA, SU32]
G8 BZT D Allen, 156 Middlecotes, Tile Hill, Coventry, CV4 9AZ [IO92FJ, SP27]
G8 CA P G Cross Axe Vale ARC, Balls Farm Cottage, Musbury Rd, Axminster, EX13 5TT [IO80LS, SY29]
G8 CAA C F Broomfield, 8 Woodview Cres, Hildenborough, Tonbridge, TN11 9HD [JO01DF, TQ54]
G8 CAB J R G Sawford, 68 Harlyn Dr, Pinner, HA5 2DA [IO91TO, TQ18]
G8 CAC M G Barker, 3 Burley Cl, Desford, Leicester, LE9 9HX [IO92IO, SK40]
G8 CAH A F Parsons, 153 Denham DrIVe, Ashford, Middx, TW15 2AP [IO91SK, TQ07]
GW8 CAK P L Kenyon, The Elvins, Norton, Presteigne, LD8 2EP [IO82LH, SO26]
G8 CAM I H Foster, 28 Quilter Gdns, Orpington, BR5 4NA [JO01BI, TQ46]
G8 CAR P J Perkins Chiltern ARC, 26 Colne Rd, High Wycombe, HP13 7XN [IO91PP, SU89]
G8 CBA G A Tipler, Scots House, Chorley, Bridgnorth, Shropshire, WV16 6PR [IO82SK, SO78]
G8 CBB C J Barnes, 44 Wheatley Dr, North Wooton, North Wootton, Kings Lynn, PE30 3QQ [JO02FS, TF62]
G8 CBC R G Whitbrook, 158 Walsall Rd, Willenhall, WV13 2EB [IO82XO, SO99]
G8 CBE K W Quarman, 127 Highfield Ln, Hemel Hempstead, HP2 5JG [IO91SS, TL00]
G8 CBH P Berkeley, 1 The Beeches, Banstead, SM7 2AZ [IO91VH, TQ25]
G8 CBL J G Bruton, 10 Chestnut Dr, Bulkington, Devizes, SN10 1TB [IO81XI, ST96]
G8 CBM Details withheld at licensee's request by SSL.
G8 CBN R B Casson, Westways, Whole House Rd, Seascale, CA20 1QY [IO84GJ, NY00]
G8 CBO K N Smith, 6 Hermitage Cl, North Mundham, Chichester, PO20 6JZ [IO90OT, SU80]
GM8 CBQ N A Hendry, 52 Corndavon Terr, Mastrick, Aberdeen, AB1 5PJ [IO87TH, NJ72]
G8 CBU R C Aldous, 23 Aldhous Cl, Luton, LU3 2LZ [IO91SV, TL02]
G8 CCF S W C Hall, Knackershole Barn, West Knowle, Dulverton, Somerset, TA22 9RU [IO81FA, SS92]
G8 CCN R F W Read, The Elms, 76 School Rd, Downham, Billericay, CM11 1QN [JO01FP, TQ79]
G8 CCQ E M Peel, Chucks Corner, Deans Ln, Walton on The Hill, Tadworth, KT20 7UD [IO91VG, TQ25]
G8 CCR Details withheld at licensee's request by SSL.
G8 CCS R B Spokes, 155 Scraptoft Ln, Leicester, LE5 2FF [IO92LP, SK60]
G8 CCV M E ODonnell, 40 Mercers Dr, Bradville, Milton Keynes, MK13 7AY [IO92OB, SP84]
G8 CDB P R Strudwick, 40 Fifth Ave, Chelmsford, CM1 4HD [JO01FR, TL70]
G8 CDC Capt E P A Jones, Tudor House, Stoneleigh Rd, Blackdown, Leamington Spa, Warks, CV32 6QR [IO92FH, SP36]
G8 CDG N A Broadbent, 2 Market Hill, Clare, Sudbury, CO10 8NN [JO02GB, TL74]
G8 CDJ Details withheld at licensee's request by SSL.
G8 CDW E H Double, 2 Ormesby Dr, Swaffham, PE37 7SL [JO02IP, TF80]
G8 CEC R D Watkins, 18 Wolvesey Pl, Chandlers Ford, Eastleigh, SO53 4EA [IO90HX, SU42]
G8 CEE Details withheld at licensee's request by SSL.
G8 CEP D Clough, 165 Pilgrims Way, Andover, SP10 5HT [IO91GF, SU34]
G8 CEQ A W Reynolds, Ashwater Village Stores, Ashwater Cottage, Ashwater, Beaworthy, NW Devon, EX21 5EJ [IO70UR, SX39]
G8 CER P M Cockroft, Flat, 6 Cheriton, West Cliff Rd, Bournemouth, BH4 8AZ [IO90BR, SZ09]
G8 CES F Shrimpton, 23 Rotherwood Rd, Putney, London, SW15 1LA [IO91VL, TQ27]
G8 CET W S Marsden, 163 Buxton Old Rd, Disley, Stockport, SK12 2AY [IO83XI, SJ98]
G8 CEX B J Turner, 225 Westborough Rd, Westcliff on Sea, SS0 9PS [JO01IN, TQ88]
GJ8 CEY A Hearne, 2 Teighmore Park, La Chevre Rue, Grouville, Jersey, JE3 9EF
G8 CEZ R A Fuller, 35 Chichester Walk, Wimborne, BH21 1SL [IO90AS, SZ09]
G8 CFI J K Downs, 11A Brookside, Coppull, Chorley, Lancs, PR7 4QP [IO83QO, SD51]
GM8 CFS D Slight, North Mains, Ormiston, Tranent, East Lothian, EH35 5NG [IO85MW, NT47]
GM8 CFW C H Macphee, 20 Rowena Ave, Glasgow, G13 2JH [IO75TV, NS57]
G8 CGK C G Dixon, Kyrles Cross, Peterstow, Ross on Wye, HR9 6LD [IO81QW, SO52]
G8 CGM P H Raybould, 115 Curlew Cres, Bedford, MK41 7HY [IO92SD, TL05]
G8 CGT A G Campbell, 27 Denton Gr, Walton on Thames, KT12 3HE [IO91TJ, TQ16]
GD8 CGV Details withheld at licensee's request by SSL.
G8 CGW J Elliott, 92 Hinckley Rd, Barwell, Leicester, LE9 8DN [IO92HN, SP49]
G8 CGX B R Hopkins, 10 Hampden Rd, Flitwick, Bedford, MK45 1HX [IO91SX, TL03]
G8 CGZ D R Moore, 17 Ashover Rd, Old Tupton, Chesterfield, S42 6HH [IO93GE, SK36]
G8 CHI A P Tidder, 3 The Oaks, Yateley, GU46 6EA [IO91OH, SU86]
G8 CHK R S King, 28 Jenkinson Rd, Towcester, NN12 6AW [IO92MC, SP64]
GM8 CHL A C James, 16 Corslet Rd, Currie, EH14 5LY [IO85IV, NT16]
G8 CHN G R Barber, 666 Bradford Rd, Birkenshaw, Bradford, BD11 2EE [IO93DR, SE22]
G8 CHO S F Humm, 235 Felmongers, Harlow, CM20 3DP [JO01BS, TL41]
G8 CHQ Details withheld at licensee's request by SSL.
G8 CHY K J Twort, 39 Mile End Ln, Stockport, SK2 6BN [IO83WJ, SJ98]
G8 CID D A L Austin, 92 St. Andrews Pl, Melton, Woodbridge, IP12 1QX [JO02QC, TM25]
GM8 CIF D Macdonald, 22 Drummie Rd, Devonside, Tillicoultry, FK13 6HT [IO86DD, NS99]
G8 CIG P E J Tester, 25 Rosemary Gdns, Blackwater, Camberley, GU17 0NE [IO91OI, SU86]
G8 CIK T F Whetstone, 36 Stotfield Rd, Nottingham, NG8 4DA [IO92JW, SK54]
G8 CIT W A M Mc Killop, 2 Moores Green, Wokingham, RG40 1QG [IO91OK, SU86]
G8 CJ F Ellesmere, 70 Circle Pine Dr, Mankato, Minnesota 56001, USA
G8 CJA Dr M Dowson, The Granary, St. Peters Rd, Arnesby, Leicester, LE8 5WJ [IO92KM, SP69]
G8 CJD C R Hutton, 25 Fiddlers Ln, East Bergholt, Colchester, CO7 6SJ [JO01MX, TM03]
G8 CJG R J B Kirsch, 37 Western Rd, Leigh on Sea, SS9 2PR [JO01HN, TQ88]
G8 CJH D C Fletcher, 35 Gossmore Ln, Marlow, Bucks, SL7 1QQ [IO91ON, SU88]
G8 CJL A Dorling, 4 The Pastures, Ipswich, IP4 5UQ [JO02OB, TM24]
G8 CJM A R Croft, 15 Blenheim Ave, Chatham, ME4 6UU [JO01GI, TQ76]
G8 CJQ R G Barnes, 3 Ivy Cotages, Church Ln, Mobberley, Ches, WA16 7RD [IO83UH, SJ78]
G8 CJT C Coles, 15 Somerdale Ave, Bath, BA2 2PG [IO81TI, ST76]
GM8 CJW J West of Stow, Stowmill, nr Galashiels, Selkirkshire, TD1 2RB [IO85NQ, NT44]
G8 CK W E Bartholomew, 61 Osmaston Rd, Stourbridge, DY8 2AN [IO82WK, SO88]
G8 CKA H Panagakis, 6 The Pines, 38 The Avenue, Poole, Dorset, BH13 6HJ [IO90BR, SZ09]
G8 CKB P E Ebsworth, Forlandsvag, 5395 Steinsland, Norway, ZZ9 9FO
G8 CKC C I Curgenven, 14 Harrington Ln, Exeter, EX4 8PG [IO80GR, SX99]
G8 CKD G P Webb, 14 Old Rd South, Kempsey, Worcester, WR5 3NJ [IO82VD, SO84]
GW8 CKJ A H Williams, 54 St. Augustine Rd, Griffithstown, Pontypool, NP4 5EZ [IO81LQ, ST29]
G8 CKK A C Zerafa, 2 Furnwood, St. George, Bristol, BS5 8ST [IO81RK, ST67]
G8 CKM P J Parker, The Birches, Wem Rd, Shawbury, Shrewsbury, Shropshire, SY4 4NW [IO82QS, SJ52]
G8 CKN R G Powers, The Dell, Hussell Ln, Medstead, Alton, GU34 5PF [IO91LD, SU63]
G8 CKS J C Sargent, The Coach House, Speltham Hill, Hambledon, Waterlooville, PO7 4RU [IO90LW, SU61]
G8 CKT Details withheld at licensee's request by SSL.
G8 CKV S A Dale, 30 Almond Rd, Dogsthorpe, Peterborough, PE1 4LT [IO92VO, TF10]
G8 CKW R F Cobbold, 80 Halstead Rd, Lexden, Colchester, CO3 5AF [JO01KV, TL92]
G8 CLF A C V Humphreys, 24 Van Dyck Ave, New Malden, KT3 5NQ [IO91UJ, TQ26]
G8 CLJ I W Richmond, 48 Broadstone Rd, Harpenden, AL5 1RF [IO91TT, TL11]
G8 CLK K R Woollven, Inglenook, Perry Hill, Worplesdon, Guildford, GU3 3RD [IO91QG, SU95]
G8 CLS W J Williams, 11 Maisemore Ave, Patchway, Bristol, BS12 6BT [IO81RM, ST68]
G8 CLV D Knaggs, 17 Reservoir Rd, Solihull, B92 8BA [IO92CK, SP18]
G8 CLW J E Griffin, 185 Eastcote Ave, West Molesey, KT8 2EX [IO91TJ, TQ16]
G8 CLY J C Lythgoe, 18 Ranleigh Walk, Harpenden, AL5 1SR [IO91TT, TL11]
G8 CMD A Ashford, 56 Guarlford Rd, Barnards Green, Malvern, WR14 3QP [IO82UC, SO74]
G8 CME M L Crawshaw, 50 Kibble Gr, Brierfield, Nelson, BB9 5EW [IO83VT, SD83]
G8 CMG R J Williams, 18 Woodford Cres, Marsh Mills, Plymouth, PL7 4QY [IO70XJ, SX55]
G8 CMJ Details withheld at licensee's request by SSL.
G8 CMK W R Blankley, 16 Charles Rd, St. Leonards on Sea, TN38 0QA [JO00GU, TQ80]
G8 CMP C D Heymans, 59 Hill View, Widnes, WA8 9BN [IO83PJ, SJ58]
GW8 CMU M J Adcock, 7 Channel Cl, Rhoose, Barry, CF62 3EH [IO81HJ, ST06]
G8 CNB M N Farrington, Grange Villa, 218 Derby Rd, Butterley, Ripley, DE5 8HX [IO93HA, SK44]
G8 CNE Details withheld at licensee's request by SSL.
GW8 CNF S Biddiscombe, 20 Arlington Cl, Newport, NP9 6QF [IO81LO, ST39]
GW8 CNG Details withheld at licensee's request by SSL.[Op: G Edwards. Station located near Merthyr Tydfil.]
G8 CNI A Smith, Kerdoon, Lyth Hill, Shrewsbury, Shropshire, SY3 0BT [IO82OP, SJ40]
G8 CNN J E Price, Oxted Pl East, Broadham Green Rd, Oxted, RH8 9PF [IO91XF, TQ35]
GW8 CNS W J Mathias, Grenan Bungalow, Highland Ave, Bryncethin, Bridgend, CF32 9YH [IO81FN, SS98]
GW8 COE Details withheld at licensee's request by SSL.
G8 COF Details withheld at licensee's request by SSL.
GD8 COH S T Dimmock, 13 Dumbells Terr, Laxey, IM4 7NY
GW8 COJ A J Roberts, 7 Barnfield, Ponthir, Newport, NP6 1TN [IO81MO, ST39]
G8 COM G E Brutnall, 66 Arkwright Rd, Irchester, Wellingborough, NN29 7EF [IO92QG, SP96]
G8 CON J E Beith, Beechcroft, 18 Ave Rd, New Milton, BH25 5JP [IO90ES, SZ29]
G8 COR G Peters, 17 Washington Ln, Euxton, Chorley, PR7 6DE [IO83QP, SD51]
G8 CPA J A Vizor, Glenview, Old Shaw Ln, Shaw, Swindon, SN5 9PH [IO91CN, SU18]
G8 CPB J J Kozminski, Heronsforde, Park View Rd, Woldingham, Caterham, CR3 7DL [IO91XG, TQ35]
G8 CPF M I Edwards, 15 Clarendon Rd, Trowbridge, BA14 7BR [IO81VH, ST85]
G8 CPH M J Watson, The Tubbery, Henley, Ipswich, Suffolk, IP6 0BR [JO02NC, TM15]
G8 CPK Flt Lt D J Hibbin, 165 Byron St., Loughborough, LE11 5JN [IO92JS, SK52]
G8 CPQ V S Humphrey, 5 Wistow Rd, Limbury, Luton, LU3 2UR [IO91SV, TL02]
G8 CQA Details withheld at licensee's request by SSL.[Station located in Gloucester post code area.]
G8 CQG P J Cornell, 22 Ravine Rd, Bournemouth, BH5 2DU [IO90CR, SZ19]
G8 CQH Dr P J Best, 21 Greening Dr, Edgbaston, Birmingham, B15 2XA [IO92AL, SP08][CAIRO Users' enquiries to: Dr Peter J Best, EEAP, Aston University, Aston Triangle, Birmingham, B4 7ET. Tel: (0121) 359 3611 ext. 4274.]
GW8 CQK E J Allen, 37 St. Ilans Way, Caerphilly, CF8 1EW [IO81HN, ST08]
G8 CQR R E Lee, Cobnar, 83 Whitstone Rd, Shepton Mallet, BA4 5PP [IO81RE, ST64]
G8 CQV W S Hunter, 2 Green Acre, Goosnargh, Preston, PR3 2BQ [IO83PT, SD53]
G8 CQW K J Bagnall, Derbyshire Co. Raynet, 19 Banwell Cl, Mickleover, Derby, DE3 5QP [IO92FV, SK33]
G8 CQZ C T Powlesland, 61A Elmstead Ln, Chislehurst, BR7 5EQ [JO01AK, TQ47]
G8 CRB S Blunt, Blk 18 Simei St, 1, Melville Park # 10-13, 529943, Singapore
G8 CRC C R Callegari, 5 Brocks Ghyll, Eastbourne, BN20 9RQ [JO00CT, TQ50]
G8 CRM P M Watson, Tall Oak, 6 New Rd, Chevington, Bury St. Edmunds, IP29 5QL [JO02HF, TL76]
G8 CRN E T Boyd, 7 Julian Cl, Haverhill, CB9 0NN [JO02FB, TL64]
G8 CRR P J Claxton, 25 Hall Cl, Pattingham, Wolverhampton, WV6 7DE [IO82UO, SO89]
G8 CRV J A Christian, 5 Towers Way, Corfe Mullen, Wimborne, BH21 3UA [IO80XS, SY99]
G8 CRX S J Winford, 6 Summerdale, Billericay, CM12 9EL [JO01EO, TQ69]
G8 CRZ P J Hunt, 17 Selfridge Ave, Southbourne, Bournemouth, BH6 4NB [IO90CR, SZ19]
G8 CSA R J Chapman Silverthorn RC, 50 Graeme Rd, Enfield, EN1 3UT [IO91XP, TQ39]
G8 CSC J Townsend, 34 Orchard Cl, Scaynes Hill, Haywards Heath, RH17 7PQ [IO90XX, TQ32]
GM8 CSE H C Hogarth, 32 Broomhall Park, Edinburgh, EH12 7PU [IO85IW, NT17]
G8 CSK S R Browning, 12 Sunderland Cl, Woodley, Reading, RG5 4XR [IO91NK, SU77]
G8 CSN Details withheld at licensee's request by SSL.
G8 CSP W M Smith, 3 Lower Alt Rd, Hightown, Liverpool, L38 0BA [IO83LM, SD20]
G8 CSQ P H Benson, Ashbank Bungalow, Duke St, Bentham, Lancaster, LA2 7HX [IO84RC, SD66]
G8 CSY P A Thompson, 26 Carleton Ave, Blackpool, FY3 7JN [IO83LU, SD33]
G8 CTA R Denton, 60 Sandsacre Ave, Bridlington, YO16 5UG [IO94VC, TA16]
G8 CTB K A Chambers, 24 Primrose Cl, Flitwick, Bedford, MK45 1PJ [IO92RA, TL03]
G8 CTD A G Tait, Birch Glen, 71 Twemlows Ave, Higher Heath, Whitchurch, SY13 2HD [IO82QW, SJ53]
G8 CTJ M A Maxey, 28 Herald Way, Burbage, Hinckley, LE10 2NX [IO92HM, SP49]
G8 CTL M D Julian, 46 Rowlington Terr, Ashington, NE63 0LY [IO95FE, NZ28]
G8 CTO J N Walker, Sandy Vale, 13 Orchard Rd, Reigate, RH2 0PA [IO91VF, TQ25]
G8 CTX C Havercroft, 28 Anglers Way, Cambridge, CB4 1TZ [JO02BF, TL46]
G8 CUA R E Boittier, 52 Tithelands, Harlow, CM19 5NB [JO01AS, TL40]
G8 CUB R V Ray, Little Mallards, Mallard Way, Hutton, Brentwood, CM13 2NF [JO01EP, TQ69]
G8 CUG P R Cockram, 14 Langshott Cl, Woodham, Addlestone, KT15 3SE [IO91RI, TQ06]
G8 CUL M P Stevens, 67 New Rd, East Hagbourne, Didcot, OX11 9JX [IO91JO, SU58]
G8 CUN G E Rawlings, 109 The Upway, Basildon, SS14 2JD [JO01FN, TQ78]
G8 CUP Details withheld at licensee's request by SSL.
G8 CUV Details withheld at licensee's request by SSL.
G8 CVF J P Dobson, 11A Glenburn Ave, Eastham, Wirral, Merseyside, L62 8DJ [IO83MH, SJ37]
GM8 CVN J W Struthers, 7 Magdala Cres, Edinburgh, EH12 5BE [IO85JW, NT27]
G8 CVP R H Perry, 49 Harwich Rd, Little Clacton, Clacton on Sea, CO16 9NE [JO01NU, TM12]
G8 CVQ A F Parr, 8 Kingston Ave, North Cheam, Sutton, SM3 9TZ [IO91VI, TQ26]
G8 CVS J E Jenkinson, 16 Leybourne Cl, Walderslade, Chatham, ME5 9JN [JO01GI, TQ76]
G8 CVV B J Chuter, 27 Tas Combe Way, Eastbourne, BN20 9JA [JO00DT, TQ50]
G8 CVY Details withheld at licensee's request by SSL.

G8 CW H Crosland, Lydgate View, Holmfirth Rd, Newmill, Huddersfield, HD7 7LF [IO93CN, SE10]
G8 CWC C A Hudson, 11 Parkview Rd, Chapeltown, Sheffield, S6 2AU [IO93GJ, SK38]
G8 CWE T E Cook, Delwyn, 141 Station Rd, Watlington, Kings Lynn, PE33 0JG [JO02EQ, TF61]
G8 CWF R G Dubery, Chalkcroft, Green St Green Rd, Dartford, DA2 8DX [JO01DJ, TQ56]
G8 CWJ J P Abbott, 20 Highbury Ave, Salisbury, SP2 7EX [IO91CB, SU13]
G8 CWQ G I Horsfall, Lancaster New Rd, Garstang, Cabus, Preston, Lancs, PR3 1AD [IO83OV, SD44]
G8 CWU J Cragg-Sapsford, 78 Babbacombe Rd, Styvechale, Coventry, CV3 5PA [IO92FJ, SP37]
G8 CXA D W Froggatt, 2 Cobden Ave, Mexborough, S64 0AD [IO93IL, SE40]
G8 CXF J A Mould, 58 Chedworth Cres, Paulsgrove, Portsmouth, PO6 4ET [IO90KU, SU60]
G8 CXL N K Read, 86 Telford Ave, Lillington, Leamington Spa, CV32 7HP [IO92FH, SP36]
G8 CXV R G Brown, 19C Arlington Dr, Mapperley Park, Nottingham, NG3 5EN [IO92KX, SK54]
G8 CXW P D Appleby, 23 Oban Dr, Garswood, Ashton in Makerfield, Wigan, WN4 0SJ [IO83PL, SJ59]
G8 CXZ M Mills, 145 Park St., Haydock, St. Helens, WA11 0BL [IO83PL, SJ59]
G8 CYA N J Parker, 10 Lockhart Cl, Kenilworth, CV8 1RB [IO92FI, SP27]
G8 CYE S J Cook, 24 Beaufort Ct, Beaufort Rd, Ham, Richmond, TW10 7YG [IO91UK, TQ17]
G8 CYG W S Steer, 62 Wetheral Dr, Stanmore, HA7 2HL [IO91UO, TQ19]
G8 CYJ Dr D A Nicole, 53 Cobbett Rd, Southampton, SO18 1HJ [IO90HW, SU41]
G8 CYL P D Smith, 61 Waverley Dr, Chertsey, KT16 9PF [IO91RI, TQ06]
G8 CYT F M White, 12 Burcombe Rd, Kinson, Bournemouth, BH10 5JT [IO90BS, SZ09]
G8 CYU P York-Jones, Abbotsbury, 323 Old Bath Rd, Cheltenham, GL53 9AJ [IO81XV, SO91]
G8 CYV Details withheld at licensee's request by SSL.
G8 CYW S J Wisher, 20 Beweshill Crsnt, Winlaton, Blaydon-on-Tyne, Tyne & Wear, NE21 6BW [IO94DW, NZ16]
G8 CYX D W Storey, 43 Harwood Cl, Welwyn Garden City, AL8 7ST [IO91VT, TL21]
G8 CYY W G Hopkinson, 27 Nab Wood Gr, Shipley, BD18 4HR [IO93CT, SE13]
G8 CZE F W Beesley, 9 Northway Fairfield, Droylsden, Manchester, M43 6EF [IO83WL, SJ99]
G8 CZG Capt D W G Bell, 3 Pendle View, Brockhall Village, Old Langho, Blackburn, BB6 8AT [IO83ST, SD73]
G8 CZJ J M Butcher, 116 Cres Rd, Reading, RG1 5SN [IO91MK, SU77]
G8 CZM Details withheld at licensee's request by SSL.[Op: K A Jones.]
G8 CZP Details withheld at licensee's request by SSL.
G8 CZV Details withheld at licensee's request by SSL.
G8 CZZ Details withheld at licensee's request by SSL.
G8 DAB B S Smith, 18 Derry Park, Minety, Malmesbury, SN16 9RA [IO91AO, SU09]
G8 DAP M Capps, 156 Ash Cres, Eckington, Sheffield, S31 9AF [IO93GJ, SK38]
G8 DAZ W N Carrick, 10 Crosswood Cl, Loughborough, LE11 4BP [IO92JS, SK51]
G8 DBD R G Taylor, The Dairy Flat, High St., Hartley Wintney, Hook, Hants, RG27 8NZ [IO91NH, SU75]
G8 DBM Dr P A L Beaumont, Newlands, Bolton Percy, York, YO5 7AD [IO93JU, SE54]
G8 DBO K L Smith, Wilson Hall Farm, Slade Ln, Melbourne, Derby, DE73 1AG [IO92HT, SK42]
G8 DBP J S Mills, 93 Gays Rd, Hanham, Bristol, BS15 3JX [IO81RK, ST67]
G8 DBS V R H Ballard, 21 Rendermore Cl, Penkridge, Stafford, ST19 5JY [IO82WR, SJ91]
G8 DBU N C Greensted, 11 Molland Cl, Ash, Canterbury, CT3 2JG [JO01PG, TR25]
G8 DBV P G Jarrett, 3 Carisbrooke Rd, Strood, Rochester, ME2 3SN [JO01FJ, TQ77]
G8 DBY D Goodwill Derby & Dist AR, 94 Palmerston St., Derby, DE23 6PF [IO92GV, SK33]
G8 DCA M P Telkman, 23 Sheppard Rd, Basingstoke, RG21 3HT [IO91KG, SU65]
G8 DCD J G Durrant, 27 Trafford Rd, Willerby, Hull, HU10 6AJ [IO93SS, TA02]
G8 DCJ P McQuail, Myrtle Cottage, Post Office Ln, Draycott, Moreton in Marsh, GL56 9JZ [IO92DA, SP13]
G8 DCX R G Sangster, 10 Addison Rd, Banbury, OX16 9DH [IO92HB, SP43]
G8 DCZ Details withheld at licensee's request by SSL.[Op: P Metcalfe. Station located near Hurstpierpoint.]
G8 DDC C A Asquith Dunstable Dwn Rd, 36 Sunningdale, Luton, LU2 7TE [IO91TV, TL02]
G8 DDK K H Wright, 12 Bushmead Rd, Luton, LU2 7EU [IO91TV, TL02]
G8 DDY P C Thompson, 69 High St., Shanklin, PO37 6JJ [IO90JP, SZ58]
G8 DEA D E Allender, Broadshaw House, Hazelwood, Skipton, N Yorks, BD23 6JS [IO94BA, SE05]
G8 DEC A Malcolm, 68 Old Birmingham Rd, Lickey End, Bromsgrove, B60 1DG [IO82XI, SO97]
G8 DEE Dr M G Woodhouse, 24 Hurst Park Ave, Cambridge, CB4 2AE [JO02BF, TL46]
G8 DEL D R Coppen, 12 Morton Rd, London, N1 3BA [IO91WM, TQ38]
G8 DEM B T Willetts, 11 Albert Rd, Oldbury, Warley, B68 0NA [IO92AL, SP08]
G8 DET J H Bowen, Craigielea, South View Rd, Danbury, Chelmsford, CM3 4DX [JO01GR, TL70]
G8 DEU D G Holman, 35 Salterns Point, Salterns Way, Lilliput, Poole, BH14 8LN [IO90AQ, SZ08]
G8 DEW J E Hindley, 380 Cheetham Hill Rd, Dukinfield, SK16 5LB [IO83XL, SJ99]
G8 DEX J A Hosking, 15 Cornfield Cl, Macclesfield, SK10 2TY [IO83WG, SJ97]
G8 DEY D G Parr, 58 Ritson St., Toxteth, Liverpool, L8 0UF [IO83MJ, SJ38]
G8 DF A E Mitchell, 6 Charnwood Way, Lillington, Leamington Spa, CV32 7BU [IO92FH, SP36]
G8 DFA D C Hyde, Little Trematon, Polurrian Cliff Rd, Mullion, Helston, Cornwall, TR12 7EW [IO70IA, SW61]
GM8 DFC R W Cliff, 32 Lochardil Rd, Inverness, IV2 4LD [IO77VK, NH64]
G8 DFI B Oliver, 6 Catherton Rd, Cleobury Mortimer, Kidderminster, DY14 8EB [IO82SJ, SO67]
G8 DFN A T Barton, The Hackett, Thornbury, Bristol, BS12 2UA [IO81SO, ST69]
G8 DFU T A Lovelock, Los Perales 9, El Puertito Del Sauzal, Santa Cruz, de Tenerife, Islas Canarias 38360, X X
GI8 DGB B Moore, 2 Sheridan Gr, Fort Rd, Helens Bay, Bangor, BT19 1LN [IO74DQ, J48]
G8 DGH E Townsend, The Manor House, Fleckney, Leics, LE8 0AP [IO92MM, SP69]
G8 DGR R V Smallwood, The Island, Hyde End Ln, Brimpton, Reading, RG7 4TH [IO91JI, SU56]
G8 DGU J D Judge, 23 Minster Ln, Barrow in Furness, LA13 9NY [IO84JC, SD27]
G8 DHA D A Bishop, Oyston Lodge, Lynstone Rd, Bude, Cornwall, EX23 8LR [IO70RT, SS20]
G8 DHE G R Mather, 72 Cranleigh Rd, Worthing, BN14 7QW [IO90TT, TQ10]
G8 DHF S E Matthews, 213 Hucclecote Rd, Brockworth, Gloucester, GL3 3TZ [IO81VU, SO81]
G8 DHJ C R Pickering, 28 George V Ave, Margate, CT9 5QA [JO01QJ, TR37]
G8 DHL N J Ellis, Open View, New Rd Hill, Midgham, Reading, RG7 5RY [IO91JJ, SU56]
G8 DHQ D F Digby, 73 Bedford St., Crewe, CW2 6JB [IO83SC, SJ75]
G8 DHU M E Baxter, 11B The Leys, Roade, Northampton, NN7 2NR [IO92ND, SP75]
G8 DHV Details withheld at licensee's request by SSL.
GI8 DHW J A Hendron, 9 Drumahiskey Rd, Balnamore, Bendooragh, Ballymoney, BT53 7QL [IO65RB, C92]
G8 DHY Details withheld at licensee's request by SSL.
G8 DHZ W E Bardgett, 9 South Terr, Redland, Bristol, BS6 6TG [IO81QL, ST57]
G8 DIN D S Cooke, 11 The Avenue, Nunthorpe, Middlesbrough, TS7 0AA [IO94JM, NZ51]
G8 DIQ T C Hall, 7 Sweetlake Cottage, Nobold, Shrewsbury, SY5 8NH [IO82OQ, SJ41]
G8 DIR K J Walker, 12 Willow Park, Minsterley, Shrewsbury, SY5 0EH [IO82MP, SJ30]
G8 DIS D O Hawes, 65 Oakwood Cres, Winchmore Hill, London, N21 1PA [IO91WP, TQ39]
G8 DIT Details withheld at licensee's request by SSL.
G8 DIU B Cannon, 38 Sandringham Rd, Worcester Park, KT4 8UJ [IO91VI, TQ26]
G8 DIY P M Geeson, 33 High St., Sutton, Ely, CB6 2RB [JO02BJ, TL47]
G8 DJC D J Cook, 75 Windmill Ln, Castlecroft, Wolverhampton, WV3 8HN [IO82VN, SO89]
G8 DJF A D Dickson, 7 Sandford Gdns, High Wycombe, HP11 1QT [IO91PO, SU89]
G8 DJK Details withheld at licensee's request by SSL.
G8 DJL J Renaut, 47 Church Rd, Ferndown, BH22 9ES [IO90BT, SZ09]
G8 DJN J G Patrick, Appletrees, The St., Stour Provost, Gillingham, SP8 5RZ [IO80UX, ST72]
G8 DJO M V Adcock, 37 Ashpole Rd, Bocking, Braintree, CM7 5LW [JO01GV, TL72]
G8 DJU J Frisby, 2 Westfield Rd, Hoddesdon, EN11 8QX [IO91XS, TL30]
G8 DJW G J Membury, 11 York Terr, Dorchester, DT1 2DP [IO80SQ, SY68]
GM8 DKB E R Taynton, 42 Craigmount Park, Edinburgh, EH12 8EE [IO85IW, NT17]
G8 DKD C E B Weale, 156 Habberley Ln, Kidderminster, DY11 5JY [IO82UJ, SO87]
G8 DKF D Robbins, 158 Jeffcock Rd, Bradmore, Wolverhampton, WV3 7AF [IO82WN, SO89]
GM8 DKG Dr C M Pegrum, 8 Dudley Dr, Glasgow, G12 9SD [IO75UV, NS56]
G8 DKI Dr D W Lucas, The Old Barn, The St., Charlton, Malmesbury, SN16 9DL [IO81XO, ST98]
G8 DKK B J Harber, 45 Brandles Rd, Letchworth, SG6 2JA [IO91VX, TL23]
G8 DKS T Stevens, Heath Holdings Farm, Stone Heath, Leigh, Stoke on Trent, ST10 4PG [IO82XV, SJ93]
G8 DKV M C Coldicott, 78 Rectory Ln, Breadsall, Derby, DE21 5LL [IO92GW, SK33]
G8 DLF D H Griffin, 1 Starapple St, PO Box 273, Belmopan, Belize, Central America, X X
G8 DLH A J Hall, 39 St. Pauls St. North, Cheltenham, GL50 4AQ [IO81XV, SO92]
G8 DLL M S Monro, 6 Yew Tree Rd, Hayling Island, PO11 0QE [IO90MT, SU70]
G8 DLM B S Cooper, Minster House, Cross Ln, Brancaster, Kings Lynn, PE31 8AE [JO02HX, TF74]
G8 DLP R P Baker, Hallen, Wooden Rd West, Wednesbury, West Midlands, WS10 7SG [IO82XN, SO99]
G8 DLT G C Baraclough, 2 Orchard Cl, Corfe Mullen, Wimborne, BH21 3TW [IO80XS, SY99]
GM8 DLU J F M Bell, Fordeldean, Dalkeith, Midlothian, EH22 2PG [IO85MV, NT36]
G8 DLX M T Crampton, 55 Gilbert Ave, Bilton, Rugby, CV22 7BZ [IO92II, SP47]
G8 DLZ P G Lea, 7 Cressex Rd, High Wycombe, HP12 4PG [IO91OO, SU89]
G8 DMQ P A Eggett, 29 Sweet Briar Dr, Calcot, Reading, RG3 7AD [IO91LJ, SU66]
G8 DMR G W Chaloner, 9 Fairthorne Rise, Old Basing, Basingstoke, RG24 7EH [IO91LG, SU65]
G8 DMT M C Caley, 40 Spenser Way, Clacton, Jaywick, Clacton on Sea, CO15 2QT [JO01NS, TM11]
G8 DMU A D Frazer, Keld Farm, Greenhow, Pateley Bridge, Harrogate, HG3 5JL [IO94BB, SE16]
GI8 DMX J T H Moffitt, 101 Boghill Rd, Templepatrick, Ballyclare, BT39 0HS [IO64XP, J28]
G8 DNL R W K Smith, 6 Wordsworth Cl, Bishops Waltham, Southampton, SO32 1RT [IO90JW, SU51]
GW8 DOA G R Pollard, 3 Carey Walk, Cwrt Herbert, Neath, SA10 7DD [IO81CP, SS79]
G8 DOB I D Stuart, 87 Redgrove Park, Cheltenham, GL51 6QZ [IO81WV, SO92]
G8 DOF P J White, 2 Coltsfoot Cl, Huntington Park, Huntington, Chester, CH3 6DU [IO83NE, SJ46]
G8 DOH Dr A J Seeds, 114 Beaufort St., London, SW3 6BU [IO91VL, TQ27]
G8 DOR A D Barrett, 7 Laceys Dr, Hazlemere, High Wycombe, HP15 7JY [IO91PP, SU89]
G8 DOT Details withheld at licensee's request by SSL.
G8 DOW B Lee, 19 Lizard Head, Littlehampton, BN17 6RY [IO90RT, TQ00]
G8 DPE V T Brooks, 19 Malham Ave, Wigan, WN3 5PR [IO83QM, SD50]
G8 DPH T A Booth, 155 Oxford Rd, Windsor, SL4 5DX [IO91QL, SU97]
G8 DPQ D A Hendon, 2 Ellis Ave, Onslow Village, Guildford, GU2 5SR [IO91QF, SU94]
G8 DPS Dr R H Biddulph, 59 Ditton Rd, Surbiton, KT6 6RF [IO91UJ, TQ16]
GM8 DPV J F Hunting, 77 Califer Rd, Forres, IV36 0JB [IO87EO, NJ05]
G8 DPW D A Holden, 63 High St., Queenborough, ME11 5AG [JO01IJ, TQ97]
G8 DPY J C Wood, 9 High View Cl, Loughton, IG10 4EG [JO01AP, TQ49]
G8 DQF L Johnston, 9 Tunbridge Cl, Burwell, Cambridge, CB5 0EL [JO02DG, TL56]
G8 DQH S J Scott, 19 Maple Gdns, Walton Lodge, Stone, ST15 0EJ [IO82WV, SJ93]
G8 DQJ A G Negus, 17 Courtenay Gdns, Upminster, RM14 1DH [JO01DN, TQ58]
G8 DQK A G Symonds, Aircliffe, Ham Rd, Wantage, OX12 9EE [IO91GO, SU38]
G8 DQN N O Hunter, 33 Chapel Ct, Billericay, CM12 9LX [JO01FO, TQ69]
G8 DQX R A Gape, 38 Quorn Gdns, Leigh on Sea, SS9 2TB [JO01HN, TQ88]
G8 DQZ A H F Lord, 1 Grange Meadow, Elmswell, Bury St. Edmunds, IP30 9GE [JO02LF, TL96]
G8 DR D R Aston, 5 Claybrook Cl, East Finchley, London, N2 8JG [IO91VO, TQ28]
GM8 DRA R M Macleod, 22 Firthview, Dingwall, IV15 9PF [IO77SO, NH55]
G8 DRE D J Atkinson, 54 Egret Cres, Longridge Park, Colchester, CO4 3FP [JO01LV, TM02]
G8 DRQ Dr R C Cochrane, 134 Moor Ln South, Ravenfield, Rotherham, S65 4QR [IO93IK, SK49]
G8 DSG W J Jones, Elm Hurst, Station Rd, Baschurch, Shrewsbury, SY4 2BB [IO82NT, SJ42]
G8 DSM J Witherspoon, 109 Bromsgrove Rd, Redditch, B97 4RL [IO92AH, SP06]
G8 DSU R J Gill, 61 Cross Deep Gdns, Twickenham, TW1 4QZ [IO91UK, TQ17]
G8 DT F N Bedwell, 95 Warburton Rd, Canford Heath, Poole, BH17 8SD [IO90AR, SZ09]
G8 DTA A J Parsons, 20 Paddocks Ln, Prestbury, Cheltenham, GL50 4NX [IO81XV, SO92]
G8 DTE M J Pusey, 74 Carrington Rd, High Wycombe, HP12 3HT [IO91OO, SU89]
G8 DTF R E Price, 42 Worsley Rd, Worsley, Manchester, M28 2SH [IO83TM, SD70]
G8 DTM F L Partington, 21 East Rd, Wymeswold, Loughborough, LE12 6ST [IO92KT, SK62]
G8 DTQ B G Petifer, 14 Wood Ln, Caterham, CR3 5RT [IO91WG, TQ35]
G8 DTS B L Norcliffe, 2 Alexander Dr, Heswall, Wirral, L61 6XT [IO83KI, SJ28]
G8 DTT W J H Moore, 26 Richard Moon St., Crewe, CW1 3AX [IO83SC, SJ75]
G8 DUB A W Redman, 46 Tukes Ave, Bridgemary, Gosport, PO13 0SE [IO90JT, SU50]
G8 DUF R S Bird, 129 Park Rd, Formby, Liverpool, L37 6AD [IO83LN, SD20]
G8 DUI D J Cox, 18 Sawyers, Elmsett, Ipswich, IP7 6QH [JO02LC, TM04]
G8 DUM J P McGuire, 39 Cobham Rd, Halesowen, B63 3JZ [IO82XK, SO98]
G8 DUO I E Casewell, 6 Petrel Cl, Woosehill, Wokingham, RG41 3TF [IO91NJ, SU86]
GW8 DUP R G Harris, 64 Frederick Pl, Llansamlet, Swansea, SA7 9SX [IO81BP, SS69]
G8 DUT H F L Orgel, 1 Taunton Gr, Whitefield, Manchester, M45 6TJ [IO83UM, SD80]
G8 DUV Dr C C Zammit, 6 Boveney House, Segsbury Gr, Bracknell, RG12 9JX [IO91PJ, SU86]
G8 DUW I E Redfern, 8 Lilac Gr, Lickhill Lodge, Stourport on Severn, DY13 8SR [IO82UI, SO87]
GW8 DUY C L Davies, 9 Templeogue Wood, Dublin 6W, Ireland
G8 DV A P Morgan, 6 Ratcliff Lawns, Southam, Cheltenham, GL52 3PA [IO81XW, SO92]
G8 DVF T T Jones, 5 Blue Hatch, Frodsham, Warrington, WA6 7QJ [IO83PH, SJ57]
GW8 DVH R O Jones, Bronwydd, Tregaron, Dyfed, SY25 6NG [IO82AF, SN65]
G8 DVJ G F Wilks, 8 Chestnut Gr, East Barnet, Barnet, EN4 8PU [IO91WP, TQ29]
G8 DVK Details withheld at licensee's request by SSL.
G8 DVN D J Smith, 3 Woods Ln, Calverton, Nottingham, NG14 6FF [IO93KA, SK64]
G8 DVR Details withheld at licensee's request by SSL.
G8 DVS A B G Sterry, 1 Wavell Garth, Sandal Magna, Wakefield, WF2 6JP [IO93GP, SE31]
G8 DVU R J West, 55 Burney Bit, Pamber Heath, Tadley, RG26 3TL [IO91KI, SU66]
G8 DVW R Leadbeater, The Birches, Torpenhow, Carlisle, Cumbria, CA5 1JF [IO84JR, NY13]
G8 DWA M T Bowen, 232 Longford Rd, Cannock, WS11 1NE [IO82XQ, SJ90]
G8 DWI J Featherstone, 7 Highcliffe Rd, Winchester, SO23 0JE [IO91IB, SU42]
G8 DWL B C Bond The Grafton ARS, 86 Agar Gr, Camden Town, London, NW1 9TL [IO91WN, TQ28]
G8 DWP P A Lee, 223 Chelmsford Rd, Shenfield, Brentwood, CM15 8SA [JO01EP, TQ69]
G8 DWW C T Garcia, Flat D, 1 Westover Cl, Westbury on Trym, Bristol, BS9 3LR [IO81QL, ST57]
G8 DWX G H Haslip, 14 Chatsworth Ave, Telscombe Cliffs, Peacehaven, BN10 7EA [IO90XT, TQ30]
G8 DX R G Lavis, 116 Frome Rd, Bath, BA2 2PP [IO81TI, ST76]
G8 DXF C J Tarran, Woodlands, School Rd, West Wellow, Romsey, SO51 6AR [IO90FX, SU21]
G8 DXH Details withheld at licensee's request by SSL.
G8 DXI W O'Connor, 15 Southcote Ave, Feltham, TW13 4EQ [IO91TK, TQ17]
G8 DXJ C R Richardson, 110 Exning Rd, Newmarket, CB8 0AF [JO02EG, TL66]
G8 DXK G D Moore, 17 Ferrybridge Green, Hedge End, Southampton, SO30 0JX [IO90IV, SU41]
G8 DXM Details withheld at licensee's request by SSL.
G8 DXO R W Humble, 3 Abbey Gdns, Galhampton Rd, Galhampton, Yeovil, BA22 7AG [IO81RB, ST62]
G8 DXP A Cheasley, 25 Normanhurst Rd, Walton on Thames, KT12 3EQ [IO91TJ, TQ16]
G8 DXT C Beresford, 2 Moulton Cl, Swanwick, Alfreton, DE55 1ES [IO93HB, SK45]
G8 DXU B M Pollard-Wilkins, Systems Integration Electronic, 14 Seabeach Ln, Eastbourne, BN22 7NZ [JO00DT, TQ60]
G8 DXV H K King, 11 Priory Mead, Doddinghurst, Brentwood, CM15 0NB [JO01DQ, TQ59]
G8 DYA C L West, 14 Ashleigh Gdns, Wymondham, NR18 0EX [JO02NN, TG10]
G8 DYF Details withheld at licensee's request by SSL.[Station located near Orpington.]
G8 DYG M L Marshallsay, 2 Prospect Cttgs, Lime St., Eldersfield, Gloucester, GL19 4NX [IO81UX, SO83]
G8 DYI K Holdway, 18 Pennymore Cl, Hunters Oak, Trentham, Stoke on Trent, ST4 8YQ [IO82VX, SJ84]
G8 DYK J M Taylor, Millstone House, Main St., Norwell, Newark, NG23 6JN [IO93OD, SK76]
G8 DYN J H Postlethwaite, 70 Felhampton Rd, New Eltham, London, SE9 3NX [JO01AK, TQ47]
GW8 DYR Details withheld at licensee's request by SSL.
G8 DZC P J Makosz, 27 Wessex Gdns, Twyford, Reading, RG10 0BA [IO91NL, SU77]
G8 DZH J D Ray, 9 Albion Hill, Loughton, IG10 4RA [JO01AP, TQ49]
G8 DZJ G A Booth, 68 Tarragon Dr, Meir Heath, Stoke on Trent, ST3 7YE [IO82WX, SJ94]
G8 DZK K R Ash, 16 Dartmouth Rd, Ruislip, HA4 0DB [IO91TN, TQ18]
G8 DZW R L Brookes, 29 Ripley Rd, Hightown, Liversedge, WF15 6QE [IO93DR, SE12]
G8 EAD M R Hutchings, 109 Longlands Way, Heatherside, Camberley, GU15 1RU [IO91PI, SU96]
G8 EAH I B Carress, Two Hoots, 1 Riplingham Rd, Skidby, Cottingham, HU16 5TR [IO93SS, TA03]
G8 EAI P W Eversden, 20 Wheatlands, Titchfield Common, Fareham, PO14 4SL [IO90IU, SU50]
G8 EAM H J S Newton, 9 Periton Ln, Minehead, TA24 8AQ [IO81GE, SS94]
G8 EAN J M J Cunningham, 62 Kings Hill, Beech, Alton, GU34 4AN [IO91LD, SU63]
G8 EAP Details withheld at licensee's request by SSL.[Op: K Rothwell. Station located near Bury.]
G8 EAS A G Gale, 7 Glebe Cl, Lewes, BN7 1LB [IO90XU, TQ40]
G8 EAW R S Hatch, 208 Wensley Rd, Blackburn, BB2 6SS [IO83RR, SD62]
G8 EAX S A Herod, 17 Stennetts Cl, Trimley St. Mary, Trimley, Ipswich, IP10 0TZ [JO02PA, TM24]
G8 EBD G P Welch, 18 Alderdale, Wolverhampton, WV3 9JF [IO82WO, SO89]
G8 EBH Details withheld at licensee's request by SSL.
G8 EBJ J G Davies, 17 Gate Hill Ct, Notting Hill Gate, London, W11 3QT [IO91VM, TQ28]
G8 EBM S J Haseldine, Leamington House, Burland Green Ln, Weston Underwood, Derbyshire, DE6 4PF [IO92FX, SK24]
G8 EBQ R P Martin, 10 Westways, Stoneleigh, Epsom, KT19 0PQ [IO91VI, TQ26]
G8 EBT R H Lees, 5 Ridgeside, Bledlow Ridge, High Wycombe, HP14 4JN [IO91NQ, SU89]
G8 EBX P J Starling, 14 Merton Pl, Littlebury, Saffron Walden, CB11 4TH [JO02CA, TL53]
G8 ECG K Montgomery, The Old Village Post Office, High St, Cuddesdon, Oxford, OX44 9HP [IO91KR, SP50]
G8 ECI D C Brown, Qatar Gas, Ras Laffan Plant, P O Box 2666, Doha, The Arabian Gulf
G8 ECJ R D Gilks, 21 Willow Cl, Flackwell Heath, High Wycombe, HP10 9LH [IO91PO, SU98]
GW8 EDE Details withheld at licensee's request by SSL.
G8 EDH P M Clarke, 746 Alexandria Dr, Naperville, Illinois 60565, USA, X X
G8 EDJ Details withheld at licensee's request by SSL.
G8 EDL D Holding, 10 Grantham Cres, Ipswich, IP2 9PD [JO02NB, TM14]
G8 EDS Details withheld at licensee's request by SSL.
G8 EDX C E Vitiello, Manor House, North St., Rothersthorpe, Northampton, Northants, NN7 3JB [IO92ME, SP75]
G8 EEA D G Hill, 872 Oldham Rd, Thornham, Rochdale, OL11 2BN [IO83WO, SD91]
G8 EEI T W Coleman, 14 Norman Ct, Stapleton Hall Rd, London, N4 4QD [IO91WN, TQ38]
G8 EEK B C Bruce, 3 Mount Pleasant, Great Totham, Maldon, CM9 8DS [JO01IS, TL81]
G8 EEM C J Gill, 31 Airedale Dr, Garforth, Leeds, LS25 2JF [IO93HS, SE43]
G8 EEY A M Mobbs, 149 The Paddocks, Old Catton, Norwich, NR6 7HR [JO02PQ, TG21]
G8 EEZ Details withheld at licensee's request by SSL.
G8 EFK E G Carter, 44 Plattes Cl, Shaw, Swindon, SN5 9SA [IO91CN, SU18]
G8 EFP Details withheld at licensee's request by SSL.
G8 EFU C R Bloxidge, 33 Rocklands Dr, Sutton Coldfield, B75 6SP [IO92CN, SP19]
G8 EFY N C Marshall, 8 Ellesboro Rd, Harborne, Birmingham, B17 8PT [IO92AL, SP08]
G8 EGC J Alty, 84 Ralthmore Cres, Southport, PR9 8PW [IO83MQ, SD31]
G8 EGE J J Denton, 32 Highfields Mead, East Hanningfield, Chelmsford, CM3 8XA [JO01GQ, TL70]
G8 EGF L G Hicks, 9 Pine Gr, Edenbridge, TN8 5HU [JO01AE, TQ44]
G8 EGG D Hemingway, Ivanhoe, Glen Rd, Hindhead, GU26 6QE [IO91OC, SU83]
G8 EGL C Burton, 4 Ashdown Way, Misterton, Doncaster, DN10 4BP [IO93NK, SK79]
G8 EGM M J Booth, 16 Falcon Dr, Birdwell, Barnsley, S70 5SN [IO93GM, SE30]
G8 EGT Details withheld at licensee's request by SSL.
G8 EGU M J Smith, 35 Queen St., Balderton, Newark, NG24 3NS [IO93OB, SK85]
G8 EGV I Robinson, 27 Lundwood Gr, Owlthorpe, Sheffield, S19 6SR [IO93GJ, SK38]
G8 EHD P J Brenton, 40 Furneaux Rd, Milehouse, Plymouth, PL2 3ET [IO70WJ, SX45]

G8	EHF	J Healen, 12 Primrose Ln, Standish, Wigan, WN6 0NR [IO83PO, SD51]
G8	EHG	P Healen, 12 Primrose Ln, Standish, Wigan, WN6 0NR [IO83PO, SD51]
GW8	EHK	C J S Dunbar, 29 Heol y Nant, Baglan, Port Talbot, SA12 8ER [IO81CO, SS79]
GW8	EHQ	J E Brown, 106 Marlborough Rd, Penylan, Cardiff, CF2 5BY [IO81KL, ST17]
G8	EHS	Dr A M Fletcher, 35 Wimborne Ave, Ipswich, IP3 8QW [JO02OB, TM24]
G8	EHX	M P Melbourne, 32 Lake Farm Rd, Rainworth, Mansfield, NG21 0ED [IO93KC, SK55]
G8	EIE	R A K Forster, 7 Western Way, Alverstoke, Gosport, PO12 2NE [IO90KS, SZ59]
G8	EIG	Details withheld at licensee's request by SSL.
G8	EIH	Dr D W Temple, 17 Trinity Rd, Nailsea, Bristol, BS19 2NT [IO81OK, ST47]
G8	EII	M J Smith, 17 Girton Cl, Owlsmoor, Camberley, Surrey, GU15 4UP [IO91PI, SU86]
G8	EIN	N B Shephard, 166 Chaldon Way, Coulsdon, CR5 1DF [IO91WH, TQ35]
G8	EIO	G J Etheridge, 24 Lyttleton Ave, Halesowen, B62 9ED [IO82XL, SO98]
G8	EJC	R D Drew, 13 Parc Corner, Penzance, Cornwall, TR18 4SP [IO70FC, SW43]
G8	EJP	N A Killgren, 72 Lakewood Rd, Chandlers Ford, Eastleigh, SO53 5AA [IO90HX, SU42]
GM8	EJS	J S Gilmour, Barnmill, Mosspark Ave, Milngavie, Glasgow, G62 8NL [IO75UW, NS57]
GM8	EKF	F Benson, 53 Warriston Dr, Edinburgh, EH3 5NA [IO85JX, NT27]
G8	EKH	E Mann, 306 Pellon Ln, Halifax, HX1 4QD [IO93BR, SE02]
G8	EKI	G F Fuller, 8 The Avenue, West Wickham, BR4 0DY [IO91XJ, TQ36]
G8	EKN	M N Biltcliffe, 19 Kennedy Rd, Bicester, OX6 8BE [IO91KV, SP52]
G8	EKO	Details withheld at licensee's request by SSL.
G8	EKP	I H Hodgson, 38 Prospect Park, Exeter, EX4 6NA [IO80FR, SX99]
G8	EKU	R J Powe, 44 Church St., Langford, Biggleswade, SG18 9QR [IO92UB, TL14]
G8	EKW	G R Thornton, 4 Fir Tree Cl, Exmouth, EX8 4EU [IO80HP, SY08]
G8	EKZ	A Jones, 52 Elizabeth Cres, Stoke Gifford, Bristol, BS12 6NZ [IO81RM, ST67]
G8	ELG	E Joyce, 1 Dellmeadow, Abbots Langley, WD5 0BA [IO91SQ, TL00]
G8	ELH	D L E Fisher, 17 Thrushel Cl, Green Meadow, Swindon, SN2 3PP [IO91CO, SU18]
G8	ELP	A D Stockley, c/o Icom UK Ltd, Unit 9 Herne Bay, West Ind.Est. Sea S, Herne Bay Kent, CT6 8LD [JO01NI, TR16]
GW8	ELR	D S Hands, Quarre Bach, Tegryn, Llanfynrnach, Dyfed, SA35 1XX [IO71RX, SN23]
G8	ELS	M G A Jarvis, 31 The Downings, Herne Bay, CT6 7EJ [JO01NI, TR16]
G8	ELV	R D Pearson, 21 West Parade, Spalding, PE11 1HD [IO92WS, TF22]
G8	ELX	P F T Redman, 24 Mayfield Rd, Farnborough, GU14 8RS [IO91OH, SU85]
G8	EMA	D C Pedley, Higher Sutton, South Milton, Kings Bridge, Devon, TQ7 3JG [IO80CG, SX64]
G8	EMB	W H Tickell, 26 Shear Brow, Blackburn, BB1 7EX [IO83SS, SD62]
G8	EMH	D Roebuck, 7 Elm Tree Cl, Northanston, Anston, Sheffield, S31 7FG [IO93GJ, SK38]
G8	EMX	G J Hankins, 11 Cottesbrook Rd, Acocks Green, Birmingham, B27 6LE [IO92CK, SP18]
G8	ENA	E J Fellows, Richeldis, 343 Wake Green Rd, Birmingham, B13 0BH [IO92BK, SP08]
G8	ENB	R W S Whitby, 138 Browns Ln, Stanton on The Wolds, Keyworth, Nottingham, NG12 5BN [IO92LU, SK63]
G8	END	I R Bodie, 20 Arwenack Ave, Falmouth, TR11 3JW [IO70LD, SW83]
G8	ENO	G K Tong, 6 Arncliffe Rd, West Park, Leeds, LS16 5JE [IO93ET, SE23]
G8	ENW	P Baker, Top of The Hill, Post Office Ln, Cleeve Hill, Cheltenham, GL52 3PS [IO81XW, SO92]
G8	ENY	R A Hersey, 13 North Terr, Mildenhall, Bury St. Edmunds, IP28 7AA [JO02GI, TL77]
G8	EOJ	E B March, 36 Handel Cl, Brighton Hill, Basingstoke, RG22 4DL [IO91KF, SU65]
G8	EOM	D J Garrard, 48 Shorefields, Benfleet, SS7 5BQ [JO01GN, TQ78]
G8	EOP	Details withheld at licensee's request by SSL.
G8	EOW	D R Thorn, 3 Gorricks, Stony Stratford, Milton Keynes, MK11 1HB [IO92NB, SP73]
G8	EOX	D J See, 28 Wayside Mews, Maidenhead, SL6 7EJ [IO91PM, SU88]
G8	EPB	A H Baldwin, 42 Martin Ave, Hampton, Evesham, WR11 6NP [IO92AC, SP04]
G8	EPC	M J Dyke, 18 Powderham Ave, Berkeley Hunterton, Worcester, WR4 0DN [IO82VE, SO85]
G8	EPH	C W Kilvington, Ayresome House, Barrow Rd, New Holland, Barrow upon Humber, DN19 7QH [IO93TQ, TA02]
G8	EPK	D M Skye, 112 Eastmoor Park, Harpenden, AL5 1BP [IO91TT, TL11]
G8	EPQ	R Prew, 16 Stokenchurch Pl, Bradwell Common, Milton Keynes, MK13 8AT [IO92OB, SP83]
G8	EPR	D N Hicks, 17 Branches Cl, Bewdley, DY12 2HD [IO82UI, SO77]
G8	EPZ	C J Ward, 4 The Hawthorns, Charvil, Reading, RG10 9TS [IO91NL, SU77]
G8	EQA	P G D Wood, 76 Andes Cl, Ocean Village, Southampton, SO14 3HS [IO90HV, SU41]
G8	EQB	A Vickers, 3 Wingrove Ave, Fulwell, Sunderland, SR6 9HJ [IO94HW, NZ45]
G8	EQC	D T Cliffe, Common Farm, Bromley Hayes, Lichfield, Staffs, WS13 8JE [IO92CR, SK11]
G8	EQD	D Wright, 22 West Hill, Rotherham, S61 2HB [IO93HK, SK39]
GW8	EQI	J A Fellows, 8 The Links, Gwernaffield, Mold, CH7 5DZ [IO83JE, SJ26]
GW8	EQO	B Tyler, Flat A 1st Floor, 42 Caperidge Dr, Peninsula Village, Discovery Bay, Lantau Hong Kong
G8	EQX	B P Taylor, 46 Hunters Field, Stanford in The Vale, Faringdon, SN7 8LX [IO91FP, SU39]
G8	EQZ	C Reynolds, 49 Westborough Way, Anlaby Common, Hull, HU4 7SW [IO93TR, TA02]
GW8	ERA	M S Voss, 9 Chapel Cl, Garndiffaith, Pontypool, NP4 7QS [IO81LR, SO20]
G8	ERC	J F Stewart Sussex Raynet, 44 Salisbury Rd, Seaford, BN25 2DB [JO00BS, TV49]
G8	ERJ	Details withheld at licensee's request by SSL.
G8	ERN	R A Walker, 7 Moreland Croft, Minworth, Sutton Coldfield, B76 1XZ [IO92CM, SP19]
G8	ERQ	Details withheld at licensee's request by SSL.
G8	ERV	K Blackman, 131 London Rd, Apsley, Hemel Hempstead, Herts, HP3 9SQ [IO91SR, TL00]
G8	ESB	Details withheld at licensee's request by SSL.[Op: Rik Royall, The Three Tuns, Knayton, Thirsk, North Yorks, YO7 4AN.]
G8	ESH	E P Owen, Owen Motor Co Ltd, Leamilngton Rd, Ryton on Dunsmore, Coventry, CV8 3EL [IO92GI, SP37]
G8	ESK	B M Kermode, 6 Ascham Hall, Lady Park Ave, College Rd, Bingley, West Yorks, BD16 4UB [IO93CU, SE14]
G8	ESL	W P Miller, 4 Paddock Ct, Bempton Ln, Bridlington, YO16 5FW [IO94VC, TA16]
GM8	ESQ	Details withheld at licensee's request by SSL.
G8	ESW	W W T Brade, Brade End, Whitstable Rd, Herne Bay, CT6 8BN [JO01NI, TR16]
G8	ETC	Details withheld at licensee's request by SSL.
G8	ETD	T J Rumble, Santos, Tolgus Mount, Redruth, Cornwall, TR15 3TA [IO70JF, SW64]
G8	ETI	N G Foggin, 12 Linnetsdene, Covingham, Swindon, SN3 5AG [IO91DN, SU18]
GM8	ETJ	K S McCartney, Cairndhue, Walter St., Langholm, DG13 0AX [IO85LD, NY38]
G8	ETP	M R Furnival, Eaton Hey, The Dr, Cobham, Surrey, KT11 2JQ [IO91TH, TQ16]
G8	ETR	Details withheld at licensee's request by SSL.
G8	ETS	D Swale, 369 Scalby Rd, Scarborough, YO12 6TG [IO94SG, TA08]
G8	ETU	A Metcalf, 10 Manor Bend, Galmpton, Brixham, TQ5 0PB [IO80FJ, SX85]
G8	ETV	P E H Richardson, 3 Butlers Cl, Amersham, HP6 5PY [IO91QQ, SU99]
G8	ETW	A Horsfield, 37 Hereward Way, Deeping St. James, Peterborough, PE6 8QA [IO92UQ, TF10]
G8	EUD	K Fidler, Westwood House, 12 Westfield Rd, Maidenhead, SL6 5AU [IO91PM, SU88]
G8	EUE	M P Gasper, The Barn, Back Rd, Wenhaston, Halesworth, IP19 9DZ [JO02SH, TM47]
GW8	EUF	C N I Hall, Trans-Wales Trails, Cwmfforest Riding Centre, Brecon, Wales, LD3 0EU [IO81JW, SO12]
GM8	EUG	N L Robertson, 10 Warrenpark Rd, Largs, KA30 8EF [IO75NS, NS25]
G8	EUH	D J Pickard, Smith Hill Farm, Shaw Ln, Triangle, Sowerby Bridge, West Yorks, HX6 3EZ [IO93AQ, SE02]
G8	EUK	Details withheld at licensee's request by SSL.
G8	EUQ	Details withheld at licensee's request by SSL.
G8	EUT	Dr A V Barford, Salcombe Rd, Marlborough, Malborough, Kingsbridge, Devon, TQ7 3BX [IO80CF, SX73]
G8	EUV	C M Fenton-Coopland, 14 Chevril Ct, Wickersley, Rotherham, S66 0BN [IO93IK, SK49]
G8	EUX	P H Saul, 51 Windsor Cl, Towcester, NN12 6JB [IO92MD, SP64]
G8	EVD	T J Cartwright, 132 Mere Rd, Wigston, LE18 3RL [IO92KO, SP69]
GM8	EVF	A M Bryce, 23 Primrose Ave, Inverkip, Greenock, PA16 0DS [IO75NV, NS27]
G8	EVI	A S Clark, 134 Bridge St., Gainsborough, DN21 1LP [IO93OJ, SK88]
G8	EVR	K W Taylor, 39 Bowerfield Cres, Hazel Gr, Stockport, SK7 6JB [IO83WI, SJ98]
G8	EVY	W M Dunell Cambridge & Dar, 4 Orchard Rd, Haslingfield, Cambridge, CB3 7JT [JO02AD, TL45]
G8	EWC	A Rouse, 18 Clover Dr, Thorrington, Colchester, CO7 8HL [JO01MU, TM02]
G8	EWD	M Smith, 47 Salisbury Rd, Market Drayton, TF9 1AR [IO82SV, SJ63]
G8	EWF	B H Gilbert, 19 Sandown Rd, Brighton, BN2 3EH [IO90WT, TQ30]
G8	EWG	J L Fitch, 102 Morley Ave, Wood Green, London, N22 6NG [IO91WO, TQ39]
G8	EWH	Details withheld at licensee's request by SSL.
G8	EWL	C G Burgess, 125 Braybourne Cl, Uxbridge, UB8 1UL [IO91SN, TQ08]
G8	EWN	D L Edmonds, Great House Cottage, Ripponden, West Yorks, HX6 4LQ [IO93AQ, SE01]
G8	EWT	G R Diacon, Raddlebarn, High Cogges, Witney, Oxon, OX8 6UW [IO91GS, SP30]
GD8	EXI	Dr S R Baker, Ballanarran, Surby, nr Port Erin, Rushen, Isle of Man, IM9 6TE
G8	EXJ	B Jones, 38 Wyresdale Rd, Lancaster, LA1 3DU [IO84OB, SD46]
G8	EXN	C Briggs, 22 Woodlesford Cres, Moor End Rd, Halifax, HX2 0RB [IO93BR, SE02]
G8	EXQ	T J Connell, 32 Hope Ave, Stanford-le-Hope, SS17 8DH [JO01FM, TQ68]
G8	EXS	P N Atherton, 10 Cheriton Dr, Ravenshead, Nottingham, NG15 9DG [IO93KC, SK55]
G8	EXT	H D Howard, 10 Lawnside, London, SE3 9HL [JO01AL, TQ37]
G8	EYC	Details withheld at licensee's request by SSL.
GW8	EYG	Details withheld at licensee's request by SSL.
G8	EYP	A J Faulkner, 38 Lillington Rd, Leamington Spa, CV32 5YZ [IO92FH, SP36]
G8	EYQ	J Clee, 38 Natty Ln, Illingworth, Halifax, HX2 9DS [IO93BS, SE02]
G8	EYT	J A Chapman, 7 Spilsbury Croft, Solihull, B91 3UB [IO92CJ, SP17]
G8	EYY	M J Hancock, 12 Mellor Rd, Hillmorton, Rugby, CV21 4BP [IO92JI, SP57]
G8	EZB	M D Whitlock, 85 Antrobus Rd, Sutton Coldfield, B73 5EL [IO92BN, SP19]
G8	EZE	P J Swallow, 17 Pangdene Cl, Burgess Hill, RH15 9US [IO90WW, TQ21]
G8	EZG	A S Pybus, 46 Broadlands, Desborough, Kettering, NN14 2TH [IO92OK, SP88]
G8	EZL	T G Lambert, 40 Deepdale Rd, Cullercoats, North Shields, NE30 3AN [IO95GA, NZ37]
G8	EZO	Details withheld at licensee's request by SSL.
G8	EZR	K R James, Amberley, Harnham Rd, Withington, Cheltenham, Glos, GL54 4DD [IO91AU, SP01]
G8	EZT	R Elgy, 130 Stebbing House, Queensdale Cres, London, W11 4TG [IO91VM, TQ28]
G8	EZU	K G Darbyshire, 24 Neston Rd, Walshaw, Bury, BL8 3DB [IO83TO, SD71]
G8	EZV	G L White, 94 Wingate Rd, Luton, LU4 8PY [IO91SV, TL02]
G8	EZZ	R S Chambers, 15 Barnfield Cl, Braunton, EX33 2HL [IO71WC, SS43]
G8	FAB	H McIntyre obo Southampton ARC, 90 Evenlode Rd, Millbrook, Southampton, SO16 9EH [IO90GW, SU31]
G8	FAD	W P M Chown, Gable End, North Waltham, Basingstoke, Hants, RG25 2BE [IO91JF, SU54]
G8	FAE	P C Ealey, 3 Pembridge Rd, Bovingdon, Hemel Hempstead, HP3 0QJ [IO91RR, TL00]
G8	FAK	S M Sherratt, 21 Tweedale Cl, Station Rd, Mursley, Milton Keynes, MK17 0SB [IO91OW, SP82]
G8	FAR	R J C Elms, 23 Bethell Ave, Ilford, IG1 4UX [JO01AN, TQ48]
G8	FAS	S M Hotham, 36 South St., Crewkerne, TA18 8DB [IO80OV, ST40]
G8	FAT	B W Haines, 20 Westfield Gdns, Kenton, Harrow, HA3 9EJ [IO91UO, TQ18]
G8	FAW	J Cutts, Fourseasons, Straight Rd, Battisford, Stowmarket, IP14 2LZ [JO02LD, TM05]
G8	FAX	E D Bye, 10 Daws Heath Rd, Rayleigh, SS6 7QH [JO01HN, TQ89]
G8	FBB	J H Lloyd, 70 Heath Dr, Ware, SG12 0RJ [IO91XT, TL31]
G8	FBF	D P Fellows, 10 Benning Way, Wokingham, RG40 1XX [IO91OK, SU86]
G8	FBG	Details withheld at licensee's request by SSL.
G8	FBK	L J West Knights, 27 Marlborough Rd, London, W4 4EU [IO91UL, TQ27]
G8	FBM	M J Bates, 6 Bay Cl, Horley, RH6 8LF [IO91VE, TQ24]
GU8	FBO	R B Stockwell, Fleurs Des Champs, La Colline Des Bas Courtills, Saint Saviours, Guernsey, Channel Islands, GY7 9YJ
G8	FBQ	B Corker, 46 Danelaw, Great Lumley, Chester-le-Street, DH3 4LU [IO94FU, NZ24]
G8	FBV	E L Skinner, 18 Huntsmans Dr, Upminster, RM14 3YU [JO01DN, TQ58]
G8	FC	R E Finch RAF Locking Rac, 1 Cherrytree Cottage, Church Rd, Penn, High Wycombe, Bucks, HP10 8LN [IO91PP, SU99]
G8	FCA	R W Payne, 23 Roundwood Cl, Hitchin, SG4 0RD [IO91VX, TL23]
G8	FCD	M L Christieson, Tilgate, Perrymans Ln, High Hurstwood, Uckfield, TN22 4AG [JO01BA, TQ42]
GM8	FCK	M R Crayton, 15 Belleisle Pl, Kilmarnock, KA1 4UD [IO75RO, NS43]
G8	FCQ	M A Lister, 246 Wigston Ln, Leicester, LE2 8DH [IO92KO, SK50]
G8	FCT	R Chadwick, Ithaca, Heck Ln, Hensall, Goole, DN14 0RD [IO93KQ, SE52]
G8	FCW	Details withheld at licensee's request by SSL.
G8	FDC	I D Hydes, 2 Stable Ct, Martlesham Heath, Ipswich, IP5 7UQ [JO02OB, TM24]
G8	FDE	B M McManus, 6 Rowley Rd, St. Neots, Huntingdon, PE19 1UF [IO92UF, TL16]
G8	FDF	J C Bastable, Home Farm, Burrowbridge, Bridgewater, Somerset, TA7 0RF [IO81MB, ST33]
G8	FDG	W E Baxendale, Juverna, 28 Westland Ave, Darwen, BB3 2ST [IO83SQ, SD62]
G8	FDJ	Dr J S Roberts, 2 Lomas Lea, Stannington, Sheffield, S6 6EW [IO93FJ, SK38]
G8	FDR	M H Bingham, 18 Ladywell Gate, Welton, Brough, HU15 1NL [IO93RR, SE92]
G8	FEA	Details withheld at licensee's request by SSL.
G8	FEC	Details withheld at licensee's request by SSL.
G8	FEG	Details withheld at licensee's request by SSL.
G8	FEI	G D Gaunt, 2 Stanhope Ave, Hayes, Bromley, BR2 7JR [JO01AJ, TQ46]
G8	FEJ	M J Woudstra, Flat 4, 15 Amherst Rd, Bexhill on Sea, TN40 1QH [JO00FU, TQ70]
G8	FEK	E M Gawthorpe, 27 Weelsby Way, Hessle, HU13 0JN [IO93SR, TA02]
G8	FEQ	J W N Noble, 11 Sharpecroft, Harlow, CM19 4AA [JO01BS, TL40]
GW8	FEY	R B Suddaby, Tyddyn Rheinallt, Penmon, Beaumaris, Gwyned, LL58 8SP [IO73XH, SH68]
G8	FEZ	F A Stuart, 70 Pear Tree Rd, Herne Bay, CT6 7EQ [JO01NI, TR16]
G8	FF	S Southgate, 12 Pearsons Cl, Holt, NR25 6EH [JO02NV, TG03]
G8	FFA	E C D Davis, 24 Redcar Ave, Bobblestock, Hereford, HR4 9TJ [IO82PB, SO44]
G8	FFC	C R McManus, 6 Rowley Rd, St. Neots, Huntingdon, PE19 1UF [IO92UF, TL16]
G8	FFD	J G V Russell, 45 Sara Cl, Sutton Coldfield, B74 4BW [IO92BO, SP19]
G8	FFF	C J Player, 7 Petlands, Little Walden, Saffron Walden, CB10 1XF [JO02DB, TL54]
GM8	FFH	D Brown, 14 Barloan Pl, Dumbarton, G82 3QW [IO75RW, NS47]
G8	FFM	B R P Jackson, 50 Holland Rd, Ampthill, Bedford, MK45 2RS [IO92SA, TL03]
G8	FFN	A J Blakemore, 20 Derwent Rd, Whitmore Park, Coventry, CV6 2HB [IO92FK, SP38]
G8	FFU	C C Burrows, 6 Brook Way, Lower Somersham, Ipswich, IP8 4PE [JO02MC, TM04]
G8	FFW	P S Rycroft, Shore View House, 100 Pilling Ln, Preesall, Poulton-le-Fylde, FY6 0HG [IO83MW, SD34]
G8	FFZ	P E Ewington, 26 Dickens Rd, Rugby, CV22 5RW [IO92II, SP57]
G8	FGB	S T Whitehead, 74 Manchester Rd, Haslingden, Rossendale, BB4 5TE [IO83UQ, SD72]
G8	FGD	C A Jones, 148 Fouracre Cres, Downend, Bristol, BS16 6PZ [IO81SM, ST67]
G8	FGK	Details withheld at licensee's request by SSL.
G8	FGQ	H R Brittan, Meadowhurst Cottage, Woodcock Heath, Kingstone, Uttoxeter, Staffs, ST14 8QS [IO92AU, SK02]
G8	FGY	P G Griffiths, 5 Chestnut Cres, Carlton Colville, Lowestoft, NR33 8BQ [JO02UL, TM49]
G8	FGZ	C S Boon, 103 Lambley Ln, Burton Joyce, Nottingham, NG14 5BL [IO92LX, SK64]
G8	FHD	H O Williams, 29A Kingsnorth, Whiston, Prescot, L35 3TE [IO83OJ, SJ49]
G8	FHF	Details withheld at licensee's request by SSL.
G8	FHI	M Clarke, 2 The Grove, Penton Grafton, Andover, SP11 0RS [IO91FF, SU34]
GM8	FHK	J Gallacher, 23 East Ave, Carluke, ML8 5TS [IO85BR, NS85]
G8	FHN	Details withheld at licensee's request by SSL.
GM8	FHV	J A Heinrichsons, 16 Kimmeter Pl, Annan, DG12 6JU [IO84JX, NY26]
G8	FHY	M H J Castell, 5 Heyford Rd, Steeple Aston, Bicester, OX6 3SH [IO91IW, SP42]
G8	FIF	D P Howlett, 28 Mary Mead, Warfield, Bracknell, RG42 3SZ [IO91PK, SU87]
G8	FIG	C J Cole, White Gables, 157 Cherry Tree Rd, Beaconsfield, HP9 1BD [IO91QO, SU99]
G8	FIK	J L Brown, 6 Melton, Stantonbury, Milton Keynes, MK14 6BH [IO92OB, SP84]
G8	FIS	A Firth, 8 Lyndale Ave, Osbaldwick, York, YO1 3QB [IO93LW, SE65]
G8	FIW	Details withheld at licensee's request by SSL.
G8	FIZ	G Stubbings, Chelford Rd, Marthall, Ollerton, Knutsford, Ches, WA16 8SZ [IO83UG, SJ77]
G8	FJA	P W Webster, 3 Eden Ave, Bare, Morecambe, LA4 6QL [IO84OB, SD46]
G8	FJF	P O D Crocombe, 2 Brede Valley View, Icklesham, Winchelsea, TN36 4DA [JO00IW, TQ91]
G8	FJR	D A Jowett, 59 Old Rd, Thornton, Bradford, BD13 3DQ [IO93CS, SE13]
G8	FJV	Details withheld at licensee's request by SSL.
GW8	FKB	A L Williams, Hendy, Cerrigcewen, Bodorgan, Anglesey, LL62 5ED [IO73TF, SH47]
G8	FKC	Details withheld at licensee's request by SSL.
G8	FKF	C H Sargeant, Northview, South Marsh Rd, Stallingborough, Grimsby, DN37 8AN [IO93UN, TA10]
G8	FKH	D G Balharrie, 27 Norfolk Rd, Uxbridge, UB8 1BL [IO91SN, TQ08]
G8	FKL	G Twibell, Greenhill Cottage, Reading Rd, Moulsford, Oxon, OX10 9DR [IO91KO, SU68]
G8	FKP	C Simkins, Finches, Cherry Cl, Prestwood, Great Missenden, HP16 0QD [IO91PQ, SP80]
G8	FKV	G A Ash, 23 Lea Cres, Ruislip Gdns, Ruislip, HA4 6PN [IO91TN, TQ08]
G8	FKY	Details withheld at licensee's request by SSL.[Op: Beryl A Fletcher, Brambles, Hyde, Fordingbridge, Hants, SP6 2QL.]
G8	FKZ	D J Russell, Elmdale, Utterson View, Lowden, Chippenham, SN15 2RN [IO81WK, ST97]
G8	FLE	J A Ennis Carlisle&Dis AR, 30 Hillcrest Ave, Carlisle, CA1 2QJ [IO84NV, NY45]
G8	FLL	D L Roseaman, 101 Westbrook, Bromham, Chippenham, SN15 2EE [IO81XJ, ST96]
GI8	FLQ	R D Moore, 9 Vermont Ave, Conlig, Newtownards, BT23 7PF [IO74DN, J46]
G8	FLV	A H Nicholson, 48 Quaker Ln, Northallerton, DL6 1EE [IO94GI, SE39]
G8	FLX	M G Vaughan, 57 Strangeways, Watford, WD1 3SR [IO91SQ, TQ09]
G8	FMA	E Sillars, 34 Sandown Rd, Stevenage, SG1 5SF [IO91WW, TL22]
G8	FMC	D J Keston, 29 Buckingham Rd, Tring, HP23 4HG [IO91PT, SP91]
G8	FMD	C S Wells, Darwells, Talskiddy, St Columb, Cornwall, TR9 6EB [IO70MK, SW96]
G8	FME	A G Hilton, 28 Eastern Esplanade, Broadstairs, CT10 1DR [JO01RI, TR36]
G8	FMH	A C Stables, 32 Horwood Gdns, Basingstoke, RG21 3NR [IO91KG, SU65]
G8	FMI	F J Steed, 19 Chancery Ln, Debenham, Stowmarket, IP14 6RN [JO02OF, TM16]
G8	FMJ	J P Slater, 57 Freshbrook Rd, Lancing, BN15 8DE [IO90UT, TQ10]
GM8	FMR	D W Taylor, 9 Porterfield Rd, Inverness, IV2 3HW [IO77VL, NH64]
G8	FMT	P L E March, Devonholm, 2 Bedford Rd, Holwell, Hitchin, Herts, SG5 3RX [IO91UX, TL13]
G8	FMX	Details withheld at licensee's request by SSL.
G8	FMZ	P R McNamara, 86 York Rd, Teddington, TW11 8SN [IO91TK, TQ17]
G8	FNG	P H Robinson, 63 High St, Wolstamton, Newcastle, Staffs, ST5 0ER [IO83VA, SJ84]
G8	FNH	M A Nash, 62 Conbar Ave, Rustington, Littlehampton, BN16 3LY [IO90RT, TQ00]
G8	FNJ	M L Baker, 8 Wynton Rise, Stowmarket, IP14 2AB [JO02ME, TM05]
GW8	FNO	R J Gregory, 5 Bryn Castell, Radyr, Cardiff, CF4 8RA [IO81IM, ST18]
G8	FNU	A J Gilfillan Nottingham University Radio Am, 15 Willow Rd, Carlton, Nottingham, NG4 3BH [IO92LX, SK64]
G8	FNV	L P Bushby, 36 Goldsmith Rd, Worthing, BN14 8EP [IO90TT, TQ10]
GW8	FNX	W J A Huxley, No2 Bungalow, Nant Mawr Rd, Buckley, Clwyd, CH7 2BS [IO83KD, SJ26]
GW8	FOD	L R Hutchings, 7 Tegfan, Belgrano, Abergele, LL22 9YD [IO83FH, SH97]
G8	FOT	B M Butterworth, 21 Higher Dr, Purley, CR8 2HQ [IO91WH, TQ36]
G8	FOY	L H Oakes, 23 Campbell Cl, Penrhyn Bay, Llandudno, Conwy, LL30 3FD [IO83CH, SH88]
G8	FOZ	R Evans, 84 The Fairways, Leamington Spa, CV32 6PP [IO92FH, SP36]
G8	FPG	S J Banner, Oedhofstrasse 18, A-33Oo Amstetten, Austria, ZZ9 9BA
G8	FPU	R Hutton, 5 Tollemache Rd, Claughton, Birkenhead, L43 8SU [IO83LJ, SJ38]
G8	FPW	F H Brown, Oxcroft Bank, Shepeau Stow, Whaplode Drove, Spalding, Lincs, PE12 0TY [IO92XQ, TF31]
GI8	FQB	D J Allen The Queens University of Belfa, 40 Bramblewood Dr, Banbridge, BT32 4RA [IO64UI, J14]
G8	FQC	C A L Riggs, 19 Green Ln, Halesowen, B62 9LP [IO82XL, SO98]
G8	FQM	J Groom, 9 Ridgemoor Cl, Tilford Rd, Hindhead, GU26 6QX [IO91PC, SU83]
G8	FQN	R J Schneider, 15 Hope Ln, Upper Hale, Farnham, Surrey, GU9 0HY [IO91OF, SU84]
G8	FQS	Dr P W G Simpson, 17 Reynard Cl, Roffey, Horsham, RH12 4GX [IO91UB, TQ13]
G8	FQT	Details withheld at licensee's request by SSL.

G8

G8 FQV G J Thompson, 25 Meadow Ave, Codnor, Ripley, DE5 9QN [IO93HB, SK44]
G8 FQZ C A Stocker, 8 Brook Dr, Astley, Tyldesley, Manchester, M29 7HS [IO83SM, SD70]
G8 FRA Details withheld at licensee's request by SSL.
G8 FRB A Veitch, 126 Greythorn Dr, West Bridgford, Nottingham, NG2 7GA [IO92KW, SK53]
G8 FRC P W Milton, 103 Butts Hill Rd, Woodley, Reading, RG5 4NT [IO91NK, SU77]
G8 FRG J P Dunbar, Rushmede, 149 St. Neots Rd, Sandy, SG19 1BU [IO92UD, TL15]
G8 FRH P J J Lyall, 20 Horn Ln, Woodford Green, IG8 9AA [JO01AO, TQ49]
G8 FRL M A Wallace, Room 34,Flat 3, Rowan House, West Suffolk Hospital, Hardwick Ln, Bury St Edmunds, IP33 2QZ [JO02IF, TL86]
G8 FRO A P Mackay, 53 Regent St., Lancaster, LA1 1SH [IO84OB, SD46]
G8 FRS K J Gurr, The Tryst, 35 Shelley Rd, Stratford upon Avon, CV37 7JS [IO92DE, SP25]
G8 FRY Details withheld at licensee's request by SSL.
G8 FSJ R Page, 39 Carlton St., Kettering, NN16 8EB [IO92PJ, SP87]
G8 FSL A D S Benham, 15 South Lodge Dr, Southgate, London, N14 4XD [IO91WP, TQ29]
GU8 FSU V J Stockwell, -le-Chene Lodge-le-Chene Hill, Forest, Guernsey CI, Guernsey, GY9 9LE
G8 FSZ A H Othen, 33 Manor Dr, New Haw, Addlestone, KT15 3RJ [IO91RI, TQ06]
G8 FTI G R Evans, 5 Lord Knyvett Cl, Stanwell Village, Staines, TW19 7PF [IO91SK, TQ07]
G8 FTO Details withheld at licensee's request by SSL.
G8 FTP P D Jarrett, 15 Groveside Est, East Rudham, Kings Lynn, PE31 8RL [JO02IT, TF82]
G8 FTT Details withheld at licensee's request by SSL.
G8 FTW R G Goodchild, 48 Coral Dr, Ipswich, IP1 5HS [JO02NB, TM14]
G8 FTX D J Gotch, 25 Glen Rise, Baildon, Shipley, BD17 5DD [IO93CU, SE13]
G8 FUB L Jones, 52 New Ln, Aughton, Ormskirk, L39 4UD [IO83NN, SD40]
G8 FUH S R J Melling, 15 Woodbridge Hill, Gdns, Guildford, Surrey, GU2 6AR [IO91QF, SU95]
G8 FUJ P A French, Oakdene, Forward Green, Earl Stonham, Stowmarket, IP14 5HJ [JO02ME, TM06]
G8 FUL J Masterton, 29 Walnut Dr, Bishops Stortford, CM23 4JT [JO01BU, TL41]
G8 FUO R W Britton, 12 Bulkeley Ave, Windsor, SL4 3LP [IO91QL, SU97]
G8 FUQ Details withheld at licensee's request by SSL.
G8 FUT M W G Coles, 1 Church St., Taunton, TA1 3JE [IO81KA, ST22]
G8 FUV N Jenkinson, 77 Brosscroft, Hadfield, Hyde, SK14 7HE [IO93AL, SK09]
G8 FUY T E Withington, 27 Swan Meadow, Much Wenlock, TF13 6JQ [IO82RO, SO69]
G8 FVC D McLay, 55 Alexandra Rd, Sible Hedingham, Halstead, CO9 3NE [JO01HX, TL73]
GW8 FVI C C Reeves, Pennyfarthing, 6 Marine Terr, Criccieth, LL52 0EF [IO72VV, SH43]
G8 FVJ D J Still, 133 Feltham Rd, Ashford, TW15 1AB [IO91SK, TQ07]
G8 FVK P V Marks, 24 Abbey Crags Way, Knaresborough, HG5 8EF [IO94GA, SE35]
GM8 FVN G P Adams, Heath Ct, Morven Way, Monaltrie, Ballater, AB35 5SF [IO87LB, NO39]
G8 FVT D O Bainton, 86 Holywell Ave, Whitley Bay, NE26 3AD [IO95GB, NZ37]
G8 FW K E Walters, 8 Danesbower Ln, Blofield, Norwich, NR13 4LP [JO02RP, TG30]
G8 FWA J T Errington, 39 Highfield St., Fleckney, Leicester, LE8 8BD [IO92LM, SP69]
G8 FWE J H Maidment, 1 Minstead Dr, Yateley, GU46 6EH [IO91NI, SU86]
G8 FWF J A Bowers, 39 Upton Rd, Haylands, Ryde, PO33 3HP [IO90JR, SZ59]
G8 FWH J S Hill, 21 Somersby Rd, Woodthorpe, Mapperley, Nottingham, NG3 5QB [IO92KX, SK54]
G8 FWI G W Tysoe, 35 Offley Rd, Hitchin, SG5 2AZ [IO91UW, TL12]
G8 FWJ Details withheld at licensee's request by SSL.[Op: W A Rowlands-Lumb. Station located in London EC1.]
G8 FWK J Cranfield, 65 Broome Manor Ln, Swindon, SN3 1NB [IO91CN, SU18]
G8 FWM B L Phillips, 5 Wrexham Cl, North Park, Callands, Warrington, WA5 5RL [IO83QJ, SJ59]
G8 FWY P Russell, 57 Norburn Park, Witton Gilbert, Durham, DH7 6SG [IO94ET, NZ24]
G8 FXD Details withheld at licensee's request by SSL.
G8 FXG N C Lay, West Down Lodge, Bradworthy, Holsworthy, Devon, EX22 7RZ [IO70SV, SS31]
G8 FXM D A Toombs, 1 Chalgrove, Halifax Way, Welwyn Garden City, AL7 2QJ [IO91WT, TL21]
G8 FXU D R Percival, Trebakken, 11 Lamborne Cl, Sandhurst, Camberley, Surrey, GU17 8JL [IO91OH, SU85]
G8 FXV Dr M J White, 2 Mill Cl, Denmead, Waterlooville, PO7 6PE [IO90LV, SU61]
G8 FXX R M Limb, Sanfelice, Newlands Dr, Maidenhead, SL6 4LL [IO91OM, SU88]
G8 FYD F A Blake, 10 Meadow Way, Didcot, OX11 0AU [IO91IO, SU58]
G8 FYE Details withheld at licensee's request by SSL.
G8 FYH L G J Parkes, 273 Mersea Rd, Colchester, CO2 8PP [JO01KU, TM02]
GM8 FYJ M J Joyner, 1 Spey Walk, Holytown, Motherwell, ML1 4ST [IO85AT, NS76]
G8 FYK K G Payne, 39 Castleway, Pendleton, Salford, M6 7AL [IO83UL, SJ89]
GI8 FYP R H Miskelly, 29 Cherryvalley Park, Belfast, BT5 6PL [IO74BO, J37]
G8 FZA Details withheld at licensee's request by SSL.
G8 FZI M P Logsdon, Pilgrims Cottage, Langford Budville, Wellington, Somerset, TA21 0RE [IO80IX, ST12]
G8 FZT T Unsworth, Heathview, 15 Fenton Rd, Warboys, Huntingdon, PE17 2SD [IO92XJ, TL38]
G8 FZW J A Brown, 9 West St., Weedon, Northampton, NN7 4QU [IO92LF, SP65]
G8 FZZ S V Lamb, 27 Marsh Cl, Rushey Mead, Leicester, LE4 7TJ [IO92KQ, SK60]
GW8 GAB J W Richards, Sichar Isaf, Brongest, Newcastle-Emlyn, Dyfed, SA38 9ET [IO72SB, SN34]
G8 GAD Details withheld at licensee's request by SSL.
G8 GAG W K Ramsbottom, The Wagon House, Temple Grafton, Alcester, Warks, B49 6NS [IO92CE, SP15]
G8 GAJ J H N Niman, Nimans Ltd, 500 Broadway, Salford, M5 2UE [IO83UL, SJ79]
G8 GAR H C Taylor, 21 Windermere Rd, Coulsdon, CR5 2JF [IO91WH, TQ35]
G8 GAT M T E Smith, 17 Anthonys Ave, Poole, BH14 8JQ [IO90AR, SZ09]
GM8 GAX P G Howson, 1 Howetown, Fishcross, Alloa, FK10 3AW [IO86CD, NS89]
G8 GAZ T B Brady, 57 Green Ln, Great Barr, Birmingham, B43 5LE [IO92AM, SP09]
G8 GBE P A Richardson, 50 Amberley Rd, Gosport, PO12 4EW [IO90KT, SU60]
G8 GBH S W Hannah, 3 West Bank Mews, Kegworth, Derby, DE74 2TX [IO92IU, SK42]
G8 GBM R J Head, 29 Kingslea Rd, Solihull, B91 1TQ [IO92CJ, SP17]
G8 GBP C Fawdon, 21 Bevan Cl, Southampton, SO19 9PE [IO90HV, SU41]
G8 GBQ A L Fawdon, 2 The Riddings, Coventry, CV5 6AU [IO92FJ, SP37]
G8 GBU D G Barker, 311 Uttoxeter Rd, Mickleover, Derby, DE3 5AH [IO92FV, SK33]
G8 GBW J J Iggleden, Hangar Ct, Hawkley Hurst, Hawkley, Hants, GU33 6NS [IO91MB, SU73]
G8 GBY D A Robinson H.A.D.A.R.S., 2 Thornton Ave, Newstead St., Kingston upon Hull, Hull, HU5 3ND [IO93TS, TA02]
G8 GCL N E Waghorn, 39 Cumberland Ave, Broadstairs, CT10 1HU [JO01RI, TR36]
G8 GCM Details withheld at licensee's request by SSL.
G8 GCO N D C Wall, The Red House, North Cl, Ipswich, Suffolk, IP4 2TL [JO02OB, TM14]
G8 GCS C J Coker, 46 Clarendon Rd, Ipplepen, Newton Abbot, TQ12 5QS [IO80EL, SX86]
G8 GCT D M Balharrie, 17 Sampson Rd, North Walsham, NR28 9AR [JO02QT, TG22]
G8 GCU G D Drinkwater, Wrens Nook, 6 Wilcox Leys, Moreton Morrell, Warwick, CV35 9BG [IO92FE, SP35]
GM8 GCY J M Heron, 99 Queens Rd, Fraserburgh, AB43 9PT [IO87XQ, NJ96]
G8 GD H E Ward, 71 Great Tattenhams, Epsom Downs, Epsom, KT18 5RE [IO91VH, TQ25]
G8 GDC R F Laver, 243 Rundells, Harlow, CM18 7HQ [JO01BR, TL40]
G8 GDH D Brown, 56 Paddock Rd, Staincross, Mapplewell, Barnsley, S75 6LE [IO93FO, SE31]
G8 GDI R Dunn, 48 Stanley Hill, Amersham, HP7 9HL [IO91QP, SU99]
GM8 GDN M D Brunton, 2 Easter Pl, Portlethen, Aberdeen, AB1 4XL [IO87TH, NJ72]
G8 GDO Details withheld at licensee's request by SSL.
G8 GDZ R J Thompson, 23 Fox Hill, Selly Oak, Birmingham, B29 4AG [IO92AK, SP08]
G8 GEA K T Warriner, Windover, 16 The Ridgeway, Friston, Eastbourne, BN20 0EU [JO00CS, TV59]
GM8 GEC Details withheld at licensee's request by SSL.
G8 GEE R A Sherwood, 22 Sherlock Rd, Whoberley, Coventry, CV5 8EY [IO92FJ, SP37]
G8 GEF S J Edwards, Fernlea, Meathop, Grange-Over-Sands, Cumbria, LA11 6RB [IO84NF, SD48]
G8 GES A R Phillips, 39 Stonechat Rd, Billericay, CM11 2NZ [JO01FO, TQ69]
G8 GEZ L S Wooller, 4 Old Ct Cl, Brighton, BN1 8HF [IO90WU, TQ30]
G8 GF W A Higgins, 17 Ascham Pl, Meads, Eastbourne, BN20 7QQ [JO00DS, TV69]
G8 GFA R J Marshall, 1A Harrold Rd, Lavendon, Olney, MK46 4HU [IO92QE, SP95]
G8 GFF N Sanderson, 54 Kelvedon Cl, Chelmsford, CM1 4DG [JO01FS, TL70]
G8 GFH Details withheld at licensee's request by SSL.
GJ8 GFI G Dickman, -le-Valandre, Balmoral Mews, St. Helier, Jersey, JE2 4NJ
G8 GFS M F Winiberg, 5 Parc Villas, Newlyn, Penzance, TR18 5EA [IO70FC, SW42]
G8 GFW J Douglas, 1030 Shields Rd, Newcastle upon Tyne, NE6 4SR [IO94FX, NZ26]
G8 GFY D C King, 16 Oldfield, Honley, Huddersfield, HD7 2RL [IO93CO, SE11]
G8 GFZ T J Cockram, The Bungalow, Dyke Hill, South Chard, Chard, TA20 2PY [IO80MU, ST30]
G8 GG H M Fenton, 5 Cromer Rd, St. Annes, Lytham St. Annes, FY8 3HD [IO83LS, SD33]
G8 GGH J A Wood, 16 Ramptons Meadow, Tadley, RG26 3UR [IO91KI, SU66]
G8 GGI R B Geddes, 107 Dukes Ave, New Malden, KT3 4HR [IO91VJ, TQ26]
G8 GGM P A Burfoot, 18 Ember Rd, Langley, Slough, SL3 8ED [IO91RM, TQ07]
G8 GGR C M L Coleman, 18 Chester St., Coventry, CV1 4DJ [IO92FJ, SP37]
G8 GGS M Clarke, Ruskin Ashurst Dr, Boxhill, Tadworth, Surrey, KT20 7LS [IO91UG, TQ15]
GW8 GGW N J Dudman, 52 Gardden Rd, Rhosllanerchrugog, Wrexham, LL14 2EP [IO83LA, SJ24]
G8 GGZ D C Hall, The Bungalow, St. Johns Rd, Stalham, Norwich, NR12 9BQ [JO02SS, TG32]
G8 GHB A C Hunt, 21 Plumpton Gdns, Cantley, Doncaster, DN4 6SN [IO93LM, SE60]
G8 GHD Details withheld at licensee's request by SSL.
G8 GHF Details withheld at licensee's request by SSL.
G8 GHH C D Gibbs, 10 Waverley Rd, Westbrook, Margate, CT9 5QB [JO01QJ, TR37]
G8 GHK W B White, 60 Parklands, Hawkwell, Rochford, SS4 1SH [JO01IO, TQ89]
G8 GHL S J Garland, 53 The Cres, Horsham, RH12 1NA [IO91TB, TQ13]
G8 GHN J F Veaney Clifton ARS, 188 Shroffold Rd, Downham, Bromley, BR1 5NJ [JO01AK, TQ37]
G8 GHO J D Wood, 17 Yew Tree Park Rd, Cheadle Hulme, Cheadle, SK8 7EP [IO83VI, SJ88]
G8 GHQ P A Laverock, 7 Gr Gdns, Woodbridge, IP12 4LL [JO02PC, TM24]
G8 GHR B A Farey, 5 Ivel View, Sandy, SG19 1AU [IO92UD, TL14]
G8 GHT J A Sansum, 50 Farm Rd, Maidenhead, SL6 5JD [IO91OM, SU88]
GM8 GHV W Sherriffs, Hillcrest, Disblair, Newmachar, Aberdeen, AB5 0ND [IO87TH, NJ72]
G8 GIF K M Turner, 16 Waters View, Pelsall, Walsall, WS3 4HJ [IO92AP, SK00]
G8 GIG A S Patterson, Rose Cottage, Ingatestone Rd, Blackmore, Ingatestone, CM4 0RS [JO01DQ, TL60]
G8 GIH K W Foster, 52 Bottesford Ave, Scunthorpe, DN16 3EN [IO93QN, SE80]
G8 GIJ G P Eveleigh, 22 Reigate Ave, Sutton, SM1 3JL [IO91VJ, TQ26]
G8 GIK J S Hart, 28A Dunton Rd, Stewkley, Leighton Buzzard, LU7 0HZ [IO91OW, SP82]
G8 GIN J R Walker, 11 Burrett Gdns, Walsoken, Wisbech, PE13 3RP [JO02CQ, TF41]
G8 GIU J Harman, 13 Linthorpe Ct, South Shields, NE34 9BU [IO94GX, NZ36]
G8 GIY Details withheld at licensee's request by SSL.
G8 GIZ A S Warne Chester ARC, 113 Queens Rd, Vicars Cross, Chester, CH3 5HF [IO83NE, SJ46]
G8 GJA P R Reeves, 77 Cale Way, Wincanton, BA9 9BS [IO81SB, ST72]
G8 GJF D R Giltrow, 17 Brookfield Cl, Milton under Wychwood, Chipping Norton, OX7 6JQ [IO91EU, SP21]
G8 GJG N Giltrow, Bramington Farmhous, The Greens, Leafield, Oxford, OX8 5NJ [IO91FU, SP31]
G8 GJI Dr E H Smith, 6 Kingcup Dr, Bisley, Woking, GU24 9HH [IO91QH, SU96]
G8 GJM R J C Harwood, 9 Cornwall Cl, Woosehill, Wokingham, RG41 3AG [IO91NJ, SU76]
G8 GJO A D Heasman, 170 Plum Ln, Shooters Hill, London, SE18 3HF [JO01AL, TQ47]
G8 GJQ D S Grant, The Warrens, Knatts Valley Rd, Knatts Valley, Sevenoaks, TN15 6XX [JO01CH, TQ56]
G8 GJU M R Bernard, 22 Meadow Ln, Over, Cambridge, CB4 5NF [JO02AH, TL37]
G8 GJV T M England, 2 Ashmead Dr, Hardwick, Cambridge, CB3 7XT [JO02AF, TL35]
G8 GJW C J Drouet, Barn Lea, Holcot Rd, Brixworth, Northampton, NN6 9BS [IO92NH, SP77]
GI8 GJX S Bunting, 67 Church Cres, Glengormley, Newtownabbey, BT36 6ET [IO74AQ, J38]
G8 GKA Details withheld at licensee's request by SSL.
G8 GKC C J Ridley, 67 Kingfisher Rd, Larkfield, Aylesford, ME20 6RF [JO01FH, TQ65]
G8 GKH R Hadley, 36 Folly Ln, St. Pauls, Cheltenham, GL50 4BY [IO81XV, SO92]
G8 GKL C Rauch, 40 Russett Cl, Reffley Est, Kings Lynn, PE30 3HB [JO02FS, TF62]
G8 GKQ D G Crump, 24 Bower Green, Longcot, Faringdon, SN7 7TU [IO91EO, SU29]
G8 GKR M N Fellows, Richelds, 343 Wake Green Rd, Moseley, Birmingham, B13 0BH [IO92BK, SP08]
GW8 GKS J A Evans, 10 Bryn Dr, Castle Rd, Coedpoeth, Wrexham, LL11 3LJ [IO83LB, SJ25]
G8 GKT P A Peake, 34 Blackhalve Ln, Wednesfield, Wolverhampton, WV11 1BH [IO82WO, SJ90]
G8 GKU Details withheld at licensee's request by SSL.
G8 GKX D P Nicholson, Buena Vista, New Rd, Old Snydale, Pontefract, WF7 6HD [IO93HQ, SE42]
G8 GLB P R Brown, 4 King Edgar Cl, Ely, CB6 1DP [JO02DJ, TL58]
G8 GLC I S Cooper, 77A Benhill Wood Rd, Sutton, SM1 3SL [IO91VI, TQ26]
G8 GLI J A Husk, Bran Dhu, Common Moor, Liskeard, Cornwall, PL14 6EP [IO70SL, SX26]
GW8 GLO Details withheld at licensee's request by SSL.
G8 GLP B W Barnett, 21 Primrose Walk, Maldon, CM9 5JJ [JO01IR, TL80]
G8 GLQ C A Short, 175 Wellington Hill, West, Henleaze, Bristol, BS9 4QW [IO81QL, ST57]
G8 GLS G Wimlett, Unit 129 Brookfield, Pl Walton Summit, Bamber Bridge, Preston, PR5 8AE [IO83QR, SD52]
G8 GLV A A Brown, Temple Ln Cottage, Pleasant Stile, Littledean, Cinderford, GL14 3NX [IO81ST, SO61]
GW8 GLW Details withheld at licensee's request by SSL.
G8 GLX S Braidford, 25 Shepherd Leaze, Wotton under Edge, GL12 7LH [IO81TP, ST79]
G8 GLY A L Higgins, 92 Stamford Rd, Bournemouth, BH6 5DT [IO90CR, SZ19]
G8 GMA D Elliott, 56 Lincoln Ave, Willenhall, WV13 1JQ [IO82XO, SO99]
G8 GMB S H Bradshaw, 82 Arden Way, Market Harborough, LE16 7DD [IO92NL, SP78]
G8 GMC M H Harbach, 7 Beech Rd, Willenhall, WV13 3DD [IO82XO, SO99]
G8 GME R A Isaac, Pancake Hill, Lower Chedworth, Chedworth, Cheltenham, Glos, GL54 4AP [IO91BT, SP01]
G8 GMK L F Lowe, 10 Overlea Dr, Burnage, Manchester, M19 1LG [IO83VK, SJ89]
G8 GML P B Melbourne, 131 Sedgwick St., Cambridge, CB1 3AL [JO02BE, TL45]
G8 GMT R B Mainwaring, 48 Wallows Ln, Bescot Grange Estat, Walsall, WS2 9BY [IO92AN, SP09]
G8 GMU Details withheld at licensee's request by SSL.
G8 GNI Details withheld at licensee's request by SSL.
G8 GNJ S Walker, 30 Mount Ave, Heckmondwike, WF16 9PF [IO93ER, SE22]
G8 GNN M J Leeson, 14 Woodside Ave, Mansfield, NG18 4RH [IO93JD, SK55]
G8 GNX J Bartholomew, 33 Manor Way, Woodmansterne, Banstead, SM7 3PN [IO91WH, TQ25]
G8 GNZ G H Blake, 22 Cannon Leys, Galleywood, Chelmsford, CM2 8PD [JO01FQ, TL70]
GW8 GOC M G Black, 53 Cherry Orchard Rd, Lisvane, Cardiff, CF4 5UE [IO81JN, ST18]
G8 GOI Details withheld at licensee's request by SSL.
G8 GOJ A G Hobbs, 83 St. Peters St., South Croydon, CR2 7DG [IO91WI, TQ36]
G8 GOM A J G Ireson, Rose Lawn, The Avenue, Wellingborough, NN8 4ET [IO92PH, SP86]
G8 GON A W N Jefford, 37 Marions Way, Exmouth, EX8 4LF [IO80HP, SY08]
GW8 GOO P W Nelson, 15 Hill St., Bethesda, Bangor, LL57 3TD [IO73XE, SH66]
G8 GOP C Pearce, 85 Percy Rd, Hampton, TW12 2JT [IO91TK, TQ16]
G8 GOR A B Pearce, 153 Henver Rd, Newquay, TR7 3EJ [IO70LK, SW86]
G8 GOS K A Roche, Linchets, 76 Medstead Rd, Beech, Alton, GU34 4AE [IO91LD, SU63]
G8 GOV B C Ackroyd, 34 Springfield Rd, Baildon, Shipley, BD17 5LZ [IO93CU, SE13]
G8 GP E V E Neal, 34 Manor Ave, Brockley, London, SE4 1PD [IO91XL, TQ37]
G8 GPA Revd S G C Smith, The Vicarage, Cheriton Fitzpaine, Crediton, Devon, EX17 4JB [IO80EU, SS80]
G8 GPF D E Clark, 2 Cleeve Lodge Cl, Downend, Bristol, BS16 6AQ [IO81SL, ST67]
G8 GPH A R C Tutton, 34 Seward Rd, Badsey, Evesham, WR11 5HQ [IO92BC, SP04]
GW8 GQE J E Moore, Bryn Edin Dutlas, Knighton, Powys, Wales, LD7 1RW [IO82KJ, SO27]
G8 GQG J P Crow, 58 Cooden Dr, Bexhill on Sea, TN39 3AX [JO00FU, TQ70]
G8 GQJ R P Clark, 9 Conigree, Chinnor, OX9 4JY [IO91MQ, SP70]
G8 GQS B Summers, Cobwebs, 9 Prior Croft Cl, Camberley, GU15 1DE [IO91PH, SU85]
G8 GQT Details withheld at licensee's request by SSL.
G8 GRA Details withheld at licensee's request by SSL.[Op: N R Atkins. Station located near Chippenham.]
G8 GRB R L Day, 20 Linacre Rd, Watcombe, Torquay, TQ2 8LF [IO80FL, SX96]
G8 GRC J C Drakeley, 186 Conway Rd, Fordbridge, Birmingham, B37 5LD [IO92DL, SP18]
G8 GRD L M Hetherington, 5 Withey Cl West, Stoke Bishop, Bristol, BS9 3SX [IO81QL, ST57]
GD8 GRE C C R Wilkinson, The Warren, The Sloping Rd, Santon, Isle of Man, IM4 2HP
G8 GRP E A Poole, Ramillies Hall School, Ramillies Ave, Cheadle Hulme, Cheadle, SK8 7AJ [IO83VJ, SJ88]
G8 GRQ A R Plail, 7 Mornington Rd, Whiteshill, Whitehill, Bordon, GU35 9EN [IO91NC, SU73]
G8 GRT R V Oakley, 32 Windmill Cl, Ellington, Huntingdon, PE18 0AJ [IO92UH, TL17]
GW8 GRY P W Fear, 30 Beulah Rd, Rhiwbina, Cardiff, CF4 6LX [IO81JM, ST18]
G8 GRZ Details withheld at licensee's request by SSL.
G8 GS C W Farrell, 138 Broadway, Knaphill, Woking, GU21 2RL [IO91QH, SU95]
G8 GSL I D Liston-Brown, 355 Monmouth Dr, Sutton Coldfield, B73 6JX [IO92BN, SP09]
G8 GSM Details withheld at licensee's request by SSL.
G8 GSQ S R Thompson, 2399 PO Box, Reading, RG7 4FB [IO91KJ, SU66]
G8 GSU R H Wade, St John's Lodge, Church St, Crowthorne, Berks, RG11 7PD [IO91LJ, SU66]
GW8 GT F R Clare, Glenview, Newport Rd, Magor, Gwent, NP6 3BZ [IO81NO, ST48]
G8 GTC M V A E Coombs, 19 Burley Cl, Verwood, BH31 6TQ [IO90BU, SU00]
G8 GTD S C Porter, 20 Newbridge Rd, Ambergate, Belper, DE56 2GR [IO93GB, SK35]
G8 GTE W G C Austin, 33 Slingsby Gdns, Cochrane Park, Newcastle upon Tyne, NE7 7RX [IO94FX, NZ26]
G8 GTH J A Curzon, Cobs, 20 The Brow, Friston, Eastbourne, BN20 0ES [JO00CS, TV59]
G8 GTI K G Barnes, 75 Southmeade, Liverpool, L31 8EG [IO83MM, SD30]
G8 GTP J M Horrocks, Fairview, 17 Woodgrove, Whitfield, Manchester, M25 7ST [IO83UM, SD80]
G8 GTR D Murray, 27 Station Ave, Walton on Thames, KT12 1NF [IO91SI, TQ16]
G8 GTU Details withheld at licensee's request by SSL.
G8 GTV B S Raby, Hunthays, Awliscombe, Honiton, Devon, EX14 0QB [IO80JT, ST10]
G8 GTW W J L Stuart, 9 Charminster Cl, Great Sankey, Warrington, WA5 1JY [IO83QJ, SJ58]
G8 GTZ N M K Matthews, 12 Petrel Croft, Basingstoke, RG22 5JY [IO91KF, SU54]
G8 GUA R Wood, 339 Horse Rd, Hilperton, Hilperton Marsh, Trowbridge, BA14 7PE [IO81VI, ST85]
G8 GUH G C OHara, 2 Stroud Ct, Bottesford, Nottingham, NG13 0ER [IO92OW, SK83]
GM8 GUJ J Stubbs, 147 Caiyside, Edinburgh, EH10 7HR [IO85JV, NT26]
G8 GUL D J Brown, 21 Withcote Ave, Leicester, LE5 6SW [IO92LP, SK60]
G8 GUN H Parker, 67 Bustleholme Ln, Stone Cross, West Bromwich, B71 3BD [IO92AN, SP09]
G8 GUS M E Board, 99 Athelstan Rd, Bitterne, Southampton, SO19 4DE [IO90HV, SU41]
G8 GUU A Everitt, 9 Princes St., Toddington, Dunstable, LU5 6ED [IO91RW, TL02]
G8 GUV Details withheld at licensee's request by SSL.[Op: A B Ward. Station located near Harlow.]
GM8 GUX Mrth J M Thomson, 2 Wilton Hill, Hawick, TD9 8BA [IO85OK, NT51]
G8 GVG R E Johnson, 4 Ton Ln, Lowdham, Nottingham, NG14 7AR [IO93LA, SK64]
G8 GVN E F Shield, 14 Wellwood St., Amble, Morpeth, NE65 0EL [IO95FH, NU20]
G8 GVS F W Crook, 19 Scafell Dr, Norley Hall Est, Wigan, WN5 9TX [IO83PM, SD50]
G8 GVV P M Richmond, 57 The Fairway, Daventry, NN11 4NW [IO92KG, SP56]
G8 GVW P L Shillito, Little Orchard, Thorney Rd, Kingsbury Episcopi, Martock Somerset, TA12 6BG [IO80OX, ST42]
G8 GVZ L Sullivan, 89 Richmond Cres, Mossley, Ashton under Lyne, OL5 9LQ [IO83XM, SD90]
G8 GWJ J S Vincent, 12 Spelman Rd, Norwich, NR2 3NJ [JO02PO, TG20]
G8 GWK C R Cornell, 80 Walcot Ave, Round Green, Luton, LU2 0PR [IO91TV, TL12]
G8 GWL Details withheld at licensee's request by SSL.
G8 GWM N I Hay, 1 Orchard Gdns, Effingham, Leatherhead, KT24 5NR [IO91TG, TQ15]
G8 GWP G R Atkinson, 127 Davies Rd, West Bridgford, Nottingham, NG2 5HZ [IO92KW, SK53]
G8 GWR Dr S F Linney, 192 Thornhill Rd, Streetly, Sutton Coldfield, B74 2EP [IO92BN, SP09]
G8 GWT Details withheld at licensee's request by SSL.

G8 GWX R T Howells, 52 Upton Park, Upton, Chester, CH2 1DG [IO83NF, SJ46]
G8 GXF J Ashmore, 3 The Cedars, Stockwell Rd, Tettenhall, Wolverhampton, WV6 9AZ [IO82WO, SJ80]
G8 GXN K H Raynor, 17 Kirkstone Walk, Nuneaton, CV11 6EZ [IO92GM, SP39]
GM8 GXQ A C Mee, 9 Hatton Farm Gdns, Hatton, Peterhead, AB42 0QL [IO97BK, NK03]
GW8 GXW D G John, 7 Cambrian Terr, Borth, SY24 5HU [IO72XL, SN69]
G8 GYB V R Vesma, 7 Usborne Cl, Staplehurst, Tonbridge, TN12 0LD [JO01GD, TQ74]
G8 GYH A J Rice, 117 Manners Way, Southend on Sea, SS2 6QP [JO01IN, TQ88]
G8 GYI G C Stanley, 133 Park Ln, Kidderminster, DY11 6TE [IO82UJ, SO87]
G8 GYL I L Bishop, 2 Steepleton Water, Steepleton, Winterbourne Steepleton, Dorchester, DT2 9LN [IO80RR, SY69]
G8 GYM R D Claridge, 124 Pemdevon Rd, Croydon, CR0 3QP [IO91WJ, TQ36]
G8 GYO P B Hilton, 7 King George Ave, Droitwich, WR9 7BP [IO82WG, SO86]
G8 GYP V Holmes, 104 York Ave, Hayes, UB3 2TP [IO91SM, TQ08]
G8 GYQ Dr F J M Crossley, 4 Bollin Dr, Congleton, CW12 3SJ [IO83VD, SJ86]
G8 GYS P T Wright, 1 Lambourne Way, Thruxton, Andover, SP11 8NE [IO91FF, SU24]
G8 GYU A Oakes, Boskenwyn Chapel, Boskenwyn, Helston, Cornwall, TR12 7AB [IO70JC, SW62]
G8 GYZ Dr J R F Guy, 25 Station Rd, Wimborne, BH21 1RQ [IO90AT, SZ09]
G8 GZC G G Tew, Beech House, Pretty Oak, Chard, Somerset, TA20 3PT [IO80MV, ST31]
G8 GZI R M Fisher, 65 Kylemore Ave, Mossley Hill, Liverpool, L18 4PZ [IO83NI, SJ38]
G8 GZN A G Hill, 9 The Colliers, Heybridge Basin, Maldon, CM9 4SE [JO01IR, TL80]
G8 GZR R J Langdon, 42 Caldbeck Dr, Woodley, Reading, RG5 4LA [IO91NK, SU77]
G8 GZV R E Duke, 7 Headley Rd, Billericay, CM11 1BJ [JO01FP, TQ69]
G8 GZW Revd A J Davis, The Rectory, The St., Great Chart, Ashford, TN23 3AY [JO01KD, TQ94]
G8 GZX J S Eadie, 5 Silver St., Cublington, Leighton Buzzard, LU7 0LJ [IO91OV, SP82]
G8 GZZ N P Rew, 2 Wansford Green, Woking, GU21 3QH [IO91QH, SU95]
G8 HAD J V T Horsley, Fybough, Castlemaine, Co. Kerry, Ireland, XXXX XXX
G8 HAM S C Collins, 2 St. Teresas Dr, Chippenham, SN15 2BD [IO81WK, ST97]
G8 HAT J J McKenzie, 6 The Guelders, Purbrook, Waterlooville, PO7 5QT [IO90LU, SU60]
G8 HAU R F Lambarth, 38 Kirkley Park Rd, Lowestoft, NR33 0LG [JO02UL, TM59]
G8 HAV P P Fox, 5 Llandovery Cl, Winsford, CW7 1NA [IO83RE, SJ66]
G8 HAX Details withheld at licensee's request by SSL.
G8 HB L M Gunnell, 18 Borrowdale Rd, Malvern, WR14 2DS [IO82UC, SO74]
GM8 HBB J F Edwards, 58 Maxwellton Ave, East Kilbride, Glasgow, G74 3AF [IO75WS, NS65]
G8 HBE Details withheld at licensee's request by SSL.
GW8 HBP M J Bobby, Hafan, Church St., Penycae, Wrexham, LL14 2RL [IO83LA, SJ24]
G8 HBQ P A Davies, 24 Upland Gr, Leeds, LS8 2SX [IO93FT, SE33]
G8 HBR M K Cook, 5 Helmshore Rd, Haslingden, Rossendale, BB4 4BG [IO83UQ, SD72]
GM8 HBY C D Ross, 16 Glebe Cres, Airdrie, ML6 7DH [IO85AU, NS76]
G8 HBZ S J Stephenson, 6 Livingstone Cl, Rothwell, Kettering, NN14 6HT [IO92OK, SP88]
G8 HCJ A M Levett, 44 Chilkwell St., Glastonbury, BA6 8DA [IO81PD, ST53]
G8 HCK A Rutter, The Uplands, Castle Howard Rd, Malton, YO17 0NJ [IO94OD, SE77]
G8 HCL V P E Menday, Brindle Crest, Camp End Rd, St. Georges Hill, Weybridge, KT13 0NR [IO91SI, TQ06]
G8 HCO Details withheld at licensee's request by SSL.
G8 HCQ S B Watson, 6 Hope St., St. Annes, Lytham St. Annes, FY8 3SL [IO83LS, SD32]
G8 HCR G P Williams, 146 Minard Rd, Catford, London, SE6 1NJ [JO01AK, TQ37]
G8 HCU Details withheld at licensee's request by SSL.
G8 HCW C Morgan, 24 High Mead, Wootton Bassett, Swindon, SN4 8LW [IO91BN, SU08]
G8 HCZ Dr P J Iredale, 13 Meadow Cl, Panfield, Braintree, CM7 5AE [JO01GV, TL72]
G8 HDA R A O Jones, Gorswen, 29 Avon Dale, Newport, TF10 7LS [IO82TS, SJ71]
G8 HDD G Millington, 11 Unwin Cl, Fearnhill Park, Letchworth, SG6 3RS [IO91VX, TL23]
GW8 HDH J R Dowdall, 56 Goetre Bellaf Rd, Dunvant, Swansea, SA2 7RP [IO71XO, SS59]
G8 HDJ P W A Muxlow, 16 Station Rd, Grasby, Barnetby, DN38 6AP [IO93TM, TA00]
G8 HDK M L Balls, Swallow Ln, Tydd Gote, Wisbech, Cambs, PE13 5PQ [JO02BR, TF41]
G8 HDL M I Connell, 38 White Cl, High Wycombe, HP13 5NG [IO91OP, SU89]
G8 HDN Dr M A Sleightholm, 4 Bridge Ave, Otley, LS21 2AA [IO93DV, SE24]
G8 HDP R J Jenkins, 15 The Ct, Blanchmans Rd, Warlingham, CR6 9BT [IO91XH, TQ35]
G8 HDS P W Mac Kimm, 36 Links View, Rochdale, OL11 4DD [IO83VO, SD81]
G8 HDW R J Parnell, 76 St. Johns Rd, Launceston, PL15 7DE [IO70TP, SX38]
G8 HEB T B Brady, 120 Baltimore Rd, Great Barr, Birmingham, B42 1QL [IO92AM, SP09]
G8 HED Details withheld at licensee's request by SSL.
G8 HEJ N J W Arnold, Blackmore Rd, Stondon Massey, Hook End, Brentwood, Essex, CM15 0DT [JO01DQ, TQ59]
G8 HEL D E Cannell, 50 Bailey Cres, Fleetsbridge Park, Poole, BH15 3HA [IO90AR, SZ09]
G8 HEM Details withheld at licensee's request by SSL.[Op: M N Totham. Station located near Cheltenham.]
G8 HER A Lambert, 2 Huxley Cl, Locks Heath, Southampton, SO31 6RR [IO90IU, SU50]
GW8 HEZ R G Barrett, 7 St. Augustines Pl, Penarth, CF64 1BJ [IO81JK, ST17]
G8 HFL L S Caine, 25 Smallbrook Rd, Broadway, WR12 7EP [IO92BA, SP03]
G8 HFW T E Hall, 23 Burcott Gdns, Addlestone, KT15 2DE [IO91SI, TQ06]
GM8 HGD A M Jones, 22 Whin Park Circle, Portlethen, Aberdeen, AB1 4SS [IO87TH, NJ72]
G8 HGG P W Abernethy, Enfield House, Idlicote Rd, Halford, Shipstone on Stour, CV36 5DA [IO92EC, SP24]
G8 HGI M Warriner, 135 Showfields Rd, Tunbridge Wells, TN2 5UN [JO01DC, TQ53]
G8 HGL D Lambert, 4 Tamworth Rd, Bedford, MK41 8QY [IO92SD, TL05]
G8 HGM K Ellis, 11 Ringwood Cl, Eastbourne, BN22 8UH [JO00DS, TQ60]
G8 HGN R E Harrison, 59 Grange Rd, Great Burstead, Billericay, CM11 2RQ [JO01FO, TQ69]
G8 HGP W H Plucknett, 17 Skylark Corner, Poplars, Stevenage, SG2 9NL [IO91WV, TL22]
GM8 HGT Details withheld at licensee's request by SSL.
G8 HGX R Chettle, 18 Belgrave Rd, Slough, SL1 3RE [IO91QM, SU98]
G8 HHN H Workman, 4 Medway, Calf Cl Est, Jarrow, NE32 4DR [IO94GX, NZ36]
G8 HHO M J Strange, Invicta House, 141 Potton Rd, Biggleswade, SG18 0ED [IO92UC, TL14]
G8 HHP B K Ellison, 209 Woodbottom, Terr Walsden, Todmorden, Lancs, OL14 6PF [IO83WQ, SD92]
G8 HHQ M R Dyer, Flat, 12 Easter Ct, St. Johns Rd, Bournemouth, BH5 1EJ [IO90BR, SZ19]
G8 HHR J K Bardell, 239 Meadow Rd, Droitwich, WR9 9BZ [IO82WG, SO86]
G8 HHV Details withheld at licensee's request by SSL.[Op: W D Oscroft, 103 Bramcote Drive, Beeston, Notts NG9 1DU.]
G8 HHX Details withheld at licensee's request by SSL.
G8 HHZ P L Woods, 20 Ridge Way, Oadby, Leicester, LE2 5TN [IO92LO, SP69]
G8 HIK Details withheld at licensee's request by SSL.
G8 HIO T P Ellis, Hollybush House, Hawley Green, Blackwater, Camberley Surrey, GU17 9BS [IO91OH, SU85]
G8 HIP P J Peberdy, 56 Bethulie Rd, Derby, DE23 8UT [IO92GV, SK33]
G8 HIQ S Whitehouse, 2 Lindholme Dr, Littleworth Park, Rossington, Doncaster, DN11 0UR [IO93LL, SK69]
GI8 HIT Details withheld at licensee's request by SSL.
G8 HJD C J Tubis, Glovers, Tinkerpot Ln, West Kingsdown, Sevenoaks, TN15 6AB [JO01CH, TQ56]
G8 HJF C F Williams, Kingsclere House, Foxs Ln, Kingsclere, Newbury, RG20 5SL [IO91IH, SU55]
G8 HJG C A Williams, Kingsclere House, Foxs Ln, Kingsclere, Newbury, RG20 5SL [IO91IH, SU55]
G8 HJH M J Norton, 179B Kimbolton Rd, Bedford, MK41 8DR [IO92SD, TL05]
G8 HJJ Details withheld at licensee's request by SSL.
G8 HJS J G Harris, 57 Evesham Rd, Stratford upon Avon, CV37 9BA [IO92DE, SP15]
G8 HJT D J Husband, 51 High St., Fortuneswell, Portland, DT5 1JQ [IO80SN, SY67]
G8 HJV T J Lyne, 9 Bodle Cres, Bexhill on Sea, TN39 4BG [JO00FU, TQ70]
G8 HKE Details withheld at licensee's request by SSL.
G8 HKK M L North, 6 Flatwoods Cres, Claverton Down, Bath, BA2 7AH [IO81UI, ST76]
G8 HKN R D Meakins, 335 Ct Rd, Orpington, BR6 9BZ [JO01BI, TQ46]
GW8 HKY J L Gearey, y Berllan, Abergwili Rd, Carmarthen, SA31 2HH [IO71UU, SN42]
G8 HLE R E W Marshall, 54 Tudor Ave, Maidstone, ME14 5HJ [JO01GG, TQ75]
G8 HLF P Soesan, 51 Campion Way, Boughton Vale, Rugby, CV23 0UR [IO92JJ, SP57]
G8 HLH R L Wheeler, 14 Robins Ln, St. Helens, WA9 3NF [IO83PK, SJ59]
G8 HLQ E T Birch, 17 Canalside, Preston Brook, Ches, WA7 3AQ [IO83QH, SJ58]
G8 HMA R P Smith, 5 Newton Cl, Loughborough, LE11 5UU [IO92JS, SK52]
G8 HMG P J Walker, 12 Brownlow Rd, Redhill, RH1 6AW [IO91VF, TQ25]
G8 HMJ M A Kellett, Wistow Gate, Glen Rd, Newton Harcourt, Leicester, LE8 9FH [IO92LN, SP69]
G8 HMR Details withheld at licensee's request by SSL.
G8 HMV J R G Nicholas, 4 Lion Ln, Clee Hill, Ludlow, SY8 3NJ [IO82QI, SO57]
G8 HMZ P M Cheseldine, 6 Lissett Cl, Doddington Park, Lincoln, LN6 0SY [IO93QE, SK96]
G8 HNA S A Clark, 1 Roman Rd, Broadstone, BH18 9DF [IO80XS, SY99]
G8 HNB C E Clark, 1 Roman Rd, Broadstone, BH18 9DF [IO80XS, SY99]
G8 HNI A K Chennells, Copper Beech, 9 Huntercombe Lnes, Taplow, Maidenhead Berks, SL6 0PQ [IO91QM, SU98]
G8 HNK J Easteal, The Chalkers, Baydon Rd, Lambourn Woodlands, Hungerford, RG17 7TS [IO91FL, SU37]
G8 HNM R A Parker, 59 Waterloo Rd, Wellington, Somerset, TA21 8HY [IO80JX, ST12]
G8 HNO N Read, 31 Drake Ave, Teignmouth, TQ14 9NA [IO80FN, SX97]
G8 HNS Details withheld at licensee's request by SSL.
G8 HOA D W Watson, 30 Louis Way, Dunkeswell, Honiton, EX14 0XW [IO80JU, ST10]
G8 HOP M W Hopkins, 10 Spencer Cl, Church Crookham, Fleet, GU13 0EG [IO91OG, SU85]
G8 HOR E G York, Newton Villa, The Path, Great Bentley, Colchester, CO7 8PN [JO01MU, TM12]
G8 HOU H Cox, 21 North Ave, Hayes, UB3 2JE [IO91TM, TQ08]
G8 HPD G W Black, Celtic, Treligga Downs Rd, Delabole, PL33 9DL [IO70PO, SX08]
G8 HPF Dr I R McLenaghan, 145 Thorndon Gdns, Epsom, KT19 0QE [IO91UI, TQ26]
G8 HPJ P H Beaumont, 1 Byron Rd, Mexborough, S64 0DG [IO93IL, SE40]
G8 HPL Dr W G Taylor, Byewell, Chester Rd, Rossett, Wrexham, LL12 0HN [IO83MC, SJ35]
G8 HPN G J P Staniewicz, Meeting House, The Street, South Harting, Petersfield, GU31 5QB [IO90NX, SU71]
G8 HPR R D Evans, 14 Silverdale Terr, Highley, Bridgnorth, WV16 6LX [IO82TK, SO78]
G8 HPS A J Hancock, 9 Elmside, Willand Old Village, Willand, Cullompton, EX15 2RN [IO80HV, ST01]
G8 HPV D Green, 67 Coombe Park Rd, Binley, Coventry, CV3 2NW [IO92GJ, SP37]
G8 HPW M Hanaghan, 9 Goole Rd, Grindon, Sunderland, SR4 8HT [IO94GV, NZ35]
G8 HPY A M Mander, 8 Vesper Terr, Kirkstall, Leeds, LS5 3JP [IO93ET, SE23]
G8 HQA S G Phillips, 1577 Fairway Dr, Los Altos, Ca 94022, USA
G8 HQC N G Edmunds, 15 Parkfield Rd, Newbold on Avon, Rugby, CV21 1EW [IO92IJ, SP47]
GW8 HQM S M Bastow, Bryn Goleu, Rhosgadfan, Caernarfon, Gwynedd, LL54 7LB [IO73VC, SH55]
G8 HQO N Johnson, 56 Clarkson Ave, Wisbech, PE13 2EG [JO02CQ, TF41]
G8 HQP D P Kimber, 5 Rolan Dr, Shirley, Solihull, B90 1EH [IO92BJ, SP17]
G8 HQV K R Preston, 30 Bury Hill, Melton, Woodbridge, IP12 1LF [JO02PC, TM25]
G8 HQW P J Kirby, 2 Kneeton Park, Middleton Tyas, Richmond, DL10 6SB [IO94EK, NZ20]
G8 HQY S F Scanlon, 42 Trevanie Ave, Quinton, Birmingham, B32 1EX [IO92AL, SP08]
G8 HRA C Ryalls, 15 Belmont Way, South Elmsall, Pontefract, WF9 2BT [IO93IO, SE41]
G8 HRB R Orgill, 486 Littleworth Rd, Cannock, WS12 5JB [IO92AQ, SK01]
G8 HRC D L Nuttall Havering Dist Rd, Fairkytes Art Centre, 51 Billet Ln, Hornchurch Essex, RM11 1AX [JO01CN, TQ58]
G8 HRF K Dodman, 105 Pheasant Rise, Bar Hill, Cambridge, CB3 8SD [JO02AF, TL36]
G8 HRH Details withheld at licensee's request by SSL.
G8 HRI D R Darkes, 70 Braemar Rd, Lillington, Leamington Spa, CV32 7EY [IO92FH, SP36]
G8 HRK Details withheld at licensee's request by SSL.
G8 HRW S R Watkin, 9 Longden Cl, Haynes, Bedford, MK45 3PJ [IO92TB, TL14]
G8 HSG G D Cowling Goole Radio & Electronics Soci, Laissez Faire, Reedness, Goole, DN14 8ET [IO93OQ, SE72]
G8 HSH F J Chinn, 24 Kings Ct, The Esplanade, Bognor Regis, PO21 1NZ [IO90QS, SZ99]
G8 HSI J Carey, 7 Church Rd, Walton on The Naze, CO14 8DF [JO01PU, TM22]
G8 HSK D G Couzens, 195 Weald Dr, Furnace Green, Crawley, RH10 6NZ [IO91WC, TQ23]
G8 HSM T R Tubb, Ebeneezer, Hawkswood Rd, Downham, Essex, CM11 1JZ [JO01GP, TQ79]
G8 HSO Details withheld at licensee's request by SSL.
G8 HSR Details withheld at licensee's request by SSL.
G8 HST M P Sanders, 19 Brunswick Gdns, Hainault, Ilford, IG6 2QU [JO01AO, TQ49]
GM8 HSY Dr H M Reekie, 5 Golf Course Rd, Bonnyrigg, EH19 2EU [IO85KV, NT36]
G8 HTA K H C Parker, 20 River Ave, Hoddesdon, EN11 0JS [IO91XS, TL30]
G8 HTB A Barker, Attica, 116 Barnsley Rd, Wakefield, WF1 5NX [IO93GP, SE31]
G8 HTH G Calvert, 48 Lerowe Rd, Wisbech, PE13 3QL [JO02CQ, TF41]
G8 HTM J F Taylor, Perry House, 188 Walstead Rd, Walsall, WS5 4DN [IO92AN, SP09]
G8 HTN A Kettlety, 106 Denton Rd, Audenshaw, Manchester, M34 5BD [IO83WL, SJ99]
G8 HTO A G Farrell, 206 London Rd, Delarpe, Northampton, NN4 8AU [IO92NF, SP75]
G8 HTP F E Jones, 48 Birdwell Dr, Great Sankey, Warrington, WA5 1XA [IO83QJ, SJ58]
G8 HTQ S Fieldsend, 59 Lamb Ln, Monk Bretton, Barnsley, S71 2DX [IO93GN, SE30]
G8 HTR Details withheld at licensee's request by SSL.
G8 HTV M D L Brumwell, 8 Cambridge Gr, Kippax, Leeds, LS25 7JF [IO93HS, SE43]
G8 HUA Details withheld at licensee's request by SSL.
GI8 HUD T N L Huddleston, 29 North Parade, Belfast, BT7 2GF [IO74BN, J37]
G8 HUE P J Pryke, Unicorn House, 30 Burgess Pl, Martlesham Heath, Ipswich, IP5 7QZ [JO02OB, TM24]
G8 HUF S R Carpenter, 12 Rowans Cl, Farnborough, GU14 9EJ [IO91OH, SU85]
G8 HUG I Coulson, 56 Potterdale Dr, Little Weighton, Cottingham, HU20 3UX [IO93SS, SE93]
G8 HUH T M Rabbitts, Laurel Cottage, Wick Ln, Brent Knoll, Highbridge, TA9 4BU [IO81MG, ST35]
G8 HUL Details withheld at licensee's request by SSL.[Station located near Southsea.]
G8 HUO D Sharpe, 37 Oulton Ave, Bramley, Rotherham, S66 0SS [IO93IK, SK49]
GW8 HUS A J Mead, 12 Wyelands View, Mathern, Chepstow, NP6 6HN [IO81PO, ST59]
G8 HUT Details withheld at licensee's request by SSL.
G8 HUV M P Rowlands, 3 Littledown View, Great Durnford, Salisbury, SP4 6AU [IO91CD, SU13]
G8 HVA C K Gellion, Talbot Inn, Red Rd, Alton, Stoke on Trent, ST10 4BX [IO92BX, SK04]
G8 HVF C F Billson, Knotts End, Bateman Rd, East Leake, Loughborough, LE12 6LN [IO92JT, SK52]
G8 HVL F R Weston, 1 Church Ln, Rearsby, Leicester, LE7 4YE [IO92LR, SK61]
G8 HVO H T Volz, 32 Carlton Ave, Rose Green, Bognor Regis, PO21 3LR [IO90PS, SZ99]
G8 HVR Details withheld at licensee's request by SSL.
G8 HVS H J Brown, 97 Belgrave Rd, Eastwood, Leigh on Sea, SS9 4SL [JO01HN, TQ88]
G8 HVT M J Evans, 25 Walnut Cl, Nailsea, Bristol, BS19 2YH [IO81OK, ST46]
G8 HVV C M Goadby, 2 Boundary Rd, Red Lodge, Bury St. Edmunds, IP28 8JQ [JO02FH, TL77]
G8 HVX A V Staniforth, 25 Brown Ct, East Brunswick, Nj 08816, U S A, X X
G8 HVY Details withheld at licensee's request by SSL.
G8 HVZ R Anderson, 23 Callington Rd, Saltash, PL12 6DU [IO70VJ, SX45]
G8 HWA J Marshall, 145 Hykeham Rd, Lincoln, LN6 8AD [IO93RE, SK96]
G8 HWF D J Hurst, 3 Plaitford Walk, Maybush, Southampton, SO16 4EU [IO90GW, SU31]
G8 HWG Details withheld at licensee's request by SSL.
G8 HWJ E M Smith, 103 Aylesbury Rd, Wendover, Aylesbury, HP22 6JN [IO91PS, SP80]
GW8 HWL P Jenkins, 20 Dimbath Ave, Blackmill, Bridgend, CF35 6ED [IO81FN, SS98]
GW8 HWS J M S Mills, 1 Cumnock Pl, Cardiff, CF2 2AG [IO81KL, ST17]
G8 HWZ Details withheld at licensee's request by SSL.[Station located near Atherstone.]
G8 HXB Details withheld at licensee's request by SSL.[Op: G C Hibbert, Flat 5, 25 Chatham Place, Brighton, Sussex, BN1 3TN.]
G8 HXD M J Ledger, 58 Mount Pleasant Cl, Lightwater, GU18 5TR [IO91PI, SU96]
G8 HXE K M Haywood, 6 Lydney Rd, Flixton, Urmston, Manchester, M41 8RN [IO83TK, SJ79]
G8 HXR M J Brooke, 1207 Bourges Boulevard, Peterborough, PE1 2AU [IO92UO, TF10]
G8 HXT P A Randall, 22 Cunningham Ave, St. Albans, AL1 1JL [IO91UR, TL10]
G8 HXU R Longshawe, Caretakers Flat, Linacre College, St Cross Rd, Oxford, OX1 3TN [IO91JS, SP50]
GI8 HXY N Henderson, 8 Oldfort Park, Moira, Craigavon, BT67 0QD [IO64UL, J15]
G8 HYF J T Cook, 19 Southfield Cl, Rufforth, York, YO2 3RE [IO93JW, SE55]
GW8 HYI E Whitfield, 6 Lon y Berth, Cae Del, Mold, CH7 1QY [IO83KD, SJ26]
G8 HYK J B Brockwell, 1 Maxwell Cl, Lichfield, WS13 6TY [IO92CQ, SK10]
G8 HYL H J Tuff, 14 Tibbets Cl, Inner Park Rd, Wimbledon, London, SW19 6EF [IO91VK, TQ27]
G8 HYM S V Bradley, 3 Bernay Gdns, Bolbeck Park, Milton Keynes, MK15 8QD [IO92PB, SP84]
G8 HYP M R Peers, 12 Lower Icknield Way, Aston Clinton, Aylesbury, HP22 5JS [IO91PT, SP81]
G8 HYU D A Pepper, 52 Kingstylw Cl, Crick, Northants, NN6 7ST [IO92KI, SP57]
G8 HYZ R T Hunter, 1 Wetherby Cres, North Hykeham, Lincoln, LN6 8SX [IO93QE, SK96]
G8 HZJ R Ingamells, Moor View, Small Banks, Cocking Ln Adding, Moorside Ilkley, LS29 0QQ [IO93BW, SE04]
G8 HZL D Wildman, 2 Bluecoat Walk, Harmans Water, Bracknell, RG12 9NP [IO91PJ, SU86]
G8 HZN R Orchard, 26 Oswald Rd, Fetcham, Leatherhead, KT22 9TZ [IO91TH, TQ15]
G8 HZO N S Pratt, 13 Cambridge Gr, Otley, LS21 1DH [IO93DV, SE24]
G8 HZQ P J Healy, 63 Hazelwood Dr, St. Albans, AL4 0UP [IO91US, TL10]
G8 HZS T R Storey, 50 Longfield Rd, Darlington, DL3 0HX [IO94FN, NZ21]
GW8 HZW Details withheld at licensee's request by SSL.
GW8 IAD S N Bowen, 6 Penywaun Cl, St. Dials, Cwmbran, NP44 4JZ [IO81LP, ST29]
G8 IAJ J Richardson, 43 Front St., Leadgate, Consett, DH8 7SB [IO94CU, NZ15]
G8 IAK R W Thomas, 88 Parkway, West Wimbledon, London, SW20 9HG [IO91VJ, TQ26]
G8 IAM Details withheld at licensee's request by SSL.
G8 IAN M Lees, 62 Sandy Ln, Romiley, Stockport, SK6 4NH [IO83XK, SJ99]-
G8 IAR P Smith, 161 Leighton Ave, Leigh on Sea, SS9 1PX [JO01HN, TQ88]
G8 IAY Details withheld at licensee's request by SSL.
G8 IB D C Woodford, 29 Norman Ave, Abingdon, OX14 2HQ [IO91IQ, SU59]
G8 IBC D L Herke, 24 The Lawns, Cove, Farnborough, GU14 0RF [IO91OG, SU85]
G8 IBK M J Murray, Heads Nook Hall, Heads Nook, Carlisle, Cumbria, CA4 9AA [IO84OV, NY45]
G8 IBL H G Hallybone, Clifton, 24 Nightingale Rd, Godalming, GU7 3AG [IO91QE, SU94]
G8 IBO T M Gill, 10 Hawthornes, Tilehurst, Reading, RG3 6FN [IO91LJ, SU66]
G8 IBP R May, 10 Lime Cl, Wokingham, RG41 4AW [IO91NJ, SU86]
G8 IC M P Dawson, 60 Ashenhurst Rd, Todmorden, OL14 8DS [IO83WR, SD92]
G8 ICC A F Campbell, 67 Guilfords, Old Harlow, Harlow, CM17 0HX [JO01BS, TL41]
G8 ICD J M Trigg, 105 Broadfield, Harlow, CM20 3PX [JO01BS, TL41]
G8 ICJ J D Hickin, 210 Hough Rd, Pleck, Walsall, WS2 9BQ [IO82XN, SO99]
GW8 ICS D I Davies, 53 Crawshay St., Ynysybwl, Pontypridd, CF37 3EF [IO81HP, ST09]
GW8 ICT C P Hopley, Cynlas, Tabernacle St., Buckley, CH7 2JT [IO83LE, SJ26]
G8 IDE J B F Pimlott, 40 Queens Rd, St. Budeaux, Higher St. Budeaux, Plymouth, PL5 2NW [IO70VJ, SX45]
G8 IDJ I D Judd, 33 Coles Mede, Otterbourne, Winchester, SO21 2EG [IO91HA, SU42]
G8 IDK R C Voisey, 2 Chester Pl, Malvern, WR14 1RQ [IO82UD, SO74]
G8 IDL D A D Smith, The Old Forge, High St., Brinkley, Newmarket, CB8 0SE [JO02ED, TL65]
G8 IEA S C Parham, 132 Wrotham Rd, Gravesend, DA11 7LB [JO01EK, TQ67]
G8 IED Details withheld at licensee's request by SSL.
G8 IEI J K McKillop, 2 Moores Green, Wokingham, RG40 1QG [IO91OK, SU86]
G8 IEM M A Hall, Cotters, Old Lyndhurst Rd, Cadnam, Southampton, SO40 2NL [IO90FW, SU21]
G8 IER P G Nice, 41 Hillfoot Rd, Shillington, Hitchin, SG5 3NH [IO91TX, TL13]
G8 IEV B Guy, Hawthorn Folly, The-Cul-De-Sac, Stickford, Boston, Lincs, PE22 8EY [JO03AC, TF35]
G8 IEW C A Davies, Applecroft, St. Johns Rd, Slimbridge, Gloucester, GL2 7DF [IO81TR, SO70]
G8 IEZ C T Moss, 26 Somerset Ave, Wilpshire, Blackburn, BB1 9JD [IO83SS, SD63]
G8 IFF Details withheld at licensee's request by SSL.
G8 IFH K B Thomas, 1 Byways, Yateley, Camberley, Surrey, GU17 7NE [IO91OH, SU85]

G8 IFN N O Hinderwell, 3 Pine Gr, Witham, CM8 2NT [JO01HT, TL81]
G8 IFQ Details withheld at licensee's request by SSL.
G8 IFR Details withheld at licensee's request by SSL.
G8 IFT I L Gordon, 40 Grange Cres, Rubery, Rednal, Birmingham, B45 9XB [IO82XJ, SO97]
G8 IFU M G Allen, 4 Forest Way, Wimborne, BH21 7PB [IO90BT, SU00]
G8 IGE D J Russell, 90 Halleys Way, Houghton Regis, Dunstable, LU5 5HZ [IO91RV, TL02]
G8 IGM Details withheld at licensee's request by SSL.
G8 IGP Details withheld at licensee's request by SSL.
G8 IGQ D Bell Siemens Amateur Radio Club, Sports Office, Technology Dr, Beeston, Nottingham, Notts, NG9 1LA [IO92JW, SK53]
G8 IHA J E Gregory, 2 Abbey Dale Cl, Kilburn, Belper, DE56 0PY [IO93GA, SK34]
G8 IHF D F Cochrane, Bridge Farm, Draycote, Rugby, Warks, CV23 9RB [IO92HH, SP46]
GM8 IHT Details withheld at licensee's request by SSL.
GM8 IID N G Paterson, 4 Cambridge Rd, Renfrew, PA4 0SL [IO75TU, NS56]
G8 IIF Details withheld at licensee's request by SSL.
G8 IIG G W Punter, 18 Lodge Rd, Sharnbrook, Bedford, MK44 1JP [IO92RF, SP95]
GM8 IIH W Jarvie, Wester Auchinrivoch, Banton, Kilsyth, Glasgow, G65 0QZ [IO75XX, NS77]
G8 III L Roberts, 15 Henley Cl, Perrycrofts, Tamworth, B79 8TQ [IO92DP, SK20]
G8 IIK D J Hooker, Pennywood, Clarke Rd, Greatstone, New Romney, TN28 8PB [JO00LX, TR02]
G8 IIM P G Coates, 2 Gleneagles Dr, Tovil, Maidstone, ME15 6FH [JO01GG, TQ75]
GM8 IIO W I Robson, 38 Glebe Park, Duns, TD11 3EE [IO85TS, NT75]
G8 IIZ W J Rush, 17 Hagden Ln, Watford, WD1 8HQ [IO91TP, TQ19]
G8 IJC C J Phillipson, Bridge House, Lincoln Rd, Dunston, Lincoln, LN4 2EX [IO93TD, TF06]
G8 IJE B J Laxton, 1 Stoney Ln, Walsall, WS3 3RF [IO82XO, SJ90]
GI8 IJF N O Bailey, 20 Brunswick Rd, Bangor, BT20 3DU [IO74DP, J48]
G8 IJG D A J Adams, 77 Chestnut Cres, Shinfield, Reading, RG2 9HA [IO91MJ, SU76]
G8 IJI K Williamson, 4 Lynwood Dr, Sandal, Wakefield, WF2 7EF [IO93GP, SE31]
G8 IJP R J Rogers, 45 Church Rd, Westoning, Bedford, MK45 5LP [IO91SX, TL03]
G8 IJZ Details withheld at licensee's request by SSL.
G8 IK V D Morse, 42 Kingscote Rd, Dorridge, Solihull, B93 8RA [IO92CI, SP17]
G8 IKA D R Poll, 66 Southlands Ave, Orpington, BR6 9NF [JO01BI, TQ46]
G8 IKD Details withheld at licensee's request by SSL.
G8 IKG K Raper, 4 Gravel Rd, Church Crookham, Fleet, GU13 0BB [IO91OG, SU85]
G8 IKK J M Channon, 82 Lowther Rd, Bournemouth, BH8 8NS [IO90BR, SZ09]
G8 IKL Details withheld at licensee's request by SSL.
G8 IKP J S Grandshaw, 1 Melstock Ave, Preston, Weymouth, DT3 6JX [IO80SP, SY68]
G8 IKU A M G Jessop, 229 Russell Ct, Woburn Pl, London, WC1H 0ND [IO91WM, TQ38]
G8 ILB N A Allinson, 9 Aislaby Gr, Billingham, TS23 3QQ [IO94IO, NZ42]
G8 ILD R D Barrow, 6 Bean Leach Dr, Stockport, SK2 5HZ [IO83WJ, SJ98]
G8 ILG J Law, 29 Brackenwood, Orton Wistow, Peterborough, PE2 6YP [IO92UN, TL19]
G8 ILI A J Gibbings, 16 Turnberry Ave, Eaglescliffe, Stockton on Tees, TS16 9EH [IO94HM, NZ41]
G8 ILK J Bywaters, 2 Wansford Green, Woking, GU21 3QH [IO91QH, SU95]
G8 ILM R W Slater, Mozartstrasse 1, 76351 Linkenheim, Germany
G8 ILS Details withheld at licensee's request by SSL.
G8 ILU J F Parker, 6 Cedar Dr, Bourne, PE10 9SQ [IO92TS, TF02]
G8 ILW D V E Couse, First Sale Co The Boys Brigade, 6 Reading Dr, Sale, M33 5DL [IO83TK, SJ79]
G8 ILZ I R Walker, 51 Whitlock Dr, Wimbledon, London, SW19 6SJ [IO91VK, TQ27]
G8 IMB M Stubbs, Crofters, Harry Stoke Rd, Stoke Gifford, Bristol, BS12 6QH [IO81RM, ST67]
G8 IMC D N Chapman, 57 Dunster Rd, West Bridgford, Nottingham, NG2 6JE [IO92KW, SK53]
G8 IMH M H Fereday, 35 Manor House Park, Bilbrook, Codsall, Wolverhampton, WV8 1ES [IO82VO, SJ80]
G8 IMM R Keeley, 6 Standings Rise, Whitehaven, CA28 6SX [IO84FN, NX91]
G8 IMN D R W Fitch, 37 Ashfurlong Cres, Sutton Coldfield, B75 6EN [IO92CN, SP19]
G8 IMP J H Mallett, 22 Mill Ln, Shoreham By Sea, BN43 5AG [IO90UU, TQ20]
G8 IMR A C Talbot, 15 Noble Rd, Hedge End, Southampton, SO30 0PH [IO90IV, SU41]
G8 IMS M H Stroud, 9 Gloucester Rd, Guildford, GU2 6TG [IO91QG, SU95]
G8 IMX D N Jones, Borie Du Ritou, 46150 Lherm, France, X X
G8 IMZ A H Palfrey, 73 Rosedale Ave, York, YO2 5LH [IO93KW, SE55]
G8 INC K Davenport, 10 Woodend Ln, Hyde, SK14 1DT [IO83XK, SJ99]
G8 INE Details withheld at licensee's request by SSL.
G8 INI Details withheld at licensee's request by SSL.
G8 INK Details withheld at licensee's request by SSL.
G8 INL B G Miller, 1 The Meadows, Monk Fryston, Leeds, LS25 5PJ [IO93IS, SE43]
G8 INN P W Harding, 51 Hilda Wharf, Aylesbury, HP20 1RJ [IO91OT, SP81]
G8 INO A D Brown, 25 Birch Ln, Haxby, York, YO3 3RP [IO94LA, SE65]
G8 INT R M Lilley, 9 Mardale Cl, Congleton, CW12 2DQ [IO83VE, SJ86]
G8 INZ T R Prentice, 36 Ives Cl, Yateley, Camberley, Surrey, GU17 7RD [IO91OH, SU85]
G8 IOJ D G Martin, 54 The Crossway, Portchester, Fareham, PO16 8PB [IO90KU, SU60]
G8 IOK J Noden, 32 Wynter Rd, Bitterne, Southampton, SO18 6NY [IO90HW, SU41]
GM8 IOL R C Thomson, 50 Craigs Cres, Edinburgh, EH12 8HU [IO85IW, NT17]
G8 ION Dr J B Hollis, 5 Brierley Cl, Dunstable, LU6 3NB [IO91RU, TL02]
G8 IOS K W Evans, 12 Moxhull Dr, Walmley, Sutton Coldfield, B76 1LZ [IO92CM, SP19]
G8 IOW P S Wright, 70 Hardy Barn, Shipley, Heanor, DE75 7LY [IO93HA, SK44]
G8 IPF H Billingham, Tanglewood, Brookside Orchard, Ashington, West Sussex, RH20 3BD [IO90TW, TQ11]
G8 IPG A Shaw, Topaz, 330 Woodlands Rd, Woodlands, Southampton, SO40 7GF [IO90FV, SU31]
G8 IPQ A R C Badcock, 7 Heathfield Rd, Hiltingbury, Chandlers Ford, Eastleigh, SO53 5RP [IO91HA, SU42]
G8 IPS J A Walker, Great Lype Farm, Moor Ln, Charlton, Malmesbury, SN16 9DR [IO81XO, ST98]
G8 IPT P R Hughes, 27 Hemsworth Ave, Little Sutton, South Wirral, L66 4SG [IO83MG, SJ37]
GW8 IQC M J White, 5 Marlowe Cl, Rogerstone, Newport, NP1 0BT [IO81LO, ST28]
G8 IQF C M Newell, Berry House, 16 Pembroke Rd, Framlingham, Woodbridge, IP13 9HA [JO02QF, TM26]
G8 IQG D R Parkes, 4 Mayridge, Titchfield Common, Fareham, PO14 4QP [IO90IU, SU50]
G8 IQK Details withheld at licensee's request by SSL.
G8 IQM C B Sills, 38 Firs Rd, Edwalton, Nottingham, NG12 4BX [IO92KV, SK53]
G8 IQO P G Henley, 29 Springfield Cl, Westham, Pevensey, BN24 5JF [JO00DT, TQ60]
G8 IQP M R Challands, Braewinds, Shaw Ln, Holbrook, Derbys, DE5 0TG [IO93HA, SK44]
G8 IQT T C Spicer, 3 Parkers Fields, Quorn, Loughborough, LE12 8EJ [IO92JS, SK51]
G8 IQX M R Dixon, 57 Northease Dr, Hove, BN3 8PP [IO90VU, TQ20]
G8 IQY J W Chapman, 2 Greenlea Cl, Bebington, Wirral, L63 7RU [IO83MI, SJ38]
G8 IRC D L de Fraine, 31 Cobden Sq, Bedford, MK40 2JE [IO92SD, TL05]
GU8 IRF M T P Corbin, -le-Coudre Bungalow, Route Du Coudre, St Pierre Du Bois, Guernsey, Channel Islands, GY7 1HZ
G8 IRL Dr K D Brown, 56 Haydock Cl, Alton, GU34 2TL [IO91MD, SU73]
G8 IRN A Telford, 9 Fellside, Tower Wood, Windermere, LA23 3PW [IO84MH, SD39]
G8 ISE G Sharp, 46 Coronation St., Monk Bretton, Barnsley, S71 2ES [IO93GN, SE30]
G8 ISI F V Breame, 68 Church Rd, Bramshott, Liphook, GU30 7SH [IO91OC, SU83]
G8 ISM P R Goldsmith, 3 High Terr, Lenham, Kent, ME17 2QD [JO01IF, TQ85]
G8 ITB Details withheld at licensee's request by SSL.[Station located near Keston.]
GI8 ITD T R Davidson, 26 Lower Parklands, Dungannon, BT71 7JN [IO64OL, H76]
GU8 ITE D J Eaton, Glenfield-le-Foulon, St. Andrew, Guernsey, GY6 8UF
GW8 ITI J W Evans, Rosegarth, Tuckers Villas, Woodbine Rd, Blackwood Gwent, NP2 1QH [IO81JP, ST19]
GW8 ITO I A Strachan, 23 Cherry Orchard Rd, Lisvane, Cardiff, CF4 5UE [IO81JN, ST18]
G8 ITU P N Wragg, The Whey Inne Cottage, Main St., Farnsfield, Newark, NG22 8EA [IO93LC, SK65]
G8 ITW I A Seager, 183 Heathfield Rd, Hitchin, SG5 1TE [IO91UX, TL13]
G8 ITX O C Williams, Dormers, Cliff Rd, Waldringfield, Woodbridge, IP12 4QL [JO02PB, TM24]
G8 ITY P J Gittins, 55 Gloucester Rd, Tilgate, Crawley, RH10 5HR [IO91VC, TQ23]
GW8 ITZ J Austin RAF Sealand ARC, Ib4/Amw, 30 Mu, RAF Sealand, Deeside, Clwyd, CH5 2LS [IO83MF, SJ37]
G8 IUA S K H Clark, 19 Marlow Cres, Twickenham, TW1 1DD [IO91TK, TQ17]
G8 IUB D S Cottam Birmingham ARS, The School of Public Policy, J G Smith Building, University of Birmingham, Edgbaston, Birmingham, B15 2TT [IO92AK, SP08]
G8 IUC R Glover, 8 Woodberry Way, Chingford, London, E4 7DX [JO01AP, TQ39]
G8 IUD B J Sermons, 17 Wellside, Marks Tey, Colchester, CO6 1XG [JO01JV, TL92]
G8 IUK J D Baker, 16 Porthkerry Ave, Welling, DA16 2DT [JO01BL, TQ47]
G8 IUM M S Richardson, 27 Cell Farm Ave, Old Windsor, Windsor, SL4 2PD [IO91RL, SU97]
G8 IUP I Walukiewicz, Louise Cottage, Branksome Ave, Chilbolton, Stockbridge, SO20 6AH [IO91GD, SU33]
G8 IUQ M D Wareing, 106 Station Rd, Eccles, Manchester, M30 0GA [IO83TL, SJ79]
G8 IUT M G Dale ARC Nottingham, 2 Ward Ave, Mapperley, Nottingham, NG3 6EQ [IO92KX, SK54]
G8 IUV J A Johnston, 58 Sherbourne Cl, Chesterton, Cambridge, CB4 1RT [JO02BF, TL46]
G8 IVB P A Samson, 12 Queen Mary Rd, Upper Norwood, London, SE19 3NW [IO91WK, TQ37]
G8 IVC Details withheld at licensee's request by SSL.
G8 IVO R A Hartland, 44 Siddons Rd, Hampton Dene, Hereford, HR1 1XD [IO82PB, SO53]
GW8 IVT Details withheld at licensee's request by SSL.
G8 IWB A J Parker, 38 Boverton Dr, Brockworth, Gloucester, GL3 4DB [IO81WU, SO81]
GM8 IWC W H McKinlay, 36 Frogston Terr, Edinburgh, EH10 7AE [IO85JV, NT26]
G8 IWE R J Thomas, 47 Beakes Rd, Smethwick, Warley, B67 5RS [IO92AL, SP08]
G8 IWF T K Bierney, 5318 N 106 Ave, Glendale, Az 85307, USA, X X
G8 IWI J E Pearce, 34 Fleetwood Ave, Westcliff on Sea, SS0 9RA [JO01IN, TQ88]
G8 IWJ G R Strange, 12 Bronington Ave, Bromborough, Wirral, L62 6DT [IO83MH, SJ38]
G8 IWO N F Jones, 14 Salcombe Gr, Walcot, Swindon, SN3 1ER [IO91CN, SU18]
G8 IWQ A D Jacques, 20 Fairfield Dr, Codsall, Wolverhampton, WV8 2AB [IO82VP, SJ80]
G8 IWT R H Shears, 15 Hale Pit Rd, Bookham, Leatherhead, KT23 4BS [IO91TG, TQ15]
G8 IWX B S Homer, 164 Okehampton Cres, Welling, DA16 1DB [JO01BL, TQ47]
G8 IXA D F Hutchings, Northcot, 121 Chestnut Ln, Chesham Bois, Amersham, HP6 6DZ [IO91QQ, SU99]
G8 IXC L P Prior, 64 Montfort Rd, Walderslade, Chatham, ME5 9HA [JO01GI, TQ76]
G8 IXE Details withheld at licensee's request by SSL.
G8 IXK B R Owen, 105 Alexandra Ave, Luton, LU3 1HQ [IO91SV, TL02]
G8 IXL P R V Baker, Doules Mead, Heath Ln, Crondall, Farnham, GU10 5PA [IO91OF, SU84]
G8 IXP R J Lister, 8 Carlton Ave, Wilmslow, SK9 4EP [IO83VI, SJ88]
G8 IXX J A Brister, 49 Tiverton Rd, Loughborough, LE11 2RU [IO92JR, SK51]
G8 IXY M Shipton, 8 Linden Cl, Benfleet, SS7 4BD [JO01GN, TQ78]
GM8 IXZ A K Legood, 25 Frankfield Pl, Dalgety Bay, Dunfermline, KY11 5LR [IO86HA, NT18]
G8 IYA Details withheld at licensee's request by SSL.
G8 IYD D T Hancock, 12 Perrys Ln, Wroughton, Swindon, SN4 9AU [IO91CM, SU18]
G8 IYE P J Shore, Watsall House, Hopton Bank, nr Kidderminster, Worcester, DY14 0QB [IO82RJ, SO67]
G8 IYG P Gobey, Green Acre, Battle Ridge, Hopton, Stafford, ST18 0BG [IO82XT, SJ92]
G8 IYH A P Bevington, Malthouse, Hoggs Ln, Purton, Swindon, SN5 9HQ [IO91BO, SU08]
G8 IYJ C R Buckland, 40 Lineacre Cl, Grange Park, Swindon, SN5 6DB [IO91BN, SU18]
G8 IYK R J Sayers, 120 Birmingham Rd, Enfield, Redditch, B97 6EP [IO92AH, SP06]
G8 IYN C D Marsh, 6 de Burgh Hill, Dover, CT17 0BS [JO01PD, TR34]
G8 IYS J L Simkins, 18 Riding Hill, Sanderstead, South Croydon, CR2 9LN [IO91XH, TQ36]
G8 IYZ A J Barker, 8 Manor Ave, Attenborough, Beeston, Nottingham, NG9 6BP [IO92JV, SK53]
G8 IZR P I Higginson, 18 Park Meadow, Houghton, Westhoughton, Bolton, BL5 3UZ [IO83RN, SD60]
G8 IZW P R Cain, 22 Ditton Green, Luton, LU2 8RU [IO91TV, TL12]
G8 IZZ Details withheld at licensee's request by SSL.
G8 JAB A P Berriman, Meadowside, Little in Sight, St. Ives, TR26 1AX [IO70GE, SW54]
G8 JAC A J Jackson, 27 Ellesmere Dr, Hamsey Green, Sanderstead, South Croydon, CR2 9EH [IO91XH, TQ36]
G8 JAD J Townsend, 56 Seymour Rd, Northfleet, Gravesend, DA11 7BN [JO01EK, TQ67]
G8 JAG Details withheld at licensee's request by SSL.
G8 JAI A N Livesley, Beckgatehead, Barbon, via Carnforth, Lancs, LA6 2LJ [IO84RF, SD68]
G8 JAN P R Biggadike, 49 Willow Rd, Downham Market, PE38 9PG [JO02EO, TF60]
G8 JAO Details withheld at licensee's request by SSL.
G8 JAP J A Perschky, 30 Belgrave Rd, Ilford, IG1 3AW [JO01AN, TQ48]
G8 JAS Details withheld at licensee's request by SSL.
G8 JAU Details withheld at licensee's request by SSL.
GM8 JAV C M Taylor, 19/5 Damside, Dean Village, Edinburgh, EH4 3BB [IO85JW, NT27]
G8 JAW B S Heed, 3 Woodcote Green, Downley, High Wycombe, HP13 5UN [IO91OP, SU89]
G8 JAY A D Jay, 282 Hatherley Rd, Cheltenham, GL51 6HR [IO81WV, SO92]
G8 JBC C L Jervis, 15 Mercer Gr, Wednesfield, Wolverhampton, WV11 3AN [IO82XO, SJ90]
G8 JBD P A Godfrey, 3 Lowry Way, Lowestoft, NR32 4LW [JO02UL, TM59]
G8 JBE Details withheld at licensee's request by SSL.
G8 JBQ R J Hughes, Ct Church View, Pickett Ln, South Perrott, Beaminster, Dorset, DT8 3HU [IO80PU, ST40]
G8 JBT D Bellingham, 22 Princes Dr, Codsall, Wolverhampton, WV8 2DJ [IO82VO, SJ80]
G8 JBW A J Dodd, 27 Lancaster Rd, Newcastle, ST5 1DS [IO83VA, SJ84]
G8 JCB P J Pullinger, 264 Woodlands Rd, Woodland, Hants, SO40 7GH [IO90FV, SU31]
G8 JCC C M Purchase, 142 Reservoir Rd, Gloucester, GL4 6SA [IO81VU, SO81]
G8 JCD M R Northey, 132 Freemantle Rd, Bilton, Rugby, CV22 7HY [IO92II, SP47]
G8 JCL J E Essex, 40 Lincoln Walk, Heywood, OL10 3JB [IO83VO, SD81]
G8 JCN W Allen, 143 Cherry Cres, Rawtenstall, Rossendale, BB4 6DS [IO83UQ, SD82]
GM8 JCR R A Pontin, 24 Elm Ln, Perth, PH1 1EL [IO86GJ, NO02]
G8 JCT S R Lewis, 6 Greenfield Ave, Balsall Common, Coventry, West Midlands, CV7 7UG [IO92EK, SP28]
G8 JCU N W Taylor, 9 Balladine Rd, Anstey, Leicester, LE7 7BE [IO92JQ, SK50]
G8 JCV P L Hewitt, 28 Amersham Ave, Langdon Mills, Basildon, SS16 6SJ [JO01EN, TQ68]
G8 JCW Details withheld at licensee's request by SSL.
G8 JCZ E Tootle, 44 Lord St., Hindley, Wigan, WN2 3EW [IO83RM, SD60]
G8 JD Details withheld at licensee's request by SSL.
G8 JDQ K R C Few, 35 Whitton Cl, Swavesey, Cambridge, CB4 5RT [JO02AH, TL36]
G8 JDT S J Gale, 39 Thorley Park Rd, Bishops Stortford, CM23 3NG [JO01BU, TL41]
G8 JDW J D Ward, 147 Gainsborough Dr, Westcliff on Sea, SS0 0SN [JO01IN, TQ88]
G8 JEE E J Pumphrey, 103 Cambridge Rd, Raynes Park, West Wimbledon, London, SW20 0PU [IO91VJ, TQ26]
G8 JEF Details withheld at licensee's request by SSL.[Op: L Richardson. 68 Fairfield Road, Ipswich, IP3 9LB.]
G8 JEG J R Wiles, 1 Ullswater Cl, Liden, Swindon, SN3 6LH [IO91DN, SU18]
G8 JEI N A Cross, 15 Field Gate, Rossington, Doncaster, DN11 0YB [IO93LL, SK69]
G8 JEM E R Cheer, 15 Stibbs Way, Bransgore, Christchurch, BH23 8HG [IO90DS, SZ19]
G8 JET D R Higginson, 28 High St., Misterton, Doncaster, DN10 4BU [IO93NK, SK79]
G8 JFC F J Wilmott, 8 The Fairway, Offerton, Stockport, SK2 5DR [IO83WJ, SJ98]
G8 JFD Details withheld at licensee's request by SSL.
GM8 JFE T K Telfer, Deanview, Wilton Dean, Hawick, TD9 7HZ [IO85OK, NT41]
G8 JFF A G H Sibley, 43 Winterborne Rd, Abingdon, OX14 1AL [IO91IQ, SU49]
G8 JFJ Details withheld at licensee's request by SSL.[Op: C M Parry. Station located near Waterlooville.]
G8 JFL D J Crough, 32 Roundaway Rd, Ilford, IG5 0NP [JO01AO, TQ49]
G8 JFT N V Hewitt, 36 Princes Terr, Kemptown, Brighton, BN2 5JS [IO90WT, TQ30]
G8 JFX T J Simmons, 27 Blackbourne Rd, Elmswell, Bury St. Edmunds, IP30 9UH [JO02LF, TL96]
GM8 JFZ D Ferguson, 12 Hazelwood Rd, Strathaven, ML10 6HG [IO75XQ, NS64]
GM8 JGB W C Fleming, 65 Dundonald Park, Cardenden, Lochgelly, KY5 0DG [IO86ID, NT29]
G8 JGC Details withheld at licensee's request by SSL.
G8 JGE C J Newbury, 37 Johns Ave, Hendon, London, NW4 4EN [IO91VO, TQ28]
G8 JGF P C Walters, 3 Inkerman St., Selston, Nottingham, NG16 6BQ [IO93IB, SK45]
G8 JGH Details withheld at licensee's request by SSL.
GM8 JGI Details withheld at licensee's request by SSL.
G8 JGL N P Owen, 59 Fernwood Dr, Leek, ST13 8JA [IO83XC, SJ95]
G8 JGM J Martin, 17 Leys Ave, Cambridge, CB4 2AN [JO02BF, TL46]
G8 JGU Details withheld at licensee's request by SSL.
G8 JH G F Budden, 18 Mayfield Cl, Timperley, Altrincham, WA15 7TE [IO83UJ, SJ78]
G8 JHA T White, 24 Chapel St., Tingley, Wakefield, WF3 1RE [IO93FR, SE22]
G8 JHC I R Whitworth, 104 The Dormers, Highworth, Swindon, SN6 7PD [IO91DP, SU29]
G8 JHE M A Brogan, 23 High St., Aldreth, Ely, CB6 3PQ [JO02BI, TL47]
G8 JHH M P Baugh, 71 Hatch Ln, Old Basing, Basingstoke, RG24 7EF [IO91LG, SU65]
G8 JHL J M Lovell, 32 Larchwood Dr, Wilmslow, SK9 2NU [IO83VH, SJ88]
G8 JHM I J Carney, 39 Blenheim Cres, Luton, LU3 1HB [IO91SV, TL02]
G8 JHO P G Evans, 5 Hunters Cl, Bilston, WV14 7BN [IO82XN, SO99]
G8 JIC A Baker, 13 Strathville Rd, South Shields, London, SW18 4QX [IO91VK, TQ27]
G8 JIF G T Horsley, 54 Wellesley Rd, Ipswich, IP4 1PL [JO02OB, TM14]
G8 JII Details withheld at licensee's request by SSL.
G8 JIK R C Mann, 38 Hollow Wood, Olney, MK46 5LY [IO92PD, SP85]
G8 JIP G T Miller, 39 Scrivens Mead, Thatcham, RG19 4FQ [IO91JJ, SU56]
G8 JIS T Macey, 63 Church Rd, Codsall, Wolverhampton, WV8 1EH [IO82VP, SJ80]
G8 JIU P T Dunham, 19 The Lunds, Anlaby, Hull, HU10 7JJ [IO93SR, TA02]
G8 JIX Details withheld at licensee's request by SSL.
G8 JIY Details withheld at licensee's request by SSL.
G8 JJE B G C Thompson, 21 Birling Pl, Corby, NN18 0LZ [IO92PL, SP88]
GM8 JJN J A Pryde, 20 Thorndale Gdns, Allandale, Bonnybridge, FK4 2HG [IO85AX, NS77]
G8 JJR K McMahon, 5 Crossgates, Wadworth, Doncaster, DN11 9TE [IO93KL, SK59]
G8 JK R Chadbone, 7 Grafton Orchard, Chinnor, OX9 4DR [IO91NQ, SP70]
GW8 JKC J N Kendall, 23 Gardd Eryri, Dwygyfylchi, Penmaenmawr, LL34 6PW [IO83AG, SH77]
G8 JKD C M Littman, Watergate, Longbredy, Dorchester, Dorset, DT2 9HP [IO80QQ, SY58]
G8 JKF Details withheld at licensee's request by SSL.
G8 JKM A M Mears, 56 Corby Rd, Weldon, Corby, NN17 3HT [IO92QL, SP98]
G8 JKR J K Rigney, 35 Douglas Rd, Bedford, MK41 7YF [IO92SD, TL05]
G8 JKV D M Leary, The Farmhouse, Blackers Hill Farm, St Ives, Huntingdon Cambs, PE17 4NE [IO92XI, TL37]
G8 JLA K Turner, 13 Stanhope St., Saltburn By The Sea, TS12 1AL [IO94MN, NZ62]
G8 JLB B M Silver, 280 Britten Rd, Brighton Hill, Basingstoke, RG22 4HR [IO91KF, SU64]
G8 JLD J A Garters, Sun Patch, Garfield Rd, Hailsham, BN27 2BT [JO00DU, TQ50]
G8 JLE R M Langner, 113 Norton Ln, Sheffield, South Yorks, S8 8GX [IO93GH, SK38]
G8 JLM P R Higham, 56 Coopers Ave, Heybridge, Maldon, CM9 4YX [JO01IR, TL80]
G8 JLT Details withheld at licensee's request by SSL.
G8 JLV I P Gulliver, 20 Meadow Rd, Melksham, SN12 7AR [IO81WJ, ST96]
GW8 JLY L Leach, 4 Ollivant Cl, Danescourt, Llandaff, Cardiff, CF5 2RJ [IO81JM, ST17]
G8 JM W G Hall, 48 Hawkdene, Chingford, London, E4 7PF [IO91XP, TQ39]
G8 JMA R D Mackay, 40 Highpark, Shevington, Wigan, Lancs, WN6 8DF [IO83PN, SD50]
G8 JMB J M Button, 16 Meadow Rise, Broadstone, BH18 9ED [IO80XS, SY99]
GM8 JME G F Drinkwater, 11 Bankton Park Wes, Murieston, Livingston, West Lothian, EH54 9BP [IO85GV, NT06]
G8 JMG J M Gartland, 175 Talbot St., Whitwick, Coalville, LE67 5AY [IO92HR, SK41]
G8 JMJ Details withheld at licensee's request by SSL.
G8 JMK D J Butler, 9 Rye Cl, Highridge, Bishopsworth, Bristol, BS13 8DP [IO81QJ, ST56]
G8 JMP D Beech, 8 Copthorne Dr, Lightwater, GU18 5TE [IO91QI, SU96]

G8

G8 JMU J W Potter, 15 Alterton Cl, Goldsworth Park, Woking, GU21 3DD [IO91QH, SU95]
G8 JMY D A Hugman, 13 Cold Harbour, North Waltham, Basingstoke, RG25 2BH [IO91JF, SU54]
G8 JNE K B Holborow, The Granary, Rull Farm, Whimple, Nr.Exeter Devon, EX5 2NX [IO80HS, SY09]
G8 JNJ Details withheld at licensee's request by SSL.
G8 JNR R H Hedderley, 17 Linford Cl, Handsacre, Rugeley, WS15 4EF [IO92BR, SK01]
G8 JNS Details withheld at licensee's request by SSL.
G8 JNZ Details withheld at licensee's request by SSL.[Station located near Barnehurst.]
GI8 JOA D S C Thompson, 16 Lynden Gate Park, Portadown, Craigavon, BT63 5YJ [IO64SK, J05]
GI8 JOD Details withheld at licensee's request by SSL.
G8 JOJ Details withheld at licensee's request by SSL.[Op: M J Cooper. Station located near Malmesbury.]
G8 JON A D Riley, 28 Hawker Way, Woodley, Reading, RG5 4PF [IO91NK, SU77]
G8 JOR Details withheld at licensee's request by SSL.
G8 JOX J A Dobson, 14 Lawyers Cl, Evenley, Brackley, NN13 5SJ [IO92KA, SP53]
GW8 JOY T S Bowen, 7 Bedford Cl, Greenmeadow, Cwmbran, NP44 5HN [IO81LP, ST29]
G8 JPA J I Hunt, 59 Dawson Dr, Trimley St. Mary, Trimley, Ipswich, IP10 0YN [JO02PA, TM24]
G8 JPD A M R Sutton, Karena Gweek, Helston, Cornwall, TR12 6UB [IO70JC, SW72]
G8 JPJ D R Jones, 184 Harwich Rd, Little Clacton, Clacton on Sea, CO16 9PU [JO01NU, TM12]
G8 JPU D J Potts, 25 Southlands Rd, Congleton, Ches, CW12 3JY [IO83VD, SJ86]
G8 JPV M S Parnell, 1 Pope Rd, Wellingborough, NN8 3DW [IO92PH, SP86]
G8 JPW J B Abbott, 11 Red House Rd, Bodicote, Banbury, OX15 4BB [IO92IA, SP43]
G8 JQA W J Westlake, 102 Radipole Ln, Southill, Weymouth, DT4 9RT [IO80SO, SY68]
G8 JQB J M Williams, Wallop Smithy, Westbury, Shrewsbury, Salop, SY5 9RT [IO82MP, SJ30]
G8 JQG J Hough, 72 Pennine Ct, Macclesfield, SK10 2RN [IO83WG, SJ97]
G8 JQK Details withheld at licensee's request by SSL.
G8 JQS G T Greensmith, Japonica, Hawthorn Ave, Biggin Hill, Westerham, TN16 3SG [JO01AH, TQ45]
G8 JQV D C Marchant, 11 Derehams Ln, Loudwater, High Wycombe, HP10 9RH [IO91PO, SU99]
GW8 JQW R B Thomas, 121 Tyn y Tower, Baglan, Port Talbot, SA12 8YE [IO81CO, SS79]
G8 JR N P Haskins, Fen House, Castle St., Eye, IP23 7AW [JO02NH, TM17]
GI8 JRE J Donnelly, 9 Lomond Heights, Cookstown, BT80 8XW [IO64PO, H87]
G8 JRF M J Wilson, 19 The Willows, Highworth, Swindon, SN6 7PG [IO91DP, SU29]
GW8 JRL C R Jones, 21 Hallfield Cl, Flint, CH6 5HL [IO83KF, SJ27]
G8 JRN R E Stockdale, Linden Cottage, Eddington, Berks, RG17 0EU [IO91FK, SU36]
G8 JRW M J Austin, 57 Glynde Cres, Bognor Regis, PO22 8HT [IO90QT, SU90]
G8 JRZ A Mills, 42 Mora Ave, Chadderton, Oldham, OL9 0EJ [IO83WN, SD90]
G8 JSC K R Austin, 85 Sewall Highway, Wyken, Coventry, CV2 3NH [IO92GK, SP38]
G8 JSE F W Cowlin, Ululantes, 9 Zealand Cl, Hinckley, LE10 1TJ [IO92HN, SP49]
G8 JSF R H Williams, 35 Broadhurst Gr, Basing, Lychpit, Basingstoke, RG24 8SB [IO91LG, SU65]
G8 JSL P S Smith, 13 Manor Garth, Pakenham, Bury St. Edmunds, IP31 2LB [JO02JG, TL96]
G8 JSM C Wood, 240 Newton Rd, Winwick, Warrington, WA2 8QN [IO83QK, SJ69]
G8 JSN P F Bailey, 50 Amis Ave, New Haw, Addlestone, KT15 3ET [IO91RI, TQ06]
G8 JSR V J Hinksman, 23 Kings Rd, Whitley Bay, NE26 3BD [IO95GB, NZ37]
G8 JSW W J Wellington, 57 Hillcrest, Whitley Bay, NE25 9AF [IO95GB, NZ37]
G8 JTD H S Johnstone Otley ARS, 16 Riverside Cres, Otley, LS21 2RS [IO93DV, SE24]
G8 JTG E R Spanton, 41 Hillhouse Rd, Stone, Dartford, DA2 6HG [JO01DK, TQ57]
G8 JTI B A Storey, 13 West Cres, Matlock, DE4 3LB [IO93FD, SK26]
G8 JTJ Details withheld at licensee's request by SSL.
G8 JTQ J Kench, Rose Cottage, Middle Hill, Stroud, GL5 1NU [IO81VR, SO80]
G8 JTU Details withheld at licensee's request by SSL.
G8 JUC J M Wheatley, 44 Kingswood Cl, Boldon, Boldon Colliery, NE35 9LG [IO94GW, NZ36]
G8 JUF Details withheld at licensee's request by SSL.
G8 JUG N M Spenceley, 7 Cleveland Ave, Wimbledon Chase, London, SW20 9EW [IO91VJ, TQ26]
G8 JUK B T Storeton-West, Nazdar, Camps Heath, Oulton, Lowestoft, NR32 5DW [JO02UL, TM59]
G8 JUS T K Gale, 6 Falkland Rd, Wash Common, Newbury, RG14 6NY [IO91HJ, SU46]
G8 JUW Details withheld at licensee's request by SSL.
G8 JUX Details withheld at licensee's request by SSL.
GM8 JUY R S McMillan, 12 Parkthorn View, Dundonald, Kilmarnock, KA2 9EZ [IO75QN, NS33]
G8 JVD J Levy, 33 Crouchview Cl, Southend Rd, Wickford, SS11 8QB [JO01GO, TQ79]
G8 JVE M F J Rowe, 97 Old Worthing Rd, East Preston, Littlehampton, BN16 1DU [IO90ST, TQ00]
G8 JVI A J Hicks, 10 Evans Cl, Eynsham, Witney, OX8 1QY [IO91HS, SP40]
G8 JVJ K Hankinson, 12 Strathmore Rd, Doncaster, DN2 6DD [IO93KM, SE50]
G8 JVM R C Bown, Park View, Chapel St., Dawley, Telford, TF4 3DD [IO82SP, SJ60]
G8 JVS M Fairey, Boston Gates, Whinns Ln, Thorp Arch, Wetherby, LS23 7AL [IO93IV, SE44]
G8 JVU A Johnson, Clematis Cottage, Wheatlow Brook, Milwich, Stafford, ST18 0EW [IO82XV, SJ93]
G8 JVW P J Boswell, 12 Holloway Dr, Wombourne, Wolverhampton, WV5 0PA [IO82VM, SO89]
GM8 JVZ Dr M J Nimmo, The Ct, 6 Farington St., Dundee, DD2 1PJ [IO86LK, NO32]
G8 JWC D M Luscombe, 31 Tewkesbury Dr, Prestwich, Manchester, M25 0HR [IO83UM, SD80]
G8 JWD J M Rees, 4 Shalbourne Cl, Hungerford, RG17 0QH [IO91FJ, SU36]
G8 JWE J E C Hickman, 41 Field Rd, Ramsey, Huntingdon, PE17 1JP [IO92WK, TL28]
G8 JWI R Adams, 33 Lawrence Cl, Hertford, SG14 2HH [IO91WT, TL31]
G8 JWT R J Trett, East Town Barn, High St, Stoke Goldington, Newport Pagnall, Bucks, MK16 8NR [IO92OD, SP84]
G8 JX H G Fielding, 89 Chapel Lands, Alnwick, NE66 1ES [IO95DJ, NU11]
G8 JXC V W Morley, 1391 Leek Rd, Stoke on Trent, ST2 8BW [IO83WA, SJ94]
G8 JXG J M Dean, 6 Greenleas, Pembury, Tunbridge Wells, TN2 4NS [JO01DD, TQ64]
G8 JXK S J Blew, 9 Craig Lea, Taunton, TA2 7SY [IO81KA, ST22]
G8 JXP D P C McCabe, 78 Oakleigh Rd, Stratford upon Avon, CV37 0DN [IO92DE, SP15]
G8 JXV T I P Trew, Ringstone Lodge, 66 Oakwood Rd, Horley, Surrey, RH6 7BX [IO91WE, TQ24]
GM8 JYJ G G Russell, 8 Devon Gdns, Carluke, ML8 5DE [IO85BR, NS85]
G8 JYN A C Stables Basingstoke ARC, 32 Horwood Gdns, Basingstoke, RG21 3NR [IO91KG, SU65]
G8 JYR R T Gall, Lorien, Wetherby Rd, Rufforth, York, YO2 3QB [IO93JW, SE55]
G8 JYS M A Fletcher, 88 Langton Rd, Norton, Malton, YO17 9AE [IO94OD, SE77]
G8 JYV K Dumbill, 30 Caithness Dr, Crosby, Liverpool, L23 0RQ [IO83LL, SJ39]
G8 JYX P W Johnson, 42 College Gdns, Chingford, London, E4 7LG [IO91XP, TQ39]
G8 JZA Details withheld at licensee's request by SSL.
G8 JZC D G Bromby, Coombe House, Kersey Rd, Flushing, Falmouth, TR11 5TR [IO70LD, SW83]
G8 JZE Details withheld at licensee's request by SSL.
G8 JZI P C Smith, 23 Monks Dr, Withnell, Chorley, PR6 8SG [IO83RQ, SD62]
G8 JZJ A C Smith, 27 Ronald Rd, Beaconsfield, HP9 1AJ [IO91QO, SU99]
G8 JZL C W Hough, 112 Douglas Dr, Moreton, Wirral, L46 6BY [IO83KJ, SJ29]
G8 JZO J Gibbs, 81 Clements Cl, Spencers Wood, Reading, RG7 1HH [IO91MJ, SU76]
G8 JZT S R Osborn, 67 Chessington Ave, Bexleyheath, DA7 5NP [JO01BL, TQ47]
GW8 JZV G C H Edwards, 14 Fernfield, Port Talbot, SA12 8AL [IO81CO, SS79]
G8 JZZ Details withheld at licensee's request by SSL.
G8 KAE R D Bushell, 102 Winchester Gdns, Northfield, Birmingham, B31 2QB [IO92AJ, SP07]
G8 KAI Details withheld at licensee's request by SSL.
G8 KAM J Humandies, 70 Orchard Rise Wes, Sidcup, Kent, DA1 8SZ [JO01CK, TQ57]
G8 KAP D Patrick, Quarry Side, Stockdalewath, Dalston, Carlisle, Cumbria, CA5 7DP [IO84MT, NY34]
G8 KAV Details withheld at licensee's request by SSL.[Op: P M Hutchison.]
G8 KB P A Johnson, Sunnybank, 55 Rodney Hill, Loxley, Sheffield, S6 6SG [IO93FJ, SK38]
G8 KBB D A Roberts, 7 Rowanhayes Cl, Ipswich, IP2 9SX [JO02NB, TM14]
G8 KBC Details withheld at licensee's request by SSL.
G8 KBD D F Pike, 3 Oakfield Rd, West Bridgford, Nottingham, NG2 5DN [IO92KW, SK53]
G8 KBG A W Price, Botterham House, Botterham, Swindon, Dudley, DY3 4RA [IO82VM, SO89]
G8 KBH D J Ward, 3 Sherbourne Cl, Poulton-le-Fylde, FY6 7UB [IO83LU, SD34]
G8 KBM Details withheld at licensee's request by SSL.
GW8 KBO Details withheld at licensee's request by SSL.
G8 KBV Details withheld at licensee's request by SSL.[Op: D J Baxter.]
G8 KBZ Details withheld at licensee's request by SSL.
G8 KCB J A Nally, 313 Wyndhurst Rd, Stechford, Birmingham, B33 9DL [IO92CL, SP18]
G8 KCC S V Ludlow, 94 Cherry Tree Rd, Holtspur, Beaconsfield, HP9 1BH [IO91QO, SU99]
GW8 KCH K C Houston, 6 Ashgrove, Llanellen, Abergavenny, NP7 9HP [IO81LS, SO31]
GM8 KCK N G Robinson, 2 Ryehill Gr, Edinburgh, EH6 8ET [IO85KX, NT27]
G8 KCM Details withheld at licensee's request by SSL.
GM8 KCQ Details withheld at licensee's request by SSL.
GM8 KCS Details withheld at licensee's request by SSL.
G8 KCY M Bover, 27 Gainsborough Rd, Crawley, West Sussex, RH10 5LD [IO91VC, TQ23]
G8 KDD D Coton, 36 Arden Rd, Furnace Green, Crawley, RH10 6HS [IO91VC, TQ23]
G8 KDF M R Sach, 25 The Elms, Codicote, Hitchin, SG4 8XS [IO91VU, TL21]
G8 KDG Details withheld at licensee's request by SSL.
G8 KDI Details withheld at licensee's request by SSL.
G8 KDL S Whitt, Hunts Cottage, Kiln Ln, Buxhall, Stowmarket, IP14 3DU [JO02KE, TL95]
G8 KDM A Smith, 2A Chesterfield Rd, Barlborough, Chesterfield, S43 4TR [IO93IG, SK47]
G8 KDO Dr P J Topham, 13 Hawthorn Rd, Northampton, NN3 2JH [IO92NG, SP76]
G8 KDQ P F Pique, 41A Woodmansterne Rd, Coulsdon, CR5 2DJ [IO91WH, TQ25]
G8 KDU R D Eager, 45 Fleetwood Ave, Herne Bay, CT6 8QW [JO01NI, TR16]
G8 KEA M S Sutton, 178 Cole Ln, Borrowash, Derby, DE72 3GN [IO92HV, SK43]
G8 KED C Mullineaux, 27 Ashfield Ave, Lancaster, LA1 5EB [IO84OB, SD46]
G8 KEJ M W Johnson, 23 The Crest, Surbiton, KT5 8JZ [IO91UJ, TQ16]
G8 KET Details withheld at licensee's request by SSL.
G8 KFB Details withheld at licensee's request by SSL.
G8 KFD R D Gwynn, 227 Sketchley Rd, Burbage, Hinckley, LE10 2DY [IO92HM, SP49]
G8 KFF R F Parker, 17 Valley Rd, Streetly, Sutton Coldfield, B74 2JE [IO92BN, SP09]
GI8 KFG P A Douglas, 21 Hillhead Rd, Ballycarry, Carrickfergus, BT38 9HE [IO74CS, J49]
G8 KFJ D J Greig, West Manor Lodge, The St., Walberton, Arundel, BN18 0PJ [IO90QU, SU90]
G8 KFK P A Loten, 15 Hornsea Burton Rd, Hornsea, HU18 1TP [IO93WV, TA24]
G8 KFN R A Heron, 46 Bradvue Cres, Bradville, Milton Keynes, MK13 7AJ [IO92OB, SP84]
G8 KFQ M P Rudd, 3 The Cherry Pit, Littleworth Rd, Downley, High Wycombe, HP13 5FA [IO91OP, SU89]
G8 KFS B L Russell, 56 Kingsmead Ave, Tolworth, Surbiton, KT6 7PP [IO91UJ, TQ16]
G8 KGA Details withheld at licensee's request by SSL.
G8 KGC D G Barker Nunsfield Hse Rd, 311 Uttoxeter Rd, Mickleover, Derby, DE3 5AH [IO92FV, SK33]
G8 KGH M Cooper, 35 Cote Lea Park, Bristol, BS9 4AH [IO81QL, ST57]
G8 KGI M K Curran Fareham&Dist AR, 2 Bridges Ave, Paulsgrove, Portsmouth, PO6 4PA [IO90KU, SU60]
G8 KGK G V Higton, Hillcrest, Cowbrow, Lupton, Carnforth, LA6 1PJ [IO84PF, SD58]
G8 KGR R C Tidswell, Helloplane, Clubhurn Ln, Surfleet, Spalding, PE11 4BQ [IO92WU, TF22]
G8 KGV P M Jessop, 12 Highfield Ave, Harpenden, AL5 5UA [IO91TT, TL11]
G8 KHH C F C Young, 572 Falmer Rd, Brighton, BN2 6NA [IO90XU, TQ30]
G8 KHI R W Partridge, 34 Milestone Cl, Stevenage, SG2 9RR [IO91WV, TL22]
G8 KHJ Details withheld at licensee's request by SSL.
G8 KHN F J Vinnicombe, Birches, Whinwhistle Rd, East Wellow, Romsey, SO51 6BN [IO90FX, SU31]
G8 KHR Details withheld at licensee's request by SSL.
G8 KHT J W Clarkson, Ashfield Farm, Ashfield Rd, Castleton, Whitby, YO21 2EN [IO94ML, NZ60]
G8 KHU D M Fielding, 216 Andover Rd, Newbury, RG14 6PY [IO91HJ, SU46]
G8 KHV R E S Evans, 6 Park End, Lichfield, WS14 9US [IO92CQ, SK10]
G8 KIG P W Winwood, 2 The Warren, Abingdon, OX14 3XB [IO91IQ, SU59]
G8 KIH J C Sargent, 9 Lee Woottens Ln, Kingswood, Basildon, SS16 5HD [JO01FN, TQ78]
G8 KIK D W Bland, 17 Knowles Cl, Kirklevington, Yarm, TS15 9NL [IO94HL, NZ40]
G8 KIL G F L Redman, 2 Ridings Mead, Chippenham, SN15 1PG [IO81WL, ST97]
G8 KIO Details withheld at licensee's request by SSL.
G8 KIT Details withheld at licensee's request by SSL.
G8 KIW Details withheld at licensee's request by SSL.
G8 KJ A W Wright, The Old Dairy, Thames St., Sonning, Reading, RG4 6UR [IO91NL, SU77]
G8 KJA Details withheld at licensee's request by SSL.
G8 KJC Details withheld at licensee's request by SSL.
G8 KJJ L C Haywood, 9 Canberra Cres, Wilford Hill, West Bridgford, Nottingham, NG2 7FL [IO92KV, SK53]
GW8 KJK G Park, 34 Delafield Rd, Abergavenny, NP7 7AW [IO81LT, SO21]
G8 KJP P D King, 38 St. Bedes Gdns, Cherry Hinton, Cambridge, CB1 3UF [JO02CE, TL45]
G8 KJU P J Lynch, 6 Laburnum Cl, Marlow, SL7 3LF [IO91ON, SU88]
G8 KKA B D Stevens, 2 Hawthorn Cres, Shepton Mallet, BA4 5XR [IO81RE, ST64]
G8 KKD D R L Jones, Garsdon Mill, Gardsdon, Malmesbury, Wilts, SN16 9NR [IO81XO, ST98]
G8 KKF C R Hills, 8 Blackdale, Cheshunt, Waltham Cross, EN7 6DF [IO91XR, TL30]
G8 KKN D K McFarlane, 10 Green Ln, Vicars Cross, Chester, CH3 5LA [IO83NE, SJ46]
G8 KKS Details withheld at licensee's request by SSL.
G8 KKU J C Walker, 21 Garden Hedge, Leighton Buzzard, LU7 8DU [IO91QW, SP92]
G8 KKZ M Dickinson, Stoneleigh, 63 Davenport Ave, Hessle, HU13 0RN [IO93SR, TA02]
G8 KLA P G Webber, 60 Trowley Hill Rd, Flamstead, St. Albans, AL3 8EE [IO91ST, TL01]
G8 KLD Details withheld at licensee's request by SSL.
G8 KLE Details withheld at licensee's request by SSL.
G8 KLG R T Miflin, Lynwood, Beechfield Rd, Corsham, SN13 9DW [IO81VK, ST87]
G8 KLH G L Robotham, 38 Ennerdale Ave, Stanmore, HA7 2LD [IO91UO, TQ18]
G8 KLQ Details withheld at licensee's request by SSL.
G8 KLR Details withheld at licensee's request by SSL.
G8 KLX D J L Clowes, 54 Heathview Gordon, House Rd, London, NW5 1LR [IO91WN, TQ28]
G8 KMG L H Hipkin, 92 Webb Rise, Stevenage, SG1 5PD [IO91VV, TL22]
G8 KMH L P D Kellett, 79A Lower Icknield Way, Chinnor, OX9 4EA [IO91NR, SP70]
G8 KMK F B Craven Denby Dale ARS, 170 Lindley Moor Rd, Mount, Huddersfield, HD3 3UE [IO93BP, SE11][Correspondence to the Secretary: Eric Stewart, G0DBU.]
G8 KMM J M Bryant, 12 Dale Tree Rd, Barrow, Bury St. Edmunds, IP29 5AD [JO02HF, TL76]
G8 KMP M P Pollock, 25 Meadow Ln, Burgess Hill, RH15 9HZ [IO90WW, TQ31]
G8 KMR M J Davis, 8 Mead Cl, Leckampton, Cheltenham, GL53 7DX [IO81XV, SO92]
G8 KNF D R Hawkins, 189 Brettenham Rd, London, E17 5AX [IO91XO, TQ39]
GW8 KNG W J ONeil, 94 Bishops Rd, Whitchurch, Cardiff, CF4 1LY [IO81JM, ST17]
G8 KNJ T Blinco, 9 Powell Cl, Forest Hill, Oxford, OX33 1EN [IO91KS, SP50]
G8 KNM A Broomfield, 20 Beaumont Lawns, Marlbrook, Bromsgrove, B60 1HZ [IO82XI, SO97]
G8 KNN J Bigwood, 29 Harvey Goodwin Ave, Cambridge, CB4 3EX [JO02BF, TL45]
G8 KNQ S R Bacon, 59 Birchwood Ln, South Normanton, Alfreton, DE55 3DB [IO93HC, SK45]
G8 KNS M B R Jelfs, Adams Acre, Chapel Ln, Corfe Mullen, Wimborne, Dorset, BH21 3SL [IO80XS, SY99]
G8 KNU R M Jacobs, 20 Wilford Ave, Wakes Meadow, Northampton, NN3 9UQ [IO92OF, SP76]
GJ8 KNV Details withheld at licensee's request by SSL.
G8 KNX B S Harmer, 6 Greenside, Yarnfield, Stone, ST15 0NA [IO82VV, SJ83]
G8 KOC R F Backham, 15 Rushmead Cl, South Wootton, Kings Lynn, PE30 3LY [JO02FS, TF62]
G8 KOD R J Adams, Swinneys, Station Rd, Brize Norton, Carterton, OX18 3QA [IO91FS, SP30]
G8 KOE M Newell, 105 Oakcroft Gdns, Littlehampton, BN17 6LU [IO90RT, TQ00]
GM8 KOF D M McNaughton, Dunbar Rd, Haddington, EH41 3PJ [IO85OX, NT57]
G8 KOL D G Slocombe, 7 Talbot Ave, Studd Hill, Herne Bay, CT6 8AD [JO01NI, TR16]
G8 KOM D J Hanson, 42 Choseley Rd, Knowl Hill, Reading, RG10 9YT [IO91OM, SU87]
G8 KOQ N Morris, 88 Tynesbank, Little Hulton, Worsley, Manchester, M28 0SL [IO83TM, SD70]
G8 KOS S R Head, 3 Ripon Gdns, Waterlooville, PO7 8ND [IO90LV, SU61]
G8 KOV D W Dunn, Wayside Lodge, 8 Dursley Rd, Woodfield, Dursley, GL11 6PE [IO81TQ, ST79]
G8 KOX Details withheld at licensee's request by SSL.
G8 KPD B Fothergill, 53 Meadow Ct, Ponteland, Newcastle upon Tyne, NE20 9RA [IO95DB, NZ17]
G8 KPE E Howard, 15 Amherst Rd, Bexhill on Sea, TN40 1QH [JO00FU, TQ70]
G8 KPF S H Bandy, 5 Barrons Row, Harpenden, AL5 1SD [IO91TT, TL11]
G8 KPG G W Wright, 58 Lifton Croft, Kingswinford, DY6 8RZ [IO82WL, SO88]
GM8 KPH M R Hobson, 17 Well Brae, Pitlochry, PH16 5HH [IO86DQ, NN95]
G8 KPI G D Dann, 40 Marham Rd, Lowestoft, NR32 2SU [JO02UL, TM59]
G8 KPL Details withheld at licensee's request by SSL.
G8 KPM Dr H J Cummings, 4 Evelyn Pl, Bradville, Milton Keynes, MK13 7UG [IO92OB, SP83]
G8 KPN M E J Cummings, 121 Cleveland Rd, Ealing, London, W13 0EN [IO91UM, TQ18]
G8 KPS Details withheld at licensee's request by SSL.
G8 KPV G C Hickman, Calverton Rd, Blidworth Bottoms, Blidworth, Mansfield, Notts, NG21 0NW [IO93KC, SK55]
G8 KPY D P Pratt, 77 Hayfield Rd, St. Mary Cray, Orpington, BR5 2DL [JO01BJ, TQ46]
G8 KQA R J Laslett, Dinnages, St. End Ln, Broad Oak, Heathfield, TN21 8SA [JO00DX, TQ62]
GD8 KQF R N Ferguson, Moaney Moar House, Corlea Rd, Ballasalla, IM9 3BA
G8 KQH O C Harvey, 33 Copthall Way, New Haw, Addlestone, KT15 3TU [IO91RI, TQ06]
G8 KQV Dr S T Evans, 4 Holcot Ln, Anchorage Park, Portsmouth, PO3 5TR [IO90LT, SU60]
G8 KQW Details withheld at licensee's request by SSL.[Op: I G Lamb. Station located near Newbury.]
G8 KQZ G E J Dawkins, 8 Chancery Ln, Eye, Peterborough, PE6 7YF [IO92VO, TF20]
G8 KRA Details withheld at licensee's request by SSL.
G8 KRD S E Pike, 12 Rowan Dr, Wootton Bassett, Swindon, SN4 7ES [IO91BM, SU08]
G8 KRG Dr C G Harrison, 53 Peveril Cl, Whitefield, Manchester, M45 6NS [IO83UM, SD80]
G8 KRR N P Foster, 40 Landseer Ave, Tingley, Wakefield, WF3 1UE [IO93FR, SE22]
G8 KRU P A Horne, Hayne Ln, Weston, Gittisham, Honiton, Devon, EX14 0PD [IO80JS, SY19]
G8 KRV J B Cottier, 83 Elizabeth Dr, Leyfields, Tamworth, B79 8DE [IO92DP, SK20]
G8 KRY T E Hall, 2 Sheridan Cl, Narborough, Leicester, LE9 5QW [IO92JO, SP59]
G8 KSA W G Hall, 67 Selwyn Dr, Bishopsgarth, Stockton on Tees, TS19 8XF [IO94HN, NZ42]
G8 KSC D Goodwin, 41 Newpool Rd, Knypersley, Stoke on Trent, ST8 6NT [IO83VC, SJ85]
G8 KSD A D Hewett, 1 Mountside Cottage, Westfield Ln, Etchinghill, Folkestone, Kent, CT18 8BY [JO01NC, TR13]
GW8 KSE W C Salisbury, 28 Dyke St., Brymbo, Wrexham, LL11 5AH [IO83LB, SJ25]
GW8 KSF A P Salisbury, 28 Dyke St., Brymbo, Wrexham, LL11 5AH [IO83LB, SJ25]
G8 KSH A A Wilkins, 66 Waltham Gdns, Banbury, Oxon, OX16 8FD [IO92IB, SP44]
GM8 KSJ D M Cowie, 8 Centre St., Kelty, KY4 0EQ [IO86HD, NT19]
GM8 KSQ A C McElroy, Craiglea, Liquorstane, Falkland, Fife, KY7 7DQ [IO86KE, NO20]
G8 KST T J Mayer, 61 Rawley Cres, New Duston, Northampton, NN5 6PU [IO92MG, SP76]
G8 KSW J Wood, 38 Beech Ln, West Hallam, Ilkeston, DE7 6GU [IO92HX, SK44]
G8 KSX A P Thompson, 5 Rose Terr, Addingham, Ilkley, LS29 0NE [IO93BW, SE04]
G8 KSZ I M Newbold, 40 Heath Cl, Stonnall, Staffs, WS9 9HU [IO92BP, SK00]
G8 KTA P R Thomas, 28 The Green, Braunston, Daventry, NN11 7HW [IO92JH, SP56]
G8 KTB A G Platt, Weare St Gill, Ockley, Surrey, RH5 5NW [IO91TD, TQ13]
G8 KTC M A Rhys, 2 Sun Ln, Teignmouth, TQ14 8EF [IO80GN, SX97]
G8 KTE C Price, 4 Greenway Cl, Helsby, Warrington, WA6 0QX [IO83OG, SJ47]
G8 KTG D J P Smith, 76 Reigate Rd, Brighton, BN1 5AG [IO90WU, TQ20]
GW8 KTQ J F Read, The Thatch, Radyr, Cardiff, Sth Glam, CF4 8EA [IO81JM, ST17]
G8 KTV D C Adams, Pennington Cross Co, 1 North St., Lymington, SO41 8FY [IO90FS, SZ39]

G8 KTX M D Butler, 7 Bassett Rd, Coundon, Coventry, CV6 1LF [IO92FK, SP38]
G8 KUA C I Bridgland, 10 Eastlands Gr, Stafford, ST17 9BE [IO82WT, SJ92]
G8 KUC Dr K L Smith University of Kent A.R.S, Staple Farm House, Durlock Rd, Staple, Canterbury, CT3 1JX [JO01PG, TR25].
G8 KUE Details withheld at licensee's request by SSL.
G8 KUH Details withheld at licensee's request by SSL.
G8 KUL Details withheld at licensee's request by SSL.
GI8 KUO N P S Hetherington, The Dell, Nightingale Rd, Farncombe, Godalming, GU7 2HU [IO91QE, SU94]
G8 KUV A L Simonds, 12 Raymond Cl, Seaford, BN25 3DW [JO00BS, TV49]
G8 KUZ J Wiggins, 35 Downing Ave, Newcastle, ST5 0LB [IO83VA, SJ84]
G8 KVN A P Nelson, 29 Coxford Rd, Maybush, Southampton, SO16 5FG [IO90GW, SU31]
G8 KVO C G Miller, Broomwood, South Park, Sevenoaks, TN13 1EL [JO01CG, TQ55]
G8 KVP J Parkins, Bardon Manor, Washford, Watchet, Somerset, TA23 0PY [IO81HD, ST04]
G8 KVU C M Smith, 48 Sherbourne Cres, Coundon, Coventry, CV5 8LE [IO92FJ, SP37]
G8 KW R G Shears, 3 Windmill Park, Wrotham Heath, Sevenoaks, TN15 7SY [JO01EG, TQ65]
G8 KWD G A Bettley, 1 Dovetrees, Covingham Park, Swindon, SN3 5AX [IO91DN, SU18]
G8 KWI Details withheld at licensee's request by SSL.
G8 KWJ D R Barnwell, Bernagh, Duncombe St., Kingsbridge, TQ7 1LR [IO80CG, SX74]
GD8 KWM R D Corkill, 8 Hildesley Rd, Douglas, IM2 5AZ
G8 KWN R C Bryant, 63 Maple Way, Burnham on Crouch, CM0 8DN [JO01JP, TQ99]
G8 KWP A C H Darragh, 8 The Stables, Station Ln, Guilden Sutton, Chester, CH3 7SY [IO83OF, SJ46]
G8 KWR Details withheld at licensee's request by SSL.
G8 KWU J Austin, 5 Cranwell Rd, Greasby, Wirral, L49 3PP [IO83KJ, SJ28]
G8 KWV J W Bailey, 27 West Mead, Ewell Ct, Epsom, KT19 0BJ [IO91UI, TQ26]
G8 KWW G W Kennion, 7 Brayfield Rd, Littleover, Derby, DE23 6LD [IO92GV, SK33]
GM8 KXF G I R Robb, 82 Samson Ave, Kilmarnock, KA1 3ED [IO75SO, NS43]
G8 KXI J W Sargent, 26 Stratford Way, Watford, WD1 3DJ [IO91TP, TQ19]
G8 KXJ F G Vincent, 8 Carfax Cl, Sidley, Bexhill on Sea, TN39 5EG [JO00FU, TQ70]
G8 KXM M S Mollatt, 2A Field Ave, Baddeley Green, Stoke on Trent, ST2 7AS [IO83WB, SJ95]
G8 KXO B J Gamble, 62 Waine House, High St., Brownhills, Walsall, WS8 6DL [IO92AP, SK00]
G8 KXW J R Watts, 117 Beverley Rd, Ruislip Manor, Ruislip, HA4 9AN [IO91TN, TQ18]
G8 KYC Details withheld at licensee's request by SSL.
GI8 KYI T J Carlisle, 55 North Rd, Carrickfergus, BT38 8NA [IO74CR, J48]
G8 KYK C A Keens, Highcliffe, Old Hill, Woking, Surrey, GU22 0DF [IO91RH, SU95]
G8 KYM A A Paget, 19 Carlton Ave, Streetly, Sutton Coldfield, B74 3JF [IO92BO, SP09]
G8 KYT M L Hayward, 43 Jacobs Cl, Waterlooville, PO8 0PA [IO90LW, SU71]
G8 KYX A G Bond, Maple Lodge, Eyhurst Cl, Kingswood, Tadworth, KT20 6NR [IO91VG, TQ25]
GW8 KZA J P Wells, 30 St. Andrews Rd, Barry, CF62 8BR [IO81IJ, ST16]
G8 KZB O M Ward, Hartley, Green Ln, Milford, Godalming, GU8 5BG [IO91QE, SU94]
G8 KZG P M Delaney, 6 Eastview Cl, Wargrave, Reading, RG10 8BJ [IO91NM, SU77]
G8 KZJ E J Lockyear, 142 Cres Dr, Petts Wood, Orpington, BR5 1BE [JO01AJ, TQ46]
GW8 KZN W Clinton, 26 Aston Cres, Newport, NP9 5RA [IO81MO, ST38]
G8 KZO R J Edgeley, 17 Barleycroft, Cowfold, Horsham, RH13 8DP [IO91UB, TQ13]
G8 KZP F W Tuck, Whalebone Cottage, Whalebone Yard, Wells Next The Sea, NR23 1EH [JO02KW, TF94]
G8 KZS N S Smith, 9 The Bowley, Diseworth, Derby, DE74 2QL [IO92HT, SK42]
G8 KZY C J Denison, 40 Leysholme Dr, Leeds, LS12 4HQ [IO93ES, SE23]
G8 KZZ K A Farrow, 3 Stirling Way, Horsham, RH13 5RX [IO91UA, TQ22]
G8 LAB R E Harste, 7 Stewards Cl, Epping, CM16 7BU [JO01BQ, TL40]
G8 LAK A J Owen, 2 Walkham Terr, Horrabridge, Yelverton, PL20 7TR [IO70WM, SX57]
G8 LAM R J G Lambley, 31 Ridgeway Rd, Redhill, RH1 6PQ [IO91VF, TQ25]
G8 LAN R M Garner, 14 Copse Rd, Clevedon, BS21 7QL [IO81NK, ST47]
G8 LAX R S R Worton, 1 Kingswell Ride, Cuffley, Potters Bar, EN6 4LH [IO91WQ, TL30]
G8 LAY E R Hibbett, 5 Qualitas, Roman Hill, Bracknell, RG12 7QG [IO91OJ, SU86]
GM8 LBC C Dalziel, 9 Dunlop Ct, Low Waters, Hamilton, ML3 7YJ [IO75XS, NS75]
G8 LBG J J L Cook, Highlands, Littledown, Shaftesbury, SP7 9HD [IO81VA, ST82]
G8 LBI Details withheld at licensee's request by SSL.
G8 LBJ K Banks, 21 St. Marys Ave, Billinge, Wigan, WN5 7QL [IO83PL, SJ59]
G8 LBM P G Matthews, 15 Tennyson Way, Melton Mowbray, LE13 1LJ [IO92NS, SK72]
G8 LBS C P Ranson, 100 Stone Lodge Ln, West Chantry, Ipswich, Suffolk, IP2 9HR [JO02NB, TM14]
G8 LBV Details withheld at licensee's request by SSL.
G8 LCA J F Scott, 123 Cotswold Way, Tilehurst, Reading, RG3 6SR [IO91LJ, SU66]
G8 LCE M G Perrett, 23 Mancroft Rd, Caddington, Luton, LU1 4EJ [IO91SU, TL01]
GI8 LCJ D A Craig, 40 Chilton Rd, Carrickfergus, BT38 7JT [IO74CR, J48]
G8 LCL S R Tames, 21 Lind Cl, Earley, Reading, RG6 5QX [IO91MK, SU77]
G8 LCM K G Day, Powys Lodge, 6 Ct Rd, Strensham, Worcester, WR8 9LP [IO82WB, SO93]
G8 LCO Details withheld at licensee's request by SSL.
G8 LCP N D F Jamieson, 68 Knowles Ln, Holmewood, Bradford, BD4 9AP [IO93DS, SE13]
G8 LCS J F Monte, 11 Woodfield Ave, Hyde, SK14 5BB [IO83XK, SJ99]
G8 LCV D K Frankling, 1O Old Shipyard Centre, West Bay, Bridport, DT6 4HG [IO80OR, SY49]
G8 LCZ J F C Sellick, 9 Sandhurst Ave, Lytham St. Annes, FY8 2DA [IO83LS, SD32]
G8 LDB K A Oldham, 165 Mountsorrel Ln, Rothley, Leicester, LE7 7PU [IO92KR, SK51]
G8 LDC Dr J A Salthouse, 10 Ramillies Ave, Cheadle Hulme, Cheadle, SK8 7AL [IO83VJ, SJ88]
G8 LDJ C W Douglas, 22 Connaught Rd, Sittingbourne, ME10 1EH [JO01II, TQ96]
GI8 LDM Dr E E Richey, Lyndenlodge, Algeo Dr, Derrychara, Enniskillen, BT74 6JL [IO64EI, H24]
G8 LDP Details withheld at licensee's request by SSL.
G8 LDV B B Harrad, 32 Woodfield Ave, Northfleet, Gravesend, DA11 7QG [JO01EK, TQ67]
G8 LDW P Harness, 7 Castlegate, Gipsey Bridge, Boston, PE22 7BS [IO93WA, TF24]
G8 LDY R W Tompkins, 16 Garden Cl, Watford, WD1 3DP [IO91TP, TQ09]
GM8 LEA N J Adam, Bridaig Villa, Gladstone Ave, Dingwall, IV15 9PG [IO77SO, NH55]
G8 LEB R Hill, Rose Lodge, Colne Fields, Somersham, Huntingdon, PE17 3DL [JO02AJ, TL37]
G8 LED R W Guppy Northampton ARC, 4A Church St, Helmdon, Brackley, Northants, NN13 5QJ [IO92KB, SP54]
G8 LEI G M Buckley, 53 Wannock Ave, Willingdon, Eastbourne, BN20 9RH [JO00CT, TQ50]
G8 LEM R O Griffith, 9 Devonshire Rd, West Kirby, Wirral, L48 7HR [IO83JI, SJ28]
G8 LEO J G Ransome, 25 Hurst Rise Rd, Oxford, OX2 9HE [IO91IR, SP40]
G8 LER D T Steer, 56 Compton Ave, Mannamead, Plymouth, PL3 5DA [IO70WJ, SX45]
G8 LES M J Sanders, 39 Telegraph Ln, Four Marks, Alton, GU34 5AX [IO91LC, SU63]
G8 LF E M Byrne, 40 Wentworth Ave, Ascot, SL5 8HQ [IO91PJ, SU96]
GM8 LFB J M Rabbitts, Keepers House, Clythness Lighthouse, Lybster, Caithness, KW3 6BA [IO88JH, ND23]
G8 LFC K R King, 121 Godstone Rd, Whyteleafe, CR3 0EH [IO91XH, TQ35]
G8 LFN J Smith, 193 Armley Ridge Rd, Armley, Leeds, LS12 2QY [IO93ET, SE23]
G8 LFY A Penn, 46 Ladbrooke Dr, Potters Bar, EN6 1QR [IO91VQ, TL20]
G8 LGA R A Ward, 1 Horton, Downswood, Maidstone, Kent, ME15 8TN [JO01GG, TQ75]
G8 LGB J L Stephens, Preachers View, 16 Church St., Wells Next The Sea, NR23 1JA [JO02KW, TF94]
G8 LGC J C Williams, 24 Hilltop Gdns, Denaby Main, Doncaster, DN12 4SB [IO93IL, SK49]
G8 LGE P E Devine, 3 The Hawthorns, Outwood, Wakefield, WF1 3TL [IO93GR, SE32]
G8 LGI Details withheld at licensee's request by SSL.
G8 LGM R W Field, Gedney, 20 Hill Rd, Watlington, OX9 5AD [IO91MP, SU69]
G8 LGS P R Chitty, 109 Bannings Vale, Saltdean, Brighton, BN2 8DH [IO90XT, TQ30]
G8 LGT D J Blakemore, 20 Derwent Rd, Whitmore Park, Coventry, CV6 2HB [IO92FK, SP38]
G8 LGU R B Milliken, 15 Lee Gr, Chigwell, IG7 6AD [JO01AO, TQ49]
G8 LGX Details withheld at licensee's request by SSL.
G8 LGY R M Tyson, 19 Wood Ln, Gedling, Nottingham, NG4 4AD [IO92LX, SK64]
G8 LHD D A J Allen, 21 Goldings Cl, Haverhill, CB9 0EQ [JO02FC, TL64]
G8 LHF P P Earl, 38 Gerard Rd, Clacton on Sea, CO16 8FP [JO01NT, TM11]
GW8 LHO D C Hopkins, 31 The Links, Trevethin, Pontypool, NP4 8DG [IO81LR, SO20]
G8 LHP Dr A J Milne, 49 Cleevemount Rd, Cheltenham, GL52 3HF [IO81XV, SO92]
G8 LHQ M W Tuffrey, 50 Lynette Ave, London, SW4 9HD [IO91WK, TQ27]
G8 LHS D M Waters, 84 Littlehaven Ln, Horsham, RH12 4JB [IO91UB, TQ13]
G8 LHT I T Harwood, 38 Spring Cres, Sprotbrough, Sprotbrough, Doncaster, DN5 7QF [IO93JM, SE50]
G8 LHW P M Cunnington, 4 Hilltop Cl, Rayleigh, SS6 7TD [JO01HN, TQ89]
G8 LHZ P A Avon, 17 Moggs Mead, Petersfield, GU31 4NX [IO91MA, SU72]
G8 LIE N Borrell, 5 Cawthorn Cl, Hemlington, Middlesbrough, TS8 9RF [IO94JM, NZ51]
G8 LIH G B Storey, 27 Dyche Rd, Jordanthorpe, Sheffield, S8 8DQ [IO93GH, SK38]
G8 LII J W Lee, 225 Ave Rd, Rushden, NN10 0SN [IO92RG, SP96]
G8 LIK S Hurst, 16 Shaw Ln, Holmfirth, Huddersfield, HD7 1PY [IO93CN, SE10]
G8 LIL P Hurren, 4 Glyn Way, Threemilestone, Truro, TR3 6DT [IO70KG, SW74]
G8 LIP B P Greenbeck, 10 Campbell Ave, Bottesford, Scunthorpe, DN16 3SA [IO93QN, SE80]
G8 LIR A Rennison, 294 Bolton Rd, Hawkshaw, Bury, BL8 2PP [IO83UO, SD71]
G8 LIT Details withheld at licensee's request by SSL.
G8 LIU N R Clyne, 78 Halford Rd, Ickenham, Uxbridge, UB10 8QA [IO91SN, TQ08]
G8 LIX R D Keates, 35 Walsh Gr, New Oscott, Birmingham, B23 5XE [IO92BM, SP09]
GM8 LJ J Marshall, 36 James Sq, Crieff, PH7 3EY [IO86BI, NN82]
G8 LJC Details withheld at licensee's request by SSL.
GW8 LJJ E R Edwards, 11 Old Village Rd, Barry, CF62 6RA [IO81IJ, ST16]
G8 LJQ C P Asquith, 24 Greenacre Park, Hornsea, HU18 1UW [IO93VV, TA24]
G8 LJU J A Spicer, 42 Westminster Rd, Malvern Wells, Malvern, WR14 4ES [IO82TC, SO74]
G8 LJY A S Griffiths, 17 Ferenberge Cl, Farmborough, Bath, BA3 1DH [IO81SI, ST66]
G8 LKA S A N Whitehead, Plummers Cottage, Sally Deards Ln, Rabley Heath, Welwyn, AL6 9UE [IO91VU, TL21]
G8 LKI Details withheld at licensee's request by SSL.
GM8 LKL A J Hogg, Fairfield, 43 Muir Wood Rd, Currie, EH14 5JN [IO85IV, NT16]
G8 LKM Details withheld at licensee's request by SSL.
G8 LKP J L Duchscherer, 36 Hamdon Cl, Stoke Sub Hamdon, TA14 6QN [IO80PW, ST41]
G8 LKQ D R Falkner, 97 Foxhill, Olney, MK46 5HE [IO92PD, SP85]
G8 LKS D D Burton, 48 West Beeches Rd, Crowborough, TN6 2AG [JO01CB, TQ53]
G8 LKW H Colville, 7 Ockam Croft, Northfield, Birmingham, B31 3XP [IO92AJ, SP07]
GW8 LKX M T Corrigan, 3 Heathway, Heath, Cardiff, CF4 4JQ [IO81JM, ST18]
G8 LKZ G W Spray, 34 Manor Park, Dousland, Yelverton, PL20 6LX [IO70XM, SX56]
G8 LLB Details withheld at licensee's request by SSL.
G8 LLC G Riding, 30 Kendal Ave, Cullercoats, North Shields, NE30 3AQ [IO95GA, NZ37]
G8 LLD P E Pritchard, 74 Harcourt Ave, Sidcup, DA15 9LN [JO01BK, TQ47]
G8 LLJ M J Tutt, 52 School Ln, Dewsbury Moor, Dewsbury, WF13 4DU [IO93EQ, SE22]
G8 LLK Details withheld at licensee's request by SSL.
G8 LLU K P Austen, 101 Ufton Ln, Sittingbourne, ME10 1JA [JO01II, TQ96]
GM8 LLY Dr W J Muir, The Chestnuts, Eshiels, Peebles, EH45 8NA [IO85KP, NT24]
G8 LM J R Jennings, Leics Metre Wave Cg, Millside, Ullesthorpe, Lutterworth Leics, LE17 5DE [IO92IL, SP58]
G8 LMD Details withheld at licensee's request by SSL.
G8 LMF P A Rigby, 92 Albany Rd, Ansdell, Lytham St. Annes, FY8 4AR [IO83MR, SD32]
G8 LMI D J Morgan, 31 Raglan Precinct, Town End, Caterham, CR3 5UG [IO91WG, TQ35]
G8 LMN S Dodic, 22 Launceston Dr, Horeston Grange, Nuneaton, CV11 6GN [IO92GM, SP39]
G8 LMO M G Moody, 59 Helmton Rd, Sheffield, S8 8QJ [IO93GI, SK38]
G8 LMS L P Geering, 43 Croftlands, Hanging Heaton, Batley, WF17 6DG [IO93EQ, SE22]
G8 LMW C M Smith, 28 Fern Cres, Groby, Leicester, LE6 0BE [IO92JP, SK50]
G8 LMX S R Smith, 2 Acle Gdns, Ludham Ave, Bulwell, Nottingham, NG6 8NY [IO93JA, SK54]
G8 LMY D J Sweetland, 15 Wasdale Cl, Owlsmoor, Camberley, Surrey, GU15 4YQ [IO91PI, SU86]
G8 LNC D C Golding, 27 Wesermarsch Rd, Cowplain, Waterlooville, PO8 8JJ [IO90LV, SU61]
G8 LNE Details withheld at licensee's request by SSL.
G8 LNG D Severn, 20 Somerton Ave, Silverdale Est, Wilford, Nottingham, NG11 7FD [IO92KW, SK53]
GM8 LNH R A Pascal, 19 Clach Na Strom, Whiteness, Shetland, ZE2 9LG [IP90IF, HU34]
G8 LNQ C B Tindill, Hunters Moon, Station Rd, Newton-le-Willows, Bedale, DL8 1SX [IO94DH, SE28]
G8 LNS D G Wigens, 25 High View, Birchanger, Bishops Stortford, CM23 5QG [JO01CV, TL52]
G8 LNT Details withheld at licensee's request by SSL.
G8 LNU L A J Tucker, 23 Mill Rd, Denmead, Waterlooville, PO7 6PA [IO90LV, SU61]
G8 LOC J G F Locke, 64 Braemar Rd, Olton, Solihull, B92 8BS [IO92CK, SP18]
G8 LOE Details withheld at licensee's request by SSL.
G8 LOJ S Dorrington-Ward, Higher Dairy, Stoke Abbott, Beaminster, Dorset, DT8 3JT [IO80OT, ST40]
G8 LOK L E Currington, 14 Homerfield, Welwyn Garden City, AL8 6QZ [IO91VT, TL21]
G8 LOL Details withheld at licensee's request by SSL.
GM8 LON R B Bruce, 36 Mallaig Ave, Gowrie Park, Dundee, DD2 4TW [IO86LL, NO33]
G8 LOP P A Coomber, 10 Streeton Way, Earls Barton, Northampton, NN6 0HX [IO92PG, SP86]
G8 LOT A S White, 5 Langdon Way, Bermondsey, London, SE1 5QN [IO91XL, TQ37]
G8 LOU P G Mattos, Croft Cottage, Newham Bottom, Ruardean Woodside, Ruardean, GL17 9UB [IO81RU, SO61]
G8 LOW R McLachlan Lowe Electronic, Arc.Chesterfield Rd, Matlock, Derbyshire, DE4 5LE [IO93FD, SK36]
G8 LOZ J B Ramsay, Strathmore, 5 Parkhurst Rd, Guildford, GU2 6AP [IO91QF, SU95]
G8 LPA N J E Hilbery, 16 Albert Rd, Ashford, TW15 2LU [IO91SK, TQ07]
G8 LPC R E Cawley, 59 The Horseshoe, Hemel Hempstead, HP3 8QS [IO91SR, TL00]
G8 LPI R J Bray, 2 Hill Park, Headlands, Walsall, WS9 9RD [IO92AP, SK00]
G8 LPN K R Edwards, Stable Cottage, Main St., Poundon, Bicester, OX6 0BB [IO91LW, SP62]
G8 LPX C J Morgan, 43 Ferndown Rd, Brooklands, Manchester, M23 9AW [IO83UJ, SJ78]
G8 LQB W J Morrison, 92 Bury Hill, Melton, Woodbridge, IP12 1JD [JO02PC, TM25]
G8 LQF J A Pettifor, 12 Windmill Rd, Atherstone, CV9 1HP [IO92FN, SP39]
G8 LQL W J Cowell, 44 South Dr, Fulwood, Preston, PR2 9SR [IO83PT, SD53]
G8 LQM P P Green, 2 High St., Flitton, Bedford, MK45 5DU [IO92SA, TL03]
G8 LQO C McKenzie, 6 Pasturefield Cl, Sale, M33 2LD [IO83UJ, SJ89]
G8 LQP R Lines, 2 Salisbury Cl, Birmingham, B13 8JX [IO92BK, SP08]
G8 LQV Revd P A Barlow, St Georges House, Jumpers Rd, Christchurch, Dorset, BH23 2JR [IO90CR, SZ19]
G8 LQZ R H Banfield, 2 Laleham Cl, Eastbourne, BN21 2LQ [JO00DS, TV69]
G8 LRD P D Hutchings, 59 Braemor Rd, Calne, SN11 9DU [IO81XK, ST97]
GM8 LRI J A Graham, Thistledrum Cottage, Gowkhall, Dunfermline, Fife, KY12 9NX [IO86FB, NT08]
GW8 LRO A P Williams, 163 Garth Ave, Glyncoch, Pontypridd, CF37 3AD [IO81IO, ST09]
G8 LRS D R Massey, 63 West End, Brampton, Huntingdon, PE18 8SG [IO92VH, TL17]
GM8 LRV W S Read, 5 Caiystane Dr, Edinburgh, EH10 6SP [IO85JV, NT26]
G8 LSC P J Wheeler, 3 Oatfield Rd, Orpington, BR6 0ER [JO01BJ, TQ46]
G8 LSD A C Wyatt, 75 Millbrook Rd, Crowborough, TN6 2SB [JO01CB, TQ53]
G8 LSE K R Robertson, 82 Beddington Gdns, Carshalton, SM5 3HQ [IO91WI, TQ26]
G8 LSG B L Kaye, 37 Pleasant View, Darfield Rd, Cudworth, Barnsley, S72 8RZ [IO93HN, SE30]
G8 LSI R A N Dungan, 2 Lamorna Cl, Orpington, BR6 0TD [JO01BJ, TQ46]
G8 LSN D G E Harding, 68 Lagham Rd, South Godstone, Godstone, RH9 8HB [IO91XF, TQ34]
G8 LSS A D Tompson, 38 The Cres, Caddington, Luton, LU1 4JA [IO91SU, TL01]
G8 LSV Details withheld at licensee's request by SSL.
G8 LSZ J W Isham, 15 Pinner Park Gdns, Harrow, HA2 6LQ [IO91TO, TQ19]
G8 LTD S R Vaslet, 4 Coniston Cres, Redmarshall, Stockton on Tees, TS21 1HT [IO94HO, NZ32]
G8 LTK Details withheld at licensee's request by SSL.
G8 LTN A A T Brown, Casita, The Ridge, Cold Ash, Thatcham, RG18 9HT [IO91IK, SU56]
GW8 LTV G K Snellgrove, 142 Arail St., Six Bells, Abertillery, NP3 2NQ [IO81KR, SO20]
G8 LTW R A Warner, 60 Manor Rd, Streetly, Sutton Coldfield, B74 3NF [IO92BN, SP09]
G8 LTY A C W Harman, 6 Chaplin Walk, Gainsborough Meadow, Great Cornard, Sudbury, CO10 0YT [JO02JA, TL84]
G8 LU W A Hoskins, 53 Woodside Rd, South Norwood, London, SE25 5DP [IO91XJ, TQ36]
G8 LUB D J T Burrell, 5 Tregarthen, Treverbyn Rd, St. Ives, TR26 1HA [IO70GF, SW54]
G8 LUK N J L Macassey, 58 Stone St., Faversham, ME13 8PS [JO01KH, TR06]
G8 LUL R Myers, 33 Withenfield Rd, Manchester, M23 9BT [IO83UJ, SJ89]
G8 LUN D J Greenaway, Pentire, Stanley Rd, Bulphan, Upminster, RM14 3RX [JO01EN, TQ68]
G8 LUP A K Semark, 22 Morris Ave, Billericay, CM11 2LB [JO01FO, TQ69]
GI8 LUR A T Hewitt, 18 Knockview Ave, Doagh Rd, Newtownabbey, BT36 6TZ [IO74AQ, J38]
G8 LVB R J Hook, Heatherside, 8 Chalkpit Rd, Portsmouth, PO6 4EX [IO90KU, SU60]
G8 LVC P B Johnson, 54 Beechwood Cl, Chandlers Ford, Eastleigh, SO53 5PB [IO90HX, SU42]
G8 LVF A Sierota, 20 Marder Rd, Northfields, London, W13 9EN [IO91UM, TQ17]
G8 LVI A C Hilton, 16 Grafton Rd, St. Peters, Broadstairs, CT10 3DP [JO01RI, TR36]
GM8 LVJ W R Brunsdon, 8 Findlay Cttgs, Edinburgh, EH7 6HE [IO85KX, NT27]
G8 LVK Details withheld at licensee's request by SSL.
G8 LVL D R Holmes, 30 Roydale Cl, Loughborough, LE11 5UW [IO92JS, SK52]
G8 LVQ M D Wilson obo White Rose ARS, 73 PO Box, Leeds, LS1 7WN [IO93FT, SE33]
G8 LVU E R Brown, 30A Colleton Dr, Twyford, Reading, RG10 0AX [IO91NL, SU77]
G8 LVV M T Fletcher, 33 The Knoll, Tansley, Matlock, DE4 5FP [IO93FD, SK36]
G8 LVW C T Snell, Ranworth, 138 Main Rd, Great Leighs, Chelmsford, CM3 1NP [JO01GT, TL71]
G8 LWA D S Tyler, 7 Stephens Cres, Horndon on The Hill, Stanford-le-Hope, SS17 8LZ [JO01EM, TQ68]
G8 LWC Dr J P Stuart, 4 Pine Gr, Havant, PO9 2RW [IO90MU, SU70]
G8 LWK H L Millard, 32 Rosehill Rd, Burnley, BB11 2JS [IO83US, SD83]
G8 LWO F W Merritt, 23 Manor Cl, Wickham, Fareham, PO17 5BZ [IO90JV, SU51]
G8 LWQ S Wood, Lucerne, Berrycroft, Soham, Ely, CB7 5BL [JO02EI, TL57]
G8 LWS G Rowlands Ariel RA Gp Lws, c/o Gareth Rowlands, Engineering Pigeon Holes, BBC Outside Broadcast Base, Kendal Avenue, Acton, London, W3 0RP [IO91UM, TQ18]
G8 LWY E Batts, 2 Andover Cl, Feltham, TW14 9XG [IO91SK, TQ07]
G8 LXI Details withheld at licensee's request by SSL.
G8 LXN W S Askey, 32 Hurst Rise, Matlock, DE4 3EP [IO93FD, SK36]
G8 LXS G H Pascoe, Broadhampston, Totnes, Devon, TQ9 6DB [IO80DL, SX86]
G8 LXY S C Clarke, 126 Putteridge Rd, Luton, LU2 8HQ [IO91TV, TL12]
G8 LY C R Hall, Restawhile, 10 Clanwilliam Rd, Lee on The Solent, PO13 9HX [IO90JT, SU50]
G8 LYA Details withheld at licensee's request by SSL.
G8 LYB S A Tompsett, 9 Ashlawn Rd, Rugby, CV22 5ET [IO92JI, SP57]
G8 LYG W P Leach, 15 Beech Lea, Blunsdon, Swindon, SN2 4DE [IO91CO, SU19]
GM8 LYO P Mahood, The Briars, Woodside, Northmuir, Kirriemuir, DD8 4PG [IO86LQ, NO35]
GM8 LYQ I A B Lindsay, 10 Mertoun Pl, Edinburgh, EH11 1JZ [IO85JW, NT27]
GM8 LYS I Langmead, 15 Sunnyside Dr, Drumoak, Banchory, AB31 5EW [IO87TB, NO79]
G8 LYV K Kearns, 79 Church Rd, Hatfield Peverel, Chelmsford, CM3 2LB [JO01HS, TL71]
G8 LYW B J Theedom, 83 Caulfield Rd, Shoeburyness, Southend on Sea, SS3 9LP [JO01JM, TQ98]
G8 LZA P A Spenceley, Little Plot, Silver St., Misterton, Crewkerne, TA18 8NG [IO80OU, ST40]
G8 LZG G W Allen, 16 Headlands Dr, Hessle, HU13 0JR [IO93SR, TA02]
G8 LZK M J Ball, Off Daniels Cres, Long Sutton, Spalding Lincs, PE12 9DR [JO02BS, TF42]
G8 LZL M E Cook, 41 Beacon Dr, Loughborough, LE11 2BD [IO92JS, SK51]
G8 LZO J K Hibbert, 42 Margery Ave, Scholar Green, Stoke on Trent, ST7 3HU [IO83VC, SJ85]
G8 LZR Details withheld at licensee's request by SSL.
G8 LZS P J Martin, 3 Pankhurst Dr, Harmanswater, Bracknell, RG12 9PS [IO91PJ, SU86]

G8 LZU S M Tomlinson, 72 Enborne Rd, Newbury, RG14 6AJ [IO91HJ, SU46]
G8 LZV Details withheld at licensee's request by SSL.
GW8 LZY S H Brown, Maes yr Haidd, 8 Glanceulan, Penrhyncoch, Aberystwyth, SY23 3HF [IO82AK, SN68]
G8 MAA G S Chaplin, 8 Manor House Dr, Northwood, HA6 2UJ [IO91SO, TQ09]
G8 MAF T L Beckham, 33 Eleanor Rd, Bowes Park, London, N11 2QS [IO91WO, TQ39]
G8 MAG S G Blake, 26 Nightingale Dr, Towcester, NN12 6RA [IO92MC, SP64]
G8 MAN K Cox, 55 Reddicap Heath Rd, Sutton Coldfield, B75 7DX [IO92CN, SP19]
G8 MAR M J N Sibley, 10 Ainley Cl, Birchencliffe, Huddersfield, HD3 3RJ [IO93CQ, SE11]
G8 MAV P A Lewis, Westbank, 46 Weyside Rd, Guildford, GU1 1HX [IO91RF, SU95]
G8 MBB D J Maud, 32 Curlew Glebe, Dunnington, York, YO1 5PQ [IO93MX, SE65]
G8 MBI Details withheld at licensee's request by SSL.
G8 MBJ J Fisher, 16 Chestnut Cl, Watlington, Kings Lynn, PE33 0HX [JO02EQ, TF61]
G8 MBK P M Bland, 17 Knowles Cl, Kirklevington, Yarm, TS15 9NL [IO94HL, NZ40]
G8 MBM C J Proctor, 15 Chiltern St., Aylesbury, HP21 8BN [IO91OT, SP81]
G8 MBO R Stead, Dew House, Marton, Marton Cum Grafton, York, YO5 9QY [IO94HB, SE46]
G8 MBQ R J Jones, 46 Wilmington Cl, Woodley, Reading, RG5 4LR [IO91NK, SU77]
G8 MBU R F Williams, Picardy, 14 Coronation Ave, Northwood, Cowes, PO31 8PN [IO90IR, SZ49]
G8 MBV I P Wood, Tessian Lodge, Lydden Rd, Swingfield, nr Folkestone Kent, CT15 7HE [JO01OD, TR24]
G8 MCA G M Bryan, 34 Shelbury Cl, Sidcup, DA14 4BE [JO01BK, TQ47]
G8 MCC C E Divall, 22 Knightstone Rise, St. Andrews Rd, Bridport, DT6 3DR [IO80PR, SY49]
G8 MCD Details withheld at licensee's request by SSL.
G8 MCG Details withheld at licensee's request by SSL.
G8 MCI T J Ostley, 39 Farm Cl Rd, Wheatley, Oxford, OX33 1XJ [IO91KR, SP50]
G8 MCJ B C Pritchard, 14 Rugby Way, Croxley Green, Rickmansworth, WD3 3PH [IO91SP, TQ09]
GW8 MCL P Rowles, 51 Cowbridge Rd West, Ely, Cardiff, CF5 5BQ [IO81JL, ST17]
G8 MCR V H Eagles, 3 Church Rd, Buckhurst Hill, IG9 5RU [JO01AP, TQ49]
G8 MCS B W Bugden, 63 The Fairway, Burnham, Slough, SL1 8DY [IO91QM, SU98]
G8 MCU A M Birt, 190 Epsom Rd, Guildford, GU1 2RR [IO91RF, TQ05]
G8 MCV Details withheld at licensee's request by SSL.
G8 MCW P J Elkins, 615 Blandford Rd, Upton, Poole, BH16 5ED [IO80XR, SY99]
G8 MCY M W Dannatt, 46 Laburnham Rd, Biggleswade, SG18 0NX [IO92UC, TL14]
G8 MCZ Details withheld at licensee's request by SSL.
G8 MDA P E Packer, 165 Stourton Ave, Hanworth, Feltham, TW13 6LD [IO91TK, TQ17]
G8 MDC S J Vickers, 13 Hipkins, Bishops Stortford, Herts, CM23 4DY [JO01BU, TL41]
G8 MDM C W Nock, 67 Woodland Rd, Northfield, Birmingham, B31 2HZ [IO92AJ, SP07]
G8 MEA C J Wilson, 27 Cedarwood Dr, St. Albans, AL4 0DN [IO91US, TL10]
G8 MEC D Uttley, 1 Edgeside, Great Harwood, Blackburn, BB6 7JS [IO83TS, SD73]
G8 MED P G Shirtliff, 2 Birch Ave, Newton, Preston, PR4 3TX [IO83NS, SD43]
G8 MEE K Patman, 45 Postland Rd, Crowland, Peterborough, PE6 0JB [IO92WQ, TF21]
G8 MEH Details withheld at licensee's request by SSL.
G8 MEI R T Whitby, 24 Macaulay Ave, Great Shelford, Cambridge, CB2 5AE [JO02BD, TL45]
G8 MEJ Details withheld at licensee's request by SSL.
G8 MEM A Lillywhite, 78 Studland Rd, Hanwell, London, W7 3QX [IO91TM, TQ18]
GW8 MER M L Busson, 15 Nant y Milwr Cl, Henllys, Cwmbran, NP44 6JW [IO81LP, ST29]
G8 MES Details withheld at licensee's request by SSL.[Op: L E Johnson. Station located near Rotherham.]
G8 MEX I G Glenn, 257 Wimpole Rd, Barton, Cambridge, CB3 7AE [JO02AE, TL45]
G8 MFE Details withheld at licensee's request by SSL.
G8 MFF R Hedley, 6 Allnatt Ave, Winnersh, Wokingham, RG41 5AU [IO91NK, SU77]
G8 MFH R Lake, 3 Pembridge Chase, Bovingdon, Hemel Hempstead, HP3 0QR [IO91RR, TL00]
G8 MFI S C M McGuigan, 194 Westmount Rd, Eltham Park, London, SE9 1XQ [JO01AL, TQ47]
G8 MFM R F Wood, 36 New England Rd, Haywards Heath, RH16 3JS [IO90WX, TQ32]
G8 MFO T C Sorensen, 19 Manor Rd, Lancing, BN15 0PH [IO90UU, TQ10]
G8 MFP Mp C J Reed, 28 Adkinson Ave, Dunchurch, Rugby, CV22 6RG [IO92II, SP47]
GW8 MFQ A John, 79 Harding Cl, Boverton, Llantwit Major, CF61 1GX [IO81GJ, SS96]
G8 MFR R M Irwin, Copperfield, 97 Offerton Ln, Stockport, SK2 5BS [IO83WJ, SJ98]
G8 MFU D J Parry, 19 Norton Ln, Great Wyrley, Walsall, WS6 6PE [IO82XP, SJ90]
G8 MFV R Hickmott, Brisley Cottage, Canterbury Rd, Challock, Ashford Kent, TN25 4DW [JO01KF, TQ95]
GM8 MFZ Dr N S Kennedy, Blackpots Farmhouse, Whitehills, Banff, AB4 2NS [IO87TH, NJ72]
G8 MGD D S Marshall, Shelwyn, Aesops Orchard, Woodmancote, Cheltenham, GL52 4TZ [IO81XW, SO92]
G8 MGG W R Whiteside, Dawstone, Flookburgh Rd, Allithwaite, Grange Over Sands, LA11 7RJ [IO84ME, SD37]
G8 MGK J T Dosher, 40 Bromfield Rd, Southcrest, Redditch, B97 4PN [IO92AH, SP03]
G8 MGP A D Hill, 5 Lilac Walk, Kempston, Bedford, MK42 7PE [IO92SC, TL04]
G8 MGQ D R Garwood, 135 Bath Rd, Bradford on Avon, BA15 1SS [IO81UI, ST86]
G8 MGS Details withheld at licensee's request by SSL.
G8 MGY N J Hulmston, 28 Axholme Rd, Thingwall, Wirral, L61 1BJ [IO83KI, SJ28]
G8 MGZ P E Haynes, 2 The Chase, Furnace Green, Crawley, RH10 6HW [IO91VC, TQ23]
G8 MHA L D Humphrey, 9 The Swallows, Old Harlow, Harlow, CM17 0AR [JO01BS, TL41]
G8 MHD C H Cooper, 16 Paulton Dr, Bishopston, Bristol, BS7 8JJ [IO81QL, ST57]
G8 MHE G Cross, 117 Broadway, Eccleston, St. Helens, WA10 5PB [IO83OL, SJ49]
GW8 MHG Dr R Dewsberry, 24 Bronllwyn, Pentyrch, Cardiff, CF4 8QL [IO81IM, ST08]
G8 MHM Details withheld at licensee's request by SSL.
G8 MHN S T Scrase, 5 Clinton Rd, Leatherhead, KT22 8NU [IO91UG, TQ15]
G8 MHO A S D Fraser, 184 Old Rd, Old Harlow, Harlow, CM17 0HQ [JO01BT, TL41]
G8 MHR Details withheld at licensee's request by SSL.
GM8 MHU J I Fraser, 12 Auchlea Pl, Mastrick, Aberdeen, AB1 6PD [IO87TH, NJ72]
G8 MIA A N Malbon, 9 Send Rd, Caversham, Reading, RG4 8EH [IO91ML, SU77]
G8 MIC M P Williams, 2 High Point, North Hill, Highgate, London, N6 4BA [IO91WN, TQ28]
G8 MIE K G Croucher, 140 Dane Rd, Coventry, CV2 4JW [IO92GJ, SP37]
G8 MIF F R Golding, 16 Lessness Park, Belvedere, DA17 5BG [JO01BL, TQ47]
G8 MIH R J Green, 33 Bulkington Ave, Worthing, BN14 7HH [IO90TT, TQ10]
G8 MIN R C Welsh, 142 Claremont Rd, Hextable, Swanley, BR8 7QT [JO01CJ, TQ57]
G8 MIT C M Wyatt, 273 Nuthurst Rd, Weston Heath, Birmingham, B31 4TQ [IO92AJ, SP07]
GI8 MIV G A Hutchinson, 40 Oldstone Hill, Muckamore, Antrim, BT41 4SB [IO64VQ, J18]
G8 MIW J A West, 21 Gardenia Cres, Mapperley, Nottingham, NG3 6JA [IO92KX, SK54]
G8 MJE Details withheld at licensee's request by SSL.
G8 MJF K Bottomley, The Cottage, Meadow Bank Farm, Daleside Park, Darley Harrogate, HG3 2PX [IO94DA, SE25]
G8 MJH P R Harrison, 154 Cherrydown Ave, Chingford, London, E4 8DZ [IO91XO, TQ39]
G8 MJK S G Pretty, Edwal,Nayland Dr, Mill Rd, Battisford, Stowmarket, Suffolk, IP14 2LS [JO02LD, TM05]
G8 MJT J R Howard, 16 Howards Meadow, Kings Cliffe, Peterborough, PE8 6YJ [IO92RN, TL09]
G8 MJX D Comber, 1 Brympton Rd, Stoke, Coventry, CV3 1GW [IO92GJ, SP37]
G8 MKC V C Webley Milton Keynes & District AR So, 817 PO Box, Milton Keynes, MK6 3LE [IO92PA, SP83]
G8 MKE C Rose, 45 Clent Rd, Oldbury, Warley, B68 9ES [IO92AL, SP08]
G8 MKK Details withheld at licensee's request by SSL.
G8 MKM Details withheld at licensee's request by SSL.
G8 MKO R K Pocock, 3 Brewery Cttgs, Netherley Rd, Tarbock Green, Prescot, Merseyside, L35 1QG [IO83OJ, SJ48]
G8 MKS P Moore, 34 Roseneath Rd, Great Lever, Bolton, BL3 3AX [IO83SN, SD70]
G8 MKT R C Maxwell, 24 Jensen, Tamworth, B77 2RH [IO92DO, SK20]
G8 MKW J O Green, Huntley, Chesham Rd, Wigginton, Tring, HP23 6HH [IO91QS, SP90]
G8 MLA P H Richardson, 26 Foxs Way, Comberton, Cambridge, CB3 7DL [JO02AE, TL35]
G8 MLB N J Bourner, Woodpeckers, 11 Richborough Rd, Sandwich, CT13 9JE [JO01QG, TR35]
G8 MLC I M Thrippleton, 161 York Ave, East Cowes, PO32 6BD [IO90IS, SZ59]
G8 MLD M J Warren, 17 Bolehill Park, Hove Edge, Brighouse, HD6 2RS [IO93CR, SE12]
G8 MLF A S Douglas, 10 Bexley Dr, Worsley, Little Hulton, Manchester, M38 9WL [IO83TM, SD70]
GM8 MLH J W Martindale, Alt-Na-Feidh, Dalmally, Argyll, PA33 1AA [IO76MJ, NN12]
G8 MLK J E Owen, The Old Coach House, Callow Hill, Virginia Water, GU25 4LD [IO91RJ, SU96]
G8 MLP D J Packard, 93 Elstree Gdns, Belvedere, DA17 5DR [JO01BL, TQ47]
G8 MLQ Details withheld at licensee's request by SSL.
G8 MLW D A Carr, 39 Fallowfield Rd, Walsall, WS5 3DH [IO92AN, SP09]
GM8 MMA W J Williamson, Leeskol, Camb Yell, Shetland, ZE2 9DA [IP90LO, HU59]
G8 MMF P W F Dorrington, 6 Thurstons, Binsted, Hants, GU34 4PD [IO91ND, SU74]
G8 MMG D A Bentley, 55 Saddlers Rd, Quedgeley, Gloucester, GL2 4SY [IO81UT, SO71]
G8 MMH Details withheld at licensee's request by SSL.
G8 MML Details withheld at licensee's request by SSL.
G8 MMM G A Nicholas, Greenbank, Chester High Rd, Neston, South Wirral, L64 7TR [IO83LH, SJ37]
G8 MMN Dr M W Holmes, Old Orchard House, High St., Norley, Warrington, WA6 8JS [IO83QF, SJ57]
G8 MMO K J Krelle, 4 Heath Dr, Brookwood, Woking, GU24 0HG [IO91QH, SU95]
G8 MMP M Swain, 38 Longdale Ln, Ravenshead, Nottingham, NG15 9AD [IO93JC, SK55]
GM8 MMW W I Dick, Glencairn, 58 Kirkland Rd, Glengarnock, Beith, KA14 3AJ [IO75PR, NS35]
G8 MNC M W Bilkey, Gargus Farm, Tregony, Truro, Cornwall, TR2 5SQ [IO70NG, SW94]
GM8 MNG C J Raine, Broomhill Edgehead, Pathhead, Midlothian, Scotland, EH37 5RN [IO85MV, NT36]
G8 MNL P E Carruthers, 16 Wivenhoe Cl, Rainham, Gillingham, ME8 7QB [JO01HI, TQ86]
G8 MNO W E B Stewart, 9 Ashley Rd, Marnhull, Sturminster Newton, DT10 1LQ [IO80UX, ST71]
G8 MNR D L Jenkins, 41 Station Rd, Bishops Cleeve, Cheltenham, GL52 4HH [IO81XW, SO92]
G8 MNT D S Readings, Flat 4 Malvern Hill House, East Approach Dr, Pittville, Cheltenham, Glos, GL52 3JE [IO81XV, SO92]
G8 MNY J D H Stockley, 27 Campden Rd, South Croydon, CR2 7ER [IO91XI, TQ36]
G8 MNZ Details withheld at licensee's request by SSL.
GI8 MOA Details withheld at licensee's request by SSL.
G8 MOB M J O'Beirne, 7 Hillside, St. Marys Rd, Ditton Hill, Surbiton, KT6 5HB [IO91UJ, TQ16]
G8 MOF F W H Bellamy, 3 Manor Rd, Crowle, Scunthorpe, DN17 4ET [IO93NO, SE71]
G8 MOH L R H Jarrett, Egremont, Newport Rd, Stafford, ST16 1DH [IO82WT, SJ92]
GM8 MOI C G Stirling, 2 St. Thomas Well, Cambusbarron, Stirling, FK7 9PR [IO86AC, NS79]
G8 MOK G McKay, 20 Blandford Rd, Winton, Eccles, Manchester, M30 8WA [IO83TL, SJ79]
G8 MOL P J Marshall, 134 Gladbeck Way, Enfield, EN2 7EN [IO91WP, TQ39]
GI8 MOV F Warwick, 44 Beech Green, Doagh, Ballyclare, Co Antrim, N Ireland, BT39 0QB [IO64XR, J28]
GW8 MOZ G T Elliott, 9 Hove Ave, St. Julians, Newport, NP9 7QP [IO81MO, ST38]
GD8 MPF A J Willmott, 33 Cannan Ave, Kirk Michael, IM6 1HG
G8 MPG G A Rigby, 1 Route D'Anton, Petit Caudos, 33380, M105, France
G8 MPM W C Brock, 15 Picketleaze, Frogwell, Chippenham, SN14 0DN [IO81WL, ST97]
G8 MPS I G D Macdonald, The Old Smithy, Old Cornsay, Durham, DH7 9EL [IO94CS, NZ14]
GW8 MQ J Millie, yr Hafod, 32 Rhos Las, Tregynwr, Carmarthen, SA31 2DY [IO71UU, SN41]
G8 MQF M J Cooper, 70 Furnace Dr, Furnace Green, Crawley, RH10 6JE [IO91VC, TQ23]
G8 MQK J Lindley, 3 Daisy Green, Linthwaite, Huddersfield, HD7 5PJ [IO93BO, SE01]
G8 MQN Details withheld at licensee's request by SSL.
G8 MQX R B Eccles, 6 Queens Dr, Barnsley, S75 2QJ [IO93GN, SE30]
G8 MQY B Densham, 8 Faugere Cl, Brackley, NN13 6LR [IO92KA, SP53]
G8 MRA Details withheld at licensee's request by SSL.
G8 MRB Details withheld at licensee's request by SSL.
G8 MRI R J C Davey, Woodstock, 1 The Limes, Chesham Bois, Amersham, HP6 5NW [IO91QQ, SU99]
G8 MRJ Details withheld at licensee's request by SSL.
G8 MRM R S Hardee, Long Acres, Kedington Rd, Sturmer, Haverhill, CB9 7XR [JO02FB, TL64]
G8 MRN M B Watch, 9 High Dr, Rowner, Gosport, PO13 0QS [IO90JT, SU50]
G8 MRQ Dr P W Lawrence, 7 The Woodlands, Corton, Lowestoft, NR32 5EZ [JO02UM, TM59]
G8 MRS A P Radley Southend&Dis AR, 16 Kingsley Ln, Thundersley, Benfleet, SS7 3TU [JO01HN, TQ78]
GM8 MRW S S C McQueen, 123 Greenwood Rd, Clarkston, Glasgow, G76 7LL [IO75US, NS55]
G8 MRY Details withheld at licensee's request by SSL.
G8 MRZ L M Overton, Brackenwood, Brook Ln, Brocton, Stafford, ST17 0TZ [IO82XS, SJ91]
G8 MSA Details withheld at licensee's request by SSL.
G8 MSE Details withheld at licensee's request by SSL.
G8 MSM R F Hudson, 4 Lake Dr, Peacehaven, BN10 7QD [IO90XT, TQ40]
G8 MSN A B Mellows, 791 Coastland Dr, Palo Alto, Ca 94303, USA, ZZ7 9CO
G8 MSQ P J Robinson, 40 Lynchmere Ave, Lancing, BN15 0PB [IO90UU, TQ10]
G8 MSR R J Tebboth, Laurel Bank, Newlands Ave, Thames Ditton, KT7 0HF [IO91TJ, TQ16]
GM8 MST Dr G R Kelly, 36 Craigleith Dr, Edinburgh, EH4 3JU [IO85JW, NT27]
G8 MSW Details withheld at licensee's request by SSL.
G8 MSY J O Wilkinson, 11 Wigmore Rd, Tadley, RG26 4HH [IO91KI, SU56]
G8 MTA B D C Haylett, 5 Riverside Cl, Whittlesey, Peterborough, PE7 1DL [IO92WN, TL29]
G8 MTB M Greenfield, 8 The Spinney, Clayton, Newcastle, ST5 4DA [IO82VX, SJ84]
G8 MTI M J Dibsdall, 28 Ct Farm Ave, Ewell, Epsom, KT19 0HF [IO91UI, TQ26]
GW8 MTJ C T Davies, 11 Beaumaris Way, Gr Park, Blackwood, NP2 1DF [IO81JQ, ST19]
G8 MTK S L Crouch, 13 Windermere Cl, Cove, Farnborough, GU14 0JZ [IO91OG, SU85]
G8 MTL D J Epton Mtl Amateur Radio Club, 61 Cartmel Dr, Dunstable, LU6 3PT [IO91RV, TL02]
G8 MTM Details withheld at licensee's request by SSL.
G8 MTO D M Cox, 130 Maidstone Rd, Paddock Wood, Tonbridge, TN12 6DY [JO01EE, TQ64]
G8 MTQ Details withheld at licensee's request by SSL.
G8 MTR P Pointer, 9 Cavendish Cl, Horsham, RH12 5HX [IO91UC, TQ13]
G8 MTV J M Wood, Spa Cottage, Dalton Rd, Croft-on-Tees, Darlington, Co Durham, DL2 2SQ [IO94FL, NZ20]
GM8 MUE Dr B Ray, Hillwood, Woodleigh Rd, Ledbury, HR8 2BG [IO82SA, SO73]
G8 MUF J R W Ames, 16 Vere Gdns, Henley Rd, Ipswich, IP1 4NZ [JO02NB, TM14]
G8 MUX J S Mottram, 13 Derwent Rd, High Ln, Stockport, SK6 8AT [IO83XI, SJ98]
G8 MUY G Steele, 101 Victoria Rd, Padiham, Burnley, BB12 8TA [IO83UT, SD83]
G8 MUZ E G Cressey, Seefeld, 5 Heronsgate, Frinton on Sea, CO13 0AW [JO01OU, TM22]
G8 MVC R W Westlake, 41 Coningham Rd, London, W12 8BP [IO91VM, TQ27]
G8 MVD K Wilks, 72 Grasmere Rd, Bradford, BD2 4HX [IO93DT, SE13]
G8 MVE G A Pluck, Meadowside, Water Ln, Garboldisham, Diss, IP22 2SB [JO02LJ, TM08]
G8 MVJ C R Chambers, Hollybank, Back St., Langtoft, Driffield, YO25 0TD [IO94SC, TA06]
G8 MVO J Cappleman, 160 Main St., Cayton, Scarborough, YO11 3TF [IO94TF, TA08]
G8 MVP M Fogden, 12 Jireh Ct, Perrymount Rd, Haywards Heath, RH16 3BH [IO91WA, TQ32]
G8 MVS N J Fuller, 11 Hayes Mead Rd, Bromley, BR2 7HR [JO01AJ, TQ36]
G8 MVV I Smith, 8 Crowndale Pl, Pipers Croft, Packmoor, Stoke on Trent, ST6 6XL [IO83VB, SJ85]
G8 MVX A E Rogers, 114 Heathbank Ave, Irby, Wirral, L61 4YG [IO83KI, SJ28]
G8 MVY E H Phillips, 2 Clarks Farm Cttgs, Bunglers Hill Rd, Farley Hill, Reading, RG7 1TW [IO91MJ, SU76]
G8 MWA J Hale Medway A.R.T., obo Medway A R T, 136 Bush Rd, Cuxton, Rochester, ME2 1HB [JO01FJ, TQ76]
G8 MWC R A Mewse, 3 Butterwick Rd, Freiston, Boston, PE22 0LF [JO02AX, TF34]
G8 MWD D R Lewing, 94 Carville Cres, Brentford, TW8 9RD [IO91UL, TQ17]
G8 MWF Details withheld at licensee's request by SSL.
G8 MWN K C W Harris, Bella Vista, Station Rd, Bere Ferrers, Yelverton, PL20 7JS [IO70VK, SX46]
G8 MWQ Details withheld at licensee's request by SSL.
G8 MWR L W Ross, 81 Ringwood Highway, Coventry, CV2 2GT [IO92GK, SP38]
G8 MWU P M Stafford, 5 Westmead Dr, Newbury, RG14 7DJ [IO91IJ, SU46]
G8 MWW W H A Westlake, West Park, Clawton, Holsworthy, Devon, EX22 6QN [IO70TS, SX39]
G8 MWX A A Priestley, 55 Derwent Ave, Garforth, Leeds, LS25 1HN [IO93HS, SE43]
G8 MXB A Woodhouse, 21 Lincoln Ave, Levenshulme, Manchester, M19 3LA [IO83VK, SJ89]
G8 MXD G S York, 41 Stroud Rd, Patchway, Bristol, BS12 5EN [IO81RM, ST58]
G8 MXE Details withheld at licensee's request by SSL.[Op: W G G Scott.]
GW8 MXG P N Berry, 1 Broadway Cttgs, Laleston, Bridgend, CF32 0HY [IO81EM, SS88]
G8 MXI K J Nicholls, Flexbury, Bradworthy, Holsworthy, Devon, EX22 7TQ [IO70TV, SS31]
G8 MXQ Details withheld at licensee's request by SSL.
G8 MXV K J Ayriss, 6 Langstons, Trimley St. Mary, Trimley, Ipswich, IP10 0XL [JO02PA, TM24]
G8 MXW C D Down, 100 Lynwood Dr, Merley, Wimborne, BH21 1UQ [IO90AS, SZ09]
G8 MXZ S P Roper, 39 Lipscomb Cl, Hermitage, Thatcham, RG18 9SZ [IO91IK, SU57]
G8 MY R W Parfitt, 29 Manor Rd, Farnborough, GU14 7EX [IO91PG, SU85]
G8 MYE R N Preston, 13 Boulters Cl, Stowmarket, IP14 1SQ [JO02LE, TM05]
G8 MYF M L Johnson, Brookfield, 42 Marlborough Rd, Ryde, PO33 1AB [IO90KR, SZ69]
G8 MYG C J Hunt, Rosings, Sandrock Hill, Crowhurst, near Battle, E Sussex, TN33 9AY [JO00GU, TQ71]
G8 MYI G A Harris, 1 School Villas, Great Easton, Dunmow, CM6 2HA [JO01EV, TL62]
G8 MYJ C D Drewe, 37 Baker St., Chelmsford, CM2 0SA [JO01FR, TL70]
G8 MYK A Rowley, Holly Cottage, 368 Highters Heath Ln, Hollywood, Birmingham, B14 4TE [IO92BJ, SP07]
G8 MYO C M Tyler, 22 Temeside, Ludlow, SY8 1PB [IO82PI, SO57]
G8 MYV D V Webster, 8 Rowlheys Pl, Lancers Mews, West Drayton, UB7 9NQ [IO91SM, TQ07]
G8 MZA D G Garrett, Brookside Farm, Tonge, Melbourne, Derbyshire, DE7 1BD [IO92IX, SK44]
G8 MZD P J Diggins, 8 Gloucester Gdns, Bagshot, GU19 5NU [IO91PI, SU96]
G8 MZF Details withheld at licensee's request by SSL.
G8 MZG J R Adams, 7 Ragley Mews, Caversham Park, Caversham, Reading, RG4 6SG [IO91ML, SU77]
GW8 MZR R W Harris, 15 Quarry Rise, Undy, Magor, Newport, NP6 3JU [IO81OO, ST48]
G8 MZV P I Clayton, 217 Prestbury Rd, Cheltenham, GL52 3ES [IO81XV, SO92]
G8 MZW S Adams, 29 Rothbury Gr, Bingham, Nottingham, NG13 8TG [IO92MW, SK64]
G8 MZX R A Barker, 79 South Parade, Boston, PE21 7PN [IO92XX, TF34]
G8 MZY Prof D A Cushman, 50 St. Peters St., Syston, Leicester, LE7 1HJ [IO92LQ, SK61]
G8 MZZ P R Boam, 36 Copeland Dr, Heritage Walk, Stone, ST15 8YP [IO82WV, SJ93]
GW8 NAC K D Davies, 45 Castle View, Simpson Cross, Haverfordwest, Dyfed, SA62 6EN [IO71LT, SM81]
G8 NAG M T Smith, 18 Manor Ln, Verwood, BH31 6HX [IO90BU, SU00]
G8 NAI J Lazzari, 3 Terson Way, Parkhall Est, Weston Coyney, Stoke on Trent, ST3 5RQ [IO82WX, SJ94]
G8 NAL P Corbishley, 15 High St., Hardingstone, Northampton, NN4 7BT [IO92NF, SP75]
G8 NAM P W Buttress, 150 Littlecroft, South Woodham Ferre, South Woodham Ferrers, Chelmsford, CM3 5GF [JO01HP, TQ89]
G8 NAP P D Beacon, 67 St. Helena Rd, Polesworth, Tamworth, B78 1NJ [IO92EO, SK20]
G8 NAV Details withheld at licensee's request by SSL.[Op: N C D Vernon. Station located near Whitstable.]
G8 NAX Details withheld at licensee's request by SSL.
G8 NBB Details withheld at licensee's request by SSL.
G8 NBO L C Phillips, 14 Heal Park Cres, Fremington, Barnstaple, EX31 3AP [IO71WB, SS53]
G8 NBR Details withheld at licensee's request by SSL.
GI8 NBW S A Robinson, 51 Moyan Rd, Stranocum, Ballymoney, BT53 8LD [IO65TC, D02]
GW8 NCF Details withheld at licensee's request by SSL.
G8 NCK N E Brown, 9 Redhill Cl, Coton Ln, Tamworth, B79 8EJ [IO92DP, SK20]
GM8 NCM Details withheld at licensee's request by SSL.
G8 NCS M F Green, 21 Hill View Rise, Northwich, Ches, CW8 4XA [IO83RG, SJ67]
G8 NCT R J Smith, 4 Halfway Houses, Main Rd, Deeping St Nicholas, Spalding, PE11 3DH [IO92VR, TF21]
G8 NCW J W Tranmer, 149 Ashwood Rd, Potters Bar, EN6 2QD [IO91VQ, TL20]
GI8 NCX M T Sinclair, 43 Edgcumbe Gdns, Belfast, BT4 2EH [IO74BO, J37]
G8 NCZ I P Ruddock, 294 Willowfield, Harlow, CM18 6SD [JO01BS, TL40]
G8 ND N E Dalby, Newlandene, St. Cyrils Rd, Stonehouse, GL10 2QG [IO81UR, SO80]

G8 NDB Dr G R Jarrett, 480 Warwick Rd, Solihull, B91 1AG [IO92CK, SP18]
G8 NDD R J Welch, 16 Eddisbury Rd, West Kirby, Wirral, L48 5DS [IO83JJ, SJ28]
G8 NDE J C Turner, 9 Clifton Ave, Culcheth, Warrington, WA3 4PD [IO83RK, SJ69]
G8 NDF D Simpson, 10 Buckingham Way, Byram Cum Sutton, Byram, Knottingley, WF11 9NN [IO93IR, SE42]
G8 NDK K Lindley, 29 Wharfedale Dr, Bridlington, YO16 5FB [IO94VC, TA16]
G8 NDL K R Ginn, 8 Kettlewell Ct, Swanley, BR8 7BP [JO01CJ, TQ56]
G8 NDN C F Keens, Toad Hall, 69 Lillywhite Cres, Andover, SP10 5NA [IO91GF, SU34]
G8 NDP R Callan, 99 Lingfield Rd, Stevenage, SG1 5SQ [IO91WW, TL22]
G8 NDQ A Kaye, 68 Pontefract Rd, Ferrybridge, Knottingley, WF11 8PW [IO93IR, SE42]
G8 NDT Details withheld at licensee's request by SSL.
G8 NDV P R Fay, 42 Roberts Rd, Salisbury, SP2 9BY [IO91CB, SU13]
G8 NED L N Fennelow Wisbech Rec, 39 Clarence Rd, Wisbech, PE13 2ED [JO02CQ, TF41]
G8 NEF Dr R M A Peel, 76 Cypress Gr, Ash Vale, Aldershot, GU12 5QW [IO91PG, SU85]
G8 NEH C J Nunn, 29 Wheatland Cl, Winchester, SO22 4QL [IO91HB, SU42]
G8 NEI K I Marsh, 1 Parr Cl, Exeter, EX1 2BG [IO80FR, SX99]
G8 NEM Rev R T Moll, 79 Moor Ln, Crosby, Liverpool, L23 2SQ [IO83LL, SD30]
G8 NEO D G Edwards, 9 Compit Hills, Cromer, NR27 9LJ [JO02PV, TG24]
GM8 NET A W Fraser, 69 Benbecula, East Kilbride, Glasgow, G74 2BS [IO75WS, NS65]
G8 NEU J N Wright, Beckside, Newbiggin, Heads Nook, Carlise, Cumbria, CA4 9DH [IO84PU, NY54]
G8 NEY D Millard, 12 Kidston Way, Rudloe, Corsham, SN13 0JZ [IO81VK, ST87]
G8 NFB I R Fisher, Gallopers Cottage, Sarratt Rd, Croxley Green, Rickmansworth, WD3 3JF [IO91SP, TQ09]
G8 NFD K R Gardiner, 8 Foxlands Dr, Sutton Coldfield, B72 1YZ [IO92CM, SP19]
GM8 NFG J R Aitken, Hamabo, 10 Lynn Park, Kirkwall, KW15 1SL [IO88MX, HY40]
G8 NFM F Turner, 46 Main St., Kings Newton, Melbourne, Derby, DE73 1BX [IO92GT, SK32]
G8 NFO T Davies, 59 Seaford Rd, Eastbourne, BN22 7JS [JO00DS, TQ60]
G8 NFP A J Crockett, 57 Upland Rd, Sutton, SM2 5HW [IO91VI, TQ26]
G8 NFU Details withheld at licensee's request by SSL.
G8 NFZ S Sims, 71 Green St., Eastbourne, BN21 1QZ [JO00DS, TV59]
G8 NGF D R Stone, 7 The Lindens, Prospect Hill, Walthamstow, London, E17 3EJ [IO91XO, TQ38]
GM8 NGG C B Wilson, 8 Larch Gr, Silverton Hill, Hamilton, ML3 7NF [IO75XS, NS75]
G8 NGJ P A J Richardson, Brembridge Farm, Shillingford, Tiverton, Devon, EX16 9BT [IO81GA, ST02]
G8 NGK R J Brannon, Glendarvel, 14 Digby Rd, Ipswich, IP4 3ND [JO02OB, TM14]
G8 NGM N C King, 42 Constance Cl, Witham, CM8 1XY [JO01HS, TL81]
G8 NGZ V A Edwards, 33 Eyrescroft, Bretton, Peterborough, PE3 8ES [IO92UO, TF10]
G8 NHD P A Mart, 84 Castle Green, Westbrook, Warrington, WA5 5XA [IO83QJ, SJ59]
G8 NHF D Goldshaker, 41 Peak Hill, London, SE26 4LS [IO91XK, TQ37]
G8 NHG R Wilkins, 135 Walton Way, Newbury, RG14 2LL [IO91IJ, SU46]
G8 NHM J E Graves, Willses, Upper Ln, Brighstone, Newport, PO30 4BA [IO90HP, SZ48]
G8 NHN Details withheld at licensee's request by SSL.[Op: B Lewis. Station located near Lytham St Annes.]
G8 NHO J Austin, 5 Mercia Rd, Baldock, SG7 6RZ [IO91VX, TL23]
G8 NIC M D Toms, Tower Hill, 9 Church St., Beaminster, DT8 3BA [IO80PT, ST40]
G8 NIE D J Sharpe, 5 Drydales, Kirkella, Anlaby, Hull, HU10 7JU [IO93SR, TA02]
G8 NIK M Carena, Armorel, Shire Ln, Chorleywood, Rickmansworth, WD3 5NH [IO91RP, TQ09]
G8 NIL D E Bales, 18 The Grove, Magazine Ln, Wisbech, PE13 1LF [JO02BP, TF40]
GU8 NIS Details withheld at licensee's request by SSL.[Correspondence to D Eaton, obo Guernsey ARS, PO Box 100, Guernsey.]
G8 NIU R C F Whiting, 1 Chestnut Dr, Egham Hill, Egham, TW20 0BJ [IO91RK, SU97]
G8 NJA D Webber Torbay ARS, 43 Lime Tree Walk, Milber, Newton Abbot, TQ12 4LF [IO80EM, SX87]
G8 NJD Details withheld at licensee's request by SSL.[Station located in London SE13.]
G8 NJE G Brier, 28 McCartney Walk, Brighton Hill, Basingstoke, RG22 4NZ [IO91KF, SU64]
G8 NJF A J Cox, 21 Torvale Rd, Wightwick, Wolverhampton, WV6 8NL [IO82VO, SO89]
G8 NJO G A Vallely RAF Henlow ARC, 2 Kayser Ct, Biggleswade, SG18 8BG [IO92UC, TL14]
G8 NJQ V E Green, 7 Raglis Cl, Webheath, Redditch, B97 5RN [IO92AH, SP06]
GW8 NJW Details withheld at licensee's request by SSL.
G8 NJY Details withheld at licensee's request by SSL.
G8 NKA E R Gaze, Bank View, The Green, Stapleton, Darlington, DL2 2QQ [IO94EM, NZ21]
G8 NKD Details withheld at licensee's request by SSL.
G8 NKJ L Reid, 26 Mansion Ave, Whitefield, Manchester, M45 7SS [IO83UN, SD80]
G8 NKM A D ODonovan, 2 Mackenzie Rd, Beckenham, BR3 4RU [IO91XJ, TQ36]
G8 NKN S B Gorwits, 13 Parkin Ln, Apperley Bridge, Bradford, BD10 0NF [IO93DU, SE13]
G8 NKQ M J Brightwell, 4 Geneva Rd, Ipswich, IP1 3NP [JO02NB, TM14]
G8 NKW P B Walker, Rosewood, Pony Cart Ln, Stelling Minnis, Canterbury, CT4 6AU [JO01ME, TR14]
G8 NKY P S Wiles, 15 Harbour View Cl, Parkstone, Poole, BH14 0PF [IO90AR, SZ09]
G8 NLF I R Munro, 30 Willow Cl, Bordon, GU35 0TH [IO91NC, SU83]
G8 NLP C H Saunders, 215 Uttoxeter New Rd, Derby, DE22 3LJ [IO92GW, SK33]
G8 NLS S O Brien, 4 Clinton Pl, East Herrington, Sunderland, SR3 3SN [IO94GU, NZ35]
G8 NLY Details withheld at licensee's request by SSL.
G8 NMA K Brunning, 2 Frog Hall, Silverstone, Towcester, NN12 8TT [IO92LC, SP64]
G8 NME A A A King, 24 Leckwith Ave, Bexleyheath, DA7 5RD [JO01BL, TQ47]
G8 NMH B F O'Regan, 10 School Hill, Little Sandhurst, Sandhurst, Camberley, Surrey, GU17 8LD [IO91OH, SU85]
G8 NMK C C Eccles, 21 Verdon Pl, Barford, Warwick, CV35 8BT [IO92EF, SP26]
G8 NML G V Plant, 5 Lyon St., Queensbury, Bradford, BD13 1AY [IO93BS, SE13]
GM8 NMM C D Reid, Carpenters Croft, Nethermuir, Maud, Peterhead, Aberdeenshire, AB42 5RE [IO87WL, NJ94]
G8 NMO D Pechey, Jays Lodge, Crays Pond, Reading, RG8 7QG [IO91KM, SU68]
G8 NMQ I J P Wooller, 6 Riseley Rd, Maidenhead, SL6 6EP [IO91PM, SU88]
G8 NMT J Hicks, 14 Oakwood, Flackwell Heath, High Wycombe, HP10 9DW [IO91PO, SU89]
G8 NMU Details withheld at licensee's request by SSL.
G8 NMW Details withheld at licensee's request by SSL.
GW8 NNF R J Galpin, 23 Heol y Delyn, Lisvane, Cardiff, CF4 5SR [IO81JM, ST18]
G8 NNJ D J Purkiss, 9 Clydesdale Rd, Hornchurch, RM11 1AG [JO01CN, TQ58]
G8 NNQ P A Gray, 23 Gainsford Rd, Southampton, SO19 7AS [IO90HV, SU41]
G8 NNS G D Stamp, 41 Willoughby Rd, Liscard, Wallasey, L44 3DZ [IO83LK, SJ29]
G8 NNT Details withheld at licensee's request by SSL.
G8 NNU T E A Rowe, 68 Cobourg Rd, Montpelier, Bristol, BS6 5HX [IO81RL, ST57]
G8 NNX M I Cohen, 20 Scafell Lawn, Woodlands Est, Gateacre, Liverpool, L27 5RH [IO83OJ, SJ48]
G8 NOB N P Bean, 33 Badger Cl, Guildford, GU2 6PJ [IO91QG, SU95]
G8 NOC Details withheld at licensee's request by SSL.
G8 NOF R G Holt, Tile House, Vicarage Hill, Tanworth in Arden, Solihull, B94 5EB [IO92BI, SP17]
G8 NOS A D Swallow, 2 Wheatfield Cl, Bredbury, Stockport, SK6 1EW [IO83WK, SJ99]
G8 NPF E C Harrison, 5 Nelson Gdns, Braintree, CM7 9TG [JO01GV, TL72]
G8 NPK J A Cattermole, 6 Grafton House, The Farmlands, Northolt, UB5 5ER [IO91TN, TQ18]
G8 NPM G A Clark, 19 Stewards Cl, Epping, CM16 7BU [JO01BQ, TL40]
G8 NPO P G Fynn, 14 Milton Gdns, Summerleys Rd, Princes Risborough, HP27 9DD [IO91NR, SP80]
G8 NPR J E Blackshaw, 23 Cherry Orchard, Oakington, Cambridge, CB4 5AY [JO02AG, TL46]
G8 NPV Details withheld at licensee's request by SSL.
G8 NPY R H Worsfold, The Forge, Cucklington, Wincanton, Somerset, BA9 9PT [IO81TB, ST72]
G8 NPZ P J Whiteman, 22 Hartsbourne Rd, Earley, Reading, RG6 5PY [IO91MK, SU77]
G8 NQA Details withheld at licensee's request by SSL.
G8 NQC P A Manser, 61 Galsworthy Dr, Caversham Park, Caversham, Reading, RG4 6QB [IO91ML, SU77]
G8 NQF D A Kiley, 2 Thorn Lodge, Spencer Rd, Eastbourne, BN21 4PA [JO00DS, TV69]
G8 NQI J Gartside, 12 Starfield Ave, Hollingworth Lake, Littleborough, OL15 0NG [IO83WP, SD91]
G8 NQN M F Bancroft, 31 Edward Cl, Southowram, Halifax, HX3 9SP [IO93CQ, SE12]
G8 NQO A B Whyatt, 11 The Perrings, Nailsea, Bristol, BS19 2YD [IO81OK, ST46]
G8 NQR F J Rendell, 11 Alawn Ln, Sutton, Ely, Cambs, CB6 2RE [JO02BJ, TL47]
G8 NQW Details withheld at licensee's request by SSL.
G8 NQY W J W Lea, 20 Gloucester Rd, Walsall, WS5 3PN [IO92AN, SP09]
G8 NRB A E Green, 156 Union St., Dunstable, LU6 1HB [IO91RV, TL02]
G8 NRC S Deighton, 23 Winston Way, Thatcham, RG19 3TY [IO91IJ, SU56]
G8 NRF G Wood, 3 Cleveleys Rd, Great Sankey, Warrington, WA5 2SR [IO83QJ, SJ58]
G8 NRI A R Wheeler, 21 The Spinnaker, Ferrers, South Woodham Ferrers, Chelmsford, CM3 5GL [JO01HP, TQ89]
G8 NRL E M C Stevens, 6 Coxwell Gdns, Coxwell Rd, Faringdon, SN7 7HB [IO91EP, SU29]
G8 NRP M J Andrew, 80 Hamble Dr, Abingdon, OX14 3TE [IO91IQ, SU59]
G8 NRR R J Bambrook, 26 Croft Rd, Thame, OX9 3JF [IO91MR, SP70]
G8 NRS P E Smith N.A.R.S.A., 52 Grantham Dr, Woodhill, Bury, BL8 1XW [IO83UO, SD71]
G8 NRT Details withheld at licensee's request by SSL.
G8 NRU D Carr, 189 Lloyd St., Heaton Norris, Stockport, SK4 1NH [IO83WK, SJ89]
G8 NRV M Crossfield, 59 The Cres, Arden Park, Bredbury, Stockport, SK6 2DY [IO83WK, SJ99]
G8 NRY P S Martin, Howden Lodge, 18 Cooper Ln, Shelf, Halifax, HX3 7RD [IO93CS, SE12]
G8 NS W D Johnson, 5 South Lodge, Bicester House, Kings End, Bicester, Oxon, OX6 7AH [IO91KV, SP52]
G8 NSD F J Taylor, 96 Elvaston Rd, North Wingfield, Chesterfield, S42 5HH [IO93HE, SK46]
G8 NSE F E Wood, 96 Manchester Rd, Astley, Tyldesley, Manchester, M29 7EJ [IO83SL, SJ69]
G8 NSK J W Barnes, 23 Spenser Rd, Kings Lynn, PE30 3DP [JO02FS, TF62]
G8 NSP J H Miller, Forder Villa, Forder, Saltash, Cornwall, PL12 4QR [IO70VJ, SX45]
G8 NSS P Leach, 5 Capesthorne Cl, Ashfields Park Est, Werrington, Stoke on Trent, ST9 0PF [IO83WA, SJ94]
G8 NST J W Leek, 13 First Ave, Enfield, EN1 1BL [IO91XP, TQ39]
GM8 NSU Details withheld at licensee's request by SSL.
G8 NSV Details withheld at licensee's request by SSL.
GM8 NSZ M F Stanway, 30 Durham Ave, Edinburgh, EH15 1PA [IO85KW, NT27]
G8 NT R F Stanbridge, 3 Smyth Ct, High St., Leiston, IP16 4BZ [JO02SE, TM46]
G8 NTD K P Johnson, 24 Capers Cl, Enderby, Leicester, LE9 5QD [IO92JO, SP59]
G8 NTG W W Howell, 6 Unity Ave, Sneyd Green, Hanley, Stoke on Trent, ST1 6DE [IO83VA, SJ84]
G8 NTH Details withheld at licensee's request by SSL.
G8 NTI Details withheld at licensee's request by SSL.
G8 NTJ K P Hand, 290 Hednesford Rd, Heath Hayes, Cannock, WS12 5DS [IO92AQ, SK01]
G8 NTQ M J Roper, 25 Harvest Green, Newbury, RG14 6DW [IO91HJ, SU46]
G8 NTR Dr J K Williams, Wild Rose Cottage, 133A Wilts Ln, Pinner, HA5 2NB [IO91TO, TQ08]
G8 NTS J A Smart, Greystone, High St., Blunsdon, Swindon, SN2 4AR [IO91CO, SU19]
G8 NTU Details withheld at licensee's request by SSL.
G8 NTY C D Mallows, 31 Mayfield Rd, Sutton Coldfield, B73 5QJ [IO92CN, SP19]
G8 NTZ D T Kowalczyk, 56 Fountains Garth, Bracknell, RG12 7RH [IO91OJ, SU86]
G8 NUF Details withheld at licensee's request by SSL.
G8 NUJ D A Brooker, 207 Eden Park Ave, Beckenham, BR3 3JW [IO91XJ, TQ36]
G8 NUT W Thomas, 14 Phoebe Bell Ct, The Flats, Bromsgrove, Worcs, B61 8LF [IO82XI, SO97]
GW8 NV Details withheld at licensee's request by SSL.
G8 NVB N V Brown, 32 Trinity Cl, Pound Hill, Crawley, RH10 3TW [IO91WD, TQ23]
G8 NVC B J Ellis, 7 Highmoor Cl, Corfe Mullen, Wimborne, BH21 3PU [IO80XS, SY99]
GM8 NVE D Watters, 111 Bankton Park East, Livingston, EH54 9BN [IO85GV, NT06]
G8 NVF J M Withey, 44 Ross Way, Caddington, Slip End, Luton, LU1 4DD [IO91SU, TL01]
GM8 NVG A J Wilson, Lochend, Beith, Ayrshire, KA15 2LN [IO75RS, NS45]
G8 NVH S P Reynolds, 242 Butchers Ln, Mereworth, Maidstone, ME18 5QH [JO01EG, TQ65]
G8 NVI A E Stevens, 67 New Rd, East Hagbourne, Didcot, OX11 9JX [IO91JO, SU58]
G8 NVQ M J Golder, 11 Angler Rd, Ramleaze, Shaw, Swindon, SN5 9SX [IO91BN, SU18]
G8 NVT R C Hatfield, 1 Slade Cl, Ottery St. Mary, EX11 1SY [IO80IS, SY19]
G8 NVW A Pearce, 670 Chorley Old Rd, Bolton, BL1 5QE [IO83SO, SD61]
G8 NVX M I K E Moss, 24 Magna Ln, Dalton, Rotherham, S65 4HH [IO93IK, SK49]
G8 NVY J M Coombs, 306 NW 7th Ave, Mineral Wells, Texas 76067, USA, X X
G8 NVZ Dr G C Evans, 4 Holcot Ln, Anchorage Park, Portsmouth, PO3 5TR [IO90LT, SU60]
G8 NWC G L Boor, 27 Welbeck Dr, Spalding, PE11 1PD [IO92WT, TF22]
G8 NWI J J Vine, 117 Betterton Rd, Rainham, RM13 8ND [JO01CM, TQ58]
G8 NWK G Milner, 3 Briggs Villas, Queensbury, Bradford, BD13 2EP [IO93BS, SE13]
G8 NWL J A Mason, 47 Rivenhall Gdns, Woodford, London, E18 2BU [JO01AO, TQ38]
G8 NWM V B Maxfield, 50 Hanthorpe Rd, Morton, Bourne, PE10 0NT [IO92TT, TF02]
G8 NWP O Kurlak, 54 Manton Dr, Luton, LU2 7DJ [IO91TV, TL02]
G8 NWR A R Neath, 2 Harpley Rd, Defford, Worcester, WR8 9BL [IO82WC, SO94]
G8 NWU Maj M W Wright, 69 Wroxham Dr, Wollaton, Nottingham, NG8 2QR [IO92JW, SK53]
G8 NWZ M G Percy, 37 Albert Rd, Wellingborough, NN8 1EL [IO92PH, SP86]
G8 NXB N M Borrett, High Timbers, 12 Belmont Rise, Cheam, Sutton, SM2 6EQ [IO91VI, TQ26]
GW8 NXC S A Nelson, 37 Brook Way, Romsey, SO51 7JZ [IO91GA, SU32]
G8 NXD M A Waterfall, 12 Boskenna Rd, Four Lanes, Redruth, Cornwall, TR16 6LS [IO70JE, SW63]
G8 NXG Details withheld at licensee's request by SSL.[Op: R A Burn, Semi-dormant QRP/ATV station located Hinckley, Leics. Pse QSL via bureau.]
G8 NXH M B Booth, 5 Westwinn View, Whinmoor, Leeds, LS14 2HY [IO93GU, SE33]
G8 NXI Details withheld at licensee's request by SSL.
G8 NXJ I D Livesey, 26 Hilltop Rd, Twyford, Reading, RG10 9BN [IO91NL, SU77]
GW8 NXK G J Garner, 31 Clare St., Manselton, Swansea, SA5 9PG [IO81AP, SS69]
G8 NXQ W T Povey, 31 Baddlesmere Rd, Whitstable, CT5 2LB [JO01MI, TR16]
G8 NXS D Stevenson, 86 Kingston Rd, Luton, LU2 7SA [IO91TV, TL02]
G8 NXU C H Matthews, 117 Cliff Rd, Felixstowe, IP11 9SA [JO01QX, TM33]
G8 NXV Details withheld at licensee's request by SSL.
G8 NYB D L Reed, 59 Cowley Ave, Chertsey, KT16 9JJ [IO91RJ, TQ06]
G8 NYC J Primmer, 46 Grantham Cres, Ipswich, IP2 9PD [JO02NB, TM14]
G8 NYD M Perry, 23 Victors Cres, Hutton, Brentwood, CM13 2HZ [JO01EO, TQ69]
G8 NYH R N Adams, 2 Longwill Ave, Melton Mowbray, LE13 1UR [IO92NS, SK71]
G8 NYJ I Gibbs, 32 Beswick Ave, Bournemouth, BH10 4EY [IO90BR, SZ09]
G8 NYK M N Nicholson, 14 Whyke Ct, Whyke Cl, Chichester, PO19 2TP [IO90OT, SU80]
G8 NYM M D Lister, 97 Hightown Rd, Liversedge, WF15 8DG [IO93DR, SE12]
G8 NYR B L Rabey, 36 Park Way, St. Austell, PL25 4HR [IO70OI, SX05]
GW8 NYS G E Sweeney, 106 Pontygwindy Rd, Caerphilly, CF8 3HF [IO81HN, ST08]
G8 NYW Details withheld at licensee's request by SSL.[Op: M J Sargent. Station located near Aldershot.]
G8 NYZ N J Cooper, 52 Peckforton View, Kidsgrove, Stoke on Trent, ST7 4TA [IO83VB, SJ85]
G8 NZB B C Durrant, Brymar, 16 Merrymeet, Whitestone, Exeter, EX4 2JP [IO80ER, SX89]
G8 NZC K A Edmunds, 8 Belmont Dr, Pensby, Wirral, L61 9NB [IO83KI, SJ28]
G8 NZD C H Atkinson, 8 Southwood Rd, Dunstable, LU5 4EA [IO91RU, TL02]
GW8 NZE M R Bartlett, 29 Delafield Rd, Abergavenny, NP7 7AW [IO81LT, SO21]
G8 NZK N G OHagan, 5 Bankside, Nine Mile Ride, Finchampstead, Wokingham, RG40 3QB [IO91NJ, SU76]
GM8 NZL E M Hogg, Fairfield, 43 Muir Wood Rd, Currie, EH14 5JN [IO85IV, NT16]
GW8 NZN D F Roberts, 54 South Dr, Rhyl, LL18 4ST [IO83GH, SJ07]
G8 NZO J W Crozier, 30 Wendover Rd, Perry Common, Birmingham, B23 5JE [IO92BM, SP09]
G8 NZP D W Murcott, 8 Plainview Cl, Aldridge, Walsall, WS9 0YY [IO92BO, SP09]
G8 NZR M K Pullan, 18 Heathfield, Mirfield, WF14 9BJ [IO93DQ, SE12]
G8 OAD G R Baxter, 4 Deeping Rd, Baston, Peterborough, PE6 9NP [IO92TR, TF11]
GM8 OAH W R Easton, 21 Cameron Ave, Bishopton, PA7 5ES [IO75SV, NS47]
G8 OAQ M J Brown, 29 The Osiers, Buckden, St. Neots, Huntingdon, PE18 9UX [IO92UG, TL16]
G8 OBB T J Hooker, Inglewood, Woodside Rd, Lower Woodside, Luton, LU1 4DJ [IO91SU, TL01]
G8 OBH G E Goodfellow, 5 Laleham Gdns, Cliftonville, Margate, CT9 3PN [JO01QJ, TR37]
G8 OBK M J Bruce-Smith, 28 Belmont Rd, Bramhall, Stockport, SK7 1LE [IO83WI, SJ88]
G8 OBP Details withheld at licensee's request by SSL.
GW8 OBW J L McMinn, 108 Smithfield Rd, Wrexham, LL13 8ES [IO83MB, SJ35]
G8 OCA J E Astle, 40 Stotfold Rd, Arlesey, SG15 6XT [IO92UA, TL13]
G8 OCE J B Hardy, 79 Cinderhill Ln, Norton, Sheffield, S8 8JA [IO93GH, SK38]
G8 OCM E A Dubbins, 2 Elizabeth Ave, Rose Green, Bognor Regis, PO21 3EL [IO90PS, SZ99]
G8 OCN B T Dubbins, 2 Elizabeth Ave, Rose Green, Bognor Regis, PO21 3EL [IO90PS, SZ99]
G8 OCO M J Hughes, 198 Wellington St., Long Eaton, Nottingham, NG10 4JN [IO92IV, SK43]
GI8 OCR J T J McIlveen, 31 Edenaveys Cres, Armagh, BT60 1NT [IO64QH, H94]
G8 OCS D C Simpson, 6 St. Martins Cl, The Willows, Stratford upon Avon, CV37 9QW [IO92DE, SP15]
G8 OCV C Smart, 23 Diban Ave, Elm Park, Hornchurch, RM12 4YF [JO01CN, TQ58]
G8 OCW M Chadwick, Ithaca, Heck Ln, Hensall, Goole, DN14 0RD [IO93KQ, SE52]
G8 ODA W Brown, 48 Woodkirk Gdns, Dewsbury, WF12 7JA [IO93ER, SE22]
G8 ODE Details withheld at licensee's request by SSL.
G8 ODK R Varley, 41 Lang Ln, West Kirby, Wirral, L48 5HQ [IO83JJ, SJ28]
G8 ODM P J Yeates, 69 Hilltop Way, Salisbury, SP1 3QQ [IO91CC, SU13]
G8 ODP D A Liddle, Holly Lodge, 16 Winkfield Rd, Windsor, SL4 4BG [IO91QL, SU97]
G8 ODT J S Laing, 138 Hillside Rd, Great Barr, Birmingham, B43 6NQ [IO92AN, SP09]
G8 ODV L Armstrong, 15 Aged Miners Homes, Fleming Field, Shotton Colliery, Durham, DH6 2JG [IO94HS, NZ34]
G8 ODX S E Smythe, 44 Askew Gr, Repton, Derby, DE65 6GR [IO92FU, SK32]
G8 OEF A C Roberts, 14 Mill Cl, Shepshed, Loughborough, LE12 9UA [IO92IS, SK42]
G8 OEJ E M Bray, Rothesay, 6 Empshott Rd, Southsea, PO4 8AU [IO90LT, SZ69]
G8 OEK P C Brown, The Well House, Stoke Rd, Westbury Sub Mendip, Wells, BA5 1HD [IO81PF, ST44]
G8 OEL Details withheld at licensee's request by SSL.
G8 OEM Details withheld at licensee's request by SSL.
G8 OEN A C Willgoose, 6 Skegby Ln, Mansfield, NG19 6QR [IO93JD, SK56]
G8 OER Details withheld at licensee's request by SSL.
G8 OEU T R Hipwood, 3 Camview, Paulton, Bristol, BS18 5XA [IO81RH, ST65]
G8 OEY S M Carpenter, 12 Rowans Cl, Farnborough, GU14 9EJ [IO91OH, SU85]
G8 OFA M P Cranage, 67 Church Rd, Laverstock, Salisbury, SP1 1QZ [IO91CB, SU13]
G8 OFJ Dr M J Turner, 23 Clarence Walk, Meadvale, Redhill, RH1 6NF [IO91VF, TQ24]
G8 OFN R Pashley, 50 Cherry Bank Rd, Sheffield, S8 8RD [IO93GI, SK38]
G8 OFO R P M Short, 10 Burgate Fields, Fordingbridge, SP6 1LR [IO90CW, SU11]
G8 OFR R J Coole, 31 Calcutt St., Cricklade, Swindon, SN6 6BA [IO91BP, SU19]
G8 OFX A R Nelson, 37 Brook Way, Romsey, SO51 7JZ [IO91GA, SU32]
G8 OFZ I McGowan, Meld House, Hawthorn Rd, Belle Vue, Shrewsbury, SY3 7NB [IO82PQ, SJ41]
G8 OGE E P Woolley, Carmel, Sampsons Ln, Crowhurst, Battle, TN33 9AU [JO00GV, TQ71]
G8 OGJ Details withheld at licensee's request by SSL.[Op: I P Beeby.]
G8 OGO Details withheld at licensee's request by SSL.[Op: Mike Wade, Watts Palace Cott, Broad Oak, nr Rye, E Sussex TN31 6EX. (01424) 882283. Interests: history and philosophy of communications, improving understanding via amateur radio, multi-interest networks.]
G8 OGP S G Martin, 60 Bridge Down, Bridge, Canterbury, CT4 5BA [JO01NF, TR15]
G8 OGR J C Holton, Sussex House, 24 Great Austins, Farnham, GU9 8JQ [IO91OE, SU84]
G8 OHA J G Wade, The Trees, Higher Rd, Yealmpton, Plymouth, Devon, PL8 2HR [IO80AI, SX55]
G8 OHC G Scholes, 14 Braemar Rd, Bulwell, Nottingham, NG6 9HN [IO93JA, SK54]
G8 OHG J N Myall, 52 Princethorpe Way, Binley, Coventry, CV3 2HF [IO92GJ, SP37]
G8 OHL Details withheld at licensee's request by SSL.

G8 OHM Details withheld at licensee's request by SSL.[Correspondence to: N Gutteridge, G8BHE, obo The South Birmingham Radio Society, c/o Hampstead House, Fairfax Road, West Heath, Birmingham, B31 3QY.]
G8 OHP C Cainsford-Betty, 84 Eastwick Dr, Great Bookham, Bookham, Leatherhead, KT23 3NX [IO91TG, TQ15]
G8 OHR Details withheld at licensee's request by SSL.
G8 OHW I G M Roberts, 13 Fearnley Dr, Ossett, WF5 9EU [IO93FQ, SE22]
G8 OID C R Vaslet, Little Copse Farm, Heath End, East Woodhay, Newbury, Berks, RG20 0AT [IO91HI, SU46]
GW8 OIG C D Osborn, Lowfield, Station Rd, Raglan, NP5 2EP [IO81NS, SO40]
GW8 OIJ A F Stark, 5 Ladyhill Rd, Alway, Newport, NP9 9RY [IO81MO, ST38]
G8 OIM D Langford, Chancton Hollow, Chestnut Cl, Storrington, Pulborough, RH20 3PA [IO90SW, TQ11]
GM8 OIO J H N Baxter, 14 Davidson Pl, Aberdeen, AB1 7RL [IO87TH, NJ72]
G8 OIS Details withheld at licensee's request by SSL.
G8 OIV C Merrell, 40 Fanton Walk, Wickford, SS11 8QT [JO01GO, TQ79]
G8 OIY Dr S A Robertson, 109 Camden Rd, London, NW1 9HA [IO91WN, TQ28]
G8 OJK V Willett, 20 The Green, Sharlston, Sharlston Common, Wakefield, WF4 1EF [IO93HP, SE31]
G8 OJQ A Hopkinson, Springfield, Neston Rd, Ness, South Wirral, L64 4AR [IO83LG, SJ37]
G8 OJR W J Pearce, 160 Philip Ln, Tottenham, London, N15 4JN [IO91XO, TQ38]
G8 OKB R J McCann, 14 Sandyacre Way, Stourbridge, DY8 1JD [IO82WK, SO98]
G8 OKD M R Bailey, 37 Dane Rd, Dane Bank, Denton, Manchester, M34 2HZ [IO83WK, SJ99]
G8 OKE R A Brown, 8 Grassmere Way, Waterlooville, PO7 8QD [IO90LV, SU71]
G8 OKI L J Mather, 69 Wadnall Way, Knebworth, SG3 6DT [IO91VU, TL21]
G8 OKN M T Gallagher, Sharmer House, Fosse Way, Radford Semele, Leamington Spa, CV31 1XH [IO92GG, SP36]
GW8 OKR B Kirkpatrick, Amroth, 74 Heol Don, Whitchurch, Cardiff, CF4 2AT [IO81JM, ST18]
G8 OKS B Dawson, 20 Carthorpe Dr, Billingham, Cleveland, TS23 3DJ [IO94IO, NZ42]
G8 OKZ D P Shillington, 6 Moss Cl, Willaston, South Wirral, L64 2XQ [IO83LG, SJ37]
G8 OLD Details withheld at licensee's request by SSL.
GI8 OLH T S Lavery, 2 Seapark, Castlerock, Coleraine, BT51 4TH [IO65OD, C73]
G8 OLI R T W Walker Brunel University ARS, 161 Long Ln, Hillingdon, Uxbridge, UB10 9JN [IO91SM, TQ08]
G8 OLK P C Smith, 64 Church Rd, Worle, Weston Super Mare, BS22 9DE [IO81MI, ST36]
G8 OLL R J Porter, 1 Knockley Cttgs, Parkend Rd, Bream, Lydney, GL15 6JZ [IO81RR, SO60]
G8 OLY D C Curwell, 9 St. Georges Rd, Aldershot, GU12 4LD [IO91OF, SU85]
G8 OMB D G Parker, 146 Merlin Ave, Nuneaton, CV10 9QJ [IO92FM, SP39]
G8 OMC D Smith, 71 Ashbourne Ave, Newsprings, Aspull, Wigan, WN2 1HW [IO83QN, SD60]
G8 OMI N J Barnacle, 1 Henleydale, Stratford Rd, Shirley, Solihull, B90 4AT [IO92CJ, SP17]
G8 OMQ D N Bliss, 11 Bubblestone Rd, Otford, Sevenoaks, TN14 5PN [JO01CH, TQ55]
G8 OMR P J Fellingham Brighton & Dist, 26 Fitch Dr, Bevendean, Brighton, BN2 4HX [IO90WU, TQ30]
G8 OMU Details withheld at licensee's request by SSL.
G8 OMW F H Rowan, 91 St. Nicholas Rd, Littlemore, Oxford, OX4 4PW [IO91JR, SP50]
G8 ONF L Tonge, 33 Pendlebury Rd, Gatley, Cheadle, SK8 4BU [IO83VJ, SJ88]
GW8 ONP J C Eastwood, 52 Glanceulan, Penrhyncoch, Aberystwyth, SY23 3HF [IO82AK, SN68]
G8 ONR M J Loader, 20 Chollacott Cl, Whitchurch, Tavistock, PL19 9BW [IO70WM, SX47]
G8 ONS K E Creighton, 10 Oram Cl, Allery Banks, Morpeth, NE61 1XF [IO95DD, NZ28]
G8 ONX Details withheld at licensee's request by SSL.
G8 ONY B W Goodhew, 101 Brier Rd, Sittingbourne, ME10 1YL [JO01II, TQ86]
G8 OOF Dr G T H Ellison, Sch of Bio & Chem Sciences, University of Greenwich, Wellington St, Woolwich, SE18 6PF [JO01AL, TQ47]
G8 OOQ Dr M H Barton, 23 Caledonia Pl, Bristol, BS8 4DL [IO81QK, ST57]
G8 OOS M J Reeson, 19 Southlands Ave, Louth, LN11 8EW [JO03AI, TF38]
G8 OOV D G M Platten, Boscarhyn Syra Cl, St Kew Highyway, Bodmin, Cornwall, PL30 3ED [IO70ON, SX07]
G8 OPA P C Barry, 32 Rutland Ave, Sidcup, DA15 9DZ [JO01BK, TQ47]
G8 OPC D B Crawley, 9 Gwynns Walk, Hertford, SG13 8AD [IO91XT, TL31]
G8 OPG A A Foss, 3 Rowan Cl, Wincanton, BA9 9SG [IO81TB, ST72]
G8 OPI J A S Spooner, 30 Counting House L, Great Dunmow, Essex, CM6 1BX [JO01EV, TL62]
G8 OPO G J Bartels, 43 Willowbank, Chippenham, Wilts, SN14 6QG [IO81WK, ST97]
G8 OPP J H Birkett, 13 The Strait, Lincoln, LN2 1JD [IO93RF, SK97]
G8 OPS G L O'Connor, 14 The Knoll, Great Gonerby, Grantham, NG31 8JY [IO92QW, SK93]
G8 OPV R J Staley, 237 Grange Rd, Kings Heath, Birmingham, B14 7RT [IO92BK, SP08]
G8 OPY G Winston, 3 Bowbank Cl, Shoeburyness, Southend on Sea, SS3 9NU [JO01JM, TQ98]
G8 OQG P Jobbins, 35 Keys Ave, Horfield, Bristol, BS7 0HQ [IO81RL, ST57]
G8 OQN Details withheld at licensee's request by SSL.
G8 OQO S Spacagna, Salterns, 52 Cupernham Ln, Romsey, SO51 7LG [IO90GX, SU32]
G8 OQP T A Spacagna, 2 Crook Cttgs, Crook Hill, Braishfield, Romsey, SO51 0QB [IO91GA, SU32]
G8 OQR M Crossman, 31A Eastcote Gr, Southend on Sea, SS2 4QA [JO01IN, TQ88]
G8 OQT Details withheld at licensee's request by SSL.[Op: J W Lambert. Station located near Rickmansworth.]
GW8 OQV W D Jackson, Pleasant Cottage, Brockweir Common, Brockweir, Gwent, NP6 7NT [IO81QR, SO50]
G8 OQW D P Barber, 205 Meadgate Ave, Chelmsford, CM2 7NJ [JO01FR, TL70]
G8 ORE Details withheld at licensee's request by SSL.
G8 ORG I McLuskie, 9 Holly Ln, Alsager, Stoke on Trent, ST7 2RS [IO83UC, SJ85]
GJ8 ORH Details withheld at licensee's request by SSL.
G8 ORM M E Baguley, 42 Kendall Ave, Shipley, BD18 4DY [IO93CT, SE13]
G8 ORO D Coulter, 5 High Gr, Whitehaven, CA28 6TA [IO84FN, NX91]
G8 ORQ J W Brock, 78 Capel Gdns, Ilford, IG3 9DG [JO01BN, TQ48]
G8 ORR Details withheld at licensee's request by SSL.
G8 ORU S Capper, 128 Wardrew Rd, St. Thomas, Exeter, EX4 1EZ [IO80FR, SX99]
G8 ORV Details withheld at licensee's request by SSL.[Op: F C Thorogood. Station located near Southend-on-Sea.]
G8 ORX K C Rashleigh, Menabilly, 43 Oxshott Way, Cobham, KT11 2RU [IO91TH, TQ16]
G8 ORY H M Sandler, 1 Beech Lawns, Torrington Park, London, N12 9PP [IO91VO, TQ29]
G8 ORZ D J Thawaites, 12 Waldeck Rd, Chiswick, London, W4 3NP [IO91UL, TQ17]
G8 OSC M S Saunders, 15 Church Farm Rd, Heacham, Kings Lynn, PE31 7JB [JO02GV, TF63]
G8 OSD Details withheld at licensee's request by SSL.
G8 OSE Details withheld at licensee's request by SSL.
G8 OSF C Doggett, Quest Cottage, 66 St. Leonards Rd, Chesham Bois, Amersham, HP6 6DR [IO91QQ, SU99]
G8 OSG N P McAlpine, 52 Sparrows Herne, Basildon, SS16 5HL [JO01FN, TQ78]
G8 OSH N G Hubbard, 16 Chipperfield Cl, New Bradwell, Milton Keynes, MK13 0EP [IO92OB, SP84]
G8 OSJ D M Halliwell, 46 Parklands, Kidsgrove, Stoke on Trent, ST7 4US [IO83VC, SJ85]
G8 OSN Details withheld at licensee's request by SSL.
G8 OSX K R Dawson, Mayfield House, 3 The Green, Bonehill, Tamworth, B78 3HW [IO92DO, SK10]
G8 OSZ S Ashley, 23 Blenheim Rd, Wellingborough, NN8 5YJ [IO92PH, SP86]
G8 OTA H D C C O Tani, 23 Bennett St., Bath, BA1 2QL [IO81TJ, ST76]
G8 OTC C I Anderson, 1 Winchester Cl, Lichfield, WS13 7SL [IO92CQ, SK11]
G8 OTD S P Ballard, 20 Gainsborough Cl, Welland, Malvern, WR13 6SH [IO82UB, SO74]
G8 OTE Details withheld at licensee's request by SSL.
G8 OTG R A Cannon, 111 Brangbourne Rd, Bromley, BR1 4LP [IO91XK, TQ37]
G8 OTH C Churchill, 87 Bradley Cres, Shirehampton, Bristol, BS11 9SR [IO81PL, ST57]
GM8 OTI Dr J A Cooke, 6 Greenbank Terr, Edinburgh, EH10 6ER [IO85JW, NT27]
G8 OTJ Details withheld at licensee's request by SSL.
G8 OTP K Owens, 124 Orchard Way, Addlestone, KT15 1LW [IO91SI, TQ06]
G8 OTS T R Ellinor Ariel Radio Group, 5 Parkwood Rd, Banstead, SM7 1JJ [IO91VH, TQ25]
G8 OTU A E Hall, 39 West Rd, West Heath, Congleton, CW12 4HH [IO83VD, SJ86]
G8 OTV S A Morse, 3 Lydiates Cl, Brownswall Est, Sedgley, Dudley, DY3 3ND [IO82WM, SO99]
G8 OTZ D R Logan, 283 Sandy Ln, Droylsden, Manchester, M43 7JU [IO83WL, SJ99]
G8 OUD M R Janaway, 18 Anson Dr, Sholing, Southampton, SO19 8RP [IO90IV, SU41]
G8 OUH I R Harfield, White Gates, Crofton Ave, Lee on The Solent, Hants, PO13 9NJ [IO90JT, SU50]
G8 OUI D K Baines, 1 Carole Cl, Sutton Leach, St. Helens, WA9 4PW [IO83PK, SJ59]
GW8 OUM D J Briggs, 69 Kimberley Villas, Tredegar, Gwent, NP2 3LD [IO81JS, SO10]
G8 OUS S W Greendale, 2 Elfleda Rd, Cambridge, CB5 8LZ [JO02BF, TL45]
G8 OUT B Horrocks, 17 Wood Gr, Whitefield, Manchester, M45 7ST [IO83UN, SD80]
G8 OUU Details withheld at licensee's request by SSL.
G8 OUX Details withheld at licensee's request by SSL.
G8 OVA L J Boddington, 33 Sorrel House, Tyburn Rd, Birmingham, B24 0TQ [IO92CM, SP19]
G8 OVH N J Dobson, 38 Crail Cl, Wokingham, RG41 2PZ [IO91NJ, SU86]
G8 OVO N M Lihou, 6 St. Laurence Ave, Warwick, CV34 6AR [IO92EG, SP26]
G8 OVS Details withheld at licensee's request by SSL.
G8 OVV B Higton, Brydon House, 20 Lydstep Cl, Oakwood, Derby, DE21 2RY [IO92GW, SK33]
G8 OWA Dr R Lewin, 180 Ladybank Rd, Mickleover, Derby, DE3 5RR [IO92FV, SK33]
G8 OWB I Morris, 14 Storth Bank, Glossop, SK13 9UX [IO93AK, SK09]
G8 OWL Details withheld at licensee's request by SSL.
G8 OWM I R Mountford, 6 Windley Cres, Darley Abbey, Derby, DE22 1BZ [IO92GW, SK33]
G8 OWN Details withheld at licensee's request by SSL.
G8 OWO K D Metcalf, 58 Forest Gate, Evesham, WR11 6XY [IO92AB, SP04]
G8 OWS J D Greenall, 6 Neasham Dr, Darlington, DL1 4LG [IO94FM, NZ31]
G8 OWZ O W Cockram, 446 Holdenhurst Rd, Bournemouth, BH8 9AE [IO90BR, SZ19]
G8 OXA S A Clarey, 34 Venetian Cres, Darfield, Barnsley, S73 9PL [IO93HM, SE40]
G8 OXD P A Brown, 4 Elm Ct, Walnut Walk, Polegate, BN26 5AG [JO00CT, TQ50]
G8 OXE M P Brooks, 1 Church St., Wimblington, March, PE15 0QS [JO02BM, TL49]
G8 OXG N C Powell, 42 Sheraton Dr, Kidderminster, DY10 3QR [IO82VJ, SO87]
G8 OXI E D Mannix, 16 Cromer Villas Rd, London, SW18 1PN [IO91VK, TQ27]
G8 OXR C J Drewitt, 33 The Roundway, Kingskerswell, Newton Abbot, TQ12 5BN [IO80FM, SX86]
G8 OXS A M Lambert, Tudor House, Alma Rd, Headley Down, Bordon, GU35 8JR [IO91OC, SU83]
G8 OXU C W Madge, 13 Blackwood Chine, South Woodham Ferrers, Chelmsford, CM3 5FZ [JO01HP, TQ89]
G8 OXX P A Bailey, 236 Sandy Ln, Droylsden, Manchester, M43 7JX [IO83WL, SJ99]
GW8 OXZ V E Mander, 4 Withy Ave, Forden, Welshpool, SY21 8NJ [IO82KO, SJ20]
GI8 OYA Details withheld at licensee's request by SSL.
G8 OYB K R J Armstrong, 2 Blackthorn Gr, Shawbirch, Telford, TF5 0LL [IO82RR, SJ61]
G8 OYF J F Popplewell, 25 Bowling Green Ln, The Groves, York, YO3 7NZ [IO93LX, SE65]
G8 OYL W E Shave, 26 Hessle Ave, Boston, PE21 8DA [IO92XX, TF34]
G8 OYM P R Taylor, 59 Penlands Vale, Steyning, BN44 3PL [IO90UV, TQ11]
G8 OYQ M F Everitt, 48 Rant Meadow, Hemel Hempstead, HP3 8EQ [IO91SR, TL00]
G8 OYS Details withheld at licensee's request by SSL.
GW8 OYT B N Jones, 6 Rhodfa Maes Hir, Rhyl, LL18 4JF [IO83GH, SJ08]
G8 OYY J D Bishop, 89 Milton Cres, East Grinstead, RH19 1TQ [IO91XC, TQ33]
G8 OZD A J Batty, 15 Bracken Dr, Baguley, Manchester, M23 1LT [IO83UJ, SJ88]
G8 OZH J A Burrell, 6 Blenheim Croft, Brackley, NN13 7ET [IO92KA, SP53]
G8 OZJ C T Dodd, 122 Rampton Rd, Willingham By Stow, Willingham, Cambridge, CB4 5JF [JO02AH, TL46]
GW8 OZO R H Owen, Gongl Gam, Aberffraw, Tycroes, Gwynedd, LL63 5RJ [IO73SF, SH37]
G8 OZP R Platts, 220 Rolleston Rd, Burton on Trent, DE13 0AY [IO92EU, SK22]
G8 OZQ S C Pallett, 6 Lancaster Cl, Agar Nook, Coalville, LE67 4TG [IO92IR, SK41]
G8 OZT N Morley, 155 Fitton Rd, St Germans, Kings Lynn, Norfolk, PE34 3AY [JO02EQ, TF51]
GM8 OZW Details withheld at licensee's request by SSL.
G8 OZY P F E Harrison, 91 Obelisk Rise, Northampton, NN2 8QU [IO92NG, SP76]
G8 OZZ Details withheld at licensee's request by SSL.
G8 PAD Details withheld at licensee's request by SSL.
G8 PAG M R Rose, 20 Broad Piece, Soham, Ely, CB7 5EL [JO02DI, TL57]
GM8 PAH D J Schofield, 3 Craiglockhart Gr, Edinburgh, EH14 1ET [IO85JV, NT26]
G8 PAI D J Rout, Two Akers, Wrabness Rd, Ramsey, Harwich, CO12 5NE [JO01OW, TM13]
G8 PAL P Hankinson, 5 Britain St., Bury, BL9 9PD [IO83UN, SD80]
G8 PAN S W Day, 16 Blenheim Way, Market Harborough, LE16 7LQ [IO92ML, SP78]
G8 PAT P G McGuinness, 9 Farmdale Rd, Carshalton, SM5 3NG [IO91VI, TQ26]
G8 PBE W Parkin, 2 Winchester Cl, Moor Ln, Wilmslow, SK9 6BZ [IO83UH, SJ88]
G8 PBH Dr A J Kent, 54 Mayfield Dr, Stapleford, Nottingham, NG9 8JG [IO92IW, SK43]
G8 PBI I A Murphy, 1 Park Ln, St. Clement, Truro, TR1 1SX [IO70LG, SW84]
G8 PBJ T D Davis, Russett House, Goddards Green Rd, Benenden, Cranbrook, TN17 4AN [JO01HB, TQ83]
GW8 PBM A J D Royston, 12 Hurford Cres, Graigwen, Pontypridd, CF37 2LD [IO81HO, ST09]
G8 PBV W K Smiles, 4 Norham Rd, Newton Hall, Durham, DH1 5NW [IO94FT, NZ24]
GW8 PBX D E Jones, Pen-y-Gorlan, 56 Wern, Llanfairpwllgwyngyl, Anglesey Gwynedd, LL61 5AQ [IO73VF, SH57]
G8 PCA F D Armstrong, 30 Church St., Great Baddow, Chelmsford, CM2 7HY [JO01GR, TL70]
G8 PCB F J Mitchell, 18 Castle Mount, Tisbury, Salisbury, SP3 6PP [IO81WB, ST92]
G8 PCS C M Ladds, 4 Tower Ln, Alnwick, NE66 1XR [IO95DJ, NU11]
GJ8 PCY P J Falle, 2 Greystones, Gorey Village Main R, Grouville, Jersey, JE3 9EP
G8 PDE W Burin, 40 Cheltenham Dr, The Cotswolds, Boldon, Boldon Colliery, NE35 9HE [IO94GW, NZ36]
GI8 PDK Dr D A Courtney, Fort House, 79 Fort Rd, Ballylesson, Belfast, BT8 8LX [IO74AM, J36]
G8 PDM C A Commander, 8 Cannon Pl, Hampstead, London, NW3 1EJ [IO91VN, TQ28]
G8 PDP R Hinchliffe, 34 Oaklea, Ash Vale, Aldershot, GU12 5HP [IO91PG, SU85]
G8 PDY S Procter, 8 Pond End Rd, Sonning Common, Reading, RG4 9SA [IO91MM, SU78]
G8 PEA K Wibberley, 5 Marston Rd, Croft, Leicester, LE9 3GX [IO92JN, SP59]
G8 PEE N C Pike, 201 Bredon Ave, Binley, Coventry, CV3 2FD [IO92GJ, SP37]
G8 PEN C A Vernon, 48 Long Beach, Hemsby, Great Yarmouth, Norfolk, NR29 4JD [JO02UQ, TG51]
G8 PEO A W Rose, 11 Ploughmans Dr, Shepshed, Loughborough, LE12 9SG [IO92IS, SK42]
G8 PEU D R Dudley, 23 Davies Ave, Heald Green, Cheadle, SK8 3PF [IO83VI, SJ88]
GM8 PEV R M Evans, Glen Park, 2 Main St., Saline, Dunfermline, KY12 9TL [IO86FC, NT09]
GI8 PFB R J Rosborough, 9 Windrush Park, Belfast, BT8 4LZ [IO74BN, J36]
G8 PFE Details withheld at licensee's request by SSL.
G8 PFG Details withheld at licensee's request by SSL.[Station located near Corsham.]
G8 PFO Details withheld at licensee's request by SSL.
G8 PFP Dr P J Buttery, 25 de Verdun Ave, Belton, Loughborough, LE12 9TY [IO92HS, SK42]
G8 PFR M B Gibson, 64 The Chase, Penn, High Wycombe, HP10 8BA [IO91PP, SU99]
G8 PFS Details withheld at licensee's request by SSL.
G8 PFT L P Hinson, 47 Saxon Wood Rd, Shirley, Solihull, B90 4JR [IO92CJ, SP17]
G8 PFZ R Harrison, 19 Southey Way, Larkfield, Aylesford, ME20 6TS [JO01FH, TQ65]
G8 PG Details withheld at licensee's request by SSL.[Op: A D Taylor, ERD, FISTC, AECS Cert. 37 Pickerill Road, Greasby, Merseyside, L49 3ND.]
G8 PGD Details withheld at licensee's request by SSL.
G8 PGF A L G Price, 11 Gatcombe Gdns, Titchfield, Fareham, PO14 3DR [IO90JU, SU50]
G8 PGH K C James, 44 The Oakfield, Littledean Hill Rd, Cinderford, Glos, GL14 X [IO81VU, SO81]
G8 PGI P M Lord, 32 Scotland Way, Countesthorpe, Leicester, LE8 5QZ [IO92KN, SP59]
GI8 PGJ D H Campbell, 74 Richmond Ct, Lisburn, BT27 4QX [IO64XM, J26]
G8 PGU L Waterfall, 14 Gloweth View, Highertown, Truro, TR1 3JZ [IO70KG, SW74]
G8 PHB P W McKenzie, 21 Arnside Walk, Chapel House, Newcastle upon Tyne, NE5 1BT [IO94DX, NZ16]
G8 PHG Dr B M Cook, Hinsley Mill House, Hinsley Mill Ln, Market Drayton, TF9 1HP [IO82SV, SJ63]
G8 PHJ M J Palmer, 16 Trulock Rd, Tottenham, London, N17 0PH [IO91XO, TQ39]
G8 PHM Details withheld at licensee's request by SSL.
G8 PHN Details withheld at licensee's request by SSL.
G8 PHS E S Campbell, 2 Russell Ave, March, PE15 8EL [JO02AN, TL49]
GM8 PHU J Savage-Lowden, Carn Bheag Glaichbea, Kiltarlity By Beauly, Inverness-shire, IV4 7HR [IO77SK, NH53]
G8 PHV R N Wetton, 8 St. Moritz Cl, Old Northwick Ln, Worcester, WR3 7ND [IO82VF, SO85]
G8 PIB Details withheld at licensee's request by SSL.[Station located near Bexley.]
G8 PIC C M Pomphrett, 113 Prospect Rd, Scarborough, YO12 7LF [IO94TG, TA08]
G8 PID R Noble, 389 Birmingham Rd, Walsall, WS5 3NU [IO92AN, SP09]
G8 PIL T Summerhayes, 36 Hesketh Cres, Swindon, SN3 1RY [IO91CN, SU18]
G8 PIP P Elwell, 4 Richmond Gr, Wollaston, Stourbridge, DY8 4SF [IO82WL, SO88]
G8 PIQ P T Tagg, 22 Hambledon Rd, Waterlooville, PO7 7UB [IO90LV, SU60]
GM8 PIV E F Souter, 10 James Gray St., Shawlands, Glasgow, G41 3BS [IO75UT, NS56]
GM8 PIW D K Smith, 26 Park Ave, Oakville, Ontario, Canada L6J 3X8
G8 PIY D C Clifton, 10 Scotney Rd, Basingstoke, RG21 5SR [IO91KG, SU65]
G8 PJC J P McDonald, 60 Deanfield Rd, Henley on Thames, RG9 1UU [IO91NM, SU78]
G8 PJD P J Deffee, 18 Poplar Rd, Kensworth, Dunstable, LU6 3RS [IO91RU, TL01]
G8 PJE P J Evans, Flat 11, 45 Kappadokias, 2028 Strovolos, Nicosia, Cyprus, X X
G8 PJF E D Summers, 7 Nursery Ave, Bexleyheath, DA7 4JX [JO01BL, TQ47]
G8 PJH P J Hazelton, 20 Greenways, Chelmsford, CM1 4EF [JO01FS, TL70]
G8 PJL J P A Thomson, The Strip, Shepherds Green, Rotherfield Greys, Oxon, RG9 4QW [IO91MN, SU78]
GM8 PJM R J Brockie, Fisherie Post Offic, Turriff, Aberdeen, AB5 7SP [IO87TH, NJ72]
G8 PJQ C D Cole, 70 Throgmorton Rd, Yateley, Camberley, Surrey, GU17 7FA [IO91OH, SU85]
G8 PJX H Miles, 63 North View, Chilton Moor, Houghton Spring, Tyne & Wear, DH4 5NW [IO94GU, NZ34]
G8 PK B Wilson, 15 Tannery Rd, Sawston, Cambridge, CB2 4UW [JO02CD, TL45]
GW8 PKB L D Rudge, Eirianfa, 8 Penrallt Est, Llanystumdwy, Criccieth, LL52 0SR [IO72UW, SH43]
G8 PKG I J Bosworth, 6 Busbys Cl, Stonesfield, Witney, OX8 8EU [IO91GU, SP31]
G8 PKJ G G Rowland, 18 Heights Way, Est, Leeds, LS12 3SN [IO93ET, SE23]
GM8 PKL Details withheld at licensee's request by SSL.
G8 PKM C E Mitchell, 6 Morden Rd, Papworth Everard, Cambridge, Cambs, CB3 8UN [IO92WG, TL26]
G8 PKN J D Hoare, 67 Upthorpe Dr, Wantage, OX12 7DG [IO91HO, SU48]
G8 PKP J P Lydiate, 11 Avondale Rd, Pitsea, Basildon, SS16 4TT [JO01FN, TQ78]
G8 PKV M S James, 4 Humber Cl, Woosehill, Wokingham, RG41 3UA [IO91NJ, SU76]
G8 PLJ A Waterson, 93 Darfield Rd, Cudworth, Barnsley, S72 8HG [IO93HN, SE30]
G8 PLL Details withheld at licensee's request by SSL.
G8 PLO R A Clubley, Church Hill House, High St, Wethersfield, Essex, CM7 4BY [JO01FW, TL73]
G8 PLQ A F S Cook, 84 Maple Cp490, Hudson Heights, Quebel, Canada, JOP 1JO
G8 PLR P N Paterson, 1 The Bakery, Northend, Leamington Spa, CV33 0TY [IO92GD, SP35]
G8 PLW Details withheld at licensee's request by SSL.
G8 PLZ D A Peck, 2 Cameron Rd, Cambridge, CB4 2LY [JO02BF, TL46]
G8 PMA L K Pennell, Mill House, Shrewton, Salisbury, Wilts, SP3 4JU [IO91BE, SU04]
G8 PMB P M Blackford, 11 Hendras Parc, Carbis Bay, St. Ives, TR26 2TT [IO70GE, SW53]
G8 PME B M P Toon, 24 Holly Rd, Weymouth, DT4 0BB [IO80SO, SY67]
GW8 PMJ D A Hughes, 70 Grovers Field, Abercynon, Mountain Ash, CF45 4PQ [IO81IP, ST09]
G8 PMQ R Whittaker, 77 Church Rd, Owlsmoor, Camberley, Surrey, GU15 4TP [IO91PI, SU86]
G8 PMR L A W Morris, 17 Kestrel Cl, Hornchurch, RM12 5LS [JO01CM, TQ58]
G8 PMT M J Tribe, 17 Warwick Rd, Anerley, London, SE20 7YN [IO91XJ, TQ36]
G8 PMU Details withheld at licensee's request by SSL.
G8 PNC Details withheld at licensee's request by SSL.
GW8 PNE P K Griffiths, 5 Dumfries St., Treherbert, Treorchy, CF42 5PL [IO81FQ, SS99]
G8 PNM N S J Cocking, Ridgeway, 18 Delmont Gr, Stroud, Glos, GL5 1UN [IO81VR, SO80]
G8 PNN G Emmerson, 72 The Gables, Widdrington, Morpeth, NE61 5RB [IO95EF, NZ29]
G8 PNO Details withheld at licensee's request by SSL.
G8 PNQ G R Baptiste, 3A Westfield Rd, Cheam, Sutton, SM1 2JY [IO91VI, TQ26]

GW8 PNR K H Snow, 9 Dyffryn Rd, Gorseinon, Swansea, West Glam, SA4 6BB [IO71XP, SS59]
G8 PNX Details withheld at licensee's request by SSL.[Op: M E Slattery.]
G8 PNZ K Green, 17 Cedar Gr, Blurton, Stoke on Trent, ST3 2AU [IO82WX, SJ84]
G8 PO Cdr J E Ironmonger OBE, 15 Monks Way, Hill Head, Fareham, PO14 3LU [IO90JT, SU50]
G8 POC C F Lorton, 102 Oxford Rd, Cumnor, Oxford, OX2 9PQ [IO91IR, SP40]
G8 POE J A Phillips, 235 Barn Mead, Harlow, CM18 6ST [JO01BS, TL40]
G8 POG P I Wood, 23 Shipley Ave, Newcastle upon Tyne, NE4 9QY [IO94EX, NZ26]
G8 POH Details withheld at licensee's request by SSL.
G8 POK G West, 6 Willerton Cl, Chidswell, Dewsbury, WF12 7SQ [IO93EQ, SE22]
G8 POL M T W Williams, 22 Charlecote Dr, Nottingham, NG8 2SB [IO92JW, SK53]
G8 PON A J Morgan, 3 Clarence Ct, George Trollope Rd, Watton, Thetford, IP25 6AS [JO02JN, TF90]
G8 POO S G Robinson, 23 Jameson Dr, Cragside, Corbridge, NE45 5EX [IO84XX, NY96]
G8 POR Details withheld at licensee's request by SSL.
G8 POS A Axon, Ashton Lodge, Tudor Gr, Groby, Leicester, LE6 0YL [IO92JP, SK50]
G8 POZ D J Warner, Rondor, Colchester Rd, Thorpe-le-Soken, Clacton on Sea, CO16 0LA [JO01NU, TM12]
G8 PP L A G Parnell, 17 Beehive Ct, Gubbins Ln, Harold Wood, Romford, RM3 0RS [JO01CO, TQ59]
G8 PPA S Ornstein, 19 Colvin Gdns, Barkingside, Ilford, IG6 2LH [JO01AO, TQ49]
G8 PPD A C Saunders, Suffolk House, Main Rd, Chelmondiston, Ipswich, IP9 1DX [JO01OX, TM23]
G8 PPF R M E Rogers, 24 Treza Rd, Tolponds, Porthleven, Helston, TR13 9NB [IO70IC, SW62]
G8 PPG P B Rawlinson, 15 Elmbourne Dr, Belvedere, DA17 6JE [JO01BL, TQ47]
G8 PPN A Osmond, 275 London Rd, Bishops Stortford, CM23 3LS [JO01BU, TL41]
G8 PPQ G T Boakes, 8 Clarendon St., Herne Bay, CT6 8JX [JO01NI, TR16]
G8 PPR Details withheld at licensee's request by SSL.
G8 PPS Details withheld at licensee's request by SSL.[Op: S Larkin, The Old Forge, Brampton, Beccles, Suffolk, NR34 8EL.]
G8 PPZ A E Davis, 5 Bycullah Rd, Enfield, EN2 8EE [IO91WP, TQ39]
G8 PQ D C Derry, 38 Green Park Way, Chillington, Kingsbridge, TQ7 2HY [IO80DG, SX74]
G8 PQA A D Gapper, 56 Hollyguest Rd, Mount Hill, Hanham, Bristol, BS15 3RW [IO81RK, ST67]
G8 PQB G H Grantham, Dunroamin, Badby Rd West, Daventry, NN11 4HJ [IO92JG, SP56]
G8 PQJ J N Robinson, Rose Cottage, Wavering Ln, Gillingham, SP8 4NR [IO81UA, ST72]
G8 PQM P Tregear, 65 Sea Ln Gdns, Ferring, Worthing, BN12 5EG [IO90ST, TQ00]
G8 PQY I R Bailey, 2 Spring Hill, Leeds, LS16 8EA [IO93FU, SE23]
G8 PQZ G M Collier, 6 Copse Cl, Tilehurst, Reading, RG31 6RH [IO91LL, SU67]
G8 PRB L W Evans, 14 Blyth Cl, St. Catherines Hill, Christchurch, BH23 2TE [IO90CS, SZ19]
G8 PRC D T Hind Plymouth Radio Club, 4 Thornyville Villas, Plymouth, PL9 7LA [IO70WI, SX55]
G8 PRH A Hartley, 3 Rotherhead Cl, Horwich, Bolton, BL6 5UG [IO83RO, SD61]
G8 PRI Details withheld at licensee's request by SSL.
G8 PRJ S L Sanders, 19 Brunswick Gdns, Hainault, Ilford, IG6 2QU [JO01AO, TQ49]
G8 PRN R L Morley, 21 Meadow View, Skelmanthorpe, Huddersfield, HD8 9ET [IO93EO, SE21]
G8 PRP B D Youster, 24 Sunningdale Rd, Weston Super Mare, BS22 0XP [IO81MI, ST36]
G8 PRR Details withheld at licensee's request by SSL.
G8 PRU D K Egan Prudential ARS, 19 Sycamore Cl, Longmeadow, Dinas Powys, The Vale of Glam, CF64 4TG [IO81JK, ST17]
G8 PSC J W Benoy, 57 Bowers Park Dr, Woolwell, Plymouth, PL6 7SH [IO70WK, SX56]
G8 PSE N J Boid, Elmtree House, Dunswell Rd, Cottingham, HU16 4JB [IO93TS, TA03]
G8 PSF A P Ball, 20 Inverness Ave, Enfield, EN1 3NT [IO91XP, TQ39]
G8 PSJ Details withheld at licensee's request by SSL.
G8 PSL M A Jervis, 15 Mercer Gr, Wednesfield, Wolverhampton, WV11 3AN [IO82XO, SJ90]
G8 PSO R J E Gould, 20 Southwood Dr, Coombe Dingle, Bristol, BS9 2QU [IO81QL, ST57]
G8 PSP Details withheld at licensee's request by SSL.
G8 PSS J Meldrum, 7 Kerryhill Dr, Pity Me, Durham, DH1 5FN [IO94ET, NZ24]
G8 PST A Murray, 2 Scratton Rd, Southend on Sea, SS1 1EN [JO01IM, TQ88]
GM8 PSV B J Thomson, 48/7 North Gyle Gr, Edinburgh, EH12 8LF [IO85IW, NT17]
G8 PSZ C J Wood, Wudum Wic, Farm Cl, Farcroft Meadows, Market Drayton, TF9 3UH [IO82SV, SJ63]
G8 PTD Details withheld at licensee's request by SSL.
G8 PTF R A Duddin, 16 Gateley Rd, Oldbury, Warley, B68 0NU [IO92AL, SP08]
G8 PTH A N Emmerson, 71 Falcutt Way, Northampton, NN2 8PH [IO92NG, SP76]
G8 PTI H H Frowen, 16 Wyebank Way, Tutshill, Chepstow, NP6 7DN [IO81PP, ST59]
G8 PTJ Details withheld at licensee's request by SSL.
G8 PTK G W Fisher, 14 Newhall Dr, Bradford, BD6 1DG [IO93DS, SE12]
G8 PTL M J Fleming, 20 Church St., Chasetown, Burntwood, WS7 8QL [IO92AQ, SK00]
G8 PTN D P Stoney, 9 Mackinley Ave, Stapleford, Nottingham, NG9 8HU [IO92IW, SK43]
G8 PTP Details withheld at licensee's request by SSL.
G8 PTR Details withheld at licensee's request by SSL.
GW8 PTS W E Leddington, 4 Cherry Walk, Monmouth, NP5 4DE [IO81PT, SO41]
G8 PTW C K Wallace, Windy Ridge, Moor Ln, Diseworth, Leics, DE74 2QQ [IO92HT, SK42]
G8 PTY P Thornton-Evison, Greyfriars, Townsend, Gr, Wantage, OX12 0AT [IO91GO, SU39]
G8 PUB S R Lucas, 84 Woodman Rd, Warley, Brentwood, CM14 5AZ [JO01DO, TQ59]
G8 PUE J Taylor, The Jays, 5 Watling Cl, Bourne, PE10 9XL [IO92TS, TF02]
G8 PUH B R Merrell, 16 Box Cl, Broadfield, Crawley, RH11 9QT [IO91VC, TQ23]
G8 PUJ W J Mac Donald, 40 Latchett Rd, South Woodford, London, E18 1DJ [JO01AO, TQ49]
G8 PUK S P Mann, 1 Blackthorn Ave, Bramley, Rotherham, S66 0LU [IO93IK, SK49]
G8 PUN J M Keleher, 9 Broadwell Dr, Pennington, Leigh, WN7 3NE [IO83RL, SJ69]
G8 PUT C I Townsend, 2 Fieldhouse, Holmfirth, Huddersfield, HD7 1EN [IO93CN, SE10]
G8 PUX J N Tubis, Glovers, Tinkerpot Ln, West Kingsdown, Sevenoaks, TN15 6AB [JO01CH, TQ56]
G8 PUY N C Dowsett, 162 Brentwood Rd, Herongate, Brentwood, CM13 3PF [JO01EO, TQ69]
GW8 PVD Details withheld at licensee's request by SSL.
G8 PVG D D Hobbs, 46 Gloucester Rd, Bridgwater, TA6 6DZ [IO81LC, ST23]
G8 PVK R S Still, The Manor House, 5 Beechwood Ave, Boscombe, Bournemouth, BH5 1LY [IO90CR, SZ19]
GJ8 PVL Details withheld at licensee's request by SSL.
G8 PVM G W Hancox, 30 Ferndale Rd, Coal Aston, Sheffield, S18 6BU [IO93GH, SK37]
GW8 PVN J Hamilton, 39 Caestory Cres, Raglan, NP5 2EQ [IO81NS, SO40]
G8 PVR J R Riggs, Lee Wood, Wotter, Plympton, Plymouth, PL7 5EG [IO70XK, SX56]
G8 PW W D Manson, 12 Sheep Gate Dr, Tottington, Bury, BL8 3JZ [IO83TO, SD71]
G8 PWA D Lee, 49 Cockerell Cl, Wimborne, BH21 1XR [IO90AS, SZ09]
G8 PWD S M Geary, 6 Windles Row, Lyppard Woodgreen, Worcester, WR4 0RS [IO82VE, SO85]
G8 PWE Details withheld at licensee's request by SSL.
G8 PWG P F Byrne, 4 Brandwood, Staghills Rd, Newchurch, Rossendale, BB4 7UH [IO83UQ, SD82]
G8 PWK M Forsey, 84 Garner Rd, Walthamstow, London, E17 4HH [IO91XO, TQ39]
G8 PWO J C Thwaites, 15 Spring Head Rd, Kemsing, Sevenoaks, TN15 6QL [JO01CH, TQ55]
G8 PWT L S Cook, 16 Florence Rd, Maidstone, ME16 8EN [JO01GG, TQ75]
G8 PWU J A A Crossland, 1 Carter Ln, Flamborough, Bridlington, YO15 1LW [IO94WC, TA27]
G8 PWX A R Fraser, 43 Edith St., Tynemouth, North Shields, NE30 2PN [IO95GA, NZ36]
G8 PWY Details withheld at licensee's request by SSL.
G8 PX F A Jefferies, 1 Lovelace Rd, Oxford, OX2 8LP [IO91IS, SP51]
G8 PXA R W Lucyk, 106 Peet St., Derby, DE22 3RG [IO92GW, SK33]
G8 PXB Details withheld at licensee's request by SSL.[Op: S J Hopkins.]
G8 PXD Details withheld at licensee's request by SSL.
G8 PXG W I E Halls, 4 Hillside, Pembroke Rd, Erith, DA8 1DA [JO01CL, TQ57]
G8 PXO G Murray, Glenelg, 25 Prowses, Hemyock, Cullompton, EX15 3QG [IO80JV, ST11]
G8 PXU B J Gascoigne, 108 Blandford Ave, Castle Bromwich, Birmingham, B36 9JD [IO92CM, SP19]
G8 PXW Details withheld at licensee's request by SSL.
G8 PY D M Bell East Midlands Communications, Spindrift, The Green, Westborough, Newark, Notts, NG23 4HQ [IO93OC, SK85]
G8 PYD G W Farrell, 95 Washington Rd, Maldon, CM9 6JF [JO01IR, TL80]
G8 PYE P J Barrett, 9 Waverley Ave, Surbiton, KT5 9HD [IO91UJ, TQ16]
G8 PYG Details withheld at licensee's request by SSL.
G8 PYP Details withheld at licensee's request by SSL.[Operator: S J Damon. Correspondence c/o PO Box 703, West Moors, Ferndown, Dorset, BH22 0YB.]
G8 PYT Details withheld at licensee's request by SSL.
G8 PYV R E R Baldwin, 8 Adams Cl, Ampthill, Bedford, MK45 2UB [IO92SA, TL03]
G8 PYX P K Hunt, 25 Pike Pursue Ln, Richmond, DL10 4PS [IO94DJ, NZ10]
G8 PZA T R Cassidy, Firbank, The St., Guston, Dover, CT15 5ET [JO01PD, TR34]
G8 PZD T D Wills, 66 Kipling Rd, St. Marks, Cheltenham, GL51 7DQ [IO81WV, SO92]
G8 PZE D W Southern, 54 Hillcrest Dr, Southdown, Bath, BA2 1HE [IO81TI, ST76]
G8 PZF B Simpson, 1 Hazelhurst Rd, Heaton, Bradford, BD9 6BJ [IO93CT, SE13]
G8 PZI W Nolan, South Lawn, 77 Reigate Rd, Reigate, RH2 0RE [IO91VF, TQ25]
G8 PZJ Details withheld at licensee's request by SSL.[Op: D Billingham. Station located near Purley.]
G8 PZL Details withheld at licensee's request by SSL.
G8 PZP Details withheld at licensee's request by SSL.
G8 PZR N W Clarke, Barn Cottage, Three Mile Pond Farm, Sawbridgeworth, Herts, CM21 9BZ [JO01BT, TL41]
G8 PZT Details withheld at licensee's request by SSL.[Station located near Kidderminster.]
G8 PZU J E Graham, 2 Hillside Cl, Brierfield, Nelson, BB9 5DS [IO83VT, SD83]
G8 PZW Details withheld at licensee's request by SSL.
G8 PZX F M A Gunn, 8 College Gdns, Hornsea, HU18 1EF [IO93VV, TA14]
G8 QM V J Flowers, 9 Laburnum Gr, Sunniside, Newcastle upon Tyne, NE16 5LY [IO94DW, NZ25]
G8 QR R F C Brake, 315 Unthank Rd, Norwich, NR4 7QA [JO02PO, TG20]
G8 QX K Hopkinson, 33 Everard Ave, Sheffield, S17 4LY [IO93FH, SK38]
G8 QZ H O Sills, 29 Briar Gate, Long Eaton, Nottingham, NG10 4BN [IO92IV, SK43]
G8 RAA D T Keight, 74 Moor View Dr, Coombe Valley, Teignmouth, TQ14 9UR [IO80FN, SX97]
G8 RAC J P Maines, Harlequins, Harlequin Ln, Crowborough, TN6 1HU [JO01BB, TQ52]
G8 RAF R E Finch RAF Locking ARC, 1 Cherrytree Cottage, Church Rd, Penn, High Wycombe, Bucks, HP10 8LN [IO91PP, SU99]
G8 RAJ R J Shore, 42 King George Ave, Moordown, Bournemouth, BH9 1TX [IO90BS, SZ09]
G8 RAL Details withheld at licensee's request by SSL.
G8 RAN K B Reeman, Zennor, Alfreda Ave, Hullbridge, Essex, SS5 6LT [JO01HP, TQ89]
G8 RAO A C Yates, 87 Princess Rd, Oldbury, Warley, B68 9PW [IO92AL, SP08]
GW8 RAS R Neville, 11 Heol Urban, Danescourt, Llandaff, Cardiff, CF5 2QP [IO81JL, ST17]
G8 RAU B R Lewis, 4 Davie Ln, Whittlesey, Peterborough, PE7 1YZ [IO92WN, TL29]
G8 RAX C N Hill, 183 Manchester Rd, Swinton, Manchester, M27 4FA [IO83UM, SD70]
G8 RBG J Loveday, 3 Chester Grange, Glebe Rd, Grimsby, DN33 2HW [IO93WM, TA20]
G8 RBI C L Allen, 8 Shoulbard, Fleckney, Leicester, LE8 8TX [IO92LM, SP69]
GW8 RBJ S Barnes, 72 Cwmavon Rd, Port Talbot, SA12 8RF [IO81CO, SS79]
G8 RBK I R Brown, 94 Darley Abbey Dr, Darley Abbey, Derby, DE22 1EF [IO92GW, SK33]
G8 RBQ L Edwards, 5 Windmill Rise, York, YO2 4TU [IO93KW, SE55]
G8 RBR W J Egerton, 26 Poplar Dr, Alsager, Stoke on Trent, ST7 2RW [IO83UC, SJ85]
G8 RBS P H Bickersteth, Tregarth, Fernsplat, Chacewater, Truro, TR4 8RJ [IO70KF, SW74]
G8 RBU T A Dewey, Dorket House, 93 Calverton Rd, Arnold, Nottingham, NG5 8FQ [IO93KA, SK54]
G8 RBV D Deighton, Stanhill Works, Tennyson Ave, Oswaldtwistle, Lancs, BB5 4QZ [IO83TR, SD72]
G8 RBW C D Ellison, 156 Dale Rd, Matlock Bath, Matlock, DE4 3PS [IO93FD, SK25]
G8 RBX L J C Fitzpatrick-Browne, 24 Beechmount Ave, Hanwell, London, W7 3AG [IO91TM, TQ18]
G8 RBY P V Hodson, 43 Thorpe Rd, Melton Mowbray, LE13 1SE [IO92NS, SK71]
G8 RCE K R Shergold, 102 Vicarage Cres, Redditch, B97 4RP [IO92AH, SP06]
G8 RCF D E Sayer, 15 Ashcroft, Chard, TA20 2JH [IO80MU, ST30]
G8 RCG M Arnfield, Cleabarrow, Plumley Moor Rd, Plumley, Knutsford, WA16 0TU [IO83SG, SJ77]
G8 RCL G A Whiston, 32 Ladywood Rd, Old Hall, Warrington, WA5 5QR [IO83QJ, SJ59]
G8 RCO D M Russell, 53 The Campions, Borehamwood, WD6 5QE [IO91UQ, TQ19]
G8 RCP Details withheld at licensee's request by SSL.
G8 RCU B Alsop, 10 Romney Cl, Houghton, Philadelphia, Houghton-le-Spring, DH4 4XH [IO94GU, NZ35]
G8 RCZ G Fermor, 26 Byron Rd, Exeter, EX2 5QN [IO80GR, SX99]
G8 RDA K J Forster, 10 Springfield Oval, Witney, OX8 5EG [IO91GT, SP31]
G8 RDB R George, Juniper Cottage, Hillesden Hamlet, nr Buckingham, Bucks, MK18 4BX [IO91LX, SP63]
G8 RDE Details withheld at licensee's request by SSL.
G8 RDG R G W Maltby, 21 Long Ln, Tilehurst, Reading, RG3 6YQ [IO91LJ, SU66]
G8 RDH Details withheld at licensee's request by SSL.
GW8 RDI R C Colclough, 6 Hill St., Mumbles, Swansea, SA3 4EF [IO81AN, SS68]
G8 RDJ J E Davies, 11 Bromley Rd, Broken Cross, Macclesfield, SK10 3LN [IO83WG, SJ87]
G8 RDK L H Mayhew, 47 Beeches Ave, Worthing, BN14 9JE [IO90TU, TQ10]
G8 RDL B A Obee, 1 Grange Gdns, Rayleigh, SS6 9BD [JO01HO, TQ79]
G8 RDN T N Sale, 20 Redwood Dr, Chase Terr, Walsall, West Midlands, WS7 8AS [IO92AQ, SK00]
G8 RDP J Webb, 6 Chatsworth Ave, Fleetwood, FY7 8EG [IO83LV, SD34]
G8 RDQ Details withheld at licensee's request by SSL.
G8 RDT K J Williams, Corn Acres, Dodwell, Stratford on Avon, Warks, CV37 9ST [IO92CE, SP15]
G8 RDX Details withheld at licensee's request by SSL.
G8 RDY P A Kilbride, 47 Arundel Dr, Battenhall, Worcester, WR5 2HU [IO82VE, SO85]
G8 RDZ Details withheld at licensee's request by SSL.
G8 REF P R Ellis, 15 Alexander Cl, Barrack Ln, Aldwick, Bognor Regis, PO21 4PS [IO90PS, SZ99]
GM8 REG R Bell, Fairview, Main St., Aberchirder, Huntly, AB54 7SY [IO87QN, NJ65]
G8 REM Details withheld at licensee's request by SSL.
G8 REO R Mitchell, Dial House, 4 Friendly Fold Rd, Ovenden, Halifax, HX3 5QF [IO93BR, SE02]
G8 REQ F E Robinson, 13 Dorset Dr, Pensby, Wirral, L61 8SX [IO83KI, SJ28]
G8 RER J Fothergill, 53 Meadow Ct, Ponteland, Newcastle upon Tyne, NE20 9RA [IO95DB, NZ17]
G8 RES M C Howard, Rosia, 22 Downham Rd, Watlington, Kings Lynn, PE33 0HS [JO02EP, TF61]
G8 REU Details withheld at licensee's request by SSL.[Op: R M Guy. Station located near Bournemouth.]
GW8 REV C I Marsh, 1 Beale Cl, Llandaff, Cardiff, CF5 2RU [IO81JM, ST17]
G8 REY Details withheld at licensee's request by SSL.
G8 REZ J A G Lavender, 59 Ferndale Rd, New Milton, BH25 5EX [IO90ES, SZ29]
G8 RF J R Raby, 38 Broadway, Codsall, Wolverhampton, WV8 2EL [IO82VP, SJ80]
G8 RFC R F Cassell, 1 St. Saviour Cl, Colchester, CO4 4PW [JO01LV, TM02]
G8 RFD P E Short, 193 Conygre Gr, Bristol, BS12 7HZ [IO81RM, ST67]
G8 RFE M L Wallace, 26 Parsons Dr, Glen Parva, Leicester, LE2 9NS [IO92JO, SP59]
G8 RFF K Richardson, 46 Kilnhurst Rd, Todmorden, OL14 6AX [IO83WR, SD92]
G8 RFK D J Gardner, 122 All Saints Ave, Maidenhead, SL6 6LT [IO91PM, SU88]
G8 RFL D A Robinson, 25 Angelica, Amington, Tamworth, B77 3JZ [IO92EP, SK20]
G8 RFN W J Seeney, 20 Dovehouse Cl, Eynsham, Witney, OX8 1EX [IO91HS, SP41]
GM8 RFO N Baxter, 24 Hillview Cres, Cults, Aberdeen, AB1 9RT [IO87TH, NJ72]
G8 RFP D Clarke, 29 Haugh Ln, Ecclesall, Sheffield, S11 9SB [IO93FI, SK38]
G8 RFQ Details withheld at licensee's request by SSL.
G8 RFV Details withheld at licensee's request by SSL.
G8 RFZ D A Cooke, 33 Wayside Ave, Bushey, Watford, WD2 3SH [IO91TP, TQ19]
G8 RGO M Robson, 8 Grange Rd, Burley in Whafedale, Burley in Wharfedale, Ilkley, LS29 7NF [IO93DV, SE14]
G8 RGU M Burt, Hartcliff Farm, Okeford Fitzpaine, Blandford Forum, Dorset, DT11 0EF [IO80UV, ST81]
G8 RHC J D Cranage, 67 Church Rd, Laverstock, Salisbury, SP1 1QZ [IO91CB, SU13]
G8 RHL J M Carver, 46 Orchard St., Wombwell, Barnsley, S73 8HQ [IO93HM, SE30]
G8 RHM K Hoggett, 14 Wyld Ct, Allesley Park, Coventry, CV5 9LQ [IO92FK, SP28]
G8 RHN M J Kirkham, Greyfriars, Hall Dr, Canwick, Lincoln, LN4 2RG [IO93RF, SK96]
G8 RHO L G Cole, St Petrocks, Lower Dingle, West Lmalvern, Worcs, WR14 4BQ [IO82TC, SO74]
GW8 RHP Rev J M Williams, The Vicarage, Brynymaen, Colwyn Bay, LL28 5EW [IO83DG, SH87]
G8 RHQ S Ormondroyd, 64 Witney Green, Lowestoft, NR33 7AP [JO02UK, TM59]
G8 RHT Details withheld at licensee's request by SSL.
G8 RHU E Carvill, 42 Park Dr Cl, Newhaven, BN9 0RR [JO00AT, TQ40]
G8 RIB P L Fallon, 17 Blundell Rd, Widnes, WA8 8SS [IO83OI, SJ48]
G8 RIC K J Murphy, 5 Wenlock Cl, Offerton, Stockport, SK2 5XP [IO83WJ, SJ98]
GW8 RIE Details withheld at licensee's request by SSL.
G8 RII P J Garnett, 2 The Rookery, Newton-le-Willows, WA12 9PW [IO83QK, SJ59]
G8 RIK R G T Milner, Lyndene, Holyhead Rd, Montford Bridge, Shrewsbury, SY4 1EE [IO82NR, SJ41]
G8 RIN R G Boardman, Ivy Dene, Redditch Rd, Alvechurch, Birmingham, B48 7TL [IO92AI, SP07]
G8 RIP M Walmsley, 3 Methuen Ave, Hoghton, Preston, PR5 0JN [IO83QR, SD52]
G8 RIT M F Coleman, West Cottage, Hill View Rd, Michelmersh, Romsey Hants, SO5 0NN [IO90HW, SU41]
G8 RIW B Harvey, 56 Oakwood Dr, Wybers Wood, Grimsby, DN37 9RN [IO93WN, TA20]
G8 RJB R J Bridgwater, 31 Pembroke Ave, Worthing, BN11 5QS [IO90TT, TQ10]
GM8 RJD Details withheld at licensee's request by SSL.[Op: J R Addinall, 93 Eskhill, Penicuik, Midlothian, EH26 8DE.]
G8 RJF K Freer, 54A High Ln East, West Hallam, Ilkeston, DE7 6HW [IO92HX, SK44]
G8 RJH J N E Rogers, 263 Eastbourne Rd, Polegate, BN26 5DL [JO00CT, TQ50]
G8 RJM S A Reap, Woodlands, Station Rd, Market Bosworth, Nuneaton, Warks, CV13 0NP [IO92HO, SK40]
G8 RJO D A Shaw, 31 Windwhistle Circle, Weston Super Mare, BS23 3TU [IO81MH, ST35]
GW8 RJU Details withheld at licensee's request by SSL.
G8 RJY D E Riches, 92 Barons Rd, Bury St. Edmunds, IP33 2LY [JO02IF, TL86]
G8 RJZ M Wills, 9 Allerdale Cl, Thirsk, YO7 1FW [IO94HF, SE48]
GI8 RKC R F Bowring, 6 Ferngrove Meadows, Aghagallon, Craigavon, BT67 0GF [IO64VL, J16]
G8 RKG D A Peck, 14 Woodside Way, Hedge End, Southampton, SO30 4BH [IO90IV, SU41]
G8 RKH L J M Hunt, 15 Oxford St., Northwood, Cowes, PO31 8PT [IO90IR, SZ49]
G8 RKO J M Butler, 36 Park Rd, Bracknell, RG12 2LU [IO91PK, SU86]
G8 RKV R Morcom, 9 Newburgh St., Winchester, SO23 8UY [IO91IB, SU42]
G8 RKX A Titley, 6 Spring View, Luddenden Foot, Luddendenfoot, Halifax, HX2 6EX [IO93AR, SE02]
G8 RKZ K A Stimpson, 4 Moors Ct, Ditchfield Ln, Finchampstead, Wokingham, RG40 4HP [IO91NJ, SU76]
G8 RLD R L Dowdell, 11 Woods Rd, Tuxedo, New York 10987, USA
GI8 RLE J R Ashe, 49 Deans Walk, Sleepy Valley, Richhill, Armagh, BT61 9LD [IO64RJ, H94]
G8 RLF R G Dickerson, 49 Greygoose Park, Harlow, CM19 4JW [JO01AS, TL40]
GI8 RLG H G Emerson, Little Castle Dillo, Armagh, Co Armagh, N Ireland, BT61 7DF [IO64QI, H84]
G8 RLH T Alston, Colt Park, Chapel-le-Dale, Carnforth, Lancs, LA6 3JF [IO84TE, SD77]
G8 RLJ G N Clinch, 36 Tyzack Rd, High Wycombe, HP13 7PU [IO91PP, SU89]
G8 RLN J E Barnett, Fairview Villa, 11 Ridge St., Stourbridge, DY8 4QF [IO82VL, SO88]
GW8 RLV G T Williams, Cefn Grug, 37 Ffordd y Llan, Cilcain, Mold, CH7 5NH [IO83JE, SJ16]
G8 RLW G N Woodward, 20 The Vale, Skelton, York, YO3 6YH [IO94KA, SE55]
G8 RMC K W Tolman, 60 Norfolk Ave, Sanderstead, South Croydon, CR2 8BP [IO91XI, TQ36]
G8 RMI S C Blake, 10 Dimore Cl, Hardwicke, Gloucester, GL2 4QQ [IO81UT, SO71]
G8 RMJ M A B Juby, Linden House, 16 Sunnyside, Diss, IP22 3DS [JO02NJ, TM18]
G8 RMP J R Bond, 19 Compton Ave, Mannamead, Plymouth, PL3 5DA [IO70WJ, SX45]
GM8 RMR E M Scott, 'Beechview', Enzie Slackhead, Buckie, Banffshire, AB56 2BR [IO87TH, NJ72]
G8 RMX A A Hewson, 20 Bawhead Rd, Earby, Colne, BB8 6PE [IO83WV, SD94]
GI8 RNG W R Smyth, 11 Legacorry Rd, Richhill, Armagh, BT61 9QB [IO64RI, H94]

G8 RNH M F Williams, Flat 2, 103 Albert Rd, London, SE25 4JE [IO91XJ, TQ36]
G8 RNJ Details withheld at licensee's request by SSL.
G8 RNM I M Lucking, 32 Nolton Pl, Edgware, HA8 6DL [IO91UO, TQ19]
G8 RNT P S Walkling, 36 Highlands House, Whamcliffe Rd, Southampton, SO19 7GG [IO90HV, SU41]
G8 RNU I H Strange, Holly Ln, Tansley, Matlock, Derby, DE4 5FF [IO93FD, SK35]
G8 ROD Details withheld at licensee's request by SSL.
G8 ROG Details withheld at licensee's request by SSL.
G8 RON R N Eyes, 6 Bakers Ln, Southport, PR9 9RN [IO83MP, SD31]
G8 ROO Rev J Wylam, Alwinton Vicarage, Harbottle, Morpeth, Northd., NE65 7BE [IO85WI, NT90]
G8 ROR T R Lees, 6 Far Cross, Cavendish Park, Matlock, DE4 3HG [IO93FD, SK36]
G8 ROS R A Platt, 40 Solent Dr, Darcy Lever, Bolton, BL3 1RN [IO83TN, SD70]
G8 ROU D M Hardy, Thorntree House, Main Rd, Wensley, Matlock, DE4 2LL [IO93ED, SK26]
G8 ROZ A Cope, 4 Room Cttgs, Chilbolton, Stockbridge, SO20 6BG [IO91GD, SU34]
G8 RPA K Mendum, 23 Eton Ave, East Barnet, Barnet, EN4 8TU [IO91WP, TQ29]
GM8 RPE J Robinson, 18 Craigshannoch Rd, Wormit, Newport on Tay, DD6 8ND [IO86MK, NO32]
G8 RPH Details withheld at licensee's request by SSL.
G8 RPI G R Atkinson, Flat 1, 1 Heather Cl, Walkford, Christchurch, BH23 5RP [IO90DR, SZ29]
GI8 RPP M E Elder, 16 Cregg Rd, Claudy, Londonderry, N Ireland, BT47 GHX [IO74BO, J37]
GI8 RPT N D Copeland, 34 Glenkyle Park, Carnmoney, Newtownabbey, BT36 6SP [IO74AQ, J38]
G8 RPV Details withheld at licensee's request by SSL.
G8 RQA M P Munn, Tolenhof 5, 6443 BC Brunssum, Zuid Limburg, The Netherlands, ZZ2 8NE
G8 RQB C I Porter, 66 Downlands, Royston, SG8 5BY [IO92XB, TL34]
G8 RQF J Duffy, 5 Birch Ct, Prudhoe, NE42 6PZ [IO94BX, NZ06]
GI8 RQI D J Allen, 40 Bramblewood Dr, Banbridge, BT32 4RA [IO64UI, J14]
G8 RQN P Needham, 2 Woodridge Cl, Bracknell, RG12 9QX [IO91PJ, SU86]
G8 RQP J Strike, Dunningwell Hall, The Green, Millom, Cumbria, LA18 4NZ [IO84IF, SD18]
G8 RQQ W R Reeve, 96 Essex Rd, Mawneys, Romford, RM7 8AX [JO01BO, TQ48]
G8 RQY Details withheld at licensee's request by SSL.
G8 RRA K A Hammersley, 28 Green Ln, Coventry, CV3 6DF [IO92FJ, SP37]
G8 RRC P S Sharpe, 59 Davison Dr, Cheshunt, Waltham Cross, EN8 0SX [IO91XR, TL30]
G8 RRL F J A Giles, 40 Jaywick Ln, Clacton on Sea, CO16 8BD [JO01NT, TM11]
G8 RRN M D Jones, 6 Shortway, Amersham, HP6 6AQ [IO91QQ, SU99]
GJ8 RRP J K Parry, No2 Thornley, Bagatelle Rd, St Saviour, Jersey, Channel Islands, JE2 7TZ
G8 RRR H R Potter, 7 Baugh Rd, Bristol, BS16 6PL [IO81SL, ST67]
G8 RRS M Ellison, 22 Cotebrook Dr, Upton, Chester, CH2 1RD [IO83NF, SJ46]
G8 RRU G C Manley, 23 Hyburn Cl, Bricket Wood, St. Albans, AL2 3QX [IO91TR, TL10]
G8 RRW Details withheld at licensee's request by SSL.
G8 RSA Dr S M Hasko, 105 High St., Brampton, Huntingdon, PE18 8TQ [IO92VH, TL27]
G8 RSB D Clarke, 67 Freemantle Rd, Bagshot, GU19 5LY [IO91PI, SU96]
GM8 RSC J E Chinnock, 3 Dundee St., Letham, Forfar, DD8 2PQ [IO86OP, NO54]
G8 RSD Details withheld at licensee's request by SSL.
G8 RSK P C Tyrell, 14 Park Farm Rd, Horsham, RH12 5EW [IO91UB, TQ13]
G8 RSL R J Carter, 50 Mansion Ln, Iver, SL0 9RN [IO91RM, TQ08]
G8 RSQ K D Williamson, 17 Ray Bond Way, Aylsham, Norwich, NR11 6UT [JO02OS, TG12]
G8 RSV S Staniforth, 30 Woodland Dr, Sheffield, S12 3HW [IO93GI, SK38]
G8 RSX T J Beck, 10 Rookery Cl, Hatfield Peverel, Chelmsford, CM3 2DF [JO01HS, TL71]
G8 RTA Me W W Evans, 1 Gurnard Heights, Gurnard, Cowes, PO31 8EF [IO90IS, SZ49]
G8 RTB R D Breeze, 119 Sundorne Rd, Shrewsbury, SY1 4RP [IO82PR, SJ51]
G8 RTC Details withheld at licensee's request by SSL.
GM8 RTI J S Grieve, Elhanan, Myrtlefield Ln, Westhill, Inverness, IV1 2UE [IO77WL, NH74]
G8 RTK L J Man, 29 Westfield Ave, Yeadon, Leeds, LS19 7NU [IO93DU, SE14]
G8 RTN G Smith, 1 Abbey Rd, Goldington, Bedford, MK41 9LG [IO92SD, TL05]
G8 RTO D L Hoyle, 59 Lower Manor Ln, Burnley, BB12 0EF [IO83VT, SD83]
G8 RTV Details withheld at licensee's request by SSL.
G8 RTW Details withheld at licensee's request by SSL.
GW8 RTZ Details withheld at licensee's request by SSL.
GW8 RUA S E Cleal, 34 Meadows View, Marford, Wrexham, LL12 8LS [IO83MC, SJ35]
G8 RUN D A Wood, 38 Henley Rd, Chester, CH4 8DY [IO83NE, SJ36]
G8 RUP A McConachie, 1 Ploughed Paddock, Nailsea, Bristol, BS19 2NB [IO81OK, ST47]
G8 RUR R E W Braggins, 9 Leigh Ave, Loose, Maidstone, ME15 9UY [JO01GF, TQ75]
G8 RUX R B Gough, 1 Silvertrees, Emsworth, PO10 7ST [IO90MU, SU70]
GM8 RVC R J Mackay, 9 Haining Pl, Grangemouth, FK3 9DR [IO86DA, NS98]
G8 RVG Details withheld at licensee's request by SSL.
G8 RVO Details withheld at licensee's request by SSL.
GJ8 RVT M J Turner Jersey ARS, 338 PO Box, St. Helier, Jersey, JE4 9YG
G8 RVV D C Dawson, Trekkers, Farm Ln, Send, Woking, GU23 7AT [IO91RG, TQ05]
G8 RVY P W Lee, 40 Thanet Lee Cl, Cliviger, Burnley, BB10 4UE [IO83VS, SD83]
G8 RVZ I Martin, Rectory Gate, Harlaston, Tamworth, Staffs, B79 9JX [IO92DQ, SK21]
G8 RW R W Standley, 47 Crest Rd, Hayes, Bromley, BR2 7JA [JO01AJ, TQ36]
G8 RWE K K Ambler, 20 Clover Way, Hedge End, Southampton, SO30 4RP [IO90IV, SU41]
G8 RWG N F Montanana, 324 Yorktown Rd, College Town, Camberley, Surrey, GU15 4PZ [IO91OI, SU86]
G8 RWH I R Jackson, 5 Vivien Cl, Chessington, KT9 2DE [IO91UI, TQ16]
G8 RWM F R Box, 11 Cook Ave, Newport, PO30 2LL [IO90IQ, SZ58]
G8 RWN M McKenzie, 9 Broomhouse Cl, Denby Dale, Huddersfield, HD8 8UX [IO93EN, SE20]
G8 RWS J F Carrick, 306 Chester Rd, Ellesmere Port, Whitby, South Wirral, L66 2NY [IO83NG, SJ37]
G8 RWT A J Coote, 28 Woolslope Rd, West Moors, Ferndown, BH22 0PD [IO90BT, SU00]
G8 RWX Details withheld at licensee's request by SSL.
G8 RXA C G Hampson, 7 Merryfield Cl, Bransgore, Christchurch, BH23 8BS [IO90DS, SZ19]
G8 RXB E M Hampson, 21 Marlowe Rd, Wallasey, L44 3DA [IO83LK, SJ39]
G8 RXG D E Gulvin, 9 Somerset Gdns, Pitsea, Basildon, SS13 3JJ [JO01GN, TQ78]
G8 RXH T F Parker, 2D Hubbards Chase, Hornchurch, RM11 3DJ [JO01CN, TQ58]
G8 RXJ F Poulton, 40 Queen St., Briercliffe, Burnley, BB10 2HE [IO83VT, SD83]
G8 RXL Details withheld at licensee's request by SSL.
G8 RXM Details withheld at licensee's request by SSL.
G8 RXP Details withheld at licensee's request by SSL.
G8 RXU Details withheld at licensee's request by SSL.
G8 RXY G H Alcock, 61 Henshall Hall Dr, Congleton, CW12 3TY [IO83VD, SJ86]
G8 RXZ P L Allgood, 83 Glovers Way, Telford, TF5 0NY [IO82RR, SJ61]
G8 RY Details withheld at licensee's request by SSL.
G8 RYD Details withheld at licensee's request by SSL.
G8 RYE D M Cope, 55 Castle Rd, Mountsorrel, Loughborough, LE12 7ET [IO92KR, SK51]
G8 RYJ P E L Tegg, Glendale, 36 Wrecclesham Hill, Wrecclesham, Farnham, GU10 4JW [IO91OE, SU84]
G8 RYL I D Smith, 12 Windmill Ln, Fulbourn, Cambridge, CB1 5DT [JO02CE, TL55]
G8 RYO P Sargent, The Old School House, Burgh on Bain, Lincoln, LN3 6JY [IO93TF, TF07]
G8 RYP D J Riggs, 33 Morley Rd, Southville, Bristol, BS3 1DT [IO81QK, ST57]
G8 RYR C R Harris, The Beeches, Coles Ln, Capel, Dorking, RH5 5HS [IO91UD, TQ14]
G8 RYW J F Hicks, 22 Courthouse Rd, Maidenhead, SL6 6JB [IO91PM, SU88]
G8 RZ H R Fox, 8 Mill Yard, Harrington, Workington, CA14 5QG [IO84FO, NX92]
G8 RZA Details withheld at licensee's request by SSL.
G8 RZD J R M Keeble, 22 Pishiobury Dr, Sawbridgeworth, CM21 0AE [JO01BT, TL41]
G8 RZJ D G Brazier, 44 Letchworth Ave, Feltham, TW14 9RY [IO91SK, TQ07]
G8 RZL T S Claydon, Hill Farm House, South St, Litlington, Royston, Herts, SG8 0QS [IO92WB, TL34]
G8 RZN D Dunn, Plot4 Pigeons Corne, Front Rd, Murrow, Wisbech Cambs, PE13 4JU [JO02AP, TF30]
G8 RZS E A R Humpston, 2 The Glebe, Hildersley, Ross on Wye, HR9 5BL [IO81RV, SO62]
G8 SAD P C Good obo Stevenage & District ARS, 80 Meredith Rd, Stevenage, SG1 5QS [IO91VV, TL22]
G8 SAE Details withheld at licensee's request by SSL.
G8 SAL K Hale Saltash Dist AR, 58 St. Stephens Rd, Saltash, PL12 4BJ [IO70VJ, SX45]
G8 SAN Dr R Charlton, The Cottage, Marsh Ln, Hampton in Arden, Solihull, B92 0AH [IO92DK, SP28]
GM8 SAP D J Cooper, 3 Garvel Rd, Milngavie, Glasgow, G62 7JD [IO75TW, NS57]
G8 SAR M Elliott, 54 Bankhouse Rd, Trentham, Stoke on Trent, ST4 8EL [IO82VX, SJ84]
G8 SAS R J Enright, 45 Gorham Dr, Tonbridge, TN9 2DU [JO01DE, TQ64]
G8 SAU Details withheld at licensee's request by SSL.
G8 SAX P J Wilkinson, 60 Whalley Dr, Aughton, Ormskirk, L39 6RF [IO83NN, SD40]
G8 SBA R A Winkworth, 23 Birkbeck Rd, North Finchley, London, N12 8DZ [IO91VO, TQ29]
G8 SBB J C Irlam, 13 Ashmeads Cl, Colehill, Wimborne, BH21 2LG [IO90AT, SU00]
G8 SBE Details withheld at licensee's request by SSL.
G8 SBF S Larkins, 47 Westcotts Green, Warfield, Bracknell, RG42 3SG [IO91PK, SU87]
G8 SBJ T M Blankley, 16 Charles Rd, St. Leonards on Sea, TN38 0QA [JO00GU, TQ80]
GW8 SBK L Cleak, 71 Pillmawr Rd, Malpas, Newport, NP9 6WG [IO81LO, ST39]
GW8 SBN J T Kemp, Poldhu 259 Delfford, Rhos, Pontardawe, Swansea, SA8 3EP [IO81CR, SN70]
G8 SBO P J Sibert, Sunnybrook, Cotton Row, Holmbury St. Mary, Dorking, RH5 6NB [IO91TD, TQ14]
G8 SBU D F Thompson, Four Winds, 131 St. Johns Rd, Exmouth, EX8 4EW [IO80HP, SY08]
G8 SBV E O Thompson, Little Pippin, 11 Lodge Rd, Sharnbrook, Bedford, MK44 1JP [IO92RF, SP95]
G8 SBZ A J Middleton, 60 Broadway, Oldbury, Warley, B68 9DL [IO92AL, SP08]
G8 SC C Collins, Calumet New Rd, Ridgewood, Uckfield, E Sussex, TN22 5SX [JO00BW, TQ41]
G8 SCH R Brown, Chant House, North St., Rotherfield, Crowborough, TN6 3JU [JO01CB, TQ52]
GW8 SCR C E Burnell, 22 Delafield Rd, Abergavenny, NP7 7AW [IO81LT, SO21]
G8 SCY C T Rosewall, Suhaili, 12 Treloggan Ln, Newquay, TR7 2JN [IO70LJ, SW86]
G8 SCZ K B Roberts, 8 Maymills Cttgs, Mill Ln, Eastry, Sandwich, Kent, CT13 0LB [JO01PF, TR35]
G8 SD R A Simpson, Little Chalet, 27 Wannock Ln, Lower Willingdon, Eastbourne, BN20 9SB [JO00CT, TQ50]
G8 SDC Details withheld at licensee's request by SSL.
G8 SDD A A Austin, 15 Brockholme Rd, Mossley Hill, Liverpool, L18 4QG [IO83NI, SJ38]
G8 SDE R Pitts, 84 Prospect Ave, Pye Nest, Halifax, HX2 7HP [IO93BR, SE02]
G8 SDN E A McIver, 31 Harts Hill, Bedford, MK41 9AL [IO92SD, TL05]
G8 SDS W J G Barton South Dorset RS, 13 Paynes Cl, Piddlehinton, Dorchester, DT2 7TF [IO80TS, SY79]
G8 SDT Details withheld at licensee's request by SSL.
G8 SDU R O Clayton, 1 Raymond Rd, Norwich, NR6 6PL [JO02PQ, TG21]
G8 SDX C V Dale, Stonehouse, Rudyard, Leek, Staffs, ST13 8RX [IO83XC, SJ95]
G8 SED P J G Starling, 5 Ash Cl, Bacton, Stowmarket, IP14 4NR [JO02MG, TM06]
G8 SEE R C Stone, 10 Rosemullion Gdns, Tolvaddon, Camborne, TR14 0EY [IO70IF, SW64]
G8 SEJ Details withheld at licensee's request by SSL.
G8 SEK C J Watts, 63 The Vineries, Colehill, Wimborne, BH21 2PY [IO90AT, SU00]
G8 SEQ J E Beech, 124 Belgrave Rd, Wyken, Coventry, CV2 5BH [IO92GJ, SP37]
G8 SER D H Bentley, Bowland, Caythorpe Rd, Lowdham, Nottingham, NG14 7EA [IO93MA, SK64]
G8 SEV P Matthews, 6 Grange View, Eastwood, Nottingham, NG16 3DE [IO93IA, SK44]
G8 SEY A Graver, 8 Ave Rd, Bishops Stortford, CM23 5NU [JO01CU, TL42]
G8 SEZ Details withheld at licensee's request by SSL.[Op: B D Morrison. Station located near Bedford.]
G8 SFA S C Milsom, 5 Maple Gr, Prudhoe, NE42 6PU [IO94BX, NZ06]
G8 SFB D A Munro, 6 The Spinney, Leamington Spa, CV32 6ED [IO92FG, SP36]
G8 SFD C P Williams, 14 Milton Pl, Bideford, EX39 3BN [IO71VA, SS42]
G8 SFF M A Watson, 7 Grange Ln, Willingham By Stow, Gainsborough, DN21 5LB [IO93PI, SK88]
G8 SFI S E Firth, 8 Lyndale Ave, Osbaldwick, York, YO1 3QB [IO93LW, SE65]
G8 SFM K A Saunders, Tamarisk, Tetbury Ln, Leighterton, Tetbury, GL8 8UP [IO81UO, ST89]
G8 SFQ T J McNamara, 12 Scarsdale Rd, Great Barr, Birmingham, B42 2JW [IO92BM, SP09]
G8 SFR S A Morton, 2 The Coppice, Pembury, Tunbridge Wells, TN2 4EY [JO01DD, TQ64]
G8 SFS Dr J M Holmes, Old Orchard House, High St., Norley, Warrington, WA6 8JS [IO83QF, SJ57]
GW8 SFT D J Mansell, 6 Penrheidol, Penparcau, Aberystwyth, SY23 1QW [IO72XJ, SN58]
G8 SFU D J Kirkham, 2 Thames Meadow Dr, Hogsthorpe, Skegness, Lincs, PE24 5PU [JO03DF, TF57]
G8 SGB P H J Houseago, 11 Arnstones Cl, Colchester, CO4 3AS [JO01LV, TM02]
G8 SGF P J Gilliland, 43 Oyster Row, Cambridge, CB5 8LJ [JO02BF, TL45]
G8 SGH P A Marshall, 123 Rochford Gdn Wa, Rochford, Essex, SS4 1QJ [JO01IO, TQ89]
G8 SGI S M Pascoe, 59 Woodvale Ave, Doddington Park, Lincoln, LN6 3RD [IO93QE, SK96]
G8 SGK A P Stannard, 70 Primose Hill, Haverhill, Suffolk, CB9 7LS [JO02FC, TL64]
G8 SGM A R Boyce, 26 All Saints Cl, Doddinghurst, Brentwood, CM15 0NH [JO01DQ, TQ59]
G8 SGP G L Wheeler, 6 Landress Ln, Beverley, HU17 8HA [IO93SU, TA03]
G8 SGT S P Williams-Conley, 120 Westward Rd, Ebley, Stroud, GL5 4ST [IO81VR, SO80]
G8 SGV M J Williams, 30 Park Dr, Belgrave, Victoria, Austria, 3160
G8 SGW R N Young, 27 Thornford Dr, Westlea, Swindon, SN5 7BB [IO91CN, SU18]
G8 SGX S P Saltmer, 35 The Green, Newby, Scarborough, YO12 5JA [IO94SH, TA08]
G8 SGY M A Law, 195 Overdown Rd, Tilehurst, Reading, RG3 6NU [IO91LJ, SU66]
G8 SH Dr P N Nield, 9 Lea Green Ln, Wythall, Birmingham, B47 6HE [IO92BJ, SP07]
G8 SHC P H Hammond, 35 West Green, Barrington, Cambridge, CB2 5RZ [JO02AD, TL34]
G8 SHE Details withheld at licensee's request by SSL.[Op: Richard J H Shears, 7 Shakespeare Close, Caversham Park Village, Caversham, Reading, Berks, RG4 0QE.]
G8 SHF C A Scrase, 37 Cypress Cl, Honiton, EX14 8YW [IO80JS, SY19]
G8 SIC B M Harris, 13 Barn Park Rd, Peverell, Plymouth, PL3 4LP [IO70WJ, SX45]
GW8 SIE R J Stark, Roneragh, Llanrheadr, Denbigh, Clwyd, LL16 4NN [IO83HD, SJ06]
G8 SIG A F Jeffery, 14 Holly Mount, Shavington, Crewe, CW2 5AZ [IO83SB, SJ75]
G8 SIK A H J Sturt, 1 Sandringham Gdns, Finchley, London, N12 0NY [IO91VO, TQ29]
G8 SIM J E Green, 7 Russell Rd, Runcorn, WA7 4BG [IO83PH, SJ58]
GM8 SIQ R S Steyn, 8 Harrison Hey, Liverpool, L36 5YR [IO83NJ, SJ49]
GW8 SIT M W Shewring, 2 Glan Hafan, Trefechan, Aberystwyth, Dyfed, SY23 1BE [IO72WJ, SN58]
G8 SIU D Stillwell, 27 Lesley Owen Way, Shrewsbury, SY1 4RB [IO82PR, SJ51]
G8 SJA P A Farrar, 8 Eagle Cl, Fareham, PO16 8QX [IO90KU, SU50]
G8 SJD J L Day, 5 Broom Cl, Belper, DE56 2TZ [IO93GA, SK34]
GW8 SJN H Davies, 42 Heol-y-Ffynon, Efail Isaf, Pontypridd, M Glam, CF38 1AU [IO81IN, ST08]
G8 SJO S J Ootam, 9 Harewood Rd, Isleworth, TW7 5HB [IO91UL, TQ17]
G8 SJP I Phillipps, 24 Acres End, Amersham, HP7 9DZ [IO91QP, SU99]
G8 SJR B J Vincent, 28 Tintagel Rd, Orpington, BR5 4LQ [JO01BI, TQ46]
GI8 SJS R A E Hoey, 28 Hanwood Heights, Dundonald, Belfast, BT16 0XU [IO74CO, J47]
GI8 SJZ M E Craig, 8 Connsbrook Dr, Belfast, BT4 1LU [IO74BO, J37]
G8 SKA R K Holden, 47 Heston Ave, Great Barr, Birmingham, B42 2NT [IO92AM, SP09]
G8 SKI Details withheld at licensee's request by SSL.
G8 SKK G F de Voil, Glenley, 9 Carroll Ave, Merrow, Guildford, GU1 2QJ [IO91RF, TQ05]
GI8 SKR G J Bannister, 65 Osborne Dr, Belfast, BT9 6LJ [IO74AN, J37]
G8 SKU Details withheld at licensee's request by SSL.
G8 SLB P W Lockwood, 36 Davington Rd, Dagenham, RM8 2LR [JO01BN, TQ48]
G8 SLC M J Truman, The Leatings, 5 Cotwood, Kennall Vale, Ponsanooth Truro, TR3 7HJ [IO70KE, SW73]
GW8 SLG Details withheld at licensee's request by SSL.
G8 SLM E A M Santer, 0B0 Sth London M.C, Zoe Lucky 3 Barnpark, Summerhill Liverton, Newton Abbot, TQ12 6HE [IO80DN, SX87]
G8 SLP J H D Barry, Mount Lodge, 7 Sandy Ln, Chester, Ches, CH3 5UL [IO83NE, SJ46]
G8 SLU M B Hack, Anmee The Ride, Ifold Loxwood, Billinghurst, Sussex, RH14 0TF [IO91RB, TQ03]
G8 SMA C B Ward, 25 Blewbury Dr, Tilehurst, Reading, RG3 5HJ [IO91LJ, SU66]
G8 SMC R Baines SMC Northern RC, 327 Langer Ln, Wingerworth, Chesterfield, S42 6TY [IO93GE, SK36]
G8 SME J A Crimes, 1276 New Chester Rd, Eastham, Wirral, L62 9AF [IO83MH, SJ37]
G8 SMG J F Pacey, White Lodge, Golf Dr, Camberley, GU15 1JG [IO91PH, SU85]
G8 SMH K A Hempsall, 69 Wantage Rd, Didcot, OX11 0AE [IO91IO, SU58]
G8 SMQ N C G Guilford, White House Cottage, Petersons Ln, Aylsham, Norwich, NR11 6HD [JO02PT, TG12]
G8 SMR J B Heath Sth Mnchester Rd, 19 Anson Rd, Swinton, Manchester, M27 5GZ [IO83TM, SD70]
G8 SMZ C B Shaw, 1 Guilford Cttgs, East Langdon, Dover, CT15 5JD [JO01QD, TR34]
GM8 SNB G Allan, 13 Mitchell Dr, Rutherglen, Glasgow, G73 3QP [IO75VT, NS66]
GM8 SNE P J Walter, Allt Beag, Pitconnochie Rd, Crossford, Dunfermline, KY12 8QD [IO86GB, NT08]
G8 SNF I A Hewitt, 26 Outwoods Dr, Loughborough, LE11 3LT [IO92JS, SK51]
GW8 SNG Details withheld at licensee's request by SSL.
G8 SNH J W Pumfrey, 61 The Ridings, Burgess Hill, RH15 0PL [IO90WW, TQ31]
G8 SNI A Sellwood, 44 Fairhurst St., Leigh, WN7 4EE [IO83RL, SD60]
G8 SNJ P Townsend, Moss Nook, Fieldend Ln, Holmbridge, Holmfirth West York, HD7 1NH [IO93CN, SE10]
G8 SNM D B I Wright, 6 Weeks House, 2 Hardwicke Rd, Ham, Richmond, TW10 7TY [IO91UK, TQ17]
G8 SNQ R G Knock, 18 The Hawthorns, Eccleston, Chorley, PR7 5QW [IO83PP, SD51]
G8 SNT Details withheld at licensee's request by SSL.
G8 SNV Dr M N Dey, 18 Ripley Rd, Hampton, TW12 2JH [IO91TK, TQ17]
G8 SOG L E Ayres, 2 Purbeck Rd, Hornchurch, RM11 1NA [JO01CN, TQ58]
G8 SOI D Carter, 35 Upland Rd, West Mersea, Colchester, CO5 8DR [JO01LS, TM01]
G8 SOK J A Sturrock, 4 Ann St., Edinburgh, EH4 1PJ [IO85JW, NT27]
G8 SOL R A Simpkins, 100 Sturgeon Ave, Clifton Est, Nottingham, NG11 8HF [IO92JV, SK53]
G8 SOU R Topping, 47 Celtic Rd, Deal, CT14 9EF [JO01QE, TR35]
G8 SOZ Details withheld at licensee's request by SSL.
G8 SPC R T White, 3 Robin Ln, Clevedon, BS21 7EX [IO81NK, ST47]
G8 SPD M S Beevers, 2 Pool House Cotts, Astley, Stourport on Severn, Worcs, DY13 0RH [IO82UH, SO76]
G8 SPE R Armstrong, 159 Gunnersbury Ln, Acton, London, W3 8HP [IO91UM, TQ17]
GW8 SPH D E Phillips, Rhos-y-Coed, Rosebush, Clynderwen, Dyfed, SA66 7QY [IO71OW, SN02]
G8 SPI E W J Nash, 18 Salisbury Gdns, Downend, Bristol, BS16 5RE [IO81RL, ST67]
GW8 SPM P R Armitage, 24 Fields Rd, Tredegar, NP2 4LW [IO81JS, SO10]
G8 SPP C W Parkinson, 77 Lime Gr, Doddinghurst, Brentwood, CM15 0QX [JO01DP, TQ59]
G8 SPU R T Doughty, 47 Red Lion Cl, Tividale, Warley, B69 1TP [IO82XM, SO99]
GM8 SQ G Proctor, 6 Roscobie Park, Banchory, AB31 5RE [IO87RB, NO69]
G8 SQC R F M Hawkins, 229 Sutton Rd, Maidstone, ME15 9BJ [JO01GG, TQ75]
G8 SQF Details withheld at licensee's request by SSL.
G8 SQH D J C Hutchinson, 250 Lyttleton Ave, Charford, Bromsgrove, B60 3LD [IO82XH, SO96]
G8 SQN J O Bower, 13 Downs Cl, Bradford on Avon, BA15 1PR [IO81UI, ST86]
G8 SQY S J Cade, 1 Trinity Cl, Great Paxton, St. Neots, Huntingdon, PE19 4YL [IO92TH, TL16]
G8 SQZ S P Westlake, 11 Mount Rd, Evesham, WR11 6BE [IO92AC, SP04]
G8 SRC D C Forrest Swindon ARC, 19 Burns Way, Swindon, SN2 6LP [IO91CO, SU18]
G8 SRD Details withheld at licensee's request by SSL.
G8 SRG L L Williams obo Solihull Raynet Group, 25 Streetsbrook Rd, Shirley, Solihull, B90 3PB [IO92CK, SP18]
G8 SRH J C Wigley, 52 Worlebury Hill Rd, Worlebury, Weston Super Mare, BS22 9SZ [IO81MI, ST36]
G8 SRK A E Trigg, 9 Threshers Walk, East Goscote, Leicester, LE7 3ZW [IO92LR, SK61]
G8 SRN E V Day, 10 Carlrayne Ln, Menston, Ilkley, LS29 6HH [IO93DV, SE14]
G8 SRS B Naylor Stockport RS, 47 Chester Rd, Poynton, Stockport, SK12 1HA [IO83WI, SJ98]
GW8 SRW Dr R C V Macario, 3 Ashburnham Dr, Mayals, Swansea, SA3 5DS [IO71XO, SS69]
G8 SSE K G Lawrence, Pak House, 7 Canada Way, Lower Wick, Worcester, WR2 4DJ [IO82VE, SO85]

G8 SSI M L Ecott, 37 The Ridgeway, Waddon, Croydon, CR0 4AD [IO91WI, TQ36]
G8 SSL A T Marwood, 10 Eastham Rd, Plains Est, Arnold, Nottingham, NG5 6QX [IO92KX, SK64]
G8 SSP C J Horswell, 9 Bosham Walk, Gosport, PO13 0QJ [IO90JT, SU50]
G8 SSX D O Baker, 99 Repton Rd, Wigston, LE18 1GD [IO92KO, SP59]
G8 SSY E Davies, 5 Cheapside, Horsell, Woking, GU21 4JG [IO91RH, SU96]
G8 STD E C John St Dunstans ARS, obo St. Dunstans Ars, 52 Broadway Ave, Wallasey, L45 6TD [IO83LK, SJ39]
G8 STE D C Barber, 12 Hulton Rd, Gaywood, Kings Lynn, PE30 4QE [JO02FS, TF62]
G8 STF T G Woods, 46 New St., St. Helens, WA9 3XL [IO83PK, SJ59]
G8 STI J H Maiden, 42 Timberdine Ave, Worcester, WR5 2BD [IO82VE, SO85]
G8 STJ D A Carter, 4 Sandale Cl, Gamston, Nottingham, NG2 6QG [IO92KW, SK63]
G8 STM Details withheld at licensee's request by SSL.
G8 STO R D Hills, 5 Larkfield Ave, Gillingham, ME7 2LN [JO01GJ, TQ76]
G8 STR B H B Beestin, Lynwood House, 16 Grant Rd, Crowthorne, RG45 7JG [IO91OI, SU86]
G8 STW Details withheld at licensee's request by SSL.
G8 STY J G Holmes, 45 College Ave, Gillingham, ME7 5HY [JO01GJ, TQ76]
G8 SUG G Peterson, 51 Springfield Rd, Harrow, HA1 1QF [IO91TN, TQ18]
G8 SUM K R Smith, 11 Church St., Earl Shilton, Leicester, LE9 7DA [IO92IN, SP49]
G8 SUN S A Williams, 11 Cotman Dr, Hinckley, Leics, LE10 0GB [IO92HN, SP49]
G8 SUP Details withheld at licensee's request by SSL.
G8 SUQ J C Corbidge, 11 Berkeley Cl, Folkestone, CT19 5NA [JO01NC, TR23]
G8 SUT Details withheld at licensee's request by SSL.
G8 SUV B C Pont, 56 Ravenhill Rd, Bristol, BS3 5BT [IO81QK, ST57]
G8 SUW N R Pont, Maisemoor, 17 Vicarage Ln, Shapwick, Bridgwater, TA7 9LR [IO81OD, ST43]
G8 SUY Details withheld at licensee's request by SSL.
GM8 SVB A R Duncan, Drumduan, Bellevue Rd, Banff, AB45 1BJ [IO87RP, NJ66]
G8 SVC Details withheld at licensee's request by SSL.
G8 SVE Details withheld at licensee's request by SSL.
G8 SVF Details withheld at licensee's request by SSL.
GW8 SVN J Iliffe, Ellesmere House, 2 Stow Park Ave, Newport, NP9 4FH [IO81LN, ST38]
G8 SVO N H Kendall, Chittlebirch, Cripps Corner, Robertsbridge, Sussex, TN32 5SA [JO00GX, TQ72]
G8 SVR J F Allart, 16 Front St., Sherburn Hill, Durham, DH6 1PA [IO94GS, NZ34]
G8 SVT T Ellis, 58 Greenland Dr, Sheffield, S9 5GJ [IO93HJ, SK38]
G8 SVX K Hartley, 242 Huddersfield Rd, Mirfield, WF14 9PY [IO93DQ, SE11]
G8 SVZ F J Keeble Buckle, 4 Croft Cl, Meeting Green, Wickhambrook, Newmarket, CB8 8YG [JO02GE, TL75]
G8 SWC H J Moyle, 9 Park Approach, Welling, DA16 2AW [JO01BL, TQ47]
G8 SWF T Holborn, 81 Hurlfield Rd, Gleadless, Sheffield, S12 2SF [IO93GI, SK38]
G8 SWG Details withheld at licensee's request by SSL.
G8 SWK D Taylor, 19 Armley Grange Oval, Leeds, LS12 3QJ [IO93ET, SE23]
G8 SWL E T Theodorson, Firclose, Orchard St, Drayton / Daventry, Northants, NN11 5EX [IO92JG, SP56]
G8 SWR Details withheld at licensee's request by SSL.
G8 SWY Details withheld at licensee's request by SSL.
G8 SWZ G I Davies, 167 Aldersley Rd, Tettenhall, Wolverhampton, WV6 9NJ [IO82WO, SJ80]
G8 SX H S Wood, 11 Bronshill Gr, Allerton, Bradford, BD15 7AJ [IO93CT, SE13]
G8 SXA J M Davies, Ballards Piece, Forest Hill, Marlborough, SN8 3HN [IO91DJ, SU26]
G8 SXB D J Mullenger, 6 Churchfields, Kingsley, Bordon, GU35 9PJ [IO91ND, SU73]
G8 SXC R L Davies, 13 Stapleford Cl, Woodley, Romsey, Hants, SO51 7HU [IO91GA, SU32]
G8 SXD B L Davies, Ballards Piece, Forest Hill, Marlborough, SN8 3HN [IO91DJ, SU26]
G8 SXI G Howells, The Rectory, Stonehill, Rackheath, Norwich, NR13 6NG [JO02QQ, TG21]
G8 SXJ F H Hutchings, 21 School Ln, St. Ives, Ringwood, BH24 2PF [IO90CU, SU10]
G8 SXQ A J Leigh, 12 Bowmens Lea, Aynho, Banbury, OX17 3AG [IO91IX, SP53]
G8 SXR Details withheld at licensee's request by SSL.
G8 SXS S R N Chenery, 230 Queens Rd, Clarendon Park, Leicester, LE2 3FT [IO92KO, SK50]
G8 SXU J M Simmons, 167 Bourne Vale, Hayes, Bromley, BR2 7LX [JO01AJ, TQ46]
G8 SXV J Thompson, 7 Windsor Dr, Wingerworth, Chesterfield, S42 6TG [IO93GE, SK36]
G8 SXY Details withheld at licensee's request by SSL.
G8 SYC A P Harris, 17 Fir Tree Cl, Patchway, Bristol, BS12 5ER [IO81RM, ST58]
G8 SYD M J Thomson, 2 Bencroft Rd, Hemel Hempstead, HP2 5UY [IO91SS, TL00]
G8 SYG Details withheld at licensee's request by SSL.
G8 SYM D Whittle, 16 Garner Dr, Astley, Tyldesley, Manchester, M29 7RT [IO83SM, SD70]
G8 SYR Details withheld at licensee's request by SSL.
G8 SYS P V Evans, 5 Compton Dr, Streetly, Sutton Coldfield, B74 2DA [IO92BN, SP09]
G8 SYT P F Fleet, 33 Spital Terr, Gainsborough, DN21 2HD [IO93OJ, SK89]
G8 SYU Details withheld at licensee's request by SSL.
G8 SYV J A Morgan, Linden Lea, Fakes Rd, Hemsby, Great Yarmouth, NR29 4JL [JO02UQ, TG51]
G8 SZB P J Marten, 2 Marlborough Ct, Wokingham, RG40 1TA [IO91OJ, SU86]
GW8 SZC P R Henry, Ael-y-Bryn, New Rd, Llanmorlias, Swansea, SA4 3TY [IO71WO, SS59]
G8 SZR M Matthews, 183 Parsonage Ln, Enfield, EN1 3UH [IO91XP, TQ39]
GM8 SZS B A McCaffrey, 2A James Gr, Kirkcaldy, KY1 1TN [IO86KC, NT29]
G8 SZX D A Towers, 20 Valiant Cl, Glenfield, Leicester, LE3 8JH [IO92JP, SK50]
G8 SZZ S A Jackson, 18 Blakesware Gdns, Bush Hill Park, London, N9 9HU [IO91XP, TQ39]
G8 TA K Atack Wolverhampton Rd, 29 High Hill, Essington, Wolverhampton, WV11 2DW [IO82XO, SJ90]
G8 TAA Details withheld at licensee's request by SSL.
G8 TAE G B Wooltorton, 13 Almond Rd, Gorleston, Great Yarmouth, NR31 8EJ [JO02UN, TG50]
G8 TAO P W Davis, Woodend Farm, Lit Warley Hall Ln, Brentwood, Essex, CM13 3EX [JO01DN, TQ68]
G8 TAQ A J Dyce, 26 Forest Rd, Winford, Sandown, PO36 0JY [IO90JP, SZ58]
G8 TAU A J Fisher, 2 Hillside Mns s, Barnet Hill, Barnet, EN5 5RH [IO91VP, TQ29]
G8 TAY R F Manning, 3 Palgrave Cl, Taverham, Norwich, NR8 6LP [JO02OQ, TG11]
G8 TB B W Wynn, 67 Old Lodge Ln, Purley, CR8 4DN [IO91WH, TQ36]
G8 TBB J W O'Meara, 117 Little Sutton Ln, Sutton Coldfield, B75 6SN [IO92CN, SP19]
G8 TBF R A Jenkins, 13 Baulk Ln, Worksop, S81 7DF [IO93KH, SK58]
GW8 TBG M Terry, 265 Delffordd, Rhos, Pontardawe, Swansea, SA8 3EP [IO81CR, SN70]
G8 TBH Details withheld at licensee's request by SSL.
G8 TBI Details withheld at licensee's request by SSL.
G8 TBK Details withheld at licensee's request by SSL.
G8 TBL N P Mosedale, 22 Mada Rd, Orpington, BR6 8HQ [JO01AI, TQ46]
G8 TBR R W S Richardson, Nevan, Queens Rd, Crowborough, TN6 1EJ [JO01CB, TQ53]
G8 TBU N R Doe, 1 Blackbrook Cttgs, Blackbrook, Dorking, RH5 4DS [IO91UE, TQ14]
G8 TBV A S Kyle, 6 Mill Hill Dr, Halesworth, IP19 8DB [JO02RI, TM37]
G8 TBW R C Crathorne, 340 Farnborough Rd, Castle Vale Est, Birmingham, B35 7PD [IO92CM, SP19]
G8 TBX S W Pybus, Primrose Cottage, Copt Hewick, Ripon, N Yorks, HG4 5BY [IO94GD, SE37]
G8 TCH M Bell, 15 The Green, Greatham, Hartlepool, TS25 2HG [IO94JP, NZ42]
G8 TCS R J Dutton, Broad Oak Cottage, Hand Ln, Heath Hill, Sherrifhales, Salop, TF11 5RR [IO82TP, SJ70]
G8 TDB Details withheld at licensee's request by SSL.
G8 TDG J L Shuker, 1 Rydal Cl, Burlish Park, Stourport on Severn, DY13 8JX [IO82UI, SO87]
G8 TDP D J Cooke, 19 St. Aldwyn Rd, Seaham, SR7 0AN [IO94HU, NZ44]
G8 TDW Details withheld at licensee's request by SSL.
G8 TEB D B Clarke, 31 Grafton Way, New Duston, Northampton, NN5 6NG [IO92MG, SP76]
G8 TEC G E Cook, The Coach House, 2 Abbey Hill Rd, Netley Abbey, Southampton, SO31 5FB [IO90HV, SU40]
G8 TEE Details withheld at licensee's request by SSL.
G8 TEF A Crute, 1 The Causeway, Partridge Green, Horsham, RH13 8JH [IO91UB, TQ12]
G8 TEK K B Worley, 33 Lynbrook Cl, Netherton, Dudley, DY2 9HE [IO82XL, SO98]
G8 TEL P J Hynes, 3 Holt Park Gdns, Leeds, LS16 7RB [IO93EU, SE24]
G8 TEO B A Jay, Brookhouse, 125 Tower Rd South, Bristol, BS15 5BT [IO81SK, ST67]
G8 TEQ Dr D K Linsdall, 2B Linkswood Rd, Burnham, Slough, SL1 8AT [IO91QM, SU98]
G8 TET Details withheld at licensee's request by SSL.
G8 TEX T A W Eley, 20 Elsham Rd, Leytonstone, London, E11 3JH [JO01AN, TQ38]
G8 TFB S R Haywood, 12 Elm Terr, Tividale, Warley, B69 1UD [IO82XM, SO99]
G8 TFF L W Dewhurst, 8 Stirling Ave, Leamington Spa, CV32 7HN [IO92FH, SP36]
G8 TFI Details withheld at licensee's request by SSL.
G8 TFO J V Buckley, 5 Churchill Ave, Droitwich, WR9 8NP [IO82WG, SO86]
G8 TFT D Alsop, 6 Haydon, Fatfield, Washington, NE38 8PF [IO94FV, NZ35]
G8 TFU P Simpson, 22 School Ln, Arkwright Town, Chesterfield, S44 5BZ [IO93HF, SK47]
G8 TFW L A Stamp, 41 Willoughby Rd, Liscard, Wallasey, L44 3DZ [IO83LK, SJ29]
G8 TFY N M Richards, 38 Parsons Rd, Irchester, Wellingborough, NN29 7EA [IO92QG, SP96]
G8 TGB M Verrall, 1 Speedwell Ave, Weedswood Est, Chatham, ME5 0SB [JO01GI, TQ76]
G8 TGD D W Troop, 10 Mellowdew Rd, Coventry, CV2 5GL [IO92GJ, SP37]
G8 TGH B N Wilmott, 27 Apple Gr, Bognor Regis, PO21 4NB [IO90PS, SZ89]
G8 TGR A R Sutcliffe, 9 Wilfred Owen Cl, Wimbledon, London, SW19 8SW [IO91VK, TQ27]
G8 TGS Dr W R Williams, Room 422, British Embassy, Bfpo 2, X X
G8 TGY R H Arey, 622 Scott Hall Rd, Chapel Allerton, Leeds, LS7 3QJ [IO93FT, SE33]
G8 THE Dr R J Hill, 41 North Down, Staplehurst, Tonbridge, TN12 0PQ [JO01GD, TQ74]
G8 THF D Hodgkins, 86 Granby Ct, Bletchley, Milton Keynes, MK1 1NF [IO92PA, SP83]
G8 THG F R Humphries, 169 Bloxham Rd, Banbury, OX16 9JU [IO92HB, SP43]
G8 THH D J Baker, 5 Larkspur Cl, Bishops Stortford, CM23 4LL [JO01BU, TL42]
G8 THK Details withheld at licensee's request by SSL.[Station located in Twickenham, Middlesex, near London Heathrow Airport. Op: David.]
GW8 THM M W Griffin, 3 Pritchard Cl, Danescourt, Llandaff, Cardiff, CF5 2QS [IO81JM, ST17]
G8 THN A R Gascoigne, 110 Wymondley Rd, Hitchin, SG4 9PX [IO91UW, TL12]
G8 THS Details withheld at licensee's request by SSL.
G8 THZ A D Tipper, 24 Waverley Rd, Hoylake, Wirral, L47 3DD [IO83JJ, SJ28]
G8 TIA D Trickett, 25 Spring St., Halesowen, B63 2SY [IO82XL, SO98]
G8 TIC Details withheld at licensee's request by SSL.
G8 TIJ Details withheld at licensee's request by SSL.[Station located in London SE16.]
G8 TIM T J Timms, Greenways, 14 Windmill Rd, North Leigh, Witney, OX8 6RQ [IO91GT, SP31]
G8 TIR R J Bates, 51 Boyton Rd, Ipswich, IP3 9PD [JO02OA, TM14]
G8 TIS P M Bond, 5 Greenacres Cl, Emley, Huddersfield, West Yorks, HD8 9RA [IO93EO, SE21]
GW8 TIX G C May, 19 McLaren Cttgs, Abertysswg, Rhymney, NP2 5BH [IO81IR, SO10]
G8 TIY J A Sheardown, 5 Winteringham Ln, West Halton, Scunthorpe, DN15 9AX [IO93QQ, SE92]
G8 TJB Details withheld at licensee's request by SSL.
G8 TJD C J Turner, Shade Oak, 14 Penny Ln, Guarlford, Malvern, WR13 6PG [IO82UC, SO84]
G8 TJF G M Swetman, Little Copse, Yallands Hill, Monkton Heathfield, Taunton, TA2 8NA [IO81LA, ST22]
G8 TJG F Starkey, 13 Thorncliffe Dr, Darwen, BB3 3QA [IO83SQ, SD72]
G8 TJH W A Reynolds, Rose Cottage, Rainton, near Thirsk, North Yorks, YO7 3PH [IO94GE, SE37]
G8 TJI M J Oldfield, Willows, Stablebridge Rd, Aston Clinton, Bucks, HP22 5ND [IO91PT, SP81]
G8 TJQ Details withheld at licensee's request by SSL.
G8 TJR C G Colebrook, 14 Yiewsley Dr, Darlington, DL3 9XS [IO94EM, NZ21]
G8 TKD D Hensby, 28 Moorland Cres, Whitworth, Rochdale, OL12 8SU [IO83VP, SD81]
G8 TKO Details withheld at licensee's request by SSL.
G8 TKQ J F Ackerley, 24 Macaulay Rd, Lutterworth, LE17 4XB [IO92JL, SP58]
G8 TKR Details withheld at licensee's request by SSL.
G8 TKY T F Bootyman, 6 Stable Ct, Welbeck, Worksop, S80 3LP [IO93JG, SK57]
G8 TKZ Details withheld at licensee's request by SSL.
G8 TLA M Robinson, 19 St. Wilfrids Cres, Brayton, Selby, YO8 9EU [IO93KS, SE53]
G8 TLH R D Rogers, 14 Coningsby Dr, Franche, Kidderminster, DY11 5LU [IO82UJ, SO87]
G8 TLI I H W White, 11 Keith Ave, Great Sankey, Warrington, WA5 3NZ [IO83QJ, SJ58]
G8 TLL L G Stewart, The Spinney, Holmes Ln, Winterton, Scunthorpe, DN15 9QY [IO93RP, SE91]
G8 TLP R L Pearce, Wisteria House, 68 North St., Barming, Maidstone, ME16 9HF [JO01FG, TQ75]
G8 TLU F G Laird, 7 Meadow Rd, Toddington, Dunstable, LU5 6BB [IO91RW, TL02]
G8 TLZ Details withheld at licensee's request by SSL.
G8 TMA H N S Colborn, 180 Elm Hill, Warminster, BA12 0AS [IO81WF, ST84]
G8 TMB R J Cook, 31 Butley Rd, Felixstowe, IP11 8NY [JO01QX, TM23]
G8 TMC E A L Clibbon, 39 Praze Rd, Newquay, TR7 3AF [IO70LK, SW86]
G8 TMD T Clint, 11 Homelea, Rothwell, Leeds, LS26 0PP [IO93GS, SE32]
GI8 TME Details withheld at licensee's request by SSL.
G8 TMJ P J Faulkner, 8 Parkfield Rd, Cheadle Hulme, Cheadle, SK8 6EX [IO83VI, SJ88]
G8 TML Details withheld at licensee's request by SSL.
G8 TMM E C Gilbert, 34 School Ln, Harpole, Northampton, NN7 4DR [IO92MF, SP66]
G8 TMP V J Tyers, 19 King Richards Hill, Whitwick, Coalville, LE67 5BT [IO92HR, SK41]
G8 TMQ P Stevens, 49 Ridge St., Stourbridge, DY8 4QF [IO82VL, SO88]
G8 TMV C Tuckley, 15 Charmandean Rd, Worthing, BN14 9LQ [IO90TT, TQ10]
G8 TMY Details withheld at licensee's request by SSL.
G8 TNA S J Thompson, Hollies, Chapel Hill, Sticker, St. Austell, PL26 7HG [IO70NH, SW95]
G8 TNB P Thompson, Lyndhurst Cottage, Main St., Weston, Newark, NG23 6ST [IO93NE, SK76]
G8 TNC Details withheld at licensee's request by SSL.
G8 TND C B Schiffman, 55 Bawdsey Ave, Newbury Park, Ilford, IG2 7TN [JO01BN, TQ48]
G8 TNE D J Pickford, 80 Hollowood Ave, Littleover, Derby, DE23 6JD [IO92FV, SK33]
G8 TNH P A Jeffries, 22 Ingrams Way, Hailsham, BN27 3NP [JO00CU, TQ50]
G8 TNK B W Godwin North Kent Radio Soc, 20 Pembury Rd, Bexleyheath, DA7 5NB [JO01BL, TQ47]
G8 TNQ F J Cousins, 6 Paddocks Cl, Orpington, BR5 4PP [JO01BI, TQ46]
G8 TNS S C Ward, 7 Redwing Cl, Oakham, LE15 6DA [IO92PQ, SK80]
G8 TNU A D Lambert, 16 Mentone Rd, Ashley Cross, Parkstone, Poole, BH14 8AU [IO90AR, SZ09]
G8 TNZ S M James, 72 Parkway, Wickham Market, Woodbridge, IP13 0SS [JO02QD, TM35]
G8 TOD A A G Woodyatt, Bochym, Lowdilow Ln, Elmstone Hardwicke, Cheltenham, GL51 9TH [IO81XW, SO92]
G8 TOF Details withheld at licensee's request by SSL.
G8 TOI R Hempstead, 21 Lymington Ave, Clacton on Sea, Essex, CO15 4PJ [JO01OT, TM11]
G8 TOK Details withheld at licensee's request by SSL.[QSL via Bureau or direct via G4NSY.]
G8 TOM W A Hitchcock, 400 Lordship Ln, East Dulwich, London, SE22 8ND [IO91XK, TQ37]
G8 TOQ J B Jackson, 1 Dolly Garth, Arkengarthdale Rd, Reeth, Richmond, DL11 6QX [IO94AJ, NZ00]
GW8 TOX K Taylor, Swn-y-Don, Beaumaris, Anglesey, North Wales, LL58 8RW [IO73XH, SH67]
G8 TPA Details withheld at licensee's request by SSL.
G8 TPC B W Taylor, 161 Sidegate Ln, Ipswich, IP4 4JN [JO02OB, TM14]
G8 TPK D A Redfern, 57A Queen St., Waingroves, Ripley, DE5 9TJ [IO93HA, SK44]
G8 TPM N Wellsbury, Elgar, 15 Woodlands Rd, Cookley, Kidderminster, DY10 3TL [IO82VJ, SO87]
G8 TPP M F S Strudwick, 65 Neave Cres, Haroldhill, Romford, RM3 8HN [JO01CO, TQ59]
G8 TPR Details withheld at licensee's request by SSL.[Op: G J Voller.]
G8 TPT M N W Male, 47 Broadway, Jaywick, Clacton on Sea, CO15 2ET [JO01NS, TM11]
G8 TPX J Bates, 60 Blenheim Rd, Northolt, UB5 4TP [IO91TN, TQ18]
G8 TPY Details withheld at licensee's request by SSL.
G8 TQH A I McMullin, Fair View, Rickham, East Portlemouth, Salcombe, Devon, TQ8 8PJ [IO80CF, SX73]
G8 TQI P R Herod, 4 St. James Rd, Little Paxton, St. Neots, Huntingdon, PE19 4QW [IO92UG, TL16]
G8 TQJ R Markfort, 105 Woodlands Way, Southwater, Horsham, RH13 7TF [IO91TA, TQ12]
G8 TQK A F Mayhew, 51 Upland Rd, Sutton, SM2 5HW [IO91VI, TQ26]
G8 TQP R W Healey, 35 Tirlebank Way, Newtown, Tewkesbury, GL20 8ES [IO81WX, SO93]
G8 TQV R S Tuckett, 89 Hillbrook Rd, Upper Tooting, London, SW17 8SF [IO91WK, TQ27]
G8 TQY Details withheld at licensee's request by SSL.
G8 TQZ B W Woods, 84 Beauly Way, Rise Park, Romford, RM1 4XR [JO01CO, TQ59]
G8 TRB Details withheld at licensee's request by SSL.
G8 TRF K J Maskell Maidstone YMCA, Ars.YMCA Sprts Ctre, Melrose Cl, Cripple St Maidston, ME15 6BD [JO01GG, TQ75]
G8 TRG R E Green Tower Radio Gp, 2 Ragley Walk, Rowley Regis, Warley, B65 9NT [IO82XL, SO98]
G8 TRK A K Wilson, 65 Beacon Rd, Boldmere, Sutton Coldfield, B73 5SX [IO92BN, SP19]
GW8 TRO K Prosser, 51 Highmead, Pontllanfraith, Blackwood, NP2 2PF [IO81JP, ST19]
G8 TRP Details withheld at licensee's request by SSL.
G8 TRQ M F Walker, 20 Littlewood Ln, Cheslyn Hay, Walsall, WS6 7EJ [IO82XQ, SJ90]
G8 TRR W J Pickard, 19 Canham Cl, Kimpton, Hitchin, SG4 8SD [IO91UU, TL11]
G8 TRS R E Deakin Tamworth ARC, 12 Henley Cl, Perrycrofts, Tamworth, B79 8TQ [IO92DP, SK20]
G8 TRU S P Lynch, 40 Copthall Dr, London, NW7 2NB [IO91VO, TQ29]
G8 TRY G G E Scott, 19 Penkett Rd, Wallasey, L45 7QF [IO83LK, SJ39]
G8 TSD Details withheld at licensee's request by SSL.
G8 TSE Details withheld at licensee's request by SSL.
G8 TSG D Johansen, 45 Marfords Ave, Bromborough, Wirral, L63 0JJ [IO83MH, SJ38]
GM8 TSI I J Raine, 131 Lisburn Rd, Saintfield, Ballynahinch, BT24 7BX [IO74BL, J35]
G8 TSJ Details withheld at licensee's request by SSL.
G8 TSN M M Gascoigne, 110 Wymondley Rd, Hitchin, SG4 9PX [IO91UW, TL12]
G8 TSQ A D Goodson, 6 Sandhills Cl, Belton, Loughborough, LE12 9TT [IO92HS, SK42]
GI8 TST F J Mimna, 96 Meelmore Dr, Strathroy, Omagh, Co. Tyrone, BT79 7XD [IO64IO, H47]
G8 TSV G H Rowlands, 8 Walbury Cl, Brant Rd, Lincoln, LN5 9TN [IO93RE, SK96]
G8 TSY Details withheld at licensee's request by SSL.[Op: P J Turner.]
G8 TSZ A R Twyford, 62 Crookham Rd, Church Crookham, Fleet, GU13 0NH [IO91NG, SU85]
G8 TTD Dr P H Palin, Meadow Vale, Beech Rd, Elswick, Preston, PR4 3YB [IO83NU, SD43]
G8 TTE Dr E C Thomas, Fairfield, St. Marys Rd, Manton, Oakham, Leics, LE15 8SU [IO92PP, SK80]
G8 TTI D J Kearns, 37 High St., Sutton Benger, Chippenham, SN15 4RQ [IO81XM, ST97]
G8 TTK Details withheld at licensee's request by SSL.
G8 TTP R . S . Nicholls, 57 Mandalay Ct, London Rd, Patcham, Brighton, BN1 8QW [IO90WU, TQ20]
G8 TTQ A H Kirk, 4/5 Dairy Cttgs, Low Row, Brampton, Cumbria, CA8 2LF [IO84QX, NY56]
G8 TTU C Smithson, 4 Calder Ave, Littleborough, OL15 9JE [IO83WP, SD91]
G8 TTX K A Sugg, 28 Well Cl, Winscombe, BS25 1HQ [IO81OH, ST45]
G8 TTY R F Spiers, 9 Orchard Way, Charlton Horethorne, Sherborne, DT9 4PJ [IO81SA, ST62]
G8 TUB M G Schwarzer, Whitelands, Back Ln, Charnock Richard, Chorley, PR7 5JT [IO83PP, SD51]
G8 TUH D George, East Winch Rd, Blackborough Gnd, Blackborough End, Kings Lynn, Norfolk, PE32 1SF [JO02FQ, TF61]
G8 TUL H R J Dowson, 11 Willaston Ave, Black0, Blacko, Nelson, BB9 6LU [IO83VU, SD84]
G8 TUN C J Denton, 34 Brook Ln, Ormskirk, L39 4RE [IO83NN, SD40]
G8 TUT E V Brett, 38 Shenstone Rd, Maypole, Birmingham, B14 4TJ [IO92BJ, SP07]
G8 TUU R J Boyce, 22 Mistley Cl, Bexhill on Sea, TN40 2TD [JO00FU, TQ70]
G8 TVB D J Travett, 9 Milbury Cl, Exminster, Exeter, EX6 8AF [IO80GQ, SX98]
G8 TVC T M Webb, 9 Granary Ct, Ramsey, Huntingdon, PE17 1HY [IO92WK, TL28]
G8 TVH Details withheld at licensee's request by SSL.
G8 TVI M W N Ashton, c/o Bonas USA Inc, 2231 Gateway Blvd, Charlotte Nc28208, USA
G8 TVL D C Bathurst, 40 High Oaks, St. Albans, AL3 6DN [IO91TS, TL10]
G8 TVM K W Biggs, 33 Blanford Gdns, West Bridgford, Nottingham, NG2 7UQ [IO92KW, SK53]
G8 TVO D H Bunch, 2 Canal Cottage, Huntworth, Bridgwater, Somerset, TA6 6LR [IO81MC, ST33]

G8 TVT T L Collins, 11 Rosedale, Leven, Beverley, HU17 5NE [IO93UV, TA14]
GM8 TVV N J Coote, 130 Castle Gdns, Paisley, PA2 9RD [IO75SU, NS46]
G8 TVW D H Young, 58 Furzefield Rd, Welwyn Garden City, AL7 3RJ [IO91VT, TL21]
GW8 TVX R P Hope, 75 Priors Way, Dunvant, Swansea, SA2 7UH [IO71XO, SS59]
G8 TVZ R J Midgeley, 2 Digswell Park Cttgs, Digswell Park Rd, Welwyn Garden City, AL8 7NN [IO91VT, TL21]
G8 TWA G H N Jasper, Clann Farm, Clann Ln, Lanivet, Bodmin, PL30 5HD [IO70OK, SX06]
GI8 TWB B H Mitchell, 7 Crea Rd, Randalstown, Antrim, N Ireland, N Ireland, BT41 3DX [IO64TT, J09]
G8 TWC H L Mather, 63 Woodhall Ln, Welwyn Garden City, AL7 3TG [IO91VT, TL21]
G8 TWD T B Manning, 21 Skeffington Rd, East Ham, London, E6 2NA [JO01AM, TQ48]
G8 TWE N P Mallet, Gwynant, 31 Belmont Rd, Maidenhead, SL6 6JL [IO91PM, SU88]
G8 TWL D N J Webster, 56 Greenhaugh, West Moor, Newcastle upon Tyne, NE12 0WA [IO95FB, NZ27]
G8 TWR J F F Evans, 49 Inverness Ave, Enfield, EN1 3NU [IO91XP, TQ39]
G8 TWS M G Corbett, Jacobs Ladder, Leys Hill, Walford, Ross on Wye Hfds, HR9 5QS [IO81QV, SO52]
G8 TWT B M Fisher, 24 Vessey Rd, Worksop, S81 7PG [IO93KH, SK58]
G8 TXA R Heeley, Rosemary Cottage, 4 Cherry Tree Ln, Halesowen, B63 1DU [IO82XK, SO98]
GM8 TXC J F D Hedley, Flat 3/2, White St., Partick, Glasgow, G11 5RT [IO75KU, NS56]
G8 TXF P W Hand, 20 Thornley Cl, Semele, Radford Semele, Leamington Spa, CV31 1UL [IO92GG, SP36]
GM8 TXK P W A Shulver, Falcove, Finlaystone Rd, Kilmacolm, PA13 4RE [IO75QV, NS37]
G8 TXM P Simm, 28 Beech Ave, Huddersfield, HD5 8DZ [IO93DP, SE11]
G8 TXN Details withheld at licensee's request by SSL.
G8 TXO Details withheld at licensee's request by SSL.
G8 TXQ P J Naylor, 24 Castle View Dr, Cromford, Matlock, DE4 3RL [IO93FC, SK25]
G8 TXW G F Sutcliffe, 41 Rose Ave, Irlam, Manchester, M44 6AQ [IO83SK, SJ79]
G8 TXX J Taylor, The Jays, 5 Watling Cl, Bourne, PE10 9XL [IO92TS, TF02]
G8 TYC N F Pritchard, 222 Ryde Park Rd, Rednal, Birmingham, B45 8RJ [IO92AJ, SP07]
G8 TYD D E Prouse, 1 Eastdown, Hartland, Bideford, Devon, EX39 6AQ [IO70SX, SS22]
G8 TYF H G Sasse, 358 Western Rd, Leicester, LE3 0ED [IO92KO, SK50]
G8 TYG P Fuller, 133A Furtherwick Rd, Canvey Island, SS8 7AT [JO01HM, TQ88]
G8 TYH R Marsh, 58 Statham Ave, Lymm, WA13 9NL [IO83SJ, SJ68]
G8 TYN C J Elliott, 4 Ivel View, Sandy, SG19 1AU [IO92UD, TL14]
G8 TYP C T Ferguson, Rocklands, 88 Ridgeway Rd, Long Ashton, Bristol, BS18 9HA [IO81QK, ST57]
GW8 TYS T Griffin Thomas, Pen-y-Banc, 3 Blue Anchor Rd, Swansea, SA4 3JQ [IO71WP, SS59]
G8 TYV Details withheld at licensee's request by SSL.
G8 TYX J E Hodgson, 26 Highleys Dr, Oadby, Leicester, LE2 5TL [IO92LO, SP69]
G8 TYY Dr T P Hopkins, 5 Rochester Cl, Bacup, OL13 8RN [IO83VR, SD82]
G8 TZE D S Pritt, 387 London Rd, Clanfield, Waterlooville, PO8 0PJ [IO90MW, SU71]
G8 TZJ A Sellers, 2 Dunkenshaw Cres, Lancaster, LA1 4LQ [IO84OA, SD45]
G8 TZK J D Sherbourne, 4 Chelston Terr, West Buckland, Chelston, Wellington, TA21 9HT [IO80JX, ST12]
G8 TZL J M Wright, 32 St. Annes Rd, Marshside, Southport, PR9 9TQ [IO83MQ, SD31]
G8 TZN R S North, 5 George Rd, Guildford, GU1 4NP [IO91RF, SU95]
G8 TZT C S Camm, 151 Hebden Rd, Haworth, Keighley, BD22 8RE [IO93AT, SE03]
G8 TZW G M Dunn, 29 Sundridge Rd, Kingstanding, Birmingham, B44 9NY [IO92BN, SP09]
G8 TZY M R D Goldsworthy, 17 Horseman Ln, Copmanthorpe, York, YO2 3UD [IO93KV, SE54]
G8 UA H Tee, 25 Norfolk Ave, Burnley, BB12 6DG [IO83UT, SD83]
G8 UAD R G White, 72 Green Ln, Bournemouth, BH10 5LF [IO90BS, SZ09]
G8 UAI D J Lockwood, 61 Green Ln, Tickton, Beverley, HU17 9RH [IO93TU, TA04]
G8 UAP T E Williamson, Thie-Ny-Keyll, Off Maurys Ln, West Wellow, Hants, SO5 0DB [IO90HW, SU41]
G8 UAQ Details withheld at licensee's request by SSL.
G8 UAS E H Wood, Nykoping, 5 Delphfields Rd, Appleton, Warrington, WA4 5BY [IO83RI, SJ68]
G8 UAV Details withheld at licensee's request by SSL.
G8 UAW D A Hallgarten, 14 Antrim Gr, London, NW3 4XR [IO91WN, TQ28]
G8 UAY J D Earley, 8 Oatlands Rd, Boorley Green, Botley, Southampton, SO32 2DE [IO90IW, SU51]
G8 UAZ A G Dipple, 47 Tamar Dr, Aveley, South Ockendon, RM15 4NA [JO01DM, TQ58]
G8 UBF G G R Beadle, 81 Halford Rd, Uxbridge, UB10 8QA [IO91SN, TQ08]
G8 UBH B S Capon, 21 Apperley Way, Homer Hill, Halesowen, B63 2PN [IO82WL, SO98]
G8 UBJ R A Lester, 48 Stanford Ave, Hassocks, BN6 8JJ [IO90WW, TQ31]
G8 UBN P G Hodgson, 19 Mongers Piece, Chineham, Basingstoke, RG24 8RL [IO91LG, SU65]
G8 UBP A S Jacketts, 18 Tatton Ct, Egerton Rd, Fallowfield, Manchester, M14 6XH [IO83VK, SJ89]
G8 UBU R I Jarvis, 39 Moy Rd, Monkwick Est, Colchester, CO2 8NZ [JO01KU, TM02]
G8 UBX R L Manning, 3 Lawns Cl, Melbourn, Royston, SG8 6DR [JO02AB, TL34]
G8 UCB Details withheld at licensee's request by SSL.
G8 UCC I S Bradley, 8 Hunt Ave, Heanor, DE75 7QB [IO93HA, SK44]
G8 UCH C Burton, 7 Springhill Dr, Slack Ln, Crofton, Wakefield, WF4 1EX [IO93GP, SE31]
G8 UCM J H Corkhill, 15 Woolton Hill Rd, Liverpool, L25 6HU [IO83NJ, SJ48]
G8 UCN A G Crookes, 23 Helliwell Ln, Deepcar, Sheffield, S30 5QH [IO93GJ, SK38]
G8 UCP M J Culling, 101 Orchard Dr, Park St., St. Albans, AL2 2QL [IO91TR, TL10]
G8 UCR D J Davis, 58 Abney Rd, Sheffield, S14 1PD [IO93GI, SK38]
GI8 UCS A J Edwards, 126 Merville Garden Village, Newtownabbey, BT37 9TJ [IO74BP, J38]
G8 UCW C R Tilling, 54 The Leys, Bidford on Avon, Alcester, B50 4DN [IO92BE, SP15]
G8 UCY D Walker, Bessbrook, 43 Wimborne Rd, Corfe Mullen, Wimborne, BH21 3DS [IO80XS, SY99]
G8 UCZ C J Wren, 24 Willow Way, Martham, Great Yarmouth, NR29 4SH [JO02TQ, TG41]
G8 UDA B Watson, 3 Anderton Rise, Millbrook, Torpoint, PL10 1DA [IO70VI, SX45]
G8 UDB R G A Youard, 12 Northampton Park, London, N1 2PJ [IO91WN, TQ38]
G8 UDD S W James, 24 Pocklington Cl, Chelmer Village, Chelmsford, CM2 6SQ [JO01GR, TL70]
G8 UDG D E Roberts, 40 Hanson Ct, Woburn Cl, Cambridge, CB4 2SE [JO02BF, TL46]
G8 UDH M Radford, 32 Millfields, Writtle, Chelmsford, CM1 3LP [JO01FR, TL60]
G8 UDI P S Newman, 66 Western Way, Barnet, EN5 2BT [IO91VP, TQ29]
G8 UDJ M J Loach, 82 Honey Bottom Ln, Dry Sandford, Abingdon, OX13 6BX [IO91IQ, SP40]
G8 UDS J E Farrant, 589 Galleywood Rd, Chelmsford, CM2 8BS [JO01FR, TL70]
G8 UDT C G Flint, Mildenhall, 13 Fellow Green, West End, Woking, GU24 9LL [IO91QI, SU96]
G8 UDV A J Frost, 10 Ramsden Sq, Cambridge, CB4 2BJ [JO02BF, TL46]
G8 UEA R Glass, 103 Greenleach Ln, Roe Green, Worsley, Manchester, M28 2RT [IO83TM, SD70]
G8 UEC Details withheld at licensee's request by SSL.
G8 UEE S Melvin, 2 Salters Ct, Newcastle upon Tyne, NE3 5BH [IO95EA, NZ26]
G8 UEF P J Mobberley, The Willows, Sladd Ln, Wolverley, Kidderminster Worcs, DY11 5SX [IO82UK, SO88]
G8 UEI A W Howells, 20 Dighton Ct, John Ruskin St., London, SE5 0PR [IO91WL, TQ37]
G8 UEK N A Hosker, 17 Morgan Cl, Blacon, Chester, CH1 5XH [IO83MF, SJ36]
G8 UEM C J Hollebon, 5 Evelyn Ave, Aldershot, GU11 3QB [IO91OF, SU84]
G8 UEP Details withheld at licensee's request by SSL.
G8 UEQ Details withheld at licensee's request by SSL.
G8 UEZ C A Southall, 40 Tathall End, Hanslope, Milton Keynes, MK19 7NF [IO92OC, SP84]
G8 UFF A D Patis, 3 Grosvenor Pines, Grosvenor Rd, Bournemouth, BH4 8BQ [IO90BR, SZ09]
G8 UFN A D Billington, 8 Wilcox Rd, Chipping Norton, OX7 5LE [IO91FW, SP32]
G8 UFR Details withheld at licensee's request by SSL.
G8 UFX I R Fowler, 1 Mayfields, Shefford, SG17 5AU [IO92UA, TL13]
G8 UG J K Coomber, 23 Bramcote Ave, Mitcham, CR4 4LW [IO91WJ, TQ26]
G8 UGB Details withheld at licensee's request by SSL.
G8 UGD Details withheld at licensee's request by SSL.
G8 UGH J Yates, Forest House, 102 Holborn Hill, Ormskirk, L39 3LJ [IO83NN, SD40]
G8 UGK P D Warburton, 4384 Henneberry Rd, Manlius, New York 13104, USA
G8 UGL J R Wakenell, 15 Cuckoo Oak Green, Madeley, Telford, TF7 4HT [IO82SP, SJ70]
G8 UGM A D Warriner, 26 Alderbrook Dr, Attleborough Gdns, Nuneaton, CV11 6PL [IO92GM, SP39]
G8 UGN C R Wood, 2 Stivers Way, Harlington, Dunstable, LU5 6PH [IO91SX, TL03]
GM8 UGO W A Kay, 22 Linton Terr, Perth, PH1 1LE [IO86GJ, NO02]
G8 UGR Details withheld at licensee's request by SSL.
G8 UGS P Marks, 106 Darlton Dr, Arnold, Nottingham, NG5 7LW [IO93KA, SK54]
G8 UHF J S Lathnury, 45 Midway Rd, Midway, Swadlincote, DE11 7NT [IO92FS, SK32]
G8 UHJ P Q Couch, 6 Quantock Gdns, Ramsgate, CT12 6SW [JO01QI, TR36]
G8 UHK A M Rolls, 9 Mareschal Rd, Guildford, GU2 5JF [IO91RF, SU94]
G8 UHM W S C Rosser, Fanshawgate House, Fanshaw, Holmesfield, Sheffield, S18 5WA [IO93GH, SK37]
G8 UHO D A Reay, 78 Wyresdale Rd, Lancaster, LA1 3DY [IO84OB, SD46]
G8 UHS J Spinks, 29 New Rd, Abbey Wood, London, SE2 0QH [JO01BL, TQ47]
G8 UHT P Shaw, Pool View, 33 Pool Ln, Winterley, Sandbach, CW11 4RZ [IO83TC, SJ75]
G8 UHV G B Watson, 22 School Rd, Laughton, Sheffield, S31 7YP [IO93GJ, SK38]
G8 UHW C I Mobbs, 5 Garth Ave, Leeds, LS17 5BH [IO93FU, SE23]
G8 UIB Details withheld at licensee's request by SSL.[Op: J F Saltmarsh, 74 Lansdowne Road, Bayston Hill, Shrewsbury, Shropshire, SY3 0JG.]
G8 UIG M E G Chedzoy, Helena, Union Drove, Picts Hill, Langport, TA10 9EY [IO81OB, ST42]
G8 UIH Details withheld at licensee's request by SSL.
G8 UIO D R Salter, 9 Old Milverton Rd, Milverton, Leamington Spa, CV32 6BA [IO92FH, SP36]
G8 UIQ P Willgress, 20 Onslow Rd, Leagrave, Luton, LU4 9AJ [IO91SV, TL02]
GI8 UIU Dr P R J Moore, 13 Ballygallum Rd, Ballygallum, Downpatrick, BT30 7DA [IO74DH, J54]
G8 UIV P F Morton-Thurtle, Trenley Hall, Stodmarsh Rd, Canterbury, CT3 4AH [JO01NG, TR15]
G8 UIW S G Threlfall-Rogers, 43 Nanpantan Rd, Loughborough, LE11 3ST [IO92JS, SK51]
G8 UIY G J Platt, 78 Harvest Ln, Moreton, Wirral, L46 7UE [IO83KJ, SJ29]
G8 UJF D J Headland, Hazelwood, Haywards Ln, Cheltenham, GL52 6RF [IO81XV, SO92]
G8 UJG R C Hawkins, 101 Tobyfield Rd, Bishops Cleeve, Cheltenham, GL52 4NZ [IO81XW, SO92]
G8 UJO B C Leveton, 12 Rugge Dr, Eaton, Norwich, NR4 7NJ [JO02PO, TG20]
G8 UJS Details withheld at licensee's request by SSL.
G8 UJV M Johnson, 9 South Rd, Brampton, Huntingdon, PE18 8PX [IO92VH, TL27]
G8 UKB F A Northwood, 3 Kempton Cl, Spalding, PE11 3BX [IO92VS, TF22]
G8 UKG R Moody, 47 The Valley, Comberton, Cambridge, CB3 7DF [JO02AE, TL35]
G8 UKH L G Luck, 157 Shakespeare Ave, Hayes, UB4 0BQ [IO91TM, TQ18]
G8 UKI T P H Gawn, 15 Barradon Cl, Watcombe Park, Torquay, TQ2 8QE [IO80FM, SX96]
G8 UKN Details withheld at licensee's request by SSL.
G8 UKT Details withheld at licensee's request by SSL.
G8 UKV M Vincent, 9 Sleapford, Longlane, Wellington, Telford Shropshire, TF6 6HQ [IO82RR, SJ61]
GW8 UKW A Case, 2 Cranog Cl, Trefil, Trefin, Haverfordwest, SA62 5AT [IO71KW, SM83]
G8 UKY R M F Mills, 3 Hallam Moor, Liden, Swindon, SN3 6LS [IO91DN, SU18]
G8 UKZ Dr A G Walker, 1 Church Ln, Milcombe, Banbury, OX15 4SA [IO92HA, SP43]
G8 ULG Details withheld at licensee's request by SSL.
G8 ULH J Wilson, 2 Reston Ct, Cleethorpes, DN35 0JQ [IO93XN, TA30]
G8 ULJ G T Sowter, 21 Seawell Rd, Poughill, Bude, EX23 8PD [IO70RT, SS20]
G8 ULL M R W Williams, 9 Fir Tree Dr, West Winch, Kings Lynn, PE33 0PR [JO02EQ, TF61]
G8 ULO Details withheld at licensee's request by SSL.
G8 ULQ C P Copsey, 19 Windsor Dr, Solihull, B92 8HS [IO92CK, SP18]
G8 ULV Details withheld at licensee's request by SSL.
G8 UMA J E H Spencer, 16 Beacon Cl, Boundstone, Wrecclesham, Farnham, GU10 4PA [IO91OE, SU84]
G8 UMB R C Kennedy, 15 Cradley Rd, New Eltham, London, SE9 2HD [JO01AK, TQ47]
G8 UMG G D Wyche, 17 Clifton Gr, Mansfield, NG18 4HY [IO93JD, SK55]
G8 UML M G B Spencer, 79 Salisbury Cl, Alton, GU34 2TP [IO91MD, SU73]
G8 UMM G Scott, Middleton Hall, Nursing Home, Middleton St George, Darlington, Co. Durham, DL2 1HA [IO94GM, NZ31]
G8 UMX Details withheld at licensee's request by SSL.
G8 UMY P F Mahoney, 2 Elm Lodge, Elm Ave, Eastcote, Ruislip, HA4 8PH [IO91TN, TQ18]
G8 UN P W Harrad, 488 Brandlesholme Rd, Bury, BL8 1JH [IO83UO, SD71]
G8 UNO R A Clarke, 58 Orpin Rd, Redhill, RH1 3EY [IO91WG, TQ25]
G8 UNP R E A Burningham, 5 Norman Way, Steyning, BN44 3SE [IO90UV, TQ11]
G8 UNU J McFarlane, Crofton House, Church Ln, Wainfleet St Mary, PE24 4HP [JO03CC, TF45]
G8 UNZ Details withheld at licensee's request by SSL.
G8 UOD Details withheld at licensee's request by SSL.
G8 UOJ A M Abrahams, 69 Culverhouse Rd, Luton, LU3 1PY [IO91SV, TL02]
G8 UOL W Cooper, 4 Pan Walk, Chelmsford, CM1 2HD [JO01FR, TL60]
G8 UOP Details withheld at licensee's request by SSL.
G8 UOW M T Buckle, 3 Whiston Dr, Filey, YO14 0DB [IO94UF, TA18]
G8 UOZ M K Freestone, 12 St. Martins Approach, Ruislip, HA4 7QD [IO91SN, TQ08]
G8 UPF K J T Hutchinson, 8 Innage Cres, Bridgnorth, WV16 4HU [IO82SM, SO79]
GM8 UPI Details withheld at licensee's request by SSL.
G8 UPJ P B Nunn, 2 Mildred Ave, Prestwich, Manchester, M25 0LR [IO83UM, SD80]
G8 UPK D H Akester, 38 Birchdale, Oakwood Park, Bingley, BD16 4SE [IO93BU, SE14]
G8 UPM F W M Richards, 114 Crowther Rd, Newbridge, Wolverhampton, WV6 0HY [IO82WO, SO89]
G8 UPO D Steele, High Lee Hall, St Marys Rd, New Mills, Stockport, SK12 3BW [IO83XI, SJ98]
G8 UQC D R Bolton, 12 Falterley Rd, Wythenshawe, Manchester, M23 9BR [IO83UJ, SJ88]
G8 UQI Details withheld at licensee's request by SSL.
G8 UQJ J G McMillan, Pipers Meadow, Chain Ln, Battle, TN33 0HG [JO00FW, TQ71]
G8 UQP R Wassell, 215 Pasture Rd, Stapleford, Nottingham, NG9 8JB [IO92IW, SK43]
G8 UQQ Details withheld at licensee's request by SSL.
G8 UQR T D Riley, 74 Marshall Hill Dr, Mapperley, Nottingham, NG3 6FP [IO92KX, SK54]
G8 UQV F Hall, 478 Darwen Rd, Bromley Cross, Bolton, BL7 9DX [IO83SO, SD71]
G8 URB F C McGilp, Beechfield Ln, Frilsham, Hermitage, Newbury, Berks, RG16 9XD [IO91LJ, SU66]
G8 URC D K Matheson, 42 Wolseley Ave, Studd Hill, Herne Bay, CT6 8AL [JO01NI, TR16]
G8 URH Details withheld at licensee's request by SSL.
G8 URI G Cross, 11 Highfield App/Ch, Billericay, Essex, CM11 2PD [JO01FO, TQ69]
G8 URO Details withheld at licensee's request by SSL.
G8 URP G C Nye, 16 Sedgewick Gdns, Up Hatherley, Cheltenham, GL51 5QD [IO81WV, SO92]
G8 URU R J Drew, Clattering Ford, Rd Head, Carlisle, Cumbria, CA6 6NT [IO85OB, NY57]
G8 URZ N P Bird, 15 Bramley Cl, Colletts Green, Powick, Worcester, WR2 4SR [IO82UD, SO85]
G8 USF R L Grimsdale, 21 Friar Rd, Brighton, BN1 6NG [IO90WU, TQ30]
G8 USG Details withheld at licensee's request by SSL.
G8 UST M C Freeman, Sunnyside Long Ln, Shirebrook, Mansfield, Notts, NG20 8AZ [IO93JE, SK56]
G8 USU Details withheld at licensee's request by SSL.
G8 USV C D Holland, 29 Woodgarth Ln, Worsley, Manchester, M28 2PS [IO83TL, SD70]
G8 USX A P Birchall, 7 Southview Cottage, Liverpool Rd West, Church Lawton, Stoke on Trent, ST7 3DL [IO83UC, SJ85]
GW8 UTK B J Davies, Rhosyr, Llanfair Pg, Gwynedd, LL61 5JB [IO73VF, SH57]
G8 UTN M A Redman, 19 Richmond Rd, Rugby, CV21 3AB [IO92JI, SP57]
G8 UTQ W Vander Byl, 126 Newtown Rd, Carlisle, CA2 7LN [IO84MV, NY35]
G8 UTW H C Mirams, 58 Ing Head Terr, Shelf, Halifax, HX3 7LB [IO93CR, SE12]
G8 UUG A Lenton, 37 Hawkley Dr, Tadley, RG26 3YH [IO91KI, SU66]
G8 UUL M R Browne, 13 Latchingdon Rd, Cold Norton, Chelmsford, CM3 6JG [JO01IQ, TL80]
G8 UUM C D Barnett, 8 Mount Ave, Hednesford, Cannock, WS12 4DA [IO82XR, SJ91]
G8 UUN G P J Curtis-Smith, 16 Mersea Rd, Colchester, CO2 7EX [JO01KV, TL92]
G8 UUS P Owen, 2 Plantation Rd, Nottingham, NG8 2ER [IO92JW, SK53]
G8 UUV M W B Copsey, 53 Boyd Ave, Dereham, NR19 1LU [JO02LP, TF91]
GM8 UUW C D Fyfe, 2 Marmion Rd, Galashiels, TD1 2DE [IO85OO, NT53]
G8 UVB S J Bartlett, 48 Barrymore Walk, Rayleigh, SS6 8YF [JO01HO, TQ89]
G8 UVC Details withheld at licensee's request by SSL.
G8 UVE Details withheld at licensee's request by SSL.
G8 UVG C C Parker-Larkin, 2 Mitchells Rarm Cttgs, Maldon Rd, Langford, Maldon, Essex, CM9 4SP [JO01IR, TL80]
G8 UVM A W Read, Yew Tree House, Edge, Glos, GL6 6NR [IO81VS, SO81]
G8 UVN B Rigby, 76 Woodland Rd, Rode Heath, Stoke on Trent, ST7 3TL [IO83UC, SJ85]
G8 UVQ B M A Smith, 161 Sharp St., Newland Ave, Hull, HU5 2AE [IO93TS, TA03]
G8 UVR Details withheld at licensee's request by SSL.[Station located near Borehamwood.]
G8 UVS A R Sillence, 74 Atherley Rd, Shirley, Southampton, SO15 5DS [IO90GV, SU41]
G8 UVU M A Soble, 1 The Coverts, Tadley, RG26 3TS [IO91KI, SU56]
G8 UVZ B J Hart, 63 Newcastle Rd, Congleton, CW12 4HL [IO83VD, SJ86]
G8 UWD R Hornby, 63 Shearwater Dr, Bicester, OX6 0YR [IO91KV, SP52]
G8 UWE M Jefford, 37 Marions Way, Exmouth, EX8 4LF [IO80HP, SY08]
G8 UWF Details withheld at licensee's request by SSL.
G8 UWG A P Kay, 19 Chesnut Gr, Higher Tranmere, Birkenhead, L42 0LB [IO83LJ, SJ38]
G8 UWH G A Breed, 39 Woodlow, Thundersley, Benfleet, SS7 3RQ [JO01HN, TQ78]
G8 UWI A F Downing, 21 Firfield Rd, Thundersley, Benfleet, SS7 3UU [JO01HN, TQ88]
G8 UWJ S Coleyshaw, 58 Suffolk Ave, Derby, Derbyshire, DE21 6ER [IO92GW, SK33]
G8 UWL D Cooper, 52 Derby Rd, Birkenhead, L42 7HB [IO83LJ, SJ38]
G8 UWM M J Crossley, 3 Derby St., Stockport, SK3 9HF [IO83WJ, SJ88]
G8 UWO G K Earp, 39 Southbourne Rd, Wallasey, L45 8QA [IO83LK, SJ29]
G8 UWS J Stopford, Bracken, The St., West Hougham, Dover, CT15 7BH [JO01OC, TR24]
G8 UWU Details withheld at licensee's request by SSL.
G8 UWY Details withheld at licensee's request by SSL.
G8 UXB B E Gilbert, 45 Kellaway Ave, Westbury Park, Bristol, BS6 7XS [IO81QL, ST57]
G8 UXE Details withheld at licensee's request by SSL.
GW8 UXI M W Griffiths, 19 Cromwell Rd, Risca, Newport, NP1 7AF [IO81KO, ST29]
G8 UXL P L Nicholson, 14 Boyne Rd, Budleigh Salterton, EX9 6SE [IO80IP, SY08]
G8 UXT Details withheld at licensee's request by SSL.
G8 UXU Details withheld at licensee's request by SSL.
G8 UXW J C Benton, 43C Stakes Hill Rd, Waterlooville, Hants, PO7 7LA [IO90LV, SU60]
G8 UXY K F Bone, 69 Tone Hill, Tonedale, Wellington, TA21 0AY [IO80JX, ST12]
G8 UYB K H Reece, 26 George St., Barnton, Northwich, CW8 4JQ [IO83RG, SJ67]
G8 UYD G R Ridgeway, 6 Rosewood Ave, Blackburn, BB1 9SZ [IO83SS, SD62]
G8 UYH D J W Robinson, 47 Oak Walk, Benfleet, SS7 4NS [JO01GN, TQ78]
G8 UYI D H Rodgers, 138 Coburg Cres, London, SW2 3HU [IO91WK, TQ37]
G8 UYK F M Rowntree, 1 Howards Thicket, Gerrards Cross, SL9 7NT [IO91RN, SU98]
G8 UYL R M Rumbelow, The Chase, Knott Park, Oxshott, Leatherhead, Surrey, KT22 0HR [IO91TH, TQ15]
G8 UYM J E N N Sanderson, 5 Babbacombe Dr, Rudd Hill Est, Ferryhill, DL17 8DA [IO94FQ, NZ33]
G8 UYN Details withheld at licensee's request by SSL.[Op: A Sharpe. Station located near Saffron Walden.]
G8 UYR B Smith, 3 Harwin Cl, Aldersley, Wolverhampton, WV6 9LF [IO82WO, SJ80]
G8 UYW T K Carrig, 12 Longmoor Dr, Liphook, GU30 7XA [IO91OB, SU83]
G8 UYY M J Daish, 27 Westbourne Rd, Portsmouth, PO2 7LB [IO90LT, SU60]
G8 UYZ D R Coomber, 14 Francis Green Ln, Penkridge, Stafford, ST19 5HF [IO82WR, SJ91]
G8 UZJ Details withheld at licensee's request by SSL.[Station located near Cheltenham.]
GW8 UZL J H Lewis, Trian, 14 Carreg y Gad, Llanfairpwll, Anglesey, LL61 5QF [IO73VF, SH57]
G8 UZM Dr R L Jefferson, 1 Carters Cttgs, Heddon on The Wall, Newcastle upon Tyne, NE15 0DW [IO94CX, NZ16]

G8 UZO Details withheld at licensee's request by SSL.
G8 UZQ S Haywood, 19 Crich Way, Newhall, Swadlincote, DE11 0UU [IO92FS, SK22]
G8 UZU D A McCoye, 26 Holcombe View Cl, Oldham, OL4 2QD [IO83XN, SD90]
G8 UZV J A Dowie, 19 Brooklands Pk, Wolverley Park, Kidderminster, DY11 5EB [IO82UJ, SO87]
G8 UZW J K Durrant, 114 Rosebank Ave, Hornchurch, RM12 5QS [JO01CN, TQ58]
G8 UZY D J Fisher, 26 Beach Rd, Burton Bradstock, Bridport, Dorset, DT6 4RF [IO80PQ, SY48]
G8 UZZ J R Fogg, 13 Barnhey Cres, Meols, Wirral, L47 9RN [IO83KJ, SJ28]
G8 VA A J Martin, 180 Clarendon Park Rd, Leicester, LE2 3AF [IO92KO, SK50]
G8 VAD J P Goodings, 133 Lache Ln, Chester, CH4 7LU [IO83NE, SJ36]
G8 VAE A T Griffiths, 11 The Dreys, Sewards End, Saffron Walden, CB10 2LL [JO02DA, TL53]
G8 VAF K R Chittenden, Collinus, Boxford Ln, Boxford, Sudbury, CO10 5JX [JO02KA, TL94]
GM8 VAM G A Brazier, 117 East Clyde St., Helensburgh, Dunbartonshire, G84 7PL [IO76PA, NS38]
G8 VAT G Denton, 46 Woollin Cres, West Ardsley, Tingley, Wakefield, WF3 1ET [IO93FR, SE22]
G8 VAU Details withheld at licensee's request by SSL.
G8 VBA R M Webb, 78 Station Rd, Rolleston on Dove, Burton on Trent, DE13 9AB [IO92EU, SK22]
G8 VBC R J Timms, 1 Butt Ln, Blackfordby, Swadlincote, DE11 8BG [IO92FS, SK31]
G8 VBE C A Thomas, 152 Bristol Rd, Edgbaston, Birmingham, B5 7XH [IO92BL, SP08]
G8 VBK R F Penver, 56 Cottesmore Ave, Ilford, IG5 0TG [JO01AO, TQ49]
G8 VBO Details withheld at licensee's request by SSL.
GM8 VBP E M Sampson, 47 Muirend Rd, Perth, PH1 1JD [IO86GJ, NO02]
G8 VBW M Corbett, 11 Leivers House, Derwent Cres, Arnold, Nottingham, NG5 6TF [IO92KX, SK54]
GM8 VBX D Coulthart, 23 Larchfield Rd, Dumfries, DG1 4HU [IO85EB, NX97]
GW8 VCA D J Dyer, Rhianfa, 24C Fforest Hill, Aberdulais, Neath, SA10 8HD [IO81CQ, SS79]
G8 VCB Details withheld at licensee's request by SSL.
G8 VCH J Grevatt, 7 Shelley Rd, East Grinstead, RH19 1SX [IO91XD, TQ33]
G8 VCI G Gwynne, 5 Stanstead Ave, Nottingham, NG5 5BL [IO93JA, SK54]
G8 VCJ C W Gunn, Kaduna House, Trispen, Truro, Cornwall, TR4 9BA [IO70LH, SW85]
G8 VCL R G German, 29 Glenthorne Gdns, Sutton, SM3 9NL [IO91VJ, TQ26]
G8 VCN M J Newport, 64 The Limes, Harston, Cambridge, CB2 5QT [JO02BD, TL45]
G8 VCQ W J Norman, 27 Newport Rd, Barnstaple, EX32 9BG [IO71XB, SS53]
G8 VCU S M Morgan, 24 Hall Farm Cl, Melton, Woodbridge, IP12 1RL [JO02QC, TM25]
G8 VCW Details withheld at licensee's request by SSL.
G8 VDA R D Maile, 8 Red House Cl, Newton Longville, Milton Keynes, MK17 0AH [IO91OX, SP83]
G8 VDF L C Perry, 5 Molly Cl, Temple Cloud, Bristol, BS18 5AE [IO81RH, ST65]
G8 VDG K H Player, 307 Birchanger Ln, Birchanger, Bishops Stortford, CM23 5QP [JO01CV, TL52]
G8 VDJ R W Pauley, Lamplight, Casterton Ln, Tinwell, Stamford,Lincs, PE9 3UQ [IO92RP, TF00]
G8 VDO Details withheld at licensee's request by SSL.
G8 VDP K W Roberts, 35A Rockley Ave, Birdwell, Barnsley, South Yorkshire, S70 5QY [IO93GM, SE30]
G8 VDQ C C S Parnell, 213 Northfield Ave, West Ealing, London, W13 9QU [IO91UL, TQ17]
G8 VDS N P Barchha, 90 King St., Wallasey, L44 8AN [IO83LK, SJ39]
G8 VDU K R Dexter, Plain-An-Gwarry Farm, Plain-An-Gwarry, Marazion, Cornwall, TR17 0DR [IO70GD, SW53]
G8 VDX P J Astle, 1 Beech Gr, Newhall, Swadlincote, DE11 0NH [IO92FS, SK22]
GW8 VEE G N Blore, Apsley House, 18 Kings Oak Ct, Wrexham, LL13 8QH [IO83MA, SJ34]
G8 VEH R M Bray, 14 Hadlow Way, Lancing, BN15 9DE [IO90UT, TQ10]
G8 VEL E G Cawkwell, 9 King St., Winterton, Scunthorpe, DN15 9RN [IO93QP, SE91]
G8 VEN H E Chapman, 24 Croft Rd, Cosby, Leicester, LE9 1SE [IO92JN, SP59]
G8 VEQ A R Stone, 47 Oakford Villas, North Molton, South Molton, EX36 3HJ [IO81CB, SS72]
G8 VER R M M Heath Verulam ARC, 26 Lancaster Ave, Hadley Wood, Barnet, EN4 0EX [IO91VQ, TQ29]
G8 VEZ T D Wagg, 15 Barncroft Way, Havant, PO9 3AA [IO90MU, SU70]
G8 VF A A H Moss, 1 Clarendon Cres, Eccles, Manchester, M30 9AU [IO83TL, SJ79]
G8 VFE P R Wilson, 33 Norton Ave, Norton, Stockton on Tees, TS20 2JJ [IO94IO, NZ42]
GW8 VFF A Wilkins, 11 Redhouse Rd, Ely, Cardiff, CF5 4FG [IO81JL, ST17]
G8 VFH Details withheld at licensee's request by SSL.
G8 VFI D J Franklin, 49 Hope Rd, Benfleet, SS7 5JQ [JO01GN, TQ78]
G8 VFL E G Bury, 26 Northfield Ave, Hanham, Bristol, BS15 3RB [IO81RK, ST67]
G8 VFM J Callaghan, 271 Belvoir Rd, Coalville, LE67 3PL [IO92HR, SK41]
G8 VFN Details withheld at licensee's request by SSL.
GW8 VFQ R Elliott, 19 Pencoed, Killay, Dunvant, Swansea, SA2 7PQ [IO71XO, SS69]
G8 VFR P A Dunning, 250 Glynswood, Chard, TA20 1BX [IO80MV, ST30]
G8 VFS Details withheld at licensee's request by SSL.
G8 VFU B A Croft, 9 Beverley St., Burnley, BB11 4LE [IO83US, SD83]
G8 VG A P W Windle, 8 Ricardo Rd, Minchinhampton, Stroud, GL6 9BY [IO81VQ, SO80]
GW8 VGB R J Morgan, Puffin House, 4 Underhill Ln, Horton, Gower, Swansea, SA3 1LB [IO71VN, SS48]
G8 VGC Details withheld at licensee's request by SSL.
G8 VGF J V Middleton, Portland House, 14 Toronto St., Lincoln, LN2 5NN [IO93RF, SK97]
GW8 VGG F D Lord, Redwick, 47 Glasllwch Cres, Newport, NP9 3SF [IO81LO, ST28]
G8 VGI A S J Lilly, 47 Horton St., Frome, BA11 3DP [IO81UF, ST74]
G8 VGO G A Grieve, 68 Pelican Rd, Pamber Heath, Tadley, RG26 3EL [IO91KI, SU66]
G8 VGQ P J Andrews, 14 Cotswold Community, Ashton Keynes, Swindon, SN6 6QT [IO91AP, SU09]
G8 VGU T R A Burgess, 14 Shrubcote, Appledore Rd, Tenterden, TN30 7BA [JO01IB, TQ83]
G8 VGY E K Brown, 26 Foxfield Way, Oakham, LE15 6PR [IO92PQ, SK80]
G8 VH F M Trier, Fairlawn, Little Cranmore Ln, West Horsley, Leatherhead, KT24 6HZ [IO91SG, TQ05]
G8 VHF J Goodier Bewlay Bros ARC, 20 Poleacre Ln, Woodley, Stockport, SK6 1PG [IO83WK, SJ99]
G8 VHG I W Gower, 10 Homethorpe, Hull, HU6 9EU [IO93TS, TA03]
GW8 VHI R G Woolley, 7 Old Rd, Baglan, Port Talbot, SA12 8TR [IO81CO, SS79]
G8 VHK M Stanford, 64 Forest Rd, Worthing, BN14 9LY [IO90TT, TQ10]
G8 VHL S P Price, 35 Western Rd, Goole, DN14 6QW [IO93NR, SE72]
G8 VHN G E Prentice, The Willow, Barretts Ln, Needham Market, Ipswich, IP6 8RZ [JO02MD, TM05]
G8 VHP N G Pittaway, Danford, Fullers Rd, Rowledge, Farnham, GU10 4DF [IO91OE, SU84]
G8 VIB J M Sim, 22 Dene View, Ashington, NE63 8JT [IO95EE, NZ28]
G8 VIC M E Rump, 24 Stoneleigh Ave, Brighton, BN1 8NP [IO90WU, TQ30]
G8 VIK K J Smith, 11 Windmill Rd, Chalfont St. Peter, Gerrards Cross, SL9 9PW [IO91RO, SU99]
G8 VIQ Details withheld at licensee's request by SSL.
G8 VIU M W Bennett, 89 Froxfield Rd, Havant, PO9 5PW [IO90MU, SU70]
G8 VIV K F Bilke, 44 Oaktree Dr, Hook, RG27 9RN [IO91MG, SU75]
GM8 VIW Details withheld at licensee's request by SSL.
G8 VJ Details withheld at licensee's request by SSL.[Op: D J Evans. Station located in Surrey.]
G8 VJF J E Parsons, 104 Bringhurst, Orton Goldhay, Peterborough, PE2 5RZ [IO92UN, TL19]
G8 VJG K W Halls, Little Rema, Cray Rd, Crockenhill, Swanley, BR8 8LP [JO01BJ, TQ56]
GW8 VJM Details withheld at licensee's request by SSL.
G8 VJO C Green, 12 Spenser Gr, Great Harwood, Blackburn, BB6 7JU [IO83TS, SD73]
G8 VJR D J Fowles, 529 Lexington Building, Fairfield Rd, Bow, London, E3 2UF [IO91XM, TQ38]
G8 VJS M S Epton, 78 Haigh Moor Rd, West Ardsley, Tingley, Wakefield, WF3 1EE [IO93FR, SE22]
G8 VJU K S Earl, 210 Churchill Ave, Chatham, ME5 0JS [JO01GI, TQ76]
GM8 VJV A R Devereux, South Fell, 24 Kirkhouse Rd, Blanefield, Glasgow, G63 9BX [IO75UX, NS57]
G8 VJW G P Davey, 147 Deeds Gr, High Wycombe, HP12 3PA [IO91OO, SU89]
G8 VJY A R F Crafer, 155 Upham Rd, Swindon, SN3 1DR [IO91CN, SU18]
GI8 VKA R J Coulter, 34 Toberdowney Valley, Lismenary Rd, Ballynure, Ballyclare, BT39 9TS [IO74AS, J39]
G8 VKD Details withheld at licensee's request by SSL.
G8 VKI R I Tetchner, 101 Saxon Way, Bradley Stoke, Bristol, BS12 9AR [IO81RM, ST68]
GM8 VKN M Tarr, Westholme, 1 Methven Dr, Dunfermline, KY12 0AH [IO86GB, NT08]
G8 VKO G A Tandy, 33 Anthony Way, Slough, SL1 5PQ [IO91QM, SU98]
G8 VKS J D Williams, 7 Saville Cl, Bishopstoke, Eastleigh, SO50 6NU [IO90IX, SU42]
GM8 VKT Details withheld at licensee's request by SSL.
G8 VKV Details withheld at licensee's request by SSL.
G8 VL J I M Sinclair, 65 Oatlands Dr, Weybridge, KT13 9LR [IO91SJ, TQ06]
GI8 VLB N G Harmon, 51 Bryansburn Rd, Bangor, BT20 3SD [IO74DP, J48]
GW8 VLD H F Hartwright, The Nook, Dark Ln, Rhayader, LD6 5DB [IO82FH, SN96]
G8 VLJ R V Jackson, 13 Watton Cl, Thelwall, Warrington, WA4 2HH [IO83RJ, SJ68]
G8 VLL A R Kett, 476 Earlham Rd, Norwich, NR4 7HP [JO02PP, TG20]
G8 VLN M Kremer, 24 Hutchings Walk, Hampstead Garden Sub, London, NW11 6LT [IO91VO, TQ28]
G8 VLP M E Clark, 6 Shalcross Dr, Cheshunt, Waltham Cross, EN8 8UX [IO91XQ, TL30]
G8 VLR R A Law, 19 Central Dr, Bramhall, Stockport, SK7 3JU [IO83WJ, SJ88]
G8 VLS D P Leeder, 15 King Edward Rd, Heaton, Newcastle upon Tyne, NE6 5RE [IO94FX, NZ26]
G8 VLY R C Macbeth, 9 Woodside, Summer St., Stroud, GL5 1PL [IO81VR, SO80]
G8 VLZ R Manning, 1 Homer Park, Hooe, Plymstock, Plymouth, PL9 9NN [IO70WI, SX55]
G8 VME S Clarke, 129 Greenham Wood, North Lake, Bracknell, RG12 7WH [IO91PJ, SU86]
G8 VMF M L Clayton, 54 Banks Rd, Golcar, Huddersfield, HD7 4RE [IO93BP, SE01]
G8 VML L J Gibson, 57A Heritage Park, Hatch Warren, Basingstoke, RG22 4XT [IO91KF, SU64]
G8 VMO Dr A J Thomas, 21 Hunts Field, Clayton-le-Woods, Chorley, PR6 7TT [IO83QQ, SD52]
G8 VMP K N B Webster, 164 Station Rd, Finchley, London, N3 2SG [IO91VO, TQ29]
G8 VMQ Dr A J Parker, 78 Whitbarrow Rd, Lymm, WA13 9BA [IO83SJ, SJ68]
G8 VMS A T P Leigh, 5A Latimer Rd, Wimbledon, London, SW19 1EW [IO91VK, TQ27]
G8 VMY D B Whitfield, Framingham, Manor Rd, Hayling Island, PO11 0QR [IO90MT, SZ79]
G8 VNA Details withheld at licensee's request by SSL.
G8 VND Details withheld at licensee's request by SSL.
G8 VNF B Benson, 31 Helston Cl, Sutton Park, Brookvale, Runcorn, WA7 6AA [IO83PH, SJ58]
G8 VNL J R Darlington, 111 Maas Rd, Northfield, Birmingham, B31 2PP [IO92AJ, SP07]
GW8 VNN A F J Dowsett, 70 Warren Dr, Broughton, Chester, CH4 0PT [IO83MD, SJ36]
G8 VNO G R Edmonds, 37 Milton Ct, Ickenham, Uxbridge, UB10 8NB [IO91SN, TQ08]
G8 VNP R A Elden, 124 Larchcroft Rd, Ipswich, IP1 6PQ [JO02NB, TM14]
G8 VNU N A Coote, Deepwood, Farm Ln, Ashtead, KT21 1LR [IO91UH, TQ15]
G8 VNX J G Dixon, 19 Pheasant Dr, Wincham, Northwich, CW9 6PX [IO83SG, SJ67]
G8 VOB K P Fisher, 50 Queen St., Henley on Thames, RG9 1AP [IO91NM, SU78]
G8 VOC V F Prank, 1 Daisy Cottage, Long Green, Bedfield, Woodbridge, IP13 7JD [JO02PG, TM26]
G8 VOH Dr P J Renshaw, 2 The Anchorage, Parkgate, South Wirral, L64 6TS [IO83LG, SJ27]
G8 VOI R K Reeves, 13 Gloucester Rd, Waterlooville, PO7 7BJ [IO90LU, SU60]
G8 VOJ J V Read, 2 Corton Rd, Lowestoft, NR32 4PH [JO02VL, TM59]
G8 VOK C F Ransome, 48 Chatsworth Dr, Rushmere Park, Ipswich, IP4 5XD [JO02OB, TM24]
G8 VOP Details withheld at licensee's request by SSL.
G8 VOT H Roberts, The Mount, Bog Height Rd, Darwen, BB3 0LF [IO83SR, SD62]
GW8 VOV Details withheld at licensee's request by SSL.
G8 VPD I A Morley, 2 Livingstone Walk, Parkwood Est, Maidstone, ME15 9JB [JO01GF, TQ75]
G8 VPE J Noy, 14 Poplar Dr, Filby, Great Yarmouth, NR29 3HU [JO02TP, TG41]
G8 VPG S P OSullivan, 15 Witney Cl, Saltford, Bristol, BS18 3DX [IO81SJ, ST66]
G8 VPH B K Aveling, 6 Brambling, Wilnecote, Tamworth, B77 5PQ [IO92EO, SK20]
G8 VPO J Chalmers, Hedsor Cottage, Maidenhead Park Ct, Maidenhead, Berks, SL6 8HU [IO91PN, SU88]
G8 VPQ Details withheld at licensee's request by SSL.
G8 VPR Details withheld at licensee's request by SSL.
G8 VPS Details withheld at licensee's request by SSL.[Station located near St. Leonards-on-Sea.]
GM8 VPT R McMurray, 36 Low Rd, Perth, PH2 0NF [IO86GJ, NO12]
G8 VPX A Davies, 14 Primrose Gr, Keighley, BD21 4NP [IO93BU, SE04]
G8 VQA S D Foulser, 9 Oak Coppice Cl, Eastleigh, Hants, SO50 8PH [IO90IX, SU41]
G8 VQC A Aldcroft, 4 Allington St., Liverpool, L17 7AD [IO83MJ, SJ38]
G8 VQE B A Haylett, 160 Hookfield, Harlow, CM18 6QN [JO01BS, TL40]
G8 VQH M Russ, 71 Farriers Cl, Martlesham Heath, Ipswich, IP5 7SN [JO02OB, TM24]
G8 VQK M J Nightingale, 17 Tavistock Cl, Thorney, Peterborough, PE6 0SP [IO92WO, TF20]
G8 VQM Details withheld at licensee's request by SSL.
G8 VQO D S Lowndes, 8 Winster Rd, Chaddesden, Derby, DE21 4JX [IO92GW, SK33]
G8 VQQ J Locke, 2 Norton Cl, Daventry, NN11 5HY [IO92KG, SP56]
G8 VQS R J Kugler, 96 Sanforth St., Whittington Moor, Chesterfield, S41 8RU [IO93GG, SK37]
G8 VQU P Kajtowski, 4 Charnock Gr, Gleadless, Sheffield, S12 3HE [IO93GI, SK38]
GW8 VQY M I Luke, 33 Maiden St., Maesteg, CF34 9HP [IO81EO, SS88]
G8 VR K E V Willis, 6 Lerryn Gdns, Broadstairs, CT10 3BH [JO01RJ, TR36]
G8 VRE R J Waring, 61 Staddiscombe Rd, Staddiscombe, Plymouth, PL9 9LU [IO70WI, SX55]
G8 VRG G B Turnbull, 39 Northern Woods, Flackwell Heath, High Wycombe, HP10 9JL [IO91PO, SU98]
G8 VRH R D Tucker, 6 Cartmel Dr, Ulverston, LA12 9PU [IO84KE, SD27]
G8 VRI Details withheld at licensee's request by SSL.
G8 VRN D Sutton, 14 Brocklesby Cl, Gainsborough, DN21 1TT [IO93OJ, SK88]
G8 VRO D G Shelford, 148 Abbots Rd, Abbots Langley, WD5 0BL [IO91SQ, TL00]
GW8 VRS D Fone, 64 Chapel Rd, Abergavenny, NP7 7DS [IO81LT, SO21]
G8 VRV G M Dyer, Chez Nous, 11 Fore St., St. Stephen, St. Austell, PL26 7NN [IO70NI, SW95]
G8 VRW M J Davis, 4 Joel Cl, Earley, Reading, RG6 5SN [IO91MK, SU77]
G8 VSF J A Williams, Staithe Marsh House, The Staithe, Stalham, Norwich, NR12 9DA [JO02SS, TG32]
G8 VSH P S Taylor, 62 Westfield Ave, Ashchurch Gdns, Tewkesbury, GL20 8QP [IO82WA, SO93]
G8 VSI M S Sutcliffe, 27 Bowling Hall Rd, East Bowling, Bradford, BD4 7LE [IO93DS, SE13]
G8 VSJ Details withheld at licensee's request by SSL.
G8 VSK Details withheld at licensee's request by SSL.[Op: P C Watts.]
G8 VSN G S Tennant, 85 Coronation Dr, South Normanton, Alfreton, DE55 2HS [IO93IC, SK45]
G8 VSR J S Rowley, 30 St. Edmunds Ave, Porthill, Newcastle, ST5 0AB [IO83VA, SJ84]
G8 VST R H Round, 12 Wheal Gerry, Ln, Camborne, TR14 8TY [IO70IF, SW64]
G8 VSV D F Petty, 7 Luscombe Cl, Ipplepen, Newton Abbot, TQ12 5QJ [IO80EL, SX86]
G8 VSX R S Hammond, 55 Claremont Rd, Grimsby, DN32 8NS [IO93XN, TA20]
G8 VTB R Burgess, Robins, Beaconsfield Rd, Chelwood Gate, Haywards Heath, RH17 7LG [JO01AB, TQ42]
G8 VTF Details withheld at licensee's request by SSL.
GI8 VTK A J Boston, 55 Stranmillis Gdns, Stranmillis Rd, Belfast, BT9 5AT [IO74AN, J37]
G8 VTN Details withheld at licensee's request by SSL.
G8 VTX I R King, The Limes, The St., Erpingham, Norwich, NR11 7QD [JO02PU, TG23]
G8 VU D G Blair, 121 Longstomps Ave, Chelmsford, CM2 9BZ [JO01FR, TL70]
GW8 VUG I Wilkinson, 6 Cwm Teg, Old Colwyn, Colwyn Bay, Wales, LL29 8ZA [IO83DG, SH87]
G8 VUK A L Palmer, Corner Holme, The Greens, Culgaith, Penrith, Cumbria, CA10 1QT [IO84QP, NY62]
G8 VUM P L Reade, 35 The Gastons, Lawrence Weston, Bristol, BS11 0QZ [IO81QL, ST57]
G8 VUN R Roberts, 5480 Laburnum Ave., Powell River, British Columbia, Canada V8A 4M8
G8 VUR D J Bartlett, The Corner House, 2 Stamford Rd, Colsterworth, Grantham, NG33 5JD [IO92QT, SK92]
G8 VUS A P Branton, 20 Sling Ln, Malvern, WR14 2TU [IO82UC, SO74]
GW8 VUV A Gravell, 49 Rehoboth Rd, Five Roads, Llanelli, SA15 5DJ [IO71VR, SN40]
G8 VVB C W Heath, 3 Trebellan Dr, Hemel Hempstead, HP2 5EL [IO91SS, TL00]
G8 VVC C F Haver, 31 Edenham Rd, Hanthorpe, Bourne, PE10 0RB [IO92TT, TF02]
G8 VVG S C Hackett, 184 Beccles Rd, Bradwell, Great Yarmouth, NR31 8QD [JO02UN, TG50]
G8 VVK Details withheld at licensee's request by SSL.
G8 VVM D J Merrick, The Conifers, 259 Wigmore Rd, Wigmore, Gillingham, ME8 0LZ [JO01GI, TQ76]
G8 VVP P J North, 84A Park Rd, Great Sankey, Warrington, WA5 3ET [IO83QJ, SJ58]
G8 VVQ Details withheld at licensee's request by SSL.
G8 VVR C R Key, 23 Oxford Rd, Kesgrave, Ipswich, IP5 7EL [JO02OB, TM24]
GW8 VVX R L Williams, The Basement, 9 Charlton St., Llandudno, LL30 2AA [IO83CH, SH78]
G8 VVY R C A Shelley, 22 Stonecote Ridge, Chalford, Bussage, Stroud, GL6 8JY [IO81WR, SO80]
G8 VVZ A M Stephens, 12 Sutherland Walk, Aylesbury, HP21 7NS [IO91OT, SP81]
GM8 VWC J Welsh, 7 South Cathkin Cottage, Rutherglen, Glasgow, G73 5RG [IO75VT, NS65]
GW8 VWD C E J Wilde, 13 Beechwood Dr, Heolgerrig, Merthyr Tydfil, CF48 1TH [IO81HR, SO00]
G8 VWH S M Hall, 31 Somerton Gdns, Earley, Reading, RG6 5XG [IO91NK, SU77]
G8 VWJ M A Hoare, 45 Tilehurst Rd, Reading, RG1 7TT [IO91MK, SU77]
G8 VWN Details withheld at licensee's request by SSL.
G8 VWP J B A Jones, 27 St. Davids Cl, Iver Heath, Iver, SL0 0RS [IO91RN, TQ08]
G8 VWU J Marriott, 9 Albany Walk, Woodston, Peterborough, PE2 9JN [IO92UN, TL19]
G8 VWV A R Ball, 250 Peppard Rd, Emmer Green, Caversham, Reading, RG4 8UA [IO91ML, SU77]
G8 VXH Details withheld at licensee's request by SSL.[Station located near Sevenoaks.]
G8 VXQ B P Hayward, 35 Dorchester Rd, Solihull, B91 1LW [IO92CJ, SP17]
G8 VXR C J Hunt, 41 Maylands Way, Harold Park, Romford, RM3 0BQ [JO01DO, TQ59]
G8 VXU Llewelyn, White House, Old Beer Rd, Seaton, EX12 2PX [IO80KQ, SY28]
G8 VXV B Machen, 59 Park Ave, New Lodge, Barnsley, S71 3TT [IO93GN, SE30]
G8 VXX Details withheld at licensee's request by SSL.
G8 VXZ I P Brownlee, Crispin Cottage, 21 Marksbury, Bath, BA2 9HS [IO81SI, ST66]
G8 VYF P A V Maynard, 14 Tilling Dr, Walton, Stone, ST15 0AA [IO82WV, SJ93]
G8 VYI Details withheld at licensee's request by SSL.
G8 VYJ R E Daly, 82 Derwent Rd, Thatcham, RG19 3UR [IO91IJ, SU56]
G8 VYK P J Shepherd GEC Sen Clb ARS, GEC Sports & Soc Cl, Gardiners Ln, Basildon Essex, SS14 3AP [JO01FO, TQ79]
G8 VYP C J Syms, 24 Warmdene Rd, Patcham, Brighton, BN1 8NL [IO90WU, TQ30]
G8 VYQ G E Todd, 100 Avebury Dr, Washington, NE38 7DB [IO94FV, NZ35]
G8 VYT D J Tombs, 24 Ferndale Rd, Northville, Bristol, BS7 0RP [IO81RM, ST67]
GM8 VYZ A Raine, Broomhill, Edgehead Pathhead, Midlothian, Scotland, EH37 5RN [IO85MV, NT36]
G8 VZA Details withheld at licensee's request by SSL.
G8 VZB A R Poole, 6 Rutland Ave, Willsbridge, Bristol, BS15 6EZ [IO81SK, ST67]
G8 VZD C A Ramsey, 10 Lindley Rd, London, E10 6QT [IO91XN, TQ38]
G8 VZE Details withheld at licensee's request by SSL.
G8 VZI M D Warren, 407 Locking Rd, Weston Super Mare, BS22 8NW [IO81MI, ST36]
G8 VZJ M R Webb, 24 College Ave, Grays, RM17 5UW [JO01EL, TQ67]
G8 VZL S M Rawlings, 16 Ebden Lodge, 308 High St, Worle, Weston Super Mare, BS22 0JP [IO81MI, ST36]
G8 VZR N P Giddings, Primrose Cottage, Birchin Cross Rd, Knatts Valley, Sevenoaks, TN15 6XJ [JO01CH, TQ56]
G8 VZS I D Chapman, 188 Goodhart Way, West Wickham, BR4 0HA [JO01AJ, TQ36]
G8 VZT D B Hall, 4 Steventon Rd, Wellington, Telford, TF1 2AS [IO82RQ, SJ61]
G8 VZZ P J Marks, Flat 3, 47 The Thoroughfare, Woodbridge, IP12 1AH [JO02PC, TM24]
G8 WAH Dr R H West, 70 Werneth Rd, Simmondley, Glossop, SK13 9NJ [IO93AK, SK09]
G8 WAJ N A Treanor, 23 Norton Ave, Penketh, Warrington, WA5 2RB [IO83QJ, SJ58]
G8 WAM G J Weeks, 36 Australian Ave, Salisbury, SP2 7JT [IO91CB, SU13]
GW8 WAO F C Woodhouse, Brynhyfryd, Meen-y-Groes, Newquay, Dyfed, SA45 9RL [IO72TE, SN35]
G8 WAV C G Jacobs, 133 Fordham Rd, Isleham, Ely, CB7 5QX [JO02EH, TL67]
G8 WBC Details withheld at licensee's request by SSL.
G8 WBG S G Netherton, 33 Bethel Rd, St. Austell, PL25 3HB [IO70OI, SX05]
G8 WBH Details withheld at licensee's request by SSL.
G8 WBI Details withheld at licensee's request by SSL.
G8 WBK A P B Maufe, 11 Calder Way, Silsden, Keighley, BD20 0QU [IO93AV, SE04]

G8 WBL T H Mole, 205 Wood Ln, Handsworth Wood, Handsworth, Birmingham, B20 2AA [IO92AM, SP09]
G8 WBN D G Neale, 34 Liverpool St., Southampton, SO14 6FZ [IO90HV, SU41]
G8 WBO S M Holley, 37 Bouverie Ave, Salisbury, SP2 8DU [IO91CB, SU12]
G8 WBP P R Humphreys, 910 High Ln, Chell, Stoke on Trent, ST6 6HE [IO83VB, SJ85]
G8 WBT A P Farnborough, 9 Mitchelmore Rd, Yeovil, BA21 4BA [IO80QW, ST51]
G8 WBU A Greenall, 44 Hardy Rd, Blackheath, London, SE3 7NN [JO01AL, TQ37]
GI8 WBZ A G Smith, 27A Cairn Rd, Carrickfergus, BT38 9AP [IO74BS, J39]
GW8 WCA K C Winter, Derwen, Hillside Rd, Redbrook, nr Monmouth Gwent, NP5 4LY [IO81PS, SO50]
G8 WCH R Shepherd, 299 West Wycombe Rd, High Wycombe, HP12 4AA [IO91OP, SU89]
G8 WCL D G Dean Liverpool & District Amateur R, 24 Hathaway Rd, Gateacre, Liverpool, L25 4ST [IO83NJ, SJ48]
G8 WCN Details withheld at licensee's request by SSL.
G8 WCQ V W McClure, 43 Roman Way, Seaton, EX12 2NT [IO80LR, SY29]
G8 WCT A M Grindrod, 7 Styveton Way, Steeton, Keighley, BD20 6TP [IO93AV, SE04]
G8 WCX A H Essex, Highwood Elms, Highwood Hill, Mill Hill, London, NW7 4HB [IO91VP, TQ29]
G8 WCZ D F Stowell, 23 Granville St., Leamington Spa, CV32 5XW [IO92FH, SP36]
G8 WDC G G E Scott Wirral&Dist ARC, 19 Penkett Rd, Wallasey, L45 7QF [IO83LK, SJ39]
G8 WDT I A Baillie, 11 Carpenters Ln, Stanton Lacy, Ludlow, SY8 2AD [IO82PJ, SO57]
G8 WDV J A Brinsley, c/o GPT Spec Proj Mgt Ltd, PO Box 9854, Riyadh 11423, Saudi Arabia
G8 WDX R M Lamkin, 22 Hampton Ct Parade, East Molesey, KT8 9HE [IO91TJ, TQ16]
G8 WEG K T Woolley, 3 Harwood St., West Bromwich, B70 9JF [IO82XM, SO99]
G8 WEI D Kostryca, 9 Cherry Tree Rd, Gainsborough, DN21 1RG [IO93OJ, SK89]
G8 WEM M P O'Neill, Coolrake, Moone, Athy, Co Kildare Ireland
G8 WEN Details withheld at licensee's request by SSL.
G8 WET E A Matthews, 18 Runnalow, Letchworth, SG6 4DT [IO91VX, TL23]
G8 WEV A R Haward, Grey Shingles, Pottersheath Rd, Welwyn, AL6 9ST [IO91VU, TL21]
GW8 WEY T H Jones, 18 Wimborne Cres, Sully, Penarth, CF64 5SR [IO81JJ, ST16]
G8 WFP C N Kershaw, 13 Glen Terr, Halifax, HX1 2YN [IO93BR, SE02]
GW8 WFS J Lawson-Reay, The Nook, Conway Rd, Llanrhos, Llandudno, LL30 1PY [IO83CH, SH78]
G8 WFU Details withheld at licensee's request by SSL.
G8 WGD P D Randall-Cook, 3 Wellmeadow, Staunton, Coleford, GL16 8PQ [IO81QT, SO51]
G8 WGE I A Robinson, 26 Wick Rd, Teddington, TW11 9DW [IO91UK, TQ17]
G8 WGM Dr A J McLeod, 5 Albemarle Rd, Bournemouth, BH3 7LZ [IO90BR, SZ09]
G8 WGN J P Marks, Stam 69, 1275 Cg Huizen, The Netherlands
G8 WGQ D Onione, 19 Chapman Cl, Kempston, Bedford, MK42 8RU [IO92RD, TL04]
G8 WGU A M Irving, 23 Woodlea Park, Sauchie, Alloa, FK10 3BG [IO86CD, NS89]
G8 WHB H M P Couchman, Pond Cottage, Woodside Green, Lenham, Maidstone, ME17 2EU [JO01IG, TQ95]
G8 WHD P G Whittington, 10 Talbot Terr, Lewes, BN7 2DS [JO00AU, TQ41]
G8 WHE Details withheld at licensee's request by SSL.
G8 WHN A James, 12 Cedar Rd, Weybridge, KT13 8NY [IO91SI, TQ06]
GI8 WHP S A Craig, 46 Coast Rd, Larne, BT40 1UZ [IO74CU, D40]
G8 WHR S A Wood, 90 Plymyard Ave, Bromborough, Wirral, L62 6BR [IO83MH, SJ38]
G8 WHX R C Clarkson, 3 Debden Cl, Kingston upon Thames, KT2 5GD [IO91UK, TQ17]
G8 WHZ Details withheld at licensee's request by SSL.
G8 WID Details withheld at licensee's request by SSL.
G8 WIG Details withheld at licensee's request by SSL.
G8 WIM H Spooner Wimbledon&Dis Rd, 71 Templecombe Way, Morden, SM4 4JF [IO91VJ, TQ26]
G8 WIR J V Vousden, 44 Castle Rd, Tankerton, Whitstable, CT5 2DY [JO01MI, TR16]
GI8 WIU S Douthart, 75 Market St., Ballycastle, BT54 6DS [IO65UE, D14]
G8 WJB J A C Geer, 31 The Beeches, Salisbury, SP1 2JH [IO91CB, SU12]
GM8 WJK J D Nicholson, Upper Barswick, St Margarets Hope, Orkney, Scotland, KW17 2RN [IO88MS, ND48]
G8 WJQ D J Pipe, 72 Hall Rd, Rolleston, Rolleston on Dove, Burton on Trent, DE13 9BY [IO92EU, SK22]
G8 WJS Dr D J Mahy, Ferry Ln Cottage, Ferry Ln, Uckinghall, Tewkesbury, GL20 6ER [IO82VA, SO83]
G8 WJY M G Garton, 58 Mayfield Cl, Catshill, Bromsgrove, B61 0NP [IO82XI, SO97]
G8 WJZ Details withheld at licensee's request by SSL.
G8 WKI S J Willisson, 15 Glenalmond Rd, Kenton, Harrow, HA3 9JY [IO91UO, TQ18]
G8 WKK M G Daniels, 6 Middlemead, Stratton on The Fosse, Bath, BA3 4QH [IO81SG, ST65]
G8 WKL M G Daniels Downside Scl AR, 6 Middlemead, Stratton on The Fosse, Bath, BA3 4QH [IO81SG, ST65]
G8 WKO D C Crouchley, Brambledown, The St., Barney, Fakenham, NR21 0AD [JO02LU, TF93][Op: D C Crouchley, Montana, Shire Lane, Cholesbury, nr Tring, Herts, HP23 6NA.]
G8 WKT J A Larkins, The Smithy, High St., Tormarton, Badminton, GL9 1HU [IO81TM, ST77]
G8 WKX R D Denton, 18 Sealand Ct, Esplanade, Rochester, ME1 1QH [JO01GJ, TQ76]
G8 WKZ K J Spragg, 88 Low Ln, Brookfield, Middlesbrough, TS5 8EB [IO94JM, NZ41]
G8 WLB S V Austen, Shiralee, The Plain, Smeeth, Ashford, Kent, TN25 6RA [JO01LC, TR04]
G8 WLL S D Lown, 50 Fall Birch Rd, Lostock, Bolton, BL6 4LG [IO83RO, SD61]
G8 WLO R G Best, 31 Farnham Rd, Loughborough, LE11 2LH [IO92JS, SK51]
G8 WLV R J Barber, 12 The Dawneys, Crudwell, Malmesbury, SN16 9HE [IO81XP, ST99]
G8 WMC E P A Holman, Weavers Cottage, The Shoe, North Wraxall, Chippenham Wilts, SN14 8SA [IO81UL, ST87]
G8 WMG J A Bassnett, 105 Edgemoor Dr, Thornton, Crosby, Liverpool, L23 9UF [IO83LL, SD30]
G8 WMJ C B Hayles, 55 Frogmore Park Dr, Blackwater, Camberley, GU17 0PJ [IO91OH, SU85]
G8 WMK G C Bessant, 145 Ballens Rd, Lords Wood, Chatham, ME5 8PG [JO01GI, TQ76]
G8 WMQ Details withheld at licensee's request by SSL.
G8 WMS J F J Seal, 74 Ridgacre Ln, Birmingham, B32 1EN [IO92AL, SP08]
GW8 WNB K Phillips, Lluest y Coed, 39 Llwyn Ynn, Talybont, LL43 2AG [IO72WS, SH52]
G8 WNF Details withheld at licensee's request by SSL.
G8 WNJ R K Simmons, 1 Clare Corner, New Eltham, London, SE9 2AE [JO01AK, TQ47]
G8 WNK J Davies, 6 Grendale Ave, Stockport, SK1 4BL [IO83WJ, SJ99]
G8 WOO Details withheld at licensee's request by SSL.
G8 WOW Details withheld at licensee's request by SSL.
G8 WOX A F Hartland, 22 Granville Crest, Kidderminster, DY10 3QS [IO82VJ, SO87]
G8 WOY Details withheld at licensee's request by SSL.
G8 WPD Details withheld at licensee's request by SSL.
G8 WPE K F S Marshall, 5 Trindehay, Basildon, SS15 5DL [JO01FN, TQ68]
G8 WPF A J Middleton, 125 Pickersleigh Rd, Malvern, WR14 2LF [IO82UC, SO74]
G8 WPK Details withheld at licensee's request by SSL.
G8 WPL D F Hughes, 12 Spencer St., Reddish, Stockport, SK5 6UH [IO83WK, SJ89]
G8 WPO S B Pateman, 28 Stubbs Ln, Braintree, CM7 3NR [JO01GU, TL72]
G8 WPU I G Rivett, 30 Millside Cl, Kingsthorpe, Northampton, NN2 7TR [IO92NG, SP76]
G8 WPV A V Reason, 71 Cavendish Rd, Hazel Gr, Stockport, SK7 6HU [IO83WI, SJ98]
G8 WQB Details withheld at licensee's request by SSL.
G8 WQG A J Warren, Clifden Farm, Quenchwell Rd, Carnon Downs, Truro, Cornwall, TR3 6LN [IO70LF, SW84]
G8 WQH P E Wood, 29 Buchan Cl, Galley Common, Nuneaton, CV10 9RR [IO92FM, SP39]
G8 WQK G E Smith, 11 Hare Ln, Hatfield, AL10 8PW [IO91VS, TL20]
G8 WQM F T Trice, 122 Scrapsgate Rd, Minster on Sea, Sheerness, ME12 2DJ [JO01JK, TQ97]
G8 WQT T N Rickard, 13 Cecilia Rd, Ramsgate, CT11 7DY [JO01RI, TR36]
G8 WQW J N Shergold, 35 Orchard Gr, New Milton, Hants, BH25 6NZ [IO90ER, SZ29]
G8 WQZ D M Mead, 9 Abraham Dr, Silver End, Witham, CM8 3SP [JO01HU, TL81]
G8 WRA D M Mann, 23 School Green Ln, North Weald, Epping, CM16 6EH [JO01CR, TL50]
G8 WRC T M Harston, 16 Money Ln, West Drayton, UB7 7NU [IO91SM, TQ07]
G8 WRG N K Read Warks Raynet Gr, 86 Telford Ave, Lillington, Leamington Spa, CV32 7HP [IO92FH, SP36]
G8 WRI W J Lawrence, 26 Cedarwood Dr, Tuffley, Gloucester, GL4 0AG [IO81VU, SO81]
G8 WRL R L Williamson, 35 Villiers Ave, Twickenham, TW2 6BL [IO91TK, TQ17]
G8 WRT Details withheld at licensee's request by SSL.[Station located near Lichfield.]
G8 WRV R Bygrave, 35 East St., St. Neots, Huntingdon, PE19 1JU [IO92UF, TL16]
G8 WRY G G Brock, 54 Lord Haddon Rd, Ilkeston, DE7 8AW [IO92IX, SK44]
G8 WSC R J Burg, 20 Rowan Way, Witham, CM8 2LJ [JO01HT, TL81]
G8 WSF F A Price, 26 Teviot Gdns, Pensnett, Brierley Hill, DY5 4QL [IO82WL, SO98]
G8 WSG Details withheld at licensee's request by SSL.
G8 WSH M J French, 17 Upland Rd, West Mersea, Colchester, CO5 8DX [JO01LS, TM01]
G8 WSM Details withheld at licensee's request by SSL.[Weston-Super-Mare RS - details c/o G4ZUX, 36 Tormynton Road, Worle, Weston-Super-Mare, BS22 9HT.]
G8 WSP P M Arup, Alma House, Broadway Rd, Windlesham, GU20 6BU [IO91QI, SU96]
G8 WSQ A V Beeston, 69 High St., Repton, Derby, DE65 6GF [IO92FU, SK32]
G8 WSS M A Blair, 12 Medoc Cl, Felmore End, Pitsea, Basildon, SS13 1NR [JO01GN, TQ78]
G8 WST Details withheld at licensee's request by SSL.
G8 WSU J B Hoggarth, Cotherstone, Rockingham Paddocks, Kettering, NN16 9JR [IO92PJ, SP88]
G8 WSV M A Cartwright, 9 Montgomery Cl, Kettering, NN15 5BY [IO92PJ, SP87]
G8 WSW R F Carter, 46 Arterial Rd, Leigh on Sea, SS9 4DA [JO01HN, TQ88]
G8 WSX G H Goodyer Chichester ARC, Flat, 54 Wyndham Rd, Petworth, GU28 0EQ [IO90QX, SU92]
G8 WSY P Bloor, 216 Waterloo St., Burton on Trent, DE14 2NB [IO92ET, SK22]
G8 WSZ J M Foster, 1 Thorn Ct, Four Marks, Alton, GU34 5BY [IO91LC, SU63]
G8 WT Details withheld at licensee's request by SSL.[Station located at Whalley.]
G8 WTA E P Williams, 24 Jays Mead, Wotton under Edge, GL12 7JF [IO81TP, ST79]
G8 WTB D H Crowe, 54 Carters Mead, Harlow, CM17 9ER [JO01BS, TL40]
GW8 WTJ S A Ellis, 56 Lynton Walk, Rhyl, LL18 3RP [IO83GH, SJ08]
G8 WTM R D Britt, 2 Lindisfarne Ct, Maldon, CM9 6UQ [JO01IR, TL80]
G8 WTN J B Capon, 24 Furness Cl, Chadwell St. Mary, Grays, RM16 4JB [JO01EL, TQ67]
G8 WTR S P Gudgeon, 10 The Sq, Hillsborough, BT26 6AG [IO64XL, J25]
G8 WUD I B Kallar, 23 Felmore Ct, Pitsea, Basildon, SS13 1PW [JO01GN, TQ78]
G8 WUF D J Legg, 2 Birkbeck Rd, Wimbledon, London, SW19 8NZ [IO91VK, TQ27]
G8 WUG R W J Spence, 8 Stoneleigh, Sawbridgeworth, CM21 0BT [JO01BT, TL41]
G8 WUJ C M Bracher, 29 Bungalow Park, Holders Rd, Amesbury, Salisbury, SP4 7PJ [IO91CE, SU14]
G8 WUO K J Baker, 57 Hedingham Rd, Hornchurch, RM11 3QH [JO01CN, TQ58]
G8 WUR S J Browning, 55 Curlew Ave, Manor Park, Chatteris, PE16 6PL [JO02AL, TL38]
G8 WUU J D Cooper, 156 Church Rd, Benfleet, SS7 4EN [JO01GN, TQ78]
G8 WUY R P Dowthwaite, 13 Lynton Cl, Knutsford, WA16 8BH [IO83TH, SJ77]
G8 WUZ P C Dineen, 5 Reynolds Dr, Little Paxton, St. Neots, Huntingdon, PE19 4QB [IO92UF, TL16]
G8 WV N H Sedgwick, 77 Lakes Ln, Newport Pagnell, MK16 8HT [IO92PC, SP84]
G8 WVB S Ayer, 335 Ings Rd, Kingston upon Hull, Sutton on Hull, Hull, HU7 4UY [IO93US, TA13]
G8 WVH J L Bull, 12 Eastfield Cres, Merthen, Laughton, Sheffield, S31 7YT [IO93GJ, SK38]
G8 WVO P L Dawson, 35 Crofton Rd, Ipswich, IP4 4QP [JO02OB, TM14]
G8 WVR Details withheld at licensee's request by SSL.
G8 WVZ C Edwards, 9 Bradworth Cl, Scarborough, YO11 3PZ [IO94TF, TA08]
G8 WW L B Carter, 65 Parry Rd, Wyken, Coventry, CV2 3LW [IO92GK, SP38]
G8 WWC G M Ludlow, 4 Cypress Ct, Waterloo St., Cheltenham, GL51 9BY [IO81WV, SO92]
G8 WWD G Hunter, 14 Kempsford Cl, Oakenshaw South, Redditch, B98 7YS [IO92AG, SP06]
G8 WWF P C O'Ryan, 12 Minton Cl, Congleton, CW12 3TD [IO83VD, SJ86]
G8 WWI P D Leverington, 28 Burymead, Old Town, Stevenage, SG1 4AY [IO91VW, TL22]
G8 WWJ J E Kirton, 13 Saltersford Rd, Grantham, NG31 7HH [IO92QV, SK93]
G8 WWM A P Morgan, 316 Middle Rd, Sholing, Southampton, SO19 8NT [IO90HV, SU41]
G8 WWN M D Hayward, 8 Fry Cl, Isle of Grain, Rochester, ME3 0EE [JO01IK, TQ87]
G8 WWW M A Harrington, Rose Cottage, 17 Church Rd, Penponds, Camborne, TR14 0QE [IO70IE, SW63]
G8 WWY W P Kemp, 3 Border Way, Vicars Cross, Chester, CH3 5PQ [IO83NE, SJ46]
G8 WXB R C Orr, 24 Wentworth Rd, Hertford, SG13 8JP [IO91XS, TL31]
G8 WXC I J McGeachy, 1 Marina Gdns, Rodwell, Weymouth, DT4 9QZ [IO80SO, SY67]
G8 WXH Details withheld at licensee's request by SSL.[Op: G S Boreham. Station located in London E17.]
G8 WXL E E Chesworth, 46 Darlington Cres, Saughall, Chester, CH1 6DB [IO83MF, SJ37]
G8 WXP R C Hadland, 32 Oxford Rd, Carshalton Beeches, Carshalton, SM5 3QY [IO91VI, TQ26]
G8 WXQ M J Corby, 15 Farmers Cl, Maidenhead, SL6 3PZ [IO91OM, SU87]
G8 WXR J A Edwards, 5 Brindley Rd, Silsden, Keighley, BD20 0LD [IO93AV, SE04]
G8 WXU G Evans, 24 Beaufort Rd, Billericay, CM12 9JL [JO01EP, TQ69]
G8 WYB M L Jefferson, 16 Edge Dell, Stoney Haggs, Scarborough, YO12 4LL [IO94SG, TA08]
G8 WYI P Herring, 36 Rensburg Rd, Walthamstow, London, E17 7HN [IO91XN, TQ38]
G8 WYJ Details withheld at licensee's request by SSL.
G8 WYO Details withheld at licensee's request by SSL.
G8 WYP D W Allan West Yorkshire Police ARC, 283 Cliffe Ln, Gomersal, Cleckheaton, BD19 4SB [IO93DR, SE22]
G8 WYR M Howes Leeds & Dis ARS, Yarnbury RUFC, Brownberrie Ln, Horsforth, Leeds, West Yorks, LS18 5HB [IO93EU, SE23]
G8 WYS R T Pinches, 119 Ridge Rd, Kingswinford, DY6 9RG [IO82VL, SO88]
G8 WYT P T Phillips, 155 Franklands Village, Haywards Heath, RH16 3RF [IO90WX, TQ32]
GW8 WYW C M Burn, Wood View Cottage, Tainant, Penycae, Wrexham Clwyd, LL14 1UG [IO83LA, SJ24]
G8 WYX J O A Brown, 7 Ct Cl, Bitterne, Southampton, Hants, SO18 5EJ [IO90HV, SU41]
GW8 WYY E M Brennan, 24 Gareth Cl, Thornhill, Cardiff, CF4 9AF [IO81JM, ST18]
G8 WYZ Details withheld at licensee's request by SSL.
GW8 WZC K D Farrant, 79 Eustace Dr, Bryncethin, Bridgend, CF32 9EX [IO81FN, SS98]
G8 WZJ A J Collier, 44 Cockington Cl, Leigham, Plymouth, PL6 8RQ [IO70WJ, SX55]
G8 WZK D J Collins, 71 Trench Rd, Tonbridge, TN10 3HG [JO01DF, TQ54]
G8 WZN W G Dimery, 81 Sandbeck House, Gr Pl, Doncaster, DN1 3AT [IO93KM, SE50]
G8 WZO P J Evans, 35 Shelley Rd, East Grinstead, RH19 1SX [IO91XD, TQ33]
GW8 WZR D A T Gale, 9 Fenner Brockway Cl, Newport, NP9 9EQ [IO81MO, ST38]
G8 WZT G W Greenfield, 21 Laburnum Way, Gillingham, SP8 4RU [IO81UA, ST82]
G8 WZW K Aspden, Langriggs Farm, Goosehouse Ln, Darwen, Lancs, BB3 0EH [IO83SQ, SD62]
G8 WZX R G Bishop, Jorol, 14 Detling Ave, Broadstairs, CT10 1SL [JO01RI, TR36]
G8 XAA A B Williams Bristol Raynet, obo Bristol Raynet Group, 38 Seneca St., St. George, Bristol, BS5 8DX [IO81RK, ST67]
G8 XAC Details withheld at licensee's request by SSL.
GW8 XAH C W L Sargent, Nethania, 26 Cheriton Dr, Thornhill, Cardiff, CF4 9DF [IO81JM, ST18]
G8 XAJ T R Sherman, 5 Lansdown Rd, Canterbury, CT1 3JP [JO01NG, TR15]
G8 XAN R B Woods, 4A Market Pl, Long Eaton, Nottingham, NG10 1LS [IO92IV, SK43]
G8 XAO G P Woodman, 10 Wrayfield Rd, North Cheam, Sutton, SM3 9TH [IO91VI, TQ26]
GW8 XAS G J Evans, Wynona, Esplanade, Penmaenmawr, LL34 6LY [IO83AG, SH77]
G8 XAU Details withheld at licensee's request by SSL.
G8 XAX K Tully, 1 The Ridgeway, Dovercourt, Harwich, CO12 4AT [JO01PW, TM23]
G8 XBA Details withheld at licensee's request by SSL.
G8 XBC Details withheld at licensee's request by SSL.
G8 XBE P E Read, 2 Priors Cl, Kingsclere, Newbury, RG20 5QT [IO91JH, SU55]
G8 XBF T F Parker Barking Rad, Elec Soc, 2D Hubbards Chase, Hornchurch, RM11 3DJ [JO01CN, TQ58]
G8 XBG Details withheld at licensee's request by SSL.
G8 XBI Details withheld at licensee's request by SSL.
G8 XBP Details withheld at licensee's request by SSL.
G8 XBW Details withheld at licensee's request by SSL.
GW8 XBY P G Allwood, 14 Cotman Cl, St. Julians Est, Newport, NP9 7PU [IO81MO, ST38]
G8 XBZ D J Atkins, 86 Dereham Rd, New Costessey, Norwich, NR5 0SY [JO02OP, TG10]
G8 XCD N Burrows, 1 Oak St., Weedon, Northampton, NN7 4RQ [IO92KF, SP65]
G8 XCE D Baker, 48 Elmwood St., Burnley, BB11 4BP [IO83US, SD83]
G8 XCH Details withheld at licensee's request by SSL.[Station located near Hayes.]
G8 XCJ I Coton, 77 Lockesfield Pl, London, E14 3AJ [IO91XL, TQ37]
G8 XCL I M Davis, 28 Sycamore Cl, Lydd, Romney Marsh, TN29 9LE [JO00KW, TR02]
G8 XCQ A B Gates, 53 Church Rise, Chessington, KT9 2HA [IO91UI, TQ16]
G8 XCR C H Girling, 20 Fore St., Praze An Beeble, Praze, Camborne, TR14 0JX [IO70IE, SW63]
G8 XCX Details withheld at licensee's request by SSL.
G8 XCY A C Worsfold, 5 Turner Cl, Langley, Eastbourne, BN23 7PF [JO00DT, TQ60]
G8 XDL R Medcalf, 19 All Saints Rd, Warwick, CV34 5NL [IO92FG, SP26]
G8 XDM P W Mutter, 7 Rochford Way, Croydon, CR0 3AG [IO91WJ, TQ36]
G8 XDQ J P Coleman, 48 Stoughton Ln, Stoughton, Leicester, LE2 2FH [IO92LO, SK60]
G8 XDR C A Johnstone, 24 Elibank Rd, Eltham, London, SE9 1QH [JO01AL, TQ47]
G8 XDS N D Lightowler, 236 Amersham Rd, Hazelmere, Hazlemere, High Wycombe, HP15 7QN [IO91PP, SU89]
G8 XDV Details withheld at licensee's request by SSL.
G8 XDX M R King, 52 Falkland Rd, Evesham, WR11 6XS [IO92AB, SP04]
G8 XDZ P A Long, 11 Bailey Dale, Stanway, Colchester, CO3 5LB [JO01KV, TL92]
G8 XEC G M Murray, 3 Domoney Cl, Thatcham, RG19 4DY [IO91IJ, SU56]
G8 XEF M J McIver, 18 West End Rd, Kempston, Silsoe, Bedford, MK45 4DU [IO92SA, TL03]
G8 XEI V L Nolan, 127 Martins Ln, Blakehall, Skelmersdale, WN8 9BQ [IO83OM, SD40]
G8 XEJ H R Hillerby, 29 Huddersfield Rd, Skelmanthorpe, Huddersfield, HD8 9AR [IO93EO, SE21]
G8 XER J A Smith, 47 Lower Shelton Rd, Marston Moretaine, Marston Moreteyne, Bedford, MK43 0LN [IO92RB, SP94]
G8 XET C H Street, Russets, Isle Brewers, Taunton, Somerset, TA3 6QN [IO80NX, ST32]
G8 XEU R E Stephens, 21 St. James Ave, Lancing, BN15 0NN [IO90UU, TQ10]
G8 XEZ M D Ward, 11 Rogate Gdns, Portchester, Fareham, PO16 8DS [IO90KU, SU60]
G8 XFA Details withheld at licensee's request by SSL.
G8 XFK R V Young, Chapel House, 4 Ratten Row, North Newbold, York, YO4 3SF [IO93QT, SE93]
G8 XFO F J Bartels, Crouchmead, Lower Rd, Little Hallingbury, Bishops Stortford, CM22 7RA [JO01CT, TL51]
G8 XFT J D Carp, 33Su Det Akrotiri, Bfpo 57, X X
G8 XFU E Carver, 46 Orchard St., Wombwell, Barnsley, S73 8HQ [IO93HM, SE30]
G8 XFX R D Curry, 6 Merriville Rd, Arle, Cheltenham, GL51 8JF [IO81WV, SO92]
G8 XFY I A Downie, 120 South View, Broughton, Brigg, DN20 0EY [IO93RN, SE90]
G8 XGA Details withheld at licensee's request by SSL.
G8 XGB K Dickson, 29 Sunnyfields Dr, Minster on Sea, Sheerness, ME12 3DH [JO01JJ, TQ97]
G8 XGG S J Gwilliam, 22 Manning Rd, Droitwich, WR9 8HW [IO82WG, SO86]
GM8 XGI Details withheld at licensee's request by SSL.
G8 XGK P R F Manford, Braemar Church Rd', Stanmore, Middx, HA7 4AG [IO91UO, TQ19]
G8 XGO P R McKellow, 155 Pittmans Field, Harlow, CM20 3LE [JO01BS, TL41]
GM8 XGP Details withheld at licensee's request by SSL.
G8 XGQ A H S A Saunders, 19 Beauchamp Cl, Eaton Socon, St. Neots, Huntingdon, PE19 3BU [IO92UF, TL15]
G8 XGR Details withheld at licensee's request by SSL.
G8 XGS J S Hindmarsh, Roseworth Cottage, West Hexham Rd, Throckley, Newcastle upon Tyne, NE15 9EB [IO94CX, NZ16]
G8 XGT Details withheld at licensee's request by SSL.
G8 XGW N Shearing, 45 St. Michaels Ave, Gedling, Nottingham, NG4 3NN [IO92KX, SK64]
G8 XHD P A Riebold, 8 Ash Lodge, The Woodlands, Shoeburyness, Southend on Sea, SS3 9RZ [JO01JM, TQ98]
G8 XHH P A Read, The Mount, Beech Hill, Haxham, Hexham, NE46 3AG [IO84WX, NY96]
G8 XHK K M Prior, 9 Snowdrop Cl, Broadfield, Crawley, RH11 9EG [IO91VC, TQ23]
G8 XHN R J Harman, 17 Coldharbour Ln, Bushey, Watford, Herts, WD2 3NU [IO91TP, TQ19]
G8 XHO Details withheld at licensee's request by SSL.

G8 XHU G E Arrowsmith, The Island, Ford St, Wellington, Somerset, TA21 9PE [IO80JX, ST11]
G8 XHW D J Barclay, 13 Avondale Terr, Chester-le-Street, DH3 3ED [IO94FU, NZ25]
G8 XHX K Barker, 183 Macaulay St., Grimsby, DN31 2EL [IO93WN, TA20]
G8 XIG E Bailey, 49 Brander St., Leeds, LS9 6QH [IO93GT, SE33]
G8 XII C L Boddy, Que Sera, Church Ln, Bledlow Ridge, High Wycombe, HP14 4AX [IO91NQ, SU79]
G8 XIJ J Bryant, 10 Kingsland Rd, Farnworth, Bolton, BL4 0HW [IO83SN, SD70]
G8 XIM I M Churchill, 12 Wyedale Ave, Coombe Dingle, Bristol, BS9 2QQ [IO81QL, ST57]
G8 XIN M Chapman, 4 Amberley Ct, Sidcup Hill, Sidcup, DA14 6JT [JO01BK, TQ47]
G8 XIR K L Church, 90 Cruden Rd, Gravesend, DA12 4HR [JO01EK, TQ67]
G8 XIU Details withheld at licensee's request by SSL.
G8 XIX Details withheld at licensee's request by SSL.
G8 XIY A J Tee, 136 Burstellars, St. Ives, Huntingdon, PE17 6YJ [IO92XH, TL37]
G8 XIZ H T Tillotson, 30 St. Laurence Rd, Northfield, Birmingham, B31 2AX [IO92AK, SP08]
G8 XJB B E Simmons, 88 Wellcome Ave, Dartford, DA1 5JW [JO01CK, TQ57]
G8 XJE J A K Williams, 298 Brampton Rd, Bexleyheath, DA7 5SE [JO01BL, TQ47]
G8 XJK D P Hamilton, 4 Lilley Cl, Bury St. Edmunds, IP33 2HZ [JO02IF, TL86]
G8 XJL M P Halford, 35 The Limes, Stony Stratford, Milton Keynes, MK11 1ET [IO92NB, SP73]
G8 XJN W H Hefferman, 74 Balmoral Dr, Borehamwood, WD6 2RB [IO91UP, TQ29]
G8 XJO S R Hedicker, 1 Hares Cl Cttgs, Greatham, Liss, Hants, GU33 6HG [IO91NB, SU73]
G8 XJU P R Curtis, 24 Aragon Dr, Warwick, CV34 6LR [IO92FG, SP36]
G8 XKB Details withheld at licensee's request by SSL.
G8 XKH W E Flood, 29 Northampton Cl, Bracknell, RG12 9EF [IO91PJ, SU86]
G8 XKT D K Last, 77 Brunswick Rd, Ipswich, IP4 4BS [JO02OB, TM14]
G8 XKV Details withheld at licensee's request by SSL.
G8 XLB J H Martin, Thatched Cottage, Thaxted Rd, Saffron Walden, CB11 3BJ [JO02DA, TL53]
G8 XLC Details withheld at licensee's request by SSL.
G8 XLE W S Metcalf, 1 Macfarlane Cl, Impington, Cambridge, CB4 4LZ [JO02BF, TL46]
G8 XLG C J Proctor, 309 Chester Rd, New Oscott, Sutton Coldfield, B73 5BJ [IO92BN, SP19]
G8 XLH A D Ralph, 15 Portchester Cl, Peterborough, PE2 8UP [IO92VN, TL29]
G8 XLI J S N Rigby, 53 Standish St., St. Helens, WA10 1HY [IO83PK, SJ59]
GW8 XLL R Stubbs, Glen Dene, Church St., Rhuddlan, Rhyl, LL18 2YA [IO83GG, SJ07]
G8 XLZ K J Riley, 38 Belle Vue Bank, Low Fell, Gateshead, NE9 6BS [IO94EW, NZ26]
G8 XMG C A R Poulter, 279 Aragon Rd, Morden, SM4 4QP [IO91VJ, TQ26]
G8 XMH D Higgins, 80 Hill Morton Rd, Sutton Coldfield, B74 4SG [IO92BO, SP19]
G8 XML J M T Hopper, Chapel Farm, Normanby Rd, Nettleton, Market Rasen, LN7 6TB [IO93UL, TF19]
G8 XMO H J Houghton, 21 John Gwynn House, Newport St., Worcester, WR1 3NY [IO82VE, SO85]
G8 XMS L M Sellar, 4 Barrowcrofts, Histon, Cambridge, CB4 4EU [JO02BG, TL46]
G8 XMU E Jones, 3 Byland Cl, Boston Spa, Wetherby, LS23 6PU [IO93HV, SE44]
GW8 XMW D O Jones, 79 Heol Waunyclun, Trimsaran, Kidwelly, SA17 4BS [IO71VR, SN40]
G8 XMZ R E Linton, Birch Cottage, Heyrose Ln, Over Tablay, Knutsford Ches, WA16 0HY [IO83SH, SJ77]
G8 XNA J E Lane, 12 Penarwyn Woods, St. Blaze Gate, Par, PL24 2DG [IO70PI, SX05]
G8 XNB R P Lelliott, Smugglers Cottage, Oreham Common, Henfield, BN5 9SB [IO90UV, TQ21]
G8 XNC R Lacey, 12 Melville Ave, Frimley, Camberley, GU16 5NA [IO91PH, SU85]
G8 XND D J F Lucas, 3 Woodlands Ave, Burghfield Common, Reading, Berks, RG7 3HU [IO91LJ, SU66]
G8 XNG R W Porter, 13 Martinfield, Covingham Park, Swindon, SN3 5BA [IO91DN, SU18]
G8 XNH H C Pearce, 7 Coronation Way, Doublestiles, Newquay, TR7 3JL [IO70LK, SW86]
G8 XNI Details withheld at licensee's request by SSL.[Op: A J Peachment, Bridgwater, Somerset.]
G8 XNJ Details withheld at licensee's request by SSL.
G8 XNL J G Rigby, 43 Corser St., Stourbridge, DY8 2DE [IO82WK, SO98]
G8 XNN H Vadgama, 20 Hollies Walk, Wootton, Bedford, MK43 9LB [IO92RC, TL04]
G8 XNO P Lambert, 92 Winterslow Dr, Leigh Park, Havant, PO9 5DZ [IO90MV, SU70]
G8 XNQ Details withheld at licensee's request by SSL.
GW8 XNV Details withheld at licensee's request by SSL.
G8 XNY I C Scrimshaw, 70 Hawerby Rd, Laceby, Grimsby, DN37 7BE [IO93VN, TA20]
G8 XOB P M Ashcroft, Fendley Corner, Sauncey Wood, Harpenden, AL5 5DW [IO91UT, TL11]
GM8 XOC D G Bird, 1 Harbour St., Gardenstown, Banff, AB45 3YT [IO87TQ, NJ86]
G8 XOD C J Burns, 3 Ellenbrook Rd, Boothstown, Worsley, Manchester, M28 1FX [IO83TM, SD70]
G8 XOE B J Baker, Linden Lea, Greenway Ln, Fivehead, Taunton, TA3 6PU [IO81MA, ST32]
G8 XOH Details withheld at licensee's request by SSL.
G8 XOM P F Cook, 37 Castle Ln, Haverhill, CB9 9NG [JO02FC, TL64]
G8 XOQ R F Smith, 100 Crawford Rd, Hatfield, AL10 0PE [IO91VS, TL20]
G8 XOR D J Sparrow, 23 Tranmere Gr, Ipswich, IP1 6DU [JO02NB, TM14]
G8 XOS P H Stoner, 6 Tintern Dr, Formby, Liverpool, L37 6DT [IO83LN, SD30]
G8 XOT M S Suckling, 105 Havenbaulk Ln, Littleover, Derby, DE23 7AD [IO92FV, SK33]
G8 XOV L Sedgwick, 28 Fairhaven Rd, Redhill, RH1 2LA [IO91WG, TQ25]
G8 XOX R W Sneath, Dragonhold, No 3 Field, Freathy, Torpoint, Cornwall, PL10 1JP [IO70VI, SX35]
G8 XPB S J Curtis, 1 Stile End, Wickle Trafford, Mickle Trafford, Chester, CH2 4QR [IO83OF, SJ46]
G8 XPC J M Caswell, 16 Birchwood Dr, Rushmere Village, Rushmere, Ipswich, IP5 7EB [JO02OB, TM24]
G8 XPF Details withheld at licensee's request by SSL.
G8 XPL Details withheld at licensee's request by SSL.
G8 XPM Details withheld at licensee's request by SSL.
G8 XPO B J Walker, 20 Delvin Rd, Henleaze, Bristol, BS10 5EJ [IO81QL, ST57]
G8 XPQ B H Whitehead, 63 Swinford Hollow, The Grange, Little Billing, Northampton, NN3 9HP [IO92OF, SP86]
G8 XPV Details withheld at licensee's request by SSL.
G8 XPZ S Lovell, 98B Baker Rd, Newthorpe, Nottingham, NG16 2DP [IO93IA, SK44]
G8 XQA P D Lineham, 10 Streetsbrook Rd, Shirley, Solihull, B90 3PL [IO92CK, SP18]
G8 XQF Details withheld at licensee's request by SSL.
G8 XQH E W Massey, 21 Arlington Dr, Macclesfield, SK11 8QL [IO83WG, SJ97]
G8 XQI P T Nightingale, 53 Philips Ave, Farnworth, Bolton, BL4 9BJ [IO83TN, SD70]
G8 XQL J R Alcock, Welland Rd, Upton on Severn, Upton upon Severn, Worcester, Worcs, WR8 0SJ [IO82VB, SO83]
G8 XQN A D Cleave, 42 The Silver, Birches, Kempston, Beds, MK42 7TS [IO92SC, TL04]
G8 XQP Details withheld at licensee's request by SSL.
G8 XQQ Details withheld at licensee's request by SSL.
G8 XQS M V Chapple, 34 Deepdale Way, Darlington, DL1 2TA [IO94FM, NZ31]
G8 XQT C Dodds, 5 Cascadia Cl, Loudwater, High Wycombe, HP11 1JW [IO91PO, SU89]
G8 XQZ Dr G A Farmer, 39 Plough Rise, Upminster, RM14 1XR [JO01DN, TQ58]
G8 XRE P A I McKnight, 5 Malvern Dr, Dibden Purlieu, Southampton, SO45 5QY [IO90GU, SU40]
G8 XRG R L Margetts, Mowbray, Arbor Rd, Croft, Leicester, LE9 3GE [IO92JN, SP59]
G8 XRL R D Mills, 131 High Rd East, Felixstowe, IP11 9PS [JO01QX, TM33]
G8 XRP R S Pryor, 27 Hollickwood Ave, Friern Barnet, London, N12 0LS [IO91WO, TQ29]
G8 XRR J R Nicholson, 14 Boyne Rd, Budleigh Salterton, EX9 6SE [IO80IP, SY08]
G8 XRS G A Nuttall, 28 Medoc Cl, Wymans Brooks, Cheltenham, GL50 4SP [IO81WV, SO92]
GW8 XRU T Nutbeem, 6 Morris Rise, Blaenavon, NP4 9PA [IO81LS, SO20]
G8 XRW D S Owen, 18 Bushey Cl, Capel, Ipswich, IP9 2HW [JO02MA, TM03]
G8 XSA W P Ash, Cornerways, 53 Waxland Rd, Halesowen, B63 3DN [IO82XK, SO98]
GI8 XSB F D J Aughey, 239 Bridge St., Portadown, Craigavon, BT63 5AR [IO64SK, J05]
G8 XSD J Atkinson, 168 Whipperley Way, Luton, LU1 5LJ [IO91SU, TL02]
G8 XSF M E Ainley, 152 Bourne View Rd, Huddersfield, HD4 7JS [IO93CO, SE11]
G8 XSI J Brinham, 16 Glynn Rd, Padstow, PL28 8EF [IO70MM, SW97]
G8 XSP T E Britton, 17 Richmond Rd, Mangotsfield, Bristol, BS17 3EZ [IO81SL, ST67]
G8 XST W J Butchers, 12 Church Rd, St. Marychurch, Torquay, TQ1 4QY [IO80FL, SX96]
G8 XSU M J Bond, Sunnyhurst, Golden Soney, Tockholes, Darwen, BB3 0NL [IO83RQ, SD62]
GI8 XSY K A Steenson, 108 Morgans Hill Rd, Cookstown, BT80 8BW [IO64OP, H87]
G8 XTC P A Holcroft, Hillcroft, Newbury Rd, Headley, Thatcham, RG19 8LA [IO91II, SU56]
G8 XTD R W Cavendish, 14 Coopers Cl, Chigwell Row, Chigwell, IG7 6EU [JO01BO, TQ49]
G8 XTE P N Connor, 20 Longfield, Lutton, Ivybridge, PL21 9SN [IO80AJ, SX55]
G8 XTF R O Cooper, 8 Bigwood Ct, Bigwood Rd, London, NW11 6SS [IO91VN, TQ28]
G8 XTG Details withheld at licensee's request by SSL.
G8 XTJ J P Fitzgerald, 21 Honor Rd, Prestwood, Great Missenden, HP16 0NJ [IO91PQ, SP80]
G8 XTK Details withheld at licensee's request by SSL.[Op: J S Discombe.]
G8 XTO R G Evans, 8 Pines Way, Radstock, Bath, BA3 3EZ [IO81SH, ST65]
G8 XTQ F J O Eite, 439 Yardley Rd, Yardley, Birmingham, B25 8NB [IO92CL, SP18]
G8 XTR P R A Emmans, 8 Bonington House, Ayley Croft, Bush Hill Park, Enfield Middx, EN1 1XT [IO91XP, TQ39]
G8 XTU M J L Fowler, 33 St. Ursulas Rd, Bell Vue, Doncaster, DN4 5ED [IO93KM, SE50]
G8 XTW P G Seaford, 14 Nevis Cl, Leighton Buzzard, LU7 7XD [IO91PW, SP92]
G8 XTX S J Sillitoe, 12 Howden Cl, London, SE28 8HD [JO01BM, TQ48]
G8 XUA Details withheld at licensee's request by SSL.
G8 XUE L M Radcliffe, 25 Oakleigh Dr, Codsall, Wolverhampton, WV8 1JP [IO82VP, SJ80]
G8 XUH J W Pearson, 8 Alder Gr, Darfield, Barnsley, S73 9JL [IO93HM, SE40]
G8 XUJ Details withheld at licensee's request by SSL.
GW8 XUM P D Jeavons, Manora, Penisarwaun, Caernarfon Gwynedd, LL55 3PW [IO73WD, SH56]
G8 XUN M I Hickman, 24 Calverley Rd, Kings Norton, Birmingham, B38 8PW [IO92AJ, SP07]
G8 XUU Details withheld at licensee's request by SSL.
G8 XUW D J Shields, 54 Wildmoor Ln, Catshill, Bromsgrove, B61 0PA [IO82XI, SO97]
G8 XVC Details withheld at licensee's request by SSL.
G8 XVD Details withheld at licensee's request by SSL.
G8 XVI V S Wake, 30 Moss Gdns, Alwoodley, Leeds, LS17 7BH [IO93FU, SE24]
G8 XVJ Details withheld at licensee's request by SSL.
G8 XVK J R Gray, The Pheasantries, Huddersfield Rd, Bretton, Wakefield, WF4 4JX [IO93FO, SE21]
G8 XVO C R Hetherington, 23 Falkland Ct, Braintree, CM7 9LL [JO01GV, TL72]
GM8 XVU W J Halliday, 35 Main Rd, Gatehead, Kilmarnock, KA2 0AR [IO75RO, NS33]
G8 XVV Details withheld at licensee's request by SSL.
G8 XVX A J Honeywell, 17 Whittaker St., Radcliffe, Manchester, M26 2TD [IO83UN, SD70]
G8 XWG Details withheld at licensee's request by SSL.
G8 XWI R C Marsh, 15 Beacon Cl, Rubery, Rednal, Birmingham, B45 9DA [IO82XJ, SO97]
GI8 XWJ Details withheld at licensee's request by SSL.
G8 XWK K L Morgan, 8 Pines Way, Radstock, Bath, BA3 3EZ [IO81SH, ST65]
G8 XWR M L Izzard, 17 Jack Jarvis Cl, Newmarket, CB8 8HY [JO02EF, TL66]
G8 XWU G B McLean, 53 Bardon View Rd, Dordon, Tamworth, B78 1QL [IO92EO, SK20]
GW8 XWW D P Moseley, Holm Leigh, 42 Severn Cres, Garden City, Chepstow, NP6 5EA [IO81PP, ST59]
G8 XXA J S Harrison, 10 Gaia Ln, Lichfield, Staffs, WS13 7LW [IO92CQ, SK11]
G8 XXF S F Rushowski, 150 Eskdale Ave, Chesham, HP5 3BE [IO91QR, SP90]
G8 XXG Details withheld at licensee's request by SSL.
G8 XXI J C Akines, Yarrow Cottage, Cottagers Plot, Laceby, Grimsby, DN37 7DX [IO93WN, TA20]
G8 XXJ J W Allchin, 40 Vale Rd, Seaford, BN25 3EZ [JO00BS, TV49]
G8 XXK Details withheld at licensee's request by SSL.
G8 XXL D C Alexander-Pye, 11 Clinton Cl, East Hanningfield, East Hanningfield, Chelmsford, CM3 8AZ [JO01GQ, TL70]
G8 XXM C Beecher, 23 Burhill Way, St. Leonards on Sea, TN38 0XP [JO00GU, TQ80]
G8 XXQ T F Bell, 9 Westover Ave, Warton, Carnforth, LA5 9QP [IO84OD, SD57]
G8 XXR W Bodicoat, 22 Brookdale Rd, Sutton in Ashfield, NG17 4LP [IO93JD, SK55]
G8 XXU M A Caulton, 115 Delves Green Rd, Walsall, WS5 4NH [IO92AN, SP09]
G8 XXV G L Clarke, 28 Little Potters, Bushey Heath, Bushey, Watford, WD2 3QT [IO91TP, TQ19]
G8 XXY B D Dawson, The Mythes, Alfrick, Worcester, WR6 5HH [IO82TE, SO75]
G8 XXZ P R Grace, 6 Davis Gr, Yardley, Birmingham, B25 8LQ [IO92CL, SP18]
G8 XYA N D Southorn, 20 Bratton Ave, Devizes, SN10 5BA [IO91AI, SU06]
G8 XYL Details withheld at licensee's request by SSL.
G8 XYQ D I Stanford, 7 Norman Cl, Melton, Woodbridge, Suffolk, IP12 1JT [JO02PC, TM25]
G8 XYR R E Tiller, Wayside, Ockley Ln, Keymer, Hassocks, BN6 8NU [IO90WW, TQ31]
G8 XYS R J Travett, 39 Amwell Rd, Kings Hedges, Cambridge, CB4 2UH [JO02BF, TL46]
G8 XYY P E J Redmile, 26 Wood Ln, Quorn, Loughborough, LE12 8DB [IO92KR, SK51]
G8 XZA Details withheld at licensee's request by SSL.
G8 XZB J F Payne, 25 Ringwood Rd, Oldfield Park, Bath, BA2 3JL [IO81TJ, ST76]
G8 XZC A N Pinder, 2 Eleanor Rd, Woodlands, Harrogate, HG2 7AJ [IO93FX, SE35]
G8 XZD Details withheld at licensee's request by SSL.
GW8 XZJ S W Rees, Bryngwyn House, 6 West Ave, Griffithstown, Pontypool, NP4 5AJ [IO81LQ, ST29]
G8 XZO Details withheld at licensee's request by SSL.
G8 XZQ M J M Fowler, 1 Mayfields, Shefford, SG17 5AU [IO92UA, TL13]
G8 XZX J E Tyler, 49 Bosworth Rd, Barlestone, Nuneaton, CV13 0JE [IO92HP, SK40]
GM8 XZY P C Tilbrook, 32 Bow Butts, Crail, Anstruther, KY10 3UR [IO86QG, NO60]
G8 YAA S R A Stigant, 81 Sandown Rd, Benfleet, SS7 3SH [JO01HN, TQ78]
G8 YAE C J Wenn, 191A Fullwell Ave, Clayhall, Ilford, IG5 0XA [JO01AO, TQ49]
G8 YAG M A Walsh, 202 Oldham Rd, Grasscroft, Oldham, OL4 4DW [IO83XM, SD90]
G8 YAO D S Whitelock-Wainwright, 1 Axbridge Ave, Sutton Leach, St. Helens, WA9 4NZ [IO83PK, SJ59]
GM8 YAQ R K Wroblewski, 9 The Woodlands, Rosyth, Dunfermline, KY11 2JD [IO86GB, NT18]
GW8 YAS A R Miller, 137 Marine Dr, Rhos on Sea, Colwyn Bay, LL28 4HY [IO83CH, SH88]
G8 YAT I V Naylor, The Old Rectory, Gratwich, Uttoxeter, Staffs, ST14 8SE [IO92AV, SK03]
G8 YAU R W Newton, Cascades, Top Rd, Worlaby, Brigg, DN20 0NN [IO93SO, TA01]
G8 YAV M Nicholls, 51 Lancaster Rd, Westville, Hucknall, Nottingham, NG15 6FN [IO93JA, SK54]
G8 YAY P Nicholls, 6 Welholme Ave, Grimsby, DN32 0HP [IO93WN, TA20]
G8 YAZ G A Oates, 21 Churchill Mns s, Cooper St., Runcorn, WA7 1DH [IO83PI, SJ58]
G8 YBD J W Crawshaw, 108 Warbreck Dr, Bispham, Blackpool, FY2 9PL [IO83LU, SD33]
G8 YBH A J Bristow, 2 Nursery Cttgs, Staplehurst Rd, Marden, Tonbridge, TN12 9BS [JO01GE, TQ74]
G8 YBL Details withheld at licensee's request by SSL.[Station located near Banstead.]
G8 YBO R C Colebrook, 29 Tunstall Terr, Darlington, DL1 4XH [IO94FM, NZ21]
G8 YBR I Davidson, Appledore, 1 Mooracre Ln, Bolsover, Chesterfield, S44 6ER [IO93IF, SK47]
G8 YBT Details withheld at licensee's request by SSL.
G8 YBW M P Furnival, 5 Hoghton Gr, Southport, PR9 0PW [IO83MP, SD31]
G8 YBY E A M Santer, Zoe Lucky, 3 Barn Park, Liverton, Newton Abbot, TQ12 6HE [IO80DN, SX87]
G8 YBZ M J Hampson, 7 Merryfield Cl, Bransgore, Christchurch, BH23 8BS [IO90DS, SZ19]
G8 YCG H W King, 2 Oddcroft, Colne Engaine, Colchester, CO6 2ET [JO01IW, TL83]
G8 YCJ G T Vickery, 33 Edgehill Rd, Chislehurst, BR7 6LA [JO01AK, TQ47]
G8 YCK K W Tomlinson, 27 Brackens Ln, Alvaston, Derby, DE24 0AQ [IO92GV, SK33]
G8 YCN E J Smith, Innisfree, Heatherton Park, Bradford on Tone, Taunton, TA4 1ET [IO80JX, ST12]
G8 YCP J R A Sergeant, 5 Jedburgh Cl, North Shields, NE29 9NU [IO95GA, NZ36]
G8 YCQ N A Storey, 15 Tower Ave, Kimberley, Upton, Pontefract, WF9 1ED [IO93IO, SE41]
G8 YCR Details withheld at licensee's request by SSL.
G8 YCV Details withheld at licensee's request by SSL.
G8 YDA R K Knight, 45 Bullimore Gr, Kenilworth, CV8 2QF [IO92FH, SP27]
G8 YDC J A Jebb, 30 Runnymede, Nunthorpe, Middlesbrough, TS7 0QL [IO94JM, NZ51]
G8 YDI D W Belcher, 53 Rock Rd, Oundle, Peterborough, PE8 4LN [IO92SL, TL08]
G8 YDQ K M Bromage, 16 Calrofold Dr, Waterhayes, Chesterton, Newcastle, ST5 7SZ [IO83VA, SJ84]
G8 YDU Dr S H Cole, Lower Mead, Liphook Rd, Whitehill, Bordon, GU35 9AF [IO91NC, SU83]
GM8 YEC P W Eunson, Houll, 11 Twageos Rd, Lerwick, ZE1 0BB [IP90KD, HU44]
G8 YEF A G Eaton, 2 The Firs, Sandy Ln, Guildford, GU3 1HQ [IO91QF, SU94]
G8 YEG L Fairhurst, 64 Beresford Cres, Newcastle, ST5 3RH [IO82VX, SJ84]
G8 YEJ J P O Glover, 5 Meadow Rise, Wymondham, Melton Mowbray, LE14 2AP [IO92PS, SK81]
G8 YEN M L Stevens, Parkhill Lodge, Parkhill Cross, Totnes Rd, Ipplepen, Newton Abbot, TQ12 5TT [IO80EL, SX86]
G8 YEO M R Smith Yeovil ARC, The Knap, Milborne Port, Sherborne, Dorset, DT9 5AR [IO80SX, ST61]
G8 YEP B A Meyer, 42 Sandcross Ln, Reigate, RH2 8EL [IO91VF, TQ24]
G8 YEQ N S Littleboy, 33 The Vale, Oakley, Basingstoke, RG23 7LD [IO91JF, SU55]
G8 YEV C Sykes, 3 Hants Rd, Chadderton, Oldham, OL9 7RX [IO83WM, SD90]
GW8 YEY D O Westmoreland, 27 Palmerston Rd, Barry, CF63 2NR [IO81JJ, ST16]
GM8 YFA A C Regnart, Lyndhurst, Corstorphine Rd, Thornhill, DG3 5NB [IO85CF, NX89]
G8 YFF R Waygood, 16 Marston Cl, Sherford, Taunton, TA1 4HZ [IO81KA, ST22]
G8 YFG Details withheld at licensee's request by SSL.
G8 YFH D Oliver, 15 Dowden Gr, Alton, GU34 2HH [IO91MD, SU74]
G8 YFK J S B Mason, 1 Royston Ave, Boyatt Wood, Eastleigh, SO50 4NH [IO90HX, SU42]
GI8 YFM Details withheld at licensee's request by SSL.
GM8 YGA A T McIntosh, 97 Ballindean Rd, Dundee, DD4 8NY [IO86NL, NO43]
GM8 YGB A McNicoll, Strathfiddich, 17 Scotston Gdns, Dundee, DD4 7UN [IO86NL, NO43]
G8 YGD J B Malyon, 24 Avondale Rd, Aldershot, GU11 3HQ [IO91OF, SU84]
GI8 YGG P Foley, 5 Woodland Dr, Cookstown, BT80 8PL [IO64PP, H87]
GM8 YGI P R Sime, 15 Birchwood Pl, Carrickwood Est, Mount Vernon, Glasgow, G32 0NX [IO75WU, NS66]
G8 YGK W J Standing, 72 Ivydore Ave, Durrington, Worthing, BN13 3JD [IO90TU, TQ10]
G8 YGM D P Southward, 3A Carnoustie Cl, West Derby, Liverpool, L12 9NE [IO83NK, SJ49]
G8 YGO G D T Tarr, 40 The Garth, Coniston, LA21 8EQ [IO84LI, SD39]
G8 YGS J A Yates, 13 Langton Rd, Worthing, Sussex, BN14 7BY [IO90TT, TQ10]
G8 YGT B Senior, 1 Bedale Cl, Coalville, LE67 3BE [IO92HR, SK41]
G8 YGX Details withheld at licensee's request by SSL.
G8 YHF S M Kenyon, 8 Dunedin Gdns, Ferndown, BH22 9EQ [IO90BT, SZ09]
G8 YHH R N Taylor, 25 Daven Rd, Grangeland Park, Congleton, CW12 3RA [IO83VD, SJ86]
G8 YHI A N Kaye, 73 Woodthorpe Park Dr, Sandal, Wakefield, WF2 6SU [IO93GP, SE31]
G8 YHJ E B M J Jones, 34 Overlea Dr, Hawarden, Deeside, CH5 3HS [IO83LE, SJ36]
G8 YHM Details withheld at licensee's request by SSL.
G8 YHS Details withheld at licensee's request by SSL.
G8 YHX A E Briggs, 49 Milldale Ave, Buxton, SK17 9BG [IO93AG, SK07]
G8 YIB Dr G R Barber, 52 Braemar Rd, Sutton Coldfield, B73 6LS [IO92BN, SP19]
G8 YIG C E Fawcett, 24 Quarry Rise, Stalybridge, SK15 1US [IO83XL, SJ99]
G8 YIH Details withheld at licensee's request by SSL.
GM8 YIJ Details withheld at licensee's request by SSL.
GM8 YIK Dr A G Robson, Flat 5, 23 Queen Charlotte St., Edinburgh, EH6 6AX [IO85KX, NT27]
G8 YIN S L Wood, 246 Rush Green Rd, Romford, RM7 0LA [JO01CN, TQ58]
GI8 YJD R J Perver, 12 Orchardville Ave, Bangor, BT19 1LP [IO74DP, J48]
G8 YJE Details withheld at licensee's request by SSL.
GI8 YJF D Roxburgh, 5 Forestbrook Park, Rostrevor, Newry, BT34 3DX [IO64VC, J11]
G8 YJM N J Patrick Gleed, 8 Linnet Cl, Patchway, Bristol, BS12 5RL [IO81QM, ST58]
GW8 YJN A D Price, 147 Fleming Cres, Haverfordwest, SA61 2SQ [IO71MT, SM91]
G8 YJQ P J Holt, 13 Norbiton Hall, London Rd, Kingston upon Thames, KT2 6RA [IO91UJ, TQ16]
G8 YJS G P Hammond, South View, Cawston Rd, Reepham, Norwich, NR10 4LU [JO02NS, TG12]
G8 YJT Details withheld at licensee's request by SSL.
GI8 YJV P D Lloyd, 18 Demesne Rd, Holywood, BT18 9NB [IO74CP, J47]

G8 YJZ P G Rayson, 26 Leys Rd, Pattishall, Towcester, NN12 8JZ [IO92LE, SP65]
G8 YKE C P E Andrew, 17 St. James Cl, Kettering, NN15 5HB [IO92PJ, SP87]
G8 YKF Details withheld at licensee's request by SSL.
G8 YKG M F Armour, 15 Lime Gr, Thornton Cleveleys, FY5 4DE [IO83LU, SD34]
G8 YKO S Bardsley, 73 Highlands, Royton, Oldham, OL2 5HL [IO83WN, SD90]
GM8 YKT E G Brumby, 141 Morriston Rd, Elgin, IV30 2NB [IO87IP, NJ26]
G8 YKV A A Cragg, 11 Windmill Ave, Hassocks, BN6 8LH [IO90WW, TQ31]
G8 YKW Details withheld at licensee's request by SSL.
G8 YKY D Canham, 82 Rugby Rd, Binleywoods, Binley Woods, Coventry, CV3 2AX [IO92GJ, SP37]
G8 YKZ T L B Carroll, Rose Villa, North Rd, Whitemoor, St Austell Cornwal, PL26 7XL [IO70NJ, SW95]
G8 YLA R Cato, 9 Lime Kiln Rd, Mannings Heath, Horsham, RH13 6JH [IO91UB, TQ22]
G8 YLB P J Clampin, 22 Gr Rd, Oldbury, Warley, B68 9JL [IO92AL, SP08]
GM8 YLH Details withheld at licensee's request by SSL.
GW8 YLK B J Evans, Mynyddmelyn, Pontfaen, Fishguard, SA65 9SL [IO71NX, SN03]
G8 YLM M A Farnworth, 109 Avondale Rd, Darwen, BB3 1NT [IO83SQ, SD62]
G8 YLU Details withheld at licensee's request by SSL.
G8 YMB M B Button, 29 The Dawneys, Crudwell, Malmesbury, SN16 9HE [IO81XP, ST99]
G8 YMD B J Joyner Sek YMCA ARC, Brimar, Nelson Park Rd, St. Margarets At Cliffe, Dover, CT15 6HL [JO01QD, TR34]
G8 YML A P Shearing, 28 Woodside Rd, Downend, Bristol, BS16 2SL [IO81RL, ST67]
G8 YMM P G Stevenson, 14 Camelford Rd, Greenbank, Bristol, BS5 6HW [IO81RL, ST67]
G8 YMN M Shorter, Silverlea, 53 The St., Adisham, Canterbury, CT3 3JN [JO01OF, TR25]
G8 YMO C J Seeney, 91 Dovehouse Cl, Badgers Walk, Eynsham, Witney, OX8 1EW [IO91HS, SP40]
G8 YMQ A R P Simpson, Parsonage Farm House, Lake Ln, Barnham, Bognor Regis, PO22 0JD [IO90QT, SU90]
G8 YMR A B Snow, 45 Knightsway, Tewkesbury, GL20 8DY [IO81WX, SO93]
G8 YMS P Swarbrook, 6 Hazel Gr, Leek, ST13 8UU [IO83XC, SJ95]
G8 YMU L B Shaw, 108 Brookvale Rd, Olton, Solihull, B92 7JA [IO92CK, SP18]
G8 YMV Details withheld at licensee's request by SSL.
G8 YMW A N Sneath, 80 South St. North, New Whittington, Chesterfield, S43 2AB [IO93HG, SK37]
G8 YMZ J Trent, The Hollies, Bourn Rd, West Bergholt, Colchester, CO6 3EP [JO01KV, TL92]
G8 YNB A G Taylor, 21 Cubbington Cl, Barton Hills, Luton, LU3 3XJ [IO91SW, TL02]
G8 YNC P D Tuck, 30 Brownlow Rd, New Southgate, London, N11 2DE [IO91WO, TQ39]
G8 YNF G S Holman, 62 The Ridge, Kennington, Ashford, TN24 9EU [JO01KD, TR04]
G8 YNG A P Hall, 33 Deanwood Rd, Dover, CT17 0NT [JO01PD, TR24]
G8 YNH M R J Hall, 22 Tyson Rd, Folkestone, CT19 6JR [JO01OC, TR23]
GW8 YNJ Details withheld at licensee's request by SSL.
G8 YNK M Higton, 120 Station Rd, Mickleover, Derby, DE3 5FN [IO92FV, SK33]
G8 YNO R S Hayward, 32 Beck Ct, Beck Ln, Beckenham, BR3 4RB [IO91XJ, TQ36]
G8 YNP J D Hill, Coach House Cottage, 15 Pike Ln, Armitage, Rugeley, WS15 4AF [IO92BR, SK01]
G8 YOA A A Lake, The Granary, Zeal Monachorum, Crediton, Devon, EX17 6DH [IO80CT, SS70]
G8 YOC M J W Witchard, 110 Bradley Rd, Huddersfield, HD2 1QY [IO93DQ, SE12]
G8 YOE C V Victory, 88 Abbotts Dr, Stanford-le-Hope, SS17 7BS [JO01FM, TQ68]
G8 YOF M C Warren, 14 Birchwood Rd, St. Annes, Bristol, BS4 4QH [IO81RK, ST67]
G8 YOG J R Woodard, 213 Leicester Rd, Ibstock, LE67 6HP [IO92HQ, SK41]
G8 YOJ G N White, 6 Westlands Ave, Tetney, Grimsby, DN36 5LP [IO93XL, TA30]
G8 YOK J S Ward, 3 Sherbourne Cl, Poulton-le-Fylde, FY6 7UB [IO83LU, SD34]
G8 YOL Details withheld at licensee's request by SSL.
G8 YOX A F Munday, 2 Broadway Cl, Bourne, PE10 9BN [IO92TS, TF02]
G8 YOY M R Maxwell, 962 Bury Rd, Breightmet, Bolton, BL2 6NX [IO83TN, SD70]
G8 YPC Details withheld at licensee's request by SSL.
G8 YPF Details withheld at licensee's request by SSL.
GI8 YPG E Muldoon, 29 Eglish Rd, Dungannon, BT70 1LA [IO64OL, H76]
G8 YPH T McKnight, 39 Victoria Rd, Hope, Salford, M6 8FZ [IO83UL, SJ79]
G8 YPJ Details withheld at licensee's request by SSL.
G8 YPK V L Maddex, 140A Kents Hill Rd, Benfleet, SS7 5PH [JO01GN, TQ78]
G8 YPL P H Martin, 23 Molyneux Rd, Maghull, Liverpool, L31 3DX [IO83MM, SD30]
G8 YPN P D Lutman, Underbanks Farm East, Reeth Rd, Richmond, DL10 4SE [IO94CJ, NZ10]
G8 YPQ M Waring, 17 Arundel Dr, Carlton in Lindrick, Worksop, S81 9DL [IO93KI, SK58]
GW8 YPR R B Williams, 54 Woodlands Ave, Talgarth, Brecon, LD3 0AT [IO81JX, SO13]
G8 YPT R J Wiggett, 171 Kingshurst Way, Birmingham, B37 6EA [IO92CL, SP18]
G8 YPV G D Williams, 54 Greenacre, Wembdon, Bridgwater, TA6 7RF [IO81LD, ST23]
G8 YPY D W Wilson, 35 Darbishire Rd, Fleetwood, FY7 6QA [IO83LW, SD34]
G8 YQA D C Arnold, Belhaven, 12 Moorfields, Leek, ST13 5LU [IO83XC, SJ95]
G8 YQB K Batterham, 17 Selba Dr, Bevendean, Brighton, BN2 4RG [IO90WU, TQ30]
G8 YQD Details withheld at licensee's request by SSL.
G8 YQH V G Carter, 69 Angela Cres, Horsford, Norwich, NR10 3HE [JO02OQ, TG11]
G8 YQJ N J Dutton, Little Farden, 17 Brickendon Ln, Brickendon, Hertford, SG13 8NU [IO91XS, TL30]
G8 YQN Details withheld at licensee's request by SSL.
G8 YQO D V Henderson, Reverie Pennys Ln, Margaretting, Ingatestone, Essex, CM4 0HA [JO01EQ, TL60]
G8 YQQ Details withheld at licensee's request by SSL.[QSL via Bureau or direct via G4NSY.]
G8 YQY L A Brown, The Haven, High Rd, Saddlebow, Kings Lynn, PE34 3AW [JO02ER, TF61]
G8 YRC J W H Cross, 12 The Deans, Portishead, Bristol, BS20 8BG [IO81OL, ST47]
GM8 YRE J A Firth, 6 Upper Burnside Dr, Thurso, KW14 7XB [IO88FO, ND16]
G8 YRF R S Foxley, 160 Lyndhurst Rd, Worthing, BN11 2DW [IO90TT, TQ10]
G8 YRG Details withheld at licensee's request by SSL.
G8 YRI Details withheld at licensee's request by SSL.
G8 YRJ Details withheld at licensee's request by SSL.
G8 YRL B J Trim, 27 Wilts Rd, Wokingham, RG40 1TS [IO91OJ, SU86]
G8 YRM Details withheld at licensee's request by SSL.
G8 YRN J T Polding, 17 Josephine Rd, Cowlersley, Huddersfield, HD4 5UD [IO93CP, SE11]
GM8 YRT W N Stewart, Hillside Cottage, Glendoick, Glencarse, Perthshire, PH2 7NS [IO86IJ, NO22]
GM8 YRX E R Saxon, 73 Upper Burnside Dr, Thurso, KW14 7XB [IO88FO, ND16]
G8 YRY C P Rourke, 35 Chestnut Manor Cl, Staines, TW18 1AG [IO91SK, TQ07]
G8 YSJ W R Bannerman, Gate House, Eaton Bank, Duffield, Belper, DE56 4BH [IO92GX, SK34]
GM8 YSN P W Coe, 1A Carleton Ave, Skipton, BD23 2TE [IO83XW, SD95]
G8 YSO Details withheld at licensee's request by SSL.
G8 YSQ N N Sidgwick, 27 Meldon Cl, Manor Park, Darlington, DL1 2BB [IO94FM, NZ31]
G8 YSV C E Snow, 34 Inglewood, St. Johns, Woking, GU21 3HX [IO91RH, SU95]
G8 YSX B T Twist, High View, The Lizard, Wymondham, Norfolk, NR18 9BH [JO02NN, TG10]
G8 YTF Details withheld at licensee's request by SSL.
GI8 YTH S G Moore, 7 Cyprus Ave, Belfast, BT5 5NT [IO74BO, J37]
GW8 YTO A L Ham, 15 Denbigh Dr, Boverton, Llantwit Major, CF61 2GQ [IO81GJ, SS96]
G8 YTP S C Holgate, 91 Valley Rd, Heaton Mersey, Stockport, SK4 2DB [IO83VJ, SJ89]
G8 YTR S P Higgs, 5 Lawnswood Cl, Cowplain, Waterlooville, PO8 8RU [IO90LV, SU61]
G8 YTU F J Adams, 27 Challenger Cl, Malvern, WR14 2NN [IO82UC, SO74]
G8 YTX K Bagshaw, 36 St. Peters Rd, Buxton, SK17 7DX [IO93BG, SK07]
G8 YTY Details withheld at licensee's request by SSL.
G8 YTZ Details withheld at licensee's request by SSL.
GM8 YUI G McClintock, 13 St. Andrews Dr, Gourock, PA19 1HY [IO75NW, NS27]
GW8 YUJ J E Milburn, Orme View, Marianglas, Anglesey, LL73 8PE [IO73VI, SH58]
G8 YUK A White, 10 Stott Dr, Flixton, Urmston, Manchester, M41 6WA [IO83TK, SJ79]
GM8 YUM G W J Walker, 19 Barassie Dr, Kirkcaldy, KY2 6HW [IO86JD, NT29]
G8 YUP B Stevens, 17 Ridge Cres, Marple, Stockport, SK6 7JA [IO83XJ, SJ98]
G8 YUR M C Robelou, 27 Albany Park Ave, Enfield, EN3 5NT [IO91XP, TQ39]
G8 YVA H L Pearson, 66 Slater Rd, Bentley Heath, Solihull, B93 8AL [IO92CJ, SP17]
G8 YVC M J Smith, 31 Burringham Rd, Scunthorpe, DN17 2BD [IO93QN, SE80]
G8 YVF H D A Mountjoy, 16 Riverholme Dr, West Ewell, Epsom, KT19 9TQ [IO91UI, TQ26]
G8 YVP M Nicholson, The Old Rectory, Nether Denton, nr Brampton, Cumbria, CA8 2LY [IO84QX, NY56]
G8 YVQ C R Harper, 3 Maze Hill Lodge, Park Vista, Greenwich, London, SE10 9LY [JO01AL, TQ37]
G8 YVT J C Howe, 28 Rarey Dr, Weaverthorpe, Malton, YO17 8HA [IO94RC, SE97]
G8 YVU P Kirkup, 337 Wheatley Ln Rd, Fence, Burnley, BB12 9QA [IO83VU, SD83]
G8 YVW C Stacey, 157 Ormond Rd, Sheffield, S8 8FT [IO93GH, SK38]
G8 YWA W G Waller, 6 Barton Cl, Cupernham, Romsey, SO51 7QE [IO90GX, SU32]
G8 YWH G K Davies, 19 Poplar Dr, Woodside Park, Marchwood, Southampton, SO40 4XH [IO90GV, SU31]
G8 YWK W Gleave, 8 Abbey Rd, Durham, DH1 5DQ [IO94ET, NZ24]
GI8 YWR P S C Pollock, 7 Cossack Ct, Townparks North, Antrim, BT41 4HN [IO64VR, J18]
G8 YWT B I Richardson, Brockwood, Gr Rd, Ryde, PO33 3LH [IO90JR, SZ59]
G8 YWU K Stubbins, 61 Goosemoor Ln, Erdington, Birmingham, B23 5PW [IO92BM, SP19]
G8 YXE Details withheld at licensee's request by SSL.
G8 YXI D P Shemeld, 128 Bradley St., Crookes, Sheffield, S10 1PB [IO93GJ, SK38]
G8 YXJ R A Skells, 31 Perry Rd, Leverington, Wisbech, PE13 5AE [JO02BQ, TF41]
G8 YXQ D I Chatterton, 13 Salisbury Rd, Dover, CT16 1EX [JO01PD, TR34]
GM8 YXR E V Ferris, 16 Grosvenor Terr, Glasgow, G12 0TB [IO75UV, NS56]
G8 YXX Details withheld at licensee's request by SSL.
G8 YXZ R P Dominy, 8 Meadow Rd, Claygate, Esher, KT10 0RZ [IO91TI, TQ16]
G8 YYA H Duesbury, 4 Harbour View Cl, Poole, BH14 0PF [IO90AR, SZ09]
G8 YYB Gray, 68 Sixth Cross Rd, Twickenham, TW2 5PD [IO91TK, TQ17]
G8 YYC G E Miller, 25 Penscroft Gdns, Borehamwood, WD6 2QZ [IO91UP, TQ29]
GW8 YYF K Jones, 3 Penffordd, Parc St. Catwg, Pentyrch, Cardiff, CF4 8TJ [IO81IM, ST18]
G8 YYL Lady G Johnson, Yeldfield, Chorleywood, Herts, WD3 5SB [IO91SP, TQ09][The Lady Johnson of Marle.]
GI8 YYM I L Ferris, 48 Abbey Gdns, Stormont, Belfast, BT5 7HL [IO74BO, J37]
G8 YYR S Hill, Fourways, Mile End, Coleford, Glos, GL16 7QE [IO81QT, SO51]
G8 YYW M A Kimber, 90 Moorland Rd, Goole, DN14 5TX [IO93NQ, SE72]
G8 YYX A H Layton, 7 Higher Saxifield, Harle Syke, Burnley, BB10 2HB [IO83VT, SD83]
G8 YZC R A Smith, 86 Manor Rd, Borrowash, Derby, DE72 3LN [IO92HV, SK43]
G8 YZD Details withheld at licensee's request by SSL.
G8 YZF M Bishop, 6 Tiverton Cl, Kingswinford, DY6 8PD [IO82WL, SO98]
G8 YZH Details withheld at licensee's request by SSL.
G8 YZL P A Thackeray, 20 Oaklands Cl, Verwood, BH31 6NZ [IO90BV, SU00]
G8 YZS M C Gardner, 137 Castle Rd, Carisbrooke, Newport, PO30 1DP [IO90IQ, SZ48]
G8 YZU A W Gibson, 14 Cassino Rd, Melbourne Park Est, Chelmsford, CM1 2EW [JO01FS, TL60]
G8 YZY D B Spencer, 28 Watery Ln, Minehead, TA24 5NZ [IO81GE, SS94]
G8 ZAC E P McCormick, 526 Watling St. Rd, Ribbleton, Preston, PR2 6TU [IO83QS, SD53]
G8 ZAD R Mantle, 37 Willis Rd, Stockport, SK3 8HQ [IO83WJ, SJ88]
G8 ZAJ C J French, 26 Wood St., Ash Vale, Aldershot, GU12 5JG [IO91PG, SU85]
G8 ZAU D J Hoodless, 21 Meadow Cl, Eastwood, Nottingham, NG16 3DQ [IO93IA, SK44]
G8 ZAX R E Rees, 19 Beechwood Rise, West End, Southampton, SO18 3PW [IO90HW, SU41]
G8 ZAZ Details withheld at licensee's request by SSL.
GW8 ZBC C P Lucas, 8 Hawker Cl, Broughton, Chester, CH4 0SQ [IO83MD, SJ36]
G8 ZBJ Details withheld at licensee's request by SSL.
G8 ZBN T R Nye, 18 Kingsway, Chandlers Ford, Eastleigh, SO53 2FE [IO90HX, SU42]
G8 ZBZ G W Pallister, Half Acre, High Kelling, Holt, Norfolk, NR25 6RD [JO02NV, TG13]
G8 ZCJ J C M Skidmore, 55 Elmsleigh Rd, Heald Green, Cheadle, SK8 3UD [IO83VJ, SJ88]
G8 ZCS A S Westerman, 133 Western Rd, Goole, DN14 6RF [IO93NR, SE72]
GW8 ZCV G M Byars, 31 Roman Reach, Caerleon, Newport, NP6 1SQ [IO81MO, ST39]
G8 ZDF Details withheld at licensee's request by SSL.
G8 ZDK Details withheld at licensee's request by SSL.[Op: R G Guttridge. Station located near Aldershot.]
GM8 ZDQ Details withheld at licensee's request by SSL.
G8 ZDS P A Hocking, 42 Rosevean Ave, Camborne, TR14 8UG [IO70IF, SW64]
G8 ZDT P F Langford, 46 Elmer Gdns, Rainham, RM13 7BS [JO01CM, TQ58]
G8 ZDU R C Arnold, 12 Wilts Ave, Crowthorne, RG45 6NG [IO91OI, SU86]
G8 ZEE A Hudson, 1 Laburnum Ct, Elm Farm Est, Cheltenham, GL51 0XE [IO81WV, SO92]
G8 ZEF J A Haworth, 117 Rutherford Dr, Over Hulton, Bolton, BL5 1DW [IO83SN, SD60]
GW8 ZEI E Whitham, Thistle Bank, Maes Llydan, Benllech, Tyn y Gongl, LL74 8RD [IO73VH, SH58]
GM8 ZEJ Dr J C Borland, 4 Shanter Pl, Kilmarnock, KA3 7JB [IO75SO, NS43]
G8 ZEK P A Jacobi, Highbury, Furzehill, Wimborne, Dorset, BH21 4HD [IO90AT, SU00]
G8 ZEN R Dyson, 14 Lonsdale Ave, Kingsway, Rochdale, OL16 5HP [IO83WO, SD91]
G8 ZEO Details withheld at licensee's request by SSL.
GM8 ZEQ M Smith, Haremuir Bungalow, Benholm, Inverbervie, Montrose Angus, DD10 0HX [IO86TT, NO77]
G8 ZES P Street, 20 Hollinshead Ave, Cross Heath, Newcastle, ST5 9DD [IO83VA, SJ84]
G8 ZEV C A Hartt, 8 Singer Cl, Paignton, TQ3 3JU [IO80FK, SX86]
G8 ZEW A K Joy, 16 Willow Cres, Great Houghton, Northampton, NN4 7AP [IO92NF, SP75]
G8 ZFD C Askin, 54 York Rd, Greenwood Ave, Hull, HU6 9RA [IO93TS, TA03]
G8 ZFI P M Bryant, 21 Devonshire Cl, Stevenage, SG2 8RY [IO91VV, TL22]
G8 ZFL A F Butcher, 4 Maple Cl, Oldland Common, Bristol, BS15 6PX [IO81SK, ST67]
G8 ZFQ M Kanelis, 57 Ringwood Ave, Redhill, RH1 2DY [IO91WG, TQ25]
G8 ZFS P F Wiley, 9 Simpson Ave, Hunmanby, Filey, YO14 0LB [IO94UE, TA07]
G8 ZFT R M Thompson, 329 Prestbury Rd, Prestbury, Cheltenham, GL52 3DF [IO81XV, SO92]
GM8 ZFW J A Morris, Wealthyon Cottage, Keig, Alford, Aberdeenshire, AB3 8BH [IO87TH, NJ72]
G8 ZFX P A Snelling, 4 Chapel Rd, Carleton Rode, Norwich, NR16 1RN [JO02NL, TM19]
GM8 ZGC C S Dowers, 6 Masonhill Pl, Masonhill, Ayr, KA7 3PA [IO75QK, NS32]
G8 ZGF R P Mackrell, 17 Townfield Ave, Worsthorne, Burnley, BB10 3JG [IO83VS, SD83]
G8 ZGI R R L Noquet, 15 Pearce Manor, Chelmsford, CM2 9XH [JO01FR, TL60]
G8 ZGO G C Bywater, 27 Shirley Rd, Maidenhead, SL6 4PH [IO91PM, SU88]
G8 ZGQ A J Longuet, 10 Severnmead, Grovehill, Hemel Hempstead, HP2 6DX [IO91SS, TL00]
G8 ZGS J Holden, 128 Greenways, Norwich, NR4 6HA [JO02PO, TG20]
G8 ZGU R R C Smith, 31 Eriswell Rd, Worthing, BN11 3HP [IO90TT, TQ10]
G8 ZHA R A Morrall, 61 Archer Rd, Leamore, Walsall, WS3 1AW [IO92AO, SK00]
G8 ZHC K Moth, 38 Annesley Rd, New Moston, Manchester, M40 3PB [IO83WM, SD80]
G8 ZHD Details withheld at licensee's request by SSL.[Station located near Stocksfield.]
G8 ZHM Details withheld at licensee's request by SSL.
GW8 ZHN P G Gibbons, 40 Grange Gdns, Llantwit Major, CF61 2XB [IO81GJ, SS96]
G8 ZHP Details withheld at licensee's request by SSL.[Correspondence to: Five Bells Group, c/o B K Tatnall, 73 Acacia Avenue, Spalding, Lincs, PE11 2LW.]
G8 ZHR N P Lawes, 87 Glebelands, Crayford, Dartford, DA1 4RY [JO01CK, TQ57]
G8 ZHS P C Lester, 44 Overton Dr, Wanstead, London, E11 2NJ [JO01AN, TQ48]
GI8 ZHW J F McDonnell, 6 Sandhurst Park, Bangor, BT20 5NU [IO74EP, J58]
G8 ZIA A M Bowman, Evergreen, Durham Rd, Stockton on Tees, TS21 3LT [IO94HO, NZ42]
G8 ZIC C H Harrison, 2 Bridgemere Cl, Radcliffe, Manchester, M26 4FS [IO83TN, SD70]
G8 ZID M J D Sisley, 6 Spearhill, Lichfield, WS14 9UD [IO92CQ, SK10]
G8 ZIH J A H Eady, Pytchley Lodge, Kettering, Northants, NN14 1EE [IO92PI, SP87]
G8 ZIK E Serwa, 93 Long Knowle Ln, Wednesfield, Wolverhampton, WV11 1JG [IO82WO, SJ90]
GW8 ZIL I F Bell, 102 Ewenny Rd, Bridgend, CF31 3LN [IO81FL, SS97]
G8 ZIP K Lake, 22 Chapmans Cl, Aqueduct, Stirchley, Telford, TF3 1ED [IO82SP, SJ60]
G8 ZIR B J Down, 12 Coach Rd, Henlow, SG16 6BT [IO92UA, TL13]
G8 ZIW G A Ludar-Smith, 8 Spencer Rd, Sandy, SG19 1AT [IO92UC, TL14]
G8 ZIY P G Eyre, 27 Holborn View, Codnor, Ripley, DE5 9RB [IO93HB, SK45]
G8 ZJD Details withheld at licensee's request by SSL.
G8 ZJE D J Webb, 51 Garden Rd, Walton on The Naze, CO14 8RR [JO01PU, TM22]
G8 ZJH B McCourt, 97 Arrowe Rd, Greasby, Wirral, L49 1RY [IO83KJ, SJ28]
G8 ZJK R H Cole, 15 Ashwood Cl, Hayling Island, PO11 9AX [IO90MS, SZ79]
GM8 ZJL R A McCann, Ardbroilach, Ardbroilach Rd, Kingussie, PH21 1LD [IO77XC, NH70]
G8 ZJM I Macalindin, The Orchards, Idridgehay, Derby, DE56 2SJ [IO93FB, SK24]
G8 ZJO S V A Tomschey, 21 Momus Boulevard, Coventry, CV2 5LL [IO92GJ, SP37]
G8 ZJP R S Powell, 57 Staplehurst, Wooden Hill, Bracknell, RG12 8DB [IO91OJ, SU86]
GM8 ZJS J W Thomson, 23 Douglas Rd, Longniddry, EH32 0LQ [IO85NX, NT47]
G8 ZJU K N Henry, Breedons Ct, Breedons Hill, Pangbourne, Reading, Berks, RG8 7AT [IO91KL, SU67]
G8 ZJX R S Green, 56 Kitchener Rd, Strood, Rochester, ME2 3AP [JO01FJ, TQ76]
G8 ZK C J Archer Siemens Amateur Radio Club, Sports Office, Technology Dr, Beeston, Nottingham, Notts, NG9 1LA [IO92JW, SK53]
G8 ZKE D C Rigby Aes Soc, 145 Knightlow Rd, Harborne, Birmingham, B17 8PY [IO92AL, SP08]
GM8 ZKF D M Robson, 6 Ladywood, Moor Rd, Milngavie, Glasgow, G62 8AT [IO75UW, NS57]
G8 ZKG R G Roberts, 93 Newtown Rd, Malvern, WR14 1PD [IO82UC, SO74]
G8 ZKH J C Barry, 14 Parklands Rd, Hassocks, BN6 8JZ [IO90WW, TQ31]
G8 ZKK N H S Collins, 34 Wellfield Rd, Alrewas, Burton on Trent, DE13 7EZ [IO92DR, SK11]
G8 ZKN I P Diment, 16 Riverside, Isleworth, Middx, TW7 6HW [IO91TL, TQ17]
G8 ZKO R E Harsent, 3 The Dr, Barking, IG11 9JB [JO01BM, TQ48]
G8 ZKV Details withheld at licensee's request by SSL.
G8 ZKW G A Powell, 37 The Lizard, Wymondham, NR18 9BH [JO02NN, TG10]
G8 ZKZ P C Weedon, 31 Fitzilian Ave, Romford, RM3 0QU [JO01CO, TQ59]
G8 ZLF T B Gilleard, 3 Paul Cres, Humberston, Grimsby, DN36 4DF [IO93XM, TA30]
G8 ZLG R E Green, 14 Birches Cl, Downs Rd, Epsom, KT18 5JG [IO91UH, TQ25]
GM8 ZLI W H G Gordon, 5 Glennie Gdns, Tranent, EH33 2DN [IO85MW, NT47]
G8 ZLK S S V Cook, 90 Chatham Gr, Chatham, ME4 6LY [JO01GI, TQ76]
G8 ZLL I Thomas, 62 Castlebar Park, London, W5 1BU [IO91UM, TQ18]
G8 ZLT M P Chambers, 6 Swan Ln, Kings Lynn, PE30 4HE [JO02FS, TF62]
G8 ZLU M J Wright, 2 Pensarn Gr, Reddish, Stockport, SK5 7LE [IO83WK, SJ89]
G8 ZLV Details withheld at licensee's request by SSL.
G8 ZMA Details withheld at licensee's request by SSL.
G8 ZMB F W Burns, 40 Thornbridge Ave, Sheffield, S12 3AB [IO93HI, SK38]
G8 ZMC A J McCalden, 127 Kings Rd, Godalming, GU7 3EU [IO91QE, SU94]
G8 ZME M J O'Toole, Forest Green, Fryern Ct Rd, Burgate Cross, Fordingbridge, Hants, SP6 1LZ [IO90CW, SU11]
G8 ZMH Details withheld at licensee's request by SSL.
G8 ZMI D S Adie, 73 Kensington Dr, Woodford Green, IG8 8LN [JO01AO, TQ49]
G8 ZMJ T G Brown, Greenhow, Cliff Ln, Wilton, Pickering N Yorks, YO18 7LE [IO94PF, SE88]
G8 ZML B G Ewart, 36 Sycamore Rise, Wooldale, Holmfirth, Huddersfield, HD7 2TJ [IO93CN, SE10]
G8 ZMM R W Bunney, 33 Cherville St., Romsey, SO51 8FB [IO90GX, SU32]
G8 ZMN B D Carr, 17 Fir Tree Ln, Thorpe Willoughby, Selby, YO8 9PG [IO93KS, SE53]
G8 ZMQ P H Burnley, 45 Ashwell Rd, Heaton, Bradford, BD9 4AX [IO93CT, SE13]
GW8 ZMU M T Goodall, 91 Uzmaston Rd, Haverfordwest, SA61 1UA [IO71MT, SM91]
G8 ZMV R E Toms, 9 Upper Gordon Rd, Highcliffe, Christchurch, BH23 5ND [IO90DR, SZ29]
G8 ZNB A L S Harris, New Skeeby Grange, Skeeby, Richmond, North Yorks, DL10 5ED [IO94DK, NZ20]

G8 ZNE T L Tucker, 20 Hainault Rd, Leytonstone, London, E11 1EE [IO91XN, TQ38]
G8 ZNK G M Barnes, 80 Boxgrove, Goring By Sea, Worthing, BN12 6AR [IO90ST, TQ10]
G8 ZNL Dr M D Speight, Newlands, Rainton, Thirsk, North Yorks, YO7 3PX [IO94GE, SE37]
G8 ZNU V A Arnold, 143 Orchard Rd, Burgess Hill, RH15 9PJ [IO90WW, TQ31]
G8 ZNW Details withheld at licensee's request by SSL.
G8 ZOB J T Neary, 1 Ashling Ct, Tyldesley, Manchester, M29 8QS [IO83SM, SD70]
G8 ZOE S J Trott, 8 St. Brelades Gr, St. Annes, Bristol, BS4 4QJ [IO81RK, ST67]
G8 ZOH Details withheld at licensee's request by SSL.
G8 ZOJ G P Barrett, 27 Rogers Rd, London, SW17 0EB [IO91VK, TQ27]
G8 ZON K Maddocks, 70 Kings Rd, Southsea, PO5 4DN [IO90KS, SZ69]
G8 ZOP J Martin, 3 Chamberlain Ave, Maidstone, ME16 8NR [JO01FG, TQ75]
G8 ZOU A W Nicholson, 31 Charles Bennett Ct, Reed Pond Walk, Haywards Heath, RH16 3SS [IO90XX, TQ32]
G8 ZOV Dr R A Nicholson, 24 Barnmead, Haywards Heath, RH16 1UZ [IO91WA, TQ32]
G8 ZOW P J Oram, 2 Newbolt Cl, Paulerspury, Towcester, NN12 7NH [IO92MC, SP74]
G8 ZOY G P Page, 1 Montagu Gdns, Wallington, SM6 8EP [IO91WI, TQ26]
G8 ZPE P R Cooper, 2 Upper Steeping, Desborough, Kettering, NN14 2SQ [IO92OK, SP88]
G8 ZPF S G Cook, Far Green Cottage, Far Green, Coaley, Dursley, GL11 5EL [IO81UQ, SO70]
G8 ZPK R J Birtles, 49 Greenfield Ave, Shavington, Crewe, CW2 5HE [IO83SB, SJ65]
G8 ZPL Details withheld at licensee's request by SSL.
G8 ZPO R E Blackwell, 46 Wyatts Dr, Thorpe Bay, Southend on Sea, SS1 3DG [JO01IM, TQ98]
G8 ZPU S Penny, 23 Humphrey Burtons, Rd, Coventry, CV3 6HW [IO92FJ, SP37]
G8 ZPW A W T Martin, 23 Portfield Rd, Christchurch, BH23 2AF [IO90CR, SZ19]
G8 ZQA P Stonebridge, Bridge House, 207 Henley Rd, Ipswich, IP1 6RL [JO02NB, TM14]
G8 ZQB J A Smith, 7 Mill Hill Cl, Whetstone, Leicester, LE8 6NF [IO92JN, SP59]
G8 ZQC T J Walsh, The Hobbits, South Molton St., Chulmleigh, EX18 7BW [IO80BV, SS61]
G8 ZQG S R Wood, 152 Letchworth Rd, Leicester, LE3 6FH [IO92JP, SK50]
G8 ZQJ D J Young, 9 Larchfield House, Highbury Est, Highbury, London, N5 2DE [IO91WN, TQ38]
G8 ZQK C A Marshall, 157 High St., Northchurch, Berkhamsted, HP4 3QT [IO91QS, SP90]
G8 ZQM K E Pascoe, 21 Cotswold Ave, Sticker, St. Austell, PL26 7ER [IO70NH, SW95]
G8 ZQO S P Potter, 23 Covent Garden, Cambridge, CB1 2HS [JO02BE, TL45]
G8 ZQR Details withheld at licensee's request by SSL.
GM8 ZQY S L Frey, 2 Balgeddie Gdns, Glenrothes, KY6 3QR [IO86JF, NO20]
G8 ZRD I R Gilzean, 13 Jubilee Cl, Waterbeach, Cambridge, CB5 9NY [JO02CG, TL46]
G8 ZRE D C W Hewitt, 31 Broadmead, Vicars Cross, Chester, CH3 5PT [IO83NE, SJ46]
G8 ZRG B J Hawes, 129 Wycombe Ln, Woodburn Green, Wooburn Green, High Wycombe, HP10 0HJ [IO91PO, SU98]
G8 ZRL Details withheld at licensee's request by SSL.
G8 ZRM R F Myers, 9 Romney Rd, Rottingdean, Brighton, BN2 7GG [IO90XT, TQ30]
G8 ZRN G W John, 29 Park Rd, Northville, Bristol, BS7 0RH [IO81RM, ST67]
G8 ZRQ R P Knight, 9 Crispin Rd, Strood, Rochester, ME2 3TW [JO01FJ, TQ76]
G8 ZRU D J Moger, 82 Waller Rd, Banbury, OX16 9NR [IO92HB, SP44]
G8 ZRV G C Sargant, 9 Orchard Way, Reigate, RH2 8DS [IO91VF, TQ24]
G8 ZSD I M Worthington, 7 Bowness Cl, Gamston, Nottingham, NG2 6PE [IO92KW, SK63]
G8 ZSH Details withheld at licensee's request by SSL.
G8 ZSI Details withheld at licensee's request by SSL.
G8 ZSK A J Allcock, 30 Clyde Gr, Crewe, CW2 8NA [IO83SC, SJ65]
G8 ZSM L Barlow, 30 North Dr, Thornton Cleveleys, FY5 3AQ [IO83LU, SD34]
G8 ZSO R J Boorman, 1 Hill Barton Farm, Sidmouth Rd, Clyst St. Mary, Exeter, Devon, EX5 1DR [IO80HQ, SY09]
G8 ZSP A M Blanchard, 41 Deane Dr, Galmington, Taunton, TA1 5PQ [IO81KA, ST22]
G8 ZSZ I Dickinson, 16 Heathfield Gr, Beeston, Nottingham, NG9 5EB [IO92JV, SK53]
G8 ZTB S L Fenn, 21 Waarem Ave, Canvey Island, SS8 9DS [JO01HM, TQ78]
G8 ZTD J M Francis, 9 Holland Cl, Bognor Regis, PO21 5TW [IO90PT, SU90]
G8 ZTF J W Hargraves, 321 Northway, Maghull, Liverpool, L31 0BW [IO83MM, SD30]
G8 ZTG J Harman, 20 Sunview Ave, Peacehaven, BN10 8PJ [JO00AS, TQ40]
G8 ZTM N E J Ledeux, 14 Jubilee Cl, Cam, Dursley, GL11 5JQ [IO81TQ, SO70]
G8 ZTO K Maw, Yews Mill Cottage, The Yews, Firbeck, Worksop, S81 8JW [IO93KJ, SK58]
GW8 ZTP B W Male, 58 The Avenue, Mountain Ash, CF45 4DU [IO81HQ, SO00]
G8 ZTQ Details withheld at licensee's request by SSL.
G8 ZTR J A Macdonald, 74 Bradford Rd, Boston, PE21 8BJ [IO92XX, TF34]
G8 ZTS S E Mack, 3 Blacksmiths Ln, Dovercourt, Harwich, CO12 4HY [JO01PW, TM23]
G8 ZTT P P Fox Mid Cheshire AR, 5 Llandovery Cl, Winsford, CW7 1NA [IO83RE, SJ66]
GM8 ZTV F R Millar, 13 Edzell Park, Kirkcaldy, KY2 6YB [IO86JD, NT29]
G8 ZTY K G Orlowski, 1 Fortune Cottage, Park Rd, Rickmansworth, WD3 1HT [IO91SP, TQ09]
G8 ZUD Details withheld at licensee's request by SSL.[Op: J D Rose. Station located near Carshalton.]
G8 ZUF K W Rogers, 36 Goodacre Rd, Ullesthorpe, Lutterworth, LE17 5DL [IO92JL, SP58]
G8 ZUI G C Shaw, 8 Nightingale Pl, Buckingham, MK18 1UF [IO92MA, SP73]
G8 ZUW P Seddon, 41 Radbourne Rd, Shirley, Solihull, B90 3RS [IO92CJ, SP17]
G8 ZUZ D Unwin, 57 Sutton Rd, Kirkby in Ashfield, Nottingham, NG17 8GY [IO93IC, SK45]
G8 ZVC M A Tungate, Estar Cottage, Gorsethorn Way, Fairlight, Hastings, E Sussex, TN35 4BQ [JO00HU, TQ81]
G8 ZVE Details withheld at licensee's request by SSL.
G8 ZVI L M Hart, 28A Dunton Rd, Stewkley, Leighton Buzzard, LU7 0HZ [IO91OW, SP82]
G8 ZVK B E Ackroyd, Flat 3, 104 Mantle St., Wellington, TA21 8BD [IO80JX, ST12]
G8 ZVM M R Atkinson, Menamber Farm, Releath, Helston, TR13 0HE [IO70JD, SW63]
G8 ZVS R M Bird, 80 Clearmount Rd, Rodwell, Weymouth, DT4 9LE [IO80SO, SY67]
G8 ZVV S N Brookes, 29 Nithsdale Rd, Weston Super Mare, BS23 4JP [IO81MH, ST35]
G8 ZVX A P Breeds, The Grange, 26 Heighton Rd, South Heighton, Newhaven, BN9 0JU [JO00AT, TQ40]
G8 ZWA P J Collins, 40 Shacklegate Ln, Teddington, TW11 8SH [IO91TK, TQ17]
G8 ZWC L J Curtis, 34 Gaisford Rd, Worthing, BN14 7HW [IO90TT, TQ10]
GW8 ZWD D C Carnell, 11 Carlyon Rd, Newbridge, Newport, NP1 5DH [IO81KQ, ST29]
G8 ZWE D Casey, 26 Riders Way, Chinnor, OX9 4TT [IO91NQ, SP70]
G8 ZWF R W Cowling, 3 Braidway Ct, Upper Battlefield, Shrewsbury, SY4 4AB [IO82PS, SJ51]
G8 ZWH E J Cooper, 2 Moorgate Dr, Astley, Tyldesley, Manchester, M29 7DG [IO83SM, SD70]
G8 ZWL O R Dale, Stone House, Heath House Ln, Horton, Leek, ST13 8RX [IO83XC, SJ95]
G8 ZWQ Details withheld at licensee's request by SSL.
G8 ZWS Details withheld at licensee's request by SSL.
G8 ZWU K P Graham, 670 Stafford Rd, Fordhouses, Wolverhampton, WV10 6NW [IO82WO, SJ90]
G8 ZXA Details withheld at licensee's request by SSL.
G8 ZXG Details withheld at licensee's request by SSL.
G8 ZXI Details withheld at licensee's request by SSL.
G8 ZXJ Details withheld at licensee's request by SSL.
G8 ZXL Details withheld at licensee's request by SSL.
G8 ZXN Details withheld at licensee's request by SSL.
G8 ZXS Details withheld at licensee's request by SSL.
G8 ZXT J E Marshall, 58 Sandbed Ct, Manston, Leeds, LS15 8JJ [IO93GT, SE33]
G8 ZXU P McGuinness, Poultry Farm, Spout Ln, Little Wenlock, Telford, TF6 5BL [IO82RP, SJ60]
G8 ZXY W D Mason, 365 Heath Rd South, Birmingham, B31 2BJ [IO92AJ, SP07]
G8 ZYC M I Sneap Zycomm Elect Lt, 51 Nottingham Rd, Ripley, DE5 3AS [IO93HB, SK45]
G8 ZYH E G Hitch, 35 Hawthorndene Rd, Hayes, Bromley, BR2 7DY [JO01AI, TQ46]
G8 ZYI N R Hitch, Aspen Lodge, Greenlands, St Mary's Platt, Sevenoaks, Kent, TN15 8LL [JO01DG, TQ65]
G8 ZYM I P J Hammond, 1 Old Rectory Cl, Barham, Ipswich, IP6 0PY [JO02NC, TM15]
G8 ZYR P C Hodgkinson, 25 Polisken Way, Trevisphan Park, St. Erme, Truro, TR4 9RB [IO70LH, SW85]
G8 ZYT S R T Higlett, 65 Nursery Rd, Hoddesdon, EN11 9LD [IO91XS, TL31]
G8 ZYZ B J Joyner, Brimar, Nelson Park Rd, St. Margarets At Cliffe, Dover, CT15 6HL [JO01QD, TR34]
GW8 ZZD J K Kenchington, 36 Lando Rd, Pembrey, Burry Port, SA16 0UR [IO71UQ, SN40]
G8 ZZE Details withheld at licensee's request by SSL.
G8 ZZG T J Lock, 40 Chertsey Rd, Ashford Common, Ashford, TW15 1SQ [IO91SK, TQ07]
G8 ZZM J H Hewitt, 71 Cres Rd, Bolton, BL3 2JP [IO83TN, SD70]
G8 ZZR P C Vince, Two Hoots, 2 Little Park Dr, Hanworth, Feltham, TW13 5HZ [IO91TK, TQ17]
G8 ZZS D G Vaughan, 12 Old Lyndale, 3 Hillside Rd, St Albans, Herts, AL1 3QZ [IO91US, TL10]
G8 ZZT J T Tonks, Flat, 3 Greystone Passage, Dudley, DY1 1SL [IO82WM, SO98]
G8 ZZV A P Tye, 3 Parkwood Ct, Forest Park, Bulwell, Nottingham, NG6 9FB [IO92JX, SK54]
G8 ZZY A W Smart, 101 Bardon Rd, Coalville, LE67 4BF [IO92HR, SK41]

M0

M0 AAA P Sayer Reading Novices Amateur Aerial, 90 Pitcroft Ave, Earley, Reading, RG6 1NN [IO91MK, SU77]
M0 AAB M Austin, 33 Totternhoe Rd, Dunstable, LU6 2AF [IO91RV, TL02]
M0 AAC P Bergin, 15 Monks Way, Harmondsworth Ln, Harmondsworth, West Drayton, UB7 0LE [IO91SL, TQ07]
M0 AAD M Stockton, 37 Ney St., Ashton under Lyne, OL7 9NL [IO83WM, SD90]
M0 AAE N M Carr Lancashire Six Metre Club, 15 Westlands, Leyland, Preston, PR5 3XT [IO83PQ, SD52]
M0 AAF D M Hodgson, 48 The Greenway, The Wells, Epsom, KT18 7HZ [IO91UH, TQ16]
M0 AAG Details withheld at licensee's request by SSL.
MI0 AAH R J Stinson Gi Contest Group, 51 Cloncarrish Rd, Birches, Portadown, Craigavon, BT62 1RN [IO64RL, H95]
M0 AAI Details withheld at licensee's request by SSL.
M0 AAJ T B Mansfield, 18 Station Rd, Histon, Cambridge, CB4 4LQ [JO02BF, TL46]
M0 AAK M J Pearson, 56 Parkwood Green, Parkwood, Gillingham, ME8 9PP [JO01HI, TQ86]
M0 AAL C Caito, Knowle Farm House, Church Knowle, Wareham, Dorset, BH20 5NG [IO80XP, SY98]
M0 AAM R Armstrong, 16 Ford, Queensbury, Bradford, BD13 2BH [IO93BS, SE02]
M0 AAN W D Glover, 21 West End Way, Lancing, BN15 8RL [IO90UT, TQ10]
M0 AAO Details withheld at licensee's request by SSL.
MW0 AAP I G Parker, 104 Campbell Dr, Windsor Quay, Cardiff, CF1 7TQ [IO81JL, ST17]
M0 AAQ G L Neilson, 104A Edleston Rd, Crewe, CW2 7HD [IO83SC, SJ75]
M0 AAR J E Kemp, 394 Great Thornton St., Hull, HU3 2LT [IO93TR, TA02]
M0 AAS J Whittaker, 9 Renown Cl, Birchwood, Warrington, WA3 6NG [IO83RK, SJ69]
M0 AAT Details withheld at licensee's request by SSL.
M0 AAU C M Davies, 27 Foxley Gr, Bicton Heath, Shrewsbury, SY3 5DF [IO82OR, SJ41]
M0 AAV V M Morris The Friday Nighters, 21 Cranhill Rd, St, BA16 0BY [IO81PC, ST43]
MI0 AAW BEM S J Blakley, 123 Mount Merrion Ave, Belfast, BT6 0FN [IO74BN, J37]
M0 AAX Details withheld at licensee's request by SSL.
M0 AAY Details withheld at licensee's request by SSL.
MI0 AAZ J M Anderson, 1 Loguestown Ct, Coleraine, BT52 2HS [IO65QD, C83]
M0 ABA T W Hackett, 81 Whitmore Ct, Whitmore Way, Basildon, SS14 2TN [JO01FN, TQ78]
M0 ABC A Holdsworth Mimram Contest Group, 26 Chelveston, Welwyn Garden City, AL7 2PW [IO91WT, TL21]
MI0 ABD J Mc Carrison, 11 Boretree Island Park, Newtownards, BT23 7BW [IO74DN, J46]
M0 ABE S Blood, 14 Duchy Cl, Chelveston, Wellingborough, NN9 6AW [IO92RH, SP96]
M0 ABF K C Molyneux, 162 Curbar Rd, Great Barr, Birmingham, B42 2AX [IO92BM, SP09]
M0 ABG L Borde, Kernanderry, Faringdon Rd, Frifford Heath, Abingdon, Oxon, OX13 6QJ [IO91HQ, SU49]
M0 ABH Details withheld at licensee's request by SSL.
M0 ABI M Lennon, 15 Worcester Gdns, Greenford, UB6 0BH [IO91TN, TQ18]
MM0 ABJ C I Ewart, 13 Princes St., Innerleithen, EH44 6JT [IO85LO, NT33]
M0 ABK M Gray, 19 Marsh View, Newton, Preston, PR4 3SX [IO83NS, SD43]
MM0 ABL J F Clarkson, 21 Snowdon Terr, Seamill, West Kilbride, KA23 9HN [IO75NQ, NS24]
MM0 ABM A B McKeeman, Cottage No2, Hauplands Farm, Ardrossan, Ayrshire, KA22 8PL [IO75OQ, NS24]
MI0 ABN N J Crawford, 10 White Mountain Rd, Lisburn, BT28 3QY [IO64WN, J26]
M0 ABO C Knowles, 24 Northern Gr, Bolton, BL1 4JZ [IO83SO, SD71]
M0 ABP J Barker, 6 Acredykes, Bempton, Bridlington, YO15 1LY [IO94VD, TA17]
MM0 ABQ W R Couse, The Old Station Hse, Wigtown, Wigtownshire, DG8 9ED [IO74SU, NX45]
M0 ABS Details withheld at licensee's request by SSL.
M0 ABT Details withheld at licensee's request by SSL.
M0 ABU D J York Dowdales School Radio Club, 47 Kirkstone Cres, Barrow in Furness, LA14 4ND [IO84JD, SD27]
MW0 ABV Details withheld at licensee's request by SSL.
M0 ABW W Johnson, 74 High Meadows, Romiley, Stockport, SK6 4QE [IO83WK, SJ99]
M0 ABX E A Carter, 7 Kirkstall Cl, Willindon Trees, Eastbourne, BN22 0UQ [JO00DT, TQ50]
M0 ABY A M Soane, 24 Nurseries Rd, Wheathampstead, St. Albans, AL4 8TP [IO91UT, TL11]
M0 ABZ E Martin, 20 Easters Gr, Stoke on Trent, ST2 7PF [IO83WB, SJ95]
M0 ACA E H Morley, 36 Hazeldene Rd, Stoke on Trent, ST4 8DN [IO82VX, SJ84]
M0 ACB E A McDonald, 32 Butterwick Rd, Messingham, Scunthorpe, DN17 3PB [IO93QM, SE80]
M0 ACD P A Hillman, 119 Porter Rd, Basingstoke, RG22 4JT [IO91KF, SU65]
M0 ACE Details withheld at licensee's request by SSL.
M0 ACF H L Philps, 29 Foresters Park Rd, Melksham, SN12 7RW [IO81WI, ST96]
M0 ACG R D Briggs A1 Contest Group, 11 PO Box, Rugeley, WS15 4YR
M0 ACI Details withheld at licensee's request by SSL.
M0 ACJ C Kenyon, Anchorage, St Martins Green, Helston, Cornwall, TR12 6BU [IO70KB, SW72]
M0 ACK M K Jackson, 121 Kiln Ln, Eccleston, St. Helens, WA10 4RH [IO83OL, SJ49]
M0 ACL E L L Jones, 47 Pine Cres, Chandlers Ford, Eastleigh, SO53 1LN [IO90HX, SU42]
M0 ACM D C Forrest, 19 Burns Way, Swindon, SN2 6LP [IO91CO, SU18]
M0 ACN K M Kidd, 9 Bulstrode Pl, Kegworth, Derby, DE74 2DS [IO92IU, SK42]
M0 ACO A Kidd, 9 Bulstrode Pl, Kegworth, Derby, DE74 2DS [IO92IU, SK42]
M0 ACP E R Fish, 2 Chudleigh Rd, Harrogate, HG1 5NP [IO93FX, SE35]
M0 ACQ W D Arnold, Fir Cottage, 9 Mornington Rd, Whitehill, Bordon, Hants, GU35 9EN [IO91NC, SU73]
MM0 ACR L Skinner, 1A Wingate Rd, Folkestone, CT19 5QE [JO01OC, TR23]
MM0 ACS M Lovatt 5th West Lothian Scout Group, 21 Crathes Gdns, Livingston, EH54 9EN [IO85FU, NT06]
MM0 ACT R Skinner, 1A Wingate Rd, Folkestone, CT19 5QE [JO01OC, TR23]
M0 ACU M C Eddyvean, 41 Liddell Rd, Cowley, Oxford, OX4 3QU [IO91JR, SP50]
M0 ACV T O Bevan, 6 Buttermere Gr, West Auckland, Bishop Auckland, DL14 9LG [IO94DP, NZ12]
M0 ACW R J S Williams Over The Hill DX Group, Dyffryn Coed, 25 Peghouse Rise, Stroud, GL5 1RU [IO81VS, SO80]
M0 ACY A R D Michaelides, 17 Craven Lodge, Craven Hill, London, W2 3ER [IO91VM, TQ28]
M0 ACZ D M Cox, 130 Maidstone Rd, Paddock Wood, Tonbridge, TN12 6DY [JO01EE, TQ64]
M0 ADA C A Rule, The Workshop, Meaver Rd, Mullion, Helston, TR12 7DN [IO70JA, SW61]
M0 ADB N Pringle, 21 Petersmiths Dr, New Ollerton, Newark, NG22 9RZ [IO93LF, SK66]
M0 ADC Details withheld at licensee's request by SSL.
MD0 ADD V G Wilson, 129 Ballabrooie Dr, Douglas, IM1 4HH
M0 ADE Details withheld at licensee's request by SSL.
M0 ADF B G Ibison, 21 Douglas House, Scholes, Wigan, WN1 1YE [IO83QN, SD50]
M0 ADG D Morris, 86 Richardson St., Carlisle, CA2 6AG [IO84MV, NY35]
M0 ADH M L Mettam, 1 Inglewood Ct, Sothall, Sheffield, S19 6PA [IO93GJ, SK38]
M0 ADJ W A F Davidson The Squarebashers Expedition G, 23 PO Box, Tewkesbury, GL20 5RN [IO81WX, SO83]
M0 ADK B C A Simister, 3 Beech Tree Rd, Featherstone, Pontefract, WF7 5EB [IO93HQ, SE41]
M0 ADL A Jermaks, 52 Laburnum Cres, Allestree, Derby, DE22 2GR [IO92GW, SK33]
M0 ADM D S Powis, Farlingaye High School, Ransom Rd, Woodbridge, Suffolk, IP12 1DJ [JO02PC, TM24]
M0 ADN H B G Epps, Pikes Cottage, Stowhill, Childrey, Wantage, Oxon, OX12 9XQ [IO91GO, SU38]
M0 ADO D A J Kennard, 16 Cantilupe Cres, Aston, Sheffield, S31 0AT [IO93GJ, SK38]
M0 ADP Details withheld at licensee's request by SSL.
M0 ADQ Details withheld at licensee's request by SSL.
M0 ADR G S Galbraith, 24 Airedale, Hadrian Lodge West, Wallsend, NE28 8TL [IO95FA, NZ26]
M0 ADS Details withheld at licensee's request by SSL.
M0 ADT G R C Herbert 485 (Harbourne & Quinton) Sqn, 421 Redditch Rd, Kings Norton, Birmingham, B38 8ND [IO92AJ, SP07]
M0 ADU S P Hodgson Exiles Contest Group, Mill Ln Farmhouse, Mill Ln, Irby in The Marsh, Skegness, PE24 5BB [JO03CD, TF46]
M0 ADW R N Latham, 47 Oldfield Park, Westbury, BA13 3LQ [IO81VG, ST85]
M0 ADY A M Grundy, 16 Mill Rise, Westdene, Brighton, BN1 5GD [IO90WU, TQ20]
M0 ADZ A J Gall, 116 Admiralty Rd, Great Yarmouth, NR30 3DG [JO02UO, TG50]
M0 AEA Capt A Madeira, 28 Lacey Ct, Wilmslow, SK9 4BH [IO83VI, SJ88]
M0 AEB E A Borley, 18 Greenside Walk, The Dales, Nottingham, NG3 7HJ [IO92KX, SK64]
MI0 AEC Dr S N Roper, 73 St. Johns Park, Belfast, BT7 3JG [IO74BN, J37]
M0 AED D P Linden, 11 Beechfield Ave, Harrowbeer Ln, Yelverton, PL20 6DU [IO70WL, SX56]
M0 AEE M J Clarke, Okanagan, 20 Eden Cl, Barugh Green, Barnsley, S75 1RA [IO93FN, SE30]
M0 AEF Details withheld at licensee's request by SSL.
M0 AEG Details withheld at licensee's request by SSL.
MI0 AEH Details withheld at licensee's request by SSL.
M0 AEI J C Burns, 13 Balmoral Ave, St. Helens, Merseyside, WA9 3TU [IO83PK, SJ59]
M0 AEJ V Trend, 64 Shutlock Ln, Moseley, Birmingham, B13 8NZ [IO92BK, SP08]
M0 AEK J C Sloan, 120 Pear Tree Ln, Bexhill on Sea, TN39 4NR [JO00FU, TQ70]
MW0 AEL S W Townsend, 42 Burns Cres, Bridgend, CF31 4PY [IO81EM, SS88]
M0 AEM M C Howden, 2 Glebelands Rd, Knutsford, Ches, WA16 9DZ [IO83TH, SJ77]
M0 AEN M D Austen, 11 Corn Avill Cl, Abingdon, OX14 2ND [IO91JQ, SU59]
M0 AEO E G Oliver, 47 Bailey Rd, Cowley, Oxford, OX4 3HU [IO91JR, SP50]
M0 AEP G S Dawes, 50 Highfield Rd, North Thoresby, Grimsby, DN36 5RT [IO93XL, TF29]
M0 AEQ M A Bardell, 15 Stantonbury Cl, New Bradwell, Milton Keynes, MK13 0EY [IO92OB, SP84]
MW0 AER T P Anziani Ynys Mon Amateur Radio Users G, Ty-Coch, Penrhyd, Amlwch, Anglesey, Gwynedd, LL68 1AA [IO73TJ, SH49]
MW0 AES S Hill, 1A Rugby Ave, Neath, SA11 1YT [IO81CP, SS79]
M0 AET K E Jones, Ferny Hoolet, Pale Ln, Winchfield, Basingstoke, Hants, RG27 8SW [IO91NG, SU75]
M0 AEU F J Heritage, 50 Laurel Cl, North Warnborough, Hook, RG29 1BH [IO91MG, SU75]
MW0 AEV E L L Jones, 18 Madryn Terr, Llanbedrog, Pwllheli, LL53 7PF [IO72SU, SH33]
MI0 AEX J D Smith, 144 Bangor Rd, Holywood, BT18 0EY [IO74CP, J48]
M0 AEZ Details withheld at licensee's request by SSL.
M0 AFA D Clifton, 12 School Field, Kingsley, Bordon, Hants, GU35 9ND [IO91ND, SU73]
M0 AFB P B Stevens, 11 Nutwood Ave, Brockham, Betchworth, RH3 7LT [IO91UF, TQ24]
M0 AFC T Boon, 27 Meadowside Ave, Clayton-le-Moors, Accrington, BB5 5XF [IO83TS, SD73]
MW0 AFD S Edwards, 59 St. Andrews Rd, Colwyn Bay, LL29 6DL [IO83DG, SH87]
M0 AFE A C Mould, 95 Stanton Rd, Southampton, SO15 4HU [IO90GW, SU31]
M0 AFF F W Hallsworth, 11 Tiree Cl, Hazel Gr, Stockport, SK7 6AY [IO83WI, SJ98]
M0 AFG A B Kneller, 34 Redbridge Gr, Havant, PO9 3DE [IO90MU, SU70]
M0 AFH L G Walden, Serendipity Kennels, Brightstone Ln, Farringdon, Alton, GU34 3EU [IO91LC, SU63]
M0 AFJ T Hague, 27 Bridge Rd, Cosgrove, Milton Keynes, MK19 7JH [IO92NB, SP74]
M0 AFK G A Patten, 20 Laslett St., Worcester, WR3 8JS [IO82VE, SO85]
M0 AFL Details withheld at licensee's request by SSL.
M0 AFM D Corden, Whisper Corner, Filance Ln, Penkridge, Stafford, ST19 5HQ [IO82WR, SJ91]
M0 AFN H W Williams, Laurels, 9 Edward Rd, Biggin Hill, Westerham, TN16 3HN [JO01AH, TQ45]
M0 AFO P T Butler, 45 Hall Rd, Manor Est, Alton, GU34 2NX [IO91MD, SU74]
M0 AFP Details withheld at licensee's request by SSL.

M0 AFQ B K Eagleton, 3 Coldridge Cl, Pendeford, Wolverhampton, WV8 1XZ [IO82WO, SJ80]
M0 AFR P M K Walker, 3 St. Johns Terr, Lewes, BN7 2DL [JO00AU, TQ41]
M0 AFS P Whiteley, 53 Sharp Ln, Almondsbury, Huddersfield, HD4 6SS [IO93DP, SE11]
MI0 AFT J Stewart, 60 Liberty Rd, Carrickfergus, BT38 9DJ [IO74BR, J39]
M0 AFU B J Moore, 17 Underley St., Burnley, BB10 2BX [IO83VT, SD83]
M0 AFV R G Rippin, Gaverne, Welford Rd, Long Marston, Stratford upon Avon, CV37 8RA [IO92CD, SP14]
M0 AFW C Parkinson, 4 Campion Dr, Killamarsh, Sheffield, S31 8TG [IO93GJ, SK38]
M0 AFX D P Waters, Meadowcroft, Steam Mill Rd, Bradfield, Manningtree, CO11 2QY [JO01NW, TM12]
M0 AFY R Ford, 70 Jubilee Rd, Darnall, Sheffield, S9 5EH [IO93HJ, SK38]
M0 AFZ P T D Nairne, 137 Barden Rd, Tonbridge, TN9 1UX [JO01DE, TQ54]
M0 AGA K L Gunstone, 8 Edward Ave, Sutton in Ashfield, NG17 4BW [IO93ID, SK45]
M0 AGB R W Levey, 516 Ashingdon Rd, Ashingdon, Rochford, SS4 3HZ [JO01IO, TQ89]
M0 AGC Details withheld at licensee's request by SSL.
MW0 AGE J A C Chinnock, 22 Mill Rd, Pyle, Bridgend, CF33 6AP [IO81DM, SS88]
M0 AGF G J Smith, Kane Farm, Boston Rd, Heckington, Sleaford, NG34 9JQ [IO92UX, TF14]
M0 AGG Details withheld at licensee's request by SSL.
M0 AGH R A Abbott, 10 Greendale Dr, Middlewich, CW10 0PH [IO83SE, SJ66]
M0 AGI G C S Ingram, 7 Cheriswood Ave, River View, Exmouth, EX8 4HG [IO80HP, SY08]
M0 AGJ A Bowker, 196 Springwell Ln, Doncaster, DN4 9AY [IO93KL, SE50]
M0 AGK Details withheld at licensee's request by SSL.
M0 AGL J Monks, 2 Low Hutton Park, Huttons Ambo, York, YO6 7HH [IO94NC, SE76]
M0 AGM P A Allen, 170 Claremount Rd, Claremount, Halifax, HX3 6JP [IO93BR, SE02]
M0 AGN G Stewart, 2 Clifton House, Crowedge, Stocksbridge, Sheffield, S36 4HF [IO93DM, SE10]
M0 AGO H Wray, 22 Askew Dale, Guisborough, TS14 8JG [IO94LM, NZ51]
M0 AGP Details withheld at licensee's request by SSL.
M0 AGQ M A Shepherd, 17 Friar Cres, Brighton, BN1 6NL [IO90WU, TQ30]
M0 AGR M R Bray, 2 Cambourne Dr, Fixby, Huddersfield, HD2 2NF [IO93CQ, SE11]
M0 AGS E Smeaton, 27 Sandringham Ave, Burton on Trent, DE15 9BJ [IO92ET, SK22]
M0 AGT R Markham, 11 Queens Cres, Stoke Sub Hamdon, TA14 6QX [IO80PW, ST41]
M0 AGU J Shorthouse, 84 Mount Pleasant, Ackworth, Pontefract, WF7 7HU [IO93HP, SE41]
M0 AGV T McGuigan, 14 Ash Gr, Heald Green, Cheadle, SK8 3JA [IO83VI, SJ88]
M0 AGW W Mason, 104 Chester Rd, Poynton, Stockport, SK12 1HG [IO83WI, SJ98]
M0 AGX K M Harvey, 77 Innes Gdns, London, SW15 3AD [IO91VK, TQ27]
M0 AGY M Griffin, 15 Bank St., St. Columb, TR9 6AT [IO70MK, SW96]
M0 AGZ A C Roberts, 19 Essex Ave, Slough, SL2 1DP [IO91QM, SU98]
M0 AHA A Hodgson, 1 Grand Ave, Seaford, BN25 2QY [JO00BS, TQ40]
M0 AHB D J Pisani, 19 Tudor Rd, Broadstone, BH18 8AP [IO90AS, SZ09]
M0 AHC M C Collins, 17 Puttenham Rd, Chineham, Basingstoke, RG24 8RB [IO91LH, SU65]
M0 AHD M Carter, 5 Broadwell Rd, Middlesbrough, TS4 3PP [IO94JN, NZ51]
M0 AHF G May, 14 Tennyson Ave, Dukinfield, SK16 5DP [IO83XL, SJ99]
MW0 AHG P G Bennett, 13 Thornbury Cl, Baglan, Port Talbot, SA12 8EU [IO81CO, SS79]
MI0 AHH C Doris, 92 Coolnafranky Park, Cookstown, BT80 8PW [IO64PP, H87]
MI0 AHI J Doris, 92 Coolnafranky Park, Cookstown, BT80 8PW [IO64PP, H87]
M0 AHJ C L John, 5 Highfield Gdns, Aldershot, GU11 3DB [IO91OF, SU84]
M0 AHK P Hayes, 1C Leopold Rd, London, NW10 9LN [IO91VN, TQ28]
M0 AHL A J Hall, Crest Cottage, Chapel Ln, Grundisburgh, Woodbridge, Suffolk, IP13 6TS [JO02PC, TM25]
M0 AHM A F Eyres, 59 Charlesworth Dr, Waterlooville, PO7 6AZ [IO90LV, SU61]
M0 AHN Details withheld at licensee's request by SSL.
M0 AHO C A Turland, 2 Ludlow Cl, Beeston, Nottingham, NG9 3BY [IO92JW, SK53]
M0 AHP Details withheld at licensee's request by SSL.
M0 AHQ J Turner, 27 Woodlands, Long Sutton, Spalding, PE12 9LY [JO02BS, TF42]
M0 AHS M J Nicholas, 4 Chesterfield Mews, Chesterfield Rd, Ashford, TW15 3PF [IO91SK, TQ07]
M0 AHT W G Burt, 3 Edward St., Hetton-le-Hole, Houghton-le-Spring, DH5 9EL [IO94GT, NZ34]
M0 AHU H Challis, 5 Blair Park, Knaresborough, HG5 0TH [IO94GA, SE35]
M0 AHV H Banks, 104 Viking Rd, Bridlington, YO16 5TB [IO94VC, TA16]
M0 AHW E J Horner, 75 Barcombe Rd, Preston, Paignton, TQ3 1QB [IO80FK, SX86]
M0 AHX Details withheld at licensee's request by SSL.
M0 AHY Details withheld at licensee's request by SSL.
M0 AHZ R Brown, 17 Ridgeway, North Seaton, Ashington, NE63 9TJ [IO95FE, NZ28]
M0 AIA P Knapton, 1 School Rd, Scapegoat Hill, Golcar, Huddersfield, HD7 4NU [IO93BP, SE01]
M0 AIB S J P Budd, 4043 PO Box, Worthing, BN13 3YL [IO90SU, TQ10]
M0 AIC S Deary, 35 Milwain Dr, Heaton Chapel, Stockport, SK4 5AT [IO83VK, SJ89]
M0 AID K J Marsh, 40 Haines Park, Taunton, TA1 4RG [IO81KA, ST22]
MW0 AIE R D Duncombe, 10 Maidenwells, Pembroke, Dyfed, SA71 5ET [IO71MP, SR99]
M0 AIF J E Stone, 223 Burgess Rd, Southampton, SO17 1TU [IO90HW, SU41]
M0 AIG P K Gardner, 49 Rounton Rd, London, E3 4HA [IO91XM, TQ38]
MI0 AIH D G Martin, 28 Crossglebe, Cookstown, BT80 9DD [IO64OO, H77]
M0 AII Details withheld at licensee's request by SSL.
M0 AIJ C E Blake, 100 Penryn Ave, Fishermead, Milton Keynes, MK6 2BE [IO92PA, SP83]
MM0 AIK B G Devlin Scottish DX Contest Club, Borrodale, Main St., Thornhill, Stirling, FK8 3PW [IO76WE, NN60]
M0 AIL J W Wenglaryck, 16 Langford Dr, Luton, Beds, LU2 9AJ [IO91TV, TL12]
M0 AIM J R Linford Camberra Contest Club, Camberra Lodge, Heath Ride, Finchampstead, Wokingham, RG40 3QJ [IO91NI, SU86]
M0 AIN Y H Murad, PO Box 8525, Amman, 11121, Jordan
M0 AIO M A Faoury, PO Box 8525, Amman, 11121, Jordan
M0 AIP P G Richards, 189 Shakespeare Rd, St. Marks, Cheltenham, GL51 7HS [IO81WV, SO92]
M0 AIQ Details withheld at licensee's request by SSL.
M0 AIR S Meynell, Tawa, 9 Teachers Terr, Rake Rd, Liss, GU33 7ED [IO91NB, SU72]
M0 AIS A D Benns, 7 Brooklands Rd, Burnley, BB11 3PR [IO83VS, SD83]
M0 AIT R P Holt, 41 Garden Ave, Ilkeston, DE7 4DF [IO92IX, SK44]
M0 AIU G H J Geurts, 2 The Cottage, The Cedars, The Common, Stanmore, Middx, HA7 3HR [IO91UP, TQ19]
M0 AIV J E Freeze, 4 Alston Gdns, Maidenhead, SL6 6DY [IO91PM, SU88]
MM0 AIW B J Humphrey, 10 McDowall Dr, Stranraer, DG9 7NA [IO74LV, NX06]
M0 AIX L Hutchinson, 16 Willow Glen, Branton, Doncaster, DN3 3JD [IO93LM, SE60]
M0 AIY R C Carter, 16 Holts Ln, Clayton, Bradford, BD14 6BL [IO93CS, SE13]
M0 AIZ R E Ramm, 19 Riddell Cl, Alcester, B49 6QP [IO92BF, SP05]
M0 AJA J Hodgson, 1 Grand Ave, Seaford, BN25 2QY [JO00BS, TQ40]
M0 AJB A J Birch The North West 320 DX Club, 17 The Stakes, Castlemeadow, Moreton, Wirral, L46 3SW [IO83KJ, SJ29]
M0 AJC M J McInally, Flat 7, 32-33 Edgar Rd, Cliftonville, Margate, CT9 2EJ [JO01QJ, TR37]
M0 AJD M R Saxton, Cartref, Church Ln, High Toynton, Horncastle, LN9 6NN [IO93XF, TF26]
M0 AJE G K Howell, 2404 Kennon Ave, Norfolk, Virginia 23513-4317, USA
M0 AJF R J Wolfe, 56 High St., Langford, Biggleswade, SG18 9RU [IO92UB, TL14]
MI0 AJG I Davies, 46 Edenaveys Cres, Armagh, BT60 1NT [IO64QH, H94]
MW0 AJH J F Donnell, 42 Wentworth Cres, Mayals, Swansea, SA3 5HT [IO71XO, SS69]
M0 AJI S K G Nursey, Crown House, Burston, Diss, Norfolk, IP22 3TW [JO02NJ, TM18]
M0 AJJ P D Olson, 10 Euston St., Liverpool, L4 5PR [IO83MK, SJ39]
M0 AJL Details withheld at licensee's request by SSL.
M0 AJM A Martin, 75 Gosling Rd, Slough, SL3 7TN [IO91RM, TQ07]
M0 AJN Details withheld at licensee's request by SSL.
M0 AJO Details withheld at licensee's request by SSL.
M0 AJP P F Aston, 10 Browning Cl, Colchester, CO3 4JJ [JO01KV, TL92]
MM0 AJQ J J Stone, 39 Connaught Pl, Edinburgh, EH6 4RN [IO85JX, NT27]
M0 AJR M A Binns, 22 Rydings Dr, Brighouse, HD6 2DA [IO93CQ, SE12]
M0 AJS R Davey, 7 Birch St., Church Warsop, Mansfield, NG20 0TA [IO93KF, SK56]
M0 AJT C Towle, 116 Stainton Dr, Grimsby, DN33 1JB [IO93WN, TA20]
M0 AJV K P Darton St Swithuns Amatuer Radio Club, 8 Foster Gr, Sandy, SG19 1HP [IO92UD, TL14]
M0 AJW Details withheld at licensee's request by SSL.
M0 AJX G Jones, 7 Hardwick View, Skegby, Sutton in Ashfield, NG17 3BW [IO93JD, SK56]
M0 AJY F A Weidema, Middachtensingel 67, 6825 Hh Arnhem, Holland
M0 AJZ R J Dyer, 76 Sandy Ln, Farnborough, GU14 9HJ [IO91OH, SU85]
M0 AKA G R Powell, 203 Queenborough Rd, Minster on Sea, Sheerness, ME12 3EL [JO01JK, TQ97]
M0 AKB D Vickers, 69 Hallowes Ln, Dronfield, Sheffield, S18 6ST [IO93GH, SK37]
M0 AKC P A Temlund, Box 1368, RAF Menwith Hill, Harrogate, N Yorks, HG3 2RF [IO94DA, SE25]
M0 AKD Dr G P N Dublon, 25 Carr Ln, Sandal, Wakefield, WF2 6HJ [IO93GP, SE31]
M0 AKE R S Johnson, 24 Balmoral Ave, Stanford-le-Hope, SS17 7BD [JO01FM, TQ68]
M0 AKF M H Temblett, 42 Westward Rd, Bristol, BS13 8DB [IO81QK, ST56]
M0 AKG P Dingley, 95 Culm Lea, Cullompton, EX15 1NJ [IO80HU, ST00]
M0 AKH R N R March, 214 Kenyon St, Stratford, Ct 06497 2556, USA
M0 AKI E F Woollen, 6 Back Ln, Kington Magna, Gillingham, SP8 5EL [IO81TA, ST72]
M0 AKJ R H Hunt, 8 Spicer Cl, Cullompton, EX15 1QD [IO80HU, ST00]
M0 AKK S M Elden, 124 Larchcroft Rd, Ipswich, IP1 6PQ [JO02NB, TM14]
M0 AKL K Lewis, 39 Timberwood Dr, Hilton Ny 14668, USA
MM0 AKM J A Hood, 15 Silverknowes Cres, Edinburgh, EH4 5JE [IO85IX, NT27]
M0 AKN G Leach, 115 Churchill Ave, Lakeview, Northampton, NN3 6PF [IO92NG, SP76]
M0 AKO Details withheld at licensee's request by SSL.
M0 AKP Details withheld at licensee's request by SSL.
M0 AKQ R Gawan, 39 The Filberts, Fulwood, Preston, PR2 3YS [IO83PS, SD53]
M0 AKR K Daniels, 5 Hunts Ln, Grappenhall, Stockton Heath, Warrington, WA4 2DU [IO83RJ, SJ68]
M0 AKS R Lusty, 16 Kirk View, Rossendale, BB4 9UQ [IO83VQ, SD82]
M0 AKT S R Major, 26 North St., Pewsey, SN9 5EX [IO91CI, SU16]
MI0 AKU R J Kilgore The Foyle & District ARC, 29 Duncastle Park, New Buildings, Londonderry, BT47 2QL [IO64HW, C41]
M0 AKV B P McMahon, 14001 El Camino Real, Ocean Springs, Ms 39564, USA
MM0 AKW Details withheld at licensee's request by SSL.
MM0 AKX J C Ramsay ATC Scotland & N.Ireland Regio, 78 Wheatlands Ave, Bonnybridge, FK4 1PL [IO86BA, NS88]
M0 AKY T D Money, 119 Twyford Way, Canford Heath, Poole, BH17 8SR [IO90AR, SZ09]
M0 AKZ R L Taylor, 46 Cres Rd, Netherton, Dudley, DY2 0NW [IO82WL, SO98]
M0 ALA Details withheld at licensee's request by SSL.
M0 ALB N F Hixson, 24 Springfield Rd, Poole, BH14 0LQ [IO90AR, SZ09]
MM0 ALC A J Barth, 21 Jamieson Rd, Darvel, KA17 0BT [IO75UO, NS53]
M0 ALD J Britten, 64 Beechwood Mount, Leeds, LS4 2NQ [IO93FT, SE23]
M0 ALE P R Johnson, 91 Highlands Rd, Andover, SP10 2PZ [IO91GE, SU34]
M0 ALF R Faithfull, 5 Hadleigh Rd, Portsmouth, PO6 3RD [IO90LU, SU60]
MW0 ALG D A Burge, Ucheldir, Maen-y-Groes, New Quay, Ceredigion, Wales, SA45 9TH [IO72TE, SN35]
M0 ALH S A Case, 5 Haldon Gr, Birmingham, B31 4LN [IO92AJ, SP07]
MW0 ALI Details withheld at licensee's request by SSL.
MM0 ALJ Details withheld at licensee's request by SSL.
M0 ALK R Cook, 3 Mill Cl, Hartford, Huntingdon, PE18 7YL [IO92WI, TL27]
M0 ALL H R Gregory, Chevaliers, High Cross, Constantine, Falmouth, TR11 5RT [IO70KC, SW72]
MM0 ALM Details withheld at licensee's request by SSL.
MM0 ALN Details withheld at licensee's request by SSL.
M0 ALO D Hooper, 21 High St., Great Linford, Milton Keynes, MK14 5AX [IO92OB, SP84]
M0 ALP I E Blunt, Burleigh, 9 Lindrick Cl, Macclesfield, SK10 2UG [IO83WG, SJ97]
M0 ALQ G S Denby, 9 Courtney House, Mulberry Cl, London, NW4 1QN [IO91VO, TQ28]
M0 ALR T Knight, 6 The Close, Great Horwood, Milton Keynes, MK17 0QG [IO91NX, SP73]
MI0 ALS E F Stanford, 33 Glenview Gdns, Belfast, BT5 7LY [IO74BN, J37]
M0 ALT I A Halliwell, Claremont House, 61 Cliffe Rd, Shepley, Huddersfield, HD8 8AG [IO93DN, SE10]
MW0 ALU D H Green, 116 Crogen, Chirk, Wrexham, LL14 5BJ [IO82LW, SJ23]
M0 ALV I C Haddow, 152 Maldon Rd, Colchester, CO3 3AY [JO01KV, TL92]
M0 ALW M S Adams, 41 Lower Keyford, Frome, BA11 4AR [IO81UF, ST74]
M0 ALX Details withheld at licensee's request by SSL.[Station located in Wandsworth, London.]
MM0 ALY A F Brown, 1 Stewart Park Cour, Rosehill, Aberdeen, AB2 2GB [IO87TH, NJ72]
M0 ALZ T E Thompson, 78 Charlton Park, Midsomer Norton, Bath, BA3 4BW [IO81SG, ST65]
M0 AMA J S Greathead, 16 Thackley Old Rd, Windhill, Shipley, BD18 1DD [IO93CU, SE13]
M0 AMB B Metcalfe, 5 Oakdale Ave, Bradford, BD6 1RP [IO93CS, SE13]
M0 AMC R Vernon, 92 Saltmarsh, Orton Malborne, Peterborough, PE2 5NN [IO92UN, TL19]
M0 AMD Details withheld at licensee's request by SSL.
M0 AME D J F Draper, 81 Chambercombe Rd, Ilfracombe, EX34 9PQ [IO71WE, SS54]
M0 AMF R E Jefferies, 38 Towbury Cl, Redditch, B98 7YZ [IO92AG, SP06]
MW0 AMG D J Hindley, 87 Norton Rd, Penygroes, Llanelli, SA14 7RU [IO71XT, SN51]
M0 AMH R W Turner Bradford & Ilkley Comm College, 7 Highfield Cres, Baildon, Shipley, BD17 5NR [IO93CU, SE13]
MW0 AMI R L Hall, 33 Heol y Garreg Las, Llandeilo, SA19 6EB [IO81AV, SN62]
MW0 AMJ L S Carter, 13 Maesdolau, Idole, Carmarthen, Dyfed, SA32 8DQ [IO71UT, SN41]
M0 AMK G J Cusick, 2 Axis Rd, Watchfield, Swindon, SN6 8SQ [IO91EO, SU29]
MW0 AML G R Farmer, 14 Garw Fechan Rd, Pontyrhyl, Bridgend, CF32 8BX [IO81FO, SS98]
M0 AMM G Smith, 174 Whitby Way, Darlington, DL3 9UQ [IO94EM, NZ21]
MW0 AMN G R A H Thomas, Ffaldwen, Heol y March, Cardiff, CF5 6TS [IO81HL, ST07]
MW0 AMO P Roberts, 13 Ty Derwen, Nantyffyllon, Maesteg, CF34 0QA [IO81EO, SS89]
M0 AMP A S Davies, 27 Foxley Gr, Bicton Heath, Shrewsbury, SY3 5DF [IO82OR, SJ41]
MW0 AMQ G Thomas, Ffaldwen, Heol y March, Welsh Saint Donats, Cardiff, CF5 6TS [IO81HL, ST07]
M0 AMR H Dick, 32 Great Harlings, Shotley Gate, Ipswich, IP9 1NY [JO01PX, TM23]
M0 AMS M Burke, 236 Pooltown Rd, Ellesmere Port, South Wirral, L65 8HZ [IO83NG, SJ37]
MW0 AMT A D Lamb, 7 Bollam Cl, Deeside, CH5 4JH [IO83LF, SJ26]
M0 AMU J F Couvaras, 3 Ct Cl, Boydell Ct, St. Johns Wood Park, London, NW8 6NN [IO91VM, TQ28]
MM0 AMV R W Moodie, 20 Gosford Rd, Port Seton, Prestonpans, EH32 0HF [IO85MX, NT47]
MM0 AMW D B Gillies, 10 Killeonan, Campbeltown, PA28 6PL [IO75EJ, NR61]
M0 AMX J S Howell, 1 East Terr, Blennerhasset, Carlisle, CA5 3QY [IO84IS, NY14]
MM0 AMY I N T Gillespie, Flat 15, 20 Kensington Rd, Glasgow, G12 9NX [IO75UV, NS56]
M0 AMZ J G Williams, 61 Longfield Rd, South Woodham Ferrers, Chelmsford, CM3 5JJ [JO01HP, TQ89]
MM0 ANA W G Shand Turriff and District Radio Clu, 4 Sunnyhill Pl, Turriff, AB53 4EU [IO87SN, NJ75]
M0 ANB R B Palmer, 115 Francis Ave, Ilford, IG1 1TT [JO01BN, TQ48]
M0 ANC R Jones, 15 Derby Rd, Sandiacre, Nottingham, NG10 5HW [IO92IW, SK43]
M0 AND S D Jones, 15 Derby Rd, Sandiacre, Nottingham, NG10 5HW [IO92IW, SK43]
M0 ANE Details withheld at licensee's request by SSL.
M0 ANF Dr A M Nicolson, 21 Groves Cl, Bourne End, SL8 5JP [IO91PN, SU98]
M0 ANG D N Kirkham, 26 Britannia Gate, London, E16 1SB [JO01AM, TQ48]
M0 ANH J Waller, 56 Daventry Rd, Dunchurch, Rugby, CV22 6NS [IO92II, SP47]
M0 ANI V A Ford, Romney, 103 Old Rd East, Gravesend, DA12 1PB [JO01EK, TQ67]
M0 ANJ J N Smith, 21 Acacia Cres, Carlton, Nottingham, NG4 3JH [IO92KX, SK64]
M0 ANK Details withheld at licensee's request by SSL.
M0 ANL A King, 79 Spring Hill Rd, Accrington, BB5 0EX [IO83TR, SD72]
M0 ANM J Blakeley, 90 Park Rd, Great Sankey, Warrington, WA5 3ET [IO83QJ, SJ58]
M0 ANN G T Wardale, 25 The Cres, Huyton, Liverpool, L36 6ER [IO83OJ, SJ49]
M0 ANO R J Spencer, 2 New Houses Whiteway Farm, Whiteway, Cirencester, GL7 7BA [IO91AR, SP00]
M0 ANP N D Crooks, 1 Dent St., Colne, BB8 8JG [IO83VU, SD83]
M0 ANQ J Chadwick, 4 Toronto St., Bolton, BL2 6PE [IO83TN, SD70]
M0 ANR J Rowlinson, 24 Hillingdon Rd, Stretford, Manchester, M32 8PB [IO83UK, SJ89]
M0 ANS A J Rawlings, 57 High St., Nash, Milton Keynes, MK17 0EP [IO92NA, SP73]
MM0 ANT R Veal, 5 George Walk, Tranent, EH33 2EN [IO85MW, NT47]
M0 ANU G A Coolledge, 49A Enfield Ave, New Waltham, Grimsby, DN36 4RB [IO93XM, TA20]
MW0 ANV R H Davies, 55 Maes Padarn, Llanberis, Caernarvon, LL55 4TE [IO73WC, SH55]
M0 ANW Details withheld at licensee's request by SSL.
MW0 ANX J C H Jensen, Pistyll Canol Farm, Llandeilo Rd, Llandybie, Ammanford, Dyfed, SA18 2LQ [IO81AU, SN61]
M0 ANY Details withheld at licensee's request by SSL.
M0 ANZ Dr R E Turnbull, Taman, 38 Elsdon Rd, Gosforth, Newcastle upon Tyne, NE3 1HY [IO95EA, NZ26]
M0 AOA D Young, 196A Goddard Ave, Hull, HU5 2BY [IO93TS, TA03]
M0 AOB J W G Allen, 149 Penistone Rd, Waterloo, Huddersfield, HD5 8RP [IO93DP, SE11]
M0 AOC F J Hart, 18 Hopcott Cl, Minehead, TA24 5HB [IO81GE, SS94]
M0 AOD D Kay-Newman, Kay Spray, Pottery Rd, Horton, Ilminster, Somerset, TA19 9QN [IO80MW, ST31]
M0 AOE W C Stacey, 10 Copse Cl, Watchet, TA23 0HW [IO81HE, ST04]
MM0 AOF D S Henry, 25 Claremont St., Aberdeen, AB10 6QQ [IO87WD, NJ90]
M0 AOG G E Dyson, 60 Westbrook Park Rd, Peterborough, PE2 9JG [IO92UN, TL19]
M0 AOH J M Barber, 9 Chiswick St., Carlisle, CA1 1HQ [IO84MV, NY45]
M0 AOI J Russell, 46 Eastleigh Dr, Tingley, Wakefield, WF3 1PF [IO93FR, SE22]
M0 AOJ A C Elliott, Alrikitra, 26 Watery Ln, Minehead, TA24 5NZ [IO81GE, SS94]
M0 AOK S C Millar, 4 Broomfield, Benfleet, SS7 2ST [JO01HN, TQ88]
MM0 AOL R S Bloomfield, Torphin, Princes St., California, Falkirk, FK1 2BX [IO85DX, NS97]
M0 AOM M C Goodrich, 24 Sunstar Ln, Polegate, BN26 5HS [JO00CT, TQ50]
M0 AON J H Edwards, 10 Bridgwater Rd, Taunton, TA1 2DS [IO81LA, ST22]
M0 AOO W Cummins, 234 Overpool Rd, Great Sutton, South Wirral, L66 2JG [IO83MG, SJ37]
M0 AOP D V Gaudier, 3 Hall Pl Dr, Weybridge, KT13 0AJ [IO91SI, TQ06]
MM0 AOQ C D Greig, 5 Mitchell Pl, Stuartfield, Peterhead, AB42 5WE [IO87XM, NJ94]
M0 AOR K M Morgan, 21 Western Rise, Ketley, Telford, TF1 4BS [IO82SQ, SJ61]
M0 AOS P A Williams, 68 Old Crown Rd, Lupset, Wakefield, WF2 8UF [IO93FQ, SE31]
M0 AOT M J Stanley, Low Moor, Fenton, Keswick, Cumbria, CA12 4AZ [IO84KO, NY22]
M0 AOU M A Newton, 35 West St., Blaby, Leicester, LE8 4GY [IO92JN, SP59]
M0 AOV Details withheld at licensee's request by SSL.
M0 AOW A Piachaud, 184 Tufnell Park Rd, London, N7 0EE [IO91WN, TQ28]
M0 AOX Details withheld at licensee's request by SSL.
MM0 AOY D C Stephen, 16 The Sq, Portlethen, Aberdeen, AB12 4QA [IO87WB, NO99]
M0 AOZ Details withheld at licensee's request by SSL.
M0 APA J W Smith Iii, PO Box 1553, RAF Menwith Hill, Harrogate, HG3 2RF [IO94DA, SE25]
M0 APB H K Lansley, 24 Old Manor Cl, Holne Cross, Ashburton, Newton Abbot, TQ13 7JF [IO80CM, SX77]
M0 APC M I Brown, 61 Valley Rd, Pudsey, LS28 9EU [IO93ES, SE23]
M0 APD J Udall, 4 Church Ln, Chilcote, Caldwell, Swadlincote, DE12 6RT [IO92ES, SK21]
MI0 APE Details withheld at licensee's request by SSL.
MM0 APF J N Fisher Inverclyde Contest Group, 23 Ranfurly Rd, Bridge of Weir, PA11 3EL [IO75RU, NS36]
M0 APH A P Gilbert, 3 Wheatlands Ln, Baslow, Bakewell, DE45 1RF [IO93EF, SK27]
M0 API P A Schranz, 604 Hotham Rd South, Hull, HU5 5LE [IO93TS, TA03]
M0 APJ Details withheld at licensee's request by SSL.
M0 APK D E Allen, 162 Wood Ln, Newhall, Swadlincote, Derbyshire, DE11 0LY [IO92FS, SK22]

M0 APL B Tucker, 15 Southgate Way, Barrow Hill, Chesterfield, S43 2NR [IO93HG, SK47]
M0 APM Details withheld at licensee's request by SSL.
M0 APN A P Nelson, 29 Coxford Rd, Maybush, Southampton, SO16 5FG [IO90GW, SU31]
M0 APO Details withheld at licensee's request by SSL.[Station located in Bradford, West Yorkshire, IO93BS.]
M0 APP LtCl R E Mathis, 53 Queensway, Mildenhall, Bury St. Edmunds, Suffolk, IP28 7JY [JO02GI, TL77]
MW0 APQ P Quick, Tollgate Cottage, Golden Gr, Carmarthen, South Wales, SA32 8NH [IO71XU, SN52]
M0 APR Details withheld at licensee's request by SSL.
M0 APS Details withheld at licensee's request by SSL.
M0 APT Details withheld at licensee's request by SSL.
M0 APU ATC R E Hayes, Nato Mews, Rnas Yeovilton, Yeovil, Somerset, BA22 [IO80QX, ST51]
M0 APV J H A Collier, 8 Princes Gdns, Margate, CT9 3AP [JO01QJ, TR37]
M0 APW E N Kemp, 118 Marine Cres, Goring By Sea, Worthing, BN12 4HR [IO90ST, TQ10]
M0 APX Details withheld at licensee's request by SSL.
M0 APY R H Arey, 622 Scott Hall Rd, Chapel Allerton, Leeds, LS7 3QJ [IO93FT, SE33]
M0 APZ F Piper, Wicheves, Haysbrook Ave, Worsley, Manchester, M28 0AY [IO83SM, SD70]
M0 AQA G Shaw, 6 Bromstone Rd, Broadstairs, CT10 2HA [JO01RI, TR36]
M0 AQB Details withheld at licensee's request by SSL.
M0 AQC Details withheld at licensee's request by SSL.
MW0 AQD R J Hallett, 151 Trealaw Rd, Tonypandy, CF40 2NX [IO81GO, SS99]
M0 AQE E Entwistle, 50 Chorley Rd, Heath Charnock, Chorley, PR6 9JS [IO83QO, SD61]
M0 AQF T R Davies, 20 The Coppice, Impington, Cambridge, CB4 4PP [JO02BF, TL46]
MW0 AQG K T Dorney, 15 Gr Park Ave, Rhyl, LL18 3RG [IO83GH, SJ08]
M0 AQH E Blackburn, 2 Stockwell Dr, Knaresborough, HG5 0LW [IO94GA, SE35]
M0 AQI Details withheld at licensee's request by SSL.
MJ0 AQJ N Jones, 7 Avon House, Mont Cochon, St.Helier, Jersey, JE2 3JU
M0 AQK K Hesketh, 7 Southey Rd, St. Helens, WA10 3SN [IO83OK, SJ49]
M0 AQL Details withheld at licensee's request by SSL.
M0 AQM A C Stubbs, 5 Jankyns Croft, Old Buxton Rd, Disley, Stockport, SK12 2DH [IO83XI, SJ98]
M0 AQN R Vinju, PO Box 3183, 2601 Dd Delft, The Netherlands
M0 AQO G D H Willson, 40 Grace Gdns, Twyford Park Est, Bishops Stortford, CM23 3EX [JO01BU, TL41]
M0 AQP A S Bellamy, 23 Naseby Rd, Kettering, NN16 0LQ [IO92PJ, SP87]
M0 AQQ K G Evans, 5 Garswood Ave, Rainford, St. Helens, WA11 8JW [IO83OM, SD40]
M0 AQR M Ruiz, 11 Crimicar Dr, Sheffield, S10 4EF [IO93FI, SK38]
M0 AQS A Jeffs, 789 Borough Rd, Birkenhead, L42 6QN [IO83LI, SJ38]
MW0 AQT E F Lucocq, 96 Carisbrooke Way, Cyncoed, Cardiff, CF3 7HX [IO81KM, ST17]
M0 AQU H Thurgood, 17 Oatland Rd, Sandsacre, Bridlington, YO16 5UJ [IO94VC, TA16]
MI0 AQV B J McDaid, 41 Crannog Park, Strathfoyle, Londonderry, BT47 1NF [IO65IA, C42]
M0 AQW M J Storkey Processed Audio Group, 9 Snatchup, Redbourn, St. Albans, AL3 7HD [IO91TT, TL11]
MI0 AQX J E May, 8 Oak Vale Ave, Newry, BT34 2BQ [IO64UE, J02]
M0 AQY Details withheld at licensee's request by SSL.
MW0 AQZ Details withheld at licensee's request by SSL.
M0 ARA J A Layton, 6 Granby Rd, Cheadle Hulme, Cheadle, SK8 6LS [IO83VI, SJ88]
MW0 ARB C P Chatwin, New St Post Office, Pontnewydd, Cwmbran, Gwent, NP44 1EE [IO81LP, ST29]
M0 ARC J Dunnington East Yorks Contest Group, 73 West Hall Garth, South Cave, Brough, East Yorks, HU15 2HA [IO93QS, SE93]
MW0 ARD A R Davies, 2 Carnhedryn, Solva, Haverfordwest, SA62 6XT [IO71JV, SM72]
MW0 ARE J M Carey, Mount Pleasant, New Mills, Whitebrook, Monmouth, Gwent, NP5 4TY [IO81PS, SO50]
M0 ARF Details withheld at licensee's request by SSL.
MM0 ARG A B McKeeman Dalry Amateur Radio Group, Cottage No2, Hauplands Farm, Ardrossan, Ayrshire, KA22 8PL [IO75OQ, NS24]
M0 ARH N Ravilious, 5 Wellington Rd, Stevenage, SG2 9HR [IO91WV, TL22]
M0 ARI Details withheld at licensee's request by SSL.
M0 ARJ Details withheld at licensee's request by SSL.
M0 ARK B M Shepherd, 178 Rhodesway, Bradford, BD8 0DB [IO93CT, SE13]
MW0 ARL H P Davies, Garth Wen, 1 Bryn Siriol, Coedpoeth, Wrexham, LL11 3PZ [IO83LB, SJ25]
M0 ARM L J Hill, 65 The Boundary, Oldbrook, Milton Keynes, MK6 2QS [IO92PA, SP83]
MI0 ARN J H Kornreich, 35 Charlotte Dr, Spring Valley, New York 10977, USA
M0 ARO J M Parsons, 36 Gainsborough, Milborne Port, Sherborne, DT9 5BD [IO80SX, ST61]
M0 ARP Details withheld at licensee's request by SSL.
M0 ARQ J A Churchill, 59 Highfield, Letchworth, SG6 3PY [IO91VX, TL23]
M0 ARR J E Arregger, 55 Lebanon Park, Twickenham, TW1 3DH [IO91UK, TQ17]
M0 ARS Details withheld at licensee's request by SSL.
M0 ART B Addis, 30 Deneside, Seghill, Cramlington, NE23 7ER [IO95FB, NZ27]
M0 ARU Details withheld at licensee's request by SSL.
MW0 ARV M Thomas, 12 School Rd, Rhosllanerchrugog, Wrexham, LL14 1BB [IO83LA, SJ24]
M0 ARW Details withheld at licensee's request by SSL.
M0 ARX M B Richardson, 138 Northd. Ave, Welling, DA16 2PY [JO01BL, TQ47]
M0 ARY M E O'Rourke, Valley Farm, Laundry Ln, Huntingfield, Halesworth, IP19 0PY [JO02RH, TM37]
M0 ARZ S R Hurst, 25 Florence Rd, Abington, Northampton, NN1 4NA [IO92NF, SP76]
M0 ASA Details withheld at licensee's request by SSL.
MM0 ASB R L Barbour, 40 Mannerston Holdings, Linlithgow, EH49 7ND [IO85FX, NT07]
M0 ASC A S Clayton, Woodhurst, Salisbury Ave, Broadstairs, Kent, CT10 2DT [JO01RI, TR36]
M0 ASD A Gallichan, 4 Wigston Rd, Hillmorton, Rugby, CV21 4LT [IO92JI, SP57]
M0 ASE C Humphreys, 40 Baffins Rd, Copnor, Portsmouth, PO3 6BG [IO90LT, SU60]
M0 ASF U Nehmzow, 2 Heyworth Ave, Romiley, Stockport, SK6 4NF [IO83XK, SJ99]
M0 ASG C Nehmzow, 2 Heyworth Ave, Romiley, Stockport, SK6 4NF [IO83XK, SJ99]
M0 ASH N E Ash, Media Services (Bche), Polhill Ave, Bedford, Beds, MK41 9EA [IO92SD, TL05]
M0 ASI N W Johns, 12 Meadow Park, Plymouth, PL9 9NT [IO70WI, SX55]
M0 ASJ S B Griggs, 6 The Willows, Elgin Ave, Chelmsford, CM1 1TN [JO01FR, TL70]
M0 ASK G A Burrows, 158 Leybourne Ave, Bournemouth, BH10 6EY [IO90BS, SZ09]
MW0 ASL J C Phillips, 57 Ffordd Llanerch, Penycae, Wrexham, LL14 2ND [IO83LA, SJ24]
M0 ASM J Percival, 9 Gipsy Ln, Irchester, Wellingborough, NN29 7DJ [IO92QG, SP96]
M0 ASN J C Marron, 70 Cotswold Cres, Billingham, TS23 2QB [IO94IO, NZ42]
M0 ASO P H J Bluthner, 31 Westway, Garforth, Leeds, LS25 1DA [IO93HS, SE33]
MU0 ASP Details withheld at licensee's request by SSL.
M0 ASQ T L Lindsey, 20 Downham Rd, Runcton Holme, Kings Lynn, PE33 0AD [JO02EP, TF60]
M0 ASR D Campanario, 3 Foxearth, Leek Rd, Werrington, Stoke on Trent, ST9 0DG [IO83XA, SJ94]
M0 AST A G Stanley, 22 Brookhouse Rd, Walsall, WS5 3AD [IO92AN, SP09]
M0 ASU K A Hamer, 5 Town Head, Haverigg, Millom, LA18 4HF [IO84IE, SD17]
MI0 ASV G Best, 1 Bensons Rd, Lisburn, BT28 3QX [IO64WM, J26]
M0 ASW Details withheld at licensee's request by SSL.
M0 ASX R C Kautz, 42 Marlborough Rd, Pilgrims Hatch, Brentwood, CM15 9LN [JO01DP, TQ59]
M0 ASY Lt S M Werner, 4225 Pl Ste.Helene, Chomeday, Laval, Quebec, Canada H7W 1P3
M0 ASZ D Clark, 16 Roman Way, Welwyn, AL6 9RJ [IO91VU, TL21]
M0 ATA Details withheld at licensee's request by SSL.
M0 ATB R J Hutton, 57 Sandy Ln, Upton, Poole, BH16 5EJ [IO80XR, SY99]
M0 ATC C J Hoare Air Training Corps, 16 Shrivenham Rd, Highworth, Swindon, SN6 7BZ [IO91DP, SU29]
M0 ATD A H Holzapfel, Surrey Ct Wey 2/11, University of Surrey, Guildford, Surrey, GU2 5XH [IO91QF, SU95]
M0 ATE Details withheld at licensee's request by SSL.
MW0 ATF J L Phelps, 10 Sunnycrest, Newbridge, Newport, NP1 5FZ [IO81KQ, ST29]
MW0 ATG H Thomas, 15 Coronation Terr, Pontypridd, CF37 4DP [IO81IO, ST09]
MW0 ATH Details withheld at licensee's request by SSL.
MW0 ATI G C Roberts, 23 Heol Dirion, Coedpoeth, Wrexham, LL11 3HL [IO83LB, SJ25]
M0 ATJ D J Doland, 54 Ducklington Ln, Witney, OX8 7JB [IO91GS, SP30]
MW0 ATK S J Brewer, 12 Oakfield St., Llanbradach, Caerphilly, CF83 3NX [IO81JO, ST19]
M0 ATL P G Nash, 110 Cranborne Rd, Potters Bar, EN6 3AJ [IO91VQ, TL20]
M0 ATM Details withheld at licensee's request by SSL.
M0 ATN L S J Ham, 2 Heathfield Cl, Bingley, BD16 4EQ [IO93CU, SE13]
MM0 ATO Details withheld at licensee's request by SSL.
MW0 ATP S F Tongue, Rose Dale, Ruthin Rd, Bwlchgwyn, Wrexham, LL11 5UR [IO83KB, SJ25]
M0 ATQ J M W Torry, Tamarisk Cottage, 41 Nevill Rd, Rottingdean, Brighton, E Sussex, BN2 7HH [IO90XT, TQ30]
MW0 ATR D R Williams, 3 y Waun, Llanelli, Dyfed, SA14 8NH [IO71WQ, SN50]
M0 ATS J G Ammundsen, 7 Bosville Rd, Sevenoaks, TN13 3JD [JO01CG, TQ55]
MW0 ATT E G Cooke, 36 Coed Mor Est, Penyffordd, Holywell, CH8 9HY [IO83IH, SJ18]
M0 ATU G H McQuire, 23 Clarence Ave, Knott End on Sea, Poulton-le-Fylde, FY6 0AH [IO83MW, SD34]
M0 ATV Details withheld at licensee's request by SSL.
M0 ATW Details withheld at licensee's request by SSL.
M0 ATX E P Williams, Dyffryn Coed, 25 Peghouse Rise, Stroud, GL5 1RU [IO81VS, SO80]
M0 ATY C G J Kirkland, 9 Holland Way, Newport Pagnell, MK16 0LL [IO92PB, SP84]
M0 ATZ C D Hardy, 32 Stanhope Dr, Wirral, L62 2DG [IO83MI, SJ38]
M0 AUA D Hazel, 1 Rhodesway, Wirral, L60 2UA [IO83KH, SJ28]
MM0 AUB A Taylor, 38 Carrick Pl, Dunure, Ayr, KA7 4LU [IO75OJ, NS21]
M0 AUC R Tapson, 3 The Quadrant, Trelander, Truro, TR1 1FX [IO70LG, SW84]
M0 AUD J Bedford, 29 Burstock Rd, London, SW15 2PW [IO91VL, TQ27]
M0 AUE Details withheld at licensee's request by SSL.
M0 AUF N E Handforth, 6 Queensbury Ave, Bromborough, Wirral, L62 7HB [IO83MH, SJ38]
M0 AUG G E Ashton, 53 Dorset Ave, High Crompton, Shaw, Oldham, OL2 7EG [IO83WN, SD90]
MW0 AUH J J Everson, 35 Brynwern, Pontypool, NP4 6HH [IO81LQ, SO20]
M0 AUI R W Moore, 36 Higher Wood, Bovington, Wareham, BH20 6NG [IO80VQ, SY88]
M0 AUJ Details withheld at licensee's request by SSL.
M0 AUK J H Sporton, 199 Glaisdale Dr, Bilborough, Nottingham, NG8 4GY [IO92JX, SK54]
M0 AUL A Bewley, 40 Whiteacres, Morpeth, NE61 2UT [IO95DD, NZ28]
M0 AUM D B Tordoff, 40 Whiteacres, Morpeth, NE61 2UT [IO95DD, NZ28]
M0 AUN G Northall, 108 Wilkinson Ave, Broseley, TF12 5EF [IO82SO, SJ60]
M0 AUO Details withheld at licensee's request by SSL.
MM0 AUP P J Laird, Kanlee, Carness Rd, St. Ola, Kirkwall, Isle of Orkney, KW15 1TB [IO88MX, HY41]
M0 AUQ H Wright, 25 Seegar Dr, Morphett Vale 5162, South Australia
M0 AUR A G Taylor, 4 Lower St., Stutton, Ipswich, IP9 2SQ [JO01NX, TM13]
M0 AUS A W Dowie, 31 St. James Rd, Scawby, Brigg, North Lincs, DN20 9BD [IO93RM, SE90]
M0 AUU Details withheld at licensee's request by SSL.
M0 AUV Details withheld at licensee's request by SSL.
M0 AUW R Hull, 10 Hambrook St., Charlton Kings, Cheltenham, GL52 6LW [IO81XV, SO92]
MW0 AUX D Johnson, Garreg Lwyd, 2 Garreg Wen Est, Rhosybol, Ynys Mon, Gwynedd, LL68 9RL [IO73TI, SH48]
M0 AUY L K Jeffries, 1 Pear Tree Cl, Bransgore, Christchurch, BH23 8NH [IO90DS, SZ19]
M0 AUZ G K Vick, 5 Mirrool St, Narrabeen, Nsw 2101, Australia
M0 AVA D A Salsbury, 38 Laxey Cres, Leigh, WN7 5HF [IO83RM, SD60]
M0 AVB C Proffitt West Pennine Packer User Group, 17 Tempest St., Bolton, BL3 4HR [IO83SN, SD60]
M0 AVC I B Clifton, 12 School Fields, Kingsley, Bordon, GU35 9PQ [IO91ND, SU73]
M0 AVD Details withheld at licensee's request by SSL.
M0 AVE Details withheld at licensee's request by SSL.
M0 AVF A F de Araujo, 119 King Edwards Dr, Harrogate, HG1 4HW [IO94FA, SE35]
M0 AVG Details withheld at licensee's request by SSL.
M0 AVH D H Eaton-Watts, 129 Blake Rd, West Bridgford, Nottingham, NG2 5LA [IO92KW, SK53]
MI0 AVI G S Millar Newry High School Radio Club, 23 Ashgrove Rd, Newry, BT34 1QN [IO64UE, J02]
M0 AVJ R T Shillabeer, 134 Redhouse Park, Hogmoor Rd, Bordon, Hants, GU34 1HS [IO91MD, SU73]
M0 AVK D A Swift, 8 Gr Ln, Buxton, SK17 9HG [IO93BF, SK07]
M0 AVL D H Meakin, 47 Sheridan St., Pleck, Walsall, WS2 9QX [IO82XN, SO99]
M0 AVM Details withheld at licensee's request by SSL.
M0 AVN A H Oatey, 42 Rue St. Pierre, Ivybridge, PL21 0HZ [IO80BJ, SX65]
MU0 AVO Details withheld at licensee's request by SSL.
M0 AVP A P Baughan, 9 Kyl Cober Parc, Stoke Climsland, Callington, PL17 8PH [IO70UN, SX37]
MW0 AVQ J B Worthington, 69 Sandy Ln, Deeside, Clwyd, CH5 2JE [IO83LF, SJ36]
M0 AVS V G Saundercock, Eastmoor View, 40 Stephens Rd, Liskeard, PL14 3SX [IO70SL, SX26]
MW0 AVT Details withheld at licensee's request by SSL.
M0 AVU M Scott, 36 Glebe Cres, Forest Hall, Newcastle upon Tyne, NE12 0JR [IO95FA, NZ27]
MM0 AVV S Macdonald, 24 Birch Terr, Girvan, KA26 0DB [IO75NF, NX19]
M0 AVW C W Spence, 32 Woodford Walk, Harewood Park, Thornaby, Stockton on Tees, TS17 0LT [IO94IM, NZ41]
M0 AVX P Cook, Westway Boundary Hs, Shiney Row, Houghton-le-Spring, Tyne & Wear, DH4 4PZ [IO94FU, NZ35]
M0 AVY A H Johnson, 131 Rylands Rd, Kennington, Ashford, TN24 9LU [JO01KD, TR04]
M0 AVZ D Clutterbuck, 2 Spring Valley Dr, Leeds, LS13 4RN [IO93ET, SE23]
M0 AWA I W Morphett, 196 Luddington Village, Stratford upon Avon, Warks, CV37 9SJ [IO92CE, SP15]
M0 AWB A W Boom, 108 Sheepwalk, Paston, Peterborough, PE4 7BL [IO92VO, TF10]
MW0 AWC L G Gagnon, 2 Ashley Ct, Field Ln, Appleton, Warrington, WA4 5JR [IO83RI, SJ68]
M0 AWD M W Mansfield, Little Ash, St. Ln, Lower Whitley, Warrington, WA4 4EN [IO83RH, SJ67]
M0 AWE K W Capewell, 9 High Croft, Claverdon, Warwick, CV35 8LH [IO92DG, SP16]
M0 AWF D Bonshor, 1315 Lakeside Dr, Wilson, North Carolina, USA 27896
M0 AWG Details withheld at licensee's request by SSL.
M0 AWH P S Bush, 144 Stoke Ln, Westbury on Trym, Bristol, BS9 3RN [IO81QL, ST57]
M0 AWI J W Ross, 18 Bernay Gdns, Bolbeck Park, Milton Keynes, MK15 8QD [IO92PB, SP84]
M0 AWJ K A Gray, 137 Astaire Ave, Eastbourne, BN22 8UU [JO00DS, TQ60]
M0 AWK R T Suggitt, 10 Rectory Ave, Guisborough, TS14 6QL [IO94LM, NZ61]
MI0 AWL A G Smith, 27A Cairn Rd, Carrickfergus, BT38 9AP [IO74BS, J39]
MW0 AWM M T Pepperell, 4 Greenway, Hook, Haverfordwest, SA62 4LL [IO71MS, SM91]
M0 AWN C D Gladman, 24 Priory Rd, Chessington, KT9 1EF [IO91UI, TQ16]
MW0 AWO S Jones, 6 Heol Will Hopkin, Llangynwyd, Maesteg, CF34 9ST [IO81EO, SS88]
M0 AWP P J Oliver, 45 Charlotte Cl, Mount Hawke, Truro, TR4 8TS [IO70JG, SW74]
M0 AWQ A B Rayner, Mittel, St. End Ln, Broad Oak, Heathfield, TN21 8RY [JO00DX, TQ62]
MJ0 AWR R L Gelber, 205 West End Ave, New York, Ny 10023, USA
M0 AWS M D Tatum, 11 Kempton Cl, Ipswich, IP1 6QZ [JO02NB, TM14]
M0 AWT K P Bates, 1 Sea View Terr, Sennen, Penzance, TR19 7AR [IO70DB, SW32]
MM0 AWU G Moffat, 4 Crosswood Ave, Balerno, EH14 7HT [IO85HV, NT16]
M0 AWV L J Hrycan, 40 Marina Dr, Marple, Stockport, SK6 6JL [IO83WJ, SJ98]
M0 AWW Details withheld at licensee's request by SSL.
M0 AWX G H Schoof, 5 Canal Row, Haigh, Wigan, WN2 1NA [IO83QO, SD50]
M0 AWY D J Ashdown, Cartwheels, 4 Honeysuckle Cl, Hailsham, BN27 3TP [JO00DU, TQ50]
M0 AWZ Details withheld at licensee's request by SSL.
MW0 AXA W S Townsend, 133 Hazeldene Ave, Brackla, Bridgend, CF31 2JR [IO81FM, SS97]
M0 AXB Details withheld at licensee's request by SSL.
M0 AXC Details withheld at licensee's request by SSL.
M0 AXD P F J Eames, 6 Sirius Cl, Seaview, Isle of Wight, PO34 5LH [IO90KR, SZ69]
M0 AXE D W Marshall, 59 Kirkstone Rd, Sheffield, S6 2PN [IO93FJ, SK38]
M0 AXF Details withheld at licensee's request by SSL.
M0 AXG K A Wheeler, 26 Melverton Ave, Wolverhampton, WV10 9HN [IO82WO, SJ90]
M0 AXH K Jasinski, 35B Friars Pl Ln, London, W3 7AQ [IO91UM, TQ28]
M0 AXI Capt L A Rayner, 4 Elysium Ct, 19 Aldridge Rd, Ferndown, Dorset, BH22 8LT [IO90BT, SZ09]
M0 AXJ A Clay, 52 Berrylands Rd, Wirral, L46 7UA [IO83KJ, SJ29]
M0 AXK Details withheld at licensee's request by SSL.
MM0 AXL J J Cook, 2 Springwells, St. Anns, Lockerbie, DG11 1HG [IO85GF, NY09]
MW0 AXM Details withheld at licensee's request by SSL.
M0 AXN J S H Davis, 60 West Bar St., Banbury, OX16 9RZ [IO92HB, SP44]
M0 AXO G P Morris, 16 Madin Rd, Tipton, DY4 8JT [IO82XM, SO99]
M0 AXP J Faxholm, 21 St. Peters Gr, London, W6 9AY [IO91VL, TQ27]
M0 AXQ Details withheld at licensee's request by SSL.
MM0 AXR T Rees, Laurel House, 7 Leewood Park, Dunblane, Perthshire, FK15 0NX [IO86AE, NN70]
MW0 AXS Details withheld at licensee's request by SSL.
M0 AXT Details withheld at licensee's request by SSL.
M0 AXU A D Russell, 126 Boothferry Rd, Hessle, HU13 9AX [IO93SR, TA02]
M0 AXV M Amplett, 44A Darby Rd, Coalbrookdale, Telford, TF8 7EW [IO82SP, SJ60]
M0 AXW J G Davies, 243A Bradford Rd, Winsley, Bradford on Avon, BA15 2HL [IO81UI, ST86]
M0 AXX E Moody, The Apiaries, Sawmill Cttgs, Rufford, Newark, Notts, NG22 9DG [IO93LE, SK66]
M0 AXY Details withheld at licensee's request by SSL.
M0 AXZ P Morgan, 20 Bishops Way, Buckden, St. Neots, Huntingdon, PE18 9TZ [IO92VH, TL16]
M0 AYA S E Sellman, Ireland Farm, Banbury Rd, Gaydon, Warwick, CV35 0HH [IO92GE, SP35]
M0 AYB T P Davies, Brooklands, 95 High Brigham, Brigham, Cockermouth, CA13 0TJ [IO84HP, NY03]
M0 AYC J D Soakell, 47 Lentune Way, Lymington, SO41 3PE [IO90FS, SZ39]
MM0 AYD Details withheld at licensee's request by SSL.
MM0 AYE J Welsh, 7 South Cathkin Cottage, Rutherglen, Glasgow, G73 5RG [IO75VT, NS65]
M0 AYF D Kostryca, 9 Cherry Tree Rd, Gainsborough, DN21 1RG [IO93OJ, SK89]
M0 AYG M R Wood, 12 Ashfield Rd, Compton, Wolverhampton, WV3 9DP [IO82VO, SO89]
M0 AYH Details withheld at licensee's request by SSL.
M0 AYI G E Waring, 7 Tynedale Terr, Stanley, DH9 7TZ [IO94DU, NZ15]
M0 AYJ Details withheld at licensee's request by SSL.
M0 AYK G G Yunnie, St Marys Lodge, Battenhall Rd, Worcester, Worcs, WR5 2HP [IO82VE, SO85]
M0 AYL D R Hill, 3 Morcar Rd, Stamford Bridge, York, YO4 1PR [IO93NX, SE75]
MW0 AYM D R Pearson, Warren Cottage, Pontfadog, Llangollen, Clwyd, LL20 7AT [IO82KW, SJ23]
M0 AYN Details withheld at licensee's request by SSL.
M0 AYO H W Parker, 21 Mayfield St., Spring Bank, Hull, HU3 1NS [IO93TS, TA02]
M0 AYP M A Greenwood, 14 Laurel Terr, Stanningley, Pudsey, LS28 7QJ [IO93ET, SE23]
M0 AYQ D W Smith, Box No 1069, Menwith Hill, Harrogate, N.Yorkshire, HG3 2RF [IO94DA, SE25]
MI0 AYR Details withheld at licensee's request by SSL.
M0 AYS C F Pocock, 4 Broad Fields, Harpenden, AL5 2HJ [IO91TT, TL11]
M0 AYT J T Pallonen, 7 Butterfields, Camberley, GU15 3EU [IO91OH, SU85]
M0 AYU I J Gibson, 7 Peverells Wood Cl, Chandlers Ford, Eastleigh, SO53 2FY [IO90HX, SU42]
M0 AYV A R Eyles, 35 Drayton St., Winchester, SO22 4BJ [IO91HB, SU42]
M0 AYW P J Garden, 58 Minerva Ct, Eatons Hill, Queensland, Australia 4037
M0 AYX A N King, 49 Drift Way, Cirencester, GL7 1WN [IO91AQ, SP00]
M0 AYY R J Corfield, 18 South Rd, Kingsclere, Newbury, RG20 5RY [IO91JH, SU55]
MI0 AYZ I J Kyle, 1 Portulla Dr, Pond Park Rd, Lisburn, BT28 3JS [IO64XM, J26]
M0 AZA Details withheld at licensee's request by SSL.
M0 AZB R Goddard, 6 Upper Ley Dell, Greenside Ct, Chapeltown, Sheffield, S30 4AL [IO93GJ, SK38]
M0 AZC P Niel, 19 Fountains Cl, Whitby, YO21 1JS [IO94QL, NZ81]
M0 AZD A K Watts, 3 Laburnum Rd, Hedge End, Southampton, SO30 0QA [IO90IV, SU41]

M0 AZE M L Surplice, 43A Cremorne Rd, Four Oaks, Sutton Coldfield, B75 5AQ [IO92CO, SP19]
M0 AZF C K Jordan St Marys College Radio Club, 31 Rocky Bank Rd, Tranmere, Birkenhead, L42 7LB [IO83LJ, SJ38]
M0 AZG J T Kisiel, Wayside Cottage, South Stoke Rd, Woodcote, Reading, RG8 0PL [IO91LM, SU68]
M0 AZH Details withheld at licensee's request by SSL.
M0 AZI P C J Warren, 5 St. Georges Ct, Bracken Rd, North Baddesley, Southampton, SO52 9AT [IO90GX, SU31]
M0 AZJ G P Gould, 32 Archer Rd, Kenilworth, CV8 1DJ [IO92EI, SP27]
M0 AZK D D Hill, 61 Blandford Ave, Birmingham, B36 9JB [IO92CM, SP19]
M0 AZL T Wrede, 145 Coldershaw Rd, London, W13 9DU [IO91UM, TQ17]
M0 AZM I D Norton, The Marina Hotel, 324 Marine Rd, Morecambe, Lancs, LA4 5AA [IO84NB, SD46]
MW0 AZN R A Cullis, 18 Church Cl, Croesyceiliog, Cwmbran, NP44 2EJ [IO81LP, ST39]
M0 AZO Details withheld at licensee's request by SSL.
M0 AZP P R Lemasonry, 7 Eastwood Rd, Sittingbourne, ME10 2LZ [JO01II, TQ86]
M0 AZQ C Sheridan, 3 Edward Cl, Aylesbury, HP21 9YQ [IO91OT, SP81]
M0 AZR S J Gale, 39 Thorley Park Rd, Bishops Stortford, CM23 3NG [JO01BU, TL41]
M0 AZS R F Buckle, 25 Portsmouth Cl, Strood, Rochester, ME2 2QY [JO01FJ, TQ76]
M0 AZT M P Thomas, 36 Seaview Ave, Peacehaven, BN10 8SA [JO00AS, TQ40]
M0 AZU Details withheld at licensee's request by SSL.
M0 AZV N P Devine, Bfbs, Royal Air Force, Akrotiri, Cyprus, Bfpo 57
M0 AZW C L Coe, 19 Seton Rd, Taverham, Norwich, NR8 6QE [JO02OQ, TG11]
M0 AZX Details withheld at licensee's request by SSL.
M0 AZY F J Willis, 99 Kenilworth Ct, Asthill Gr, Coventry, CV3 6JB [IO92FJ, SP37]
M0 AZZ A C Bond, 45 Cherry Cl, Knebworth, SG3 6DS [IO91VU, TL21]
M0 BAA Details withheld at licensee's request by SSL.
MM0 BAB M J Richard, Rowland Cottage, Fordell Est, Hillend, Dunfermline, Fife, KY11 5HA [IO86HB, NT18]
MM0 BAC C I Mackay, 26 Abbey Grange, Newtongrange, Dalkeith, EH22 4RJ [IO85LU, NT36]
M0 BAD Details withheld at licensee's request by SSL.
M0 BAE A A Radford, 33 Priory Ave, Kirkby in Ashfield, Nottingham, NG17 9BU [IO93JB, SK55]
M0 BAF E Richardson, Military Cottage, Irthington, Carlisle, Cumbria, CA6 4NE [IO84OW, NY45]
MM0 BAG G B Craig, 9 Green Dr, Inverness, IV2 4EX [IO77VK, NH64]
M0 BAH A J Tyler, Christmas Tree Cottage, 92 Crawley Rd, Roffey, Horsham, RH12 4DT [IO91UB, TQ13]
M0 BAI P Byrne, 303B Brodie Ave, Liverpool, L19 7ND [IO83NI, SJ48]
MW0 BAJ D H Nelson, 8 Willow Park, Gladstone Way, Queensferry, Deeside, CH5 2TX [IO83LE, SJ36]
M0 BAK K Williams, Cranesbie, 6 Dore Rd, Sheffield, South Yorks, S17 3NB [IO93FH, SK38]
M0 BAL F Johnson, 7 Pharos Ct, Pharos St., Fleetwood, FY7 6BG [IO83LW, SD34]
M0 BAM A Tobin, 22 Brookway Dr, Charlton Kings, Cheltenham, GL53 8AJ [IO81XV, SO92]
M0 BAN Details withheld at licensee's request by SSL.
M0 BAO A J Edwards, The Old House, North St., Haselbury Plucknett, Crewkerne, TA18 7RJ [IO80PV, ST41]
M0 BAP W C L Stewart, Hill Farm, Edge, Stroud, Glos, GL6 6PH [IO81VS, SO80]
M0 BAQ N T Wright Deben High School Amateur Radi, Karachi, 1 Garfield Rd, Felixstowe, IP11 7PU [JO01QX, TM23]
M0 BAR B Bartley, 5 Cookes Wood, Deerness Vale, Broompark, Durham, DH7 7RL [IO94ES, NZ24]
M0 BAS Details withheld at licensee's request by SSL.
MI0 BAT S J Gilmore, 8 Fortfield, Dromore, BT25 1DD [IO64WK, J15]
M0 BAU G Hoyle, 39 Randle Meadow, Great Sutton, South Wirral, L66 2SE [IO83MG, SJ37]
M0 BAV L N R Evans, 16 Kynaston Dr, Wem, Shrewsbury, SY4 5DE [IO82PU, SJ52]
M0 BAW Details withheld at licensee's request by SSL.
M0 BAX B A Pope, 12 Wilton Cl, Christchurch, BH23 2PL [IO90CR, SZ19]
M0 BAY G A Bilson, 101 Birkinstyle Ln, Shirland, Alfreton, DE55 6BT [IO93HD, SK45]
M0 BAZ J A Waterfield, 287 Turves Green, Birmingham, B31 4BS [IO92AJ, SP07]
M0 BBA D A Whitfield, 18 Yates Cl, Great Sankey, Warrington, WA5 1XH [IO83QJ, SJ58]
M0 BBB S Tsiakkouris, 7 Isidorou St, Strovolos, Nicosia, Cyprus
M0 BBC G D Eddowes Ariel Radio Group (Bbc Club), Flat 1, 47 The Avenue, Ealing, London, W13 8JR [IO91UM, TQ18]
M0 BBD Details withheld at licensee's request by SSL.
M0 BBE J S Hayward, 32 Brendon Gr, Bingham, Nottingham, NG13 8TN [IO92MW, SK63]
MI0 BBF D J Doherty, 175 Bridge Rd, Glarryford, Ballymena, BT44 9QA [IO64TX, D01]
M0 BBG D R Langeheine, 37 Hayman Rd, Brackley, NN13 6JA [IO92KA, SP53]
M0 BBH M A Redman, 19 Richmond Rd, Rugby, CV21 3AB [IO92JI, SP57]
M0 BBI Details withheld at licensee's request by SSL.
MW0 BBJ G R Ottaway, Garden Cottage, Abermarlais, Llangadog, Dyfed, Wales, SA19 9NG [IO81BW, SN62]
M0 BBK J E Meakin, White House Farm, Osmotherley, North Allerton, North Yorks, DL6 3AQ [IO94II, SE49]
MW0 BBL Details withheld at licensee's request by SSL.
MW0 BBM B A Meredith, 27 Hyde Pl, Llanhilleth, Abertillery, NP3 2RT [IO81KQ, SO20]
M0 BBN K F Corser, 85 Coleshill Rd, Marston Green, Birmingham, B37 7HT [IO92DL, SP18]
M0 BBO S Woodford, 31 Seaborough View, Crewkerne, TA18 8JB [IO80OV, ST40]
M0 BBP Details withheld at licensee's request by SSL.
M0 BBQ K D Taylor, 241 Daventry Rd, Cheylesmore, Coventry, CV3 5HH [IO92FJ, SP37]
M0 BBR R C Rogers, 11 Broomfield, Octavia Way, Staines, TW18 2QD [IO91SK, TQ07]
M0 BBS Maj F W Mooney, 10 Blencowe Cl, Bicester, OX6 9UJ [IO91KW, SP52]
M0 BBT T M Pirrie, Walnut Thatch, Radway, Warks, CV35 0UE [IO92GD, SP34]
MW0 BBU S K A Lloyd, 41 Coombs Dr, Milford Haven, SA73 2NU [IO71LR, SM90]
M0 BBV R Lyford, 4 Wrentham Est, Old Tiverton Rd, Exeter, EX4 6ND [IO80FR, SX99]
M0 BBW S J Gleadall, 59 Old Chapel Rd, Smethwick, Warley, B67 6HU [IO92AL, SP08]
M0 BBX Details withheld at licensee's request by SSL.
M0 BBY W J Deegan Iii, 11121 Lakeshore Dr East, Carmel, Indiana 46033-4403, USA
M0 BBZ Details withheld at licensee's request by SSL.
M0 BCA Details withheld at licensee's request by SSL.
M0 BCB A J Norris, 6 Bawtry Walk, Carlton Rd, Nottingham, NG3 2GF [IO92KX, SK54]
M0 BCC R M Clapp, 20 Wychwood Dr, Trowell, Nottingham, NG9 3RB [IO92IW, SK43]
M0 BCD M J Reynolds, 110 Bilborough Rd, Bilborough Est, Nottingham, NG8 4DN [IO92JX, SK54]
M0 BCE Dr W A Johnstone, 67 Station Ln, Birkenshaw, Bradford, BD11 2JE [IO93DS, SE22]
M0 BCF R J Cranwell, 39 Rivergreen, Nottingham, NG11 8FW [IO92JV, SK53]
M0 BCG I A Williams, Alma Cottage, South Marston, Swindon, SN3 4SN [IO91DO, SU18]
M0 BCH C R Chadburn, 31 Darwin Cl, Top Valley, Nottingham, NG5 9LN [IO93JA, SK54]
M0 BCI N Armstrong, 112 Chandos St., Netherfield, Nottingham, NG4 2LW [IO92LX, SK64]
M0 BCJ G Lewis, 42 Ladywood Rd, Ilkeston, DE7 4NE [IO92IX, SK44]
M0 BCK K Bell, 71 Wheatfield Rd, Stanway, Colchester, CO3 5YA [JO01KV, TL92]
M0 BCL P R Williams, 3 Upper Holway Rd, Taunton, TA1 2HF [IO81LA, ST22]
M0 BCM M L Vest Jr, 417 Elm St, Morgantown, Wv 26505, USA
M0 BCN D J James, 82 Overbury Cl, Birmingham, B31 2HD [IO92AJ, SP07]
M0 BCO C P Sardeson, 106 Main St., Willoughby on The Wolds, Loughborough, LE12 6SZ [IO92LT, SK62]
M0 BCQ A L Jackson Craven Radio Group, 14 Earls View, Sutton in Craven, Keighley, BD20 7PR [IO93AV, SE04]
MM0 BCR L S Haynes, 29 Invercauld Rd, Aberdeen, AB16 5RP [IO87WD, NJ90]
M0 BCS N Saxvoisin, 2 South Entrance, Saxmundham, IP17 1DQ [JO02RF, TM36]
M0 BCT M D Danfer, 72 Warwick Ave, Woodbridge, IP12 1JY [JO02PC, TM24]
M0 BCU J F Beckett, 23 Newton Ave, Wakefield, WF1 2PX [IO93FQ, SE32]
M0 BCV S Graham, 4 Oakland Ave, Ellenborough, Maryport, CA15 7BU [IO84GR, NY03]
M0 BCW P W Mason, 15 Granton Ave, Clifton Est, Nottingham, NG11 9AL [IO92JV, SK53]
M0 BCX Details withheld at licensee's request by SSL.
M0 BCY A W Lees, 71 Sara House, Larner Rd, Erith, DA8 3RF [JO01CL, TQ57]
M0 BCZ R Burton, 103 Ingrave Rd, Brentwood, CM15 8BA [JO01DO, TQ69]
MM0 BDA Dr R A August, Smiddyhill House, Stracathro, By Brechin, Angus, DD9 7QE [IO86QS, NO66]
M0 BDB R H W Taylor, 88 Hillside Cres, Leigh on Sea, SS9 1HQ [JO01IM, TQ88]
M0 BDC J Wright, 17 Claremont View, Woodlesford, Leeds, LS26 8TA [IO93GS, SE32]
M0 BDD D Webster, 210 Walesby Ln, New Ollerton, Newark, NG22 9UU [IO93LE, SK66]
M0 BDE D Oakes, 44 Sarum Ave, West Moors, Ferndown, BH22 0ND [IO90BT, SU00]
M0 BDF J D Reid, The Bailiwick, Woodlands, Wimborne, Dorset, BH21 8LN [IO90AV, SU00]
M0 BDG A J King, 19 Greenhayes, Cheddar, BS27 3HZ [IO81OG, ST45]
M0 BDH P A Fisher, 21 Charlotte Cl, Mount Hawke, Truro, TR4 8TS [IO70JG, SW74]
M0 BDI R P Lansdale, Harbour House, Spinney Ln, Itchenor, Chichester, PO20 7DJ [IO90NT, SU80]
M0 BDJ R W Hawkins, 7 Consort Rd, Cowes, PO31 7SQ [IO90IS, SZ49]
MM0 BDK I M Hutchison, Burnbrae, Aultnaskiach, Inverness, IV2 4BB [IO77VL, NH64]
M0 BDL D Ferris, 167 Lonsdale Ave, Doncaster, DN2 6HF [IO93KM, SE60]
M0 BDM K Fujita, PO Box 4724, General Post Office, Hong Kong
M0 BDN B Lowe, 393 St George St E, Fergus, N1M 1K6, Canada
MI0 BDO J T Christie, 3 Victoria Dr, Sydenham, Belfast, BT4 1QT [IO74BO, J37]
M0 BDP Details withheld at licensee's request by SSL.
M0 BDQ K Kisselev, 162 Parsloes Ave, Dagenham, RM9 5PX [JO01BN, TQ48]
M0 BDR D W Kearton, Skellcote, Woodhall, Askrigg, Leyburn, DL8 3LA [IO84XH, SD99]
M0 BDS Dr G Butler, 51 Leamington Dr, Beeston, Nottingham, NG9 5LN [IO92JV, SK53]
M0 BDT P J Knight, 16 Pembroke Cl, Marston Moreteyne, Bedford, MK43 0JX [IO92RB, SP94]
M0 BDU D D Goodwin, 15 Tennyson Rd, Bentley, Doncaster, DN5 0EG [IO93KN, SE50]
M0 BDV Details withheld at licensee's request by SSL.
M0 BDW P J A Hayes, 131 High St., Henlow, SG16 6AE [IO92UA, TL13]
MI0 BDX A Patterson, 33 Marlborough Park, Carryduff, Belfast, BT8 8NL [IO74BM, J36]
M0 BDY P Andrews, 52 Trenchard Cres, Springfield, Chelmsford, CM1 6FG [JO01FS, TL70]
MI0 BDZ M S Chancellor, 55 Brae Hill Park, Belfast, BT14 8FP [IO74AP, J37]
MW0 BEA Details withheld at licensee's request by SSL.
M0 BEB M Skull, 15 Hambleton Sq, Billingham, TS23 2RZ [IO94IO, NZ42]
M0 BEC R E Millerchip, 16 Kennedy Cres, Gosport, PO12 2NN [IO90KS, SZ59]
MM0 BED J Macdonald, Somerled, Craigowan Rd, Campbeltown, PA28 6QH [IO75EK, NR72]
M0 BEE D Martin Whitehaven Amateur Radio Club, Bayshore, Gilcrux, Carlisle, Cumbria, CA5 2QD [IO84HR, NY13]
MM0 BEF B C Sim, 23 Willow Dr, Girvan, KA26 0DE [IO75NF, NX19]
M0 BEG Details withheld at licensee's request by SSL.
M0 BEH P W Mutter, 7 Rochford Way, Croydon, CR0 3AG [IO91WJ, TQ36]
MI0 BEI R S Shilliday, 106 Lisburn Rd, Ballynahinch, BT24 8TS [IO74AK, J35]
M0 BEJ G Moody, 25 Norbiton Common Rd, Kingston upon Thames, KT1 3QB [IO91UJ, TQ16]
M0 BEK C T Dunn, 11 Thames Ave, Burnley, BB10 2PZ [IO83VT, SD83]
M0 BEL A G Owen, 116 Bridgetown Rd, Stratford upon Avon, CV37 7JA [IO92DE, SP25]
M0 BEM M R Taperell, 16 Parkhall Croft, Birmingham, B34 7BU [IO92CM, SP18]
MW0 BEN P R Hughes, 6 Coed y Nant, Wrexham, LL13 7QH [IO83MA, SJ34]
M0 BEO K J Anderson, 53 Spruce Hill, Harlow, CM18 7SS [JO01BR, TL40]
MW0 BEP Details withheld at licensee's request by SSL.
M0 BEQ H Walsh, 38 Potter Hill, Greasbrough, Rotherham, S61 4PA [IO93HK, SK49]
MW0 BER D J Jones, 3 Clwch, Rhosmeirch, Llangefni, LL77 7SJ [IO73UG, SH47]
MI0 BES J L May, 8 Oak Vale Ave, Newry, BT34 2BQ [IO64UE, J02]
MW0 BET V T Hughes, Manley, 1 Garden Dr, Llandudno, LL30 3LL [IO83CH, SH88]
M0 BEU G R Wilson, 33 Church Rd, Illogan, Redruth, TR16 4SR [IO70IF, SW64]
M0 BEV R Knapp, 85 Eastern Ave, Liskeard, PL14 3TD [IO70SK, SX26]
M0 BEW Details withheld at licensee's request by SSL.
M0 BEX J M Hrycan, 40 Marina Dr, Marple, Stockport, SK6 6JL [IO83WJ, SJ98]
MW0 BEY M T J Beynon, Sunrise, Kilgetty-Lane, Stepaside, Narberth, Pembrokeshire, SA67 8JL [IO71PR, SN10]
M0 BEZ N M Clissold, 2 Whitebeam Cl, Gloucester, GL2 0UG [IO81VV, SO82]
M0 BFA D Wilson, 4 Downton Rd, Swindon, Wilts, SN2 5JN [IO91CO, SU18]
M0 BFB K R Francks, 63 Parc Godrevy, Pentire, Newquay, TR7 1TY [IO70KJ, SW86]
M0 BFC Details withheld at licensee's request by SSL.
MW0 BFD N A Hutchinson, 4 Penycae Rd, Port Talbot, SA13 2EL [IO81CO, SS78]
M0 BFE C B Cooper Iii, 4 Hilltop Rd, Caversham, Reading, RG4 7HR [IO91ML, SU67]
MM0 BFF I Macaulay, 3A Melbost, Point, Isle of Lewis, HS2 0BG [IO68UF, NB43]
M0 BFG N O Bourne, Exiner Str 22G, 29303 Bergen, Germany
M0 BFJ D E F J Cusack, 7 Battledown Priors, Cheltenham, GL52 6RB [IO81XV, SO92]
M0 BFK S J Bertrand, 15 Flag Walk, Waterlooville, PO8 8LE [IO90LV, SU61]
M0 BFL B J Jayne, 38 Townsfields, Lichfield, WS13 8AA [IO92BQ, SK10]
M0 BFM S Jones, 186 Earle Rd, Liverpool, L7 6HH [IO83MJ, SJ38]
M0 BFN N Y C Kan, 195A Chiswick High Rd, London, W4 2DR [IO91UL, TQ27]
M0 BFO I G M Roberts, 13 Fearnley Dr, Ossett, WF5 9EU [IO93FQ, SE22]
M0 BFP D W Phillipson National Radiofone Amateur Rad, 9 Austins Cl, Market Harborough, LE16 9BJ [IO92ML, SP78]
M0 BFQ T Norton, 5 Cullen Gr, Manchester, Lancs, M9 6LJ [IO83VM, SD80]
M0 BFR G J Vinke, 48 Dunraven Dr, Plymouth, PL6 6AR [IO70WK, SX46]
M0 BFS P Barnes, 100 Money Bank, Wisbech, PE13 2JF [JO02CP, TF40]
M0 BFT S D Smith, 1 Field Ave, Tydd St. Giles, Wisbech, PE13 5LJ [JO02BR, TF41]
M0 BFU A R Crowder, 19 Leys Rd, Blackpool, FY2 0SH [IO83LU, SD33]
M0 BFV L Papazoglou, 45 Mosslands Dr, Wallasey, L45 8PF [IO83LK, SJ29]
M0 BFW D J S Taylor, 45 Ardsley Rd, Chesterfield, S40 4DG [IO93GF, SK37]
M0 BFX A S Daniels, 9 Moss Fold, Astley, Tyldesley, Manchester, M29 7FP [IO83SM, SD70]
MW0 BFY J K Lester, 1 Maes Hwylfa, Llanrhaeadr, Denbigh, LL16 4PG [IO83HD, SJ06]
M0 BFZ I M Stuart, 20 Beresford St., Nelson, Lancs, BB9 0JB [IO83VT, SD83]
M0 BGA Details withheld at licensee's request by SSL.
M0 BGB S A Fross, 12 Lambscroft Way, Chalfont St. Peter, Gerrards Cross, Bucks, SL9 9AX [IO91RO, TQ09]
M0 BGC Details withheld at licensee's request by SSL.
M0 BGD D G Kelly, 80 High St., Hail Weston, St. Neots, Huntingdon, Cambs, PE19 4JW [IO92UF, TL16]
M0 BGE T G Parker, 24 Burrows Cl, Lawford, Manningtree, CO11 2HE [JO01MW, TM03]
M0 BGF G R Reid, 13 Wynmore Dr, Bramhope, Leeds, West Yorks, LS16 9DQ [IO93EV, SE24]
M0 BGG D Rosher, 40 Sycamore Rd, Stowupland, Stowmarket, IP14 4DR [JO02ME, TM05]
M0 BGI R Buttery, 96 Rupert St., Lower Pilsley, Chesterfield, S45 8DE [IO93HD, SK46]
M0 BGJ L A Thorne, 41 Ingleby Rd, Long Eaton, Nottingham, NG10 3DG [IO92IV, SK43]
M0 BGK Details withheld at licensee's request by SSL.
MW0 BGL Details withheld at licensee's request by SSL.
M0 BGM A S K Murphy, 10 Byron St., Ilkeston, DE7 5JG [IO92IX, SK44]
M0 BGN J P M Stooker, 36 Norfolk Farm Rd, Woking, Surrey, GU22 8LF [IO91RH, TQ05]
MM0 BGO R Herd, Kilrymont, Newtown, Ceres, Cupar, Fife, KY15 5LY [IO86MH, NO31]
M0 BGP Details withheld at licensee's request by SSL.

M1

M1 AAB R J Monksummers The, 29 Cloverfields, Peacemarsh, Gillingham, SP8 4UP [IO81UA, ST82]
MM1 AAC L R McCartney, Cairndhue, Walter St., Langholm, DG13 0AX [IO85LD, NY38]
M1 AAD Details withheld at licensee's request by SSL.
M1 AAE G W Edwards, 36 Patrick Stirling Ct, Barnstone St., Doncaster, DN4 0EU [IO93KM, SE50]
M1 AAF R J Maskill 115 (Peterborough) Sqn ATC A.R, 21 Clayton, Orton Goldhay, Peterborough, PE2 5SB [IO92UN, TL19]
M1 AAG J P Ingrey, 8 Redman Gdns, Biggleswade, SG18 0DF [IO92UC, TL14]
M1 AAH D G Creber, 8 George Manning Way, Gowerton, Swansea, SA4 3HB [IO71XP, SS59]
M1 AAI Details withheld at licensee's request by SSL.
MW1 AAK Details withheld at licensee's request by SSL.
MM1 AAL M N Stuart, 3 Arran View, Largs, KA30 9ER [IO75NT, NS25]
M1 AAM Details withheld at licensee's request by SSL.
M1 AAR T A Parkin, 39 Clifford Ave, Wakefield, WF2 7LF [IO93GP, SE31]
M1 AAS A Jackson, 29 Bramble Ave, Birkenhead, L41 0AX [IO83LJ, SJ28]
MW1 AAT Details withheld at licensee's request by SSL.
M1 AAU Details withheld at licensee's request by SSL.
M1 AAV R C Blinco, 32 Beverley Pl, Cloverdale, Western Australia, 6105
M1 AAW Details withheld at licensee's request by SSL.
M1 AAX D J Brown Leicestershire Ip, 29 Marlborough Way, Ashby de La Zouch, LE65 2NN [IO92GR, SK31]
MM1 AAY N A Thomson, 10 Edgemoor Park, Balloch, Inverness, IV1 2RB [IO77WL, NH74]
M1 AAZ A D Joslin, Holland Cottage, Kirby Rd, Great Holland, Frinton on Sea, CO13 0HZ [JO01OU, TM22]
MM1 ABA I Hopley, 6 Ninian Pl, Portlethen, Aberdeen, AB12 4QW [IO87WB, NO99]
M1 ABB Details withheld at licensee's request by SSL.
M1 ABC B Johnson, 20 Valleyside, Hemel Hempstead, HP1 2LN [IO91SS, TL00]
M1 ABD Details withheld at licensee's request by SSL.[Dorset Police Packet & Contest Gp, correspondence via G0ROZ.]
M1 ABE Details withheld at licensee's request by SSL.
M1 ABF K M Perkin, 25 Rownall View, Leek, ST13 8JN [IO83XC, SJ95]
M1 ABI P Lindsay, 6 Rothley Cl, Ponteland, Newcastle upon Tyne, NE20 9TD [IO95CB, NZ17]
M1 ABJ R A Pascoe The Kanga Gang, Seaview House, Crete Rd East, Folkestone, Kent, CT18 7EG [JO01OC, TR23]
M1 ABK Details withheld at licensee's request by SSL.
M1 ABL Details withheld at licensee's request by SSL.
M1 ABM M M Chapman, Woodcroft, Windmill Green, Stone Cross, Pevensey, BN24 5DY [JO00DT, TQ60]
M1 ABO S M Rutter Cromer High School Amateur Rad, Old Stores, The St., Sutton, Norwich, NR12 9RF [JO02SS, TG32]
M1 ABP Details withheld at licensee's request by SSL.
M1 ABQ Details withheld at licensee's request by SSL.
M1 ABR T Robinson, 35 Stoneham Ln, Southampton, SO16 2NU [IO90HW, SU41]
M1 ABS B Whitehouse, Hill House, Colburn Village, Catterick Garrison, North Yorks, DL9 4PD [IO94DJ, SE19]
MW1 ABT R A Macleod, 24 Heol Powis, Gungrog Hill, Welshpool, SY21 7TP [IO82KP, SJ20]
M1 ABU K Cackett, 162 Hart Dyke Rd, Swanley, BR8 7EF [JO01BJ, TQ56]
M1 ABV B C Breet, 23 Mitchell St., Eccles, Manchester, M30 8AJ [IO83TL, SJ79]
M1 ABX S Painting, Claytons, Inkpen, Newbury, Berks, RG17 9QE [IO91GJ, SU36]
M1 ABY T Highams, 16 Lumley Rd, Sutton, SM3 8NN [IO91VI, TQ26]
M1 ACA G Barnett, 63 Sandcroft, Sutton Hill, Telford, TF7 4AB [IO82SP, SJ70]
M1 ACB S G Thomas, 49 Worcester Rd, Ipswich, IP3 0RS [JO02OA, TM14]
M1 ACC J R Chambers, 9 Farnborough Rd, Swindon, SN3 2DR [IO91DN, SU18]
M1 ACE P D Musk, 38 Glenwood, Welwyn Garden City, AL7 2JS [IO91WT, TL21]
M1 ACF A J Jackson Acf/Ccf National Net Club, 27 Ellesmere Dr, Hamsey Green, Sanderstead, South Croydon, CR2 9EH [IO91XH, TQ36]
M1 ACG R J Horn, 8 Hullbridge Mews, London, N1 3QU [IO91WM, TQ38]
M1 ACH Details withheld at licensee's request by SSL.
M1 ACI M J Smith, 28 Repton Cl, Luton, LU3 3UL [IO91SV, TL02]
M1 ACJ S R Shearing, 42 Meadow Park, Wesham, Preston, PR4 3DN [IO83NS, SD43]
M1 ACK R C Mackay, Conifers, The St., West Clandon, Guildford, GU4 7TJ [IO91RG, TQ05]
M1 ACL Details withheld at licensee's request by SSL.
M1 ACM A Iannetta, Heathfield Cottage, Heathfield, Alkington, Berkeley, Glos, GL13 9PL [IO81SQ, ST79]
M1 ACN M D I Goom, 7 Hillbrook Rd, Thornbury, Bristol, BS12 2EZ [IO81RO, ST68]
M1 ACO R S Hankin, The Croft, Portinscale, Keswick, CA12 5TX [IO84KO, NY22]

M1 ACP S R Stevens, 7 Pembroke Dr, Ponteland, Newcastle upon Tyne, NE20 9HS [IO95CA, NZ17]
M1 ACQ S D Thomas, 111 Jersey Ave, Bristol, BS4 4QX [IO81RK, ST67]
M1 ACR G D Wilks, 154 Woodland Dr, Anlaby, Hull, HU10 7HT [IO93SR, TA02]
M1 ACS T D Girdler, 8 Westmorland Ave, Loughborough, LE11 3RY [IO92JS, SK51]
M1 ACT C J Leeds, 23 Fairfax Rd, Norwich, NR4 7EZ [JO02PO, TG20]
M1 ACU Details withheld at licensee's request by SSL.
M1 ACV R S S King, 5 Havenwood Rise, Nottingham, NG11 9HD [IO92JV, SK53]
M1 ACY Details withheld at licensee's request by SSL.
M1 ACZ Details withheld at licensee's request by SSL.
M1 ADA M N Pritchard, 15 Comet Dr, Ditherington, Shrewsbury, SY1 4AY [IO82PR, SJ51]
M1 ADB C Pritchard, 120 Whitchurch Rd, Harlescott, Shrewsbury, SY1 4ED [IO82PR, SJ51]
M1 ADC K B Dagger, 46 Hamilton Rd, Chorley, PR7 2DL [IO83QP, SD51]
M1 ADE Details withheld at licensee's request by SSL.
M1 ADG Details withheld at licensee's request by SSL.
M1 ADI R A Bradbrook, 48 Springfield Green, Chelmsford, CM1 7HS [JO01FR, TL70]
M1 ADJ J J Moore, 34 Alder Gr, Poulton-le-Fylde, FY6 8EH [IO83MU, SD33]
M1 ADK G B Pratt, 2 Houghton Rd, Newbottle, Houghton-le-Spring, DH4 4EF [IO94GU, NZ35]
M1 ADL S M Barlow, 48 Albert St., Featherstone, Pontefract, WF7 5EX [IO93HQ, SE42]
M1 ADM G Snook, 3 Harehills Park Ave, Leeds, LS9 6BP [IO93FT, SE33]
M1 ADN T E Whiting, 28 Legarde Ave, Hull, HU4 6AP [IO93TR, TA02]
M1 ADO Details withheld at licensee's request by SSL.
M1 ADP K Eastwood, 120 The Stour, Daventry, NN11 4PT [IO92JG, SP56]
M1 ADQ K M Wells, 42 Eggesford Rd, Sinfin, Derby, DE24 3BH [IO92GV, SK33]
M1 ADR S J Thompson, 19 Cavalier Cl, Norwich, NR7 0TE [JO02QP, TG20]
M1 ADS P Peterson, 8 Croftacres, Ramsbottom, Bury, BL0 0LX [IO83UP, SD71]
M1 ADT R Vickerstaff, 16 Sewell Wontner Cl, Kesgrave, Ipswich, IP5 2GB [JO02OB, TM24]
M1 ADU M F Thompson, 96 Carlingford Rd, Hucknall, Nottingham, NG15 7AG [IO93JA, SK54]
M1 ADV I G Owen, 11 Thatchers Walk, Stowmarket, IP14 2DR [JO02LE, TM05]
M1 ADW Details withheld at licensee's request by SSL.
M1 ADX J Towns, 5 Padgate, Thorpe End, Norwich, NR13 5DG [JO02QP, TG21]
M1 ADZ N W Davis, 395 Chingford Rd, London, E17 5AF [IO91XO, TQ39]
M1 AEA Details withheld at licensee's request by SSL.
M1 AEB G J Haughie, 227 Inglewhite, Skelmersdale, WN8 6JQ [IO83ON, SD40]
M1 AEC Details withheld at licensee's request by SSL.
M1 AED M Cattell, 12 Fairway Rd, Oldbury, Warley, B68 8BE [IO82XL, SO98]
M1 AEE A E Fugler, 9 Westover Rd, Fleet, GU13 9DG [IO91OG, SU85]
M1 AEG A Green, Michigane, Whitemoor, Nanpean, St. Austell, Cornwall, PL26 7XN [IO70NJ, SW95]
M1 AEH D F Cossey, 11 Halden Ave, Norwich, NR6 6UX [JO02PQ, TG21]
M1 AEI D C Bowyer, East Foldhay, Zeal Monachorum, Crediton, Devon, EX17 6DH [IO80CT, SS70]
M1 AEJ A S Benjamin, 4 Girton Cl, Owlsmoor, Sandhurst, GU47 0UP [IO91OI, SU86]
M1 AEK D G Mulliner, 106 Matson Rd, Bridlington, YO16 4TP [IO94VB, TA16]
MM1 AEL C Haswell, 6 Lochlann Rd, Culloden, Inverness, IV1 2HB [IO77WL, NH74]
M1 AEM Details withheld at licensee's request by SSL.
M1 AEN R M Salt, 1 Weaver Cttgs, Rue Hill, Cauldon Low, Stoke on Trent, ST10 3HD [IO93BA, SK04]
M1 AEO R D East, 4 Brewer Rd, Barnstaple, EX32 8EX [IO71XB, SS53]
M1 AEP M E Griffin, 76 Waylands, Swanley, BR8 8TN [JO01CJ, TQ56]
M1 AEQ F R Allenby, 9 Church View, Holme on Spalding Moor, York, YO4 4BG [IO93OU, SE83]
M1 AER Details withheld at licensee's request by SSL.
M1 AES Details withheld at licensee's request by SSL.
M1 AET M E Stephens, 26 Hill St., Wigan, WN6 7EQ [IO83QN, SD50]
M1 AEU J Edmunds, 52 Cotswold Rd, Chipping Sodbury, Bristol, BS17 6DP [IO81TM, ST78]
M1 AEV Details withheld at licensee's request by SSL.
M1 AEW Details withheld at licensee's request by SSL.
M1 AEX R D Pyman, Broumfield, North St., Barming, Maidstone, ME16 9HF [JO01FG, TQ75]
M1 AEY R A Wyeth, 112 Main Rd, Crockenhill, Swanley, BR8 8JL [JO01BJ, TQ56]
M1 AEZ R D Barrett, Kennoway, Hay Ln, Fulmer, Slough, SL3 6HJ [IO91RN, SU98]
M1 AFA Details withheld at licensee's request by SSL.
M1 AFB N M Davison, 1 Retford Cl, Derby, DE21 4DX [IO92GW, SK33]
M1 AFC Details withheld at licensee's request by SSL.
M1 AFD Details withheld at licensee's request by SSL.
M1 AFE M Read, Greenacres, North Rew Ln, Martinstown, Dorchester, Dorset, DT2 9DU [IO80RR, SY69]
M1 AFF R C Dyer, Highbank Cottage, Underhill, Moulsford, Wallingford, Oxon, OX10 9JH [IO91KN, SU58]
M1 AFH D Morris, 8 Falcon Cl, Bristol, BS9 3NH [IO81QL, ST57]
M1 AFK D Dore, 132 Cortis Rd, London, SW15 3AQ [IO91VK, TQ27]
M1 AFL J Hinde, 40 Reynolds Cl, Flanderwell, Rotherham, S66 0XL [IO93IK, SK49]
M1 AFM A Mori, 35 Oliver Rd, Southampton, SO18 2JQ [IO90HW, SU41]
M1 AFN N Wright, 50 Ledbury Rd, Portsmouth, PO6 4BS [IO90KU, SU60]
M1 AFP P E Jefford, 61 Willow Way, Flitwick, Bedford, MK45 1LN [IO91SX, TL03]
M1 AFQ A P Brooks, 86 Violet Rd, Norwich, NR3 4TS [JO02PP, TG21]
M1 AFR Details withheld at licensee's request by SSL.
M1 AFU S Derwin, 5 Hawthorne Gr, Yarm, TS15 9EZ [IO94HM, NZ41]
M1 AFV S K Wells, 44 Darkwood Cl, Shadwell, Leeds, LS17 8BH [IO93FU, SE33]
MW1 AFW C E Jones, 41 Trosnant Cres, Penybryn, Hengoed, CF82 7FW [IO81JP, ST19]
M1 AFY Details withheld at licensee's request by SSL.
M1 AFZ C M Grizzell, 11 Quarry Way, Stapleton, Bristol, BS16 1UP [IO81RL, ST67]
M1 AGA R S Taylor, 5 Thirlmere Dr, Bury, BL9 9QE [IO83UN, SD80]
M1 AGB Details withheld at licensee's request by SSL.
M1 AGC D J Fahy, 57 Thornway, Bramhall, Stockport, SK7 2AH [IO83VI, SJ88]
M1 AGD Details withheld at licensee's request by SSL.
M1 AGE D N Thorley, Stokeleigh, Wyson Ln, Brimfield, Ludlow, Salop, SY8 4NW [IO82PH, SO56]
M1 AGF Details withheld at licensee's request by SSL.
M1 AGG D J Mapeley, 75 Ormonde, Stantonbury, Milton Keynes, MK14 6DH [IO92OB, SP84]
M1 AGH D Hirst, 10 Orchard Gdns, Purton, Swindon, SN5 9EJ [IO91BO, SU08]
M1 AGI S G Hirst, 10 Orchard Gdns, Purton, Swindon, SN5 9EJ [IO91BO, SU08]
M1 AGJ D J Walsh, 13 Shaws Ave, Warrington, WA2 8AU [IO83QJ, SJ68]
M1 AGK R M Large, 7 The Hedgerow, Longlevens, Gloucester, GL2 9JE [IO81VV, SO81]
M1 AGL P Bennett, 31 Westbury Rd, Westgate on Sea, CT8 8QX [JO01QJ, TR36]
M1 AGM D S Munro, 11 Crane Lodge Rd, Hounslow, TW5 9PG [IO91TL, TQ17]
M1 AGN D J Goodchild, 17 Dale Rd, Buxton, SK17 6LN [IO93BG, SK07]
M1 AGO G J Robertson, 22 Carlton Villas, Hatt, Saltash, PL12 6PS [IO70UK, SX36]
M1 AGP J G Davies, 10 Gorselands, Hollesley, Woodbridge, IP12 3QL [JO02RB, TM34]
M1 AGR M D Skinner, 81 Verdon St., Sheffield, S3 9QN [IO93GJ, SK38]
M1 AGT S J Hall, Crest Cottage, Chapel Ln, Grunisburgh, Woodbridge, Suffolk, IP13 6TS [JO02PC, TM25]
M1 AGU G R Ridgway, 27 Naverne Meadows, Woodbridge, IP12 1HU [JO02PC, TM24]
M1 AGW S J Whitehouse, 47 Mulberry Rd, Walsall, WS3 2NG [IO82XO, SJ90]
M1 AHA S D Lefevre, 9 Old Barn Cres, Hambledon, Waterlooville, PO7 4SW [IO90KW, SU61]
M1 AHB Details withheld at licensee's request by SSL.
M1 AHC J H Pryde, 16 Chantry Rd, Worthing, BN13 1QN [IO90TT, TQ10]
M1 AHD Details withheld at licensee's request by SSL.
M1 AHE Details withheld at licensee's request by SSL.
M1 AHF M J Byatt, 54 St. Leonards Rd, Plymouth, PL4 9NE [IO70WI, SX45]
M1 AHG Details withheld at licensee's request by SSL.
M1 AHH S J Tombs, 42 Parkfield Rd, Pucklechurch, Bristol, BS17 3PS [IO81SL, ST67]
M1 AHJ Details withheld at licensee's request by SSL.
M1 AHK B D Tutty, 39 Mere Gate, Margate, CT9 5TR [JO01QJ, TR37]
MM1 AHL D McArthur, 117 Dumbuck Cres, Dumbarton, G82 1EH [IO75RW, NS47]
M1 AHM S Williamson, 5 Orchard Dr, South Hiendley, Barnsley, S72 9BQ [IO93HO, SE31]
M1 AHN G R Boyce, 3 Woodview, Four Oaks, Newent, GL18 1LU [IO81SW, SO62]
M1 AHO Details withheld at licensee's request by SSL.
M1 AHQ N R Grigsby, 32 South Ham Rd, Basingstoke, RG22 6AD [IO91KG, SU65]
M1 AHR K J Peters, 82 Blackmoor Rd, Moortown, Leeds, LS17 5JP [IO93FU, SE23]
M1 AHS S T Murphy, 6 Lynne Way, London, NW10 9JP [IO91UN, TQ28]
M1 AHT L V Russell, 106 Stambridge Rd, Rochford, SS4 1DP [JO01IO, TQ89]
MW1 AHU Dr G C Armstrong, Plas Porth Uchaf, Conwy, Gwynedd, LL32 8RL [IO83CG, SH77]
M1 AHW B F Cannon, 7 Stopham Rd, Pulborough, RH20 1DP [IO90RW, TQ01]
M1 AHX W Taylor, 28 Jeremy Rd, Wolverhampton, WV4 5BZ [IO82WN, SO99]
M1 AHY M R Rhodes, 1 Chetwode, Overthorpe, Banbury, OX17 2AB [IO92IB, SP44]
M1 AHZ S A Murray, 41 Springfield Ct, Leek, ST13 6LZ [IO83XC, SJ95]
M1 AIB P Lewis, 2 Hopgoods Green, Bucklebury, Reading, RG7 6TB [IO91JK, SU57]
MW1 AID M E Mickels, 3 Gorslas, North Cornelly, Bridgend, CF33 4NG [IO81DM, SS88]
M1 AIE I A Elliott, 8 Crowland Rd, Hartlepool, TS25 2JJ [IO94JP, NZ42]
M1 AIF Details withheld at licensee's request by SSL.
M1 AIG Details withheld at licensee's request by SSL.
M1 AIH N G Bullough, 29 Redfern Rd, Stone, ST15 0LF [IO82WV, SJ83]
M1 AIJ R D Walker, 38 Wheatley St., West Bromwich, B70 9TJ [IO82XM, SO99]
M1 AIK T D Bardgett, 49 St. James St., South Petherton, TA13 5BN [IO80OW, ST41]
M1 AIL Details withheld at licensee's request by SSL.
M1 AIM A Moore, 45 Carters Way, Wisborough Green, Billingshurst, West Sussex, RH14 0BX [IO91RA, TQ02]
M1 AIN D J White, Hollingworth Gate, Hollingworth Ln, Walsden, Todmorden, OL14 6QY [IO83WQ, SD92]
M1 AIO J K R Kyle, 2 Castle Rd, Ellon, AB41 9EY [IO87XI, NJ93]
M1 AIP R C Bray, 10 Upwell Rd, March, PE15 9DT [JO02BN, TL49]
M1 AIQ Details withheld at licensee's request by SSL.
MM1 AIR Details withheld at licensee's request by SSL.
M1 AIS J C White, Pathways Down, Barton Rd, St Nicholas At Wade, Birchington, Kent, CT7 0PY [JO01PI, TR26]
M1 AIT Details withheld at licensee's request by SSL.
M1 AIU G J Griffiths, 10 Teesdale Rd, Long Eaton, Nottingham, NG10 3PG [IO92IV, SK43]
M1 AIV G L Tomlinson, 92 Gale St., Rochdale, OL12 0BG [IO83WP, SD81]
M1 AIW B J Reeves, 143A Chantry Gdns, Southwick, Trowbridge, BA14 9QP [IO81VH, ST85]
M1 AIX W Greenall, 356 Warrington Rd, Abram, Wigan, WN2 5XA [IO83QM, SD60]
M1 AIY M I Bastin, 14 Golvers Hill Rd, Kingsteignton, Newton Abbot, TQ12 3BP [IO80EN, SX87]
M1 AIZ P Jarvis, 6A Britannia Rd, Bilston, West Midlands, WV14 8DS [IO82XN, SO99]
M1 AJA L A Mason, Oxley Hall, Weetwood Ln, Leeds, UK, LS16 8HL [IO93EU, SE23]
M1 AJB A Bloor, 1 Arthur Rd, Biggin Hill, Westerham, TN16 3DD [JO01AH, TQ45]
M1 AJF D Hill, 9 Grosvenor Gdns, Newton-le-Willows, WA12 8LY [IO83QK, SJ59]
M1 AJG M A Ramskill, 7 Hobart Rd, Dewsbury, WF12 7LS [IO93EQ, SE22]
MW1 AJI J Williams, 1 Glam err, Penrhiwfer, Tonypandy, Mid Glam, CF40 1SA [IO81GO, ST08]
M1 AJJ Details withheld at licensee's request by SSL.
M1 AJK D J Attwood, 16 Raleigh St., Barrow in Furness, LA14 5RH [IO84JC, SD16]
M1 AJL K J Day, Lazy Days, 13 Guildford Rd, Southport, PR8 4JU [IO83LO, SD31]
M1 AJM S J Hoskins, 46 Allington Way, Maidstone, ME16 0HN [JO01FG, TQ75]
M1 AJN A J Balls, Jonmer, Swallow Ln, Tydd Gote, Wisbech, PE13 5PQ [JO02BR, TF41]
M1 AJO R E Pearce, Kolner, 86 Thrupp Ln, Thrupp, Stroud, GL5 2DG [IO81VR, SO80]
MW1 AJP G B Morrison, 16 Russell Dr, Prestatyn, LL19 7YR [IO83GH, SJ08]
M1 AJQ W J Clarke, 10 Athol Cl, Sinfin, Derby, DE24 9LZ [IO92GV, SK33]
M1 AJR Details withheld at licensee's request by SSL.
M1 AJS Details withheld at licensee's request by SSL.
M1 AJT P Amos, 16 Eastry Rd, Erith, DA8 1NN [JO01BL, TQ47]
M1 AJU S C J Bradford, 49 Ave Rd, Ramsgate, CT11 8EP [JO01RI, TR36]
M1 AJV G Elsmore, 7 Thorn Gdns, Dumpton Park, Ramsgate, CT11 7AS [JO01RI, TR36]
M1 AJW I W Peters, 7 Rougemont Cl, Plymouth, PL3 6QY [IO70WJ, SX45]
MW1 AJX M J Clarke, 41 Min Awel, Flint, CH6 5TG [IO83KF, SJ27]
M1 AJY Details withheld at licensee's request by SSL.
M1 AJZ A Brewer, 28 West Ln, Edwinstowe, Mansfield, NG21 9QT [IO93LE, SK66]
M1 AKA R T Schuring, 21A Cokeham Rd, Sompting, Lancing, BN15 0AE [IO90TT, TQ10]
M1 AKB Details withheld at licensee's request by SSL.
M1 AKD Details withheld at licensee's request by SSL.
M1 AKF P B Wilson, 7 The Cedars, New Longton, Preston, PR4 4AF [IO83PR, SD52]
M1 AKG Dr J G Robertson, 1 Padley Cl, Dodworth, Barnsley, S75 3SE [IO93FN, SE30]
M1 AKH D Mulvana, 43 Saltmarsh, Orton Malborne, Peterborough, PE2 5NL [IO92UN, TL19]
M1 AKI S D Smith, 1 Field Ave, Tydd St. Giles, Wisbech, PE13 5LJ [JO02BR, TF41]
M1 AKJ Details withheld at licensee's request by SSL.
M1 AKL V Parsons, Gull Cottage, Briar Cl, Fairlight, Hastings, TN35 4DP [JO00IV, TQ81]
M1 AKN R J Day, 53 Out Westgate, Bury St. Edmunds, IP33 3NX [JO02IF, TL86]
M1 AKO J P O'Neill, Shafto Cottage, Low Ousterley, Craghead, Stanley, Co. Durham, DH9 6DW [IO94DU, NZ25]
M1 AKP J R Kirkman, 32 Broad Oak Way, Cheltenham, GL51 5LG [IO81WV, SO92]
M1 AKQ S J Hutton, 11 College Gdns, Hornsea, HU18 1EF [IO93VV, TA14]
M1 AKS S M Shields, 139 Longcroft Ln, Welwyn Garden City, AL8 6EL [IO91VT, TL21]
M1 AKT D Thomas, 82 Fir Tree Est, Thurgoland, Sheffield, S30 7BG [IO93GJ, SK38]
M1 AKU D Leonard, 63 Little Coates Rd, Grimsby, DN34 4NN [IO93WN, TA20]
M1 AKV R Kowalski, 2 Newcross Park, Kingsteignton, Newton Abbot, TQ12 3TJ [IO80EN, SX87]
M1 AKW S R Crossland, 63 Temple St., Half Acres, Castleford, WF10 5RH [IO93HR, SE42]
M1 AKX N J Williams, 20 Hillside Dr, St. Catherines Hill, Christchurch, BH23 2RU [IO90CS, SZ19]
M1 AKZ Details withheld at licensee's request by SSL.
M1 ALA Details withheld at licensee's request by SSL.
M1 ALB D I Isherwood, 14A Chaucer Gr, Leigh, WN7 5JZ [IO83RM, SD60]
MM1 ALC Details withheld at licensee's request by SSL.
M1 ALD B A C Moran, 36 Yew Tree Cl, Chatham, ME5 8XN [JO01GH, TQ76]
M1 ALE A W Ratcliff, 71 Spring Gdns, Leek, ST13 8DD [IO83XC, SJ95]
M1 ALF R B Coston, 22 Leagate, Urmston, Manchester, M41 9LD [IO83TK, SJ79]
M1 ALG J A Sindall, 9 James St., Rotherham, S60 1JU [IO93HK, SK49]
M1 ALH K I Whinney, 24 Saxon Park, Barrets Ln, Needham Market, Ipswich, IP6 8SA [JO02MD, TM05]
M1 ALI Details withheld at licensee's request by SSL.
M1 ALK J W Newsome, 241 Skellow Rd, Skellow, Doncaster, DN6 8JL [IO93JO, SE51]
M1 ALM R J Hodgkins, 38 Byron Rd, Gillingham, ME7 5QH [JO01GJ, TQ76]
M1 ALN S Dring, 23 Hardstoft Rd, Pilsley, Chesterfield, S45 8BL [IO93HD, SK46]
M1 ALO Details withheld at licensee's request by SSL.
M1 ALP M L W Clarke, 40 Penrice Cl, Greenstead Est, Colchester, CO4 3XN [JO01LV, TM02]
M1 ALQ Details withheld at licensee's request by SSL.
M1 ALR C D Moore, 28 Hillside View, Peasedown St. John, Bath, BA2 8ES [IO81SH, ST75]
M1 ALT Details withheld at licensee's request by SSL.
M1 ALU T A Summers, 14 Ollerton Rd, Edwinstowe, Mansfield, NG21 9QG [IO93LE, SK66]
MM1 ALV C F C Cregan, 31A Nelson St., Edinburgh, EH3 6LJ [IO85JW, NT27]
M1 ALX D A Philip, 9 Wallace Cl, Uxbridge, UB10 0SB [IO91SM, TQ08]
M1 ALY S Ford, 18 Woodlands Rd, Newton Abbot, TQ12 4ER [IO80FM, SX87]
M1 ALZ W Allsey, 5 Dudley Rd, Cadishead, Manchester, M44 5UA [IO83SK, SJ79]
M1 AMA A C Day, 9 Arundel Rd, Tewkesbury, GL20 8AS [IO82WA, SO93]
M1 AMB D J Snow, 7 Aynsley Cl, Cheadle, Stoke on Trent, ST10 1DP [IO92AX, SK04]
M1 AMC W C Hand, 19 Burns St., Narborough, Leicester, LE9 5EA [IO92JN, SP59]
MM1 AMD J D Gauson, 112A High St., New Pitsligo, Fraserburgh, AB43 6NN [IO87VO, NJ85]
M1 AMF J M Barnham, 35 Chatsworth Dr, Market Harborough, Leics, LE16 8BS [IO92NL, SP78]
M1 AMG Details withheld at licensee's request by SSL.
M1 AMH Details withheld at licensee's request by SSL.
M1 AMI D F Chambers, 17 Field Way, Wivenhoe, Colchester, CO7 9HG [JO01LU, TM02]
M1 AMJ D R Bonnett, 254 Norwich Rd, Wisbech, PE13 3UT [JO02CQ, TF40]
M1 AMK Details withheld at licensee's request by SSL.
MW1 AML D A Cornick, 36 Pinewood Sq, St. Athan, Barry, CF62 4JR [IO81HJ, ST06]
M1 AMM K Longley, 94 The Elms, Colwick, Nottingham, NG4 2GW [IO92KW, SK64]
M1 AMN Details withheld at licensee's request by SSL.
M1 AMP H Withers, 23 Fernie Rd, Guisborough, TS14 7LZ [IO94LM, NZ61]
M1 AMQ N O Bourne, Exiner Str 22G, 29303 Bergen, Germany
M1 AMR S J Westwood, 118 Abbey Ln, Leigh, WN7 5NU [IO83RM, SD60]
MW1 AMS D H W Williams, 3 Maes Hyfryd, Garndolbenmaen, LL51 9SX [IO72VX, SH44]
M1 AMT Details withheld at licensee's request by SSL.
M1 AMW C A Whitehead, 78 Cop Ln, Penwortham, Preston, PR1 0UR [IO83PR, SD52]
M1 AMY Details withheld at licensee's request by SSL.
M1 AMZ K Birch, 16 Brentwood Ave, Thornton Cleveleys, FY5 3QR [IO83LU, SD34]
M1 ANA D W Anger, Plantation, Mill St., Polstead, Colchester, CO6 5AD [JO02KA, TL93]
M1 ANC C A McLean, 18 Chatfield Rd, Gosport, PO13 0TN [IO90JT, SU50]
MW1 AND M J Griffiths, 32 Hill St., Aberaman, Aberdare, CF44 6YG [IO81GQ, SO00]
M1 ANE P J Brennan, 5 Westover Rd, Southampton, SO16 9BJ [IO90GW, SU31]
M1 ANG A D Reynolds, 18 Laburnum Gr, Irby, Wirral, L61 4UT [IO83KI, SJ28]
M1 ANH Details withheld at licensee's request by SSL.
M1 ANI Details withheld at licensee's request by SSL.
M1 ANJ P Herbert, 2 Wheatsheaf Cttgs, Alconbury Hill, Alconbury Weston, Huntingdon, PE17 5JH [IO92UJ, TL17]
M1 ANK R H W Taylor, 88 Hillside Cres, Leigh on Sea, SS9 1HQ [JO01IM, TQ88]
M1 ANL D V Clapp, 1 Berks Rd, Bristol, BS7 8EX [IO81QL, ST57]
M1 ANM A N Maylin, 221 Branksome Ave, Stanford-le-Hope, SS17 8DD [JO01FM, TQ68]
M1 ANN A J Webb, 53 Highland Way, Redditch, B98 7RH [IO92AG, SP06]
M1 ANO P G Worledge, 1 Rushlade Cl, Paignton, TQ4 7BZ [IO80FK, SX85]
MM1 ANP Dr J A Tobias, 15 Rozelle Dr, Newton Mearns, Glasgow, G77 6YU [IO75TS, NS55]
M1 ANQ D J Hirst, 66 Turncroft Ln, Stockport, SK1 4AB [IO83WJ, SJ99]
M1 ANR R Hall, 12 West Ln, Edwinstowe, Mansfield, NG21 9QT [IO93LE, SK66]
M1 ANS Details withheld at licensee's request by SSL.
M1 ANT D J Simcock, 51 Broadway, Stockport, SK2 5SF [IO83WJ, SJ98]
MW1 ANU G C Roberts, 23 Heol Dirion, Coedpoeth, Wrexham, LL11 3HL [IO83LB, SJ25]
M1 ANV Details withheld at licensee's request by SSL.
M1 ANW M Briddon, 5 Spencer St., Chesterfield, S40 4SD [IO93GF, SK37]
M1 ANX M Fearnley, 23 Sycamore Rd, Eccles, Manchester, M30 8LH [IO83TL, SJ79]
M1 ANY G A King, Perrymill, Smiths Green, Mathon, Malvern, WR13 5PE [IO82TC, SO74]
M1 ANZ P D Dongray, 353 Dereham Rd, Norwich, NR2 3UT [JO02PP, TG20]
M1 AOA Details withheld at licensee's request by SSL.
M1 AOB R C Pentney, 4 Caley Rd, Tunbridge Wells, TN2 3BL [JO01DD, TQ54]

M1 AOC Details withheld at licensee's request by SSL.
M1 AOD V L Bolger, Little Annaside, Bootle Station, Millom, Cumbria, LA19 5XL [IO84HG, SD08]
MM1 AOE I W Davidson, 61 Darley Rd, Cumbernauld, Glasgow, G68 0JR [IO85AX, NS77]
M1 AOF A S Wainwright, 6 The Cres, Royal Hospital School, Ipswich, Suffolk, IP9 2RT [JO01NX, TM13]
M1 AOG S P Moriarty, 39 Chetwode, Banbury, OX16 7QW [IO92HB, SP44]
M1 AOH Details withheld at licensee's request by SSL.
M1 AOI N J Pollard, 5 The Stackfield, Wirral, L48 9XS [IO83KJ, SJ28]
M1 AOJ Details withheld at licensee's request by SSL.
M1 AOK R J Cranwell, 39 Rivergreen, Nottingham, NG11 8FW [IO92JV, SK53]
M1 AOL J M Pepper, 9 Hornbeam Cl, Owlsmoor, Camberley, Surrey, GU15 4UE [IO91PI, SU86]
M1 AON A R Albinson, 16 Teesdale Rd, Long Eaton, Nottingham, Notts, NG10 3PG [IO92IV, SK43]
MI1 AOO H P McGoldrick, 11 Burnbank, Cookstown, BT80 8DX [IO64PP, H87]
M1 AOP Details withheld at licensee's request by SSL.
M1 AOQ Details withheld at licensee's request by SSL.
M1 AOR S E Darrigan, 32 Brownlow Rd, Wirral, L62 1AU [IO83MI, SJ38]
M1 AOS A Jeffs, 789 Borough Rd, Birkenhead, L42 6QN [IO83LI, SJ38]
M1 AOT L Cranwell, 39 Rivergreen, Nottingham, NG11 8FW [IO92JV, SK53]
M1 AOU R R Pinchen, 9 Orwell Cl, Swindon, SN2 3LZ [IO91CO, SU18]
M1 AOV H J Delafield, 205 South Ave, Abingdon, OX14 1QU [IO91IQ, SU49]
M1 AOW Details withheld at licensee's request by SSL.
M1 AOX J P O'Neill, 1 Links Ave, Little Sutton, South Wirral, L66 1QS [IO83MG, SJ37]
M1 AOY Details withheld at licensee's request by SSL.
M1 AOZ H B Cartwright, 405 Marion House, Fitzroy Rd, London, NW1 8UD [IO91WM, TQ28]
M1 APB S E Sanders, 9 Kyl Cober Parc, Stoke Climsland, Callington, PL17 8PH [IO70UN, SX37]
MM1 APC J W Hulme, 180 Crosslet Rd, Dumbarton, G82 2LH [IO75RW, NS47]
M1 APD P C Lincoln, Providence Cottage, Great Glemham, Saxmundham, Suffolk, IP17 2DN [JO02RE, TM36]
M1 APE M Brooks, 20 Watergate Ln, Leicester, LE3 2XP [IO92JO, SK50]
M1 APF J L Thompson, 97 King Ln, Leeds, LS17 5AX [IO93FU, SE33]
MM1 APG A W McTaggart, 131 Huron Ave, Livingston, EH54 6LQ [IO85FV, NT06]
M1 APH P Hildebrand, 1 Lloyd Cl, Heslington, York, YO1 5EU [IO93LW, SE65]
M1 API D Wilson, 50 Langstone Rd, Portsmouth, PO3 6BX [IO90LT, SU60]
M1 APJ J J Coombe, 20 Victoria Rd, Mount Charles, St. Austell, PL25 4QD [IO70OI, SX05]
M1 APK P A Henson, 29 Station Rd, Rainworth, Mansfield, NG21 0AH [IO93KC, SK55]
M1 APL A G Speakman, 12 Allerton Ave, Leeds, LS17 6RF [IO93FU, SE33]
M1 APN Details withheld at licensee's request by SSL.
MM1 APO G J Davidson, 2 Chapel Ct, Wigtown, Newton Stewart, DG8 9ET [IO74SU, NX45]
M1 APP D J James, 82 Overbury Cl, Birmingham, B31 2HD [IO92AJ, SP07]
M1 APQ M R Hodson, 93 Deer Park Rd, Fazeley, Tamworth, B78 3SZ [IO92DO, SK10]
MW1 APR Details withheld at licensee's request by SSL.
MM1 APS C A Stuart, 16 Ladylands Terr, Selkirk, TD7 4BB [IO85ON, NT42]
M1 APT I Waterhouse, Little Bracken, 9 Willingdon Drove, Eastbourne, BN23 8AL [JO00DT, TQ60]
M1 APU R Buston, 36 Hawthorn Rd, Exeter, EX2 6EA [IO80FR, SX99]
M1 APV T Graham, 59 Hillside Rd, Ramsbottom, Bury, BL0 9NJ [IO83UP, SD71]
M1 APW Details withheld at licensee's request by SSL.
M1 APX M Simpson, 23 Main St., Marston Trussell, Market Harborough, LE16 9TY [IO92ML, SP68]
M1 APY A P Young, 41 Lewis Rd, Radford Semele, Leamington Spa, CV31 1UQ [IO92GG, SP36]
MM1 APZ M Metz, 31C Logie Ave, Aberdeen, AB16 7TQ [IO87WE, NJ90]
M1 AQA W M Huddleston, 17 Moorside Rd, Brookhouse, Lancaster, LA2 9PJ [IO84PB, SD56]
M1 AQB F E Strike, 66 Bad Bargain Ln, Burnholme, York, YO3 0LW [IO93LX, SE65]
M1 AQC J K Jedrzejewski, 7 Shedfield Way, East Hunsbury, Northampton, NN4 0SD [IO92NE, SP75]
M1 AQD G Lewis, 42 Ladywood Rd, Ilkeston, DE7 4NE [IO92IX, SK44]
M1 AQE M Clarke, 8 Winship Terr, Newcastle upon Tyne, NE6 2JY [IO94FX, NZ26]
M1 AQF E B Bennett, 57 Vicarage Ln, Coventry, CV7 9AD [IO92FL, SP38]
M1 AQG F J Dearling, 5 William Booth House, Queen St., Portsmouth, PO1 3JB [IO90KT, SU60]
M1 AQH Details withheld at licensee's request by SSL.
M1 AQI M A Sanderson, 14 Hazelwood Ave, York, YO1 3PD [IO93LW, SE65]
M1 AQJ P Jackson, 18 Avon St., Stockport, SK3 8DR [IO83WJ, SJ88]
M1 AQK Details withheld at licensee's request by SSL.
MM1 AQL Details withheld at licensee's request by SSL.
MM1 AQM Details withheld at licensee's request by SSL.
M1 AQN P S Wilson, 14 Jerome Way, Shipton on Cherwell, Kidlington, OX5 1JT [IO91IU, SP41]
M1 AQO G B Nutsey, 5 Highfield Ave, Worsley, Manchester, M28 1AL [IO83SM, SD70]
M1 AQP C J Chapman, 9 Edinburgh Ave, Sawston, Cambridge, CB2 4DW [JO02CD, TL45]
M1 AQQ Details withheld at licensee's request by SSL.
M1 AQR A J Cooper, Farley Cottage, Farley Ln, Great Haywood, Stafford, Staffs, ST18 0RA [IO92AT, SK02]
M1 AQS R J Watson, 12 Orchard Cres, Penkridge, Stafford, ST19 5BT [IO82WR, SJ91]
M1 AQT A Morris, 2 Brancote Mount, Birkenhead, L43 6XS [IO83LI, SJ38]
M1 AQU P Mulliner, 9 Pleasant Ave, Great Houghton, Barnsley, S72 0BU [IO93HN, SE40]
M1 AQV T M Larkman, 207 Watford Rd, Croxley Green, Rickmansworth, WD3 3RY [IO91SP, TQ09]
M1 AQW Details withheld at licensee's request by SSL.
M1 AQX A M Pell, 1 Oak Gr, Daventry, NN11 5XG [IO92KG, SP56]
M1 AQY G M Cottam, 38 Pinewood Dr, Morpeth, NE61 3SX [IO95DE, NZ18]
M1 AQZ M D G Stead, 14 Norma Cres, Whitley Bay, NE26 2PD [IO95GA, NZ37]
M1 ARA Details withheld at licensee's request by SSL.
M1 ARB M Thomas, Pen Barri, 3 Uplands Vean, Truro, TR1 1NH [IO70LG, SW84]
M1 ARC Details withheld at licensee's request by SSL.
M1 ARE Details withheld at licensee's request by SSL.
M1 ARF D Riley, 9 Century Ave, Mansfield, NG18 5EE [IO93JD, SK56]
M1 ARG P Wakelam, 6 Stockton Rd, Durham, DH1 3DX [IO94FS, NZ24]
M1 ARH W H Mountford, 3 Spurstow Cl, Birkenhead, L43 2NQ [IO83LJ, SJ28]
M1 ARI M F Axford, The Rectory, Englands Ln, Queen Camel, Yeovil, BA22 7NN [IO81RA, ST52]
M1 ARJ Details withheld at licensee's request by SSL.
M1 ARK Details withheld at licensee's request by SSL.
M1 ARL D Edwards, 28 Solingen Est, Blyth, NE24 3EP [IO95GC, NZ38]
MW1 ARM D J Rees, y Coed, Tan Lan Hill, Tan Lan, Holywell, Flintshire, CH8 9JB [IO83II, SJ18]
M1 ARN Details withheld at licensee's request by SSL.
M1 ARO P J Cusson, 18 Austerfield Ave, Doncaster, DN5 9QU [IO93KM, SE50]
M1 ARP A Davies, 6 Lees Meadow, Talaton, Exeter, EX5 2SG [IO80IS, SY09]
M1 ARQ J D Campbell, 23 Rowan Ave, High Wycombe, HP13 6JA [IO91PP, SU89]
M1 ARR M S Moss, 42 Burton Bank Ln, Moss Pit, Stafford, ST17 9JW [IO82WS, SJ92]
M1 ARS H Cawley, 11 Cleveland Way, Winsford, CW7 1QL [IO83RE, SJ66]
M1 ART F Melhuish, Alverdean, Mile End Rd, Coleford, GL16 7QD [IO81QT, SO51]
M1 ARU R Stanley, 18 Southdown Rd, Minster on Sea, Sheerness, ME12 3BG [JO01JJ, TQ97]
M1 ARV J M Hrycan, 40 Marina Dr, Marple, Stockport, SK6 6JL [IO83WJ, SJ98]
M1 ARW Details withheld at licensee's request by SSL.
M1 ARX S Russell, 13 Burdett Rd, Crowborough, TN6 2EN [JO01CB, TQ53]
MI1 ARY P Carson, 10 Belgravia Ave, Bangor, BT19 6XA [IO74EP, J58]
M1 ARZ Details withheld at licensee's request by SSL.
M1 ASA K Wrack, 18 Carrs Rd, Cheadle, SK8 2EE [IO83VJ, SJ88]
M1 ASB I R Smith, 12 Arnside Cl, High Ln, Stockport, SK6 8AN [IO83XI, SJ98]
M1 ASC D A Whitfield, 18 Yates Cl, Great Sankey, Warrington, WA5 1XH [IO83QJ, SJ58]
M1 ASD L J Hrycan, 40 Marina Dr, Marple, Stockport, SK6 6JL [IO83WJ, SJ98]
M1 ASE Details withheld at licensee's request by SSL.
M1 ASF Details withheld at licensee's request by SSL.
MW1 ASG J Johnson, 42 Winchester Way, Gresford Park, Gresford, Wrexham, LL12 8HL [IO83MC, SJ35]
M1 ASH A Tobin, 22 Brookway Dr, Charlton Kings, Cheltenham, GL53 8AJ [IO81XV, SO92]
MM1 ASI R D Smith, 13 Clerics Hill, Kirkliston, EH29 9DP [IO85HW, NT17]
MI1 ASJ Details withheld at licensee's request by SSL.
M1 ASK W H Booker, 3 Hollybank Ave, Sheffield, S12 2BL [IO93HI, SK38]
M1 ASM R J Guscott, 14 Polwhele Rd, Tregurra, Truro, TR1 1RF [IO70LG, SW84]
MI1 ASN P B McMahon, 26 Ballycraigy Rd, Newtownabbey, Co Antrim, BT36 8ST [IO74AQ, J38]
M1 ASO Details withheld at licensee's request by SSL.
M1 ASP R B I Rutherford, 3 Stevenson St., Oban, PA34 5NA [IO76FL, NM83]
M1 ASQ J R J Stevens, Fourwinds, Trencreek, Newquay, Cornwall, TR8 4NP [IO70LJ, SW86]
M1 ASR G L Jefferies, 34 Marshall Terr, Leeds, LS15 8EA [IO93GT, SE33]
M1 ASS Details withheld at licensee's request by SSL.
M1 ASU R Birch, Hill View, Little Rissington, Cheltenham, Glos, GL54 2ND [IO91DV, SP11]
M1 ASV A S Evans, 45 Parkway, Romford, RM2 5PL [JO01CO, TQ59]
M1 ASW Details withheld at licensee's request by SSL.
M1 ASX A J F Gasking, 48 Eastlea Ave, Watford, WD2 4RH [IO91TQ, TQ19]
M1 ASY Details withheld at licensee's request by SSL.
M1 ASZ E Willox, 3 Pound Gate, Hassocks, BN6 9LU [IO90WW, TQ31]
M1 ATA A T N Betts, 5 Glebelands, Biddenden, Ashford, TN27 8EA [JO01HC, TQ83]
M1 ATB G S Gale, 107 Parkway Vale, Leeds, LS14 6XE [IO93GT, SE33]
M1 ATC Flt Lt R B Courtney Air Training Corps, 5 Bute Cl, Highworth, Swindon, SN6 7HN [IO91DP, SU29]
M1 ATD R T Burden, 78 High Kingsdown, Bristol, BS2 8EP [IO81QL, ST57]
M1 ATF Details withheld at licensee's request by SSL.
M1 ATG P Robinson, 59 Bouverie Rd West, Folkestone, CT20 2RN [JO01OB, TR23]
M1 ATH D Parnell, 9 Mayfield Rd, Carlton Colville, Lowestoft, NR33 8RF [JO02UK, TM59]
M1 ATI R J Proctor, 11 Redfern Rd, Winsford, CW7 1NE [IO83RE, SJ66]
M1 ATJ P E Whitby, 90 Manor Rd, Martlesham Heath, Ipswich, IP5 7SY [JO02OB, TM24]
M1 ATK Details withheld at licensee's request by SSL.
M1 ATL R T Neilson, 13 Meadow Head Ave, Whitworth, Rochdale, OL12 8TH [IO83VP, SD81]
M1 ATM Details withheld at licensee's request by SSL.
M1 ATN S E Sellman, Ireland Farm, Banbury Rd, Gaydon, Warwick, CV35 0HH [IO92GE, SP35]
MM1 ATO A G Dixon, 9 Riverford Cres, Conon Bridge, Dingwall, IV7 8HL [IO77SN, NH55]
M1 ATP A Plitsch, 125 Bridge Rd, Oulton Broad, Lowestoft, NR33 9JU [JO02UL, TM59]
M1 ATQ R H Drummond, Harrogate Ladies College, Clarence Dr, Harrogate, HG1 2QG [IO93FX, SE25]
MM1 ATR C Robinson, 5 Camsail Rd, Rosneath, Helensburgh, G84 0RZ [IO76OA, NS28]
MM1 ATS J D Reid, Station House, 41 Alloway, Murdoch Lone, Ayr, Ayrshire, KA7 4PY [IO75QK, NS31]
M1 ATT A J Hutchinson, 12 Hawksnest Gdns East, Leeds, LS17 7JQ [IO93FU, SE33]
M1 ATU M C Bloss, 20 Barry Walk, Brighton, BN2 2HP [IO90WT, TQ30]
M1 ATV R J Ballard, 42 Milton Ave, Alfreton, DE55 7LA [IO93HC, SK45]
M1 ATW Details withheld at licensee's request by SSL.
MM1 ATX A M Finlayson, 39 Paisley Ave, Edinburgh, EH8 7LG [IO85KW, NT27]
MM1 ATY D H Shirley, 15 Seggarsdean Cres, Haddington, EH41 4RH [IO85OW, NT57]
M1 ATZ I Graham, 6 Paulsons Cl, Riseley, Bedford, MK44 1DG [IO92SG, TL06]
M1 AUA M B Turnbull, 10 Neasham Hill, Neasham, Darlington, Co. Durham, DL2 1QY [IO94GL, NZ31]
M1 AUB L V Richards, 10 Blackthorn Way, Alcester, B49 6BW [IO92BF, SP05]
M1 AUC A M Scott, 31 Ravenscourt Rd, Orpington, BR5 2PN [JO01BJ, TQ46]
MM1 AUD F M Baird, 3 Sportsman Walk, Dunrubin, Golspie, Sutherland, KW10 6SH [IO87AX, NC80]
M1 AUE E E Marcus, 140 Limes Rd, Hardwick, Cambridge, CB3 7XX [JO02AF, TL35]
MM1 AUF C F McClintock, 13 St. Andrews Dr, Gourock, PA19 1HY [IO75NW, NS27]
MM1 AUG M R McClintock, 13 St. Andrews Dr, Gourock, PA19 1HY [IO75NW, NS27]
M1 AUH A C Easton, 58 Ellesmere Rd, Stockton Heath, Warrington, WA4 6DZ [IO83QI, SJ68]
MI1 AUI V W Hughes, 7 Craiglands Manor, Newtownabbey, BT36 5FG [IO74AQ, J38]
M1 AUJ Details withheld at licensee's request by SSL.
M1 AUK B J Pittaway, 66 Montrose Ave, Leamington Spa, CV32 7DY [IO92FH, SP36]
M1 AUL Details withheld at licensee's request by SSL.
M1 AUM Details withheld at licensee's request by SSL.
M1 AUN J K Yarnall, 85 Wombourne Park, Wombourne, Wolverhampton, WV5 0LX [IO82VM, SO89]
M1 AUO Rev D R Eady, Little Gables, Hayes Ln, Woodmancote, Cirencester, GL7 7EE [IO91AS, SP00]
M1 AUP A Bailey, 3 Francis St., Stoke on Trent, Staffs, ST6 6LP [IO83VB, SJ85]
M1 AUQ D A Birch, 22 Queensland Rd, Weymouth, DT4 0LR [IO80SO, SY67]
M1 AUR M S Garlick, 212 South Parkway, Leeds, LS14 6EL [IO93GT, SE33]
M1 AUS Details withheld at licensee's request by SSL.
MI1 AUU Details withheld at licensee's request by SSL.
MW1 AUV P B Davies, 4 Caradog Pl, Townhill, Swansea, SA1 6NH [IO81AO, SS69]
M1 AUW T D Roberts, 109 Gordon Ave, Norwich, Norfolk, NR7 0DS [JO02QP, TG20]
M1 AUX J Chadwick, 4 Toronto St., Bolton, BL2 6PE [IO83TN, SD70]
M1 AUY G R West, 15 Hildens Dr, Tilehurst, Reading, RG31 5HW [IO91LK, SU67]
M1 AUZ G Wright, 15 Stretton Dr, Southport, PR9 7DR [IO83MP, SD31]
MM1 AVA C Stevenson, 8 Carlaverock Gr, Tranent, EH33 2EB [IO85MW, NT47]
M1 AVB J Pidgeon, 5 Brook Terr, Axmouth, Seaton, EX12 4AG [IO80LR, SY29]
M1 AVC Dr T M C Moore, Ship To Shore, PO Box 400, Winchester, Hants, SO22 4RU [IO91HB, SU42]
MM1 AVD A E Bishop, Glenlair Mill, Knockvennie, Castle Douglas, Kirkcudbrightshire, DG7 3NX [IO85AA, NX77]
M1 AVE Details withheld at licensee's request by SSL.
MM1 AVG Details withheld at licensee's request by SSL.
MI1 AVH T E Allingham, 17 Coagh Rd, Cookstown, BT80 8RL [IO64PP, H87]
M1 AVI Details withheld at licensee's request by SSL.
M1 AVJ Details withheld at licensee's request by SSL.
M1 AVK D A Parker, 852 Tong Rd, Leeds, LS12 5HE [IO93ES, SE23]
M1 AVL S G Ridgeon, 7 Southlands, Haxby, York, YO3 3PB [IO94LA, SE65]
M1 AVM S J Cooper, 93 Langton Rd, Norton, Malton, YO17 9AE [IO94OD, SE77]
M1 AVN Details withheld at licensee's request by SSL.
M1 AVP Details withheld at licensee's request by SSL.
M1 AVQ Details withheld at licensee's request by SSL.
M1 AVR S McIver, 52 Lorn Rd, Dunbeg, Oban, PA37 1QQ [IO76GK, NM83]
M1 AVS J Banks, 407 Poulton Rd, Wallasey, Merseyside, L44 4DF [IO83LJ, SJ39]
M1 AVT C O Deacon, 27 Gadloch View, Kirkintilloch, Glasgow, G66 5NS [IO75WV, NS67]
M1 AVU G Purrier, Archways, Forge Hill, Lydbrook, GL17 9QS [IO81RU, SO61]
M1 AVV S Linney, 5 Elliscales Ave, Dalton in Furness, LA15 8BW [IO84JD, SD27]
M1 AVW R V Hedges, 2 Sand Hutton Ct, Sand Hutton, York, YO4 1LU [IO94MA, SE65]
M1 AVX D Faunt, 6405 Regent St, Oakland, Ca 94618 1313, USA
M1 AVY Details withheld at licensee's request by SSL.
M1 AVZ K M Goodacre, 6 Kirkland Ave, Birkenhead, L42 6QF [IO83LI, SJ38]
MM1 AWA I B Carruthers, 49 High St., Brydekirk, Annan, DG12 5LY [IO85IA, NY17]
M1 AWB S P Webb, 29 Markfield Cres, Islands Brow, St. Helens, WA11 9LD [IO83PL, SJ59]
M1 AWC M Worrall, 15 Whitegate Dr, Bolton, BL1 8SF [IO83SO, SD71]
M1 AWD Details withheld at licensee's request by SSL.
MI1 AWE D Holland, 35 Ashfield Rd, Clogher, BT76 0HJ [IO64JJ, H54]
M1 AWF R Gould, 22 Vereker Dr, Sunbury on Thames, TW16 6HF [IO91TJ, TQ16]
M1 AWG J M Adams, 12 The Birches, Benfleet, SS7 4NT [JO01GN, TQ78]
M1 AWH Details withheld at licensee's request by SSL.
M1 AWJ Details withheld at licensee's request by SSL.
M1 AWK R P Jackson, 4 Hornbrook Gdns, Plymouth, PL6 6LS [IO70WL, SX46]
M1 AWL D Mitchell, 28C Fraser Rd, Aberdeen, AB25 3UH [IO87WD, NJ90]
MI1 AWM D Reid, 179 Melmount Rd, Sion Mills, Strabane, BT82 9LA [IO64GS, H39]
M1 AWN J R J Saunders, 60 Bowhill, Kettering, NN16 8TU [IO92PJ, SP87]
M1 AWO Details withheld at licensee's request by SSL.
M1 AWP Details withheld at licensee's request by SSL.
M1 AWQ Details withheld at licensee's request by SSL.
M1 AWR C E Harden, 13 Greenfield Rd, Coleford, GL16 8BY [IO81QT, SO51]
M1 AWS A Jones, 35 St. Marys Cl, Aspull, Wigan, WN2 1RL [IO83RN, SD60]
MW1 AWT P McCarthy, 43 Bodnant Rd, Llandudno, LL30 1LT [IO83CH, SH78]
MI1 AWU P P Connolly, 94 North Parade, Belfast, BT7 2GJ [IO74BN, J37]
MM1 AWV R Lynch, 21 Carnoustie Ave, Gourock, PA19 1HF [IO75NW, NS27]
M1 AWX S H Yendell, 35 Chester Rd, Newquay, TR7 2RH [IO70LJ, SW86]
M1 AWY Details withheld at licensee's request by SSL.
M1 AWZ Details withheld at licensee's request by SSL.
MM1 AXA Details withheld at licensee's request by SSL.
M1 AXB Details withheld at licensee's request by SSL.
M1 AXC P A Higson, 38 Oakfield Rd, Keynsham, Bristol, BS18 1JH [IO81SJ, ST66]
M1 AXD P J Gartell, 20 Crouch Ct, Forresters Rd, Tewkesbury, GL20 5TH [IO81WX, SO83]
M1 AXE G J Ecclestone, 45 Lytham Rd, Rugby, CV22 7PG [IO92II, SP47]
M1 AXF Details withheld at licensee's request by SSL.
M1 AXG D J C Tucker, 2 Meyrick Park Mns s, Bodorgan Rd, Bournemouth, BH2 6NH [IO90BR, SZ09]
MI1 AXH Details withheld at licensee's request by SSL.
M1 AXI Details withheld at licensee's request by SSL.
M1 AXJ S J Gover, 13 Alan Gr, Fareham, PO15 5HQ [IO90JU, SU50]
MM1 AXK G Ritchie, 4 Ravenswood Rise, Dedridge West, Livingston, EH54 6PE [IO85FV, NT06]
M1 AXL Details withheld at licensee's request by SSL.
M1 AXM D Russell, 16 Eyewell Green, Seaton, EX12 2BW [IO80LR, SY29]
M1 AXN Details withheld at licensee's request by SSL.
M1 AXO Details withheld at licensee's request by SSL.
M1 AXP Details withheld at licensee's request by SSL.
M1 AXQ J McKinney, 7 Bramleys, Kingston, Lewes, BN7 3LF [IO90XU, TQ30]
M1 AXR Details withheld at licensee's request by SSL.
MW1 AXS J L Phelps, 10 Sunnycrest, Newbridge, Newport, NP1 5FZ [IO81KQ, ST29]
M1 AXT D Wallis, 12 Fitzroy Dr, Leeds, LS8 1RW [IO93FT, SE33]
M1 AXU G H Jesson, 13 Strathmore Rd, Hinckley, LE10 0LW [IO92HM, SP49]
MW1 AXV Details withheld at licensee's request by SSL.
M1 AXW J R M Bumford Salop Amateur Radio Soc, 19 Bewdley Ave, Telford Est, Monkmoor, Shrewsbury, SY2 5UQ [IO82PR, SJ51]
M1 AXX R W Brotherton, Richanchor, Mill Ln, Acaster Malbis, York, YO2 1UJ [IO93KV, SE54]
M1 AXY Details withheld at licensee's request by SSL.
M1 AXZ Details withheld at licensee's request by SSL.
M1 AYA P R Booth, 61 Coalpit Ln, Rugeley, WS15 1EW [IO92AR, SK01]
M1 AYB Details withheld at licensee's request by SSL.
M1 AYC A J Booth, 35 Gillamore Dr, Whitwick, Coalville, LE67 5PA [IO92HR, SK41]
M1 AYD Details withheld at licensee's request by SSL.
MI1 AYE Details withheld at licensee's request by SSL.

MI1 AYF J P L Jones, 107 Belfast Rd, Whitehead, Carrickfergus, BT38 9SU [IO74DR, J49]
M1 AYG R S Sanders, Fern Cottage, 4 Railway Terr, Burford Rd, Lechlade, GL7 3EP [IO91DQ, SP20]
M1 AYH R Harrison, Terendak, Thorn Ln, Lower Burraton, Saltash, Cornwall, PL12 4JN [IO70VJ, SX45]
M1 AYI Details withheld at licensee's request by SSL.
M1 AYK G Baker, 17 Gr Farm Cl, Leeds, LS16 6DA [IO93EU, SE23]
MI1 AYL B J P Byrne, Suncrest, 6 Holymount Rd, Gilford, Craigavon, BT63 6AT [IO64UJ, J04]
M1 AYM Details withheld at licensee's request by SSL.
M1 AYN J R Hewlett, 28 Coombs Rd, Coleford, GL16 8AY [IO81QT, SO51]
M1 AYP B Jones, 5 South Walk, Birmingham, B31 3HY [IO92AJ, SP07]
M1 AYR K J Miller, 15A Holly Cl, Cherry Willingham, Lincoln, LN3 4BH [IO93SF, TF07]
MW1 AYS N Smith, 2 Fitzhamon Rd, Porthcawl, CF36 3JA [IO81DL, SS87]
M1 AYT R Daines, 49 Billet Rd, London, E17 5DL [IO91XO, TQ39]
M1 AYU Details withheld at licensee's request by SSL.
M1 AYV T W Owens, 31 Oakfields, Burnopfield, Newcastle upon Tyne, NE16 6PQ [IO94DV, NZ15]
MM1 AYW Details withheld at licensee's request by SSL.
M1 AYX Details withheld at licensee's request by SSL.
M1 AYY L Rosen, 363 Somercotes, Basildon, SS15 5UG [JO01FN, TQ68]
M1 AYZ K Walker, 14 Birch Ave, Shildon, DL4 2EE [IO94EP, NZ22]
M1 AZA M A Gould, 36 Wistaria Rd, Wisbech, PE13 3RH [JO02CQ, TF41]
M1 AZB J Titterton, 33 Jersey Rd, Crawley, RH11 9QB [IO91VC, TQ23]
M1 AZC G A Taylor, 12 Repton Dr, Newcastle, ST5 3JF [IO82VX, SJ84]
M1 AZD A J Clayton, 15 Linley Cl, Aldridge, Walsall, WS9 0ES [IO92AO, SK00]
M1 AZE Details withheld at licensee's request by SSL.
M1 AZF S W Arbuckle, 8 Hungerford Rd, Calne, SN11 9BG [IO91AK, SU07]
M1 AZG A J Ruston, 68 Cotwall End Rd, Dudley, DY3 3EN [IO82WM, SO99]
M1 AZH L J Hill, 65 The Boundary, Oldbrook, Milton Keynes, MK6 2QS [IO92PA, SP83]
MW1 AZI S Dunlop, 13 Queen St., Nantyglo, Brynmawr, NP3 4LZ [IO81KS, SO10]
M1 AZJ A Bottrell, 10 Harbour Rd, Par, PL24 2BB [IO70PI, SX05]
M1 AZK Details withheld at licensee's request by SSL.
MI1 AZL R M Lyttle, 62 Alveston Park, Carryduff, Belfast, BT8 8RP [IO74BM, J36]
M1 AZM P Jefferson, 20 Buckstone Gr, Leeds, LS17 5HW [IO93FU, SE23]
M1 AZN R S P Freemantle, 2 Hamilton Dr, Guildford, GU2 6PL [IO91QG, SU95]
M1 AZO K R Austen, 12B Downs Rd, Folkestone, CT19 5PW [JO01OC, TR23]
M1 AZP A P Hunt, 21 Park View, New Malden, KT3 4AY [IO91VJ, TQ26]
M1 AZQ A P Beale, 1 Rubens Rd, Northolt, UB5 5JH [IO91TM, TQ18]
MW1 AZR R K Snelling Gwent Raynet Group, 91 Oakfield Rd, Newport, South Wales, NP9 4LP [IO81LO, ST28]
MI1 AZS P J Milliken, 15 Windslow Green, Carrickfergus, BT38 9BA [IO74CR, J48]
M1 AZT Details withheld at licensee's request by SSL.
M1 AZU Details withheld at licensee's request by SSL.
M1 AZV R S Aston, 88 Cotgrave Ln, Tollerton, Nottingham, NG12 4FY [IO92LV, SK63]
M1 AZW N P Bolt, Esc 2 73 Rue Pereire, 78100 St Germain En Lame, France
M1 AZX Details withheld at licensee's request by SSL.
M1 AZY J P Drummond, 60 Park Ln, Exeter, EX4 9HP [IO80GR, SX99]
M1 AZZ Details withheld at licensee's request by SSL.
M1 BAA Details withheld at licensee's request by SSL.
MM1 BAB M Paterson, 9 Lochinvar Pl, Bonnybridge, FK4 2BL [IO85BX, NS87]
M1 BAC J A Booth, 35 Gillamore Dr, Whitwick, Coalville, LE67 5PA [IO92HR, SK41]
M1 BAD D R Redding, 53 Cadwell Dr, Maidenhead, SL6 3YS [IO91PM, SU87]
MM1 BAE A W Craig, 90 Leslie Terr, Aberdeen, AB25 3XB [IO87WD, NJ90]
M1 BAF J R Mellor, 38 New Acres, Newburgh, Wigan, WN8 7TU [IO83OO, SD41]
MM1 BAH R J Hunter, 42 Grosvenor Pl, Aberdeen, AB25 2RE [IO87WD, NJ90]
M1 BAI A J Saunders, 128 Foxcroft Dr, Wimborne, BH21 2LA [IO90AT, SU00]
MW1 BAJ J S Alexander, 34 Trefelin St., Port Talbot, SA13 1DQ [IO81CO, SS79]
M1 BAL J D Barber, 14B Cliff Dr, Radcliffe on Trent, Nottingham, NG12 1AX [IO92LW, SK64]
M1 BAM Details withheld at licensee's request by SSL.
M1 BAN T D Baldwin, 51 Queens Rd, Broadstairs, CT10 1PG [JO01RI, TR36]
M1 BAO D P A Legg, 20 Beechwood Cres, Eastbourne, BN20 8AE [JO00DS, TV59]
MW1 BAP R E Faulkner, The Coppins, Cefn Gorwydd, Llangammarch Wells, Powys, LD4 4DP [IO82EC, SN84]
M1 BAQ W Worsley, 7 West Dr, Swinton, Manchester, M27 4ED [IO83UM, SD70]
M1 BAR G S Bleads Bar-Packers Contest Group, 11 Surrey Way, Brinnington, Stockport, SK5 8DE [IO83WK, SJ99]
M1 BAS C L Bastin, 42 Peterborough Rd, Exeter, EX4 2EG [IO80FR, SX99]
M1 BAT C J Grimsditch, 48 Sterndale Rd, Stockport, SK3 8QU [IO83WJ, SJ88]
M1 BAU Details withheld at licensee's request by SSL.
M1 BAV G Noble, 96 Foxroyd Ln Est, Dewsbury, WF12 0BD [IO93EP, SE21]
M1 BAW I C Beith, Beechcroft, 18 Ave Rd, New Milton, BH25 5JP [IO90ES, SZ29]
M1 BAX E S Owen, 4 Madin Rd, Tipton, DY4 8JT [IO82XM, SO99]
MM1 BAY Details withheld at licensee's request by SSL.
M1 BAZ Details withheld at licensee's request by SSL.
M1 BBA I P Randles, 29 St. Peters Mews, Birkenhead, L42 1RT [IO83MI, SJ38]
M1 BBB P S Marshall, Thrushel View, Tinhay, Lifton, Devon, PL16 0AJ [IO70UP, SX38]
M1 BBC G D Eddowes Ariel Radio Group (Bbc Group), Flat 1, 47 The Avenue, Ealing, London, W13 8JR [IO91UM, TQ18]
M1 BBD N Escreet, 55 Grammar School Rd, Hull, HU5 4NX [IO93TS, TA03]
MI1 BBE S Calvert, 2 Wanstead Gdns, Dundonald, Belfast, BT16 0ET [IO74CN, J47]
M1 BBF M V Lonsdale, 5 Brookvale Rd, Langley Mill, Nottingham, NG16 4AX [IO93IA, SK44]
M1 BBG Details withheld at licensee's request by SSL.
M1 BBH C M Tan, 28B Staghill Ct, University of Surrey, Guildford, Surrey, GU2 5XH [IO91QF, SU95]
M1 BBI F A Purvis, 38 Manners Gdns, Seaton Delaval, Whitley Bay, NE25 0DR [IO95FB, NZ37]
M1 BBJ Details withheld at licensee's request by SSL.
M1 BBK C J Ellis, 2 Little Fawsley, Fawsley, Daventry, NN11 3BU [IO92JE, SP55]
M1 BBM Details withheld at licensee's request by SSL.
M1 BBN Details withheld at licensee's request by SSL.
M1 BBO E Y O Chung, Harrogate Ladies College, Clarence Dr, Harrogate, HG1 2QG [IO93FX, SE25]
M1 BBP J Mossman Trafford Radio Group, 31 Marlborough Rd, Stretford, Manchester, M32 0AW [IO83UK, SJ79]
MI1 BBQ Details withheld at licensee's request by SSL.
M1 BBR M E Deglos, Saint Etaine, Crellow Ln, Stithians, Cornwall, TR3 7BA [IO70JE, SW73]
M1 BBS D P W Hawkes, 23 Gloucester Ave, Margate, CT9 3NN [JO01QJ, TR37]
M1 BBT Details withheld at licensee's request by SSL.
M1 BBU G Price, 118 Broadstone Rd, Stockport, SK4 5HS [IO83WK, SJ89]
M1 BBV R A Hooks, 23 Eccleston Rd, Kirk Sandall, Doncaster, DN3 1NX [IO93LN, SE60]
M1 BBW L T A Parnell, Deer Park, Stevenstone, Torrington, Devon, EX38 7HY [IO70WW, SS51]
M1 BBX Details withheld at licensee's request by SSL.
M1 BBY I Booley, 24 Calderbrook Rd, Littleborough, OL15 9HL [IO83WP, SD91]
M1 BBZ J A Welsh, 151 St. Stephens Rd, Saltash, PL12 4NH [IO70VJ, SX45]
M1 BCA P A Van Den Bossche, 22 St. James Pl West, Plymouth, PL1 3AT [IO70WI, SX45]
M1 BCB D Ball, 20 Easton Gdns, Borehamwood, WD6 2PJ [IO91UP, TQ29]
M1 BCC P K Jones, 7 Greyswood Rd, Stoke on Trent, ST4 6LF [IO82VX, SJ84]
M1 BCD A W Inglis Walton High School Amateur Rad, 15 Morris Dr, Kingston Hill, Stafford, ST16 3YE [IO82WT, SJ92]
M1 BCE M R Holland Communications Club Anstey Jun, 31 The Avenue, Andover, SP10 3EP [IO91GF, SU34]
M1 BCF Details withheld at licensee's request by SSL.
M1 BCG R W Thompson Bristol Contest Group, 179 Newbridge Hill, Bath, BA1 3PY [IO81TJ, ST76]
M1 BCH N H Macan, Parkside, 2 Dry Hill Park Cres, Tonbridge, Kent, TN10 3BG [JO01DE, TQ54]
MM1 BCI Details withheld at licensee's request by SSL.
M1 BCJ Details withheld at licensee's request by SSL.
M1 BCK Details withheld at licensee's request by SSL.
M1 BCL Details withheld at licensee's request by SSL.
M1 BCM J G Worthing, 11 Copley Cl, Redhill, RH1 2BE [IO91VF, TQ25]
M1 BCN G Lee, 14 Beadnell Way, Newcastle upon Tyne, NE3 3HB [IO95EA, NZ26]
M1 BCO R Burton, 103 Ingrave Rd, Brentwood, CM15 8BA [JO01DO, TQ69]
M1 BCP G Hasman, 23A Hollingwood Cres, Hollingwood, Chesterfield, Derbyshire, S43 2HD [IO93HG, SK47]
MW1 BCQ R J Cobb, 107 High St., Neyland, Milford Haven, SA73 1TR [IO71MR, SM90]
M1 BCR Details withheld at licensee's request by SSL.
M1 BCS A H Holzapfel, Surrey Ct Wey 2/11, University of Surrey, Guildford, Surrey, GU2 5XH [IO91QF, SU95]
M1 BCT Details withheld at licensee's request by SSL.
M1 BCU A K Howe, 29 Constable Cl, Houghton Regis, Dunstable, LU5 5ST [IO91RV, TL02]
M1 BCV P L Johnson, Squirrels, Nepcote Ln, Findon, Worthing, West Sussex, BN14 0SF [IO90TU, TQ10]
M1 BCW Details withheld at licensee's request by SSL.
M1 BCX Details withheld at licensee's request by SSL.
M1 BCY T M Keeler, 72 Grafton Rd, Selsey, Chichester, PO20 0JB [IO90OR, SZ89]
M1 BCZ R M Craggs, 78 Croft Rd, Portland, DT5 2EP [IO80SN, SY67]
M1 BDA Details withheld at licensee's request by SSL.
MM1 BDB M Biagi, Barcosh, Dalry Rd, Beith, KA15 1JJ [IO75QR, NS35]
M1 BDC J F Loy, 3 Westfield Bridge Ct, Workington, CA14 5AW [IO84FP, NX92]
M1 BDD E Reynolds, 2 Blackford Rd, Birmingham, B11 3SH [IO92BK, SP08]
M1 BDE Details withheld at licensee's request by SSL.
M1 BDF Details withheld at licensee's request by SSL.
M1 BDG R Culshaw, 1 Coe Ln, Tarleton, Preston, PR4 6HH [IO83OQ, SD42]
M1 BDH T C E Woods, 13A Brocas St, Eton, Windsor, Berks, SL4 6BW [IO91QL, SU97]
M1 BDI Details withheld at licensee's request by SSL.
M1 BDJ G D W Hamlin, Down Farm Bungalow, Middle Wallop, Stockbridge, Hants, SO20 8EA [IO91FD, SU33]
M1 BDK Details withheld at licensee's request by SSL.
M1 BDL A G Statham, 24 Fulton Cl, High Wycombe, HP13 5SP [IO91OP, SU89]
M1 BDM Details withheld at licensee's request by SSL.
M1 BDN Details withheld at licensee's request by SSL.
M1 BDO Details withheld at licensee's request by SSL.
M1 BDP D R Lewis, 52 Redfern Rd, Stone, ST15 0LG [IO82WV, SJ83]
M1 BDQ J A Tabbal, 40 Drayton Ct, Drayton Gdns, London, SW10 9RH [IO91VL, TQ27]
M1 BDR J C Sargent Essex Raynet(Basildon District, 9 Lee Woottens Ln, Kingswood, Basildon, SS16 5HD [JO01FN, TQ78]
M1 BDS P L Colwell, 16 River View, Gillingham, ME8 6XW [JO01HJ, TQ86]
M1 BDT G M Bhattie, 108 St Thomas Ct, Manchester Rd, Huddersfield, West Yorks, HD1 3HU [IO93CP, SE11]
M1 BDU Details withheld at licensee's request by SSL.
M1 BDV W E Davis, The Nutshell, 1 Poolway Rise, Coleford, Glos, GL16 8DG [IO81QT, SO51]
M1 BDW U J D Rose, 45 Ringstead Cres, Weymouth, DT3 6PT [IO80SP, SY68]
M1 BDX N J Pratt, 11 Rayleigh Rd, Bristol, BS9 2AU [IO81QL, ST57]
M1 BDY J Beard, 1 Mitcham Rd, Ilford, IG3 8QW [JO01BN, TQ48]
M1 BDZ N A Drumm, 1 Beech St., Harrogate, HG2 7PL [IO93FX, SE35]
M1 BEA Details withheld at licensee's request by SSL.
M1 BEB T G Parker, 24 Burrows Cl, Lawford, Manningtree, CO11 2HE [JO01MW, TM03]
M1 BEC P Schofield Eton College Amateur Radio Soc, Fives Ct Flat, Eton College, Windsor, Berks, SL4 6DU [IO91QL, SU97]
M1 BED S S Gould Bedford & District Amateur Rad, 87 Wentworth Dr, Bedford, MK41 8QD [IO92SD, TL05]
M1 BEE N Williams Whitehaven Amateur Radio Club, 1 Dorset Cl, Whitehaven, CA28 8JP [IO84FM, NX91]
M1 BEF A Marzuki, 52A Filey St., Sheffield, South Yorks, S10 2FG [IO93GJ, SK38]
M1 BEG Details withheld at licensee's request by SSL.
M1 BEH I D Norton, The Marina Hotel, 324 Marine Rd, Morecambe, Lancs, LA4 5AA [IO84NB, SD46]
M1 BEI R M White, 75 Fernley Rd, Birmingham, B11 3NP [IO92BK, SP08]
M1 BEJ E H Walker, 104 Fairmile Rd, Christchurch, BH23 2LN [IO90CR, SZ19]
M1 BEK Details withheld at licensee's request by SSL.
M1 BEL M Kirushnamoorthe, 11 Heaton Rd, Manchester, M20 4PX [IO83VK, SJ89]
MM1 BEM Details withheld at licensee's request by SSL.
M1 BEN H Dickson, 40 Montclair Dr, Liverpool, L18 0HB [IO83NJ, SJ48]
M1 BEO D E Tatlow, Merriford House, Cricklade St, Poulton, Cirencester, Glos, GL7 5HT [IO91BQ, SP10]
M1 BEP A M Amos, 41 Gorse Hill, Fishponds, Bristol, BS16 4HW [IO81RL, ST67]
M1 BEQ A K Strange, 24 Stourton Dr, Barrs Ct, Bristol, BS15 7AL [IO81SK, ST67]
M1 BER A Martin, 11 The Mount, Worcester Park, KT4 8UD [IO91VI, TQ26]
M1 BES Details withheld at licensee's request by SSL.
M1 BET A Cullen, Hollytree Farmhouse, Town Drove, Quadring, Spalding, Lincs, PE11 4PU [IO92VV, TF23]
M1 BEV D Ward, 6 Riverton Cl, Lincoln, LN1 3QZ [IO93RG, SK97]
M1 BEW C Sampson, Brookvale, Nooklands, Fulwood, Preston, Lancs, PR2 8XN [IO83PS, SD53]
M1 BEX Details withheld at licensee's request by SSL.
M1 BEY R C Henton, 1 Roker Cl, Nottingham, NG8 5RA [IO92JX, SK54]
M1 BEZ S Vaudrey, 309 Frederick St., Oldham, OL8 4HG [IO83WM, SD90]
M1 BFA S J Goodfield, 47 The Martins, Stroud, GL5 4PG [IO81VR, SO80]
M1 BFB Details withheld at licensee's request by SSL.
M1 BFC D M Eade, 44 Belmont St., Southsea, PO5 1ND [IO90KS, SZ69]
M1 BFD P M Germaney, 1 Oak Pl, Rue Du Temple, St John, Jersey, JE3 4BH
MM1 BFE J Murray, 11 Hazel Rd, Banknock, Bonnybridge, FK4 1LQ [IO85AX, NS77]
M1 BFF A J Buckley, 6 Munro St., Stoke on Trent, ST4 5HA [IO82VX, SJ84]
M1 BFG H C Tribe, The Paddock, Wix Hill, West Horsley, Leatherhead, KT24 6ED [IO91SG, TQ05]
M1 BFH S Howard, 5 York Rd, St. Leonards on Sea, TN37 6PU [JO00GU, TQ81]
M1 BFI Z A Billington, 63 Westfield Rd, Dunstable, LU6 1DN [IO91RV, TL02]
M1 BFJ R D Keay, 16 Willoughby Cl, Penkridge, Stafford, ST19 5QT [IO82WR, SJ91]
M1 BFK C A J Baker, 37 Boulter Cl, Roborough, Plymouth, PL6 7AY [IO70WK, SX56]
M1 BFL S Shenstone, Jubilee Cottage, Marlow Rd, Ln End, High Wycombe, Bucks, HP14 3JP [IO91OO, SU89]
M1 BFM P Barnes, 100 Money Bank, Wisbech, PE13 2JF [JO02CP, TF40]
M1 BFN S Keightley, 11 Sandringham Ave, Wisbech, PE13 3ED [JO02CQ, TF41]
M1 BFO P Aplin, 30 Cheviot Dr, Charvil, Reading, RG10 9QD [IO91NL, SU77]
M1 BFP Details withheld at licensee's request by SSL.
M1 BFQ G P Birch, 12 Solway Ct, Hawthorn Ave, Colchester, CO4 3JW [JO01LV, TM02]
M1 BFR A J Day, 46 Beatrice Ave, Saltash, PL12 4NG [IO70VJ, SX45]
M1 BFS J D Housego, 16 Ligo Ave, Stoke Mandeville, Aylesbury, HP22 5TX [IO91OS, SP81]
MM1 BFT Details withheld at licensee's request by SSL.
M1 BFU C Blandford, 18 Worthing Rd, Patchway, Bristol, BS12 5HX [IO81RM, ST58]
M1 BFV F Richardson, 239 Riverway, Measham, Swadlincote, DE12 7NT [IO92FQ, SK31]
M1 BFW Details withheld at licensee's request by SSL.
M1 BFX J R Belcher, 101 Colne Dr, Romford, RM3 9LA [JO01CO, TQ59]
M1 BFY A N C Davey, Woodstock, 1 The Limes, Amersham, Bucks, HP6 5NW [IO91QQ, SU99]
M1 BFZ Details withheld at licensee's request by SSL.
M1 BGA G Ward, 4 Quay St., Wivenhoe, Colchester, CO7 9DD [JO01LU, TM02]
M1 BGB J P Andrews, 20 Laburnum Cl, Frome, BA11 2UB [IO81UF, ST74]
M1 BGC W M Andrews, 20 Laburnum Cl, Frome, BA11 2UB [IO81UF, ST74]
M1 BGD S Woodford, 31 Seaborough View, Crewkerne, TA18 8JB [IO80OV, ST40]
MW1 BGE R W Roberts, Cefn Triol, Saron, Llandysul, Carmarthenshire, SA44 5HB [IO72TA, SN33]
M1 BGF M K Shearman, 4 McDonough Cl, Fitton Hill, Oldham, Lancs, OL8 2PD [IO83WM, SD90]
M1 BGG A W Lees, 71 Sara House, Larner Rd, Erith, DA8 3RF [JO01CL, TQ57]
MI1 BGH D J Christie, 5 Moneydig Park, Garvagh, Coleraine, BT51 5JP [IO64QX, C81]
MM1 BGI J Martin, 3 Lismore Ave, Edinburgh, EH8 7DW [IO85KW, NT27]
M1 BGJ K J Prior, 74 Walden Way, Frinton on Sea, CO13 0BQ [JO01PU, TM22]
M1 BGK J E Lewis, 12 Eastleigh Rd, Staple Hill, Bristol, BS16 4SQ [IO81SL, ST67]
M1 BGL G L K Langford, 105 King St., Felixstowe, IP11 9DY [JO01QX, TM23]
MW1 BGM S Jones, 6 Heol Will Hopkin, Llangynwyd, Maesteg, CF34 9ST [IO81EO, SS88]
M1 BGN N A Grose, 6 Innage Gdns, Bridgnorth, WV16 4HW [IO82SM, SO79]
M1 BGO O J Stinchcombe, 47 Liddington Rd, Longlevens, Gloucester, GL2 0HL [IO81VU, SO81]
MW1 BGP S K A Lloyd, 41 Coombs Dr, Milford Haven, SA73 2NU [IO71LR, SM90]
M1 BGQ Details withheld at licensee's request by SSL.
M1 BGR D J Wrigley, 3 Whitstable Cl, Chadderton, Oldham, OL9 9LX [IO83WM, SD90]
M1 BGS M T Elvers, 27 Tunbury Ave, Chatham, ME5 9EH [JO01GI, TQ76]
M1 BGT R J Williams, 6 Pennine Way, Charvil, Reading, RG10 9QH [IO91NL, SU77]
M1 BGU Details withheld at licensee's request by SSL.
M1 BGV Details withheld at licensee's request by SSL.
M1 BGW M A Stokes, 70 The Dr, Totton, Southampton, SO40 9EN [IO90GV, SU31]
MM1 BGX A Cromack, 4 Millpark Ave, Oban, PA34 4JN [IO76GJ, NM82]
M1 BGY Details withheld at licensee's request by SSL.
MM1 BGZ J Galloway, 144 Strathkinnes Rd, Kirkcaldy, KY2 5PZ [IO86JC, NT29]
M1 BHA A H Ayres, Stanloe, 3 Balsall St. East, Balsall Common, Coventry, West Midlands, CV7 7FQ [IO92EJ, SP27]
M1 BHB N T Armer, 17 Keswick Rd, Lancaster, LA1 3HJ [IO84OB, SD46]
M1 BHC M L R Lee, 11 Sturrocks, Basildon, SS16 4PQ [JO01FN, TQ78]
M1 BHD Details withheld at licensee's request by SSL.
M1 BHE B R Vickers, 10 Beckett Cres, Dewsbury, WF13 3PW [IO93EQ, SE22]
M1 BHF Details withheld at licensee's request by SSL.
M1 BHG K Bell, 71 Wheatfield Rd, Stanway, Colchester, CO3 5YA [JO01KV, TL92]
M1 BHH B G Firth, Bingley Cottage, Bingley Gr, Woodley, Reading, Berks, RG5 4TT [IO91NK, SU77]
M1 BHI M P Thomas, 36 Seaview Ave, Peacehaven, BN10 8SA [JO00AS, TQ40]
M1 BHJ G E Payne, 2 Little Birch Croft, Bristol, BS14 0JB [IO81RJ, ST66]
M1 BHK Details withheld at licensee's request by SSL.
M1 BHL A E Clack, 174 Rowley Ave, Sidcup, DA15 9LG [JO01BK, TQ47]
M1 BHM Details withheld at licensee's request by SSL.
M1 BHN J D A Chambers, 2 Appleton Way, Hucclecote, Gloucester, GL3 3RP [IO81VU, SO81]
MM1 BHO R A Hopkins, Westmost Cottage, Knockbrex Stables, Borgue, Kirkcudbright, DG6 4UE [IO74VT, NX54]
M1 BHP D P Hartley, 112 Woolmer Rd, Nottingham, NG2 2FD [IO92KW, SK53]
M1 BHQ G L Pridmore, 63 Rotherham Baulk, Carlton in Lindrick, Worksop, S81 9LE [IO93KI, SK58]
M1 BHR Details withheld at licensee's request by SSL.
M1 BHS M C Montanaro, 29 Northd. Rd, Leamington Spa, CV32 6HE [IO92FH, SP36]
M1 BHT Details withheld at licensee's request by SSL.
M1 BHU R P Mullen, 112 Ramsden St., Barrow in Furness, LA14 2BU [IO84JC, SD26]
M1 BHV Details withheld at licensee's request by SSL.
M1 BHW P W Frier, 58 Parklands, Rochford, SS4 1SH [JO01IO, TQ89]
M1 BHX K J Hayden, 41 Millers Way, Honiton, EX14 8JB [IO80JT, ST10]
MW1 BHY Dr I M Jones, 5 Greenhall Park, Johnston, Haverfordwest, SA62 3PT [IO71MS, SM91]
M1 BHZ T J Shepherd, 40 Pheasant Way, Cirencester, GL7 1BL [IO91AR, SP00]
M1 BIA Details withheld at licensee's request by SSL.

M1 BIB P L Brooke, 179 Pinza Cl, Newmarket, CB8 7AR [JO02EG, TL66]
M1 BIC Details withheld at licensee's request by SSL.
MM1 BID J Cameron, Tipperton, 407 Smerclate, Lochboisdale, Isle of South Uist, HS8 5TU [IO67HC, NF71]
MW1 BIE J R Birks, 4 Awel y Mor, Pen-y-Maes, Holywell, Flintshire, CH8 7HA [IO83JG, SJ17]
MM1 BIF T Cassidy, 14 Hillshaw Green, Bourtreehill South, Irvine, KA11 1EQ [IO75QO, NS33]
M1 BIG T.W Read, 57 Ollard Ave, Wisbech, PE13 3HF [JO02CQ, TF41]
M1 BIH Details withheld at licensee's request by SSL.
M1 BII R Jaffar, Blackwater 0/9, Surrey Ct, University of Surrey, Guildford, Surrey, GU2 5XH [IO91QF, SU95]
M1 BIJ Details withheld at licensee's request by SSL.
M1 BIK Details withheld at licensee's request by SSL.
M1 BIL C D Pugh, 16 Park Gate, Somerhill Rd, Hove, BN3 1RL [IO90WT, TQ20]
MM1 BIM M K Long, 51 Fairways, Dunfermline, KY12 0DX [IO86GC, NT08]
M1 BIN Details withheld at licensee's request by SSL.
M1 BIO C Shackleton Heather Grove First School Rad, 159 Halifax Rd, Brighouse, HD6 2EQ [IO93CQ, SE12]
M1 BIP M A Shepherd Brighton Tcp/Ip Group, 17 Friar Cres, Brighton, BN1 6NL [IO90WU, TQ30]
MM1 BIQ R H Harman, 2 Dornoch Ct, Kilwinning, KA13 6QN [IO75PP, NS24]
M1 BIR M Moore, 27 The Moorlands, Bacup, OL13 8BT [IO83VQ, SD82]
M1 BIS J Mallen, 9 Firbeck Cres, Langold, Worksop, S81 9SA [IO93KJ, SK58]
MM1 BIT A I Stewart, 141 Main St., Bellshill, ML4 3DZ [IO75XT, NS76]
M1 BIU M W Redstall, 56 Westmorland Rd, Felixstowe, IP11 9TJ [JO01QX, TM33]
MW1 BIV W T Eldridge, Pontartamddwr, Tregaron, Dyfed, SY25 6LW [IO82AG, SN66]
MI1 BIW R H Robinson, 31 Brantwood Gdns, Antrim, BT41 1HP [IO64VR, J18]
M1 BIX G Perrins, Iona, Cornhill Gdns, Leek, Staffs, ST13 5PZ [IO83XC, SJ95]
M1 BIY J A Laker, 9 Monterey Dr, Havant, PO9 5TQ [IO90MU, SU70]
M1 BIZ Details withheld at licensee's request by SSL.
M1 BJA I J Moth, 38 Annesley Rd, New Moston, Manchester, Lancs, M40 3PB [IO83WM, SD80]
MW1 BJB S K Mandal, 168 Wrexham Rd, Rhostyllen, Wrexham, LL14 4DN [IO83LA, SJ34]
M1 BJC P Marshall, 75 Drewstead Rd, London, SW16 1AA [IO91WK, TQ37]
MW1 BJD A J Jones, 63 North Rd, Loughor, Swansea, SA4 6QF [IO71XP, SS59]
M1 BJE S J Robinson, 140 The St., Kirtling, Newmarket, CB8 9PD [JO02FE, TL65]
MW1 BJF V T Hughes, Manley, 1 Garden Dr, Llandudno, LL30 3LL [IO83CH, SH88]
M1 BJG O Ferula, 28 Allison St., Carstairs Junction, Lanark, ML11 8RG [IO85EQ, NS94]
M1 BJH Details withheld at licensee's request by SSL.
M1 BJI Details withheld at licensee's request by SSL.
MM1 BJJ G McKenna, 54 Taylor St., Ayr, KA8 8AU [IO75QL, NS32]
M1 BJK J F Park, 15 Duncan Gdns, Bath, BA1 4NQ [IO81TJ, ST76]
M1 BJL J Holland, 4 The Avenue, Newton Abbot, TQ12 2BY [IO80EM, SX87]
MI1 BJM R Jones, 155 Ulsterville Park, Portadown, Craigavon, BT63 5HD [IO64SK, J05]
M1 BJN J Bax, 17 Caroland Cl, Smeeth, Ashford, TN25 6RY [JO01LC, TR04]
MM1 BJO R A Lowe, 81 Paradykes Ave, Loanhead, EH20 9LF [IO85KV, NT26]
MM1 BJP A McDermid, 96 Strathleven Dr, Alexandria, G83 9PQ [IO75RX, NS37]
M1 BJQ Details withheld at licensee's request by SSL.
M1 BJR Details withheld at licensee's request by SSL.
M1 BJS G J Ausher, 94 New Rd, Ditton, Aylesford, ME20 6AE [JO01FH, TQ75]
MM1 BJT D Smith, 1076 Aikenhead Rd, Glasgow, G44 4TJ [IO75VT, NS56]
MM1 BJU Details withheld at licensee's request by SSL.
MW1 BJV D J Jones, 3 Clwch, Rhosmeirch, Llangefni, LL77 7SJ [IO73UG, SH47]
MI1 BJW D A Stanley, 57 Schomberg Ave, Belmont Rd, Belfast, BT4 2JR [IO74BO, J37]
MI1 BJX P Bell, 3 Ashton Park, Coleraine, BT52 1NH [IO65QC, C83]
M1 BJY H Y E Quah, University of Surrey, Battersea Ct, Guildford, Surrey, GU2 5XH [IO91QF, SU95]
MM1 BJZ P D Fraser, 28B Afton Rd, Cumbernauld, Glasgow, Lanarkshire, G67 2DR [IO85AW, NS77]
M1 BKA P B Longley, 4 Walkers Yard, Radcliffe on Trent, Nottingham, NG12 2FF [IO92LW, SK63]
M1 BKB W L Geldert, 25 Albert Rd, Morecambe, LA4 4HE [IO84NB, SD46]
M1 BKC Details withheld at licensee's request by SSL.
MI1 BKD Details withheld at licensee's request by SSL.
M1 BKE P J Hewlett, 28 Coombs Rd, Coleford, GL16 8AY [IO81QT, SO51]
M1 BKF W B Hill, 492 Earlham Rd, Norwich, NR4 7HP [JO02PP, TG20]
M1 BKG A D Whiteside, South Lodge, Browhead, St Annes School, Windermere, Cumbria, LA23 1NW [IO84NJ, NY40]
M1 BKH R H Clements, 33 James Carter Rd, Colchester, CO3 5XN [JO01KU, TL92]
M1 BKI J V R Cairns, 26 Roman Way, St. Margarets At Cliffe, Dover, CT15 6AH [JO01QD, TR34]
M1 BKJ J M Colles, The Old Rectory, Church Ln, Bromeswell, Woodbridge, Suffolk, IP12 2PJ [JO02QC, TM35]
MM1 BKK Details withheld at licensee's request by SSL.
M1 BKL P Coddington, Southerly House, 14B North St, Wareham, Dorset, BH20 4AG [IO80WQ, SY98]
MW1 BKM Details withheld at licensee's request by SSL.
M1 BKN J C Robertson, 60 Warburton Rd, Poole, BH17 8SF [IO90AR, SZ09]
MI1 BKO J P Cosgrove, 9 Burnreagh Dr, Newtownards, BT23 8UF [IO74DN, J46]
M1 BKP I Zainal-Abidin, 18 Barnwood Cl, Guildford, GU2 6GG [IO91QG, SU95]
M1 BKQ T B Beadman, Cottage Farm, Shackerstone, Nuneaton, Warks, CV13 6NL [IO92GP, SK30]
MI1 BKR S J Toner, 25 Garnock Hill, Belfast, BT10 0AW [IO74AN, J36]
M1 BKS M Kevern, Wheal Bal, Trewellard, Pendeen, Penzance, TR19 7SS [IO70DD, SW33]
MM1 BKT A G Andrews, 48 Woodburn Terr, Dalkeith, EH22 2HT [IO85LV, NT36]
M1 BKU C S Milton, 28 Hilton Rd, Newton Abbot, TQ12 1BJ [IO80EM, SX87]
M1 BKV D Oakes, 44 Sarum Ave, West Moors, Ferndown, BH22 0ND [IO90BT, SU00]
M1 BKW S W Plant, 17 New Rd, Driffield, YO25 7DJ [IO94SA, TA05]
M1 BKX Details withheld at licensee's request by SSL.
M1 BKY C N Sturt, 22 Cardigan Cl, St. Johns, Woking, GU21 1YP [IO91QH, SU95]
M1 BKZ C R Ransome, 42 Gypsey Rd, Bridlington, YO16 4AZ [IO94VC, TA16]
MI1 BLA M S Chancellor, 55 Brae Hill Park, Belfast, BT14 8FP [IO74AP, J37]
M1 BLB J Woodburn, 4 Dodding Holme, Skelsmergh, Mealbank, Kendal, LA8 9DH [IO84PI, SD59]
MW1 BLC L Chesters, 45 Briar Dr, Buckley, CH7 2AP [IO83LD, SJ26]
MW1 BLD C Smith, Ty-Uchaf, Llanbadrig, Cemaes Bay, Anglesey, LL67 0LN [IO73TK, SH39]
M1 BLE C G Beech, 5 Jenner Way, Weymouth, DT3 6RW [IO80SP, SY68]
M1 BLF Details withheld at licensee's request by SSL.
MW1 BLG A W Williams, 5 Pont y Crychddwr, Llanllyfni, Caernarvon, LL54 6DH [IO73UA, SH45]
M1 BLH J Escreet, The Paddock, Carlton Ln, Aldbrough, Hull, North Humberside, HU11 4RA [IO93WT, TA23]
M1 BLI D Brissenden, 29 Ethel Rd, Norwich, NR1 4DB [JO02PP, TG20]
M1 BLJ S Brion, 165 Kings Head Hill, London, E4 7JG [IO91XP, TQ39]
M1 BLK S G Tott Reading Central Scout District, 8 Cheddington Cl, Tilehurst, Reading, RG3 4HA [IO91LJ, SU66]
MM1 BLL Details withheld at licensee's request by SSL.
M1 BLM P J D Brockett, 220 Goring Rd, Goring By Sea, Worthing, BN12 4PG [IO90TT, TQ10]
M1 BLN A J Hill, 150 Claughton Ave, Crewe, CW2 6ET [IO83SB, SJ75]
M1 BLO P M Hoggard, 41 Helvellyn Cl, Bransholme, Hull, HU7 5AX [IO93UT, TA13]
M1 BLP Details withheld at licensee's request by SSL.
M1 BLQ Details withheld at licensee's request by SSL.
M1 BLR R C Andreang Raywell Park Scout ARS, 6 Beech Ave, Bilton, Hull, HU11 4EN [IO93VS, TA13]
MW1 BLS Details withheld at licensee's request by SSL.
MW1 BLT C E Thomas, 29 Heol Offa, Johnstown, Wrexham, LL14 2BA [IO83LA, SJ34]
M1 BLU Details withheld at licensee's request by SSL.
M1 BLV Details withheld at licensee's request by SSL.
M1 BLW E Banks, 165 Burstall Hill, Bridlington, YO16 5NH [IO94VC, TA16]
M1 BLX Details withheld at licensee's request by SSL.
M1 BLY C R Scoffield, 41 All Saints Way, Aston, Sheffield, S31 0FJ [IO93GJ, SK38]
MI1 BLZ D Kyle, Sea Breezes, Rathlin Island, Ballycastle, Co. Antrim, BT54 6RT [IO65VH, D15]
M1 BMA I M Al-Bulooshi, Benbow Block, Military Rd, HMS Sultan, Gosport, Hants, PO12 3BY [IO90KT, SU50]
M1 BMB I P Goodhew, 38 St. Kilda Rd, London, W13 9DE [IO91UM, TQ18]
M1 BMC K S Termie, 1 Brocks Hill Cl, Oadby, Leicester, LE2 5RB [IO92LO, SP69]
M1 BMD W H Curtis, Yacht Doublit, Capitainerie Cap Monastir, Bp60, 5000 Monastir, Tunisia
M1 BME L W Champion, 46 Kendal Rise Rd, Rubery, Birmingham, West Midlands, B45 9PX [IO92AJ, SP07]
M1 BMF R A Clarke, 52 Lydate Rd, Halesowen, B62 0DW [IO82XK, SO98]
M1 BMG W F Griffin, 13 Hillside Ave, Halesowen, B63 2SR [IO82XL, SO98]
M1 BMH Details withheld at licensee's request by SSL.
M1 BMI G D Fretwell, 17 Cross Ln, Stocksbridge, Sheffield, S30 5AY [IO93GJ, SK38]
MM1 BMJ Details withheld at licensee's request by SSL.
MM1 BMK E W Mitchell, North Gowanwell Cottage, Methlick, Ellon, Aberdeenshire, AB41 7JL [IO87VL, NJ84]
M1 BML M F Carroll, 66 Cambridge Rd, Kings Heath, Birmingham, B13 9UD [IO92BK, SP08]
M1 BMM L Papazoglou, 45 Mosslands Dr, Wallasey, L45 8PF [IO83LK, SJ29]
M1 BMN Details withheld at licensee's request by SSL.
MI1 BMO P M Houston, 37 Bell Doo, Strabane, BT82 9PG [IO64GT, H39]
M1 BMP Details withheld at licensee's request by SSL.
M1 BMQ J Waters, 1 Borrowdale Ave, Blyth, NE24 5LX [IO95FD, NZ28]
M1 BMR C J Robinson, 3 Blacksmiths Copse, Hailsham, BN27 3XB [JO00CU, TQ51]
MM1 BMS T Campbell, 66 Antonine Gdns, Clydebank, G81 6BJ [IO75TW, NS47]
M1 BMT Details withheld at licensee's request by SSL.
M1 BMU E A Woodward, 20 The Vale, Skelton, York, YO3 6YH [IO94KA, SE55]
M1 BMV J Fox, Stonecroft, Horley, Banbury, Oxon, OX15 6BJ [IO92HC, SP44]
M1 BMW J E Burrill, 3 Godfrey Ave, Gosberton, Spalding, PE11 4HF [IO92WU, TF23]
M1 BMX P A Edmunds, Beech Gr, 28A Station Rd, Corton, Lowestoft, NR32 5HF [JO02UM, TM59]
M1 BMY G R Tuby, 128 Knightsbridge Rd, Solihull, B92 8RB [IO92CK, SP18]
M1 BMZ A K Caswell, 42 Medina Rd, Birmingham, B11 3SA [IO92BK, SP18]
M1 BNA C J Thompson, 80 Aston Rd, Willerby, Hull, HU10 6SG [IO93SS, TA02]
M1 BNB Details withheld at licensee's request by SSL.
M1 BNC Details withheld at licensee's request by SSL.
M1 BND P Finlay, 12 Harbottle Ave, Newcastle upon Tyne, NE3 3HR [IO95EA, NZ26]
MI1 BNE J H Grob, 18 Chestnut Glen, Glenavy, Co Antrim, BT29 4GJ [IO64VO, J17]
M1 BNF J Pennell Salisbury Amateur Radio Club, 12 Mons Ave, Bulford Barracks, Salisbury, SP4 9NN [IO91DE, SU14]
M1 BNG R A Smith, 6 Hermitage Cl, North Mundham, Chichester, PO20 6JZ [IO90OT, SU80]
M1 BNH P J Walton, 4 Mansion Ave, Whitefield, Manchester, M45 7SS [IO83UN, SD80]
M1 BNI Details withheld at licensee's request by SSL.
M1 BNJ Details withheld at licensee's request by SSL.
M1 BNK A S Wood, 45 Earl St., Hartlepool, TS24 0DS [IO94JQ, NZ53]
M1 BNL Details withheld at licensee's request by SSL.
M1 BNM Details withheld at licensee's request by SSL.
M1 BNN Details withheld at licensee's request by SSL.
MI1 BNO B Bonnar, 49 Cullycapple Rd, Aghadowey, Coleraine, BT51 4AR [IO65QA, C82]
M1 BNP Details withheld at licensee's request by SSL.
M1 BNQ Details withheld at licensee's request by SSL.
M1 BNR G B Forster Holsworthy Community College, Treloyhan, Deer Valley Rd, Holsworthy, Devon, EX22 6DA [IO70TT, SS30]
MM1 BNS P Cerutti, 5 Durham Pl, Bonnyrigg, EH19 3EX [IO85KU, NT36]
M1 BNT Details withheld at licensee's request by SSL.
M1 BNU A Robertson, 3 Singleton Rd, Weeton, Preston, PR4 3PA [IO83MT, SD33]
M1 BNV C Taylor, 7 Merryfields, Rochester, ME2 3ND [JO01FJ, TQ77]
M1 BNW Details withheld at licensee's request by SSL.
MM1 BNX Details withheld at licensee's request by SSL.
MW1 BNY B J Fitzpatrick, 21 Heol Coed Leyshon, Coytrahene, Coytrahen, Bridgend, CF32 0DT [IO81EN, SS88]
MI1 BNZ L W Wright, 5 Woodview Park, The Donahies, Ireland
M1 BOA G V Heard, The White House, Thurgarton, Norwich, Norfolk, NR11 7PD [JO02OU, TG13]
M1 BOB R Allen, 43 Vowell Cl, Bristol, BS13 9HS [IO81QJ, ST56]
M1 BOC Details withheld at licensee's request by SSL.
M1 BOD P C Hanfrey, 49 Allotment Rd, Niton, Ventnor, PO38 2DZ [IO90IO, SZ57]
MI1 BOE A M Prenter, 156 Greenville Rd, Belfast, BT5 5JY [IO74BO, J37]
M1 BOF L W Blount, 61 New John St., Halesowen, B62 8HH [IO82XL, SO98]
MW1 BOG Details withheld at licensee's request by SSL.
MM1 BOH K A Dearing, Flat 1, 2 New Bells Ct, Edinburgh, Midlothian, EH6 6RY [IO85KX, NT27]
M1 BOI M S Zuppone, via Sebino 32, Roma 00199, Italy
M1 BOJ D A Kinghorn, 36 Brookside Ave, Rainford, St. Helens, WA11 8DF [IO83OM, SD40]
MM1 BOK I Cowie, 29 Invercauld Rd, Aberdeen, AB16 5RP [IO87WD, NJ90]
M1 BOL D Rowlands, 6 Avocet Way, Banbury, OX16 9YA [IO92IB, SP43]
M1 BOM Details withheld at licensee's request by SSL.
M1 BON P E Jeffries, 5 Darnick Rd, Sutton Coldfield, B73 6PE [IO92BN, SP09]
MW1 BOO S Singh, 32 Beatty Ave, Roath Park, Cardiff, CF2 5QT [IO81JM, ST17]
M1 BOP M Riley, 63 Clifford Rd, Ipswich, IP4 1PJ [JO02OB, TM14]
MI1 BOQ Details withheld at licensee's request by SSL.
MW1 BOR C L Hiscocks, 20 Brynderwen Rd, Newport, NP9 8LQ [IO81MO, ST38]
M1 BOS W A Thomas, 11 Arran Green, Rochester, ME2 2ND [JO01FJ, TQ76]
MM1 BOT K Lindsay, 27 Buccleuch Rd, Selkirk, TD7 5DL [IO85NN, NT42]
M1 BOU N A Leach, 17 Keswick Rd, Lancaster, LA1 3HJ [IO84OB, SD46]
M1 BOV O M Parker, 37 Springbank Cres, Gildersome, Morley, Leeds, LS27 7DN [IO93ES, SE22]
M1 BOW G Scothorn, School House, Kirk Balk, Hoyland, Barnsley, S74 9HU [IO93GM, SE30]
M1 BOX R Frostick, 27 Marsh St., Wombwell, Barnsley, S73 0AD [IO93HM, SE40]
M1 BOY P W Smith, 20 Fairfield Gdns, Eastwood, Leigh on Sea, SS9 5SF [JO01HN, TQ88]
M1 BOZ A Thompson, 26 Balmoral Ave, Clitheroe, BB7 2QH [IO83TU, SD74]
M1 BPA R R Hill, 32 Ravensdale Ave, Long Eaton, Nottingham, NG10 4GG [IO92IV, SK43]
M1 BPB C F C Ballam, Merryfield, Mill Rd, Waldringfield, Woodbridge, IP12 4PY [JO02PB, TM24]
M1 BPC O E Ballam, Merryfield, Mill Rd, Waldringfield, Woodbridge, IP12 4PY [JO02PB, TM24]
M1 BPD A J Collins, Chestnuts, 2 St. Teresas Dr, Chippenham, SN15 2BD [IO81WK, ST97]
M1 BPE D Fitzpatrick, 21 Stambridge Rd, Clacton on Sea, CO15 3JR [JO01NT, TM11]
M1 BPF Details withheld at licensee's request by SSL.
M1 BPG Details withheld at licensee's request by SSL.
M1 BPH C S Graham, 22 Tennyson Rd, Colne, BB8 9SD [IO83VU, SD84]
M1 BPI S J Nicholson, 53 Violet Ave, Edlington, Doncaster, DN12 1NW [IO93JL, SK59]
M1 BPJ Details withheld at licensee's request by SSL.
M1 BPK J Bloor, 1 Arthur Rd, Biggin Hill, Westerham, TN16 3DD [JO01AH, TQ45]
M1 BPL A W White, 1 Redhouse, Gallamore Ln, Middle Rasen, Market Rasen, LN8 3UB [IO93TJ, TF08]
M1 BPM J W Richardson, Newton House, East Hauxwell, Leyburn, North Yorks, DL8 5LS [IO94DI, SE19]
M1 BPN A D Burchell, 25 Cherbury Cl, London, SE28 8PG [JO01BM, TQ48]
M1 BPO R W Thornton, 2 Sceptre Gr, New Rossington, Doncaster, DN11 0RW [IO93LL, SK69]
M1 BPP P J Brechany, 24 Teignmouth Ave, Mansfield, NG18 3JQ [IO93KD, SK56]
M1 BPQ D I Fower, 52 North St., Leek, ST13 8DQ [IO83XC, SJ95]
M1 BPR P A Curran North Notts Amateur Radio Soci, 29 Wingfield Ave, Worksop, S81 0SY [IO93KH, SK58]
M1 BPS A P Wellman, Pendil Gr, Market Pl, Northleach, Cheltenham, Glos, GL54 3EJ [IO91BT, SP11]
M1 BPT Details withheld at licensee's request by SSL.
M1 BPU B Lake, Brymore, 1 Eunice Gr, Chesham, Bucks, HP5 1RL [IO91QQ, SP90]
M1 BPV Details withheld at licensee's request by SSL.
M1 BPW P Williams, 26 Downs View Rd, Westbury, BA13 3AQ [IO81VG, ST85]
M1 BPX K Hunt, 81 Front St., Frosterley, Bishop Auckland, DL13 2RH [IO94AR, NZ03]
M1 BPY D Dixey, 102 Blackberry Rd, Stanway, Colchester, CO3 5RZ [JO01KV, TL92]
M1 BPZ S A L McPherson, 7 Burford Ave, Boothville, Northampton, NN3 6AF [IO92NG, SP76]
M1 BQA P R Kilpin, 28 Ash Hayes Rd, Nailsea, Bristol, BS19 2LW [IO81OK, ST47]
M1 BQB G C Stokes, 9 The Haven, Harwich, CO12 4LA [JO01PW, TM23]
M1 BQC S C Sparks, 36 Tormynton Rd, Weston Super Mare, North Somerset, BS22 9HT [IO81MI, ST36]
M1 BQD K J Sparks, 36 Tormynton Rd, Weston Super Mare, North Somerset, BS22 9HT [IO81MI, ST36]
M1 BQE C L Sparks, 36 Tormynton Rd, Weston Super Mare, North Somerset, BS22 9HT [IO81MI, ST36]
M1 BQF J R Stewart, 45 The Cross, Wivenhoe, Colchester, CO7 9QH [JO01LU, TM02]
M1 BQG S K Norman, 18 Everest Pl, Swanley, BR8 7BX [JO01CJ, TQ56]
M1 BQH M J Perry, 2 Riseholme Ave, Wollaton, Nottingham, NG8 2TE [IO92JW, SK53]
M1 BQI B A Bosson, 43A Compton Bassett, Calne, SN11 8RG [IO91AK, SU07]
MI1 BQJ P J Letters, 24 Old Grange Ave, Carrickfergus, BT38 7UE [IO74CR, J48]
M1 BQK D J Viney, Longcroft, The Ridgeway, Fernhurst, Haslemere, GU27 3JU [IO91PB, SU82]
MI1 BQL J Iznerowicz, 12 Cappy St., Ravenhill Rd, Belfast, BT6 8ER [IO74BO, J37]
M1 BQM G W Preedy, 6 Grange Cl, Prescot Rd, Stourbridge, DY9 7LA [IO82WK, SO98]
M1 BQN M P Ballard, 43 Eyewell Green, Seaton, EX12 2BN [IO80LR, SY29]
MW1 BQO E Buckley, 3 Cae Eithin, Minffordd, Penrhyndeudraeth, LL48 6EF [IO72WW, SH53]
M1 BQP I F Wingate, 56 West Ave, Rudheath, Northwich, CW9 7ES [IO83SG, SJ67]
M1 BQQ A J Gardiner, 24 Backbury Rd, Hereford, HR1 1SD [IO82PB, SO54]
M1 BQR S Hall, 58 Lower Meadow Ct, Northampton, NN3 8AX [IO92NG, SP76]
M1 BQS F T Gibson, 27 Flaxlands Ct, Northampton, NN3 8LX [IO92OG, SP76]
M1 BQT F Y Steel, 2 Larkfield, Cholsey, Wallingford, OX10 9QT [IO91KN, SU58]
M1 BQU S P Daniels, 71 Firsby Ave, Croydon, CR0 8TP [IO91XJ, TQ36]
M1 BQV E G Poole, 13 Welford Gdns, Abingdon, OX14 2BN [IO91IQ, SU59]
M1 BQW V Edwards, 232 Earlham Rd, Norwich, NR2 3RH [JO02PP, TG20]
M1 BQX A G Franklin, 71 Parkside, Revesby, Boston, PE22 7NH [IO93XD, TF36]
M1 BQY I L Carter Trowbridge & District Amateur, 12 Bobbin Ln, Westwood, Bradford on Avon, BA15 2DL [IO81UI, ST85]
M1 BQZ W J Rowley, 6 Sea King Cres, Colchester, CO4 4RJ [JO01LV, TM02]
MI1 BRA N E Moore Belfast Royal Academy Amateur, 164 Ardenlee Ave, Belfast, BT6 0AE [IO74BN, J37]
M1 BRB W Strickland, 159 North Circular Rd, London, N13 5EL [IO91WO, TQ39]
M1 BRC R Strickland, 159 North Circular Rd, London, N13 5EL [IO91WO, TQ39]
M1 BRD A Strickland, 159 North Circular Rd, London, N13 5EL [IO91WO, TQ39]
M1 BRE J B Scott, 14 Baddeley Hall Rd, Stoke on Trent, ST2 7JZ [IO83WB, SJ95]
M1 BRF B Bull Furness Amateur Radio Soc, 2 Maylands Gr, Barrow in Furness, LA13 0AN [IO84JD, SD27]
M1 BRG L C Bacon Bridlington Raynet Group, 100 Etherington Dr, Hull, HU6 7JT [IO93TS, TA03]
MM1 BRH R E Mason, 11 Stobbs Terr, Kilwinning, KA13 6JB [IO75PP, NS34]
M1 BRI J P Billingham Brighton College Amateur Radio, 22 Lawday Pl Ln, Farnham, GU9 0BT [IO91OF, SU84]
M1 BRJ O M Suleyman, 61 Victoria Rd, London, N9 9SU [IO91XO, TQ39]
M1 BRK Details withheld at licensee's request by SSL.
M1 BRL Details withheld at licensee's request by SSL.
MM1 BRM J Webster, 40 Greenhill Terr, Knockentiber, Kilmarnock, KA2 0BZ [IO75RO, NS33]
M1 BRN J Dann, 127 Ladybrook Ln, Mansfield, NG18 5JH [IO93JD, SK56]
MI1 BRO S R McCormick, 71 Greyabbey Rd, Ballywalter, Newtownards, BT22 2NY [IO74GM, J66]
M1 BRP R C I Gifford, 107 Cardinal Ave, Kingston upon Thames, KT2 5RZ [IO91UK, TQ17]
M1 BRQ Details withheld at licensee's request by SSL.

M1 BRR P Joslin, Holland Cottage, Kirby Rd, Great Holland, Frinton on Sea, CO13 0HZ [JO01OU, TM22]
MI1 BRS R S Dickey, 8 Coachmans Way, Hillsborough, BT26 6HQ [IO64WL, J26]
MM1 BRT B Connelly, 2 Faside Cres, Wallyford, Musselburgh, EH21 8AH [IO85LW, NT37]
M1 BRU D N Pope, 27 Cross Rd, Cholsey, Wallingford, OX10 9PE [IO91KN, SU58]
M1 BRV Details withheld at licensee's request by SSL.
M1 BRW P H Wells, 7 Kings Meadow, Overton, Basingstoke, RG25 3HP [IO91IF, SU54]
M1 BRX D Seddon, 22 Newton Rd, Lowton St Marys, Warrington, WA3 1EB [IO83RL, SJ69]
M1 BRY J W Fisher, 18 The Smooting, Tealby, Market Rasen, LN8 3XZ [IO93UJ, TF19]
M1 BRZ D C Lee, 2 Ancaster Ave, Lincoln, LN2 4AY [IO93RF, SK97]
MM1 BSA Details withheld at licensee's request by SSL.
M1 BSB W M Charman, 23 The Vineries, Wimborne, BH21 2PU [IO90AT, SU00]
M1 BSC B E James, 10 The Hyde, New Milton, BH25 5GA [IO90ES, SZ29]
M1 BSD C J Bootz, 10 Davenport Ave, Nantwich, CW5 5QJ [IO83RB, SJ65]
M1 BSE J V Wharton, 12 Brent Ct, Watergate Ln, Leicester, LE3 2XQ [IO92JO, SK50]
M1 BSF Details withheld at licensee's request by SSL.
M1 BSG W J Billingham, 30 Ash Gr, Conisbrough, Doncaster, DN12 2HH [IO93JL, SK59]
MW1 BSH S Quick, 19 Edward St., Port Talbot, SA13 1YG [IO81CO, SS79]
M1 BSI C A Bowen, 45 Morris Rd, Nottingham, NG8 6NE [IO92JX, SK54]
MI1 BSJ R Blythe, 159 Victoria Rd, Bready, Strabane, BT82 0DZ [IO64HW, C30]
MI1 BSK R D Robinson, 30 Trasnagh Dr, Newtownards, BT23 4PD [IO74DO, J47]
M1 BSL D G Hughes, 121 Felixstowe Rd, Ipswich, IP3 8EA [JO02OB, TM14]
M1 BSM G R Meyer, 447-449 Manchester Rd, Stockport, SK4 5DJ [IO83VK, SJ89]
M1 BSN J D C Riley, Peacehaven, Great St., Norton Sub Hamdon, Stoke Sub Hamdon, TA14 6SH [IO80OW, ST41]
M1 BSO B J Ford, 7 Courtwick Rd, Wick, Littlehampton, BN17 7NE [IO90RT, TQ00]
M1 BSP M D Ford, 7 Courtwick Rd, Wick, Littlehampton, BN17 7NE [IO90RT, TQ00]
M1 BSQ Details withheld at licensee's request by SSL.
M1 BSR Details withheld at licensee's request by SSL.
M1 BSS Details withheld at licensee's request by SSL.
M1 BST Capt D Ellwood, 41 Southampton Rd, Portsmouth, PO6 4SA [IO90KU, SU60]
M1 BSU Details withheld at licensee's request by SSL.
M1 BSV M Black, 9 New St., Nelson, Lancs, BB9 8JW [IO83VU, SD83]
M1 BSW Details withheld at licensee's request by SSL.
M1 BSX A R May, 18 Pennant Hills, Havant, PO9 3JZ [IO90LU, SU60]
M1 BSY W Ginger, Wenick House, 152 Hawks Rd, Hailsham, E Sussex, BN27 1NA [JO00DV, TQ51]
M1 BSZ Details withheld at licensee's request by SSL.
MW1 BTA Details withheld at licensee's request by SSL.
M1 BTB P J Brown, 46 Greygoose Park, Harlow, CM19 4JW [JO01AS, TL40]
M1 BTC V Felgate, 36 Milburn Cres, Stockton on Tees, TS20 2DN [IO94IN, NZ42]
M1 BTD D Wilkinson, 3 Grosvenor Ave, Liverpool, L23 0SB [IO83LL, SJ39]
M1 BTE C G Taylor, 11 Alexandra Cl, Swanley, BR8 7BS [JO01CJ, TQ56]
M1 BTF Details withheld at licensee's request by SSL.
M1 BTG G A Lee, 16 Phoenix Chase, North Shields, NE29 8SS [IO95GA, NZ36]
M1 BTH J V Warren, 2 St. Julien Cres, Weymouth, DT3 5DT [IO80SP, SY68]
M1 BTI D W Piggin, 10 Snowden Ave, Urmston, Manchester, M41 6EL [IO83TK, SJ79]
MM1 BTJ D W Gonella, 34 Lochy Pl, Erskine, PA8 6AY [IO75SV, NS47]
M1 BTK J W T Watson, 24 Peregrine Dr, Benfleet, SS7 5EJ [JO01GN, TQ78]
M1 BTL D Surman, 27 Stanley Rd, Hinckley, LE10 0HP [IO92HN, SP49]
MW1 BTM C T Jones, 17 Gr House Ct, Pontygwaith, Ferndale, CF43 3LJ [IO81GP, ST09]
M1 BTN K Hughes, 18 Red Lion Cl, Talke, Stoke on Trent, ST7 1SZ [IO83UB, SJ85]
M1 BTO N Martin, 28 Churchmead Cl, Lavant, Chichester, PO18 0AY [IO90OU, SU80]
M1 BTP B A Pickup, 9 Darwin St., Sunderland, SR5 2EJ [IO94HW, NZ35]
M1 BTQ D J Tysoe, 21 Burnt Cl, Luton, LU3 3SU [IO91SV, TL02]
M1 BTR J M Charles, Ash Tree, Priory Ln, Grimoldby, Louth, LN11 8SP [JO03BI, TF38]
M1 BTS B E Rutlidge, Myrtle Cottage, Poughill, Bude, Cornwall, EX23 9EW [IO70RU, SS20]
M1 BTT F S Pealing, 7 Hemyock Rd, Birmingham, B29 4DG [IO92AK, SP08]
M1 BTU P Mason, 24 Elborough Rd, Swindon, SN2 2LR [IO91CO, SU18]
MW1 BTV D Ellis, 26 Morfa Lodge Est, Porthmadog, LL49 9PF [IO72WW, SH53]
M1 BTW R Smith, 131 Livingstone St., Birkenhead, L41 4HQ [IO83LJ, SJ38]
M1 BTX Details withheld at licensee's request by SSL.
M1 BTY Details withheld at licensee's request by SSL.
M1 BTZ A Harradine, 2 Balmoral Ave, Northwich, CW9 8BH [IO83RF, SJ67]
M1 BUA J Bell, 5 Burntwood Cl, London, SW18 3JU [IO91VK, TQ27]
M1 BUB Details withheld at licensee's request by SSL.
M1 BUC A J Benson, 12 Longfellow Rd, Caister on Sea, Great Yarmouth, NR30 5RH [JO02UP, TG51]
M1 BUD Details withheld at licensee's request by SSL.
MM1 BUE C S Addison, South Lothian, Crimond, Fraserburgh, AB43 8QU [IO97AO, NK05]
M1 BUF R G Hanney, 74 Avon Rd, Bournemouth, BH8 8SF [IO90BR, SZ19]
M1 BUG M R Dugdale, 57 Macauley Ave, Blackpool, FY4 4YF [IO83LT, SD33]
M1 BUH A Parkinson, 20 Brookhill Rd, Darton, Barnsley, S75 5EL [IO93FO, SE20]
M1 BUI G R H Chance, 111 Pencarn Parc, Four Lanes, Redruth, TR16 6LQ [IO70JE, SW63]
M1 BUJ I C Lewis, Tarntoft, Raynsway Marina, Thurmaston, Leicester, LE4 8BA [IO92KQ, SK60]
M1 BUK A J Hall, 4 Stockton Rd, Leicester, LE4 9DS [IO92KP, SK60]
M1 BUL D L Atkins, Gr Cottage, 79 Roe Ln, Southport, PR9 7HR [IO83MP, SD31]
MW1 BUN D C Luke, 56 Maerdy Park, Pencoed, Bridgend, CF35 5HX [IO81FM, SS98]
MM1 BUO D W Gerrie, 43 Balgownie Way, Bridge of Don, Aberdeen, AB22 8XR [IO87WE, NJ91]
M1 BUP N Foyen, 72 Bunces Cl, Eton Wick, Windsor, SL4 6PL [IO91QL, SU97]
M1 BUQ I D Houghton, 14 Windfield Gdns, Little Sutton, South Wirral, L66 1JJ [IO83MG, SJ37]
M1 BUR H C Craven, 4 Amanda Dr, Louth, LN11 0AZ [JO03AJ, TF38]
M1 BUS M W Edmunds, Beech Gr, 28A Station Rd, Corton, Lowestoft, NR32 5HF [JO02UM, TM59]
M1 BUT P Nixon, 18 Queens Dr, Middlewich, CW10 0DG [IO83SE, SJ76]
M1 BUU C D Evans, Sunnybank, Stanbury, Keighley, West Yorks, BD22 0HA [IO93AT, SE03]
M1 BUV Details withheld at licensee's request by SSL.
M1 BUW R F Hill, Jarrah, 68 Shallowford Rd, Plymouth, PL6 5TD [IO70WJ, SX55]
M1 BUX S G Leaker, 166 Beckett Rd, Doncaster, DN2 4BB [IO93KM, SE50]
M1 BUY Details withheld at licensee's request by SSL.
M1 BUZ Details withheld at licensee's request by SSL.
M1 BVA M Allen, 18 Clittaford Rd, Plymouth, PL6 6DT [IO70WK, SX46]
M1 BVB T S Duley, 96 Steventon Rd, Drayton, Abingdon, OX14 4LD [IO91IP, SU49]
M1 BVC D Tucker, 137 Seaford Rd, London, W13 9HS [IO91UM, TQ18]
M1 BVD S Brosnan, 4 Black Rd Cl, Hayes, UB3 4QJ [IO91SM, TQ07]
M1 BVE Details withheld at licensee's request by SSL.
MW1 BVF B Hunter, Minffordd, Sun St., Ffestiniog, Blaenau Ffestiniog, LL41 4NE [IO82AX, SH74]
M1 BVG C R Worbey, 61 Stotfold Rd, Hitchin, SG4 0QW [IO91VX, TL23]
M1 BVH Details withheld at licensee's request by SSL.
M1 BVI K Worrall, 66 Elm St., Hollingwood, Chesterfield, S43 2LH [IO93HG, SK47]
M1 BVJ B G Muizelaar, 13 Belsay, Oxclose, Washington, NE38 0NE [IO94FV, NZ25]
M1 BVK N S Anderson, 23 Banbury, Washington, NE37 3AY [IO94FV, NZ35]
M1 BVL R Clark, 36 Southfields, Stanley, DH9 7PH [IO94DU, NZ15]
M1 BVM S A Turner, 16 Standish St., South Moor, Stanley, DH9 7AD [IO94DU, NZ15]
M1 BVN J K Taylor, 169 Meadow Way, Leighton Buzzard, LU7 8XP [IO91QW, SP92]
MI1 BVO R G Graham, 21 Meadowvale Cres, Bangor, BT19 1HQ [IO74DP, J47]
M1 BVP M A Taylor, 56 Newgate Ln, Mansfield, NG18 2LQ [IO93JD, SK56]
M1 BVQ K Broxup, 71 Meadway, Streethouse, Pontefract, WF7 6DT [IO93HQ, SE32]
M1 BVR R D Barrett Buckinghamshire Volunteer Resc, Kennoway, Hay Ln, Fulmer, Slough, SL3 6HJ [IO91RN, SU98]
M1 BVS C W Bunn, 311 Stourbridge Rd, Dudley, DY1 2EF [IO82WL, SO98]
M1 BVT M J Beardsley, 121 Wood Rd, Dudley, DY3 2LR [IO82WM, SO99]
M1 BVU D A Parker, 2 Burnett Cl, Saltash, PL12 4LL [IO70VJ, SX45]
M1 BVV P M Dullingham, 5 Turnberry Walk, Bedford, MK41 8AZ [IO92SD, TL05]
MM1 BVW A R Gordon, 84 Riverside Dr, Haddington, EH41 3QP [IO85OX, NT57]
M1 BVX S J Simpson, 125 Woodhouse Rd, Keighley, BD21 5NP [IO93BU, SE03]
M1 BVY I F Butcher, 8 Glebe Ave, Flitwick, Bedford, MK45 1HS [IO91SX, TL03]
MM1 BVZ Details withheld at licensee's request by SSL.
M1 BWA K Moore, 44 Napier St., Dalton in Furness, LA15 8HR [IO84JD, SD27]
M1 BWB Details withheld at licensee's request by SSL.
MM1 BWC J R Ganson, 1 Beechwood Terr West, Newport on Tay, DD6 8JH [IO86MK, NO42]
M1 BWD B J Brain, 11 Dickens Rd, Flitwick, Bedford, MK45 1QB [IO92RA, TL03]
M1 BWE P J Davis, 8 Norfolk Pl, Welling, DA16 3HR [JO01BL, TQ47]
M1 BWF P Roberts, 28 Central Dr, Shirebrook, Mansfield, NG20 8BQ [IO93JE, SK56]
M1 BWG Details withheld at licensee's request by SSL.
M1 BWH F Laycock, 8 Melling Way, Liverpool, L32 1TP [IO83NL, SJ49]
MW1 BWI J Hayhurst, Angorfa, Cymyran Rd, Caergeiliog, Anglesey, LL65 3HN [IO73RG, SH37]
M1 BWJ J W Bottle, 2 Manor Cttgs, East Sutton Hill, East Sutton, Maidstone, ME17 3DJ [JO01HF, TQ84]
M1 BWK P F Cox, 19 Gobions, Basildon, SS16 5AY [JO01FN, TQ78]
M1 BWL A N Mann, 10 Hawkins Way, Wootton, Abingdon, OX13 6LB [IO91IQ, SP40]
M1 BWM J H Wilson, 30 Tennyson Rd, Dartford, DA1 5DJ [JO01CK, TQ57]
M1 BWN S A Jarrett, 17 Wolmers Hey, Great Waltham, Chelmsford, CM3 1DA [JO01FT, TL61]
M1 BWO Details withheld at licensee's request by SSL.
M1 BWP R Bond, 219 Bexhill Rd, St. Leonards on Sea, TN38 8BH [JO00GU, TQ71]
M1 BWQ J P R Burn, 9 Birbeck Way, Frettenham, Norwich, NR12 7LG [JO02PR, TG21]
M1 BWR Dr E H N Oakley, Brooklands Lodge, Park View Cl, Wroxall, Ventnor, Isle of Wight, PO38 3EQ [IO90JO, SZ58]
M1 BWS Dr A K A Kerr, 14 Glam I, St. Helens, WA10 3XT [IO83PK, SJ59]
M1 BWT S R C Chudley, Orchard Lea, Oxford St., Lee Common, Great Missenden, Bucks, HP16 9JT [IO91PR, SP90]
M1 BWU S Constable, 9 Ridgeway Cl, Heathfield, TN21 8NS [JO00DX, TQ52]
M1 BWV A A Newton, 11 The Highlands, Bexhill on Sea, TN39 5HL [JO00FU, TQ70]
M1 BWW Details withheld at licensee's request by SSL.
M1 BWX Details withheld at licensee's request by SSL.
M1 BWY K G Kaminskyj, 23 Cathkin Cl, Leicester, LE3 6PW [IO92JP, SK50]
M1 BWZ P Quick, 70 Trent Ave, Maghull, Liverpool, L31 9DE [IO83MM, SD30]
M1 BXA K D Fisher, 180 Richmond Rd, Sheffield, S13 8TG [IO93HI, SK38]
M1 BXB Dr P C Smith, 11 Sunningdale Rd, Saltash, PL12 4BN [IO70VJ, SX45]
M1 BXC A C Blakeney, 2 Dixon Ave, Grimsby, DN32 0AJ [IO93XN, TA20]
M1 BXD M Cross, Overmist, 1 Grenville Cl, Stokenham, Kingsbridge, TQ7 2SY [IO80DG, SX84]
M1 BXE P P Hyde, 10 Highfield Cres, Comeytrowe Ln, Taunton, TA1 5JH [IO81KA, ST22]
MM1 BXF G R Nesbitt, 34 Abercromby Cres, Helensburgh, G84 9DX [IO76PA, NS38]
M1 BXG Details withheld at licensee's request by SSL.
M1 BXH G J Timms, 74 Park Gwyn, St. Stephen, St. Austell, PL26 7PN [IO70NI, SW95]
M1 BXI Details withheld at licensee's request by SSL.
M1 BXJ M W Ellis, 62 Peterborough Rd, Crowland, Peterborough, PE6 0BA [IO92WQ, TF20]
M1 BXK I Armstrong, 19 Nash Ave, South Shields, NE34 8NS [IO94GX, NZ36]
M1 BXL G Barrett, 114 William St., Long Eaton, Nottingham, NG10 4GD [IO92IV, SK43]
M1 BXM M Forster, 38 Barlow Rd, Moulton, Northwich, CW9 8QS [IO83RF, SJ66]
M1 BXN A R Cubitt, 132 Cauldwell Hall Rd, Ipswich, IP4 5BP [JO02OB, TM14]
M1 BXO D R Ellis, 67 Ambersham Cres, East Preston, Littlehampton, BN16 1AJ [IO90ST, TQ00]
M1 BXP K R Singleton, 28 Cuckoo Ln, Ashford, TN23 5DB [JO01KD, TQ94]
M1 BXQ J W Squire, 57 The Avenue, Chinnor, OX9 4PE [IO91NQ, SP70]
M1 BXR T A Fray, 38 Gunners Gr, London, E4 9SS [IO91XO, TQ39]
M1 BXS M D Stainsby-Tron, 26 Moorfield Rd, Liverpool, L23 9UD [IO83LL, SD30]
M1 BXT S T Kerr, 3 Gilfil Rd, Nuneaton, CV10 7BU [IO92GM, SP39]
M1 BXU D Napper, 47 Mallard Walk, Sidcup, DA14 6SG [JO01BK, TQ47]
M1 BXV A F R Bain, 32 Long Meadow, Wirral, L60 8QQ [IO83KH, SJ28]
M1 BXW Details withheld at licensee's request by SSL.
M1 BXX M R Broxton, Harding House, 3 Chapel Rd, Llanreath, Pembroke Dock, Dyfed, SA72 6TL [IO71MQ, SM90]
M1 BXY Details withheld at licensee's request by SSL.
MM1 BXZ N J Fowler, 4 Cairnhill Dr, Rosehearty, Fraserburgh, AB43 7JU [IO87WQ, NJ96]
M1 BYA Details withheld at licensee's request by SSL.
M1 BYB Details withheld at licensee's request by SSL.
M1 BYC B A Cockerill, 79 Eastlands Rd, Rugby, CV21 3RR [IO92JI, SP57]
M1 BYD T M Lee, 91 Old Vicarage Park, Narborough, Kings Lynn, PE32 1TG [JO02HQ, TF71]
M1 BYE Details withheld at licensee's request by SSL.
MW1 BYF D Davies, 47 Woodland Ave, Pencoed, Bridgend, CF35 6UW [IO81FM, SS98]
MD1 BYG A R Chatel, 22 Cooil Dr, Douglas, Isle of Man, IM2 2HA
M1 BYH A K Moss, 14 Nelson St., Macclesfield, SK11 6UN [IO83WG, SJ97]
M1 BYI P C Stockley, 41 Fairway Ct, Cleethorpes, DN35 0NN [IO93XM, TA30]
MI1 BYJ A G Cobb, 7 Kinallen Cttgs, Dromara, Dromore, BT25 2PW [IO64XJ, J24]
MW1 BYK Details withheld at licensee's request by SSL.
MI1 BYL W L R Norton, 25 Rathvarna Ave, Lisburn, BT28 2UZ [IO64XM, J26]
M1 BYM S J Mennie, 21 Washbrook Cl, Barton-le-Clay, Bedford, MK45 4LF [IO91SX, TL03]
M1 BYN S M Roberts, Sunnycroft, Tillington, Hereford, HR4 8LH [IO82OC, SO44]
M1 BYO I A Rogers, Yew Tree Cottage, Shucknall, Hereford, HR1 3SJ [IO82QB, SO54]
M1 BYP B T Seaby, 4 Parkside, Wollaton, Nottingham, NG8 2NN [IO92JW, SK53]
M1 BYQ R J Josephs, 9 Park View, Cleethorpes, DN35 7TG [IO93XN, TA20]
M1 BYR M C Haywood, 4 Wentworth Gate, Birmingham, B17 9EB [IO92AL, SP08]
M1 BYS Details withheld at licensee's request by SSL.
M1 BYT H Bloomfield, 49 Oak Cres, Garforth, Leeds, LS25 1PW [IO93HT, SE43]
M1 BYU Details withheld at licensee's request by SSL.
MI1 BYV B Hawkins, 53 Hillside Dr, Kilkeel, Newry, BT34 4JF [IO64XB, J31]
M1 BYW Details withheld at licensee's request by SSL.
MM1 BYX E G McDonald, 3 Kent Rd, Alloa, FK10 2JN [IO86CD, NS89]
MM1 BYY R A M Allan, Torguish House, Daviot, Inverness, Inverness Shire, IV1 2XQ [IO77WK, NH73]
M1 BYZ E M Hedley, 17 Chowdene Bank, Gateshead, NE9 6JJ [IO94EW, NZ25]
M1 BZA R W Stoddart, 163 Flatts Ln, Middlesbrough, TS6 0PP [IO94KN, NZ51]
M1 BZB A Warner, Lamesley Vicarage, Gateshead, Tyne & Wear, NE11 0EU [IO94EV, NZ25]
M1 BZC E J Gray, Room 7, 2 King Sq, Bristol, BS2 8JD [IO81QL, ST57]
M1 BZD D V Atkins, 21 Greystones Dr, Keighley, BD22 7AL [IO93AU, SE03]
M1 BZE J D Page, 38 Hunderton Rd, Hereford, HR2 7AE [IO82PB, SO53]
M1 BZF Details withheld at licensee's request by SSL.
M1 BZG L Raybould, Flat 2, 309 Stourbridge Rd, Halesowen, B63 3QT [IO82XK, SO98]
M1 BZH J M Rose, 80 March End Rd, Wolverhampton, WV11 3QU [IO82XO, SJ90]
M1 BZI F M Lee, 26 Lache Hall Cres, Chester, CH4 7NF [IO83NE, SJ36]
M1 BZJ P L Buer, 7 Lonsdale Walk, Orrell, Wigan, WN5 0DZ [IO83PN, SD50]
M1 BZK D J Burt, 54 St. Leonards Rd, Plymouth, PL4 9NE [IO70WI, SX45]
M1 BZL S J Cape, Penford Mill, Black Dog, Crediton, Devon, EX17 4QJ [IO80DV, SS71]
M1 BZM Details withheld at licensee's request by SSL.
M1 BZN C J Bartlett, Mount Pleasant, Ridgeway Cross, Cradley, Malvern, WR13 5JD [IO82TC, SO74]
M1 BZO W M Southwell, 33 High View, Portishead, North Somerset, BS20 8RF [IO81OL, ST47]
M1 BZP D A M Hillyard, 27 Harveys Ln, Winchcombe, Cheltenham, GL54 5QU [IO91AW, SP02]
M1 BZQ M Warner, 6 Vine Tree Cl, Withington, Hereford, HR1 3QW [IO82QC, SO54]
M1 BZR D J Wright, 2 Cawder Hall Farm, Southend Rd, Corringham, Stanford-le-Hope, Essex, SS17 9NQ [JO01FM, TQ78]
M1 BZS Details withheld at licensee's request by SSL.
M1 BZT R W Swannell, 4 Sycamore Cl, Kettering, NN16 9ST [IO92PJ, SP87]
M1 BZU D M Pye, 8 Bishops Rd, Hove, BN3 6PQ [IO90WU, TQ20]
M1 BZV J Pye, 8 Town House Rd, Nelson, BB9 9LW [IO83VU, SD83]
M1 BZW Details withheld at licensee's request by SSL.
M1 BZX W H Thompson, 77 Springfield Rd, Etwall, Derby, DE65 6LA [IO92EV, SK23]
M1 BZY R L Allpress, 48 St. Chads Ave, Portsmouth, PO2 0SB [IO90LT, SU60]
M1 BZZ P M Lonsdale, 14 Donne Cl, Wirral, L63 9YJ [IO83LI, SJ38]
M1 CAA T G Wells, 55 Holt Park Rd, Leeds, LS16 7QS [IO93EU, SE24]
M1 CAB Details withheld at licensee's request by SSL.
MM1 CAC G G Mathers, 46 Castle St., Fraserburgh, AB43 9DH [IO87XQ, NJ96]
M1 CAD K L Sapsed, 12 Brookmead Way, Havant, PO9 1RT [IO90MU, SU70]
M1 CAE R H Naylor, 93 Woodland Rd, Halton, Leeds, LS15 7DN [IO93GT, SE33]
M1 CAF Details withheld at licensee's request by SSL.
M1 CAG N T Howe, 45 Kettering Rd, Islip, Kettering, NN14 3JT [IO92RJ, SP97]
M1 CAH C Bennet, Kesgrave High School, Main Rd, Kesgrave, Ipswich, IP5 7PB [JO02OB, TM24]
M1 CAI J A Mortimer, 4 Nethercliffe Cres, Guiseley, Leeds, LS20 9HN [IO93DV, SE14]
M1 CAJ A J Burton, Mill House Farm, Walesby, Market Rasen, Lincs, LN8 3UR [IO93UJ, TF19]
M1 CAK P J Prior, 17 Layton Ave, Malvern, WR14 2ND [IO82UC, SO74]
M1 CAL S Casey Red Rose A.R.G, 102 Dunkirk Ln, Leyland, Preston, PR5 3SQ [IO83PQ, SD52]
M1 CAM Details withheld at licensee's request by SSL.
MW1 CAN P A Owen, 6 Caer Efail, Dwyran, Llanfairpwllgwyngyll, LL61 6JA [IO73UD, SH46]
M1 CAO R D Cameron, 16 Falkland Pl, Chatham, ME5 9HR [JO01GI, TQ76]
M1 CAP R J Taylor, 223 Sarehole Rd, Birmingham, B28 8HA [IO92BK, SP18]
M1 CAQ G T Humphreys, 45 High Meadows, Romiley, Stockport, SK6 4QE [IO83WK, SJ99]
M1 CAR R A Griffiths, 22 Quarry Rd, Hereford, HR1 1SS [IO82PB, SO54]
MI1 CAS A Mulholland, 38 Antrim Rd, Lisburn, BT28 3DN [IO64XM, J26]
M1 CAT J C Bunting North Downs Amateur Radio Cont, 4 Springfield Rd, Sittingbourne, ME10 2NB [JO01II, TQ86]
M1 CAU A M N Shelswell, Waterways, Dock Ln, Melton, Woodbridge, Suffolk, IP12 1PE [JO02PC, TM25]
M1 CAV J D Morrison, Arthurton House, 818 Holderness Rd, Hull, North Humberside, HU9 3LP [IO93US, TA13]
M1 CAW S C Dennis, 74 Connegar Leys, Blisworth, Northampton, NN7 3DF [IO92ME, SP75]
M1 CAX K Biggs, 22 Wallingford Cl, Bracknell, RG12 9JE [IO91PJ, SU86]
M1 CAY M D Harris, Weathercock Cottage, East Mersea Rd, West Mersea, Colchester, Essex, CO5 8SL [JO01LT, TM01]
M1 CAZ P G Poore, 42 Kibblewhite Cres, Twyford, Reading, RG10 9AX [IO91NL, SU77]
M1 CBA S E Noble, 24 Newfield Rd, Coventry, CV1 4EA [IO92FK, SP38]
M1 CBB M W Wade, 42 New Rd, Burnham on Crouch, CM0 8EH [JO01JP, TQ99]
M1 CBC Details withheld at licensee's request by SSL.
M1 CBD Details withheld at licensee's request by SSL.
M1 CBE L Mills, 37 Ambleside Ave, Telscombe Cliffs, Peacehaven, BN10 7LS [IO90XT, TQ40]
MW1 CBF A H David, 62 The Uplands, Brecon, LD3 9HT [IO81HW, SO02]
M1 CBG Details withheld at licensee's request by SSL.
M1 CBH J D Tomlins, 23 Riber Cres, Old Tupton, Chesterfield, S42 6HU [IO93GE, SK36]

M1	**CBI**	R H Bird, 72 Charterfield Dr, Kingswinford, DY6 7RD [IO82VM, SO88]
M1	**CBJ**	J A Butler, 4 Vittery Cl, Brixham, TQ5 8LJ [IO80FJ, SX95]
M1	**CBK**	R Scott, 47 Masham House, Kale Rd, Erith, DA18 4BN [JO01BL, TQ47]
M1	**CBL**	A Baxter, 408 Leeds Rd, Wakefield, WF1 2JB [IO93FQ, SE32]
MW1	**CBM**	K A Stoddart, Plas Newydd, Pendref St, Newborough, Llanfairpwllgwyngyll, Gwynedd, LL61 6TA [IO73TD, SH46]
MI1	**CBN**	Details withheld at licensee's request by SSL.
M1	**CBO**	R Appleby, 3 St. Judes Way, Burton on Trent, DE13 0LR [IO92ET, SK22]
M1	**CBP**	B Fegan, 85 Cavendish Cl, Old Hall, Warrington, WA5 5PS [IO83QJ, SJ58]
M1	**CBQ**	Details withheld at licensee's request by SSL.
M1	**CBR**	A T Hutton, 29 Manor Rd, Ashford, TW15 2SL [IO91SK, TQ07]
M1	**CBS**	M Beale, 83 Sandgate, Swindon, SN3 4HH [IO91DN, SU18]
M1	**CBT**	C B Taylor, 48 Northdown Park Rd, Cliftonville, Margate, CT9 3PT [JO01QJ, TR37]
M1	**CBU**	R A Stansfield, Sundene, 157 Hollin Ln, Crigglestone, Wakefield, WF4 3EG [IO93FP, SE31]
M1	**CBV**	G M Leeder, 89 Chesterton Ave, Harpenden, AL5 5ST [IO91TT, TL11]
M1	**CBW**	Details withheld at licensee's request by SSL.
MI1	**CBX**	P Doris, 92 Coolnafranky Park, Cookstown, BT80 8PW [IO64PP, H87]
M1	**CBY**	D P Howse, 24 Sandown Rd, Bishops Cleeve, Cheltenham, GL52 4BY [IO81XW, SO92]
M1	**CBZ**	A K Howse, 24 Sandown Rd, Bishops Cleeve, Cheltenham, GL52 4BY [IO81XW, SO92]
M1	**CCA**	B A Albutt, 48 Rushfield Rd, Liss, GU33 7LP [IO91NA, SU72]
MI1	**CCB**	Details withheld at licensee's request by SSL.
M1	**CCC**	Details withheld at licensee's request by SSL.
M1	**CCD**	D J Bell, 56 Moorland Ave, Lincoln, LN6 7RD [IO93RF, SK96]
M1	**CCE**	D K Soutter, Cranleigh, Bossiney Rd, Tintagel, Cornwall, PL34 0AG [IO70PP, SX08]
M1	**CCF**	M J Buckley, Springfield, 12 Ranmore Ave, Croydon, CR0 5QA [IO91XI, TQ36]
M1	**CCG**	G Smales, 6 Chestercourt Cttgs, Camblesforth, Selby, YO8 8HZ [IO93LR, SE62]
M1	**CCH**	J Clarey, 2 Sebastopol Cttgs, Plantation Farm Rd, Redmere, Ely, CB7 4SS [JO02EK, TL68]
MI1	**CCI**	J N Barr, 58 Glenbank Rd, Londonderry, BT48 0BD [IO65HA, C41]
M1	**CCJ**	E L Huntley, 12 Ventnor Rd, Portland, DT5 1JE [IO80SN, SY67]
M1	**CCK**	J Illsley, Flat 1, 12 Ventnor Rd, Portland, DT5 1JE [IO80SN, SY67]
M1	**CCL**	R G Chantler, 68 Chilton Ln, Ramsgate, CT11 0LQ [JO01QH, TR36]
MW1	**CCM**	Details withheld at licensee's request by SSL.
M1	**CCN**	P D Terry, 13 Lamsey Ln, Heacham, Kings Lynn, PE31 7LA [JO02FV, TF63]
M1	**CCO**	Details withheld at licensee's request by SSL.
M1	**CCP**	Details withheld at licensee's request by SSL.
M1	**CCQ**	A Bantoft, 25 Montgomery Way, Kings Lynn, PE30 4YH [JO02FS, TF62]
MM1	**CCR**	A G Annan, Easter Cottage, Blairlogie, Stirling, FK9 5PX [IO86BD, NS89]
MI1	**CCS**	Details withheld at licensee's request by SSL.
MI1	**CCT**	Details withheld at licensee's request by SSL.
M1	**CCU**	Details withheld at licensee's request by SSL.
M1	**CCV**	Details withheld at licensee's request by SSL.
M1	**CCW**	C Connolly, 47 Northfield Ave, Rothwell, Leeds, LS26 0SN [IO93GR, SE32]
M1	**CCX**	P G Tully, 9 Beechcroft, Rothbury, Morpeth, NE65 7RA [IO95BH, NU00]
M1	**CCY**	R G Coxon, 2 Bewick Folly, Old Bewick, Alnwick, Northd., NE66 4EA [IO95BL, NU02]
M1	**CCZ**	Details withheld at licensee's request by SSL.
M1	**CDA**	L R Hoare, 51 Kings Rd, Birchington, CT7 0DU [JO01PI, TR36]
M1	**CDB**	L G Mahon, 367 Whitton Ave East, Greenford, UB6 0JT [IO91UN, TQ18]
MM1	**CDC**	J Town, 11 Edinburgh Rd, Greenlaw, Duns, TD10 6XF [IO85SQ, NT74]
M1	**CDD**	Details withheld at licensee's request by SSL.
M1	**CDE**	T D Cooke, 1 Sarah Gdns, Margate, CT9 3XB [JO01QJ, TR36]
M1	**CDF**	Details withheld at licensee's request by SSL.
M1	**CDG**	Details withheld at licensee's request by SSL.
M1	**CDH**	Details withheld at licensee's request by SSL.
M1	**CDI**	Details withheld at licensee's request by SSL.
M1	**CDJ**	F Waite, 91 Priors Hill, Wroughton, Swindon, SN4 0RL [IO91CM, SU18]
M1	**CDK**	R B Woods Danpac(Derbyshire & Nottingham, 4A Market Pl, Long Eaton, Nottingham, NG10 1LS [IO92IV, SK43]
M1	**CDL**	D J Hall, 4 Burns Cl, Peterborough, PE1 3JJ [IO92VO, TF10]
M1	**CDM**	Details withheld at licensee's request by SSL.
M1	**CDN**	C L Niemann, 60 Spring Gdns, London, N5 2DT [IO91WN, TQ38]
M1	**CDO**	E J Reilly, Vine Cottage, Bourton on The Hill, Moreton in Marsh, Gloucester, GL56 9AG [IO91DX, SP13]
M1	**CDP**	A T McDade, 1 Bradford Walk, Corby, NN18 0PF [IO92PL, SP88]
M1	**CDQ**	R B Damm, 81 Sidney St., Cleethorpes, DN35 7NQ [IO93XN, TA20]
M1	**CDR**	Details withheld at licensee's request by SSL.
M1	**CDS**	Details withheld at licensee's request by SSL.
M1	**CDT**	C A Behan, 19 St. Chads Cl, Burton on Trent, DE13 0ND [IO92ET, SK22]
M1	**CDU**	G B Temple, 7 Bank Top, Chillingham, Alnwick, Northd., NE66 5NG [IO95BM, NU02]
M1	**CDV**	A McEwen, 23 Cobholm Rd, Great Yarmouth, NR31 0BU [JO02UO, TG50]
M1	**CDW**	Details withheld at licensee's request by SSL.
M1	**CDX**	K P Leach, 82 Powell Ave, Blackpool, FY4 3HH [IO83LT, SD33]
M1	**CDY**	Details withheld at licensee's request by SSL.
M1	**CDZ**	Details withheld at licensee's request by SSL.
M1	**CEA**	T V Gladman, 43 Queens Rd, New Malden, KT3 6BY [IO91VJ, TQ26]
M1	**CEB**	K D White, 30 Hastings Rd, Coventry, West Midlands, CV2 4JD [IO92GJ, SP37]
M1	**CEC**	C D Thompson, 23 Cowling Brow, Chorley, Lancs, PR6 0QE [IO83QP, SD51]
M1	**CED**	P S Hughson, 9 Pykestone Cl, Oakwood, Derby, Derbyshire, DE21 2JW [IO92GW, SK33]
M1	**CEE**	Details withheld at licensee's request by SSL.
MI1	**CEF**	E J Fagan, 78 Malone Rd, Belfast, BT9 5BU [IO74AN, J37]
MI1	**CEG**	G McWilliams, 65 Callan St., Armagh, BT61 7RG [IO64QI, H84]
M1	**CEH**	Details withheld at licensee's request by SSL.
M1	**CEI**	D L Philpotts, 33 St. Guthlac St., Hereford, Herefordshire, HR1 2EY [IO82PB, SO53]
M1	**CEJ**	J D Riley, The Rafters, Doncaster Rd, Wragby, Wakefield, WF4 1QX [IO93HP, SE41]

UK Novice Callsigns

2E0

2E0 AAK R J Berridge, Bracklyn, St. Clare Rd, Walmer, Deal, CT14 7QB [JO01QE, TR35]
2E0 AAL M B Bullock, 9 Weirfield Rd, Darley Abbey, Derby, DE22 1DH [IO92GW, SK33]
2E0 AAM W R Smith, 23 Beacon View, Walsall, WS2 0DY [IO82XN, SO99]
2E0 AAP Details withheld at licensee's request by SSL.
2M0 AAR G Morrison, 4 Invercauld Rd, Aberdeen, AB1 5RT [IO87TH, NJ72]
2E0 AAS A Evans, 25 Offerton Ave, Derby, DE23 8DU [IO92GV, SK33]
2E0 AAV P B O'Connor, Flat 12, 39 Varden Croft, Edgbaston, Birmingham, B5 7LR [IO92BL, SP08]
2M0 AAW C M Wylie, 3 Kings Cres, Elderslie, Johnstone, PA5 9AD [IO75SU, NS46]
2E0 AAX E V Wills, 23 Falcons Way, Salisbury, SP2 8NR [IO91CB, SU12]
2M0 AAY D I H Herron, 21 Southfield Ave, Paisley, PA2 8BY [IO75ST, NS46]
2E0 AAZ W C Elston, 49 Langdale Rd, Kingsthorpe, Northampton, NN2 7QQ [IO92NG, SP76]
2E0 ABB B Pitch, Hayne Cottage, 27 Church Rd, Winscombe, BS25 1BJ [IO81NH, ST45]
2E0 ABD R A Mold, 179 Braeside Ave, Patcham, Brighton, BN1 8SP [IO90WU, TQ30]
2E0 ABI P J Earnshaw, Dunelm, Ayton Rd, Irton, Scarborough, YO12 4RQ [IO94SF, TA08]
2E0 ABJ A Percy, 76 Windsor Rd, Co. Rd, Hull, HU5 4HQ [IO93TS, TA03]
2E0 ABL F Hayes, 88 Johns Rd, Fareham, PO16 0RX [IO90JU, SU50]
2M0 ABN R Kelsall, Parkview, Kinnoir, Huntly, Aberdeenshire, AB54 5YU [IO87PK, NJ54]
2E0 ABQ S P Wells, 15 Beaumont Caravan Park, Mill Ln, Bradwell, Great Yarmouth, NR31 8HP [JO02UN, TG50]
2E0 ABS J S Loader, 21 Canford View Dr, Colehill, Wimborne, BH21 2UW [IO90AT, SU00]
2E0 ABT L Simons, Westwood, Faris Ln, Woodham, Addlestone, KT15 3DJ [IO91RI, TQ06]
2E0 ABU Details withheld at licensee's request by SSL.
2E0 ABW Details withheld at licensee's request by SSL.
2E0 ACA I V F Craig, Partridge Cottage, Redpale, Dallington, Heathfield, TN21 9NR [JO00EW, TQ61]
2W0 ACD P G Bennett, 13 Thornbury Cl, Baglan, Port Talbot, SA12 8EU [IO81CO, SS79]
2E0 ACE A P Backhouse, 113 Bucklesham Rd, Kirton, Ipswich, IP10 0PF [JO02PA, TM24]
2E0 ACJ S G Lydiate, 11 Avondale Rd, Pitsea, Basildon, SS16 4TT [JO01FN, TQ78]
2E0 ACK J E N Field, 14 Regent Rd, Harborne, Birmingham, B17 9JU [IO92AL, SP08]
2E0 ACO Details withheld at licensee's request by SSL.
2E0 ACP J C Everitt, 55 Risborough Rd, Bedford, MK41 9QR [IO92SD, TL05]
2E0 ACQ H L Barnes, 24 Burleigh Pl, Oakley, Bedford, MK43 7SG [IO92RE, TL05]
2E0 ACR H P Anese, 13 Dennis Rd, Kempston, Bedford, MK42 7HF [IO92RC, TL04]
2E0 ACS Details withheld at licensee's request by SSL.
2M0 ACT K W Goodwin, 5 Firth Dr, Macduff, AB44 1XY [IO87SP, NJ76]
2E0 ACU A C Rose, 68 Wheatley Ln, Wins Hill, Burton on Trent, DE15 0DX [IO92ET, SK22]
2E0 ACV E L Barnes, 3 Layton Rd, Ashton on Ribble, Preston, PR2 1PB [IO83OS, SD53]
2E0 ACY K L Cannon, 35 Loddon Bridge Rd, Woodley, Reading, RG5 4AP [IO91NK, SU77]
2E0 ADA R K Gaskell, 18 Woodcroft, Kennington, Oxford, OX1 5NH [IO91JR, SP50]
2E0 ADE Details withheld at licensee's request by SSL.
2E0 ADF R L Phillips, 57 Hollytrees, Bar Hill, Cambridge, CB3 8SF [JO02AG, TL36]
2E0 ADL J P Wresdell, Bracey Bridge Farm, Harpham, Driffield, East Yorks, YO25 0DE [IO94TA, TA06]
2E0 ADM L T Shooter, 9 Albert Rd, Corfe Mullen, Wimborne, BH21 3QB [IO80XS, SY99]
2E0 ADN M C Hulme, 71 Victoria Gdns, Ferndown, BH22 9JQ [IO90BT, SU00]
2E0 ADO D J Whetstone, 70 Heanor Rd, Smalley, Ilkeston, DE7 6DX [IO93HA, SK44]
2E0 ADR S N Watson, 109 Union St., Middlesbrough, TS1 4ED [IO94JN, NZ41]
2E0 ADU Details withheld at licensee's request by SSL.
2E0 ADZ A D Powis, Fircroft, Pound Ln, Dallinghoo, Woodbridge, IP13 0LN [JO02PD, TM25]
2E0 AED A D Sumner, 20 Woodlands Way, Southwater, Horsham, RH13 7HZ [IO91TA, TQ12]
2E0 AEG R P Holt, 41 Garden Ave, Ilkeston, DE7 4DF [IO92IX, SK44]
2M0 AEL A Batty, 31 Brae Cres, Mintlaw, Peterhead, AB42 5FD [IO87XM, NJ94]
2E0 AEM Details withheld at licensee's request by SSL.
2W0 AEQ G R Perry, 25 Moira Terr, Adamsdown, Cardiff, CF2 1EJ [IO81KL, ST17]
2E0 AER Details withheld at licensee's request by SSL.
2E0 AES B R Hyde, 54 The Byway, Darlington, DL1 1EQ [IO94FM, NZ31]
2E0 AEX V Lee, Clayton Lodge, Sunnyside, Edgerton Rd, Huddersfield, HD3 3AD [IO93CP, SE11]
2E0 AEY Details withheld at licensee's request by SSL.
2E0 AEZ J C P McCarthy, 29 Harefield Rd, Swaythling, Southampton, SO17 3TG [IO90HW, SU41]
2E0 AFD D I Spillett, 56 North Ln, Rustington, Littlehampton, BN16 3PW [IO90RT, TQ00]
2E0 AFI D W Shallcross, 523 New St., Hilcote, Alfreton, DE55 5HU [IO93IC, SK45]
2E0 AFL M J Coxhead, Baronsway, Parker Ln, Whitestake, Preston, PR4 4JX [IO83PR, SD52]
2E0 AFO D E Goodall, 181 Farhalls Cres, Horsham, RH12 4BX [IO91UB, TQ13]
2E0 AFP R Forrest-Webb, 1 Trelasdee Cttgs, St. Weonards, Hereford, HR2 8PU [IO81PV, SO52]
2E0 AFZ A J Wilson, 10 Duchy Ave, Fulwood, Preston, PR2 8DH [IO83PS, SD53]
2E0 AGA J S Ingram, 2 Sylvan Cres, Skegby, Sutton in Ashfield, NG17 3DL [IO93JD, SK56]
2E0 AGB J L Oldham, 82 Mersey Bank Rd, Hadfield, Hyde, SK14 7PN [IO93AL, SK09]
2E0 AGE K L Phillips, 3 Lang Ln, Wirral, L48 5HE [IO83JJ, SJ28]
2E0 AGG F R Lugg, 4 Newbury Cl, Walsall, WS6 6DF [IO82XP, SJ90]
2E0 AGI E G C McLusky, 11 Ripon Rd, Killinghall, Harrogate, HG3 2DG [IO94FA, SE25]
2E0 AGJ S R Warren, Clifden Farm, Quenchwell Rd, Carnon Downs, Truro, Cornwall, TR3 6LN [IO70LF, SW84]
2E0 AGL J G C McLusky, 11 Ripon Rd, Killinghall, Harrogate, HG3 2DG [IO94FA, SE25]
2E0 AGM M J Ritchie, 173 St. Saviours Rd, Coley Park, Reading, RG1 6EY [IO91MK, SU77]
2E0 AGN Details withheld at licensee's request by SSL.
2E0 AGP Details withheld at licensee's request by SSL.
2E0 AGR D W Kearton, Skellcote, Woodhall, Askrigg, Leyburn, DL8 3LA [IO84XH, SD99]
2E0 AGV S J Gore, 79 Park Rd, Spixworth, Norwich, NR10 3NP [JO02PQ, TG21]
2E0 AGW M D Brown, 49 West Ave, Mayland, Chelmsford, CM3 6AE [JO01JQ, TL90]
2M0 AHB E A Shearer, Acham, Losset Rd, Alyth, Blairgowrie, PH11 8BU [IO86JO, NO24]
2E0 AHD T H Best, 3 Gordon Rd, Topsham, Exeter, EX3 0LJ [IO80GQ, SX98]
2E0 AHF A H Skinner, Bryony, 55 Gilbert St., Enfield, EN3 6PE [IO91XQ, TQ39]
2E0 AHH C G L Speight, 1 Lyndene Ave, Roe Green, Worsley, Manchester, M28 2RJ [IO83TM, SD70]
2E0 AHI G E Smith, 38 Elizabeth Ave, Enfield, EN2 8DP [IO91WP, TQ39]
2E0 AHJ D I Smith, 102 Rosebarn Ln, Pennsylvania, Exeter, EX4 5DU [IO80FR, SX99]
2E0 AHM A T Tavender, 38 Hesley Gr, Chapeltown, Sheffield, S30 4TX [IO93GJ, SK38]
2E0 AHN B W Smith, 31 Essex Gdns, Market Harborough, LE16 9JS [IO92ML, SP78]
2I0 AHO J S Barron, 400 Upper Rd, Trooperslane, Carrickfergus, BT38 8PW [IO74BR, J38]
2E0 AHP Details withheld at licensee's request by SSL.
2E0 AHS D R McNeil-Watson, 5 Bakers Rd, Halvergate, Norwich, NR13 3PY [JO02SO, TG40]
2E0 AHT K S F Dossett, 18 Carrington Ave, Hornsea, HU18 1JQ [IO93WW, TA24]
2E0 AHW H Hamilton, Windrush, 58 Kingswood Way, Selsdon, South Croydon, CR2 8QQ [IO91XI, TQ36]
2E0 AHX Details withheld at licensee's request by SSL.
2E0 AHY Details withheld at licensee's request by SSL.
2E0 AID J R Barnett, 13 Kendals Cl, Radlett, WD7 8NQ [IO91UQ, TQ19]
2E0 AIE M J J Hooks, 1 Dove House Cl, Bromham, Bedford, MK43 8PS [IO92RD, TL05]
2E0 AIH A J Hodkin, 24 The Heights, Market Harborough, LE16 8BQ [IO92NL, SP78]
2E0 AIK M Footring, 26 Ernest Rd, Wivenhoe, Colchester, CO7 9LG [JO01LU, TM02]
2E0 AIL G Harper, 48 Norlands Ln, Widnes, WA8 5AS [IO83PJ, SJ58]
2E0 AIM J E Cowley, 39 Alpine Way, Tow Law, Bishop Auckland, DL13 4DS [IO94CR, NZ13]
2E0 AIQ Dr J T B Moyle, 35 Midland Rd, Olney, MK46 4BL [IO92PD, SP85]
2E0 AIS Details withheld at licensee's request by SSL.
2E0 AIT K J Lloyd, 2 Bishopstone Dr, Beltinge, Herne Bay, CT6 6RE [JO01OI, TR26]
2E0 AIW S J Bilbie, 1 Windmill Ln, Mansfield, NG18 2AL [IO93JD, SK56]
2E0 AIZ T J Stevens, 7 Keates Rd, Cherry Hinton, Cambridge, CB1 4ER [JO02CE, TL45]
2E0 AJC M Jessop, 29 Lichfield Ave, Mansfield, NG18 4SY [IO93JC, SK55]
2W0 AJD R G Meal, 4 Sparrow Cl, Brampton, Huntingdon, PE18 8PY [IO92VH, TL26]
2E0 AJG S D Snow, 2 Park Ln, Lower Froyle, Alton, GU34 4LU [IO91NE, SU74]
2E0 AJH A J Walsh, 79 Greencroft, Penwortham, Preston, PR1 9LB [IO83PR, SD52]
2W0 AJI Details withheld at licensee's request by SSL.
2E0 AJJ Details withheld at licensee's request by SSL.
2E0 AJK D D Forsyth, 20 Chapel View, Rowlands Gill, NE39 2PN [IO94DW, NZ15]
2W0 AJM S Singh, 32 Beatty Ave, Roath Park, Cardiff, CF2 5QT [IO81JM, ST17]
2E0 AJO D L Turton, 29B Pitt St., Low Valley, Wombwell, Barnsley, S73 8AS [IO93HM, SE40]
2E0 AJP L Humphrey, 4 Bluebell Rd, Bassett, Southampton, SO16 3LQ [IO90HW, SU41]
2E0 AJQ A P Nelson, 29 Coxford Rd, Maybush, Southampton, SO16 5FG [IO90GW, SU31]
2E0 AJR Details withheld at licensee's request by SSL.
2E0 AJS A J King, 19 Greenhayes, Cheddar, BS27 3HZ [IO81OG, ST45]
2E0 AJT W R Atkins, 55 Park Grange Croft, Lower Norfolk Park, Sheffield, S2 3QJ [IO93GJ, SK38]
2M0 AJW J J Cook, 2 Springwells, St. Anns, Lockerbie, DG11 1HG [IO85GF, NY09]
2E0 AJZ M J Papworth, 7 Edwin Cl, Siege Cross, Thatcham, RG19 4GW [IO91JJ, SU56]
2E0 AKD M I Hall, 6 Amor Pl, Taunton, TA1 4SG [IO81KA, ST22]
2E0 AKE A Heald, 94 Haylings Rd, Leiston, IP16 4DT [JO02SE, TM46]
2E0 AKF C Heald, 94 Haylings Rd, Leiston, IP16 4DT [JO02SE, TM46]
2E0 AKP R J Jarvis, 82 Muxton Ln, Muxton, Telford, Salop, TF2 8PE [IO82SR, SJ71]
2M0 ALG B J Humphrey, 10 McDowall Dr, Stranraer, DG9 7NA [IO74LV, NX06]
2E0 ALO K M Andrews, Birchwood House, Cross Rd, Albrighton, Wolverhampton, WV7 3RB [IO82UP, SJ80]
2E0 ALP Details withheld at licensee's request by SSL.
2E0 ALQ P J L Godfrey, 10 Studland Ave, Rugby, CV21 4HN [IO92JI, SP57]
2E0 ALR M Clark, Cherry Garden, Windsor Rd, Crowborough, TN6 2HR [JO01CB, TQ52]
2M0 ALS M Gill, Easter Templand, Fortrose, Ross-shire, IV10 8RA [IO77WN, NH75]
2E0 ALT T Bailey, 21 Gargrave Pl, Lupset, Wakefield, WF2 8AR [IO93FQ, SE32]
2E0 ALW R J Hatcher, 61 Holland Rd, Holland, Oxted, RH8 9AU [JO01AF, TQ45]
2E0 ALX Details withheld at licensee's request by SSL.
2W0 ALZ D G Gunning, 19 Gordon Ave, Prestatyn, LL19 8RU [IO83HH, SJ08]
2W0 AMB J Lloyd-Owen, Gorwel, Maesdu Ave, Llandudno, LL30 1NR [IO83CH, SH78]
2E0 AME D Lockwood, 24 Enfield Dr, Batley, WF17 8DY [IO93ER, SE22]
2E0 AMH J W H Ward, 11 Hants Rd, Canterbury, CT1 1SJ [JO01NG, TR15]
2E0 AML C E Hydes, 2 Stable Ct, Martlesham Heath, Ipswich, IP5 7UQ [JO02OB, TM24]
2E0 AMM Master C Jolliffe, Hylton Cottage, Friday St, Pebworth, Stratford on Avon, Warks, CV37 8XW [IO92CC, SP14]
2E0 AMN G Morrish, 27 Lawns Wood, Telford, TF3 2HR [IO82SQ, SJ60]
2E0 AMO R K Potts, 7 Crewe Green Ave, Haslington, Crewe, CW1 5NT [IO83TC, SJ75]
2E0 AMR Details withheld at licensee's request by SSL.
2E0 AMS D R P Drew, 34 Church St., Eastwood, Nottingham, NG16 3HS [IO93IA, SK44]
2E0 AMT Details withheld at licensee's request by SSL.
2E0 AMU R Tattersall, 56 Larch Rd, New Ollerton, Newark, NG22 9SX [IO93LE, SK66]
2E0 AMV K M Morgan, 21 Western Rise, Ketley, Telford, TF1 4BS [IO82SQ, SJ61]
2E0 AMW G J Fowle, 12 Lytham Rd, Broadstone, BH18 8JS [IO90AS, SZ09]
2W0 AMX J H Morris, Hafod, Dob, Tregarth, Bangor, Gwynedd, LL57 4PN [IO73WE, SH66]
2W0 AMY P C Cash, 7 Wordsworth Ave, Bridgend, CF31 4SB [IO81EM, SS88]
2E0 AMZ Details withheld at licensee's request by SSL.
2E0 ANA Details withheld at licensee's request by SSL.
2E0 ANB D M Berry, 11 Woodlands Ave, Wellington, Telford, TF1 2AR [IO82RQ, SJ61]
2E0 ANC D G Pearson, 157 Great Elms Rd, Hemel Hempstead, HP3 9UL [IO91SR, TL00]
2E0 ANE G W D Steele, 77 Ashtree Rd, Frome, BA11 2SE [IO81UF, ST74]
2E0 ANF Details withheld at licensee's request by SSL.
2E0 ANG P Andrews, 52 Trenchard Cres, Springfield, Chelmsford, CM1 6FG [JO01FS, TL70]
2E0 ANI R D Goulding, 117 Greystone Rd, Broadgreen, Liverpool, L14 6UF [IO83NJ, SJ49]
2E0 ANK J D Gossington, 121 Elm Rd, Thetford, IP24 3HL [JO02IJ, TL88]
2E0 ANL Master A J W Hardie, Tana Merah, Church Ln, Clarborough, Retford, DN22 9NQ [IO93NI, SK78]
2E0 ANM Master R A W Hardie, Tana Merah, Church Ln, Clarborough, Retford, DN22 9NQ [IO93NI, SK78]
2E0 ANN K Grimshaw, 38 Pilling Ave, Lytham St. Annes, FY8 3QG [IO83LS, SD32]
2E0 ANO A C Foad, 7 Boundary Rd, Newark, NG24 4DZ [IO93OB, SK75]
2M0 ANP Details withheld at licensee's request by SSL.
2E0 ANS M J Baldwin, 22 Banbury Villas, Hook Green Rd, Southfleet, Gravesend, DA13 9NF [JO01DJ, TQ67]
2E0 ANU F K Campbell, 1 Meadow Cl, Trench, Telford, Salop, TF2 6PA [IO82SR, SJ61]
2E0 ANV M W Wheaton, 2 Grouch View, Rettendon, nr Chelmsford, Essex, CM3 5DS [JO01HP, TQ89]
2E0 ANW Details withheld at licensee's request by SSL.
2E0 ANX Details withheld at licensee's request by SSL.
2E0 ANY P J A Hayes, 131 High St., Henlow, SG16 6AE [IO92UA, TL13]
2E0 ANZ J V W Constance, 22 Gorse Cres, Ditton, Aylesford, ME20 6EU [JO01FG, TQ75]
2E0 AOC B Bass, 13 Albert Rd, Wellington, Telford, TF1 3AR [IO82RQ, SJ61]
2E0 AOD C E Pooler, 18 Johnstone Cl, Wrockwardine Wood, Telford, TF2 7DA [IO82SR, SJ61]
2M0 AOF E Skea, Craigard, Craigton, North Kessock, Ross-shire, IV1 1YG [IO77VM, NH64]
2E0 AOG G Ryder, 11 Claremont Gdns, Farsley, Pudsey, LS28 5BF [IO93ET, SE23]
2E0 AOH A Camm, 6 Lamb Cres, Ripley, DE5 3EX [IO93HB, SK45]
2E0 AOI Dr G Butler, 51 Leamington Dr, Beeston, Nottingham, NG9 5LN [IO92JV, SK53]
2E0 AOK P D Godolphin, 3 Knipe View, Bampton, Penrith, CA10 2RF [IO84ON, NY51]
2E0 AON L M Taylor, 18 Folly Ln, North Crawley, Newport Pagnell, MK16 9LW [IO92QC, SP94]
2E0 AOO R G Walsh, 76 Yew Tree Ln, Dukinfield, SK16 5DX [IO83XL, SJ99]
2E0 AOP A Warner, Lamesley Vicarage, Gateshead, Tyne & Wear, NE11 0EU [IO94EV, NZ25]
2E0 AOR M A Jolley, 30 Oban Dr, Blackburn, BB1 2HY [IO83SR, SD72]
2E0 AOS E M Hedley, 17 Chowdene Bank, Gateshead, NE9 6JJ [IO94EW, NZ25]
2E0 AOT M Labourn, 6 Healey Dr, Ossett, WF5 8NA [IO93FQ, SE21]
2E0 AOU S A Tilly, 16 Mitchell Dr, Milburn Park, Ashington, NE63 9JT [IO95FE, NZ28]
2E0 AOV L M Blair, 6 Haycocks Cl, Telford, TF1 3NN [IO82RQ, SJ61]
2E0 AOW C Warhurst, 12 Potters Way, Ilkeston, DE7 5EX [IO92IX, SK44]
2E0 AOX J E Carlisle, 24 Loweswater Ave, Astley, Tyldesley, Manchester, M29 7EG [IO83SL, SJ69]
2E0 AOY Details withheld at licensee's request by SSL.
2E0 AOZ I F Marshall, 14 Laythorpe Ct, Pontefract, WF8 2TW [IO93IQ, SE42]
2E0 APA M Ashdown, 42 Alpine Ave, Tolworth, Surbiton, KT5 9RJ [IO91UJ, TQ26]
2M0 APB J Macdonald, 63 Perceval Rd South, Stornoway, Isle of Lewis, HS1 2TL [IO68TF, NB43]
2M0 APC G J Grant, 125 Middlefield Terr, Aberdeen, AB2 4PD [IO87TH, NJ72]
2E0 APD A Hosking, 28 Nimrod, The Concourse, London, NW9 5TU [IO91VO, TQ29]
2E0 APE Details withheld at licensee's request by SSL.
2W0 APF Details withheld at licensee's request by SSL.
2E0 APG G Wilson, 15 Wigmores, Telford, TF7 5NA [IO82SP, SJ60]
2E0 APH M S Haynes, 34 Pear Tree Mead, Harlow, CM18 7BY [JO01BS, TL40]
2W0 API Details withheld at licensee's request by SSL.
2E0 APJ M D France, 52 St. Oswald Rd, Lupset, Wakefield, WF2 8EH [IO93FQ, SE32]
2E0 APK J R Griffin, 93 Oldmixon Rd, Hutton, Weston Super Mare, BS24 9QA [IO81MH, ST35]
2E0 APL J M Whitton, 299 Spital Rd, Wirral, L62 2AQ [IO83MI, SJ38]
2E0 APM V G Andrews, 34 Forest Rd, Barnsley, S71 3BG [IO93GN, SE30]
2E0 APN T P Humble, 43 Whiteley Cres, Bletchley, Milton Keynes, MK3 5DQ [IO91OX, SP83]
2W0 APO G Rowlands, Dalar Wen, Rhosmeirch, Llangefni, Gwynedd, LL77 7SJ [IO73UG, SH47]
2E0 APP N D F Jamieson, 68 Knowles Ln, Holmewood, Bradford, BD4 9AP [IO93DS, SE13]
2E0 APQ D Cooksey, 5 Wellswood Ave, Ketley Bank, Telford, TF2 0BD [IO82SQ, SJ60]
2E0 APR A A Lake, The Granary, Zeal Monachorum, Crediton, Devon, EX17 6DH [IO80CT, SS70]
2E0 APS R T Thorne, 28 Delamere Rd, Colchester, CO4 4NH [JO01LV, TM02]
2W0 APT G E Davies, 27 Twyniago, Pontarddulais, Swansea, SA4 1HX [IO71XR, SN50]
2E0 APU L A Jones, 86 The Grove, Bournemouth, BH9 2TY [IO90BS, SZ09]
2E0 APV J B Isaacs, 11 Houndsmill, Horsington, Templecombe, Somerset, BA8 0ED [IO81SA, ST72]
2E0 APW P G Smith, 155 Grosvenor Rd, Dalton, Huddersfield, HD5 9UA [IO93DP, SE11]
2M0 APX S Barbour, 40 Mannerston, Linlithgow, West Lothian, EH49 7ND [IO85FX, NT07]
2E0 APY J I Rayson, 127 Brewsters Rd, Nottingham, NG3 3BY [IO92KX, SK54]
2E0 APZ O M Parker, 37 Springbank Cres, Gildersome, Morley, Leeds, LS27 7DN [IO93ES, SE22]
2E0 AQA D A Lewis, Burley Cottage, Shortwood, Standon, Staffs, ST21 6RG [IO82UW, SJ73]
2E0 AQB J A Young, 9 Caincross Rd, Nottingham, NG8 4AZ [IO92JX, SK54]
2E0 AQC E Smith, 14 St. Crispin Cl, North Killingholme, Grimsby, DN40 3JN [IO93UP, TA11]
2W0 AQD T G Rogers, The Willows, 48 Hillock Ln, Gresford, Wrexham, LL12 8YL [IO83MC, SJ35]
2E0 AQE J T Brown, 119 Borrowdale Rd, Birmingham, B31 5QL [IO92AJ, SP07]
2E0 AQF C E Sutton, Bridge Farm, Wold Newton, Driffield, East Yorks, YO25 0YN [IO94TD, TA07]
2E0 AQG J Springett, 1 Sycamore Lodge, Paynes Rd, Freemantle, Southampton, SO15 3SE [IO90GV, SU41]

2E1

2E1 AAB Details withheld at licensee's request by SSL.
2E1 AAE Details withheld at licensee's request by SSL.
2E1 AAH Details withheld at licensee's request by SSL.
2E1 AAI S T Weir, 59 Kennington Rd, Kennington, Oxford, OX1 5PB [IO91JR, SP50]
2E1 AAR D J Stears, 10 Lady Winefrides Walk, Great Billing, Northampton, NN3 9EE [IO92OG, SP86]
2E1 ABE H E J Forder, 157 Kennington Rd, Kennington, Oxford, OX1 5PE [IO91JR, SP50]
2E1 ABI A D Sampson, 47 Falcutt Way, Kingsthorpe, Northampton, NN2 8NR [IO92NG, SP76]
2E1 ABK Details withheld at licensee's request by SSL.
2E1 ABP C D Joyner, 19 Roman Cl, Deal, Kent, CT14 9XJ [JO01QF, TR35]

2E1 ABR S J Williamson, 120 Warbreck Hill Rd, Blackpool, FY2 0TR [IO83LU, SD33]
2E1 ABS M S Williamson, 120 Warbreck Hill Rd, Blackpool, FY2 0TR [IO83LU, SD33]
2E1 ABW S T Minnock, 32 Sandwood Rd, Sandwich, CT13 0AQ [JO01QG, TR35]
2E1 ABY M J OBrien, 14 Westdean Cl, River, Dover, CT17 0NP [JO01PD, TR24]
2E1 ACC S M Yates, 74 Olton Rd, Shirley, Solihull, B90 3NN [IO92CK, SP18]
2E1 ACF E W Harvey, 36 Seaview Ave, Little Oakley, Harwich, CO12 5JB [JO01OW, TM22]
2E1 ACG V J Hammonds, 28 Kingshurst Rd, Shirley, Solihull, B90 2QP [IO92BJ, SP17]
2E1 ACH D A Garrard, 20 Knowsley Ct, Tudor Grange, Kenton Bank Foot, Newcastle upon Tyne, NE3 2FJ [IO95DA, NZ26]
2E1 ACK D A R Nye, 5 Charles Rd, Deal, CT14 9AT [JO01QF, TR35]
2E1 ACP J W Williams, 754 Old Lode Ln, Solihull, B92 8NH [IO92CK, SP18]
2E1 ACR Details withheld at licensee's request by SSL.
2E1 ACU M A Brooks, 2 Long Gair, Snook Hill, Blaydon on Tyne, NE21 6QY [IO94DW, NZ16]
2E1 ACV Details withheld at licensee's request by SSL.
2E1 ACW C E Pooler, 18 Johnstone Cl, Wrockwardine Wood, Telford, TF2 7DA [IO82SR, SJ61]
2E1 ADJ J S Bridgman, 5 Drayton Ave, Mackworth Est, Derby, DE22 4JU [IO92FW, SK33]
2E1 ADP T Thompson, 19 Park End, Summer Ln Caravan, Banwell, Weston Super Mare, BS24 6JD [IO81NH, ST35]
2E1 ADQ C K Hammett, Rosehill, Ladock, Truro, Cornwall, TR2 4PQ [IO70MH, SW85]
2E1 ADR D J J Palmer, 133 Victoria Rd East, Thornton Cleveleys, FY5 5HH [IO83LU, SD34]
2E1 ADT K D Barbery, 17 Polbreen Ave, St. Agnes, TR5 0TR [IO70JH, SW75]
2E1 ADV M D Harris, 73 Whitecross Rd, Weston Super Mare, BS23 1EH [IO81MI, ST36]
2E1 AEC C J Vincent, 9 Sleapford, Long Ln, Telford, Shropshire, TF6 6HQ [IO82RR, SJ61]
2E1 AEE J P Hurrell, Sunrise, Museum Hill, Haslemere, GU27 2JR [IO91PC, SU93]
2E1 AEF P M Hibbs, 17 Orchard Rd, Shalford, Guildford, GU4 8ER [IO91RF, TQ04]
2E1 AEG Details withheld at licensee's request by SSL.
2E1 AEI A C Mason, 22 Forehill Cl, Preston, Weymouth, DT3 6DS [IO80SP, SY68]
2E1 AEJ E Jones, 19 Foxhollow, Bar Hill, Cambridge, CB3 8EP [JO02AG, TL36]
2E1 AEK R L Phillips, 57 Hollytrees, Bar Hill, Cambridge, CB3 8SF [JO02AG, TL36]
2E1 AEM N M Fortune, 12 Wignalls Meadow, Hightown, Liverpool, L38 9EN [IO83LM, SD20]
2E1 AER Details withheld at licensee's request by SSL.
2E1 AES G Smith, 11 Broadway Gdns, Stevensons Cl, Wimborne, BH21 1LS [IO90AT, SZ09]
2E1 AFA J F Davis, 30 Bonny Wood Rd, Hassocks, BN6 8HR [IO90WW, TQ31]
2E1 AFD G M Dorman, Carmine, North Rd, South Kilworth, Lutterworth, LE17 6DU [IO92KK, SP68]
2E1 AFF P R Mundy, Welbeck House, 19 Manor Rd, Burgess Hill, RH15 0NW [IO90WX, TQ31]
2E1 AFH G B Tibbett, 16 The Dingle, Fulwood, Preston, PR2 3EX [IO83PS, SD53]
2E1 AFI J Charnley, 30 Dunkirk Ave, Fulwood, Preston, PR2 3RY [IO83PS, SD53]
2E1 AFK R J Field, 3 Waveney Dr, Belton, Great Yarmouth, NR31 9JU [JO02TN, TG40]
2E1 AFN E P Williams, Dyffryn Coed, 25 Peghouse Rise, Stroud, GL5 1RU [IO81VS, SO80]
2E1 AFR F G Batty, 26 Kingsmead Park, Elstead, Godalming, GU8 6DZ [IO91PE, SU94]
2E1 AFS L Jenkins, 49 Harts Gr, Chiddingfold, Godalming, GU8 4RG [IO91QC, SU93]
2E1 AFX Details withheld at licensee's request by SSL.
2E1 AFY Details withheld at licensee's request by SSL.
2E1 AGD I P Mills, 18 Holland Ave, Walton-le-Dale, Preston, PR5 4RJ [IO83QR, SD52]
2E1 AGF E J Berry, Roseneath, Walcote Rd, South Kilworth, Lutterworth, LE17 6EQ [IO92KK, SP58]
2E1 AGG K Berry, Roseneath, Walcote Rd, South Kilworth, Lutterworth, LE17 6EQ [IO92KK, SP58]
2E1 AGJ C D Hawkins, 54 Coltham Rd, Short Heath, Willenhall, WV12 5QF [IO82XO, SJ90]
2E1 AGK G A Charles, 8 The Birches, Cheadle, Stoke on Trent, ST10 1EJ [IO92AX, SK04]
2E1 AGL V L Bancroft, 7 Moorlands Ct, Greetland, Halifax, HX4 8LF [IO93BQ, SE02]
2M1 AGP A Spence, 6 Woodend Terr, Aberdeen, AB1 6YG [IO87TH, NJ72]
2E1 AGQ J M Collins, 61 Albemarle Rd, Gorleston, Great Yarmouth, NR31 7AS [JO02UN, TG50]
2E1 AGV E Muircroft, 34 Newman Ct, Newman Rd, Sheffield, S9 1LQ [IO93GK, SK39]
2M1 AGY Details withheld at licensee's request by SSL.
2E1 AHC J V Warren, 2 St. Julien Cres, Weymouth, DT3 5DT [IO80SP, SY68]
2E1 AHO P A Clay, Ranville, Billington Ln, Derrington, Stafford, ST18 9LR [IO82WS, SJ82]
2M1 AHZ R S Rogers, 4 Deans Rd, Fortrose, IV10 8TJ [IO77WN, NH75]
2W1 AID S J Williams, 62 Turberville St., Maesteg, CF34 0LU [IO81EO, SS89]
2E1 AII D R Swann, 31 Eastgate Gdns, Newcastle upon Tyne, NE4 8DR [IO94EX, NZ26]
2M1 AIR A Ross, 16 Croft Rd, Kiltarlity By, Kiltarlity, Beauly, IV4 7HZ [IO77SK, NH54]
2E1 AIT J A Tonks, 295 Quinton Rd West, Quinton, Birmingham, B32 1PG [IO92AK, SP08]
2E1 AIV Details withheld at licensee's request by SSL.
2E1 AIY B E Whalley, 1 Lees Farm Dr, Madeley, Telford, TF7 5SU [IO82SP, SJ60]
2E1 AIZ A E Blackwell, 5 Tollgate Rd, Culham, Abingdon, OX14 4NL [IO91IP, SU59]
2I1 AJA E McGall, 20 Sandyknowes Cres, Newtownabbey, BT36 5DJ [IO74AQ, J38]
2E1 AJC M Williamson, 7 Hanbury Hill, Stourbridge, DY8 1BE [IO82WK, SO98]
2E1 AJD S J Deakin, The Woodlands, 170 Bewdley Rd, Stourport on Severn, DY13 8PJ [IO82UI, SO87]
2E1 AJE S L Jewell, 56 Meadowlands, Kirton, Ipswich, IP10 0PP [JO02PA, TM23]
2E1 AJK Details withheld at licensee's request by SSL.
2E1 AJL Details withheld at licensee's request by SSL.
2E1 AJO A J Rose, 68 Wheatley Ln, Winskill, Burton on Trent, DE15 0DX [IO92ET, SK22]
2E1 AJU Jnr A Williamson, 45 Glenconner Rd, Liverpool, L16 3NJ [IO83NJ, SJ49]
2E1 AJV N J Clasper, 7 Longmynd Way, Stourport on Severn, DY13 0BA [IO82UH, SO77]
2E1 AJW D J Blackburn, 20 Clive Ave, Hastings, TN35 5LW [JO00HU, TQ81]
2W1 AJZ D J Heward, 4 Llanover Cl, Blaenavon, NP4 9LT [IO81LS, SO20]
2W1 AKD C B Beedle, 2 Chestnut Gr, Maesteg, CF34 0NT [IO81EO, SS89]
2E1 AKK T Goodwin, 26 Turner Rd, Long Eaton, Nottingham, NG10 3GP [IO92IV, SK43]
2E1 AKL N Murray, 17 Windmill Ct, Newcastle upon Tyne, NE2 4BA [IO94EX, NZ26]
2E1 AKN A T Hilton, 24 Mount St., Derby, DE1 2HH [IO92GV, SK33]
2E1 AKO A Williamson, 45 Glenconner Rd, Liverpool, L16 3NJ [IO83NJ, SJ49]
2E1 AKT Details withheld at licensee's request by SSL.
2E1 AKV R Davis, 38 Glen Ave, Herne Bay, CT6 6HY [JO01NI, TR16]
2E1 AKW Details withheld at licensee's request by SSL.
2E1 AKX G J Gatusch, 64 Starle Cl, Canterbury, CT1 1XJ [JO01NG, TR15]
2W1 AKZ A L Evans, 4 Elm Gr, Rhyl, LL18 3PE [IO83GH, SJ08]
2E1 ALB D J Twittey, 5 Maxstoke Ln, Meriden, Coventry, CV7 7ND [IO92EK, SP28]
2W1 ALD I A Evans, 4 Elm Gr, Rhyl, LL18 3PE [IO83GH, SJ08]
2I1 ALE D Auld, 9 Chilton Rd, Carrickfergus, BT38 7JT [IO74CR, J48]
2E1 ALF Details withheld at licensee's request by SSL.
2E1 ALG M C Donnelly, 58 Bucklesham Rd, Kirton, Ipswich, IP10 0PB [JO02PA, TM23]
2E1 ALK Details withheld at licensee's request by SSL.
2E1 ALO R Godfrey, 225 Tentelow Ln, Southall, Middx, UB2 4LP [IO91TM, TQ17]
2E1 ALP C A Langley, 21 Kelvin Gr, Chessington, KT9 1DP [IO91UI, TQ16]
2E1 ALQ D J Powis, Fircroft Pound Ln, Dallinghoo, nr Woodbridge, Suffolk, IP13 0LN [JO02PD, TM25]
2E1 ALV E D Christie, 15 Bockhampton Rd, Kingston upon Thames, KT2 5JU [IO91UK, TQ17]
2E1 AMB A C Collins, 141 Downside Ave, Findon Valley, Worthing, BN14 0EY [IO90TU, TQ10]
2E1 AMJ Details withheld at licensee's request by SSL.
2E1 AMK Details withheld at licensee's request by SSL.
2E1 AMT C T Nock, 14 Warrens Hall Rd, Dudley, DY2 8DJ [IO82XL, SO98]
2E1 AMX T P Munn, 25 Wye Rd, Newcastle, ST5 4AZ [IO82VX, SJ84]
2E1 ANA D J Bentley, 4 Highway, Edgcumbe Park, Crowthorne, RG45 6HE [IO91OI, SU86]
2E1 ANE A Taylor, 113 Queensway, Grantham, NG31 9RG [IO92QW, SK93]
2E1 ANJ Details withheld at licensee's request by SSL.
2E1 ANK I V F Craig, Partridge Cottage, Redpale, Dallington, Heathfield, TN21 9NR [JO00EW, TQ61]
2E1 ANN M V Kearney, 18 Wayside Mews, Maidenhead, SL6 7EJ [IO91PM, SU88]
2E1 ANP Details withheld at licensee's request by SSL.
2E1 ANQ A O T Bell, 8 Silk Mill Green, Leeds, LS16 6DU [IO93EU, SE23]
2M1 ANT J L Miles, 58 Fogralea, Lerwick, ZE1 0SE [IP90JD, HU44]
2M1 ANY L T R Waterall, 3 Wavell St., Grangemouth, FK3 8TG [IO86DA, NS98]
2E1 ANZ Dr R E Turnbull, Taman, 38 Elsdon Rd, Gosforth, Newcastle upon Tyne, NE3 1HY [IO95EA, NZ26]
2E1 AOB E P Lattka, South Fen Lodge, 9 The Row, Sutton, Ely, CB6 2PD [JO02BJ, TL47]
2W1 AOD D E Wanklyn, 21 Upper St., Maesteg, CF34 9DU [IO81EO, SS89]
2E1 AOF G A Tutt, 46 Heathcroft Ave, Sunbury on Thames, TW16 7TL [IO91SK, TQ17]
2E1 AOG J R Menday, 3 Ash Gr, Guildford, GU2 5UT [IO91QF, SU95]
2W1 AOK R S Gill, 45 Biggin Ln, Ramsey, Huntingdon, PE17 1NB [IO92WK, TL28]
2M1 AOL G A Sibbald, 30 Langton View, East Calder, Livingston, EH53 0LE [IO85GV, NT06]
2E1 AOQ Details withheld at licensee's request by SSL.
2E1 AOR E A E Bullivant, 93 Sutherland Rd, Cheslyn Hay, Walsall, WS6 7BT [IO82XP, SJ90]
2E1 AOU M J Larcombe, 52 Orchard Rd, Burgess Hill, RH15 9PL [IO90WW, TQ31]
2E1 AOX E K J Fox, 249 Warminster Rd, Norton, Sheffield, S8 8PR [IO93GI, SK38]
2E1 AOY A R Vann, 3 Mile Planting, Richmond, DL10 5DB [IO94DK, NZ10]
2E1 APC Details withheld at licensee's request by SSL.
2E1 APQ M J Hawkins, 38 Victoria Ave, Burgess Hill, RH15 9PX [IO90WW, TQ31]
2E1 APR P C Hawkins, 38 Victoria Ave, Burgess Hill, RH15 9PX [IO90WW, TQ31]
2E1 APT G M K Houlden, 29 Ct Barton, Weston, Portland, DT5 2HJ [IO80SN, SY67]
2E1 APW D Jenkinson, 72 Poole Rd, Salterbeck, Workington, CA14 5DW [IO84FO, NX92]
2E1 APX D Johnson, 3 Plantation Ave, Swalwell, Newcastle upon Tyne, NE16 3JN [IO94DW, NZ26]
2E1 APY D E Johnston, 11 Granville St., Deal, CT14 7EZ [JO01QF, TR35]
2E1 AQC Details withheld at licensee's request by SSL.
2E1 AQE Z Levett, 5 Park Rd, Yapton, Arundel, BN18 0JE [IO90QT, SU90]
2E1 AQF M D Lewin, Flat Above, 280 Narborough Rd, Leicester, LE3 2AQ [IO92KO, SK50]
2E1 AQG P M Lewis, 20 Hermitage Way, Stourport on Severn, DY13 0DA [IO82UH, SO87]
2E1 AQH S A Allgood, 53 The Avenue, Leighton Bromswold, Huntingdon, PE18 0SH [IO92TI, TL17]
2E1 AQI J R Turner, 35 Oakfield Rd, Harpenden, AL5 2NP [IO91TT, TL11]
2E1 AQR S D Daley, 16 Bassett Rd, Sheffield, S2 5EZ [IO93GJ, SK38]
2E1 AQS M L F Snary, 12 Borden Ave, Enfield, EN1 2BZ [IO91XP, TQ39]
2E1 AQU M Sezen, 29 Coronation Ave, Moordown, Bournemouth, BH9 1TW [IO90BR, SZ09]
2M1 AQV Details withheld at licensee's request by SSL.
2E1 AQX N H Storbeck, Cockerham House, 33 Huddersfield Rd, Barnsley, S75 1DN [IO93GN, SE30]
2E1 AQY N A Stretch, 5 Ledwych Rd, Droitwich, WR9 9LA [IO82VG, SO86]
2E1 ARA A J Stretch, 5 Ledwych Rd, Droitwich, WR9 9LA [IO82VG, SO86]
2E1 ARF A T Townley, 11 Walton Rd, Hartlebury, Kidderminster, DY10 4JA [IO82VI, SO87]
2E1 ARG M Trigg, 41 Veasey Rd, Hartford, Huntingdon, PE18 7TA [IO92WI, TL27]
2E1 ARS W D Hornby, 14 Essex Rd, Stevenage, SG1 3EZ [IO91VV, TL22]
2E1 ARU R D Dilley, Thauma, 6 Watton Rd, Knebworth, SG3 6AH [IO91VU, TL22]
2E1 ARW R Whitney, Gr View, Upthorpe Rd, Stanton, Bury St. Edmunds, IP31 2AP [JO02KH, TL97]
2E1 ASC J Berke, 190 Hall Rd, Handsworth, Sheffield, S13 9AN [IO93HJ, SK48]
2E1 ASF T M G Stevens, 33 Langham Rd, Hastings, TN34 2JE [JO00HU, TQ81]
2E1 ASI H M Bird, 15 Ludlow Dr, Stirchley, Telford, TF3 1EG [IO82SP, SJ60]
2E1 ASN J C Everitt, 55 Risborough Rd, Bedford, MK41 9QR [IO92SD, TL05]
2E1 ASP E E Gill, 120 High St., Offord Cluny, St. Neots, Huntingdon, PE18 9RQ [IO92VG, TL26]
2M1 ASQ K A Gray, 34 Murrayston, Lerwick, ZE1 0RE [IP90JD, HU44]
2E1 ASU J R Parry, 38 Milford Dr, Bearcross, Bournemouth, BH11 9HJ [IO90AS, SZ09]
2E1 ASV Details withheld at licensee's request by SSL.
2E1 ATB Details withheld at licensee's request by SSL.
2E1 ATD D J Roden, 12 Pimbury Rd, Short Heath, Willenhall, WV12 5QN [IO82XO, SJ90]
2E1 ATE S E Rolinson, 14 Waters View, Lichfield Rd, Pelsall, Walsall, WS3 4HJ [IO92AP, SK00]
2E1 ATF J Rowsell, North Lodge, 61 Barrack Rd, Bexhill on Sea, TN40 2AZ [JO00FU, TQ70]
2E1 ATH J Minnock, 32 Sandwood Rd, Sandwich, CT13 0AQ [JO01QG, TR35]
2E1 ATL M Chamberlain, 54 Henray Ave, Glen Parva, Leicester, LE2 9QJ [IO92KO, SP59]
2M1 ATN Mstr D J Robertson, 13 Swanston View, Fairmilehead, Edinburgh, EH10 7DG [IO85JV, NT26]
2E1 ATR T Ralph, 15 Portchester Cl, Peterborough, PE2 8UP [IO92VN, TL29]
2E1 ATS A W J Cundall, Hawthorn House High Levels, nr Thorne, Doncaster, DN8 4SW [IO93MP, SE61]
2E1 ATT M R Anthony, 77 Brayfield Rd, Littleover, Derby, DE23 6GT [IO92FV, SK33]
2E1 ATV R J I Corden, 24 Lavington Ave, Cheadle, SK8 2HH [IO83VJ, SJ88]
2E1 ATY Details withheld at licensee's request by SSL.
2E1 AUA R M Hope, 3 Farm Cres, Sittingbourne, ME10 4QD [JO01IH, TQ96]
2E1 AUB R G Jeffs, 28 Ludlow Ave, Luton, LU1 3RW [IO91TU, TL01]
2E1 AUG J M Poole, 18 Grosvenor Ave, Kidderminster, DY10 1SS [IO82VJ, SO87]
2E1 AUL Details withheld at licensee's request by SSL.
2E1 AUM R E Raymer, New House, Clacton Rd, Wix, Manningtree, CO11 2RU [JO01NV, TM12]
2E1 AUN D Austin, 29 Salisbury Cl, Sittingbourne, ME10 3BP [JO01JI, TQ96]
2E1 AUQ E W Harding, 17 Summerfield Cl, Wokingham, RG41 1PH [IO91NK, SU87]
2E1 AUR C T Saxton, Rock Farm House, 3 Acres Rd, Bebington, Wirral, L63 7QD [IO83LI, SJ38]
2E1 AUS J E Simmonds, 3 Mallings Ln, Bearsted, Maidstone, ME14 4EY [JO01HG, TQ85]
2E1 AUU A Bowman, 7 Wystan Ct, Repton, Derby, DE65 6SA [IO92FU, SK32]
2W1 AVC Details withheld at licensee's request by SSL.
2E1 AVM S L Peacock, Gkn Defence, 106 PO Box, Hadley Castle Works, Telford, TF1 4QW [IO82SQ, SJ61]
2E1 AVS A S Jewell, 56 Meadowlands, Kirton, Ipswich, IP10 0PP [JO02PA, TM23]
2E1 AVT J A Hoggan, 25 Clingan Rd, Boscombe East, Bournemouth, BH6 5PY [IO90CR, SZ19]
2E1 AVU C A Howgego, 12 Bettespol Meadows, Redbourn, St. Albans, AL3 7EW [IO91TT, TL11]
2E1 AVX J Preece, 8 Gregory Rd, Hedgerley, Slough, SL2 3XL [IO91QN, SU98]
2M1 AVY I M Sinclair, 3 Ben More Dr, Paisley, PA2 7NU [IO75TT, NS56]
2M1 AVZ N K Sinclair, 3 Ben More Dr, Paisley, PA2 7NU [IO75TT, NS56]
2I1 AWD Details withheld at licensee's request by SSL.
2E1 AWF M J Hampson, 7 Merryfield Cl, Bransgore, Christchurch, BH23 8BS [IO90DS, SZ19]
2E1 AWG P D Howgego, 12 Bettespol Meadows, Redbourn, St. Albans, AL3 7EW [IO91TT, TL11]
2E1 AWH Details withheld at licensee's request by SSL.
2E1 AWI J H Wheeldon, 11 Stathern Walk, Grantham, NG31 7XG [IO92QW, SK93]
2E1 AWJ Details withheld at licensee's request by SSL.
2E1 AWK E E Matthews, Thisledo, Dale Bank Rd, Cheadle, Stoke on Trent, ST10 1RE [IO92AX, SK04]
2E1 AWM Details withheld at licensee's request by SSL.
2E1 AWO P I Hobbis, 680 Chester Rd, Castle Bromwich, Kingshurst, Birmingham, B36 0LJ [IO92DL, SP18]
2E1 AWQ D J Tropman, 91 Reindeer Rd, Deer Park, Fazeley, Tamworth, B78 3SW [IO92DO, SK10]
2E1 AWR A D Stean, 31 Lanesfield Park, Greenhill, Evesham, WR11 4NU [IO92AC, SP04]
2E1 AWS N D Cook, 30 Osborne Gdns, Herne Bay, CT6 6SH [JO01NI, TR16]
2E1 AWW Details withheld at licensee's request by SSL.
2E1 AWX Details withheld at licensee's request by SSL.
2E1 AXD A P Richmond, 57 The Fairway, Daventry, NN11 4NW [IO92KG, SP56]
2E1 AXE L M Richmond, 57 The Fairway, Daventry, NN11 4NW [IO92KG, SP56]
2E1 AXF R Mizon, 78 Finchfield, Parnell, Peterborough, PE1 4YQ [IO92VO, TF20]
2I1 AXG D L Cosgrove, 18 Lynn Cres, Dromore, BT25 1PY [IO64WJ, J25]
2I1 AXH K E Bird, 115 Halftown Rd, Lisburn, BT27 5RF [IO64WL, J26]
2E1 AXI F R Preece, 8 Gregory Rd, Hedgerley, Slough, SL2 3XL [IO91QN, SU98]
2E1 AXK S A Griffin, 121 Home Ct, Railway Terr, Feltham, TW13 4AW [IO91TK, TQ17]
2E1 AXL Details withheld at licensee's request by SSL.
2E1 AXN Details withheld at licensee's request by SSL.
2E1 AXO S A Sugden, 21 Pembroke, Hanworth, Bracknell, RG12 7RD [IO91OJ, SU86]
2E1 AXP B T Lyon, 28 Welbeck, Bracknell, RG12 8UQ [IO91OJ, SU86]
2E1 AXR Details withheld at licensee's request by SSL.
2I1 AXS Details withheld at licensee's request by SSL.
2E1 AXT Details withheld at licensee's request by SSL.
2E1 AXU Details withheld at licensee's request by SSL.
2E1 AXV Details withheld at licensee's request by SSL.
2E1 AXX Details withheld at licensee's request by SSL.
2E1 AXY Details withheld at licensee's request by SSL.
2E1 AXZ R J Aley, 39 Westwood Ave, March, PE15 8AX [JO02BN, TL49]
2E1 AYB J J Richardson, 107 Pinehurst Rd, West Moors, Ferndown, BH22 0AL [IO90BT, SU00]
2E1 AYC Details withheld at licensee's request by SSL.
2E1 AYE S Fisher, 8 Tubbenden Ln, Orpington, BR6 9PN [JO01BI, TQ46]
2E1 AYK Details withheld at licensee's request by SSL.
2E1 AYL Details withheld at licensee's request by SSL.
2W1 AYO E J Phillips, 2 Oak St., Newport, NP9 7HW [IO81MO, ST38]
2W1 AYP A W Norris, 2 Oak St., Newport, NP9 7HW [IO81MO, ST38]
2E1 AYS P A Cartwright, 41 Sandgate Dr, Kippax, Leeds, West Yorks, LS25 7EX [IO93HS, SE43]
2E1 AYX R J Wallis, 26 Heather Bank, Osbaldwick, York, YO1 3QH [IO93LW, SE65]
2W1 AYZ D A Maxwell, Cae Fron, Nant Bychan, Moelfre, LL72 8HE [IO73VI, SH58]
2E1 AZA E H Williams, 37 Danesby Cres, Denby, Ripley, DE5 8RF [IO93GA, SK34]
2E1 AZD Details withheld at licensee's request by SSL.
2E1 AZF J G F Heaton, 39 Goldieslie Rd, Sutton Coldfield, B73 5PE [IO92CN, SP19]
2E1 AZG S A Mayfield, 184 Wharf Rd, Pinxton, Nottingham, NG16 6LQ [IO93IC, SK45]
2E1 AZI M E Whitehouse, 102 Osmaston Rd, Harborne, Birmingham, B17 0TN [IO92AK, SP08]
2M1 AZJ J R Gordon, 140 Mains Hill, Erskine, PA8 7JE [IO75SV, NS47]
2E1 AZL Details withheld at licensee's request by SSL.
2E1 AZM R A Bratley, 17 Bruce Walk, Openshaw, Manchester, M11 1EE [IO83VL, SJ89]
2E1 AZN R W Smith, Greenacres, Main St., Sawdon, Scarborough, YO13 9DY [IO94RG, SE98]
2E1 AZO W Sykes, The Lilacs, Harewood End, Hereford, HR2 8JT [IO81PW, SO52]
2E1 AZP C J Vincent, 10 Laburnum Dr, Earl Shilton, Leicester, LE9 7HU [IO92IN, SP49]
2E1 AZQ J H Perkins, Highfield House, Newtown, Longnor, Buxton, SK17 0NF [IO93BD, SK06]
2E1 AZS M Thomas, Flat 2, 34 Black Walk, Upper Tything, Worcester, Worcs, WR1 1JZ [IO82VE, SO85]
2W1 AZU D Williams, 75 Queens Ave, Maesgeirchen, Bangor, LL57 1NH [IO73WF, SH57]
2E1 AZV B Tancock, 12 Belgrave St., Nelson, BB9 9HR [IO83VU, SD83]
2E1 AZW R M Roberts, Rose Cottage, Castle Hill, Middleham, Leyburn, DL8 4QN [IO94CG, SE18]
2E1 BAC T M Trailor, 3 Welford Rd, Shearsby, Lutterworth, LE17 6PE [IO92LM, SP69]
2E1 BAD B Rowland, Deacons Cottage, Brideway, Off Broughton Rd, Croft, Leics, LE9 6EE [IO92JM, SP59]
2E1 BAE T R Ladley, 3 Quarry Wood, Aldington, Ashford, TN25 7EY [JO01LC, TR03]
2E1 BAI M J Martin, 130 Heath St., Hednesford, Cannock, WS12 4BP [IO82XR, SJ91]
2E1 BAM K A Worthington, 48 South Dr, Chorltonville, Manchester, M21 8EG [IO83UK, SJ89]
2E1 BAP J E Huddleston, Milford House, 9 Gallowfields Rd, Richmond, DL10 4DB [IO94DJ, NZ10]
2E1 BAQ Details withheld at licensee's request by SSL.
2M1 BAR R McKenzie, 25 Glen Cres, Yoker, Glasgow, G13 4EF [IO75TV, NS56]
2E1 BAU N Morgan, 5 Sealy Cl, Spital Bebbington, Wirral, L63 9LP [IO83MH, SJ38]
2E1 BAW S E Fechcer, 1 Hungerford Dr, Maidenhead, SL6 7UT [IO91PM, SU88]
2E1 BAY S E Jinks, Nyali, Main St., South Littleton, Evesham, WR11 5TJ [IO92BC, SP04]

2E1 BBA Details withheld at licensee's request by SSL.
2E1 BBC B Bicknell, 30 Alloe Field View, Halifax, HX2 9EP [IO93BS, SE02]
2E1 BBG M V Vowles, 14 Frobisher Rd, Ashton, Bristol, BS3 2AU [IO81QK, ST57]
2E1 BBH O W P Cowell, 38 Brompton Dr, Pinkneys Green, Maidenhead, SL6 6SP [IO91PM, SU88]
2E1 BBN Details withheld at licensee's request by SSL.
2E1 BBO P N H Kent, Old Cottage, Hermitage Ln, Detling, Maidstone, ME14 3HP [JO01GH, TQ75]
2E1 BBP K Uytendhal, 76 Gautby Rd, Birkenhead, L41 7DR [IO83LJ, SJ29]
2E1 BBV E H Walker, 104 Fairmile Rd, Christchurch, BH23 2LN [IO90CR, SZ19]
2E1 BBX S E Smy, 62 Mayfield Rd, Ipswich, IP4 3NG [JO02OB, TM14]
2M1 BBY J Duncan, 132 John St., Penicuik, EH26 8NJ [IO85JT, NT26]
2M1 BCA S F Andrew, Fereneze View, Gateside Rd, Barrhead, Glasgow, G78 1TT [IO75TT, NS45]
2E1 BCC M J Kluger-Langer, 23 Vernon Walk, Northampton, NN1 5ST [IO92NF, SP76]
2E1 BCD J D Rail, 16 The Cres, Truro, TR1 3ES [IO70LG, SW84]
2E1 BCF Details withheld at licensee's request by SSL.
2E1 BCJ M E Exell, Baregains, Callow, Hereford, HR2 8DE [IO82PA, SO43]
2E1 BCL Details withheld at licensee's request by SSL.
2E1 BCN Details withheld at licensee's request by SSL.
2W1 BCP G Rowlands, Dalar Wen, Rhosmeirch, Llangefni, Gwynedd, LL77 7SJ [IO73UG, SH47]
2E1 BCR D M Gaston, 8 Linden Cl, Chelsfield, Orpington, BR6 6JJ [JO01BI, TQ46]
2E1 BCV Details withheld at licensee's request by SSL.
2E1 BCW C P Whittaker, Ashurst, 42 Barnehurst Ave, Barnehurst, Bexleyheath, DA7 6QB [JO01CL, TQ57]
2E1 BCX A Whittaker, Ashurst, 42 Barnehurst Ave, Barnehurst, Bexleyheath, DA7 6QB [JO01CL, TQ57]
2E1 BCZ M A Ward, 20 Dalmeny Rd, Erith, DA8 1JX [JO01BL, TQ47]
2E1 BDA M J Austin, 1 Sandhurst Terr, Higher Trehaverne, Truro, TR1 3RL [IO70LG, SW84]
2E1 BDB P Hudson, 1 Dean Moore Cl, St. Albans, AL1 1DW [IO91UR, TL10]
2E1 BDC P J Kennedy, 71 Wilbert Ln, Beverley, HU17 0AJ [IO93SU, TA03]
2E1 BDG S J Crook, 38 Coulton Ave, Northfleet, Gravesend, DA11 8DY [JO01EK, TQ67]
2E1 BDJ Details withheld at licensee's request by SSL.
2E1 BDL Details withheld at licensee's request by SSL.
2E1 BDM Details withheld at licensee's request by SSL.
2E1 BDN J A K Reid, 4 Harles Acres, Hickling, Melton Mowbray, LE14 3AF [IO92MU, SK62]
2E1 BDO E R R Milton, 35 Holybrook Rd, Reading, RG1 6DG [IO91MK, SU77]
2E1 BDR M A Colwell, 24 Berkeley Ave, Reading, RG1 6JE [IO91MK, SU77]
2E1 BDT D P Palmer, 6 Victoria Rd, South Woodham Ferrers, Chelmsford, CM3 5LR [JO01HP, TQ89]
2E1 BDU A C Swain, 32 College Rd, Harrogate, HG2 0AQ [IO93FX, SE25]
2E1 BDV P J McLusky, 11 Ripon Rd, Killinghall, Harrogate, HG3 2DG [IO94FA, SE25]
2E1 BDW W Berry, 17 St. Pauls Rd, St. Leonards on Sea, TN37 6RS [JO00GU, TQ81]
2E1 BDX S Whittle, 48 Kitchener Rd, Great Yarmouth, NR30 4HU [JO02UO, TG50]
2E1 BDY Details withheld at licensee's request by SSL.
2E1 BDZ Details withheld at licensee's request by SSL.
2E1 BEB J A Wright, 18 Anderida Ct, Mansell Cl, Little Common, Bexhill on Sea, TN39 4XD [JO00FU, TQ70]
2E1 BEF P A Hill, 35 Ave Rd, Stratford upon Avon, CV37 6UW [IO92DE, SP25]
2E1 BEG C M Hill, 35 Ave Rd, Stratford upon Avon, CV37 6UW [IO92DE, SP25]
2E1 BEI M D Geldart, Thrapston House, 42 Collington Ave, Bexhill on Sea, TN39 3NE [JO00FU, TQ70]
2E1 BEL J Lisk, 37 Ashton Rd, Enfield, EN3 6DQ [IO91XQ, TQ39]
2E1 BEM A J Ransome, 48 Chatsworth Dr, Ipswich, IP4 5XD [JO02OB, TM24]
2E1 BEO S E Elliott, 4 Ivel View, Sandy, SG19 1AU [IO92UD, TL14]
2E1 BET Details withheld at licensee's request by SSL.
2M1 BEU M A Horne, 10 Blair Pl, Kirkcaldy, KY2 5SQ [IO86JC, NT29]
2E1 BEV M Norman, 124 Porthcawl Green, Tattenhoe, Milton Keynes, MK4 3AL [IO91OX, SP83]
2E1 BFA S E Lawrence, 4 Dale Park Rise, Leeds, LS16 7PP [IO93EU, SE23]
2E1 BFD P N Costello, 8 Manor Cl, Clifton, Shefford, SG17 5EJ [IO92UA, TL13]
2E1 BFF R G Greer, 159 Lucas Ave, Chelmsford, CM2 9JR [JO01FR, TL70]
2E1 BFG Details withheld at licensee's request by SSL.
2E1 BFI Details withheld at licensee's request by SSL.
2E1 BFP J C Philpot, 17 Jervis Ct, Cotmanhay, Ilkeston, DE7 8PX [IO92IX, SK44]
2E1 BFT J E Foster, 30 Chapel Rd, Grassmoor, Chesterfield, S42 5EL [IO93HE, SK46]
2E1 BFW P A Hyde, 10 Highfield Cres, Comeytrowe Ln, Taunton, TA1 5JH [IO81KA, ST22]
2E1 BFX Details withheld at licensee's request by SSL.
2E1 BFY Details withheld at licensee's request by SSL.
2E1 BGG H J Faulkner, 1 Westland, Martlesham Heath, Ipswich, IP5 7SU [JO02OB, TM24]
2E1 BGH T Warhurst, 12 Potters Way, Ilkeston, DE7 5EX [IO92IX, SK44]
2E1 BGM Details withheld at licensee's request by SSL.
2E1 BGN A B Smith, Flat 1, 29 Wensley Cres, Doncaster, South Yorks, DN4 6QL [IO93LM, SE60]
2E1 BGO M A Radforth, 4 Walcote Rd, Southkilworth, Lutterworth, Leics, LE17 6EE [IO92KK, SP68]
2E1 BGS A J Overton, 34 Featherstone, Blindley Heath, Lingfield, RH7 6JY [IO91XE, TQ34]
2E1 BGU S J J Lindley, 10 Fleet St., Scissett, Huddersfield, HD8 9JJ [IO93EO, SE21]
2E1 BGV R Issatt, 20 Priory Way, Snaith, Goole, DN14 9HB [IO93LQ, SE62]
2E1 BGX A C Clark, 43 Harts Ln, Barking, IG11 8NA [JO01AM, TQ48]
2E1 BGY K L Elliott, 48 Weymouth Ave, Parr, St. Helens, WA9 3QX [IO83PK, SJ59]
2E1 BHB A M J Comis, 178 Lordswood Rd, Birmingham, B17 8QH [IO92AL, SP08]
2E1 BHC P G D Comis, 178 Lordswood Rd, Birmingham, B17 8QH [IO92AL, SP08]
2E1 BHD Details withheld at licensee's request by SSL.
2E1 BHF J E Clifford, 70 Strathmore Ave, Luton, LU1 3NZ [IO91TU, TL02]
2E1 BHH T A Easting, Newlyn, Honey Tye, Leavenheath, Colchester, CO6 4NY [JO01KX, TL93]
2E1 BHL Details withheld at licensee's request by SSL.
2E1 BHQ A P Murphy, 2 Walnut Tree Cotages, Walcote Rd, South Kilworth, Lutterworth, LE17 6EG [IO92KK, SP68]
2E1 BHR R L Feay, 19 Dorset Ave, Diggle, Oldham, OL3 5PL [IO93AN, SE00]
2E1 BHS Details withheld at licensee's request by SSL.
2E1 BHT L Wiseman, 76 Haywood Rd, Mapperley, Nottingham, NG3 6AE [IO92KX, SK54]
2E1 BHU P W Gosling, 10 Prospect Rd, Carlton, Nottingham, NG4 1LY [IO92KX, SK54]
2I1 BHV R V Simpson, 26 Oak Grange, Waringstown, Craigavon, BT66 7SU [IO64UK, J15]
2W1 BHX D M Collins, 24 Glade Cl, Coed Eva, Cwmbran, NP44 4TF [IO81LP, ST29]
2E1 BHZ S J Moss, 10 Thorngate Ct, Penwortham, Preston, PR1 0XN [IO83PR, SD52]
2E1 BID M C Shaw, 1 Castleton Ave, Wembley, HA9 7QH [IO91UN, TQ18]
2E1 BIF R J Baines, 327 Langer Ln, Wingerworth, Chesterfield, S42 6TY [IO93GE, SK36]
2E1 BIG J D Baines, 327 Langer Ln, Wingerworth, Chesterfield, S42 6TY [IO93GE, SK36]
2E1 BIH A H Sims, 4 Fawn Cl, Wingerworth, Chesterfield, S42 6PZ [IO93GE, SK36]
2E1 BII A M Hall, 11 School Ln, Dronfield, Sheffield, S18 6RY [IO93GH, SK37]
2I1 BIJ R E Hammond, 75 Deramore Dr, Portadown, Craigavon, BT62 3HJ [IO64SK, J05]
2E1 BIM C J Faulkner, 1 Westland, Martlesham Heath, Ipswich, IP5 7SU [JO02OB, TM24]
2E1 BIR Details withheld at licensee's request by SSL.
2E1 BIU Details withheld at licensee's request by SSL.
2W1 BIV G Houghton, 47 Marion Rd, Prestatyn, LL19 7DA [IO83GI, SJ08]
2W1 BIY L A Butler, Moyddin Fach, Gorsgoch, Llanybydder, Ceredigion, SA40 9TN [IO72VD, SN45]
2E1 BJA Details withheld at licensee's request by SSL.
2I1 BJB A Whale, 22 Mill Hill, Waringstown, Craigavon, BT66 7QL [IO64UK, J15]
2E1 BJD R D Saunders, 31 Greenwood Rd, High Green, Sheffield, S30 4GU [IO93GJ, SK38]
2E1 BJE A D Cowdell, 6 Pearl St., Bedminster, Bristol, BS3 3EA [IO81QK, ST57]
2E1 BJF Details withheld at licensee's request by SSL.
2E1 BJG S J Benson, 7 Crofton Cl, Attenborough, Beeston, Nottingham, NG9 5HX [IO92JV, SK53]
2E1 BJL B J Lawrence, 4 Dale Park Rise, Cookridge, Leeds, LS16 7PP [IO93EU, SE23]
2E1 BJM A Orchard, 68 Simpson Rd, Bletchley, Milton Keynes, MK1 1BA [IO92PA, SP83]
2E1 BJN B J Orchard, 68 Simpson Rd, Bletchley, Milton Keynes, MK1 1BA [IO92PA, SP83]
2E1 BJO A Sporton, 14 Harles Acres, Hickling, Melton Mowbray, LE14 3AF [IO92MU, SK62]
2E1 BJP Details withheld at licensee's request by SSL.
2E1 BJS D Childs, 26 Granby Gr, Highfield, Southampton, SO17 3RZ [IO90HW, SU41]
2E1 BJU A Rychlinski, 82 Marine Parade, Seaford, BN25 2QR [JO00BS, TV49]
2E1 BJW R C Ng, 1 Station Rd, Hailsham, BN27 2BE [JO00DU, TQ50]
2E1 BKA D S Clarke, 28 Wallett Ave, Beeston, Nottingham, NG9 2QR [IO92JW, SK53]
2E1 BKD R Roychoudhuri, 62 Parkway, Eastbourne, BN20 9DY [JO00DT, TQ50]
2E1 BKF G E Muircroft, 34 Newman Ct, Newman Rd, Sheffield, S9 1LQ [IO93GK, SK39]
2E1 BKK C M Berry, Roseneath, Walcote Rd, South Kilworth, Lutterworth, LE17 6EQ [IO92KK, SP58]
2E1 BKM L Martin, Bayshore, Gilcrux, Carlisle, Cumbria, CA5 2QD [IO84HR, NY13]
2E1 BKP S Martin, Bayshore, Gilcrux, Carlisle, Cumbria, CA5 2QD [IO84HR, NY13]
2E1 BKT D J Jump, 29 Ainslie Rd, Smithills, Bolton, BL1 5LU [IO83SO, SD61]
2E1 BKU Details withheld at licensee's request by SSL.
2E1 BKW Details withheld at licensee's request by SSL.
2E1 BKX D P Scott, Wayside Stables, Uddens Dr, Uddens, Wimborne, Dorset, BH21 7BQ [IO90AT, SU00]
2E1 BLA D Burgess, 67 Fair Cl, Beccles, NR34 9QT [JO02SK, TM49]
2E1 BLG M L M Fish, 44 Billing Ave, Finchampstead, Wokingham, RG40 4JE [IO91NJ, SU76]
2E1 BLJ M J Peters, 28 The St., Geldeston, Beccles, NR34 0LB [JO02SL, TM39]
2E1 BLK Details withheld at licensee's request by SSL.
2E1 BLN P J Robinson, Windrush, 7 Kenderdine Cl, Bednall, Stafford, ST17 0YS [IO82WS, SJ91]
2E1 BLO T W Waterworth, Stonethwaite, Darley, Harrogate, North Yorks, HG3 2PP [IO94DA, SE25]
2E1 BLP B O W Mulley, Woodlands, Old Stowmarket Rd, Woolpit, Bury St. Edmunds, IP30 9QS [JO02KF, TL96]
2E1 BLS D W Woolger, 24 Weald Dr, Furnace Green, Crawley, RH10 6JU [IO91VC, TQ23]
2E1 BLT Details withheld at licensee's request by SSL.
2E1 BLV C J Pitchford, 9 Pickhurst Ln, Bromley, BR2 7JE [JO01AJ, TQ46]
2E1 BLZ E M Sully, 10 The Paddock, Pound Hill, Crawley, RH10 7RQ [IO91WC, TQ23]
2E1 BMA Details withheld at licensee's request by SSL.
2E1 BMC P J Read, 2 Corton Rd, Lowestoft, NR32 4PH [JO02VL, TM59]
2E1 BMF D M Carslake, 38 Loppets Rd, Tilgate, Crawley, RH10 5DW [IO91VC, TQ23]
2E1 BMI L M Mc Clusky, 11 Ripon Rd, Killinghall, Harrogate, HG3 2DG [IO94FA, SE25]
2E1 BMJ D M Peters, 28 The St., Geldeston, Beccles, NR34 0LB [JO02SL, TM39]
2E1 BMK F E Peters, 28 The St., Geldeston, Beccles, NR34 0LB [JO02SL, TM39]
2E1 BML P M Phillips, 9 Moat Walk, Crawley, RH10 7ED [IO91WC, TQ23]
2E1 BMP H V Eyre, St. Michael Mead, The Common, Barton Turf, Norwich, NR12 8BA [JO02RR, TG32]
2E1 BMU T J Quantrill, 7 Clare Rd, Kessingland, Lowestoft, NR33 7PS [JO02UK, TM58]
2E1 BMV B E Mycock, 69 Bentley Rd, Uttoxeter, ST14 7EN [IO92BV, SK03]
2E1 BNA S C Eyre, St. Michael Mead, The Common, Barton Turf, Norwich, NR12 8BA [JO02RR, TG32]
2E1 BND D J Witham, 129 Upper Moor Green Rd, Cowes, PO31 7LF [IO90IS, SZ49]
2E1 BNE Details withheld at licensee's request by SSL.
2W1 BNI B Curtis, Porth-y-Felin Rd, Holyhead, Gwynedd, LL65 1PL [IO73QH, SH28]
2E1 BNL D T Saunders, 2 Abbey Cl, Pyrford, Woking, GU22 8RY [IO91RH, TQ05]
2E1 BNN C J Mansfield, 10 Stanley Rd, Lordswood, Chatham, ME5 8LN [JO01GJ, TQ76]
2E1 BNP Details withheld at licensee's request by SSL.
2E1 BNQ Details withheld at licensee's request by SSL.
2E1 BNR Details withheld at licensee's request by SSL.
2E1 BNU S J Williams, 70 Linden Way, Sendmarsh, Ripley, Woking, GU23 6LP [IO91RG, TQ05]
2E1 BNV P D Longthorne, 130 Spashett Rd, Lowestoft, NR32 4DH [JO02UL, TM59]
2E1 BNX K W Gibson, 6 Beaumont Caravan Park, Mill Ln, Bradwell, Great Yarmouth, NR31 8HP [JO02UN, TG50]
2E1 BOA P J Radford, Northmead, 111 Mount Ambrose, Redruth, Cornwall, TR15 1NW [IO70JF, SW74]
2E1 BOB M R D Haines, Park Cottage, Beech Ln, Grasscroft, Oldham, OL4 4EP [IO83XM, SD90]
2E1 BOC M Freeman, 5 Northfield Rd, Beeston, Nottingham, NG9 5GS [IO92IV, SK53]
2E1 BOE C S Greenbank, 114 Wignals Gate, Holbeach, Spalding, PE12 7HR [IO92XT, TF32]
2E1 BOF Details withheld at licensee's request by SSL.
2E1 BOG Details withheld at licensee's request by SSL.
2E1 BOJ M J Beeley, Bridport House, Cilcennin, Lampeter, Dyfed, SA48 8RL [IO72WF, SN56]
2E1 BOM M D Love, 72A Hart Plain Ave, Waterlooville, Hants, PO8 8RX [IO90LV, SU61]
2E1 BOO M R Constantine, The Old Exchange, Burnley Rd, Mytholmroyd, Hebden Bridge, HX7 5PD [IO93AR, SE02]
2E1 BOR Details withheld at licensee's request by SSL.
2E1 BOS Details withheld at licensee's request by SSL.
2E1 BOV C J O'Beirne, 4 Duncan Way, Hartford, Huntingdon, PE18 7SZ [IO92WI, TL27]
2E1 BOX Details withheld at licensee's request by SSL.
2E1 BOZ S Razzaq, 37 Newport St., Nelson, BB9 7RW [IO83VU, SD83]
2E1 BPA B L Callaway, 53 Fearon Rd, North End, Portsmouth, PO2 0NJ [IO90LT, SU60]
2E1 BPB D A Guy, 14 Lower New Rd, West End, Southampton, SO30 3FL [IO90IW, SU41]
2W1 BPC E D Greenhalgh, 6 Clifton Gr, Rhyl, LL18 4AF [IO83GH, SJ08]
2E1 BPD Details withheld at licensee's request by SSL.
2W1 BPG Rev J W Binny, The Rectory, Rectory Dr, St. Athan, Barry, CF62 4PD [IO81HJ, ST06]
2W1 BPH Rev E D D Lewis, 5 Oak Gr, St. Anthan, St. Athan, Barry, CF62 4JN [IO81HJ, ST06]
2W1 BPJ M Butler, 178 Squirrel Walk, Pontarddulais, Swansea, SA4 1UG [IO71XR, SN50]
2E1 BPL J Whitney, Gr View, Upthorpe Rd, Stanton, Bury St. Edmunds, IP31 2AP [JO02KH, TL97]
2E1 BPN A J Collins, 4 Beckham Rd, Lowestoft, NR32 2BY [JO02UL, TM59]
2E1 BPP D L J Fuller, 11 Homefield Ave, Beccles, NR34 9UB [JO02SK, TM49]
2E1 BPQ J A Painter, 20 Beech Rise, Sleaford, NG34 8BJ [IO93TA, TF04]
2E1 BPR J A Painter, 20 Beech Rise, Sleaford, NG34 8BJ [IO93TA, TF04]
2W1 BPS E A Mainwaring, Tyle Bach, Manordeilo, Llandeilo, Dyfed, SA19 7BA [IO81AW, SN62]
2E1 BPT Details withheld at licensee's request by SSL.
2E1 BPV D E J Roberts, 98 Pinkneys Rd, Maidenhead, SL6 5DN [IO91OM, SU88]
2E1 BQD Details withheld at licensee's request by SSL.
2E1 BQE L A M D Wall, 12 Elm Cl, Ryde, PO33 1ED [IO90KR, SZ59]
2E1 BQF L Maughan, 151 Locks Rd, Locks Heath, Southampton, SO31 6LF [IO90IU, SU50]
2E1 BQJ G M Blue, 96 Upper Shaftesbury Ave, Highfield, Southampton, SO17 3RT [IO90HW, SU41]
2W1 BQO Details withheld at licensee's request by SSL.
2E1 BQS J B Benton, 24A Eggington Rd, Wollaston, Stourbridge, DY8 4QJ [IO82WL, SO88]
2W1 BQW M A Bollingham, 2 Bendrick Rd, Barry, CF63 3RE [IO81JJ, ST16]
2E1 BRA M J Scotton, 15 Gr Rd, Walton, Stone, ST15 0DW [IO82WV, SJ83]
2E1 BRC G Thornsby, 20 Stowupland Rd, Stowmarket, IP14 5AG [JO02ME, TM05]
2E1 BRD M A Larcombe, 52 Orchard Rd, Burgess Hill, RH15 9PL [IO90WW, TQ31]
2E1 BRE Details withheld at licensee's request by SSL.
2E1 BRG C E Sanderson, 14 Hazelwood Ave, Osbaldwick, York, YO1 3PD [IO93LW, SE65]
2E1 BRJ O W Roberts, 15 St. Andrews Rd, Enfield, EN1 3UA [IO91XP, TQ39]
2E1 BRM J D Sargent, 15 Wilton Rd, Balsall Common, Coventry, CV7 7QW [IO92EJ, SP27]
2E1 BRR Details withheld at licensee's request by SSL.
2E1 BRT A J Cooper, 5 Roman Ave South, Stamford Bridge, York, YO4 1EZ [IO93NX, SE75]
2E1 BRW B M Bonnard, 250 Gibbon Rd, Newhaven, BN9 9EX [JO00AS, TQ40]
2E1 BRY S E Feay, 19 Dorset Ave, Diggle, Oldham, OL3 5PL [IO93AN, SE00]
2E1 BRZ T M Richards, 59 Abshot Rd, Titchfield Common, Fareham, PO14 4NB [IO90IU, SU50]
2E1 BSC C A Castle, 26 Chestnut Walk, Pulborough, RH20 1AW [IO90RX, TQ01]
2E1 BSF S Laepong, Harpenden, Whatling Rd, Battle, E Sussex, TN33 0NA [JO00FW, TQ71]
2E1 BSH N J Thomas, 10 Crespin Way, Brighton, BN1 7FG [IO90WU, TQ30]
2E1 BSI M D Shenton, 25 Cherry Cl, Arnold, Nottingham, NG5 8GS [IO93KA, SK54]
2E1 BSJ M A E Dolby, 36 Gregory St., Ilkeston, DE7 8AE [IO92IX, SK44]
2E1 BSQ T J Watkins, 51, Alciston, Polegate, E Sussex, BN26 6UW [JO00BT, TQ50]
2E1 BSS Details withheld at licensee's request by SSL.
2E1 BST Details withheld at licensee's request by SSL.
2E1 BSU M D Walsh, 1 Mereside Cl, Longton, Preston, PR4 5ZY [IO83OR, SD42]
2E1 BSV J E England, 17 Barnfield, Much Hoole, Preston, PR4 4GE [IO83OQ, SD42]
2E1 BSW K L Leeman, 5 Serlby Rise, Off Gordon Rd, Nottingham, NG3 2LS [IO92KX, SK54]
2E1 BSY D C Toone, 39 Bramcote Rd, Wigston, LE18 1DB [IO92KO, SP69]
2E1 BSZ M R Curtis, 20 Alder Rd, Folkestone, CT19 5BZ [JO01OC, TR23]
2E1 BTD Details withheld at licensee's request by SSL.
2E1 BTG M J J Joyner, Brimar, Nelson Park Rd, St. Margarets At Cliffe, Dover, CT15 6HL [JO01QD, TR34]
2E1 BTH Details withheld at licensee's request by SSL.
2E1 BTJ P Head, 91 Bedford Rd, Cranfield, Bedford, MK43 0HA [IO92QB, SP94]
2E1 BTK K A Carson, 99 Sandwich Rd, Whitfield, Dover, CT16 3LU [JO01PD, TR34]
2E1 BTN Details withheld at licensee's request by SSL.
2E1 BTR R J Musson, 14 Alfreton Rd, South Normanton, Alfreton, DE55 2AS [IO93HC, SK45]
2E1 BTS A M Chapman, 15 Norwood Rd, Somersham, Huntingdon, PE17 3EY [IO02AJ, TL37]
2W1 BTT S E Thomas, 2 Morlais Ct, Hendredenny, Caerphilly, CF8 2UG [IO81HN, ST08]
2W1 BTU H A Fowler, 19 Monmouth Ct, Hendredenny, Caerphilly, CF8 2TG [IO81HN, ST08]
2W1 BTV K S Robinson, Tor Gwyrdd, Heol Clyd, Ty Isaf, Caerphilly, CF8 2AL [IO81HN, ST08]
2E1 BTX D A Reeve, 7 All Saints Ave, Westbrook, Margate, CT9 5QW [JO01QJ, TR37]
2E1 BUA R S Thompson-Pettitt, 34 Pheasant Way, Kingsthorpe, Northampton, NN2 8BJ [IO92NG, SP76]
2E1 BUC Details withheld at licensee's request by SSL.
2E1 BUE C R Dangerfield, 70 Singledge Ln, Whitfield, Dover, CT16 3EW [JO01PD, TR24]
2E1 BUJ P A Stott, Wellview, 12 Castle View, Ovingham, Prudhoe, NE42 6AT [IO94BX, NZ06]
2E1 BUM F A Stone, 18 New Rd, Hilton, Derby, DE65 5FH [IO92EU, SK23]
2E1 BUN J M E Holdsworth, 57 Westland Rd, Faringdon, SN7 7EY [IO91FP, SU29]
2E1 BUO M S Thorne, 119 Burgess Rd, Bassett, Southampton, SO16 7AE [IO90HW, SU41]
2E1 BUR J A Izzard, 4 Bankside Ct, Bancroft Rd, Bexhill on Sea, TN39 4AG [JO00FU, TQ70]
2E1 BUX M J Elliott, 6 Masefield Cl, Dukinfield, SK16 5DY [IO83XL, SJ99]
2E1 BVC S A Ash, 71 Bridgewater Dr, Northampton, NN3 3AF [IO92NF, SP76]
2E1 BVD Details withheld at licensee's request by SSL.
2E1 BVG G A Barlow, 105 Fitzworth Ave, Poole, BH16 5BA [IO80XR, SY99]
2E1 BVH P E Mostyn, 3 Woodlands Ave, Cheadle Hulme, Cheadle, SK8 5DD [IO83VJ, SJ88]
2E1 BVJ E A E Constantine, The Old Exchange, Burnley Rd, Mytholmroyd, Hebden Bridge, HX7 5PD [IO93AR, SE02]
2E1 BVK J B North, 22 Foxland Cl, Cheswick Green, Shirley, Solihull, B90 4HL [IO92CJ, SP17]
2E1 BVL Details withheld at licensee's request by SSL.
2E1 BVM Details withheld at licensee's request by SSL.
2E1 BVN A F Haughey, 153 Mount Pleasant Ln, Bricket Wood, St. Albans, AL2 3XH [IO91TQ, TL10]
2E1 BVO H A Smith, 1 Parsons Dr, Glen Parva, Leicester, Leics, LE2 9NS [IO92JO, SP59]
2E1 BVQ N S Brown, 17 Western Dr, Leyland, Preston, PR5 3JB [IO83PQ, SD52]
2E1 BVS E Elliott, 6 Masefield Cl, Dukinfield, SK16 5DY [IO83XL, SJ99]
2E1 BVU E J Parker, 10 Priory Rd, Wollaston, Wellingborough, NN29 7PW [IO92PG, SP96]
2E1 BVV J C Elliott, 6 Masefield Cl, Dukinfield, SK16 5DY [IO83XL, SJ99]
2E1 BVX Details withheld at licensee's request by SSL.
2E1 BVY D Andrews, 16 Scafell Way, Clifton Est, Nottingham, NG11 9FW [IO92JV, SK53]
2E1 BWA W H Giles, 336 London Rd, St. Albans, AL1 1EA [IO91UR, TL10]
2E1 BWF R J Heathcote, The Lane use, Top Rd, Acton Trussell, Stafford, ST17 0RQ [IO82WS, SJ91]

2E1 BWJ E A R Starkie, 169 Saunders Ln, Mayford, Woking, GU22 0NT [IO91QH, SU95]
2E1 BWM G L Smith, Ais17D-Commcen, Hqptc, RAF Innsworth, Gloucester, UK, GL3 1EZ [IO81VV, SO82]
2E1 BWZ Details withheld at licensee's request by SSL.
2E1 BXB T J Chick, 27 Stoney Stile Rd, Alveston, Bristol, BS12 2NG [IO81RO, ST68]
2E1 BXC R E George, 85 Devonshire Way, Croydon, CR0 8BW [IO91XI, TQ36]
2I1 BXH A E W Ardill, 2 Lonsdale Ct, Shore Rd, Newtownabbey, BT37 0FA [IO74BQ, J38]
2E1 BYB J A Hayles, Beamans, Chale St., Chale Green, Ventnor, PO38 2JQ [IO90IO, SZ47]
2E1 BYC J A Flavill, 5 Brockenhurst Cl, Canterbury, CT2 7RX [JO01MG, TR15]
2W1 BYD R G Williams, 8 Heol Gomer, Abergele, LL22 7UG [IO83EG, SH97]
2W1 BYK A M Sellors, 12 Morfa View, Bodelwyddan, Rhyl, LL18 5TT [IO83FG, SH97]
2E1 BYL Details withheld at licensee's request by SSL.
2E1 BYN Details withheld at licensee's request by SSL.
2E1 BYO N Walker, 79 Arklow Dr, Hale Village, Liverpool, L24 5RR [IO83OI, SJ48]
2E1 BYP Details withheld at licensee's request by SSL.
2E1 BYQ R G Craven, 16 Alexandra Rd, Stafford, ST17 4DE [IO82WT, SJ92]
2E1 BYR Details withheld at licensee's request by SSL.
2E1 BYU S R Lambert, 31 Orchard Way, Burgess Hill, RH15 9PB [IO90WW, TQ31]
2M1 BYW T N Conlan, 12 Rowantree Rd, Mayfield, Dalkeith, EH22 5ER [IO85LU, NT36]
2E1 BYY W Booth, 26 The Cricket Gr, Hardy Ln, Chorlton Cum Hardy, Manchester, M21 7LZ [IO83UK, SJ89]
2E1 BYZ Details withheld at licensee's request by SSL.
2E1 BZA Details withheld at licensee's request by SSL.
2E1 BZB D J Peter, 41 Coleswood Rd, Harpenden, AL5 1EF [IO91TT, TL11]
2E1 BZE M J Papworth, 7 Edwin Cl, Siege Cross, Thatcham, RG19 4GW [IO91JJ, SU56]
2E1 BZG C Jones, 378 London Rd, Earley, Reading, RG6 1BA [IO91MK, SU77]
2E1 BZH G W Low, 40 Lilac Cres, Runcorn, WA7 5JX [IO83PH, SJ58]
2E1 BZI D E Low, 40 Lilac Cres, Runcorn, WA7 5JX [IO83PH, SJ58]
2E1 BZJ O H J Milton, 35 Holybrook Rd, Reading, RG1 6DG [IO91MK, SU77]
2E1 BZL S Brooks, 15 Tarn Cl, Winsford, CW7 2SA [IO83RE, SJ66]
2E1 BZQ Details withheld at licensee's request by SSL.
2E1 BZR Details withheld at licensee's request by SSL.
2E1 BZS D R Newton, 9 Calcott, Stirchley, Telford, TF3 1YG [IO82SP, SJ70]
2E1 BZU Details withheld at licensee's request by SSL.
2E1 CAA A T Masson, Appletree Cottage, Neath Gdns, Tilehurst, Reading, RG3 4UL [IO91LJ, SU66]
2E1 CAB A J Hind, 16 Dawney Dr, Four Oaks, Sutton Coldfield, B75 5JA [IO92CO, SP19]
2E1 CAD D J Balharrie, 27 Norfolk Rd, Uxbridge, UB8 1BL [IO91SN, TQ08]
2E1 CAE C S Backshell, 26 The Tithings, Halton Brook, Runcorn, WA7 2DT [IO83PI, SJ58]
2E1 CAG P Stemp, 3 Loxwood, East Preston, Littlehampton, BN16 1DT [IO90ST, TQ00]
2E1 CAH R S Richards, 39 North Holme Ct, Northampton, NN3 8UX [IO92NG, SP76]
2E1 CAI S J Wersby, Perrandale, Merstone, Newport, Isle of Wight, PO30 3DF [IO90IP, SZ58]
2E1 CAM T J Capon, 5 Spinney Ave, Goostrey, Crewe, CW4 8JE [IO83TF, SJ77]
2E1 CAN Details withheld at licensee's request by SSL.
2E1 CAQ M R Porter, 7 Long Rd, Framingham Earl, Norwich, NR14 7RY [JO02QN, TG20]
2E1 CAR A T McMahon, 6 Curzon Park North, Chester, CH4 8AR [IO83NE, SJ36]
2E1 CAT L P ORyan, 12 Minton Cl, Congleton, CW12 3TD [IO83VD, SJ86]
2E1 CAW R F Marshall, 15 Whisby Ct, Holton-le-Clay, Grimsby, DN36 5BG [IO93XM, TA20]
2E1 CAX K E R Hales, Brook View, 15 Meadow Bank Rd, Hereford, HR1 2ST [IO82PB, SO53]
2E1 CAZ T A Wong, 16 Fraser Rd, Thorplands Brook, Northampton, NN3 8YL [IO92NG, SP76]
2E1 CBA A J Clayton, 15 Linley Cl, Aldridge, Walsall, WS9 0ES [IO92AO, SK00]
2E1 CBD W R Schofield, Birch Trees, New Platt Ln, Goostrey, Crewe, CW4 8NJ [IO83TF, SJ77]
2E1 CBE R W J Crabtree, 46 Hill Village Rd, Four Oaks, Sutton Coldfield, B75 5BD [IO92CO, SP19]
2E1 CBG R B Newton, 9 Calcott, Stirchley, Telford, TF3 1YG [IO82SP, SJ70]
2E1 CBH S M Stretch, 5 Ledwych Rd, Droitwich, WR9 9LA [IO82VG, SO86]
2E1 CBK J Porter, 7 Long Rd, Framingham Earl, Norwich, NR14 7RY [JO02QN, TG20]
2E1 CBL Details withheld at licensee's request by SSL.
2E1 CBQ W Holland, 4 Woodhouse Ln, Norden, Rochdale, OL12 7RG [IO83VP, SD81]
2E1 CBU C M Richards, 39 North Holme Ct, Northampton, NN3 8UX [IO92NG, SP76]
2E1 CCB C Burrows, 40 Fairmile Rd, Christchurch, BH23 2LL [IO90CR, SZ19]
2E1 CCC D J Eaton, 49 Farnley Cl, Norton Brow, Windmill Hill, Runcorn, WA7 6NN [IO83QI, SJ58]
2E1 CCE S A L McPherson, 7 Burford Ave, Boothville, Northampton, NN3 6AF [IO92NG, SP76]
2E1 CCF P E Izzard, 7 Yardley Dr, Kingsthorpe, Northampton, NN2 8PE [IO92NG, SP76]
2E1 CCG R E Winship, 32 Lytes Cary Rd, Keynsham, Bristol, BS18 1XD [IO81SJ, ST66]
2E1 CCI A D Z Murphy, 34 Hawkenbury Way, Lewes, BN7 1LT [IO90XU, TQ41]
2E1 CCJ F Martin, 18 The Hawthornes, John Ogaunts Way, Belper, DE56 0DH [IO93GA, SK34]
2E1 CCL Details withheld at licensee's request by SSL.
2E1 CCN D G Hill, 1 Gleneagles Ct, Edwalton, Nottingham, NG12 4DN [IO92KV, SK63]
2E1 CCO Details withheld at licensee's request by SSL.
2E1 CCQ Details withheld at licensee's request by SSL.
2E1 CCT J O Gale, 17 Cheltenham Rd, Parkstone, Poole, BH12 2ND [IO90AR, SZ09]
2E1 CCU R J Hudson, 1 Cherry Clack, North St., Punnetts Town, Heathfield, TN21 9DT [JO00DX, TQ62]
2E1 CCW D G Stoodley, Sedbury Branches Ln, Sherfield English, Romsey, Hants, SO51 6JW [IO91FA, SU22]
2E1 CCZ J C Murray, 34 Brackendale, Halton Brook, Runcorn, WA7 2EF [IO83PH, SJ58]
2E1 CDB S A Reilly, 137 Lockton Ave, Heanor, DE75 7ER [IO93HA, SK44]
2M1 CDC Details withheld at licensee's request by SSL.
2E1 CDD R A Smith, 6 Hermitage Cl, North Mundham, Chichester, PO20 6JZ [IO90OT, SU80]
2E1 CDH J Lavery, Cadnant, Newby West, Carlisle, Cumbria, CA2 6QU [IO84MV, NY35]
2E1 CDJ Details withheld at licensee's request by SSL.
2E1 CDK G P Langdon, 43 Daniel St., Ryde, PO33 2BH [IO90KR, SZ59]
2E1 CDL Details withheld at licensee's request by SSL.
2E1 CDO Details withheld at licensee's request by SSL.
2M1 CDP C R C Proctor, 17 Ninians Rise, Kirkintilloch, Glasgow, G66 3HU [IO75WW, NS67]
2E1 CDQ A M Pitt, 3 Moat Hill, Birstall, Batley, WF17 0DX [IO93ER, SE22]
2E1 CDR R P Crofts, Little Isle, Woodgate Green, Knighton on Teme, Tenbury Wells, WR15 8LX [IO82RH, SO67]
2E1 CDS B Allen, 78 Bargates, Christchurch, BH23 1QL [IO90CR, SZ19]
2E1 CDT K H Smith, 52 Grantham Dr, Bury, BL8 1XW [IO83UO, SD71]
2E1 CDU C P Smith, 52 Grantham Dr, Bury, BL8 1XW [IO83UO, SD71]
2E1 CDZ A M Woods, 90 Winchester Rd, Sandy, SG19 1DP [IO92UD, TL14]
2E1 CEA L Royle, 14 Lindinis Ave, Salford, M6 5AB [IO83UL, SJ89]
2E1 CEB Details withheld at licensee's request by SSL.
2E1 CEC K A Asquith, 31 Cawley Ln, Heckmondwike, WF16 0BN [IO93EQ, SE22]
2W1 CEE D Whish, 62 Marion Rd, Prestatyn, LL19 7DF [IO83GH, SJ08]
2M1 CEG R K Hopkins, The Elms, Brechin Rd, Kirriemuir, DD8 4DE [IO86MQ, NO35]
2E1 CEH J M R Baldock, Knightsbridge House, St Teath, Bodmin, Cornwall, PL30 3JE [IO70PO, SX08]
2M1 CEJ A Troup, Drumdruils, Glen Rd, Bridge of Allan, Stirling, FK9 4LZ [IO86AE, NS79]
2E1 CEK N C Braddon, 45 Booth Bed Ln, Goostrey, Crewe, CW4 8NB [IO83TF, SJ77]
2E1 CEL Details withheld at licensee's request by SSL.
2M1 CEN D C Boyd, 15 Pladda Ave, Port Glasgow, PA14 6EW [IO75QW, NS37]
2M1 CEO T A E Boyd, 15 Pladda Ave, Port Glasgow, PA14 6EW [IO75QW, NS37]
2E1 CEP J H Rotherham, 10 Townfield View, Windmill Hill, Runcorn, WA7 6QD [IO83PI, SJ58]
2E1 CEQ M A Sayers, Flat 6, 24 Knole Rd, Boscombe, Bournemouth, Dorset, BH1 4DH [IO90BR, SZ19]
2E1 CER Details withheld at licensee's request by SSL.
2E1 CES K J Nicholas, Greenbank, Chester High Rd, Neston, South Wirral, L64 7TR [IO83LH, SJ37]
2E1 CEU J A Columbine, West Lodge, 166 Tollerton Ln, Tollerton, Nottingham, NG12 4FW [IO92KV, SK63]
2E1 CEV D J Plunkett, 150 Oaks Ave, Garden Village, Stocksbridge, Sheffield, S30 5EN [IO93GJ, SK38]
2E1 CEW Details withheld at licensee's request by SSL.
2E1 CEX D M Peake, 71 Sutton Rd, Huthwaite, Sutton in Ashfield, NG17 2NZ [IO93ID, SK45]
2E1 CEY J Hyde, 46 Oak Ln, Whitefield, Manchester, M45 8ET [IO83UN, SD80]
2W1 CEZ J E Brown, Kingsdown Cottage, Fron, Montgomery, Powys, SY15 6SB [IO82JN, SO19]
2E1 CFA Details withheld at licensee's request by SSL.
2E1 CFB Details withheld at licensee's request by SSL.
2E1 CFC S M R Clayton, 6 Albert Rd, Bunny, Nottingham, NG11 6QE [IO92KU, SK53]
2E1 CFE R D Snodin, Ruan, 7 Earlswood Cl, Breaston, Derby, DE72 3UF [IO92IV, SK43]
2E1 CFF P M Bird, 132 Braemar Rd, Sutton Coldfield, B73 6LZ [IO92BN, SP19]
2E1 CFG A J Huddleston, Milford House, 9 Gallowfields Rd, Richmond, DL10 4DB [IO94DJ, NZ10]
2E1 CFH F P Fairclough, 3 Cyprus Terr, Sandfield Rd, Wallasey, L45 1JL [IO83LK, SJ39]
2E1 CFJ J M Bisset, 42 Brandon St., Leicester, LE4 6AW [IO92KP, SK50]
2E1 CFQ D L Sedgley, 10 South View Rd, Christchurch, BH23 1JH [IO90CR, SZ19]
2E1 CFR A Phillips, Upsdell, Orchard Rd, Pratts Bottom, Orpington, BR6 7NS [JO01BI, TQ46]
2M1 CFS C J Laing, 7 Vandeleur Gr, Edinburgh, EH7 6UE [IO85KW, NT27]
2E1 CFT G J Ellis, Batson, 1 Highfield Park, Marlow, SL7 2DE [IO91ON, SU88]
2E1 CFU Details withheld at licensee's request by SSL.
2E1 CFW Details withheld at licensee's request by SSL.
2W1 CGC E J Harrison, 19 Eldon Dr, Abergele, LL22 7DA [IO83EG, SH97]
2E1 CGE Details withheld at licensee's request by SSL.
2M1 CGG I D B Sleigh, 43 Murieston Rd, Livingston, EH54 9AX [IO85GU, NT06]
2E1 CGH Details withheld at licensee's request by SSL.
2E1 CGI Details withheld at licensee's request by SSL.
2E1 CGP B R Thomas, 2 Fieldhouse, Cinderhills, Holmfirth, Huddersfield, HD7 1EN [IO93CN, SE10]
2E1 CGR G O C Harrison, Hindleap Corner, Priory Rd, Forest Row, E Sussex, RH18 5JF [JO01AC, TQ43]
2E1 CGT Details withheld at licensee's request by SSL.
2E1 CGW T D Girdler, 8 Westmorland Ave, Loughborough, LE11 3RY [IO92JS, SK51]
2E1 CGZ M I Thomas, 11 Beaufort Ave, Ferndale, Kidderminster, DY11 5NH [IO82UJ, SO87]
2E1 CHF S J Orton, 76 High St., Earls Barton, Northampton, NN6 0JG [IO92PG, SP86]
2E1 CHG Details withheld at licensee's request by SSL.
2E1 CHI Details withheld at licensee's request by SSL.
2E1 CHK I R Gifford, 16 Newstead Walk, Carshalton, SM5 1AW [IO91VJ, TQ26]
2W1 CHM S G Middleton, 64 Cog Rd, Sully, Penarth, CF64 5TE [IO81JJ, ST16]
2W1 CHO Details withheld at licensee's request by SSL.
2W1 CHP Details withheld at licensee's request by SSL.
2E1 CHR J J S Mulira, 31 Ashleigh Rd, Solihull, B91 1AF [IO92CK, SP17]
2E1 CHS Details withheld at licensee's request by SSL.
2E1 CHX M S Reavill, 11 Clarence Rd, Attenborough, Beeston, Nottingham, NG9 5HY [IO92JV, SK53]
2M1 CHZ C W Vincent, 133 Maxwell Dr, Glasgow, G41 5AE [IO75UU, NS56]
2E1 CIB J D Shattock, Three Ways Cottage, Ripe, Lewes, Sussex, BN8 6AW [JO00BU, TQ51]
2E1 CIC Details withheld at licensee's request by SSL.
2E1 CIH Details withheld at licensee's request by SSL.
2E1 CII Details withheld at licensee's request by SSL.
2E1 CIK D A Gillatt, 10 Marlborough Gr, York, YO1 4AY [IO93LW, SE65]
2E1 CIN Details withheld at licensee's request by SSL.
2E1 CIO C A Brooks, 34 Wentworth Cres, New Marske, Redcar, TS11 8DB [IO94LN, NZ62]
2W1 CIP R Bufton, 7 Laburnum Cl, Rassau, Ebbw Vale, NP3 5TS [IO81JT, SO11]
2E1 CIQ M S Begg, 11 Lilac Rd, Normanby, Middlesbrough, TS6 0BS [IO94KN, NZ51]
2E1 CIT J J Watkins-Field, Sharlions, 27 Bosvigo Rd, Truro, TR1 3DG [IO70LG, SW84]
2E1 CIX W Scott, 18 Manor Gdns, Killinghall, Harrogate, HG3 2DS [IO94FA, SE25]
2E1 CIY B Harratt, 49 Northfield St., South Kirkby, Pontefract, WF9 3NG [IO93IO, SE41]
2E1 CJB K Jordan, 11 Sandringham Pl, Hucknall, Nottingham, NG15 8EU [IO93JA, SK54]
2E1 CJC R P Richardson, Moorside, 64 Moor Cl, Killinghall, Harrogate, HG3 2DZ [IO94FA, SE25]
2E1 CJD P Taylor, 13 Mackenzie Cres, Chapeltown, Burncross, Sheffield, S30 4UR [IO93GJ, SK38]
2E1 CJE Details withheld at licensee's request by SSL.
2E1 CJF S D J Curtis, Jo53C, John Lester Ct, Meyrick Rd, Salford, Lancs, M6 5EN [IO83UL, SJ89]
2M1 CJG K J Potter, 34 Norman Rise, Dedridge, Livingston, EH54 6LY [IO85FV, NT06]
2E1 CJH J J Russell, 57 Norburn Park, Witton Gilbert, Durham, DH7 6SG [IO94ET, NZ24]
2E1 CJI Details withheld at licensee's request by SSL.
2E1 CJK Details withheld at licensee's request by SSL.
2E1 CJM Details withheld at licensee's request by SSL.
2E1 CJN H M Hughes, 46 The Boundary, Oldbrook, Milton Keynes, MK6 2HT [IO92PA, SP83]
2M1 CJO C Haswell, 6 Lochlann Rd, Culloden, Inverness, IV1 2HB [IO77WL, NH74]
2E1 CJR L D Pearce, 18 Pier Rd, North Woolwich, London, E16 2LH [JO01AL, TQ47]
2E1 CJT Details withheld at licensee's request by SSL.
2E1 CJU K D Boulton, 110 Southgate, Sutton Hill, Telford, TF7 4HQ [IO82SP, SJ70]
2E1 CJV Details withheld at licensee's request by SSL.
2E1 CJW R Hewitt, 31 Broadmead, Vicars Cross, Chester, CH3 5PT [IO83NE, SJ46]
2E1 CJX Details withheld at licensee's request by SSL.
2E1 CJY Details withheld at licensee's request by SSL.
2E1 CJZ Z E Hodges, 12 Linwal Ave, Houghton on The Hill, Leicester, LE7 9HD [IO92LP, SK60]
2E1 CKA M A Evans, 15 Melrose Ave, Vicars Cross, Chester, CH3 5JA [IO83NE, SJ46]
2M1 CKB D A Campbell, 6 Station Rd, Ardersier, Inverness, IV1 2ST [IO77XN, NH75]
2E1 CKC Details withheld at licensee's request by SSL.
2M1 CKE Details withheld at licensee's request by SSL.
2E1 CKH K Riley, 16 King St., Westhoughton, Bolton, BL5 3AX [IO83RN, SD60]
2E1 CKI J Eccleston, 426 St. Helens Rd, Leigh, WN7 3QG [IO83RL, SJ69]
2E1 CKJ N A Quiruga, 3 Condor Gr, Heath Hayes, Cannock, WS12 5YB [IO92AQ, SK01]
2E1 CKK S R Woolley, 12 Princess Rd, Uttoxeter, ST14 7DN [IO92BV, SK03]
2E1 CKM V Beardsley, 2 King Edward St., Sandiacre, Nottingham, NG10 5BS [IO92IW, SK43]
2E1 CKN J P Hayter, 1 Winchester Cl, Newport, PO30 1DR [IO90IQ, SZ48]
2E1 CKQ E S Swain, 11 Blackdown, Fullers Slade, Milton Keynes, MK11 2AA [IO92NB, SP73]
2E1 CKT Details withheld at licensee's request by SSL.
2E1 CKV C M Schollick, Greencroft, 55 Darlington Rd, Richmond, DL10 7BG [IO94DJ, NZ10]
2E1 CLB M J L Hance, Old Swinford Hospital, Hagley Rd, Stourbridge, West Midlands, DY8 1QX [IO82WK, SO98]
2E1 CLC M R Holmes, 86 Portland Ave, Murston, Sittingbourne, ME10 3QY [JO01JI, TQ96]
2E1 CLG G Harvey, 15 Cheyne Walk, Meopham, Gravesend, DA13 0PF [JO01EI, TQ66]
2E1 CLH D R J Simpkins, 158 Golden Dr, Eaglestone, Milton Keynes, MK6 5BN [IO92PA, SP83]
2E1 CLL C Gandhi, 36 Warrender Cl, Bramcote Hills, Beeston, Nottingham, NG9 3EB [IO92JW, SK53]
2E1 CLM E J Woolley, 82 Pennycroft Rd, Uttoxeter, ST14 7ET [IO92BV, SK03]
2E1 CLN J P Hudson, 10 Canberra Rd, Bexleyheath, DA7 5SG [JO01BL, TQ47]
2E1 CLQ G Lawrence, 20 Keats Rd, Walsall, WS3 1DT [IO92AO, SK00]
2E1 CLT C M Sheffield, 167 Hawkswell Dr, Willenhall, WV13 3EL [IO82XN, SO99]
2E1 CLV D W Gray, 81 Rectory Rd, Markfield, LE67 9WN [IO92IQ, SK40]
2M1 CLY J M Moss, Ptarmigan Cottage, Laggan Bridge, nr Newtonmore, Inverness Shire, PH20 1BT [IO77VA, NN69]
2E1 CMB Details withheld at licensee's request by SSL.
2E1 CMC T S Bull, 35 Coach Dr, Eastwood, Nottingham, NG16 3DR [IO93IA, SK44]
2E1 CMD S G Bull, 35 Coach Dr, Eastwood, Nottingham, NG16 3DR [IO93IA, SK44]
2E1 CME L Williams, 37 Mickledales Dr, Marske By The Sea, Redcar, TS11 6DF [IO94LO, NZ62]
2E1 CMG Details withheld at licensee's request by SSL.
2E1 CMH Details withheld at licensee's request by SSL.
2E1 CMJ L S Holyer, 39 Lenacre Ave, Whitfield, Dover, CT16 3HH [JO01PD, TR24]
2E1 CML B M Norris, 28 Fir Tree Rd, Bellfields Est, Guildford, GU1 1JJ [IO91RG, SU95]
2I1 CMM J Iznerowicz, 12 Cappy St., Ravenhill Rd, Belfast, BT6 8ER [IO74BO, J37]
2E1 CMP Details withheld at licensee's request by SSL.
2E1 CMS D C Aitchison, 649 Chorley New Rd, Horwich, Bolton, BL6 6LH [IO83RO, SD61]
2E1 CMZ D E Bracher, 29 Bungalow Park, Holders Rd, Amesbury, Salisbury, SP4 7PJ [IO91CE, SU14]
2E1 CNA M C Dibdin, 6 St. Johns Cl, Newport, PO30 1LU [IO90IQ, SZ48]
2E1 CND A P Davies, 35 Doddington, Hollinswood, Telford, TF3 2DJ [IO82SQ, SJ70]
2E1 CNG S H Robertson, 28 Frith Rd, Bognor Regis, PO21 5LL [IO90PS, SZ99]
2E1 CNJ C Melia, 4 Skellington Fold, Netherley, Liverpool, L27 8YA [IO83NJ, SJ48]
2E1 CNM A G Raxworthy, 327 Fareham Rd, Gosport, PO13 0AB [IO90JT, SU50]
2W1 CNN A Gray, 36 Heol Pentre Felen, Morriston, Swansea, SA6 6BY [IO81AQ, SS69]
2E1 CNO L Call, 32 Hunger Hills Dr, Horsforth, Leeds, LS18 5JU [IO93EU, SE23]
2E1 CNP J R Lawrence, 4 Dale Park Rise, Leeds, LS16 7PP [IO93EU, SE23]
2E1 CNQ M C H Hicks, The Old Coach House, 16 Avon Carrow, Avon Dassett, Leamington Spa, CV33 0AR [IO92HD, SP44]
2E1 CNT Details withheld at licensee's request by SSL.
2E1 CNU Z G Sznober, 7 Botany Bay Cl, Aqueduct, Telford, TF4 3RJ [IO82SP, SJ60]
2E1 CNX Y Eaton, 49 Farnley Cl, Norton Brow, Windmill Hill, Runcorn, WA7 6NN [IO83QI, SJ58]
2E1 CNY K A Wilson, 12 New St., Elworth, Sandbach, CW11 3JF [IO83TD, SJ76]
2E1 COB R C Bishop, 21 Trimnel Green, Chawson, Droitwich, WR9 8ST [IO82WG, SO86]
2E1 COC J F H Taylor-Cram, 7 Hart Plain Ave, Cowplain, Waterlooville, PO8 8RP [IO90LV, SU61]
2E1 COD B Call, 32 Hunger Hills Dr, Horsforth, Leeds, LS18 5JU [IO93EU, SE23]
2E1 COF J Mullan, 68 Wordsworth Rd, Daybrook, Nottingham, NG5 6HJ [IO93KA, SK54]
2E1 COG D Oakes, 2 Hillcrest, Scotton, Catterick Garrison, DL9 3NJ [IO94DI, SE19]
2E1 COI L Brown, 62 Brickfields Rd, Worcester, WR4 9TW [IO82VE, SO85]
2E1 COJ Details withheld at licensee's request by SSL.
2E1 COL C B Peel, 45 Garden Rd, Brighouse, HD6 2DH [IO93CQ, SE12]
2E1 COM M J Holt, 20 Lingfield Mount, Moortown, Leeds, LS17 7EP [IO93FU, SE33]
2E1 CON C I Lowe, 28 Alpine Dr, Leigh, WN7 5HT [IO83RM, SD60]
2E1 COQ J A Gardner, 165 Park Cl, Ashley Park, Walton on Thames, KT12 1EW [IO91SJ, TQ06]
2W1 COT A J Outram, 36 Midland Rd, Llansamlet, Swansea, SA7 9QY [IO81BP, SS69]
2E1 COV I D Cockshoot, 72 Princess Margaret Ave, Cliftonville, Margate, CT9 3EF [JO01RJ, TR37]
2E1 COX Details withheld at licensee's request by SSL.
2E1 CPA G A Porter, 65 Bartlett St., Wavertree, Liverpool, L15 0HN [IO83MJ, SJ38]
2E1 CPB A T Cadey, The Millstone, 45 St. Mildreds Rd, Westgate on Sea, CT8 8RJ [JO01QJ, TR37]
2E1 CPC M J A Sheppard, 37 Oakfield Rd, Kingswood, Bristol, BS15 2NT [IO81RK, ST67]
2E1 CPD R S Hankin, The Croft, Portinscale, Keswick, CA12 5TX [IO84KO, NY22]
2E1 CPE A J Porter, 7 Long Rd, Framingham Earl, Norwich, NR14 7RY [JO02QN, TG20]
2E1 CPF R E Kensall, 40 Eskdale Ave, Ramsgate, CT11 0PB [JO01QI, TR36]
2E1 CPH M P Buckle, 1 Clipstone Rd West, Forest Town, Mansfield, NG19 0EF [IO93KD, SK56]
2E1 CPI B G Rowley, School House, Aldbrough St John, Richmond, North Yorks, DL11 7SU [IO94DL, NZ21]
2E1 CPL M G Cockshoot, 17 Muir Rd, Dumpton, Ramsgate, CT11 8AX [JO01RI, TR36]
2E1 CPM M P Carter, 191 Westgate, Almondbury, Huddersfield, HD5 8XN [IO93DP, SE11]
2E1 CPP N P Newman, 89 Sea Pl, Goring By Sea, Worthing, BN12 4BH [IO90TT, TQ10]
2E1 CPQ P Goodwin, 60 Dale Cres, Congleton, CW12 3EP [IO83VD, SJ86]
2E1 CPR S D Pollard, 94 Crosland Hill Rd, Crosland Hill, Huddersfield, HD4 5NZ [IO93CP, SE11]
2E1 CPV P Porter, 7 Long Rd, Framingham Earl, Norwich, NR14 7RY [JO02QN, TG20]

2E1 CPW J E Mawson, 38 Springbank Rd, Gildersome, Morley, Leeds, LS27 7DJ [IO93ES, SE22]
2E1 CQA H M R Wall, 53 Carr Hill Rd, Calverley, Pudsey, LS28 5PZ [IO93DT, SE23]
2E1 CQE D J Easton, 2 Milton Terr, Milborne St. Andrew, Blandford Forum, DT11 0LJ [IO80US, SY89]
2E1 CQG S R Crofts, 7 Cowleaze Rd, Broadmayne, Dorchester, DT2 8EW [IO80TQ, SY78]
2E1 CQH P T Coldwell, 12 Stileham Bank, Milborne St. Andrew, Blandford Forum, DT11 0LE [IO80US, SY89]
2E1 CQI G Guild, 7 Coronation Rd, Preston Brook, Runcorn, WA7 3AR [IO83QH, SJ58]
2E1 CQJ J D Mellor, 4 Dockery, Swan Ln, Lockwood, Huddersfield, HD1 3TP [IO93CP, SE11]
2E1 CQL R A Hounslow, 46 Garrick Rd, Abington, Northampton, NN1 5ND [IO92NF, SP76]
2E1 CQM I Hurst, 50 Church Hill, Kirkby in Ashfield, Nottingham, NG17 8LJ [IO93IC, SK45]
2E1 CQO A D Wall, 53 Carr Hill Rd, Calverley, Pudsey, LS28 5PZ [IO93DT, SE23]
2E1 CQP V J Brightwell, 4 Buckingham Rd, Margate, CT9 5SS [JO01QJ, TR37]
2E1 CQQ E A Whelan, 47 Romanby Rd, Northallerton, DL7 8NG [IO94GI, SE39]
2E1 CQR R A Shaw, The Anchorage, South Parade, Croft on Tees, Darlington, DL2 2SN [IO94FL, NZ20]
2E1 CQS N D Milne, 1 Hampden Cl, Stoke Poges, Slough, SL2 4JF [IO91QM, SU98]
2E1 CQT G D M Vellino, East House, Cranbrook Rd, Goudhurst, Cranbrook, TN17 1DR [JO01FC, TQ73]
2E1 CQU Details withheld at licensee's request by SSL.
2E1 CQX E D Scholes, 7 Abbotsbury, Orton Malborne, Peterborough, PE2 5PS [IO92UN, TL19]
2E1 CQY J P Chapman, 34 Grange Cl, Hitchin, SG4 9HD [IO91UW, TL12]
2E1 CRA M K Lewis, 15 Highcliffe Ave, Saughall, Chester, CH1 5DP [IO83NE, SJ36]
2E1 CRB J T Levick, 63 Sussex Gdns, Chessington, KT9 2PU [IO91UI, TQ16]
2E1 CRC A J Perry, The School House, Menin Way, Farnham, Surrey, GU9 8DY [IO91OF, SU84]
2E1 CRE K P Blanshard, 30 Torquay Cres, Symonds Green, Stevenage, SG1 2RS [IO91VV, TL22]
2E1 CRF B Willimott, 1 Wells Croft, Meanwood, Leeds, LS6 4LA [IO93FT, SE23]
2E1 CRH R E Doull, 148 The Maples, Harlow, CM19 4RD [JO01BR, TL40]
2E1 CRI J Mosby, 1 School Rd, Scapegoat Hill, Golcar, Huddersfield, HD7 4NU [IO93BP, SE01]
2E1 CRK P W Bell, 122 Howard Dr, Letchworth, SG6 2DE [IO91VX, TL23]
2E1 CRP Details withheld at licensee's request by SSL.
2W1 CRR N C Martin, Beavers Meadow, Prince of Wales Cl, Houghton, Milford Haven, SA73 1NR [IO71MR, SM90]
2E1 CRW P G Smith, 155 Grosvenor Rd, Dalton, Huddersfield, HD5 9UA [IO93DP, SE11]
2E1 CRX C A Hannah, 5 Main Rd, Orby, Skegness, Lincs, PE24 5HT [JO03CE, TF46]
2E1 CRY Details withheld at licensee's request by SSL.
2E1 CRZ N P Massey, 105 Blakedown Rd, Halesowen, B63 4NR [IO82XK, SO98]
2E1 CSA N G Dent, 28 Lightsfield, Oakley, Basingstoke, RG23 7BY [IO91JG, SU55]
2E1 CSB C N Sturt, 22 Cardigan Cl, St. Johns, Woking, GU21 1YP [IO91QH, SU95]
2E1 CSC A L Robinson, 5 Hazel Gr, Welton, Lincoln, LN2 3JX [IO93SH, TF07]
2E1 CSD A E H Smith, 30 Lime Gr, Grantham, NG31 9JD [IO92QW, SK93]
2E1 CSE Details withheld at licensee's request by SSL.
2E1 CSH Q Y Sim, 5 Balmoral Rd, Worcester Park, KT4 8SR [IO91VJ, TQ26]
2E1 CSI G P G Harmer, 81 Borodin Cl, Basingstoke, RG22 4EW [IO91KF, SU64]
2E1 CSN Details withheld at licensee's request by SSL.
2E1 CSQ Details withheld at licensee's request by SSL.
2E1 CSW Details withheld at licensee's request by SSL.
2E1 CSY A L H Brown, 35 Peterborough Rd, Castor, Peterborough, PE5 7AX [IO92TN, TL19]
2E1 CSZ F Fisher, 93 Harborne Rd, Edgbaston, Birmingham, B15 3HG [IO92AL, SP08]
2E1 CTC S A Whatmore, 14 Iden Cres, Staplehurst, Tonbridge, TN12 0NU [JO01GD, TQ74]
2E1 CTF C I Dixon, 70 Sussex Rd, Lowestoft, NR32 4HF [JO02VL, TM59]
2E1 CTG T D Cooke, 1 Sarah Gdns, Margate, CT9 3XB [JO01QJ, TR36]
2E1 CTL Details withheld at licensee's request by SSL.
2E1 CTQ D M Joyner, 31 Brookfield Rd, Dover, CT16 2AU [JO01PD, TR34]
2E1 CTR Details withheld at licensee's request by SSL.
2E1 CTS M L Broderick, 192 Brownside Rd, Worsthorne, Burnley, BB10 3JW [IO83VS, SD83]
2E1 CTT M J Broderick, 192 Brownside Rd, Worsthorne, Burnley, BB10 3JW [IO83VS, SD83]
2E1 CTU S Crawshaw, 98 Grassington Dr, Burnley, BB10 2SP [IO83VT, SD83]
2E1 CTV Details withheld at licensee's request by SSL.
2E1 CTW D H Pickles, 44 Healdwood Dr, Reedley, Burnley, BB12 0EA [IO83VT, SD83]
2E1 CTZ E L Setterfield, 54 Hallam Rd, Nelson, BB9 8AB [IO83VU, SD83]
2E1 CUA B D Hewson, 2 Ribchester Way, Brierfield, Nelson, BB9 0YH [IO83VT, SD83]
2E1 CUF Details withheld at licensee's request by SSL.
2E1 CUG I A Burnley, 8 Wren Hill, Batley, WF17 0QL [IO93ER, SE22]
2E1 CUL R E Bucknell, 29 Haverholt Cl, Colne, BB8 9SN [IO83VU, SD84]
2W1 CUN R Nosw0Rthy, Nethway, Victoria Rd, Llandrindod Wells, LD1 6AP [IO82HF, SO06]
2W1 CUP S Morris, Boderw, 9 Glen View, Rhyd y Foel, Abergele, Clwyd, LL22 8EB [IO83EG, SH97]
2E1 CUS R J Higgins, Mary Hare Grammar School, Arlington Manor, Snelsmore Common, Newbury, RG14 3BQ [IO91IK, SU47]
2E1 CUX M J Bettesworth, 3 Glen Leigh, Redruth, TR15 1BU [IO70JF, SW64]
2E1 CUY M P B Swain, 13 Gatley Dr, Guildford, Surrey, GU4 7JJ [IO91RG, TQ05]
2E1 CUZ K B Hardcastle, 3 St. Johns Gr, Kirk Hammerton, York, YO5 8DE [IO93IX, SE45]
2E1 CVA C G Schofield, Pilgrims, 8 Broomfield Ride, Oxshott, Leatherhead, KT22 0LW [IO91TI, TQ16]
2E1 CVD D G Hazeldine, 30 Oldford Walk, Stourport on Severn, DY13 0DW [IO82UH, SO86]
2E1 CVE R J Trow, 5 Cranberry Dr, Stourport on Severn, DY13 8TH [IO82UI, SO87]
2E1 CVG Details withheld at licensee's request by SSL.
2W1 CVH D V Evans, The Oaks, Victoria Rd, Llandrindod Wells, LD1 6AP [IO82HF, SO06]
2E1 CVJ D A Johnson, 12 Meadway, Harpenden, AL5 1JL [IO91UT, TL11]
2E1 CVM C Burgess, 8 Mayfair Cl, Dukinfield, SK16 5HR [IO83XL, SJ99]
2E1 CVO P S Veitch, 244 Grange Rd, Guildford, GU2 6QY [IO91RG, SU95]
2E1 CVS J R Gwynne, 30 Ryder Cres, Hillside, Southport, PR8 3AE [IO83LO, SD31]
2E1 CVU J D Parker, 19 Mayfair Cl, Dukinfield, SK16 5HR [IO83XL, SJ99]
2E1 CVV A N Mann, 10 Hawkins Way, Wootton, Abingdon, OX13 6LB [IO91IQ, SP40]
2J1 CVX R M Whittaker, Coeur Joyeux, La Rue Des Sapins, St. Peter, Jersey, JE3 7AD
2E1 CWA P J Burton, 14 Gorsey Rd, Mapperley Park, Nottingham, NG3 4JL [IO92KX, SK54]
2E1 CWE M Skewes, 47 Pentrevah Rd, Penwithick, St. Austell, PL26 8UA [IO70OI, SX05]
2J1 CWG J Totty, 34 Clos Paumelle, St Saviour, Jersey, Channel Islands, JE2 7TW
2J1 CWH C Totty, 34 Clos Paumelle, St Saviour, Jersey, Channel Islands, JE2 7TW
2E1 CWI C M Burton, 14 Gorsey Rd, Mapperley Park, Nottingham, NG3 4JL [IO92KX, SK54]
2E1 CWJ D C Thatcher, 6 Ivel View, Sandy, SG19 1AU [IO92UD, TL14]
2E1 CWK Details withheld at licensee's request by SSL.
2E1 CWL S L West, 37 Hallam Cl, Doncaster, DN4 7RT [IO93KM, SE60]
2E1 CWN P Kemble, 88 Mayfield Rd, Ipswich, IP4 3NG [JO02OB, TM14]
2E1 CWP J R Parker, 19 Mayfair Cl, Dukinfield, SK16 5HR [IO83XL, SJ99]
2E1 CWQ P J D Millward, 28 Olive Gr, Burton Joyce, Nottingham, NG14 5FG [IO92LX, SK64]
2E1 CWX E Parker, 19 Mayfair Cl, Dukinfield, SK16 5HR [IO83XL, SJ99]
2E1 CXB Details withheld at licensee's request by SSL.
2E1 CXC M L Kneebone, 34 Henver Rd, Newquay, TR7 3BN [IO70LK, SW86]
2E1 CXD C J Phillips, 34 Crich Ave, Littleover, Derby, DE23 6ES [IO92FV, SK33]
2E1 CXE J E Mortimer, 19 Greenland Dr, Humberstone, Leicester, LE5 1AB [IO92LP, SK60]
2E1 CXF A Whittaker, 9 Ingham St., Padiham, Burnley, BB12 8DR [IO83UT, SD73]
2E1 CXG J P Williams, 31 Coniston Ave, Wigan, WN1 2EY [IO83QN, SD50]
2W1 CXJ D C Williams, 43 St. Margarets Rd, Whitchurch, Cardiff, CF4 7AB [IO81JM, ST18]
2E1 CXK S J Stephens, 26 Hill St., Wigan, WN6 7EQ [IO83QN, SD50]
2E1 CXO K M Skidmore, 239 Alfreton Rd, Blackwell, Alfreton, DE55 5JN [IO93HC, SK45]
2E1 CXP E Bradshaw, 41 Sherwood Rd, Woodley, Stockport, SK6 1LH [IO83WK, SJ99]
2E1 CXS N W Bell, 228 Birchover Way, Allestree, Derby, DE22 2RT [IO92GW, SK33]
2E1 CXW A T Shall, 76 Townsend Ln, Harpenden, AL5 2RQ [IO91TT, TL11]
2I1 CXY A Bird, 198 Ashmount Gdns, Hill Hall Est, Lisburn, BT27 5DB [IO64XM, J26]
2E1 CXZ A P R Musson, 110 Marples Ave, Mansfield Woodhouse, Mansfield, NG19 9DW [IO93JE, SK56]
2E1 CYA M J Ball, 23 Caroline Cl, Alvaston, Derby, DE24 0QX [IO92HV, SK33]
2W1 CYC D G Tiltman, 16 St. Georges Rd, Heath, Cardiff, CF4 4AQ [IO81JM, ST17]
2E1 CYE R J E Stenhouse, Oak Lodge, Lower Holbrook, Ipswich, Suffolk, IP9 2RJ [JO01OX, TM13]
2M1 CYF M J Dearness, 16 Papdale Rd, Kirkwall, KW15 1JT [IO88MX, HY41]
2E1 CYH Details withheld at licensee's request by SSL.
2E1 CYP J Bick, 45 Gloucester Rd, Almondsbury, Bristol, BS12 4HH [IO81RN, ST68]
2I1 CYQ Details withheld at licensee's request by SSL.
2E1 CYR Details withheld at licensee's request by SSL.
2E1 CYS I S Limbert, 739 Queens Dr, Stoney Croft, Stoneycroft, Liverpool, L13 4BS [IO83NJ, SJ49]
2E1 CYT S A Willmer, Flat 1, 107 London Rd, Bexhill on Sea, E Sussex, TN39 3LB [JO00FU, TQ70]
2E1 CYZ M K Baxter, 8 Birch Cl, Romsey, SO51 5SQ [IO90GX, SU32]
2E1 CZA C E Baxter, 341 Woodlands Rd, Woodlands, Southampton, SO40 7GE [IO90FV, SU31]
2E1 CZB B Fido, 1 Bramley Gr, Scotter, Gainsborough, DN21 3UJ [IO93QL, SE80]
2E1 CZF A J Pink, 15 Primrose Hill, Daventry, NN11 5BX [IO92KG, SP56]
2E1 CZH Details withheld at licensee's request by SSL.
2E1 CZI Details withheld at licensee's request by SSL.
2E1 CZJ J C Bosworth, 57 Livingstone Rd, Derby, DE23 6PS [IO92GV, SK33]
2E1 CZK M P Dixon, 6 Paddock Rise, Verney Fields, Stonehouse, GL10 2RD [IO81UR, SO80]
2E1 CZO B J Coombs, 10 Horseshoe Walk, Widcombe, Bath, BA2 6DE [IO81TJ, ST76]
2E1 CZQ C L Sharman, 21 Ferrers Green, Churston Village, Churston Ferrers, Brixham, TQ5 0LF [IO80FJ, SX85]
2E1 CZR D Mills, 11 Northfield Rd, Dagenham, RM9 5XH [JO01BN, TQ48]
2E1 CZS S A Backhouse, 113 Bucklesham Rd, Kirton, Ipswich, IP10 0PF [JO02PA, TM24]
2E1 CZT Details withheld at licensee's request by SSL.
2E1 CZU W Dawes, 30 Trapstyle Rd, Ware, SG12 0BB [IO91XT, TL31]
2E1 CZV E T McIntosh, 69 Southbank Rd, Southport, PR8 6QN [IO83MP, SD31]
2E1 CZW Details withheld at licensee's request by SSL.
2E1 CZY C D J Rayfield, Falcondale, Ash Ln, Almondsbury, Bristol, BS12 4DB [IO81QN, ST58]
2E1 CZZ J E Craker, 7 Burnt House Ln, Stubbington, Fareham, PO14 2LF [IO90JT, SU50]
2E1 DAA Details withheld at licensee's request by SSL.
2E1 DAC S E Coombs, 10 Horseshoe Walk, Widcombe, Bath, BA2 6DE [IO81TJ, ST76]
2E1 DAE J R Simmons, 6 Sporton Cl, South Normanton, Alfreton, Derbyshire, DE55 2HH [IO93HC, SK45]
2E1 DAG Details withheld at licensee's request by SSL.
2E1 DAH Details withheld at licensee's request by SSL.
2E1 DAI Details withheld at licensee's request by SSL.
2E1 DAK C M Wilderspin, 59 Underwood Pl, Oldbrook, Milton Keynes, MK6 2NU [IO92OA, SP83]
2E1 DAL Details withheld at licensee's request by SSL.
2W1 DAM A W Patrick, 52 Huntsmans Corner, Borras Park, Wrexham, LL12 7UH [IO83MB, SJ35]
2E1 DAN P E Dawson, 286 Denewood Cres, Bilborough, Nottingham, NG8 3DD [IO92JX, SK54]
2W1 DAO J Patrick, 52 Huntsmans Corner, Borras Park, Wrexham, LL12 7UH [IO83MB, SJ35]
2E1 DAP D I Featherstone, 39 Cavendish Rd, Bolsover, Chesterfield, S44 6HN [IO93IF, SK46]
2E1 DAQ J W Codd, 12 Broome Cl, Horsham, RH12 5XG [IO91UB, TQ13]
2E1 DAR A V Cottle, 1 Bathite Cttgs, Shaft Rd, Monkton Combe, Bath, BA2 7HN [IO81TI, ST76]
2E1 DAS J Fishwick, 65 Josephs Rd, Guildford, GU1 1DN [IO91RF, SU95]
2E1 DAT J G E Porter, The Bungalow, Holy Island, Northd., TD15 2SE [IO95CQ, NU14]
2E1 DAV D M Dennison, 10 Cornmead, Welwyn Garden City, AL8 7QR [IO91VT, TL21]
2E1 DBA M J Loader, 3 Nursery Gdns, Romsey, SO51 5UU [IO90GX, SU32]
2E1 DBH G J Willoughby, 59 Chasewater Ave, Copnor, Portsmouth, PO3 6JB [IO90LT, SU60]
2E1 DBI P G Dennison, 10 Cornmead, Welwyn Garden City, AL8 7QR [IO91VT, TL21]
2E1 DBJ A E Chebil, 70 Whitley Ct Rd, Quinton, Birmingham, B32 1EY [IO92AL, SP08]
2E1 DBK A P Davenport, 7 Nether Cl, Duffield, Belper, DE56 4DR [IO92GX, SK34]
2E1 DBL M D Bradley, 56 Guildford Rd, Birkdale, Southport, PR8 4JX [IO83LO, SD31]
2E1 DBM E C Bradley, 56 Guildford Rd, Birkdale, Southport, PR8 4JX [IO83LO, SD31]
2E1 DBN H R Hardware, 59 Baulk Ln, Harworth, Doncaster, DN11 8PF [IO93LK, SK69]
2E1 DBP A R Gener, 41 Chace Ave, Potters Bar, EN6 5LZ [IO91WQ, TL20]
2E1 DBR P F Harvey, Rose Cottage, Eppleby, nr Richmond, N Yorks, DL11 7AR [IO94DM, NZ11]
2E1 DBT Dr R Verma, 43 Farley Rd, Derby, DE23 6BW [IO92GV, SK33]
2E1 DBV S J H Payne, 4 Stevens Way, March, PE15 8SL [JO02BN, TL49]
2E1 DBW J D Bishton, 23 Mountjoy Cl, Wimborne, BH21 3AX [IO90AS, SZ09]
2E1 DBZ S R Issatt, 20 Priory Way, Snaith, Goole, DN14 9HB [IO93LQ, SE62]
2E1 DCC I Chapman, 2 Friars Bank, Friars Hill, Guestling, Hastings, TN35 4ET [JO00HV, TQ81]
2E1 DCF R M Brown, Highbury, Carrs Hill, Badingham, Woodbridge, IP13 8NE [JO02QG, TM36]
2M1 DCH Details withheld at licensee's request by SSL.
2E1 DCJ Details withheld at licensee's request by SSL.
2E1 DCL Details withheld at licensee's request by SSL.
2E1 DCM B B Frost, Dale House, 39 Doxey, Stafford, ST16 1EB [IO82WT, SJ92]
2E1 DCN J A Austen, 11 Corn Avill Cl, Abingdon, OX14 2ND [IO91JQ, SU59]
2E1 DCO R J Austen, 11 Corn Avill Cl, Abingdon, OX14 2ND [IO91JQ, SU59]
2E1 DCP C D E Childs, 17 Gladstone Rd, Burgess Hill, RH15 0QQ [IO90WX, TQ32]
2E1 DCQ Details withheld at licensee's request by SSL.
2E1 DCR J B Moss, 137 Upper Batley Ln, Batley, WF17 0QT [IO93ER, SE22]
2E1 DCS J J Padgett, 52 Hillhead Dr, Birstall, Batley, WF17 0PJ [IO93ER, SE22]
2E1 DCU A D Birks, 56 George St., South Normanton, Alfreton, DE55 2AY [IO93HC, SK45]
2E1 DCV B A Shields, 20 Gresley Ct, Grantham, NG31 7RH [IO92QV, SK93]
2E1 DCW J Avery, 8 Birch Cl, Romsey, SO51 5SQ [IO90GX, SU32]
2E1 DCX M R Hinson, 35 The Pines, Roffey, Horsham, RH12 4UF [IO91UB, TQ23]
2E1 DCY J C Mason, 1 Bramble Dr, Stoke Bishop, Bristol, BS9 1RE [IO81QL, ST57]
2I1 DDF C R McBrien, 11 Beechfield Ave, Glenfield, Carrickfergus, BT38 7SN [IO74CR, J48]
2E1 DDG M G Beaver, 8 Watergate Ln, Woolton, Liverpool, L25 8QJ [IO83NI, SJ48]
2E1 DDH M P Wiffin, Lenhill, Bury Rd, Stuston, Diss, IP21 4AD [JO02NI, TM17]
2E1 DDL Details withheld at licensee's request by SSL.
2E1 DDN J C Boardman, 215 Church Green Rd, Fishtoft, Boston, PE21 0RP [JO02AX, TF34]
2E1 DDO C Armstrong, 26 Castle View, Ovingham, Prudhoe, NE42 6AU [IO94BX, NZ06]
2E1 DDQ M Smith, 67 Cornwall Way, Ainsdale, Southport, PR8 3SH [IO83LO, SD31]
2E1 DDR C Smith, 67 Cornwall Way, Ainsdale, Southport, PR8 3SH [IO83LO, SD31]
2E1 DDS W J L Colthorpe, 37 Hill Rd, Eastbourne, BN20 8SN [JO00CS, TQ50]
2E1 DDT Details withheld at licensee's request by SSL.
2E1 DDU Details withheld at licensee's request by SSL.
2E1 DDV T A Smith, 2A Chesterfield Rd, Barlborough, Chesterfield, S43 4TR [IO93IG, SK47]
2E1 DDW R J D Young, 17 Upper Kings Dr, Eastbourne, BN20 9AN [JO00DT, TQ50]
2E1 DDX Details withheld at licensee's request by SSL.
2E1 DDZ S J Arter, 18 Essex Rd, Westgate on Sea, CT8 8AP [JO01QJ, TR36]
2W1 DEA P Smith, 51 St. Cadocs Rd, Trevethin, Pontypool, NP4 8JW [IO81LR, SO20]
2E1 DEC H J S Scott-Telford, 9 Squires Cl, Rochester, ME2 2TZ [JO01FJ, TQ76]
2E1 DED P R Davis, Four Winds, Waters Upton, Telford, Shropshire, TF6 6NP [IO82RS, SJ61]
2E1 DEG Details withheld at licensee's request by SSL.
2E1 DEI E Pugsley, 156 Thistledown Rd, Clifton Est, Nottingham, NG11 9ED [IO92JV, SK53]
2E1 DEJ Details withheld at licensee's request by SSL.
2E1 DEK P W Moss, 22 Battersby St., Ince, Wigan, WN2 2NA [IO83QN, SD60]
2E1 DEL S J Peck, 16 Greenacres, Mile End, Colchester, CO4 5DX [JO01KV, TL92]
2E1 DEM N Humphreys, 90 Wedgewood Cres, Ketley, Telford, TF1 4BN [IO82SQ, SJ61]
2E1 DEN P W Hart, 82 Bagslate Moor Rd, Rochdale, OL11 5YH [IO83VO, SD81]
2E1 DEP L K Darton, 8 Foster Gr, Sandy, SG19 1HP [IO92UD, TL14]
2E1 DEQ Details withheld at licensee's request by SSL.
2E1 DES G Johnson, 6 Paynell, Dunholme, Lincoln, LN2 3SW [IO93SI, TF08]
2E1 DET A E Whyman, 8 Staplers Cl, Great Totham, Maldon, CM9 8UN [JO01IS, TL81]
2E1 DEV Details withheld at licensee's request by SSL.
2E1 DEW Details withheld at licensee's request by SSL.
2E1 DEX J R Constable, 9 Foxglove Cl, Clacton on Sea, CO15 4TY [JO01NT, TM11]
2E1 DEY Details withheld at licensee's request by SSL.
2E1 DEZ Details withheld at licensee's request by SSL.
2E1 DFA Details withheld at licensee's request by SSL.
2E1 DFB B R Williams, Highview, Dover Rd, Guston, Dover, CT15 5EH [JO01PD, TR34]
2E1 DFC N Herbert, 26 Knolls Way, Clifton, Shefford, SG17 5QZ [IO92UA, TL13]
2E1 DFD A R Ellis, 52 Phyllis St., Passmonds, Rochdale, OL12 7NA [IO83VO, SD81]
2E1 DFE R J Scott, 69 Baker St., Poolstock, Wigan, WN3 5HG [IO83QM, SD50]
2E1 DFF T D Pole, The Vicarage, Greenhill Rd, Alveston, Bristol, BS12 2QT [IO81RO, ST68]
2M1 DFG Details withheld at licensee's request by SSL.
2E1 DFH M Keyser, Rosemount, Church Whitfield Rd, Whitfield, Dover, CT16 3HZ [JO01PD, TR34]
2E1 DFI Details withheld at licensee's request by SSL.
2E1 DFJ A J Murray, 56 Malthouse Ln, Kenilworth, CV8 1AD [IO92EI, SP27]
2E1 DFK N Charlton, 94 Cromarty, Ouston, Chester-le-Street, DH2 1JU [IO94EV, NZ25]
2E1 DFL R J Reeves, 73 Connaught Rd, Cromer, NR27 0DB [JO02PW, TG24]
2E1 DFM D Ankers, 2 Dashwood Cl, Slough, SL3 7NB [IO91RL, SU97]
2E1 DFO Details withheld at licensee's request by SSL.
2E1 DFQ Master M P Ginnever, 7 Passmonds Way, Rochdale, OL11 5AN [IO83VO, SD81]
2W1 DFR Details withheld at licensee's request by SSL.
2E1 DFS R E Bearcroft, 11 Blakes Hill, North Littleton, Evesham, WR11 5QN [IO92BC, SP04]
2W1 DFT P D Donovan, 30 The Alders, Llanyravon, Cwmbran, NP44 8JE [IO81LP, ST39]
2E1 DFU Master C A Harrison, 22 Sherburn Gate, Chapeltown, Sheffield, S30 4EU [IO93GJ, SK38]
2E1 DFV K Harrison, 22 Sherburn Gate, Chapeltown, Sheffield, S30 4EU [IO93GJ, SK38]
2W1 DFW S A Mosey, 49 Alban Rd, Llanelli, SA15 1EP [IO71WQ, SN50]
2W1 DFX P W Mosey, 49 Alban Rd, Llanelli, SA15 1EP [IO71WQ, SN50]
2E1 DFZ M G Axon, 48 Cowslip Rd, Poole, Dorset, BH17 7QZ [IO90AR, SZ09]
2E1 DGA M Kennedy, 15 Wallwork St., Openshaw, Manchester, M11 1FY [IO83WL, SJ89]
2E1 DGD D J Carter, 30 Swift Way, Sandal, Wakefield, WF2 6SR [IO93GP, SE31]
2E1 DGF R J M Omielan, 2 Cypress Gr, Henleaze, Bristol, BS9 4RX [IO81QL, ST57]
2E1 DGG N J Parnell, Four J's, Fernihough Ave, Honeybourne, Evesham, Worcs, WR11 5XS [IO92BC, SP14]
2E1 DGH T G Francis, 24 Waterside, Evesham, WR11 6BU [IO92AC, SP04]
2J1 DGI D P Mahrer, 4 Fairfield Ave, La Pouquelaye, St. Helier, Jersey, JE2 3FT
2E1 DGJ J A Rogers, 58 Broadwater Cres, Welwyn Garden City, AL7 3TU [IO91VT, TL21]
2W1 DGK N D J Booth, 30 Treowen Rd, Pennar, Pembroke Dock, SA72 6NZ [IO71MQ, SM90]
2W1 DGL P I Lewis, 17 South Meadows, Pembroke, SA71 4EW [IO71MQ, SM90]
2W1 DGM J H Foster, 1 Carr Terr, Pennar, Pembroke Dock, SA72 6RJ [IO71MQ, SM90]
2E1 DGN Details withheld at licensee's request by SSL.
2E1 DGP G Harrison, 22 Sherburn Gate, Chapeltown, Sheffield, S30 4EU [IO93GJ, SK38]
2E1 DGR M B M Chebil, 70 Whitley Ct Rd, Quinton, Birmingham, B32 1EY [IO92AL, SP08]
2E1 DGS G Bradshaw, 3 Falmouth Ave, Haslingden, Rossendale, BB4 6QN [IO83UQ, SD72]

2E1 DGT R J Clothier, 20 Southdown Rd, Rodwell, Weymouth, DT4 9LJ [IO80SO, SY67]
2M1 DGU J D Ferguson, 21 Pentland Dr, Edinburgh, Midlothian, EH10 6PU [IO85JV, NT26]
2E1 DGV P H Close, Carosa, Hervines Rd, Amersham, HP6 5HS [IO91QQ, SU99]
2E1 DGW D W Williams, 53 Derby Way, Stevenage, SG1 5TR [IO91VV, TL22]
2E1 DGX D G Ford, Stonecroft, London Rd, Dorking, RH4 1TA [IO91UF, TQ15]
2E1 DGY T Nokes, 10 Elsworth Pl, Cambridge, CB2 2RG [JO02BE, TL45]
2E1 DGZ J A Bloxam, 73 Rock Rd, Cambridge, CB1 4UG [JO02BE, TL45]
2E1 DHA B L Clarkson, 231 Overdown Rd, Tilehurst, Reading, RG3 6NX [IO91LJ, SU66]
2E1 DHB Details withheld at licensee's request by SSL.
2E1 DHD T E Stean, 31 Lanesfield Park, Greenhill, Evesham, WR11 4NU [IO92AC, SP04]
2W1 DHE Details withheld at licensee's request by SSL.
2M1 DHG G D D Russell, 32 Pulpit Dr, Oban, PA34 4LE [IO76GJ, NM82]
2E1 DHH Details withheld at licensee's request by SSL.
2M1 DHI A M Cowley, 7 Fernie Brae, Gardenstown, Banff, AB45 3YL [IO87TQ, NJ76]
2E1 DHJ D D E Jones, 11 Kylemilne Way, Stourport on Severn, DY13 9NA [IO82UI, SO87]
2E1 DHK J C Rees, 29 Tuckton Rd, Southbourne, Bournemouth, BH6 3HR [IO90CR, SZ19]
2E1 DHL P J Westwood, 62 Blackbrook Ln, Bickley, Bromley, BR2 8AY [JO01AJ, TQ46]
2W1 DHM P A Faraday, Hillcrest, Twyn Ln, Glascoed, nr Pontypool, Gwent, NP4 0UB [IO81MQ, SO30]
2E1 DHN E E Pomeroy, 110 Millstrood Rd, Whitstable, CT5 1PT [JO01MI, TR16]
2E1 DHO J R Howell, 4 West End, Stokesley, Middlesbrough, TS9 5BN [IO94JL, NZ50]
2E1 DHQ S J L Williams, 15 The Meadows, Broomfield, Herne Bay, CT6 7XB [JO01NI, TR16]
2E1 DHS Details withheld at licensee's request by SSL.
2E1 DHU Details withheld at licensee's request by SSL.
2W1 DHV J Tuite, 44 Gorlan, Conwy, LL32 8RS [IO83BG, SH77]
2E1 DHW Details withheld at licensee's request by SSL.
2E1 DHX D Walker, 8 Wenlock Ct, Hunt Rd, Somerford, Christchurch, BH23 3BY [IO90DR, SZ19]
2E1 DHY C J Houghton, 25 Woodberry Dr, Sittingbourne, ME10 3AT [JO01JI, TQ96]
2W1 DHZ R G Williams, 103 Victoria Rd, Prestatyn, LL19 7SR [IO83HI, SJ08]
2E1 DIA J Laffin, 154 Blenheim Dr, Allestree, Derby, DE22 2GN [IO92FW, SK33]
2E1 DIB D A Coleman, 21 Lone Pine Dr, West Parley, Ferndown, BH22 8LW [IO90BS, SZ09]
2E1 DIC Details withheld at licensee's request by SSL.
2E1 DID I C Beith, Beechcroft, 18 Ave Rd, New Milton, BH25 5JP [IO90ES, SZ29]
2W1 DIG G M Williams, 103 Victoria Rd, Prestatyn, LL19 7SR [IO83HI, SJ08]
2E1 DIH J J Bentley, 4 Highway, Edgcumbe Park, Crowthorne, RG45 6HE [IO91OI, SU86]
2W1 DIK P Watson, Eirianfa, Cwmdad, Carmarthen, Dyfed, SA33 6XJ [IO71TW, SN33]
2M1 DIL R J M Gibbons, 4 Elm Gdns, Bearsden, Glasgow, G61 3BH [IO75UW, NS57]
2E1 DIM R J Steed, 42 Lynch Rd, Farnham, GU9 8BY [IO91OF, SU84]
2M1 DIN I P Mallinson, 15 Rowallan Gdns, Glasgow, G11 7LH [IO75UV, NS56]
2E1 DIO J Kubiesa, 103 Fifers Ln, Hellesdon, Norwich, NR6 6EF [JO02PP, TG21]
2M1 DIQ K C Gordon, 31 Roselea Dr, Milngavie, Glasgow, G62 8HE [IO75UW, NS57]
2E1 DIR Details withheld at licensee's request by SSL.
2M1 DIT G J Davidson, 2 Chapel Ct, Wigtown, Newton Stewart, DG8 9ET [IO74SU, NX45]
2E1 DIX Details withheld at licensee's request by SSL.
2E1 DJC S E Bradshaw, 20 Parkside Rd, Chaddesden, Derby, DE21 6QQ [IO92GW, SK33]
2E1 DJD A Neville, 1 Bethel Ave, Blackpool, FY2 9NA [IO83LU, SD33]
2E1 DJE Details withheld at licensee's request by SSL.
2E1 DJF C M H Hornby, 14 Essex Rd, Stevenage, SG1 3EZ [IO91VV, TL22]
2E1 DJI G M Johns, 6 Wakefield Ave, Northbourne, Bournemouth, BH10 6DS [IO90BS, SZ09]
2M1 DJL Details withheld at licensee's request by SSL.
2E1 DJM M E Seeby, 59 Dallamoor, Telford, TF3 2EE [IO82SQ, SJ70]
2E1 DJO Details withheld at licensee's request by SSL.
2E1 DJQ P L R Nicolson, Norton Hall, Pebworth, Warks, CV37 8XH [IO92CC, SP14]
2E1 DJR M Fletcher, 21 Lincoln Way, Thetford, IP24 1DG [JO02IK, TL88]
2E1 DJS K A Fletcher, 21 Lincoln Way, Thetford, IP24 1DG [JO02IK, TL88]
2E1 DJT J L Marshall, 37 Whin Bank, Scarborough, YO12 5LD [IO94SG, TA08]
2E1 DJZ Details withheld at licensee's request by SSL.
2E1 DKA Details withheld at licensee's request by SSL.
2E1 DKD A M McVey, 2 Rathmore Rd, Cambridge, CB1 4AD [JO02BE, TL45]
2E1 DKE T B Jeffs, 34 Hayhurst Rd, Whalley, Clitheroe, BB7 9RL [IO83TT, SD73]
2E1 DKF Details withheld at licensee's request by SSL.
2E1 DKG Details withheld at licensee's request by SSL.
2E1 DKI Details withheld at licensee's request by SSL.
2E1 DKK Details withheld at licensee's request by SSL.
2E1 DKM R J Hawkes, 23 Gloucester Ave, Cliftonville, Margate, CT9 3NN [JO01QJ, TR37]
2E1 DKN C M Thomas, 4 Sunningdale Walk, Herne Bay, CT6 7TR [JO01NI, TR16]
2E1 DKO G J Daniell, Meadow View, Broad Marston Rd, Pebworth, Stratford upon Avon, CV37 8XR [IO92CC, SP14]
2E1 DKP Details withheld at licensee's request by SSL.
2E1 DKQ A Jackson, 1 Belvedere Ave, Atherton, Manchester, M46 9LQ [IO83SM, SD60]
2E1 DKR Details withheld at licensee's request by SSL.
2E1 DKU A Whittle, 9 Dale View, Littleborough, OL15 0BP [IO83WP, SD91]
2E1 DKX L J Clarkson, 231 Overdown Rd, Tilehurst, Reading, RG3 6NX [IO91LJ, SU66]
2E1 DKY Details withheld at licensee's request by SSL.
2E1 DKZ A J Lovell, 98B Baker Rd, Newthorpe, Nottingham, NG16 2DP [IO93IA, SK44]
2E1 DLA P R S Craig, Partridge Cottage, Redpale, Dallington, Heathfield, TN21 9NR [JO00EW, TQ61]
2E1 DLC M F Sadler, Hillview, Forest Mill Ln, Horton, Ilminster, Somerset, TA19 9QU [IO80MW, ST31]
2E1 DLD E A Harrison, 1 Winnipeg Cl, Lower Wick, Worcester, WR2 4XT [IO82VE, SO85]
2E1 DLF G A Stilton, 30 City Mills, Skeldergate, York, North Yorks, YO1 1DB [IO93LW, SE65]
2M1 DLH R F Faulks, 23 Bentinck Dr, Troon, Ayrshire, KA10 6HX [IO75QN, NS33]
2E1 DLK S J Jeremy, Sheen Cottage, Thorpe Ln, Fylingthorpe, Whitby, YO22 4TH [IO94RK, NZ90]
2E1 DLN I C Dellbridge, 19 Cleeve Cl, Stourport on Severn, DY13 0NY [IO82UH, SO86]
2E1 DLO B Bush, 19 Fanns Rise, Purfleet, RM19 1GP [JO01CL, TQ57]
2E1 DLP C J Chittock, Drackenhill, 8 Castle Rise, West Ayton, Scarborough, YO13 9JY [IO94SG, SE98]
2E1 DLQ A M Langford, 31 Rosehall Cl, Oakenshaw, Redditch, B98 7YD [IO92AG, SP06]
2E1 DLR R J Diaper, 30 Holmcroft Rd, Kidderminster, DY10 3AG [IO82VJ, SO87]
2E1 DLS J F Field, 27 Lovelace Rd, Barnet, EN4 8EA [IO91WP, TQ29]
2E1 DLT P Lilley, 24 Brook Cl, Bulwell, Nottingham, NG6 8NL [IO92JX, SK54]
2E1 DLU Details withheld at licensee's request by SSL.
2E1 DLW M M Burford, 26 Shrubbery Rd, Bromsgrove, B61 7BH [IO82XH, SO97]
2E1 DLX P G Thackray, 20 Darfield St., Leeds, LS8 5DB [IO93FT, SE33]
2E1 DLZ C Edis, 70 Holly Rd, Watnall, Nottingham, NG16 1HP [IO93IA, SK54]
2E1 DMA S Whitney, Gr View, Upthorpe Rd, Stanton, Bury St. Edmunds, IP31 2AP [JO02KH, TL97]
2E1 DME D A Priestley, 25 Meadow Cl, Eastwood, Nottingham, NG16 3DQ [IO93IA, SK44]
2E1 DMH M T Rippin, Gaverne, Welford Rd, Long Marston, Stratford upon Avon, CV37 8RA [IO92CD, SP14]
2E1 DMI R W Dixon, 97 Sunny Blunts, Peterlee, SR8 1LN [IO94HR, NZ43]
2E1 DMK Details withheld at licensee's request by SSL.
2E1 DML M P Bailey, 44 Fishponds Dr, Crigglestone, Wakefield, WF4 3PB [IO93FP, SE31]
2E1 DMM Details withheld at licensee's request by SSL.
2E1 DMO V P Kitson, 23 Mountbatten Ave, Sandal, Wakefield, WF2 6EY [IO93GP, SE31]
2E1 DMP J G Kitson, 23 Mountbatten Ave, Sandal Magna, Sandal, Wakefield, WF2 6EY [IO93GP, SE31]
2E1 DMQ J A Mortimer, 4 Nethercliffe Cres, Guiseley, Leeds, LS20 9HN [IO93DV, SE14]
2E1 DMR J A Walker, Stornoway, Isington Rd, Isington, Alton, GU34 4PP [IO91NE, SU74]
2E1 DMS T G Elliott, Manor House, Bewholme, Driffield, E Yorks, YO25 8DX [IO93VW, TA14]
2E1 DMT G M Jones, 62 Mallings Dr, Bearsted, Maidstone, ME14 4HG [JO01HG, TQ85]
2E1 DMU S A Wilson, 3 Pinewood Dr, Chatham, ME5 8XU [JO01GH, TQ76]
2E1 DMW Details withheld at licensee's request by SSL.
2E1 DMY R G Fairchild, 412 Livingstone Rd, Bradford, BD2 1QD [IO93DT, SE13]
2E1 DMZ R Laverick, 12 Greenlands Rd, Redcar, TS10 2DG [IO94LO, NZ62]
2E1 DNB P J Lomas, 15 Norman Pl, Leeds, LS8 2AW [IO93FU, SE33]
2E1 DND J P U Ursell, 7 Eton Rd, Frinton on Sea, CO13 9JA [JO01PU, TM22]
2E1 DNE Details withheld at licensee's request by SSL.
2E1 DNH Details withheld at licensee's request by SSL.
2E1 DNJ A P Scott, 11 Redgrave Rise, Winstanley, Wigan, WN3 6HG [IO83PM, SD50]
2E1 DNM M R Lixenberg, 41 Essex Park, Finchley, London, N3 1ND [IO91VO, TQ29]
2I1 DNO D Davey, 15 Heathfield, Culmore Rd, Londonderry, BT48 8JD [IO65IA, C42]
2E1 DNP Details withheld at licensee's request by SSL.
2E1 DNR B J Darton, 3 Holwell Rd, Welwyn Garden City, AL7 3RA [IO91VT, TL21]
2E1 DNS Details withheld at licensee's request by SSL.
2E1 DNT D J Hardiman, 43 Nettlecombe, Shaftesbury, SP7 8PR [IO81VA, ST82]
2E1 DNU Details withheld at licensee's request by SSL.
2E1 DNY Details withheld at licensee's request by SSL.
2E1 DNZ D T Harrop, 85 Bewick Rd, Gateshead, NE8 1RR [IO94EW, NZ26]
2E1 DOA P J Dalby, 76 Amery Gdns, Gidea Park, Romford, RM2 6RU [JO01CO, TQ58]
2E1 DOB J Palmer, Reeds Farm, Beaford, nr Winkleigh, North Devon, EX19 8LR [IO70XV, SS51]
2E1 DOC A C McInnes, West View Cottage, Dalton, Thirsk, North Yorks, YO7 3HS [IO94HE, SE47]
2E1 DOD R U Dodd, 27 Lancaster Rd, Newcastle, ST5 1DS [IO83VA, SJ84]
2E1 DOE I A Elliott, 8 Crowland Rd, Hartlepool, TS25 2JJ [IO94JP, NZ42]
2E1 DOH D T Hart, Prospect House, 25 The Spain, Petersfield, GU32 3JZ [IO91MA, SU72]
2E1 DOL D Silvers, 32 Walesby Ct, Leeds, LS16 6RX [IO93EU, SE23]
2E1 DOM D R Beal, 57 Milton Rd, Eastbourne, BN21 1SN [JO00DS, TQ50]
2E1 DOO M R Hauxwell, 65 Harleston Way, Heworth, Gateshead, NE10 9BQ [IO94FW, NZ26]
2E1 DOQ Details withheld at licensee's request by SSL.
2E1 DOR E E Gerrard, 39 Lade Fort Cres, Lydd on Sea, Romney Marsh, TN29 9YG [JO00LW, TR02]
2E1 DOU P E Jeffries, 5 Darnick Rd, Sutton Coldfield, B73 6PE [IO92BN, SP09]
2E1 DOW C Milburn, 16 Dorset Cres, Moorside, Consett, DH8 8HX [IO94BU, NZ04]
2E1 DOX Details withheld at licensee's request by SSL.
2E1 DOY C A Scholes, 14 Braemar Rd, Nottingham, NG6 9HN [IO93JA, SK54]
2E1 DOZ S R Brenchley, 19 Hoopers Way, Oakley, Basingstoke, RG23 7DE [IO91JF, SU55]
2E1 DPA T A Yarrow, Hall View, Barton, Richmond, North Yorks, DL10 6JP [IO94EL, NZ20]
2E1 DPC B Barnes, 10 Cranbourne Rd, Rochdale, OL11 5JD [IO83VO, SD81]
2E1 DPD C M Pearson, 56 Parkwood Green, Parkwood, Gillingham, ME8 9PP [JO01HI, TQ86]
2E1 DPF A J Timmins, 42 Owen Ave, Long Eaton, Nottingham, NG10 2FS [IO92IV, SK53]
2E1 DPG D Stone, Bridor, 12 Robertson Ave, Leasingham, Sleaford, NG34 8NJ [IO93SA, TF04]
2E1 DPH S L Wood, 4 Colenutts Rd, Ryde, PO33 3HS [IO90JR, SZ59]
2E1 DPJ Details withheld at licensee's request by SSL.
2E1 DPK S F B Hill, 22 Sandy Point Rd, Hayling Island, PO11 9RP [IO90MS, SZ79]
2E1 DPL J W C Walliker, Southside, The Green, Newsham, Richmond, N Yorks, DL11 7RD [IO94BL, NZ10]
2E1 DPO T Tennant, The Bungalow, Dyson Ln, Newsham, Richmond, North Yorks, DL11 7RA [IO94BL, NZ11]
2E1 DPS D T Place, 34 Holcroft, Orton Malborne, Peterborough, PE2 5SL [IO92UN, TL19]
2E1 DPT C L E King, 6 Woodgate Rd, Whalley Range, Manchester, M16 8LX [IO83UK, SJ89]
2E1 DPU J Towle, 36 Old Orchard, Haxby, York, YO3 3DT [IO94LA, SE65]
2E1 DPV B W Woods, 64 Yarningale Rd, Weeford Est, Coventry, CV3 3EQ [IO92GJ, SP37]
2E1 DPW B T Seaby, 4 Parkside, Wollaton, Nottingham, NG8 2NN [IO92JW, SK53]
2E1 DPX A J Grizzell, 11 Quarry Way, Stapleton, Bristol, BS16 1UP [IO81RL, ST67]
2E1 DQA K G M Hobbs, 20 Holding St., Rainham, Gillingham, ME8 7JP [JO01HI, TQ86]
2E1 DQB Details withheld at licensee's request by SSL.
2E1 DQC Details withheld at licensee's request by SSL.
2E1 DQD I G Evans, 14 St. Georges Cttgs, Tethering Drove, Hale, Fordingbridge, SP6 2NJ [IO90DX, SU11]
2E1 DQE C Williams, 7 Potters Dr, Hopton, Great Yarmouth, NR31 9RW [JO02UN, TG50]
2E1 DQF M F McKay, 98 de Lacy Ct, New Ollerton, Newark, NG22 9RW [IO93LE, SK66]
2E1 DQG D M Templeton, 59 Oaklands Way, Fareham, PO14 4LF [IO90IU, SU50]
2E1 DQH L A Templeton, 59 Oaklands Way, Fareham, PO14 4LF [IO90IU, SU50]
2W1 DQI J A Adcock, Bleake House, Cefn Coch, Welshpool, Powys, SY21 0AE [IO82HO, SJ00]
2E1 DQK G J Horsley, 5 Edwards Gdns, Swanley, BR8 8HP [JO01CJ, TQ56]
2E1 DQL M W Dunne, 40 Egmont Rd, Hamworthy, Poole, BH16 5BZ [IO80XR, SY99]
2E1 DQM N D Edwards, 609 Upper Richmond Rd West, Richmond, TW10 5DU [IO91UL, TQ17]
2E1 DQN C P J Langton, 10 Little Comptons, Horsham, RH13 5UW [IO91UA, TQ22]
2E1 DQQ D C Horsley, 5 Edwards Gdns, Swanley, BR8 8HP [JO01CJ, TQ56]
2E1 DQS M B Garnett-Frizelle, 17 Bridport Ave, New Moston, Manchester, M40 3WP [IO83VM, SD80]
2E1 DQT A H M Charles, 6 Bridewell St., Wymondham, NR18 0AR [JO02NN, TG10]
2E1 DQU Details withheld at licensee's request by SSL.
2E1 DQV S Jackson, 1 Bow St., Bridlington, YO15 3DU [IO94VB, TA16]
2E1 DQW Y Javaheri, 66 Bifield, Orton Goldhay, Peterborough, PE2 5SW [IO92UN, TL19]
2E1 DQY P S Hurren, 64 Ship Rd, Pakefield, Lowestoft, NR33 7DP [JO02UK, TM59]
2E1 DQZ C M Houlden, 29 Ct Barton, Weston, Portland, DT5 2HJ [IO80SN, SY67]
2W1 DRB P A Waller, 4 Rose Ct, Ty Canol, Cwmbran, NP44 6JH [IO81LP, ST29]
2E1 DRC P J Hatcher, 32 Slough Rd, Iver Heath, Bucks, SL0 0DT [IO91RM, TQ08]
2M1 DRD G J A Costa, 54 High St., Dollar, FK14 7BA [IO86EE, NS99]
2W1 DRE T Nowell, Pigeon House, Llangathen, Carmarthen, Dyfed, SA32 8QH [IO71XV, SN52]
2E1 DRI I D Brenkley, The Alms House, Easby, Richmond, North Yorks, DL10 7EX [IO94DJ, NZ10]
2E1 DRJ K T Silliman, 12 Cleardene, Dorking, RH4 2BY [IO91UF, TQ14]
2W1 DRL N Evans, Ty-Newydd, Tregynon, Newtown, Powys, SY16 3ER [IO82IN, SO19]
2W1 DRM L Fryer, Pantycrai, Adfa, Newtown, Powys, SY16 3DB [IO82HO, SJ00]
2E1 DRN M D Fairchild, 89 Park Rd, Congresbury, Bristol, BS19 5HE [IO81OI, ST46]
2E1 DRO S R Croall, 71 Frederick Rd, Malvern, WR14 1RS [IO82UD, SO74]
2W1 DRP G Mainwaring, 3 Elias St., Neath, SA11 1PP [IO81CP, SS79]
2E1 DRR B E Naylor, 16 Dorchester Cl, Basingstoke, RG23 8EX [IO91KG, SU65]
2E1 DRS R S Pamment, 5 New Captains Rd, West Mersea, Colchester, CO5 8QP [JO01KS, TM01]
2E1 DRT S R Davies, 11 Gravel Pits Cl, Bredon, Tewkesbury, GL20 7QL [IO82WA, SO93]
2E1 DRU K L McCann, Treverven, Back Ln, Hemingbrough, Selby, YO8 7QP [IO93MS, SE63]
2E1 DRV J F Wohlgemuth, 36 Gr Rd, Bexleyheath, DA7 6AX [JO01BK, TQ57]
2E1 DRW H W May, 18 Pennant Hills, Bedhampton, Havant, PO9 3JZ [IO90LU, SU60]
2E1 DRX R L Hauxwell, 65 Harleston Way, Heworth, Gateshead, NE10 9BQ [IO94FW, NZ26]
2E1 DRY G Symonds, Green Roofs, Thorpe Rd, Haddiscoe, Norwich, NR14 6PP [JO02TM, TM49]
2E1 DSA A R May, 18 Pennant Hills, Havant, PO9 3JZ [IO90LU, SU60]
2E1 DSB B R Cooper, 2 Heather Way, Great Moulton, Norwich, NR15 2HP [JO02OL, TM18]
2E1 DSD A Dickins, 4 Downside, Lewes, BN7 1EE [IO90XU, TQ40]
2E1 DSE R C Symonds, Green Roofs, Thorpe Rd, Haddiscoe, Norwich, NR14 6PP [JO02TM, TM49]
2E1 DSF A K Parker, 103 Nortonwood Ln, Windmill Hill, Runcorn, WA7 6QQ [IO83PI, SJ58]
2E1 DSG Details withheld at licensee's request by SSL.
2E1 DSH K G Southworth, 188 Bispham Rd, Southport, PR9 7BP [IO83MP, SD31]
2E1 DSK P L Kelly, 5 Cardiff Rd, Hanwell, London, W7 2BW [IO91UL, TQ17]
2E1 DSL M J Perry, 2 Riseholme Ave, Wollaton, Nottingham, NG8 2TE [IO92JW, SK53]
2E1 DSM S J Ainsworth, 25 Grangeway Ct, Runcorn, WA7 5FA [IO83PH, SJ58]
2E1 DSN R Marsden, 26 Lincombe Rise, Leeds, LS8 1QH [IO93FT, SE33]
2E1 DSP B Polwarth, 101 Hartside, Lemington, Newcastle upon Tyne, NE15 8BZ [IO94DX, NZ16]
2E1 DSR Details withheld at licensee's request by SSL.
2M1 DST W J M Ross, 10 Sorleys Brae, Dollar, FK14 7AS [IO86DD, NS99]
2E1 DSU J J Hewitt, 5 Oakmount Rd, Streetly, Sutton Coldfield, B74 2EG [IO92BN, SP09]
2E1 DSV Details withheld at licensee's request by SSL.
2E1 DSW M C Porter, 29 Cedar Way, Pucklechurch, Bristol, BS17 3RN [IO81SL, ST67]
2E1 DSY Details withheld at licensee's request by SSL.
2E1 DTB S J Haddon, 1 Victoria Pl, Easton, Portland, DT5 2AA [IO80SN, SY67]
2E1 DTC Details withheld at licensee's request by SSL.
2E1 DTD P S Keeler, 72 Grafton Rd, Selsey, Chichester, PO20 0JB [IO90OR, SZ89]
2E1 DTE C D Folkerd, The Old Manse, Ealing Green, London, W5 5QT [IO91UM, TQ18]
2E1 DTF O R Folkerd, The Old Manse, Ealing Green, London, W5 5QT [IO91UM, TQ18]
2E1 DTH P N Hacker, Wallnook Cottage, Rendham, Suffolk, IP17 2AU [JO02RF, TM36]
2E1 DTI Details withheld at licensee's request by SSL.
2E1 DTJ T D Hacker, Wallnook Cottage, Rendham, Suffolk, IP17 2AU [JO02RF, TM36]
2E1 DTK R H Cleary, 5 Gregson Rd, Halton View, Widnes, WA8 0BX [IO83PI, SJ58]
2E1 DTL Details withheld at licensee's request by SSL.
2E1 DTM M R Bishton, 23 Mountjoy Cl, Wimborne, BH21 3AX [IO90AS, SZ09]
2E1 DTN J V Bennett, Rectory Cottage, Broad St., Wrington, Bristol, BS18 7LD [IO81OI, ST46]
2W1 DTO Details withheld at licensee's request by SSL.
2E1 DTP S D Harland, 3 Waterfall Terr, Barton, Richmond, DL10 6LZ [IO94EL, NZ20]
2E1 DTQ Details withheld at licensee's request by SSL.
2E1 DTR B G Ellis, 70 Leaside, Halton Brook, Runcorn, WA7 2NH [IO83PH, SJ58]
2E1 DTS M Poulter, 26 West Cres, Duckmanton, Chesterfield, S44 5HE [IO93HF, SK47]
2E1 DTT C R Rainbow, 65 Eleanor Rd, Waltham Cross, EN8 7DW [IO91XQ, TL30]
2E1 DTW Details withheld at licensee's request by SSL.
2W1 DTX N E Farthing, Bay View, Bull Bay, Amlwch, Anglesey, Gwynedd, LL68 9ST [IO73TK, SH49]
2E1 DTY Details withheld at licensee's request by SSL.
2E1 DTZ Details withheld at licensee's request by SSL.
2E1 DUB M T Rogers, 14 Yewdale, Shevington, Wigan, WN6 8DE [IO83PN, SD50]
2E1 DUC Details withheld at licensee's request by SSL.
2E1 DUD Details withheld at licensee's request by SSL.
2E1 DUE Details withheld at licensee's request by SSL.
2E1 DUF B J Solway, 2 Valley View Dr, Newbridge, Truro, TR1 3UL [IO70LG, SW84]
2E1 DUG M A Spensley, Strands Holme, Gunnerside, Richmond, North Yorks, DL11 6LF [IO84XJ, SD99]
2E1 DUI G J Tunley, Ashfields, 5 Sawyers Mill, Shillingford, Tiverton, EX16 9RY [IO81GA, SS92]
2E1 DUL K L Robins, 4 Scotstown Cres, Peterhead, AB42 1GU [IO97CM, NK14]
2E1 DUM L Griffiths, 91 Worrall Rd, Sheffield, S6 4BA [IO93FJ, SK39]
2E1 DUN Details withheld at licensee's request by SSL.
2E1 DUP Details withheld at licensee's request by SSL.
2E1 DUQ J Wright, 31 Sandpit Ln, St. Albans, AL1 4EW [IO91US, TL10]
2E1 DUS B T Legg, Mardi Gras, Cann, Shaftesbury, Dorset, SP7 0DF [IO80VX, ST82]
2E1 DUT M A Bryan, 3 Wheatfield Lea, Cranbrook, TN17 3ND [JO01GC, TQ73]
2E1 DUU Details withheld at licensee's request by SSL.
2E1 DUV A J Martindale, The Beeches, Raby Dr, Bromborough, Wirral, L63 0NL [IO83MH, SJ38]
2E1 DUW M R Goldby, Waylands Gate, St. Johns Rd, Bashley, New Milton, BH25 5SD [IO90ES, SZ29]
2E1 DVB E T Southam, 16 Parkside Ave, Littlehampton, BN17 6BG [IO90RT, TQ00]
2E1 DVC Details withheld at licensee's request by SSL.

2E1 DVD M S Ball, 3 The Chesters, Middleton Tyas, North Yorks, DL10 6PP [IO94EK, NZ20]
2E1 DVE Details withheld at licensee's request by SSL.
2E1 DVG J S Moore, Bon Accord, 122 Fairmead Ave, Westcliff on Sea, SS0 9SB [JO01IN, TQ88]
2E1 DVK J E Sheppard, 45 Alma Rd, Birkdale, Southport, PR8 4AN [IO83LP, SD31]
2E1 DVM G S Thomas, 11 Beaufort Ave, Kidderminster, DY11 5NH [IO82UJ, SO87]
2M1 DVQ G Russell, Leslie Lark, Headswood, Denny, FK6 6BW [IO86BA, NS88]
2E1 DVR P J Savage, Laneside, Croasdale Dr, Parbold, Wigan, WN8 7HR [IO83OO, SD41]
2M1 DVS R D Guyan, 2A Southerton Rd, Kirkcaldy, KY2 5NA [IO86JC, NT29]
2E1 DVY J P Davies, 36 Arundel Rd, Southport, PR8 3DQ [IO83LO, SD31]
2E1 DWD L R Dominy, 26 Sweetmans Rd, Shaftesbury, SP7 8EH [IO81VA, ST82]
2E1 DWF J H Mansfield, 14 Rowan Way, Langford, Bristol, BS18 7HE [IO81OI, ST46]
2M1 DWG E C Anthony, 26 Fishers Green, Bridge of Allan, Stirling, FK9 4PU [IO86AD, NS79]
2E1 DWK Details withheld at licensee's request by SSL.
2E1 DWM J Bush, 19 Fanns Rise, Purfleet, RM19 1GP [JO01CL, TQ57]
2E1 DWW Details withheld at licensee's request by SSL.
2E1 DWY Details withheld at licensee's request by SSL.
2E1 DWZ J Bradbury, 276 College St., Long Eaton, Nottingham, NG10 4GW [IO92IV, SK43]
2E1 DXB I R Humberstone, 20 Kingswood Rd, Colchester, CO4 5JX [JO01KV, TL92]
2E1 DXF Details withheld at licensee's request by SSL.
2E1 DXK A M D Bures, 80 Clifford Rd, Barnet, EN5 5NY [IO91VP, TQ29]
2E1 DXL Details withheld at licensee's request by SSL.
2E1 DXP Details withheld at licensee's request by SSL.
2E1 DXV A M Powers, Sheringham House, Elm Rd, Evesham, Worcs, WR11 5DL [IO92AC, SP04]
2E1 DXW D W Anger, Plantation, Mill St., Polstead, Colchester, CO6 5AD [JO02KA, TL93]
2E1 DYB Details withheld at licensee's request by SSL.
2E1 DYC M I Reid, 41 Allhallowgate, Ripon, HG4 1LF [IO94FD, SE37]
2E1 DYF R Firth, 8 Lyndale Ave, York, YO1 3QB [IO93LW, SE65]
2E1 DYG D Firth, 8 Lyndale Ave, York, YO1 3QB [IO93LW, SE65]
2E1 DYI Details withheld at licensee's request by SSL.
2E1 DYM D G Sznober, 7 Botany Bay Cl, Telford, TF4 3RJ [IO82SP, SJ60]
2E1 DYN W J Sznober, 7 Botany Bay Cl, Telford, TF4 3RJ [IO82SP, SJ60]
2E1 DYO Details withheld at licensee's request by SSL.
2E1 DYR Details withheld at licensee's request by SSL.
2E1 DYS K A Eastman, 23 Haughgate Cl, Woodbridge, IP12 1LQ [JO02PC, TM25]
2E1 DYT C T Block, 4 Christchurch Dr, Woodbridge, IP12 4TJ [JO02PC, TM24]
2E1 DYU E Poppel, 23 Pembroke Ave, Woodbridge, IP12 4JB [JO02PC, TM24]
2E1 DYV H F D Eastman, 23 Haughgate Cl, Woodbridge, IP12 1LQ [JO02PC, TM25]
2E1 DZA Details withheld at licensee's request by SSL.
2E1 DZC P R Thompson, 28 The Common, Quarndon, Derby, DE22 5JY [IO92FX, SK34]
2E1 DZD S Marshall, 15 Searby Rd, Sutton in Ashfield, NG17 5JQ [IO93JC, SK55]
2E1 DZE Details withheld at licensee's request by SSL.
2E1 DZF H R Bennet, 48 Fairway, Copthorne, Crawley, RH10 3QA [IO91WD, TQ33]
2E1 DZG J M Sanderson, 393 Chesterfield Rd North, Pleasley, Mansfield, NG19 7RA [IO93JD, SK56]
2E1 DZH J L Richards, Little Piece, Stocks Mead, Washington, Pulborough, RH20 4AU [IO90TV, TQ11]
2E1 DZI M A Bures, 80 Clifford Rd, Barnet, EN5 5NY [IO91VP, TQ29]
2E1 DZJ R A L Morris, 7 Chapmans Cl, Stirchley, Telford, TF3 1ED [IO82SP, SJ60]
2E1 DZK S J Kirby, 2 Kneeton Park, Middleton Tyas, Richmond, DL10 6SB [IO94EK, NZ20]
2E1 DZL R K Neville, 1 Bethel Ave, Blackpool, FY2 9NA [IO83LU, SD33]
2M1 DZM S J Murdoch, 27 Keystone Quadrant, Milngavie, Glasgow, G62 6LW [IO75UW, NS57]
2E1 DZP A M Cannon, 7 Stopham Rd, Pulborough, RH20 1DP [IO90RW, TQ01]
2E1 DZR Details withheld at licensee's request by SSL.
2M1 DZS C Clark, 3 Old Cttgs, Seton Mains, Longniddry, East Lothian, EH32 0PG [IO85MX, NT47]
2E1 DZT Details withheld at licensee's request by SSL.
2M1 DZW B Scott, 87 Hercus Loan, Musselburgh, EH21 6BA [IO85LW, NT37]
2M1 DZX R J Stuart, The Pink Lodge, Balavil, Conon Bridge, Dingwall, IV7 8AJ [IO77SN, NH55]
2E1 DZY Details withheld at licensee's request by SSL.
2E1 DZZ J R Poole, Brae Side, Snape, Bedale, North Yorks, DL8 2TQ [IO94EG, SE28]
2E1 EAA A P Cowley, 14 Monrtpelier, Quarndon, Derbyshire, DE22 5JW [IO92GX, SK34]
2E1 EAC S P Eager, 46 Forest Rd, Horsham, RH12 4HJ [IO91UB, TQ13]
2E1 EAG J R Oxley, 97 Defoe Cres, Colchester, CO4 5LQ [JO01KV, TL92]
2E1 EAH D M Daker, 13 Honeypot Rd, Brompton on Swale, Richmond, DL10 7HT [IO94EJ, SE29]
2E1 EAI S P Taylor, 5 Bexhill Ave, Timperley, Altrincham, WA15 7RT [IO83TJ, SJ78]
2E1 EAJ Details withheld at licensee's request by SSL.
2E1 EAK Y L Wood, Ashfield, New Mill Rd, Holmfirth, Huddersfield, HD7 2SQ [IO93CN, SE10]
2E1 EAM P R Beattie, 25 Springfield Cl, Buckden, St. Neots, Huntingdon, PE18 9UR [IO92UH, TL16]
2E1 EAP D P Hughes, 163 Belmont Rd, Bolton, BL1 7AW [IO83SO, SD71]
2E1 EAQ K H Hicks, 47 The Causeway, March, PE15 9NU [JO02BN, TL49]
2E1 EAR Details withheld at licensee's request by SSL.
2E1 EAS H J Southgate, 1 Rolfe Dr, Burgess Hill, RH15 0LA [IO90WW, TQ31]
2E1 EAT R Ghuman, 291 Wakefield Rd, Huddersfield, HD5 8AG [IO93CP, SE11]
2M1 EAU B C Haswell, 6 Lochlann Rd, Culloden, Inverness, IV1 2HB [IO77WL, NH74]
2E1 EAV Details withheld at licensee's request by SSL.
2E1 EAW M J Barnett, Fairview Villa, 11 Ridge St., Stourbridge, DY8 4QF [IO82VL, SO88]
2E1 EAX R J Edwards, Manor Cottage, Manor Rd, Elmsett, Ipswich, IP7 6PN [JO02LC, TM04]
2E1 EAY J R Hall, 58 Lower Meadow Ct, Thorplands, Northampton, NN3 8AX [IO92NG, SP76]
2E1 EAZ D Finch, 30 St. Pauls Ave, Nottingham, NG7 5EB [IO92JX, SK54]
2E1 EBA Details withheld at licensee's request by SSL.
2E1 EBB Details withheld at licensee's request by SSL.
2W1 EBC S L Faraday, Hillcrest, Glascoed, Pontypool, Gwent, NP4 0UB [IO81MQ, SO30]
2W1 EBD J M Faraday, Hillcrest, Glascoed, Pontypool, Gwent, NP4 0UB [IO81MQ, SO30]
2E1 EBE Details withheld at licensee's request by SSL.
2E1 EBF G P Weaver, 17 Grasmere Ave, Harpenden, AL5 5PT [IO91TT, TL11]
2E1 EBG Details withheld at licensee's request by SSL.
2E1 EBH P Polley, 5 Ct Rd, Weymouth, DT3 5DQ [IO80SP, SY68]
2M1 EBJ D Martin, 117 Drumossie Ave, Inverness, IV2 3SQ [IO77VL, NH64]
2E1 EBK B D Dexter, 43 Anglers Ln, Spondon, Derby, DE21 7NT [IO92HV, SK43]
2E1 EBL K P Dignall, 19 Hilary Cl, Widnes, WA8 3HT [IO83PJ, SJ58]
2E1 EBM A J Morris, 41 Cumberland Rd, Sale, M33 3QT [IO83UJ, SJ79]
2E1 EBN S J Valvona, Flat, 5 Somerville, East Hill Rd, Ryde, PO33 1LU [IO90RK, SZ69]
2E1 EBO C G Branch, 50 California Rd, Mistley, Manningtree, CO11 1JQ [JO01NW, TM13]
2E1 EBR C M Smith, 142 Victoria St., Grantham, NG31 7BW [IO92QV, SK93]
2E1 EBS D Dexter, 43 Anglers Ln, Spondon, Derby, DE21 7NT [IO92HV, SK43]
2E1 EBV P H Dyson, 54 Woodlands Dr, Skelmanthorpe, Huddersfield, HD8 9DB [IO93EO, SE21]
2E1 EBX L S Collinson, 12 Victoria Ave, Hunstanton, PE36 6BX [JO02FW, TF64]
2E1 EBZ M P Watkins, 10 Rowan Dr, Billingshurst, RH14 9NF [IO91SA, TQ02]
2E1 ECA R G Parkinson, 79 Redford Ave, Horsham, RH12 2HW [IO91UB, TQ13]
2E1 ECB G D Long, 2 Riversmeade, Leigh, WN7 1JA [IO83RM, SD60]
2M1 ECF J Mackenzie, Hazelgrove, Inverfarigaig, Inverness, IV1 2XR [IO77SG, NH52]
2E1 ECG A J Topping, 30 St. Pauls Ave, Nottingham, NG7 5EB [IO92JX, SK54]
2J1 ECH F C Whittaker, Coeur Joyeux, La Rue Des Sapins, St. Peter, Jersey, JE3 7AD
2E1 ECI Details withheld at licensee's request by SSL.
2M1 ECJ P S Thackery, Leachkin Lodge, Upper Leachkin, Inverness, IV3 6PN [IO77UL, NH64]
2E1 ECL C Gray, 19 Marsh View, Newton, Preston, PR4 3SX [IO83NS, SD43]
2E1 ECM P S Fletcher, 171 Obelisk Rise, Northampton, NN2 8TX [IO92NG, SP76]
2E1 ECN M T Chapman, 15 Norwood Rd, Somersham, Huntingdon, PE17 3EY [JO02AJ, TL37]
2E1 ECQ Details withheld at licensee's request by SSL.
2E1 ECR C Gregson, 11 Coupe Green, Hoghton, Preston, PR5 0JR [IO83QR, SD52]
2E1 ECV P J Elsey, 129 Kingsway, Chandlers Ford, Eastleigh, SO53 5BX [IO90HX, SU42]
2E1 ECW R Starkie, 2 Oak Hill Cl, Wigan, WN1 2QL [IO83QN, SD50]
2E1 ECX R D Keay, 16 Willoughby Cl, Penkridge, Stafford, ST19 5QT [IO82WR, SJ91]
2E1 ECY S R Cowley, 14 Montpelier Quarndon, Derby, Derbyshire, DE22 5JW [IO92GX, SK34]
2W1 EDA R E Kurtz, 5 Tenby Ct, Caerphilly, Mid Glam, CF83 2UE [IO81JN, ST18]
2E1 EDB M Bradwell, 6 Moorfoot Gdns, Gateshead, NE11 9LA [IO94EW, NZ26]
2M1 EDC J P McFadden, 208 Carmunnock Rd, Glasgow, G44 5AP [IO75VT, NS56]
2E1 EDD S Lansdell, 42 Marylebone Cres, Derby, DE22 4JX [IO92FW, SK33]
2E1 EDE R E Williams, 8 Marjorie Rd, Chaddesden, Derby, DE21 4HQ [IO92GW, SK33]
2W1 EDF Details withheld at licensee's request by SSL.
2W1 EDJ F H Parry, The Bungalow, Hillside Country Club, Capel Hill, Tonyrefail, Porth, CF39 9YU [IO81GO, ST09]
2E1 EDK D P Watkins, Dolauhirion, Cilycwm Rd, Llandovery, Dyfed, SA20 0TU [IO82CA, SN73]
2W1 EDL M J Churcher, 71 Twyn Rd, Abercarn, Newport, NP1 5JY [IO81KP, ST29]
2M1 EDM D Martin, 7 Eddington Dr, Newton Mearns, Glasgow, G77 5AX [IO75TS, NS55]
2E1 EDO Details withheld at licensee's request by SSL.
2E1 EDP C K Affleck, 36 Kings Rd, Horsham, RH13 5PR [IO91UA, TQ22]
2E1 EDQ P A Mackie, 33 Mentmore Gdns, Appleton, Warrington, WA4 3HF [IO83RI, SJ68]
2J1 EDR A R Price, Chataignier House, Rue de La Croix, St Owen, Jersey, Channel Islands, JE3 2HA
2M1 EDT J Wilson, 4F Langside St., Clydebank, G81 5HJ [IO75TW, NS57]
2E1 EDU R H Felds, 93 Bancroft Ln, Mansfield, NG18 5LL [IO93JD, SK56]
2E1 EDV J E Blanche, 31 Gandalfs Ride, South Woodham Ferrers, Chelmsford, CM3 5WX [JO01HP, TQ89]
2E1 EDW J C Blanche, 31 Gandalfs Ride, South Woodham Ferrers, Chelmsford, CM3 5WX [JO01HP, TQ89]
2E1 EDX R C Earp, 4 Buriton Rd, Winchester, SO22 6HX [IO91HC, SU43]
2E1 EDY B J Pitty, 12 St. Leonards Rd, Horsham, RH13 6EJ [IO91UB, TQ12]
2E1 EDZ Details withheld at licensee's request by SSL.
2E1 EEB Details withheld at licensee's request by SSL.
2E1 EED C A Smith, 24 Fron Uchaf, Colwyn Bay, Clwyd, LL29 6DS [IO83DG, SH87]
2E1 EEF I D Houghton, 14 Windfield Gdns, Little Sutton, South Wirral, L66 1JJ [IO83MG, SJ37]
2E1 EEG H T Rhodes, 38 Whitmore Dr, Ribbleton, Preston, PR2 6LA [IO83QS, SD53]
2E1 EEI Details withheld at licensee's request by SSL.
2E1 EEK S P Paffett, 14 Western Rd, Chandlers Ford, Eastleigh, SO53 5DA [IO90HX, SU42]
2E1 EEM D I Peacock, 35 Heathfield Dr, Tyldesley, Manchester, M29 8PW [IO83SM, SD70]
2W1 EEN C Cater, 40 Frances Ave, Wrexham, LL12 8BN [IO83MB, SJ35]
2W1 EEP J Jones, 10 Wernfadog, Pontfadog, Llangollen, Clwyd, LL20 7AR [IO82KW, SJ23]
2E1 EER Details withheld at licensee's request by SSL.
2E1 EET L Edwards, 39 Foskitt Ct, Northampton, NN3 9AX [IO92OF, SP86]
2E1 EEU M C Neil, 208 Stonelow Rd, Dronfield, Sheffield, S18 6ER [IO93GH, SK37]
2E1 EEV Details withheld at licensee's request by SSL.
2E1 EEW P J Brechany, 24 Teignmouth Ave, Mansfield, NG18 3JQ [IO93KD, SK56]
2E1 EEX A P Cree, 58 Lime Gr, Forest Town, Mansfield, NG19 0HP [IO93KD, SK56]
2E1 EEY Details withheld at licensee's request by SSL.
2E1 EEZ J M Balfe, Rift Valley, Townhouse Rd, Costessey, Norwich, NR8 5BX [JO02OP, TG11]
2E1 EFA S J Balfe, Rift Valley, Townhouse Rd, Costessey, Norwich, NR8 5BX [JO02OP, TG11]
2E1 EFB Details withheld at licensee's request by SSL.
2E1 EFC Details withheld at licensee's request by SSL.[letter 96]
2E1 EFD C B Brown, Highbury, Carrs Hill, Badingham, Woodbridge, IP13 8NE [JO02QG, TM36]
2E1 EFF Details withheld at licensee's request by SSL.
2E1 EFG S P Philpot, 93 Princess Dr, Grantham, NG31 9QA [IO92QW, SK93]
2E1 EFH Details withheld at licensee's request by SSL.
2E1 EFI R Birch, 1 Paignton Ave, Portsmouth, PO3 6LL [IO90LT, SU60]
2E1 EFJ Details withheld at licensee's request by SSL.
2E1 EFK P G Tinkler, 27 Cavendish Dr, Carlton, Nottingham, NG4 3DX [IO92KX, SK64]
2E1 EFL J D Rayson, 127 Brewsters Rd, St. Anns, Nottingham, NG3 3BY [IO92KX, SK54]
2E1 EFQ D J Bushby, 66 Sandy Rd, Everton, Sandy, SG19 2JU [IO92VD, TL25]
2E1 EFR E Gudgin, 22 Clifton Rd, Shefford, SG17 5AE [IO92UA, TL13]
2E1 EFS C L Hill, 9 Pine Rd, Glenfield, Leicester, LE3 8DH [IO92JP, SK50]
2E1 EFT L Froggatt, 255 Rushton Rd, Desborough, Kettering, NN14 2QB [IO92OK, SP88]
2E1 EFU Details withheld at licensee's request by SSL.
2E1 EFX T J Laundon, 55 Parkway, Eastbourne, BN20 9DY [JO00DT, TQ50]
2E1 EFY P J Prior, 17 Layton Ave, Malvern, WR14 2ND [IO82UC, SO74]
2E1 EFZ Details withheld at licensee's request by SSL.
2E1 EGA B M Holyoake, Oakebury Cottage, Churches Green, Dallington, Heathfield, TN21 9NX [JO00EW, TQ61]
2E1 EGC A G Fox, 31 Pierson Rd, Windsor, SL4 5RE [IO91QL, SU97]
2E1 EGD C J Bartlett, Mount Pleasant, Ridgeway Cross, Cradley, Malvern, WR13 5JD [IO82TC, SO74]
2E1 EGE E P R Orr, The Old Corner House, 43 High St., Westham, Pevensey, BN24 5LJ [JO00DT, TQ60]
2E1 EGF T W E Davies, 15 Hoo Rd, Meppershall, Shefford, SG17 5LP [IO92UA, TL13]
2E1 EGG J T Westlake, 33 Querns Rd, Canterbury, CT1 1PX [JO01NG, TR15]
2E1 EGH M P Moore, 164 Ardenlee Ave, Belfast, BT6 0AE [IO74BN, J37]
2E1 EGI G C Durrant, 51 Raglan Ave, Waltham Cross, EN8 8DA [IO91XQ, TL30]
2E1 EGJ S J Braithwaite, 33 Somers Park Ave, Malvern, WR14 1SE [IO82UD, SO74]
2E1 EGL I P Thetford, 5 Wellington Pl, Albion Rd, Great Yarmouth, NR30 2HS [JO02UO, TG50]
2E1 EGM G W D Steele, 77 Ashtree Rd, Frome, BA11 2SE [IO81UF, ST74]
2E1 EGN N S Ralph, 12 Chestnut Cl, Great Waldinfield, Great Waldingfield, Sudbury, CO10 0RU [JO02JB, TL94]
2E1 EGO Details withheld at licensee's request by SSL.
2E1 EGQ Details withheld at licensee's request by SSL.
2E1 EGR N J Sponer, 27 Romsley Cl, Halesowen, B63 3DP [IO82XK, SO98]
2E1 EGS J C White, Pathways Down, Barton Rd, St Nicholas At Wade, Birchington, Kent, CT7 0PY [JO01OI, TR26]
2E1 EGT J D Strand, 129 Malmesbury Rd, Chippenham, SN15 1PZ [IO81WL, ST97]
2E1 EGU W R Lupton, Eaton Cottage, Shorts Green Ln, Motcombe, Shaftesbury, SP7 9PA [IO81VA, ST82]
2E1 EGV I Townson, 53 Brompton Rd, Bradford, BD4 7JD [IO93DS, SE13]
2E1 EGW M W C Robinson, 5 Elmley Cl, Malvern, WR14 2QT [IO82UC, SO74]
2E1 EGY W J Holland, 2 Gate Farm Cttgs, Elmsett, Ipswich, IP7 6NX [JO02MB, TM04]
2E1 EHA J T Smith, 25 Poyntz Rd, Overton, Basingstoke, RG25 3HJ [IO91IF, SU54]
2E1 EHB I T Greenall, 356 Warrington Rd, Abram, Wigan, WN2 5XA [IO83QM, SD60]
2E1 EHC D Greenall, 356 Warrington Rd, Abram, Wigan, WN2 5XA [IO83QM, SD60]
2E1 EHD D J Kingshott, 17 Ellington Dr, Basingstoke, RG22 4EZ [IO91KF, SU64]
2E1 EHE I A Thornton, Four Winds, Glenmore Rd East, Crowborough, TN6 1RE [JO01BB, TQ53]
2E1 EHF J P Goodman, 21 Loxley Rd, Smethwick, Warley, B67 5BL [IO92AL, SP08]
2E1 EHG P F Goodman, 21 Loxley Rd, Smethwick, Warley, B67 5BL [IO92AL, SP08]
2E1 EHH A F Goodman, 21 Loxley Rd, Bearwood, Smethwick, Warley, B67 5BL [IO92AL, SP08]
2E1 EHI P M Trickey, 59 Shelley Dr, Sutton Coldfield, B74 4YD [IO92BO, SK10]
2E1 EHJ D Van'T Riet, 65 Selangor Ave, Emsworth, PO10 7LR [IO90MU, SU70]
2E1 EHL E K Jones, 23 Cotswold Way, Risca, Newport, NP1 6QT [IO81LO, ST29]
2E1 EHM D P Whittaker, 68 Querns Rd, Canterbury, CT1 1PZ [JO01NG, TR15]
2E1 EHP D A Moses, 121 Badger Ave, Crewe, CW1 3JN [IO83SC, SJ65]
2E1 EHQ B Fure, 28 Bonner Hill Rd, Kingston upon Thames, KT1 3HE [IO91UJ, TQ16]
2E1 EHR M A Blackwell, 5 Tollgate Rd, Culham, Abingdon, OX14 4NL [IO91IP, SU59]
2E1 EHS J E Megone, 16 Mercer Cl, Basingstoke, RG22 6NZ [IO91KG, SU65]
2E1 EHT S C Burtenshaw, Rivendell, Yapton Rd, Barnham, Bognor Regis, PO22 0BA [IO90QT, SU90]
2E1 EHV P H Wells, 7 Kings Meadow, Overton, Basingstoke, RG25 3HP [IO91IF, SU54]
2E1 EHW Z E Sheffield, 167 Hawkswell Dr, Willenhall, WV13 3EL [IO82XN, SO99]
2E1 EHX J D Riley, The Rafters, Doncaster Rd, Wragby, Wakefield, WF4 1QX [IO93HP, SE41]
2E1 EHY N A Waters, 15 Laureate Cl, Margate, CT9 2TJ [JO01QJ, TR37]
2E1 EHZ J P R Burn, 9 Birbeck Way, Frettenham, Norwich, NR12 7LG [JO02PR, TG21]
2E1 EIA Details withheld at licensee's request by SSL.
2E1 EIB T P Woolrych, 20 Meadow Dr, Devizes, SN10 3BJ [IO91AI, SU06]
2E1 EIC M D Ross, 3 Little Ln, Clophill, Bedford, MK45 4BG [IO92TA, TL03]
2W1 EID R J Owens, 8 Lambourne Walk, Bettws, Newport, NP9 6UB [IO81LO, ST29]
2E1 EIE S A Lambert, 66 Horsley Rd, Kilburn, Belper, DE56 0NE [IO93GA, SK34]
2E1 EIF Details withheld at licensee's request by SSL.
2E1 EIG C T Pearse, 10 Stanley Dr, Farnborough, Hants, GU14 0PL [IO91OG, SU85]
2E1 EIH L P Pilkington, Purbeck, 10 Stanley Dr, Farnborough, GU14 0PL [IO91OG, SU85]
2E1 EII B Kearns, 37 High St., Sutton Benger, Chippenham, SN15 4RQ [IO81XM, ST97]
2E1 EIJ D A Veale, 5 Heathfield Cl, Dronfield, Sheffield, S18 6RJ [IO93GH, SK37]
2E1 EIK N F Smith, 20 The Glebe, Cossall, Nottingham, NG16 2SG [IO92IX, SK44]
2E1 EIL Details withheld at licensee's request by SSL.
2W1 EIN C G D Bodley, 34 Claremont Rd, Newbridge, Newport, NP1 5DL [IO81KQ, ST29]
2E1 EIO Details withheld at licensee's request by SSL.
2E1 EIP L Smith, 45 Station Rd, Harpenden, AL5 4XE [IO91TT, TL11]
2E1 EIR G H Davis, 3 Norton Cl, Southwick, Fareham, PO17 6HD [IO90KU, SU60]
2E1 EIS Details withheld at licensee's request by SSL.
2E1 EIU D V Meddings, 42A Argyle Rd, Poulton-le-Fylde, FY6 7EW [IO83MU, SD33]
2E1 EIV A R Eyre, St. Michael Mead, The Common, Barton Turf, Norwich, NR12 8BA [JO02RR, TG32]
2E1 EIW A Howard, 23 Vicarage Cl, Shillington, Hitchin, SG5 3LS [IO91TX, TL13]
2E1 EIX A Chruscinski, 104 Brick Kiln Ln, Mansfield, NG18 5JT [IO93JD, SK56]
2E1 EIY G W Milner-Smith, 44 Tomline Rd, Felixstowe, IP11 7PA [JO01QX, TM33]
2E1 EJC T E Bain, 23 Salisbury Cres, Blandford Forum, DT11 7LX [IO80WU, ST80]
2E1 EJD W F Ling, Valley Farm Equestrian Centre, Wickham Market, Woodbridge, Suffolk, IP13 0ND [JO02QD, TM25]
2E1 EJE I Clarke, 11 Hartland Dr, Birtley, Chester-le-Street, DH3 2LZ [IO94FV, NZ25]
2U1 EJF A R Scheffer, Keukenhof, Route de Carteret, Castel, Guernsey, GY5 7YS
2M1 EJI R A Lynch, 21 Carnoustie Ave, Gourock, PA19 1HF [IO75NW, NS27]
2M1 EJK L Lewis, 181 Kent Dr, Helensburgh, G84 9RX [IO76PA, NS38]
2E1 EJM G Alcock, 8 Ivel Rd, Sandy, SG19 1AX [IO92UD, TL14]
2E1 EJO C J Short, 92 Livermore Green, Peterborough, PE4 5DQ [IO92UO, TF10]
2E1 EJP Details withheld at licensee's request by SSL.
2E1 EJR S K Watson, 6 Chatsworth Pl, Peterborough, PE3 9NP [IO92UN, TL19]
2E1 EJU D B Lumley, 3 Ribble Cl, St. Ives, Huntingdon, PE17 6HU [IO92XI, TL37]
2E1 EJV Details withheld at licensee's request by SSL.
2E1 EJX M R Tullett, 50 Beaconsfield Rd, Burton on Trent, DE13 0NP [IO92ET, SK22]
2M1 EJY D B Smith, 13 Clerics Hill, Kirkliston, EH29 9DP [IO85HW, NT17]
2E1 EKA Details withheld at licensee's request by SSL.
2E1 EKC D Donnelly, 15 Antwerp Rd, Sunderland, SR3 3JQ [IO94GV, NZ35]
2E1 EKD M J Finch, 4 Fountain Dale Ct, Nottingham, NG3 4QT [IO92KX, SK54]
2U1 EKE C D Ayres, Rousay, Bailiffs Cross Rd, St. Andrew, Guernsey, GY6 8RY
2E1 EKF D A I Rouse, 141 High St., Garlinge, Margate, CT9 5LY [JO01QJ, TR36]

2E1 EKG S E Lincoln, Providence Cottage, Gt Glemham, Saxmundham, Suffolk, IP17 2DN [JO02RE, TM36]
2U1 EKH K B Johnson, 14 Les Genats, Cobo, Castel, Guernsey, Channel Islands, GY5 7YQ
2M1 EKI K Simpson, 53 Jedworth Ave, Glasgow, G15 7QE [IO75TV, NS57]
2E1 EKJ P A Holt, The Dovecote, 29 High St., Haslingfield, Cambridge, CB3 7JW [JO02AD, TL45]
2E1 EKL D R Tudor, 13 Woodlands Rd, Stalybridge, SK15 2SG [IO83XL, SJ99]
2E1 EKM D Baldwin, 51 Queens Rd, Broadstairs, CT10 1PG [JO01RI, TR36]
2E1 EKN P T Baldwin, 51 Queens Rd, Broadstairs, CT10 1PG [JO01RI, TR36]
2E1 EKQ M S Eades, 5 Mallard Cl, Skellingthorpe, Lincoln, LN6 5SE [IO93QF, SK97]
2W1 EKR D R Pollard, 19 The Dr, Bargoed, CF81 8JX [IO81JQ, ST19]
2E1 EKU D H Stroud, 9 Gloucester Rd, Guildford, GU2 6TG [IO91QG, SU95]
2W1 EKW Details withheld at licensee's request by SSL.
2W1 EKX Details withheld at licensee's request by SSL.
2E1 EKY K L Palmer, Warwick School, Myton Rd, Warwick, CV34 6PP [IO92FG, SP26]
2E1 ELA Details withheld at licensee's request by SSL.
2E1 ELB Details withheld at licensee's request by SSL.
2E1 ELC Details withheld at licensee's request by SSL.
2E1 ELE P J Williams, 20 Elm Cl, Great Haywood, Stafford, ST18 0SP [IO82XT, SJ92]
2W1 ELF R S Williams, 43 St. Margarets Rd, Whitchurch, Cardiff, CF4 7AB [IO81JM, ST18]
2E1 ELH D J Green, The Abbey, Warwick Rd, Southam, Warks, CV33 0HN [IO92HG, SP46]
2E1 ELI D J Liddicoat, 45 Randall Rd, Chandlers Ford, Eastleigh, SO53 5AJ [IO91HA, SU42]
2E1 ELJ P J Hoare, 81 Sheepfold Rd, Guildford, GU2 6TU [IO91QG, SU95]
2E1 ELK Details withheld at licensee's request by SSL.
2E1 ELL R Appleby, 3 St. Judes Way, Burton on Trent, DE13 0LR [IO92ET, SK22]
2E1 ELM G W D Appleby, 3 St. Judes Way, Burton on Trent, DE13 0LR [IO92ET, SK22]
2U1 ELN S P Johnson, Killarney, Vale Rd, St. Sampson, Guernsey, GY2 4DN
2E1 ELO M Cornell, 22 Ravine Rd, Bournemouth, BH5 2DU [IO90CR, SZ19]
2E1 ELP Details withheld at licensee's request by SSL.
2E1 ELQ M L Bentley, 55 Stirling Ct Rd, Burgess Hill, RH15 0PS [IO90WX, TQ31]
2E1 ELR Details withheld at licensee's request by SSL.
2E1 ELS Details withheld at licensee's request by SSL.
2E1 ELT C R Worbey, 61 Stotfold Rd, Hitchin, SG4 0QW [IO91VX, TL23]
2M1 ELU I Sinclair, Airdanair, Kilchrenan, Taynuilt, Argyll, PA35 1HG [IO76JI, NN02]
2E1 ELV G R Watson, Harvest Cottage, West St., Stoke Sub Hamdon, Somerset, TA14 6PZ [IO80OW, ST41]
2E1 ELX Details withheld at licensee's request by SSL.
2E1 ELY Details withheld at licensee's request by SSL.
2E1 ELZ A B Siddall, 6 Delside Ave, Manchester, M40 9LF [IO83VM, SD80]
2E1 EMB Details withheld at licensee's request by SSL.
2E1 EMD A J Ore, 25 Melton Rd, Wymondham, NR18 0DB [JO02NN, TG10]
2E1 EME N M Ore, 25 Melton Rd, Wymondham, NR18 0DB [JO02NN, TG10]
2E1 EMF A G Speakman, 12 Allerton Ave, Leeds, LS17 6RF [IO93FU, SE33]
2E1 EMG D Rust, 34 Marlingford Way, Easton, Norwich, NR9 5HB [JO02NP, TG11]
2E1 EMH V Newton, 60 The Lynch, Winscombe, BS25 1AR [IO81OH, ST45]
2E1 EMI G M Paterson, 4 Rowallan Dr, Bedford, MK41 8AW [IO92SD, TL05]
2E1 EMJ I D Hydes, 2 Stable Ct, Martlesham Heath, Ipswich, IP5 7UQ [JO02OB, TM24]
2E1 EMK J C Roff, 6 Canal Cl, Wilcott, Wilcot, Pewsey, SN9 5NW [IO91CI, SU16]
2E1 EML G P James, 3 Lathom Rd, Irlam, Manchester, M44 6ZD [IO83SK, SJ79]
2E1 EMM Details withheld at licensee's request by SSL.
2E1 EMN J L Thompson, 97 King Ln, Leeds, LS17 5AX [IO93FU, SE33]
2E1 EMO Details withheld at licensee's request by SSL.
2E1 EMP T R Draycott, 32 Irwin Dr, Nottingham, NG6 7BH [IO92JX, SK54]
2E1 EMQ T Draycott, 32 Irwin Dr, Nottingham, NG6 7BH [IO92JX, SK54]
2E1 EMR A J Watson, The Spinney, Youngsbury Ln, Wadesmill, Ware, SG12 0TX [IO91XU, TL31]
2E1 EMS E J Jones, 47 Pine Cres, Chandlers Ford, Eastleigh, SO53 1LN [IO90HX, SU42]
2E1 EMT Details withheld at licensee's request by SSL.
2E1 EMU Details withheld at licensee's request by SSL.
2E1 EMV Details withheld at licensee's request by SSL.
2E1 EMW T J Gale, 33 Watson Cl, Upavon, Pewsey, SN9 6AF [IO91CH, SU15]
2E1 EMX K M Messenger, Alga Lodge, Alga Terr, Scarborough, North Yorks, YO11 2DG [IO94TG, TA08]
2E1 EMY Details withheld at licensee's request by SSL.
2E1 ENA I Clark, 146 York Rd, Haxby, York, YO3 3EL [IO94LA, SE65]
2E1 ENB D Wallis, 12 Fitzroy Dr, Leeds, LS8 1RW [IO93FT, SE33]
2E1 ENC Details withheld at licensee's request by SSL.
2E1 END R J Gale, 9 Spruce, Tamworth, B77 4ES [IO92EP, SK20]
2E1 ENE A Dykes, 149 Mayfield Rd, Chaddesden, Derby, DE21 6FZ [IO92GW, SK33]
2E1 ENF Details withheld at licensee's request by SSL.
2E1 ENH Details withheld at licensee's request by SSL.
2M1 ENI D R Paterson, Leuchlands Croft, Whitecairns, Aberdeen, Aberdeenshire, AB23 8UT [IO87WF, NJ91]
2M1 ENK D Paterson, Leuchlands Croft, Whitecairns, Aberdeen, Aberdeenshire, AB23 8UT [IO87WF, NJ91]
2E1 ENL J Bell, 122 Howard Dr, Letchworth, SG6 2DE [IO91VX, TL23]
2E1 ENM R G Morris, 14 Storrs Hill Rd, Ossett, WF5 0DL [IO93FQ, SE21]
2E1 ENN G E Cattle, 50 Oakland Ave, York, YO3 0DF [IO93LX, SE65]
2E1 ENO C Sutton, 32 Queensway, Euxton, Chorley, PR7 6PW [IO83QP, SD51]
2E1 ENP J B Langford, 35 Wellington Gdns, Selsey, Chichester, PO20 0RF [IO90OR, SZ89]
2E1 ENQ P B Crook, 1 Hillpark Ave, Fulwood, Preston, PR2 3QQ [IO83PS, SD53]
2E1 ENR S Taylor, Flat 1, 17A Commercial St., Batley, WF17 5HJ [IO93ER, SE22]
2E1 ENS Details withheld at licensee's request by SSL.
2E1 ENU A Thomson, 15 Maplewood, Newcastle upon Tyne, NE6 4NP [IO94FX, NZ26]
2E1 ENW J E Hobson, 11 Lime Ave, Swinton, Manchester, M27 0GF [IO83UM, SD70]
2E1 ENX W A Thomas, 11 Arran Green, Rochester, ME2 2ND [JO01FJ, TQ76]
2E1 ENY R H Chenery, 2 Hazelton Green, Stafford, ST17 9NH [IO82WS, SJ92]
2E1 ENZ R E Baxter, 16 Mortomley Cl, High Green, Sheffield, S30 4HZ [IO93GJ, SK38]
2E1 EOA L Knight, 60 Leys Rd, Hemel Hempstead, HP3 9LE [IO91SR, TL00]
2E1 EOB R J Whalley, Runways Farm, Bourne End Ln, Hemel Hempstead, HP1 2RR [IO91RR, TL00]
2E1 EOC A J Bridges, Acomb, Station Rd, Eckington, Pershore, WR10 3BB [IO82WB, SO94]
2E1 EOD N J Bridges, Acomb, Station Rd, Eckington, Pershore, WR10 3BB [IO82WB, SO94]
2E1 EOE D M Clarke, 31 Ashfield Rd, Huthwaite, Sutton in Ashfield, NG17 2NX [IO93ID, SK45]
2E1 EOG Details withheld at licensee's request by SSL.
2W1 EOH Details withheld at licensee's request by SSL.
2E1 EOI J M Allum, 122 Long Chaulden, Hemel Hempstead, HP1 2HY [IO91RS, TL00]
2E1 EOK C M Blackman, 131A London Rd, Hemel Hempstead, Herts, HP3 9SQ [IO91SR, TL00]
2E1 EOL S Carrington, 135 Richmond Park Rd, Bournemouth, BH8 8UA [IO90BR, SZ19]
2E1 EOM Details withheld at licensee's request by SSL.
2E1 EON A F Knowles, Blackthorns, Treskinnick Cross, Poundstock, Bude, EX23 0DT [IO70RS, SX29]
2E1 EOO D G Webb, 31 Ridding Cl, Shirley, Southampton, SO15 5PJ [IO90GW, SU41]
2E1 EOP R F Lickman, 192 Redbridge Hill, Southampton, SO16 4LZ [IO90GW, SU31]
2E1 EOQ D G Horton, Glen View, New Rd, Bush, Bude, EX23 [IO70RT, SS20]
2E1 EOR K J Horton, Glen View, New Rd, Bush, Bude, EX23 9LE [IO70RU, SS20]
2E1 EOS J P Titterall, 12 Youngsbury Ln, Wadesmill, Ware, SG12 0TY [IO91XU, TL31]
2E1 EOT L G Taylor, 18 Folly Ln, North Crawley, Newport Pagnell, MK16 9LW [IO92QC, SP94]
2E1 EOU S R Turner, 36A West End, Whittlesey, Peterborough, PE7 1LS [IO92WN, TL29]
2M1 EOV M Macleod, 4 Portnaguran, Isle of Lewis, Scotland, HS2 0HD [IO68WG, NB53]
2E1 EOW Details withheld at licensee's request by SSL.
2E1 EOX M J Homer, 63 Forest Ln, Harrogate, HG2 7HB [IO94GA, SE35]
2E1 EOY Details withheld at licensee's request by SSL.
2E1 EOZ R A Cannon, 148 Whitmore Way, Basildon, SS14 2PA [JO01FN, TQ78]
2E1 EPA P Allaker, 3 Eden Cttgs, Watling St., Consett, DH8 6HZ [IO94CU, NZ15]
2E1 EPC R A Hoban, 3 Lake Lock Gr, Stanley, Wakefield, WF3 4JJ [IO93GR, SE32]
2E1 EPD M B Freedman, Rivermeade, Irwell Vale, Ramsbottom, Bury, BL0 0QA [IO83UQ, SD72]
2E1 EPE P D M Freedman, Rivermeade, Irwell Vale, Ramsbottom, Bury, BL0 0QA [IO83UQ, SD72]
2E1 EPF M L Cogan, 6 Birch Cl, Locking, Weston Super Mare, BS24 8BP [IO81NH, ST35]
2E1 EPG B R Winch, 52 Palm Rd, Southampton, SO16 5HF [IO90GW, SU31]
2E1 EPH D J Lingwood, 114 Onslow Gdns, Wallington, SM6 9QG [IO91WI, TQ26]
2W1 EPL G W Morris, 10 Orton Gr, Rhyl, LL18 1DG [IO83GH, SJ08]
2E1 EPM D J Edwards, 2 Intake Cl, Stanley, Wakefield, WF3 4HY [IO93GR, SE32]
2W1 EPN D E Richardson, 5 Glan y Mor, The Knap, Barry, CF62 6FF [IO81IJ, ST16]
2E1 EPP Details withheld at licensee's request by SSL.
2E1 EPQ T J Fanning, 23 Santa Maria Way, Stourport on Severn, DY13 9RX [IO82UH, SO87]
2E1 EPR Details withheld at licensee's request by SSL.
2E1 EPS R J Taylor, 7 Ashline Gr, Whittlesey, Peterborough, PE7 1DW [IO92WN, TL29]
2E1 EPT B T Leahy, 31 Harrow Rd, Carshalton, SM5 3QH [IO91VI, TQ26]
2M1 EPV D M Stewart, 19 Arthur St., Blairgowrie, PH10 6PF [IO86HO, NO14]
2E1 EPX A E Richardson, 12 Catcheside Cl, Whickham, Newcastle upon Tyne, NE16 5RX [IO94DW, NZ26]
2E1 EPY J A Young, 9 Caincross Rd, Nottingham, NG8 4AZ [IO92JX, SK54]
2E1 EPZ A J Morris, 7 The Elms, Hertford, SG13 7UY [IO91XT, TL31]
2E1 EQB C Palmer, 85 Weymouth Rd, Hayes, UB4 8NH [IO91SM, TQ08]
2E1 EQC C J Marshall, Thistledome, First Ave, Garston, Watford, WD2 6PS [IO91TQ, TQ19]
2E1 EQD E H Antonsen, 73 Manor Rd, Dawley, Telford, TF4 3EB [IO82SP, SJ60]
2E1 EQE D J Evans, 5 Compton Dr, Streetly, Sutton Coldfield, B74 2DA [IO92BN, SP09]
2E1 EQF J R J Tennant, Greengables, Trench Rd, Trench, Telford, TF2 6NU [IO82SR, SJ61]
2E1 EQG C J Wooff, 57 Alexandra Rd, Ashton in Makerfield, Wigan, WN4 8QS [IO83QL, SJ59]
2E1 EQI V A Collins, 85 Malvern Rd, Swindon, SN2 1AU [IO91CN, SU18]
2E1 EQJ D K Russell, 80 Hayward Ave, Donnington, Telford, Salop, TF2 8DD [IO82SR, SJ71]
2E1 EQK P J Gibson, 43 First Ave, Ketley Bank, Telford, TF2 0AJ [IO82SQ, SJ61]
2W1 EQM Details withheld at licensee's request by SSL.
2E1 EQN D D Reavill, 11 Clarence Rd, Beeston, Nottingham, NG9 5HY [IO92JV, SK53]
2E1 EQQ K S Wood, 52 Ashfield Ave, Beeston, Nottingham, NG9 1PY [IO92JW, SK53]
2E1 EQS R J Williams, 6 Pennine Way, Charvil, Reading, RG10 9QH [IO91NL, SU77]
2E1 EQT B Bass, 13 Albert Rd, Wellington, Telford, TF1 3AR [IO82RQ, SJ61]
2E1 EQU D Cooksey, 5 Wellswood Ave, Ketley Bank, Telford, TF2 0BD [IO82SQ, SJ60]
2E1 EQW C M Goodwin, 8 Elderberry Cl, The Rock, Telford, TF3 5EN [IO82SQ, SJ60]
2E1 EQY S R Heard, 42 Hallowell Down, South Woodham Ferrers, Chelmsford, CM3 5FS [JO01HP, TQ79]
2E1 EQZ N J Collicott, 71 Blenheim Dr, Bredon, Tewkesbury, GL20 7QQ [IO82WA, SO93]
2E1 ERA Details withheld at licensee's request by SSL.
2M1 ERB J Galloway, 144 Strathkinnes Rd, Kirkcaldy, KY2 5PZ [IO86JC, NT29]
2E1 ERD M A Pennington, 6 Salvin Cl, Ashton in Makerfield, Wigan, WN4 8XS [IO83QL, SJ59]
2E1 ERE R Draycott, 32 Irwin Dr, Nottingham, NG6 7BH [IO92JX, SK54]
2E1 ERF M Riley, 63 Clifford Rd, Ipswich, IP4 1PJ [JO02OB, TM14]
2E1 ERG T R Dobbs, 11 Plantation Cres, Bredon, Tewkesbury, GL20 7QG [IO82WA, SO93]
2E1 ERI D A Hambly, 193 The Avenue, Lowestoft, NR33 7LJ [JO02UL, TM59]
2E1 ERJ D Lusty, 104 Polstain Rd, Threemilestone, Truro, TR3 6DB [IO70KG, SW74]
2E1 ERK M D Lusty, 104 Polstain Rd, Threemilestone, Truro, TR3 6DB [IO70KG, SW74]
2E1 ERL Details withheld at licensee's request by SSL.
2E1 ERM G F Tepper, 144 Doncaster Rd, Mexborough, S64 0JW [IO93IL, SK49]
2E1 ERN M S Haynes, 34 Pear Tree Mead, Harlow, CM18 7BY [JO01BS, TL40]
2M1 ERO A J Rucklidge, Southbogs, Leslie Insch, Aberdeenshire, Scotland, AB52 6PQ [IO87QH, NJ62]
2E1 ERP J Youle, 42 Melbourne Ave, Dronfield Woodhouse, Sheffield, S18 5YW [IO93GH, SK37]
2E1 ERQ Details withheld at licensee's request by SSL.
2E1 ERR B C I Kearney, 18 Wayside Mews, Maidenhead, SL6 7EJ [IO91PM, SU88]
2E1 ERS A Swindlehurst, Mount St Marys College, Spinkhill, Sheffield, South Yorks, S31 9YL [IO93GJ, SK38]
2E1 ERT M M Reynolds, Barborough Hall School, Barlborough, Chesterfield, S43 4TJ [IO93IH, SK47]
2E1 ERU Details withheld at licensee's request by SSL.
2E1 ERW M J Bilson, 101 Birkinstyle Ln, Shirland, Alfreton, DE55 6BT [IO93HD, SK45]
2W1 ERX D H Hodges, 71 Open Hearth Cl, Griffithstown, Pontypool, NP4 5LU [IO81LQ, ST29]
2W1 ERY E J Waldron, 100 Porthmawr Rd, Cwmbran, NP44 1NB [IO81LP, ST29]
2E1 ERZ J S Waldron, 100 Porthmawr Rd, Cwmbran, NP44 1NB [IO81LP, ST29]
2E1 ESA Details withheld at licensee's request by SSL.
2E1 ESB Details withheld at licensee's request by SSL.
2E1 ESC R F Marsden, 26 Lincombe Rise, Leeds, LS8 1QH [IO93FT, SE33]
2E1 ESD G J M Tench, Bracken Lodge, Brook Ln, Brocton, Stafford, ST17 0TZ [IO82XS, SJ91]
2E1 ESE N S Garland, 3 Penwood Heights, Penwood, Burghclere, Newbury, RG20 9EY [IO91HI, SU46]
2E1 ESH P R James, 36 Lemon Hill, Mylor Bridge, Falmouth, TR11 5NA [IO70LE, SW83]
2E1 ESJ N Evans, 38 Cockster Rd, Stoke on Trent, Staffs, ST3 2EG [IO82WX, SJ84]
2E1 ESK L P C Hodge, 11 Glebelands, Bampton, OX18 2LH [IO91FR, SP30]
2E1 ESL K H Risdale, 14 Carmarthen Way, Rushden, NN10 0TN [IO92QG, SP96]
2E1 ESM K R James, 36 Lemon Hill, Mylor Bridge, Falmouth, TR11 5NA [IO70LE, SW83]
2E1 ESN C McLean, 18 Chatfield Rd, Gosport, PO13 0TN [IO90JT, SU50]
2E1 ESO P A J McLean, 18 Chatfield Rd, Gosport, PO13 0TN [IO90JT, SU50]
2E1 ESP R P Francis, Downsview, Weavers Ln, Halland, near Lewes, E Sussex, BN8 6PR [JO00BW, TQ41]
2E1 ESQ A C Mould, 95 Stanton Rd, Southampton, SO15 4HU [IO90GW, SU31]
2E1 ESR K L Gater, 110 Byrds Ln, Uttoxeter, ST14 7NB [IO92BV, SK03]
2E1 ESS R Whittaker, 29 Blenheim Dr, Bredon, Tewkesbury, GL20 7NQ [IO82WA, SO93]
2W1 EST G E Davies, 27 Twyniago, Pontarddulais, Swansea, SA4 1HX [IO71XR, SN50]
2E1 ESU Details withheld at licensee's request by SSL.
2E1 ESV Details withheld at licensee's request by SSL.
2E1 ESW D J Wilkinson, 139 Church Rd, Jackfield, Telford, TF8 7ND [IO82SO, SJ60]
2E1 ESX Details withheld at licensee's request by SSL.
2E1 ESY Details withheld at licensee's request by SSL.
2E1 ESZ W Wilsdon, 2 Long Catlis Rd, Gillingham, ME8 9SR [JO01HI, TQ86]
2E1 ETA S E Guest, 3A Waterloo Terr, Bridgnorth, Salop, WV16 4EG [IO82SM, SO79]
2E1 ETB R T Moore, 13 Flanderwell Ln, Bramley, Rotherham, S66 0QJ [IO93IK, SK49]
2E1 ETC A J Mackie, 33 Mentmore Gdns, Appleton, Warrington, WA4 3HF [IO83RI, SJ68]
2W1 ETE Details withheld at licensee's request by SSL.
2E1 ETF S J Price, 86 Woodside, Gosport, PO13 0YU [IO90JU, SU50]
2E1 ETG Details withheld at licensee's request by SSL.
2E1 ETH M J Shaw, 14 Ashtead Cl, Fareham, PO16 9TP [IO90KU, SU60]
2M1 ETI I McLeary, 146 Captains Rd, Edinburgh, EH17 8DX [IO85KV, NT26]
2E1 ETJ A Willis, Kilncroft, Broad Layings, Woolton Hill, Newbury, Berks, RG20 9TS [IO91HI, SU46]
2E1 ETK Details withheld at licensee's request by SSL.
2E1 ETL Details withheld at licensee's request by SSL.
2M1 ETM S T H Mackie, 25 Carlaverock Dr, Tranent, EH33 2EE [IO85MW, NT47]
2W1 ETN D L Jorgensen, 4 Coleridge Cres, Killay, Swansea, SA2 7DJ [IO71XO, SS59]
2E1 ETO Details withheld at licensee's request by SSL.
2E1 ETP Details withheld at licensee's request by SSL.
2E1 ETQ Details withheld at licensee's request by SSL.
2E1 ETR S E Carter, 242 Thorney Leys, Witney, OX8 7YP [IO91GS, SP30]
2E1 ETS P A Strange, 24 Stourton Dr, Longwell Green, Barrs Ct, Bristol, BS15 7AL [IO81SK, ST67]
2W1 ETT C D Charles, 13 King Georges Ave, Llanelli, SA15 1LY [IO71WQ, SN50]
2W1 ETU K J Morris, 15 Christopher St., Llanelli, SA15 1DF [IO71WQ, SS59]
2W1 ETV L D Thomas, 45 Dillwyn St., Llanelli, SA15 1BT [IO71WQ, SS59]
2W1 ETW R D H Thomas, 24 Heol Innes, Llanelli, SA15 4LA [IO71WQ, SN50]
2E1 EUB Details withheld at licensee's request by SSL.
2E1 EUD Details withheld at licensee's request by SSL.
2E1 EUE C M Troy, 90 Evenlode Rd, Southampton, SO16 9EH [IO90GW, SU31]
2E1 EUF Details withheld at licensee's request by SSL.
2E1 EUG Details withheld at licensee's request by SSL.
2E1 EUI J R Grant, 3 Maytree Rd, Chandlers Ford, Eastleigh, SO53 5RT [IO91HA, SU42]
2E1 EUK R Strickland, 159 North Circular Rd, London, N13 5EL [IO91WO, TQ39]
2E1 EUL A Strickland, 159 North Circular Rd, London, N13 5EL [IO91WO, TQ39]
2E1 EUM V Ollis, 36C Belfast Sq, Brize Norton, Carterton, OX18 3TA [IO91ES, SP20]
2M1 EUO D A A Storey, 1 Drummond St., Greenock, PA16 9DN [IO75OW, NS27]
2E1 EUQ Details withheld at licensee's request by SSL.
2W1 EUR A K J Joynes, 25 St. Annes Gdns, Maesycwmmer, Hengoed, CF8 7QQ [IO81HN, ST08]
2E1 EUS Details withheld at licensee's request by SSL.
2E1 EUT Details withheld at licensee's request by SSL.
2E1 EUU Details withheld at licensee's request by SSL.
2M1 EUV E Clark, 3 Old Cttgs, Seton Mains, Longniddry, East Lothian, EH32 0PG [IO85MX, NT47]
2E1 EUW Details withheld at licensee's request by SSL.
2E1 EUY W E Partridge, 15 Cranbourne Ave, Wolverhampton, WV4 6RJ [IO82WN, SO99]
2E1 EUZ Details withheld at licensee's request by SSL.
2E1 EVA D Guest, 22 City Rd, Nottingham, NG7 2JJ [IO92JW, SK53]
2E1 EVB Details withheld at licensee's request by SSL.
2E1 EVE A M Vines, 30 Catalina Cl, Christchurch, BH23 4JG [IO90DR, SZ19]
2E1 EVG M W E Seddon, 3 Badgers Ct, Foxlands Cl, Leavesden, Watford, WD2 7LY [IO91TQ, TL10]
2E1 EVH F Stevenson, 56 Barrows Hill Ln, Westwood, Nottingham, NG16 5HJ [IO93IB, SK45]
2E1 EVI N J Ward, 69 Hamelin Rd, Gillingham, ME7 3EX [JO01GI, TQ76]
2E1 EVJ R M Chipperfield, 5 Lullingstone Cl, Hempstead, Gillingham, ME7 3TS [JO01GI, TQ76]
2E1 EVK R P Chipperfield, 131 Hempstead Rd, Hempstead, Gillingham, ME7 3QE [JO01GI, TQ76]
2E1 EVL Details withheld at licensee's request by SSL.
2E1 EVM J A Merrick, The Conifers, 259 Wigmore Rd, Gillingham, ME8 0LZ [JO01GI, TQ76]
2E1 EVN A Hoyle, York House, 2 Minster Cl, Greetland, Halifax, West Yorks, HX4 8QW [IO93BQ, SE02]
2E1 EVO C E H Black, 206 Bradford Rd, Tingley, Wakefield, WF3 1RX [IO93FR, SE22]
2E1 EVQ G W E Knight, 232 PO Box, Southampton, SO15 4ZA [IO90GW, SU31]
2E1 EVS G Odgers, 1 Southgate St., Redruth, TR15 2LY [IO70JF, SW74]
2E1 EVT Details withheld at licensee's request by SSL.
2E1 EVV D J Elsey, 129 Kingsway, Chandlers Ford, Eastleigh, SO53 5BX [IO90HX, SU42]
2E1 EVW J L Best, Longview, Central Rd, Dearham, Maryport, CA15 7ER [IO84GQ, NY03]
2E1 EVX T Bostock, 85 Ashdown Rd, Chandlers Ford, Eastleigh, SO53 5QH [IO91HA, SU42]
2E1 EVY T J Pattle, 2 Scotts Ct, Station Rd, Alderholt, Fordingbridge, SP6 3RB [IO90CW, SU11]
2E1 EVZ Details withheld at licensee's request by SSL.
2M1 EWA D A Cromar, 30 Ardgowan St., Greenock, PA16 8EH [IO75OW, NS27]
2E1 EWB J E Best, Longview, Central Rd, Dearham, Maryport, CA15 7ER [IO84GQ, NY03]
2E1 EWC R C I Gifford, 107 Cardinal Ave, Kingston upon Thames, KT2 5RZ [IO91UK, TQ17]

2E1 EWD Details withheld at licensee's request by SSL.
2E1 EWG P Joslin, Holland Cottage, Kirby Rd, Great Holland, Frinton on Sea, CO13 0HZ [JO01OU, TM22]
2E1 EWH S A Jarrett, 17 Wolmers Hey, Great Waltham, Chelmsford, CM3 1DA [JO01FT, TL61]
2E1 EWI M J Panton, Lavers, Preston Rd, Lavenham, Sudbury, Suffolk, CO10 9QD [JO02JC, TL94]
2E1 EWJ Details withheld at licensee's request by SSL.
2E1 EWK B K Ashman, 108 Eastwood Dr, Highwoods, Colchester, CO4 4SL [JO01LV, TM02]
2E1 EWL W Jones, 13 The Coppice, Enfield, EN2 7BY [IO91WP, TQ39]
2E1 EWM C Joyce, 21 The Boulevard, Pevensey Bay, Pevensey, BN24 6SB [JO00ET, TQ60]
2E1 EWN B P Storkey, 9 Snatchup, Redbourn, St. Albans, AL3 7HD [IO91TT, TL11]
2E1 EWO Details withheld at licensee's request by SSL.
2E1 EWR D Fitzpatrick, 21 Stambridge Rd, Clacton on Sea, CO15 3JR [JO01NT, TM11]
2E1 EWT J A F Jackson, 14 Glebe Rd, Welwyn, AL6 9PB [IO91VT, TL21]
2E1 EWU T Bailey, The Cottage, Alderton Rd, Grittleton, Chippenham, Wilts, SN14 6AN [IO81VM, ST88]
2E1 EWV J G Davies, 10 Gorselands, Hollesley, Woodbridge, IP12 3QL [JO02RB, TM34]
2E1 EWW Details withheld at licensee's request by SSL.
2E1 EWX E M Lawrence, 468 Bexhill Rd, St. Leonards on Sea, TN38 8AU [JO00GU, TQ70]
2E1 EWY C S J Gladman, 43 Queens Rd, New Malden, KT3 6BY [IO91VJ, TQ26]
2E1 EWZ I R Williams, 3 Windham Cres, Wawne, Hull, HU7 5XW [IO93TT, TA03]
2E1 EXA P A J Johnson, 24 Balmoral Ave, Stanford-le-Hope, SS17 7BD [JO01FM, TQ68]
2E1 EXB Loweth, 3 Kirkwick Ave, Harpenden, AL5 2QH [IO91TT, TL11]
2E1 EXC S C Curlis, 76 Barnacres Rd, Hemel Hempstead, HP3 8JQ [IO91SR, TL00]
2E1 EXD T V Gladman, 43 Queens Rd, New Malden, KT3 6BY [IO91VJ, TQ26]
2E1 EXE B Allington, 87 Leybourne Ave, Bournemouth, Dorset, BH10 6ET [IO90BS, SZ09]
2E1 EXG R P Dennis, 17 Knoll Rise, Weymouth, DT3 6AP [IO80SP, SY68]
2E1 EXI D B Taylor, 188 Walstead Rd, Walsall, WS5 4DN [IO92AN, SP09]
2E1 EXJ P T Parsonage, 14 Firbeck Gdns, Stafford, ST17 4QQ [IO82WS, SJ92]
2E1 EXK J E Wilkes, 47 Greenwood Park, Cannock, WS12 4DQ [IO82XR, SJ91]
2E1 EXL L J Taylor, Southgate, South St., Wincanton, Somerset, BA9 9DW [IO81TB, ST72]
2E1 EXN G S D Lambley, 21 Sancroft Ct, Battersea Bridge Roa, London, SW11 3AJ [IO91WL, TQ27]
2I1 EXO M B J Ardill, 2 Lonsdale Ct, Shore Rd, Newtownabbey, BT37 0FA [IO74BQ, J38]
2E1 EXP R Hall, 91 Nethershire Ln, Shiregreen, Sheffield, S5 0DB [IO93GK, SK39]
2E1 EXR Details withheld at licensee's request by SSL.
2E1 EXS Details withheld at licensee's request by SSL.
2E1 EXT J Snow, 41 Ludlow Rd, Feltham, TW13 7JE [IO91TK, TQ17]
2I1 EXU P A Robinson, 20 Harwood Park, Carrickfergus, BT38 7LZ [IO74CR, J48]
2E1 EXV T J Burslem, Honeysuckle Cottage, School Rd, Waldringfield, Woodbridge, IP12 4QR [JO02PB, TM24]
2E1 EXW H J A Burslem, Honeysuckle Cottage, School Rd, Waldringfield, Woodbridge, IP12 4QR [JO02PB, TM24]
2E1 EXX G M A Burslem, Honeysuckle Cottage, School Rd, Waldringfield, Woodbridge, IP12 4QR [JO02PB, TM24]
2E1 EXY B P Eckersley, 10 Pytches Cl, Melton, Woodbridge, IP12 1SE [JO02PC, TM24]
2E1 EXZ C J Thornton, 12 Montagus Harrier, Guisborough, Cleveland, TS14 8PB [IO94LM, NZ51]
2E1 EYA Details withheld at licensee's request by SSL.
2E1 EYC M K Davies, 31 Long Meadow, Mansfield Woodhouse, Mansfield, Notts, NG19 9QW [IO93JE, SK56]
2E1 EYD D de Aston, 18 Lime Gr, Scunthorpe, DN16 2HL [IO93QN, SE80]
2E1 EYE N C Butler, 1 Old Brewery Pl, Kimpton, Hitchin, SG4 8RX [IO91UU, TL11]
2E1 EYF E M Hedley, 17 Chowdene Bank, Gateshead, NE9 6JJ [IO94EW, NZ25]
2E1 EYG R M H Walters, 11 Grouse Cl, Stratford upon Avon, CV37 9FE [IO92DE, SP15]
2E1 EYH M C Montanaro, 29 Northd. Rd, Leamington Spa, CV32 6HE [IO92FH, SP36]
2E1 EYI S J Riddle, 57 Frome Rd, Trowbridge, BA14 0DN [IO81VH, ST85]
2E1 EYJ Details withheld at licensee's request by SSL.
2E1 EYK O Overton, Eastrigg, Scorers Ln, Great Lumley, Chester-le-Street, DH3 4JH [IO94FU, NZ24]
2E1 EYL M A Harris, 7 Severn Rd, Melksham, SN12 8BQ [IO81WJ, ST96]
2W1 EYM D Davies, 47 Woodland Ave, Pencoed, Bridgend, CF35 6UW [IO81FM, SS98]
2W1 EYN D C Luke, 56 Maerdy Park, Pencoed, Bridgend, CF35 5HX [IO81FM, SS98]
2E1 EYO Details withheld at licensee's request by SSL.
2W1 EYP W J Evans, 26 Station Rd, Llangynwyd, Maesteg, CF34 9TF [IO81EO, SS88]
2E1 EYQ M Brooks, 38 California, Blaydon on Tyne, NE21 6LZ [IO94DW, NZ16]
2W1 EYR Details withheld at licensee's request by SSL.
2E1 EYS B A Egglestone, 4 Lancaster Cl, Etherley Dene, Bishop Auckland, DL14 0RP [IO94DP, NZ12]
2E1 EYT M A Jolley, 30 Oban Dr, Blackburn, BB1 2HY [IO83SR, SD72]
2E1 EYU A M L Bone, 23 Lincoln St., Gateshead, NE8 4EE [IO94EW, NZ26]
2E1 EYV A C Wilson, 14 Glentress Mews, Bolton, BL1 5JS [IO83SO, SD61]
2E1 EYW C J Kelly, 4 Ambrose Ln, Harpenden, AL5 4AX [IO91TT, TL11]
2E1 EYX Details withheld at licensee's request by SSL.
2E1 EYY Details withheld at licensee's request by SSL.
2W1 EYZ S E Gray, 36 Heol Pentre Felen, Morriston, Swansea, SA6 6BY [IO81AQ, SS69]
2M1 EZA J Boyle, 33 St. Mungo Ave, Glasgow, G4 0PH [IO75VU, NS56]
2I1 EZB C Morris, 115 Glenarm Rd, Larne, BT40 1EE [IO74CU, D40]
2E1 EZD Details withheld at licensee's request by SSL.
2E1 EZE Details withheld at licensee's request by SSL.
2E1 EZF Details withheld at licensee's request by SSL.
2E1 EZG Details withheld at licensee's request by SSL.
2E1 EZH C S Pearson, 14 Mill Croft, Richmond, DL10 4TR [IO94CJ, NZ10]
2E1 EZJ C S Norris, 68 Northcote Rd, Bournemouth, BH1 4SQ [IO90BR, SZ19]
2E1 EZM A J Chittock, Drackenhill, 8 Castle Rise, West Ayton, Scarborough, YO13 9JY [IO94SG, SE98]
2E1 EZN C J Turk, 33 Eskdale Ave, Rochdale, OL11 3JX [IO83VO, SD81]
2E1 EZO Details withheld at licensee's request by SSL.
2E1 EZP M Lewis, The Manor House, Warmington, Banbury, Oxon, 0X17 1BU
2E1 EZQ Details withheld at licensee's request by SSL.
2E1 EZR T Morrison, 16 Russell Dr, Prestatyn, LL19 7YR [IO83GH, SJ08]
2E1 EZT C M Palmer, South Cross, Weeping Cross, Bodicote, Banbury, OX15 4ED [IO92IA, SP43]
2E1 EZU Details withheld at licensee's request by SSL.
2E1 EZV Details withheld at licensee's request by SSL.
2E1 EZX C W Hattam, 10 Burywick, Harpenden, AL5 2AE [IO91TT, TL11]
2E1 EZY L M Harrison, 7 Romsley Cl, Mickleover, Derby, DE3 5SD [IO92FV, SK33]
2E1 FAC R Rose, 77 Glebe Ln, Maidstone, ME16 9BA [JO01FG, TQ75]
2E1 FAD C Parrott, 55 Brown St., Macclesfield, SK11 6RY [IO83WG, SJ97]
2E1 FAE P Rose, 77 Glebe Ln, Maidstone, ME16 9BA [JO01FG, TQ75]
2E1 FAF M J Muriel, 13 York Rd, Sale, M33 6EZ [IO83UK, SJ79]
2E1 FAG Details withheld at licensee's request by SSL.
2E1 FAH R F R McKinnell, 17 Coronation Ave, Long Clawson, Melton Mowbray, LE14 4NF [IO92MT, SK72]
2E1 FAI J McKinnell, 17 Coronation Ave, Long Clawson, Melton Mowbray, LE14 4NF [IO92MT, SK72]
2E1 FAJ P H Rice, 15 Towy Ave, Llandovery, SA20 0EH [IO81CX, SN73]
2E1 FAL A I Johnson, 28 Hillside Rd, Blidworth, Mansfield, NG21 0TR [IO93KC, SK55]
2E1 FAN K Blanchard, 27 Willow Ct, Sleaford, NG34 7GJ [IO92TX, TF04]
2E1 FAO Details withheld at licensee's request by SSL.
2E1 FAP V Constantine, 34 Ladysmith Rd, Ashton under Lyne, OL6 9DJ [IO83XM, SD90]
2E1 FAS D J Russell, 15 Hereford Rd, Seaforth, Liverpool, L21 1EG [IO83LL, SJ39]
2E1 FAT S J Hallsworth, 27 Westfield Ave, Heanor, DE75 7BN [IO93HA, SK44]
2E1 FAU B D B Love, 21 Ripon Dr, Blaby, Leicester, LE8 4AU [IO92KN, SP59]
2E1 FAV Details withheld at licensee's request by SSL.
2E1 FAX Details withheld at licensee's request by SSL.
2E1 FAY D J Swaine, 61 Denham Way, Camber, Rye, TN31 7XR [JO00JW, TQ91]
2E1 FAZ J K Kilgannon, Orchard House, Ave Rd, Stratford upon Avon, CV37 6UN [IO92DE, SP25]
2E1 FBA R Locke, 3 Marigold Cl, Lincoln, LN2 4SZ [IO93RG, SK97]
2W1 FBD S G Tuppen, 78 Heol Dewi, Brynna, Pontyclun, CF72 9SQ [IO81GM, SS98]
2E1 FBE Details withheld at licensee's request by SSL.
2E1 FBF Details withheld at licensee's request by SSL.
2E1 FBG S S Fahey, 277 Mount Pleasant Ave, St. Helens, WA9 2PU [IO83PK, SJ59]
2E1 FBH Details withheld at licensee's request by SSL.
2E1 FBI Details withheld at licensee's request by SSL.
2E1 FBK C M Strange, 17 Elmwood, Chippenham, SN15 1AP [IO81WL, ST97]
2E1 FBL Details withheld at licensee's request by SSL.
2E1 FBN M J Gardiner, 2 Coronation Cttgs, Ipswich Rd, Helmingham, Stowmarket, IP14 6EL [JO02OD, TM15]
2E1 FBP W R Trevitt, Laundry Cottage, Helmingham, Stowmarket, Suffolk, IP14 6EH [JO02OD, TM15]
2E1 FBR C E Sutton, Bridge Farm, Wold Newton, Driffield, East Yorks, YO25 0YN [IO94TD, TA07]
2E1 FBS D C Seabridge, 31 Charlestown Dr, Allestree, Derby, DE22 2HA [IO92GW, SK33]
2E1 FBT Details withheld at licensee's request by SSL.
2E1 FBU A Knight, 156 Alwyn Rd, Bilton, Rugby, CV22 7RA [IO92II, SP47]
2E1 FBV D J Egglestone, 4 Castle View, Ushaw Moor, Durham, DH7 7NT [IO94ES, NZ24]
2E1 FBW I P Colley, 225 Botley Rd, Burridge, Southampton, SO31 1BJ [IO90IV, SU50]
2W1 FBX R Henderson, 1107 Llangyfelach Rd, Tirdeunaw, Swansea, SA5 7HY [IO81AP, SS69]
2E1 FBY I Roper, 1 Hedge Side, Bradford, BD8 0AL [IO93CT, SE13]
2E1 FBZ G Dunne, 164 Durham Rd, Bradford, BD8 9HU [IO93CT, SE13]
2E1 FCA S D Rogers, 7 Buckleigh Rd, Wath upon Dearne, Rotherham, S63 7JB [IO93HL, SK49]
2E1 FCC M J Williams, Corn Acres, Dodwell, Stratford upon Avon, Warks, CV37 9ST [IO92CE, SP15]
2E1 FCD M Williams, Corn Acres, Dodwell, Stratford upon Avon, Warks, CV37 9ST [IO92CE, SP15]
2E1 FCE J Lindsay, 10 New Station Rd, Swinton, Mexborough, S64 8AH [IO93IL, SK49]
2E1 FCF A P Rickman, 35 Cedar Way, Basingstoke, RG23 8NG [IO91KG, SU65]
2E1 FCG M Hardy, 18A High St., Hoyland, Barnsley, S74 9AB [IO93GM, SE30]
2E1 FCH A C Mott-Gotobed, 29A Attwood Cl, Basingstoke, RG21 8YY [IO91KG, SU65]
2E1 FCI D Houghton, 160 Mason Lathe Rd, Sheffield, S5 0TQ [IO93GK, SK39]
2E1 FCJ B Webb, 100 Rother View Rd, Rotherham, S60 2UR [IO93HK, SK49]
2E1 FCK M J M Jelinek, 3 Holbrook Cl, Eastbourne, BN20 7JT [JO00DS, TV69]
2E1 FCL Details withheld at licensee's request by SSL.
2E1 FCM A K Humphries, 254 Sprotborough Rd, Doncaster, DN5 8BY [IO93KM, SE50]
2E1 FCN G P Johnson, 28 Hillside Rd, Blidworth, Mansfield, NG21 0TR [IO93KC, SK55]
2E1 FCO I C Brady, 119 Newport Rd, West Cowes, PO31 7PT [IO90IS, SZ49]
2E1 FCP Details withheld at licensee's request by SSL.
2E1 FCS T Furby, 14 Coupland Rd, Rotherham, S65 3LJ [IO93IK, SK49]
2E1 FCT M Graseley, 6 Wroxham Cl, Bramley, Rotherham, S66 0XB [IO93IK, SK49]
2E1 FCV Details withheld at licensee's request by SSL.
2E1 FCW R J Taylor, 164 Ulverley Green Rd, Solihull, B92 8AB [IO92CK, SP18]
2E1 FCX T A D Gardiner, 7 The Rydes, Bodicote, Banbury, OX15 4EJ [IO92IA, SP43]
2E1 FCY C Frank, 36 Melksham Rd, Nottingham, NG5 5RX [IO93KA, SK54]
2E1 FCZ Details withheld at licensee's request by SSL.
2E1 FDA R A Wilkins, 6 Leng Cres, Norwich, NR4 7NX [JO02PO, TG20]
2E1 FDB E M Braithwaite, 33 Somers Park Ave, Malvern, WR14 1SE [IO82UD, SO74]
2E1 FDC L K Peck, 17 Mill Ln, Barton-le-Clay, Bedford, MK45 4LN [IO91SX, TL03]
2E1 FDD M K Warren, 145 Shirehall Rd, Sheffield, S5 0JL [IO93GK, SK39]
2E1 FDE J Pesani, 5 Japonica Dr, Leegomery, Telford, TF1 4XD [IO82SR, SJ61]
2E1 FDF G Bilson, 38 Woodhouse Ln, Bolsover, Chesterfield, S44 6BL [IO93IF, SK47]
2E1 FDG E J Townsend, 14 Birch Gr, Sandy, SG19 1NG [IO92UD, TL14]
2E1 FDI S A Brock, The Firs, The Butts, Norwich, Norfolk, NR16 2EQ [JO02LK, TM08]
2E1 FDJ K G Newnam, 1 Wheatlands Cl, Maulden, Bedford, MK45 2AQ [IO92SA, TL03]
2E1 FDK L J Rule, The Kiteshop, Meaver Rd, Mullion, Helston, Cornwall, TR12 7DN [IO70JA, SW61]
2E1 FDL C J Rule, The Kiteshop, Meaver Rd, Mullion, Helston, Cornwall, TR12 7DN [IO70JA, SW61]
2E1 FDM J C Mew, Tehig, Back St., Garboldisham, Diss, IP22 2SD [JO02LJ, TM08]
2E1 FDP G P Westwood, 62 Blackbrook Ln, Bromley, BR2 8AY [JO01AJ, TQ46]
2E1 FDQ A J Burton, Mill House Farm, Walesby, Market Rasen, Lincs, LN8 3UR [IO93UJ, TF19]
2E1 FDR P J Dover, 11 Roman Rd, Barton-le-Clay, Bedford, MK45 4QJ [IO91SX, TL03]
2E1 FDS S J Dover, 11 Roman Rd, Barton-le-Clay, Bedford, MK45 4QJ [IO91SX, TL03]
2E1 FDT P Brown, 14 Hanslope Cres, Nottingham, NG8 4BE [IO92JX, SK54]
2E1 FDU M L Darton, 8 Foster Gr, Sandy, SG19 1HP [IO92UD, TL14]
2E1 FDV P R McNamara, 39 Waveney Rd, Bungay, NR35 1LH [JO02RK, TM38]
2E1 FDW H Shemming, 6 Smiths Pl, Kesgrave, Ipswich, IP5 7YR [JO02OB, TM24]
2E1 FDX E E J Rowland, The Rectory, High St., Sandy, Beds, SG19 1AQ [IO92UD, TL14]
2E1 FDY Details withheld at licensee's request by SSL.
2E1 FDZ J Gill, 71 Waarden Rd, Canvey Island, SS8 9AB [JO01HM, TQ78]
2E1 FEB D C Wilkes, 36 Selbourne, Sutton Hill, Telford, TF7 4AY [IO82SP, SJ70]
2E1 FEC F W Dunmore, 36 Dove Rise, Oadby, Leicester, LE2 4NY [IO92LO, SK60]
2E1 FED D A Tomlinson, 3 Holgate Rd, Nottingham, NG2 2EB [IO92KW, SK53]
2E1 FEE Details withheld at licensee's request by SSL.
2E1 FEF K S Rhodes, Jubilee House, Kent Ave, Theddlethorpe, Mablethorpe, LN12 1QE [JO03CI, TF48]
2E1 FEG B Rhodes, Jubilee House, Kent Ave, Theddlethorpe, Mablethorpe, LN12 1QE [JO03CI, TF48]
2W1 FEH T D Coombe, 37 Chapel Ln, Croesyceiliog, Cwmbran, NP44 2PW [IO81LP, ST39]
2E1 FEI R L Frost, North View, Long Green, Bedfield, Woodbridge, IP13 7JQ [JO02PG, TM26]
2E1 FEJ D Keene, Sweetnaps, 4 Blackberry Ln, Four Marks, Alton, GU34 5BN [IO91LC, SU63]
2E1 FEK Details withheld at licensee's request by SSL.
2E1 FEM P D Higginson, 4 Haywood Ln, Cheswardine, Market Drayton, TF9 2RP [IO82SU, SJ72]
2E1 FEO P D Meigh, White Ash, Mickfield, Stowmarket, Suffolk, IP14 5LR [JO02NF, TM16]
2E1 FEP R J Giddings, 7 Foster Gr, Sandy, SG19 1HP [IO92UD, TL14]
2E1 FEQ A S Daniels, 9 Moss Fold, Astley, Tyldesley, Manchester, M29 7FP [IO83SM, SD70]
2E1 FER A J Ruskin, 15 Arlington Ave, Mansfield Woodhouse, Mansfield, NG19 9DH [IO93JD, SK56]
2E1 FES L R O'Neill, 3 Barlow Cl, Tamworth, B77 3ES [IO92DP, SK20]
2E1 FET N J Moore, 84 Franklynn Rd, Haywards Heath, RH16 4DH [IO90WX, TQ32]
2E1 FEU C Bagshaw, 35 School Ln, Ollerton, Newark, Notts, NG22 9AS [IO93LE, SK66]
2E1 FEV F A D Pragnell, 32 Alfred St., East Cowes, PO32 6SA [IO90IS, SZ59]
2E1 FEW A McDonald, 3 Wallington Walk, Billingham, TS23 3XJ [IO94IO, NZ42]
2E1 FEX R D Ibbotson, 10 Cawthorne Rd, Wakefield, WF2 7HW [IO93GP, SE31]
2E1 FEY S J Pinder, 13 Brisbane Cl, Mansfield Woodhouse, Mansfield, NG19 8QZ [IO93JE, SK56]
2E1 FEZ E W P Perkins, 10 St. Marys Cl, Sudbury, CO10 0PN [JO02IB, TL84]
2E1 FFA Details withheld at licensee's request by SSL.
2E1 FFB O M Suleyman, 61 Victoria Rd, London, N9 9SU [IO91XO, TQ39]
2E1 FFC Details withheld at licensee's request by SSL.
2E1 FFD D P Bull, 8 Mayfield Ln, Martlesham Heath, Ipswich, IP5 7TZ [JO02OB, TM24]
2E1 FFE G J Timms, 74 Park Gwyn, St. Stephen, St. Austell, PL26 7PN [IO70NI, SW95]
2E1 FFF J L Parry, 23 Hawksworth, Tamworth, B77 2HH [IO92EO, SK20]
2E1 FFG D Faraday, Top Farm, Church Rd, Streatley, Luton, LU3 3PN [IO91SW, TL02]
2E1 FFH Details withheld at licensee's request by SSL.
2E1 FFI Details withheld at licensee's request by SSL.
2E1 FFJ R Dennis, The Old Chapel House, Farlesthorpe, Alford, Lincs, LN13 9PH [JO03CF, TF47]
2E1 FFK Details withheld at licensee's request by SSL.
2E1 FFL P Harbinson, 19 Pennine Rd, Dewsbury, WF12 7AW [IO93EQ, SE22]
2E1 FFM C M Hepplestone, 14 Sike Ln, Totties, Holmfirth, Huddersfield, West Yorks, HD7 1UL [IO93CN, SE10]
2E1 FFN G Burgess, Stickle-Fuen, Kerrys Gate, Hereford, HR2 0AH [IO81NX, SO33]
2E1 FFO I A Rogers, Yew Tree Cottage, Shucknall, Hereford, HR1 3SJ [IO82QB, SO54]
2E1 FFP R Hall, 57 Sixth Ave, Edwinstowe, Mansfield, NG21 9PW [IO93LE, SK66]
2E1 FFQ S H J Booth, 20 Edna Rd, Leigh, WN7 5ES [IO83RM, SD60]
2E1 FFS M E Williams, Timbers, Barrack Ln, Aldwick, Bognor Regis, West Sussex, PO21 4BZ [IO90PS, SZ99]
2E1 FFT T Hindley, 14 High St., Astley, Tyldesley, Manchester, M29 7DD [IO83SL, SD60]
2E1 FFU S J Chapman, 9 Edinburgh Ave, Sawston, Cambridge, CB2 4DW [JO02CD, TL45]
2E1 FFV S T Hindley, 14 High St., Astley, Tyldesley, Manchester, M29 7DD [IO83SL, SD60]
2E1 FFW Details withheld at licensee's request by SSL.
2E1 FFX P R Fryer, 17 Tilia Rd, Tamworth, B77 3BE [IO92EP, SK20]
2E1 FFY M P Sadler, 30 Holbeck Ave, Scarborough, YO11 2XQ [IO94TG, TA08]
2E1 FFZ D Fawcett, 6 Wand Hill, Boosbeck, Saltburn By The Sea, TS12 3AW [IO94MN, NZ61]
2E1 FGA O J Marshall, 9 Losinga Cres, Norwich, NR3 2RR [JO02PP, TG21]
2E1 FGB N Horn, 12 Melbourne Rd, Chichester, PO19 4NE [IO90OU, SU80]
2E1 FGC A M Jones, 70 Devon Rd, Smethwick, Warley, B67 5EJ [IO92AL, SP08]
2E1 FGD P J Rogers, 20 Clayfield Cl, Nottingham, NG6 8DG [IO92JX, SK54]
2E1 FGE R W J Wild, The Form, 9 Augusta Rd, Portland, DT5 1DE [IO80SN, SY67]
2E1 FGF S M Roberts, Sunnycroft, Tillington, Hereford, HR4 8LH [IO82OC, SO44]
2E1 FGG T G Jones, 354 Bridgeman St., Bolton, BL3 6SJ [IO83SN, SD70]
2E1 FGH Details withheld at licensee's request by SSL.
2E1 FGI Details withheld at licensee's request by SSL.
2E1 FGJ J E Parker, 76 Elm Rd, Grays, RM17 6LD [JO01EL, TQ67]
2W1 FGK C J Hogan, West Hill, Llanasa Rd, Gwespyr, Holywell, Clwyd, CH8 9LT [IO83HH, SJ18]
2E1 FGL D P Hogan, West Hill, Llanasa Rd, Gwespyr, Holywell, Clwyd, CH8 9LT [IO83HH, SJ18]
2E1 FGM Details withheld at licensee's request by SSL.
2E1 FGN J A Sutton, Karena, Gweek, Helston, Cornwall, TR12 6UB [IO70JC, SW72]
2E1 FGO S J Ely, 20 Tenterleas, St. Ives, Huntingdon, PE17 4QP [IO92XH, TL37]
2E1 FGP D D Stead, 18 Cornwall Ave, Mansfield, NG18 3JG [IO93KD, SK56]
2E1 FGR J L Clark, 22 Heol yr Wylan, Cwmrhydyceirw, Swansea, SA6 6TB [IO81AQ, SN60]
2E1 FGT A G Noden, 9 Abbotsbury Cl, Saltdean, Brighton, BN2 8SR [IO90XT, TQ30]
2M1 FGU M McCrum, 1 Jamieson St, Forfar, Angus, DD8 2HY [IO86NP, NO45]
2E1 FGV Details withheld at licensee's request by SSL.
2M1 FGW J Leith, 6/7 Sparrowcoft, Forfar, Angus, DD8 2AP [IO86NP, NO45]
2E1 FGX C J Adams, 20 Welbeck Dr, Burgess Hill, RH15 0BB [IO90WX, TQ32]
2E1 FGY Details withheld at licensee's request by SSL.
2E1 FGZ Details withheld at licensee's request by SSL.
2W1 FHA S D Jefferson, 7 Haddon Cl, Rhyl, LL18 2JP [IO83GH, SJ08]
2E1 FHB Details withheld at licensee's request by SSL.
2E1 FHC J F Beckett, 23 Newton Ave, Wakefield, WF1 2PX [IO93FQ, SE32]
2E1 FHD K Glover, 14 Crawley Cres, Eastbourne, BN22 9RN [JO00DT, TQ60]
2E1 FHE Details withheld at licensee's request by SSL.
2W1 FHF C A Cartwright, 14 Clement Dr, Rhyl, LL18 4HU [IO83GH, SJ08]
2E1 FHG Details withheld at licensee's request by SSL.
2E1 FHH C E Rose, Monksfield, Monk Soham, Woodbridge, Suffolk, IP13 7EX [JO02OF, TM26]
2E1 FHI C H Anderson, 11 Langmead Rd, Crewkerne, TA18 8DY [IO80OU, ST40]
2E1 FHJ E Telenius-Lowe, 27 Hertford Rd, Stevenage, SG2 8RZ [IO91VV, TL22]
2E1 FHK C J Green, 24 Leasown, Burghill, Hereford, HR4 7SA [IO82OC, SO44]
2W1 FHL J R Lightfoot, 29 Clement Dr, Rhyl, LL18 4HU [IO83GH, SJ08]

2E1 FHM Details withheld at licensee's request by SSL.
2E1 FHN Details withheld at licensee's request by SSL.
2E1 FHO Details withheld at licensee's request by SSL.
2E1 FHP S J Mennie, 21 Washbrook Cl, Barton-le-Clay, Bedford, MK45 4LF [IO91SX, TL03]
2E1 FHQ C S Ward, 1 Horton Downs, Downswood, Maidstone, ME15 8TN [JO01GG, TQ75]
2M1 FHR E G McDonald, 3 Kent Rd, Alloa, FK10 2JN [IO86CD, NS89]
2M1 FHS R C Blain, 39 Redwell Pl, Alloa, FK10 2BT [IO86CC, NS89]
2M1 FHT D McKay, 13 Dorothy Terr, Tillicoultry, FK13 6EU [IO86DD, NS99]
2E1 FHU R M Townsend, 56 Seymour Rd, Northfleet, Gravesend, DA11 7BN [JO01EK, TQ67]
2E1 FHV C J Gould-Fellows, 7 New Rd, Pebworth, Stratford upon Avon, CV37 8XS [IO92CC, SP14]
2E1 FHW J Seedle, 3 Knaresborough Cl, Poulton-le-Fylde, FY6 7SJ [IO83MU, SD34]
2E1 FHX D E Kirk, 23 Oak Ave, Uxbridge, UB10 8LP [IO91SN, TQ08]
2E1 FHY S A Kirk, 37 Glebe Ave, Ruislip, HA4 6QZ [IO91TN, TQ18]
2E1 FHZ D Hebb, 45 Marklew Ave, Grimsby, DN34 4AD [IO93WN, TA20]
2E1 FIA J J B Bishop, 43 Darwin Rd, Ipswich, IP4 1QE [JO02OB, TM14]
2M1 FIB J McGhie, 16 Boyach Cres, Isle of Whithorn, Newton Stewart, DG8 8LD [IO74TQ, NX43]
2E1 FIC C F C Ballam, Merryfield, Mill Rd, Waldringfield, Woodbridge, IP12 4PY [JO02PB, TM24]
2E1 FID O E Ballam, Merryfield, Mill Rd, Waldringfield, Woodbridge, IP12 4PY [JO02PB, TM24]
2E1 FIE M P Robertson, 28 Frith Rd, Bognor Regis, PO21 5LL [IO90PS, SZ99]
2E1 FIF M A Lawrance, 10 Clay Pit Piece, Saffron Walden, CB11 4DR [JO02CA, TL53]
2E1 FIG C Bennet, Kesgrave High School, Main Rd, Kesgrave, Ipswich, IP5 7PB [JO02OB, TM24]
2E1 FIH C L Shapland, 120 Church Rd, Enfield, EN3 4NY [IO91XP, TQ39]
2E1 FII J Owen, 70 Churston Dr, Morden, SM4 4JQ [IO91VJ, TQ26]
2E1 FIJ R Owen, 70 Churston Dr, Morden, SM4 4JQ [IO91VJ, TQ26]
2E1 FIK J A Wilson, 6 St. Pauls Rd, Scunthorpe, DN16 3DL [IO93QN, SE80]
2E1 FIL P A Ridsdale, Pebsham Hall Farm, Bodle St Green, Hailsham, E Sussex, BN27 4RD [JO00EV, TQ61]
2E1 FIM R B Damm, 81 Sidney St., Cleethorpes, DN35 7NQ [IO93XN, TA20]
2E1 FIN K A Wilson, 6 St. Pauls Rd, Scunthorpe, DN16 3DL [IO93QN, SE80]
2E1 FIO B Knott, 76 New Barns Ave, Mitcham, CR4 1LF [IO91WJ, TQ26]
2E1 FIP A Handley, 4 Southwood Dr, Thorne, Doncaster, DN8 5QS [IO93MO, SE61]
2E1 FIQ H C Pook, 2 Andridge Cttgs, Sprigs Holly Ln, Radnage, High Wycombe, HP14 4DZ [IO91NQ, SU79]
2E1 FIR A J Higton, 18 Moseley St., Ripley, DE5 3DA [IO93HB, SK35]
2E1 FIS P C Duffin, 44 Legion Cl, Poole, BH15 4EA [IO80XR, SY99]
2E1 FIT A J Morris, 133 Shuttlewood Rd, Bolsover, Chesterfield, S44 6NX [IO93IF, SK47]
2E1 FIU N D Bayliss, 12 New St., Elworth, Sandbach, CW11 3JF [IO83TD, SJ76]
2E1 FIV D J Dalzell, 9 Pyms Ln, Crewe, CW1 3PJ [IO83SC, SJ65]
2E1 FIW P Webb, 29 Markfield Cres, St. Helens, WA11 9LD [IO83PL, SJ59]
2E1 FIX J D Shaw, 50 Elmpark Way, Rochdale, OL12 7JQ [IO83VP, SD81]
2E1 FIY Details withheld at licensee's request by SSL.
2E1 FIZ J J Barber-Lomax, Ruscombe, Spring Rd, Kinsbourne Green, Harpenden, Herts, AL5 3PP [IO91TT, TL11]
2E1 FJA J A Campbell, Rattlecombe Hollow, Rattlecombe Rd, Shennington, Banbury, Oxon, OX15 6LZ [IO92GB, SP34]
2E1 FJB S J Elliott, 16 Valentine Rd, Leicester, LE5 2GH [IO92LP, SK60]
2E1 FJC J A Read, 22-24 Stoney Rd, Grundisburgh, Woodbridge, Suffolk, IP13 6RF [JO02OC, TM25]
2E1 FJD J Needham, 122 Alfreton Rd, Newton, Alfreton, DE55 5TR [IO93HC, SK45]
2E1 FJE D Tyson, 7 Hazel Rd, Filey, YO14 9NE [IO94UF, TA18]
2E1 FJF Details withheld at licensee's request by SSL.
2W1 FJG A J Heigh, Janneen, Henry St., Rhostyllen, Wrexham, Clwyd, LL14 4DA [IO83LA, SJ34]
2W1 FJH T G Rogers, The Willows, 48 Hillock Ln, Gresford, Wrexham, LL12 8YL [IO83MC, SJ35]
2E1 FJI K Biggs, 22 Wallingford Cl, Bracknell, RG12 9JE [IO91PJ, SU86]
2E1 FJJ C Reddington, 2 South St., Newton, Alfreton, DE55 5TT [IO93HC, SK45]
2E1 FJK J Gray, 19 Marsh View, Newton, Preston, PR4 3SX [IO83NS, SD43]
2E1 FJL M Gray, 19 Marsh View, Newton, Preston, PR4 3SX [IO83NS, SD43]
2E1 FJM L L Stockwell, 167 Hathaway Rd, Grays, RM17 5LW [JO01DL, TQ67]
2W1 FJN I Pearson, Warren Cottage, Pontfadog, Llangollen, Clwyd, LL20 7AT [IO82KW, SJ23]
2E1 FJO B R Vickers, 10 Beckett Cres, Dewsbury, WF13 3PW [IO93EQ, SE22]
2E1 FJP B Melling, 68 Westfield Dr, Ribbleton, Preston, PR2 6TH [IO83PS, SD53]
2E1 FJQ V G Druce, 7 Buckingham Rd, London, N22 4SR [IO91WO, TQ39]
2M1 FJR Details withheld at licensee's request by SSL.
2E1 FJS Details withheld at licensee's request by SSL.
2W1 FJT C M Mitchell, 52 Snowdon Dr, Wrexham, LL11 2YA [IO83MB, SJ35]
2E1 FJU C W Peggram, Cherry Trees, Broad Ln, Bracknell, RG12 9BY [IO91PJ, SU86]
2E1 FJV I J Page, 7 Riverdale Cl, Swindon, SN1 4ED [IO91CN, SU18]
2E1 FJW N V Taylor, 4 Lower St., Stutton, Ipswich, IP9 2SQ [JO01NX, TM13]
2E1 FJX J W Jordan, 31 Rotherham Rd, Dinnington, Sheffield, S31 7RG [IO93GJ, SK38]
2E1 FJY K J Prior, 74 Walden Way, Frinton on Sea, CO13 0BQ [JO01PU, TM22]
2W1 FJZ P Edwards, Cam Or Afon, Dolywern, Pontfadog, Llangollen, Clwyd, LL20 7AD [IO82KW, SJ23]
2E1 FKA Details withheld at licensee's request by SSL.
2M1 FKB P J Hatton, 64 Abercromby Cres, Helensburgh, G84 9DN [IO76PA, NS38]
2E1 FKC Details withheld at licensee's request by SSL.
2E1 FKD E J Merrington, Cartref, Ball Ln, Kingsley, Warrington, Ches, WA6 8HP [IO83QG, SJ57]
2E1 FKE G Riley, 5 Lune Way, Widnes, WA8 8YQ [IO83OI, SJ48]
2E1 FKF L D Bentley, 55 Stirling Ct Rd, Burgess Hill, RH15 0PS [IO90WX, TQ31]
2E1 FKG M E Burdis, 6 Gordon Rd South, Branksome, Poole, BH12 1EF [IO90BR, SZ09]
2E1 FKH L B James, 107 Spendmore Ln, Coppull, Chorley, PR7 4PY [IO83QP, SD51]
2M1 FKI K H Gillen, 1 Ashton Dr, Helensburgh, G84 7JT [IO75PX, NS38]
2E1 FKJ J Ewing, 12 Linford Cl, Rugeley, WS15 4EF [IO92BR, SK01]
2E1 FKK S G Jenkins, 31 Ashbrook Cres, Rochdale, OL12 9AJ [IO83WP, SD91]
2E1 FKL T R Jenkins, 31 Ashbrook Cres, Rochdale, OL12 9AJ [IO83WP, SD91]
2E1 FKM C Andrews, 29 Hallgate Ln, Pilsley, Chesterfield, S45 8HN [IO93HD, SK46]
2E1 FKN Details withheld at licensee's request by SSL.
2E1 FKO Details withheld at licensee's request by SSL.
2E1 FKP J R Skarratt, 39 Andover Ave, Middleton, Manchester, M24 1JG [IO83VM, SD80]
2E1 FKQ Details withheld at licensee's request by SSL.
2E1 FKR Details withheld at licensee's request by SSL.
2E1 FKS Details withheld at licensee's request by SSL.
2E1 FKT M S Boyes, 19 Rowe Ashe Way, Locks Heath, Southampton, SO31 7EY [IO90IU, SU50]
2E1 FKU T D Lord, The Fairfields, 15 Blackthorn Cl, Melbourne, Derby, Derbyshire, DE73 1LY [IO92GT, SK32]
2E1 FKV K D Williams, 22 West St., Tamworth, B79 7JE [IO92DP, SK20]
2E1 FKW R G Furniss, 91 Nethershire Ln, Sheffield, S5 0DB [IO93GK, SK39]
2W1 FKX I W Taylor, 75 St. Margarets Dr, Rhyl, LL18 2LA [IO83GH, SJ08]
2E1 FKY M J Purcell, 14 Adelaide Rd, Blacon, Chester, CH1 5SY [IO83ME, SJ36]
2E1 FKZ A Woodward, 19 Hazel Gr, Winchester, SO22 4PQ [IO91IB, SU42]
2E1 FLA K Woodward, 19 Hazel Gr, Winchester, SO22 4PQ [IO91IB, SU42]
2E1 FLB K Williams, 14 Bordon Pl, Stratford upon Avon, CV37 9AU [IO92DE, SP15]
2E1 FLC A C Williamson, 39 Lordswell Rd, Burton on Trent, DE14 2TA [IO92ET, SK22]
2E1 FLD P Whittaker, 26 Station Rd, Woodville, Swadlincote, DE11 7DX [IO92FS, SK31]
2E1 FLE Details withheld at licensee's request by SSL.
2E1 FLF N Thornton, 13 Cross Rd, Dewsbury, WF12 0EB [IO93EP, SE21]
2M1 FLG A McDermid, 96 Strathleven Dr, Alexandria, G83 9PQ [IO75RX, NS37]
2M1 FLH Details withheld at licensee's request by SSL.
2E1 FLI J A Moriarty, 5 Fell View, Grimsargh, Preston, PR2 5LN [IO83QT, SD53]
2E1 FLJ K G Kaminskyj, 23 Cathkin Cl, Leicester, LE3 6PW [IO92JP, SK50]
2E1 FLK A J Forshaw, 6 Nixon Rd, Cuddington, Northwich, CW8 2QL [IO83QF, SJ57]
2E1 FLL G N Forshaw, 6 Nixon Rd, Cuddington, Northwich, CW8 2QL [IO83QF, SJ57]
2E1 FLM J D Brunlees, 91 Blake Ln, Sandiway, Northwich, CW8 2NW [IO83QF, SJ67]
2E1 FLN G E McMillan, 10 Thornton Ct, Girton, Cambridge, CB3 0NS [JO02BF, TL46]
2E1 FLO J M Coupe, 9 Weaver Cl, Crich, Matlock, DE4 5ET [IO93FD, SK36]
2E1 FLP P J Smith, 55 Countess Way, Euxton, Chorley, Lancs, PR7 6PT [IO83QP, SD51]
2E1 FLQ A Evans, 92 Rozel Ct, Southampton, SO16 9QE [IO90GW, SU31]
2E1 FLR P J Walker, 54 Burnage Ln, Manchester, M19 2NL [IO83VK, SJ89]
2E1 FLS S E Rose, 77 Glebe Ln, Maidstone, ME16 9BA [JO01FG, TQ75]
2E1 FLT Dr L Holland, 18 Anneforde Pl, Bracknell, RG42 2ES [IO91OK, SU87]
2E1 FLU D V Griffiths, 8 Oakfield Ave, Chester, CH2 1LQ [IO83NF, SJ46]
2E1 FLV M R Clulow, 60 Gilders, Sawbridgeworth, CM21 0EH [JO01BT, TL41]
2E1 FLW S H Burling, Ongar Cottage, 28 Main Rd, Langley, Macclesfield, Ches, SK11 0BU [IO83XF, SJ97]
2E1 FLX P J Dix, 2 Collett Ave, Shepton Mallet, BA4 5PL [IO81RE, ST64]
2E1 FLY S Archer, 80 Melbourne Rd, Clacton on Sea, CO15 3JA [JO01NT, TM11]
2E1 FLZ J M Watkins, 28 Bath Rd, Wells, BA5 3LG [IO81QF, ST54]
2M1 FMA Details withheld at licensee's request by SSL.
2E1 FMB T H Huntriss, 36 Woodfield, Bamber Bridge, Preston, PR5 8EB [IO83QR, SD52]
2E1 FMC R A Sanderson, 2 High St., Thornhill, Dewsbury, WF12 0PS [IO93EP, SE21]
2E1 FMD B Harrison, Beverley, Paice Ln, Medstead, Alton, Hants, GU34 5PT [IO91LC, SU63]
2E1 FME A D Knott, 11 Amhurst Cl, Leicester, LE3 9NA [IO92JP, SK50]
2E1 FMF L McGregor, 93 Heath End Rd, Flackwell Heath, High Wycombe, HP10 9ES [IO91PO, SU89]
2W1 FMG R D Saunders, 1 Taliesin, Forgeside, Cwmbran, NP44 3NR [IO81LP, ST29]
2E1 FMH C H Walker, 54 Burnage Ln, Manchester, M19 2NL [IO83VK, SJ89]
2M1 FMI Details withheld at licensee's request by SSL.
2E1 FMJ Details withheld at licensee's request by SSL.
2E1 FMK Details withheld at licensee's request by SSL.
2E1 FML S L S Bond, Southamfields Farm, Coventry Rd, Southam, Leamington Spa, Warks, CV33 0BG [IO92HG, SP46]
2E1 FMM G Barrett, 114 William St., Long Eaton, Nottingham, NG10 4GD [IO92IV, SK43]
2E1 FMN Details withheld at licensee's request by SSL.
2E1 FMO D M Tink, 96 Theobald Rd, Norwich, NR1 2NX [JO02PO, TG20]
2E1 FMP R J Robinson, 18 Trafalgar Ave, Audenshaw, Manchester, M34 5GH [IO83WL, SJ99]
2E1 FMQ C A Bowen, 45 Morris Rd, Nottingham, NG8 6NE [IO92JX, SK54]
2E1 FMR D Pearce, 55 Salesbury Way, Wigan, WN3 5QQ [IO83QM, SD50]
2E1 FMS Details withheld at licensee's request by SSL.
2E1 FMT R Cockings, 4 Freeman Gdns, High Green, Sheffield, S30 4NT [IO93GJ, SK38]
2E1 FMU R Liversidge, 17 Millbank Cl, High Green, Sheffield, S30 4NS [IO93GJ, SK38]
2E1 FMV P S Burgess, 27 Watergate St., Ellesmere, SY12 0EX [IO82NV, SJ43]
2E1 FMW A W Oughton, 176 South Lodge Dr, London, N14 4XN [IO91WP, TQ39]
2M1 FMX S Fraser, Hopefield Cottage, Gladsmuir, Tranent, East Lothian, EH33 2AL [IO85NW, NT47]
2E1 FMY Details withheld at licensee's request by SSL.
2W1 FMZ V E Lee, 23 Forest Cl, Coed Eva, Cwmbran, NP44 4TE [IO81LP, ST29]
2E1 FNA K A Johnson, 64 Chandlers Way, Hertford, SG14 2EF [IO91WT, TL31]
2E1 FNB J Morgan, 171 Town Rd, London, N9 0HJ [IO91XO, TQ39]
2E1 FNC C I Fletcher, 10 Moor Side, Boston Spa, Wetherby, LS23 6PD [IO93HV, SE44]
2E1 FND J Dann, 127 Ladybrook Ln, Mansfield, NG18 5JH [IO93JD, SK56]
2E1 FNE G Scott, 19 Witton Gdns, Jarrow, NE32 5YJ [IO94GX, NZ36]
2E1 FNF J McCarthy, 19 West Bank, North End Rd, Yapton, Arundel, BN18 0DW [IO90QT, SU90]
2E1 FNG Details withheld at licensee's request by SSL.
2E1 FNH Details withheld at licensee's request by SSL.
2E1 FNI G Ramsey, 9 Pevensey Cl, Southampton, SO16 9HF [IO90GW, SU31]
2E1 FNJ L C Carr, 10 Bonds Rd, Hemblington, Norwich, NR13 4QF [JO02RP, TG31]
2E1 FNK A S James, 36 Lemon Hill, Mylor Bridge, Falmouth, TR11 5NA [IO70LE, SW83]
2E1 FNL K Harvey, 29 The Hobbins, Bridgnorth, WV15 5HH [IO82TM, SO79]
2E1 FNM R A Fletcher, 27 Myrtle Ave, Long Eaton, Nottingham, NG10 3LZ [IO92IV, SK43]
2E1 FNN C A Newman, Newland, Seldom Seen Ln, Silecroft, Millom, Cumbria, LA18 4NX [IO84HF, SD18]
2E1 FNO M J Warren, 5 St. Georges Ct, Bracken Rd, North Baddesley, Southampton, SO52 9AT [IO90GX, SU31]
2E1 FNP G S Cullum, 4 Morland Rd, Ipswich, IP3 0LE [JO02OA, TM14]
2U1 FNQ J T Le Page, Heathwick, Les Martins, St Martins, Guernsey, GY4 6QJ
2E1 FNR D W Rowland, Wortham, Hugus Rd, Threemilestone, Truro, TR3 6DD [IO70KG, SW74]
2W1 FNS A Ellis, 44 Winchester Way, Gresford, Wrexham, LL12 8HL [IO83MC, SJ35]
2E1 FNT A S Garland, 40 Skiers View Rd, Hoyland, Barnsley, S74 0BS [IO93GL, SE30]
2E1 FNU E W Lockitt, Flat 5, 629 Wilbraham Rd, Manchester, M21 9JT [IO83UK, SJ89]
2E1 FNV D W Summerwill, 52 Lanmor Est, Lanner, Redruth, TR16 6HN [IO70JF, SW73]
2E1 FNW Details withheld at licensee's request by SSL.
2E1 FNX J W E Meredith, 2 Hamilton Rd, Dawley, Telford, TF4 3NG [IO82SP, SJ60]
2E1 FNY C B Warren, Clifden Farm, Quenchwell, Carnon Downs, Truro, Cornwall, TR3 6LN [IO70LF, SW84]
2E1 FNZ G W Browne, 30 Dereham Rd, Easton, Norwich, NR9 5EJ [JO02NP, TG11]
2U1 FOA C Fallaize, Lorbert, Pleinheaume Rd, Vale, Guernsey, GY6 8NR
2E1 FOB M G Kowalski, 47 Graveney Pl, Springfield, Milton Keynes, MK6 3LU [IO92PA, SP83]
2E1 FOC A J Rixon, Chiltern View Cottage, 7 Bishops Meadow, Bierton, Aylesbury, Bucks, HP22 5EF [IO91OT, SP81]
2E1 FOD P Chapman, 8 Bristow Cl, Bletchley, Milton Keynes, MK2 2XP [IO91PX, SP83]
2E1 FOE S P Woodward, 8 Hobart Rd, Ilford, IG6 2EB [JO01BO, TQ49]
2E1 FOF S M Simpson, 588 Dereham Rd, Norwich, NR5 8TE [JO02OP, TG10]
2E1 FOG R R Hill, 32 Ravensdale Ave, Long Eaton, Nottingham, NG10 4GG [IO92IV, SK43]
2E1 FOH F V Webley, 1 Bates Cl, Willen, Milton Keynes, MK15 9HZ [IO92PB, SP84]
2E1 FOI P J Sanders, 20 Moor Park Rd, Hereford, HR4 0RR [IO82PB, SO44]
2E1 FOJ I M Wilkinson, 1 Homefield Cl, Swanley, BR8 7JH [JO01CJ, TQ56]
2E1 FOK J A Coxhead, Barronsway, Parker Ln, Whitestake, Preston, Lancs, PR4 4JX [IO83PR, SD52]
2E1 FOL P M Evans, 27 Glentrammon Rd, Orpington, BR6 6DE [JO01BI, TQ46]
2E1 FOM L Moore, 16 Sutton Rd, Bolton, BL3 4QR [IO83SN, SD60]
2E1 FON S J Simpson, 125 Woodhouse Rd, Keighley, BD21 5NP [IO93BU, SE03]
2E1 FOO C D Meakin, 11 Lorimer Ave, Gedling, Nottingham, NG4 4BS [IO92LX, SK64]
2E1 FOP Details withheld at licensee's request by SSL.
2M1 FOQ A M Pryde, 11 Barleyknowe Pl, Gorebridge, EH23 4HD [IO85LU, NT36]
2E1 FOR S J Crowder, Sunny Croft, Bleasby Rd, Thurgarton, Nottingham, Notts, NG14 7FW [IO93MA, SK64]
2E1 FOS G D Barrington, 856 Burnley Rd, Bury, BL9 5JT [IO83UO, SD81]
2E1 FOT C A Pearson, 24 Broadlands, Netherfield, Milton Keynes, MK6 4HL [IO92PA, SP83]
2E1 FOU Details withheld at licensee's request by SSL.
2E1 FOV P J Halford, 39 Deepdale, Telford, TF3 2EJ [IO82SQ, SJ70]
2E1 FOW Details withheld at licensee's request by SSL.
2E1 FOX J R King, 48 Northfield Rd, Harpenden, AL5 5HZ [IO91TT, TL11]
2M1 FOY Dr D J Floyd, 84 Pentland Terr, Edinburgh, EH10 6HF [IO85JV, NT26]
2E1 FOZ L Dimmock, 26 Foxford Walk, Manchester, M22 5QN [IO83UJ, SJ88]
2E1 FPA Details withheld at licensee's request by SSL.
2E1 FPB N A France, 86 Palatine Dr, Bury, BL9 6RR [IO83UO, SD81]
2E1 FPC Details withheld at licensee's request by SSL.
2E1 FPD Details withheld at licensee's request by SSL.
2E1 FPE L Haines, 26 Maydman Sq, Portsmouth, PO3 6HT [IO90LT, SU60]
2E1 FPF K J Roberts, 42 Birdwood Gr, Fareham, PO16 8AF [IO90KU, SU50]
2E1 FPG C M Roberts, 42 Birdwood Gr, Fareham, PO16 8AF [IO90KU, SU50]
2E1 FPH A R Lockwood, 14 Milton Park Ave, Southsea, PO4 8JG [IO90LT, SZ69]
2E1 FPI G I Goodearl, 4 South Terr, High St., Farningham, Dartford, DA4 0DF [JO01CJ, TQ56]
2E1 FPJ S I Anders, 34 Lazenby Cres, Ashton in Makerfield, Wigan, WN4 9NJ [IO83QL, SJ59]
2W1 FPK A Cartwright, 14 Clement Dr, Rhyl, LL18 4HU [IO83GH, SJ08]
2U1 FPL J P Cooper, La Tonnellerie, Barras Ln, Vale, Guernsey, Channel Islands, GY6 8EJ
2E1 FPM J G Noel, 48 Burnside, Telford, TF3 1SX [IO82SP, SJ60]
2E1 FPN J Cocks, 26 Britten Way, Waterlooville, PO7 5XB [IO90LU, SU60]
2E1 FPO Details withheld at licensee's request by SSL.
2E1 FPP Details withheld at licensee's request by SSL.
2E1 FPQ R M Walton, 78 St. Lawrence Rd, North Wingfield, Chesterfield, S42 5LL [IO93HE, SK46]
2E1 FPR Details withheld at licensee's request by SSL.
2E1 FPS G A Williams, 22 West St., Tamworth, B79 7JE [IO92DP, SK20]
2E1 FPT J S Heffernan, 10 The Copperfields, Tupman Cl, Rochester, ME1 1RY [JO01GJ, TQ76]
2W1 FPU J C Overland, Mon Abri, Wrexham Rd, Caergwrle, Wrexham, Clwyd, LL12 9HN [IO83LC, SJ35]
2E1 FPV J J Stringer, The Cottage, High St., Oakhill, Bath, Avon, BA3 5AL [IO81RF, ST64]
2E1 FPW J K Green, 2 Broadmeadow, Kingswinford, DY6 7HG [IO82WM, SO88]
2W1 FPX G M Parry, 23 Heol Dirion, Coedpoeth, Wrexham, LL11 3HL [IO83LB, SJ25]
2E1 FPY R J Westgarth, 11 Beckford Ave, Bracknell, RG12 7ND [IO91OJ, SU86]
2E1 FPZ T Watkinson, 29 Ash Cres, Nuthall, Nottingham, NG16 1FL [IO92JX, SK54]
2W1 FQA I C Tuite, 44 Gorlan, Conwy, LL32 8RS [IO83BG, SH77]
2E1 FQB P D Elcombe, 16 Blenheim Ave, Martham, Great Yarmouth, NR29 4TW [JO02TQ, TG41]
2E1 FQC Details withheld at licensee's request by SSL.
2W1 FQD Details withheld at licensee's request by SSL.
2E1 FQE C Carter, 331A Ordnance Rd, Enfield, EN3 6HE [IO91XQ, TQ39]
2W1 FQF R G Williams, Plas Iolyn, Picton, Holywell, Flintshire, CH8 9JQ [IO83IH, SJ18]
2W1 FQG M E Fray, 6 Meirwen Dr, Cardiff, CF5 4ND [IO81IL, ST17]
2W1 FQH A J Fray, 6 Meirwen Dr, Cardiff, CF5 4ND [IO81IL, ST17]
2M1 FQI G Robinson, 12 Hannahston Ave, Drongan, Ayr, KA6 7AU [IO75SK, NS41]
2M1 FQJ J M Clark, 16 Bonnyton Ave, Drongan, Ayr, KA6 7DG [IO75SK, NS41]
2M1 FQK P M Smith, 9 Glenmount Pl, Ayr, KA7 4JE [IO75QK, NS31]
2W1 FQL C H Bowen, 69 Dylan, Llanelli, SA14 9AS [IO71WQ, SN50]
2W1 FQM Details withheld at licensee's request by SSL.
2E1 FQN E A Costa, 5 Webb Est, Clapton Common, London, E5 9BB [IO91XN, TQ38]
2E1 FQO M E Dunthorne, 56 Nottingham Rd, Lowdham, Nottingham, NG14 7AP [IO93LA, SK64]
2E1 FQP M Dunthorne, 56 Nottingham Rd, Lowdham, Nottingham, NG14 7AP [IO93LA, SK64]
2E1 FQQ P Wilson, 46 Moor Ave, Penwortham, Preston, PR1 0NE [IO83PR, SD52]
2E1 FQR C R D Livens, 122 Fairmead Ave, Westcliff on Sea, SS0 9SB [JO01IN, TQ88]
2E1 FQS C Murray, 1 Throckmorton, Warboys, Huntingdon, PE17 2RY [IO92XJ, TL38]
2E1 FQT Details withheld at licensee's request by SSL.
2E1 FQU Details withheld at licensee's request by SSL.
2E1 FQV Y J Young, 41 Park Rd, Donnington, Telford, TF2 8BP [IO82SR, SJ71]
2E1 FQW A J Young, 41 Park Rd, Donnington, Telford, TF2 8BP [IO82SR, SJ71]
2E1 FQX J W Griffin, 4 Cundishall Cl, Whitstable, CT5 4DA [JO01MI, TR16]
2E1 FQY O G James, 24 Fryer Ave, Leamington Spa, CV32 6HY [IO92FH, SP36]
2E1 FQZ Details withheld at licensee's request by SSL.
2E1 FRA G E Millard, 20 Tor O Moor Rd, Woodhall Spa, LN10 6TD [IO93VD, TF26]
2E1 FRB M R Howard, 36 Thornton Cres, Horncastle, LN9 6JP [IO93WE, TF26]

2E1 FRC Mast H W Doyle, Hurst House, Stratford Rd, Henley in Arden, Solihull, West Midlands, B95 6AB [IO92CG, SP16]
2E1 FRD J P Montanaro, 29 Northd. Rd, Leamington Spa, CV32 6HE [IO92FH, SP36]
2E1 FRE A G Franklin, 71 Parkside, Revesby, Boston, PE22 7NH [IO93XD, TF36]
2E1 FRF W Bradbury, 192 Greenwood Rd, Nottingham, NG3 7FY [IO92KW, SK64]
2E1 FRG J C E Bishop, 6 Old Budbrooke Rd, Hampton Magna, Warwick, CV35 8RS [IO92EG, SP26]
2E1 FRH G L Parker, 420 Meadow Ln, Nottingham, NG2 3GD [IO92KW, SK53]
2E1 FRI J J Wood, 17 St. Peters Cl, Henley, Ipswich, IP6 0RH [JO02NC, TM15]
2E1 FRJ J C Taylor, Willow Cottage, Bulls Rd, Hemingstone, Ipswich, IP6 9RF [JO02ND, TM15]
2E1 FRK D P Barrett, 5 Swift Cl, Flitwick, Bedford, MK45 1RG [IO91RX, TL03]
2E1 FRL J D Mageehan, 37 Gosbecks Rd, Colchester, CO2 9JR [JO01KU, TL92]
2E1 FRM B A Smith, 38 Drury Rd, Colchester, CO2 7UX [JO01KV, TL92]
2E1 FRN C J Dalton-Moore, 12 Nightingale Cl, Great Alne, Alcester, B49 6PE [IO92BF, SP05]
2E1 FRO D J Mageehan, 37 Gosbecks Rd, Colchester, CO2 9JR [JO01KU, TL92]
2E1 FRP A Brotherhood, 23 St. James Rd, Shepshed, Loughborough, LE12 9JB [IO92IS, SK41]
2E1 FRQ A O Storry, 99 Swineshead Rd, Wyberton Fen, Boston, PE21 7JG [IO92XX, TF24]
2E1 FRR K Sylvester, 8 Beacon Park Cl, Skegness, PE25 1HQ [JO03DD, TF56]
2E1 FRS Details withheld at licensee's request by SSL.
2E1 FRT M White, 100 Burnham Rd, Coventry, CV3 4BQ [IO92GJ, SP37]
2E1 FRU D Totty, Sycamore Farm, Church Ln, Morton, Alfreton, DE55 6GU [IO93HD, SK46]
2E1 FRV B Pearce, 23 Furnival St., Manton, Worksop, S80 2NF [IO93KH, SK57]
2E1 FRW T J Depledge, 16 Pennington Walk, Retford, DN22 6LR [IO93MH, SK78]
2E1 FRX A J Coundley, Vale Cottage, Fritham, Lyndhurst, Hants, SO43 7HJ [IO90EW, SU21]
2E1 FRY E A Young, 7 The Quadrant, Fordingbridge, SP6 1BW [IO90CW, SU11]
2E1 FRZ T Jennings, 4 Holly Rise, New Ollerton, Newark, NG22 9UZ [IO93LF, SK66]
2E1 FSA Details withheld at licensee's request by SSL.
2E1 FSB A Ashfield, 14 Achilles Ave, Warrington, WA2 9RW [IO83QJ, SJ69]
2E1 FSC P A Bush, 3 Smarts Cttgs, The Green, Bearsted, Maidstone, ME14 4EA [JO01HG, TQ85]
2E1 FSD A D Marston, 10 Chestnut Cl, Queniborough, Leicester, LE7 3DW [IO92LQ, SK61]
2E1 FSE J T Dixon, 11 Holden Dr, Burgh-le-Marsh, Skegness, PE24 5LZ [JO03CD, TF56]
2E1 FSF J A D Kinch, 3 Hamble Ct, Basingstoke, RG21 4DA [IO91LG, SU65]
2E1 FSG R T Kinch, 3 Hamble Ct, Basingstoke, RG21 4DA [IO91LG, SU65]
2E1 FSH C D Forber, 111 Sefton St., Newton-le-Willows, WA12 9LF [IO83QK, SJ59]
2E1 FSI R P G Almond, 6 Bowen Rd, Rugby, CV22 5LF [IO92JI, SP57]
2E1 FSJ T R Taylor, 46 Bullards Ln, Woodbridge, IP12 4HE [JO02PC, TM24]
2E1 FSK A C Williamson, 5 Orchard Dr, South Hiendley, Barnsley, S72 9BQ [IO93HO, SE31]
2E1 FSL D A Johnston, 25 Haughgate Cl, Woodbridge, IP12 1LQ [JO02PC, TM25]
2E1 FSM J W Ford, 36 Westland, Martlesham Heath, Ipswich, IP5 7SU [JO02OB, TM24]
2E1 FSN K M A Stooke, 29 Tweed Cl, Swindon, SN2 3PU [IO91CO, SU18]
2E1 FSO I P M Johnson, 61 Wickdown Ave, Swindon, SN2 3DY [IO91CO, SU18]
2E1 FSP M C Johnson, 41 Westland, Martlesham Heath, Ipswich, IP5 7SU [JO02OB, TM24]
2E1 FSQ Details withheld at licensee's request by SSL.
2E1 FSR G P H Hammond, 50 Fernhill Cl, Woodbridge, IP12 1LB [JO02PC, TM25]
2E1 FSS P J Bloom, 12 Pytches Cl, Melton, Woodbridge, IP12 1SE [JO02PC, TM24]
2E1 FST J Cohen, 42 Millway, London, NW7 3RA [IO91VO, TQ29]
2E1 FSU A Harradine, 2 Balmoral Ave, Northwich, CW9 8BH [IO83RF, SJ67]
2E1 FSV D Feetenby, 32 Hawkins Cl, Daventry, NN11 4JQ [IO92KG, SP56]
2E1 FSW Details withheld at licensee's request by SSL.
2E1 FSX D J Charlton, 20 Bailey Cres, South Elmsall, Pontefract, WF9 2TL [IO93IO, SE41]
2E1 FSY M J Fey, 37 Winnards Cl, West Parley, Ferndown, BH22 8PA [IO90BS, SZ09]
2E1 FSZ T Bashford, 198 Uplands Rd, West Moors, Ferndown, BH22 0EY [IO90BT, SU00]
2E1 FTA S Goodall, 37 Woodfield Rd, Bournemouth, BH11 9EU [IO90BS, SZ09]
2E1 FTB Details withheld at licensee's request by SSL.
2E1 FTC B J Hunt, 2A Golf Rd, Radcliffe on Trent, Nottingham, Notts, NG12 2GA [IO92LW, SK63]
2E1 FTD Details withheld at licensee's request by SSL.
2E1 FTE C Bingham, 56 Newgate Ln, Mansfield, NG18 2LQ [IO93JD, SK56]
2E1 FTF E J Sheppard, 4 Lindrick Ave, Swinton, Mexborough, S64 8TE [IO93IL, SK49]
2E1 FTG H J Smith, Restormel House, Squires Ln, Martlesham Heath, Ipswich, Suffolk, IP5 7UG [JO02OB, TM24]
2E1 FTH R C Megone, 16 Mercer Cl, Basingstoke, RG22 6NZ [IO91KG, SU65]
2E1 FTI M Burrows, 40 Fairmile Rd, Christchurch, BH23 2LL [IO90CR, SZ19]
2E1 FTJ D L Lacy, 30 Gr Rd East, Christchurch, BH23 2DQ [IO90CR, SZ19]
2E1 FTK N L Brewer, 31 Dale St., Rawmarsh, Rotherham, S62 7BZ [IO93HL, SK49]
2E1 FTL D Johnson, 85 Cavendish Cl, Old Hall, Warrington, WA5 5PS [IO83QJ, SJ58]
2E1 FTM Details withheld at licensee's request by SSL.
2E1 FTN G Reeds, 26 Holme Leaze, Steeple Ashton, Trowbridge, BA14 6EH [IO81WH, ST95]
2E1 FTO B E Geall, 129 Jewell Rd, Bournemouth, BH8 0JP [IO90CR, SZ19]
2E1 FTP Details withheld at licensee's request by SSL.
2E1 FTQ Details withheld at licensee's request by SSL.
2E1 FTR P A White, 61 North St., Pewsey, SN9 5ES [IO91CI, SU16]
2E1 FTS T Anderson, 1 Thames Cl, Ferndown, BH22 8XA [IO90BT, SU00]
2E1 FTT M Anderson, 1 Thames Cl, Ferndown, BH22 8XA [IO90BT, SU00]
2E1 FTU P Kennedy, 22 Kirkwick Ave, Harpenden, AL5 2QX [IO91TT, TL11]
2E1 FTV J D Bailey, 13 Newark Rd, Mexborough, S64 9EZ [IO93IL, SE40]
2E1 FTW L Cohen, 42 Millway, London, NW7 3RA [IO91VO, TQ29]
2E1 FTX M A Gillen, 170 Rushton Gr, Harlow, CM17 9PT [JO01BS, TL40]
2E1 FTY T C Everitt, 39 Keats Rd, Larkfield, Aylesford, ME20 6TP [JO01FH, TQ75]
2E1 FTZ T J Kay, 5 Church St., Brighouse, HD6 3NF [IO93CQ, SE12]
2E1 FUA P A McCormack, 119 Lockgate West, Windmill Hill, Runcorn, WA7 6LE [IO83PI, SJ58]
2E1 FUB B R W Bailey, The Cottage, Alderton Rd, Grittleton, Chippenham, SN14 6AN [IO81VM, ST88]
2W1 FUD D Green, 23 Litchard Park, Bridgend, CF31 1PF [IO81FM, SS98]
2E1 FUE T W Sherratt, 28 Church St., Bollington, Macclesfield, SK10 5PY [IO83WH, SJ97]
2E1 FUF Details withheld at licensee's request by SSL.
2E1 FUG A B Crowther, 3 Andrew Cres, Hill Tree Park, Crosland Hill, Huddersfield, HD4 7AF [IO93CO, SE11]
2E1 FUH D J Hartley, 99 Carr Rd, Fleetwood, FY7 6QQ [IO83LW, SD34]
2E1 FUI A J Taylor, 46 Bullards Ln, Woodbridge, IP12 4HE [JO02PC, TM24]
2E1 FUJ J W R Hartley, 99 Carr Rd, Fleetwood, FY7 6QQ [IO83LW, SD34]
2W1 FUL S S Luke, 33 Maiden St., Maesteg, CF34 9HP [IO81EO, SS88]
2E1 FUM L A Bennett, 7 Parsons Dr, Ellington, Huntingdon, PE18 0AU [IO92UH, TL17]
2E1 FUN Details withheld at licensee's request by SSL.
2E1 FUO Details withheld at licensee's request by SSL.
2E1 FUP Details withheld at licensee's request by SSL.
2E1 FUQ J Quartermaine, 41 Althorp Rd, Northampton, NN5 5EQ [IO92NF, SP76]
2E1 FUR Details withheld at licensee's request by SSL.
2E1 FUS A Sherratt, 28 Church St., Bollington, Macclesfield, SK10 5PY [IO83WH, SJ97]
2E1 FUT G Ashfield, 14 Achilles Ave, Warrington, WA2 9RW [IO83QJ, SJ69]
2E1 FUU Details withheld at licensee's request by SSL.
2W1 FUV G M E Thomas, 2 Heol Fach, Pencoed, Bridgend, CF35 6UT [IO81FM, SS98]
2W1 FUW G D Thomas, 2 Heol Fach, Pencoed, Bridgend, CF35 6UT [IO81FM, SS98]
2E1 FUY F S Overland, 1 Horse Shoe Cl, Kingsley, Warrington, WA6 8DY [IO83PG, SJ57]
2E1 FUZ C G Hope, 40 Cavendish Rd, Blackpool, FY2 9JR [IO83LU, SD33]
2E1 FVA Details withheld at licensee's request by SSL.
2E1 FVB P C Hadwen, 54 Fairfield Ave, Felixstowe, IP11 9JJ [JO01QX, TM33]
2E1 FVC Details withheld at licensee's request by SSL.
2E1 FVD C A Ballam, Meeryfield, Mill Rd, Waldringfield, Woodbridge, Suffolk, IP12 4PY [JO02PB, TM24]
2E1 FVE D Bragg, 2 The Croft, Fleetwood, FY7 8DY [IO83LV, SD34]
2W1 FVF Details withheld at licensee's request by SSL.
2E1 FVG Details withheld at licensee's request by SSL.
2W1 FVH R Hallett, Llamedos, 151 Trealaw Rd, Tonypandy, CF40 2NX [IO81GO, SS99]
2E1 FVI S K Bhattie, 43 Spire Ct, Marsh, Huddersfield, HD1 4NW [IO93CP, SE11]
2E1 FVJ R E Halford, 104 Gladstone Ave, London, N22 6LH [IO91WO, TQ39]
2E1 FVK T Kiely, 192 Morley Ave, London, N22 6NT [IO91WO, TQ39]
2E1 FVL G J Abbotts, Lanterns, Gullivers Cl, Horley, Banbury, OX15 6DY [IO92HC, SP44]
2E1 FVM P Woods, 28 Henniker Rd, Debenham, Stowmarket, IP14 6PY [JO02OF, TM16]
2I1 FVN Details withheld at licensee's request by SSL.
2E1 FVO M T Lyne, Yew Tree Cottage, Pitmans Corner, Wetheringsett, Stowmarket, Suffolk, IP14 5PX [JO02NG, TM16]
2E1 FVP Details withheld at licensee's request by SSL.
2E1 FVQ S Worthington, 11 Plantagenet Ct, Nottingham, NG3 1HJ [IO92KX, SK54]
2E1 FVR P A Parsons, 142 Heath Park Rd, Romford, RM2 5XL [JO01CN, TQ58]
2E1 FVS M P Hotchin, 15 Stocks Dr, Shepley, Huddersfield, HD8 8EY [IO93DO, SE10]
2E1 FVT L H Green, 48 Clydesdale Rd, Hornchurch, RM11 1AG [JO01CN, TQ58]
2E1 FVU S G Papworth, 4 Queensway, Sturton By Stow, Lincoln, LN1 2AD [IO93QH, SK88]
2E1 FVV A G Hanner, 10 St. Edmunds Cl, Crawley, RH11 7SR [IO91VD, TQ23]
2I1 FVW S K Johnston, 20 Dorchester Gdns, Newtownabbey, BT36 5JJ [IO74AQ, J38]
2E1 FVX Details withheld at licensee's request by SSL.
2E1 FVY D M Ripley, 5 Rope Walk, Cranbrook, TN17 3DZ [JO01GC, TQ73]
2E1 FVZ F Robertson, 44 Shellbeach, Leysdown on Sea, Sheerness, ME12 4RL [JO01LJ, TR06]
2E1 FWA C Walker, 12 Bradshaw Cres, Honley, Huddersfield, HD7 2EG [IO93CO, SE11]
2W1 FWB G J Edwards, Camorafon, Dolywern, Pontfadog, Llangollen, Clwyd, LL20 7AD [IO82KW, SJ23]
2E1 FWC Details withheld at licensee's request by SSL.
2E1 FWD C L Booth, 8 Heathfield Mews, Martlesham Heath, Ipswich, IP5 7UF [JO02OB, TM24]
2E1 FWE M M Brown, Barn Cottage, Coton, Chacombe, Banbury, OX17 2JX [IO92IC, SP44]
2E1 FWF Details withheld at licensee's request by SSL.
2E1 FWG B A Bosson, 43A Compton Bassett, Calne, SN11 8RG [IO91AK, SU07]
2E1 FWH D P Crossland, 19 Hawker Cl, Wimborne, BH21 1XW [IO90AS, SZ09]
2W1 FWI O G Doak, 27 Hill St., Gilfach Goch, Porth, CF39 8TW [IO81GO, SS98]
2W1 FWJ T G Doak, 27 Hill St., Gilfach Goch, Porth, CF39 8TW [IO81GO, SS98]
2E1 FWK Details withheld at licensee's request by SSL.
2E1 FWL Details withheld at licensee's request by SSL.
2E1 FWM M E Foister, 38 Nine Acres, Kennington, Ashford, TN24 9JW [JO01KD, TR04]
2E1 FWN Details withheld at licensee's request by SSL.
2E1 FWO A Scott, Shrublands Cttgs, Brockford Green, Stowmarket, Suffolk, IP14 5NL [JO02NF, TM16]
2E1 FWP P C Curtis, 18 Dawlish Cres, Weymouth, DT4 9JN [IO80SO, SY67]
2E1 FWQ R Matthews, 18 Hawkins Cl, Daventry, NN11 4JQ [IO92KG, SP56]
2E1 FWR Details withheld at licensee's request by SSL.
2W1 FWS Details withheld at licensee's request by SSL.
2E1 FWT B A Tichler, Timbers, Hatching Green, Harpenden, Herts, AL5 2JP [IO91TT, TL11]
2E1 FWU M Pinnock, Myrtle Lodge, 55 Homewood Rd, St. Albans, Herts, AL1 4BG [IO91US, TL10]
2E1 FWV Details withheld at licensee's request by SSL.
2E1 FWW D W Matthews, 14 Frobisher Cl, Daventry, NN11 4JH [IO92KG, SP56]
2E1 FWX G J Sessions, 10B Victoria Rd, Diss, Norfolk, IP22 3HE [JO02NI, TM17]
2E1 FWY C J Whiting, 21 Ducksen Rd, Mendlesham, Stowmarket, IP14 5SD [JO02MG, TM16]
2E1 FWZ S Strobel, 16 Barley Ponds Rd, Ware, SG12 7EZ [IO91XT, TL31]
2E1 FXA Details withheld at licensee's request by SSL.
2E1 FXB Details withheld at licensee's request by SSL.
2E1 FXC M Taylor, Kerensa, High St., Thorndon, Eye, Suffolk, IP23 7LX [JO02NG, TM16]
2I1 FXD S J Clarke, 53 Lisnamuck Rd, Black Hill, Blackhill, Coleraine, BT51 4HN [IO65QB, C82]
2E1 FXE Details withheld at licensee's request by SSL.
2M1 FXF N Hamilton, 19 Abbots Cres, Doonfoot, Ayr, KA7 4JR [IO75QK, NS31]
2E1 FXG P B Sebborn, 32 Wroxham Way, Harpenden, AL5 4PP [IO91TT, TL11]
2M1 FXH B Newcombe, 9 Calder Rd, Bellsquarry, Livingston, EH54 9AA [IO85FU, NT06]
2W1 FXI J A Horton, 60 Pentwyn, Radyr, Cardiff, CF4 8RE [IO81IM, ST18]
2E1 FXJ B Guyver, 5 Whitehead Cl, Lychpit, Basingstoke, RG24 8SG [IO91LG, SU65]
2E1 FXK R Guyver, 5 Whitehead Cl, Lychpit, Basingstoke, RG24 8SG [IO91LG, SU65]
2E1 FXL R C de La Rue, 80 Keable Rd, Marks Tey, Colchester, CO6 1XR [JO01JU, TL92]
2E1 FXM E B Birch, 8 Eleanor Way, Warley, Brentwood, CM14 5AQ [JO01DO, TQ59]
2E1 FXN D E Boland, 242 Longley Ave West, Sheffield, S5 8UH [IO93GJ, SK39]
2E1 FXO P Large, 1 Long Ing Cttgs, Shaw Ln, Holmfirth, Huddersfield, HD7 1PP [IO93CN, SE10]
2E1 FXP A Baxter, 408 Leeds Rd, Wakefield, WF1 2JB [IO93FQ, SE32]
2M1 FXQ I R Street, 5 Calder Rd, Bellsquarry, Livingston, EH54 9AA [IO85FU, NT06]
2E1 FXR G T A Hall, Greenbank The Green, Hardstoft, Pilsley, Chesterfield, Derbyshire, S45 8AE [IO93HD, SK46]
2E1 FXS E L Breeze, 9 Yew Tree Cl, Bradwell, Great Yarmouth, NR31 8NZ [JO02UN, TG50]
2E1 FXT S J Leak, 3 Stelfox Ln, Audenshaw, Manchester, M34 5HE [IO83WL, SJ99]
2E1 FXU J G Schofield, 6 Naburn Cl, Stockport, SK5 8JQ [IO83WK, SJ99]
2E1 FXV R G Heath, 7 Albany Rd, Lymm, WA13 9LU [IO83SJ, SJ68]
2E1 FXW S J Buckley, 44 Stevenage Rd, Knebworth, SG3 6NN [IO91VU, TL22]
2W1 FXX A P Wood, 18 Priory Oak, Brackla, Bridgend, CF31 2HY [IO81FM, SS98]
2E1 FXY A C Capon, Bramleys, Station Rd, Cantley, Norfolk, NR13 3SH [JO02SN, TG30]
2E1 FXZ U D Reddy, 221 Kings Dr, Eastbourne, E Sussex, BN21 2UJ [JO00DS, TQ50]
2E1 FYA V R I Edirisinghe, 53 Upper Ratton Dr, Eastbourne, BN20 9BY [JO00CS, TQ50]
2E1 FYB C Waller, 32 Mill Green, Eastry, Sandwich, CT13 0LE [JO01PF, TR35]
2E1 FYC D C Martyr, 2 de Roos Rd, Eastbourne, BN21 2QA [JO00DS, TV69]
2E1 FYD T Hawkins, 114 Arthur St., Derby, DE1 3EH [IO92GW, SK33]
2W1 FYE Details withheld at licensee's request by SSL.
2E1 FYF B T Barrett, Garden Flat, 16 Surbiton Gr, Ryde, PO33 1EB [IO90KR, SZ59]
2W1 FYG Details withheld at licensee's request by SSL.
2E1 FYH M Essex, 6 Chapel Cl, Westwoodside, Doncaster, DN9 2PD [IO93NL, SK79]
2E1 FYI P M Wootton, 74 White Lodge Park, Shawbury, Shrewsbury, SY4 4NU [IO82QS, SJ52]
2E1 FYJ Details withheld at licensee's request by SSL.
2E1 FYK J Evans, 16 Kynaston Dr, Wem, Shrewsbury, SY4 5DE [IO82PU, SJ52]
2E1 FYL G Lyon, 47 Chapel Ln, Spondon, Derby, DE21 7JT [IO92HW, SK43]
2E1 FYM A A Harris, 11 Sea View Rd, Cliffsend, Ramsgate, CT12 5EH [JO01QI, TR36]
2E1 FYN Details withheld at licensee's request by SSL.
2E1 FYO Details withheld at licensee's request by SSL.
2E1 FYP S D Dunn, 33 Southbrook Cl, Poole, BH17 8BG [IO90AS, SZ09]
2E1 FYQ C D Chapman, 376 Wallisdown Rd, Bournemouth, BH11 8PS [IO90BR, SZ09]
2E1 FYR P D Geilern, 20 New Borough, Wimborne, BH21 1RA [IO90AT, SZ09]
2E1 FYS K McGregor, 75 Sutton Ave, Chellaston, Derby, DE73 1RJ [IO92GV, SK33]
2E1 FYT D Goodman, 52 Crayford Rd, Alvaston, Derby, DE24 0HN [IO92GV, SK33]
2E1 FYU H Horwood, Willow Lodge, Ipswich Rd, Yaxley, Eye, Suffolk, IP23 8BX [JO02NH, TM17]
2E1 FYV M N Jones, 35 Pendle Way, Meole Brace, Shrewsbury, SY3 9QS [IO82OQ, SJ41]
2E1 FYW R A Bright, Sandiacre, Orchard Ln, Hanwood, Shrewsbury, SY5 8LE [IO82OQ, SJ40]
2E1 FYX B E Fowler, Old Bakery, 4 Langley St., Derby, Derbyshire, DE22 3GL [IO92GW, SK33]
2E1 FYY W J White, 1 Red House, Gallamore Ln, Middle Rasen, Market Rasen, LN8 3UB [IO93TJ, TF08]
2E1 FYZ T R Wilkie, Bramcote, Grange Rd, Sandilands, Sutton on Sea, LN12 2RE [JO03DH, TF58]
2E1 FZA M A Crossley, 111 Skagen Ct, Bolton, BL1 2JD [IO83SO, SD71]
2E1 FZB G J Turner, 35 Horncastle Rd, Wragby, Market Rasen, LN8 5RB [IO93UG, TF17]
2E1 FZC J A Charter, 36 Northd. Ave, London, E12 5HD [JO01AN, TQ48]
2E1 FZD M G Roberts, 13 Fearnley Dr, Ossett, WF5 9EU [IO93FQ, SE22]
2E1 FZE H L Glover, 14 Crawley Cres, Eastbourne, BN22 9RN [JO00DT, TQ60]
2E1 FZF P R Ferris, 116 Capel Rd, Forest Gate, London, E7 0JS [JO01AN, TQ48]
2E1 FZG Details withheld at licensee's request by SSL.
2E1 FZH M R Saunders, 6 Hodson Cl, Bury St. Edmunds, IP32 6RS [JO02IG, TL86]
2E1 FZI M L Gardiner, 24 Backbury Rd, Hereford, HR1 1SD [IO82PB, SO54]
2E1 FZJ Details withheld at licensee's request by SSL.
2E1 FZK B P Barnett, 7 Holts Ln, Tutbury, Burton on Trent, DE13 9LE [IO92DU, SK22]
2E1 FZL A M Barnett, 7 Holts Ln, Tutbury, Burton on Trent, DE13 9LE [IO92DU, SK22]
2E1 FZM S W Chandler, 33 Kingfisher Dr, Birmingham, B36 0RD [IO92DM, SP18]
2E1 FZN G W Admans, 1 Oak Tree Cl, Hertford, SG13 7RG [IO91XS, TL31]
2E1 FZO P A J Geier, 24 Deirdre Ave, Wickford, SS12 0AX [JO01GO, TQ79]
2E1 FZP P Chesterton, Richmond House, Rodington Heath, Shrewsbury, SY4 4QX [IO82QR, SJ51]
2E1 FZQ M R Birch, 67 Edgeworth Dr, Manchester, M14 6RS [IO83VK, SJ89]
2E1 FZR G B Phillips, 34 Chambercombe Rd, Ilfracombe, EX34 9PH [IO71WE, SS54]
2E1 FZS A H Dyche, 29 Snailbeach, Minsterley, Shrewsbury, Salop, SY5 0NS [IO82NO, SJ30]
2E1 FZT M A Tait, William Cottage, Hinderclay, Diss, IP22 1HX [JO02LI, TM07]
2E1 FZU D A Peters, 25 Corndon Cres, Shrewsbury, SY1 4LD [IO82PR, SJ51]
2E1 FZV D C Burke, 6 St. Farm Ln, Ixworth, Bury St. Edmunds, IP31 2JE [JO02JH, TL97]
2E1 FZW L Baker, Brickworks, Old Stowmarket Rd, Woolpit, Bury St. Edmunds, IP30 9QS [JO02KF, TL96]
2E1 FZX M R Johnson, 24 Balmoral Ave, Stanford-le-Hope, SS17 7BD [JO01FM, TQ68]
2E1 FZY J M Doyle, 33 Bodenham Rd, Birmingham, B31 5DP [IO92AJ, SP07]
2E1 FZZ M E Clarke, 78 Woking Rd, Cheadle Hulme, Cheadle, SK8 6NU [IO83VI, SJ88]
2E1 GAA F N Dixon, 18 Lansdowne Dr, Sutton on Sea, Mablethorpe, LN12 2JD [JO03DH, TF58]
2E1 GAB A J Rowe, 51 Cranwell Dr, Manchester, M19 1NE [IO83VK, SJ89]
2W1 GAC W T Hicks, 2 Second Ave, Morriston, Clase, Swansea, SA6 7LN [IO81AP, SS69]
2E1 GAD C L Lee, 2 Edward Rd, Romford, RM6 6UH [JO01BN, TQ48]
2E1 GAE M S Ewart, 27 Fairfield Rd, Heckmondwike, WF16 9NP [IO93ER, SE22]
2E1 GAF S S Ewart, 27 Fairfield Rd, Heckmondwike, WF16 9NP [IO93ER, SE22]
2E1 GAG L Raine, 3 Hainworth, Keighley, BD21 5QH [IO93BU, SE03]
2E1 GAH J Tildesley, 26 Vale Mill Ln, Haworth, Keighley, BD22 0EF [IO93AU, SE03]
2E1 GAI Details withheld at licensee's request by SSL.
2E1 GAJ Details withheld at licensee's request by SSL.
2E1 GAK L M Williams, 1 Home Farm Cottage, Leebotwood, Church Stretton, Salop, SY6 6LX [IO82ON, SO49]
2E1 GAL Details withheld at licensee's request by SSL.
2E1 GAM Details withheld at licensee's request by SSL.
2E1 GAN B D Lea, 49 Hewitt Ave, Hereford, HR4 0QR [IO82PB, SO44]
2E1 GAO Details withheld at licensee's request by SSL.
2E1 GAP Details withheld at licensee's request by SSL.
2E1 GAQ M Chapman, 15 Smithies Moor Ln, Birstall, Batley, WF17 9AT [IO93ER, SE22]
2E1 GAR M R Szybut, 27 Middleton Ave, Littleover, Derby, DE23 6DN [IO92FV, SK33]
2E1 GAS F A Sadler, 12 Yokecliffe Dr, Wirksworth, Matlock, DE4 4EX [IO93FB, SK25]
2E1 GAT N J Godbold, 16 Mariners Way, Aldeburgh, IP15 5QH [JO02TD, TM45]
2W1 GAU D Conde, 13 Broxton Rd, Wrexham, LL13 9BA [IO83MB, SJ35]

2E1 GAV L W Clayton, 40 School Ln, Ramsgate, CT11 8QX [JO01RI, TR36]
2W1 GAW Details withheld at licensee's request by SSL.
2E1 GAX K Brassington, 35 Harpur Ave, Littleover, Derby, DE23 7EJ [IO92FV, SK33]
2E1 GAY J H Burch, 1 The Bungalows, Red House Ln, Leiston, IP16 4LS [JO02TE, TM46]
2W1 GAZ M R Woosnam, 8 Hillside, Risca, Newport, NP1 6QD [IO81KO, ST29]
2E1 GBA P R B Bones, 1 Hawthorn Way, Burwell, Cambridge, CB5 0DQ [JO02DG, TL56]
2E1 GBB Details withheld at licensee's request by SSL.
2E1 GBC J K Houlden, Queach Cottage, Leys Hill, Walford, Ross-on-Wye, HR9 5QU [IO81QV, SO52]
2E1 GBD P Houghton, 19 Gilthwaites Ln, Denby Dale, Huddersfield, HD8 8SG [IO93EN, SE20]
2E1 GBE D H Thorne, Willow Lodge, The St., Barrow, Bury St. Edmunds, Suffolk, IP29 5AP [JO02GF, TL76]
2E1 GBF J Thorne, Willow Lodge, The St., Barrow, Bury St. Edmunds, Suffolk, IP29 5AP [JO02GF, TL76]
2M1 GBG M Thomson, 194 East Main St., Broxburn, EH52 5HQ [IO85GW, NT07]
2E1 GBH D Hales, Brookview, 15 Meadow Bank Rd, Hereford, Herefordshire, HR1 2ST [IO82PB, SO53]
2E1 GBI K J Suddes, 10 Tilecroft, Welwyn Garden City, AL8 7QY [IO91VT, TL21]
2E1 GBJ Details withheld at licensee's request by SSL.
2E1 GBK J M Glanfield, 21 Rogers Cl, Felixstowe, Suffolk, IP11 9DG [JO01QX, TM23]
2E1 GBL J H S Taylor, 3 Foxgrove Ln, Felixstowe, Suffolk, IP11 7JS [JO01QX, TM33]
2E1 GBM J E Lake, 55 Colneis Rd, Felixstowe, Suffolk, IP11 9HH [JO01QX, TM33]
2E1 GBN A P Vincent, 9 Broad Park Ave, Ilfracombe, EX34 8DZ [IO71WE, SS54]
2E1 GBO D J McArdell, 9 Station Rd, Mablethorpe, Lincs, LN12 1HA [JO03DI, TF58]
2E1 GBP G Mattock, 4 Yew Tree Ln, Thulston, Derby, Derbyshire, DE72 3FG [IO92HV, SK43]
2E1 GBQ Details withheld at licensee's request by SSL.
2E1 GBR O G Spevack, Fircroft, Orestan Ln, Effingham, Surrey, KT24 5SN [IO91TG, TQ15]
2E1 GBS V A Barnes, 16 Aldingbourne Park, Hook Ln, Aldingbourne, Chichester, PO20 6YR [IO90PT, SU90]

Irish Republic

This list includes new call signs and amendments notified up to June 1997. The Irish Radio Transmitters Society is indebted to the Office of the Director of Telecommunications Regulation for supplying much of the data in this listing.

Every effort is made to ensure that the information is accurate and complete, but no responsibility can be accepted for errors or omissions.

Call signs with a three letter suffix ending in 'B' indicate a Class B VHF/UHF licence. The EJ prefix is used instead of EI when operating from offshore islands.

As Ireland has adopted CEPT Recommendation T/R 61-01, visitors from the UK may use their own call sign preceded by 'EI/' and followed by '/P' or '/M' as appropriate.

While in Ireland, visitors must, of course, comply with the local licensing regulations. These regulations are broadly similar to the UK Amateur Radio regulations.The regulatory authority for radio may be contacted at the following address:

Office of the Director of Telecommunications Regulation
Radio Section
Abbey Court Irish Life Centre
Lower Abbey Street
Dublin 1

Tel: 00 353 1 804 9600
Radio Section direct line: 00 353 1 804 9621
Fax: 00 353 1 804 9680

The Irish Radio Transmitters Society (IRTS) is the national member society for Ireland of the IARU. The IRTS may be contacted at PO Box 462, Dublin 9. E-mail: jryan@iol.ie

EI0

EI0 CF Finbar O'Connor, Malin Head Radio, Ballygorman, Lifford, Co. Donegal
EI0 CG D.J. Fitzpatrick, 6 Ballymace Green, Templeogue, Dublin 6
EI0 CH Gerry Butler, 130 Glenageary Avenue, Dun Laoghaire, Co. Dublin
EI0 CI Bernard Kelly, 42 Walnut Avenue, Dublin 9
EI0 CK Thomas McManus, 513 South Circular Road, Kilmainham, Dublin 8
EI0 CL Michael J. Higgins, Roevehagh, Clarenbridge, Co. Galway
EI0 CM Roy Edwards, 52 St. Canices Road, Ballymun, Dublin 11
EI0 CN Brendan Coughlan, 254 Glenwood Estate, Dublin Road, Dundalk, Co. Louth
EI0 CP Sean Taaffe, 'San Marino', 56 Muirhevna, Dublin Road, Dundalk, Co. Louth
EI0 CR Colm J. Headon, 59 The Stiles Road, Clontarf, Dublin 3
EI0 CT William C. Nolan, 25 Beech Park, Athlone, Co. Westmeath
EI0 CV Michael H. Smye, Herons Gate, Manorial Road, Parkgate, Wirral L64 6QW, England
EI0 CX Kenneth M. Bishop, 12 Oakfield Drive, Kempsley, Worcs., England
EI0 CY George H. O'Reilly, 97 St. Assams Avenue, Raheny, Dublin 5
EI0 CZ Brendan Kilmartin, 47 Shannon Banks, Corbally, Co. Clare, via Limerick
EI0 DA Vincent Rafter, Hillquarter, Coosan, Athlone, Co. Westmeath
EI0 DB David V. Aldridge, The QTH, Illeigh, Thurles, Co. Tipperary
EI0 DC Colm Ward, 7 Sydenham Terrace, Ballinacurra, Limerick
EI0 DD Thomas Hurley, 112 Ballindrum, Athy, Co. Kildare
EI0 DE Charles P. Lynch, 21 Oaklands, Salthill, Galway
EI0 DG Irvine Ferris, Kilclone, Co. Meath
EI0 DH Patrick Dunne, 25 Riversfield, Midleton, Co. Cork
EI0 DI Dr. John J.V. Caraher, Swan Park, Monaghan
EI0 DJ Breda Condon, Golf Links Road, Castletroy, Co. Limerick
EI0 DK John Hill, 19 Stapolin Lawns, Baldoyle, Dublin 13
EI0 ARC Apollo Radio Club, 14 Clonfert Road, Kimmage, Dublin 12
EI0 ARS A.R.S.I, P.O. Box 938, Dublin 6
EI0 DMF Dundalk Maytime Festival, Dundalk Amateur Radio Soc, 113 Castletown Road, Dundalk, Co. Louth
EI0 NDR North Dublin Radio Club, Chanel College, Coolock, Dublin 5
EI0 RTS Irish Radio Transmitters Society, P.O. Box 462, Dublin 9

EI1

EJ1 D Dalkey Island Contest Group, c/o Daniel Coughlan EI5HD, 157 Shanganagh Cliffs, Shankill, Co. Dublin
EI1 AA Irish Leprechaun Contest Group, QSL via EI2BB
EI1 CH Patrick Donnelly, 28 Ailesbury Park, Newbridge, Co. Kildare
EI1 CI Tom Fitzgerald, Fermoy Road, Ballyhooly, Co. Cork
EI1 CK Robert W. Semple, "Algonquin", 16 Cairn Hill, Foxrock, Dublin 18
EI1 CL C.D. Rutter, 82 Willow Park Avenue, Ballymun, Dublin 11
EI1 CM Marilynn S. Stockwell, Ardgannon, Tuam, Co. Galway
EI1 CP James A. Butler, Depot Signals, Ceannt Barracks, Curragh, Co. Kildare
EI1 CR John R. Williams, 39 Ashton Park, Monkstown, Co. Dublin
EI1 CS Rev. Fr. Finbarr Buckley, 'Curraghmore', Cherry Grove, Model Farm Road, Cork
EI1 CT Thomas J. Gorman, Ivy Villa, Borris, Co. Carlow
EI1 CV Richard A. Smye, Herons Gate, Manorial Road, Parkgate, Wirral L64 6QW
EI1 CW Thomas Byrne, 35 Montrose Close, Artane, Dublin 5
EI1 CX Richard F. Chambers, 43 Fairways, Rathfarnham, Dublin 14
EI1 CY John Moloney, 162 Vernon Avenue, Clontarf, Dublin 3
EI1 DB James C. O'Mahony, 838 Portola Drive, San Francisco, California, USA
EI1 DC David G. Fitzgibbon, 18 Gilford Park, Sandymount, Dublin 4
EI1 DD Blackrock Radio Scouts, QSL via EI2CA
EI1 DE Michael R. O'Rourke, "Kiltanna", Knockaderry, Newcastle West, Co. Limerick
EI1 DF Eugene A. Ryan, The Lough, Tuckmill, Baltinglass, Co. Wicklow
EI1 DG Patrick McGrath, 15 Castleknock Way, Laurel Lodge, Castleknock, Dublin 15
EI1 DI John Feely, "Churchmount", Manorhamilton, Co. Leitrim
EI1 DJ Brendan Griffin, 20 Sarto Park, Naas, Co. Kildare
EI1 DK William E. Boles, 205 Emmet Road, Inchicore, Dublin 8
EI1 EM John Owen-Jones, Delamore House, Carrigeen, Bordhill, Co. Tipperary

EI2

EI2 E Cormac McHenry, 31 Rathdown Park, Terenure, Dublin 6
EI2 P Roderic Mooney
EI2 V Irish Air Corps Signals Amateur Radio Club, Air Support Signals, Casement Aerodrome, Baldonnel, Dublin 22
EI2 X Michael Beazley, 165 Orchard Street, Yonkers, N.Y. 10703, USA
EI2 AA John Smith, 33 Laurel Park, Clondalkin, Dublin 22
EI2 AB D. O'Neill, 11 St. Patricks Avenue, Crossmolina, Co. Mayo
EI2 AF Dermot K. Donnelly, 43 Park Drive, Ranelagh, Dublin 6
EI2 AG J.R. Murphy, 235 Marian Park, Drogheda, Co. Louth
EI2 AH Raymond S. Jordan, 35 Bryanstown Village, Drogheda, Co. Louth
EI2 AI Dermot J. Ryan, 75 Ballinteer Crescent, Ballinteer, Dublin 16
EI2 AJ Tom Kelly, 32 Rushbrook, Blanchardstown, Dublin 15
EI2 AK Dermot Cowley, Fieldstown, Monasterboice, Co. Louth
EI2 AR Patrick R. McCabe, Bishop Street, Tuam, Co. Galway
EI2 AW Anthony Condon, Golf Links Road, Castletroy, Co. Limerick
EI2 BA Collins Amateur Radio Club, c/o C. Healy, Collins Barracks, Cork
EI2 BB James R. Bartlett, "Chickamauga", Deansgrange Road, Blackrock, Co. Dublin
EI2 BD Bernard J. Bland, 9088E Captain Dreyfuss Av, Scottsdale, Arizona 85260, USA
EI2 BL Fr. Pacificus Jennings OFM Cap., Ard Mhuire, Creeslough, Co. Donegal
EI2 BM William Morrissey, 23 Ballinacurra Gardens, Limerick
EI2 BW E. Tully, Largy, Killybegs, Co. Donegal
EI2 BY Amateur Radio Society of Ireland, P.O. Box 938, Terenure, Dublin 6
EI2 CA Paul Martin, "Sitka", Cronroe, Ashford, Co. Wicklow
EI2 CC Donal Lonergan, "Ormond", 47 Hazelbrook Drive, Terenure, Dublin 6W
EI2 CD J. Misstear, 80 Rathgar Road, Dublin 6
EI2 CE Owen McArdle, 15 The Laurels, Dundalk, Co. Louth
EI2 CF Gerald O'Reilly, Monavally, Tralee, Co. Kerry
EI2 CG David O'Leary, 117 Cappaghmore, Clondalkin, Dublin 22
EI2 CH Gerard P. Morgan, N6CBX, Bunatubber, Corrandulla, Co. Galway
EI2 CI Joseph M. Purfield, 12 Wolseley Street, South Circular Road, Dublin 8
EI2 CJ Patrick J. Doran, Leighlinbridge, Co. Carlow
EI2 CK Dan Byrne, 35 Endsleigh Estate, Douglas Road, Cork
EI2 CL Michael McNamara, 92 Griffith Court, Dublin 3
EI2 CM Robert L. Williams, K6EMN
EI2 CN Douglas Turnbull, Deforest House, Coolfore, Monasterboice, Co. Louth
EI2 CP Peter J. Gleeson, 'Tinarana', Killaloe, Co. Clare
EI2 CR Sean Carvin, Apt. 9 "Millbrook", Old Lucan Road, Palmerstown Village, Dublin 20
EI2 CS Robert Harrison, 32 Coolatree Road, Dublin 9
EI2 CT Philip Connell, Knightstown, Wilkinstown, Navan, Co. Meath
EI2 CV Tony Bourke, 'Hillcrest', Spur Hill, Togher, Cork
EI2 CW Ed McIntyre, 22 Olympic Drive, Strabane, Co. Tyrone
EI2 CX John V.G. O'Donovan, G3RPI, 199 Court Road, Orpington, Kent BR6 0PX, England
EI2 CY Rev. Michael Breen, The Presbytery, Sutton, Dublin 13
EI2 CZ Patrick Doyle, "Shalom", Hawthorn Drive, Hill View, Waterford
EI2 DA Declan Howard, 11 Island View Estate, Sea Park, Malahide, Co. Dublin
EI2 DB N. Mulhall, Knockroe Lane, Ballyragget, Co. Kilkenny
EI2 DC Stuart W. Wallbridge, "Lissadel", Ulverton Close, Dalkey, Co. Dublin
EI2 DD Sean M. Reilly, Cratloekeel, Co. Clare
EI2 DE Paul J. Blood W1KYY, 7333 Yahley Mill Road, Richmond, VA (218) 23231, USA
EI2 DF Patrick Rohan, 139 Ballyroan Road, Templeogue, Dublin 16
EI2 DG Declan J. Graham, "Carnadoon", Cullen Upper, Kilbride, Co. Wicklow
EI2 DH P. Gillespie, 46 Limekiln Road, Dublin 12
EI2 DI Allan McMurty, 20 Tower View Crescent, Bangor, Co. Down
EI2 DJ Michael Wright, 5 Woodview Park, The Donahies, Dublin 13
EI2 DL Jim Stacey, Monang, Abbeyside, Dungarvan, Co. Waterford
EI2 DN A. Ryan, Telecomms & Radio Div., Dept. of Communications, Findlater House, Dublin 1
EI2 DP Evelyn A. Robinson, 26 Frascati Park, Blackrock, Co. Dublin
EI2 DQ Edward O'Connor, Templemichael, Glanmire, Co. Cork
EI2 DR Details withheld at licensee's request
EI2 DS J.B. O'Dwyer, St. Joseph's Road, Naas, Co. Kildare
EI2 DT Gerald P. McGorman, "Teaghlach", Legnakelly, Clones, Co. Monaghan
EI2 DU Kevin Murray, c/o 56 Crestwood Estate, Galway
EI2 DV R. Austin, The Eire Farm, Timoleague, Bandon, Co. Cork
EI2 DW Karen G. Wright, 5 Woodview Park, The Donahies, Dublin 13
EI2 DY Sean Donohoe, Proudstown Lodge, Navan, Co. Meath
EI2 DZ Peter W. Harris, 18 Vera Road, Coventry, RI 02816, USA
EI2 EA Matthew J.P. Murtagh, 84 St. Patricks Road, Greenhills Road, Dublin 12
EI2 EB Harry McMullan, Mill Road, Bunclody, Co. Wexford
EI2 EC Lawrence Grehan, 87 Ballyedmonduff Road, Stepaside, Co. Dublin
EI2 ED John Hosty, 17 Ardilaun Road, Newcastle, Galway
EI2 EE Sean O'Hara, Tully Hill, Rathcormac, Sligo
EI2 EF Cornelius Guiney, Ballyhearney, Valentia, Co. Kerry
EI2 EG Frederick Owens, 14 McCurtin Street, Cork
EI2 EH Sylvester O'Farrell, 73 Fifth Avenue, Ottawa, Ontario K1S 2M3, Canada
EI2 EI R.A. Penn, 46 Ladbrooke Drive, Potters Bar, Herts EN6 1QR
EI2 EJ James McQuaid, Mullaghvadden, Dungannon, Co. Tyrone
EI2 EK John P. McCafferty, Kilsallagh, Ballinlaw, Slieverue, Co. Kilkenny
EI2 EL Robert W. Shaw, "Ronomia", Glack, Dublin Road, Longford
EI2 EM Charles Lyons, 24 Oakwood Avenue, Swords, Co. Dublin
EI2 EN Edmund J. Burke, 33 Ailesbury Grove, Dublin 16
EI2 EO M.F. Hayes, Curry Cummor, Tuam, Co. Galway
EI2 ER Details withheld at licensee's request
EI2 ES G. Shedwell, 7 St. Mel's Road, Longford
EI2 ET Manfred Lauterborn, DK2PZ
EI2 EU Andrew Doyle, 52 The Faythe, Wexford
EI2 EV David Warton, 6644 23rd Avenue, Wyattsville, MD 20782, USA
EI2 EW Jorg M. Eibner, Emoring 3, 469 Herne, Germany
EI2 EX Padraic Cawley, Ballybeg, Knockmore, Ballina, Co. Mayo
EI2 EY Jeremiah J. O'Mahony, "St. Anthonys", 61 Westcourt, Ballincollig, Co. Cork
EI2 EZ A.J. Moore, G4RHX, 5 Longbeck Lane, New Marske, Redcar, Cleveland TS11 8AT, England
EI2 FA William G. Ryan, 115 Fortview Drive, Ballinacurra Gardens, Limerick
EI2 FB Liam Lyons, 'Cuan Oir', The Hill, Crosshaven, Co. Cork
EI2 FC J.T. O'Sullivan, Lahinch, Co. Clare
EI2 FD John R. Masterson, Rincoola, Granard, Co. Longford
EI2 FE Patrick Keelan, 59 Dean Cogan Place, Navan, Co. Meath
EI2 FF Paul M. Doherty, Allen Street, Cappoquin, Co. Waterford
EI2 FG John Hearne, The Quay, Fethard-On-Sea, Co. Wexford
EI2 FH Aidan Q. McDermott, Mullaghroe (via Boyle), Co. Sligo
EI2 FI Killian Harford, 4 Fancourt Road, Balbriggan, Co. Dublin
EI2 FJ Michael C. Griffin, Cahirfilane, Castlemaine, Co. Kerry
EI2 FK Robert J.V. Patterson, Murvagh, Laghey, Co. Donegal
EI2 FL Liam R. Wynne, 30 Allen Park Road, Stillorgan, Co. Dublin
EI2 FM Denis Cadogan, 53 Bancroft Park, Tallaght, Dublin 24
EI2 FN John McGowan, 15 Lr. Kindlestown, Delgany, Co. Wicklow
EI2 FO Piaras N. O'Donnchadha, Foxtown, Summerhill, Co. Meath
EI2 FP Joseph McGloughlin, 20 Tymon Close, Old Bawn, Tallaght, Dublin 24
EI2 FQ Flor Lynch, 'Nephin', Coronea, Skibbereen, Co. Cork
EI2 FR Adrian T. O'Gorman, 15 Gibbons Terrace, Balbriggan, Co. Dublin
EI2 FS Edward Taylor, 26 Lough Mahon Road, Blackrock, Cork
EI2 FT Mary T. Lyons, 24 Oakwood Avenue, Swords, Co. Dublin
EI2 FV Richard F. Meredith, Beech Grove, Athy, Co. Kildare
EI2 FW Garret P. O'Sullivan, 1 Coolrua Drive, Dublin 9
EI2 FX Richard J. Cullen, 25 Moyclare Park, Baldoyle, Dublin 13
EI2 FY Patrick J. McDonnell, 5 Ballyboughal Square, Ballyboughal, Co. Dublin
EI2 FZ Patrick Griffin, Duagh Camp, Tralee, Co. Kerry
EI2 GA Gerald Breslin, 37 Carnhill Estate, Derry
EI2 GB David Redmond, 89 Beechdale, Kilcoole, Co. Wicklow
EI2 GC Seamus McGiff, Main Street, Buttevant, Co. Cork
EI2 GD John E. Radcliffe, 26 Lanahrone Avenue, Corbally, Limerick
EI2 GE Declan Collison, 17 Main Street, Moneygall, Birr, Co. Offaly
EI2 GF Michael Byrne, 117 Wicklow Heights, Wicklow
EI2 GG John Campbell, 4 Bollinbarn, Macclesfield, Cheshire SK10 3DL

EI2	GH	Hans Tyhuis, 338 Beechmount Drive, Dundalk, Co. Louth
EI2	GI	Peter Bluett, 50 Dukesmeadow Avenue, Kilkenny
EI2	GJ	Peter A. McGorman, Teaghlach, Legnakelly, Clones, Co. Monaghan
EI2	GK	Michael Behan, Coolagad Cottage, Blacklion, Greystones, Co. Wicklow
EI2	GM	John J. Hill, Decomade, Lissycasey, Co. Clare
EI2	GN	John P. Ketch, 9 Rockcliffe Terrace, Blackrock Road, Cork
EI2	GO	William A. Fahy, 25 Melbourne Avenue, Bishopstown, Cork
EI2	GP	Thomas Rea, Bridge Street, Headford, Co. Galway
EI2	GQ	Alfred Berger, "Cuil na Sionnach", Farrandau, Castletownsend, Co. Cork
EI2	GR	Adrian W. O'Leary, 34 Lotabeg Estate, Mayfield, Cork
EI2	GS	Frank O'Brien, Stonehurst, Monastery, Enniskerry, Co. Wicklow
EI2	GT	Dermot Gleeson, 77 Quarry Road, Thomond Gate, Limerick
EI2	GU	John Kelly, Cavan Road, Ballindrait, Co. Donegal
EI2	GV	Robert J. Pegritz, P.O. Box 7921, Newark, Delaware 19714, USA
EI2	GW	Fergus McDonald, 14 Glendoher Avenue, Rathfarnham, Dublin 16
EI2	GX	Tony Stack, 162 St. Peter's Road, Greenhills, Dublin 12
EI2	GY	John McSwiney, 22 Victoria Terrace, Glenbrook, Passage West, Co. Cork
EI2	GZ	Dr. Robert Finlay, 86 Crodaun Forest Park, Celbridge, Co. Kildare
EI2	HA	Michael Boyce, Ambrosetown, Duncormick, Co. Wexford
EI2	HB	Tipperary Radio Club, c/o Mike Hoare, EI9FE, "Glencoe", Ballykisteen, Tipperary
EI2	HC	Dick Bean, K1HC, 422 Everett Street, Westwood, MA 02090, USA
EI2	HD	Dermot P. Miley, 26 Slievebloom Park, Walkinstown, Dublin 12
EI2	HE	Joseph S. Johnson N1FEY, 27 Parkvale, Sandyford Road, Dundrum, Dublin 16
EI2	HF	Pat McGrath, Ballintemple, Castlegar, Co. Galway
EI2	HG	Damien Commins, 5 Cruachan Park, Rahoon, Galway
EI2	HH	Andy Linton, 75 Farran Park, Waterford
EI2	HI	Hugh O'Donnell, Kilmalogue, Cahir, Co. Tipperary
EI2	HK	Patrick A V Walsh, "Rockfield", Hillcrest Road, Sandyford, Co. Dublin
EI2	HL	Mark J. Doyle, 28 Ashton Avenue, Knocklyon, Dublin 16
EI2	HM	Kevin M.J. Dillon, 4 Dargle Road, Blackrock, Co. Dublin
EI2	HN	Patrick McGrath, Carriglawn, Ballysimon Road, Limerick
EI2	HO	John Hagin-Meade, 26 Broadford Walk, Ballinteer, Dublin 16
EI2	HP	David Waugh, 16 Seaview Avenue, Millisle BT22 2BN, Northern Ireland
EI2	HQ	Joseph Quigley, Newtown Kells, Co. Kilkenny
EI2	HR	David Hooper, 14 Corbally Way, Westbrook Lawns, Tallaght, Dublin 24
EI2	HS	Robert Hyde, 12 Avondale Drive, Kilcohan, Waterford
EI2	HT	Paddy Twomey, "Rosanore", Pinewood Crescent, Loughrea, Co. Galway
EI2	HV	Patrick K. Keogh, 27 St Joseph's Square, Fermoy, Co. Cork
EI2	HW	John M Forristal, 86 Roselawn, Tramore, Co. Waterford
EI2	HX	Patrick J. Fitzpatrick, 24 Ascail A Do, Yellow Batter, Drogheda, Co. Louth
EI2	HY	Anthony O'Rourke, 13 Hazel Road, Togher, Cork
EI2	HZ	Trevor Wiseman, Kent View, Nurses Cottage, Foxbush, Hildenborough, Kent TN11 9HT, England
EI2	IA	Eoghan O hUallachain, 171 Martello, Po Mearnog, Co. Atha Cliath
EI2	IB	Michael Kiely, Ballinrush Lower, Kilworth, Co. Cork
EI2	IC	David Norton, c/o Nurse B Doherty, The Cottages, Glentogher, Carndonagh, Co. Donegal
EI2	ID	Richard P. Ebbs, Flat 5, 14 McDermott Street, Midleton, Co. Cork
EI2	IE	Michael Conaghan, 5 Mont Pelier View, Tallaght, Dublin 24
EI2	IF	Patrick P. Rosney, 221 Sundays Well, Blessington Road, Naas, Co. Kildare
EI2	IG	Ivan O'Sullivan, Rathoneigue, Bartlemy, Fermoy, Co. Cork
EI2	IH	Hugh Galt, 9 Corbawn Avenue, Shankill, Co. Dublin
EI2	II	Enda Broderick, Tullahill, Loughrea, Co. Galway
EI2	IJ	Mary T. Daly Scanlon, Meehan, Coosan, Athlone, Co. Westmeath
EI2	IK	Liam J. Thunder, 47 Walnut Lawn, Courtlands Estate, Dublin 9
EI2	IL	Paul F. O'Toole, 8 Novara Park, Bray, Co. Wicklow
EI2	IM	Details withheld at licensee's request
EI2	IN	Brendan F. Logue, Whitecross, Julianstown, Co. Meath
EI2	MC	Jack Quinn, 1210 San Mateo Drive, Menlo Park, California 94025, USA
EI2	WW	Wicklow Wireless Society, c/o Sean Donelan EI4GK, "Inniscarra", Ballybride Road, Shankill, Co. Dublin
EI2	ADB	Thomas J. Boland, 8 Hillsbrook Grove, Perrystown, Dublin 12
EI2	AFB	John Bergin, Mount Elland House, Ballyragget, Co. Kilkenny
EI2	AHB	Charles G. Duncan, Woodroad House, Lisnagry, Co. Limerick
EI2	AIB	Patrick D. Kelly, 95 Butterfield Avenue, Rathfarnham, Dublin 14
EI2	ALB	Donal Heffernan, 103 Mayorstone Park, Limerick
EI2	AMB	D. Tocher, 2 Mount Shannon Road, Lisnagry, Co. Limerick
EI2	ANB	Timothy Hall, 7 New Westfields, North Circular Road, Limerick
EI2	APB	Matthew Gavigan, 41 Watermeadow Drive, Old Bawn, Dublin 24
EI2	AQB	Andrew Dodswell, Rathward, Ballysheedy, Co. Limerick
EI2	ARB	J.F. Sherry, 6 Monastery Road, Clondalkin, Dublin 22
EI2	ASB	Liam J. Convey, 8 Hillcrest Road, Lucan, Co. Dublin
EI2	ATB	Seamus McGabhann, The Orchard, Monastery Road, Clondalkin, Dublin 22
EI2	AVB	Michael Goss, 41 Churchview Avenue, Killiney, Co. Dublin
EI2	AYB	Raymond J. Rafferty, 11 Chestnut Close, Lisbeg Lawn, Renmore, Co. Galway
EI2	AZB	Desmond Coyne, 79 Allen Park Road, Stillorgan, Co. Dublin
EI2	BCB	D. Healy, Drompeach, Lombardstown, Mallow, Co. Cork
EI2	BJB	Anthony Skinner, 123 Lr. Kilmacud Road, Stillorgan, Co. Dublin
EI2	BRB	Robert E. Cloherty, St. Judes, Mincloon, Raheen, Galway
EI2	BRC	City of Dublin Amateur Radio Club c/o E.D. Berkenheier, 'Southern Comfort', Glasnarget, Rathdrum, Co. Wicklow
EI2	BUB	Martin C. Bradley, Clonroadbeg, Ennis, Co. Clare
EI2	BWB	William J. Dundon, Ardshanbally, Adare, Co. Limerick
EI2	BXB	A. Smyth, 167 Glenview Park, Tallaght, Dublin 24
EI2	BYB	A. Lewis, 'Janaleema', 3 Mackerel Hall, Royston, Herts S68 5BS
EI2	CBB	Sean O Briain, Ballyegan, Lisseton, Co. Kerry
EI2	CCB	Joseph R. Elliott, Harlockstown, Ashbourne, Co. Meath
EI2	CEB	George Coyle, Ballyare, Ramelton, Letterkenny, Co. Donegal
EI2	CGB	Bryan C. Jones, 115 Clarke Road, Cork Road, Waterford
EI2	CHB	Eamonn O'Connor, 23 Watermill Drive, Raheny, Dublin 5
EI2	CIB	Kerril J. Curran, 11 Fortfield Drive, Terenure, Dublin 6
EI2	CKB	Donal Maher, 30 McDara Road, Shantalla, Galway
EI2	CLH	Lahinch Radio Club, c/o T. O'Sullivan, Station Road, Lahinch, Co. Clare
EI2	CMB	Karen Tobin, 8 Ferndale Road, Glasnevin, Dublin 11
EI2	CNM	Co-Operation North Maracycle, Dundalk Amateur Radio Soc, 113 Castletown Road, Dundalk, Co. Louth
EI2	COB	John O'Donoghue, 2 Rosehall, Templeogue, Dublin 6
EI2	CPB	George Adjaye, 245 Lr. Kimmage Road, Dublin 6
EI2	CRB	Stephen A. Murray, 10 Sarah Place, Islandbridge, Dublin 8
EI2	CRG	Carlow & District Amateur Radio Group, c/o Pat Hutton, EI6HF, 50 Ash Grove, Tullow Road, Carlow
EI2	CTB	Aidan V. Grant, 19 Chestnut Close, Viewmount, Waterford
EI2	CWB	Thomas J. Cahill, 43 Rialto Buildings, South Circular Road, Dublin 8
EI2	CYB	Alan Murray, 128 Carrickhill Rise, Portmarnock, Co. Dublin
EI2	DBB	Details withheld at licensee's request
EI2	DDB	Details withheld at licensee's request
EI2	DFB	Michael B. Quinn, Glenane, Killeagh, Co. Cork
EI2	DGB	Michael J. Power, Tinternfarm, Saltmills, Co. Wexford
EI2	DHB	Ronald Griffin, 60 Curryhill Road, Strabane, Co. Tyrone
EI2	DJB	Richard Humphreys, 34 Sycamore Road, Mount Merrion, Co. Dublin
EI2	DKB	Stephen P. Fitzpatrick, White's Flats, The Mall, Ballyshannon, Co. Donegal
EI2	DLB	Keith Chadwick, 32 Deanarath Road, Clondalkin, Dublin 22
EI2	DMB	Brian Winters, St. Judes, Cartown, Termonfeckin, Drogheda, Co. Louth
EI2	DNB	Desmond Behan, 257 Oakley Park, Celbridge, Co. Kildare
EI2	DRG	Dublin Experimental Repeater Group
EI2	DSB	Aileen O'Neill, "Parknasilla", Old Blackrock Road, Cork
EI2	DSJ	Dundalk Scout Jamboree, Dundalk Amateur Radio Soc, 113 Castletown Road, Dundalk, Co. Louth
EI2	DTB	Brian Duffy, 'Oakfield', Hettyfield, Douglas Road, Cork
EI2	DUB	William S. Wigham, "Sitka", Clogheen, Milcon, Blarney, Co. Cork
EI2	DWB	Charles McHugh, Donegal Street, Ballybofey, Lifford, Co. Donegal
EI2	DYB	Francis T. Shortt, 14 McKelvey Avenue, Finglas, Dublin 11
EI2	DZB	Donal Jevens, 3 Oaktree Rise, Newlands, Wexford
EI2	EAB	John J. Godkin, 9 Ballinteer Gardens, Dundrum, Dublin 16
EI2	EBB	John A. McGennis, 4 College Drive, Terenure, Dublin 6W
EI2	ECB	Kenneth W. Gall, 109 Kimmage Road West, Terenure, Dublin 12
EI2	EDB	Colm Donelan, "Inniscarra", Ballybride Road, Shankill, Co. Dublin
EI2	EGB	Nicholas Mallon, 67 Kimmage Road West, Dublin 12
EI2	EHB	Maurice Wilson, 26 Castle Avenue, Clontarf, Dublin 3
EI2	EIB	James F. Linehan, 35 Maulbawn Estate, Passage West, Co. Cork
EI2	EJB	Wilfred J. Higgins, 23 Nutgrove Park, Clonskeagh, Dublin 14
EI2	ELB	Niall J Coveney, 7 Bellevue Park, Greystones, Co. Wicklow

EI2	EMB	David B. Meehan, Ballincurrig, Leamlara, Midleton, Co. Cork
EI2	EQB	Patrick Paul Loughnane, 52 Glenwood Estate, Carrigaline, Co. Cork
EI2	ERB	W. H. Raitt, Main Street, Stranorlar, Ballybofey P.O., Lifford, Co. Donegal
EI2	ESB	Stanley Raitt, Main Street, Stranorlar, Ballybofey P.O., Lifford, Co. Donegal
EI2	ETB	Mark Patrick Henry, Main Street, Wicklow Town
EI2	EUB	Owen J. O'Sullivan, 7 Donscourt, Bishopstown, Cork
EI2	EVB	Paul J Fahy, 16 Drumacrin Avenue, Bundoran, Co. Donegal
EI2	EWB	Neil S. Mac Parthalain, 17 Pairc an Charraigin, Baile an Locha, Corcaigh
EI2	EXB	Robert J. Byrne, 27 Aylsbury, Ballincollig, Co. Cork
EI2	EYB	Robert Lester, Glen Cottage, Kilcully, Whitescross, Co. Cork
EI2	EZB	Paul F Murphy, 56 Cabinteely Avenue, Dublin 18
EI2	FAB	Paul C. Ivers, 87 Rutland Grove, Crumlin, Dublin 12
EI2	FBB	Cormac Moore, 25 The Close, Orlynn Park, Lusk, Co. Dublin
EI2	FCB	Terry McEvoy, The Cinema, Monasterevan, Co. Kildare
EI2	FDB	Sean Lynch, Killeenacoff, Westport, Co. Mayo
EI2	FEB	John Lyons, 24 Oakwood Avenue, Swords, Co. Dublin
EI2	FFB	Gareth Martin, 8 Springfield, Blackhorse Avenue, Dublin 7
EI2	FHB	Charles Kinsella, Johnstown, Sea Road, Arklow, Co. Wicklow
EI2	FIB	P. J. O'Reilly, Balgeeth, Kilskyre, Kells, Co. Meath
EI2	FJB	Martin Hanley, 22 St Asicus Villas, Athlone, Co. Westmeath
EI2	FKB	Sean O Suilleabhain, 141 Brookwood Avenue, Artane, Dublin 5
EI2	FLB	Padraic O Cathain, c/o Kate Kelly's, Ballintubber, Claremorris, Co. Mayo
EI2	FMB	Jerry Walsh, Shippool, Innishannon, Co. Cork
EI2	FNB	John Francis McDonnell, Burrin, Ballyglass, Claremorris, Co. Mayo
EI2	FOB	John Corless, Coolaght, Claremorris, Co. Mayo
EI2	FPB	Noel Aidan McIntyre, The Diamond, Lifford, Co. Donegal
EI2	FQB	Rev Fr Martin Gerard MacEntee OP
EI2	FRB	Donogh Roche, Shalom, Blarney Road, Shanakiel, Cork
EI2	FRC	Fingal Radio Club, c/o Chris Yeates, EI7AAB, 75 Georgian Village, Castleknock, Dublin 15
EI2	FSB	Anthony Baldwin, Rathlin, Dromnea, Kilcrohane, Co. Cork
EI2	FTB	Dermot Wall, Cabra, Dublin 7
EI2	FUB	Michael Murphy, "Grove Lodge", Turra, Ballickmoyler, Co. Laois
EI2	FVB	Tomas Kelly
EI2	FWB	Dermott Finegan, Apartment 3, James McSweeney House, Berkley Street, Dublin 7
EI2	FXB	Paul Hurley, 17 Turlough Gardens, Fairview, Dublin 3
EI2	FYB	Aidan Kinane, Golf Links Road, Tipperary
EI2	FZB	Alan Geoghegan, 106 O'Molloy Street, Tullamore, Co. Offaly
EI2	GRC	Regional Tech College Radio Club, c/o Tom Frawley EI3ER, R.T.C, Dublin Road, Galway
EI2	IHE	Plassey Amateur Radio Club, c/o David Tocher, University of Limerick, Limerick
EI2	IPA	International Police Assocn. ARC, c/o George Moran EI7EC, 13 Iona Drive, Drumcondra, Dublin 9
EI2	KSS	2nd Kerry Sea Scouts, c/o Brian O'Daly, 38 Derrylea, Tralee, Co. Kerry
EI2	MRC	Mostrim Radio Club, c/o Tom McLoughlin, Cranley More, Mostrim, Co. Longford
EI2	NCR	North County Radio Club, c/o G. Fitzgerald, EI8FE, 56 Strand Street, Skerries, Co. Dublin
EI2	NKR	North Kildare Radio Club, c/o Fr. Robert Swinburne, Salesian College, Celbridge, Co. Kildare
EI2	PAR	Phoenix Amateur Radio Club, 30 Woodview Grove, Blanchardstown, Dublin 15
EI2	PKT	Dublin Packet Repeater, c/o EI6FZ
EI2	QRP	QRP Club of Ireland, c/o Marino Institute of Education, Griffith Avenue, Dublin 9
EI2	RTC	Eugene A. Hanley, Cork Regional Tech., Rossa Avenue, Bishopstown, Cork
EI2	RTE	RTE Amateur Radio Society, c/o Michael Wright, Radio Centre, R T E, Donnybrook, Dublin 4
EI2	SDR	South Dublin Radio Club, Ballyroan Community Centre, Marian Road, Rathfarnham, Dublin 14
EI2	SNR	St. Raphaels College ARC, Loughrea, Co. Galway
EI2	SRC	Sligo Amateur Radio Club, c/o David Dillon EI8BEB, 84 Knocknaganny Park, Sligo
EI2	TCD	Micro Electronics and Elec. Engineering Dept., Trinity College, Dublin 2
EI2	TRC	Thomond Radio Club, c/o William G. Ryan EI2FA, 115 Fortview Drive, Ballinacurra Gardens, Limerick
EI2	UCD	U.C.D. Radio Club, Newman House, 86 St. Stephens Green, Dublin 2
EI2	VBV	Frank M. Lewis Jnr. W3JGM, 2461 Mullinix Mill Road, RFD3. Mount Airy, MD 21771, USA
EI2	VHF	Brian Credico, 1 Sylvan Road, Fairlands Park, Newcastle, Co. Galway
EI2	WRB	Waterford Radio Beacon, c/o Eamonn Phelan, 14 Ursuline Crescent, Waterford
EI2	WRC	South Eastern Amateur Radio Group, c/o Eamonn Phelan, 14 Ursuline Crescent, Waterford

EI3

EI3	A	John Llewellyn G4JTM, 5 The Rowans, St. Mary's Park, Portishead, Bristol
EI3	C	Kevin Kilduff, Swellan Lower, Cavan
EI3	H	Colm Ardiff, 113 Sutton Park, Sutton, Dublin 13
EI3	I	J.F. Brezina, 33 Clonmore Road, Mount Merrion, Dublin 4
EI3	K	Eugene Larkin, Thomastown Cross, Donaghmore, Dundalk, Co. Louth
EI3	M	Sean Fanning, "Lissadell", 42 Redesdale Road, Mount Merrion, Dublin 4
EI3	U	Victor Thorne, Dublin Inst of Technology, Kevin Street, Dublin 8
EI3	V	James J. Moore, 8 Meadow Mount, Churchtown, Dublin 16
EI3	X	Rev. Bro. Leo P. O'Gorman, De La Salle, Hospital, Co. Limerick
EI3	Y	Ian Clarke, Deerpark, Kilcullen, Co. Kildare
EI3	AC	K. O'Beirne, 149 Tritonville Road, Sandymount, Dublin 4
EI3	AE	Dan Lloyd, 78 Park Road, Navan Road, Dublin 7
EI3	AF	J. O'Connell, 16 Seafield Drive, Dublin 3
EI3	AG	William A. Thompson, Main Street, Moville, Co. Donegal
EI3	AK	3rd Field Signal Company, Sarsfield Barracks, Limerick
EI3	AL	Tom O'Sullivan, Knockroe, Kildimo, Co. Limerick
EI3	AR	Patrick O'Floinn, "The Moorings", Ballymahon Road, Athlone, Co. Westmeath
EI3	AS	Thomas J. Moran, Clonmoney West, Newmarket-on-Fergus, Co. Clare
EI3	AU	T.P. O'Brien, 43 Manor Street, Waterford
EI3	AV	Patrick Maher, "Knockearl", Cloughjordan, Co. Tipperary
EI3	AX	William E. Curristan, Quay Street, Donegal
EI3	BA	Charles Coughlan, 14 Clonfert Road, Kimmage, Dublin 14
EI3	BB	Robert Bolton, "Avalon", Central Avenue, Firgrove Estate, Cork
EI3	BD	J.J. Drudy, Strandhill, Co. Sligo
EI3	BF	John J. Hickey, Ard Mo Chroi, Cullen, Mallow, Co. Cork
EI3	BK	Jeremiah O'Sullivan, 70 Uam Var Drive, Bishopstown, Co. Cork
EI3	BL	Michael L. Barrett, 25 Morrisons Terrace, Ballina, Co. Mayo
EI3	BN	Michael O'Connor, Main Street, Foynes, Co. Limerick
EI3	BT	T.K. Dempsey, "Loughill", Moylough, Co. Galway
EI3	BV	James Shortland, 24 Inchvale Drive, Shamrock Lawn, Douglas, Co. Cork
EI3	BW	Bernard Flynn, c/o Irish Net Ltd., Castletownbere, Co. Cork
EI3	CA	Brian Fox, Rosebank, Farranlea Park, Model Farm Road, Cork
EI3	CD	Fergal Holmes, Mill Cottage, Shankill, Co. Dublin
EI3	CE	Michael Meegan Jr., 46 McSwiney Street, Dundalk, Co. Louth
EI3	CG	Rev. Alan Malone, WA7KBN, Eyrecourt, Co. Galway
EI3	CH	Ultan Rice, 32 Emer Terrace, Castletown Road, Dundalk, Co. Louth
EI3	CI	J.F.X. Hickey, 51 Fortfield Road, Dublin 6
EI3	CJ	M. Keeney, 128 North Road, Finglas, Dublin 11
EI3	CK	Rev. Fr. M. Gallagher, Curraun, Westport, Co. Mayo
EI3	CL	M.G. Rogers, 131 Cromwellsfort Road, Dublin 12
EI3	CN	Michael Lawlor, 16 Wainsfort Grove, Terenure, Dublin 6
EI3	CP	Colum P. Clarke, 5 Glen Garth, The Park, Cabinteely, Dublin 18
EI3	CR	Joseph McCormack, 17 Victoria Avenue, Newtownards, Co. Down
EI3	CS	Gerald Walsh, 51 Seskin View Road, Tallaght, Dublin 24
EI3	CT	Frank Cox, "The Orchards", Piltown, Co. Kilkenny
EI3	CV	John G. Daly, "St. Kierans", Rope Walk, Cork
EI3	CW	Thomas Moroney, 19 Emmet Street, North Circular Road, Dublin 1
EI3	CX	Thomas P. Dwyer, 25 Summer Street, North Circular Road, Dublin 1
EI3	CY	Ted Crowley, "Cuilin", New Long Hill, Killough Lower, Kilmacanogue, Bray, Co. Wicklow
EI3	CZ	Roderick Power, Corballis Cottage, Mount Seskin, Saggart, Co. Dublin
EI3	DA	Philip O'Donnell, 112 River Valley, Swords, Co. Dublin
EI3	DB	Michael P. Mee, 27 Oaklawns, North Road, Drogheda, Co. Louth
EI3	DC	John A. Hayne, 46 Glasthule Road, Sandycove, Co. Dublin
EI3	DD	Sean O'Rourke, Gallowstown, Roscommon, Co. Roscommon
EI3	DE	Noel F. Martin, 14 Fernvale Drive, Crumlin, Dublin 12
EI3	DF	James McAndrew, 19 Dromartin Park, Kilmacud, Dublin 14
EI3	DG	Patrick E. Griffin, 17 Lambeecher Estate, Bath Road, Balbriggan, Co. Dublin
EI3	DH	Norbert Payne, 3 Lakelands Grove, Stillorgan, Co. Dublin
EI3	DI	Brendan Lehane, 36 Commons Road, Clondalkin, Dublin 22
EI3	DJ	Colin Lafferty, "Cartymore", Athenry, Co. Galway
EI3	DK	Fredrick R. Elder, Tirkeeran, Cregg, Claudy, Co. Derry
EI3	DL	M. Van Der Vlist, Jahon Vam Limbeeck Laan 4, 3971 BZ Driebergen, Holland
EI3	DM	W.M. Foley, 128 Ranelagh, Dublin 6
EI3	DN	Peter O'Sullivan, 21 The Drive, Woodpark, Ballinteer, Dublin 16

EI3 DO Patrick A. Moran, 398 North Circular Road, Dublin 7
EI3 DP Jim Ryan, 11 Knockgriffin, Midleton, Co. Cork
EI3 DQ Stephen H. Nankivell, 16 Corbawn Court, Shankill, Co. Dublin
EI3 DR Details withheld at licensee's request
EI3 DT Sean N. Walsh, Feenagh, Quin, Co. Clare
EI3 DU Rev. Fr. Seamus A. Cullen, Garty Lough, Arva, Co. Cavan
EI3 DV Gregory M. Murphy, 15 Rose Avenue, Bayworth Mead, Abingdon, Oxon. OX14 1XX, England
EI3 DW F.A. Stoltz, Analog Devices B.V., Raheen Industrial Park, Limerick
EI3 DY Michael Staunton, 'Glenina', Enniskerry Road, Sandyford, Co. Dublin
EI3 DZ Stewart C. Crampton, 4 Brooklands Park, Whitehead, Carrickfergus, Co. Antrim
EI3 EA Gerard A. O'Sullivan, 28 O'Malley Park, Limerick
EI3 EB John Kealy, 6 Red Houses, The Swan, Athy, Co. Kildare
EI3 EC John K. O'Sullivan, 'Dunhallow', 4 Woodvale Road, Beaumont, Cork
EI3 ED Edward H. Brooks, GD4HOX, Elmwood, Somerset Road, Douglas, Isle of Man, IM2 5AE
EI3 EE Liam P. McGuire, 95 Lye Avenue, Birmingham BB2 3UG, England
EI3 EF Details withheld at licensee's request
EI3 EG A. O'Meara, 42 Halldene Drive, Bishopstown, Cork
EI3 EH John Kelly, Templeton Glebe, Killashee, Co. Longford
EI3 EI Phyllis M. MacArthur, 11 Woodlawn, Upper Churchtown Road, Dublin 14
EI3 EJ Seamus McQuaid, Mullaghvadden, Dungannon, Co. Tyrone
EI3 EK Joseph Flynn, Dunsany, Co. Meath
EI3 EL M. Rosney, 19 Thornhill Gardens, Celbridge, Co. Kildare
EI3 EM M.D. Lynch, Plush, Cloverhill, Belturbet, Co. Cavan
EI3 EN Rev. Fr. Henry Houlihan, 15 Fortfield Drive, Dublin 6
EI3 EP John Cronin, 7 Kinahan Street, (off Infirmary Road), Dublin 7
EI3 EQ Malcolm Bowden, Rosseragh, Ramelton, Letterkenny, Co. Donegal
EI3 ER Thomas Frawley, Engineering Department, Regional Tech. College, Dublin Road, Galway
EI3 ES Michael Mulhall, The Pike, (Co. Laoise), via Athy, Co. Kildare
EI3 ET William Borland, Glencar Scotch, Letterkenny, Co. Donegal
EI3 EU Kevin B. Clancy, 11 Raheen Court, Tallaght, Dublin 24
EI3 EV Thomas P. Gogarty, 29 Abbey Avenue, Corbally, Limerick
EI3 EW Roger A. Adair, 16 Sparsholt Road, London N19
EI3 EX Anthony P. McCullion, 7 Kennedy Street, Strabane, Co. Tyrone
EI3 EY Eamon O'Hara, Lisdamiet, Quay Street, Donegal
EI3 EZ Olan O'Brien, 68 Uam Var Drive, Bishopstown, Cork
EI3 FA Colm F.X. Mooney, 5 Brookwood Road, Artane, Dublin 5
EI3 FB Raymond K. Preston, 16 Marcella Park, Scrabo Road, Newtownards, Co. Down
EI3 FC Peter E. Harris, Dunbeg, 50B Avondale Lawn, Blackrock, Co. Dublin
EI3 FD Timothy J. Sinnott, Broomville, Ardattin, Carlow
EI3 FE Pierce Meagher, 20 Raglan Lane, Dublin 4
EI3 FF Hugh Mellerick, 6 Foxford Avenue, Melbourn Estate, Bishopstown, Cork
EI3 FG Jack Keane, Branchfield, Ballymote, Co. Sligo
EI3 FH Sean Malone, Clogherevagh, Sligo
EI3 FI Martin F. O'Dea, Ard-Na-Glass, Grange, Co. Sligo
EI3 FJ Laurence Murphy, 2 Windmill Lane, New Ross, Co. Wexford
EI3 FK John F. Grace, 3 Carrickhill Road, Portmarnock, Co. Dublin
EI3 FM Fintan Muldoon, N4AYW, 4800 Truesdale Place, Charlotte, NC 28277-8649, USA
EI3 FN Isaac F. Wheelock, Beech Heights, Monart, Enniscorthy, Co. Wexford
EI3 FO James Fitton, Ballinara Cross, Waterfall, Cork
EI3 FP Denis Maloney, 32 Chalfont Place, Malahide, Co. Dublin
EI3 FQ Joseph P. Walsh, 39a Ulverton Close, Dalkey, Co. Dublin
EI3 FR Tom Doherty, 262 Beechdale, Dunboyne, Co. Meath
EI3 FS Andrew Connor, 5 Kerrcourt, Girvan, Ayrshire KA22 0BP, Scotland
EI3 FT Midland Radio Experimenters Club, c/o George Shedwell, 7 St. Mels Road, Longford
EI3 FU John Lawlor, Main Street, Castletownroche, Mallow, Co. Cork
EI3 FV Hugh R. Duffy, 10 Glenaboy Gardens, Salthill, Galway
EI3 FW Craig L. Robinson, 'Enniscree', Templecarrig, Delgany, Co. Wicklow
EI3 FX Henry Harty, 'Faraday', Killeen, Patrickswell, Co. Limerick
EI3 FY Eamonn G. Douglas, 94 Shangan Avenue, Ballymun, Dublin 11
EI3 FZ James Mannix, Keelclogherane, Faha, Killarney, Co. Kerry
EI3 GA John O'Grady, 4 McDermott Villas, Navan, Co. Meath
EI3 GC Liam Curran, 3 Rosemount Road, North Circular Road, Dublin 7
EI3 GD Raymond A. Percival, 7 Meadow Mount, Churchtown, Dublin 16
EI3 GE Jim Echlin, Stylebawn Cottage, Delgany, Co. Wicklow
EI3 GF Michael C. Quinn, 13 Seafield, Wicklow
EI3 GG Gerard Elliott GI4OWA, 4 Fernbrae Gardens, Kilfennan, Derry, BT47 1XS, Northern Ireland
EI3 GH Liam Field, 'Ittledy', Portrane Road, Donabate, Co. Dublin
EI3 GJ William J. Byrne, Ballynerrin Lower, Wicklow
EI3 GK Martin P. Behan, Ballymotey, Enniscorthy, Co. Wexford
EI3 GM Christina Connor, 5 Kerrcourt, Girvan, Ayrshire KA22 0BP, Scotland
EI3 GN Timothy J. McCarthy, Ballintrim, Rostellan, Midleton, Co. Cork
EI3 GO Martin J. Ffrench, Donanore, New Ross, Co. Wexford
EI3 GP Michael Crosbie, Skehana, Peterswell, Co. Galway
EI3 GQ Richard K. Mansfield, "Woodlands", Green Hill, Fermoy, Co. Cork
EI3 GR Lonan Mag Fhogartai, 11 Liskey Brae, Fintona, Co. Tyrone
EI3 GS John P. Cowman, 29 Beaumont Lawn, Ballintemple, Cork
EI3 GT David Sherwood, Maudlintown, Wexford
EI3 GU Patrick A. Murtagh, 31 Seaview Park, Shankill, Co. Dublin
EI3 GV Brendan M. De hOra, 7 Lakelands Lawn, Stillorgan, Blackrock, Co. Dublin
EI3 GW William A. Wilson, 10 Kickham Street, Mullinahone, Thurles, Co. Tipperary
EI3 GY Luke Conroy Sr., 407 Carnlough Road, Cabra West, Dublin 7
EI3 GZ John J. O'Sullivan
EI3 HA Anthony Casey, 54 Beladd Park, Dublin Road, Portlaoise, Co. Laois
EI3 HB Kinsale Radio Club, Vocational School, Kinsale, Co. Cork
EI3 HC Fr. Walter P. McNamara, Holy Ghost Missionaries, Ardbraccan, Navan, Co. Meath
EI3 HD Patrick J. Mangan, 87 South Avenue, Mount Merrion, Co. Dublin
EI3 HE Eamonn Doyle, Kilcavan, Carnew, Co. Wicklow
EI3 HF Sonia M. Malone, 7 Clonard Drive, Dundrum, Dublin 16
EI3 HG Andy Green, 20 Greenbank, Ashley Court, Waterford
EI3 HH Tom Jeffery, Meadowcroft, Dromstrasna Collins, Abbeyfeale, Co. Limerick
EI3 HI Aidan Duffy, St. Brendans Street, Portumna, Co. Galway
EI3 HJ John Harte, Oldtown, Moycullen, Co. Galway
EI3 HK Liam Murphy, 67 Gracepark Meadows, Drumcondra, Dublin 9
EI3 HM Seosamh O hIarnain, Na Haille Thiar, Indreabhan, Conamara, Co. na Gaillimhe
EI3 HN Charles Patrick Nolan, 95 Strodes Crescent, Staines, Middlesex TW18 1DG, England
EI3 HO William T Heaslip, 10 Rockpark Avenue, Tralee, Co. Kerry
EI3 HP Massimo Fussi, 31 Foxrock Mount, Foxrock Park, Dublin 18
EI3 HQ Dr. Barry Breslin, 6 Glendoher Park, Rathfarnham, Dublin 16
EI3 HR Humphrey Joseph Ryan, Ballycurreen, Farmers Cross, Cork 4
EI3 HS Austin Grogan, 25 Townparks, Skerries, Co. Dublin
EI3 HT James A. Duggan, 'Ard na Mara', Friars Hill, Wicklow
EI3 HV Thomas Curwen, 14 Marian Park, Patrickswell, Co. Limerick
EI3 HW James M. Hogan, 5 Eastlands, Tramore, Co. Waterford
EI3 HX David G Lafferty, Carrickshandrum, Killygordon, Lifford, Co. Donegal
EI3 HY Michael Hentschel, Ballure, Port Salon P.O., Letterkenny, Co. Donegal
EI3 HZ Noel Moore, Ballynacragga, Newmarket-on-Fergus, Co. Clare
EI3 IA David Heale, Apt 2, Ranelagh Court, 12/13 Ranelagh Road, Dublin 6
EI3 IB William A. Whiddett, 43 James Road, Hatboro, PA 19040, USA
EI3 IC Martin J O'Connor, Rooska Farm, Rooska, Bantry, Co. Cork
EI3 ID Naish Kelly, Grace Dieu, Ballyboughal, Co. Dublin
EI3 IE Liam P. O'Riordan, Ballintubbrid East, Carrigtwohill, Co. Cork
EI3 IF Patrick J O'Doherty, Cromogue, Dromcollogher, Charleville, Co. Cork
EI3 IG Michael Everett Clarke, Ballindrimley, Castlerea, Co. Roscommon
EI3 IH Tony McNamara, Ashfield, Old Portmarnock, Co. Dublin
EI3 II Raymond Cassidy, 64 College Park, Corbally, Limerick
EI3 IJ Patrick C. Corkery
EI3 IK Brian Gundry, Beherna, Ryefield, Virginia, Co. Cavan
EI3 IL Gerry Coyne, "San Giovanni", Hazelwood Avenue, Sligo
EI3 IM Paul Reilly, 253 Ballsgrove, Drogheda, Co. Louth
EI3 IN Patrick Augustine Jennings, 43 Sutton Park, Sutton, Dublin 13
EI3 IO D. I. Court, 9 Greenwood Close, Petts Wood, Kent, BR5 1QG, England
EJ3 CB Fr. Tim Vaughan, SPS, CC, Bere Island, Co. Cork
EI3 ADB James G. Lacy, An Cuan, Fairbrook Lawn, Rathfarnham, Dublin 14
EI3 AEB Michael M. Byrne, 46 Chanel Road, Coolock, Dublin 5
EI3 AHB Paul C.P. Culbin, 51 Marian Grove, Rathfarnham, Dublin 14
EI3 AIB Samuel J. Sherrard, "Ailsa Craig", Ballyfoyle, Co. Kilkenny
EI3 AJB James G. Gough, Shron, Greencastle, Co. Donegal
EI3 AKB Gerard S.M. McCarthy, 71 Weston Road, Churchtown, Dublin 14
EI3 ANB Michael J. Dunne, 27 Greenhills, Athy, Co. Kildare
EI3 AOB Michael A. Adams, Palmerstown, Oldtown, Co. Dublin
EI3 APB Robert R. Hunter, 4 Eden Vale Road, Dublin 6
EI3 AUB Martin J.P. Crowe, 49 Ailesbury Road, Ballsbridge, Dublin 4
EI3 AWB Michael F. McGoldrick, Commons Road, Navan, Co. Meath
EI3 AYB John G. Gilmour, Cloneen, Drumcliff, Co. Sligo
EI3 AZB Thomas J. Murray, Warren, Boyle, Co. Roscommon
EI3 BAB F. Field, 1 Sonesta, Malahide, Co. Dublin
EI3 BCB James P. Keane, Birchgrove, Ballinasloe, Co. Roscommon
EI3 BGB Eoin Moran, 75a Braemore Road, Churchtown, Dublin 14
EI3 BHB Mark Jevins, 17 Braemore Avenue, Churchtown, Dublin 14
EI3 BIB Gerry Kelly, Goff Street, Roscommon
EI3 BJB Frank Joynt, St. Raphaels College, Loughrea, Co. Galway
EI3 BKB Seamus Barrett, 6 Greenfields, Rosbrien, Limerick
EI3 BLB John F. McCourt, The Square, Roscommon
EI3 BOB Malcolm F. Granville, "Greenbanks", Balleigham, Manor Cunningham, Co. Donegal
EI3 BPB Arthur F. Sherwin, 1 Shanakiel Lawn, Shanakiel Road, Sundayswell, Co. Cork
EI3 BQB Alfie Coyle, 10 Old Road, Cashel, Co. Tipperary
EI3 BRB T. Coyle
EI3 BSB Ruiadhe St.John Murphy, 7 Coolmine Close, Clonsilla, Co. Dublin
EI3 BTB Charlie Andrews, Dromore, Killygordan, Lifford, Co. Donegal
EI3 BVB Peter J. Towey, Knockroe, Castlerea, Co. Roscommon
EI3 BXB J.T. Cromie, R.P. Manse, Stranorlar, Lifford, Co. Donegal
EI3 BYB J. Kelly, 38 Shandon Park, Kilkenny
EI3 BZB Martin F. Kearney, 466 Mourne Road, Drimnagh, Dublin 12
EI3 CAB Douglas Port, 8 Betterton Drive, Sidcup, Kent DA14 4PS, England
EI3 CIB Rory J. Higgins, 55 Wierview Drive, Stillorgan, Co. Dublin
EI3 CJB Peter F. Cahill, 80 Dunston Road, Battersea, London
EI3 CKB Asta Keil, P.O. Box 6, Avoca, Co. Wicklow
EI3 COB Declan Mullally, Mount Prospect, Raheen, Limerick
EI3 CQB Details withheld at licensee's request
EI3 CSB Alan Brocklebank, 46 Oaklands Avenue, Swords, Co. Dublin
EI3 CTB Justin Behan, 25 Birchdale Road, Kinsealy Court, Kinsealy, Co. Dublin
EI3 CUB Timothy A. Cronin, Kinsealy Road, Old Portmarnock, Co. Dublin
EI3 CVB Lawrence B. Reardon, 21 Redesdale Road, Mount Merrion, Co. Dublin
EI3 CXB Fiachra R. Furlong, 4 Raheen Heights, Arklow, Co. Wicklow
EI3 CYB James Allen, 40 Victoria Street, South Circular Road, Dublin 8
EI3 CZB David P. Corcoran, 22 Glebemount, Friars Hill, Wicklow
EI3 DCS Donahies Community School, c/o R. Savage, Donahies Community School, Dublin 13
EI3 DDB Mark Whyte, 'Lawndale', Rocky Road, Wicklow
EI3 DEB Margaret M. Tobin, 78 St. Brendans Avenue, Coolock, Dublin 5
EI3 DHB Terence N. McCafferty, 32 Railway Road, Strabane, Co. Tyrone, Northern Ireland
EI3 DIB John A. Lofthouse, Monroe East, Ardfinnan, Clonmel, Co. Tipperary
EI3 DKB Robert J. Rutbotham, 58 Philipsburg Terrace, Marino, Dublin 3
EI3 DLB John P. McCarthy, 'Villa Nova', Douglas Road, Cork
EI3 DMB William P. Lanigan, 191 Monread Heights, Naas, Co. Kildare
EI3 DNB Richard F. Moore, 43 Upper Galliagh Road, Derry City, Northern Ireland
EI3 DQB Raymond Elgy, 2 Hillcrest Grove, Kilkishen, Co. Clare
EI3 DRB John Furlong, 3 Wellmount Court, Finglas West, Dublin 11
EI3 DSB Frederick O'Keeffe, 2 Grand View Place, Dillons Cross, Cork
EI3 DUB Details withheld at licensee's request
EI3 DWB Finbarr Sheehy, 35 Oatfield Park, Clane, Co. Kildare
EI3 DYB P.J. Tallon, Ballinree, Mostrim, Co. Longford
EI3 DZB Denise M. Lyons, 24 Oakwood Avenue, Swords, Co. Dublin
EI3 EAB Graham Bryce, 29 Knocklyon Avenue, Dublin 16
EI3 EBB Alan D Foley, "Windsor Cottage", Model Farm Road, Cork
EI3 ECB Details withheld at licensee's request
EI3 EDB Enda P. Dalton, 48 Cill Eanna, Raheny, Dublin 5
EI3 EEB John O'Hea, Myrtle Hill, Ballygarvan, Co. Cork
EI3 EFB Michael A Fitzgerald, Broomfield House, Midleton, Co. Cork
EI3 EHB Brian Canning, Aughnaglace House, Cloone P.O., Co. Leitrim
EI3 EIB David R. Doyle, 20 Dalysfort Road, Salthill, Galway
EI3 EKB Joe McCormack, 18 Mount Carmel Drive, Moate, Co. Westmeath
EI3 EMB Enda L Murphy, 61 Tiffany Downs, Bishopstown, Cork
EI3 ENB Paul Norris, Clogga, Mooncoin, Co. Kilkenny
EI3 EPB John J O'Rourke, 31 South Parade, Waterford
EI3 EQB Wilfred Ferguson, 12 Ashdale Road, Terenure, Dublin 6W
EI3 ESB Joseph Fadden, Knockthomas, Castlebar, Co. Mayo
EI3 ETB Matthew J Byrne, 60 College Rise, Drogheda, Co. Louth
EI3 EVB Garrett O'Hanlon, 17 Alderbury Grove, Earls Court, Waterford
EI3 EXB Daniel Doran, Convent View, Mooncoin, Co. Kilkenny
EI3 EYB John Flavin, 13 Kimmage Road West, Terenure, Dublin 12
EI3 EZB Brendan Meehan, 40 Holly Road, Dublin 9
EI3 FAB Peter McDaid, "Cloniff", Adelaide Road, Glenageary, Co. Dublin
EI3 FBB Paul McEntagart, St. Martin's, Dublin Road, Navan, Co. Meath
EI3 FCB Kevin Craig, Rathconnell, Mullingar, Westmeath
EI3 FDB Patrick Kearney, Curraclough, Bandon, Co. Cork
EI3 FEB Andrew McCormack, Clonkill, Mullingar, Co. Westmeath
EI3 FFB Eamonn Kavanagh, Ballyverane, Bansha, Co. Tipperary
EI3 FHB Richard J. Cullinan, Ballingaddy, Ennistymon, Co. Clare
EI3 FIB Joseph Desbonnet, Roscam, Galway
EI3 FJB Noel Clarke, Smithstown, Maynooth, Co. Kildare
EI3 FKB Paul McQuaid, 84 Sutton Park, Sutton, Dublin 13
EI3 FLB Mark Wall, 22 Sunshine Crescent, Waterford
EI3 FMB Rudy Stoklosa, 16 Winters Hill, Sunday's Well, Cork
EI3 FNB Eoin O'Connor, 4 Seaview Close, Carrigaline, Co. Cork
EI3 FOB Noel Lane, 53 Dundanion Road, Beaumont, Cork
EI3 FPB Brian Crowley, 88 Blackwater Heights, Youghal, Co. Cork
EI3 FQB Kieran Groeger, Ballyclamasy, Youghal, Co. Cork
EI3 FRB Barry Prendergast, 4 Prospect Farm, Windmill Hill, Youghal, Co. Cork
EI3 FSB Kilian Kelly, Redbarn, Youghal, Co. Cork
EI3 FTB Hugh O'Leary, Stumphill, Midleton, Co. Cork
EI3 FUB Barry Bridgeman, 65 Fairhill Drive, Fairhill, Cork
EI3 FVB Miriam P. Curtin, 18 Central Avenue, Bishopstown, Cork
EI3 FWB Stephen O'Leary, 35 Alderwood Est., South Douglas Road, Cork
EI3 FXB Ciaran Ferry, Brinalack P.O., Gweedore, Letterkenny, Co. Donegal
EI3 FYB James Ryan, 10 MacCurtin Villas, College Road, Cork
EI3 FZB Michael Walsh, 437 St. John's Park, Waterford
EI3 PKT Digi-Repeater, c/o Fr. F. Buckley EI1CS, Cork Radio Club, Wilton Park House, Wilton, Cork
EI3 VNO John E. Lee, N2DFP, 180 Van Cortlandt Pk. Sth, Bronx, NY 10463 USA

EI4

EI4 A D. K. McCrossan, 33 Ashgrove Drive, Naas, Co. Kildare
EI4 E Galway Radio Experimenters Club (see EI4GRC)
EI4 H Donal P. Fitzmaurice, 93 Willow Park Grove, Glasnevin North, Dublin 11
EI4 J Gavin G. Halpin, "Dookinella", Rathmullen Road, Drogheda, Co. Louth
EI4 L John E. Scanlon, 7 St. John's Road, Wexford
EI4 T Ronald E. McCrea, Tullyhommon, Co. Fermanagh, Northern Ireland
EI4 W Peadar Seoighe, 15 Pairc Leana Bui, Gaillimh
EI4 Y Eric F. Matthews, 9 Pine Copse Road, Dublin 16
EI4 Z Tom O'Donnell, 26 Glasnevin Avenue, Ballymun, Dublin 11
EI4 AB Christopher Connolly, 29 Marian Park, Waterford
EI4 AD Analog Devices Radio Club, c/o Gerard Dowling, Analog Devices bv, Raheen, Limerick
EI4 AE T.J. Martin, Banba, Killybegs, Co. Donegal
EI4 AG John Maher, 6032 N 28th Street, Arlington, VA 22207, USA
EI4 AH D.W. Ray, 13 Renmore Road, Galway
EI4 AJ Louis B. D'Alton
EI4 AL Michael G. Bourke, 4th Field Sigs., Custume Barracks, Athlone, Co. Westmeath
EI4 AN Jack Love, "The Stone Wall", Carnalynch, Bailieboro, Co. Cavan
EI4 AR Ronald G. Hall, 32 Marino Green, Marino, Dublin 3
EI4 AV J.J. Canavan, Gulladuff House, Moville, Co. Donegal
EI4 AZ Neal Doherty, "Limefield Bungalow", Ballynally, Moville, Co. Donegal
EI4 BA Owen Mooney, 184 Forrest Hills, Rathcoole, Co. Dublin

EI4 BB Brendan Daly, 28 Templeogue Wood, Templeogue, Dublin 6W
EI4 BC A.D. Patterson, GI3KYP
EI4 BK T. Deegan, 27 Oakland Drive, Greystones, Limerick
EI4 BN D.P. Hegarty, Growtown, Dunshaughlin, Co. Meath
EI4 BS Patrick Trant, 38 Fernhill Park, Terenure, Dublin 12
EI4 BT J.S. Craig, "Cloonagh", Ballymote, Co. Sligo
EI4 BV Aer Lingus A.R.C., Dublin Airport, Co. Dublin
EI4 BX Jim Bellew, Long Avenue, Dundalk, Co. Louth
EI4 BY Trevor Campbell Davis, 9 Cloister Road, North Acton, London W3 0DE, England
EI4 BZ David Moore, Dooneen, Carrigtwohill, Co. Cork
EI4 CA Comdt. Des Butler, Limcarra, Ballincar, Sligo
EI4 CB Declan F. Keane, 11 Springfield Park, Templeogue, Dublin 6
EI4 CC J.B. Eddy, 5 Mossgrove Avenue, Caherdavin Heights, Limerick
EI4 CD Basil W. Aldwell, GU2HML
EI4 CE Con O'Callaghan, 14 Lower Main Street, Newmarket, Co. Cork
EI4 CF Rev. Fr. Niall Foley, St. Joseph's College, Garbally Park, Ballinasloe, Co. Galway
EI4 CG Mike Babe, Lisrenny, Tallanstown, Dundalk, Co. Louth
EI4 CH Brendan Derrane, 23 The Crescent, Lifford, Ennis, Co. Clare
EI4 CI Pierce O'Brien, Clavinstown, Drumree, Co. Meath
EI4 CJ Rev. Fr. A. Costello, Terenure College, Dublin 6
EI4 CK John M. McCormack, Dispensary Residence, Ravensdale, Dundalk, Co. Louth
EI4 CL Patrick Robinson, 26 Frascati Park, Blackrock, Co. Dublin
EI4 CM Paul O'Brien, 7 Old Coach Road, East Weymouth, Mass. 02189, USA
EI4 CN Michael P. Brown, 5A The Woodlands, Rathfarnham Castle, Rathfarnham, Dublin 14
EI4 CP James C. Smith
EI4 CR Rev. Fr. Thomas G. Lorigan, Leachten Lodge, Waterfall Road, Cork
EI4 CS Rev. Bro. Francis P. Crummey, Christian Brothers, St. Mary's College, Mullingar, Co. Westmeath
EI4 CT Harry Dunleavy, Tavanaugh, Cloghans Hill, Co. Mayo
EI4 CV James I. Menton, 190 St. Donaghs Road, Kilbarrack, Dublin 5
EI4 CW Robert J. Brown, 26 Bawnmore Road, Belfast BT9 6LA
EI4 CX Michael Grace, 1 Lishin, Tullyglass, Shannon Airport, Co. Clare
EI4 CY Ernie Berkenheier, 'Southern Comfort', Glasnarget, Rathdrum, Co. Wicklow
EI4 CZ Bernard J. Monaghan, 829 Greenwood Avenue, Chicago, Illinois 60027, USA
EI4 DA John P. Hurley, 10 Makedonias Street, Ayios Andreas, Nicosia, Cyprus
EI4 DB Desmond P. Gallagher, Cummeen, Strandhill Road, Sligo
EI4 DC Patrick Tuohy, "Mill View", Ballinagough, Whitegate, Co. Clare
EI4 DD Martin Rossiter, 205 Sundrive Road, Dublin 12
EI4 DE Martin Lynch, 30 Woodview Grove, Blanchardstown, Dublin 15
EI4 DG Dr. J.K. McDarby, "Parkview", Neale Road, Ballinrobe, Co. Mayo
EI4 DH Denis Rooney, Cornagilla, Manorhamilton, Co. Leitrim
EI4 DI Owen Furniss, 40 Broadford Crescent, Ballinteer, Dublin 16
EI4 DJ Dr. William D. Hutchinson, GI4FUM
EI4 DK Dominic Cafolla, GI4DOM
EI4 DL Roy J. Fleming, 'Basin View', Tralee, Co. Kerry
EI4 DM Mike Fogarty, 11 Willow Place, Athlone, Co. Westmeath
EI4 DN J. Kirk, 15 Victoria Road, Chichester, West Sussex PO19 4HY
EI4 DO William J. Harvey, 20 Eaton Square, Terenure, Dublin 6W
EI4 DP Ian James McMullan, 48 Merton Drive, Ranelagh, Dublin 6
EI4 DQ Tom Cocking, 'Scartlea', Saleen, Midleton, Co. Cork
EI4 DR Michael Foley, Woodbine, Strandhill Road, Sligo
EI4 DS John A. O'Riordan, CT1DHG
EI4 DT James Cullinan, Ballycallan, Kilkenny
EI4 DU Leslie W. Long, 79 Hawthorn Heights, Letterkenny, Co. Donegal
EI4 DV Michael B. Rooney, c/o The Post Office, Glencar, Sligo
EI4 DW Ken McDermott, Curraghamone, Ballybofey, Co. Donegal
EI4 DY Andrew P. Henry, Main Street, Wicklow Town
EI4 DZ Noel Cameron, 16 St. Mary's Crescent, Westport, Co. Mayo
EI4 EA Kevin McElhatten, 25 Cherrywood Road, Loughlinstown, Co. Dublin
EI4 EB Peadair P. Kinsella, Carrigduff, Bunclody, Co. Wexford
EI4 ED Tony Whelan, 5 Allenagh, Longford, Co. Longford
EI4 EF George F. Peterson, 152 Clonkeen Crescent, Kill-'o-the-Grange, Co. Dublin
EI4 EG Details withheld at licensee's request
EI4 EH Tom Kiely, 28 Martins Avenue, Sligo
EI4 EI H. Convery, 6 Carnhill Avenue, Newtownabbey, Co. Antrim
EI4 EJ R. Walsh, 87 Brian Road, Marino, Dublin 3
EI4 EK William McCauley, Manorcunningham, Letterkenny, Co. Donegal
EI4 EL Christopher Flanagan, Kilnameela Cottage, Ahiohill, Enniskeane, Co. Cork
EI4 EM Luke Conroy Jr., 407 Carnlough Road, Cabra West, Dublin 7
EI4 EO G.F. Williams, Francis Street, Ballina, Co. Mayo
EI4 EP E. Quinn, Glenahilt, Burtonport, Letterkenny, Co. Donegal
EI4 EQ Danny Campbell, GI4NKD, 109 Drumgor Park, Craigavon, Co. Armagh BT65 4AH, Northern Ireland
EI4 ER H. Gallagher, Garda Station, Sligo
EI4 ES Bernard C. Leavey, 29 Farmhill Park, Goatstown, Dublin 14
EI4 ET Anthony A. Campbell, Abbeylands, Ballyshannon, Co. Donegal
EI4 EU Patrick J. Clifford, 75 Newark Street, Lindenhurst, NY 11757, USA
EI4 EV Thomas Walshe, 2 Clonkeen Crescent, Dun Laoghaire, Co. Dublin
EI4 EW William H. Kelly, 7 St. Donaghs Road, Kilbarrack, Dublin 13
EI4 EY John Phelan, 72 Meadow Vale, Raheen, Limerick
EI4 EZ Howard R. Spier, Kaieteur-Ville, Knocklyon Avenue, Knocklyon Woods, Templeogue, Dublin 16
EI4 FA John Garner G3KEC, Portuan Road, Hannafore, West Looe, Cornwall PL13 2DN
EI4 FB Anthony Walsh, Killamaster, Carlow
EI4 FC John Kenny, Ballynaglearagh, Lattin, Co. Tipperary
EI4 FD Michael Keane, 59 Mornington Heights, Trim, Co. Meath
EI4 FE Details withheld at licensee's request
EI4 FF Rev. Donal Kilduff CC, Kill, Cootehill, Co. Cavan
EI4 FG Stephen Staunton, 6 Shamrock Close, Douglas, Co. Cork
EI4 FI Frank Rima, Tully, Renvyle, Co. Galway
EI4 FJ Patrick Lohan, Ruane, Ballygar, Co. Galway
EI4 FK Hugh McNulty, College Farm Road, Letterkenny, Co. Donegal
EI4 FL John J. Farrell, 79 Brookside, Bettystown, Co. Meath
EI4 FN Fintan Meyler, Springfield House, Hill Street, Wexford
EI4 FP Ambrose Owens, Deanhill, Navan, Co. Meath
EI4 FR Todd Harvey KA6QOJ, 19 Woodview Close, Moylish, Limerick
EI4 FS Waldemar Borrmann, 30 Mount Ievers, Sixmilebridge, Co. Clare
EI4 FT Joseph A. Bugg, 'Marianvilla', Kinvara, Co. Galway
EI4 FV Joseph F. Dillon, 5 Verbena Lawn, Sutton, Dublin 13
EI4 FW Laurance Byrne, Glenfin Road, Ballybofey, Co. Donegal
EI4 FX Liam Fitzgerald, 'Melrose', Dwyers Road, Midleton, Co. Cork
EI4 FY John V. Porter, 9 Sutton Gardens, Waterside, Derry
EI4 FZ Brendan Murphy, Curra Cottage, Curraheen, Tralee, Co. Kerry
EI4 GA Paul Smith, Sheepgrange, Drogheda, Co. Louth
EI4 GB Liam Mangan, "Mount Ievers", Sixmilebridge, Co. Clare
EI4 GC John C. Farrell, Curravarahane, Bandon, Co. Cork
EI4 GD Gerry Cregg, Killaraght, via Boyle, Co. Sligo
EI4 GE Shane Halpin, c/o Amateur Radio EI4J, 'Dookinella', Rathmullen Road, Drogheda, Co. Louth
EI4 GF Thomas O'Rourke, 10 Hazelbrook Drive, Terenure, Dublin 6
EI4 GG Ronan O'Gorman, 7 St. Senan's Road, Lifford, Ennis, Co. Clare
EI4 GH John M. Bruce, 30 Ballymenoch Road, Holywood, Co. Down
EI4 GI Brian Short, 10 Castleview, Tallanstown, Dundalk, Co. Louth
EI4 GJ John Slocum, Shanakill, Ballymacoda, Co. Cork
EI4 GK John J. Donelan, "Inniscarra", Ballybride Road, Shankill, Co. Dublin
EI4 GL Brian O'Daly, Tralee
EI4 GM Alfred C. Deeney, 43 Coolnevaun, Stillorgan, Dublin 14
EI4 GN Details withheld at licensee's request
EI4 GO John H. Dalton, 28 Rosmeen Gardens, Dun Laoghaire, Co. Dublin
EI4 GP Garrett Sinnott, Ballycadden, Bunclody, Co. Wexford
EI4 GQ John D. Crichton GI4YWT, 10 Bann Drive, Lisnagelvin, Londonderry, Northern Ireland
EI4 GR Michael Dorgan, Ballinasare, Annascuil, Co. Kerry
EI4 GT Brendan P. Norton, 58 Oxmantown Road, Dublin 7
EI4 GU Kevin P. Harris, 21 River Valley Court, Swords, Co. Dublin
EI4 GV Peter Vekinis, Ballyvogue Cottage, Goleen, Co. Cork
EI4 GW Robert Browning, 53 Caulside Park, Muckamore, Antrim, Northern Ireland
EI4 GX Joseph Earley, 3 Whitworth Terrace, Drumcondra, Dublin 3
EI4 GY John Coleman, Station Road, Cootehill, Co. Cavan
EI4 GZ James Gallagher, 58 St. Jarlath Road, Cabra, Dublin 7
EI4 HA Brian T. Cullen, 109 Grange Heights, Waterford
EI4 HC Fr. William Kingston 9L4WK, Catholic Mission, Koindu, P.O.Box 200, Kenema, Sierra Leone, West Africa
EI4 HD James W. Monaghan Sr. N7HKO, 1206 9th Avenue, S.E. Puyallup, WA 98372, USA
EI4 HE Robert McGrogan, Clonnagapple House, Delvin, Co. Westmeath
EI4 HF Michael J. Lee, 'Helvellyn', Knockaverry, Youghal, Co. Cork
EI4 HG Dze Stefan Zalewski, Marian Villa, Kinvara, Co. Galway
EI4 HH James Holohan, 7 Hilton Gardens, Balinteer Avenue, Balinteer, Dublin 16
EI4 HI Details withheld at licensee's request
EI4 HJ Christopher Holt, 92 Hazlewood Park, Artane, Dublin 5
EI4 HK Claire Gardner, 75 Farran Park, Waterford
EI4 HL Tim Jones, 405 Grand Prom, Dianella 6062, Perth, Western Australia
EI4 HM D.F.J. Walmsley, 2 St. Margaret's Court, Uttoxeter Road, Draycott, Stoke-on-Trent, Staffs ST11 9SL, England
EI4 HN Daniel Taggart, 12 Oak Rise, Dublin Road, Omagh, Co. Tyrone BT78 1TN
EI4 HO Liam Maher, "Hillside", 55A St Patrick's Avenue, Tipperary
EI4 HP Ivan William McCaffrey, Shanaghy, Ballina, Co. Mayo
EI4 HQ Cormac Gebruers, Springfield, Cobh, Co. Cork
EI4 HR Ray Martin, 96 College Rise, Drogheda, Co. Louth
EI4 HS John Kelly, 25 Abbey Court, Abbey Farm, Celbridge, Co. Kildare
EI4 HT David Ryan, 11 Knockgriffin, Midleton, Co. Cork
EI4 HU Donal Hurley, Shanagarry, Midleton, Co. Cork
EI4 HV Jimmy Hamill, 67 Windsor Avenue, Coleraine, Co. Derry, BT52 2DR, Northern Ireland
EI4 HW Frankie McEvoy, 12 Rousseau Grove, Norwood, Waterford
EI4 HX Peter B. Grant, 37 Glenmore Park, Dundalk, Co. Louth
EI4 HY Prakash Madhavan, 14 The Village, Porterstown Road, Clonsilla, Dublin 15
EI4 HZ Anne C. Clear Vekinis, Ballyvogue Cottage, Goleen, Co. Cork
EI4 IA Eugene J Drumm, Mullingar, Killygordan P.O., Co. Donegal
EI4 IB Nicholas Jordan, Tristernagh, Ballynacargy, Mullingar, Co. Westmeath
EI4 ID Matthew L. Merry, 705 Manor Terrace, Moorestown NJ 08057, USA
EI4 IE Hilary Moore, Dooneen, Carrigtwohill, Co. Cork
EI4 IF James Duggan, Cahirkeem Strand, Eyeries, Bantry, Co. Cork
EI4 IG Aidan Brodigan, 7 Fairway Lawns, Bettystown, Co. Meath
EI4 IH Joseph O'Neill, Hermitage, Collon Road, Slane, Co. Meath
EI4 II Bernard Gondard, 2 Farnogue Heights, Newlands, Wexford
EI4 IJ Martin Michael Scanlon, Meehan, Coosan, Athlone, Co. Westmeath
EI4 IK Shane McKeever, Kilkenny City
EI4 IL Cyril W. Forde, 5 Faiche Seannaigh, Rath Eanaigh, Baile Atha Cliath 5
EI4 IM John W. Spendlove, Tullaghanrock, Edmonstown, Ballaghadereen, Co. Roscommon
EI4 IN Ben Gaughran, 9 Hillside, Skerries, Co. Dublin
EI4 RF 4 Metre Beacon, c/o EI6DN
EI4 ABB Aengus Cullinan
EI4 ACB Mark G. Davis, 44 Keatingstown, Wicklow, Co. Wicklow
EI4 ADB Timothy J. Crawford, 24 Main Street, Ballywalter, Co. Down, Northern Ireland
EI4 AEB R. O'Brien, 3 Batterstown, Dunboyne, Co. Meath
EI4 AFB Kenneth W. McAllister, "Lermoos", Demesne Road, Malahide, Co. Dublin
EI4 AGB Cyril Moriarty, Zonnedauw 10, Valkenswaard, The Netherlands
EI4 AHB Fredrick B. Edmondson, c/o H. Rattery, Marble Hill, Port-na-Blagh, Co. Donegal
EI4 AJB John R Ashe, 49 Dean's Walk, Sleepy Valley, Richill, Co. Armagh BT61 9LD, Northern Ireland
EI4 AKB Marie B. Enright, Crossfield, Firhouse Road, Templeogue, Dublin 16
EI4 ALE Galway VHF Group, c/o Steve Wright, Blood Bank, Regional Hospital, Galway
EI4 ANB Kevin Connolly, 37 Haven Hill, Summercove, Kinsale, Co. Cork
EI4 AQB Gerard M. Garvey, 135 Inishannagh Park, Newcastle, Galway
EI4 ARL Listowel Radio Club, c/o M. Crowley, Church Street, Listowel, Co. Kerry
EI4 ASB George Emerson, c/o Willie Long, Dunkineely, Co. Donegal
EI4 ATB Gerard O'Brien, 5 Willow Avenue, Caherdavin Heights, Limerick
EI4 AXB J. McDermott, Celtic Avenue, Roscommon
EI4 AYB P. Coyle, Knocknacarra, Salthill, Co. Galway
EI4 AZB Michael Jordan, The Walk, Roscommon
EI4 BAB Patrick P. Kearney, 466 Mourne Road, Drimnagh, Dublin 12
EI4 BBB Donal Cussen, Tinraheen, The Ballagh, Enniscorthy, Co. Wexford
EI4 BGB Brendan Finn, 6 Ashbrook Avenue Road, Dundalk, Co. Louth
EI4 BHB Patrick T. Power, Blenheim Hill, Grantstown, Waterford
EI4 BKB David Branigan, "Dunrath", 9 Elton Park, Sandycove, Co. Dublin
EI4 BLB Paul Baker, 10 Monaloe Avenue, Blackrock, Co. Dublin
EI4 BNB Timothy O'Riordan, Curraleigh House, Curraheen Road, Bishopstown, Cork
EI4 BOB John B. Dynes, 30 Breagh Road, Portadown, Co. Armagh, Northern Ireland
EI4 BPB Peter O'Farrelly Jnr., Ardshanbally, Adare, Co. Limerick
EI4 BSB Sarah A. Boyle, Loughanure, Annagry P.O., Letterkenny, Co. Donegal
EI4 BST 132nd Bayside Scouts Radio Group, c/o Sean O Suilleabhain, 141 Brookwood Avenue, Artane, Dublin 5
EI4 BVB James A. Challen, 66 Abbey Park, Ferrybank, Waterford
EI4 BWB David Johnston VK6BWB, 65 Benara Road, Noranda, Western Australia 6062
EI4 BXB Francis J. Lawrence, Mothel Road, Carrick-on-Suir, Co. Tipperary
EI4 BYB William J. Bourne, 25 St. Egnies Terrace, Buncrana, Co. Donegal
EI4 BZB John V. Goff, 14 Hillcrest Court, Lucan, Co. Dublin
EI4 CAB Peter Wilson, 9 Navan Road, Castleknock, Dublin 15
EI4 CCB Michael R. Kennedy, 342 Tymon Heights, Firhouse, Dublin 24
EI4 CDB Martin J. Murphy, Convent Road, Claremorris, Co. Mayo
EI4 CHB Rachel Doyle, Parknashaw House, Avoca, Arklow, Co. Wicklow
EI4 CIB Thomas R. Doyle, Parknashaw House, Avoca, Arklow, Co. Wicklow
EI4 CJB Keith R. Roberts, 31 Cybi Place, Holyhead, Gwynedd LL6S 1DT, Wales
EI4 CKB Michael Martin, Tullycanna, Ballymitty, Co. Wexford
EI4 CLB Ronan B. O'Neill, 57 Hazelbrook Road, Terenure, Dublin 6
EI4 CMB John P. Doyle, 95 Blarney Park, Kimmage, Dublin 12
EI4 COB Michael Dargan
EI4 CQB Francis G. Halford, 478 Mourne Road, Drimnagh, Dublin 12
EI4 CRB James Keppel, Tobinstown, Tullow, Co. Carlow
EI4 CSB John P. Whelan, Ballyclough, Kilworth, Co. Cork
EI4 CTB Brendan J. Fidgeon, "Crannog", St. Patrick's Road, Wicklow
EI4 CUB Richard Ryan, Claremont, Killeshin Road, Carlow
EI4 CVB Joseph E. May, The Lighthouse, Mornington, Drogheda, Co. Louth
EI4 CWB Victor Bradshaw, 8 Woodview, Ashford, The Murrough, Wicklow
EI4 CXB Alan F. Sheane, Ballycullen, Ashford, Co. Wicklow
EI4 CYB Brendan J. O'Kane, 79 Martello Court, Portmarnock, Co. Dublin
EI4 DBB David M. O'Sullivan, 5 Woodlawn, Lahinch Road, Ennis, Co. Clare
EI4 DCB Dan Gallagher, Rocklawn, Carrigohane, Co. Cork
EI4 DDB James McDaid, 10 De Burgh Terrace, Derry BT48 7LQ
EI4 DEB Kenneth Connolly, Craughwell, Co. Galway
EI4 DFB Kevin W. King, 284 Rathmullen Park, Drogheda, Co. Louth
EI4 DGB Nigel Gamble, 309 Coneyburrow Road, Lifford, Co. Donegal
EI4 DHB Philip Masterson, 15 Kilcross Grove, Sandyford, Co. Dublin
EI4 DIB Anthony Allen, Co. Louth
EI4 DJB Rory A.J. Hinchy, 'Waverley', 299 Navan Road, Dublin 7
EI4 DLB John Madden, The Cottage, Clarendon Street, Derry
EI4 DMB Paul M. Browne, 8 Casino Park, Malahide Road, Dublin 3
EI4 DMR William F. Hurley, Regional Tech. College, Moylish, Limerick
EI4 DNB D.R. Rowkins, The Grove, Kilteenan, Roscommon
EI4 DOB Kevin Barry, The Square, Passage West, Co. Cork
EI4 DPB Brendan Minish, Inismhara, Killadangan, Westport, Co. Mayo
EI4 DRB Patrick Lyons, Crinaloo, Rathcoole, Mallow, Co. Cork
EI4 DSB John Kirwan, 256 Navan Road, Dublin 7
EI4 DUB Tony O'Regan, Ballincrokig, Dublin Pike, Cork
EI4 DVB C.J. Colgan, 11 St. Johns Park, Moira, Craigavon BT67 0NL, Northern Ireland
EI4 DWB Trevor T. Jones, 7 Dove Close, Thorely, Bishops Stortford, Herts CM23 4JD, England
EI4 EAB Brendan Power, 12 Faughart Road, Crumlin, Dublin 12
EI4 ECB Paul O'Brien, 32 The Strand, Donabate, Co. Dublin
EI4 EDB Peter Green, Castlerea, Co. Roscommon
EI4 EEB Martin Sweeney, Lisaleen, Tuam, Co. Galway
EI4 EFB Thomas O'Dea, Kilcornan, Clarinbridge, Co. Galway
EI4 EHB Anthony M. Rogers, 60 Poleberry, Waterford City
EI4 EIB Michael Sweeney, Lisaleen, Kilconly, Tuam, Co. Galway
EI4 EKB Jakob Conrady, Rockwell House, Kilconly, Tuam, Co. Galway
EI4 ELB Bartholomew O'Leary, 1 Mellows Park, Renmore, Galway
EI4 ENB Patrick Casey, 64 Summerstown Drive, Wilton, Cork
EI4 EOB Nigel Barlow, Heathstown, Coralstown, Mullingar, Co. Westmeath
EI4 EPB Brendan L. Kelly, 145 Swords Road, Whitehall, Dublin 9
EI4 EQB Paul R. Delany, 17 Wheatfield, Martello, Portmarnock, Co. Dublin
EI4 ERB Andrew Earley, c/o IRTS, P.O. Box 462, Dublin 9
EI4 ESB Dermot Madsen, 7 Gracefield Avenue, Artane, Dublin 5

EI4 EUB John F. Malone, Greenfield House, 1 Santa Sabina Manor, Greenfield Road, Sutton, Co. Dublin
EI4 EVB William G. Cooney, Loughaderra, Castlemartyr, Co. Cork
EI4 EXB Conor O'Neill, Reencaheragh, Portmagee, Co. Kerry
EI4 EYB John A. Bracken, Cork
EI4 FAB Vincent Long, 22 Carberry Grove, Knocknaheeny, Cork City
EI4 FBB Keith Martin, "Sitka", Cronroe, Ashford, Co. Wicklow
EI4 FCB Deaglan O Meachair, 8 Cnoc na Manach, Cill Mhantain
EI4 FDB William Keyes, 76 Captain's Avenue, Crumlin, Dublin 12
EI4 FEB Details withheld at licensee's request
EI4 FFB Details withheld at licensee's request
EI4 FGB Cathal T. Stockdale, 21 Woodview, Ashford, Co. Wicklow
EI4 FHB Details withheld at licensee's request
EI4 FIB Stuart Keyes, 76 Captain's Avenue, Crumlin, Dublin 12
EI4 FJB David Corbett, 28 Woodview, Killygoan, Monaghan
EI4 FKB Stephen Hand, 12 Church Street, Rosslea, Enniskillen, Co. Fermanagh, BT92 7DD, Northern Ireland
EI4 FLB Danny Traynor, Three Mile House, Co. Monaghan
EI4 FNB Mark Kilmartin, 8 Trim Road, Kilmore West, Dublin 5
EI4 FOB Details withheld at licensee's request
EI4 FPB Laurence Wright, 5 Woodview Park, The Donahies, Dublin 13
EI4 FQB Gerard Kelly, 20 Oakview Avenue, Clonsilla, Dublin 15
EI4 FRB William Michael Ankers, 175 Delwood Park, Castlenock, Dublin 15
EI4 FSB Peter Thomas Kelly, 93 Tonlegee Road, Raheny, Dublin 5
EI4 GRC Galway Radio Experimenters Club, c/o Joe Bugg, EI4FT, 'Marianvilla', Kinvara, Co. Galway
EI4 GST Greystones Scout Troop, c/o John McGowan, EI2FN, 15 Lr. Kindlestown, Delgany, Co. Wicklow
EI4 IMD Galway Radio Experimenters Club (see EI4GRC)
EI4 KRC Kilkenny Radio Club, c/o Michael Drennan EI7GH, Ballyroberts, Cuffesgrange, Co. Kilkenny
EI4 LRC Limerick Radio Club, c/o Anthony Condon, EI2AW, Golf Links Road, Castletroy, Co. Limerick
EI4 PKT Digi-Repeater, Limerick Radio Club, c/o EI3FX
EI4 VIV Robert B. Block, K6LX, P.O. Box 1259, Santa Barbara, CA 93102, USA
EI4 VMI Bill Rothwell, 215 Dovercourt Avenue, Ottawa, Ontario KIZ 7H3, Canada

EI5

EI5 D Thomas M. Cronin, "Castlebuoy", Strand Road, Baldoyle, Dublin 13
EI5 H William P. Kennedy, 7 Morley Terrace, Waterford
EI5 I Redmond Burke, 'Beechwood', Laburnum Park, Model Farm Road, Cork
EI5 M James E. Corcoran, 31 Shantalla Drive, Beaumont, Dublin 9
EI5 Q Eileen E. Corry, Barkhall, Port Road, Letterkenny, Co. Donegal
EI5 R Frank Colbert, 'Edel', 200 Viewmount Park, Waterford
EI5 V Signals Amateur Radio Club, 2nd Field Signals, Collins Barracks, Dublin 7
EI5 Z Donal O'Brien, 90 Sutton Park, Sutton, Dublin 13
EI5 AC Patrick F. McHugh, 74 Ard Connell, Glenties, Co. Donegal
EI5 AE Rev. Eric W. Stanley, St. Marys Rectory, Nenagh, Co. Tipperary
EI5 AG John V. Paul, Kingsbury Mews, Morehampton Lane, Herbert Park, Ballsbridge, Dublin 4
EI5 AH M. Meldrum, Finisklin, Sligo
EI5 AJ Eamon Cassidy, 64 College Park, Corbally, Limerick
EI5 AK R.M. Sloan, Shancurragh, Athlone, Co. Westmeath
EI5 AL Paul Charles Layton, 189 Old Youghal Road, Cork
EI5 AR J. Lysaght, 47 Mount Farren, Assumption Road, Co. Cork
EI5 BB William Hughes, 55 Road "A", Hartstown, Clonsilla, Dublin 15
EI5 BC C.F. Morgan, Monastery Park, Clondalkin, Dublin 22
EI5 BD Adrian Hopkins, 1 Carrig Orchard, Killincarrig, Delgany, Co. Wicklow
EI5 BF Details withheld at licensee's request
EI5 BH Paul Quast, 1 Auburn Villas, Athlone, Co. Westmeath
EI5 BK Ken Broughton, "Havenside", Castletown-Berehaven, Co. Cork
EI5 BT P.I. Dawson, 33 Ardlea Road, Artane, Dublin 5
EI5 BW Peter Farrell, "Greenacres", Ballindoolin, Edenderry, Co. Offaly
EI5 BX J. Casey, 138c Ballincurra Gardens, Limerick
EI5 BY Thomas J. Kilcline, Carrigeen, Rahara, Co. Roscommon
EI5 CA T.J. O'Brien, Silverbridge, Ballycar, Newmarket-on-Fergus, Co. Clare
EI5 CB Matthew Fahey, "Athina", 26 Chestnut Grove, Caherdavin Lawn, Limerick
EI5 CD Desmond J. Walsh, 17 The Rise, Owenabue Heights, Carrigaline, Co. Cork
EI5 CE Patrick Timmons, Kilberry, Wilkinstown, Co. Meath
EI5 CF George Twist, "Atlantic Lodge", Carhoo East, Dingle, Co. Kerry
EI5 CG B. Babe, Lisrenny, Tallanstown, Dundalk, Co. Louth
EI5 CH John A Ferguson, Drumbee-More, Armagh BT60 1HP, Northern Ireland
EI5 CI Joseph B. Clarke, 58 Cowper Road, Rathmines, Dublin 6
EI5 CJ Rev. Fr. Thomas Fives, Terenure College, Terenure, Dublin 6
EI5 CK Robert O'Donnell, 4 Ballyman Road, Enniskerry, Co. Wicklow
EI5 CL William B. Mannion, "An Culaun", Galway Road, Tuam, Co. Galway
EI5 CM Reginald J. Scarff, 4 Seapark Drive, Clontarf, Dublin 3
EI5 CN James D. Chadwick, 6 The Orchard, Monkstown Valley, Monkstown, Co. Dublin
EI5 CP Nicholas A. Lambert, Ballycross, Bridgetown, Co. Wexford
EI5 CS Daniel B. Walter, 29 Marian Crescent, Rathfarnham, Dublin 14
EI5 CT George T. Kilishek, 367 Teddy Avenue, Lancaster, PA 17601, USA
EI5 CV John C. Byrne, 21 Fairfield Avenue, East Wall, Dublin 3
EI5 CW Liam Moran, 15 Avonmore Lawn, Blackrock, Co. Dublin
EI5 CX Eric K.H. Nagel, Room 1522, 53rd W. Jackson Boulevard, Chicago, Illinois, USA
EI5 CY Timothy P. O'Connell, Tully, Glenroe, Co. Limerick
EI5 CZ Patrick A. Dillon, Depot Sigs, Ceannt Barracks, Curragh, Co. Kildare
EI5 DA Rev. Fr. P.J. O'Kelly, Parochial House, Tullyallen, Drogheda, Co. Louth
EI5 DB Andrew C. Nissipmarov, P.O. Box 113, Limassol, Cyprus
EI5 DC Michael M.A. Moore, Depot Sigs, Ceannt Barracks, Curragh, Co. Kildare
EI5 DD Stephen Wright, Blood Bank, Regional Hospital, Galway
EI5 DE Edward Tuthill, Green Briar, Loughanure, Clane, Co. Kildare
EI5 DG Thomas A. McGuinn, 13 Avondale Lawn Extn., Blackrock, Co. Dublin
EI5 DH Thomas Moore, Ballynacragga, Newmarket-on-Fergus, Co. Clare
EI5 DI Paul O'Kane, 36 Coolkill, Sandyford, Dublin 18
EI5 DJ Seamus E. Regan, 20 Sandfield Gardens, Blackrock, Dundalk, Co. Louth
EI5 DK Michael Kinsella, 9 Shannon Grove, Corbally, Limerick
EI5 DL George C. Black, Finart, South Douglas Road, Cork
EI5 DN John M. Smith, 33 Laurel Park, Clondalkin, Dublin 22
EI5 DO Albert G. Brown, 19 James Connolly Park, Clondalkin, Dublin 22
EI5 DQ William J.C. Curtis, 8 Station Road, Kesh, Enniskillen, Co. Fermanagh, Northern Ireland
EI5 DR Edward F. Kelly, Cregganavar, Breaffy, Castlebar, Co. Mayo
EI5 DS Edward Collins, 20 Balgriffin Park, Raheny, Dublin 13
EI5 DT Patrick Keeney, Legandara, Lifford, Co. Donegal
EI5 DU Michael Bourke, Midland Experimental R.C., 107 Retreat Park, Athlone, Co. Westmeath
EI5 DV John Brady, 22 Handle Road, Billerica, Mass. 01821, USA
EI5 DW Niall F. Kelly, 2 Rhebogue Avenue, Corbally, Limerick
EI5 DY William A. Murphy, 98 Langton Park, Newbridge, Co. Kildare
EI5 DZ Brian Duggan, 62 Glenwood Road, Dublin 5
EI5 EA John Rogers, 77 Sandy Lane, Woking, Surrey, England
EI5 EB Victor Cromien, Ardmore, Church Road, Greystones, Co. Wicklow
EI5 EC Louis Fleming, Main Street, Roscommon
EI5 ED Gareth McComb, 5 Fairhill Crescent, Carnmoney, Newtownabbey, Co. Antrim
EI5 EE J. Harvey Makin, 'Rowson Heights', Goggins Hill, Ballinhassig, Co. Cork
EI5 EF Thomas Russell, 1 Redwood Avenue, Kilnamanagh, Tallaght, Dublin 24
EI5 EG Howard Morris, Garronboy, Killaloe, Co. Clare
EI5 EH Patrick Egan, NR2N, 146 Andover Lane, Aberdeen Tnp, NJ 07747, USA
EI5 EI Gerry O'Sullivan, 7 Gortbeg Avenue, Finglas, Dublin 11
EI5 EJ R.P. Gorman, 70 Bloomfield Drive, Athlone, Co. Westmeath
EI5 EK Nigel Lyon-Bowie, 20 Butterfield Close, Rathfarnham, Dublin 14
EI5 EL Dominic McLaughlin, 24 Josephine Avenue, Limavady, Co. Derry BT49 9BA
EI5 EM Tony Breathnach, 159 Ceide Ard Mor, Baile Atha Cliath 5
EI5 EN Romano Morelli, 12 Capel Street, Dublin 1
EI5 EO Ballinteer Radio Club, Ballinteer Comm. School, Dublin 16
EI5 EP Lucien F. Dienick, Ardnafleming, Glenbowers, Coolbawn, Nenagh, Co. Tipperary
EI5 EQ David Clegg
EI5 ER Frank Nolan, 92 Roselawn, Tramore, Co. Waterford
EI5 ES J. Maidment, Derreencoosare, Keadue, Co. Roscommon
EI5 ET R.H. Allen, c/o C. Remedial Clinic, Vernon Avenue, Clontarf, Dublin 3
EI5 EU Godfrey O'Donnell, c/o B. O'Domhnaill, Supermarket, Main Street, Ardara, Co. Donegal
EI5 EV Joseph Murphy, Carriganurra, Slieverue, Co. Kilkenny
EI5 EW G. Singleton, 13 Compostella, Milford Grange, Castletroy, Co. Limerick
EI5 EX George Norton, 4 Newtown Park, Tallaght, Dublin 24
EI5 EY Denis Moran, Carhoonaknock, Shrowown, Listowel, Co. Kerry
EI5 EZ Declan J. McLoughlin, 5 Beechwood Street, Derry BT48 9JN, Northern Ireland
EI5 FA Graham P. Clarke, 2 Chestnut Court, Collinswood, Collins Avenue, Dublin 9
EI5 FB Timothy O'Sullivan, 51 Mitchels Avenue, Tralee, Co. Kerry
EI5 FC John A. Coakley, Glenn-na-Smol, Galway's Place, Douglas West, Co. Cork
EI5 FD Peter Keane, 76 Wyattville Park, Loughlinstown, Co. Dublin
EI5 FE Brendan Somers, 37 Carricklawn, Coolcotts, Wexford
EI5 FF Dieter Keil, Newbawn, Rathdrum, Co. Wicklow
EI5 FG Daniel D. Burke, Post Office, Ballybunion, Co. Kerry
EI5 FH Alphonsus Ward, c/o George Houston, 4 Albert Place, Lifford, Co. Donegal
EI5 FI Diarmuid J. O'Sullivan, 1 Greenlea Avenue, Terenure, Dublin 6W
EI5 FJ J.P. Hallisey, 68 Park Road Estate, Killarney, Co. Kerry
EI5 FK Charles Coughlan, 12 Forest Ridge Crescent, Wilton, Cork
EI5 FL Patrick F. O'Keeffe, Rathcooney, Glanmire, Co. Cork
EI5 FM Phil Bond, GV Assendelfstaat 3C, NH 1961 Heemskerk, Netherlands
EI5 FN Denis P. Hughes, 63 Teffia Park, Longford
EI5 FO Terence J. Devlin, 4 Legion Terrace, Park Road, Longford
EI5 FP John F. Greenan, Deer Park, Boyle, Co. Roscommon
EI5 FQ Owen A. Daunt, The Hollies, Gleann Rua, Fountainstown, Co. Cork
EI5 FR John F. Joynt, Moanmore West, Masonbrook, Loughrea, Co. Galway
EI5 FS Heinz Glocker, 10 Monaloe Avenue, Blackrock, Co. Dublin
EI5 FT Colm D. Leahy, Trabolgan, Whitegate, Midleton, Co. Cork
EI5 FV John Twamley, 30 Grange Park Crescent, Raheny, Dublin 5
EI5 FW Ian Platt, Beach Cottage, Dromcloc, Bantry, Co. Cork
EI5 FX Michael J. Flynn, St. Mary's College, Rathmines, Dublin 6
EI5 FY William Rice, 'Oakwall', 16A Presentation Road, Galway
EI5 FZ Michael Hurley, 84 St. Brendan's Park, Tralee, Co. Kerry
EI5 GA Joseph H. Cleves, Braystown House, College Hill, Slane, Co. Meath
EI5 GB Martin Whyte, 79 The Paddocks, Naas, Co. Kildare
EI5 GC Richard Glynn, 17 Oatlands, Salthill, Galway
EI5 GD Br. Terence Flynn, "San Antone", Wheatfield, Boghall Road, Bray, Co. Wicklow
EI5 GE Joseph T. Leahy, 'Galtymore', Bansha, Barnlough, Co. Tipperary
EI5 GF Brendan P. Porter, 237 Culmore Road, Derry
EI5 GG Michael J. Murphy, 87 Ard Easmuinn, Dundalk, Co. Louth
EI5 GH Sean Cooney, 59 Sandford Road, Ranelagh, Dublin 6
EI5 GI Daniel J. Kelleher, 249 Wedgewood, Dundrum, Dublin 16
EI5 GJ Thomas Mortell, 36 Foxborough Drive, Lucan, Dublin 22
EI5 GL Brendan J. Martin, The Flat, 917 The Crescent, Droichead Nua, Co. Kildare
EI5 GM Jeremy M. Sheehan, 'Villa Maria', Cork Road, Kinsale, Co. Cork
EI5 GN Iain Fisher, 21 Rathdown Park, Greystones, Co. Wicklow
EI5 GO Patrick Bergin, 77 Hazelwood, Shankill, Co. Dublin
EI5 GP Nick McDonagh-Greaves, Flat 10, Cramhurst House, Sutton Grove, Sutton, Surrey SM1 4TH, England
EI5 GQ William Mitchell, Coolenearl, Avoca, Co. Wicklow
EI5 GR Gavin B. Stewart GI4TVV, 68 Kingsdale Park, Knock, Belfast BT5 7BZ
EI5 GS International Police Association, c/o Jim Jeffers, 73 Beech Grove, Lucan, Co. Dublin
EI5 GT Desmond Chambers, Burrishoole, Newport, Co. Mayo
EI5 GU Bridget B. Harris, 21 River Valley Court, Swords, Co. Dublin
EI5 GV John M. Heenan, 8 Wynnefield House, 26 Charleville Road, Rathmines, Dublin 6
EI5 GW Cathal E. O Murchadha, 11 Woodford Road, Watford, Herts WD1 1PB, England
EI5 GX Eamonn G. Roddy, 7 Marlborough Avenue, Derry City BT48 9BQ, Northern Ireland
EI5 GY Alex Charampopoulos, 1 Upper Pembroke Street, Dublin 2
EI5 GZ Joe O'Brien WG2C, 255 Fieldston Terrace, The Bronx, NY 10471, USA
EI5 HA Patrick F. Browne, 63 Oakland Park, Dundalk, Co. Louth
EI5 HB Mannix McAlister, 'Kintyre', Iona Road, Mayfield, Cork
EI5 HC W. Leslie Clarke, 2 Glencregagh park, Belfast BT6 0NT
EI5 HD Daniel Coughlan, 157 Shanganagh Cliffs, Shankill, Co. Dublin
EI5 HE Sean Corcoran, Baltydaniel East, Mallow, Co. Cork
EI5 HF Conor McGlynn, 24 Knocklyon Road, Templeogue, Dublin 16
EI5 HG J. Eugene O'Malley, 11 Silvercourt, Tivoli, Cork
EI5 HH John Madden, The Cottage, 53 Clarendon Street, Derry
EI5 HI Helen O'Sullivan, 'Sundown', Mount Oval, Rochestown, Co. Cork
EI5 HJ Gerard Molloy, Mahanagh, Cummer, Tuam, Co. Galway
EI5 HK Paul Jenkinson, The Squirrells, Urraghry, Aughrim, Ballinasloe, Co. Galway
EI5 HL John Rogers, 35 Glen Ellen Crescent, Swords, Co. Dublin
EI5 HM Joseph Harding, 13 Blackhill Crescent, Donecarney, Co. Meath
EI5 HN Rainer Allraun, Trautenauer Str. 12, 8700 Wurzburg, Germany
EI5 HO Beverly McLaughlin, 1A Taylor Park, Limavady, Co. Derry BT49 ONT
EI5 HP Anthony C Marston, Granagh (Liskennet West), Kilmallock, Co. Limerick
EI5 HQ Batt O'Brien, Sparrowsland, Bree, Co. Wexford
EI5 HR Patrick Hennessy, 35/36 The Square, Castlecomer, Co. Kilkenny
EI5 HT Dermot Fagan, 56 Dunmore Lawn, Ballymount, Kingswood, Dublin 24
EI5 HU Martin Sheridan
EI5 HV Brian Tansey, 12 Clonmult Terrace, Midleton, Co. Cork
EI5 HW Aidan Murphy, c/o I.R.T.S., P.O. Box 462, Dublin 9
EI5 HX Walter E. Roberts, 24 Leeds Road, Barwick in Elmet, Leeds LS15 4JD, England
EI5 HY Alan Adams, 57 Abbey Drive, Ferrybank, Waterford
EI5 HZ Paul D. Reilly, 44 The Drive, Castletown, Celbridge, Co. Kildare
EI5 IA Donald Geoffrey Gibbons, Patrick's Chair, Boheh, Liscarney, Westport, Co. Mayo
EI5 IB David Halton, 35 Delacy Park, Shannon, Co. Clare
EI5 IC Eddie Murphy, 12 Limewood Park, Raheny, Dublin 5
EI5 ID James Mc Manus, 10 The Orchard, Slieverue, Waterford
EI5 IE Patrick A. Mulreany, Tulach na Greine, 51 Millar Ridge Road, Wellington NV 89444, USA
EI5 IF Patrick Molloy, 71 Bannow Road, Cabra West, Dublin 7
EI5 IG John Wilson, 19 Seapark Road, Clontarf, Dublin 3
EI5 IH Gerard Richardson, 104 Kildonan Road, Finglas West, Dublin 11
EI5 II Thomas Walsh, 28 Clonshaugh Park, Clonshaugh, Dublin 17
EI5 IJ William J. Hillick, 22 Lower Ormond Quay, Dublin 1
EI5 IK Conor C. Daly, 9 Saint David's, Artane, Dublin 5
EI5 IL Patrick Gaffney, 32 St Anne's Drive, Montenotte Park, Cork
EI5 IM Luke E. Fleming, Scariff, Ogonnolloe, Co. Clare
EI5 IN Keith Nolan, Loughanstown, Knockdrin, Mullingar, Co. Westmeath
EI5 RV Louis Varney, 82 Folders Lane, Burgess Hill, W. Sussex RH15 0DX, England
EI5 WL Wang Laboratories, Plassey Technological Prk, Limerick, c/o Todd Harvey EI4FR
EI5 AAB D. McDonnell, "Brickfield", Newry Road, Dundalk, Co. Louth
EI5 ABB Timothy P. O'Leary, 81 Corbawn Drive, Shankill, Co. Dublin
EI5 ACB P. Buckley, Main Street, Banteer, Co. Cork
EI5 AGB Allistair Edwards, 11 Arranmore Road, Dublin 14
EI5 AIB Maurice Boles, 2 Merton Road, Rathmines, Dublin 6
EI5 AJB A.G. Smith, 26 Garnerville Park, Belfast BT4 2WY
EI5 AKB Michael Phelan, Acragar, 1 The Folly, Co. Waterford
EI5 ALB Ann McCourt, Drumline, Newmarket-on-Fergus, Co. Clare
EI5 AOB Thomas Lane, Manor, Tulsk, Castlerea, Co. Roscommon
EI5 APB Thomas Lane Jr., Manor, Tulsk, Castlerea, Co. Roscommon
EI5 AQB Patrick Gibbons, Granayhan, Scramore, Co. Roscommon
EI5 ASB T. O'Neill, 13 Pondsfield, New Ross, Co. Wexford
EI5 AUB Vera Ffrench, Donanore, New Ross, Co. Wexford
EI5 AXB Leo Hilliard, Cahirdown, Listowel, Co. Kerry
EI5 AZB James Hennessy, Lower Ballinacroona, Knocklong, Co. Limerick
EI5 BAB Rory B. O'Doherty, 53 Violet Street, Waterside, Derry
EI5 BBB Cecil E. Fairman, Finn View House, Trennamullin, Ballybofey, Co. Donegal
EI5 BDB Fr. Seamus Reid, GI6EWM, 30 Ashwood, Co. Armagh BT66 8PF, Northern Ireland
EI5 BEB Patrick J. Bowe, 197 Gloucester Avenue, Chelmsford, Essex CM2 9DX, England
EI5 BFB Patrick G. O'Rourke, 30 Michael Collins Road, Mervue, Galway
EI5 BGB Details withheld at licensee's request
EI5 BHB Kieran O'Carroll, Cathay Pacific Airways, Flight Ops. Dept. 10/F, Block T, C.P. Building, Hong Kong Int. Airport
EI5 BIB James O'Driscoll, 211 Decies Road, Ballyfermot, Dublin 11
EI5 BKB Patrick Harnett, Charleville View, Tullamore, Co. Offaly
EI5 BLB Maurice E. Doherty, Eagle Lodge, Beechlawn, Ballinasloe, Co. Galway
EI5 BMB George Hayes, Barefield P.O., Ennis, Co. Clare
EI5 BNB Peter McNally, 17 Duncarrig, Sutton, Dublin 13
EI5 BPB Michael J. McGarry, 20 Ballyraine Park, Letterkenny, Co. Donegal
EI5 BTB Hugh Drumm, 18 Serpentine Park, Ballsbridge, Dublin 4
EI5 BUB Graham Wilson, Raford, Kiltulla, Athenry, Co. Galway
EI5 BVB Peter J. Mathews, Creevelea, Emyvale, Co. Monaghan
EI5 BWB John J. Crerand, 23 Melrose Avenue, Fairview, Dublin 3
EI5 BYB Benjamin Rodriguez, 'Ballyvartry', Calary, nr Bray, Co. Wicklow

EI5 BZB Vincent Cantwell, "Villa Maria", Trim, Co. Meath
EI5 CAB Alan C. Hildebrand, "Estoil - Sol", Rooslee Road, Dunboyne, Co. Meath
EI5 CBB Alex W. Frazer, 40 Raphoe Road, Crumlin, Dublin 12
EI5 CEB Frank Bourke, Graigue Lower, Cuffesgrange, Co. Kilkenny
EI5 CFB Peter J. Crimmings, 7 The Rise, Friars Hill, Wicklow
EI5 CGB Sean Molloy, Main Street, Ballingarry, Thurles, Co. Tipperary
EI5 CHB Patrick J. Martin, 24 Tamarisk Avenue, Kilnamanagh, Tallaght, Dublin 24
EI5 CJB John Roche, 323 Mill Street, Callan, Co. Kilkenny
EI5 CLB Francis McAuley, 61 Templeroan Park, Templeogue, Dublin 16
EI5 CMB William P. Polion, 39 Glasilawn Road, Glasnevin, Dublin 11
EI5 COB Dominic B. Doyle, 21 Cherrywood, Celbridge, Co. Kildare
EI5 CQB Patrick O'Shaughnessy, Lackandarragh Lower, Enniskerry, Co. Wicklow
EI5 CRC Cork Radio Club, c/o Anthony Stack, 33 Park Court, Ballyvolane, Cork
EI5 CTB Trevor C. Plowman, 5 Mackie's Place, Dublin 2
EI5 CVB Frank W. Upton, 3 Raheen Crescent, Tallaght, Dublin 24
EI5 CXB Micheal O'Sullivan, 41 Grange Park, Rathfarnham, Dublin 14
EI5 CYB Joe Ivory, 15 Lower Killmagig, Avoca, Co. Wicklow
EI5 DBB Gerard Dowling, 2A Elm Park, Ennis Road, Limerick
EI5 DCB Michael Hoban, 63 Avondale, Waterford City
EI5 DDB Brendan H. Cornyn, Dowra, via Carrick-on-Shannon, Co. Leitrim
EI5 DGB Claire B. Whelan, 126 Farranshoneen, Waterford City
EI5 DHB Teresa A. Foley, Coolgrange, Granges Road, Kilkenny
EI5 DIB Brian Cremin, 'Glencairn', Boherboy Road, Lotabeg, Mayfield, Cork
EI5 DJB Lawrence C. Hoey, Ballinderry, Four-Mile-House, Roscommon
EI5 DKB James Walsh, 117 Kincora Avenue, Clontarf, Dublin 3
EI5 DLB Kevin Foley, 18 Pine Grove, Ashling Heights, Raheen, Limerick
EI5 DNB Edmund McCrystal, 33 Richmond Park, Omagh, Co. Tyrone, Northern Ireland
EI5 DPB Karl P. Madden, Lagore Road, Dunshaughlin, Co. Meath
EI5 DRB Donal Caulfield, 10 Beechcourt, Killiney, Co. Dublin
EI5 DSB Eamonn Sheeran, Boyerstown, Navan, Co. Meath
EI5 DTB Adrian D. Davis, 20 Harlech Downs, Goatstown Road, Dublin 14
EI5 DWB Brendan J. Flanagan, 11 Churchview Park, Killiney, Co. Dublin
EI5 DYB Joe Chester, 199 Ardilaun, Portmarnock, Co. Dublin
EI5 EAB Michael White, 6 Tormey Villas, Athlone, Co. Westmeath
EI5 EBB Thomas W. Vickery, 4 Marino Street, Bantry, Co. Cork
EI5 ECB Leo Power, Ballynoe, Ballinhassig, Co. Cork
EI5 EDB Mark Dennehy, 'Pine Copse', 167 Applewood Heights, Greystones, Co. Wicklow
EI5 EEB William White, Knockglass, Ladysbridge, Co. Cork
EI5 EFB Mark A. O'Leary, Seamount, Carrigaline, Co. Cork
EI5 EGB James A. Kavanagh, 62 Deerpark Road, Castlenock, Dublin 15
EI5 EHB John P. Somers, Perrin House, Glengariff, Co. Cork
EI5 EIB Mike Mulvihill, 7 Willbrook Downs, Rathfarnham, Dublin 14
EI5 EJB Kieran A. Jefferies, 1 Park Road, Muskerry Estate, Ballincollig, Co. Cork
EI5 EKB Thomas R. Fitzmaurice, Ridge Road, Moate, Co. Westmeath
EI5 EMB M. E. Grant, The School House, Church Road, Wittering, Cambs, PE8 6AF, England
EI5 EOB John King, 24 Willow Park, Raheen, Limerick
EI5 EQB Nathaniel Panday, 53 Cherryfield Road, Walkinstown, Dublin 12
EI5 ERB John P Ward, Station Road, Glenties, Co. Donegal
EI5 ESB John Gartlan, Carrickedmond, Kilcurry, Dundalk, Co. Louth
EI5 EUB Paul C Dolan, 583 Ashling Park, Dundalk, Co. Louth
EI5 EVB Morgan J. Cohen
EI5 EXB Philip Bartlett, "Chickamauga", Deansgrange Road, Blackrock, Co. Dublin
EI5 EYB Jeremiah C. Forde, 8 Newbrook Grove, Mullingar, Co. Westmeath
EI5 FSC 5th Field Signals Coy FCA, c/o Joe Bannon, EI9FB, The Mall, Sligo
EI5 HAM Irish Amateur Radio Association for Blind and Disabled, c/o John Walsh, 26 Verbena Grove, Kilbarrack, Dublin 13
EI5 LID Galway Radio Experimenters Club (see EI4GRC)
EI5 OMD Order of Malta, Drogheda, Dundalk Amateur Radio Soc, 113 Castletown Road, Dundalk, Co. Louth
EI5 TCR Donegal (Tir Conaill) Amateur Radio Society, c/o Ken McDermott, EI4DW, Curraghamone, Ballybofey, Co. Donegal
EI5 VFH Robert B. Church, N3RYX, 342 Sonora Drive, Camarillo, CA 93010 - 6030, USA
EI5 WAR Wicklow Amateur Radio Club, c/o Padraig O Meachair, 8 Cnoc na Manach, Cill Mhantain

EI6

EI6 C John J. Moriarty, 7 Shrewsbury Lawn, Cabinteely, Dublin 18
EI6 D Leo Purcell, "Gleann-na-Greine", Naas, Co. Kildare
EI6 K Thomas Fay, 45 Chanel Road, Artane, Dublin 5
EI6 Q Louis J. Robinson, 117 Mangerton Road, Drimnagh, Dublin 12
EI6 S George McClarey, "Rosemount", Mountnugent, Co. Cavan
EI6 U Ian Morris Amateur Radio Group, c/o Derek Peyton, 123 Springhill Avenue, Blackrock, Co. Dublin
EI6 Y Patrick O'Doherty, 82 Cherrywood Park, Clondalkin, Dublin 22
EI6 AD David Connolly, The Pier, Ballycotton, Co. Cork
EI6 AE J.P. Horgan, "Dunleer", Woodleigh Park, Bishopstown, Co. Cork
EI6 AG Alexander F. Barrett, Annmount House, Glounthane, Co. Cork
EI6 AH John J. O'Carroll, Lower Main Street, Ballybunion, Co. Kerry
EI6 AI William Long, Dunkineely, Co. Donegal
EI6 AK John A. Mooney, 'The Cottage', South Douglas Road, Cork
EI6 AL D.J. O'Connor, 33 Whitehall Road, Terenure, Dublin 6
EI6 AN Anthony Darcy, Coneyburrow Road, Lifford, Co. Donegal
EI6 AS Albert Latham, 226 Belgard Heights, Tallaght, Dublin 24
EI6 AU M.F. Whelan, 7 Lorcan Drive, Santry, Dublin 9
EI6 AV Very Rev. Patrick Dean Deegan, St. Eunan's, Raphoe, Co. Donegal
EI6 AX Bernard O'Sullivan, Cahermore, Bantry, Co. Cork
EI6 AY J.C. Boyce, 6 Westland Street, Derry, Northern Ireland
EI6 AZ Cedric Rourke, GI3IVJ
EI6 BA Thomas J. Foley, 40 Hillcourt, Donnybrook, Douglas, Cork
EI6 BL Frank H. Eccleston, 52 Charnwood, Vevay Road, Bray, Co. Wicklow
EI6 BP A.E. McAlpine, GI3IIF
EI6 BR Dr. T. Pilsworth, GD4GDK
EI6 BS School of Electrical Engineering, Regional Tech. College, Moylish, Limerick
EI6 BT Jerry Cahill, Killea, Broomfield East, Midleton, Co. Cork
EI6 BY H. McGivern, 3 St. Helens Park, Dundalk, Co. Louth
EI6 BZ W. Hourigan, 68 Rathgar Road, Dublin 6
EI6 CB Cornelius J. Connolly, The Square, Skibbereen, Co. Cork
EI6 CE Sean Grant, Clonmacken Road, Caherdavin, Limerick
EI6 CF Details withheld at licensee's request
EI6 CH Bernard Smith, GW4ANO
EI6 CJ James M. Doherty, 58 Glengoland Gardens, Suffolk, Co. Antrim, Northern Ireland
EI6 CK Guy Jean-Francois Poinboeuf, Sheshoon, The Old Road, The Curragh, Co. Kildare
EI6 CL T. Lehane, 10 Pinewood Park, Rathfarnham, Dublin 6
EI6 CM Dr. Barry J. Field, 1 Rathclaren, Killarney Road, Bray, Co. Wicklow
EI6 CP Patrick Daly, 12 Stella Avenue, Glasnevin, Dublin 9
EI6 CR Derek Doyle, College Street, Ballyshannon, Co. Donegal
EI6 CS William J. Ormond, 9260 224th Street, Queens Village, New York 11424, USA
EI6 CV Patrick E.C. Ronaghan, Moynehall, Cavan
EI6 CW John Jeffrey Walsh, 6 Fancourt Heights, Balbriggan, Co. Dublin
EI6 CX Brendan Jordan, Dunsink Observatory, Castleknock, Dublin 15
EI6 CY Patrick J. Manning, Rosegreen, Clare Road, Ennis, Co. Clare
EI6 CZ Roland E. Hall, 'Pinewood Lodge', 16 Tullyvarraga Hill, Shannon, Co. Clare
EI6 DA Patrick Mooney, 57 Johnstown Road, Dun Laoghaire, Co. Dublin
EI6 DB Millstreet A.R.C., c/o Daniel Cowman, Drishane Road, Millstreet, Co. Cork
EI6 DC Azim Jina, Glenarm House, 184 Lr. Glanmire Road, Tivoli, Cork
EI6 DD Dundalk Field Day & Competitions, Dundalk Amateur Radio Soc, 113 Castletown Road, Dundalk, Co. Louth
EI6 DE Kate Whyte, Ballycarney, Clarina, Co. Limerick
EI6 DF Rev. Fr. Michael McCarthy, St. Athanasius, 21 Mount Stewart Street, Carluke, Lanarkshire, Scotland ML8 5EB
EI6 DG John Ryan, 23 Dollymount Grove, Clontarf, Dublin 3
EI6 DH Colin Kennedy, 74 Dunmore Park, Ballymount, Clondalkin, Dublin 22
EI6 DJ Robert Hickson, 1 Watermill Close, Oldbawn, Tallaght, Dublin 24
EI6 DK Victor P. Moran, 32 Anne Devlin Avenue, Rathfarnham, Dublin 14
EI6 DL Anthony Magliocca, Custume Place, Athlone, Co. Westmeath
EI6 DM Eugene McSherry
EI6 DN John Molloy, Kilmoon, Ashbourne, Co. Meath
EI6 DO A.W. Boles, 2 Merton Road, Rathmines, Dublin 6
EI6 DP Gerard P. Fitzgerald, 43 Upper William Street, Limerick
EI6 DQ Patrick M. O'Shea, Graigue, Inchigeela, Macroom, Co. Cork
EI6 DR Bill Kelsey, 2716 C.R. 26, Mt. Cory, Ohio 45868, USA
EI6 DS John A. Stringer, GI3KDR
EI6 DT Anthony Enright, "Crossfield", Firhouse Road, Templeogue, Dublin 16
EI6 DU William Guiry, Sunville, Woodpark, Castleconnell, Co. Limerick
EI6 DV Desmond J. Doyle, Microelectronics & Elect., Engineering, Trinity College, Dublin 2
EI6 DW Patrick Lynch, Malin Head Radio Station, Lifford, Co. Donegal
EI6 DY John Gilligan, 14 Arden Vale, Tullamore, Co. Offaly
EI6 DZ Dermot F. Kenny, 6 Pinewood Grove, Renmore, Galway
EI6 EA Kevin D. Hayes, 'Lima', Plassy Avenue, Corbally, Limerick
EI6 EB R. McKinty, 3 Rhanbuoy Park, Craigavad, Co. Down, Northern Ireland
EI6 EC Fr. Michael Reaume, Marianist Community, St. Columba's, Church Av, Ballybrack, Co. Dublin
EI6 ED Daniel Boyce, 35 Ashburn Avenue, Derry, Northern Ireland
EI6 EE Peter J. Gillen
EI6 EF Frank Malone, Elmhill, Grove Road, Malahide, Co. Dublin
EI6 EG Joseph Cosgrave, 103 McIntosh Park, Pottery Road, Dun Laoghaire, Co. Dublin
EI6 EH Tom Clarke, Balrath, Kells, Co. Meath
EI6 EI Frank McCarron, Demense, Raphoe, Lifford, Co. Donegal
EI6 EJ Thomas P. McKay, 10 Kylecare Road, Dublin 4
EI6 EK Colin G. McGuire, 402 Sth. Franklin, St. Louis, MI 48880, USA
EI6 EL Andrew Forde, 57 Hazel Park, Newcastle, Galway
EI6 EM E.J. Cooke, Ovidstown, Straffan, Co. Kildare
EI6 EN David Lamb, Kilcoleman Farm, Enniskeane, Co. Cork
EI6 EQ Daniel Costello, Ballyfin Road, Portlaoise, Co. Laois
EI6 ER Michael P. O'Sullivan, 14 Pleasant Drive, Mount Pleasant, Waterford
EI6 ES T. Bluett, Convent Road, Clonakilty, Co. Cork
EI6 ET John F. Murphy, Cooleanig, Beaufort, Killarney, Co. Kerry
EI6 EU Carl McGowan, 22 Durham Road, Sandymount, Dublin 4
EI6 EV Donal O hUallachain, 171 Martello, Port Mearnog, Co. Atha Cliath
EI6 EW Anthony Baker, 10 Monaloe Avenue, Blackrock, Co. Dublin
EI6 EX Sean Keevey, Golden Grove Road, Roscrea, Co. Tipperary
EI6 EY Michael J. McElligott, Skehenerin, Listowel, Co. Kerry
EI6 EZ J.P. Martin, 15 Corduff Grove, Blanchardstown, Dublin 15
EI6 FB Niall Syms, 71 Hazelwood, Shankill, Dublin 18
EI6 FC Noel P. Keller, Circular Road, Kilkee, Co. Clare
EI6 FD Thomas Lohan, Ballincurry, Glinsk, Castlerea, Co. Roscommon
EI6 FE Paul Kirkby, 93 Aidan Park, Shannon, Co. Clare
EI6 FF John B. McClintock, Moneygreggan, Newtowncunningham, Lifford, Co. Donegal
EI6 FG Anthony McGarry, Ballydoogan, Magheraboy, Sligo Town
EI6 FH Thomas Rock, Pearse Road, Ballymote, Co. Sligo
EI6 FI Thomas J. Keely, 1 St. Ann's Terrace, Sligo
EI6 FJ John Higgins, 124 Cromwell Road, London SW7 4ET
EI6 FL William B. Chapple, 3 Knapton Terrace, Knapton Road, Dun Laoghaire, Co. Dublin
EI6 FM Dermot P. O'Connell, Ballinvarrig, Whitechurch, Co. Cork
EI6 FN Patricia Brennan, West End, Bundoran, Co. Donegal
EI6 FO J.S. Robinson, 54 High Street, Comber, Co. Down, Northern Ireland
EI6 FP Paul McGlinchey, Dunwiley, Stranorlar, Co. Donegal
EI6 FQ Dr. Michael A. Radix, 63 Highfield Park, Dundrum, Dublin 14
EI6 FR Declan P. Craig, 167 St. James's Road, Greenhills, Dublin 12
EI6 FT Paul M.J. Wallace, 24 Rosemount Avenue, Artane, Dublin 5
EI6 FV Colum Bruce, Carrig, Lacken Road, Blessington, Co. Wicklow
EI6 FW John J. O'Kelly, 6 The Palms, Roebuck, Clonskeagh, Dublin 14
EI6 FX Liam J. Daly, 1 Elm Park, Blackrock, Dundalk, Co. Louth
EI6 FY Richard Prendergast, Mogeely Road, Castlemartyr, Co. Cork
EI6 FZ Dermot Flanagan, 132 Upr. Kilmacud Road, Stillorgan, Co. Dublin
EI6 GA Brendan Lynch, 37 Hazelwood, Shankill, Co. Dublin
EI6 GB Thomas Hughes, 'Inisfree', Newrath, Rathnew, Co. Wicklow
EI6 GC Alfred L. Sammon, 11 Drumclay Road, Enniskillen, Co. Fermanagh
EI6 GD Edward Dalton, 14 Garvans Terrace, Dungarvan, Co. Waterford
EI6 GE Terence Caulfield, 39 Jamestown Avenue, Inchicore, Dublin 8
EI6 GF Michael J. McLoughlin, Spencerstown, Murrintown, Co. Wexford
EI6 GG Paul Cotter, Derryvillane, Glanworth, Co. Cork
EI6 GH Padraig F. Barry, 21 Belgrave Road, Monkstown, Co. Dublin
EI6 GI John D. Hickey, c/o Murphy Barracks, Ballincollig, Co. Cork
EI6 GJ Edmond Fitzgerald, 'Sea Winds', Ballinamona, Shanagarry, Midleton, Co. Cork
EI6 GK Bruce R. Williams, Kernan Cottage, Derreen, Gort, Co. Galway
EI6 GL Ronan W. Lynch, 37 Hazelwood, Shankill, Co. Dublin
EI6 GM Karl M. Reddy, 8 The Rise, Woodpark, Ballinteer, Dublin 16
EI6 GN Dan R. Nelson, Chinook, 6 Kindlestown Heights, Delgany, Co. Wicklow
EI6 GO Ronald P. McGrath, Hill Grove, Prior Park Road, Clonmel, Co. Tipperary
EI6 GP Aengus O'Fearghail, The Reask, Dunshaughlin, Co. Meath
EI6 GQ Bernard Jouaux, 52 rue des Grovantes, 54770 Bouxieres Aux Chene, France
EI6 GR Patrick Kielty, 3 Kiltalown Court, Jobstown, Tallaght, Dublin 24
EI6 GS Domhnall O Chnaimhsi, Clochbhaile, Leitir-Mhic-a-Bhaird, Dun na nGall, Co. Dhun na nGall
EI6 GT Michael P. Larkin, Malin Road, Moville, Co. Donegal
EI6 GU Edward Lawlor, 8 Primrose Street, Broadstone, Dublin 7
EI6 GV Padraig McCormack, Main Street, Kilnaleck, Co. Cavan
EI6 GW Tony Kirby, Greenfields, Ballincollig, Co. Cork
EI6 GX James A. Whelan, 26 Main Street, Kilcoole, Co. Wicklow
EI6 GY Ian C. White, 79 Hawthorn Drive, Waterford
EI6 GZ Brendan Wall, Riverstown, Rathfeigh, Tara, Co. Meath
EI6 HA Michael Troy, 3 Farleigh Place, Middle Glanmire Road, Montenotte, Cork
EI6 HB Denis O'Flynn, Ladysbridge P.O., Ladysbridge, Castlemartyr, Co. Cork
EI6 HC George B. Berrich, 19 Craigmaddie Gardens, Balmore Park, Torrance, Glasgow G64 4LW, Scotland
EI6 HD Patrick O'Keeffe, 8 Clonard Grove, Sandyford Road, Dublin 16
EI6 HE Steve Canavan, Station Road, Corofin, Co. Clare
EI6 HF Patrick N. Hutton, 50 Ash Grove, Tullow Road, Carlow
EI6 HG Aidan J. Cornyn, Rockfort Farm, Formil, Castleblaney, Co. Monaghan
EI6 HH Richard Bermingham, 17 Fremont Drive, Melbourn, Bishopstown, Cork
EI6 HJ Donal Skelly, Bredagh Glen, Moville, Co. Donegal
EI6 HK John Geoghegan, 120 St James Road, Walkinstown, Dublin 12
EI6 HL John Walsh, 26 Verbena Grove, Kilbarrack, Dublin 13
EI6 HM Robert G. Alexander, KO4AO, Tynagh, Loughrea, Co. Galway
EI6 HN Denis Bernard Rowe, 23 Glenwood Drive, Onslow Gardens, Commons Road, Cork
EI6 HO Francis E. Harkin, RT 1 Box 155A, Cushing MN 56443, USA
EI6 HP David Sellars, Driney, Drumcong, Carrick-on-Shannon, Co. Leitrim
EI6 HQ D M Brosnan, Killarney Road, Castleisland, Co. Kerry
EI6 HR Michael O'Brien, 50 Beech Park, Tramore, Co. Waterford
EI6 HS Paul Breen, 21 Sli an Aifrinn, Athlone, Co. Westmeath
EI6 HT Meine H. Rouwhof, Tigh-na-Mara, Durrus, Co. Cork
EI6 HU Guido Junold, 15 Elm Park, Lisbeg Lawn, Renmore, Galway
EI6 HV Kenneth Mooney, 46 St John's Court, Artane, Dublin 5
EI6 HW Noel Mulvihill, Hillquarter, Coosan, Athlone, Co. Westmeath
EI6 HX Christopher Hannigan, 14 Conthem Road, Bearny, Strabane, Co. Tyrone BT82 8NY, Northern Ireland
EI6 HY Michael J. Bonar, Galwolie, Cloghan P.O., Lifford, Co. Donegal
EI6 HZ Richard Coleman, 20 Clontarf Estate, Skehard Road, Blackrock, Cork
EI6 IA Christopher Maverley, 3 St. Stephens Place, Friars Street, Cork
EI6 IB Fergus Millar, Dromore, Carrick on Shannon, Co. Leitrim
EI6 IC Francis G. Fahy, Athenry Road, Loughrea, Co. Galway
EI6 ID Patrick Serridge, 21 Lissara Heights, Warrenpoint, Co. Down BT34 3PG, Northern Ireland
EI6 IE Eugene Horgan
EI6 IF Denis Collins, The Moorings, Robertstown, Naas, Co. Kildare
EI6 IG Martin Leonard, 14 Temple View Avenue, Clarehall, Dublin 17
EI6 IH Janet Louisa Serridge, 21 Lissara Heights, Warrenpoint, Co. Down BT34 3PG, Northern Ireland
EI6 II Walter Baumann
EI6 IJ John B. Riordan, Kinnego, Buncrana, Co. Donegal
EI6 IK Anthony O. Hodgins, 31 Cherrybrook Drive, Drogheda, Co. Louth
EI6 IL Denis F. Brennan, Ramsfort Lodge, Fort Road, Gorey, Co. Wexford
EI6 IM Michael J. Grifferty, Barnacogue, Swinford, Co. Mayo
EI6 IN Sean McMorrow, 'San Juda', Convent Avenue, Bray, Co. Wicklow
EI6 AAB Maurice Donworth, Main Street, Bruff, Co. Limerick
EI6 ADB Michael Larkin, 3 The Oval, Gouldavoher, Limerick
EI6 AFB Colin D. Heymans, 48 Belton Park Avenue, Dublin 9
EI6 AGB John Greensmith, 2 Bridge Street, Tipperary
EI6 AJB Michael R. Forde, 22 Maywood Crescent, Raheny, Dublin 5
EI6 AKB Francis Matthews, 38 Balkill Park, Howth, Co. Dublin
EI6 ALB Leo O'Leary, 70 Mourne Road, Drimnagh, Dublin 12

EI6 AMB John Melvin, Derreen Upper, Kilkerrin, Ballinasloe, Co. Galway
EI6 ANB Paul Kinney, 4 Merton Road, Rathmines, Dublin 6
EI6 AOB Robert Cullen, 35 Aylmer Road, Newcastle, Co. Dublin
EI6 AQB Sean L. Forde, 'Gougane Barra', 603 Howth Road, Raheny, Dublin 5
EI6 ARB John C. O'Sullivan, 142 Lr. Kilmacud Road, Stillorgan, Co. Dublin
EI6 AVB Cecil Coyne, 1 Clonsilla Close, Blanchardstown, Co. Dublin
EI6 AXB Michael Kingston, 5 Merval Crescent, Clareview, Limerick
EI6 AYB Eugene Cullen, Upper Strandhill, Co. Sligo
EI6 AZB Philip Maguire, Mullaghadeen, Monaghan, Co. Monaghan
EI6 BAB Christopher Cummins, 1 Columba Terrace, Kells, Co. Meath
EI6 BBB Anthony Farrelly, Georges Cross, Castletown K.P., Navan, Co. Meath
EI6 BCB Declan Kerr, Craigs Road, Dunmoe, Navan, Co. Meath
EI6 BDB Martin Kerr, Dunmoe, Navan, Co. Meath
EI6 BEB John Murtagh, Proudstown, Navan, Co. Meath
EI6 BHB John M. Walsh, Birmingham Road, Tuam, Co. Galway
EI6 BKB Donnachadha O'Shea
EI6 BOB Francis X. Gilmore, 32 Lr. Salthill, Galway
EI6 BPB Sean Corry, 56 Beechwood Drive, Rathnapish, Carlow
EI6 BRB James V. Harte, "Thorndale", Upper Kilmoney Road, Carrigaline, Co. Cork
EI6 BSB John P. Kelly, 83 Fairways, Rathfarnham, Dublin 14
EI6 BTB Richard P. Murnane, 93 Mayorstone Drive, Limerick
EI6 BWB Tony Archer, "Redthorn", Rosebank, Douglas Road, Cork
EI6 BXB Shane E. Finnegan, Meteorological Office, Dublin Airport, Co. Dublin
EI6 BZB Thomas Moore, Carragarry, Drum, Co. Monaghan
EI6 CBB Declan Garvey, 55 Lansdowne Park, Ennis Road, Limerick
EI6 CEB Eamon J. Fidgeon, 'Crannog', St. Patricks Road, Wicklow
EI6 CFB Michael J. Gleeson, 2 Ballyronan, Kilquade, Co. Wicklow
EI6 CGB Michael Kennedy, 35 Cherrington Road, Shankill, Co. Dublin
EI6 CKB Francis J. Mulligan, 50 Courtown Park, Kilcock, Co. Kildare
EI6 CLB Charles J. Dowdall, 20 Gransha Drive, Glen Road, Belfast
EI6 CMB John P. Connolly, The Square, Skibbereen, Co. Cork
EI6 CNB Patrick Timmons, Moneystown, Bray, Co. Wicklow
EI6 CPB Hugh D. Forde, Railway House, Albert Walk, Bray, Co. Wicklow
EI6 CQB Patrick Hanley, 107 Applewood Heights, Greystones, Co. Wicklow
EI6 CRB Kenneth Duffy, 16 Brookhaven Drive, Blanchardstown, Dublin 15
EI6 CTB Ian Hurley, Ardbrae, Ennis Road, Limerick
EI6 CUB Patrick J. White, 12 Cypress Grove South, Templeogue, Dublin 6
EI6 CWB Michael J. Tinkler, 1 Fatima Terrace, Bray, Co. Wicklow
EI6 CXB John P. Ward, 5 Lacknalocha, Mallow, Co. Cork
EI6 CZB Edmond M. Power, Touraneena, Ballinamult, Co. Waterford
EI6 DAB Kieran McGann, 1 Riverside Close, Shannon Banks, Corbally, Limerick
EI6 DBB Sean Kinghan, Batterstown, Co. Meath
EI6 DCB Michael McGowan, 40A Dargle Wood, Knocklyon Road, Templeogue, Dublin 16
EI6 DFB John P. Cronin, Tinegeragh, Watergrasshill, Co. Cork
EI6 DGB James McCole, 33 College Rise, Drogheda, Co. Louth
EI6 DJB Gary Hawkins, 33 St. Brendans Crescent, Greenhills, Dublin 12
EI6 DKB James M. McCabe, 4 Domville Green, Templeogue, Dublin 16
EI6 DLB Alan F. Laffan, 14 Walkinstown Park, Dublin 12
EI6 DMB Patrick J. McEvoy, 36 Idrone Park, Knocklyon Woods, Dublin 16
EI6 DNB Dermot A. McNally, Caulstown, Dunboyne, Co. Meath
EI6 DOB Frank O'Neill, 67 Pearse Gardens, Sallynoggin, Co. Dublin
EI6 DRB Joseph W. Roche, 42 Greenlea Road, Dublin 6W
EI6 DSB James A. Hyland, 224 Road 1, Balrothery Estate, Tallaght, Dublin 24
EI6 DTB Pat Monahan, 7 Whitethorn Avenue, Inniscarra View, Ballincollig, Co. Cork
EI6 DVB Breiffni F. Hogan, Bawnfone, Butlerstown, Co. Waterford
EI6 DWB Patrick J. Haughney, 15 Meadow Road, Riverview, Knockboy, Waterford
EI6 DYB James E. Farrell, 29 Lismore Drive, Waterford
EI6 DZB Ann Farrell, 29 Lismore Park, Waterford
EI6 EBB James M. Twomey, Clonkerdon House, Cappagh, Dungarvan, Co. Waterford
EI6 EDB Edward Lyons, 41 Kennington Road, Templeogue, Dublin 6W
EI6 EEB John Lynn, 81 Pine Valley Avenue, Rathfarnham, Dublin 16
EI6 EFB Brendan Beasley, 145 St Peter's Road, Walkinstown, Dublin 12
EI6 EGB Fr. Robert Swinburne, Salesian College, Maynooth Road, Celbridge, Co. Kildare
EI6 EIB Philip C. McCarthy, Kilgobbin Cross, Ballinadee, Bandon, Co. Cork
EI6 EJB Thomas Allen, Six Cross Roads, Kilbarry, Waterford
EI6 ENB David P. O'Rourke, 29 Aspen Road, Kinsealy Court, Swords, Co. Dublin
EI6 EPB Dolores T. O'Byrne, 31 Shenick Avenue, Skerries, Co. Dublin
EI6 EQB Anthony F. Fay, 48 Ardlea Road, Artane, Dublin 5
EI6 ESB Robert A. Stack, 33 Park Court, Ballyvolane, Cork
EI6 ETB Warren Daly, Kilmoney South, Carrigaline, Co. Cork
EI6 EVB Frank Mason, 27 Dundanion Court, Blackrock Road, Blackrock, Cork
EI6 EWB Shane R. Moloney, 6 Rockboro Road, Old Blackrock Road, Cork
EI6 EXB Paul G. Mannion, 75 Crestwood, Coolough Road, Galway
EI6 EYB Aidan McDonald, "Briarleas", Knockarourke, Donoughmore, Co. Cork
EI6 EZB Stephen G. Walmsley, 7 Mosley Park, Kilfennan, Derry BT47 1HR, Northern Ireland

EI7

EI7 A Thomas J. McCrossan, Newtowncunningham, Co. Donegal
EI7 D Comdt. Michael Ryan, "Polimicus", 21 Strand Road, Baldoyle, Dublin 13
EI7 H Roy Cookman, 31 Weston Road, Churchtown, Dublin 14
EI7 M East Cork Amateur Radio Group, c/o Denis O'Flynn, EI6HB, Ladysbridge P.O., Castlemartyr, Co. Cork
EI7 R Seamus C. Rossiter, 34 Celtic Park Avenue, Whitehall, Dublin 9
EI7 V John Cahill, 10 Strawberry Hill, Sundays Well, Cork
EI7 Y Kieran F. Williams, 35 The Rise, Malahide, Co. Dublin
EI7 AC Thomas H. Perrott, 204 Ballyroan Road, Dublin 16
EI7 AE D. Bourke, 8 Pearse Street, Clonakilty, Co. Cork
EI7 AF Robert Williams, Sanmartineus, Mountain Road, Clonmel, Co. Tipperary
EI7 AG Stafford Charles McConnell, Killaloe, Co. Clare
EI7 AH Michael P. Power, "Glendale", 130 Lismore Park, Waterford
EI7 AK Details withheld at licensee's request
EI7 AL P. Twomey, The Rock, Carrigaline, Co. Cork
EI7 AN L. Muller, 4 St. Thomas Mead, Mount Merrion, Dublin 4
EI7 AV Tomas O Canainn, Ard Barra, Gleann Maighair, Co. Chorcaigh
EI7 BA John Tait, Ballykennefick, Whitegate, Co. Cork
EI7 BB Geoffrey Dean, Lohart, Kenmare, Co. Kerry
EI7 BD J. J. Quilter, Ballinvownig, Dingle, Co. Kerry
EI7 BF A.F. Fogarty, 36 Monastery Walk, Clondalkin, Dublin 22
EI7 BK Brian Cullen, 67 College Green, Ennis, Co. Clare
EI7 BN Jeff Miles, 84 Goatstown Road, Dublin 14
EI7 BR David Fitzgerald, 114 Willow Park Avenue, Ballymun, Dublin 11
EI7 BS Patrick J. McGorman, 'Teaghlach', Legnakelly, Clones, Co. Monaghan
EI7 BV John Breen, 10 Fortfield Park, Terenure, Dublin 6
EI7 BW W. Meredith, 29 Leybourne Road, Strood, Rochester, Kent ME2 3QF, England
EI7 BY M.C. Kenny, "Clareview", 16 Aisling Drive, Ennis Road, Limerick
EI7 BZ George Young, 36 Rathdown Drive, Terenure, Dublin 6
EI7 CB G. O'Sullivan, 227 Templeogue Road, Dublin 6
EI7 CC Peter R. Ball, 21 Doonamana Road, Dun Laoghaire, Co. Dublin
EI7 CD Sean Nolan, 12 Little Meadow, Pottery Road, Dun Laoghaire, Co. Dublin
EI7 CE John E. Murtagh, 2075 Tully Estate, Maddenstown, Curragh, Co. Kildare
EI7 CF T.E. Sloan, "Kincora", Newtownbutler, Clones, Co. Monaghan
EI7 CH J.J. Brown, 2 Tullyglass Hill, Shannon, Co. Clare
EI7 CI T.P. O'Gorman, 129 Kimmage Road West, Dublin 6
EI7 CJ Laurence Loughran, 434 8th Avenue, Menlo Park, CA 32051, USA
EI7 CK Henry Kennedy, 50 Larkfield Gardens, Terenure, Dublin 6
EI7 CL Michael F. North, 135 Downpatrick Road, Dublin 12
EI7 CN Patrick J. Ryan, 787 Sharavogue, Borris Great, Portlaoise, Co. Laois
EI7 CP Robert W. Gallagher, The Old House, Achill Sound, Co. Mayo
EI7 CR Michael Mullins, 124 Lurgan Park, Murrough, Renmore, Galway
EI7 CS Brendan Rooney, Lower Road, Glencar, via Sligo, Co. Leitrim
EI7 CT J.G. O'Sullivan, 109 Main Street, Mallow, Co. Cork
EI7 CV Sean Linehan, 2 College Grove, Dunshaughlin, Co. Meath
EI7 CW Clare Dixon, The Point, Crosshaven, Co. Cork
EI7 CX Ian McStay, 37 Clonkeen Drive, Foxrock, Co. Dublin
EI7 CY Joseph P. Lawless, 45 Coolameer Drive, Rathcoole, Co. Dublin
EI7 CZ Thomas May, 41 Ballygall Road East, Dublin 11
EI7 DA Fergus McGee, 5E College Grove, Castleknock, Dublin 15
EI7 DB James McMahon, 43 Laurence Road, Maynooth, Co. Kildare
EI7 DC James Bergin, 3 Corbawn Grove, Shankill, Co. Dublin
EI7 DD Desmond Ryan, "Polimicus", 21 Strand Road, Baldoyle, Dublin 13
EI7 DE Michael Cawley, 12 Thornhill Road, Mount Merrion, Dublin 4
EI7 DF Roderick Walsh
EI7 DG Robert Loftus, 2 Glenwood Road, Raheny, Dublin 5
EI7 DH Michael Ennis, 47 Singleton Road, Scarborough, Ontario M1R 1H8, Canada
EI7 DI Thomas Mollen, 7 St. Synocks Terrace, Ferbane, Co. Offaly
EI7 DJ Cork Radio Club DX Group, QSL via W O'Reilly, EI8AU, Mount Oval, Rochestown, Co. Cork
EI7 DK Raymond Walsh, 628 Howth Road, Dublin 13
EI7 DL David Tighe, "Westward", Shore Road, Strandhill, Co. Sligo
EI7 DM R. Walsh, 2 Auburn Villas, Athlone, Co. Westmeath
EI7 DN Patrick Treanor, 113 Seafield Road, Clontarf, Dublin 3
EI7 DO Eugene G. McArdle, "Emyvale", Little Borough, Merton, Lincs, England
EI7 DP Gerhard Lerch, Main Street, Lahinch, Co. Clare
EI7 DR Paul P. Farrell, 43 Glenmore Drive, Drogheda, Co. Louth
EI7 DS Richard White, c/o Willie Long, Dunkineely, Co. Donegal
EI7 DT Stanley W. Bell, 21 Plantation Avenue, Lisburn, Co. Antrim
EI7 DU William E. Williamson, 18 Hinton Park, Londonderry
EI7 DV Edward O'Loughlin, Mountrice, Monasterevan, Co. Kildare
EI7 DW Tony O'Connor, 41 Wesley Heights, Sandyford Road, Dublin 16
EI7 DY Michael Conroy, West End, Bundoran, Co. Donegal
EI7 DZ Michael Ronan, Symphony House, Kilmacanogue, Bray, Co. Wicklow
EI7 EA H.B. Mutter, 8800 Alton Parkway, Silver Spring, MD 20910, USA
EI7 EC George Moran, 56 Rivervalley Grove, Swords, Co. Dublin
EI7 ED Details withheld at licensee's request
EI7 EE John A. Murtagh, 345 Orwell Park, Templeogue, Dublin 12
EI7 EF Details withheld at licensee's request
EI7 EG Michael J. Murphy, 5163 W. 88th Place, Oak Lawn, IL 60453, USA
EI7 EH Alan Shattock, Scurlock's Leap, Manor Kilbride, Blessington, Co. Wicklow
EI7 EI P.J. Casey, Gort, Kilgarvan, Co. Kerry
EI7 EJ June Dunne, 26 Duncreggan Road, Derry BT48 0AD, Co. Derry, Northern Ireland
EI7 EK Luigi Infante, Dublin Street, Monasterevin, Co.Kildare
EI7 EL Thomas Lande, 13 Haldene Villas, Bishopstown, Cork
EI7 EM John Fitzgerald, The Old Schoolhouse, Inch, Whitegate, Co. Cork
EI7 EN Anthony A. Carroll, 100 Johnstown Avenue, Dun Laoghaire, Co. Dublin
EI7 EO Tom Kelly
EI7 EP Cormac P. Holmes, 9 Lansdowne Park, Ennis Road, Limerick
EI7 EQ Manus McClafferty, "Errarooey", Falcarragh, Co. Donegal
EI7 ER Padraig McGowan, 22 Durham Road, Sandymount, Dublin 4
EI7 ES Arthur J. Owens, 192 Mount Agnes Road, Fair Hill, Cork
EI7 ET Michael Ryan, 43 Firgreen, Ballysimon Road, Limerick
EI7 EU Thomas A. Buckley, 6 Cherry Drive, Listowel, Co. Kerry
EI7 EV Thomas Barnes, "Lyndhurst", Ballybrown, Clarina, Co. Limerick
EI7 EW Jonathan H. Falkner, 52 Shrewsbury Lawn, Cabinteely, Dublin 18
EI7 EX Isaac H.V. Stewart GI4POV, 164 Ballymoney Road, Ballymena, Co. Antrim BT43 5BZ
EI7 EY Kenneth Jordan, 25 Vanessa Close, Celbridge, Co. Kildare
EI7 EZ Brendan Joyce, Pillar Park, Buncrana, Co. Donegal
EI7 FA Noel Meehan, Rougey View, Bundoran, Co. Donegal
EI7 FB Brian P. Cusack, 23 Leinster Square, Dublin 6
EI7 FC Stewart Greer, Joppa Lodge, Rosses Point, Co. Sligo
EI7 FD John M. Cashman, 'Glenview', Crush, Glanmire, Co. Cork
EI7 FE Liam O'Brien, 33 Heywood Heights, Clonmel, Co. Tipperary
EI7 FF George R.H. Northridge, 9 Stillorgan Park, Blackrock, Co. Dublin
EI7 FG Aileen Finucane, 7 Rectory Slopes, Bray, Co. Wicklow
EI7 FH Declan Joyce, 105 Tudor Lawn, Dangan Upper, Galway
EI7 FI Morgan H. Evans, 'Annacura', Blacklion, Greystones, Co. Wicklow
EI7 FJ William McLoughlin, Spencerstown, Murrintown, Co. Wexford
EI7 FL Bernard McMahon, 34 Rose Park, Kill Avenue, Dun Laoghaire, Co. Dublin
EI7 FM Derek Peyton, 123 Springhill Avenue, Blackrock, Co. Dublin
EI7 FN John A. Hegarty, "The Nook", 1 Cookstown Road, Moneymore, Derry, Northern Ireland
EI7 FO Tobias Stapleton, Swiss Cottage, Knocknagore, Crosshaven, Co. Cork
EI7 FP Declan Merriman, 83 Balally Park, Dundrum, Dublin 14
EI7 FQ William K. Walsh, 8 Lower Pouladuff Road, Cork
EI7 FR Harry McGrath, WB2EZM, 50 Somerset Street, Huntington Station, Long Island, NY 11746-8422, USA
EI7 FS Gerard Ronan, 12 Marian Park, Patrickswell, Co. Limerick
EI7 FT Patrick Tobin, 78 St. Brendans Avenue, Coolock, Dublin 5
EI7 FV Vincent McGettrick, "Creggan", Greenfield Road, Sutton Cross, Dublin 13
EI7 FW Christoph Weritz, DL9YEL, Shelmalier Commons, Barntown, Co. Wexford
EI7 FX Thomas Lambe, Gray Acre Road, Newtownbalregan, Dundalk, Co. Louth
EI7 FY Patrick Geary, Kilbeg, Ladysbridge, Co. Cork
EI7 FZ Donal F. Egan, 25 Ballymace Green, Templeogue, Dublin 14
EI7 GA Partick J. McLoughlin, Aughadrina, Castlebar, Co. Mayo
EI7 GB Wilfred Bryans, 6 Slieve Bloom Park, Drimnagh, Dublin 12
EI7 GD Stephen Fitzgerald, Rossmoney, Carrowholly, Westport, Co. Mayo
EI7 GE Michael A. Kelly, 29 Old Cabra Road, North Circular Road, Dublin 7
EI7 GF Patrick M. O'Neill, 8 Abbey Drive, Rathculliheen, Ferrybank, Co. Waterford
EI7 GG Charles Harkin, 19 Slievenamore Park, Derry, BT 488NJ, Northern Ireland
EI7 GH Michael Drennan, Ballyroberts, Cuffesgrange, Co. Kilkenny
EI7 GI Tony McNally, Caulstown, Dunboyne, Co. Meath
EI7 GJ Alan Hackett G0DMM, Lugboy, Elphin, Co. Roscommon
EI7 GK Padraig O Meachair, 8 Cnoc na Manach, Cill Mhantain
EI7 GL John M. Desmond, 4 Rathmore Lawn, South Douglas Road, Cork
EI7 GM Paul Kearney, Dublin 9
EI7 GN John Sherwood, Newtown, Ballinderry, Mullingar, Co. Westmeath
EI7 GO William F. Cantwell, Kickham Street, Mullinahone, Thurles, Co. Tipperary
EI7 GP Michael F. Foley, 67 Park Road, Strabane, Co. Tyrone, Northern Ireland
EI7 GQ Michael P. Ryan, VE7FWY, 8882 - 214 B Street, Langley, B.C., Canada V3A 6X5
EI7 GR Daniel O'Driscoll, 15 Ardagh Park, Blackrock, Co. Dublin
EI7 GS George M.K. Browne, 58A Monastery Rise, Clondalkin, Dublin 22
EI7 GT Patrick Weldon, 215 Cedarwood Park, Dundalk, Co. Louth
EI7 GU Thomas Walk, Sandgasse 25/27, P.O. Box 110355, D-8750 Achaffenburg, Germany
EI7 GV Peter Clandillon, 108 Abbey Park, Baldoyle, Dublin 13
EI7 GW Joseph Breen, 7 Watermill Road, Raheny, Dublin 5
EI7 GX Thomas Verling, 6 Corrovorrin Avenue, Ennis, Co. Clare
EI7 GY Joe Ryan, 34 Watson Road, Killiney, Co. Dublin
EI7 GZ John Gibney, 1 Stoneylea, Vevay Road, Bray, Co. Wicklow
EI7 HA Charles J. Reason, 91 Fortfield Road, Terenure, Dublin 6
EI7 HB Ian Mellows, Saoirse Nursing Home, Meath Road, Bray, Co. Wicklow
EI7 HC Pauric Brophy, 70 Woodbrook, Mountrath, Co. Laois
EI7 HD Michael Webb, Motor Barge 'Snipe', The Watergate, Athlone, Co. Westmeath
EI7 HE Francis Fitzpatrick, 45 Oaklands, Arklow, Co. Wicklow
EI7 HF Emmet A. Caulfield, 10 Beechcourt, Killiney, Co. Dublin
EI7 HG Charles Farnan, 22 Lavarna Road, Terenure, Dublin 6W
EI7 HH Aidan F. Murray, 9 Whitestrand Avenue, Lower Salthill, Galway
EI7 HI Details withheld at licensee's request
EI7 HJ Daniel Gillespie, 81 Lisfannon Park, Derry BT48, Northern Ireland
EI7 HK Patrick J. Murray, 4 Dunboy, Brighton Road, Foxrock, Dublin 18
EI7 HL Harry F. Malthouse, 10 Leoville, Dunmore Road, Waterford
EI7 HM Brendan A. Rooney, 194 Glasnevin Avenue, Dublin 11
EI7 HN Vincent J. Neff, 14 Westgate Road, Bishopstown, Cork
EI7 HO John P. O'Connell, 11 Suffolk Avenue, Shirley, Southampton, Hampshire SO1 5EF, England
EI7 HP Brian Hoffmann, 27 Ursuline Court, Waterford
EI7 HQ Kay Eyman, WA0WOF, R.R.2, Box 366, Garnett, Kansas 66032, USA
EI7 HS Dr. Kibon M. Aboud, Somerville, Rostrevor Road, Rathgar, Dublin 6
EI7 HT Tom McGrath, Rose Cottage, Piperstown, Dublin 24
EI7 HU Michael G. McCarthy, Ballintrim, Upper Aghada, Midleton, Co. Cork
EI7 HV Bro. Bernard P. Vallocheril; Mount St Mary's School, 75 Parade Road, Delhi Cantt P.O., New Delhi 110010 INDIA
EI7 HW Joe McEvoy, 11 Valleycourt, Athlone, Co. Westmeath
EI7 HX Jack Fenlon
EI7 HY Eugene Harrington, 429 Maple Hill Drive, Hackensack, NJ 07601, USA
EI7 HZ Christopher Kenneth Youens, Derrynaneane, Kilmactranny, Boyle, Co. Roscommon
EI7 IA Michael Doyle, 2 Geraldine Street, Phibsboro, Dublin 7
EI7 IB Joseph Carr, Leitir, Kilcar, Co. Donegal

EI7 ID Con Murphy, P.O. Box 88, Yarloop, Western Australia 6218
EI7 IE Nicholas J. Cummins, 25 Pinewood Park, Rathfarnham, Dublin 14
EI7 IF John A. Muhl, Mulroe, Dunnbeacon, Durrus, Bantry, Co. Cork
EI7 IG John Ronan, P.O Box 20, Cahir, Co. Tipperary
EI7 IH Janet Serridge, 21 Lissara Heights, Warrenpoint, Co. Down BT34 3PG, Northern Ireland
EI7 II Albert A. Kleyn, Clashadoo, Durrus, near Bantry, Co. Cork
EI7 IJ Thomas Kenny, Aughnadrung, Virginia, Co. Cavan
EI7 IK Bill Igoe, 28 The Paddocks, Naas, Co. Kildare
EI7 IL Joseph Hernon, 5 Heatherton Park, South Douglas Road, Cork
EI7 IM Kevin O'Herlihy, 285 St. James Road, Greenhills, Dublin 12
EI7 IN John Pirollo, 25 Park Lodge, Castleknock, Dublin 15
EI7 AAB Chris Yeates, 75 Georgian Village, Castleknock, Dublin 15
EI7 ADB George H. McCourt, 8 Market Square, Roscommon
EI7 AEB Daniel F. Campbell, 109 Drumgor Park, Craigavon, Co. Armagh, Northern Ireland
EI7 AFB Paul Murtagh, 345 Orwell Park, Templeogue, Dublin 12
EI7 AGB John Tynan, Main Street, Banteer, Co. Cork
EI7 AHB Ray McCabe, 14 Botanic Park, Dublin 9
EI7 AIB Michael A. Malone, 30 Cloontuskert, Lanesboro, Co. Longford
EI7 AKB Patrick Grady, Killult, Falcarragh, Co. Donegal
EI7 ALB Simon Kenny, 1 Tullyglass Court Lr., Shannon, Co. Clare
EI7 AOB Patrick A. Marry, Boyne Road, Navan, Co. Meath
EI7 APB Anthony Harris, G8ZNB, c/o Raby Estates Office, Staindrop, Nr. Darlington, Co. Durham DL2 3NF, England
EI7 AQB Paul A. Cleary, 50 Pine Copse Road, Dublin 16
EI7 ARB J. Garvey, 57 Grange Court, Rathfarnham, Dublin 14
EI7 ASB Details withheld at licensee's request
EI7 ATB Karen Horan, 69 Trimbleston Gardens, Rock Road, Blackrock, Co. Dublin
EI7 AVB Paul W. Byrne, 41 Broadford Close, Ballinteer Avenue, Dublin 16
EI7 AXB Enda P. Folan, 7 Carragh Close, Knocknacarra, Galway
EI7 AYB Andrew MacMahon, Mt. Gabriel Radar Station, Schull P.O., Co. Cork
EI7 AZB Raymond G. Magauran, Glenstal Abbey, Murroe, Co. Limerick
EI7 BBB Daniel J. O'Callaghan, Coolgreen, Whites Cross, Co. Cork
EI7 BCB John J. O'Kelly, 6 The Palms, Roebuck, Clonskeagh, Dublin 14
EI7 BDB John S. Steele, 42 St. Finbar's Terrace, Bohermore, Galway
EI7 BFB David D. Tobin, 8 Ferndale Road, Glasnevin, Dublin 11
EI7 BIB Peter J. Stewart, Illistrin, Letterkenny, Co. Donegal
EI7 BJB Jim Spain, Johnstown Bridge, Co. Kildare, via Enfield, Co. Meath
EI7 BLB Thomas O'Shea, 29 Drumcairn Road, Armagh BT61 8DQ, Northern Ireland
EI7 BMB Anthony J. Moore, 9 Glendown Green, Templeogue, Dublin 12
EI7 BYB Margaret E. Elder, 16 Cregg Road, Claudy, Co. Derry, Northern Ireland
EI7 CBB Terry J. Upton, 238 Huntstown Road, Mulhuddart, Dublin 15
EI7 CDB Details withheld at licensee's request
EI7 CEB Fintan Sheerin, 733 Hartstown, Clonsilla, Dublin 15
EI7 CGB William F. Brown, 101 Colmcille Road, Gurranabraher, Cork
EI7 CHB Derek McGonagle, North Strand, Skerries, Co. Dublin
EI7 CIB Eugene McKenna, St. Martins, 2 Upper North Road, Drogheda, Co. Louth
EI7 CMB Joseph Moore, "Chez Nous", Sandyhall, Julianstown, Co. Meath
EI7 CNB Frederick W. Markham, Knocknagow, Menloe Gardens, Blackrock, Cork
EI7 COB The Science Society, c/o Dominic B. Doyle, St. Patrick's College, Maynooth, Co. Kildare
EI7 CQB Billy O'Connor, Killarney Road, Castleisland, Co. Kerry
EI7 CRB Jeremiah J. Driscoll, 8 Dromnanane Park, Beaumont, Dublin 9
EI7 CSB Michael S. Scannell, "Prague", Carrigeen Park, Ballinlough, Cork
EI7 CSG Michael Kelly, 20 Barrack Street, Castlecomer, Co. Kilkenny
EI7 CTB Brian Menton, 190 St. Donagh's Road, Donaghmede, Dublin 13
EI7 CXB Ronan Doyle, 20 Tullyvarraga Crescent, Shannon, Co. Clare
EI7 CYB Stephen F. Webster, 15 Meadow Grove, Dundrum, Dublin 16
EI7 CZB Patrick Brown, 189 Brandon Road, Drimnagh, Dublin 12
EI7 DAB Thomas Rogers, 11 Griffith Place, Waterford City
EI7 DAR Dundalk Amateur Radio Society, 113 Castletown Road, Dundalk, Co. Louth
EI7 DBB Teresa M. Buckley, 11 Shanglas Road, Whitehall, Dublin 9
EI7 DCB S. Timothy Jones, 71 Kilcohan Park, Waterford
EI7 DDB Peter J. Donohoe, 6 Rafters Lane, Crumlin, Dublin 12
EI7 DEB Gillian Quinlan, 39 Forest Hills, Rathcoole, Co. Dublin
EI7 DGB Stan O'Reilly, 12 Ocean Wave, Dr. Colohan Road, Salthill, Galway
EI7 DJB Con Mac Parthalain, 17 Pairc an Charraigin, Baile an Locha, Corcaigh
EI7 DKB Ken FitzGerald Smith, Passage West, Co. Cork
EI7 DMB Adrian Jackson, Skehana, Menlough, Ballinasloe, Co. Galway
EI7 DNB John Barry, Windsor Hill, Glounthane, Co. Cork
EI7 DOB Michael Munnelly, Gibbstown, Navan, Co. Meath
EI7 DPB Andrew Gilligan, 6 Thornberry, Windsor Hill, Glounthanne, Co. Cork
EI7 DQB Gerard M. Freaney, Oranbeg, Oranmore, Co. Galway
EI7 DRB William J. Glynn, Cloonemore, Tuam, Co. Galway
EI7 DSB Liam Rainford, 16 Sandyford Hall Cres., Kilgobbin Road, Dublin 18
EI7 DZB Thomas Farrell, 63 Monacurragh, Carlow
EI7 EAB Richard Lawless, Menlo Park, Menlo, Galway
EI7 EBB Ailbe Kinane, Ardlamon, Tipperary, Co. Tipperary
EI7 ECB Gerard McGinley, Woodtown, Ramelton, Co. Donegal
EI7 EDB Sean Martin Curtin, 6 St. Stephen's Place, Watergrasshill, Co. Cork
EI7 EEB Peter J. Duggan
EI7 EGB Margaret Lane, Ballincurrig, Leamlara, Midleton, Co. Cork
EI7 EHB Helen Meehan, Ballincurrig, Leamlara, Midleton, Co. Cork
EI7 EIB Andrew Boyle, 42 Monivea Park, Galway
EI7 EJB Joseph O'Callaghan, Main Street, Ballyclough, Mallow, Co. Cork
EI7 EKB Sean Doherty, Cromogue, Dromcollogher, Charleville, Co. Cork
EI7 ENB Aidan Noone, "The Hermitage", Deerpark Road, Ravensdale, Dundalk, Co. Louth
EI7 EOB Donal Ward, Welchtown Road, Curraghamone, Ballybofey, Lifford, Co. Donegal
EI7 EPB Robert J McGowan, 42 Grange Erin, Douglas, Cork
EI7 EQB Fergal Purcell, 17878 Preston Road #370, Dallas, Texas 75252, USA
EI7 ERB Desmond Murphy, 13 Forest Park, Drogheda, Co. Louth
EI7 ESB Sean Santry, 2 Wood Brook, East Cliffe, Glanmire, Co. Cork
EI7 EUB Michael Joseph McArdle, Big Ash, Knockbridge, Dundalk, Co. Louth
EI7 EVB Patrick J. Quinn
EI7 EWB Walter Keating, Ballydrehid, Cahir, Co. Tipperary
EI7 EXB Enda McDonnell, Moore Street, Loughrea, Co. Galway
EI7 EYB Martin Hynes, 3 Lynnderry Court, Mullingar, Co. Westmeath
EI7 EZB David Elliott, 7 Anglesea Terrace, Anglesea Street, Cork
EI7 KRC Kells Radio Club, c/o Fitzgerald Kitchens, Bective Street, Kells, Co. Meath
EI7 LPD Galway Radio Experimenters Club (see EI4GRC)
EI7 MPD Galway Radio Experimenters Club (see EI4GRC)
EI7 NET Westnet DX Group, c/o Declan Craig, EI6FR, 167 St. James's Road, Greenhills, Dublin 12
EI7 RPD Galway Radio Experimenters Club (see EI4GRC)
EI7 RSD St. David's School Amateur Radio Club, Greystones, Co. Wicklow
EI7 SPD Galway Radio Experimenters Club (see EI4GRC)
EI7 SPM Galway Radio Experimenters Club (see EI4GRC)
EI7 TRG Tipperary Amateur Radio Group, 33 Heywood Heights, Clonmel, Co. Tipperary
EJ 7NET Westnet DX Group (see EI7NET)

EI8

EI8 A Ambrose E. MacNamara, 11 Shanowen Drive, Whitehall, Dublin 9
EI8 B Donal F. O'Dwyer, 119 St. Mobhi Road, Glasnevin, Dublin 9
EI8 C Fintan J. Murphy, 6 Palace Street, Drogheda, Co. Louth
EI8 D H. McElligott, 10 Eglington Park, Dublin 4
EI8 H Patrick J. Fagan, Cloughernal, Granard, Co. Longford
EI8 I Larry Duggan, Cashellachan, Ballyshannon, Co. Donegal
EI8 K C.S. O'Rourke, Skerries Road, Lusk, Co. Dublin
EI8 Q P.N. Staunton, Woodlands, Glanmire, Co. Cork
EI8 R Patrick Duggan, 'Walford', 24 Shrewsbury Road, Dublin 4
EI8 S James Newman, 39 Hillcrest Park, Dublin 11
EI8 U 11th Field Signals Company, H.Q., Cathal Brugha Barracks, Rathmines, Dublin 6
EI8 W Jim Gaffney, 14 Abbey Meadows, Cahir Road, Clonmel, Co. Tipperary
EI8 Z James D. Upton, 11 Cardiffcastle Road, Finglas, Dublin 11
EI8 AC William J. Doherty, 11 Cedarwood Road, Glasnevin North, Dublin 11
EI8 AH Seamus McShane, Dunkineely, Co. Donegal
EI8 AI Cavan Amateur Radio Club, c/o Kevin Kilduff EI3C, Swellan Lower, Cavan
EI8 AJ John Reddington, 28 Abbey Park, Baldoyle, Dublin 13
EI8 AP G. Cassidy, 9 Halldene Estate, Waterfall Road, Bishopstown, Cork
EI8 AR Rev. Bro. John Shortall, De La Salle Brothers, Dundalk, Co. Louth
EI8 AT J.S. Murray, 6 Glen Garth, The Park, Cabinteely, Co. Dublin
EI8 AU W.C. O'Reilly, Mount Oval, Rochestown, Co. Cork
EI8 AV W. Schmitt, Tierenane Lodge, Ballickmoyler, Co. Carlow
EI8 BC William K. Ryan, 11 Wendell Avenue, Martello, Portmarnock, Co. Dublin
EI8 BD John J. Gallagher, "Calderstones", Link Road, Brownshill, Carlow
EI8 BN Leo Dorrington, 9 Church Street, Tralee, Co. Kerry
EI8 BP Seamus McCague, 10 Dromartin Close, Goatstown, Dublin 14
EI8 BR Leo J. McHugh, 1 Glenageary Woods, Upper Glenageary Road, Dun Laoghaire, Co. Dublin
EI8 BS L. Byrne
EI8 BX John J. Keely, Walshestown, Lusk, Co. Dublin
EI8 CA C.J. Martin, 8 Ashgrove Drive, Naas, Co. Kildare
EI8 CC Ger Gervin, 61 Johnstown Road, Dun Laoghaire, Co. Dublin
EI8 CE Aidan McGrath, "Tinhalla", Carrick-on-Suir, Co. Waterford
EI8 CG Rev. Fr. J. Griffin, St. Senans, Miltown Malbay, Co. Clare
EI8 CH Rev. Fr. D. Coyle, c/o 63 Nutgrove Avenue, Dublin 4
EI8 CI Shane Kenny, "Kilbrae", Glen o' the Downs, Co. Wicklow
EI8 CJ Thomas Naughton, 9 College Park, Corbally, Limerick
EI8 CK T.J. King, 4 Bavan Villas, Mayorbridge, Newry, Co. Down, Northern Ireland
EI8 CL P.J. Griffin, 64 Tymon Crescent, Oldbawn, Dublin 24
EI8 CM Patrick E. Mahon, 16 Shanliss Grove, Santry, Dublin 9
EI8 CN Patrick E. Flynn, 127 Pine Valley Avenue, Grange Road, Rathfarnham, Dublin 16
EI8 CP Rev. Fr. Patrick M. Kelly, The Presbytery, Brittas Bay, Co. Wicklow
EI8 CR Bryan A. Yeomans, Bayview House, Front Strand, Youghal, Co. Cork
EI8 CS Daniel J. O'Sullivan, Derragh, Cullen, Co. Cork
EI8 CT Sean R. Reilly, 46 Sycamore Road, Dundrum, Dublin 16
EI8 CV Alan Cooney, 96 Kincora Avenue, Clontarf, Dublin 3
EI8 CX Rev. Fr. James F. Madigan, Cragmore, Askeaton, Co. Limerick
EI8 CY Daniel O'Regan, 23 Shanliss Walk, Santry, Dublin 9
EI8 CZ Patrick J. O'Leary, 30 Grattan Park, Greystones, Co. Wicklow
EI8 DA Patrick Brennan, 22 Highfield, Drogheda, Co. Louth
EI8 DB James Butler, 6 Weavers Square, Dublin 8
EI8 DC Lillian Higgins, Roevahagh, Kilcolgan, Co. Galway
EI8 DD Tom McNamara, 24 Oakley Crescent, Highfield Park, Galway
EI8 DE Dr. John B. Kearns, 126 Glenageary Avenue, Dun Laoghaire, Co. Dublin
EI8 DF Robert J. Semple, "Lisieux", 99 Rathfarnham Road, Dublin 14
EI8 DG Maurice Byrne, 75 New Ireland Road, Rialto, Dublin 8
EI8 DH Bernard Walsh, 10 Park View, Athboy, Co. Meath
EI8 DJ Donal Kelly, Olivette, Camden Road, Crosshaven, Co. Cork
EI8 DK Jim Jeffers, 73 Beech Grove, Lucan, Co. Dublin
EI8 DL John F. Moran, 1 Connolly Crescent, Longford
EI8 DN Liam McNulty, Town Parks, Raphoe, Co. Donegal
EI8 DO Longford Amateur Radio Club, c/o Secretary, John F. Moran, 1 Connolly Crescent, Longford
EI8 DP Patrick J. McGowan, Quay Street, Donegal
EI8 DQ Aemar Higgins, 1 Cairnshill Park, Cairnshill Road, Belfast BT8 4RG
EI8 DR John F. McKinney, 11 Curlew Way, Waterside, Derry BT47 1LQ
EI8 DS Maurice Quinlan, 11 St. Colman's Square, Cobh, Co. Cork
EI8 DT William S. Barker, 18 Main Street, Seskinore, Omagh, Co. Tyrone
EI8 DU Bernadette Smith, 33 Laurel Park, Clondalkin, Dublin 22
EI8 DV Michael B. Bennett, Forenoughts, Naas, Co. Kildare
EI8 DW James P. Keenan, 36 Newtownpark Avenue, Blackrock, Co. Dublin
EI8 DY Joseph Donnelly, 20 Harmony Avenue, Donnybrook, Dublin 4
EI8 DZ P. Conboy, 29 Lisbrak, Longford
EI8 EA Keith Burnside, 21 Dermott Walk, Comber, Co. Down
EI8 EB Thomas Schewe, Barkenkoppel 37, D-2000 Hamburg 65, Germany
EI8 EC Thomas Finucane, 7 Rectory Slopes, Bray, Co. Wicklow
EI8 ED J. West, 28 Shenick Avenue, Skerries, Co. Dublin
EI8 EE Kevin Fitzsimons, 4 Station Road Cottages, Sutton, Dublin 13
EI8 EF Details withheld at licensee's request
EI8 EG Joseph Deegan, 6 McLoone Terrace, Letterkenny, Co. Donegal
EI8 EJ Werner Thore, Upper Uggon, Tullow, Co. Clare
EI8 EK Sean Ward, Sessiaghoneill, Ballybofey, Co. Donegal
EI8 EL Bernard P. Curtin, 42 Lime Trees Road East, Maryborough Estate, Douglas, Cork
EI8 EM Alan Cronin, College View, Clonroadmore, Ennis, Co. Clare
EI8 EN S.W. Kinnear, 25 Beech Lawn, Dundrum, Dublin 16
EI8 EO James R. Moloney, 17 Beechbrook, Delgany, Co. Wicklow
EI8 EP Joseph Pasieka, 550 Howth Road, Raheny, Dublin 5
EI8 EQ Ben Croly, Kilshanroe, Via Enfield, Co. Kildare
EI8 ER Marvin Wallis, Leamore House, Greystones, Newcastle, Co. Wicklow
EI8 ET Noel Grier, "Mistra Bungalow", Aughnish, Ramelton, Letterkenny, Co. Donegal
EI8 EU John Sullivan, Moneymore, Oranmore, Co. Galway
EI8 EV James O'Hara, Binghamstown, Belmullet, Co. Mayo
EI8 EW Frank Devitor, Lower Grange, Goresbridge, Co. Kilkenny
EI8 EX John J. O'Sullivan, 23 Avon Court, Ballincollig, Co. Cork
EI8 EY John Cooley, 18 McDermot Avenue, Mervue, Galway
EI8 EZ Alan P. Robinson, 26 Frascati Park, Blackrock, Co. Dublin
EI8 FA David K. Hall, Cherry Lawn, Killarney Road, Bray, Co. Wicklow
EI8 FB Jim Fitzgerald, "Dun Eochaill", 78 Pearse Park, Tipperary
EI8 FC James Ryan, Tullovin Bridge, Croom, Co. Limerick
EI8 FD Walter B. Albert, 82 Ailesbury Grove, Ballinteer, Dublin 16
EI8 FE Gerald FitzGerald, 56 Strand Street, Skerries, Co. Dublin
EI8 FF William Cullen, 35 St. Manntan's Road, Wicklow
EI8 FG Michael Mulcahy, "Stagmount", Knockchapple, Co. Cork
EI8 FH Malcolm Joyce, "Nigella", Eyrecourt, Ballinasloe, Co. Galway
EI8 FI Kevin M. Keane, 2 Doon Road, Ballybunion, Co. Kerry
EI8 FJ Stephen Patterson, Regional Tech. College, Port Road, Letterkenny, Co. Donegal
EI8 FL L.H. Kirk, 9 Allen Park Road, Stillorgan, Co. Dublin
EI8 FM William N. Campbell, GI4NKY, 68 Richmond Court, Lisburn, Co. Antrim BT27 4QX, Northern Ireland
EI8 FN Michael J. Beirne, 3 Belgrove Lawn, Chapelizod, Dublin 20
EI8 FO Gerard McCallion, c/o Christopher Gildea, Liscooley, Castlefin, Lifford, Co. Donegal
EI8 FP John P. Flood, 43 Hillside, Dalkey, Co. Dublin
EI8 FQ Patrick C. Farrell, Loughanstown, Rathowen, Co. Westmeath
EI8 FR Margaret McAnnalan, Chapel Hill, Ballycotton, Co. Cork
EI8 FS Terence P. Walsh, Cork
EI8 FT William J. Cousins, 35 Knocknamana Park, Letterkenny, Co. Donegal
EI8 FV Martin Hughes, 37 Parkview, Church Hill, Passage West, Co. Cork
EI8 FW Padraig Moroney, Main Street, Broadford, Co. Clare
EI8 FX Joseph A. O'Connor, 9 John Street, Ardee, Co. Louth
EI8 FY James McKinney
EI8 GA Matthew J. Hurley, Park, Youghal, Co. Cork
EI8 GB Ben Adjaye, 245 Lr. Kimmage Road, Dublin 6
EI8 GC Fintan Van Ommen Kloeke
EI8 GD James Tighe, 1 Avondale Drive, Hanover, Co. Carlow
EI8 GE Dr. John Malone, Greenfield House, 1 Santa Sabina Manor, Greenfield Road, Sutton, Co. Dublin
EI8 GF Paul Normoyle, 51 Ferndale Avenue, Glasnevin, Dublin 11
EI8 GG James Ryan, Skehanagh, Watergrass Hill, Co. Cork
EI8 GH Thomas P. Finn, Kyleclonherbert, Portlaoise, Co. Laois
EI8 GI Helen McNally, Caulstown, Dunboyne, Co. Meath
EI8 GJ Gerard McGrane, 2 Ferndale, Navan, Co. Meath
EI8 GK Phelim Blake, 16 Glenville Lawn, Clonsilla, Dublin 15
EI8 GL Colm D. McConnell, 67 Woodbine Park, Raheny, Dublin 5
EI8 GM Peter D. White, 1 Coliemore Road, Dalkey, Co. Dublin
EI8 GN David R. Vizard, Abbey Road, Kilcrea, Ovens, Co. Cork
EI8 GO Thomas Molloy, 2 Derrynane Close, Powerscourt Lawns, Waterford
EI8 GP Martin J. Gillespie, Dreenan, Ballybofey, Co. Donegal
EI8 GQ Garry Wilson, 3 Church Bay, Crosshaven, Co. Cork
EI8 GR Aidan Riordan, 1 Portmarnock Crescent, Carrick Hill, Portmarnock, Co. Dublin
EI8 GS Jim Barry, Windsor Hill, Glounthaune, Co. Cork
EI8 GT Joseph D. Duffin W2ORA, 4 West Central Avenue, Moorestown, NJ 08057-2415, USA
EI8 GU Thomas Mooney, 'Oaklands', Lagore Little, Dunshaughlin, Co. Meath
EI8 GV Eugene O'Connor, Knocknafinchy, Castleisland, Co. Kerry
EI8 GW David A. Perris, 44 Westpark, Tallaght, Dublin 24
EI8 GX Kevin O'Sullivan, 'Sundown', Mount Oval, Rochestown, Co. Cork
EI8 GY Details withheld at licensee's request
EI8 GZ Finbar Moloney, 21 Clonard Park, Sandyford Road, Dundrum, Dublin 16

EI8 HA Jim Murphy, Kilcarrig, Bagenalstown, Co. Carlow
EI8 HB James Clarke, G0DTK, 66 Harefield Road, Swaythling, Southampton SO17 3TH, England
EI8 HC Eoin Savage, 8 Wellington Close, Chelmsford, Essex CM1 2EE, England
EI8 HD Paul A. Corbett, 3 Seapoint, Barna, Co. Galway
EI8 HE Denis D. Pavis, 269 Lr. Braniel Road, Belfast BT5 7NR, Northern Ireland
EI8 HF J. Coughlan, 57 Butterfield Close, Rathfarnham, Dublin 14
EI8 HG N. John Hooper KK4YL, P.O. Box 3, Irvington, VA 22480, USA
EI8 HH Andreas Imse, Hinter der Kirche 31, D-6500 Mainz 41, Germany
EI8 HJ Ronan Coyne, 31 Lakeshore Drive, Renmore, Galway
EI8 HK Joseph Gray, "Caravelle", Hextable (near Shanley), Kent BR8 7LS, England
EI8 HL Robert G. Barry, 386 St. Johns Park, Waterford
EI8 HM Denis Ryan, 30 Lanndale Lawns, Dublin 24
EI8 HN Philip D. Murphy, 8 Buttercup Park, Darndale, Dublin 17
EI8 HO Gerard Dykes, 30 St Benildus Avenue, Ballyshannon, Co. Donegal
EI8 HP Michael Adrian Fitzgerald, "Roskeen", Kerry Pike, Co. Cork
EI8 HQ John Quinn, Newtown, Kilcolgan, Co. Galway
EI8 HR William Power, 25 Doyle Street, Waterford
EI8 HS John Kelleher, 40 Rosewood Estate, Ballincollig, Co. Cork
EI8 HT Gerald Kenneally, 23 Knockaverry, Youghal, Co. Cork
EI8 HU Patrick J O'Mahony, Ballintotas, Castlemartyr, Co. Cork
EI8 HV Michael Regan, 66 Silverheights Avenue, Silverheights, Cork
EI8 HW Details withheld at licensee's request
EI8 HX John P Maher, 44 Grange Heights, Waterford
EI8 HY W.P. Hughes, 155 Shanliss Road, Santry, Dublin 9
EI8 HZ John Edmundson, Drumbuoy, Lifford, Co. Donegal
EI8 IA Seamus Campbell, Reynoldstown, Clogherhead, Co. Louth
EI8 IB Brendan Kavanagh, 26 Ely Drive, Tallaght, Dublin 24
EI8 IC Tim Makins, Coolmeen, Ballyfarnon, Co. Sligo
EI8 ID Darko Volfer, 68 Huntstown Wood, Clonsilla, Dublin 15
EI8 IE Michael McCann, 36 Quarry Road, Cabra, Dublin 7
EI8 IF Michael O'Connor, 34 O'Brien Street, Tipperary
EI8 IG James A. Farrell, 12 Byrneville Estate, Dungarvan, Co. Waterford
EI8 IH Ciaran McCarthy, 35 Dun na Mara Drive, Renmore, Galway
EI8 II David Cowman
EI8 IJ Tony Kennedy, 76 Westgate Road, Bishopstown, Cork
EI8 IK Robert Michael Wright, College Street, Ballyshannon, Co. Donegal
EI8 IL Tom Quinn, Moross, Tamney, Co. Donegal
EI8 IM Details withheld at licensee's request
EI8 IN Larry Hess, 212 Malahide Marina Vill., Malahide, Co. Dublin
EI8 AAB Michael Pettigrew, 3 Willow Mews, St. Albans Park, Sandymount, Dublin 4
EI8 ABB James Christie
EI8 ACB David H. Kerr, 64 Balrothery Estate, Tallaght, Dublin 24
EI8 AEB Ian Smith, 33 Laurel Park, Clondalkin, Dublin 22
EI8 AGB Noel Bell, 81 Ballinclea Heights, Killiney, Co. Dublin
EI8 AHB Thomas J. Trimble, Cloonadra, Lanesboro, Co. Longford
EI8 AKB J. Chaffer, New Bungalow, Ullid, Kilmacow, Via Waterford
EI8 APB Patrick J. Kinahan, Main Street, Castleknock, Dublin 15
EI8 AQB Barry Moore, Flat 4, Oakhurst, Woodlands Road, Guildford, Surrey, England
EI8 ARB Terence Lynch, 77a Landscape Park, Churchtown, Dublin 14
EI8 ASB P. Higgins, Sweeneys Oughterard, House Hotel, Co. Galway
EI8 AUB Frank Joynt, Masonbrook, Loughrea, Co. Galway
EI8 AVB Maurice J. Fitzgerald, 188 Upper Salthill Road, Salthill, Co. Galway
EI8 AXB Barbara Martin, "Sitka", Cronroe, Ashford, Co. Wicklow
EI8 BAB Martin O'Rourke, Miltowngrange, Doristown, Castlebellingham, Co. Louth
EI8 BCB John A. Nolan, 19 Burrin Street, Carlow
EI8 BDB Patrick J. McCann, 24 Woodview Court, Glenalbyn Road, Stillorgan, Co. Dublin
EI8 BEB David J. Dillon, 84 Knocknaganny Park, Sligo
EI8 BFB Patrick J. Geoghegan, Kilmolin, Enniskerry, Co. Wicklow
EI8 BGB James Proudfoot, Nobber, Co. Meath
EI8 BHB Details withheld at licensee's request
EI8 BLB William Grant, 209 Ballybeg Square, Waterford
EI8 BNB Pat McCarthy, 14 Raheen Court, Tallaght, Dublin 24
EI8 BRB Ivan Sproule, Bella, Collooney, Co. Sligo
EI8 BVB Ciaran M. Dorgan, 130 Hillside, Dalkey, Co. Dublin
EI8 BWB Finbarr J. Leach, c/o Ministry of Works, Architects Branch, P.O. Box 330, Maseru, Lesotho, Africa
EI8 BXB Kevin M. Prendergast, 18 Northbrook Road, Leeson Park, Dublin 6
EI8 BZB James P. Fitzsimons, Scotchmans Road, Monkstown, Co. Cork
EI8 CAB F.A.L. Gardner, 44 Munster Street, Phibsborough, Dublin 7
EI8 CBB Patrick O'Haire, 15 Culhaine Street, Dundalk, Co. Louth
EI8 CCB David R. Connor, Carnagore, Carrigart, Latterkenny, Co. Donegal
EI8 CDB Patrick F. Moran, 18 Clybaun Heights, Knocknacarra, Galway
EI8 CEB Noel L. Byrne, 5 Millmount Place, Drumcondra, Dublin 9
EI8 CGB Jackie Tierney, 384 Collins Avenue, Whitehall, Dublin 9
EI8 CHB David Kearney, Ballybeg, Currow, Killarney, Co. Kerry
EI8 CIB Neil J. Murphy, "Raymur", Holmston Avenue, Glenageary, Co. Dublin
EI8 CKB Gerard C. Scullion, 13 Orritor Crescent, Cookstown, Co. Tyrone
EI8 CLB George D. O'Reilly, 96 Boulevard North, Bayside, Sutton, Dublin 13
EI8 CMB Joseph Clarke, Balrath, Kells, Co. Meath
EI8 CNB James Meehan, Killaloonty, Tuam, Co. Galway
EI8 CQB Peter J. Glennane, 15 Rochestown Avenue, Dun Laoghaire, Co. Dublin
EI8 CTB Ian Webster, 6 Grange Park Rise, Raheny, Dublin 5
EI8 CUB Kenneth Purcell, 67 Woodview Grove, Blanchardstown, Dublin 15
EI8 CYB Thomas Cooney, 14 Cairns Drive, Sligo
EI8 DBB John Timoney, Glen Road, Glenties, Co. Donegal
EI8 DDB Alexander P. Walsh, 79 Castleview Road, Clondalkin, Dublin 22
EI8 DEB Martin O'Donnell, 154 The Grove, Celbridge, Co. Kildare
EI8 DFB Lesley O'Donnell, 154 The Grove, Celbridge, Co. Kildare
EI8 DGB Sean Sheehy, 48 Murmont Road, Montenotte, Cork
EI8 DIB Kenneth Foley, Coolgrange, Granges Road, Kilkenny
EI8 DLB Patrick J. Kyles, Burtonport, Letterkenny, Co. Donegal
EI8 DMB A. T. Dalton, 42 Ramlyn Park, Ballybone, Galway
EI8 DQB Kieran Corrigan, Parsonstown, Togher, Drogheda, Co. Louth
EI8 DRB Gerard Kavanagh, Oranmore, Co. Galway
EI8 DSB Michael Barry, Orlagh, Templeogue, Dublin 16
EI8 DUB Hugh Boyle, Devlinmore, Carrigart, Letterkenny, Co. Donegal
EI8 DWB Pat Walsh, Lower Main Street, Ballintra, Co. Donegal
EI8 DYB Tom Maher, 44 Grange Heights, Waterford
EI8 DZB Martin Burke, Moneygreggan, Newtowncunningham, Lifford, Co. Donegal
EI8 EEB Shane Tohill, 46 Carnmeen Park, Warrenpoint, Co. Down, Northern Ireland
EI8 EFB Mark Fergus Cusack, 54 Foxwood, Swords, Co. Dublin
EI8 EIB Donal Patrick Leader, Marino Inst. of Education, Griffith Avenue, Dublin 9
EI8 EJB Brian Gerard Whelan, 19 Aulden Grange, Santry, Dublin 17
EI8 EKB Paul Callan, 29 Adare Green, Coolock, Dublin 17
EI8 EOB Sean Kennedy, 114B Newfield Estate, Drogheda, Co. Louth
EI8 EPB Seamus A. Ryan, Modeshill, Mullinahone, Co. Tipperary
EI8 ERB Thomas J Harte, Stonestown, Robinstown, Navan, Co. Meath
EI8 ESB Robbie Phelan, 24 Glenfield, Kilmallock, Co. Limerick
EI8 ETB Thomas McCoy, 96 Lower Friars Walk, Ballyphehane, Cork
EI8 EUB Tony Cooke, Shean House, Athy, Co. Kildare
EI8 EXB Gerry Ormond, Gainstown, Navan, Co. Meath
EI8 EYB George W. Quinlan, 16 Plunkett Terrace, Cobh, Co. Cork
EI8 EZB Details withheld at licensee's request
EI8 LSS Louth Sea Scouts, Dundalk Amateur Radio Soc, 113 Castletown Road, Dundalk, Co. Louth
EI8 VHF Drogheda & Dist. VHF Radio Club, c/o Anthony Allen, EI4DIB, Co. Louth

EI9

EI9 A Desmond F. Cornwall, 14 Mountain View Park, Greystones, Co. Wicklow
EI9 C J. O'Mahony, 89 Ballymun Road, Glasnevin, Dublin 9
EI9 D J.G. Carroll, 41 Barton Road East, Churchtown, Dublin 14
EI9 G Terry Tierney, 55 Moyne Road, Ranelagh, Dublin 6
EI9 K William Curtis, 30 Hillside Gardens, Skerries, Co. Dublin
EI9 P Phil Cantwell, 'Villa Maria', Manorland, Trim, Co. Meath
EI9 S Thomas J. Sheerin, 24 Goatstown Road, Dublin 14
EI9 V Con Hunter, 30 Coolgarrif Road, Beaumont, Dublin 9
EI9 AB Ken E. Dixon, The Point, Crosshaven, Co. Cork
EI9 AC Daniel P. Meehan, Castle Street, Donegal
EI9 AD John J. Dolan, 'Naomh Mog', 26 Shelbourne Park, Limerick
EI9 AE Norman Miller, G3MVV, "Oak Tree", Ashwood, Arklow, Co. Wicklow
EI9 AL Brian Toner, 19 Springfield Drive, Dooradoyle, Limerick
EI9 AR Francis T. Crosbie, "Chez Nous", Ballinasloe, Co. Galway
EI9 AX Denis M. O'Sullivan, 69 St. Columbas Rise, Swords, Co. Dublin
EI9 BA Liam o Tuathalain, 72 Bothar Bhinn Eadair, Cluain Tarbh, Baile Atha Cliath 3
EI9 BC Alf Bradshaw, 70 Rail Park, Maynooth, Co. Kildare
EI9 BD Jim Naughton, Crannagh, Cloghans P.O., Ballina, Co. Mayo
EI9 BG Tom Donnellan, Rosmadda, Parteen, Co. Clare
EI9 BL R. Casey, Foxford, Bansha, Co. Tipperary
EI9 BS Hugh Pollard, 7 Waltham Terrace, Blackrock, Co. Dublin
EI9 BT Dermot Cullen, 1 Upper Beechwood Avenue, Ranelagh, Dublin 6
EI9 BV John G. Mahon, 48 St. Brendan's Avenue, Coolock, Dublin 5
EI9 BW Henry Boyle, 120 Pinebrook Road, Artane, Dublin 5
EI9 BX W.F. Hurley, "La Plata", 20 Chestnut Grove, Caherdavin Lawn, Limerick
EI9 BZ Denis T. Walsh, Charleville Road, Tullamore, Co. Offaly
EI9 CA Cathal O'Reilly, 16 O'Connell Avenue, Berkeley Road, Dublin 7
EI9 CB Limerick City S.W. Radio Club, 'Santa Lucia', Ballykeelaun, Parteen, Limerick
EI9 CC Richard K. Wilson, 9 Navan Road, Castleknock, Dublin 15
EI9 CD J.W. Welch, Deer Park Lodge, Ravensdale, Dundalk, Co. Louth
EI9 CE John McGorman, 8 MacCurtain Street, Legnakelly, Clones, Co. Monaghan
EI9 CF Seamus O'Dea, 27 Millmount, Mullingar, Co. Westmeath
EI9 CG Vincent H. Brady, 21 St. Assams Park, Raheny, Dublin 5
EI9 CI Timothy O'Mahony, Roxtown, Fedamore, Kilmallock, Co. Limerick
EI9 CJ Tom McDermott, Rockmarshall, Jenkinstown, Dundalk, Co. Louth
EI9 CK Rev. Fr. Nicholas O'Grady, St. Paul's Retreat, Mount Argus, Dublin 6W
EI9 CL Rev Fr. D. McKenna S.J., Woodlands, Raheny, Dublin 5
EI9 CN Lawrence McGriskin, 42 Ard Easmuinn, Dundalk, Co. Louth
EI9 CP Peter Clancy, 9 Chanel Road, Dublin 5
EI9 CS Donald Riordan, Lombardstown, Pallasgreen, Co. Limerick
EI9 CT Douglas C. Morris, GW2FVZ, 48 Pen y Cefn Road, Caerwys, Mold, Clwyd, CH7 5BH, Wales
EI9 CW Raymond Bonar, 28 Cherry Grove, Naas, Co. Kildare
EI9 CX Corp. William Buckley, 1 H Block, Married Quarters, Arbour Hill, Dublin 7
EI9 CY James A. Roche, Ballyclamasy, Youghal, Co. Cork
EI9 DA Kieran Daly, Granarogue, Carrickmacross, Co. Monaghan
EI9 DB Peter F. McGovern, Barran, Blacklion, Co. Cavan
EI9 DC Dermot Cronin, Derhill, Church Road, Killiney, Co. Dublin
EI9 DD Army Apprentice School Radio Club, Devoy Barracks, Naas, Co. Kildare
EI9 DE Charles P. Gallagher, 1 Airfield Drive, Churchtown, Dublin 14
EI9 DF Alan Aston, c/o Air Corp Sigs., Baldonnel, Co. Dublin
EI9 DH Brian Beary, Main Street, Oola, Co. Limerick
EI9 DI Thomas A. Drake, P.O. Box 4574, Nicosia, Cyprus
EI9 DJ Dermot E. O'Dwyer, Dublin Road, Navan, Co. Meath
EI9 DK Sean Dunne, 7 Allee des Messiers, 78220 Viroflay, France
EI9 DL Thomas McLoughlin, Cranley More, Mostrim, Co. Longford
EI9 DM R.J. Long, Dunkineely, Co. Donegal
EI9 DO Patrick Kennedy, 8 Elm Park, Celbridge, Co. Kildare
EI9 DP Edward L.A. Davy-Thomas, 51 La Touche Park, Greystones, Co. Wicklow
EI9 DQ Raymond McAteer, GI4MFM, "Four Winds", Glenullin Road, Garvagh, Co. Derry BT51 5DQ, Northern Ireland
EI9 DR Gerard M. O'Doherty, 22 Meadowvale Close, Raheen, Limerick
EI9 DS James Waters, Clonsilla, Gorey, Co. Wexford
EI9 DT Robert J. Armstrong, 95 Craighill, Ballycrakiay, Mullamore, Co. Antrim
EI9 DU J. Gartlan, Rathbraughan, Co. Sligo
EI9 DV Bert S.J. Clarke, 38 Teignmouth Road, London NW2 4HN
EI9 DW Dr. David W. Hughes, 17 Rochester Court, Coleraine, Co. Derry
EI9 DY Matthew Rafter, "Crockroy", St. Catherines Road, Killybegs, Co. Donegal
EI9 DZ Gerald E. Birkhead, 103 Roselawn Road, Castleknock, Dublin 15
EI9 EA J.C. McCabe, 8 Robin Villas, Palmerstown, Dublin 20
EI9 EB Edward McCourt, Drumline, Newmarket-on-Fergus, Co. Clare
EI9 EC Stanley Dornan, 9 Clonallon Gardens, Belfast BT4 2BY
EI9 ED Ronald F. McGrane, Cavan Road, Kells, Co. Meath
EI9 EE Michael Grady, Fycorrenagh, Cullion Road, Letterkenny, Co. Donegal
EI9 EF Patrick J. Kenny, Slieve Ban, Malin Head, Co. Donegal
EI9 EG James J. Lysaght, 32 Hyde Avenue, Rosbrien, Limerick
EI9 EH Patrick Murphy, 22 Rosmore Close, Templeogue, Dublin 12
EI9 EI M. Murray, Lisdaulan, Rahara, Roscommon
EI9 EJ Jackie Buckley, Finuge, Lixnaw, Co. Kerry
EI9 EK William Wallace, 108 Shanliss Avenue, Santry, Dublin 9
EI9 EL Ken O'Brien, 2 Dartry Park, Dublin 6
EI9 EM Maurice H. McFadden, 121 Greystown Avenue, Belfast 9
EI9 EN T.E. Radisic, 54 St. Albans Road, London NW5 1RH, England
EI9 EO Michael Mulkerrin, 9 Cypress Grove North, Dublin 6
EI9 EP Jim Duffy, 69 Fitzroy Avenue, Dublin 3
EI9 EQ Egidio A. Giani, 15 Pinewood Avenue, Hillview, Waterford
EI9 ER Dermot Ryan, 53 Culmore Road, Palmerstown, Dublin 20
EI9 ES Peter Morrison, 9 Fortfield Avenue, Terenure, Dublin 6
EI9 ET Richard Craig, 9 Brookvale Terrace, Townsend Street, Skibbereen, Co. Cork
EI9 EU George F. Jackson, 7412 Leahy Road, New Carrollton, MD 20784, USA
EI9 EV Donald Yeomans, Cnoc-Na-Geaorad, Quay Road, Dungloe, Co. Donegal
EI9 EW William Furlong, Ballycoheir, New Ross, Co. Wexford
EI9 EX Eugene A. Hanley, "Cronomore", Durtaheen Road, Fermoy, Co. Cork
EI9 EY W.C. Foley, Coolgrange, Granges Road, Kilkenny
EI9 EZ Patrick Geoghegan, 24 Shanbally, Cappoquin, Co. Waterford
EI9 FA Details withheld at licensee's request
EI9 FB Joseph Bannon, Carrigeenroe, Boyle, Co. Roscommon
EI9 FC Frank Cullen, 28 Gosworth Park, Dalkey, Co. Dublin
EI9 FD Edward Navagh, Knochharley, Brownstown, Navan, Co. Meath
EI9 FE Mike Hoare, "Glencoe", Ballykisteen, Tipperary
EI9 FF Thady Kennedy, 45 Erne Dale Heights, Ballyshannon, Co. Donegal
EI9 FG Henry J. Cassidy, 12 Mahon Crescent, Blackrock, Cork
EI9 FH Philip McGettigan, Gortmacall, Milford, Co. Donegal
EI9 FI Rev. H.L.F. Bolster, 24 Limavady Road, Londonderry BT47 1GD
EI9 FJ Paul Sumner, 10 Shanrath Road, Santry, Dublin 9
EI9 FK William Somerville-Large, 'Vallombrosa', Thornhill Road, Bray, Co. Wicklow
EI9 FL George C. Sheehan, 6 Whitepoint Drive, Cobh, Co. Cork
EI9 FM Noel Lafferty, Carrickshandrum, Killygordon, Lifford, Co. Donegal
EI9 FN Percy Masters, Knockatee, Dunmore, Co. Galway
EI9 FO Thomas Devine, c/o D. Doherty, Coolatree, Ballindrait, Lifford, Co. Donegal
EI9 FP Thomas P. Gaffney, 14 Maretimo Gardens East, Blackrock, Co. Dublin
EI9 FQ James Fitzgerald, 21 St. Aidans Avenue, Darnen, Lancs. BB3 2BS, England
EI9 FR Brian J. Lamb, 179 Ryevale Lawns, Leixlip, Co. Kildare
EI9 FS J.R. Cole, 7886 Sarahurst Drive, Dublin, OH 43017, USA
EI9 FT James F. Buckley, 'Avoca', Sleveen Park, Kinsale, Co. Cork
EI9 FV Gerry A. Lawlor
EI9 FW Tina Keil, Newbawn, Rathdrum, Co. Wicklow
EI9 FX Mark O'Rourke, Derrycammagh, Castlebellingham, Dundalk, Co. Louth
EI9 FY Patrick Devine, Begrath, Collon, Drogheda, Co. Louth
EI9 FZ Albert Millar, Braganstown, Castlebellingham, Dundalk, Co. Louth
EI9 GA Declan Goggin, Desert, Clonakilty, Co. Cork
EI9 GB John Doherty, 3 Rockfield Terrace, Buncrana, Co. Donegal
EI9 GC Comdt. James Roche, McKee Barracks, Dublin 7
EI9 GE Patrick J. Kane, 11 Ashlawn Court, Upper Dargle Road, Bray, Co. Wicklow
EI9 GF Kevin J. Boyd, 40 Killyman Street, Moy, Dungannon, Co. Tyrone BT1 7SJ, Northern Ireland
EI9 GG William Walsh, 6 Dolphin Terrace, Crosshaven, Co. Cork
EI9 GH Jacqueline A. Bluett, Convent Road, Clonakilty, Co. Cork
EI9 GI Christopher Costello, Murphy Barracks, Ballincollig, Co. Cork
EI9 GJ Wexford VHF Group Radio Club, c/o William McLoughlin, Spencerstown, Murrintown, Co. Wexford
EI9 GK Michael Kirwan, 15 Ashbrook Grove, Ennis Road, Limerick
EI9 GL Paul Healy, 20 Blackheath Drive, Clontarf, Dublin 3
EI9 GM Kieran McElhinney, c/o G. Houston, 4 Albert Place, Lifford, Co. Donegal
EI9 GN Harry Cheetham, 14 Castleton Avenue, Carlton, Nottingham NG4 3NZ
EI9 GO Eamonn G. Phelan, 14 Ursuline Crescent, Waterford
EI9 GQ Eamon Shelton, 10 Belgrave Place, Wellington Road, Cork
EI9 GR Hugh Cumming, 55 Barrmill Road, Mansewood, Glasgow, Scotland
EI9 GS James A. Mackessy, Caherdale, Knockaderry, Newcastle West, Co. Limerick

EI9 GT Peter J. McNally, 'Lorna', 57 Holmpatrick, Skerries, Co. Dublin
EI9 GU Sean Brennan, Mylerspark, New Ross, Co. Wexford
EI9 GV Seamus F. Holland, 21 Valentia Road, Drumcondra, Dublin 9
EI9 GW Michael J. Flynn, Cullentragh, Mayo Abbey, Claremorris, Co. Mayo
EI9 GX Jesse G. Lawrence, Port Roe, Nenagh, Co. Tipperary
EI9 GY Patricia Mangan, "Mount Ievers", Sixmilebridge, Co. Clare
EI9 GZ Gerald V. Gavin, 14 Parslicktown Avenue, Mulhuddart, Dublin 15
EI9 HC Stephen Nolan, 1 Millfield Lawn, Lower Dublin Hill, Cork
EI9 HD James McGrory, 69 Woodbine Park, Raheny, Dublin 5
EI9 HE John M. Cotter, 34 Ennafort Park, Raheny, Dublin 5
EI9 HF Donal Mahon, 30 McDara Road, Shantalla, Galway
EI9 HG Joe Ryan, Kylebeg, Newtown, Nenagh, Co. Tipperary
EI9 HH Marie Wiseman, 19 Foxbush, Hildenborough, Kent TN11 9HT, England
EI9 HI Vincent O'Connor, 63 Ardilaun Road, Newcastle, Galway
EI9 HJ Jeremiah J. Kelleher, 1 Glencairn Park, Rossa Avenue, Bishopstown, Cork
EI9 HK Tony Clifford, 'Bofeenaun', Brookfield, Rochestown Road, Cork
EI9 HL Mark McNulty, Town Parks, Raphoe, Co. Donegal
EI9 HM Finbarr Carroll, Ballybeg, Dingle, Co. Kerry
EI9 HN John B Kelly, 'Tara', 32 Blean, Athenry, Co. Galway
EI9 HO Joseph P Clarke, 61 Whitebridge Manor, Killarney, Co. Kerry
EI9 HP William P O'Neill, 120 Route 208, New Paltz, New York 12515, USA
EI9 HQ Declan Lennon, 45 Pearse Park, Sallynoggin, Dun Laoghaire, Co. Dublin
EI9 HR Richard Ryan, 22 Convent Hill, Waterford
EI9 HS Michael Farrell, Sunnyside House, Castle Street, Roscommon
EI9 HT Trevor L. Grant, The School House, Church Road, Wittering, Cambs, PE8 6AF, England
EI9 HU Brian McCan, 17 Castleowen, Blarney, Co. Cork
EI9 HV Maurice Keating, 20 Kincora Avenue, Clontarf, Dublin 3
EI9 HW John Fitzgerald, Killua, Clonmellon, Navan, Co. Meath
EI9 HX Patrick G. O'Connor, Togher, Ballinasloe, Co. Roscommon
EI9 HY Brian P. Lynch, 47 Roselawn, Ballydowd, Lucan, Co. Dublin
EI9 HZ Eoin O'Cleary
EI9 IA Bruce T. Marshall, 52 Cornell Street, Roslindale, MA 02131, USA
EI9 IB Mark J Boothmann, Norton Place, Ballysax, The Curragh, Co. Kildare
EI9 IC Anthony D. McDermott, 262 Crodaun Forest Park, Celbridge, Co. Kildare
EI9 ID Anthony Stack, 33 Park Court, Ballyvolane, Cork
EI9 IE Niall McLoughlin, 5 Sente des jardins du couchant, 27610 Romilly sur Andelle, France
EI9 IF Alan Dean, 15 Delaford Avenue, Knocklyon, Dublin 16
EI9 IG Sean Doran, 11 Hillbrook Estate, Tullow, Co. Carlow
EI9 IH Liam Carlos, Ballindrimley, Castlerea, Co. Roscommon
EI9 II Graham O'Sullivan, 20 Kingsford Park, Douglas, Cork
EI9 IJ Frank J. Minto, 26 Edenmore Grove, Raheny, Dublin 5
EI9 IK Dominic J. Nolan, 7 Glenfield Close, Clondalkin, Dublin 22
EI9 IL Anthony Liddy, 26 Sarsfield Avenue, Garryowen, Limerick
EI9 IM Derry Lawlor, 29 Grove Park Avenue, Finglas, Dublin 11
EI9 AAB Francis J. Gordon, 4 Henley Villas, Churchtown, Dublin 14
EI9 ADB Gerard Connolly, Patrick Street, Tullamore, Co. Offaly
EI9 AEB Anthony T. Cullen, 188 Glenview Park, Tallaght, Dublin 24
EI9 AHB William Dumpleton, The Lodge, Dunsink Observatory, Finglas, Dublin 11
EI9 AIB Adrian Swanton, Gould's Hill, Mallow, Co. Cork
EI9 AKB James Travers, 59 Mondalea Woods, Firhouse Road, Dublin 16
EI9 ALB Details withheld at licensee's request
EI9 AMB Stephen Hanlon, 13 Fortfield Park, Mountview, Blanchardstown, Dublin 15
EI9 AQB Patrick J. O'Loughlin, Recess, Connemara, Co. Galway
EI9 AWB Details withheld at licensee's request
EI9 BAB Keith Garland, 32 New Park Road, Blackrock, Co. Dublin
EI9 BDB David T. Deane, 12 Chelmsford Road, Ranelagh, Dublin 6
EI9 BEB Patrick J. Costelloe, Edenvale, Ennis, Co. Clare
EI9 BFB John Healy, 6 Westerton Rise, Ballinteer Road, Ballinteer, Dublin 16
EI9 BGB Robert Bolton, 29 Ballinaspig Lawn, Bishopstown, Co. Cork
EI9 BHB Brian T. Cullen, 47 Rice Park, Waterford
EI9 BIB John C. Roche, 56 Ballybeg Court, Waterford
EI9 BJB James Bustard, Clarcam, Donegal
EI9 BKB John E. Lamb, 179 Ryevale Lawns, Leixlip, Co. Kildare
EI9 BNB John B. Kelly, Buttersland, New Ross, Co. Wexford
EI9 BOB Chris Nagle, "Leenane", Marian Park, Pouladuff Road, Cork
EI9 BPB John J. McBride, 35 Main Street, Newtownstewart, Co. Tyrone
EI9 BRB Alan Keane, Turnings, Straffan, Co. Kildare
EI9 BSB Dermot V. O'Neill, Eadiestown, Naas, Co. Kildare
EI9 BTB Thomas G. Griffin, 82 Orchardstown Drive, Templeogue, Dublin 14
EI9 BWB John L. Reilly, "Terra Nova", 56 Curraheen Road, Bishopstown, Cork
EI9 BXB Louis O'Toole, 2 Seapark, Malahide, Co. Dublin
EI9 BYB Noel Fay, 27 Ratoath Drive, Finglas West, Dublin 11
EI9 BZB Alexander Palatianos, 39 Glen Lawn Drive, The Park, Cabinteely, Dublin 18
EI9 CBB John F. Marron, Briarless, Julianstown, Co. Meath
EI9 CCB John P. McCarthy, 45 Sunday's Well Road, Cork
EI9 CEB Tony M.D. Proudfoot, 6 Larch Grove, Clonsilla, Dublin 15
EI9 CHB Geoffrey Curtis-Smith, 11 Mersea Road, Colchester, England
EI9 CKB Jan Ritsma, "Kaladan", Hazelthatch Road, Celbridge, Co. Kildare
EI9 CMB David R. Coleman, Drutamon, Canningstown, Cootehill, Co. Cavan
EI9 CNB Francis R. Small, Kilkenny Abbey, The Pigeons, Athlone, Co. Westmeath
EI9 COB Padraig J. Crowley, No. 31 Platoon, Devoy Barracks, Naas, Co. Kildare
EI9 CPB John V. Gardner, 44 Munster Street, Phibsborough, Dublin 7
EI9 CQB Donal G. Terry, 36 Sunday's Well Road, Cork
EI9 CRB Martin J. Whyte, 66 Rafters Road, Drimnagh, Dublin 12
EI9 CSB Paul Thim, Paskelijegatan 3, 216 27 Malmo, Sweden
EI9 CTB Thomas J. Fahy, "San Michele", Porterstown Road, Dublin 15
EI9 CUB Patrick McCabe, Carrickacromin, Mountain Lodge, Cootehill, Co. Cavan
EI9 CXB Vincent P. Gallagher, 17 Avoca Avenue, Bray, Co. Wicklow
EI9 CZB Christopher Mann, Woodford, Listowel, Co. Kerry
EI9 DBB John A. McCarthy, 24 Sunrise Crescent, Waterford
EI9 DDB Derek O'Hanlon, 3 Cherry Court, Granstown Village, Waterford
EI9 DFB Gerard Barron, Kildermody, Kilmeaden, Co. Waterford
EI9 DGB Alan Adams, Ballinattin, Tramore, Co. Waterford
EI9 DHB Kieran Howley, 7 Corcoran Terrace, Kells Road, Kilkenny
EI9 DIB Nicholas M. Madigan, 176 Hennessy's Road, Waterford
EI9 DJB Edward F. Steadman, 21 Pleasant Avenue, Mount Pleasant, Waterford
EI9 DLB Luke Mooney, "Knocklaun", 5 Bernadette Place, Western Road, Cork
EI9 DNB Martin Fitzgerald, 31 Mounlaun, Tramore, Co. Waterford
EI9 DPB David McCabe, 26 North Summer Street, North Circular Road, Dublin 1
EI9 DQB Patrick Crawley, Townparks, Ardee, Co. Louth
EI9 DSB Thomas F Crawley
EI9 DUB Eamonn Gibson, 64 Wellington Drive, Templeogue, Dublin 6W
EI9 DVB Liam Brady, "Stonehaven", Eaton Brae, Shankill, Co. Dublin
EI9 DYB Alan M. Brown, 9 Casana View, Howth, Co. Dublin
EI9 DZB Stephen J. Curran, 70 Gilford Road, Sandymount, Dublin 4
EI9 EAB Tony Kenny, 130 Ard O'Donnell, Letterkenny, Co. Donegal
EI9 EEB Martin McElwee, Dromfad, Kerrykell, Letterkenny, Co. Donegal
EI9 EFB John F. Sinclair, Ardeskin, Donegal Town
EI9 EGB John J. Nee, Carrygawley, Letterkenny, Co. Donegal
EI9 EHB Kevin Muldoon, 33 Ardeskin, Donegal Town
EI9 EIB Roy Tallon, 1 Blackwater Abbey, Navan, Co. Meath
EI9 EJB Leslie Ferguson, 101 Meadowbrook, Mill Road, Corbally, Co. Limerick
EI9 EKB Conor Keegan, Rathfern House, Windgates Lower, Greystones, Co. Wicklow
EI9 ELB Hazel Leahy, 'Galtymore', Bansha, Barnlough, Co. Tipperary
EI9 EMB Details withheld at licensee's request
EI9 ENB Michael J. Shevlin, Bomany, Letterkenny, Co. Donegal
EI9 EOB Joseph Gallagher, Chapel Road, Dungloe, Co. Donegal
EI9 EPB Michael Considine, Ard Muire, Fr. Russell Road, Ballykeefe, Co. Limerick
EI9 EQB Bruno Nardone, 187 Delwood Park, Castleknock, Dublin 15
EI9 ERB Brian Canning, Aughnaglace House, Cloone P.O., Co. Leitrim
EI9 ESB John J. Costello, "April Rise", Gurrane West, Killorglin, Co. Kerry
EI9 ETB Johnathan Kilpatrick, Carrickbrack, Convoy, Lifford, Co. Donegal
EI9 EUB David Gilmartin, 171 Charlemont, Griffith Avenue, Dublin 9
EI9 EVB John Paul Oxley, Wolfe Tone Street, Mountmellick, Co. Laois
EI9 EXB David Eames, 40 The Chantries, Balrothery, Balbriggan, Co. Dublin
EI9 EYB Revere Richardson, Barronsland, Thomastown, Co. Kilkenny
EI9 EZB Thomas Hallinan, P.O. Box 20, Cahir, Co. Tipperary
EI9 GRC St. Joseph's College Radio Club, Garbally College, Ballinasloe, Co. Galway
EI9 RMR Royal Meath Radio Club, c/o Phil Cantwell EI9P, 'Villa Maria', Manorland, Trim, Co. Meath

---

UK Surname Index

A

Aanestad, T ... G0LVA
Aaron, K G ... G4KCI
Abberley, F C ... GM4ZFG
Abberstein, K A ... G0LZM
Abbey, A F ... G3OVH
Abbey, G S ... G1YWU
Abbey, J R ... G7KLN
Abbishaw, J B ... G6CQH
Abbot, A M ... G4NPA
Abbot, S D ... G4NPB
Abbott, A D ... G6GBL
Abbott, A J ... G1GOP
Abbott, A L ... G6PXA
Abbott, C W ... G4YJV
Abbott, D G ... G4MPT
Abbott, D L ... G4RFU
Abbott, G ... G7DLI
Abbott, H N E ... G7LGY
Abbott, J B ... G8JPW
Abbott, J P ... G8CWJ
Abbott, J R ... G4PBA
Abbott, M ... G6SGQ
Abbott, M G ... G7WIR
Abbott, N ... G6DSQ
Abbott, P C ... G0FUI
Abbott, R A ... M0AGH
Abbott, R J ... G7RCL
Abbott, S ... G1GOQ
Abbotts, G J ... 2E1FVL
Abbruscato, J V ... G0AOH
Abebe, B A ... G7VXJ
Abel, A ... GM4GYR
Abel, I ... G3ZHI
Abel, R M ... G4FKX
Abel, S J ... G7ETC
Abela, C ... G0ATP
Abell, B ... G1SKV
Abercrombie, A M ... GM4XMH
Abernethy, P W ... G8HGG
Abeynayake, K ... G7RFS
Abraham, D A ... GM4UTC
Abraham, M J ... G1DDK
Abrahams, A I ... G7VZP
Abrahams, A M ... G8UOJ
Abrahams, L M ... G6GEG
Abrahams, S J ... G1ESA
Abram, J ... G1PJH
Abram, J N ... G6CZZ
Abram, M P ... G1CZU
Abram, W P ... GI6KJC
Abramczyk, R S ... GW7CKR
Abrams, B R ... G0BII
Abrey, C E ... G3RZY
Abson, T ... G6MLS
Aburrow, M B ... G3KWE
Aburrow, P J ... G4BQY
Ace, P H ... GW4SPL
Aceves Ii, W E ... G0TWY
Acheson, B A ... G1JFQ
Acke, P R ... G3FYF
Ackerley, G D ... G3VUN
Ackerley, J F ... G8TKQ
Ackerley, N ... G3RIR
Ackerman, B M ... G0GNP
Ackland, N W J A ... G0IIK
Ackley, P N ... G3LRP
Ackrill, D J ... G0DJA
Ackroyd, A D ... G0LGD
Ackroyd, A F ... G4UJX
Ackroyd, B ... G8GOV
Ackroyd, B E ... G8ZVK
Ackroyd, F ... G6SYV
Ackroyd, R ... G4SYV
Ackroyd, S B ... G1GOR
Acomb, H ... G4CCV
Acott, J M ... G4ILH
Acott, L F ... G1SEM
Acquier, J F ... G3EQJ
Acres, B D ... G4MXQ
Acton, E R ... G6CYX
Acton, J D ... G0NFH
Acton, M ... G3MBJ
Acton, M J ... G4STL
Adair, D R J ... G3BVB
Adair, T M ... GD4YON
Adalian, D Y ... G2ACG
Adam, A S ... GM0VFD
Adam, A W ... GM3DAP
Adam, B S ... GM0HBM
Adam, N J ... GM8LEA
Adam, P G ... G7KXS
Adam, R A ... G7JAQ
Adam, T ... GM0NBA
Adam, W L ... G1TJT
Adams, A ... G1EGZ
Adams, A E ... G0TKI
Adams, A J ... G3DXQ
Adams, A L ... G7OPB
Adams, A P ... GW0KZG
Adams, A R ... G3YOA
Adams, B N C ... G4RFV
Adams, B W ... G0CPZ
Adams, C ... G3URL
Adams, C A G ... G1RAM
Adams, C F ... G4KZG
Adams, C J ... 2E1FGX
Adams, C J ... G0YCA
Adams, C J ... G3YNC
Adams, D A ... G3JUU
Adams, D A ... G4YLK
Adams, D A ... G4ZEW
Adams, D A J ... G8IJG
Adams, D B ... G0GIE
Adams, D C ... G8KTV
Adams, D H ... GW3VBP
Adams, D M ... G4NWA
Adams, F J ... G8YTU
Adams, G H ... G3ICA
Adams, G L ... G1LEQ
Adams, G L ... G3LEQ
Adams, G L ... G4LEQ
Adams, G P ... GM8FVN
Adams, H W ... G6PYR
Adams, I W ... GW4EJG
Adams, J A ... GI7HYU
Adams, J H ... G3VZF
Adams, J J ... G4JZL
Adams, J L ... GW6LZH
Adams, J M ... G6AFK
Adams, J M ... M1AWG
Adams, J P ... GM1JHU
Adams, J R ... G0OFS
Adams, J R ... G0WSY
Adams, J R ... G8MZG
Adams, J S ... G3TPZ
Adams, J S ... GM1BWV
Adams, K ... G4JIH
Adams, K J ... G7OAH
Adams, K W ... G0VDK
Adams, L E ... G4RKV
Adams, M ... G1ICH
Adams, M ... GU7OMJ
Adams, M D ... G0AMO
Adams, M D ... G0ARC
Adams, M D ... G0RBL
Adams, M D ... G4RBS
Adams, M E ... G0PPZ
Adams, M E ... G4GAJ
Adams, M J ... G3ZLQ
Adams, M J ... G4LOF
Adams, M P ... G4IYA
Adams, M P ... G6ZLJ
Adams, M S ... M0ALW
Adams, N ... G1ZUG
Adams, P D ... G6LZB
Adams, P D ... G7EGU
Adams, P E ... G0HWY
Adams, P E R ... G0NOV
Adams, P J ... G1YLE
Adams, P J ... G4XKA
Adams, P K ... G7CHJ
Adams, P W ... G3UKE
Adams, R ... G8JWI
Adams, R A ... G6NYG
Adams, R B ... G8BYP
Adams, R F ... G0JLE
Adams, R J ... G0UMK
Adams, R J ... G4LRM
Adams, R J ... G8KOD
Adams, R N ... G8NYH
Adams, R R ... G3CDJ
Adams, S ... G8MZW
Adams, S A ... G0FRV
Adams, S J ... G0KVZ
Adams, S M ... G6VSC
Adams, S P ... G0WCF
Adams, S R ... G0ULF
Adams, S T ... G0OMM
Adams, T C ... G1ZSK
Adams, T E ... G4CHD
Adamson, A T ... G4SPV
Adamson, G H ... GM0RDA
Adamson, K ... GI6BPF
Adamson, M ... G4UTQ
Adamson, R M ... GW6KWU
Adamson, W ... GI4CPP
Adcock, G G M ... G4EUK
Adcock, H W ... G4STX
Adcock, J A ... 2W1DQI
Adcock, J L ... G0FWD
Adcock, M J ... GW8CMU
Adcock, M V ... G8DJO
Adcock, P M ... GW1JNR
Adderley, A G ... G1JCH
Adderley, M ... G4OAW
Addey, S ... G4DUK
Addicott, M J ... G7SSA
Addidle, V M ... GI3VHM
Addis, B ... M0ART
Addis, C J ... G6DSP
Addis, G R J ... G3TEB
Addison, C S ... MM1BUE
Addison, J ... G4PKD
Addison, J L ... G6YOZ
Addison, M J ... G6MJA
Addison, M P ... G7SVZ
Addison, M R B ... G4TQY
Addison, P J M ... G7UVZ
Addison-Lees, C J ... G4OHV
Addlesee, M D ... G1TFK
Addy, J ... G3YYP
Addy, R D K ... GI0FCC
Addy, W T ... G3KRX
Adey, A G ... G7BCO
Adey, A J ... G0HDD
Adey, T F ... G3TLF
Adie, D S ... G8ZMI
Adie, W L ... GM7LDU
Adkin, F W ... GM3LAU
Adkins, A J W ... G3MVU
Adkins, G ... G1IQA
Adlam, M J ... G7DZI
Adlard, M P ... G7FBD
Adlem, S N ... G7RZU
Admans, G W ... 2E1FZN
Adrain, S J ... GI4SQL
Adshead, J E ... G4UJA
Adshead, P ... G7OCP
Aedy, A E ... G4GMT
Affleck, C K ... 2E1EDP
Affleck, L J ... G4MUD
Affolter, A R ... G0CKH
Afford, A N ... G6AFG
Afford, B D ... G0AGM
Afford, B H ... G6AKF
Afford, L ... G4SHY
Agacy, R N ... G1EQM
Agar, D W ... G0EBW
Agar, J ... G4KKQ
Agar, J E ... G8AZA
Agass, C A ... G7SQX
Agass, J M ... G6ZCH
Ager, D ... G7RUY
Ager, J G ... G1VRJ
Ager, T R A ... G4UXJ
Aggus, R B ... G4CZZ
Agness, G D ... G4ODC
Agnew, J T ... GI6IRL
Agnew, R E ... G1ZNX
Agnew, W H ... G4PZD
Agombar, K C ... G8BOJ
Aguilar, E C ... G4KBJ
Aherne, T ... G6UGT
Ahmed, M ... G0RNH
Aigeldinger, H P ... G0PTI
Aiken, D J W ... G4CBO
Aiken, J C ... GM0AZU
Ailsby, E J ... G4MZL
Aindow, J D ... G4GUV
Ainge, R C ... G4MRA
Ainge, R W ... G6GA
Ainger, A W S ... G1ZYJ
Ainley, C M ... G7JWY
Ainley, M E ... G8XSF
Ainley, R D ... G7WIZ
Ainley-Smith, B ... G1EHA
Ainscough, D ... G1OMY
Ainslie, D W ... G6FXR
Ainsworth, A W ... G0AXF
Ainsworth, B ... G3WOR
Ainsworth, B D ... G4GPW
Ainsworth, J ... G7USB
Ainsworth, P ... G6GEN
Ainsworth, R G ... G4UPU
Ainsworth, R H ... G1JMP
Ainsworth, S ... G0HTP
Ainsworth, S J ... 2E1DSM
Aird, A M ... GM0UGG
Aird, D W ... G3MFE
Airey, R ... G1NDV
Airs, D P ... G7VIB
Airs, M D ... G7VIA
Aisher, J H ... G4YPA
Aisthorpe-Buckley, P G ... G4TMF
Aitchison, C ... G4JDG
Aitchison, D ... G3BSA
Aitchison, D C ... 2E1CMS
Aitchison, M R ... G4YFK
Aitchison, P J ... G3LSQ
Aitchison, W W ... G6CZX
Aithison, J H ... G0JFC
Aitken, B A L ... GM0CBA
Aitken, J D W ... G3UAC
Aitken, J R ... GM8NFG
Aitken, J T ... GM0HZM
Aitken, M T ... G2ACK
Aitken, R ... G4VCT
Aitken, R ... GM4SUR
Aitken, R I ... GM1JVU
Aitken, S ... GM0IOA
Aitkenhead, A F ... GM0PGV
Aitkenhead, D A ... GM4BHU
Aitkenhead, R J ... GM0STB
Aitkenhead, R J ... GM4UQG
Aizlewood, I Y ... G4ZGJ
Aizlewood, J D ... G4WZV
Aizlewood, J R ... G0DLT
Akass, P C ... G3DJY
Akehurst, B W ... G0OPF
Akehurst, M ... G3MIQ
Akerman, D A W ... G4YDA
Akers, T F L ... GM6HWZ
Akester, D H ... G8UPK
Akhurst, W ... G1HDK
Akiki, M E ... G7LBP
Akines, J C ... G8XXI
Akines, W J ... G4PCH
Akrill, A ... G1GGJ
Akse, G H ... G0UVR
Alban, G P ... G0GPA
Alban, R F C ... GW3SPA
Albee, T K ... G0UTW
Albers, B A ... G0KGT
Albers, F R ... G4YGN
Albinson, A R ... M1AON
Alblas, J G J ... G4XNL
Albon, K G ... G0CVB
Albon, M R ... G7ULT
Albone, G J ... G7OME
Albrighton, G P ... G4ZUR
Albury, D J ... G7OZH
Albutt, B A ... M1CCA
Alcock, A D ... G6SBW
Alcock, G ... 2E1EJM
Alcock, G H ... G8RXY
Alcock, J L ... G4CJM
Alcock, J R ... G8XQL
Alcock, N ... G4SPU
Alcock, R A ... G6DSJ
Alcock, R C ... G6DKY
Aldcroft, A ... G8VQC
Aldcroft, W N ... G1VGC
Alden, L H ... G7VML
Alder, J F ... G4GMZ
Alder, M J ... G4UTR
Alder, N G ... G3DYO
Alder, S ... G0HTS
Alder, S ... G0PWK
Alderman, A E ... G4LQD
Alderman, G W ... G3BNE
Alderman, G W ... G3JXA
Alderman, J B ... G2HV
Alderman, J B ... G4JBA
Alderman, W J ... G7PHH
Aldersey, B J ... G0DPE
Aldersey, J ... G1AQI
Alderson, B ... G0JQA
Alderson, B ... G3KJX
Alderson, F D ... G0LRU
Alderson, H B ... G3BKJ
Alderson, K C ... G0LJN
Alderson, R H ... G4ZQC
Alderton, A F ... G7GUB
Alderton, I R J ... G6IZA
Alderton, R P ... G4OQK
Aldhous, L R G ... G4UYF
Aldous, R C ... G8CBU
Aldred, G W ... G4YGD
Aldred, J R ... G1UEO
Aldred, T L ... G7VVM
Aldridge, A E G ... G3PJQ
Aldridge, A W ... G0CIG
Aldridge, D J ... G3VGR
Aldridge, D J ... G6FOV
Aldridge, E V ... G0FJI
Aldridge, F L ... G4FLA
Aldridge, I C ... G4AJU
Aldridge, T ... G4DAJ
Aldridge, T J ... G4GJR
Aldus, K F C ... G0TAH
Aldworth, E W B ... G3FHN
Alecio, A ... G0RKC
Aleemayehu, B ... G7VXG
Alefs, C R ... G1MPG
Alemayehu, T ... G7VXE
Alesbury, D E ... G3HSV
Alexander, C I ... G0TID
Alexander, D G ... G3KLH
Alexander, D R R ... G4GVM
Alexander, E ... G4NLN
Alexander, G ... G0JGE
Alexander, G ... G0OSD
Alexander, I M ... GM7OTT
Alexander, I R ... G4AKD
Alexander, J ... GM0KWW
Alexander, J A ... GM7OJJ
Alexander, J L ... GW6CJJ
Alexander, J M ... G1HEJ
Alexander, J S ... MW1BAJ
Alexander, K ... G8BVT
Alexander, L R ... GM0LVK
Alexander, L R ... GM0MGV
Alexander, L R ... GM4TRC
Alexander, M J ... G4VMN
Alexander, M J ... GM7IHJ
Alexander, P ... G1OSK
Alexander, P R G ... GW4RXO
Alexander, R ... G6UGO
Alexander, R A ... G4YED
Alexander, R C A ... GM0DEQ
Alexander, R L ... G4ZPB
Alexander, R S ... G4WJN
Alexander, S E ... G3TLY
Alexander, S M ... G7ENT
Alexander, T J ... G1AKV
Alexander-Pye, D C ... G8XXL
Alexandre, A C ... GJ4YBM
Alexandrou, A ... G4UKS
Alexandrou, V M L ... G6SRQ
Aley, R J ... 2E1AXZ
Aley, R J ... G0VOH
Alford, C G J ... G6SRU
Alford, J C ... G4DOE
Alford, J L ... G4NBW
Alford, K D ... G1YWF
Alger, J K S ... G7NGB
Alison, N ... G0RPL
Allack, K S G ... G0WNE
Allaker, P ... 2E1EPA
Allan, A G ... G3BEH
Allan, A M ... G4BRN
Allan, C F ... GM7JMO
Allan, D W ... G0RZP
Allan, D W ... G3WYP
Allan, D W ... G8WYP
Allan, E J ... G7RGO
Allan, G ... GM4TPE
Allan, G ... GM6FDQ
Allan, G ... GM8SNB
Allan, G M ... GM4HYF
Allan, G S ... G7DFX
Allan, G T ... GM4SBP
Allan, G W D ... G0BGA
Allan, H ... G1RSP
Allan, J ... G3IJA
Allan, J B ... G4LTH
Allan, J L ... GM6UWF
Allan, J M ... GM3JVX
Allan, P ... G1NPA
Allan, P ... G3OJA
Allan, P ... G4NTA
Allan, P D ... G4RYO
Allan, P M ... G1NBP
Allan, R A M ... MM1BYY
Allan, R D ... G3TQZ
Allan, S ... GI3LFH
Allan, W ... GM1XIN
Allan, W L ... G4ODS
Allanson, P M ... G0CEW
Allanson, S ... G0TYS
Allard, M E ... G3WFC
Allardyce, D R ... G4YWL
Allardyce, J W ... GM6PKP
Allart, J F ... G8SVR
Allaway, E J ... G3FKM
Allbright, R J S ... G3RCE
Allbright, R V A ... G2JL
Allbutt, S J ... G3WXU
Allchin, A B ... G4JGH
Allchin, J W ... G8XXJ
Allchin, N R ... G7PGZ
Allcock, A J ... G8ZSK
Allcock, A R ... G4XDN
Allcock, G A ... G3ION
Allcock, P J ... G1EHB
Allcock, R ... G7JWO
Allcroft, J H ... G0TMN
Alldred, J P ... G0SVL
Alldrick, D F ... G3MZP
Alldridge, D A ... G6LKS
Alldridge, D F ... G4UXA
Alldridge, L A ... G6ZWG
Allely, E J ... GW0PZT
Allely, P E W ... GW3KJW
Allen, A ... G3GBU
Allen, A ... G4XEJ
Allen, A C ... G4HRH
Allen, A D ... G3BNU
Allen, A J ... G7JOG
Allen, A J ... GI0UZC
Allen, A R B ... G7IEX
Allen, B ... 2E1CDS
Allen, B ... G7GKA
Allen, B R ... G7BVZ
Allen, B W ... G7UIL
Allen, C ... G4XFL
Allen, C J ... G4SYA
Allen, C L ... G8RBI
Allen, D ... G0BWE
Allen, D ... G3SDT
Allen, D ... G6YAY
Allen, D ... G7OWX
Allen, D ... G8BZT
Allen, D A ... G6FDS
Allen, D A J ... G8LHD
Allen, D E ... G6PCJ
Allen, D E ... M0APK
Allen, D G ... G4CNZ
Allen, D J ... G6TLN
Allen, D J ... GI8FQB
Allen, D J ... GI8RQI
Allen, D M ... GW7LZO
Allen, D W ... G7TPM
Allen, E G ... G3DRN
Allen, E J ... GW8CQK
Allen, E W G ... G3JHP
Allen, G ... G0OCC
Allen, G ... G6SCZ
Allen, G A ... GI4OZJ
Allen, G K ... G3HST
Allen, G L A ... G0KNX
Allen, G T ... G3JTK
Allen, G W ... G3IUO
Allen, G W ... G8LZG
Allen, H ... G4TLW
Allen, I ... G6VGH
Allen, J ... G4PDP
Allen, J ... G4WDR
Allen, J ... G6PKM
Allen, J ... G7NUX
Allen, J A ... G4VHX
Allen, J C ... G0CPG
Allen, J C W ... G0SGV
Allen, J M ... GW3TUD
Allen, J P ... G6EBG
Allen, J R ... G0NIE
Allen, J R ... G6YPE
Allen, J S ... G1PVU
Allen, J S ... G3DOT
Allen, J W G ... G7TTD
Allen, J W G ... M0AOB
Allen, K B ... G7IZF
Allen, K G ... G6CJK
Allen, K G A ... GI4RSI
Allen, L ... G0EIB
Allen, L ... G3AAH
Allen, L F L ... G3CJD
Allen, M ... M1BVA
Allen, M A ... G1IPP
Allen, M F ... GW0RBZ
Allen, M G ... G8IFU
Allen, M H ... G6RBR
Allen, M R ... G1ZKK
Allen, N B ... G4MIS
Allen, N J ... G4IHR
Allen, N T ... G0RCN
Allen, N T ... G4VVF
Allen, P ... G1RJE
Allen, P A ... M0AGM
Allen, P F ... G3USH
Allen, P H ... G0IAP
Allen, P J ... G1DYT
Allen, P M ... G1GVJ
Allen, R ... G2DSP
Allen, R ... G4VLI
Allen, R ... G4WOI
Allen, R ... M1BOB
Allen, R C ... G6JTV
Allen, R F ... G3NVL
Allen, R K ... G0NBE
Allen, R M ... G3JKK
Allen, R N ... G1IPO
Allen, S C ... G1SCA
Allen, S R ... G4CYR
Allen, S R ... G7LRU
Allen, S W ... GM0FHE
Allen, T K ... G7HKZ
Allen, T M ... G4ETU
Allen, V A ... GW1ZXN
Allen, V P ... G1HRD
Allen, W ... G8JCN
Allen, W K ... G3ZVD
Allen F.B.S, C W ... G6JYO
Allenby, F R ... M1AEQ
Allender, D E ... G8DEA
Allenet, R F ... GJ3XZE
Allenson, M ... G3TGD
Allerston, D H ... G5PQ
Allerton, J G ... G4PDW
Allerton Austin, M R D ... G3ZCJ
Allgood, I F ... G1NOO
Allgood, P L ... G8RXZ
Allgood, R ... G0KSD
Allgood, S A ... 2E1AQH
Allibone, D J ... G0SZT
Allies, K ... G0LDU
Alliker, D G ... G6WNA
Allin, G E ... G4ZUC
Allin, J S ... G3VLN
Allin, M D ... G4HWI
Allingham, T E ... MI1AVH
Allington, B ... 2E1EXE
Allington, M ... GW1AKT
Allinson, J R ... G0UUE
Allinson, N A ... G8ILB
Allis, G V ... G0LRS
Allisett, W J ... GU3NDX
Allisette, M H ... GU4EON
Allisette, R W ... GU4CHY
Allison, C ... G4YIQ
Allison, D N ... G7BGM
Allison, D S ... G3IZA
Allison, E J E ... G4JNQ
Allison, E W ... G7TMD
Allison, G ... G0CGZ
Allison, H J ... G3XSE
Allison, J B ... G4URW
Allison, J H G ... G0LYY
Allison, M ... G4JJM
Allison, M A ... G6UGS
Allison, R ... GD4FJI
Allison, T J ... G0TYM
Allison, V ... G3TNX
Allman, B J ... G6HLL
Allman, D ... G0FYC
Allman, M ... G4TWV
Allnutt, A P ... G6DTH
Allport, A R ... G7UCN
Allport, B ... G1RKD
Allpress, R L ... M1BZY
Allsebrook, D ... G1VAC
Allsey, W ... M1ALZ
Allsop, G ... G4ZMC
Allsop, G P ... G4PYM
Allsop, J H G ... G3OGX
Allsopp, D F ... G0DFA
Allsopp, D J ... G4UHW
Allsopp, F ... G3IFA
Allsopp, J T ... G4YDM
Allsopp, P J ... G4LCM
Allsopp, R M ... G1YFT
Allt, B R ... G0OQJ
Allum, J M ... 2E1EOI
Allwood, L ... G4GTT
Allwood, L M ... G3VQO
Allwood, P G ... GW8XBY
Allwood, R J ... G0NGC
Allwood, S A ... G0ALH
Almey, C B ... G7STZ
Almond, D M ... G7VJP
Almond, R P G ... 2E1FSI
Almond, S C ... G7JSE
Almond, T V ... G4WNW
Alperowicz, B ... G0LPN
Alpine, S J ... G4LKH
Alsop, B ... G8RCU
Alsop, D ... G8TFT
Alston, D R ... G4DAL
Alston, R N M ... G6PVS
Alston, S R ... G4VSR
Alston, T ... G8RLH
Alston-Pottinger, B R ... G4YZF
Alston-Pottinger, S A ... G6YZF
Altman, B ... G7MNE
Altoft, J B ... G0HWH
Alton, R J ... G3ZYD
Altschul, A B ... G3JDP
Alty, J ... G8EGC
Alvey, J ... G1ESE
Alwyn-Clark, T J ... G7JRD
Al-Bulooshi, I M ... M1BMA
Al-Katan, A W ... G6SFE
Amare, B ... G7VXK
Ambler, A E ... G3SNG
Ambler, D N ... G1BKL
Ambler, K K ... G8RWE
Amblin, R J ... G3LYN
Ambridge, N C ... G4FRL
Ambrose, C R ... G7CPQ
Ambrose, J R ... G3LNT
Ambrose, P J ... G0RRT
Amer, D ... G3SIV
Amer, J ... G0ALQ
Amery, C B ... G8AXN
Ames, A M ... G4SVD
Ames, C J ... G4SVI
Ames, J E ... G4XVI
Ames, J R W ... G8MUF
Ames, M ... G1IQF
Ames, R L ... G8BQS
Ames, S A ... G4OAV
Ameson, R ... GW1URD
Amey, P N ... G6BIF
Amies, A G ... G6NKD
Ammundsen, J G ... M0ATS
Amor, B C ... G1XHM
Amos, A M ... M1BEP
Amos, F ... G4WUM
Amos, J L ... G1KGU
Amos, K ... G4YRF
Amos, K S ... G2DPQ
Amos, M A ... G4LAH
Amos, M J ... G0ACD
Amos, P ... M1AJT
Amos, R C W ... G6EJF
Amplett, M ... M0AXV
Anders, S I ... 2E1FPJ
Anderson, A ... GM4VIR
Anderson, A A ... G7SFG
Anderson, A F ... G7WJP
Anderson, A G ... GM3BCL
Anderson, A J ... G0BFM
Anderson, A L ... G4CAV
Anderson, A M ... G1MOE
Anderson, A W ... GD3HQR
Anderson, A W A ... GM6AWA
Anderson, B T ... G7UHX
Anderson, C A ... G1DQD
Anderson, C H ... 2E1FHI
Anderson, C I ... G8OTC
Anderson, C M ... G4TBN
Anderson, C R ... G4LIT
Anderson, D A ... G0HUC
Anderson, D G L ... GM4JJJ
Anderson, D H A ... GW0GFN
Anderson, D M ... GM4SQM
Anderson, D P ... G6YBC
Anderson, D S ... GM6BIG
Anderson, F ... G1OPZ
Anderson, F E ... GI4ZAH
Anderson, G ... G1VGP
Anderson, G A ... GM0VGI
Anderson, G H ... GM0SYU
Anderson, G H I ... GI4TPI
Anderson, G W ... G3NPA
Anderson, H F M ... GW1NED
Anderson, H W ... G7VOV
Anderson, I F ... G3OAX
Anderson, I R ... G7VVO
Anderson, J ... G0TDK
Anderson, J ... GM6KKL
Anderson, J ... G0UMB
Anderson, J A ... G6ZCI
Anderson, J B ... G4DBP
Anderson, J C ... G0MPP
Anderson, J D ... G1OZD
Anderson, J E ... G3BIO
Anderson, J G ... G0GVO
Anderson, J L ... G4ZYL
Anderson, J M ... MI0AAZ
Anderson, J M S ... GW0AKV
Anderson, K J ... M0BEO
Anderson, L J ... GM3ZXH
Anderson, M ... 2E1FTT
Anderson, M ... G4VMA
Anderson, M ... GI3WWY
Anderson, M C ... G7STL
Anderson, M C D ... G0GNI
Anderson, N S ... M1BVK
Anderson, P R ... G7DQC
Anderson, R ... G6LYZ
Anderson, R ... G8HVZ
Anderson, R E ... G0WBB
Anderson, R G ... G3SXL
Anderson, R H ... G4RHA
Anderson, R H ... GI0UAG
Anderson, R J ... G4AEV
Anderson, R J ... GM0SCW
Anderson, S R ... G0EAT
Anderson, S W ... G1OXD
Anderson, T ... 2E1FTS
Anderson, T J ... G4ZBE
Anderson, T W ... G6DLA
Anderson, W ... G0NYF
Anderson, W ... GM3OIV
Anderson-Mochrie, I H ... G3VCM
Andersz, C ... GM0PEO
Anderton, C ... G1TAR
Anderton, C J ... G0DNQ
Anderton, D ... G0DYC
Anderton, G ... G6WIT
Anderton, M J ... G6ZLG
Anderton, N ... G0KLF
Anderton, R H ... G6RSQ
Anderton, T D ... G4YYL
Andre, R J ... G0SSG
Andreang, K R ... G4GZN
Andreang, R C ... G0VRM
Andreang, R C ... G4CMT
Andreang, R C ... M1BLR
Andreoli, D E ... G4MAQ
Andres, P D ... G0VJN
Andress, J C ... G0JCA
Andrew, A C ... G6ATL
Andrew, A D ... G7WKP
Andrew, C P E ... G8YKE
Andrew, D ... G6OUT
Andrew, D ... G7NLR
Andrew, D H ... G6TRA
Andrew, D W M ... G4GZH
Andrew, I ... G1FXU
Andrew, J ... G4RPM
Andrew, L ... G0GWV
Andrew, L ... G0KBV
Andrew, L H ... G4VZS
Andrew, M J ... G8NRP
Andrew, P ... G7GCW
Andrew, P C R ... G7VVB
Andrew, R S T J ... GM3WFJ
Andrew, S F ... 2M1BCA
Andrew, S F ... GM7WGM
Andrew, S J ... G3SNA
Andrew, W ... GM1MRY
Andrews, A ... G1VAO
Andrews, A G ... MM1BKT
Andrews, A J ... G4LKI
Andrews, B H ... G4CTS
Andrews, W J ... 2E1FKM
Andrews, C C ... G0SZE
Andrews, C R ... G4RVE

Andrews, D ... 2E1BVY
Andrews, D B ... G4EZZ
Andrews, D C ... G4CWB
Andrews, D J ... G3MXJ
Andrews, D J ... G4FOC
Andrews, D L ... G4BRQ
Andrews, D N ... G4ZOG
Andrews, G ... G6IRH
Andrews, G L ... G1XWN
Andrews, G M E ... G3PNV
Andrews, H ... G7KHX
Andrews, H ... G7UXU
Andrews, J ... G7NNC
Andrews, J B ... G4TEH
Andrews, J C M ... G0PUR
Andrews, J G ... G1HUL
Andrews, J M ... G3RMY
Andrews, J P ... M1BGB
Andrews, J R ... G4IWN
Andrews, J R ... G4PTX
Andrews, J R ... G7NVT
Andrews, J T A ... G3GEF
Andrews, J W ... G0REF
Andrews, K ... G1NCF
Andrews, K M ... 2E0ALO
Andrews, L F ... G0NZF
Andrews, L J ... G0WMA
Andrews, L R ... G7IQU
Andrews, M ... G6IRG
Andrews, M J ... G0PHR
Andrews, M R ... G0KBY
Andrews, P ... 2E0ANG
Andrews, P ... M0BDY
Andrews, P C ... G4TOI
Andrews, P J ... G8VGQ
Andrews, P L ... G1KUQ
Andrews, P L ... G4XKM
Andrews, P W ... G6MNJ
Andrews, R ... G1JMN
Andrews, R ... G1JZJ
Andrews, R D ... G6SRV
Andrews, R E ... GW3NLN
Andrews, R G ... G6TQZ
Andrews, R J ... G1DYW
Andrews, R L ... G6MNI
Andrews, R S ... G3BNG
Andrews, R S ... GW1XUD
Andrews, R V ... G3UZW
Andrews, R W ... G4BWB
Andrews, S I ... G6RPW
Andrews, V G ... 2E0APM
Andrews, W D ... GW2DHM
Andrews, W M ... M1BGC
Andreyev, V A ... G0SZZ
Andronov, G ... G6IRJ
Andronov, I ... G1VVX
Anese, H P ... 2E0ACR
Anese, O R ... G0DKE
Angel, D R ... G4ZXU
Angel, J M ... G6YJR
Angel, R ... G4ZUP
Angell, P M ... G3ZTX
Angell, R V ... G4CCE
Angelou, G ... G7GXO
Anger, D W ... 2E1DXW
Anger, D W ... M1ANA
Angier, T J ... G0HMK
Angiolini, J M ... GM0GFV
Angiolini, M ... GM1SBD
Anglin, J T ... G4GZ
Angold, P C ... G3WTQ
Angove, A W ... G6ZWI
Angove, C ... G4NCS
Angus, H ... G0EVS
Angus, H A ... GM4MUZ
Angus, J ... G1ZDR
Angus, R ... G0CHP
Angwin, F T ... G1IQE
Ankcorn, C T ... G0LYD
Ankers, D ... 2E1DFM
Annan, A G ... MM1CCR
Anness, P E ... G0WCO
Annett, G T ... G7RFY
Annis, H G H ... G6TRD
Ansell, C E ... G6NBD
Ansell, D N ... G3ZWY
Ansell, I T ... G4YBN
Ansell, I T ... G6UT
Ansell, M L ... G1WUA
Ansell, P J ... G4MZK
Ansell, P W ... G4MZJ
Anslow, N G V ... G4GD
Anson, M J ... G4TQC
Anstead, L J ... G4HOU
Anstie, D A ... G7UVB
Anstiss, D J ... G0GWX
Anstock, D A ... G6DZT
Anstock, P A ... G6OJZ
Anthoney, G ... GM0AAX
Anthony, E C ... 2M1DWG
Anthony, J ... G3ERD
Anthony, J ... G3KQF
Anthony, M G ... G4THN
Anthony, M R ... 2E1ATT
Anthony, R J ... GW0DFY
Antill, C A ... G0JFV
Antley, A ... GW3UTG
Antliff, P J ... G0SOO
Antmony, A M ... G1JFR
Antonelos, P ... G0IHS
Antonsen, E H ... 2E1EQD
Antrobus, C L ... G3JCJ
Anyon, T H ... G3YEI
Anziani, T P ... GW4ZWN
Anziani, T P ... MW0AER
Aplin, O T ... G0CSX
Aplin, P ... M1BFO
Apperly, M B ... G6FSU
Applebee, W F J ... G4ZKJ
Appleby, D T ... G7FMU
Appleby, F B ... G4RYH
Appleby, G W D ... 2E1ELM
Appleby, M S ... G3ZNU
Appleby, N J ... G0WIW
Appleby, P D ... G8CXW
Appleby, P W ... G4BLS
Appleby, R ... 2E1ELL
Appleby, R ... M1CBO
Appleby, R J ... G3INU
Appleby, T A ... G3RZ
Appleby, T S ... G4TLN
Appleton, A A ... G1HDO
Appleton, C D ... G4GBK
Appleton, D A ... G4KCP
Appleton, J ... G3XPZ
Appleton, J K ... G4SVA
Appleton, M ... G1EHE
Appleton, V ... G6CHC
Appleton, W G ... G4HEZ
Appleyard, A ... G4ZBS
Appleyard, A C ... G6PVU
Appleyard, J M ... G3JMA
Appleyard, P A E ... G0AMX
Appleyard, S F ... G3PND
Appleyard, T N ... G0LNV
Appleyard, W J ... G7VGF
Apps, N C ... G1FTK
Aquilina, M F C ... G1RQM
Arak, R G ... G4EEJ
Arakawa, T ... GW0RTA
Aram, G D ... G3SET
Aram, P V ... G1HDM
Aram, S J ... G6SRY
Arbon, M ... G4SAW
Arbuckle, S W ... M1AZF
Arcari, D ... GM0ISA
Arcari, D ... GM0JWK
Arcari, D ... GM4AAF
Archard, T N J ... G3ZDB
Archer, C J ... G4VFK
Archer, C J ... G8ZK
Archer, C W ... GI7ONC
Archer, D J ... G4GKE
Archer, J ... G0EQD
Archer, K A ... G4CMZ
Archer, P ... G6ODI
Archer, P A ... G3VVI
Archer, P S ... G6AKK
Archer, R ... G0RHA
Archer, R C ... G6WXS
Archer, S ... 2E1FLY
Archer, S ... G2DWZ
Archer, S ... G4ZFF
Archer, S D ... G1KFT
Archibald, D D ... GM4WJL
Archibald, D J ... GM0LEW
Ardern, R V ... G0SKX
Ardern, T H ... G3LJF
Ardill, A E W ... 2I1BXH
Ardill, A E W ... GI0XBK
Ardill, D A S ... GI7TDF
Ardill, M B J ... 2I1EXO
Ardley, W F ... G0BOF
Arey, J R ... G7LUS
Arey, R H ... G8TGY
Arey, R H ... M0APY
Argall, R S K ... G7TMX
Argue, S A ... GI1PXX
Argument, H G ... G0UGI
Argyle, M ... G7LIL
Arigho, D G ... G3NVM
Aris, F A ... G0LYG
Arkell, R S ... G0UYP
Arkless, S M ... G4HCJ
Arkwright, N B ... G6BKY
Arlette, D T ... G0AEW
Arliss, M R ... G7GEI
Armatage, G A ... G0KGQ
Armer, N T ... M1BHB
Armfield, M ... G7MVO
Armistead, B T ... GM4LBN
Armistead, C J ... G4DMI
Armitage, C E ... G0GFA
Armitage, D J ... G0AOU
Armitage, D J ... G3FVA
Armitage, J T ... G0OHR
Armitage, P R ... GW8SPM
Armitage, R ... G1RFC
Armitage, R D ... G3ZSA
Armitage, R G ... G3TZE
Armour, M F ... G8YKG
Armour, T D J ... GM6FOT
Armstrong, A ... G0FBW
Armstrong, A ... G1PAL
Armstrong, A K ... G1GQY
Armstrong, A S ... G3YZW
Armstrong, B ... G6SRW
Armstrong, B D A ... G3EDD
Armstrong, C ... 2E1DDO
Armstrong, C A R M G4TYG
Armstrong, C M ... G1GQZ
Armstrong, D ... G0FYQ
Armstrong, D L ... G6XAT
Armstrong, D P ... G6SRT
Armstrong, F ... G4NHC
Armstrong, F D ... G8PCA
Armstrong, F J ... G3JRL
Armstrong, G ... G4JUB
Armstrong, G ... G7TFR
Armstrong, G ... GI4XGO
Armstrong, G ... GM4DMK
Armstrong, G C ... MW1AHU
Armstrong, G F ... G0LIU
Armstrong, H ... G3RII
Armstrong, I ... M1BXK
Armstrong, J ... G7ROC
Armstrong, J B ... GW3EJR
Armstrong, J E ... G3GDA
Armstrong, J F ... G4VPU
Armstrong, J V ... G6YPF
Armstrong, K B ... G1AQX
Armstrong, K R J ... G8OYB
Armstrong, L ... G8ODV
Armstrong, M ... G0BMQ
Armstrong, M D ... G7PLA
Armstrong, N ... G6AWY
Armstrong, N ... G7PNX
Armstrong, N ... M0BCI
Armstrong, N G ... G3KKN
Armstrong, R ... G8SPE
Armstrong, R ... M0AAM
Armstrong, R A ... G4JYT
Armstrong, R I ... G0BIA
Armstrong, R I ... G3PSG
Armstrong, R J ... GI3HHN
Armstrong, R L ... GI4GOW
Armstrong, R W ... G0NWM
Armstrong, R W ... G3PGC
Armstrong, S A ... G1EZI
Armstrong, T A B ... GM0EHL
Armstrong, T P A ... GW6ZWH
Armstrong, W ... G3PRE
Armstrong, W ... GI6DSH
Armstrong, W M ... GI4XTC
Armstrong, W T ... G4TMX
Armstrong-Bednall, A M ... G7MLT
Arnfield, H P ... G3LX
Arnfield, M ... G8RCG
Arnison, M J ... G4THC
Arnold, B R ... G3FP
Arnold, B W ... G3HDO
Arnold, C J ... G3PJC
Arnold, D A ... G6HLH
Arnold, D C ... G8YQA
Arnold, D J ... G7OGN
Arnold, D J ... GM0JPG
Arnold, D M ... G0BID
Arnold, D M J ... G6UGP
Arnold, E G ... G3YBF
Arnold, E R ... G6AKE
Arnold, F I ... G0ATY
Arnold, H T ... GI1DTH
Arnold, I W ... G4IJM
Arnold, J L ... G3WXH
Arnold, J P ... G4NPH
Arnold, J W ... G7SOO
Arnold, K F ... G3XNP
Arnold, K M ... G0CAY
Arnold, K R ... G4DIQ
Arnold, L ... G8AHE
Arnold, M J ... G4IZK
Arnold, N J W ... G8HEJ
Arnold, P A ... G4CFG
Arnold, P J ... GW1AQJ
Arnold, R C ... G8ZDU
Arnold, R E ... G1VTU
Arnold, R T ... G7GBC
Arnold, R W D ... G1GRB
Arnold, S ... G1HSA
Arnold, T M ... G6UGX
Arnold, V ... G4FXA
Arnold, V A ... G8ZNU
Arnold, W D ... M0ACQ
Arnold, W E ... G4AVZ
Arnold, W M ... G3YET
Arnott, J ... G1WKK
Arregger, J E ... M0ARR
Arris, T M ... G4OSB
Arrowsmith, B A ... G0KOU
Arrowsmith, F B ... GD3HFC
Arrowsmith, G E ... G8XHU
Arrowsmith, H ... G7KNK
Arrowsmith, J R ... G4IWA
Arrowsmith, J R ... G6NWR
Arrowsmith, K G ... GW4VVL
Arrowsmith, K M ... G6JOS
Arscott, D C ... G6YJO
Arscott, J A ... G3VSL
Arscott, P J ... G6GFA
Arter, D J ... G1NLQ
Arter, J ... G7MEZ
Arter, S J ... 2E1DDZ
Arthur, D G B ... G0DLN
Arthur, J F ... GM7DTC
Arthur, J J ... G1CPC
Arthur, J W ... GJ4JVP
Arthur, M J ... G0BVS
Arthur, M S H ... GW2FYV
Arthurs, D J ... G7OGR
Artingstall, B N ... G1BBK
Artingstoll, T M ... G0JOE
Arts, D ... G4IHE
Artus, S ... G0EBS
Artym, R ... G7EXM
Arumugam, T V ... G0TAV
Arundel, C ... G4KDX
Arundel, C J ... G1YNH
Arundel, J ... G3HCX
Arundel, W E M ... G0LXH
Arup, P M ... G8WSP
Ascough, S W ... G4DZQ
Ash, A ... G3PZB
Ash, A ... G3SKY
Ash, D F ... G1BWW
Ash, G A ... G8FKV
Ash, K R ... G8DZK
Ash, N D ... G6ASH
Ash, N E ... M0ASH
Ash, P ... G7APA
Ash, S A ... 2E1BVC
Ash, W P ... G8XSA
Ashall, N P ... G6KRS
Ashall, R ... G4AHF
Ashbee, G D ... G6LKV
Ashbee, G D ... G6ZLS
Ashbee, H P ... G6ZLT
Ashbee, J Q ... G7JOW
Ashbee, R A ... G1HDP
Ashbee, T G ... G6LKW
Ashberry, R J ... G6RTM
Ashburner, E ... G0EHV
Ashby, A E ... G3HCW
Ashby, D E ... G3SUV
Ashby, D L ... G0BOT
Ashby, J E ... G1DPB
Ashby, M S ... G0UUH
Ashby, P S ... G1ZLC
Ashby, P S ... G6UZG
Ashby, S G ... G6XCE
Ashby, T R ... G3NBR
Ashcombe, A P ... G4APA
Ashcombe, J C ... G0IGO
Ashcroft, C O ... G1KGV
Ashcroft, F ... G4CJL
Ashcroft, G ... G0CGW
Ashcroft, G A ... G8AKL
Ashcroft, H D U V ... G4CCM
Ashcroft, K ... G3MSW
Ashcroft, N R ... G4IRU
Ashcroft, P D U V ... G0RLS
Ashcroft, P H ... G4CTI
Ashcroft, P M ... G8XOB
Ashdown, B ... G0AGR
Ashdown, B ... G4KZT
Ashdown, D J ... M0AWY
Ashdown, M ... 2E0APA
Ashdown, M ... G0BQV
Ashdown, M ... G3KIN
Ashdown, N A J ... G4NAJ
Ashdown, S E ... G1SDO
Ashe, J R ... GI0ADD
Ashe, J R ... GI8RLE
Ashfield, A ... 2E1FSB
Ashfield, G ... 2E1FUT
Ashfield, M ... G7HCF
Ashfield, N J ... G0OYR
Ashfield, S ... G4XMO
Ashford, A ... G8CMD
Ashford, G W F ... G2AOZ
Ashford, H V ... G3WGY
Ashford, P ... G7TKT
Ashley, A J ... G7SJW
Ashley, C D R ... G4UMN
Ashley, E P ... G0BIN
Ashley, S ... G8OSZ
Ashlin, C P ... G0GRX
Ashlin, C P ... G4MUQ
Ashman, B K ... 2E1EWK
Ashman, N ... G0KVJ
Ashman, N ... G3NSI
Ashman, N E ... G3FIT
Ashman, P ... G4VPO
Ashman, R P ... G6NYC
Ashmore, D M ... G3SXI
Ashmore, J ... G7KVZ
Ashmore, J ... G8GXF
Ashmore, M ... G7BSE
Ashmore, S B ... G0UAU
Ashton, A ... G1NAQ
Ashton, C K ... G3HNY
Ashton, D A ... G4HRV
Ashton, D R ... G6GEV
Ashton, D W H ... G3THA
Ashton, G ... G1ARD
Ashton, G E ... M0AUG
Ashton, J ... G4NTS
Ashton, K R ... G7IBH
Ashton, L R ... G4TDQ
Ashton, M W N ... G8TVI
Ashton, P ... G0DCS
Ashton, P ... G4CHG
Ashton, R H ... G4SAN
Ashton, R L ... G0AXR
Ashton, S P ... G7PCA
Ashton, T A ... G4PBZ
Ashton-Jones, J ... G7LVG
Ashurst, J ... G1SGW
Ashwood, D G ... G3TVX
Ashworth, A ... G4ZCG
Ashworth, D ... G3SYA
Ashworth, E J ... G0PFM
Ashworth, E J ... G7DNM
Ashworth, E V ... G6XCD
Ashworth, H ... G3CUF
Ashworth, J B ... G6AXB
Ashworth, J V ... G3BXC
Ashworth, L E ... G0KSF
Ashworth, R ... G4UMH
Askam, W B ... G0VNQ
Asker, B J ... G1NJG
Asker, T B ... G1KZD
Askew, A ... G4BPC
Askew, C J E ... G1NTE
Askew, D M E ... G3PCG
Askew, F ... G4NLG
Askew, G ... GM1OPO
Askew, J P ... G7TIV
Askew, J W ... G3ZMK
Askew, K ... G7AGA
Askey, J D ... G6SWJ
Askey, W S ... G8LXN
Askham, P H ... G6NJO
Askin, C ... G8ZFD
Askin, J ... G6JEN
Aslaksen, N ... G4KHO
Aslam, M ... G0KMK
Aslan, J J ... G6LFJ
Aslett, J T C ... G4HRN
Aslin, D W ... G3WGN
Asmussen, H ... G0WAZ
Aspden, G ... G0AMV
Aspden, K ... G8WZW
Aspey, S M ... G1ZEU
Aspinal, A ... G0NDM
Aspinall, D J ... G3XIP
Aspinall, H A ... G3RXH
Aspinall, J B D B ... G4ZAW
Aspinall, P ... G4ZLJ
Aspinwall, B A ... G3MBO
Aspland, J A ... G4PFZ
Aspland, S ... G7WAK
Asquith, C A ... G4ENB
Asquith, C A ... G8DDC
Asquith, C P ... G8LJQ
Asquith, D S ... G4UVD
Asquith, K A ... 2E1CEC
Asquith, N K G ... G0UZW
Asquith, P M ... G4ENA
Assenheim, J G ... G4BOK
Astbury, T ... GM0GMD
Astfalck, P C ... G6DKZ
Astington, G ... G4ZYX
Astle, J E ... G8OCA
Astle, P J ... G8VDX
Astle, R N ... G7PXY
Astles, F ... G1GJM
Astley, A J ... G0AJA
Astley, M J ... G7CAH
Aston, A ... GW0IWD
Aston, D R ... G8DR
Aston, E ... G4TYF
Aston, J ... G7OBF
Aston, J J F ... G6EQT
Aston, K F ... G0WZV
Aston, M ... G6HXB
Aston, P F ... M0AJP
Aston, R P ... G3WND
Aston, R S ... M1AZV
Atack, K ... G8TA
Atack, K M ... G4WAS
Atchinson, G ... G0DGI
Athawes, A P ... G4RUZ
Atherfold, J T ... G0FZB
Atherley, A N ... G4ORW
Athersmith, D A ... G1KRR
Atherton, A ... G4TZS
Atherton, J ... G0LJT
Atherton, J ... G7FGO
Atherton, M J ... G3ZAY
Atherton, P O ... G8EXS
Atherton, R S ... G0JIT
Athey, R J T ... G3GRC
Atkin, C C ... G4YAM
Atkin, F ... G4VHH
Atkins, A D W ... G3RRK
Atkins, A S ... G1VNZ
Atkins, C H ... G7MPZ
Atkins, C P ... G6OOT
Atkins, D E ... G7PIX
Atkins, D J ... G7USV
Atkins, D J ... G8XBZ
Atkins, D L ... M1BUL
Atkins, D V ... M1BZD
Atkins, G ... G3HQG
Atkins, G T P ... G0ATK
Atkins, I P ... G0HOX
Atkins, J ... G0WGB
Atkins, J R ... G0SDF
Atkins, N ... GI4CUV
Atkins, P J ... G3RJU
Atkins, P J ... G4DOL
Atkins, P R ... G4CKK
Atkins, R J W ... G0MWH
Atkins, R W H ... GW0FYF
Atkins, T B J ... G4ABN
Atkins, T F A ... G0JJK
Atkins, W H ... G0KTF
Atkins, W R ... 2E0AJT
Atkinson, A ... G1DKP
Atkinson, A ... G4BYD
Atkinson, B ... G0SWO
Atkinson, B ... G3JDY
Atkinson, B ... G3TEP
Atkinson, B S ... G3GSI
Atkinson, C G ... G0BQT
Atkinson, C H ... G8NZD
Atkinson, D ... G4ACL
Atkinson, D J ... G8DRE
Atkinson, D T ... G0WOU
Atkinson, G R ... G8GWP
Atkinson, G R ... G8RPI
Atkinson, I ... G6REM
Atkinson, J ... G0COI
Atkinson, J ... G8XSD
Atkinson, J B ... G0VHD
Atkinson, J B ... G4XBU
Atkinson, J W ... G0BVJ
Atkinson, K ... G1OGH
Atkinson, K ... G4NFX
Atkinson, L O ... G4FAA
Atkinson, M ... G0WQM
Atkinson, M C ... G7RIM
Atkinson, M R ... G8ZVM
Atkinson, P ... G0WXD
Atkinson, P ... G7EWU
Atkinson, P ... G7NRG
Atkinson, R C ... G3OPH
Atkinson, R V ... G4YJZ
Atkinson, S ... G0IQM
Atkinson, S ... G3YPS
Atkinson, S M ... G1PTY
Atkinson, S N ... G1PTZ
Atkinson, T E ... G7TTV
Atkinson, W E ... G3AFV
Atkiss, B A ... G3VYA
Atrill, N J ... G6GFO
Attack, A C ... G7IYN
Attenborough, K B ... G3WKD
Atter, D ... G3GRO
Atter, D ... G3WSC
Atter, K F ... GM3PTI
Atterbury, B A ... G7HCB
Atterbury, R S ... G4NQI
Attew, P T ... G0GRB
Attle, A ... G0RNY
Attlee, J W ... G0ATT
Attreed, D C ... G0CQE
Attwater, R A ... GU4WQP
Attwood, D J ... M1AJK
Attwood, P E ... G0RNZ
Attwood, R ... G7EYD
Attwood, R F ... G0PTA
Atwell, S R ... G6IRF
Au, S Y ... G0NYJ
Aubin, J ... G0RKP
Aubury, A E ... G4XQN
Auckland, R G ... G2PA
Auckland, S ... G4NSO
Aucote, R G ... G0BYF
Aucott, G L ... G7EKT
Audcent, A E ... G4MCE
Audsley, A ... G3NDV
Audus, L J ... G0DDM
Aughey, D ... GI7VBS
Aughey, F D J ... GI8XSB
Aughey, R ... GI3VPV
Augood, E G ... G3MML
August, R A ... GM1KQK
August, R A ... MM0BDA
Auker-Howlett, A ... G0FLT
Auker-Howlett, W R G0FML
Auld, C L ... GM0SUY
Auld, D ... 2I1ALE
Auld, D ... GI0UTE
Auld, P R ... G1JNQ
Aulsebrook, J H ... G0CQT
Ault, D J ... G1ZEK
Ault, J T ... G3KTU
Aunger, F J ... G6FJA
Aungiers, G ... G4PNH
Aungiers, G P ... G0KMP
Aungles, J F ... G0ABF
Aungles, J F ... G3NMD
Ausher, G J ... M1BJS
Aust, P A ... G6EBI
Aust, S H ... G4YBI
Austen, J A ... 2E1DCN
Austen, J M ... G6UZJ
Austen, K P ... G4GJA
Austen, K P ... G8LLU
Austen, K R ... M1AZO
Austen, M D ... M0AEN
Austen, R J ... 2E1DCO
Austen, S V ... G8WLB
Austin, A A ... G8SDD
Austin, A C ... G1JZK
Austin, B A ... G0GSF
Austin, B J ... G1IKV
Austin, D ... 2E1AUN
Austin, D ... G0WYW
Austin, D ... GW1XHG
Austin, D A L ... G8CID
Austin, D E ... G4BLX
Austin, D G W ... G1XUW
Austin, D K ... G0VGX
Austin, D M ... G4GTP
Austin, F E ... G4NFA
Austin, G ... G0VUS
Austin, G ... G4OAU
Austin, G E ... G4DPA
Austin, G J ... G1LQV
Austin, G J ... G6NYH
Austin, J ... G6EIZ
Austin, J ... G8KWU
Austin, J ... G8NHO
Austin, J ... GW8ITZ
Austin, K ... G7ABF
Austin, K R ... G6ATK
Austin, K R ... G8JSC
Austin, L ... G3BXF
Austin, L ... G3CYH
Austin, L A ... G0WYX
Austin, L E ... G0NMD
Austin, M ... G0VPH
Austin, M ... G3REM
Austin, M ... M0AAB
Austin, M B ... G1GDA
Austin, M J ... 2E1BDA
Austin, M J ... G8JRW
Austin, N ... G6DSG
Austin, N A ... G6WAO
Austin, N W ... G2FQR
Austin, P E ... G0AXQ
Austin, P G ... G7BXA
Austin, P G ... G7UEG
Austin, R F G ... G4JTT
Austin, S ... G1XXR
Austin, T O ... G3RCA
Austin, W G ... G7HIO
Austin, W G C ... G8GTE
Austwick, J ... GM0RIP
Authers, S W ... G4SWA
Auty, C ... GM3STU
Auty, C ... GM4GPP
Auty, E ... G4DYM
Auty, J ... G3TFO
Auty, P ... G4FKS
Auty, P ... G4RDL
Auty, S ... G3JRY
Auty, W L ... G0CVM
Aveling, B K ... G8VPH
Aveling, D ... G3IZU
Aveling, D J ... G4REQ
Avenell, M J ... G1YOS
Avenia, F ... G4MWM
Averill, N ... GI0SRP
Averill-Elias, C E J ... G4VHB
Avern, J H W ... G6HHE
Avery, E D ... G3WBB
Avery, J ... 2E1DCW
Avery, L J ... G3GAO
Avery, R L ... G3TQD
Avery, R V G ... G4XRA
Avery, S J ... G7JKW
Avill, P ... G3TPX
Avis, C L H ... G4CBN
Aviss, H W ... G7JUD
Avon, P A ... G8LHZ
Avory, L J ... G2FQP
Awbery, R E ... G3YZN
Axe, J C ... G4EHN
Axford, D ... G4LHU
Axford, M F ... M1ARI
Axon, A ... G8POS
Axon, H ... G4PM
Axon, M G ... 2E1DFZ
Axtell, T S ... G4SVC
Ayer, S ... G8WVB
Ayers, D ... G4SXT
Ayers, D W I ... G6GBC
Ayers, J H ... G4IXP
Ayers, J R ... G4SBN
Ayers, R F ... G4ZWB
Ayers, S E ... G7OQQ
Ayers, W H ... G0NOU
Ayers-Hunt, P R ... G6NMA
Aykroyd, L R ... G0NAS
Aykroyd, P J B ... G7DCQ
Ayles, P F ... G1JAS
Aylett, M ... G7NZK
Ayley, R W ... G6AKG
Ayling, C R ... G4HSU
Ayling, M C ... G4BNO
Ayling, R F ... G0UZA
Ayling, R F ... G3YUH
Ayling, S G ... G4ASL
Aylmer-Kelly, J R E G4SCZ
Aylott, M R ... G7ARM
Aylward, B R ... G6ODE
Aylward, G R ... G0XAN
Aylward, K ... G6NYF
Aylwin, K G ... G4SEK
Aynge, M J ... G6FXN
Aynge, R G ... G6IRE
Ayre, G J ... G0MFR
Ayre, J H ... G3AT
Ayre, L K ... G3DPR
Ayre, P ... G4TFF
Ayres, A H ... M1BHA
Ayres, A W ... G3UNP
Ayres, B P ... GU1HTY
Ayres, C D ... 2U1EKE
Ayres, C M ... G0KTC
Ayres, E A ... G4SKT
Ayres, J ... G3DQT
Ayres, L E ... G8SOG
Ayres, M L ... G4OQG
Ayres, N E ... G4ADR
Ayres, R A ... GU4ASO
Ayres, W J ... GU1WJA
Ayris, D ... G4GZA
Ayriss, K J ... G8MXV
Ayton, A A ... G4YEJ

B

Baal, R G ... G7HZV
Babbage, A D ... G0WWL
Babbage, A K ... G4YKK
Babbage, D S ... G4CI
Babbage, N J ... G1DZB
Baber, J L ... G7ARJ
Bache, J P ... G7JMZ
Bache, P F ... G1JUR
Back, C D ... G6HCH
Backham, R F ... G8KOC
Backhouse, A P ... 2E0ACE
Backhouse, C I ... G7VNN
Backhouse, K J ... G4RHR
Backhouse, S A ... 2E1CZS
Backhouse, W A ... G4HZI
Backshell, C S ... 2E1CAE
Backshell, C S ... G7WFS
Backus, J H ... G4PFJ
Bacon, G R ... G7INC
Bacon, J D ... G0TXV
Bacon, J D ... G3YLA
Bacon, L C ... G6JFJ
Bacon, L C ... M1BRG
Bacon, M K ... G3WMB
Bacon, P W ... G3ZSS
Bacon, R A ... G4ARK
Bacon, R J ... G6OBY
Bacon, R R ... G3WRJ
Bacon, S R ... G8KNQ
Badcock, A R C ... G8IPQ
Badcock, C M ... G1NPI
Badcock, R F W ... G4GEK
Badcock, W F ... G2BAP
Baddeley, J ... G7DOY
Baddeley, J E ... G6YCO
Baddeley, W J ... G4SVR
Badger, E W ... G3OZN
Badger, G C ... G3OHC
Badger, J F ... G4YZO
Badger, P D ... G4OZN
Badham, P H ... G0WXJ
Badham, R H ... G4PKE
Badman, F L ... G1NVG
Badz, W ... G4CGF
Baggaley, D M ... G7MMK
Baggaley, S ... G4RQG
Baggett, P M ... G4ORX
Baggott, G G ... G4GUW
Bagley, A ... GW3LJP
Bagley, A ... GW4EKE
Bagley, A W ... G3XPY
Bagley, D P ... G1NFB
Bagley, G C ... G3FHL
Bagley, J ... G0SNR
Bagley, J ... G4BDW
Bagley, J ... G7UCC
Bagley, R G ... G4HQD
Bagnall, A L ... G4CWD
Bagnall, D ... G6JWZ
Bagnall, J ... G4FSH
Bagnall, K J ... G8CQW
Bagnall, M A ... G0ICW
Bagnall, R I ... G1HGR
Bagnall, V A ... G3AFK
Bagshaw, C ... 2E1FEU
Bagshaw, C R ... G0JNI
Bagshaw, D T ... G3FNQ
Bagshaw, J A ... G1YBT
Bagshaw, J M ... G1GQB
Bagshaw, K ... G8YTX
Bagshaw, P ... G3NEO
Baguant, J ... GM4JGO
Baguley, C ... G0CEJ
Baguley, M E ... G8ORM
Baguley, M V ... G6ZTT
Baguley, M V ... G7LQD
Baguley, P E ... G4HUF
Baguley, P E ... G4HVP
Bagwell, R B ... G7WLY
Bagwell, R C ... G4HZV
Bagwell, W J ... G3YDE
Bagworth, A G ... G1YUU
Bahlke, A J ... G7OQA
Bahlke, G R ... G1LZW
Baier, C C ... G0FYG
Bailes, C J ... G7PAE
Bailes, J H ... G7NQX
Bailes, J L ... G0MVA
Bailey, A ... G0ELV
Bailey, A ... G4XSN
Bailey, A ... M1AUP
Bailey, A H W ... G3SFM
Bailey, A R ... G3IBN
Bailey, A W ... G3YNI
Bailey, A W T ... G1LCB
Bailey, B E ... G3MAH
Bailey, B M ... G3ZGF
Bailey, B R W ... 2E1FUB
Bailey, B V ... G1UFA
Bailey, C A ... G6CQR
Bailey, C F ... G7HUV
Bailey, C W R ... G3TRX
Bailey, D ... G4GUC
Bailey, D A ... G3ZNR

Bailey, D J G4NKT
Bailey, D J G6YB
Bailey, D J C G4FFM
Bailey, E G4LUE
Bailey, E G6AJ
Bailey, E G7LRG
Bailey, E G8XIG
Bailey, E GM8BBA
Bailey, E J G3SFT
Bailey, E J G4NNI
Bailey, E R G0MWM
Bailey, G J G1ZTJ
Bailey, I R G8PQY
Bailey, J G0PTC
Bailey, J A G6KAE
Bailey, J C E G4RNE
Bailey, J D 2E1FTV
Bailey, J E G3BNW
Bailey, J E G6JEB
Bailey, J E G7NMP
Bailey, J H G6PCN
Bailey, J S G0FJB
Bailey, J W G8KWV
Bailey, K G G0GZV
Bailey, K N H G4DFD
Bailey, L G0IDZ
Bailey, L B G4JLJ
Bailey, L F G0KXN
Bailey, L N G0BRB
Bailey, L S G0NSB
Bailey, L W G4HME
Bailey, M J G4RHB
Bailey, M P 2E1DML
Bailey, M P G3ZEK
Bailey, M R G7KSG
Bailey, M R G8OKD
Bailey, N G4GPJ
Bailey, N O GI8IJF
Bailey, P G7DPB
Bailey, P A G8OXX
Bailey, P D G7VHM
Bailey, P F G8JSN
Bailey, P J G3PJB
Bailey, P J G7BHY
Bailey, P L G6IUH
Bailey, P R G3TFP
Bailey, R G0FBY
Bailey, R G4OEE
Bailey, R G4PPP
Bailey, R A B G0IAX
Bailey, R E K G7NFN
Bailey, R G G0VFS
Bailey, R G G3WCQ
Bailey, R H G6WLE
Bailey, R J G7UTN
Bailey, R L G6GZJ
Bailey, R M G4PJL
Bailey, R V G1CJK
Bailey, S G3VDK
Bailey, S J G4MCQ
Bailey, S J GW7RHV
Bailey, T 2E0ALT
Bailey, T 2E1EWU
Bailey, T G0CRF
Bailey, T G0OMT
Bailey, T G1BRR
Bailey, T G6CRF
Bailey, T B G6KKM
Bailey, W G4ZXV
Bailey, W G6UKN
Bailey, W D G7PZF
Bailey, W G G2CHI
Bailie, J A GI3XEQ
Bailie, J B GI4PGN
Bailie, R W GI4TUV
Baillie, I A G8WDT
Baillie-Searle, C G GD4EIP
Baily, A C G7GPI
Baily, C J G7KID
Baily, D G1SQM
Baily, J W G4MDG
Bain, A GM1VSR
Bain, A F R M1BXV
Bain, G L G1IDV
Bain, J G4NAA
Bain, J GM0KXJ
Bain, J C GM3KAI
Bain, J S G7CRY
Bain, R F G6UEX
Bain, T E 2E1EJC
Bainbridge, D G0EPM
Bainbridge, M J R G4GSY
Baines, A D G7TDN
Baines, D G7RCP
Baines, D K G8OUI
Baines, G G7VDT
Baines, J D 2E1BIG
Baines, J G G4TUX
Baines, J H GW7LXI
Baines, K R G0DNL
Baines, M I G7NCY
Baines, R G3VKK
Baines, R G3YBO
Baines, R G8SMC
Baines, R J 2E1BIF
Baines, R V G1PHK
Baines, T J G4AEB
Bainton, D O G8FVT
Baird, A M G0PTM
Baird, C L GM7ASN
Baird, D P GM3OYH
Baird, F M MM1AUD
Baird, J G7EHN
Baird, N A GI7TTJ
Baird, N D GM0RIV
Baird, N D GM4JNB
Baird, P H G7FER
Baird, P K G7GBN
Baird, S A GM0FHS
Baird, S J L GI1XLK
Bairstow, A G3NBS
Bairstow, A C G4RSW
Baister, M G0UDZ
Baizley, P J G6ZXV
Bajjon, A S G6BWT
Baker, A G6TVK
Baker, A J G3PFM
Baker, A J G G0CKB
Baker, A K G4GNX
Baker, A M R G0VOD
Baker, A N G0TQR
Baker, A R G3KFN
Baker, A W G3JSF
Baker, A W E G3UXY
Baker, B G1WWZ
Baker, B J G8XOE
Baker, B R G4XYO
Baker, C G1SLA
Baker, C A G4FFN
Baker, C A J M1BFK
Baker, C F G7EDX
Baker, C F J G7ADP
Baker, C J G0NOL
Baker, C J G3HQS
Baker, C J G4HAY
Baker, C J G4WUV
Baker, C L G6ZDQ
Baker, C M G0HMJ
Baker, C M G7DIS
Baker, C S G7OSB
Baker, C W G4LDS
Baker, D G8XCE
Baker, D A G0BSM
Baker, D D G0GWL
Baker, D F G6HNI
Baker, D H E G3IYF
Baker, D J G8THH
Baker, D O G8SSX
Baker, D R G3XMD
Baker, E G7VLJ
Baker, E A G2FMW
Baker, E A G3EVK
Baker, E T G1POM
Baker, F G4VTR
Baker, F T G2FTB
Baker, G G6YAT
Baker, G M1AYK
Baker, G D G0CQI
Baker, G R G4ZRZ
Baker, H G G3EBL
Baker, I M GM6HFH
Baker, J G7ACG
Baker, J B G0MTQ
Baker, J D G3AJP
Baker, J D G8IUK
Baker, J E G7STT
Baker, J F GI0OXG
Baker, J I G3YHB
Baker, J L G4XIP
Baker, J M G6LIB
Baker, J M G7JMB
Baker, J S G0HQQ
Baker, J W B GW2SB
Baker, K G4RPV
Baker, K J G3WTV
Baker, K J G8WUO
Baker, L 2E1FZW
Baker, L A G0TEO
Baker, L A G8JIC
Baker, L F G4RZY
Baker, L F G4WAW
Baker, M A G1JUP
Baker, M A G4HGR
Baker, M B G0GMB
Baker, M C G3ZBP
Baker, M G G4YZR
Baker, M J G4OTL
Baker, M L G8FNJ
Baker, M P G4CEI
Baker, M R G4SHM
Baker, M W G4ASN
Baker, M W G4WUW
Baker, N C G7MZL
Baker, N J G4PHF
Baker, N J G4VYH
Baker, P G8ENW
Baker, P GW6VZW
Baker, P A E G4HSO
Baker, P G G7IPH
Baker, P H G6ESQ
Baker, P J G1LUN
Baker, P M G4PBF
Baker, P R G0GKS
Baker, P R V G8IXL
Baker, R G0AIH
Baker, R G0BLB
Baker, R G0LKO
Baker, R G1BZU
Baker, R G1LTQ
Baker, R G3CRS
Baker, R A G7PWD
Baker, R E G4SFY
Baker, R G G8AIF
Baker, R J G4DJC
Baker, R J G6FVB
Baker, R J GW3OVD
Baker, R K G1RBX
Baker, R R G8DLP
Baker, R R G3PKZ
Baker, R S GD4HPN
Baker, S D G7HRM
Baker, S J G3RXQ
Baker, S R GD8EXI
Baker, T C G3WNP
Baker, T E G0RCB
Baker, T J G4CLE
Baker, T R G3XEB
Baker, V J G4PJO
Baker, W G3HDQ
Baker, W C G0MGS
Baker, W M G1GRZ
Baker, W P G1ZBW
Baker, W R GW4RGI
Baker, W T G7GFH
Baker Munton, A G0KPY
Bakewell, J A GM6ZDW
Bakin, S C G6FSW
Bakrania, P G0MHA
Balaam, P G4HMS
Balaam, P G7HMS
Balaam, P E G4LNA
Balcombe, K M G6ODD
Balderson, R A G6VYR
Balderson, R E G6IVW
Balderson, S W G7WLG
Balderston, C R G4XMP
Balding, A J R G6JAS
Balding, S A G0ZAA
Balding, S G G7VRK
Baldock, D G3XNB
Baldock, D W G6NKL
Baldock, J M R 2E1CEH
Baldock, K A P G3UID
Baldock, K D G7HLP
Baldock, R J G0FFQ
Baldock, W A G7ENY
Baldry, K G G7OZE
Baldry, M C G6MCB
Baldry, M J E G7BYI
Baldry, V W G6IVU
Baldwin, A G6CAR
Baldwin, A H G8EPB
Baldwin, A L G7HMK
Baldwin, D 2E1EKM
Baldwin, D G4IGV
Baldwin, D C G4BFR
Baldwin, D F G0WTU
Baldwin, D V T G4KLC
Baldwin, G J G4KDU
Baldwin, J E C G3UHK
Baldwin, J H G4RYB
Baldwin, J M G7HNF
Baldwin, J V R G4KGE
Baldwin, L R G6WDC
Baldwin, L V G6GTJ
Baldwin, M J 2E0ANS
Baldwin, O D G0RCL
Baldwin, P C G3LCF
Baldwin, P I G4CQV
Baldwin, P S G7HHG
Baldwin, P T 2E1EKN
Baldwin, R D G6ZDS
Baldwin, R E R G8PYV
Baldwin, R J H G3WZ
Baldwin, S A G0FRD
Baldwin, T D M1BAN
Baldwin, T V G6EQY
Bale, B J G2ACN
Bale, C G G0MPH
Bale, E S G1XMV
Bale, I J G4SOL
Balen, H G4MHB
Bales, D E G8NIL
Bales, J C G0HAT
Balfe, J M 2E1EEZ
Balfe, S J 2E1EFA
Balfour, G R GM0SHD
Balfour, G R GM3GBZ
Balfour, J W GM0RRU
Balfour, J W GM4PYJ
Balharrie, D G G8FKH
Balharrie, D J 2E1CAD
Balharrie, D M G8GCT
Balister, R G3KMA
Balkham, S J G7KMA
Balkwell, R G1WPL
Ball, A G3RIT
Ball, A F P G4EDK
Ball, A G G3UQW
Ball, A P G8PSF
Ball, A R G8VWV
Ball, A R GW0VWD
Ball, A T G7HNR
Ball, D M1BCB
Ball, D G G7ANB
Ball, D J G6EJR
Ball, D L G1ZLL
Ball, E A G0PFI
Ball, E C G7STN
Ball, E G G4MYY
Ball, E H G4UOZ
Ball, F G7VHY
Ball, G A G4OJF
Ball, G J G6WXI
Ball, J G4OOS
Ball, J E G4CEY
Ball, J E G4KIP
Ball, J H G6HNR
Ball, J P G4DPI
Ball, J R G3UYX
Ball, J S G4WBE
Ball, J S G4XUO
Ball, K A G3XVW
Ball, K S GJ3FKW
Ball, L D G1DZD
Ball, L M G4YCE
Ball, M J 2E1CYA
Ball, M J G8LZK
Ball, M R G7MGQ
Ball, M S 2E1DVD
Ball, N T G1HHU
Ball, P J G3HQT
Ball, P W G0URC
Ball, R G4UXB
Ball, R G7MYJ
Ball, R A G0INZ
Ball, R D G3BDY
Ball, R J G6HNS
Ball, R O G4IRS
Ball, S G1VGM
Ball, S P G4BBZ
Ball, T GM1AXI
Ball, V J G7RXI
Ball, W L G3ZKD
Ballam, B C G0WDH
Ballam, C A 2E1FVD
Ballam, C F C 2E1FIC
Ballam, C F C M1BPB
Ballam, O E 2E1FID
Ballam, O E M1BPC
Ballance, J H W G1YBG
Ballance, K G3KNB
Ballance, M N G8AQB
Ballance, P W G6HEB
Ballantyne, G A R G0NXG
Ballantyne, I G7JXT
Ballantyne, J R G0AJS
Ballantyne, M A G1MOV
Ballantyne, R G3LDU
Ballard, J G0SMR
Ballard, J E G0AOT
Ballard, M J G7WKK
Ballard, M P M1BQN
Ballard, M S G1XGG
Ballard, P F G2BPF
Ballard, R A G3TWB
Ballard, R J M1ATV
Ballard, S A C G8BFK
Ballard, S P G8OTD
Ballard, V R H G8DBS
Ballard, W H G7IGV
Balley, E A G4KLQ
Ballinger, T J G0RYR
Balloch, I A GM3UTQ
Balls, A J M1AJN
Balls, J A G7UYP
Balls, J B G4ALC
Balls, L R G3YYQ
Balls, M L G8HDK
Balm, P L G7PDZ
Balment, T M G0ERX
Balmer, S G4OPY
Balmford, J J W G6DAP
Balmforth, A J G3RKQ
Balmforth, N S G4TEL
Balon, S F G1HAW
Balsdon, C J G1BOB
Balsdon, J E G4DEY
Balson, A K G0CRV
Bamber, B J G4RKU
Bamber, D J G4ZZS
Bamber, M D G0SHY
Bamber, R L G6JIY
Bambrey, R R GW7SWB
Bambridge, P J G7VRY
Bambrook, R J G8NRR
Bamford, C R B G4FQP
Bamford, H M G0WFK
Bamford, I M G0BGH
Bamford, J M G7MOF
Bamford, W A G1STO
Bampton, H G G7CEH
Banahan, M F G4BUH
Banaszak, L G G0PSZ
Bance, B R G1XFR
Bancil, N S G6KKD
Bancroft, J R GW4XXP
Bancroft, M F G8NQN
Bancroft, V L 2E1AGL
Band, K S G3WSB
Bandara Mieee, G G7NLF
Bandy, S H G8KPF
Banester, J H G4CKB
Banfield, J D G7PCF
Banfield, J M G1SSS
Banfield, R H G8LQZ
Bangle, M G4WMP
Banham, G A G2BBN
Banham, G I G6ZWF
Banham, N B G4YFV
Banham, P G1XAR
Banham, T E G6WWA
Banister, C J G7TQC
Banister, D F GW7TDQ
Banks, A G7LNZ
Banks, A A J G4XKS
Banks, A D G4GZZ
Banks, B J G4POR
Banks, B J G6JJB
Banks, D A G7SIQ
Banks, D F G0VFB
Banks, D K G7EKA
Banks, D R G6KIE
Banks, E GJ2CNC
Banks, E M1BLW
Banks, E J G4SAR
Banks, G J G0PMB
Banks, G V GM7HNU
Banks, H M0AHV
Banks, J M1AVS
Banks, J K G1WDJ
Banks, J S G1XLN
Banks, J T G7LKP
Banks, K G8LBJ
Banks, K F G0DAY
Banks, P A GW7NTP
Banks, R C G7GPJ
Banks, R M G4WND
Banks, R M G4YNL
Banks, R S G7UTH
Banks, S G1YAX
Banks, S E G1MZG
Banks, S F G6VVE
Banner, A D G7UMW
Banner, C E R G7UBA
Banner, G H G3AHX
Banner, L J G7SFS
Banner, M J G6YCL
Banner, S J G8FPG
Bannerman, M W GM3ZXE
Bannerman, R S GM4LUD
Bannerman, W R G8YSJ
Bannier, P J GU3HFN
Bannier, P J GU4SXM
Bannister, A G0OVQ
Bannister, D E G0DEB
Bannister, D J G1PKR
Bannister, D J G4FNZ
Bannister, E W G0RBA
Bannister, G J GI8SKR
Bannister, M E G0FTY
Bannister, P E G1HNH
Bannister, R E G4GPX
Bannister, S R G4HNZ
Bannister, W G0RPT
Bant, L F G7DCM
Banthorpe, A M G1SJO
Bantoft, A M1CCQ
Banwatt, P G6VVB
Baptiste, G J G8PNQ
Baraclough, G C G8DLT
Barber, B G0OYP
Barber, B M G7TPG
Barber, B W G0EUR
Barber, B W G1DPH
Barber, C A G4ZWD
Barber, C H G0JZM
Barber, C S G0JPQ
Barber, D G0LSX
Barber, D G2AKR
Barber, D C G8STE
Barber, D H G0UFS
Barber, D P G8OQW
Barber, D R G4JBW
Barber, G G4KYO
Barber, G R G8CHN
Barber, G R G8YIB
Barber, J G7NEM
Barber, J A G0LAU
Barber, J A G1JRD
Barber, J D M1BAL
Barber, J G G3TTJ
Barber, J M M0AOH
Barber, J S F GM4NNH
Barber, K S G0UZH
Barber, M G7EKU
Barber, M G G0DVZ
Barber, P G6SFH
Barber, R E G3NEF
Barber, R J G8WLV
Barber, R W G7HYG
Barber, T A G3TRB
Barbery, K D 2E1ADT
Barber-Lomax, J J 2E1FIZ
Barbour, R D G7VVP
Barbour, R L MM0ASB
Barbour, S 2M0APX
Barbour, W J GM1GES
Barchha, N P G8VDS
Barclay, A P G6ZLU
Barclay, D J G8XHW
Barclay, E M GM0KVI
Barclay, J GM4SDQ
Barclay, L W G3HTF
Barclay, P J GM0PXV
Barcroft, P A G7MOF
Bardell, J K G8HHR
Bardell, M A M0AEQ
Barden, J P G6UWK
Barden, P R G0JUY
Bardfield, C R G4EZH
Bardgett, T D M1AIK
Bardgett, W E G8DHZ
Bardsley, E G0RST
Bardsley, S G8YKO
Bardy, A E G6ODA
Bareham, A F G7FWG
Bareham, K G1RRR
Barfield, A H G6NXR
Barfield, M G7MRV
Barfield, T A G0NSO
Barfoot, C J G1TJH
Barfoot, N A G7GPY
Barford, A V G8EUT
Barham, C G G4MYB
Barham, D C G0DGW
Barker, A G8HTB
Barker, A J G6XVZ
Barker, A J G8IYZ
Barker, A K G3NXQ
Barker, A R G4YKC
Barker, B G4VRT
Barker, B C G3NIJ
Barker, C G6BZV
Barker, C G G4USG
Barker, C J G1EZJ
Barker, C M G7WDO
Barker, C S G6JJE
Barker, D B G7SGB
Barker, D E G4GZL
Barker, D F G1ATN
Barker, D G G8GBU
Barker, D G G8KGC
Barker, D J G1DOT
Barker, D R G1ISJ
Barker, E G G3OTO
Barker, F G6YAQ
Barker, F A G0BKE
Barker, F W G4AUQ
Barker, G G6SFY
Barker, G R A H G6LQM
Barker, H F G4BXY
Barker, H H G3SYB
Barker, J G3SAZ
Barker, J M0ABP
Barker, J F G0UPB
Barker, J M G3ZLD
Barker, J P G3PAX
Barker, J P G4JOB
Barker, J R GM3PDX
Barker, K G7PEX
Barker, K G8XHX
Barker, K C G0OBA
Barker, K E G6MLV
Barker, K H G0KLW
Barker, L F G1TWF
Barker, L M G4TJY
Barker, M G7JXU
Barker, M G G8CAC
Barker, N G4LRN
Barker, N A H G4NBU
Barker, P G0DZU
Barker, P G4HPS
Barker, P G8BBZ
Barker, P A G1MQB
Barker, P J G0RVV
Barker, P M G1ALC
Barker, P R G4BPV
Barker, R G0FDU
Barker, R G0KDV
Barker, R G4JNH
Barker, R A G4IDE
Barker, R A G7RCN
Barker, R A G8MZX
Barker, R G G7ARB
Barker, R G G7FOX
Barker, R G G7MRG
Barker, R J GW3WVV
Barker, R N GW3UTL
Barker, R P A G0VAZ
Barker, R S G0NUN
Barker, S G8AXX
Barker, S J G0WYP
Barker, S R G0JUM
Barker, S R G3CHD
Barker, T G G6XVY
Barker, T J G4AZT
Barker, T M G0HLA
Barker, W R G4JIQ
Barker, W S G0MVW
Barker, W S G6PRP
Barker, W S E GI4SXV
Barker Read, G R G0FSX
Barkley, D H G0DPI
Barkley, D H G0HRW
Barkley, I H G7EDK
Barkley, R A G1GGN
Barkley, R D G7IXC
Barksfield, F A G0IHH
Barling, R G4TGP
Barlow, A J G1GKR
Barlow, A S G3PYM
Barlow, B W G0ADL
Barlow, C G1ORL
Barlow, C G7GYN
Barlow, C F G4MWI
Barlow, C H G3XTL
Barlow, C R G7SXC
Barlow, D G0UJU
Barlow, D G3FKU
Barlow, D H G3PLE
Barlow, D J G1MZD
Barlow, E P G3KLO
Barlow, G A 2E1BVG
Barlow, G S G1DZF
Barlow, J G0LQQ
Barlow, J G7BJN
Barlow, J A G0LBW
Barlow, J A GW6PQT
Barlow, J C G6SUV
Barlow, J S G1SEC
Barlow, J S G3VOU
Barlow, K G0BZU
Barlow, K J GU7PMC
Barlow, L G8ZSM
Barlow, L G G3JMR
Barlow, P A G8LQV
Barlow, R A G4ZTL
Barlow, R K G3VFQ
Barlow, S G1RZR
Barlow, S M M1ADL
Barlow, T R G3QX
Barnacle, N J G8OMI
Barnard, A R G6OEJ
Barnard, C J G6GFQ
Barnard, F G4FB
Barnard, G E G0BMJ
Barnard, J B G7RPJ
Barnard, K G7HAJ
Barnard, K F G4MMA
Barnes, A G0EVB
Barnes, A G3LTB
Barnes, A G6OOY
Barnes, B 2E1DPC
Barnes, B J G7IBB
Barnes, B M G4IJA
Barnes, C G4OVM
Barnes, C B G8CBB
Barnes, C G G1BCC
Barnes, C N G4KQK
Barnes, D G7MGM
Barnes, D A G0RIF
Barnes, D E G0MTL
Barnes, D E G6VLP
Barnes, D F G3UVB
Barnes, D F G4VJF
Barnes, D L G6IVE
Barnes, E G0WHL
Barnes, E L 2E0ACV
Barnes, F G4LDE
Barnes, F G G0WOV
Barnes, G H G6FPF
Barnes, G M G4SGA
Barnes, G M G8ZNK
Barnes, H G3CCC
Barnes, H G6SZG
Barnes, H L 2E0ACQ
Barnes, H R G0HRB
Barnes, I N G1SHC
Barnes, J G4DVH
Barnes, J G7VTK
Barnes, J A F G0FQJ
Barnes, J B G3KVH
Barnes, J E G0GPV
Barnes, J G G3AOS
Barnes, J H G3RDH
Barnes, J K GM4HAA
Barnes, J K GM4JKB
Barnes, J P G7DMP
Barnes, J P G8AHN
Barnes, J R G0DDT
Barnes, J T GI3USS
Barnes, J T GI6YM
Barnes, J T GI6YMC
Barnes, J W G8NSK
Barnes, K E G4MKQ
Barnes, K G G8GTI
Barnes, L A G0FAJ
Barnes, L W G0OYZ
Barnes, M A G7DUA
Barnes, M C G6BQQ
Barnes, M J G0FVE
Barnes, M M G4MMB
Barnes, M W G6OCA
Barnes, N A G7EIF
Barnes, N R G1SGP
Barnes, P M0BFS
Barnes, P M1BFM
Barnes, P J G7KNP
Barnes, P W G4YNO
Barnes, R D F G0PCE
Barnes, R F G0DKS
Barnes, R G G8CJQ
Barnes, R G S G0VFZ
Barnes, R G S G8AZN
Barnes, R K G4YGV
Barnes, R W G4YLI
Barnes, S GW8RBJ
Barnes, V A 2E1GBS
Barnes, W B G0RAT
Barnes, W J G6SJA
Barnes, W J GI1HEK
Barnes, W M G4CDK
Barnes, W T R G6XML
Barnes, W W G0PQZ
Barnet, W A G0JID
Barnetson, I D GM0ONN
Barnett, A B G7OLH
Barnett, A J G2BHQ
Barnett, A M 2E1FZL
Barnett, B P 2E1FZK
Barnett, B W G8GLP
Barnett, C G1FFV
Barnett, C D G8UUM
Barnett, D A G0RHH
Barnett, D R G1LVH
Barnett, D W G7UGR
Barnett, G M1ACA
Barnett, G W G6LZV
Barnett, H M G4OPM
Barnett, J G7GIJ
Barnett, J C GI6GRV
Barnett, J E G8RLN
Barnett, J G G0JGB
Barnett, J R 2E0AID
Barnett, J W GM7IFX
Barnett, K L G0HOF
Barnett, K V G7CED
Barnett, M J 2E1EAW
Barnett, M J G0HCO
Barnett, M J G7JCC
Barnett, N F J G0JUR
Barnett, P G0EYC
Barnett, P G1YBZ
Barnett, P J G4GSK
Barnett, P N G4TMC
Barnett, P W G8BIZ
Barnett, R J G6NBL
Barnett, R M G4XCV
Barnett, S J G1DEA
Barnett, T S G4HVD
Barnett, W D G0PAK
Barnett-Bone, M R G3VOO
Barney, D G3VIC
Barnfather, S J G0MZV
Barnham, J M M1AMF
Barnish, G G4OVY
Barnish, R G1OJT
Barns, F J B G3AGP
Barnsley, M G3HZM
Barnwell, D R G8KWJ
Barnwell, L G0HZF
Barnwell, M G4FKQ
Baron, C G0HXQ
Baron, C G6PWY
Baron, J C G6NBF
Baron, P G0WWF
Baron, W F G0ELK
Barr, A G4LLZ
Barr, A E G0VBD
Barr, C D G3PFO
Barr, C J G0SLU
Barr, E GI7FFF
Barr, G S GI4NBO
Barr, J N MI1CCI
Barr, J S GI1CET
Barr, L G7KOF
Barr, M G4OQN
Barr, M S GI6IBL
Barr, O S GM1MCA
Barr, P R GI4GOV
Barr, S T G0CLV
Barr, W D G7VZA
Barraclough, S G3VQS
Barraclough, S J G0SJB
Barraclough, T G3VSR
Barraclough, V G3XVG
Barrall, E A L G2BCB
Barras, J P G4JTV
Barrasford, J T A G6WML
Barrass, B G7DGP
Barratt, A A G1DPI
Barratt, A J G2AA
Barratt, A J G4JDI
Barratt, G A G4VEO
Barratt, J F G3WVQ
Barratt, M D G6VXM
Barratt, M S G3RYN
Barratt, R G4VEP
Barratt, R E G4WJB
Barrell, D G G4BMC
Barrell, F A G3DCN
Barrell, G T G3TKQ
Barrett, A G2BWW
Barrett, A G7IIF
Barrett, A D G8DOR
Barrett, B D G0PWB
Barrett, B T 2E1FYF
Barrett, C R GW7UFH
Barrett, C W G3APX
Barrett, D A G3ZQH
Barrett, D M P G1ZRQ
Barrett, D P 2E1FRK
Barrett, G 2E1FMM
Barrett, G G4FWH
Barrett, G M1BXL
Barrett, G P G8ZOJ
Barrett, J G4GQV
Barrett, J G0XRN
Barrett, J M GW7OZP
Barrett, J R G0SGF
Barrett, J S G GW4AMX
Barrett, K G6PRK
Barrett, K GW0SIS
Barrett, K GW4NBY
Barrett, M H G6VGO
Barrett, P G6EJI
Barrett, P D G4WQU
Barrett, P J G8PYE
Barrett, P L G4CTM
Barrett, P R G1DOS
Barrett, R G0TDE
Barrett, R G4CWC

Barrett, R D M1AEZ
Barrett, R D M1BVR
Barrett, R E G3YCY
Barrett, R F G4ZIZ
Barrett, R G GW8HEZ
Barrett, S J G4IVH
Barrett, T R G4SPM
Barrett, V G1NXR
Barrett, V P G3KDD
Barrick, R G7FVH
Barrington, B J G3ZQW
Barrington, G D 2E1FOS
Barrington, J E G4ZUV
Barrington, N H D G4PUZ
Barrington, S L G1KBC
Barron, B G GI6EBY
Barron, C J G4JNZ
Barron, J S 2I0AHO
Barron, S R G7JYF
Barron, W M GI4JUA
Barrett, D J G4GVN
Barrow, A E G7UXN
Barrow, K E G4ZNT
Barrow, R D G8ILD
Barrow, T G0VUA
Barrowman, N W GM0LTQ
Barry, A W G3XJH
Barry, D A G3ONU
Barry, D G G4WWP
Barry, E G J G7JQT
Barry, J C G8ZKH
Barry, J H D G8SLP
Barry, P C G8OPA
Barry, P F G3RJS
Barry, P M G4VVC
Barry, P M N GW7RFP
Barry, P T G3KFU
Barry, T T G0CEM
Barson, D G G3UEH
Barson, J G0DWW
Barson, J A G3FJN
Barson, M D G6ETA
Bartels, F J G8XFO
Bartels, G J G8OPO
Barter, A G8ATD
Barth, A J MM0ALC
Barth, K A G0KSW
Bartholomew, E L G3FCV
Bartholomew, J G8GNX
Bartholomew, P R G0PQW
Bartholomew, T N G0JXY
Bartholomew, W E G8CK
Barthorpe, S H G0UVM
Bartlam, R N G0HDF
Bartle, G E G3JEB
Bartle, M A G6IBK
Bartle, M R G0UJD
Bartle, W G3GQF
Bartlett, A GW3YSA
Bartlett, C J 2E1EGD
Bartlett, C J M1BZN
Bartlett, D J G0VHF
Bartlett, D J G4VIX
Bartlett, D J G8VUR
Bartlett, E F J G7JSU
Bartlett, J G0GZB
Bartlett, J G6DLT
Bartlett, J R GD3ZND
Bartlett, L R G6TPQ
Bartlett, M J G3PXH
Bartlett, M R GW8NZE
Bartlett, P J G7VSS
Bartlett, R G G4LJN
Bartlett, S J G8UVB
Bartlett, S M G1JBJ
Bartlett, T H G3ITB
Bartlett, W E H G4KIH
Bartley, B G7SSF
Bartley, B M0BAR
Bartley, N A G7UFO
Bartnik, C A G1RYO
Bartolo, J J GM7FGH
Bartoloni, V I G7GKH
Barton, A N G3JQI
Barton, A T G8DFN
Barton, A W G6GKK
Barton, D I GW1JPF
Barton, G J G6DAW
Barton, I G3TYE
Barton, J E G0RPD
Barton, J H G6SZB
Barton, J M G7AGY
Barton, K R G3KKM
Barton, K R W G1ZFB
Barton, M C G6TRS
Barton, M H G8OOQ
Barton, M J G4FIS
Barton, N G4BZV
Barton, P H G0IFY
Barton, R A G4JEE
Barton, R J C G6SPN
Barton, S A G0FSB
Barton, W H G1ZNT
Barton, W J G G7JIM
Barton, W J G G8SDS
Bartram, A A G7ELS
Bartram, A A G4TKZ
Bartram, J A G0JYL
Bartram, P J G3GPX
Barugh, H C G1ZVE
Barville, P G3XJS
Barwell, F G6AKS
Barwell, G W G0EEB
Barwell, J H G7VBF
Barwick, B S G0HRF
Barwick, M G1BPE
Barwick, P J G4SUO
Barwise, R V G1JUH
Barwood, D A W G7VPS
Barwood, D E G4EGR
Basden, F T G4TUW
Basford, D M G0OUT
Basford, D M G0UGJ
Basford, R J G3VKM
Bashford, B G G7PIK
Bashford, T 2E1FSZ

Bashir, A M G0TVM
Basilio, E G3HVH
Baskerville, E E G1ZSU
Baskerville, N P G0UQQ
Baskerville, S G1JTC
Baskeyfield, M C G7SEJ
Baskeyfield, S G3HVI
Bason, M G G1ZBG
Bass, B 2E0AOC
Bass, B 2E1EQT
Bass, E M G4IPG
Bass, F G0EGK
Bass, J J G0DKY
Bass, M D G3OJE
Bass, R G4IPH
Bass, T W G3KMD
Bassam, S J G4XJV
Bassett, A J G0LLX
Bassett, A J G3EXP
Bassett, C M G1LRK
Bassett, K B G7NBU
Bassett, K R G3YIU
Bassett-Smith, I M G3XPR
Bassford, B W G4YNX
Bassford, J A G3HHT
Bassford, R K G2BZR
Bassford, S H G3YZB
Bassil, M J G4AEU
Bassnett, J A G8WMG
Bastable, J C G8FDF
Bastable, R R GW6IUK
Bastin, C L M1BAS
Bastin, D E G4XVY
Bastin, M I M1AIY
Bastin, R L G3LHA
Bastin, R R GW3JEZ
Baston, P GW0PJA
Bastow, R A G3BAC
Bastow, S M GW8HQM
Batchelor, D W G4RBD
Batchelor, G P G1NPM
Batchelor, I G GW0NLY
Batchelor, J H V G1YTR
Batchelor, M G4ZJU
Batchelor, M D G0RBB
Batchelor, R G1ALK
Batchelor, T P S G4PXU
Batchelor, T R G0JUW
Bate, D G4XEE
Bate, D G6LQG
Bate, E W G3LUC
Bate, G G0IFV
Bate, G G4CRC
Bate, G A F G0FHT
Bate, H L G8AMD
Bate, J G0HQO
Bate, J G7OAE
Bate, S D G4XVO
Bate, S I G7NZO
Bateman, A R G7HZU
Bateman, D G6EHN
Bateman, D J G0CGJ
Bateman, G P G0RCS
Bateman, G P G3LCG
Bateman, I G3ZKH
Bateman, I J GW0BAZ
Bateman, L G4KBO
Bateman, M J G7JPN
Bateman, N D G3UUB
Bateman, P F GW0BAH
Bateman, P M G1ZOV
Bateman, R F G6SLE
Bateman, R W G7HJZ
Bateman, S I G6RQP
Bateman, W M G1GSJ
Bates, A G1XYD
Bates, A GM0KNT
Bates, A R G6MOI
Bates, A S G3AZW
Bates, C E G4STY
Bates, C F G7GKP
Bates, C J G4ORY
Bates, C K GM6KAY
Bates, D G0LZL
Bates, D C J G1OLE
Bates, D L R G1NCK
Bates, G G6HFF
Bates, H G3AFR
Bates, H G3TRU
Bates, J G0BZP
Bates, J G4PLU
Bates, J G8TPX
Bates, J P G6YSB
Bates, J R G3OJK
Bates, K A GW3KGV
Bates, K C W G7KUU
Bates, K P M0AWT
Bates, M J G8FBM
Bates, M N G6BYT
Bates, P G0IYD
Bates, P GM1NZD
Bates, P A G1OFF
Bates, P C G0KWQ
Bates, P J GM3HAM
Bates, P J GM4BYF
Bates, R D G0IZE
Bates, R J G8TIR
Bates, R M GW0YDX
Bates, S G3IEZ
Bates, S G G1IGE
Bates, S J G4XUH
Bates, S J G6MNZ
Bates, S P G1YJW
Bates, V A G6MML
Bates, V W G6SSH
Bateson, W M G4RTS
Batey, A D G7LBL
Batey, J W G1TIF
Batey, W A G0DEO
Bath, D A G1YRP
Bath, G N G3NMZ
Bath, M E G3YAB
Bath, M J G4EEZ
Batham, D W G1JIY
Batham, J A G3LNC
Bathe, D E G6TPO

Batho, G J G7SOF
Bathurst, A J E G4PNB
Bathurst, A W G0WCB
Bathurst, D C G8TVL
Batiste, G C G7CCN
Batkin, F D G0OCR
Batley, J I G0IID
Batson, P C G0TDB
Batt, M S G3SJI
Battell, D B G7TVY
Batten, D G6DEN
Batten, E W G3BKN
Batten, I G G1FVC
Batten, J F G8BIJ
Batten, M D G1OUR
Batterham, K G8YQB
Battersby, M K G3ZZH
Battersby, R G0IMB
Battershill, P S G0IUH
Battle-Welch, J G4NYZ
Batts, E G8LWY
Batty, A 2M0AEL
Batty, A J G8OZD
Batty, C J G1MBE
Batty, F G 2E1AFR
Batty, G E G0THB
Batty, K G6LBO
Batty, P D G1HMY
Batty, R G0LBB
Baty, L G3DVY
Bauer, A D G7NGQ
Bauers, C N G4JUV
Baugh, M F G0SRR
Baugh, M P G8JHH
Baugh, P G6CQT
Baugh, S J G4AUC
Baughan, A P M0AVP
Bauly, B J G6SYW
Baum, K G6ZDP
Baum, K G G1VNS
Bautista, J J G4JTC
Baverstock, B H G4WCJ
Baverstock, C A G0PRS
Baverstock, C A G4WCK
Baverstock, J E G6ANI
Baverstock, S C G6OAI
Bavin, J G GM0GMI
Bavin, S L GM1TFZ
Bavister, R C G1WLE
Bawden, A B G1FJF
Bawden, A C G7UGH
Bawley, R G7VCN
Bax, G J G4SBD
Bax, J M1BJN
Baxendale, G G4IBS
Baxendale, H D G4OJK
Baxendale, J K G0UCV
Baxendale, W E G8FDG
Baxter, A 2E1FXP
Baxter, A M1CBL
Baxter, A W G0GTM
Baxter, A W M G1MPP
Baxter, C D G0FIV
Baxter, C J 2E1CZA
Baxter, C J G7KJR
Baxter, E J G0TII
Baxter, E T G7KTC
Baxter, F GM3VEY
Baxter, G R G8OAD
Baxter, H E G4ORD
Baxter, J H G4NMV
Baxter, J H N GM8OIO
Baxter, K M G7IRI
Baxter, M A G0KEA
Baxter, M E G8DHU
Baxter, M J G1PVW
Baxter, M J G6MAE
Baxter, M K 2E1CYZ
Baxter, N GM8RFO
Baxter, O R G0WRQ
Baxter, P G0VYJ
Baxter, P G4ZVN
Baxter, P G7EYG
Baxter, P G G1DEZ
Baxter, P J G4EOW
Baxter, P J G6IVR
Baxter, P J G6RSV
Baxter, R G1YZH
Baxter, R E 2E1ENZ
Baxter, R J M GI6KKG
Baxter, R J T G3WTB
Baxter, R N G7SYS
Baxter, W A G6SRZ
Bayes, M W G4DZC
Bayfield, R G3IAV
Bayley, P G G4ASG
Bayling, A E G4AZS
Baylis, A D G3LCL
Baylis, B J G1HOJ
Baylis, D G4EKQ
Baylis, F H W GM1RII
Baylis, J E G3UXX
Baylis, J F G4LMA
Baylis, M E G1GSK
Baylis, P J G6NYL
Bayliss, C R G3WKZ
Bayliss, D G4FJJ
Bayliss, D L G4OEF
Bayliss, F J G6AXB
Bayliss, J G1SKL
Bayliss, M D G4JHK
Bayliss, M H A G1YLV
Bayliss, M P G3PQ
Bayliss, N D 2E1FIU
Bayliss, T H G2HHH
Bayliss-Blomley, C GW1XDY
Baynes, N W E G4OFO
Baynham, D H G3DHB
Baynham, M A G3VCW
Bazley, J G3HCT
Bazley, L W G2FIF
Bazley, N I G6AFB
Bazyk, J G4WOB
Beacall, G H G4RPB
Beach, J G1KJH
Beach, J F G7PMA

Beach, K J G0UBJ
Beach, M J G7NMT
Beacham, I G G0GHR
Beacham, W H G4FMH
Beacher, D GM0SRQ
Beachill, R G7JKL
Beacon, J G1JBG
Beacon, P D G8NAP
Beadle, D G7JQZ
Beadle, G G R G8UBF
Beadle, R C G3VQG
Beadle, R G G0KGR
Beadle, R L G7ODG
Beadle, R W G1HGQ
Beadle, S G0DWC
Beadman, T B M1BKQ
Beakhust, D J G3OSQ
Beakhust, M D G0DGF
Beal, A S G6CSK
Beal, D R 2E1DOM
Beal, E A G4HNX
Beal, G G4ELU
Beal, L J G4YVJ
Beal, M A G0MDT
Beal, M R J G0AOB
Beal, S J G3WZK
Beale, A P M1AZQ
Beale, A Q GM1FML
Beale, D G0DBX
Beale, D S G1KII
Beale, K C G3YTF
Beale, M M1CBS
Beales, A J G6ZDV
Beales, D A G3MWO
Beales, M D G3YAW
Bealing, L L G6PJP
Beals, A GW4AGV
Beals, I GW0SYZ
Beamer, E L G4TKY
Beames, R F E GM4XMF
Beamish, S J G7JCF
Beamond, T W G3VLF
Bean, B A M G6AGO
Bean, B J G6PKS
Bean, C J G3VWD
Bean, D G4OUS
Bean, L G4UBL
Bean, M R G7UTC
Bean, N K G6DGR
Bean, N P G8NOB
Beane, D G0TAG
Beanland, C J G3BVU
Bearchell, R J G6YRY
Bearcroft, R E 2E1DFS
Beard, A B G0GJR
Beard, F J G4IRR
Beard, J M1BDY
Beard, R V G4MDS
Beardall, J H G6ITY
Beardmore, B G G4VPP
Beardmore, E G7PFR
Beardmore, E R I C G4LCL
Beardmore, J D G6DZX
Beards, P H G4IZX
Beardshaw, P D G1UGL
Beardsley, B G0FVZ
Beardsley, M J M1BVT
Beardsley, V 2E1CKM
Beardsmore, J E M G3SIO
Beardsmore, R G0KYK
Beardwell, C E J G1OTQ
Beare, D G3UPT
Bearne, R B G4DUA
Bearpark, K GD4TOW
Bearpark, T G4KFA
Beasley, A G0CXJ
Beasley, C D G1DPJ
Beasley, G G3LNS
Beasley, M E G4SHI
Beasley, R C G1CSN
Beastall, D G7VHU
Beatrup, C G7IIS
Beatrup, M A G7DIZ
Beattie, A G0TNL
Beattie, C M G3YSR
Beattie, D F G3OZF
Beattie, D W GI0URN
Beattie, D W GI4XFR
Beattie, G H H GI1NAV
Beattie, J GI3NQH
Beattie, P R 2E1EAM
Beattie, R A G0SKV
Beattie, R J G6CKW
Beattie, R J GI4FME
Beattie, S GI0LTT
Beattie, W M GM8AT
Beauchamp, B J G0JEL
Beauchamp, S H G3SYD
Beaugie, R C GW1JFT
Beaumont, A G J G7FWF
Beaumont, G G1YUQ
Beaumont, G G4OY
Beaumont, G G7LHV
Beaumont, J G4EIM
Beaumont, J P G4JPB
Beaumont, M J G4VCX
Beaumont, P A L G8DBM
Beaumont, P H G8HPJ
Beaumont, R C G4SEB
Beaumont, R C G7FNU
Beaumont, R E G G6AXC
Beaumont, R S G4HFV
Beaumont, S P G7IBD
Beavan, B S A G7GHB
Beavan, F GW1SMJ
Beaven, B O W G4BZU
Beaven, P H G1TUN
Beaver, A J G0SVI
Beaver, D J G0GUK
Beaver, G G4CLD
Beaver, M G 2E1DDG
Beaver, P C G1LQZ
Beavis, M J G7USH
Beavon, J R G G3PPR
Beazley, A L G3WEF
Beazley, S J G7BIM

Bebbington, R A G7MXP
Beck, B C G1MRT
Beck, G W G4MIF
Beck, N C G0DMJ
Beck, P G G6JJO
Beck, R G0KQM
Beck, R G6IBI
Beck, R A G6NOW
Beck, T J G8RSX
Beckerleg, R G3ZMP
Beckers, A E G7UKH
Beckers, B H G6EBO
Beckers, R J GW4VBV
Becket, D J G3SCH
Beckett, D A G4UQW
Beckett, E F G4CMR
Beckett, G G7JHE
Beckett, J F 2E1FHC
Beckett, J F G7WHS
Beckett, J F M0BCU
Beckett, N S G3MWP
Beckett, P A H GW4YCU
Beckett, R G4PIR
Beckett, S B G1JZL
Beckham, T L G8MAF
Beckingham, G D G7GQC
Beckingham, J A G7JUP
Beckinsale, R J GW1AXG
Beckley, D A E G0KCB
Beckley, D H G0FLC
Beckley, M J G3OTR
Beckly, D E G0PGI
Beckman, H G4YNI
Beddington, G G6LZM
Beddoe, K S G3YOM
Beddoes, J H G7EPM
Beddow, D A G6CBB
Beddow, G G0CYO
Beddows, J G G3NQK
Bedell, M D G0IDL
Bedford, D G4ABS
Bedford, D A G1UVL
Bedford, D F G6YRV
Bedford, D J G3ZVH
Bedford, D J M0AUD
Bedford, M G0DXK
Bedford, M D G4AEE
Bedford, P R G7BSL
Bedford, R G0CUX
Bedford-White, M G7JOS
Bedson, D F G1KVI
Bedwell, F N G8DT
Bedwell, P G3KCD
Bedwell, R J G1VBB
Bee, G S G6UCO
Bee, P K G4ZNN
Bee, R G6PXN
Bee, R J G3SZS
Beebe, R W GU4YOX
Beech, C G M1BLE
Beech, D G8JMP
Beech, E H S T G6ORQ
Beech, G G4WUF
Beech, G M G0KBN
Beech, G M G0WXQ
Beech, J G4YVG
Beech, J E G8SEQ
Beech, K J G4WRB
Beech, R G1HUM
Beech, R A GM1BXG
Beech, R H G0KCP
Beech, R J G4ZIS
Beecham, C R G0CMU
Beecham, J G G1LAP
Beecham, P R G6PZ
Beecham, R G4OQV
Beecher, A F G4MMG
Beecher, C G8XXM
Beecher, T G1XBE
Beeching, A J G0WIX
Beeching, B W G M G4MXZ
Beeching, T A G7FHV
Beecroft, R G6SDY
Beedan, D G0PDN
Beedle, C B 2W1AKD
Beedle, L GW0VMS
Beedle, T H GW0TOM
Beedles, R GW4RBA
Beehlar, J N P G3ZCT
Beekar, R V G3WY
Beeley, M J 2E1BOJ
Beeney, J M G7VAE
Beeney, M J G7VAD
Beer, A G G0GPO
Beer, M S G3OGZ
Beer, N H G0NIN
Beer, R E F G1UTF
Beer, R J G0FAE
Beesley, C F G4NHT
Beesley, C F G4OUG
Beesley, F W G8CZE
Beesley, K G3XUE
Beesley, K J G6KWC
Beesley, M A G4RUN
Beesley, P C G4PGG
Beesley, P J G1HNG
Beesley, S T C G7PHB
Beeson, M H G7PIJ
Beeson, P D G7RCE
Beeson, P J G1CNN
Beestin, B H B G8STR
Beeston, A V G8WSQ
Beeston, C S G4YCG
Beeston, P A G0WTV
Beeston, P C G0OCY
Beeton, D G R G1RLD
Beever, P F G4PKU
Beever, P R G6HNP
Beever, R N G6CKR
Beevers, B W G3VOI
Beevers, J S G1GMP
Beevers, M S G8SPD
Beevers Jnr, P R G1GCV
Beezley, C J G4FEA
Begg, C R G4MCF
Begg, D G3YXJ

Begg, D M GM6GFL
Begg, M S 2E1CIQ
Beggs, B J GM3YEH
Beggs, M T GI4TMB
Behan, C A M1CDT
Behan, S G0MUL
Behan, T C G0NON
Beharie, R A G7VKN
Beighton, T A G4JVJ
Beilby, W M G1OSG
Beir, E V G1KOG
Beirne, M G0DCO
Beith, I C 2E1DID
Beith, I C M1BAW
Beith, J E G8CON
Beith, M A GM0OXS
Beith, S J G1ZVC
Bekenn, W J F G3WIU
Belcher, D F J G0MCO
Belcher, D W G8YDI
Belcher, J R M1BFX
Belcher, M M J G1SSL
Belcher, R A GW3XIS
Belcher, R V GW4PCJ
Belcher, S A G0NIJ
Beldon, N J G4KGM
Beldon, S A G6UKB
Belfield, F A G G4YAG
Belfield, J E G0NNZ
Belfield, J E G7ECR
Belger, L W A G3JLB
Belham, N D N G2BKO
Bell, A G1AQP
Bell, A G3FBH
Bell, A G4IAB
Bell, A G4MHQ
Bell, A F G4YWX
Bell, A J GW4JJW
Bell, A M GM4MPY
Bell, A O T 2E1ANQ
Bell, A P G1YVZ
Bell, A P G6AXO
Bell, B D GM0HQT
Bell, B S G0NUD
Bell, C A G0PXQ
Bell, C D GM3RWP
Bell, C F G8AKC
Bell, C J G1ZSG
Bell, C R G3NIE
Bell, D G4NLL
Bell, D G4TYN
Bell, D G8IGQ
Bell, D E GM6TMH
Bell, D H GW0MXV
Bell, D J M1CCD
Bell, D M G0SGE
Bell, D M G8PY
Bell, D W C GM4IQU
Bell, D W G G8CZG
Bell, F G1TIH
Bell, F A G7DSP
Bell, F J G7BAI
Bell, F J G7CND
Bell, G G3NPT
Bell, G GM4XLI
Bell, G C H G7FIF
Bell, G G G0MXR
Bell, G J G4PMY
Bell, G S G6LLD
Bell, H J GI4STJ
Bell, H V G3MAZ
Bell, I F GW8ZIL
Bell, I M GI6PLO
Bell, J 2E1ENL
Bell, J G3DII
Bell, J G3JON
Bell, J GM0BBU
Bell, J GM0JMO
Bell, J GM4UDX
Bell, J M1BUA
Bell, J A G4XBL
Bell, J C GM0EDR
Bell, J F GM4SLY
Bell, J F M GM8DLU
Bell, J L G4FOL
Bell, J M GI7TMQ
Bell, J P G4LSA
Bell, J T GM1KMH
Bell, K M0BCK
Bell, K M1BHG
Bell, K F G1VWM
Bell, K W G0KWB
Bell, M G1WXK
Bell, M G8TCH
Bell, M A GW4JJV
Bell, M C C G4PPE
Bell, M D G0ECM
Bell, M J G6UGW
Bell, M R C G4CXT
Bell, M T GI6ETD
Bell, N S G4WLJ
Bell, N W 2E1CXS
Bell, N W GI4OHW
Bell, P MI1BJX
Bell, P C G0KFB
Bell, P F G0DWR
Bell, P J G4KVR
Bell, P W 2E1CRK
Bell, P W G0GWI
Bell, R G0WYY
Bell, R G7NNN
Bell, R GM8REG
Bell, R A G1ILO
Bell, R A G4VNA
Bell, R J G7BRX
Bell, R J I G7THG
Bell, R M G7KDX
Bell, R W G7IAS
Bell, S A G0SBI
Bell, S K G7JTK
Bell, S L G4RHS
Bell, S R G1EWH
Bell, S W GI4JER
Bell, T G0LEF
Bell, T F G8XXQ
Bell, T H G3RFE

Bell, W GM0JNB
Bell, W A R GI3MUS
Bell, W M G4WMB
Bell, W T G0JAL
Bell, W W GM1ZTB
Bellaby, M J G7BGY
Bellamy, A S M0AQP
Bellamy, B F G7IIO
Bellamy, F W H G8MOF
Bellamy, J G3TRD
Bellamy, K G7MAQ
Bellamy, M G G1TPC
Bellamy, M J G7DTX
Bellamy, M J G7MAP
Bellamy, R G0BIQ
Bellamy, R P G4JZV
Bellamy, R W G0NRB
Bellas, M W G0MDV
Bellenot, R P G0TEL
Bellerby, R G3ZYE
Bellfield, A R G4GLN
Bellfield, H L W G3SBV
Belling, J H G0RHO
Bellinger, D J G7DRR
Bellinger, J A C Q G0RGT
Bellingham, D G8JBT
Bellis, G GW7PCX
Bellis, J G7SRM
Bellis, J GD0TFO
Bellringer, B G3JYF
Bellwood, P C G4BHB
Belsham, C H G3DUH
Belshaw, C G0FXV
Belshaw, M R GI6PJD
Belshaw, T W R G3UTS
Belshaw, W J GI7UFX
Belt, G G4ZPO
Belt, G N G0SCV
Benbow, B R G4ASJ
Benbow, G L G3HB
Benbow, R D G4YDI
Bence, G G4RXF
Bendall, A F G0EXA
Bendall, D F G1RAX
Bendall, P J S G3NBU
Bender, M L G4WQL
Bendermacher, P GM0PEX
Bendrey, D S G7BYN
Benfield, A G G0NZE
Benfold, K M G1NQN
Benford, H G6ZB
Benham, A D S G8FSL
Benham, C W R G7GLM
Benham, D J G3YJP
Benham, R G3NSH
Benjamin, A S M1AEJ
Benjamin, H J G3MNB
Benjamin, T P G7KRE
Benn, A G1WWH
Benn, A D G8AFV
Benn, D G4GOP
Bennellick, R P G4WPN
Bennellick, V R G3WPN
Bennet, G 2E1FIG
Bennet, C M1CAH
Bennet, H R 2E1DZF
Bennett, A G1JHX
Bennett, A G4VVM
Bennett, A G6ORS
Bennett, A G7ASF
Bennett, A C G0OPI
Bennett, A C G1WLQ
Bennett, A J G6YTW
Bennett, A J G7WCB
Bennett, A R G4KNX
Bennett, A T G4WRZ
Bennett, A W G0AHB
Bennett, A W G6YCG
Bennett, A W GW3RTZ
Bennett, B GI7HEW
Bennett, B R G7HOJ
Bennett, C G4VXN
Bennett, C G7WFR
Bennett, C A G0KUW
Bennett, C A G7JDR
Bennett, C L G6WXN
Bennett, D B G0VPS
Bennett, D E K G0WQQ
Bennett, D G G4GLH
Bennett, D J G0CYF
Bennett, E G3ZJO
Bennett, E GI0OHG
Bennett, E A G G6SCP
Bennett, E A T G0EAB
Bennett, E B M1AQF
Bennett, E L G0PKV
Bennett, E M G1HFT
Bennett, G G0SHU
Bennett, G E G7GZW
Bennett, G G E G5BZ
Bennett, G H G1FRD
Bennett, G J G0IZG
Bennett, G J G3CYL
Bennett, G J G3DNF
Bennett, G J G4LJO
Bennett, G S G6WNB
Bennett, G T G1DOR
Bennett, H G4LPV
Bennett, I G7SEZ
Bennett, I A G6HNJ
Bennett, I F G6TVJ
Bennett, I G G4PYG
Bennett, J G0PUA
Bennett, J G3FWA
Bennett, J G3PVG
Bennett, J F G4XCT
Bennett, J I G7TLY
Bennett, J S G3KLC
Bennett, J V 2E1DTN
Bennett, K G0DUK
Bennett, K G7NMB
Bennett, K H G8AVV
Bennett, K M G4WDZ
Bennett, L A 2E1FUM
Bennett, L C G4HLN
Bennett, L P G4GXA

Bennett, M G6XIR
Bennett, M D G7PSV
Bennett, M D G7TRF
Bennett, M F G3UKL
Bennett, M J G4HUO
Bennett, M P G4XPU
Bennett, M W G8VIU
Bennett, N E G4IGG
Bennett, P G4NMY
Bennett, P M1AGL
Bennett, P E G1YQI
Bennett, P G 2W0ACD
Bennett, P G G7RHT
Bennett, P G MW0AHG
Bennett, P J G3VDU
Bennett, P J G6KFY
Bennett, P J G6LLF
Bennett, P L E G3HEH
Bennett, P R G3ZEG
Bennett, R G0FMJ
Bennett, R G0SHZ
Bennett, R G4IXK
Bennett, R G4KZQ
Bennett, R GM0PTP
Bennett, R GW4GSS
Bennett, R B G0AOI
Bennett, R D G0OOJ
Bennett, R E GW0GFH
Bennett, R F C G3SIH
Bennett, R J M G6GLT
Bennett, R T G4DIY
Bennett, S G G3TMX
Bennett, S J G0UPP
Bennett, T A G3DEB
Bennett, T B G7MST
Bennett, T E G1URZ
Bennett, W G1VSD
Bennett, W G4WWB
Bennett, W D G3LMS
Bennett, W J G6RBO
Bennetts, S J G6LKY
Bennewitz, F O M G6BPH
Bennewitz, R M G6BPG
Benney, S P G1WZV
Bennie, S J G GM4PTQ
Bennington, V A W G3JFU
Bennion, R C G7UOL
Bennison, G R G4PHV
Bennison, R C G3BRV
Bennison, R J G6FXO
Benns, A D M0AIS
Benou, E V G0BBD
Benoy, J W G8PSC
Bensley, A G3PTZ
Benson, A G0MZZ
Benson, A H G3XDM
Benson, A J M1BUC
Benson, B G8VNF
Benson, C E H G3MUX
Benson, C S G6DLO
Benson, D GW1LFO
Benson, D GW3TQI
Benson, D GW4LFO
Benson, D S GI1PUM
Benson, E G4RVX
Benson, F G7RBN
Benson, F GM8EKF
Benson, G G4ESK
Benson, J G4CAR
Benson, J G4GBH
Benson, P GW6ZDM
Benson, P A G0PVF
Benson, P A G0SPA
Benson, P H G8CSQ
Benson, R J G7CTH
Benson, R N G7SOQ
Benson, S J 2E1BJG
Benson, W E J GI0NYI
Benstead, A G3IFC
Benstead, F J G4YZD
Benstead, R H G4FKB
Benstead, S J G3ZAE
Benstock, A L G1HHT
Bent, A G7IER
Bent, H G0EZW
Bent, P R W G6ZLD
Bentham, B M G7NRX
Bentham, R S G4SHC
Bentley, B G1ELQ
Bentley, B W G4GQS
Bentley, D A G8MMG
Bentley, D H G8SER
Bentley, D J 2E1ANA
Bentley, D J G4RVK
Bentley, D J G6LKZ
Bentley, E I G7ECY
Bentley, J J 2E1DIH
Bentley, K R G1KRB
Bentley, L D 2E1FKF
Bentley, M J G6AKN
Bentley, M K G0CBW
Bentley, M L 2E1ELQ
Bentley, P G3VUD
Bentley, P G T G6BQM
Bentley-Beard, J M G3IYB
Benton, C D G6JWX
Benton, D J G7TLC
Benton, J B 2E1BQS
Benton, J C G8UXW
Benton, J F G4FCQ
Benton, K F GU0NHD
Benton, M D G0TTV
Benton, N G4KBS
Benton, T G G0OEL
Benton, W A G1OIO
Benwell, G T G4BHP
Benyon, A G3FXG
Benyon, R P G4KSK
Benzie, E M M G1PCN
Benzie, S P GM4JQA
Beresford, A C G4GGO
Beresford, C G8DXT
Beresford, G G7KQW
Beresford, G G7WFG
Beresford, J M G6PCX
Beresford, W L G4XOS

Beresford-Pym, I G3AIV
Berg, A R G4SCR
Berg, J R G4LON
Bergelin, F G3DDS
Bergin, P M0AAC
Bergman, A G4NBN
Bergna, L G3EIE
Bergstrom-Allen, N G0AWH
Berke, J 2E1ASC
Berkeley, A R G1COX
Berkeley, P G8CBH
Berkeley, R A G0WUO
Berkerey, A R G1HGT
Bernard, M G3RE
Bernard, M G4AKQ
Bernard, M G6JRL
Bernard, M R G8GJU
Bernard, P H G7ORI
Bernier, J G0CVS
Bernstein, A T G4BHV
Berridge, H J J G3ZCW
Berridge, J C G0OXX
Berridge, J C G0UJR
Berridge, L A G1PEI
Berridge, R J 2E0AAK
Berriman, A J G4CII
Berriman, A P G8JAB
Berrisford, R A G0CQP
Berrisford, T G0UPV
Berrow, A E G4MVB
Berry, A G0DKT
Berry, A G G4VRM
Berry, A H G6SLG
Berry, A J G7VPN
Berry, B A G4JSB
Berry, B M G6FFR
Berry, C G4YJJ
Berry, C G6RIM
Berry, C M 2E1BKK
Berry, D F G3SFG
Berry, D F G4DFB
Berry, D M 2E0ANB
Berry, D M G0FQO
Berry, E J 2E1AGF
Berry, G G0JFJ
Berry, J H G0JMU
Berry, I M J G7IAN
Berry, I R G1VIR
Berry, J G1WOS
Berry, J D G4DDW
Berry, J H G7LCK
Berry, K 2E1AGG
Berry, M G8BHX
Berry, N G8BYH
Berry, N M GW0EVG
Berry, P G4ZJC
Berry, P B G6SOY
Berry, P N GW8MXG
Berry, R G4AVD
Berry, R F G4GMU
Berry, R I GM6JJN
Berry, R M J G6NBK
Berry, S A G4IWR
Berry, S C G0KIK
Berry, S L G4LRT
Berry, V E GW0EVE
Berry, W 2E1BDW
Berry, W W G1RHE
Bers, J V G6EKA
Bertos, G G GW4YVN
Bertram, J GM0GMN
Bertram, J C R G7HOV
Bertrand, S J M0BFK
Berwick, P G G4BOP
Beschizza, D G4POP
Besford, A D G3NHU
Besford, A D G3YRC
Bessant, D J T G1YUE
Bessant, G C G8WMK
Bessell, R A G0WHV
Bessent, S G1JJA
Best, C A G0ATZ
Best, D G3SQO
Best, D J G6TRN
Best, G MI0ASV
Best, H W GI7IIJ
Best, J E 2E1EWB
Best, J L 2E1EVW
Best, L P G3THM
Best, M G0SFS
Best, M C G0IKR
Best, M W G0NAU
Best, P J G8CQH
Best, P W G8BLS
Best, R C GI3VAF
Best, R G G8WLO
Best, T H 2E0AHD
Best, T H G0WUB
Best, W E G0SCY
Beswarick, E T G6BTP
Beswick, B E G1OHD
Beswick, W G4GQW
Betambeau, B J G0AFV
Bethell, M J G7NUK
Bethell, P R G0RRX
Bethell, P R G6NBE
Bethell, R J G6ZDY
Bethune, N C GM4IUS
Bettany, D G1DPN
Bettesworth, M J 2E1CUX
Bettles, E W G3XKE
Bettley, A D G0RRG
Bettley, A D G4LDL
Bettley, G A G8KWD
Bettney, N A G7EKY
Betts, A G4FPH
Betts, A J G0VLC
Betts, A J G0VSI
Betts, A T N M1ATA
Betts, J G4HMG
Betts, J W G4TTM
Betts, M D G4MVV
Betts, M E G4FFW
Betts, M J G4EAK
Betts, P A G0PAB
Betts, P S G7VDC

Betts, R L C G0TRB
Betts, W E G7HFQ
Bettyes, G R G7NDQ
Bevan, A A G3GZZ
Bevan, A J G4FAV
Bevan, D A G3ZTT
Bevan, D A G4XUV
Bevan, D D N G4WPO
Bevan, D G GW4DMR
Bevan, I A P G0YAP
Bevan, J G1YFF
Bevan, J E G4JGQ
Bevan, K H GW3XKB
Bevan, M W H G0BRL
Bevan, N H GW1HAX
Bevan, R W GD3XPA
Bevan, R W GW0BNH
Bevan, S G4GFX
Bevan, S A G3MIZ
Bevan, T M G4TWW
Bevan, T O G4TTF
Bevan, T O M0ACV
Beveridge, J GM4MTV
Beveridge, J R N L G6IVQ
Beverstock, K G G3YZZ
Bevin, D G4WIV
Bevington, A C G5KS
Bevington, A J G7UNB
Bevington, A P G8IYH
Bevington, B C G0KJG
Bevington, P A G4ZUI
Bevins, A B G7CVZ
Bevis, A J G0OGA
Bew, H GW4VEI
Bewley, A M0AUL
Bewley, C A GM1RHX
Bewley, J W G1SQI
Bewley, J W G7GYY
Bewley, K C G0JGV
Bewley, K C G0NEA
Bewley, M A G7LLD
Bexley, A J G4JKK
Bexon, I P G4OCS
Beyfus, C A R G4AVW
Beynon, C S GW3VKL
Beynon, C S GW3WSU
Beynon, C S GW4BRS
Beynon, C S GW6BRC
Beynon, D GW7VGB
Beynon, J G4ILL
Beynon, M GW4GSH
Beynon, M T J GW7UWM
Beynon, M T J MW0BEY
Beynon, P M G1KDG
Bhamra, R S G0IND
Bhattie, G M M1BDT
Bhattie, S K 2E1FVI
Bhogal, J S G7LOE
Biagi, M MM1BDB
Biart, M D G4FDB
Bibb, C E G0EAN
Bibb, M G7PWK
Bibb, R S G0RVD
Bibby, F P G4PDD
Bibby, J A G3YQQ
Bibby, J C G6WOC
Bibby, J W G6JBY
Bibby, M M GW3NJY
Bibby, R D G1PIX
Bibby, W A G4SYJ
Bichard, A L GU4XGB
Bick, J 2E1CYP
Bickell, T G0UIO
Bicker, R GI0WJI
Bickersteth, P H G8RBS
Bickerton, D L G4VSN
Bickerton, H S G1XTX
Bickham, W B G3TJH
Bickley, R G4MZQ
Bickley, R F G0EBK
Bicknell, B 2E1BBC
Bicknell, C N G6FDX
Bicknell, K W G0NYT
Bicknell, P K G7UHH
Bicknell-Thompson, R J G1OGV
Biddell, A C W GM3GNM
Biddiscombe, M GW3YKZ
Biddiscombe, S GW8CNF
Biddle, M E G4UEK
Biddle, P E G4LUX
Biddle, R R G0ROY
Biddle, R S G7RAY
Biddle, T F G6LPS
Biddlecombe, J P G1COB
Biddlecombe, K W J G1YVI
Biddlecombe, R G4THV
Biddlecombe, T F G3WAO
Biddles, C O G6EJU
Biddles, M D G7OBR
Biddulph, R H G8DPS
Bidmead, W E G4EUV
Bidwell, M G0LDM
Bieber, D A G4AIR
Bieber, S W G7RNB
Bielawski, E R GW6JTX
Bierney, T K G8IWF
Bierton, K A G6KHD
Biggadike, P R G8JAN
Bigger, C J G4COU
Bigger, J M G7HDW
Biggin, A G7DGE
Biggs, G G0OSF
Biggs, J G0FGC
Biggs, J R G7BRP
Biggs, K 2E1FJH
Biggs, K M1CAX
Biggs, K W G8TVM
Biggs, S I G2FWZ
Biggs, T K G6GJN
Bigg-Wither, C M G4TGX
Biginton, D J G0WYG
Biginton, D J G1WYG
Bignall, J E G6YRX
Bignell, M A G1MZM
Bignell, M J G1ZLD

Bignell, R G0BGL
Bigwood, J G8KNN
Bigwood, P C J G3WYW
Bilbie, S J 2E0AIW
Biles, R M L G0GHX
Bilke, F G3UZD
Bilke, K F G8VIV
Bilkey, M W G8MNC
Billam, J E G7KWA
Billett, J E G1SEU
Billingham, D J G3KFD
Billingham, H G8IPF
Billingham, J P G0BRI
Billingham, J P M1BRI
Billingham, N G6JJA
Billingham, W J M1BSG
Billings, N C G0LSN
Billington, A D G8UFN
Billington, C A GI3WSS
Billington, C J G0TNO
Billington, G A G3EAE
Billington, Z A M1BFI
Bills, A J G3KZG
Billson, C F G8HVF
Billups, R S G7DAY
Bilmen, J G G7RFZ
Bilsland, R C G7WIG
Bilson, G 2E1FDF
Bilson, G A M0BAY
Bilson, M J 2E1ERW
Biltcliffe, D N G6DB
Biltcliffe, D N G6NB
Biltcliffe, M N G8EKN
Biltcliffe, R J G2BSJ
Bilton, F G6FFQ
Bilton, K H G0JIY
Bilton, K H G6YSK
Binding, I T S G4RVG
Bindon, D P A G3RSU
Bindon, G J G1VVE
Biner, P M G4FSE
Bingham, C 2E1FTE
Bingham, D G G3VBD
Bingham, G A GW0VEN
Bingham, J T G4TAJ
Bingham, M H G8FDR
Bingham, M J G0AGI
Bingham, N G GI6ECV
Bingham, W A G4WUS
Bingham, W L G3NYB
Bingley, A G1SIU
Binks, M S G0LJF
Binks, P J G4WDM
Binnell, C J G0TVR
Binning, A J GM3XIJ
Binning, J G3AJS
Binning, J G4BCV
Binning, J G7ORE
Binnington, A J G6WEX
Binnington, D R G6JFK
Binnington, F J G0SCM
Binns, A R J G0NJF
Binns, D G3MGI
Binns, E H G0PAJ
Binns, E V G4OSO
Binns, J C G6HJU
Binns, L G3RGN
Binns, M A G7PDA
Binns, M A M0AJR
Binns, M O G3YYU
Binns, M R G4OSP
Binns, M S G0TVU
Binns, R G3OTE
Binns, R A G6ELO
Binns, T G1SRA
Binns, T G7KRG
Binny, J W 2W1BPG
Birbeck, S G7WEC
Birch, A G4EAP
Birch, A J G4NXG
Birch, A J M0AJB
Birch, B J G7CMI
Birch, B T G0CGT
Birch, C T G0SGJ
Birch, D G0GKH
Birch, D G0ORM
Birch, D G3VRE
Birch, D A M1AUQ
Birch, D J GW4HDZ
Birch, D K GM1EHK
Birch, D N G0UDB
Birch, D R G1XHT
Birch, E B 2E1FXM
Birch, E M G0WTO
Birch, E T G8HLQ
Birch, G G0GWY
Birch, G G7VNX
Birch, G P M1BFQ
Birch, J A G0PZW
Birch, J F G7FUW
Birch, J R G3YZP
Birch, J R G4GGZ
Birch, K M1AMZ
Birch, M A G3KMO
Birch, M R 2E1FZQ
Birch, R 2E1EFI
Birch, R M1ASU
Birch, S C G0AXS
Birch, W T G6ETI
Birchall, A P G8USX
Birchall, D R G4MAU
Birchall, R N G3KMV
Birchall, T G4ERQ
Bircham, J W GW1JBF
Birchenough, A S C GD0KEO
Bircher, A W G4LFU
Bird, A 2I1CXY
Bird, A G3MJX
Bird, A C G0WQL
Bird, A H G6TMF
Bird, A J G6HOC
Bird, A R G6FSD
Bird, A R J G1IEM
Bird, B J G7WDD
Bird, C A G0OBW
Bird, C F G0SDA

Bird, D G G4JIK
Bird, D G GM8XOC
Bird, D H G4DHY
Bird, D J G4NEL
Bird, D L G6EJD
Bird, D P G0FJK
Bird, D S G4OOY
Bird, E C G7JTO
Bird, G A G3GDB
Bird, G D G4FRI
Bird, H B G3OUQ
Bird, H M 2E1ASI
Bird, J C G3GIH
Bird, J F P G1ALA
Bird, J H GW4BDV
Bird, J L G0PIS
Bird, J P G4CPN
Bird, J R G4CEK
Bird, K E 2I1AXH
Bird, K L G4JED
Bird, K W G3ZAW
Bird, L V G4ZYE
Bird, M L G4KFB
Bird, M P G1DZO
Bird, N P G8URZ
Bird, N S G6LQI
Bird, P A G1EHM
Bird, P A GI4RYD
Bird, P M 2E1CFF
Bird, R G1WDF
Bird, R G4ALY
Bird, R A GU4XIT
Bird, R H M1CBI
Bird, R I G0ILS
Bird, R J G7URS
Bird, R M G8ZVS
Bird, R S G8DUF
Bird, R W G1AXK
Bird, S J G0UFE
Bird, S J GI6EBX
Bird, V C G3ZLY
Bird, V J GW0DST
Bird, W C G4JAV
Bird, W D GI4XIR
Bird, W T G4NBF
Birk, G J G0DKX
Birkbeck, J W G3IGV
Birkby, B G0NEI
Birkby, G T L G1TTB
Birkby, J G0KMU
Birkenshaw, I G4UWK
Birkenshaw, S M G0TVL
Birkett, I B GM0FTX
Birkett, J H G8OPP
Birkett, M R G3OBZ
Birkett, N J G3EKX
Birkett, R G0UQC
Birkhead, G E G4KOQ
Birkill, S J G8AKQ
Birkinshaw, A J G3DMC
Birkmyre, H A G1HAB
Birkmyre, J A G1DNA
Birks, A D 2E1DCU
Birks, C E G7FNV
Birks, D J G7KHB
Birks, D L G4SFD
Birks, E G G3XXO
Birks, J R MW1BIE
Birley, J A G3PYN
Birrell, W W G0BYT
Birse, J C G4ZVD
Birt, A M G8MCU
Birt, A W G3NR
Birt, D P G6IAW
Birt, D P G7FEP
Birt, D W G3GIW
Birt, E J G7SAI
Birt, N J G7UOQ
Birt, S G G3ZNG
Birtles, R J G8ZPK
Birtwell, E G7RNO
Birtwhistle, S G7TYO
Birtwistle, A G4YYD
Birtwistle, J W G3UQU
Biscombe, P J W G0IXV
Bishop, A D G3MSV
Bishop, A E G4PFM
Bishop, A E MM1AVD
Bishop, B J G1LGY
Bishop, D A G8DHA
Bishop, E A F G0KZA
Bishop, G L G6MAA
Bishop, G M G0WUG
Bishop, I G R G7GFP
Bishop, I L G8GYL
Bishop, J G7VPQ
Bishop, J C E 2E1FRG
Bishop, J D G8OYY
Bishop, J J B 2E1FIA
Bishop, J K P G1TFM
Bishop, K G1DZP
Bishop, K R G1ATL
Bishop, K W G0SWK
Bishop, M G8YZF
Bishop, M D G0GQT
Bishop, N H G4LQE
Bishop, P G0WJO
Bishop, P J G1TQY
Bishop, R A G3GGG
Bishop, R A G4PNI
Bishop, R A G7GCB
Bishop, R C 2E1COB
Bishop, R C G6ZLQ
Bishop, R E G0SQL
Bishop, R E G7JSX
Bishop, R G G4DLM
Bishop, R G G8WZX
Bishop, R P G0IBJ
Bishop, R W G2AHC
Bishop, S J G6XCK
Bishop, S M G1NXU
Bishton, J D 2E1DBW
Bishton, M R 2E1DTM
Bisley, B A G3OFI
Biss, R C G7SUS
Bisset, J M 2E1CFJ

Bissett, J G3YMO
Bisson, J L G0MHF
Bittan, C J J G1TBO
Bjart, E G0AZY
Blabey, K A GM0FZM
Black, A R G6YTV
Black, C C GD3SKH
Black, C C GI4MBQ
Black, C E H 2E1EVO
Black, C J B G4LOJ
Black, D B GM2BLC
Black, D J G3HAB
Black, G GM7UMG
Black, G H GM3XPQ
Black, G W G8HPD
Black, J GM0FQV
Black, J B H GM0MFU
Black, J C G0UKA
Black, M GI4OYG
Black, M M1BSV
Black, M G GW8GOC
Black, M P G4HJY
Black, M T P GM0PIV
Black, N G4RYS
Black, N R H G4NOC
Black, P M G7KOG
Black, S E G3VSY
Black, S N G0MXA
Black, T J GI4IKF
Black, W J G4TME
Black, W R GI0LXN
Blackadder, J G4SDV
Blackadder, J GM1PGL
Blackaller, L F G6OUG
Blackaller, S G G4NLM
Blackburn, A H G0REP
Blackburn, A M G7RNT
Blackburn, D G G1JBE
Blackburn, D J 2E1AJW
Blackburn, E M0AQH
Blackburn, E M G6KYK
Blackburn, G R G7IWK
Blackburn, H S G7NDO
Blackburn, J G1GSM
Blackburn, J G4ACI
Blackburn, J G4EAB
Blackburn, K C G0LUU
Blackburn, K W G6HNQ
Blackburn, M G7UKR
Blackburn, M GD7IEH
Blackburn, M W G0YNM
Blacker, R G4GBE
Blackett, J R G4IOG
Blackett, K G4IGU
Blackett, P M G7BDK
Blackford, D W G0RFC
Blackford, D W G3NPB
Blackford, P M G8PMB
Blacklaw, P F GM7DPI
Blackley, M E G1YWT
Blacklock, M G7VZM
Blackman, A M G6TME
Blackman, C M 2E1EOK
Blackman, D G7PWP
Blackman, D I G6TNR
Blackman, I G4OCQ
Blackman, J C GM1AQV
Blackman, K G8ERV
Blackman, M G1YJH
Blackmoor, C I G0HAW
Blackmoor, G J G0BLT
Blackmore, M G G0VBK
Blackmore, R G0CYN
Blackmore, S P G4TIO
Blacksell, G G6CLA
Blackshaw, J E G8NPR
Blackwell, A G1WJO
Blackwell, A E 2E1AIZ
Blackwell, C H G6BXO
Blackwell, D G4UFX
Blackwell, D C G4XOH
Blackwell, D J G4PWP
Blackwell, J D G7VWN
Blackwell, M A 2E1EHR
Blackwell, M J G3VUH
Blackwell, R E G8ZPO
Blackwell, R P G4PMK
Blackwood, G G0JAQ
Blades, F G0BPN
Blades, G D G6OTL
Blades, J C G4ZFX
Blades, J N G0AAU
Bladon, A E G3GZX
Blagburn, D J G6DAD
Blagg, K R G1UGG
Blaikie, R G4OGR
Blain, D A G0IBQ
Blain, F G G3JLN
Blain, J M G7LSF
Blain, R G3NTI
Blain, R C 2M1FHS
Blair, A J G6KXW
Blair, D G G8VU
Blair, G G0RRQ
Blair, L M 2E0AOV
Blair, M A G8WSS
Blair, N GW7ASZ
Blair, P K G3LTF
Blake, A E G0UPM
Blake, A K G1BCB
Blake, A S N G3VEZ
Blake, B M GW1XOT
Blake, B P GW0XAP
Blake, C E M0AIJ
Blake, D M G7JVN
Blake, F A G8FYD
Blake, F B G8AAF
Blake, F R G3YLR
Blake, G G4WMF
Blake, G H G8GNZ
Blake, G L G3MHH
Blake, H A C G3OFW
Blake, H S G0BUQ
Blake, I M G4TWU
Blake, J W G7DSD
Blake, L G1KNB

Blake, M J G7ART
Blake, P J G0GEK
Blake, P V G7OWR
Blake, R L G6NOY
Blake, R N GM6YRZ
Blake, S C G8RMI
Blake, S G G8MAG
Blakeborough, P G3PYB
Blakeley, D W G3GRS
Blakeley, D W G3KZN
Blakeley, G G4CBM
Blakeley, J M0ANM
Blakeley, M E G4YKX
Blakely, H G7OIL
Blakeman, T J G6QN
Blakemore, A J G8FFN
Blakemore, D J G8LGT
Blakemore, G J G7AMD
Blakemore, P G7ACR
Blakemore, R G1IKF
Blakeney, A C M1BXC
Blakeney, P R G8BLB
Blakeston, A M G7MTG
Blakeway, R G1PXM
Blakey, J G3LRI
Blakley, S J MI0AAW
Blamey, K M G4KKB
Blampied, D C G4GJO
Blampied, P L GU7DSB
Blanchard, A M G8ZSP
Blanchard, D R GW3VBC
Blanchard, J S G4VVS
Blanchard, K 2E1FAN
Blanchard, W F G3JKV
Blanchard, W G H G3LHB
Blanchard, W J G3KOG
Blanche, J C 2E1EDW
Blanche, J E 2E1EDV
Bland, A M G6SLH
Bland, D W G8KIK
Bland, J D G1WSC
Bland, K R G7RGK
Bland, L G1RPX
Bland, M N G4WPE
Bland, M S G7NEX
Bland, P J G7MNQ
Bland, P M G8MBK
Bland, R G3BKL
Blandford, C M1BFU
Blandford, R P G1PCA
Blandford, S J G7GSF
Blanking, P J G6YUB
Blankley, T M G8SBJ
Blankley, W R G8CMK
Blanning, R G0NZU
Blanshard, K P 2E1CRE
Blasdell, R J G3ZNW
Blatchford, R B G4JVW
Blatchford, R J G4ZGX
Blaxall, G G7LIT
Blaxland, F J G4WSH
Blay, P J G6AFL
Blay, P R G6HCL
Blay, W G0GSK
Blayer, S G4STV
Blayney, A W G4WUL
Blayney, C R G6XVN
Blayney, K L G6XMI
Blayney, R M GW4NPC
Bleach, A R G4WNJ
Bleach, R G7VWC
Bleads, G S M1BAR
Bleaney, J C B G4UOO
Bleaney, S L GW3VPL
Blears, A G4PDX
Blease, P G6NBP
Bledowski, J A GM4YWU
Bleek, N G4WKT
Blemings, R E G4YYH
Blenkin, J G4OXZ
Blessin, G G7ANP
Blest, T G6OEI
Blew, S J G8JXK
Blewett, M G3IGQ
Blewett, M J G4VRN
Blewett, M J G8AHK
Blewett, S W G0HHB
Blewitt, C M J G4JXJ
Blezard, C R GI4RNC
Blezard, T W G1BBJ
Blichfeldt, J F G0OEQ
Bligh, F G G0MTA
Blight, B M G8AAD
Blight, D J G0PGL
Blight, J J G4SOF
Blight, T J C G1FNQ
Blinco, R C M1AAV
Blinco, T G8KNJ
Blinkhorn, S G1XGP
Blinman, G A G1ZBM
Bliss, D N G8OMQ
Bliss, F H G3IFB
Bliss, M W G4AQS
Blissett, A C G4OPD
Blizzard, P G0WLG
Block, C T 2E1DYT
Block, J D G7PSF
Blofield, W L G4MBV
Bloice, J F G0FJN
Blomeley, G G0NWJ
Blomley, G GW0HUS
Blood, S M0ABE
Bloodworth, A W G1IQG
Bloodworth, J G4VWO
Bloodworth, R P G4VWP
Bloom, P J 2E1FSS
Bloomer, B G4KES
Bloomer, N J G4OPZ
Bloomer, P G6DAK
Bloomfield, D G4ATL
Bloomfield, D J G0KUC
Bloomfield, D W G3UZK
Bloomfield, G A G4PXY
Bloomfield, G D G6ZLM
Bloomfield, H M1BYT
Bloomfield, R S MM0AOL

Bloor, A M1AJB
Bloor, G G3UD
Bloor, J M1BPK
Bloor, P G8WSY
Bloor, R N G7RSM
Bloor, R N G7UQG
Bloor, V J G4VJB
Blore, G N GW8VEE
Blore, H G GW0EMB
Blore, T G G0TGB
Bloss, M C M1ATU
Blott, R F G7OXV
Blount, A J G7RWG
Blount, B G6BTH
Blount, C J G0DJL
Blount, L W M1BOF
Blower, G D G4MQR
Blower, W A G1RJD
Blowers, D H F G0MBF
Blowers, J I H G4DWU
Bloxam, J A 2E1DGZ
Bloxam, T W GW3LJS
Bloxham, M K G1KXQ
Bloxidge, C R G8EFU
Bloy, P F G1KBH
Bloyce, G M G0UHU
Bluck, C GW0BEW
Blue, G M 2E1BQJ
Bluer, E G7WKM
Bluer, H G3UUZ
Bluff, J W G3ASR
Bluff, J W G3SJE
Blumfield, C E G0ALV
Blumfield, S M G0SST
Blumson, S G GW0WNB
Blundell, D R GW0TPL
Blundell, D W G0IWG
Blundell, E C G3RXI
Blundell, J A G3DBM
Blundell, J W G4UQS
Blundell, M W G6ITW
Blunden, L G G2CNO
Blunden, M G3PFH
Blunden, P P G7NSK
Blunn, P R G4JHC
Blunsdon, T GW7NTA
Blunt, D I G4PFU
Blunt, I E M0ALP
Blunt, P J G0UXG
Blunt, R I G1SGA
Blunt, S G8CRB
Bluthner, P H J M0ASO
Blyth, A GM4TAL
Blyth, B V G6ZHC
Blyth, P S G1LOK
Blythe, A A G3LOJ
Blythe, D W G3KCT
Blythe, J R G4TJX
Blythe, M S G4HFO
Blythe, N A G0SQI
Blythe, R G4MWH
Blythe, R MI1BSJ
Blythe, W S G0PPH
Blything, J C G1LEX
Boag, K B GI4TPY
Boakes, G T G8PPQ
Boal, R J G3AXI
Boalch, D R A G0BEP
Boaler, P J G0EJL
Boam, P R G8MZZ
Board, M E G8GUS
Boardall, R S G8AJZ
Boardman, A S G7ROM
Boardman, F H G3XUB
Boardman, G D G4CDI
Boardman, J C 2E1DDN
Boardman, R G G8RIN
Boardman, R M G7HCU
Boast, D G G1AKX
Boast, D W G G3HPZ
Boast, J S G3WDL
Bobbett, J A G0MSL
Bobby, M J GW0ZWZ
Bobby, M J GW8HBP
Bobin, V J G1FBH
Bockhoefer, E G4TBY
Bocking, C R G1YRK
Bocock, C F P G0SXU
Bodaly, G L W G0PYI
Boddington, J F W B G1MRD
Boddington, L J G4JDC
Boddington, L J G8OVA
Boddy, C L G8XII
Boddy, H G4MGP
Boddy, J W G0OMX
Boddy, K G3JRX
Boddy, R G6AKL
Boddy, R N G4MGQ
Bodecott, M I G1ZDX
Boden, B W G0HGI
Boden, J B G4CDZ
Boden, P M G6MOD
Boden, P S G6KHC
Boden, S A G4XCK
Boden, S A G6ZLV
Bodenham, D C G6XKV
Bodicoat, W G8XXR
Bodie, I R G8END
Bodill, M T G6IBN
Bodle, J G7WHX
Bodle, J H R GM7AWY
Bodle, M R G0HCI
Bodley, C G D 2W1EIN
Bodley, G H GW7RZC
Bodman, D J G1UGV
Bodman, D R G G4UJJ
Body, B H G0RIZ
Body, D G0HTJ
Body, J E G6FPC
Boehner, H G4YUV
Boele, F G7TIJ
Bohan, H P GM0FIQ
Boid, N J G6CCV
Boid, N J G8PSE
Boittier, R E G0SRB
Boittier, R E G8CUA

Bokor, R G1CLT
Bolam, S J G0OKF
Boland, C F G0RBM
Boland, D E 2E1FXN
Boland, J H G0NEO
Bolas, P G1IXJ
Bold, R H G4CNW
Bolderson, P G1OUX
Bolderston, B GU1HYN
Bolger, V L M1AOD
Bolland, J G G6OEM
Bollingham, M A 2W1BQW
Bolsover, J E G7IHD
Bolster, G J GW1USQ
Bolster, H L F GI4SVO
Bolt, B L G0FGE
Bolt, C G0ACX
Bolt, C G3NN
Bolt, D T G1HHC
Bolt, D T G1RCD
Bolt, E H G1HHD
Bolt, N P M1AZW
Bolt, S M G4SUI
Bolter, J H W G3NNY
Bolton, A G1EAB
Bolton, A G4VSQ
Bolton, A GM3BMI
Bolton, D R G8UQC
Bolton, F J G1JFZ
Bolton, G L G3UDZ
Bolton, I G3JES
Bolton, J G6LZZ
Bolton, J D G4XPP
Bolton, J E G4WEL
Bolton, J H G3YYG
Bolton, J R G3HBN
Bolton, M E G0OVV
Bolton, M I K GM0DBW
Bolton, N A M G3HMV
Bolton, P G3CVK
Bolton, P G4CXE
Bolton, P R GI4UWT
Bolton, R F GI4BBE
Bolton, R W G4SSJ
Bolton, S I G0DTW
Bolton, T W G4OUM
Bolton, W J E G3FBN
Bolwell, H W G0JGU
Bona, R E G3SGX
Bond, A G6ZOI
Bond, A C G7JAW
Bond, A C M0AZZ
Bond, A G G8KYX
Bond, A R G8BJP
Bond, B C G3AFT
Bond, B C G3ZKE
Bond, B C G8DWL
Bond, B G G7PNW
Bond, C G1ODQ
Bond, C J G8BLP
Bond, D J G1NZR
Bond, G J G4GJB
Bond, H A S G0PWM
Bond, J A G1ZHH
Bond, J P G8APY
Bond, J R G8RMP
Bond, J S G7IIH
Bond, L G3LDT
Bond, M A GM4HYR
Bond, M F C G1ALL
Bond, M J G8XSU
Bond, N J G3IHX
Bond, P C G3BEG
Bond, P M G8TIS
Bond, R M1BWP
Bond, S F G7CAF
Bond, S L S 2E1FML
Bondar, P G6FTE
Bonds, P E G6GUH
Bondy, D N G4NRT
Bone, A M L 2E1EYU
Bone, D M G4MOI
Bone, H E G3EHQ
Bone, J S G7OOE
Bone, K G6RTN
Bone, K F G8UXY
Bone, S D G0SZS
Bone, S H G4PPJ
Bonehill, B L G3LHC
Bones, D P G4RSD
Bones, F L G3KSG
Bones, J C E G0WNI
Bones, K GI4ERM
Bones, P R B 2E1GBA
Bones, W H G4CFP
Bonfield, C G1ILN
Bonfield, D G G7UDM
Bonfield, D M G4JXK
Bonfield, P J G3ZZB
Bongers, W G0RZD
Bonham, S N G7APL
Boniface, A C G4XSA
Boniface, O G4DSC
Boniface, R F G0THE
Bonnar, B MI1BNO
Bonnard, B M 2E1BRW
Bonner, A P G2BHY
Bonner, C R G3TGF
Bonner, J F G0GKP
Bonnett, D R M1AMJ
Bonnett, E A G4XER
Bonnett, J A G6BNJ
Bonney, A K G0KTD
Bonney, R W G0KFI
Bonney, S A G7VGA
Bonnick, K H G3BIF
Bonning, J G7HUU
Bonnor, H R G4LYL
Bonsall, C G3UWR
Bonser, M J G7KXN
Bonser, W J G0PBB
Bonser, W J G0SNB
Bonshor, D M0AWF
Bonson, P D G4FUY
Bonter, G D GW0AGA
Boocock, A G4REG

Boocock, F A G4IRP
Booer, A K G3ZYR
Booker, M G1MQF
Booker, W H M1ASK
Booley, I M1BBY
Boom, A W M0AWB
Boon, B S G4TRE
Boon, C S G8FGZ
Boon, I E G4TRF
Boon, J G1DZR
Boon, K G0AWE
Boon, M J G0JMB
Boon, T M0AFC
Boone, E J G6DLF
Boonham, A Y G6DAQ
Boor, A C G1EWI
Boor, G L G8NWC
Boorman, L J G7FHB
Boorman, P A G6NLM
Boorman, P E G0JBA
Boorman, R J G8ZSO
Boorman, T D GW1FBL
Boot, A J G1OQW
Boot, A J G4YSL
Boot, D A G1UCC
Boot, J A G1KIG
Boot, K G4WUN
Boote, K G1SQC
Boote, S R G6BER
Booth, A J G3TQE
Booth, A J G4JBX
Booth, A J M1AYC
Booth, A M G6BSF
Booth, A V G4YQQ
Booth, B G3NXU
Booth, B B G1KMC
Booth, B K G3SYC
Booth, C L 2E1FWD
Booth, C W G3HKA
Booth, D G1UCJ
Booth, D G6FWK
Booth, G A G8DZJ
Booth, H V G2AS
Booth, I N G7HRP
Booth, J A M1BAC
Booth, J D G3IDQ
Booth, J G G6ZHB
Booth, J T G0BLR
Booth, M B G8NXH
Booth, M D G3YNO
Booth, M J G8EGM
Booth, M M G4RSF
Booth, M P G7JVP
Booth, M W G4DCF
Booth, N G2DSF
Booth, N D J 2W1DGK
Booth, N S G1MOB
Booth, P G4PAA
Booth, P K G3XEM
Booth, P R M1AYA
Booth, R G4RSG
Booth, R G7NWC
Booth, R G8BIW
Booth, R A G0TTL
Booth, R F G1WQH
Booth, R O G G7VKM
Booth, S H J 2E1FFQ
Booth, S W G4DFS
Booth, T G4YTD
Booth, T A G8DPH
Booth, T J G6AAU
Booth, V H G0RVA
Booth, W 2E1BYY
Booth, W G7JQF
Boothby, B G6PWZ
Boothman, W A G3SWP
Boothroyd, G G0GHK
Boothroyd, G G4AWT
Boothroyd, G G7HAH
Boothroyd, J G0MTJ
Boothroyd, K G1FYS
Booth-Isherwood, N G4SNN
Bootles, C P G6NOZ
Bootman, D E G3MWG
Booty, D S G3KKQ
Bootyman, T F G0VKJ
Bootyman, T F G7TTW
Bootyman, T F G8TKY
Bootz, C J M1BSD
Boraston, J G3YLJ
Borde, L M0ABG
Boreham, A J G1DZQ
Boreham, S M G7AMP
Boreham, T D G4YMP
Borer, M P G0NYM
Borg, J C G0AIS
Borkowski, C A G4ILP
Borland, A M G4DBD
Borland, B M S GM1FBM
Borland, I A C GW0KOD
Borland, J GM4YZU
Borland, J C GM8ZEJ
Borland, W H G3EFS
Borley, D R G4CAX
Borley, E A M0AEB
Borley, L R G4LIK
Borley, P J M G0VEF
Borne, D F G4CYW
Borrell, N G8LIE
Borrer, B W G1JRC
Borrett, G G G1NLS
Borrett, M R G8NXB
Borrett, P J R G3XTC
Borrington, J G G6XMG
Borrow, M J G6GKL
Borrowdale, G G0OFR
Borrows, B GM0LLJ
Borsay, O C G4DPS
Borthwick, M GM0HLK
Bortowski, J P GW0FPY
Bosanquet-Bryant, N F G6TNQ
Bosanquet-Bryant, P W G6DLZ
Bosansko, A G6DMY
Bosher, T M G0UVH

Boskett, G J G6UWM
Boskett, S W GI7IFW
Bosley, M J G0FGS
Bosley, S S G3ORB
Bosley, S S G4SSB
Boss, I G1SNQ
Boss, R M G4ZDE
Bossino, C J GW0IQT
Bosson, B A 2E1FWQ
Bosson, B A M1BQI
Bostock, A J G7VMB
Bostock, H N B G1SVW
Bostock, T 2E1EVX
Boston, A J GI8VTK
Boston, L L G1ONK
Boston, S J G3MYY
Boswell, A G G4ZYQ
Boswell, A G P G3NOQ
Boswell, G E G0TMW
Boswell, P J G8JVW
Boswell, P L G4AEK
Bosworth, A G3GUD
Bosworth, D J G0ECD
Bosworth, D J G4NAC
Bosworth, I J G8PKG
Bosworth, J C 2E1CZJ
Bosworth, J D G8BAV
Bosworth, P G7UPS
Botfield, J P G4UOY
Botham, B T GW0BTB
Botherway, A G0AJB
Botrel, J A GJ3YUL
Bott, D D G7VHV
Bott, V M G4AFZ
Bottle, J W M1BWJ
Bottom, J J G3SDG
Bottomley, D D G3GAQ
Bottomley, E F G1MCZ
Bottomley, E J GM0HNP
Bottomley, H G4WCY
Bottomley, H F G8BCL
Bottomley, H M G6JRM
Bottomley, J D G3TQQ
Bottomley, J P G7RKE
Bottomley, K G8MJF
Bottomley, M G0WNJ
Bottomley, M GM0RXP
Bottomley-Mason, J G1CFZ
Bottoms, C H G4PIP
Bottrell, A M1AZJ
Bottrell, W J G6DZY
Bottrill, A G1RTV
Boucher, K G4KBA
Boucher, T G3OLB
Boucher, W T G1GBC
Boughton, C G G0VUC
Boughton, D C G7PER
Boughton, K W G1YJK
Boull, G P G4NVH
Boult, B G G7HMQ
Boult, D G7HCE
Boult, J G4LOM
Boultbee, G S G3VHI
Boulter, C G4UXY
Boulter, H R GW1GJS
Boulton, A R G1JJG
Boulton, C M G4JQS
Boulton, D M W G4DGY
Boulton, K D 2E1CJU
Boulton, R F G4VXV
Boundey, G T G0VII
Bounds, A G G3KDP
Boundy, S J G0EDH
Bourke, C J G4UOR
Bourke, N P G6ESR
Bourn, R J G7CUL
Bourne, B G4DLC
Bourne, C J G0RFF
Bourne, C J G3LNK
Bourne, C J G4ETK
Bourne, C P G4RPG
Bourne, C S G4EJD
Bourne, E J S G7ARV
Bourne, F A G G3YJQ
Bourne, F R G0VTB
Bourne, J G3MLX
Bourne, J G G6JJK
Bourne, J R G3VYD
Bourne, M G7MTV
Bourne, N O M0BFG
Bourne, N O M1AMQ
Bourner, H A A G3NCB
Bourner, J L G7JXV
Bourner, N J G4JYU
Bourner, N J G8MLB
Bournes, A K GW6YTY
Bousfield, M C G0AFU
Bousfield, T S G0SJU
Boutell, C G7HMN
Bovenizer, G T G4ELB
Bover, M G8KCY
Bovey, C J G0AFT
Bowden, B A G0ENN
Bowden, C G3OCB
Bowden, C E G4MFB
Bowden, C L G7IPX
Bowden, D G4VVR
Bowden, D A G3PNF
Bowden, D G G1MPT
Bowden, G G3YVR
Bowden, G G4JBY
Bowden, I P D G0RUE
Bowden, J H G6NKJ
Bowden, K D G4SBE
Bowden, M A D G4NTQ
Bowden, M G G4ZXE
Bowden, O G7GQK
Bowden, P J G4MKW
Bowden, R G7FKZ
Bowden, R J G4JOU
Bowden, R M G7IND
Bowden, R T G3IXZ
Bowden, S G5GW
Bowden, S B G4KIV
Bowden, S L G4TDP
Bowden, W M G0MAL

Bowditch, A W M G4WSB
Bowdler, K G0KMB
Bowe, P J G1MVI
Bowell, E R G0OTE
Bowell, R G3LRL
Bowen, C A 2E1FMQ
Bowen, C A M1BSI
Bowen, C H 2W1FQL
Bowen, D C G0MIU
Bowen, D J G6ATS
Bowen, E GW1IRL
Bowen, G C GW0UMC
Bowen, I J G0JJY
Bowen, J D GW3TSQ
Bowen, J H G8DET
Bowen, J J G1YOY
Bowen, J L GW6MMM
Bowen, M T G8DWA
Bowen, M T GW3KGI
Bowen, M T GW3UWS
Bowen, P G1BMC
Bowen, P A G3TZL
Bowen, S N GW8IAD
Bowen, T G4JKQ
Bowen, T S GW8JOY
Bowen, W T G4AAU
Bowen-Lock, F C G4HLS
Bower, A F G3MKU
Bower, A H B G3COJ
Bower, D J G7UHQ
Bower, F GM0JFB
Bower, G A P GM1PGO
Bower, I P G6RAJ
Bower, J G0KED
Bower, J O G8SQN
Bower, L R G4HKY
Bower, P G GM3OFT
Bower, P W G0LNK
Bower, R GM4DLG
Bower, T H G4VVG
Bowering, A C G0JKU
Bowerman, S G0OGM
Bowers, A B G1TKT
Bowers, B R G0VAX
Bowers, C O J G4YSH
Bowers, D G3GKK
Bowers, D S G7SYU
Bowers, D W GW4AVC
Bowers, J G0NLL
Bowers, J A G8FWF
Bowers, J M G6BIM
Bowers, J R G0FAD
Bowers, P G0HES
Bowers, R C G G0MQX
Bowes, A E G4XTW
Bowes, B J G6CRG
Bowes, J G0HUE
Bowes, J L G4MB
Bowes, J S G4ZFY
Bowgen, R E G0BOW
Bowhill, A J G4FTI
Bowhill, F R G1SSZ
Bowie, A J GM4YWQ
Bowker, A M0AGJ
Bowker, B G7MRO
Bowker, D G4ZBM
Bowker, J G1ZWQ
Bowlas, D R G6VGT
Bowler, R S G3ZHV
Bowler, S G G7ABV
Bowles, B G7JMX
Bowles, D A G6IBD
Bowles, D C G1CFK
Bowles, D J G0CRK
Bowles, F K GM4KAV
Bowles, J A G0GMJ
Bowles, J H G7HOS
Bowles, P W G3ECM
Bowles, S M G0UWF
Bowles, W G C G0KCZ
Bowley, A J G6CHI
Bowley, C J G4OQL
Bowmaker, A G0REV
Bowmaker, A S P G0EBP
Bowmaker, R G0NRD
Bowman, A 2E1AUU
Bowman, A K G4NJN
Bowman, A M G8ZIA
Bowman, C GW3NTR
Bowman, D G0MRF
Bowman, D J G0BOH
Bowman, D R G3LUB
Bowman, F GI7EZF
Bowman, F J G4SNU
Bowman, I P G7ESY
Bowman, J D G0WNK
Bowman, M D GM4LVW
Bowman, M R G7VHE
Bowman, R A G0CIS
Bowman, R G G4DDV
Bowman, S J G7ESX
Bowmer, J L G4BHX
Bown, P K G7PVM
Bown, R C G8JVM
Bowring, R F GI8RKC
Bowron, P J G6UCT
Bowry, N J GM6OUL
Bowser, G P G4NRU
Bowskill, R W G0LKU
Bowthorpe, M J T G0CVZ
Bowyer, A M GW1KQN
Bowyer, A P G4OYH
Bowyer, D C M1AEI
Bowyer, D E G3NHB
Bowyer, J A G4KGS
Bowyer, M J G6CJN
Bowyer, P C C G4MJS
Box, A R G0HAD
Box, F G6XOB
Box, F R G8RWM
Box, M S G3RZG
Box, S G1PUZ
Boxall, I F S G1COT
Boxwell, L M G4RET
Boyce, A R G8SGM
Boyce, C S G4XWP
Boyce, E J G1LAW

Boyce, G R M1AHN
Boyce, J GI3TJJ
Boyce, M GM0RRK
Boyce, P J GW4NHB
Boyce, R G3ZVO
Boyce, R G6PXQ
Boyce, R G6YDD
Boyce, R J G8TUU
Boyce, S T E A G3VBV
Boycott, T G1ODJ
Boyd, A G4LRP
Boyd, D GM1SZH
Boyd, D C 2M1CEN
Boyd, E T G8CRN
Boyd, I F GI7GKC
Boyd, J GM7RAK
Boyd, J M GM3YAV
Boyd, K J GI4SLQ
Boyd, P G4TUZ
Boyd, P A GM6PCW
Boyd, R GI7THC
Boyd, S J GI4KCE
Boyd, T A E 2M1CEO
Boyd, W G4BID
Boyd, W GI0BDT
Boyd, W GI4ZOS
Boydell, J C G3TAX
Boyden, P G M GJ4YAD
Boydon, E G G1SJG
Boydon, M G G1UDT
Boyd-Livingston, D J G1KLC
Boyer, J A G0WRX
Boyes, A S G0PWO
Boyes, A S G7HCV
Boyes, J G4YJX
Boyes, M S 2E1FKT
Boyes, P G0TYH
Boyes, R M G1JTD
Boylan, T J G7MFQ
Boyle, G A G3OWC
Boyle, J 2M1EZA
Boyle, P G0NVT
Boyle, R H M GI0VTS
Boylett, J E G3OLY
Boyne, A C G3YVH
Boys, C A E G8BCO
Boyton, J A G0HKE
Bozac, P B G0GOE
Bozman, B J G6OUJ
Bpophy, I R G1OBA
Brabbins, C J G3MPZ
Brabbins, S W G6KJT
Brabon, K A G6FXG
Brace, J GW3JBZ
Brace, R GW3IHN
Brace, R G G3WOO
Bracegirdle, R C GM4UGB
Bracewell, B G3GED
Bracewell, C J G4HVF
Bracewell, S P G7GOQ
Bracey, A G7VOK
Bracey, E M GW4UTS
Bracey, M G4NFB
Bracey, R H GM4BZI
Bracher, C M G4SXR
Bracher, C M G8WUJ
Bracher, D E 2E1CMZ
Bracken, R L G6EQZ
Brackenborough, A R G4ZVC
Brackenbury, S G0WJQ
Brackenridge, J M .. G1YKL
Brackley, K J G3WHV
Brackley, S J G0RFI
Brackstone, L H G7UFF
Bradberry, D J G4OZM
Bradbrook, R A M1ADI
Bradbury, A G3OMU
Bradbury, D W G0KDK
Bradbury, G G6DAO
Bradbury, J 2E1DWZ
Bradbury, J G G6LFG
Bradbury, M J R G4RDF
Bradbury, P G4EXK
Bradbury, P J G7MKG
Bradbury, S M G0URT
Bradbury, S W G1BDT
Bradbury, W 2E1FRF
Bradbury-Harrison, R G G4GOT
Braddock, I G0RUN
Braddon, N C 2E1CEK
Brade, J T K G4CDH
Brade, R G G0VES
Brade, R G G3VIR
Brade, W W T G8ESW
Brades, D E G0IZW
Bradfield, H G G0VTL
Bradfield, P A G1GSN
Bradfield, R G G4XIM
Bradford, C J N G7EQZ
Bradford, D G3LCK
Bradford, I J GW4ZQV
Bradford, M S G3HLI
Bradford, P F G0DNO
Bradford, S C J M1AJU
Bradford, W G4ZGH
Bradford, W J G1IEK
Bradley, A G1NBT
Bradley, A G7SPS
Bradley, A G G1XZJ
Bradley, A H G4YPS
Bradley, C G7KCB
Bradley, C J G1FYF
Bradley, C W W G0RFS
Bradley, D G0PBU
Bradley, D G7REK
Bradley, D C G0GZC
Bradley, D E G3VUY
Bradley, D K G6HV
Bradley, D R G2DVA
Bradley, D W G7HFC
Bradley, D W G7SVM
Bradley, E C 2E1DBM
Bradley, F A G6HNY
Bradley, G G0IQQ

Bradley, G V G0FJZ
Bradley, I S G8UCC
Bradley, J G0ILR
Bradley, J G3MAV
Bradley, J F G6EJH
Bradley, J R G1OSD
Bradley, L GW7NYN
Bradley, L A G4RZD
Bradley, M A G7KEM
Bradley, M D 2E1DBL
Bradley, M D G4WCB
Bradley, M S G7SPR
Bradley, O P GJ3NCJ
Bradley, P G3NEP
Bradley, P G3UJO
Bradley, P G6XQE
Bradley, P J G4BZE
Bradley, P J G7VVK
Bradley, R GW4KGD
Bradley, R E G0UOJ
Bradley, R I G1DQL
Bradley, R M G0OKD
Bradley, S L G1GFC
Bradley, S V G8HYM
Bradley, T H G7KEL
Bradley-Feary, E C S G0PQN
Bradnam, S E G3XXX
Bradnock, J G0IZQ
Bradshaw, A G4FJD
Bradshaw, A J G1BDU
Bradshaw, B I G6CJT
Bradshaw, B P G0LVJ
Bradshaw, D G3TKI
Bradshaw, D J G3SUX
Bradshaw, D M G3HGW
Bradshaw, E 2E1CXP
Bradshaw, E C P G1YLM
Bradshaw, G 2E1DGS
Bradshaw, G G7WGO
Bradshaw, I S G1RLK
Bradshaw, J G0ELR
Bradshaw, J G3BZC
Bradshaw, L G F G0MRL
Bradshaw, M G6UWI
Bradshaw, P J G4CTE
Bradshaw, R E G0CTS
Bradshaw, R H G4PDE
Bradshaw, R R G0GXJ
Bradshaw, S E 2E1DJC
Bradshaw, S H G1WVR
Bradshaw, S H G8GMB
Bradshaw, S J G4OUK
Bradshaw, S N G4FLJ
Bradshaw, S P G4UHX
Bradshaw, W H G4SKS
Bradwell, M 2E1EDB
Bradwell, R H G0FGP
Brady, G M G0UFI
Brady, I C 2E1FCO
Brady, K F G0MKN
Brady, T B G8GAZ
Brady, T B G8HEB
Brady, W J G0JWB
Braeman, N M G4FUP
Bragg, D 2E1FVE
Bragg, G H G4XSK
Braggins, R E W G8RUR
Braham, H L G3OJS
Braham, P A GW4BYA
Braidford, S G8GLX
Brailey, K D G6FXH
Brailsford, C D G7FHA
Brailsford, D P G6YUD
Brailsford, E G3PAF
Brain, B J M1BWD
Brain, C H G4GUO
Brain, D F G7RUJ
Brain, J R B G3VVO
Brain, M G0VNB
Braisher, A W F G0LLR
Braithwaite, E M 2E1FDB
Braithwaite, E R G7WHG
Braithwaite, I G4COL
Braithwaite, J B W G3PMH
Braithwaite, J B W G3PWK
Braithwaite, R G3DAQ
Braithwaite, S J 2E1EGJ
Brake, A J G6EBZ
Brake, R F C G8QR
Bramall, A D G3TJT
Bramall, J R G4HUD
Brambley, J M G0JWD
Bramhill, J G2BMI
Bramhill, M C G4MJL
Bramley, A N G4NDU
Bramley, D G0CYR
Bramley, P J G4WAI
Bramley, R G G4NDT
Brammer, A G6RFR
Bran, K N G3UDC
Branagan, G G6PKV
Branagh, D W GI7JEM
Branagh, J H GI3YRL
Branagh, K R GI4SBA
Branch, C D G0UUR
Branch, C G 2E1EBO
Branch, H E G7RVY
Branch, M L G1OWI
Branch, Y N G7AKP
Brand, J G0MGN
Brand, J J W GM4OWR
Brand, J W G3SSN
Brand, R A G0SJR
Brand, R H G6CEZ
Brand, V E G3JNB
Brander, D GM7OWU
Brandhuber, J A G4PDY
Brandon, B J G4TDU
Brandon, D G4UXD
Brandon, J C G4WXL
Brandram, J P G7OXY
Brandwood, G W G3YJF
Branegan, J GM4IHJ
Branham, D G4PWT
Braniff, B M GI1XTK

Branigan, J O G1MXV
Brannigan, E P GI7TPX
Brannigan, J R G3FHM
Brannon, D G0AVE
Brannon, P M G4IQJ
Brannon, R J G8NGK
Bransby, D A P G7RTJ
Branton, A P G8VUS
Branton, P M G2CNN
Branton, S J G6XWK
Brasenell, D H G6ZLY
Brash, G M GM4FBR
Brash, H M GM3RVL
Brash, J K G0RRP
Brash, P M GM4PGM
Brasier, A S G3XMP
Brasier, J A C G3YLX
Brass, M G4YMB
Brassington, K 2E1GAX
Brassington, M G4UMG
Brassington, P E G4MDM
Brassington, R G6YCN
Bratley, R A 2E1AZM
Bratt, I J G7DXM
Bratt, P J G7DXL
Brattle, D H GM7RHP
Braund, G S G3ZPI
Bravery, R A G3SKI
Brawn, D J G4XJE
Bray, A G1YAN
Bray, A F G0FXX
Bray, B G0ELB
Bray, D S G4MKI
Bray, E M G8OEJ
Bray, F G0NHA
Bray, G G4ZSJ
Bray, J GM6FPD
Bray, M R M0AGR
Bray, M S G0JLP
Bray, N V F GW0MOJ
Bray, R G3KEL
Bray, R C M1AIP
Bray, R J G8LPI
Bray, R M G1WOR
Bray, R M G8VEH
Bray, R P G3NSB
Bray, T J G0JNZ
Bray, T M G7UBQ
Braybrook, M J G1XRM
Braybrooke, P A G6XVQ
Braybrooke, S P G1EOH
Brayshaw, C D G7VOL
Brayshaw, J A G0FNS
Brayshaw, J S G6GRU
Brayshaw, P G7GNU
Brazenall, P R G0NXD
Brazier, D G G8RZJ
Brazier, F S G0SMQ
Brazier, G A GM8VAM
Brazier, L P G4ZBC
Brazier, P A G0OCW
Brazier, P C G6JFN
Brazier, R G G0OSG
Brazington, K J G4LZV
Brazzill, J H G3WP
Breach, J F GI3OJO
Breaden, B G G6OUM
Breadon, L M GI7NMK
Breakspear, R M G1DOL
Breakwell, K J G6JJG
Brealy, D J G3TSE
Breame, F V G8ISI
Brean, G D T G7CSP
Brear, J H G4ITS
Brebner, D G6HOB
Brechany, P J 2E1EEW
Brechany, P J M1BPP
Breck, P A G7EWS
Breckell, D G6FAZ
Breckell, N A G7GOK
Breckons, C M W G6XWD
Bredael, M J G3TYU
Breed, D C G0EGJ
Breed, G A G8UWH
Breedon, F B G3MZU
Breedon, K G1HMZ
Breeds, A P G8ZVX
Breen, D J G0FQI
Breen, S R G7HZQ
Breese, P A G7PZQ
Breet, B C M1ABV
Breeze, A J G0UUZ
Breeze, E L 2E1FXS
Breeze, M D G G3NGA
Breeze, R D G8RTB
Breingan, J R G0RRO
Brejnak, T A GM0VEQ
Brelsford, I C G7CRA
Bremerman, K H G0PZG
Bremner, H J GM3SER
Bremner, S I GM0NOZ
Bremner, S S GM6FYY
Brenchley, J R G6PKT
Brenchley, S R 2E1DOZ
Brend, A G G4PXE
Brenig-Jones, M J G3ZEQ
Brenkley, I D 2E1DRI
Brennan, A E B G1CYD
Brennan, C R S GI4CXH
Brennan, E M GW8WYY
Brennan, F G0BSW
Brennan, J G4TXC
Brennan, J G G7HYO
Brennan, J J G7WJQ
Brennan, K C J GI6OCC
Brennan, P J M1ANE
Brennan, S J GW3ZXI
Brennan, W GI3FTT
Brennan, W M G3CQE
Brent, J H G3FZS
Brent, P J J G4LEG
Brentnall, A M G0THW
Brentnall, D G1JNT
Brenton, P J G8EHD
Brereton, M G8ALE
Breslin, G M A GI4ZLD

Brett, E V G8TUT
Brett, J A G0TFP
Brett, J L G6NBI
Brett, J W G0WWG
Brett, M D G7AHR
Brett, M E G3YDC
Brett, M J G4XVM
Brett, R F G4NDQ
Brett, S P G3MRD
Brett, S P G4COT
Brettell, J B G1LMR
Brettell, K D G0OYK
Brettell, P J G0PYK
Brettle, P D GW7JHK
Brewer, A M1AJZ
Brewer, A J G1TDK
Brewer, A K G1GMV
Brewer, G D G4LJ
Brewer, K G4WYO
Brewer, M D G7SHQ
Brewer, N L 2E1FTK
Brewer, P G G4OSJ
Brewer, S J GW7VKI
Brewer, S J MW0ATK
Brewer, W J G1XHW
Brewerton, P D GW1CLZ
Brewitt, R G4SQY
Brewster, C E GW6RNV
Brice, D F T G0UIL
Brice-Stevens, P N G0WAT
Brickham, K D G3RQY
Brickley, M R G6VVC
Bricknall, K G1DZY
Bricknell, R D R G0SCT
Brickstock, E J G3IVQ
Brickwood, J M G1TTG
Brickwood, M K G1PTO
Brickwood, N M G6DAI
Briddon, M M1ANW
Bridge, J G3ZVQ
Bridge, J H C G0RGC
Bridge, M H F G3VC
Bridge, R J G4WMV
Bridgeland, A P G6DUQ
Bridgeland-Taylor, T R G0JWJ
Bridgeman, E F G0OKH
Bridgeman, P J G3SUY
Bridgen, W G4XSB
Bridger, C J G6CYL
Bridger, P R G1KIV
Bridges, A J 2E1EOC
Bridges, C W GM4NGJ
Bridges, F R G3WPM
Bridges, M J G0LLC
Bridges, N J 2E1EOD
Bridges, P B GM3OBG
Bridges, P C G6DLJ
Bridges, R K G7RFW
Bridgewood, J G1UCP
Bridgland, C I G8KUA
Bridgland, R G G4ALZ
Bridgman, J S 2E1ADJ
Bridgnell, D K G4VPJ
Bridgwater, R J G8RJB
Bridle, G G1GSY
Bridle, K E G1DZZ
Bridle, P G G6CHD
Bridson, R E G3VEB
Brien, T P B G1CYY
Brier, G G8NJE
Brierley, D K G3YGJ
Brierley, G T G4SRP
Brierley, J R G3VO
Brierley, N GW2DNJ
Brierley, S G0UMQ
Brierley, S W G7ONZ
Brierley, W F G0XBB
Briers, A G0KZT
Briers, D G GM1BUL
Brigden, N G1UBV
Briggs, A D G6SJQ
Briggs, A E G0JND
Briggs, A E G8YHX
Briggs, A P G1YOI
Briggs, C G8EXN
Briggs, D F G0LQC
Briggs, D G G0BDM
Briggs, D J G0INF
Briggs, D J GW8OUM
Briggs, E GI3IJU
Briggs, F G4NEW
Briggs, F G4RD
Briggs, H D G1XVR
Briggs, J G7PIR
Briggs, J D G6GWP
Briggs, J L S G7LYH
Briggs, M R G4SMB
Briggs, N I G3WGL
Briggs, R G7SVL
Briggs, R D M0ACG
Briggs, R G G4AMY
Briggs, R J G4YJB
Briggs, S G7VZK
Brigham, G N L G3LEO
Bright, E G3OYS
Bright, E C G7NFQ
Bright, J E G3TJW
Bright, J G G4HJI
Bright, L A G4KEB
Bright, L V G4BHQ
Bright, R A 2E1FYW
Bright, R F G3PQI
Brighting, J A G3UWY
Brightman, G M G7DPF
Brightman, J E G0JXN
Brightman, M B G4NHD
Brighton, A W G7BQY
Brighton, D G4ISK
Brighton, M D G6KAI
Brightwell, M J G8NKQ
Brightwell, R G0TBU
Brightwell, W J 2E1CQP
Brigstocke, A A GW0NUV
Brigstocke, C L GW0MOI
Briley, T A G0EIH

Brill, D H G4USD
Brimacombe, M J G4DZP
Brimley, M G0WSB
Brind, C R G0CDS
Brindle, A W G0BGK
Brindle, G F G3VXE
Brindle, H G3YZX
Brindle, H D G3YZY
Brindle, J G0DVT
Brindle, J G4SHX
Brindle, S P G3RDV
Brindley, E T G4CZH
Brindley, J R G4CWJ
Brindley, M D G6VTX
Brindley, P G2TO
Brindley, P G6BSE
Brindley, P D G0HEV
Brindley, W D G0MVT
Brinham, J G8XSI
Brinkley, R C G0UPD
Brinkworth, N G G3UFB
Brinnen, D N G7BUK
Brinnen, N R F G3VDV
Brinsley, J A G8WDV
Brinton, A W G7DIR
Brion, C J G4BTN
Brion, S M1BLJ
Brisbar, G P GW3LWU
Briscoe, A J G4YUB
Briscoe, A L G4GKR
Briscoe, C G3YHR
Briscoe, J M GM8AOB
Briscoe, M G0WJC
Briscoe, S R G1YHV
Brisley, J R G4MPN
Brislin, A R G6WPO
Brissenden, D M1BLI
Bristeir, R G G0ADG
Brister, J A G8IXX
Bristow, A J G8YBH
Bristow, B M G4KBB
Bristow, B M G4MDF
Bristow, C G GI3PSQ
Bristow, C J G7TGB
Bristow, D E G0EOH
Bristow, E G0EPK
Bristow, G E G1HGW
Bristow, K G3XCY
Britt, R D G8WTM
Brittain, D K G0UQB
Brittain, J G7TVS
Brittain, J V G G1LEO
Brittain, R E G6IUF
Brittan, H R G8FGQ
Britten, B E G1MOF
Britten, J M0ALD
Britten, P N G6BPJ
Britton, A J GM1JKJ
Britton, D C G0SCK
Britton, J G6CQO
Britton, J R GW6DGU
Britton, L D G0BNG
Britton, R G4PQG
Britton, R W G8FUO
Britton, T E G8XSP
Brixton, A G G1DOK
Broad, B J G6LZX
Broad, C D G7OQL
Broad, D M G1PVT
Broad, G R G6AXE
Broad, M D G1BDP
Broad, M R G6MNN
Broad, P G G0SWU
Broad, S GD1AQY
Broadbent, A G3PYH
Broadbent, A E G3HVE
Broadbent, A F G0NCG
Broadbent, C G G1RPP
Broadbent, C M G4UTT
Broadbent, I G1AXE
Broadbent, I D G0KMT
Broadbent, J L G4MBK
Broadbent, J M G3UYH
Broadbent, N A G8CDG
Broadbent, R G7GTD
Broadbent, R H G2AND
Broadbent, R J C G0AUK
Broadbent, W H G1YUV
Broadbent MBE, R J C G3AAJ
Broadberry, R G1PNF
Broadbridge, D G3RNT
Broadfoot, B J G1KCY
Broadhead, P A G7OLC
Broadhead, S E G0HQU
Broadhurst, C R H G3PH
Broadhurst, F G4KVG
Broadhurst, G G G0BVC
Broadhurst, G G G0SVP
Broadhurst, J J G4JJB
Broadhurst, P A G4NFD
Broadley, D L G1XFP
Broadley, J D G3WBP
Broadley, P J G0HSN
Broadwater, J W G6HHS
Broadway, M G6LLG
Broadway, M V L G4GFI
Brobin, P G G7MHX
Brock, A C G4BOZ
Brock, B T N GW6NKG
Brock, C H G6DGV
Brock, C J G3ISB
Brock, E A G4HQK
Brock, G F G1ZBH
Brock, G G G8WRY
Brock, H T G3FD
Brock, H T G7SRC
Brock, J C D G3KDQ
Brock, J W G8ORQ
Brock, L W G3WXF
Brock, P G1CXK
Brock, S A 2E1FDI
Brock, S H G7SZM
Brock, W C G4VNV
Brock, W C G8MPM
Brock, W W G7VIM

Brockbank, C J G3RCD
Brockett, J W G4KXP
Brockett, P J D M1BLM
Brockett, P S G1LSB
Brockie, E H GM4EHB
Brockie, R J GM8PJM
Brockis, J G G0SGA
Brocklehurst, D G4VDB
Brocklehurst, S G6VMV
Brocklesby, D E G7RLH
Brockman, T H G4TPH
Brocks, R A W G3WHR
Brockway, J G4OUH
Brockway, M J G4OUI
Brockwell, J B G8HYK
Broder, P D G1NIW
Broderick, M J 2E1CTT
Broderick, M L 2E1CTS
Brodie, B S G4LUU
Brodie, D W G4ZDT
Brodie, D W G7TXQ
Brodie, D W G7UBC
Brodie, E A D G6HTB
Brodie, J F G3RYH
Brodie, P K G1WQL
Brodie, R A GM7IHR
Brodie, S J G0PSY
Brodie, S J G7SNR
Brodie, T G GM0KZU
Brodribb, B C G1EHS
Brodribb, P B G3ONL
Brodrick, C J G7VUN
Brodrick, R J G4LAF
Brodrick, T D G1DOJ
Brodzky, J G3HQX
Brogan, J C GW3NIN
Brogan, M A G8JHE
Brogden, G K G1IGX
Brolan, P G1MPU
Bromage, K G1YAO
Bromage, K M G8YDQ
Bromage, M G7ABZ
Bromby, D G G8JZC
Bromfield, A J G6JJI
Bromfield, D A GW4KFI
Bromfield, G G0MIT
Bromiley, J G2ANC
Bromley, G G4NID
Bromley, G G4UTN
Bromley, N H GW3IVR
Bromley, P M J G6MMJ
Bromley, R C G1TJR
Bromley, T G1WPR
Bromley, V J G0ROQ
Bromsgrove, B G0AJL
Bromsgrove, T H O M G4STZ
Brook, A D G0JKA
Brook, A J G7WGP
Brook, D G1YCV
Brook, J R G7VDH
Brook, K G4ENR
Brook, K F G0LIO
Brook, M A G7NVL
Brook, M A W G0DIJ
Brook, N G0KIY
Brook Foster, D G3NRU
Brooke, A G0UEF
Brooke, A S G6YCE
Brooke, C W G4NHY
Brooke, D A G6GZH
Brooke, D L G7SCO
Brooke, D R G0TRH
Brooke, H G3GJV
Brooke, H N G6ERB
Brooke, M G4SKO
Brooke, M J G8HXR
Brooke, P L M1BIB
Brooke, P W G7TBV
Brooke, S G1GKK
Brooker, A D G3NAT
Brooker, A D G4WGZ
Brooker, D A G8NUJ
Brooker, G P G4XVJ
Brooker, H W G3FAM
Brooker, J G3JMB
Brooker, J H R G3VOR
Brooker, K W G3OXH
Brooker, M G G0WVM
Brooker, P G G3WXC
Brooker, R J G3HBI
Brooker, T W G1FZQ
Brooker, V J G7NHY
Brooker-Carey, A G3OGH
Brookes, C W G1XMP
Brookes, F J G3PGZ
Brookes, G B G3NSO
Brookes, G H G7OEB
Brookes, G J G6SYX
Brookes, G R G0SER
Brookes, I G4TQA
Brookes, J A G4YMW
Brookes, K G1NOS
Brookes, K H G0UZE
Brookes, L E G0HPG
Brookes, P G G0HPH
Brookes, P J G0DFK
Brookes, R G0DVI
Brookes, R G6YCM
Brookes, R L G8DZW
Brookes, R L W G4PUM
Brookes, S E G1IGC
Brookes, S M G1WLU
Brookes, S N G8ZVV
Brookes, T W G0JRZ
Brookfield, B G4ZDD
Brookfield, D J G1PJL
Brooking, D G1KHX
Brooking, P G4SHH
Brookman, B M GW7KIS
Brooks, A A G4VAV
Brooks, A D G1KAC
Brooks, A D G3IDB
Brooks, A H G0RBX
Brooks, A J G4KIQ
Brooks, A N G7MKP

Brooks, A P M1AFQ
Brooks, A W G3VLI
Brooks, C A 2E1CIO
Brooks, C R G1IGA
Brooks, D E G1PGN
Brooks, D J P G7VLM
Brooks, D M G0VIE
Brooks, D R G4IAR
Brooks, D R G4KXH
Brooks, E F G3HFW
Brooks, E H GD4HOX
Brooks, G A G1YMN
Brooks, G D G4RHT
Brooks, G G GM4NHX
Brooks, G R G3LVB
Brooks, G T G4KGF
Brooks, H G GM4FXX
Brooks, I B G0AAY
Brooks, I D G6IAN
Brooks, I N G1RVK
Brooks, J G4IAQ
Brooks, J A G0GDJ
Brooks, K G0SPH
Brooks, K R G3XSJ
Brooks, M 2E1EYQ
Brooks, M M1APE
Brooks, M A 2E1ACU
Brooks, M E G6EBQ
Brooks, M P G4IWB
Brooks, M P G8OXE
Brooks, P G4NZQ
Brooks, P G4UMI
Brooks, P M GM0SIA
Brooks, R A G7JGZ
Brooks, R F G3YLL
Brooks, R L G3FGP
Brooks, R W D GW1JFV
Brooks, S 2E1BZL
Brooks, S C GW6JFM
Brooks, T G8DPE
Brooks, V T G4JOA
Brooks, W A G2DDS
Brooks, W J G4DTT
Brooksbank, E GW6YUC
Brooksbank, V J G1EHR
Brookson, C B G4GBA
Broom, C J G7BFE
Broom, I D G1PMD
Broom, J I G0IVY
Broom, P J G5DQ
Broom, R J G0ONT
Broom, R M G4LFE
Broom, R W G4GWW
Brooman, M G G0LGE
Broome, J D G0CCF
Broome, J D GW7LUB
Broome, K F G3RBW
Broome, R D G1YKO
Broomfield, A G8KNM
Broomfield, C F G8CAA
Broomfield, F W J G3JTI
Broomhall, I K G0JYZ
Broomhead, J H W G3NCX
Brophy, M D G1MGZ
Brosnan, S M1BVD
Brosnan, T P G4DRO
Brotherhood, A 2E1FRP
Brotherton, I D G1BRS
Brotherton, I D G2BDV
Brotherton, I D G2BRS
Brotherton, R P G7JSC
Brotherton, R W M1AXX
Brotherton, W D W GW3WWB
Brothwell, I R G4EAN
Brothwood, R D G6ICH
Brouder, P J G3ZJH
Brough, B J G1BCE
Brough, B W GD4PTV
Brough, C A G0VKZ
Brough, D W G3HUR
Brough, G E G7RIN
Brough, S D G1UUL
Brough, S R G4NQK
Brough, W G0GLY
Broughton, A J G1EAE
Broughton, C K G1RSK
Broughton, D A G1IEX
Broughton, G G7TZV
Broughton, J H G4ZSV
Broughton, P N G3ZJF
Broughton, R E G6YTR
Broughton, R P G4LZK
Broughton, R W G0MRB
Browell, J T G4PUI
Browell, R G G4ZMJ
Brown, A G0GWD
Brown, A G1XYS
Brown, A G2WQ
Brown, A GI7URC
Brown, A GM4ZIL
Brown, A GM7VCV
Brown, A GW7ERI
Brown, A A G1ALD
Brown, A A G8GLV
Brown, A A J G4WSI
Brown, A A T G8LTN
Brown, A D G4UIM
Brown, A D G8INO
Brown, A F MM0ALY
Brown, A G G0NBD
Brown, A J G0LXE
Brown, A J G3TMO
Brown, A J G4GPV
Brown, A L H 2E1CSY
Brown, A M G6UZR
Brown, A N GW7WEE
Brown, A P G4EIQ
Brown, A R G0AFJ
Brown, A R G4VSB
Brown, A S G0FKI
Brown, A S G4WWG
Brown, A T G1OPJ
Brown, A W G4CCB
Brown, B J C G3JFD
Brown, B P G1MLP
Brown, B R G3FMZ
Brown, C G0UNJ

Brown, C G1KIE
Brown, C G4LIL
Brown, C A G7VNO
Brown, C B 2E1EFD
Brown, C D GW4NQJ
Brown, C G G7SOH
Brown, C H GM0GZX
Brown, C J G1ZDC
Brown, C J GM0RLZ
Brown, C L G4CLB
Brown, C M GM4WEW
Brown, C R G6PRI
Brown, C R G7GJZ
Brown, C T G0JRM
Brown, C T G6ZWC
Brown, C W G0MGU
Brown, C W G1UKG
Brown, C W W G3CEI
Brown, D G0EMF
Brown, D G0SUX
Brown, D G6MMG
Brown, D G6XMH
Brown, D G7SRE
Brown, D G7UQH
Brown, D G8GDH
Brown, D GM8FFH
Brown, D A G4MUI
Brown, D A G6LQP
Brown, D A GM6JFP
Brown, D A T GM6JUA
Brown, D C G0NWV
Brown, D C G6PRL
Brown, D C G8ECI
Brown, D E GW0ROL
Brown, D G G4PJN
Brown, D H G3YGD
Brown, D H G7LLP
Brown, D H J G0HBB
Brown, D J G4KFN
Brown, D J G7KDP
Brown, D J G8GUL
Brown, D J M1AAX
Brown, D M G7OVE
Brown, D P G6UZT
Brown, D R D G3UOC
Brown, D S G0CKQ
Brown, D S G0LYX
Brown, D S G4LNM
Brown, D S G4UNS
Brown, D T GM6RAK
Brown, D W GM4RWE
Brown, E G0WDO
Brown, E G3CSP
Brown, E G4STK
Brown, E G4YXE
Brown, E A G1YYD
Brown, E A GM0ALU
Brown, E C G3AXF
Brown, E C GM0BRS
Brown, E C D GM0BPO
Brown, E F GI0AHZ
Brown, E K G8VGY
Brown, E M E G3ZIE
Brown, E R G8LVU
Brown, F H G8FPW
Brown, F J J G4MWN
Brown, G G1VCY
Brown, G G3MZV
Brown, G GD0KWM
Brown, G GJ4HXJ
Brown, G A G3HLP
Brown, G B GW0PUP
Brown, G D G4VRX
Brown, G E G0BHW
Brown, G E G2DBP
Brown, G F G0PWU
Brown, G F G1IEJ
Brown, G F G2BJK
Brown, G H GM4PVH
Brown, G S G0JSL
Brown, G S G4WUA
Brown, G S D G6NKM
Brown, H J G8HVS
Brown, I G4XTG
Brown, I A G4YSN
Brown, I D G0NPO
Brown, I D G3TLH
Brown, I D G3TVU
Brown, I J G7PLP
Brown, I K G7UWV
Brown, I L GM0ILB
Brown, I L GM0VFA
Brown, I R G8RBK
Brown, J G0DPX
Brown, J G0OGG
Brown, J G0PJU
Brown, J G3JOE
Brown, J G4EJE
Brown, J G4UVL
Brown, J G6ETC
Brown, J G7GOX
Brown, J A G0CZS
Brown, J A G0GJB
Brown, J A G0JSM
Brown, J A G1MBG
Brown, J A G1XMI
Brown, J A G4NUK
Brown, J A G8FZW
Brown, J A GM7SPA
Brown, J A GW3RNC
Brown, J A H GM4HVU
Brown, J C G6PCP
Brown, J C G7FGD
Brown, J D G0RBR
Brown, J D G0UFY
Brown, J D G4PHE
Brown, J E 2W1CEZ
Brown, J E G0KJW
Brown, J E G0SOU
Brown, J E G3ECP
Brown, J E GW8EHQ
Brown, J F G3KTN
Brown, J G G6UV
Brown, J L G8FIK
Brown, J L GM4ZIT
Brown, J M G0PIA

Brown, J N G0PBF
Brown, J O G3DVV
Brown, J O A G8WYX
Brown, J P G4XSS
Brown, J P G7GMY
Brown, J P GM7HHB
Brown, J R D G4UBB
Brown, J S GM4VLX
Brown, J T 2E0AQE
Brown, J V E G4PXV
Brown, K G0IKB
Brown, K G4ZNY
Brown, K G7CXK
Brown, K GM4WEX
Brown, K C G3PHT
Brown, K D G3XQE
Brown, K D G8IRL
Brown, K E G0PSW
Brown, K E G6NKI
Brown, K T G7EXO
Brown, L 2E1COI
Brown, L G4ZQS
Brown, L A G8YQY
Brown, L B G7LQO
Brown, L G G4WSN
Brown, L P G3JBF
Brown, L S G7LSB
Brown, M G0MNH
Brown, M G3ZXM
Brown, M A G7WMB
Brown, M B G7SQE
Brown, M C G3OVE
Brown, M C GM0SGH
Brown, M D 2E0AGW
Brown, M D D G3ZPW
Brown, M E G0DCG
Brown, M E B G3XMW
Brown, M I G7VED
Brown, M I M0APC
Brown, M J G0JLS
Brown, M J G3UDP
Brown, M J G8OAQ
Brown, M K G6XVO
Brown, M L B G4ZPN
Brown, M M 2E1FWE
Brown, M P B G4RAA
Brown, M R G3HUD
Brown, N GM4VHZ
Brown, N D F G0GDA
Brown, N E G8NCK
Brown, N I G5NB
Brown, N J G7MVU
Brown, N S 2E1BVQ
Brown, N V G8NVB
Brown, P 2E1FDT
Brown, P G4EYP
Brown, P G4WWY
Brown, P G6LPX
Brown, P A G0HIH
Brown, P A G3PZL
Brown, P A G3WRI
Brown, P A G8OXD
Brown, P C G8OEK
Brown, P D G0DGU
Brown, P H G3WUZ
Brown, P J G4ZJW
Brown, P J GI0KOV
Brown, P J M1BTB
Brown, P R G8GLB
Brown, P W G4AJE
Brown, P W G6PJC
Brown, P W G7ILD
Brown, P W GM4IDV
Brown, R G1LPL
Brown, R G3LQP
Brown, R G3XNO
Brown, R G3ZNT
Brown, R G4LZT
Brown, R G6PRO
Brown, R G7LYB
Brown, R G8SCH
Brown, R GI6IVJ
Brown, R GM4IKU
Brown, R M0AHZ
Brown, R A G0NGG
Brown, R A G3SCZ
Brown, R A G8OKE
Brown, R B GW6JQT
Brown, R C G3JRC
Brown, R C G4JTZ
Brown, R C GW6WRP
Brown, R E G0PBW
Brown, R E G1NLZ
Brown, R E G3GZH
Brown, R G G8CXV
Brown, R G GM7RGC
Brown, R I G4VKY
Brown, R I G4ZTH
Brown, R J G0KZV
Brown, R J G7JFS
Brown, R J G7PKD
Brown, R J G7SPZ
Brown, R J GI4BXB
Brown, R L G1ZOB
Brown, R M 2E1DCF
Brown, R M G4NAG
Brown, R N G3WPT
Brown, R S G7BZM
Brown, R T G1LCY
Brown, R T G1SCO
Brown, R T A G3TOF
Brown, R W G1FFA
Brown, S G4HPV
Brown, S G4ISB
Brown, S G6CRD
Brown, S G8BCU
Brown, S F G4LU
Brown, S H GW8LZY
Brown, S I G4PRF
Brown, S J G0NZQ
Brown, S J G1UEV
Brown, S J G1ZBB
Brown, S J G4ATU
Brown, S J G4NYL
Brown, S J GD4ELI
Brown, T G0NSA

Brown, T G4TZB
Brown, T G G2HMK
Brown, T G G8ZMJ
Brown, T J G7TQE
Brown, T M G7NUG
Brown, T R G7FJK
Brown, W G4MGK
Brown, W G7VDO
Brown, W G8ODA
Brown, W GW7FAE
Brown, W B G6QY
Brown, W C G3FBU
Brown, W D G0HZD
Brown, W D G GD4XTT
Brown, W F G1GCS
Brown, W G GM0UEQ
Brown, W H G3NQX
Brown, W I GM0IYA
Brown, W J GI1JJC
Brown, W L GW7DWE
Brown, W N G4WNB
Brown, W S GI1TVH
Brown, W T G1BBY
Brown, W T G3NTM
Browne, A O G0HWL
Browne, D F G4XKF
Browne, D F G7HCQ
Browne, D G W G7PRH
Browne, G GI0MXT
Browne, G R G3MMJ
Browne, G W 2E1FNZ
Browne, H W R J GI7REP
Browne, J R G3XZG
Browne, L W G0VCD
Browne, M R G8UUL
Browne, M W G8AOG
Browne, R G G0NUU
Browne, S G0GWA
Browne, S R G4SJU
Brownell, G A G4HRA
Brownett, G S J G4FAZ
Brownhill, W H GM0KJX
Browning, D M G3UEY
Browning, K D G7NUV
Browning, L W G1FBO
Browning, R GI0MQN
Browning, R C G4BEB
Browning, R E W G4UMW
Browning, S J G8WUR
Browning, S R G8CSK
Browning, W L C GW4ISF
Brownjohn, K R G4MTS
Brownlee, H W G3ZGR
Brownlee, I P G8VXZ
Brownlees, J L GI6VCG
Brownlie, I GM4EGD
Brownlow, M L G4LCU
Brownlow, P J G6OCP
Brownlow, P N G4ESC
Brownsett, J F C G6PAA
Brownsett, P J G7RDA
Brownson, W H G3NYI
Broxton, M R M1BXX
Broxup, B J G7KWC
Broxup, K M1BVQ
Broxup, W G4OPN
Bruce, A G4NSM
Bruce, A GM7JGH
Bruce, A J G6AKP
Bruce, A J G6OCO
Bruce, A W H G6MON
Bruce, B C G8EEK
Bruce, C D G7UPV
Bruce, D J G2XD
Bruce, D W G3IGZ
Bruce, J GM1IEL
Bruce, M G G G3TUY
Bruce, N G6XVM
Bruce, P K G1FJH
Bruce, P S G4WPB
Bruce, R B GM8LON
Bruce, T N G6IAT
Bruce-Smith, M J G8OBK
Bruckshaw, D A G1JRF
Brudenell, D A G4BUL
Brugsch, H I G0GKU
Bruin, T A C G0PRN
Brumby, E G GM8YKT
Brumby, G J G0KWX
Brumby, P G0RRM
Brumwell, M D L G8HTV
Brundle, M J G6EBL
Brunlees, J D 2E1FLM
Brunning, A N G4PTW
Brunning, J G8NMA
Brunnschweiler, P G7SON
Brunsch, D A G0UOG
Brunsden, D C G4ZLQ
Brunsdon, M K G6UZO
Brunsdon, W R GM8LVJ
Brunskill, R GM3MIE
Brunt, D J G6KTE
Brunt, D L G7UWS
Brunt, J W G0TPY
Bruntlett, K D G4MXI
Bruntnell, V D G7GEU
Brunton, A GM6VGL
Brunton, M D GM8GDN
Brunton, P S G0RBV
Brunton, R L G4TUT
Brunton, S J G0EXI
Bruring, A G7MJD
Brusch, D J G0SCQ
Brush, N J GW0MOQ
Brushwood, D J G3ZJB
Brushwood, P D G4VIQ
Brutnall, G E G4PAV
Brutnall, G E G8COM
Bruton, J G G8CBL
Bryan, C G4EHG
Bryan, C G7ANK
Bryan, C J G6TPI
Bryan, C W G6OAN
Bryan, D G4JCL
Bryan, G M G8MCA
Bryan, H G6TMN
Bryan, M G4DTB
Bryan, M GW6ATT
Bryan, M A 2E1DUT
Bryan, N A G0DXC
Bryan, P E G0DQD
Bryan, R M A G0NLA
Bryan, S A G7VHG
Bryan, S G G0SGB
Bryan, T E G4AOH
Bryan, T W G3GSO
Bryan, W G4BMV
Bryant, A J G1FTV
Bryant, A M G3AGO
Bryant, A R G3EFE
Bryant, C J G3WIE
Bryant, D G7JLS
Bryant, D D G1JFU
Bryant, D G G0NES
Bryant, G H G0TZV
Bryant, H G4TTG
Bryant, J G4LTR
Bryant, J G8XIJ
Bryant, J A G1XAM
Bryant, J M G4CLF
Bryant, J M G8KMM
Bryant, K G0BSK
Bryant, K D G7JLT
Bryant, L R G3WJO
Bryant, M J S GW6NLP
Bryant, P F G0JWV
Bryant, P M G8ZFI
Bryant, P S G7HGY
Bryant, R C G8KWN
Bryant, R D G4UBM
Bryant, R E G3WBC
Bryant, S F G3YSX
Bryant, T C GW3SB
Bryant, T C GW4LZP
Bryce, A M GM8EVF
Bryce, G R GM3JOB
Bryce, J G4LWY
Bryce, J GM3GBY
Bryden, G T GM4CAC
Bryden, J F G4HNQ
Bryder, J F G0JOP
Brydon, I G0PMZ
Brynes, M G G0IBB
Bryson, G F GM4DYZ
Brzenczek, P M GW0TTN
Bubb, G M G7KNS
Bubez, J T G0SQF
Bubloz, G G4FNL
Buchan, A J GM0EFH
Buchan, A V G6ZOJ
Buchan, N J G0MAR
Buchan, P B B G3INR
Buchan, R G7PQM
Buchanan, G I GM3KCY
Buchanan, J GM4VGR
Buchanan, N B GI0TJU
Buchanan, N W L G0KUF
Buchanan, W S GM1EAH
Buck, A G G7MBD
Buck, C J G1TZF
Buck, D A G4HAS
Buck, E G4SGC
Buck, F D G3JNO
Buck, G G0RFQ
Buck, J G7VNZ
Buck, J L H G7RWF
Buck, L N W G0DLR
Buck, M A G0FKZ
Buck, M F GW4NHH
Buck, M J G6YCI
Buck, M R GW6KGR
Buck, P G8AUL
Buck, P W G3LWT
Buck, R L G4RXQ
Buck, T G G6VXL
Buckby, R I G2DJ
Buckby, R I G3VGW
Buckenham, H A G3PGN
Buckenham, P G0DKB
Buckett, H J G3ARL
Buckett, W G3ODO
Buckie, I E G6WXK
Buckingham, D R G0ENJ
Buckingham, J L G4BCS
Buckingham, P J G1JBI
Buckingham, P R G6ECS
Buckingham, R F G6ECR
Buckingham, S M G7TZN
Buckland, C R G8IYJ
Buckland, C V G3WJU
Buckland, D G3JKM
Buckland, E M G0JWN
Buckland, J G7MJF
Buckland, J E G4SLL
Buckland, M J G0XAE
Buckland, M J G7UDN
Buckle, D GW0FUR
Buckle, I G0MIF
Buckle, J L G1LES
Buckle, M E J G4RMV
Buckle, M P 2E1CPH
Buckle, M T G8UOW
Buckle, P M GW1VBA
Buckle, R F M0AZS
Buckle, R K G6HGM
Buckle, T G0EWV
Buckler, C A GD4HIT
Buckler, J M G0HNI
Buckler, M D GD0EEM
Buckley, A G4STW
Buckley, A J M1BFF
Buckley, C G7VKJ
Buckley, C F GW4HHO
Buckley, D G1XRY
Buckley, D G3VLX
Buckley, D G4OQD
Buckley, D G7ORT
Buckley, D C G0EQV
Buckley, E G0CVC
Buckley, E MW1BQO
Buckley, E A G4KVS
Buckley, F GW3JRJ
Buckley, G G4OTM
Buckley, G J G0PNG
Buckley, G M G8LEI
Buckley, H G3CZO
Buckley, J H H G7WLN
Buckley, J M G4HGL
Buckley, J V G8TFO
Buckley, L G3PTX
Buckley, M J M1CCF
Buckley, M W G7RME
Buckley, N M G7EQM
Buckley, P G0APB
Buckley, R D GI3HCP
Buckley, R E G1GIE
Buckley, S J 2E1FXW
Buckman, R N G3CZL
Buckmaster, J M G0BYQ
Buckmaster, S F G6VGN
Bucknall, B A G6PJM
Bucknell, R E 2E1CUL
Bucknell, W C G0ULQ
Buckstone, R J C G5JR
Buckwell, R L G0MBV
Budas, M GM4VTB
Budas, V T GM3VTB
Budd, C G0LOJ
Budd, C J G4NBG
Budd, D G6NLS
Budd, D J G6DAH
Budd, J A G4WMW
Budd, M A G7CSS
Budd, S J P M0AIB
Budden, G A G3WZP
Budden, G F G8JH
Budden, J C G0PCW
Buddery, D J M G3OEP
Buddery, D L G3SEP
Budding, I T G4GWK
Buddle, R W G4UOX
Budge, G P GW0MGQ
Budgen, P G7MAG
Buer, P L M1BZJ
Bues, M J G8AAI
Buffham, I G3TMA
Bufton, N T R GW0MNO
Bufton, R 2W1CIP
Buga, O H G0MLD
Bugby, M J G1HVA
Bugden, B W G8MCS
Bugg, J E G1PAJ
Bugg, J P G6NLR
Bugg, M G6HNN
Bugg, T D G6XMM
Buggs, D G G1YRF
Buick, A S G4TZZ
Buik, D A G0DAB
Buksh, K G3YWZ
Bulbeck, M G4YOI
Bulbrook, J S G6MMB
Bulcock, M J G4KQI
Bulcock, M J G7CYC
Bulger, G V G3WIP
Bull, A C G1MLK
Bull, A J G7NSM
Bull, A P G3ICB
Bull, B G4AGB
Bull, B G4ARF
Bull, B M1BRF
Bull, C J G6TLX
Bull, D J G0LKZ
Bull, D P 2E1FFD
Bull, G G7HMB
Bull, H E G3ABM
Bull, J L G8WVH
Bull, M A G4MIK
Bull, M D G4BHZ
Bull, M J GM1MLY
Bull, R G G1KAJ
Bull, S G 2E1CMD
Bull, T S 2E1CMC
Bullard, D P G7TWA
Bullard, J W G0FTW
Bullen, R A G1GDH
Buller, B J G4JUM
Buller, D S G3LQT
Bullers, S G1VKL
Bullett, V D G3EAO
Bulleyment, G G G3XIV
Bullimore, R E G4VOT
Bullivant, C H G3DIC
Bullivant, E A E 2E1AOR
Bullock, D C G6UWO
Bullock, G G G0SED
Bullock, M B 2E0AAL
Bullock, N H G3TAQ
Bullock, P L G4WUO
Bullock, R G4XRV
Bullock, R GW1FJI
Bullock, R K G0EML
Bullock, S R G6DOT
Bullock, W D G4CVG
Bullough, M L G1OVS
Bullough, N G M1AIH
Bullough, P G1CWD
Bullough, R P G0TBB
Bullwinkle, J A G1BPI
Bulman, J G1PWM
Bulmer, H S G4FZS
Bulmer, M N G6MNB
Bulmer, P D G0TTS
Bulpin, J GW0BNN
Bulteel, J B G0UHT
Bulteel, J B G4PII
Bultitude, D A G7SVT
Bultitude, I A G6KHW
Bumford, J R M G0GTN
Bumford, J R M M1AXW
Bumstead, D E G4FHV
Bunce, F C D G0GAJ
Bunce, M G G3YFO
Bunce, P A G7VBZ
Bunce, T J L GW4RRR
Bunch, D H G8TVO
Bundell, R A G0KQR
Bundey, D A G3JQQ
Bundle, D C G0CDJ
Bundle, N J G4LRV
Bundock, P A J G7LBT
Bundy, M W G4WVD
Bundy, W H G7AKZ
Bunker, D G G0MDC
Bunker, K R G2DFY
Bunkum, C B B G1JXS
Bunn, C W M1BVS
Bunn, G D G7MLC
Bunn, M C G6ITU
Bunn, P J G6UZL
Bunney, I A G3LFR
Bunney, R W G8ZMM
Bunston, K W C G3AAK
Buntin, J W G1STA
Bunting, H G4BUN
Bunting, J C G7MHZ
Bunting, J C M1CAT
Bunting, M C G6BZ
Bunting, M D G1EAJ
Bunting, S GI8GJX
Bunting, S C J G7ACQ
Bunyan, A S G3XLR
Bunyan, R J G0OHQ
Burbage, R G7AQL
Burbanks, J C G3SJJ
Burbeck, P G0OMH
Burbeck, R D G4NOB
Burbidge, N G6ZLP
Burbidge, R K G3GAI
Burbridge, B G0FMA
Burbridge, M J G0VPU
Burbury, P R G7CIU
Burch, G D G7VQI
Burch, G R G3CQO
Burch, J H 2E1GAY
Burchell, A D M1BPN
Burchell, C R G3NKQ
Burchell, D J T G6TPT
Burchell, D W G0ANV
Burchell, G P G7VXS
Burchell, L J G0DJB
Burchell, P GW0VSO
Burchell, R P G0WBV
Burchell, S G1DAZ
Burchett, P G0LMG
Burchmore, A R G1RCV
Burchmore, A R G3RCV
Burchmore, A R L G4BWV
Burchmore, M G0ARQ
Burden, C G G4AXC
Burden, I P G0RRI
Burden, J P H G3UBX
Burden, P R G1WFO
Burden, R T M1ATD
Burden, W H C G3EAT
Burden, W R G3XCJ
Burdess, R G4OFU
Burdett, D J G7MFW
Burdett, J G0SUT
Burdett, J M G1KNA
Burdett, R A G0DLB
Burdis, B G0WKQ
Burdis, B G0WZB
Burdis, B G7OXN
Burdis, M E 2E1FKG
Burdon, I R T G1TKQ
Burdon, K A G4NHZ
Burdon, M A G7LXN
Burdon, R GW0PZQ
Burdon, T H G3PVR
Bures, A M D 2E1DXK
Bures, M A 2E1DZI
Burfield, H E R G3IIX
Burfield, R A G6MNK
Burfoot, P A G8GGM
Burford, A R G3VTI
Burford, J F G4OAZ
Burford, J F G4ZOZ
Burford, M M 2E1DLW
Burg, R J G8WSC
Burge, A J GI3OTU
Burge, D A MW0ALG
Burge, D A S G4ILV
Burge, J C G3VHN
Burge, N W J G0BTQ
Burge, P G G1BDK
Burgess, A G4GLV
Burgess, A B G3LPO
Burgess, C 2E1CVM
Burgess, C G G8EWL
Burgess, C M GM0EZH
Burgess, C S G0RKE
Burgess, D 2E1BLA
Burgess, D P G4IYS
Burgess, G 2E1FFN
Burgess, G M G4HDY
Burgess, I T G0KQJ
Burgess, J G3KKP
Burgess, J E G0TAZ
Burgess, J N G7CII
Burgess, J W G4BNP
Burgess, K G1HFS
Burgess, M G7HID
Burgess, M J G6UCQ
Burgess, O W G6SSM
Burgess, P G3VPT
Burgess, P D G4BCH
Burgess, P M G6OUO
Burgess, P S 2E1FMV
Burgess, R G8VTB
Burgess, R P G0XGM
Burgess, R P G7VRB
Burgess, R W G G3RXG
Burgess, S F P G4TRM
Burgess, S J G7CPN
Burgess, S R G4CQO
Burgess, S W G4NMS
Burgess, T G6ZDT
Burgess, T R A G8VGU
Burgess, W F G3VTN
Burgess-Lee, R E G3NAL
Burgin, D F G7MII
Burgin, K N G4CRG
Burgon, J G3TII
Burhouse, G B G4MVA
Burin, W G8PDE
Burke, A J G6BEN
Burke, A M G4HIY
Burke, B P G0SFR
Burke, D C 2E1FZV
Burke, J G3MVX
Burke, J GM4TNP
Burke, J GM7FRC
Burke, J A G1JAB
Burke, J U G3HEA
Burke, M G7VHR
Burke, M M0AMS
Burke, T G4UGR
Burke, T J G1LXU
Burke, T R G3UPM
Burke, W M GM3TQH
Burkett, L R G1PUC
Burkitt, A R G1VLD
Burkitt, C J G3PZE
Burkitt, E T G3TB
Burkitt, N D G4FXT
Burleton, A S G6GHD
Burley, A L G0MBM
Burley, T R A GW0EGF
Burling, J D G4CKA
Burling, M J G1HND
Burling, R G G0CMB
Burling, S H 2E1FLW
Burlington, G M G4HXQ
Burman, R A H G4RSN
Burn, C M GW8WYW
Burn, D R J GI0AQD
Burn, G P W G7HLW
Burn, H G0RKN
Burn, J F G4VSF
Burn, J P R 2E1EHZ
Burn, J P R M1BWQ
Burn, P S GM4PSW
Burn, T R G7KVH
Burnand, P B G0VET
Burnard, A E G2FCA
Burndred, E F G0KBJ
Burndred, W T G0KBI
Burnell, C E GW8SCR
Burnet, J Y G1LDN
Burnet, P G3YQJ
Burnet, W G3WZA
Burnett, A H G4LSU
Burnett, A J G1SPU
Burnett, A M G6IVT
Burnett, A W G0IFC
Burnett, C G G6GCI
Burnett, C M G7LYX
Burnett, D J G0MXB
Burnett, F G G4CYB
Burnett, F S C G3RSM
Burnett, J G4NDP
Burnett, J D G0HCU
Burnett, J G G3OLW
Burnett, J W G6GCJ
Burnett, P G4BLL
Burnett, P D G7SKA
Burnett, P W G1DAT
Burnett, R C G6YTX
Burnett, S J GM4LHW
Burnham, H A G6TMK
Burnham, J W G3UJK
Burnie, J G3ZXO
Burningham, R E A G8UNP
Burnley, I A 2E1CUG
Burnley, P H G8ZMQ
Burns, C J G8XOD
Burns, D A G3GLV
Burns, D A GI0UXD
Burns, F W G8ZMB
Burns, I T G0AFH
Burns, J A GM4ZXJ
Burns, J C G7HQA
Burns, J C M0AEI
Burns, M C G7NDJ
Burns, M J C G3TCA
Burns, N F GM4DIN
Burns, P H GI4NLQ
Burns, R GM1VAX
Burns, R F G3OOU
Burns, R F G3VCP
Burns, R I G4VDT
Burns, R T G3LHQ
Burns, R W G7FMB
Burns, S B G1WUS
Burns, T G0FZL
Burns, T S G0LVX
Burns, W R G7SVP
Burnside, A G7OKZ
Burnside, I G G0WHH
Burnside, K J G GI4IYO
Burnside, R J G GI4RXS
Burr, E P G0JDV
Burr, J H G3CTI
Burr, J W G4TEU
Burrage, D H G4GUU
Burrell, A G G0PKI
Burrell, D G4KTR
Burrell, D J T G8LUB
Burrell, J A G8OZH
Burrell, W G G0EEW
Burridge, P R G3CQR
Burrill, C H B G4OSC
Burrill, J E M1BMW
Burroughs, M E G0IWC
Burrow, F H GW8BME
Burrow, J D G0NYD
Burrow, P W G0GGE
Burrow, R H G1LOP
Burrows, A J G1HHS
Burrows, C 2E1CCB
Burrows, C G8FFU
Burrows, C R G0GFI
Burrows, D G6LLL
Burrows, D R G7OBY
Burrows, G A M0ASK
Burrows, G D G4IGW
Burrows, J E G6GFC
Burrows, J F G0GPK
Burrows, J M G1LCS
Burrows, J W G3SUI
Burrows, K S G0OZK
Burrows, M 2E1FTI
Burrows, N G8XCD
Burrows, R A G0BUR
Burrows, R A G6DUN
Burrows, R W G6WEH
Burrows, R W G8BYI
Burrows, S J G6MMD
Burrows, S M G7BXB
Burrows, T G4PHQ
Burrows, T R G6EJM
Burrows, W G0ANE
Burslem, G M A 2E1EXX
Burslem, H J A 2E1EXW
Burslem, T J 2E1EXV
Bursnall, M GM3NUB
Burson, P S G3ORE
Burston, J C G4XKK
Burston, S J G6RSB
Burt, A P G6WXM
Burt, C G6PVV
Burt, D G G0DAX
Burt, D J M1BZK
Burt, F R G4NBX
Burt, G R W GM3OXX
Burt, J G GM7WLO
Burt, J W G0RJB
Burt, M G8RGU
Burt, M H GW7MHB
Burt, P T G3NBQ
Burt, R G1YKZ
Burt, R GM3NTX
Burt, W G M0AHT
Burtenshaw, D E G0AYV
Burtenshaw, P D G4ZWE
Burtenshaw, S C 2E1EHT
Burton, A G0NTZ
Burton, A J 2E1FDQ
Burton, A J M1CAJ
Burton, A K G4VUA
Burton, A S G1GAV
Burton, B R G1PHU
Burton, C G8EGL
Burton, C G8UCH
Burton, C M 2E1CWI
Burton, C R G1XET
Burton, C W G1OBR
Burton, D G0GKY
Burton, D G4XXG
Burton, D D G8LKS
Burton, D J G0SFV
Burton, E G0MPY
Burton, G A G3GQR
Burton, G N G7HCI
Burton, H B G2JR
Burton, H R G0WWQ
Burton, J G0CAP
Burton, J G1LUU
Burton, J G4XQI
Burton, J GW2DXQ
Burton, J D G3TJR
Burton, J F G6IVP
Burton, J G G4TQE
Burton, J L G4ZHX
Burton, K A G4JRW
Burton, K M G6SSN
Burton, N C H G3CFW
Burton, N F G4BLM
Burton, N J G6BBH
Burton, P H G4CUR
Burton, P J 2E1CWA
Burton, P L G3GEX
Burton, P L A G3ZPB
Burton, P N GW0JTK
Burton, R M0BCZ
Burton, R M1BCO
Burton, R A G1XEH
Burton, R B G6OBW
Burton, R F G1RAY
Burton, R J G3NOD
Burton, R W G4KPX
Burton, S M G0ETD
Burton, T M G4UDA
Burton, W G4CWA
Burton, W C GW3TUG
Burwell, S G0NLK
Bury, E G G8VFL
Bury, P R G1LNQ
Busby, D T G4HFL
Busby, G C G4ORB
Busby, P W G0JCD
Busby, T A G4HDM
Bush, A S G1EAM
Bush, B 2E1DLO
Bush, B G3IVM
Bush, B E G4WEY
Bush, D B G4SVS
Bush, J 2E1DWM
Bush, M A G3LZM
Bush, N A G7NYV
Bush, P A 2E1FSC
Bush, P S M0AWH
Bush, R A G7RVH
Bush, R B G6PKY
Bushby, D J 2E1EFQ
Bushby, J R G3WLV
Bushby, L P G8FNV
Bushell, A G0NUR
Bushell, A G7RTM
Bushell, J P O GW3OSV
Bushell, P N G7CEU
Bushell, R D G8KAE
Bushell, R G G0HTM
Bushell, R J G4UNM
Bushnell, M A G0TZH
Bushnell, M V G0UXP
Bushnell, M V G4AZH
Buss, D G1IBO
Bussell, D G4EOT
Bussey, D J G0IXL
Busson, M L GW8MER
Buston, R M1APU
Buswell, A T G0WEJ
Butcher, A A G4SIB
Butcher, A C G3FSN
Butcher, A D G0ROU
Butcher, A F G8ZFL
Butcher, A J G7VKC
Butcher, A V G0ILL
Butcher, A W G3KPJ
Butcher, B J G4HKG
Butcher, C G4RLA
Butcher, C B G4VGT
Butcher, C J G1NQY
Butcher, E J B G5XG
Butcher, G J G0JFD
Butcher, H J GM0SUF
Butcher, I F M1BVY
Butcher, J G0RWX
Butcher, J A G4HKZ
Butcher, J B G3LAS
Butcher, J G G0BRG
Butcher, J M G4GWJ
Butcher, J M G8CZJ
Butcher, K A G4NLO
Butcher, K F G3XCQ
Butcher, M E G3SLZ
Butcher, M T G1SLG
Butcher, P G4PMZ
Butcher, P A G3UDH
Butcher, P L G4GXB
Butcher, P R G3XPB
Butcher, R J G3UDI
Butcher, R P G4NWH
Butcher, S M G1MMA
Butchers, W J G8XST
Butland, R C G6RKJ
Butler, A A G1EAN
Butler, A J G0JCG
Butler, A N G7JWH
Butler, A R G6MNL
Butler, A W G8BIT
Butler, C G6ZGI
Butler, C C G2DYF
Butler, D A G0PFO
Butler, D J G4ASR
Butler, D J G8JMK
Butler, D K G4ZMP
Butler, F G0LWI
Butler, G 2E0AOI
Butler, G M0BDS
Butler, G C G4JVA
Butler, G W G4BXU
Butler, H G3XHR
Butler, H C G4JSW
Butler, J GW0NFN
Butler, J A M1CBJ
Butler, J M G0NRK
Butler, J M G8RKO
Butler, J M GM3ZMA
Butler, L A 2W1BIY
Butler, L H T G3DXS
Butler, M 2W1BPJ
Butler, M D G8KTX
Butler, M G GW0MNP
Butler, M J G4UXC
Butler, M J G4WET
Butler, M J E G0HCX
Butler, M R G0LMD
Butler, M S G0GXZ
Butler, N C 2E1EYE
Butler, P B G7KMO
Butler, P P G1GTA
Butler, P P J G3DIF
Butler, P S G6AXK
Butler, P T M0AFO
Butler, P W G4URM
Butler, R A G0VKL
Butler, R G G3ZAC
Butler, R J G4JXC
Butler, S G0RFJ
Butler, S G6VUE
Butler, S E GI7ISX
Butler, S J B G0CJW
Butler, S L G6XMA
Butler, T O G6VVZ
Butlin, R G1HFR
Butlin, R G G3YKS
Butlin, S J G3MRP
Butson, I R G4HKC
Butt, A G3WZF
Butt, D J G1BDS
Butt, J A B G0JJG
Butt, L G G4KON
Butt, S J G7NZC
Butterfield, G A G4FPE
Butterfield, J GW7DWR
Butterfield, P N G4AAQ
Butterfield, R A G3VYB
Butterfield, R A G4UOI
Butters, C D GW3UAY
Butters, J R G6FBA
Butters, L G1EBU
Butterwick, J K G0AOO
Butterworth, A G7DPW
Butterworth, B M G8FOT
Butterworth, D G7VDQ
Butterworth, D B G7JMU
Butterworth, G G6IBH
Butterworth, G C G4VDQ
Butterworth, H G4BIH
Butterworth, J G3PQX
Butterworth, J A GM0WHF
Butterworth, J E G7OJB
Butterworth, J K G0UOS
Butterworth, J L G0LHN
Butterworth, M J G6HNF
Butterworth, P G0GPH
Butterworth, R C G0FYH
Butterworth, R C G6FDU
Butterworth, R E G4IJB
Buttery, C G1KAK
Buttery, J A G7OPJ
Buttery, P J G8PFP
Buttery, R G7SMD
Buttery, R M0BGI
Buttimore, D G1NMF
Buttle, S A G0EZM
Button, A J G3YSK
Button, C A G0RPY
Button, I G1KOK
Button, J F G1WQQ

Button, J M G8JMB
Button, L J G1LNR
Button, M B G8YMB
Button, N P G4IRX
Buttress, H G3VHL
Buttress, P W G8NAM
Buxcey, K G4BUX
Buxcey, K G8BUX
Buxton, B G4FRP
Buxton, C A G1KGQ
Buxton, D A G7MLR
Buxton, D H G4PHY
Buxton, E G6DLX
Buxton, J S G7JSB
Buxton, S G6NLU
Buzzing, B R G4BYM
Buzzing, P B G4EFS
Byars, G M GW8ZCV
Byatt, M J M1AHF
Bye, A S B G3TCI
Bye, E D G8FAX
Bye, R A G7DOR
Byers, A G4CUF
Byers, D J G4IZU
Byers, D M G6OCB
Byers, W J G3XIU
Byford, R N G4MKR
Byford, S C G6UZM
Bygate, R G1UUT
Bygrave, R G8WRV
Bygrave, R W G0KNJ
Byham, D R G7NKM
Byles, G R G3KRY
Byles, M P G6UWS
Bylo, C G1CIH
Bylo, N G4CSN
Byne, D L G3MRQ
Byng, P A G7WCK
Byrer, D L G0POH
Byrne, A J G6SOZ
Byrne, B C E G0NSW
Byrne, B E GM0NYB
Byrne, B J P MI1AYL
Byrne, B M G6HCI
Byrne, C G6JJF
Byrne, C M G7DRK
Byrne, D G0LVN
Byrne, D G3KPO
Byrne, D M G2HNU
Byrne, E J H G1RUZ
Byrne, E M G8LF
Byrne, F E G0EHH
Byrne, M G G3RYZ
Byrne, P M0BAI
Byrne, P F G8PWG
Byrne, T W G3NDQ
Byrom, D T G6LQE
Byrom, W R G3UCR
Byron, G G7MAO
Byron, N H G0PZM
Bysshe, P J S G3WKX
Bysshe, P J S G3WYK
Bytheway, J G4FBH
Bywater, G C G8ZGO
Bywaters, J G8ILK

C

Cabban, E A G0ETU
Cabban, M D G0XDI
Cabban, M D G1WPF
Cabban, P G K G4ABC
Cabban, P G K G4OST
Cable, R E G4WSL
Cachart, E B G0FDR
Cackett, K M1ABU
Caddick, J C G4VZL
Caddick, J M G4TCP
Caddy, R A G0VBM
Cade, C A G0DMB
Cade, S J G1EBV
Cade, S J G8SQY
Cadey, A T 2E1CPB
Cadey, G G0CEY
Cadman, D G6XGF
Cadman, P J G4JCP
Cadman, P S G G3PCC
Cadman, S C G1DIG
Cadman, T M G0HFE
Cadogan, C K G3XWB
Cadwallader, R M C GM1WMS
Cady, D N G0HGO
Cafolla, D G GI4DOM
Cage, H W L G0APS
Cage, I M G4CTZ
Cahill, D M G3TGE
Cahill, G G4GXW
Cahill, K M G6IZF
Cahill, P C G0LBZ
Cahill, T J G0UXN
Cain, A G3YCX
Cain, A H G3DVF
Cain, C C G7KZG
Cain, J M C G GM4GDF
Cain, L W G4IKO
Cain, P R G8IZW
Cain, S J G6KZU
Caine, C R G4IWS
Caine, L S G8HFL
Caine, L W G4XVQ
Caine, N P C GM1REZ
Caine, R GM1REY
Caine, R S G0LGW
Caine, R W G6UFP
Caine, S P G6NPW
Caine, W T G6JNZ
Caines, B A G4PAY
Caines, C F G1OQF
Caines, P M G4PDA
Caines, R R J G3ORC
Caines, R J G7GLW
Cainsford-Betty, C G8OHP
Caira, R A G4URD
Cairney, T M G1DPT
Cairns, E P W G0INA
Cairns, J B GI4ALM
Cairns, J S GW3ITT
Cairns, J V R M1BKI
Cairns, P G3ISP
Cairns, R G3HFA
Cairns, W E G7OPN
Caito, C M0AAL
Cake, A C G3CNO
Cakebread, S J V G3IDI
Calder, A R P G3BEU
Calder, D GM4WHD
Calder, I A M GM0WNS
Calder, J A G7HSN
Calder, J J G4CQC
Calder, L GM1BKF
Calder, N GM0ERB
Calderwood, A A GM7FGF
Calderwood, D W GW4VHO
Caldicott, B C G7NME
Caldwell, A B GM0NGJ
Caldwell, P A G4PAC
Caldwell, T B GM4NDO
Caley, A G3OLZ
Caley, A S G0RUK
Caley, D G0PTL
Caley, J A G2FSS
Caley, M C G8DMT
Caligari, E GM0RM
Caligari, E G6TKY
Calkin, G A G4RTO
Calkin, R A G4MNT
Call, B 2E1COD
Call, C L G4JGT
Call, L 2E1CNO
Callaghan, G W G4SPE
Callaghan, H G4OQH
Callaghan, J G8VFM
Callaghan, J GM4ZNS
Callaghan, P G4WHM
Callaghan, P W G4WHL
Callaghan, R J G6VXC
Callaghan, T E GM6WTH
Callaghan, V G3JMH
Callan, R G8NDP
Callanan, T C G0HNO
Callaway, B B G4FIG
Callaway, B L 2E1BPA
Callaway, T GM0WJY
Callegari, A G G3OMD
Callegari, C R G8CRC
Callen, D GI4TEA
Callender, A C G7VCO
Callender, D C G3GEA
Callow, K GM3END
Callow, N G0TUP
Callow, P A F G4RNO
Calloway, H R G6CSR
Callum, J A G3ZMO
Calpin, P C G6HAT
Calter, P R G0RMD
Calthorpe, E G0HXL
Calver, S P G6TXV
Calvert, A G7MZW
Calvert, B G3OHA
Calvert, E I G4EIC
Calvert, G G8HTH
Calvert, I D G0PCM
Calvert, M G4VBW
Calvert, R G6BXR
Calvert, R A G0DSO
Calvert, S MI1BBE
Calvert, T R G4GBS
Calvert-Toulmin, B M G4YZH
Calvin, A G GI4XFE
Calvin, W GI4PMF
Camac, D R G1IOO
Camber, L G G0XAM
Cambery, V P G0IGN
Cambridge, W A G1CRT
Cameron, A J M G4STT
Cameron, A M G3OGJ
Cameron, C GM0HIG
Cameron, C A C G1EHX
Cameron, D G M GM0AJK
Cameron, D M GM1BZR
Cameron, H GM4VIS
Cameron, H A GM0GSG
Cameron, J G3PQK
Cameron, J MM1BID
Cameron, J G G4YIM
Cameron, J R G4XNF
Cameron, R D M1CAO
Cameron, R J GM4OHY
Cameron, S GM4UTP
Camley, M G E GI6ECD
Camm, A 2E0AOH
Camm, C S G8TZT
Cammell, M W G1PYL
Cammies, S C G3VNI
Cammish, M T G1FJJ
Camp, M G3MFK
Camp, N J G7KFQ
Camp, N J G7RIH
Campanario, D M0ASR
Campbell, A GM3NKG
Campbell, A A G3LMB
Campbell, A A GI6ETQ
Campbell, A F G8ICC
Campbell, A G G8CGT
Campbell, A G GM0WNR
Campbell, A J G6DTT
Campbell, A J M GM3VAR
Campbell, A T GM1JNC
Campbell, B GM0FQQ
Campbell, C GM1LUZ
Campbell, C G G7KDM
Campbell, C J A G4TDN
Campbell, C R G6TXW
Campbell, D G3AJL
Campbell, D A 2M1CKB
Campbell, D E G5SD
Campbell, D F GI3VFW
Campbell, D F GI4NKD
Campbell, D G G4ELG
Campbell, D H GI8PGJ
Campbell, D M GM0VYM
Campbell, D V G0MWV
Campbell, E S G8PHS
Campbell, F K 2E0ANU
Campbell, G G4RTG
Campbell, G GM3HNE
Campbell, H A G7HNX
Campbell, I G3URK
Campbell, I A GM6TIB
Campbell, I D GM4FFP
Campbell, J GM4RUP
Campbell, J GM4WQQ
Campbell, J A 2E1FJA
Campbell, J A GJ7DPH
Campbell, J A GM1YKE
Campbell, J C G7PUA
Campbell, J D M1ARQ
Campbell, J F GI1PHF
Campbell, J G GM0AMB
Campbell, J G GM4UWO
Campbell, J G GM4YEQ
Campbell, J J G3GR
Campbell, J J G4IUA
Campbell, J P G7OKR
Campbell, J P GM4LHJ
Campbell, J P C G4SGJ
Campbell, J R GI6UFU
Campbell, J R GM1TCP
Campbell, J R M GM8BCB
Campbell, J T G7EWF
Campbell, K A G4NOX
Campbell, L M GI1CDZ
Campbell, M G G3MRA
Campbell, N B G0PZI
Campbell, P A GI0CMJ
Campbell, P V G0PGF
Campbell, Q G G4OEU
Campbell, R GI0CXR
Campbell, R C G6ABO
Campbell, R M GM1NET
Campbell, R M GM6OQN
Campbell, S B GM4OSS
Campbell, S C GM1ZUN
Campbell, S G G3HDM
Campbell, S J G1BMP
Campbell, T G1LZK
Campbell, T G1RUY
Campbell, T MM1BMS
Campbell, T J GI1XGA
Campbell, W GI3PQW
Campbell, Y M G7HER
Campbell Davis, T F G3YMM
Campden, A G4CCR
Campion, A G0GND
Campion, W G6IZG
Canale, H R S GM3XFC
Canavan, J J GI3TZC
Cancellor, J H G3DVA
Cane, H G G8BGG
Cane, R F G4KRH
Cangir, F G0SPJ
Canham, D G8YKY
Cann, D J G6VCM
Cann, R P G4WHK
Cann, T G4CTC
Cannard, L G G1UES
Cannell, D E G8HEL
Cannell, J E G7OAI
Cannell, R G0RIC
Canning, A J G0OPB
Canning, D D G0VIF
Canning, G W G4GBN
Canning, J R G1OMI
Canning, R G G0ARF
Cannings, D W G4DWC
Cannon, A M 2E1DZP
Cannon, B G8DIU
Cannon, B F M1AHW
Cannon, D I G1OKF
Cannon, D J G4IWQ
Cannon, D M G4ZHE
Cannon, D P GD4MCR
Cannon, K L 2E0ACY
Cannon, R A 2E1EOZ
Cannon, R A G8OTG
Cannon, T G0VQR
Cannon, T C G6YLW
Cannon, T P G0HNL
Cannon, T W G7MUT
Cansfield, D J G0LDJ
Cant, A G G1XWV
Cant, D H G1PJE
Cant, G T G0CQB
Cantrill, H N G3BJY
Cantwell, J A G1ZRS
Cantwell, T G0PVZ
Canty, D H G4NNT
Cape, S J M1BZL
Capel, A W J G4ROX
Capell, V E G3JOR
Capewell, A E G0JVZ
Capewell, C G7FLR
Capewell, K W M0AWE
Capewell, M W G3FZR
Capewell, P G0MQC
Capindale, J W G0CSV
Caplan, D H G3PXC
Capocci, F R GM6YPQ
Capon, B S G8UBH
Capon, C A 2E1FXY
Capon, I J G0KRL
Capon, J B G8WTN
Capon, N D G1URJ
Capon, T J 2E1CAM
Capon, T J G7VQT
Capp, D A G3CPT
Capper, B G G4DBC
Capper, S G8ORU
Cappleman, J G8MVO
Capps, E T G0KRF
Capps, M G8DAP
Capstick, E G GM7OAF
Capstick, M H GW4RCE
Capstick, M T G4RCD
Capstick, S G7STU
Capstick, W B G3JYP
Carberry, T M G4SBW
Carbutt, P G2AFV
Carby, I GM4XXO
Carby, I C D G4YCS
Card, K L G0NJG
Card, M G7PGH
Cardell, D M G0RHL
Cardell, J S G0RHM
Carden, D G3RIK
Carden, J E G7VKO
Carder, A V G0HHK
Carder, R O G7SRK
Cardiff, J D G0TBJ
Carding, D J G6SGD
Cardno, W D GM0NRT
Cardoo, G GM2CRV
Cardwell, A GI7IUH
Cardwell, C D G4COV
Cardwell, P A G7UEX
Cardwell, P H GW3FXI
Cardwell, R GW0CCR
Cardwell, R GW1RCC
Cardwell, R GW4PUX
Cardy, P F W G4WPC
Cardy, R S G4PRZ
Care, C M G3WDM
Care, M R G4MEX
Care, R H G4MEY
Care, W G7RIO
Careless, D H G3HJD
Carena, M G8NIK
Carey, J G8HSI
Carey, J M GW7SWN
Carey, J M MW0ARE
Carey, M R GW0VJS
Carey, M R GW4VSE
Carey, P J E G3PBY
Carey, P J E G3UXH
Carey, P J E G6GEC
Carey, S R G4MJW
Carey, W C A G6BLB
Cargill, D J G0EPE
Cargill, W R GM0UZV
Cariss, R J G7ACD
Carlile, A K G0MNI
Carlin, J D GM0NAE
Carline, J M G4JJY
Carline, L G1HQC
Carling, B J G3XLQ
Carling, E D G0CGL
Carlisle, G R G3SRJ
Carlisle, J E 2E0AOX
Carlisle, T J GI8KYI
Carlisle, T R GI0WAA
Carlsen, D C G3XRC
Carlson, N R G6YME
Carlton, A P G7KBD
Carlton, G D G4TOA
Carman, J G1URW
Carment, P M G5WW
Carmichael, J C GM3MSG
Carmichael, K M GM0MNW
Carmichael, R M GM0HYW
Carmichael, W G4BQR
Carnall, A J GM6FXZ
Carnegie, P A G GM1CMF
Carnell, D C GW8ZWD
Carney, G C G4DRZ
Carney, I J G8JHM
Carp, J D G0XFT
Carp, J D G8XFT
Carpenter, A B G3YWW
Carpenter, A W GM0SNT
Carpenter, C G1NQB
Carpenter, H F G7DEH
Carpenter, I W G6CSW
Carpenter, I W G7AED
Carpenter, J P G1YDQ
Carpenter, L J G4CNH
Carpenter, L J G4YRN
Carpenter, P J G1TMD
Carpenter, R B P G4BAH
Carpenter, S M G8OEY
Carpenter, S R G8HUF
Carpenter, S V G3ZQF
Carpenter, W C A G4KQJ
Carpenter, W G G3OYQ
Carr, B G4LOY
Carr, B D G8ZMN
Carr, B E G4UHQ
Carr, C D G7UVY
Carr, D G6UFT
Carr, D G8NRU
Carr, D A G8MLW
Carr, D H G6PZS
Carr, D T G0WGY
Carr, F E G4VCF
Carr, G G G3KEK
Carr, I C GM6SEV
Carr, J GM0RSF
Carr, J A G6KTF
Carr, J C G4WQJ
Carr, K A G1WUH
Carr, L A G6LFR
Carr, L C 2E1FNJ
Carr, N M G0JHC
Carr, N M M0AAE
Carr, N R G1SHU
Carr, P E G4CMC
Carr, P H G0BWL
Carr, T E F G0NAL
Carre, P R GU4XEA
Carress, I B G8EAH
Carress, W J G0WYK
Carrett, D N G4OPK
Carrick, D G1VIG
Carrick, G W G3MRV
Carrick, J F G8RWS
Carrick, W N G8DAZ
Carrick Smith, J P G6CFA
Carrick-Smith, L G6GOR
Carrig, T K G8UYW
Carrigan, S A G4OUJ
Carrington, E J G0RCF
Carrington, G E G0GFX
Carrington, J G G0VNW
Carrington, M G G8BEG
Carrington, P G7VKZ
Carrington, R K G0RRZ
Carrington, S 2E1EOL
Carrington, W G7VCA
Carroll, C J G3WWC
Carroll, J G0KQP
Carroll, J A G3NFW
Carroll, J G GD0GBA
Carroll, J P G0MZU
Carroll, M A G1BOP
Carroll, M F M1BML
Carroll, R G0OCD
Carroll, R V G7MQW
Carroll, T G0DCC
Carroll, T G0HCD
Carroll, T J G7NHD
Carroll, T L B G8YKZ
Carroll, W J GM0PKP
Carrotta, D F G0PMM
Carruthers, D E G6RVZ
Carruthers, D K G0MGT
Carruthers, G P GW4HGJ
Carruthers, H G0NPQ
Carruthers, I B MM1AWA
Carruthers, P E G8MNL
Carruthers, T G4XLA
Carruthers, W S G3MPY
Carsboult, F J G1MYD
Carslake, D M 2E1BMF
Carslake, P A G1UMJ
Carslake, R E R G4UMJ
Carslake, R E R G6RC
Carslaw, A GM0BCY
Carslaw, J C GM2ACY
Carson, B H GM0GCO
Carson, C L G7OVK
Carson, D G4IHO
Carson, J GM3OXK
Carson, J E GW0DSL
Carson, K A 2E1BTK
Carson, M G4OTQ
Carson, M GI0XPE
Carson, M B G2AIV
Carson, P MI1ARY
Carswell, V M B G0FNU
Carter, A G7IJD
Carter, A A E G4TBR
Carter, A N G6XPZ
Carter, A N G7BSG
Carter, A T G6FSF
Carter, B G1AVA
Carter, B A G0TXA
Carter, B A GW0TPR
Carter, B P G3ODU
Carter, C 2E1FQE
Carter, C G4TYA
Carter, C D G4UPQ
Carter, C W G1KWX
Carter, D G1KPC
Carter, D G8SOI
Carter, D A G8STJ
Carter, D A P G3VYW
Carter, D F G0VFH
Carter, D J 2E1DGD
Carter, D J G4DJD
Carter, D J G6BIU
Carter, E A M0ABX
Carter, E G G8EFK
Carter, G F G4SCI
Carter, G P G0ISE
Carter, I L G0GRI
Carter, I L M1BQY
Carter, I P G0NQX
Carter, J G3SFZ
Carter, J A G6GQL
Carter, J D G7JTH
Carter, J G G0GIL
Carter, J H G3PHR
Carter, J H G4LPY
Carter, J J G3WMN
Carter, J J G4CFO
Carter, J M G0LHZ
Carter, J M G7TFB
Carter, J P G0CHF
Carter, J S G4GEY
Carter, J T G0FUD
Carter, K E G7FPG
Carter, L B G8WW
Carter, L C G7DNV
Carter, L J G6HCF
Carter, L S MW0AMJ
Carter, L V G7NHP
Carter, M M0AHD
Carter, M A G0NEV
Carter, M A G6ETX
Carter, M E G1VRP
Carter, M E B G6JNV
Carter, M J G1UUS
Carter, M J G6GCE
Carter, M J G7AJT
Carter, M J G7HNL
Carter, M P 2E1CPM
Carter, M P G6CHJ
Carter, N G6SZP
Carter, N J G3PEC
Carter, N J G6VCH
Carter, P G1BPB
Carter, P G6FTJ
Carter, P A F G3VAI
Carter, P J G3XUL
Carter, P J G6CLK
Carter, P L G6HRF
Carter, R C M0AIY
Carter, R F G8WSW
Carter, R J G6XQB
Carter, R J G7ECO
Carter, R J G8RSL
Carter, R M G4VSO
Carter, S G0GPT
Carter, S G6PPY
Carter, S E 2E1ETR
Carter, S R G6JNW
Carter, T G1HJP
Carter, V G G8YQH
Carter, W G2NJ
Carter, W G6IJG
Carter, W G6WSX
Carter, W H G0LVD
Carthey, P E G4AGA
Cartledge, R G G4IHV
Cartledge, W J G6XXC
Cartlidge, F S GW4LWO
Cartlidge, J A G0URU
Cartlidge, S N G0MJG
Cartmel, C G4EST
Cartwright, A 2W1FPK
Cartwright, A M G7MNS
Cartwright, C A 2W1FHF
Cartwright, D J G7ODR
Cartwright, F A G7NHK
Cartwright, G A G4OYJ
Cartwright, H B M1AOZ
Cartwright, J E GW1WZI
Cartwright, J J G7GAP
Cartwright, J N GW0KRQ
Cartwright, L G0UVL
Cartwright, L C G6KJY
Cartwright, M A G8WSV
Cartwright, M W G4DVM
Cartwright, N E G4DKX
Cartwright, P A 2E1AYS
Cartwright, P G G1GAN
Cartwright, P J G4YWA
Cartwright, P O G3POC
Cartwright, T G0VUB
Cartwright, T J G8EVD
Cartwright, W A G0DQO
Cartwright, W H G6IJQ
Carvell, C J G0UBK
Carvell, M J G6DNH
Carvell, P J G0UZI
Carvell, R G0PSG
Carvell, T G G0TFV
Carver, C S GW4EYO
Carver, D J G0URY
Carver, E G4MXS
Carver, E G8XFU
Carver, J D GM0IZC
Carver, J M G8RHL
Carver, P S GW4WWE
Carvill, E G8RHU
Carville, R G1XZQ
Carvin, S G0DRD
Carwood, S G6SPI
Carwood, W J G0WVT
Case, A GW8UKW
Case, D A G4BTI
Case, E J GW4HWR
Case, L G8BJQ
Case, P J G4JHD
Case, S A M0ALH
Case, T G4ZVR
Casemore, P J G3SGF
Casemore, R J G3WVW
Casewell, I E G8DUO
Casey, D G4MVE
Casey, D G8ZWE
Casey, I J G7DNG
Casey, S G4JQK
Casey, S G7VOQ
Casey, S M1CAL
Casey, T D G3KXW
Casey, V M G7VTA
Cash, C R G0SMO
Cash, D P G7MEG
Cash, G G7LNO
Cash, J N G7AZA
Cash, P C 2W0AMY
Cashmore, M R G1IWC
Cashmore, R F GW3CKB
Cashmore, R F GW4SRO
Cashmore, V A GW4MOK
Caslake, S G0DIR
Casling, J D G3MWZ
Casperd, S G G3XON
Caspersz, A B G1TMW
Cass, G H G4HEV
Cass, K G G3GLO
Cass, K R G3HWW
Cass, K R G3WVO
Cassar, V G7CXO
Cassell, K S G7GJP
Cassell, R F G8RFC
Cassells, R GM4WLN
Cassere, D A G3TAF
Cassidy, F G4HBI
Cassidy, F J GM6JEP
Cassidy, J G3HSW
Cassidy, J T G4XXT
Cassidy, T MM1BIF
Cassidy, T G GM4RKM
Cassidy, T R G8PZA
Cassidy, W N G0WUI
Cassling, R N L G4RPC
Casson, J G1KNC
Casson, R B G8CBN
Cast, D W A G4XQD
Castell, J P G4AXK
Castell, M H J G8FHY
Castle, B A G3ZJX
Castle, C A 2E1BSC
Castle, C E G7HFP
Castle, D H G1PGJ
Castle, E A GM0WRH
Castle, F J G0FTP
Castle, G E G6TXL
Castle, I R G6DUI
Castle, P R G7PTX
Castle, W D G6OQJ
Castledine, N M G4ALB
Castleton, I M G0HLY
Castley, K B G0FDJ
Castling, W G0MZE
Castree, J D G0AUM
Caswell, A K M1BMZ
Caswell, J J E G7JTV
Caswell, J M G8XPC
Caswell, P G6PPV
Caswell, R G4GMN
Catchpole, D J G0PFN
Catchpoole, B P G1WMV
Cater, A M A G0EVX
Cater, C 2W1EEN
Cater, C C G6MIJ
Cater, D B G4WHZ
Cater, G M N G7GBK
Cates, L V G4AVE
Catherall, E GW0SLM
Catherall, L C G0MMT
Catherwood, D F G3ZRN
Cathmoir, R S F G0DCL
Catling, D J G4ECU
Catling, G E J G8ABX
Catling, P O G4FVA
Cato, R G8YLA
Caton, A G0OQP
Caton, C G G0JPS
Caton, P A G1BWX
Caton, R G0LHD
Cator, B G4AED
Cattanach, P H G7RGA
Cattani, A L G4RMJ
Cattell, D W G1KDO
Cattell, M M1AED
Catterall, B G4IWC
Catterall, D A G0NZR
Catterall, P J G4OBK
Catterall-Annal, R G G4SDH
Cattermole, J A G8NPK
Cattermole, T J G6DNA
Catterson, J L G0GIF
Cattle, G E 2E1ENN
Cattle, S G1XSA
Cattley, T J H G0CWZ
Catton, C S G1BSN
Cattrall, C G4UGK
Cattran, D B G0AJG
Catts, A H J GU3LPV
Caudy, C GW1KQV
Caughey, S GI0SSW
Caulfield, J J GW3ISJ
Caulton, D G7NPW
Caulton, M A G8XXU
Caunce, C R G3ZUG
Caunce, K G0PTT
Causer, W E D G1ZTK
Cavanagh, P R G3KVM
Cave, C A G1HXZ
Cave, C J R G6ABP
Cave, C T G7VNC
Cave, H G G4RLX
Cave, J W G0WJM
Cave, M R S G0JRW
Cave, T R G0JRX
Cavendish, R W G8XTD
Cavers, W B GM4KHS
Cavers, W B GM4UIB
Cavie, G R G7SJF
Cavill, R W G3GN
Cawkwell, E G G8VEL
Cawkwell, G H G3ULD
Cawley, H M1ARS
Cawley, J A G3NFQ
Cawley, M R G4JZT
Cawley, R E G8LPC
Cawood, K R G0DIL
Cawser, D C G0FSF
Cawson, F H P G2ART
Cawthorne, B M G6VPF
Cawthorne, C J G4KPY
Cawthorne, D G6HQS
Cawthorne, N S G3TXF
Cawthorne, S A GM4KOO
Cawthron, V G4ODG
Cecil, W G GM3KHH
Cedar, B L G8BMQ
Centanni, J D G0CKE
Ceresole, J P G8BSD
Cerutti, P MM1BNS
Cervini, P C G1RPQ
Cesnavicius, P A G6SPG
Chace, M H G6DHU
Chace, P H G6DHT
Chadbone, R G8JK
Chadburn, C C G6WHL
Chadburn, C R M0BCH
Chaddock, A P G6GNW
Chadwick, A G7KDJ
Chadwick, A T G4ZVJ
Chadwick, F G0HFC
Chadwick, G G6CSN
Chadwick, H C G4GKG
Chadwick, J G6CLL
Chadwick, J M0ANQ
Chadwick, J M1AUX
Chadwick, K G4EVC
Chadwick, L E G0TTO
Chadwick, L M G0OJB
Chadwick, L S G0GAB
Chadwick, M G8OCW
Chadwick, N D G3VMK
Chadwick, P G0TAC
Chadwick, P C G4HGF
Chadwick, P E G3RZP
Chadwick, R G4BPK
Chadwick, R G8FCT
Chadwick, T G0CSU
Chadwick, T G G0TBF
Chadwick, T W G4ZVU
Chadwick, W A G1IBS
Chaffe, W R G2DLJ
Chainey, R G0INY
Chaldecott, J A G1XDO
Chalker, R G1JRR
Chalkley, K A G4BXR
Challacombe, N J G0LGG
Challands, M R G8IQP
Challen, A S G6DTW
Challen, P J G1EBW
Challen, S M G1EBX
Challenger, J G7PIB
Challenger, J GW4FPX
Challenger, J H G4EIB
Challinor, A J G1ELX
Challinor, C J G0EXD
Challinor, P W G4YVD
Challis, A D G0VRE

Challis, D A W G0PQC
Challis, H M0AHU
Challis, J O GM1USN
Challis, J W G1DLU
Challis, M G G1RYB
Challis, M S G6BTR
Challis, R O G6EEA
Challis, S G6AUD
Challoner, S P G6VUX
Challoner, T C GW4KBG
Challons, E M G2CXR
Chalmers, C G6ETZ
Chalmers, D F G3WQG
Chalmers, D J H G0IYE
Chalmers, G GM0ALW
Chalmers, J G8VPO
Chalmers, M A GM0ALX
Chaloner, D G3VSJ
Chaloner, G W G8DMR
Chaloner, M J G0UGR
Chalwin, G W K G3CZM
Chamberlain, A J G4ROA
Chamberlain, A J G7CFC
Chamberlain, A W E G7VGN
Chamberlain, B K G0WLS
Chamberlain, D J G6NUI
Chamberlain, E W G6TED
Chamberlain, F J G3XBN
Chamberlain, G G4NPD
Chamberlain, I F G7TIE
Chamberlain, M 2E1ATL
Chamberlain, M J G3WPH
Chamberlain, P C G4XHF
Chamberlain, P R G4GQO
Chamberlain, R G G3VYU
Chamberlain, R J G4WYM
Chamberlain, S G0JQZ
Chamberlain, W B GM6FSG
Chamberlin, D M C G7NLS
Chamberlin, S B G0UYA
Chambers, A G1XKN
Chambers, A G GJ5NO
Chambers, B G8AGN
Chambers, B B G1IGQ
Chambers, B L G4TAM
Chambers, C R G8MVJ
Chambers, C W G1TQU
Chambers, D A GI4OYI
Chambers, D F M1AMI
Chambers, D J G4GIK
Chambers, D S G4SYT
Chambers, E G6YLS
Chambers, E K G0HKC
Chambers, J D G7OLX
Chambers, J D A M1BHN
Chambers, J P G4TQW
Chambers, J R M1ACC
Chambers, K A G8CTB
Chambers, K G GI0USS
Chambers, K L G0LUP
Chambers, M G1AJE
Chambers, M P G8ZLT
Chambers, M S G3SBW
Chambers, N F G0IRM
Chambers, P G6DTX
Chambers, P A G7HIT
Chambers, P D G4ZUU
Chambers, P M G1LQC
Chambers, R A GI4IOO
Chambers, R H G8BCA
Chambers, R J G4BAQ
Chambers, R K G0IIP
Chambers, R S G8EZZ
Chambers, S A G6WYD
Chambers, S M G7CEV
Chambers, T A G0KQK
Chambers, W H GI3ONZ
Chamings, P R G6USO
Chamley, S T G0TOU
Champ, C G1PLP
Champ, C J G1VSO
Champion, A G7LBH
Champion, I S G3PUX
Champion, L W M1BME
Champion, M C G1MPD
Champion, P G3KKZ
Champion, R S G6KHG
Chan, C K G0DXS
Chance, A P G1HYX
Chance, C M G7EYS
Chance, G R H M1BUI
Chance, M J G0TFJ
Chance, R J G4NEC
Chancellor, M S MI0BDZ
Chancellor, M S MI1BLA
Chandler, B J G4PIO
Chandler, E R G4ERC
Chandler, F J S G3HHM
Chandler, J G6XWU
Chandler, J G7RSR
Chandler, J K G0NVM
Chandler, K J G0ORH
Chandler, M A G4ZYP
Chandler, M W M GW0TZG
Chandler, N E G1UFS
Chandler, P R G3PID
Chandler, R W G4XUZ
Chandler, S A G G3UDD
Chandler, S W 2E1FZM
Chandler, T W G3EBJ
Chandler, V L G6IIO
Chandler, W G J G0TAA
Chandler, W H G0VBH
Chandler, W L G0VBG
Chandless, L C G6PLR
Chaney, R J G1CQA
Channon, J M G8IKK
Chant, E A G4MZE
Chantler, C D G7VEP
Chantler, E N G0ORD
Chantler, R G M1CCL
Chantry, G A G0VEU
Chapaton, D GM0CWL
Chaplin, B F G G0TNB
Chaplin, D G4CSM
Chaplin, G S G8MAA
Chaplin, G T G3UTW
Chaplin, K C G4NWO
Chaplin, T G G1UGH
Chapman, A J G3RIY
Chapman, A M 2E1BTS
Chapman, B G M G0KKR
Chapman, C G4LZF
Chapman, C A G8BUV
Chapman, C D 2E1FYQ
Chapman, C J M1AQP
Chapman, C N G2HDR
Chapman, D G4PPN
Chapman, D C G3NGK
Chapman, D G G4NLP
Chapman, D G GM4NVI
Chapman, D J G4SJV
Chapman, D J K G3RLN
Chapman, D N G8IMC
Chapman, D W M GI6DNI
Chapman, E J G7SJV
Chapman, E S GJ2FMV
Chapman, G K J G7RMG
Chapman, H GW0TSL
Chapman, H E G8VEN
Chapman, H S G3NZL
Chapman, I 2E1DCC
Chapman, I G0MTK
Chapman, I D G8VZS
Chapman, J G7GJY
Chapman, J GI4LVC
Chapman, J A G7UUP
Chapman, J A G8EYT
Chapman, J D G4FIT
Chapman, J G H G6PGT
Chapman, J L J G4FMG
Chapman, J P 2E1CQY
Chapman, J W G4CCI
Chapman, J W G8IQY
Chapman, K D G7CSJ
Chapman, K J G4YZN
Chapman, K M GM1FYW
Chapman, L C G4NVJ
Chapman, L E G4ZID
Chapman, M 2E1GAQ
Chapman, M G8XIN
Chapman, M A G0BQI
Chapman, M B G4ZKE
Chapman, M I G0EMT
Chapman, M J G4NGF
Chapman, M M M1ABM
Chapman, M T 2E1ECN
Chapman, N D G6PXV
Chapman, P 2E1FOD
Chapman, P G3LBL
Chapman, P A G0SOX
Chapman, P C G7BHN
Chapman, P E G6ZNR
Chapman, P J G0NMP
Chapman, P J G1TOW
Chapman, P J G4JCG
Chapman, P J G4VBS
Chapman, P L G3JSR
Chapman, P M G1PAT
Chapman, P M G7FPY
Chapman, P P G0BYI
Chapman, P R G4HUH
Chapman, P R G5VH
Chapman, P S G6AXW
Chapman, R G7KEA
Chapman, R A G7PEU
Chapman, R D G3XPC
Chapman, R H G1LPC
Chapman, R J G0NLG
Chapman, R J G2HR
Chapman, R J G3SRA
Chapman, R J G8CSA
Chapman, R M GM4IGS
Chapman, R R G4HVH
Chapman, R W C G3LFF
Chapman, S G4UDD
Chapman, S J 2E1FFU
Chapman, T G0MKA
Chapman, T G1AXR
Chapman, T G4GWU
Chapman, T C G0OOD
Chapman, T J G3PTQ
Chapman, T M G7BAD
Chapman, V W G0KAT
Chapman, W B G4FDA
Chappell, A E G3HVJ
Chappell, C G4MEK
Chappell, H G G6PPU
Chappell, J G6KJK
Chappell, J D G0BWB
Chappell, J F G0JZT
Chappell, J W H G4HUC
Chappell, K C G1OPG
Chappell, K J G1DOA
Chappell, L G2AAY
Chappell, M J G0GQX
Chappell, R G6XPY
Chappell, S C G0LOQ
Chappell, S J G3XRJ
Chappell, S J G4ZTQ
Chappell, W G0NCV
Chappelle, A G E G0OVR
Chapple, A G G4KKD
Chapple, C L G4KBX
Chapple, M V G8XQS
Chapple, N G1LMX
Chapple, S R G6SC
Chapple, T A M G3SGZ
Chard, P E W G6NPZ
Chardin, D W G6VTY
Charles, A G4ORE
Charles, A H M 2E1DQT
Charles, C D 2W1ETT
Charles, C J G0LWA
Charles, G A 2E1AGK
Charles, H J G7VZI
Charles, J G4YIR
Charles, J M M1BTR
Charles, J R GW6NPD
Charles, J S G3KVG
Charles, S G G3XYA
Charleston, M A G0TKA
Charlesworth, C G4PTU
Charlesworth, F R G1GBE
Charlesworth, H S G4FMQ
Charlesworth, M G G0WWZ
Charlesworth, P G0FRQ
Charlesworth, P H G6GVJ
Charlesworth, R A G0TSN
Charlesworth, R A G4UNL
Charlesworth, S J G0DSC
Charleton, M J GD4UQV
Charlton, A B G6NUZ
Charlton, D J 2E1FSX
Charlton, F C GW3KKG
Charlton, G G4MLF
Charlton, G A G1DKV
Charlton, J G3VRF
Charlton, M G0RMC
Charlton, M G6OXZ
Charlton, M P G6ZQS
Charlton, N 2E1DFK
Charlton, N G7WCD
Charlton, N A G0KXB
Charlton, R G8SAN
Charlton, R D G3CPC
Charlton, R I G1EBZ
Charlton, R L G7DAR
Charman, G I G0OMN
Charman, M S GW4UEJ
Charman, T J G6HBJ
Charman, W M M1BSB
Charnley, F N G0LCR
Charnley, F N G4RPW
Charnley, F N G7CLR
Charnley, J 2E1AFI
Charnock, D G0BCU
Charnock, J L G4WXX
Charrett, F W G3COO
Charter, J A 2E1FZC
Charteris, R R G0NFO
Chase, B F G4MYE
Chase, C W C G7MTJ
Chastell, R E G4SXF
Chaston, R J G1HOP
Chatel, A R MD1BYG
Chater, J F G0EFU
Chater-Lea, D J G4EPX
Chatfield, D G G3JXU
Chatfield, E J G3BLG
Chattenton, K W G4KIR
Chatterton, D I G8YXQ
Chatterton, D K G7FFT
Chatterton, D R G4PLE
Chatterton, F G3WYV
Chatwin, C N G7IVR
Chatwin, C P MW0ARB
Chatwin, R GW0VAW
Chatwin, R GW7SIT
Chaudhry, J K G4TGZ
Chaudhry, M Y G4UTE
Chaudri, E H G3DCS
Chavasse, G B B GW4URJ
Chawner, D R A G0NTG
Chawner, M A G0SPC
Chaytor, R G G7SNJ
Cheadle, E N G3NUG
Cheasley, A G8DXP
Cheasley, M P GM4RDB
Cheatle, C J G3MYC
Chebil, A E 2E1DBJ
Chebil, M B M 2E1DGR
Checketts, R H G G0FBA
Checkley, B G7SGL
Chedzoy, M E G G8UIG
Cheek, I G1ECA
Cheer, A G1IOP
Cheer, E R G8JEM
Cheese, B R G6TCQ
Cheeseman, G L G0DCN
Cheeseman, I R G1DOG
Cheeseman, M E G1IPQ
Cheeseman, P I G4XST
Cheeseman, P R G3KDE
Cheeseman, V H G3AOK
Cheesewright, N L G7GJS
Cheesman, N J G7AUS
Cheesman, V C G3VDC
Cheetham, A H G4AFI
Cheetham, D G G0HEX
Cheetham, E M G6PLT
Cheetham, G G G0EHK
Cheetham, G M G7UVF
Cheetham, J G6ALM
Cheetham, K J G4RWD
Cheetham, R G0FLQ
Cheetham, R A G3JZT
Cheetham, R J G0NYR
Cheetham, S G7NEV
Chell, A G6AYG
Chell, R G0MKL
Chell, S H G7EEI
Cheneler, E G6JUG
Chenery, A B G6PLU
Chenery, D G0PPI
Chenery, G A G4BVI
Chenery, R H 2E1ENY
Chenery, S R N G8SXS
Cheney, C J G3RSE
Chennell, D GW4DUY
Chennells, A K G8HNI
Chennells, J M G4CIJ
Chenoweth, D A G1HYQ
Chenoweth, W R G6PJS
Cheriton, D G G6VGZ
Cherrett, A G3BEJ
Cherrie, H GM0NHT
Cherrington, D A G4FIF
Cherrington, D K G4WRX
Cherrington, J G G1BUN
Cherrington, M O G0KIE
Cherry, A R GM0TEA
Cherry, M F G1JYH
Cherry, N G G0NHN
Cherry, R M GI0BMR
Cherry, S M G3SJK
Cheseldine, P M GM8HMZ
Chesney, W R GI4DCC
Chessell, B C G7KZT
Chester, B F G4DXB
Chester, K J G4IIP
Chester, M F G6MHU
Chester, P S G3YPD
Chester, R B G6USN
Chester, S G4DRA
Chesterman, J A G6HRA
Chesters, C A G1GIF
Chesters, E J GM7FLZ
Chesters, K B G4NXW
Chesters, L MW1BLC
Chesters, M GM0OYU
Chesterton, P 2E1FZP
Chesterton, W D G0AIZ
Chesterton, W H G0WHC
Chester-Brown, P C G7KXA
Chesworth, E E G8WXL
Chesworth, G H GM7OKX
Chesworth, P G1ZYM
Chetcuti, J GW3PYX
Chettle, A C G4DOW
Chettle, J P G8ATA
Chettle, R G8HGX
Chettle, S G8ATB
Chetwood, N P G0UNQ
Chetwynd, J G1RCW
Chew, G G G7SQH
Chew, K G0AWC
Chewter, W A G0IQK
Chiappi, L G1MYE
Chibnell-Smith, T I N G7IFR
Chick, C J V G0DWA
Chick, J C G4NWJ
Chick, M E J GW1AYA
Chick, P J GW1AYB
Chick, T J 2E1BXB
Chick, W L G3LUF
Chicken, E G3BIK
Chiddick, J D G6PWJ
Chiddick, R H A G8AUN
Chidgey, C R G3YHV
Chidgey, C R G4AHG
Chidgey, R A K G6FBB
Chidlon, F G4BBX
Chidlow, F G3WCM
Chidwick, A E J G4XDW
Chilcott, M J G7KII
Child, D J G7JIJ
Child, G W GM0EUA
Child, P I G6WVS
Child, P R G7UFQ
Childe, G G0CPO
Childs, C D E 2E1DCP
Childs, D 2E1BJS
Childs, D G7VGY
Childs, D F G3GVC
Childs, G L G3XEW
Childs, J L G4RUS
Chilinski, A G0UOB
Chilles, A A GM4OSQ
Chilton, C G1GTG
Chilton, D G6LIJ
Chilton, E G1GTF
Chilton, F E G7IZW
Chilton, G J G1JPS
Chilton, N J G7SIB
Chilton, R G3WJC
Chilton, R H G3IKQ
Chilvers, A G3SZ
Chilvers, B G0CUL
Chilvers, F G G3JCK
Chilvers, O S G3JOC
Chilvers, O S G4ANT
Chimber, P S G3ZZW
Chin, A K L GI0XAC
Chin, I K C GI0TWX
Chin, J GI7NIL
Chinn, C D G0KKX
Chinn, E G4SWK
Chinn, E W G6OXY
Chinn, F J G8HSH
Chinn, R G3VQZ
Chinnery, J E P G0ONS
Chinnock, J A C MW0AGE
Chinnock, J E GM8RSC
Chipman, P W C G0GEF
Chippendale, D R G1ECC
Chipper, H R G0UWI
Chipperfield, R H C G6AGF
Chipperfield, R M 2E1EVJ
Chipperfield, R P 2E1EVK
Chipperfield, T T G3VFC
Chisholm, A GM3PIL
Chisholm, A A G3INL
Chisholm, J G7KGP
Chisholm, R E G0MQT
Chisholm, T G0RTC
Chislett, C G0CAM
Chislett, D J G4XDU
Chisman, J C G4ADS
Chittenden, K R G8VAF
Chittock, A J 2E1EZM
Chittock, C J 2E1DLP
Chitty, M D G1SMB
Chitty, P R G8LGS
Chivers, A G4BIO
Chivers, D C G3XNX
Chivers, I A G6CEK
Chivers, J B G0AKK
Chivers, J W H G4DJP
Chivers, L L G1DJD
Chivers, O J G7JXZ
Chivers, R A G1YOD
Chiverton, N W G4SFO
Chmielewski, J J G4GQA
Cholerton, F G1PIY
Chomer, J G0ODM
Choong, L S G0MFY
Chorley, A G4BKH
Chorley, A G4HSN
Chorley, B G0RWA
Chorley, D J G6AYD
Chorley, H E G0OEZ
Chorley, H J G0PWV
Chorley, P A G7XPC
Chorley, P J G4YMG
Chouings, M J G3XAW
Choules, S E G6PWF
Chowaniec, Z T G3PTN
Chown, C J G4RJT
Chown, J E G0BQE
Chown, M L G0AYA
Chown, W P M G8FAD
Chrees, S T G3DZW
Chrismas, P M G1GDD
Christian, J A G8CRV
Christian, R G G3GKS
Christie, D J MI1BGH
Christie, D W GI0ISQ
Christie, E D 2E1ALV
Christie, G GI4VWC
Christie, G M GM7GMC
Christie, H H GM4SNP
Christie, J GM4WWT
Christie, J D GM1ASY
Christie, J T G3NKY
Christie, J T GI7RAM
Christie, J T MI0BDO
Christie, L A GM6RWW
Christie, S G3WHB
Christie, T A G7JHC
Christie, T N GM6RQW
Christieson, M L G8FCD
Christison, G GM0BKS
Christlo, W G0KNL
Christmas, E J G0JOS
Christmas, J S G4OZO
Christmas, M A GM1SRR
Christmas, S J G0BEC
Christmas, T R G1IUT
Christodoulou, P G0VBI
Christofi, G K G0JKZ
Christopher, C H G4PZJ
Christopher, H R G3VBI
Christopher, R J G0ISI
Christy, J A G1ECI
Chruscinski, A 2E1EIX
Chrysostomou, P G6JEU
Chrzanowski, M K G7IAM
Chubb, D G G0IWR
Chubb, D G G4LUY
Chubb, F W G3ARE
Chubb, I F A G1ECD
Chubb, S J G1SQH
Chudley, S R C M1BWT
Chung, E Y O M1BBO
Chung, T F G6FFX
Church, D S G0SYB
Church, F J G3HCH
Church, I T G1HQH
Church, J C G4INI
Church, K J G1HYU
Church, K L G8XIR
Church, M A G4ENZ
Church, R GW4FSY
Church, R A G3KJC
Church, R C G3RRS
Church, S N G1LIK
Churchard, T D G0UFQ
Churcher, M J 2W1EDL
Churchill, C G8OTH
Churchill, C E L G1PBX
Churchill, I M G8XIM
Churchill, J A M0ARQ
Churchill, J C G7SVH
Churchill, J E G7PWJ
Churchill, J H A L GD3MBC
Churchill, J M G7ONI
Churchley, A R G4EAQ
Churchman, E R G0SHA
Churchman, H T G0CTH
Churchman, M A G1SVN
Churchyard, E G G1SVR
Churchyard, E G G3SVR
Churchyard, E G G3TVR
Churms, M G4NQL
Chuter, B J G8CVV
Chuter, R J G7HYA
Cilia, R P G1PLV
Ciotti, P R G3XBZ
Civil, A J G0CIV
Civil, G I G0LKY
Civil, R A G6ION
Civita, L K A G0USA
Civita, V G G0UFO
Clabon, I R G0OFN
Clacher, N G4XQF
Clack, A E M1BHL
Clack, S L G6NUX
Clack, S P G4YEK
Clafton, R A G7EIX
Clague, R GW6WGY
Clamp, D W L G6GNY
Clamp, G G3YTX
Clamp, K W G3ZOW
Clamp, R A G0IMD
Clampin, G K G0YKC
Clampin, P J G8YLB
Clanachan, D R GM4TYU
Clancy, J G4TKS
Clancy, M P G1MWT
Clapham, M J G4PIT
Clapp, D V M1ANL
Clapp, R M M0BCC
Clapp, W H G0GDT
Clapperton, M F J G0OYA
Clappison, R G7JLE
Clapton, K H G7NMR
Clare, F R GW3NWS
Clare, F R GW8GT
Clare, M J G7TLD
Clare, N J G6LNF
Clarence, D R G0UPJ
Clarey, J M1CCH
Clarey, S A G8OXA
Clare-Noon, R G4EJQ
Claridge, M J G0UWU
Claridge, R D G7DBT
Claridge, R D G8GYM
Claridge, V J G0WUW
Clark, A G7VFW
Clark, A GM4ZJV
Clark, A A GM3DIN
Clark, A C 2E1BGX
Clark, A F G0OPD
Clark, A J G7AJC
Clark, A L G4WIY
Clark, A M G3UBY
Clark, A M GM0EJY
Clark, A P G7ABR
Clark, A R G0ATX
Clark, A R G3BII
Clark, A S G7LEN
Clark, A S G8EVI
Clark, A T D GW4SKP
Clark, A W GM3EXS
Clark, B G3VCL
Clark, B G4KZI
Clark, B GW3HGL
Clark, B E G3IMC
Clark, B J G3BEC
Clark, B P G1ECE
Clark, B S D G3PHU
Clark, C 2M1DZS
Clark, C GM2ASU
Clark, C B G1GTH
Clark, C E G8HNB
Clark, C M G1GQJ
Clark, D G6FTH
Clark, D G6SSQ
Clark, D G7BIK
Clark, D M0ASZ
Clark, D A G0HVB
Clark, D A H GM1MSO
Clark, D A W G4JT
Clark, D C G0WDC
Clark, D E G6XYA
Clark, D E G8GPF
Clark, D J G7UDE
Clark, D J GW0KWA
Clark, D L G0KQV
Clark, D M GM7OSQ
Clark, D R G4ELJ
Clark, E 2M1EUV
Clark, E G0WIC
Clark, E G0WNF
Clark, E R G6CSX
Clark, G G0MGC
Clark, G GM4HFD
Clark, G A G8NPM
Clark, G G G0BKR
Clark, G H G4WDD
Clark, G M G0TYQ
Clark, G S G3ZHU
Clark, G S G6AGA
Clark, H D L G3YOY
Clark, I 2E1ENA
Clark, J G0NOA
Clark, J G1OOB
Clark, J GM4BRD
Clark, J G G4AHX
Clark, J H G4DBE
Clark, J L 2E1FGR
Clark, J M 2M1FQJ
Clark, J M G7SGD
Clark, J N G6IIZ
Clark, J P G3GII
Clark, J S GM0LBN
Clark, J V G6YIN
Clark, J W G7NWB
Clark, K G2DOT
Clark, K P G4MPQ
Clark, K R G0CQO
Clark, K R G7KYJ
Clark, L G6LIK
Clark, M 2E0ALR
Clark, M GM6AES
Clark, M C GM6OFO
Clark, M E G8VLP
Clark, M J G0EHR
Clark, N G0BXE
Clark, N A K G3ZQY
Clark, N R G6YMA
Clark, P G0BTH
Clark, P G0VDT
Clark, P D G7PQI
Clark, P E G0VCY
Clark, P F G4PGS
Clark, P J G4FUG
Clark, P J G6RCD
Clark, P J G8BZR
Clark, P L G4IUV
Clark, P N G0TBO
Clark, P R G7WKH
Clark, P R S G7WHN
Clark, R G0CXV
Clark, R G0FXR
Clark, R G0PBR
Clark, R GM1BXI
Clark, R GM1BVL
Clark, R A GM4JCW
Clark, R C G0MOU
Clark, R E G4DDP
Clark, R J G4KYF
Clark, R J G7FCK
Clark, R P G7AGR
Clark, R P G7PRJ
Clark, R P G8GQJ
Clark, R S G0IGA
Clark, R T G3LFE
Clark, S A G8HNA
Clark, S D G4XGR
Clark, S H G6NUO
Clark, S K H G8IUA
Clark, T G4ZLU
Clark, T GM0WFA
Clark, T H G0UVC
Clark, V G GM3WSR
Clark, W C T G0WTC
Clark, W F GM4XND
Clark, W J GM1MSN
Clark OBE, C G G8BKQ
Clarke, A G4NFY
Clarke, A A G6CBU
Clarke, A E W G4GUD
Clarke, A G G1GTQ
Clarke, A O G6XXN
Clarke, A T G7RFB
Clarke, B G0ECZ
Clarke, B C G0MXX
Clarke, B D G4ICB
Clarke, B J GW6RDV
Clarke, B L G6GKO
Clarke, C B G1ATG
Clarke, C E C L G4LPW
Clarke, C G G6BWA
Clarke, C J G4XCX
Clarke, C S G3SQU
Clarke, D G0IDD
Clarke, D G8RFP
Clarke, D G8RSB
Clarke, D B G6XWY
Clarke, D B G8TEB
Clarke, D F G4KYG
Clarke, D G G0IMJ
Clarke, D J G3KYZ
Clarke, D K G3JSW
Clarke, D L GW0MVS
Clarke, D M 2E1EOE
Clarke, D M G7JGD
Clarke, D P G7KAO
Clarke, D S 2E1BKA
Clarke, E A G6HWB
Clarke, E M GI7CNS
Clarke, E P GI3UOY
Clarke, E T G3UYD
Clarke, F G E G3WII
Clarke, F J G3XWZ
Clarke, F K G7BXG
Clarke, F M G7CBE
Clarke, G G G7GYH
Clarke, G H G3CXT
Clarke, G H G3YTW
Clarke, G L G8XXV
Clarke, G M GM0SZL
Clarke, G P G1DNZ
Clarke, G R G4LKM
Clarke, G T S G0PBK
Clarke, H E J GW0PYU
Clarke, H J GI7CNT
Clarke, I 2E1EJE
Clarke, I G1XTD
Clarke, I M G0RTF
Clarke, J G0DTK
Clarke, J G4FFD
Clarke, J A G1IHT
Clarke, J A G3TIS
Clarke, J A G7BWY
Clarke, J B G1HIU
Clarke, J B G4XMT
Clarke, J C G4FMU
Clarke, J D GI6XOV
Clarke, J E G0OBE
Clarke, J G S G3GZI
Clarke, J H G2AAN
Clarke, J K G6IIP
Clarke, J L G7INK
Clarke, J M GI4GUH
Clarke, J P G0YTF
Clarke, J P GI3FFF
Clarke, J P GI4HCN
Clarke, J R G0JDL
Clarke, J R G3OWQ
Clarke, L J GW7RSE
Clarke, L T G1LQB
Clarke, M G8FHI
Clarke, M G8GGS
Clarke, M M1AQE
Clarke, M E 2E1FZZ
Clarke, M E G4OVZ
Clarke, M J G4ULW
Clarke, M J M0AEE
Clarke, M J MW1AJX
Clarke, M L W M1ALP
Clarke, M R G7DUK
Clarke, M R G7RHD
Clarke, M S G3CQL
Clarke, N A G0CAS
Clarke, N V GM3PAE
Clarke, N W G8PZR
Clarke, P G7WJZ
Clarke, P A G1GTR
Clarke, P F L G3LST
Clarke, P I G6DTO
Clarke, P J G7FML
Clarke, P M G8EDH
Clarke, R GW3CCF
Clarke, R A G1IPX
Clarke, R A G8UNO
Clarke, R A M1BMF
Clarke, R C GM0VUJ
Clarke, R J G1GTS
Clarke, R J G6IJN
Clarke, S G8VME
Clarke, S C G8LXY
Clarke, S J 2I1FXD
Clarke, S R G7UID
Clarke, T C G4IVR
Clarke, T E G2FLH
Clarke, T E G4RKD
Clarke, T S G4YKF
Clarke, V G G0SCD
Clarke, W J M1AJQ
Clarke, W L GI0KAN
Clarke, W M G0NGE
Clarkson, A G2CJK
Clarkson, B L 2E1DHA
Clarkson, B L G7WHO
Clarkson, C J GM3ZEU
Clarkson, D A G4JLP
Clarkson, F G G7TWU
Clarkson, G D G3RHM
Clarkson, G F G7ITM
Clarkson, J F M0ABL
Clarkson, J W G8KHT
Clarkson, L J 2E1DKX
Clarkson, M H G4UCM
Clarkson, R C G8WHX
Clarkson, R W G4IHY
Clarkson, W F G4XYR
Clark-Booth, C N G4YQN
Clary, E W G3DTG
Clasby, F W G3KFC

Name

Clasper, N J 2E1AJV
Clasper, R N GM0LOT
Claughton, J L G0MPM
Claxton, J F G4SCM
Claxton, M J G7TYT
Claxton, P J G8CRR
Claxton, R E G0TOT
Clay, A M0AXJ
Clay, B E G1MWS
Clay, B E G6ECN
Clay, E J G0BTT
Clay, G P G4GQJ
Clay, M F G4PJP
Clay, P A 2E1AHO
Clay, P A G7THV
Clay, R A G0VJI
Clay, R A G1DNY
Clay, T H G3TSV
Clayden, K D E G0KXH
Clayden, M J G6MIC
Claydon, C J G4ITC
Claydon, C J C GM4ZJI
Claydon, F J G7SMT
Claydon, J G3JBP
Claydon, T S G8RZL
Clayfield, D H G7WHD
Clayphon, A E G6IJK
Clayson, E C G3IIY
Clayson, E R GW4PQE
Clayton, A G1YTU
Clayton, A G4SEG
Clayton, A J 2E1CBA
Clayton, A J M1AZD
Clayton, A R G7HZZ
Clayton, A S M0ASC
Clayton, C P G4UUU
Clayton, C R I G3VCY
Clayton, E G0KZH
Clayton, F G0JGY
Clayton, G C G0VXD
Clayton, I D G6DHW
Clayton, J G0KJC
Clayton, J G4PDQ
Clayton, J J G5BK
Clayton, J R G0JRC
Clayton, K G G0KEU
Clayton, L W 2E1GAV
Clayton, M L G8VMF
Clayton, N G0GVS
Clayton, N K G0IFS
Clayton, N R G4MCI
Clayton, P I G8MZV
Clayton, P J G4ANQ
Clayton, P J G7LYQ
Clayton, R G0OOO
Clayton, R G3ZJT
Clayton, R G4SSH
Clayton, R G7OOO
Clayton, R G7ROY
Clayton, R H G0DAM
Clayton, R O G8SDU
Clayton, S A G6XGG
Clayton, S J G7FTM
Clayton, S M R 2E1CFC
Clayton, T G0TKJ
Claytonsmith, F GM3JKS
Claytonsmith, M H G4JKS
Clayworth, W E GM4WFN
Cleak, J P GW4JBQ
Cleak, L GW8SBK
Cleal, S E GW8RUA
Cleall, A F G3BSU
Cleall, P E G8AFN
Clear, D P G0KNU
Clear, R S G4FFX
Cleary, K M G4ATZ
Cleary, M L G7RRP
Cleary, P G G3WZE
Cleary, R H 2E1DTK
Cleaton, J G4GHA
Cleave, A D G8XQN
Cleaver, A I G0LCB
Cleaver, D R G0JVF
Cleaver, N J G4HOW
Cleaver, T J G0FFI
Cleaver, T J G0IOD
Cleaver, T J G6ZZZ
Clee, J G8EYQ
Cleeton, G G3LBS
Cleeve, J E G3JVC
Clegg, B J G3VQH
Clegg, E H G3CXR
Clegg, G G4FIQ
Clegg, G G G1LCN
Clegg, J J G3UYG
Clegg, J T G4BIC
Clegg, K G1DNX
Clegg, N G3FCD
Clegg, P B G3TTB
Clegg, P J G3MGH
Clegg, R B L G7RER
Clegg, R K G2PB
Clegg, W T G3EFK
Cleghorn, T G4ZTC
Cleghorn, T J G0VYB
Cleland, C GI0DHW
Cleland, M W G3PWU
Clelland, A S G3UUQ
Clem, G F G7IUE
Clemence, J R G3YHK
Clemens, D T M G3VXM
Clemens, P A G4RAR
Clement, G A GW0OLN
Clements, A B G0PIK
Clements, A J G3AOJ
Clements, A W S G4KDZ
Clements, D R G6KTG
Clements, F W G4ZGO
Clements, J H G3SGS
Clements, J T G6ECJ
Clements, L H G7DXK
Clements, M G G7IPA
Clements, P M G6IZP
Clements, R C G0FGW
Clements, R H M1BKH
Clements, S G6NQB
Clements, S D G1YBB
Clements, T G G0IKP
Clements, T R G0STF
Clements, W A G6RMJ
Clementson, G T G4TIM
Clemmetsen, A R G3VZJ
Clemons, A E G4ZRY
Clench, D J G0BVV
Cleveland, R W G0AKO
Cleverley, M G7PNF
Cleverley, M GW6MIH
Cleverley, R A GW4RKZ
Cleverley, R C G0GJE
Cleverley, R C G4DMC
Clewer, A R G4CTA
Clewes, B G7GXR
Clewes, P D A G6TXR
Clewley, I D G7KAK
Clews, D J G6GLZ
Clews, M J G1VVF
Clews, R G G0SMZ
Clibbon, E A L G8TMC
Cliff, D G1IGW
Cliff, J G6LDH
Cliff, J M G0WXU
Cliff, N H G1VIT
Cliff, N H J G3GRB
Cliff, R W GM0SXP
Cliff, R W GM8DFC
Cliffe, D H G0JWE
Cliffe, D T G8EQC
Cliffe, S G1PDS
Cliffe, W M G4YYV
Clifford, B G7GFR
Clifford, B B G0EYQ
Clifford, B J G7SOJ
Clifford, C J G4IIC
Clifford, G G4SHK
Clifford, J D G4BVE
Clifford, J E 2E1BHF
Clifft, R G4VRJ
Clifft, R K G4DES
Clift, D B G0RVL
Clift, G C G1KIM
Clift, K H W G0ECL
Clift, M G3UNV
Clift, M G3VDM
Clift, M J G7FDL
Clifton, A C G4JCZ
Clifton, A M G4UXU
Clifton, B D G7OWQ
Clifton, C G0UDS
Clifton, D M0AFA
Clifton, D C G8PIY
Clifton, D L B G3WOK
Clifton, D N G7ENA
Clifton, G M G7NZM
Clifton, I B M0AVC
Clifton, J E G6LAE
Clifton, J H G0UIU
Clifton, J S G3XYE
Clifton, P G3UQJ
Clifton, R S G0MYC
Clifton, S P GW4WBT
Clift-Jones, A R G4YCJ
Clinch, A M G1XNR
Clinch, G G3RPD
Clinch, G N G8RLJ
Clinch, J G G3MJK
Cline, D G G4VGY
Cline, S P G4MDZ
Clingan, J F G3TNI
Clink, S G7VSR
Clink, S GM1VBE
Clint, T G8TMD
Clinton, G D G3OYT
Clinton, R H G0BUX
Clinton, W GW8KZN
Clissold, N M M0BEZ
Clitheroe, R G G3TGT
Clode, C J G0CJC
Clode, G K G4AXP
Cloke, D J G4BMO
Cloke, F W G0EON
Cloke, R J G1EXU
Cloke, T G G0JYX
Close, B G0VLQ
Close, C R G1HQE
Close, D G1VKC
Close, D A G3MFG
Close, P H 2E1DGV
Close, P W G7TCU
Clothier, R J 2E1DGT
Clotworthy, C J GI7TEU
Clough, B G6YUX
Clough, D G8CEP
Clough, F G1GCF
Clough, F D G5FD
Clough, J E GM0MDD
Clough, J E GM0VKG
Clough, T J G4PHR
Clough, T J G4WMS
Clough, T J G7CSV
Clough, W G3USW
Cloutman, R W G G6PWL
Clover, I G4UGD
Clover, R E G0RMU
Clowes, B C GW4HBZ
Clowes, D J L G8KLX
Clowes, K G4WKC
Clowes, M J G0FOC
Clowes, P G6KKN
Clowes, W J G3ICZ
Clubley, D A G6XXJ
Clubley, R A G8PLO
Cluer, G G4AVV
Clues, B J G0KOY
Clulee, B D G0LXG
Clulee, B D G3UDN
Clulee, B D G6WAR
Cluley, G S G7FGR
Cluley, J C G4YIG
Clulow, M R 2E1FLV
Clune, P J G0LUL
Clutson, R M G0WHO
Clutterbuck, D M0AVZ
Clutterbuck, P G4FTQ
Clutterham, R J G0AGS
Clutton, M G4OAB
Clutton, M G G4VQH
Clyne, N R G8LIU
Cmoch, S W G1DPW
Coad, C P G6ELZ
Coad, J R G6IZQ
Coaker, R M G0LME
Coakes, R J T G7CFG
Coalston, L A G7TDJ
Coan, B GM0BWR
Coan, M J G4EOL
Coate, D J G G1YHE
Coates, A G0EWT
Coates, A C L G3TVV
Coates, A M G7DJN
Coates, A P G6NPE
Coates, B G0SHM
Coates, D G G1HEN
Coates, G P G0COA
Coates, I E G1MSS
Coates, J A G1WJG
Coates, J M G0MDS
Coates, J M G4GYU
Coates, J N G0SYE
Coates, K H G3IGU
Coates, P G G8IIM
Coates, P R G4VMC
Coates, P V G6PVC
Coates, R G1KFF
Coates, R F G4DIH
Coates, W A G3WAC
Coates, W C G1OFX
Coathup, N F G0HZZ
Coatman, R A G1ZME
Coatsworth, J G3HRB
Cobb, A G0WJK
Cobb, A G MI1BYJ
Cobb, B J G1DBH
Cobb, E J GU4WRO
Cobb, F A GW3YIH
Cobb, G G4UZU
Cobb, G R G3IXG
Cobb, J H G3VHS
Cobb, L L N G3UI
Cobb, R J MW1BCQ
Cobb, R M C D G3HZE
Cobbe, I M G3ZRZ
Cobbett, F C A G3OGL
Cobbett, W E G6ALK
Cobbledick, D G G4VRL
Cobbold, R F G8CKW
Coben, S G1KGO
Cobley, J A G4RMD
Cobley, P G7MZT
Coburn, J S GW0NOO
Coburn, M J G0NMA
Coburn, M G4DUL
Coburn, P GW0NOP
Coburn, S J GW7COB
Cochrane, A J G4UIO
Cochrane, D F G8IHF
Cochrane, H GM0LYH
Cochrane, M J G1IXN
Cochrane, P G3RVC
Cochrane, R G4IZO
Cochrane, R A G7IHF
Cochrane, R C G8DRQ
Cock, D W G3DON
Cock, J W W G3HN
Cockayne, P L G7GJA
Cockbill, R G0KTP
Cockburn, A J G7FBF
Cockburn, D M GM0VWQ
Cockburn, J GM3GKJ
Cockburn, K J G0SKJ
Cockburn, S GM0VXB
Cockcroft, G M G6WEI
Cockell, B D G8BJK
Cocker, A J G7LFM
Cocker, D J G4RYE
Cocker, P D G0TYW
Cocker, S H G7PHM
Cocker, W J F G0PZK
Cockerell, W G G0LKI
Cockerill, A S G4UTV
Cockerill, B A M1BYC
Cockerill, E G4GOZ
Cockerill, F G G2AMQ
Cockfield, B G4ETW
Cockfield, B G G0KRK
Cocking, N S J G8PNM
Cocking, R I G1BEB
Cockings, D R G3WF
Cockings, R 2E1FMT
Cockman, A P G4UTF
Cockman, P M G4UAI
Cockman, R B G1HNN
Cockram, A G1CSZ
Cockram, O W G8OWZ
Cockram, P R G8CUG
Cockram, T J G8GFZ
Cockrill, A R GM4OPJ
Cockrill, J R G4CZB
Cockroft, A G4OWF
Cockroft, J W C G1MSR
Cockroft, M W G4YLM
Cockroft, P M G8CER
Cocks, D E T G0FWO
Cocks, J 2E1FPN
Cocks, K V GM6AXZ
Cocks, S J G4ZUL
Cockshaw, S W G4NBH
Cockshoot, I D 2E1COV
Cockshoot, M G 2E1CPL
Cockshoot, S J G1WWR
Codd, J W 2E1DAQ
Codd, M J G3NNA
Coddington, P M1BKL
Codling, M R G7FEA
Codling, R G4FRC
Codling, T G3UPI
Codrai, T A G3WQY
Coe, A J G4PNL
Coe, A J GW1MKV
Coe, A K G6UHS
Coe, C L G7VUU
Coe, C L M0AZW
Coe, C R G7SWA
Coe, D A G4PZQ
Coe, D L G7SDC
Coe, E J G0COE
Coe, P W GM8YSN
Coe, S G G6NUR
Coe, S J G3FCT
Coe, S V G0ECA
Coenraats, P S G0CEC
Coffey, D G1SFU
Coffin, B V G7EPX
Coffin, G G3XFN
Coffin, S J G7JJW
Cogan, M L 2E1EPF
Cogan, T W G7USJ
Coggan, J M G4RHF
Cogger, G I G4YRT
Coggin, R T G3KMC
Coggin, W E G1SXJ
Coggins, J F G3TFC
Coggon, G F G4FYE
Coghill, F W GM0PXR
Coghlan, P G G6IOB
Cogman, D M H G3ZOZ
Cognet, P J G1PDY
Cogzell, R F G G3OTY
Cogzell, R F G G4ACF
Cohen, A D G6YIP
Cohen, B G7ITZ
Cohen, H G G4GHS
Cohen, J 2E1FST
Cohen, L 2E1FTW
Cohen, L J G0EFL
Cohen, M H G4XWA
Cohen, M I G8NNX
Cohen, P GM3LKY
Cokayne, F G0VFU
Coker, A G G3WHM
Coker, C J G0SBM
Coker, C J G4FCN
Coker, C J G7MNL
Coker, C J G8GCS
Coker, G C G6CLD
Colbeck, C D G4IER
Colbeck, P G4PHK
Colborn, H N S G8TMA
Colbourn, N T G0HOD
Colclough, A G7CYZ
Colclough, G G6ALN
Colclough, J D G4CWN
Colclough, R G7FSA
Colclough, R C GW8RDI
Colclough, W G1FXW
Colclough, W G1MAD
Coldbeck, D G6ABG
Coldham, F W G1GGK
Coldicott, M C G8DKV
Coldwell, D J G0RKY
Coldwell, P T 2E1CQH
Cole, A B G7VGJ
Cole, A D G4XAE
Cole, B G0WLC
Cole, B D G3PQJ
Cole, B J G1HBO
Cole, B S G7WWW
Cole, C D G8PJQ
Cole, C J G8FIG
Cole, C M G4NVW
Cole, D L G3RCQ
Cole, E R C G7MRH
Cole, G G4AWI
Cole, H A G3VJE
Cole, H M G3OHK
Cole, I D G6GBT
Cole, L G G8RHO
Cole, M G4ZRC
Cole, M G G0DKR
Cole, M J M G6FPH
Cole, P G6PLV
Cole, P J G0XAO
Cole, P J G6JVP
Cole, P T G0THA
Cole, P W GW7SSQ
Cole, R G7MPH
Cole, R A G3REB
Cole, R H G8ZJK
Cole, R J G3XKC
Cole, R J J G G4SJZ
Cole, S H G8YDU
Cole, S J G0KEY
Cole, S K G3YOL
Cole, S R G3ZDG
Cole, S R G7ITD
Cole, S R GW4BLE
Cole, T A G7EEG
Cole, T J G4RQH
Cole, T J GW7JPC
Cole, V A J G3MHC
Cole, V B GW7HOM
Cole, W C G0KFW
Cole, W E GW0PDA
Cole, W G GW4JUW
Colebourne, M K G4WRM
Colebrook, C G G8TJR
Colebrook, J L G3BJD
Colebrook, R C G8YBO
Colegate, J G3RVX
Colegate, P G3RVY
Coleman, A G0PBO
Coleman, A L G0LUE
Coleman, B R G4NNS
Coleman, C M L G8GGR
Coleman, D A 2E1DIB
Coleman, D R N G0UTR
Coleman, E W G1LGO
Coleman, G A G7PYT
Coleman, G C G4IVT
Coleman, G N G3ZEZ
Coleman, G V G7RLX
Coleman, I J G1YVV
Coleman, I T G4GBT
Coleman, J C G0KGI
Coleman, J C H G3YGB
Coleman, J M G1GTU
Coleman, J P G8XDQ
Coleman, K G0TNV
Coleman, K M G6SSX
Coleman, L G0WJA
Coleman, L J G6KGA
Coleman, M G6TID
Coleman, M F G8RIT
Coleman, M H G5BH
Coleman, R G3KDV
Coleman, R G4LQX
Coleman, R A G1JKP
Coleman, R C G0BXP
Coleman, R G G3PYV
Coleman, R J G4RJC
Coleman, R S G0WYD
Coleman, S G0BTV
Coleman, S A G4YFB
Coleman, T W G8EEI
Coles, A G3SAQ
Coles, B J G0EHW
Coles, C G8CJT
Coles, D E G G4NZY
Coles, D E J G3NYD
Coles, D H E G7GZC
Coles, D J G6PJW
Coles, F E G7TKB
Coles, F P J E G3PZC
Coles, G D G0BKU
Coles, G G G0CCA
Coles, J D R G4IJU
Coles, J F G0VWL
Coles, J F G1ANA
Coles, M G0VXC
Coles, M G G0AJX
Coles, M W G G8FUT
Coles, P G0XAF
Coles, V B G4EGN
Coles-Macgregor, J V G4FYW
Coley, A N G1HQG
Coley, L W G0BEO
Coley, R A G3IFF
Coleyshaw, S G8UWJ
Colgan, M G7SQI
Colgan, N GI1FSJ
Colhoun, C J GI0BZM
Collar, A E G4ZMX
Collar, M D GM8AGM
Collar, P G G3TGN
Collard, H G G2CVA
Collerton, K W G4BDC
Colles, J M M1BKJ
Collett, A J G4NBS
Collett, A J G4NXO
Collett, C G7FPU
Collett, C A G1PZP
Collett, F G3OVT
Collett, J G7BJL
Collett, J G7HGR
Collett, J F G3BUR
Collett, J T GI3AMY
Collett, P G0BQF
Collett, P G4LYC
Collett, R G4LRQ
Collett, R A G4AUN
Collett, R H G0KSV
Collett, R W G3EGS
Collette, R G3CUR
Colley, I P 2E1FBW
Colley, J A G4XJC
Colley, L D G3AGX
Colley, M G1JGE
Colley, R P G1DPX
Colley, S G1ILC
Collicott, M T G0BNT
Collicott, N J 2E1EQZ
Collie, F J H G4YQJ
Collier, A J G6IZK
Collier, A J G6WXZ
Collier, A J G8WZJ
Collier, C P G6SXM
Collier, D I G7TRU
Collier, E G GW1KJE
Collier, G L GM0LOD
Collier, G L GM0SNG
Collier, G M G8PQZ
Collier, H A G4JXI
Collier, J H A M0APV
Collier, J R G0OSU
Collier, J W G7DVD
Collier, P P GM4ZPC
Collier, Q G G3WRR
Collier, Q G G4ALE
Collier, S A G4WCS
Collier, S W P G3LLD
Collier, T G7UKA
Collier, W G0AYF
Collier, W R G0TGU
Collier Webb, N R GJ3DVC
Collier-Webb, N R GJ0VJP
Collinge, B P G7NYD
Collings, J G3BIX
Collings, L G7AVZ
Collings, S G4SGI
Collings, S GM4YYF
Collings, S A G4SAC
Collingwood, J G0VBX
Collins, A G4SOR
Collins, A C 2E1AMB
Collins, A D G1ZYB
Collins, A E G7IUV
Collins, A G G7IKO
Collins, A J 2E1BPN
Collins, A J G6RNW
Collins, A J M1BPD
Collins, A M G0MUN
Collins, B A G1YPL
Collins, B E G0FMB
Collins, B J G4YQR
Collins, B S G3MXA
Collins, C G8SC
Collins, C GW3WEQ
Collins, C A G0PUX
Collins, C A G3THX
Collins, C E G0JBC
Collins, C J G7PVJ
Collins, C J G7PWS
Collins, C R T G0JAP
Collins, D G1XZG
Collins, D G4NHW
Collins, D A W G7NGH
Collins, D H G4ZYF
Collins, D J G0DJC
Collins, D J G0TSM
Collins, D J G7TYR
Collins, D J G8WZK
Collins, D J GW0WVF
Collins, D J A GW4YNP
Collins, D M 2W1BHX
Collins, D S W G7MEY
Collins, E G4ZME
Collins, E H G4UPS
Collins, F C G4UTG
Collins, F W G0RXO
Collins, G G4DKG
Collins, G M G6XUJ
Collins, G S G6EMB
Collins, J G0ILH
Collins, J G0OKL
Collins, J G0POW
Collins, J E G3MBW
Collins, J M 2E1AGQ
Collins, J M G0DHU
Collins, J R G0EPV
Collins, J T G G7GDT
Collins, K G4FIP
Collins, L R G0IPB
Collins, L V G1DNQ
Collins, L W G4LWC
Collins, M C M0AHC
Collins, M G G0GCA
Collins, M J G4WGJ
Collins, M J G7JDN
Collins, N H S G8ZKK
Collins, N J G3SPN
Collins, P J G8ZWA
Collins, P S G4GHZ
Collins, R G2AQJ
Collins, R G4PCE
Collins, R G7ROQ
Collins, R A G1DNP
Collins, R A G6VUG
Collins, R D G0VTA
Collins, R E G3ROC
Collins, R T G3TRC
Collins, R T G5LK
Collins, R T G7RAT
Collins, R W G1OQK
Collins, S G0SLF
Collins, S G0WDQ
Collins, S G7UCE
Collins, S C G8HAM
Collins, S H G7TBS
Collins, S J G6SXJ
Collins, T G1PNC
Collins, T G6ALJ
Collins, T L G8TVT
Collins, V A 2E1EQI
Collins, V F G1THA
Collins, W G4FLF
Collins, W A G7MDY
Collins, W J G0PVV
Collins, W T GI6JXG
Collinson, A G4BNS
Collinson, A R G1WJQ
Collinson, D G0OLO
Collinson, E G0VWX
Collinson, H G3FKV
Collinson, H I G6WZM
Collinson, J D G7ULH
Collinson, L S 2E1EBX
Collinson, M G0LFD
Collinson, R G3ZWZ
Collinson, R C G7SZO
Collinson, T G G4RVO
Collis, A E G7UYK
Collis, A W G3NWH
Collis, G G4YBA
Collis, G H G4GHC
Collis, M A G7AGC
Collis Bird, N G G0WZK
Colliton, J E G0IIE
Collop, C G G3AXN
Collopy, M F G3XVC
Colls, J S G6PKA
Collyer, G F G0LUO
Collyer, I R P G1SAQ
Collyer, S G G3PZK
Colman, D G6VLV
Colman, R B G1YJJ
Colmer, E A G4KMF
Colson, J D G4XMY
Colson, R H G4GYN
Coltart, D R G3SYM
Colthorpe, W J L 2E1DDS
Coltman, H D G3PVJ
Coltman, T S G3TTH
Colton, A G0UYE
Colton, D P G7SQY
Columbine, C G4ANU
Columbine, J A 2E1CEU
Colville, A G7UOU
Colville, H G8LKW
Colvin, A J G0UYD
Colvin, F C G6BNM
Colvin, R G0BXQ
Colwell, L W G4SYY
Colwell, M A 2E1BDR
Colwell, P L M1BDS
Colwill, J L G0DQH
Colyton-Smith, L J G0MYT
Comben, P E G7AAR
Comber, A J G7BAU
Comer, G M G6USG
Comer, P H G0CTU
Comerford, J J W GW0ENT
Comfort, A G0DBL
Comis, A M J 2E1BHB
Comis, P G D 2E1BHC
Comis, S D G4XDM
Comley, A T G1FPF
Commander, C A G8PDM
Common, B K G4SUA
Compton, A G4IDW
Compton, A P G1KAR
Compton, A P G1SHH
Compton, C A G1CHV
Compton, C W G7RJO
Compton, J R G4COM
Compton, L G0HKI
Compton, R W G1ZPU
Conant, P E M G0MLX
Concannon, W J G4EDM
Conce, C V G7SHI
Conde, D 2W1GAU
Condliffe, K G4AYY
Condliffe, W J G1XUA
Condron, J GW7KDZ
Conduit, C P G4KCZ
Cone, R A G7JBZ
Congrave, S R G6SHZ
Conibear, I P G4TAH
Conlan, T N 2M1BYW
Conley, M G1IWA
Conlon, J GI7NEB
Conlon, J N G4RKB
Conlon, K A G0RLO
Conlon, K A G7KRC
Conlon, L D B G0BBE
Conlon, L D B G0KRG
Conlon, M G GI6NPF
Conlon, M V G0ZMC
Conn, G E G7HDT
Conn, J N GM6SXK
Conneely, R H J G0HKB
Connell, J G0DRJ
Connell, L G0DGB
Connell, M I G8HDL
Connell, R G4PZS
Connell, R J G4JQY
Connell, T J G3UWI
Connell, T J G8EXQ
Connelly, B MM1BRT
Connelly, J J GM3RGU
Conner, S GM0TET
Connery, L D GW4ZBN
Connett, R G G0WSC
Connolly, C M1CCW
Connolly, D L G1REH
Connolly, D P G3KHK
Connolly, E L GI6FXY
Connolly, L J G4WPL
Connolly, M J G0NKC
Connolly, M J G7VJF
Connolly, P A G3ZYO
Connolly, P M G0CUN
Connolly, P P MI1AWU
Connolly, R A GI7IVX
Connolly, R C G4LLN
Connolly, T M G7BYV
Connolly, W N GM4ZET
Connor, D J G7LKI
Connor, E G4DDB
Connor, F G4FMI
Connor, I L G7PHD
Connor, J W G1OQV
Connor, L G7AXI
Connor, M G4PIX
Connor, P N G8XTE
Connor, W G0GSO
Connors, P F G4PLZ
Consitt, J P G4OBR
Consolante, R C G1FSW
Constable, A G4WFI
Constable, J R 2E1DEX
Constable, M A G4FMC
Constable, S M1BWU
Constable, S A G7GOA
Constance, A N GM1OXE
Constance, J V W 2E0ANZ
Constance, J V W G0VGD
Constantine, A J G7OOP
Constantine, E A E 2E1BVJ
Constantine, L K G6CQG
Constantine, M I G0MIC
Constantine, M R 2E1BOO
Constantine, R J G3UGF
Constantine, V 2E1FAP
Convery, F GI3ZTL
Convery, W G6WQN
Conway, A C G3WQL
Conway, B G4MCG
Conway, B M G6BLC
Conway, G I G1VSJ
Conway, G P G0GPC
Conway, P G4LAN
Conway, P G4RPX
Conway, P A G3UFI
Conway, R D G7HRH
Conway, R H G3ZBC
Cooch, J P G4MRX
Coogan, J P G6LMW
Coogan, P J G6FFU
Cook, A G4PIQ
Cook, A C G7VQV
Cook, A F S G8PLQ
Cook, A J G0LGM
Cook, A J G6HQX
Cook, A J G7GFD
Cook, A J G7IEO
Cook, A M G6HIA
Cook, A R J G4MRS
Cook, B A G0RIX
Cook, B A G7TYP
Cook, B M G8PHG
Cook, B T G4BWJ
Cook, C A R G3OTH
Cook, C E GM0OMC
Cook, C G GM0VHR
Cook, C R G0GMC
Cook, C R G1ZMS
Cook, C R G3ZMS
Cook, D A G6XXB
Cook, D J G1BWE
Cook, D J G6DNJ
Cook, D J G8DJC

Name

Cook, D P G4YNJ
Cook, E G0VZE
Cook, E D G1HYT
Cook, E R G3NAV
Cook, F G4MTW
Cook, F A G3UZL
Cook, G C G4RKX
Cook, G E G8TEC
Cook, G G G4CUI
Cook, G H G4EXJ
Cook, G M G3IXB
Cook, G T G4WXH
Cook, H S G4SYD
Cook, H T G1JRL
Cook, J G0AFQ
Cook, J G0DPC
Cook, J G3OPW
Cook, J GW1AXU
Cook, J A G4OYC
Cook, J H G0EQM
Cook, J H G4GAR
Cook, J I GW6GCK
Cook, J J 2M0AJW
Cook, J J GM7POK
Cook, J J MM0AXL
Cook, J J L G8LBG
Cook, J R G7VUL
Cook, J R E G1AZA
Cook, J T G8HYF
Cook, L S G8PWT
Cook, M G1NQH
Cook, M A G7SED
Cook, M C G4OTN
Cook, M E G8LZL
Cook, M G G0TPO
Cook, M J G1ELZ
Cook, M J G7RIA
Cook, M K G8HBR
Cook, M W G0PXE
Cook, N D 2E1AWS
Cook, N D G0MQV
Cook, N M G7JVE
Cook, P G1FCY
Cook, P M0AVX
Cook, P C G4AVP
Cook, P C G6IZO
Cook, P F G7PFM
Cook, P F G8XOM
Cook, P J G0PEQ
Cook, P J G4NCA
Cook, P R G6GFG
Cook, R G0JVK
Cook, R GM4OFZ
Cook, R M0ALK
Cook, R H GM4BYT
Cook, R J G8TMB
Cook, R L G4XHE
Cook, R M GM1MYR
Cook, R N G7MMA
Cook, R R G6WYF
Cook, R V G7IQD
Cook, S G1CPF
Cook, S G6BPK
Cook, S E G0ATJ
Cook, S F G1HOL
Cook, S G G3YBR
Cook, S G G8ZPF
Cook, S H G6OQC
Cook, S J G8CYE
Cook, S S V G8ZLK
Cook, T E G8CWE
Cook, W A GW0KQX
Cook, W H G4ZPV
Cook, W R GM3MLW
Cook, W S G0GUI
Cooke, A G4UCT
Cooke, A J D G8APK
Cooke, A R G3IFX
Cooke, B G G0FCO
Cooke, B K GW6MHV
Cooke, B O GW0ION
Cooke, D A G8RFZ
Cooke, D G G7GRA
Cooke, D J G8TDP
Cooke, D S G8DIN
Cooke, E G MW0ATT
Cooke, E J G0HMO
Cooke, E J GW7MMH
Cooke, F G3XLC
Cooke, G A G6LFT
Cooke, G F G0IUB
Cooke, G F G4NIX
Cooke, G J G4WDC
Cooke, H A C G4RCV
Cooke, J G1MPW
Cooke, J A GM8OTI
Cooke, J C GW0SYN
Cooke, J E G6TYB
Cooke, K D G6AYH
Cooke, L A G4FMJ
Cooke, M J G4DYC
Cooke, P G4HVZ
Cooke, R G3WTL
Cooke, R G7PMB
Cooke, R G GM1MQA
Cooke, R J G3LDI
Cooke, S C G3VJK
Cooke, S J G4NIY
Cooke, T D 2E1CTG
Cooke, T D M1CDE
Cooke-Sanderson, A G1DPY
Cooknell, D A G3DPM
Cooknell, F G2CO
Cooksey, D 2E0APQ
Cooksey, D 2E1EQU
Cookson, D A G0DRM
Cookson, J E G4XWD
Cookson, J H G6ETP
Cookson, R I G7CUA
Coole, G E G0LHH
Coole, R J G8OFR
Cooley, G G GI0LEK
Cooley, M A G3XOC
Cooling, T R G4XMQ
Coolledge, G A M0ANU
Coombe, C M G7TGK
Coombe, C R G3LPA
Coombe, J J M1APJ
Coombe, T D 2W1FEH
Coomber, C J E G4NVK
Coomber, D G8MJX
Coomber, D R G8UYZ
Coomber, G K G4YOF
Coomber, J K G8UG
Coomber, P A G8LOP
Coombes, D S G7BHR
Coombes, J E G0CQK
Coombes, P J G6AXY
Coombes, R G6WWR
Coombes, R J G3ZNH
Coombes, R J G4IGL
Coombes, R S G4WWR
Coombes, R S G4ZEJ
Coombes, W C G4ERV
Coombs, B J 2E1CZO
Coombs, G G1JYA
Coombs, J M G8NVY
Coombs, M G G4YBB
Coombs, M P G3VTO
Coombs, M V A E G8GTC
Coombs, M W G1KOM
Coombs, N B G6XQG
Coombs, N J G4IZV
Coombs, P G1KBN
Coombs, P A GW4XYI
Coombs, S E 2E1DAC
Coombs, T J G1IWE
Coombs, V J W G4DJG
Cooney, M J G3TYH
Cooney, P B G0UXK
Coope, B A G4ADA
Cooper, A G4VMY
Cooper, A G4YSG
Cooper, A C GW1ZBE
Cooper, A E G1OOG
Cooper, A J 2E1BRT
Cooper, A J M1AQR
Cooper, A M GW3TKD
Cooper, A P GD4PNY
Cooper, A R G0OWJ
Cooper, B G6GNO
Cooper, B D G4RKO
Cooper, B P G0TUM
Cooper, B R 2E1DSB
Cooper, B S G8DLM
Cooper, C GM3JKC
Cooper, C C G4XTX
Cooper, C H G8MHD
Cooper, C N G1SDJ
Cooper, D G8UWL
Cooper, D A G4LQY
Cooper, D A W G0HVP
Cooper, D G G4NZX
Cooper, D G G7DBD
Cooper, D G GM1TGA
Cooper, D J G1EXV
Cooper, D J GM8SAP
Cooper, D R G6DRC
Cooper, D W G0KYR
Cooper, E G0ZEC
Cooper, E G3YBA
Cooper, E C G3GEG
Cooper, E D G4HRT
Cooper, E J G8ZWH
Cooper, E M G3UGO
Cooper, E T G4VAS
Cooper, F G G0BYU
Cooper, F H G2QT
Cooper, F O G3RIQ
Cooper, G G3HJP
Cooper, G G4OQX
Cooper, G GM4LFA
Cooper, G E G1PII
Cooper, G J G4YXA
Cooper, G W G0KVK
Cooper, G W G3VJB
Cooper, G W G4XXS
Cooper, H A G3SWW
Cooper, H G G0BLQ
Cooper, I G7AQN
Cooper, I G G3XYV
Cooper, I J G0BMS
Cooper, I S G8GLC
Cooper, J G3DPS
Cooper, J G7RCO
Cooper, J G7VIE
Cooper, J GM1TGS
Cooper, J GM4OYK
Cooper, J D G8WUU
Cooper, J F G4RAC
Cooper, J J G3XEV
Cooper, J M G2UD
Cooper, J P 2U1FPL
Cooper, J P G0MNK
Cooper, J S GM0FRI
Cooper, K G G4AQN
Cooper, M G1ANI
Cooper, M G1BWU
Cooper, M GM6MGH
Cooper, M G8KGH
Cooper, M B G1VUG
Cooper, M E G6JEY
Cooper, M G G0NLV
Cooper, M H G4SMP
Cooper, M J G0KOI
Cooper, M J G8MQF
Cooper, M J GJ0SVZ
Cooper, M W G6UIC
Cooper, N A B G6FPK
Cooper, N J G8NYZ
Cooper, P G7TTF
Cooper, P D G0KDA
Cooper, P F H GU0SUP
Cooper, P J G1JKN
Cooper, P J G3CXI
Cooper, P J G4MEN
Cooper, P R G0VUM
Cooper, P R G8ZPE
Cooper, R A G4GPB
Cooper, R A GI0WOW
Cooper, R D G1DBZ
Cooper, R F G4TSB
Cooper, R H G0MSM
Cooper, R I G0WFR
Cooper, R I G4WFR
Cooper, R I G6IIU
Cooper, R J G0AMM
Cooper, R O G8XTF
Cooper, R W G1JBN
Cooper, S G GM4AFF
Cooper, S J G3YTI
Cooper, S J G6MGA
Cooper, S J M1AVM
Cooper, S M G0EPI
Cooper, S M G0PZR
Cooper, S M G0WZC
Cooper, S S G7RXX
Cooper, T A G4CBY
Cooper, T J G1ASW
Cooper, T J G4UZT
Cooper, T J G4XOP
Cooper, T M G6XWZ
Cooper, T R G0RNT
Cooper, T S G7SLR
Cooper, T W G3YHA
Cooper, W G8UOL
Cooper, W C GW0OTY
Cooper, W E G4CIA
Cooper, W G G4PIY
Cooper, W H G7TYU
Cooper, W J M G4WJM
Cooper, W L G7FDD
Cooper, W M C K G0KDL
Cooper, W T GM4LNH
Cooper Iii, C B M0BFE
Cooper-Bland, H G0ABR
Coopland, C R G1NBG
Coopman, W M G0XBD
Coote, A J G8RWT
Coote, G W G1IPU
Coote, J A C G0THV
Coote, L G G3AHB
Coote, M G7MTE
Coote, N A G8VNU
Coote, N J GM8TVV
Coots, H G7WJE
Cope, A G8ROZ
Cope, A R G7UVC
Cope, D G4NXY
Cope, D M G8RYE
Cope, I D G4IUZ
Cope, J W G4DYJ
Cope, P G7ASK
Cope, R A G0RLX
Cope, R S G4KMX
Cope, S T G4VAX
Copeland, D J G0JXJ
Copeland, G GI4DRY
Copeland, J M G6NPJ
Copeland, N D GI8RPT
Copestake, A GW4YSV
Copestake, R R GW3LFC
Copland, A I GM1SXX
Copland, I M G1HJL
Copley, S W G0UEM
Coppack, N P G6PGV
Coppen, D R G8DEL
Copper, A P G6MEE
Copperwaite, A E G8AQO
Copping, C E W G4OC
Coppins, P J G3GET
Copplestone, J H G0RFM
Copsey, A G1OTE
Copsey, A J G1OWJ
Copsey, C P G8ULQ
Copsey, M W B G8UUV
Copsey, R A G4TDF
Copsey, S D G1ZMY
Copson, J M G3TUL
Corallini, A R G0DCW
Corallini, D E G4YOS
Corallo, J C T G0HZN
Coram, N W G G0SIK
Corben, J B G4EXT
Corbett, A J G0WFV
Corbett, G T G0JZS
Corbett, J R G G3TWS
Corbett, M G4MPW
Corbett, M G8VBW
Corbett, M G G8TWS
Corbidge, J C G8SUQ
Corbin, M T P GU8IRF
Corbishley, P G8NAL
Corbu, J A J G1BTU
Corby, M J G8WXQ
Corcoran, M G GI4TTL
Cordell, A J G0FPL
Corden, D G7UIX
Corden, D M0AFM
Corden, R J I 2E1ATV
Corder, D J G6SPB
Corderoy, C J T GI4CZW
Corderoy, J H G6LFD
Cording, G R K G6RVB
Cordingley, I G4AQR
Cordingley, R G3BAP
Cordrey, P W G0UTB
Corduroy, A J G1PRM
Cordwell, A G0NFY
Cordwell, D G7CWR
Core, A G G0AGC
Corey, J GI6UFS
Corey, R A G0ANY
Corfield, D G7NBJ
Corfield, J A G4ZKG
Corfield, R J M0AYY
Cork, K G4ZNL
Corker, A G4NJI
Corker, B G6GLR
Corker, B G8FBQ
Corkhill, J H G8UCM
Corkill, D M G4DJK
Corkill, R D GD8KWM
Corkin, P G G0PXN
Corkish, W C GD0IFU
Corless, A G3ANE
Corlett, W D GD1GHK
Corley, P G G1NSW
Cormack, D S G4VZR
Cormack, J G GM0HBI
Cormack, J G GM0IYP
Cormack, J G GM4JUE
Cornall, J M G0PES
Cornell, C R G8GWK
Cornell, J A S G6ZKM
Cornell, M 2E1ELO
Cornell, P J G4TBI
Cornell, P J G8CQG
Cornell, R E G7HBO
Cornell, R P G0GOH
Corner, C C G1MHA
Corner, D W G3ZXF
Corner, M G7MYK
Corner, P W G3SPK
Cornes, D C GW7KWB
Cornes, G G G7SXP
Cornes, I L G4OUT
Cornes, K J G4THY
Cornes, P G6YLZ
Corney, D G4UPT
Corney, J G0ICQ
Cornick, D GW0ALM
Cornick, D A MW1AML
Cornish, F J G0UOW
Cornish, R D G0DOK
Cornish, R M G3AZY
Cornish, W M G7OKI
Cornish, D G4SAX
Cornthwaite, R G7SLQ
Cornwall, B R G3ZFX
Cornwell, A V G4XUK
Cornwell, R H G1WEF
Corp, A W G1ZTM
Corps, A A G7CDO
Corps, R J G3FOR
Corr, R C G4GXM
Corrall, R E GW1EWQ
Corrigan, J GM0FXJ
Corrigan, M T GW8LKX
Corrigan, P GM0DQR
Corrigan, P P G4JNN
Corrigan, S V G1TSV
Corris, S W G7JWU
Corsellis, T G1KTE
Corser, K F G7BSA
Corser, K F M0BBN
Corsi, D GW6FED
Corsi, L D J G3IKG
Corson, J C G3WQJ
Corston, J E G2BCY
Cort-Wright, P J G3SEM
Cory, C N G3MEV
Cosford, H G8ACL
Cosgrif, C T G0DZC
Cosgrove, D L 2I1AXG
Cosgrove, F P G0DMW
Cosgrove, F P G0IOQ
Cosgrove, I W G4UOW
Cosgrove, J J GI3HXH
Cosgrove, J P MI1BKO
Cosgrove, W J G1EXR
Cosham, G R G6AUE
Cossar, D L GM3WIL
Cossens, A H C G7LQV
Cossey, A D G7WEW
Cossey, D F M1AEH
Cossins, M G0TBV
Costa, E A 2E1FQN
Costa, G J A 2M1DRD
Costello, C G1LQM
Costello, J M G4PIW
Costello, M E G3YPP
Costello, M G G1TQT
Costello, P J G1IAV
Costello, P N 2E1BFD
Costello, R J G6BUU
Costelloe, S T P G0NIH
Coster, J H GM3SHR
Costford, R W G GM3ITE
Costford, T G GM1XOI
Costford, T G GM7WKQ
Costigan, C J G1AXS
Costigan, P P L G1DAX
Costin, G H G4GFU
Costis, C G0WQU
Coston, R B M1ALF
Coton, D G8KDD
Coton, I G8XCJ
Coton, T A G1AUX
Cotsford, R C G0SAV
Cott, J G6SSV
Cott, P J G3BDV
Cottage, D J G7UWL
Cottam, D S G0HVN
Cottam, D S G3IUB
Cottam, D S G8IUB
Cottam, G M M1AQY
Cottee, B S G7SMQ
Cotter, T S G7LRO
Cotterill, D W G7DWC
Cotterill, K J G7NHQ
Cotterill, R E S G7AKQ
Cottham, T W G4KTB
Cottier, J B G8KRV
Cottingham, J S G3YHS
Cottis, H G G3OA
Cottis, S M G4KMH
Cottle, A V 2E1DAR
Cottle, D C G4XWQ
Cottle, E G7OOF
Cotton, D G7EHD
Cotton, D W G1ATA
Cotton, E G4PVO
Cotton, F G0LFI
Cotton, L J G7CBY
Cotton, M S G7MVE
Cotton, M W G3NFC
Cotton, M W G4HBY
Cotton, R G6YLT
Cotton, S G7CBZ
Cotton, S J H GW4FKJ
Cotton, V F G3BCI
Cotton, V F G4PRS
Cotton, V P G4HGD
Cotton, W E G4PQZ
Cottrell, D G G7GLQ
Cottrell, J G G3PSY
Cottrell, J R G1OKY
Cottrell, N B G3JFR
Cottrell, P F G0LKF
Cottrell, P F G4CLV
Cottrell, R B G3VOS
Cottrell, R C G3SHY
Cottrill, C G0EIJ
Couch, M G GW4ZQY
Couch, N B G7OPM
Couch, P Q G8UHJ
Couch, R A G6ERI
Couchman, B T G4IYC
Couchman, H M P G8WHB
Couchy, D T G8BJA
Coughlan, C G1HQD
Coughlan, S G G6FTI
Coughlin, A GW3TOB
Coughtrie, J B GM0RUW
Coull, P G3XVY
Coull, P M GM4GLD
Coulman, J T G3HJC
Coulson, B A G4NZZ
Coulson, D K G0OAP
Coulson, I G8HUG
Coulson, I GM0KKE
Coulson, J R G6KGB
Coulson, R G3OBJ
Coulson, R G G1YAW
Coulson, W A G0MFW
Coulstock, A J G6TXZ
Coulstock, B J G6TXY
Coultas, M A G0SLP
Coulter, D G8ORO
Coulter, D A GI0TIE
Coulter, M A L G0MEY
Coulter, R J GI8VKA
Coulthard, B G7WHT
Coulthard, W J G0FPO
Coulthart, D GM8VBX
Coundley, A J 2E1FRX
Coupar, D W GM3YVX
Coupe, B G4RHZ
Coupe, D P G4ZML
Coupe, D W G7DGF
Coupe, E G1WYB
Coupe, E J G0MUP
Coupe, J J G0ASH
Coupe, J M 2E1FLO
Coupe, J N G4MWC
Coupe, J N G4NRO
Coupe, J N G6FSA
Coupe, M G1EUT
Coupe, N G3KBS
Courcoux, M S G1BLT
Courcoux, P E R G3EBP
Course, A P G4HND
Court, A J G7END
Court, D C G4RPA
Court, D I G3SDL
Court, K L G4GSZ
Court, M L G4AKB
Court, P G G1ROK
Court, T J G4IAG
Courtenay, B G7UFI
Courteney, M N G4VWE
Courtier-Dutton, D L G3FPQ
Courtnell, E G G7CQZ
Courtney, D A GI8PDK
Courtney, G R D G1LHE
Courtney, R B M1ATC
Courts, D L G7VQF
Couse, D V E G8ILW
Couse, W R MM0ABQ
Cousens, T A G3KXU
Cousins, A J G0HXP
Cousins, C W G4DEL
Cousins, D J G3NCC
Cousins, E M G0DRG
Cousins, F J G8TNQ
Cousins, G H F G4PRU
Cousins, P R G1LVR
Cousins, P W G4NJJ
Cousins, W GM0DTJ
Coutts, A G GM0MZD
Coutts, A M GM3KPD
Coutts, D S GM3VTH
Couvaras, J F M0AMU
Couzens, D G G8HSK
Couzens, E J G0PGO
Couzens, G A G3NTA
Couzins, J D G1KMJ
Covel, J G7JOF
Covell-London, V M G0APV
Coventon, E H G4LHY
Coventry, K G0VDM
Coverdale, M G4LTI
Covey, T C G4ARO
Cowan, A W GM0UDL
Cowan, C A GM0UIG
Cowan, G J M GM7NOA
Cowan, G M GM7GXI
Cowan, J H GM7OIN
Cowan, J I GM0HNJ
Cowan, R M GM0SEP
Cowan, R M GM4SRL
Coward, E A G4EYQ
Coward, M A G3NFJ
Coward, M J G7JKD
Coward, M R G3PRH
Coward, R M G4XKR
Coward, S G6NUQ
Cowburn, P A G7CYM
Cowdell, A D 2E1BJE
Cowdell, K R G8BDZ
Cowdell, S V G4XED
Cowden, R R B GI2BZV
Cowdery, R A G3UKB
Cowee, J W G6FYC
Cowell, B J G6USL
Cowell, F O G6OFU
Cowell, K G0OKV
Cowell, M S A G1XNX
Cowell, O W P 2E1BBH
Cowell, R A G1IVY
Cowell, W C A G0WCC
Cowell, W J G8LQL
Cowell, W S G0OPL
Cowgill, M S G1VOX
Cowgill, R G4YMO
Cowie, D M GM8KSJ
Cowie, G A GM7SBB
Cowie, I MM1BOK
Cowie, J GM3RNZ
Cowie, J M GM6KJD
Cowie, S J G0EZB
Cowles, R C G7OED
Cowley, A G1VTV
Cowley, A G3FCM
Cowley, A J G3ETK
Cowley, A J G4CXA
Cowley, A M 2M1DHI
Cowley, A P 2E1EAA
Cowley, B J G4WDH
Cowley, D J G0DAC
Cowley, J E 2E0AIM
Cowley, L G G6VBR
Cowley, M G0GAG
Cowley, M J G1YNP
Cowley, M K G7RTQ
Cowley, N A G6MFU
Cowley, S I G4TNZ
Cowley, S R 2E1ECY
Cowlin, A T GM7SUW
Cowlin, F W G8JSE
Cowling, C J G4VTS
Cowling, G D G0FRX
Cowling, G D G8HSG
Cowling, J F G3HWM
Cowling, L F G3IGX
Cowling, O M G0MHU
Cowling, P G8BMZ
Cowling, R W G8ZWF
Cowling, W E G0LCW
Cowper, C N L GM0OYV
Cowperthwaite, E G4UJI
Cowsill, A G4MBA
Cowsill, J S G4EDJ
Cowtan, D J G1DQB
Cowtan, M E G1GUL
Cox, A G4PCB
Cox, A J G0OBG
Cox, A J G8NJF
Cox, B E G0DMH
Cox, B L G7TOL
Cox, B N G4KWX
Cox, C G G4JEC
Cox, D G3KHZ
Cox, D G3ZAT
Cox, D G G0LOM
Cox, D H R G0NYS
Cox, D J G8DUI
Cox, D J B G0EWA
Cox, D M G8MTO
Cox, D M M0ACZ
Cox, D V G0RRJ
Cox, E H G1YFH
Cox, F V G1OPW
Cox, G G4SWB
Cox, G R G4AL
Cox, H G8HOU
Cox, H E G3YEB
Cox, H W G0OBH
Cox, I C GM6JNQ
Cox, I F G6ECK
Cox, J P G4WZK
Cox, K G0UUL
Cox, K G8MAN
Cox, K R G7JXB
Cox, L F G3PRN
Cox, L R G1JGF
Cox, M C G3RWR
Cox, M J B G1XWM
Cox, N A G4LJP
Cox, N D G0JZA
Cox, P GM0WOB
Cox, P E G1MEG
Cox, P F M1BWK
Cox, P J G0UQY
Cox, P R G3RYV
Cox, P W G4BUB
Cox, R G4AEL
Cox, R J C G0FOK
Cox, R P G1MKE
Cox, R R G1OVO
Cox, R S G1ION
Cox, R V G3LQJ
Cox, R W G3PLP
Cox, R W G3VIS
Cox, S G1DRY
Cox, S D G7EKJ
Cox, S J G0JGF
Cox, S J G7MXG
Cox, T G0PXP
Cox, T H G G0FYV
Cox, T J E G7JUH
Cox, T W G3NKN
Cox, W F G3IWC
Coxhead, J A 2E1FOK
Coxhead, M J 2E0AFL
Coxon, D G0GHM
Coxon, D T G0DTC
Coxon, K C G0HDV
Coxon, L A GM1MUY
Coxon, R G M1CCY
Coxon, W I L L G4IFH
Coxshall, R A G1EWR
Coy, D N G3PYI
Coyle, A R G0HTX
Coyle, D C GI4DGI
Coyle, E J GI4EPK
Coyle, M GI0NWN
Coyle, R GM4ZMK
Coyne, B D G3DCO
Coyne, G J G3YJR
Coyne, J B G4ODV
Coyne, J J GW6GKP
Cozens, M V G0JDQ
Crabb, C J G0OBC
Crabbe, G S C G4ZFN
Crabbe, J G4ECT
Crabbe, J G6CRC
Crabbe, J J N G3WFM
Crabbe, L W G3CON
Crabtree, J M G7ISB
Crabtree, M P G1NAA
Crabtree, P J G7TOO
Crabtree, R W J 2E1CBE
Crabtree, R W J G7UCJ
Cracknell, M P G0CPU
Cracknell, P G G0KDT
Cracknell, R G G2AHU
Cracknell, R G G6YTP
Cracknell, V G4KPZ
Craddock, R E GW4SLK
Craddock, S C G8AGR
Cradock, W G G0HHH
Cradock-Hartopp, K A G4PZR
Crafer, A R F G8VJY
Craft, A P G7OGJ
Craft, M R G4NSN
Craft, P M GM0RKU
Cragg, A A G8YKV
Cragg, C W G2HDU
Craggs, A GW8ARC
Craggs, D R G3RYP
Craggs, R M M1BCZ
Craggs, S G0BAU
Cragg-Sapsford, J G8CWU
Craib, G R G GM1YGW
Craib, J R W GM1LKD
Craig, A W MM1BAE
Craig, C C GM0EWU
Craig, D A GI8LCJ
Craig, D C G0URX
Craig, D C G6KGU
Craig, D F GM3AWF
Craig, D G G4YYC
Craig, D R GM6GFH
Craig, G G4IUT
Craig, G B MM0BAG
Craig, G S G4IWD
Craig, I V F 2E0ACA
Craig, I V F 2E1ANK
Craig, J GI6OQL
Craig, J GM3HVK
Craig, J G H GM7UPY
Craig, J J G4DJL
Craig, J S J G0UNA
Craig, J S J G3SGR
Craig, K G3UIR
Craig, K H G3ZSX
Craig, L M GM0DPU
Craig, M E GI8SJZ
Craig, P M G7BOH
Craig, P M G7ULL
Craig, P R S 2E1DLA
Craig, S A GI8WHP
Craig, S F GI4SSF
Craig, W N G6JJ
Craigen, P G7MEI
Craigen, W S G4GTX
Craighead, G G6IJL
Craigie, J W GM0JAV
Craine, J F G3XNU
Craine, W F G3PMF
Crake, M J G4HUQ
Craker, D J G0CEP
Craker, J E 2E1CZZ
Crallan, R A G7NSQ
Cram, D GM3NIG
Cram, N GM4RIN
Cramond, J R GM4NHI
Cramp, A W G7HSA
Cramp, G G4JZP
Cramp, N A G1HYA
Crampton, A G0JDK
Crampton, I M G0EJH
Crampton, M T G8DLX
Crampton, S C GI3XDD
Cranage, J D G8RHC
Cranage, M P G8OFA
Crane, A B G3GOX
Crane, B E G3ABH
Crane, C A G1UKW
Crane, C L G4YFN
Crane, H J G4JFY
Crane, J G G4ZGL
Crane, K G G1VGA
Crane, L A G0RGH
Crane, L A G3PED
Crane, M G0KHZ
Crane, M A G0FLU
Crane, P L G7IPI
Crane, R J G7FXH
Crane, S G0CUH
Crane, S G0KUY
Crane, T P G7WHZ
Cranfield, J G1LSF
Cranfield, J G8FWK
Crankshaw, P F GM7VXR
Cranmer, P A G4TFP
Cranwell, L M1AOT
Cranwell, M P G0PNA
Cranwell, R J M0BCF
Cranwell, R J M1AOK
Crapper, G E R G3HOK
Crask, S A G7AHP
Craske, S A G3ZLS
Cratchley, G W G4ILI
Cratchley, L J G3IXC
Crathorne, R C G8TBW
Craven, F B G0DDB
Craven, F B G8KMK
Craven, F J G4LAW
Craven, H C M1BUR
Craven, J G3JCU
Craven, K G1HIR
Craven, K W E G4LKP
Craven, M S G4KMC
Craven, R G 2E1BYQ
Crawford, C G0GKI
Crawford, D R G1JBO
Crawford, E C GM4GUQ

Crawford, G I G6IOE
Crawford, G L K G3XPX
Crawford, H M G0UMU
Crawford, I GI0PUZ
Crawford, J G0VNY
Crawford, J GI4PGH
Crawford, M G3KDU
Crawford, N J MI0ABN
Crawford, N P G7SQM
Crawford, R C G0KEQ
Crawford, R C G6NUL
Crawford, S K C G6HCQ
Crawford, T J GI4OPH
Crawford-Baker, J R G0HWO
Crawley, D B G8OPC
Crawley, G S G0IVW
Crawley, J L R GM3LBX
Crawley, J S G6UHQ
Crawley, K J G1BRP
Crawshaw, D M G1YHL
Crawshaw, G D G0ENF
Crawshaw, G D G7AEF
Crawshaw, J W G8YBD
Crawshaw, M G4BLH
Crawshaw, M G4CPS
Crawshaw, M L G8CME
Crawshaw, P G3UFV
Crawshaw, S 2E1CTU
Craxton, R T G3IKL
Crayden, P M G6NQF
Crayton, M R GM8FCK
Craze, T L G3VNN
Creasey, J G4RIP
Creasey, J A G4FSD
Creasy, E L G4FBI
Creber, D G M1AAH
Cree, A P 2E1EEX
Cree, J D G3TBK
Cree, K S J GI0VFT
Creed, P J M G1LZT
Creedy, R G G1ONF
Creek, A J G7BNL
Creek, D G0GEZ
Creek, D P G1HJJ
Creek, L N G1PAB
Cregan, C F C MM1ALV
Cregan, J H GM4CRV
Creighton, J K GI6WFI
Creighton, K E G8ONS
Creighton, P J G4TGQ
Creissen, P G0UOQ
Crellin, J K G0JSZ
Crellin, J R G3LYX
Crellin, S G1TIQ
Crespel, P GJ0NSG
Cressey, D G4MQM
Cressey, E G G8MUZ
Cressey, J M G0VBN
Cressey, M D J G1WQA
Cressey, P R G0SXW
Cresswell, A R G7ADS
Cresswell, C C G3ZZY
Cresswell, H E G4ZCJ
Cresswell, J G4AMF
Cresswell, J S G4UKQ
Cresswell, P N G0VVM
Cresswell, S H G6VWV
Cretney, R GW6HRG
Crewe, R G0WJU
Crewson, R W G4NDS
Crichton, J GM4EFL
Crichton, J D GI4YWT
Crick, C A G4CJR
Crick, D G GW0WIM
Crick, M J R G1YSA
Crickett, A J G4WIP
Cridland, N S G4FTK
Cridland, R D G3ZGP
Crimes, J A G8SME
Crimes, M J G0RCY
Crimes, M P G7JCE
Crimlisk, A T G0BLW
Crines, R C G3VVZ
Cripps, B J G6RQZ
Cripps, G J G7WJW
Cripps, J R G3XWL
Cripps, P K G3SKT
Cripps, R P GM7FZC
Cripsey, S G T G4CHZ
Cripwell, I G GW0WGG
Crisp, A D G7DFW
Crisp, B G5PW
Crisp, D J G4OAE
Crisp, F J G3GZJ
Crisp, G P G7MLX
Crissell, R P G1DXH
Criswick, V J G7NQN
Critchley, A G3SXC
Critchley, C J G0GUN
Critchley, H W G6NUK
Critchley, J O G1LRJ
Critchley, J R G1DSA
Critchley, L G3EEL
Critchley, P J I G1OJB
Critchlow, D G3PTV
Critchlow, D G4YRD
Critchlow, H D G3SHU
Critoph, D E G1JXZ
Crittenden, K A G0CGB
Croall, S R 2E1DRO
Croasdale, C G7LTL
Crocker, G G1NUA
Crocker, J G4BQB
Crocker, J C G4RGO
Crocker, J G G4RLQ
Crocker, R G0KAU
Crocker, R H G0KNV
Crockett, A J G8NFP
Crockett, H B G8ACA
Crockett, R J G4EDL
Crockford, I GM1AUZ
Crockford, S GM1KBZ
Crocombe, P O D G8FJF
Croft, A B G6XEX
Croft, A R G8CJM
Croft, A V G6NUS
Croft, B A G8VFU
Croft, I P G7GZK
Croft, J R G3SXA
Croft, M J G1ASX
Croft, P G0WSP
Croft, P D G6MID
Croft, R J G7VRX
Crofts, A J G4GKB
Crofts, F E G4FLM
Crofts, M G4DYW
Crofts, M A G4JAQ
Crofts, P J G4ZHD
Crofts, R P 2E1CDR
Crofts, R W L G7RVW
Crofts, S R 2E1CQG
Croft-Smith, D A GM0OAD
Croker, J J G3WCL
Cromack, A MM1BGX
Cromack, H T H G0FGI
Cromack, J T G6YLV
Cromar, C G6FYZ
Cromar, D A 2M1EWA
Cromie, D GI0WCE
Crompton, A J G4VCV
Crompton, C R G0JIQ
Crompton, D P G4IAD
Crompton, D W G4DXX
Crompton, F A G1BOO
Crompton, J P G0CUV
Crompton, L C G6MCC
Crompton, L D E GW6UHY
Crompton, L L GM7HBT
Crompton, W GW7AEL
Cronin, C S C GW1ZKM
Cronin, E G3ZKX
Cronin, E P G4TZV
Cronin, P M D G4GRO
Cronk, J E GW3MEO
Cronshaw, R G7VNK
Crook, A E G4AYS
Crook, A J G1GFD
Crook, B M G4AZN
Crook, B M G5LO
Crook, C H G3YOG
Crook, D A G6ZQU
Crook, F W G8GVS
Crook, J W G6RAD
Crook, M G1ZRP
Crook, M G G4THA
Crook, N G4ZLP
Crook, N A G4OKA
Crook, P G0NDL
Crook, P B 2E1ENQ
Crook, R P G6SZS
Crook, S J 2E1BDG
Crook, T D E G6GFR
Crookall, E J G7BBN
Crookall, M L G4TEX
Crookbain, J C C G6SPH
Crooke, D N GM0RHP
Crookes, A G G8UCN
Crookes, K D G0NHO
Crookes, T F G4RBU
Crooks, A P G7EFL
Crooks, J W A G4CEW
Crooks, N D M0ANP
Crooks, P A G4KGG
Crooks, R G4RRU
Crooks, S J G0LMA
Crooks, W G GW4XWC
Croot, D M G7UVL
Croot, S G G0WXH
Cropley, L I G0DFC
Crosby, A G1TDR
Crosby, J M G0POU
Crosby, L G0OIP
Crosby, L F G3FGT
Crosby, M G1PSJ
Crosby, R A W G4PQJ
Crosby-Clarke, D W G4SWM
Crosfill, M L G7TJD
Crosland, H G8CW
Crosland, P L G1EME
Crosland, P L G6JNS
Crosland, T D G4PNK
Cross, A G0HKG
Cross, A D G0SAC
Cross, A D G4WGE
Cross, A R G3WEA
Cross, A R G4XNI
Cross, B G3ZBZ
Cross, B M G0POR
Cross, D G4SWZ
Cross, D A G0KLQ
Cross, D L G7TIB
Cross, E J G4ZEG
Cross, G G4IEN
Cross, G G8MHE
Cross, G G8URI
Cross, H W G3HCY
Cross, I A G7UVN
Cross, J B G4GOR
Cross, J W G1DXG
Cross, J W H G8YRC
Cross, M M1BXD
Cross, M W G3OKU
Cross, N A G8JEI
Cross, O L G4DFI
Cross, P G G0GHH
Cross, P G G7AXE
Cross, P G G8CA
Cross, P H G3RWI
Cross, P M G3OZD
Cross, P S G7MWH
Cross, R G G0TRW
Cross, R J G3TMC
Cross, S G1VWB
Cross, S A G0TPJ
Cross, T GW4LFW
Cross, T R G6RMC
Cross, V M GW0HYJ
Cross, W H G0ELZ
Cross, W H G0SJW
Crossan, J M G3OJW
Crossan, W J G3BWY
Crossfield, J E G3SEQ
Crossfield, M G8NRV
Crossfield, T G2DML
Crosskey, S S G4XDE
Crossland, D P 2E1FWH
Crossland, G H G4CWG
Crossland, H N G4SBG
Crossland, J A A G8PWU
Crossland, S G6OFV
Crossland, S R M1AKW
Crossley, D N G6BIT
Crossley, F J M G8GYQ
Crossley, J G1IWG
Crossley, M G0DNP
Crossley, M A 2E1FZA
Crossley, M J G8UWM
Crossley, N I G0WPC
Crossley, R M G4CZP
Crossley, W B G7BMO
Crossman, M G8OQR
Crosson Smith, S A G4LMX
Croston, W M G1JCQ
Crotty, P V G4YLX
Crouch, A W G1JKL
Crouch, H J G0KVC
Crouch, P D A G4ZGC
Crouch, S L G8MTK
Croucher, A R G6ERJ
Croucher, C G4BLD
Croucher, K G G8MIE
Croucher, P G4YPC
Croucher, P G5RS
Croucher, R G3XRP
Croucher, R S G0KVF
Croucher, R S M G0NRJ
Croucher, V G G3AFY
Crouchley, D C G8WKO
Crough, D J G8JFL
Crow, B G4UYJ
Crow, B GW4LFV
Crow, C M GW6IOA
Crow, J G1YNQ
Crow, J P G8GQG
Crow, L J G0TNH
Crow, S G G3DFH
Crow, V L G4YDN
Crowden, J GM1VGZ
Crowder, A R M0BFU
Crowder, J D G0GDU
Crowder, S J 2E1FOR
Crowe, A GI1FWK
Crowe, D H G8WTB
Crowe, D M G1OYG
Crowe, J D G4NHU
Crowe, J T G6ZXO
Crowe, P A G4MQP
Crowe, R H G3RVA
Crowe, T P G1ANF
Crowe-Haylett, B E G G0OKK
Crowhurst, G M G4ZPY
Crowhurst, J G4YKN
Crowley, B J G4CWO
Crowley, G S GM7SJC
Crowley, P W G2AZP
Crowley, R C GW4UGI
Crowley-Milling, M C G6MX
Crownshaw, H S C G5NV
Crowsley, M E G0JIW
Crowther, A G6ZOB
Crowther, A GD0MWL
Crowther, A B 2E1FUG
Crowther, B G1HYG
Crowther, B G6VRF
Crowther, I H G3KLF
Crowther, J L G3KMM
Crowther, P D G0ABU
Crowther, S G6XXE
Crowther, S D G G6BXK
Crowther-Watson, M G3IAR
Crowton, G J G6WIG
Croxall, W G2FRT
Croxford, C F G4YCW
Croxford, I L G3OIC
Croxford, M W G7HQH
Croydon, A D G0VAT
Croydon, A J G4VTC
Croysdale, J H G3OZV
Crozier, A D GI7VXC
Crozier, J W G8NZO
Crozier, R D G0ZAP
Crucefix, D J G0HTF
Cruddas, C R G GW0LKA
Cruickshank, D GM6EUC
Cruise, S R G6OFT
Crump, D G G8GKQ
Crump, S J J G7BYH
Cruse, T P G0JUE
Crust, P W G3XYC
Crutchley, D G G3RLW
Crutchley, M A G3MRZ
Crutchley, R J C R G6WI
Crutchley, S J G7HRF
Crute, A G8TEF
Crute, T G4DGB
Crux, J A G3JAG
Cruz, M T D G7HSP
Crye, D G6BSK
Crymble, N J GI4MCH
Csapo, P G6RFU
Cubberley, F J G3OCW
Cubitt, A R M1BXN
Cubitt, C C G4MQK
Cubitt, O L G0TGQ
Cuddington, R C G4MQL
Cudmore, C A G0FIY
Cuff, B L G3YMD
Cuff, B L G4SAU
Cuff, D G0YBU
Cuffe, P P G0IEP
Cuff, R V G0SEG
Culkin, M J G6ELW
Cull, C G G1BDZ
Cull, J G8ALR
Cull, M S GW7DRN
Cull, T G1SUM
Cullen, A GW0DJX
Cullen, A M1BET
Cullen, B H G4UIU
Cullen, D P G7ORB
Cullen, K D G G1EWT
Cullen, W J G4WNO
Culling, J C G0SNF
Culling, M J G8UCP
Cullington, C J P J G0WIA
Cullingworth, C GM1XLH
Cullingworth, S A G7CCL
Cullis, N T C G1JGD
Cullis, R A GW7UKJ
Cullis, R A MW0AZN
Cullum, G P G4EXM
Cullum, G S 2E1FNP
Cullup, A N G4FCI
Cullup, T W G0RCH
Culpan, R M G4GND
Culpan, S D G4GPZ
Culshaw, L G G0LGC
Culshaw, R M1BDG
Cumberland, B R GI0USB
Cumiskey, D GW0UXX
Cumiskey, P J G1OMX
Cumming, A G GM4HMN
Cumming, H S GM0HSC
Cumming, J D GM0GUJ
Cumming, J W G3VQY
Cumming, R G3NOI
Cummings, A G8BIO
Cummings, C G4BOH
Cummings, F G3UQK
Cummings, G W G8ANA
Cummings, K G0VON
Cummings, M E J G8KPN
Cummings, R G GI7VIQ
Cummings, R W C GI0KOW
Cummins, H J G8KPM
Cummins, J G4ERR
Cummins, J R G0OTJ
Cummins, S F G3VKX
Cummins, W M0AOO
Cundall, A W J 2E1ATS
Cundall, P G7LNT
Cundall, V C G3FAU
Cundell, G K G4EYF
Cundy, A F G0RCO
Cunliffe, A G4EII
Cunliffe, A W G6ERK
Cunliffe, J G0IZO
Cunliffe, J R G3ZOC
Cunliffe, J W G3EXU
Cunliffe, J W D G6LNV
Cunliffe, N P G4OWS
Cunliffe, P J G1KKH
Cunliffe, R GI0HVJ
Cunliffe, R T G3BZB
Cunliffe, W H G4RIG
Cunnah, G D G3OFP
Cunningham, A J G0PET
Cunningham, A J GM0NWI
Cunningham, B S GI1DNK
Cunningham, D P G0EQE
Cunningham, I J G3XKA
Cunningham, J G G0LIQ
Cunningham, J J P GI1BSJ
Cunningham, J M J G8EAN
Cunningham, K GI6VCL
Cunningham, M GI7UPQ
Cunningham, M P G4CUC
Cunningham, M P G6IOM
Cunningham, P G3MDL
Cunningham, P M G0NXH
Cunningham, R T G6DHN
Cunningham, V D G4FDF
Cunnington, A P G7VKB
Cunnington, P M G8LHW
Cupples, C G A GI4EQN
Curd, E G7DPZ
Curgenven, C I G8CKC
Curley, A R G6SDE
Curlis, S C 2E1EXC
Curnow, B J G3UKI
Curnow, E G7CVA
Curnow, J G0ETQ
Curphey, W A G3AGC
Curr, J GM7AOM
Curran, A M G4UMM
Curran, B A G4OQE
Curran, D GM7SWX
Curran, G E GI4JJD
Curran, M K G3VEF
Curran, M K G8KGI
Curran, M W B G4YTT
Curran, P A G6TLA
Curran, P A G6TLB
Curran, P A M1BPR
Curran, T GM1TCN
Curran, W GM1KCH
Currell, A J G4VBX
Currell, I G3WBA
Currell, R A G4EIK
Currey, B G3LYZ
Currey, C G0FKJ
Currey, N A G0COX
Currey, N A G3SSM
Currie, B G7HQW
Currie, C G1IHS
Currie, D R G4WGO
Currie, G A C GM3VIO
Currie, G A L GM7CPJ
Currie, G P GM0JCR
Currie, K M G4EDN
Currie, L C G3UKX
Currie, S A W GI3NYJ
Currie, S J G1VPH
Currie, T H GM0FRH
Currie, W GM3ZTA
Currigan, P F G6IIN
Currington, L E G8LOK
Curry, A S G3DMQ
Curry, G H GI4XLB
Curry, G H GI6ATZ
Curry, H GI0ABH
Curry, J S G3UVU
Curry, R D G8XFX
Curry, R G G6IZE
Curry, S F G7HHI
Curryer, C H G4XWY
Curson, C W G4SCG
Curson, D C G1HYC
Curson, K E G0EQB
Curtis, A M G4ZKH
Curtis, A S G0ASC
Curtis, A S G4JYH
Curtis, B 2W1BNI
Curtis, B C G0KEK
Curtis, B R G4UEY
Curtis, C G0JRR
Curtis, C G3MKV
Curtis, D W G0ULB
Curtis, E W G3PPX
Curtis, F L G3SVK
Curtis, H G G3DLG
Curtis, J A G6VGY
Curtis, J E G0SEC
Curtis, J F G0HVA
Curtis, J G G0JWY
Curtis, J G G4VMW
Curtis, L J G8ZWC
Curtis, M H G7PSR
Curtis, M R 2E1BSZ
Curtis, M R G3IAB
Curtis, P G0UTX
Curtis, P C 2E1FWP
Curtis, P J GW3EPF
Curtis, P R G8XJU
Curtis, R G4MGZ
Curtis, R GW4AZE
Curtis, R E G0CYC
Curtis, R W G3EUK
Curtis, S D J 2E1CJF
Curtis, S J G8XPB
Curtis, S M G0XAK
Curtis, W G0PNJ
Curtis, W D G8BGR
Curtis, W H M1BMD
Curtis, W J G1HHW
Curtis, W J C GI3GRD
Curtis, W P G6KTB
Curtis-Smith, G P J G8UUN
Curwell, D C G8OLY
Curwen, J M G4ZEZ
Curwen, L N G6THP
Curwen, R H G3PDC
Curwen, T G0BZQ
Curzen, P G3XXI
Curzon, J A G8GTH
Curzon, Q G0BVW
Curzon, R F G0HBA
Cusack, D E F J M0BFJ
Cushing, J S G3KHC
Cushion, C C G3RHW
Cushman, D A G8MZY
Cushnahan, J A GI4ELQ
Cusick, D G0BQH
Cusick, G A G1BTT
Cusick, G J M0AMK
Cusiter, G W GM4FVS
Cussans, A A G7VFI
Cusson, P J M1ARO
Cutbush, R G4ADK
Cutcliffe, A G G0CGM
Cuthbert, C R G4FJT
Cuthbert, J GI4OYL
Cuthbert, J C G3YYZ
Cuthbert, P E G0AHR
Cuthbert, R J GM4GLE
Cuthbert, S M G6MGE
Cuthbertson, J F G0RQY
Cuthill, J G0MMC
Cutland, A P GJ7RWT
Cutler, P J G3MXF
Cutler, W M G6RCH
Cutmore, N A F G6ALG
Cutmore, T B G3KHS
Cutt, J G GM0VHC
Cutt, J K GM0LDT
Cutter, A C G0MBB
Cutter, D J G3UNA
Cutting, G C G3GNQ
Cutting, L S G3KAA
Cutts, A B G6BUV
Cutts, D A GW4FGC
Cutts, D C G4YJQ
Cutts, D G G4FAW
Cutts, H B G3RDP
Cutts, J G8FAW
Cutts, R G3RJM
Czarnota, S P G7LNI

D

Da Silva Curiel, R A G7GLY
Dabbs, H G G4HMU
Dabbs, R E T G2RD
Dabbs, S N G4GFN
Dabell, P G0WOP
Dabhi, H T G0KSN
Dabhi, M T G0RRY
Dabinett, D R GW4DEP
Dables, C K G7VVA
Daborn, A F G6CMF
Dabrowski, S G4KSJ
Dack, R J G0OMU
Dackham, J J G1BBT
Da-Costa, V V G1ZHQ
Dadak, H E P G0MMQ
Daddy, P W G0PSL
Dade, D G3XCT
Dadswell, J H G0BKP
Daft, G L G7ERB
Daft, R J E G4TRD
Dagger, K B M1ADC
Dagnall, A K G0MNY
D'Agostino, A G1EBA
Dahle, E GM3UWO
Dailey, A H G3UMH
Dailey, A J GM0REZ
Dailey, K M G0RVH
Daily, C E G1PXH
Daines, G L G7DLF
Daines, P R G6PHJ
Daines, R M1AYT
Dainotto, G G0VCF
Dainty, M J G6JAM
Dainty, P A G0JQU
Dairon, L J T GM6JLM
Daish, M J G8UYY
Daish, R P G3SLD
Daker, D M 2E1EAH
Dakin, C C G0NUX
Dakin, N E G4ZSO
Dakin, P R G4PRD
Dalby, L G G1ULZ
Dalby, N E G8ND
Dalby, P J 2E1DOA
Dale, A C G1GVT
Dale, A J G1PDD
Dale, C G0DOA
Dale, C V G8SDX
Dale, E G1VKT
Dale, E G4KTW
Dale, G G3MFH
Dale, J H G6NET
Dale, J L G2DSY
Dale, J S D G1VDY
Dale, K R G0JPC
Dale, M G6AGI
Dale, M G G6ABU
Dale, M G G8IUT
Dale, O R G8ZWL
Dale, R G3RLS
Dale, R S G0OHL
Dale, S A G8CKV
Dales, J R G4DFH
Daley, G G0DAL
Daley, S D 2E1AQR
Dale-Green, D J G4GLP
Dallas, W G4HEE
Dalley, M A M G4VYI
Dalley, P A G6JAC
Dalling, D W GW4PHT
Dally, K R G4FZR
Dally, M A G4PCD
Dalrymple, D W G3OLK
Dalrymple, J A J GM3JSX
Dalrymple, M W GM0AYR
Dalrymple, M W GM4SUC
Dalton, D G4ZTY
Dalton, E D G3ZLJ
Dalton, H G G4GZK
Dalton, K J G4UEF
Dalton, L D G7RRA
Dalton, R G3PWS
Dalton, W GI0VLE
Dalton-Kirby, A G4DDS
Dalton-Moore, C J 2E1FRN
Daly, E G0TXF
Daly, H J G0AFG
Daly, M K G0AUX
Daly, P G0GTE
Daly, R E G0UKV
Daly, R E G7SDH
Daly, R E G8VYJ
Daly, T J G0TIM
Daly, W G0SXA
Dalzell, D J 2E1FIV
Dalzell, W G G7UJE
Dalziel, A M GM4FGD
Dalziel, C GM8LBC
Damm, R B 2E1FIM
Damm, R B M1CDQ
Damon, L G3CPG
Danagher, P J G1RMW
Danby, A G0KGA
Danby, C M G0DWV
Dance, A J G7KWN
Dance, D GM4CXP
Dance, M L G3IPP
Dancer, K E GW7WIX
Dancer, K G GW7VEU
Dancock, W D G0LJK
Dancy, R G G3JRD
Dandy, E G3BJB
Dane, P G G4PWA
Danenberg, D T G0CRM
Danfer, M D G7LBJ
Danfer, M D M0BCT
Dangerfield, A A G1XAL
Dangerfield, C R 2E1BUE
Daniel, A J G4PND
Daniel, B G7DZY
Daniel, B J G0BAB
Daniel, C G7NPL
Daniel, K S G3ZJE
Daniel, R F G4RUW
Daniell, G J 2E1DKO
Daniells, P R G4CBQ
Daniels, A H G1ZOD
Daniels, A J G7LCS
Daniels, A S 2E1FEQ
Daniels, A S M0BFX
Daniels, B A G0IIB
Daniels, B J G6DAN
Daniels, C G4OQI
Daniels, D G4XHP
Daniels, D M GW0SRF
Daniels, F T G1ORN
Daniels, G J G0DGH
Daniels, I G4VUR
Daniels, I B G4VTD
Daniels, K G0PRE
Daniels, K M0AKR
Daniels, K A W G6WHY
Daniels, M G4IFJ
Daniels, M G G1SOM
Daniels, M G G7GZS
Daniels, M G G8WKK
Daniels, M G G8WKL
Daniels, P R G1USW
Daniels, R B G0GZL
Daniels, R F G0PWD
Daniels, R G G6ABW
Daniels, S B G6UIM
Daniels, S P G6BVN
Daniels, S P M1BQU
Daniels, S R G0DKM
Daniels, T J GW7KTP
Daniels, W H G1XAK
Daniels, W J S G4HUG
Danielson, G L GD4AM
Daniels-Galey, A G0JZZ
Danks, E R G8BKL
Danks, J L G5DS
Danks, K G0DBI
Dann, C G1AXW
Dann, G D G8KPI
Dann, J 2E1FND
Dann, J M1BRN
Dann, J R G3PYO
Dann, M G3NHE
Dann, P J G7PJW
Dann, V H G4PPD
Dannatt, M W G8MCY
Danner, J W G7LAW
Dansey, T J G0BIX
Danton, A G0SGP
Danton, J E GM6NYT
Daramy, J A G0EWI
Darben, E A G8AIJ
Darby, B G1XBL
Darby, B D G6ALW
Darby, C E G4NPI
Darby, D G6AGN
Darby, D M G4SRV
Darby, G M G7GJU
Darby, G R G7RTC
Darby, J G6OGF
Darby, J C G4TVC
Darby, J G G6XSA
Darby, K W G3MLD
Darby, R G7TZQ
Darby, R L G0UIS
Darbyshire, J A G4GIS
Darbyshire, K G G8EZU
Darbyshire, M S G7ING
Darbyshire, S G3TVF
Darbyshire, W A G3FFR
D'Arcy, S A G6VYS
Dare, E G G7FBE
Dare, K G6JUI
Dare, L M G7GEB
Dark, J G4AHQ
Dark, J F H G3XXK
Darke, D F H G0KFN
Darke, P D G1DXD
Darke, T L GM3VQJ
Darker, C J G0ULJ
Darkes, D R G8HRI
Darkes, L G7MFL
Darkin, M J G3KTH
Darley, A V G6ZQH
Darley, J R G1AOE
Darling, J H G1DSB
Darling, M W M G0NUH
Darling, W G G1DNI
Darlington, E G3YWD
Darlington, J R G8VNL
Darlington, K G3MXE
Darlington, S J G1UTP
Darragh, A W G8KWP
Darragh, P W F G3MNV
Darrell, G J GW0CTG
Darrigan, S E M1AOR
Darrington, J G3WHL
Darroch, R GM7PGM
Darroch, R W O GM0OVD
Dart, J C GW0WHU
Dart, J M G3KZJ
Dart, P M G0VZX
Darton, B J 2E1DNR
Darton, K P M0AJV
Darton, K S G4LSE
Darton, L K 2E1DEP
Darton, M L 2E1FDU
Darwent, R R G0UHF
Darwin, N G0SDK
Darwood, D A G3YKO
Darwood, M C G0HZT
Darwood, S G4SBZ
Das, D N G7ONV
Das Neves Pedro, C G6FYE
Dasilva-Hill, G M A G1HER
Dasilva-Hill, K G C G6KEN
Date, A P G G0JEP
Date, A R W D G0FQU
Daulman, A J G4KQL
Daunt, C D G4BQM
Davage, J E B G6LNP
Davenport, A P 2E1DBK
Davenport, F G G8BOE
Davenport, J T G4OQU
Davenport, K G8INC
Davenport, M A G6NZA
Davenport, M T G0AXE
Davenport, P A G0GLQ
Davenport, P J GW0GZQ
Davenport, R G3IQI
Davenport, R G3ROD
Davenport, R H GW4ANK
Davenport, T M G1HRM
Davey, A G6YXG
Davey, A E G1DGV
Davey, A N C M1BFY
Davey, A R G6WLX
Davey, B G G4ITG
Davey, B J G0LCP
Davey, C G7GLO
Davey, D 2I1DNO
Davey, E S G1JBT
Davey, G G6TBT
Davey, G A G4XSM
Davey, G P G8VJW
Davey, G W G3NGI
Davey, H R G0CLD
Davey, J G1PKA
Davey, J H G G3AID
Davey, J R G3FPN
Davey, M R G4FZM
Davey, R G3ZSI
Davey, R G4VLW
Davey, R M0AJS

Davey, R J C G8MRI
Davey, R T GI0IIZ
Davey, R W G4KEM
Davey, T E G0GTD
Davey, W E S G0THI
Davey-Thomas, E L D G3AGA
Davey-Thomas, E L D G3MPD
David, A H MW1CBF
David, E G4LQI
David, E L G7EOE
David, M J G4MEM
David, S E GW0NCU
David, W M GW4WMD
Davidson, A F GM3FAO
Davidson, A J G0UVQ
Davidson, A J B G4CDG
Davidson, B G1YXG
Davidson, D G GM4ZGV
Davidson, G B G3YSL
Davidson, G I GM6TYL
Davidson, G J 2M1DIT
Davidson, G J MM1APO
Davidson, G L GM0JKS
Davidson, G L GM4TPF
Davidson, G L GM4XKG
Davidson, H W B GI4BTG
Davidson, I G4KDW
Davidson, I G6EMH
Davidson, I G8YBR
Davidson, I W MM1AOE
Davidson, J GI3FJX
Davidson, J C G7PVA
Davidson, J C GM0IGF
Davidson, J D G6CTA
Davidson, J L G4BVZ
Davidson, J R GM3UAG
Davidson, J S G0VJD
Davidson, J W T GM0CPI
Davidson, M G0GVF
Davidson, M G4WRU
Davidson, M C G8ABU
Davidson, M I G3YSM
Davidson, P G4RTU
Davidson, P G4TZW
Davidson, P C G0DOR
Davidson, R GM4DOF
Davidson, R R W G1AVB
Davidson, R W G4DAT
Davidson, T R GI8ITD
Davidson, W A F M0ADJ
Davidson, W M GM4NXT
Davidson, W P G4YDD
Davie, E A G2XG
Davie, M W M G0NJS
Davie, P GM0VEK
Davie, R K G4KTY
Davie, W R N G4MQD
Davies, A G0TZY
Davies, A G3IIV
Davies, A G7RZW
Davies, A G8VPX
Davies, A GW3INW
Davies, A GW4BIS
Davies, A M1ARP
Davies, A C GW3MSY
Davies, A E G7NIH
Davies, A E GW6JLH
Davies, A G G0EZU
Davies, A G G4NUZ
Davies, A H G4JY
Davies, A J G3PBI
Davies, A L M G4EVR
Davies, A M G0HDB
Davies, A P 2E1CND
Davies, A R MW0ARD
Davies, A S M0AMP
Davies, A W R G6ILH
Davies, B G0PWI
Davies, B G1DEO
Davies, B G4UCE
Davies, B F J G3PHL
Davies, B J GW8UTK
Davies, B J R G3OYU
Davies, B L G8SXD
Davies, B M G0VGP
Davies, B M G6RPV
Davies, B V GW4KAZ
Davies, B W G1AZE
Davies, C GW3YBN
Davies, C A G4VWS
Davies, C A G8IEW
Davies, C E G4PAJ
Davies, C H GI3HNM
Davies, C J G4JNI
Davies, C J G7GZB
Davies, C J GW4VFE
Davies, C K GW3MOM
Davies, C L G4FVP
Davies, C L GW8DUY
Davies, C M GM6YQA
Davies, C M GW7HAE
Davies, C M M0AAU
Davies, C N G4GLF
Davies, C R G0HRQ
Davies, C R G3JAU
Davies, C T GW8MTJ
Davies, D 2W1EYM
Davies, D G8AUM
Davies, D GW3YUC
Davies, D GW4MVY
Davies, D MW1BYF
Davies, D A G4EYJ
Davies, D A GW6MYY
Davies, D C G4CEN
Davies, D E GW0FEU
Davies, D G GW4RML
Davies, D G GW7CEI
Davies, D H G0TLT
Davies, D H G4YKT
Davies, D I GW8ICS
Davies, D J G6GCO
Davies, D J GW0CYG
Davies, D J T G6EDB
Davies, D M G4WVK
Davies, D P G7CKS
Davies, D R G0BVU
Davies, D R GW0VEW
Davies, D R GW1CDH
Davies, D T GW0NKJ
Davies, D W GW6KHH
Davies, D W GW7SXN
Davies, E G3PGM
Davies, E G4XVV
Davies, E G8SSY
Davies, E A G GM4APK
Davies, E G GW4TAU
Davies, E M GW2HIY
Davies, E S G3LIZ
Davies, F G3HXG
Davies, G G1DSC
Davies, G G4VEW
Davies, G GW0NDZ
Davies, G GW6SBD
Davies, G A G7MJS
Davies, G C G4OWV
Davies, G E 2W0APT
Davies, G E 2W1EST
Davies, G H GW7VST
Davies, G I G8SWZ
Davies, G J G7HTS
Davies, G K G8YWH
Davies, G L G7SDM
Davies, G M GW4TGO
Davies, G R G1ESL
Davies, G T D G7FLK
Davies, H G0OSH
Davies, H G0WXY
Davies, H G4DZH
Davies, H G6DOQ
Davies, H G7IXI
Davies, H GE7OSH
Davies, H GW8SJN
Davies, H C G0BOK
Davies, H P MW0ARL
Davies, H T G7KZN
Davies, I MI0AJG
Davies, I C GW6BUW
Davies, I E G3IZD
Davies, I E G3TQP
Davies, I E GW1RQF
Davies, I R C G7SBD
Davies, I S G3KZR
Davies, J G1ZFD
Davies, J G6GQJ
Davies, J G8WNK
Davies, J GM4YBJ
Davies, J GW4IOI
Davies, J A G7ENI
Davies, J A GW0ETM
Davies, J A GW0VMZ
Davies, J B GW4GFL
Davies, J B GW4XQH
Davies, J C G7OAA
Davies, J D G0PVR
Davies, J D GW3JVW
Davies, J D G G3KZE
Davies, J E G0ISY
Davies, J E G1PGB
Davies, J E G4UGQ
Davies, J E G8RDJ
Davies, J F G1YJR
Davies, J G 2E1EWV
Davies, J G G7WJX
Davies, J G G8EBJ
Davies, J G M0AXW
Davies, J G M1AGP
Davies, J H G3YJD
Davies, J H GW0DZL
Davies, J J G1TRV
Davies, J J G3PAG
Davies, J J H G3LJD
Davies, J M G4ETQ
Davies, J M G8SXA
Davies, J N G0PTU
Davies, J P 2E1DVY
Davies, J W G0VXK
Davies, J W GW6JWD
Davies, K G0USY
Davies, K D GW8NAC
Davies, K F G4CGR
Davies, K R G1YHI
Davies, L GM1IKQ
Davies, L J GM1YPJ
Davies, L J L G0RDV
Davies, L J L G7KRS
Davies, L R G0OEN
Davies, L R G3MVK
Davies, M G0WZY
Davies, M G7TNS
Davies, M D G6IPN
Davies, M D L G1CGJ
Davies, M G G0VPC
Davies, M G G4AUZ
Davies, M J G1YDA
Davies, M J G4CGH
Davies, M K 2E1EYC
Davies, M K G4HFS
Davies, M M G0VWB
Davies, M R GW3ITD
Davies, M R GW4GNY
Davies, M R GW4HVN
Davies, M T G0CBF
Davies, N G7MBI
Davies, N GW0MAW
Davies, N GW6GW
Davies, N D G7UNU
Davies, N J H G0AXI
Davies, N V G0OOX
Davies, N W G GW1JRM
Davies, O G4VPF
Davies, P G0WAO
Davies, P G1XCB
Davies, P A G0KQA
Davies, P A G8HBQ
Davies, P B MW1AUV
Davies, P G N G0SXY
Davies, P J G4EYX
Davies, P J G4MIO
Davies, P M G0PBL
Davies, P W G0BHI
Davies, R G0JUU
Davies, R G1ZEX
Davies, R G7JLZ
Davies, R GW0AJY
Davies, R GW1EHI
Davies, R A G0WJX
Davies, R A G7MCP
Davies, R B GW3KYA
Davies, R D GW4BCC
Davies, R E G1HWP
Davies, R G G3TAZ
Davies, R H G4NDL
Davies, R H MW0ANV
Davies, R J GW7HLZ
Davies, R K GW0JSX
Davies, R L G8SXC
Davies, R P G7JGL
Davies, R R G4PDK
Davies, R T G0MJP
Davies, R T E G0UTZ
Davies, R W G GW0APL
Davies, S G1SLF
Davies, S B G1LWP
Davies, S D G0LQP
Davies, S E G3HCJ
Davies, S H G6YPV
Davies, S H G6ZTR
Davies, S J G4KNZ
Davies, S J GW0AJI
Davies, S M GW1EAV
Davies, S R 2E1DRT
Davies, S R G1HRI
Davies, S W G4AEC
Davies, T G8NFO
Davies, T GW4ADL
Davies, T D J GW3YAF
Davies, T E G4ZKW
Davies, T G G0JIX
Davies, T J G0JLI
Davies, T J G7PPQ
Davies, T P M0AYB
Davies, T R GW4BVN
Davies, T R M0AQF
Davies, T W E 2E1EGF
Davies, V A GW3OCD
Davies, V M GW1PMQ
Davies, W E GW0OQH
Davies, W E GW4HLO
Davies, W G GW4XPN
Davies, W J G4YWD
Davies-Jones, A G0EXU
Davin, A K G7NKI
Davis, A G7BDR
Davis, A B GW6NYR
Davis, A D G6ALZ
Davis, A D G6JGT
Davis, A D H GU4TFM
Davis, A E G8PPZ
Davis, A J G0TPE
Davis, A J G1HRY
Davis, A J G8GZW
Davis, A P G1NZK
Davis, A R G3FSA
Davis, A R GW4UMX
Davis, A V H G3MGL
Davis, A W G3VTR
Davis, B C GI6EWO
Davis, B J G0XIT
Davis, B T G3UJB
Davis, C G3ZFC
Davis, C A G3SZR
Davis, C E G0EGR
Davis, C J G3VMU
Davis, C J G7VQM
Davis, C L G7KIF
Davis, C M G1LLA
Davis, C P G1JRP
Davis, C R G3TNQ
Davis, D G4AQK
Davis, D G4GJE
Davis, D H G0JXQ
Davis, D H G3SVI
Davis, D H G4VVY
Davis, D J G8UCR
Davis, E C G7WBR
Davis, E C D G8FFA
Davis, E G G1ZCS
Davis, E N G3FTP
Davis, G G0TBW
Davis, G G6FBH
Davis, G GW6BAH
Davis, G B G3ISL
Davis, G F G6MJW
Davis, G H 2E1EIR
Davis, G R G4MFX
Davis, G W G3CMH
Davis, G W G3ICO
Davis, H G0SNQ
Davis, H B G4MX
Davis, H E G4OYA
Davis, I J G4LPL
Davis, I M G8XCL
Davis, J G0OWS
Davis, J G7IIQ
Davis, J A G6XGJ
Davis, J D G3MER
Davis, J D G3PAQ
Davis, J D T G4HGK
Davis, J F 2E1AFA
Davis, J J G0RVI
Davis, J L G6CTK
Davis, J M G0JMD
Davis, J M G6VUJ
Davis, J R G3WTD
Davis, J R K G6DID
Davis, J S G3SQV
Davis, J S H M0AXN
Davis, K A G1FJS
Davis, K C G3TYR
Davis, K L G7TJX
Davis, M G0BGB
Davis, M G1UCN
Davis, M G4NXS
Davis, M E G7MBH
Davis, M J G0ROT
Davis, M J G1EKC
Davis, M J G4KRT
Davis, M J G4LFG
Davis, M J G8KMR
Davis, M J G8VRW
Davis, M M E G6KYE
Davis, M P G4PQW
Davis, M S G0KHB
Davis, M S G1UUZ
Davis, N W M1ADZ
Davis, P G0RFZ
Davis, P G7DIL
Davis, P H G4XYU
Davis, P J G0RIU
Davis, P J M1BWE
Davis, P R 2E1DED
Davis, P R M G7KNU
Davis, P W G8TAO
Davis, R 2E1AKV
Davis, R G0WMU
Davis, R G6RAE
Davis, R A G0MEO
Davis, R A G3RLO
Davis, R C G3TDL
Davis, R G G4PTT
Davis, R M G1UFJ
Davis, R P G1HXR
Davis, S G8BFA
Davis, S C G0IAW
Davis, S F J G3KVR
Davis, S H G4DHZ
Davis, S J G4SJD
Davis, S M G6YPY
Davis, S M G7AUR
Davis, S N G7WKX
Davis, T D G8PBJ
Davis, T T GW7DJL
Davis, W E M1BDV
Davis, W H G3WMO
Davison, A E R G0PFD
Davison, B J G0VEI
Davison, B J GI1WZA
Davison, D N G3VFX
Davison, H M G3TVW
Davison, I J G0LNX
Davison, J G3JKD
Davison, J M G1SBN
Davison, J M G7RDL
Davison, K J G7RZB
Davison, M J G3ZUB
Davison, N M M1AFB
Davison, R H G3NVX
Davy, A F G0MJB
Davy, D W G6EWP
Davy, K G6PZT
Davy, M D G6SWZ
Davy-Jones, J G0USE
Daw, A M G1DSF
Daw, A P G1APD
Daw, B G7KNW
Daw, R L M G1DSE
Daw, R M G0MCE
Daw BEM, J G0UUW
Dawber, J H G7GVB
Dawber, S G6OGD
Dawe, B A G0PWC
Dawe, I D G3SPI
Dawe, R G G4XAF
Dawes, A G6VZF
Dawes, G S M0AEP
Dawes, J A G6HVL
Dawes, R G3SEN
Dawes, W 2E1CZU
Dawkes, D G G0ICJ
Dawkes, D G G1WAC
Dawkes, D G G4WAC
Dawkes, L P GM8BFE
Dawkins, F O A G3AXK
Dawkins, G E J G8KQZ
Dawkins, M J G7CXZ
Dawkins, R G G1EAX
Dawkins, R G GW4FRH
Daws, A G4PVX
Daws, D J G4WNR
Daws, D R G0RRD
Dawson, A G3ZVU
Dawson, A A G0NEU
Dawson, B G8OKS
Dawson, B D G8XXY
Dawson, B R G4OQZ
Dawson, C G4UZS
Dawson, C G8BRD
Dawson, D G1NEV
Dawson, D C G7UZN
Dawson, D C G8RVV
Dawson, D E G0CIW
Dawson, D J G1UCB
Dawson, D P E G0ELJ
Dawson, E G0AID
Dawson, F W P G1HCM
Dawson, G GM3GBX
Dawson, G W H G0VBT
Dawson, H G6KSE
Dawson, H G G0OQZ
Dawson, J D G4XGT
Dawson, J R G3BJJ
Dawson, K H E G3XSK
Dawson, K R G8OSX
Dawson, K W G3KWO
Dawson, L A G7EQQ
Dawson, M D G4HWJ
Dawson, M D R G0GYJ
Dawson, M F GM4OKG
Dawson, M J G G3TCL
Dawson, M P G8IC
Dawson, M W G0VYC
Dawson, N P G7GGA
Dawson, P 2E1DAN
Dawson, P L G8WVO
Dawson, P M J G7TGN
Dawson, P W G6YIU
Dawson, R G0PEV
Dawson, R G0XAB
Dawson, R G G6TJE
Dawson, R H GI6CMA
Dawson, R J G6LXJ
Dawson, R J GW0EYH
Dawson, R O G4ELA
Dawson, S G0HXU
Dawson, S H GI4OVN
Day, A C G4RIM
Day, A C G6OXQ
Day, A C M1AMA
Day, A J M1BFR
Day, A M G7CGT
Day, B G3WIS
Day, C G4MAS
Day, C A J G7FDN
Day, D I G6TJI
Day, D J G4TDI
Day, D S G6XXO
Day, E J G4OBV
Day, E V G8SRN
Day, F A G4PZU
Day, G A G3GAD
Day, G K G4DED
Day, G W G1PHA
Day, I M G1YCJ
Day, J G3NWP
Day, J F GW1YDN
Day, J H G0WBG
Day, J H G4WXD
Day, J H G6OFZ
Day, J L G8SJD
Day, K G3LDJ
Day, K G G8LCM
Day, K J M1AJL
Day, L J G4OUZ
Day, M H G4ZKI
Day, M T G0PUL
Day, N P G4OBT
Day, P E H G3PHO
Day, P L G4KYY
Day, R G0AOM
Day, R G8ADV
Day, R E G1LOA
Day, R G G1CEO
Day, R J M1AKN
Day, R L G8GRB
Day, R P G7RBF
Day, S W G8PAN
Day, T S G0VSM
Day, T S G3ZYY
Day, T W G0SLI
Day, W H G6FZV
Daykin, J A G3WHW
Daymond, D S GW7AVB
Daymond, P A J G0LRJ
Daynes, R C G0MJB
D'Cruz, A J G1TWG
de Almeida, R F A G0NQF
de Araujo, A F M0AVF
de Aston, D 2E1EYD
de Ath, R F GM0UCB
de Bank, J G1XHA
de Banks, M J G7ITO
de Bass, F W J G4LXD
de Bertodand, G A G0AJC
de Buriatte, A H G0PYF
de Cadenet, P G4ZOW
de Carte, P G4PAZ
de Castillo, D J R G0SUO
de Faubert Maunder, L F G4TIS
de Fraine, D L G8IRC
de Frece, J G0TRN
de La Bertauche, L C G3RCO
de La Haye, D C G6BAL
de La Haye, K G GU0RAG
de La Mothe, P D G3VIE
de La Rue, R C 2E1FXL
de Lannoy, M G0WBI
de Maio, A F G4WBZ
de Muth, R D G4JRD
de Putron, T J GU3LYC
de Renzi, J M G1PUY
de Rose, V B G0CLO
de Silva, M A G0WMD
de Voil, G F G8SKK
Deacon, A G6DOW
Deacon, C E G7AJW
Deacon, C J G1LUX
Deacon, C J G4IFX
Deacon, C O M1AVT
Deacon, D G3BCM
Deacon, J R G1TRL
Deacon, K I G6DWW
Deacon, N H G7KMK
Deacon, R H G1DLA
Deacon, S C G6JAK
Deacon, S J G0HLS
Deacon, T R G0AHI
Deadman, W G G7LJU
Deak, B G4PNP
Deakin, A G0EOM
Deakin, A D G1ENR
Deakin, A E G3JXT
Deakin, C G0HRX
Deakin, C J F G0CBV
Deakin, D G1CAX
Deakin, F J W G0OFO
Deakin, P A GD0LQA
Deakin, R E G0HYR
Deakin, R E G4FWC
Deakin, R E G8TRS
Deakin, R N G6LIY
Deakin, R P G6IPQ
Deakin, S G G0JYF
Deakin, S J 2E1AJD
Deakin, T E G6KZA
Deal, R C G7EOB
Deamer, C E G3NDC
Dean, A G3MRJ
Dean, A A G1RSF
Dean, A W G0NRQ
Dean, B G G4KCD
Dean, B H G6NZC
Dean, D G G3AHD
Dean, D G G7SXG
Dean, D G G8WCL
Dean, D J G3JSK
Dean, H F G3ANH
Dean, J G1AVC
Dean, J C S GM4FBP
Dean, J M G8JXG
Dean, J T G6NEA
Dean, J W G6NEP
Dean, M G0CQU
Dean, M C G4RSX
Dean, N G4JHM
Dean, N S G0OQS
Dean, P G3FNT
Dean, P G G0IAI
Dean, P H G0UFN
Dean, R G1AOF
Dean, R J G7UEL
Dean, S R G7ELI
Dean, W C W G4MFD
Deane, A P G7KCG
Deane, C G G6DHY
Deane, D J G3ZOI
Deane, R A J G3VYQ
Deane, R E G4FQQ
Deans, B Q GM4TQN
Deans, D S GM4VZY
Deans, R D G0CFI
Dear, N C G4DFQ
Dearden, F G4IYP
Dearden, G R G1YPR
Dearing, B M G4HMM
Dearing, D A G7PUI
Dearing, K A MM1BOH
Dearing, M C A G G0KBP
Dearling, F J M1AQG
Dearman, A J G1NEB
Dearness, M J 2M1CYF
Dearsley, R G G1EMW
Dearsley, R G G3MOZ
Deary, M D G7UEN
Deary, S M0AIC
Deas, G L GM7SCJ
Deas, J T GM7OHB
Deasington, R J GM1PKY
Death, G M G1BEK
Debney, C W G3ZYW
Debono, V J G3NJK
Dedman, R T G4DFY
Dee, H J G4LRZ
Dee, R W G7DIG
Deegan, J P G4DQC
Deegan Iii, W J M0BBY
Deeley, M G4VZO
Deeley, M J G0OQC
Deeley, W G G4KZJ
Deeprose, R J G4VBK
Deffee, P J G8PJD
Defries, D G1TXW
Degerdon, M E G1DSG
Degg, R G0JOD
Degg, R G1ATC
Degg, R G3ATC
Deglos, M E M1BBR
Deighton, A G4PTA
Deighton, D G8RBV
Deighton, S G8NRC
Deione, E S G3VTP
Delacassa, D L G0NPF
Delafield, H J M1AOV
Delaforce, M P G1OKT
Delaforce, R G G6UJB
Delamare, R G G6ALR
Delamere, P J G6HBK
Delaney, F G4GKT
Delaney, J P G1GBF
Delaney, J W G6ZBY
Delaney, P A GW0HPQ
Delaney, P M G8KZG
Delanoy, D B L G3FOQ
Delarue, R A G4FJC
Delfosse, D C GW4PYK
Delhaye, C R G3NDJ
Dell, D G4WLA
Dell, D H G3PQF
Dell, E A J G7RBH
Dell, F A H G4GFA
Dellbridge, G R G0PMF
Dellbridge, G R G0SSR
Dellbridge, I C 2E1DLN
Dellbridge, R G G0PMG
Dellett, P GI0STC
Delmonte, E J G4CCJ
Delmonte, H W G3UPD
Delves, A N N G0HXM
Delves, J L G3VHH
Delves, R W J G7MJJ
Demain, R A G0MFK
Demchak, R E G0MXK
Demeza, E C G1FTX
Dempsey, L G6PXX
Dempster, B G1RVF
Dempster, C J G3OAK
Dempster, D GI7LAI
Dempster, D GM4FZT
Dempster, I B G4RCZ
Dempster, K J G0ENO
Dempster, P A G0GBC
Dempster, S GM0POF
Dempster, W M GM0MDX
Denby, A W G0UZB
Denby, G G3FCW
Denby, G S M0ALQ
Denby, J S G3TSA
Denby, M M G4GYL
Dence, R L G4DAM
Dench, E G4NPE
Dench, M R G1TWS
Denecker, V G0LMX
Denford, D M G0HSH
Denford, J W G G0GFK
Denham, C G G4VLL
Denham, G J G1SWU
Denham, J A G7JUY
Dening, A C G4JBH
Denison, A J G4ICF
Denison, C J G8KZY
Denison, D M G1GYR
Denison, I J G6XSC
Denison, M D G1JLB
Denley, R F G4HRG
Denman, C B G7SYT
Denman, E T G0PXT
Denman, E T G0SEV
Denman, J C G3OND
Denman, K R G3VOD
Denmead, J S G7JAH
Dennehy, M S G0IZK
Dennett, C GM7MYF
Dennett, E G0EDM
Denney, A R G0TIH
Denney, D J G0XAJ
Denney, F A G4LVE
Denney, I C G3JPZ
Denney, S A G3CIM
Denney, T A G3VLD
Dennick, G M G4MFK
Denning, P G1WJP
Denning, P M G0MAO
Dennis, A C G0PZX
Dennis, A P G7GUA
Dennis, B J G4UTM
Dennis, B J GW4MTU
Dennis, C E G7ELF
Dennis, J F G7RAG
Dennis, J R G4ZOQ
Dennis, K J G7PHT
Dennis, M GM7RNJ
Dennis, M C G8BTY
Dennis, M C GM6KIW
Dennis, M J G7FEK
Dennis, P M G1SQG
Dennis, R 2E1FFJ
Dennis, R H E G0OFA
Dennis, R P 2E1EXG
Dennis, S C M1CAW
Dennis, T GM7RMV
Dennison, D M 2E1DAV
Dennison, P G 2E1DBI
Dennison, W P G6USX
Denniss, J W G0NMJ
Denniss, K J G7DEE
Denny, A G0JNJ
Denny, D A G4DAD
Denny, D E G3ZQE
Denny, M O G3FGW
Denovan, J G0DYN
Denscombe, C F G4HAC
Densem, J H G4KJV
Densham, B G8MQY
Densham, T F G4GOG
Dent, A G G0FVW
Dent, A H J G4KJN
Dent, B G6GLV
Dent, F M G0EZA
Dent, J E G1CSO
Dent, J S G1YFE
Dent, M C G6PHF
Dent, N G 2E1CSA
Dent, R J G7CSL
Dentamaro, N D G0WUC
Denton, B F G4GAT
Denton, C J G8TUN
Denton, D W G0FUE
Denton, E R G3WLO
Denton, G G8VAT
Denton, G B G1EAZ
Denton, H E G1KEV
Denton, J G4DUU
Denton, J J G8EGE
Denton, J W V G7CWM
Denton, P L G6CGF
Denton, R G4YRZ
Denton, R G8CTA
Denton, R A G1HKD
Denton, R D G8WKX
Denton, R E G7TQT
Denton-Powell, C G0MRR
Denyer, A K G4MLG
Denyer, L M G0HPN
Depledge, J C G6LKJ
Depledge, N W G0TSR
Depledge, T G1VIY
Depledge, T J 2E1FRW
Deravi, F GW6WDS
Derbyshire, N G1CKV
Derbyshire, T GW0SSD
Derbyshire, T S G6USU
Derham, D C G3EXL
Dermont, A A G8BGT
Derrick, A J G4DEQ
Derrick, C B G0PQH
Derrick, H C F G7JQW
Derrick, J A G4PKM
Derrick, J W G7KAC
Derricott, R T G4VPE
Derry, D C G8PQ
Derry, F I G1RIP
Dervin, C G4RBZ
Derwin, S M1AFU
Desborough, C L G3NNG
Desborough, C L G3PIA
Desoer, A P G7VSI
Dessau, N G6ZIO
Deutsch, M E G3VJG
Devaney, J P G1YEK
Devenish, J J G1AYL
Devenish, W J J L G6GML
Devenney, G S GI6NDZ
Deverell, F H G1WBG
Deverell, S A G2FVX
Devereux, A R GM8VJV
Devereux, E M G3SED
Devereux, E M G4PYS
Devereux, G P GM7NHU
Devereux, J G6VXD
Deville, P S G4KWM
Deville, S G6TJC
Devine, A M GM6BAO
Devine, B A G1XZP
Devine, E C G7RIJ
Devine, M W G1HZJ
Devine, N P M0AZV
Devine, P G0HJX
Devine, P E G8LGE
Devine, S J G0TKD

Devine, T J ... GI4XGQ
Devlin, B G ... GM0EGI
Devlin, B G ... MM0AIK
Devlin, D S ... G7RMD
Devlin, T B ... G4SPW
Devonshire, J E ... GW4LFF
Devries, M J ... GM0TQB
Dew, H M S ... G0DEW
Dew, J E ... G3FPY
Dew, J E ... G4UXG
Dewar, D H ... G1UCZ
Dewar, I M ... GM1BLX
Dewe, J ... G4EHM
Dewey, D I ... G3ZDD
Dewey, T A ... G8RBU
Dewhurst, C J ... G4KLD
Dewhurst, K ... G0CHL
Dewhurst, L W ... G8TFF
Dewhurst, R J ... G1JYJ
Dewhurst, W ... G6RZG
Dewick, P W ... G6PHC
Dewing, K J ... G0ULG
Dews, F ... G3HPD
Dews, K W ... G3PMW
Dewsberry, R ... GW8MHG
Dewsbery, R P ... G7PZB
Dexter, A R ... G4CDT
Dexter, B D ... 2E1EBK
Dexter, D ... 2E1EBS
Dexter, I R ... G0ITP
Dexter, K R ... G8VDU
Dexter, P ... G6GZR
Dey, M N ... G8SNV
Dharas, M R ... G1ZHL
Di Duca, A A ... G1EBB
Di Ponio, G G ... GM7OLN
Diacakis, A ... G7VVR
Diacon, G R ... G8EWT
Diamond, A G ... GM0IKY
Diamond, D ... G3UEE
Diaper, C J ... G1IBJ
Diaper, G R ... G1IUW
Diaper, R J ... 2E1DLR
Dias, B A C ... G0HMH
Diaz Aguilar, J ... G4ZHS
Dibbins, A J ... G7OTQ
Dibden, F J ... G4SHO
Dibden, W P ... G6OXW
Dibdin, M C ... 2E1CNA
Dibdin, M C ... G7UYX
Dibsdall, M J ... G8MTI
Dick, D K D ... G6PWQ
Dick, H ... M0AMR
Dick, P J ... GM4DTH
Dick, W E ... G4VUU
Dick, W I ... GM8MMW
Dickason, A W ... G0EKD
Dicken, A K ... G4WGH
Dicken, P C ... G1PCD
Dickens, A ... G4YQM
Dickens, K ... G4OCH
Dickens, P H ... G6PHH
Dickenson, B W ... G7MIF
Dickenson, C R ... G3GAR
Dicker, A H ... GW3VEN
Dickerson, C H ... G0KFT
Dickerson, R G ... G8RLF
Dickey, A R ... G6NZB
Dickey, R S ... MI1BRS
Dickie, G K ... G7JDE
Dickie, J ... GM3DGD
Dickie, R J ... GM0IPW
Dickie, W ... GM3ZEA
Dickins, G D ... G0VWS
Dickins, A ... 2E1DSD
Dickins, A D ... G3HLR
Dickinson, A F S ... G0EAV
Dickinson, A M W ... G0GTI
Dickinson, B ... G0KBD
Dickinson, B W ... G1LZF
Dickinson, C ... G0CVT
Dickinson, E P ... G1MFX
Dickinson, I ... G8ZSZ
Dickinson, I H ... G3OUI
Dickinson, J R ... G0WUV
Dickinson, K ... GM7UKL
Dickinson, M ... G8KKZ
Dickinson, M H ... G7DRT
Dickinson, M S ... GD3XJR
Dickinson, N ... G0FZA
Dickinson, P ... G0VEX
Dickinson, P C ... G7RKU
Dickinson, R ... G4LYA
Dickinson, R ... GM0FCP
Dickinson, S ... G7PCZ
Dickman, G ... GJ8GFI
Dicks, C R ... G7OLO
Dicks, R W ... G0TES
Dicks, T H ... G7BLK
Dickson, A D ... G8DJF
Dickson, A M ... GM0KVE
Dickson, D J R ... GM1JTJ
Dickson, H ... M1BEN
Dickson, J F ... G4WDN
Dickson, J J ... GM0IGJ
Dickson, J M ... G4UPR
Dickson, K ... G8XGB
Dickson, R D D ... G7BEA
Dickson, R F ... G1WFU
Dickson, T W ... GM3DIE
Dickson, W H ... GI0HSB
Dickson-Smith, A W ... G4VBY
Didcott, C D ... G2FHF
Didmom, B F ... G4RIS
Dievendorff, R ... G0MFO
Diffey, H ... G4TLQ
Digby, A ... G0JLX
Digby, C L ... G4NEO
Digby, D F ... G8DHQ
Digby, P ... G4NKX
Diggins, P J ... G8MZD
Dighton, M ... G3TEJ
Dighton, P ... G1YTC
Digman, E ... G3BVA
Dignall, K P ... 2E1EBL
Dignan, J J ... G7NQB
Dignum, B J ... G0DDE
Dignum, R P ... G0LPD
Digweed, C E ... GW1EMZ
Dilks, J M ... G7JGI
Dilks, J P ... G7OVS
Dilley, R D ... 2E1ARU
Dillon, C R ... G3WCD
Dillon, J T H ... GM3YQK
Dillow, E H ... G1HRL
Dilworth, I J ... G3WRT
Dilworth, J ... GI4TBV
Dimbleby, N J ... G0TRE
Diment, I P ... G8ZKN
Diment, J E ... G4LTC
Dimery, W G ... G8WZN
Dimes, D C J ... G4HRP
Dimmick, A J ... GM0USI
Dimmock, A D ... G1SUK
Dimmock, F A ... G0CFD
Dimmock, J P ... G8AZR
Dimmock, L ... 2E1FOZ
Dimmock, R M ... G1HIJ
Dimmock, S T ... GD8COH
Dimmock, W R C ... G0NRO
Dimon, R W ... G7AQI
Dineen, P C ... G8WUZ
Dinger, F ... GM0CSZ
Dingle, B ... G4ITV
Dingle, N J ... G1XNI
Dingle, P L ... GM1WBT
Dingle, R ... G7VAY
Dingle, R J ... G0OCB
Dingley, P ... M0AKG
Dingley, W J ... G0UCS
Dingwall, F B ... G0SGX
Dingwall, J A ... G4ILW
Dinning, R J ... GM0GOV
Diplock, A J ... G4NRV
Diplock, J D ... GW3UZS
Diplock, O ... G3NXK
Diplock, O ... G3OCQ
Dipple, A G ... G8UAZ
Diprose, M K ... G4AKA
Disley, R J ... G3KQY
Disney, R ... G0HNZ
Diss, D S ... G4YAX
Diss, P E ... G6NEK
Dissanayake, M B ... G0CUA
Distin, R V ... G6IPH
Ditchburn, W ... G7TSR
Ditchfield, C ... G0JQX
Ditchman, D P ... G0WIU
Divall, C E ... G8MCC
Divall, J R ... G4MDC
Diver, J ... G1IJT
Dix, B G ... G0LPW
Dix, C M ... G6ALS
Dix, D L ... G4JZS
Dix, P J ... 2E1FLX
Dix, W A ... G5IX
Dix, W H ... G4ZEU
Dixey, D ... M1BPY
Dixey, M P ... G4OSU
Dixey, P ... G6JLI
Dixon, A ... G0GSU
Dixon, A ... G8BQF
Dixon, A A ... G1KKZ
Dixon, A B ... G0VPK
Dixon, A G ... MM1ATO
Dixon, A J ... G4IVU
Dixon, A R ... G1SPO
Dixon, B ... G0GZT
Dixon, B ... G0TOK
Dixon, B ... G1UWV
Dixon, B ... G6APN
Dixon, B G ... G1YXA
Dixon, C G ... G8CGK
Dixon, C I ... 2E1CTF
Dixon, D A ... G0BXV
Dixon, D P ... G7BPG
Dixon, D W ... G0AYD
Dixon, D W D ... G1DCG
Dixon, F G ... G7OJU
Dixon, F N ... 2E1GAA
Dixon, G W ... G6SXN
Dixon, H H ... G7HRI
Dixon, I R ... G1BEJ
Dixon, I R ... G4BVY
Dixon, J ... GM4RSJ
Dixon, J A ... G6YIQ
Dixon, J G ... G8VNX
Dixon, J S ... G7UCL
Dixon, J T ... 2E1FSE
Dixon, J T ... GW4YLF
Dixon, K ... G4XKD
Dixon, K A ... G7UCM
Dixon, K J ... G7GWZ
Dixon, M E K ... G0AKE
Dixon, M J ... G7EYL
Dixon, M P ... 2E1CZK
Dixon, M P ... G7VEK
Dixon, M R ... G8IQX
Dixon, M W ... G3PFR
Dixon, M W ... G4ZTT
Dixon, P ... G0HHA
Dixon, P ... G0WQB
Dixon, P ... G1UWR
Dixon, P ... G4JBR
Dixon, P A ... G7JRK
Dixon, R ... G4YAV
Dixon, R A ... GM3ZDH
Dixon, R J ... GI0BFO
Dixon, R J ... GM0CUY
Dixon, R N ... G3SNT
Dixon, R W ... 2E1DMI
Dixon, R W ... GW7DTB
Dixon, S ... G4IYK
Dixon, T C ... G4NHL
Dixon, T S ... G1YUI
Dixon, U J N ... GW4JIY
Dixon, W ... G4PUE
Dixon, W R ... G3XIH
Djali, P K ... G4JTE
Doak, O G ... 2W1FWI
Doak, T G ... 2W1FWJ
Dobbs, C T ... G0BOX
Dobbs, G C ... G3RJV
Dobbs, G G ... G4LAY
Dobbs, G H ... G3RGO
Dobbs, J L ... G0OWH
Dobbs, R L ... G7FGK
Dobbs, T R ... 2E1ERE
Dobby, I R ... GW4LDP
Dobbyn, A R ... G0IAD
Dobdinson, R G ... G3RGD
Dobinson, C J ... G4YAK
Doble, P J ... G7RVG
Dobson, B ... G4XEB
Dobson, C C ... G3ICK
Dobson, D S ... G7PHY
Dobson, I A ... G4XVW
Dobson, I R ... G6LNL
Dobson, J ... G6WJD
Dobson, J A ... G0SJZ
Dobson, J A ... G8JOX
Dobson, J P ... G8CVF
Dobson, J S ... G1YZQ
Dobson, L ... G4WPP
Dobson, L J ... G7WCW
Dobson, N J ... G8OVH
Dobson, P H C ... G6ABA
Dobson, R G R ... G3JDD
Dobson, R T ... G4OBX
Dobson, S ... G7PHW
Docherty, A W ... GM4FXL
Docherty, H A ... G4ZJO
Dockar, D R ... G4IDD
Docker, M F ... G3OOW
Dockerill, J R ... G3JYK
Dockerty, M ... G7WFD
Dockerty, P ... G3YYI
Dockery, D ... G4IBH
Dockray, K ... G0EMM
Dockray, M M ... G1PEN
Dodd, A J ... G4YTY
Dodd, A J ... G8JBW
Dodd, A S ... G0SXK
Dodd, C A ... G3XMZ
Dodd, C T ... G8OZJ
Dodd, D ... G4DKZ
Dodd, D C ... GD3RFK
Dodd, D M ... G6DOX
Dodd, G B ... G6UIP
Dodd, G H ... G0HIP
Dodd, H A ... G4GRA
Dodd, J A ... G1FNU
Dodd, J F ... G4FJB
Dodd, K ... G6GWY
Dodd, L A ... G7VTC
Dodd, M A ... G0IXA
Dodd, M E ... GD4RFK
Dodd, N F ... G1NST
Dodd, P D ... G6SFW
Dodd, P G ... G3LDO
Dodd, P I ... G7ODL
Dodd, R ... G0UDO
Dodd, R E ... G1BWG
Dodd, R U ... 2E1DOD
Dodd, S J ... G0CIM
Dodding, M J ... G0LMI
Dodds, B ... G3YRH
Dodds, B ... G6XXQ
Dodds, C ... G8XQT
Dodds, D A ... GM4WLL
Dodds, G D ... G4ZND
Dodds, J A ... G0FOR
Dodds, M ... G0DOD
Dodds, R T ... G7IYX
Dodds, S D ... G7TFL
Dodds, T ... G0NJZ
Dodge, B J ... G3PCX
Dodge, C I D ... G4UWB
Dodge, J W ... G6ILN
Dodge, R ... G1AYC
Dodgshon, G H ... G4DNC
Dodgson, M S ... G0EKX
Dodic, S ... G8LMN
Dodman, K ... G8HRF
Dodman, P C ... G7RBQ
Dodshon, C C ... G7HBB
Dodson, A T ... G3MGU
Dodson, C ... G7KZV
Dodson, J A ... G0LSY
Dodson, J B ... G3CQK
Dodson, L R ... G0IKE
Dodson, L S ... G6HVQ
Dodson, M B ... G6RII
Dodson, M G ... G7MAB
Dodson, R J ... G4RNK
Dodson, R S ... G3PPD
Dodsworth, G W ... G0PHS
Dodsworth, J A ... G0GFU
Dodsworth, M H W ... G0URF
Dodwell, G D ... GM4CFS
Doe, M A ... G6YYA
Doe, N R ... G8TBU
Doe, R A C ... G3PJX
Doel, B J ... G3SXK
Doggett, C ... G8OSF
Doherty, D J ... MI0BBF
Doherty, D J D ... GW4HZH
Doherty, J ... G7HIK
Doherty, J ... GI4XJD
Doherty, J G ... GI4AXV
Doherty, K J ... GI4TED
Doherty, M ... GI4TAV
Doherty, N C ... GI4HPZ
Doherty, T A ... GI0OTC
Doig, A ... GM1JTK
Doig, G W ... G1DBI
Doig, M P ... GW4CQZ
Dolan, M B ... G3KZU
Doland, D J ... M0ATJ
Dolby, A ... G3TZH
Dolby, A M P ... G0SYH
Dolby, J ... G3PDD
Dolby, J M ... G7IRT
Dolby, M A E ... 2E1BSJ
Dolby, M F ... G6KKO
Dolden, T W ... G1BUB
Doley, O C ... G3BRA
Dollery, C T ... G3GAF
Dollery, C T ... G4HXX
Dollery, P N ... G4TNB
Dollimore, P J ... G1LLQ
Dolling, D S ... G0FVH
Dolling, P G ... G0RBN
Dolling, P G ... G4LQZ
Dolman, L J ... G4EXN
Dolphin, C ... GW7VRZ
Dolphin, D ... G0AQF
Dolphin, P R A ... G3ELH
Dolton, R H ... G3HTO
Domachowski, P S ... G7JSQ
Doman, C G ... G4EZQ
Dombrowski, P D ... GW1NYO
Dominy, L R ... 2E1DWD
Dominy, R P ... G8YXZ
Domville, R ... G6RKS
Donachie, F ... G4XWT
Donachie, G K ... G1LMQ
Donachie, I B ... G0WJZ
Donachie, V ... G0CWD
Donacie, J M ... GM3PXG
Donaghy, E J ... G7WAA
Donald, A K ... G1ROX
Donald, C ... GM6MJY
Donald, C B ... G0OEB
Donald, G M ... GM4JCR
Donald, I W ... G1YUP
Donald, L ... G1WWY
Donald, S ... G6EDD
Donald, V A ... G7RXE
Donald, W K ... G7ENQ
Donaldson, A ... GM7CQQ
Donaldson, A G ... GM0LYM
Donaldson, A L ... GM0DGK
Donaldson, I A ... GM4SZA
Donaldson, J ... GM3ZSH
Donaldson, J H ... G0RBT
Donaldson, L R M ... GM4AJR
Donaldson, M C ... G1HZL
Donaldson, P E K ... G7JQY
Donaldson, R A J ... GM0SRD
Donaldson, R B ... GM4RIW
Donaldson, W ... GM3FFQ
Donaldson, W C ... GM1MXW
Donaldson, W M J ... GI4SYM
Donati, D ... G8BAD
Donbavand, E J ... G6BIX
Donders, B ... G3UXJ
Done, G B ... G7PLR
Dongray, P D ... M1ANZ
Donin, J A ... G4YJD
Donington, G R ... G4LNO
Donkin, B R ... G7OUZ
Donley, T ... G0SQX
Donn, B ... G3XSN
Donn, G ... G4IHS
Donn, I ... G6FUT
Donnachie, F R ... G1RON
Donnan, J B ... GM0WUX
Donnan, O J P ... GI0WYJ
Donnan, R ... GI7WJY
Donnell, J F ... MW0AJH
Donnelly, A ... G4PWJ
Donnelly, A A M ... GI4NSV
Donnelly, D ... 2E1EKC
Donnelly, J ... GI0VVJ
Donnelly, J ... GI8JRE
Donnelly, M C ... 2E1ALG
Donnelly, M C ... GI3NSV
Donnelly, P J ... GI4VCZ
Donnet, R ... GM7KXJ
Donnett, J G ... GM0MVY
Donnison, A L ... G4IOX
Donnison, D C ... G4BDY
Donno, R J ... G3YBK
Donoghue, M ... G1NGZ
Donoghue, W T ... G6DIA
Donoughue, G ... G4THX
Donovan, F P ... G4ALD
Donovan, G A ... GW8BNL
Donovan, J ... G4GHK
Donovan, P D ... 2W1DFT
Donovan, P D ... G6ILM
Dons, E K ... GM0AXY
Dons, E M C ... GM0SYL
Dons, E M C ... GM4YMM
Donson, R ... G7MTU
Doody, D C ... G6OXP
Dooks, B E ... G0RHI
Dooley, J ... G4MYG
Dooley, R ... GW0AZW
Doores, J W ... GW0WEY
Doran, A ... GM0HVD
Doran, C J ... G3VZH
Doran, P ... GI4WYE
Doran, R ... G0DFS
Doran, R ... G4LPZ
Doran, R ... G7DFS
Doran, R J ... G4VRC
Dore, D ... M1AFK
Dore, J R ... GW3XPK
Dore, R C ... GW7GAH
Dore, W A ... G4CGE
Dorey, B P ... G7OYX
Dorian, I R ... G6EME
Doris, C ... MI0AHH
Doris, J ... MI0AHI
Doris, P ... MI1CBX
Dorling, A ... G8CJL
Dorling, R A A ... G0PJZ
Dorman, A M ... G0AMD
Dorman, G M ... 2E1AFD
Dorman, J A ... G1LVY
Dorman, S R C ... G3ABP
Dormann, D ... G7IXG
Dormer, C ... GW6EWQ
Dornan, M P ... G4SOY
Dornan, S ... GI3TNK
Dornan, S ... GI7GXZ
Dorney, K T ... MW0AQG
Dorning, J A ... G0NEB
Dorrance, V S ... G6ZQL
Dorrell, I A A ... G0WZU
Dorrell, I A A ... G6TJH
Dorrell, K J ... G4AZO
Dorrian, A J ... GU7NHX
Dorrian, C J ... GI7VGR
Dorricott, B D ... G4SDL
Dorrill, A A ... G0BLL
Dorrington, N P ... G4VAF
Dorrington, P W F ... G8MMF
Dorrington, S ... G1KGL
Dorrington, S M ... G1KWH
Dorrington-Ward, S ... G8LOJ
Dorsett, A ... GM4PSX
Dosher, J T ... G8MGK
Dossett, K S F ... 2E0AHT
Doswell, J ... G3VYE
Doswell, K R ... G1OET
Dotchin, R M ... G3WEP
Dotchon, P H ... GW1TKO
Double, E H ... G8CDW
Doubleday, G J E ... G6BZQ
Douce, I R ... G0JDD
Dougan, M J ... GI0DWN
Dougherty, G ... G4ZSK
Dougherty, J J ... G4FUT
Doughty, A M ... G6ZQJ
Doughty, K F W ... GM0IJV
Doughty, P R ... G3TKK
Doughty, R ... G0FHM
Doughty, R C ... G1AOC
Doughty, R T ... G8SPU
Douglas, A ... GM4WDO
Douglas, A C ... G0HDJ
Douglas, A C ... G0PCS
Douglas, A L ... GM4HVM
Douglas, A S ... G0DVW
Douglas, A S ... G8MLF
Douglas, C B ... G0OGW
Douglas, C G ... GD3ZEX
Douglas, C W ... G8LDJ
Douglas, D A ... G0HKO
Douglas, G J ... G3SZY
Douglas, I ... GM4FGS
Douglas, I A M ... G3NID
Douglas, J ... G8GFW
Douglas, J E ... G4ZXR
Douglas, J N ... GM4DVG
Douglas, L C ... GI0RJO
Douglas, N A ... G4SHJ
Douglas, P A ... GI8KFG
Douglas, R G ... G0FUH
Douglas, S E ... G6JLL
Douglas, W ... G4NTW
Douglass, J M ... G6EZA
Douglass, M J ... G6MKD
Doull, J A ... GM6TJD
Doull, R E ... 2E1CRH
Douse, A E H ... G3XYK
Douthart, S ... GI8WIU
Douthwaite, J D ... G6OQO
Dovaston, N G ... G4ODE
Dove, D C ... G3DOV
Dove, F G ... G0IAV
Dove, R J ... G6VYM
Dover, G W ... G4AFJ
Dover, G W ... G4RMS
Dover, J F ... G0KOH
Dover, P J ... 2E1FDR
Dover, S J ... 2E1FDS
Doveton, S ... G0JNR
Dovey, I J ... G7PBX
Dowd, R ... G0RPO
Dowdall, C J ... GI1LGM
Dowdall, J R ... GW8HDH
Dowdell, A ... G4UJE
Dowdell, R L ... G8RLD
Dowdeswell, R E ... G1WIW
Dowding, M J ... GU0PSP
Dowell, A M ... G6TJK
Dowell, C M ... G0TJQ
Dowell, J ... G4NFZ
Dower, K J ... G0BDJ
Dower, M J ... G6XSB
Dowers, C S ... GM8ZGC
Dowie, A W ... M0AUS
Dowie, J A ... G8UZV
Dowkes, W M ... G1SZT
Dowler, P ... G6EAR
Dowles, P C F ... G3VNP
Dowling, A F ... G3GUE
Dowling, B L ... G4IJV
Dowling, J A ... G1PET
Dowling, J G ... GD0TFG
Dowling, R ... G3NKH
Dowling, R M ... G3WMT
Down, A J ... G4KQD
Down, B J ... G8ZIR
Down, C D ... G8MXW
Down, E I J ... GW0DDK
Down, G M ... G6ZTP
Down, J D ... G7TLS
Down, J T ... G7NWS
Down, M J ... G4ALR
Down, N E ... G3SRX
Down, S ... G7IJE
Down, S J ... G3USE
Downer, B D ... G3ZQI
Downer, J ... GW6TYJ
Downes, D J ... G0KRD
Downes, I A ... G6IPC
Downes, L P ... G4TMD
Downes, M ... G7JZD
Downes, W E ... G6HHD
Downey, M E ... G4AHJ
Downey, P ... G3WDK
Downham, P S ... G3WIB
Downham, R G ... G6TJJ
Downie, I A ... G8XFY
Downing, A ... G0JQS
Downing, A ... G3ZES
Downing, A F ... G8UWI
Downing, J E ... G7IXT
Downing, M K ... G7WIY
Downing, T E ... G3MXH
Downs, G K ... GI4MTZ
Downs, G W ... G3UCK
Downs, J K ... G8CFI
Downs, N ... G0DND
Downs, N J ... G7KHN
Downs, P J ... G0PFF
Downs, R ... GM0JSW
Downs, R D ... G3OEB
Downton, B V ... G3ZQR
Dowse, G J ... G6HHH
Dowse, J R J G ... G0GKN
Dowse, V J ... G4WGP
Dowsett, A F J ... GW8VNN
Dowsett, E F G ... G4BDD
Dowsett, G P ... G6FBI
Dowsett, N C ... G8PUY
Dowsett, P H ... G0HWS
Dowson, D ... G3BYX
Dowson, H R J ... G8TUL
Dowson, J R ... G8BTU
Dowson, M ... G8CJA
Dowson, R J ... G4TDG
Dowthwaite, M D ... G0LAQ
Dowthwaite, R P ... G8WUY
Doxey, A J ... G1EKH
Doxey, J ... G0WOC
Doy, J ... G7OCH
Doyle, A J ... G0FAU
Doyle, B ... G6IPB
Doyle, B P ... G7DFQ
Doyle, D ... G3RDK
Doyle, D J ... G0DYG
Doyle, E J ... G6RRZ
Doyle, F ... GM0SXD
Doyle, G ... G1FLA
Doyle, G J ... G7CJJ
Doyle, H W ... 2E1FRC
Doyle, J F A ... G4TIK
Doyle, J H L ... G3DID
Doyle, J J ... GW4FOI
Doyle, J M ... 2E1FZY
Doyle, M J ... G7JZJ
Doyle, M R ... G4FRJ
Doyle, P ... G1MBN
Doyle, R A ... G0JDE
Doyle, S J ... GM1YLB
Drabble, M J ... G7DRF
Drabble, R ... G1SLE
Drackley, E G ... G3HTP
Drage, A J ... G1AZD
Drage, A J ... G7KDS
Drage, D B ... G2BNI
Drage, P M ... G7DMK
Drage, R G T ... G0GOB
Drain, A ... G7RGZ
Drain, G S ... GM1FUD
Drain, R J ... GI4POC
Drake, D H ... G0EZJ
Drake, F J ... G1NQQ
Drake, G ... G3SJD
Drake, G L ... G0MFT
Drake, J ... GW1RHQ
Drake, M J ... G6JAR
Drake Brockman, R M ... GM4UPL
Drakeford, K ... G4VSJ
Drakeley, J C ... G8GRC
Drake-Brockman, L M ... GM4UOD
Draklely, J E ... G4FGF
Dransfield, W ... G4DCY
Draper, D J F ... M0AME
Draper, D W ... GW4JUI
Draper, J G ... G6BAM
Draper, K ... G1SUW
Draper, M ... G4BSA
Draper, N L ... G3FLN
Draper, P C ... G4DSG
Draper, P R ... G0LUI
Draper, P R ... G1TKY
Draper, R H ... G4BU
Drawmer, P G ... G4YFY
Dray, L R ... G3SHD
Draycott, C D ... G7DNB
Draycott, D J ... G4LVF
Draycott, D S ... G0LNN
Draycott, K A ... G3UQT
Draycott, K V ... G2BOI
Draycott, M J ... G3XTQ
Draycott, P R ... G0OXB
Draycott, R ... 2E1ERE
Draycott, S ... G4JFH
Draycott, T ... 2E1EMQ
Draycott, T R ... 2E1EMP
Drayton, C N H ... G1IBX
Drea, W P ... G0CBU
Dredge, I A R ... G4DIE
Dredge, R G ... G6JUJ
Dredge, W C ... G4HJF
Dreiling, G ... GW7KBI
Drennan, M ... GM7PZH
Dress, R ... G4KQM
Dressel, R K ... G0WNY
Dresser, A M ... G7MFP
Dresser, P ... G4VOK
Dresser, R ... G0FDA
Drever, M R ... G0WKA
Drew, B G ... G0MBG
Drew, D R P ... 2E0AMS
Drew, D R P ... G7BTX
Drew, G P ... G0CZE
Drew, J G ... G7UIY
Drew, M P ... G0HWK
Drew, P ... GW6IPR
Drew, P W ... G1OPV
Drew, R D ... G8EJC
Drew, R H ... G6JYX
Drew, R J ... G8URU
Drew, T J ... G6FBF
Drewe, C D ... G8MYJ
Drewett, R J ... G0GXV
Drewitt, C J ... G8OXR
Drewitt, T C ... G0BMU
Drewry, J R ... G1TAU
Dring, L S ... G7HHM
Dring, R A ... G0BXG
Dring, S ... M1ALN
Drinkall, M L ... G6XGK
Drinkwater, A J ... G0EHY
Drinkwater, C ... G3FNK
Drinkwater, E J ... G6SXD
Drinkwater, G D ... G8GCU
Drinkwater, G F ... GM8JME
Drinkwater, J ... G0JYD
Drinkwater, K ... G3RHR
Drinkwater, M B ... G1YHF
Driscoll, D P ... G1MVT
Driscoll, J C W ... GM3NZJ
Driscoll, J G ... GI0TDP
Driscoll, M ... GW4OKI
Driver, A L ... G1VWU
Driver, C M ... G6CMD
Driver, D P ... G7VQC
Driver, E V ... G1VMX
Driver, G C ... G3NDE
Driver, G P ... G4ZSM
Driver, K ... G6KGW
Drobnica, C ... G4VWN
Drohan, G ... G6DIE
Drohan, M B ... G4NQM
Dronfield, M W ... G1WWI
Dronfield, P J ... G4RNA
Drouet, C J ... G8GJW
Druce, B ... G3ZGT
Druce, V G ... 2E1FJQ
Drumm, N A ... M1BDZ
Drummond, A ... G6KAM
Drummond, A J ... G7PSU
Drummond, I R ... GM0PCH
Drummond, J M ... G0RGO
Drummond, J M ... G0TWD
Drummond, J P ... M1AZY
Drummond, R H ... M1ATQ
Drummond, T ... G0ILT
Drury, A J ... G4FZP
Drury, B W ... G1UMK
Drury, H D ... G4HMD
Drury, I M ... G0FXQ
Drury, J E E ... G3XTG
Drury, N E ... G4SCO
Drury, S A ... G6ALU
Drury, S J ... G7RLK
Drury, S N ... G4ZPQ
Drybrough, D A S ... G0AUO
Dryburgh, G ... GM7CNW
Dryden, J L ... G4DSN
Dryer, G R C ... G0IKO
Drysdale, I G ... GM3TYS
Drysdale, N ... GM0KMD
Drysdale, P H ... G4VCW
Du Cros, G M ... G7UTI
Du Feu, G P ... G0NUO
Du Feu, G P ... G1MQC
Du Heaume, J C ... G0SNP
Du Plessis, P G ... G7FZN
Dubbins, B T ... G8OCN
Dubbins, E A ... G8OCM
Dubery, M D ... G4EZR
Dubery, R G ... G8CWF
Dublon, C F R ... GW0KZE
Dublon, G P N ... M0AKD
Dublon, H A M ... GW7SQP
Dubourg, T D ... G0TDD
Duce, A ... G1VOJ
Duce, M ... G4BQF
Duce, W R ... G0SZC
Duchscherer, J L ... G8LKP
Duck, A J ... GW0DQT
Duck, I A ... G3DTX
Duckett, G R ... G0KGW
Duckett, W F ... G8BKF
Duckfield, K ... GW0BZA
Duckles, C K ... G6KIA
Duckling, C ... G3SVL
Duckling, S C ... G7TAJ
Duckworth, A ... G4BG
Duckworth, C S ... G0PUY
Duckworth, D K ... GM4UGN
Duckworth, J ... G0DTO
Duckworth, J ... G3FM
Duckworth, R ... G0WMZ
Dudbridge, C A ... G7TJO
Dudbridge, J W ... G3UUO
Duddin, R A ... G8PTF
Duddington, J E ... G4ATH
Duddington, J E ... G4BFH
Duddridge, J E ... G4NVM
Dudek, M P ... G6BWW
Dudeney, J F ... G8BFS
Dudhill, B ... G4NMP
Dudley, A ... G0IMF
Dudley, A J ... G0HLI
Dudley, D R ... G8PEU
Dudley, E J ... G0NED
Dudley, E P ... G1ILA
Dudley, G M D ... G0DGQ
Dudley, I C ... G3YRP
Dudman, N J ... GW8GGW
Dudman, P ... GW6GTS
Duell, A R ... G1XPF
Duell, J F ... G6CHO
Duell, P D ... G0TLG
Duerden, M D ... G1KWK
Duesbury, E J ... G3EYO
Duesbury, H ... G8YYA
Duesbury, P G ... G6KIB
Dufeu, M A J ... G4GRQ
Duff, A D ... G7FTH
Duff, C I ... G0SHS
Duff, D A ... G3VYV
Duff, D W ... GM4UGF
Duff, M C ... G4CGM
Duff, W ... G0GWS
Duffell, B ... G3VGZ
Duffett-Smith, P J ... G3XJE
Duffield, A G ... G0BCO
Duffield, S T ... G7CBW
Duffill, D H ... G0UBY
Duffin, G ... G4TUI
Duffin, G E ... G2FTY
Duffin, P C ... 2E1FIS
Duffner, W G ... G6KGG
Duffus, J L ... G4EWB
Duffy, A F ... G1JCW
Duffy, B J ... G7UWK
Duffy, B M ... G1VKJ
Duffy, C F ... G0PVP

Duffy, C J G1NFE
Duffy, C K G4MPO
Duffy, G S GI7LQI
Duffy, J G8RQF
Duffy, J GM0DZE
Duffy, J P GM0LIM
Duffy, L S G3TXP
Duffy, M GM0GYL
Duffy, P J G4NPG
Duffy, V L G4CJP
Dufour, I G G3PWB
Dufrane, J C G E G4MBW
Dufton, W E G3WUH
Dugdale, M R M1BUG
Duggan, A G0LAX
Duggan, D E GW0JBN
Duggan, G A G6VNI
Duggan, H E GW0MSY
Duggan, J B GW0LXD
Duggan, J C GW1XVM
Duggan, L N G7AVD
Duggan, P E G7POL
Duggan-Keen, J E GW1FWE
Duggins, A F G4SNO
Duggins, F H G7RGN
Duguid, S GM4WFV
Duguid, W K R GM4RLV
Duignan, C GI4UKH
Duke, A J G0EKV
Duke, G R G4GKH
Duke, P G4YJK
Duke, R E G8GZV
Dukes, C R G6UJC
Dukesell, D A G0RKT
Duley, P D G6LXF
Duley, R J G6TYF
Duley, T S M1BVB
Dullingham, P M M1BVV
Dumbill, K G8JYV
Dumbrille, C C G3OGA
Dumont, L M G F G4AML
Dun, A C GM6KRD
Dunbar, C J S GW8EHK
Dunbar, I GM0KDP
Dunbar, J P G8FRG
Dunbar, M R G6RRY
Dunbar, P R L GW3WCA
Dunbar, W F G4PUK
Duncan, A GM0PYC
Duncan, A GM3DZB
Duncan, A G G0FIF
Duncan, A R GM8SVB
Duncan, B S G1LJL
Duncan, C D D G7FOP
Duncan, C G GM6MUZ
Duncan, C M GM0EKM
Duncan, E G4KSL
Duncan, G G H GM4ZEX
Duncan, J 2M1BBY
Duncan, J H G4IZM
Duncan, J K G6SLD
Duncan, J R G3KUD
Duncan, P G4HIX
Duncan, P S G3TKA
Duncan, R A GM7UDI
Duncan, S E G0OFT
Duncombe, H G3XJN
Duncombe, R D MW0AIE
Dundas, J M GM0OPS
Dunell, W G2XV
Dunell, W M G3BYW
Dunell, W M G8EVY
Dunford, A G3XOF
Dunford, D J G0PTK
Dunford, G D G7AAF
Dunford, N C G6HVD
Dunford, P J G3YXW
Dungan, R A N G8LSI
Dunglinson, J F G4CGW
Dunham, A L G6OHM
Dunham, L E S G6SXB
Dunham, M R G1OQX
Dunham, P T G8JIU
Dunham, R G3ZSQ
Dunhill, J J G1DEP
Dunkerley, C G3FBI
Dunkerley, S R GW4GFZ
Dunkley, J E G0ITL
Dunkley, M P G0TTA
Dunlop, A J G7IET
Dunlop, C J G6LKH
Dunlop, D GI7TTO
Dunlop, J A GM3GVD
Dunlop, J A GM3KBZ
Dunlop, J I GI3YDM
Dunlop, J J GM0WDF
Dunlop, P R G7NHR
Dunlop, R H GM3NMN
Dunlop, S MW1AZI
Dunlop, T GM4YMA
Dunmore, F W 2E1FEC
Dunn, A L GM6MD
Dunn, B G4FQW
Dunn, B G7JUO
Dunn, B P GM3XTR
Dunn, C G0AMW
Dunn, C GM4OSV
Dunn, C L G0MTR
Dunn, C L G4KVI
Dunn, C T G8BEK
Dunn, C T M0BEK
Dunn, D G0WLK
Dunn, D G7TKJ
Dunn, D G8RZN
Dunn, D B G3SCD
Dunn, D C L G4NKU
Dunn, D J GW3XRM
Dunn, D W G8KOV
Dunn, E G1UBT
Dunn, F G0KZS
Dunn, F A G4KPV
Dunn, G G0PGW
Dunn, G A N G4MZI
Dunn, G M G8TZW
Dunn, G R G4DYH
Dunn, G S G3YAA
Dunn, G S G6RIQ
Dunn, H G6VCN
Dunn, H G4XCO
Dunn, I B GM7IQG
Dunn, J G7HGB
Dunn, J G7IYB
Dunn, J GM0VIV
Dunn, J G GM4ZNG
Dunn, J W B G1HEX
Dunn, K A R G1NDK
Dunn, L E G6DOV
Dunn, M G7ELA
Dunn, M J G7IZN
Dunn, N R G4UWG
Dunn, P L G0NLQ
Dunn, P N G4XOC
Dunn, R G7LQK
Dunn, R G8GDI
Dunn, R J G4WBG
Dunn, R L G4PMW
Dunn, S C G0BIF
Dunn, S C P G4KCR
Dunn, S D 2E1FYP
Dunn, S P GM4PMH
Dunn, S T G0KHN
Dunn, W V G0MKC
Dunnachie, N GM6VVX
Dunne, E J G6CTH
Dunne, G 2E1FBZ
Dunne, H T G0EFN
Dunne, J M G0UPG
Dunne, J R G3AGR
Dunne, K G4VLB
Dunne, K P G0GKD
Dunne, M J GI4MJD
Dunne, M W 2E1DQL
Dunne, P D G0LZA
Dunne, R F G0MGM
Dunnett, A J S GM6PYD
Dunnett, J M G4RGA
Dunning, J G1DBK
Dunning, P A G4DEA
Dunning, P A G8VFR
Dunnington, J G3LZQ
Dunnington, J M0ARC
Dunnington, J P GM4EIW
Dunphy, D G4JOX
Dunstan, K G4WKD
Dunstan, P C GW7UHM
Dunstan, R G4MFQ
Dunstan, W H G0CSY
Dunster, A G0RSK
Dunster, G J G0MFQ
Dunster, J M G7RUT
Dunster, S B G7RUV
Dunthorne, M 2E1FQP
Dunthorne, M E 2E1FQO
Dunthorne, P J G4NTT
Dunwell, J G1PCG
Dunwell, K G4WLG
Dunworth, I G4SNL
Dupree, B C G4INB
Durance, D B G4EXX
Durant, C G4CEX
Durban, J G6LNU
Durbridge, A P G1OSO
Durbridge, C J G4EML
Durdin, P R GW0WBP
Durell, C A G3PNT
Durell, D G6UIL
Durell, D A G1OGM
Durell, R M G3LRX
Durey, E R G4MOV
Durey, M S G1NMN
Durham, D R G3SIR
Durham, P J G3ZOX
Durling, A R C G3CPD
Durnall, C C G4DHL
Durnford, A B G4WJY
Durno, G S GW7NIW
Durrand, J GM4XLN
Durrans, J E GW4AYQ
Durrant, A M G7VHF
Durrant, B C G8NZB
Durrant, B M G0SIU
Durrant, B R G4LVD
Durrant, D A J G6DOR
Durrant, G C 2E1EGI
Durrant, J E G3WBM
Durrant, J G G8DCD
Durrant, J K G8UZW
Durrant, J W G0SWF
Durrant, K F G4UBC
Durrant, M L GM7CTI
Durrant, N H G0TNJ
Durrant, P T G G1UNB
Durrant, S G1HCO
Durrant, S P G1LTZ
Durrell, J R G0HJZ
Durston-Wyatt, J R G6MBD
Duschek, W G G4VUY
Dussart, J J C G4DGQ
Dussold, I G G7TMS
Dutfield, P V G3OBD
Duthie, G GM7GEF
Duthie, I F M G0DUT
Dutson, K G0LOH
Dutton, A G1KLK
Dutton, A G F G3TIE
Dutton, G F G3FJV
Dutton, J J G0ITM
Dutton, L A G6HFK
Dutton, N J G8YQJ
Dutton, R D L G6QQ
Dutton, R J G8TCS
Dutton, V R G4IIS
Duval, R G0JAG
Duxbury, J M G6LNS
Duxbury, P S G7OSC
Dwight, D A G1DQQ
Dwight, J N G1DWT
Dwyer, A G0HBU
Dwyer, C L T G3FMR
Dwyer, D S G1PWU
Dwyer, K I GW7EQC
Dwyer, P J G0UXM
Dwyer, S G6RXP
Dwyer, S E G1CNI
Dyball, H H G4FOJ
Dyce, A J G8TAQ
Dyche, A H 2E1FZS
Dyde, M R G6BAA
Dye, D C G3WPG
Dye, D J G4TRT
Dye, P S G0WNX
Dyer, A G4ODI
Dyer, B GW3HNC
Dyer, C J G7CJD
Dyer, D G4WUK
Dyer, D D G0PRU
Dyer, D D G4DNX
Dyer, D J GW8VCA
Dyer, D L G4CXQ
Dyer, E J G4CPT
Dyer, G M G8VRV
Dyer, G W GW0NUS
Dyer, K G GW0RHC
Dyer, K G GW7JVS
Dyer, K W G0KWD
Dyer, M R G8HHQ
Dyer, P A G1FZL
Dyer, P E G4POU
Dyer, P J G4BUV
Dyer, R C M1AFF
Dyer, R J G1XIE
Dyer, R J M0AJZ
Dyer, W N GM7JUX
Dyke, J R G4NVA
Dyke, M J G8EPC
Dyke, P R G0LUC
Dyke, S G3ROZ
Dykes, A 2E1ENE
Dykes, A D GW1MNC
Dykes, J R G0NDU
Dymock, J R G7DSW
Dymond, G R G4VQL
Dymond, L J G4PEK
Dymott, A R G4OAG
Dynes, J B GI6FTM
Dynes, P L GI3OZW
Dyson, A G0PAV
Dyson, A I G0HUW
Dyson, A P G0BXT
Dyson, C G0WAD
Dyson, D H G7DHD
Dyson, G E M0AOG
Dyson, H G4JLO
Dyson, J G7OJE
Dyson, J A G7RXJ
Dyson, J F G0BXX
Dyson, J H G6LXI
Dyson, K W G3DDA
Dyson, M G4SKY
Dyson, P G0UEE
Dyson, P H 2E1EBV
Dyson, R G8ZEN
Dyson, T J G0RLL
Dyson, T J G1RIV
Dyson-Bawley, G G G7UWJ

E

Eade, D M M1BFC
Eade, S A G4RFS
Eades, A J G1VIO
Eades, G D G1LTE
Eades, M S 2E1EKQ
Eadie, A J G1KFO
Eadie, J S G8GZX
Eady, D R M1AUO
Eady, J A H G8ZIH
Eager, R D G8KDU
Eager, S P 2E1EAC
Eagle, G G G4UTX
Eagles, N D G0CPS
Eagles, V H G8MCR
Eagleton, B K M0AFQ
Eagling, C J G6PMD
Eales, M G1YEU
Ealey, P C G8FAE
Eames, A A G4UNP
Eames, B V G3SBF
Eames, D GI6MZL
Eames, P F J G7VHL
Eames, P F J M0AXD
Eamus, I J G3KLT
Eamus, I J G4VRS
Eardley, A J N G3UXO
Eardley, G G6LMJ
Earl, A D G7RZV
Earl, C A G3OXV
Earl, K S G4OTS
Earl, P P G8LHF
Earland, R J A G3AJK
Earle, D A G4FTA
Earle, J A P GI6VLY
Earle, R A G4HTJ
Earle, S G0SWG
Earle, S D G7FQP
Earley, J D G8UAY
Early, D G3DGW
Early, R G7KJA
Earnshaw, C G3DMO
Earnshaw, D G3LHP
Earnshaw, D G4MZF
Earnshaw, D A G8BAA
Earnshaw, D J G0FNJ
Earnshaw, D J G3MEZ
Earnshaw, D J W G7RQU
Earnshaw, G W G3ZXC
Earnshaw, H L G0UFG
Earnshaw, J D G4YSS
Earnshaw, P J 2E0ABI
Earnshaw, P J G0UUU
Earnshaw, R A G1FUI
Earnshaw, R J G4FVF
Earp, A T GW7IBT
Earp, D G4SGG
Earp, G K G8UWO
Earp, K E G0MZJ
Earp, R C 2E1EDX
Easdon, B G0RZI
Easdown, J N G4HIZ
Easey, B G1JGM
Easey, J D G4XBE
Easom, A G4OPI
Eason, J G G3RYQ
Eason, J M G3BBC
Eason, W E E G4MQN
East, A R G0TLH
East, B W GM3NNZ
East, D A G4PWM
East, E R G1ROO
East, J G0OQX
East, J A G7HSS
East, J B G0HIC
East, K G4UNF
East, M G4IOF
East, R D M1AEO
East, R P G0GEB
East, S G4BMU
Easteal, J G8HNK
Easteal, J R G4DNH
Easter, K G3ISK
Easter, S J G7URR
Easterbrook, M A GI0JPW
Eastgate, G M G4SNV
Eastham, F J G7PZE
Eastham, J G1BWI
Eastham, T J G1AVD
Eastick, B G1AVF
Eastick, W E R G1RSO
Easting, R G7NZV
Easting, T A 2E1BHH
Eastman, H F D 2E1DYV
Eastman, K A 2E1DYS
Eastman, M D G0DDZ
Eastment, J D GW4LXO
Easton, A C M1AUH
Easton, B W M G4XMA
Easton, C G0FSG
Easton, D F GM7AWK
Easton, D J 2E1CQE
Easton, J A H G0CYE
Easton, P GM1RCP
Easton, W G3VGN
Easton, W R GM8OAH
Eastty, K F G3LVP
Eastwood, D E G1CBX
Eastwood, E G1WCQ
Eastwood, H G0BGV
Eastwood, J C GW8ONP
Eastwood, K M1ADP
Eastwood, S A G7POT
Eaton, A G G8YEF
Eaton, C C G4TDY
Eaton, D J 2E1CCC
Eaton, D J GU8ITE
Eaton, E G0WKS
Eaton, G E G3SMK
Eaton, J G3EZZ
Eaton, J H G0AOV
Eaton, J R GM4LBV
Eaton, K J G0RTX
Eaton, K J GW1FKY
Eaton, P G4EDW
Eaton, P M G7PME
Eaton, P R G0SVW
Eaton, P R G1BWH
Eaton, R G6MZJ
Eaton, W D G3TAO
Eaton, W J H GM3KIG
Eaton, Y 2E1CNX
Eaton-Watts, D H G7THU
Eaton-Watts, D H M0AVH
Eaton-Williams, R H G3CGB
Eatwell, C P G3AJW
Eatwell, W A G7LZY
Eaves, A G6FZW
Eaves, T C G4GUY
Eavis, M E C G0AKI
Eavis, P J G7CTP
Ebbetts, R F G7JWV
Ebden, R C G0JOZ
Eborall, M J G0JUQ
Ebsworth, P E G8CKB
Eccles, A W G3KNZ
Eccles, C C G8NMK
Eccles, D G G4LAG
Eccles, D J G6YXU
Eccles, D L G7VKR
Eccles, F C GI3TIJ
Eccles, G G1JNG
Eccles, M J GM3PPE
Eccles, R B G8MQX
Eccles, W A G4UFY
Eccleston, E C H G1DQR
Eccleston, G G4OTS
Eccleston, J 2E1CKI
Eccleston, M D G0BSB
Eccleston, P G3JIK
Ecclestone, G J M1AXE
Eckersall, G G4HFG
Eckersley, B P 2E1EXY
Eckersley, J A G1RRE
Eckersley, R J G4FTJ
Eckersley, W G3GAG
Eckles, G A H G5GC
Eckley, D W E G3UFQ
Ecott, M L G8SSI
Eddowes, G D G3AYC
Eddowes, G D G3NOH
Eddowes, G D M0BBC
Eddowes, G D M1BBC
Eddy, G C G7PVE
Eddy, G C G7TCC
Eddy, R L G7SEQ
Eddyvean, M C M0ACU
Eden, A R G4DOK
Eden, G G0BCX
Eden, G R G1HXT
Eden, H G6VCR
Eden, J GM0EXN
Eden, J P F G6NFJ
Eden, J T G3VJP
Eden, L A G1WPH
Eden, T G1JBV
Edeson, R E G4FBA
Edgar, A E G7HBR
Edgar, A H G3IOE
Edgar, G GI4MCW
Edgar, J S GI4FVM
Edgar, R A G0KYS
Edgar, R K G1ANV
Edgcombe, A J G6TOI
Edge, M G G7GCU
Edge, R G4EMD
Edgecock, A G G4AZD
Edgecombe, R H G8BXD
Edgecumbe, L W G0NXI
Edgeler, N G0VQN
Edgeley, R J G8KZO
Edgerton, J D M G0EDG
Edgington, J E G0KTV
Edgington, J R G3KJT
Edginton, R L G3AGF
Edgley, B W G7KEI
Edib, M S G3YTY
Edinborough, R J G0BAJ
Edinburgh, G G3SDY
Edinburgh, P D G1DEN
Edirisinghe, V R I 2E1FYA
Edis, C 2E1DLZ
Edis, M J E G4RPT
Edis, N L G0VQC
Edis, P M G4RLI
Edis, R G4RLJ
Edisbury, D R G0OKR
Ediss, R C G6XYF
Edlin, G E G7EIK
Edmett, K A G1ITJ
Edmonds, A J G0HIF
Edmonds, C W G4CFR
Edmonds, D J G0NAX
Edmonds, D L G8EWN
Edmonds, G B G6HIG
Edmonds, G R G8VNO
Edmonds, K A R G3XCB
Edmonds, P J G4OFN
Edmonds, R G0KVB
Edmonds, R J G4HHX
Edmonds, S E G7ORV
Edmonds, W K GW0BMI
Edmondson, B G3HCZ
Edmondson, D G7KDH
Edmondson, D E G4IPP
Edmondson, D J G0RGL
Edmondson, F G8AHW
Edmondson, L G7UYM
Edmondson, P G7KDG
Edmondson, R H G3YEC
Edmondson, R J G6BPN
Edmondson, S G G4CIC
Edmunds, A C G1DQS
Edmunds, C B G7AIK
Edmunds, C L G6OHY
Edmunds, D G3MJW
Edmunds, G GW0KTL
Edmunds, J M1AEU
Edmunds, K A G8NZC
Edmunds, M V G0ISO
Edmunds, M W G0JKI
Edmunds, M W M1BUS
Edmunds, N G G8HQC
Edmunds, P A M1BMX
Edson, J G4SAA
Edson, S K G6NZG
Edward, B B G6ILX
Edwardes, J C G0WHN
Edwards, A G0NWS
Edwards, A G0SUA
Edwards, A GW0HIR
Edwards, A C G1NTN
Edwards, A C G3KGN
Edwards, A D G4ZON
Edwards, A E G1ZNK
Edwards, A F GM0URZ
Edwards, A G G3HNP
Edwards, A G G3MBL
Edwards, A G G7KXU
Edwards, A J G6DXD
Edwards, A J G7HHU
Edwards, A J G7WFL
Edwards, A J GI8UCS
Edwards, A J M0BAO
Edwards, A M G4GBI
Edwards, A N G0HWU
Edwards, A R G1FJM
Edwards, A R G3DAC
Edwards, A S G0HDG
Edwards, A S G1IDF
Edwards, A S G3OMX
Edwards, A S G7JLC
Edwards, A W G7GGJ
Edwards, B H G4SPQ
Edwards, B J G1HSC
Edwards, B L G3WCE
Edwards, B P G6HIE
Edwards, B R G3RJB
Edwards, B W G4JFR
Edwards, C G7MJP
Edwards, C G8WVZ
Edwards, C D G G4GLG
Edwards, C J G6CTY
Edwards, C J G6YXY
Edwards, C J G7UDJ
Edwards, C T GW3KUY
Edwards, C W GM7UPD
Edwards, D GW0CVY
Edwards, D GW1YZF
Edwards, D M1ARL
Edwards, D A G7RAU
Edwards, D C GW4XMI
Edwards, D F G0OWP
Edwards, D G G8NEO
Edwards, D J 2E1EPM
Edwards, D J G4HXC
Edwards, D J G4ZWR
Edwards, D J G6SJX
Edwards, D N GW0TMU
Edwards, D R G4KMJ
Edwards, D R G7ERC
Edwards, D R G8BFV
Edwards, E J G3DHY
Edwards, E J G7UAC
Edwards, E R GW3RUE
Edwards, E R GW8LJJ
Edwards, F A G0FIE
Edwards, F B GI3BUP
Edwards, G G0NZH
Edwards, G G2FLY
Edwards, G A G6IPU
Edwards, G C H GW8JZV
Edwards, G D G7WFT
Edwards, G D GW2ABJ
Edwards, G F G6BFE
Edwards, G J 2W1FWB
Edwards, G J G0XDL
Edwards, G S G1ENA
Edwards, G W GW0LTC
Edwards, G W M1AAE
Edwards, H G7NKW
Edwards, H A G3AGW
Edwards, I O GM0GUM
Edwards, I R T G4WRK
Edwards, J G0MJZ
Edwards, J G4NFE
Edwards, J G4XAM
Edwards, J G7PEB
Edwards, J GW0ONY
Edwards, J A G4BVA
Edwards, J A G7SSD
Edwards, J A G8WXR
Edwards, J A W G3ERR
Edwards, J F GM8HBB
Edwards, J G G0WGH
Edwards, J H GW3TCV
Edwards, J H M0AON
Edwards, J L G7IKD
Edwards, J M G6EWV
Edwards, J S G0JSE
Edwards, J W G1LCC
Edwards, K G0IVI
Edwards, K G1ZOE
Edwards, K GW1EWW
Edwards, K A G4RYQ
Edwards, K A V G6BWE
Edwards, K J G4WPZ
Edwards, K R G6GSF
Edwards, K R G8LPN
Edwards, K V G3XUO
Edwards, K W GW4LWL
Edwards, L 2E1EET
Edwards, L G8RBQ
Edwards, L M G6KHM
Edwards, L M GW0GLI
Edwards, M G1HLQ
Edwards, M G4OJM
Edwards, M G6KJO
Edwards, M GW4LHL
Edwards, M H G0MQQ
Edwards, M I G8CPF
Edwards, M J G4BZM
Edwards, M J GW4SJO
Edwards, M R G4GDY
Edwards, M S G4KHY
Edwards, M W G3WYT
Edwards, N A F G5SG
Edwards, N D 2E1DQM
Edwards, N I G1WFF
Edwards, N L G4FCB
Edwards, N M G3XZB
Edwards, O H GW0ALF
Edwards, P 2W1FJZ
Edwards, P G0NFI
Edwards, P G4WFW
Edwards, P G7MLQ
Edwards, P A G4IKJ
Edwards, P A G7LJP
Edwards, P D G7VUX
Edwards, P F G7TWJ
Edwards, P F G7VLP
Edwards, P I GW6LMF
Edwards, P J G1MNA
Edwards, P J GW4ESL
Edwards, P J GW8ARR
Edwards, P M G1KTU
Edwards, P S G7FPO
Edwards, R A G0NQK
Edwards, R J 2E1EAX
Edwards, R J G0FIU
Edwards, R J G4DBU
Edwards, R J G4FGQ
Edwards, R L G6OHR
Edwards, R M GW0GWE
Edwards, R O G4TSD
Edwards, R S MW0AFD
Edwards, S C G6EDJ
Edwards, S F GW0CNJ
Edwards, S J G0DMY
Edwards, S J G8GEF
Edwards, S J R G4GPY
Edwards, S M GW4TVE
Edwards, T GW0PZS
Edwards, T K G7JEJ
Edwards, T W G0PWW
Edwards, V G4XPA
Edwards, V M1BQW
Edwards, V A G8NGZ
Edwards, W G G6ZUE
Edwards, W H G4YXC
Edwards, W M GW1SBO
Edwards-Hanham, W G G3FQC
Edworthy, R M G3URU
Edy, G D G4AXD
Eeles, R A G0SWC
Eeles, R I G1WWB
Eesemann, H L H G4BYJ
Egan, D G0KZM
Egan, D K G0PPS
Egan, D K G8PRU
Egan, D K GW4XKE
Egan, G G0NWE
Egan, P T G1WTX
Egan, R G4XOM
Egar, G G0WIR
Egerton, M J G6UAW
Egerton, R C G1MMO
Egerton, S E G4BWX
Egerton, W J G8RBR
Eggett, D G R G7RAJ
Eggett, P A G8DMQ
Egglestone, B A 2E1EYS
Egglestone, D J 2E1FBV
Egglestone, J G7WBC
Eggleton, P G3XMQ
Eggs, E J G6CTV
Egleton, D M G7FFB
Eglin, T G4TMU
Eglington, P A GW6JUL
Eisenberg, A G7GPT
Eite, F J O G8XTQ
Eke, R G0UQS
Elcoate, A G0RTH
Elcock, T D G1KKD
Elcocks, B G G3RJX
Elcombe, P D 2E1FQB
Elden, R A G8VNP
Elden, S M M0AKK
Elder, F R GI4AHD
Elder, H GM4HRW
Elder, M E GI8RPP
Eldrett, R P G0HZQ
Eldridge, G R GI0POB
Eldridge, K H G6SHS
Eldridge, R A G3RAE
Eldridge, W T MW1BIV
Element, M G0EBD
Elesmore, G M1AJV
Eley, A T G3GHB
Eley, C A E G3LRA
Eley, J G4VCY
Eley, J K G3LMR
Eley, K G7HST
Eley, P A G6VCY
Eley, R S G7FDG
Eley, T A W G8TEX
Elford, A T G0JZW
Elford, J G6YIJ
Elford, R T G0XAY
Elgar, P A J G1PPW
Elgin, J K GI7SOB
Elgy, R G8EZT
Elie, C F G4ELE
Elkington, D F G0PAN
Elkins, J G G6SJY
Elkins, P J G8MCW
Ellam, E B G3MUN
Ellams, L M G0TJR
Ellard, D W J G1ZJK
Ellefsen, A O G3FJO
Ellenger, F W G2ZU
Elleray, H J G1KCW
Ellerby, J GM3LHV
Ellerby, M F G1HSH
Ellerton, J A G3NCN
Ellery, B W G0COJ
Ellery, C J G4EAS
Ellery, E I T G3YIN
Ellery, F G3VCE
Ellery, G S G3BIT
Ellesmere, F G8CJ
Ellett, M C GW6XYE
Ellin, S J G6SWW
Ellingworth, D F G6ZDE
Ellinor, T R G0ARG
Ellinor, T R G4DFA
Ellinor, T R G8OTS
Ellins, D P G0GGL
Elliot, I E G3HMB
Elliot, J A G3KIQ
Elliot, N D M GM0ELL
Elliot, P J G3MFO
Elliot, R D GM3ZRA
Elliot, R J G7CXV
Elliot, W G4BIU
Elliott, A C G3GBI
Elliott, A C M0AOJ
Elliott, A J G3ZOG
Elliott, A P G6GEK
Elliott, B G1SLI
Elliott, B GM1RRJ
Elliott, B P GM4JTA
Elliott, B R G6AMF
Elliott, B T G3FJE
Elliott, B T G4MEO
Elliott, C G1NZL
Elliott, C G4UJW
Elliott, C G G4MBS
Elliott, C J G8TYN
Elliott, D G1YAS
Elliott, D G4ZOY
Elliott, D G8GMA
Elliott, D G G7ULF
Elliott, D L G4DUT
Elliott, E 2E1BVS
Elliott, E G1NZP
Elliott, E A G3YGC
Elliott, E W G3BYY
Elliott, F R A G4PDZ
Elliott, G G0KQO
Elliott, G G3FMO
Elliott, G G7NSU
Elliott, G A GI4OWA
Elliott, G D GM3BXD
Elliott, G T GW8MOZ
Elliott, H G4GSO
Elliott, I A 2E1DOE
Elliott, I A M1AIE
Elliott, J G8CGW
Elliott, J C 2E1BVV
Elliott, J G G0LOS
Elliott, J M G0KYH
Elliott, J R G4XYE
Elliott, K L 2E1BGY
Elliott, K M GM4NTX
Elliott, L G7NZU
Elliott, L W G4OGB
Elliott, M G1DQU
Elliott, M G8SAR
Elliott, M A G6EDE
Elliott, M J 2E1BUX
Elliott, M J G4SRC
Elliott, M J G4VEC

Elliott, M J G6SRC
Elliott, N R G1XAX
Elliott, P C G0TXL
Elliott, P R G4MQS
Elliott, R GW8VFQ
Elliott, R A G4ERX
Elliott, R D G7UFT
Elliott, R L C G4NJK
Elliott, R M G0RVJ
Elliott, R P G7PTW
Elliott, R W G7JLK
Elliott, S E 2E1BEO
Elliott, S J 2E1FJB
Elliott, S J G0WEX
Elliott, S J G7VLU
Elliott, S P G1IKT
Elliott, T G 2E1DMS
Elliott, T W G0EHX
Elliott, V G0IGB
Elliott, W GI4OYM
Elliott, W V G4HGZ
Elliot-West, P J G0KFY
Ellis, A 2W1FNS
Ellis, A G3PJR
Ellis, A GW2HFR
Ellis, A K G0NXS
Ellis, A R 2E1DFD
Ellis, A R G7OGK
Ellis, A S G7HEP
Ellis, A W G3WAW
Ellis, B E G3VXF
Ellis, B G 2E1DTR
Ellis, B G G3NSU
Ellis, B J G8NVC
Ellis, C J M1BBK
Ellis, C R G4NVX
Ellis, D MW1BTV
Ellis, D J G4FBB
Ellis, D J G6LXL
Ellis, D R M1BXO
Ellis, D T G6GUC
Ellis, D V N G4RAB
Ellis, E S GD3LSF
Ellis, G C GW6OHX
Ellis, G G E G4TEG
Ellis, G J 2E1CFT
Ellis, G J G7RWO
Ellis, G J G7VMS
Ellis, H E A G7FKS
Ellis, H T G3XCI
Ellis, H V G7BUN
Ellis, J G0ELO
Ellis, J G3HRD
Ellis, J G3HWZ
Ellis, J G4ABE
Ellis, J A G0MVU
Ellis, J H G2FNK
Ellis, J R G7PFY
Ellis, J V G1VSQ
Ellis, K G1ZLQ
Ellis, K G8HGM
Ellis, K B G0DPJ
Ellis, K E G5KW
Ellis, M G1ORA
Ellis, M G4TTO
Ellis, M I G4GDL
Ellis, M J G6RIC
Ellis, M M G0OAR
Ellis, M M G4ROM
Ellis, M S G2CAZ
Ellis, M W G4UDE
Ellis, M W G7UKF
Ellis, M W M1BXJ
Ellis, M W T G4VXB
Ellis, N G1OIS
Ellis, N J G8DHL
Ellis, N P D G6MKJ
Ellis, P G0ANL
Ellis, P G4AEM
Ellis, P G4FVJ
Ellis, P A G3YAS
Ellis, P G G1ILF
Ellis, P I G0VMQ
Ellis, P R G0UVG
Ellis, P R G8REF
Ellis, P S G0SSX
Ellis, R G0MBJ
Ellis, R A G0AGB
Ellis, R D G6MKL
Ellis, R J G3FSX
Ellis, R P G3SN
Ellis, R S G1WKV
Ellis, S G1YMP
Ellis, S G3JNY
Ellis, S A GW8WTJ
Ellis, S D G4MPP
Ellis, T G1WRQ
Ellis, T G7PST
Ellis, T G8SVT
Ellis, T P G8HIO
Ellis, T R G7AJG
Ellis, W B G0UUX
Ellis, W G G0IOV
Ellis, W J G4IFK
Ellis, W P B G4DQZ
Ellis, W R GW0MMY
Ellison, B G1JBW
Ellison, B F G0SIW
Ellison, B K G8HHP
Ellison, C D G8RBW
Ellison, C H G3DTJ
Ellison, C R G4WYF
Ellison, D E G6BXS
Ellison, D H G0JTA
Ellison, G J R G3LZN
Ellison, G T H G8OOF
Ellison, J G0KPA
Ellison, J C H G2PK
Ellison, J D G1XXF
Ellison, K G0NRE
Ellison, L C GD0PNK
Ellison, M G8RRS
Ellison, P G0WIG
Ellison, S G7APS
Ellison, T S G4UFM
Elliston, E E G6ILT
Elliston, M W GU7CNI
Ellis-Brown, E G6DJO
Ellner, J T G6OHT
Ellory, F R G3CUI
Ellsmore, J W G1BEG
Ellson, D A G0GDG
Ellwood, D M1BST
Ellwood, J GW4JLK
Ellwood, J F T GW0RLQ
Ellwood, P G0RHF
Elmer, P M G4LSQ
Elmer, R J G6PHQ
Elmes, A J G7VNG
Elmore, S P GW4MBL
Elms, P R G0IJU
Elms, R J C G8FAR
Elms, R L GW0TVX
Elphick, C D G7HFL
Elphick, R H G4IPT
Elsdon, J G3JRM
Elsdon, J G4RLS
Elsdon, S T G4TUH
Else, A G4RPD
Else, J J G6KJF
Else, M T G7TGL
Else, N G4JGP
Else, R G O G1BFK
Elsey, D J 2E1EVV
Elsey, M G4YME
Elsey, P J 2E1ECV
Elsey, R D G4NCO
Elsley, E G3YUQ
Elsley, M L G6IEE
Elsom, C J G1POC
Elsom, P E G1OPD
Elston, I C G3YRX
Elston, W C 2E0AAZ
Elsworth, D C G0ADJ
Elsworth, I C G1NQS
Elsworth, K G G6XYD
Elsworth, W J G0ADI
Elsworthy, H W G3GMN
Elton, W J GW3RIH
Elvers, M T M1BGS
Elvin, D G G0INO
Elvin, K J G1ANS
Elvins, I G3WUG
Elvins, F A G3VTC
Elvy, M T G3BD
Elwell, D R G4MUS
Elwell, P G8PIP
Elwell-Sutton, S A G3ZTP
Elwood, J C G1XXA
Elwood, P E G0POC
Elworthy, S E GW1ZNC
Ely, B W G3TGB
Ely, S J 2E1FGO
Emanuel, S GW0VFF
Emary, B H GW1DQV
Emblem English, T P G7ITX
Emblen, C F G0FQZ
Emblen, K A G0VCJ
Embleton, A G4MBH
Embleton, A G G3BNF
Embling, A R G7TLG
Embling, M G G0BSA
Embrey, A F G3KNG
Emeney, T G3RIM
Emeny, R J G4JAC
Emerson, D T G3SYS
Emerson, E G0DLX
Emerson, H G GI8RLG
Emerson, J G0BAN
Emerson, R G6NEZ
Emery, C H GW4UGP
Emery, D L G4FEB
Emery, F D S G4ASI
Emery, J H G1HSE
Emery, R E G0WDT
Emery, R W G3FYX
Emery, S G4HHP
Emery, T W F G4ZRF
Emery, W G3BYN
Emes, R D G3EPV
Emes, R D G3YDD
Emlyn-Jones, S GW4BKG
Emm, M J G0TQZ
Emmans, P R A G8XTR
Emmanuel, C W G0DXF
Emmerson, A G1RHO
Emmerson, A N G8PTH
Emmerson, G G0UQD
Emmerson, G G4AAX
Emmerson, G G8PNN
Emmerson, M H G3OQD
Emmerson, R S G3NOG
Emmerton, P G G4IOV
Emmett, D A G3TMR
Emmett, D J G8AED
Emmett, L G3VKO
Emmett, R G G0FZX
Emmett, S J G0TZT
Emmett, S R G7RGG
Emmott, J W G3ANG
Emms, S W A G0RDU
Emms, V H G3IUS
Emons, E G1MYM
Empringham, A G1MFW
Empringham, P G6GZS
Emson, N J G4UXO
Endacott, A J G3TLK
Endall, J R G2HOS
Endean, C M G G7VQK
Endean, D S G0UBG
Endean, M M G7VQL
Enderby, D GM0FMW
Endersby, C R G7HTB
Endersby, E R G4DTA
Endersby, H F G0LLO
Endersby, J H GW4VIB
Endersby, R E G4SPI
Endersby, R L G1AZJ
Endicott, J E G1WZG
Endicott, J Q G6FBJ
Endicott, J W G3UWH
Engel, G F M G4MVF
England, A GW0OEJ
England, C R G7SDR
England, G H G1PJX
England, J E 2E1BSV
England, J E G4XYY
England, P G4GRK
England, P C G1MYQ
England, P S J G7UPF
England, R A G7EFC
England, R H G4REH
England, T M G8GJV
Englehard, E P G0DNB
English, A G1OAP
English, B J G6DPA
English, C L G6ZOE
English, D G7LFB
English, D J G6LXP
English, J H G2DZF
English, J J G7TYX
English, K S G4FYD
English, O C G4HST
English, R J GM3YKE
English, R S G4NVV
English, V J G1OAU
Ennis, J A G3XWA
Ennis, J A G4ARS
Ennis, J A G8FLE
Ennis, M J G7VBD
Enoch, D G G3KLZ
Enright, A K G4KSP
Enright, R J G0RJE
Enright, R J G8SAS
Enticknap, C J G6XSF
Enticknap, D R J G6YIG
Entwisle, G V G3MXT
Entwistle, E M0AQE
Entwistle, G E G0HGG
Entwistle, M D G6VHE
Entwistle, N J G0BRM
Entwistle, P L G8AFC
Entwistle, Y D G1SXY
Epps, H B G M0ADN
Epton, D J G4EKB
Epton, D J G4MTL
Epton, D J G8MTL
Epton, M S G8VJS
Erber, M E G7HJQ
Ereaut, G M G6HBN
Erents, S K G8AGY
Ernest, A M GW3LQE
Ernster, P A G4NYY
Erratt, C D G0MXY
Erridge, H C G6CUA
Errington, J T G8FWA
Errington, M G4YGF
Errington, S C G0UUF
Errock, G A G3HCO
Erskine, C A GW1USX
Erskine, F W G4LPK
Erskine, N C GW0GDI
Erskine, W D GM1BTL
Ersser, E T G1KRK
Erwood, A F G7AFE
Escreet, B A G4SPC
Escreet, J M1BLH
Escreet, N M1BBD
Escreet, P A G1FGI
Esdale, D J G0RMX
Eskelson, A G0POY
Esler, G W GI4VBZ
Esler, S D G0VHE
Esplin, D J G3VTG
Esposito, S J G4MGG
Esser, M E G6MZN
Essex, A H G8WCX
Essex, J E G8JCL
Essex, M 2E1FYH
Esslemont, G B GM3FRZ
Etchells, J G1LWE
Etchells, P R G6DXH
Etchells, W F J G0GJQ
Ethell, H G G0DBG
Etheridge, A G0HXF
Etheridge, G J G8EIO
Etheridge, J G G4DKP
Etheridge, P G4ERG
Etheridge, W H G4HTS
Etherington, B G1ISP
Etherington, W R G0JZU
Etherton-Scott, W F G1HLS
Eunson, P W GM8YEC
Eustace, G T G2FXD
Evans, A 2E0AAS
Evans, A 2E1FLQ
Evans, A G3WSJ
Evans, A G4BPE
Evans, A GW0CWG
Evans, A C GW4YBE
Evans, A D G1PKZ
Evans, A E G7FKG
Evans, A E GW3VEP
Evans, A E GW3XXB
Evans, A G G1GJR
Evans, A J G0PBP
Evans, A J G3PXI
Evans, A J G4RVN
Evans, A J G6RIB
Evans, A J GW1TJK
Evans, A L 2W1AKZ
Evans, A L G6WLZ
Evans, A R GW4ARC
Evans, A R GW4HDR
Evans, A S M1ASV
Evans, A V G3ZZX
Evans, B G0GOO
Evans, B G7IJY
Evans, B J G6YXT
Evans, B J GW8YLK
Evans, B L G1ILG
Evans, C G3LZV
Evans, C G4NHE
Evans, C D G4EYA
Evans, C D GW0IRP
Evans, C D M1BUU
Evans, C H G3LUO
Evans, C J G6CKE
Evans, C J GM4FVO
Evans, C R GW7RCR
Evans, C S G3XKE
Evans, C T E G6IZZ
Evans, D G3ZWL
Evans, D G6EDM
Evans, D G7NYW
Evans, D A G7RAB
Evans, D A J G0EDL
Evans, D B GW1XFB
Evans, D C G7CJS
Evans, D E G4HFR
Evans, D E GW4GTE
Evans, D E J G6EDF
Evans, D J 2E1EQE
Evans, D J G1GBV
Evans, D J G1JRU
Evans, D J G4EQR
Evans, D J G4YND
Evans, D J G7WLC
Evans, D K GW1ANW
Evans, D M G7UCZ
Evans, D O G0EVA
Evans, D P T GW3IVK
Evans, D R G4AMJ
Evans, D R G4YNE
Evans, D R GW4JCD
Evans, D S G3YNK
Evans, D V 2W1CVH
Evans, D W GW0BVN
Evans, D W GW0SZW
Evans, E GW1TDX
Evans, E D G3JAH
Evans, E E GM6JAG
Evans, E I G1PDA
Evans, E J GW0TCN
Evans, E O GW1OSQ
Evans, F W GW8AWM
Evans, G G0HOP
Evans, G G4IQK
Evans, G G4JDE
Evans, G G6AUK
Evans, G G8WXU
Evans, G GM4CAI
Evans, G GW0HPC
Evans, G GW1MLE
Evans, G A G4SDW
Evans, G B GW0UJJ
Evans, G B GW1JVB
Evans, G C G8NVZ
Evans, G E G3TAV
Evans, G E R G0HDC
Evans, G J G1WKZ
Evans, G J GW8XAS
Evans, G K G6MKQ
Evans, G M G1SCR
Evans, G M G1YJB
Evans, G P G7NOI
Evans, G R G8FTI
Evans, G T G3ZZV
Evans, G W G1CIY
Evans, G W GW3WWN
Evans, H G7RRZ
Evans, H I GI0AZB
Evans, I A 2W1ALD
Evans, I D G7IDE
Evans, I G 2E1DQD
Evans, I L A M G1OBC
Evans, I P G6JAF
Evans, J 2E1FYK
Evans, J GW3BAZ
Evans, J GW5BI
Evans, J GW6LMI
Evans, J A GW8GKS
Evans, J D G0AJE
Evans, J E G6NNO
Evans, J E G7ACC
Evans, J F F G8TWR
Evans, J G G3WET
Evans, J H G6FBC
Evans, J H G8AGJ
Evans, J H G4NXB
Evans, J L G7CEC
Evans, J L GM3VJY
Evans, J M G0BME
Evans, J M GW0IAU
Evans, J R G0OGY
Evans, J S G1HCR
Evans, J T G3VDB
Evans, J W GW8ITI
Evans, K G0XAA
Evans, K G3VKW
Evans, K G M0AQQ
Evans, K W G8IOS
Evans, L G7IGU
Evans, L J S G7MML
Evans, L N R M0BAV
Evans, L P G7CHN
Evans, L W G8PRB
Evans, M G1HSF
Evans, M G6HWM
Evans, M GW4TPG
Evans, M A 2E1CKA
Evans, M A GI0AYG
Evans, M C G7ULU
Evans, M C E G4VLN
Evans, M D G7EBI
Evans, M D G7NNT
Evans, M J G4CFT
Evans, M J G7OLI
Evans, M J G8HVT
Evans, M J P GW3UCJ
Evans, M S G0TZL
Evans, M W G0KFV
Evans, M W G4MMH
Evans, N 2E1ESJ
Evans, N 2W1DRL
Evans, N G0AOE
Evans, N G4IIN
Evans, N G7RYG
Evans, N D G1HSG
Evans, N E GI4BDR
Evans, N J G6EWX
Evans, P G0VPN
Evans, P G3DLH
Evans, P G4RUJ
Evans, P GW0WRW
Evans, P B G0RBJ
Evans, P B GW0MMB
Evans, P G G8JHO
Evans, P H G4BKI
Evans, P J G8PJE
Evans, P J G8WZO
Evans, P M 2E1FOL
Evans, P R G4RAV
Evans, P R G7REH
Evans, P V G8SYS
Evans, P W G1TAN
Evans, R G0IGS
Evans, R G0NDB
Evans, R G7VWG
Evans, R G8FOZ
Evans, R GM0CDV
Evans, R GW0HXB
Evans, R GW0WLQ
Evans, R A G0HLB
Evans, R A G3VHE
Evans, R C G3LQC
Evans, R C G0RPX
Evans, R D G3PRO
Evans, R D G8HPR
Evans, R E G0DHB
Evans, R E S G7CRS
Evans, R E S G8KHV
Evans, R G G4XAT
Evans, R G G8XTO
Evans, R J G7ODB
Evans, R J GW3NNB
Evans, R M GD0PLT
Evans, R M GM8PEV
Evans, R N G4AGE
Evans, R R G4RSB
Evans, R R GW6MKI
Evans, R W G0VCW
Evans, R W G4GEZ
Evans, R W GM0AYW
Evans, R W GW4CXK
Evans, R W GW6PMC
Evans, S A G4XEL
Evans, S D GW7CVF
Evans, S J GW7GWO
Evans, S J H G1EKM
Evans, S K G1WLD
Evans, S P G6AHX
Evans, S R G7IFE
Evans, S T G8KQV
Evans, S T S G3VGO
Evans, S W G0EVJ
Evans, T G1CNY
Evans, T B G0WIF
Evans, T L G7VZU
Evans, V S G4AVT
Evans, W G6ILY
Evans, W GW3AGB
Evans, W D GW3CDP
Evans, W E G4DIR
Evans, W G K GW4PNZ
Evans, W I GW0OBB
Evans, W J 2W1EYP
Evans, W J GW4LKS
Evans, W R G4EQM
Evans, W T GW4PWZ
Evans, W W G8RTA
Eve, C R GJ7AOG
Eveleigh, D G1VYM
Eveleigh, G P G8GIJ
Evely, N K G0CEL
Evenden, S J G1OJL
Everett, R A G3AGZ
Evennett, T A G0LGF
Everall, M R G6FTA
Everall, T G4VMH
Everard, A G0ARZ
Everard, A J G4IEC
Everard, B G3ODL
Everard, D G7CEW
Everard, J C G0SZO
Everard, K L G0NKZ
Everard, P G4UNW
Everard, P G7GFX
Everest, G C G4TZX
Everest, G R J G3XUP
Everest, H J G3NND
Everett, A C G0VRB
Everett, A K G1LJJ
Everett, C G0GXP
Everett, E F G7SEU
Everett, J G4CDP
Everett, J L G0HJK
Everett, P C P G3SFE
Everett, R C G1DWC
Everett, S P G6MKO
Everett, S R G7GRR
Everett, W C G0ODK
Everingham, J N G4TRN
Everington, K G6YII
Everist, J W G4CVC
Everitt, A G0LMO
Everitt, A G8GUU
Everitt, A J G7LTH
Everitt, B N G6OQV
Everitt, C E H G7OQS
Everitt, H E G1AZM
Everitt, J C 2E0ACP
Everitt, J C 2E1ASN
Everitt, J C G0VCE
Everitt, J L G7OMG
Everitt, L J G0TIB
Everitt, L J G7MOA
Everitt, M C G6KRN
Everitt, M F G8OYQ
Everitt, R G G1JZT
Everitt, R H O G4ZFE
Everitt, S G0WXA
Everitt, T C 2E1FTY
Everley, C M G4PPK
Everley, G A D G4UQH
Everley, M G6GUD
Evers, R GM0BVK
Eversden, P W G8EAI
Everson, J J MW0AUH
Everson, R A G6DIO
Everton, L E G4NBI
Everton, W R G7GJT
Eves, A G0JJM
Eves, C G0RXV
Eves, D G G0RNP
Eves, G W G4CWT
Eves, T R E G6DIM
Evetts, D A N G3RKP
Evill, B R G3SSC
Evill, J M G6HJV
Evison, R D G7JFU
Evison, R V G4SEE
Ewald, B L G0NXL
Ewan, J GW0UEO
Ewart, B G G8ZML
Ewart, C I MM0ABJ
Ewart, M S 2E1GAE
Ewart, S S 2E1GAF
Ewen, J A G3HGM
Ewen, J A G6RGA
Ewen-Smith, B M G3URZ
Ewen-Smith, J D G4JKA
Ewer, N G3ZUS
Ewing, C R G7DBR
Ewing, F J GM4LHM
Ewing, H D F G7GRB
Ewing, J 2E1FKJ
Ewing, P A G0WEZ
Ewing, P E GM1FNX
Ewington, P E G8FFZ
Excell, A A G4DII
Exell, M E 2E1BCJ
Exell, S G G1VNU
Exeter, F M J G3BNV
Exley, H H G4FOT
Exley, R A G4IUE
Exton, M J G7NUM
Eyers, M J G0IHI
Eyers, S L G1VNM
Eyers, T S G0UJY
Eyes, J D G7OMN
Eyes, R B G8RON
Eyles, A R M0AYV
Eyles, C J G3SJH
Eynon, C GW6HJO
Eyre, A R 2E1EIV
Eyre, B E G3TPP
Eyre, C I G1DDI
Eyre, D B G0UQZ
Eyre, D J G0TFD
Eyre, H H G5KM
Eyre, H V 2E1BMP
Eyre, J M G7TZZ
Eyre, K G4WRN
Eyre, L E G7UIR
Eyre, P G G8ZIY
Eyre, S G2FZU
Eyre, S C 2E1BNA
Eyre, T R G7ONM
Eyre, W A G7HNW
Eyres, A F M0AHM
Ezra, R J G3KOJ

F

Fabb, M W R G3ZCS
Facer, D A G1OJD
Facer, D H G3WBZ
Fadil, M J L G4CCA
Fagan, E J MI1CEF
Fagan, W F G7IUI
Fagg, J E G6RCJ
Fagg, M N G3SRC
Fagg, M N G4DDY
Fagg, P M G4CCY
Fagg, R C G3FVV
Fahey, S S 2E1FBG
Fahy, D J M1AGC
Faichney, K D G4ZJE
Fails, V M GI4WWF
Fairall, C R M G6ZTV
Fairbairn, I GM4VJV
Fairbairn, I GM4XZZ
Fairbank, D J G0AII
Fairbotham, K N G0UDP
Fairbrass, G E G4GAC
Fairchild, C T G3YY
Fairchild, D T G0DDF
Fairchild, M D 2E1DRN
Fairchild, R G 2E1DMY
Fairclough, F P 2E1CFH
Fairclough, F W G3OEI
Fairey, A G G0OIN
Fairey, M G8JVS
Fairfull, S GM7PVV
Fairgrieve, O J GM0PKW
Fairhead, R S G3UNB
Fairholm, R GM4VBE
Fairhurst, A P G6ZIY
Fairhurst, D F G4GTS
Fairhurst, G K G4WGF
Fairhurst, J T G6DBJ
Fairhurst, L G8YEG
Fairhurst, P G6XFQ
Fairhurst, P A G0KLU
Fairhurst, V G6EWZ
Fairman, F J G0NLV
Fairmington, P G3PLJ
Fairweather, A S G7GAR
Fairweather, S J G6BEL
Faithfull, B A G3KOS
Faithfull, J H G0DCU
Faithfull, M J G6UBH
Faithfull, R M0ALF
Faiz, N A G7VLF
Falamerzi, A G0VQV
Falconer, E A G7HRU
Falconer, K G1BWJ
Falconer, K J G3YOI
Falconer, S D G7GUO
Falding, G B G3LGF
Falkner, D R G8LKQ
Fallaize, C 2U1FOA
Falle, P J GJ8PCY
Fallick, T M G4FYI
Fallon, J B G0PTB
Fallon, J K G0OBU
Fallon, J P G3SGV
Fallon, M P G0UOH
Fallon, P J G0CDK
Fallon, P L G8RIB
Fallowfield, R A G6CUI
Fallows, A G G4UCL
Fallows, G J G1BMB
Fallows, I G0UAA
Fallows, J W GI4OWB
Faloon, K M GM0WPU
Falstein, D D G6BAT
Falstein, H T G3KUF
Fambely, P G0BHP
Fambely, P G0TRC
Fanner, J E C G0ITR
Fanning, P I E G1RDU
Fanning, T J 2E1EPQ
Fantham, A J G3TGL
Fantom, D GM4UGG
Faoury, M A M0AIO
Faraday, D 2E1FFG
Faraday, D M GW7TLU
Faraday, J M 2W1EBD
Faraday, P A 2W1DHM
Faraday, S L 2W1EBC
Faragher, L J G7AYI
Faram, A J G1XIO
Fardell, P J G0LQU
Fardoe, G R G1NXB
Fare, G N G3OGQ
Farey, B A G8GHR
Farey, R E G6LLP
Farish, J E G4WJF
Farleigh, B G4RJ
Farler, D S G3WJA
Farley, C G4WVF
Farley, C G6GEX
Farley, H C G4SNI
Farley, M C G3DFD
Farley, P G3TDF
Farley, P G4YPU
Farley, R B G4LOG
Farley, S F G4NNQ
Farline, G K G6AIB
Farman, D J G3TWZ
Farman, P A G7SCE
Farman, R A G7EFD
Farmar, A N G1GZI
Farmer, A H G G4ZQI
Farmer, A J P G0PDP
Farmer, A M G4MDR
Farmer, C B B F G1LDP
Farmer, F T G6JY
Farmer, G A G8XQZ
Farmer, G R MW0AML
Farmer, K W G4VJT
Farmer, M G7MRF
Farmer, N C B G7CTE
Farmer, R G1KEB
Farmer, R G4AOC
Farmer, S E G7BSP
Farmer, T A G4RBH
Farmer, T A G4UGO
Farmer, T M G6PPA
Farn, D R G4HRY
Farnborough, A P G8WBT
Farndon, G G0IYW
Farnell, P H G6CNL
Farnham, D F G0BPL
Farnie, G H G4FXM
Farnley, R G0KTR
Farnley, R G0SQU
Farnworth, M A G8YLM
Farnworth, M J GM0TOF
Farnworth, R A G1FTT
Farquhar, A M GM0JOV
Farquhar, J G G GM6KJQ
Farr, C E G4XPH
Farr, D S G4WUB
Farr, G G3UTC
Farr, G D G1ILH
Farr, J R G3CJG
Farr, K G G6WFM
Farr, M B G4CAJ
Farr, S N G4STE
Farr, T R G7OUF
Farraday, D J E GW3VWT
Farrall, G A G3MNT
Farrance, G V G3KPT
Farrance, K G0AKF
Farrance, R G3TRH
Farrant, E M P GD4BEG
Farrant, J E G8UDS
Farrant, K D GW8WZC
Farrar, E A G7LOV
Farrar, J R G4KXF
Farrar, K G1HZR
Farrar, P A G8SJA
Farrar, P J G0JMZ
Farrar, W G3ESP
Farrell, A G G8HTO
Farrell, C W G8GS
Farrell, F J G0VQI
Farrell, G W G8PYD
Farrell, J J G0HUR
Farrell, J J GI4JOR
Farrell, L P GM8BOW
Farrell, M D G0PFT
Farrell, M D G8ASG
Farrell, S K G6UFL
Farrell, T G6XSI
Farrelly, B A G4MLE
Farrery, J G3XHZ
Farrey, M W G0SMF
Farrier, R GI0TIU
Farrington, B G7SML
Farrington, J R G0VAV
Farrington, M N G8CNB
Farrington, P G4SFC
Farrow, A G0TAM
Farrow, B S G3FIR
Farrow, I E G4BME
Farrow, J G7SXJ

Name

Farrow, K A G8KZZ
Farrow, P A G1HTN
Farrow, R G1OQY
Farrow, S C G0PCD
Farrugia, M G0WSW
Farthing, N E 2W1DTX
Farthing, S J G0XAR
Fasham, D A G7FBO
Fasham, M D G7ITS
Fauchon, P W C G0JSP
Faulconbridge, J H G0GWN
Faulconbridge, K G0HCF
Faulconbridge, W J G0ITF
Faulkner, A H G0SKG
Faulkner, A H G6GWI
Faulkner, A J G1SCV
Faulkner, A J G8EYP
Faulkner, C J 2E1BIM
Faulkner, D W G4DWF
Faulkner, D W G4ODF
Faulkner, H J 2E1BGG
Faulkner, J G0CYX
Faulkner, K M G6YXV
Faulkner, M J G0IFK
Faulkner, M J G3IZJ
Faulkner, P J G3IMF
Faulkner, P J G8TMJ
Faulkner, P J GI7FGQ
Faulkner, R E MW1BAP
Faulkner, R G G1OXT
Faulkner, R P G0CAL
Faulkner, S A G4HUW
Faulkner, S A GU3MLR
Faulkner, S J G0CQS
Faulkner, V K G4EQZ
Faulkner-Court, J D G0CDO
Faulkner-Court, J D G0SDX
Faulks, R F 2M1DLH
Faunt, D M1AVX
Fautley, P H G0ASG
Fautley, R F G3ASG
Faux, K D G3BSF
Faversham, R M G0ISM
Fawbert, T H G6ZXF
Fawcett, A G0VGN
Fawcett, A N GI0PMO
Fawcett, B G4FEJ
Fawcett, C E G8YIG
Fawcett, D 2E1FFZ
Fawcett, P L G0FBK
Fawcett, S G G7VMO
Fawdon, A L G8GBQ
Fawdon, C G8GBP
Fawke, R J G4YGY
Fawkes, B J G3VQW
Fawkes, C S G4UDG
Fawkes, P D G4MOC
Fawkes, R G0HUV
Faxholm, J M0AXP
Fay, C J G4UDB
Fay, C W GW4SBB
Fay, I D G1HLT
Fay, K E G0AMZ
Fay, P R G8NDV
Fayerman, A L G3BWX
Fayers, D J G3YKC
Fazal, M R G7VVT
Fazey, J N G6XYO
Feakes, N E G0NKY
Fear, P W GW8GRY
Fear, W J G7MJM
Fearn, I V GW6WMB
Fearn, J R G4EHX
Fearnley, A C G1SEF
Fearnley, D C G3XYI
Fearnley, M M1ANX
Fearns, J G7RFH
Fearnside, G I G0BWY
Feary, G E G4MDH
Feast, J K G6SEF
Feast, Z I D G6SEE
Featherstone, C G1GZJ
Featherstone, D G4JFD
Featherstone, D I 2E1DAP
Featherstone, E A G0GPW
Featherstone, J G8DWI
Featherstone, J G G4OVO
Featherstone, P E G3XOP
Featherstone, R D G4MWJ
Featherstone, S A G6IPW
Feaviour, T P G0CFT
Feay, J V G1KFG
Feay, K G1GZK
Feay, R L 2E1BHR
Feay, S E 2E1BRY
Fechcer, S E 2E1BAW
Fedyk, J G6DBL
Feeley, D G6DIZ
Feeley, J F G4MRB
Feetenby, D 2E1FSV
Feetham, N A G7WDM
Fegan, B M1CBP
Fegen, A M C L GM0LYT
Feilen, R D GM7WED
Feist, A J G3PMV
Feldman, S H G3GBN
Felds, R H 2E1EDU
Felgate, J E G4YVK
Felgate, V M1BTC
Fell, A J G7MJC
Fell, A L G7JDW
Fell, E G3LIQ
Fell, J M G0API
Fell, M J G4TRP
Fell, S G7MHO
Fell, T G7DGW
Fell, T G7KCR
Fellingham, P J G7FJC
Fellingham, P J G8OMR
Fellows, A E J G7UGC
Fellows, B G0ONH
Fellows, D P G8FBF
Fellows, E J G8ENA
Fellows, G F G7GHP
Fellows, J A GW8EQI
Fellows, J W E G0RIS
Fellows, L G4HCZ
Fellows, M N G8GKR
Fellows, T J G7KYF
Fells, C R G0LUW
Felstead, H J G7UPI
Felton, G W GW0FEM
Felton, M J G0AGT
Felton, M J G1LSN
Felton, P J G3ZUJ
Felton, P M G1PMF
Felton, R J G0ENB
Felton, W G3XZF
Fenby, R G G3PLS
Fenelon, M G4WJE
Fenn, C V H G0VQE
Fenn, J C G4UJV
Fenn, P M G6FVF
Fenn, S L G8ZTB
Fennah, A J GW1JNI
Fennah, F M GW0EHS
Fennah, H D GW4FLZ
Fennell, A S G0HRS
Fennell, A S G0JIF
Fennell, F G3HGQ
Fennell, F G0FPM
Fennell, W J C G3LYT
Fennelly, D L G7ITW
Fennelow, L N G0DWI
Fennelow, L N G1WRC
Fennelow, L N G4ODH
Fennelow, L N G4PQL
Fennelow, L N G8NED
Fenner, G E G3VMO
Fenner, N K G1NMP
Fensome, E J G6ZJD
Fentham, A M G3TON
Fenton, C A P G0MSA
Fenton, C A P G0NAR
Fenton, H M G8GG
Fenton, J R G0KCM
Fenton, R C G3NQF
Fenton, R R G4WIH
Fenton, W N G3ZJP
Fenton-Coopland, C M G8EUV
Fenton-Coopland, J P G0TQF
Fenwick, B G8BTC
Fenwick, F G0AOJ
Fenwick, J E G3VCK
Fenwick, S G1WKS
Fenwick, S G3AIO
Fenwick, T H G4MXX
Fereday, B A G4TDO
Fereday, M H G8IMH
Fereday, M J G3VOW
Ferentiuk, M G1CIA
Ferguson, A GM7IIL
Ferguson, A F GM3TRI
Ferguson, C I GM4HNK
Ferguson, C T G8TYP
Ferguson, D G4PHM
Ferguson, D GM8JFZ
Ferguson, D E GM3YMX
Ferguson, D M G0TUQ
Ferguson, D M G6FS
Ferguson, E J G6JJT
Ferguson, I B F GM0ILQ
Ferguson, I C GM4YXK
Ferguson, J A GI4GPC
Ferguson, J A GI4RYP
Ferguson, J D 2M1DGU
Ferguson, J W GM4ARJ
Ferguson, M T G0JPU
Ferguson, N P G0BPK
Ferguson, P H G7TUH
Ferguson, R G3XCG
Ferguson, R GD0IOM
Ferguson, R J GI7ILU
Ferguson, R N GD4GNH
Ferguson, R N GD8KQF
Ferguson, R T G4KQO
Ferguson, R W GM0XCW
Ferguson, R W GM3YTS
Ferguson, S GI0OKQ
Ferguson, W GM6VCV
Ferguson, W H GM3KMH
Ferguson, W R GM4AGL
Fergusson, R G4POE
Fergusson, S W GW0DYH
Fergusson, W W G3BLR
Ferigan, D M G3XRE
Ferigan, D M G3ZYV
Fermor, G G8RCZ
Fern, M G6HWR
Fern, T J G4LOH
Fernandes, E N J G4FIH
Fernandez, A G0IAF
Fernandez, A J T M G4XPT
Fernandez, R G G3NMT
Ferne, M A L G6GGZ
Ferneyhough, A W G0EVH
Ferneyhough, S D G8AVH
Fernie, C E G0RDB
Fernie, D J G4EYB
Fernie, S G6CNO
Ferns, D A G7ABQ
Ferns, D B G7ABP
Fernyhough, V J G6JUO
Ferrand, R G7VBR
Ferrari, P J G4IHI
Ferriday, M W G6CGD
Ferrie, T GM4KHI
Ferrier, J A G0ATW
Ferris, A P GM0LKT
Ferris, B H G7HAR
Ferris, D M0BDL
Ferris, E V GM8YXR
Ferris, I L GI8YYM
Ferris, L H G4YJF
Ferris, M E G1JTM
Ferris, P R 2E1FZF
Ferris, P R G0LLE
Ferris, R A R G0VZB
Ferris, R T GI0OUM
Ferris, S GW0RMB
Ferris, W J GD1CRZ
Ferry, M J G7DDI
Ferryman, R C G4BBH
Ferula, O M1BJG
Few, K R C G8JDQ
Fewings, G G7SUM
Fewkes, A F G4WAF
Fewtrell, R G4HQI
Fey, M J 2E1FSY
Fiander, D G4APS
Fidler, G G3TDV
Fidler, K G8EUD
Fidler, R B G4EXZ
Fido, B 2E1CZB
Fidoe, J A G0LCV
Fiedler, C G0GYP
Field, B J G3TTY
Field, C H G0TRG
Field, C H G4ZRU
Field, D A G7NKC
Field, D C G3WNV
Field, D I G3XTT
Field, F J G0CEN
Field, G W R G2GC
Field, H M G4XFD
Field, J C G7JSH
Field, J E N 2E0ACK
Field, J F 2E1DLS
Field, K C B G3BKG
Field, K F G0KFF
Field, K F G0LTR
Field, K J E G3PVT
Field, M J G0VXF
Field, N J G4LQF
Field, R E G0OOU
Field, R J 2E1AFK
Field, R L G3ODY
Field, R T G4JQJ
Field, R W G8LGM
Field, S G G4SYR
Field, S J G0MTS
Field, T J G0MQD
Fielden, D G4LFX
Fielder, A A G0CPT
Fielder, D A G7IGQ
Fielder, F A G6XFR
Fieldhouse, P G G7CQI
Fielding, A C G7GGV
Fielding, A P G4CPM
Fielding, B A G7CMP
Fielding, D M G8KHU
Fielding, E G3WDN
Fielding, E G4IHF
Fielding, G R G1DFP
Fielding, H G G8JX
Fielding, J G4VIL
Fielding, M G4IOJ
Fielding, M E G3JPO
Fielding, P R G4TMA
Fields, D G4SKX
Fields, I G G7IGF
Fieldsend, D L G6HCW
Fieldsend, N T G1BWZ
Fieldsend, S G8HTQ
Fifield, B E G7MFH
Figg, A D G4XIG
Figures, P W GM4CAH
Filby, E G G4AQ
Filby, R J G0HJR
Fildes, G T G4ZYY
Fildes, J V M G0HHG
Fill, D W G3UBB
Fillingham, D G4OVR
Filmer, A T G6EMP
Filmer, K G G3XPO
Final, M R G4TOO
Finbow, A G4MPV
Finbow, N D G7OBX
Finbow, P J G0DEH
Finch, A G3XQM
Finch, B R G6YGW
Finch, C G0LEB
Finch, C W G3AHO
Finch, D 2E1EAZ
Finch, F E G4FEF
Finch, G M G7CDH
Finch, H J G3FML
Finch, J A G3VNU
Finch, J E G4YVB
Finch, J T G6PQI
Finch, K G3XIQ
Finch, M J 2E1EKD
Finch, M J G0VXJ
Finch, P R G7PKU
Finch, P R E G4XVG
Finch, R G3RAF
Finch, R D G0HYZ
Finch, R E G4DDM
Finch, R E G8FC
Finch, R E G8RAF
Finch, R G G3JYS
Finch, R J G0IFL
Finch, R P G4AJO
Finch, R T G4PNE
Finch, S N G0HMG
Finch, T B G0VSA
Finch, T M J G1LWF
Findlay, A S G0FIN
Findlay, I G GM7VYR
Findlay, I P G6BSO
Findlay, R G3PRF
Findlay, R W R G3MFU
Findlay, T GM4DOZ
Findlay, W GM0LEG
Findlay, W GM4ZNC
Findon, G D G3TQF
Finegan, P R G7CHH
Fineman, N R G4JHU
Fineman, R E G4BOL
Fingerhut, G A G0ENW
Finlay, A D GM3NEQ
Finlay, B J G4USK
Finlay, J H G6IMG
Finlay, P M1BND
Finlay, R GI0WVN
Finlay, T R GI6BVQ
Finlay, W J GM7HLI
Finlayson, A M MM1ATX
Finlayson, I G GM0WZO
Finlayson, I G GM1MBT
Finlayson, J K GM4IXH
Finlayson, V G7IBX
Finlay-Maxwell, D G3BGA
Finnegan, C J G7CQT
Finnegan, J A GI4FFL
Finnegan, S E G7EYE
Finnemore, D G G3WJJ
Finneran, T F G0PKN
Finney, I K G7DYA
Finney, M S G6NZO
Finnigan, F GI0JGQ
Finnigan, M J GW4UJF
Finnis, C A G4TIL
Finnis, J L E G4TJT
Finnis, R P J GW0HDY
Finon, A G7DIB
Firby, I A G7VCG
Firby, J C G3UOI
Firew, D G7VXF
Firks, D R G0SYT
Firmager, D G6IQE
Firmin, G Y GM4GQM
Firmin, J L G4GUS
Firmin, P S G0FUU
Firmstone, J W GW6DBP
Firmstone, P J GW0MDQ
Firth, A G0FUY
Firth, A G4ETD
Firth, A G8FIS
Firth, A G G7UKK
Firth, A J GM6WOF
Firth, B G M1BHH
Firth, B H G4KCT
Firth, D 2E1DYG
Firth, D G G7VBU
Firth, D R G3WLT
Firth, E A G6XSK
Firth, G F G3MFJ
Firth, H M GM4ZZH
Firth, I R G3WRS
Firth, I R G3WWF
Firth, J A GM8YRE
Firth, J F G1TMF
Firth, K L R G7WGL
Firth, M G3MMK
Firth, M J G3ZJV
Firth, M L G6EDU
Firth, M R G4JMT
Firth, M S G7PMX
Firth, R 2E1DYF
Firth, R L G6IMH
Firth, S E G8SFI
Firth, T S G1FFH
Fischer, P S G4USL
Fish, A F GW1CLA
Fish, A G G4GPL
Fish, D G0KFE
Fish, E R M0ACP
Fish, F B G4YFD
Fish, G L G3ADJ
Fish, G L G3IHH
Fish, J GM1ZJI
Fish, J G GM3NYG
Fish, J H G4MH
Fish, J M GM0ANG
Fish, J P G4NRP
Fish, K J G1ZVZ
Fish, L N G0LQN
Fish, W L M 2E1BLG
Fish, W M G4FBO
Fishburn, M E G7VOC
Fisher, A G1DZH
Fisher, A G6CUK
Fisher, A G G0MIN
Fisher, A G G4VBH
Fisher, A J G8TAU
Fisher, A W G3WSD
Fisher, B G7RTU
Fisher, B G7URZ
Fisher, B M G8TWT
Fisher, C E G3OQV
Fisher, C J G7EIT
Fisher, D A G4EXU
Fisher, D B G6YQJ
Fisher, D C GM1FNY
Fisher, D I G4IGN
Fisher, D J G0DNF
Fisher, D J G8UZY
Fisher, D L E G8ELH
Fisher, E C G0FFN
Fisher, E J G7UKM
Fisher, F 2E1CSZ
Fisher, F J G3LXJ
Fisher, G G0WTL
Fisher, G G4MJX
Fisher, G G4OZQ
Fisher, G G7VWK
Fisher, G J G6UTK
Fisher, G P G1NXX
Fisher, G S G0DSU
Fisher, G W G8PTK
Fisher, H N GM0EFQ
Fisher, I R G8NFB
Fisher, J G4PGK
Fisher, J G4XQZ
Fisher, J G8MBJ
Fisher, J N GM0NAI
Fisher, J N MM0APF
Fisher, J P G0IVZ
Fisher, J S G3VNA
Fisher, J W M1BRY
Fisher, K G6ENK
Fisher, K G6LMR
Fisher, K A M G3WSN
Fisher, K D M1BXA
Fisher, K E G0LKX
Fisher, K J G4ITR
Fisher, K P G8VOB
Fisher, L J G7LOA
Fisher, M G1HZQ
Fisher, M G1IML
Fisher, M GW1LXD
Fisher, M J G1MBX
Fisher, M J G3UBI
Fisher, M R G4OKH
Fisher, M S G6VTA
Fisher, M W G4WDU
Fisher, P G0LFV
Fisher, P A G0VKB
Fisher, P A G7WER
Fisher, P A M0BDH
Fisher, P E G7OJD
Fisher, P H G4PPL
Fisher, P M G6MKZ
Fisher, P R G4GHQ
Fisher, R G4KJU
Fisher, R D K G1YLH
Fisher, R E G0UEB
Fisher, R M G8GZI
Fisher, R S G4KBK
Fisher, R W G3PWJ
Fisher, R W G7IQA
Fisher, S 2E1AYE
Fisher, S G3YWU
Fisher, S G4AKT
Fisher, S G7RPV
Fisher, S C G2FFN
Fisher, S K G1YTV
Fisher, S M G1SMY
Fisher, S W G6YXO
Fisher, T J G0UCU
Fishlock, T W F G1BFF
Fishpool, A F G4WIF
Fishwick, J 2E1DAS
Fishwick, R G G6FUY
Fisk, G G7RVX
Fisk, R C G4CPV
Fisk, R D G7AVU
Fisk, S H G0UWW
Fitch, C J G4LVJ
Fitch, D R W G8IMN
Fitch, G L G4WKZ
Fitch, J L G8EWG
Fitch, N A S G3FPK
Fitches, H G7BCW
Fitchett, F B S G0UNP
Fitt, G V G4ILN
Fitton, G L G4MDT
Fitton, S E G0WWE
Fitz Patrick, J T G7DFP
Fitzgerald, A N G0MEN
Fitzgerald, D E G4CKS
Fitzgerald, E G3XUX
Fitzgerald, E C G6NZT
Fitzgerald, J GW6UFO
Fitzgerald, J G G3EUS
Fitzgerald, J G G4JGF
Fitzgerald, J P G8XTJ
Fitzgerald, R G0HBW
Fitzgerald, W A GW4YMJ
Fitzgerald, W J G0BOY
Fitzgerald, W M GI4EEB
Fitzherbert, H B G4GAP
Fitzhugh, S C G6PHT
Fitzjohn, L G7TBT
Fitzjohn, S T G7TBU
Fitzmaurice, A T G0KXW
Fitzpatrick, B J MW1BNY
Fitzpatrick, D 2E1EWR
Fitzpatrick, D M1BPE
Fitzpatrick, G C G1CGU
Fitzpatrick, S J G1YMC
Fitzpatrick-Browne, L J C G8RBX
Fitzsimmons, B W G0GGM
Fitzsimons, J G4JFS
Fitzsimons, J G6WEW
Fitzsimons, R GI0HHZ
Fitzsimons, W T GI4OXO
Fitzwater, J F G4HVO
Fitzwater, L J G6GGY
Fitz-Patrick, J S G3OKF
Flack, B M G4AMP
Flack, J W G0UBZ
Flack, M D G7VCE
Flaherty, I J G4RPJ
Flanagan, C G7NRO
Flanagan, D GW4ZAR
Flanagan, D M C GM6KXP
Flanagan, G W G7FPR
Flanagan, P J GI3HJA
Flanders, R G4PEA
Flanner, F C P G3AVE
Flannigan, E P G4SZB
Flather, A S G3XMK
Flatman, C J G4WTD
Flatman, M P G0FDT
Flatman, N C G0EBQ
Flatman, P B G6BJC
Flatt, T L G4AZF
Flatters, D M G7ENC
Flatters, D R G7IQO
Flattley, J G1TNK
Flaum, R R G3BDH
Flavell, R G G3LTP
Flavell, W H G1ZJJ
Flavill, J A 2E1BYC
Fleet, C J G0PVN
Fleet, D A G0TBK
Fleet, P F G8SYT
Fleet, R G1FZS
Fleet, T D G0FSL
Fleetwood, D G1ZDF
Fleetwood, D E G0IXZ
Fleetwood, D M G6WPR
Fleetwood, J D G0UJP
Fleetwood, M F G0TMF
Fleetwood, R V G4RAL
Flegg, J D G4OYY
Flello, D C G4WOS
Fleming, A K G4JAH
Fleming, E D GM2BWF
Fleming, G G0IIG
Fleming, G G0ODY
Fleming, G M G1MHF
Fleming, H L G3JNW
Fleming, M J G8PTL
Fleming, P G7KXG
Fleming, S T GI7FCM
Fleming, W C GM8JGB
Flemming, I G3ZDQ
Fletcher, A G4AVF
Fletcher, A M G8EHS
Fletcher, B G4CWL
Fletcher, B G G6HCV
Fletcher, B P M G4MFW
Fletcher, B R G0SWH
Fletcher, C F G3DXZ
Fletcher, C I 2E1FNC
Fletcher, D G0MZL
Fletcher, D C G8CJH
Fletcher, D R GU4WRP
Fletcher, D W G0IDF
Fletcher, D W G4FUS
Fletcher, E G0UAY
Fletcher, E G6YQI
Fletcher, G K G7TEG
Fletcher, G O G7TES
Fletcher, H D G3TVM
Fletcher, J G0AWX
Fletcher, J G4EDX
Fletcher, J C GM0LYO
Fletcher, J E G4EGB
Fletcher, J G G4EDD
Fletcher, K A 2E1DJS
Fletcher, L G3AYY
Fletcher, L A G4SXH
Fletcher, M 2E1DJR
Fletcher, M G6EDT
Fletcher, M A G8JYS
Fletcher, M K G0NVS
Fletcher, M L G1RRX
Fletcher, M N G3EPG
Fletcher, M T G8LVV
Fletcher, N G4YNK
Fletcher, P A G6NZL
Fletcher, P J G4KTS
Fletcher, P S 2E1ECM
Fletcher, R A 2E1FNM
Fletcher, R A G7PCC
Fletcher, R A D G1YQO
Fletcher, R C G6RIJ
Fletcher, R L G6EDR
Fletcher, R M G7EXD
Fletcher, R R G0NWU
Fletcher, S G1SVF
Fletcher, S A GW6MKV
Fletcher, S C G4RFC
Fletcher, S H G0SZJ
Fletcher, S R G7GVJ
Fletcher, T D G0MZR
Fletcher, T L G3CNW
Fletcher, V E GW0NEC
Fletcher, W G3DRP
Fletcher, W H G3NXT
Fletcher, W S G4JNG
Flett, A W GM0HTT
Flett, G W GM0HQG
Flett, J GM4MKU
Fleury, B G0WYC
Flewitt, J R G0MFJ
Flewitt, M H G1WSW
Flicos, P G3ZWD
Flindall, W A G4IPN
Flindell, W J G0FAA
Flinn, B J G6DPH
Flint, A G4VLV
Flint, C G G8UDT
Flint, G N G6WOI
Flint, P E G0NXB
Flisher, N E G1NQX
Flitterman, D D M G0PZC
Flood, A G0SNZ
Flood, G H G0APY
Flood, R B G3TSI
Flood, W E G8XKH
Florence, A J G7CDK
Florentini, M A M G7HUC
Flounders, B R GW1VJB
Flower, E R G0GCN
Flower, G M A G1AVJ
Flower, J C G1LAO
Flowerday, R P G3ZHH
Flowers, D E G3EJC
Flowers, J C G1CUW
Flowers, J L G0JLF
Flowers, J L G0RTL
Flowers, V J G8QM
Floyd, A G3PNQ
Floyd, A G4GVB
Floyd, D J 2M1FOY
Floyd, S A GM3KXQ
Floyd, W F GW5AF
Floyde, M N G4AOB
Flux, C J G7IVF
Flynn, B W GM8BJF
Flynn, C D G7EOG
Flynn, J W G0VVN
Flynn, L J G4UWP
Flynn, R G GW1XXL
Flynn, R H G4OBW
Foad, A C 2E0ANO
Foad, J A G0UGF
Fochtmann, M K G4BHJ
Foden, W R G7BAB
Fogarty, C J GI1RSR
Fogden, C F G7ESG
Fogden, J R A G4WSX
Fogden, M G8MVP
Fogell, G L G0RMI
Fogg, E G G0HVF
Fogg, J C G0SWN
Fogg, J C G3PHZ
Fogg, J R G8UZZ
Fogg, M A G0RVK
Fogg, R G3ZCD
Foggin, G R G3GRF
Foggin, N G G8ETI
Foister, M E 2E1FWM
Foley, C GI6PME
Foley, C W G0XCF
Foley, M GI3SOO
Foley, N A G7PKV
Foley, P GI8YGG
Foley, T B G4FYO
Foley, T E J G7MTF
Folkerd, C D 2E1DTE
Folkerd, O R 2E1DTF
Folland, M E G7PVF
Follant, J GW0CBL
Follett, J GW1MBV
Follows, S R G4UQX
Fone, C J G4LAI
Fone, D GW8VRS
Fonseca, P G7VBA
Fooks, G P G4UTY
Foord, E P G0RJJ
Foot, M R G4WHY
Foot, N E G4WHO
Foote, C V GI0OKU
Foote, G G7NCR
Foote, M D G6PXZ
Foote, N GI4MNF
Foote, R G4GQP
Foote, S B G4FOH
Foote, S W G1EUM
Footring, L M G4SDI
Footring, M 2E0AIK
Forber, C D 2E1FSH
Forbes, A GM7RJG
Forbes, F G3TOM
Forbes, J J G3JKU
Forbes, P G0NZJ
Forbes, S G6YIE
Forbes, S K G6YQT
Forbes, T N G2BFC
Forbes, T W G0JWO
Forbes, W S G0NNN
Ford, A A GM4UEH
Ford, A J G6HBQ
Ford, A J M G6FPP
Ford, A M G7EQY
Ford, B J M1BSO
Ford, B S G7VFH
Ford, B T G0EYF
Ford, C F G3FDS
Ford, C F G4ZVS
Ford, D A G4UKP
Ford, D A G6PHU
Ford, D E G0IDP
Ford, D G 2E1DGX
Ford, D R G G6PHV
Ford, E J G4UYQ
Ford, E T G3DEY
Ford, E W K G7AHZ
Ford, G G0MHC
Ford, G G0PEJ
Ford, G G4APN
Ford, G A E G4RSK
Ford, I G1GZM
Ford, I H G4DKS
Ford, I J G1ODX
Ford, J G0NHQ
Ford, J A G7UZF
Ford, J C GM2DWW
Ford, J P G6LFW
Ford, J W 2E1FSM
Ford, K G7CNZ
Ford, K A G4ZTG
Ford, K J G1SLU
Ford, K R G1BBI
Ford, K S G4ZKF
Ford, M A G1IFS
Ford, M D M1BSP
Ford, P C G0RHK
Ford, P E G3VRU
Ford, P E G4WFL
Ford, R G0ENY
Ford, R M0AFY
Ford, R G G4GTD
Ford, R H GJ4JVI
Ford, R J GW0DRS
Ford, R S G3NKO
Ford, S G4BEM
Ford, S G4DPV
Ford, S M1ALY
Ford, T A G7WFH
Ford, T W GW7ESF
Ford, V A M0ANI
Ford, W G4AQP
Ford, W J G0TGP
Forde, G D G0VXY
Forde, J E G2FSP
Forde, L S J G4ZLF
Forder, H E J 2E1ABE
Forder, M J G0SPS
Forder, M J G7KGH
Fordham, G W A G1SKY
Fordham, M D G0LQV
Fordham, S E G4HOP
Fordyce, A A G1WLW
Foreman, C M G1IMI
Foreman, G A G4NEE
Foreman, E B G0EXS
Foreman, M G4HBT
Foreman, M G7LSZ
Foreman, M G G1TWX
Foreman, M J G4HCI
Forge, P H G6AER
Forhead, B R G7TUQ
Forknell, M L G7JXF
Forrest, A D GM0TEX
Forrest, A E G4OQC
Forrest, D A GM7UTD
Forrest, D C G8SRC
Forrest, D C M0ACM
Forrest, J E G4WJV
Forrest, J L G3VVW
Forrest, K G1BTI
Forrest, P G3GIL
Forrester, C R G7LWJ
Forrester, M A G4KUA
Forrester, W J D G3XAN
Forrest-Webb, R 2E0AFP
Forryan, A M G4OKD
Forse, I G4DHA
Forsey, D J G1DAV
Forsey, M G8PWK
Forshaw, A J 2E1FLK
Forshaw, G N 2E1FLL

Forshaw, P D G0JJI
Forshaw, P J G4HSS
Forshaw, S H G4XSV
Forster, A G G4NNJ
Forster, A J G0UYG
Forster, A R G1YDD
Forster, B G0SND
Forster, C G4FFU
Forster, D G3NWY
Forster, D I G3KZZ
Forster, D J G7VZL
Forster, D L G0VSZ
Forster, D S G0NUL
Forster, G G0CPD
Forster, G B G7VTE
Forster, G B M1BNR
Forster, G C G0VSY
Forster, I G4FFV
Forster, J G G1SKQ
Forster, K J G8RDA
Forster, M G2AGG
Forster, M M1BXM
Forster, M P A G7WGE
Forster, P G0CBN
Forster, P M G1PGG
Forster, P W F G3VWQ
Forster, R A K G8EIE
Forster, T C G4DQR
Forster, W J G6TOC
Forster-Pearson, R G6GGD
Forsyth, C M G0TQS
Forsyth, D D 2E0AJK
Forsyth, E N G0KHR
Forsyth, J A GM4OOU
Forsyth, J I G0PAU
Forsyth, M W J G1OWD
Forsyth, W GM1OVJ
Forsythe, J T GI0PGC
Fort, P T G3ZCX
Fortescue, M T G7IMH
Fortescue, R J G1HTO
Forth, D G1CDY
Fortnum, C A M G1JGR
Fortt, S G0OBT
Fortune, D GM7SKB
Fortune, N M 2E1AEM
Fortune, R G GM4WTK
Fortune, S A GM7NNH
Fortune, V GI0BDU
Forward, D N G7MSS
Forward, J D G3HTA
Forward, K N G0FAK
Forward, K N G1URG
Forward, R H G4AV
Fosbraey, R A G7BWV
Fosbrook, C J G0KCF
Fosbrook, T J G6DXN
Fosh, D R G0URO
Foskett, K E R G4UEN
Foss, A A G8OPG
Fossey, J W G0TUX
Fost, B J G3BUF
Foster, A E G7OJQ
Foster, A J G6VHG
Foster, A J G7IGJ
Foster, A S GM3OXA
Foster, A W GW3GAH
Foster, C G7BJB
Foster, C G7DEX
Foster, C A GW6MKR
Foster, C N G0ONQ
Foster, D G4PHP
Foster, D G7ESZ
Foster, D E G7URL
Foster, D H G4IPI
Foster, D M GW3HGJ
Foster, D V G3KQR
Foster, F W G3LD
Foster, G G3ZOQ
Foster, G E GW0EHB
Foster, G E GW0GZR
Foster, G H G3POD
Foster, G R G1DRG
Foster, H G1DOE
Foster, H C G3SVC
Foster, H C G4EZS
Foster, I G6NFU
Foster, I H G8CAM
Foster, J B G3TKB
Foster, J E 2E1BFT
Foster, J E G0FZZ
Foster, J E G7NCF
Foster, J F G4YVP
Foster, J H 2W1DGM
Foster, J J G1PJK
Foster, J M G1HTL
Foster, J M G8WSZ
Foster, J R G7SZC
Foster, J S G3KGF
Foster, J T GM4PWQ
Foster, J V G0MYH
Foster, J W G7NEN
Foster, K P G0EYN
Foster, K W G8GIH
Foster, M E G1SWE
Foster, M G G3VOF
Foster, M G G8AMG
Foster, M S G4KLE
Foster, N P G8KRR
Foster, N R G0NBJ
Foster, N R G0ULA
Foster, P A G1PAF
Foster, P A G7TMO
Foster, P L G0OSV
Foster, P W G1LOV
Foster, R G0GRU
Foster, R C G1NJI
Foster, R P G1FVN
Foster, R T G3AOV
Foster, S G4EEF
Foster, S G4MPK
Foster, T G7TMF
Foster, T H G3PQP
Foster, W J G3JWF
Foster-Jones, P A G1KEO
Fothergill, B G8KPD
Fothergill, J G8RER
Fouche, D W G0DWF
Fougere, T J G4MHK
Foulds, D H G4GFE
Foulds, P M G1SRD
Foulds, T K G6SFC
Foulis, A R GM7PGT
Foulkes, C H G3UFZ
Foulkes, F C G3KOM
Foulkes, P M G7OEA
Foulser, S D G8VQA
Foulser, S R G6UTL
Foulsham, C A G6YOY
Foulsham, W G0MWF
Found, A G7BYG
Found, R H E G4GDQ
Fountain, G L G1ODZ
Fountain, N A G7ELC
Fountaine, A A G7AFL
Fountaine, E G G0CGC
Fountaine, E G G0MKR
Fountaine, E G G0OUR
Fountaine, S E G1CBY
Fower, D I M1BPQ
Fowle, C A G7HKT
Fowle, G J 2E0AMW
Fowler, A G0TYP
Fowler, A J G0GFP
Fowler, A J G1IPA
Fowler, A J G4IPA
Fowler, B E 2E1FYX
Fowler, D A G4MDN
Fowler, D W G4YWG
Fowler, E A C G7MUF
Fowler, G C E G6NZN
Fowler, G D K G1EKP
Fowler, H A 2W1BTU
Fowler, H H G4RXH
Fowler, I R G8UFX
Fowler, J A G4VHG
Fowler, J C G7MHT
Fowler, J V P G7KLZ
Fowler, K E P G0NOM
Fowler, M D GW3GKZ
Fowler, M J L G8XTU
Fowler, M J M G8XZQ
Fowler, N J MM1BXZ
Fowler, P G4TQO
Fowler, R A G3IQF
Fowler, R H G0JJP
Fowler, S C G0NKU
Fowler, S D G0TGM
Fowler, W C G0JHJ
Fowler, W R L G0VFX
Fowles, D J G8VJR
Fowles, G G3XGV
Fox, A G 2E1EGC
Fox, A J G6JHD
Fox, B J G0OJR
Fox, B J H G7PVG
Fox, C A G3PKL
Fox, C J G7UEV
Fox, C M G0WYR
Fox, D G0NRR
Fox, D G0XRC
Fox, D J GI4CYK
Fox, D J W G4SSZ
Fox, E J G3AVJ
Fox, E K J 2E1AOX
Fox, G C G3AEX
Fox, G S G4NVU
Fox, H R G8RZ
Fox, I A G4UDF
Fox, J M1BMV
Fox, J C G0OAZ
Fox, J E G7PTE
Fox, J F G4CUD
Fox, J W G3KHR
Fox, J W F G0OWB
Fox, K M G4MDQ
Fox, M C G1NOM
Fox, M E G3ZUZ
Fox, M P G0EYT
Fox, M P G3SID
Fox, N G6YQN
Fox, P G4MCK
Fox, P P G0IRA
Fox, P P G7CRG
Fox, P P G8HAV
Fox, P P G8ZTT
Fox, R A G4HFN
Fox, R J G0UOI
Fox, S G4HGU
Fox, S G6WFK
Fox, S A G0MZI
Fox, S J G4FAB
Fox, S K G1NON
Fox, S W G4GVV
Fox, T B G3HU
Foxall, A F G6NPK
Foxall, B R G0PCF
Foxall, P T G0ICM
Foxley, E G6UFM
Foxley, R S G8YRF
Foxton, T G0KOE
Fox-Roberts, P G0TZU
Fox-Roberts, P A GI0USQ
Foy, D G0FWM
Foy, D J G4WCO
Foy, E J G6HIJ
Foy, J A G0JAF
Foy, M J G0SKI
Foyen, N M1BUP
Fozzard, M C W G3IWB
Fradgley, K G0WBA
Fradley, T A G6NLQ
Frakes, W G0BWF
Fraley, D M G1JYR
Frame, K R GM6IZU
Frame, T F GM0PPE
Frame, W GM3ZWG
Frampton, D J G6ACJ
Frampton, J A G4YQH
Frampton, J R G6CUE
Frampton, M B G6HWO
Frampton, P G4SGH
Frampton, P W G6NNK
France, A G0VUH
France, D W G4HGB
France, J F G0VRC
France, J F G3KAF
France, L A GW3PEX
France, L A GW0ARA
France, L A GW3SON
France, M D 2E0APJ
France, M E G4FDL
France, N A 2E1FPB
France, R G3ZOU
Francey, C GM7LQR
Francis, B J GW1KQY
Francis, D H G6PWT
Francis, G W G0JWQ
Francis, J D F G3LWI
Francis, J E G4BKO
Francis, J G G4XVE
Francis, J M G8ZTD
Francis, M J G3LOV
Francis, M T G0GBY
Francis, P E G4AKK
Francis, P J G7MMW
Francis, R G0TJT
Francis, R H O G4EIE
Francis, R N G3RWU
Francis, R P 2E1ESP
Francis, R T G0EYP
Francis, T G 2E1DGH
Francis, W D G0RZO
Francks, C G4WMX
Francks, K R G7JOE
Francks, K R M0BFB
Francl, V L A D G4OQM
Frank, A S G0TDY
Frank, C 2E1FCY
Frank, J G0MSX
Frank, S G G7SOP
Frankcom, K G3ZBI
Frankham, L W J G4GFJ
Frankl1N, K V G3JKF
Frankland, T W G4INM
Franklin, A C G6LMU
Franklin, A G 2E1FRE
Franklin, A G M1BQX
Franklin, A J G6GAF
Franklin, B A G0OAE
Franklin, C E G1FLI
Franklin, D J G0MQL
Franklin, D J G8VFI
Franklin, G GM4BXG
Franklin, J H G1XAJ
Franklin, K G7NWU
Franklin, L G7VSN
Franklin, L R G3LWF
Franklin, M A G3VYI
Franklin, M H G4LJA
Franklin, M N G1WSM
Franklin, P G7UVD
Franklin, P J G1FOA
Franklin, R G4XHH
Franklin, R D G3ITH
Franklin, S M G3XVH
Frankling, D K G8LCV
Franks, C G0OXR
Franks, J W G3SQQ
Franks, K S G3URI
Franks, M E G7UXP
Franks, M K G4MKF
Franks, R W G8BIQ
Frankum, S P G0WZH
Fraser, A GM0HAQ
Fraser, A M GM3AXX
Fraser, A M G4PEC
Fraser, A R G8PWX
Fraser, A S D G8MHO
Fraser, A W G1BBR
Fraser, A W GM8NET
Fraser, C F GM0HBF
Fraser, D J GM6JRX
Fraser, D S S G0BDL
Fraser, E P G1SIB
Fraser, G E G4MGF
Fraser, I M G3TVT
Fraser, I R G3BWN
Fraser, J GM0OKJ
Fraser, J C GM4WJA
Fraser, J G GU0VNF
Fraser, J G GU7PMD
Fraser, J I GM8MHU
Fraser, K J G4FMA
Fraser, L J G4YUC
Fraser, P D MM1BJZ
Fraser, P S G3HZT
Fraser, R G4PHH
Fraser, R R B G GM0NTL
Fraser, S 2M1FMX
Fraser, S G1FBQ
Frater, C S H G0KIH
Frati, J B GM0MYQ
Fray, A J 2W1FQH
Fray, H A GW6RCK
Fray, M E 2W1FQG
Fray, R A GW0OYD
Fray, T A M1BXR
Frayne, G C GW0GMX
Frazer, A D G8DMU
Frazer, E L G7OBU
Frazer, G E GI4SJQ
Frear, M E G0MEF
Frear, M E G4YPT
Frearson, J A G3OVK
Frearson, T W G0COY
Frederick, D J G4XXM
Frederick, G A G4NZO
Frederick, N G6YXX
Free, A J G4EYE
Free, M GW0OGE
Freeborough, D J G1JLE
Freeburn, R A GI6WHZ
Freedman, D G3VSH
Freedman, M G0WHW
Freedman, M B 2E1EPD
Freedman, P D M 2E1EPE
Freegard, S M G7UEM
Freeland, J J GM4SZG
Freeman, A H G4TKX
Freeman, A L G0PIT
Freeman, B S G3ITF
Freeman, C J G6HWT
Freeman, D J G7RTR
Freeman, D M G6NZS
Freeman, D R G0LTP
Freeman, D T G4EUN
Freeman, G C A G4FCC
Freeman, J G4XQQ
Freeman, J E G4MGX
Freeman, J J G6ZJJ
Freeman, J R G1OHU
Freeman, M 2E1BOC
Freeman, M C G8UST
Freeman, P J G4DMS
Freeman, P M G4PXS
Freeman, R T G G4SDJ
Freeman, R W G3TCZ
Freeman, S J W G3LQR
Freemantle, R S P M1AZN
Freer, A C G3ZKW
Freer, J G1EBL
Freer, J A GM7LJE
Freer, K G8RJF
Freer, R A G0PBY
Freer, R A F G4MFF
Freeston, D J G4DBF
Freestone, J S G4TEM
Freestone, M K G8UOZ
Freestone, P J GD4PFL
Freeta, M S GW6AHY
Freeth, G R G4HFQ
Freeze, J E M0AIV
French, A J G3XQP
French, C J G8ZAJ
French, C R G0PAF
French, D R G3TIK
French, D R G4PKO
French, D W R G4RTF
French, G E G0TMI
French, I H G0WGV
French, J E G3LGL
French, J R G4IET
French, M J G8WSH
French, P A G8FUJ
French, P R G0PKF
French, R M G0BMF
French, S C G0NNJ
French, S J G6SWT
French, S P G6ZTZ
French, T J G7HEZ
French, W J D G1CYR
French-St-George, A G0CGY
Frend, P E GI0FZT
Frenzel, C D G4SNJ
Frenzel, J T G4SNK
Frettsome, K G4ABV
Frettsome, R G4WPW
Fretwell, G D M1BMI
Fretwell, M C G4OET
Fretwell, P H G4UFC
Frew, J F GM3YLD
Frew, R I G3SEF
Frewen, W E L G0FLA
Frey, M J G3VXZ
Frey, N P G1BKV
Frey, S L GM8ZQY
Fricker, T D G0NQN
Friedman, M R G1END
Friel, A G1GWE
Friel, C D G3EFX
Friel, C D G4AUF
Friend, A J G1EUN
Friend, C A GI3KSY
Friend, G R GW1NGD
Friend, J D G4HLI
Friend, P G3SQR
Friend, R R J G7VPD
Frier, P W M1BHW
Friis, A S G4LNC
Frings, J A C K G3FFH
Fripp, D M G0GIY
Frisby, B L G4WDJ
Frisby, F M G3ZNY
Frisby, J G8DJU
Frisby, R G4OAA
Frisby, R G G2CFC
Frisby, S A GW4NFF
Friston, R J G0FMI
Frith, D J G1DMC
Frith, I C G4GIR
Frith, W H G3FRE
Fritsch, D H G G0CKZ
Frizell, J W G1WSE
Frizzell, J G G0LEO
Frizzell, S G7PUW
Froggatt, B G3YXR
Froggatt, B G3DUZ
Froggatt, D J G4KKE
Froggatt, D W G8CXA
Froggatt, H G3HQH
Froggatt, L 2E1EFT
Froggatt, S R G1LMN
Froggatt, T J G7JGF
Froggett, B J G0LPV
Froggett, P G0LVV
Fromm, S A G0BMD
Fronius, R A E G3MCW
Fross, S A M0BGB
Frost, A G3FTQ
Frost, A A G0MVM
Frost, A D G4FUB
Frost, A J G8UDV
Frost, B B 2E1DCM
Frost, B J G6UTN
Frost, C G1XRO
Frost, C J G6ACD
Frost, C L G0KEB
Frost, D A G7VQE
Frost, J A G0DCR
Frost, J R GW0NDA
Frost, J S G4SYL
Frost, M R G0IDH
Frost, P A G3VYK
Frost, P A GW4NLD
Frost, P L G6YXH
Frost, R L 2E1FEI
Frost, R R C G3SZF
Frost, R S G3UUV
Frost, R T GM6FT
Frost, S L G4VNM
Frost, S O GW0MYK
Frostick, R M1BOX
Froud, J H G3YHH
Frow, R D G0VAI
Frowd, K W GW4ZCM
Frowen, H H G8PTI
Fry, A C GW3NWC
Fry, A F G0VXX
Fry, A R G4WBV
Fry, A R G4XRS
Fry, B E G7PBA
Fry, C R G3NDI
Fry, D J G4JSZ
Fry, D K G1MJI
Fry, E W G6LKQ
Fry, M M G0GLU
Fry, M R G7NDS
Fry, P G3TZV
Fry, P C G4AKG
Fry, P G G0FUS
Fry, P W G4SBF
Fry, R E GW7TIH
Fry, R F G7SMF
Fry, S M GW6UFH
Fry, S W G0WRZ
Fry, W R P G1YRR
Fryer, A G7APV
Fryer, D M G1OQG
Fryer, D W G6DBQ
Fryer, L 2W1DRM
Fryer, L G4KMU
Fryer, P R 2E1FFX
Fryer, S E C G3ERO
Frykman, G N G0GNF
Frykman, G N G7WKS
Fudge, H W G3DZS
Fudge, M W G G1THP
Fugler, A E M1AEE
Fugler, I N G4IIY
Fujita, K M0BDM
Fulcher, K G G7IYQ
Fuller, A J G4OYN
Fuller, A N G4YUT
Fuller, D H M G1CQR
Fuller, D K G1OZO
Fuller, D L J 2E1BPP
Fuller, D P G4XPY
Fuller, E E G6XYP
Fuller, F G0TCK
Fuller, F G4GCJ
Fuller, G G4VFH
Fuller, G F G8EKI
Fuller, G H G4DRF
Fuller, G W G0KRS
Fuller, G W G3TFF
Fuller, J E G3IQE
Fuller, J N G0OIO
Fuller, J T G4WPI
Fuller, K R G0VOB
Fuller, L G0ULN
Fuller, L G7MYY
Fuller, N J G8MVS
Fuller, P G4FNJ
Fuller, P G8TYG
Fuller, P T E G0PVQ
Fuller, R A G0HHJ
Fuller, R A G8CEZ
Fuller, R G G0EBG
Fuller, R J G0EUQ
Fuller, R J G6PWS
Fuller, R W G0HFK
Fuller, R W G6YQU
Fuller, S G G7PQO
Fullerton, I R G6ILZ
Fulton, A C GM3CSO
Fulton, D I GI3CFH
Fulton, D I GI4OUN
Fulton, J G4ZOJ
Fulton, P M GW3MMU
Funnell, C G7IZV
Funnell, M E G3YQW
Funnell, P C G8AFI
Funnell, P J G4BWN
Furby, D J G1MCO
Furby, T 2E1FCS
Furby, W G6BAF
Fure, B 2E1EHQ
Furmage, G G0NAQ
Furminger, S J G7TYH
Furmston, A B G0TCW
Furness, A N P G7SKQ
Furness, D W G4XKW
Furness, K G7TRA
Furness, M J G3LNF
Furness, R H GD4IHC
Furness, T G4WNG
Furness, W M G3SMM
Furniss, D G0IFR
Furniss, R G7FBY
Furniss, R G 2E1FKW
Furnival, M P G8YBW
Furnival, M R G8ETP
Furze, W L G0UIQ
Furze, W L G7AOX
Fusco, T GI4NKZ
Fussey, R J R G3BQE
Futcher, H G G8BTL
Fyall, G GM0NXO
Fyfe, C D GM8UUW
Fyffe, A F G4ENF
Fyles, D A G1RVV
Fynn, P G G8NPO
Fyrth, J M G0DPS
Fysh, J GM0JEF
Fyson, J G1XDS
Fyson, J H G1YBK

G

Gabbatiss, J G G7UNY
Gabbitas, R W G3KHU
Gabel, P G1TAI
Gabell, F C G4LDJ
Gabriel, N A G4JUZ
Gabriel, P R G4IKI
Gabriel, S G3HCQ
Gadd, C M G4JZZ
Gadeberg, E G0CMT
Gadney, R N GW7KIV
Gadsby, J G1TEW
Gadsden, B G1YHQ
Gadsden, D M G4NXV
Gadsden, H W G3JZW
Gadsden, P G G3MTP
Gaffin, A G0MWO
Gaffney, E M G7GZZ
Gaffney, J P G4UAA
Gage, B R G4PGA
Gage, T C G1MPJ
Gagen, P R G4OTC
Gagg, J A G4XRB
Gaggs, D B G3DFA
Gagnon, A C G4DGW
Gagnon, L G MW0AWC
Gail, R O N I G7LYG
Gain, W A G1IFW
Gainey, P R G0DZM
Gainswin, S M G1GMG
Gair, D M G3YEY
Gair, K G A G3SCE
Gait, P J G6XQO
Galbraith, G G1NOR
Galbraith, G S M0ADR
Galbraith, K R GI6GDM
Gale, A G3GWF
Gale, A G G8EAS
Gale, A S G7NUR
Gale, B B G3EBK
Gale, B D R G3UJE
Gale, D A T GW8WZR
Gale, D K G1GVM
Gale, E A G4TMV
Gale, G S M1ATB
Gale, J A G3LLK
Gale, J F G7IYO
Gale, J M G3JMG
Gale, J O 2E1CCT
Gale, J T G3WIM
Gale, M G GM4PXB
Gale, M O G6XYT
Gale, P F G3OJG
Gale, R J 2E1END
Gale, S J G8JDT
Gale, S J M0AZR
Gale, T J 2E1EMW
Gale, T K G8JUS
Galea, M E G7PDO
Galer, P J G6AZN
Gall, A J M0ADZ
Gall, D A G0AOK
Gall, I J GM8BNH
Gall, R B GM4UFD
Gall, R T G8JYR
Gallacher, C I G4JCX
Gallacher, D GM4DNS
Gallacher, I GM8FHK
Gallacher, W J G0IRY
Gallagher, A J R G0TOE
Gallagher, C D GW4ZTW
Gallagher, E G7DYL
Gallagher, G G3SNV
Gallagher, K G0KEV
Gallagher, M J GM4AHB
Gallagher, M T G8OKN
Gallagher, P J G1WSS
Gallagher-Daggitt, G E G0UZS
Gallamore, G G8BRU
Gallant, L A J G0CPE
Galley, G N G0JZL
Gallichan, A M0ASD
Gallienne, J F GU4WMG
Gallimore, F G4PKX
Galliver, G E L G7SOX
Gallon, E A G7UTF
Gallon, M G6URX
Gallop, D L G3LXQ
Gallop, J E G3YIW
Gallop, R G0KNQ
Galloway, B R G0IFX
Galloway, C A G3RNV
Galloway, G A G4KSC
Galloway, J 2M1ERB
Galloway, J MM1BGZ
Galloway, K D G4ZVE
Galloway, W J GI3GSB
Galpin, A J G1AOL
Galpin, I A N G1SMD
Galpin, R J GW8NNF
Galsworthy, B E GW0UIZ
Galsworthy, P D G0TDU
Galvin, A E G0IKD
Gamble, B J G8KXO
Gamble, E G G2DAU
Gamble, E J G0ASZ
Gamble, J M G6MDB
Gamble, N GI7FJY
Gamble, P GM3UYR
Gamble, P R G1SGZ
Gamble, R B G6GFI
Gamble, T G0UNX
Gamble, T K G1MTG
Gamblen, J G G4ERS
Gambles, G F G6ZEA
Gambles, R J G6DBU
Game, P B G6TQF
Gammage, R W S G6BVR
Gammage, T A G4TAG
Gammage, T W G3YOV
Gammans, D C R G7EQR
Gammon, A P G1EUQ
Gammon, C D G3NHG
Gammon, H V G4YIU
Gammon, I G G4SCV
Gander, S D G7TGY
Gandhi, C 2E1CLL
Gandy, R G GM0MNV
Gane, E J G3LWX
Gane, G A T GM6PTX
Gane, P M GM4SUF
Gane, R A G1ESO
Ganley, P D G4YWJ
Gannaway, J N G3YGF
Ganson, J R MM1BWC
Gant, E B G4DV
Gant, R C G0LXP
Gape, R A G8DQX
Gapper, A D G8PQA
Gapper, A G G4JYX
Garbett, C A G0RCX
Garbett, S C T G4XDX
Garbutt, B S G0LPX
Garbutt, K G0PBA
Garbutt, M G0OEJ
Garbutt, N I B G4PJJ
Garcia, C T G8DWW
Garcia, F R G0VUF
Garcia-Rodriguez, J G6BNW
Gard, A E N G4LWA
Garde, P G F G6MCE
Garden, G G4LJR
Garden, J G4DXZ
Garden, L Y G0EJI
Garden, P J M0AYW
Gardener, D G1EPD
Gardener, J H G1ECV
Gardiner, A J M1BQQ
Gardiner, B J V G4SXL
Gardiner, C R G7VJW
Gardiner, D G8BAS
Gardiner, F W W G0MXQ
Gardiner, G H G0EJJ
Gardiner, G S G0OLM
Gardiner, G W G3WEB
Gardiner, I M G3PHD
Gardiner, J G4OIT
Gardiner, J G7BVV
Gardiner, K G6YDQ
Gardiner, K R G8NFD
Gardiner, M J 2E1FBN
Gardiner, M J G4MVX
Gardiner, M L 2E1FZI
Gardiner, P G1MRI
Gardiner, T A D 2E1FCX
Gardiner, T N GI1CKU
Gardner, A GW0OZB
Gardner, A H P G0NTH
Gardner, A R G4OKC
Gardner, C A G4WEO
Gardner, D J G1LTL
Gardner, D J G8RFK
Gardner, D J GM4JZB
Gardner, D N G7TQW
Gardner, G G1CFJ
Gardner, G M G0HEM
Gardner, G W G4ZEN
Gardner, I W G3CDM
Gardner, I W G4ZVH
Gardner, J G3URP
Gardner, J GM4MF
Gardner, J A 2E1COQ
Gardner, J A G7RUX
Gardner, J B G3WDV
Gardner, J C G6ULI
Gardner, J F J GM3VPN
Gardner, J M G6MCO
Gardner, J P GU7CQN
Gardner, K G7SZG
Gardner, K E G0VFL
Gardner, M C G8YZS
Gardner, P A G0IPE
Gardner, P B G1DCU
Gardner, P K M0AIG
Gardner, P W G1PWY
Gardner, R G7NLJ
Gardner, R E G0KZQ
Gardner, R J G4WKN
Gardner, R J GM4KVJ
Gardner, R W C G3CGE
Gardner, S H G4PSP
Gardner, T G3YVZ
Gardner, W E G1HMW
Gardner, W E G3FYR
Gardner, W J G0PDO
Gardner, W K G6DXU
Gare, C S G3WOS
Garfirth, S G6ANR
Garford, J D G3SVS
Garforth, A G3IGC
Gargett, S L G7UYN
Garland, A S 2E1FNT
Garland, B W GW0LDZ
Garland, C M G3RJT
Garland, H D G0NFJ
Garland, M K G1JGS
Garland, N S 2E1ESE
Garland, P N J G6MCX
Garland, R G G3EBO
Garland, S J G8GHL
Garlick, D G3LCR
Garlick, M P G4YNG
Garlick, M S M1AUR
Garlick, S E G7API
Garlinge, A W G0HHI
Garman, R G G3NCG
Garmany, H GM0GYQ
Garment, R M G6NMI
Garner, G G1OWK
Garner, G D J G6XQP
Garner, G J GW8NXK
Garner, J D G3ZJG
Garner, J F G4ZKQ
Garner, J M G3KEC
Garner, J S H G2BGG
Garner, M J G1WUC
Garner, P R GW0INN
Garner, R M G8LAN
Garner, R P G7HHL
Garner, S H G4SGZ

Garner, S J ... G3WSL
Garner, T ... G1ULP
Garner, T ... GW0DYT
Garner, T R ... G3XZY
Garner, W ... GM3UHT
Garnett, A P ... G0JBB
Garnett, J ... G1YUX
Garnett, P A ... G4LZJ
Garnett, P J ... G8RII
Garnett-Frizelle, M B ... 2E1DQS
Garnett-Frizelle, R E ... G7ORW
Garnham, C J ... G6MCG
Garnham, J A ... G0PMX
Garnham, T A ... G1IVK
Garrad, D H ... G2FAB
Garrard, D A ... 2E1ACH
Garrard, D J ... G8EOM
Garratt, D H ... G1IFX
Garratt, D J ... G0FXL
Garratt, D J ... G0UHP
Garratt, D J ... G1WSD
Garratt, F ... G4HOM
Garratt, G M ... G7LVU
Garrbutt, S F ... G0LLM
Garrett, B M ... G6ENO
Garrett, D G ... G8MZA
Garrett, F E ... G3MVZ
Garrett, G C ... G1KXY
Garrett, G C H ... G7GLN
Garrett, G S ... G3IJW
Garrett, J ... G3RHP
Garrett, K ... G7VLQ
Garrett, R ... G0MUR
Garrett, R M ... G3BP
Garrett, S G E ... G4EVN
Garrett, W R ... G1WRG
Garrick, C D ... G0KOQ
Garrington, D E ... GM3RFA
Garrington, M J ... GM7SPB
Garrod, N J ... G0OQK
Garrott, E T ... G0LMJ
Garry, R ... G8ALG
Garside, D H ... G3SSH
Garside, K S ... G4LGH
Garside, N G ... GW6FBM
Garstang, F K ... G0VVH
Garston, P ... GW4SII
Gartell, P J ... M1AXD
Garters, J A ... G8JLD
Garth, R N ... G7ELK
Garthwaite, P S ... G3OXR
Gartland, J M ... G8JMG
Gartley, R ... G4MYH
Garton, M G ... G8WJY
Gartside, J ... G8NQI
Garvey, T P ... G0BML
Garwood, D R ... G8MGQ
Garwood, M R ... G4DLD
Garwood, R ... G6FBR
Garwood, S J ... G7USP
Gascoigne, A R ... G8THN
Gascoigne, B J ... G8PXU
Gascoigne, M M ... G8TSN
Gascoigne, P G ... G4IMB
Gascoigne, S A T ... G6UUZ
Gascoyne, R W ... G1PWJ
Gash, P T ... G7AOA
Gash, R A H ... G1UTG
Gaskell, D C ... G8BGW
Gaskell, D R ... G0REL
Gaskell, E A ... G0RJX
Gaskell, P D ... G1DDD
Gaskell, P D ... G4MWO
Gaskell, R K ... 2E0ADA
Gaskell, R K ... G0RKG
Gasken, R ... G4RXD
Gaskin, P T ... G8AYY
Gasking, A J F ... M1ASX
Gasper, M P ... G8EUE
Gass, J D ... G4NGL
Gass, P R ... G4XZC
Gasser, D M J ... G4KWY
Gasson, K S J ... G2BGU
Gaston, A E ... GM0HPK
Gaston, A E ... GM4RIV
Gaston, C D T ... G4KEI
Gaston, D M ... 2E1BCR
Gaston, D M ... G7RPS
Gateley, A J ... G1NAN
Gater, K L ... 2E1ESR
Gater, K T ... G7STC
Gater, M J E ... G4ICC
Gater, P ... G0UCQ
Gater, P D ... G7PEH
Gater, P M ... G7TNY
Gates, A ... G0ARV
Gates, A B ... G8XCQ
Gates, W E ... G3ENB
Gathergood, M C ... G0OIE
Gathergood, R F S ... G4LUA
Gatland, J P ... G6ANW
Gatrell, A E ... G4SVB
Gatusch, G J ... 2E1AKX
Gaudier, D V ... M0AOP
Gaughan, J G ... G4FEO
Gaukroger, C A ... G7CLO
Gauld, A ... G0KFG
Gault, A R ... GI6PYP
Gault, J ... GM0LPB
Gaunt, B P M A ... G0LPG
Gaunt, C J ... G7BRZ
Gaunt, E J ... G7UXH
Gaunt, G D ... G8FEI
Gaunt, G P ... G4IJO
Gaunt, J P ... G1TAG
Gaunt, J W ... G0JWG
Gaunt, K I ... G7CIY
Gaunt, L ... G4MLV
Gaunt, N ... G3MEK
Gauntlett, B W G ... G3WQK
Gauntlett, B W G ... G4LYU
Gauntlett, G R ... G3VLL
Gauson, J D ... MM1AMD
Gaut, J W ... G0CCV
Gautrey, N ... G6GGW
Gavin, J P G ... G0XAS
Gaw, S J ... GM4XWL
Gawan, R ... M0AKQ
Gawn, T P H ... G8UKI
Gawne, A J ... GD7LAV
Gawthorne, E H ... G4KFV
Gawthorpe, E M ... G8FEK
Gawthorpe, A S ... G0RVM
Gay, A M ... G4HMK
Gay, F T ... G3CFV
Gay, M ... G3OCL
Gaygan, T R ... G4AFV
Gaylard, M R ... G6IUQ
Gayler, M J ... G4SDZ
Gayne, A R ... G7KPF
Gayther, B ... GW6KFH
Gayther, J F ... G0TPD
Gayton, G A ... G4CJU
Gayton, R J ... G8ATC
Gaze, E R ... G8NKA
Gealer, B J ... G3DEF
Geall, B E ... 2E1FTO
Gealy, R H ... G3PTG
Gear, J P ... G0BYG
Gearey, J L ... GW8HKY
Gearing, G D ... GM3ALB
Gearing, G F ... G3JJG
Geary, A ... G4XVL
Geary, N ... G0COT
Geary, S J ... G3MO
Geary, S L ... G4JZA
Geary, S M ... G8PWD
Gebhardt, K D ... G7AUF
Geddes, D R ... G4RKR
Geddes, R B ... G8GGI
Geduldig, P ... G0TYE
Gee, A C ... G2UK
Gee, A E ... G3HAV
Gee, A J ... G1IMM
Gee, A S ... G1AGA
Gee, B E ... G3LDG
Gee, B W ... G0VRU
Gee, C S ... G4ZUN
Gee, C W D ... G0CKM
Gee, D H ... G4KYX
Gee, D M ... G1AGB
Gee, G ... G0TKW
Gee, G N ... G0ORW
Gee, J ... G1IRC
Gee, J ... G4BAV
Gee, J ... G4IRC
Gee, J M ... G7GEE
Gee, J N ... G0RHW
Gee, M ... G0BDR
Gee, M S M S ... G6EXG
Gee, R ... G0PQE
Geen, K R C ... G4LRB
Geer, J A C ... G8WJB
Geere, D H ... G3UON
Geering, A R ... G4LMS
Geering, L P ... G8LMS
Geeson, B L ... G4KHJ
Geeson, P M ... G8DIY
Geeson, R D ... G3NJX
Gehammar, A U ... G4YHN
Geier, P A J ... 2E1FZO
Geiger, P G ... G4CNI
Geilern, P D ... 2E1FYR
Geis, R O ... G0ASF
Gelber, R L ... MJ0AWR
Geldart, D ... G7NCB
Geldart, D ... G8BRK
Geldart, M D ... 2E1BEI
Geldart, T G ... G4PXR
Gelder, W ... G7HUW
Geldert, W L ... M1BKB
Gell, B P ... G6XSS
Gell, F E ... G3JTO
Gell, J ... G4EAX
Gell, N ... G7PYQ
Gellatly, J C ... G3ZVV
Gellion, C K ... G8HVA
Gellion, P ... G0AVI
Gelsthorpe, H R ... G3THO
Gemmell, A ... GM0DVO
Gemmell, F S ... G4GGR
Gemmell, T K ... GM4RPO
Gemmill, A D ... G1DFR
Gener, A R ... 2E1DBP
Genes, T M ... G6CNQ
Genon, H P ... GM4JPJ
Gent, C J ... G4AKE
Gent, D ... G0GCS
Gent, M J ... G0RUV
Gent, N W ... GM4HQU
Gentle, B L ... G0GWB
Gentle, D A ... G4RVL
Gentles, J D ... GM3IDS
Gentles, J D ... GM4WZP
Gentry, M A ... G6DXP
Gentzler, J ... G3XRA
George, A ... GM0EIT
George, A T ... G7TNX
George, B P ... G7TFU
George, B R ... G3ZOH
George, C M ... G7MYN
George, C R H ... G4GBZ
George, D ... G8TUH
George, D A T ... GW0VGW
George, D E ... GW1OUP
George, E J ... G6AIH
George, G ... GM0DQV
George, G S ... G0OTF
George, J ... G6VTQ
George, K C ... G3VZZ
George, K J ... G3XPJ
George, M ... G3XYG
George, M A I ... G0VKF
George, M J ... GM3DNV
George, M L ... G0NFL
George, P ... GW1YHA
George, R ... G8RDB
George, R E ... 2E1BXC
George, R E ... G7VRM
George, R P ... G4VTU
George, R P ... G6AII
George, T M ... GW3NJG
George, T R ... G4AMT
George, T S ... GW3RDB
George, T S ... GW4ZRW
George, W J S ... G6TQC
Geraghty, D J ... G4UIT
Geraghty, J J ... G0PJG
Gerard, C J A ... G4PKW
Gerard, D M ... G7IMT
Gerard, K J ... GM4TPX
Gerhardi, V I ... G6GDI
Gericke, C S ... G4MKX
German, R A E ... G3OZT
German, R G ... G8VCL
Germaney, D J ... G7HYS
Germaney, P M ... M1BFD
Gerolemou, N ... G1NEK
Gerrard, A E ... G7JSF
Gerrard, A F ... G0FTS
Gerrard, A F ... G4TFU
Gerrard, C ... G3YOS
Gerrard, C ... G4AXL
Gerrard, C A ... G7CAQ
Gerrard, C S ... G3LSJ
Gerrard, D E ... G4NJL
Gerrard, D T ... G3BCC
Gerrard, E E ... 2E1DOR
Gerrard, I E ... G0EFZ
Gerrard, M ... GW6OKC
Gerrard, P ... G7JJC
Gerrard, R ... G3NOM
Gerrard, R L ... G3LAZ
Gerrard, R T ... G7HVO
Gerrard, T ... G7MGC
Gerrard, W A ... G4ZRB
Gerrie, D W ... MM1BUO
Gerrity, J W ... G4OQP
Gershon, M R ... G4HIM
Gervais, D W ... GI7JHV
Gethin, E ... G6HWD
Gething, D B ... G3XZK
Gething, P ... G1JNM
Gethings, J T ... GU0ELF
Gethings, J T ... GU0JTG
Getty, R J ... GI6JRY
Geurts, G H J ... M0AIU
Gevers, W G J ... GM0DTA
Ghani, N ... G3WZH
Ghassempoory, M ... GW7JDX
Ghetti, G A ... G1CPD
Ghillyer, K C ... G4YGZ
Ghiotti, C S ... G7OCG
Ghuman, R ... 2E1EAT
Gibb, D J ... G8BME
Gibb, J M ... G0GVU
Gibb, J R ... G0GVT
Gibb, M J ... GM0GIB
Gibb, P E ... GM1CBQ
Gibbard, B F ... G0AZF
Gibbard, J W ... G1BES
Gibbard, J W ... G1KOP
Gibbens, K G B ... G1JYT
Gibbings, A A J ... G6SWD
Gibbings, A J ... G8ILI
Gibbings, M ... G3BW
Gibbings, M ... G3FDW
Gibbins, A M ... G4GNC
Gibbins, E G ... G3JTG
Gibbon, D ... GW4DTQ
Gibbon, R J ... G3XAG
Gibbon, J V ... G1GVP
Gibbons, D ... G0DUM
Gibbons, D A ... G1DVX
Gibbons, E T ... G7BWE
Gibbons, F ... G0MPR
Gibbons, F H ... G3TIP
Gibbons, F J ... G3TBU
Gibbons, G P ... G1PEU
Gibbons, H C ... G6CVY
Gibbons, J R ... G0IQY
Gibbons, M G W ... GM7ODA
Gibbons, P A ... G6MRT
Gibbons, P G ... GW8ZHN
Gibbons, R A G ... GW0AIY
Gibbons, R J M ... 2M1DIL
Gibbons, R S ... G0FOT
Gibbons, W ... G0EQN
Gibbs, A D ... G0RSY
Gibbs, A E ... G0RGP
Gibbs, A H ... G1SPJ
Gibbs, A M ... G0VSB
Gibbs, B C ... G3MBN
Gibbs, B D ... G4CSG
Gibbs, C D ... G8GHH
Gibbs, D ... GW7OVV
Gibbs, D E ... G7CON
Gibbs, F W ... G7UUB
Gibbs, G G ... G3AAZ
Gibbs, I ... G4GWB
Gibbs, I ... G8NYJ
Gibbs, J ... G3ZZZ
Gibbs, J ... G8JZO
Gibbs, J A L ... G3LIO
Gibbs, J B ... G4SHU
Gibbs, J R ... G4UQR
Gibbs, K B ... G7PPX
Gibbs, L G ... G1NMQ
Gibbs, M J ... G3PSR
Gibbs, P A G ... G4DFG
Gibbs, R J ... G0UBA
Gibbs, R N ... G0RBQ
Gibbs, R P ... G7IWW
Gibbs, R R J ... GM3SVE
Gibbs, R S ... G8BNB
Gibbs, S ... G1JQK
Gibbs, S H ... GU3MBS
Gibbs, S R ... G4YYR
Gibbs, W E ... G4EBO
Gibson, A ... G0RCI
Gibson, A ... G0TFT
Gibson, A G ... GI0TJV
Gibson, A J ... G7NMZ
Gibson, A J ... G7VGU
Gibson, A K ... G7TGF
Gibson, A W ... G8YZU
Gibson, B ... G0ADU
Gibson, B ... G4UKD
Gibson, C C ... GW4UNY
Gibson, C J F ... GM0UKZ
Gibson, C S ... GW6VMB
Gibson, D ... GI3OQR
Gibson, D I ... G6ZUO
Gibson, D L ... G3JDG
Gibson, D M ... G4RGN
Gibson, D N ... G3CVM
Gibson, D R ... G4LXA
Gibson, D S ... G4VOQ
Gibson, E M ... G0JUI
Gibson, E M ... G1JTQ
Gibson, F T ... M1BQS
Gibson, G ... G3ZFZ
Gibson, H J E ... G6AIG
Gibson, H K ... G4WYD
Gibson, H L ... G2BUP
Gibson, I H ... GI4MDD
Gibson, I J ... M0AYU
Gibson, J ... G1ZUF
Gibson, J ... G7TGE
Gibson, J K ... G3WYN
Gibson, K F ... GI7JAM
Gibson, K J ... G4MIV
Gibson, K W ... 2E1BNX
Gibson, L ... G1JTR
Gibson, L ... G6UMN
Gibson, L H ... G3RCX
Gibson, L J ... G8VML
Gibson, M ... G0EEK
Gibson, M ... GI7JEB
Gibson, M A ... G7VKG
Gibson, M B ... G8PFR
Gibson, M R ... G7SOM
Gibson, N I ... G4XOR
Gibson, P ... G0EOK
Gibson, P A ... G7LJA
Gibson, P D ... G4UVB
Gibson, P F ... G1WVS
Gibson, P J ... 2E1EQK
Gibson, P M ... G1LDC
Gibson, P M ... GM0GYN
Gibson, P S ... GW1ENG
Gibson, R ... G0GFF
Gibson, R ... GI3NPP
Gibson, R ... GW4WFP
Gibson, R J M ... GM7MTK
Gibson, R S ... G3JAX
Gibson, R S ... G4WGR
Gibson, R W ... G3RWG
Gibson, T H M ... G4LQC
Gibson, W S M ... GM0KMG
Gibson-Ford, K R ... G6XDY
Giddens, D A ... G3IKB
Giddings, B J ... G1JLG
Giddings, K G ... G6YDS
Giddings, M R ... GM3XLB
Giddings, N P ... G8VZR
Giddings, R C ... GW0RCG
Giddings, R J ... 2E1FEP
Giffen, I ... GM4MIG
Gifford, B J ... G0MPZ
Gifford, G M ... G1ACB
Gifford, G V ... G4PFK
Gifford, I R ... 2E1CHK
Gifford, P W ... G3AWP
Gifford, R C I ... 2E1EWC
Gifford, R C I ... M1BRP
Gifford, R M ... G7SGM
Gigg, B M ... GW1UVM
Gilbert, A G ... G0OMD
Gilbert, A P ... M0APH
Gilbert, A R ... G4ENW
Gilbert, B ... G3NZU
Gilbert, B E ... G8UXB
Gilbert, B H ... G0BOO
Gilbert, B H ... G8EWF
Gilbert, B J ... G7OYF
Gilbert, B J D ... G6IUS
Gilbert, C A ... G6HVZ
Gilbert, C N ... G7DYU
Gilbert, D ... G0NFA
Gilbert, D ... G0VZS
Gilbert, D B ... G1MHP
Gilbert, D J ... GW3OYL
Gilbert, E C ... G8TMM
Gilbert, E J ... G3YBE
Gilbert, I ... G0FNF
Gilbert, J M ... G0OFD
Gilbert, L ... G1CJC
Gilbert, P J ... G7PXR
Gilbert, R A ... G3YVI
Gilbert, R C ... G0ROB
Gilbert, R J ... GW3HDR
Gilbertson, G R ... G4SGU
Gilbertson, V W R ... G1ZPZ
Gilbey, A W ... G4YTG
Gilbey, D G ... G1ITL
Gilbey, D L J ... G1DEQ
Gilbody, A J ... GI1HZX
Gilbody, C M ... GI4XFS
Gilbody, H B ... GI4WVN
Gilbourne, N R ... G0WPM
Gilbraith, B S ... G7MXQ
Gilby, P B ... G8AFU
Gilchrist, A J ... G8BVJ
Gilchrist, D A ... G1JRW
Gilchrist, G E ... GI0JQH
Gilchrist, H ... GM0EWK
Gildersleve, I R ... G3YAR
Gilding, A R ... G3KSH
Giles, A W ... G4OJH
Giles, B H ... G7SSH
Giles, D W ... G7OXA
Giles, F J A ... G8RRL
Giles, G W ... G6VRU
Giles, H E E ... G7UES
Giles, J R ... G0IIM
Giles, M ... G4FBF
Giles, M ... G7KNM
Giles, P J ... G0GRY
Giles, R ... G0VWV
Giles, R W ... G4LBH
Giles, S C ... G4FDI
Giles, T G ... G4CDY
Giles, W H ... 2E1BWA
Giles-Holmes, M P ... G4IML
Gilfillan, A J ... G0FVI
Gilfillan, A J ... G8FNU
Gilfillan, J B S ... GM3BQN
Gilham, D E P ... G7LNM
Gilham, R ... G6OKB
Gilhooly, F A ... GM0AXX
Gilks, R D ... G8ECJ
Gill, A J ... G0SNK
Gill, B M ... G7HSV
Gill, C ... G3XKL
Gill, C ... G4CXU
Gill, C A ... GD7ESU
Gill, C J ... G0CCB
Gill, C J ... G8EEM
Gill, D ... G6IIK
Gill, D ... GW4YCO
Gill, D J ... G4ABU
Gill, D W ... GM0HLV
Gill, E E ... 2E1ASP
Gill, J ... 2E1FDZ
Gill, J ... G4YEW
Gill, J D ... G4FBE
Gill, J F ... G0RQG
Gill, J O ... G3UAE
Gill, J V L ... G6AIK
Gill, K ... G7OMF
Gill, K C ... G3EEQ
Gill, K J ... G1DRI
Gill, K T ... GW4YDX
Gill, M ... 2M0ALS
Gill, M ... G0PLL
Gill, M H ... G3VJX
Gill, M I ... G1EUV
Gill, M J ... G0WAN
Gill, P C ... G3YTE
Gill, P D ... G4IEV
Gill, P J ... G0WID
Gill, R A ... G3CXP
Gill, R A ... G3MQI
Gill, R A ... G3NKJ
Gill, R A ... G3ROQ
Gill, R E ... G7BHG
Gill, R H ... G4KOY
Gill, R J ... G8DSU
Gill, R S ... 2W1AOK
Gill, T C ... GM7SLZ
Gill, T H ... GW0KSZ
Gill, T M ... G8IBO
Gill, T V ... GD4TVG
Gill, W ... G0PPK
Gill, W ... G1PFZ
Gill, W ... GI1ZYY
Gillain, L J ... G4YEO
Gillam, G F ... G3ZHA
Gillard, A W ... G1CGP
Gillard, B N ... G4VVP
Gillard, F W ... G7CBL
Gillard, J H ... GM4FNV
Gillard, M J ... G4RGJ
Gillatt, D A ... 2E1CIK
Gille, J M ... G0FZV
Gilleard, T B ... G8ZLF
Gillen, K ... G0VRS
Gillen, K H ... 2M1FKI
Gillen, M A ... 2E1FTX
Gillen, P J ... G4GVW
Gillen, S ... G0WYZ
Giller, J ... G1JGT
Gillespie, A W ... G6MCN
Gillespie, D A ... GI0OZQ
Gillespie, F ... G7THI
Gillespie, F W ... G7UPU
Gillespie, I N T ... MM0AMY
Gillet, E ... GW7PFK
Gillett, B J ... G1SWI
Gillett, D J ... G3WAG
Gillett, M S ... G6RKV
Gillham, A J ... G3ING
Gillham, F W H ... G0KUU
Gilliatt, R H ... G8BGL
Gillies, A B ... GW4XSP
Gillies, D B ... GM1PKN
Gillies, D B ... MM0AMW
Gillies, D W ... GM4HSR
Gillies, N ... G7BCK
Gilligan, P J ... G4UXR
Gilliland, F M ... GI1MJJ
Gilliland, P J ... G8SGF
Gilling, R D ... G0AHV
Gillingham, I J ... G0FLB
Gillingham, R J ... G6TCJ
Gilliver, J P ... G6JPG
Gillmore, D J ... G0DER
Gillon, A S ... G1GWJ
Gillott, D A ... G4TMZ
Gillott, J ... G6YOR
Gillott, M I ... G1KPZ
Gillott, W ... G7BZE
Gillson, I D ... G1MZW
Gilman, A J ... G4GFD
Gilmartin, J P ... G7WAZ
Gilmore, D L ... GI4KIX
Gilmore, J S T ... G6MRU
Gilmore, S J ... MI0BAT
Gilmore, T M ... G0UXR
Gilmour, J S ... GM8EJS
Gilmour, T C D ... GM1VZG
Gilowski, E A ... G7VHZ
Gilpin, J A ... G6REA
Gilroy, W K S ... G6YIW
Gilruth, J ... GM1VWA
Gilson, P ... G3WSZ
Giltrow, D R ... G8GJF
Giltrow, N ... G8GJG
Gilzean, I R ... G8ZRD
Gimber, G L ... G7PMV
Gingell, R M ... G6BUY
Ginger, N J J ... G1IFV
Ginger, W ... M1BSY
Ginn, K R ... G8NDL
Ginn, R ... G7CQA
Ginnever, M P ... 2E1DFQ
Ginsberg, D W ... G0JGX
Ginsburg, B ... G1INI
Ginsburg, R ... G1INJ
Ginty, W J ... G6FTP
Gipp, D E ... G4OCU
Girdham, M H ... G6RML
Girdler, T D ... 2E1CGW
Girdler, T D ... M1ACS
Girgis, M S ... G1JRX
Girling, A R ... G0UDK
Girling, C H ... G8XCR
Girling, D P ... G4POT
Girling, R A ... G4FCD
Girt, J L ... G1HSL
Girvan, W ... GM2FVV
Gisborne, G G ... G1YLC
Gisby, A M ... G0UZD
Gittins, D A ... G6NVP
Gittins, E W ... GW6YDT
Gittins, P J ... G8ITY
Gittoes, T S ... GW7JYJ
Giudice, G ... GW6TQH
Giuliani, G G ... G0WMX
Giuliano, C ... G7UQC
Givens, A B ... GM3YOR
Gizzi, F ... G6XQR
Gladden, R R ... G4IRY
Gladdis, L V ... G1EUW
Gladman, C D ... M0AWN
Gladman, C S J ... 2E1EWY
Gladman, T V ... 2E1EXD
Gladman, T V ... M1CEA
Gladwell, R ... G3DXB
Gladwin, D ... G4WLV
Gladwin, S J ... G0BFQ
Gladwin, T W ... G3UFA
Gladwish, D J ... G6HVX
Glaisher, P E ... G4RWW
Glaisher, R L ... G6LX
Glanfield, J M ... 2E1GBK
Glanville, H C ... G6MCT
Glanville, J H E ... G3TZG
Glanville, P ... GM7JFR
Glanville, S J ... G0OGL
Glaser, A N ... G3ZEN
Glasgow, R E ... GM4UYZ
Glashan, A G ... GM4JCM
Glasper, B F ... G0BKC
Glass, J E ... G7DPE
Glass, J V ... G4OJG
Glass, R ... G8UEA
Glass, S F ... G4HSK
Glasscock, L J ... G6IGK
Glasscock, P E ... G7HMC
Glasscott, E M ... G3TSF
Glassey, R B ... G0VTQ
Glassford, B ... G0EQJ
Glaze, F B ... G6MSD
Glazebrook, K R ... G0DPO
Glazier, M S ... G1HSI
Glazzard, S D ... G1XYO
Gleadall, S J ... G7WIM
Gleadall, S J ... M0BBW
Gleave, B M ... G1JPT
Gleave, C ... G6SLN
Gleave, D ... G6JPT
Gleave, P ... G6MCS
Gleave, W ... G8YWK
Gledhill, A J ... G6MCI
Gledhill, C D ... G4NWF
Gledhill, P ... G0LFS
Gledhill, R A ... G3ZYN
Gledhill, V M ... G4JIN
Gleed, B F ... G0IOU
Gleed, J B ... G1STL
Gleek, D A ... G0MOX
Gleek, V ... G4WIS
Gleeson, R ... G6TCD
Glen, A W ... G0OQR
Glen, M A ... G0DQS
Glendinning, G ... GW0EJP
Glendinning, K B ... GM4EZJ
Glendinning, M ... GM7GIS
Glendinning, S A ... GI7JKM
Glenister, A ... G0VNN
Glenister, L C H ... G4GLX
Glenn, C S ... G4ZCR
Glenn, I G ... G8MEX
Glenn, J D ... G7RIE
Glenn, W ... GI4KUM
Glennon, J ... GM0ZAM
Glennon, M J ... G4JVZ
Glenwright, I L ... G1IUY
Glew, W J ... G4NEG
Gliddon, W F ... G4NGB
Gloistein, M E P ... GM0HCQ
Glossop, E ... G4LJQ
Glossop, I ... G0SFH
Glotham, G A J ... G0CLX
Glover, C B ... G6OKA
Glover, D ... G6RMA
Glover, D ... G7TRV
Glover, D J ... G6AZP
Glover, F W ... G6HWA
Glover, G E ... G4KPB
Glover, G N ... G3AAV
Glover, H L ... 2E1FZE
Glover, J ... G3FIC
Glover, J P O ... G8YEJ
Glover, J S ... G4TOX
Glover, K ... 2E1FHD
Glover, K R ... G7HHN
Glover, M ... GW0BKJ
Glover, M J ... G0PDM
Glover, M W ... G0ISK
Glover, P ... G1ITP
Glover, R ... G8IUC
Glover, R ... G0WGP
Glover, R A ... G4OWW
Glover, R M ... G7SMB
Glover, T W ... G6MSC
Glover, W B ... G4BQW
Glover, W D ... M0AAN
Glover, W L ... G0EOR
Glozier, A E ... G3CRR
Glynn, A J ... G4WZS
Glynn, M D ... G3AAS
Glynn, R ... G0NFR
Goacher, A T ... G1OBQ
Goacher, D J ... G3LLZ
Goad, G ... G0SIO
Goad, R W ... G4EFA
Goadby, C M ... G8HVV
Goadby, D V ... G8BZN
Goadby, P G ... G3MCP
Goan, S D J ... G1AUU
Goatcher, F ... G4KNL
Goben, P ... G4BVV
Goben, P ... G4SKM
Gobey, P ... G8IYG
Goble, C F ... G4OZX
Godber, P V ... G4YTF
Godbold, N J ... 2E1GAT
Godbold, P L ... G4UDU
Godbold, S A ... G0NRX
Godbold, S M ... G0WQZ
Goddard, A ... G7TEA
Goddard, A B ... G6NUD
Goddard, A G ... G0FLH
Goddard, A G ... G3NQR
Goddard, A M ... G1STK
Goddard, B A ... G7IYI
Goddard, B L ... G4FRG
Goddard, C ... G4XAN
Goddard, C A ... G4LAA
Goddard, D ... G3UQH
Goddard, E C ... G7BYD
Goddard, F H P B ... G4OVS
Goddard, G H ... G0GNW
Goddard, G P ... G6ERQ
Goddard, G W ... G6DDU
Goddard, H ... G7GGY
Goddard, J M ... G0JOM
Goddard, L J ... G0MVB
Goddard, M ... G0OJU
Goddard, M J ... G6MCY
Goddard, M R ... G0GBU
Goddard, N D ... G7UWU
Goddard, N L ... G0OAS
Goddard, N R ... G3UXR
Goddard, R ... G7SAM
Goddard, R ... M0AZB
Goddard-Watson, H A ... G7JNN
Godden, C F ... G4BXI
Godden, I G ... G4CZX
Godden, L ... G1HSJ
Godden, M ... G0ACQ
Godden, N ... G7GSC
Godden, W R ... G4GCE
Godding, D ... G0VHK
Godfrey, B J ... G0OVC
Godfrey, D ... G1KWS
Godfrey, D J ... G0KIU
Godfrey, D P ... G4FYL
Godfrey, E H ... G3GC
Godfrey, E L ... GM3IRV
Godfrey, E R ... G6ERG
Godfrey, F I H ... G1NSK
Godfrey, J ... G0WOS
Godfrey, K E ... GW3VEW
Godfrey, M P ... G4YLC
Godfrey, P A ... G8JBD
Godfrey, P F ... G4BAN
Godfrey, P J L ... 2E0ALQ
Godfrey, R ... 2E1ALO
Godfrey, S ... G7AJR
Godley, R H ... G3WCK
Godlieb, E R ... G4XRG
Godolphin, P D ... 2E0AOK
Godolphin, P D ... G4XTA
Godsave, M E ... G1IHI
Godward, C J ... G1GDJ
Godwin, A E ... G0WYN
Godwin, A E ... G6SVZ
Godwin, B W ... G8AOL
Godwin, B W ... G8TNK
Godwin, D ... G2DOJ
Godwin, D J ... G7NJG
Godwin, D R ... G4YIX
Godwin, E J ... G0PCB
Godwin, J P ... G0NHG
Godwin, K T ... G0PCA
Godwin, N S ... G6XQT
Godwin, P S ... G4III
Godwin, R C ... G0IXF
Godwin, R L G ... G3VDH
Godwin, S ... G6YWI
Goff, R C ... G4FON
Goff, R E ... GW6NMO
Goff, R F G ... G1EDA
Goffin, L C ... G0VDR
Goffin, L C ... G7RPK
Gohl, M J G ... G7BAC
Gold, A R ... G3SKR
Gold, J F ... G6SVY
Gold, J P ... G1PFY
Goldberg, J L ... G3ETH
Goldbey, J ... G4DUW
Goldblatt, R ... G0ABK
Goldby, M R ... 2E1DUW
Golder, M J ... G8NVQ
Goldie, A M ... GM0DEX
Goldie, J ... G0UVT
Goldie, W ... GM4GIH
Golding, A R ... G3UKD
Golding, B G ... G6AUR
Golding, D C ... G8LNC
Golding, F R ... G8MIF
Golding, H J ... GI4SPT
Golding, J L ... G7VHX
Golding, M M ... G0MMG
Golding, N A ... G6RIG
Golding, P A L ... G1KLW
Golding, R N ... G3SRT
Golding, R N ... G3VZG
Golding, S ... G0NBK
Golding, S H D ... G3FXV
Goldingay, C J ... G4DFC
Goldman, G A ... G6CXQ
Goldman, M H ... G4LCB
Golds, P R ... G1ZZC
Goldsbrough, M A ... G3WOP
Goldshaker, D ... G8NHF

Goldsmith, A F G3VIG
Goldsmith, E H G3GRW
Goldsmith, J B GW0DLW
Goldsmith, J O G4KTX
Goldsmith, J P J G4CJG
Goldsmith, P R G8ISM
Goldstraw, W N G0DTQ
Goldsworthy, M R D G8TZY
Goldsworthy, T H G4BHD
Goldthorpe, P A G0AQS
Goligher, R T G G GI4LIF
Golightly, G G G0IGH
Golightly, J F G0IGC
Golightly, J W G4FOQ
Golledge, D J G4WVL
Golledge, P R G3EDW
Golledge, R W G6GCU
Golley, C R L G4JYF
Golton, J W G7VMJ
Gomer, L D G0MER
Gomes, G E V G7BFT
Gomez, C D A G6BYF
Gomm, F A S G6CNR
Gonella, D W MM1BTJ
Gonsalves, T H G0OYJ
Gooch, B G6IRR
Gooch, L J G6PTI
Good, B G G1NYB
Good, H M G0NAT
Good, M J G8AVA
Good, P G7HCL
Good, P C G8SAD
Good, W R G7VQO
Goodacre, D R G7EXG
Goodacre, J H M G G6GO
Goodacre, K M M1AVZ
Goodall, A L G4RFP
Goodall, B L G6NRA
Goodall, B S G8BUB
Goodall, D G0GYH
Goodall, D E 2E0AFO
Goodall, D E G7UEC
Goodall, D J G6VZS
Goodall, E F GW4WJU
Goodall, J G0SKR
Goodall, J GI4OCV
Goodall, J D G0BOR
Goodall, M J G0MGI
Goodall, M T GW8ZMU
Goodall, N J D GM7GNO
Goodall, R G3ONQ
Goodall, R F GM0OGZ
Goodall, R W G4XZD
Goodall, S 2E1FTA
Goodall, S H G3PTT
Goodayle, P J G7TOI
Goodchild, D J M1AGN
Goodchild, K H G1FBU
Goodchild, K T G6EFO
Goodchild, R G4CLL
Goodchild, R G G8FTW
Goode, A G2DTQ
Goode, A M G4TXE
Goode, C N S GM4HHY
Goode, J G GM1ENI
Goode, J P G0BLF
Goode, M G G4SMA
Goode, P G7HOE
Goodearl, G I 2E1FPI
Goodearl, M J G4XNO
Gooden, R C G0CWW
Goodenough, P D G3TJS
Goodes, F M G7IBN
Goodes, M C G1LHD
Goodey, J A G6XSY
Goodey, M G3RPC
Goodey, M G4BRA
Goodey, M J G0GJV
Goodfellow, G E G8OBH
Goodfield, G G GW4CNL
Goodfield, S J G0GNM
Goodfield, S J M1BFA
Goodge, R G6SVX
Goodger, B F G3DGR
Goodger, F A G0GOX
Goodger, P J G0BAI
Goodger, R C G0CFQ
Goodhand, C J G6MWD
Goodhead, C J G6NMR
Goodhew, A D G0BAR
Goodhew, A D G0ISF
Goodhew, A D G7BAR
Goodhew, B W G8ONY
Goodhew, L P M1BMB
Goodier, B G1SYV
Goodier, B D GW0PRM
Goodier, G C G6PMW
Goodier, I G0UWK
Goodier, J G4KUC
Goodier, J G6AZV
Goodier, J G8VHF
Gooding, D M G0RKK
Gooding, E P G3WYG
Gooding, G L G7JRN
Goodings, A S G1GER
Goodings, J P G8VAD
Goodison, A G1GAI
Goodkin, N M G0WKE
Goodkin, N R G0WOQ
Goodlad, T A GM3ZET
Goodlad, T A GM4LER
Goodliffe, J G7PLE
Goodman, A D G7PCH
Goodman, A F 2E1EHH
Goodman, D 2E1FYT
Goodman, D A GI4JFP
Goodman, E S G4LEM
Goodman, F A G6YWK
Goodman, H H G3PGW
Goodman, H K G7RVD
Goodman, J C G4PIJ
Goodman, J C G6SFR
Goodman, J P 2E1EHF
Goodman, K S G7TXE
Goodman, M J G4UQA
Goodman, N K G4XFG
Goodman, P B G4LKT
Goodman, P L 2E1EHG
Goodman, P L G7ALR
Goodman, R V G3KOB
Goodrich, M C M0AOM
Goodrich, M J G3YAD
Goodridge, M H GW0VND
Goodrum, M G3ZQU
Goodship, D A G0DIO
Goodson, A D G8TSQ
Goodson, J E G4CFX
Goodson, J G G6VET
Goodson, P W G4PCF
Goodwill, D G8DBY
Goodwill, D W G1VAB
Goodwill, M C G4ZUY
Goodwin, A F G3XDR
Goodwin, A J G7ABL
Goodwin, C E G7NKB
Goodwin, C M 2E1EQW
Goodwin, D G8KSC
Goodwin, D C GW0LJW
Goodwin, D D M0BDU
Goodwin, D G G4SOT
Goodwin, E D GM0LTJ
Goodwin, E M G1XHR
Goodwin, E S J GW0MSW
Goodwin, G S G0IHA
Goodwin, G W G0LNB
Goodwin, H S G1NSG
Goodwin, J GW7KZS
Goodwin, J A G0PRF
Goodwin, J M G6CNX
Goodwin, J W G0NBH
Goodwin, K W 2M0ACT
Goodwin, L P G6ANO
Goodwin, M B G3WKR
Goodwin, M J G7NBE
Goodwin, P 2E1CPQ
Goodwin, S G7NAR
Goodwin, S M G1VWP
Goodwin, T 2E1AKK
Goodwin, T G0HOA
Goodwin, T M G7TBM
Goodworth, K G0RFV
Goody, B D G7VGC
Goody, T S GM0MXZ
Goody, T S GM0PSS
Goody, T S GM7NSS
Goodyear, B G6AUP
Goodyer, G H G6NMQ
Goodyer, G H G8WSX
Goodyer, T G0ATG
Gookey, A A F G6SBU
Goolding, B C G0UVN
Gooman, M D I M1ACN
Goom, V W G4AMW
Goozee, H B G6GFJ
Gordon, A G4BCT
Gordon, A G4TTB
Gordon, A G7GJV
Gordon, A GM4PCT
Gordon, A M G3XOI
Gordon, A R MM1BVW
Gordon, A R B GM6RXQ
Gordon, B J G4GHP
Gordon, D G0MRD
Gordon, D C G4OG
Gordon, D J F GW6ZUQ
Gordon, D T G6ENN
Gordon, E G1SWP
Gordon, E G4VMU
Gordon, G G4SAL
Gordon, G GM6KFO
Gordon, H M E GM7KRQ
Gordon, I L G8IFT
Gordon, I P G6ENU
Gordon, J A G4LIA
Gordon, J C D G4XUI
Gordon, J L G7WGI
Gordon, J R 2M1AZJ
Gordon, K C 2M1DIQ
Gordon, K W G4AQJ
Gordon, M J G4BRW
Gordon, N D GW1AUT
Gordon, N H G4CDQ
Gordon, P T GD3GCE
Gordon, R A G3UHJ
Gordon, R B G4HCG
Gordon, R L G6FYW
Gordon, R R G0NMG
Gordon, S J G6ENS
Gordon, S J G7GKN
Gordon, S J GI7GRY
Gordon, T J G6MWB
Gordon, W A G3SEG
Gordon, W H G GM8ZLI
Gordon, W J GI4DXK
Gordon-Laycock, W A G3XYD
Gordon-Smith, D G3UUR
Gore, S G1INL
Gore, S J 2E0AGV
Goring, R G6VYT
Gorman, B G1VTQ
Gorman, D D G0OOG
Gorman, D J G6GEY
Gorman, H G0MVG
Gorman, N A G7VYJ
Gorman, P K G1FXE
Gormley, V G4GVG
Gornall, A G6IVB
Gornall, R C G7DME
Gorny, V A G4HHL
Gorrill, D F GM4UJZ
Gorse, D G G7HRQ
Gorse, R G4AOM
Gorsuch, I G1AOQ
Gortmans, N A G0LPZ
Gorton, A G0HHC
Gorton, J M G4JMG
Gorton, J R G4UTJ
Gorton, R J G6VPH
Gorton, R L G3NIQ
Gorvett, D P G6YJA
Gorwill, J A G0CEE
Gorwits, S B G8NKN
Gosbee, K H G6NTW
Gosden, A J G7GDC
Gosling, B G6KVI
Gosling, D J G0NEZ
Gosling, P W 2E1BHU
Gosling, R J G0NNI
Gosnell, P C G0PLC
Gosney, C L G0AAP
Goss, G F G6DEA
Goss, G R G0RKS
Goss, R J G0VTP
Gossage, J G7TIU
Gossington, J D 2E0ANK
Gostick, P F G0SJV
Gotch, D J G8FTX
Gotch, M S G0IMG
Gott, G F G3MUO
Gott, G W J GM6PFJ
Gott, I F G1KKS
Gotts, M S G7HHT
Gottschlich, G F H GM4XUC
Goudge, L G G6UOU
Goudie, W GM4WXQ
Gough, A J GM7VUJ
Gough, C P G1URR
Gough, D M G0DZH
Gough, E W G4ZEY
Gough, J GW3WXA
Gough, J G GI0TQD
Gough, K M GW1SRB
Gough, L F GI1BEU
Gough, L W G0CWO
Gough, M S G7IJQ
Gough, N J G6EFR
Gough, R G0BET
Gough, R B G8RUX
Gough, R F G3AWK
Gough, S L G0SLG
Gough, W B G4DDJ
Goulborn, D G G0VBB
Goulbourne, D J G4EHK
Goulbourne, D V E G0NPK
Gould, A J G3GHN
Gould, A J G3JKY
Gould, A M G1AMG
Gould, A T G4UAM
Gould, B F A G1VKA
Gould, B P G0FZE
Gould, D I G3UEG
Gould, E J G7KGI
Gould, G P G6UMP
Gould, G P M0AZJ
Gould, H M G1FZE
Gould, J G6JPQ
Gould, J GI3SUM
Gould, J GW7NIQ
Gould, J W G4POD
Gould, M A G4OKE
Gould, M A M1AZA
Gould, P A G1DVW
Gould, P A G6DBY
Gould, P A G7PAG
Gould, P D G0EDY
Gould, P L G4FVZ
Gould, R M1AWF
Gould, R A G0VKM
Gould, R J E G8PSO
Gould, S P GM0ULK
Gould, S S G0VJJ
Gould, S S G3WTP
Gould, S S M1BED
Goulden, D G1JZU
Goulden, J S G0TBM
Goulden, W G6XQW
Goulding, L H G4EPW
Goulding, R D 2E0ANI
Goulding, R G GW3GWA
Goulding, S G G0WMO
Gouldsbrough, E G1VVN
Gouldstone, R H J G3TAG
Gould-Fellows, C J 2E1FHV
Goulsbra, D C G4UHZ
Goulty, J A G8BBV
Gourlay, A C GM0OYE
Gourlay, B F GM0LBR
Gourley, D J G0MJY
Gourley, G D G0OZJ
Gove, C A GM4LCJ
Gover, S J M1AXJ
Govier, J G6YJD
Gow, H B F G3LAG
Gow, P K J G7VHJ
Gowans, J L GM4JRG
Gowen, P J A G3IOR
Gower, C G6CNZ
Gower, D C G1ROH
Gower, D O G4GRJ
Gower, G A G0VAY
Gower, I W G8VHG
Gower, R H G1XUP
Gowers, D A G0IZV
Gowland, D G4LGA
Gowland, D G4PFQ
Gowland, G L G1GCY
Goy, A M G4HJD
Goy, S A G1PLU
Grace, D GW4OUU
Grace, J G3VVR
Grace, P D G4MPG
Grace, P R G8XXZ
Gracey, V I GI3WEM
Gracie, P A C G1FBS
Graffham, A J G1SWY
Graffham, M F G6AUO
Graffham, O P G1JDE
Grafton, E C G2CGL
Graham, A H G7OCC
Graham, B P G4LGB
Graham, C S M1BPH
Graham, D G3XHQ
Graham, D G6XSZ
Graham, D C G1JPE
Graham, E C G0CTM
Graham, F R GM4IEO
Graham, G GI3JCD
Graham, G GM4ZHL
Graham, H GM3VVM
Graham, I M1ATZ
Graham, I R G7BZH
Graham, J G6HFW
Graham, J GW6CNS
Graham, J A G3TXL
Graham, J A G4LIC
Graham, J A GM8LRI
Graham, J C G0GHB
Graham, J D G3HDT
Graham, J E G8PZU
Graham, J R GM1XPE
Graham, J W GM7UVS
Graham, K G7IVG
Graham, K G GM0AVB
Graham, K P G8ZWU
Graham, M B G3XMG
Graham, M J G6EXE
Graham, M T G7EUG
Graham, N M GI3RXV
Graham, N T G6ENY
Graham, P G0PJY
Graham, P G7WGB
Graham, P J G6JZE
Graham, P R G3GLK
Graham, R G0ESW
Graham, R G7PTH
Graham, R GM2DTB
Graham, R G MI1BVO
Graham, R N G3OAY
Graham, S G4YQE
Graham, S G7UPX
Graham, S GM4ZFS
Graham, S M0BCV
Graham, S J G0THS
Graham, S R GM7DRY
Graham, T GM7UFN
Graham, T M1APV
Graham, V E GI1BEW
Graham, W C GM1SYC
Graham, W J GM3GDS
Graham Cbe, D G G0VGJ
Graham Cbe, D G G7PLQ
Graham-Brown, E C GW7VMT
Graham-Kerr, W A G7FTJ
Grainge, B F G3JPM
Grainger, B A G4TOG
Grainger, D A G4UQM
Grainger, J H G3OZE
Grainger, M A G0VQB
Grainger, P G0SLN
Grainger, P W G4XWR
Grainger, T G4GGL
Grainger, W J G3JYO
Granby, P W GW4OKF
Grandfield, D E G7TCH
Grandfield, H G0DOU
Grandshaw, J S G8IKP
Grandshaw, R H G7WBA
Grane, J A G6MCQ
Grange, A W G7AYD
Grange, M A G4JJX
Grange, R J GI3TDY
Grange, T S R G7MXL
Grangel, C G3ELI
Grannell, P K G3UOK
Grannell, P K G4TQB
Granshaw, A G6AZR
Grant, A G1VAG
Grant, A I GW4KPD
Grant, A J G0LBH
Grant, A J G7BMW
Grant, C A GM4XEP
Grant, D G3UWL
Grant, D G4UAY
Grant, D J G4EZK
Grant, D S G8GJQ
Grant, F GM0PKQ
Grant, G J 2M0APC
Grant, G M GM3UKG
Grant, H G1NOZ
Grant, I R G4MWG
Grant, J B G3TYA
Grant, J M G4OVT
Grant, J M L GM7UGV
Grant, J R 2E1EUI
Grant, J R G1XTP
Grant, J W GM0CTY
Grant, J W R G3SQN
Grant, L D G3XNG
Grant, L P G3TJU
Grant, M G0TZE
Grant, M A GM0JVC
Grant, M G G4DCS
Grant, P G4YDW
Grant, P A G6MRW
Grant, R G4UAC
Grant, R M C P GM4DQJ
Grant, R S G3TEI
Grant, R S G3XPH
Grant, S G0HBY
Grant, S GM7KHA
Grant, S D G1TSY
Grant, S P G6ENR
Grant, S W GM4YHS
Grant, T L G4IXI
Grant, T M GM3UDJ
Grant, T P G4VKJ
Grant, W A G4SMZ
Grantham, E J G7AKV
Grantham, G H G8PQB
Graseley, M 2E1FCT
Grassby, N W G4CPY
Grassi, M G0PRH
Grattan, H O C G3XKS
Grattan, K R GD4RGR
Gratton, E W GW4OCN
Gratton, G G1HCU
Gratton, H J G6GN
Gratton, T G GM3TGG
Graupner, L M GM0GNY
Gravell, A GW8VUV
Gravenor, W G G0KVM
Graver, A G8SEY
Graver, J F G7VYT
Graves, A G1KRU
Graves, D P GW0AYP
Graves, H A A G3LXW
Graves, J A G7CJO
Graves, J E G8NHM
Graves, J L G0KSJ
Graves, L A G4BCP
Graves, M R G7RUN
Graves, N P G1WKH
Graves, S J G1HCV
Gray, G8YYB
Gray, A 2W1CNN
Gray, A G0MYA
Gray, A GM0ART
Gray, A A C G4UEV
Gray, A C D G4SRI
Gray, A D J G7HRY
Gray, A G G1XGW
Gray, A G G3XQU
Gray, A K G4DJX
Gray, A L G7BWF
Gray, A L GW1BCI
Gray, A T G7GRQ
Gray, B G0GRR
Gray, B G4TFB
Gray, B G6WME
Gray, C 2E1ECL
Gray, C GM7IKA
Gray, C E G0HZS
Gray, D G4TFC
Gray, D G7CNC
Gray, D GM7NTJ
Gray, D F GJ3XOJ
Gray, D H K G1JDF
Gray, D J G0FLX
Gray, D J G0LEA
Gray, D L GM6KRO
Gray, D M G3YPL
Gray, D M G4FQV
Gray, D S J G0BOI
Gray, D W 2E1CLV
Gray, E C G3CPS
Gray, E H G0CTZ
Gray, E J M1BZC
Gray, F M G3TBH
Gray, G W G1HMT
Gray, H G7KOT
Gray, H A G3LTT
Gray, H D G1MNP
Gray, I K G0SNU
Gray, I L M G3VAJ
Gray, J 2E1FJK
Gray, J G0RQE
Gray, J G4GSE
Gray, J GM3PLO
Gray, J E GW6ZUS
Gray, J E GW7RLS
Gray, J E GW7SKC
Gray, J F GM3LRG
Gray, J F GM3ZRC
Gray, J L G0WIJ
Gray, J M G0ASL
Gray, J R G8XVK
Gray, K G4BDG
Gray, K A 2M1ASQ
Gray, K A M0AWJ
Gray, K P G0LFE
Gray, L A GJ0NAC
Gray, L C G3FTK
Gray, L W G1HTT
Gray, M 2E1FJL
Gray, M G6CKY
Gray, M G8AWO
Gray, M M0ABK
Gray, M A G1RIY
Gray, M F G0OXY
Gray, M J G4EPU
Gray, M J G6ZUT
Gray, M R G1HNU
Gray, N J G4OOW
Gray, O G7PEG
Gray, P A G8NNQ
Gray, P D GM0FWY
Gray, P J G0HYT
Gray, P J G1FLL
Gray, R G0DOB
Gray, R G4TQG
Gray, R G6SVV
Gray, R D GW1NWF
Gray, R H G7HIX
Gray, R H R G1AIV
Gray, R R G4AWO
Gray, R R G6VUK
Gray, S G0ASK
Gray, S G7FUV
Gray, S G7LHS
Gray, S GM1HTI
Gray, S E 2W1EYZ
Gray, S J S GM0NTW
Gray, S T GM6XQX
Gray, T W G6VEG
Gray, W H G1LSL
Gray, W H GW4IUY
Gray, W R G7MGP
Gray, W S GM6GQT
Graydon, D C G1EDE
Grayer, G H G3NAQ
Grayshon, P G G1AOR
Grayson, A K G1JTY
Grayson, D G1UDE
Grayson, E A G6OJX
Grayson, G W H G3YWI
Grayson, I J G3RYK
Grayson, M B G4OTE
Grayson, V R G0HYH
Gready, J L E Q GJ6ENP
Greatbatch, D G3YTV
Greatbatch, D G4KCU
Greathead, A P G3ZID
Greathead, J S M0AMA
Greatorex, D J GM0AEG
Greatorex, J M R G4PIM
Greatorex, K G0THF
Greatorex, P M G4FEM
Greatrex, A GW4OQB
Greatrex, M GW4HDB
Greatrix, B J G4ICZ
Greatrix, G P G7HNM
Greaves, A B G3JOX
Greaves, B W G4BJO
Greaves, C G7DYO
Greaves, F H G7POA
Greaves, G G3DZV
Greaves, G W G7VOS
Greaves, J G3UXM
Greaves, K G0PVE
Greaves, M G4VET
Greaves, R G4LUH
Greaves, R F G0MOH
Greaves, R H S G4JVV
Greaves, R L G4ZEO
Grech-Cini, H J G1PZB
Grech-Cini, W J G1PZA
Greed, G T G0LCQ
Greed, P K G1HDG
Greed, P T G3MQD
Greed, P T G3PZV
Greed, W G4GBX
Greed, W G G0MZQ
Green, A G6BFM
Green, A M1AEG
Green, A C G1GNA
Green, A E G8NRB
Green, A F G1TIE
Green, A J GW4JGU
Green, A J GW4UNV
Green, A P G7HSB
Green, A R G0CRE
Green, A R G7KXR
Green, A R G8BNG
Green, A T G3TRL
Green, A W G0OJJ
Green, A W GW4ZWO
Green, B C G1BFQ
Green, B D GW4HYZ
Green, B F G3PMI
Green, B H GW2FLZ
Green, B N G3KCB
Green, C G3TUJ
Green, C G4VUG
Green, C G8VJO
Green, C GW1WTZ
Green, C A G3CYI
Green, C A G4SAJ
Green, C J 2E1FHK
Green, D 2W1FUD
Green, D G1COV
Green, D G6ZBT
Green, D G8HPV
Green, D A G1DVV
Green, D F G0LJG
Green, D F G7DKX
Green, D H GW7NTZ
Green, D H MW0ALU
Green, D J 2E1ELH
Green, D J G0SZI
Green, D L G6XEB
Green, D R G4ZFV
Green, D R H G1PBU
Green, D W G0VAQ
Green, D W G4OTV
Green, E G0EAD
Green, E M R G4EZM
Green, E W G0ATS
Green, F G3OS
Green, F E A G3GMY
Green, F E V G4UVO
Green, G G6TQD
Green, G F G4KGV
Green, H G3AMH
Green, H G4PIK
Green, H G D C GW4VAG
Green, H R G0DXJ
Green, I B G1IXF
Green, I J G7CWI
Green, J G4UPI
Green, J A G0SXZ
Green, J A G1GWO
Green, J C G0IIF
Green, J D G4WSO
Green, J E G8SIM
Green, J G G3ZNV
Green, J K 2E1FPW
Green, J L G3PYF
Green, J L G3WLX
Green, J M G1DVU
Green, J O G8MKW
Green, J R G3WVR
Green, J T G0PBT
Green, J T G1EKX
Green, K G4CYC
Green, K G8PNZ
Green, K A G0SEW
Green, K H G1NAK
Green, K J G0PHP
Green, L G7VWE
Green, L A G6BZG
Green, L H 2E1FVT
Green, L P G6DPL
Green, M G1HYO
Green, M G4JGE
Green, M G4JHE
Green, M G4WSP
Green, M G6MDC
Green, M G7DYD
Green, M G7PFI
Green, M A J G6PVA
Green, M D G7IPN
Green, M E GJ1EXC
Green, M F G8NCS
Green, M J G0LPI
Green, M L G0SGQ
Green, M P G4PMG
Green, M R G0BMG
Green, M S GW0GAI
Green, N J G4SDY
Green, N M G7LGS
Green, P A G0ABI
Green, P A G6VTN
Green, P D G6RFV
Green, P D G6TCK
Green, P F G3EWM
Green, P J G4VZT
Green, P J GM7LAC
Green, P M G7VTS
Green, P P G8LQM
Green, P R G0ELM
Green, P R G4LWF
Green, P R G4MEB
Green, P R G6MEB
Green, P T G4PHL
Green, P W G2DLO
Green, R G0DCF
Green, R G1AUQ
Green, R G1PYU
Green, R G3ENO
Green, R G4JII
Green, R A G1BHV
Green, R A G4UDV
Green, R E G3TRG
Green, R E G8TRG
Green, R E G8ZLG
Green, R J G6PAJ
Green, R J G8MIH
Green, R S G8ZJX
Green, R W G1EDD
Green, S G0SRG
Green, S G1DVT
Green, S G4EKM
Green, S GU3EJL
Green, S B G4YZM
Green, S C H GM4JXP
Green, S E G3ISG
Green, S J G1INK
Green, S M GW4SMG
Green, S N G6JPM
Green, S R GM4HDE
Green, S R C G4DNA
Green, S W G4WER
Green, T G1LYO
Green, T B G7AJS
Green, T D G4NQN
Green, T G G6PMX
Green, T G G G7AHB
Green, T M G7ODW
Green, T N G3GLL
Green, V A G1IXE
Green, V E G8NJQ
Green, W G6HWF
Green, W D G8BCQ
Green, W E G4DMB
Green, W T G4KGX
Greenacre, D G0TGR
Greenacre, J G7NLP
Greenall, A G8WBU
Greenall, D 2E1EHC
Greenall, F B G0PMN
Greenall, I T 2E1EHB
Greenall, J D G8OWS
Greenall, J R GW0UDR
Greenall, J R GW4EIR
Greenall, N B G3BWR
Greenall, S J G7MXX
Greenall, W M1AIX
Greenan, P P GI3RNO
Greenaway, B F G3THQ
Greenaway, D J G8LUN
Greenbank, A G3ZVM
Greenbank, C S 2E1BOE
Greenbank, J M G6AUQ
Greenbeck, B P G8LIP
Greendale, S W G8OUS
Greene, B J G1ZYD
Greene, J O G0SZG
Greene, J O G1OWR
Greener, P G0ABS
Greener, P G4ULK
Greenfield, A H G4TPB
Greenfield, G W G6POF
Greenfield, G W G8WZT
Greenfield, J E G0VPZ
Greenfield, K J G7RQB
Greenfield, M G8MTB
Greenfield, P C G7DNW
Greenfield, S A G3PFP
Greenfield, S M A G1TWH
Greengrass, R E G4NRG
Greenhalgh, A G7FEO
Greenhalgh, A H G0JAB
Greenhalgh, D A G0IWN
Greenhalgh, D H G0KDB
Greenhalgh, E D 2W1BPC
Greenhalgh, E L P G0AQI
Greenhalgh, G GW0MOF
Greenhalgh, K M G7VMI
Greenhalgh, P N G0TON
Greenhalgh, P N G3XGE
Greenhalgh, R G4HRI
Greenhough, J E G4UIQ
Greenhough, R G1FYQ
Greenhough, R G4KMW
Greenland, C G4SEZ
Greenleaf, A C G6IGU
Greenless, G S GM4NSL
Greenley, J V G6OJV
Greenough, F G4EHY
Greenough, K S G8BEQ
Greenow, M G G6PTY
Greenshields, I K G4FSU
Greenslade, G E G4HPJ
Greensmith, G T G8JQS
Greensted, N C G8DBU
Greenstreet, N E K G4BOJ
Greenway, B F G7UAF
Greenway, D A G7UAG
Greenway, P GM0PRO
Greenwood, A G0IPN
Greenwood, A J G3SIQ
Greenwood, A S G1IYA
Greenwood, C G4KAM
Greenwood, C G4TFT
Greenwood, D A G7JXQ
Greenwood, D W G3FCS
Greenwood, G G3OAR
Greenwood, G H G3BMX
Greenwood, G L G4LIX
Greenwood, G W G4VTP
Greenwood, H G0VKK

Greenwood, I ... GI0AIJ
Greenwood, I B ... G4SCU
Greenwood, J ... G3KRZ
Greenwood, J B ... G3TUN
Greenwood, J D ... G0KNH
Greenwood, J F ... G3ZJY
Greenwood, L I ... G4LIW
Greenwood, M A ... M0AYP
Greenwood, M D ... G0KHM
Greenwood, M D ... G3YPE
Greenwood, N ... G7VRL
Greenwood, P H ... G2BUJ
Greenwood, R ... G4YOR
Greenwood, R B ... G0IFO
Greenwood, R C ... G4UFZ
Greenwood, R T ... G3LBA
Greenwood, R W ... G4EOO
Greenwood, S ... G0LRR
Greenwood, S A ... G4TWG
Greenwood, T R R H ... G4AYR
Greenwood, W H C ... G3GVH
Greer, B ... G0JEE
Greer, D D ... G4EEH
Greer, M ... G4UJO
Greer, N D ... GI4TBP
Greer, P W ... GI7WLS
Greer, R G ... 2E1BFF
Greer, T ... GI4TGR
Greevy, J B ... G6JVA
Gregg, I K ... G6PMS
Gregg, J R W ... G4MAK
Gregg, M ... G0KNN
Gregg, P N ... G0AHM
Gregg, T F ... G1WQU
Gregor, A G R ... GM4KIA
Gregor, G R ... G4OWH
Gregory, A ... G7AQF
Gregory, A ... G7AYP
Gregory, A K ... G4RJU
Gregory, A R ... G1ZDT
Gregory, C P ... G1NJJ
Gregory, D B ... G0SLV
Gregory, D E H ... G7SVJ
Gregory, E J ... G3ORW
Gregory, F J ... G3AQM
Gregory, H ... G3GIY
Gregory, H ... G3VDF
Gregory, H R ... M0ALL
Gregory, J ... G4PFO
Gregory, J ... G7ESV
Gregory, J A ... G4JWF
Gregory, J E ... G8IHA
Gregory, K H ... G3WEU
Gregory, L J ... G6HWH
Gregory, L R K ... G2AVI
Gregory, M ... G4HGM
Gregory, M P G ... G0JYQ
Gregory, M R ... G1HRH
Gregory, N ... G0HIK
Gregory, P ... G1MRZ
Gregory, P ... G6MRZ
Gregory, P C ... G4HXV
Gregory, P J ... G0BHH
Gregory, P L ... G7DSZ
Gregory, R ... G7VOE
Gregory, R D ... G4FQT
Gregory, R J ... GW8FNO
Gregory, R W ... G6KZI
Gregory, S ... G7ESI
Gregson, C ... 2E1ECR
Gregson, D W ... G6IGV
Gregson, P ... G1WHP
Grehan, K ... GM1JYV
Greig, A S ... G0CLM
Greig, C ... GM1OXC
Greig, C D ... MM0AOQ
Greig, D J ... G8KFJ
Greig, M ... GM0MMN
Grellis, C A ... G1YIL
Gresham, E ... G0JOU
Gresswell, C J ... G0WFH
Gresswell, D M ... G3PWY
Gresty, S ... G0FRB
Grevatt, D ... G7CEK
Grevatt, D M ... G7AIF
Grevatt, J ... G8VCH
Grevett, D ... G0BCW
Greville, B E ... G3JCW
Greville-Smith, L G ... G4SUJ
Grey, D R ... G0JCP
Grey, K ... G3WNR
Grey, M J ... G1IVI
Grice, J R ... G4BBO
Grice, N ... GW0MKP
Grice, P G ... G4INA
Grice, T ... G4PSL
Gridley, A B ... G4XPJ
Gridley, E J ... G4XUY
Grier, E P ... GM4YHA
Grierson, C ... GM4YLN
Grierson, M J ... G3TSO
Grieve, A D ... G7JSK
Grieve, G A ... G8VGO
Grieve, J F ... G4ARZ
Grieve, J G ... G6JHG
Grieve, J L ... GM6ENX
Grieve, J R ... GM1MWK
Grieve, J S ... GM0HTH
Grieve, J S ... GM0OTI
Grieve, J S ... GM8RTI
Grieve, L J ... GM0NTI
Grieve, W ... GM1TXE
Grieveson, G W ... G4GXU
Griffen, D J ... G3EGQ
Griffin, A ... G7NAN
Griffin, A P ... G7VMD
Griffin, B J ... G0CGE
Griffin, B J ... G0HEA
Griffin, C E ... G3NTG
Griffin, D H ... G8DLF
Griffin, D K ... G3XRK
Griffin, E E ... G1EUX
Griffin, F A ... G0IBX
Griffin, G R ... G0AQH
Griffin, I ... G4RZZ
Griffin, J A ... G6XQY
Griffin, J E ... G8CLW
Griffin, J R ... 2E0APK
Griffin, J S ... G6ZUV
Griffin, J W ... 2E1FQX
Griffin, K ... G4HDP
Griffin, M ... M0AGY
Griffin, M E ... M1AEP
Griffin, M J ... G3IIN
Griffin, M J ... G6ZEH
Griffin, M W ... GW8THM
Griffin, P ... G4IFU
Griffin, P A ... G0DJF
Griffin, R ... G4XFZ
Griffin, R ... GI7BET
Griffin, R J ... G0KLH
Griffin, S A ... 2E1AXK
Griffin, S M ... G1OHH
Griffin, S M ... G6DBZ
Griffin, W F ... M1BMG
Griffin Thomas, T ... GW8TYS
Griffith, C ... G1PLE
Griffith, D ... GW0OPY
Griffith, D G ... G0OAB
Griffith, J M ... GW0UKF
Griffith, R O ... G8LEM
Griffith, S J N ... G0WWP
Griffith, W G ... G0NZX
Griffiths, A ... G3PTM
Griffiths, A A ... GW7DZA
Griffiths, A C ... GW7DZC
Griffiths, A P ... G1XYP
Griffiths, A S ... G8LJY
Griffiths, A T ... G8VAE
Griffiths, B E ... G6NTY
Griffiths, C ... G7UHT
Griffiths, C P ... GW1HCW
Griffiths, C S ... G0HHU
Griffiths, D ... G0CEQ
Griffiths, D ... GW0JUJ
Griffiths, D C ... G3RDQ
Griffiths, D G ... G0LZR
Griffiths, D G ... G4PKV
Griffiths, D G ... GW7FGL
Griffiths, D H ... GW0LML
Griffiths, D J ... G7UMF
Griffiths, D L ... GW3XHG
Griffiths, D M ... G4DMG
Griffiths, D V ... 2E1FLU
Griffiths, E ... GW1TIU
Griffiths, E H ... G2ASX
Griffiths, E J ... G3TFG
Griffiths, F A ... G3MED
Griffiths, G ... G3ZIL
Griffiths, G ... G7IZA
Griffiths, G A ... G3STG
Griffiths, G A ... G4FOX
Griffiths, G A ... G4NRC
Griffiths, G B ... G0KQS
Griffiths, G C ... GW3ONN
Griffiths, G D ... G3POX
Griffiths, G J ... M1AIU
Griffiths, G P ... GW0PDB
Griffiths, G R ... GW4PXQ
Griffiths, G T ... GW3LHK
Griffiths, H ... G3GCI
Griffiths, H ... GW0OUH
Griffiths, H ... GW0WPT
Griffiths, H A ... GW3YCD
Griffiths, H D ... G7MQQ
Griffiths, J ... G1KFI
Griffiths, J ... G4ZZM
Griffiths, J ... GW7CBU
Griffiths, K D ... G1VDF
Griffiths, K E ... G3WIC
Griffiths, K R ... G0EVP
Griffiths, L ... 2E1DUM
Griffiths, L ... GW6PFK
Griffiths, M ... G6IVC
Griffiths, M A ... G1ZHN
Griffiths, M A ... GW0NNY
Griffiths, M E ... G6KIZ
Griffiths, M J ... G3WLG
Griffiths, M J ... G7OGP
Griffiths, M J ... MW1AND
Griffiths, M W ... GW7TUU
Griffiths, M W ... GW8UXI
Griffiths, N ... G0WPO
Griffiths, N ... G7UUG
Griffiths, P ... G0KEP
Griffiths, P G ... G8FGY
Griffiths, P K ... GW8PNE
Griffiths, P N ... G7RRR
Griffiths, R ... GW0DIV
Griffiths, R A ... M1CAR
Griffiths, R C ... G3HPO
Griffiths, R D G ... G0AUF
Griffiths, R F ... G3JTQ
Griffiths, R G ... G0ISB
Griffiths, R O ... GW0IXK
Griffiths, R S ... GW6JVB
Griffiths, R W ... G7NHB
Griffiths, S G ... G2CGF
Griffiths, S R ... G4SWN
Griffiths, S R ... G4SWO
Griffiths, T E ... GW0ULX
Griffiths, T G ... G4GRN
Griffiths, T J ... G3NPZ
Griffiths, T J ... G7PEE
Griffiths, W ... GW0UHJ
Griffiths, W A ... GW4IEU
Griffiths, W E ... G6CVW
Griffiths, W F ... GW1AUO
Griffths, J ... G1UJV
Griggs, G H ... G0MAI
Griggs, A C ... G4KMB
Griggs, D A ... G0IPT
Griggs, M J ... G4YJN
Griggs, S B ... M0ASJ
Griggs, S G ... G7BKM
Griggs, W G W ... G7EYO
Grigor, A B ... G7TPW
Grigsby, J E ... G3WNG
Grigsby, N R ... M1AHQ
Grigson, P J W ... G0TLE
Grimbleby, T C ... G4MPL
Grime, A ... G7EQK
Grime, D J ... G1CFC
Grime, K W ... G4PMV
Grime, M ... G4PSE
Grimes, A G R ... G0GVZ
Grimes, B R ... G0LGZ
Grimes, M R ... G6NMK
Grimes, S I ... G1VIS
Grimmett, A ... G6HEE
Grimsdale, R L ... G8USF
Grimsditch, C J ... M1BAT
Grimshaw, D ... G6JPN
Grimshaw, G ... G3TQX
Grimshaw, G T ... G3PWN
Grimshaw, K ... 2E0ANN
Grimshaw, K ... G7RVF
Grimshaw, P ... G4FYS
Grindel, E ... GD4VGM
Grindell, D W ... G3RLL
Grindle, M ... GW1ZKE
Grindrod, A ... G4EXF
Grindrod, A M ... G8WCT
Grindrod, M P ... G4EDY
Grint, E S ... G4XLY
Grint, W G ... G1FGK
Grisdale, G L ... G5GZ
Grisley, E J ... G7TQX
Grist, E B ... G3GJX
Grizzell, A J ... 2E1DPX
Grizzell, C M ... M1AFZ
Grob, J H ... MI1BNE
Groeber, N E ... G1FDD
Groeger, J R ... G4XXW
Grogan, W ... G4APP
Gronbech, P D ... G6ULP
Groom, C M ... G1GDT
Groom, D M ... G1VJT
Groom, E L ... G3ONE
Groom, G L ... G3YLC
Groom, I C ... G0FCA
Groom, J ... G8FQM
Groom, J C ... G0CMW
Groom, K K ... G0VTV
Groom, M J D ... G3WME
Groom, P B ... G4FIE
Groom, P B ... G6YWN
Groom, R J ... G4RKP
Groome, B L ... G1WPG
Grose, N A ... M1BGN
Grosjean, P ... G4SFS
Grossart, C D ... GM0KLO
Grossmith, E ... G3WOH
Grosvenor, R I H ... G6KSF
Grounsell, I ... G1UUF
Grove, G H ... G3PXU
Grove, M ... G0GUG
Grove, R G ... GU4XGG
Grove, S P ... G4BSM
Grover, A M ... G6DBX
Grover, K G ... G3KIP
Grover, K W ... G7THK
Groves, C ... G6FPQ
Groves, F G ... G0AUB
Groves, H J ... G3UYM
Groves, J ... G4GRY
Groves, J R ... G0VQA
Groves, K ... GW3HDF
Groves, L K ... G7LKG
Groves, P J ... G3MWQ
Groves, P J ... G7UWP
Groves, R S E ... G4HMQ
Groves, T J ... G0HCC
Groves, T J ... G4KUJ
Groves, T M ... GI4ORG
Grubb, M P ... G0WMV
Grubb, M R ... G7PQT
Grubb, R N ... G3FNL
Grudzinski, A E ... G7TIZ
Gruffydd, L D ... GW4CFC
Grundey, J H ... GD7ESM
Grundy, A M ... M0ADY
Grundy, A R ... G7DEC
Grundy, D A ... G4WZF
Grundy, G ... G0EDS
Grundy, G ... G3XEC
Grundy, G ... G4YJW
Grundy, J A ... G7ULN
Grundy, N ... G4GLJ
Grunewald, U ... G0BBB
Grylls, B ... G4ZCN
Gubbins, C W ... G3LHU
Gubbins, R A ... G4BKQ
Gudgeon, J R ... G4MDU
Gudgeon, S P ... G8WTR
Gudgin, E ... 2E1EFR
Gudonis, M ... G0SFD
Guerrero, C ... G0GSX
Guest, A L ... G1ITN
Guest, A P ... G0MSP
Guest, C B M ... G7VXZ
Guest, D ... 2E1EVA
Guest, D H ... GM3TFY
Guest, G S ... G6HBS
Guest, I P ... G7OIT
Guest, J M ... G4GON
Guest, J S ... G4YVA
Guest, M G ... G0VYN
Guest, R ... G1HTW
Guest, R W ... G4RWG
Guest, S E ... 2E1ETA
Guest, W A ... G4IYB
Guffick, I ... G7JXY
Guffogg, J P V ... G4UAL
Guggiari, F R ... G0GJO
Guilbert, G H ... GU4YZV
Guilbert, P G ... GU0DXX
Guild, G ... 2E1CQI
Guild, G M ... G0VQL
Guilford, N C G ... G8SMQ
Guilfoyle, W T ... G3UBC
Guinan, C T ... G0FZF
Guinnessy, B J ... G4MXL
Guite, J R ... G4FNP
Gulley, J V P ... GW4TQD
Gullick, D ... G0DGA
Gullick, J E ... G0DGE
Gulliford, B O ... G6ANV
Gulliford, G ... G4SQI
Gulliford, G A ... G0DVP
Gulliver, B J ... GW7DHX
Gulliver, I P ... G8JLV
Gully, D J ... G4YOC
Gully, P E ... G2BQP
Gulvin, D E ... G8RXG
Gulyas, I ... GW0CKL
Gumb, J A G ... G4RDC
Gumbrell, G ... G1AUM
Gumbrell, M A ... G6CVV
Gumbrill, L S ... G2BAH
Gummow, B R ... G1PDH
Gunbie, A R ... G6LAJ
Gunby, M T I ... G4HOD
Gundry, B H C ... G4SBU
Gundry, G P J ... G1LGB
Gunia, J R ... G7HIJ
Gunn, C W ... G8VCJ
Gunn, F M A ... G8PZX
Gunn, G T ... G8ACT
Gunn, J N ... G4FCT
Gunn, S J ... G6JRZ
Gunn, S J ... G7VID
Gunn, W D ... GM6JHH
Gunnell, L M ... G8HB
Gunnill, G ... G3AVV
Gunning, D G ... 2W0ALZ
Gunstone, K L ... M0AGA
Guppy, J E ... G6VHL
Guppy, P F ... G0BCH
Guppy, R W ... G0TNM
Guppy, R W ... G3GWB
Guppy, R W ... G8LED
Gurbutt, A R ... G0NQA
Gurden, G R ... G4RWB
Gurney, A P ... G4FWF
Gurney, C W ... G0UTA
Gurney, I K ... G7RPP
Gurney, L W ... G4LBJ
Gurney, S F ... G0ETZ
Gurney-Smith, F A ... G0JIV
Gurnhill, A G ... G0ROW
Gurnhill, C D ... G6ZUW
Gurowich, R ... G6ENM
Gurr, J H ... G4JG
Gurr, K J ... G8FRS
Gurr, M A ... G0MAZ
Gurr, M P ... G7BVH
Gurton, I R G ... G0CPN
Guscin, S C W ... G7ONG
Guscott, R J ... M1ASM
Gustar, G J ... G4USO
Gutkowski, M S ... G7UHB
Gutten, N J ... G6IGW
Gutteridge, J M ... G3PEZ
Gutteridge, N ... G3MAR
Gutteridge, N L ... G1MAR
Gutteridge, N L ... G8BHE
Guttridge, C ... G7GIV
Guttridge, J ... G3JQS
Guttridge, P R ... G3TCU
Guttridge, R ... G6EKT
Guttridge, R J W ... G4YTV
Guy, B ... G8IEV
Guy, C ... G4NOQ
Guy, C ... G6MRY
Guy, C J ... G4DDI
Guy, D A ... 2E1BPB
Guy, D P ... G0ANP
Guy, H W ... G4TFW
Guy, J H ... G7GDQ
Guy, J M ... G1SYR
Guy, J R F ... G8GYZ
Guy, M J ... G0VNJ
Guy, N W ... G6YEA
Guy, P ... GI0HCJ
Guy, W ... G4ZSD
Guyan, R D ... 2M1DVS
Guymer, C A ... G1ZRT
Guyver, B ... 2E1FXJ
Guyver, R ... 2E1FXK
Gwillam, J A ... G6NMH
Gwillam, J A ... G7GXX
Gwilliam, E M ... G7EXX
Gwilliam, G S ... G4FJO
Gwilliam, S J ... G8XGG
Gwilt, J ... GI4MUE
Gwinnett, J P C ... G0WEU
Gwinnutt, D F ... GW4DZD
Gwynn, R D ... G8KFD
Gwynne, A E ... GW3LNR
Gwynne, G ... G8VCI
Gwynne, J R ... 2E1CVS
Gwynne, R ... G4CKT
Gynane, M ... G7AWW
Gynn, R G ... G3SBP
Gyurgyak, L ... G0BMP

H

Haase, R ... G4VHE
Habens, I D ... G3WXG
Haberman, A ... G0NMB
Habib, P A ... G0RNJ
Hack, M B ... G8SLU
Hack, S E ... G7CVK
Hacker, J R ... G1DER
Hacker, P N ... 2E1DTH
Hacker, T D ... 2E1DTJ
Hackett, D ... G0GBV
Hackett, D M ... G0WDK
Hackett, J F ... G0OAF
Hackett, J F ... G0TLF
Hackett, J J ... G6VZZ
Hackett, J T ... G6MTH
Hackett, M R ... G7BRV
Hackett, P B ... G6XEE
Hackett, P F ... G3WCJ
Hackett, R A ... G4LAJ
Hackett, R G ... G0CUC
Hackett, S C ... G8VVG
Hackett, T W ... M0ABA
Hackford, M J ... G0IBZ
Hacking, F A ... G0IGP
Hacking, I ... G3VDO
Hackney, C E ... G4VMD
Hackney, M J ... G6UPH
Hackwell, K L ... G4VFR
Hackwill, T M ... G4MUT
Hackworth, T J ... G0IXR
Hadaway, D J ... G4MCM
Haddad, C M ... G6RQY
Haddaway, K ... G7MPB
Hadden, W R ... GI0EBT
Haddington, K A ... G1OBU
Haddock, A E ... G6VBQ
Haddock, C P ... G3UZM
Haddock, P ... G3ITK
Haddon, J A ... G4FCA
Haddon, M J ... G0VOP
Haddon, M J ... G4ZIY
Haddon, S J ... 2E1DTB
Haddon, T J ... G4KMA
Haddon, T J ... G6AEN
Haddow, I C ... M0ALV
Haddrell, C J ... G4OPO
Haden, B W ... G1NTP
Haden, H ... G4GWF
Haden, J N ... G0MWI
Haden, W ... G1SKM
Hadfield, G W ... G1JYW
Hadfield, R ... G4ANN
Hadjidakis, D ... G1GIJ
Hadjigeorgiou, C D ... G7DWM
Hadland, R C ... G8WXP
Hadler, P A ... G4CZU
Hadley, A ... G0ETH
Hadley, B G W ... G0JKY
Hadley, K C ... G4HDK
Hadley, N R ... G4BSW
Hadley, P ... G8GKH
Hadley, T ... G4NNO
Hadlow, K A ... G0KFA
Hadwen, P C ... 2E1FVB
Haffenden, O P ... G0PVA
Haffenden, P J ... G4SVQ
Hagan, C W ... GI6IES
Hagan, E B ... GI0EUG
Hagan, V ... GI6NNP
Hagar, R J ... G1CDJ
Hagen, J ... G0LPQ
Haggart, J S ... G3JQL
Haggarty, A R ... G3MJH
Hagger, D E ... G1DEU
Hagger, E ... G6TSQ
Hagger, L R ... G6HSW
Hague, G S ... G4VJO
Hague, J ... G4GOY
Hague, J D ... GM3JIJ
Hague, M N ... G0PYD
Hague, P ... GM0HTG
Hague, R ... G3ZQV
Hague, R ... G4XOU
Hague, T ... M0AFJ
Hagues, J A ... G1JAH
Hagues, R M ... G7GAB
Hahn, M B ... G4JRB
Hahn, S ... G0VMJ
Haig, J E ... G4VXU
Haigh, A J ... G4PLS
Haigh, G F ... G1GYA
Haigh, S A ... G0VQK
Haigh, S C ... G3SJW
Haighton, F R E ... GD4INU
Haighton, M ... GW4JMN
Haile, J O ... G8ADC
Hailes, N S J ... G4OTB
Hailey, K R ... G7EDU
Hailstone, B K ... G4BEO
Haines, A ... GM7VLC
Haines, A W ... G3OSH
Haines, B L ... G1IYB
Haines, B W ... G8FAT
Haines, C ... G0SKQ
Haines, F W ... GW1PZZ
Haines, G E ... G4SXY
Haines, J R ... G6WNG
Haines, L ... 2E1FPE
Haines, M R D ... 2E1BOB
Haines, M R D ... G7UJL
Haines, R ... G0DBN
Hainesborough, M S H ... G0PYV
Hainsworth, D ... G3ZNK
Hainsworth, G W ... G4JFC
Hainsworth, P G ... G4JVD
Hair, J ... G0LSU
Haith, P ... G0KUD
Hajdukiewicz, M ... G0OVU
Haji-Michael, C J ... G6WDJ
Hakes, J ... G0HSK
Hakes, J C ... G4KWJ
Hakes, K T ... G4KWK
Hakes, L J ... G8BLV
Hakes, N ... G6JXP
Halberg, D M ... G4XUU
Halbert, C M ... G7RWC
Halbertsma, S P ... G7CRK
Halden, M P ... G1SES
Hale, A R ... G4ZKT
Hale, C T ... G3SFB
Hale, J ... G5MW
Hale, J ... G8MWA
Hale, J G ... G7GTP
Hale, J S ... G3FTH
Hale, K ... G0AKH
Hale, K ... G4GXK
Hale, K ... G8SAL
Hale, M ... G4EQK
Hale, M A ... GW0POA
Hale, P E ... G2HS
Hale, P J ... G4OAD
Hale, R ... G0ISG
Hale, R ... GW0SKO
Hale, S G ... G7RLG
Hale, S M ... G6PAP
Hales, A J H ... G4NEX
Hales, D ... 2E1GBH
Hales, K E R ... 2E1CAX
Hales, K G ... G0RQF
Hales, R G ... G4XGI
Hales, S ... G7IKL
Hales, V C ... G1ZGC
Haley, A P ... G4MMT
Haley, M J ... GD1XMA
Halfhide, D E J ... G3CJP
Halford, J R ... GM4EWZ
Halford, J T ... G4CVE
Halford, M P ... G8XJL
Halford, N ... GW3UNH
Halford, P J ... 2E1FOV
Halford, R E ... 2E1FVJ
Halfpenny, J B ... G6PFU
Halfyard, S ... G4EIS
Haliburton, J ... GM3ULG
Haliburton, J ... GM4GRC
Haliburton, J H ... GM4AQO
Halifax, P ... G1JAU
Hall, A ... G3FWN
Hall, A ... G3UWA
Hall, A C ... G1MGF
Hall, A E ... G0ROV
Hall, A E ... G4TQP
Hall, A E ... G8OTU
Hall, A E ... G8DLH
Hall, A J ... M0AHL
Hall, A J ... M1BUK
Hall, A M ... 2E1BII
Hall, A M ... G4TLJ
Hall, A P ... G8YNG
Hall, A P L ... G0THT
Hall, A P L ... G4CYE
Hall, A R ... G4CRN
Hall, A S ... G3WOX
Hall, B ... GW0EDC
Hall, B A ... G1DDY
Hall, B B ... G0SUS
Hall, B J ... G0BXN
Hall, B R ... G7MHK
Hall, C A ... G4HJB
Hall, C A ... G6LWA
Hall, C A ... GW7GPD
Hall, C B ... G1ISX
Hall, C B ... G4HNB
Hall, C E ... G3ETA
Hall, C J ... GM4EMX
Hall, C M J ... G6UVQ
Hall, C N ... G4SET
Hall, C N I ... GW8EUF
Hall, C R ... G8LY
Hall, C R ... GM4JPZ
Hall, C S ... G4FGW
Hall, C T ... G6HTH
Hall, D ... G1XLX
Hall, D ... G3KAH
Hall, D ... G4TID
Hall, D ... G6FZC
Hall, D B ... G8VZT
Hall, D C ... G0LSF
Hall, D C ... G5JJ
Hall, D C ... G8GGZ
Hall, D D ... G7NOF
Hall, D H ... G4DHH
Hall, D H ... G6TSZ
Hall, D J ... M1CDL
Hall, D M ... G0KCG
Hall, D M ... G0OVW
Hall, D M ... G1MSX
Hall, D M ... G4XTL
Hall, D R ... G7EDF
Hall, D S ... G4EME
Hall, D V ... G7UFD
Hall, E P ... G4RGP
Hall, E P ... G6EXN
Hall, F ... G4ZTX
Hall, F ... G8UQV
Hall, F J ... G3NSY
Hall, G ... G6LWE
Hall, G ... G6UVL
Hall, G C ... G6VPJ
Hall, G F S ... G3RZJ
Hall, G J ... G4XZI
Hall, G M ... G1KOH
Hall, G T A ... 2E1FXR
Hall, G W ... G6HTL
Hall, H G ... G3LZT
Hall, I J ... G3IWH
Hall, J ... G0ODQ
Hall, J ... G3FJL
Hall, J A ... G4BOQ
Hall, J C ... G3KVA
Hall, J D A ... G4DUP
Hall, J E ... G3ZFQ
Hall, J H ... G0HKK
Hall, J H ... GD1JJH
Hall, J L ... G1ZHZ
Hall, J L ... G3TOK
Hall, J M ... G4WTZ
Hall, J R ... 2E1EAY
Hall, J R ... G0UYM
Hall, J R A ... G4LGX
Hall, J W ... G3WLD
Hall, K ... G4OSK
Hall, K ... GM1FSZ
Hall, L ... G6XAR
Hall, L ... GM7CHX
Hall, L C ... G1NKZ
Hall, L D ... G4IGC
Hall, L E R ... G3IGI
Hall, M A ... G3USC
Hall, M A ... G8IEM
Hall, M C ... G0ODN
Hall, M C ... G4GSB
Hall, M D ... G1IMD
Hall, M E ... G7RAO
Hall, M F ... G0BQK
Hall, M F ... G0WZT
Hall, M F ... G7NFO
Hall, M I ... 2E0AKD
Hall, M R ... G3VQQ
Hall, M R J ... G8YNH
Hall, M W ... G0AWA
Hall, N ... G7CLQ
Hall, N C ... G4DQL
Hall, N F ... G0DWJ
Hall, N L ... G7RTY
Hall, P ... G0ELC
Hall, P D ... G4AQA
Hall, P F ... G1JMD
Hall, P J ... G4YEE
Hall, P S ... G6SDI
Hall, R ... 2E1EXP
Hall, R ... 2E1FFP
Hall, R ... G3TUK
Hall, R ... GW4VRH
Hall, R ... M1ANR
Hall, R B ... G4DVJ
Hall, R D ... G3UWB
Hall, R D ... GW7WXM
Hall, R E ... GI4CFV
Hall, R G ... G0OGN
Hall, R G ... G4WGA
Hall, R J ... G0GSA
Hall, R L ... MW0AMI
Hall, R M ... G0DXP
Hall, R N ... G4XDV
Hall, R O ... G7UPP
Hall, R S J ... G4TER
Hall, R W ... G3XIY
Hall, R W ... G4UGU
Hall, S ... G0BPQ
Hall, S ... M1BQR
Hall, S A ... G1KAP
Hall, S A ... G7EUW
Hall, S H ... G4VEG
Hall, S J ... G1RJN
Hall, S J ... M1AGT
Hall, S K ... G6VRM
Hall, S M ... G8VWH
Hall, S R ... G4FSR
Hall, S T ... G3BR
Hall, S W ... G3NQA
Hall, S W ... G7GPZ
Hall, S W C ... G8CCF
Hall, T ... GM3HBT
Hall, T C ... G8DIQ
Hall, T E ... G8HFW
Hall, T E ... G8KRY
Hall, T E ... GI0MSJ
Hall, T G R ... G1TOB
Hall, W ... G6IXK
Hall, W G ... G8JM
Hall, W G ... G8KSA
Hall, W H ... G3RMX
Hall, W M ... G4WMH
Hall, W S ... GM2AOL
Hall, W T S ... G4FRN
Hallam, J ... G0DME
Hallam, K D ... G3KKB
Hallam, K W ... G0UKH
Hallam, M ... G6XWF
Hallam, P J ... GI0IBC
Hallam, P J ... GI4GVS
Hallam, R F ... G4HLB
Hallam, S T ... G7BMM
Hallam, T ... G6ADD
Hallard, P ... G6UKV
Hallatt, J S ... G3DBY
Haller, D A ... G1KJY
Halleron, S J ... G6YJI
Hallet, M H ... G3MDR
Hallett, G R ... G4RQP
Hallett, J E ... G6CAC
Hallett, P H ... G7FXY
Hallett, R ... 2W1FVH
Hallett, R J ... MW0AQD
Hallgarten, D A ... G8UAW
Halliday, A J ... G0IZN
Halliday, C T ... G0FNA
Halliday, E C ... G3JMY
Halliday, J ... G7SWZ
Halliday, J R ... G4HMX
Halliday, J W ... G6TCV
Halliday, T ... G0FEG
Halliday, W J ... GM8XVU
Halligan, T M ... GM0WPW
Hallin, J A ... G7DQA
Halling, C W ... G3RPQ
Halliwell, D M ... G8OSJ
Halliwell, I A ... M0ALT
Halliwell, J K ... G7HHO
Halliwell, R E ... G4SVT
Halliwell, W J ... G4VYX
Halliwell, W K ... GW0LKJ
Hallows, B ... G3XPI
Halls, D P ... G4LVG
Halls, K W ... G8VJG
Halls, P J ... G4CRY
Halls, R L ... G3EIW
Halls, W I E ... G8PXG
Hallsworth, F W ... M0AFF
Hallsworth, M J ... G1GYC
Hallsworth, R I ... G6XTD
Hallsworth, S J ... 2E1FAT
Hallybone, H G ... G8IBL
Hall-Brooks, K ... G3ONB
Hall-Osman, R ... G1OFL
Halman, J H ... G3JNP
Halmshaw, B H ... G0JSF
Halsall, L E ... G1LFG
Halsall, P ... G0HXD
Halsall, P ... GD1GJB
Halsall, R S ... G1TGC
Halse, G L ... G3GRV
Halsey, D T ... G7LAN
Halsey, I S J ... G1JBZ
Halson, I D ... G6MSN
Halstead, R L ... G7HSR
Halton, J M ... G3ZDZ
Ham, A J ... GW1UOV
Ham, A L ... GW8YTO
Ham, B T ... G0NPV
Ham, D E ... G3ZUO
Ham, L S J ... M0ATN
Hambidge, D F A ... G0DZO
Hamblen, E D ... G3MZA
Hambleton, J L ... G4KIB
Hambleton, S E ... G7ETS
Hamblett, P K ... G0TKT
Hamblett, R ... G0DNU
Hamblin, R J ... G0TKZ

Hambly, D A 2E1ERI
Hambly, P C G6YJJ
Hambrook, J W GM1RJS
Hamer, H J G G1JHP
Hamer, J G4ZQD
Hamer, J T G3LMQ
Hamer, K A M0ASU
Hamer, K W G0FSR
Hamer, N A G7NYO
Hamer, R G0MOK
Hames, D G7FCL
Hamill, I B G3SMF
Hamill, J GI4NRQ
Hamill, J GI4ORI
Hamill, W J GI4KUZ
Hamilton, A GI0TJJ
Hamilton, A D G G4ERD
Hamilton, A J G4UCW
Hamilton, A T GI4HVI
Hamilton, B G4SZD
Hamilton, B A GM6RQU
Hamilton, B G GI3VYY
Hamilton, C M G3XWV
Hamilton, D GM4CAM
Hamilton, D E G4TKW
Hamilton, D G G4GLC
Hamilton, D P G8XJK
Hamilton, D W G1LGW
Hamilton, F R G0DYB
Hamilton, H 2E0AHW
Hamilton, H G7VKT
Hamilton, H S GM4JRF
Hamilton, I C G0TCD
Hamilton, I C G7TTZ
Hamilton, J G0TBZ
Hamilton, J GW8PVN
Hamilton, J A G4JSD
Hamilton, L GM3ITN
Hamilton, M C G3KEV
Hamilton, M J G7VKS
Hamilton, M J W GM3TAL
Hamilton, N 2M1FXF
Hamilton, N C G4TXG
Hamilton, N S GM0ARY
Hamilton, R GI0SRU
Hamilton, R H G1YIC
Hamilton, S A G4UIY
Hamilton, T G0HIN
Hamilton, T E G0OBF
Hamilton-Cooper, S N G1MVQ
Hamilton-Sturdy, W G1TRZ
Hamilton-Wedgwood, K R G3XKW
Hamlett, J G3EOO
Hamlin, G D W M1BDJ
Hamlyn, K G4WQB
Hamlyn, S E GW7IPS
Hamm, A D G4EBI
Hamm, D G6KHP
Hammersley, G E G3ZCL
Hammersley, K A G8RRA
Hammersley, N P G4BCW
Hammersley, P G7IXP
Hammett, A H G3VWK
Hammett, C D G3AWR
Hammett, C K 2E1ADQ
Hammett, H W G3OVX
Hammett, R J G0DNV
Hammett, R W G0KYV
Hammett, R W G4ARE
Hammon, A J G0HNA
Hammond, A D G7AJJ
Hammond, A J G7PSQ
Hammond, A M J G3PGA
Hammond, B G4SIJ
Hammond, D A W G4TVL
Hammond, D G G3KAR
Hammond, G P G8YJS
Hammond, G P H 2E1FSR
Hammond, I P J G8ZYM
Hammond, J GW4XTY
Hammond, J H G3ZUU
Hammond, J S GW3JBH
Hammond, J T G0FLP
Hammond, M J G4HIE
Hammond, P F G7CNX
Hammond, P H G8SHC
Hammond, P L G1EDK
Hammond, P V G6XEF
Hammond, R E 2I1BIJ
Hammond, R E G4FKR
Hammond, R K G1NFN
Hammond, R S G8VSX
Hammond, R W G4DBW
Hammond, S A G7PJL
Hammond, S C G6PNJ
Hammond, W G6BRD
Hammond, W A E G4SOB
Hammonds, S V G4KUR
Hammonds, V J 2E1ACG
Hamnett, F E G2CFH
Hamon, A P GU4WTN
Hamon, M C G4UIC
Hampshire, J A G6ZUZ
Hampson, C G G7RDT
Hampson, C G G8RXA
Hampson, E M G1MCR
Hampson, E M G8RXB
Hampson, G M G0JIL
Hampson, G M G0RER
Hampson, G R G7KSP
Hampson, H GW0AWN
Hampson, I J G1DFT
Hampson, J G0VXH
Hampson, J H G3PJL
Hampson, J R G4LPO
Hampson, K J GM4DUX
Hampson, K M G3WFW
Hampson, M G0TSD
Hampson, M J 2E1AWF
Hampson, M J G8YBZ
Hampson, R A GW4MGH
Hampson, R S G7RYC
Hampson, T G6DEG
Hampton, A D G7EWA
Hampton, D M G3UHU
Hampton, I L G3JLH
Hampton, M W G0BJO
Hampton, R C G4OSA
Hamriding, J R G1FOD
Hamson, R E G0AXC
Hamson, R E G0NOC
Hamstead, R T G4RTH
Hanaghan, M G8HPW
Hance, M J L 2E1CLB
Hancock, A GM4IEF
Hancock, A J G8HPS
Hancock, B A G4NPM
Hancock, D A G3OMY
Hancock, D T G8IYD
Hancock, E J G3BHW
Hancock, E W G3GWY
Hancock, G G4HEW
Hancock, G A G0UNC
Hancock, G F G6YEC
Hancock, I R G6BJG
Hancock, J G4TAK
Hancock, J G6ISM
Hancock, J C G1ZBK
Hancock, K G0MGJ
Hancock, K H G4KIY
Hancock, L J GW0LTH
Hancock, M G6ISN
Hancock, M J G4URU
Hancock, M J G8EYY
Hancock, N H G7SBP
Hancock, P A G0EFS
Hancock, P J G6ISG
Hancock, P N GU3WOW
Hancock, R G6BBD
Hancock, R GW4SCK
Hancock, R A G3GEI
Hancock, R A G4BBT
Hancock, R C GW6MXB
Hancock, S D GU6RWD
Hancock, S R G3GBD
Hancocks, N S J G4XTF
Hancock-Baker, F G1WTT
Hancox, E B G0MNJ
Hancox, G W G8PVM
Hancox, L A G0LBG
Hancox, R J G7UWG
Hancox, W G7JZK
Hand, K P G8NTJ
Hand, M S G0WCI
Hand, P W G8TXF
Hand, R B G7TPH
Hand, S GI7UIM
Hand, W G0BHT
Hand, W C M1AMC
Handcocks, E M G5HN
Handford, T J G4YTH
Handforth, N E M0AUF
Handley, A 2E1FIP
Handley, C G6KTK
Handley, J G4DJR
Handley, J A G0EAA
Handley, J A G4RNF
Handley, L T G7IMO
Handley, M J G1AOV
Handley, P W G4VAJ
Handley, R GW3GJQ
Handley, W N G2FRZ
Hands, C D G4KQV
Hands, D C G1PGS
Hands, D S GW8ELR
Hands, G G0FBG
Hands, R W G4EJU
Handscombe, F C G4BWP
Handscombe, F C G4MBC
Handstock, R G4TOY
Handy, C J G6OJO
Handy, E J G4RTI
Handy, F C G4XNE
Handy, J P G6UVU
Handyside, R GM0NUQ
Hanfrey, P C M1BOD
Hankin, G A G6YWT
Hankin, J E G G4NEH
Hankin, R S 2E1CPD
Hankin, R S M1ACO
Hanking, N G0DSX
Hankins, G J G8EMX
Hankins, T R F G7RVI
Hankinson, K G8JVJ
Hankinson, P G8PAL
Hanks, C H G0MWX
Hanley, D G3YVY
Hanman, T R G1ITY
Hanmer, P J GW0WMY
Hanna, A G GI0SMU
Hanna, D K GI4UXW
Hanna, L GI3WHA
Hannaby, E G0AEX
Hannaford, D G G0BYY
Hannaford, J L G7KCE
Hannaford, R H G3VIV
Hannah, C A 2E1CRX
Hannah, C E G4LOP
Hannah, S W G8GBH
Hannam, D S G6OLR
Hannam, P D G3ZNB
Hannam, P J G6TSJ
Hannan, G D G1JVF
Hannan, G R GW0UXU
Hannan, M R T G4FVG
Hannan, P P G1XVT
Hannan, R J G4RQJ
Hannant, D W G4YDQ
Hannaway, P GI7MSJ
Hannell, C W G0INQ
Hannemann, R B G7GQH
Hanner, A G 2E1FVV
Hanney, J F G8BDF
Hanney, N F G4YPE
Hanney, R G M1BUF
Hannigan, C GI0VJE
Hannigan, R N G7RGY
Hanraads, M G0ILZ
Hanrahan, R J G6BJQ
Hanratty, T G0JRT
Hansell, B G G0BXH
Hansell, J A G4COS
Hansell, J J G7VUD
Hansen, A D A G3VLJ
Hansen, L C G7MWI
Hansford, C J G7VCJ
Hansford, V G G4WTX
Hansley, A J G0JRN
Hanslow, S G0IHP
Hanson, B C G0SRA
Hanson, B R M G3RYC
Hanson, D G7MJX
Hanson, D J G8KOM
Hanson, F I G3RBD
Hanson, G R G0UUS
Hanson, G S G4CPA
Hanson, J G0FJT
Hanson, N G6VNO
Hanson, P G0NVY
Hanson, P D G0WDP
Hanson, P M G4IFR
Hanton, J M G4MCA
Hanton, K G4SGS
Harada, A E G4INX
Harbach, M H G8GMC
Harber, B J G8DKK
Harber, J W G3JWH
Harber, P W G4LGY
Harber, R P G6LHQ
Harber, T T G4SZS
Harbinson, P 2E1FFL
Harbinson, P J GI4NEZ
Harbison, A S GI0SRL
Harbison, R B GI3PDN
Harbord, R H G4YDY
Harbottle, J W G0OWV
Harbour, D W G0EID
Harbron, D B G7PHG
Harcourt, B G8BEA
Harcourt, G P G4EBE
Harcourt, R A G7MUB
Hard, C J J GW4ITJ
Hardacre, G GM1PZT
Hardacre, K G1BTG
Hardaker, I P G1YBA
Hardaker, M L G4HJH
Hardcastle, A G G0STK
Hardcastle, D G G3DGH
Hardcastle, J G0JUA
Hardcastle, J A G3JIR
Hardcastle, K B 2E1CUZ
Hardcastle, P R G7SLP
Hardcastle, W A G3XQX
Hardee, R S G8MRM
Harden, C E M1AWR
Harden, R J G4DUB
Harder, W F G0AQW
Harders, J B C G0VNK
Hardes, S C R G7ICV
Hardie, A J W 2E0ANL
Hardie, A W G0VUL
Hardie, B G S G0WYL
Hardie, C B GW3YQP
Hardie, D W G4ITY
Hardie, K G4SNU
Hardie, L GM2FHH
Hardie, N J GM7GRH
Hardie, R A W 2E0ANM
Hardie, R H G7JVK
Hardie, R W GM4HKH
Hardie, W GM6RGY
Hardiman, A M G0FUM
Hardiman, D J 2E1DNT
Hardiman, M R J G6FBS
Hardiman, P G0SOC
Hardiman, P G4LBM
Hardiman, P J G6HSS
Hardiman, R D G1RPV
Harding, A G0HVT
Harding, A G W G1JHM
Harding, A P G0VHQ
Harding, A R G1VGH
Harding, A T G1OOX
Harding, C G1XOZ
Harding, C G7XOZ
Harding, C A G6XAK
Harding, C P G6HHF
Harding, D G7UWW
Harding, D G1PMA
Harding, D G6IHH
Harding, D A G0DQI
Harding, D C G6YWU
Harding, D G E G8LSN
Harding, D R G3YHG
Harding, E W 2E1AUQ
Harding, H G0HZY
Harding, J E G3TZU
Harding, J F G4PFR
Harding, J K G3XFL
Harding, J N G0ABW
Harding, K G6WOV
Harding, K E G7FFI
Harding, M E G7DHJ
Harding, M R G4WOP
Harding, N J G0NZL
Harding, P GW4LJS
Harding, P R G6YOP
Harding, P W G8INN
Harding, R A G3AKU
Harding, R F G1HNW
Harding, R H G4SBM
Harding, R J G3RJH
Harding, R P H G6BJL
Harding, S G4PYU
Harding, S A J G4OBN
Harding, S D G7RTA
Harding, S J G6XAN
Harding, S W G4JGS
Harding, S W G7LDJ
Harding, T E GI3ZKT
Harding, T T G3RGQ
Harding, V E G0KCE
Harding, W A G7LJR
Hardinges, D S G7HSO
Hardingham, B W G0BAL
Hardingham, R W G1OAY
Hardisty, C H G0HDL
Hardman, D V G0VLV
Hardman, J M G0DQG
Hardman, M J G0PXG
Hardman, R E G3LGV
Hardman, R W G6LWC
Hards, D G G0UOL
Hardware, D R G0MLK
Hardware, H R 2E1DBN
Hardwick, E H G1DEV
Hardwick, R G1NUH
Hardy, A H G1BTF
Hardy, C G0ROZ
Hardy, C D M0ATZ
Hardy, C J G1FMT
Hardy, D G4BXH
Hardy, D G6APZ
Hardy, D H G8AJA
Hardy, D M G8ROU
Hardy, E B G3BSL
Hardy, G G4HBL
Hardy, G GM1FLO
Hardy, J B G8OCE
Hardy, J S P G3KND
Hardy, K G0HPV
Hardy, L A G1UDA
Hardy, M 2E1FCG
Hardy, P GM3VNH
Hardy, R G3TIX
Hardy, R GM1NSU
Hardy, T G0LKK
Hare, D A G3NHV
Hare, G G7SVX
Hare, J P G1EXG
Haresign, B G0TUT
Harfield, I R G8OUH
Hargan, R A A GI3TME
Hargrave, A G G0AXA
Hargraves, J W G8ZTF
Hargreaves, A A G4SYW
Hargreaves, B G4WUZ
Hargreaves, C E G6NRL
Hargreaves, H G4ZSI
Hargreaves, J G1EDM
Hargreaves, J H G5VO
Hargreaves, M J G6EXP
Hargreaves, P G3TEY
Hargreaves, P J G4ZEM
Hargreaves, R G3EJV
Hargreaves, R G3XZQ
Hargreaves, R G4IVO
Hargreaves, R A G3OHH
Hargreaves, R D G1IHN
Hargreaves, R W G4YVQ
Hargreaves, W G0PLG
Haria, H J G0CZV
Haring, G GW3EQL
Harker, R A G4CMK
Harkess, R D GM3THI
Harkin, R G3MID
Harkness, D R J GM4NNK
Harkness, G F G0IGR
Harkness, I G0UED
Harkness, R B GM3PSJ
Harkness, W S G0CQG
Harknett, J E G3TVH
Harknett, R F G3KZC
Harland, A M G6BBG
Harland, E J G3VPF
Harland, L T G6AIU
Harland, S D 2E1DTP
Harland, W E G7WDY
Harle, S G3MEA
Harley, E G0UUV
Harley, G R G6TGI
Harley, I D G6BJJ
Harley, P A G4UDH
Harling, I K G7HFS
Harling, P G4LDD
Harlow, C W G8BTK
Harlow, J G3SHL
Harman, A C W G8LTY
Harman, G F G0HSL
Harman, J G8GIU
Harman, J G8ZTG
Harman, M W G G3NZP
Harman, P G1FWR
Harman, P G3WIR
Harman, P G6WIR
Harman, P H J G4XGD
Harman, R G0BAP
Harman, R G7GQO
Harman, R C G0BSH
Harman, R H MM1BIQ
Harmer, A E G8XHN
Harmer, B S G8KNX
Harmer, D G4AOL
Harmer, D G G7PSX
Harmer, E R B G3EWJ
Harmer, G P G 2E1CSI
Harmer, K M G0UPK
Harmer-Knight, R G4LBT
Harmon, N G GI8VLB
Harmsworth, H J G3ACQ
Harness, A G1HPI
Harness, P G7LSP
Harness, P G8LDW
Harnett, R G4WXJ
Harney, M W G4YMH
Harper, A J G1BFG
Harper, A J G7VIY
Harper, A L J G6MSH
Harper, A M G0PQG
Harper, B C GM0JFL
Harper, B D G0TOS
Harper, C G6GJD
Harper, C G G4UJR
Harper, C J G7DIO
Harper, C R G8YVQ
Harper, C S G1VJH
Harper, C T GM0JFK
Harper, E G0WXL
Harper, E J M GI0AZA
Harper, G 2E0AAI
Harper, G G4WHA
Harper, G W G4IVZ
Harper, G W G4NVN
Harper, J E G0AHQ
Harper, J E G3RLJ
Harper, J M G0WZS
Harper, J V G7EQB
Harper, J W G0MJX
Harper, K D G1HVZ
Harper, K J G0EKH
Harper, L G4FNC
Harper, L G6EHV
Harper, L E G0CJA
Harper, M C G0MBH
Harper, M S T J G7SRH
Harper, N G6RZY
Harper, N F G3ZCV
Harper, P G4MZM
Harper, P J G1TJX
Harper, R E G6BZH
Harper, R E D G3KSF
Harper, R I C GW4VGB
Harper, R J GW0KPU
Harper, T W G0KIN
Harper, V G4UWZ
Harper Bill, J S G3IZM
Harpham, D R G1EZU
Harpur, P J G3ZWF
Harrad, B B G8LDV
Harrad, P W G8UN
Harradine, A 2E1FSU
Harradine, A M1BTZ
Harradine, C G0NDW
Harrap, C P G1SWX
Harratt, B 2E1CIY
Harratt, C W G4RJH
Harries, A G6BCS
Harries, B J GW7FXX
Harries, D G G4WYN
Harries, I W GW7SXU
Harries, M G4PXF
Harrigan, J GI4JRA
Harrigan, R G0JUZ
Harriman, M J G4NLS
Harrington, A G1XLW
Harrington, A V G7JJD
Harrington, B G4SQF
Harrington, C A G4TPC
Harrington, C D G3LUL
Harrington, D M G3VQM
Harrington, J A G0ERH
Harrington, M G7BNF
Harrington, M G7REF
Harrington, M A G8WWW
Harriott, D R G3IIO
Harris, A A 2E1FYM
Harris, A C G3EWF
Harris, A G G0CZV
Harris, A L S G8ZNB
Harris, A P G8SYC
Harris, B G3HZR
Harris, B A G3XGY
Harris, B G G1JMC
Harris, B J G0DRH
Harris, B M G8SIC
Harris, B P G0NPN
Harris, B W N G3GTF
Harris, C A G4TBO
Harris, C A G7PJC
Harris, C E G3ORH
Harris, C E G7UDX
Harris, C E G7URV
Harris, C R G3XAO
Harris, C R G8RYR
Harris, D G4LSB
Harris, D GM0TPI
Harris, D GW3NDR
Harris, D GW7VBE
Harris, D A G6DEV
Harris, D D G1POK
Harris, D F G7MXT
Harris, D G G6XTK
Harris, D K G0CER
Harris, D L G6FEI
Harris, D P GW0ONU
Harris, E C G3PKA
Harris, E J G2FXO
Harris, E J G6PNB
Harris, E L G3LQO
Harris, F J T G4IEY
Harris, F J T G6QM
Harris, G A G8MYI
Harris, G H G4RNU
Harris, G J G3TPQ
Harris, G J G4KWZ
Harris, G P G7NJB
Harris, G W J G0LXZ
Harris, I G3WAE
Harris, I G6PFX
Harris, I D G4IDH
Harris, J G1OEP
Harris, J G2ATZ
Harris, J G2FPY
Harris, J G3RWC
Harris, J G4PFT
Harris, J GW0KKN
Harris, J GW3RYE
Harris, J A G0JVD
Harris, J A G6OWD
Harris, J B G4EKO
Harris, J D G3LWM
Harris, J D G3PFJ
Harris, J D GM0PNS
Harris, J F G4VMO
Harris, J F V GW4LNK
Harris, J G G0OZO
Harris, J G G8HJS
Harris, J G GM4WKO
Harris, J J G4AJZ
Harris, J P G4RXI
Harris, J W G1TQR
Harris, J W G4GOA
Harris, K G6DRM
Harris, K C G0IEE
Harris, K C G1FMU
Harris, K C G1XIC
Harris, K C W G8MWN
Harris, K P G G7WSN
Harris, K S G4NSP
Harris, K W G6YLR
Harris, K W H G1GEL
Harris, L C G0ULH
Harris, L N G4YUD
Harris, M A 2E1EYL
Harris, M A G6ZLE
Harris, M D 2E1ADV
Harris, M D M1CAY
Harris, M E G0IFJ
Harris, M H G3RUQ
Harris, M J G0HOC
Harris, M J G7AIL
Harris, M L G6VEX
Harris, M P GW6DEP
Harris, M R G3VUI
Harris, M R G3YIA
Harris, M W E G G0UUG
Harris, N C G3TXC
Harris, N J G6FYL
Harris, N R G0GZU
Harris, N R GM1IUA
Harris, P C G G0LSE
Harris, P G G4ZOB
Harris, P H G3OBV
Harris, P J G0WUA
Harris, P N G4SPZ
Harris, P R G0POM
Harris, R G0SBC
Harris, R G3UVW
Harris, R A G4FFA
Harris, R G G3SPY
Harris, R G G3ZFR
Harris, R G G4APV
Harris, R G GW8DUP
Harris, R J G0EER
Harris, R J G0NCL
Harris, R J G1AFW
Harris, R J G3OTK
Harris, R J G4GIY
Harris, R J G8BIR
Harris, R J GW0MOW
Harris, R N G0KIA
Harris, R P G6IQF
Harris, R S G6XTJ
Harris, R S G6XTM
Harris, R W GW8MZR
Harris, S J G0SJH
Harris, S J G2AUK
Harris, S L S G4WQG
Harris, T C G0OIB
Harris, T J G3XIN
Harris, T R G7FIU
Harris, U M G4RDA
Harris, V A G0EXG
Harris, V H G4GXF
Harris, V H G6MWY
Harris, W E G3HLM
Harris, W J G1DKA
Harris, W J G1EMR
Harris, W R G3CCM
Harrison, A G0PXD
Harrison, A G6PXJ
Harrison, A GM0GON
Harrison, A C G7MXS
Harrison, A H G1NRM
Harrison, A P G1XUR
Harrison, A P G7SEG
Harrison, A R G0FPG
Harrison, A T G4SAI
Harrison, B 2E1FMD
Harrison, B G1YUB
Harrison, B A G7PQD
Harrison, B D R G0ILD
Harrison, B E G0SNS
Harrison, B J G4NWI
Harrison, B L G7VEJ
Harrison, C A 2E1DFU
Harrison, C A G4MZU
Harrison, C D GW0TWR
Harrison, C G G8KRG
Harrison, C H G4TTS
Harrison, C H G8ZIC
Harrison, D G4PGQ
Harrison, D G6EHO
Harrison, D G7NQZ
Harrison, D C G0FXD
Harrison, D G G6WVM
Harrison, D I G4JSG
Harrison, D J G4ROR
Harrison, D M G1LCR
Harrison, D M G1SPA
Harrison, D W G1MDE
Harrison, E G4ITI
Harrison, E A 2E1DLD
Harrison, E C G8NPF
Harrison, E J 2W1CGC
Harrison, F G3SFL
Harrison, F G3XII
Harrison, F R G4MJT
Harrison, G 2E1DGP
Harrison, G A G2BHG
Harrison, G O C 2E1CGR
Harrison, G R G0NRN
Harrison, H G3RIF
Harrison, H C G3ACR
Harrison, H H G0IVX
Harrison, I G1XUQ
Harrison, I G4JPX
Harrison, I A G0UNZ
Harrison, I B G G0RTJ
Harrison, J G0NTR
Harrison, J G2FZN
Harrison, J A G1ZDA
Harrison, J E G4HCB
Harrison, J E GD4XWF
Harrison, J J GW6VBO
Harrison, J L G4CLP
Harrison, J S G8XXA
Harrison, K 2E1DFV
Harrison, K G0KBZ
Harrison, K H G1NDL
Harrison, L G4RDY
Harrison, L M 2E1EZY
Harrison, M G3USF
Harrison, M G4SRH
Harrison, M D G7SZD
Harrison, M D L G0IYK
Harrison, M E J G3HUB
Harrison, M G G6YWV
Harrison, M J F G3HKH
Harrison, M L G6DPS
Harrison, M R G0LFX
Harrison, N G3NJU
Harrison, N G3WVU
Harrison, N G4OFK
Harrison, N P G6XAC
Harrison, P G1PQK
Harrison, P G4VAM
Harrison, P G8ADY
Harrison, P B G0MYM
Harrison, P F E G8OZY
Harrison, P J G6NVS
Harrison, P J G6WVC
Harrison, P J C GM3CFK
Harrison, P K G7TEW
Harrison, P R G4ZWX
Harrison, P R G8MJH
Harrison, P W G3WAB
Harrison, R G0SHJ
Harrison, R G3VPR
Harrison, R G8PFZ
Harrison, R GD4VBA
Harrison, R M1AYH
Harrison, R A G4LMF
Harrison, R E G8HGN
Harrison, R G G6UTC
Harrison, R J G1YXH
Harrison, R J G3TMQ
Harrison, R L S G3EPK
Harrison, R P G4UJS
Harrison, S A G1KSC
Harrison, S D G7FPS
Harrison, S J G0NKN
Harrison, S J G0UTN
Harrison, S J G4JJS
Harrison, S M G0AJF
Harrison, S W G0TOQ
Harrison, S W H G3KYV
Harrison, T G1EDO
Harrison, T GM3NHQ
Harrison, W A G3SNH
Harrison, W F GW1ACV
Harriss, S A G0RTI
Harrisson, C K G3OPJ
Harrod, R G3RWN
Harrod, R G4JUR
Harrold, C H G4RRN
Harrold, C H G7RNN
Harrold, G D G7EOA
Harrold, M J G3VXA
Harrold, M N G0NSN
Harron, R GI6ISQ
Harrop, D G G6YLQ
Harrop, D S G4BMQ
Harrop, D T 2E1DNZ
Harrop, F G3DVL
Harrop, H D G0REI
Harrop, I J G7EGQ
Harrop, W A G4VPG
Harrower, A R G3ESF
Harrower, D M K GM6NX
Harry, P H J G4BOF
Harry, R G G0EWC
Harsant, J C GM4RJE
Harsent, R E G8ZKO
Harste, R E G8LAB
Harston, J T GW6MWG
Harston, T M G8WRC
Hart, A G1TWQ
Hart, A H G0BOL
Hart, A L G4BLI
Hart, B D G7VHO
Hart, B J G8UVZ
Hart, D A G1JEA
Hart, D A GI4DAV
Hart, D B G4YGH
Hart, D G G4XXK
Hart, D I G1DRW
Hart, D T 2E1DOH
Hart, E G G0SRZ
Hart, F J G1HCX
Hart, F J M0AOC
Hart, H J G4TWK
Hart, J C G4SQV
Hart, J E G3ZGA
Hart, J L G7VHN
Hart, J P G4POF
Hart, J R G0OXN
Hart, J S G8GIK
Hart, L A L G6RSI
Hart, L M G8ZVI
Hart, M C G0EZE
Hart, M J G1ENN
Hart, M L G6REC
Hart, P G4JSM
Hart, P G5RR
Hart, P B G7HUK
Hart, P D G0THD
Hart, P J G3SJX
Hart, P L G4XVP
Hart, P W 2E1DEN
Hart, R G4EJL
Hart, R G M G1XAG
Hart, R W G0NNE
Hart, T G0HEO
Hart, T P G4KPF
Hart, T W G3VFO
Hartas, P J G6HTA
Harte, R J G4VBI
Hartgroves, S G4OIA
Hartigan, C G7KLP
Hartigan, J J G0UWH
Hartigan, P F G4FYX
Hartin, J B GI0EWP
Hartland, A F G6KRC
Hartland, A F G7CHW
Hartland, A F G8WOX
Hartland, R A G8IVO
Hartley, A G6URF
Hartley, A G8PRH
Hartley, B G0GGT

Hartley, B M G4SXI
Hartley, C G0MUB
Hartley, C P G3VJV
Hartley, C S G7BBD
Hartley, C W G7DTY
Hartley, C W B G0FLZ
Hartley, D J 2E1FUH
Hartley, D P M1BHP
Hartley, E J G3DDK
Hartley, G G1EZW
Hartley, J G4ING
Hartley, J R G3WGQ
Hartley, J W R 2E1FUJ
Hartley, K G8SVX
Hartley, K J G7MZN
Hartley, N J G7SZF
Hartley, P W G0SKN
Hartley, R D G3ZQZ
Hartley, R J G0NCY
Hartley, R S G0HDX
Hartley, S G0FUW
Hartley, S R G1GHO
Hartley, T R G0WBU
Hartman, L S G6MFT
Hartnell, A G G7SVV
Hartnell, J H P G6YWX
Hartnell, R G4NYE
Hartshorn, D G7SFF
Hartshorn, G G1YPT
Hartshorn, R G3YSY
Hartshorn, T H G4HMW
Hartshorn, V GW3VZO
Hartshorne, W G4TEC
Hartstone, G R G3LOA
Hartt, C A G8ZEV
Hartwell, D S G0VOV
Hartwell, H S GW0PNC
Hartwell, J G3UOJ
Hartwright, H F GW8VLD
Harvatt, P G0CPY
Harverson, E F G3OEG
Harvey, A G1XFX
Harvey, A F GI7VIW
Harvey, A G G7JTF
Harvey, A G GU7DHI
Harvey, A M G4MUR
Harvey, A M G7RRC
Harvey, B G8RIW
Harvey, C G6LVT
Harvey, C H G3ZIO
Harvey, C J G3XWM
Harvey, D C G1GYF
Harvey, D F G3XBY
Harvey, D J G4BJN
Harvey, D J G6DJH
Harvey, E W 2E1ACF
Harvey, E W G4TNQ
Harvey, F G4YOA
Harvey, F W G7FEE
Harvey, G 2E1CLG
Harvey, I D G4COR
Harvey, J C G2CQJ
Harvey, J K G4IVJ
Harvey, J S G0VMY
Harvey, J W G4RBE
Harvey, K 2E1FNL
Harvey, K A G1WMN
Harvey, K M M0AGX
Harvey, L A G3MJN
Harvey, L G G4OOX
Harvey, L M G4XJU
Harvey, L W G3ONT
Harvey, M J G1XKW
Harvey, N GM0NLU
Harvey, N A G4WTQ
Harvey, N R G0UFX
Harvey, O C G8KQH
Harvey, P G0RLA
Harvey, P G3FLG
Harvey, P GI3VFK
Harvey, P F 2E1DBR
Harvey, P G G1OPT
Harvey, R G0FQD
Harvey, R G0TBY
Harvey, R A G3YHM
Harvey, R C G4BBR
Harvey, R D G3IJV
Harvey, R G G4HTR
Harvey, R G G4YKZ
Harvey, R K GU4JHH
Harvey, R L G2FSA
Harvey, R S G7IPB
Harvey, S G0OYF
Harvey, S G6KHN
Harvey, S C G4NMR
Harvey, S C G6ACT
Harvey, S J G6SVJ
Harvey, S N G6TGD
Harvey, S R G7IHN
Harvey, W A GM4VYQ
Harvey, W J G0GWW
Harwood, A G4PQU
Harwood, A J G4HHZ
Harwood, B G0MLW
Harwood, F A G4GKW
Harwood, H W G1PVJ
Harwood, I T G8LHT
Harwood, J G7PIC
Harwood, J H G3WLY
Harwood, J H G6YWW
Harwood, J L G0PWJ
Harwood, R E G1IMG
Harwood, R E GM0HRT
Harwood, R J C G8GJM
Harwood, R W G6YLP
Harwood, S R G4OWT
Haselden, F C G1CCW
Haseldine, S J G8EBM
Haselgrove, P S G7UFZ
Haselup, M R G6WID
Haskett, E G1XYZ
Haskett, E G3XYZ
Haskett, E G4OZG
Haskins, N P G8JR
Hasko, S M G8RSA
Haslam, A G7NZR
Haslam, J H G4KBQ
Haslam, M E F G3UJU
Haslam, P J G4WMA
Haslam, T G3XSI
Haslehurst, A G6JCV
Haslehurst, D G4VLK
Haslehurst, H E G6JCT
Haslewood, C S G7TCW
Haslip, G H G8DWX
Hasman, G M1BCP
Hassall, G G0WHD
Hassall, K P GW1PND
Hassall, T A G7KLT
Hassel, H GM4SSA
Hassell Bennett, R G3WJN
Hassmann, P R GW4REX
Hasted, A S G1FGS
Hasted, E C G3BHF
Hasted, T W G7WKC
Hastelow, P B G7HRK
Hastie, J A G4IRV
Hastie, K A G4DKH
Hastilow, P W G4NHA
Hastings, D J G6WMG
Hastings, E C G4YZZ
Hastings, F R GW0BPV
Hastings, J E H G6EQS
Hastings, M C G8ASI
Hastings, T J G0CFN
Hastry, B A G1OND
Haswell, A W G6JSI
Haswell, B C 2M1EAU
Haswell, C 2M1CJO
Haswell, C MM1AEL
Haswell, C K GM2CWL
Haswell, G G7MWU
Haswell, M J G4KAX
Hatch, C GW6NNR
Hatch, E J G3ISD
Hatch, J G0IYG
Hatch, J G0WRA
Hatch, J G3OOL
Hatch, J G7WRA
Hatch, R J G0UHS
Hatch, R J B G6CGQ
Hatch, R S G8EAW
Hatcher, D J G7NGF
Hatcher, P J 2E1DRC
Hatcher, R G G7EYA
Hatcher, R J 2E0ALW
Hate, M L G7JUA
Hately, J N GM4EQY
Hately, M C GM3BSQ
Hately, M C GM3HAT
Hatfield, D C G4WBP
Hatfield, J H G6JCM
Hatfield, J P G7EAT
Hatfield, R C G8NVT
Hatfull, C L G3HZI
Hathaway, R G G6CWU
Hathaway, R J G3YIG
Hathaway, R L S G3JHI
Hathway, R G7ANF
Hathway, T W G4WVH
Hatley, B G4JPE
Hatt, B M G6SMZ
Hatt, F H G3CPH
Hatt, J D G1CBS
Hatt, M G G3YTR
Hattam, C W 2E1EZX
Hattam, M T G4KGA
Hatter, J E G4JPY
Hatter, P G6PYL
Hattersley, A F G0CWS
Hattersley, M E G1EVA
Hattersley, R G3PJN
Hattersley, W W G7BWH
Hattie, W M GM4LYV
Hattley, G C W G7GBF
Hatton, D G6VBK
Hatton, D C G1BYQ
Hatton, D F G4VWI
Hatton, D J G1BYB
Hatton, J W GM4RJX
Hatton, K J G4IZW
Hatton, L G4WPG
Hatton, P J 2M1FKB
Hatton, R G0VVP
Hatton, S G4VZX
Hatton, S J G7IHC
Hatwood, M T GW4VLU
Haughey, A F 2E1BVN
Haughey, P P G3JXR
Haughie, G J M1AEB
Hauton, H D G0TNS
Hauxwell, M R 2E1DOO
Hauxwell, R L 2E1DRX
Havard, F J G0WNU
Havard, F J G4VEY
Havard, M G6UYT
Havard, S E G4USN
Havell, G R G7IHV
Havenhand, B G G3OOP
Havenhand, P G0EXF
Haver, C D G1IXV
Haver, C F G8VVC
Haver, D F G3YFS
Haver, I M G6VEY
Havercroft, C G8CTX
Haverson, R F G4DHT
Havran, V GD4VGN
Haw, A J G7KHT
Haward, A K G4NLF
Haward, A R G8WEV
Hawes, A F G1JPP
Hawes, B A G3YLY
Hawes, B J G8ZRG
Hawes, C H G0UCH
Hawes, D O G8DIS
Hawes, I E G1IYE
Hawes, J A G4JHP
Hawes, J B G0DWD
Hawes, J G G4UAZ
Hawes, M J G7TAF
Hawes, P F G0BQY
Hawes, P R G4CKW
Hawke, F J GM3FUT
Hawke, R D G4FPG
Hawken, D J G6NQO
Hawker, E GW4JCE
Hawker, G P G1REL
Hawker, J P G3VA
Hawkes, A G1ACJ
Hawkes, C G0BVQ
Hawkes, C G6PNI
Hawkes, C G7RSA
Hawkes, D P W M1BBS
Hawkes, D W G4FOR
Hawkes, R A G6FEJ
Hawkes, R H G2CCH
Hawkes, R J 2E1DKM
Hawkesford, N A F G0UYI
Hawkesworth, R J G6FBT
Hawkesworth, W J G1WXR
Hawkes-Bayliss, J G6EHP
Hawkey, M G6NWW
Hawkings, J A G4GVE
Hawkings, K F G4RXB
Hawkins, A C G4GKK
Hawkins, A C G4OPA
Hawkins, A C G6NRF
Hawkins, A J G4TCC
Hawkins, B G4YBH
Hawkins, B MI1BYV
Hawkins, C G4RND
Hawkins, C C G3VLC
Hawkins, C D 2E1AGJ
Hawkins, C D G0CEU
Hawkins, D J G3PVS
Hawkins, D J G7LEL
Hawkins, D M G7AAS
Hawkins, D R G8KNF
Hawkins, E J GW4ZEA
Hawkins, G G1RWI
Hawkins, G J G1YOA
Hawkins, G R G4ATQ
Hawkins, G W A G0PSK
Hawkins, H G1JMG
Hawkins, H R G7JCU
Hawkins, I J G3PNO
Hawkins, I R G1FXG
Hawkins, J G3HWQ
Hawkins, J GW4NLE
Hawkins, J C G1CJH
Hawkins, J L G3LXD
Hawkins, K R G4OFL
Hawkins, M G0GFT
Hawkins, M G0HZL
Hawkins, M G7OKF
Hawkins, M C G1GOO
Hawkins, M J 2E1APQ
Hawkins, M J G3ZNI
Hawkins, P C 2E1APR
Hawkins, P C G3YUD
Hawkins, P J G0DOQ
Hawkins, P J G4JAA
Hawkins, P M G4KHU
Hawkins, R C G8UJG
Hawkins, R F M G8SQC
Hawkins, R W M0BDJ
Hawkins, S M G4TVR
Hawkins, T 2E1FYD
Hawkridge, A T G0OWI
Hawkridge, C J G4UKA
Hawkridge, C J D G4RBC
Hawkridge, P G H G6IEQ
Hawkridge, W G4MXM
Hawkshaw, M R G7AQA
Hawksworth, G W G3JQC
Hawkyard, L N G G5HD
Hawley, D G4KIE
Hawley, D G6AUY
Hawley, J G7TYB
Haworth, A G0PYW
Haworth, A F G7PZU
Haworth, B G0MBE
Haworth, B G1ZED
Haworth, J G4IMQ
Haworth, J A G8ZEF
Haworth, K G4XUL
Haworth, P G6OWI
Haworth, P G G4SCQ
Hawthorn, R W G4TCA
Hawthorne, H B GI6JCL
Hawthorne, M D G7SXL
Hawthorne, W R G3MCS
Hawthorn-Slater, G M GW7KIO
Hawxby, A G1KAO
Haxton, A R GM0ARH
Hay, B J G4KRO
Hay, D G0TGW
Hay, J N G1GYH
Hay, N A G3ZWR
Hay, N I G8GWM
Hay, S P G3HYH
Hay, W W GM6AOJ
Hayball, W G G6HST
Haycock, A GD2HCX
Hayden, E A G4LZU
Hayden, K J M1BHX
Hayden, T GW0HCN
Hayden, T GW0IIW
Haydock, A R G0SVQ
Haydock, C H G4OIB
Haydock, G A G6LVO
Haydock, S G4KZW
Haydon, A J G7MID
Haydon, D R G4NLH
Haydon, G R G4XQT
Haydon, M P G1KVO
Haydon, S G4WBI
Haydu Jones, T C G3OAD
Haye, T G7DNF
Hayes, B G4WXR
Hayes, C J G G7ACN
Hayes, C M G1AUI
Hayes, D E G6AOI
Hayes, D P G4AKY
Hayes, F 2E0ABL
Hayes, G G4UHO
Hayes, G B G4SEC
Hayes, G M GM7IHZ
Hayes, H W G4RLD
Hayes, J R GM0NBM
Hayes, J W GW3FPH
Hayes, M B G4BMD
Hayes, P M0AHK
Hayes, P D G3POQ
Hayes, P J A 2E0ANY
Hayes, P J A M0BDW
Hayes, R E M0APU
Haygarth, C J G1CPO
Hayhurst, F G0BYO
Hayhurst, J MW1BWI
Hayhurst, M G0ECB
Hayler, K L G4KYC
Hayler, P G0URL
Hayler, P J G1ITE
Hayles, C B G8WMJ
Hayles, J A 2E1BYB
Haylett, B A G8VQE
Haylett, B D C G8MTA
Haylett, P W G3IPV
Haylock, A G0BXA
Haylock, D W J G3ADZ
Haylock, P N G7KJW
Haylor, P G6DRN
Hayman, B W G6TDB
Haymes, M E G1KJQ
Haynes, A B G0JMX
Haynes, A J G4URA
Haynes, B G0IOE
Haynes, G R M G4FLY
Haynes, K L G3WRO
Haynes, L C W G0BAW
Haynes, L M G4URL
Haynes, L S MM0BCR
Haynes, M P G6NVU
Haynes, M S 2E0APH
Haynes, M S 2E1ERN
Haynes, N D G0VQO
Haynes, P E G8MGZ
Haynes, S G7HRZ
Haynes, S F G0CLG
Haynes, S J G1LSJ
Haynes, W G3STT
Haynes, W L GW1KEU
Hays, E A GW3RGL
Hayselden, R J G7USI
Hayter, A H G4UWJ
Hayter, D G4HZU
Hayter, D T G3JHM
Hayter, D W G3OCI
Hayter, G R G0AZD
Hayter, J P 2E1CKN
Hayter, R G4OAC
Hayter, R C G8BFO
Hayter, S W G6RAQ
Hayton, T G G7REA
Hayward, A A G1GYQ
Hayward, B G0VMB
Hayward, B J G0KIC
Hayward, B P G8VXQ
Hayward, C G4AHH
Hayward, D C V G0KBX
Hayward, D F S G3OMH
Hayward, D M G2UH
Hayward, D R C G0ODF
Hayward, E A H G0IRD
Hayward, J S G7NZJ
Hayward, J S M0BBE
Hayward, M D G8WWN
Hayward, M G G3LGA
Hayward, M L G8KYT
Hayward, M R G6JSF
Hayward, P C G0TNG
Hayward, P C G3JMX
Hayward, P C G4PGB
Hayward, R G3UCC
Hayward, R G G4OPR
Hayward, R M G4RTY
Hayward, R S G8YNO
Hayward, T J G3HHD
Hayward, W M G4LJT
Hayward, W R G7NBR
Haywood, A G7CRQ
Haywood, G A G0DPA
Haywood, J P G6PNM
Haywood, J R G7KPM
Haywood, J T G7SSW
Haywood, J W G1URL
Haywood, K M G8HXE
Haywood, L C G8KJJ
Haywood, M C M1BYR
Haywood, M W G4HLP
Haywood, R G6GES
Haywood, S G8UZQ
Haywood, S G G7CXT
Haywood, S R G8TFB
Hayzen, D I G7LXH
Hazel, D M0AUA
Hazeldine, D G 2E1CVD
Hazell, B T G1JHK
Hazell, C H H G6CWF
Hazell, E J G0HAU
Hazell, J R G7VWM
Hazell, M V G1EDP
Hazell, O M G0SOH
Hazelton, P J G8PJH
Hazelwood, P G G4KZB
Hazlewood, B L G4DYU
Hazlewood, J R G0CGQ
Hazzard, D R G4HUM
Hazzledine, M G0MRY
Head, A E G4BLG
Head, A J G3YPY
Head, D J G7PNE
Head, J D G4VUD
Head, J P G7MZI
Head, J S G0JCH
Head, M J G1WLO
Head, N E G6BRB
Head, N M G7LUO
Head, N P G4ZAL
Head, P 2E1BTJ
Head, P J G4FYY
Head, P W D G4LKW
Head, R G4XBW
Head, R J G8GBM
Head, R M G1DVP
Head, S R G8KOS
Head, T A G1KIZ
Head, V G6VYV
Head, W F G6WDF
Headey, M C G0VVV
Headland, D J G8UJF
Headland, J N G3BFP
Headland, R B G4XRX
Heal, A T G1FGT
Heald, A 2E0AKE
Heald, C 2E0AKF
Heald, D H G3SHJ
Heald, F R G4YDB
Heald, G J G0VLJ
Heald, J G0UEA
Heale, D C G6HGE
Healen, J G8EHF
Healen, P G8EHG
Healey, C R J G0NCS
Healey, D R G1HPJ
Healey, M T G3TNO
Healey, P G0OON
Healey, R G6HY
Healey, R W G8TQP
Healey, S J G1NNQ
Healey, T L G7BNS
Healless, A J G0IYH
Healy, C GM0JUB
Healy, P J G8HZQ
Healy, R J G6PGD
Heaney, G GI4DAH
Heaney, G A H G3MDQ
Heaney, G F S G7ROL
Heaney, J L G6RUY
Heaney, P M GI7UGJ
Heap, A G1OCS
Heap, H R G5HF
Heape, B G3SZH
Heaps, C H D G0DQE
Heaps, C H D G4NOR
Heaps, W H GD7VNS
Heapy, G A G3RGV
Heard, C F G7HBV
Heard, D W G6YEK
Heard, G V M1BOA
Heard, K A G0KAH
Heard, S R 2E1EQY
Heard, V E G0BFI
Hearl, P G6IXT
Hearn, A H R G3UEQ
Hearn, C C G3UIB
Hearn, C F G3REU
Hearn, D G1VAY
Hearn, L C G3WER
Hearn, M G4AMI
Hearn, N J A G4XTR
Hearn, S W G7UUW
Hearn, W S G4JLW
Hearne, A GJ8CEY
Hearne, F H G G1OFA
Hearne, N C G0IRW
Hearsey, M G6FRS
Hearsey, M G8ATK
Hearson, P G3SIU
Hearsum, D R G2HLP
Heartfield, T A G7URM
Heasley, J GI4GID
Heasman, A D G8GJO
Heasman, D E G4LMG
Heasman, F J G3EEW
Heasman, N G4XDK
Heasman, T J G4LMH
Heater, C K G6TSX
Heath, A G4GDR
Heath, A E G0FYY
Heath, A E G1MTC
Heath, A J G8BHY
Heath, C J G3PUT
Heath, C W G8VVB
Heath, D A G7MAY
Heath, D J G7UVU
Heath, H B G3HOI
Heath, J G7HIA
Heath, J B G4IRB
Heath, J B G8SMR
Heath, L R G0IDR
Heath, M C B G0BAO
Heath, P D G0CZY
Heath, P R G3NAU
Heath, R A G4HWN
Heath, R G 2E1FXV
Heath, R G G4XIZ
Heath, R M M G3UJV
Heath, R M M G3VER
Heath, R M M G8VER
Heath, R T G6FQL
Heath, W G4IAF
Heath, W D G3ABS
Heath, W H G0MRQ
Heathcote, A L G0RAJ
Heathcote, D G3KQE
Heathcote, D P G0AOD
Heathcote, F M G6BCZ
Heathcote, R J 2E1BWF
Heathcote, R J G7SNN
Heather, A S G0PQA
Heather, D G6SNA
Heather, S A G6BUS
Heathershaw, D F G3TLI
Heathershaw, D F G4EKT
Heathershaw, J G4CHH
Heathershaw, O G4USH
Heathfield, J C G1YKI
Heathfield, K A G3SDO
Heatley, A J R G0OPU
Heatley, R K G0OPV
Heaton, D G3YSV
Heaton, D C G1WZK
Heaton, J G1YYH
Heaton, J G F 2E1AZF
Heaton, R B G3UGX
Heaton, R J G7JER
Heaton, R V G3JIS
Heaton-Jones, E H G3CJ
Heaven, S H G3MGF
Heavens, C G G4AMD
Heavingham, T J G6KEQ
Heaviside, J W G3NYX
Heaviside, K O G0NPG
Heaysman, A G G1IBP
Hebb, D 2E1FHZ
Hebborn, J GM4TJL
Hebden, A G G8BYB
Hebden, C S G3GRQ
Hebden, D G0MMX
Hebden, D J G8BLC
Hebden, J G6NVV
Hebdige, R G1IOR
Hebel, S A G4UCU
Hebenton, D GM4API
Heckley, A G4AVY
Hector, G G4TVD
Hector, W J G G3JTH
Hedderley, R H G8JNR
Hedge, B G6JCY
Hedgecock, B D G6HTM
Hedgecock, J B G1EDS
Hedgeland, P M S G2DBA
Hedges, A K G6PYM
Hedges, D G G7TKD
Hedges, J A G3MMQ
Hedges, J E G7ANQ
Hedges, K W G3XMR
Hedges, M S G0JHK
Hedges, R V M1AVW
Hedges, S A G7UXK
Hedicker, P N G1SED
Hedicker, S R G8XJO
Hedison, P G4NLC
Hedley, D J G0OYN
Hedley, E M 2E0AOS
Hedley, E M 2E1EYF
Hedley, E M M1BYZ
Hedley, G G0SOF
Hedley, G G8BDQ
Hedley, G H G4WOX
Hedley, J A G4WVJ
Hedley, J F D GM8TXC
Hedley, R G8MFF
Hedley, W B G0FBU
Hedtmann, J GM4YRI
Heed, B S G8JAW
Heed, R H D G4FAK
Heeley, A N G3PFT
Heeley, R G8TXA
Heeley, T G3SWU
Heerma Van Voss, S F C GM0UMJ
Heesom, J T G4TQH
Heesom, R A G0VYK
Hefferman, W H G8XJN
Hefferman, J S 2E1FPT
Hefford, A J G1NEN
Heftel, D W G0CNF
Hegarty, J A GI4NFW
Hegarty, J F G3HKZ
Heggerty, N C G0PTN
Heggie, A G GM0NAZ
Heggie, W K GM3NHW
Heggs, B E G3VDQ
Heggs, O G3NLR
Heigh, A J 2W1FJG
Heilbron, S G3MIP
Hein, J G B GM1YME
Heinrichsons, J A GM8FHV
Heley, J F G1IUQ
Helgesen, K H G7CVY
Helie, I GM7VHQ
Heller, L F G1HSM
Hellewell, A G4PNT
Hellier, A R G4ZZD
Helliwell, J G0ESP
Helliwell, P M G0KAO
Helm, A G B G4BCX
Helm, G G6VEZ
Helm, J H G1OVB
Helm, P G4MZT
Helm, P G6AMX
Helm, P G8AEN
Helman, B G4TIC
Helsdon, B A G6XRE
Hely, M D G2CYN
Hembery, A M G6JKX
Hembery, C L G1IMF
Hemenway, A J G1VIZ
Heming, P R G3WBW
Hemingway, D G8EGG
Hemingway, N S G7BPM
Hemmens, H C GW4KUS
Hemming, A H G1MJO
Hemming, J A G0UYT
Hemming, L G6GTU
Hemmings, A E G7LNX
Hemmings, A P G0PHT
Hemmings, M J G0NCQ
Hemmings, N P G4NQQ
Hemmings, R K G0TVT
Hemmings, R K G3VCT
Hemmings, S J G7FLE
Hemmins, D N G6DRP
Hemphill, J E GM3CTG
Hemphill, P R G0HUG
Hempsall, K A G8SMH
Hempstead, R G8TOI
Hemsil, K I G7TLK
Hemsworth, S P G0JTT
Hencher, J T G1UVB
Hender, C H G4HYV
Hender, C H G4IAP
Henderson, A G4MBT
Henderson, A K G0EJF
Henderson, A M G1RRW
Henderson, A P G3SXH
Henderson, B L G0EET
Henderson, C G7GSH
Henderson, C A P G4FAM
Henderson, D G7FKP
Henderson, D C G0PVT
Henderson, D H G4NTC
Henderson, D V G8YQO
Henderson, E E G3LYD
Henderson, G G7RGX
Henderson, G R GM0LRA
Henderson, G R GM3RTJ
Henderson, H G GM3AEY
Henderson, I G4RVC
Henderson, J GM4HKV
Henderson, J A G0KKH
Henderson, J F G4NZD
Henderson, J I G6PXK
Henderson, J W GM4HKW
Henderson, K G6SVH
Henderson, M R GM6JKU
Henderson, N GI8HXY
Henderson, P I GM4VRE
Henderson, P J G4LAZ
Henderson, P J GI4KBW
Henderson, R 2W1FBX
Henderson, R G3ZEM
Henderson, R GM4DTJ
Henderson, R G GM0UET
Henderson, R P G1ISW
Henderson, R W G3NAN
Henderson, R W W G0LNA
Henderson, S F GI4OUP
Henderson, S J G1BNV
Henderson, S N G0EES
Henderson, S P G6GZB
Henderson, T GM4ITH
Henderson, T M GI7GVI
Henderson, W D G3KOZ
Henderson, W J GM0VIT
Hendon, D A G8DPQ
Hendricks, G L G1OKV
Hendriksen, G F G3EHA
Hendron, J A GI8DHW
Hendry, C M G0NNS
Hendry, C M G0ODR
Hendry, K J G0BBN
Hendry, K J G4RSE
Hendry, K J G6RSE
Hendry, N A GM8CBQ
Hendry, R J G6OLM
Hendry, S J G0ASN
Hendy, E G0IUC
Hendy, P G1OFY
Henery, C A GM0HBT
Henery, R F G7KIW
Henk, A J G4XVF
Henley, D M G3OQO
Henley, P G G8IQO
Henly, H R G3IHR
Henman, B J G3UNT
Henman, E R A G6HM
Henman, M G6UVN
Henne, G G1XUE
Henneman, R J G6JKV
Henney, W J G7NCU
Hennigan, T G4GNW
Henning, A J G0DPQ
Henny, P F G4NJC
Henretty, D F G0KOF
Henry, A G4DHE
Henry, A E GI4CRL
Henry, D GU0HRY
Henry, D S MM0AOF
Henry, J C GI0DVU
Henry, J C GI4GTY
Henry, K N G8ZJU
Henry, K P G4CFB
Henry, M G4PWK
Henry, M J G3LZE
Henry, N P G3OFK
Henry, P R GW8SZC
Henry, R GM7AON
Henry, R J GW4JQQ
Henry, S GM3ZNM
Henry, S A G7VDJ
Henry, S T GU4GNS
Henry, W C GM3WCH
Hensby, D G8TKD
Hensford, L A GW2BHS
Henshall, R G4NIL
Henshall, R A G0VJR
Henshaw, A G3VQR
Henshaw, A G6KSH
Henshaw, G G6CDT
Henshaw, R G3IPO
Henshaw, R G0GBN
Henshaw, J P T G6AOF
Henson, A I G4TNG
Henson, D F J G1XBY
Henson, D J G0DOH
Henson, P G6AUX
Henson, P A M1APK
Henson, P B G4UYA
Henson, P J G4IIH
Henson, R E G1WBM
Henson, S N G6IXS
Henson, W T G6JCI
Henstock, A G4XIN
Henstock, G A G0HEN
Henstridge, G N G4UED
Henton, R C M1BEY
Henville, R J G3TPH
Henwood, P N G3RWF
Henwood, S A G4PGR
Hepburn, M W GM0SXQ
Hepburn, M W GM7NXI
Hepke, K R G4NJB
Heppel, G H GM2DRB
Heppenstall, B GW4CWU
Hepple, T J G4UNI
Hepplestone, C M 2E1FFM
Hepplestone, D J G4KUL
Heptinstall, I P G0OPH
Heptonstall, C J G1HEP
Heptonstall, J G4EFU
Hepworth, A G G0BZG
Hepworth, A G G4VOG
Hepworth, C GM1DVO
Hepworth, D C G4LKX
Hepworth, F A G4SAV
Hepworth, G O G1LFI
Hepworth, S G G0KMN

Herbert, D G0VXE
Herbert, D W G4PPS
Herbert, G R C M0ADT
Herbert, I H G7BYA
Herbert, J J G4JJH
Herbert, L G6SVF
Herbert, M J G7TNH
Herbert, N 2E1DFC
Herbert, P M1ANJ
Herbert, R M G2KU
Herbert, R M G2TV
Herbert, V V G0ATB
Hercombe, B F G4JCH
Herd, D J P G4SOZ
Herd, R MM0BGO
Herd, T G0TJY
Herdman, T L G6HD
Heredge, P C G4PWI
Herf, L G0EOP
Heritage, A V G4EOG
Heritage, C F L G6UVO
Heritage, F J M0AEU
Heritage, H J GM0PUN
Herke, D L G8IBC
Hermes, W R G3YHC
Hern, A G1UFX
Herod, F C G4DZV
Herod, P R G8TQI
Herod, S A G8EAX
Herold, J G7RPC
Heron, H G1HTF
Heron, J M GM8GCY
Heron, R G4UDO
Heron, R A G8KFN
Heron, R M G4FBC
Herr, R V G7TER
Herrett, B G0DGD
Herrett, C G4LSV
Herrick, R B G0SZY
Herridge, D J G1VSE
Herridge, F A G3IDG
Herries, J W G0MKY
Herring, A N GM0CQQ
Herring, P G8WYI
Herring, P R G4FQF
Herring, R N G4BNE
Herringshaw, R E G4HTH
Herrington, B W G4KLY
Herrington, J A G0VJH
Herrits, P R R GW1GVG
Herrmann, P J G0SVB
Herron, D I H 2M0AAY
Herron, I M G0VHB
Herron, W H GM4LHQ
Hersee, G C G1CRP
Hersey, P J G4UDW
Hersey, R A G8ENY
Hersom, J G7BYE
Herwig, C T G4YVE
Herwig, E G G4YEP
Heselwood, R J G4UYI
Hesford, J R G4UAE
Hesketh, D A G6MTB
Hesketh, J C G1JZX
Hesketh, J G G1OLM
Hesketh, K M0AQK
Hesketh, P A J G4LIG
Heslop, R G G3KMQ
Hesman, J M G4EUI
Hespley, R H G2DXH
Hessom, D R G4GFM
Hester, P A A G7IVW
Hetherington, A G0CDH
Hetherington, C C G4ZOH
Hetherington, C R G8XVO
Hetherington, J G G0NGS
Hetherington, J G G4YGU
Hetherington, K W G1YUN
Hetherington, L M G8GRD
Hetherington, N P S GI8KUO
Hetherington, R J GM1OVW
Hetherington, S A G7DMH
Hetherington, W A G7DMG
Hettiarratchi, D G1VOP
Heuser, S G B G4SUG
Heward, A G0PVY
Heward, C M G0CMH
Heward, D J 2W1AJZ
Heward, G J G4RJM
Heward, J M G6UKZ
Hewer, I G4CLR
Hewes, A H E G1VCZ
Hewes, C G6SFF
Hewes, R S G3TDR
Hewes, R S G3UES
Hewett, A D G8KSD
Hewett, E T G G7KUA
Hewett, J G4SVE
Hewett, L R G4PBE
Hewett, N P G0DDV
Hewett, P J G1ENP
Hewett, R W S G3XLU
Hewins, E E GW3GSJ
Hewins, F GW1HFW
Hewins, M F W G3WKI
Hewison, R G7VYM
Hewitson, T G3EWH
Hewitt, A G6PFN
Hewitt, A E GW3YNM
Hewitt, A P G3SVD
Hewitt, A T GI8LUR
Hewitt, B R G0AHC
Hewitt, C G7LPF
Hewitt, C J G3UKN
Hewitt, C R G0PAE
Hewitt, D C W G8ZRE
Hewitt, D G G0CAD
Hewitt, E R G0PVJ
Hewitt, E Y G0KSX
Hewitt, F W G6JCX
Hewitt, G M G4IVN
Hewitt, G P G0MCB
Hewitt, G S G6IXN
Hewitt, I G4SVL
Hewitt, I A G7RAL
Hewitt, I A G8SNF
Hewitt, J G0CUO
Hewitt, J A G7IFM
Hewitt, J C G0BNZ
Hewitt, J H G8ZZM
Hewitt, J J 2E1DSU
Hewitt, J P G3IWT
Hewitt, J R G7TUV
Hewitt, J R GM3ZOT
Hewitt, K G6DER
Hewitt, L G0PVO
Hewitt, L J G2HNI
Hewitt, M A G0SQS
Hewitt, M A G7JVC
Hewitt, M D G4NCU
Hewitt, M D G7IIN
Hewitt, M J G4AYO
Hewitt, M J G7TKM
Hewitt, M J G7TKP
Hewitt, M R GW3DMV
Hewitt, N V G8JFT
Hewitt, P G0NUE
Hewitt, P L G8JCV
Hewitt, R 2E1CJW
Hewitt, R C G4MHJ
Hewitt, S W GI4MDO
Hewitt, S W GI7PIZ
Hewitt, T D G7WEM
Hewitt, T I G7WEL
Hewitt, W G4WEP
Hewitt, W A G1EDT
Hewitt, W J J G3UQS
Hewitt, W T G3YFD
Hewlett, C G GM1PFU
Hewlett, J R M1AYN
Hewlett, P J M1BKE
Hewson, A A G8RMX
Hewson, B G0TVO
Hewson, B D 2E1CUA
Hewson, D A G6YXB
Hewson, R G4KFF
Hewson, R G4NJA
Hey, J R G3TDZ
Hey, P G4JHS
Heyburn, W G GI4OYE
Heyes, A G1VSK
Heyes, A G3ZHE
Heyes, M R G0EIM
Heyes, R G4SPN
Heymans, C D G8CMP
Heys, A G1OLF
Heys, J D G3BDQ
Heys, S G4KJO
Heywood, C G7FVR
Heywood, F G4IDT
Heywood, J G4IAL
Heywood, J M G0GZA
Heywood, P G0UIT
Heywood, R R G1MET
Heywood, W H G2ALZ
Heywood-Bell, T G GW7RNC
Heyworth, A G0VAH
Hibbard, A J G8AQN
Hibberd, D G0ODH
Hibberd, D G0REQ
Hibberd, G F G3PYP
Hibberd, G W G4OTX
Hibberd, J W G3YMU
Hibberd, T D G4PZH
Hibberd, W D J G7TYW
Hibberson, A G3EVX
Hibbert, C M G1APA
Hibbert, E W G3UUX
Hibbert, J G3YCV
Hibbert, J G6UMX
Hibbert, J K G8LZO
Hibbert, S G GW7OTU
Hibbett, E R G8LAY
Hibbin, D J G4AOP
Hibbin, D J G8CPK
Hibbin, R C G4IKL
Hibbitt, M G3ULN
Hibbs, C F G0UPZ
Hibbs, P M 2E1AEF
Hick, A B G0LBY
Hickey, K Q G4UFF
Hickey, P A G3WDX
Hickey, R G P G6LVJ
Hickin, A J G3PXL
Hickin, J D G8ICJ
Hickinbottom, T W G3LCZ
Hickingbottom, J E G7LHU
Hickling, T E G7HBU
Hickman, A R G0IAS
Hickman, E G7JWJ
Hickman, G C G8KPV
Hickman, J E C G8JWE
Hickman, M F G3VGE
Hickman, M I G8XUN
Hickman, M W G7BGZ
Hickman, S R G1DKW
Hickmott, R G8MFV
Hicks, A G G1EDU
Hicks, A J G0TXY
Hicks, A J G8JVI
Hicks, A W T G1DSM
Hicks, B C G0SAA
Hicks, C G0WTB
Hicks, D G7MYR
Hicks, D A B G0IZY
Hicks, D B J G3JOF
Hicks, D G C G3GIZ
Hicks, D G C G6IFA
Hicks, D N G8EPR
Hicks, J G8NMT
Hicks, J A G6XRF
Hicks, J F G8RYW
Hicks, J L G4VTM
Hicks, J W G4XRU
Hicks, K H 2E1EAQ
Hicks, L G G8EGF
Hicks, L J G4GMS
Hicks, M C H 2E1CNQ
Hicks, P A G4KCX
Hicks, P T G4DVP
Hicks, R T G0GJF
Hicks, S W G6DYK
Hicks, W A E G1FBZ
Hicks, W T 2W1GAC
Hicks-Arnold, E A G0CDZ
Hicks-Arnold, F G6MB
Hickton, D R G6DRH
Hide, A G1ZKT
Hide, R E G0LFF
Hiendl, M J G0PCV
Hier, W H GM3HSF
Higbee, D G0SQH
Higbee, K M G6VPK
Higgin, M G0ECG
Higgin, M G7ELR
Higginbotham, J B GM4GXR
Higgins, A GM0PTY
Higgins, A L G8GLY
Higgins, A R G1TRN
Higgins, C D G3NRQ
Higgins, D G3XVR
Higgins, D G8XMH
Higgins, E F G8AZC
Higgins, F G3YNG
Higgins, G GM1TKI
Higgins, G M G3UBD
Higgins, J G4UAF
Higgins, J GM3ZXG
Higgins, J A G6VLT
Higgins, J K G4IGX
Higgins, K W G1HRV
Higgins, L A GD3FOC
Higgins, M A GI3YMT
Higgins, M J G7SRU
Higgins, N C G7VIK
Higgins, N R G4ZQL
Higgins, R G6ISX
Higgins, R J 2E1CUS
Higgins, S P G3UPX
Higgins, W G3PNR
Higgins, W A G8GF
Higgins, W J G1MCW
Higginson, D R G8JET
Higginson, P D 2E1FEM
Higginson, P I G8IZR
Higginson, T GW3AHN
Higginson, W G4EGG
Higgott, B G6WQY
Higgs, A J F G6XAG
Higgs, G R G4AWG
Higgs, J C G1AFT
Higgs, P B GW0WXW
Higgs, P B GW4IGF
Higgs, S A GI7UBV
Higgs, S P G8YTR
Higgs, T G4TUA
Higgs, V W G3WVJ
High, R D G7VXP
Higham, A M G6ZBV
Higham, A R G1YQG
Higham, C G4IBE
Higham, P R G8JLM
Highams, T M1ABY
Highley, K W GW4MOL
Highley, P J G0GXA
Higlett, M A G6WTM
Higlett, S R T G8ZYT
Hignett, J GM4LVV
Higson, J R G4NTY
Higson, P A M1AXC
Higson, R N G2HFP
Higton, A J 2E1FIR
Higton, B G8OVV
Higton, B GM0VBE
Higton, B A G0JWA
Higton, G V G8KGK
Higton, M G8YNK
Higton, P R G3XXR
Hilbery, N J E G8LPA
Hilbourne, A C G6NVY
Hildebrand, P M1APH
Hildebrand, P G G3VJO
Hildebrand, R J G6EAZ
Hildich, M A G4BWR
Hildreth, S L G7FVP
Hiles, R G4SXZ
Hill, A G0JDZ
Hill, A G4PYQ
Hill, A A G4VPZ
Hill, A D G0XBA
Hill, A D G8MGP
Hill, A E G4BQJ
Hill, A G G8GZN
Hill, A H G0IZF
Hill, A H G0SPM
Hill, A H G7NWH
Hill, A J G0JTP
Hill, A J G3AUU
Hill, A J M1BLN
Hill, A P G1HAF
Hill, B G0GFJ
Hill, B F G4PHZ
Hill, B H W G4RNB
Hill, B J G0GRT
Hill, B N G7KGN
Hill, B W G0LMQ
Hill, C G6COE
Hill, C A G6EFY
Hill, C B C G3LGS
Hill, C D G6BFN
Hill, C G G2KG
Hill, C L 2E1EFS
Hill, C M 2E1BEG
Hill, C M G7SIS
Hill, C N G8RAX
Hill, D GM0UEL
Hill, D M1AJF
Hill, D B G6OLJ
Hill, D D M0AZK
Hill, D G 2E1CCN
Hill, D G G8EEA
Hill, D J G7MMQ
Hill, D L G4IQM
Hill, D M G3ZWH
Hill, D P G4LKU
Hill, D R G3AWQ
Hill, D R M0AYL
Hill, D W G3ZCH
Hill, D W G8BEH
Hill, E C L G4KGW
Hill, E J G4YEH
Hill, E P G4HZE
Hill, E T G7DFE
Hill, F G3YWH
Hill, F J G3EVV
Hill, F W G6APQ
Hill, G C G3DFL
Hill, G F G6EOA
Hill, G K H G0NPC
Hill, H A G1EML
Hill, H E G0NIZ
Hill, I GM0CNB
Hill, I E G6HL
Hill, I M G4SWR
Hill, I R G6ZVE
Hill, J G7CLY
Hill, J GM3YAU
Hill, J C G0WEH
Hill, J C G3XYH
Hill, J D G8YNP
Hill, J R G4NBR
Hill, J R A G4CFH
Hill, J S G8FWH
Hill, J W G3JIP
Hill, K A G3CSY
Hill, K J G1EMM
Hill, K P G0VSL
Hill, L F G0MUA
Hill, L J M0ARM
Hill, L J M1AZH
Hill, L P G6ISY
Hill, L W G6COK
Hill, M G6IFC
Hill, M A G1PPH
Hill, M A GW1NRS
Hill, M A GW4EZW
Hill, M A GW4SUE
Hill, M C G7GOV
Hill, M C T G7ECI
Hill, M J G0JAC
Hill, M J G4MJF
Hill, M L G4KUY
Hill, M M G0GFG
Hill, M N S G0BEV
Hill, M N S G3ZQM
Hill, N J G4UKO
Hill, P G1APC
Hill, P G6PNO
Hill, P A 2E1BEF
Hill, P A G7SDT
Hill, P J G0OCP
Hill, P J G1NRX
Hill, P M G0DRX
Hill, P N G4IOA
Hill, P S G6PNG
Hill, P T B G1SGU
Hill, P V G4FRM
Hill, R G3UIP
Hill, R G4OWU
Hill, R G8LEB
Hill, R A G6XEN
Hill, R F M1BUW
Hill, R J G0IUW
Hill, R J G8THE
Hill, R L G0IMV
Hill, R M G4DLT
Hill, R R 2E1FOG
Hill, R R M1BPA
Hill, R T A G3SMZ
Hill, R V C G7VRW
Hill, R Y H GM0FTG
Hill, S G8YYR
Hill, S MW0AES
Hill, S B G0OUD
Hill, S C G0CAZ
Hill, S F G6FPX
Hill, S F B 2E1DPK
Hill, S G G4XRO
Hill, S G G6PFP
Hill, S J G1HRU
Hill, T C G7VJY
Hill, T J G4YFC
Hill, T P G0SPF
Hill, T S GM4NWK
Hill, W A G4UKI
Hill, W B M1BKF
Hill Esq B.E.M., R C G3SGQ
Hillard, B F G4PDG
Hillary, M J G7KZI
Hilleard, D G4CQM
Hillerby, H R G8XEJ
Hillerby, J G0KDU
Hilliard, J G G4YRX
Hilliard, R J G7LPK
Hillier, B R W G7JKI
Hillier, F S GW4HDW
Hillier, M C G0MQM
Hillier, M D G0DDW
Hillier, P G1MUQ
Hillier, R C G0HBE
Hilling, J C G3HRX
Hillman, B G0UXO
Hillman, C J G1OCH
Hillman, C M G7UWR
Hillman, G E G3OKH
Hillman, J G7LIO
Hillman, N M G G1NMH
Hillman, P A M0ACD
Hills, A J R G1RCE
Hills, B C G8AOO
Hills, C R G8KKH
Hills, D G6PYF
Hills, D R G6MFM
Hills, D W G7OEF
Hills, F L G0KUQ
Hills, J V G0NUG
Hills, J W C G4APO
Hills, M J S GW1VMA
Hills, P L G7JRV
Hills, R C G3HRH
Hills, R D G8STO
Hills, R F G0BDA
Hills, W F J G4LGU
Hillson, R A E GW4OWX
Hillum, R A G6PAE
Hillyard, D A M M1BZP
Hill-Smith, C E R G7RKX
Hilsley, R W G3PBT
Hilton, A C G8LVI
Hilton, A G G8FME
Hilton, A T 2E1AKN
Hilton, D B G6VXE
Hilton, G G4NKW
Hilton, I T G7OND
Hilton, J G1HAC
Hilton, J G3JMZ
Hilton, J A G0IPC
Hilton, N J G7KMD
Hilton, P B G8GYO
Hilton, P J G7MYT
Hilton, R G4WZI
Hilton, R M G7DLD
Hilton, R W G1LKH
Hilton, W G7JZI
Hilton, W J G0NRI
Hilton-Jones, D G4YTL
Himmo, O D G1KGA
Himsworth, P J G4AUX
Hinch, K G2BXZ
Hinchcliffe, J G4CDD
Hinchcliffe, N G7SVU
Hinchliffe, J H G0JTL
Hinchliffe, K B G0HIY
Hinchliffe, M H G0LQK
Hinchliffe, R G8PDP
Hincks, S E G3ASM
Hind, A J 2E1CAB
Hind, A J G7RUF
Hind, D T G3PRC
Hind, D T G3VNG
Hind, D T G8PRC
Hind, G G GM1FMV
Hind, J C G7RUE
Hind, J E G6BMG
Hind, J P G3XKN
Hind, M A G7DQG
Hind, W F C GM4IWK
Hinde, G J G4LSP
Hinde, H G3KRS
Hinde, J G4PED
Hinde, J M1AFL
Hinderwell, N O G8IFN
Hindes, R G G3IGM
Hindle, A G4DTO
Hindle, H J R G3WBG
Hindle, L G0HVZ
Hindle, N J G7VEX
Hindle, P J G0LYC
Hindle, P J G3SBL
Hindley, D J MW0AMG
Hindley, F GM3VBY
Hindley, J E G8DEW
Hindley, M R H G4VHM
Hindley, R S G1NEE
Hindley, S T 2E1FFV
Hindley, T 2E1FFT
Hindmarsh, C N G7RKJ
Hindmarsh, J S G8XGS
Hinds, G H G1GKH
Hinds, W R G6PTZ
Hine, C J G2FZO
Hine, D J G3VFF
Hine, D M G4DIG
Hine, L G6CAT
Hine, M D G6VFA
Hine, S O G3RUB
Hines, B G6DYX
Hines, G A G6SAQ
Hines, M V G0TXN
Hingley, N K G4JSV
Hingston, B E G6UPR
Hinken, M G0SHG
Hinks, P R G4LTK
Hinksman, V J G8JSR
Hinson, D C G0JKH
Hinson, G T G4IFB
Hinson, L P G8PFT
Hinson, M R 2E1DCX
Hinton, C D G7JRM
Hinton, G D G7NCW
Hinton, J C G4EZE
Hinton, J G G0OUK
Hinton, K G7MAT
Hinton, M G7EWZ
Hipkin, I H L GM7TWM
Hipkin, L H G8KMG
Hipkin, S J G0OFF
Hipwell, G D G7KGS
Hipwell, J A G0LQM
Hipwell, W F G3HTX
Hipwood, T R G8OEU
Hircock, K G4KFE
Hird, D M M G4RVH
Hird, G G4VVH
Hird, M G6TSS
Hirons, B W G4JPU
Hirons, G S G4JPT
Hirons, J E G6TGJ
Hirst, A G8BRF
Hirst, A D G3TOW
Hirst, C G0EPY
Hirst, D M1AGH
Hirst, D J M1ANQ
Hirst, D P G0HFO
Hirst, I D R G7FGY
Hirst, J G M M G0POJ
Hirst, M T G0RHD
Hirst, P G7NDC
Hirst, P A G0PHI
Hirst, P G G0GXF
Hirst, R G6MSW
Hirst, R G G4XJG
Hirst, R I G4APO
Hirst, S G M1AGI
Hirst, W G4SZR
Hirst, W P G6TSM
Hiscock, J P K G4RLM
Hiscocks, C L MW1BOR
Hiscoe, G A G4GYF
Hislop, J D G7OHO
Hislop, J D G7UPG
Hislop, R G0UEH
Hislop, R H G4GPI
Hitch, E G G8ZYH
Hitch, N R G8ZYI
Hitcham, C J G1MPL
Hitchcock, A D G3ESB
Hitchcock, B M G1NMY
Hitchcock, W A G8TOM
Hitchen, B G1VZW
Hitchens, B J G3MHN
Hitchens, D J G0BXI
Hitchens, S G4VXI
Hitchin, L M G7OET
Hitchings, W L G3HWL
Hitchins, B P C G4CTU
Hitchins, B P C G4GXP
Hitchins, D E J G4GMB
Hitchman, M J G3HAN
Hixson, N F M0ALB
Hizzey, P A G6YLO
Ho, C M GM7TJV
Ho, P P K G7RJD
Hoad, R A G1ZMG
Hoare, A V G3WBY
Hoare, B E G0VEC
Hoare, C J G4AJA
Hoare, C J G4BPO
Hoare, C J G7BPO
Hoare, C J M0ATC
Hoare, D G1DFW
Hoare, H A G4PJD
Hoare, J D G8PKN
Hoare, J E G6XRH
Hoare, K J G0XRO
Hoare, L R M1CDA
Hoare, M A G4NBC
Hoare, M A G8VWJ
Hoare, P J 2E1ELJ
Hoare, R G G0DUF
Hoath, P J G7JBW
Hoban, J M G0EVT
Hoban, J V G3EGC
Hoban, R A 2E1EPC
Hobbis, A J G7ONO
Hobbis, P I 2E1AWO
Hobbs, A G G8GOJ
Hobbs, A J G1TQN
Hobbs, A J G3OJX
Hobbs, B S G1KJX
Hobbs, D D G8PVG
Hobbs, D J GW0LYF
Hobbs, D P G3HEO
Hobbs, G C G1JRZ
Hobbs, G J G1EHT
Hobbs, J H G1FTD
Hobbs, J N G1CTT
Hobbs, J W D G3HFY
Hobbs, J W D G3JQN
Hobbs, K G G7NCV
Hobbs, K G M 2E1DQA
Hobbs, L A G7IYH
Hobbs, M B G6YLN
Hobbs, N A J G7IYG
Hobbs, N D G7LXV
Hobbs, P A F G3LET
Hobbs, R C G7KHZ
Hobbs, R L G6IRY
Hobbs, S A G6XRI
Hobday, S J G3SKV
Hobden, D J GM3XMY
Hobden, S A G0VAJ
Hobin, J E G3OJ
Hobin, J E G3XIX
Hobkirk, A G P G4VZH
Hobley, J H G4FBQ
Hoblin, R J P G6AOH
Hobro, D G G4IDF
Hobro, D G G4MHC
Hobson, A G G2AGH
Hobson, C A G0MDK
Hobson, C A GM7JGE
Hobson, D G4YMN
Hobson, J E 2E1ENW
Hobson, L G0CUI
Hobson, L G4JQO
Hobson, M R GM8KPH
Hobson, P G4HOJ
Hobson, P R G6OLU
Hobson, S J G4KXE
Hobson, W C G7PAA
Hockey, A M G4DZW
Hockey, J G4YJH
Hockey, P G7VOO
Hockin, D E G4UGT
Hocking, A B G0FHX
Hocking, D W G4FSS
Hocking, E S G4LKV
Hocking, G A G4LCO
Hocking, J G0FHY
Hocking, P A G8ZDS
Hockley, A S GW4ZUW
Hockley, J D G0ANW
Hockley, J I GW0XYL
Hodby, E G G6ZIP
Hodder, E G G6PUA
Hodder, R G7IFD
Hodds, R H G0KLG
Hoddy, M G G0JXX
Hodge, A N G4UPY
Hodge, D E R G4FFS
Hodge, G A G4NTB
Hodge, K B GW4NEI
Hodge, K R GM3JIG
Hodge, L P C 2E1ESK
Hodge, M P G7OLG
Hodge, R E G3YJN
Hodge, R G G1IOF
Hodge, R M GM4PPT
Hodge, R T G4MMI
Hodgekins, D G1UFV
Hodges, A F GW1NGX
Hodges, B G4JKF
Hodges, C A G4JZI
Hodges, C N H G4PQN
Hodges, D G G6IXH
Hodges, D H 2W1ERX
Hodges, E G4YUO
Hodges, G J G7OEY
Hodges, G L D G3KRT
Hodges, J G0WRN
Hodges, J R G0URH
Hodges, J S G7DLE
Hodges, K G G0CHI
Hodges, K L G6TDG
Hodges, L G0EJV
Hodges, M G G6WZN
Hodges, M J G4OPE
Hodges, M L GW7TAA
Hodges, P G7AZK
Hodges, R B G0RYL
Hodges, R J G3SUR
Hodges, R J G3YVO
Hodges, R M B M G4DDH
Hodges, R S G0NFZ
Hodges, S G7BMT
Hodges, V E GW4VLQ
Hodges, Z E 2E1CJZ
Hodgetts, B A G4YQG
Hodgetts, D T G1BJJ
Hodgetts, D W G7AWU
Hodgetts, E G3WPY
Hodgetts, G E G7FXZ
Hodgetts, J G4NTE
Hodgetts, J B G2FXZ
Hodgetts, L W G3OCK
Hodgetts, P G4JTL
Hodgetts, S A G1LMU
Hodgetts, T I D G4USQ
Hodgetts, T J G4AYD
Hodgkins, D G8THF
Hodgkins, I G6AJC
Hodgkins, J G3JZP
Hodgkins, J E G3EJF
Hodgkins, P G0FDK
Hodgkins, R J M1ALM
Hodgkinson, A G G1SAM
Hodgkinson, A J G3LLJ
Hodgkinson, B GW7BOY
Hodgkinson, G G3SUN
Hodgkinson, M G4HCC
Hodgkinson, P C G8ZYR
Hodgkinson, P J G4IPB
Hodgkinson, R G3OGU
Hodgkinson, R C G3ZMM
Hodgkiss, I G0FMP
Hodgkiss, P G3WTY
Hodgkisson, D R G6DRO
Hodgson, A G0ADO
Hodgson, A M0AHA
Hodgson, A J G6KSK
Hodgson, C G0OBN
Hodgson, C W R G0JYJ
Hodgson, D M M0AAF
Hodgson, E A G0SBO
Hodgson, E D G3RAR
Hodgson, F T G3JXG
Hodgson, G G0BZH
Hodgson, H G4BOC
Hodgson, I H G8EKP
Hodgson, J M0AJA
Hodgson, J E G8TYX
Hodgson, J W G0EGS
Hodgson, N L G3RSS
Hodgson, N L G3WAH
Hodgson, P G0JYC
Hodgson, P G G0CLT
Hodgson, P G G8UBN
Hodgson, P M G0HAA
Hodgson, R G3DUW
Hodgson, R J G4KBH
Hodgson, S P G0LII
Hodgson, S P M0ADU
Hodgson, W G G4WVR
Hodgson, W G G1OVZ
Hodkin, A J 2E0AIH
Hodkin, A R G1YLG
Hodkinson, B W G3XAE
Hodkinson, D G4YSU
Hodkinson, H B GU0BFE
Hodkinson, J W G6COB
Hodkinson, K E G4MDE
Hodkinson, T P GU6JSC
Hodskinson, V G G0GHV
Hodson, H L B G7TLL
Hodson, H R G3MFX
Hodson, M C G6VUN
Hodson, M R M1APQ
Hodson, P E D G1DRX
Hodson, P V G8RBY
Hodson, R G7MOM
Hodson, S G1XTA
Hodson, S M GW4ZXD
Hodson, W L G6GKG
Hoey, G E G4JCF
Hoey, J GM0ARD
Hoey, R A E GI8SJS
Hoffman, R R G0CBO
Hogan, C J 2W1FGK
Hogan, C J G0SKZ
Hogan, D G3PUZ
Hogan, D P 2E1FGL
Hogan, E M G6ZVJ
Hogan, J R G4RLR
Hogan, M F G4RMN
Hogan, R J G4VCQ
Hogan, W G6VFB
Hogarth, H C GM8CSE
Hogarth, I G0IXE
Hogarth, J D GM0BPH
Hogarth, S G7EWL
Hogben, G S G0JUL
Hogg, A G0AUR
Hogg, A J GM8LKL
Hogg, C A G3NRZ
Hogg, D G7KIT
Hogg, D J G4CAF
Hogg, E M GM8NZL
Hogg, F G4BNH
Hogg, G D G1EVC
Hogg, H GM3PPJ
Hogg, H E G0ANF

Hogg, H M	G3NGX
Hogg, J	G3NUA
Hogg, J	GW4AKO
Hogg, J M	G1YHP
Hogg, J M	G2OG
Hogg, K M	G0UZJ
Hogg, S M	G0RMJ
Hoggan, A	G8ASX
Hoggan, J A	2E1AVT
Hoggan, T	GM6FBZ
Hoggard, P M	M1BLO
Hoggard, R A	G7PUL
Hoggart, D W	G0PMJ
Hoggarth, J B	G8WSU
Hoggett, K	G8RHM
Holbert, T H	G3DXJ
Holborn, R E	G6OWG
Holborn, T	G8SWF
Holborow, K B	G8JNE
Holbrook, A R	GW3HOJ
Holbrook, H	G4LPH
Holbrook, J W	G6VWF
Holbrook, V E B	G6CWW
Holbrough, M F H	G0IKI
Holburn, D M	G3XZP
Holburn, K L	G4AIK
Holcroft, P A	G8XTC
Holdaway, F J	G3OQL
Holdaway, M P	G6ZEL
Holdaway, P	G0DCP
Holden, A C	G1EVD
Holden, A J	G0LHY
Holden, A P	G7SMN
Holden, B	G4SXE
Holden, C C	G7BLD
Holden, D A	G8DPW
Holden, D M	G3WUN
Holden, E	G3MAJ
Holden, G	G0OYI
Holden, J	G8ZGS
Holden, J A	G7IKU
Holden, J P	G1IOQ
Holden, L	G4XDO
Holden, M L	G4HOL
Holden, P J	G0JSG
Holden, R K	G8SKA
Holden, W T	G4WTH
Holden OBE, G G P	G2HIX
Holder, A F	G4ZBH
Holder, W G	G7ARI
Holderness, C J	G6LVM
Holderness, R J	G3XDA
Holding, B	G0DMQ
Holding, D	G8EDL
Holding, E	G0DVS
Holding, J R	G4ASP
Holding, N W	G4CSH
Holding, W	G6PUV
Holdom, G L	G4SVU
Holdsworth, A	G0SAH
Holdsworth, A	M0ABC
Holdsworth, C L	G1DLO
Holdsworth, D	G1VGO
Holdsworth, D W	G6COG
Holdsworth, J B	G4XFF
Holdsworth, J M E	2E1BUN
Holdsworth, J S	GM4SQS
Holdsworth, K	GM4SQT
Holdsworth, M D	G7NFG
Holdway, A R	G0KQT
Holdway, K	G8DYI
Holdway, P	G0WXK
Hole, M J H	G0GJC
Hole, R	G0LAR
Hole, R S	G0IKC
Hole, S K	G7WIK
Holford, E G	G0EJZ
Holgate, M J	G4ZLL
Holgate, R F	G6XTT
Holgate, S C	G8YTP
Holker, P M	G3OIP
Holland, A J	G4VFL
Holland, A R	G6FVH
Holland, C D	G8USV
Holland, C G	G0FYP
Holland, D	MI1AWE
Holland, D C	G3WFT
Holland, D J	G4LDT
Holland, D V	G1XXJ
Holland, F J	GI0AIQ
Holland, G A	G3CMC
Holland, G E	G1IOU
Holland, G J	G0VHJ
Holland, G J	G7HJD
Holland, G W	G7UWO
Holland, J	G4COQ
Holland, J	M1BJL
Holland, J G	G3GHS
Holland, K O	G3MCD
Holland, L	2E1FLT
Holland, M R	G7VRJ
Holland, M R	M1BCE
Holland, P	G4TCB
Holland, P A M	G7EXC
Holland, P F	G6IFE
Holland, P J	G3TZO
Holland, P J	GI0EWE
Holland, P T	G6CHX
Holland, P W	G3XLI
Holland, R G	G2BQY
Holland, R G	G3BPE
Holland, R J	G1AUH
Holland, R W	G4DTZ
Holland, S	G7VOH
Holland, T	G0VWT
Holland, W	2E1CBQ
Holland, W J	2E1EGY
Holland, W J	G0EFV
Holland Carter, J N	G3OWB
Hollands, P A	G0VYA
Hollands, P M	G1PXW
Hollebon, C J	G8UEM
Hollebon, F W	G0HQS
Hollebon, G	G4UEL
Holley, K	G4IMU
Holley, K G	G6CNB
Holley, M A	G3CIL
Holley, M J T	G4EHQ
Holley, M V	G1ACY
Holley, S M	G8WBO
Hollick, R C	G6ZAX
Holliday, M R	G4DCK
Holliday, T F	G7KWQ
Hollidge, G T	G6BOF
Holliman, P A	G4EAZ
Hollinger, W R	GI7DBZ
Hollinghurst, M R C	G4NOE
Hollingsbee, D A H	G3TDT
Hollingsbee, I M C	G4NVC
Hollingshurst, S C	G3GEV
Hollingsworth, D	G3VKU
Hollingsworth, K	G7AFO
Hollingworth, B	G6GGV
Hollingworth, L J	G4NCV
Hollinshead, K	G7NSR
Hollinshead, M I	G4IDZ
Hollinshead, N	G6IFS
Hollis, B J	G0LCI
Hollis, F M	G4LNG
Hollis, J B	G8ION
Hollis, P J	G7BMC
Hollis, P R	G8BPY
Hollis, S	G7OYP
Hollister, C R	G4SQQ
Hollom, J H	G3RBQ
Hollow, D S	G4HRE
Hollow, R K	G4SOK
Holloway, A J	G7MWJ
Holloway, B	G0GFR
Holloway, C	G7FKJ
Holloway, C G	G4WIE
Holloway, C P	G0GGF
Holloway, D G	G3BBX
Holloway, D W L	G6UPQ
Holloway, E R	G0NHX
Holloway, J W	G4LFQ
Holloway, L A R	G4YTX
Holloway, M A	G4YRY
Holloway, M H	G0MHH
Holloway, R E	G3HUA
Hollowell, E E	GW0GUY
Hollows, R L	G0MCA
Hollyoake, A	G0CWC
Hollyoake, A	G1SAN
Holman, D F	G7SHC
Holman, D G	G8DEU
Holman, E P A	G8WMC
Holman, G	G6TGE
Holman, G S	G8YNF
Holman, H M	G7HJJ
Holman, I R	G7VYQ
Holmden, H M	G4KCC
Holme, G M	G2FFK
Holmes, A J	G4CRW
Holmes, A J	G4ISN
Holmes, C	G0LJH
Holmes, C	G1NCL
Holmes, C A	G0JHG
Holmes, C A	G0LUR
Holmes, D	G1ACG
Holmes, D A	G4KIZ
Holmes, D J	G4ZAO
Holmes, D L	G1EVG
Holmes, D L M	G4XEO
Holmes, D R	G8LVL
Holmes, D W J	G4FZZ
Holmes, E C	G4TLY
Holmes, E J	G3ALK
Holmes, G	G0PZD
Holmes, G M	G6ENZ
Holmes, H W	G0RBG
Holmes, H W M	G4TWT
Holmes, J	G3UEU
Holmes, J	G4VMG
Holmes, J G	G8STY
Holmes, J H	G1GNY
Holmes, J M	G8SFS
Holmes, J R	G3PKQ
Holmes, J W	G4GMG
Holmes, K D	G6BTX
Holmes, K F	G6IRW
Holmes, M A	G7HQE
Holmes, M D	G6AIZ
Holmes, M J	G4YGX
Holmes, M O	G0TYN
Holmes, M R	2E1CLC
Holmes, M W	G8MMN
Holmes, P	G0MAY
Holmes, R A	G7SVQ
Holmes, R S	GM0WRU
Holmes, S	G0HBO
Holmes, S	GW3GA
Holmes, S R	G4TWS
Holmes, S T	G7ECE
Holmes, V	G8GYP
Holmshaw, R K	G0KMF
Holohan, A L	G7OAV
Holroyd, A R	G0NUF
Holroyd, J	G7JOO
Holroyd, T P	G3YHD
Holroyd, W H	G7JGW
Holstead, J	G3OZC
Holt, C	G3ALA
Holt, C S	G6IRX
Holt, D E	G1BTV
Holt, D J	G6EXQ
Holt, D M	G4LRD
Holt, E	GM0WED
Holt, E W	G3MHQ
Holt, F A	G0WVY
Holt, G M	G3PTS
Holt, J E	G3KYL
Holt, K	GD3IBQ
Holt, M J	2E1COM
Holt, P	G4AIB
Holt, P A	2E1EKJ
Holt, P G	G1FOE
Holt, P J	G8YJQ
Holt, P M	G4LPP
Holt, R	G3TTU
Holt, R G	G8NOF
Holt, R J	GM1BNP
Holt, R P	2E0AEG
Holt, R P	M0AIT
Holt, S G	G0GEU
Holt, S H	G3ZLB
Holt, W H	G7DHM
Holt, W J	GW0SGG
Holtam, M	G1EDX
Holtham, G W	G4DBV
Holtham, M	G0EIG
Holtham, P N	G3ZXY
Holtham, R E	G4EKS
Holtom, R G	G4MMF
Holton, J C	G8OGR
Holton, J G J	G3PSC
Holtum, G F	GW4SLZ
Holyer, L S	2E1CMJ
Holyoake, B M	2E1EGA
Holyoake, R M	G4WAY
Holyoake, V A	G4OJJ
Holyoake, V A	G7OJY
Holzapfel, A H	M0ATD
Holzapfel, A H	M1BCS
Homan, J A	G6ZEN
Homans, S W	GM4BNM
Homden, R W	G0RMF
Homer, A M	G1IHJ
Homer, B S	G8IWX
Homer, M A	G6IFT
Homer, M J	2E1EOX
Homer, M J	G6AIQ
Homer, M R	G1YMH
Homer, P	G4BXT
Homer, P J	G4KQU
Homer, R R	G7UTV
Homer, S G	G6HFZ
Homer, W	G1JLZ
Homewood, E A	G4KBY
Hone, D J	G7OQH
Honer, R G	G4KTL
Honey, W A J	G6GAB
Honeybone, P W	G7AVF
Honeyman, R W	GM0NUI
Honeysett, D J	GW4TEJ
Honeysett, R A	G1UBW
Honeywell, A J	G8XVX
Honeywell, M J	G0ABB
Honeywell, P J	G7FCJ
Honeywood, E R	G3GKF
Honnor, F B	G6XUL
Honnor, R W	G0ILM
Honour, P	G4IHX
Hood, A	GM4UXX
Hood, A J	GM7GDE
Hood, B J	G6APV
Hood, D F	G4PNC
Hood, J A	MM0AKM
Hood, J H	G3MZI
Hood, J R	GM4COX
Hood, L A	G3LCW
Hood, M	G7BEE
Hood, R	G1IXU
Hood, R A	G4BIA
Hood, T J	GM4LRU
Hoodless, D J	G8ZAU
Hook, C D	G4FJW
Hook, D W	G7VUE
Hook, F U	G0BOP
Hook, G A	G2CIL
Hook, L R	G3PPO
Hook, M	G0CIO
Hook, R J	G8LVB
Hook, R N	G3ZLM
Hook, S H	G6TCP
Hooke, A H	G3CMT
Hooker, D J	G7OHK
Hooker, D J	G8IIK
Hooker, G A	G4OEM
Hooker, K H	G7PQB
Hooker, T J	G8OBB
Hookham, R A	G4GWV
Hooks, M J	G7IXM
Hooks, M J J	2E0AIE
Hooks, M M	GW3MKT
Hooks, R A	M1BBV
Hoole, R J	G1APH
Hoon, K J	G7DHP
Hoopet, A J	G1JMF
Hooper, B W	G0VUU
Hooper, D	G7VNW
Hooper, D	M0ALO
Hooper, D A	G4FVW
Hooper, D G	G0CYU
Hooper, I A	G4PRQ
Hooper, J C	G3XEI
Hooper, J L	G3FYS
Hooper, J R	G3PCA
Hooper, J R	G3XRT
Hooper, M W	G7UUK
Hooper, P O	G3KSP
Hooper, R	G4COF
Hooper, R A	G4LXR
Hooper, R C R	G4ZIX
Hooper, R G	G6YEX
Hooper, R S G	G6HTS
Hooppell, A R	G4TKV
Hoose, D	G4BSD
Hoose, D	G4UEW
Hoose, J A	G0NYK
Hooton, D	G6VFC
Hooton, G G	G0RUA
Hope, A	G0ACZ
Hope, A W	GM3MGT
Hope, B C	G4ZPP
Hope, B D	GW4EXE
Hope, C G	2E1FUZ
Hope, C J	G7WFF
Hope, C W	G3IPY
Hope, G	G1DLJ
Hope, G M	G4GVY
Hope, G P	G6YLD
Hope, J	G7AUL
Hope, J B	G3RXM
Hope, J J	G7JDF
Hope, J W	G7KYG
Hope, K D	G4HAZ
Hope, R	G6VMF
Hope, R E	G6ZAY
Hope, R M	2E1AUA
Hope, R M	G0PNB
Hope, R M	G6YEY
Hope, R P	G0DRQ
Hope, R P	GW1SVG
Hope, R P	GW8TVX
Hope, R W	G6LVN
Hopewell, F	G4PGC
Hopewell, J	G4JPQ
Hopewell, P	G4DCI
Hopkins, A C	G4TPD
Hopkins, A T	GW4YKW
Hopkins, B L	G7KDR
Hopkins, B R	G8CGX
Hopkins, C B E	G7RHX
Hopkins, C G	GW6MWN
Hopkins, C J	G7EOC
Hopkins, D A	G0MXI
Hopkins, D C	GW8LHO
Hopkins, D G	G3BJR
Hopkins, G C	G4CST
Hopkins, J	GM1XJE
Hopkins, J A	G4PWX
Hopkins, J T	G1EMT
Hopkins, J W	GM4LPT
Hopkins, K G R	G7NQQ
Hopkins, M B	G0OEI
Hopkins, M J	G0HFR
Hopkins, M W	G8HOP
Hopkins, N E	G1IME
Hopkins, P A	G0LIW
Hopkins, R	G4PUG
Hopkins, R	G4VMV
Hopkins, R	G4XUE
Hopkins, R	GW4NOS
Hopkins, R A	MM1BHO
Hopkins, R F	G1APL
Hopkins, R J	G3NLX
Hopkins, R K	2M1CEG
Hopkins, R M T	G3IWW
Hopkins, S J	G1IHL
Hopkins, T	G1HBC
Hopkins, T G	GW4EVL
Hopkins, T H	GI7CBD
Hopkins, T P	G8TYY
Hopkins, W B	G3FPI
Hopkins, W J	G3XNI
Hopkinson, A	G0VKI
Hopkinson, A	G8OJQ
Hopkinson, A S	G0SIY
Hopkinson, J	G6GWV
Hopkinson, J E	G6VPL
Hopkinson, K	G8QX
Hopkinson, M	G0AUH
Hopkinson, M	G4JDK
Hopkinson, P J	G4EBX
Hopkinson, T A	G0AUG
Hopkinson, W G	G8CYY
Hopley, A E	G7WBU
Hopley, C P	GW8ICT
Hopley, I	M1ABA
Hopley, R W	G0NDH
Hopley, S W	G4YTK
Hopley, W	G4ETA
Hoppe, D A	G0NVC
Hoppe, M A	G0GLA
Hopper, B	G0CRN
Hopper, C D	G4BKC
Hopper, D A	G0NFG
Hopper, J	G6KES
Hopper, J M T	G8XML
Hopper, L P M	G1ZGO
Hopps, K	G7TZS
Hopson, L S	G6SKF
Hopton, D J	G1NYJ
Hopton, J	G3WMP
Hopton, S R	G0VZT
Hopwood, A T	G1AJA
Hopwood, F R	GW8BIA
Hopwood, J I	G0EDT
Hopwood, P A	G3UKH
Hopwood, P J	G4JZD
Horabin, T W	G0LNW
Horbaczewskyj, P S	G4ZXO
Horder, A D J	G4RBW
Horder, D A	G7HDR
Hordley, T N	G8BXQ
Hore, C F	G6GWX
Hore, K J	G0UHD
Horley, J H R	G4KME
Horn, L	G6DYU
Horn, M G	G6PSC
Horn, N	2E1FGB
Horn, P S	G3GGH
Horn, R J	M1ACG
Horn, T A T	G6MTF
Horn, W A	G4GPD
Hornby, A C	G1HBD
Hornby, C M H	2E1DJF
Hornby, D	G0WCK
Hornby, D G	G7UCP
Hornby, E J	G7CYQ
Hornby, J C	G4RAK
Hornby, R	G8UWD
Hornby, W D	2E1ARS
Horne, A J	G4CKQ
Horne, A J	G4GJV
Horne, A J	G4SGL
Horne, A R	G0TPH
Horne, B M	G4NLB
Horne, D R	G0JUT
Horne, E J	G3YCQ
Horne, H	G0KKF
Horne, H J	G4AIF
Horne, J I	GM0EAS
Horne, K D	GM3YBQ
Horne, K R	GM4FSF
Horne, L W	GW0JTE
Horne, M A	2M1BEU
Horne, M A	GM7UQM
Horne, M J	G8KRU
Horne, P R	G3JRH
Horne, R J	G0JVN
Horne, R W	G4HBA
Horne, R W T	GM3KJA
Horner, B H	G0CFG
Horner, E J	M0AHW
Horner, E L	G4LCX
Horner, G	G1XDW
Horner, J D	G6RUP
Horning, B J	GM4TOE
Hornsby, A R	G0KTX
Hornsby, C	G0CVL
Hornsby, G	G0IKL
Hornsby, J S	G0AJH
Hornsey, D	G3MAN
Hornsey, D J	G0IVE
Horobin, P J	G6KJH
Horoszko, M W	G4BDX
Horrabin, C W	G3SBI
Horrobin, T H	G3WDD
Horrocks, B	G8OUT
Horrocks, J K	G0OEV
Horrocks, J M	G8GTP
Horsburgh, D I	G3UOM
Horsburgh, G	GM4XHV
Horsburgh, J J	GM1RDG
Horscroft, W T	G0JLD
Horscroft, W T	G0OWX
Horsefield, I	G4OPP
Horseman, D J	G4TNE
Horseman, L	G4IPF
Horsepool, P J	GU0JCI
Horsfall, A	G4CBW
Horsfall, A	G7DCT
Horsfall, G B	G3GKG
Horsfall, G I	G8CWQ
Horsfall, M	G7DMS
Horsfall, R	G7KEK
Horsfall, W S	G3GXX
Horsfield, A	G8ETW
Horsfield, C D S	G7CGV
Horsfield, E	G4UFR
Horsfield, G D	G4SKQ
Horsfield, J M	G0UBM
Horsfield, K	G1HIP
Horsfield, M	G1HIO
Horsfield, M	G6LWK
Horsfield, R	G0ONO
Horsley, D C	2E1DQQ
Horsley, G J	2E1DQK
Horsley, G T	G8JIF
Horsley, J V T	G8HAD
Horsman, A R	G0MBA
Horsman, A R	G0PKT
Horsman, B J	G4MZC
Horsman, R	G3XOD
Horswell, C J	G8SSP
Horti, A I	G0FDI
Horton, A	G0LKG
Horton, A B	GM0BUL
Horton, A E	G4BOV
Horton, A J	G0GWP
Horton, A M	G0JAH
Horton, B A	G0HNG
Horton, C B	GM1KJF
Horton, D A	G3RZF
Horton, D D	G4XIB
Horton, D E	G6BCV
Horton, D G	2E1EOQ
Horton, D M	G7SWK
Horton, D R	G0RFY
Horton, D R	G4ENG
Horton, G A	G1IUB
Horton, J A	2W1FXI
Horton, J R	G4WCC
Horton, K D	G0AEL
Horton, K F	G4TZT
Horton, K J	2E1EOR
Horton, L R	G4GRM
Horton, M E	G1TOT
Horton, M E	G4DMH
Horton, P G	G7IOO
Horton, R	G0HCA
Horton, R	G3XWH
Horton, R	G7XWH
Horton, R P	G4AOJ
Horton, R P	G7NXQ
Horton, S	G7CGO
Horwood, D J	G7HGQ
Horwood, G M	G6IFR
Horwood, H	2E1FYU
Horwood, W G	G7OCQ
Hosfield, J	G0SVK
Hosker, N A	G8UEK
Hoskin, I S	G4GDU
Hoskin, M E	G6GGN
Hosking, A	2E0APD
Hosking, C	G0PGB
Hosking, J A	G8DEX
Hosking, J R	GW0JTF
Hoskins, J	G4XYM
Hoskins, L F	G3VN
Hoskins, S J	M1AJM
Hoskins, V R	G1LUI
Hoskins, W A	G8LU
Hossack, J	GM3DKW
Hossack, W S	GM3UBJ
Hostekens, M	G1EHU
Hotchen, K C	G6VVL
Hotchin, J M	G4ATA
Hotchin, M P	2E1FVS
Hotchkiss, E S M	GW3VLU
Hotchkiss, G D	G4BEQ
Hotham, S M	G8FAS
Hother, J A	G7NCC
Hotson, N R	G0TDG
Houchen, M B	G4AIZ
Hough, B R	G0KKF
Hough, C W	G8JZL
Hough, D	G6XTV
Hough, H G	G0AVS
Hough, J	G8JQG
Hough, J R	G0KRR
Hough, J T	G3MQK
Hough, R	G3XTN
Hough, W H	G1RIH
Hough, W J	G0IHF
Hougham, E H	G0GBT
Hought, C G	GI4SFZ
Houghting, M F	G0MAX
Houghton, A	G0WWH
Houghton, A E	G0WMB
Houghton, B	G1MRU
Houghton, B	G4BCO
Houghton, C A	G0GVD
Houghton, C J	2E1DHY
Houghton, C J	G4JNE
Houghton, C P	G1THE
Houghton, D	2E1FCI
Houghton, D	G0MXW
Houghton, D	G3UPY
Houghton, D E	G0TQG
Houghton, F	G3VZM
Houghton, G	2W1BIV
Houghton, H J	G8XMO
Houghton, I A	G0ADT
Houghton, I D	2E1EEF
Houghton, I D	M1BUQ
Houghton, J G	G1KEP
Houghton, J P	G4ZNE
Houghton, J W	G4NIA
Houghton, K J	G7PYW
Houghton, N	G0GHD
Houghton, P	2E1GBD
Houghton, R D	G7UYR
Houghton, R G	G3SCL
Houghton, S A	G6TGQ
Houghton, S D	G7TAV
Houghton, T C	G1KPI
Houghton, T C	G6PTJ
Houghton, V	G7ERH
Houghton, W C	G4WCH
Houghton, W G	G3AMO
Houlden, C M	2E1DQZ
Houlden, D K	G7VQP
Houlden, G M K	2E1APT
Houlden, J K	2E1GBC
Houldershaw, H G	G7GZV
Houlding, S T	G0BYA
Houldridge, I M	G6GKT
Houlihan, J	G4HVV
Houlihane, T S	G7BIX
Houlston, R W	G4PVB
Hoult, D	G1DSP
Hoult, D	G4DSP
Hoult, D	G4OO
Houltby, A	G3ZSF
Houltby, C E	G3UHS
Houlton, A D	G7HCJ
Houlton, S D	G7LXQ
Hounslow, N I	G4TMY
Hounslow, P M	G4YFE
Hounslow, R A	2E1CQL
Hourston, G M	G7IZD
Hourston, R C	G7GLZ
Housden, B K	G4WZZ
Housden, J	G4CPI
House, E J	G7JAI
House, R	G0ULU
Houseago, P H J	G8SGB
Housego, J D	M1BFS
Houston, J B	GM4OPU
Houston, K	GM0RMW
Houston, K C	GW8KCH
Houston, P M	MI1BMO
Houston, S R	GM7OHQ
Houtby, R A	G3HJY
Hoverd, R E	G0ERR
How, C J	G4JCA
How, D I	G0PAR
How, K S	G1XXP
Howard, A	2E1EIW
Howard, B A	G1SVP
Howard, B J	G0LJD
Howard, C	G1EMU
Howard, C	G4JMW
Howard, C J	G0OWR
Howard, C P	GD0HWA
Howard, D J L	G0TVA
Howard, D M	G6LVS
Howard, D P	G4IZA
Howard, D W	G1AJB
Howard, D W	G7GQX
Howard, E	G8KPE
Howard, F N	GW3DEX
Howard, G	G0BBH
Howard, G A	G7SUR
Howard, G W	G0LEM
Howard, G W	G4ZLW
Howard, H D	G8EXT
Howard, H L	G0KSY
Howard, I B	G2DUS
Howard, J	G0KKU
Howard, J	G0NJO
Howard, J	G6YLA
Howard, J A	G0BUV
Howard, J L	G3KQI
Howard, J R	G8MJT
Howard, J S	G0UFZ
Howard, J W	G4EVI
Howard, K	G0FEQ
Howard, K	G7HEF
Howard, K A	G0PDD
Howard, K J	G7VLD
Howard, K K	G4HNR
Howard, K T	G0DUV
Howard, M C	G8RES
Howard, M E	G7UGB
Howard, M J	G7SOE
Howard, M J	GW7HGU
Howard, M K J	G7PXC
Howard, M R	2E1FRB
Howard, M W D	G1OSJ
Howard, N J	G4OVV
Howard, P	G0WPH
Howard, P A	G4UMB
Howard, P G	G0AFN
Howard, P K	G1OFG
Howard, P R	G4JKC
Howard, R	G7EKX
Howard, R A	GM4EGX
Howard, S	M1BFH
Howard, S J	G1VON
Howard, S J	G7DNQ
Howard, T	G3NVC
Howard, T L	G7LBM
Howard, W J	G6TGM
Howarth, A	G0RHY
Howarth, B A	G0YRT
Howarth, C E	G7SOV
Howarth, D	G6TCW
Howarth, E D	G0OBD
Howarth, G	G0DRL
Howarth, G W H	G4YVO
Howarth, J	G7NOQ
Howarth, J A	G4WNI
Howarth, J H	GM1VKG
Howarth, J R	GW0AEZ
Howarth, K	G6ZVO
Howarth, L J	G0GEH
Howarth, N J	G4YAC
Howarth, P	G0CCL
Howarth, P A	G3YAC
Howarth, P D	G7ULM
Howarth, P S	G1HVL
Howarth, R L	G4GIV
Howarth, S J	G4YGA
Howarth, S J	G7SYY
Howarth, T E	G4BKF
Howat, A R	G7LZB
Howat, G D	G7FCO
Howat, J N	G3CFQ
Howchen, T N	G1FBW
Howcroft, S	G1EEA
Howcroft, S	G6IFQ
Howden, M	G6JKP
Howden, M C	M0AEM
Howdon, E	G0IBU
Howe, A K	M1BCU
Howe, A R L	G4KBC
Howe, C	G4RTJ
Howe, F R	G3CO
Howe, F R	G3FIJ
Howe, J C	G8YVT
Howe, J W	G3NXZ
Howe, K D	G4KDH
Howe, L C	G4DXJ
Howe, N T	M1CAG
Howe, P	G4IIO
Howe, P J	G4CHL
Howe, R T	G4LGE
Howe, R W	G3PLB
Howe, T	G1VID
Howe, T A	G3VID
Howe, T D	G0ETP
Howe, T J	G1AJC
Howell, A	G7SWH
Howell, A D	G0VJM
Howell, A G	G3SQK
Howell, B	G4LOI
Howell, B N	G6IXQ
Howell, C A	G0VWF
Howell, C P	G4LJU
Howell, D	GW3WCV
Howell, E R	G0ONJ
Howell, E T	G3GUP
Howell, G A	G4JVT
Howell, G B	G0HGB
Howell, G H	G6KVK
Howell, G K	M0AJE
Howell, J D	G0ICB
Howell, J G	GM4ZQH
Howell, J M	G4BXZ
Howell, J R	2E1DHO
Howell, J S	M0AMX
Howell, M A	G1UBH
Howell, M R	G0ZMH
Howell, P J	G3ZTZ
Howell, P N	G0NAP
Howell, R	G4WHQ
Howell, R A	G0HZE
Howell, T D	G6FBV
Howell, W S	G0TMK
Howell, W W	G8NTG
Howells, A W	G8UEI
Howells, C E P	G4DXP
Howells, D E	GW4IHN
Howells, D V	GW0WBQ
Howells, D W	G4YWP
Howells, E	G0GAL
Howells, E J	GW7WFO
Howells, G	G8SXI
Howells, G D	GW1FKL
Howells, J	GW4BUZ
Howells, J J W	G6BJS
Howells, M J	G1GIG
Howells, M J	GW7IAT
Howells, R T	G4FFY
Howells, R T	G8GWX
Howell-Pryce, J B	G4ZHI
Howell-Walmsley, A M J	GW7IBZ
Howes, A T J	G0HOV
Howes, C	G0MVV
Howes, D A	G4KQH
Howes, G T	G4UOA
Howes, J L	G0NMS
Howes, M	G4LAD
Howes, M	G8WYR
Howes, M G	G0BPW
Howes, M L	G7VMP
Howes, P	G0CUD
Howes, P	G4TTC
Howes, R J	G4OWY
Howes, R M	G7PJD
Howes, S E	G7FTW
Howes, T	G1KRF
Howett, C J	G4ILR
Howett, P A	G1UXJ
Howett, P R	G4MD
Howgate, A I	G7WHM
Howgego, C A	2E1AVU
Howgego, P D	2E1AWG
Howgego, R J	G4DTC
Howie, D	G0GSJ
Howie, J	GM7JGR
Howie, J B	GM4DIJ
Howkins, M	G4CPG
Howland, A M	G4ZKS
Howland, M J	G4MIX
Howland, P I	G0WJN
Howlett, A J	G1HBE

Howlett, B G0VZA
Howlett, B J P G3JAM
Howlett, D P G8FIF
Howlett, R M G1OSE
Howls, D A GW4KDD
Howman, A G0FVF
Howorth, D W G4IFT
Howorth, M G7DPC
Howorth, N G0DEI
Howorth, W A G4LNE
Howroyd, D T E G4YPN
Howse, A K M1CBZ
Howse, D P M1CBY
Howse, G A G6YLB
Howse, M J G7PTV
Howsham, I G0CVN
Howson, I G GW6RSP
Howson, P G GM8GAX
Hoy, H L G4ZWM
Hoy, J L GW6NQU
Hoy, S W G7JPU
Hoyland, J W G0HOJ
Hoyle, A 2E1EVN
Hoyle, A G0KQG
Hoyle, D H G7MET
Hoyle, D L G8RTO
Hoyle, D W G4HGN
Hoyle, G M0BAU
Hoyle, H G3MCI
Hoyle, S G7MAJ
Hoyle, W D G0DRV
Hruza, P G4SGQ
Hrycan, J M M0BEX
Hrycan, J M M1ARV
Hrycan, L J M0AWV
Hrycan, L J M1ASD
Hubball, M N G1JHL
Hubbard, A D G7TTY
Hubbard, A W G0HUB
Hubbard, B G4KEX
Hubbard, F G0IBY
Hubbard, J E G1EEC
Hubbard, J V G4PWG
Hubbard, K C G0VFJ
Hubbard, M G3OVL
Hubbard, M F G4IND
Hubbard, N G G8OSH
Hubbard, N W G4ZYC
Hubbard, P J G4DKB
Hubbard, S T G6RYD
Hubber, G W G3NVJ
Hubber, P W H GW4KWV
Hubbert, L J G0FTB
Hubert, P J G3YWM
Hubert, R D G6ZEQ
Hubner, J G4YHG
Huckfield, R L GW4ZWC
Huckle, R C GW1AHU
Hucklebridge, J F G3ENR
Huddart, M J G4OQR
Hudders, H G0THH
Hudders, K R G6SPR
Huddleston, A J 2E1CFG
Huddleston, A J G3WZZ
Huddleston, D C G0SGT
Huddleston, J E 2E1BAP
Huddleston, T N L GI8HUD
Huddleston, W M M1AQA
Huddlestone, J A G1UJX
Hudgell, G G6DYM
Hudman, R G G7OBC
Hudsmith, G M G4LTM
Hudson, A G8ZEE
Hudson, A J G3KIC
Hudson, A R G6BBE
Hudson, B R G4MMV
Hudson, B W G0TUI
Hudson, C A G8CWC
Hudson, D G3BDD
Hudson, D J G6OVO
Hudson, D R G4XUW
Hudson, D W G3BOR
Hudson, E A G6YMQ
Hudson, E G G4HSI
Hudson, G G7IPP
Hudson, G J GM4SVM
Hudson, G P G6DRW
Hudson, H G G0JQW
Hudson, H J G3FVL
Hudson, J G0DBC
Hudson, J G4NS
Hudson, J E G3PEW
Hudson, J J G7RBZ
Hudson, J L G0IMN
Hudson, J O G4ABQ
Hudson, J P 2E1CLN
Hudson, J R G7DMM
Hudson, J W GM3LQH
Hudson, K B G0OBP
Hudson, M G G3LEJ
Hudson, M S G7VDS
Hudson, P 2E1BDB
Hudson, P A G4ANV
Hudson, P A G4VAH
Hudson, P H GW3IEQ
Hudson, P M G7UCB
Hudson, P R G8BHD
Hudson, P T G3UMM
Hudson, R G1XZW
Hudson, R G4JFN
Hudson, R A G1IVL
Hudson, R C G0TUU
Hudson, R C G7BIV
Hudson, R D G4YUL
Hudson, R F G1RUR
Hudson, R F G8MSM
Hudson, R J 2E1CCU
Hudson, S A GW0NQQ
Hudson, S W G7MIE
Hudson, T G1JAX
Hudson, W D G0CSW
Hudspeth, K GW0ARK
Hueck, A S G6MWS
Huff, A D G4WYI
Huffadine, J R G3VXH
Huffer, L F G0JTX
Huggett, J G3JKT
Huggett, L G G3ZAL
Huggett, M W G0LEJ
Huggett, V J G1DSQ
Huggins, I A G1RGV
Huggins, J G0DZX
Huggins, J B G0DIP
Huggins, M C G6XRK
Hughes, A G4ZOF
Hughes, A GW7UUH
Hughes, A E G0WMH
Hughes, A E G1CKQ
Hughes, A J G4KOR
Hughes, A L GW1TFB
Hughes, A N G1ZSY
Hughes, A S GW0LBA
Hughes, A T GW1YSM
Hughes, A W GW3YGH
Hughes, B G0TCI
Hughes, B A G4XVR
Hughes, B C G4SGE
Hughes, B C G7NNR
Hughes, B R G4CUQ
Hughes, C G0DQW
Hughes, C G6YMH
Hughes, C E GW0FBT
Hughes, C J G4MEP
Hughes, C M G7IAK
Hughes, C R G7INJ
Hughes, D G0HIZ
Hughes, D G0RQH
Hughes, D G4EZI
Hughes, D G7VOA
Hughes, D GW0JZQ
Hughes, D A GW8PMJ
Hughes, D F G8WPL
Hughes, D G M1BSL
Hughes, D J G4PDR
Hughes, D J G7LFC
Hughes, D J G7MEU
Hughes, D L G0RVW
Hughes, D L GW3VNO
Hughes, D P 2E1EAP
Hughes, D R J G7EWH
Hughes, D S GW6APR
Hughes, D T N G1NBJ
Hughes, D W GI4JNS
Hughes, E D C GW7SUC
Hughes, G A G0MYD
Hughes, G A G6DYI
Hughes, G C G0LUJ
Hughes, G E GW0DQY
Hughes, G J G0GJH
Hughes, G J G6YFB
Hughes, G J G7SHU
Hughes, G R GW6PXF
Hughes, H G0ECJ
Hughes, H G4HSC
Hughes, H G A GW6KAV
Hughes, H J 2E1CJN
Hughes, H O G0OQG
Hughes, I GW4YCT
Hughes, I W GM0KXF
Hughes, J G0JQK
Hughes, J G3RRM
Hughes, J A G4KGT
Hughes, J A R GM3LCP
Hughes, J D G4VVI
Hughes, J D G7CJC
Hughes, J D GW7UIZ
Hughes, J G GW1FBI
Hughes, J P E G1NSP
Hughes, J R G7VRH
Hughes, J R GW7OIV
Hughes, K G1RAO
Hughes, K GW3SUH
Hughes, K M1BTN
Hughes, K A G4XVS
Hughes, K F G1XUC
Hughes, K J G4CJT
Hughes, L A F G0MPL
Hughes, L J G7LSY
Hughes, L R G7TEE
Hughes, L T G4MVJ
Hughes, M G1HBF
Hughes, M G1KAS
Hughes, M GM3RIJ
Hughes, M GM4ISM
Hughes, M GW3VFZ
Hughes, M E G4MEH
Hughes, M J G8OCO
Hughes, M P G3KBH
Hughes, M P GM0NBO
Hughes, M R G6EAY
Hughes, M V G0PJM
Hughes, N D GW0WWY
Hughes, N H G0EFP
Hughes, N M C K GI4SZP
Hughes, P G8AOZ
Hughes, P GI0VHG
Hughes, P A G1PAH
Hughes, P E G0KHQ
Hughes, P F G3OSR
Hughes, P G GM1ZCD
Hughes, P G GW0ABE
Hughes, P H G4VQI
Hughes, P R G8IPT
Hughes, P R MW0BEN
Hughes, P W G0PWH
Hughes, P W V G0BXC
Hughes, R G0TWM
Hughes, R B GW0RBH
Hughes, R F G0JER
Hughes, R G G0AGJ
Hughes, R G G4PBD
Hughes, R G G4PLH
Hughes, R J G3GVV
Hughes, R J G4AJS
Hughes, R J G8JBQ
Hughes, R O K G7TTP
Hughes, R P G3HAG
Hughes, R P G7UOI
Hughes, R P GW3PWA
Hughes, R W G0FTI
Hughes, R W G1MKR
Hughes, R W GW6AZX
Hughes, S G6ZVU
Hughes, S P GW0XAQ
Hughes, S R G0WBL
Hughes, T G0SCP
Hughes, T GM4DSO
Hughes, T P GM3EDZ
Hughes, T R G4WKJ
Hughes, T S G4KST
Hughes, T W GW4TOD
Hughes, V G0CVH
Hughes, V T MW0BET
Hughes, V T MW1BJF
Hughes, V W MI1AUI
Hughes, W G0LYJ
Hughes, W G4LVY
Hughes, W G6JXS
Hughes, W D G4UCH
Hughes, W D I GW4ZXL
Hughes, W T R G0MYN
Hughesdon, R G0NCX
Hughesdon, S M G4WGY
Hughson, P S M1CED
Hugill, T J S G4FJK
Hugman, D A G8JMY
Huish, M A G3VRV
Huke, J C G7PRE
Huke, M G7SGI
Hulands, M R G4BHT
Hulands, R W G6ZER
Hulatt, P G0BDD
Hulbert, K H G6BJR
Hulcoop, G S G0LDK
Hulett, J G3OCH
Hulett, M A G1HNX
Hull, B C GW0SFI
Hull, E G1XOP
Hull, F A GI4WME
Hull, L A J G3ROK
Hull, N G G7KUJ
Hull, N J G6ZVV
Hull, N J G7IRC
Hull, N J G7RNL
Hull, P D G4DCP
Hull, P J G1NTK
Hull, R M0AUW
Hull, R A G7HIW
Hull, W F G0VKN
Hulme, A G7OEM
Hulme, A M B G0CDY
Hulme, D G4OZW
Hulme, D A G6GVK
Hulme, E G3BQT
Hulme, F G3WI
Hulme, G C G0DLK
Hulme, J W MM1APC
Hulme, M C 2E0ADN
Hulme, P G G0NPE
Hulme, S G3SRM
Hulmes, T A GW7KYT
Hulmston, N J G8MGY
Hulse, P F G0WAR
Hulse, P J G7KAT
Human, M J G1MPC
Humberstone, I R 2E1DXB
Humberstone, S G0ORK
Humble, R W G8DXO
Humble, T P 2E0APN
Humby, P A G6AUS
Hume, H GM1VUH
Hume, J F G1XAS
Humes, E H G1KNG
Humes, I L G1KNH
Humes, S M G1KNF
Hume-Spry, P H G0FOE
Humm, A J G4CWE
Humm, S F G8CHO
Hummerstone, B G3HBR
Humphrey, B J 2M0ALG
Humphrey, B J MM0AIW
Humphrey, G C V G3BMQ
Humphrey, G W G0MAH
Humphrey, J H GW7VEH
Humphrey, L 2E0AJP
Humphrey, L D G8MHA
Humphrey, L S G6BFP
Humphrey, M I G0SWY
Humphrey, N R G3VHW
Humphrey, V S G8CPQ
Humphreys, A GW6WFW
Humphreys, A C V G8CLF
Humphreys, A J G6RXZ
Humphreys, C M0ASE
Humphreys, D G4MXT
Humphreys, D G7JKC
Humphreys, D F R G4GHR
Humphreys, D G G4XDG
Humphreys, D G G4XDH
Humphreys, F R G0CWU
Humphreys, G T M1CAQ
Humphreys, H M GI3EVU
Humphreys, J E G1LSX
Humphreys, L B G0JBM
Humphreys, M G0LHL
Humphreys, N 2E1DEM
Humphreys, P G0UTU
Humphreys, P A D G0JXC
Humphreys, P R G8WBP
Humphreys, P R GW6YMS
Humphreys, S M G0FQW
Humphries, A K 2E1FCM
Humphries, B L H G0AGU
Humphries, C D G4UXE
Humphries, D B G4ETG
Humphries, E W G0EZT
Humphries, F R G8THG
Humphries, F W F G3DCE
Humphries, J G7FFZ
Humphries, J G7MKB
Humphries, J A G4KLS
Humphries, J W GW0EAW
Humphries, M J G3LRQ
Humphries, P N G6BJK
Humphries, R W J G4UKL
Humphries, S G4UXF
Humphries, T M G0OLS
Humphries, W J F G7IHL
Humphries, W S G0OOI
Humphrys, R C G6ULD
Humpoletz, J E G3ITL
Humpston, E A R G8RZS
Humpston, G G4GYO
Hundy, P E J G7OPD
Hunniford, H E GI6DEO
Hunnisett, I N G0RNF
Hunnisett, I N G7GQS
Hunsdale, D G4ZOI
Hunt, A G1YLJ
Hunt, A C G8GHB
Hunt, A D G3CHU
Hunt, A F K G7TMH
Hunt, A G G4TYT
Hunt, A J G0NFV
Hunt, A J G6ISB
Hunt, A M G3HHV
Hunt, A P M1AZP
Hunt, A S G6NRK
Hunt, B G8BII
Hunt, B J 2E1FTC
Hunt, B T G4SJN
Hunt, C G0GYB
Hunt, C A GW7KIL
Hunt, C J G8MYG
Hunt, C J G8VXR
Hunt, C P G6MSQ
Hunt, C P G6VZU
Hunt, D G0WKY
Hunt, D G4VNE
Hunt, D J G7SBK
Hunt, D R G7VLV
Hunt, D R M G6MFR
Hunt, F G2HOJ
Hunt, G G1EFM
Hunt, G L G7RFM
Hunt, G T G1GEI
Hunt, H G2FUM
Hunt, I M G1MXM
Hunt, J G6XRL
Hunt, J A G2FSR
Hunt, J A G6IFV
Hunt, J D G3RJE
Hunt, J H G7GFM
Hunt, J I G8JPA
Hunt, J P G3LPN
Hunt, J R G3PVU
Hunt, K M1BPX
Hunt, K A G4UKE
Hunt, L A G4NQ
Hunt, L J M G8RKH
Hunt, M D GW4CBR
Hunt, M R G6AJA
Hunt, M R G6DEK
Hunt, M R GW6VBN
Hunt, P G4AEO
Hunt, P E G1ZHM
Hunt, P J G8CRZ
Hunt, P K G8PYX
Hunt, P L G0REC
Hunt, P L G3FWB
Hunt, P M G3TVL
Hunt, R A G0CLR
Hunt, R D G3EKI
Hunt, R D G4XAB
Hunt, R H M0AKJ
Hunt, R M G4MIJ
Hunt, R T G2MJ
Hunt, S G1VUK
Hunt, S G3YQ
Hunt, S E G3TXQ
Hunt, S K GM4YDC
Hunt, T E G1RAB
Hunt, T S G4DWM
Hunt, W E G0SZX
Hunter, A C G3WBF
Hunter, A F GM3LTW
Hunter, A S G6LBQ
Hunter, B MW1BVF
Hunter, B C G4ZUQ
Hunter, C S G7IOT
Hunter, D S G4WRT
Hunter, E G0OTQ
Hunter, F A G3KTT
Hunter, F T GI4NKB
Hunter, G G8WWD
Hunter, G A G6YFF
Hunter, G A GM3ULP
Hunter, G T G1OSL
Hunter, H L GM0LOO
Hunter, I G G4UQP
Hunter, J G3BPI
Hunter, J G3IMV
Hunter, J GM0EEH
Hunter, J GM7AYW
Hunter, J GW4GZX
Hunter, J C Q GM7BTL
Hunter, J E G6HU
Hunter, J G H G1OPK
Hunter, J J G0WOZ
Hunter, J J G3AZ
Hunter, J J G7VQX
Hunter, J P GM3KAK
Hunter, J T GI7JLD
Hunter, K A GM0JHE
Hunter, K M G4ENJ
Hunter, M G7TFV
Hunter, M T J G7IOS
Hunter, N O G8DQN
Hunter, P G0GSZ
Hunter, P C G3OFF
Hunter, P R G7MJN
Hunter, R C GW4OFQ
Hunter, R E G3LUI
Hunter, R J MM1BAH
Hunter, R T G8HYZ
Hunter, S C GM3WZV
Hunter, S V G7UAK
Hunter, W G0WKP
Hunter, W GM1SXZ
Hunter, W F GM3HUN
Hunter, W M GM7BAS
Hunter, W S G8CQV
Hunting, J F GM8DPV
Huntley, A R G7PGG
Huntley, E L M1CCJ
Huntley, G F T G1NNC
Huntley, J C G3PPI
Huntley, L H B G4LW
Huntley, R C L G1VWZ
Hunton, I G0EOX
Hunton, W G7HAF
Huntriss, T H 2E1FMB
Huntsman, M P G4MZN
Huntsman, P H G3KBQ
Huntsman, R A G3KBR
Hunt-Duke, A T G4IOT
Hupton, D G0OMF
Hurd, A C G7PUP
Hurel, C M J GU5BVQ
Hurley, C W G4EBW
Hurley, R P G0CGH
Hurley, T D G7KRH
Hurley, T F G1NSR
Hurn, D G G1EOC
Hurnandies, J G8KAM
Hurp, P J G1RBY
Hurr, D A G6MFH
Hurrell, B B G6UPI
Hurrell, D L G3JMK
Hurrell, G G7GGH
Hurrell, J P 2E1AEE
Hurrell, K A V G3NBC
Hurrell, M J G4UGV
Hurrell, R F G4HSM
Hurren, P G8LIL
Hurren, P S 2E1DQY
Hurst, A M G7VGO
Hurst, B P G4XIL
Hurst, C L G7MER
Hurst, D J G1LCI
Hurst, D J G8HWF
Hurst, G D G3UFC
Hurst, G M G0RTZ
Hurst, G M G4MIT
Hurst, I 2E1CQM
Hurst, J B S G4EFY
Hurst, K W G7ULS
Hurst, M G0CCM
Hurst, M J G4ASZ
Hurst, P L G0CCN
Hurst, R E G3JJU
Hurst, S G8LIK
Hurst, S K G6WTA
Hurst, S R G7REV
Hurst, S R M0ARZ
Hurst, T R GW0GEV
Hurt, A C G0OJY
Hurt, R C G0HDS
Hurton, T W G6SJG
Husband, D J G8HJT
Husk, J A G6RLS
Husk, J A G8GLI
Hussain, A G1YMY
Hussey, A S G4KUN
Hussey, D R G0MIB
Hussey, G W S G4PGZ
Hutchence, A B G3IKA
Hutchens, J B GM0SYY
Hutcheson-Collins, R A G3AXS
Hutchings, C F G3LCA
Hutchings, D G6YMU
Hutchings, D F G8IXA
Hutchings, F H G8SXJ
Hutchings, G H G7VGD
Hutchings, L R GW8FOD
Hutchings, M E G4ZIW
Hutchings, M P G1LLJ
Hutchings, M R G8EAD
Hutchings, N W G4YQO
Hutchings, P D G8LRD
Hutchings, R J G4MEA
Hutchings, S A G0DVE
Hutchings, S P K G6YMV
Hutchings, T S G4POB
Hutchings, W G0UUA
Hutchins, K S G0BMN
Hutchins, W E G0HVC
Hutchinson, A J M1ATT
Hutchinson, A N GM6ARB
Hutchinson, B R G G3VGH
Hutchinson, C W G0HYX
Hutchinson, D J G1XGE
Hutchinson, D J C G8SQH
Hutchinson, D K GW4NMQ
Hutchinson, D M GW0DHQ
Hutchinson, E G0CEB
Hutchinson, G G0NDE
Hutchinson, G A GI8MIV
Hutchinson, J D GW4HHD
Hutchinson, J E G3OJJ
Hutchinson, K J T G8UPF
Hutchinson, L M0AIX
Hutchinson, M GI4UHP
Hutchinson, N A MW0BFD
Hutchinson, P M G0OXT
Hutchinson, P R GI7EXN
Hutchinson, S G0HYS
Hutchinson, S J G7REJ
Hutchinson, S M GM3WPA
Hutchinson, T GI4YRP
Hutchinson, V A G GI0EJU
Hutchinson, W D GI4FUM
Hutchinson, W D GI4SIW
Hutchison, C H G0DLI
Hutchison, D I GM7RPT
Hutchison, I M MM0BDK
Hutchison, J GM0OYT
Hutchison, J D G0SLE
Hutchman, W A GI0WAH
Hutley, A J G6SPS
Hutley, K L J G0VDP
Hutson, D H GW0PNE
Hutson, J E G4RBI
Hutt, G N G0HZC
Hutt, K J G0TSH
Hutt, L A G7AYO
Hutt, P G0DYS
Hutt, R A GJ0RKM
Hutton, A R GM4EAF
Hutton, A R GM4ZRH
Hutton, A T M1CBR
Hutton, C R G8CJD
Hutton, D R G1XBO
Hutton, H C G6WJW
Hutton, L G G0ODD
Hutton, L K G1AJG
Hutton, P G G4UAH
Hutton, R G3WYH
Hutton, R G8FPU
Hutton, R J G7OGB
Hutton, R J M0ATB
Hutton, S G6ZES
Hutton, S G G1MAB
Hutton, S J M1AKQ
Hutton, T G1WDQ
Hutton, T H L GW0HUT
Hutton, T N GI1ACN
Hutty, P A G0UME
Hutty, P R G0UMA
Huxham, M J G4UXP
Huxley, R G1BNX
Huxley, W G1OMV
Huxley, W J A GW8FNX
Huxtable, G T G7HFU
Hyams, P J GW4OZU
Hyams, R C G4NYA
Hyatt, C G0FBC
Hyatt, K T I GW4TES
Hyatt, R F G4WOC
Hyde, A C GM1CHT
Hyde, A F GW4ILF
Hyde, A M G4URI
Hyde, B G0MHE
Hyde, B R 2E0AES
Hyde, B V G0SFA
Hyde, B V G1SFA
Hyde, C G G7JKH
Hyde, D G0PDH
Hyde, D C G8DFA
Hyde, D W G1OKI
Hyde, E G0RFG
Hyde, G K G7LXK
Hyde, J 2E1CEY
Hyde, J H G0OQY
Hyde, J L G0GDS
Hyde, M J G4IHZ
Hyde, N C G4LTL
Hyde, P G1EFO
Hyde, P A 2E1BFW
Hyde, P A G4CSD
Hyde, P P G7KIQ
Hyde, P P M1BXE
Hyde, R G3TCQ
Hyde, R G3ZDW
Hyde, R G6RAF
Hyde, T A G7ANO
Hyde, T J G6KHA
Hydes, A F G3XSV
Hydes, C E 2E0AML
Hydes, I D 2E1EMJ
Hydes, I D G8FDC
Hydes, N S G4OPB
Hyett, R E G4MBJ
Hyland, N J G1ESS
Hyman, H G3IZQ
Hyman, S N G4CCT
Hyndman, A J G7GTG
Hynd-Smith, A G1FBY
Hynes, B GM0CQV
Hynes, M A G1ZRN
Hynes, P J G8TEL
Hynes, T J G4DUV
Hyslop, A GM0UDY
Hyslop, K T G1NPJ

I

Iannetta, A M1ACM
I'Anson, P W G3YFV
I'Anson, R G4CGG
I'Anson-Holton, J C R G0BST
Ibbetson, A L G3XAQ
Ibbitson, T G0VTI
Ibbitson, T G1RHW
Ibbotson, A G1OKB
Ibbotson, D A G0BCK
Ibbotson, T G1WYE
Ibbotson, R D 2E1FEX
Ibbotson, R O G0DVY
Ibison, B G M0ADF
Ibrahim, E R G6HYJ
Iceton, J L G0IIL
Icke, K P G4XQG
Icke, S W G4ZWY
Ickringill, J C W G3HHU
Ide, R P G0GYI
Iggleden, J J G8GBW
Igglesden, H J G4TPW
Igo, P GI4NJQ
Ikin, E S G4VPC
Ikin, K A G1RWX
Ikonomou, V G0VAS
Iles, C A G0VOE
Iles, D J GM4XGA
Iles, S E F G6YFG
Iles, S H G3BWQ
Iliffe, J GW8SVN
Illidge, S A GW4HBS
Illing, H G0RNU
Illingworth, T H H G4DOG
Illman, D E G7CVM
Illman, S M G6VF
Illsley, J M1CCK
Illsley, S G0BES
Illston, M A G1YYP
Ilott, J E G4KWW
Ilott, T K G4EOJ
Ilsley, D A G4ZVW
Ilston, J G1OXO
Ilston, J F G0JAN
Imber, D M G0VIS
Imianowski, A C G4KKU
Imianowski, E G GW0FWZ
Imperato, J F GW6ITB
Import, D A G3INI
Import, S G7ITT
Ince, A G4YUK
Ince, A M G0BZS
Ince, D P G4UPO
Ince, E J G4UPP
Ince, S L G4UWY
Inch, B S G0PAX
Inch, C R G4INC
Inch, M J G6TGY
Indri, L G6VHO
Ineson, D B G4ZAZ
Ineson, E G0PHF
Ingall, R W G3USZ
Ingamells, R G8HZJ
Ingarfill, E W G6LGC
Ingerslev, L C F G4RZC
Ingham, D I G4RPL
Ingham, D J G7LTR
Ingham, G I G4VSV
Ingham, J T G3RMQ
Ingham, P G6HDD
Ingham, W B G4DWO
Ingle, N P G3ZNE
Ingle, P D I G7CUP
Ingle, P M G6HYI
Ingle, R M G0MQI
Ingle, R S G6YFH
Ingle, T R G7MZE
Ingleby, W G G4TQV
Inglis, A W G6OAS
Inglis, A W M1BCD
Inglis, D A GM7CPM
Inglis, F GM3NLB
Inglis-Smith, P B G3LAC
Ingman, N G4AJF
Ingmire, G L G7BNW
Ingram, A P G1OYM
Ingram, D M GM1OIN
Ingram, E H G3TDX
Ingram, G C S G7RRW
Ingram, G C S M0AGI
Ingram, J A G7BSK
Ingram, J A C G4FDS
Ingram, J A C G4THW
Ingram, J D G0DOO
Ingram, J S 2E0AGA
Ingram, K C G7SVF
Ingram, L GM1CFL
Ingram, M C G7VOX
Ingram, P J G3GYC
Ingram, P M J G4OZL
Ingram, R A G3YIY
Ingram, S C G6ZY
Ingram, V J G3UUK
Ingram, W G4PEF
Ingram, W E G4ZKM
Ingrey, J P M1AAG
Inman, C G6JOX
Inman, K G6GEL
Inman, R M G4KZS
Inman, R M F G3MYG
Innes, D G GM7GVD
Innes, G A G4YDH
Innes, J GM7OLQ
Innes, J R GI1JQP
Innes, M G6MYG
Inness, M J GW6GMF
Innocenti, R GM7TRH
Inns, D J GM0RTY
Inskip, D J G6HQE
Inskip, D J G6ZVR
Insole, B E GW6TTK
Insole, L J G1FSM
Insole, S M GW7KIP
Instone, G C G6KAW
Instone, P E W G4SOA
Instone, S W J GW0NPL
Insull, B G3ESW
Iona, X G4RTC
Iredale, P J G8HCZ
Ireland, C E G0SSN
Ireland, D F G4KPO
Ireland, G GM3DUM
Ireland, G H G3CCL
Ireland, I F G0FRS
Ireland, I F G4BJQ
Ireland, J G0AQB
Ireland, J G GD7HSX
Ireland, L J GW0JHH
Ireland, O H G0TYJ
Ireland, P D E G0TGX
Ireland, R J G3YXQ
Ireland, W GM0PSQ
Ireson, A J G G8GOM
Ireson, K G7CGW
Ireson, M T G3OKB
Ireson, R A G6HGG
Irish, C J G6PDE
Irish, R T G4LUF
Irlam, I J G7WLL
Irlam, J G3JBT
Irlam, J C G8SBB
Ironmonger, C S G6HYF
Ironmonger OBE, J E G8PO
Irons, K H G1FVA
Ironside, A G4PGE
Irvin, H G1DEX
Irvine, A H G7WOS
Irvine, A H GM7OAW
Irvine, D S G0RZT
Irvine, G A GI4OHI
Irvine, H GI0PGU
Irvine, H GI3XRQ
Irvine, H M GI3TLT
Irvine, R B GM3EWC
Irvine, R J L G7KHV
Irving, A M G8WGU
Irving, I A G0BAQ
Irving, M R G3ZHY
Irving, R D G0IIN
Irving, R N A G3SYX
Irving, T C G2FJT
Irwin, A R GI5TK

Irwin, B G4DSR
Irwin, B A W G4XZF
Irwin, D B G0SPG
Irwin, H G4KHC
Irwin, H J GI0LTF
Irwin, P A G7UNL
Irwin, P J G0KLR
Irwin, R D GI7KPJ
Irwin, R H G7TUP
Irwin, R M G8MFR
Irwin, T F R GM4VIK
Isaac, A J G3JII
Isaac, R A G8GME
Isaac, R F R G0HAE
Isaac, R P G7HBI
Isaac, S A G3OVS
Isaacs, A M G3SGL
Isaacs, C A G3PIY
Isaacs, D R G7WGC
Isaacs, J B 2E0APV
Isaacson, B J G7MAZ
Isbill, R J G3GFR
Isham, C J G3OEC
Isham, C J G5YC
Isham, J W G8LSZ
Isherwood, D I M1ALB
Isherwood, M E G0WRS
Isherwood, M E G4VSS
Ishmael, D G7CWE
Isles, G G6HBY
Isles, J L G3NEH
Isom, D A G0DAI
Isom, T O G4RLC
Ison, A T G7BNM
Ison, C R G4DAP
Ison, D H G1LYV
Ison, J H G3YMS
Issatt, R 2E1BGV
Issatt, S R 2E1DBZ
Isted, B G G7TYI
Ives, A F G4DDT
Ives, E G7CTG
Ives, P C W G3ASQ
Ives, P S G3NIW
Ives, R J G3MSL
Iveson, D G6UYU
Ives-Whitaker, J A G0MBX
Ives-Whitaker, P M G7FPW
Izatt, R GM1IXW
Iznerowicz, J 2I1CMM
Iznerowicz, J M1BQL
Izzard, A G0NDJ
Izzard, A H G0NDI
Izzard, J A 2E1BUR
Izzard, M L G8XWR
Izzard, P E 2E1CCF

J

Jablonski, A P G7PDP
Jack, A G G0VQW
Jack, G J G6EBC
Jack, J W GM2AJW
Jack, R D GM3IXW
Jack, R L G3PNC
Jackaman, K J G1HBR
Jackett, R J G0JSJ
Jacketts, A S G8UBP
Jacklin, D G8BJV
Jacklin, D A GW3VNZ
Jacklin, H E G1XAA
Jacklin, J C G1JZN
Jackman, D R G4WPT
Jackman, W H G1IYY
Jackson, A 2E1DKQ
Jackson, A G7EVN
Jackson, A G7UXF
Jackson, A M1AAS
Jackson, A C J GI4TCR
Jackson, A D G0UEU
Jackson, A J G8JAC
Jackson, A J M1ACF
Jackson, A L G0VJL
Jackson, A L M0BCQ
Jackson, A M G4FNK
Jackson, A V G0STY
Jackson, A W G7SVE
Jackson, B G0NYL
Jackson, B GW4UFQ
Jackson, B C G4MKT
Jackson, B E G0GII
Jackson, B R P G8FFM
Jackson, C D G7UPN
Jackson, C F G0EUD
Jackson, D G1KYN
Jackson, D A G4ESY
Jackson, D B G0EGG
Jackson, D D GI4NAE
Jackson, D I G4WXS
Jackson, D J G1FQI
Jackson, D M G0WJD
Jackson, D P G4PYH
Jackson, D R G0GQP
Jackson, D R GD3RVQ
Jackson, D W G6ZET
Jackson, E G3DTP
Jackson, E E G6WJX
Jackson, E G A G3CDE
Jackson, E M G0OAO
Jackson, F G4XHC
Jackson, F E G3ZMX
Jackson, F J D GM4PDB
Jackson, F S G4LWM
Jackson, F S G6FTB
Jackson, G G4CKH
Jackson, G J G4FZK
Jackson, G L G3EEO
Jackson, G L G3OZ
Jackson, G S G1HWH
Jackson, H D G3MDN
Jackson, I G3OHX
Jackson, I A G3TYP
Jackson, I R G1RMC
Jackson, I R G7GLR
Jackson, I R G8RWH
Jackson, J G0NDD
Jackson, J G3ZWB
Jackson, J G4OPV
Jackson, J GI6RTB
Jackson, J GM3DDL
Jackson, J A G0FVS
Jackson, J A F 2E1EWT
Jackson, J B G8TOQ
Jackson, J F R GW0BTW
Jackson, J L G1XQP
Jackson, J W E G3TZZ
Jackson, K G4NEJ
Jackson, K B G6YMW
Jackson, K D G3KJ
Jackson, K G G4WEG
Jackson, K P G4KXG
Jackson, L G0NPJ
Jackson, L G4HZJ
Jackson, L C G1DMM
Jackson, L D G6MYH
Jackson, L H GJ1JTF
Jackson, L J G0GCY
Jackson, M G2FMU
Jackson, M G3CRC
Jackson, M G6THB
Jackson, M F G0SMJ
Jackson, M G G1LLZ
Jackson, M J G4SDR
Jackson, M K M0ACK
Jackson, M N G0FDE
Jackson, M P G0BYK
Jackson, M S G6FEP
Jackson, O GM4VYU
Jackson, P G0PPQ
Jackson, P G3ADV
Jackson, P G3KNU
Jackson, P G4WBH
Jackson, P G6MYI
Jackson, P M1AQJ
Jackson, P B G3WQ
Jackson, P C G7AMW
Jackson, P R G8BEB
Jackson, R G6VVS
Jackson, R A G3ASH
Jackson, R A G3YCW
Jackson, R P G1ZGF
Jackson, R P M1AWK
Jackson, R T G3YZS
Jackson, R V G8VLJ
Jackson, S 2E1DQV
Jackson, S G0UQL
Jackson, S G1PEE
Jackson, S G3CVG
Jackson, S GW3ZBB
Jackson, S A G8SZZ
Jackson, S L G6ONW
Jackson, S M G1TKN
Jackson, S P G0UGY
Jackson, T G0TJA
Jackson, T D G4HYY
Jackson, T D G7HRJ
Jackson, V T G4UPG
Jackson, W G0IIU
Jackson, W G7VPL
Jackson, W A G0DLL
Jackson, W D GW8OQV
Jackson, W J GI4TCS
Jackson, J G4CSV
Jacob, C K G3VPG
Jacob, C W G3RXN
Jacob, F G GW0OET
Jacob, P B H G0FKE
Jacob, S G G0HSV
Jacobi, P A G8ZEK
Jacobs, A D G0PAD
Jacobs, C A G4LEP
Jacobs, C G G8WAV
Jacobs, C R R G4KUW
Jacobs, E T G6HQI
Jacobs, J G1GFW
Jacobs, J G4VYA
Jacobs, M J G7EID
Jacobs, R M G7ONE
Jacobs, R M G8KNU
Jacobs, S B G3SUS
Jacobs, T J G7EIE
Jacobs, T J G7SCC
Jacobsen, C GW4ITO
Jacobsen, M G1AJD
Jaconelli, A GM1FTE
Jacques, A D G8IWQ
Jacques, D M G7VTW
Jacques, J G4AXF
Jacques, J L G1NQO
Jacques, P J G6YMY
Jacques, S G0PSJ
Jaffar, R M1BII
Jagdev, P S G7IVN
Jaggard, W C GW1OIK
Jagger, D K GW3KAJ
Jagger, S B GM3BGB
Jagger, W J G4JSC
Jaggs, P J G7IQM
Jago, R T GW4TIE
Jakins, A J G7GWA
Jakowuik, A GM7GPG
Jakusz-Gostomski, A K G6YFJ
Jallands, D M M G7FDF
Jallands, S L G7HFE
James, A G1VOC
James, A G7VIR
James, A G8WHN
James, A A G0KLK
James, A C GM8CHL
James, A E G7FFK
James, A J P GW3KZT
James, A L G1VNY
James, A L G6FVJ
James, A N G4IVD
James, A R G7WDS
James, A S 2E1FNK
James, A T G3RUV
James, A T A G1CHN
James, B G0AQJ
James, B G3NXJ
James, B E M1BSC
James, B G GW4TFX
James, B H GW0WGW
James, C B G0SDD
James, C B G6MUJ
James, C E R G7SUT
James, C F G0GFY
James, C L G0EXY
James, D G0ENM
James, D G GW4VDP
James, D H GW6KRQ
James, D I G0OYO
James, D J M0BCN
James, D J M1APP
James, D M G4DOC
James, D M GW4OZH
James, D R GM4UTK
James, D S G6PYH
James, D T G4IJP
James, D W G0FCQ
James, D W GW6DFX
James, E D G3KQG
James, E O E GW0NKF
James, E R G4PQM
James, F G6NW
James, F D C G4YVF
James, F H G0LOF
James, F L G0FLJ
James, G P 2E1EML
James, G T G7EQU
James, H G3MCN
James, H D G3HZP
James, H W G3UPZ
James, I G0JXZ
James, I J P G5IJ
James, I M G4OQJ
James, J GW0KPD
James, J S G7NLY
James, J W G4MXO
James, K G4XQA
James, K C G8PGH
James, K F G0BFK
James, K R 2E1ESM
James, K R G8EZR
James, L B 2E1FKH
James, L D G0NMN
James, L H GW4JJR
James, M G6BZE
James, M GW4YID
James, M D G3OZK
James, M D G6IXE
James, M G G3GVY
James, M J G7UTS
James, M L GW1FOF
James, M S G6SFX
James, M S G8PKV
James, O G 2E1FQY
James, P A G6NSQ
James, P D G0IQI
James, P J G1HPU
James, P J G4NTP
James, P R 2E1ESH
James, P R G6HYN
James, P W G6PZF
James, R G0LIA
James, R G0LRH
James, R G3ZSZ
James, R E A G0REA
James, R F G4TOT
James, R P G0WUK
James, R P GB7SDX
James, R P GM4CXM
James, R W G3AHE
James, S A G1DAK
James, S I W G3NZW
James, S M G8TNZ
James, S W G8UDD
James, T C G0VAU
James, T R G4EWW
James, W D G3SKK
James, W E G4JAZ
Jameson, B C G6VPN
Jameson, C J G0UFU
Jameson, G G7SMC
James-Robertson, R E G4ELL
Jamieson, J N GM1POA
Jamieson, M M G0BHU
Jamieson, M T GI4VIZ
Jamieson, N D F 2E0APP
Jamieson, N D F G8LCP
Jamieson, R GM1JPJ
Jamieson, R G G6RVH
Jamieson, W GM7ANE
Jamil, M L G0RAN
Jamison, W A G3XGH
Janaway, A D G4VFD
Janaway, M R G8OUD
Jandrell, A P G4HWH
Janes, B J G3XYL
Janes, H C G0TIG
Janes, P A GW1SXU
Janes, R A G0JNA
Janes, R J G4XKQ
Janes, R K GW4IUN
Jappy, D A G6STW
Jaques, N W G0HFU
Jaramillo, R G1OSN
Jardine, A J GM7SAK
Jardine, D M G0FDV
Jardine, J C G7UYJ
Jardine, J R G3XJY
Jardine, T D GM2BMJ
Jardine, W H G1YHS
Jarman, G T D G0GBS
Jarman, M A G7VYN
Jarman, S B G1YGP
Jarmyn, S W G6FQN
Jarrard, D G0DEP
Jarratt, B F G1ATY
Jarratt, K T G6WLA
Jarrett, A E G4FRZ
Jarrett, D C G4DCJ
Jarrett, D C G4LSF
Jarrett, E M G6EMJ
Jarrett, G R G8NDB
Jarrett, G V G4PPV
Jarrett, L R H G8MOH
Jarrett, P A G7DPK
Jarrett, P D G8FTP
Jarrett, P E G4CDJ
Jarrett, P G G8DBV
Jarrett, P S G1UZC
Jarrett, S A 2E1EWH
Jarrett, S A M1BWN
Jarvie, J C G3XTI
Jarvie, W GM8IIH
Jarvill, W M G7SYC
Jarvis, A G6TTL
Jarvis, A H G0NTA
Jarvis, B C G0NTB
Jarvis, C W G6XTY
Jarvis, D A G4PGF
Jarvis, D J G4CEU
Jarvis, D R G4TFZ
Jarvis, E R G7JLL
Jarvis, G M G0WGJ
Jarvis, H A G7EXT
Jarvis, I M G6CYO
Jarvis, J J G3SUG
Jarvis, J W G4NEY
Jarvis, M G A G8ELS
Jarvis, N J G4SPD
Jarvis, P M1AIZ
Jarvis, P N G3OWJ
Jarvis, R A G7OHM
Jarvis, R A W G6DRX
Jarvis, R C G4OBA
Jarvis, R F G4JPA
Jarvis, R I G8UBU
Jarvis, R J 2E0AKP
Jarvis, S A G1RXB
Jarvis, S C G4RRI
Jarvis, S D G6DSB
Jarvis, S J G4TIA
Jarvis, T G6XTZ
Jarvis, W A G6SND
Jarvis, W H GM8APX
Jasinski, K G0ZHP
Jasinski, K M0AXH
Jasper, G H N G8TWA
Jasper, L A G0GZN
Jasper, R G0CIR
Javaheri, Y 2E1DQW
Javes, S G1BGC
Jay, A D G1BPM
Jay, B A G8TEO
Jay, C G4WWO
Jay, C E G4TDR
Jayne, B J M0BFL
Jeacock, T G G0EZY
Jeans, C G4IPJ
Jeans, W P G3GAA
Jeavons, A P G3OZL
Jeavons, P D GW8XUM
Jebb, J A G8YDC
Jebb, S B G6AJF
Jeckells, G S G1MJE
Jeckells, H G1OMT
Jedrzejewski, J K M1AQC
Jeeves, A J G0RTW
Jeeves, M L G6CBY
Jeeves, R S G7EPR
Jeffcoate, S E G7IMB
Jeffels, J G7VGQ
Jefferies, F A G8PX
Jefferies, G L M1ASR
Jefferies, L E GW0BXZ
Jefferies, R E M0AMF
Jefferies, T G G1JPK
Jeffers, J G0UNB
Jeffers, M G1VDC
Jefferson, D M G6PZE
Jefferson, I P G1JCC
Jefferson, I P G2LO
Jefferson, I P G4IXT
Jefferson, M L G8WYB
Jefferson, P M1AZM
Jefferson, R L G8UZM
Jefferson, S D 2W1FHA
Jefferson, T M G4IAJ
Jeffery, A C G4YRL
Jeffery, A F G8SIG
Jeffery, D R N G4HWM
Jeffery, G T G7BRM
Jeffery, M A G7ORR
Jeffery, M J G7HRN
Jeffery, M W G6JVK
Jeffery, P G0MID
Jeffery, R A G6RBM
Jeffery, R P G0GZI
Jeffery, S D G1FAA
Jeffery, S W G0LAP
Jeffery, W F G4NFH
Jefferys, A B GU4RUK
Jefferys, D J G6IWZ
Jefferys, J F G0SPO
Jefferys, S J L G0IDB
Jefford, A W N G8GON
Jefford, M G8UWE
Jefford, P E M1AFP
Jefford, T J G7HIY
Jeffrey, A E G3HWD
Jeffrey, H B G0VJZ
Jeffrey, J D G3RDF
Jeffrey, S G0UYS
Jeffrey, S B G3JSB
Jeffries, A L G4MOJ
Jeffries, E A G4GBV
Jeffries, L K M0AUY
Jeffries, P A G8TNH
Jeffries, P E 2E1DOU
Jeffries, P E M1BON
Jeffries, R P G4KAR
Jeffs, A G0RJM
Jeffs, A M0AQS
Jeffs, A M1AOS
Jeffs, M R G7LJN
Jeffs, P H G4AYA
Jeffs, R G 2E1AUB
Jeffs, T B 2E1DKE
Jeffs, W A G3OAF
Jehle, A E B G4VHD
Jelfs, M B R G8KNS
Jelinek, M J M 2E1FCK
Jellett, R G4OAY
Jelley, G M G7DFV
Jelley, N D G0CRT
Jelley, R H C G1OYT
Jelly, S J G6URJ
Jemmison, M B G0VAC
Jempson, B G4OVW
Jenkin, D R G0PNZ
Jenkins, A G6CYZ
Jenkins, A T G7CYD
Jenkins, B J G7PKG
Jenkins, C G0VRX
Jenkins, C G7RQL
Jenkins, C J GM0NQP
Jenkins, D G GW4RIX
Jenkins, D J GW7LMW
Jenkins, D L G8MNR
Jenkins, D P GW0PUM
Jenkins, E GW0TSX
Jenkins, F E G G1CKB
Jenkins, G G7GMV
Jenkins, H V G4UXM
Jenkins, J A GW0JKB
Jenkins, J G4LJW
Jenkins, J O D GW3JUN
Jenkins, J P GW0ADS
Jenkins, J R G0KHY
Jenkins, L 2E1AFS
Jenkins, L GW3HXX
Jenkins, L J G0BTS
Jenkins, L J G0LTG
Jenkins, M GW1WGR
Jenkins, M GW3OMN
Jenkins, M G GW7EYP
Jenkins, M R G0SBK
Jenkins, N GW7RQV
Jenkins, P G0ECK
Jenkins, P GW8HWL
Jenkins, P A GW4MII
Jenkins, P E GW7LYD
Jenkins, P H GW7HTU
Jenkins, R G4XEV
Jenkins, R GW4ATY
Jenkins, R A G0WTK
Jenkins, R A G7WNN
Jenkins, R A G8TBF
Jenkins, R A F G1BPM
Jenkins, R B GW4UYT
Jenkins, R E G0VHZ
Jenkins, R H GW6VFH
Jenkins, R J G8HDP
Jenkins, R M G6CYR
Jenkins, R N GI4RVT
Jenkins, S F G4CHO
Jenkins, S G 2E1FKK
Jenkins, T F G6AJG
Jenkins, T R 2E1FKL
Jenkins, W R G4USW
Jenkinson, A E G3OMC
Jenkinson, B G3JHC
Jenkinson, B G7BBJ
Jenkinson, D 2E1APW
Jenkinson, J A G7GDI
Jenkinson, J E G8CVS
Jenkinson, J N G0HGM
Jenkinson, K J G0LDY
Jenkinson, M J G0RGE
Jenkinson, N G8FUV
Jenkinson, P G0VGK
Jenkinson, R J G4SEF
Jenkinson, S R A G6HBZ
Jenkinson, W L G0OJT
Jenks, D J G1RFQ
Jenman, H R G0CGO
Jenner, A B G0GQH
Jenner, G W G3KIW
Jenner, J E G3ETJ
Jenner, J M G0PEG
Jenner, P L G3NCA
Jennings, A G4DZT
Jennings, A G4PZF
Jennings, B G4TLM
Jennings, D G0NZV
Jennings, D T G3HBV
Jennings, F G G0MXE
Jennings, G T G0MJL
Jennings, G W G3MWH
Jennings, J M G0EOG
Jennings, J R G4VOZ
Jennings, J R G8LM
Jennings, J W E G6URK
Jennings, J W J G1KIC
Jennings, M G6ZEW
Jennings, P L GW3RMJ
Jennings, R GI4NFH
Jennings, R B GI4WRJ
Jennings, R H G3NXV
Jennings, R H G3SOE
Jennings, R P G0JHD
Jennings, R W H G4BY
Jennings, S E GI4RKC
Jennings, S F G3XMU
Jennings, S H G0TEQ
Jennings, T 2E1FRZ
Jennings, W G G0IEC
Jennings, W H F G4JWY
Jennison, D G4UEE
Jennison, R C G2AJV
Jennison, R L G4HXM
Jenno, A C G7LTZ
Jensen, F F G1HQQ
Jensen, J C H MW0ANX
Jensen, P G7BTP
Jensen, P M G4GZT
Jensen, T E GI7GHM
Jenson, R L G4FCE
Jephcott, A GU6JPE
Jephcott, R E G3XNN
Jepson, J J G0JYR
Jeremy, M L G3VEO
Jeremy, S J 2E1DLK
Jermaks, A M0ADL
Jermany, C E G1EBP
Jermey, K E G7WCM
Jermy, R J G0UVP
Jerram, N G4EAH
Jerrome-Jones, M J GD4WBY
Jervis, C L G8JBC
Jervis, F N GJ3UQM
Jervis, I H G0LCY
Jervis, M A G8PSL
Jessemey, R E G3UOG
Jesson, G H M1AXU
Jessop, A M G G8IKU
Jessop, G R G6JP
Jessop, K F G3TAA
Jessop, M 2E0AJC
Jessop, P M G8KGV
Jessup, G C G4BKB
Jessup, G G G3AMG
Jessup, G P G7AYA
Jessup, P G7VQW
Jewell, A S 2E1AVS
Jewell, B G7VPB
Jewell, B E G7WKV
Jewell, D S G0LUA
Jewell, E R G4ELM
Jewell, G A G0TAD
Jewell, L J G4OCX
Jewell, S L 2E1AJE
Jewell, S T G4DDK
Jewers, G M G1MKT
Jewitt, P G4ENL
Jewkes, M V G0TWH
Jewson, D F G3XFB
Jex, A B G0OOR
Jillings, K P G3OIT
Jinks, F A GW3XVQ
Jinks, I G4TKM
Jinks, J V G1WRU
Jinks, M W G0GIT
Jinks, R I G7RVM
Jinks, S E 2E1BAY
Jobber, A R G6EXU
Jobbins, P G8OQG
Jobbins, R H G1DFZ
Jobes, R G4XYP
Jobling, C H G4YHP
Jobling, E G4SLN
Jobling, J B G3ZUP
Jobson, A K G7SUQ
Jobson, C B G0UTI
Jocys, J B G4WQD
Johansen, D G8TSG
Johanssen, N G1ZPA
Johansson, P GW1NGN
John, A GW8MFQ
John, C L M0AHJ
John, D G3WCB
John, D G GW8GXW
John, E C G3SEJ
John, E C G3STD
John, E C G8STD
John, G S GW1EOI
John, G W G8ZRN
John, H T GW1MGR
John, H T GW6GWK
John, H T GW7MGR
John, L GW3MMT
John, R G4DDD
John, S I G0OJS
John, W M GW1YKT
John, W O GW4WOJ
Johns, A R G1MSH
Johns, A R G4TVJ
Johns, A S G6HAA
Johns, D A GW4XES
Johns, D J G0VNO
Johns, E G0RWI
Johns, F D GW0CXK
Johns, G M 2E1DJI
Johns, I M G7URH
Johns, J A GW4ZBU
Johns, M S G0GZM
Johns, N W M0ASI
Johns, R R G0NMW
Johns, R S GW4NXD
Johns, R W G4FVX
Johns, W D GW4TOU
Johns, W M G7MTT
Johnsen, H A G0DHZ
Johnson, A G0PUK
Johnson, A G1NTB
Johnson, A G1OPS
Johnson, A G3ENZ
Johnson, A G4FWR
Johnson, A G8JVU
Johnson, A E G0AUV
Johnson, A E G4EFH
Johnson, A G G4DUC
Johnson, A H G7WBQ
Johnson, A H M0AVY
Johnson, A I 2E1FAL
Johnson, A J G4UFG
Johnson, A J J G6BLK
Johnson, A L R G0KCJ
Johnson, A R G3WHJ
Johnson, A T G4RQK
Johnson, A T G4TEI
Johnson, B G0GAQ
Johnson, B G1GTY
Johnson, B G1NUP
Johnson, B G6URM
Johnson, B G7WBV
Johnson, B GJ6FTU
Johnson, B M1ABC
Johnson, B A G3XIB
Johnson, B A G7IUT
Johnson, B E G6EAL
Johnson, B J G1YNX
Johnson, B J G7SXZ
Johnson, B M G3LOX
Johnson, B P G1SPK
Johnson, B W G6MYO
Johnson, C G4BFT
Johnson, C G4PYD
Johnson, C A G0JQI
Johnson, C A E G6MUS
Johnson, C G G7RBL
Johnson, C O L A G6LRT
Johnson, C P G0BZN
Johnson, C R G7HJV
Johnson, D 2E1APX
Johnson, D 2E1FTL
Johnson, D G1GNS
Johnson, D G4DHF
Johnson, D G7GJE
Johnson, D G7TZX
Johnson, D MW0AUX
Johnson, D A 2E1CVJ
Johnson, D A G4DPZ
Johnson, D A G8AZM
Johnson, D A J G4AON
Johnson, D C W G1IHO
Johnson, D E G3HLG
Johnson, D E G3MPN
Johnson, D E G7SMX
Johnson, D J G0BUS
Johnson, D J G4UIA
Johnson, D J M G1YQS
Johnson, D M G1PNL
Johnson, D M G6RTY
Johnson, D R G0IES
Johnson, D R G0NGI
Johnson, D S G0KIQ
Johnson, D W G8AYK
Johnson, E G6YMZ
Johnson, E J G4LUW
Johnson, E R G1IUM
Johnson, F M0BAL
Johnson, F D G0GSR
Johnson, G 2E1DES
Johnson, G G8YYL
Johnson, G A G0SSK
Johnson, G D G6BSX
Johnson, G M G4ZWA
Johnson, G P 2E1FCN
Johnson, G P G0AAK
Johnson, G P G7ATW
Johnson, H G1YMV
Johnson, H G4XNK
Johnson, H A G0ITK
Johnson, H P G0KGE
Johnson, H R W G1PHJ
Johnson, I J G1DGW
Johnson, I L G0SUQ
Johnson, I M G1IMJ
Johnson, I M G7VLR
Johnson, I P M 2E1FSO
Johnson, J G0HJJ
Johnson, J G4XTE
Johnson, J MW1ASG
Johnson, J B G0NDS
Johnson, J B G3JJW
Johnson, J C G3GSC
Johnson, J C G6VZM
Johnson, J C G7AOQ
Johnson, J D G0AHJ
Johnson, J D G0FDS
Johnson, J D G0VIJ
Johnson, J D G7UIJ
Johnson, J G G7UAT
Johnson, J J G3JJJ
Johnson, J L G0KSC
Johnson, J R G1JER
Johnson, J R G1PVQ
Johnson, K G4SVN
Johnson, K G7NHV
Johnson, K A 2E1FNA
Johnson, K B 2U1EKH
Johnson, K E G1PQW
Johnson, K E G7SKM
Johnson, K P G8NTD
Johnson, K P GW7DZB
Johnson, K R G7GJD
Johnson, K R GM4WBO
Johnson, L C G1SGO
Johnson, M G3YQZ
Johnson, M G3ZUI
Johnson, M G6THC
Johnson, M G8UJV
Johnson, M A G0GCK
Johnson, M C 2E1FSP
Johnson, M J G7APH
Johnson, M L G8MYF
Johnson, M P GU6AJE
Johnson, M R 2E1FZX
Johnson, M R G0BPU
Johnson, M S G1RJA
Johnson, M T G4ZRT
Johnson, M W G8KEJ
Johnson, N G0JOW
Johnson, N G0RGQ
Johnson, N G7UIO
Johnson, N G8HQO
Johnson, N B G6EID
Johnson, N C G6OWO
Johnson, N W G0SNX
Johnson, P G0DIG
Johnson, P G0LEU
Johnson, P G0PPJ
Johnson, P G1FHH
Johnson, P G4YFX
Johnson, P G6DFC
Johnson, P G7RFD
Johnson, P A G1JST
Johnson, P A G4LXC
Johnson, P A G4TLO
Johnson, P A G8KB
Johnson, P A J 2E1EXA
Johnson, P B G0UMV
Johnson, P B G3UMV
Johnson, P B G4UMV
Johnson, P B G8LVC
Johnson, P C G8BFC
Johnson, P E G3JPE
Johnson, P E G7DDF
Johnson, P F G4EQQ
Johnson, P F G4UCX
Johnson, P I G4EMV
Johnson, P L M1BCV
Johnson, P M G1LJQ
Johnson, P R G4RMT
Johnson, P R G7JJZ
Johnson, P R M0ALE

Johnson, P T G0UPW
Johnson, P W G0SHX
Johnson, P W G4TCT
Johnson, P W G8JYX
Johnson, R G0MAT
Johnson, R G0UYR
Johnson, R G2FFO
Johnson, R GM0VYI
Johnson, R C GW0BNP
Johnson, R D G1ZUC
Johnson, R D G4BWF
Johnson, R E G8GVG
Johnson, R E D G7RFE
Johnson, R G G7HHK
Johnson, R K G3VZT
Johnson, R K G7COA
Johnson, R M G3XOV
Johnson, R P G0OOV
Johnson, R P C GM7NHS
Johnson, R S G0GFC
Johnson, R S GW4UJT
Johnson, R S M0AKE
Johnson, R T G0BCZ
Johnson, R W GM7CZC
Johnson, S G0USJ
Johnson, S A P G0IJJ
Johnson, S C G1PYC
Johnson, S C G7GJC
Johnson, S N GI5SJ
Johnson, S P 2U1ELN
Johnson, T E W G7TUF
Johnson, T J G0ABM
Johnson, T L G3BJN
Johnson, T M G0WBR
Johnson, W G4CNK
Johnson, W M0ABW
Johnson, W A G4ZXZ
Johnson, W D G0AXZ
Johnson, W D G8NS
Johnson, W W G1ZLF
Johnson-Roberts, B G3RGG
Johnston, A R G4DTE
Johnston, A R G6YNA
Johnston, B G1PYR
Johnston, D A 2E1FSL
Johnston, D E 2E1APY
Johnston, D H GI6RMO
Johnston, G L G0RDN
Johnston, G R GI0CUR
Johnston, G W E G7OZA
Johnston, I G7STO
Johnston, J G4DPN
Johnston, J GM1HDF
Johnston, J A G8IUV
Johnston, J G G3PHJ
Johnston, J P GM4ENP
Johnston, J T A GM3LYY
Johnston, K L GM1KKI
Johnston, K R GW4BCB
Johnston, L G8DQF
Johnston, M R GM0MRJ
Johnston, P G0FVN
Johnston, R GI6WFX
Johnston, R GM3XUW
Johnston, R J GI7BON
Johnston, S G4WZM
Johnston, S G4XSJ
Johnston, S GM7VOZ
Johnston, S G GI4IBV
Johnston, S K 2I1FVW
Johnston, V G3KXV
Johnston, V H G1PKS
Johnston, V M G3NUL
Johnstone, A W G1EBS
Johnstone, C A G8XDR
Johnstone, D GM1JZM
Johnstone, D N GM4EVS
Johnstone, F E G3IDC
Johnstone, H S G1RRG
Johnstone, H S G8JTD
Johnstone, J T GM0CBC
Johnstone, J W G1WSH
Johnstone, P G6RAU
Johnstone, P G GM0KMJ
Johnstone, R G3XQL
Johnstone, R T S GM3EZA
Johnstone, R W GM0WWX
Johnstone, R W E GM1YGV
Johnstone, S G G6SGJ
Johnstone, S N GI0HHV
Johnstone, W G0BPF
Johnstone, W A G7WKL
Johnstone, W A M0BCE
Joiner, M C G3ZYZ
Joiner, W M G4OAX
Joines, A W G4LVH
Joll, J W G0TQT
Joll, R H G3OSY
Jolley, D P G6RDZ
Jolley, D V G1IQU
Jolley, F G4XHZ
Jolley, G E G G4BUF
Jolley, M A 2E0AOR
Jolley, M A 2E1EYT
Jolley, M E G3URN
Jolliff, M L G7ILQ
Jolliffe, C 2E0AMM
Jolliffe, E G G3IMX
Jolliffe, E M G6JXW
Jolliffe, R T G3ZGC
Jolly, A J GW4JCO
Jolly, D R H G3TJY
Jolly, I B G4BTW
Jolly, J W G3STB
Jolly, K J G3LOC
Jolly, M J G1OZV
Jolly, R G4VMJ
Joly, G C G6DFY
Joly, N F G3FNJ
Jonas, R T G6EPL
Jones, A G3CTZ
Jones, A G4TDZ
Jones, A G4UGA
Jones, A G8EKZ
Jones, A GW4RUX
Jones, A M1AWS
Jones, A A G6JS
Jones, A B G4ICU
Jones, A B G4ORJ
Jones, A B G4OUY
Jones, A D G3SGA
Jones, A E G1ENS
Jones, A E GW1HBU
Jones, A E H G0EEI
Jones, A F G0JFA
Jones, A F M G1EBT
Jones, A G G1WNY
Jones, A G GW4OGC
Jones, A G GW4TFS
Jones, A H GW4RYW
Jones, A I G4VMZ
Jones, A J G7UEK
Jones, A J GW1CEV
Jones, A J GW4VPX
Jones, A J MW1BJD
Jones, A L G1UWN
Jones, A L G3XJO
Jones, A M 2E1FGC
Jones, A M G0PJC
Jones, A M G7LNP
Jones, A M GM8HGD
Jones, A M D G0PZN
Jones, A N G7APQ
Jones, A N GW0SCN
Jones, A P G0YSS
Jones, A P GW7VHD
Jones, A R G0MKW
Jones, A R E G6DGA
Jones, A S G0UIW
Jones, A S G4NHF
Jones, A S G4RWY
Jones, A T G1YBI
Jones, A V GW6NSK
Jones, A W G3CRF
Jones, A W GW7RTK
Jones, B G0PLW
Jones, B G1BDI
Jones, B G4ZQG
Jones, B G8EXJ
Jones, B GW1BDH
Jones, B GW3WRE
Jones, B GW4OPW
Jones, B GW7KNN
Jones, B M1AYP
Jones, B A G4PBY
Jones, B A G4YFZ
Jones, B A G8ASO
Jones, B E G0UKB
Jones, B E GW7LJG
Jones, B G G1FUJ
Jones, B G G7IFI
Jones, B G GW6ZYI
Jones, B G A G3GCW
Jones, B H J G0UQE
Jones, B M G1VJN
Jones, B M G7CMU
Jones, B N GW8OYT
Jones, B S G4ISQ
Jones, C 2E1BZG
Jones, C G0PIO
Jones, C G7WCJ
Jones, C A G6ZEZ
Jones, C A G7SPM
Jones, C A G8FGD
Jones, C A GW7PBR
Jones, C B G3KOP
Jones, C D G6ITH
Jones, C E G0NOR
Jones, C E G1ARR
Jones, C E MW1AFW
Jones, C F G4DHV
Jones, C H GW3NKM
Jones, C H P GW6IXA
Jones, C J G0IQN
Jones, C J G1PKO
Jones, C J G4HHU
Jones, C J G6OAV
Jones, C J GW6OWQ
Jones, C L C G0FIJ
Jones, C M G0LSJ
Jones, C P G4PUH
Jones, C R GW8JRL
Jones, C T MW1BTM
Jones, C W G0FTU
Jones, D G0DSR
Jones, D G0IBW
Jones, D G1XJN
Jones, D G3UVR
Jones, D G4FAH
Jones, D G4MGR
Jones, D G4RVJ
Jones, D G6EPJ
Jones, D A G4GDS
Jones, D A GW7FNQ
Jones, D B GW0NNB
Jones, D B GW4EIN
Jones, D C G4YBQ
Jones, D C G7DRI
Jones, D D GM0CIA
Jones, D D E 2E1DHJ
Jones, D D G GW4WPJ
Jones, D E G4LXH
Jones, D E G4SXD
Jones, D E G7SSB
Jones, D E GW6WAG
Jones, D E GW8PBX
Jones, D G G0KSG
Jones, D G G4GRU
Jones, D G G G6FEO
Jones, D H G7VNR
Jones, D H GW4XMU
Jones, D I G0WVV
Jones, D J G4NBK
Jones, D J GW4YUX
Jones, D J MW0BER
Jones, D J MW1BJV
Jones, D L G0EYW
Jones, D M G0HHP
Jones, D M G1DMJ
Jones, D M G4FQR
Jones, D M G6PGG
Jones, D M G7VSJ
Jones, D M GI3KVD
Jones, D N G0REE
Jones, D N G7HCC
Jones, D N G8IMX
Jones, D O GW8XMW
Jones, D P G4FAQ
Jones, D P G7UIV
Jones, D R G8JPJ
Jones, D R L G8KKD
Jones, D S GW3XYW
Jones, D T GW1SQT
Jones, D W GW0UDJ
Jones, D W K G4LIH
Jones, E 2E1AEJ
Jones, E G0FWR
Jones, E G0SGU
Jones, E G8XMU
Jones, E GW6HQA
Jones, E GW7NFM
Jones, E GW7UNV
Jones, E A G1VRA
Jones, E A G7TSB
Jones, E B M J G8YHJ
Jones, E C GW4JPP
Jones, E D G0RLV
Jones, E F G3EUE
Jones, E G G3NQE
Jones, E G G3ZLX
Jones, E I G4YML
Jones, E J 2E1EMS
Jones, E K 2E1EHL
Jones, E L GW3FJI
Jones, E L L M0ACL
Jones, E L L MW0AEV
Jones, E P A G8CDC
Jones, E R B G1HBV
Jones, E V G3JQK
Jones, E W G0TMJ
Jones, E W GW0MLN
Jones, E W P G3TVK
Jones, F G2ADA
Jones, F G3RVN
Jones, F GW4SLS
Jones, F E G8HTP
Jones, F G GW3CF
Jones, F M G1HBW
Jones, G G3VAS
Jones, G G4TPV
Jones, G GW0DDL
Jones, G GW1HPP
Jones, G GW1ZTP
Jones, G M0AJX
Jones, G B GW0KAM
Jones, G B J G6ERZ
Jones, G D G1PQY
Jones, G H S G3VKV
Jones, G J GW6STS
Jones, G J GW6TGR
Jones, G L GW6PVK
Jones, G L GW7UAE
Jones, G M 2E1DMT
Jones, G M G4OEX
Jones, G M GW7GDH
Jones, G O G3VSB
Jones, G O GW0ANA
Jones, G P G3UZZ
Jones, G P GW4RUY
Jones, G P GW7CMM
Jones, G R G7TIM
Jones, G R GW4UCK
Jones, G S G1FEO
Jones, G T GW0SEO
Jones, G W F G4DPH
Jones, H G0LLD
Jones, H G0TDZ
Jones, H A G0OKO
Jones, H G G0GQW
Jones, H G G4VSU
Jones, H I GW0WSU
Jones, H J G4VQS
Jones, H M G0MGX
Jones, H M GW4XZJ
Jones, H N G6YFL
Jones, H R GW4GFS
Jones, H T GW3NQP
Jones, H V GW0SAJ
Jones, H W G1NCN
Jones, I GW3TLP
Jones, I B GW1IQS
Jones, I D G0GPS
Jones, I D GW4RDW
Jones, I F G0HSE
Jones, I F G4MLW
Jones, I G GW7SDE
Jones, I M MW1BHY
Jones, I P GW7DKI
Jones, I R G4JVC
Jones, I W GW1OXJ
Jones, I W GW4FQU
Jones, J 2W1EEP
Jones, J G1RXR
Jones, J G4PKP
Jones, J GW4UYU
Jones, J A G1CUJ
Jones, J A G3NPJ
Jones, J A G4XTU
Jones, J A G6CYN
Jones, J A G7EZD
Jones, J A G8AZT
Jones, J A GD4WOW
Jones, J B G2CUJ
Jones, J B G4JYI
Jones, J B A G8VWP
Jones, J C GW7NJT
Jones, J C V G1RAG
Jones, J C V G1ZUU
Jones, J D GW4LPU
Jones, J D GW6NSG
Jones, J F GW1UVN
Jones, J H G3TQL
Jones, J H T G4TLL
Jones, J K GW0PND
Jones, J M G3UED
Jones, J M GW0APJ
Jones, J M GW0PTX
Jones, J M A GW4KYK
Jones, J P G GW3IGG
Jones, J P L MI1AYF
Jones, J R G3FTU
Jones, J R GW0KJZ
Jones, J S G0REO
Jones, J S G1WZX
Jones, J S J G0HQK
Jones, J T J G3JTJ
Jones, J W G4DGO
Jones, K GW1BDF
Jones, K GW8YYF
Jones, K A G0IUI
Jones, K A G0RJA
Jones, K B G4SGV
Jones, K B GM1YUH
Jones, K E G3RRN
Jones, K E M0AET
Jones, K G G3PSZ
Jones, K G GW7TSO
Jones, K M G0HMU
Jones, K M G4UXL
Jones, K M GW4AHO
Jones, K M GW4SUD
Jones, K R G1HPQ
Jones, K T G4FPY
Jones, L G8FUB
Jones, L A 2E0APU
Jones, L A GW4WKQ
Jones, L B G7NIR
Jones, L D W GW7BDG
Jones, L E G6XRT
Jones, L F G3LBZ
Jones, L H G4DLK
Jones, L H G4KLT
Jones, M G1YWY
Jones, M GW0WVL
Jones, M GW1SGG
Jones, M GW3WTZ
Jones, M GW4NNL
Jones, M GW4ZNU
Jones, M GW6VRN
Jones, M A GW7ODP
Jones, M A T G6DFZ
Jones, M C G1JHR
Jones, M C GM0BCP
Jones, M D G8RRN
Jones, M G G0RPQ
Jones, M G G0TQE
Jones, M G GW7UQN
Jones, M H G3GRU
Jones, M H GW4NBM
Jones, M I GW7OLL
Jones, M J G0MIX
Jones, M J G3PQT
Jones, M J G7BBY
Jones, M J G7JCD
Jones, M J G7UWI
Jones, M K G0ERE
Jones, M L G0MLJ
Jones, M L G4TIF
Jones, M N 2E1FYV
Jones, M P GW0ADY
Jones, M R G4MFN
Jones, M R G6OAU
Jones, M R G7VGZ
Jones, M V G4WTA
Jones, M W G0OQT
Jones, M W G4NKC
Jones, N GW0VQZ
Jones, N MJ0AQJ
Jones, N C G7RWQ
Jones, N F G8IWO
Jones, N G G6HYP
Jones, N J G0NJJ
Jones, N J GW7SDB
Jones, N P E G6HQD
Jones, N R V GW4XXJ
Jones, N W G1UDP
Jones, O GW1BDG
Jones, O D GW3DRV
Jones, P G2JT
Jones, P G3PYU
Jones, P G3YLV
Jones, P G6IIM
Jones, P G7RKT
Jones, P G7TIA
Jones, P GI4VSC
Jones, P GW4ZCL
Jones, P A G4UGW
Jones, P A GW4HAT
Jones, P A GW4UKU
Jones, P D G4DXO
Jones, P D G4EQV
Jones, P F GW3FPF
Jones, P G G3TZS
Jones, P G GW7HDX
Jones, P J G0JCT
Jones, P K M1BCC
Jones, P M GW1ESU
Jones, P R G1XTJ
Jones, P R G4GNK
Jones, P R G7TZD
Jones, P W G4ERF
Jones, P W GW7ASL
Jones, P W F G3ESY
Jones, R G0EPJ
Jones, R G1HST
Jones, R G1LRU
Jones, R G4OEK
Jones, R G4SWH
Jones, R G6NSM
Jones, R G7RGR
Jones, R G7SMJ
Jones, R GW3MDK
Jones, R GW6TM
Jones, R M0ANC
Jones, R MI1BJM
Jones, R A G0UEW
Jones, R A G4NMU
Jones, R A G6DSD
Jones, R A GW0BYZ
Jones, R A GW0MHY
Jones, R A O G0UCI
Jones, R A O G8HDA
Jones, R B G6LRU
Jones, R B G6RBJ
Jones, R B GW0LOI
Jones, R C G1WVW
Jones, R C G4SAS
Jones, R C G7DGH
Jones, R D G6TGW
Jones, R D GM1MYF
Jones, R E G8AXK
Jones, R E GW4FCV
Jones, R E GW7KAX
Jones, R G G4AIJ
Jones, R G G6UMK
Jones, R H G3SFO
Jones, R J G6XEV
Jones, R J G7VJU
Jones, R J G8MBQ
Jones, R J G G3OFZ
Jones, R L G4KQQ
Jones, R L G4NWG
Jones, R M G0BGR
Jones, R M G3NKL
Jones, R M GW0KLY
Jones, R M G G7SWW
Jones, R M P G0EUZ
Jones, R O GW8DVH
Jones, R P GW7CMF
Jones, R R GW4HOQ
Jones, R S G0RMG
Jones, R T GW0BZE
Jones, R T H G0INX
Jones, R W G3YIQ
Jones, R W G4RIU
Jones, R W R G3YMK
Jones, S G0HAB
Jones, S G6FES
Jones, S GW0RHE
Jones, S GW3HHF
Jones, S M0BFM
Jones, S MW0AWO
Jones, S MW1BGM
Jones, S A G0MKV
Jones, S A GW0COU
Jones, S C G4GVD
Jones, S C G4MQQ
Jones, S D M0AND
Jones, S E G4ZAX
Jones, S G G0VZP
Jones, S G G6LRL
Jones, S G B GW1YKY
Jones, S K G4KQP
Jones, S K G4TQF
Jones, S M G4DSF
Jones, S M G4YYS
Jones, S R G4AXW
Jones, S R G7SRJ
Jones, S R GW0GEI
Jones, S T G1KNX
Jones, S W G3LXB
Jones, T G0IYV
Jones, T G6MUQ
Jones, T G7VMQ
Jones, T GW0IVT
Jones, T GW4RQQ
Jones, T A B G7IRD
Jones, T B G0RBW
Jones, T C GW7AWO
Jones, T D G4IPR
Jones, T E GW0HGN
Jones, T G 2E1FGG
Jones, T G G1HPS
Jones, T G GW0JAI
Jones, T G GW7TGZ
Jones, T H G3UUL
Jones, T H GW8WEY
Jones, T J G1MGN
Jones, T J G1NOL
Jones, T L G7JUB
Jones, T M G1BQZ
Jones, T M J G0NFT
Jones, T O V GW6OAW
Jones, T R G1HPV
Jones, T R G4KRB
Jones, T R GW7WLD
Jones, T T G8DVF
Jones, T W G7BXN
Jones, T W GW0VYG
Jones, T W GW4TNF
Jones, T W GW4TWJ
Jones, T W GW4VEQ
Jones, T W GW7GWM
Jones, V G GW4TEE
Jones, V H GW0GSW
Jones, W 2E1EWL
Jones, W G0PWP
Jones, W GD4XOD
Jones, W A G0GBX
Jones, W D G7MOX
Jones, W E G0TYV
Jones, W E J G1PXE
Jones, W F G7PUJ
Jones, W G G0GOR
Jones, W G G0NIX
Jones, W G G7CZW
Jones, W G G7WHP
Jones, W G H GW4KJW
Jones, W H GW1DLP
Jones, W J G0WPK
Jones, W J G8DSG
Jones, W J GW7CEQ
Jones, W J E GW0IVG
Jones, W L GW6MUP
Jones, W S A GW0KZW
Jones, W W G3CSL
Jopling, J A G0CXX
Jopson, B J G0UKP
Jopson, L V G6QA
Jordan, A S G0HAS
Jordan, B G7UNK
Jordan, B E G6NRZ
Jordan, B J G7SNT
Jordan, B W G4EWJ
Jordan, C K G0PZO
Jordan, C K M0AZF
Jordan, D A G3OVI
Jordan, D J G1TNR
Jordan, F A G4BZR
Jordan, I A G4GET
Jordan, J E G7HLV
Jordan, J W 2E1FJX
Jordan, K 2E1CJB
Jordan, K G7SNP
Jordan, L L G6GXE
Jordan, M G7NKE
Jordan, M E G4VAO
Jordan, M J R G7EVQ
Jordan, M P G4ASQ
Jordan, M V G1FVH
Jordan, N G0VSX
Jordan, P F G7NMJ
Jordan, S M G4GES
Jordan, T G4WCQ
Jordon, J R G7PPW
Jorgensen, D L 2W1ETN
Joseph, J L GW0UMO
Josephs, R J M1BYQ
Josko, S K G4ZVZ
Joslin, A D M1AAZ
Joslin, J G3NPY
Joslin, P 2E1EWG
Joslin, P M1BRR
Jouault, M L GJ4ZFM
Jow, H G4PZT
Jowett, D A G8FJR
Jowett, J H G3CFR
Joy, A K G8ZEW
Joy, B W F G0DJV
Joy, H M G4MUB
Joy, R J G0DRC
Joy, R J G0LSG
Joyce, A P G6REG
Joyce, C 2E1EWM
Joyce, C A G1NSQ
Joyce, D A G0FGJ
Joyce, E G8ELG
Joyce, H J G0IFH
Joyce, J T G4JTJ
Joyce, L J G0SZK
Joyce, R A G3WLM
Joyce, R A G4DDC
Joyner, B J G8YMD
Joyner, B J G8ZYZ
Joyner, C D 2E1ABP
Joyner, D M 2E1CTQ
Joyner, G R F G1PPG
Joyner, J G4DRH
Joyner, M J GM8FYJ
Joyner, M J J 2E1BTG
Joynes, A K J 2W1EUR
Joynes, H D GW7SNF
Joynson, K GM0GCF
Joynt, J F G4WVY
Jubb, A H G0WIN
Jubb, A H G3PMR
Juby, M A B G8RML
Juby, N R GW0TVK
Judd, B P G0RZM
Judd, G P G4ULQ
Judd, I D G8IDJ
Jude, S R G7SOZ
Juden, S G4PKK
Judge, A C G0NCW
Judge, A J G0PQF
Judge, A J G5ZG
Judge, G P GW7VKX
Judge, H F G7IWU
Judge, J D G8DGU
Judge, W K GW4KEV
Judges, F L G3TDU
Judkins, E G3OMJ
Judkins, P E G4DSM
Juett, D A G1TPV
Juffs, M R G4NVY
Juggins, J W G7VFA
Jukes, G W E G1HWZ
Jukes, I R G1MNU
Jukes, P G0CVO
Julian, H G G3UFX
Julian, M D G8CTL
Julians, M G G6ZBO
Jul-Christensen, F G4MJC
Jump, D J 2E1BKT
Jump, H K G7RGV
Jupp, B C G0SDE
Jupp, D V G6ITM
Jupp, L A G4MBB
Jupp, M R G1HWY
Jupp, P G G6UYY
Justice, J C G6ZFA
Justice, M E G1JMK
Justin, P G4AZL
Justin, S G6XUD

K

Kaberry, N G7EVW
Kaczmarek, J J G7GBJ
Kaine, J R G4RPK
Kaiser, R A G4MFE
Kajtowski, P G8VQU
Kalas, P A G3VCN
Kalawsky, R S G7NII
Kaliski, M P G0ULI
Kallar, I B G8WUD
Kaminski, H G1NBX
Kaminskyj, K G 2E1FLJ
Kaminskyj, K G M1BWY
Kamm, D P G7RFV
Kan, N Y C M0BFN
Kane, E K GM0MJR
Kane, H GI4ORK
Kane, J GM1MVK
Kane, J G GM0ODB
Kane, P G7ITF
Kane, R W GM4UWN
Kane, W GI7MBP
Kanelis, M G8ZFQ
Kapoutsis, C G6URT
Karande, A N G1TNP
Karazy-Kulin, M B G6PGQ
Karklins, E C GW6KRK
Karkoszka, J P G0RLY
Kasser, J E G3ZCZ
Kastner-Walmsley, J O G3HUI
Kathrens, M V G3WBX
Kathuria, V P G0KUA
Katoh, Y O S I G0GRV
Katz, J S G0TZX
Katzmann, M G G4NYV
Kautz, R C M0ASX
Kavanagh, B C G1RVT
Kavanagh, M R GM4VKI
Kay, A G1FOG
Kay, A G4SPY
Kay, A A G1RUG
Kay, A P G8UWG
Kay, A S GM6LSG
Kay, B G4VBJ
Kay, C R G4TIH
Kay, D E G0MXH
Kay, D J G4WWQ
Kay, G G4WMY
Kay, H G0FAB
Kay, H G4KIC
Kay, J A G4RJK
Kay, J D G3AAE
Kay, J J G6OBG
Kay, L G G0KBS
Kay, M P G1EOJ
Kay, P J D G3KOD
Kay, P N G0KUX
Kay, P N G1YJI
Kay, R G0LHV
Kay, R G3OQF
Kay, R E G3NSW
Kay, R S G0MZP
Kay, R V G4VBC
Kay, S G3OMA
Kay, T J 2E1FTZ
Kay, T S G4ZTK
Kay, W A GM8UGO
Kaye, A G8NDQ
Kaye, A K G4UUZ
Kaye, A N G8YHI
Kaye, B L G8LSG
Kaye, D R G6PGM
Kaye, M A G3WPQ
Kaye, R G0JEA
Kaye, R C G6RO
Kaylor, D J G4OBB
Kay-Newman, D M0AOD
Kaznowski, M G6OBA
Keable, D A G1ICA
Keal, B C G4HDU
Keal, J M G7HQQ
Keal, R F J G7LJM
Kealey, J P G3YJY
Keane, D J G0OFP
Keane, P A G4DUQ
Kear, C S J G4JHQ
Kearnes, R J G7HJK
Kearney, A G0UZU
Kearney, A G1IBM
Kearney, B C I 2E1ERR
Kearney, J J G0NHC
Kearney, M V 2E1ANN
Kearney, N F GI7TSA
Kearney, R J G3XOK
Kearns, B 2E1EII
Kearns, D E G0HVS
Kearns, D J G8TTI
Kearns, G G4MYA
Kearns, K G8LYV
Kearns, M A G4XAR
Kearns, T M G4GIX
Kearsey, H J G3WDR
Kearsley, B G0FOW
Kearsley, T D G4WFT
Kearton, D W 2E0AGR
Kearton, D W M0BDR
Keasley, P J G0JXR
Keat, G P G0GKB
Keates, A J G4TWN
Keates, D J G0DJK
Keates, R D G8LIX
Keating, J G G0ODO
Keating, M G0VKA
Keating, M R G4WVM
Keats, T R G4CCN
Keavey, P J G1UTW
Keay, D S GM1DSK
Keay, G S G4ELC
Keay, R D 2E1ECX
Keay, R D M1BFJ
Keay, W A B G3YXB
Kebbell, M E G4UWF
Keddie, D GM1RGM
Keddie, J W GM7LUN
Keeble, A B G4RUI
Keeble, A C G4HPU
Keeble, A P G0MJC
Keeble, C R G3TUU
Keeble, D G7WAF
Keeble, E G1EOK
Keeble, G G G6CDU
Keeble, J R M G8RZD
Keeble, K J G6ZYM
Keeble, P J G4ETC
Keeble Buckle, F J G8SVZ
Keech, A G4PPW
Keech, N G7VDV
Keechan, B E G0GFE
Keedy, B G6LIC
Keefe, R C G4SIS
Keegan, G J G6DGK
Keeler, C G0SCC
Keeler, D E G3KXI
Keeler, E G G4FPM
Keeler, J H G4EZN
Keeler, J H G6UW
Keeler, P S 2E1DTD
Keeler, T M M1BCY
Keeley, J G0MFB
Keeley, J J G0FYS
Keeley, R G8IMM
Keeley, R J G6RAV
Keeley-Osgood, R A G0GIA
Keeling, B M G4EUW

Keeling, C D G7JDZ
Keeling, J M G0WQC
Keeling, T G G6OBD
Keely, D T GW0OGI
Keen, A G7DWN
Keen, A M G7CSX
Keen, C F D G8BYC
Keen, D B J G3ONK
Keen, D J G7DQZ
Keen, D S G3JVN
Keen, F A H G3MYE
Keen, K J G7FIM
Keen, M G G3PBQ
Keen, N E G7OPO
Keen, P G0VGY
Keen, R D G7OOW
Keen, S M G4PWS
Keen, W A G4XPB
Keenan, A GM4SFA
Keenan, J GM0WFB
Keenan, J H GI4WAH
Keenan, J J GM7VET
Keenan, K G0XKK
Keenan, K G7NXA
Keenan, M P J GI4SZW
Keenan, P A T R GI0PTQ
Keenan, S C GD4MDY
Keene, D 2E1FEJ
Keene, G E G7EVF
Keene, G W G6BZP
Keene, R G7CDP
Keens, C A G8KYK
Keens, C F G8NDN
Keep, R E G7SPE
Keepin, K J GW7UMS
Keeping, C G4PTF
Keeping, M W G8AVZ
Keers, R G0VSF
Keeton, L G0JNT
Keeys, W G7LAX
Keggen, G E G4VBR
Keighley, L G3FLV
Keighley, P J G0KPH
Keighley, R G4ARL
Keight, D T G8RAA
Keightley, N G0BNR
Keightley, S M1BFN
Keightley, S J G3ZZL
Keir, A G4KZO
Keir, D A G0BVB
Keir, J T GM4ISY
Keiser, B E G0PAA
Keitch, K P G0IFA
Keith, D C G3RQF
Keleher, J M G8PUN
Kelk, J J G4JMP
Kelk, V W G1FDL
Kell, P E G7GCF
Kell, S J G4KEL
Kellagher, J J GI1SLZ
Kelland, C L G0JEK
Kellaway, G J G3RTE
Kellaway, H S C J GW3CBA
Kellaway, J O G4NUJ
Kellaway, T A G4KHL
Kelle, A H G4AYB
Kelle, H J G1PAS
Kelleher, A G0WXE
Kelleher, M B G7VHK
Kellett, L P D G8KMH
Kellett, M G4RHC
Kellett, M A G8HMJ
Kellett, T G3EGF
Kellett, T A G6DSW
Kelleway, M F G4XIU
Kelley, C I G1GAS
Kelley, C J G4YIE
Kelley, G W G5KC
Kellingley, P C G7HOK
Kellow, T G G3ZHK
Kells, C G7UYU
Kelly, A G4LVK
Kelly, A N G4TYD
Kelly, C J 2E1EYW
Kelly, C P G4KPP
Kelly, D G M0BGD
Kelly, D M G4HXN
Kelly, E GM0RJG
Kelly, E I GD4RFW
Kelly, F G4TFV
Kelly, G GD4NTR
Kelly, G A R G6OWT
Kelly, G F G3CBF
Kelly, G R GM8MST
Kelly, G T G4FQN
Kelly, J GM0SYV
Kelly, J A G4JAK
Kelly, J A GI6MYQ
Kelly, J C G0HMZ
Kelly, J C G0KPT
Kelly, J C G1OGC
Kelly, J E G3YGG
Kelly, J F GM3TCW
Kelly, J J G0SQJ
Kelly, J P G1NDU
Kelly, J W G0MRK
Kelly, J W R GM1VKI
Kelly, K G3VKF
Kelly, L A G7OPY
Kelly, M B G6EIG
Kelly, M I GI4MEQ
Kelly, M M G1NMR
Kelly, M M G4XGP
Kelly, M V G0PHN
Kelly, M V G G0WON
Kelly, N A G4UUF
Kelly, P GM4ENK
Kelly, P A G6ITO
Kelly, P B GI4BGB
Kelly, P D G3SDH
Kelly, P G GW7TZG
Kelly, P J G6VWZ
Kelly, P L 2E1DSK
Kelly, R GM0GRD
Kelly, R P G6GAD
Kelly, S GI7TOX
Kelly, S A G4USM
Kelly, S E GD3HDL
Kelly, S P G6HDF
Kelly, S P GD7DUZ
Kelly, T G3MOK
Kelly, T G G3MNN
Kelly, V B GW4ACO
Kelly, W M GD0JBL
Kelsall, G G7UTR
Kelsall, P I G0CYB
Kelsall, R 2M0ABN
Kelsall, R G G0BDW
Kelsall, R H G4FM
Kelsall, R J G1VBL
Kelsall, S L G1IUE
Kelsey, B G1AFK
Kelsey, P G1MLV
Kelsey, T G G3ASJ
Kelsey-Stead, W A G0COQ
Kelson, P M GW0DXQ
Kemble, J P G0XJK
Kemble, M G0JMK
Kemble, P 2E1CWN
Kemble, P R G3UYK
Kemley, J G7HYH
Kemmis, P B G4MGI
Kemp, B B C G1VTO
Kemp, C J G7VCM
Kemp, C W G7SSS
Kemp, D E G0POQ
Kemp, E N M0APW
Kemp, F W G4JEO
Kemp, G G4DSA
Kemp, G C G0EIR
Kemp, G C G4WGK
Kemp, I A G3MGK
Kemp, I P G4AEG
Kemp, J G4OWD
Kemp, J A G0WJH
Kemp, J E M0AAR
Kemp, J H G7FCY
Kemp, J T GW8SBN
Kemp, L A W E G4DXL
Kemp, L R G7BHH
Kemp, M J G1CWI
Kemp, O W G4TLK
Kemp, P K GW4JPN
Kemp, P T G4XBJ
Kemp, R A G4VUW
Kemp, R E G3VWL
Kemp, R G G0SEL
Kemp, R J G4ZEV
Kemp, R L G G3YYF
Kemp, S G6XAU
Kemp, S J G1KID
Kemp, S P G7IMY
Kemp, T G6JDC
Kemp, W L GM0BBR
Kemp, W P G8WWY
Kempe, C W G4MLX
Kemplen, D F G1NSV
Kempson, H E G3BHM
Kempster, A W G7HUZ
Kempster, R G0MML
Kempton, A C G1BYS
Kempton, D C GM0DGL
Kempton, R C G6CYT
Kemsley, R E G3FUN
Kench, J G8JTQ
Kenchington, J K GW8ZZD
Kendal, W B G3GDU
Kendal, W M G4MKD
Kendall, D S G3ILB
Kendall, I J G6ARO
Kendall, J G4HLK
Kendall, J A G7WFV
Kendall, J N GW8JKC
Kendall, M G G0EMK
Kendall, M G G4JXG
Kendall, N H G8SVO
Kendall, R G3KYS
Kendall, R E G G8BNE
Kendall, R L G0UPU
Kendall, S G0MOM
Kendall, T R G4LDB
Kendall M.Ed, N J G4JNK
Kendrick, A W G3RDW
Kendrick, C R G0STW
Kendrick, M S G6CYU
Kendrick, R N G0SJT
Kendrick, S G0POE
Kendrick-Finn, M G6BOI
Kenington, P B G1ISR
Kennard, D A J M0ADO
Kennard, D P G3YEN
Kennard, H E G G4TLE
Kennard, J I G6BDH
Kennard, L J G0IVR
Kennard, L J G3ABA
Kennaugh, A M GD6HCB
Kenneally, B G J G6AJT
Kennedy, B G3ZUL
Kennedy, B M GI7ALP
Kennedy, D G3VZE
Kennedy, D G6DHI
Kennedy, D A G7GWF
Kennedy, E G6YES
Kennedy, E D GM0VMV
Kennedy, G C G0IEQ
Kennedy, G R G3OGK
Kennedy, I T GI7ALQ
Kennedy, J B G3CLE
Kennedy, J E G1HSW
Kennedy, J E G4YEX
Kennedy, J S G4DND
Kennedy, L A G4TEP
Kennedy, M 2E1DGA
Kennedy, M J H G1YHG
Kennedy, M O D G1KSQ
Kennedy, M W G6ADG
Kennedy, N S GM8MFZ
Kennedy, O H G3OCS
Kennedy, P 2E1FTU
Kennedy, P B G0EWD
Kennedy, P B G1CCX
Kennedy, P J 2E1BDC
Kennedy, P J G4BVQ
Kennedy, P J GI1YSG
Kennedy, P S G1CLJ
Kennedy, R G7RKO
Kennedy, R A GI1WYZ
Kennedy, R C G8UMB
Kennedy, S H G0FCU
Kennedy, T GI7HVC
Kennedy, T A G0PLK
Kennedy, W J G3MCX
Kennett, D A G7CMB
Kennett, M G4OVX
Kenney, R G4LYO
Kenney, T P G1FAD
Kennion, G W G8KWW
Kenny, J M G7RKH
Kenny, R A G1NNF
Kensall, M J G7FMI
Kensall, R E 2E1CPF
Kent, A F G0FMD
Kent, A F C G3DSZ
Kent, A J G3THG
Kent, A J G8PBH
Kent, A M G4ZSC
Kent, A N G6EXZ
Kent, B G G6MUW
Kent, D A G7LIW
Kent, D W G7PJH
Kent, E G1HCC
Kent, F R G3PIH
Kent, G T G G0VNA
Kent, K M G1SCQ
Kent, K M G7RNA
Kent, M A G7GXF
Kent, M J G3PSS
Kent, M J G4SUK
Kent, N M G4ABG
Kent, P N H 2E1BBO
Kent, R C G6MUV
Kent, R F G7TNP
Kent, R J G0ROS
Kent, R M G3KCF
Kent, R S G4POY
Kent, W E B G3YCN
Kentch, C S G0FJY
Kentfield, H D G1UID
Kentish, R GW4JRK
Kent-Woolsey, P G G6ULS
Kenward, A B K G4WPD
Kenward, R A G6WTD
Kenward, R C G4APU
Kenworthy, J G4YWO
Kenworthy, N J G0WEF
Kenyon, A V GW4DOO
Kenyon, C M0ACJ
Kenyon, E G GW0KTE
Kenyon, G G G3HMF
Kenyon, I G4VAP
Kenyon, P L GW8CAK
Kenyon, S M G8YHF
Kenyon, W P G0EIZ
Kenyon, W R G6TTX
Kenzie, B W G4PDI
Keohane, A G3XUC
Keon, A J G0DLF
Ker, J GM0ICP
Kerby, R G0CHK
Kerins, J GM1NEW
Kermode, B M G8ESK
Kernaghan, H D GI3USK
Kernahan, N W G1EKU
Kernohan, H GI0JEV
Kernohan, J N GI0RBO
Kernohan, W GI4OGQ
Kerr, A G G6GOG
Kerr, A G GI4IVI
Kerr, A G GM3KBP
Kerr, A J M G4WJZ
Kerr, A K A M1BWS
Kerr, A P G0IMC
Kerr, D B G0DBK
Kerr, I A G1RPY
Kerr, J G4DJN
Kerr, J A G6RLG
Kerr, K M GM4YXI
Kerr, P I G1TAL
Kerr, R B GM4FDT
Kerr, R B GM4MFL
Kerr, R E GI0RUC
Kerr, R H GM4OZZ
Kerr, R W GM0CBX
Kerr, S T M1BXT
Kerr, T W G4BDN
Kerr, W A GI1HQU
Kerridge, D G7JXL
Kerridge, R G G1FPK
Kerrigan, J G0TXW
Kerrison, A R G0OEY
Kerrison, B J G7GAZ
Kerrison, S A G3MFQ
Kerry, M J G4ATG
Kerry, M J G4BMK
Kerry, M J GW1SXT
Kerry, P G6LSD
Kerry, P L G3SFK
Kerr-Munslow, D C G7IEA
Kersey, A G0IBN
Kersey, E V G4LBU
Kershaw, C N G8WFP
Kershaw, M L G4GUI
Kershaw, N G0CBH
Kershaw, P S G6EYA
Kershaw, R G4PJE
Kershaw, T G7MMM
Kershaw, V F G3GKI
Kershenbaum, E G0ERC
Kerslake, J M GW0OJK
Kerstein, N G4BPN
Kerton, A K G3CJR
Kerton, P G G0EOZ
Kessel, J E G7RWH
Kessel, M J G4WJX
Kessler, A S G3XEP
Kesterton, E G D G6HGK
Keston, D J G8FMC
Ketley, H G1JGY
Kett, A R G8VLL
Kett, K B G4NVP
Kett, N S G6ARM
Kettlety, A G8HTN
Kettlewell, T T G4TWR
Kevern, M M1BKS
Kewell, T S D G0JDM
Kewn, J A G7TBJ
Key, B J G6PGO
Key, C G4TOF
Key, C R G8VVR
Key, J P G4SXX
Key, L G0FQX
Key, M D G3WHG
Key, M J G4JQF
Key, N G G6LSB
Keye, A G6ZYP
Keyes, R S GW4IED
Keyes, D G GI0LDI
Keyser, I H G3ROO
Keyser, M 2E1DFH
Keyte, B H G3SIA
Keyte, P F G1AFJ
Keyworth, A G4MWL
Khachaturian, A G7GJN
Khalaf, M S G4KRD
Kharbanda, S R R G2PU
Kibblewhite, K C G0HER
Kidd, A M0ACO
Kidd, C J C G3YTQ
Kidd, C S G4GTW
Kidd, D R G3XNK
Kidd, J A G1MMX
Kidd, K M M0ACN
Kidd, M N G6JVO
Kidd, P L GM0LPK
Kidder, G D G3NZO
Kiddle, A R G4HVC
Kiddle, A R G4WRS
Kiddy, A E G6THH
Kiddy, M G7OEI
Kidgell, J W G0FKV
Kidman, H M G1ARZ
Kidman, M R G7LAF
Kidman, M R GW3SDK
Kielthy, M J G7CIK
Kiely, D P G0RBD
Kiely, T 2E1FVK
Kier, E J G1DTS
Kier, T G0OOP
Kierman, J G G1HQW
Kiernan, S G4ROI
Kiff, H G0DOC
Kiff, P G7AAP
Kift, G W J GW0LDQ
Kightley, A H G3MZZ
Kilbride, P A G8RDY
Kilburn, W F G1PMG
Kiley, D A G8NQF
Kilgannon, J K 2E1FAZ
Kilgore, R J GI0WYO
Kilgore, R J MI0AKU
Kilgour, G A GM7KVU
Kilkenny, I G7FEG
Kilkenny, M V G1IUF
Kill, C E G1OQI
Killeen, J R G3KPV
Killeen, W G1VHW
Killgren, N A G8EJP
Killick, B J G3GPQ
Killick, H P G1TXV
Killick, K D G0SNM
Killigrew, K S G6DZH
Killip, E L G0APZ
Kilminster, J H G7TPB
Kilminster, R G3JHK
Kilmister, J G0CQJ
Kilmister, S R G6MOT
Kilner, A E GW4EJT
Kilner, E G1KZA
Kilpatrick, H GM1ASA
Kilpin, P R M1BQA
Kilroy, J A G4PBC
Kilvington, C W G8EPH
Kimbell, G F G3TCT
Kimber, B L M G1SGS
Kimber, B W G1NVO
Kimber, D P G8HQP
Kimber, H G G3HAC
Kimber, K G G4ZXF
Kimber, M A G8YYW
Kimber, N J G7CWZ
Kimber, P G1HSX
Kimber, R J G1BZW
Kimberley, R G4VHT
Kimble, A N G4BFM
Kimblin, S G6OBE
Kimm, A J G4YMQ
Kimmings, A G1HCE
Kimmitt, M H G G4GOO
Kimoto, Y G0TDX
Kimpton, J N G4WAO
Kimpton, R P G6GUN
Kin, R G7BOB
Kinal, G V G0GZZ
Kincaid, A A GI4TOR
Kincaid, J GM4WSY
Kinch, J A D 2E1FSF
Kinch, N L G3XEE
Kinch, R T 2E1FSG
Kind, S S G4AYP
Kinder, G C G0XGK
Kinder, M E G3NVV
Kinder, M G G0CZD
Kinder, M R G0KYB
King, A G0IAG
King, A GM0OTU
King, A M0ANL
King, A A A G8NME
King, A B G7ANY
King, A D G6KTX
King, A E G0DDJ
King, A H G G3YEZ
King, A J 2E0AJS
King, A J M0BDG
King, A K G0IRL
King, A N GM4GJG
King, A N M0AYX
King, A R G3SNE
King, A R J GI6BDI
King, B D G0AQE
King, B E G8ARA
King, B R H G3SGK
King, C A G7LQN
King, C B G3CEG
King, C J G6MYT
King, C J G6PGN
King, C K GW1VRW
King, C L E 2E1DPT
King, C M G7NBL
King, C R G7JAO
King, D A G0NNP
King, D C G1PWF
King, D C G8GFY
King, D G G3EON
King, D G GM1FPD
King, D G N G3PFS
King, D H E G3TQN
King, D I G4EOD
King, D M G0DMK
King, D M G6KWA
King, D M GM6OOA
King, D R G0IJE
King, D R G0RWJ
King, D S G1EEL
King, D W G4BBQ
King, E F G4HWC
King, E J G3DCC
King, E M GM1ATW
King, G A G1VQH
King, G A G3UVC
King, G A G6WIM
King, G A M1ANY
King, G C G4FTL
King, G E G1HXN
King, G J G1TCF
King, G M G3MY
King, G M G8BJB
King, G R A G3XTH
King, G W G7UAR
King, G W H GW3ENN
King, H G0KMV
King, H G C G4HKA
King, H K G8DXV
King, H L G1TMQ
King, H S G3ASE
King, H W G8YCG
King, I A G6VIK
King, I R G8VTX
King, I W G3VVU
King, J G4EMC
King, J GW0CTK
King, J A G0RSA
King, J A G7VTT
King, J A F G4AND
King, J R 2E1FOX
King, J W G0JIM
King, J W G6JIM
King, K G G3RGE
King, K G G4RS
King, K P J G7ANJ
King, K R G8LFC
King, K W G3ACB
King, L G0VTC
King, L H G3CML
King, M G3XKD
King, M G7SFD
King, M A G6OWX
King, M J G4GLI
King, M J G6FTC
King, M P G0RAM
King, M R G8XDX
King, M W J G1XPW
King, N C G3DNS
King, N C G8NGM
King, N D G0OHK
King, P G1KFQ
King, P G3WKP
King, P A G2RSA
King, P B G6BOK
King, P D G8KJP
King, P E G6RYF
King, P G G0OIK
King, P J G1XXW
King, P J G3PVA
King, P J A G7IFL
King, P L G4GFY
King, P M G4JXE
King, P R G0RTV
King, P W G1EEM
King, R G0AKL
King, R G4XJK
King, R G6CYA
King, R A G4JYA
King, R G G4RLK
King, R H G4NIZ
King, R H G7AJX
King, R H G7MQS
King, R J G0TDV
King, R J G0THQ
King, R L G4VXD
King, R L G6XWM
King, R M GM7BOW
King, R R G0VSS
King, R S G7IMV
King, R S G8CHK
King, R S S M1ACV
King, S L G1LHQ
King, T G4ZUM
King, T R G1FDO
King, W G1FYN
King, W J G0WIY
Kingdon, G K G4RTX
Kingdon, G R G6RNT
Kingdon, M G G4XYB
Kinger, M E G4UMS
Kinghan, A J GI4OZI
Kinghorn, D A M1BOJ
Kinghorn, H W G1EOM
Kinghorn, J T GM4DQK
Kinghorn, P G1UDL
Kings, A GW0HBD
Kingshott, D J 2E1EHD
Kingshott, R J G4CFN
Kingsley, N G3RCB
Kingsley-Williams, P C G1PKW
Kingston, R D G4RHM
Kingston, R K G0PRB
Kingstone, R A G4HHB
Kinnell, G GM7NQP
Kinnell, J T GM1RZB
Kinnersley, A G7OSO
Kinnes, R GM4ALK
Kinrade, D H GD4EBA
Kinrade, R E G6POC
Kinsella, B S G7KIN
Kinsella, J M G0UGX
Kinsella, M P G1RNL
Kinselley, N E G1BYT
Kinsey, P G6ONZ
Kinson, A A G0KOC
Kinton, S N G6PGP
Kinvig, L M GD0ADV
Kipp, M A G4FBK
Kippax, A E G1JXU
Kippen, K F G3KFK
Kipping, D M G0ACL
Kipping, M A C G1GAR
Kirby, D GW0PLP
Kirby, D G G6ZYQ
Kirby, D P GW4SZV
Kirby, G L G0EEL
Kirby, G M G0PXM
Kirby, H E G1HXO
Kirby, I S G6JPI
Kirby, J G3JYG
Kirby, J C G4FZN
Kirby, J D G0GXB
Kirby, K H G0ECR
Kirby, K H G4VKK
Kirby, L G0SRC
Kirby, L G4CRT
Kirby, L GW0ULC
Kirby, M J G1EEO
Kirby, P G3XUD
Kirby, P J G8HQW
Kirby, R J G7ODX
Kirby, S J 2E1DZK
Kirby, T H G4VXE
Kirby, W E G6FGE
Kirby-Parkinson, G G0NEL
Kirk, A H G8TTQ
Kirk, C G1SCX
Kirk, C H G0CKY
Kirk, D G3GTW
Kirk, D A G4NXA
Kirk, D E 2E1FHX
Kirk, D F G7RIU
Kirk, G C G4FKG
Kirk, H G1ADE
Kirk, H N G3JDK
Kirk, I G6URR
Kirk, J J G3ZDF
Kirk, M A G0JPZ
Kirk, M A G1LBK
Kirk, M B G0MYZ
Kirk, M E G4UCZ
Kirk, P J G0IYU
Kirk, S A 2E1FHY
Kirk, S M G7DCK
Kirk, T G3OMK
Kirk, W G1VWH
Kirkby, D J N G7DVI
Kirkby, J R G0DSH
Kirkham, A G7TMM
Kirkham, A H G7MQF
Kirkham, D J G8SFU
Kirkham, D N M0ANG
Kirkham, G V G7OAP
Kirkham, M J G8RHN
Kirkham, P W G6CYV
Kirkham, W C G4WTP
Kirkland, A G G4HDO
Kirkland, C F G1VIN
Kirkland, C G J M0ATY
Kirkland, G C GM0KDO
Kirkland, J G0UNM
Kirkland, K H G7BOR
Kirkland, K L GM4HCE
Kirkman, C G7VPW
Kirkman, C P G1JVH
Kirkman, J R G3RDI
Kirkman, J R M1AKP
Kirkman, M E G7VPV
Kirkman, N A G7KIE
Kirkpatrick, B GW8OKR
Kirkpatrick, G C G1PXG
Kirkup, P G8YVU
Kirkup, P N G0RTU
Kirkwood, D M G3YQO
Kirkwood, R I G4RIK
Kirk-Bayley, K GJ0KKB
Kirsch, R J B G8CJG
Kirsop, P J G0BAA
Kirsop, P J G4WCE
Kirtley, N L G3RQR
Kirton, D G G4EHR
Kirton, J E G8WWJ
Kirushnamoorthe, M M1BEL
Kirwan, S J G0WFE
Kisiel, J T M0AZG
Kissack, A C GD0TEP
Kissack, A C GD7MAN
Kissack, B V G3MTD
Kissack, G A GD1ASB
Kisselev, K M0BDQ
Kissick, R L J G3RJK
Kitchen, A G7COD
Kitchen, B G4GHB
Kitchen, D G3KVP
Kitchen, D F G0FCL
Kitchen, I G0WZM
Kitchen, S P G7UEJ
Kitchener, C J G4LYB
Kitchener, J H G7UYW
Kitchener, M G G6LSC
Kitchener, R W M G4IKQ
Kitchener, S M G6DZJ
Kitchin, J B GW3JUV
Kitching, A G0DHS
Kitching, L G3LEK
Kitching, R C G3XVS
Kitching, W F G4FBZ
Kiteley, D F G7SCZ
Kiteley, M P G7WDC
Kitson, D G3TRK
Kitson, G L G4ZAD
Kitson, J G 2E1DMP
Kitson, K M G0BWQ
Kitson, M J G1MBM
Kitson, P D G7GXE
Kitson, V P 2E1DMO
Kittle, A D G4GWT
Kittrick, E D G0FEV
Kitzmann, R P G1OXL
Klatzko, L S G0LZU
Kleeman, H G4BHY
Kleijn, J W G0VLP
Klein, H GD0VKS
Kliffen, J A G0ACA
Klose, H GI0RJY
Kluger-Langer, M J 2E1BCC
Klunder, J R G7KTQ
Knaggs, C M G0LYZ
Knaggs, C M G0UDD
Knaggs, D G8CLV
Knapp, F GM4JRP
Knapp, G C C G3NMJ
Knapp, R M0BEV
Knapp, S G4JRQ
Knapper, P M G7OAL
Knapton, P M0AIA
Knatchbull, H N G4VLP
Kneale, F E G4DBG
Kneale, H O G7MIP
Kneale, J D GD0BFN
Kneale, J R GD1MAN
Kneebone, A F G6CEP
Kneebone, B L G6FQP
Kneebone, M L 2E1CXC
Kneebone, P G G7EKD
Knell, A J G0BNE
Knell, M P G7GPA
Knell, S R G0CKP
Kneller, A B M0AFG
Kneller, C I G0UPH
Knibb, E C G4ZJR
Knibbs, D GW0IRC
Knibbs, K E G1GHG
Knibbs, V D G3JNZ
Knight, A 2E1FBU
Knight, A M B G7GDJ
Knight, A R G2HKQ
Knight, B G0ODT
Knight, B G7DMZ
Knight, B L G7MSC
Knight, C S GM0CFK
Knight, D G6MTG
Knight, D G B G2DFL
Knight, E G G4NVD
Knight, E J G7UBB
Knight, F H GD3VEM
Knight, F J G4GAN
Knight, G A G3XRD
Knight, G J G4ZXK
Knight, G W E 2E1EVQ
Knight, H T G0EXP
Knight, J C G3OUP
Knight, J C G3TPB
Knight, J F G3MTK
Knight, J H G1PPK
Knight, J R G3JRK
Knight, J R GW0CYI
Knight, K G1ASD
Knight, K G G4DFZ
Knight, L 2E1EOA
Knight, L G2DXK
Knight, M A R G7OZU
Knight, M B G4BNW
Knight, M J G7FWL
Knight, M M I G6YET
Knight, P G4BMM
Knight, P G4RDD
Knight, P A G6EPN
Knight, P J M0BDT
Knight, P T G0SDG
Knight, P V G4CEC
Knight, R G0AXP
Knight, R A G3NOK
Knight, R H G4VML
Knight, R H G6JPJ
Knight, R K G8YDA
Knight, R L G3DPW
Knight, R M G3VUK
Knight, R P G8ZRQ
Knight, S A G1UBN
Knight, S J G0BGI
Knight, T G2FUU
Knight, T G7VHA
Knight, T M0ALR
Knight, T A G4FHK
Knight, V G1REU
Knight, W G7HMW
Knighton, J D G1DVH
Knighton, R G0GER
Knighton, R J G4NYB
Knights, B C M G1XUG
Knights, D J G0MPK
Knights, K D P G7VNP
Knights, M G3NJF
Knights, M R G0WBH
Knights, M T G3TQY
Kniveton, K V G4IDU
Knock, G J G4FTX
Knock, R G G8SNQ
Knorr, R D GM4MYL
Knott, A D 2E1FME
Knott, A R G6KSN
Knott, B 2E1FIO
Knott, C G3WMX
Knott, D S G3UYL
Knott, H F G3CU
Knott, J K G6POE

Knott, J R GW7BEY
Knott, K V T G6POD
Knott, M J G0WCR
Knott, M P G0ICE
Knott, P S GI1KGZ
Knott, R E GW4VZJ
Knott, T G7LXU
Knowler, A J G0LNP
Knowler, D J G0XXX
Knowler, D J G4SFB
Knowler, W A G7AZM
Knowles, A F 2E1EON
Knowles, A G G3LUA
Knowles, C G6FWT
Knowles, C M0ABO
Knowles, C J G G4TBZ
Knowles, D E GW3UVA
Knowles, E G2XK
Knowles, F G3AKI
Knowles, J G0KWE
Knowles, J C G1ASE
Knowles, J D G3RPA
Knowles, J H G2FXS
Knowles, J H G7BWD
Knowles, J S G4OFP
Knowles, K G4FPN
Knowles, M P G1ZOY
Knowles, M W G7GPB
Knowles, N J G6SDG
Knowles, P G G4YPK
Knowles, P J G7DEY
Knowles, R G0LZX
Knowles, R J G4HQA
Knowles, S V G3UFY
Knowlson, C R V G0OPG
Knowlson, G T G0GBF
Knowlson, M B G7KHE
Knox, D E G4UFN
Knox, G F G0IGK
Knox, J B GM0MHV
Knox, J W G3GMT
Knox, R GI3WEL
Knox, W J G7BIL
Knutton, K G6KZR
Kobiela, G D G6XRY
Koch, A C G1DGY
Koenen, B H N G0MWQ
Koenig, J T G1XRQ
Koester, M C G0JEX
Koeze, T H G0THK
Kok, F C S G7NSP
Koker, P C G3YPU
Kolbe, G GM4LPJ
Kolker, A M J G0EZF
Koncz, S GM0WRY
Konemann, B G7VFF
Konos, W G GI4TUJ
Konowicz, R J G0YYY
Koopman, D J G1TLH
Koops, J B G0VZO
Korda, A G3WKS
Korda, A G4FDC
Kornreich, J H MI0ARN
Kostryca, D G8WEI
Kostryca, D M0AYF
Kostryca, W G4VUE
Koszegi, G Y G7EUV
Kotowicz, A J G6UZA
Kowalczyk, D T G8NTZ
Kowalczyk, Z G4GCU
Kowalski, G J G0RDG
Kowalski, M G 2E1FOB
Kowalski, R M1AKV
Kowalsky, M D G0VYH
Kozma, D E G1WQI
Kozma, K P G4VKA
Kozminski, J J G8CPB
Kraft, E G4FTP
Krarup, A C G3TAP
Kratzer, C G G0SLY
Kraus, H G G1PZK
Kravchenko, V G0KBO
Kraven, I J G0ICK
Kraven, I J G4JIJ
Krelle, K J G8MMO
Kremer, M G8VLN
Kressman, R I G3SIT
Kreuchen, K D O G0PER
Kroll, A R G G1LPN
Krom, J G G1RLZ
Krzymuski, J G4DQW
Kubiesa, J 2E1DIO
Kubiesa, S G7SVD
Kugler, R J G8VQS
Kuik, M G1SHI
Kuipers, J H G4GUX
Kunzler, P A G0TJX
Kurian, P P G4HYT
Kurlak, O G8NWP
Kurnatowski, A R G G4XTK
Kurtz, R E 2W1EDA
Kusin, M E GM0PDQ
Kusin, V J GM4HCO
Kuss, C W G0DZI
Kutscherauer, H C G3EYC
Kyle, A S G8TBV
Kyle, D MI1BLZ
Kyle, I J GI8AYZ
Kyle, I J MI0AYZ
Kyle, J D G6PGJ
Kyle, J K R M1AIO
Kyle, N S R GI4CQL
Kynaston, J A G4AHZ
Kynaston, J M G1ZLB

L

Labourn, M 2E0AOT
Labron, P G0DWO
Lacey, C G0IMX
Lacey, D J G4JBE
Lacey, F A G4NSW
Lacey, G A G4JFZ
Lacey, J E G3GLB
Lacey, R G8XNC
Lack, M S G7KYL
Lacy, A G4AUD
Lacy, D L 2E1FTJ
Lacy, G D GW6JJV
Lacy, K M G3ZON
Laczko, I GM0BQQ
Ladd, L J G7LUQ
Laddiman, S M G7GTQ
Ladds, C M G8PCS
Ladley, C E G1PRS
Ladley, T R 2E1BAE
Ladner, L G G1IVO
Laepong, S 2E1BSF
Lafferty, C G4KDS
Lafferty, P B GM0HWB
Laffin, J 2E1DIA
Lagar, R G0RFT
Laidler, W G0WLA
Lainchbury, J A G4XIQ
Lainchbury, J B G1IJJ
Laing, A W T GM1MFD
Laing, C J 2M1CFS
Laing, E P G0OAI
Laing, J G6KLQ
Laing, J S G8ODT
Laing, R T G3TXT
Laing, S G4MQC
Laird, C J G4UCY
Laird, F G G8TLU
Laird, M R G1ZMV
Laird, P J MM0AUP
Laister, D A H G0RTT
Lait, P E G0IFQ
Lake, A A 2E0APR
Lake, A A G8YOA
Lake, A E G4DVW
Lake, B M1BPU
Lake, C GW0LBJ
Lake, D G3ZCA
Lake, F I G4YXS
Lake, J E 2E1GBM
Lake, K G8ZIP
Lake, R G8MFH
Lake, S G G1WFG
Laker, J A M1BIY
Laker, P J E G1INB
Lakey, B L G1FWZ
Lakhaney, H G0DXI
Lalley, A E G4ZFK
Lam, T P G0REU
Lam, W G4KPG
Lamb, A D MW0AMT
Lamb, D H G1IZA
Lamb, D T G0ACK
Lamb, E W G0HGS
Lamb, G J G7GMU
Lamb, G W G0NDG
Lamb, I C I G6LD
Lamb, J E G1GUW
Lamb, J F G1FQX
Lamb, J G G3GXW
Lamb, J P G0WXG
Lamb, J R G1IVP
Lamb, K G G4BUW
Lamb, M A G7PSL
Lamb, M J GW0DSP
Lamb, N G1BMN
Lamb, P R G3VRW
Lamb, R G4TVX
Lamb, S J G0VOA
Lamb, S V G8FZZ
Lamb, V GW7AIY
Lambarth, R F G8HAU
Lambe, A G4BRU
Lambe, J V G7MNK
Lambert, A G8HER
Lambert, A D G7JAF
Lambert, A D G8TNU
Lambert, A F G0UIX
Lambert, A M G8OXS
Lambert, B M G4YJA
Lambert, B W G6FNA
Lambert, C E G7VNL
Lambert, C J G3TA
Lambert, C J G4PCN
Lambert, D G8HGL
Lambert, D M G1ABA
Lambert, D R G4POI
Lambert, D W G7UCV
Lambert, E C G3FKI
Lambert, E P G4AYL
Lambert, G G0IXT
Lambert, G R G3TUO
Lambert, H W G0TLV
Lambert, I R G4LWG
Lambert, J A ' G3FNZ
Lambert, K G7BVL
Lambert, K A G4SLE
Lambert, L GM0ASY
Lambert, L GM0COD
Lambert, L J G1XEP
Lambert, M M G3XGZ
Lambert, N J G7HCO
Lambert, N M G7BFH
Lambert, P G8XNO
Lambert, P F P G3CYX
Lambert, P J G1LSZ
Lambert, P M G0TGK
Lambert, R G3OFQ
Lambert, R S G6GND
Lambert, S A 2E1EIE
Lambert, S R 2E1BYU
Lambert, T G G8EZL
Lambert, W E G1LUR
Lamberton, R G4LAM
Lambert-Hutchinson, C N R G7SVI
Lambeth, C P G1YDI
Lambeth, M E G2AIW
Lamble, J R G6NGA
Lamble, R M G6YGR
Lambley, G S D 2E1EXN
Lambley, R J G G8LAM
Lambourn, E H G0SHB
Lambourne, E T G0OLG
Lambourne, R W G1FXC
Lamden, D P G7RJW
Lamford, K J G6ODT
Laming, M G0FFG
Lamkin, R M G8WDX
Lamming, P A G3NXL
Lamont, A G4KDE
Lamont, D G3WPV
Lamont, J G4MLQ
Lamont, R I GM4LYQ
Lamont, R S GM4KVY
Lampard, S G6AJX
Lancashire, W E G4SSL
Lancaster, B R G6YCW
Lancaster, N H G6NII
Lancaster, P G7NVI
Lancaster, R G0CFH
Lancaster, R S G3CFG
Lancaster, T G G1EOO
Lancastle, M R G7VYZ
Lancefield, G G3DWQ
Lancefield, G G3KUE
Landen-Turner, G L G0OXA
Lander, C D G4DDA
Lander, D H G4LQL
Lander, K H G3LCX
Landin, D M GW0VFQ
Landon, E J G3MHT
Landon, K H G3IXI
Landon, L B G3IBE
Landor, C P GJ4YLP
Landor, E C G0IPO
Landor, P M GJ3AME
Landrebe, K G G4YDJ
Landricombe, L H G0KYE
Lane, D G3POZ
Lane, D G6AVY
Lane, D A G3VOM
Lane, G E G4PZM
Lane, J E G8XNA
Lane, J F G4WVW
Lane, J R G2CJL
Lane, M G0SHC
Lane, M A G0TMP
Lane, M A G3VOV
Lane, M T G6YGV
Lane, N GI6FOR
Lane, P G1MNX
Lane, P G3MQX
Lane, R G4AWU
Lane, R D G7JPP
Lane, S F GW4PRP
Lane, S G G1HRF
Lane, S R G3REZ
Lane, S R G4GSP
Lane, V W G4FRA
Lane, W M G3MWL
Laney, R E R G4RAE
Lanfear, A D G4CQI
Lang, D H A G6NAG
Lang, G G1BGK
Lang, G H G3ETY
Lang, G M G0OBQ
Lang, M J G4DVK
Lang, R J G3KAY
Langan, J G0KKO
Langdale, S C G1JYK
Langdon, B T G4PUD
Langdon, C N G0VNH
Langdon, D S G6CKM
Langdon, G P 2E1CDK
Langdon, M M G4WHV
Langdon, P G4OLE
Langdon, P J G0EXB
Langdon, R H G4RHL
Langdon, R J G8GZR
Langdon, T G G3MHV
Lange, A GJ7DTA
Langeheine, D R M0BBG
Langer, A S G6EII
Langer, R F G4RQF
Langfield, A B G3IOA
Langfield, P A G7EBF
Langford, A C G4IQW
Langford, A K G4ARY
Langford, A M 2E1DLQ
Langford, A M G7WHV
Langford, A S G0PQY
Langford, B A G4GWP
Langford, D G8OIM
Langford, E G6BKQ
Langford, G G G4LNZ
Langford, G L K M1BGL
Langford, G W G0MKU
Langford, J G1GIT
Langford, J B 2E1ENP
Langford, M G7TDP
Langford, P F G8ZDT
Langford, P L G1MRE
Langford, R J G4FAD
Langford, R P G4RLT
Langford, T B G4VHL
Langham, C E G7URT
Langham, S T G0DMN
Langham, T W G4ZDF
Langley, A A G3KHQ
Langley, C A 2E1ALP
Langley, C J G3XGK
Langley, F G0UMR
Langley, P M G7KCK
Langley, T G G1HZH
Langlois, S C GJ4ODX
Langmaid, F C G3RAM
Langmead, A R N G4OOE
Langmead, I GM8LYS
Langmuir, I G7NAO
Langner, R M G8JLE
Langridge, D J G6GYG
Langridge, J G0SCB
Langston, R G3MDT
Langston, S T L G1JNX
Langton, A GM4HTU
Langton, C P J 2E1DQN
Langton, S G6ART
Langton, V A G7PBN
Langtree, P G0KBA
Langtry, W H GI4AAM
Langwade, M C G3VET
Langworthy, J W G7TXP
Langworthy, W A G7TXO
Lanham, A E G0TVC
Lanham, J G6HXV
Lanham, K G1XFL
Laniosh, B V G4NZK
Lankester, M R G8ATL
Lankshear, W A W G0HFB
Lannon, R W GW6ZHM
Lanoe, P L GU7TSI
Lansdale, R P M0BDI
Lansdell, S 2E1EDD
Lansdowne, J D G0HFA
Lansdowne, K F L G3DUN
Lansley, C R G0LOX
Lansley, H K M0APB
Lansley, P S L G3GUN
Lansown, S G GW4SDT
Lapham, R P G7UXS
Lappage, D L W G7MQZ
Lapper, M N GW2DUR
Lappin, J GI0OND
Lapsley, R A GM0BWT
Lapthorn, R G3XBH
Larbalestier, P R G4TEB
Larcombe, D R G0JLA
Larcombe, K T GW0PUH
Larcombe, K T GW3RNH
Larcombe, M A 2E1BRD
Larcombe, M A G7OIA
Larcombe, M J 2E1AOU
Larcombe, M J G7ZMS
Large, J D GM1ZIV
Large, J D J G7MLO
Large, J D J G7SRA
Large, L H T G4CYZ
Large, P 2E1FXO
Large, R M M1AGK
Large, S J G0VCC
Large, W T G6LBV
Largent, S G4VBQ
Larimer, R F GI7THY
Lark, E C G3CWC
Larke, R W GI6BDN
Larkin, A R G1JNY
Larkins, J A G8WKT
Larkins, S G8SBF
Larkman, T M M1AQV
Larmour, A GI4SWS
Larn, J O G7MUG
Larsen, D V G4FMY
Larsen, T R G7PHI
Larson, J E G3NBL
Larson, N J G7RXB
Larssen, J . O G6TBJ
Larter, R S G4YFI
Lasbury, W E G6DQT
Lascelles, C V G6FHJ
Lascelles, P S G3YLW
Lascelles, R G G3AKX
Lash, P W G6KSO
Lasham, K GW0NDP
Lasher, N G6HIU
Laslett, R R G8KQA
Lassman, A R G7PHE
Lassman, H G G3JZX
Last, D K G8XKT
Last, D S GM4THP
Last, D W G6LEU
Last, E E G7HMF
Last, J A G0DID
Last, J D GW3MZY
Last, P A G3THS
Last, R A G0FCH
Last, T W G1UBA
Laszkiewicz, A F G1GHY
Laszkiewicz, A L G1PSS
Laszkiewicz, L S G3KAU
La Traille, L J GW7LPM
Latham, A G7VFQ
Latham, A C G0DIM
Latham, D G6HXL
Latham, D G6LYK
Latham, F E G1DMW
Latham, R N M0ADW
Latham, S R G0NQG
Latham, S R G7NUN
Lathnury, J S G8UHF
Lathrope, J G3OQX
Lathwood, F C G3MUL
Latimer, D E G3XBB
Latimer, D W G3VUS
Latimer-Sufit, M L G8ACK
Latta, A GM6RVL
Latter, E T G0UMI
Lattin, G P G6ZHO
Lattka, E P 2E1AOB
Lattka, G P P G0PPL
Laud, N B G0MMO
Lauder, D M G0SNO
Lauder, D M G4WTB
Lauder, D M G6BOB
Lauder, J GM3PZG
Laugher, S J G7LYN
Laughlan, A N G1YCM
Laughton, D A G1CUG
Laughton, K M G1ESW
Laughton, P G3LBO
Laundon, T J 2E1EFX
Laurence, A A G3GGI
Laurence, A S G7PSZ
Lauwers, S G7JEF
Lavell, M G G4XDQ
Lavelle, L S G3SSZ
Lavender, J A G G8REZ
Lavender, P E G3RAN
Laver, A G7RQE
Laver, R F G8GDC
Laver, R J G6DNO
Laverack, B G4NAP
Laverick, J W G4PFE
Laverick, R 2E1DMZ
Laverock, P A G8GHQ
Laverty, S J GI3RQU
Lavery, J 2E1CDH
Lavery, T S GI8OLH
Lavey, A T G0MTG
Lavin, H F G4HAP
Lavin, K C G7LMR
Lavis, C J G6HIQ
Lavis, J J G4LYG
Lavis, R G G8DX
Law, B F G0GSC
Law, D J G0GUD
Law, D J G0LBK
Law, E K G3WGE
Law, E W G4CVW
Law, F A G6RHP
Law, J G8ILG
Law, K G4ALF
Law, K H G4WMZ
Law, M A G8SGY
Law, M R G6OKU
Law, N R G0ADA
Law, R A G8VLR
Law, T J G0WCT
Law, T S W G6TQL
Lawbuary, L V G3WMZ
Lawes, A J G1KKF
Lawes, A N GW3GFM
Lawes, D J P G6CLU
Lawes, G H F G3PLT
Lawes, L F W G3ZTM
Lawes, N P G8ZHR
Lawford, P T G6PIM
Lawford, T W G4KGY
Lawless, K P G7PHV
Lawless, P J GM4PGV
Lawless, T GM6JOD
Lawley, C E G6GTA
Lawley, D J G4BUO
Lawley, E R E G4ERL
Lawley, E W G8ADX
Lawlor, N J G6AJY
Lawlor, P J G3MUA
Lawman, S P G0UIH
Lawrance, A A G3RZV
Lawrance, B M G4ZYO
Lawrance, G D G4EOB
Lawrance, M A 2E1FIF
Lawrence, A GM3IQL
Lawrence, A G G6TQM
Lawrence, A P G6TPH
Lawrence, A W J G6JGP
Lawrence, B C G0SHO
Lawrence, B H G3KGX
Lawrence, B J 2E1BJL
Lawrence, B R G4LSL
Lawrence, C J G1KAT
Lawrence, C J R G1JHB
Lawrence, D G0PGK
Lawrence, D D G4FYT
Lawrence, D F G4VKC
Lawrence, D J G0RTQ
Lawrence, D J L G6XAW
Lawrence, D W J G6HXR
Lawrence, E M 2E1EWX
Lawrence, E V G1EOQ
Lawrence, F H G2LW
Lawrence, G 2E1CLQ
Lawrence, G J G6XAV
Lawrence, G R G0XGL
Lawrence, H G3YYE
Lawrence, J G3MEY
Lawrence, J G3PQY
Lawrence, J B G4VYN
Lawrence, J C GW3WEZ
Lawrence, J E K G0KYL
Lawrence, J E T GW3JGA
Lawrence, J L G4VII
Lawrence, J L GW3YJL
Lawrence, J R 2E1CNP
Lawrence, K R G8SSE
Lawrence, K R G6FAH
Lawrence, L W G0HRD
Lawrence, L W G4VJJ
Lawrence, M GW0SZN
Lawrence, M A G0VCR
Lawrence, M A G4JXO
Lawrence, P G1MUT
Lawrence, P D GM7JYW
Lawrence, P G G4XAL
Lawrence, P W G8MRQ
Lawrence, R D G4LDA
Lawrence, R I G6VLE
Lawrence, R I G0IED
Lawrence, R L G0RQZ
Lawrence, R M G8AWB
Lawrence, R W G1YCR
Lawrence, S C G0SCL
Lawrence, S E 2E1BFA
Lawrence, S J G6GOL
Lawrence, S M G7PMU
Lawrence, S R G4EOF
Lawrence, T B G4TDE
Lawrence, W J G8WRI
Lawrence, W S G6RZQ
Lawrenson, C M GM3SJY
Lawrie, R GM7SFE
Lawrie, S E GM0VWZ
Lawrie, W W GM4VZI
Lawry, T R G4VTL
Laws, C J G7SCM
Laws, C J G GW0CUM
Laws, K I G0CZR
Laws, R F G4IHH
Lawson, C K G3JCL
Lawson, D G4JHO
Lawson, D F G0FBM
Lawson, F M G0VXI
Lawson, F R G1AET
Lawson, J G0GSV
Lawson, J A G0FDX
Lawson, J A G0GVA
Lawson, J C G3WSV
Lawson, K G4AAH
Lawson, K GW0RKH
Lawson, M L G0OHN
Lawson, M W G4YQA
Lawson, P G4FCL
Lawson, R G1RLB
Lawson, R E A G3ZUA
Lawson, R N G5ZK
Lawson, S G0TBC
Lawson, S J GW6WPL
Lawson, T S GM6DCU
Lawson, W F L G0GAX
Lawson-Reay, J GW8WFS
Lawton, B W G0HWV
Lawton, D G4SSQ
Lawton, D G6NSZ
Lawton, D B G0ANO
Lawton, D B G0FQL
Lawton, G G1XKD
Lawton, G G7EVY
Lawton, J GW4YAW
Lawton, J M G0LHU
Lawton, J M G4ZOC
Lawton, K G4YQW
Lawton, K A G1AYP
Lawton, M P GW4IQP
Lawton, N G7VIJ
Lawton, P E G4XZG
Lawton, P J G7IXH
Lawton, R J G3XZR
Lax, D R G4AHN
Lax, K C G3VVL
Laxton, B J G8IJE
Lay, K C G5LY
Lay, M H G3VKP
Lay, N C G8FXG
Laycock, F M1BWH
Laycock, G B G3XWN
Laycock, G R D G7ADW
Laycock, G W G3WZW
Laycock, J G6KQF
Laycock, R G1IMY
Layland, A J G4NUS
Layne, D I G0AHK
Layne, D I G0MUD
Layne, D I G7MUD
Layphries, T R G0XTL
Layton, A H G8YYX
Layton, A I G6BTC
Layton, E G1XJZ
Layton, J A M0ARA
Layton, J F G3VGG
Layton, J F G4AAL
Layton, W G2BLL
Layzell, R J G0OVP
Lazarus, S G3TUA
Lazenby, C A G0PZH
Lazzari, J G8NAI
Le Bon, J O N G6VAH
Le Boutillier, K D GU6EFB
Le Boutillier, P P GU3UOQ
Le Brocq, D H GJ4YCR
Le Carpentier, B A GW4SUN
Le Couteur Bisson, A G4WZH
Le Feuvre, P L G0DBS
Le Fevre, P E G8AWZ
Le Good, A S G4BNV
Le Good, G P G4GUN
Le Gresley, N D G4SEV
Le Grove, D W G6JWM
Le Grys, B W G3GOT
Le Jehan, R E GJ0FYB
Le Lievre, B F GU4LJC
Le Maistre, G W G0HLG
Le Page, J T 2U1FNQ
Le Page, L M E GU4SYQ
Le Page, N K GU4NYT
Le Piez, R J G7UAX
Le Quesne, F P GJ4HSW
Le Sbirel, R J G0ITQ
Le Sueur, E G G4CEQ
Le Ves Conte, M T G6YCT
Le Vine, D T G1FKP
Lea, B D 2E1GAN
Lea, P G G8DLZ
Lea, W J W G8NQY
Leach, A J G1VAJ
Leach, A L G4WOQ
Leach, A S G3WIW
Leach, B O G3DXY
Leach, D G1LHZ
Leach, D J G6RDL
Leach, F A G2HNF
Leach, G G7ALK
Leach, G M0AKN
Leach, H G4ANE
Leach, I G1NRY
Leach, J G0FVO
Leach, J F G7OXK
Leach, K P M1CDX
Leach, L GW8JLY
Leach, L C G1WZO
Leach, M J G4HGV
Leach, N A M1BOU
Leach, P G8NSS
Leach, P A G G0OSP
Leach, P R G7GBZ
Leach, R G7HBH
Leach, T G6YCV
Leach, V G4GFT
Leach, W P G8LYG
Leack, M J G6ZHL
Leadbeater, R G8DVW
Leadbetter, G A G4VJE
Leader, A F G0CZZ
Leader, C G3LJQ
Leader, J G7RVS
Leader, J E G0KLJ
Leader, W J G3BRR
Leader Chew, B A G4TXI
Leader-Chew, T G6LEB
Leah, D R G3FGH
Leah, R C K G0VZR
Leah, R K G0TZD
Leahy, B T 2E1EPT
Leak, H D G0RJT
Leak, J G0BXO
Leak, M J G4VNC
Leak, S G0LSM
Leak, S D G0TGH
Leak, S J 2E1FXT
Leake, A P G0NAA
Leaker, S G M1BUX
Leal, C J G3ISX
Lean, G D G3WJG
Leaney, N G G1JKF
Lear, V C G3TKN
Learmonth, W GM4MKY
Learmonth, W G G4YZE
Learoyd, C J G6SKS
Learoyd, W A J G6SKT
Leary, D M G8JKV
Leary, J G6ZOT
Leary, J R G4YSB
Leary, T D GW4WPA
Leask, D GM7VPT
Leask, E GM6UNQ
Leask, G J G3IYX
Leask, L W J G2BIM
Leask, R I G4CEO
Leather, J D G0JIU
Leather, P G7KEJ
Leather, P K G4JYK
Leatherbarrow, G G4KQC
Leaver, D J F G1NKP
Leaver, E W G7GWH
Leaver, I C G7DBO
Leavey, A A G0SCG
Leavold, L I G0AJJ
Leavold, R L G6VOV
Leburn, R H G6JAZ
Leckie, D J C GM4NFI
Leckie, T A GM0HZO
Leddington, R J G6SKU
Leddington, W E GW8PTS
Ledeux, N E J G8ZTM
Ledger, A A G6LBR
Ledger, A D G4PYA
Ledger, C L K G3UBL
Ledger, D J G3XTM
Ledger, M J G8HXD
Ledger, R M B G7SWU
Ledger, R W G7OEV
Lee, G4EQJ
Lee, A GM1VLA
Lee, A A G1ULE
Lee, A J G4TNL
Lee, A P G6LWO
Lee, A R G0EDK
Lee, B G4VDJ
Lee, B G8DOW
Lee, B J G0TWT
Lee, B M G1IIX
Lee, B T L G7FTF
Lee, C B G0ORS
Lee, C C G0FZO
Lee, C D G7DQX
Lee, C F G6BQJ
Lee, C J T GW7VEL
Lee, C L 2E1GAD
Lee, C M G3GXG
Lee, C R G0BNY
Lee, C W G4GQY
Lee, D G1OTA
Lee, D G8PWA
Lee, D GW0DJU
Lee, D C M1BRZ
Lee, D G G4NIF
Lee, D J G4UHJ
Lee, D P G0ROX
Lee, D P G0SUK
Lee, D R G6XUV
Lee, D R G6YGP
Lee, D W G1AEY
Lee, E N G1OKU
Lee, F G3ZDP
Lee, F F G3JVO
Lee, F M M1BZI
Lee, F W G3YCC
Lee, G M1BCN
Lee, G A M1BTG
Lee, G S G4XXI
Lee, G W G4IVQ
Lee, H M G6FLY
Lee, I M G0SBR
Lee, J G1IVH
Lee, J A V G6SLZ
Lee, J C G4LHE
Lee, J D G4TTJ
Lee, J J G0CGV
Lee, J J GM7ORX
Lee, J M G0INT
Lee, J M G7AGO
Lee, J R G4JTK
Lee, J T G1HUE
Lee, J V F G4AEH
Lee, J W G0TGN
Lee, J W G1HLV
Lee, J W G4TSN
Lee, J W G8LII
Lee, K A G1RWR
Lee, K J GM6KDB
Lee, K Y G0EDB
Lee, L G1VGE
Lee, L G7BQI
Lee, L GW4OWZ
Lee, L H G5FH
Lee, M G3MCE
Lee, M GM6OPK
Lee, M A G4BAL
Lee, M B R G6FKN
Lee, M J G3XUQ
Lee, M J G6BKR
Lee, M J G8BGM
Lee, M L R M1BHC
Lee, M R G3VYF
Lee, M W G0LXV
Lee, N G0PWN
Lee, N GM0GLD
Lee, N D G4WKY
Lee, P G0MFH
Lee, P G1IWX
Lee, P A G8DWP
Lee, P D G3SPL
Lee, P F G3ZKO
Lee, P G G4UVX
Lee, P G G7HMR

Lee, P J G0DLP
Lee, P L GW0MHK
Lee, P W G4GEW
Lee, P W G8RVY
Lee, R GW7ARD
Lee, R A G0BWK
Lee, R E G8CQR
Lee, R G G0LEE
Lee, R J G0GVL
Lee, R J G0VQQ
Lee, R N G3XLM
Lee, R V G4GRV
Lee, S G GW7GCD
Lee, S J G0JPF
Lee, T J G4PGT
Lee, T M M1BYD
Lee, T W G4TWL
Lee, V 2E0AEX
Lee, V E 2W1FMZ
Lee, W G7FUR
Lee, W GW0ESU
Lee, W E G3MHR
Lee, W M GW3MFY
Leech, A C G6ADK
Leech, D C G7DIU
Leech, D C G7JPS
Leech, M A L G0KCD
Leech, P R G0GQV
Leech, S F G4WEE
Leeder, D P G8VLS
Leeder, G M M1CBV
Leeder, K P W G7HIM
Leeder, M G6AVF
Leedham, J G0BKA
Leedham, R V G0MDU
Leedham, V C G0BKB
Leeds, C J M1ACT
Leeds, R J G0WPS
Leeds, R J G4RZN
Leek, J W G8NST
Leek, L G0NOB
Leek, T G0HQP
Leeman, K L 2E1BSW
Leeman, P G1ZBA
Leeman, T R G0MLM
Leeming, A J G4LLQ
Lees, A G1YUS
Lees, A C D G1UTJ
Lees, A W M0BCY
Lees, A W M1BGG
Lees, B D G0JBO
Lees, C G0UXQ
Lees, C M G4WXG
Lees, D I G7NCT
Lees, E C G1UZS
Lees, F J G0JDR
Lees, G G0MGL
Lees, G G4WXB
Lees, G G6CXM
Lees, J D G6AJU
Lees, J M G0WLJ
Lees, K G0UFT
Lees, L J G1DMH
Lees, M G4KGL
Lees, M G8IAN
Lees, N F G0ONW
Lees, P A G4WIG
Lees, P D G1JSK
Lees, R G1GLS
Lees, R H G8EBT
Lees, R N G0PFE
Lees, T E G0ICG
Lees, T R G8ROR
Leese, A G3UP
Leese, C W G7ORU
Leese, D G3RKN
Leese, D J G6ZJM
Leese, G J G7OJM
Leese, P T V G6UXX
Leesley, G G4WEC
Leeson, M J G8GNN
Leeson, R G1VNR
Lees-Oakes, R G1JAA
Leete, R B G7SIR
Leetham, P I G4YVV
Leeves, R J G2LV
Lefever, J C GM0SCO
Lefever, J C GM4CAZ
Lefevre, S D M1AHA
Legan, P A G7ELB
Legate, C J G1HZD
Legg, A W G6JWO
Legg, B T 2E1DUS
Legg, D J G8WUF
Legg, D P A M1BAO
Legg, D T G3TFZ
Legg, G K G0EEF
Legg, M R G7OMO
Legg, P W G4BRX
Leggat, J F GM0LOK
Leggat, S GM7CPY
Leggatt, W R P G0UNN
Legge, D G7TQA
Legge, D J G G3MP
Legge, V M G4ZXP
Leggett, A J C G4KNN
Leggett, G D G7JYD
Leggett, J V G1JVL
Leggett, K N G4JKZ
Legood, A K GM8IXZ
Leigh, A B G4JZC
Leigh, A E H G0FBX
Leigh, A J G4DEN
Leigh, A J G8SXQ
Leigh, A S G0CHX
Leigh, A T P G8VMS
Leigh, C V G0VVR
Leigh, J R G4ILK
Leigh, R F E G4EWY
Leigh, R J G0HKN
Leigh, W N G1YNU
Leighfield, T G3KEU
Leighs, A R G4SXK
Leighton, A J GJ6SNQ
Leighton, B C G3JWP
Leighton, C GW4VBM
Leighton, F GJ4WRR
Leighton, J GW4RQS
Leighton, L G6HXW
Leighton, M D G3UKM
Leighton, R S G4OWG
Leiper, G R GM4VQY
Leitch, D M G6FIU
Leitch, I T G0PAI
Leitch, J GM0RFK
Leitch, P W GI1EOS
Leitch, W D G6HKE
Leith, J 2M1FGW
Leith, J L GM6GJW
Leithead, J M G0VTK
Lekesys, J G4BYW
Lelliott, R P G8XNB
Lemasonry, P R G1CQX
Lemasonry, P R M0AZP
Lemay, J K G4ZTR
Lemin, D G4TDL
Lemin, M H G4UUB
Lemon, J J G4FYJ
Lempriere, D G4SXQ
Lengyel, F J G3TUC
Lenihan, M C G0JVX
Lennard, D G4ZHK
Lennard, J C G1HXP
Lennard, P J G3VPS
Lennon, J B G4CRM
Lennon, M M0ABI
Lennon, P L G4CMP
Lennon, T R G7IIP
Lennox, B J GW0ZIG
Lennox, C H G4LXU
Lennox, E M T G4JKY
Lennox, J C GW3NSP
Lennox, M N G6VZB
Lennox, T GM0EWQ
Lennox, V J G7PMF
Lenthall, R C G0RUS
Lenton, A G8UUG
Lenzi, M J G7HNY
Leon, C H G6IQA
Leonard, C C G6FHK
Leonard, C G G0GSB
Leonard, C S G4ERO
Leonard, D M1AKU
Leonard, D H G4HKU
Leonard, D R G0ORT
Leonard, G G4DDN
Leonard, G R G1GNX
Leonard, I E J G4UKV
Leonard, J G3CUN
Leonard, J D G4YUU
Leonard, J F G4YWE
Leonard, M S G4YJM
Leonard, N A G7GOR
Leonard, P C GI0CAH
Leonard, S E C G6GTH
Leong, R G W G6ODU
Lepage, R E G0ADE
Lepper, J H G3JHL
Lerner, D J G4XWZ
Leslie, C J G0JWT
Leslie, D S G6CBL
Leslie, J S GM6UNL
Leslie, R A G4VHK
Leslie-Reed, P A G4DPW
Lester, E M G6NAP
Lester, G A G4JBF
Lester, J A G4DZA
Lester, J G M GI1DAO
Lester, J K MW0BFY
Lester, P C G8ZHS
Lester, R A G8UBJ
Lesurf, J T G6OFA
Lethbridge, L P J G3SXE
Lether, A M G4UQI
Lett, D R G4IYT
Letters, P J M1BQJ
Letts, J B G3URQ
Letts, R G0LGA
Letts, R J G1POG
Leung, E W Y G7TTM
Leung, W G4ZCF
Levay, P J G3SPB
Levens, M H G7TBQ
Leventhall, P E G3CJJ
Lever, D R G7VWA
Lever, J J E G4TEZ
Lever, R C G2DXU
Leverington, P D G8WWI
Levesconte, R G4MSS
Levesley, J M G0HJL
Leveton, B C G8UJO
Levett, A M G8HCJ
Levett, B G G3TXH
Levett, C G4WYL
Levett, J K G3VTL
Levett, M J G4TSG
Levett, Z 2E1AQE
Levey, R W M0AGB
Levi, R H J G3NQT
Levick, J T 2E1CRB
Levick, W G G7OLF
Levin, N D G1EOU
Levinas, A A G1EEY
Levingston, C G0XBQ
Leviston, J T G3NFB
Levitt, A G4EFX
Levitt, K B G1LMA
Levitt, P G4HAI
Levoir, L G6VJB
Levring, E H G0VZL
Levy, A G7UET
Levy, A S G0UQA
Levy, J G8JVD
Levy, M G0RMK
Levy, M E G4ACU
Levy, Eur Ing, M G0OWO
Lewarne, J O H G0MTU
Lewcock, S G GM1ARK
Lewczenko, G G6AKC
Lewer, J A G6EZW
Lewer, S K G6LJ
Lewin, D S G4PKT
Lewin, G P G0NEN
Lewin, M D 2E1AQF
Lewin, P M G0JKW
Lewin, R G8OWA
Lewing, D R G8MWD
Lewington, F E B G6JGM
Lewis, A G1ZSF
Lewis, A G4ELK
Lewis, A G6FIT
Lewis, A C G3XHM
Lewis, A C G6ACL
Lewis, A J A GW0JTU
Lewis, A L J G0TNQ
Lewis, A M G7BSF
Lewis, A V G7CAV
Lewis, B R G8RAU
Lewis, B T GW0VPB
Lewis, B W G4VPS
Lewis, B W D G4UWL
Lewis, C G7UTY
Lewis, C D H GU3NHL
Lewis, C E G4RQC
Lewis, C J GW3YTL
Lewis, C V G4CYY
Lewis, D G1AEQ
Lewis, D G4KPH
Lewis, D GW0UKE
Lewis, D GW1PQE
Lewis, D GW7CEP
Lewis, D GW7TDK
Lewis, D A 2E0AQA
Lewis, D A GW7VSO
Lewis, D D GW4HBK
Lewis, D E G1FMW
Lewis, D J G4CLC
Lewis, D M G1YCN
Lewis, D M G6SLY
Lewis, D R G4NIP
Lewis, D R G7TXT
Lewis, D R M1BDP
Lewis, E D D 2W1BPH
Lewis, G M0BCJ
Lewis, G M1AQD
Lewis, G E G1TTK
Lewis, G E G3NFS
Lewis, G J GW4NKR
Lewis, G M G7CEM
Lewis, G W G7KWP
Lewis, H F G3GIQ
Lewis, I G3XTY
Lewis, I C G7KGV
Lewis, I C M1BUJ
Lewis, J G0UUN
Lewis, J G8AYV
Lewis, J A G7KWO
Lewis, J C G3TCY
Lewis, J E G7OUQ
Lewis, J E M1BGK
Lewis, J E P G4HAE
Lewis, J F G0CKU
Lewis, J F G0UUM
Lewis, J F G3FBR
Lewis, J H GW8UZL
Lewis, J L GW0RAD
Lewis, J R GW1EEZ
Lewis, K G4GGK
Lewis, K M0AKL
Lewis, K C GI1KDS
Lewis, L 2M1EJK
Lewis, L GW7OTI
Lewis, L A G7LBD
Lewis, M 2E1EZP
Lewis, M D G1LQH
Lewis, M E G4CSE
Lewis, M G G1MDS
Lewis, M J G0TWL
Lewis, M J G7DGC
Lewis, M J GW7KDU
Lewis, M K 2E1CRA
Lewis, M M GW7DVB
Lewis, M S G0JDY
Lewis, M S G4ZAC
Lewis, M T G0VGH
Lewis, N G4EPM
Lewis, N L G0WZI
Lewis, N T P GW6JGE
Lewis, P M1AIB
Lewis, P A G8MAV
Lewis, P F A G4VFG
Lewis, P G G0LAW
Lewis, P G G3EMF
Lewis, P I 2W1DGL
Lewis, P J G3WBI
Lewis, P J G4ZFP
Lewis, P M 2E1AQG
Lewis, P N F A G0SCR
Lewis, P N F A G4APL
Lewis, P R G3RQX
Lewis, P W GW7RQW
Lewis, R GW0WRI
Lewis, R GW4UIL
Lewis, R F G6WKI
Lewis, R G G1OXF
Lewis, R J G3YCO
Lewis, R K G6UNN
Lewis, S G4PLM
Lewis, S K G4PLK
Lewis, S R G8JCT
Lewis, T A G4ZQO
Lewis, T W GW4AZW
Lewis, V A G4DQP
Lewis, W J GW3RTA
Lewis, W S G0RIR
Lewkowicz, S P G1CMZ
Lexton, S M G1UGO
Leybourne, P M G7ESK
Leyden, C G4RYY
Leyland, T J B G6NAJ
Le-Mottee, A O T G4HLM
Li, K C G0TOY
Libby, F L G7REL
Liberson, T C G0MKO
Licence, B H G4DCG
Lichfield, H W G7EPW
Lickley, A G7POV
Lickman, R F 2E1EOP
Lidbetter, P E C G4NER
Liddell, A J G7JWE
Liddell, S GM4BGS
Liddiard, N A G1MJV
Liddiard, R D G1EFA
Liddicoat, D J 2E1ELI
Liddicoat, S E G0IGD
Liddle, D A G8ODP
Liddle, D W GM1PSZ
Liddle, G F G4OAS
Liddle, K G7TRG
Liddle, K G7TWV
Lidster, S G0ROR
Lidstone, G C G4OQ
Liedberg, R B GM0BCI
Liepziger, R W G1OCK
Liffchak, L B G6ODW
Lifton, A M G0PEH
Liggett, N GI0PFK
Liggins, T P G0FYU
Light, D A F G7DNY
Light, F J G1JOT
Light, G F G0IIC
Light, R J G4NXI
Lightbody, A A G4CTY
Lightfoot, A K G4ONJ
Lightfoot, J GW7UCA
Lightfoot, J R 2W1FHL
Lightfoot, P A G0OFW
Lightfoot, P B G6ZHU
Lightfoot, R W GW3MZN
Lightfoot, S G0GOF
Lightly, A A GW3VFL
Lightly, A L GW6UXY
Lightowler, N D G8XDS
Lihou, N M G8OVO
Lill, N G4YWR
Lilley, B G W G7UZE
Lilley, C P G6YGQ
Lilley, D G4IAU
Lilley, D G4NOK
Lilley, G K G4TYO
Lilley, N J G0TMZ
Lilley, N S G3INN
Lilley, P 2E1DLT
Lilley, R M G8INT
Lilley, T G1YAA
Lillingstone, K L G4SOP
Lillington, M J I G3JFY
Lilly, A S J G8VGI
Lillycrop, R M G1ZJO
Lillywhite, A G8MEM
Limb, L W G2DTD
Limb, R C G4PVY
Limb, R M G8FXX
Limbert, I S 2E1CYS
Limbert, R P G4IUP
Limebear, R W L G3RWL
Limehouse, R W G3WTN
Limehouse, W F E G2FDF
Limond, F B G0AHX
Linacre, T G4VKV
Lincoln, A D G4CIG
Lincoln, E W G3OAL
Lincoln, J GM0JOL
Lincoln, P C M1APD
Lincoln, S E 2E1EKG
Linda, M J G4GTH
Lindell, S G G4IEH
Linden, D P M0AED
Linden, R G4RFI
Linden, S J G0DUA
Lindenberg, M F L G1DKI
Lindgren, B E A G0BMZ
Lindgren, V A G4BYG
Lindley, A G4ITQ
Lindley, B T G7VDR
Lindley, D J GW0BNO
Lindley, J G8MQK
Lindley, K G8NDK
Lindley, L GW3KFA
Lindley, M G0HGL
Lindley, M H G0UNY
Lindley, P G0BAD
Lindley, R J G7IAE
Lindley, S J J 2E1BGU
Lindop, B R P G3WUA
Lindop, D A A G6XDN
Lindsay, A C G0ERA
Lindsay, A C G4NRD
Lindsay, B G G0BEG
Lindsay, C G3KTZ
Lindsay, C W G4VJI
Lindsay, C W GM1MLW
Lindsay, D GM4HQF
Lindsay, D J G0PLZ
Lindsay, D R GM1AHC
Lindsay, G G0KGL
Lindsay, G G4CDB
Lindsay, G G4KOT
Lindsay, G G8BZL
Lindsay, G C GM0BEL
Lindsay, I A B GM8LYQ
Lindsay, I F GM0TWB
Lindsay, J 2E1FCE
Lindsay, J G GM4ZRX
Lindsay, K G0JWL
Lindsay, K MM1BOT
Lindsay, M J A G0IYY
Lindsay, P M1ABI
Lindsay, P S G4CLA
Lindsay, R G4HUX
Lindsay, S W G0KDS
Lindsay Smith, R G3YBS
Lindsay-Smith, W A G3WNI
Lindsell, T J G1BMK
Lindseth, W D G7CVS
Lindsey, R J G6DCV
Lindsey, T L M0ASQ
Lindsley, P E G3UDV
Line, D J G4ZRM
Line, S J G3XYO
Lineham, P D G8XQA
Linehan, B K R G0UCK
Lines, F E G1YHX
Lines, G I G7ISR
Lines, G T G0ROH
Lines, J G6BEB
Lines, J H G6XBG
Lines, M J G1SEO
Lines, R G8LQP
Lines, R J GW4KOE
Lines, S C G6TPB
Linfoot, A P G1UVK
Linfoot, J S G0CPP
Linford, D T G1INC
Linford, J R G3WGV
Linford, J R M0AIM
Linford, R S G0UOP
Linford, R S G3YQF
Ling, H G4XGE
Ling, H A G4CCH
Ling, L R D G3UHY
Ling, R W G4BDU
Ling, W F 2E1EJD
Lingard, C A G4GQC
Lingard, D L J G0CLH
Lingard, E F G3WNQ
Lingard, P I G7TSM
Lingwood, A J G3NKA
Lingwood, D J 2E1EPH
Link, I P G0LCL
Linn, B G1JTZ
Linnegar, G W G0OVB
Linnell, D J G0MJK
Linney, A G7RNX
Linney, K A G3UDA
Linney, R S G6CJW
Linney, S M1AVV
Linney, S F G8GWR
Linsdall, D K G8TEQ
Linskill, D G0AGF
Linsley, D J G7RAW
Linsley, K T G0PXF
Linsley, P F G3PDL
Linstead, V J G1NNL
Lintern, D T GW4VEB
Lintern, G R G0DOV
Linton, D GI0ITJ
Linton, D GI3TEN
Linton, M G0IYM
Linton, P E G4DVY
Linton, P J G4LPX
Linton, R E G8XMZ
Lintorn, R L G6RLH
Lintott, R G0KDR
Linzey, R N G6KXB
Lipman, N H G6ASA
Lipscomb, R A G0JBP
Lipscomb, S N G0JBQ
Liptrott, S J G0HKZ
Liptrott, S J G4EGY
Liquorice, D A G6ILE
Lisbona, D G0MNS
Lish, R C G6ENC
Lishman, T A G0WVD
Lishman, W G2AKK
Lishman, W G4JS
Lisk, J 2E1BEL
Lisle, D G7VWO
Lisle, J G1DUB
Lisle, M E T G4FPZ
Lisle, R J G0SLR
Lisney, D L G3MNO
Lister, D G0ZDL
Lister, D J G6HXX
Lister, H G7AJN
Lister, M A G8FCQ
Lister, M D G1JUI
Lister, M D G8NYM
Lister, M H G4FHA
Lister, R G4BDE
Lister, R J G8IXP
Lister, T G0IEH
Liston-Brown, I D G8GSL
Liston-Smith, I L G4JQT
Litchfield, D M G0NQV
Litchman, M G0TOC
Litherland, B H G4IMT
Little, A B G7EYV
Little, A R G4PSO
Little, B J GI4PID
Little, D F G7PQP
Little, F G0HJB
Little, F G G6OPL
Little, H G4SUL
Little, J W GM3KEZ
Little, M K G6LED
Little, N J G GW3YVN
Little, P D G0WYV
Little, P W GI7JOJ
Little, R G0ACR
Little, R G0NDC
Little, R S G4CIM
Little, R T G0MZK
Little, S D G7KSQ
Little, S D G7RFC
Little, W T G0UTG
Littleboy, N S G8YEQ
Littleboy, R A G1FFB
Littlechild, P E G0HWT
Littlefield, R J G4GYJ
Littlejohns, C S GW0TQM
Littleproud, B G3AMK
Littler, A G0LRM
Littler, C B G4YKH
Littler, D A G4RIE
Littler, J G0LVY
Littler, T R G4MNZ
Littlewood, A G5LW
Littlewood, D G6DCT
Littlewood, M J G0CIJ
Littlewood, M T G3OYI
Littlewood, R E G4YET
Littlewood, R K G3XLX
Littlewood, W G G6EEV
Littlewood, W G G7AEE
Littman, C M G8JKD
Litton, D S G4MGC
Lively, G D G3KII
Livens, C R D 2E1FQP
Livens, W G2CKB
Livermore, E T F G7LQP
Livermore, H E G3XEG
Liversidge, D J G4JMY
Liversidge, G S G0KYJ
Liversidge, R 2E1FMU
Liversidge, T J G1GOG
Livesey, C J G0LJC
Livesey, D G3THD
Livesey, G A G3FIB
Livesey, I D G8NXJ
Livesey, J G0JJL
Livesey, S G7JTL
Livesey, W G6MIU
Livesley, A N G8JAI
Livesley, J A G4YAB
Livingston, D M GM0WNM
Livingston, G E GW3LAI
Livingston, H A GM0FTH
Livingston, J A G4FDD
Livingston, M J G6PIB
Livingston, M P GM0FTL
Livingstone, D GI0DQJ
Livingstone, D A P G0PDE
Livingstone, F D M G3IXA
Livsey, D W G4BQH
Livsey, R GW3SMY
Livsey, W G GD0BCM
Lixenberg, J G3MOL
Lixenberg, M R 2E1DNM
Llewellin, B C S G4DEZ
Llewellin, B C S G4ZDA
Llewellyn, C N GW0RTP
Llewellyn, D H GW4EEO
Llewellyn, J N G4JTM
Llewellyn, M V G0IVB
Llewellyn, S A G0ESO
Llewellyn-Jones, J P G0TXQ
Llewelyn, G8VXU
Llewelyn, G GW3RXD
Lloyd, A G4OLW
Lloyd, A J G1TLW
Lloyd, A L G7RFF
Lloyd, A W G6BTO
Lloyd, B G6CLW
Lloyd, B D G1NNA
Lloyd, C F G4GKD
Lloyd, C K G4VTT
Lloyd, D G6MJB
Lloyd, D C G6CLX
Lloyd, D J G6CXO
Lloyd, D J G7HII
Lloyd, D J GW0BVY
Lloyd, D K G4HJT
Lloyd, D S G1HRA
Lloyd, E G0HEJ
Lloyd, G GW4FNO
Lloyd, G B G1NNB
Lloyd, G B G7LEY
Lloyd, G D G7PHF
Lloyd, G J G3RBL
Lloyd, I G6LES
Lloyd, J B GM1ZTA
Lloyd, J C G0PWE
Lloyd, J C G4OLS
Lloyd, J H G8FBB
Lloyd, J R G4NDD
Lloyd, K F G7BIQ
Lloyd, K J 2E0AIT
Lloyd, K J G0JNH
Lloyd, K J G3XTP
Lloyd, L D G0NTT
Lloyd, M J G1PKP
Lloyd, M J G3KLY
Lloyd, P G4PHD
Lloyd, P D GI8YJV
Lloyd, P S G1BLQ
Lloyd, R C GW4IQA
Lloyd, R M G0SAT
Lloyd, R M G1OKP
Lloyd, S K A MW0BBU
Lloyd, S K A MW1BGP
Lloyd, T D GW0TWQ
Lloyd, T S G6UED
Lloyd-Owen, J 2W0AMB
Loach, E R G4ZXS
Loach, F A G4JUD
Loach, G A G4PVZ
Loach, M J G8UDJ
Loader, C M G7HXU
Loader, E J G6HXU
Loader, J E D G0SRX
Loader, J S 2E0ABS
Loader, J S G0VBV
Loader, M J 2E1DBA
Loader, M J G8ONR
Loades, G Y G3PVD
Loake, G K G0GBI
Loakes, M B G0DBR
Lobb, M T G4EKV
Lobban, F J G4JQW
Loch, P G0TCF
Lock, A A G4TSS
Lock, A A G4ZHR
Lock, A D G3NWL
Lock, C GW7EYQ
Lock, D G7UZO
Lock, D C G7JTR
Lock, K C G4ZZP
Lock, M T G7UOZ
Lock, N J G1EUJ
Lock, P G4STB
Lock, P C G7JUR
Lock, R J G4JZU
Lock, S G7CTT
Lock, T J G8ZZG
Locke, D D G3LKV
Locke, D J GW3TKG
Locke, J G8VQQ
Locke, J G F G8LOC
Locke, R 2E1FBA
Locke, R V G1LOC
Locke, S L GW0SGL
Locker, J W G7MIZ
Lockett, D J G7JTD
Lockett, N J L G4EMB
Lockett, T J G1KQS
Lockey, B V G3HYR
Lockey, F G0RXQ
Lockhart, B L G6UNX
Lockhart, G B G3NQZ
Lockhart, J L G0CXG
Lockhart, V G0BVL
Lockitt, E W 2E1FNU
Lockitt, M F G7RXW
Lockley, M G4WAM
Lockton, A S G0VDC
Lockwood, A L G1YOF
Lockwood, A R 2E1FPH
Lockwood, D 2E0AME
Lockwood, D G3ZTR
Lockwood, D A G0FBR
Lockwood, D A G0OLE
Lockwood, D J G8UAI
Lockwood, J H G0MEI
Lockwood, J L G3XLL
Lockwood, L E G0HMT
Lockwood, M J G6SCG
Lockwood, P G7ELV
Lockwood, P W G8SLB
Lockwood, T G4HZN
Lockyear, E J G8KZJ
Lockyer, C G G0IRV
Lockyer, D E C GW3HCL
Lockyer, J K G0MUZ
Lockyer, J W K G0MQB
Lockyer, L F G4VRI
Locock, D G G7CQB
Lodge, C F G4IIK
Lodge, G J G4JJC
Lodge, H I G7PPS
Lodge, P A G0TIA
Lodge, S G0KWG
Lody, B R G0PCZ
Loewenthal, L S G4FGX
Loffler, H H G4YAO
Lofthouse, K G0BVP
Lofthouse, N G1DJQ
Loftus, C G4IFI
Logan, B G0VDN
Logan, B G1YNF
Logan, D R G8OTZ
Logan, G GI7MCY
Logan, G H G0AVU
Logan, L G G4ILG
Logan, L G G6RRB
Logan, P GM0NKX
Logan, R P G4WXF
Logan, T R GM3VBT
Logan, W D G4EZF
Logsdon, M P G8FZI
Lokuge, R M N G0TAO
Lole, A J G7SGA
Lomas, A G0VVG
Lomas, D A G4XOW
Lomas, E G0KZO
Lomas, G G4SYC
Lomas, J H G3HEQ
Lomas, J S GW0UTC
Lomas, K A G4ONE
Lomas, N B G3OGN
Lomas, P J 2E1DNB
Lomas, R G G6REN
Lomas, R T G0OMZ
Lomas, S G1NUN
Lomas, W G1YOH
Lomax, G D GW0FXA
London, D J G0VGB
Loney, S G G1XBR
Long, A GM4BRM
Long, A D G1SJT
Long, A E G3XDL
Long, A P G6CXI
Long, B W G4ULI
Long, D W G1YZL
Long, D W G7AEB
Long, E C G3MJS
Long, F A G3TUF
Long, G D 2E1ECB
Long, J G4HAG
Long, J H G4IQZ
Long, K G7JZY
Long, L K GW1MGI
Long, M K MM1BIM
Long, N R W G4BIN
Long, P A G8XDZ
Long, P J G0PHE
Long, S W G1VVJ
Long, V G4YRC
Long, V G G4DUM
Long, V G G4RNV
Long, W D G0PRJ
Long, W M GM0VCN
Longden, E G0IPX
Longden, E G3ZQS
Longhurst, D W G7UEI
Longhurst, J F G3VLH
Longhurst, P E G3ZVI
Longhurst, P W G7WBM
Longhurst, W H G3AAO
Longley, C H G1JYD
Longley, H G7OZZ
Longley, K M1AMM
Longley, M J G6GVL
Longley, P B M1BKA
Longley, R S GM4XJX
Longman, F J G3HOH
Longman, G C G6WQU
Longshawe, R G8HXU
Longson, F G3SSA
Longson, M G0NMY
Longson, R J S G4LGI
Longstaff, B G1NTI
Longstaff, D G4WCD
Longstaff, J B G3VJR
Longstaff, P R G6NFZ
Longstaffe, E B G3SQE
Longstaffe, E W G1GFQ
Longthorne, P D 2E1BNV
Longton, J G G4WFE
Longuet, A J G8ZGQ
Longworth, C L G4FWP
Lonnon, B L G3ZUM
Lonsdale, C GW0FJH

Lonsdale, M V M1BBF
Lonsdale, P M M1BZZ
Lonsdale, S A G7FFW
Loon, D C G1PMJ
Looney, S A G4NIE
Loose, J C G4BJS
Loosley, G G1HZA
Loosley, L C G6HLP
Loosmore, C GW0BBQ
Lord, A G4KHT
Lord, A GM7DAP
Lord, A GW7RFA
Lord, A H F G8DQZ
Lord, A J G4PXC
Lord, A K G3JID
Lord, A M G0PCK
Lord, A S G1YEZ
Lord, B G7UCT
Lord, C J G3YUU
Lord, C J G7BGP
Lord, E M G3ZAU
Lord, F G4YWK
Lord, F D GW8VGG
Lord, J A G4XRK
Lord, M J G4HXT
Lord, M P G0PAS
Lord, P D G0UYJ
Lord, P H J G4UOG
Lord, P M G8PGI
Lord, R G0DTN
Lord, R G6YCU
Lord, R A G3DSK
Lord, T D 2E1FKU
Lord, W B H GM5NU
Lorenzen, J P GM7ORS
Lorimer, T GM0GTL
Lorton, C F G8POC
Lorton, J A G4EKL
Losekoot, M G1NBR
Loss, R J G4XUT
Loten, P A G8KFK
Lothian, W D GM4JZJ
Lott, A G1AEU
Lott, T M G2CIN
Lotz, J C G3VUL
Louca, J R G0PHZ
Loucks, W W G4MNI
Lough, P J L G1BJK
Loughlin, J C G4DKQ
Loughran, E G GI4WNH
Loughran, V P GI6TBC
Loughrey, N GI6JJR
Loughrey, N P GM4TTD
Loukes, A T G7VTY
Loukes, R K G7MZA
Lovatt, B W G7LIE
Lovatt, D C G7UHK
Lovatt, M GM0JCN
Lovatt, M MM0ACS
Lovatt, T H G1XII
Love, A M G0EJR
Love, B D B 2E1FAU
Love, D G1LFR
Love, D J G7ORK
Love, D R G4RBQ
Love, D R A G0DRA
Love, J GM4OBG
Love, J P G0EPL
Love, K G1ESX
Love, K G G0GPB
Love, M D 2E1BOM
Love, N A GW0VCK
Love, P J G0KOK
Love, W J G0NXQ
Loveday, G S G3IUV
Loveday, J G8RBG
Loveday, R C H G3LLG
Lovegreen, A T GM4FLX
Lovegrove, G N G7KLV
Lovegrove, K D G7GWI
Lovejoy, M M G3IXN
Lovejoy, P E S G7PJT
Lovelady, P G3LNL
Loveland, L J G3KZX
Loveland, P J G4SYB
Loveland, R A G2ARU
Loveland, R M J G0PVI
Lovelass, S M G7OSM
Lovell, A G0GRG
Lovell, A G1HZG
Lovell, A J 2E1DKZ
Lovell, C L G3JUW
Lovell, D G0MGD
Lovell, G G0KJQ
Lovell, J M G8JHL
Lovell, R J G4FHN
Lovell, S G8XPZ
Lovell, S J G1BLJ
Lovelock, B I D G4XXU
Lovelock, G P B S G3III
Lovelock, J P G0BYJ
Lovelock, P H G6ZOS
Lovelock, T A G8DFU
Lovelock, T J G7DNS
Lovely, N P G4RGE
Loveridge, A G0JGH
Loveridge, D R G0IQL
Loveridge, S H G1HRE
Lovesey, D G G3ONP
Lovesey, S R G0DBM
Lovett, C J G1MMN
Lovett, G C G7OCL
Lovett, P J G6HXZ
Low, A S S GM4UZP
Low, D E 2E1BZI
Low, G E G0RLF
Low, G M GM4LXM
Low, G W 2E1BZH
Low, J A GM0IDJ
Low, J E GM4UZR
Low, K M G1XYM
Low, P G0CIP
Low, R C G GM0ECU
Low, R M GM3BHY
Low, S A S GM4KGZ
Lowden, A D G3FIA
Lowden, J A G6NAN
Lowe, A G7VJX
Lowe, A J G1INA
Lowe, A J G7DTS
Lowe, A R G0WCM
Lowe, B G0RDR
Lowe, B M0BDN
Lowe, B A G1UGJ
Lowe, B H L G0WRG
Lowe, B H L G4PPC
Lowe, B H L G7PPC
Lowe, B H L G7WRG
Lowe, C G0UMY
Lowe, C D G1IVG
Lowe, C I 2E1CON
Lowe, C R G6SKZ
Lowe, D G1IVF
Lowe, D G3YER
Lowe, D A G4NMI
Lowe, D B G4UOF
Lowe, G W G0NRA
Lowe, G W G1GQC
Lowe, G W G3GQC
Lowe, J A G3XZX
Lowe, J B G3UBO
Lowe, J E G1ARJ
Lowe, J M G2DFP
Lowe, J R G4LOS
Lowe, J T G1WHY
Lowe, K G4NZN
Lowe, K P G1BAI
Lowe, L F G8GMK
Lowe, M C G4KHB
Lowe, M D G0NZA
Lowe, M J G1NNU
Lowe, M S G4YCD
Lowe, N G2HFC
Lowe, R G0FRL
Lowe, R A G4ZAM
Lowe, R A MM1BJO
Lowe, R F G3XVZ
Lowe, S G0OYQ
Lowe, S W G4MSZ
Lowe, T G3XWO
Lowe, T G4ZOO
Lowe, V G E G1IND
Lowe, W B G4ONG
Lowe, W D G0MLC
Lowes, G R R G4NJW
Loweth, ... 2E1EXB
Loweth, P H P G1IVN
Lowings, E A G1OEF
Lowman, E G0UXS
Lown, K R G4PTE
Lown, S D G8WLL
Lowndes, D S G8VQO
Lowrie, D B G6ARU
Lowrie, P S GI7JYK
Lowrie, P S GM6TBE
Lowry, E GI0NYO
Lowsley, G G7WAR
Lowson, M GM4XRF
Lowson, R O G2HKG
Lowson, W L L GM2BFV
Lowther, J R G4ZJT
Loxley, W E G4LMV
Loxton-Gear, R R G7LMY
Loy, J F M1BDC
Lubbock, P E GW6UAK
Lubrani, A G6SBG
Lucas, A T G4LVA
Lucas, B D G0TAR
Lucas, C H G7TDL
Lucas, C P GW8ZBC
Lucas, D G0BIE
Lucas, D H G4MAG
Lucas, D J F G8XND
Lucas, D L G6AJW
Lucas, D W G8DKI
Lucas, F J G7IZM
Lucas, G GM4EJI
Lucas, G F G0MJO
Lucas, J G1EVS
Lucas, J C G4XRN
Lucas, J F G3ISU
Lucas, K G7RYZ
Lucas, R A W G2BJW
Lucas, S R G8APZ
Lucas, S R G8PUB
Lucas, W J G0TBS
Lucas-Davis, E D G4YAS
Luck, J E G7TPZ
Luck, L G G8UKH
Luckcuck, P D G7SNA
Luckett, R J G6JAY
Luckhaus, S GD4VGL
Luckhurst, D A G4UCI
Lucking, I M G8RNM
Lucking, R R G0NIB
Luckman, A J G6FGO
Luckman, R F P G3CNG
Lucock, B A G0LCJ
Lucocq, E F GW7THM
Lucocq, E F MW0AQT
Lucyk, R W G8PXA
Ludar-Smith, G A G8ZIW
Ludford, S R G1OIW
Ludgate, K G1MSY
Ludgate, R W G4VOC
Ludlam, K E G0CKT
Ludlam, R N G0CTX
Ludlow, G M G8WWC
Ludlow, J D V GW3ZTH
Ludlow, S V G8KCC
Ludlow, V J G3JLZ
Ludwell, R B G3ZZQ
Luff, L A G3XAK
Luff, R E G4UMQ
Luft, P M G3SEZ
Lugard, C R G4LZE
Lugg, F R 2E0AGG
Lugg, F R G7JCR
Lugmayer, E F G3KIJ
Luhman, G K G0ETA
Luing, D W G4OLT
Luke, B S GW3XJC
Luke, C A G4JEH
Luke, D C 2W1EYN
Luke, D C MW1BUN
Luke, K G0HNE
Luke, M I GW8VQY
Luke, S S 2W1FUL
Luker, R B G6NAK
Luker, R J G1RKR
Lumb, J T G0ARU
Lumb, P G3IRM
Lumb, S R G0TEH
Lumbard, R V G4YIF
Lumbard, S F G0IXB
Lumkin, J E G4DSH
Lumley, D B 2E1EJU
Lumley, G E G3DJE
Lumley, T H G1SBZ
Lummis, K R G6JZV
Lumsden, G G0GKO
Lund, R G1EOY
Lund, T R G0FUG
Lundberg, O I G0CKV
Lundean, B H G3ZHT
Lundegard, T I G3GJW
Lundegard, T I G3ZRX
Lundegard, T I G4BRC
Lundie, H G3XTU
Lunn, A C G0EYE
Lunn, A H G5LL
Lunn, A J G4JAX
Lunn, D I G7TNO
Lunn, J D G3BRD
Lunnon, C GI7UBY
Lunnon, R F G1MIY
Lunt, G H G0EDF
Lupton, F G0MDR
Lupton, I A G1ROD
Lupton, K J G6ZHS
Lupton, R G1ARH
Lupton, R W G7RJI
Lupton, S N GW7UOH
Lupton, W R 2E1EGU
Lurcook, D G G4ERW
Luscombe, D M G8JWC
Luscombe, J L G1GED
Luscombe, P B G0TDQ
Luscombe, R D G3MEP
Lusty, D 2E1ERJ
Lusty, E D G3YIE
Lusty, G G3DWI
Lusty, I D G4RTR
Lusty, M D 2E1ERK
Lusty, R M0AKS
Lutas, P W G6VDK
Lutman, P D G8YPN
Lutterot, J D G0LUT
Lux, L J G4OHA
Luxton, K C G0AYM
Lyall, P J J G8FRH
Lycett, D C G6BYL
Lycett, J E G0MSZ
Lycett, S P G1YFD
Lycett, W G G4EQG
Lydall, H GM4DIZ
Lydall, K G4RIA
Lyddon, R F G4ETJ
Lyder, R D G3NBB
Lydiate, J P G8PKP
Lydiate, S G 2E0ACJ
Lydiate, S G G3SCT
Lydiate, S G G4IFD
Lyford, B J G0BPA
Lyford, R M0BBV
Lyle, G A GI0IUP
Lyle-Hodges, A G4KFR
Lymer, A GM0DHD
Lymer, J W G3BMM
Lynam, A G4VQU
Lynas, P G0EJA
Lynch, D E G6VJM
Lynch, F J GW4SKC
Lynch, J A G0NXX
Lynch, J J GI1ZJF
Lynch, P G G3KWP
Lynch, P J G8KJU
Lynch, R MM1AWV
Lynch, R A 2M1EJI
Lynch, S G6IKE
Lynch, S P G8TRU
Lyne, J R G0PEW
Lyne, M T 2E1FVO
Lyne, R H G0JTD
Lyne, T J G8HJV
Lynes, F W G3JKZ
Lynn, E GW3REY
Lynn, M F G1KOT
Lynn, W A G1OUK
Lyon, B T 2E1AXP
Lyon, C S S GW3EIZ
Lyon, G 2E1FYL
Lyon, G G0TRY
Lyon, M G0WOA
Lyon, P G8BIS
Lyon, P D GW4JUF
Lyon, T G0LHW
Lyons, B G1ISS
Lyons, B W G4SOQ
Lyons, D A W G1TYK
Lyons, E GI6DCX
Lyons, J A G6EFC
Lyons, J D G6NTE
Lyons, N T GW4JWV
Lyons, R A G4ZZL
Lyons, T GI0PVG
Lythaby, A G G6KLF
Lythall, R G0TLA
Lythgoe, J C G8CLY
Lythgoe, J R H G7VIC
Lyttle, A M GM4VGU
Lyttle, H GM7UKS
Lyttle, H C GI3SSR
Lyttle, J D GI0SQV
Lyttle, P V G1FOM
Lyttle, R M MI1AZL

M

Mabbutt, C J G4WIN
Maby, C D G4NAQ
Mac Donald, W J G8PUJ
Mac Kimm, P W G8HDS
Mac Walter, A R G3TSZ
Macadam, A I GM0TDO
Macadie, D S GM0EUY
Macalindin, I G8ZJM
Macan, N H M1BCH
Macario, R C V GW8SRW
Macarthur, I D G3NUQ
Macarthur, I M GM6FCW
Macarthy, J E G3PHM
Macartney, M C G1MYX
Macartney, N F G0ELI
Macassey, N J L G8LUK
Macaulay, A G7KLS
Macaulay, B M G4ABX
Macaulay, I MM0BFF
Macbeth, A K G1MAK
Macbeth, M P G1MAC
Macbeth, R C G8VLY
Macconnachie, A G4ECL
Macconnell, D G GM0VYY
Maccormick, P J G0VRY
Maccourt, G J G4YZX
Macculloch, L E G7WCZ
Macdiarmid, I G6BGH
Macdiarmid, W S GM4YMD
Macdonald, A D G3NPM
Macdonald, A R GM4TRH
Macdonald, C S G4ZGM
Macdonald, D GM0BNQ
Macdonald, D GM8CIF
Macdonald, D J GM0MGO
Macdonald, D L G0FSY
Macdonald, E N G4TTY
Macdonald, F G GI4VRF
Macdonald, F R G1ABQ
Macdonald, G H G4AZG
Macdonald, I G1ABM
Macdonald, I GM8AVM
Macdonald, I G D G8MPS
Macdonald, I M GM7JED
Macdonald, J 2M0APB
Macdonald, J G4AUS
Macdonald, J GM7REY
Macdonald, J MM0BED
Macdonald, J A G8ZTR
Macdonald, J G G1OTZ
Macdonald, J L GM4XZN
Macdonald, J N GM4LQS
Macdonald, M D G0VME
Macdonald, N A GM4BVU
Macdonald, P G0BOQ
Macdonald, S MM0AVV
Macdonald, S F GM4GUL
Macdonald, S J G4AQB
Macdonald, S N G0ELN
Macdonald, S T GM0HUU
Macdonald, W M C GM4CKP
Macdonell, R R GM0CJT
Macdonnell, M G7AZU
Macdougall, H GM4FBU
Macduff, R A G4AWB
Mace, H G4ZJB
Mace, J R G3ZTU
Mace, M G6HDG
Macey, C P G6SGM
Macey, D G6STD
Macey, I D G3RVD
Macey, T G8JIS
Macfadyen, A D G3ZHZ
Macfadyen, R I G4JNA
Macfarlan, W A GM3GJB
Macfarlane, A G4CF
Macfarlane, A W C GM0MAC
Macfarlane, I G7JPL
Macfarlane, J G4URO
Macfarlane, J L G7NOY
Macfarlane, N GM7NYB
Macfarlane, R GM3EAK
Macfarlane, S D GM4ZWJ
Macfie, R J G4FAX
Macgillivray, K G GM4XKP
Macgilp, N S GM0GDH
Macgovern, P F G4KHN
Macgregor, A E G3LYG
Macgregor, A G G3HMG
Macgregor, G L GM7GBD
Macham, C G0NLI
Macham, J G0GOI
Machardie, I G M G3YMV
Machen, B G8VXV
Machin, F G3LUZ
Machin, J P G6TUI
Machin, R G6EOO
Machniak, F G4ZPH
Macinnes, W J G3CLW
Macintosh, J M GM3XMS
Maciver, A E G7LKR
Maciver, D G1SJU
Mack, E F G6YJK
Mack, S E G8ZTS
Mackay, A G G4CRS
Mackay, A G G4OLK
Mackay, A L GW0SQT
Mackay, A L GW7LCP
Mackay, A P G8FRO
Mackay, C GM0KVD
Mackay, C I MM0BAC
Mackay, D J GM3SYO
Mackay, F A GM0WXX
Mackay, J C D G4ZKY
Mackay, J S GI4OCK
Mackay, J W M G2BYP
Mackay, P G G0VRA
Mackay, R C M1ACK
Mackay, R D G8JMA
Mackay, R J GM8RVC
Mackean, R M GM4HAO
Mackellar, J C G3ETZ
Macken, D G1ISB
Mackender, M A G1IZP
Mackenny, D R G0LJJ
Mackenzie, A D GM0DKK
Mackenzie, A L GM7BOZ
Mackenzie, A N G4NUO
Mackenzie, D I GM0TNK
Mackenzie, D M GM4HJQ
Mackenzie, G W A GM4NUN
Mackenzie, I GM7GAE
Mackenzie, J 2M1ECF
Mackenzie, J G0JEG
Mackenzie, J A GM0HJU
Mackenzie, J A GM0MOC
Mackenzie, J D GM1XIA
Mackenzie, L A GM0TGG
Mackenzie, L J G1XUO
Mackenzie, M G4MFO
Mackenzie, N A GM3WIJ
Mackenzie, P J GM0HNV
Mackenzie, S G4XIW
Mackenzie, W GM0FSH
Mackey, J F G6SKX
Mackey, R G3SEY
Mackie, A J 2E1ETC
Mackie, G C G4NNB
Mackie, J R G0REY
Mackie, P A 2E1EDQ
Mackie, S T H 2M1ETM
Mackie, W J GM4AIE
Mackinder, D J G4DWP
Mackinlay, A J G6IHW
Mackinnon, D I GM0ADF
Mackinnon, D I GM0WSR
Mackinnon, D M G1TFY
Mackinnon, J GM0EUM
Mackinnon, J M GM4EKC
Mackinnon, J N G0HXX
Mackinnon, M J GM4AJV
Mackinnon, N G4WAZ
Mackinnon, W H B GM7GIO
Mackinven, P J G4TFH
Mackley, P D G1UTY
Macklin, B W G4BHE
Mackmin, M J G6PUE
Mackney, R A G4RAM
Mackney, V C G6NGK
Macknish, J G0TZR
Mackrell, G E G3KAX
Mackrell, R P G8ZGF
Maclachlan, I GM0HLP
Maclacklan, W G3TYT
Maclaganwedderburn, J H G4JOV
Maclagan-Wedderburn, J H G0OMY
Maclaine, E W GI4IWP
Maclauchlan, G A GM3NVU
Maclean, D R GM4ODW
Maclean, H G0FEP
Maclean, I A G0SRO
Maclean, J GM7TKA
Maclean, K D GM7JFN
Maclennan, C J G1HKF
Maclennan, D G3KGM
Maclennan, I M GM0DRU
Maclennan, M J GM4TJD
Maclennan, S J GM4UMA
Maclennan, S T G0TDN
Macleod, A E G4EZL
Macleod, C GM6TYX
Macleod, C GM7JGP
Macleod, C S G1NTO
Macleod, J GM0CBQ
Macleod, J GM7DXT
Macleod, J V G4ELS
Macleod, K U GM4VST
Macleod, M 2M1EOV
Macleod, N F C GM4DHN
Macleod, R A MW1ABT
Macleod, R F GM4DZX
Macleod, R M GM4EWL
Macleod, R M GM8DRA
Macleod, W G1PVS
Macliesh, C J G1FKS
Macliver, D J GM0ONV
Macmanus, E G4VVE
Macmillan, D G3GNA
Macmillan, K P G4JHY
Macmillen, P GW6SIX
Macnamara, D J G1DON
Macolive, P G0LQW
Maconochie, J J GM3GIG
Macphail, P J G0PMA
Macphee, C H GM8CFW
Macphee, J B GM3VNW
Macpherson, A C G3WLA
Macpherson, C J GM0EWX
Macpherson, D A GM7OBM
Macpherson, I A GM3RXU
Macrae, J T G4DXI
Macrae, J T G4LBS
Macrobbie, W E GM1BQP
Macrory, R GI4JTS
Macvean, L P G4LES
Madagan, P M G0ALE
Madagan, P M G3RQZ
Maddams, A W G7WGU
Madden, D G4ZPI
Madden, D GM1OGZ
Madden, J H P GI6KCX
Madden, J K GI0PXS
Madder, R G G0BEH
Maddex, C G0RAS
Maddex, V L G8YPK
Maddison, D G4MJB
Maddison, J W G3KAW
Maddison, M F G1YWM
Maddison, M H G4HIC
Maddison, R L G0GNE
Maddock, K J G4JDD
Maddock, W N G2AJS
Maddocks, J A G0OZE
Maddocks, K G8ZON
Maddox, P F G6PHZ
Maddrell, J W G4JND
Madeira, A M0AEA
Madeley, F G0VBP
Madeley, J R G0BYM
Madge, C W G8OXU
Madore, B J G1HRQ
Magee, A R GI0BFD
Magee, D J G4HAJ
Magee, J GI7KMC
Magee, T GI0BEB
Magee, W R GI3XLK
Mageehan, D J 2E1FRO
Mageehan, J D 2E1FRL
Magfhogartai, L G M GI0HGP
Maggs, P E G0OVY
Maggs, P E G0PKD
Magill, B N R G3RMF
Magill, H C G0LGV
Magill, I A GI4HCX
Magill, S A N G3ONW
Magill, S H S GI0LWO
Magnus, A U G3ZFT
Magnuszewski, E R G6UAP
Magnus-Watson, P A G6FCJ
Magowan, D J GI7OOM
Magri, R G0NCH
Maguire, A J G4GYQ
Maguire, A J O G7MYG
Maguire, D A GI0BQX
Maguire, J GM0PQV
Maguire, J J GI0LEC
Maguire, J J GI4UHA
Maguire, L P G4HXI
Maguire, T L G0NHP
Magwood, R A GW1XZI
Mahady, C J GM0LDF
Mahany, B F G4XAG
Mahany, D J G4XAH
Maher, R P G7NFX
Mahon, J K G0VVQ
Mahon, L G M1CDB
Mahoney, G GW6NHB
Mahoney, J C G6FCL
Mahoney, J C G7OTG
Mahoney, P F G8UMY
Mahoney, T J G1BAL
Mahony, C D G6INI
Mahood, K R G0OXV
Mahood, P GM8LYO
Mahrer, D P 2J1DGI
Mahrer, P R GJ0KYZ
Mahy, D J G8WJS
Maiden, J H G8STI
Maiden, J R G0MEP
Maidens, M W G6HGN
Maidment, J H G8FWE
Maidment, M A G1GYI
Maile, R D G8VDA
Mailey, R L G3BEY
Main, A A G0SPP
Main, A E GM0XAV
Main, I D G7VBT
Main, R GM1DTJ
Main, R A GM1BNA
Main, T G GM4DCL
Maines, J P G8RAC
Mainhood, D C G3HZW
Mains, W T GM3NMA
Mainwaring, E A 2W1BPS
Mainwaring, G 2W1DRP
Mainwaring, L T G4ZXQ
Mainwaring, P GW4NFJ
Mainwaring, R B G8GMT
Mainwaring, W J GW8AWT
Mair, C J GM7NUQ
Maires, A I G4YPI
Maisey, P J G4XPV
Maish, A D G4ADM
Maitland, G G3XYB
Maitland, P M G1ONC
Major, A M G4LOB
Major, A R A G3TWQ
Major, B G6VNW
Major, C L G7UYB
Major, G A R G7HZP
Major, M GI7PVI
Major, S R M0AKT
Major, T L G1GHD
Mak, K W G7HYK
Makeham, B G G4BQC
Makepeace, R L G0TRJ
Makin, C G4JZM
Makin, H G3FDC
Makin, J E G1ARF
Makins, T S G G4DNV
Makinson, T G GW3RVT
Makosz, P J G8DZC
Malbon, A N G8MIA
Malcher, A G G4TPM
Malcolm, A G8DEC
Malcolm, C GM3BXW
Malcolm, C E G3UYN
Malcolm, I M GM3VWY
Malcolm, J G6SYB
Malcolm, R GM4SXJ
Malcolm, R E G0DBT
Male, A L G0RIW
Male, B W GW8ZTP
Male, J L M G6BKT
Male, M N W G8TPT
Male, R G GM1JWJ
Malekout, D G6EEF
Males, J J G4DEW
Males, S G0EVZ
Maley, B P G7JWX
Malhi, A C G1TBN
Malin, T C GM3UVO
Mallen, J M1BIS
Mallender, T R G3WFD
Mallet, D L G4HIF
Mallet, N P G8TWE
Mallett, D M G3HUL
Mallett, J H G8IMP
Mallett, K G2AXU
Mallett, R G0NJD
Mallett, T G G6LCL
Mallin, D G1MOW
Mallinder, A G3JUY
Mallinson, B G4MYW
Mallinson, G G3NAK
Mallinson, I P 2M1DIN
Mallinson, M S G0LDB
Mallinson, R W G0EQI
Mallion, G C G0TZQ
Mallows, C D G8NTY
Mallows, F G1GYJ
Mallows, V J G3TSM
Malme, J M G4PQQ
Malme, P S G4PQP
Malon, P G4BLF
Malone, B L G6KLE
Malone, E W F G4MRT
Malone, G G4IVG
Malone, J GM7DZK
Malone, K GM0WHM
Malone, M G7WBZ
Maloney, E W G7NEE
Maloney, R A G0OKN
Maloney, S G0NYZ
Malpas, E A G0IPF
Malpas, R GW0DKG
Malpass, A G7CAG
Malpass, F W G3ELF
Malpass, P M G7FFM
Malpass, S J G0OGS
Malson, T L G0TZJ
Maltby, C R G2HLB
Maltby, J K G4FGY
Maltby, R G W G8RDG
Malynicz, G L G7RGF
Malyon, J B G8YGD
Malyon, M G6NHA
Mammatt, S C G6HKK
Man, J H G0NSI
Man, L J G8RTK
Mance, F E G0BIT
Manchett, B R D G4NZC
Mancini, G D V G6XUQ
Mandal, S K MW1BJB
Mandall, C E G1ARL
Mander, A M G8HPY
Mander, R W GW4DYY
Mander, V E GW8OXZ
Manekshaw, M J E GM4UKG
Manford, P R F G8XGK
Mangan, J G0DPV
Mangan, M M G1FON
Mangan, S M G1WQN
Manger, N D W GW0EHN
Mangles, D G6LCM
Mangles, T G4STP
Manklow, C S G4CGV
Manktelow, K G GD3SKZ
Manley, G C G8RRU
Manley, M R G4SMM
Manley, W L G1ISA
Mann, A A G4CMY
Mann, A N 2E1CVV
Mann, A N M1BWL
Mann, D G8ADM
Mann, D J G0HXN
Mann, D M G8WRA
Mann, D P G7TDM
Mann, E G8EKH
Mann, F GM0KUP
Mann, G J G0VHH
Mann, J N G7KYH
Mann, K P G7OGO
Mann, L W G0HUA
Mann, M C D G8BBW
Mann, P G7OJA
Mann, P G G4IVF
Mann, P J G6WWV
Mann, P O G4HTP
Mann, R A G1GKF
Mann, R C G8JIK
Mann, S J K G6XID
Mann, S P G8PUK
Mann, T G7MSK
Mann, T J G7IYM
Manning, A E G1XBX
Manning, A J G1ZBJ
Manning, C S G1SMT
Manning, D C G0KBM
Manning, D G G4VXR
Manning, F G0IQH
Manning, G L G4GLM
Manning, J G G6KWZ
Manning, M R G0IGM
Manning, P A D G1LKJ
Manning, P J G6ZPL
Manning, R G8VLZ
Manning, R F G8TAY
Manning, R L G8UBX
Manning, S J G1IRG
Manning, T B G8TWD
Mannion, R G3SWM
Mannion, R B G3XFD
Mannix, D G3XER
Mannix, E D G8OXI
Mannix, R K G6FJT
Mansel, R R G6AWO
Mansell, D GW8SFT
Mansell, J D G7GEL
Mansell, M R G4MAN
Mansell, P G3VKN
Mansell, R G3TZD
Mansell, R T J G0OVK
Mansell, W B G2CPM
Manser, P A G8NQC
Manser, R K G1DMR
Mansfield, A F G4YIZ
Mansfield, C J 2E1BNN
Mansfield, C L G0EFA
Mansfield, D G1HUX
Mansfield, D J G1DJM
Mansfield, E G1ZJR
Mansfield, J H 2E1DWF
Mansfield, L G0WQY
Mansfield, L M G0TVQ
Mansfield, L M G2SP
Mansfield, M W M0AWD
Mansfield, N D G6ZPV
Mansfield, T B M0AAJ

Name

Mansfield, T G R G4KAU
Mansfield, W A G3UWU
Manson, I A GM1PSU
Manson, R J GM1PWS
Manson, W D G8PW
Mant, I G G4WWX
Mantell, I G4NRY
Mantell, S G4NRX
Manthri, S G7UNS
Mantle, G G G7ACJ
Mantle, G J G0OYY
Mantle, J W G1CGQ
Mantle, R G8ZAD
Mantle, R A G6FMW
Manton, B G G1WSL
Mantovani, G G4ZVB
Manwaring, J W G7UXQ
Mape, G A G7PXS
Mapeley, D J M1AGG
Maple, G W G0BZO
Maple, P A G0BPZ
Maplesden, K F G4EMZ
Maplestone, G E G7EHT
Mapp, B K G3NVP
Mapp, C G1NTG
Mappin, D E G4EDR
Mapplebeck, R G3GXN
Mapson, M H G3UUI
Mapstone, R G0BMK
Maras, F G6XKZ
Marbus, W G0ERL
Marcer, A S G7ASM
March, D J G4DYT
March, E B G8EOJ
March, P L E G8FMT
March, R N R M0AKH
March, S M G4TRI
Marchant, B C G0SMH
Marchant, D C G8JQV
Marchant, L M G7EPS
Marchant, P J G3UWM
Marchant, P S G4OIM
Marchant, R F G6YHF
Marchant, R T G3TAJ
Marchington, R S G4MRQ
Marchington, R S G4SPA
Marcus, E C GM4YRE
Marcus, E E M1AUE
Marden, R G G3MWF
Marden, R W G1RND
Mardle, D P G6WCX
Mardle, P R G0HWI
Mardlin, D P GM6JBF
Mardo, A C G0OED
Margetts, C R G7VJM
Margetts, E J GM4BOA
Margetts, R L G8XRG
Margolis, L S G3UML
Margolis, R J G4TTZ
Margrave, F W G4VRG
Marinho, A N G0XBG
Maris, A J G3XDK
Maris, P J GW1WQM
Marjoram, D J G0JVT
Markeson, G G0IOK
Markey, B G0NMH
Markfort, R G8TQJ
Markham, G C G0OZR
Markham, J H F GW6INF
Markham, P D G0GYY
Markham, P J J G0OSO
Markham, R M0AGT
Markham, R A G0KYN
Marks, A G7DPV
Marks, C G4ZPJ
Marks, J A G6GVH
Marks, J F C G0OWT
Marks, J P G8WGN
Marks, P G8UGS
Marks, P J G8VZZ
Marks, P V G8FVK
Marks, R H G4ZFC
Marks, W D M G0JJZ
Markwell, J M G4SMV
Markwell, J W G1ABO
Markwick, A E G2YK
Marland, A G7GLT
Marland, F R E D G0ONE
Marlett, J G G1WDO
Marley, C T G4YXO
Marley, G R G0VFV
Marlow, C G G6NVJ
Marlow, D J G4OMV
Marlow, J F G0LDR
Marlow, J H GW1ORP
Marlow, K A G7AFQ
Marlow, M J G3IAF
Marlow, P J C G8BTV
Marlow, S D G7ITU
Marment, M A W G4MAW
Marmont, D W G3RRD
Marobin, L G0OOS
Marquis, I J G0FPV
Marr, J G4WUI
Marriner, R A G0HBG
Marriott, A C G3VWC
Marriott, A D G1EFF
Marriott, A M M G0GFL
Marriott, A M M G4RBV
Marriott, A V G7GTH
Marriott, C J G4RCP
Marriott, D G6IQP
Marriott, G G4ZSP
Marriott, H V G4ERT
Marriott, H W G0OWG
Marriott, J G8VWU
Marriott, J F G8BFH
Marriott, K G6NHY
Marriott, K W G7PND
Marriott, L P G4FFE
Marriott, M G0OPC
Marriott, P A GM0VXA
Marriott, P G G0RUD
Marriott, R G3LTN
Marriott, S G3MMM
Marris, R Q G2BZQ
Marrison, P G G6DYA
Marritt, D M G7SAT
Marron, J C M0ASN
Marron, J H G7WEN
Marrow, J G4IME
Marrows, A C G4REC
Marsden, A G3NTD
Marsden, A G G4FEI
Marsden, D G1ZQE
Marsden, D G3ZHP
Marsden, D H G0DUG
Marsden, D N G4HCH
Marsden, D R G4RMC
Marsden, G G4JME
Marsden, K G7CLX
Marsden, M G4AXX
Marsden, M J G8BQH
Marsden, N J GI4BQN
Marsden, P D G3LGQ
Marsden, R 2E1DSN
Marsden, R A G0BZK
Marsden, R F 2E1ESC
Marsden, R S G1YSY
Marsden, W G0PDK
Marsden, W F G4YJY
Marsden, W S G8CET
Marsdon, J A G4SNY
Marsh, B J G1JOA
Marsh, C C G0PXB
Marsh, C D G8IYN
Marsh, C E G4HKQ
Marsh, C I GW8REV
Marsh, C R G1PHV
Marsh, D J G7YEW
Marsh, E G1XKY
Marsh, E G3ILE
Marsh, I G4INK
Marsh, J G0IPK
Marsh, J B G7BRB
Marsh, J T G3SYG
Marsh, K I G8NEI
Marsh, K J G7JUC
Marsh, K J M0AID
Marsh, L E G6DKF
Marsh, M J G0SWI
Marsh, M J G4GGC
Marsh, M J M G1IAR
Marsh, N J G1SGR
Marsh, N J G1ZBL
Marsh, P G4WFZ
Marsh, P J G7EYT
Marsh, R G4YZK
Marsh, R G8TYH
Marsh, R C G8XWI
Marsh, R S G7UZQ
Marsh, R T G4GRZ
Marsh, S G4BWG
Marsh, S L C GW7VCH
Marsh, S M G1BNG
Marsh, W F G4GBB
Marsh, W R G4IUO
Marshall, A G0WPN
Marshall, A G6FCN
Marshall, A GM4GIL
Marshall, A GW0FJQ
Marshall, A D G3MJM
Marshall, B R G1DVD
Marshall, B R G4BJF
Marshall, B T G0WSO
Marshall, B V G3RUX
Marshall, C G1NPU
Marshall, C G4IOK
Marshall, C A G7JJJ
Marshall, C A G8ZQK
Marshall, C C GI0EZD
Marshall, C J 2E1EQC
Marshall, C T G3YRN
Marshall, D E G0HED
Marshall, D F G0NQW
Marshall, D J G3RHG
Marshall, D S G8MGD
Marshall, D W M0AXE
Marshall, E G0TZF
Marshall, E T G4PPB
Marshall, F G0RWV
Marshall, F E G2XQ
Marshall, F G G1MVG
Marshall, G G0FEJ
Marshall, G G1WRH
Marshall, G G3XGM
Marshall, G B G6HLR
Marshall, G C G3ZDU
Marshall, G F G3RJW
Marshall, G L G8BJW
Marshall, G N GM4GVJ
Marshall, G R G3HWS
Marshall, G R G4IQC
Marshall, H G0JMN
Marshall, H D W G6JTW
Marshall, I F 2E0AOZ
Marshall, J G4MHF
Marshall, J G8HWA
Marshall, J GM8LJ
Marshall, J B GM7UOJ
Marshall, J E G0NPU
Marshall, J E G8ZXT
Marshall, J E S G4RVR
Marshall, J L 2E1DJT
Marshall, J L G3RKH
Marshall, J M GM1FAF
Marshall, J R G4KFP
Marshall, J R E G3UBM
Marshall, J T G4JIA
Marshall, J W G4MFV
Marshall, K G4IIB
Marshall, K G4LNQ
Marshall, K C G7NDB
Marshall, K E G0OGJ
Marshall, K F S G8WPE
Marshall, K R G7TLR
Marshall, K W G3ZTI
Marshall, L G1EPF
Marshall, L C G3IGN
Marshall, N A G0XTM
Marshall, N C G8EFY
Marshall, N J G1NTW
Marshall, O J 2E1FGA
Marshall, P G0MLF
Marshall, P M1BJC
Marshall, P A G7CBG
Marshall, P A G8SGH
Marshall, P J G8MOL
Marshall, P S M1BBB
Marshall, R C G3SBA
Marshall, R D G4GIQ
Marshall, R D G4KEW
Marshall, R E W G8HLE
Marshall, R F 2E1CAW
Marshall, R J G7PHL
Marshall, R J G8GFA
Marshall, R J GM4JJG
Marshall, R K GM3RXZ
Marshall, R M GM4GIO
Marshall, R W G0PFY
Marshall, R W G4ERP
Marshall, S 2E1DZD
Marshall, S G0JKJ
Marshall, S B G3YXH
Marshall, S D W G0UAI
Marshall, S E G6NHG
Marshall, S G G1RYQ
Marshall, S M G4ZXM
Marshall, T G3OPM
Marshall, W G7LER
Marshall, W GM4IMX
Marshall, W F G4IOD
Marshallsay, M L G8DYG
Marshland, B G7SWD
Marshman, J D G4LIO
Marshman, M J G6GYF
Marsland, L R G0DBE
Marsters, D A G1KIJ
Marston, A G1FTH
Marston, A C G7VDU
Marston, A D 2E1FSD
Marston, F W G G0ZIP
Marston, F W G G7VIP
Marston, J R F G1JSP
Marston, M E G7EFG
Marston, N E G0ASM
Marston, R G7OJC
Marston, R GW1WRV
Mart, P A G8NHD
Martell, P G7NPY
Marten, P J G8SZB
Marten, T G0AOA
Marten, W J G3ARO
Martich, K GW4YKM
Martin, A GM7ONJ
Martin, A M0AJM
Martin, A M1BER
Martin, A G G4XBY
Martin, A G GJ3ECC
Martin, A J G3HCU
Martin, A J G4HBV
Martin, A J G8VA
Martin, A R G8AFZ
Martin, A W T G8ZPW
Martin, C G0RYP
Martin, C D G6FHE
Martin, C J T G4JWX
Martin, C P G0UAC
Martin, C R G0XAU
Martin, C S G7KNQ
Martin, D 2M1EBJ
Martin, D 2M1EDM
Martin, D G0ORO
Martin, D G1YEV
Martin, D G6OZH
Martin, D M0BEE
Martin, D A G G3ODC
Martin, D G G3IER
Martin, D G G8IOJ
Martin, D G MI0AIH
Martin, D H G3RUZ
Martin, D J G6HKL
Martin, D M G1IJQ
Martin, D P G0OKA
Martin, D P G7AEY
Martin, D R G1ZIU
Martin, D R G4RST
Martin, D R G7HOL
Martin, D S G4DBZ
Martin, D W G3XSF
Martin, E M0ABZ
Martin, E C G0UBO
Martin, E G G0LAA
Martin, E G G0WYU
Martin, F 2E1CCJ
Martin, F D GW4LMU
Martin, F H G2CDT
Martin, F S G4UMF
Martin, G GM3NVQ
Martin, G A GI3XCZ
Martin, G D G4KOU
Martin, G E G0IQW
Martin, G E G3GWH
Martin, G R G6GQF
Martin, G W G0HXR
Martin, H G3JDO
Martin, H G3VES
Martin, H G3XCD
Martin, H G6JTI
Martin, H D W GM4TMS
Martin, H D W GM4UYE
Martin, H W G7RUK
Martin, I G6DQO
Martin, I G8RVZ
Martin, I A M G6CKL
Martin, I D G6ZNO
Martin, J G4IAS
Martin, J G4ONN
Martin, J G4TMQ
Martin, J G7VEY
Martin, J G8JGM
Martin, J G8ZOP
Martin, J MM1BGI
Martin, J A G4ULM
Martin, J C G0VKV
Martin, J C G6MC
Martin, J C G7EWR
Martin, J E G7PDR
Martin, J E GD4RAG
Martin, J E GU3YIZ
Martin, J G GM4NLJ
Martin, J H G0IXC
Martin, J H G6YHD
Martin, J H G8XLB
Martin, J N P G3YRF
Martin, J R D GI7LAA
Martin, J S G1KIB
Martin, K G1IKE
Martin, K G4UBK
Martin, K J G0GFQ
Martin, K J GM7APK
Martin, K L G0PQO
Martin, K S G3FHG
Martin, L 2E1BKM
Martin, L G7DXB
Martin, L G G4EKY
Martin, L L GM0FVJ
Martin, M J 2E1BAI
Martin, M P J G0JCZ
Martin, N G4KGQ
Martin, N G6HVE
Martin, N M1BTO
Martin, N C 2W1CRR
Martin, N J G6NHK
Martin, N S B G4GAX
Martin, N T G6OPO
Martin, P G4KHK
Martin, P A G6EON
Martin, P B R G0BUW
Martin, P C G3VLW
Martin, P C G6OPP
Martin, P C G7MSL
Martin, P H G8YPL
Martin, P I G4AZC
Martin, P J G0CTR
Martin, P J G8LZS
Martin, P R G3PSU
Martin, P S G8NRY
Martin, P S GW7LDP
Martin, P T G6GVM
Martin, R G0HNU
Martin, R G7IED
Martin, R D G7LEG
Martin, R E G0UNW
Martin, R E G6NAV
Martin, R J G4PPX
Martin, R J G6LRA
Martin, R J G7SDG
Martin, R K A G4WXY
Martin, R L G0WKL
Martin, R L G7TIN
Martin, R P G8EBQ
Martin, R W A G0RWM
Martin, S 2E1BKP
Martin, S B G0JPI
Martin, S G G8OGP
Martin, S J D G6WHH
Martin, S K GM0HMR
Martin, S M G1KNI
Martin, S S G7HON
Martin, T G1CKJ
Martin, T B G7CEN
Martindale, A G3MYA
Martindale, A 2E1DUV
Martindale, C W GM4PSF
Martindale, J G0EYB
Martindale, J H GM4VPA
Martindale, J W GM8MLH
Martindale, P T G7OHD
Martindale, S G4XFO
Martindale, T G G7RXQ
Martinelli, V C G0NNT
Martinez, J P G3PLX
Martland, A G1AHM
Martland, D V G0PXL
Martland, F G6RHQ
Martlew, B G G1NXS
Martorano, G G4NNZ
Martyn, J C G7JAN
Martyn, L N G7PUZ
Martyn, M C G7EVE
Martyn Clark, J E G6KTO
Martyr, D C 2E1FYC
Martyr, E B G6HKM
Martyr, J R D G0MWT
Martyr, J R D G3PMX
Marvelley, S E GW4TGA
Marven, A G3PWQ
Marvill, J K G0UFW
Marvin, D P G7JIQ
Marwick, D S GM1RQD
Marwood, A T G8SSL
Marwood, P G7TCY
Marzuki, A M1BEF
Marzuki, H J G7LNY
Mase, B J G3GLA
Maskell, K J G3YSC
Maskell, K J G4YTU
Maskell, K J G8TRF
Maskell, P C G0PGM
Maskell, T E G0IJS
Maskelyne, K F G1ISE
Maskew, R G4TDW
Maskill, R J G0PUV
Maskill, R J G3CCA
Maskill, R J M1AAF
Maskort, G H G0DKO
Maskrey, S G6FDK
Maslen, I A G4BYR
Maslen, W A R GM4LPG
Mason, A G4TGU
Mason, A G6JSR
Mason, A C 2E1AEI
Mason, A D G3NNQ
Mason, A D GW4TQU
Mason, A F G6GNH
Mason, A F GW0TKX
Mason, A G G1NKN
Mason, B R G4RKQ
Mason, C C G0HRJ
Mason, C J W G4OXM
Mason, C N G4UZE
Mason, D B G0PBH
Mason, D G G3USD
Mason, D I G3ZPR
Mason, D J G0SOV
Mason, D J G3RXP
Mason, D L G G4NDC
Mason, D N G4AZK
Mason, D W G0HPJ
Mason, D W G4EHW
Mason, E G0PXX
Mason, E G G3PNB
Mason, E N G0ASP
Mason, F G1SBP
Mason, F G3TWN
Mason, G G4AIT
Mason, G G4VCA
Mason, G C G4PTK
Mason, G D G7VIL
Mason, G E G3YJG
Mason, G F G4IWF
Mason, G F G5BR
Mason, G G R GI3CKF
Mason, G H G0OSC
Mason, H T G4AOA
Mason, I M G0EWZ
Mason, J A G8NWL
Mason, J C 2E1DCY
Mason, J C GW0VSW
Mason, J L G7PBV
Mason, J P G3ZJZ
Mason, J S B G8YFK
Mason, K C G4JIO
Mason, L A M1AJA
Mason, L E G4HTD
Mason, L R G1CBL
Mason, M L G4OTY
Mason, M N G0DHL
Mason, N W G0EAZ
Mason, P M1BTU
Mason, P G G0VAG
Mason, P L G1RCI
Mason, P W M0BCW
Mason, R A G4SIE
Mason, R A G6HKS
Mason, R D G3TDM
Mason, R E G4GVR
Mason, R E MM1BRH
Mason, R F G0AQU
Mason, R G M G4YPG
Mason, R I GW6WQF
Mason, R S G0VAF
Mason, R T G1GKA
Mason, S G4EWT
Mason, S G4KNR
Mason, S G6WPE
Mason, S A GW0WQE
Mason, S M G0JMY
Mason, S P G6IX
Mason, T G3WVA
Mason, V GM4GGF
Mason, W M0AGW
Mason, W D G8ZXY
Massen, M G G4YDZ
Massett, J P G0UMN
Massey, A J G0EDV
Massey, B G0UNG
Massey, D R G8LRS
Massey, E W G8XQH
Massey, G W J G3BZR
Massey, H G4PYC
Massey, H W I GI7FNP
Massey, J G6LBE
Massey, J M G6YCZ
Massey, N P 2E1CRZ
Massey, N P G7VJV
Massey, P L G4YCA
Massey, R H G0NEQ
Massey, S D G0AAV
Massheder, P M G0EIF
Massingham, G W G6GPL
Massingham, W E G3DWS
Masson, A J G3PSP
Masson, A T 2E1CAA
Masson, R G7TBR
Masson, T D G7SJK
Masterman, M B G4XXY
Masterman, M J G7PTA
Masterman, P H G3RHH
Masterman, S F G4YEI
Masters, C S G3XED
Masters, E L G0KRT
Masters, J D G3MBM
Masters, J V G0IXD
Masterson, J R G0RGS
Masterson, M R G4GGT
Masterton, J G8FUL
Matcham, C G0ALB
Matcham, W C J G4HQJ
Matheny, M E G0FMH
Mather, A G0OHY
Mather, A G1OAJ
Mather, A G3ZIK
Mather, C G1PCU
Mather, D C GM3KAM
Mather, G R G8DHE
Mather, H L G8TWC
Mather, J E G6UNC
Mather, K J G3EHE
Mather, L J G8OKI
Mather, R W G4ZDG
Mather, S C GM4OGM
Mathers, D J G4PQB
Mathers, F N GW3JBJ
Mathers, G G MM1CAC
Mathers, J G6JBG
Mathers, J C GM4CNF
Mathers, J F GW1XPP
Mathers, J M M GI7TEB
Mathers, J R GM4EAW
Mathers, M J G4ODD
Matheson, D C G4CKO
Matheson, D K G8URC
Matheson, I R G3RRA
Matheson, P R GW0HCB
Matheson, R D GM6UYC
Mathews, F G8ACJ
Mathews, P G0BLM
Mathews, P L G4WJH
Mathews, R C G3RLH
Mathias, A C GW0TLJ
Mathias, D F G0ALY
Mathias, W J GW8CNS
Mathieson, D GM1AFD
Mathis, R E M0APP
Mathur, M N G0JMM
Matkin, P G0DDI
Matkin, P L G7ASY
Matley, H G3YGU
Maton, K L G6NHU
Maton, S G0KXL
Matson, B J H G3RDO
Matsuoka, T G7RMR
Mattacks, H F G3EKJ
Mattei, S G0LGY
Matthew, J G6NID
Matthew, K G0WYS
Matthewman, A GD4GWQ
Matthewman, C GD4FWQ
Matthews, A G6YHS
Matthews, A D G4DGF
Matthews, A J G3UNM
Matthews, A P G1AQF
Matthews, A R G3VFB
Matthews, B GW0JWF
Matthews, B J G4UZF
Matthews, C A G1LQT
Matthews, C H G8NXU
Matthews, D G3ZZP
Matthews, D G G0ICD
Matthews, D J GW0CKX
Matthews, D T G0OWE
Matthews, D W 2E1FWW
Matthews, E A G3FZW
Matthews, E A G8WET
Matthews, E E 2E1AWK
Matthews, E H G3NPL
Matthews, F G4VPN
Matthews, G M GM0UUB
Matthews, G P G4CGY
Matthews, G P G4LLI
Matthews, I D G7EXP
Matthews, I J G3URM
Matthews, J G6ASK
Matthews, J A G7RCW
Matthews, J H G3LTG
Matthews, J L G3CLA
Matthews, J R G3WZT
Matthews, J R G4BFW
Matthews, J R G4HRS
Matthews, J V G3HUX
Matthews, K R G7UUD
Matthews, L N G3LHS
Matthews, M G6OYF
Matthews, M G8SZR
Matthews, M A G6MTY
Matthews, M J G3ZBF
Matthews, M J G6USD
Matthews, N M K G8GTZ
Matthews, P G4AWZ
Matthews, P G7UUA
Matthews, P G8SEV
Matthews, P G G8LBM
Matthews, P H G1TWN
Matthews, P J G6HMA
Matthews, P J G6KUA
Matthews, P J H G3BPM
Matthews, P R GM7CDI
Matthews, R 2E1FWQ
Matthews, R G4KAQ
Matthews, R A G0NTM
Matthews, R A M G4KGO
Matthews, R J G0ECI
Matthews, R J G3SAH
Matthews, R W E G0PKG
Matthews, S A G1BHB
Matthews, S A G1VKN
Matthews, S D M G1IAB
Matthews, S E G8DHF
Matthews, T G3RGC
Matthews, W GW4YWM
Matthiae, D G0VVZ
Mattingly, D G4EBC
Mattinson, H D GM1CUC
Mattison, C F K G0SLB
Mattison, J L M G0PTG
Mattock, G 2E1GBP
Mattock, J J G4VBO
Mattocks, A M G1IWS
Mattocks, J R GW4TEQ
Mattos, P G G8LOU
Mattravers, S W GM7DUG
Mauchel, D J G3OVC
Maud, D J G8MBB
Maude, A G4JBK
Maude, H W D G4YDG
Maude, J G4PVS
Maude, R G0SWM
Maude, R G0VHU
Maudsley, R H G1PHB
Maudsley, W A G4ROU
Maufe, A P B G8WBK
Maughan, J G3ZMG
Maughan, L 2E1BQF
Maughan, L G7SOR
Maughan, S G0FFB
Maule, J D G7LVM
Maule, R M G G3OEF
Maund, L E G3HNB
Maunder, A G G7IWA
Maunder, J L G4RNR
Maunder, J M G6KBI
Maunder, L C G4KRX
Maunder, T D G0IOI
Maver, P GM0VYL
Mavin, D G4OIV
Mavin, M F G0KGY
Maw, J S G0DFT
Maw, K G8ZTO
Mawby, M A G4CQN
Mawdsley, D V G3POG
Mawdsley, P M G7VQR
Mawhinney, D GI4KSO
Mawhinney, D J M GI6BNI
Mawn, B R G0BKD
Mawson, A G0JVI
Mawson, A G0UXH
Mawson, D J G0TKG
Mawson, H T G1VVT
Mawson, J E 2E1CPW
Mawson, J S GM7KZL
Mawson, S A J G7OWN
Maxey, M A G8CTJ
Maxfield, C G GW3HAI
Maxfield, D G3ZRQ
Maxfield, F R G4GQG
Maxfield, V B G8NWM
Maxted, D G7DVJ
Maxted, K J GM4JMU
Maxwell, B G1GWF
Maxwell, D A 2W1AYZ
Maxwell, D C GM0ELP
Maxwell, G D A GM4BAE
Maxwell, J G6TQS
Maxwell, J J G6UUY
Maxwell, J J W G7DXC
Maxwell, J S G6XPU
Maxwell, K W G6XPT
Maxwell, M R G8YOY
Maxwell, N A GW3UMD
Maxwell, R C G8MKT
Maxwell, S G7UJO
Maxwell, S L G6IHD
Maxwell, T GM0TKC
May, A B G5AL
May, A R 2E1DSA
May, A R M1BSX
May, B G4EXW
May, C A G3KMP
May, C H G1GWG
May, D O GW4REI
May, E J G0OZI
May, E J R G6UQN
May, F A G8AAR
May, G G1UTS
May, G G4RZF
May, G M0AHF
May, G C GW8TIX
May, H W 2E1DRW
May, H W G7VDN
May, J E MI0AQX
May, J L MI0BES
May, J P G4YMZ
May, J R G4JFG
May, J S G0CDB
May, J S G4SSD
May, J S G7FDC
May, K R G4APB
May, K W G3GMK
May, L G4HHS
May, M A G7VXD
May, M I G0UPO
May, N A G0MIG
May, N A G4FWN
May, P D G0UUQ
May, P D G7BYF
May, P R GM7KTY
May, R G4YLQ
May, R G8IBP
May, R C G G4CEH
May, R D G3KTF
May, S G3ENL
May, S T G4CTQ
Mayall, A G GW6OJK
Mayall, J G3VPH
Maycey, W D G4YIH
Maycock, B A G3JWQ
Mayer, P W G0KKL
Mayer, S J G0HKT
Mayer, S S G0HQH
Mayer, T J G8KST
Mayers, A A G3UOB
Mayers, D R G6VKL
Mayes, A C G1GXX
Mayes, A J G4ZQJ
Mayes, D W G3MMA
Mayes, J R G1PQT
Mayes, K P G0EBL
Mayes, N GI4MAY
Mayes, N B G0JUK
Mayfield, A P G7TLH
Mayfield, J P G0LXX
Mayfield, L G0IGZ
Mayfield, S A 2E1AZG
Mayfield, T M G4YQD
Mayfield, T P G7BHU
Mayhead, L V G3AQC
Mayhead, L V G4FRS
Mayhew, A F G8TQK
Mayhew, C E W G1UAR
Mayhew, L H G8RDK
Maylin, A N M1ANM
Mayman, A S G6JFU
Mayman, C L G2ABR
Maynard, B M G6LUO
Maynard, G A G3XZJ
Maynard, J G4EJA
Maynard, K G7OML
Maynard, P A V G8VYF
Maynard, P C G0TCP
Maynard, R F G1XRC
Maynard, R F G4YRM
Maynard, W H D G0DEN
Mayne, G F G4IPV
Mayo, C F G0KVR
Mayo, G E G4EUF
Mayo, J G GW7TVM
Mayor, H C G4MGB
Mayou, R G1TOK
Mayson, B J G1LHL
Maytum, R H G7BLJ
Mazura, I M G7LAL
Mc Afee, G A GI7SLN
Mc Allister, K G7OOH
Mc Anerney, C R GM3XNJ
Mc Ardle, J S GM6OFB
Mc Ateer, R GI0MRI
Mc Ateer, R GI4MFM
Mc Bride, J J GI1CAI
Mc Caldon, P R E G0DPK
Mc Call, C S G G6VGA

Mc Carrison, J ... MI0ABD
Mc Chrystal, J ... G1ETA
Mc Clusky, L M ... 2E1BMI
Mc Cormack, J ... GI4CSO
Mc Culloch, H L C ... GM4CBV
Mc Culloch, J D ... G0IUM
Mc Cullough, R D ... GI6IHX
Mc Dowall, W J ... GW6ZMN
Mc Fall, J L ... G4HFX
Mc Farland, M S D ... GW0GLX
Mc Faul, W A ... GI4FHB
Mc Ghie, A M ... GM3AUE
Mc Glasson, D J ... G6NVF
Mc Guinness, J G ... GI7KMJ
Mc Kavanagh, J F ... GI4SIP
Mc Keown, F W ... G4WXI
Mc Killop, W A M ... G8CIT
Mc Menamin, W P ... GI7MAC
Mc Mullan, I ... GI0RDJ
Mc Murtry, A ... GI3MBB
Mc Pherson, M ... G0GYO
Mc Quaid, S A ... GI4LDN
Mc Taggart, P M ... G6BJO
McAdam, S L ... G4VBD
McAdams, S C J ... GJ7DNI
McAleer, B W ... GI3MMF
McAleer, R J ... G0VLF
McAlister, C A N ... G7OJI
McAlister, P J ... G3YFK
McAlister, R C ... GI0DFD
McAllister, C ... G4KKP
McAllister, J D ... GM1AYT
McAllister, N ... GM0FSW
McAlonan, D ... GM4SFT
McAlpine, D ... GI0YES
McAlpine, I T ... G7FTO
McAlpine, N P ... G8OSG
McAlpine, P ... GI3WFP
McAlroy, H R ... G0PCN
McAndrew, D ... G3RYX
McAndrew, I W ... G4DSI
McAnespie, B P ... GI0JRD
McAra, C V ... G1EFG
McArdell, D J ... 2E1GBO
McArdle, P ... G0DAG
McArdle, P ... G0OBR
McArdle, P ... G4MSM
McAreavey, W J ... G7JSV
McArthur, D ... MM1AHL
McArthur, D R ... GM3TNT
McArthur, E J B ... G3VVA
McArthur, S ... G0CBI
McArthur, W ... GM7PVT
McAsey, A ... G0OWY
McAtee, D A ... G0KKM
McAteer, P J ... GI7IRJ
McAteer, S F ... GI0NOX
McAulay, I C ... GM6CZM
McAulay, J I ... GM6NIC
McAuley, P G ... GI4JIC
McAuliffe, P D ... G4XFB
McAuslan, D N ... G3ZMH
McAvoy, G C ... G4NFT
McAvoy, I ... G0RPA
McAvoy, I L ... G0NKA
McAvoy, J ... GM3RPM
McBean, C D ... GM1OWV
McBride, A N M ... GI4UIV
McBride, A T ... GM0IMW
McBride, J V ... G0PJI
McBrien, C R ... 2I1DDF
McBrien, G ... G6SHD
McBrien, S F ... GI7UJI
McBurney, J ... G4AUR
McBurney, R ... GI3HJH
McBurney, W ... GM0TTY
McCabe, A ... GI0VWU
McCabe, A E G ... G7PZY
McCabe, A P ... G0MCC
McCabe, D P C ... G8JXP
McCabe, I R ... G0FYD
McCabe, J ... GM4RPE
McCabe, S F ... GI4TAP
McCabe, T C ... GM4XGW
McCafferty, A J ... GM0PHM
McCaffery, K R ... G7FRW
McCaffrey, B A ... GM8SZS
McCagherty, B ... GI4GNZ
McCaig, J ... G3LQG
McCaig, J S ... GM3BQA
McCalden, A J ... G8ZMC
McCaldin, A D ... GI0HXH
McCall, D A ... GM6NIA
McCall, H G ... G4WYA
McCall, J ... GM3HGA
McCall, J S ... G3ZBS
McCall, P ... G4RGF
McCallan, M M ... GI4RYL
McCallion, A P ... GI4VKS
McCallum, A ... GW6NHL
McCallum, D N ... GW6CWZ
McCallum, G D ... GM3UCI
McCallum, J R ... G4YMC
McCallum, R J ... G4VNG
McCallum, W E ... GM0POD
McCandlish, W ... GM1FMX
McCann, A ... G3AZI
McCann, A ... G3PS
McCann, A B ... GW4GAF
McCann, B ... GI4OTZ
McCann, C D ... GI4XAA
McCann, F ... GI0IMY
McCann, I G ... G4RFJ
McCann, J ... GW4PFC
McCann, J D ... G0MCN
McCann, J H ... GI3YBZ
McCann, J H ... GI8BNC
McCann, K L ... 2E1DRU
McCann, K W ... G6YYN
McCann, L ... G7TVC
McCann, M D ... GI0DES
McCann, R A ... GM8ZJL
McCann, R J ... G8OKB
McCann, S L ... G6UBU
McCann, S M ... G4HFZ
McCarlie, G ... GM4XEQ
McCart, R ... GM0JBE
McCarthy, C J ... G3XVL
McCarthy, D M ... G0OTD
McCarthy, I D ... G3FEC
McCarthy, I D ... G3YBY
McCarthy, J ... 2E1FNF
McCarthy, J C P ... 2E0AEZ
McCarthy, J C P ... G7JTT
McCarthy, L J ... GW6KQC
McCarthy, M A ... GM4FGI
McCarthy, M P ... G0KCH
McCarthy, P ... MW1AWT
McCarthy, T B ... G0WBF
McCarthy, T E ... G1HWK
McCarthy-Stewart, C J ... G6PBS
McCartney, B J ... G4DYO
McCartney, D K ... G4TXA
McCartney, G G ... G4VYR
McCartney, K S ... GM8ETJ
McCartney, L R ... MM1AAC
McCartney, R B ... GM4BDJ
McCartney, W A ... G3SOA
McCarty, D ... G0GWZ
McCarty, R A ... G3OEM
McCaughey, R ... GI4NTO
McCaughey, W J ... GI4XJJ
McCaulay, P ... G7VTN
McCausland, B J ... GI0KPF
McClea, O ... G7UDB
McClean, S S ... GI0PFL
McClean, S T ... GI4ESI
McClelland, A ... GM0BFW
McClelland, A W W ... G0DKN
McClelland, C ... G6AG
McClelland, P ... G1OVY
McClelland, W T ... G0LGQ
McCleverty, A J ... GW0VEM
McClew, D G ... G7IJL
McClintock, C F ... MM1AUF
McClintock, G ... GM8YUI
McClintock, J T L ... GI4MAJ
McClintock, M R ... MM1AUG
McClintock, W J ... G3VPK
McClory, B A ... G3UJA
McCloskey, J M ... G1JLM
McCloud, C J ... G4EFB
McCluney, D ... GI4MVQ
McClure, C A ... GM0HJV
McClure, I F ... GM4ZRQ
McClure, K W ... GM1XBK
McClure, V W ... G8WCQ
McClurg, C R ... GI4ISR
McClurg, S M ... GI0NMV
McCluskey, P P ... G7DLC
McColl, A ... G7BZU
McCollin, J H ... G7JHM
McComb, J E ... G6ZIC
McCombe, B H ... G3ZJW
McCombe, S ... G4WLP
McConachie, A ... G8RUP
McConnachie, A ... G7RRJ
McConnell, G P ... GW0DIB
McConnell, K H ... GI7DQI
McConnell, R ... G6IQT
McConnell, S ... GI4MBM
McConnell, W H ... G6PAZ
McConville, D M C ... GI6FQT
McCorkell, D ... GI0SFX
McCormack, A P ... GM1AHF
McCormack, J R ... G7HBS
McCormack, N ... GM0KBC
McCormack, N M C G ... GM1AHG
McCormack, P ... G7VFS
McCormack, P A ... 2E1FUA
McCormack, P J ... GI4EQW
McCormack, W J ... G6RQL
McCormick, D J ... G0VCB
McCormick, E P ... G8ZAC
McCormick, G J ... GI7PWQ
McCormick, I N ... GW0JJF
McCormick, J B ... GI7FCP
McCormick, J N ... GM4ZNA
McCormick, K W ... GM1RHA
McCormick, L ... GM0LDE
McCormick, R E ... G4ZCA
McCormick, S ... GI7TFK
McCormick, S J ... G7KOS
McCormick, S R ... MI1BRO
McCormick, W D ... G1JHG
McCourt, B ... G8ZJH
McCowatt, R A ... G3WPK
McCoy, L P ... G1KON
McCoye, D A ... G8UZU
McCracken, M ... G0VCO
McCracken, P J ... G1SMX
McCracken, R D ... GM4CEA
McCracken, R L ... G6EIH
McCrandles, D J ... GM4SNZ
McCrea, J W ... GI4MRN
McCrea, R E ... GI3WBR
McCreadie, A M ... GM0BPY
McCreadie, R A H ... G0FGX
McCreery, M M ... GM0IBM
McCreery, M M ... GM0RAY
McCreight, J F ... GM3DJS
McCrimmon, T ... G4LQM
McCrory, P F ... GI7RTB
McCrum, D ... G8BLF
McCrum, M ... 2M1FGU
McCrystal, E I ... GI7FHZ
McCubbin, R W ... G4OLA
McCudden, A T ... GM4DLU
McCue, W J ... G6HKN
McCullagh, J B F ... GI4UDI
McCullagh, J K ... GI4BWM
McCullagh, J K ... GI4KKK
McCullagh, J M ... GI7JKA
McCullagh, S J ... GI6EEH
McCullagh, S J ... GI6NAQ
McCulloch, A A ... G4HTL
McCulloch, I ... G1RYS
McCulloch, J C ... GM1SRP
McCulloch, J C ... GM3FAH
McCulloch, M ... GM1XEB
McCulloch, R E A ... G0RJQ
McCulloch, S J ... G4TPO
McCulloch, T W ... G0SQD
McCulloch, W K ... G4IDB
McCullough, G M ... G1RNU
McCullough, J S ... G1WGF
McCullough, J T ... GI4SFE
McCullough, L ... GI4RMA
McCullough, T ... GI3SCM
McCullough, W J ... GI4BQI
McCurrach, R D ... G4ASF
McCurrie, P ... G4ADP
McCurry, J A ... GI4SZU
McCurry, R T ... GI4OCL
McCurry, T W ... G3VSK
McCutcheon, D B ... GI3OAU
McCutcheon, G C ... GI1WLJ
McCutcheon, M ... GI6MTL
McCutcheon, R P ... G1WVZ
McCutcheon, T L ... GM0HPL
McDade, A T ... M1CDP
McDaid, B J ... MI0AQV
McDermid, A ... 2M1FLG
McDermid, A ... MM1BJP
McDermott, D B ... G4HXU
McDermott, J ... GM6LEZ
McDermott, J C ... GI4NMZ
McDermott, J M J ... GM4NTL
McDermott, M ... G6NAD
McDermott, M A ... GM0WIB
McDermott, M J ... G7UHW
McDermott, M P ... G7TOJ
McDermott, R S ... G6TDR
McDiarmid, D ... G3FMU
McDicken, W S ... GM4XMD
McDonald, A ... 2E1FEW
McDonald, D T ... G0AQZ
McDonald, E A ... M0ACB
McDonald, E G ... 2M1FHR
McDonald, E G ... MM1BYX
McDonald, G ... GM4LZO
McDonald, G A ... G1BXN
McDonald, J ... GI0THR
McDonald, J ... GM1VYG
McDonald, J K ... GM4VEJ
McDonald, J P ... G8PJC
McDonald, M F ... G3RZM
McDonald, P G ... GI0USW
McDonald, P P ... G0SFT
McDonald, P P ... GI7KVR
McDonald, R A ... G0IZP
McDonald, R G ... G3DCZ
McDonald, S O ... G0IKN
McDonald, T S ... G7OGT
McDonald, W M ... GM0TKM
McDonnell, B ... G6MAC
McDonnell, D ... G0RZB
McDonnell, J F ... GI8ZHW
McDonnell, P J C ... G0AFS
McDonnell, P M ... G1FIM
McDonnell, T P ... G0MVN
McDonough, R ... G6EYH
McDougall, A H ... GM0AYT
McDougall, L J ... GM3CIX
McDowell, A ... G0KOO
McDowell, G G ... GI3XDX
McDowell, M ... GI4FNU
McDowell, R J ... GI6IHM
McEachen, A ... GM3JUF
McEachran, J ... GM3XGX
McElhatton, K ... GI3NFM
McElhinney, A C ... GI7GAQ
McElroy, A C ... GM8KSQ
McElroy, D P ... GI0MSH
McElroy, E A ... GI0SRN
McElroy, R J A ... GI0EEC
McElvanna, J J ... GI4OVE
McElvenney, J A ... G3LLV
McEntee, W ... G0KHI
McEvoy, A C ... G0SSL
McEvoy, H V ... G4MRR
McEwan, I ... GM0IMZ
McEwan, R ... GM3PXK
McEwan, R ... GM4VWV
McEwan, R A ... G4CHM
McEwan, W ... GM3TLN
McEwan Reid, R ... G4GTO
McEwen, A ... GM3PGY
McEwen, A ... M1CDV
McEwen, A W ... GM4RQW
McEwen, C D ... G3VKQ
McEwen, P ... G1GYM
McEwen, P J ... G4PUQ
McEwen, W J ... GI0SZH
McFadden, J P ... 2M1EDC
McFadden, M H ... GI3VCI
McFadyen, A ... G0JZE
McFall, M E ... GI4NXJ
McFarland, E ... G3GMM
McFarland, V S ... GI4RNP
McFarlane, D ... G0DMF
McFarlane, D K ... G8KKN
McFarlane, D S ... G7OFJ
McFarlane, J ... G3JUX
McFarlane, J ... G8UNU
McFarlane, J C ... GM0GAK
McFarlane, K S P ... G3ICG
McFerran, D G ... GM0OPX
McFetridge, S R A ... G0UPX
McForsyth, M ... GM4LDX
McGahon, P J ... G1IKG
McGall, E ... 2I1AJA
McGarrigle, I J ... G4JIU
McGarry, B G ... GI4HDJ
McGarry, G F ... GI3ECQ
McGarry, J W ... G0KDJ
McGarry, M J ... G0VYT
McGarry, P E ... G1NEG
McGarvey, P A ... G0IVD
McGarvie, I E ... GM4JDU
McGarvie, R ... G0END
McGeachy, I J ... G8WXC
McGee, C G ... G4CPL
McGee, G J ... G3MDM
McGee, J H ... G4HBR
McGennity, B P ... G4DBM
McGeough, K C ... G7EKM
McGeown, T J ... GI7IEZ
McGhie, A ... GM0TFF
McGhie, D I ... G6CZO
McGhie, J ... 2M1FIB
McGibbon, M ... G1SUI
McGifford, J W ... GM0KUJ
McGill, A ... GM4XOI
McGill, D ... G0VHX
McGill, J ... GM3MTH
McGill, M ... G3LBC
McGill, S P ... G0RFA
McGill, W ... G0DXB
McGill, W ... GM3LGM
McGillian, J J ... GI4NNM
McGillivray, J J ... G7MNY
McGilloway, D ... GI0BSO
McGilp, F C ... G8URB
McGinty, J ... G4GZQ
McGivern, P ... G4JXH
McGlen, E G ... G6INK
McGlynn, J ... G0HOB
McGoff, J ... G0TGF
McGoldrick, H P ... MI1AOO
McGonigal, K A ... GI3ZSC
McGookin, E J ... GI4NKQ
McGough, P M ... G1TBR
McGowan, C F ... G0KPE
McGowan, D J ... G1NAT
McGowan, G S ... G4FGJ
McGowan, I ... G8OFZ
McGowan, I J ... GM1RIG
McGowan, J ... GM3WIV
McGowan, L B ... G0DLG
McGowan, P ... G7VDD
McGowan, R D ... GM0DUX
McGrady, A ... G0NVU
McGrath, N ... G7AQK
McGrath, P J ... GD0BCJ
McGredy, D H ... G3PKX
McGregor, A G ... G0GOC
McGregor, J ... GM4LGM
McGregor, K ... 2E1FYS
McGregor, L ... 2E1FMF
McGregor, R C ... GM6PMG
McGregor, S J D ... GM4ZOA
McGrogan, L ... G4UUE
McGrory, S J ... G4GNP
McGuckian, K ... G3ZHR
McGuckin, D ... GI4OMO
McGuckin, J P ... GI6VJI
McGuckin, K M ... GI0RDM
McGuffie, W G ... G7BYU
McGugan, A ... GM4JYZ
McGuigan, J A ... GI4FCW
McGuigan, N ... G4XKX
McGuigan, P J ... GI3UYZ
McGuigan, S C M ... G8MFI
McGuigan, T ... M0AGV
McGuigan, T J ... G7CIT
McGuinness, E ... G0EXE
McGuinness, E ... GI1XBI
McGuinness, H F ... G0RPK
McGuinness, P ... G8ZXU
McGuinness, P G ... G4FDN
McGuinness, P G ... G8PAT
McGuire, A N ... G0RLE
McGuire, C ... G4WFF
McGuire, J ... G3LNW
McGuire, J P ... G8DUM
McGuire, L J ... G7NBG
McGuire, W S Y ... G0TYC
McGurgan, A ... GI6OHA
McHale, J ... G7TGP
McHale, J G ... G0HEW
McHale, M J ... G4PWD
McHardy, A R ... G6AWP
McHardy, I W T D ... GM3JFG
McHardy, J N ... GW6AEO
McHenry, R E ... G3NSM
McHenry, S J ... GM0MYS
McHugh, B ... G3THF
McHugh, B A ... G6WIO
McHugh, G ... GI7IIU
McHugh, W G ... GI0RPS
McHugh, W G ... GI4SRQ
McIlroy, H ... GI4CYU
McIlroy, J A ... G4EQX
McIlroy, J S ... GI4VJC
McIlroy, K A ... GI4JJF
McIlveen, J T J ... GI8OCR
McIlvenna, S D J ... GI7MDP
McIlwraith, J ... GM4ZTO
McInally, M J ... M0AJC
McInarlin, W ... G8AZF
McInerney, D J ... GM6AMP
McInerney, T J ... G0VYQ
McInnes, A ... GM3VCZ
McInnes, A C ... 2E1DOC
McInnes, B G R ... G7SKW
McInnes, G A ... GM4XHQ
McInnes, K D ... G3FTE
McInnes, K E N N ... GM6KDN
McInnes, T A ... G7CXB
McIntosh, A J ... G0JBV
McIntosh, A T ... GM8YGA
McIntosh, B ... G4WZG
McIntosh, E T ... 2E1CZV
McIntosh, M ... GW4IEQ
McIntosh, N J ... GM3RKO
McIntosh, W A ... GM0OTS
McIntyre, A I ... GM7BYB
McIntyre, B S ... G7KBE
McIntyre, D ... GW4YNR
McIntyre, H ... G3FLJ
McIntyre, H ... G3SOU
McIntyre, H ... G8FAB
McIntyre, J ... GM4ARU
McIntyre, M G ... GI3YDH
McIver, C E ... G7MYO
McIver, D ... G4LLS
McIver, E A ... G8SDN
McIver, I ... G6MBR
McIver, I M ... G0BKN
McIver, I M ... G0MBR
McIver, M J ... G8XEF
McIver, S ... M1AVR
McKae, W J ... G4ILA
McKaig, S H ... GI3SHI
McKane, K P ... G7NHW
McKavney, M ... G3COA
McKay, A G ... G0FQE
McKay, A J ... GM3OZB
McKay, D ... 2M1FHT
McKay, D ... G0NUT
McKay, D ... GM7WLE
McKay, D C ... G1RHH
McKay, D J ... G1JWG
McKay, D W ... G3ECI
McKay, G ... G8MOK
McKay, G ... GM3SPT
McKay, G ... GM4YWS
McKay, G K ... G3AVO
McKay, J A ... G4HOK
McKay, M F ... 2E1DQF
McKay, M J ... G6LWY
McKay, P ... G4KSN
McKay, P L ... G3WQU
McKay, R D ... G7NFE
McKay, R W ... G0WLO
McKay, S J ... G6XJJ
McKay, W A ... G1UFU
McKdavidson, B ... GM3ALZ
McKeand, E J ... G4XRQ
McKechnie, A ... G6JGL
McKechnie, A M ... G4XFV
McKee, D ... GI0GPG
McKee, H ... GW7FRD
McKee, P A ... GI0MHB
McKee, R T ... G4LJK
McKeeman, A B ... MM0ABM
McKeeman, A B ... MM0ARG
McKeever, A ... G1LIB
McKeever, J ... GI7TDA
McKeever, R ... G0HEF
McKeever, W ... GI0DSG
McKellar, C ... GM4KKW
McKellow, P R ... G8XGO
McKenna, G ... MM1BJJ
McKenzie, A A ... G3OSS
McKenzie, A B ... G0SMN
McKenzie, C ... G8LQO
McKenzie, D ... GM0TWK
McKenzie, D C ... G7OQC
McKenzie, I ... GI0LIX
McKenzie, I ... GM3KMN
McKenzie, I T ... GI0JPR
McKenzie, J ... G0IRU
McKenzie, J D ... G0OAY
McKenzie, J J ... G8HAT
McKenzie, J S ... G0CNU
McKenzie, M ... G8RWN
McKenzie, P W ... G8PHB
McKenzie, R ... 2M1BAR
McKenzie, R ... G0FFK
McKenzie, R M ... GM0AUL
McKeown, R ... GI0PFB
McKeown, W H A ... GI1RBI
McKeracher, F M A J ... G3RSI
McKernan, J D ... G1JDM
McKersie, M R ... GI1SMZ
McKersie, R I A ... GI0BFA
McKie, D J ... G4ZPW
McKie, F ... G7ATZ
McKillop, J K ... G8IEI
McKinlay, G ... GM0IVQ
McKinlay, J R ... G1SBJ
McKinlay, W H ... GM8IWC
McKinley, J S ... GI3NUM
McKinnell, J ... 2E1FAI
McKinnell, R F R ... 2E1FAH
McKinney, D J ... GI4MXW
McKinney, J ... M1AXQ
McKinney, M S ... GI4MAC
McKinney, R M ... GI4LQU
McKinney, W J M ... GI3TZB
McKinnon, R ... G0TXK
McKinnon, S J ... G0TBI
McKinty, R B ... GI3GTR
McKittrick, B ... GI7OKJ
McKnight, P ... G0RTM
McKnight, P A I ... G8XRE
McKnight, T ... G8YPH
McKone, D J ... G4DRX
McKown, L ... G3WXN
McKown, T ... G1UTR
McKune, I ... GM1OXQ
McLachlan, D A ... G1BGF
McLachlan, D B ... G4KOW
McLachlan, R ... G8LOW
McLachlan, R W ... G7EBR
McLaren, D M ... GM0OLD
McLaren, F A ... GM1SLW
McLaren, K B ... GM0EMC
McLaren, L ... G4LMC
McLaren, N B ... G4OAR
McLaren, N B ... G4WCG
McLaren, P A B ... GM6TUE
McLaren, R ... GM4FQG
McLaren, W ... GM4SVW
McLaughlan, J K ... GM7CZU
McLaughlin, C S ... GW7VHW
McLaughlin, D J ... G4RGH
McLaughlin, D J ... GM0PHG
McLaughlin, H B ... GI4PMP
McLaughlin, H D J ... GI4DQO
McLaughlin, P R ... G7TUZ
McLaverty, J ... GM0AAJ
McLay, D ... G8FVC
McLay, F J ... GM0HJS
McLean, A ... G0RFW
McLean, C ... 2E1ESN
McLean, C A ... M1ANC
McLean, C M ... G6FOH
McLean, D ... G6GXS
McLean, D ... GM3SUZ
McLean, D ... GM4VNQ
McLean, D J ... G7IZS
McLean, D L ... G3NOF
McLean, E W ... GM4EWM
McLean, G B ... G8XWU
McLean, J I ... G0BMH
McLean, L N ... GM4LFK
McLean, P A J ... 2E1ESO
McLean, R T ... G0JVE
McLean, W ... GM4REF
McLean, W A ... GM0PKB
McLeary, I ... 2M1ETI
McLellan, A S ... GM4MMM
McLellan, J ... G7VLN
McLellan, R A ... G6XKO
McLenaghan, I R ... G8HPF
McLennan, R ... GM8BQY
McLennan, S J ... G7JNS
McLeod, A J ... G8WGM
McLeod, J ... GM4INE
McLeod, J G ... G4MWW
McLeod, J H ... GW4KZC
McLeod, N J ... G4PAQ
McLeod, P A ... G0LJP
McLeod, T G ... GM4TRZ
McLeod-Stangroom, F ... G6CRX
McLewee, C H ... G3OML
McLintock, R W ... G1TGZ
McLocklin, A S ... G7RHI
McLoughlin, D A ... G4XEI
McLoughlin, F J ... G1GAD
McLoughlin, J G ... G3NNR
McLoughlin, M ... G4YSW
McLoughlin, M A ... G6PZN
McLoughlin, T J ... G4RFM
McLuckie, J C ... GM0TKE
McLuckie, S A ... GM4HML
McLuskie, I ... G0UVJ
McLuskie, I ... G8ORG
McLusky, E G C ... 2E0AGI
McLusky, E G C ... G1ZBO
McLusky, J G C ... 2E0AGL
McLusky, J G C ... G7RWZ
McLusky, P J ... 2E1BDV
McMackin, A P ... G4NHQ
McMahon, A R ... 2E1CAR
McMahon, B J ... GI4KEQ
McMahon, B P ... M0AKV
McMahon, C J ... G6FCI
McMahon, D ... GM0EOO
McMahon, J ... G4YZP
McMahon, J P ... G4KVD
McMahon, K ... G8JJR
McMahon, P B ... MI1ASN
McMahon, R ... G7KJP
McMahon, R R ... G0CBP
McManus, A ... G0CNV
McManus, B M ... G8FDE
McManus, C R ... G8FFC
McManus, D K ... GI1SZC
McManus, W ... G0WWC
McMaster, J D ... GM0TBH
McMaster, J M A ... GI7NBB
McMaster, R A ... GI1TFC
McMaster, R G ... GI7NOW
McMaster, S ... G0IDS
McMath, A M ... G6BRM
McMaw, G W A ... GI4JYJ
McMillan, A ... GM0GDD
McMillan, A H ... G4GZM
McMillan, A J ... G4MKM
McMillan, A K ... G4SSO
McMillan, G E ... 2E1FLN
McMillan, I J ... GM6PER
McMillan, J G ... G8UQJ
McMillan, P ... G0AQK
McMillan, R ... GM3SAE
McMillan, R D ... G3IUC
McMillan, R S ... GM8JUY
McMillan, T ... G1CCM
McMillan, W P ... GM4DAE
McMillan, W S ... GM4WMM
McMillen, W ... GI4RCK
McMinn, A ... GM6VYY
McMinn, D ... GM4XJY
McMinn, J L ... GW8OBW
McMinn, R Q ... G6REY
McMinn, W S ... GM6VYZ
McMullan, J K ... GW0KZF
McMullan, P ... G4RXK
McMullan, R ... G1GYN
McMullen, J P ... G7WCA
McMullen, S J ... G0WEV
McMullen, W ... G6VDL
McMullin, A I ... G7JEQ
McMullin, A I ... G8TQH
McMullon, A S ... G0SDJ
McMurdo, D V ... G3WZL
McMurray, H N ... GW4MTD
McMurray, R ... GM1EAF
McMurray, R ... GM8VPT
McNab, D ... G0PRY
McNab, D D ... G0GDM
McNab, J G ... G7PGC
McNab, W ... G0WMN
McNair, R ... G3UBZ
McNally, S W ... GI7UHJ
McNally, W J ... GI3YDO
McNamara, D P ... G0VVO
McNamara, F D B ... G3LZY
McNamara, J P ... G0EBA
McNamara, M A ... G0IFW
McNamara, P R ... 2E1FDV
McNamara, P R ... G8FMZ
McNamara, T J ... G8SFQ
McNamara, W H ... G0AWG
McNamee, M T ... GI7JGT
McNaney, P S ... G1DKG
McNaught, J G ... G3UJZ
McNaught, R E ... G6MHC
McNaughton, D M ... GM8KOF
McNeice, A H ... GI4OTG
McNeil, G J ... G3PLY
McNeil, N A J ... G1MTP
McNeil, R J ... G0RON
McNeil Greig, J C M ... G2FBU
McNeill, A W ... G3FCK
McNeill, J A ... G6LCS
McNeil-Watson, D R ... 2E0AHS
McNerlan, H ... GI4HQP
McNerlin, J L ... GI4EBS
McNicol, A S ... GM0MIW
McNicol, A S ... GM3UU
McNicoll, A ... GM8YGB
McNiel, D A C ... G4LQW
McNiff, J ... GM4HRJ
McNinch, M S ... GI6JGB
McNulty, J ... G4EUY
McNulty, J ... G7LGQ
McNulty, N ... GM0DVH
McParland, J N ... GI0VWK
McPartland, A M ... GI6FHD
McPartland, C O N ... G4JVE
McPhail, A ... G4WYH
McPheat, E ... G4OEC
McPhedran, A I ... GM3GTQ
McPhee, S W ... G0MWS
McPherson, A J C ... GM7FPN
McPherson, P H ... G3TEL
McPherson, S A L ... 2E1CCE
McPherson, S A L ... M1BPZ
McQuade, E M ... G4FLU
McQuade, P J ... G8AKP
McQuaid, J J ... GI4LDO
McQuail, P ... G8DCJ
McQuarrie, A ... G0FMN
McQue, D W ... G4NJU
McQueen, C S ... G7TIK
McQueen, J ... GM1MON
McQueen, S S C ... GM8MRW
McQuillan, P ... G0DKU
McQuillan, R S ... G0HVX
McQuillian, S A ... GM7VYG
McQuire, G H ... M0ATU
McQuire, L E ... G0ETV
McRae, C R ... GM3WRN
McRae, D ... GM4LDU
McShane, J ... G1NTL
McShea, B J ... G7MCS
McShea, P B L ... G1YQN
McSherry, J ... G4VIA
McSoley, J S ... G1YAH
McSweeney, J G ... GI4CFQ
McTaggart, A W ... MM1APG
McTaggart, J J ... G4VLM
McTaggart, W ... GM0MZB
McTait, R A ... G2BKZ
McTomney, J R ... G4VEU
McVea, R S ... GI6OZI
McVeigh, M ... GI5MV
McVety, A ... G0IXU
McVey, A M ... 2E1DKD
McVey, P ... G0PXJ
McVey, R N ... G3GMC
McVicar, J C ... GM6MCV
McVicar, M W ... GM6MYA
McVicker, T ... G4JPK
McVittie, J G ... GM0NBG
McVittie, M ... G1UIO
McWatters, A ... G3ZRL
McWhinnie, C J R ... G0MQW
McWhinnie, E ... GM0GLV
McWhinnie, M A ... GM0DYU
McWhirter, D ... GI3ZTG
McWilliam, A J ... G7UZG
McWilliam, L ... G4BTO
McWilliams, G ... MI1CEG
McWilliams, L A ... GM0LWD
McWilliams, R A ... G1WOJ
Mc'Donnell, T ... GI4WTT
Meachen, E E ... G3SFV
Meacock, D J ... G0UFC
Mead, A ... G0KNB
Mead, A C ... G4KQE
Mead, A E ... G4FQX
Mead, A J ... G6EEE
Mead, A J ... GW8HUS
Mead, A P ... G4DOA
Mead, C W ... G1YRM
Mead, D M ... G8WQZ
Mead, J ... G7MYQ
Mead, J C ... G6TDW
Mead, J C ... GW6IGY
Mead, L H ... G3ZGQ
Mead, M C ... G3TZT
Mead, P R ... G6RJZ
Meade, B A ... G3VOQ
Meade, I G ... GJ7DNJ
Meaden, B G ... G3BHT
Meadowcroft, J J J ... G4UGX
Meadows, C E ... G6KMQ
Meadows, C P ... G4KWH
Meadows, D ... G4TGB
Meadows, M H ... G6MHM
Meadows, M J ... G4GUG
Meadows, S E ... G0MHQ
Meads, M R ... G3WOT
Meads, V R ... G7HLU
Meadwell, S G ... G6LEI
Meager, S B ... G7REE
Meaker, J B ... G3RKM
Meakin, C D ... 2E1FOO
Meakin, D H ... G7NLL
Meakin, D H ... M0AVL
Meakin, J E ... M0BBK
Meakins, D ... G4SCJ
Meakins, R D ... G8HKN
Meal, R G ... 2W0AJD
Meal, R G ... GW6TAM
Meale, L D ... G4MNA
Mealey, J M ... G0LBF
Meaney, G A ... GI3KOT
Meanley, G R G ... G0GRM
Means, G J ... G1EFK
Meares, G H ... G0JGM
Mearns, D F ... GM6RXJ
Mears, A M ... G8JKM
Mears, C B ... G0AFR
Mears, D L ... G4ONI
Measom, J M ... G6IHU
Measom, P D ... G0ZZZ
Measures, A E ... G3WUC
Measures, D B ... G0HKR
Measures, L ... G7GKQ
Meaton, D G ... G1IKD
Medcalf, A ... G1IPE

Name	
Medcalf, A J	G0FRO
Medcalf, M N	G1EFL
Medcalf, R	G7RMW
Medcalf, R	G8XDL
Medcalf, R C F	G4RUA
Medcraft, R A	G4JUJ
Medcraft, R H	G3JVM
Meddick, S R	G0BQD
Meddings, D V	2E1EIU
Meddings, J G	G4DGM
Medhurst, D A C	G4XCZ
Medici, V	G3TTW
Medley, C	G0OYX
Medley, D R	G7KKJ
Medlicott, T A	G7UJK
Medlock, F	G6WLI
Medway, A P J	G6OHD
Mee, A A	G7KPN
Mee, A C	GM8GXQ
Mee, A J	G0TEC
Mee, C W	G4JQV
Mee, H	G5MY
Mee, J V	G3PJK
Mee, M	GW7NFY
Mee, M J	G7UQJ
Mee, N B	GW7EXH
Mee, R D N	G1ZGZ
Mee, S S	GW4CTV
Meech, D H	G4PBK
Meecham, W N	G0GAP
Meehan, M A C	GW0SWD
Meek, D D	GM7IZO
Meek, J M	G6BGY
Meekers, E	G4SNR
Meerman, M J M	G7DQE
Meers, H R	G3RTY
Meffan, D	G1JYE
Megone, J E	2E1EHS
Megone, R C	2E1FTH
Mehaffey, C A	G0MON
Mehaffey, J H	G0MOL
Mehta, T	GM0HSG
Meigh, P D	2E1FEO
Meigh, S J	G6ING
Meikle, H P	GM0DQC
Meikle, L F	G3TTI
Meikle, P	GM7GHL
Meikle, W S	GM0MWR
Meinerts-Hahn, W F	G3UOL
Meiring, P D	G0BSX
Mekka, R A	G4AWY
Melbourne, M P	G0UYQ
Melbourne, M P	G8EHX
Melbourne, P B	G8GML
Meldrum, B A J	GM4VHU
Meldrum, D S	G3XCO
Meldrum, E R	GM6BKK
Meldrum, J	G8PSS
Meldrum, M K	G7SQU
Melham, A	G7VNM
Melhuish, D J	GW0WLZ
Melhuish, F	M1ART
Melhuish, J W K	G6EYT
Melhuish, S J	G4TJC
Melia, A	G0KNZ
Melia, A	G5FS
Melia, A J	G3NYK
Melia, C	2E1CNJ
Melia, C L P M	G1VHX
Melia, G B	G0OPM
Melia, R S	G0CMR
Mellett, G	G4MVS
Mellett, P E	G3PIJ
Mellin, D J	G4NKP
Melling, B	2E1FJP
Melling, S R J	G8FUH
Mellings, C I	G7KYW
Mellings, D	G0WJS
Mellings, F	G0SOG
Mellor, B	G0UJF
Mellor, B	G4CQJ
Mellor, D J	G0GXT
Mellor, D J	G7IRS
Mellor, D R	G4EWK
Mellor, I	G7EMZ
Mellor, J	G0IQV
Mellor, J D	2E1CQJ
Mellor, J R	G0VEA
Mellor, J R	M1BAF
Mellor, K W	G4RTA
Mellor, P B	G4BIK
Mellor, P M	G4JNC
Mellor, R C	G0EHO
Mellor, S	G7BHX
Mellor, T A	G3TVO
Mellor, T A	G6TZD
Mellors, A J	G4GOK
Mellors, C A	G0JDX
Mellors, G W	G1KNK
Mellors, P J	G0RCP
Mellows, A B	G8MSN
Melnyczuk, T A	G0REM
Melnyczuk, T A	G0SEA
Melton, D E W	G3ZJU
Melton, J D	G0ORX
Melton, K G	GM3WKM
Melton, R J	G0OJP
Melton, T R	G7BFX
Melville, C	G6OPU
Melville, D H	G0IOP
Melville, E D	G3FCB
Melville, I T	G4EZP
Melville, I T	G4JAR
Melville, J	GM6HLT
Melvin, C E	G4VCB
Melvin, J	G3LIV
Melvin, S	G8UEE
Membury, G J	G8DJW
Memory, M F	G6XDP
Menday, J R	2E1AOG
Menday, R J	G4CXL
Menday, V P E	G8HCL
Mendham, N E W	G1DME
Mendoza, G C	G4EUC
Mendum, K	G8RPA
Menguy, J	G4VDX
Menhams, J M	G0EBN
Mennie, S J	2E1FHP
Mennie, S J	M1BYM
Menown, P A	GI4FZD
Menzel, K R	G0FIT
Menzies, D A J	GM3GNE
Menzies, G T	G4IQV
Menzies, I J	GM1FSU
Menzies, R L	GM0FHJ
Menzies, T W G	GM1GEQ
Mepham, A J	G4CBZ
Mepham, D W	G4ERA
Mepham, F E	G4CDL
Mepham, R W	G4TPJ
Mercer, A	G4CZE
Mercer, A	GM3EGU
Mercer, A D	G6WIA
Mercer, A J	GW4CZK
Mercer, B	G3KLL
Mercer, D	G3YHQ
Mercer, G H	GM4BES
Mercer, I C	G3ZER
Mercer, J D	G0THN
Mercer, N	G1ZWJ
Mercer, N	G3UVQ
Mercer, P	G0TPM
Mercer, P G	GI4VIV
Mercer, S G	G2DPY
Merchant, F V	G7CWN
Meredith, B	G4EBG
Meredith, B A	GW6ZJG
Meredith, B A	MW0BBM
Meredith, E W S	GW0VPR
Meredith, E W S	GW4XLK
Meredith, G M	G0KXV
Meredith, J R	GM0SWX
Meredith, J W E	2E1FNX
Meredith, M	GW0XLK
Meredith, M	GW4WCM
Meredith, M J	G6EEU
Meredith, W G	GW3ZNN
Merewether, H C H	G0LFR
Merrifield, J D	G0FKY
Merrell, B R	G8PUH
Merrell, C	G8OIV
Merrell, R P	G0GEC
Merrell, R P	G4GUJ
Merrett, C	G3JGH
Merrick, D J	G8VVM
Merrick, J A	2E1EVM
Merrick, L	G0VIG
Merriday, J F G	G1WEQ
Merrifield, S L	G7SFI
Merrilees, C H	GM3EOB
Merrils, G	G0UQF
Merriman, C J F	G7UHS
Merriman, J M	G7JWB
Merriman, P R	G3YJE
Merriman, R A	G3SIP
Merrin, L G	G0WAX
Merrin, M A	G0WGA
Merrington, D T	G6HKP
Merrington, E J	2E1FKD
Merrington, G	G1IVV
Merritt, F W	G8LWO
Merritt, G A	G7DRC
Merritt, R	G0WVO
Merritt, T D M	GM4WTJ
Merrix, A	G0TEM
Merrix, A	G7BRG
Merry, C D	G4CDM
Merry, D A W	G1NTV
Merry, D R	G1DRM
Merrylees, A W	G0DCI
Mersi, B	G7TKG
Merson, J Y	GM4MOA
Merz, W	G1DTE
Mesny, H R	GJ3LFJ
Messenger, K M	2E1EMX
Messenger, M S	G6XLQ
Messer, R L	G3KIL
Metcalf, A	G8ETU
Metcalf, E	G4YDE
Metcalf, H J	G6VXR
Metcalf, K D	G8OWO
Metcalf, K J	G4YSP
Metcalf, W S	G8XLE
Metcalfe, B	M0AMB
Metcalfe, E	G4XLC
Metcalfe, E	GM4SMH
Metcalfe, G	G3JDC
Metcalfe, I R	G6TUG
Metcalfe, J	G0VQJ
Metcalfe, J	G4YOV
Metcalfe, P	G6BKL
Metcalfe, R	G1LRN
Metcalfe, R J F	G1JUL
Metcalfe, S J	G7CZV
Metcalfe, T A	G1SVI
Metcalfe, T R	G7LTW
Metcalfe, W G	G0WGZ
Metcalfe, W H G	G6HKY
Metcalfe, W H G	G6VS
Methven, J D	G1IPC
Metson, P	G6RRP
Metson, S	G7FWV
Mettam, M L	M0ADH
Mettam, R W	G6XUX
Metz, M	MM1APZ
Mew, J C	2E1FDM
Mew, R P	G6HKF
Mewse, R A	G8MWC
Meyer, B A	G8YEP
Meyer, G R	M1BSM
Meyers, H A	G3CMU
Meynell, S	M0AIR
Michael, D	G0ASA
Michael, D A	GM0KCY
Michael, E D	GM0PKX
Michaelides, A R D	M0ACY
Michaelson, K B	G3RDG
Michie, G D	GM0IXO
Michie, L I	GM7PXJ
Mickels, M E	MW1AID
Mickleburgh, R	G4BRF
Micklewright, R W	G3MYM
Middel, E H R	G7IXS
Middle, C C	G4CM
Middlehurst, P T	G1DVA
Middlemass, T	G7IXQ
Middleton, A J	G1DTF
Middleton, A J	G8SBZ
Middleton, A J	G8WPF
Middleton, B	G0CML
Middleton, D F	G0CZG
Middleton, D J	G3VSV
Middleton, E A D	G7MOH
Middleton, G	G4SQK
Middleton, G	G7LBV
Middleton, G H	G6EER
Middleton, H J S	G7RZZ
Middleton, H O	G0OSR
Middleton, I W	G4BNX
Middleton, J	G4TXO
Middleton, J C R	G6BJP
Middleton, J V	G8VGF
Middleton, J W	G6MGZ
Middleton, K E	G4EJH
Middleton, L	G6OPD
Middleton, M	G2FXV
Middleton, P M	G1CFA
Middleton, R A	G7LLG
Middleton, R C	G7RDJ
Middleton, S G	2W1CHM
Middleton, S P	G7VYY
Middleton, V M	G1LFD
Midgaff, D	G4IWG
Midgeley, R J	G8TVZ
Midmore, J G	G3ZXW
Midwood, J M R	G7PTD
Midwood, P S	G7SLO
Midworth, N C	G0WTA
Mifflin, R T	G8KLG
Milan, P S	G0VLM
Milburn, C	2E1DOW
Milburn, C	G7NLA
Milburn, C A	G1CHM
Milburn, J E	GW8YUJ
Mildren, R K	G3FVD
Milenkovic, R A T O	GM4TNJ
Miles, A C	G1YDG
Miles, A C	G6RJS
Miles, A L	G7CXA
Miles, A M	G6WJT
Miles, B L	G6VDU
Miles, C A G	G3NRC
Miles, C J	G7JTN
Miles, C P	G7PUR
Miles, C R	G4TYR
Miles, D F	G3SPZ
Miles, D L	G7LED
Miles, E A	G7BAY
Miles, E G	G0RLH
Miles, F G	G3VBE
Miles, G	G2CXO
Miles, G E	G3NIR
Miles, G H	G4IWE
Miles, G P	G3TOV
Miles, G P	G7CZL
Miles, H	G8PJX
Miles, I	G6DCJ
Miles, J	G7KIA
Miles, J L	2M1ANT
Miles, L	G7KHY
Miles, L A	G4LNR
Miles, L A	G7OWF
Miles, L G	G4TWP
Miles, L J	G6LWZ
Miles, P A	GW3KDB
Miles, P R	G1GOY
Miles, R	G4XXH
Miles, R A	G6FTY
Miles, R E	G1WUM
Miles, R G	GM4CAQ
Miles, S R	G0MSV
Miles, T D J	GW3NXR
Miles-Williams, W A	G0LZG
Milford, B W	G7UOM
Milham, A A	G3OPL
Millar, A G	G6JLU
Millar, B	GI0RYU
Millar, D M	GM3JJQ
Millar, F R	GM8ZTV
Millar, G B	GM4FSB
Millar, G P	GM3UM
Millar, G S	GI0KVQ
Millar, G S	MI0AVI
Millar, I C	G1KMS
Millar, I C	GM7RKD
Millar, K C	G4VFW
Millar, L	GM7MBB
Millar, N	G0GBZ
Millar, R T	G3WLW
Millar, S C	M0AOK
Millar, T	GI7WET
Millar, W V G	GI4UPC
Millard, D	G8NEY
Millard, D J	G3SBJ
Millard, G E	2E1FRA
Millard, H L	G8LWK
Millbank, F R	G1XWZ
Millen, D R	G7RUC
Miller, A	G1FKF
Miller, A	GI4SFV
Miller, A B	GM3SOM
Miller, A B	G7PLV
Miller, A C	GM3BKC
Miller, A C	GM4ACM
Miller, A E	G0KTU
Miller, A G	G1BAQ
Miller, A K	GM1FAI
Miller, A R	GW8YAS
Miller, B E	G7BQT
Miller, B G	G8INL
Miller, C	G6YHL
Miller, C G	G8KVO
Miller, C P	G0DHV
Miller, C S	GM3NEC
Miller, D	GM1MJK
Miller, D	G4MPR
Miller, D C R	G6LEY
Miller, D D	G0EHM
Miller, D J	G6LYM
Miller, D J	G7TIY
Miller, D L	GW6JMC
Miller, D N	G4JHI
Miller, D P	G4NMJ
Miller, G B	G6WWY
Miller, G C J	G3ONG
Miller, G E	G8YYC
Miller, G G	GM4NQT
Miller, G H	GM6OA
Miller, G J	G6UCI
Miller, G T	G8JIP
Miller, G W	G1DHM
Miller, H A	G1YVY
Miller, H R	G0IIY
Miller, I	G3NXX
Miller, I M G	GM4JAE
Miller, J	GW6VDR
Miller, J A	G1OBM
Miller, J A	G6SGW
Miller, J C	GM6NHF
Miller, J H	G8NSP
Miller, J H H	GW6XII
Miller, J J	G6CLP
Miller, J M	G4MM
Miller, J M	G7SFQ
Miller, J P	G3LRU
Miller, J R	G3RUH
Miller, J S	GM4AGS
Miller, J W	G1DKY
Miller, J W	G4DJM
Miller, K D	G7AAQ
Miller, K E	G0BWJ
Miller, K E	G6TYZ
Miller, K J	M1AYR
Miller, L E	G3YEQ
Miller, M F	G0TGS
Miller, M G	G0RMO
Miller, M J	G7EGX
Miller, N L	G6CDW
Miller, N O	G3MVV
Miller, N O	G4KF
Miller, N O	G6BCL
Miller, P K	G1PKM
Miller, P L	G4AAW
Miller, P M	G4REE
Miller, R A H	GI3TJM
Miller, R B	G3CJI
Miller, R B	G4ZCK
Miller, R C	G4DSY
Miller, R E	G3AQS
Miller, R E	G4FDP
Miller, R F	G1RPA
Miller, R H	G4UKX
Miller, S A	G4YXT
Miller, S W	GI7POP
Miller, T D W	G1HWO
Miller, T G	G4BYE
Miller, T I	G1CKR
Miller, T J	G6HLU
Miller, T S	G6YHK
Miller, W	G7BVX
Miller, W B	GM3PMB
Miller, W B	GM5VG
Miller, W B L	GM6SAG
Miller, W C	G3OTW
Miller, W P	G8ESL
Millerchip, P J	G7NBZ
Millerchip, R E	M0BEC
Millership, A I	G1XDF
Millership, E T	G7FND
Millett, R E	G0HBJ
Millican, B	G4OFA
Millichamp, K	G6NHX
Millie, J	GW8MQ
Milligan, A S	GM0XAL
Milligan, G J	G1CPU
Milligan, J	G4DLQ
Milligan, R S	G4SPK
Milligan, W	GM4TPQ
Milliken, P J	MI1AZS
Milliken, R B	G8LGU
Milliken, S P C	GI7SAH
Millin, D	G4LXY
Millington, D J	G6GSI
Millington, G	G8HDD
Millington, H S	GW7MVG
Millington, M R	G6GNN
Millington, R A	G3WDQ
Millington, R P	G4AZV
Millington, T A J	G0TAJ
Millington, R G H	G3PJY
Millner, A J	G1ANN
Millner, C T W	G4NXQ
Mills, A	G0SVM
Mills, A	G3KDO
Mills, A	G8JRZ
Mills, A A	G1CAQ
Mills, A G	G3UKF
Mills, A K	G0EGX
Mills, A K	G4GPR
Mills, B F D	G3VMP
Mills, B W	G7THJ
Mills, C	G7EVT
Mills, C J	G4BYZ
Mills, D	2E1CZR
Mills, D	G1AFI
Mills, D	G4PDM
Mills, D	G7UVW
Mills, D H	G0BKH
Mills, D E	G3LEI
Mills, D J	G1UEA
Mills, E	G4XXB
Mills, E L	G4DGL
Mills, F J	G4XCY
Mills, F W	G3ZAZ
Mills, F W	G4XXA
Mills, G	G0SSC
Mills, G	G7TEI
Mills, G E	G6ZMG
Mills, G J	G3TWY
Mills, G J	GW4UZH
Mills, I P	2E1AGD
Mills, J	G4NOY
Mills, J	G4XRJ
Mills, J	G6XLC
Mills, J C	G7BRJ
Mills, J D	G6IHY
Mills, J F	G0FUV
Mills, J H	G0DJR
Mills, J M S	GW8HWS
Mills, J S	G7BDS
Mills, J S	G8DBP
Mills, J T	GI0GQG
Mills, K G	G4DIS
Mills, K L	G0GIN
Mills, K M	G4ZOK
Mills, L	M1CBE
Mills, L R N	G1HWR
Mills, M	G0HMD
Mills, M	G8CXZ
Mills, M F	G6SYA
Mills, M J	G0FQM
Mills, M J	G3TEV
Mills, R B	G4LPD
Mills, R D	G8XRL
Mills, R H	G6AMY
Mills, R J	G3ZAV
Mills, R M	G4ZES
Mills, R M F	G8UKY
Mills, R S	GW4MZB
Mills, S	G7KJT
Mills, T	G7UEO
Mills, T A	G3YAI
Mills, T E	G7ILE
Mills, T G	G3UNS
Mills, T L	G4RQA
Mills, W F	G4AYW
Millward, A	G7WBW
Millward, B	G0MDN
Millward, J	G4YNY
Millward, L	G3VDI
Millward, M G A	GM0LNT
Millward, P J D	2E1CWQ
Milman, T	G1XTN
Milne, A B	GW4BFX
Milne, A J	G8LHP
Milne, B C	G4HIV
Milne, C	GM7ILO
Milne, D J	G0WXZ
Milne, D O	GM6WLJ
Milne, D S	G6VMI
Milne, D S	GM4FUL
Milne, G M M	GM1ZQF
Milne, G S	G3UMI
Milne, J M	GM7AUW
Milne, J P	G1CIK
Milne, N D	2E1CQS
Milne, R	G3AKN
Milne, R	G6LKG
Milne, R	G6WWZ
Milne, R	GM0IDY
Milne, S G	GM4KOI
Milne, T G	G4CMG
Milner, C J L	G3ZJK
Milner, D	G0VSE
Milner, G	G3PNK
Milner, G	G8NWK
Milner, J M	G7VUP
Milner, L	G4GJH
Milner, P A	G1HWM
Milner, P S	G1HWJ
Milner, R G T	G8RIK
Milner-Smith, G W	2E1EIY
Milnes, J	G0SZV
Milns, M D	G7LTO
Milosevic, J	GW1ZFX
Milroy, R W	G7LUU
Milsom, H J	G1WYP
Milsom, K J	G0TFM
Milsom, P J	G4GSA
Milsom, R	G0UAB
Milsom, S C	G8SFA
Milton, A E	G0LGX
Milton, C D	G6XIF
Milton, C S	M1BKU
Milton, A	G3YBX
Milton, E R R	2E1BDO
Milton, G D	G0CUQ
Milton, J P	G7PLX
Milton, O H J	2E1BZJ
Milton, P W	G8FRC
Milton, R W	GW0NIY
Mimna, F J	GI8TST
Minard, C	GW0WUL
Minaudo, J D	GM1EOA
Minchin, P D	G6NHW
Mindel, S	G6NVI
Minett, A	G3SPP
Minett, D C H	G3WPP
Minett, T B	G0TNR
Minihane, K J	G1OCY
Minihane, M J	G0UHQ
Minks, J O	G3RMC
Minnis, P G	G0IRX
Minnock, J	2E1ATH
Minnock, S T	2E1ABW
Minns, J H	G0IOG
Minns, T F	G6CZQ
Minshall, J G	G0BED
Minshull, M G J	G1XYV
Minson, G R	G3EYZ
Minson, S R	G2FGB
Minter, R F	G0HVO
Mintern, K E	G8ATP
Minton, B	G0BKH
Minton, D E	G1MCG
Minton, E A M	G4EAM
Mirams, D R	G4SFU
Mirams, F N	G6YDO
Mirams, H C	G8UTW
Mirams, J A	G6KJM
Mirtle, P G R	GM1YFO
Miskelly, R H	GI8FYP
Miskimmin, R	GI0DUP
Mister, P W	G0YEF
Mistofsky, M	GM4KLO
Mitchell, A	GM0CPQ
Mitchell, A E	G8DF
Mitchell, A I	G4BZJ
Mitchell, A J	G0TPK
Mitchell, A J	G4ICE
Mitchell, A P	G1AHS
Mitchell, A P	G3YJZ
Mitchell, A R	G6EIO
Mitchell, A S E	G0EEJ
Mitchell, A W	G7SIV
Mitchell, B H	GI8TWB
Mitchell, B J	G3HJK
Mitchell, B J	G4MLI
Mitchell, B W	G0NXN
Mitchell, B W	G4YHZ
Mitchell, C A	G3UVS
Mitchell, C E	G8PKM
Mitchell, C M	2W1FJT
Mitchell, C R	G0SKA
Mitchell, D	G7MZK
Mitchell, D	GM3PAB
Mitchell, D	M1AWL
Mitchell, D J	G7FGZ
Mitchell, D J S	GM6WNX
Mitchell, D R	G2DLX
Mitchell, D W	GM7RFN
Mitchell, E P	G1POD
Mitchell, E W	MM1BMK
Mitchell, F	G3UZX
Mitchell, F J	G8PCB
Mitchell, G C	GM7FMJ
Mitchell, G J	G4AIQ
Mitchell, G J	G7BJR
Mitchell, G J	G7INA
Mitchell, G K	G4AUL
Mitchell, G P	G3OFJ
Mitchell, H	G4DYB
Mitchell, H J	G6IHG
Mitchell, H W	G2AMG
Mitchell, I E M	G4NSD
Mitchell, I K	G0BUK
Mitchell, I M	GW4XAZ
Mitchell, I S	G7MZJ
Mitchell, J	GM4TPR
Mitchell, J	GM7SBR
Mitchell, J D	G7IEG
Mitchell, J G	GW0GRQ
Mitchell, J H	G4DMX
Mitchell, J M	GM3DJT
Mitchell, J P	G0JPM
Mitchell, J P	G3KEQ
Mitchell, J R	GW3OKM
Mitchell, J W	GM6EOP
Mitchell, K M	G7NHL
Mitchell, L R V	G3BHK
Mitchell, M H	G1MDQ
Mitchell, N	G7TUG
Mitchell, N S J	GI0LRZ
Mitchell, P	G4XYK
Mitchell, P A	GW6YDA
Mitchell, P J	G1PJM
Mitchell, P W	G1WAW
Mitchell, R	G8REO
Mitchell, R C	GM3ZTW
Mitchell, R H	G3YBM
Mitchell, R K	G1PDO
Mitchell, R L H	G1KRN
Mitchell, R S	G1IZN
Mitchell, R W	G4KVC
Mitchell, R W	GM4HJK
Mitchell, S	G0JNQ
Mitchell, T	G1JZY
Mitchell, T	GM0GLH
Mitchell, T A T	G4OHT
Mitchell, T M	GM0JHF
Mitchell, T W	G3LMX
Mitchell, V	GW1GZB
Mitchell, V A	G6BCQ
Mitchell, V N	G7LLE
Mitchell, W A	G7PMK
Mitchell, W A	GM3FRI
Mitchell, W A L	G4IBI
Mitchell, W D	G4LYH
Mitchell, W D	GI4NAT
Mitchell/Watson, W P	G0LUM
Mitchelmore, S G	G3XHC
Mitchelson, N R	G0IYQ
Mitchener, G	G7SNO
Mitchener, J P	G0DVJ
Mitchinson, D	G3XID
Mitchinson, D I	G4KGN
Mitchinson, J A	G7GLA
Mitchinson, N	GM0FTJ
Mitton, D	G7PUC
Mizon, R	2E1AXF
Moakes, K A	G0TWP
Moakes, R B	G1MWI
Moan, T	G4MSJ
Moar, J L	GM4EFR
Moat, R F	G1PQX
Moate, K A	G0CRL
Moate, R F	G0CHG
Mobberley, P J	G8UEF
Mobbs, A M	G8EEY
Mobbs, C I	G8UHW
Mobbs, D A	G1IPD
Mobbs, D P	G4MEE
Mobbs, J R	G0ONG
Mobbs, L A	G6XIK
Moberley, G	G7UMK
Mobley, D W	G1AHQ
Mobley, L G H	G2HPH
Mobley, S R	G7JVF
Mock, C H	GW6FLU
Mock, W C	G1FAJ
Mockett, D E	G7LZF
Mockett, I T	G1DSI
Mockford, C E	G7APM
Mockridge, N	G6EED
Model, J S	G0WOE
Moffat, D	G6MGY
Moffat, G	G7VPP
Moffat, G	MM0AWU
Moffat, I G	G0OZS
Moffat, K L	GM0CFW
Moffat, S K	G4LHF
Moffatt, D S	G3RAU
Moffatt, F	G4DTI
Moffatt, G W	G4ZRA
Moffitt, J T H	GI8DMX
Moffitt, T J	GI4KQA
Moger, D J	G8ZRU
Mogford, P L A	GW4PCO
Mogford, R C	G0BQA
Moggeridge, J G	G1YHB
Moggridge, A E	G0OEA
Mohammed, K L	G0OVE
Mohammed, N	GM0WPY
Moir, J B	GM4VEE
Moir, N G	GM7RVR
Moir, P	G0KMM
Moisy, S	G2FXJ
Mold, J J	G1GMX
Mold, R A	2E0ABD
Mold, R A	G7PTZ
Mole, S M	G1ERU
Mole, S T	G1CBK
Mole, T H	G8WBL
Moll, R T	G8NEM
Moll, V R	G4GGV
Molland, R E	G3CNC
Mollart, J W	G4IMV
Mollatt, M S	G8KXM
Moller, G	GW4WFM
Moller, J G	GI6JMD
Molloy, M D	G0PGQ
Molloy, M J	G4DWR
Molyneaux, D F	G3YUT
Molyneux, J	G4SNG
Molyneux, J A	G6DCH
Molyneux, K C	M0ABF
Molyneux, L C	G0DBH
Monaghan, J	G1WIR
Monaghan, T	G1VJQ
Monaghan, T B	G6GUU
Monahan, R S	GM1FLQ
Monckton, C W	G7APO
Monckton, D C	G6UBW
Money, D C	G3HKD
Money, P R	G4UVA
Money, T D	M0AKY
Monk, A C	G7LFT
Monk, A E J	G0UGP
Monk, B C	G0OYB
Monk, B E	G3NAR
Monk, D L	G6EET
Monk, K D	G4UQJ
Monk, K J	G3KJM
Monk, R	G0HUQ
Monk, S F	G0EVI
Monk, W R	G1TBI
Monk, W R	G6WWW
Monk, W R	GW7SSI
Monkhouse, A D	G2DTS
Monkman, N	G7DJY
Monks, H B	G0NKM
Monks, J	M0AGL
Monksummers, C D	G0YLO
Monksummers, C D	G0ZEE
Monksummers, R J	G6TER
Monksummers, R J	M1AAB
Monnery, D C	G4VFC
Monro, M S	G8DLL
Monson, F D	G4MHY
Montagu, W D	G4YTP
Montague, R H P	G7CHB
Montanana, N F	G8RWG
Montanaro, J P	2E1FRD
Montanaro, M C	2E1EYH
Montanaro, M C	M1BHS
Monte, D G	G0OTZ
Monte, J F	G8LCS
Monteil, M J	GJ6WDK
Monteiro, G S	G0WIO
Montford, R L	G6RPD
Montford, W K	G3JZL
Montgomery, C G G	G0EBH
Montgomery, D	G6AWM
Montgomery, J	GM0HCW
Montgomery, J Y C	GM0SUH
Montgomery, K	G8ECG
Montgomery, P	G1KKA
Montgomery, R H	G0IQO
Monument, N	G0RWQ
Monument, N R	G4BTX
Moodie, D M	GM4FOZ
Moodie, R W	MM0AMV
Moodie, W T	G6EEB
Moody, B	G0UGA
Moody, D R	G0HVQ
Moody, E	M0AXX
Moody, F	G4HVW
Moody, G	G4LIM
Moody, G	M0BEJ
Moody, I D A	G6FDO
Moody, K W	G4NRZ
Moody, L W	G1EAP
Moody, M G M	G8LMO
Moody, N	G0OMG
Moody, R	G8UKG
Moon, A	G3RGB
Moon, D	G7NPG
Moon, H L	G6KQJ
Moon, J O	G4WAX
Moon, R H	G0CLF
Mooney, E C	GI4EQA
Mooney, F W	M0BBS
Mooney, P	GI4FWK
Mooney, P R	G7SPV
Mooney, P W	G0VKU
Mooney, R A D	G3WZX
Mooney, W	G3VZU
Mooney, W W	G4LOR
Moorcraft, N	G4XCS
Moore, A	G0EAM
Moore, A	G0RFE
Moore, A	M1AIM
Moore, A	G7LGH
Moore, A J	G4RHX
Moore, A J B	G1KXZ
Moore, A O	G4AFX
Moore, A R F	G3VSU
Moore, A T	G1PRP
Moore, A W	GD3GBG
Moore, B	GI8DGB
Moore, B J	M0AFU
Moore, C D	M1ALR

Moore, C J G1MTK
Moore, C J G1ZGT
Moore, C J G6CZS
Moore, C R G7DAH
Moore, C S G4ZOX
Moore, D G1THG
Moore, D G3RDS
Moore, D G6IHV
Moore, D G7SBV
Moore, D A B GM4HPK
Moore, D H K G0DHM
Moore, D R G3LSA
Moore, D R G8CGZ
Moore, D R R G0FZH
Moore, D S GI0TSA
Moore, D S GM0GRL
Moore, D T G7WJG
Moore, D W GW6DQH
Moore, E G6PAY
Moore, E B G4DVS
Moore, E J GD1RHT
Moore, F E J G3VST
Moore, G G3JQJ
Moore, G A G8BBN
Moore, G C G6JBL
Moore, G D G3YIJ
Moore, G D G8DXK
Moore, G R G0IOF
Moore, G R G4DML
Moore, H A G0VOR
Moore, I GM4KLN
Moore, I H G0CAX
Moore, J G3PHX
Moore, J G4SWT
Moore, J A G7BMY
Moore, J C G3OGM
Moore, J D G0MFI
Moore, J E GW8GQE
Moore, J G I G0BIP
Moore, J H G1MPZ
Moore, J I G3CST
Moore, J J M1ADJ
Moore, J L G1BYI
Moore, J M GI7IUR
Moore, J P G3IKR
Moore, J R G0AXB
Moore, J R G7BNE
Moore, J R G7TUM
Moore, J S 2E1DVG
Moore, K G6NVE
Moore, K GI4SNC
Moore, K GW0SFO
Moore, K M1BWA
Moore, K E G7HCT
Moore, K J G1PRE
Moore, L 2E1FOM
Moore, L M G4UXS
Moore, L R G3VFH
Moore, M M1BIR
Moore, M P 2E1EGH
Moore, M R G3NRM
Moore, N E GI7CMC
Moore, N E MI1BRA
Moore, N J 2E1FET
Moore, P G8MKS
Moore, P GI0VAB
Moore, P GW0PLN
Moore, P L GM0DNH
Moore, P R J GI8UIU
Moore, R G3YOC
Moore, R G4CEJ
Moore, R C G3SPU
Moore, R C J G1IWT
Moore, R D GI8FLQ
Moore, R K G0KPG
Moore, R K G3YPM
Moore, R P G3PLL
Moore, R S G4KRF
Moore, R T 2E1ETB
Moore, R W G3YUX
Moore, R W G4LWU
Moore, R W G6YHV
Moore, R W GD4XJR
Moore, R W M0AUI
Moore, S A G3KPB
Moore, S G GI8YTH
Moore, S J G1NNN
Moore, S J GW6FKP
Moore, S P G0GTV
Moore, S R G0OGV
Moore, T G3AJD
Moore, T D G7RHG
Moore, T J G0CXD
Moore, T M G7LZZ
Moore, T M C M1AVC
Moore, W A GW4THK
Moore, W J H G8DTT
Moore, Y E G6OZJ
Moorecroft, B G4UDY
Moorecroft, M C G1WXT
Moores, R G3GZT
Moores, R H G0BEM
Moorey, B A G7KVT
Moorfield, G A G3ZBG
Moorhead, T F G0VMT
Moorhouse, G M G6YWC
Moorhouse, J A G3VQF
Moorhouse, J V G4KVB
Moorhouse, J W G5XJ
Moorhouse, K M G4CEV
Moorhouse, M J G7RAI
Moorhouse, R J G7UUQ
Moorwood, W C L G3CAQ
Moran, B A C M1ALD
Moran, J G1MIH
Moran, J W G3XUM
Moran, P G0JIB
Moran, P G0LKR
Moran, P J G0PYQ
Moran, P J F GW0WER
Morcom, C J G3VEH
Morcom, R G8RKV
Mordas, E J G4TJM
Mordue, R G1SDN
More, J A GM6SDV
More, K L G1JLS
Moreau, B G7WJD
Morecroft, A W G4DFP
Morecroft, M K G1AEV
Morecroft, S G0TEF
Moreland, D M G7FGA
Moreman, D E G3UNY
Morency, A S G0CFF
Moreton, D G G1THO
Moreton, G J G1UWM
Moreton, I S G4XSZ
Moreton, J R G6PEJ
Moreton, L K G0RLM
Moreton, R J E G6GQP
Moreton, R S G3EMY
Morey, K G G4NCD
Morey, R P G4NAK
Morgan, A G0MWY
Morgan, A J G6JGF
Morgan, A J G8PON
Morgan, A J GW1SGE
Morgan, A M G4SUP
Morgan, A P G8DV
Morgan, A P G8WWM
Morgan, A R G6PPD
Morgan, A W G1MME
Morgan, A W G8ADS
Morgan, B A GW0GQC
Morgan, C G8HCW
Morgan, C GW3RYR
Morgan, C G D G4OLY
Morgan, C J G8LPX
Morgan, C J A G3ZTJ
Morgan, D G4NSA
Morgan, D E G6FOO
Morgan, D J G8LMI
Morgan, D J GW0POZ
Morgan, D R G7BWO
Morgan, D S G3ZKN
Morgan, D S GW4KYZ
Morgan, E A G4AIU
Morgan, E F G4YQT
Morgan, E F G4ZGS
Morgan, E G GM3DPL
Morgan, E P G1BGJ
Morgan, F T G4NWU
Morgan, G G1TBV
Morgan, G G3SNR
Morgan, G G6DDH
Morgan, G H G3LMM
Morgan, G J G3ROG
Morgan, G W G3FNO
Morgan, G W L GW3POM
Morgan, H A GI4ZTU
Morgan, H K GW7VTI
Morgan, H R G1FWF
Morgan, I D P G6VKS
Morgan, I G M GM3OZJ
Morgan, I P G1YOM
Morgan, J 2E1FNB
Morgan, J GM3LYA
Morgan, J A G8SYV
Morgan, J D GW1PLJ
Morgan, J G G3ZHL
Morgan, J K GW4UEP
Morgan, J R G3HAA
Morgan, J R W G0PXO
Morgan, J U G0KPQ
Morgan, K A G3NWX
Morgan, K H G1KHM
Morgan, K L G8XWK
Morgan, K M 2E0AMV
Morgan, K M M0AOR
Morgan, K T GI0XBH
Morgan, L GM0ATQ
Morgan, L A G0WDG
Morgan, L H G4MSW
Morgan, M GW0DOX
Morgan, M D J G0JMH
Morgan, M L G4WLK
Morgan, M L GW6RGT
Morgan, N 2E1BAU
Morgan, P G7SKF
Morgan, P M0AXZ
Morgan, P N G0KYX
Morgan, P N G0ONM
Morgan, P Q G1DSJ
Morgan, R G4BFK
Morgan, R A GW6EOL
Morgan, R H G0BMB
Morgan, R H G0GCM
Morgan, R J GW8VGB
Morgan, R J B G3KGC
Morgan, R J T G0WDX
Morgan, R S G6NTQ
Morgan, S G6LUZ
Morgan, S D GW0SQY
Morgan, S F G4SUS
Morgan, S M G8VCU
Morgan, S M GW4UGH
Morgan, T A G6CAF
Morgan, T D W G3UAS
Morgan, T G G4SYP
Morgan, T J G0CAJ
Morgan, T J G0HSM
Morgan, T J G0KFR
Morgan, T J G0VOQ
Morgan, T W G3XMM
Morgan, T W GW4OXB
Morgan, V C GW3FRK
Morgans, C R M G4LUO
Morhall, V F G6LUP
Mori, A M1AFM
Moriarty, A G GM4TRY
Moriarty, J G0ENP
Moriarty, J A 2E1FLI
Moriarty, R L G0IFN
Moriarty, S P M1AOG
Moring, S P C G6NAX
Morison, I G0DMU
Morle, C W G4HQE
Morley, A N G7FNN
Morley, A S C G7OZI
Morley, D W G3OXQ
Morley, E H M0ACA
Morley, G F G0OXH
Morley, H W GW4IUK
Morley, I A G8VPD
Morley, I C B G0XBU
Morley, J G7LFU
Morley, J P G4FSQ
Morley, M J G4OII
Morley, M L G1PHN
Morley, N G8OZT
Morley, O M G7WHA
Morley, P H G4FIV
Morley, R D G4BNL
Morley, R L G8PRN
Morley, S A G7OZJ
Morley, S J G7SMO
Morley, S M G0MCV
Morley, T R G4DUJ
Morley, V W G8JXC
Morley-Joel, L G0KFJ
Morling, A G G6CDV
Morne, A C G6EYS
Morpeth, K B G1UMF
Morphett, I W M0AWA
Morrall, B A G6EOX
Morrall, P F G4TMK
Morrall, R A G8ZHA
Morrell, B E G7HLG
Morrell, F G3SRU
Morrell, P G6HLW
Morrell-Cross, L A G7HEY
Morrey, S M G0MOR
Morrice, J C G0OAJ
Morrin, B S G0FYW
Morris, A GM1RKI
Morris, A M1AQT
Morris, A A GW0COV
Morris, A C G3SWT
Morris, A C G6ZPR
Morris, A D G6VQN
Morris, A E G1VKB
Morris, A G D GW6KLC
Morris, A H G0BRQ
Morris, A J 2E1EBM
Morris, A J 2E1EPZ
Morris, A J 2E1FIT
Morris, A J G0UOZ
Morris, A P G7TIW
Morris, A R G4ENS
Morris, B A G3VBG
Morris, B A E GW4XXF
Morris, B G GW0BOD
Morris, B J G4KSQ
Morris, B T G6VIF
Morris, C 2I1EZB
Morris, C G6CGN
Morris, C D G0CUZ
Morris, C J GW6FOF
Morris, C N G0TVJ
Morris, D G3LSY
Morris, D G4GVZ
Morris, D G7WGQ
Morris, D M0ADG
Morris, D M1AFH
Morris, D C GW2FVZ
Morris, D E GM3YEW
Morris, D H G0CGS
Morris, D H G3WFH
Morris, D H GW4HMR
Morris, D J G0VID
Morris, D J G6CKJ
Morris, D J GW7RJC
Morris, D L G6RHN
Morris, D R G3REW
Morris, D S G6FNB
Morris, D S G7DOA
Morris, G B G0TQU
Morris, G D G6KQD
Morris, G E G4INO
Morris, G E GW7KIU
Morris, G J G4FAG
Morris, G J GW1ZHX
Morris, G L G1GCJ
Morris, G P M0AXO
Morris, G R G4CEP
Morris, G R GW1ATZ
Morris, G T G4YWN
Morris, G W 2W1EPL
Morris, G W G3SGC
Morris, G W G7VND
Morris, H G GW7VFP
Morris, H R G3MUJ
Morris, I G7UAV
Morris, I G8OWB
Morris, I GD3CKO
Morris, J G0CDF
Morris, J A G4LMK
Morris, J A GM8ZFW
Morris, J A GW6DDF
Morris, J A G G6PEP
Morris, J D G0DYQ
Morris, J E G3XHW
Morris, J F G4TEY
Morris, J H 2W0AMX
Morris, J H D G6LEL
Morris, J L G0DVB
Morris, J O G4BXS
Morris, J R GM4ANB
Morris, K D G1MMD
Morris, K J 2W1ETU
Morris, K J G1BSY
Morris, K J G1WIK
Morris, L A W G8PMR
Morris, L D G0UYX
Morris, M G4WFC
Morris, M G7CSI
Morris, M I G4LDW
Morris, M P G6MQF
Morris, M W G7DNX
Morris, N G8KOQ
Morris, N E G6VXS
Morris, N L G1ISY
Morris, P G1IAG
Morris, P G4XNR
Morris, P A G6EES
Morris, P D G3ISZ
Morris, P H G0APW
Morris, P J G7GIN
Morris, P P G7NRR
Morris, P R G7GJO
Morris, R G0VZH
Morris, R G1UHJ
Morris, R G3MHY
Morris, R A L 2E1DZJ
Morris, R E GM4RRP
Morris, R G 2E1ENM
Morris, R J G0OKI
Morris, R J G6XLB
Morris, R J G7EXQ
Morris, R P G4CTR
Morris, R W G0JZH
Morris, R W G0SVU
Morris, S 2W1CUP
Morris, S G7MFE
Morris, S J G1KEM
Morris, T A G4XTM
Morris, T C G6JFV
Morris, V M G0HTO
Morris, V M G4PLY
Morris, V M M0AAV
Morrish, G 2E0AMN
Morrish, M D G6HJD
Morrison, A GM4HQZ
Morrison, A R G3KGA
Morrison, A R G4YZG
Morrison, C GI4FUE
Morrison, C G GW1DRQ
Morrison, D G0LLA
Morrison, D GM0BCQ
Morrison, D C G4JHW
Morrison, D G GM7OUR
Morrison, D J G4GCW
Morrison, D J GM0LZE
Morrison, D L GM4JMZ
Morrison, D R GM1BAN
Morrison, E M F GI4ZIQ
Morrison, G 2M0AAR
Morrison, G B MW1AJP
Morrison, H G T G7IKK
Morrison, I C GM4MIM
Morrison, J G1BSX
Morrison, J A M G0JAM
Morrison, J D G0NVR
Morrison, J D M1CAV
Morrison, J D M G0ICT
Morrison, J M GM3MUZ
Morrison, K E G7CFW
Morrison, K F M GM6KTP
Morrison, L F GM1SZN
Morrison, L M GM7ADU
Morrison, M GM7ADY
Morrison, M E GM1XSS
Morrison, N J G7TQJ
Morrison, P E G3OXO
Morrison, P L G3ZDT
Morrison, P W G0VHT
Morrison, R G G6CKN
Morrison, R M G7PKT
Morrison, R R GW7VWY
Morrison, R T G3VZP
Morrison, S J GW4XNT
Morrison, T 2E1EZR
Morrison, W J G8LQB
Morrison-Smith, D J G7VBY
Morriss, A R G4GEN
Morrissey, A D GJ3YLI
Morrissey, B D G7OSL
Morrissey, B M G4YK
Morrissey, M A G0TNX
Morrissey, M P G3HUK
Morrow, A B G7KRP
Morrow, H J GI4KSH
Morse, S A G8OTV
Morse, V D G8IK
Morstatt, J G G7AMQ
Mort, A M G4AXY
Mort, N K G3ODK
Mortimer, C E G0WBC
Mortimer, C R G4OXR
Mortimer, F J G0ULS
Mortimer, G GM3GG
Mortimer, G T G3DQG
Mortimer, J G0GGJ
Mortimer, J A 2E1DMQ
Mortimer, J A M1CAI
Mortimer, J E 2E1CXE
Mortimer, P G4DCB
Mortimer, P W G0ORN
Mortimer, R A G4AUG
Mortimer-Hampson, E J G3XSG
Mortimore, E H G3DKH
Mortimore, R S GW4BVJ
Mortlock, J S V G3UGY
Morton, A T GM8BJJ
Morton, B J G4HWA
Morton, C R GM3RUP
Morton, D G4LQV
Morton, D A G4LOX
Morton, D M G1IAD
Morton, D P G6EYJ
Morton, E A G4LFD
Morton, E H G4CDC
Morton, H B G3AUA
Morton, I H G4KNT
Morton, J E G1EGU
Morton, J W GM1GDO
Morton, N J G0SSE
Morton, N J G1KRD
Morton, R G1YRC
Morton, R G6XIH
Morton, R F G7WJJ
Morton, S A G8SFR
Morton, T G7TPD
Morton, T J G4HZT
Morton, T J GM4RTN
Morton-Thurtle, P F G8UIV
Mosby, J 2E1CRI
Mosedale, N P G8TBL
Mosedale, R T G4PJK
Moseley, A D C G7VOT
Moseley, A W G1ERY
Moseley, D P GW8XWW
Moseley, J G1SOY
Moseley, J B G0NYH
Moseley, J F G2CIW
Moseley, K J G7RWP
Moseley, R A G7DRG
Moseley, R M G4JUY
Moseley, T P G4HAD
Moser, G B G3HMR
Moses, D A 2E1EHP
Moses, R G6HKZ
Mosey, P W 2W1DFX
Mosey, S A 2W1DFW
Moss, A G3YIT
Moss, A G7OSW
Moss, A A H G3GXI
Moss, A A H G8VF
Moss, A D G6DDA
Moss, A E G0MJU
Moss, A G J G0ORY
Moss, A J G3URJ
Moss, A K M1BYH
Moss, A M G1PWD
Moss, A P G7WAQ
Moss, A S G4VVT
Moss, B G0FYE
Moss, C G3TXK
Moss, C T G8IEZ
Moss, D A G4YCQ
Moss, D L G0DLM
Moss, G W G0LCT
Moss, H J T G4OYT
Moss, J G2AGO
Moss, J B 2E1DCR
Moss, J D G1KIW
Moss, J E A G0KTW
Moss, J M 2M1CLY
Moss, L G7POI
Moss, M G G4WBK
Moss, M I K E G8NVX
Moss, M P G3RZI
Moss, M R G0EJD
Moss, M S M1ARR
Moss, P G0UYF
Moss, P A G0AHE
Moss, P A G1UUX
Moss, P A GW7JUJ
Moss, P B G3SQA
Moss, P M G4BUP
Moss, P W 2E1DEK
Moss, R G7MWC
Moss, R G7UVO
Moss, S J 2E1BHZ
Moss, T J G3UFH
Moss, W J F GW4GWS
Mossman, J G7JKK
Mossman, J M1BBP
Mossop, F G A G0DUB
Mossop, F G A G0RCW
Mossop, J E G1GWS
Moston, I D G6GQG
Mostyn, P E 2E1BVH
Moth, I H G4MBD
Moth, I J M1BJA
Moth, K G8ZHC
Moth, W D G7IRH
Mothersdale, A A G3RVP
Mothew, A P G0LWM
Mothew, A P G7EEE
Mothew, N G J G7NGM
Mott, A G G6EYD
Mott, G L G1FRL
Mott, K G6BUT
Mott, K W G0HRR
Mott, R B G0ECX
Mottart, E J GM0JLJ
Mottart, E W B GM0FHD
Mottram, G G4SLF
Mottram, J S G8MUX
Mott-Gotobed, A C 2E1FCH
Mott-Gotobed, C G3TCR
Mott-Gotobed, C G4ODM
Mott-Gotobed, C E G7LJB
Mott-Gotobed, J A M G6JDP
Moughtin, R Q GD1XMW
Mould, A B G7FAZ
Mould, A C 2E1ESQ
Mould, A C M0AFE
Mould, A P J GW3DJA
Mould, G G7VCT
Mould, J A G8CXF
Mould, K G4HSL
Moulder, A A G0PBN
Moulding, C J G4HYG
Moulding, F G6KUJ
Moulds, J M GM6GVE
Moule, D C G3YUS
Moule, J P G4ZKD
Moules, A J G1ERZ
Moulson, D G G1AWK
Moult, J F G0BGY
Moulton, B G7BKL
Moulton, P A G0CHY
Mount, B G4JFX
Mount, C A G4YJU
Mount, K E G7DOF
Mount, R D G7DOE
Mountain, C O G1ESC
Mountain, D J G1WIS
Mountain, D R G4KSA
Mountain, F C G4RUB
Mountain, G G4HJJ
Mountain, J D M G6JMB
Mountain, J R G6YDN
Mountain, L G7LZM
Mountcastle, A J G4HEN
Mounter, D J G4GRT
Mountford, I R G8OWM
Mountford, K G1USH
Mountford, M H G7POG
Mountford, W H M1ARH
Mountifield, A G4CJO
Mountjoy, H D A G8YVF
Mountney, G G0JAZ
Mountstevens, P J G0IFI
Mourant, A J GJ7HTV
Mourant, F V GU2RS
Mowat, J W GM7RDY
Mowbray, I G7TET
Mowbray, J G0EVF
Mowbray, R B GW4BDS
Mowbray, T R G3VUE
Mowle, B M G7KKE
Mowthorpe, H C G7MEQ
Moxey, J H G3MOE
Moxham, P G O G4GJU
Moxham, R C G7AXN
Moxon, H G3FOD
Moxon, L A G6XN
Moxon, R D G1UIB
Moxon, R E G4SUT
Moye, B A G7MYB
Moyle, H J G8SWC
Moyle, J T B 2E0AIQ
Moyle, J T B G1AWJ
Moyle, R R G0UWB
Moyler, B E G3LTM
Moyse, B M G4IJR
Moyse, J A J G7FLI
Moyses, S T G0GJL
Moysey, P G1TPK
Mozolowski, M H GM4HJO
Mrzyglod, M G7TNQ
Muchamore, K D G0AKM
Mucklow, M G4FIA
Mudd, A C G0HOI
Mudd, S J G0FNB
Muddell, J R G2AOY
Muddimer, C S G0PAO
Muddimer, R J G1PBD
Mudge, B S G3MDD
Mudie, A R G6ZPJ
Mugele, S B G1RXV
Mugford, R M GW3SFQ
Muggeridge, G C G4LXT
Muggeridge, G R H G1AEX
Muggeridge, N T G4ZNM
Muggeridge, R A Y M G0UEY
Muggleton, R F G0HZK
Muir, A W G0IJI
Muir, C G GM0UOU
Muir, D GM4XNQ
Muir, D R G4HMO
Muir, D W M GM1VTY
Muir, I H G6FMU
Muir, J G6CKH
Muir, M S G3WOM
Muir, N R GM1PHD
Muir, R D G3LHN
Muir, R N GM4BQD
Muir, W J GM8LLY
Muircroft, E 2E1AGV
Muircroft, G E 2E1BKF
Muirhead, D G A GM0LBM
Muizelaar, B G M1BVJ
Mulcahy, A H G3LBM
Muldoon, D G0HDM
Muldoon, E GI8YPG
Mulford, L F G0RHB
Mulhall, B G G0GGV
Mulheron, J GM3ZNC
Mulheron, P GM0VXQ
Mulholland, A MI1CAS
Mulholland, G D G7EAK
Mulholland, J M GI4AID
Mulholland, R G GM0ATA
Mulira, J J S 2E1CHR
Mullan, D J GI6EIR
Mullan, J 2E1COF
Mullaney, D P G0WJJ
Mullany, J E G4GIG
Mullard, R K L G4BQV
Mullarkey, W J G3RSW
Mulleady, B B GM0KWL
Mullee, R G6MLM
Mullen, C GW0LHO
Mullen, R P M1BHU
Mullen, S F G7NJX
Mullender, G G4NAO
Mullenger, D J G8SXB
Mullenger, H M G4OJO
Muller, C M G6JZW
Mullett, K P G7HOP
Mulley, A J G8AUB
Mulley, B O W 2E1BLP
Mulligan, S G4CBA
Mullin, E G G1KSK
Mullin, G H G4XFH
Mullin, J A G0TEV
Mullineaux, C G8KED
Mullineaux, P C G0UJE
Mullineaux, P C G3XEN
Mulliner, D G M1AEK
Mulliner, P M1AQU
Mullins, D R G3RGM
Mullins, M J G0WMP
Mullins, S J G7GGT
Mullis, P A G7MTX
Mulloch, B G4OXH
Mullock, D M A G7GFC
Mullock, E B G4OCJ
Mullock, P J G3HPM
Mullord, A J G7VFU
Mulloy, T A G1YYL
Mulroy, J J G3LGY
Mulvana, D M1AKH
Mulvaney, A J G6MAJ
Mulvany, C M G0JRY
Mulvany, P D G6IQU
Mulye, J E G0VEH
Mumby, S W G7OWE
Mumford, A M G4RXG
Mumford, B C G6NVL
Mumford, W H G7LFZ
Mummery, N A G4EPG
Muncey, N G7VPA
Munday, A F G8YOX
Munday, G R G0GXM
Munday, J G G7VBL
Munday, R A G1DYR
Munden, G C G3NIL
Munden, R G G7TUS
Mundy, F H G3XSZ
Mundy, J F G4XYS
Mundy, M G0GNV
Mundy, P G6UXW
Mundy, P R 2E1AFF
Munir, A G0MNA
Munir, N A G0MNB
Munn, A A G4HOS
Munn, J G R G7DQF
Munn, M G8RQA
Munn, M R G1JCL
Munn, P R G7PRM
Munn, R P G2ALO
Munn, T P 2E1AMX
Munns, R V G6FVZ
Munro, A J G3GBB
Munro, A J GM7GIL
Munro, B N G7HOG
Munro, D A G8SFB
Munro, D G GM4XWS
Munro, D J GM3TCM
Munro, D L G1INH
Munro, D S M1AGM
Munro, E S GM4YCY
Munro, H R GM0MCJ
Munro, H S GM4NGY
Munro, I GM4GVK
Munro, I R G8NLF
Munro, I R GM4VXM
Munro, J M B G0MIZ
Munro, N G4KGK
Munro, N A S G6YYQ
Munro, S W GM0UKD
Munro, Y V G7HOI
Munroe, E F G3TQU
Munt, P G G3KTA
Munt, R A G7POF
Munton, C G F G4XMS
Munton, D G3ORD
Muraca, F M GW0VUP
Murad, Y H M0AIN
Murakami, M GM0NJP
Murch, J A G1MDJ
Murchie, P G G4FSG
Murcott, D W G8NZP
Murcott, R M G3HVY
Murden, D W G3MWM
Murdoch, D G0KYP
Murdoch, J G3YSD
Murdoch, J GM1VFQ
Murdoch, J M GM0RED
Murdoch, J M GM3JMM
Murdoch, P R W GW0TWO
Murdoch, S GI7ULG
Murdoch, S A GW4VHP
Murdoch, S J 2M1DZM
Murfin, A G7HMZ
Murfitt, P S G3YQL
Murgatroyd, T W G4HAW
Muriel, M J 2E1FAF
Muriel, R C G3ZDM
Murison, R F GM0EZG
Murly, G D G6YHW
Murphy, A GI6ANC
Murphy, A A GI1RXM
Murphy, A C G6IHF
Murphy, A D Z 2E1CCI
Murphy, A P 2E1BHQ
Murphy, A S G4ASM
Murphy, A S K G7UAD
Murphy, A S K M0BGM
Murphy, B M GW6MLL
Murphy, C G4IBM
Murphy, C R G1SVJ
Murphy, E G1EPJ
Murphy, E A G6NTM
Murphy, E G D GW0IJC
Murphy, E J G4ZWZ
Murphy, E T J GM3SBC
Murphy, G M G0LNL
Murphy, I A G8PBI
Murphy, J A G0UDI
Murphy, J A GI7DWF
Murphy, K A GI4SZQ
Murphy, K B G4XBG
Murphy, K J G8RIC
Murphy, L J G7RUQ
Murphy, M GI0CTI
Murphy, M GM0SZQ
Murphy, M GM7CMR
Murphy, M D G0CDQ
Murphy, M F G0LQI
Murphy, P GI4OMK
Murphy, P GI4VIP
Murphy, R G1EVW
Murphy, R E G3NRX
Murphy, R J G6UAM
Murphy, R J G7AEQ
Murphy, S H G4LOT
Murphy, S J G7SNK
Murphy, S P C G6PMJ
Murphy, S T M1AHS
Murphy, T GM4OBJ
Murphy, T M G4WJW
Murphy, W G3YAE
Murray, A G8PST
Murray, A J 2E1DFJ
Murray, A J L GM4FIZ
Murray, A M GM3DOD
Murray, A R GM4EJX
Murray, A T L GM0BPT
Murray, B G4VVK
Murray, C 2E1FQS
Murray, C L GM4EAU
Murray, C P G4UBV
Murray, C W G0MIA
Murray, D G1GPE
Murray, D G8GTR
Murray, D M G0NCO
Murray, G G6KYW
Murray, G G8PXO
Murray, G E G4RDG
Murray, G M G8XEC
Murray, G M GM7OQE
Murray, G O G7SKH
Murray, G T GI1JXE
Murray, H GM7MWL
Murray, H A G GM1PVD

Murray, I C G0SMV
Murray, I H G4FHT
Murray, J MM1BFE
Murray, J A GM4PUS
Murray, J C 2E1CCZ
Murray, K GI0JRI
Murray, K C G0PLB
Murray, K P GI0STM
Murray, M A K GM0CMO
Murray, M J G8IBK
Murray, M R A G0USR
Murray, N 2E1AKL
Murray, N E G7ECQ
Murray, P S G4WYQ
Murray, R G7IWZ
Murray, R D GM1SXW
Murray, R H G7DDR
Murray, S G3ZQL
Murray, S A M1AHZ
Murray, S B G1FHI
Murray, S M G7BYJ
Murray, S Y G7MLU
Murray, T A H GM6RGD
Murray, W S GM3AWW
Murray Cbe, A R D G4KEO
Murray-Shelley, R GW4KCV
Murrell, A J G7UEH
Murrell, D G0KQY
Murrell, P J G0DUH
Murt, C J G0MWW
Murtha, P G3WZG
Murton, S E G7LJL
Murton, V J G0MQK
Muscat, A C G0AFB
Musgrave, J M G6KBS
Musgrave, M J G0ESL
Musgrave, M J G4NVT
Musgrove, D M G1VKY
Musicer, R W G0SQO
Musk, A W G1FWA
Musk, P D M1ACE
Mussard, M J G7TOU
Mussell, A G3ZNP
Mussell, B G4CXJ
Mussell, S M G7SZA
Musson, A P R 2E1CXZ
Musson, C J G1UAZ
Musson, E G P G1WTB
Musson, R J 2E1BTR
Mustard, A GM3NCO
Mustard, E G0BLU
Mustard, M N D G4WBX
Mustchin, P R G0LFH
Muster, S G1UOL
Muster, S G4UOL
Musther, A P G7UZY
Musto, R N M M GM4MHE
Musto, S P G1EEQ
Mustoe, P A G4BPD
Mutch, A I GM6JYC
Mutch, R GW0BWX
Muten, C E G0FUZ
Mutimer, A H G6YYU
Mutter, H B G0AAD
Mutter, P W G8XDM
Mutter, P W M0BEH
Mutton, M V G0DEF
Mutton, R J G3EVT
Mutton, R J G4ACZ
Muxlow, P W A G8HDJ
Muzyka, J G4OOC
Muzyka, J G4RCG
Myall, J N G8OHG
Myatt, A J GW7MDH
Myatt, B L G7CUT
Myatt, G N G3FRN
Myatt, R G0VEO
Myatt, T E G4WRL
Mycock, B E 2E1BMV
Mycock, R G0DTT
Mydat, D M G1NUG
Myers, A G1TLC
Myers, A C G7EPU
Myers, A C G7PTT
Myers, D H G8AYG
Myers, P W G3UWT
Myers, R G6LCH
Myers, R G8LUL
Myers, R F G8ZRM
Myerscough, P D G0VYO
Myerscough, R N G0SIQ
Myerscough, R N G3PXO
Myford, I A G4HTO
Myhal, P J G0DDG
Myhan, A G0VBL
Myhan, C A G0URA
Myhill, M J G0UEI
Myint, K M G0DSQ
Myland, A H G0JIS
Mylchreest, C G1EVV
Myler, C J G0JLQ
Myles, M F G7DGK
Mynett, A L G3HBW
Mynett, L J G4DRD

N

Nabb, B G4EFZ
Nadauld, A J GM3RFQ
Nagle, C G3KPZ
Nagle, M J G4RZI
Nailer, A J G4CFY
Nailor, G W G3FLB
Nairne, P T D M0AFZ
Naish, E GW2FRB
Naish, N H G4MHS
Naish, P J G3EIX
Najman, R A P G0NXM
Nakisa, G R G0WBJ
Naldoken, C C G0WUN
Nally, J A G8KCB
Nancarrow, D J G3RID
Nance, A G G0KKS
Nap, J GW0MFN
Napier, A K GM1TBW
Napp, P M G1USF
Napper, D M1BXU
Napper, R J G4FXU
Nappin, D G3MLS
Narroway, C G6FAF
Nasey, D G3ATM
Nash, A P G4ZHZ
Nash, D R G0NZT
Nash, E G3YNR
Nash, E G4KOA
Nash, E W J G8SPI
Nash, F W G3RIS
Nash, G A H G4XBD
Nash, G L GW0OQN
Nash, I R G0MVD
Nash, J F G4SFP
Nash, L C G0DSN
Nash, L C G0NWT
Nash, M A G8FNH
Nash, M A J G0IHC
Nash, M J S G0LCH
Nash, P G M0ATL
Nash, R J G4GEE
Nash, R S G0DVL
Nash, R S G1BSZ
Nash, S G4PBI
Nash, S G G0UQT
Nason, R E G1HGE
Nattrass, G A G0OGD
Naughton, J GM4WQH
Naughton, P GM4WGC
Navier, B L G0OVT
Navier, L H G3UYV
Navin, T J G3GTN
Nayler, L R G4BNY
Naylor, B G3SHF
Naylor, B G6UQ
Naylor, B G8SRS
Naylor, B E 2E1DRR
Naylor, D G0JEC
Naylor, D G3SKN
Naylor, D A R G3GHI
Naylor, D S B G4JWA
Naylor, D T G3YXS
Naylor, I V G8YAT
Naylor, J G6ZQC
Naylor, J G7ROI
Naylor, J GW3SCX
Naylor, J A G4OGY
Naylor, J S G0NIQ
Naylor, J S G4KLX
Naylor, K J G7ILP
Naylor, P A G1JJK
Naylor, P J G8TXQ
Naylor, R G1MJT
Naylor, R G4MSY
Naylor, R H M1CAE
Naylor, W G7WBJ
Naylor, W E G0RQW
Neades, P W G1YFC
Neal, A D G0BYR
Neal, B J G0IUR
Neal, D G G0SPK
Neal, D J G4ESG
Neal, E V E G8GP
Neal, G F G3LON
Neal, J E G4NQC
Neal, M R G7NJP
Neal, R A G0IEB
Neal, R A G0KVG
Neal, T B G0CQH
Neal, T M G0NMC
Neal, W W G0DED
Neale, B M G4FBN
Neale, C A G3TMU
Neale, D A G7DHW
Neale, D G G8WBN
Neale, F W J G8AQT
Neale, H G3REH
Neale, J A G7OJZ
Neale, M G7NVD
Neale, M P G0HLW
Neale, P A G3UHN
Neale, P J G0VVK
Neale, R G4OIE
Neame, N I G2AUB
Neary, B G3VHZ
Neary, C M G0TMH
Neary, H M G6CKG
Neary, J GM0XFK
Neary, J A G0NAJ
Neary, J K G4NRM
Neary, J M J G4ZQM
Neary, J T G8ZOB
Neate, B J G1XKQ
Neate, J A G1WUU
Neath, A R G8NWR
Neathey, B C T GW0DPM
Neave, R P G4DAN
Neaves, A J G4BKA
Neech, D N J G1OXV
Needham, J 2E1FJD
Needham, M W G6AWV
Needham, P G8RQN
Needham, P L G6AMM
Needham, P M N GW4MSI
Needham, R G0WXC
Needham, W H E GW6AMK
Needs, R A G4NUG
Neely, D G6ZSF
Neely, J GI4ZFA
Neenan, A L G0MPO
Neep, B H G0USV
Neeson, L M G1WRP
Negri, B G3LXN
Negus, A G G8DQJ
Negus, N R G7AEG
Negus, N R G7NYP
Nehan, A G4HUE
Nehmzow, C M0ASG
Nehmzow, U M0ASF
Neil, C J G0MVC
Neil, G R GM1VDZ
Neil, M C 2E1EEU
Neil, S P G6AEB
Neill, B A GI7UQW
Neill, D GI0EZS
Neill, F C G6NJT
Neill, J GI0CDM
Neilson, A V G4CVZ
Neilson, B G4TFA
Neilson, D R G4ZNZ
Neilson, G L M0AAQ
Neilson, R GM0EFT
Neilson, R A G1PTT
Neilson, R T M1ATL
Nelhams, B L G7HFX
Nell, A W G3VYS
Nell, K J G1KDH
Nellis, J T GM4PLI
Nelmes, A B G4PLF
Nelmes, B E A G4FQH
Nelmes, E V GW1ABB
Nelson, A P 2E0AJQ
Nelson, A P G8KVN
Nelson, A P M0APN
Nelson, A R G8OFX
Nelson, A R GM4IIR
Nelson, A S GM3LLB
Nelson, B B GJ4KBM
Nelson, C G1UMA
Nelson, D G7GUK
Nelson, D H MW0BAJ
Nelson, F G W G0RQV
Nelson, H A W GM0OGC
Nelson, J H GW4FRX
Nelson, J M G4KLA
Nelson, J O GI4TRX
Nelson, N G3DHE
Nelson, N D G4PRR
Nelson, P C GM4XAW
Nelson, P W GW8GOO
Nelson, R G0CJO
Nelson, R G3ZLF
Nelson, R A G1NGE
Nelson, R J G0TZO
Nelson, R J G1CKT
Nelson, R L G8BBK
Nelson, R L GW6MPW
Nelson, R M G7TMR
Nelson, S A G6BBK
Nelson, S A GW8NXC
Nelson, W D G4PXM
Nelson-Jones, L G4JDW
Nepean, E Y G5YN
Nerurkar, V GM0VOL
Nesbit, J S A G1ZRJ
Nesbitt, A R G0CWQ
Nesbitt, A R G3SGY
Nesbitt, G R MM1BXF
Nesbitt, K R GI1XPV
Nesbitt, S R E GI0MSI
Nesbitt, W H GI3TZX
Ness, F GD3ESV
Ness, J R GM1JLU
Nethercott, F J G0RXU
Netherton, S G G8WBG
Netherway, R T G0PDV
Nettleship, M D G4VLZ
Nettleship, T N G4WBN
Nettleton, R J G3YED
Neufville, F L G4YQZ
Neve, A R G6AED
Nevey, H M G0IBF
Nevill, A J G7JDH
Neville, A 2E1DJD
Neville, A R G7WCC
Neville, G J E G3IDZ
Neville, J F G3XSP
Neville, J R G1HGB
Neville, K G0EUJ
Neville, M L G4JUK
Neville, P J G4NZB
Neville, R GW8RAS
Neville, R H GW3KWB
Neville, R K 2E1DZL
Nevin, J G4ZHG
Nevison, A G4OSH
Nevison, J G G0VEG
New, E R G0BSJ
Newall, P F G4GEI
Newberry, P A G0VQD
Newberry, P S G4WQW
Newbold, I M G8KSZ
Newbold, J K G1GNL
Newbold, M G0VVA
Newbold, M J G1MJN
Newbold, P G G0SSU
Newbold, S M G3VQN
Newbold, S T G1RDX
Newbould, K G4VRW
Newbury, C J G8JGE
Newbury, D J G0ENR
Newbury, L G6CKD
Newby, G C G3EBH
Newby, G H G1ZBU
Newby, J G6FDB
Newby, J R G7EIC
Newby, N C G0VDZ
Newby, P L G4AVL
Newby, R G6PEI
Newby Robson, C F G1YVS
Newby-Robson, G A F W G0PFV
Newcombe, B 2M1FXH
Newcombe, P G0YZC
Newcombe, P G6YZC
Newcombe, P W G4SXO
Newell, C M G8IQF
Newell, D A G0AKS
Newell, H G6VHW
Newell, H J G7IBA
Newell, J F GW4UHK
Newell, M G8KOE
Newell, M F G1HGD
Newell, M W G3IUE
Newell, N S GI3YMY
Newell, R G6XKX
Newell, R A G6KBJ
Newell, S A M G1EFS
Newey, D G G3NDN
Newey, J E R GW6RUE
Newey, M C G4SND
Newgas, J R G7LTQ
Newham, R H G7SEK
Newing, P L K G1YEY
Newland, H E G3TEK
Newland, J R G1ETD
Newland, O E G7SNB
Newland, R H G3VW
Newland, T J G0TJN
Newland, T M G3TMN
Newlands, A GM4VAY
Newlands, A J GM4TXN
Newlands, M J G4FYG
Newman, A G4AVX
Newman, A C A G2FIX
Newman, A P G7BAX
Newman, B J G3MMN
Newman, B J G4TGN
Newman, B V G7OXG
Newman, C G6WIC
Newman, C A 2E1FNN
Newman, C G G0VXS
Newman, C J G4JCJ
Newman, D J G7SAL
Newman, D R G4CGZ
Newman, G A G4YCP
Newman, H G2BNY
Newman, J G1VPU
Newman, J G3RPY
Newman, J A G0FVC
Newman, J B H G4MGO
Newman, J E G0VDU
Newman, J K G6XBS
Newman, J P C G0PGT
Newman, K J N G4CZA
Newman, L G G4ZDO
Newman, L P V G3UUU
Newman, M G3UWZ
Newman, N P 2E1CPP
Newman, P G4KIM
Newman, P L G4INP
Newman, P S G8UDI
Newman, R G0SOI
Newman, R A G7OYW
Newman, R B G1OSM
Newman, R B G2CH
Newman, R B G3VZL
Newman, R H G7SAX
Newman, R J G6XO
Newman, R J GW0UZN
Newman, R J GW4BCF
Newman, R L G4JIE
Newman, S V E G6WQI
Newman, V H G0EFI
Newman, V H G7AIC
Newman, W A G0JLH
Newman, W G G7NLE
Newnam, K G 2E1FDJ
Newnham, G L G7EOH
Newnham, M A G6NZ
Newns, H S G0BHV
Newport, C R G4JCI
Newport, J A G4FVD
Newport, M J G1GZG
Newport, M J G8VCN
Newport, R C G0UUI
Newport, S L G4DEV
Newsham, E R G4ERN
Newsham, S G7KKC
Newsome, J W M1ALK
Newsome, M O G6GUW
Newstead, J W G4UJN
Newstead, R G3CWI
Newstead, S W G7PRC
Newstead, T J G0LQX
Newstead, T J G0WJW
Newton, A G4AHI
Newton, A A M1BWV
Newton, A V G0LPU
Newton, B G4ZAK
Newton, B GM1HGC
Newton, C E G2FKZ
Newton, C J P G0FGZ
Newton, D J S G3JJZ
Newton, D M G0GQQ
Newton, D R 2E1BZS
Newton, E W GM0LDC
Newton, G G4UOS
Newton, G A G7SMH
Newton, G D G4VUP
Newton, G P GM4LSD
Newton, H J S G8EAM
Newton, J A G6KQS
Newton, J M G1HGG
Newton, J R G0BZJ
Newton, L G G3YJU
Newton, M A G7JMV
Newton, M A G7KOL
Newton, M A M0AOU
Newton, M E G7SBZ
Newton, M J G3UKW
Newton, M R G7FOV
Newton, P K G7PRI
Newton, P W GM0EZR
Newton, R G0NAH
Newton, R B 2E1CBG
Newton, R H G7LDD
Newton, R J G0EWH
Newton, R W G8YAU
Newton, S A G7RDE
Newton, S J G3OPD
Newton, V 2E1EMH
Newton, W F G7GMZ
Newton of Brauncenell, A G4PSC
Newton-Goverd, D R GW6HKT
Ney, G R G1KEY
Ney, M J G7AZW
Neyland, T A G3RPL
Neyman, M W G6CCE
Ng, R C 2E1BJW
Nias, J D G3VRB
Niblett, M C G3RGR
Niblock, A J GI7USA
Niblock, P A G3INB
Nice, I G6RUM
Nice, K J G7TZC
Nice, P G G8IER
Nichol, J G0EUN
Nichol, R W N G0FVD
Nicholas, A E G3ZUE
Nicholas, C S J G0LZV
Nicholas, D J G7RJX
Nicholas, D M G0WWA
Nicholas, G A G8MMM
Nicholas, G A GW7EVG
Nicholas, J G GW3OIN
Nicholas, J R G G8HMV
Nicholas, K J 2E1CES
Nicholas, M J M0AHS
Nicholas, P G4OLV
Nicholas, S A G0NYY
Nicholas, S E G6ADO
Nicholas, T A GW4RVA
Nicholas, T J G7VOI
Nicholas-Letch, J C G3PRU
Nicholl, B G0DFB
Nicholl, F G0CVD
Nicholl, J H G4WZD
Nicholl, S A GI4TPN
Nicholl, W R GI7PBE
Nicholls, A C G0HEL
Nicholls, A J T G7KXP
Nicholls, B G G7FWD
Nicholls, B J G6BLU
Nicholls, C B G4NPR
Nicholls, D G0GTC
Nicholls, D A G3VMZ
Nicholls, D A G7IBU
Nicholls, D C J G6AEC
Nicholls, D E G6RQA
Nicholls, D P GI3ZVZ
Nicholls, F N G3MAX
Nicholls, G G1PPX
Nicholls, G A G1FOO
Nicholls, G J P G4JXS
Nicholls, J E G0KXG
Nicholls, K J G8MXI
Nicholls, L M G4HIB
Nicholls, M G8YAV
Nicholls, M A G3LQX
Nicholls, M J G4HIA
Nicholls, N L G4TCK
Nicholls, P G0ONA
Nicholls, P G8YAY
Nicholls, P C G0VMK
Nicholls, P D G0VXV
Nicholls, P G GW7RIB
Nicholls, P N G1BGH
Nicholls, P T G7VEG
Nicholls, R A G8BKY
Nicholls, R H GD1ETE
Nicholls, R J T G4ZHY
Nicholls, R K G0TNA
Nicholls, R . S . G8TTP
Nicholls, S G0JFM
Nicholls, S G3NJB
Nicholls, S J G7VCI
Nicholls, T W G1KSV
Nichols, A G G1XJT
Nichols, B G7TKF
Nichols, C B G0FZD
Nichols, D G3BCE
Nichols, E W G0KMR
Nichols, K J G4FNI
Nichols, L J G3GMW
Nichols, M G4PVR
Nichols, P G H G1SDP
Nichols, S W G0KYA
Nichols, T A G7TKH
Nichols, W H C G4TCF
Nicholson, A G6LVE
Nicholson, A H G8FLV
Nicholson, A W G8ZOU
Nicholson, B G1MWF
Nicholson, B G7LNL
Nicholson, B A G0MZF
Nicholson, C E G3JSY
Nicholson, D G0OPZ
Nicholson, D G4ZHW
Nicholson, D P G8GKX
Nicholson, D T G4GGE
Nicholson, F G4AFR
Nicholson, G C G1IBG
Nicholson, J G0CZP
Nicholson, J G6DPV
Nicholson, J D GM8WJK
Nicholson, J R G8XRR
Nicholson, J R GM0MFE
Nicholson, K G1TRF
Nicholson, M G1LEH
Nicholson, M G4FRW
Nicholson, M G8YVP
Nicholson, M N G8NYK
Nicholson, P G0MCK
Nicholson, P G3MYZ
Nicholson, P F G6LHI
Nicholson, P L G8UXL
Nicholson, P S G0EKR
Nicholson, P S G3VJF
Nicholson, P W G1EFT
Nicholson, R G4SQG
Nicholson, R A G8ZOV
Nicholson, R D G7EFM
Nicholson, R F H G2DOH
Nicholson, S J M1BPI
Nicholson, T GM0LNQ
Nicholson, T E G1VRC
Nickells, A B G4NOD
Nickells, B F G4NOA
Nicklin, D G4FHO
Nicklin, D N G0REK
Nickolls, K G7VLI
Nickson, K GW1OTI
Nicol, A T G8BXA
Nicol, H G S G0XTT
Nicol, I M J GM4YVR
Nicol, M J G4LLT
Nicol, R I G1NZZ
Nicol, R I GD4YMY
Nicolaides, K G4WQX
Nicolaou, D A G6KMY
Nicolaou, M G0WWW
Nicole, D A G8CYJ
Nicoll, P G5ZN
Nicoll, P J G4HQQ
Nicoll, W C G3KWN
Nicoll, W L T GM4LFZ
Nicolle, S Q G4MUV
Nicolson, A M M0ANF
Nicolson, D G4WJG
Nicolson, D B GM4RGU
Nicolson, G D B G0FBS
Nicolson, J GM3KGT
Nicolson, P L R 2E1DJQ
Nicolson, R J GM6RGU
Niebuhr, K W G6OM
Niel, P M0AZC
Nield, E W GW3ARP
Nield, P N G8SH
Niemann, C L M1CDN
Nieschalk, J G0WZQ
Nieto, R L V GM1BVA
Nightingale, E G3UUG
Nightingale, I A G0VZM
Nightingale, J G A G8AEU
Nightingale, J T G8BUJ
Nightingale, M J G8VQK
Nightingale, P T G8XQI
Nightingale, R P G7MNO
Nilan, P G0LVG
Nilan, W L G3PRD
Niles, C P G0SWL
Nilski, Z J G3OKD
Nilsson, A P G6FCX
Niman, J H N G8GAJ
Niman, M A G3LGN
Niman, M J G6JYB
Nimmo, M J GM8JVZ
Nimmo, S G6OJB
Niner, T A W G1RFS
Nisbet, J M G3OGO
Niven, A P G6SUP
Niven, I H G6EGY
Niven, M F G4AHT
Nix, D G4TBK
Nixon, A G1EFU
Nixon, A M G4DPO
Nixon, B M G7WBY
Nixon, B W G6EGU
Nixon, C R G7LYL
Nixon, J S GI0LCZ
Nixon, L G6YZU
Nixon, M G4YSO
Nixon, M J G0NTF
Nixon, P G4BBI
Nixon, P M1BUT
Nixon, P W G0GYX
Nixon, R D G7LBA
Nixon, R W G0XVC
Nixon, S S G0NBY
Noake, M P G6WJH
Noakes, F A G2ASF
Noakes, F A G2FTK
Noakes, M G4ZJQ
Noakes, R G0RND
Noakhes, G T G2AUI
Nobbs, K G0KSS
Nobbs, W R D G3XKI
Noble, A J G7BSO
Noble, B A G4JAJ
Noble, D G3MAW
Noble, D H M GM3NCS
Noble, D J GM0DJN
Noble, D M G4VQX
Noble, G M1BAV
Noble, G G G0EJX
Noble, G S G7CWK
Noble, J G3JZJ
Noble, J A S G3NHF
Noble, J G G0EVD
Noble, J S G4VWQ
Noble, J W N G8FEQ
Noble, M G4BVP
Noble, R G8PID
Noble, R J G4UYR
Noble, R W G4TZH
Noble, S C G3WWD
Noble, S M M1CBA
Noble, S M C GI4SAM
Nobles, D G4OMH
Noblet, T G3TN
Nocera, S G1VNE
Nock, A L G0BFZ
Nock, C T 2E1AMT
Nock, C W G8MDM
Nock, D G7KHW
Nock, D J G0SVA
Nock, G D G0EJO
Nock, M P G7WHU
Nock, R W G4LWN
Noden, A G 2E1FGT
Noden, C H G3JPB
Noden, J G8IOK
Noel, J G 2E1FPM
Noke, S D J G4MOE
Nokes, A R G4WZA
Nokes, D J G3JVR
Nokes, M D P G0TDT
Nokes, R W G6JTK
Nokes, T 2E1DGY
Nolan, B G0NBW
Nolan, C G0MIJ
Nolan, G G6ZQA
Nolan, J G1AIP
Nolan, J A G0HHQ
Nolan, J M G7LUR
Nolan, K J G3ZIV
Nolan, R W G3KWK
Nolan, V L G8XEI
Nolan, W G8PZI
Nolson, R E G0PMU
Noon, C T G0CTN
Noon, D GI3MMG
Noon, G S C G1BYJ
Noquet, R R L G8ZGI
Norbury, T C G1FWU
Norcliffe, B L G8DTS
Norcott, B A G7GES
Norfolk, D W E G4AWM
Norfolk, J G4MED
Norman, A E G0GGB
Norman, C J G4MHR
Norman, D G1MGT
Norman, D A G0BTX
Norman, D C G1UAE
Norman, D C G7SSR
Norman, D J G3TJB
Norman, D M G3WOQ
Norman, F L G0NCT
Norman, G A G1FLT
Norman, I A G6UEG
Norman, J G0AZR
Norman, J A G7KGL
Norman, J C G3OCF
Norman, J L G0LRE
Norman, M 2E1BEV
Norman, M G G4MGN
Norman, M N G4XDL
Norman, P G6UEI
Norman, P A J G1DUI
Norman, P J G0WMC
Norman, R A G4JPO
Norman, R E G3NAI
Norman, R G G0AFY
Norman, R P G7IQZ
Norman, S K M1BQG
Norman, S L G8BDO
Norman, W J G8VCQ
Normandale, S J G6LVG
Normington, F H G4YLJ
Norridge, D A G0BPX
Norrie, A M GM1MQE
Norrie, D G GM1RPJ
Norris, A G1WQE
Norris, A J G4KSR
Norris, A J M0BCB
Norris, A W 2W1AYP
Norris, B A G1BAR
Norris, B M 2E1CML
Norris, C S 2E1EZJ
Norris, D W G4TUP
Norris, G G6DCS
Norris, G A G1RBZ
Norris, G W G4KAG
Norris, J A G1KVC
Norris, J J G0DDU
Norris, J S G4MLR
Norris, K G0WZG
Norris, P F G4VUN
Norris, P S G3XVN
Norris, S GM3ZAS
Norris, T S G6FKY
Nortcliffe, H G7CVX
North, D P GM0MDB
North, J B 2E1BVK
North, J M G1RGJ
North, M J G1RGK
North, M L G8HKK
North, M R G4GMK
North, P J G8VVP
North, R G3WAR
North, R B A G4KHR
North, R C G3YKK
North, R S G8TZN
North, S J G4JRJ
Northall, G M0AUN
Northam, J W N GW1XBF
Northcott, R J G1YPM
Northcott, W J GW3KRD
Northeast, D J G7LWY
Northey, M R G8JCD
Northfield, J E G7ELX
Northover, P W G4OIF
Northover, T W G0TOD
Northrop, C H G0DZF
Northrop, S H G6MQG
Northway, B J G0BNJ
Northwood, D J G4VQJ
Northwood, E J G3WRL
Northwood, F A G8UKB
Norton, A B J G1ABJ
Norton, C R G4FTG
Norton, D K G0UHI
Norton, D K G7OCY
Norton, F A G6IHB
Norton, F J G0GJX
Norton, F W G0ICH
Norton, H W G0TSI
Norton, I D M0AZM
Norton, I D M1BEH
Norton, J G0GNU
Norton, J A G4TLS
Norton, J E G3PLW
Norton, L G4JNW
Norton, M J G7VVQ
Norton, M J G8HJH
Norton, R J H G1JOD
Norton, T M0BFQ
Norton, W L R MI1BYL
Norval, B L G3VHT
Norvall, K F G3IFN
Norwood, J G1CMH
NoswORthy, R 2W1CUN
Noszkay, T E G0TCH
Nothard, J P G0NVD
Notman, M A G0JBS
Notschild, A F G3RSF
Nott, P C G7JCP
Nottingham, E A G0VGA
Nottingham, G F G3DTA
Nottingham, M R G1XIY
Nottle, P C G0UOA
Nowell, J L G4FUO
Nowell, T 2W1DRE
Nowikow, C V G8BUI
Nowland, P G G4NVS
Noy, J G8VPE
Noyce, W F G0OBM
Noyes, T G7AWP
Nunn, A J G8AXO
Nunn, C J G8NEH
Nunn, D E G3JMJ

Nunn, J C ... G0MZN
Nunn, J C ... G3ULT
Nunn, J D ... G7TPV
Nunn, K R ... G0NFC
Nunn, L ... G6YZB
Nunn, P B ... G8UPJ
Nunn, R I ... G7OOV
Nunn, R W ... G6WMV
Nunneley, A D ... G0RIT
Nunns, V J ... G0NKQ
Nurse, J E ... G3UUC
Nurse, M E ... G4ZFT
Nurse, P ... G0IFT
Nursey, S K G ... M0AJI
Nussey, A ... G7VIV
Nutbeem, T ... GW8XRU
Nutbrown, L E C ... G7JTS
Nuthall, D G ... G4ZOU
Nutkins, C R A ... G1EFX
Nutkins, P ... G0HET
Nutsey, G B ... M1AQO
Nutt, D J ... GM3YDN
Nutt, G K ... G4WLI
Nutt, R S ... G4LIJ
Nutt, W A ... G1ZSP
Nuttall, D L ... G4HRC
Nuttall, D L ... G4ZST
Nuttall, D L ... G8HRC
Nuttall, D R ... G4VFQ
Nuttall, G A ... G8XRS
Nuttall, N ... G4AUM
Nye, B R ... G6SYN
Nye, C J ... G0EKN
Nye, D A R ... 2E1ACK
Nye, G C ... G8URP
Nye, T R ... G8ZBN
Nyman, M ... G6WMR
Nyman, M A ... G4OMP

O

O Brien, S ... G8NLS
O Brien, T ... G0PIL
O Connor, N ... G6LGO
O Garr, J ... G0DIE
O Grady, N ... GM3NEX
O Hagan, J A ... G6HIX
O Kane, P G ... GI3OTV
O Neill, J J ... GI4XKI
O Rourke, J ... G7NJZ
O Sullivan, C F ... G0PUO
O Tani, H D C C ... G8OTA
OBrien, A ... G6AOB
OBrien, A ... G7IPK
OBrien, B M ... G0UCT
OBrien, E M ... G3WIO
OBrien, K J ... G0NKO
OBrien, M J ... 2E1ABY
OConnell, J P ... G0PZF
OConnor, J ... G4GSM
OConnor, K C V ... G4SCH
OConnor, P R ... G4SFG
OConnor, S ... G4ECC
OConnor, S M ... G1YKK
ODea, P J ... G4XPI
ODell, J F ... G4RCF
ODell, L R ... G4ROC
ODell, P S ... GM3MUM
ODell, R J ... G1SIW
ODell, S A ... G1SIV
ODonnell, M E ... G8CCV
ODonoghue, B ... G0KQH
ODonovan, A D ... G8NKM
OFarrell, J C ... G4YIS
OHagan, N G ... G8NZK
OHara, A C ... G0KEX
OHara, G C ... G8GUH
OHara, J ... G6WSZ
OHare, H M ... G0PUQ
OHare, J ... GM3WNB
OKane, O ... GI6OJC
OKelly, P J ... GI3PKY
OMara, J J ... G0PPX
ONeil, T E ... GM4PRO
ONeil, W J ... GW8KNG
OReilly, I J ... G4MAD
OReilly, K J ... G6INM
ORyan, L P ... 2E1CAT
OShaughnessy, A F ... G0ITO
OShea, J ... G4GDP
OShea, J T ... G4IVM
OSullivan, P ... G1IKL
OSullivan, S P ... G8VPG
OToole, I ... G6JLP
Oakden, R ... G6DDP
Oakes, A ... G8GYU
Oakes, A H ... G6AHO
Oakes, A T A ... G1RPG
Oakes, D ... 2E1COG
Oakes, D ... M0BDE
Oakes, D ... M1BKV
Oakes, D N E ... G0FRN
Oakes, E J ... G7JWL
Oakes, F ... G0UOC
Oakes, I C A ... GI0HDK
Oakes, J H ... G3ESA
Oakes, L H ... G8FOY
Oakes, W J ... G1YQY
Oakley, A R ... G4JMO
Oakley, B A ... G1DUJ
Oakley, B M ... G4PBJ
Oakley, C G O ... G4YGE
Oakley, C S ... G0AEA
Oakley, C W ... G3IPD
Oakley, D H ... G0DFI
Oakley, D M ... G4PDO
Oakley, E H N ... M1BWR
Oakley, E L ... G2AZM
Oakley, H M ... G0GOJ
Oakley, J W ... G4VQZ
Oakley, N A ... G0NAO
Oakley, P J ... G0BVD
Oakley, R ... G1DDR
Oakley, R H ... G6BRJ
Oakley, R L ... G0MRO
Oakley, R V ... G8GRT
Oakton, F R ... G0IWF
Oates, D R ... GW0DXO
Oates, G A ... G8YAZ
Oates, J A ... G3LZI
Oates, J A ... G4MOU
Oates, J D L ... GM0VIY
Oates, K H ... G0DUY
Oates, P ... G7VNE
Oatey, A H ... M0AVN
Oatley, D W ... G6NJW
Oatway, G ... GW8BQK
Obee, B A ... G8RDL
Obermaier, G J E ... G4LRH
Obey, G R ... G0MSF
Ockenden, M S ... G3MHF
Ockendon, C S ... G3TPO
Odam, M E B ... G7LVN
Odd, H G ... G7JYG
Oddy, B C ... G3FEX
Oddy, G ... G4ATS
Oddy, G ... G8BVR
Odegaard, P S ... G0DJT
Odgers, G ... 2E1EVS
Odlum, K ... G4WBW
Offer, I C ... G4FDX
Offer, R E ... G1ZJP
Officer, G A ... G1NVL
Offler, F ... GM3YAO
Offord, R P ... G0JAR
Offord, R S ... G3ZYX
Ogden, A ... G5OD
Ogden, A ... G6JZN
Ogden, A J ... G1FZV
Ogden, B P ... G0GML
Ogden, E W ... G0EWO
Ogden, E W ... G0NWH
Ogden, G J ... G6XKY
Ogden, J ... G0TRK
Ogden, L L ... GW0RQM
Ogden, S M ... G4UJM
Ogden, T L ... G0UCA
Ogg, A J ... G4FPP
Ogg, G D ... GM1BOT
Ogg, M M ... GM1AIH
Ogier, J D ... G0SUU
Ogilby, J A H ... GI0GQA
Ogilvie, G R ... GM1VBG
Ogle, K ... G1YBU
Ogle, M ... G1DLH
Ogle, R W ... G1TUX
Oglesby, R A ... G6FNJ
Ogston, K J ... GM6AMI
Ohta, M ... G0CEO
Okas, J ... G3YGE
Okubo, A ... G0LHB
Olbrien, J ... G4ZKR
Oldbury, B A ... G1PAA
Oldbury, E M ... G1SJJ
Oldfield, F F ... G3DVK
Oldfield, G C ... G1TRC
Oldfield, G F ... G3GFT
Oldfield, J ... G4VJL
Oldfield, M J ... G8TJI
Oldfield, S R ... G1BHF
Oldham, J L ... 2E0AGB
Oldham, K A ... G8LDB
Oldham, P ... G0NVO
Oldham, P M ... G7HRW
Olding, R N ... G6KSQ
Oldroyd, R D ... G4JFV
Olesen, G K ... GM3MQO
Oliphant, A J ... GM3SFH
Oliphant, J ... G7OWP
Oliphant, S C ... G7IIC
Oliva, V E ... G3KPM
Oliver, A ... G4FGO
Oliver, B ... G8DFI
Oliver, D ... G6UQO
Oliver, D ... G8YFH
Oliver, D A E ... G4XQX
Oliver, D B ... G4HPT
Oliver, D C ... G4KJK
Oliver, D K ... G7BRF
Oliver, E ... G1AIA
Oliver, E G ... M0AEO
Oliver, G ... G0BJR
Oliver, G ... G1ORC
Oliver, G ... G4ORC
Oliver, G ... G7CCS
Oliver, H W ... G0IZD
Oliver, J ... GI1VMF
Oliver, J D ... G4HMC
Oliver, J H ... G7PBK
Oliver, J R ... G6CND
Oliver, K ... G4SLG
Oliver, K ... G7TKN
Oliver, K A ... GW6JFX
Oliver, K D ... G0SJY
Oliver, M E ... G4NOO
Oliver, M V ... G4HOC
Oliver, M W J ... G7TXW
Oliver, P ... G1JDO
Oliver, P ... G4ABT
Oliver, P C ... G0PEM
Oliver, P D ... G7CGC
Oliver, P J ... G4TSO
Oliver, P J ... G7NLI
Oliver, P J ... M0AWP
Oliver, R ... G1AHZ
Oliver, R B ... G3NDS
Oliver, S A ... G0NLJ
Oliver, S A ... G1LMM
Ollerhead, D ... G4JMF
Ollerton, M B ... G7GMD
Ollett, C H ... G2OR
Olley, C C ... G3AIZ
Olley, H E ... G3XDC
Olliffe, P N ... G0CEI
Ollis, V ... 2E1EUM
Olone, A ... GM3EXX
Olsen, J M ... G4ZBK
Olsen, T H ... GM0EQW
Olson, P D ... G7FMW
Olson, P D ... M0AJJ
Olver, I V ... G1LLU
Olway, J P ... G3RMA
Omeje, C N ... G0TRQ
Omer, W J ... G3DOJ
Omielan, R J M ... 2E1DGF
Oneill, C J ... G0CGD
Oneill, G ... G0LUY
Oneill, P N ... GW0EJE
Oneill, P N ... GW4UZL
Oneill, P R ... G3SPO
Oneill, T A ... G3ZGI
Onion, G A ... G6FHO
Onion, P J ... G0DZB
Onione, D ... G8WGQ
Onions, G R ... G4JVH
Onions, J T ... G4YZW
Onions, S ... G0RNX
Onn, H R ... G3WUP
Oord, E ... G7THB
Ootam, S J ... G8SJO
Openshaw, F A ... GW4TWB
Openshaw, G ... G2BTO
Openshaw, J G ... G2AYG
Opie, P G ... G4MKG
Opie, T D ... G0FHW
Opitt, L W ... G1NEJ
Opitz, H ... G0KEC
Oram, E T ... G1NUO
Oram, M A ... G6GIF
Oram, P J ... G8ZOW
Oram, R J ... G0FXI
Oram, S D ... G0SYK
Orange, G W ... G0DLQ
Orbell, E R J ... GM4TGC
Orbell, J J ... G0WGR
Orchard, A ... 2E1BJM
Orchard, A C ... G6RXK
Orchard, A J ... G1OEB
Orchard, B J ... 2E1BJN
Orchard, C H ... G0WQN
Orchard, F P ... G4GBC
Orchard, G R ... GW0CNK
Orchard, G R ... GW1ARC
Orchard, G D ... G6JKK
Orchard, H ... G0AXH
Orchard, K M ... G3TTC
Orchard, P A ... G0EYZ
Orchard, P A ... G0SMK
Orchard, R ... G7ARP
Orchard, R ... G8HZN
Orchard, T A G ... GW0DSW
Orchel, H M ... G7TFG
Orchiston, A R ... G7SYQ
Ord, P J ... G7CLG
Ord, W F ... G3GDY
Ore, A J ... 2E1EMD
Ore, N M ... 2E1EME
Oren, A ... G0UFH
Orengo, L B ... G0ECF
Orford, G ... G4FRO
Orford, J R ... G3PBF
Organ, J A ... G4ULH
Orgee, A R ... G1ICQ
Orgel, H F L ... G8DUT
Orgill, D Y ... G0ABP
Orgill, N J ... G6LBG
Orgill, R ... G8HRB
Orlowski, K G ... G8ZTY
Orman, C D ... G0SJC
Orme, D W ... G6DQD
Orme, K W ... G1VVA
Ormerod, D R ... G3SLT
Ormerod, E ... G6JWU
Ormerod, S ... G0WCS
Ormett, F J ... G1LZM
Ormond, R T ... G0EDO
Ormondroyd, S ... G8RHQ
Ormsby-Rymer, J D ... G1JOR
Orstein, S ... G8PPA
Orr, A S ... GM7IGS
Orr, C ... GM1CCN
Orr, C ... GM2CPC
Orr, E P R ... 2E1EGE
Orr, F E ... GI4XHO
Orr, J ... G1HPG
Orr, J ... G3KYE
Orr, J ... G3VKB
Orr, N ... GI6GAG
Orr, R C ... G8WXB
Orr, W ... G4IYW
Orrells, J T ... G6DQY
Orrin, J N ... G3BBK
Orr-Ewing, C I ... G5OG
Orson, J W ... G4TJO
Ortmayer, S P ... G2UG
Ortmayer, S P ... G4RAW
Orton, A S ... G0GQI
Orton, D G ... G1JOY
Orton, R S ... G3VGX
Orton, S J ... 2E1CHF
Osbaldeston, H ... G3DUI
Osbaldeston, H ... G3HOX
Osborn, B L ... G0BLO
Osborn, C D ... GW8OIG
Osborn, C J ... G3XIZ
Osborn, C J ... G4UXV
Osborn, D A ... GD4HOZ
Osborn, F H ... G2CVO
Osborn, G B ... G3EJP
Osborn, K F ... G7GSD
Osborn, P M ... G1OHI
Osborn, S R ... G8JZT
Osborne, A ... G0BSQ
Osborne, A C ... G3SLI
Osborne, B E ... GW0SZU
Osborne, B J ... G0VMI
Osborne, B T ... G0MPJ
Osborne, B T ... G0NKL
Osborne, B T ... G7DOL
Osborne, D ... G1SVV
Osborne, D A ... G7GOH
Osborne, D G ... GW0DHE
Osborne, D J ... G1XEF
Osborne, I F ... G0FIW
Osborne, J D ... G4URS
Osborne, J H ... G4GSC
Osborne, J J ... G7WGW
Osborne, J M ... G3HMO
Osborne, K ... G7NMI
Osborne, K H ... GM6KOR
Osborne, L J ... G4KLZ
Osborne, M ... G4XOL
Osborne, M A ... GW3YGM
Osborne, M R ... G1BUV
Osborne, M R ... G1JAC
Osborne, N R ... G4JEI
Osborne, P J ... G4RPF
Osborne, P R ... G7HBN
Osborne, R K ... G1JWD
Osborne, R N ... GW4BVT
Osborne, S ... G1YHT
Osborne, T ... G1FAL
Osborn-Jones, C J ... G4CSI
Osbourn, C R ... G0FRU
Osbourne, B A ... G4HWU
Osbourne, I L D ... G0KKT
Osgathorpe, R W ... G4VWB
Oskis, D J ... G0DOM
Osler, G A ... G0WXI
Osler, G A ... G1RTF
Osmond, A ... G8PPN
Ostcliffe, P M ... G0VGU
Ostley, T J ... G8MCI
Ostwind, P L ... G4LOU
Oswald, C B ... G0GBW
Oswald, D ... GM3COQ
Oswald, M B ... G0LHC
Oswald, R S ... G7PIP
Oswick, H G H ... G0SSM
Oswin, J C ... G6HDJ
Otero, G ... G1YFK
Othen, A H ... G8FSZ
Othen, R B ... G7UNW
Otley, J C ... G4CYA
Ott, F A ... G0SQP
Ottaway, A C E ... G7KPB
Ottaway, G R ... G7PIV
Ottaway, G R ... MW0BBJ
Otten, E H ... G4GEV
Otter, B ... G4CGC
Otter, J B ... G0WWO
Otter, J B ... G7UAM
Otterson, W J R ... GI4GPA
Ottewell, P J ... G3ZRE
Ottley, R ... G3WIA
Otto, P ... G7LIR
Ottolini, R ... G6INO
Ottrey, K J ... G3ECS
Ottrey, K J ... G4GCT
Ottway, R W ... G4ULZ
Otty, B S ... G0SRV
Otty, E J ... G4XRL
Otway, G K ... G8AGT
Oubridge, B L ... G7AUE
Oubridge, D ... G7LHO
Oubridge, M R ... G1SYU
Oughtibridge, G ... G4EIL
Oughton, A M ... G6VAL
Oughton, A W ... 2E1FMW
Oughton, B W ... G4AEZ
Oultram, D A ... G6GYC
Oura, I J ... G0WAW
Oura, M J ... G7STQ
Ousbey, C H ... G0CHO
Outen, J C ... G3VGU
Outen, S M T ... GW0DWQ
Outhwaite, R H M ... GU3YVV
Outram, A J ... 2W1COT
Outram, D H ... GM1XVI
Outterside, S W ... G0CDH
Overbury, F G ... G7DPR
Overell, P H ... G4FXI
Overend, B ... G7OHE
Overend, G E ... G4YFH
Overend, M ... G0PMP
Overend, S K ... G4BVS
Overett, W F ... G7LIP
Overington, H F ... G0IGI
Overington, J T ... G4UYB
Overland, F S ... 2E1FUY
Overland, J C ... 2W1FPU
Overton, A J ... 2E1BGS
Overton, G C ... G4BJD
Overton, J G ... G4GUA
Overton, L M ... G8MRZ
Overton, M J B ... G1JDP
Overton, O ... 2E1EYK
Overton, O T ... G4EWE
Overton, P C ... G0MHD
Overton, S T ... G1LQF
Overy, G A ... G1AJN
Ovey, P ... G6SEK
Owen, A ... G0GMT
Owen, A ... G4POW
Owen, A ... G6FDC
Owen, A D ... G0MTP
Owen, A G ... M0BEL
Owen, A J ... G8LAK
Owen, A W ... G2FUD
Owen, B R ... G8IXK
Owen, C B D ... GW6ZSK
Owen, C W ... G4ITP
Owen, C W ... GW3PIO
Owen, D ... G6TFV
Owen, D E E ... G4GGB
Owen, D F ... G3MCA
Owen, D J ... G7UMR
Owen, D J M ... G1OXB
Owen, D L ... G0OES
Owen, D M ... G1CIV
Owen, D R ... G0ESR
Owen, D S ... G8XRW
Owen, D W ... G4XNV
Owen, D W E ... G3ZBD
Owen, E ... G7UML
Owen, E ... G8ESH
Owen, E S ... M1BAX
Owen, F R ... G3ACT
Owen, G A H ... G6DDK
Owen, G G ... GW0KPV
Owen, H ... G2HLU
Owen, H B ... GW1PIH
Owen, H C ... G7EBA
Owen, H M ... GW4OVH
Owen, I G ... M1ADV
Owen, J ... 2E1FII
Owen, J ... G4VWL
Owen, J C ... G7GUW
Owen, J E ... G8MLK
Owen, J G ... G1UFD
Owen, J H ... G7JKN
Owen, J H ... GW6OSS
Owen, J R ... G4IJG
Owen, J S ... GW3QN
Owen, K L ... G3AUX
Owen, L ... G0AZE
Owen, L J ... GW0TSE
Owen, L R G ... G1DBL
Owen, L S ... G3MDP
Owen, M ... G3ZML
Owen, M ... G6VNZ
Owen, M C F ... G4JSX
Owen, M I ... GW0PZZ
Owen, M J ... G0FRR
Owen, M J ... G4RFR
Owen, M J ... G4YTA
Owen, M J ... G6UZY
Owen, N P ... G8JGL
Owen, P ... G1RAS
Owen, P ... G8UUS
Owen, P A ... MW1CAN
Owen, P J ... GW0KAX
Owen, P O ... G0HTY
Owen, P R ... GW0TSF
Owen, R ... 2E1FIJ
Owen, R A ... G6DDO
Owen, R H ... GW8OZO
Owen, R J ... G7PNG
Owen, S M ... G6XBH
Owen, T H ... G4PSH
Owen, V ... G3JGE
Owen, W K ... G3CSS
Owen Smith, G ... G1RJG
Owens, A I ... G0AIO
Owens, C F ... GW1ZHI
Owens, D G ... GW0DGO
Owens, G J ... G0FQP
Owens, I ... GM7WEF
Owens, J ... G4TIU
Owens, J J ... G3MUF
Owens, J S ... GW4GDM
Owens, K ... G8OTP
Owens, N ... G7AJU
Owens, R H ... GW0OPP
Owens, R J ... 2W1EID
Owens, T W ... M1AYV
Owens, W C ... G4GJS
Owers, P K W ... G1EFY
Owles, L R ... G4IRJ
Oxlade, R N ... G7RFX
Oxley, G H ... G3YYM
Oxley, I I ... G1CMT
Oxley, I M ... G7EJO
Oxley, J R ... 2E1EAG
Oxley, J W T ... G0JTO
Oxley, R E ... G3WWI
Oxley, R T ... G0FYM
Oxley, W J ... G4CRB
Oxtoby, H ... GI0JHR
Ozanne, D J ... GU3UMX
O,Donnell, S J ... G4MSV
O,Sullivan, J ... G4PPG
O'Beirne, C J ... 2E1BOV
O'Beirne, M J ... G8MOB
O'Boyle, G D ... GI6NDM
O'Brien, B ... G2AMV
O'Brien, C K ... GW6FXB
O'Brien, E J ... G3KBX
O'Brien, M M ... G7TOS
O'Brien, P ... G3DNR
O'Brien, P L J ... GW1SXN
O'Brien, R E ... G3ZCI
O'Callaghan, M R G ... G1RYF
O'Connell, M R ... G4ZIU
O'Connor, D A ... G1ROM
O'Connor, G L ... G8OPS
O'Connor, P B ... 2E0AAV
O'Connor, P B ... G1ZCY
O'Connor, P J ... G3WLK
O'Connor, S ... G1ROL
O'Connor, W ... G8DXI
O'Dell, M D ... G0AUJ
O'Dell, P A D ... G3YUI
O'Donnell, H W ... G3ALP
O'Donnell, J ... G7VCF
O'Donnell, S ... G4NYU
O'Donnell, S M ... G4RVP
O'Donnell, T R ... G4DNN
O'Flaherty, L M ... GI0UYY
O'Flaherty, L M ... GI1YEA
O'Flanagan, A P ... G4ZKX
O'Gara, M B T ... G6SEM
O'Grady, F ... G6OJI
O'Hagan, J ... G4PFY
O'Hagan, S A ... G2CR
O'Halloran, J M ... G4JRE
O'Hanlon, M G ... G4GYD
O'Hara, J ... GI0EFW
O'Hara, J B ... G4PCR
O'Hare, C J ... GM0DYF
O'Herlihy, V J ... G6IYE
O'Keeffe, R W ... G0FFL
O'Keeffe-Wilson, G Y ... G4MIA
O'Leary, S T ... G6DCM
O'Malley, S A ... G7ANV
O'Meara, J W ... G8TBB
O'Neill, G P ... G7RES
O'Neill, J P ... M1AKO
O'Neill, J P ... M1AOX
O'Neill, L R ... 2E1FES
O'Neill, M J ... G1ZJM
O'Neill, M P ... G8WEM
O'Neill, P ... GI0NNM
O'Neill, P V ... G0FCG
O'Neill, T E ... G4AHC
O'Nion, J R ... G0SMM
O'Nion, P S ... G0PSO
O'Nions, J M ... GW1LHV
O'Regan, B F ... G8NMH
O'Reilly, E ... GI4LXL
O'Reilly, K ... GI7UIP
O'Reilly, K P ... GW0KIG
O'Reilly, M ... GW7EMV
O'Reilly, W R ... G4UHT
O'Riordan, J L H ... G0UUO
O'Riordan, J P ... G0SBQ
O'Rourke, M E ... G7REM
O'Rourke, M E ... M0ARY
O'Rourke, M O R ... GI0AZL
O'Rourke, P ... G7SCX
O'Ryan, P C ... G8WWF
O'Shea, B K ... G6LHG
O'Shea, T ... GI0MSG
O'Sullivan, D J ... G1JXR
O'Sullivan, J ... G4VYJ
O'Toole, J J T ... G7UYT
O'Toole, M J ... G8ZME
O'Toole, P F ... GI7FNJ

P

Pace, R S ... G3SOI
Pacewicz, A T E ... G4YQF
Pacey, J F ... G8SMG
Pacitto, A L ... G0HKH
Pack, S E ... G7WIQ
Packard, D J ... G8MLP
Packard, K J ... G0MLO
Packer, G B ... G3UUS
Packer, J E P ... G3NRD
Packer, K ... G7WCN
Packer, M C ... G4FFC
Packer, P E ... G8MDA
Packham, D I J ... G1XEX
Packham, G ... G7CEL
Packham, W P ... G3IJH
Packington, B J ... G4LTS
Packington, H ... G4PFH
Packington, S R ... G4KWC
Packman, C J ... G6XDI
Packman, K ... G4LQN
Packman, V D ... G4VYC
Packwood, D ... G4VGG
Padbury, R M ... G4GAB
Paddock, A ... G0CRB
Paddock, D W ... G7TEX
Paddock, E W ... G3SGB
Paddock, G C ... G3ZGY
Paddock, R ... G7RMA
Paddon, A J ... G1DUL
Padfield, D N ... G6HOR
Padfield, G J ... GW4RGT
Padgett, J J ... 2E1DCS
Padgett, M ... G0TPN
Padgett, R ... G7IME
Padley, N D ... G7PWA
Padmore, R ... G4MLB
Paffett, S P ... 2E1EEK
Paganuzzi, R ... GW4YEG
Page, A ... G7LZE
Page, A C ... G7OZL
Page, A M ... G3UUM
Page, A M ... G7ILI
Page, A R ... G1SBG
Page, A S ... G6UYJ
Page, A V ... G7LPT
Page, B ... G3LMP
Page, B ... G4YFT
Page, B H ... G4JOC
Page, C F ... G3HKJ
Page, C H ... G0TRM
Page, C H ... G0WJV
Page, C J ... G4BUE
Page, C S E ... G4MTN
Page, D L ... G7WLI
Page, D R ... G0TIJ
Page, D R ... G3KWC
Page, E ... G0NID
Page, E H ... GU3HKV
Page, E K ... G4UKM
Page, G H ... G1EGB
Page, G P ... G8ZOY
Page, I J ... 2E1FJV
Page, I M ... G0CRG
Page, I M ... G4WIR
Page, J ... G3DEM
Page, J D ... M1BZE
Page, J F ... G6XDG
Page, J J ... G1POS
Page, J R ... G7KPC
Page, K R ... G4RNZ
Page, L G ... G4RHP
Page, M D J ... G4XCW
Page, M N ... G0UBU
Page, N C ... G0PAG
Page, P D ... G0EMR
Page, P E C ... G3KLK
Page, P M ... G0BRE
Page, P M ... G7FIA
Page, R ... G8FSJ
Page, R A ... G4FTY
Page, R B ... G4YAN
Page, R M ... G6TUS
Page, R W G ... G7DMU
Page, T ... GW0VSC
Paget, A A ... G8KYM
Paget, L J ... GM0ONX
Pagett, J M ... G4YTJ
Page-Jones, M A ... G7VZY
Page-Jones, R M ... G3JWI
Paice, D R ... G3MXK
Pain, H ... G3ATH
Pain, R F ... G6NAL
Pain, S J ... G4TBD
Paine, D P ... G0DAV
Paine, D R S ... G3ASX
Paine, J F ... G8BUZ
Painter, A R ... G3BPF
Painter, D G ... G4PNX
Painter, D J ... G0LEV
Painter, J A ... 2E1BPQ
Painter, J A ... 2E1BPR
Painter, K H ... G4SZO
Painter, L A ... G0KQI
Painter, M S ... G7AHE
Painter, N ... G0FFA
Painter, P ... G3TEX
Painter, P V ... G4PVP
Painter, T L ... G3NEU
Painting, P ... G3OUC
Painting, R J ... G4FAP
Painting, S ... G0LTX
Painting, S ... M1ABX
Painton, R E ... G4YSZ
Painz, W F ... G0WFP
Pairman, A S ... GM3UA
Paisnel, N A ... GJ1YOT
Pajo, O ... G0UPF
Palace, A W ... G4ALL
Palastanga, G J ... G1RPR
Paley, A S ... G3XAX
Paley, F T ... G7MTI
Paley, G W ... G1LVN
Palfreeman, A E ... G0VLK
Palfreman, D J ... G4XXZ
Palfrey, A H ... G8IMZ
Palfrey, B J ... G6YZK
Palfrey, J A ... G3MXP
Palfrey, J L ... G4XEN
Palfrey, J R ... G4CYI
Palfrey, J S ... G4XEM
Palfrey, L E ... G4CQX
Palgrave Brown, I ... G4IXD
Palin, B J ... G4AHK
Palin, D E ... G6DP
Palin, P H ... G8TTD
Paling, P R ... G4RWV
Palk, R B R ... G3RDU
Palk, S W ... G0OUO
Pallant, F W ... G3RNM
Pallester, R ... GM3VCO
Pallett, S C ... G8OZQ
Pallett, S J ... G4JDP
Pallister, D H ... G0KVO
Pallister, G W ... G8ZBZ
Pallister, J W ... G1YXF
Pallister, J W ... G6GLO
Pallonen, J T ... M0AYT
Palmer, A E ... G1YXJ
Palmer, A E ... G3OOE
Palmer, A F ... G6BHI
Palmer, A J ... G1WBL
Palmer, A L ... G8VUK
Palmer, A M ... G4VDF
Palmer, B ... G4ECO
Palmer, B A ... G1BHO
Palmer, C ... 2E1EQB
Palmer, C M ... 2E1EZT
Palmer, C M ... G4FMO
Palmer, D ... G1DHQ
Palmer, D ... G6BHH
Palmer, D A ... G4LYD
Palmer, D A ... G6CMV
Palmer, D A ... G7EME
Palmer, D A V ... G3HYV
Palmer, D E ... GM0AEY
Palmer, D G ... G3ZDY
Palmer, D J ... G4PFX
Palmer, D J ... G6TAH
Palmer, D J ... GW4XMV
Palmer, D J J ... 2E1ADR
Palmer, D M ... G1ICX
Palmer, D P ... 2E1BDT
Palmer, D R ... G0LUK
Palmer, E ... G6AOW
Palmer, E C ... G3FVC
Palmer, E M ... GW7MGW
Palmer, G A M ... GW7PVD
Palmer, G B ... G1JDT
Palmer, G C ... G0FZC
Palmer, I M ... GM0NVQ
Palmer, I R ... G7SNC
Palmer, J ... 2E1DOB
Palmer, J ... G1CXE
Palmer, J H ... G2DJA
Palmer, J N ... G7RLN
Palmer, J W C ... G0EOE
Palmer, K ... G1BHL
Palmer, K L ... 2E1EKY
Palmer, L S ... G7UPW
Palmer, M ... G6HOT
Palmer, M A ... G3KGP
Palmer, M I ... G1OQM
Palmer, M J ... G8BOP
Palmer, M J ... G8PHJ
Palmer, M P T ... G0OIW
Palmer, N T ... G4GCI
Palmer, P ... G7UMB
Palmer, P E ... G1GPS
Palmer, R B ... M0ANB
Palmer, R D ... G1RFB
Palmer, R T ... G3YJJ
Palmer, S C ... G0EQS
Palmer, S M ... G7UIU
Palmer, T ... G1NRT
Palmer, T G ... G4UPH
Palmer, W J ... GW3TKZ
Palmeri, J S ... GU0MNQ
Paloschi, J R ... G0WZD
Pamment, A P ... G0APP
Pamment, L J ... G6RPF
Pamment, R S ... 2E1DRS
Pampling, A J ... G3RSP
Pampling, J W G ... G3YMJ
Panagakis, H ... G8CKA
Pangborn, R S ... G0KLS
Pankhurst, R E ... G1YUG
Pannell, A J ... G0AHA
Panting, R C ... G4ELY
Panting, R W ... G6BPB
Panton, A J ... G6UGG
Panton, B J ... G1TWY
Panton, M J ... 2E1EWI
Pantony, D E ... G3KXB
Pantrey, D ... G7LMT
Papadopoulos, G ... G0PIU
Papaioannou, C ... G7TWW
Papazoglou, L ... M0BFV
Papazoglou, L ... M1BMM

Pape, J G0NYQ
Papps, A GW4KOS
Papworth, A R G3WUW
Papworth, G G G8AUJ
Papworth, H A G6XKE
Papworth, J S G6LUM
Papworth, M J 2E0AJZ
Papworth, M J 2E1BZE
Papworth, S G 2E1FVU
Paramor, R C G0UUD
Parcell, A J G8BIX
Pardington, I G7TOF
Pardivalla, A P G7PPL
Pardoe, A H GM0HUO
Pardoe, K C G0EXC
Pardoe, R C G0MHZ
Pardy, A E G4DAU
Pardy, F F R GW3DZJ
Parfett, J R G0LAD
Parfitt, A R G6XKF
Parfitt, G D W G7SIU
Parfitt, G O J G3BRT
Parfitt, R W G8MY
Parfitt, T S G6DFR
Pargeter, A S G1ODT
Parham, J H G3PFF
Parham, S C G8IEA
Parish, C E G0VVU
Parish, G P J G7VPX
Parish, J N G0RHV
Parish, J T P G4UAU
Parish, M R G0RHU
Parish, P G0BUP
Parish, S D G3TGK
Park, A J C G3NZV
Park, B G7AWC
Park, B J G0AJZ
Park, B M G0TME
Park, C J C G7NVG
Park, D A G3PSV
Park, E M K G3ZVS
Park, G GW8KJK
Park, H P GM4NUU
Park, H U G H G4UME
Park, J D GM0OFM
Park, J F M1BJK
Park, P E G4KVK
Park, P I GM3PIP
Park, R M L GM0MXP
Parker, A G3KAG
Parker, A G3TXE
Parker, A G4RUC
Parker, A F G4AXN
Parker, A G G1WKE
Parker, A G L G4BKP
Parker, A H G3KH
Parker, A J G0KKD
Parker, A J G3SYK
Parker, A J G8IWB
Parker, A J G8VMQ
Parker, A K 2E1DSF
Parker, A S G6DFV
Parker, B G0USM
Parker, B G3KOQ
Parker, B A G0FLG
Parker, B F G0IKX
Parker, C G4OOI
Parker, C E G0WWK
Parker, C E G1SJR
Parker, C E G7SSC
Parker, C J G1PAK
Parker, C J G4CAY
Parker, C M GW6BDM
Parker, D G3ELY
Parker, D G4DZU
Parker, D A GM0AIR
Parker, D A M1AVK
Parker, D A M1BVU
Parker, D C G4IKK
Parker, D G G8OMB
Parker, D J G7OGL
Parker, D R G6FJF
Parker, E 2E1CWX
Parker, E A GW6BOQ
Parker, E J 2E1BVU
Parker, F V G7OGM
Parker, G C L G4EMK
Parker, G E S G4ONV
Parker, G H G4JTP
Parker, G J G7PGQ
Parker, G L 2E1FRH
Parker, H G6SMG
Parker, H G8GUN
Parker, H W G7UHI
Parker, H W GW2ADZ
Parker, H W M0AYO
Parker, I G4IWV
Parker, I C G4KSS
Parker, I G MW0AAP
Parker, I T G6PMO
Parker, J G3FKY
Parker, J G3IWV
Parker, J B G G3SOL
Parker, J C G G3OLX
Parker, J D 2E1CVU
Parker, J D G0UID
Parker, J E 2E1FGJ
Parker, J F G8ILU
Parker, J H O G3FFL
Parker, J I G0PLX
Parker, J R 2E1CWP
Parker, J R G1ZLS
Parker, J T G3BXZ
Parker, J T G3ITP
Parker, K G7MOO
Parker, K E G3PKR
Parker, K G G4LTE
Parker, K H C G8HTA
Parker, L G0ING
Parker, L G5LP
Parker, L G6ITV
Parker, M C G4YFU
Parker, M G G6KXD
Parker, M H G4IUF
Parker, M P G6NWK
Parker, M R G0UMP
Parker, M R G6SNJ
Parker, N A G7VQA
Parker, N E G4VMP
Parker, N J G8CYA
Parker, N P G1AJS
Parker, O M 2E0APZ
Parker, O M M1BOV
Parker, P J G1LVW
Parker, P J G8CKM
Parker, P R GD4UQO
Parker, P W G3AVN
Parker, R G4KAK
Parker, R G4ZBO
Parker, R A G8HNM
Parker, R E G4OLP
Parker, R F G8KFF
Parker, R H G4VVO
Parker, R H GM4AWP
Parker, R R G3JPG
Parker, S G0BAY
Parker, S G0TLW
Parker, S G7TPL
Parker, S C G4OOH
Parker, S K G4HQH
Parker, S L G0LYK
Parker, S M G6MJM
Parker, S R G1WAR
Parker, T F G8RXH
Parker, T F G8XBF
Parker, T G G0TIW
Parker, T G G0WIH
Parker, T G M0BGE
Parker, T G M1BEB
Parker, T K G4NXN
Parker-Larkin, C C G8UVG
Parkes, A W G7AXW
Parkes, B F G0KCI
Parkes, B J G1WTY
Parkes, D R G8IQG
Parkes, E G1PEY
Parkes, G H G0IMI
Parkes, G W G3NL
Parkes, H W G3NZS
Parkes, J R G6BYK
Parkes, K H G3EHM
Parkes, L G J G8FYH
Parkes, R E G3REP
Parkes, R J G7MFO
Parkes, S D G0WXP
Parkes, T B G3XDH
Parkes, W K GI8AIR
Parkhurst, G F G3TOZ
Parkhurst, J C G4KFU
Parkhurst, R A P G1NYS
Parkin, A G G1OER
Parkin, A S G1NVN
Parkin, G F G0ISJ
Parkin, J G3HOY
Parkin, L G G3UVY
Parkin, M C G0JMI
Parkin, M J G0UQV
Parkin, P R G7BUF
Parkin, R J G7VEF
Parkin, R K G1SGM
Parkin, R W G7RUS
Parkin, S C G0EVN
Parkin, T G6SKK
Parkin, T A M1AAR
Parkin, T R G4YUE
Parkin, W G7EBX
Parkin, W G8PBE
Parkins, J G8KVP
Parkins, R J G0GGA
Parkinson, A M1BUH
Parkinson, A E G6HDI
Parkinson, B G1YHO
Parkinson, C M0AFW
Parkinson, C W G6WAI
Parkinson, C W G8SPP
Parkinson, D W G7LKY
Parkinson, E G4LTG
Parkinson, E G4OYW
Parkinson, F G G0SBP
Parkinson, H J G0RQU
Parkinson, I J G3YRQ
Parkinson, J P G1PIC
Parkinson, J S G7BCP
Parkinson, N J G0TTR
Parkinson, N J G6DBC
Parkinson, R G 2E1ECA
Parkinson, T J G6MXW
Parkinson, T N G1KCJ
Parkinson, W A GD0EXM
Parkinson, W R G3FNM
Parkin-Coates, G G0FHA
Parkman, F W G G0PWF
Parkman, F W G G8AHF
Parks, C E G6GXO
Parks, M J G4UPD
Parks, M J G6JTL
Parkyn, M K G7EYM
Parlett, G G3JFQ
Parmenter, J R G0TIL
Parmiter, G E G4ZFB
Parnaby, P B G4GBR
Parnell, A J G7SHM
Parnell, C G0HFX
Parnell, C C S G8VDQ
Parnell, C D G0FKQ
Parnell, D M1ATH
Parnell, J GD3YUM
Parnell, J H I G3WJP
Parnell, L A G G8PP
Parnell, L T A M1BBW
Parnell, M G6DOE
Parnell, M J H GD7BMG
Parnell, M S G8JPV
Parnell, N G4ZXI
Parnell, N J 2E1DGG
Parnell, R J G8HDW
Parnham, K G7KLQ
Parr, A G0NHS
Parr, A G3IWP
Parr, A F G8CVQ
Parr, B S G4TML
Parr, D E G6OSO
Parr, D G G8DEY
Parr, D H G3MIR
Parr, E A G0EMX
Parr, E T G6CGO
Parr, F G3KHF
Parr, F A G4CWV
Parr, G D G4AXU
Parr, J G6WVL
Parr, M S G1FLV
Parr, N G6MRN
Parr, N R G1OTJ
Parr M.B.E., C G7MTA
Parradine, F W J G7IIV
Parrett, S A J G7EUM
Parrett, W R G7VGK
Parris, G A G4GW
Parris, G L G3XFW
Parrish, B F G4XWI
Parrish, C G4RPI
Parrish, H B G7PBH
Parrish, P G7MWS
Parrish, R C G0DMV
Parrish, R M G0FKK
Parrish, S J G0IVP
Parrott, A J G1PBY
Parrott, C 2E1FAD
Parrott, D A GW4LOD
Parrott, D W G6XKC
Parrott, G C GW3WHU
Parrott, H G6XKK
Parrott, L G0AMU
Parrott, L G4MWS
Parrott, R A G3HAL
Parry, A G3YYR
Parry, A G G4XDB
Parry, A M T G4SZI
Parry, D J G4SZT
Parry, D J G8MFU
Parry, E E G4EVD
Parry, F H 2W1EDJ
Parry, G M 2W1FPX
Parry, J A G7NFT
Parry, J E GW3VVC
Parry, J E GW4TTA
Parry, J K GJ8RRP
Parry, J L 2E1FFF
Parry, J L G3KWA
Parry, J R 2E1ASU
Parry, J R G8BMH
Parry, L J G8AMK
Parry, M GW7HGJ
Parry, P J G1HGY
Parry, R GW6IYP
Parry, R O G6NWF
Parry, R P I G4VCL
Parry, R P I G7AEC
Parry, R Y G5XV
Parry, S G0OIT
Parry, S I G7GVH
Parry, T J G2FKT
Parsley, G G4EMH
Parslow, D E G4XBI
Parslow, J B GD4UHB
Parslow, S A G1DGM
Parsonage, P T 2E1EXJ
Parsons, A E G1GPT
Parsons, A F G8CAH
Parsons, A F M G6PS
Parsons, A J G8DTA
Parsons, A R G6UGH
Parsons, B G4YJS
Parsons, B W G7FIJ
Parsons, B W GW0KZK
Parsons, C G G3WTU
Parsons, C W GW3OGC
Parsons, D G1OOZ
Parsons, D A G2NM
Parsons, D A G4BZB
Parsons, D H G0LWN
Parsons, D J G7VTQ
Parsons, D R G0TCL
Parsons, D W G3SLJ
Parsons, E G3TMD
Parsons, E A G2PS
Parsons, G G0AOL
Parsons, G G1WXU
Parsons, G G7FLQ
Parsons, G A GW1RJU
Parsons, H T G3DEP
Parsons, J A G1AJU
Parsons, J E G8VJF
Parsons, J M M0ARO
Parsons, K J G4VZA
Parsons, L G0ULP
Parsons, L H G3OSJ
Parsons, M S G4ORP
Parsons, P A 2E1FVR
Parsons, P O GW4VRO
Parsons, R A G0TLQ
Parsons, R J G3RBP
Parsons, R R GI3HXV
Parsons, S E G4DTW
Parsons, S G G4UHM
Parsons, S J G4NZG
Parsons, V M1AKL
Partington, C L G0CLP
Partington, D A G6EHL
Partington, F L G4BZP
Partington, F L G8DTM
Partington, J G1KLO
Partington, K G6EOS
Partington, K J G6PBI
Partner, A R G3HKT
Partner, E C G3OMB
Partner, R G6XPB
Parton, A W G1PPZ
Parton, P N G6JMG
Partridge, B C G4KID
Partridge, B L G0VUD
Partridge, C A G0GWG
Partridge, C C G2RT
Partridge, C G G8AUU
Partridge, D H G1LMW
Partridge, D J G7PHN
Partridge, D N G0VVW
Partridge, G A G3RJD
Partridge, G S G6HZK
Partridge, H G7DTV
Partridge, I S G3PRR
Partridge, J K G0VCV
Partridge, K A G4VJQ
Partridge, N L G7PLS
Partridge, P G1FLW
Partridge, R M GW1XRV
Partridge, R W G8KHI
Partridge, W E 2E1EUY
Parvin, E G2ADR
Pascal, R A GM8LNH
Pascoe, A G G7IIV
Pascoe, E J G4DKD
Pascoe, G H G8LXS
Pascoe, J G0WBK
Pascoe, J J G4ELZ
Pascoe, K E G8ZQM
Pascoe, N E G4USB
Pascoe, N R G3IOI
Pascoe, R A G0BPS
Pascoe, R A G0ROO
Pascoe, R A M1ABJ
Pascoe, S M G8SGI
Pasek, M J G4BZS
Pasfield, J C G1JCP
Pash, B W G4VLH
Pashley, R G8OFN
Paskin, P G6AFA
Paskins, J A G4MVP
Pasquet, P G4RRA
Pasquill, S G7IEI
Pass, R J G4MGV
Passey, C A G1ABX
Passfield, J E G1JIJ
Passmore, B GW4XEF
Passmore, B W G0BWP
Passmore, G A GW4HGS
Passmore, H L P G0LHX
Passmore, J M G4IZ
Passmore, S G7UYO
Patchett, B G4VBP
Patching, W R G6WYS
Patchitt, R G0TOZ
Pate, A J GM4BYC
Patel, A S G7DNP
Patel, J G7OMI
Patel, K G7TUI
Patel, M GM0VPM
Patel, R M G7VYB
Pateman, J A G2DBI
Pateman, S B G8WPO
Paternoster, W R G4YOH
Paterson, A G6CMZ
Paterson, A GM0PGD
Paterson, A J G0HAL
Paterson, D 2M1ENK
Paterson, D A GM0KAE
Paterson, D M G3PDE
Paterson, D R 2M1ENI
Paterson, G GM3KJZ
Paterson, G F GM0PZV
Paterson, G M 2E1EMI
Paterson, I A G4TZM
Paterson, I R GM4XTI
Paterson, J GM3COB
Paterson, J H GM6KLL
Paterson, M G1RIE
Paterson, M MM1BAB
Paterson, N G GM8IID
Paterson, P J G1JDQ
Paterson, P N G0SCS
Paterson, P N G7NOC
Paterson, P N G8PLR
Paterson, R H GM4CUB
Paterson, R H G G1VIW
Patey, L E A GM0PIQ
Patis, A D G8UFF
Patman, F G G6KSR
Patman, K G8MEE
Patmore, P W G3WYD
Paton, C J GM7TFN
Paton, D H C G1WHV
Paton, J R GM4LGR
Paton, J S G6RAZ
Patrick, A J B P G6XOX
Patrick, A R G0RJD
Patrick, A W 2W1DAM
Patrick, D G8KAP
Patrick, D I C G6EGM
Patrick, I G0TAW
Patrick, J 2W1DAO
Patrick, J G G8DJN
Patrick, M E G1LWL
Patrick, P J G3TWG
Patrick, R G1GWX
Patrick, S GI6DFU
Patrick, S D G1VVB
Patrick Gleed, N J G8YJM
Patrickson, P A G3TQS
Patrick-Gleed, A M G6ENH
Pattemore, J M G4NHO
Patten, G A M0AFK
Patten, J G3PIN
Pattenden, B G1TLA
Patterson, A GM4MCV
Patterson, A MI0BDX
Patterson, A D GI3KYP
Patterson, A G G6AGP
Patterson, A S G8GIG
Patterson, D J G4THH
Patterson, E G0SLK
Patterson, E A G6PUH
Patterson, G G4XUG
Patterson, H S G1KPY
Patterson, J G0BPJ
Patterson, J G7SUV
Patterson, J D GW4UIR
Patterson, K GI0OEH
Patterson, K J GI4BJK
Patterson, M J GW0TWF
Patterson, N M G7OFU
Patterson, P G6MJR
Patterson, R J V GI0APH
Patterson, S G1CLD
Patterson, S J G4UNG
Patterson, T G6WUX
Patterson, W P GM4YRO
Pattinson, A G6TVW
Pattinson, F M G7ANH
Pattinson, N G G0JVU
Pattinson, R GW3KVX
Pattinson, T G3AOT
Pattinson, W G0JCK
Pattison, D K G0CPW
Pattison, M G1PZD
Pattison, P G6KEZ
Pattle, T J 2E1EVY
Pattman, M A G0SDW
Patton, A J G1BRB
Pattullo, I R GM4ZQQ
Paul, A H G3RJI
Paul, D G1NRE
Paul, D J G6YFZ
Paul, D J GI4NGP
Paul, E F G3CUY
Paul, M G GU6XCM
Paul, N R G3AUB
Paul, R J G7KMZ
Pauley, M K G1MKP
Pauley, O R G0BGX
Pauley, P D G1DGS
Pauley, R W G8VDJ
Pauline, W E G0WEP
Paull, M A G4JKN
Pauwels, J B K GW4XLS
Pavelin, P E G4WWH
Pavely, M C G3GWD
Pavey, C W E G3JHU
Pavey, F J G1GLF
Pavey, P M E G3NFT
Pavey, R A K G3PXM
Pavia, J G7RBR
Pavier, C G0BYX
Pavis, D D GI0BDZ
Pawley, M A G8AWV
Pawlik, L S G0LSP
Pawlowski, M A G0UDT
Pawson, I W N G0FCT
Pawson, I W N G6BRA
Paxton, A A GW6MXG
Paxton, A D H G4BIZ
Paxton, J G0FXK
Paxton, N D G0ANZ
Paxton, R J G6RGV
Paxton, S D G6VLC
Pay, D J G4WTU
Pay, K J G1EGE
Pay, P G4ZAV
Payas, L B G0SEU
Payas, S G G4YNH
Payea, D G7AFZ
Payne, A N G3RBJ
Payne, A R G7SDL
Payne, B G3XTS
Payne, B A G4MLN
Payne, B J G4CJY
Payne, B R G1FOP
Payne, C D G6KVE
Payne, C J G8AJM
Payne, D G4OAT
Payne, D GI0AWK
Payne, D J G1UYZ
Payne, D K G4UUG
Payne, D L G7OWZ
Payne, D W G3KCR
Payne, E E G4FHE
Payne, E J G6UQI
Payne, E R G6HZW
Payne, G G4EUG
Payne, G B G7BUS
Payne, G E M1BHJ
Payne, G F G4GNN
Payne, J A G6CNF
Payne, J F G8XZB
Payne, J S G7VZD
Payne, K G G8FYK
Payne, L G0NNU
Payne, L A G4MWX
Payne, M G3ZRM
Payne, M J G4ITX
Payne, M W G1PER
Payne, R E G6LOY
Payne, R E H G4SUX
Payne, R G G7SXW
Payne, R J G0NSF
Payne, R T GM4AWA
Payne, R W G8FCA
Payne, S J H 2E1DBV
Payne, S M G0UEK
Paynter, D S G0FJR
Payton, P J G7COP
Payton, P P G6ZSU
Peabody, H G3KUG
Peace, D T G4KKM
Peace, K B G3KJP
Peacey, B M G4UFV
Peacey, K GW0VMD
Peach, A J G1VGN
Peach, D M G3VXS
Peach, D R G7OBE
Peach, G L J G7SLL
Peach, M J G4BQO
Peach, M J G7OBD
Peach, P L G3GOS
Peachey, A W G0BXJ
Peachey, D G0LSI
Peachey, D G1UWD
Peachey, D C G6IOW
Peachey, M D G1UWE
Peacock, B G3TXU
Peacock, C D G6CZE
Peacock, C E G1VTN
Peacock, D E G0BJI
Peacock, D I 2E1EEM
Peacock, D J G3NOP
Peacock, F S G4ZVT
Peacock, G A G3NHP
Peacock, H E G7SDF
Peacock, K G7OQT
Peacock, M G1NQU
Peacock, M R G1XDU
Peacock, N D G4KIU
Peacock, S L 2E1AVM
Peak, W E G4VDH
Peake, A E GW3SRG
Peake, C D G0NZI
Peake, C G G4UDN
Peake, D M 2E1CEX
Peake, G A G0KLZ
Peake, J A G4LTZ
Peake, J B GW1WWE
Peake, M J G3UIJ
Peake, P A G8GKT
Peake, V R T G4GEP
Pealing, F S M1BTT
Pealing, J G3WCU
Pearce, A G8NVW
Pearce, A GM7VLZ
Pearce, A B G8GOR
Pearce, A G G6IOX
Pearce, A J G0AZQ
Pearce, A K GW4LLC
Pearce, B 2E1FRV
Pearce, B G7ACA
Pearce, B J G6MGX
Pearce, C G8GOP
Pearce, C R G4EFK
Pearce, D 2E1FMR
Pearce, D G7GQD
Pearce, D I G6WMU
Pearce, D W G1PBF
Pearce, E G0WLP
Pearce, E J G0EDD
Pearce, E J T G6GNG
Pearce, F W G1ONQ
Pearce, G M G0LFJ
Pearce, G R G4LSX
Pearce, H A G4BXC
Pearce, H C G8XNH
Pearce, I GM7STI
Pearce, J G6FRB
Pearce, J E G8IWI
Pearce, J F G1OOU
Pearce, J G H G3IGP
Pearce, J S G4NSU
Pearce, J S E G3MEC
Pearce, L G0FTA
Pearce, L D 2E1CJR
Pearce, M A G7SNX
Pearce, M J G4KQY
Pearce, M J G6GVO
Pearce, N G G4WLC
Pearce, O W G8BKX
Pearce, P A G0DIS
Pearce, P E G7SOY
Pearce, P G G0KTZ
Pearce, P R G0IMA
Pearce, R G0GHT
Pearce, R G0SYF
Pearce, R G4VKE
Pearce, R G7UTT
Pearce, R B G3ZTC
Pearce, R E M1AJO
Pearce, R F G6RVQ
Pearce, R L G8TLP
Pearce, R M G0PNY
Pearce, R M G3LUK
Pearce, R V G0LTV
Pearce, T G0HMY
Pearce, T G7BLB
Pearce, W G G0WTQ
Pearce, W J G8OJR
Pearl, B F G4FCX
Pearl, P M GW7MVQ
Pearless, C A G3PGK
Pearn, G W G6NKS
Pearn, I R G6TRP
Pears, J A G0FSP
Pears, J W G7UAQ
Pears, R A G0ECC
Pears, R A G4TRV
Pears, V G1OHX
Pearsall, T M G4OWK
Pearse, C T 2E1EIG
Pearse, L V G7LMX
Pearsey, R B G6RBP
Pearson, A G0WDJ
Pearson, A B GM1MLS
Pearson, A H G0MLV
Pearson, B E G0CBY
Pearson, B R G4CVS
Pearson, B S G0NSF
Pearson, C A 2E1FOT
Pearson, C J G4VFO
Pearson, C M 2E1DPD
Pearson, C S 2E1EZH
Pearson, D GM3TLA
Pearson, D F G4UFS
Pearson, D G 2E0ANC
Pearson, D J G4PMA
Pearson, D J G7AKM
Pearson, D R G1OJO
Pearson, D R G3ZOM
Pearson, D R GW7GHE
Pearson, D R GW7RQI
Pearson, D R MW0AYM
Pearson, E F G0JJH
Pearson, E G G1EWB
Pearson, F N G7FSH
Pearson, F T G4JVM
Pearson, G S GM8BHR
Pearson, H G0SET
Pearson, H L G8YVA
Pearson, H M G4LBF
Pearson, I 2W1FJN
Pearson, J G1FTU
Pearson, J G6SRN
Pearson, J G7VPY
Pearson, J A G0VTM
Pearson, J D G3KOC
Pearson, J L G7UBO
Pearson, J L GW0PEA
Pearson, J M G4KDM
Pearson, J R G1WXF
Pearson, J V G3SNY
Pearson, J W G4EYZ
Pearson, J W G8XUH
Pearson, K R G0CRX
Pearson, L J G3VNT
Pearson, M A G4CRU
Pearson, M J G0BRC
Pearson, M J G7BRC
Pearson, M J M0AAK
Pearson, M N G1VAL
Pearson, M S G1GBI
Pearson, P G0MTB
Pearson, P H G4GXI
Pearson, R G0PEY
Pearson, R G3VRP
Pearson, R G3XWU
Pearson, R G4GOX
Pearson, R D G8ELV
Pearson, R E G3XUH
Pearson, R H G3CAG
Pearson, R H G4FHU
Pearson, S E G0NLX
Pearson, S W G4AGH
Pearson, T J G0DXT
Pearson, W H G4MWU
Pearsons, C D G0DWM
Peart, B A C G4RKF
Peart, C G1FHK
Peart, R H G0FHK
Pease, E G G7TBE
Pease, G D G6HOW
Peasey, E G0UGM
Peasey, P A G0UIR
Peat, D G0RDP
Peat, D F G1GMH
Peat, S J G0CBT
Peat, S J G3XBF
Peat, W W W GM3AVA
Peaty, A R G1GTL
Peberdy, E W G0DKV
Peberdy, P J G8HIP
Pebody, B M G0MWU
Pechey, D G8NMO
Pechey, J R G3HOO
Pechey, W L G3TRY
Pechey, W L G4CUE
Peck, D A G8PLZ
Peck, D A G8RKG
Peck, G M G3ZVN
Peck, G W G4OIG
Peck, J W G3YCE
Peck, L K 2E1FDC
Peck, N C G7ODY
Peck, S J 2E1DEL
Peck, S J G4WSR
Peck, W P G6OTS
Peckett, N L G4KUX
Pedder, C G3VBL
Pedder, D A G G3LFX
Peddie, J A GM4RZJ
Peddie, S R G0WEI
Peden Esq, J A G3ZQQ
Pedley, C J G3YHN
Pedley, D C G8EMA
Pedley, J C G1YBM
Pedley, J M G1UBC
Pedley, P A G G1RLR
Pedley, R E G0LWF
Pedreschi, M GM3MZX
Pedretti, E P G7VRU
Pedro, R G7HIH
Peel, C B 2E1COL
Peel, C J G G4NFL
Peel, C J G G4RNS
Peel, E M G8CCQ
Peel, J S G6TAI
Peel, R M A G8NEF
Peel, R W G4TKP
Peel, W D G7BTW
Peel, W E G4URV
Peeling, R W G6MXE
Peerless, J C G1SLQ
Peers, C F G3REA
Peers, J D F G2MT
Peers, M G6FMS
Peers, M R G8HYP
Peet, A G7TOY
Peet, J G4PXJ
Peet, M J G3ZJJ
Pegg, D D GM7FLG
Pegg, F G0HXC
Pegg, R K G4CSL
Peggram, C W 2E1FJU
Peggram, R A G7RUH
Pegler, A J R G3ENI
Pegler, A J R G3TVS
Pegram, P H G7JOD
Pegrum, C M GM8DKG
Pegrum, J S G3XCK
Peiperl, M A G0UHK
Peirce, J N M G0MQE
Peirson, A G7VZB
Peirson, F G3HSG
Peirson, J G3UYC
Pell, A M M1AQX
Pell, C G3WLH
Pell, J E G6CMX
Pell, M A G7RTL
Pell, M I F G1FCR
Pell, R E G4EXP
Pellatt, R A G4LJI
Pellett, R D G3RZC
Pelling, R F G0BUJ
Pelling, R J G1GFZ
Pellowe, D G4IGZ
Pellowe, V G4VAL
Pells, D T G4DTP
Pemberton, A S G0NGH
Pemberton, C G4GNG
Pemberton, C I G4RRJ
Pemberton, D J G7VGL
Pemberton, G A G7NEH
Pemberton, G M G1BVV
Pemberton, I R GW0EHA
Pemberton, J G6WJC
Pemberton, J S G4WXO
Pemberton, L A G1TKA
Pemberton, M E G4ZMQ
Pemberton, M J G4DDL
Pemberton, M J G6DAY

Penaluna, K S G6BEH
Penberthy, R G3ZFP
Pendlebury, W E G0FQR
Pendleton, J K G4RYG
Pendleton, M A O G7PPK
Pendleton, T J G0OAU
Pendleton, T J G4IRH
Pendrick, G D G6BEI
Penfold, A G0BEX
Penfold, C F G7GSX
Penfold, R A G1OKD
Penfold, R H W G1COW
Penistone, N F G0FLI
Penketh, N G3RYY
Penlington, A GW0LAL
Penman, H C T G7BEB
Penman, R W G4BYB
Penn, A G8LFY
Penn, E M G7PEN
Penn, M J G7JYY
Penn, R A G4GEL
Penn, R I G7HKW
Penn, R P G1HLO
Penna, C S GM3POI
Pennell, J M1BNF
Pennell, J C G3EFP
Pennell, L K G8PMA
Pennells, A K G6TGB
Penney, A P G4WEM
Penney, B J G1JQA
Penney, C D G7IAU
Penney, G G4ECF
Penney, I G1JUO
Penney, J E G3JEP
Penney, J J T G0LAG
Pennington, J G3RTP
Pennington, J J G7TYJ
Pennington, J N G0OGX
Pennington, M A 2E1ERD
Pennington, P J G4EGQ
Pennington, R J G1NVS
Pennington, S G3PUW
Pennock, J A G4MLK
Pennock, R M G0FLV
Penny, A G1YYY
Penny, D J G3PEN
Penny, D T G0AIL
Penny, H GW1IKN
Penny, J GM0EUK
Penny, K A GM1WWD
Penny, M D G7DTR
Penny, P A G6TAF
Penny, R E G4XYA
Penny, S G8ZPU
Pennycook, R T GM1PGQ
Pennycook, V A GM0BWS
Pennycook, W GM4WZY
Penrose, D J G1CWZ
Pentecost, K G G7BPN
Pentecost, P D GW0LNM
Pentecost, S G6YNT
Penticost, R L G4MSH
Pentland, A GM7VOP
Pentland, J S GM0KLP
Pentland, W D GM0CWR
Pentney, R C M1AOB
Penton, C S G0BHX
Penver, R F G8VBK
Peperell, A H G3TOP
Pepper, D A G8HYU
Pepper, D J G1XQG
Pepper, D S G0SLJ
Pepper, D S G2FA
Pepper, E S G3YWA
Pepper, H R G0LZP
Pepper, J B G4EPA
Pepper, J C G6ZSQ
Pepper, J E G0CHQ
Pepper, J M M1AOL
Pepper, M S G4WEH
Pepper, R B G4MVZ
Pepperell, M T MW0AWM
Peppert, F B G3HSL
Peppiatt, M J G7TKW
Perch, N W H G3HWF
Perch, N W H G4DNB
Percival, D R G8FXU
Percival, G M GM0CFC
Percival, J M0ASM
Percival, K H G7GFK
Percival, R G4VCJ
Percival, R H G1EQU
Percival, R H G4DBA
Percival, S D G3BGR
Percy, A 2E0ABJ
Percy, H G0IBT
Percy, M G G8NWZ
Perera, I M G4FJX
Perera, P E G4AJG
Perera, P V P G0USK
Perera, P V P G0WOX
Perez, J A G4SVY
Perfect, K W G3FIK
Perfect, K W G4KWP
Perkin, F G G1TAZ
Perkin, K M M1ABF
Perkins, A G3POJ
Perkins, A G6KDY
Perkins, A J G1ADS
Perkins, A K G7RBB
Perkins, A S G0FGR
Perkins, C J G4VFX
Perkins, D K G4VZK
Perkins, D W G8ANR
Perkins, E A G3MA
Perkins, E W P 2E1FEZ
Perkins, F G2BXH
Perkins, G W G3VIJ
Perkins, H GW3NMH
Perkins, J G4IVV
Perkins, J H 2E1AZQ
Perkins, J M G3PPB
Perkins, K G0CSD
Perkins, K G G3EDS
Perkins, K M G7JDV
Perkins, M G7DRQ
Perkins, N G4EJW
Perkins, N G6IYM
Perkins, P J G3CAR
Perkins, P J G7DQL
Perkins, P J G8CAR
Perkins, R H G0TFN
Perkins, R S G4NGW
Perkins, R V G8ADG
Perkins, S J G4FPV
Perkis, R H GM5UI
Perks, A E G1FCQ
Perks, E H M G3VWX
Perks, G E T G6VAA
Perks, I D G0JRE
Perks, R A G0LBQ
Perks, R S G4ICI
Perks, R S C G6EKM
Perrat, R J G1XLT
Perrett, A H R J G4ZDR
Perrett, J P R J G6KPD
Perrett, L G R J G4XWE
Perrett, M G G8LCE
Perrett, R F G3HWE
Perrey, D R G3WUE
Perrin, B N G3NRH
Perrin, B W W G7CIV
Perrin, C I G7VQG
Perrin, R H G4AFY
Perrin, R K G7VCU
Perrin, R W G3SHS
Perrin, R W G4DFT
Perrins, G M1BIX
Perrins, M N G3EDX
Perrins, P G4AYK
Perrins, P S G4NQW
Perrisset, F E G3IMN
Perrott, C G G6ANJ
Perrott, N R GJ4TAW
Perrow, R W G3ZZT
Perry, A G7LPO
Perry, A G G3ATX
Perry, A J 2E1CRC
Perry, A J G4YUW
Perry, A P G0AYY
Perry, B G G4ZUD
Perry, C A GM0EFD
Perry, C N G3AWI
Perry, C P G6XPM
Perry, C R G4TZI
Perry, D J G1CQG
Perry, D J G7BBM
Perry, E R G4RWP
Perry, F J W G0HAX
Perry, G G1IDJ
Perry, G A G1UUK
Perry, G C G4OED
Perry, G E G0GEP
Perry, G M GM0EFC
Perry, G R 2W0AEQ
Perry, G R GW6OSE
Perry, I M G7UVV
Perry, J A G6HGQ
Perry, J C G3OHS
Perry, J F G4FJP
Perry, J H G0GOD
Perry, J P G4ERH
Perry, K M G1WGM
Perry, L C G8VDF
Perry, M G8NYD
Perry, M E G0WIT
Perry, M J 2E1DSL
Perry, M J M1BQH
Perry, M M G7DVC
Perry, M R G8AKX
Perry, N A G6DQZ
Perry, O A G0DRR
Perry, O V G4ASX
Perry, R C G3ZRP
Perry, R C G7CQD
Perry, R G G6LUJ
Perry, R H G8CVP
Perry, R P G6KXN
Perry, R T K G6YNL
Perry, S G G0BFG
Perry, T J GW0SYG
Perry, T J GW4XQK
Perryman, D C G7NMA
Perryman, H L G4KKJ
Perryman, K R GW4RKI
Perschky, J A G8JAP
Person, A D G0TJP
Pert, A J GM7HIR
Pert, J S GM0HIM
Perver, R J GI8YJD
Pesani, J 2E1FDE
Pesarini, G G0VPV
Pescod, C R G4BMW
Pescod, R W G3UCD
Peskett, G D G3LOF
Peskett, M D G7FVN
Petch, B C G1DSY
Peter, D J 2E1BZB
Peterkin, A D G4NZF
Peterkin, I F G3WDU
Peters, A G1AIO
Peters, A G1UTZ
Peters, C R G0NDF
Peters, D A 2E1FZU
Peters, D J L G6GAW
Peters, D M 2E1BMJ
Peters, F E 2E1BMK
Peters, G G8COR
Peters, G G GW0EGQ
Peters, I D G0NWY
Peters, I W M1AJW
Peters, J D G3VMJ
Peters, J H G1BAX
Peters, J H G3YDU
Peters, K J M1AHR
Peters, M J 2E1BLJ
Peters, M L G4EFE
Peters, O R G7TKY
Peters, R B G3ZUF
Peters, R S G1YRV
Peters, R W G3JXV
Peterson, F R G3ELZ
Peterson, G G8SUG
Peterson, P I W G0HUF
Peterson, W G G4EZU
Pethard, J W G G3PNJ
Pethen, M S G6MDF
Pether, E D P G4VEZ
Pether, J P G4YKG
Pether, J W H G4JGG
Petherick, K E G0ODU
Petifer, B G G0SYR
Petifer, B G G8DTQ
Petit, C J A GU0PYO
Petit, G M GU0BGP
Petit, M Y G0LVS
Peto, A R G G6KYR
Petraitis, S J G4IFM
Petre, K G4NMK
Petri, G G0PXA
Petri, R E G G0OAT
Petrie, A A G4DJZ
Petrie, D M GW1MVZ
Petrie, I C GM4EHP
Petrie, J C GM3FDN
Petrie, R S G0UMS
Petrie Baker, J W G1CKK
Pett, R G G3SHK
Petter, K R G0GAO
Petters, J D G3YPZ
Petterson, P M1ADS
Pettett, R M G7TKI
Pettett, W D GM1BVT
Pettican, D A G1WSF
Pettican, G V G4MYQ
Pettican, R A G1OKJ
Pettifor, J A G8LQF
Pettigrew, A J G0FPT
Pettigrew, D GM0EWF
Pettigrew, M W G0WLR
Pettinger, I G4YTM
Pettinger, R G1SRN
Pettinger, W G0TDW
Pettit, B A G3VD
Pettit, G T G1VSH
Pettit, H W G3XHP
Pettit, M D G3UCW
Pettit, N D G6LOJ
Pettitt, C A G0EYO
Pettitt, C A G6SL
Pettitt, T E G7THA
Pettman, P B G7FKY
Petts, A G3PXF
Petts, L G G7TNV
Petty, C J G0CJP
Petty, D F G8VSV
Petty, P M G4CVD
Pevy, A G6GS
Pevy, A G G4VIY
Pevy, A P G4XYW
Pevy, W R H G4CWP
Phaff, J J G G4KDN
Phanco, G M GM4KHE
Pharaoh, M GM3LCH
Phelps, D J G1VZC
Phelps, D S G4WJT
Phelps, J C G1XHJ
Phelps, J F G4RMX
Phelps, J L MW0ATF
Phelps, J L MW1AXS
Phelps, J V G6DFM
Phelps, M A G1KGE
Phelps, P H G6UDF
Phelps, P I G6IOV
Phelps, R A G1AJW
Philbrick, V J G1EXZ
Philip, D A M1ALX
Philip, E M GM4JLZ
Philipp, B H G4PUP
Philipp, J B G4SVK
Philipps, I G0RDI
Philipps, S A G0VRV
Philips, J W G3NGT
Philipson, J A M G4LBQ
Phillimore, B J G4VXT
Phillimore, W A G3PGI
Phillippo, S J G6SXO
Phillipps, I G8SJP
Phillips, A 2E1CFR
Phillips, A J G7UIE
Phillips, A R G8GES
Phillips, A W G3NBX
Phillips, B J G0GRO
Phillips, B L G7HEV
Phillips, B L G8FWM
Phillips, B N E G4ZLN
Phillips, B W G0GJG
Phillips, B W G4DUF
Phillips, C G3RLA
Phillips, C D G2FNT
Phillips, C J 2E1CXD
Phillips, C J G4LXJ
Phillips, C V GW0CVP
Phillips, C V GW4WUR
Phillips, D G0YAE
Phillips, D C GW1VVK
Phillips, D E GW8SPH
Phillips, D E F G0WAE
Phillips, D E F G7RVZ
Phillips, D G G8AAE
Phillips, D H GW4KQ
Phillips, D H GW4LPC
Phillips, D I W G0OOW
Phillips, D K G0DKP
Phillips, D M GM0OLF
Phillips, D M GW4SDO
Phillips, D P G1JDU
Phillips, D S G7SXB
Phillips, E H G8MVY
Phillips, E J 2W1AYO
Phillips, F B GW4KBT
Phillips, G G4KWO
Phillips, G B 2E1FZR
Phillips, G B G3XTZ
Phillips, G C GW6UGC
Phillips, G D G7HHW
Phillips, G H G3GAZ
Phillips, G J G7ISG
Phillips, G J G7VPG
Phillips, G J GI4AFH
Phillips, G J B G1BAW
Phillips, G M G0KRB
Phillips, H GW0BLZ
Phillips, H J M GW3PRA
Phillips, H S G1MUM
Phillips, I R G7CQK
Phillips, J G3VWE
Phillips, J A G8POE
Phillips, J A GW6ZCR
Phillips, J C MW0ASL
Phillips, J G G0FGF
Phillips, J G GW3VOL
Phillips, J J G3KSK
Phillips, J J M G3PXX
Phillips, J K G3VHC
Phillips, J M G4JKH
Phillips, J P C GW0TWS
Phillips, J R G1DSZ
Phillips, J T GW3LDC
Phillips, K GW8BOQ
Phillips, K GW8WNB
Phillips, K L 2E0AGE
Phillips, K S G1POJ
Phillips, L C G8NBO
Phillips, L R B G1FNA
Phillips, M G0MBZ
Phillips, M G6WPJ
Phillips, M C G3RFX
Phillips, M D K G1JDR
Phillips, M E G4CIO
Phillips, M G G7EUK
Phillips, M J G0CBB
Phillips, M L G4GTZ
Phillips, M T G3NJP
Phillips, N F C GW4JVE
Phillips, P G4MIP
Phillips, P F G4TPK
Phillips, P M 2E1BML
Phillips, P S G7KBR
Phillips, P T G8WYT
Phillips, R G6OIH
Phillips, R GM0VRP
Phillips, R A G6VPZ
Phillips, R L 2E0ADF
Phillips, R L 2E1AEK
Phillips, R L G0TYL
Phillips, R M G1GGB
Phillips, R O G4IQQ
Phillips, R R D G7IZC
Phillips, R W G0DXN
Phillips, R W G0PVS
Phillips, R W G4SBS
Phillips, S C G7PJN
Phillips, S G G8HQA
Phillips, S H G4EYR
Phillips, S M G0LOE
Phillips, S P G6UDI
Phillips, S P G7PNB
Phillips, T G0IBH
Phillips, W C G0HVG
Phillipson, B G7OCK
Phillipson, C J G8IJC
Phillipson, D W M0BFP
Phillipson, J G4BEZ
Phillipson, R G7OOI
Phillpott, A G G4IMP
Philp, A P GM3WAP
Philp, J E P G3NJM
Philp, L D G4RL
Philpot, C J G6KBC
Philpot, G A GW0CZJ
Philpot, J A G4OYU
Philpot, J C 2E1BFP
Philpot, R F G4TTQ
Philpot, R F G4TTR
Philpot, S P 2E1EFG
Philpott, A S G1JYZ
Philpott, C R G4YNC
Philpott, L J G4BI
Philpott, M M B G3MJP
Philpott, R G G0AUW
Philpott, R P G3VCH
Philpotts, D L M1CEI
Philpotts, G A F G0KRC
Philpotts, G A F G0RJP
Phips, H L M0ACF
Phin, R F G7CRN
Phippin, P J G6FRA
Phipps, G H G3IPG
Phipps, J A G4GHX
Phipps, J H G4HVG
Phipps, R G4DIC
Phipps, R J G7AYE
Piachaud, A M0AOW
Pibworth, D B G0VPE
Pibworth, D B G4KWT
Picco, S M D G6VDT
Pick, D A G2BBC
Pick, D A G3YXM
Pick, D A G6BBC
Pickard, A R GW4OES
Pickard, B A G4TFD
Pickard, D J G8EUH
Pickard, F H G2HLL
Pickard, F W G3KNI
Pickard, W J G8TRR
Pickerill, D B G7HNO
Pickerill, E G0IUK
Pickering, A H G0GKK
Pickering, C R G8DHJ
Pickering, D K G7UAY
Pickering, E G3LPS
Pickering, G G4RAY
Pickering, H H G3DUL
Pickering, J H G4OTF
Pickering, J R G0WZJ
Pickering, K G0GNL
Pickering, M W G4PLA
Pickering, N V G7HJH
Pickering, P J G3ORP
Pickering, P J G3TRF
Pickering, R G3SLK
Pickering, R C G3UWP
Pickering, R J G0BCF
Pickering, W G4DRB
Pickersgill, C M J G0UMX
Pickersgill, F A G3XXN
Pickersgill, J H G0DZG
Pickett, C E A G6TZG
Pickett, R A G7BZC
Pickett, R E G1GGU
Pickford, B H J G4DUS
Pickford, D J G8TNE
Pickford, K G7PKZ
Pickhaver, P G4JDR
Pickles, A E G0RHR
Pickles, A J G1AWU
Pickles, A J G6UZZ
Pickles, C G R G7KPS
Pickles, D A G3XVA
Pickles, D H 2E1CTW
Pickles, J H G4CWM
Pickles, J T G3ZMQ
Pickles, K J G4XZM
Pickles, R G0RAF
Pickles, R G3VCA
Pickles, R G7FAR
Pickrell, B G G8ARM
Pickstock, M A G0WMM
Pickstone, R G6NCL
Pickstone, S J G1FLX
Pickstone, T D G6NCM
Pickup, B A M1BTP
Pickup, E G4KWF
Pickup, J W G4RYT
Pickworth, M C G4VMI
Picot, A J G6LGR
Picton, A S G4OJV
Piddock, C R G1WTH
Pidgeon, A B G6CBP
Pidgeon, D G3CVF
Pidgeon, J M1AVB
Piekarski, J G GM4ZGB
Pierce, A D G0GZF
Pierce, A P G0RVE
Pierce, A R GM4TRS
Pierce, G P GW0JMF
Pierce, L E J G3TSW
Piercy, L F G0JFU
Piercy, M F G4VEA
Pieroni, E P G0ITZ
Pierson, H V G3MXV
Pierson, K J GW1APU
Pierson, P B K G3MVM
Pieters, C G6KRY
Pigg, R G A G4MHW
Piggin, D W M1BTI
Piggott, B R G0NSH
Piggott, J G2PT
Piggott, J C G4LMY
Piggott, J F GW7WFI
Pigott, G J G4XKV
Pigott, H G0MYL
Pigou, R J G3CRP
Pike, A R G7FFS
Pike, D G3WDY
Pike, D A G3VMI
Pike, D F G8KBD
Pike, D H G1LPF
Pike, D J G4UVV
Pike, J W G3IPC
Pike, N G0IPL
Pike, N C G8PEE
Pike, P H GW7MMG
Pike, S C G4ZLK
Pike, S D G7RPW
Pike, S E G8KRD
Pilbeam, S V G6AOS
Pile, J A GW4EPF
Pile, K N G0JXP
Pile, P F G7ZZY
Pilfold, E J G0JGO
Pilgrim, J D G7NFD
Pilkington, A G0KAB
Pilkington, B G0OCK
Pilkington, B A G7BAP
Pilkington, F E G0OCL
Pilkington, L H A GU3MAT
Pilkington, L P 2E1EIH
Pilkington, M S G7AWV
Pilkington, S J G3NNT
Pill, A H G0NRZ
Pill, M A J G4MQB
Pilling, A G0DCM
Pilling, F B G4HWK
Pilling, H G4SOO
Pilling, J M G7DRM
Pilling, J T G4DPU
Pilling, N G4LYE
Pillinger, S G6DDJ
Pilot, M J GW1DTA
Pimblett, P G0TPP
Pimblott, P V G3XVP
Pimlott, J B F G8IDE
Pinch, T G4ETP
Pinchen, R R M1AOU
Pinches, R T G8WYS
Pinchin, H S G3VPE
Pinchin, R G6FDD
Pinckston, M J G1OLL
Pinder, A N G8XZC
Pinder, C F G4IKD
Pinder, L B P G0CGU
Pinder, M B G4STI
Pinder, S J 2E1FEY
Pine, C S F G4OIW
Pine, R C G3RRP
Pine, R D G0RWT
Pink, A J 2E1CZF
Pink, A S G3RMZ
Pink, D C G6EGO
Pink, D R G6IAB
Pink, J J G3OQB
Pinkard, I M G6XKJ
Pinkard, K M G1LED
Pinkerton, J W D GI6GIE
Pinkerton, R GI0NCA
Pinkney, A G4VOU
Pinkney, M J G6UGA
Pinkney, N A B G6FJG
Pinkney, S M G7ACM
Pinna, J G7GLS
Pinnell, M J G4MAE
Pinnell, M V G G4VQT
Pinnell, R W G4KNJ
Pinnell, W J C G3XWK
Pinnock, A G G0PIN
Pinnock, D G G3HVA
Pinnock, M 2E1FWU
Pinsent, E J G6BMQ
Pinson, J J G6JJP
Pinson, J T G6VJN
Pinto, L V G1LRR
Piotrowski, W G0BOE
Pipe, D J G8WJQ
Pipe, H C S G7SLH
Piper, B A G1KRX
Piper, D G4JSQ
Piper, E J G0BUZ
Piper, F M0APZ
Piper, F R G3BGO
Piper, G W P G4LLU
Piper, L G4ZIP
Piper, M T G6UEV
Piper, P R G0DDY
Piper, R G6XCO
Piper, R E G3MEH
Piper, S E G1YRE
Piper, T G1HAO
Pipes, M A L G4DKV
Pipkin, J D G7FEF
Pipping, W H G0IWP
Pique, P F G8KDQ
Pirie, C G GM0TCU
Pirie, G GM4YZI
Pirie, R J G4NQE
Pirrie, T M G7OCV
Pirrie, T M M0BBT
Pisani, D J M0AHB
Pitch, B 2E0ABB
Pitcher, J A K G3JNI
Pitcher, M J G3SNP
Pitchford, C J 2E1BLV
Pitchford, C J G7USR
Pitchford, C L G7RLV
Pitchford, J S G1ZPQ
Pitfield, I P G6YFY
Pither, J P G0JME
Pithers, C J G0DFN
Pitkin, D J GW0PNI
Pitkin, I M G4KJD
Pitkin, J A G7WCR
Pitman, E L G0BOC
Pitman, K H G0FLD
Pitt, A J G4YIC
Pitt, A M 2E1CDQ
Pitt, C J G1FWY
Pitt, J E G3VRY
Pitt, N G G1GFA
Pitt, P J G3JAA
Pitt, P N G3ICH
Pitt, W D G7VDE
Pitt, W G G4IES
Pittam, G W G0LVP
Pittaway, B J M1AUK
Pittaway, N G G8VHP
Pitter, R G0CXU
Pitts, G I G0EGT
Pitts, J W G4SHN
Pitts, M A G6SMJ
Pitts, P T G3GYE
Pitts, R G8SDE
Pitty, B J 2E1EDY
Pitty, J R G4PEO
Pivac, M A GM0JML
Place, D G G7UQE
Place, D T 2E1DPS
Place, T G6DRG
Place, T F G7JWD
Plaice, A E G4MKE
Plail, A R G8GRQ
Planck, K GI4NKK
Plant, A B G3NXC
Plant, A J G1IAL
Plant, A P G0MBL
Plant, D G G3JPU
Plant, D J G1HLP
Plant, D W G4DLW
Plant, E D A G7PKY
Plant, G G7PAO
Plant, G V G8NML
Plant, J E G7RQZ
Plant, K G3NIC
Plant, M G1JHD
Plant, N D G4WUJ
Plant, P R GU7IOJ
Plant, R C G6FHR
Plant, R J G6HZV
Plant, R J G7SEO
Plant, S W M1BKW
Plant, T A G3SFA
Plant, W G0UYU
Plaskitt, M G0MAF
Plaster, M W G3OJL
Plastow, B G4DRU
Plater, D J G4MZY
Platt, A G G8KTB
Platt, D A G3JNJ
Platt, D W G8BMG
Platt, G J G6SNI
Platt, G J G8UIY
Platt, J J G2VO
Platt, J R G3FEV
Platt, M J G4XUM
Platt, R A G6DRF
Platt, R A G8ROS
Platt, S V G4WSZ
Platt, S V G7ORL
Platten, C I G7WAT
Platten, D G M G8OOV
Platten, L G0KLB
Platten, M R G1BOX
Platts, E G0OLL
Platts, R G8OZP
Platts, S G0NXT
Player, C J G8FFF
Player, K H G8VDG
Player, R K G7PTB
Playfair, D I G1FRH
Playford, C D G6HOS
Playford, K G6MRP
Playle, P P G4SGN
Playle, R E G3ESL
Pleass, L K E G3RNN
Pledger, P V G4TZO
Plenderleith, J G3OOK
Pleshkevich, V G0WKW
Plested, J A G4GYS
Plested, M J G6IYJ
Plested, P J G6DFL
Plested, R C G6AFE
Plewa, L A G6FJE
Pleydell, P G7UBX
Pleydell, P J E G0PHU
Plimmer, R P V GW3UEP
Plitsch, A M1ATP
Plowden, B H G6PHY
Plowman, J A G3AST
Pluck, A G7KPD
Pluck, G A G8MVE
Pluck, K G7LTE
Pluck, R G G7JTB
Pluck, T J G7OZQ
Plucknett, G P J G4FKA
Plucknett, W H G8HGP
Pluckrose, B G G4VOW
Plumb, N J G0PBV
Plumb, R W G3IRP
Plumbley, A J GW0SZB
Plume, R G7DJX
Plumley, J C GW4SLI
Plummer, C D G8APB
Plummer, C F G7LYS
Plummer, J Q R G0GPM
Plummer, M F G3MDI
Plumridge, D H G3KMG
Plumridge, K F G0MMA
Plumridge, K J GW4BYY
Plumtree, B G6MQU
Plumtree, J G7EAM
Plumtree, R G0RKQ
Plumtree, S E G3OSP
Plunkett, D J 2E1CEV
Plunkett, J L GM4VAC
Plunkett, P C G8BQZ
Plyer, D R P G1AIT
Pocock, C F M0AYS
Pocock, G J W GM1SQZ
Pocock, R C G0JZY
Pocock, R K G8MKO
Pocock, S G4GTU
Pocock, S J G0CPV
Podmore, B J G0VNU
Podmore, G D J G7VAG
Podmore, G S G8BCF
Podvoiskis, J G0NPI
Poffley, R M G6CZB
Pogoda, G G4YXV
Pogorzelski, A G1XVW
Pogson, T G0PLD
Poingdestre, F H GJ0LRV
Pointer, P G8MTR
Pointon, A C G3MTX
Pointon, G G6MQI
Pointon, K P G4IBN
Pokusinski, Z P G4JQU
Polain, T A G7TBW
Polding, A P G6FMQ
Polding, J T G8YRN
Pole, T D 2E1DFF
Poll, D R G8IKA
Pollard, A D G0NHZ
Pollard, A G G0BTZ
Pollard, A J G0LXW
Pollard, A S G6LGW
Pollard, C E G3DPX
Pollard, C J GM6PSK
Pollard, D G0UWX
Pollard, D R 2W1EKR
Pollard, E R G4PUN
Pollard, E G G6LHB
Pollard, E W GM3KJI
Pollard, G F G4RWJ
Pollard, G J GM7RYK
Pollard, G R GW8DOA
Pollard, J G3IY
Pollard, J B G1RQH
Pollard, J B G7TFY
Pollard, J P R G7BYS
Pollard, K G G0MHR
Pollard, K G G3WGC
Pollard, K G G4KGP
Pollard, M A G3FJY
Pollard, M D G8BWA
Pollard, N J M1AOI
Pollard, R J P G7BXO
Pollard, S D 2E1CPR
Pollard-Wilkins, B M G8DXU
Polley, D A G0JPY
Polley, D J G7VZQ
Polley, P 2E1EBH
Polley, R C G3NAA
Polley, R J G3PYC
Polley, T G1YDJ
Pollitt, G G4NPQ
Pollitt, J G0FAG
Pollock, A N GM0PEI
Pollock, C GM0PIY
Pollock, M E GM6IOU
Pollock, M P G8KMP
Pollock, P S C GI8YWR
Pollock, R D GI0WZW
Pollock, T GM4PQV
Pollock, W R GI3NVW
Polson, N G4UFW
Polwarth, B 2E1DSP
Pomeroy, E 2E1DHN
Pomeroy, P L G4ATX
Pomfret, A M G3LZZ
Pomfret, D N S G7SPL
Pomfret, I G6MHO
Pomphrett, C M G8PIC

Name

Pomroy, G J ... G0ILI
Pomroy, M ... G1ZUH
Pond, C R ... G0VDO
Pond, C W ... G7FGI
Pond, D M G ... G0IRH
Pond, E ... G7HVG
Pond, W A ... G1YJL
Pont, B C ... G8SUV
Pont, N R ... G8SUW
Pontet, D R A ... G3MRS
Pontiero, A ... GM0EUU
Pontiero, A P ... GM4UBF
Pontin, R A ... GM8JCR
Ponton, J W ... GM0RWU
Pook, H C ... 2E1FIQ
Pool, D W ... G6MAY
Pool, G ... G0DKQ
Pool, G ... G3RCW
Pool, S D ... G4LFY
Poole, A R ... G8VZB
Poole, B C J ... G3MLP
Poole, B J ... G3MRC
Poole, B J F ... G4UJL
Poole, B M ... GW3JAZ
Poole, C S ... G0DPP
Poole, D ... G1HLK
Poole, D W H ... G1MGH
Poole, E A ... G8GRP
Poole, E G ... M1BQV
Poole, H R ... GW3PKO
Poole, I ... G1RTX
Poole, I D ... G3YWX
Poole, J G ... GJ1TJP
Poole, J G ... GJ7DGJ
Poole, J M ... 2E1AUG
Poole, J R ... 2E1DZZ
Poole, J W ... G4WSK
Poole, K ... G6ZSR
Poole, M P ... G7UJY
Poole, P H ... G3ENV
Poole, P J ... G4EVY
Poole, T J ... G3VMT
Pooler, C E ... 2E0AOD
Pooler, C E ... 2E1ACW
Pooler, G E ... G0WQH
Pooley, D J ... G7UWZ
Pooley, L D C ... G7KLR
Pooley, N ... G1VII
Pooley, T P ... G0RFL
Poore, P G ... M1CAZ
Poore, R ... G4OHC
Poore, V A ... G3ZSB
Pope, B A ... G7NUB
Pope, B A ... M0BAX
Pope, D N ... M1BRU
Pope, G J ... G3ASV
Pope, G V ... G4XRD
Pope, I R ... G4GUE
Pope, K J ... G4TNA
Pope, N ... G0GSN
Pope, N G ... G4AXA
Pope, N S ... G7BNI
Pope, R L ... G0PZJ
Pope, R L ... G4HXH
Pope, W J ... G4TIG
Popek, S J N ... G4BJP
Popely, D C ... G6GYM
Pople, D G ... GM4UND
Pople, G D ... G4AVJ
Pople, G W A ... G4DKL
Poppel, E ... 2E1DYU
Popple, E ... G4YOP
Popplewell, E W ... G3IQX
Popplewell, J ... G7NUO
Popplewell, J F ... G8OYF
Porch, A ... G8AOK
Pordum, I C ... G0DTR
Porritt, S J ... G7PZP
Port, D G ... G6INU
Portch, D A ... G0KPZ
Porteous, A M ... G7GDV
Porter, A B ... G1POR
Porter, A J ... 2E1CPE
Porter, A J ... G0BZW
Porter, A J ... G0OQD
Porter, C ... GM0TAE
Porter, C A ... G6IYR
Porter, C I ... G8RQB
Porter, C J ... G0AYC
Porter, D ... G4NFO
Porter, D ... G4OYX
Porter, D E ... G1JZZ
Porter, D W ... G0GBP
Porter, E H ... G3PFZ
Porter, F J ... G0OQE
Porter, F W ... G3VVP
Porter, G ... G4GOJ
Porter, G A ... 2E1CPA
Porter, G R ... G0RPZ
Porter, G R ... G4TXM
Porter, H H ... GI6KVS
Porter, I R ... G6IYS
Porter, J ... 2E1CBK
Porter, J ... G0HPX
Porter, J ... G4OHJ
Porter, J ... GM6OML
Porter, J A ... GI3GGY
Porter, J D ... G0OSS
Porter, J F J ... G4AGN
Porter, J G E ... 2E1DAT
Porter, J P ... G3YZR
Porter, J T ... GM4XRP
Porter, K ... G7OKV
Porter, K M ... G4WEN
Porter, K R ... G6UCY
Porter, L ... G4FYR
Porter, M A ... G7PHC
Porter, M C ... 2E1DSW
Porter, M C ... G4TJK
Porter, M C G ... G4OKS
Porter, M J ... G8BVL
Porter, M R ... 2E1CAQ
Porter, M S ... G7CVC
Porter, N B ... G0IRK
Porter, N D ... G4AFA
Porter, N D ... G4NYM
Porter, N R ... G6SEJ
Porter, P ... 2E1CPV
Porter, P R ... G1LIJ
Porter, R A ... G3NII
Porter, R B ... G3VXK
Porter, R E ... G4VRP
Porter, R J ... G8OLL
Porter, R W ... G8XNG
Porter, S C ... G0UOV
Porter, S C ... G8GTD
Porter, S E G ... G4NHP
Porter, S H ... G4FGR
Porteus, R ... G6YAR
Portlock, G A ... G1FKJ
Portnoy, L R ... G4SBV
Portrey, G L ... G4BNG
Posen, M C J ... G6EYY
Posner, B H ... G3TFT
Post, A W ... G3BOI
Postel, H F ... GW6UWV
Postill, A J ... G7TFJ
Postle, O ... G3EFZ
Postle, P A ... G1DXQ
Postlethwaite, J H ... G8DYN
Potgieter, C ... G0HKJ
Pothecary, M W ... G4FUU
Pottage, J T ... G1XVF
Potter, A ... G0JIR
Potter, A M ... G1JHY
Potter, A T ... GM7GLJ
Potter, C G ... G6FQZ
Potter, D M ... G0DMP
Potter, F C ... G3CNA
Potter, G A ... G0HUI
Potter, G E ... G0EOF
Potter, H R ... G8RRR
Potter, J M ... G4PFF
Potter, J N ... G1NCC
Potter, J W ... G7WGA
Potter, J W ... G8JMU
Potter, K J ... 2M1CJG
Potter, L A ... G3ESK
Potter, N J ... G7GEX
Potter, S M ... G1JHZ
Potter, S P ... G8ZQO
Potter, W ... G1VDS
Pottinger, J M ... G1FIZ
Potts, A J ... GW1TDV
Potts, B ... G4OLC
Potts, D J ... G8JPU
Potts, D M J ... GW6KBD
Potts, G ... GM0CJY
Potts, I W A ... GI7AUY
Potts, J J ... GM4IYZ
Potts, K H ... G6HGR
Potts, R A ... G4PEN
Potts, R K ... 2E0AMO
Pougher, T D ... G0MKZ
Poulet, D G A ... G7NUC
Poulson, G F ... G7LOB
Poulston, A R ... G4OVB
Poulter, C A R ... G8XMG
Poulter, D A R ... G3WHK
Poulter, F G ... G0TCM
Poulter, J B ... G3KQP
Poulter, M ... 2E1DTS
Poulter, R G ... G0CHJ
Poulter, S F ... G0PNT
Poulton, D M ... G7NVS
Poulton, F ... G8RXJ
Poulton, G H ... G4HNG
Poulton, H J ... G4WKB
Pounder, A ... G1WFM
Pounder, F ... G0GRZ
Pounder, R H ... G3DVQ
Pountain, S M ... G0SMP
Pountain, S M ... G7KRZ
Poupard, A D ... G6MXV
Povall, N G ... GD6ZNL
Pover, B S ... GD3EUI
Pover, F J ... G1FVF
Povey, E V ... G6CHA
Povey, S R ... G7AWR
Povey, W B ... G1WBP
Povey, W C ... G3MWB
Povey, W T ... G8NXQ
Povoas, L B ... G4FZL
Pow, D J ... G6POW
Powe, R J ... G8EKU
Powell, A C ... G1VXX
Powell, A G ... GW0UKG
Powell, A J ... G0IAJ
Powell, A J F ... G3GYF
Powell, A R ... GW0PFZ
Powell, B ... G4DIA
Powell, B J R ... G4DZN
Powell, B L ... G4ZJJ
Powell, B S ... G6UDE
Powell, C D ... G6JID
Powell, C J ... G3YNJ
Powell, D A ... G3RLB
Powell, D A R ... G0OSE
Powell, D G ... G3XLW
Powell, D J ... G4FJH
Powell, D J ... G6EZY
Powell, D K ... G4JVX
Powell, D P ... G4TWC
Powell, E ... GW1TDW
Powell, E F ... G0UYW
Powell, F ... G3SEL
Powell, G A ... G8ZKW
Powell, G R ... M0AKA
Powell, H ... G1MRP
Powell, H G G ... G6SMK
Powell, J D ... G3OQU
Powell, J G ... G1OXM
Powell, J L ... G4MZZ
Powell, J R ... G3ZOL
Powell, K G ... G1JAG
Powell, K G ... G1NCG
Powell, K M G ... G0PPM
Powell, M J ... G3IJE
Powell, N C ... G8OXG
Powell, N G ... G4XPK
Powell, N G A ... G7NSZ
Powell, N J ... G0HPU
Powell, P G T ... G1PYJ
Powell, P J ... G3EFI
Powell, P J ... G4EDP
Powell, R ... G3RUJ
Powell, R ... G4URP
Powell, R ... G7MTW
Powell, R A ... G6JME
Powell, R C ... G4VAA
Powell, R D ... G0AOZ
Powell, R J ... G3OGP
Powell, R J ... G7TPA
Powell, R K ... G6CCQ
Powell, R M ... G7TRT
Powell, R M ... GW4CMW
Powell, R S ... G8ZJP
Powell, R W ... G1HLY
Powell, R W ... G4VOY
Powell, R W ... G7MON
Powell, S D ... G4WYC
Powell, S M ... GJ7VQZ
Powell, S V ... G7KZW
Powell, S W ... G3WRA
Powell, T R ... G1OVG
Powell, Y J ... G0OGP
Power, C H ... G6UGE
Power, D M ... G6KDP
Power, D W ... G3SCJ
Power, J J ... G1WVK
Power, J P ... GM0KTO
Power, M E ... G0NKV
Power, N J ... G1YOB
Power, P ... G8BVB
Power, S M ... G7CXU
Power, S W ... G0DQQ
Power, W ... G6EOR
Powers, A J ... G0JNK
Powers, A M ... 2E1DXV
Powers, M ... G4XHX
Powers, P B ... G4XHY
Powers, R G ... G8CKN
Powis, A D ... 2E0ADZ
Powis, D J ... 2E1ALQ
Powis, D S ... G4HUP
Powis, D S ... M0ADM
Powis, R ... G1DTB
Powles, C ... GW4UBQ
Powlesland, C T ... G8CQZ
Pownall, R J ... G0ISH
Pownall, R J ... G3RAC
Powney, A J ... G0TUN
Powrie, G H ... GW4PNV
Powrie, M I ... G4EVZ
Poxon, J W ... G4UPA
Poynter, A F G ... G0AKR
Poynter, F J ... G0PFJ
Poynter, I R ... G0VBS
Poyser, I ... G6NWN
Pragnell, F A D ... 2E1FEV
Pragnell, J K ... G1EPO
Prall, V H ... G7FFC
Prank, V F ... G8VOC
Prater, G ... G4TZK
Prater, G V ... G4AEI
Prater, I J ... G7VIN
Prater, W A ... G0EOL
Pratley, D C ... G6INV
Pratley, G K ... G0OBJ
Pratley, T C ... G7JRP
Pratt, A J ... G7AJB
Pratt, A M ... G0SHP
Pratt, A P ... G3YSQ
Pratt, D M ... G4DMP
Pratt, D P ... G8KPY
Pratt, D W ... G1VSM
Pratt, E G ... G0PJQ
Pratt, E I ... G4MID
Pratt, F ... G4CNJ
Pratt, F D ... G4FXJ
Pratt, G B ... M1ADK
Pratt, I P ... G0JYW
Pratt, M A ... G4RHY
Pratt, M D L ... G7MRY
Pratt, N ... G1BYK
Pratt, N J ... G0PXK
Pratt, N J ... M1BDX
Pratt, N S ... G8HZO
Pratt, N W F ... G3RTO
Pratt, R ... G0WWU
Pratt, R G ... G3RGP
Pratt, R J L ... G4DDX
Pratt, R R ... G1WQC
Precious, P M ... G7PKH
Preece, A W ... G3TCO
Preece, B A ... G4BYL
Preece, D L ... G4TCO
Preece, F G ... G4FGP
Preece, F R ... 2E1AXI
Preece, G M ... G3RSX
Preece, J ... 2E1AVX
Preece, J C M ... G1UMS
Preece, J M ... G1EGK
Preece, L ... G1ILS
Preece, L ... G1KUG
Preece, T F ... G0IAH
Preece, W J ... G7KAG
Preedy, A R ... G3LNP
Preedy, G W ... M1BQM
Preedy, N ... G7GNY
Prendergast, M J ... G0RDD
Prendiville, D J P ... G0LCN
Prenter, A M ... MI1BOE
Prentice, C D ... GM6GWB
Prentice, G E ... G8VHN
Prentice, M A ... G7SWR
Prentice, T R ... G8INZ
Prescott, C ... G4OMG
Prescott, M J ... G4WYZ
Prescott, P K ... G1WXW
Presland, J ... G3IUY
Presley, D E ... G0RQT
Presley, A ... G4BXQ
Presswood, R E ... G4AEW
Prestage, B C P ... G0WCH
Prestidge, M J ... G2BXP
Preston, B ... G7OXF
Preston, B J ... G4KJA
Preston, C G ... G7TGG
Preston, D ... G6XDH
Preston, D J ... G4WDP
Preston, D V ... G3KFS
Preston, J ... G4WDQ
Preston, J H ... G7JQJ
Preston, K P ... G0RCE
Preston, K R ... G8HQV
Preston, L J ... G7UUF
Preston, M J ... G0THY
Preston, P J ... G7LUK
Preston, R ... G4UFO
Preston, R G ... G0RAX
Preston, R G ... G1EGL
Preston, R L ... G6YGB
Preston, R N ... G8MYE
Preston, R T ... G4IVK
Preston, S ... G0EGD
Preston, S ... G1VDO
Preston, T J G ... G1GQE
Preston, W H ... G3EFL
Prestwood, M H ... G3PDH
Pretty, S G ... G8MJK
Prettyjohns, K N ... G4KSU
Prew, B J ... G4FJA
Prew, R ... G8EPQ
Priamo, V ... GW6ZCS
Price, A ... G6UDG
Price, A ... GM0IDV
Price, A D ... GW8YJN
Price, A E ... GW1ERA
Price, A J ... G6TIQ
Price, A L G ... G8PGF
Price, A R ... 2J1EDR
Price, A R ... G7UON
Price, A W ... G8KBG
Price, B ... G4DDF
Price, B D ... G4DVB
Price, B H ... G3MTQ
Price, B R ... G4NXU
Price, C ... G0LZJ
Price, C ... G7GVP
Price, C ... G8KTE
Price, C C ... G3JDA
Price, C J ... G0FZP
Price, C J ... G0NYN
Price, C J ... G7UMC
Price, C P ... G4AGJ
Price, C T ... G6PEG
Price, D ... G1ODB
Price, D ... G7WGD
Price, D ... GW1HAN
Price, D B G ... G0DOI
Price, D C ... G7CUB
Price, D D ... G4BIX
Price, D G ... G3YZ
Price, D G ... G4MSA
Price, D J G ... G0ADZ
Price, D J W ... G3RLF
Price, D T ... G3LYU
Price, D T ... GW7TIX
Price, D W ... GW0SRE
Price, E H ... G3JPP
Price, E R ... G4FP
Price, F A ... G8WSF
Price, G ... M1BBU
Price, G C ... GW3MPP
Price, G E ... G4UZG
Price, G G W ... G0NVV
Price, G H ... GW3LXI
Price, G J ... GW7LIY
Price, I B ... GW4MZG
Price, I G ... GW0OSB
Price, J ... G0ARP
Price, J ... G0CJN
Price, J ... G3AEO
Price, J ... G7NGI
Price, J A ... G6IQY
Price, J D ... G3RNP
Price, J E ... G4OIK
Price, J E ... G4OIL
Price, J E ... G8CNN
Price, J E ... GW1DKK
Price, J N ... G0WJF
Price, J P ... G3UAL
Price, J R ... G0ILC
Price, J R ... G7JCX
Price, K C ... G0PNV
Price, M A ... G7PRW
Price, M D ... GW0UKC
Price, M J ... G1XFO
Price, M J ... G4EGS
Price, M R ... G7JRL
Price, M T ... GW0JXG
Price, N ... G7EWX
Price, O ... GI0WPV
Price, P G ... GW0TDA
Price, P J ... G6GYN
Price, R ... G3VTD
Price, R ... G4CKG
Price, R ... GW4EVX
Price, R ... GW4PCX
Price, R A ... GM0DWY
Price, R E ... G8DTF
Price, R F ... GW6JJX
Price, R H ... GW0VMW
Price, R J ... GW3ECH
Price, R M ... GW3SYL
Price, R W ... G4BSO
Price, S J ... 2E1ETF
Price, S J ... G4BWE
Price, S P ... G8VHL
Price, T D ... G6WUR
Price, T E ... G3EMK
Price, T H ... G3KDH
Price, W G ... GW0TTF
Price Hopkins, M ... GW3EMI
Price-Gore, C D ... GW6GDR
Prichard, A C ... G4TPP
Prichard, A D ... G0CPA
Prichard, B H ... G4BQS
Prickett, M L W ... G3BSK
Prickett, W C E C ... G4NCF
Priday, G R ... G2HMY
Priddy, A ... G4UBI
Priddy, J E ... GM3CIG
Priddy, M G ... GW0KPM
Pridham, R S ... G4BVB
Pridmore, G L ... M1BHQ
Priestley, A A ... G8MWX
Priestley, B A ... G7JWQ
Priestley, D A ... 2E1DME
Priestley, G D ... G7JZM
Priestley, G R ... G6RXF
Priestley, H ... G3FLR
Priestley, M B ... G7HEN
Priestley, N J ... G0CBK
Priestley, R E ... G0IHO
Priestnall, F ... G6LHA
Priestnall, W ... G3BRU
Prietzel, E W ... G4ZCV
Primack, R ... G4DBO
Prime, E W ... G6CDK
Primmer, J ... G8NYC
Prince, D J ... G6DMW
Prince, D W ... GW6MPX
Prince, E ... G3KPU
Prince, G E ... G7CDU
Prince, I R ... G6EZG
Prince, K G ... G4TQL
Prince, K J ... G0NUP
Prince, K J ... G0SIG
Prince, M ... G7EUL
Prince, M J ... G4LZA
Prince, M J ... G6CZT
Prince, N H ... G3VSI
Prince, R ... GW4GSL
Prince, R F D ... G3CAJ
Prince, S ... G4PMN
Prince, W A ... G4KRL
Prince, W R ... G3XEK
Pringle, A B ... GM3MAS
Pringle, C ... GM1SVQ
Pringle, C W ... G6BTB
Pringle, N ... M0ADB
Prior, A G ... G4UWW
Prior, D C ... G7TLM
Prior, J R ... G1BKJ
Prior, K E ... G4TRW
Prior, K J ... 2E1FJY
Prior, K J ... M1BGJ
Prior, K M ... G8XHK
Prior, L P ... G8IXC
Prior, M ... G7VMY
Prior, M E ... G1VOB
Prior, P C ... G4CWR
Prior, P J ... 2E1EFY
Prior, P J ... M1CAK
Prior, R ... G1APT
Prior, R A ... G3MTG
Prior, R E ... G0ALJ
Prior, R W ... GW0IEU
Prior, S J ... G4SJP
Prisk, S J ... G7DWO
Priston, D G ... G7NVN
Pritchard, A C ... G0RUY
Pritchard, A E ... G3ODB
Pritchard, A H ... GW3WLN
Pritchard, A S ... GW7NNW
Pritchard, A W ... GW7KUK
Pritchard, B C ... G8MCJ
Pritchard, C ... GW2BFD
Pritchard, C ... M1ADB
Pritchard, C F ... G7JAE
Pritchard, D J ... G4ULP
Pritchard, D V ... G3JSJ
Pritchard, E ... G0BJZ
Pritchard, E ... G4USX
Pritchard, E G ... G4ZGP
Pritchard, G W ... GW1BAV
Pritchard, I D ... G1XYN
Pritchard, J A V ... GW0HZJ
Pritchard, J L ... GM4DMQ
Pritchard, J W ... G7JXD
Pritchard, K I ... G3WVG
Pritchard, L A ... GW0RQS
Pritchard, L A G ... G0TRI
Pritchard, L E ... G3IUW
Pritchard, M ... G4HKI
Pritchard, M ... G7EVP
Pritchard, M G ... G3VNQ
Pritchard, M N ... M1ADA
Pritchard, N B ... G8AYM
Pritchard, N F ... G8TYC
Pritchard, P E ... G8LLD
Pritchard, P H ... G6MPT
Pritchard, P J ... G1NPP
Pritchard, R D ... G1BHQ
Pritchard, R J ... G0AZX
Pritchard, S ... G1ILT
Pritchard, T E ... G0GSM
Pritchett, L W ... G1YPZ
Pritt, D S ... G8TZE
Privett, R A ... G4MZV
Privett, R V ... G0CUU
Prizeman, D W ... G0EFB
Probert, C J ... GW3YSF
Probert, D B ... G4JBU
Probert, M J ... GW4HXO
Probert, P C ... G3GPZ
Probert, P J ... GW1PJP
Probin, D S ... G3YBD
Probyn, A ... G1JII
Probyn, G S ... G7WIC
Procter, D R ... G0UTF
Procter, J B ... G8AWN
Procter, J L ... G0FQN
Procter, K I ... G3EPO
Procter, N C ... G0IZL
Procter, R F ... G0ENA
Procter, R G ... G4LAK
Procter, S ... G8PDY
Procter, S H ... G6PDM
Proctor, C J ... G1XLG
Proctor, C J ... G8MBM
Proctor, C J ... G8XLG
Proctor, C R C ... 2M1CDP
Proctor, D ... G7RYN
Proctor, D C ... G4JYW
Proctor, D J ... G0EDE
Proctor, E O ... G7ICC
Proctor, G ... GM8SQ
Proctor, J C E ... G0DCH
Proctor, J F ... G3JFP
Proctor, L ... G0LLK
Proctor, R J ... G4PZW
Proctor, R J ... M1ATI
Proctor, T E ... G4OHR
Proffitt, C ... G0DTB
Proffitt, C ... M0AVB
Proffitt, J M ... G6RJH
Profitt, A V ... G7MBX
Profitt, R F ... G4HRU
Prosser, D M ... GU0BDI
Prosser, G ... G7EJN
Prosser, K ... GW8TRO
Prosser, M E ... GW6TUD
Prosser, N R ... G1DYQ
Prosser, N R ... G7BTI
Prosser, P C ... G4TVZ
Prosser, P E ... G4TJA
Prosser, R ... G4OJP
Prosser, R ... GW4BUS
Prosser, R A ... G1RAP
Prosser, S J ... G6HZH
Prothero, E D ... G0RNO
Protheroe, D E ... G4RCI
Protheroe-Thomas, W J ... GW4TGL
Proud, D ... G0AWR
Proud, L L ... G0LLP
Proud, T A ... G3OPK
Proudfoot, J J ... G4ISS
Proudler, A J ... G4OMI
Proudler, M J ... G0NAG
Proudlove, G ... G1GQH
Proudlove, P J ... G6NAH
Prouse, D E ... G8TYD
Prouse, T M ... G1BHM
Prouse, W F ... G4SGW
Prout, C E ... G7NKU
Prout, D C ... G0LTE
Prout, D T ... G3VCV
Provan, S T ... GW0UQH
Provins, R ... G0RGJ
Prowse, C J ... G7MDV
Prowse, P V ... G4UON
Pruden, J ... G0GEL
Pruden, S G ... G0GEO
Pruitt, R D ... G1ERB
Prust, C D ... G4TPL
Pryce, S G ... G0EIY
Pryce, S J ... G0MGA
Pryce, W ... GW4WHP
Pryde, A M ... 2M1FOQ
Pryde, J A ... GM8JJN
Pryde, J H ... M1AHC
Pryde, R I ... GM3LGU
Pryer, J B A ... G0LAZ
Pryke, G A T ... G6INX
Pryke, G R ... G6WKL
Pryke, J J ... G8BXH
Pryke, P J ... G8HUE
Pryke, S A ... G1SUE
Pryor, J E ... G0KBQ
Pryor, R S ... G8XRP
Przybyla, A P ... G0EPP
Pucknell, I D ... G7EZT
Puddephatt, M W ... G7CTN
Pudsey, A F ... G4DXH
Puffett, F J ... G0MTW
Pugh, A H ... GW8ASD
Pugh, B J ... G3SMU
Pugh, C D ... M1BIL
Pugh, D ... G0SJS
Pugh, J ... G0IKZ
Pugh, K ... GM7DHA
Pugh, K G ... G3HES
Pugh, K R ... G3SIK
Pugh, M ... G4DVE
Pugh, M A ... G4VPD
Pugh, M D ... G0WKT
Pugh, M E ... G7FBX
Pugh, P D ... G1MLO
Pugh, P V ... G2CQX
Pugh, R D ... GM0OTB
Pugh, R H G ... G7RYM
Pugh, S C ... G1UFM
Pugh, S F ... G3SFX
Pugh, W F ... G6PBJ
Pugsley, E ... 2E1DEI
Pugsley, R ... G7TYQ
Pulford, J P E C ... G4RQO
Pulfrey, B ... G4VIM
Pullan, M K ... G8NZR
Pullen, A J ... G8BXJ
Pullen, D R ... G1EGM
Pullen, F J ... G4XXX
Pullen, J W ... G4TGE
Pullen, M T ... G3EYH
Pullen, P G ... G7UZI
Pullen, R ... GW1MAX
Pullen, R A R ... G0OII
Pulley, P J ... G6XLG
Pulleyblank, E A ... G7SZI
Pullin, D S ... GW4IUL
Pulling, R E ... G3REV
Pullinger, P J ... G8JCB
Pumfrey, J W ... G8SNH
Pumphrey, E J ... G8JEE
Puncer, M ... G0BLV
Puncher, B M ... G0IJK
Pung, C W ... G6RVP
Punshon, H ... G0TOH
Punshon, K R ... G4APJ
Punter, G W ... G8IIG
Purcell, A E ... G0KFS
Purcell, C ... GW0WLN
Purcell, C J G ... G0RHZ
Purcell, C S ... G0DOG
Purcell, J G ... G0OIZ
Purcell, J G ... G0WUU
Purcell, M J ... 2E1FKY
Purcell, T D ... G0DOE
Purchase, B ... G3FWD
Purchase, C M ... G8JCC
Purchese, D E E ... G3LXP
Purchon, G ... G6CFC
Purchon, J A ... G4EMQ
Purchon, N D ... G0WKM
Purchon, N D ... G0WVJ
Purchon, N D ... G7VHH
Purdom, R ... G3NMI
Purdy, B ... G7ATY
Purdy, J J ... G4ZZJ
Purdy, J R M ... G6JIF
Purdy, P W ... G6VAD
Purdy, R ... G6HZX
Purdy, R B ... G4RBP
Purdy, V J ... G3JVP
Purdy, W T ... G4LTP
Purkins, N ... G1EIB
Purkiss, B ... G7SFY
Purkiss, D J ... G8NNJ
Purnell, I ... GW4RGL
Purnell, J T ... G3UVH
Purnell, M R ... G4XIX
Purrier, G ... M1AVU
Purse, B E ... G1RNV
Purseglove, A D ... G0JDG
Purser, G ... G4ONX
Purser, G ... G4ONY
Purser, K J ... G6HZG
Purser, M R ... G0NEM
Purser, R C ... G7WAH
Purser, S J ... G4SHF
Purser, W J ... G2AXO
Purtell, J M ... GM0IST
Purtell, N B ... GM0NYP
Purves, H ... G6UQX
Purves, M A ... G4OFF
Purves, R D ... GM4IKT
Purvess, J F ... G0FWP
Purvess, R ... G1OLH
Purvis, F A ... M1BBI
Puryer, T L W ... G3MRB
Pusey, M J ... G8DTE
Putnam, R J ... G0ILN
Puttick, D ... G1YDW
Puttick, M ... G3LIK
Puttick, N M ... G4PGW
Puttock, J C ... G6WPK
Puttock, J L ... G0BWV
Puttock, J L ... G2XP
Puttock, J L ... G7SAC
Pybus, A S ... G8EZG
Pybus, B ... G0WKZ
Pybus, S W ... G8TBX
Pye, D J ... G3EVC
Pye, D M ... M1BZU
Pye, J ... M1BZV
Pye, K J ... G1OMU
Pye, R J ... G4IUH
Pye-Smith, R E S ... G4MTM
Pyke, B C ... G1LZY
Pykett, D W ... G0BEN
Pykett, R K ... G7BGT
Pyle, M W A ... G2BLA
Pym, M J ... G4GXT
Pyman, R D ... M1AEX
Pyne, A F ... G4UNH
Pyrah, M R ... G6ENA
Pyrah, R W ... G6UDA
Pywell, D T ... G7TJH

Q

Quade, S H ... G6WYQ
Quah, H Y E ... M1BJY
Quail, T ... G3TBQ
Quaintance, R E ... G0DIZ
Quaintance, R E ... G7FAI
Quaite, G ... GI4MHD
Quantick, H R ... G3UGL
Quantrell, C F ... G1ZSX
Quantrill, G ... G6WFS
Quantrill, L C ... G0DTP
Quantrill, S ... G7POZ
Quantrill, T J ... 2E1BMU
Quarman, K W ... G8CBE
Quarmby, J H ... G3XDY
Quartermaine, J ... 2E1FUQ
Quarterman, A K ... G1DTC
Quarterman, G L ... G3NHX
Quash, J ... G4UNC
Quayle, A ... GW0MHO
Quayle, A ... GW4GTC
Quayle, F ... GD4DPK
Quayle, G D ... G4NAV
Quee, M J ... G3ZWW
Quest, A M ... G4UZN
Quest, N V ... G7VVL
Quested, P W ... G0DRT
Quick, G ... G0SUZ
Quick, P ... M1BWZ
Quick, P ... MW0APQ
Quick, R J ... G7AZV
Quick, S ... MW1BSH
Quicke, M C D ... G0MCQ
Quickfall, P B ... G4RLU
Quigg, H ... G7VGV
Quigg, J A ... GI6VWS
Quigg, R K ... GI4CRQ
Quigley, C H ... GI4NKJ
Quigley, D J ... G3PRI
Quigley, J ... GM0PYM
Quigley, J ... GM0TQA
Quilter, P D ... G6KUS
Quin, H A ... GM0DXE
Quin, J M ... G4XBV
Quince, A J ... G0UBX
Quinlan, K M ... G7DZD
Quinn, A ... G7TPK
Quinn, G M M ... GI7JPW
Quinn, H A ... GI0WLW
Quinn, J A ... GI4PCQ
Quinn, N J ... G0BAF
Quinn, N J ... GI4KJC
Quinn, P J J ... GI7FCW
Quinn, S L ... G0SLQ
Quinn, T ... GM0GYM
Quinn, T E W ... G4FQA
Quinnear, D M ... G0POK

Quinnin, C W A ... G0ECQ
Quint, G D ... G7TTH
Quirk, P E ... G0VMF
Quirk, P R ... G1RQI
Quiruga, N A ... 2E1CKJ
Quy, A J ... G0FEO

R

Rabbage, B D ... G3APN
Rabbett, M J ... GI0WGS
Rabbett, M W ... GI0IOT
Rabbitt, I A E ... G1BHR
Rabbitt, W ... G0PZP
Rabbitts, J M ... GM8LFB
Rabbitts, T M ... G8HUH
Rabe, C M ... G6NLC
Rabey, B L ... G8NYR
Rabey, C J ... G0RMR
Rabjohns, J G ... G3YBG
Rabone, D C ... G4FNR
Rabson, J ... G3PAI
Rabstaff, G M ... G3ZVT
Raby, B ... GW4ZVV
Raby, B S ... G8GTV
Raby, F J ... G3ZPE
Raby, J E ... G6DGX
Raby, J R ... G8RF
Race, M C ... G1DCX
Racher, P A ... G6MQJ
Racionzer, P H ... GM0DFW
Rack, M ... G1KBL
Rackett, J S ... G3EZB
Rackham, G H ... G8BXV
Rackham, M M C ... GW4JKV
Rackham, P M ... G3IRQ
Rackstraw, G G ... G0GGR
Rackstraw, R J ... G1MFR
Racz, S ... G1KAX
Radcliffe, A D ... GD3FXN
Radcliffe, C M ... G4SRF
Radcliffe, L M ... G8XUE
Radcliffe, P A ... G1NIO
Radcliffe, P F ... G0FNP
Radcliffe, R M ... GW0GJD
Raddy, S ... GW6YNV
Radford, A A ... M0BAE
Radford, B J ... G4RQR
Radford, D G ... G3KMY
Radford, G P ... G3YDZ
Radford, J J ... G0PCY
Radford, K ... G3SZU
Radford, M ... G8UDH
Radford, P J ... 2E1BOA
Radford, P J ... G0ADW
Radford, R J ... G3GPB
Radford, S I ... GW0CYK
Radforth, M A ... 2E1BGO
Radley, A L ... G0VXR
Radley, A P ... G0TTM
Radley, A P ... G5QK
Radley, A P ... G8MRS
Radley, C J ... G0ROF
Radley, D S ... G4ABI
Radley, L R ... G4JDS
Radtke, J M ... G7LKC
Radwell, F J ... G4CTW
Rae, A D ... GM4ENN
Rae, C A ... G6KLO
Rae, F A ... GM4TYQ
Rae, G D ... G4MLO
Rae, I D ... G8BUE
Rae, J K ... GM0FUA
Rae, J R ... G0FQF
Rae, L J ... G6YNX
Rae, P S ... G7MJH
Rae, R J ... G1XSO
Rafferty, F C ... G0HMS
Rafferty, J L ... G4YVX
Rafferty, S ... G6NXP
Rafferty, S ... GI0EJT
Raffield, K G ... G3CQU
Rafter, R ... G1XNK
Raftery, A F ... G4CDO
Ragg, W L ... G4ORS
Rail, J D ... 2E1BCD
Rail, M J ... G4IXM
Railton, C J ... G3YQV
Railton, R D ... GW6RXA
Raimbach, D T ... G3ZWK
Rainbow, C R ... 2E1DTT
Rainbow, M A J ... G6OTP
Raine, A ... GM8VYZ
Raine, C J ... GM8MNG
Raine, C R ... G6VOC
Raine, D B ... G4ZHA
Raine, I J ... GM8TSI
Raine, L ... 2E1GAG
Raine, R W ... G4RXR
Rainer, D J R ... G4VTQ
Raines, M ... G0KTQ
Rainey, J D ... GI4YCZ
Rainey, M G ... GI0KMA
Rainey, R C ... GI0RBS
Rainy Brown, G ... G1NAB
Raistrick, A S ... G7IKS
Raistrick, S A ... G6LLR
Rake, S ... GW1OPE
Rall, L J ... G4AJ
Ralls, A W ... G3PDP
Ralph, A D ... G8XLH
Ralph, C F ... G1GGC
Ralph, D ... G1UCO
Ralph, D V ... GW4UZC
Ralph, J ... G0GWK
Ralph, N S ... 2E1EGN
Ralph, R P ... G4KSG
Ralph, S F ... G7EIA
Ralph, T ... 2E1ATR
Ralston, P ... G6CIP
Ralton, R C ... G6GIG
Ramm, A E M ... G7OSJ
Ramm, D J ... G7OSK
Ramm, R E ... M0AIZ
Ramplin, R W ... G0NGW
Rampton, D ... G4ZEL
Rampton, J A ... G3VFI
Rampton, K D ... G1KAW
Rampton, T ... G0HOY
Ramsay, A J ... GM6BML
Ramsay, D W C ... G0RWY
Ramsay, E M ... GM7AUX
Ramsay, J ... G4NCZ
Ramsay, J B ... G8LOZ
Ramsay, J C ... GM3OQI
Ramsay, J C ... MM0AKX
Ramsay, J C E ... G3OZZ
Ramsay, R L ... G4BBJ
Ramsbottom, W K ... G8GAG
Ramsdale, A S ... G7GNM
Ramsdale, E J ... G3GWC
Ramsdale, K ... G4YKR
Ramsden, A ... G0VDQ
Ramsden, D ... G4YPV
Ramsden, H W ... G4YKB
Ramsden, S ... G7UXL
Ramsden, T ... G6AYS
Ramsey, A ... G4MQF
Ramsey, C A ... G8VZD
Ramsey, D A ... G3UAA
Ramsey, D H ... G6SNN
Ramsey, G ... 2E1FNI
Ramsey, P ... GM4WNQ
Ramsey, P J ... G4JQE
Ramsey, W T ... G7MJB
Ramshaw, R A ... G3RLD
Ramskill, M A ... M1AJG
Ranby, M J ... G0IOL
Rance, T L ... G4HTB
Rand, H E ... G6IFK
Rand, S W ... GD7RVP
Randall, C J ... G4RBR
Randall, G K ... G6ZMU
Randall, J ... G3OAZ
Randall, J A ... G0DAF
Randall, J M ... G6WAY
Randall, L W R ... G4ACQ
Randall, M W ... G4TZN
Randall, M W ... GW0JZN
Randall, P ... G0BDG
Randall, P ... G1SJZ
Randall, P ... G6TPC
Randall, P A ... G8HXT
Randall, P F ... G3NJV
Randall, S F ... G4GGX
Randall, T M H ... GW1VYT
Randall-Cook, P D ... G8WGD
Randell, P J ... G6HKH
Randeria, Z ... G7WCI
Randle, I K ... G6TJY
Randles, D F ... G4AFT
Randles, D J ... G7JHZ
Randles, I P ... M1BBA
Randolph, W L ... G0KCC
Rands, A H ... G6PEH
Rands, D ... G7SST
Ranger, K B ... G0KJK
Ranger, M J ... G6GWE
Rank, J H ... G4NSE
Rank, N ... G6SIR
Rankin, A ... G6WTS
Rankin, D A ... G3YQB
Rankin, F R A ... GM4PVF
Rankin, G M ... G4FVL
Rankin, J G ... GM1BKR
Rankin, J R ... GM4VDG
Rankin, R ... GM0NPS
Rankine, K J R ... GM4VZW
Ranklin, J E ... GU6AYV
Ranner, P K ... G4ZCX
Ransford, D A ... G4NNY
Ransom, N J ... G0SRJ
Ransom, R J ... G6VBE
Ransom, S D ... G6MIS
Ransom, T H ... G1HHH
Ransom, T H ... G2LL
Ransom, T H ... G4FET
Ransom, T H ... G6HH
Ransome, A J ... 2E1BEM
Ransome, C B ... G8VOK
Ransome, C R ... M1BKZ
Ransome, D J ... G1KDX
Ransome, J G ... G8LEO
Ransome, R B ... G0NFE
Ransome, R C ... G7MPF
Ranson, C P ... G8LBS
Ranson, D L ... G0IWI
Ranson, J R ... G7NND
Ranson, R G ... G3ZTB
Raper, K ... G8IKG
Raper, T O ... G6LLT
Rappolt, F M ... G1VJF
Rapson, C D ... G0OZM
Rapson, N A ... G0RJI
Rash, R M W ... G3MQU
Rashleigh, K C ... G8ORX
Raspin, C J ... G4KUE
Ratcliff, A W ... M1ALE
Ratcliff, L A C ... G4GYP
Ratcliffe, C P ... G1MXO
Ratcliffe, G ... G3HKU
Ratcliffe, G R ... G6WZO
Ratcliffe, K ... G6OMX
Ratcliffe, M R ... G6JWV
Ratcliffe, N ... G3FHW
Ratcliffe, R A ... G4LRY
Ratcliffe, R B ... G4ACY
Ratcliffe, R B ... GW3KZW
Ratcliffe, S A ... G0IUA
Rathbone, J ... G3KZY
Rathbone, M ... G3ZII
Rathbone, M P ... G3UZN
Rathbone, N A ... G4KZU
Ratigan, J S ... G7DAL
Rattenbury, J ... G1HQN
Rattenbury, P E ... G6OIA
Rattey, J K ... G6OSG
Rattigan, J F ... G0GJM
Rattray, A E ... GI6UQU
Rattray, C B ... GM3HYX
Rattray, F ... G4SPR
Rattray, H G ... GI0MSK
Rattray, W G ... GM4NDV
Rauch, C ... G8GKL
Ravelini, J P ... G1UPT
Ravelini, T L ... G1HIG
Raven, P N ... G4KLM
Raven, R R ... G4CFW
Raven, T K ... G4ARI
Ravenhill, K N ... G1HDQ
Ravenscroft, G S ... G0TMG
Ravenscroft, M R ... G0TMR
Ravenscroft, P A ... G0ULO
Ravenscroft, R W ... G0RAV
Ravilious, N ... M0ARH
Rawcliffe, S W R ... G4YXU
Rawdon, A J ... G6GHE
Rawlance, T P ... G4MVN
Rawling, H S ... G2DSB
Rawlings, A J ... M0ANS
Rawlings, D V ... G0BGG
Rawlings, F ... G4JOD
Rawlings, G E ... G8CUN
Rawlings, J M ... GW4ULG
Rawlings, K R ... G4MIU
Rawlings, L F ... G3FET
Rawlings, R J ... G3UXF
Rawlings, S A ... GW4ALG
Rawlings, S M ... G8VZL
Rawlings, W J ... G3BON
Rawlingson, I J ... G1XZA
Rawlingson, J E ... G1XZB
Rawlins, A D ... G1CGH
Rawlins, F A ... G0OFX
Rawlins, P ... G6NWO
Rawlinson, D E ... G0OME
Rawlinson, J G ... G4XET
Rawlinson, L W ... G7FQL
Rawlinson, P B ... G8PPG
Rawlinson, R D ... G7VFO
Rawlinson, S B ... G4MUO
Rawlinson, T J ... G0RNA
Rawls, A V ... G6BHC
Rawson, A ... G1MCI
Rawson, J H ... G4ZJY
Rawson, J W D ... G0BQO
Rawson, M ... G0DJM
Rawson, P L ... G6PMP
Rawson, R ... G1SQW
Rawson, S ... G0MAA
Rawson, W N ... GI3XHL
Rawsthorne, R ... G0HBR
Raxter, D A G ... G7WEB
Raxworthy, A G ... 2E1CNM
Raxworthy, K J ... G7COQ
Raxworthy, P A ... G0AVP
Ray, B ... GM8MUE
Ray, I C ... G4VZB
Ray, J D ... G8DZH
Ray, K L ... G1GXB
Ray, M A ... G4XBF
Ray, R ... G0WAC
Ray, R ... G1MDG
Ray, R ... G3MDG
Ray, R ... G3NCL
Ray, R C ... G2TA
Ray, R J ... G3ZHS
Ray, R V ... G8CUB
Ray, S A ... G0UNT
Ray, S J ... G1GXC
Ray, W S ... G4HES
Raybould, B ... G4NMG
Raybould, D R ... G7HJX
Raybould, J C ... G7PVW
Raybould, J H ... G4PQI
Raybould, L ... M1BZG
Raybould, P H ... G8CGM
Raybould, T G ... G4UEQ
Raybould, W O ... G4DFE
Rayfield, C D J ... 2E1CZY
Rayland, E G ... G4ASK
Rayment, J D ... G3XAV
Rayment, R J A ... G0TTI
Raymer, B ... G6FDI
Raymer, R E ... 2E1AUM
Raymond, C E ... GW0LIK
Raymond, J T ... GW4ZYV
Rayner, A ... G1KPV
Rayner, A B ... M0AWQ
Rayner, B S ... G1VZT
Rayner, D G ... G0AFP
Rayner, D J ... G4XNP
Rayner, D K ... G3NYR
Rayner, G J ... G7KCT
Rayner, J R ... G4YSI
Rayner, K ... G1AKI
Rayner, L A ... M0AXI
Rayner, M P ... G1VUE
Rayner, R B ... G3YHP
Raynes, J M ... G0BWG
Raynor, J C ... G6ATG
Raynor, K H ... G8GXN
Raynor, P N ... G6EUF
Raynor, R H ... GM7MWX
Raynor-Smith, R G ... G0VYU
Rayson, J D ... 2E1EFL
Rayson, J I ... 2E0APY
Rayson, P G ... G8YJZ
Rayson, P M ... G1AJZ
Razey, R C ... G0ADH
Razzaq, S ... 2E1BOZ
Razzell, A G ... G0PBM
Rea, D I ... G1INW
Rea, M ... G6IWT
Rea, R H ... G0HQX
Reacher, D T ... G7VTH
Read, A ... G6CYE
Read, A B ... G0GMS
Read, A G ... G6VQC
Read, A J ... G1AEA
Read, A S M ... G3SXR
Read, A W ... G8UVM
Read, B J ... G3JDT
Read, C E ... G0SGM
Read, C K ... G4NPY
Read, C L ... G4TZA
Read, D V ... G0OIF
Read, D W ... G7CQX
Read, G C ... G1MFG
Read, G H ... G4KCB
Read, H D ... G4MYR
Read, J A ... 2E1FJC
Read, J F ... GW8KTQ
Read, J M ... G3JIZ
Read, J N ... G3UJC
Read, J R ... G4THU
Read, J V ... G8VOJ
Read, K N ... G7UYZ
Read, M ... M1AFE
Read, N ... G7DND
Read, N ... G8HNO
Read, N A ... G6WHS
Read, N K ... G8CXL
Read, N K ... G8WRG
Read, P A ... G8XHH
Read, P E ... G4YAR
Read, P E ... G8XBE
Read, P J ... 2E1BMC
Read, P M R ... G1ZPJ
Read, R C ... G4OVJ
Read, R F W ... G8CCN
Read, S ... G2ATM
Read, T W ... M1BIG
Read, W S ... GM8LRV
Reade, C J ... G1FCT
Reade, D ... GI7BDJ
Reade, J L ... G4OHQ
Reade, P L ... G8VUM
Reade, W ... G4XTO
Reader, A ... G0CRJ
Reader, G R W ... G7ITQ
Reader, K E ... G3LUH
Reader, P S ... G3TSR
Reader, T ... G6UML
Reading, D W ... G6VJR
Reading, G ... G1IFH
Reading, G J S ... G4LZD
Reading, N E ... G4WQS
Readings, D S ... G8MNT
Readings, J ... G3KFT
Readings, J M ... G4MVO
Readings, M C ... G1YDT
Readings, P ... G4FML
Rean, A J ... G4VXH
Reaney, C ... G0ROD
Reaney, C K ... G3PMA
Reanney, R A ... G0AKU
Reap, S A ... G8RJM
Reardon, D C ... G4EJK
Reason, A V ... G8WPV
Reason, G J ... G4EBF
Reason, W R ... G4GFB
Reavill, D D ... 2E1EQN
Reavill, K ... G7USM
Reavill, M S ... 2E1CHX
Reay, D A ... G8UHO
Reay, G A L ... G0GOZ
Reay, P ... G6TAK
Recardo, A M ... G0LFZ
Recardo, D G ... G0LFY
Reckitt, W H ... G3EBG
Record, F A ... G4JVQ
Record, I J ... G6TJP
Redall, P ... G4FZV
Reddecliffe, G ... G7UBK
Redden, T J ... G1AHW
Redding, A R ... G0COM
Redding, D R ... M1BAD
Redding, G W ... G0GGQ
Redding, R J ... G3VMR
Reddington, B ... G3JWY
Reddington, C ... 2E1FJJ
Reddington, G G ... G6ZDB
Reddington, R ... G1XOA
Reddish, A ... G6MGQ
Reddish, N A ... G0ORE
Reddish, T B ... G0PLA
Reddy, U D ... 2E1FXZ
Redfearn, J W ... G4YNZ
Redfern, A ... G3MVD
Redfern, B J ... G0CGI
Redfern, C B ... G4CZR
Redfern, D A ... G4SKW
Redfern, D A ... G8TPK
Redfern, H ... G3OVR
Redfern, I E ... G8DUW
Redfern, J W ... GW4RZU
Redfern, P G ... G4CLN
Redfern, S W ... G4AEQ
Redford, J ... G3SXP
Redford, K M ... G4FRR
Redgate, A S ... G1EGO
Redgate, J S ... G4ANS
Redgewell, G A ... G4TTN
Redhead, G W ... G4KXW
Redhead, J M ... G6LGN
Redhead, R F ... G4FXG
Redman, A J ... G4KUF
Redman, A W ... G8DUB
Redman, D A ... G4IDR
Redman, D J ... G7TEC
Redman, G ... G0THL
Redman, G F L ... G8KIL
Redman, J H ... G4KXL
Redman, J H K ... G4HBP
Redman, M A ... G8UTN
Redman, M A ... M0BBH
Redman, P A ... GW0TSW
Redman, P A ... GW7HEC
Redman, P F T ... G8ELX
Redman, P L ... G1HQI
Redman-White, W ... G3XSH
Redmayne, C V ... G4GLW
Redmayne, N J ... G6PQW
Redmile, P E J ... G8XYY
Redmill, D J ... G6ZTM
Redmond, D H ... G4SWF
Redmond, J ... G0IHY
Redmond, K P ... G6FKE
Redmond, W J ... GI0LMR
Redpath, J C ... G0VLZ
Redpath, R E ... G6AHR
Redpath, W P ... GM4JEM
Redshaw, A L ... G6DVE
Redshaw, D J ... G3GAW
Redstall, M W ... M1BIU
Redway, R C ... G4TRB
Redway, S M ... G4TRA
Redwin, M D J ... G0BJS
Redwood, C ... G6MXL
Reece, G H ... G4YGL
Reece, G P ... G6HTT
Reece, K H ... G8UYB
Reed, A ... G4HCD
Reed, A ... G4HKR
Reed, A ... G4LLJ
Reed, A ... G6NLD
Reed, A J ... G0ESZ
Reed, C A R ... G6WKN
Reed, C J ... G8MFP
Reed, D A ... G4YDR
Reed, D A ... G6RCY
Reed, D A ... G7RAD
Reed, D L ... G8NYB
Reed, D M ... G1VNK
Reed, D T ... G7UCO
Reed, F E ... G4PIB
Reed, F G J ... G6LOG
Reed, G E ... G4YBD
Reed, J ... G3SIB
Reed, J A ... G0JRB
Reed, J D ... GW0UDM
Reed, J E ... G1AWI
Reed, J S ... G3ZMD
Reed, J W ... G4ZTA
Reed, J W K ... G0NOH
Reed, K J ... G4NXM
Reed, M B ... G4BZF
Reed, M J ... G0JWX
Reed, N G ... G0GUT
Reed, P L ... G4BVH
Reed, P M ... G0KEM
Reed, P M ... G4ADV
Reed, P M ... G6TKA
Reed, Q L ... G0IHE
Reed, R ... G1WKT
Reed, R ... G3ZIG
Reed, R ... G4ZIG
Reed, R D ... G4ZCW
Reed, S ... G4VQR
Reed, S J ... G0AEV
Reed, S J ... G6XJU
Reed, S P ... G7MMC
Reed, T W ... G0WAL
Reed, W J ... GI7JHA
Reeder, A B ... G3BOT
Reedman, N A ... G4WKU
Reeds, G ... 2E1FTN
Reeds, M A ... G0HOT
Reeds, M J ... G3SBU
Reekie, D C ... G4UHR
Reekie, H M ... GM8HSY
Reeley, A G ... G4OIN
Reeman, K B ... G8RAN
Rees, A B ... G3IQY
Rees, C J ... G3TUX
Rees, D A ... GW1LFN
Rees, D A J ... GW7OEE
Rees, D J ... MW1ARM
Rees, D W ... G1NNV
Rees, E G ... G0DQP
Rees, E J ... GW0BUA
Rees, F J ... GW0JRF
Rees, F J ... GW0VJO
Rees, I J M ... G8JWD
Rees, J C ... 2E1DHK
Rees, J W ... G4YBL
Rees, L ... GW1ADY
Rees, L R ... GW0RIL
Rees, M C ... GW0KYT
Rees, M V ... G3KTI
Rees, N ... G0KMS
Rees, R ... GW1EPR
Rees, R ... GW4PUC
Rees, R ... GW6HSD
Rees, R A S ... GW0FMQ
Rees, R D A ... G7IXD
Rees, R E ... G8ZAX
Rees, R M ... GW0DIX
Rees, R M ... GW7FNR
Rees, R P ... GW0GPQ
Rees, R T ... GW0RTR
Rees, R T ... GW7VPJ
Rees, S G ... G0GPN
Rees, S W ... GW8XZJ
Rees, T ... MM0AXR
Rees, T A ... GW6NXL
Rees, W D G ... GW0GUB
Rees, W L ... GW6NXH
Rees, W S ... GW0NLB
Reeson, M J ... G8OOS
Rees-Jones, D A ... G3PPN
Reeve, A B ... G6ZNJ
Reeve, A V ... G7KEP
Reeve, D ... G4NRW
Reeve, D ... G6TIW
Reeve, D A ... 2E1BTX
Reeve, D C ... G7RWI
Reeve, G F ... G0PXC
Reeve, G S ... G1HHO
Reeve, H J R ... G4ACJ
Reeve, H S ... G2HFD
Reeve, J ... G8ATS
Reeve, L R G H ... G3LLX
Reeve, P J ... G4GTN
Reeve, W R ... G8RQQ
Reeves, A E ... G4ZFQ
Reeves, A V K ... G7DTT
Reeves, B J ... M1AIW
Reeves, C C ... GW8FVI
Reeves, D A ... G1PAD
Reeves, H ... G3OAB
Reeves, J ... G4SRE
Reeves, J D ... G0MVX
Reeves, J D ... G1WCY
Reeves, J D ... G4SJM
Reeves, K A ... G3WZI
Reeves, L A ... G4CEM
Reeves, M G ... G6YNW
Reeves, P R ... G8GJA
Reeves, R J ... 2E1DFL
Reeves, R K ... G8VOI
Reeves, R M ... G1TWW
Reeves, T S ... G0GHL
Reffell, G ... G0EEA
Reffold, A D ... G3SQH
Regan, A C ... G6FOI
Regan, E ... GI7CTW
Regan, F ... G3RMD
Regnart, A C ... GM8YFA
Reid, A ... G3HIF
Reid, A ... GI7IMU
Reid, B A ... G1CUH
Reid, B W ... GM7JDS
Reid, C D ... GM8NMM
Reid, C J ... G0VQM
Reid, D ... M1AWM
Reid, D A ... G0BZF
Reid, D B ... GM4YLY
Reid, D E ... G4HJX
Reid, D J ... G1AKC
Reid, F C ... G3VMD
Reid, G ... G6VEP
Reid, G F ... G4EXG
Reid, G F ... G4OIS
Reid, G M ... GM1JWC
Reid, G P ... M0BGF
Reid, G S ... G6UEU
Reid, I A C ... GM0MYV
Reid, I D ... G4LMQ
Reid, I W ... GM6FND
Reid, J A K ... 2E1BDN
Reid, J A K ... G0TYG
Reid, J D ... M0BDF
Reid, J D ... MM1ATS
Reid, J F ... G0MLR
Reid, J I ... GM7KPE
Reid, J J ... GM0EMY
Reid, J M ... G2FKP
Reid, J P ... GM4LQR
Reid, J S ... GM3NUU
Reid, K G ... G3GNB
Reid, K M ... G2UT
Reid, L ... G8NKJ
Reid, M I ... 2E1DYC
Reid, M I ... G7VDP
Reid, M W ... GW4TWE
Reid, P ... GM0FCI
Reid, R ... G4OMM
Reid, R A ... GI3UBA
Reid, R C S ... GJ3IT
Reid, R J ... GI4XFX
Reid, R P ... GM0WBW
Reid, S ... GI6EWM
Reid, S C R ... G6KVG
Reid, S G ... G0KRM
Reid, W J ... G4HEJ
Reidy, P M ... G6INA
Reigate, S C ... G1IEY
Reilly, A ... G4BVW
Reilly, A W ... G6KOE
Reilly, D T ... G0RIE
Reilly, E J ... M1CDO
Reilly, J ... GM3HOM
Reilly, L W ... G7CFP
Reilly, N P ... G0VOK
Reilly, P T ... G4VSX
Reilly, R ... G4VYL
Reilly, S A ... 2E1CDB
Reilly, V ... GM0FWJ
Reilly-Cooper, G ... G7SUF
Rekers, G J A ... G0UQO
Remedios, Y A G ... G4UDT
Remmert, H P ... G4XEG
Remnant, D ... G7LXP
Remnant, M J ... G4YYM
Remsbury, A A ... G3OLH
Renaut, J ... G8DJL
Rendall, D ... GM0MHS
Rendall, M ... G0VXW
Rendell, F B ... G4HXK
Rendell, F J ... G8NQR
Rendell, P J ... G6TJZ
Renggli, G P ... G0GID
Renner M.I.S.M., B T ... G4YEF
Rennick, P J ... GW4IZJ
Rennie, A ... G0VYS
Rennie, E R W ... G4ZMS
Rennie, G W J ... GM3OUU
Rennie, J ... GM4UWX
Rennie, J G ... G0BSY
Rennie, J L ... GM4LFL
Rennie, J M ... G3XCP
Rennie, T R ... G0WDL
Rennison, A ... G8LIR
Rennison, A N ... G3GSL
Rennison, K J ... G0CEX
Rennison, P ... G1YZI
Rennison, R ... G1YZJ
Rennison, W G ... G3BOK
Renouf, R L ... GU0UJC
Renshaw, C ... G2AQN
Renshaw, J ... G7VDW
Renshaw, J L ... G7UMU
Renshaw, P J ... G8VOH
Renshaw, T M ... G0WOM
Renton, A ... G7VVX
Renton, G W ... G7DEU
Renton, K R ... G6JMJ
Renvoize, P J ... G0JKM
Renwick, A C ... GM0PMW
Reoch, G A ... GM4LHA
Retter, A S ... G4JBG
Retter, J W C P ... G6LCU
Revan, R L ... G6CXY
Revell, D R ... G3MGZ
Revell, P ... G3IOB
Revell, S ... G3PMJ
Revill, B ... G6GJX
Revill, P B ... G4EFV
Revill, R ... G4GKZ
Rew, N P ... G8GZZ
Rew, R ... G3HAZ
Rewaj, F T ... G4TAZ
Reynard, G ... G0EUS
Reynard, T ... G7HJT
Reynolds, A ... G6ZTF
Reynolds, A D ... M1ANG
Reynolds, A J ... G3NNK
Reynolds, A J ... G4XYC
Reynolds, A W ... G8CEQ
Reynolds, B J ... G3ONR
Reynolds, C ... G8EQZ
Reynolds, C R ... G0UFP
Reynolds, C R ... GW3JPT
Reynolds, D F ... G4BPW
Reynolds, D L ... G7DPU
Reynolds, E ... M1BDD
Reynolds, E A ... G6DKE
Reynolds, E G ... G4YFF
Reynolds, G ... GM6KBG
Reynolds, G J ... GW0GDK
Reynolds, G J R ... G1XIV
Reynolds, G R ... G7BZQ
Reynolds, G R ... G7SRG
Reynolds, I H ... G1FAN
Reynolds, I H ... G4RRD
Reynolds, J A ... G3PTO
Reynolds, J K ... G4GTG
Reynolds, J R ... G3RSD
Reynolds, J R ... G6WEL
Reynolds, K L ... G7ESO
Reynolds, M C ... G6NLZ
Reynolds, M D ... G0AOS
Reynolds, M J ... G1PWW
Reynolds, M J ... G4ZMR
Reynolds, M J ... G7OUT
Reynolds, M J ... M0BCD
Reynolds, M M ... 2E1ERT
Reynolds, M T ... G4KXO
Reynolds, P C ... G4YGM
Reynolds, P N ... G3NOA
Reynolds, R ... G3IDW
Reynolds, R F S ... G3AVL
Reynolds, R J ... G0UFF
Reynolds, R J R ... G1GQI
Reynolds, R W J ... G4OMS
Reynolds, S ... GM4CAB
Reynolds, S E ... G7BNB
Reynolds, S J ... G3DYH
Reynolds, S J ... G6AHN
Reynolds, S P ... G8NVH
Reynolds, T A W ... G7IRW
Reynolds, V ... G4MVR
Reynolds, V ... G8BBC
Reynolds, V J ... G3COY
Reynolds, V J ... G4ATC
Reynolds, W A ... G8TJH
Rhenius, C M ... G1VJG
Rhind-Tutt, M A ... G4BSK
Rhodes, A ... G1XWD
Rhodes, A K ... G1LVV
Rhodes, B ... 2E1FEG
Rhodes, B ... G4ZVP
Rhodes, D ... G3LUP
Rhodes, D ... G6LST
Rhodes, F ... G3TWO
Rhodes, F ... G6AHQ
Rhodes, F R ... G4WNF
Rhodes, G ... G0LKP
Rhodes, G D ... G7EUF
Rhodes, H T ... 2E1EEG
Rhodes, J B ... G7BRU
Rhodes, J M ... G0NGF
Rhodes, K ... G1CFB
Rhodes, K S ... 2E1FEF
Rhodes, M D O ... G3XZO
Rhodes, M F ... G0SVD
Rhodes, M J ... G0VPW
Rhodes, M R ... M1AHY
Rhodes, N H ... G3GMV
Rhodes, P ... G4DCM
Rhodes, P A ... G6PAR
Rhodes, P I ... G3XJP
Rhymes, D A ... G6AON
Rhynas, J W ... GM4NJD
Rhys, J B ... G3YMN
Rhys, M A ... G8KTC
Ricalton, A ... G6OTV
Ricalton, A ... G6OTW
Ricalton, W A ... G4ADD
Rice, A J ... G8GYH
Rice, E A ... G1SHG
Rice, F ... G7LPP
Rice, L ... G3TVC
Rice, M E ... G1ZBN
Rice, P ... G3WUB
Rice, P H ... 2E1FAJ
Rice, P H ... G6AYU
Rice, P S ... G6XPP
Rice, S C ... G4NEA
Rice, V D ... G6UER
Rice, W A ... G2AGR
Rich, A J E ... G1XWK
Rich, E P ... G4DYG
Rich, G C ... G4NIK
Rich, R M ... G7RTD
Richard, I R ... G3APO
Richard, M J ... MM0BAB
Richard, S R ... G1JAZ
Richards, A ... G8BHG
Richards, A ... GW3SFC
Richards, A B ... G7KWF
Richards, A J ... GW4RYK
Richards, A J ... GW7THP
Richards, B ... G4TLR
Richards, B J D ... G7VBN
Richards, B R ... G0VTT
Richards, C M ... 2E1CBU
Richards, C N ... G1JBB
Richards, C W C ... GW3JET
Richards, D ... G7EUT
Richards, D B ... GW0SEH
Richards, D F ... G7MBY
Richards, D J ... G1XGZ
Richards, D J ... G6UYM
Richards, D P ... G3XXM
Richards, E L ... G6ELR
Richards, F A ... G0CFA
Richards, F S ... G4TJU

Richards, F W M G8UPM
Richards, G A G0PWX
Richards, G E G1WJK
Richards, H E GI3DXU
Richards, J G0ALC
Richards, J G0KJM
Richards, J G7TEN
Richards, J A G6WKO
Richards, J A G7BZN
Richards, J C GW0AQR
Richards, J D G0OMI
Richards, J H C G3BPG
Richards, J K G6TZE
Richards, J K G7PPY
Richards, J L 2E1DZH
Richards, J W GW8GAB
Richards, K G6MGR
Richards, K C G4UZO
Richards, K J G3HSU
Richards, K J GW0HYU
Richards, L T G6XPS
Richards, L V M1AUB
Richards, M G4APF
Richards, M C G6FJI
Richards, M J G1MCX
Richards, M J G1RMN
Richards, M J G3WKF
Richards, M J G6WCI
Richards, M R G7VAB
Richards, M S G G0EIQ
Richards, N M G8TFY
Richards, P GW0AVD
Richards, P A G7WCQ
Richards, P G M0AIP
Richards, P J G0DCZ
Richards, P J G6XFC
Richards, P J G8ASC
Richards, P P G1DUO
Richards, R S 2E1CAH
Richards, R T G4IRD
Richards, S A G4LTY
Richards, S D G4OAK
Richards, S D G6DUB
Richards, S D G6SGR
Richards, S D GW0LST
Richards, S D GW0PPG
Richards, S D GW4SGR
Richards, S D GW7BTC
Richards, S J G4HFT
Richards, S M G4OAJ
Richards, S M G6XHF
Richards, S P G4HPE
Richards, T G0SII
Richards, T C L G4JWU
Richards, T M 2E1BRZ
Richards, T M G7VAP
Richards, V G0GAW
Richardson, A G4CGD
Richardson, A E 2E1EPX
Richardson, A R G2CXT
Richardson, A T G4ZEB
Richardson, A W G4WMQ
Richardson, B G0WVE
Richardson, B I G8YWT
Richardson, C G0FXT
Richardson, C A G0AHD
Richardson, C D G4OCZ
Richardson, C E G4LQG
Richardson, C J G6VKI
Richardson, C J H G4DPD
Richardson, C K G3NAE
Richardson, C R G8DXJ
Richardson, D G3WHF
Richardson, D A G1OEN
Richardson, D E 2W1EPN
Richardson, D F G4GED
Richardson, D J G0AWZ
Richardson, D J G6KUN
Richardson, D K G6MRK
Richardson, D R G6MGN
Richardson, D T G1ERC
Richardson, D W G4ZGQ
Richardson, D W E G4TST
Richardson, E G7SEP
Richardson, E M0BAF
Richardson, F M1BFV
Richardson, G G4RVY
Richardson, G A GI1VAZ
Richardson, G B G1SHN
Richardson, G C G0ESJ
Richardson, G D G1IFA
Richardson, G E G7GGG
Richardson, G J G7HCR
Richardson, G K G1MDF
Richardson, H R G7AYF
Richardson, H S GD0JWR
Richardson, H W G0CZU
Richardson, I C G1HQK
Richardson, I T G3XLP
Richardson, J G0RZC
Richardson, J G6JGR
Richardson, J G8IAJ
Richardson, J A G0WYQ
Richardson, J F G4BZQ
Richardson, J J 2E1AYB
Richardson, J J G7PFT
Richardson, J R G7OHI
Richardson, J W M1BPM
Richardson, K G8RFF
Richardson, K A G7AAY
Richardson, K D G3SPV
Richardson, K M G0PEK
Richardson, L O G3FWU
Richardson, M G7VYA
Richardson, M G7VYH
Richardson, M B G1IYN
Richardson, M B M0ARX
Richardson, M C G3LLN
Richardson, M S G8IUM
Richardson, M W G4XMR
Richardson, M W G6IMV
Richardson, N G3ZPL
Richardson, N A G0PUT
Richardson, N H M G0AES
Richardson, N M R G4SFH
Richardson, P A G8GBE
Richardson, P A J G8NGJ
Richardson, P E H G8ETV
Richardson, P H G8MLA
Richardson, P J G4IBZ
Richardson, P J G4UBR
Richardson, R G3XMB
Richardson, R G4PKZ
Richardson, R C G4MQW
Richardson, R I G3KXT
Richardson, R J G3WRD
Richardson, R P 2E1CJC
Richardson, R W S G8TBR
Richardson, S G7OXB
Richardson, S D G4WCP
Richardson, S E GW0AWT
Richardson, S J GW7NJQ
Richardson, S P G4JCC
Richardson, S R G4URG
Richens, K G G3ZGU
Richer, M H G3UYE
Riches, A G4GMD
Riches, C G3UBP
Riches, D E G0XEG
Riches, D E G8RJY
Riches, D M G7VUM
Riches, M H G4MEG
Riches, M S G1TXO
Riches, S E G7KWS
Richey, E E GI8LDM
Richings, A R G3YNY
Richman, C GM7GTS
Richmond, A P 2E1AXD
Richmond, C C S G3WPR
Richmond, I W G8CLJ
Richmond, J L G1KFH
Richmond, L M 2E1AXE
Richmond, P C G7ECK
Richmond, P M G8GVV
Richmond, R W G4NUV
Richmond, S J G1DCY
Richter, H A G4OAI
Richter, H A G4WAR
Richter, T W G7HQU
Rick, A G4NTJ
Rickaby, J E G3EDT
Rickard, D J G3BEQ
Rickard, J G0TWJ
Rickard, R D K G4KNI
Rickard, T N G8WQT
Rickard-Worth, M D G0SPZ
Rickatson, J M G6MHK
Rickerby, C G1NWA
Rickerby, C G7JOA
Rickerby, D A P G7LIC
Rickers, A R G1XSI
Rickers, D GW3HEU
Rickett, E H G3PV
Ricketts, D G4LYR
Ricketts, J C G7HLT
Ricketts, R GW7AGG
Ricketts, R H GW3VGY
Rickman, A P 2E1FCF
Rickman, S A C G0JSR
Rickson, J M G4AFB
Rickward, A T G4ZZN
Rickwood, J G3JJR
Ridd, J G8BQX
Ridd, N P G7MOK
Riddell, A GM6AQB
Riddell, A M G0TGO
Riddell, A S GM4BFC
Riddell, C R G0NIC
Riddell, G T G7IKB
Riddell, P G7HIN
Riddell, S GM3YCB
Ridden, D G G6GXG
Ridden, M L G1JOG
Riddick, D J G0LZW
Riddington, R G4IHT
Riddiough, R GM4DWS
Riddiough, R J GM4SQO
Riddle, D G7WKB
Riddle, D P G4BMR
Riddle, S J 2E1EYI
Riddoch, J D G1TSL
Rideout, A B G1AKD
Rideout, R A G4OVI
Rider, A P G6GLP
Rider, B K G1KTY
Rider, J C B G4FLQ
Rider, W J G3MKW
Ridgard, S C G1JIG
Ridge, H J G3HFS
Ridgeon, S G M1AVL
Ridgeway, G M G7FZB
Ridgeway, G R G8UYD
Ridgway, A J G4VRA
Ridgway, C G0JXS
Ridgway, G R M1AGU
Ridgway, I A G0KPB
Ridgway, S G G3TZQ
Ridgwell, K E D G6XHI
Ridgwell, R C G4COW
Riding, G G8LLC
Riding, N J G6WGE
Riding, R J G3JZG
Ridler, A V G1ZJU
Ridley, A G3ZLR
Ridley, C J G0NLM
Ridley, C J G8GKC
Ridley, C S G1BLK
Ridley, D T G6OSH
Ridley, P D G7NYF
Ridley, P S G6PLB
Ridley, R A G3UTX
Ridout, P N G2DKI
Ridsdale, P A 2E1FIL
Ridyard, D N G6IMY
Riebold, P A G8XHD
Rigby, A G0SGS
Rigby, B G8UVN
Rigby, C J G1CFG
Rigby, D C G4KXV
Rigby, D C G4SKE
Rigby, D C G8ZKE
Rigby, E G6GVZ
Rigby, G A G8MPG
Rigby, G P G3KTJ
Rigby, H A G0MEQ
Rigby, J G4RXZ
Rigby, J G G8XNL
Rigby, J S N G8XLI
Rigby, K G3TRE
Rigby, M G0BHG
Rigby, M A G7JVD
Rigby, M J G4FUI
Rigby, N G7IFO
Rigby, P A G8LMF
Rigby, P W G0PXI
Rigby, R E G6IDZ
Rigby, S J G0AXV
Rigelsford, K G G3XXC
Rigg, G G6UXL
Rigg, M R G0EUP
Rigg, P F G0TVB
Rigg, P F G6TRG
Riggott, P M G4XGN
Riggott, T J G4MFU
Riggs, C A L G8FQC
Riggs, C C G0LOW
Riggs, C C G3XAS
Riggs, D J G8RYP
Riggs, J F W G4KDK
Riggs, J R G8PVR
Riggs, S P G6PDR
Rigley, M W G7NVH
Rignall, M W G3OYX
Rigney, J K G8JKR
Riley, A G6MXT
Riley, A B G7CQW
Riley, A C G4DQG
Riley, A D G8JON
Riley, C E G4JQX
Riley, C F G4RQX
Riley, D G1JDW
Riley, D M1ARF
Riley, E G0JDA
Riley, G 2E1FKE
Riley, J G0IEJ
Riley, J G3ZKG
Riley, J G6UKQ
Riley, J A G0RPG
Riley, J A G6OIB
Riley, J D 2E1EHX
Riley, J D M1CEJ
Riley, J D C M1BSN
Riley, J J A G3TZA
Riley, J N G1MSK
Riley, J P M G7BRA
Riley, J W G6ZNF
Riley, K 2E1CKH
Riley, K G3JAS
Riley, K G7UKN
Riley, K G7VPE
Riley, K J G8XLZ
Riley, M 2E1ERF
Riley, M M1BOP
Riley, M R G4CSZ
Riley, P L C G0KTT
Riley, R G4NQZ
Riley, R T G0JFX
Riley, R W N G4VPI
Riley, S G0MUH
Riley, T D G8UQR
Riley, V J GW3XET
Riley, W G0HBN
Riley, W A G7NWM
Riley, W V G0CZL
Riley-Moxon, C E G7CER
Rimell, S B G1AIG
Rimington, J P G6IFH
Riminton, P G4HGG
Rimmer, A A GM4XRY
Rimmer, B R G0JCQ
Rimmer, D G4VYP
Rimmer, H G0DVM
Rimmer, H G4INV
Rimmer, J B GM3AKK
Rimmer, J I G1KIL
Rimmer, J T G1DQZ
Rimmer, K G4TDA
Rimmer, L T G0BRZ
Rimmer, M G G3KDA
Rimmer, N F GD4HOO
Rimmer, P G G4YVI
Rimmer, P J G7REB
Rimmer, R A G3RQS
Rimmer, R C GD3YEO
Rimmer, R I G3KCE
Ring, G E G4YSE
Ring, G E G6TAL
Ringrose, B G7WLJ
Ringrose, J W G3EYU
Ringrose, R J G4KYU
Ripley, A J P G7AKD
Ripley, D M 2E1FVY
Ripley, G W GD3AHV
Ripley, S P G4AVK
Rippin, G R G0EPQ
Rippin, M T 2E1DMH
Rippin, R G M0AFV
Rippington, M D G0IEV
Risbridger, N E G1GQK
Risdale, K H 2E1ESL
Riseborough, D B G7UFV
Rish, B M GW0SFP
Risley, S G3VLR
Ristic, B GW0DHG
Ritchie, A C G4VMX
Ritchie, A L GM3WYL
Ritchie, B C G6UHL
Ritchie, C C GM7KSC
Ritchie, G MM1AXK
Ritchie, H E G0TKN
Ritchie, J GM0BTL
Ritchie, J G GM0RSI
Ritchie, J M GM0GYT
Ritchie, M D GM8ADK
Ritchie, M J 2E0AGM
Ritchie, M P G6NLG
Ritchie, P R GM3OYV
Ritchie, T M GM0GSL
Ritchie, W J G0KJJ
Ritchie, W J GM3IWX
Ritson, D G7DBV
Ritson, F J U G5RI
Ritson, J G6IFI
Ritson, K G0PKR
Ritson, M G0MIK
Ritson, M G0VRT
Ritson, V C G7FAD
Ritzema, K G0MLP
Rive, G P G3UVP
Rivers, B A G4WGX
Rivers, B L A G1FHO
Rivers, H W O G6NLE
Rivers, I A G6FDG
Rivers, J G7MAR
Rivers, J R G0GCQ
Rivers, J S G0LZF
Rivers, M J G0VIM
Rivers, M J G0WFN
Rivers, P G7PSC
Rivers, P D G4XEX
Rivers-Bland, L G3GJU
Rivett, I G G8WPU
Rivett, N B G3OXS
Rix, L G G3XJW
Rix, R E G6UEQ
Rix Cbe DL, B N R G2DQU
Rixon, A J 2E1FOC
Rixon, A J G4JIR
Rixon, B H A G3IOJ
Rixon, E J G6XHG
Rixon, J C G0COH
Rixon, R N E G6NXM
Rizzo, C A E G7ISD
Roach, A GM4FKO
Roach, J M G4ETO
Roach, M H G3TWJ
Roadnight, D M G7IEF
Roake, P W E G0KDN
Robb, D GM3ACL
Robb, G I R GM8KXF
Robb, J G4WMN
Robb, R Y A GM3BXX
Robb, W A G3THU
Robbins, B G G4TAE
Robbins, B P G0CFM
Robbins, D J G8DKF
Robbins, G E G0IRR
Robbins, G H G3LNG
Robbins, K GW3PFV
Robbins, M J G4RCS
Robbins, M J G4TXD
Robbins, R J G0MDM
Robe, W J G7HES
Robelou, M C G8YUR
Roberson, C J G7OOK
Roberson, N G0UMM
Roberts, A G3VKH
Roberts, A G4XBZ
Roberts, A G7VCC
Roberts, A GW4DTU
Roberts, A GW6BMP
Roberts, A C G1YPH
Roberts, A C G8OEF
Roberts, A C M0AGZ
Roberts, A D G0FMX
Roberts, A E G3RUR
Roberts, A E G7EMD
Roberts, A F G4ZIB
Roberts, A J G7BAV
Roberts, A J G7JDA
Roberts, A J GW8COJ
Roberts, A K G1PTH
Roberts, A P G6AZE
Roberts, B A G4DBQ
Roberts, B D G4ZCP
Roberts, B F G4VYG
Roberts, B W GW0HGC
Roberts, C GW4YJT
Roberts, C H G1CXQ
Roberts, C J G0CNG
Roberts, C J G4EVA
Roberts, C J GW4COJ
Roberts, C L G0FBO
Roberts, C L G0SFC
Roberts, C M 2E1FPG
Roberts, C M G4ZFJ
Roberts, C P GW0OLP
Roberts, C W G4EYU
Roberts, D G0TFI
Roberts, D G1DUS
Roberts, D G3NSZ
Roberts, D G4GSR
Roberts, D G4OLQ
Roberts, D GW0CSR
Roberts, D A G8KBB
Roberts, D A W GI7DIT
Roberts, D D G1UCI
Roberts, D D GW0GHG
Roberts, D E G4MFI
Roberts, D E G8UDG
Roberts, D E GW0ABL
Roberts, D E J 2E1BPV
Roberts, D F GW8NZN
Roberts, D G G0GWC
Roberts, D G G7NJF
Roberts, D G GW0VOG
Roberts, D H G0DRO
Roberts, D J G0JBT
Roberts, D K G7MYM
Roberts, D L G3KDN
Roberts, D M G1ZMJ
Roberts, D P G0RKB
Roberts, D R G3UBV
Roberts, D R G6FKR
Roberts, E G0TKL
Roberts, E G G1TXQ
Roberts, E H G7TVL
Roberts, E J G1AJV
Roberts, E J G7PHQ
Roberts, F GW3WPE
Roberts, F D G4UYM
Roberts, F J G6EVA
Roberts, F N GI4TAN
Roberts, G G0HUK
Roberts, G G0RYA
Roberts, G G6DVH
Roberts, G C MW0ATI
Roberts, G C MW1ANU
Roberts, G L GW7RLZ
Roberts, G R G6IFI
Roberts, G W GW1MPR
Roberts, G W GW4JXN
Roberts, H G8VOT
Roberts, H B G4XGF
Roberts, H G GW4CNM
Roberts, H H G6BDS
Roberts, H M GW6JWL
Roberts, H W O G6NLE
Roberts, I G4ASH
Roberts, I G6CYH
Roberts, I G M G8OHW
Roberts, I G M M0BFO
Roberts, I W G0IYX
Roberts, I W GW0IJY
Roberts, J G0LLL
Roberts, J G0PVU
Roberts, J G1DKC
Roberts, J G3OTX
Roberts, J G7OCJ
Roberts, J A G6DFH
Roberts, J A G6OIY
Roberts, J C G4ZMM
Roberts, J C GW0TBT
Roberts, J D G4UQK
Roberts, J E G0FJM
Roberts, J E G1XYT
Roberts, J F GW4WXM
Roberts, J G G0ZPV
Roberts, J I G7RZA
Roberts, J L GW0HNK
Roberts, J M J G6OIX
Roberts, J P G6OSJ
Roberts, J S G8FDJ
Roberts, J T G6NLI
Roberts, J W G3OKX
Roberts, K B G8SCZ
Roberts, K E G3DKR
Roberts, K J 2E1FPF
Roberts, K R GW1HHM
Roberts, K R GW1NGL
Roberts, K W G8VDP
Roberts, L G3EGX
Roberts, L G8III
Roberts, L E GW0MMW
Roberts, L J G4PIA
Roberts, L S G0OLR
Roberts, M E G0HZB
Roberts, M E G6MER
Roberts, M G 2E1FZD
Roberts, M J G0RUH
Roberts, M J G1ZLA
Roberts, M W R GW0HUN
Roberts, N G G4JZK
Roberts, N M G6TZO
Roberts, N P G4KZZ
Roberts, O C GW6DHC
Roberts, O P C GW4UWD
Roberts, O W 2E1BRJ
Roberts, P G0XBI
Roberts, P M1BWF
Roberts, P MW0AMO
Roberts, P B G7SJS
Roberts, P D A GW6AYM
Roberts, P E J G0JPE
Roberts, P I C GW4WSU
Roberts, P J G0OER
Roberts, P J G4DJB
Roberts, P J M GW4VPK
Roberts, P M G7FRL
Roberts, P N G6KVR
Roberts, P W G4KKN
Roberts, R G0WCU
Roberts, R G3MAK
Roberts, R G8VUN
Roberts, R GW7HEE
Roberts, R A G7HGI
Roberts, R E G3TAR
Roberts, R G G8ZKG
Roberts, R G GW4IGT
Roberts, R J G7DSQ
Roberts, R J G7SZZ
Roberts, R J GW4TYH
Roberts, R M 2E1AZW
Roberts, R S G7NQV
Roberts, R W G3CDY
Roberts, R W GW1CAV
Roberts, R W MW1BGE
Roberts, S G3AQX
Roberts, S G4OSR
Roberts, S G7UEU
Roberts, S GM4HWS
Roberts, S GW3EOP
Roberts, S GW4NZ
Roberts, S A G6ZTC
Roberts, S C G1YRY
Roberts, S D GW6BMR
Roberts, S E G0TFX
Roberts, S G G6ZMD
Roberts, S J G1SWF
Roberts, S M 2E1FGF
Roberts, S M M1BYN
Roberts, S O GW0KLC
Roberts, T G0GJN
Roberts, T G6NLH
Roberts, T GM0OAV
Roberts, T GW7PQG
Roberts, T A GW0WHT
Roberts, T D M1AUW
Roberts, T M G6IDW
Roberts, V D GW7PBP
Roberts, V E GM1ZNR
Roberts, W G3LOE
Roberts, W A G0HIJ
Roberts, W A G2RO
Roberts, W E G3GXQ
Roberts, W F G0CGA
Roberts, W G G7DDQ
Roberts, W J G1ORD
Roberts, W J G3GZQ
Roberts, W L G1TES
Robertshaw, N H G0NHM
Robertson, A GM3ZXB
Robertson, A M1BNU
Robertson, A M GM4IAO
Robertson, B GM4OIJ
Robertson, B C G4RJO
Robertson, B S G4POL
Robertson, C GM0HBK
Robertson, D G0HYN
Robertson, D GM3JDR
Robertson, D A GM1TDT
Robertson, D J 2M1ATN
Robertson, D J G0MSU
Robertson, D J GI7UGP
Robertson, D S GM0MWJ
Robertson, F 2E1FVZ
Robertson, G J G6KQN
Robertson, G J M1AGO
Robertson, G R B GM3KSD
Robertson, H GM3RQQ
Robertson, J GM0KYU
Robertson, J A G0KJU
Robertson, J A GM7REG
Robertson, J C GM6GPH
Robertson, J C M1BKN
Robertson, J G M1AKG
Robertson, J H G3ZOR
Robertson, J W G0PJV
Robertson, J W G1DUT
Robertson, K C GM1PWL
Robertson, K J G7OKC
Robertson, K R G8LSE
Robertson, K W GW1JIE
Robertson, L H H G4WTL
Robertson, M G3USX
Robertson, M A G6CMS
Robertson, M E G0WTR
Robertson, M E G1HDX
Robertson, M F G7VCK
Robertson, M J G0IOR
Robertson, M M GM7IBM
Robertson, M P 2E1FIE
Robertson, N F G0ORG
Robertson, N L GM8EUG
Robertson, P GM4RAH
Robertson, P G G0RQN
Robertson, P O G4RIZ
Robertson, R GM0CEA
Robertson, S A G8OIY
Robertson, S C G7TIO
Robertson, S H 2E1CNG
Robertson, S J GI7VCR
Robertson, W B GM1SZM
Robertson-Mudie, F GM0HMM
Robey, A L G4RFN
Robins, A G1UUP
Robins, A M G3SVT
Robins, A M G3VLG
Robins, C J G6TZI
Robins, D A G0UHW
Robins, D S G1KTZ
Robins, F L A G3GVM
Robins, I GD0AVF
Robins, K L 2E1DUL
Robins, M C G1UWH
Robins, M D G4FYQ
Robins, P G G8BSK
Robins, V G0THX
Robinson, A G0GAR
Robinson, A G2SU
Robinson, A G3TQA
Robinson, A G3XG
Robinson, A G6RJ
Robinson, A A G0RLW
Robinson, A A G7NAT
Robinson, A A G7WFM
Robinson, A C G1KSE
Robinson, A D G4ZDB
Robinson, A E G2KF
Robinson, A E G8BOB
Robinson, A G G1FMA
Robinson, A G G1ZAY
Robinson, A H G0GNJ
Robinson, A J G0WFY
Robinson, A J P G3XEY
Robinson, A L 2E1CSC
Robinson, A L G3ZYQ
Robinson, A M G0WEK
Robinson, A P G0VPO
Robinson, A R G1ILV
Robinson, B G0FEC
Robinson, B G0LRN
Robinson, B G4YIA
Robinson, B A I G6OJA
Robinson, B J G3ZEB
Robinson, B L G6ZTD
Robinson, C G1EHY
Robinson, C MM1ATR
Robinson, C A G1KJG
Robinson, C A G2DAD
Robinson, C A G3UHQ
Robinson, C A G7WGH
Robinson, C C G3IAI
Robinson, C G GI0GCR
Robinson, C J M1BMR
Robinson, C M GM0RDC
Robinson, C S GI4JDX
Robinson, C T GI6OIR
Robinson, D G0MQH
Robinson, D G0NXF
Robinson, D G7EHY
Robinson, D GI0BGE
Robinson, D GI7MDK
Robinson, D A G3UQR
Robinson, D A G6AYX
Robinson, D A G8GBY
Robinson, D A G8RFL
Robinson, D H GW0CKK
Robinson, D J G1HQJ
Robinson, D J G3MQR
Robinson, D J GI4OSG
Robinson, D J W G8UYH
Robinson, D P G1NYZ
Robinson, D R GJ3YHU
Robinson, D W G3FMT
Robinson, D W L G2BTJ
Robinson, E G0WZE
Robinson, E G0ZER
Robinson, E A G6MGP
Robinson, E G G4MET
Robinson, E J G6OSK
Robinson, E J G7VIT
Robinson, E V G3TWT
Robinson, E W C G1AIB
Robinson, F G0EVO
Robinson, F G3FLQ
Robinson, F G4EEQ
Robinson, F A G3TPV
Robinson, F E G8REQ
Robinson, F W G4YIJ
Robinson, G 2M1FQI
Robinson, G C G6TDM
Robinson, G D G0PMT
Robinson, G E G4ZMH
Robinson, G G G4IZB
Robinson, G N G4AKW
Robinson, G T G4KXU
Robinson, G W G0LNS
Robinson, G W J GI4CSP
Robinson, H G3VVE
Robinson, H G7IHE
Robinson, H M A M G7PJZ
Robinson, I G8EGV
Robinson, I A G8WGE
Robinson, J G3PTJ
Robinson, J G4AZX
Robinson, J G4TSV
Robinson, J G6GCY
Robinson, J GM8RPE
Robinson, J A G3OOS
Robinson, J A G6UEC
Robinson, J F G0KXS
Robinson, J H G4BHO
Robinson, J K G3TBX
Robinson, J L G6CIO
Robinson, J M G3VHK
Robinson, J N G8PQJ
Robinson, J R G7SCL
Robinson, K G0LCE
Robinson, K G4KKZ
Robinson, K A G4CYO
Robinson, K C G4LVN
Robinson, K S 2W1BTV
Robinson, L G G0STA
Robinson, L H G0LHR
Robinson, L J G3MPO
Robinson, L M H GI4NSS
Robinson, M G4UUA
Robinson, M G7PQL
Robinson, M G8TLA
Robinson, M G C G3ZZM
Robinson, M J GI7NFB
Robinson, M L G6DHB
Robinson, M P G7KOY
Robinson, M W C 2E1EGW
Robinson, N GI4PES
Robinson, N A G4OIR
Robinson, N G GM8KCK
Robinson, P G1PLB
Robinson, P G3MGJ
Robinson, P G3MRX
Robinson, P G4IZH
Robinson, P G4UEA
Robinson, P M1ATG
Robinson, P A 2I1EXU
Robinson, P A G0SDC
Robinson, P A G0VZI
Robinson, P G G0EYR
Robinson, P H G8FNG
Robinson, P J 2E1BLN
Robinson, P J G0MQJ
Robinson, P J G4CHI
Robinson, P J G8MSQ
Robinson, P R GI0MTE
Robinson, P S G6WZE
Robinson, P S T J G0DVC
Robinson, P W G6HGT
Robinson, R G0JEO
Robinson, R G1AIF
Robinson, R G1JUD
Robinson, R G6YGJ
Robinson, R A C G0LEX
Robinson, R D MI1BSK
Robinson, R E G8BWH
Robinson, R G G6FJH
Robinson, R H MI1BIW
Robinson, R J 2E1FMP
Robinson, R S G1FXT
Robinson, S G4HFJ
Robinson, S G4NSH
Robinson, S A GI8NBW
Robinson, S G G0BHO
Robinson, S G G6FKS
Robinson, S G G8POO
Robinson, S G GI0URI
Robinson, S J GW0WMT
Robinson, S J M1BJE
Robinson, S J R G0AIM
Robinson, S L F G0BQW
Robinson, S R G4WNV
Robinson, T G3WUX
Robinson, T M1ABR
Robinson, T E G4URX
Robinson, T J G6UUS
Robinson, T P G3KOA
Robinson, T R GI4RFH
Robinson, V K G4JTR
Robinson, W G0EDN
Robinson, W GW3XHJ
Robinson, W E G3LTV
Robley, J H G3TLG
Robotham, G L G8KLH
Robson, A G0DJO
Robson, A G GM8YIK
Robson, A J G4THI
Robson, A T G1FFL
Robson, B J GW6ZTG
Robson, B W GM6AYW
Robson, D G4UYN
Robson, D J F GM6CMQ
Robson, D M GM8ZKF
Robson, E R GD4IHA
Robson, J A W G3HMQ
Robson, J C G0NHJ

Robson, J M G4PZY
Robson, J M G7MQL
Robson, J M GM3CFS
Robson, J N G1SIO
Robson, J W G1YUL
Robson, K G0NHK
Robson, K G GM1JLP
Robson, K R G3VTY
Robson, M G8RGO
Robson, M M G1MMR
Robson, P G G3NZK
Robson, P S G3FYP
Robson, R G3IDY
Robson, R J GW8AGI
Robson, T G0VWW
Robson, T G4CGK
Robson, W I GM8IIO
Roby, G N G0CBD
Roche, A P G7WKG
Roche, F G0BPB
Roche, K A G8GOS
Roche, M J G3YJS
Rochester, K J G0LCS
Rochester, M G6ZCX
Rochester, M B G6ZCY
Rochford, T L G7RDQ
Rock, A D J G3KTR
Rock, G A G0NUM
Rock, N E G3RLY
Rock, N E G4DAR
Rock, T J G1HMI
Rocke, M J R G1YNJ
Rockett, E P G3SXQ
Rockett, R S G1WTU
Rockliffe, D J G7UMY
Rod, J G1RSE
Rodd, E W G1LTC
Rodda, J H G4VTN
Rodda, S W A G4PEM
Roddan, S J G3CSC
Rodden, K J G0ASI
Roddy, T G6OTQ
Roden, D J 2E1ATD
Roden, D S G3KXF
Roden, R G4GKO
Rodger, D C G1CHA
Rodgers, A G4ANL
Rodgers, A W G4RVS
Rodgers, C D S GM4NNC
Rodgers, C W G7OYN
Rodgers, D H G8UYI
Rodgers, E M GM1MDH
Rodgers, G GI6ATD
Rodgers, I G G7RBC
Rodgers, J G1KOD
Rodgers, K G3SRZ
Rodgers, M G0GDL
Rodgers, M G0OCG
Rodgers, M G7BQS
Rodgers, M I A G7BOJ
Rodgers, R G1UKH
Rodgers, R R GM1SMF
Rodgers, W S G1UOR
Rodgerson, K G1CBG
Rodinson, J G7UZA
Rodley, R G G1TUS
Rodman, D S G0TVW
Rodman, H D G3GJZ
Rodmell, P G3ZRS
Rodway, J E G4FRK
Rodway, P A G3UFF
Rodwell, G D G4GLL
Rodwell, I E G3TZP
Roe, A W G0TJF
Roe, B G F G4LVR
Roe, B G F G4RRM
Roe, B G F G6RRM
Roe, B J G6XFB
Roe, C E G1ZWY
Roe, D G1GNP
Roe, D H G8AFG
Roe, F A GM0ALS
Roe, G H G4HCF
Roe, N D G4ACW
Roe, N J G1DHY
Roe, P G0DLJ
Roe, W A C GW4SXN
Roe, W R G6BPY
Roebuck, D G0LJM
Roebuck, D G8EMH
Roebuck, I G0SVX
Roebuck, K G7IOF
Roedel, T C G7VZG
Roeschlaub, R F G1NBY
Roff, G A G3TJI
Roff, J C 2E1EMK
Roffey, A T G1MMZ
Roffey, D A G7OVO
Roffey, G J G7BIP
Roffey, L J G0GKE
Roger, W M GM3JDX
Rogers, A G0KHH
Rogers, A G4UJB
Rogers, A B G6HTZ
Rogers, A E G8MVX
Rogers, A J G4CZL
Rogers, A L G2FQD
Rogers, A W G G3SYZ
Rogers, B G G6YGI
Rogers, C G4JNJ
Rogers, C C G3KEG
Rogers, C C GM7TAN
Rogers, C L GW0PYN
Rogers, C R G0CNL
Rogers, C R G3LYV
Rogers, C R G4MXB
Rogers, C V R G1YAB
Rogers, C W G1WAE
Rogers, D GW3UOO
Rogers, D B G6WZP
Rogers, D F G1PIF
Rogers, D H G4XJL
Rogers, D P G4VID
Rogers, D W J G1LLW
Rogers, E D G3MWN
Rogers, E D G4BTS
Rogers, E H G4BRG
Rogers, E J H G1OMZ
Rogers, E W G0CLC
Rogers, F G0FZN
Rogers, F A G0NXE
Rogers, F R G0CXE
Rogers, G G8ABB
Rogers, G E G3TFL
Rogers, G H G6OMN
Rogers, G T GW0RJV
Rogers, H J G1ATQ
Rogers, H T G3NHR
Rogers, I G6LGM
Rogers, I A 2E1FFO
Rogers, I A G7VIX
Rogers, I A M1BYO
Rogers, I P G0BON
Rogers, J G0LAK
Rogers, J G4SME
Rogers, J A 2E1DGJ
Rogers, J F G3CIF
Rogers, J G G1PJZ
Rogers, J L G4ATR
Rogers, J N E G8RJH
Rogers, J P F O G3PQA
Rogers, J T G3EMG
Rogers, K G4TZL
Rogers, K A H G3AIU
Rogers, K M G8ZUF
Rogers, L G0NQB
Rogers, L A G1LAR
Rogers, M A R K G4RGB
Rogers, M J G0UPY
Rogers, M T 2E1DUB
Rogers, N K G1UCD
Rogers, N K W G0JZF
Rogers, N M G0GGG
Rogers, N V G6BHE
Rogers, O W G1FXD
Rogers, P G4ZMO
Rogers, P J 2E1FGD
Rogers, P J G6FMN
Rogers, P L P G0SNW
Rogers, P R G6KSU
Rogers, R B G3CWH
Rogers, R C M0BBR
Rogers, R D G8TLH
Rogers, R G G0BYV
Rogers, R J G3LIA
Rogers, R J G4LLV
Rogers, R J G8IJP
Rogers, R M E G8PPF
Rogers, R S 2M1AHZ
Rogers, S D 2E1FCA
Rogers, S G N G4TAO
Rogers, S J G1ERF
Rogers, S M G4UUH
Rogers, S P G1BHT
Rogers, S P GW4HER
Rogers, T E G0MND
Rogers, T G 2W0AQD
Rogers, T G 2W1FJH
Rogers, T J G4YSQ
Rogers, T P G6TZT
Rogers, W H A GM0CII
Rogers Jones, D G0KLT
Rogerson, F B S G3RBR
Rogerson, J G0GEQ
Rogerson, R A G3IUJ
Rogerson, W G3LXY
Rogister, F F E G0KYQ
Rohowsky, J V J G0EDI
Rohrer, C C G7VJE
Rohrlach, L G G1XPB
Rohsler, N B G4ZTM
Rolf, G L G4RTW
Rolfe, H F G0BSF
Rolfe, M C G0EXO
Rolfe, P C G4UAS
Rolfe, P G G4CBE
Rolfe, S GW0ETF
Rolinson, C P G0MLY
Rolinson, S E 2E1ATE
Rolland, G W G8AKB
Rollason, A J A G6DHD
Rollason, J A G3WCO
Rollason, J E G4VTK
Rollason, M L J G4PAH
Rollason, N G A G4NRR
Rolle, G J G7OZM
Rollin, A J GW7PUX
Rollin, P G G4AFU
Rollings, M J G4XLG
Rollins, W J G1WJR
Rollitt, A M G6HTY
Rollitt, D G3XYP
Rollo, D R GM3GRG
Rollo, L D GM4HFM
Rolls, A M G8UHK
Rolph, D G0UYC
Rolph, D D G3DXD
Romang, K J G4SKN
Romano, A F GW1PFK
Romano, C GW4TGF
Rome, J E G8BPH
Romer, H D G3CIK
Roney, D G G1RKM
Roney, J G4FYP
Roney, V G4CJK
Roobottom, E L GW0WUM
Rook, R G1OZP
Rookard, D W G1NGO
Rooke, D G0OEW
Rooke, L G4DDE
Rooker, S A G4UUI
Rooks, L D G3PUO
Roomes, D G G0RQL
Rooney, C J G3VCR
Rooney, G G3MKH
Rooney, J M GM1TDU
Rooney, P J G0AIU
Roost, B D G4PRK
Roots, R H G1FHR
Roots, T P G1PPB
Roper, B G G6WMT
Roper, C C G4TIV
Roper, C W G4FPB
Roper, D E G4ZFD
Roper, G B G8AKM
Roper, G F W G0AEP
Roper, I 2E1FBY
Roper, I M G6XHK
Roper, L G0LYN
Roper, M C G0TCO
Roper, M J G8NTQ
Roper, M P G1WTS
Roper, P A G0PRR
Roper, P A G3MII
Roper, R G G1SVD
Roper, S J G7MUE
Roper, S N MI0AEC
Roper, S P G8MXZ
Ropper, I D GM0UHC
Rosamond, P G4LHI
Rosborough, R J GI8PFB
Roscoe, B G0CZN
Roscoe, J B G4QK
Roscoe, R W G4MYZ
Roscoe, T G1UVR
Roscrow, W J G3ELW
Rose, A G1IFF
Rose, A GW6VEN
Rose, A A G0TNY
Rose, A C 2E0ACU
Rose, A D G4LXV
Rose, A J 2E1AJO
Rose, A J GM3WED
Rose, A W G8PEO
Rose, B G3WWO
Rose, C G8MKE
Rose, C E 2E1FHH
Rose, C J G7WCF
Rose, C M G1BUY
Rose, C M G4OOJ
Rose, D G G7BPF
Rose, D K G1KKE
Rose, D M G7MLK
Rose, D T G0PYS
Rose, E J G3NC
Rose, E P G1CPM
Rose, E W G G0FJL
Rose, G M G4EDH
Rose, G S G3WGF
Rose, I D G6IDU
Rose, I K G0HDZ
Rose, J G3OGE
Rose, J A G0IEW
Rose, J J G0ORU
Rose, J M M1BZH
Rose, K W G0DZV
Rose, L M G4KAB
Rose, M G7SRN
Rose, M D G3VPA
Rose, M E G3XKU
Rose, M R G8PAG
Rose, P 2E1FAE
Rose, P F G0BCS
Rose, P G G4STO
Rose, P G G5FZ
Rose, P G G6COL
Rose, P J G4YQV
Rose, P J G7NRN
Rose, P J G7TDD
Rose, P L G4NXR
Rose, P V G3ZZA
Rose, R 2E1FAC
Rose, R A G1FMC
Rose, R D G3SGM
Rose, S E 2E1FLS
Rose, S G G7PMG
Rose, S S G4HDD
Rose, T M G4OXD
Rose, U J D M1BDW
Rose, W GM0KWR
Roseaman, D L G8FLL
Rosen, L M1AYY
Rosen, P R G3YFW
Rosenberg, I G4XEW
Rosenthal, M G3NLW
Roser, G A G7LJQ
Rosevear, D C G4RLN
Rosevear, I G3GKC
Rosevear, M G7NWG
Rosewall, C T G8SCY
Rosewarn, D G1FED
Rosher, D M0BGG
Rosier, A W G1FLY
Rosier, K E G3DJK
Rosindale, J R G0GUO
Rosling, J H G4JIZ
Ross, A 2M1AIR
Ross, A G7ORJ
Ross, A C GM3VVF
Ross, A H G1SQB
Ross, A W GM4PMT
Ross, B GW0PJF
Ross, C D GM4UFP
Ross, C D GM8HBY
Ross, D G0VHO
Ross, D G7NXV
Ross, D GI4SNA
Ross, D GM0IQD
Ross, D A G1ULR
Ross, D A GM4PVQ
Ross, D B G0HIW
Ross, D B G4LOO
Ross, D J G1DBR
Ross, E G3TJC
Ross, E H GM3LWS
Ross, F S G1MYO
Ross, G A G4IEI
Ross, G W G7FWW
Ross, I A G GM1BLM
Ross, I D G1WZB
Ross, I E GI0UTV
Ross, I J GM0IJR
Ross, I J J GM0TGE
Ross, I M GM3TMK
Ross, I S G4DPF
Ross, J A G3WWG
Ross, J A GM1BSG
Ross, J D G0LBO
Ross, J W M0AWI
Ross, L GW0UHA
Ross, L W G8MWR
Ross, M D 2E1EIC
Ross, M H G0DMR
Ross, P G4KVE
Ross, R G1TJL
Ross, R N G1BIM
Ross, R P G1ZGM
Ross, S R G6HUH
Ross, T GM4YWI
Ross, W G7PRB
Ross, W J M 2M1DST
Rosser, D M GW3ZHQ
Rosser, W J R G7RRN
Rosser, W S C G8UHM
Rossi, S G0SHT
Rossiter, D J G7VVF
Rossiter, N S G6IQQ
Rossmann, W D GM0BTK
Ross-Fraser, W G1YED
Rostron, A G4ZEF
Rothera, F B G7IZY
Rotherham, J H 2E1CEP
Rothery, R G1UVD
Rothery, R J GW3RJR
Rothwell, A G3ZED
Rothwell, D J G4YPH
Rothwell, M E G4KFT
Rothwell, R A G7JDK
Rothwell, W J G0VDE
Rouget, P R G4RDM
Rough, J M L GW7GKX
Roughton, J T G4BAI
Roullier, F T G7TAX
Roulstone, R M G0SOM
Roulstone, R M G7IQV
Rounce, R H G4UKZ
Round, D M E G0BTA
Round, G G1VXT
Round, J G6LDA
Round, M J G4THE
Round, R H G8VST
Round, T D G0TDR
Rourke, B R GM0EKY
Rourke, C J GI3IVJ
Rourke, C P G8RYY
Rous, R F G0WUZ
Rouse, A G8EWC
Rouse, C R G4ESU
Rouse, D A I 2E1EKF
Rouse, D J G4TZQ
Rouse, G P G4MUP
Rouse, L A G6IFN
Rouse, M K G1YKX
Rout, D J G8PAI
Routledge, D F G4NYG
Rovardi, P M R G4HSB
Row, D S E G0EUE
Rowan, D W G4CUO
Rowan, F H G8OMW
Rowan, F J G4RWN
Rowberry, E C G1YPU
Rowbotham, F W G7SWE
Rowbotham, J G4AQT
Rowbotham, M A G1GQQ
Rowcroft, N G4NIV
Rowe, A G6AVP
Rowe, A J 2E1GAB
Rowe, A J G7PUB
Rowe, B G4WYG
Rowe, D J G6XJZ
Rowe, G G7VPC
Rowe, G T G0UAD
Rowe, I M G4WWL
Rowe, J L G3EQX
Rowe, J W K G7PBU
Rowe, K E P G1PYQ
Rowe, M G1TLE
Rowe, M F J G8JVE
Rowe, M H G0IXX
Rowe, M J G7BLX
Rowe, M L G0RXB
Rowe, N J G3EYY
Rowe, P G1DIA
Rowe, P D G3JSP
Rowe, P J G6WKQ
Rowe, T E A G8NNU
Rowell, A G7HDZ
Rowell, B A G4IOR
Rowell, B J G0MRZ
Rowell, B K G5RL
Rowell, E G1JIH
Rowland, A GW1LEL
Rowland, B 2E1BAD
Rowland, B G7UCU
Rowland, D A G0UBF
Rowland, D W 2E1FNR
Rowland, E E J 2E1FDX
Rowland, G G8PKJ
Rowland, J D R G7ATD
Rowland, M G4YUA
Rowland, M G M G4FTW
Rowland, P G4YSJ
Rowland, R D GW0FXC
Rowlands, D M1BOL
Rowlands, D C GW4JAD
Rowlands, D M G0RUJ
Rowlands, G 2W0APO
Rowlands, G 2W1BCP
Rowlands, G G8LWS
Rowlands, G H G8TSV
Rowlands, G J GW7TED
Rowlands, G W G4HIP
Rowlands, G W G7BBC
Rowlands, J G0JYV
Rowlands, J B G4OJS
Rowlands, J H G1IPY
Rowlands, M A G3NKR
Rowlands, M P G8HUV
Rowlands, R E GW4HKX
Rowlands, R W G4UHS
Rowlandson, S J G1FIP
Rowles, B G G1ZDU
Rowles, D V G0MXM
Rowles, J F G4ZUH
Rowles, M GW4WWN
Rowles, M B G2FYY
Rowles, P GW8MCL
Rowles, R GW4FOM
Rowlett, R A G0EYM
Rowley, A G8MYK
Rowley, A H G0TML
Rowley, B G 2E1CPI
Rowley, C E G0ORL
Rowley, D J G7WGK
Rowley, G G8BAG
Rowley, G G G4WKG
Rowley, H G1AKA
Rowley, J G3TQO
Rowley, J A G3KAE
Rowley, J S G8VSR
Rowley, L J G0MBC
Rowley, L J G4MIQ
Rowley, T E G6TC
Rowley, W J M1BQZ
Rowley-Guyon, J F G6WZQ
Rowlinson, J M0ANR
Rowney, R J H G4KFL
Rowntree, F M G8UYK
Rowntree, J H GW3KXC
Rowntree, R O G3ZQA
Rowsby, A G4CQS
Rowse, D M GW3LOD
Rowse, R T G3RTR
Rowsell, A J GU1JFA
Rowsell, C N G4NHN
Rowsell, J 2E1ATF
Rowsell, P J G1ULQ
Rowsell, R G4SIF
Rowsell, R W G3CMY
Rowthorn, E W F G1OOS
Roxburgh, D GI8YJF
Roxby, T G1LPS
Roy, A W G1OPL
Roy, M S G4PPU
Roy, T G7VQB
Royal, J H L G0IXH
Royal, M R G7SBN
Roychoudhuri, R 2E1BKD
Roycroft, R J G1NXV
Royle, A G3FOE
Royle, G G4FAS
Royle, J R T G3NOX
Royle, L 2E1CEA
Royle, T E G0JNE
Royston, A J D GW8PBM
Roze, J G0VMP
Rozentals, J L G6TBN
Rozentals, L M G4YGJ
Ruaux, R R G3RMK
Rubins, R F G4HAB
Rubinstein, P G3ZAR
Ruck, D E J G6RFH
Rucklidge, A J 2M1ERO
Rucklidge, P A GM4KKV
Rudcenko, S G0KBL
Rudd, B G3ELS
Rudd, J G1ODS
Rudd, J A G7OCI
Rudd, M G4LOA
Rudd, M P G8KFQ
Rudd, P D GM0CQL
Rudd, P K GU1MUP
Rudd, R F G4HWF
Rudd, T E G0AOY
Ruddell, A J G4BGH
Ruddell, F H GI4MWA
Rudderham, T S P G7HJR
Ruddle, J G GW0WLI
Ruddock, I P G8NCZ
Ruddock, J M G7KWT
Ruddock, L M G0UHM
Ruddock, M H G G6WZR
Rudd-Clarke, S G S G7MZB
Ruder, P C G0KRX
Rudge, C R G0WYF
Rudge, G C G0CDP
Rudge, L D GW8PKB
Rudge, M E GW4IDC
Rudge, S C GW7RKQ
Rudgewick-Brown, N G4LLL
Rudgley, E A G1XHL
Rudkin, B G1HMF
Rudkin, B J G4MEF
Rudkin, J W G3XHX
Rudkin, R L G6TZJ
Rudling, A T G7DYB
Rudwick, P L G3RDR
Rudwick, P L G4JMR
Ruff, J H G3MOA
Ruff, J O GI4RGV
Ruff, S GI0IQA
Ruffell, S L G7JIF
Ruffle, C D G7HQM
Ruffle, N D G6UQW
Ruffle, P J G0PJR
Ruffle, S J G4EAG
Ruffles, A E G4KTF
Rugen, G A G4OVD
Rugg, R G G2BRR
Rugg, R G G4BJC
Ruiz, E E S G7PCY
Ruiz, M M0AQR
Rule, C A M0ADA
Rule, C J 2E1FDL
Rule, E A G3FEW
Rule, G N GM7OCU
Rule, I C G4NQR
Rule, J S G0JVR
Rule, J S G0PZE
Rule, L J 2E1FDK
Rule, R C G3LDR
Rumbelow, A R G3KKC
Rumbelow, H C G4GYY
Rumbelow, M E G1BHS
Rumbelow, R M G8UYL
Rumble, N P G4OEH
Rumble, T J G8ETD
Rumble, T V G0DIU
Rumbol, N C G4WAK
Rumbold, A G G3ORX
Rumbold, D A G4RYV
Rumbold, T G G6AYY
Rumens, D J G4BOO
Rumens, M D G4UHU
Rumney, A E G6UYN
Rump, M E G8VIC
Runcie, W GM0CJK
Runciman, M P G0NAV
Runcorn, K B G4AAJ
Rundle, G R G7ULD
Runkee, R G4IBL
Rupp, J G0SIH
Rusby, G W C G0SPB
Rusby, I M G1DAE
Ruscoe, J M G3ZBW
Ruse, P T G4WYP
Rush, D L G4CRE
Rush, N E A G3HBZ
Rush, R B G7PYL
Rush, W J G8IIZ
Rushall, K H G4HVB
Rushby, A G7BSW
Rushen, P P G0XBE
Rushforth, J E G3ZPD
Rushforth, L F G4RIQ
Rushmer, D J G7OGS
Rushowski, S F G8XXF
Rushton, A G6LGH
Rushton, A J GW0UZK
Rushton, B G7IJW
Rushton, C L G0PMY
Rushton, H G0UES
Rushton, J G1AII
Rushton, J G7BRS
Rushton, J D G0WZL
Rushton, K G1NXT
Rushton, M G4SBQ
Rushton, N J G0UNR
Rushton, P F G7PMZ
Rushton, R T G4YRR
Rushworth, P W G8AHR
Ruskin, A J 2E1FER
Russ, A M G7FRH
Russ, G G7KOI
Russ, M G8VQH
Russell, A B T G3OMT
Russell, A D G6ZTI
Russell, A D M0AXU
Russell, A J G3MFL
Russell, A K G8AWS
Russell, A M G4WEV
Russell, A R G7AQV
Russell, A S G0IJO
Russell, A W G0ADM
Russell, B J GI7VCS
Russell, B L G8KFS
Russell, C G4HYU
Russell, C B G0NSL
Russell, D M1AXM
Russell, D D G3XLN
Russell, D J 2E1FAS
Russell, D J G8FKZ
Russell, D J G8IGE
Russell, D K 2E1EQJ
Russell, D M G8RCO
Russell, E G0SEY
Russell, E E M G4UUJ
Russell, F G0JZJ
Russell, F K G1YEP
Russell, F P G7LUL
Russell, F W G7VBI
Russell, G 2M1DVQ
Russell, G D GM1IDP
Russell, G D D 2M1DHG
Russell, G G GM8JYJ
Russell, G K M G6MXO
Russell, G M G0CAK
Russell, G M G6GMH
Russell, G W G6OMY
Russell, H G7NIO
Russell, I D G6PSO
Russell, I G G1HPZ
Russell, J G3IHB
Russell, J G3VRO
Russell, J G4SEU
Russell, J G6LUK
Russell, J M0AOI
Russell, J A G1BCQ
Russell, J A G3RRY
Russell, J C G0LVR
Russell, J C G7PVX
Russell, J E G4MSR
Russell, J E GM7FIS
Russell, J G V G8FFD
Russell, J H G7BGL
Russell, J H R G3OKQ
Russell, J J 2E1CJH
Russell, J R G4YMS
Russell, J S G3PMT
Russell, K G4RZQ
Russell, K GM6RFG
Russell, L V M1AHT
Russell, M G0UKM
Russell, M G1WXD
Russell, M G4ZVX
Russell, M C G0WLD
Russell, P G0PCX
Russell, P G6IDA
Russell, P G8FWY
Russell, P G G6UYK
Russell, P G G7AHO
Russell, R G0RUT
Russell, R GW3HUJ
Russell, R A G4MEU
Russell, R E B G3KLD
Russell, R G G4JLF
Russell, R P GM3DEE
Russell, R T G4BAU
Russell, R T G4TOH
Russell, S M1ARX
Russell, T G1XXI
Russell, T A G3JFH
Russell, T W G4ZRV
Russell, W H B G4EDZ
Russell-Smith, D M G0UMD
Russoff, M G3OMR
Russon, A D G0GUC
Russon, D I G7PUN
Russon, J G3APL
Rust, D 2E1EMG
Rust, D J G7MOD
Rustman, W A G0DQF
Ruston, A J M1AZG
Rutgers, K T F G0GQZ
Ruth, D K G4TAT
Ruth, T H G7TXL
Rutherford, A G0DHI
Rutherford, D A G7KUG
Rutherford, G P G7SCT
Rutherford, J G7FFR
Rutherford, K F G1KFR
Rutherford, R B I M1ASP
Rutherford, R N G0MWZ
Rutherford, T C G3TEC
Rutherford, T S GM3ZFU
Rutkowski, J M A GW4XAU
Rutland, M I G0VIX
Rutlidge, B E M1BTS
Rutt, H N G0DHR
Rutt, M P J G6YOG
Rutt, R C G0AMG
Rutt, S J G0MSR
Rutt, W S G0DEJ
Ruttenberg, M J G7TWC
Rutter, A G8HCK
Rutter, D A G0STT
Rutter, K G0FXS
Rutter, K G1EYD
Rutter, S M G7VAH
Rutter, S M M1ABO
Rutter, W E G6UYI
Rutter, W S G0GXR
Ruttle, R M G7KLU
Rutty, M J G3UPV
Ruud, S G0TSJ
Ryall, A C GW7KRY
Ryall, S D G4SIK
Ryalls, C G8HRA
Ryalls, C T J GW0KLN
Ryan, A G0DGN
Ryan, A G3VJN
Ryan, A G6AQL
Ryan, A B G3YUR
Ryan, D M G7SDP
Ryan, D W G3SSU
Ryan, F N GM6BKE
Ryan, H J G0JYH
Ryan, J A GM7RDR
Ryan, J F G0NZS
Ryan, J S G3NAW
Ryan, J R GI0EEO
Ryan, M P G0KXQ
Rybak, C P G6VEQ
Rybalka, A G0NTC
Rychlinski, A 2E1BJU
Rycroft, D H G4OKO
Rycroft, P S G8FFW
Ryder, D J G4HHM
Ryder, G 2E0AOG
Ryder, J C G0KYO
Ryder, M J G0CDA
Ryder, N T G0WKF
Ryder, P E G0TSG
Ryder, T P D G1MMI
Rye, M D G4WTE
Ryland, A G0PTR
Rylatt, R F F G3VXJ
Rylett, I D G6NLW
Ryley, J E M G3KQV
Rymer, W E G3JCZ

S

Saagi, H M G6DMR
Saagi, K R G0NQU
Saagi, L G6TMQ
Saagi, R G4AQV
Sabido, R E R GW4LKE
Sabin, D J G0MLE
Sables, M J G7NTY
Sables, M P G0PUG
Sables, M P G4ZJN
Sables, P A G0TZC
Sables, P W G4MRU
Sach, M R G8KDF
Sacharewicz, R G1XVY
Sachs, R B G2CZS
Saddington, S W G2FXQ
Sadeh, I G0UUT
Sadler, A G7NBF
Sadler, A J G7MUY
Sadler, D G0MWA
Sadler, E R G0MVK
Sadler, F A 2E1GAS
Sadler, J A G2FHM
Sadler, M F 2E1DLC
Sadler, M F G7WAL
Sadler, M P 2E1FFY
Sadler, R C G4MRP
Sadler, R E G4FAJ
Sadler, R L G4BMP
Sadler, U M G0IHM
Sadler, W R G4YSR
Sagar, A M G1FUG
Sagar, J GW3ARS
Sage, J P G7JYL
Sage, P GW0UHO
Sage, P E G1UXZ
Sage, W E G1DUU
Sager, D P G7GJI
Saggers, C G G0OSK
Saggerson, T B G3CSA
Saggerson, T B G4WSE
Saich, B L G1PRZ
Saines, S J G3RNB
Sainsbury, P J F G1BFS
Sainsbury, R P G0OSW
St Aubyn, A F P G7BZP
St Aubyn, A P W G4YNW
St.George, D L G4IOY

St John, D E A G3PQD
St John, R I G G2HBQ
St John-Murphy, T D G0HTK
St.Leger, J A C G3VDL
St Quintin, D G4PCZ
Sait, A G0OIV
Salaman, R L G4JSE
Salata, A G G4PLX
Sale, A C G0GJW
Sale, T N G8RDN
Salem, V G1DYP
Sales, B G6SGY
Sales, J J G0AZJ
Salisbury, A E GW4RYJ
Salisbury, A P GW8KSF
Salisbury, G G0KTY
Salisbury, J A G1BFZ
Salisbury, T H G2HAO
Salisbury, W C GW8KSE
Sallis, T J G6FFH
Salmon, C F G6ZGB
Salmon, C G G0REB
Salmon, D C G4KCN
Salmon, M J G3XVV
Salmon, P G7NFF
Salmon, R V G4LJX
Salmon, S G G7DIE
Salomon, P M GW3XQO
Salsbury, D A M0AVA
Salt, A G0HEE
Salt, B G0OZP
Salt, B G4ITL
Salt, B W G3VVG
Salt, D G G0JMG
Salt, F G7AJE
Salt, K E G7DXQ
Salt, P G7UQF
Salt, R M M1AEN
Salter, D J G1ERM
Salter, D R G8UIO
Salter, J L G3DQC
Salter, K G G4FKU
Salter, K J G3LFY
Salthouse, J A G8LDC
Saltmer, M G7RRY
Saltmer, S P G8SGX
Salusbury, R D GW7THE
Salvesen, T M N GM3ODP
Salzman, M J G1ZSR
Samber, D G1EIH
Sambrook, A G4JJN
Sambrook, G G4ROK
Sambrook, P O G1NCR
Samet, F C G0EVU
Sammon, A L GI4PCY
Sammons, A B G0HBC
Sammons, A L G0EAG
Sammons, R H G6XZP
Samouelle, A J G0BOM
Sample, R C L GU0TEW
Sampson, A D 2E1ABI
Sampson, A W G4WUT
Sampson, A W GM4BVD
Sampson, C G0VZK
Sampson, C M1BEW
Sampson, E M GM8VBP
Sampson, F W G G0LQB
Sampson, G P G4SRX
Sampson, K G G1MXC
Sampson, L W G3JSU
Sampson, P E G1BGQ
Samson, M J G6XMT
Samson, P A G8IVB
Samuel, D H GW4JDZ
Samuel, M R L G0VVX
Samuels, A M G0ABN
Samuels, B M G0FJA
Samuels, D H G0LTY
Samuels, E G G6NIX
Samuels, J M G4FOB
Samuels, M J G0FYL
Samuels, P J G4VSD
Samways, A J F G6LPC
Samways, R B G0GGN
Samwells, H M G7MSH
Sanby, G G3WXJ
Sanchez-Garci, F G1TDO
Sancto, V A G0LFM
Sandall, B M G3LGK
Sandall, J A G1DCZ
Sandaver, E C G4KIT
Sander, J H F GI4BUJ
Sanders, A C G6ULX
Sanders, A F G0KEJ
Sanders, A G G4VEF
Sanders, B A G0JZX
Sanders, C W G4KCM
Sanders, D G G0KRY
Sanders, G G0MTF
Sanders, G D G4YHB
Sanders, H H A G3CRH
Sanders, H R C G3IGH
Sanders, J F G1BHX
Sanders, J F G4YCL
Sanders, J T G1LMI
Sanders, L T G2HCA
Sanders, M A G3RWV
Sanders, M G G6XZK
Sanders, M J G8LES
Sanders, M P G8HST
Sanders, N R GM6WTP
Sanders, P J 2E1FOI
Sanders, R G4RSM
Sanders, R J G0MUK
Sanders, R S M1AYG
Sanders, S E M1APB
Sanders, S L G8PRJ
Sanders, T G8AEE
Sanders, T H G1ZWH
Sanderson, C E 2E1BRG
Sanderson, G L G2DBT
Sanderson, J E N N G8UYM
Sanderson, J H G7POW
Sanderson, J M 2E1DZG
Sanderson, J R G3TKS
Sanderson, J W G4MQV
Sanderson, K C G7MSF
Sanderson, K E A G4KCF
Sanderson, M A M1AQI
Sanderson, N G1CSR
Sanderson, N G4UUK
Sanderson, N G8GFF
Sanderson, R G0HYV
Sanderson, R A 2E1FMC
Sanderson, T R G4OEY
Sanderton, J A G7UOE
Sandever, D J G7MQU
Sandford, B J G4XLF
Sandford, L G6DKM
Sandham, M E G7VLA
Sandham, P H GW0VST
Sandiford, P G3STF
Sandiford, V H G0KSH
Sandilands, E GM0FSZ
Sandilands, G W G7FBU
Sandilands, T B G0HBV
Sandle, W G0SBZ
Sandler, H M G8ORY
Sandler, M C G6MVS
Sandoz, M L A G3GBS
Sands, A G G1FPC
Sands, A J G0AOX
Sands, M G3IIW
Sands, S E G0AOW
Sands, S E G4WRA
Sandum, J A G0JIO
Sandy, D H G4RSP
Sandys, E R GI2FHN
Sandys, J T G0RGX
Sangster, D M GM3FIZ
Sangster, G P GM4OBD
Sangster, G P GM4RGC
Sangster, J R GM4RAO
Sangster, R G G8DCX
Sankey, B K G7RWY
Sankey, J M G0SGI
Sansom, A J G4KIF
Sansom, I L G7UPZ
Sansom, M G0POT
Sansum, J A G8GHT
Santagata, A R G7ANG
Santer, C S G6JVT
Santer, E A M G8SLM
Santer, E A M G8YBY
Santer, L J G3PVB
Santillo, A G0XAW
Santillo, D A G7WKN
Santillo, P L G6MBW
Santos, A G4PMJ
Sanvoisin, N M0BCS
Sapsed, K L M1CAD
Sapsworth, D T G3YMW
Sardeson, C P M0BCO
Sargant, G C G8ZRV
Sargeant, A C G0KZN
Sargeant, C H G8FKF
Sargeant, A L J G0KVA
Sargent, B G0HTL
Sargent, C W L GW8XAH
Sargent, D C G4JYE
Sargent, F G G0NJB
Sargent, F G G3BLO
Sargent, J C G8CKS
Sargent, J C G8KIH
Sargent, J C M1BDR
Sargent, J D 2E1BRM
Sargent, J W G8KXI
Sargent, M J G4SCB
Sargent, P G8RYO
Sargent, P K G0BJJ
Sargent, T C G4BFS
Sargent, W C G0CTF
Sarjantson, J W GM4EHO
Sarre, R D GU4HUY
Sartin, J C G4SSE
Sartin, R D GW6JNE
Sartorius, C A G4PSR
Sass, E C GI4FHD
Sasse, H G G8TYF
Sate, A J G4LSK
Sato, R G4YHC
Satterthwaite, R G6BMY
Saueressig, J F C G0ADK
Saul, A G7DIW
Saul, D S N G4EKZ
Saul, P H G8EUX
Saunby, M G4NFP
Saundercock, V G M0AVS
Saunders, A C G8PPD
Saunders, A H G GM3VLB
Saunders, A H S A G8XGQ
Saunders, A J G6NJH
Saunders, A J M1BAI
Saunders, A R G4LLW
Saunders, A R G7VEE
Saunders, A W G1OZB
Saunders, A W G7IRF
Saunders, A W G7NRS
Saunders, B A GW4NML
Saunders, B J G1SPW
Saunders, C G7KGR
Saunders, C B G1GTO
Saunders, C D G0TVS
Saunders, C D G4ZCS
Saunders, C H G8NLP
Saunders, C P GW0VRL
Saunders, D A GM7RBW
Saunders, D E G3OWE
Saunders, D F G0WFT
Saunders, D L G3OEW
Saunders, D M N G4HZR
Saunders, D T 2E1BNL
Saunders, E S G4LMT
Saunders, G G A G7UZX
Saunders, G M J G3OYN
Saunders, H G1OUA
Saunders, J G0VJU
Saunders, J G1NQM
Saunders, J E G3KAZ
Saunders, J R J M1AWN
Saunders, J T G3OLU
Saunders, J W F G3KCV
Saunders, K G0OYC
Saunders, K A G8SFM
Saunders, M G0UVK
Saunders, M A G3ORV
Saunders, M A G6IWC
Saunders, M D G0DAU
Saunders, M J G1RNP
Saunders, M R 2E1FZH
Saunders, M S G8OSC
Saunders, N H G1PJV
Saunders, P A G0UZQ
Saunders, P J G4AQE
Saunders, P J G6HGU
Saunders, R A G7HHQ
Saunders, R B G0ERY
Saunders, R B G1LQX
Saunders, R D 2E1BJD
Saunders, R D 2W1FMG
Saunders, R F G3CVW
Saunders, R M G6VAX
Saunders, S A W G7AET
Saunders, S R G6FNE
Saunders, S W G6IKC
Saunders, W E J G6IAJ
Saunderson, R A G4FPD
Saunderson, W J GW3DKE
Savage, C E G0TCR
Savage, C J G7UII
Savage, J G3MSS
Savage, J B GI6VSK
Savage, J H G6VBD
Savage, J R G1AEB
Savage, M A G7UPT
Savage, P J 2E1DVR
Savage, P J G7LTG
Savage, R G7XYL
Savage, R L G3LTX
Savage-Lowden, J GM8PHU
Saveall, J G1FEJ
Savegar, J D G4ZSG
Saveker, C E G8AMU
Saverton, R P G6DKS
Savigar, R J G6IBW
Saville, G E G1RNZ
Saville, G H G3DNN
Saville, P G6UJU
Savin, C S G0IYT
Savin, M C G7KWD
Savin, N G0SVN
Savin, R A J G4XYN
Savory, W G GM4HUL
Saw, R J G0CJQ
Sawbridge, W GW0NSQ
Sawdy, J G G3XJM
Sawford, G R G0GRS
Sawford, J R G G8CAB
Sawford, L H G6APD
Sawkins, E J GW0BBG
Sawkins, R G3ADS
Sawkins, R E R G0GWM
Sawyer, F G G3SLN
Sawyer, P E G7LTP
Sawyer, P F G1SEG
Sawyers, B G6ZKU
Sawyers, M R G4VHI
Sawyers, P A G6ZKT
Saxby, K S G7JVO
Saxelby, R C G G4OGJ
Saxon, E R GM8YRX
Saxton, C T 2E1AUR
Saxton, D M G4WQI
Saxton, G G0BBS
Saxton, M R M0AJD
Saxton, R J S G4RRX
Saxton, T E G3LJR
Sayegh, M G0UCC
Sayer, D E G8RCF
Sayer, E W G6MVW
Sayer, J P G6XCC
Sayer, P G0VKT
Sayer, P M0AAA
Sayer, R J G0BCT
Sayers, A J G6KSV
Sayers, B J G7AZJ
Sayers, M A 2E1CEQ
Sayers, M B G1VJE
Sayers, P E G7MJI
Sayers, R J G8IYK
Sayle, T G GD6TKX
Sayles, D A G7WGX
Sayner, J G4VJS
Sayner, T F G0VWP
Saynor, J H G3EKT
Saynor, J H G4GGP
Saywell, S A G0IXW
Scadden, R G3TFM
Scaffer, C A G0WZR
Scaife, R G3RSB
Scales, N R G6HPU
Scandrett, A E G4KFC
Scanlain, S GM1VAD
Scanlin, G GW0NXW
Scanlon, M P G3JZE
Scanlon, P J G3ZKL
Scanlon, S F G8HQY
Scanlon, T A D G7VJH
Scantlebury, G F G1ODW
Scaplehorn, D G0NQJ
Scarce, R B G7RMQ
Scarfe, R S G4TUK
Scargill, D B G0UND
Scargill, R G4MBE
Scarisbrick, A L G7ELG
Scarisbrick, B H G4ACK
Scarlett, G T G0SUI
Scarlett, M N G4CAK
Scarlett, R J W G4HZF
Scarlett, W G G3RXS
Scarr, A C G0LWU
Scarr, J D G6LBL
Scarr, M L G0PUU
Scarr, M R G1XSV
Scarratt, P G0WRE
Scarsbrook, A G4AIW
Scarsbrook, B L G7TNT
Scarth, A G4IZQ
Scates, C H G1NWG
Scattergood, G GM6AAJ
Sceal, J S C G4ZBW
Schafers, G GM1MXE
Schamp, E R G1MZH
Schau, K H G3NXI
Scheffer, A R 2U1EJF
Scheffer, J GU0GUX
Schein, P S G0OMJ
Scherrer, J M G4ACP
Scherrer, J W D G1WEG
Scheuber, R A J G0BUY
Schiefer, J E G0TRR
Schiffeldrin, G G4LYM
Schiffman, C B G8TND
Schlatter, P J G0NJQ
Schmid, M S G0EDZ
Schmitz-Goertz, H T G4VDK
Schnapp, G G0WSF
Schneider, E J G0AZT
Schneider, K F G1JEH
Schneider, R J G8FQN
Schnell, F G0NLW
Schnurr, L G0AAN
Schoales, R G4NYW
Schofield, C G7NSA
Schofield, C GM7KUN
Schofield, C G 2E1CVA
Schofield, D G7OKY
Schofield, D A A G4OJI
Schofield, D J GM8PAH
Schofield, G G B G0HMQ
Schofield, H G G6JVX
Schofield, I E G7VTD
Schofield, J G3KRL
Schofield, J G 2E1FXU
Schofield, L W GM0LWW
Schofield, M L G0JBI
Schofield, M W G4WUP
Schofield, N L D GW4CAT
Schofield, P G7VME
Schofield, P M1BEC
Schofield, P D G7KNT
Schofield, S G4PJT
Schofield, S GW4VMF
Schofield, T F G4ZLI
Schofield, W E G0BAK
Schofield, W R 2E1CBD
Scholefield, P R G4TSF
Scholes, A A G6OGT
Scholes, C A 2E1DOY
Scholes, D W G7NHC
Scholes, E G4KHG
Scholes, E D 2E1CQX
Scholes, G G8OHC
Scholey, C A G0CJV
Schollick, C M 2E1CKV
Scholte, B G G1SIG
Schonborn, M G7IDJ
Schoof, G H G1SWH
Schoof, G H M0AWX
Schoolar, J R G4LYY
Schorah, P A GW3PPF
Schoth, B E G3WJM
Schou, O B G7SGG
Schrager, G R G3XKY
Schranz, P A M0API
Schrier, S J G7KQT
Schroepfer, M K G0WFU
Schulz, J P G1JJQ
Schuring, R T M1AKA
Schwartz, P S G0USZ
Schwarzer, M G G8TUB
Scivetti, A G1PXQ
Scleparis, E G4YBX
Scobbie, D GM6KDD
Scoffield, C R M1BLY
Scoffield, R J G3VSX
Scofield, D G G4UOV
Scofield, J A G1ERG
Scoles, W K G1YMA
Scothern, D G G1YFQ
Scothern, J A G6XYR
Scothorn, G M1BOW
Scotney, D G4AIV
Scotney, J P G6BKD
Scotney, P G8BKD
Scott, A 2E1FWO
Scott, A G4TNU
Scott, A E G1TWT
Scott, A G G4BYP
Scott, A J G0HNX
Scott, A J G6NIZ
Scott, A J G7MWM
Scott, A J G7TXF
Scott, A J GM8BDX
Scott, A M G7RSK
Scott, A M M1AUC
Scott, A P 2E1DNJ
Scott, A P G6MDS
Scott, B 2M1DZW
Scott, B L R G1HIE
Scott, B R G6UHD
Scott, C B GM7CPL
Scott, C N G3XRL
Scott, C S M G4CHE
Scott, D G6HPK
Scott, D GM4IZN
Scott, D A G0KJP
Scott, D G G4OWQ
Scott, D I G6LDP
Scott, D J GI1BBG
Scott, D J S G6MVR
Scott, D P 2E1BKX
Scott, E G0VSR
Scott, E G4GVF
Scott, E GM0EUO
Scott, E M GM8RMR
Scott, F R G6UDB
Scott, G 2E1FNE
Scott, G G8UMM
Scott, G E G0GLK
Scott, G G E G8TRY
Scott, G G E G8WDC
Scott, G L G6OOC
Scott, H A G3YRE
Scott, H E G7NGO
Scott, H N J G7IEQ
Scott, I D G3SYW
Scott, I M G6AQH
Scott, I R G0IRS
Scott, J G0HGH
Scott, J GM0WRR
Scott, J A G0JFP
Scott, J A G3CMI
Scott, J B M1BRE
Scott, J D GM7UJJ
Scott, J E G7BTK
Scott, J F G8LCA
Scott, J F C G1OLD
Scott, J G M G1UDR
Scott, J L G4ZTF
Scott, J M G7HOT
Scott, J M S G7IVC
Scott, J P GM3KJE
Scott, J R GM4WZL
Scott, J T C G1OWZ
Scott, K G4FOY
Scott, K GW7KVW
Scott, K A GM0PJP
Scott, K F G7WJL
Scott, K R G0MCP
Scott, K W G0KWS
Scott, M G1ERO
Scott, M G4XUX
Scott, M G6AQG
Scott, M M0AVU
Scott, M D G3LYP
Scott, M N G6XZA
Scott, M R G7IRN
Scott, N L J G6FLE
Scott, P GM0VOU
Scott, P F G1RVP
Scott, P J G7DDD
Scott, R GM0JCS
Scott, R GM1VBD
Scott, R M1CBK
Scott, R C H G4HWW
Scott, R C S G0IAL
Scott, R D G4KVQ
Scott, R J 2E1DFE
Scott, R K G4NLI
Scott, R N G7KYY
Scott, S G0NXK
Scott, S G1PJI
Scott, S G4CKR
Scott, S J G8DQH
Scott, T G7VTJ
Scott, T GM0OTX
Scott, T W G1GXF
Scott, T W G6VFN
Scott, W 2E1CIX
Scott, W G1AWP
Scott, W D G6FFM
Scott, W G G7UNZ
Scott, W J G0RWS
Scott, W K G3FUJ
Scott, W K GM4YAU
Scott, W M J G1ZVO
Scott, W T G7AYX
Scott Iverson, I P A G1MMG
Scotter, J R G0MST
Scotton, M J 2E1BRA
Scott-Dickinson, P F G0JRH
Scott-Green, A G4GWR
Scott-Green, K W G4GWQ
Scott-Stapleton, F C G4IPX
Scott-Telford, H J S 2E1DEC
Scott-Telford, H J S G7UTB
Scourfield, D GW3SVY
Scovell, A V G7DFD
Scovell, P F G1PRX
Scovell, P W G1MVV
Scrase, C A G8SHF
Scrase, C D G4MRD
Scrase, S T G8MHN
Screen, D A G4SWE
Scrimshaw, I C G8XNY
Scriven, J G4YNU
Scrivens, J GW4IWU
Scrivens, P J G0HHL
Scrivens, R G3LNM
Scroggins, K P G1TQH
Scroggs, B D G0BDS
Scrogie, N P G4CPQ
Scrutton, A G1KCS
Scullion, G C J GI1IIL
Scully, J W G3GUR
Sculthorpe, G G6TVI
Scutchings, R D G7TLP
Scutt, P A G3IBI
Seabourne, F J G0VQT
Seabourne, R G0UHN
Seabridge, C E G0VRK
Seabridge, D C 2E1FBS
Seabright, I R L G0WYT
Seabrook, A R G3ZQB
Seabrook, D J T G4LJG
Seabrook, D T G6HPY
Seabrook, M A G1BJA
Seabrook, R J G4UCO
Seaby, B T 2E1DPW
Seaby, B T M1BYP
Seaford, P G G8XTW
Seager, I A G8ITW
Seager, J G0UCP
Seager, J D G6BHB
Seager, J R G4AK
Seago, A G G4KDL
Seal, D G1RKP
Seal, E E G4HXG
Seal, G J L G4RLO
Seal, J F J G8WMS
Seal, J G G3UJG
Seal, R M P G0RMP
Seal, S W G1CUZ
Seals, A E G0TZP
Sealy, R K G4SRD
Seaman, D H G1SEI
Seaman, G P G3PWW
Seaman, G W G1MAV
Seaman, J W GI4XAP
Seaman, P J G3OTN
Seaman, P R G1TIJ
Seaman, V J G4RAP
Seaney, W C G6VOG
Sear, L G G3PPT
Searby, B S G6TUX
Searl, M J G6SBN
Searle, A D G6XYS
Searle, A L G3BDT
Searle, D J G1XYR
Searle, D W G7GOP
Searle, E G3VMY
Searle, G W G4GRR
Searle, J H G4ZAS
Searle, O E G4WCI
Searle, S J G0OWK
Searle, T J G7BAE
Searles, P A G4PAS
Searley, M G G0AVJ
Sears, G G4WXK
Sears, K E G1RHB
Sears, N J G3YEG
Sears, R F G0AHL
Seath, N P G7FUM
Seaton, G E G7MMR
Seaton, P J GM0AKJ
Seaton, R K G6VQW
Seatory, H W D G1XWS
Seatter, W J GM7UGZ
Seaward, R K G0TCA
Sebborn, P B 2E1FXG
Seccombe, D G1DDS
Seddon, B J G6XYH
Seddon, D G0TEK
Seddon, D M1BRX
Seddon, D M G4VCO
Seddon, F G3RPO
Seddon, F C G0GRC
Seddon, F C G4WFK
Seddon, H J G1VYA
Seddon, J G4IJD
Seddon, J C G4MFH
Seddon, J M G6LDY
Seddon, J P GW4GMI
Seddon, K S G0KHA
Seddon, M W E 2E1EVG
Seddon, P G8ZUW
Seddon, R N G4VXW
Seddon, T W G6TFC
Sedgbeer, A H G0MAS
Sedgebeer, D D GW4OZB
Sedgebeer, S D C GW3RVG
Sedgebeer, V R GW4TVU
Sedgley, D L 2E1CFQ
Sedgley, J M G7AII
Sedgley, J M G7HHC
Sedgley, V R G3YIC
Sedgwick, C W G0VYD
Sedgwick, C W G7RPM
Sedgwick, L G8XOV
Sedgwick, N G3VSG
Sedgwick, N G G4HDL
Sedgwick, N H G8WV
See, D J G8EOX
See, P C G4PCM
See, P H G1GQW
Seearam, C P G6XHM
Seeby, M E 2E1DJM
Seed, A G3FOO
Seed, A G3NWR
Seedhouse, W G G6ZKL
Seedle, J 2E1FHW
Seeds, A J G8DOH
Seeds, D T G7FFG
Seeley, K A G7LUV
Seeley, K E J G7LJZ
Seeney, C J G8YMO
Seeney, C R G6LDO
Seeney, W J G8RFN
Sefton, M R G0CYL
Sefton, S G3ZBA
Segal, D M G1DMS
Segal, L G6XLL
Segall, R H G7VRS
Segar, A M G0UPN
Segar, A R G0SXO
Segar, S L G0NQY
Seidner, H G0ROA
Sejwacz, A G7MFA
Selby, C D G1HMK
Selby, C M G7VXQ
Selby, I R G4DRI
Selby, K P G0BJT
Selby, R C G7GRN
Selden, O C G0RKA
Seldon, A G3LSZ
Seldon, C G3SCY
Seldon, H C G7NIA
Seldon, P A G8BSF
Self, R V G0HHX
Selkirk, J R G3NQJ
Sell, D F G1JQR
Sell, F L G0GGH
Sellar, L M G8XMS
Sellars, J GM0BLX
Sellek, R J GW0JZR
Sellek, V J GW1UIK
Sellen, D S G3YAJ
Sellens, K L G6ZRL
Seller, K G G1GJK
Sellers, A G8TZJ
Sellers, J R G3VDE
Selley, J W G7OFB
Selley, P G7FJZ
Sellick, A E G7LPB
Sellick, J F C G8LCZ
Sellman, R J G4CRK
Sellman, S E M0AYA
Sellman, S E M1ATN
Sellors, A M 2W1BYK
Sellwood, A G8SNI
Sellwood, D E G4CGA
Sellman, L H G0LYQ
Selmes, A G F G4KLF
Selvestian, F G7VWL
Selwood, J R G7WEU
Selwood, P L G3YDY
Selwyn, J G7UDU
Selwyn, M F J G3TLD
Selwyn, S M G7GZU
Selwyn-Smith, C R G1RQZ
Semark, A K G4OJN
Semark, A K G8LUP
Semmens, N S G3PUQ
Semple, J GI3OYG
Semple, R W GI3JRW
Senft, R E G0AMP
Senior, A G0TJZ
Senior, A P G7ALN
Senior, B G8YGT
Senior, D S G4MIB
Senior, G T G3HRU
Senior, J A G7RXS
Senior, J B G4LJE
Senior, J R G3ANI
Senior, M G1DCB
Senior, M F V G4EFO
Senior, M G G3PAK
Senior, P G0CSK
Senior, P J G4JNL
Senior, R E G4IEZ
Senior, S R G4XSH
Sennitt, J A G1PKV
Senter, I F G6RDD
Sephton, A F G3IJL
Sercombe, A J G7RBS
Sergeant, A H GM4HRL
Sergeant, B P GM0JPH
Sergeant, D W G3YMC
Sergeant, E G4HTE
Sergeant, J R A G8YCP
Sergent, P G G4ONF
Serlin, J H G3TLU
Sermons, A J G7LCD
Sermons, B J G8IUD
Sermons, C G1IUD
Sermons, C L G7OBN
Serplus, J C GI7LCQ
Serridge, J GI0VVB
Serridge, J L GI0VVC
Serridge, P J GI0PED
Service, N D G0TXJ
Serwa, E G8ZIK
Sesemann, M J G4WII
Sessions, G J 2E1FWX
Seton, D J G0TVE
Seton, M G4PLV
Setter, B W G4YAQ
Setterfield, D E G1SYZ
Setterfield, E L 2E1CTZ
Setterfield, W E G4XGV
Severn, D G8LNG
Severn, P R W G1NHX
Severs, M A G6UJR
Seville, D G0AKA
Seward, C W GW0PJX
Seward, G G3PBK
Sewart, C G0HHR
Sewell, A T G0EVQ
Sewell, D J G4FVK
Sewell, M A G7PBO
Sewell, P R G6GIH
Sewell, R T GM0RTS
Sewell, S L G4VCE
Sex, M R G4AZQ
Sexton, B E G1JJS
Sexton, C P G0TEB
Sexton, J H G4CNN
Sexton, J L G6ZWZ
Sexton, R D G3JQV
Sexton, R D G4IZS
Sexton, T J G1KUN
Seymour, D C G7VCY
Seymour, J G0CWK
Seymour, J S G4CPJ
Seymour, R G G1FEH
Seymour, R W G1JOO
Seymour, T R G4ZYZ
Seymour-Smith, G C G6REW
Seys, B C G3BJS
Sezen, M 2E1AQU
Shackleford, S P G2HAX
Shackleton, A D G1UVE
Shackleton, C G7UST
Shackleton, C M1BIO
Shackleton, E G0EQH
Shackleton, P A G7ALW
Shackleton, S G7VSC
Shackleton, T K G6PSZ
Shackleton, W W GM0GNT
Shackley, N J G0OSX
Shacklock, A J G0OEG
Shacklock, S H G4IP
Shadbolt, P E G6LBJ
Shaddick, E I G7EIS
Shaddick, K A G0TDS
Shaddick, R L G1YRS
Shade, B G0UOF
Shades, J J GM0EPO
Shadlock, C J G7PAC
Shadlock, R G3US
Shadwell, J G1JEM
Shail, K V G8AXV
Shakeshaft, C J G3PZH
Shakeshaft, S G4YNS
Shakespeare, F L G3HKN
Shalders, A P G6MPN
Shall, A T 2E1CXW
Shallcross, D W 2E0AFI
Shalley, C I G0FDD
Shaman, T A G4DXT
Shams-Nia, R G7JUZ
Shand, G C G7DZR
Shand, W G GM3JGS
Shand, W G MM0ANA
Shane, B J R G1JOL
Shankie, G GM3WIG
Shanklin, E J M G0UWA
Shanks, C S G4MFR
Shannon, A G3KKJ
Shannon, M G6NXV
Shapero, P G4NCK

Name

Shapland, C L 2E1FIH
Shapland, R E G4ATB
Shard, R P G4XIE
Shardlow, I G4HZB
Shardlow, J G4EYM
Shardlow, M G3SZJ
Share, J A G3OKA
Sharkey, J M GW1BBH
Sharman, A M G0UWS
Sharman, A M GW7GPU
Sharman, B G0FBL
Sharman, C L 2E1CZQ
Sharman, J G4TKO
Sharman, J R G3YCH
Sharman, K J G0TQC
Sharman, K W J G7VJA
Sharman, L J G1JTX
Sharman, M J G3TPM
Sharman, R G4YCN
Sharman, S K GU3WIY
Sharon, C C G4YUF
Sharp, A J G6AJS
Sharp, A N G0WNZ
Sharp, A R G4CWY
Sharp, A W G0JIA
Sharp, B S G1WVL
Sharp, C E D G1EIL
Sharp, C K G7VVC
Sharp, C M G0WGO
Sharp, C M GM7BBU
Sharp, E J G3RDN
Sharp, F S GM4FCP
Sharp, G G8ISE
Sharp, I P G7MZY
Sharp, J G0JYK
Sharp, J GM4DEX
Sharp, J A G0SWZ
Sharp, J F G1NWO
Sharp, J W G0WSL
Sharp, J W G7KYV
Sharp, M P G4YCH
Sharp, P G1KNU
Sharp, P A G6GAV
Sharp, R G7VTB
Sharp, R E G0TSY
Sharp, R E G1GGT
Sharp, R E G6RMV
Sharp, R E J G4VNR
Sharp, R M G3GON
Sharp, R W G4MNB
Sharpe, C G7WCP
Sharpe, C I G0WYI
Sharpe, D G8HUO
Sharpe, D J G8NIE
Sharpe, D R G3ZUN
Sharpe, J D G1WWP
Sharpe, J M G1ZGH
Sharpe, P L G0FYZ
Sharpe, P R G4LVM
Sharpe, P R G6KCG
Sharpe, P S G8RRC
Sharpe, R G G4LQH
Sharpe, R H G6IIF
Sharpe, R J G4FKY
Sharpe, W R GI4ILZ
Sharpen, P J G3XXG
Sharples, D G4ZMB
Sharples, F G4EQT
Sharples, G P G3YYC
Sharples, J G3FJ
Sharples, J D G0NIT
Sharples, R L G3PKD
Sharples, S J G4UNE
Sharples, W J G4MPI
Sharps, L G4OHP
Sharrard, S W G0HMI
Sharratt, J G3XKF
Sharratt, J C G6ZRF
Sharred, D L G3NKC
Sharred, S D G4JGV
Sharrock, A W G0AWS
Sharrott, I E G0CND
Shattock, A G G4RTP
Shattock, J D 2E1CIB
Shave, W E G8OYL
Shaw, A G3BMV
Shaw, A G8IPG
Shaw, A C B G6RDO
Shaw, A J G6OTZ
Shaw, A M G0RIB
Shaw, A R G6CMN
Shaw, B G1IQK
Shaw, B G4YKA
Shaw, B J G3CRJ
Shaw, B R G6HFS
Shaw, C B G6EUI
Shaw, C B G8SMZ
Shaw, C G G6XZC
Shaw, C M G4EKJ
Shaw, D G6OTE
Shaw, D A G0IQG
Shaw, D A G7KYZ
Shaw, D A G8RJO
Shaw, D C G1DWI
Shaw, D F G0PUD
Shaw, D G G1OAN
Shaw, D W G3SFY
Shaw, E G7DWH
Shaw, E W G6KLR
Shaw, F GM0CDC
Shaw, F A G3NXS
Shaw, F B GI4WXA
Shaw, F D G4HYJ
Shaw, F T G6NIW
Shaw, G M0AQA
Shaw, G C G8ZUI
Shaw, H GI2AFW
Shaw, H M G6SOE
Shaw, I R G4MWD
Shaw, J GM0EDQ
Shaw, J A G3CAZ
Shaw, J B G1XFM
Shaw, J D 2E1FIX
Shaw, J G G1ZPG
Shaw, J G G3ZKZ
Shaw, J J C GM3BNX
Shaw, J K G3RDD
Shaw, J M G0ERI
Shaw, J P G3YFE
Shaw, J P G4NKI
Shaw, J R G4ETI
Shaw, J S G4RAJ
Shaw, L B G8YMU
Shaw, L G G3NW
Shaw, M G4ONR
Shaw, M C 2E1BID
Shaw, M C G4EKW
Shaw, M C G6CW
Shaw, M E G3UYB
Shaw, M J 2E1ETH
Shaw, N GW0OGK
Shaw, P G1HMP
Shaw, P G6PMR
Shaw, P G8UHT
Shaw, P A G4SUY
Shaw, P B G0UFV
Shaw, P E G4MFJ
Shaw, P I A G1YSR
Shaw, P J G4VWU
Shaw, P N G0KHK
Shaw, R G1IGZ
Shaw, R G3VRD
Shaw, R G7OIW
Shaw, R A 2E1CQR
Shaw, R E G0VTN
Shaw, S G0DNK
Shaw, S A G0JPV
Shaw, T G0DKL
Shaw, V T GW4SOC
Shaw, W G3SHW
Shaw, W J G0STR
Shaw-Brookman, P M G4ZDM
Shaxted, N D GM0ONB
Shaxted, N D GM4OGI
Shayler, P R J G6TSF
Shea, D A G1CAY
Shea, J GI6PBV
Shea, J G G4OFE
Shea, T M M G0FKL
Shea, W F G4AUJ
Shead, F H G4VVQ
Shead, G H S G4WXT
Sheard, A G G0CEF
Sheard, N G4XZE
Sheard, N M G4GCP
Sheard, P R G3YCJ
Sheardown, J A G8TIY
Shearer, A J GM3SWK
Shearer, E A 2M0AHB
Shearer, M J C G4KVU
Shearer, N S L G6DWS
Shearing, A P G8YML
Shearing, C D G0KVX
Shearing, N G8XGW
Shearing, R J GW6AYR
Shearing, S R M1ACJ
Shearman, K G1XLZ
Shearman, M K M1BGF
Shearman, R F G7ARQ
Shearme, J N G2SH
Shears, L F J G1RIR
Shears, R G4BKW
Shears, R G G8KW
Shears, R H G8IWT
Sheasby, B D G3YAL
Sheather, E W G0UOD
Shedden, C C G0IZT
Sheehan, J C G7HBF
Sheen, D R G4CCW
Sheen, N G0JLL
Sheepwash, B J GI4KIS
Sheer, H A G4YAZ
Sheffield, C M 2E1CLT
Sheffield, I W GM3VEI
Sheffield, Z E 2E1EHW
Sheils, J J G0TJS
Sheils, M S GI7LCC
Sheils, T J G6VAT
Shelbourne, A G4TJJ
Sheldon, C G G6CMP
Sheldon, K G4NIJ
Sheldrake, J P G0IVV
Sheldrake, P J G0INH
Shelford, D G G8VRO
Shelford, R C G7HMI
Shell, M J G0IET
Shelley, B J GW7FFU
Shelley, C R G7HGV
Shelley, G G7MSQ
Shelley, J W G3YFZ
Shelley, M A GW3XJQ
Shelley, N F G4JYP
Shelley, R C A G8VVY
Shelley, T W G0MFV
Shelley, W R G1NLG
Shelmerdine, F A G6WZG
Shelswell, A M N M1CAU
Shelton, B L G0MEE
Shelton, C J G4PCP
Shelton, J A GM3TDS
Shelvey, H P G3WPI
Shemeld, D P G8YXI
Shemming, H 2E1FDW
Shemwell, S D G6VQV
Shenington - Gun, J D G3JSG
Shenstone, S M1BFL
Shenton, A E G1MEZ
Shenton, M D 2E1BSI
Shenton, S J G4VFJ
Shephard, E A G0OZT
Shephard, J G7TYK
Shephard, N B G8EIN
Shepherd, A GM8BSQ
Shepherd, A J G3RKK
Shepherd, B M G7TMZ
Shepherd, B M M0ARK
Shepherd, C G0COL
Shepherd, C R G4OOA
Shepherd, D R G3YNP
Shepherd, F G G1TDQ
Shepherd, F G G6RST
Shepherd, G J P G8BKH
Shepherd, H F G0MFS
Shepherd, H H G0KMW
Shepherd, I G1ZSO
Shepherd, I E G4EVK
Shepherd, I H G4LJF
Shepherd, J E G1TFT
Shepherd, J H S G1SDH
Shepherd, J M G0SEB
Shepherd, J W G0NQI
Shepherd, K M GW0ADC
Shepherd, M A M0AGQ
Shepherd, M A M1BIP
Shepherd, N W G4AWW
Shepherd, N W G8ADZ
Shepherd, P C G0KXX
Shepherd, P J G7DXV
Shepherd, P J G8VYK
Shepherd, P P G4ZMN
Shepherd, R G8WCH
Shepherd, R G G3NOB
Shepherd, R G G7PHP
Shepherd, R S G3BPJ
Shepherd, T G3HPJ
Shepherd, T J M1BHZ
Shepherd, W F G4CBC
Shepherdson, B G0TYD
Sheppard, B C G4CAA
Sheppard, B C G4CVF
Sheppard, C S G7BNO
Sheppard, E J 2E1FTF
Sheppard, J E 2E1DVK
Sheppard, J M A G0WIP
Sheppard, J M A G4WOD
Sheppard, K G7BNN
Sheppard, M J A 2E1CPC
Sheppard, P G4UWE
Sheppard, P G6UTT
Sheppard, P R G4EJP
Sheppard, R A W G4GFQ
Sheppard, R K G6CGC
Sheppard, R W G2FUF
Sheppard, S H G7BVN
Sheppard, T G0KSK
Sheppard, T G7MEX
Sheppard, W T G6MBG
Sheppeck, J G0AMY
Shepperd, P C G0CAN
Shepperd, W P G3WMA
Sher, C M G0TVZ
Sheratte, H G0KVU
Sherbourne, J D G8TZK
Sherburn, P D G7PBC
Sherdley, P G3LPL
Sherer, A G3TEU
Sheret, A G GM7RLA
Shergold, J M G6LVC
Shergold, J N G8WQW
Shergold, K R G8RCE
Shergold, N G6GNS
Shergold, R T G6JNB
Sheridan, A R G6EUW
Sheridan, C M0AZQ
Sheridan, D F G7RYL
Sheridan, M L G1CRD
Sheridan, P G4SHB
Sheriff, M G4ZVF
Sheriff, S K G7TAO
Sherlock, F W G3JPX
Sherlock, J P A GI0VGQ
Sherlock, J P A GI0WOO
Sherlock, M G4ZYN
Sherlock, R G6RQJ
Sherman, K G7VFL
Sherman, T R G8XAJ
Sherrard, R T GI3VAW
Sherratt, A 2E1FUS
Sherratt, A F G4TGM
Sherratt, L M G6IWD
Sherratt, M L G7MFZ
Sherratt, S M G8FAK
Sherratt, T W 2E1FUE
Sherriff, B G0RNV
Sherriff, B GM4SEH
Sherriff, E G1MIR
Sherriff, T M G0CHV
Sherriffs, J M GM3BCX
Sherriffs, W GM8GHV
Sherry, J GM0AZC
Shersby, B A G0ISL
Shersby, J D G3TVD
Sherwin, K G0WGI
Sherwin, P W G1JEP
Sherwood, B S G0GVX
Sherwood, C J G4TMG
Sherwood, L H G1ZST
Sherwood, P J G1KSH
Sherwood, P J G4AUY
Sherwood, R A G8GEE
Shewan, I GM4DZM
Shewan, J R G3UZB
Shewring, D GW0WLH
Shewring, M W GW8SIT
Shield, E F G8GVN
Shield, M J G0PRQ
Shield, P G8BXM
Shields, A B GM1WMO
Shields, A R G0WYM
Shields, B G3OIH
Shields, B A 2E1DCV
Shields, B A G7SJX
Shields, D J G8XUW
Shields, E A GM4SMI
Shields, G J G7VIZ
Shields, P J G4PJS
Shields, P J G7IIB
Shields, R GM6KUL
Shields, S M M1AKS
Shields, U D G4ZRI
Shiels, D C G0NWX
Shillabeer, R T M0AVJ
Shilliday, R S MI0BEI
Shillington, D P G0PKZ
Shillito, A G2FRY
Shillito, P L G8GVW
Shillitto, H G0CSS
Shilson, G M G1XOL
Shilson, J E G1IHY
Shilton, T G6NJF
Shingler, D H G4TCX
Shingles, C J G1MUC
Shinn, F R G0JKG
Shipley, S G0PQX
Shipman, D G4JFI
Shipman, R G3ZPG
Shipp, A F G7THZ
Shippen, D J G6LDM
Shipperley, G D G1EYG
Shipton, B J G7FEQ
Shipton, E GW0DSJ
Shipton, E K GW0NWR
Shipton, M G8IXY
Shipton, R G G4KYI
Shiradski, C R G6APH
Shires, N N G3BTM
Shirley, D H MM1ATY
Shirley, D W G4NVQ
Shirley, J E G0JOJ
Shirley, J V G4RCU
Shirley, R S G7MGG
Shirley, V B G1OYF
Shirley, V L G0ORC
Shirreff, F D G3BGM
Shirt, R G7FYM
Shirtliff, P G G8MED
Shirville, G P G3VZV
Shirvington, R L G0ICC
Shoesmith, C R G0SLH
Shone, A G4VWX
Shone, A M G6HAE
Shone, D P G4FBG
Shone, L P G4XME
Shonfield, P T G1CAN
Shons, A P G1PPD
Shoosmith, M G G1IJM
Shoosmith, P A L G3MDH
Shoosmith, P G G6MVO
Shooter, L T 2E0ADM
Shooter, L T G1LHW
Shooter, D G0EOJ
Shore, F A G1SDA
Shore, M J G0NNO
Shore, P J G8IYE
Shore, R J G8RAJ
Shores, J A G4ZBI
Shorey, M J G4NUH
Shorland, M G3WIK
Shorland, P E G0ARL
Short, A G G2DGB
Short, B J G3YEU
Short, C A G8GLQ
Short, C J 2E1EJO
Short, E G4NDF
Short, J A G7FHW
Short, J E G1DJI
Short, J M G4NCJ
Short, J T G4SMT
Short, P G0UFB
Short, P E G8RFD
Short, R G3AFF
Short, R G3YEE
Short, R P M G8OFO
Short, T E G6FFL
Short, W E G0LWE
Short, W J D G3BEX
Shorten, D G7SRB
Shorter, M G8YMN
Shorthouse, J M0AGU
Shorthouse, J E G4FPA
Shortland, M G G0EFO
Shortland, P M G0AGD
Shortland, R E G7OWM
Shortman, M W GW6VYD
Shotter, J J G0UZY
Shotter, M J G0CWX
Shrago, M J G1WVD
Shread, M G GM6TAN
Shrewsbury, A T G3KAN
Shrimpling, B J G7VKY
Shrimpton, F G8CES
Shrubsall, W A G0UKJ
Shubert, F G G0OHV
Shuck, R C G3NXD
Shuffell, P L G4SBX
Shufflebotham, L J G7TII
Shuker, J L G8TDG
Shurety, C R G7NIX
Shute, D G3GGN
Shute, J E G3GTA
Shutt, C A G2HGA
Shuttle, J S G1JOI
Shuttlewood, P G0INJ
Shuttleworth, B G6TFM
Shuttleworth, C G4LKZ
Shuttleworth, P E G6EPX
Siarey, G W G1PJC
Sibbald, D R G3LVN
Sibbald, G A 2M1AOL
Sibert, P J G8SBO
Sibley, P A G6BDW
Sibley, A G H G8JFF
Sibley, C B G0LMC
Sibley, M J N G8MAR
Siddall, A B 2E1ELZ
Siddall, B P G7USQ
Siddall, R G G4DIT
Siddens, R G4XVN
Siddle, A G G7TBX
Siddon, A J G7IVE
Siddon, J G4IJS
Siddons, A C R G4ZDQ
Siddons, C R G1SID
Sidgwick, G L G4FEH
Sidgwick, N N G8YSQ
Sidnell, J E G0DFF
Sidney, C J G4RJY
Sidney, C P G0LXI
Sidwell, R B G1SKW
Siebert, U G7AOW
Siemieniago, A G4LZZ
Sieroslawski, A J S G4XKC
Sierota, A G8LVF
Siertsema, W J G3KCZ
Siese, A G4CIL
Sievert, R L G1GMF
Sievey, V C G4PTC
Siggers, E A G4BEJ
Siggins, S P G6WBS
Silburn, G L G3GGP
Silcock, E T G1VOD
Silcock, L M G1HHP
Silcocks, R J G1CZW
Silcox, D B G6RWJ
Silk, T B G6NHQ
Silkstone, K A G7OLE
Sillars, D I G4IKY
Sillars, E G8FMA
Sillence, A R G4MYS
Sillence, A R G7SOU
Sillence, A R G8UVS
Sillence, C G6WZD
Sillence, D W G7HPZ
Sillifant, E W G3YTP
Silliman, K T 2E1DRJ
Sillito, R G0NXC
Sillitoe, J A G4MTG
Sillitoe, S J G8XTX
Sills, C B G8IQM
Sills, H O G8QZ
Silver, B M G8JLB
Silvera, R B W G4WDS
Silvers, D 2E1DOL
Silvers, T J G6OYV
Silverson, D J G0TNN
Silverthorne, N G1PHZ
Silverton, D M G7KME
Silvester, D D G2BFO
Silvester, L G G0SMY
Silvester, W A G4DAQ
Silveston, D M G3GDH
Silvey, C J G1VJJ
Silvey, R N K G4YQL
Sim, B C GM7VVJ
Sim, B C MM0BEF
Sim, G A GM0OTP
Sim, G J G0TLN
Sim, J M G8VIB
Sim, K G7RNG
Sim, Q Y 2E1CSH
Sim, R G0AKW
Simarpi, L J W G6SIM
Simcock, C A GW4NAW
Simcock, D J M1ANT
Simcox, G F G0ASD
Simcox, M A GI0OTL
Sime, P R GM8YGI
Simes, S F G6DKK
Simister, B C A M0ADK
Simkin, O F G3HYJ
Simkins, C G8FKP
Simkins, J L G8IYS
Simkins, M P G7OBS
Simm, P G8TXM
Simmens, M J G4YQP
Simmers, A K S GM1HNZ
Simmonds, A G G0HND
Simmonds, D E G7PCI
Simmonds, H G6PIR
Simmonds, J G1YFI
Simmonds, J G3GLX
Simmonds, J D G6MVQ
Simmonds, J E 2E1AUS
Simmonds, J K G4XNW
Simmonds, K C G0PIB
Simmonds, K R G6TWR
Simmonds, N R G4MPH
Simmonds, R G0RSS
Simmonds, S G4GJY
Simmonds, S G6WLM
Simmons, A G6UX
Simmons, A M G1THD
Simmons, B E G8XJB
Simmons, B G G6SDC
Simmons, D M G1MAL
Simmons, D T R GW7MPW
Simmons, G A G3RFO
Simmons, G D G0VXM
Simmons, H F G0HSI
Simmons, H W G0XAZ
Simmons, J A R G6MPE
Simmons, J M G8SXU
Simmons, J R 2E1DAE
Simmons, N C GI7GJX
Simmons, P A GW0VYF
Simmons, P D G3XUS
Simmons, P J G3YHJ
Simmons, R F G1BYN
Simmons, R G G0WFC
Simmons, R K G8WNJ
Simmons, S G G4HXY
Simmons, T G GW7EXW
Simmons, T J G8JFX
Simmons, T P G6YRJ
Simmons, V J G6TSC
Simms, D H G4FQZ
Simms, H J G0LJS
Simms, J P G1CQF
Simms, P R G6HPE
Simms, R F G3ZUQ
Simon, E J GM4GZW
Simon, G GU7CMH
Simon, S L G4GZV
Simonds, A L G8KUV
Simons, C S G0KEE
Simons, D A G7DKB
Simons, E G0PAQ
Simons, J G7LSW
Simons, J G G3ZVK
Simons, L 2E0ABT
Simons, P M G4CCZ
Simons, T J G0VHS
Simpkins, C P T G3MCL
Simpkins, D R G6VOE
Simpkins, D R J 2E1CLH
Simpkins, R A G8SOL
Simpkins, T G7PDS
Simpson, A F G3UMF
Simpson, A R P G8YMQ
Simpson, A T G3YRB
Simpson, B G1DWL
Simpson, B G8PZF
Simpson, B D G3PEK
Simpson, B H GW4YBV
Simpson, B L GM1KWA
Simpson, C F G3OOZ
Simpson, C J G7LCW
Simpson, D G6PII
Simpson, D G6ZGF
Simpson, D G8NDF
Simpson, D C G8OCS
Simpson, D C GI4PRH
Simpson, D J G4BZL
Simpson, D J GM3LVA
Simpson, D R G6NGN
Simpson, E G G3KQS
Simpson, E L G0LES
Simpson, E L G3GRX
Simpson, E M GM4KNH
Simpson, E S G4ITM
Simpson, G G3RIV
Simpson, G C G4LWQ
Simpson, G J G3WYI
Simpson, H C GM1FTZ
Simpson, I G7HGF
Simpson, I GM4MBG
Simpson, I F GM3YND
Simpson, J GI0WYB
Simpson, J GM4YZT
Simpson, J B G0BQP
Simpson, J D G3NJQ
Simpson, J D GM7DSC
Simpson, J H G4HSG
Simpson, J M G4BUI
Simpson, J R G3JRS
Simpson, J T G4GFV
Simpson, K 2M1EKI
Simpson, K G6XBV
Simpson, K A G4DQF
Simpson, K J G4YPQ
Simpson, K J GW4JGW
Simpson, K W G0BUC
Simpson, L R GM3UQI
Simpson, M M1APX
Simpson, M C G0ORP
Simpson, M G G8BOI
Simpson, M J G1WRD
Simpson, M R G G3UVM
Simpson, P G0RUR
Simpson, P G7SEY
Simpson, P G8TFU
Simpson, P A G3XQZ
Simpson, P J G3GGK
Simpson, P J G6IBZ
Simpson, P M G1KGC
Simpson, P W G G8FQS
Simpson, R G0SPW
Simpson, R G4NOP
Simpson, R GM7NZI
Simpson, R A G3OMS
Simpson, R A G8SD
Simpson, R B G3UWE
Simpson, R E G3SLG
Simpson, R M G0KDG
Simpson, R V 2I1BHV
Simpson, S G4SGD
Simpson, S J 2E1FON
Simpson, S J M1BVX
Simpson, S L G4ZSS
Simpson, S M 2E1FOF
Simpson, T A GI0GTJ
Simpson, T L G3NSF
Simpson, V GI4BDL
Simpson, V GM6UJG
Simpson, W G4IKW
Simpson-Fraser, A GM6FLL
Simpson-Fraser, D D GM6FLM
Sims, A F G4VHV
Sims, A H 2E1BIH
Sims, C P G1ILY
Sims, E W T G6TVD
Sims, F G S G3WQW
Sims, G G0GLO
Sims, G G4GNQ
Sims, G G4LMR
Sims, J M W G1UEQ
Sims, J O G0RMA
Sims, M C G7RBA
Sims, P J G7IBP
Sims, R J G4CVX
Sims, R W GW4VNS
Sims, S G8NFZ
Sims, M E W G4HVE
Sims-Mindry, S P G1GOZ
Sinagoca, S GM6UJO
Sinclair, A D G7HUD
Sinclair, A K GD3FLH
Sinclair, A K GD3TNS
Sinclair, A L G0NDK
Sinclair, A M G0AMS
Sinclair, B G0MBS
Sinclair, D G7VIU
Sinclair, D A G4BFV
Sinclair, F S GM4SWU
Sinclair, G G G4BWH
Sinclair, H A GI4GOS
Sinclair, I 2M1ELU
Sinclair, I G1GXH
Sinclair, I M 2M1AVY
Sinclair, I M GM6KEC
Sinclair, I R G1GIA
Sinclair, J C GM3SBE
Sinclair, J I M G8VL
Sinclair, J R G7RZT
Sinclair, J W GM4YAA
Sinclair, M J G4BVF
Sinclair, M T GI8NCX
Sinclair, N G4VWK
Sinclair, N K 2M1AVZ
Sinclair, P A GI0UQK
Sinclair, P G G3UCA
Sinclair, S J G4EKF
Sinclair, W A GM3KLA
Sindall, J A M1ALG
Singer, H R G0IAA
Singer, I G G7VDA
Singer, N A G7HYM
Singh, S 2W0AJM
Singh, S MW1BOO
Singh-Gill, M G0UQW
Singleton, C P G0FOG
Singleton, D W G3UZJ
Singleton, G R GM3HKF
Singleton, J G4WJR
Singleton, J G6EQB
Singleton, K G7MCK
Singleton, K R M1BXP
Sinkinson, J B G0PTW
Sinton, R F S GI3ONF
Sircombe, I J G0FCM
Sirignano, B G4FZG
Sirley, I G4OLZ
Sisley, M J D G8ZID
Sismey, T G G4XQV
Sissons, C G4PXN
Sitton, A M G0HGA
Sivapragasam, J A C Q G4MSE
Sivapragasam, N G4IQD
Sives, A W GM6OGN
Siviter, A A G0AZI
Siviter, D J GW7EWD
Sivyer, F J G6SQS
Sivyer, R D A G6DJV
Sixsmith, P G0JFE
Sizer, P R GW6OTD
Sizmur, S G6YRI
Skacel, G G6VWP
Skakle, B GM1INS
Skarratt, J R 2E1FKP
Skate, G R G7LTM
Skea, E 2M0AOF
Skeats, R G GI4CYJ
Skeels, W R G4HOI
Skeggs, E P C S G4EPS
Skelcher, C K G3YHF
Skelding, M G D G3VHA
Skelhorn, H R G8BPU
Skellern, A F GW0SAU
Skells, R A G8YXJ
Skells, R C G0XTA
Skelly, J G6IWB
Skelly, J D GI0OUX
Skelton, B G4VSK
Skelton, M R G7NVM
Skeoch, I T GM7NAA
Skerritt, J W G1DGN
Skerritt, P G6ZYZ
Sketch, J E GW3DDY
Sketcher, B G0PVB
Sketchley, J A G4DCE
Skett, J R G4MRJ
Skewes, M 2E1CWE
Skidmore, B G4ERB
Skidmore, C J G0NGZ
Skidmore, J C M G8ZCJ
Skidmore, K M 2E1CXO
Skidmore, S R GM7RQK
Skidmore, W N G6CSC
Skillen, J M GI4TSK
Skillett-Habin, M J GU6GDO
Skillings, C G S G0FKG
Skillington, F G4DFU
Skilton, D J G1KPU
Skinner, A G0BIR
Skinner, A H 2E0AHF
Skinner, B C GM0IJA
Skinner, D C GW0BBF
Skinner, D T G4UXT
Skinner, E L G8FBV
Skinner, F G G0EHQ
Skinner, G G3CZU
Skinner, H W G7EEU
Skinner, J G7VSM
Skinner, J A GM1FOS
Skinner, L MM0ACR
Skinner, L K G4VEH
Skinner, M D M1AGR
Skinner, M P G7URX
Skinner, P G7MLJ
Skinner, R MM0ACT
Skinner, R F G4PMR
Skinner, R H G6JQD
Skinner, R W G4ULR
Skinner, R W G6EQF
Skinner, S W G4UXN
Skipper, D E J G4WAQ
Skipper, W G G0ZAT
Skippon, H A G1EPZ
Skipworth, B A G4USA
Skipworth, D G2FFD
Skivington, P P G4UUM
Skobelski, R J G6ROQ
Skolar, P J G4EYV
Skorupinski, L G1GRN
Skoyles, E G7NFW
Skoyles, R V G3MTJ
Skrobanski, Z P G3XDZ
Skull, M M0BEB
Skulski, G Z G6TNE
Skupski, G G0VMA
Skuse, K J G0PQD
Skuse, R V G0NOQ
Skutsch, O L A F G0NFW
Skye, D A G3PLR
Skye, D M G8EPK
Skyner, M R G4GHT
Skyrme, D J G3RQP
Slack, G M G4AKR
Slack, T R G4ANW
Sladden, J T C G3MWS
Sladden, J T C G7ERN
Slade, A J G0IJN
Slade, A J G0UTT
Slade, A J F G4DDU
Slade, D H G4YTB
Slade, M G7POB
Slade, M F G4ONS

Slade, P J	G4DPP
Slade, R J	G6PCC
Sladen, P A	G8BTD
Slaney, A M	G3YZT
Slaney, J H	G0DMA
Slaney, M F	G7NBX
Slark, A N	G0GVG
Slater, A L	G0NHF
Slater, A N	G0BDN
Slater, B D	G6XYX
Slater, C	G6EUG
Slater, F	G1ONH
Slater, G	G6XMZ
Slater, G	G7PMT
Slater, G B	G7HQB
Slater, G E	G1OSH
Slater, J A	G6FIO
Slater, J D	G3FOZ
Slater, J P	G7FEL
Slater, J P	G8FMJ
Slater, J S	G0WCJ
Slater, J W	G1RFH
Slater, J W	G6EUO
Slater, M E	G3NML
Slater, N T	G4WLE
Slater, P	G1GRP
Slater, P	G1NRK
Slater, P G	G0PGS
Slater, R N	G7UVE
Slater, R W	G8ILM
Slater, S A	G0PQB
Slater, W A	G0UUC
Slater, W H	G4YWY
Slater-Walker, C M	G6ELH
Slator, R	G4MSN
Slatter, B H	G4DF
Slatter, D N	G0CJG
Slatter, D W	G6WZC
Slatter, G E F	GW0KQU
Slaughter, A J	G0KMC
Slaughter, S C	G1BYO
Sleat, W E D	GM3FJA
Sleath, M A	G1PLA
Sledden, D	G1HIH
Sledge, E	G0ICY
Slee, C M J	G7VES
Slee, D R	G7JAU
Slee, J G	G7VER
Sleeman, A L J J	G4VNI
Sleeman, G R	G7PGY
Sleeman, M C	G7SFC
Sleep, D J	G4RLZ
Slegg, P W	G0LNY
Sleigh, A J	G1OSA
Sleigh, D B	GM4BBF
Sleigh, G	G0NIA
Sleigh, I D B	2M1CGG
Sleigh, R J	G0KBK
Sleight, C F	G6EVC
Sleight, J H	G3OJI
Sleightholm, M A	G8HDN
Slenk, C P	G0APA
Slessor, G R	GM1THS
Slight, D	GM8CFS
Slight, P J	G7PMQ
Slim, R A	G4TJI
Sliman, J D R	G0XJS
Slimmon, R J	G0WAY
Slinger, A G	G1JMM
Slingsby, A T	G3ZWN
Slingsby, C G	G8ADH
Slingsby, G A	G0IWJ
Slinn, A J	G7KTR
Sliwinski, S M	G4ILX
Sloan, J C	M0AEK
Sloan, M C	GU3WHN
Sloan, P	GM6PLG
Sloan, T E	GI4AHP
Sloane, T H A	G6AAB
Slocombe, D G	G8KOL
Slone, R D	G1FVP
Slyfield, R E	G0ISC
Sm1Th, C T	G6XZM
Smale, J	G3EDV
Smale, S J	G1ORT
Smale, W J	GW3DRK
Smales, G	M1CCG
Small, B M	G0CHB
Small, C	GM3PFX
Small, D	GM6BGQ
Small, D L	G7WBH
Small, G T	G6RTD
Small, J B	G1BQC
Small, J G	G4BEU
Small, J G	GM1JIU
Small, K	G1UZR
Small, K G	GM7GTK
Small, M	G4DVI
Small, R S	G3ALI
Small, S G	G4HJE
Small, W	G0GHO
Smalley, J R	G7OOS
Smalley, R	G3LJO
Smallshaw, B J	G7CSY
Smallwood, G E	GW4TJN
Smallwood, I	G0ANK
Smallwood, J	G7KJD
Smallwood, J A	G6KDJ
Smallwood, J E	GM3AEI
Smallwood, M	G0GXX
Smallwood, R	GW6POO
Smallwood, R V	G8DGR
Smallwood, S J	GM7NEW
Smallwood, T J	G7LOG
Smart, A E	G0UYH
Smart, A G	GM0RML
Smart, A W	G8ZZY
Smart, C	G8OCV
Smart, D B	G3MGB
Smart, J A	G8NTS
Smart, J B	G0SRY
Smart, J D	G0CTP
Smart, K M	G1DWN
Smart, L E	GW0LBI
Smart, M R	G3UXQ
Smart, P K	G4SDU
Smart, R G	G1CRB
Smart, R M	G6BHA
Smart, W H	G6DJE
Smeaton, E	M0AGS
Smedley, A	G7AAW
Smedley, F A	G3VJJ
Smelt, C	G0PST
Smerdon, G A	G7JCI
Smethers, R	G3NLY
Smethers, R	G3WAS
Smethurst, E F	G4PJV
Smethurst, J	G3UGC
Smethurst, J S	G0FMO
Smethurst, K	G3GPE
Smiles, W K	G8PBV
Smiley, P J	GI7AAH
Smillie, D	GM4GM
Smillie, D A	GM4FKD
Smillie, G S	G6TRY
Smit, P C	G0LIY
Smith, A	G0PBC
Smith, A	G1DIK
Smith, A	G1HOD
Smith, A	G1YMJ
Smith, A	G3SQM
Smith, A	G4EFP
Smith, A	G4FAI
Smith, A	G4MDJ
Smith, A	G6AGY
Smith, A	G6YRC
Smith, A	G8CNI
Smith, A	G8KDM
Smith, A	GM4KQS
Smith, A	GM7CTV
Smith, A	GW0NHE
Smith, A A H	G4HEH
Smith, A B	2E1BGN
Smith, A B	G6JDY
Smith, A B	G7VIG
Smith, A B B	GI7IKP
Smith, A C	G8JZJ
Smith, A C	GM6YRH
Smith, A C F	GM4PVC
Smith, A D	G0WAS
Smith, A D	G7OKE
Smith, A D	G7PAK
Smith, A D	GI0NQC
Smith, A E	G0KQD
Smith, A E	G1CMW
Smith, A E	G4FMK
Smith, A E	G4MTH
Smith, A E	G6FNQ
Smith, A E	G6HPO
Smith, A E	G7CMX
Smith, A E H	2E1CSD
Smith, A G	G3KVT
Smith, A G	G3WPD
Smith, A G	G6XDS
Smith, A G	G7JWN
Smith, A G	GI8WBZ
Smith, A G	MI0AWL
Smith, A J	G1TKH
Smith, A J	G1VNL
Smith, A J	G3EMJ
Smith, A J	G4AUB
Smith, A J	G4OEP
Smith, A J	G7ANX
Smith, A J	GI0VIB
Smith, A J	GM0FQS
Smith, A K	G6THN
Smith, A L	G1LPJ
Smith, A L	G1MBA
Smith, A L	GW0JBH
Smith, A M	G4LSS
Smith, A M	G4PJX
Smith, A M	GU1DWO
Smith, A P	G4BBW
Smith, A P	G4XEZ
Smith, A P	G7FWE
Smith, A P	G7IZU
Smith, A R	G0RYV
Smith, A R	G3MPB
Smith, A R	G4ZSA
Smith, A R	GM0SYW
Smith, A R	GW0TXP
Smith, A R G	G0UAS
Smith, A W	G1JVY
Smith, B	G0IVF
Smith, B	G0UYV
Smith, B	G1EIO
Smith, B	G1XKL
Smith, B	G4THF
Smith, B	G6TET
Smith, B	G6UMA
Smith, B	G7DOS
Smith, B	G8UYR
Smith, B A	2E1FRM
Smith, B A	G3WCY
Smith, B A H	GM4RAZ
Smith, B B	G0IER
Smith, B D	G1ALZ
Smith, B D	G1JOJ
Smith, B E	G6RRU
Smith, B G	G4MHX
Smith, B H	G1NWU
Smith, B J	G0DAH
Smith, B J	G0OZL
Smith, B K	G4EQC
Smith, B M	G7AMU
Smith, B M A	G8UVQ
Smith, B P	G4ETN
Smith, B R	G0SUC
Smith, B R	G1GDR
Smith, B S	G4IAT
Smith, B S	G8DAB
Smith, B T	G1TJW
Smith, B W	2E0AHN
Smith, B W	G0TPB
Smith, C	2E1DDR
Smith, C	G0ICA
Smith, C	G3GHY
Smith, C	G3YA
Smith, C	G4VQP
Smith, C	G7SWV
Smith, C	MW1BLD
Smith, C A	2E1EED
Smith, C A	G0JTN
Smith, C A	G0PGA
Smith, C C	G1PGI
Smith, C C	G6PLH
Smith, C F	G1DIM
Smith, C F	G3MPF
Smith, C G	G0IHU
Smith, C G	G4WLS
Smith, C H T R	G7USE
Smith, C J	G0LIN
Smith, C J	G3UFS
Smith, C J	G4NUX
Smith, C J	G4XBS
Smith, C L	G0VAE
Smith, C L	GW3MOV
Smith, C M	2E1EBR
Smith, C M	G3OBY
Smith, C M	G7JXJ
Smith, C M	G8KVU
Smith, C M	G8LMW
Smith, C M	GM4BVO
Smith, C P	2E1CDU
Smith, C P	G1FEF
Smith, C R	GM0CLN
Smith, C S	G0VJC
Smith, C S T	G6DFB
Smith, C T	G6BFW
Smith, C T	G6THM
Smith, C T B	G4VCP
Smith, C V	G4FZH
Smith, C V S	GW6VKD
Smith, C W	G3CSE
Smith, D	G0PFC
Smith, D	G0TCS
Smith, D	G1ZJQ
Smith, D	G4COE
Smith, D	G7AJM
Smith, D	G8OMC
Smith, D	GM0KCN
Smith, D	MM1BJT
Smith, D A	G1DAN
Smith, D A	G2FHK
Smith, D A	G6SBI
Smith, D A	GM6KEV
Smith, D A D	G8IDL
Smith, D B	2M1EJY
Smith, D B	GM3PML
Smith, D C	G0PXY
Smith, D C	G6IDJ
Smith, D C	G7JRC
Smith, D C	G7NKH
Smith, D C	GW3TMS
Smith, D E	G6NZW
Smith, D F	G4TBG
Smith, D F J	GM0BFS
Smith, D G	G0GQY
Smith, D G	G0JCF
Smith, D G	G1DES
Smith, D G	G3UGJ
Smith, D G	G4EQE
Smith, D G	G7GMQ
Smith, D G	GM3WML
Smith, D H	G4YNN
Smith, D H	G6FIL
Smith, D I	2E0AHJ
Smith, D J	G0CFR
Smith, D J	G0JPL
Smith, D J	G0JYU
Smith, D J	G0SXM
Smith, D J	G1OPA
Smith, D J	G1PBB
Smith, D J	G3PZQ
Smith, D J	G4EMU
Smith, D J	G4PBO
Smith, D J	G8DVN
Smith, D J P	G8KTG
Smith, D K	GM8PIW
Smith, D K	GW7MQE
Smith, D L	G4XUR
Smith, D L	G7CLH
Smith, D M	G6XMR
Smith, D M	GM4PKJ
Smith, D N	G4RSO
Smith, D O	G6SNX
Smith, D P	G6ZFW
Smith, D P	GM0EEY
Smith, D S	G4DAX
Smith, D S	GM4DSS
Smith, D T	G7AHT
Smith, D W	G0CTE
Smith, D W	G4UQU
Smith, D W	G8ATY
Smith, D W	GM4MPC
Smith, D W	M0AYQ
Smith, D W A	G3JBV
Smith, D W J	G3XPD
Smith, E	2E0AQC
Smith, E	G3BPQ
Smith, E	GI4MRZ
Smith, E A	G0BAM
Smith, E A	G0CFS
Smith, E A	G1SKR
Smith, E A	G3JYA
Smith, E A W	G4VHQ
Smith, E B	G0BKL
Smith, E D	G7WEG
Smith, E D W	GW4VTG
Smith, E F	G4YZC
Smith, E H	G0XBO
Smith, E H	G8GJI
Smith, E J	G0VYV
Smith, E J	G3VUV
Smith, E J	G8YCN
Smith, E M	G4MJU
Smith, E M	G8HWJ
Smith, E P	G0JLY
Smith, E W	G0XBP
Smith, E W	G3LHG
Smith, F	G3YGL
Smith, F	G4GOM
Smith, F C	GW2DDX
Smith, F J	G1IZB
Smith, G	2E1AES
Smith, G	G0BZC
Smith, G	G1BQI
Smith, G	G3PLN
Smith, G	G3ZZI
Smith, G	G4PSI
Smith, G	G4RVZ
Smith, G	G6JMA
Smith, G	G6WLP
Smith, G	G6XIO
Smith, G	G6XMU
Smith, G	G7RRD
Smith, G	G8AXH
Smith, G	G8RTN
Smith, G	GM6WRY
Smith, G	GW4ZUS
Smith, G	M0AMM
Smith, G A	G3TSJ
Smith, G A	G7DOW
Smith, G A	G7IMZ
Smith, G A	G7IXK
Smith, G C	G6FLQ
Smith, G C	GM4XGY
Smith, G E	2E0AHI
Smith, G E	G2BOX
Smith, G E	G8WQK
Smith, G E R	G7UHP
Smith, G G	G6KXK
Smith, G G	GM1KUI
Smith, G H	G1EIP
Smith, G J	G1NLK
Smith, G J	G3CNX
Smith, G J	G3ZOD
Smith, G J	G4EBK
Smith, G J	G4NMD
Smith, G J	G6APJ
Smith, G J	GW4EUA
Smith, G J	M0AGF
Smith, G J A	GM4XUS
Smith, G L	2E1BWM
Smith, G L	G1JQT
Smith, G L	G4DRW
Smith, G L	G6CRM
Smith, G L	GW0OLZ
Smith, G M	G3SNO
Smith, G M	GM4GZD
Smith, G N	G0VCZ
Smith, G N	G1XAV
Smith, G N	G8KUC
Smith, G R	G3WYU
Smith, G R	G4AJJ
Smith, G R	G7ISV
Smith, G R	G7LTU
Smith, G S	G0PGJ
Smith, G S	G1LIF
Smith, G S	G4GFH
Smith, G T	G0CUB
Smith, G T	G1OEK
Smith, G W	G0HPA
Smith, G W	G3GGU
Smith, G W	G4PWB
Smith, G W	GW6TEO
Smith, H A	2E1BVO
Smith, H A	G7BFG
Smith, H B	G7OIC
Smith, H C	G0IZB
Smith, H E	G3IVF
Smith, H E	G3POY
Smith, H H	G3YSN
Smith, H J	2E1FTG
Smith, H J	G3ARU
Smith, H J	G3HSS
Smith, H J L	GM1CQC
Smith, H L	G4UIG
Smith, H M	G4KNQ
Smith, H R	G6NWM
Smith, H W	G0LQT
Smith, H W	G0OVS
Smith, I	G8MVV
Smith, I C R	G6RHV
Smith, I D	G1EYJ
Smith, I D	G8RYL
Smith, I E	G4WZQ
Smith, I G	G4GDX
Smith, I G	G7TJZ
Smith, I H	G7HMA
Smith, I H	GM1FEM
Smith, I R	G4FCY
Smith, I R	M1ASB
Smith, I W F	GW6WJM
Smith, I W K	G2BPW
Smith, J	G0JJS
Smith, J	G0MCY
Smith, J	G0UAO
Smith, J	G1DGP
Smith, J	G1VTS
Smith, J	G3ZQC
Smith, J	G4VLY
Smith, J	G4WNU
Smith, J	G7AXM
Smith, J	G8LFN
Smith, J	GM4UYP
Smith, J	GW0DRI
Smith, J A	G0AEB
Smith, J A	G0FKF
Smith, J A	G1KCR
Smith, J A	G1MRQ
Smith, J A	G8XER
Smith, J A	G8ZQB
Smith, J A	GM0REW
Smith, J A W	G4VEL
Smith, J B	G3HSR
Smith, J C	G0OIY
Smith, J C	G7UGW
Smith, J D	G0OJN
Smith, J D	G3KGW
Smith, J D	G7CKV
Smith, J D	GI7UQY
Smith, J D	MI0AEX
Smith, J E	G1SRJ
Smith, J E	G3JZF
Smith, J E	G3SMV
Smith, J E	GM0JCX
Smith, J F	G4YLT
Smith, J F	G7PFG
Smith, J G	G4XJS
Smith, J H	G4PET
Smith, J H	G4ZMA
Smith, J H	G6AAK
Smith, J H	GW0NZN
Smith, J J	GM1OSZ
Smith, J K	GM3HVT
Smith, J M	G0TQP
Smith, J M	G3OHW
Smith, J M	G3YWS
Smith, J M J	G0CCQ
Smith, J N	M0ANJ
Smith, J O L	G7WFP
Smith, J P	G3WTS
Smith, J P	G6FLH
Smith, J Q	G3SLX
Smith, J R	G0OFE
Smith, J R	G1OAW
Smith, J R	G7ATP
Smith, J R	G7MWW
Smith, J R	G7UME
Smith, J R	GI0USC
Smith, J R	GM7VFR
Smith, J R F	G0WWT
Smith, J S	G1IHE
Smith, J S	G4KJJ
Smith, J S	G7MUN
Smith, J T	2E1EHA
Smith, J T	G1BRD
Smith, J T	GM1MBZ
Smith, J W	G3ZJS
Smith, J W	GM4EOU
Smith, J W	GM4YHO
Smith, K	G4KEN
Smith, K	G6IKW
Smith, K	G7UXW
Smith, K A	G0XBN
Smith, K A	G8ACO
Smith, K B	G3IYU
Smith, K C	G8BEN
Smith, K G	G4RSZ
Smith, K H	2E1CDT
Smith, K H	G4XAQ
Smith, K J	G3JCR
Smith, K J	G8VIK
Smith, K L	G3JIX
Smith, K L	G3SRE
Smith, K L	G3UKC
Smith, K L	G8DBO
Smith, K L	G8KUC
Smith, K M	G0DHT
Smith, K N	G3RB
Smith, K N	G4PEU
Smith, K N	G4YYE
Smith, K N	G8CBO
Smith, K N	GW0BND
Smith, K R	G3TLB
Smith, K R	G8SUM
Smith, K W R	G8DNL
Smith, L	2E1EIP
Smith, L	G0EYV
Smith, L	G0GBK
Smith, L	G0RPF
Smith, L	G4VNK
Smith, L	G6AYN
Smith, L	G7UTZ
Smith, L H	G0HFT
Smith, L J	G3HDJ
Smith, L J	G3HJF
Smith, L N	G1AEJ
Smith, L R	G0RNM
Smith, L T	G0VPT
Smith, L W	G7GNA
Smith, M	2E1DDQ
Smith, M	G0WXV
Smith, M	G3TRV
Smith, M	G6STL
Smith, M	G6YZR
Smith, M	G8EWD
Smith, M	GM8ZEQ
Smith, M	GW4VUC
Smith, M A	G0FVU
Smith, M A	G1PIB
Smith, M A	G1WRO
Smith, M A	G7VGH
Smith, M A	GW7NIS
Smith, M C	G0BIW
Smith, M C	G0DPT
Smith, M C	G1MBW
Smith, M C	G7LGV
Smith, M C	GW0DIQ
Smith, M D	G1DFF
Smith, M D	G1KLI
Smith, M D	G1XAF
Smith, M D	G7TKO
Smith, M F	G4MFS
Smith, M F	G6ZGA
Smith, M G	G4FIK
Smith, M G	G6OES
Smith, M G	G6UUO
Smith, M G	GW0CES
Smith, M J	G0CHC
Smith, M J	G0TTD
Smith, M J	G3RMN
Smith, M J	G3SBR
Smith, M J	G4FQI
Smith, M J	G4OKM
Smith, M J	G4OXV
Smith, M J	G7EQP
Smith, M J	G7SDQ
Smith, M J	G8EGU
Smith, M J	G8EII
Smith, M J	G8YVC
Smith, M J	GM3WHT
Smith, M J	GW4DWX
Smith, M J	M1ACI
Smith, M J W	G4PRG
Smith, M K	G0RFD
Smith, M K	G4HMA
Smith, M O	G0FXY
Smith, M R	G0RDL
Smith, M R	G7SDD
Smith, M R	G8YEO
Smith, M R L	G4BTE
Smith, M R P	G6ZRC
Smith, M S	G3WXM
Smith, M S	GW6XZI
Smith, M T	G8NAG
Smith, M T E	G8GAT
Smith, M W	G3UAF
Smith, M W	G4XJO
Smith, M W	G6NIO
Smith, M W	G7LGG
Smith, N	G7HRD
Smith, N	G7JRI
Smith, N	MW1AYS
Smith, N A	G3HFO
Smith, N A	G3SKD
Smith, N A	G7PLL
Smith, N B C	G0ILA
Smith, N D	G0UQJ
Smith, N E	G6XZL
Smith, N F	2E1EIK
Smith, N G	G4GJC
Smith, N J	G7RMO
Smith, N J D	G3YII
Smith, N K	G1VTW
Smith, N M	G7AUQ
Smith, N M D	G7HJU
Smith, N P	G7TBF
Smith, N R	G0NIG
Smith, N R	G4DBN
Smith, N S	G8KZS
Smith, N T	G6HPS
Smith, N V	G6FVT
Smith, N W	G4EQD
Smith, P	2W1DEA
Smith, P	G0JTM
Smith, P	G3YWT
Smith, P	G4TON
Smith, P	G7JGY
Smith, P	G7OFI
Smith, P	G8IAR
Smith, P	GI0ACE
Smith, P A	G1VBP
Smith, P A	G4TNM
Smith, P A	G7LBS
Smith, P A	G7UWE
Smith, P A H	G3CJH
Smith, P A S	GW0VMR
Smith, P B	G0RRF
Smith, P C	G1BJW
Smith, P C	G1HIA
Smith, P C	G8JZI
Smith, P C	G8OLK
Smith, P C	M1BXB
Smith, P D	G1LTI
Smith, P D	G3PZZ
Smith, P D	G8CYL
Smith, P E	G0FPP
Smith, P E	G1HED
Smith, P E	G2DPL
Smith, P E	G3BRS
Smith, P E	G4JNU
Smith, P E	G6BRS
Smith, P E	G7TZM
Smith, P E	G8NRS
Smith, P F	G0VAO
Smith, P F	G1KEI
Smith, P G	2E0APW
Smith, P G	2E1CRW
Smith, P G	G3TJE
Smith, P G	G7HQF
Smith, P J	2E1FLP
Smith, P J	G0BQB
Smith, P J	G1WIY
Smith, P J	G3PPU
Smith, P J	G6XND
Smith, P J	GM0ATL
Smith, P J	GW1XBG
Smith, P J R	G4EES
Smith, P K	G0JPJ
Smith, P K	G3UPW
Smith, P M	2M1FQK
Smith, P M	G0CQR
Smith, P M	G0MEA
Smith, P M	G4LWB
Smith, P M	G4ZWQ
Smith, P P	G3YWF
Smith, P R	G0BZX
Smith, P R	G0EOS
Smith, P R	G1ALW
Smith, P R	G1OYZ
Smith, P R	G3WPB
Smith, P R	G6WBG
Smith, P R	G7NBM
Smith, P R	G7PNM
Smith, P S	G8JSL
Smith, P W	G3NEI
Smith, P W	G7PQX
Smith, P W	M1BOY
Smith, R	G0IWU
Smith, R	G0JXO
Smith, R	G0OCV
Smith, R	G0RRC
Smith, R	G0TWA
Smith, R	G1HID
Smith, R	G1TPS
Smith, R	G3DHU
Smith, R	G3NBI
Smith, R	G4DXW
Smith, R	G4IYE
Smith, R	G4LZY
Smith, R	G4MGU
Smith, R	G4NME
Smith, R	G4ZZI
Smith, R	G6EQI
Smith, R	G7JTZ
Smith, R	G8BWC
Smith, R	GM0BWU
Smith, R	GM3JPF
Smith, R	M1BTW
Smith, R A	2E1CDD
Smith, R A	G0ERS
Smith, R A	G0GIZ
Smith, R A	G0HJM
Smith, R A	G1POV
Smith, R A	G4BZG
Smith, R A	G6EKW
Smith, R A	G8YZC
Smith, R A	GW0AYQ
Smith, R A	M1BNG
Smith, R A C	GM4IOB
Smith, R A J	G3VKT
Smith, R A J	G4KZK
Smith, R B	G3YMA
Smith, R C	G0MSE
Smith, R C	G0NRU
Smith, R C	G1EIR
Smith, R C	G4ZXA
Smith, R C	G6KMD
Smith, R C	G7IAY
Smith, R D	G1JSS
Smith, R D	MM1ASI
Smith, R E	G4NCI
Smith, R E	G4YUI
Smith, R E	G6NHO
Smith, R F	G4TTX
Smith, R F	G5KN
Smith, R F	G8XOQ
Smith, R F	GM7FLM
Smith, R G	GM4RGS
Smith, R G	GW7STV
Smith, R J	G1MNY
Smith, R J	G3MYN
Smith, R J	G6NZZ
Smith, R J	G7TDR
Smith, R J	G8NCT
Smith, R J G	G0VKX
Smith, R J S	G1CJB
Smith, R K	GM0KDC
Smith, R L	G4WSF
Smith, R M	G0LLB
Smith, R M	G0UHR
Smith, R M	GM7GKT
Smith, R O	G7NEG
Smith, R O R	G0HOU
Smith, R P	G0LZY
Smith, R P	G3SVW
Smith, R P	G3XXJ
Smith, R P	G3ZEJ
Smith, R P	G8HMA
Smith, R P W	G1XDG
Smith, R R	G0FNT
Smith, R R	G6TQ
Smith, R R	GM4JHG
Smith, R R C	G8ZGU
Smith, R S	G1AEF
Smith, R S	G6TFJ
Smith, R T	G7VYI
Smith, R W	2E1AZN
Smith, R W	G1GAW
Smith, R W	G1RWS
Smith, R W	G6RNF
Smith, R W B	G3LVW
Smith, S	G1WYC
Smith, S	G4SSV
Smith, S	G7PWH
Smith, S	GM0RDZ
Smith, S A	G0BQZ
Smith, S A	G4LSG
Smith, S A	G7KHL
Smith, S A	G8AZB
Smith, S C	G0ONR
Smith, S D	G7JFM
Smith, S D	G7JKY
Smith, S D	M0BFT
Smith, S D	M1AKI
Smith, S F	G7VBW
Smith, S G	G0NWQ
Smith, S G C	G8GPA
Smith, S J	G0WUE
Smith, S L	G1KXP
Smith, S M	G6STJ
Smith, S M	G6XMP
Smith, S M	GM4LUS
Smith, S P	G3WMY
Smith, S P	G7OKT
Smith, S R	G0TDJ
Smith, S R	G4CW
Smith, S R	G6GJY
Smith, S R	G8LMX
Smith, S S	GM4GFR
Smith, S T	G0WUD
Smith, S T	G3BSI
Smith, S T	G7KMH
Smith, S W	G0MGG
Smith, S W	G4SRS
Smith, T	G0OIS
Smith, T	G3YZU
Smith, T	G4VYZ
Smith, T A	2E1DDV
Smith, T A	G1VYB
Smith, T A	G1WNW
Smith, T A	G4ZZZ
Smith, T A	G6MPK
Smith, T B	G0PSU
Smith, T D J	G7KGM
Smith, T J	GW4LZL
Smith, T K A	G4DKC
Smith, T M	G7DKS
Smith, T N	G6AQI
Smith, T R	G1GHU
Smith, T T	G7RVT
Smith, T W A	G3EFY
Smith, V A	G1JJR
Smith, V J	G6RYW
Smith, V S	G7AEH
Smith, W	G0NSE
Smith, W	G3TAS
Smith, W	G4OZC
Smith, W B	G6IM
Smith, W E	G1JMW
Smith, W E	GW4OXL
Smith, W E G	G2CBC
Smith, W F	G7PJJ
Smith, W H	G0HWE
Smith, W H	G4ROH
Smith, W J	GD0PLR
Smith, W J	GD4IOM
Smith, W J	GD7HMG
Smith, W J	GW0GIQ
Smith, W M	G8CSP
Smith, W R	2E0AAM
Smith, W T	G1VRF
Smith, W T	G4TCM
Smith, W W	G3GOV
Smith Iii, J W	M0APA
Smith Jones, A	G0OIU
Smithers, C R	G4CWH
Smithers, R A	G4HSD
Smithers, T W	G0KTN
Smithies, S R	G3OKS

Smithson, C G8TTU
Smithurst, J W G0GLS
Smithyes, K G R G0MDJ
Smith-Gauvin, S W . GJ0JSY
Smoker, J L G6PLF
Smy, J V GM4ILE
Smy, S E 2E1BBX
Smye, J R G7SSG
Smye, M H G4DHC
Smylie, J A GI4OUC
Smyth, A F G3XNE
Smyth, A G GI3POS
Smyth, A R GD3ZCN
Smyth, E A GI0UTS
Smyth, I M GI0VSG
Smyth, J GI4LZS
Smyth, J W GM3LTD
Smyth, K J G3UTA
Smyth, M D G0BXM
Smyth, M J G3YFM
Smyth, M J GI0TMS
Smyth, P J G0MQU
Smyth, P J GI1ZXM
Smyth, R A GI4CBG
Smyth, W GI7KHR
Smyth, W J G6MDM
Smyth, W R GI8RNG
Smythe, D W GI1VPA
Smythe, S B G3ODH
Smythe, S E G8ODX
Snaden, H A G4YNV
Snaith, D P G0WGQ
Snaith, G E G6KVX
Snape, D G3BPK
Snape, D G4GWG
Snape, E G0BBT
Snape, K R G3UPN
Snape, T R G7LPN
Snarey, S G0LXK
Snart, P A G7CSZ
Snary, M L F 2E1AQS
Snary, R F G4OBE
Sneap, M I G3ZYC
Sneap, M I G8ZYC
Sneath, A N G8YMW
Sneath, R W G8XOX
Snee, J J GM3JWV
Snelgrove, A GW1NCS
Snelgrove, A J G7VON
Snelgrove, J G7VOM
Snelgrove, J A GM4OFC
Snell, A P G7WEA
Snell, B J G0TSB
Snell, C T G8LVW
Snell, E E G0LDV
Snell, E W G6PRE
Snell, J A G0RDO
Snellgrove, G K GW8LTV
Snellin, K M G6DNL
Snelling, P A G8ZFX
Snelling, R E G7TXD
Snelling, R E G8ACM
Snelling, R K GW7BSC
Snelling, R K MW1AZR
Sniadowski, J R G4BRH
Sniezko-Blocki, M A G0MCM
Snitch, P A G1BYP
Snoddy, W J G4OLB
Snodin, R D 2E1CFE
Snook, C G GW4HCV
Snook, G M1ADM
Snook, T J G6AHC
Snow, A B G8YMR
Snow, A W G4FLS
Snow, C E G8YSV
Snow, C M G1XVD
Snow, D J M1AMB
Snow, D M GW3PRL
Snow, J 2E1EXT
Snow, J M S G4TSH
Snow, K H GW8PNR
Snow, M B G6MBL
Snow, P J C G7JRX
Snow, P R G1DMN
Snow, P R G4HNC
Snow, R A G0BSP
Snow, S D 2E0AJG
Snow, W C G0CYD
Snowden, L J G0UQI
Snowden, R D G0GMH
So, Y H GM0MOP
Soaft, I K G4TGV
Soakell, J D G7TSN
Soakell, J D M0AYC
Soames, C G G0TZZ
Soane, A M M0ABY
Soane, M E G0XBC
Soar, D G1DWR
Soars, A G0ALI
Soars, A J G4VCN
Soars, C E G6VAW
Soars, I J G3HGI
Soars, M J G4TCI
Sobey, R J G4AER
Sobey-Smith, R S G6OTX
Soble, A P G1ZND
Soble, M A G8UVU
Sobye, P W G0PNM
Sockett, A A G0KXZ
Soesan, P G8HLF
Softley, M W G7JDQ
Softley, R G0JUN
Sohal, G S G6DWM
Sohst, R V G6RVS
Sojkowski, D J G6DJS
Sole, A D G6XYW
Sole, D J G3AEF
Sole, J A G3NEZ
Sole, M G0PFA
Sole, M G4UQF
Sole, S W GJ6RND
Solkow, G A G6WFF
Sollom, P W G3BGL
Solly, C G7JSO
Solly, J GM3EQZ
Solman, P J G0GUL

Solomon, J E G6KNK
Solomon, S G G7SBC
Solomon, T R G0DDS
Solomons, R G6CRR
Soltysik, E G4KWR
Solway, B J 2E1DUF
Solway, G J G1PED
Somerfield, A D G4XBK
Somerfield, P J G6SYI
Somers, G M G7VFV
Somers, I E G3ZHX
Somers, O R G6RZJ
Somerville, E GM8BOM
Somerville, W J G4WJS
Somerville Roberts, R J G7BWQ
Sommerfield, D G0NHR
Sommerfield, D G4TAY
Sondhis, J S G4EIV
Sonley, J M G3XZV
Soper, M L G0VVY
Soper, R G N G3VJZ
Sorab, A G6VXZ
Sorab, P G3NDO
Sorbie, T GM3MXN
Sore, G R GM1JVI
Sorensen, T C G8MFO
Sorger, B N G4HWY
Sorrell, I B G7HQC
Sourbutts, R G GW0NEJ
Souter, E F GM8PIV
Souter, R V G1RVS
South, G G4YAP
South, K D G3SGJ
South, W G GW0HNT
South, W N G7MND
Southall, C A G8UEZ
Southall, G R G7WLV
Southall, J M G4DMT
Southall, M G3WWS
Southall, M C G7IHW
Southall, R A C G4NLK
Southam, E T 2E1DVB
Southcombe, E W G G0MKF
Southerington, R J G0IOZ
Southern, D W G8PZE
Southern, G D G3RWW
Southern, K G4PJW
Southern, R G4WAP
Southern, R A G6BDY
Southern, R V G3RST
Southey, D R J M G0EYX
Southgate, H J 2E1EAS
Southgate, K H G3XSC
Southgate, P J G6ILD
Southgate, S G8FF
Southon, E F G0MRA
Southorn, N D G8XYA
Southward, P D G8YGM
Southwart, A M G7IQC
Southwart, P D G4RVW
Southwell, B G0MQA
Southwell, B R G4VTY
Southwell, E A P G4PXH
Southwell, F S G6ZRU
Southwell, G L G4IGY
Southwell, G V GM4HVR
Southwell, J G1LIT
Southwell, L H F G3JLS
Southwell, S W G0LFN
Southwell, T G4FEU
Southwell, W M M1BZO
Southworth, D J G0IKV
Southworth, K G 2E1DSH
Southworth, W G0VYP
Soutter, D K M1CCE
Soutter, L D L G4XHK
Soutter, V G0OXW
Sowden, G W G3WGZ
Sowerbutts, J G1BQQ
Sowerby, B M G4MXY
Sowter, B C G3NAP
Sowter, G A V G2OS
Sowter, G T G8ULJ
Spacagna, S G8OQO
Spacagna, T A G8OQP
Spacey, J R G0RII
Spacey, M J G1NWZ
Spacey, R M G0LUD
Spackman, C J G3GYQ
Spafford, M G0TJG
Spain, T J GW1GXM
Spain, T R G3NSS
Spalding, D L J G1JQH
Spalding, D L J G7UMH
Spalding, I N G4RYM
Spanner, E W G0DFU
Spanswick, H J G4RHI
Spanton, E R G8JTG
Sparey, D L G0DLS
Spark, G E G7FQY
Sparke, B J G6NZY
Sparke, P G D W G7TNU
Sparkes, A C G1LZQ
Sparkes, A C G7BRN
Sparkes, C O J G0JRO
Sparkes, D A G0UJI
Sparkes, D A G0VDX
Sparkes, G F B G4UJY
Sparkes, G T G3ZTF
Sparkes, I P G7MQB
Sparkes, J S G0SDT
Sparks, A G G0VHW
Sparks, B R G4JYB
Sparks, C L M1BQE
Sparks, F M G7JVQ
Sparks, J M G0SPX
Sparks, K H G7COT
Sparks, K J M1BQD
Sparks, L G1LCH
Sparks, M G GW1ETH
Sparks, N J G4ZUX
Sparks, R J G0TEI
Sparks, S C M1BQC
Sparrey, N G0IMK
Sparrey, T M G0CRY

Sparrow, A A H G3EKD
Sparrow, B R G0DZY
Sparrow, C J G1PSL
Sparrow, D J G8XOR
Sparrow, M J G3KQJ
Sparrow, N A G7LNU
Sparry, R E G3BJC
Spashett, H A G3RK
Spashett, J M G7OYZ
Spashett, P GW1IHB
Spaven, G P G1HQO
Spavins, B V G7STG
Speak, D M G6MZW
Speak, N E G4XNS
Speak, T B G6TEX
Speake, J D G3URX
Speake, S G4WFO
Speakman, A G 2E1EMF
Speakman, A G M1APL
Speakman, B J G3UBS
Speakman, J H G4TUM
Speakman, J K G1SUH
Spear, C G3SSQ
Spear, R W G4BXM
Spearing, J J G0BUD
Spearing, T S G7DVO
Spector, E M G4GCV
Speechley, D S G4UVJ
Speed, A M M GM3JIH
Speed, D P G4ZSL
Speed, H G3BMO
Speed, P R G4YVW
Speers, J M G0BVM
Speight, A G3PYW
Speight, C G L 2E0AHH
Speight, D L GJ6WDN
Speight, H F G3XTD
Speight, M D G8ZNL
Speight, T L G0TEE
Speirs, D G1OKM
Speirs, G H G1DAU
Speirs, G W A GM0AGN
Speller, E W G3LZR
Speller, J R G0EBI
Speller, J R GJ3YLN
Spelman, P J G4EKD
Spence, A 2M1AGP
Spence, B S G4MPS
Spence, B T GM4FNE
Spence, C GM0NZM
Spence, C W M0AVW
Spence, D S M GM4MTI
Spence, E W G1OGJ
Spence, J G4OLI
Spence, J G6ZYS
Spence, J B G4BSS
Spence, J R G0AME
Spence, J S G0FPI
Spence, M G G4SOH
Spence, P G1UDS
Spence, P C G3ALC
Spence, R G0ACO
Spence, R W J G8WUG
Spence, R W R GM7RDH
Spence, S GM0RSE
Spence, S GM3YCG
Spence, S H P GM7TMT
Spence, V G2CZO
Spenceley, G G G4DQN
Spenceley, N M G8JUG
Spenceley, P A G8LZA
Spencer, A G3PMO
Spencer, A G7AUB
Spencer, A A G7VCX
Spencer, A M L GM0AQG
Spencer, B G0AYI
Spencer, B G4GVI
Spencer, B A G0BDE
Spencer, B C G4YNM
Spencer, B J C G3SMW
Spencer, B K G7UTE
Spencer, C G G6RHK
Spencer, C H G2HBA
Spencer, D B G8YZY
Spencer, D C G7MUV
Spencer, D G G3LGW
Spencer, D W G4DJA
Spencer, E G0WQK
Spencer, F G0DMS
Spencer, G G4TVT
Spencer, G GW4DRR
Spencer, G A G2KI
Spencer, G C G0KTB
Spencer, G E G4CXW
Spencer, G E G4OLO
Spencer, G S G G3ZIN
Spencer, H C G6NA
Spencer, I D G3ULO
Spencer, J G0UNE
Spencer, J G3WTO
Spencer, J E H G8UMA
Spencer, K G7GVZ
Spencer, L G G4LX
Spencer, M G4OYZ
Spencer, M G B G8UML
Spencer, M W G4PPR
Spencer, N G0HFG
Spencer, N A G7PWY
Spencer, N W GM0EWJ
Spencer, P G0LYR
Spencer, P A G3PSW
Spencer, P L G3FIY
Spencer, P S G7DUB
Spencer, R E G3TEZ
Spencer, R J M0ANO
Spencer, R S G0PMI
Spencer, W G0TUJ
Spencer, W H G4WDW
Spencer Chapman, J S G3WUK
Spender, D S G4DHU
Spendiff, A G7PNS
Spendlove, D G4CMH
Spendlove, J W G4DXY
Spensley, M A 2E1DUG
Sperry, J F G4CQH

Spershott, J H G4SIA
Spevack, O G 2E1GBR
Spicer, D J G1ORK
Spicer, G H A G6OTG
Spicer, J A G3DNH
Spicer, J A G8LJU
Spicer, J F H G1IQN
Spicer, K G3RPB
Spicer, K G6KQ
Spicer, P J G0PJS
Spicer, T C G8IQT
Spiers, E F G3THZ
Spiers, J A L GM4UIJ
Spiers, K F G1ALU
Spiers, R F G8TTY
Spillett, D I 2E0AFD
Spillett, J M G1WSN
Spillett, M H G4UAW
Spilling, R H G0DEE
Spilman, W J G3OTD
Spindler, I C G1NKF
Spink, B G6XZF
Spink, B P GM0KZX
Spink, B P GM4HEL
Spink, G G3WUI
Spink, G G4XTV
Spink, G P G4IUM
Spink, J E G1SBK
Spink, J N G1BUQ
Spink, R W GM0WRV
Spinks, G B G3UXI
Spinks, G B G8BVY
Spinks, G C G1IGP
Spinks, J G6PCA
Spinks, J G8UHS
Spinks, L A O G3YOQ
Spinks, M G1BQR
Spinner, R E G6OYU
Spinney, G M G0IFF
Spires, C E G7VHC
Spires, C M G7HVL
Spires, E W G7PXX
Spirrell, R A G7PBT
Spiteri, J G4SOM
Spiteri, N J M G6YRG
Spittle, J E G4YLS
Spittle, M R G6NJL
Spittle, R G6BLS
Spittlehouse, B G7IMD
Spittlehouse, D G7VMH
Spivey, J G2HHV
Splaine, J H G4EPH
Spoard, J E G0GZW
Spokes, R B G8CCS
Sponer, N J 2E1EGR
Spooner, D G3WDS
Spooner, D J G6TVC
Spooner, D J G7CCX
Spooner, D P G0LWL
Spooner, D R G0OEK
Spooner, H G8WIM
Spooner, H S G4KXD
Spooner, J A S G8OPI
Spooner, K J G3TNY
Spooner, M J G4PFG
Spooner, P J G1JIW
Spooner, P M G4HFU
Sporton, A 2E1BJO
Sporton, J H M0AUK
Spragg, K J G8WKZ
Spragg, R J G4HPY
Sprake, R H S G1NLJ
Spratley, K G0CJD
Spratt, A J G4IJJ
Spratt, D J G7AGZ
Spratt, H N G0JJE
Spratt, J A H G3KWG
Spray, G W G8LKZ
Spray, I W G7ANZ
Spray, R G7LKV
Spreadbury, D A G7JSD
Spreadbury, R J G3XLG
Spridgen, J H G4LQJ
Spriggs, G W G3PFE
Spriggs, J W G6AHV
Spring, K R G0SFE
Spring, N J G1LUC
Springall, P A J G1WXS
Springate, F E G3BWV
Springett, J 2E0AQG
Springett, J G7DYV
Sprint, S E G7EVR
Sprogis, E G0DSM
Spruce, G G6ZYX
Spry, A J S G0KDY
Spry, J R G1TDP
Spry, M J G1DKE
Spry, N R GW4KGR
Spurgeon, J L G4LKD
Spurgeon, M H G G0MEC
Spurling, J O N G4AQI
Spurr, M S G1EYL
Spurr, P C G3YYN
Spyrakis, P A G6TWW
Squance, E H GI4JTF
Squance, M P G3HTB
Squibb, G J G4MPA
Squibb, N J G4HZX
Squibbs, S G0OZH
Squire, A G1MZC
Squire, A B G0IYB
Squire, A J G7DXG
Squire, C J G7VUH
Squire, D A G0DAS
Squire, D J G6TAP
Squire, J G3WPZ
Squire, J E GW0AGZ
Squire, J W M1BXQ
Squire, M A G7DSN
Squire, R S G4RXE
Squire, S B G3EBV
Squire, W G4NMF
Squires, C J G3XCS
Squires, D H G4DAC
Squires, G S G7HGS
Squires, J J G0UQU

Squires, J R G7BNT
Squires, P C R G3OIF
Squires, R A G7OFM
Squires, T G4DWW
Staal, L M G1BPU
Stables, A C G8FMH
Stables, A C G8JYN
Stables, E M G3ODD
Stables, J D G3ZIJ
Stace, J H G3CCH
Stace, J H G4FUH
Stacey, A G G3BXS
Stacey, C G8YVW
Stacey, G R G1RTY
Stacey, I G4VNU
Stacey, J H G0VPJ
Stacey, J H G4XGM
Stacey, J H G6SDQ
Stacey, J H G8BXO
Stacey, R J G4XQW
Stacey, W C M0AOE
Stacey, W L G4TJZ
Stackhouse, N G1SCL
Staddon, B G G6OMH
Staddon, H F G6STI
Staddon, K G7VNQ
Stadon, J F G4TFN
Stafford, A G3CEL
Stafford, A C G1ALR
Stafford, A D G4VPM
Stafford, A F G3NYZ
Stafford, G A G6CVH
Stafford, J A G1YIQ
Stafford, J B G7TSQ
Stafford, J C GI7KEC
Stafford, P M G8MWU
Stafford, R W G4ROJ
Stafford, T GW3XZU
Stageman, J C W G4WVT
Stagg, B A G3ZQJ
Stagg, C E C G1YQL
Stagg, V A G3IKN
Stagles, A F G3RBY
Stain, C W G0IUN
Stainer, A P G4EBV
Stainer, E J G4EJS
Staines, S E G7WAS
Stainforth, W E G0VHI
Stainforth, W E G7GHI
Stainsby, F A G0NVA
Stainsby-Tron, M D M1BXS
Stainton, D E G6MBI
Stainton, G G1MQQ
Stainton, P G4WMO
Stainton, P G6ZRV
Stainton, W R A G0ELH
Staley, K E G0PWR
Staley, R J G8OPV
Stallard, W A G0NHB
Stalley, H J G1EIT
Stalley, K D G4HHA
Stallon, D A G0PKJ
Stallworthy, S G G1LZS
Stals, E G0GVJ
Stamford, M A S GW4IIA
Stamford, R J G6PCE
Stammers, K G0SXG
Stamp, G D G8NNS
Stamp, L A G8TFW
Stamp, N J G0SDI
Stamper, H G3KYM
Stamps, M G0BEZ
Stampton, W C G3ZDR
Stanborough, A J P G4PKH
Stanbridge, F B G3PZS
Stanbridge, M F G3RHU
Stanbridge, R F G8NT
Stanbrook, R A G4CBJ
Stanbury, J C GW0BBO
Stancer, C G1RVH
Stancey, G P G3MCK
Stancliffe, K J G0FKS
Standen, B L G0GUW
Standen, D C G6TFE
Standen, L G4ZQA
Standen, M Y G0JMS
Standige, M G3MBU
Standing, G H G3EZE
Standing, M G1HIB
Standing, W J G8YGK
Standley, D L F G3XSA
Standley, L G0FCZ
Standley, P E G1GSB
Standley, R W G8RW
Standring, T A G1HGF
Stanford, D I G8XYQ
Stanford, E F GI6NIY
Stanford, E F MI0ALS
Stanford, M G8VHK
Stanford, P F G0NYV
Stanford, R A G1HDR
Stanger, S T G1YLA
Stangroom, C G0VDL
Stanhope, G T G6AVT
Stanhope, K E G0WIQ
Stanhope, P GM1PST
Stanier, F A G3RIC
Stanier, J H G3APU
Staniewicz, G J P G8HPN
Staniforth, A J G4ZDX
Staniforth, A V G8HVS
Staniforth, B J G7AJP
Staniforth, D J G6CRV
Staniforth, P A G7UCQ
Staniforth, R G3EGV
Staniforth, S G8RSV
Staniland, M T G6ZKS
Stankiste, E W P G4FCH
Stanleigh, R J G4DHK
Stanley, A B G0TFC
Stanley, A G M0AST
Stanley, C H G3YRM
Stanley, C T GM1MCN
Stanley, D G8BMD
Stanley, D A MI1BJW
Stanley, D J G1ORG

Stanley, D J GW0FJP
Stanley, D L G6YZZ
Stanley, D R G4ZHU
Stanley, G B G4XJM
Stanley, G C G8GYI
Stanley, I H G6MDR
Stanley, K R G6CPE
Stanley, M J G1GSD
Stanley, M J G7NIY
Stanley, M J M0AOT
Stanley, M T G1UVI
Stanley, N G3UEC
Stanley, P D G3BSN
Stanley, P H G0VOX
Stanley, R M1ARU
Stanley, R A G1NTH
Stanley, R A G7LUW
Stanley, R C G7SQV
Stanley, R W F G6LOB
Stanley, R W L G0SSZ
Stanley, S F G1EIV
Stanley, T J G4TXK
Stanley, T P G4DBL
Stanley, W C GI0OHT
Stanley-Jackson, D B G8AU
Stanmore, K G0IYO
Stannard, A P G8SGK
Stannard, J D G1VLB
Stanners, A G0RRB
Stanners, D V M G3HEJ
Stanners, N D G8AAU
Stannett, G M G4VUX
Stansfield, A G6ZZA
Stansfield, A D G6IIA
Stansfield, D I G0EVV
Stansfield, H A G7SPP
Stansfield, J B G1ASN
Stansfield, R A M1CBU
Stansfield, R H G3UAX
Stansfield, T G4LPS
Stanton, B G7HSY
Stanton, D G G0FWF
Stanton, D H G0OLX
Stanton, G G3AJX
Stanton, G A G3SCV
Stanton, J J G6XYU
Stanton, M G0VJK
Stanton, R J G3XKG
Stanton, R K J G4KFZ
Stanton, R M G4KWD
Stanton, R M G7TNK
Stanton, S J G6JRE
Stanton, S K G4YJP
Stanway, G W G3INP
Stanway, H E GM3JOA
Stanway, M F GM4NSZ
Stanway, M F GM8NSZ
Stanway, T R G4DVA
Stanway, W G4ALP
Stanyer, K G0UKK
Stanyon, J A G4BGZ
Stapleford, D N G6SAL
Stapleford, J A G1YQU
Staplehurst, J C S G6REH
Staples, C B G GM3EZI
Staples, D G1PNX
Staples, W G G0AKY
Stapleton, C G6OCE
Stapleton, G E G0IWH
Stapleton, J R GM0GKR
Stapleton, L T GM0GEE
Stapleton, M A G4BRE
Stapleton, P A G6KPV
Stapley, A R G7VYP
Stark, A GM0BUE
Stark, A F GW8OIJ
Stark, J R M G7BHB
Stark, R J GW8SIE
Starkey, B J G4YOZ
Starkey, C D G1IAY
Starkey, F G8TJG
Starkey, G C G4ZBF
Starkey, K G0LDP
Starkey, M J G0CWB
Starkey, M W G4SCL
Starkie, D S G4AKC
Starkie, E A R 2E1BWJ
Starkie, R 2E1ECW
Starley, P J G4RCR
Starling, G S G3ALG
Starling, J A G3WJS
Starling, P J G8EBX
Starling, P J G G8SED
Starmer, D C G4TGW
Starnes, K J G7MOW
Starnes, K W G3JWK
Starr, B G6ZRS
Start, B F G1GLZ
Startup, L R GW0IXP
Staruszkiewicz, J P GM4LCP
Stasiuk, M V G1UDX
Statham, A G M1BDL
Statham, B G0HSW
Statham, C J G0HME
Stather, K G0CWP
Staton, B W G6UJI
Staton, M A G4BGT
Staton, P C G4FXY
Staton, S G1XRE
Statt, M A G7JXK
Stavert-Dobson, A J G1YZO
Stavrinides, S G0KUB
Stayt, C W G0LUN
Steabler, G D G7TSZ
Stead, D D 2E1FGP
Stead, G G0GSQ
Stead, M D G M1AQZ
Stead, R G8MBO
Stead, T A G1ALX
Steadman, B R G7IHA
Steadman, F A GW6AAG
Steadman, M R G7JUN
Steadman, P G6LMB
Steadman, R E G0KHL
Stean, A D 2E1AWR

Stean, T E 2E1DHD
Steans, R F G0KDI
Stearn, D G4PHC
Stearn, J G7DKZ
Stearn, R W G1SOG
Stearn, S C G6IIG
Stearn, W T G0MLT
Stears, D J 2E1AAR
Stears, P B G4LMM
Stebbing, A E G7MPL
Stebbings, E J G6ZKY
Stebbings, M G8BIG
Steddy, S G0WDV
Stedman, M H G3FFY
Steed, E G7NYY
Steed, F J G8FMI
Steed, M F D G3GQO
Steed, R G3YUJ
Steed, R J 2E1DIM
Steed, S A G7MVJ
Steeden, C F G2HCP
Steeden, C P G4DJY
Steedman, G R GM0IXN
Steel, F Y M1BQT
Steel, G G0RDY
Steel, G R G3TOJ
Steel, J G7TUK
Steel, J W G3VJI
Steel, L F G2BBI
Steel, M P G3TNM
Steel, P A G4TKQ
Steel, R A G0HLL
Steel, R A L G1OAE
Steel, R J GM4WWU
Steele, A D GM7IEU
Steele, A G G1HEA
Steele, A J G1HMO
Steele, A M G1ZNZ
Steele, B G G0BDK
Steele, D G8UPO
Steele, D H G0VDV
Steele, G G8MUY
Steele, G R GW3SIY
Steele, G W D 2E0ANE
Steele, G W D 2E1EGM
Steele, G W D G0WUR
Steele, H B G7NMU
Steele, J G0WUQ
Steele, J C B G3LZK
Steele, J P M G6BRU
Steele, L J G7AWL
Steele, M N G7CKX
Steele, P M G4PMS
Steele, R W G0RPC
Steele, R W G1NRG
Steele, R W G6TVB
Steele, T GW0AVW
Steele, T R G6JRH
Steele, W F A G0WUP
Steele, W W A GI1WGK
Steele Davies, N W G0CUP
Steen, B D W G4UFU
Steenson, K A GI8XSY
Steenvoorden, L G0NBC
Steer, D J G4XFM
Steer, D T G8LER
Steer, W S G8CYG
Steggles, A K G4YOD
Steiner, D G4OKQ
Stelfox, A N G6EKS
Stelfox, J G3LTS
Stellar, T A G6RCT
Stellig, R J G4CK
Stelmasiak, J GW7DRX
Stelp, R A G0TQX
Stemp, H G0DJS
Stemp, N G7KAV
Stemp, P 2E1CAG
Stenbacka, C A H G0NGD
Stenhouse, A G3GWU
Stenhouse, H F GM3MAY
Stenhouse, H W G1ZES
Stenhouse, R J E 2E1CYE
Stenner, L S J G4KPT
Stennett, W G1SOX
Stenning, M J G7WBO
Stenson, J W G3XJG
Stephen, C J G6ZFU
Stephen, D C MM0AOY
Stephen, J GM0CHM
Stephen, R C G0MYX
Stephen, R C G3ZEF
Stephen, S G G7PNK
Stephen, W E G3IVZ
Stephens, A M G8VVZ
Stephens, A S G6GIU
Stephens, C D G3MGS
Stephens, D C GW3PYD
Stephens, D G G4ANY
Stephens, D J C G0INP
Stephens, F R G1HIM
Stephens, G GW0DXZ
Stephens, G N G7SHW
Stephens, G R G6HPC
Stephens, H S G1EIX
Stephens, H V GW4ZRK
Stephens, J G1VVU
Stephens, J C G7WJK
Stephens, J D GW4UVC
Stephens, J L G8LGB
Stephens, J M G0SFN
Stephens, J T E G0CUG
Stephens, M E M1AET
Stephens, N J G1FND
Stephens, P G0BDB
Stephens, P G G3TFI
Stephens, R G6WOR
Stephens, R E G8XEU
Stephens, R M G0NPA
Stephens, S B G1LGJ
Stephens, S J 2E1CXK
Stephenson, B GM0IPV
Stephenson, B J G4VRU
Stephenson, C N G4DCD
Stephenson, D G3KUL
Stephenson, D GW3FPC

Name

Stephenson, D K A G4UNR
Stephenson, G C G0TNC
Stephenson, J C G1ZFG
Stephenson, J G G6NIS
Stephenson, J M G6CPF
Stephenson, M G0JHU
Stephenson, M G4ESJ
Stephenson, S J G8HBZ
Stephenson, T J G3EID
Stephenson, W G3AQB
Stepney, A J G8BLW
Sterland, C B G1FYE
Sterry, A B G G8DVS
Sterry, H F G1AVW
Sterry, P J G1VXU
Sterry, R C G0KNR
Sterry, R C G4BLT
Steven, A J GM4IPK
Steven, J GM0JIN
Stevens, A G1SPE
Stevens, A GM7PSH
Stevens, A E G8NVI
Stevens, A R G6FIN
Stevens, A W G0HTE
Stevens, B G8YUP
Stevens, B D G8KKA
Stevens, B E G0DZQ
Stevens, B J G6STE
Stevens, B L G0WZX
Stevens, C T G0DCE
Stevens, D H G3XDO
Stevens, D K G4FCO
Stevens, D R G7CMQ
Stevens, D R G7HBJ
Stevens, E M C G8NRL
Stevens, F A G4BHC
Stevens, F L G7LEB
Stevens, F P G4IJH
Stevens, F R G3URV
Stevens, G A G1ZEC
Stevens, G P G0GRK
Stevens, H A G6WOB
Stevens, H S G6YRQ
Stevens, H W G0NBQ
Stevens, J G3UFW
Stevens, J G4PFV
Stevens, J GM4XAV
Stevens, J C C G6SDO
Stevens, J R J M1ASQ
Stevens, K G4BVK
Stevens, M G0UUP
Stevens, M G4CFZ
Stevens, M A G0GKL
Stevens, M D G0KAS
Stevens, M E G0SWW
Stevens, M E G7MES
Stevens, M J G3CPN
Stevens, M J G7GTN
Stevens, M L G8YEN
Stevens, M P G7SFA
Stevens, M P G8CUL
Stevens, N E G4UHE
Stevens, P G4OND
Stevens, P G8TMQ
Stevens, P GW0AQC
Stevens, P GW0HMC
Stevens, P B M0AFB
Stevens, P D GI6BXY
Stevens, P G P G7LRB
Stevens, P L G3SES
Stevens, P R G4EOR
Stevens, R A W G3TVI
Stevens, R B G6IDD
Stevens, R C G0JBZ
Stevens, R C G8BBD
Stevens, R T GM1ADI
Stevens, R W G1ERQ
Stevens, R W G3MAI
Stevens, R W G3TDH
Stevens, S G0WAM
Stevens, S H G0KYI
Stevens, S J G1PQO
Stevens, S R M1ACP
Stevens, T G8DKS
Stevens, T J 2E0AIZ
Stevens, T M G 2E1ASF
Stevens, T S G3VIX
Stevens, W F G0LYI
Stevenson, A G0DDO
Stevenson, A H G4TXL
Stevenson, A W G4ZAI
Stevenson, C MM1AVA
Stevenson, D G8NXS
Stevenson, D E G7EUR
Stevenson, D R GM6FIK
Stevenson, F 2E1EVH
Stevenson, F R G4IOE
Stevenson, G GM3KC
Stevenson, I J G3YNU
Stevenson, I M GM7FIE
Stevenson, J GI0IJB
Stevenson, J C G4JCS
Stevenson, J K GI0SSA
Stevenson, J S GM0JVV
Stevenson, J W G3DKO
Stevenson, K J G4RSQ
Stevenson, L G4XBQ
Stevenson, M G8ALS
Stevenson, M I G1GPM
Stevenson, M W GI4XSF
Stevenson, N P G6TEZ
Stevenson, P G G8YMM
Stevenson, P R G1AEI
Stevenson, P R G7TRB
Stevenson, R G3VND
Stevenson, R G4HBW
Stevenson, S E G3GYW
Stevenson, W T GM4WTS
Steventon, E F G3JJA
Steventon, M K G4GWH
Steward, A E G0CHR
Steward, C A G6WWT
Steward, D D G4OLU
Steward, D D G4VYV
Steward, I C G3ZRG
Steward, R J G7JSW
Stewart, A G3ZKM
Stewart, A G7UWN
Stewart, A C GM6UHC
Stewart, A D GM6YRN
Stewart, A G G4BRB
Stewart, A H GM4AGG
Stewart, A H GM4TOQ
Stewart, A I MM1BIT
Stewart, A J GI7EBM
Stewart, A M G4PEQ
Stewart, A M GI6IXD
Stewart, C C M GM0TFE
Stewart, C J A G3GLE
Stewart, C K G0APF
Stewart, D A G3TIR
Stewart, D B G7PNA
Stewart, D G H G4UEO
Stewart, D J G0LEP
Stewart, D J G0RHG
Stewart, D J GM7NSG
Stewart, D M 2M1EPV
Stewart, D P G4UVG
Stewart, E F G0DBU
Stewart, F G7GQJ
Stewart, F C G0CSF
Stewart, F C G1HLH
Stewart, F J G0JLZ
Stewart, G M0AGN
Stewart, I H V GI4POV
Stewart, J G7NJD
Stewart, J GM0WLE
Stewart, J GM1MDO
Stewart, J MI0AFT
Stewart, J F G0CPR
Stewart, J F G6YRB
Stewart, J F G8ERC
Stewart, J G G6BWN
Stewart, J G GM3LER
Stewart, J M G0UII
Stewart, J R G7SSU
Stewart, J R M1BQF
Stewart, K G0TWV
Stewart, L G G8TLL
Stewart, M E G6OOK
Stewart, M S G4RNW
Stewart, M W G1EIZ
Stewart, N G1YKV
Stewart, N D GM1CNH
Stewart, P G7EAH
Stewart, R G0PPD
Stewart, R A G4PBP
Stewart, R A G8BHH
Stewart, R A GI7PJF
Stewart, S J G0KGX
Stewart, S M GI7MWA
Stewart, T GM0BKX
Stewart, T GM0NET
Stewart, V W GM3OWU
Stewart, W G7PHR
Stewart, W GI4EIZ
Stewart, W GI4MYT
Stewart, W C L M0BAP
Stewart, W E B G8MNO
Stewart, W J GI0PJH
Stewart, W N GM8YRT
Steyn, R S GM8SIQ
Stickland, A C R G4LUN
Stickland, P N A G0VLR
Stickland, P N A G4WTW
Stiff, B H G1NEZ
Stigant, S R A G8YAA
Stiles, E T G0BHK
Stiles, R W G3UEN
Stilgoe, B A G0RNL
Stilgoe, G P G0NRF
Stilgoe, T A G0MLH
Stiling, W G G6MBH
Still, A G G4KZX
Still, D G G0OOC
Still, D G G0WVB
Still, D J G8FVJ
Still, R J G1GSF
Still, R S G8PVK
Stiller, C L G0AUI
Stillwell, D G8SIU
Stillwell, S G7MHV
Stilton, G A 2E1DLF
Stimpson, A J G1BJE
Stimpson, B P GM3TYQ
Stimpson, J E W G4YSF
Stimpson, K A G8RKZ
Stimpson, V G3SLU
Stimson, D R G3THC
Stinchcombe, I P G4DMM
Stinchcombe, O J M1BGO
Stinson, R J GI0UJG
Stinson, R J MI0AAH
Stinton, D G G0VBZ
Stinton, H C G6IDF
Stinton, P J G4TPS
Stirk, B R G6EQD
Stirland, A E G3LZC
Stirland, A E G7VUK
Stirland, M J G1WUK
Stirling, A GM1AMC
Stirling, A J G0WVG
Stirling, C G GM8MOI
Stirling, G C GM0WDD
Stirling, I J G4ICV
Stirling, J G0SFQ
Stirling, J GM3UWX
Stirling, W D GM4DGT
Stirrat, J G4DGX
Stirrup, J G6III
Stirrup, T D G6LOC
Stitt, J GI7WCS
Stitt, T R GI4RXM
Stoakes, C D B G0BBA
Stoakes, T L G3JTS
Stoaling, A R GU7OYU
Stoate, R B G6VPW
Stobbs, S G4OOK
Stobie, D D GM3HOQ
Stock, A J G1TBL
Stock, J T G7NAI
Stock, W O G4XYH
Stockbridge, P J G4OTI
Stockbridge, R A W D G0AUU
Stockdale, C G3PCS
Stockdale, L G4SFL
Stockdale, M J G0MKK
Stockdale, R E G8JRN
Stocker, B J G4LKB
Stocker, C A G8FQZ
Stocker, D W G4JSR
Stocker, G G4DZK
Stocker, K J G1SVL
Stocker, R E G4LKC
Stocking, J S G4IWI
Stockley, A D G8ELP
Stockley, B L G7PYG
Stockley, D J G4ELP
Stockley, D J G4ICM
Stockley, H G3LLS
Stockley, J G3FMW
Stockley, J D H G8MNY
Stockley, K J G4UQN
Stockley, L A F G3EKE
Stockley, P C M1BYI
Stockley, R G7MVY
Stockley, W C G6RGN
Stocks, A G0OPQ
Stocks, D G7SCK
Stocks, D I G0VTU
Stocks, D R S G4VUM
Stocks, F G3IVG
Stocks, J B G0BFJ
Stocks, T E G4ZGE
Stockton, A M G1PKG
Stockton, D GM4ZNX
Stockton, M G4SZX
Stockton, M M0AAD
Stockwell, B A G0NOG
Stockwell, J T G0XBJ
Stockwell, L L 2E1FJM
Stockwell, M E G7DVP
Stockwell, N G G G0RIK
Stockwell, P F G G6KFD
Stockwell, R B GU8FBO
Stockwell, V J GU8FSU
Stoddart, D R G4VMB
Stoddart, J M G1JMS
Stoddart, K A MW1CBM
Stoddart, R W M1BZA
Stoddon, R L G4DLP
Stoelwinder, G J G6IJW
Stogdale, H G4FEQ
Stoker, A C G8BPW
Stoker, D G1GEY
Stokes, A H G0JWS
Stokes, A J G3ZRH
Stokes, C R G6MVF
Stokes, D R G4HTY
Stokes, E G0PRV
Stokes, G C M1BQB
Stokes, G R G0BYH
Stokes, G T G0FVB
Stokes, H GI7IPO
Stokes, H B G0VNE
Stokes, J G0IIA
Stokes, J G6VFO
Stokes, J C G0ULY
Stokes, J S G6BMV
Stokes, J W G6XMY
Stokes, M G3ZXZ
Stokes, M A M1BGW
Stokes, M D G4HCY
Stokes, P B G0KVP
Stokes, P D G4WNM
Stokes, P J G1BWO
Stokes, P J G7UXO
Stokes, R G4TDV
Stokes, R G5NZ
Stokes, R P GM1KSB
Stokes, S K G4LTX
Stokes, T G0LNE
Stokes, T G1ASG
Stokes, T J H G0WTD
Stokes-Herbst, P A G4VZC
Stokoe, H P G3WLP
Stolting, D N G0SEF
Stone, A C G0OPR
Stone, A F G1JWO
Stone, A G G4UPJ
Stone, A J G1NMW
Stone, A M G4VSI
Stone, A P G3UIS
Stone, A R G8VEQ
Stone, A W G6POQ
Stone, B G4BPJ
Stone, B J G6SRE
Stone, B M G3JFC
Stone, C E G3SGN
Stone, C G G7BYW
Stone, D 2E1DPG
Stone, D R G8NGF
Stone, F A 2E1BUM
Stone, G G1CWW
Stone, G M C G3FZL
Stone, H F G7AZI
Stone, J E M0AIF
Stone, J F G2ADM
Stone, J J MM0AJQ
Stone, J P G0DFE
Stone, J R G6OOH
Stone, J W G4ITB
Stone, J W G6SFB
Stone, K J G1YRX
Stone, K J G4UBT
Stone, M G4JGA
Stone, P G3LWD
Stone, P H G7EVC
Stone, R G4TVW
Stone, R A G1EJA
Stone, R A G4FPC
Stone, R C G8SEE
Stone, R J G7WFZ
Stonebridge, P G8ZQA
Stoneham, M W G4RVV
Stonehouse, D GW4MRM
Stonehouse, J D GW0JDS
Stonehouse, W A G3HPC
Stoneley, B G0RLB
Stoneman, N E G3SXS
Stoneman, P J G6JEM
Stoneman, R P G6FAL
Stoneman, V L G G6XFN
Stoner, P H G8XOS
Stones, G M GM7KYX
Stones, I GM0SZA
Stones, S M G0NER
Stoney, D P G8PTN
Stoney, H J G4KMN
Stonton, M W G6ONM
Stoodley, D G 2E1CCW
Stoodley, F R G3UNW
Stooke, K M A 2E1FSN
Stooker, J P M M0BGN
Stoole, D GW4RIB
Stopford, J G8UWS
Stopforth, W G4LBX
Stoppard, R G G3OAF
Storace-Rutter, W W G0WLF
Storbeck, N H 2E1AQX
Storer, A J G0DCV
Storer, F A G1INQ
Storer, J N G1WXO
Storeton-West, B T G8JUK
Storeton-West, T J G8BTX
Storeton-West, W J C G3RXF
Storey, A R G3TEH
Storey, B A G8JTI
Storey, B L G4ZGG
Storey, C G G1SBT
Storey, D A A 2M1EUO
Storey, D W G8CYX
Storey, G B G8LIH
Storey, G E G3HTC
Storey, J G0FZQ
Storey, J R G6MVP
Storey, M G4BB
Storey, N A G8YCQ
Storey, P H G3YBH
Storey, P L G0BDF
Storey, P M G3ZGG
Storey, P R G4OIQ
Storey, R G3HTD
Storey, T R G8HZS
Stork, B J G3VUU
Stork, F G3TEE
Stork, N G0TZI
Storkey, B P 2E1EWN
Storkey, L G1AXF
Storkey, M J G0MVE
Storkey, M J M0AQW
Stormes, R C G7COU
Stormont, A G G3GWR
Stormont, A H G3ZPM
Stormont, W J G0EGC
Storr, P R G0CPX
Storry, A O 2E1FRQ
Storry, B G3WSM
Storry, J G G4OID
Storry, W B GM3DUS
Stothard, A P G0JNG
Stothard, F H G0LSC
Stott, B G1WAP
Stott, G G3ZRY
Stott, J P G0CJX
Stott, M G0NEE
Stott, P A 2E1BUJ
Stott, R S G4AKF
Stott, S G4SFJ
Stout, R F G0NRM
Stout, R S GM1CEJ
Stoutt, D R GD4XXE
Stow, G J G4MCU
Stow, T J G0SWS
Stowe, C GW4ZZO
Stowell, D F G8WCZ
Stowell, J S J G0THJ
Stracey, B M G7MKF
Stracey, M A G0UBL
Stracey, P M G4HEC
Strachan, A C GM7NNS
Strachan, A J GW4IFE
Strachan, I A GW8ITO
Strachan, I S GM0TAY
Strachan, I S GM4FLP
Strachan, S C GM4CUZ
Strachan, W GM0ADX
Strachan, W GM3ZRT
Strafford, R A G3MRT
Strain, H Y G3GPG
Straker, A G1JQQ
Straker, R G G8BVQ
Strand, J D 2E1EGT
Strand, P A G3UTO
Strand, T D G6FNY
Strange, G L GM0KZJ
Strange, A H G3PUC
Strange, A K M1BEQ
Strange, C M 2E1FBK
Strange, D J G1ERS
Strange, G R G8IWJ
Strange, I H G8RNU
Strange, L D G3NYA
Strange, M J G8HHO
Strange, P A 2E1ETS
Strangeway, R A G4FOW
Stratfull, J F G3IJS
Strathdee, B GM4REN
Strathdee, K GM0LDX
Stratta, A D G0GNO
Stratton, J B G7HMU
Stratton, J W G4AHM
Stratton, R E G3XKV
Straughan, J G7SCV
Straughan, R M GW6VED
Straus, L P G4STF
Stravens, R A G1IHA
Straw, D G1RZU
Strawbridge, P O GI0LRB
Strawson, B R G7KQY
Strawson, R J G7GNE
Streatfield, E C G4UCN
Street, A R G4KSY
Street, A W R G0JYI
Street, C D G4ICS
Street, C E G1LVK
Street, C H G8XET
Street, E G4OBF
Street, I R 2M1FXQ
Street, K GM3ENJ
Street, M J G3JKX
Street, M W G0WFL
Street, P G8ZES
Street, P H G1PHS
Street, R G G3TJA
Street, S K G3VFU
Street, T G0IFD
Streeter, D W G1OOW
Streluk, S J G0PFQ
Stretch, A J 2E1ARA
Stretch, N A 2E1AQY
Stretch, S M 2E1CBH
Stretch, V G G4LVO
Strett, B J G1JEO
Strevens, A V G0MGH
Strevens, C J R G4ZHT
Stribblehill, M C G7TXZ
Strickland, A 2E1EUL
Strickland, A M1BRD
Strickland, D R G0LSD
Strickland, M R M G6WSF
Strickland, R 2E1EUK
Strickland, R M1BRC
Strickland, T F G4EOA
Strickland, W M1BRB
Stride, A E G7MYI
Stride, P L G2BUY
Strike, A J G8AUX
Strike, F E M1AQB
Strike, J G8RQP
Stringer, A J G1CWO
Stringer, A J G7KPL
Stringer, A P G1FIU
Stringer, C G3RSK
Stringer, C H G6IDG
Stringer, G M G7CJG
Stringer, J G3TYO
Stringer, J A GI3KDR
Stringer, J B G4ZDS
Stringer, J F G7WLR
Stringer, J H G0LNI
Stringer, J J 2E1FPV
Stringer, K C G0COG
Stringer, L A G4GZG
Stringer, R G3IEG
Stringer, R D G1EQB
Stringfellow, G C GW4JGL
Stringfellow, R G0WMR
Stringfellow, R G4TZR
Striplin, P C J G1FXS
Stripp, A D G7VEI
Strobel, D M G0EVY
Strobel, S 2E1FWZ
Strode, E G6XYV
Stromqvist, S J G6MPB
Stronach, J H GI3LQY
Strong, A G0UOY
Strong, A G7UOY
Strong, A C G0DVD
Strong, C J GM1MKC
Strong, M R G4UMY
Strong, N A G0CWA
Strong, P B GI3VGL
Strong, P M G0DIH
Strothard, N G1JQG
Stroud, C B S G1BJN
Stroud, D H 2E1EKU
Stroud, M H G8IMS
Stroud, R G4PSJ
Stroud, R S GM7OMM
Stroud, R W G7WJV
Stroud, W J G4CYM
Strudwick, M F S G8TPP
Strudwick, P R G8CDB
Struthers, J W GM8CVN
Strutt, B J G1XZX
Strutt, J A G4BON
Strutt, J M G4XTS
Struve, P M G1EJB
Stuart, A C G3ZHB
Stuart, C A MM1APS
Stuart, D R G0VNX
Stuart, F A G8FEZ
Stuart, G J GM0VFY
Stuart, I D G8DOB
Stuart, I M M0BFZ
Stuart, J GM7LSI
Stuart, J C GM4SFW
Stuart, J M G4SDE
Stuart, J P G8LWC
Stuart, L G3JWG
Stuart, M G6BQC
Stuart, M N MM1AAL
Stuart, P H G0JCY
Stuart, R J 2M1DZX
Stuart, S G4XSG
Stuart, W G0TKF
Stuart, W J L G8GTW
Stuart-Cole, B W G4KWN
Stuart-Turner, H N G1EYS
Stubbings, G G8FIZ
Stubbins, K G8YWU
Stubbs, A C M0AQM
Stubbs, B F G4KPN
Stubbs, C G G1TXR
Stubbs, C J G1IZD
Stubbs, D G4EFD
Stubbs, J GM8GUJ
Stubbs, L J G4DBX
Stubbs, M G8IMB
Stubbs, P G7VCP
Stubbs, R G4SMS
Stubbs, R GW8XLL
Stuckey, E J GW0DLA
Stuckey, I C G6TEQ
Stuckey, M R G0SQK
Stuckey, R J GW1NZF
Studd, G J G0WMG
Studd, J M G1MGU
Studdart, H P GW7AAU
Studdart, P K G0PNO
Studdart, S K R GW7AAV
Stump, D R G7CRM
Stumpf, W W G4PUO
Stunden, P G0MTY
Sturdy, W Y GW3RUL
Sturgeon, C R G4NOT
Sturgeon, S J G1ZJW
Sturgess, A G G1LTK
Sturgess, P G0HZG
Sturley, C J G4MVM
Sturman, A A G1CDQ
Sturman, J W G7EQG
Sturmey, T A H G4MJI
Sturrock, I GM4DEK
Sturrock, J GI3CSK
Sturrock, J A G8SOK
Sturt, A H J G8SIK
Sturt, C N 2E1CSB
Sturt, C N M1BKY
Stuttard, T A F G4YWC
Styles, J GW4HZM
Styles, P L G4PXX
Styne, M A G4UWA
Styne, N J G6OVA
Stynes, C W G1TBB
Suarez, C R G0DDN
Suarez-Fernandez, J M G0KNF
Suart, I D GM4AUP
Suckling, A G0WLX
Suckling, C W G3WDG
Suckling, M S G8XOT
Suckling, P G4KGC
Sudbury, R T G0NXU
Suddaby, R B GW8FEY
Suddes, D GW0HKQ
Suddes, K J 2E1GBI
Suffolk, A F G4WLF
Suffolk, N E G1ZYN
Sugden, D B G3MAM
Sugden, J E G6GNE
Sugden, J J G1KRZ
Sugden, R P G0GLZ
Sugden, S A 2E1AXO
Sugg, K A G8TTX
Suggate, G C G3NPI
Suggitt, R T G7VUO
Suggitt, R T M0AWK
Sugrue, J P G6TRX
Suleyman, O M 2E1FFB
Suleyman, O M M1BRJ
Sullivan, D D G4WNY
Sullivan, D H G4HUS
Sullivan, D K G0NXJ
Sullivan, H G6IJX
Sullivan, J B G4HLA
Sullivan, K G G3KYF
Sullivan, L G8GVZ
Sullivan, P E GI4NMB
Sullivan, S R G0TJE
Sully, E M 2E1BLZ
Sully, E M G0VIQ
Sully, E M G7RWK
Sumesar-Rai, A K G3RMW
Summerfield, R A G0LDO
Summerhayes, T G8PIL
Summerhill, K G G4SUM
Summers, A G4MYU
Summers, A G G4KNO
Summers, B G8GQS
Summers, B D G0SOQ
Summers, C J G0GSH
Summers, E G1TYP
Summers, E D G8PJF
Summers, H J A G0UPL
Summers, J W G1HOG
Summers, N A G3RKJ
Summers, R W G0LTO
Summers, T A M1ALU
Summerwill, D W 2E1FNV
Sumner, A D 2E0AED
Sumner, C G0POS
Sumner, D J G3PVH
Sumner, D L G4FWI
Sumner, I J G3VPX
Sumner, J G6IKS
Sumner, J G7HVR
Sumner, K M G0DTI
Sumner, M E G6AHD
Sumner, N G0ACP
Sumner, R H G3BTV
Sumner, R P G4YOY
Sumner, S K G6BRW
Sumner, T G G4TJH
Sumption, G E G3DQL
Sunderland, D G6FHM
Sunderland, J C G3TQC
Sunderland, M C G1BQV
Sunley, F C G0LEL
Sunouchi, T G0WND
Sunter, G E M G4VOB
Sunter, R E G6HMN
Surgey, T K G3ZST
Surguy, L G7AKF
Surman, D M1BTL
Surman, J G7CKP
Surooprajally, A K G0EAU
Surooprajally, A H A D G7DAC
Surplice, M L G6AHA
Surplice, M L M0AZE
Surrage, R L G G0MUQ
Surtees, B G1YNO
Sutcliffe, A R G8TGR
Sutcliffe, C B G3MZC
Sutcliffe, G F G8TXW
Sutcliffe, J G0TFK
Sutcliffe, K M G6TKB
Sutcliffe, M GM7OKA
Sutcliffe, M J G3VAK
Sutcliffe, M S G8VSI
Sutcliffe, P C K G4YYO
Sutcliffe, S R G0MPI
Sutcliffe, S W G0ECY
Suter, S A G0GXI
Sutherland, A P G0STG
Sutherland, D A G0MYJ
Sutherland, G D K G0WCZ
Sutherland, G R GM0UPE
Sutherland, J G4TBU
Sutherland, J GM0RZY
Sutherland, J A G4LOV
Sutherland, J H GD1MZJ
Sutherland, J R G7USG
Sutherland, M T S G0AIG
Sutherland, P G7RVC
Sutherland, R G GM7BCC
Sutherland, S GM4BKV
Suttenwood, D F G4PBR
Suttenwood, R S G6KNM
Sutton, A C G6VDA
Sutton, A F G3XSR
Sutton, A M R G0XPD
Sutton, A M R G8JPD
Sutton, B W G3MIH
Sutton, C 2E1ENO
Sutton, C E 2E0AQF
Sutton, C E 2E1FBR
Sutton, C J G3UHV
Sutton, D G0IPH
Sutton, D G8VRN
Sutton, D M G3ZAJ
Sutton, D R G0WTM
Sutton, E B G7GQM
Sutton, G B G4VDD
Sutton, G W G4EVW
Sutton, J G6JUP
Sutton, J G6RTG
Sutton, J A 2E1FGN
Sutton, J A G0CHN
Sutton, J G G0ANT
Sutton, J G G0TDM
Sutton, J H C G1YLX
Sutton, J M G7FMQ
Sutton, K J G6CII
Sutton, L F G3COF
Sutton, M G G1PEK
Sutton, M S G8KEA
Sutton, R G1WVV
Sutton, R H G3TXO
Sutton, R J G0ODI
Sutton, W E G3FWI
Sutton, W J G3YKP
Sutton, W V G3GLQ
Sutton-Atkins, P C G6LTD
Suzuki, M G0PSC
Swaddle, H G0GXO
Swail, W D GI4YPR
Swain, A G3KWY
Swain, A C 2E1BDU
Swain, D H G3KMS
Swain, E S 2E1CKQ
Swain, J H G0JON
Swain, M G8MMP
Swain, M P B 2E1CUY
Swain, M P B G7VNJ
Swain, M S G6RHJ
Swain, N A G4CBS
Swain, N K G7KZH
Swain, P B W G4GXQ
Swain, S G4LPF
Swain, S W G0FYX
Swain, S W G4FBS
Swain, T J G2FRI
Swaine, D J 2E1FAY
Swaine, F W G1FWS
Swaine, L G4GIT
Swainson, D G3OXN
Swale, D G8ETS
Swales, A P G1DWU
Swales, J G1DLC
Swallow, A D G8NOS
Swallow, C G3VHB
Swallow, J G7PDH
Swallow, N F S G0LFA
Swallow, P J G8EZE
Swan, G G8ASJ
Swan, G E GW4FXF
Swan, I G3MNS
Swan, I G GM8BSE
Swan, M I G4LDG
Swan, N D G1BLO
Swan, P C G0WIE
Swan, R H G0DCY
Swan, R J GW7FLA
Swan, W R GM4XYF
Swanborough, A G G0JUS
Swancutt, R P G6JQB
Swann, D R 2E1AII
Swann, G E D G0WSD
Swann, G J G4LXI
Swann, N G G1TEX
Swann, P C G3WWX
Swann, R G6GQI
Swann, R W F J G0ROE
Swannell, R W M1BZT
Swanson, D GM0BCA
Swanson, D J G3OSI
Swanson, J G G3NPC
Swanson, P A G3NNV
Swanston, J L GM3ZVF
Swanwick, J P G7UVP
Swanwick, V H G7TMU
Swarbrick, P J O G3ZGN
Swarbrook, D G4TGS
Swarbrook, P G8YMS
Swartz, J L G0UCE
Swatman, J F G4BSH
Swatton, J E G0PYJ
Swaysland, G F D G4DNE
Sweeney, A G7VTF
Sweeney, G E GW8NYS
Sweeney, H F GM0SBJ
Sweeney, P B G4VUI
Sweeney, R H G0PMS
Sweeny, P G4UOT
Sweet, B E E G7VEN
Sweet, D W G7FMV
Sweet, G R G3OZY

Sweet, H W G1KQE
Sweet, M A S G1INU
Sweet, R E GW6STK
Sweet, S A G6YRE
Sweetapple, A F G4FVU
Sweetenham, G W G0MTM
Sweeting, B GW4XYL
Sweeting, C R G6APF
Sweeting, P L G0GMA
Sweetingham, F E G4ROS
Sweetland, D J G8LMY
Sweetman, A J P G4LRI
Sweetman, B V G3ROM
Sweetman, E A G3UAZ
Swetman, G M G8TJF
Swetman, M G G1YLN
Swetmore, R G3VTE
Swift, A G6WGA
Swift, A W G1PRW
Swift, C M G3IUK
Swift, D A G6XMQ
Swift, D A M0AVK
Swift, J W G3CTP
Swift, M G6NJJ
Swift, N R G7ENS
Swift, P R G0MIH
Swift, R J G0PAY
Swift, R S G1ALN
Swift-Hook, J M G1DFI
Swinbank, P A J G7AEA
Swinbank, P A J G8AHB
Swinbourne, S R G0LJV
Swinburne, W GM4WBU
Swindells, G J G3THV
Swindells, R J G0WCP
Swinden, J GW0NSZ
Swindlehurst, A 2E1ERS
Swingewood, J G4CVU
Swingler, A R G0SIE
Swingler, P E G0DLH
Swinnerton, R G4MXE
Swinney, R G0FRZ
Swinney, R G4IXL
Swire, P W G0WJP
Sword, G C G6ILC
Swynford, P R G0PUB
Swynford-Lain, R G7AOK
Sycamore, A G3IVC
Sykes, B G2HCG
Sykes, C G8YEV
Sykes, C L G3XJZ
Sykes, D M G0JOX
Sykes, J C G6CML
Sykes, J H G3GGR
Sykes, J W G6GAN
Sykes, K G6NXW
Sykes, M P G4GAK
Sykes, R G3NFV
Sykes, W 2E1AZO
Sylvester, D C G3RED
Sylvester, K 2E1FRR
Sym, A G G6JSN
Syme, G K GM3EFH
Symes, J E G3LNN
Symes, J E G3RSC
Symes, W E M G8AIV
Symington, R GI6TFF
Symonds, A G G8DQK
Symonds, D G1VHY
Symonds, D C G0PRZ
Symonds, G 2E1DRY
Symonds, G A G3XPT
Symonds, J C G0RPU
Symonds, K J G0DLC
Symonds, K J G2FKO
Symonds, P J G1BQW
Symonds, R C 2E1DSE
Symons, A G0BQU
Symons, G S S G3DSS
Symons, G S S G3VXX
Symons, L G3UFJ
Symons, L W G0INU
Symons, M G3ZPJ
Syms, C J G8VYP
Synge, H M G3BOC
Szendzielarz, V B G0ONF
Sznober, D G 2E1DYM
Sznober, W J 2E1DYN
Sznober, Z G 2E1CNU
Szybut, M R 2E1GAR
Szymanski, G O GM4COK

T

Taaffe, S G G4DQT
Tabbal, J A M1BDQ
Tabberer, D F T G0PMH
Tabberer, H F G3YVK
Tabberer, R D G6HAJ
Tabelin, A S G0KGC
Taber, K W G7FTD
Taberner, F T G0UGS
Tabor, C D G3UGR
Tabor, E H G0TAB
Tackley, K B G3BRQ
Tadesse, K G7VYF
Taft, J M G0MSS
Tagg, P T G8PIQ
Tagg, R J G1BCU
Taggart, J GI4GNT
Taggart, R J G4UPM
Taggart, W R GI4BKR
Taggerty, P GM0IJD
Taggerty, P H G0IJP
Taha, H M G1MLC
Tahla, M GW7TTX
Tainton, H M G2BCI
Tainton, P G6UMF
Tait, A GM4LBE
Tait, A G G8CTD
Tait, B N G3DDN
Tait, C T GI0TSS
Tait, E G0WJB
Tait, E M G6MEJ
Tait, H G G4RHQ
Tait, M A 2E1FZT
Tait, V J GI4LKG
Tait, W B G4EHD
Talaber, D J G0CAG
Talbot, A C G4JNT
Talbot, A C G8IMR
Talbot, B R G1BIN
Talbot, C G0EYU
Talbot, D H G1OFO
Talbot, D N T G0JHT
Talbot, G A G3VAL
Talbot, N G4EBQ
Talbot, N H G2DZH
Talbot, P B G8BAZ
Talbot, R A G3ORK
Talbot, R G G4IWZ
Talbot, R H G0SBV
Talbot, S J G1TOL
Talbot, S R G0AGY
Talbott, M S G4NWM
Talbott, R E G3LRS
Talbott, R E G4LRO
Talbott, R E G6XRS
Talbott, W J G4NWN
Talboys, J W G3ZMT
Talkowski, L A G0SDQ
Tallarigo, M G7STR
Tallentire, J R G4AFE
Tallett, E R G1ZWK
Tallis, T H G4UUQ
Talmage, F J G0LVF
Talmage, L G G7AKO
Tam, K C G3UXL
Tambini, A D G7HNC
Tambini, P D G6GPM
Tamblin, L D G1KOW
Tames, S R G8LCL
Tamkin, C G3EWT
Tamlin, J L F G3LCY
Tamlin, S G G6VRA
Tamlyn, J G7WLM
Tamplin, A F K G4XCE
Tamplin, P J G0VGR
Tams, R G4TCG
Tams, R J G1OQB
Tan, C M M1BBH
Tancock, B 2E1AZV
Tandy, C G G6FGJ
Tandy, G A G8VKO
Tandy, L G G0SWT
Tandy, M J G0MJT
Tandy, P N G1VOQ
Tanfield, R M G6CVE
Tankard, V J G1KSN
Tankaria, D G6ESM
Tann, M G6ZGC
Tannahill, G G6JQE
Tannahill, R GM0ERT
Tanner, A J GW0NKI
Tanner, B C G6HUI
Tanner, C E G0FDY
Tanner, D G4FLR
Tanner, D A C G7KSL
Tanner, D J G0OZD
Tanner, F GI0ZAK
Tanner, G V G1LLO
Tanner, J E G8AER
Tanner, K C G0UOE
Tanner, P M G4VTO
Tanner, S G7NIU
Tanner, S H W G3LRM
Tannock, D GM2BUD
Tansley, G A G3WJY
Tansley, J G4RIT
Tansley, N A G1IRQ
Tanswell, D R G6LAU
Tant, R G G4WNP
Taperell, M R M0BEM
Taplin, M G G3WAM
Taplin, S J G8BJN
Tapp, A P G4MTQ
Tapp, G B G4ZQB
Tapp, R F G1XXH
Tappin, B A G4YXF
Tapping, C J G0OXE
Tapson, R M0AUC
Tarbatt, D A G7SKR
Tarbett, K W G0CKI
Tarbuck, D G0DTG
Tarbuck, D G6IKM
Targonski, G J G7WHB
Targonski, R G0BZT
Tarmey, P H G6AZL
Tarr, G D T G8YGO
Tarr, M GM8VKN
Tarr, R J G3PUR
Tarran, C J G8DXF
Tarrant, C P G0KRH
Tarrant, H G7KRU
Tarrant, K J G1CUB
Tarry, B J G4FKP
Tarry, D J G0WFO
Tarry, G E G4WGU
Tarry, K J G0FIC
Tarver, F G G8AIM
Tarver, L C G0LPY
Tasker, M M G0LKM
Tasker, P L G6VBJ
Tate, C G0CEZ
Tate, C O G6SUN
Tate, D G6ZFO
Tate, J G3LGT
Tate, M C G3MHX
Tate, M J G4YGQ
Tatem, S P G1SPT
Tatham, L C GW1IEB
Tatlow, C H G0FCB
Tatlow, D E M1BEO
Tatlow, M A G6FNP
Tatman, V K G4IMH
Tatnall, B K G4ODA
Tattersall, J F G0RBK
Tattersall, P R G4SYG
Tattersall, R 2E0AMU
Tattersall, W H G4WHT
Tatterson, G L G0ETL
Tatterton, A P G6TWX
Tatton, R M GM3SRV
Tatum, M D G1SYT
Tatum, M D M0AWS
Tatum, W C G4OJB
Tavender, A T 2E0AHM
Tavener, J G4XCM
Taverner, J F G7SHT
Taverner, S J G7VDX
Tavinor, A W G G7KUX
Tawn, J G4EMT
Tawney, A F G1USI
Tayler, J A GM0KCV
Taylerson, D J O G3PPC
Taylforth, C A G7WEK
Taylforth, S C G7SNQ
Taylor, A 2E1ANE
Taylor, A G0ASV
Taylor, A G0LAN
Taylor, A G4PVN
Taylor, A G4ZII
Taylor, A G6XNI
Taylor, A G7RST
Taylor, A GM4HBQ
Taylor, A GW3TMJ
Taylor, A MM0AUB
Taylor, A C G0TPA
Taylor, A G G0MWP
Taylor, A G G0UIK
Taylor, A G G7UAJ
Taylor, A G G8YNB
Taylor, A G M0AUR
Taylor, A H G4IZE
Taylor, A H G4XTZ
Taylor, A J 2E1FUI
Taylor, A J G3NYE
Taylor, A J G7BDW
Taylor, A L G3JMO
Taylor, A L G4ALT
Taylor, A L G6GOY
Taylor, A L GW4ZXH
Taylor, A M G4KJI
Taylor, A P G1THW
Taylor, A R G8BBB
Taylor, A R J G7EKC
Taylor, A S G1MSA
Taylor, A T G4KLV
Taylor, A W G4SSC
Taylor, A W J G4RZE
Taylor, B G0EIA
Taylor, B G0OYH
Taylor, B G4SSR
Taylor, B GW6TOX
Taylor, B G G3ZAG
Taylor, B G W G3AKF
Taylor, B H G0AXK
Taylor, B J G1HJD
Taylor, B J G7CKL
Taylor, B M G3TIN
Taylor, B M G4ZWV
Taylor, B P G8EQX
Taylor, B W G0TPG
Taylor, B W G8TPC
Taylor, C G0IWY
Taylor, C G1LBQ
Taylor, C M1BNV
Taylor, C A G1ETU
Taylor, C B M1CBT
Taylor, C G M1BTE
Taylor, C G D GM0IMH
Taylor, C G N G1SCN
Taylor, C J G1LBM
Taylor, C M G0GYA
Taylor, C M G7MCT
Taylor, C M GM8JAV
Taylor, C R G6EQN
Taylor, C W G3USA
Taylor, D G0VZZ
Taylor, D G3OPT
Taylor, D G4EBT
Taylor, D G4NXP
Taylor, D G7RWN
Taylor, D G8SWK
Taylor, D GW0WQP
Taylor, D A G1VDE
Taylor, D B 2E1EXI
Taylor, D B G3IZF
Taylor, D B G4HRB
Taylor, D C G6WCT
Taylor, D E G0VLI
Taylor, D F G0LSK
Taylor, D F GM4RZW
Taylor, D H G0GBL
Taylor, D J G0OZV
Taylor, D J G3YQD
Taylor, D J G6CIF
Taylor, D J G8APW
Taylor, D J GM8ARV
Taylor, D J GW0EGH
Taylor, D J C G1AMM
Taylor, D J S G7WDL
Taylor, D J S M0BFW
Taylor, D L G3ZYT
Taylor, D R G3WGB
Taylor, D R G4FMB
Taylor, D R G6EWJ
Taylor, D R GM6JWH
Taylor, D S G0RJF
Taylor, D T G1YSX
Taylor, D W GM8FMR
Taylor, E A GW6FKB
Taylor, E C GW3HDH
Taylor, E F G3SQX
Taylor, E G GW7RZN
Taylor, E W G2ALN
Taylor, E W G3FK
Taylor, E W G5VQ
Taylor, E W E G3GRT
Taylor, F C D G3RFP
Taylor, F E GM3UNJ
Taylor, F J G8NSD
Taylor, F J G G4DUO
Taylor, F R G1JMY
Taylor, G G1SDX
Taylor, G G4KPU
Taylor, G G7MHQ
Taylor, G G7UJC
Taylor, G GM6JDZ
Taylor, G A G0UNF
Taylor, G A G8AKN
Taylor, G A GM0GAV
Taylor, G A M1AZC
Taylor, G B G6JBQ
Taylor, G D G6DNV
Taylor, G D G7VDK
Taylor, G G G6APB
Taylor, G H G0MJA
Taylor, G H G3IUL
Taylor, G J G4PJY
Taylor, G R G3HRR
Taylor, G R GW4IGQ
Taylor, G S G4JZF
Taylor, G S G6TNH
Taylor, G V G3DKD
Taylor, G W G0PUW
Taylor, G W G7CFT
Taylor, H G0WRD
Taylor, H G7PBL
Taylor, H GM0CNW
Taylor, H A G0OIA
Taylor, H C G8GAR
Taylor, H C GW0RVC
Taylor, H C GW4GJT
Taylor, H G J R G3MYU
Taylor, H H G3KFG
Taylor, I G6VJA
Taylor, I E G0WXS
Taylor, I J GW6TRV
Taylor, I J C GM3IJT
Taylor, I K G3ORG
Taylor, I W 2W1FKX
Taylor, J G0OJW
Taylor, J G3EDK
Taylor, J G3PIK
Taylor, J G4ZVK
Taylor, J G6XD
Taylor, J G6XNJ
Taylor, J G7RFK
Taylor, J G7WJT
Taylor, J G8PUE
Taylor, J G8TXX
Taylor, J GW3UMB
Taylor, J B G3CVX
Taylor, J C 2E1FRJ
Taylor, J C G4KKG
Taylor, J D GW4VDW
Taylor, J D C G7OKB
Taylor, J E G3OWO
Taylor, J E G4ETM
Taylor, J E G4UVF
Taylor, J F G8HTM
Taylor, J H S 2E1GBL
Taylor, J J G4DFX
Taylor, J J G4UUP
Taylor, J K G1RVA
Taylor, J K M1BVN
Taylor, J L G1STQ
Taylor, J M G7PTM
Taylor, J M G8DYK
Taylor, J M GW7AIN
Taylor, J P G4ERU
Taylor, J R G4REU
Taylor, J R G6VJC
Taylor, J R H G6CVB
Taylor, J S D G4VHJ
Taylor, J W G0TAS
Taylor, J W G4VTA
Taylor, K G0APQ
Taylor, K G0RLN
Taylor, K G3NNW
Taylor, K G6YTC
Taylor, K GW8TOX
Taylor, K D G1NJC
Taylor, K D GW1KKJ
Taylor, K D M0BBQ
Taylor, K E G3LME
Taylor, K G G4NLV
Taylor, K J G4GDK
Taylor, K J G4RRW
Taylor, K L G4FSI
Taylor, K M G4GAI
Taylor, K P G0TQV
Taylor, K R G0XKT
Taylor, K R G4YQK
Taylor, K W G8EVR
Taylor, L G0NGB
Taylor, L G0TJC
Taylor, L GM0PGG
Taylor, L B H R G4ZAB
Taylor, L E G7PZS
Taylor, L G 2E1EOT
Taylor, L J 2E1EXL
Taylor, L J G4ZJD
Taylor, L J P G0EBF
Taylor, L M 2E0AON
Taylor, M 2E1FXC
Taylor, M G0LGJ
Taylor, M G4XRT
Taylor, M G7VRV
Taylor, M A G4OBC
Taylor, M A G8BMP
Taylor, M A M1BVP
Taylor, M C G1NDQ
Taylor, M C G4TVP
Taylor, M D GM4OMT
Taylor, M E G4YMT
Taylor, M E L V G6AGR
Taylor, M F G7UXR
Taylor, M F W G3YSG
Taylor, M G G0EAE
Taylor, M G G3UCT
Taylor, M J G4XBA
Taylor, M J G4XDC
Taylor, M P G4ZIF
Taylor, M R G1GSG
Taylor, M S G4VSW
Taylor, M V G1JPI
Taylor, N G4UFJ
Taylor, N C G1OFI
Taylor, N C A G3TOQ
Taylor, N E G0TTH
Taylor, N F G0UOM
Taylor, N H G7LLK
Taylor, N J G7DMX
Taylor, N P G4HLX
Taylor, N R G0WYA
Taylor, N R G7RMX
Taylor, N S G3BHA
Taylor, N V 2E1FJW
Taylor, N W G0RUF
Taylor, N W G8JCU
Taylor, O G0CUJ
Taylor, P 2E1CJD
Taylor, P G0HMP
Taylor, P G0MVO
Taylor, P G4KIN
Taylor, P G7SLY
Taylor, P G8BCG
Taylor, P A G4OUA
Taylor, P C G0GFZ
Taylor, P C G0VSJ
Taylor, P C G3RRG
Taylor, P D G1LZH
Taylor, P D G4MBZ
Taylor, P D G4VVD
Taylor, P L G1PLT
Taylor, P M G0WRK
Taylor, P M G1FET
Taylor, P M G4INQ
Taylor, P N G0TCE
Taylor, P N A G4OHB
Taylor, P R G7ADE
Taylor, P R G8OYM
Taylor, P R GW0MXG
Taylor, P S G0ILO
Taylor, P S G0MUX
Taylor, P S G8VSH
Taylor, R G0JSA
Taylor, R G1LHR
Taylor, R G3VGB
Taylor, R G4HZA
Taylor, R G6ERN
Taylor, R G6LRD
Taylor, R C G3JAL
Taylor, R C G4BEV
Taylor, R C GW0MVI
Taylor, R C GW2HCJ
Taylor, R D G3LDY
Taylor, R D G6TNZ
Taylor, R E G6XNQ
Taylor, R G G4CGU
Taylor, R G G6RKF
Taylor, R G G7VSL
Taylor, R G G8DBD
Taylor, R G W G4FDG
Taylor, R H G0LQO
Taylor, R H G1XRT
Taylor, R H G3KAP
Taylor, R H W M0BDB
Taylor, R H W M1ANK
Taylor, R I G3SYJ
Taylor, R J 2E1EPS
Taylor, R J 2E1FCW
Taylor, R J G3OHV
Taylor, R J G4BEL
Taylor, R J G4ROB
Taylor, R J GM0DFL
Taylor, R J M1CAP
Taylor, R K G3WWH
Taylor, R K G4KTI
Taylor, R L G0RLT
Taylor, R L G6KLH
Taylor, R L M0AKZ
Taylor, R N G6ICL
Taylor, R N G8YHH
Taylor, R P GJ6BUK
Taylor, R S G1WEX
Taylor, R S M1AGA
Taylor, R W G4CQQ
Taylor, R W M G4GXO
Taylor, S 2E1ENR
Taylor, S G1JEZ
Taylor, S GJ6HUL
Taylor, S G G8AHA
Taylor, S J G7SAP
Taylor, S J G7VGM
Taylor, S M G0INK
Taylor, S N G1NTX
Taylor, S P 2E1EAI
Taylor, S P G4EDG
Taylor, S P G6LPG
Taylor, S R G0GQO
Taylor, S R G6XNR
Taylor, S V G1BMQ
Taylor, S W G4CKX
Taylor, T G4RPR
Taylor, T G4WKP
Taylor, T A GM0GHN
Taylor, T M G0UCX
Taylor, T R 2E1FSJ
Taylor, T R G0NIP
Taylor, V G0AQT
Taylor, W G4IHP
Taylor, W GM0CZM
Taylor, W M1AHX
Taylor, W A G6SKM
Taylor, W F G0KWF
Taylor, W F G4YPF
Taylor, W G G4DLZ
Taylor, W G G4WGT
Taylor, W G G8HPL
Taylor, W J G0FOO
Taylor, W J G1MHM
Taylor, W J G6NWT
Taylor, W R G0FWH
Taylor, W R G3YTT
Taylor, W R GM4OEZ
Taylor, W T G0RUX
Taylor, Y G4RTD
Taylor-Cram, J F H 2E1COC
Taynton, B G GM1PEL
Taynton, E R GM8DKB
Teager, I J G8AWL
Teague, C P G7DUI
Teague, J G3GTJ
Teague, J J GM0TKQ
Teague, R H G7VGE
Teale, A P G3SGT
Teale, A P G3UUP
Teanby, P M G4GZC
Teanby, W S G3RRL
Teaney, H A GW0PKA
Tear, B G0CQY
Tear, S J T G4YYI
Teasdale, B G0NSP
Teasdale, C R T G3ZOP
Teasdale, J G G0BVO
Teather, P G4PLT
Tebay, D G0MHN
Tebboth, I E G6FIP
Tebboth, R J G8MSR
Tebbutt, P G4ONZ
Tebbutt, P R GM1ZBD
Tebbutt, R T G7KJX
Tebby, I W G1BQJ
Tecklenburg, D A G0XAH
Tedbury, R M H G6SNY
Tedbury, W H G4PQS
Tedd, G G0PMK
Tee, A J G8XIY
Tee, H G8UA
Tee, R G G0WPB
Tee, W H G4BYO
Teed, J M G3WWT
Teer, I V C G7IEW
Teesdale, C E G2BUV
Teesdale, R J GW4URB
Tegerdine, D R G0LRD
Tegg, P E L G8RYJ
Tehara, P S G6GBI
Telenius-Lowe, E 2E1FHJ
Telenius-Lowe, S G0WPX
Telenius-Lowe, S G4JVG
Telfer, T K GM8JFE
Telford, A G8IRN
Telford, A J G4XPO
Telford, B G6TNI
Telford, D G1FBE
Telford, J T G1REO
Telford, M H G6OET
Telford, R G7IOI
Telford, S A G7HUJ
Telkman, M P G8DCA
Temblett, M H M0AKF
Tempest, P G4HJS
Temple, D R G0OYS
Temple, D W G8EIH
Temple, G B M1CDU
Temple, J E G3XAI
Temple, P T GM0EDJ
Templeman, M G7HVN
Templeman, M J G0TCY
Templeman, P E G3EQF
Templeton, A GM0DAN
Templeton, C J G0VOT
Templeton, D M 2E1DQG
Templeton, H H GM4USY
Templeton, I C G0UVB
Templeton, I S GM0JQE
Templeton, L A 2E1DQH
Temple-Heald, J N G6BBN
Tench, G J M 2E1ESD
Tench, J D G4ZAY
Tennant, A G4VOJ
Tennant, E G GM6SYC
Tennant, F G4RFL
Tennant, G S G8VSN
Tennant, J R J 2E1EQF
Tennant, M J G7ION
Tennant, T 2E1DPO
Tenwolde, R G4WOL
Teperek, R J GM3TDI
Teperek, R J GM4HVS
Tepper, G F 2E1ERM
Termie, K S M1BMC
Ternlund, P A M0AKC
Terraneau, H G2FYO
Terrell, I M G4JXZ
Terris, I B D GM0VXZ
Terry, A G G4PZV
Terry, B W G6PEA
Terry, C G G7HUO
Terry, F H GW7EYB
Terry, J C G4GEU
Terry, J R G0VZC
Terry, K C A G3GSY
Terry, L G G7MZS
Terry, M G7HUP
Terry, M GW8TBG
Terry, N S G4IPM
Terry, P D M1CCN
Terry, R S G4FVC
Terry, W T S G4LUT
Tester, M T G4TBM
Tester, P R G8CIG
Tetchner, R I G8VKI
Tetley, M R G3RIX
Tetley, R J G0BIH
Tetlow, D G7JXC
Tetlow, D S G1GSR
Tetlow, J H G4SFN
Tett, A F G6WDR
Tett, F J G0OID
Tew, B G G3WFF
Tew, G G G8GZC
Tew, G K G7WFK
Tew, J H G4PAK
Tew, M I G4RFA
Tew, R G4JDO
Tew, R N G6DKI
Thacker, C G6MEI
Thacker, I A G6UOH
Thacker, M N G4FJF
Thacker, P M G4HSZ
Thackeray, E G3OHT
Thackeray, P A G8YZL
Thackery, A R G4BBL
Thackery, P S 2M1ECJ
Thackray, M F G0VJX
Thackray, P G 2E1DLX
Thackrey, N D GM0VPG
Thain, P G7MOB
Thain, S C G0LJE
Thaiss, C E G6NGE
Thaiss, I M G7JXX
Tham, K M G0CRU
Thane, E C G0TED
Tharme, B A G4KVL
Tharp, V G0CQY
Thatcher, B W G0JTR
Thatcher, D C 2E1CWJ
Thatcher, L C G0GES
Thatcher, M D G0AVY
Thatcher, R S G1GXP
Thawaites, D J G8ORZ
Thawley, R S G0BSU
Thayer, H S A G4DTK
Thayne, M K G3GMS
Thayne, R W G7ZRT
Theaker, B A G7RTO
Theakston, R G0UKO
Theasby, G T G8BMI
Theedom, B J G8LYW
Theedom, J S G6XNK
Theobald, J A G3EQM
Theobald, J D G4XHU
Theodorson, E T G8SWL
Thetford, I P 2E1EGL
Thexton, J W G3URE
Theyer, R G G0IBV
Thickett, D M G0FEH
Third, E B G4NZE
Thirkell, E R GM4FQE
Thirlaway, S D G7LNW
Thirlwell, T J G0VFW
Thirsk, J P G4PAT
Thirst, P A G1ANZ
Thirst, T J G4CTT
Thoennissen, H P G4AGT
Thom, D G3NKS
Thom, J S GM3WJE
Thom, P M G1NKS
Thomalla, D G7GGM
Thomas, A G4UAT
Thomas, A GW0TCV
Thomas, A C G0SFJ
Thomas, A C N G7OQO
Thomas, A E G0VSK
Thomas, A E V G6VBB
Thomas, A G G1DKX
Thomas, A G G8VMO
Thomas, A R G7GCX
Thomas, A R L G0GYW
Thomas, B G0FHL
Thomas, B G4KGU
Thomas, B G7NJH
Thomas, B C G0NNR
Thomas, B E R G0KOJ
Thomas, B J G0PID
Thomas, B J G4EIY
Thomas, B R 2E1CGP
Thomas, B S G3JLQ
Thomas, C G3YGR
Thomas, C A G8VBE
Thomas, C A GW1KTW
Thomas, C D G4IZI
Thomas, C E G7MQC
Thomas, C E MW1BLT
Thomas, C J G1ETX
Thomas, C J G3PSM
Thomas, C J G4ZCT
Thomas, C L G0VTE
Thomas, C L G6BRY
Thomas, C M 2E1DKN
Thomas, C R G1UTC
Thomas, C T G1YLQ
Thomas, C W G4OEQ
Thomas, D G0VQX
Thomas, D G4HHJ
Thomas, D G7TIR
Thomas, D GW1TFU
Thomas, D M1AKT
Thomas, D B GW4KYT
Thomas, D C J G0HAY
Thomas, D D G6TNJ
Thomas, D E G3XJL
Thomas, D E GW0IXJ
Thomas, D J G4OGW
Thomas, D J G6VAZ
Thomas, D J G7TWE
Thomas, D K GW4AZI
Thomas, D M G4FRQ
Thomas, D M GW3RWX
Thomas, D R G0BJK
Thomas, D R GW0BHJ
Thomas, D R GW4NGU
Thomas, D S G7TFZ
Thomas, D W J GW0WGE
Thomas, D W K G4IKX
Thomas, D W N G3ZLN
Thomas, E G4SLD
Thomas, E GW3PXY
Thomas, E GW7UYF
Thomas, E C G8TTE
Thomas, E G G3VBW
Thomas, E W G7UBE
Thomas, F P G1WNL
Thomas, G G4JYL
Thomas, G GW3VMX
Thomas, G MW0AMQ
Thomas, G D 2W1FUW
Thomas, G H GW6XZU
Thomas, G M E 2W1FUV
Thomas, G R G0SUB
Thomas, G R G4AWJ
Thomas, G R A H MW0AMN
Thomas, G S 2E1DVM
Thomas, G S G3SWO
Thomas, G W G3JKE
Thomas, H G0WVQ
Thomas, H GW0NPM
Thomas, H GW4SYO
Thomas, H MW0ATG
Thomas, H I G6GSC
Thomas, I G8ZLL
Thomas, I G G4PLL
Thomas, I M G7NEK
Thomas, I R G1BXT
Thomas, J G3LZO
Thomas, J G7UIA

Name

Thomas, J	GW4PAF
Thomas, J A	G1GST
Thomas, J E	G0SDZ
Thomas, J E	G4DVV
Thomas, J E	G4MYN
Thomas, J F	GW3AWC
Thomas, J H L	GW6ZRX
Thomas, J J	G4NPW
Thomas, J L	G4ZBQ
Thomas, J O	GW3EHN
Thomas, J P	G0RLI
Thomas, J P	G7NFR
Thomas, J R	GW4RWR
Thomas, K B	G8IFH
Thomas, K M	GW0GTK
Thomas, L D	2W1ETV
Thomas, L F G	GW4ZXG
Thomas, L G	G4YTN
Thomas, L W	G4WUY
Thomas, L W	GM0TKB
Thomas, M	2E1AZS
Thomas, M	G0WKH
Thomas, M	M1ARB
Thomas, M	MW0ARV
Thomas, M A	G0VQG
Thomas, M E P	GW4YDF
Thomas, M F	G4HVK
Thomas, M H	G4FCZ
Thomas, M I	2E1CGZ
Thomas, M I	G0WXN
Thomas, M P	M0AZT
Thomas, M P	M1BHI
Thomas, M R	G6BBR
Thomas, M S	G1ETW
Thomas, M S	GM4OTH
Thomas, M . R .	GW6OIO
Thomas, N J	2E1BSH
Thomas, N J	G4LGO
Thomas, N M	G4MRL
Thomas, P	GW6ZZF
Thomas, P A	G1MLR
Thomas, P C	G0IOH
Thomas, P J	G4IBO
Thomas, P J	GW0GIH
Thomas, P J A	G1FFO
Thomas, P M	GW7FSF
Thomas, P R	G8KTA
Thomas, R	G7MEA
Thomas, R B	GW8JQW
Thomas, R C	GW0DKF
Thomas, R D H	2W1ETW
Thomas, R E	G1TEY
Thomas, R E	GW0TXM
Thomas, R E W	G0RCU
Thomas, R F	GW6JBN
Thomas, R G	GW6AGS
Thomas, R G	GW7SBO
Thomas, R H	G4GCQ
Thomas, R H	G4JHA
Thomas, R J	G1RDI
Thomas, R J	G8IWE
Thomas, R J	GW0SLC
Thomas, R J	GW4TVQ
Thomas, R M	G3VNB
Thomas, R P	G4JJP
Thomas, R P	G6AGT
Thomas, R S	GW4LEU
Thomas, R S	GW4WJO
Thomas, R W	G8IAK
Thomas, S	GM7LVV
Thomas, S A	G0GVE
Thomas, S A	G0PGX
Thomas, S D	M1ACQ
Thomas, S E	2W1BTT
Thomas, S E M	G0LUG
Thomas, S G	G6EQL
Thomas, S G	M1ACB
Thomas, S J	G6HUM
Thomas, S M	G6HMU
Thomas, T A W	G7UBD
Thomas, T J	G1GGF
Thomas, T R	GW0ABT
Thomas, T R	GW3SPC
Thomas, V C	G1GFF
Thomas, V H	G6BHL
Thomas, V T	G7EWK
Thomas, V T	GW4UWR
Thomas, W	G3ORN
Thomas, W	G8NUT
Thomas, W A	2E1ENX
Thomas, W A	M1BOS
Thomas, W G	GW7VSF
Thomas, W J	GW3XHV
Thomas, W K	G4DQQ
Thomas, W L	GW1JWN
Thomas, W R	G3HWY
Thomas, W R	GW4WYX
Thomas, W V	G4AEP
Thomasson, A H	GM0GAT
Thomasson, F	G1EJH
Thompson, A	G0IEF
Thompson, A	G3KQU
Thompson, A	M1BOZ
Thompson, A C	G4CXZ
Thompson, A C	G4FIJ
Thompson, A C	G4TRU
Thompson, A E	GW0IZR
Thompson, A F	G0IVM
Thompson, A K	G6TKM
Thompson, A M	G7PFD
Thompson, A P	G8KSX
Thompson, A R	G6HUN
Thompson, B	G0KUR
Thompson, B	G1YAE
Thompson, B	G7BQX
Thompson, B E	G0CGG
Thompson, B E	G4KAL
Thompson, B G C	G4LKF
Thompson, B G C	G8JJE
Thompson, B J	G4LNT
Thompson, B J	G6DWB
Thompson, B J	G6TXB
Thompson, B N	G6HAI
Thompson, C	G6VKA
Thompson, C B	G6FLW
Thompson, C D	M1CEC
Thompson, C E	G3KMI
Thompson, C J	G0GBE
Thompson, C J	M1BNA
Thompson, C M	G1SJB
Thompson, D	G1NYG
Thompson, D	G7WAW
Thompson, D B	G0MRU
Thompson, D F	G8SBU
Thompson, D G	G3AHS
Thompson, D I	G3IDT
Thompson, D J	G4UPK
Thompson, D M	G3RKX
Thompson, D S C	GI8JOA
Thompson, D W	G3OXG
Thompson, E O	G8SBV
Thompson, E R	G3VZR
Thompson, F	G4PMB
Thompson, F	GW0UZX
Thompson, F R	G0CJS
Thompson, F W	G6PIF
Thompson, G	G0BYS
Thompson, G	GI1SYM
Thompson, G F	G4UAN
Thompson, G J	G8FQV
Thompson, G P J	G7PEZ
Thompson, G R L	G3XBH
Thompson, G T	G3RCZ
Thompson, H	G4ZAF
Thompson, H B	G6AVL
Thompson, H R	G6RMW
Thompson, I D	G0AIT
Thompson, I W	G4ZKB
Thompson, J	G0SBY
Thompson, J	G8SXV
Thompson, J	GI6UUC
Thompson, J A	G0XVS
Thompson, J A	G1SKA
Thompson, J C	GU1EYP
Thompson, J E	G3OKT
Thompson, J E	G4THG
Thompson, J H	G6DVR
Thompson, J J	G0FGG
Thompson, J K	GI3VQ
Thompson, J L	2E1EMN
Thompson, J L	M1APF
Thompson, J M C B	GI0RBC
Thompson, J P	G4YES
Thompson, J R	G0WQW
Thompson, J R	G4OVG
Thompson, J S	G6HUO
Thompson, J W	G3NWU
Thompson, J W	G3WQM
Thompson, K	G1PJO
Thompson, K	G4SLH
Thompson, K	G4ZGR
Thompson, K	G7ECG
Thompson, K C	G3VSE
Thompson, K G	G3AMF
Thompson, K G	G4ZJX
Thompson, K J	G4PAD
Thompson, L	G1NSL
Thompson, L F	G4WZU
Thompson, L W	G3VYZ
Thompson, L W	G8ARF
Thompson, M	G6HUP
Thompson, M	G7UXA
Thompson, M A	G7HYZ
Thompson, M A	GD3EFD
Thompson, M D	G7THF
Thompson, M F	M1ADU
Thompson, M J	G7UXT
Thompson, M P	G1NIG
Thompson, M R	GD3JIU
Thompson, M R	G7UZS
Thompson, N C	G1KMN
Thompson, N C	G4CLY
Thompson, P	G1MIL
Thompson, P	G6PPT
Thompson, P	G7OYD
Thompson, P	G8TNB
Thompson, P	GM1GHZ
Thompson, P A	G8CSY
Thompson, P C	G1ZUB
Thompson, P C	G8DDY
Thompson, P D	G0BFH
Thompson, P F	G0TLU
Thompson, P F	G4GQR
Thompson, P K	G3JTT
Thompson, P R	2E1DZC
Thompson, R	G0SRT
Thompson, R	G6BIA
Thompson, R	G7RXK
Thompson, R J	G4AVO
Thompson, R J	G8GDZ
Thompson, R J R	G7MRZ
Thompson, R M	G8ZFT
Thompson, R M	GM4YPL
Thompson, R V	G7HSQ
Thompson, R W	G0UAK
Thompson, R W	G3TKF
Thompson, R W	G4HTV
Thompson, R W	M1BCG
Thompson, S	G4RCH
Thompson, S B	G7UTL
Thompson, S J	G0SEN
Thompson, S J	G8TNA
Thompson, S J	GI7DZE
Thompson, S J	M1ADR
Thompson, S R	G7IFB
Thompson, S R	G8GSQ
Thompson, T	2E1ADP
Thompson, T	G4UUR
Thompson, T E	M0ALZ
Thompson, T N	G4AVN
Thompson, V M	GM4ULS
Thompson, W	G1KIK
Thompson, W	GI4UUC
Thompson, W G S	G0LAE
Thompson, W H	M1BZX
Thompson, W L	G3YAG
Thompson-Pettitt, R	G0JPD
Thompson-Pettitt, R S	2E1BUA
Thomsen, B C	G0GSY
Thomson, A	2E1ENU
Thomson, A	G4OLF
Thomson, A H	GM3YXY
Thomson, A R	GM3AHR
Thomson, B	G4SDF
Thomson, B J	GM8PSV
Thomson, C J	G6AVK
Thomson, C J W	G3PEM
Thomson, D	G3WBS
Thomson, D J	G3RGS
Thomson, D S	GM7DST
Thomson, I G F	G1OZR
Thomson, I M	GM0URD
Thomson, J M	GM4CXF
Thomson, J M	GM8GUX
Thomson, J M M	G0AXM
Thomson, J P A	G8PJL
Thomson, J S	GM4LNU
Thomson, J W	GM8ZJS
Thomson, M	2M1GBG
Thomson, M B	G0GBD
Thomson, M G S	GJ0FTZ
Thomson, M J	G6AID
Thomson, M J	G8SYD
Thomson, M J	GM4JEJ
Thomson, N A	MM1AAY
Thomson, N O	G4BQZ
Thomson, P R	GM1XEA
Thomson, R	GM3OBC
Thomson, R C	GM8IOL
Thomson, R H L	G0NXP
Thomson, R J	G7RLB
Thomson, R L	GM0KDF
Thomson, R L	GM0RNR
Thomson, R P	GJ6RFE
Thomson, S J	GM4SJL
Thomson, T M	GI4SIZ
Thomson, W C	GM7AXQ
Thomson, W N	GM8BQV
Thorburn, G G P	G4UUS
Thores, O A	GM4JKT
Thorley, D N	M1AGE
Thorley, J	G0SPY
Thorley, R	G0GDN
Thorley, R J	G7MGA
Thorman, T E B	G0TRD
Thorn, C R	G3STZ
Thorn, C S	G6OEW
Thorn, D R	G8EOW
Thorn, J	G3PQE
Thorn, J M J	G6XZS
Thornber, B P	G0LVT
Thornber, J C P	G6GNC
Thornber, S K	G6SGA
Thorndike, D K	G4OMN
Thorndike, P M	G0XOX
Thorndyke, C	G0MEZ
Thorndyke, C J	G1OUT
Thorndyke, J	G0MEV
Thorne, A C	G6KPX
Thorne, C J	G4ZTS
Thorne, D H	2E1GBE
Thorne, D K	G7UHU
Thorne, E J	GW0FCN
Thorne, F J	G4YPD
Thorne, G A	G0TIP
Thorne, G H	G0JJD
Thorne, G N	G3UIF
Thorne, J	2E1GBF
Thorne, K R	G3NBZ
Thorne, L A	G7TSY
Thorne, L A	M0BGJ
Thorne, M E	G3ZUT
Thorne, M S	2E1BUO
Thorne, R J	GW1SGH
Thorne, R T	2E0APS
Thorne, T	G4KQZ
Thornes, A	G4SJR
Thornett, J	G6YTH
Thornhill, R J	G7TGM
Thornley, C	G6CVD
Thornley, J R	G1NUS
Thornley, J R	G7WAB
Thornley, N A	G7UGY
Thorns, J A	G0AQN
Thornsby, G	2E1BRC
Thornsby, M	G6XGT
Thornsby, P	G6SUR
Thornton, A G	G0JVW
Thornton, A J	G7NZF
Thornton, A P	GM7VZF
Thornton, B	G1IIO
Thornton, B	G7ILJ
Thornton, C E	G6DVQ
Thornton, C E	GM7PWI
Thornton, C J	2E1EXZ
Thornton, E	G7BLQ
Thornton, F A	G0TYI
Thornton, G	G0BBI
Thornton, G A	G0LVO
Thornton, G A	G4YJC
Thornton, G R	G8EKW
Thornton, H R	GM3PKV
Thornton, I A	2E1EHE
Thornton, J	G1LML
Thornton, J A	G4ZJQ
Thornton, J E	G7IFK
Thornton, M	G1GDU
Thornton, M E	G1XUH
Thornton, N	2E1FLF
Thornton, P L	G4XHG
Thornton, R	G4JCY
Thornton, R	G6KCE
Thornton, R W	M1BPO
Thornton, T J	G1CNV
Thornton, W	G0NSC
Thornton, W E	G3KLJ
Thornton-Evison, P.	G8PTY
Thorogood, J B	GW4UWI
Thorogood, M J	G1DMF
Thorogood, M J	GM0JKF
Thorp, A	G7MHD
Thorp, D M	G4RCB
Thorp, I R	G6WBT
Thorp, M A	G3PQM
Thorpe, A B	G3UHX
Thorpe, A T J	G4DRK
Thorpe, C	G4LEI
Thorpe, C A	G0SAY
Thorpe, D	G0RNI
Thorpe, D	G1LPQ
Thorpe, D	G4FKI
Thorpe, D	G4ONP
Thorpe, D	G4RNT
Thorpe, H	G0UFM
Thorpe, J B E	G3YLP
Thorpe, J D	GD3YDB
Thorpe, J H	G7PIO
Thorpe, J T	G0OKZ
Thorpe, J T	G1VCE
Thorpe, S A	G4NST
Thorpe, W J	G4REB
Thow, W M	GM4GNR
Threadingham, R L	G3ADR
Threapleton, W R	G4PEL
Threlfall, T R	GW4TGT
Threlfall-Rogers, S G	G8UIW
Thrippleton, I M	G8MLC
Throne, J O	GI0AYB
Throssell, D R	G4DWA
Thrower, N	G3YSW
Thrower, P R	G7JYU
Thrussell, A N	GW1PKX
Thurbon, A D J	G4GZO
Thurgood, A H	G4LNY
Thurgood, H	M0AQU
Thurlow, A	G3WBN
Thurlow, A V	GW0KWU
Thurlow, B R	G6OVC
Thurlow, M C	G1GCT
Thurlow, P W	G3GOM
Thurman, P	G1BLB
Thurman, S I	G1PWI
Thursfield, N J	G6HUR
Thursfield, W V	G4HOH
Thurtell, D J	G1UYL
Thwaites, A L	G3HHR
Thwaites, B H	G3CVI
Thwaites, J C	G8PWO
Thwaytes, D R	G1GDB
Thwaytes, P L	G4WOH
Thynne, A	G7IKG
Tibbert, B R	G3RKZ
Tibbett, A	G4ADT
Tibbett, G B	2E1AFH
Tibbetts, G	G4OGK
Tibbetts, S P	G7VJT
Tibbit, R J	G7DWP
Tichler, B A	2E1FWT
Tickell, H	G4BJJ
Tickell, W H	G8EMB
Tickle, A	G6TWL
Tickle, I	G4ZJH
Tickle, R P	G7EPP
Tickle, S C	G1MTJ
Tickner, T A	G0CSA
Tidball, M P	GW0VBC
Tidder, A P	G8CHI
Tideswell, D A S	G0FVV
Tideswell, I S	G7HKQ
Tidey, I W	G1AML
Tidmarsh, S J	G4VMM
Tidnam, R C	G4JOI
Tidswell, A J V	GW6WQJ
Tidswell, R C	G8KGR
Tidwell, C W	G0DAE
Tierney, J D	G4JYQ
Tierney, T A	G0JSV
Tietjen, A J	G7NZH
Tietz, P	G0AEU
Tiffany, B	G3TXX
Tilbee, A R	G4HXE
Tilbrook, H H	GM6OKJ
Tilbrook, P C	GM8XZY
Tildesley, J	2E1GAH
Till, W P	G0RPV
Tiller, G D	G7UHE
Tiller, J C	G4AZU
Tiller, R D	G7JTY
Tiller, R E	G8XYR
Tillett, G H	G3KXP
Tillett, G J	G6MDN
Tillett, O S	G3TPJ
Tilley, D P	G4USI
Tilley, F W	G1XJK
Tilley, J G	G7KRI
Tilley, R E	G6HMV
Tilley, R L	G0BXF
Tilley, T W	G1BBA
Tillin, A V	G3MES
Tilling, C R	G8UCW
Tilling, D G G	G3PGT
Tillotson, H T	G8XIZ
Tills, H P	G0HPT
Tillson, G	G3TJX
Tilly, S A	2E0AOU
Tilly, S A	G1BNN
Tilsed, G H	G4SHV
Tilson, D A	G0OZF
Tilston, T J	G4YXQ
Tiltman, D G	2W1CYC
Timbrell, C J	G6JOZ
Timbrell, G G	G4STH
Timbrell, H J	G4YLO
Timbrell, L J	G7JOC
Timbrell, M J	G6JOY
Timings, R L	G0IZH
Timlett, P G	G0LWC
Timme, A W W	G3CWW
Timmins, A J	2E1DPF
Timmins, C J K	G6BRL
Timmins, F I	G4RRK
Timmins, P M	G0NDV
Timms, A R	G0FKW
Timms, B G	G7FLX
Timms, G J	2E1FFE
Timms, G J	M1BXH
Timms, M E	G7DTL
Timms, M G	G4ZYH
Timms, R G	G3JUC
Timms, R J	G8VBC
Timms, S E	G7OST
Timms, T J	G8TIM
Timms, T M	G7DTK
Timms, W R	G3BWI
Timperley, J	G1IZU
Timson, D	G6FMF
Tindal, M A	G3VLQ
Tindale, C	G1HEZ
Tindall, N M	G7UMA
Tindill, B	G4HVA
Tindill, C B	G8LNQ
Tindle, J E	GW3JXN
Tink, A D	G7DRU
Tink, D M	2E1FMO
Tinker, C L	G1GSW
Tinker, J M	G3PKC
Tinkler, M C	G0AZS
Tinkler, P G	2E1EFK
Tinkler, R F	G0IPG
Tinkler, R J	G3SMR
Tinley, D J	G7DJT
Tinning, R H	G3ZEP
Tinsley, J	G0FZU
Tinsley, J	G4TTP
Tinsley, S	G0TDC
Tinson, R E	G3XPM
Tinton, B	G3SWC
Tiplady, J	G4ELR
Tipler, G A	G8CBA
Tipp, C W	G1ZDY
Tipper, A D	G8THZ
Tipper, A R	G4KXR
Tipper, B D	G3WWL
Tipper, D P	G3JBR
Tipper, D P	G4BP
Tipper, M	G6AFZ
Tipper, T G	G1MZT
Tipping, F R	G3EHV
Tipping, N J	G4WED
Tisdale, J F	G4NRA
Tisdall, T N	G8BJL
Tite, C E	G0EYG
Tite, J R	G0HMF
Tither, P A	G1MVE
Titheridge, C	G4DYI
Titheridge, D E	G4EZX
Titherington, J F S	G4POH
Titherington, P J	G4RWM
Titherington, S K	G1GSX
Titley, A	G8RKX
Titley, C O	G3BGG
Titmarsh, B J	G7UAH
Titmuss, R K	G0AWY
Titt, R A	G3CMJ
Tittensor, M V	G4EKG
Tittensor, P J	G4PVM
Titterall, J P	2E1EOS
Titterington, R G	G3ORY
Titterington, R G	G3SDC
Titterington, S	G7LKL
Titterton, J	M1AZB
Titze, W A U	G0PLM
Tivey, D J	G0ODJ
To, D C N	G7TTL
Toas, A W	G6TRW
Toase, G V	G4NYI
Tobias, J A	MM1ANP
Tobin, A	M0BAM
Tobin, A	M1ASH
Tobin, J S	G6PBO
Toby, C	G4TZF
Tock, D I	G0SYA
Todd, A D	G7DXE
Todd, B J	G3PHW
Todd, C	G4ABM
Todd, C A	G4CLZ
Todd, D A	GW7OIK
Todd, G	G0KOG
Todd, G E	G8VYQ
Todd, H	G0HTD
Todd, J B	G1HEY
Todd, J K	G2KV
Todd, J W	G4XLM
Todd, M K	G1HJS
Todd, R	GI0STS
Todd, R K	G3WNC
Todd, R K	G4ULD
Todd, R M	GI7AQO
Todd White, B R	G3OJZ
Todorovic, S	G7RUR
Toft, R H	G4XZT
Tofts, P S	G3WBK
Togwell, B	G4FTZ
Toh, H H	G0CCG
Tohill, J	GM0FKP
Toke, L	G3ETU
Tokley, E E	G3TFV
Tokley, R F G	G4MDB
Tolcher, N J	G7UUS
Toll, P M	G7BGO
Tollefson, G H	G7OAM
Tolley, S P	G1DHA
Tolman, J	G3BTL
Tolman, K W	G8RMC
Tolman, W J M	G4RWK
Tolson, J E	G0AHU
Tomalin, A W	G3PTB
Tomalski, R	G6CQF
Tombs, C D	GW4MOG
Tombs, D J	G8VYT
Tombs, G F	G0AOR
Tombs, H W	G7NRM
Tombs, S J	M1AHH
Tomes, P W	G3YPT
Tometzki, E S A	G0TIZ
Tomkins, G B	G1IEP
Tomkins, P	G1LBH
Tomkins, S	G1UHB
Tomkins, V A	G4KEE
Tomkinson, C G	G4GEO
Tomkinson, D	G0FQG
Tomkinson, K W	G1PAG
Tomkinson, R	G7SBG
Tomkinson, S	G1UYT
Tomkys, R W	G3NOW
Tomlin, D	G7RYA
Tomlin, E A	G3YFU
Tomlins, J D	M1CBH
Tomlinson, D	G4ODO
Tomlinson, D A	2E1FED
Tomlinson, G	G3JZU
Tomlinson, G L	M1AIV
Tomlinson, G M	G1EJK
Tomlinson, G P	G6DJQ
Tomlinson, H	G7PWU
Tomlinson, H J	G3UHW
Tomlinson, J M	G3MGX
Tomlinson, J V	G3KTX
Tomlinson, K J	GW6VBC
Tomlinson, K W	G8YCK
Tomlinson, N F	G2HMB
Tomlinson, P	G4VSP
Tomlinson, P	G6TKW
Tomlinson, P A	G4CBL
Tomlinson, P D	GW7LHI
Tomlinson, R J	G6IKK
Tomlinson, R P	G4TGJ
Tomlinson, R R	G0EKG
Tomlinson, S	G3KVJ
Tomlinson, S M	G0KKK
Tomlinson, S M	G8LZU
Tommey, D G	G1MDC
Tommey, J M	G0EHT
Tompkins, B H	G6WIE
Tompkins, D S	G4MJN
Tompkins, M	G3YZA
Tompkins, R W	G8LDY
Tompsett, S A	G7APD
Tompsett, S A	G8LYB
Tompson, A D	G8LSS
Toms, M D	G8NIC
Toms, P J	G4XMZ
Toms, R E	G8ZMV
Tomschey, S V A	G8ZJO
Tomsett, D R	G1ORB
Tomson, A J	G0RRV
Tomson, I T	G0LOZ
Toner, S J	MI1BKR
Tong, C D	G7DSU
Tong, G K	G8ENO
Tonge, A G	G0BVT
Tonge, G	G1YHM
Tonge, G	G4IDG
Tonge, J R	GW7JRT
Tonge, L	G8ONF
Tonge, P	G1XPA
Tonge, P	G6YSZ
Tongue, S F	MW0ATP
Tonkin, B	G7DUC
Tonkin, G P	G5RQ
Tonkin, M C	GW7TZI
Tonks, F A	GW3WBH
Tonks, J A	2E1AIT
Tonks, J M	G0MGP
Tonks, J M	G4MBN
Tonks, J T	G8ZZT
Tonks, W T	G0PWQ
Tonner, K W	G1RFX
Toogood, A N	G4VZU
Toogood, C S C	G1MCY
Toohey, E W	G6ZFK
Tooke, B A	G4VGQ
Toolan, J	G1AVZ
Tooley, F J	G3HPB
Tooley, I C	G7EHS
Tooley, M A	G4BBD
Toombes, N C	G7WFU
Toombs, D A	G8FXM
Toombs, E	G0IVS
Toome, E J	G7JSA
Toomer, C A	G4RKE
Toon, B	G1OYU
Toon, B M P	G8PME
Toon, D J	G4IXF
Toon, I P	G4RJG
Toon, I P	G4YIT
Toon, J G	G0FNH
Toone, D C	2E1BSY
Toop, G P	G0ULZ
Toop, K E	G7PCE
Tootell, G R	G1WJC
Tooth, R W F	G4RZR
Tootill, L C	G6RGO
Tootle, E	G8JCZ
Toovey, N J	G8BWG
Tooze, L C	G7VCB
Topham, D	GM3WKB
Topham, D F	G4POK
Topham, J O	G0RCJ
Topham, P J	G8KDO
Topliss, C A	G0RFB
Topliss, J R	G4OXF
Topliss, R P	G0OTH
Topping, A J	2E1ECG
Topping, C H	GM6HGW
Topping, D A	G4PWV
Topping, D R P	G3HYG
Topping, R	G8SOU
Topping, S W	G0MTD
Tordoff, D B	M0AUM
Torence-Smith, R	G4SWQ
Torr, I D	G7GME
Torr, R C	G0KSL
Torrance, P	G4HAK
Torring, J R	G6TKH
Torry, J M W	M0ATQ
Torry, P G	G3SMT
Torunski, H L	G0GLG
Tory, J W	G0ETG
Tory, P G	G3VMQ
Toseland, C M	G7VEV
Tosh, J M M	GI0RJU
Tosler, A D	G4NXE
Tosney, V T	G1WSZ
Tostevin, L O	G4PLD
Tott, S G	M1BLK
Totten, J A	G7LWF
Totterdell, N H	G4FAL
Tottle, J	G6LDW
Tottle, J E	G4SSN
Totty, C	2J1CWH
Totty, C	GJ7UIT
Totty, D	2E1FRU
Totty, J	2J1CWG
Tough, I S	G0IHK
Toulalan, G	G8AAP
Tournant, J C	G0DEU
Tournant, P J	G6HMO
Tournier, J	G3INZ
Tout, M J	G4EUR
Tovey, M J	G4XSX
Towell, P J	G0AYX
Towers, D A	G8SZX
Towers, D J	G1GTK
Towers, E P	G4HYI
Towers, K D	G0CEV
Towers, M	G7AFW
Towers-Perkins, W	G4NIO
Towle, C	G0WKJ
Towle, C	M0AJT
Towle, J	2E1DPU
Towle, J A	G4PJZ
Towle, J G	G0WPF
Towle, R	G4IWJ
Towler, G O	G4NGS
Towler, L P	G7JJP
Towler, R	G0UKS
Towler, R	G3AMW
Towler, S	G0VTD
Towlson, M I	G0VFI
Town, C S	G3WLS
Town, J	MM1CDC
Towndrow, E V	G4JEP
Townend, A	G3DDH
Townend, D	G4GNA
Townend, D M	G0RQX
Townend, G M	G4SDX
Townend, J E	G0TOJ
Townend, J L	G3BBD
Townley, A T	2E1ARF
Townley, C P	G3BVW
Townley, E E M	GI4XFY
Townley, M E	G4JJK
Townley, N G	G4NGT
Townley, R	G7GVC
Townrow, P A	G6LTB
Towns, C H	G8BKE
Towns, H J	GM3LNI
Towns, J	M1ADX
Towns, K W	G6IKN
Townsend, A	G4NMA
Townsend, C I	G8PUT
Townsend, D F	G4SQR
Townsend, D J	G0OFC
Townsend, D J	G7NIN
Townsend, E	G8DGH
Townsend, E A	G6XNN
Townsend, E J	2E1FDG
Townsend, F H	G8BCT
Townsend, G H	G8ANN
Townsend, G M M	G3LSX
Townsend, I L	G1XRA
Townsend, J	G8CSC
Townsend, J	G8JAD
Townsend, J M	G4ILY
Townsend, K R J	G0VUG
Townsend, K R J	G7IZT
Townsend, M	G1KRI
Townsend, M E	G4EQL
Townsend, P	G8SNJ
Townsend, R M	2E1FHU
Townsend, R N	G6CIE
Townsend, S W	MW0AEL
Townsend, W S	GW7PRT
Townsend, W S	MW0AXA
Townshend, J H	G4DA
Townshend, P M	G6PMT
Townshend, R J	G6KLK
Townson, I	2E1EGV
Towse, D L	G1RFO
Toy, L	G0VEJ
Toynton, P J H	G3RGA
Tozer, J J	G3XLZ
Tozer, T R	G1WUP
Tracey, A F	G4PZX
Tracey, E J	G4YBT
Tracey, J C C	G6GTB
Tracey, W J	GM4UBJ
Tracy, J A	G1TUI
Trahearn, M E	G0OYM
Traill, K	GM4XUJ
Traill, K W	GM0TQK
Traill, T R	GM0LUF
Trailor, T M	2E1BAC
Train, G S	G4LEX
Trainer, D A	G3UPJ
Trainer, F	G7RYW
Trainer, J N	G1LMC
Tran, C W	GM2MP
Tran, C W	GM3WOJ
Tran, C W	GM4DMZ
Trangmar, R H	GW6ORE
Tranmer, J	G8ANK
Tranmer, J W	G8NCW
Tranter, A	G3PMD
Tranter, J C	G3BQQ
Tranter, J S	G4XFT
Trask, P	G0DRF
Tratt, E H J	G6TKR
Traveller, C R	G3WXW
Travers, J	GW4UVN
Travers, J M	G4VCG
Travers, J P	G0SHH
Travers, P E	G1HEW
Travett, D J	G8TVB
Travett, R J	G8XYS
Travis, D B	G4GXD
Travis, G R	G6ORD
Travis, R C	G4DNP
Trayler, A R	GM3DSD
Trayner, C	G4OKW
Traynor, A	G0FCX
Traynor, D A	G1URQ
Treacher, A L	G7MGH
Treacher, M J	G0ODS
Treadwell, C M	GW6DNZ
Treadwell, P C	G7PCT
Treanor, N A	G8WAJ
Treasure, D	G4ZBT
Treasure, K	G0SYI

Trebilcock, R J GW3ZCF
Tredgold, E G3XWI
Treece, J G G3QD
Treen, D G0RDT
Trefry, J J G0TLZ
Tregay, A T G7AYQ
Tregear, P G8PQM
Tregonning, W L G4TKT
Trelease, R C G4ROG
Trelease, V M G0CAT
Tremain, G D G6BBM
Tremain, J G3EHT
Trembath, P G7SBL
Trembeth, L O G4HBU
Tremble, C R G0PKE
Tremelling, R T G G3FWG
Trenchard, G F W G4EHU
Trenchard, M GU6JQF
Trend, A G G4BER
Trend, A T G0MKB
Trend, V M0AEJ
Treneer, R P G3UHZ
Trengove, T C F G1SZD
Trent, A G0VGT
Trent, J G8YMZ
Trepass, P H G4HBD
Tresadern, J M G7ELD
Trethewey, E G4GKQ
Trethewey, T D G4YXJ
Trett, A C G6XNP
Trett, E G0BZV
Trett, J T G6JTT
Trett, R J G8JWT
Trevelyan, G W G4UPW
Trevelyan, R S G2CKQ
Trever, J B H G4SKH
Trevethick, A D G4PKB
Trevett, J L W G4GKX
Trevitt, A L G4DHX
Trevitt, E W G3EPP
Trevitt, R G3SSE
Trevitt, V I G1HJA
Trevitt, W R 2E1FBP
Trevor, B A G6SUK
Trew, T I P G8JXV
Trezise, P C G2HAJ
Tribbeck, R G0AOG
Tribe, H C M1BFG
Tribe, H J G4SAP
Tribe, J B G0IUY
Tribe, M J G1RPT
Tribe, M J G8PMT
Tribe, M L G7NJI
Tribe, R E G4SAQ
Tribe, R W G6HEW
Tribe, S J B G0IEY
Tribute, D W G1OEQ
Trice, F T G8WQM
Trice, J M G0RGM
Tricker, S R G0AZP
Trickett, D G8TIA
Trickett, J W G4JMC
Trickett, P E G6DNX
Trickey, E F G4DCX
Trickey, G A G4CDW
Trickey, J V G7SER
Trickey, P M 2E1EHI
Trickey, R S G3DRB
Trier, F M G8VH
Trigell, A E G1JAF
Trigg, A E G8SRK
Trigg, J M G8ICD
Trigg, K J G7OJX
Trigg, M 2E1ARG
Trigger, B C G6IKQ
Trim, A M G6KAC
Trim, B J G8YRL
Trim, G E G4XSC
Trimmer, B W G0FHC
Trimmer, P G GW4KCY
Trinder, K R GM7DFI
Trinder, W G4UCC
Tring, A J G3WYB
Tringale, L G0OLT
Tringham, D N G0BSD
Tripp, A G6XTC
Tripp, C J G1PUQ
Tripp, J M B G3YWO
Trippear, P G6GLD
Trivett, B W G0TTQ
Trolan, M J G6SHF
Trolan, V A G1STP
Troll, P J G7PKQ
Trollope, N S G4FAT
Tromans, D B G4CGB
Tromans, K G0LBT
Troop, C J G4ZSN
Troop, D W G8TGD
Troop, N M G4IQR
Tropman, D J 2E1AWQ
Tropman, D J G7UHG
Trotman, C GW4LNP
Trotman, D E G4WPR
Trotman, M R G0TYA
Trotman, M R G0WNL
Trotman, M R G1FYC
Trott, A J G7MIS
Trott, P P G4BNN
Trott, S J G8ZOE
Trotter, J G0WEQ
Trotter, J G4HPX
Trotter, J M G4FLL
Trotter, S D G6KVY
Trotter, W G G0EGE
Troughton, E G0VNV
Troughton, R G4AZJ
Troughton, R H G3SFP
Troup, A 2M1CEJ
Trousdale, A R G4LXW
Trout, J G0AFC
Trow, R J 2E1CVE
Trowbridge, P M G3BLN
Trowbridge, T R W G3IBK
Trowell, D W G3RML
Trowell, E H G2HKU
Trowman, A W G4BDK
Trowsdale, R M G6HMG
Troy, A E G4KRN
Troy, C M 2E1EUE
Troy, G G3YS
Troy, M L G0WFQ
Troy, M L G1UWL
Truberg, P A GW4JOG
Truckel, G P G0LXL
Trudgen, A G4YAF
Trudgeon, T J G0ENZ
Trudgill, R C G4VLA
Trueman, A M G1IZH
Trueman, H G0BUF
Truitt, P B G4WQO
Trull, R T G3RAD
Truman, M J G8SLC
Trundle, M A G3TCG
Trunley, H J F G3RPZ
Truran, P H G4BPR
Truran, R J GW0MAV
Trusler, A D V G0FIG
Trusler, J A G1HJH
Trussler, F G G6KPR
Trussler, R W GM0CSN
Trusson, C I B G3RVM
Trusson, C I B G3WOI
Try, N W G4NWT
Trybulski, J G7SRZ
Tryhorn, L A G7GKD
Tryner, P G1ZWX
Tse, C K F G7WCX
Tsiakkouris, S M0BBB
Tubb, D R G1SNO
Tubb, T R G8HSM
Tubbs, J W G0DWS
Tubey, C J G1ZMW
Tubis, C J G8HJD
Tubis, J N G8PUX
Tubman, E J G4SIL
Tuby, G R M1BMY
Tuck, F W G8KZP
Tuck, J P G1MLH
Tuck, P D G8YNC
Tuck, P J G4TKF
Tucker, A G4PCW
Tucker, A J W G6LPD
Tucker, B G4YZQ
Tucker, B M0APL
Tucker, C G4DCH
Tucker, C E G0CRW
Tucker, C E G3TXZ
Tucker, C E G4RTV
Tucker, D G6HMX
Tucker, D M1BVC
Tucker, D J C M1AXG
Tucker, D M G1SEA
Tucker, D W G1KHS
Tucker, J C H G5TU
Tucker, J L G7LNG
Tucker, J S G0OZY
Tucker, K J G1HEQ
Tucker, L A J G8LNU
Tucker, L J G4HJN
Tucker, N R G7URW
Tucker, P D G4DWZ
Tucker, P J G3TXN
Tucker, R D G8VRH
Tucker, R E G3KTC
Tucker, R E G6AVI
Tucker, R M G4ULU
Tucker, S L G1FFR
Tucker, T L G8ZNE
Tuckett, R S G8TQV
Tuckfield, F J T G2HOX
Tuckley, C G8TMV
Tuckley, C I G1UZT
Tuddenham, E P G3XFF
Tuddenham, K W J G1WXM
Tudor, D R 2E1EKL
Tudor, E R G3INY
Tudor, M J E G0LZI
Tudor-Jones, S G3UMZ
Tuer, K H G0VRZ
Tuff, H J G8HYL
Tuff, V G G7PYR
Tuffin, D A G7WEZ
Tuffin, S J G4YAL
Tuffrey, B G0LHM
Tuffrey, M W G8LHQ
Tuffs, P G4HEB
Tufnail, B L G0HOS
Tugman, C C G4VUF
Tugwell, B L G0LTD
Tugwell, E M G0FIP
Tugwell, G L G0HPD
Tuite, A P GW0NSR
Tuite, I C 2W1FQA
Tuite, J 2W1DHV
Tuke, J B G3BST
Tulk, D J G1NLU
Tulk, P H GW1ZJC
Tulk, R J GW1GXQ
Tullett, M R 2E1EJX
Tulley, H C G0LWV
Tulloch, D J GM0DJI
Tullock, G G1IBF
Tully, C S G4CYF
Tully, C V C G7ESE
Tully, J F G4OSN
Tully, K G8XAX
Tully, P G M1CCX
Tully, W F G0ANX
Tumber, J G0SJF
Tunbridge, C A G7LIK
Tunbridge, C A G7PEC
Tunbridge, D J G1SNU
Tunbridge, D J G7MDR
Tungate, A R G0JBJ
Tungate, M A G8ZVC
Tunley, G J 2E1DUI
Tunmer, D G7SPH
Tunna, C D G4VLT
Tunnah, C GI7VPF
Tunnicliffe, D L G4BCA
Tunstall, E K G3MSO
Tunstall, M R G0SKM
Tunstall, R G0JGD
Tupman, A E G3ID
Tupman, K D G0BTU
Tuppen, S G 2W1FBD
Tuppeny, G S G4LOE
Tupper, J E G1IER
Turbert, T H G4TOM
Turgill, L G2DCU
Turk, C J 2E1EZN
Turk, C J G1RYE
Turk, E P G7BQM
Turk, P D G G3PQC
Turkington, N S GI0BEY
Turland, C A M0AHO
Turland, N A G7LNV
Turley, B J G6GUT
Turley, P M G1CQT
Turley, R H G0IYJ
Turley, S J G7JYZ
Turley, T G0TYZ
Turlington, R T G8ATE
Turnbull, A D GM1FIX
Turnbull, A P G4CUS
Turnbull, B G0OIM
Turnbull, D A GM0TUR
Turnbull, D A GM7SAR
Turnbull, G B G8VRG
Turnbull, J G6KMG
Turnbull, J C G6TNK
Turnbull, J I G7NHE
Turnbull, L R GM3HWN
Turnbull, M A GM1NDE
Turnbull, M B M1AUA
Turnbull, M D G7PWL
Turnbull, M H G4ILM
Turnbull, P W GM4CKM
Turnbull, R E 2E1ANZ
Turnbull, R E M0ANZ
Turnbull, R J G0AIF
Turnbull, R K GM0VUY
Turnbull, V M I GM7AHA
Turnell, D J G4HLE
Turnell, K J G4HLW
Turner, A G4OWN
Turner, A C G4TXV
Turner, A D G4XNY
Turner, A D G7SEH
Turner, A E G4XQE
Turner, A H G4CPE
Turner, A J G0BRR
Turner, A J G0FMU
Turner, A J G4IMJ
Turner, A J G4UUT
Turner, A J G4XBC
Turner, A P G4RUL
Turner, A R G3YPN
Turner, A R G7PYV
Turner, A S G1GTM
Turner, A W GM2HDH
Turner, B H G3RLE
Turner, B H G4WZN
Turner, B J G8CEX
Turner, B R G1SGT
Turner, B W G3YNF
Turner, C B G4SEP
Turner, C G G G4HKP
Turner, C J G7MXM
Turner, C J G8TJD
Turner, C L G3VTT
Turner, C P A G0VUT
Turner, C P A G2IC
Turner, C W G7TZO
Turner, D G4SWY
Turner, D C M G3WEI
Turner, D D G1IIY
Turner, D E G0VVF
Turner, D E GW6OZA
Turner, D G G3SBM
Turner, D G GW4WMK
Turner, D I GM7KJL
Turner, D I O G4IDX
Turner, D J G0OZG
Turner, D R G4KEY
Turner, D S G4PST
Turner, D W G4APM
Turner, E G3TTL
Turner, E A G4GLY
Turner, E A G4SRB
Turner, E H G4IRG
Turner, E J G1MSB
Turner, E J G1XES
Turner, E M G7NCG
Turner, E N GI7DBN
Turner, F G8NFM
Turner, G G4SQJ
Turner, G G7OBP
Turner, G D G1VLS
Turner, G J 2E1FZB
Turner, H G4YRH
Turner, H C A G4CZT
Turner, I J G4NHR
Turner, J G1IYX
Turner, J G7POR
Turner, J GI6KBX
Turner, J M0AHQ
Turner, J A G4KJF
Turner, J C G8NDE
Turner, J E G0KFO
Turner, J E G1IEO
Turner, J F G3AYZ
Turner, J G G2AXY
Turner, J M G4FZY
Turner, J P G4MRO
Turner, J R 2E1AQI
Turner, J R G3UST
Turner, J R A G6MEH
Turner, J W G7EJK
Turner, J W G7HKU
Turner, K G1SLO
Turner, K G4CIF
Turner, K G6KXJ
Turner, K G8JLA
Turner, K J G7RYO
Turner, K M G8GIF
Turner, K O GW6WEU
Turner, K W G4GZB
Turner, L E G4DLA
Turner, M G6ZFZ
Turner, M H G0LIP
Turner, M H G4PCS
Turner, M J G3ZMN
Turner, M J G8OFJ
Turner, M J GJ0PDJ
Turner, M J GJ8RVT
Turner, M K G7UGA
Turner, M S G3VYN
Turner, N A G0PPY
Turner, N C G4DDZ
Turner, N R GU7NCZ
Turner, N S G7TMA
Turner, O C G4KDO
Turner, P G7BXJ
Turner, P C G3TIG
Turner, P F G0UCN
Turner, P G G6EWC
Turner, P G G7MWD
Turner, P J G4HKB
Turner, P J S G0SIX
Turner, P R G1CKY
Turner, P R C G0JXD
Turner, P W G4IJE
Turner, R G0GME
Turner, R G3IMG
Turner, R G3THK
Turner, R A G6OKN
Turner, R E G7VDG
Turner, R J GM4YJL
Turner, R J G G0LXF
Turner, R L G3SMD
Turner, R M G1FZY
Turner, R P GM7SAQ
Turner, R W G3VRX
Turner, R W M0AMH
Turner, S G3JUB
Turner, S G6FJO
Turner, S A M1BVM
Turner, S H G3IST
Turner, S M G0HCR
Turner, S M G1LCE
Turner, S P G4TUR
Turner, S R 2E1EOU
Turner, S R G3UJI
Turner, T H GM4YAT
Turner, T J G7HRL
Turner, V C G0RJC
Turner, W G7PAM
Turner, W G G4LML
Turner, W G GW6MNC
Turner, W J GI4LZR
Turner, W W G2CBH
Turner-Smith, F D G3VKI
Turnham, N R G6PSA
Turnham, P R G4VLS
Turpin, E D D G3MNK
Turpin, L G G0BFU
Turquand, A J G1FXB
Turtle, D G G1OLZ
Turton, A J G0LGO
Turton, D G1JVM
Turton, D L 2E0AJO
Turton, J M G7OCX
Turton, L M J G0WLV
Turton, R G7IOH
Turton, W F G4MSG
Turvey, B D G3UMT
Turvey, G E G3DOI
Turvey, K J G6BGA
Turvey, M F G4URN
Turvey, P A G1PJJ
Turville, C L G3BSO
Tuson, I R G0MQR
Tust, M W G4LUQ
Tuthill, P J G1BMT
Tutt, B L G4ZZK
Tutt, E A G1FQK
Tutt, G A 2E1AOF
Tutt, I G G0PEC
Tutt, I T G0RNN
Tutt, M J G8LLJ
Tuttle, M J G0TMT
Tutton, A R C G8GPH
Tutty, B D M1AHK
Tuvey, C F G4IXB
Tweed, S L G7HJN
Tweedie, D P G4GQZ
Tweedie, S GI4RXX
Tweedie, T D GI7RAH
Tweedy, G G0LOP
Tweedy, J G7MPJ
Tweedy, N M G1CYN
Tweedy, W J GI4FGH
Twells, M G7DWX
Twells, P G4XWO
Tweney, P G1RCX
Twibell, G G8FKL
Twidale, D R G1FEP
Twiddy, R G0BCE
Twigg, D E G4JKM
Twigg, M V G4ZXT
Twigger, S N G1SNT
Twiggs, R C G4HIN
Twine, T R G1GDQ
Twiss, G E G3RUG
Twiss, J G3YVP
Twist, B H G3NFY
Twist, B T G8YSX
Twist, D E G6DKB
Twist, K G G6DKA
Twist, R D G4AEY
Twittey, D J 2E1ALB
Twort, K J G8CHY
Twyford, A R G8TSZ
Twyman, D P G6LJR
Twyman, R K G1ZQR
Tyas, J G3NSG
Tyblewski, J G4BZA
Tybora, G B G1HEU
Tye, A P G8ZZV
Tye, C W G1FER
Tye, D F G3ZIB
Tye, J R G4BYV
Tye, R G G1MDA
Tyerman, D J G3KCG
Tyerman, J W G7EJH
Tyers, D G0FJD
Tyers, D G6XZW
Tyers, V J G8TMP
Tylee, J E G4REK
Tyler, A J G1GKN
Tyler, A J M0BAH
Tyler, A M G1YAF
Tyler, B GW8EQO
Tyler, C M G8MYO
Tyler, D S G0WML
Tyler, D S G8LWA
Tyler, E P G0AEC
Tyler, F W G3CGQ
Tyler, I M G0KOA
Tyler, J G4GCL
Tyler, J C G0EQF
Tyler, J E G8XZX
Tyler, K G0ITI
Tyler, K J G4CRP
Tyler, M G4NUB
Tyler, R J J G0CFB
Tyler, S J G1FOU
Tyler, S R G4UDZ
Tymon, M G7ETY
Tynan, W G3SJR
Tynemouth, G G4PIF
Tyou, M C G0GUE
Tyrell, P C G8RSK
Tyrell, R A G0WKG
Tyreman, H R G3SLL
Tyreman, J L G6TVE
Tyrer, T N G6TKV
Tyrrell, J M G7VRE
Tyrrell, J R G3OBL
Tyrrell, M J G6GAK
Tysiorowski, J P G3ZSK
Tysiorowski, J P G4GLQ
Tysoe, D J M1BTQ
Tysoe, G W G8FWI
Tyson, D 2E1FJE
Tyson, D J G4ERY
Tyson, H J G3IXO
Tyson, J E G6MDG
Tyson, M J G0GAU
Tyson, P H G0RLJ
Tyson, R G4NFR
Tyson, R H GW6HUV
Tyson, R M G8LGY

U

Udall, J M0APD
Udall, R P B G2HKS
Ullathorne, E B G3PLA
Ullathorne, T P G4JXF
Ullett, J M G1ZXF
Ullman, F T G4KQT
Ullman, L A G4LUM
Underdown, D W G3MBK
Underhay, B L G1YES
Underhill, C A G0MMI
Underhill, E D G4UVW
Underhill, M J G3LHZ
Underhill, P G7JTQ
Underhill, T R G4MWP
Underseth, L H G0FVY
Underwood, A J GW0AJU
Underwood, B G6PTE
Underwood, B R G4BUD
Underwood, C C G6MNA
Underwood, C I G1WTW
Underwood, D G0HBQ
Underwood, D A G4RCJ
Underwood, F J G0DEK
Underwood, H G0VMH
Underwood, I H G6UOJ
Underwood, J B GW1IIZ
Underwood, J G G0PGY
Underwood, K J G3SDW
Underwood, N GW4WRD
Underwood, N J G4LDR
Underwood, R D G6MAW
Underwood, T F G3VDW
Underwood, V GW0HYL
Unstead, F W G3BCY
Unstead, P B G0TNP
Unsworth, A E G8BCJ
Unsworth, C G7GJM
Unsworth, M G7LVS
Unsworth, R E G6LDV
Unsworth, R S G3PJW
Unsworth, R S G3WPF
Unsworth, S J G0HWX
Unsworth, T G0ECP
Unsworth, T G8FZT
Unsworth, W G0MZO
Unterhorst, P F G0DWG
Unwin, D G8ZUZ
Unwin, F G G3KIH
Unwin, J M G4JHN
Unwin, P G4HDS
Upcott, A E G6HMJ
Uphill, J A R G2HKW
Uphill, L B G3UCE
Upsher, J J G0JJU
Upson, W J G4ZPS
Upstone, J GW4MOZ
Upton, A G3UZU
Upton, D G0UAF
Upton, G K G4SJG
Upton, P J G7CCV
Upton, R C G0TLS
Urda, E R G4XAY
Urquhart, C GM3JUD
Urquhart, D P M G4DR
Urquhart, H R GM0UTD
Ursell, C M G1JVO
Ursell, F G1JVN
Ursell, J P U 2E1DND
Urwin, C G3OPE
Usher, A C G4HZW
Usher, D J G6ZKC
Usher, E R G0NUJ
Usher, R J G1IEV
Utley, A W G4YXB
Utley, D I G0PAZ
Utteridge, A G0FFU
Utting, A B G0AUT
Utting, A D G1WZQ
Utting, A D G7MCD
Utting, N V T GJ7LJJ
Utting, S C GW7UQL
Uttley, D G8MEC
Uttley, W G3MWK
Uttridge, P R G1EQF
Uytendhal, K 2E1BBP
Uzzell, W R GW4IXC

V

Vadgama, H G8XNN
Vagars, P G0RSX
Vahl, S P G4SIG
Vaisey, L C G1LOU
Valder, R J G4XCA
Vale, F W G3CWT
Vale, R G3YHI
Vale, T G0LUQ
Valente, M E S G4EBN
Valenti, G R G6XJN
Valenti, M R G6XGV
Valentine, G K G0UVX
Valentine, M G0HIU
Valentine, M J G4ANP
Valentine, N B G3KWJ
Valentine, P S GI3RKE
Valentine, S G3ZNQ
Valentine, S F G4RWS
Valentine, S J GW4WPH
Valentine, T GM1XHZ
Valerio, P H GW4KTT
Vallard, T A W G3DKB
Valleley, P G0LQD
Vallely, G A G3ELV
Vallely, G A G8NJO
Valler, G H G6FKA
Vallins, D K G7TCK
Vallis, P J G0RAL
Vallis, R J G0COC
Vallow, P H G4FSK
Valvona, S J 2E1EBN
Van Arman, C G GM4VDL
Van Beers, L J G7RLO
Van Den Bergh, W L G6YTZ
Van Den Bossche, P A M1BCA
Van Der Steeg, M G7BEP
Van Dyke, J R GM0RYD
Van Falier, P A G1VAN
Van Haaren, D M G4XVH
Van Haaren, R L G4YVH
Van Kassel, S G R G4VWG
Van Klinkenberg, P D G7IYF
Van Praag, S S G4YFS
Van Stigt, N D G4RWA
Van Zuilen, C G7TWO
Vanbeck, D E G1GBX
Vance, C P G7HPI
Vance, D H GI3XZM
Vance, I A W G3WMS
Vandepeer, R W G1VJM
Vander Byl, W G8UTQ
Vane, B E G4SEJ
Vann, A R 2E1AOY
Vann, M S G0RYS
Vann, M S G3RLV
Vannerley, H C V K G0IZZ
Vansittart, R G6GHP
Vanson, B W G0MBP
Vanstone, S G G7FZD
Van'T Riet, D 2E1EHJ
Vardon, J H G6ESN
Vardy, K G4NEM
Vare, A D G4XZL
Varga, A P G6TEB
Varley, F L G2FCP
Varley, R G8ODK
Varnals, K D G1UAY
Varnes, N J G4YXX
Varney, D G GM3OAV
Varney, R L G5RV
Varney, T H GW4RLP
Varnham, J A G7NQU
Varns, H C G0BHB
Varrow, R J G7LZC
Varty, A A G6CQC
Vasey, J G0OHW
Vasey, M J G G0HRZ
Vaslet, B M E G6GXZ
Vaslet, C A G7IVM
Vaslet, C R G8OID
Vaslet, M R G0UMH
Vaslet, M R G4JAL
Vaslet, M R G7MHS
Vaslet, S R G8LTD
Vasper, R G3VIY
Vass, N A G4LRK
Vaudrey, S M1BEZ
Vaughan, A H G2FSH
Vaughan, B G0VJB
Vaughan, B A G3MCV
Vaughan, B W G3DQW
Vaughan, B W G3TGO
Vaughan, C D G1BHW
Vaughan, D G G8ZZS
Vaughan, D L H G0JLV
Vaughan, D P E G4FBV
Vaughan, E R G0UCZ
Vaughan, G G0HGZ
Vaughan, J G3DQY
Vaughan, J GU0FZS
Vaughan, J B GW7NGU
Vaughan, L P D G0VBO
Vaughan, M G G8FLX
Vaughan, P B G4HNU
Vaughan, P G G6KPJ
Vaughan, P J G4TCQ
Vaughan, P R G0BEA
Vaughan, R G4VYK
Vaughan, R T G0STH
Vaughan, R T G4WGB
Vaughan, S A G4WXC
Vaughton, G B G0HRH
Vaux, J F A G G3KWH
Veal, R MM0ANT
Veale, D A 2E1EIJ
Veale, D J G0XVL
Veale, D J G7UDD
Veale, F R G4LEA
Veale, P G G3ZXV
Veall, E G6HMS
Veaney, J F G8GHN
Veary, G W G7MWP
Veasey, M N J G4THT
Veitch, A G8FRB
Veitch, C F G4LEV
Veitch, J B G4MRK
Veitch, P P G7VJG
Veitch, P S 2E1CVO
Veitch, S C G7KXY
Vellacott, D J G3CFA
Vellino, G D M 2E1CQT
Venables, C G4NWY
Veness, J A G4PWE
Venison, R S G6HMF
Venn, C J G6NDA
Venn, K G0GFD
Venn, T J G0HSR
Venn, T J G3RPV
Vennard, R K GM0SEI
Vennard, R K GM3USL
Vennard, R R S GM7KSA
Venner, S B G0TAN
Venton, J M G0OKC
Venus, H G G0VKH
Verduyn, J G0BBL
Verity, J H G3ONV
Verity, M GM0RMV
Verma, R 2E1DBT
Vernalls, T P GW6IMS
Vernon, A M G4ZMU
Vernon, A T G4MMY
Vernon, C A G8PEN
Vernon, C F G4HIW
Vernon, C M G0TQJ
Vernon, J A G1WWK
Vernon, M GW6HVA
Vernon, R M0AMC
Vernon-Jones, W I G0VFE
Verrall, M G8TGB
Verrall, R B G0VEB
Vesma, V R G8GYB
Vesper, E J G3WKG
Vest, A G0EBV
Vest Jr, M L M0BCM
Vicarage, R G V G6BHY
Vicary, H M G0RJN
Vick, G K M0AUZ
Vickers, A G8EQB
Vickers, A J G7NNI
Vickers, A R G3HFM
Vickers, B R 2E1FJO
Vickers, B R M1BHE
Vickers, D G4SEQ
Vickers, D G7MOR
Vickers, D M0AKB
Vickers, J R G3ORI
Vickers, K G3RHQ
Vickers, K A G3YKI
Vickers, K M G1NAF
Vickers, M R G6NPL
Vickers, P G6ZWY
Vickers, P J G1EYT
Vickers, P J G4RHJ
Vickers, S J G8MDC
Vickers, S L V G6KML
Vickers, W J GW4PHB
Vickerstaff, J R G4EIG
Vickerstaff, N H G0VYR
Vickerstaff, R M1ADT
Vickery, B J GW4CSY
Vickery, C S G4YCV
Vickery, G T G8YCJ
Vickery, K J G3CLK
Victory, C V G8YOE
Videan, E F G4LWV
Vieira, R B G8AJF
Vigar, R A G6TAV
Viles, K W G2FUB
Villena Bota, J G4VPL
Villiers, T G1URO
Vince, C L G0TLI
Vince, P C G8ZZR
Vincent, A E W G7FUQ
Vincent, A P 2E1GBN
Vincent, B G4FVV
Vincent, B F G3SXV
Vincent, B J G8SJR
Vincent, C J 2E1AEC
Vincent, C J 2E1AZP
Vincent, C W 2M1CHZ
Vincent, D A G G3OKY
Vincent, D L G7TZB
Vincent, F G G8KXJ
Vincent, I D G4MLY
Vincent, J A G1PVZ
Vincent, J M G7NBW
Vincent, J S G8GWJ
Vincent, M G8UKV
Vincent, M I G3UKV
Vincent, M I G3ZME
Vincent, N G3NVO
Vincent, R G4TLV
Vincent, R C G4FRV
Vincent, R R G7GIG
Vincent, S J G4MQX
Vincent-Squibb, G G0BJA
Vincent-Squibb, R K G0BSN
Vine, C M G3XXF
Vine, F A G6DMV
Vine, G G3KLV
Vine, J J G8NWI

Viner, M W G4CJJ
Vines, A M 2E1EVE
Viney, B E GW4KDP
Viney, D J G3SJV
Viney, D J M1BQK
Viney, J R G3ZIC
Viney, M E G0ANN
Viney, M R G0MRV
Vining, A G0TFR
Vining, J A F G7SXX
Vinju, R M0AQN
Vinke, G J M0BFR
Vinnicombe, F J G8KHN
Vinnicombe, S W J G6UXM
Vinson, J G1GCZ
Vinson, J M G4YGP
Vinters, A E G0SQA
Vinters, A E G0WFG
Vipond, P J G1RLT
Virdee, J S G0WEC
Virtue, M L R G1NIT
Vitiello, C E G8EDX
Vivash, D S G6BJY
Vivian, J E G4PBN
Vivian, R M G1NFQ
Vivian, S G7PXV
Vize, W H GW6REQ
Vizor, J A G8CPA
Vizoso, A F G3XVB
Vlismas, T GW0TMV
Vodden, B GW3WBU
Vogan, N G GM4NKF
Voges, R P G7ILX
Voisey, N G0RLK
Voisey, R C G8IDK
Voisey, T G1ZFF
Volante, L J G0MTN
Volante, L J G0WRC
Volante, L J G7WAC
Volck, R M GW3RKV
Volkert, M GM0SZF
Voller, G C G3JUL
Voller, K J G0VZN
Voller, K J GW0CMI
Volpe, A T G0BZB
Volz, H T G8HVO
Von Fircks, N M P G4YFJ
Vosper, F R G3LQU
Voss, J A G0FMG
Voss, M S GW8ERA
Voss, S A G4DYQ
Vousden, J V G8WIR
Vowles, F D G3ZZG
Vowles, H GW3RHC
Vowles, J F G4YTW
Vowles, M V 2E1BBG
Vowles, R W G1WVM
Vukasinovic, P G0UPA
Vukasinovic, P G7SEV
Vye, J M G0ANS

W

Wackett, C G G0GVN
Waddell, A GM4LAO
Waddell, B J GM4XQJ
Waddilove, A L G4KYH
Waddingham, R G0HPM
Waddington, B A G7RBX
Waddington, D M G0IEI
Waddington, E G4GIZ
Waddington, W R G4ICX
Waddle, W H G4ZWU
Waddoups, E E G0TJW
Waddoups, L J G0GNZ
Waddoups, M W G1OBE
Wade, A G G1PPO
Wade, A G G7EVK
Wade, A I H G3NRW
Wade, A J G4AJW
Wade, D G4UGM
Wade, I L G4VKX
Wade, J E G3RIO
Wade, J G G8OHA
Wade, J L G0IRI
Wade, J T G6FLJ
Wade, K R G7GYR
Wade, L G4CAL
Wade, M W M1CBB
Wade, N C G4XOJ
Wade, O W A GW3YVC
Wade, P A G4BQX
Wade, P F G7JVB
Wade, P T G4TCE
Wade, R A G3IRW
Wade, R B G7UXD
Wade, R H G8GSU
Wade, R W G6ZWM
Wade, T P G4IDL
Wade, W G3WAD
Wade, W H G4FCF
Wadhams, P L C G0VRW
Wadley, P M GU4YBW
Wadman, I L G4KDB
Wadman, J M G0VZD
Wadsley, S M G0DVF
Wadsworth, A G4NII
Wadsworth, A C G3NPF
Wadsworth, B G3OLP
Wadsworth, J G7IGK
Wadsworth, M R G3UOF
Wadsworth, T F G4SNQ
Wadwell, G R G4ORU
Waft, G H GD1MFF
Wagenaar, A J GW0ALR
Wager, D J G7VBJ
Wager, E M G3TPI
Wager, J H G4ONO
Wager, M J G7RAZ
Wager, R A J G3VOJ
Wagg, P W G7HXN
Wagg, T D G8VEZ
Waghorn, N E G8GCL
Waghorne, D P G6DPW
Waghorne, K B G4SSP
Wagner, E S G0SNY
Wagner, F T G0IRQ
Wagoner, E A G1SBI
Wagstaff, A G1KBJ
Wagstaff, L G4XSL
Wahlgren, C G P J G0NYX
Waight, R C G0PNN
Wain, D G6PFB
Wain, D A G7SCP
Wain, E S G4NUM
Waine, D D G1THY
Wainman, C G4KBI
Wainman, N C G7RDG
Wainwright, A G0PHB
Wainwright, A S M1AOF
Wainwright, B G0HDP
Wainwright, B G1YWX
Wainwright, C A G7JRJ
Wainwright, C J G4AMN
Wainwright, D R G0HLF
Wainwright, J G1WAB
Wainwright, J G6PBW
Wainwright, P G7LSD
Wainwright, R C G3YMH
Wainwright, R G G3TEG
Wainwright, R J GW0MAP
Wainwright, S J G0FFE
Wainwright, S J G4DFO
Wainwright, T A G0NST
Wainwright, W G4MHO
Waite, B S G4IWW
Waite, F M1CDJ
Waite, I W G1COP
Waite, J P G0SLW
Waite, M G7PQS
Waite, N J G3KOX
Waite, P G0PWZ
Waites, K E G1BCX
Waitt, R J GM6LJE
Wake, A D G3GIB
Wake, J A G6VPV
Wake, R A G7BXS
Wake, V S G8XVI
Wakefield, A B G0JSK
Wakefield, B E G0MVH
Wakefield, B F GW0LKH
Wakefield, H C G6CTP
Wakefield, L H G G7IQH
Wakeford, D G0DUN
Wakeford, M J G0ASB
Wakelam, P M1ARG
Wakeley, R A G3YEP
Wakeling, A J G7MFY
Wakely, A J G4ZMY
Wakely, M C G3NWW
Wakeman, A G3EEZ
Wakeman, A E G0IUD
Wakeman, A R G3UNK
Wakenell, J R G6ZME
Wakenell, J R G8UGL
Wakes, D G1FMJ
Walch, J M G3RVI
Walden, C M G6AHH
Walden, L J M0AFH
Walden, M C G0IJZ
Walden, M G E G4XWV
Walden, P A G7NRT
Walder, P F G3EWY
Walder-Davis, I W G0KCA
Waldie, A J G3NOC
Waldren, D R G7HBC
Waldron, C B G4VFF
Waldron, C M G3ZZU
Waldron, D P G4LUB
Waldron, E J 2W1ERY
Waldron, G A G1MXD
Waldron, H G4MMD
Waldron, J S 2E1ERZ
Waldron, P I G6ZAM
Waldron, W GW0UKT
Waldron, W D G0OAW
Waldron, W J GW0FGO
Waldron, W T G4YGT
Wale, D W C G0EWR
Wale, G W G4BCG
Wales, D C G4TXT
Wales, G L G7GUG
Wales, M G3UXG
Walford, J G GM3POT
Walford, P G G0RFX
Walford, S G0WRF
Walford, T R N G3PCJ
Walker, A G4GNI
Walker, A G4UWS
Walker, A G7RCU
Walker, A G G4DIU
Walker, A G G4ORQ
Walker, A G G6XCZ
Walker, A G G8UKZ
Walker, A H G3OUT
Walker, A M G0FGA
Walker, A R GI4NKE
Walker, A S G3MPW
Walker, A T G1VYS
Walker, A V G4XDR
Walker, A W GM4RNX
Walker, B G0LCU
Walker, B G0OMB
Walker, B G6LQR
Walker, B G6OZR
Walker, B G6WZL
Walker, B C G0HDI
Walker, B J G8XPO
Walker, B M G7KRM
Walker, B M C GM0UZM
Walker, B W G4PCL
Walker, C 2E1FWA
Walker, C G4WHN
Walker, C D G1ETZ
Walker, C G G3VTS
Walker, C H 2E1FMH
Walker, C H G3AZT
Walker, C J G0VUX
Walker, C J G3USO
Walker, C J G7AVT
Walker, C J G7TTK
Walker, D 2E1DHX
Walker, D G3ULL
Walker, D G8UCY
Walker, D A G3BLS
Walker, D C G4DCW
Walker, D E G4DEM
Walker, D E G7DNE
Walker, D G G7IRE
Walker, D J G3OLM
Walker, D J G6ZAF
Walker, D J F G0AEJ
Walker, D R G1PPU
Walker, E G0FNM
Walker, E A G1VYC
Walker, E C GM4BNZ
Walker, E D G0KAQ
Walker, E D G4WKS
Walker, E G G6XCX
Walker, E H 2E1BBV
Walker, E H M1BEJ
Walker, E J G0VZF
Walker, E P GM7RMF
Walker, E S G1IRW
Walker, F D G3JWN
Walker, G G6OSW
Walker, G A G6SKR
Walker, G A M G4DAF
Walker, G D G0PXZ
Walker, G E G0IHB
Walker, G E G0TEN
Walker, G R GW6JDF
Walker, G W J GM8YUM
Walker, H G1NWH
Walker, H G3CBW
Walker, H GW4XWN
Walker, I G3RJF
Walker, I G3VNY
Walker, I G6OXN
Walker, I GM4SZJ
Walker, I J G0KAK
Walker, I R G8ILZ
Walker, J G0WMJ
Walker, J G3GNK
Walker, J GM4FAU
Walker, J A 2E1DMR
Walker, J A G3RDZ
Walker, J A G6NRY
Walker, J A G7DXX
Walker, J A G8IPS
Walker, J C G8KKU
Walker, J D G0FRY
Walker, J D G6FYU
Walker, J G G0LGB
Walker, J G G6VIN
Walker, J G GM0DJG
Walker, J H G4FHF
Walker, J N G8CTO
Walker, J R G4SSW
Walker, J R G8GIN
Walker, K G4AES
Walker, K G6IMJ
Walker, K M1AYZ
Walker, K A GM0TCC
Walker, K B G0SKW
Walker, K J G8DIR
Walker, L G1YGH
Walker, L B G6BGG
Walker, L J G1CBB
Walker, L J G3PUP
Walker, L R G4ULT
Walker, M C G4IJI
Walker, M F G8TRQ
Walker, M G G6POV
Walker, M J G3TLZ
Walker, M J G4OGZ
Walker, M J G6GEA
Walker, M J G6TKC
Walker, M J G6UOX
Walker, M K G6PFE
Walker, M S G1VIP
Walker, N 2E1BYO
Walker, N G1MDL
Walker, N A G8AYC
Walker, P G0CPJ
Walker, P G0MMH
Walker, P B G4HHH
Walker, P B G8NKW
Walker, P D G0KIW
Walker, P H G7KJY
Walker, P J 2E1FLR
Walker, P J G0JYO
Walker, P J G4PLW
Walker, P J G6KUI
Walker, P J G8HMG
Walker, P L G0RSQ
Walker, P M G0RDX
Walker, P M K M0AFR
Walker, P T G4DBY
Walker, R G0OUJ
Walker, R G0RAE
Walker, R G0TAK
Walker, R G4UTA
Walker, R G6QI
Walker, R G7NDT
Walker, R A G8ERN
Walker, R C G0UTP
Walker, R C G4AUH
Walker, R D G7SMZ
Walker, R D M1AIJ
Walker, R J G3YKW
Walker, R J G7SLV
Walker, R L G7VUB
Walker, R M G3XYJ
Walker, R S G3ZJQ
Walker, R T W G4PRI
Walker, R T W G8OLI
Walker, R W G4FNG
Walker, S G1PPQ
Walker, S G8GNJ
Walker, S E G4CPB
Walker, S J G4HTF
Walker, S M G4XMM
Walker, S M O G7HIF
Walker, S N G1BRE
Walker, S R G3IYT
Walker, T D G4ZXC
Walker, T J G0TWE
Walker, W G3BQL
Walker, W G3JWW
Walker, W A GM3LAW
Walker, W D G7KTD
Walker, W I B G3RNX
Walker, W N G3HTJ
Walker-Kier, S G7KYD
Walkley, J C G0BMT
Walkley, V T G3LHH
Walkling, P S G8RNT
Walkup, C R L G0LQZ
Wall, A D 2E1CQO
Wall, B F G1EUA
Wall, C A G6SQT
Wall, D J G6NDG
Wall, E W G0LGK
Wall, F G0CKD
Wall, G A G4XPR
Wall, G B G7PMW
Wall, H J B G4CBT
Wall, H M R 2E1CQA
Wall, I M G1HFI
Wall, J G4IZR
Wall, J A G0CRZ
Wall, K G0LRK
Wall, L A M D 2E1BQE
Wall, N D C G8GCO
Wall, N J G7MQN
Wall, S G7NPJ
Wallace, A H GI6JOP
Wallace, C K G8PTW
Wallace, D G6WZK
Wallace, D A R G3OCP
Wallace, E GM4XLU
Wallace, E R I C GM4TFJ
Wallace, G GM4MSL
Wallace, G W GM0DNG
Wallace, K G3LQW
Wallace, K A G4GZS
Wallace, M GM7LWC
Wallace, M A G8FRL
Wallace, M G G8AXA
Wallace, M L G8RFE
Wallace, N F GM4SYF
Wallace, N T GW0UIP
Wallace, P J G1OAR
Wallace, P W G0WFM
Wallace, R B GM0AOF
Wallace, R G G4MXF
Wallace, R H G1EUB
Wallace, R I G7JJX
Wallace, S G7VWH
Wallace, S A G6HSI
Wallace, U A GM1LTM
Wallbank, A R G4CIZ
Wallbank, R G0XAT
Wallbank, R G7TEJ
Wallbanks, A G7TVH
Wallen, L J G0TFB
Waller, A W H G0KRI
Waller, C 2E1FYB
Waller, D E G3SUL
Waller, D R G0FPN
Waller, E I G0ABQ
Waller, G K G0CKA
Waller, I T G4TQT
Waller, J M0ANH
Waller, J D G3DAV
Waller, K N G7BZB
Waller, M D G0PJO
Waller, P A 2W1DRB
Waller, R A GW0DHA
Waller, R G G0CJZ
Waller, R H G6HGD
Waller, R S G6ICZ
Waller, R W G7IBF
Waller, S T G6ZGD
Waller, T S G3KBI
Waller, W G G8YWA
Wallett, J J G4KNS
Walley, B M G7EHU
Walliker, J W C 2E1DPL
Wallington, P J G7BZI
Wallis, A G4DEO
Wallis, A G4TQS
Wallis, A G G4YMU
Wallis, A P G4LDC
Wallis, A S G0FQA
Wallis, C G3CWV
Wallis, C E W G0CIL
Wallis, D 2E1ENB
Wallis, D M1AXT
Wallis, G G4DQB
Wallis, J G0OGB
Wallis, J G7NIC
Wallis, J H GW4TJQ
Wallis, K R GW1UYW
Wallis, M G G0CRD
Wallis, M W G4EIA
Wallis, R G6OCF
Wallis, R J 2E1AYX
Wallis, S N G3YPK
Wallis, T G G6BWK
Walls, S J G4HNW
Wallwork, A H G4BOB
Wallwork, C J G6AHK
Walmsley, A G0UXZ
Walmsley, A G2HIO
Walmsley, A W G3ADQ
Walmsley, D F J G3HZL
Walmsley, G M G4IXE
Walmsley, J F G7FNM
Walmsley, J L G0WRB
Walmsley, J L G0ZDX
Walmsley, J L G7ZDX
Walmsley, M G0VOF
Walmsley, M G8RIP
Walmsley, P J G0NGK
Walpole, B J G4INF
Walpole, D E G0OOB
Walpole, R R G3VYG
Walsh, A G3YGZ
Walsh, A G7VZS
Walsh, A J 2E0AJH
Walsh, A T G6HSG
Walsh, C M G7IAW
Walsh, D J M1AGJ
Walsh, E G GM4FH
Walsh, G L J G6ORO
Walsh, H M0BEQ
Walsh, I GM4OLH
Walsh, I S G7WKZ
Walsh, I T G6IML
Walsh, J G7NFP
Walsh, K G G0CAE
Walsh, L G4BGL
Walsh, M A G8YAG
Walsh, M D 2E1BSU
Walsh, M D G7ATX
Walsh, M L G7DRL
Walsh, M P G7UXM
Walsh, P J G4MKS
Walsh, P J G7HKN
Walsh, R G 2E0AOO
Walsh, R H GW4UZW
Walsh, R V G4ZNK
Walsh, T G4FMM
Walsh, T D G0IYC
Walsh, T J G8ZQC
Walsh, W J G3HZJ
Walshe, L C G0EVL
Walster, B J GJ3SND
Walsworth, S N G0TAL
Walter, P G G6BRP
Walter, P J GM8SNE
Walter, R I M G1RLF
Walter, T D G4KWL
Walters, A G4RCL
Walters, B B GW3XHD
Walters, B J G4UNJ
Walters, C G A G7PES
Walters, D E GW7SOA
Walters, D J G4DFV
Walters, D J GW0ENU
Walters, D M G4RES
Walters, E S G0MLA
Walters, G N G7ISE
Walters, J A GW6EFK
Walters, J C M G6RKG
Walters, J S G4RWL
Walters, K G0GZO
Walters, K G7NEA
Walters, K E G8FW
Walters, K S G0MLB
Walters, L E G6LQY
Walters, M H G3JVL
Walters, P G1PWH
Walters, P A G0GYU
Walters, P C G8JGF
Walters, P J G3THW
Walters, P J G6YAL
Walters, R M H 2E1EYG
Walters, S C G3IMK
Walters, S I G7VFY
Walters, W B G0MIL
Walton, A F G3XBE
Walton, A R G3ZKQ
Walton, C B G6TNA
Walton, C M G6FXE
Walton, D G0MTO
Walton, D K G6DKW
Walton, D W GD7KHG
Walton, E C M G4FSN
Walton, E C M G6GMR
Walton, F R G4XWM
Walton, I A G6DMN
Walton, J G4KKO
Walton, J A G4LAV
Walton, J E G4FRD
Walton, J H G0ADQ
Walton, J P G1CSA
Walton, J R G3KQN
Walton, K E G7PMO
Walton, K G G7HQY
Walton, M A G0AZM
Walton, P G1IEC
Walton, P G4WAL
Walton, P J M1BNH
Walton, R M 2E1FPQ
Walton, R S G4XUA
Walton, R S G7SSK
Walton, T D G1JKB
Walukiewicz, I G8IUP
Wami, D L G7VYE
Wami, L G7VYD
Wand, E A G4ZOR
Wand, R E G3FNY
Wandless, K G7VKA
Wane, G G G7ODM
Wane, J E G0OFY
Wanford, A M G6LTN
Wankling, C G G6DOF
Wanklyn, D E 2W1AOD
Wann, G B D G0DNI
Wanstall, K G G7LLC
Want, D S G4OIH
Want, J P GD7LWE
Want, R F G4VXP
Waples, M H G0VRF
Waples, M I G6CPX
Wappett, J C G6YAN
Warbrick, R T GM6BHR
Warburton, A G3WPL
Warburton, D C G6LKB
Warburton, D J G7HOC
Warburton, D J GM0LVI
Warburton; D M G4IRM
Warburton, G R GD0LQE
Warburton, J G0KHJ
Warburton, P D G8UGK
Warburton, T G G3ELQ
Ward, A GW0PZU
Ward, A GW4RAF
Ward, A A G3PZX
Ward, A A G3YIR
Ward, A F G3HSP
Ward, A G GI4VQK
Ward, A J G1YIZ
Ward, A L G4FJY
Ward, B G7BQD
Ward, B A G1ZWB
Ward, B C G3SZV
Ward, B C G6BQH
Ward, B E G8BRL
Ward, B L G0TND
Ward, B R G3XKH
Ward, C B G8SMA
Ward, C C G3UHF
Ward, C C G4HON
Ward, C C GM7RWM
Ward, C F J G3TAI
Ward, C J G8EPZ
Ward, C M G6YAF
Ward, C P GI1WFA
Ward, C R G1EUC
Ward, C S 2E1FHQ
Ward, D G0MDO
Ward, D G4PGJ
Ward, D M1BEV
Ward, D A P G4NNX
Ward, D J G1HFH
Ward, D J G3ZLE
Ward, D J G6GMW
Ward, D J G8KBH
Ward, D M G1FVE
Ward, D R G4AOQ
Ward, D T G1MTU
Ward, E G0CMQ
Ward, E B G4XSW
Ward, E H G4HUR
Ward, E T G3JWC
Ward, F C G2CVV
Ward, F J S G0DGG
Ward, G M1BGA
Ward, G A G3MZB
Ward, G A GI3ZCK
Ward, G D G7TCD
Ward, G F G3TUQ
Ward, G M G3BOB
Ward, G W G0HRL
Ward, H E G8GD
Ward, J G1EYX
Ward, J G4HHT
Ward, J A G0WQA
Ward, J A G4JJ
Ward, J A G4ZJA
Ward, J A G8AWY
Ward, J D G8JDW
Ward, J E G4KXK
Ward, J G G7SNW
Ward, J J G3PNP
Ward, J J G6YVN
Ward, J M G6MMT
Ward, J S G8YOK
Ward, J W G0SUL
Ward, J W H 2E0AMH
Ward, K G4TYP
Ward, K T G4RJD
Ward, K T G6SW
Ward, K W G6YSN
Ward, L G4XGC
Ward, L GW1YBF
Ward, L J G0PRI
Ward, L R G4EPL
Ward, M G3KZB
Ward, M G4ZXN
Ward, M A 2E1BCZ
Ward, M A G4ZJP
Ward, M C L G4GHL
Ward, M D G8XEZ
Ward, M E G4MNP
Ward, M H G1JPC
Ward, M M G0KDQ
Ward, M W N G0CCK
Ward, N F GU3OPC
Ward, N G R G4OOQ
Ward, N J 2E1EVI
Ward, N P G7OVM
Ward, N V G0ASX
Ward, O M G8KZB
Ward, P G0RZR
Ward, P G4GYI
Ward, P J G1BZD
Ward, P T S G6BBI
Ward, P W G3XWX
Ward, R G1TYU
Ward, R G1VOY
Ward, R A G8LGA
Ward, R C G7BJD
Ward, R E G0FNO
Ward, R H G7SGK
Ward, R J G2BSW
Ward, R J GW5NF
Ward, R J J G4JQN
Ward, R P G0UGU
Ward, R P G4KZE
Ward, S G0JKE
Ward, S G0RBI
Ward, S C G8TNS
Ward, S D C G7NDN
Ward, S E G4VWA
Ward, S J G6BCM
Ward, S P G4MVL
Ward, S P GW1XVC
Ward, S V G0JIZ
Ward, T P G7WAE
Ward, V G0FEI
Ward, W K GM0KTJ
Wardale, G T M0ANN
Wardale, J G0DFZ
Wardale, L G G0IIJ
Wardale, P N G0EPR
Wardell, A G0DKJ
Wardell, D R G7TOD
Wardell, R G7TOB
Warden, J S GM0JZV
Wardill, C J G7CYF
Wardill, H L G4HFH
Wardle, C F G0PQR
Wardle, C W B G1LKK
Wardle, E A G4XZA
Wardle, G J G1UDB
Wardle, J G7EOX
Wardle, J A G4CVA
Wardle, J G G3MAU
Wardle, J V G7CJW
Wardle, S G4HZO
Wardle, T H G0GMZ
Wardley, S H G6JEF
Wardman, G A GW0PSV
Wardman, P R G0FBV
Wardy, G G4YFO
Ware, C H G4LIY
Ware, H E G0WGD
Ware, M J G4BJT
Ware, R F G7PRD
Ware, R J G8BSW
Wareham, J K G0CXS
Wareham, R M G0GXS
Warehand, H B G6DVO
Wareing, M D G8IUQ
Wareing, P G1LUY
Warham, J G7PZG
Warhurst, C 2E0AOW
Warhurst, T 2E1BGH
Waring, A A G7DFF
Waring, A E G6SRJ
Waring, G E M0AYI
Waring, J G G3NSN
Waring, L C GI3WUO
Waring, M G8YPQ
Waring, R J G8VRE
Waring, W E G3GGS
Waring, W M G0OYL
Wark, D GM3FRU
Warke, H GI6GAQ
Warman, B S G0EFY
Warnaby, B C G1GTP
Warncken, D J GJ4YMX
Warne, A E G3YJX
Warne, A S G4EZO
Warne, A S G8GIZ
Warne, R H G0UHG
Warner, A 2E0AOP
Warner, A GM7LTX
Warner, A M1BZB
Warner, A E G3XUF
Warner, A G G4MMZ
Warner, D C G4OER
Warner, D F GW0TXS
Warner, D J G8POZ
Warner, D W GM7PXL
Warner, E L GW4GSG
Warner, F E G4EWI
Warner, J L G6LTR
Warner, J T G1JFE
Warner, M M1BZQ
Warner, M J GW0KYY
Warner, R A G8LTW
Warner, R G F G0KJF
Warner, R J G1SAA
Warner, R M G3SAR
Warner, S J G8AQP
Warner, S P G7DCJ
Warnes, G G4YAH
Warnes, K M G0OBK
Warnett, D L GW3PDW
Warnock, N D G4JCK
Warnock, W G GI0VGL
Warr, C R G0AWM
Warr, D P G4RQI
Warr, M G0CWF
Warrell, R G G7KUB
Warren, A J G6RFM
Warren, A J G8WQG
Warren, C B 2E1FNY
Warren, D M K G0FKX
Warren, G G4BYS
Warren, H J M G3MSQ
Warren, J L G4LJY
Warren, J V 2E1AHC
Warren, J V M1BTH
Warren, M B G1JWL
Warren, M C G8YOF
Warren, M D G8VZI
Warren, M J 2E1FNO
Warren, M J G7FCE
Warren, M J G8MLD
Warren, M K 2E1FDD
Warren, P C J M0AZI
Warren, R G0FXO
Warren, S J G7LIH
Warren, S R 2E0AGJ
Warren, W S G0PNF
Warrender, R G8ASW
Warrener, P G4HOF
Warrilow, A J G4GXZ
Warrilow, I C G7ILS
Warriner, A D G8UGM
Warriner, K T G8GEA
Warriner, M G0TTG
Warriner, M G8HGI
Warriner, P G0AOP
Warriner, R G4NIQ
Warrington, E M G4EMW
Warrington, E M G4ION
Warrington, J D G1KCU
Warrington, J D G8AKE
Warrington, M F G0HIO
Warwick, A W J G0IZI
Warwick, D G4EEV
Warwick, F GI8MOV
Warwick, J G0CDR
Warwick-Oliver, E K G3YGA
Washby, A G6ZJI
Washby, J M G1KNZ
Washer, J F G3BDU
Washington, A G0PTD
Washington, I C G0VUZ
Washington, J H W G0DUZ
Washington, J W GW0HRG
Washington, J W GW4VUH
Waspe, D L G4HQM
Wassell, I J H G4KDR
Wassell, R G8UQP
Wasteney, R G7ARG
Wastnidge, D G4LYP
Watch, M B G8MRN
Waterall, L T R 2M1ANY
Waterall, S S GM0FRC
Waterall, S S GM0KBU
Waterfall, L G8PGU
Waterfall, M A G8NXD

Waterfield, D A G0WEO
Waterfield, G G0RZA
Waterfield, J A M0BAZ
Waterhouse, D J G4TDB
Waterhouse, I M1APT
Waterhouse, J H G0JHW
Waterloo, B G G6HGX
Waterman, M J G6ADX
Waterman, N M G7RZQ
Waterman, R C G4KRW
Waters, B T G4GIM
Waters, C G0BRH
Waters, C A G0ATN
Waters, C A G3TSS
Waters, D M G8LHS
Waters, D P M0AFX
Waters, D R G4ORT
Waters, F G0UQP
Waters, G GW7LLF
Waters, I M G3KKD
Waters, J G4OBS
Waters, J M1BMQ
Waters, J E G1VVY
Waters, K G G6GVF
Waters, N A 2E1EHY
Waters, P W G0PEP
Waters, P W G3OJV
Waters, R C G0RSW
Waters, R W G1SMI
Waters, S S G0AXD
Waters, T C G0GQJ
Waters, T R GW4IMC
Waters, W G3OYB
Waterson, A G8PLJ
Waterson, K J G1VGI
Waterton, W G7EFV
Waterworth, C G6AHF
Waterworth, D G4HNF
Waterworth, T W 2E1BLO
Wathen-Blower, D G0DWB
Watkin, S R G8HRW
Watkin BA Hnd, A L G7NNA
Watkins, A E GM1BID
Watkins, A J G4ULB
Watkins, A S G3EHW
Watkins, C A G3TTZ
Watkins, D P 2E1EDK
Watkins, G A G6OXH
Watkins, G H GW4YPO
Watkins, J G4VMR
Watkins, J M 2E1FLZ
Watkins, K GW6WSY
Watkins, K F G0ERF
Watkins, K N G3AIK
Watkins, K P GW6ZJO
Watkins, M P 2E1EBZ
Watkins, M R G0NBB
Watkins, P G0PDF
Watkins, R A C G6AHI
Watkins, R B G8CEC
Watkins, R P G0MYE
Watkins, T A G4VSL
Watkins, T H GW0JTJ
Watkins, T J 2E1BSQ
Watkins, W H G6YAC
Watkinson, J G G3EEH
Watkinson, K G GW0ECN
Watkinson, T 2E1FPZ
Watkins-Field, J J 2E1CIT
Watkins-Field, L G4PVV
Watkins-Field, T L G4REZ
Watling, B N G4ZPA
Watling, I R G4NYD
Watling, M G G3PCW
Watling, N P G0UML
Watmough, D B G0LRO
Watmough, J A G7NFK
Watmough, K W G3WXB
Watson, A B T GM7RYR
Watson, A C G7VBV
Watson, A D G4DZS
Watson, A J 2E1EMR
Watson, A J G1FPY
Watson, A J G4GML
Watson, A P G1KAG
Watson, B G0UTM
Watson, B G8UDA
Watson, B A G0RDH
Watson, B A G4YBS
Watson, B J G1HFY
Watson, B J G7NWR
Watson, B M M GM3LLP
Watson, C G0VLW
Watson, C D G7NEC
Watson, C F G7JSS
Watson, C J G4ABF
Watson, C L GW0PCJ
Watson, C M GM1CCI
Watson, C W G0IUL
Watson, C W G7RSW
Watson, D G4YLD
Watson, D A G4NON
Watson, D C G1FEV
Watson, D C G1TBK
Watson, D C G2BSI
Watson, D C G4WBC
Watson, D J G3YXO
Watson, D W G0DEZ
Watson, D W G8HOA
Watson, E G7VNV
Watson, F G3HRE
Watson, F G6PBQ
Watson, F E G4MSQ
Watson, G GW4EVJ
Watson, G B G8UHV
Watson, G E G0FBH
Watson, G E G3XGD
Watson, G F G0RIY
Watson, G N G0HVH
Watson, G R 2E1ELV
Watson, H A G0SHR
Watson, I G G6NRU
Watson, I M GW3OPU
Watson, J GI1RWD
Watson, J E G4JNV
Watson, J G G4ZKA
Watson, J H E GJ3EML
Watson, J M G6BHS
Watson, J W GM1OQT
Watson, J W T M1BTK
Watson, K G4EUZ
Watson, K G4GBF
Watson, K G4MOT
Watson, K G4SQW
Watson, M G1TIL
Watson, M G3JME
Watson, M G7PTC
Watson, M A G8SFF
Watson, M D G3WMQ
Watson, M J G0SAR
Watson, M J G1OOJ
Watson, M J G1SAR
Watson, M J G1WSR
Watson, M J G7IBR
Watson, M J G7MSR
Watson, M J G7RMP
Watson, M J G7RSF
Watson, M J G7RSP
Watson, M J G7RSS
Watson, M J G7RWV
Watson, M J G7SCR
Watson, M J G8CPH
Watson, M J GE7RLF
Watson, M R G6VIQ
Watson, N A G1XBV
Watson, N J G4LCE
Watson, P 2W1DIK
Watson, P G3PEJ
Watson, P B G3GJJ
Watson, P D G4UUV
Watson, P M G8CRM
Watson, R G0FPS
Watson, R G7MWZ
Watson, R GM0NJL
Watson, R A G3AYS
Watson, R A G4CVM
Watson, R A W GI3TZF
Watson, R D G0MKG
Watson, R E G7EHR
Watson, R J G0EVK
Watson, R J G4WQT
Watson, R J M1AQS
Watson, S B G8HCQ
Watson, S K 2E1EJR
Watson, S N 2E0ADR
Watson, S R G1KWF
Watson, T E G0NPP
Watson, T S G0JAA
Watson, V C G7AWG
Watson, W G4EHT
Watson, W A GM4YWV
Watson, W C GM4JNF
Watson, W G G1WGW
Watson, W J GI4NRB
Watt, A D G3NXO
Watt, A D G3WZJ
Watt, A P R G3ZBU
Watt, D GI7GUT
Watt, E GM0IOY
Watt, F L G4LCS
Watt, G G1GID
Watt, G C G4UCR
Watt, G W GM3NTL
Watt, I G GM4ZRR
Watt, J G6ZC
Watt, J GM4SRU
Watt, J G G3PFY
Watt, K G4MSF
Watt, R GM4OMW
Watt, R S G1ELE
Watters, D GM8NVE
Watton, D W J G4EFG
Watts, A C W G3MKB
Watts, A K M0AZD
Watts, A N G4HMH
Watts, A R G4UFK
Watts, A W R G4EYY
Watts, B C G4ZSZ
Watts, B W G0SXN
Watts, C J G7PVL
Watts, C J G8SEK
Watts, C S G0NVX
Watts, C T G4KLB
Watts, D G7BME
Watts, D C G3XWD
Watts, D E A G3XFH
Watts, D G A G6ZZS
Watts, D M G7WKD
Watts, F G G6IAE
Watts, F H G5BM
Watts, G R G0EVW
Watts, G R G3SDS
Watts, G R G8BCH
Watts, H B G0XBL
Watts, H P J G0BBV
Watts, J R G4MSC
Watts, J R G8KXW
Watts, J V G3ZFV
Watts, K J G7JJG
Watts, L D GW3MOP
Watts, M A G7KSS
Watts, M D G1PKU
Watts, M J G0IAK
Watts, M J D G4JZO
Watts, N J G1ZUS
Watts, P J G0FQB
Watts, R G4VIF
Watts, R G GM4ZZW
Watts, R J G6YTB
Watts, R O G3FED
Watts, S A G1GUI
Watts, S A G6YSL
Watts, T G G3PIZ
Watts, T M G4ZZY
Watts, W A G4VIW
Watts-Read, D J W G7VFN
Watts-Read, R J G7VFX
Waud, M A G6IDL
Waugh, H G B GM7PPN
Waugh, I C G0WIZ
Waugh, W GM3MYW
Waugh, W J G5BW
Waugh, W V GI4MKJ
Way, C F G3LWJ
Way, D P G0WTG
Way, D P G7VMC
Waygood, P W G4YXR
Waygood, R G8YFF
Waygood, R E G4OXK
Wayland, A J G1HJW
Waylett, N S A G3YQG
Wayman, J C G4DRS
Wayman, J L G0COZ
Wayman, S E G4JQL
Wayne, R S T B G0NDY
Waywell, C L G3ARZ
Weale, C E B G8DKD
Weale, G G3LEW
Weale, G C G4ACS
Weale, M GW7AGW
Wealleans, J G6ZJW
Wealthy, L A E G0IMM
Wear, D G4DPJ
Wearing, J J A G8BVF
Wearing, R W S G1HPB
Wears, T M G7CTX
Weatherall, P W G3MLO
Weathered, B G4SEO
Weatherer, J M GM4RJF
Weatherhead, H S G7CNP
Weatherill, D G G0LCX
Weatherley, J L G3KQL
Weatherley, M J G1SAK
Weatherley, T F G3WDI
Weatherspoon, W G4NSC
Weaver, A C G4FJQ
Weaver, A C W G7JMW
Weaver, C D G1YGY
Weaver, D J G3IZW
Weaver, D W G6BWP
Weaver, G L J G4MUW
Weaver, G P 2E1EBF
Weaver, J J L G2HNA
Weaver, J L G1RYK
Weaver, J L G3URR
Weaver, K J G1SKI
Weaver, M E G4GMW
Weaver, M R G0MJF
Weaver, P A G7VCZ
Weaver, P J G4JMB
Weaver, P T GW6JDJ
Weaver, R H G GW3KXX
Weaving, P A G0PAW
Weaving, R J G3NBN
Webb, A D G6UXG
Webb, A E G0MSO
Webb, A F G4LYF
Webb, A H G3KCJ
Webb, A J M1ANN
Webb, B 2E1FCJ
Webb, C G4NYJ
Webb, C A G4FWM
Webb, C J G4JFF
Webb, C R G6OXI
Webb, D A G4YHQ
Webb, D A G7JAK
Webb, D G 2E1EOO
Webb, D G G0SSY
Webb, D I G0WVW
Webb, D J G1KTF
Webb, D J G8ZJE
Webb, D W G6JOR
Webb, F D G2HBC
Webb, F J G7OKO
Webb, F S G4ETZ
Webb, F W G0CEK
Webb, F W G3ZKS
Webb, G A G0OVG
Webb, G P G8CKD
Webb, I T G6TNW
Webb, J G8RDP
Webb, J D G6SCM
Webb, J P G0RYO
Webb, J S G3VGI
Webb, M GW0UGQ
Webb, M A G1KVD
Webb, M J GD6ICR
Webb, M J W G3OOQ
Webb, M P G4SHA
Webb, M R G8VZJ
Webb, N D G1AMN
Webb, P 2E1FIW
Webb, P A G1RLI
Webb, P E G0KUE
Webb, P E G0WBZ
Webb, P J G7NBI
Webb, P W G6LMC
Webb, R A G0WEB
Webb, R A G3EKL
Webb, R J G1SIP
Webb, R K G3NDK
Webb, R M G0URR
Webb, R M G1WYA
Webb, R M G8VBA
Webb, S A G4GHO
Webb, S E G1UDN
Webb, S L G0AEN
Webb, S M G1SOT
Webb, S P M1AWB
Webb, S R G3TPW
Webb, S R G7CGS
Webb, T A G6EHB
Webb, T D R G1OEM
Webb, T M G8TVC
Webb, W GM3NGW
Webb, W J G1BWP
Webb, W P GW4ZUA
Webber, B G1ABW
Webber, C C G4SCD
Webber, C L G4NGR
Webber, D G3LHJ
Webber, D G3NJA
Webber, D G8NJA
Webber, D J G0PBS
Webber, D M G3ENX
Webber, H T G4ZYR
Webber, J G6FGV
Webber, J H GW0BKK
Webber, K A T G1WQY
Webber, M A G6VKX
Webber, P C G0HJW
Webber, P G G8KLC
Webber, W G7DRO
Webberley, J N G1CIR
Webberley, J P M G7MNM
Weber, R W G4FTO
Weber, S F F G8ACC
Webley, F V 2E1FOH
Webley, V C G0RKV
Webley, V C G3HIU
Webley, V C G8MKC
Webster, A G0DJQ
Webster, A J G7UHN
Webster, A P G1EWC
Webster, A S G1UAF
Webster, B J G7WJC
Webster, C J G3TBJ
Webster, D G1YQZ
Webster, D M0BDD
Webster, D J G7AYS
Webster, D J G7VSG
Webster, D N J G8TWL
Webster, D V G8MYV
Webster, E G0ERN
Webster, E E G0VDJ
Webster, E H G3HAO
Webster, E T G3JQ
Webster, F G0VBQ
Webster, F A D G6ZJK
Webster, F R A N G3YON
Webster, G G0BYP
Webster, G G7RIY
Webster, G C G3VOT
Webster, G C G6JDH
Webster, G M G6OBJ
Webster, H G3XTF
Webster, I G3YUF
Webster, J G6LJA
Webster, J MM1BRM
Webster, J D G3YOO
Webster, J L G0DRE
Webster, J T G3WDW
Webster, K G3LLE
Webster, K A G7DWV
Webster, K C G6DMM
Webster, K N B G8VMP
Webster, N GM4RXW
Webster, N R GM8AAN
Webster, N W G2DWB
Webster, P G0JCJ
Webster, P A GM7EEY
Webster, P A S G3ZTV
Webster, P K G4EVE
Webster, P M J G3WEG
Webster, P W G8FJA
Webster, R G4EGM
Webster, R GM0DZW
Webster, R GM3MOR
Webster, R GM7OBI
Webster, R H G0HKD
Webster, S A G0NWF
Webster, T I G4ZVA
Webster, T R G0WOF
Webster, W B G0RSV
Webster, W L N G3ZMB
Weddell, A GM1JFF
Wedgbury, C P G4FNQ
Wedgbury, N E G6CUQ
Wedge, J C G7WEV
Wedge, S E G6FRT
Wedgwood, A J G0PTZ
Wedgwood, A J G0TJD
Wedgwood, B J G0SBU
Weeden, A M G4MIE
Weedon, N C GW6TMW
Weedon, P C G8ZKZ
Weedon, R J G0IWW
Weedon, W G7GFY
Weeds, C A G1BZE
Weekes, R J G6FGW
Weeks, D S G1JFN
Weeks, G J G8WAM
Weeks, K G0OBO
Weetman, F G4AIH
Wegg, G W G0LPT
Weglarz, S G0AMT
Weidema, F A M0AJY
Weigh, J G6LXW
Weight, K S G0HLC
Weight, R J G1OTN
Weiner, E M G7GOZ
Weiner, J H G3YIF
Weinstock, J G0CCJ
Weir, C A G4MWQ
Weir, E M G6CEM
Weir, J G6NOL
Weir, K F GM4SEW
Weir, M H G1IRX
Weir, S GM0OPJ
Weir, S GM3SAN
Weir, S T 2E1AAI
Weir, W G4GJX
Weir, W H GI7UDV
Weiss, J G0SEQ
Weiss, L F G6OBO
Weiss, R G G0RKJ
Weiss, S W G6EFE
Welbourn, R G0VJY
Welburn, I A G4EMA
Welburn, W E G7CAS
Welch, D G7WBE
Welch, D E J G0ATD
Welch, F E W G3VQD
Welch, G J GM6WBV
Welch, G P G8EBD
Welch, J G7JJF
Welch, J D G1DXU
Welch, J W G4AQL
Welch, M G G6DIS
Welch, R H G1OGR
Welch, R J G8NDD
Welch, R P G3OFX
Welch, R W G0LQF
Welch, T G0TAX
Welchman, H G6EIY
Welding, L G4OMZ
Weldon, P G4DNI
Welford, I G4RKK
Welford, J G3WOD
Welford, P B G4YKQ
Welford, W V G4RKL
Welger, S A G7MGY
Welland, A L G0OIQ
Wellard, J A G6ZAA
Wellbeloved, L D G1NBU
Wellbeloved, R G3LMH
Weller, A W G6ORJ
Weller, C W G4ONH
Weller, D A GM7RYT
Weller, G C G0LKN
Weller, G F G3DNJ
Weller, I C G1GUB
Weller, J G0GNA
Weller, J E GW0ICU
Weller, K G3WBL
Weller, M R GI4IZF
Weller, P C GM3XOQ
Wellham, R W J G0VLX
Wellings, D C G4KLJ
Wellings, S J G0CXO
Wellington, E N D G6LAX
Wellington, T G G1YWL
Wellington, W J G8JSW
Wellman, A P M1BPS
Wellman, D D G0OBL
Wellman, E J G2HJT
Wellon, S L G6DMG
Wells, A G4ERZ
Wells, A E V G0OBI
Wells, A P G7PPF
Wells, A T G7IGG
Wells, B W G0JEZ
Wells, C G4ZZG
Wells, C G6XOG
Wells, C A G6HER
Wells, C B G3MND
Wells, C R G7NZY
Wells, C R J G4YPW
Wells, C S G8FMD
Wells, D E G6ELE
Wells, D G G0GPE
Wells, D R G7IJC
Wells, E E G3XCE
Wells, E L G7FED
Wells, F W G G3ATJ
Wells, G A G4YGS
Wells, G S G0MRE
Wells, J A G0IWB
Wells, J P GW8KZA
Wells, J R G4BSC
Wells, J W G3IZG
Wells, K M M1ADQ
Wells, M H G1CHQ
Wells, M J G4JES
Wells, M J G7TNZ
Wells, M J G7VJR
Wells, M S G3XBW
Wells, P H 2E1EHV
Wells, P H M1BRW
Wells, P J G4RKN
Wells, P M G4PAI
Wells, P S G0JEW
Wells, P S G4APD
Wells, R A G0RXH
Wells, R C G0ITS
Wells, R F C G7CFF
Wells, R G G3TPK
Wells, R J G8BNR
Wells, S J G7KRB
Wells, S K M1AFV
Wells, S P 2E0ABQ
Wells, T G M1CAA
Wells, T J G1AWD
Wells, W A G4SUZ
Wells, W H G3HVX
Wellsbury, N G8TPM
Wellspring, M J G8AWE
Welsby, J G0AJW
Welsh, D G7HVK
Welsh, D B A G7UAW
Welsh, J G0NVZ
Welsh, J GM8VWC
Welsh, J MM0AYE
Welsh, J A GI4JXM
Welsh, J A M1BBZ
Welsh, K S G0KJA
Welsh, R G4NEQ
Welsh, R C G8MIN
Welsh, R L G6WDZ
Welthy, B A G7BND
Wemyss, D J GM7FYB
Wende, M D G1CDW
Wendels, O C G4TYI
Wendes, R W G7RCC
Wendon, B J G6JMX
Wendon, L W G8AJL
Wenglaryck, J W M0AIL
Wenham, A J M G3ZXA
Wenman, K G1YTO
Wenn, C J G8YAE
Wensley, P GW4UVT
Wensley, S F G4PUB
Wensley, S F G4SVG
Wentworth, A G7PYN
Wentworth, D R J G0NLT
Wentworth, J A G0FED
Wentworth, P M G0DZA
Wentworth, P M GW0WBS
Werba, P R G7FXO
Werner, K P G7RTI
Werner, S M M0ASY
Wernham, R GD7ESR
Wernham, T D G0SBH
Wersby, S J 2E1CAI
Wersby, S J G7TAE
Weseley, A P G3XBQ
Wesil, D H G0RPJ
Wesley, A J G7FMT
Wesley, T G7PLH
Wesson, R G3MXZ
West, A E G7LNB
West, A J G4RQZ
West, A R G4XFW
West, A W G6YBN
West, B H W G4STD
West, B J GW0OSQ
West, C L G8DYA
West, D F G1PUO
West, D J G3XVO
West, D J G4SHQ
West, D J GW3TYI
West, D R G1IKI
West, E E G3KTP
West, E H G0ILE
West, F H G1LVZ
West, G G8POK
West, G E G3HOU
West, G M G3YZK
West, G R M1AUY
West, G S G7HZS
West, G W G0JAJ
West, I G G3SZC
West, J G0MMV
West, J G4AOS
West, J G4LRG
West, J G4OGG
West, J A G0KIM
West, J A G8MIW
West, J A H G6FRL
West, J F G2CMW
West, J J G0DBD
West, J T G1JKX
West, L G4FKC
West, L G4TNX
West, L F G0GNQ
West, L H G6MAO
West, M G7UUC
West, M A G1AAR
West, M F G0UCD
West, M J G1YQP
West, M S G4EJM
West, P J G6CAA
West, P J G6RHL
West, P S G4LLG
West, R G4ZIH
West, R G G3SHX
West, R H G8WAH
West, R J G8DVU
West, R S G3VSQ
West, S L 2E1CWL
West, W J G6TWD
West Knights, L J G8FBK
West of Stow, J GM8CJW
Westaby, G C G0ECS
Westall, F J G4MUU
Westall, M W G0FEF
Westall, S G G6LXU
Westbrook, B A G4WBA
Westbrook, E B G6FGY
Westbrook, H W G4ZBP
Westbrook, T H G7NKJ
Westbury, D R G3OXL
Westbury, P J G0UAP
Westbury, P S G1JAL
Westbury, T H G3TBW
Westby, D P G4UHI
Westcott, E C G4ONC
Westell, S H G3YFG
Westerby, P J G7VFZ
Westerman, A S G8ZCS
Westerman, J G4OOB
Western, R K G0AAA
Western, R K G3SXW
Western, W G G3TDW
Westgarth, R J 2E1FPY
Westgate, D R G6WSN
Westlake, A J G1DFM
Westlake, B M G1ODD
Westlake, D S H G0AIX
Westlake, J T 2E1EGG
Westlake, R W G8MVC
Westlake, S P G8SQZ
Westlake, W H A G8MWW
Westlake, W J G8JQA
Westland, M R GM0UGH
Westley, K G4WEZ
Westley, R H G0NNG
Westmeckett, R G4MRW
Westmoreland, D O GW0YEY
Westmoreland, D O GW8YEY
Westmoreland, L V G3HKQ
Westney, A E G1AEW
Weston, C G GM3VAP
Weston, C S G4WES
Weston, D G0OWD
Weston, D N G1SHJ
Weston, D W G0WWD
Weston, E E G0VPY
Weston, F R G8HVL
Weston, G G6ZGK
Weston, G P G0RSU
Weston, H F P G3AQF
Weston, I K G4UAQ
Weston, J F R G3LYW
Weston, J F R G4LYW
Weston, K M G7IHP
Weston, L H G4CBK
Weston, R A G6RRV
Weston, R B G0LIB
Weston, R C G0RXI
Weston, R G G4XZS
Weston, R I G0MLI
Weston, R J G3RGJ
Weston, S L H G7VSW
Weston, W J GW4VKG
Westripp, P F G0SLD
Westwater, A GM0BUH
Westwater, M C GM4KAT
Westwell, H G3CTQ
Westwell, P A G4HLF
Westwood, C W G3VFD
Westwood, D L G4NPN
Westwood, G P 2E1FDP
Westwood, G W G3VWJ
Westwood, H A R O G4NRF
Westwood, J G1NWM
Westwood, J W G4HCQ
Westwood, N G4NYH
Westwood, P J 2E1DHL
Westwood, P J G7VFD
Westwood, S J M1AMR
Wetherell, K A G6IMN
Wetherill, P G1PAW
Wetton, R N G8PHV
Wevill, K G4UKW
Weymouth, R W G1ELI
Wh1Ttle, D S G6ESK
Whale, A 2I1BJB
Whale, M P G7MNG
Whaling, G G0PPR
Whalley, B A G0TKH
Whalley, B E 2E1AIY
Whalley, D G4EIX
Whalley, G D G0GMM
Whalley, H G2HW
Whalley, M G1CIT
Whalley, P A G3YXN
Whalley, R H G6PEZ
Whalley, R J 2E1EOB
Whalley, S J G4DVN
Whan, D A G4OXU
Wharrie, R J G1XDT
Wharton, E G4NUY
Wharton, G B G0NLO
Wharton, J G7LLY
Wharton, J G G0AZH
Wharton, J V M1BSE
Wharton, R P G0IEN
Wharton, T J G0AZG
Whateley, M J P G1WCB
Whateley, R J S I G1TPN
Whatley, F G G3JOT
Whatley, H E G2BY
Whatley, M G G7IBL
Whatley, M J H G7FZJ
Whatley, P GW1VDT
Whatley, R E G1JXA
Whatling, B J G0BRW
Whatmore, A K G4UVZ
Whatmore, S A 2E1CTC
Whatmore, W D G4BJX
Whatmough, D L G4ORV
Whattam, R C G8ACQ
Whattingham, D G7WDE
Whatton, P E G4DCV
Wheater, A C G0GEA
Wheatley, A R G1RBA
Wheatley, C J G1GUC
Wheatley, G N G4HNJ
Wheatley, J H G0JSC
Wheatley, J H G0JST
Wheatley, J H G7JST
Wheatley, J M G8JUC
Wheatley, K J G3BBR
Wheatley, L J G1XPD
Wheatley, M P G0JJO
Wheatley, P M G0AAT
Wheatley, P P G4MQT
Wheatley, R J G7TZW
Wheaton, M W 2E0ANV
Wheaton, M W G0LKW
Wheaton, M W G4ZPE
Wheaton Smith, S W G C G3ROW
Wheeldon, B S G1CYQ
Wheeldon, J H 2E1AWI
Wheeldon, N L G0IMP
Wheeldon, R A GM7NRV
Wheele, D W E G3AKJ
Wheeler, A J C G4NWS
Wheeler, A J C G4WRC
Wheeler, A R G8NRI
Wheeler, B A J G4KYJ
Wheeler, C G6YAH
Wheeler, C H G0TOX
Wheeler, D W G0GMK
Wheeler, G L G8SGP
Wheeler, G Q G3ZOF
Wheeler, J G7WBL
Wheeler, J F GM0UYZ
Wheeler, J S G0IUE
Wheeler, J T G7CRR
Wheeler, K A M0AXG
Wheeler, K J G0SXS
Wheeler, L R S G0EFR
Wheeler, M G6DOD
Wheeler, M G G4LAL
Wheeler, O T G0NCE
Wheeler, P A C G4PFA
Wheeler, P J G0JNU
Wheeler, P J G8LSC
Wheeler, R G3MGW
Wheeler, R A G6GOW
Wheeler, R B G1LJT
Wheeler, R E G3THY
Wheeler, R L G8HLH
Wheeler, S F G3MHM
Wheeler, T M G7MIM
Wheeler, W E M G0BNW
Wheeler, W H C G3BFC
Wheelhouse, J S G6LNW
Whelan, C G0WFF
Whelan, C GW3NJW
Whelan, D F G1MLJ
Whelan, E A 2E1CQQ
Whelan, J F G6NOI
Whelan, J F G7HHZ
Whelan, K R G3KRW
Whelan, N R G7COC
Whelan, R C G3PJT
Whelan, T G1BIF
Whenham, G A G3TFA
Wherrett, C J G4IIX
Whetstone, D J 2E0ADO
Whetstone, G A G1SXB
Whetstone, G F G4YIK
Whetstone, J F G4OUB
Whetstone, T F G8CIK
Whetton, E T G3DJJ
Whetton, R L G4XKL
Wheway, B A G0RRE
Wheway, J H A G4JJQ

Name

Whibley, A E G0JKP
Whiddington, L E G7KDF
Whiffen, J R G7FQE
Whiffin, I A G0CTQ
Whiffin, K J G4RNJ
While, V R G6IQL
Whiles, G G6LYE
Whiley, S J G6ORM
Whillock, A K G4BWD
Whillock, A K G4ZLX
Whinney, K I M1ALH
Whipp, A G1ZBP
Whish, D 2W1CEE
Whistlecraft, J B G7LXT
Whistlecroft, C J G3BZS
Whiston, G A G8RCL
Whitaker, A J T G3RKL
Whitaker, A J T G3UOS
Whitaker, C H G0FLL
Whitaker, M G G3IGW
Whitaker, M J G4DNG
Whitaker, M J G7RRL
Whitburn, S A G0SWE
Whitbread, D T S G4WIQ
Whitbread, H G1MOS
Whitbread, K V G3XDU
Whitbread, R C G8AYN
Whitbrook, R G G8CBC
Whitburn, P W GW4EAI
Whitby, C J G1DJU
Whitby, E P G6XOD
Whitby, F A G6XOE
Whitby, P E M1ATJ
Whitby, R T G8MEI
Whitby, R W S G8ENB
Whitcher, A E G7JUL
Whitchurch, K A G6NQM
Whitchurch, P A G3SWH
Whitchurch, V C G4HSA
Whitcomb, J M G7DLY
Whitcombe, B G G6AQZ
Whitcombe, W J G6SIQ
White, A G0OHA
White, A G1WRY
White, A G8YUK
White, A GM6JOA
White, A B G6TOT
White, A C G3BGY
White, A C GM3HEN
White, A D G7AUP
White, A E G4YBG
White, A E G7JVG
White, A E J GW6VKY
White, A F GM0KAZ
White, A J G6OLV
White, A J G7REO
White, A L S G1DYB
White, A M G4IPY
White, A R J G7OAS
White, A S G8LOT
White, A T G6YBH
White, A W G4IOQ
White, A W M1BPL
White, C G6AOV
White, C G7NJE
White, C GI7VAF
White, C A G4ROP
White, C A G4TXF
White, C H G4FKE
White, C H G4JBL
White, C R G7VKF
White, D G1MCJ
White, D G1VUY
White, D A W G3URW
White, D B G0OKT
White, D E G0JFT
White, D G O G3KES
White, D J M1AIN
White, D O G3ZPA
White, D P G4OGH
White, D S G3OHL
White, E G GW3LAD
White, E J G0RJH
White, E R J G4JIG
White, F C M G8CYT
White, F S G3JFW
White, F W G4FLW
White, G L G0GLW
White, G L G7WID
White, G L G8EZV
White, G M G8APM
White, G N G8YOJ
White, H A G0FRM
White, H E G0WUH
White, H E G4LFB
White, H R G4UBH
White, I F G3SEK
White, I F G5RP
White, I H W G8TLI
White, J G4BCZ
White, J G4WVS
White, J GM4RDI
White, J A G0RNS
White, J A G7LFQ
White, J C 2E1EGS
White, J C G7BWI
White, J C M1AIS
White, J F A G4MNE
White, J H G4BCY
White, J L G6LJF
White, J R G1BLA
White, J T G4FQS
White, J W G4YWB
White, K G0RSM
White, K A G0RSL
White, K A G4KTU
White, K A G7HQR
White, K D M1CEB
White, K G G7VEM
White, K J G4YXP
White, K J G7TRM
White, L A G0JAO
White, L A G1KRY
White, L E G1YXY
White, L P G0EKC
White, M 2E1FRT
White, M G1MSG
White, M GW0MTI
White, M A G3WOE
White, M C G4YRV
White, M J G1KKG
White, M J G8FXV
White, M J GW8IQC
White, M R G0WRM
White, M R G4HZG
White, M S G1LUM
White, P G1LGQ
White, P G6OZT
White, P G7ULJ
White, P A 2E1FTR
White, P A P G6CJB
White, P G G0BHA
White, P G G0DDA
White, P H G6NCU
White, P J G1SXA
White, P J G8DOF
White, P M G3WJI
White, P P G0WHY
White, P R G4VQF
White, P R G4ZMT
White, P T GW6XJM
White, R G1FCK
White, R G1ITR
White, R G6GRG
White, R A G0DQB
White, R A G6XCY
White, R A G7JET
White, R B G4PGY
White, R G G8UAD
White, R H C G3ISQ
White, R J G1SAJ
White, R J G4ZJK
White, R J G6NFE
White, R L G3RWO
White, R M M1BEI
White, R T G8SPC
White, R W G3UQE
White, R W GI0RYK
White, S G4TSW
White, S A G6TYT
White, S G G4BDB
White, S J G4XXD
White, S J G7DVG
White, S J G7TSW
White, S K G0SZD
White, S P G1MTB
White, S P G3ZVW
White, T G8JHA
White, T GI7THH
White, T B K G4YQS
White, T M G0BXL
White, V R G7BCI
White, W B G8GHK
White, W E G4TFI
White, W J 2E1FYY
Whitecross, J GM0WTP
Whitehall, J G0VAD
Whitehand, J G0NFD
Whitehead, A G4JKW
Whitehead, A GM1PDL
Whitehead, A M G1JTN
Whitehead, B H G8XPQ
Whitehead, C A M1AMW
Whitehead, C E G0PHD
Whitehead, C J G3XAC
Whitehead, D G GW3FDZ
Whitehead, D M G7UYE
Whitehead, D R G4NRH
Whitehead, F J G1YEL
Whitehead, F L G0CFE
Whitehead, F L G4MLL
Whitehead, G G2ACZ
Whitehead, G H G0GHW
Whitehead, I R G4RTQ
Whitehead, J B G0OZA
Whitehead, K J G4LOW
Whitehead, L G6XJC
Whitehead, M T J G0UXI
Whitehead, N D G4HL
Whitehead, P T G4SCE
Whitehead, R G4GWZ
Whitehead, R G G3ZUK
Whitehead, R G G4DUN
Whitehead, S A G6UUR
Whitehead, S A N G8LKA
Whitehead, S J G3RDA
Whitehead, S T G8FGB
Whitehead, T J G6HRX
Whitehead, W E G1JSY
Whitehill, A C GW3IRK
Whitehouse, A F G6HOV
Whitehouse, A G G4BTK
Whitehouse, B M1ABS
Whitehouse, B C G1LWH
Whitehouse, D G0ALA
Whitehouse, D G4OSI
Whitehouse, D N F GW4URY
Whitehouse, G R G7LUF
Whitehouse, J G J G6LJU
Whitehouse, J M G3UEK
Whitehouse, K G G3OHN
Whitehouse, M E 2E1AZI
Whitehouse, P V GW4GWL
Whitehouse, R C G6BCG
Whitehouse, R C G7FYZ
Whitehouse, R T G3YJW
Whitehouse, S G8HIQ
Whitehouse, S J G7RRM
Whitehouse, S J M1AGW
Whitehurst, F A G6VSQ
Whitehurst, J G6CUT
Whitelaw, D GM0ODW
Whitelegg, L W G0CCU
Whitelegg, N G3XQB
Whiteley, C J G3NYS
Whiteley, M GM0SVS
Whiteley, M J G6JTC
Whiteley, P M0AFS
Whiteley, R A G4EUJ
Whiteley, S E G2DAN
Whitelock, P W G0GMY
Whitelock-Wainwright, D S G8YAO
Whiteman, A D G1ADW
Whiteman, P J G8NPZ
Whiteman, S J G3YQR
Whitemore, L J G0GWF
Whitenstall, R L G7AIH
Whiteside, A D M1BKG
Whiteside, L K G7TVT
Whiteside, L T G0MEW
Whiteside, N G4HUN
Whiteside, O J H G7FTS
Whiteside, W R G8MGG
Whiteway, G GW4VWY
Whitfield, A G7BXE
Whitfield, D A M0BBA
Whitfield, D A M1ASC
Whitfield, D B G8VMY
Whitfield, E GW8HYI
Whitfield, G B G3ETQ
Whitfield, G H G4KKL
Whitfield, H G0GBQ
Whitfield, H G6AUC
Whitfield, H G7AWZ
Whitfield, J D G4ZCZ
Whitfield, M R G4MPJ
Whitfield, M S W G0EBZ
Whitfield, P W G0VWE
Whitfield, S J M G3IMW
Whitford, B M G3ZNF
Whitford, P A G3MME
Whitgreave, A G6SKP
Whitham, E GW8ZEI
Whitham, E G G6CPY
Whitham, G G0EPA
Whitham, N D G4SEN
Whithorn, S G G3NAY
Whiting, B E G4DYV
Whiting, C J 2E1FWY
Whiting, G GM7MZZ
Whiting, G A G3MMS
Whiting, G D G6FYT
Whiting, J G4RKG
Whiting, J C G6JUT
Whiting, J C G7GRC
Whiting, J G G4KDQ
Whiting, M J G1XUM
Whiting, N J G4BRK
Whiting, P J G4YQC
Whiting, P J G4ZFR
Whiting, P J G7OPS
Whiting, R C F G8NIU
Whiting, R E G3POF
Whiting, R J G0FGN
Whiting, T E M1ADN
Whitington, J G4KHM
Whitlam, E M G6PBL
Whitley, M P G7ONF
Whitley, S A G7TXR
Whitling, P R G0NEP
Whitlock, A J G0VFM
Whitlock, A J G8ALQ
Whitlock, M D G8EZB
Whitmarsh, C R G0FDZ
Whitmore, I A G4WIA
Whitmore, J GD4IHB
Whitmore, K R G0WKN
Whitnear, S G0BPR
Whitney, J 2E1BPL
Whitney, K G7OOG
Whitney, M D G0VNT
Whitney, R 2E1ARW
Whitney, S 2E1DMA
Whitt, S G8KDL
Whittaker, A 2E1BCX
Whittaker, A 2E1CXF
Whittaker, A D G7THO
Whittaker, A J G8BFM
Whittaker, B C G3LUW
Whittaker, C G0UXF
Whittaker, C G3ZRW
Whittaker, C GJ7SLU
Whittaker, C P 2E1BCW
Whittaker, C V G3UK
Whittaker, D G3XAB
Whittaker, D G4FSJ
Whittaker, D G4ZER
Whittaker, D P 2E1EHM
Whittaker, E G0IZJ
Whittaker, E W D G7KVY
Whittaker, F B G4IAY
Whittaker, F C 2J1ECH
Whittaker, J G6PRA
Whittaker, J M0AAS
Whittaker, J C G7HJP
Whittaker, K B G2FXL
Whittaker, K F G7UHZ
Whittaker, K R G6ICS
Whittaker, P 2E1FLD
Whittaker, P E G1ZDG
Whittaker, R 2E1ESS
Whittaker, R G7DOB
Whittaker, R G8PMQ
Whittaker, R J G6EVV
Whittaker, R M 2J1CVX
Whittaker, S A G1OEH
Whittaker, T G1JKV
Whittaker, T R G3JNM
Whittam, T A G1HKR
Whitten, H S G4LTU
Whittering, R G G3URA
Whitters, D R G6WBR
Whitticombe, A S GW4ODN
Whittingham, A G0CBJ
Whittingham, N G4ISU
Whittingham, P G1XAP
Whittingham, S A G4CLG
Whittington, B F G7PMY
Whittington, J R G3SHZ
Whittington, P G G8WHD
Whittington, R W G3UQD
Whittle, A 2E1DKU
Whittle, B L G3YBU
Whittle, C D G4YXZ
Whittle, D G8SYM
Whittle, G G6GVR
Whittle, K L A G7RFT
Whittle, P G4BBU
Whittle, P L G1ICO
Whittle, S 2E1BDX
Whittles, B G4KUD
Whittlestone, P R M G3OAH
Whittock, B G G7IYA
Whittock, J G W G4VKO
Whitton, D J G4YEB
Whitton, J M 2E0APL
Whitton, K V G1RWT
Whitty, B J G3HWX
Whitty, D K G4FEV
Whitwam, A J G0TLP
Whitwell, R G4EBL
Whitworth, A D G1YWN
Whitworth, E C J G3OTJ
Whitworth, E S G4TUO
Whitworth, I L G3OFB
Whitworth, I R G8JHC
Whitworth, R S G4CTP
Whorton, J E G0OEU
Whorwell, R L G3CTR
Whtye, A J GM7KVB
Whyatt, A B G8NQO
Whyatt, M J GM4SKB
Whyborn, D M G4KIK
Whyborn, N D G4JNX
Whyle, B R G4YIV
Whyman, A E 2E1DET
Whysall, D G6XJD
Whysall, J G6XJE
Whysall, P J G6LYA
Whyte, I G0CPF
Whyte, W GM3OJC
Whytock, J N G7TOZ
Wibberley, K G8PEA
Wiblin, D G6XJI
Wiblin, M W G4WZJ
Wickenden, C G G4ICH
Wickenden, R J G6IQH
Wickens, D H G6WZA
Wickens, F R G3NXN
Wicker, D G0NCC
Wicker, D I G0HAV
Wickers, L F G7GBY
Wickers, P F G7EQX
Wickham, A A G3XHK
Wickham, B F G2DW
Wickham, K A W G3YWB
Wickham, M E G4IGK
Wickham, R J G4JQB
Wickins, B J A G1LIE
Wicks, A J G8BSP
Wicks, G J G0OIR
Wicks, G M G6ORN
Wicks, L J G3ZUD
Wicks, M G L G1IRV
Wicks, R J G1RJW
Wicks, T E G6FIB
Widders, R G3LFD
Widdett, S A G4DFN
Widdop, G G0PTO
Widdows, B K G0MJE
Widdows, M G7BWW
Widdowson, A G G3PET
Widdowson, R M G0CSM
Widdup, B G1BMI
Widdus, D G G7HFH
Widger, P F G0HNW
Wieland, P J G6DOM
Wieloch, D K G7PSW
Wienrich, C F G0REN
Wiewiorka, J GM3YKA
Wiffen, E G G3JTX
Wiffin, M P 2E1DDH
Wigens, D G G8LNS
Wigg, M S G3ZGJ
Wigg, S A G1KQH
Wigg, T C M G3SKF
Wigg, T C M G4IBC
Wigg, W O G5OW
Wiggett, R J G8YPT
Wiggins, H P G2CP
Wiggins, J G8KUZ
Wiggins, J L G0EIP
Wiggins, M J G7BJC
Wiggins, M W G0KBF
Wigginton, C J G6DOI
Wigglesworth, L G3ZMV
Wiggs, D G G4FYM
Wigham, C M G6TYU
Wigham, G R G1IPK
Wightman, F R GW6RQG
Wightman, J E G3DLO
Wightman, K G0KEN
Wigington, D G3ICQ
Wigley, J C G8SRH
Wigley, M G1VPC
Wigley, P J G4RVU
Wigmore, C R G1BZB
Wigmore, R A G4NOU
Wignall, B D G4EAJ
Wignall, H R GM0TFQ
Wignall, K G0PFU
Wilberforce, P W C GM4AXS
Wilbraham, L R G0ILW
Wilbraham, W G G0JGT
Wilby, E G3RZX
Wilby, E A G4EHJ
Wilby, P R G3YRU
Wilcock, P G4VPW
Wilcockson, R J G1VPE
Wilcox, B M G6TRO
Wilcox, C J G4HQC
Wilcox, D A G3JSA
Wilcox, G K G0PHQ
Wilcox, J E G0SIV
Wilcox, J G G3OYF
Wilcox, J H G4RNN
Wilcox, J J G4GBW
Wilcox, T W A GW0KNC
Wild, C J GU6TKE
Wild, G G3RFN
Wild, I D G7SWQ
Wild, J G3GYU
Wild, J H G4LJB
Wild, J P G6IMQ
Wild, L A G3KSW
Wild, L J G4OAN
Wild, P G8BMJ
Wild, P J GU6NCZ
Wild, P R G4ISW
Wild, R W J 2E1FGE
Wilde, B K G3VWH
Wilde, C E J GW8VWD
Wilde, D H E G0FGB
Wilde, D S G6UOO
Wilde, F GW4FKW
Wilde, G T G4JXR
Wilde, J R G0FOI
Wilden, D M GJ0BFF
Wilders, H R G6LTL
Wilders, S G6BCE
Wilders, S J G3ZEO
Wilderspin, C M 2E1DAK
Wilderspin, M T G0RXJ
Wildigg, W D G3JZI
Wilding, C G4SQP
Wilding, J W G4NLW
Wilding, N B GM0CQC
Wildman, D G8HZL
Wildman, L G0NUZ
Wildman, T A G8BFW
Wildsmith, A C G6UQC
Wildsmith, J H G6CUY
Wileman, C H G0SSF
Wileman, M F G4LTF
Wileman, N J G0INV
Wileman, R A G0FPZ
Wiles, C K G0KNW
Wiles, D G1EUD
Wiles, J E G0LDS
Wiles, J E G1LDS
Wiles, J E G4TVA
Wiles, J N R G4WQZ
Wiles, J R G8JEG
Wiles, P G7EYR
Wiles, P S G8NKY
Wiles, S A G4GDC
Wiley, P F G8ZFS
Wilford, J R G4NQY
Wilkerson, G S G8BPN
Wilkes, A G7KJE
Wilkes, A D G4HYW
Wilkes, A R G4NTV
Wilkes, B E G4RWQ
Wilkes, D GM4LIS
Wilkes, D C 2E1FEB
Wilkes, D F G0TUC
Wilkes, G J G4SEL
Wilkes, J A G3FRX
Wilkes, J E 2E1EXK
Wilkes, J L G7PHK
Wilkes, L E G3KJK
Wilkes, P M G7LIN
Wilkes, R G0DUQ
Wilkes, R G4TQR
Wilkes, R W J G0OWU
Wilkes, S G7LEX
Wilkie, A E C G0BXK
Wilkie, A H G0BLS
Wilkie, C R C G0CBM
Wilkie, J GM0OFL
Wilkie, L G4SBI
Wilkie, M D G1UKA
Wilkie, R D B G4JMJ
Wilkie, T J G4VIE
Wilkie, T R 2E1FYZ
Wilkie, W G4SIQ
Wilkin, R G0UKX
Wilkins, A GW8VFF
Wilkins, A A G8KSH
Wilkins, B S G7GUS
Wilkins, D A G7JAV
Wilkins, D C T G6CVP
Wilkins, D J G0MMJ
Wilkins, D M G0TKU
Wilkins, D M G5HY
Wilkins, G J J G3VED
Wilkins, G R G0TSK
Wilkins, H K G3TBF
Wilkins, I K G6OXL
Wilkins, J G0IBK
Wilkins, P G4LRL
Wilkins, R G4UWQ
Wilkins, R G7OXH
Wilkins, R G8NHG
Wilkins, R A 2E1FDA
Wilkins, R F G2ALM
Wilkins, R R T G3XFA
Wilkins, S B GM7RBP
Wilkins, S D W G0NIF
Wilkins, T G1MIN
Wilkins, T A GM6KON
Wilkins, W S GM6FVP
Wilkinson, A G4YVU
Wilkinson, A G4ZHH
Wilkinson, A C G6RIY
Wilkinson, A E G3RHZ
Wilkinson, A G G7VTZ
Wilkinson, A G S G1SHM
Wilkinson, A J G7SKX
Wilkinson, A J GW4PVU
Wilkinson, B G0PXH
Wilkinson, B H G0GBG
Wilkinson, B H G0MJH
Wilkinson, C G0OSA
Wilkinson, C G4UXH
Wilkinson, C B G0NQE
Wilkinson, C B G3FYQ
Wilkinson, C C R GD8GRE
Wilkinson, C T G7LAK
Wilkinson, C T G7OPW
Wilkinson, D G0BWO
Wilkinson, D G4LEH
Wilkinson, D M1BTD
Wilkinson, D A G7MCE
Wilkinson, D B G6JSZ
Wilkinson, D H G0WOJ
Wilkinson, D H G3YFI
Wilkinson, D J 2E1ESW
Wilkinson, D J G4KHF
Wilkinson, D J G4KNV
Wilkinson, E J G4XHT
Wilkinson, F T G6HRZ
Wilkinson, G G3XDP
Wilkinson, G R G4YKO
Wilkinson, I G4RJA
Wilkinson, I G6XDC
Wilkinson, I GW8VUG
Wilkinson, I M 2E1FOJ
Wilkinson, J G4HGT
Wilkinson, J G4TZE
Wilkinson, J G4WIL
Wilkinson, J G6NQL
Wilkinson, J G7OHS
Wilkinson, J A G3TRA
Wilkinson, J A G6PFD
Wilkinson, J C G1UGK
Wilkinson, J D G7TTE
Wilkinson, J G GI1XIB
Wilkinson, J H G0EZO
Wilkinson, J O G8MSY
Wilkinson, K G0VKW
Wilkinson, K A G4YSM
Wilkinson, M G7JPJ
Wilkinson, N G4HVT
Wilkinson, N J G4HCK
Wilkinson, P J G8SAX
Wilkinson, P R G4MWF
Wilkinson, R G3VVT
Wilkinson, R G6LDJ
Wilkinson, R E G3TXA
Wilkinson, R E G4JUH
Wilkinson, R G G4OKY
Wilkinson, R J G0WJR
Wilkinson, R J G3KAC
Wilkinson, R L G0VXG
Wilkinson, R L G7MRQ
Wilkinson, R W G3KWW
Wilkinson, S G0FMY
Wilkinson, S G0LZQ
Wilkinson, S G4IQF
Wilkinson, S A G1TLI
Wilkinson, S P G0HMN
Wilkinson, T G0KNG
Wilkinson, W G3WJH
Wilkinson, W A G3XJI
Wilkinson, W E G4MSK
Wilkinson, W F G0ESA
Wilks, D R G3VCG
Wilks, G D M1ACR
Wilks, G F G8DVJ
Wilks, K G8MVD
Will, S GM4SID
Willams, D E G0SCZ
Willan, W K G7IKM
Willard, M E G7POU
Willats, J A G6YPM
Willcocks, P W G4BWY
Willcocks, P W G8AIE
Willcox, B GM3OHQ
Willerton, V G7HZR
Willett, V G8OJK
Willetts, B T G8DEM
Willetts, G A G0DBJ
Willetts, G H H G0AAM
Willetts, P J G6YAK
Willetts, P R G0KYG
Willetts, S G G1DKF
Willey, D J G1WVO
Willey, G R G3VPO
Willey, L D G4ABW
Willey, T G G2AMM
Willford, L H G0UGZ
Willgoose, A C G8OEN
Willgress, P G8UIQ
Williams, A G0NTJ
Williams, A G0RDS
Williams, A G4PQY
Williams, A GW0GST
Williams, A GW7JLG
Williams, A B G3TAD
Williams, A B G3ZKI
Williams, A B G4JPS
Williams, A B G7TQR
Williams, A B G8XAA
Williams, A C G7KMR
Williams, A D GM7SXI
Williams, A E W G3MHD
Williams, A F G4CHJ
Williams, A G GW0BBC
Williams, A G GW0LAY
Williams, A H GW8CKJ
Williams, A H E G4WWA
Williams, A J GW4OJX
Williams, A K G7MXD
Williams, A L GW3NKZ
Williams, A L GW8FKB
Williams, A L N GW6EUT
Williams, A M G6OLY
Williams, A P GW8LRO
Williams, A R G0CSJ
Williams, A R G3KSU
Williams, A S G0MJV
Williams, A S G1YWI
Williams, A T GW0FYO
Williams, A V G3MCB
Williams, A V G6JEK
Williams, A W MW1BLG
Williams, B G6OSZ
Williams, B GW1VAW
Williams, B D GW0GHF
Williams, B J G4IAW
Williams, B M G1ORS
Williams, B M G7CMN
Williams, B M G7RMZ
Williams, B M H G7FSI
Williams, B O G7DKR
Williams, B R 2E1DFB
Williams, C 2E1DQE
Williams, C GW6DOK
Williams, C A G8HJG
Williams, C D G6AMW
Williams, C D L G4FXQ
Williams, C E GW0ESK
Williams, C F G8HJF
Williams, C J G7NBP
Williams, C P G1FAZ
Williams, C P G8SFD
Williams, C R G4EYT
Williams, C S G4GKY
Williams, D 2W1AZU
Williams, D G0NBN
Williams, D G0PWA
Williams, D G4CVN
Williams, D G4NIR
Williams, D G4UNB
Williams, D G7TXX
Williams, D GW4SAE
Williams, D A G0JMR
Williams, D A G0KUS
Williams, D A G7GQW
Williams, D A G GW3ORL
Williams, D A V G3CCO
Williams, D C 2W1CXJ
Williams, D C G0ODE
Williams, D C GU7FWO
Williams, D C GW7WLX
Williams, D C E G1PZH
Williams, D E G4BII
Williams, D E GW4VES
Williams, D E A GW4WQC
Williams, D E B G0LSQ
Williams, D F G4UUW
Williams, D F S G0ESI
Williams, D H GW4BNJ
Williams, D H GW7PYH
Williams, D H W MW1AMS
Williams, D I G1LTH
Williams, D J G6ONE
Williams, D J G6TOY
Williams, D J G7IRP
Williams, D J GW3XJA
Williams, D L G0LJQ
Williams, D L GW4TUC
Williams, D L GW6RVI
Williams, D M G0UMT
Williams, D M G6KNI
Williams, D R GW4LPA
Williams, D R MW0ATR
Williams, D W 2E1DGW
Williams, D W G7PMI
Williams, E G6ADR
Williams, E A G6VYK
Williams, E D G4LHR
Williams, E E G4JZR
Williams, E G GW4WAN
Williams, E H 2E1AZA
Williams, E J G2AKY
Williams, E J GW4VHS
Williams, E L L G7JDY
Williams, E M GW0AGL
Williams, E P 2E1AFN
Williams, E P G8WTA
Williams, E P M0ATX
Williams, E R G0WMQ
Williams, E W G0ULL
Williams, F G4UJC
Williams, F F G1XWO
Williams, G G1AGM
Williams, G G1TBX
Williams, G G1WER
Williams, G G4FKH
Williams, G G4MTF
Williams, G G4VYB
Williams, G GW4FGL
Williams, G GW4LCF
Williams, G GW4RII
Williams, G GW6VLA
Williams, G A 2E1FPS
Williams, G A GW0HHD
Williams, G A GW1MFY
Williams, G C G1RGT
Williams, G D G8YPV
Williams, G F G3RNR
Williams, G G G1YUD
Williams, G G GW4VMT
Williams, G J G1MCT
Williams, G J G4DXN
Williams, G M 2W1DIG
Williams, G M G4LZQ
Williams, G M G6JUQ
Williams, G N W G1YHJ
Williams, G P G8HCR
Williams, G R G7NLW
Williams, G T GW8RLV
Williams, G V GW2DLK
Williams, H G0JWZ
Williams, H G1JXX
Williams, H G3ROS
Williams, H A G4WNA
Williams, H G G6LJB
Williams, H L G3WZS
Williams, H O G8FHD
Williams, H W G7UDK
Williams, H W M0AFN
Williams, I GW4MEI
Williams, I A M0BCG
Williams, I C GW3GLY
Williams, I M G0PEF
Williams, I R 2E1EWZ
Williams, I W GW4TUD
Williams, J G0MNC
Williams, J G1JPD
Williams, J G6JQH
Williams, J GW4SXB
Williams, J GW4WLT
Williams, J GW6WVD
Williams, J MW1AJI
Williams, J A G6GSV
Williams, J A G8VSF
Williams, J A GW0GXQ
Williams, J A GW6VYE
Williams, J A K G8XJE
Williams, J A R G7JHX
Williams, J C G8LGC
Williams, J D G8VKS
Williams, J D GW4WLZ
Williams, J D GW6ZAN
Williams, J E GW3SSK
Williams, J E GW4XLP
Williams, J F G0KZI

Name

Williams, J F G4DYK
Williams, J F G7TDW
Williams, J F GD6OXG
Williams, J G G0TFL
Williams, J G G6EVI
Williams, J G GW0SXE
Williams, J G M0AMZ
Williams, J H G7GGN
Williams, J H GW1VRR
Williams, J I G0DSK
Williams, J J G0DQM
Williams, J J O G7LSA
Williams, J K G8NTR
Williams, J M G0VMC
Williams, J M G8JQB
Williams, J M GW8RHP
Williams, J P 2E1CXG
Williams, J R GW4TSG
Williams, J T G7VQS
Williams, J T GW4WVB
Williams, J W 2E1ACP
Williams, J W G3TOS
Williams, K 2E1FLB
Williams, K G0KGM
Williams, K GW3TMH
Williams, K GW6BHQ
Williams, K M0BAK
Williams, K C G7NHF
Williams, K D 2E1FKV
Williams, K D G7MIW
Williams, K H G4FCR
Williams, K J G7ILR
Williams, K J G8RDT
Williams, K L GW0EZQ
Williams, K L GW0RNK
Williams, K R G6XCV
Williams, K R GW0KWO
Williams, L 2E1CME
Williams, L G0RPI
Williams, L G7CNQ
Williams, L G7UXY
Williams, L A G0FLW
Williams, L L G8AVX
Williams, L L G8SRG
Williams, L M 2E1GAK
Williams, L M G0ASQ
Williams, L M G0LWZ
Williams, L W G4UUX
Williams, M 2E1FCD
Williams, M G7WEP
Williams, M GW0OUV
Williams, M GW3LCQ
Williams, M D L G0BEU
Williams, M E 2E1FFS
Williams, M F G8RNH
Williams, M G G6BMZ
Williams, M J 2E1FCC
Williams, M J G0ISX
Williams, M J G0JMW
Williams, M J G0TRP
Williams, M J G0WIL
Williams, M J G4PGX
Williams, M J G8SGV
Williams, M J GW3VXC
Williams, M P G4GRS
Williams, M P G4ZMW
Williams, M P G8MIC
Williams, M R W G8ULL
Williams, M S GW1HKY
Williams, M T W G8POL
Williams, M W G6JEL
Williams, N G7MRL
Williams, N M1BEE
Williams, N A F G3GFC
Williams, N D GW0JFQ
Williams, N H G0RPM
Williams, N J M1AKX
Williams, N J H GW1YXR
Williams, N L H G3BYG
Williams, N W GW3UTE
Williams, O C G8ITX
Williams, O E GW0IXM
Williams, O W G0PHY
Williams, P G3UDW
Williams, P GW7MYD
Williams, P M1BPW
Williams, P A G4PUA
Williams, P A M0AOS
Williams, P B W G3XRI
Williams, P D G0OJX
Williams, P E G3YPQ
Williams, P H G4LIQ
Williams, P J 2E1ELE
Williams, P J G3XXE
Williams, P J G4ZKC
Williams, P L GW1CJJ
Williams, P M G4NPU
Williams, P M GW3NUO
Williams, P R G7UYA
Williams, P R GW0IQP
Williams, P R M0BCL
Williams, P R M G6IEI
Williams, P S G4KIL
Williams, P S GM4VXA
Williams, P T G0LJB
Williams, R G0DNX
Williams, R G0RNW
Williams, R G4NLU
Williams, R G4PMM
Williams, R GW4CC
Williams, R GW4HSH
Williams, R GW4SXA
Williams, R A GW7SOK
Williams, R B GW8YPR
Williams, R C S G6OMA
Williams, R D G1ELJ
Williams, R D G4LVQ
Williams, R D G7NGN
Williams, R E 2E1EDE
Williams, R E G4IUX
Williams, R E G4RCT
Williams, R F G3KQW
Williams, R F G8MBU
Williams, R G 2W1BYD
Williams, R G 2W1DHZ
Williams, R G 2W1FQF
Williams, R G GW0PEB
Williams, R H G4VKM
Williams, R H G8JSF
Williams, R H GM7GOD
Williams, R J 2E1EQS
Williams, R J G3TVN
Williams, R J G8CMG
Williams, R J M1BGT
Williams, R J S G0OOF
Williams, R J S M0ACW
Williams, R J W G4NYK
Williams, R K G1PRL
Williams, R L GW7NYR
Williams, R L GW8VVX
Williams, R M G4UOJ
Williams, R M G6CTR
Williams, R S 2W1ELF
Williams, R S G3KXY
Williams, R T GW0CDG
Williams, S GW1FJC
Williams, S A G8SUN
Williams, S C G0VOJ
Williams, S D G0VNI
Williams, S D G4DSX
Williams, S D G4ZDP
Williams, S E G0IEG
Williams, S E G3EKW
Williams, S F G4RIO
Williams, S G G3LQI
Williams, S J 2E1BNU
Williams, S J 2W1AID
Williams, S J GW4OGO
Williams, S J L 2E1DHQ
Williams, S K G4KJX
Williams, S K W G4JLQ
Williams, S L GW1KDE
Williams, S M G0WMS
Williams, S M GW0OFH
Williams, S R G6FOL
Williams, S R GW6OZV
Williams, S T G7OQG
Williams, T G1ITS
Williams, T G1XHO
Williams, T GW1GEX
Williams, T G G4JYN
Williams, T G G4YVY
Williams, T I G6CIZ
Williams, T J G1EWE
Williams, T J G3XLS
Williams, T J P G6KBQ
Williams, T L GW0HXS
Williams, T P G1IFL
Williams, V G6XNU
Williams, W GW0IQZ
Williams, W GW4ZYM
Williams, W A G0LHQ
Williams, W C G7AQD
Williams, W D GW4PEX
Williams, W D O GW4RCM
Williams, W E G0DTF
Williams, W F G3RFJ
Williams, W G GW0JMJ
Williams, W J G8CLS
Williams, W R G4YUJ
Williams, W R G8TGS
Williamson, A 2E1AJU
Williamson, A 2E1AKO
Williamson, A G0WUY
Williamson, A B G6FVM
Williamson, A C 2E1FLC
Williamson, A C 2E1FSK
Williamson, A G GM0RAO
Williamson, A R GI0NWG
Williamson, B C G7UYI
Williamson, C G4FZJ
Williamson, C M G7VDY
Williamson, C S G4CVK
Williamson, C S G4IEB
Williamson, D G1KSS
Williamson, D L G0EGP
Williamson, D M G7VDZ
Williamson, F G4VDZ
Williamson, G GM0BUI
Williamson, I G0EQC
Williamson, I R G7ULZ
Williamson, J G4XJN
Williamson, J G7DZH
Williamson, J A GM0FET
Williamson, J C GM4WWJ
Williamson, J E G7PLM
Williamson, J E GI3JOZ
Williamson, J P A G7URN
Williamson, J R G4NTG
Williamson, J R G6WYL
Williamson, K G8IJI
Williamson, K D G8RSQ
Williamson, M 2E1AJC
Williamson, M G8ATG
Williamson, M A G0EGA
Williamson, M A G0LBL
Williamson, M D G1IJC
Williamson, M R G1OQO
Williamson, M S 2E1ABS
Williamson, N G3URG
Williamson, P J G7IFA
Williamson, P S G4WUU
Williamson, R G0MTT
Williamson, R G4GGI
Williamson, R A G4NPT
Williamson, R L G8WRL
Williamson, S G3WGU
Williamson, S M1AHM
Williamson, S J 2E1ABR
Williamson, T E G8UAP
Williamson, V R G6HSC
Williamson, W G3RUO
Williamson, W D A G0WDW
Williamson, W J GM8MMA
Williamson-Armsby, M A G1KXF
Williams-Berry, T G4SON
Williams-Conley, S P G8SGT
Williams-Davies, M P G6UWW
Willicombe, D R G0DEC
Willies, C J G6DFA
Willies, D F G3HRK
Willies, J M G3DRL
Williiams, J B G3XIE
Willimott, B 2E1CRF
Willis, A 2E1ETJ
Willis, A C G0IOO
Willis, A F G G0ATV
Willis, A K G4XNA
Willis, A M G6PFF
Willis, A M GI0PEZ
Willis, A R G4JSN
Willis, A R GW0VBU
Willis, A S G0VKE
Willis, B L GW4ZUJ
Willis, C G0ADB
Willis, C S G7JIN
Willis, D C GW0JDW
Willis, D R G4NMC
Willis, D S J G7KRO
Willis, D W G6BXV
Willis, F J G7SXK
Willis, F J M0AZY
Willis, G J G8AIQ
Willis, H A G3ZPK
Willis, H J G2CYT
Willis, J G0JOK
Willis, J G4MES
Willis, J C G0AST
Willis, J E G0GGS
Willis, J G G4DOQ
Willis, J L G6TWJ
Willis, K E V G8VR
Willis, K L G3LSW
Willis, L GW4RJW
Willis, M L G4FVI
Willis, P B E G3GLW
Willis, P R G0IYS
Willis, R C G6NPP
Willis, R C G6XCU
Willis, R C G7ERO
Willis, S J G7PDU
Willis, T L G7NCD
Willis, W D G6GJV
Williscroft, R G4GYA
Willisson, S J G8WKI
Willkins, R H G0LPF
Willmer, I P G4TRG
Willmer, S A 2E1CYT
Willmot, N H G3ZHC
Willmott, A J GD8MPF
Willmott, P C L G6KCV
Willmott, R P G6HSD
Willmott, W J G7VQJ
Willoughby, B G4TYS
Willoughby, C G G4GKC
Willoughby, C G G4HLL
Willoughby, G J 2E1DBH
Willoughby, M A G7UQV
Willoughby, W G1HFK
Willox, E M1ASZ
Willox, I A G3YFT
Wills, A G0WFJ
Wills, A G G G3KCN
Wills, A J GM4IZY
Wills, D G3NMX
Wills, D G G3XKX
Wills, D G G5UM
Wills, E L G2AZC
Wills, E V 2E0AAX
Wills, G B G3XZW
Wills, J H G4AXO
Wills, M G3OIL
Wills, M G8RJZ
Wills, M A G0TNF
Wills, N G7PQW
Wills, R G7VFE
Wills, R E G3TPT
Wills, R H G0BDP
Wills, R M G3ZZS
Wills, T D G8PZD
Wills, V W G7WGV
Wills, W J G3JXW
Wills Browne, A J GD0MAN
Wills Browne, A J GD4XWB
Willson, A D G3WNS
Willson, B F G8AKU
Willson, B R G0RNQ
Willson, G D H M0AQO
Willson, M J G8JRF
Willson, P J G6GTC
Willy, J G3CLL
Wilmington, H N GW0BOJ
Wilmot, J W G3EHP
Wilmot, M K G0TAT
Wilmot, R C GW3RRI
Wilmot, R I G0JHM
Wilmot, R J G0JHL
Wilmot, R W G0LHE
Wilmot, R W G4PEY
Wilmott, B N G8TGH
Wilmott, F J G8JFC
Wilmshurst, H J G1BIA
Wilmshurst, M B G1JZG
Wilmshurst, T H G3IBY
Wilsdon, P C G1ZTG
Wilsdon, W 2E1ESZ
Wilshaw, J H F G3MPX
Wilshaw, M G G6UOZ
Wilshaw, S P G7VRR
Wilson, A G1OGE
Wilson, A G3JHS
Wilson, A G4SXC
Wilson, A GM4FSA
Wilson, A C 2E1EYV
Wilson, A C G0IRJ
Wilson, A C G0PVX
Wilson, A C G6NDJ
Wilson, A C G7APY
Wilson, A C GM6UHE
Wilson, A D GM0UGO
Wilson, A E G3MAE
Wilson, A F G0HWP
Wilson, A J 2E0AFZ
Wilson, A J GM8NVG
Wilson, A K G6OZU
Wilson, A K G8TRK
Wilson, A M G7UUT
Wilson, A R G4BOY
Wilson, A R G4CZV
Wilson, A R C G6ZAC
Wilson, A Y GM6KPL
Wilson, B G6HWI
Wilson, B G8PK
Wilson, B GW4KFD
Wilson, B I G0KFQ
Wilson, B K G3ZUV
Wilson, B S G1XCL
Wilson, B V G0JOG
Wilson, B W G4OON
Wilson, C G0VAR
Wilson, C G4AZM
Wilson, C G6RMY
Wilson, C A G7FQB
Wilson, C B G0PHO
Wilson, C B GM8NGG
Wilson, C D G0MGW
Wilson, C D G6AQY
Wilson, C D G7UBW
Wilson, C H G0AFZ
Wilson, C J G4GPU
Wilson, C J G8MEA
Wilson, C N G4VVZ
Wilson, C N G4ZAP
Wilson, C R G0HQN
Wilson, C R GM1ZEL
Wilson, C W G0MPS
Wilson, C W M GM4UZY
Wilson, D G7HOA
Wilson, D G7OBW
Wilson, D G7TUL
Wilson, D G7UXE
Wilson, D G7VEB
Wilson, D G7WGY
Wilson, D M0BFA
Wilson, D M1API
Wilson, D A G0RPW
Wilson, D A G7LBO
Wilson, D A GI3LEG
Wilson, D E J G3OST
Wilson, D H G4OKZ
Wilson, D J G1HKS
Wilson, D J G4OMD
Wilson, D J D G3VCQ
Wilson, D R G3SZA
Wilson, D R G7RKV
Wilson, D S GM0BRJ
Wilson, D V G4AWF
Wilson, D W G7JLX
Wilson, D W G8YPY
Wilson, E C G0ECW
Wilson, E D G3OJQ
Wilson, E R G4GXS
Wilson, E R GM0MFD
Wilson, E S G5CW
Wilson, E S GI3CWY
Wilson, F G3YQA
Wilson, F E G4ISO
Wilson, F H G0OHF
Wilson, F P G7SZL
Wilson, F R G3ETX
Wilson, F R G7LHT
Wilson, G 2E0APG
Wilson, G G7LEV
Wilson, G R G7PQY
Wilson, G R M0BEU
Wilson, G R B G3APV
Wilson, H G3OGW
Wilson, H G3ZIF
Wilson, H G6OBT
Wilson, H GM4SJC
Wilson, H A G4KTG
Wilson, H L G0GDQ
Wilson, H S G4ZYI
Wilson, I G0CLL
Wilson, I G1KVW
Wilson, I GM1XOG
Wilson, I C G3YUZ
Wilson, I C G4ZCD
Wilson, I E G4BGW
Wilson, I H GM0KEL
Wilson, I H GM4UPX
Wilson, I M G6YAI
Wilson, I R G0SPQ
Wilson, I S GM4JPG
Wilson, J 2M1EDT
Wilson, J G0SDL
Wilson, J G1KBA
Wilson, J G3MCF
Wilson, J G6ODO
Wilson, J G6PTF
Wilson, J G8ULH
Wilson, J GD6NDE
Wilson, J GM3KJF
Wilson, J GM7MMI
Wilson, J A 2E1FIK
Wilson, J A G0ISV
Wilson, J A G4PZB
Wilson, J B G3PNH
Wilson, J B G4KOJ
Wilson, J D G3CYU
Wilson, J D G7BHW
Wilson, J D GM4XLZ
Wilson, J E G6BOS
Wilson, J E GI3NEB
Wilson, J F G3PYE
Wilson, J F G3UUT
Wilson, J F G5PI
Wilson, J H M1BWM
Wilson, J K G4SMX
Wilson, J M G0HOG
Wilson, J O GM6WQH
Wilson, J R G6YVS
Wilson, J S G4BCJ
Wilson, K G4SMK
Wilson, K G4WGN
Wilson, K G4WHF
Wilson, K G6RMX
Wilson, K A 2E1CNY
Wilson, K A 2E1FIN
Wilson, K A G3ZYF
Wilson, K D G6LTK
Wilson, K E G0KRE
Wilson, K J G0WHQ
Wilson, K M G4FPO
Wilson, L G4TDC
Wilson, L W GI4SZZ
Wilson, M G0JYG
Wilson, M A G1CSS
Wilson, M A I G4RQL
Wilson, M D G0WNP
Wilson, M D G4GOU
Wilson, M D G4SYE
Wilson, M D G8LVQ
Wilson, M J G7VVZ
Wilson, M N G1AAP
Wilson, M P G6CQB
Wilson, M P D G0TFU
Wilson, M S W G6LJH
Wilson, P 2E1FQQ
Wilson, P G0FVG
Wilson, P G0NGP
Wilson, P G0ULM
Wilson, P G4DUI
Wilson, P G7VIH
Wilson, P A G0TIX
Wilson, P B M1AKF
Wilson, P D G0VCA
Wilson, P D G1XGS
Wilson, P D G6SKN
Wilson, P F G4CPW
Wilson, P G G4ASB
Wilson, P J G1ELK
Wilson, P J G1PQJ
Wilson, P J G3HFX
Wilson, P J G6FNG
Wilson, P J G6MQY
Wilson, P R G8VFE
Wilson, P S G6GTZ
Wilson, P S M1AQN
Wilson, P T G1RCN
Wilson, R G0NJK
Wilson, R G0VHY
Wilson, R G4NCX
Wilson, R G4TYW
Wilson, R G6OBP
Wilson, R GM4BIT
Wilson, R A G4RW
Wilson, R A GI0BRO
Wilson, R A GI0UEG
Wilson, R B GM4FIB
Wilson, R B GM4XBR
Wilson, R C G4AVS
Wilson, R D G0FEK
Wilson, R D G4LRC
Wilson, R E G4ZPR
Wilson, R E GI4SZY
Wilson, R G G1EUF
Wilson, R G G4NZU
Wilson, R L G1BAB
Wilson, R M G0TUB
Wilson, R W G4HIH
Wilson, R W P G3DSV
Wilson, S G1KEH
Wilson, S G3VMW
Wilson, S GD1OOY
Wilson, S A 2E1DMU
Wilson, S H G1UTA
Wilson, S J G4HNO
Wilson, S J G6BOX
Wilson, S J G6HRK
Wilson, T GM4DPC
Wilson, T A G3PGG
Wilson, T A G4TKI
Wilson, T A G6MQZ
Wilson, T D GI4PBS
Wilson, T J GI4ULE
Wilson, T J GI4VJZ
Wilson, T J P GD0EOT
Wilson, T M GI4PBT
Wilson, T M GM0FNE
Wilson, T W G7CZA
Wilson, V G MD0ADD
Wilson, W E F G0NCZ
Wilson, W J GM4YLU
Wilson, W J G G0RET
Wilson-Dutton, L C G0TGV
Wilton, R F K G3ZHO
Wilton, R M G0CIX
Wilton, S N G7HCD
Wilton, V F G0ORV
Wiltshire, A G3SAD
Wiltshire, A D G0OVO
Wiltshire, D G G6ZZR
Wiltshire, L S G0IAY
Wiltshire, N B G6AQW
Wiltshire, R G1UMY
Wiltshire, R G G7SFM
Wiltshire, T R G8AKA
Wimble, W J G4TGK
Wimlett, G G8GLS
Wimpenny, J R C E G0LVH
Winch, B R 2E1EPG
Winchcombe, T GW6ZH
Winchester, A N GM0SSQ
Winchester, G R GM4CUX
Winchester, P R G4KHX
Wincott, G G7MKT
Wincott, R A GW0LNZ
Winder, R G1HFA
Windle, A P W G8VG
Windle, M G0SJP
Windsor, A B G1OQU
Windsor, P A GM4OCA
Windsor, S GM6KMK
Windwick, S GM6VQZ
Winfield, K R G0TZM
Winfield, N J G0UWO
Winfield, S G0SVV
Winfield, S P G0WRT
Winford, S J G8CRX
Wing, G G4AUV
Wingate, I F M1BQP
Wingfield, F G G4KVV
Wingfield, I C GW4IHM
Wingfield, J A G4VLF
Wingrove, A W G7UIQ
Wingrove, M G6VIO
Winiberg, M F G8GFS
Winkler, A J G0FJJ
Winkley, D A G1DYC
Winkup, R I G4ZLT
Winkworth, R G3BGF
Winkworth, R A G8SBA
Winkworth, R F G4AZA
Winlove-Smith, S R G7BPR
Winlow, B M G0WSX
Winnard, K J GW3TKH
Winnard, K J GW4WVO
Winnert, L A G4IRZ
Winnett, P T G4RSU
Winning, C D T G4YWZ
Winning, W M G1NBK
Winsford, P W G4DC
Winship, O G G4UJK
Winship, R E 2E1CCG
Winship, T A G0DSB
Winslow, B D R G3TYG
Winsor, L K G2FS
Winstanley, C M G0KAR
Winstanley, E H G3SUA
Winstanley, J G4OKP
Winstanley, K J G3ZCR
Winston, G G8OPY
Winter, A G0EJB
Winter, D J GW7CSK
Winter, F E G7UJR
Winter, G G G3XCW
Winter, H G G6HRL
Winter, I G GW3KJN
Winter, J W G1AMS
Winter, K C GW8WCA
Winter, M J G3OHP
Winter, S K G0NVJ
Winter, T G4AOK
Winter, W G G4WGW
Winterbottom, A G6ZJV
Winterbourne, J GM1KWG
Winterburn, D G3DQQ
Winterburn, D E M G0KVN
Winterburn, R E G1NSB
Winterburn, W A G4AOO
Winterflood, C G W G4NNN
Winteridge, W G G4SEM
Winters, C R G4MQG
Winters, D G6EZM
Winters, D A G3IPL
Winters, J F G4SGY
Winters, L M G4RZH
Winters, R G3NVK
Winters, R S G4IPL
Winters, S R G4XUQ
Winters, T J G4ZLH
Winters, W H GW0PQI
Winterton, P G1BMW
Winter-Kaines, M G4LAP
Winthorpe, E J G7RIR
Wintle, C D S GJ4GG
Wintle, J G W G0UJW
Winton, A C M GM3MWX
Winton, D G1ICK
Winton, D GM7RXL
Winton, T H G G7RLQ
Winton, V GW4JUN
Winwood, P W G8KIG
Winyard, G G4XWW
Wirthner, M J G4RDH
Wisbey, D H G0RIQ
Wise, C D G1YXT
Wise, M V G0GPX
Wise, R J G1IJY
Wise, S G1FHY
Wiseman, A G7CIQ
Wiseman, A G7STS
Wiseman, B M G6KQZ
Wiseman, C R G1PUV
Wiseman, D S G6UPA
Wiseman, D W G7SUA
Wiseman, F L GW3GRY
Wiseman, G M G7JXR
Wiseman, I G1LSK
Wiseman, J N G8BPQ
Wiseman, L 2E1BHT
Wiseman, R E G3PXV
Wiseman, T E G3DEJ
Wishart, R G7FXU
Wishart, R N G0FDH
Wishart, S GI3PKP
Wisher, A F G4WFS
Wisher, S J G8CYW
Wislocki, T F S G4JRY
Witchard, M J W G8YOC
Withall, P G4ZSW
Witham, D J 2E1BND
Wither, A W G0ADP
Withers, A G1AAG
Withers, D G3AIN
Withers, D E G0HUD
Withers, G J G0VAP
Withers, H M1AMP
Withers, R C S G0KXT
Withers, S F G0MBQ
Withers, T H A G3HGE
Witherspoon, J G8DSM
Withey, D W G8AZV
Withey, J M G8NVF
Withey, M R G0KZD
Withington, T E G8FUY
Withnell, S C H G0AIN
Witley, P C M G0KHF
Witney, R M G4ICP
Witson, C G0KIJ
Witt, J G6CTC
Witter, A G0VAM
Witts, A J G1DIL
Witts, D J G7OOU
Witts, J A G6BBW
Wixon, A J W G4PNM
Wodehouse, P A G4CA
Woffenden, A G3UGB
Wogden, M G4KXQ
Wohlgemuth, J F 2E1DRV
Woiwod, A J G3AWA
Wojtuszek, R T G1INY
Wolfe, B S G3MTR
Wolfe, D J G1EUG
Wolfe, P D G4EGU
Wolfe, R G0WII
Wolfe, R J M0AJF
Wolfenden, E J G7OJT
Wolfenden, J H G4HYH
Wolfenden, J M G4GBM
Wolfson, L G4VUK
Wolk, R C G7SUU
Wollaston, R G4IVB
Wollen, I D G3UZI
Wolpers, R E G3LCB
Wolstenholme, L G0RDF
Womack, P G4VZF
Womack, R G0LBV
Womersley, R V G1JAJ
Wong, T A 2E1CAZ
Wood, A G4EEE
Wood, A G7MEE
Wood, A G7NTI
Wood, A A G4LED
Wood, A C G1FJD
Wood, A C GM3LIW
Wood, A E G6EVX
Wood, A F H G3RDC
Wood, A G R G3VJM
Wood, A J G4RRT
Wood, A L G0ENV
Wood, A P 2W1FXX
Wood, A S G6VIY
Wood, A S M1BNK
Wood, A V G0GHP
Wood, B G0OQQ
Wood, B G6GOV
Wood, B D G1BMJ
Wood, B D G4RFO
Wood, B F H G0WOI
Wood, B J W G0GOQ
Wood, B W G4RDS
Wood, C G1ILJ
Wood, C G3TAW
Wood, C G8JSM
Wood, C F G7PIG
Wood, C H G3PZN
Wood, C J G8PSZ
Wood, C M G1LIQ
Wood, C N G6NOO
Wood, C N E G7FKX
Wood, C P GD6TWF
Wood, C R G8UGN
Wood, C S J GM0EZX
Wood, C W G7TXV
Wood, D G0FDW
Wood, D A G3HKO
Wood, D A G4TIW
Wood, D A G8RUN
Wood, D E G0ABV
Wood, D E GM4JEF
Wood, D G G0VIK
Wood, D G G4CQR
Wood, D H G4ZJL
Wood, D H G7BNK
Wood, D J G0MRX
Wood, D J G0WZA
Wood, D J G4KAE
Wood, D L G1LNA
Wood, D L G4GDT
Wood, D M J G3EAY
Wood, D W G3YXX
Wood, E G2DBW
Wood, E GW4XZP
Wood, E A G4LWW
Wood, E H G8UAS
Wood, E J F G0JLU
Wood, E M G4PKF
Wood, F E G8NSE
Wood, F G G0RME
Wood, F H G1GET
Wood, F J G4IVC
Wood, G G3VIP
Wood, G G6WRC
Wood, G G6YVD
Wood, G G8NRF
Wood, G T F G4XCR
Wood, G W G0LTZ
Wood, H E G7NAL
Wood, H F G6HFD
Wood, H S G8SX
Wood, H T G4MTA
Wood, I P G8MBV
Wood, J G3ERA
Wood, J G3MXR
Wood, J G3VG
Wood, J G8KSW
Wood, J A G0CFZ
Wood, J A G0PSI
Wood, J A G8GGH
Wood, J C G6HRM
Wood, J C G7JHS
Wood, J C G8DPY
Wood, J D G8GHO
Wood, J F G0JFW
Wood, J F G7NNU
Wood, J G G0KUI
Wood, J H G1VDI
Wood, J I G0MPQ
Wood, J J 2E1FRI
Wood, J L G3YQC
Wood, J M G8MTV
Wood, J R G4EAT
Wood, J R G4LZW
Wood, J R G4OFD
Wood, J T G1XPY
Wood, K G4YKV
Wood, K J G3WCS
Wood, K J G4JOA
Wood, K J G7BCS
Wood, K S 2E1EQQ
Wood, L J G0WZN
Wood, L W G4ZSB
Wood, M G4HLZ
Wood, M A G4VIT
Wood, M D G1KVR
Wood, M G S G0ITU
Wood, M H G7HVD

Wood, M J G7HMV
Wood, M J G7KPH
Wood, M R M0AYG
Wood, N G4UFL
Wood, N J GW4OXG
Wood, P G0WHS
Wood, P G4UIW
Wood, P G7PJI
Wood, P D G0FQC
Wood, P D G7UKP
Wood, P E G8WQH
Wood, P G D G8EQA
Wood, P I G8POG
Wood, P R G0HWQ
Wood, R G0BJC
Wood, R G6GHU
Wood, R G8GUA
Wood, R A G7BRR
Wood, R A B G0GSP
Wood, R D G1PXL
Wood, R E G G1PJT
Wood, R F G4UDK
Wood, R F G4XSD
Wood, R F G8MFM
Wood, R H G6RZS
Wood, R H G7UUN
Wood, R J G0AJQ
Wood, R J G1RZZ
Wood, R J G6GVS
Wood, R K G0LZD
Wood, R P GW6ZZP
Wood, R R G4CCU
Wood, R S G0UAW
Wood, R W GW4XRW
Wood, S G1HFE
Wood, S G3EZX
Wood, S G7JPB
Wood, S G8LWQ
Wood, S A G8WHR
Wood, S C G0BJD
Wood, S C G4CWS
Wood, S D G7JNX
Wood, S J G6GVU
Wood, S L 2E1DPH
Wood, S L G8YIN
Wood, S M J GW6LSL
Wood, S R G8ZQG
Wood, T G4MIZ
Wood, T GJ3JGY
Wood, T G G7FWM
Wood, T J G3UNI
Wood, T R G4KFS
Wood, V G3HRF
Wood, V S G1NVE
Wood, W J G4VNX
Wood, W M G7OLU
Wood, Y L 2E1EAK
Woodage, B A T G1OUM
Woodall, D G7RML
Woodall, K A G7MKW
Woodard, J R G8YOG
Woodberry, R E G0IEO
Woodburn, J M1BLB
Woodbury, G E G0WIS
Woodcock, A G1WYK
Woodcock, B M G4CIB
Woodcock, B R G1OOM
Woodcock, G D G4AOE
Woodcock, J G0WBT
Woodcock, L S G4RHK
Woodcock, R D G7EDA
Woodcock, R M G4GZU
Woodcock, T A H GW1GFU
Wooden, E G G3PRQ
Wooden, P J G3OWA
Wooderson, N W G3RZS
Woodfield, B W G3REL
Woodfield, P G3SFU
Woodfine, B G0MVF
Woodford, A E G1OMP
Woodford, D C G8IB
Woodford, G A G0KNM
Woodford, I M G1SHQ
Woodford, M A G0JVG
Woodford, R V G3HJS
Woodford, S M0BBO
Woodford, S M1BGD
Woodford, S R G6BLA
Woodgate, R J G6RJW
Woodhall, D G3ZGZ
Woodhall, F G0MZY
Woodhead, B A G0RGN
Woodhead, P G4UQL
Woodhead, S L G4YAJ
Woodhouse, A G1WFJ
Woodhouse, A G8MXB
Woodhouse, A R G0CMN
Woodhouse, B J G1GXW
Woodhouse, D C G3TWX
Woodhouse, D M G7VLK
Woodhouse, E M G7FBT
Woodhouse, F C GW8WAO
Woodhouse, G A G3DDX
Woodhouse, H G G3MFW
Woodhouse, J G7VLL
Woodhouse, J C G3MLA
Woodhouse, J K G6GPF
Woodhouse, M G G8DEE
Woodhouse, M H G7GQB
Woodhouse, P R G1SXF
Wooding, H W G3LPQ
Wooding, M J G6IQM
Wooding, P R G0VQY
Wooding, T J G0VUN
Woodland, A G4KVP
Woodland, F H G1ZYS
Woodland, J M GW4KHQ
Woodley, B G6PKZ
Woodley, C G3XPU
Woodley, G R G1OWM
Woodley, R A G1HKT
Woodlock, C R G0UDE
Woodman, D G4ULV
Woodman, G P G8XAO
Woodman, R G4DSD
Woodman, R K G4MLS
Woodman, R M G3JYL
Woodnoth, W E G0WOL
Woodnutt, D W G0RAU
Woodnutt, H G G6IYA
Woodnutt, H N G3DEQ
Woodnutt, J R G6XDB
Woodroffe, A C G3OVZ
Woodroffe, M K G0MUY
Woodroffe, R J G2DQX
Woodroffe, S G6FNI
Woodrow, J C G4FZB
Woodruff, E G G4ZWS
Woodruffe, D R G1AGK
Woods, A H G G0PWG
Woods, A J G0FCV
Woods, A J G7SQW
Woods, A M 2E1CDZ
Woods, B K G7MNT
Woods, B M G7FJN
Woods, B W 2E1DPV
Woods, B W G8TQZ
Woods, D G0NIL
Woods, D G7CCH
Woods, D P G3ONI
Woods, D R G1GAP
Woods, D S G3TGC
Woods, F S G1HKM
Woods, G G0AXU
Woods, G G3LPT
Woods, G H G0TJI
Woods, G O GW4JPC
Woods, G P GI7MSY
Woods, J G4RLL
Woods, J J G3TGR
Woods, J J L G7HQP
Woods, J R G1EUH
Woods, J S G1AGT
Woods, J V G0MPW
Woods, J W G3OQC
Woods, P 2E1FVM
Woods, P GM4JLD
Woods, P B G3HLN
Woods, P L G8HHZ
Woods, P M GM0LIR
Woods, R A GM7KIM
Woods, R B G8XAN
Woods, R B M1CDK
Woods, T C E M1BDH
Woods, T G G8STF
Woods, T J G1AGQ
Woods, T M G7IDP
Woods, W A GW0EWY
Woodward, A 2E1FKZ
Woodward, A G1AAL
Woodward, A G1HKQ
Woodward, B D G0KDD
Woodward, E A M1BMU
Woodward, E F G3IPJ
Woodward, E O J G2YW
Woodward, G G6WJR
Woodward, G N G8RLW
Woodward, H C G4JUC
Woodward, I G6OSV
Woodward, I A GW1IAW
Woodward, J D G6KLA
Woodward, J G G7UCG
Woodward, J S P GW6IDK
Woodward, K 2E1FLA
Woodward, K G6AAZ
Woodward, P S G4JWL
Woodward, R D G0MWE
Woodward, R W G4KUV
Woodward, S P 2E1FOE
Woodward, T J W G3XAU
Woodworth, G S GW4ZAG
Woodyatt, A A G G8TOD
Wood-Hill, G G4TWH
Woof, J P G7IUW
Woofenden, A G G2AVF
Wooff, C G3SPJ
Wooff, C J 2E1EQG
Wooffindin, K G4NCB
Wookey, P E G6ESJ
Wooldridge, L G G1HFJ
Wooldridge, M G7NGC
Wooldridge, P A G1HKP
Woolen, D J GW3SRF
Woolfenden, E A G7LDR
Woolfenden, G G0TUW
Woolford, A B G3SNN
Woolgar, D C G4UII
Woolgar, J D GW0FZY
Woolgar, S G4UIH
Woolger, D W 2E1BLS
Woolhouse, P J G7PWV
Woollams, D G4ZSR
Woollams, G J G4YZL
Woollard, A N G6TWA
Woollard, M E G7USX
Woollard, R J G0TUL
Woollen, E F M0AKI
Woollen, W B G4MFP
Woolier, I J P G8NMQ
Wooller, J G3FPB
Wooller, K A G0KFP
Wooller, L S G8GEZ
Wooller, W G G3GYZ
Woollerton, A K G4EYL
Woollett, P G3RQJ
Woolley, C P G6IMM
Woolley, C P G7HLD
Woolley, D J G3ZZF
Woolley, D L G8AMJ
Woolley, E B H GU2FRO
Woolley, E J 2E1CLM
Woolley, E P G8OGE
Woolley, K T G8WEG
Woolley, R C G4HIJ
Woolley, R E G6VGC
Woolley, R G G4OZD
Woolley, R G GW8VHI
Woolley, S R 2E1CKK
Woollin, M G4ADE
Woolliss, J D G4NPS
Woollons, J C G1OSP
Woollven, K R G8CLK
Woolmer, D E G1FFU
Woolmore, G E G7IUL
Woolnough, B E G6LKA
Woolnough, D J G0MIV
Woolnough, J M G7TXN
Woolridge, M G1JFL
Woolridge, R E J G7LNJ
Woolrych, H N G4NWR
Woolrych, H N G4TIX
Woolrych, H N G4WIK
Woolrych, H N G6EVY
Woolrych, T P 2E1EIB
Woolterton, G B G8TAE
Woolterton, M R G4TAD
Woomans, I R B G4NCY
Woosnam, M R 2W1GAZ
Wooster, D G1SHT
Wooster, S G7VSP
Wootten, F W G4XVK
Wootton, A J G0BXD
Wootton, J G0AWP
Wootton, K L G1JKR
Wootton, N R G7LPY
Wootton, P A G0LWP
Wootton, P M 2E1FYI
Wootton, P N J G1EQL
Wootton, W T G0FBQ
Worbey, C R 2E1ELT
Worbey, C R M1BVG
Worden, H H G0KXD
Wordley, C J G4LAE
Wordsworth, D M G0MYP
Wordsworth, R G0GPR
Wordsworth, T W G7OFV
Worgan, H J GW1XJJ
Worker, F H G3APZ
Workman, D J GW0OHJ
Workman, E G3RUD
Workman, H G8HHN
Worland, B D G6WJS
Worledge, P G1AAH
Worley, D G G4EHZ
Worley, H E R G4WZB
Worley, J P G4TYY
Worley, K B G8TEK
Worlledge, P G M1ANO
Wormald, A W G G3GGL
Wormald, C J G0PJW
Wormald, D A C G7BMZ
Wormwell, B J G3WGK
Worner, M J P G1YDK
Wornham, J GD4RVQ
Worrall, A J G3IWA
Worrall, K G3LIT
Worrall, K M1BVI
Worrall, K W G3MCC
Worrall, L H G0TYK
Worrall, M M1AWC
Worrall, S G G7RBT
Worrall, S J G1IJZ
Worrall, T G7VPM
Worrall, T H G0TCJ
Worsdale, I G4RUE
Worsdale, P G0CEG
Worsdale, P G0LEN
Worsell, R E G4CUG
Worsey, T G7TYS
Worsfold, A C G8XCY
Worsfold, C W G0HDH
Worsfold, M G0VPX
Worsfold, M G4PRJ
Worsfold, R H G8NPY
Worsley, A T G6TDL
Worsley, K G1FOW
Worsley, R C G0MYR
Worsley, W M1BAQ
Worsnop, J G0SNV
Worsnop, J G7DFC
Worsnop, J C G4BAO
Worswick, I G0UZO
Worswick, W G0NLP
Worth, R G4ZQF
Wortham, C R G4AGC
Worthing, J G M1BCM
Worthington, I M G8ZSD
Worthington, J GW3COI
Worthington, J B MW0AVQ
Worthington, J C G6FXL
Worthington, K A 2E1BAM
Worthington, S 2E1FVQ
Worthington, T GM4PXG
Worthington, T J G1GUT
Worthington, W D C G3ZBM
Worthy, I G6POJ
Wortley, K G0CVJ
Wortman, A S G4MLD
Worton, R G1XWR
Worton, R S R G8LAX
Worviell, B G1NWT
Worvill, M B G4CRH
Woudstra, M J G8FEJ
Wozencroft, D G6WOZ
Wozniak, J R G0WPL
Wrack, K M1ASA
Wragg, A G4ZNI
Wragg, A D G4UZZ
Wragg, A L G3WEX
Wragg, G G1FHZ
Wragg, K M G0FEZ
Wragg, K M G4WAB
Wragg, M G0HMX
Wragg, P N G8ITU
Wragg, R J G7IAG
Wragg, S G7LPE
Wraige, S G4HFF
Wraight, J G0DHJ
Wraith, I G7GHH
Wratten, D R G6XJB
Wratten, G E G6MAR
Wratten, G R G0III
Wratten, T C GM4CAU
Wratten, W F G6MQR
Wray, H M0AGO
Wray, M G4SOI
Wray, M G7KJO
Wray, M P G6DIR
Wray, T H W G6EUP
Wrede, T M0AZL
Wren, C J G8UCZ
Wren, G H G3XQJ
Wren, J P G3IRA
Wrench, W R H G7AKJ
Wresdell, J F G3XYF
Wresdell, J P 2E0ADL
Wridgway, C N G3GGO
Wright, A C G4OJY
Wright, A C G6AQV
Wright, A C GW3LDH
Wright, A G G0BJN
Wright, A J G0KRU
Wright, A J G0REG
Wright, A J G4EPN
Wright, A J G4RCC
Wright, A O GM0TLX
Wright, A P G0WTF
Wright, A P GW0LIS
Wright, A R GI0VGV
Wright, A W G8KJ
Wright, A W GM0OWM
Wright, A W GM3IBU
Wright, B A G4HJW
Wright, B A G7UWB
Wright, B J G0TVN
Wright, B P G0XAD
Wright, C G0OPE
Wright, C A G7UWC
Wright, C G G4JNM
Wright, C H G4ZFU
Wright, C J G1XRP
Wright, C J G7BPV
Wright, C J G7ODT
Wright, C J GM4HWO
Wright, C J J G0EEZ
Wright, C L G3YFL
Wright, C S G6ZWL
Wright, C S GW7FYG
Wright, C W G3XME
Wright, D G0MTV
Wright, D G3XOU
Wright, D G4ASA
Wright, D G6ORH
Wright, D G8EQD
Wright, D A G4NES
Wright, D A G7LVE
Wright, D B G7TRL
Wright, D B I G8SNM
Wright, D F G8BKG
Wright, D G G0BNU
Wright, D H G0GJA
Wright, D H G4BKE
Wright, D I GW0VML
Wright, D I GW1MVL
Wright, D J G0MXD
Wright, D J M1BZR
Wright, D K G3VBQ
Wright, D L G3HRN
Wright, D T G3WTR
Wright, D W G3UUY
Wright, E G G0SVH
Wright, E J G1VQK
Wright, F E G6FXM
Wright, F E GM4WPU
Wright, F G G6KAA
Wright, G G3VUF
Wright, G G4FUJ
Wright, G G6YBM
Wright, G G6ZAO
Wright, G GW1SXP
Wright, G M1AUZ
Wright, G R A G2YD
Wright, G S G3AER
Wright, G W G8KPG
Wright, H G0LKB
Wright, H M0AUQ
Wright, H A GI6GNA
Wright, H B G4KOV
Wright, H E G7VAR
Wright, H I G3IVA
Wright, H J GW6VDY
Wright, H R C G0EKK
Wright, H V G4TQK
Wright, I G6JRI
Wright, I H G4RRQ
Wright, I T G0PSB
Wright, J 2E1DUQ
Wright, J G0AJO
Wright, J G0OWA
Wright, J G4DMF
Wright, J G6POI
Wright, J G6TRQ
Wright, J G7JIP
Wright, J G7TCV
Wright, J GM3OXU
Wright, J M0BDC
Wright, J A 2E1BEB
Wright, J A GM7CPR
Wright, J D G0ANH
Wright, J D G1LUF
Wright, J I G3SZG
Wright, J M G8TZL
Wright, J N G8NEU
Wright, J S G3VPW
Wright, J S G6KNE
Wright, J W G0KTS
Wright, J W G4ZQT
Wright, K G4EYN
Wright, K GI4KCO
Wright, K E G3WGG
Wright, K H G8DDK
Wright, L G6ZJS
Wright, L B GW0PBJ
Wright, L S G0UWQ
Wright, L W MI1BNZ
Wright, M G6DMJ
Wright, M GW0VSH
Wright, M D G1SXO
Wright, M D G6ELG
Wright, M E G4RLF
Wright, M E G6GYK
Wright, M E J G1EUI
Wright, M J G8ZLU
Wright, M L G4GXN
Wright, M L G6FIQ
Wright, M M H G6ZZN
Wright, M P G0GCI
Wright, M S G0VWH
Wright, M W G8NWU
Wright, N M1AFN
Wright, N K G6JDO
Wright, N R G4IYI
Wright, N T M0BAQ
Wright, P G6EVK
Wright, P D G6NCP
Wright, P F G0WXF
Wright, P J G3JDM
Wright, P J G6IDQ
Wright, P L G1VBQ
Wright, P M G4CGP
Wright, P R G6PBZ
Wright, P S G8IOW
Wright, P T G8GYS
Wright, R G0TYB
Wright, R G1PKQ
Wright, R A G1OIF
Wright, R A G1SWZ
Wright, R D G3WZR
Wright, R F W G0IDC
Wright, R H GW2NF
Wright, R J G3TOY
Wright, R W G0EEN
Wright, S G7MMY
Wright, S A G4LBY
Wright, S C G4GFC
Wright, S C G6OBU
Wright, S J G0PWL
Wright, S J G7DSA
Wright, S P G6YRK
Wright, S P G7ARF
Wright, T B G6NFB
Wright, T D G0DRW
Wright, T E G1LYI
Wright, T H G0LZS
Wright, T J G3HRP
Wright, T K G3KVE
Wright, T R G0UNK
Wright, W E G7EPY
Wright, W H G4BNK
Wright, W P GM3UCH
Wright, W R G0FAH
Wright, W W G0VMO
Wrighton, H T GW7KGD
Wrightson, M W G7OQJ
Wrigley, D G6GXK
Wrigley, D J M1BGR
Wrigley, J GD7DPG
Wrigley, W H A GD7ARS
Wring, D P G4WRQ
Writer, E E G1AAQ
Wroblewski, R K GM8YAQ
Wroe, D P G0MXD
Wroe, J G G4IUJ
Wroe, P D G0KXY
Wroe, R G1WTN
Wroe, R G6TAS
Wszeborowski, J A G0DSI
Wuille, J R G3SZM
Wulwick, M A G4SIN
Wunderlich, W G R G4FXR
Wustrau, M R G0PBZ
Wyard, A R G7FSR
Wyatt, A C G8LSD
Wyatt, C M G8MIT
Wyatt, C W J G6YBO
Wyatt, D E G0VAL
Wyatt, D W G3PNW
Wyatt, E H G3PNX
Wyatt, F C C G4MQU
Wyatt, G J GW8ASA
Wyatt, H F G0UIJ
Wyatt, J M G1ZEI
Wyatt, K G6ICX
Wyatt, M B G1IFY
Wyatt, M F G7STM
Wyatt, M R G0CNA
Wyatt, P A G7WDG
Wyatt, R E G6MQN
Wyatt, R J G3KNJ
Wyatt, S G6AQC
Wyatt, T A G G7KHC
Wybrew, F C GW4TUL
Wybrow, R A G2VJ
Wyche, G D G8UMG
Wye, I D G0OKY
Wyer, W L G4CVO
Wyeth, K R G6CUV
Wyeth, R A M1AEY
Wylam, J G8ROO
Wyles, G G6IAK
Wyles, P W GW4TIZ
Wyles, V K G4YLZ
Wylie, A B GM3MTS
Wylie, C M 2M0AAW
Wylie, E M C GM0KHP
Wylie, G M GM0GMO
Wylie, J G G4LYX
Wylie, P G0GZE
Wylie, R S GM1WIB
Wylie, W K GM3KTD
Wyman, E M G0UDF
Wyman, M L G7JLI
Wyman, R H G7JLH
Wynes, G C G3TLV
Wynford-Thomas, D GW3YQM
Wynn, B W G8TB
Wynn, M G7GXM
Wynn, R A G4BNB
Wynne-Jones, T A G6ZFV
Wynters, D A J G6KCJ
Wyse, A M H G3IWE
Wysocki, N G6CPO
Wysome, R A G8BWB
Wyspianski, A T G1AWF

Y

Yale, J G3ZTY
Yallop, A H G3SVQ
Yallop, B F G4KPJ
Yallop, E G0IUG
Yallop, M E G4YNT
Yam, J G6REV
Yard, E G3KRG
Yardley, J H G G3HFZ
Yardley, P M G0INS
Yarker, A B G3TAY
Yarker, J O G3GJY
Yarnall, G G4OMJ
Yarnall, J K G1JLQ
Yarnall, J K M1AUN
Yarnold, G J G4CNG
Yarrow, D A G6TDX
Yarrow, N GM4PJR
Yarrow, T A 2E1DPA
Yates, A G3WEC
Yates, A C G6LUF
Yates, A C G8RAO
Yates, A R G6CPS
Yates, B G4TVN
Yates, B G7UOS
Yates, B A G8AMY
Yates, B C G0DKZ
Yates, D A G0PBE
Yates, D C G3PGQ
Yates, G G0CVA
Yates, G W G7BLM
Yates, J G3TDC
Yates, J G8UGH
Yates, J A G0NNF
Yates, J A G8YGS
Yates, J A GM4AGX
Yates, J C G3MNJ
Yates, J M G1SQA
Yates, J R G1UZD
Yates, J R G4AKP
Yates, K G3XGW
Yates, K L G1HGA
Yates, N A G0TDL
Yates, P G0NPY
Yates, P M G7BZD
Yates, R G8ACR
Yates, S G7ETK
Yates, S G7VKW
Yates, S C G7ENM
Yates, S J G7JXN
Yates, S M 2E1ACC
Yates, S S GW0HNS
Yates, T B G3RWE
Yates, T O G6LUE
Yates, V G M G7HBW
Yates, W G6ADY
Yates Jones, H M G7RGI
Yaxley, P G L G4YLW
Yaxley, R A G3YHO
Yea, P D G0WKU
Yeaman, D J G4ASY
Yeandel, J D G4OOL
Yearl, H G0WKI
Yearsley, P H G1UTM
Yeates, A J G0KOS
Yeates, D J G4FND
Yeates, K C G0TTW
Yeates, P F G3KLQ
Yeates, P J G1XXE
Yeates, P J G8ODM
Yeatman, D J G1AGW
Yeatman, P N G0PSF
Yeats, J R G7STX
Yeend, J J G3CGD
Yeldham, H J G6XOU
Yeldham, L C G0KUV
Yendell, S H M1AWX
Yeo, D G0IAE
Yeo, F G G6XCR
Yeo, I N G0PCQ
Yeo, K A G7UCR
Yeo, W H J G2CVY
Yeoell, D G3MLH
Yeoman, D G4SQA
Yeoman, R W GM3ZGH
Yeomans, A G0SOE
Yeomans, M J G4YTO
Yeomans, S A G7NDL
Yeomans, T C G4FOD
Yeomanson, S W G3UNF
Yetton, T A G1SOB
Yiacoumis, A M G3YAP
Yilmaz, A H G3PRK
Yim, S S G0CBR
Yirrell, M J G6RXD
Yohn, S T G7OEW
York, A D G4EQP
York, C G3XDI
York, D J M0ABU
York, E G G8HOR
York, G S G8MXD
York, J A G3KJY
York, M E GW0NKG
York, M J G1BKI
York, P S G4UAX
York, R G1USV
York, R K E G6ZGH
Yorke, A G7KUM
Yorke, D A G4JLG
Yorke, E G1AGV
Yorke, M G4ASW
Yorke, T R G1JWY
Yorke, T R G1WRN
York-Jones, P G8CYU
Youard, R G A G8UDB
Youd, N M G1AAD
Youde, J E G0GUF
Youle, J 2E1ERP
Young, A GM4LLY
Young, A GM7LUP
Young, A F G3YBP
Young, A G GM0LZC
Young, A J 2E1FQW
Young, A J G0EJC
Young, A J G1AJY
Young, A J G4AWD
Young, A N G4BVG
Young, A P M1APY
Young, A W G G6NFC
Young, C F C G8KHH
Young, C F H G0CCC
Young, C F H G4CCC
Young, C M GM0WFI
Young, D GM0DYD
Young, D M0AOA
Young, D E G4ZHN
Young, D H G8TVW
Young, D I G1NSD
Young, D J G8ZQJ
Young, D M G1IDZ
Young, D M G4RQU
Young, E G0SQE
Young, E GM0TXG
Young, E A 2E1FRY
Young, E C G0NRG
Young, E C G4MZX
Young, E H P G3ATK
Young, E J G3RKI
Young, F G7NBV
Young, G F G4BLU
Young, G I G0UCY
Young, H G3YHY
Young, H C G3HIA
Young, H F G4JTO
Young, H S G3AEU
Young, H V G3LCI
Young, I R P G7III
Young, J G0WQR
Young, J G4KZD
Young, J G6MMS
Young, J A 2E0AQB
Young, J A 2E1EPY
Young, J A GM4DQD
Young, J B G1ICP
Young, J C G4PPZ
Young, J E G0BIV
Young, J G GI0CNI
Young, J J G1XGL
Young, J R G3KLP
Young, J R F G4JVY
Young, J S G3UIK
Young, J V G1NIV
Young, J W GM4RCN
Young, K G1WZR
Young, K D G1XAW
Young, K D G6UXF
Young, K J G3ZCG
Young, K J J G3HUO
Young, L V G7APU
Young, M D G7ILN
Young, M G G4KPL
Young, N S G6OTB
Young, P G7SQC
Young, P A G0HWC
Young, P A G1MRX
Young, P J G6AHE
Young, P M GM0GBH
Young, R G1OIZ
Young, R G4GWC
Young, R G4XYD
Young, R A G6FVL
Young, R C G7RNQ
Young, R C GM0GRW
Young, R E G6JOL
Young, R F G4NQS
Young, R G GM0TUS
Young, R J GI7PBQ
Young, R J GM0NEG
Young, R J D 2E1DDW
Young, R J F G6CIT
Young, R L GM7ITG
Young, R N G8SGW
Young, R V G8XFK
Young, S G4EOV
Young, S G7LFV
Young, T A G4INT
Young, T E G1FXM
Young, W G4XQR
Young, W G4YGO
Young, W H G4KUU
Young, W J G4SJT
Young, W L G4DTL
Young, W T H G3NZR
Young, Y J 2E1FQV
Younge, E J G3IVH
Younger, A W G0RMN
Younger, K A J GM3OIB
Youngman, D G3ULY
Youngman-Smith, N S G1ZRR
Youngs, D C G3JIE
Youngs, S J G1KYV
Youster, B D G8PRP
Yoxall, G W G7DBC
Yu, J G3ZQT
Yu, J G4LWT
Yue, N Y L G6RHO
Yuen, F K Y G7WDA
Yuill, S J G7UHL
Yukawa, T G0WWM
Yule, G H M G3IED
Yunnie, G G G7WIL
Yunnie, G G M0AYK

Z

Zacharov, B G0KGD
Zaim, T Y G0WIV
Zainal-Abidin, I M1BKP
Zak, K W G0OQI
Zalicks, L R G4YOT
Zammit, C C G8DUV
Zammit, M N L G3WXD
Zara, P R J G1UFT
Zarattini, M G4HJL
Zarucki, B G7OLZ
Zeal, C G4BGM
Zeal, R W S GW7PRK
Zealand, K B G7MOJ
Zerafa, A C G8CKK
Zervas, J K G0OOT
Zielinski, R G0ELG
Ziemacki, A G1GBR
Zilberberg, L J G4ZLB
Zimmermann, P A G0UPS
Zollman, P M G4DSE
Zrobok, P G0UZP
Zulawski, D C G0WOH
Zuppone, M S M1BOI

UK Local Authority Index

These pages show amateurs whose details have been released by SSL, listed by local authority district and then by postcode.

The format has been changed from that used in previous years due to the changes in local government introduced in 1996 and 1997. (Further changes are planned to come into effect in 1998.)

The essence of these changes are the introduction of unitary authorities which combine the roles of former district and county councils. In Scotland and Wales, all local authorities are now unitary and this has brought back some old familiar names such as Clackmananshire and Flintshire, but some new ones still remain eg Highland and Powys.

Unfortunately the position in England is much more confused. Relatively few district authorities have converted to unitary status and those that have have/are being phased in during 1996, 7 and 8. This means that a "county" map of England now shows not only the traditional counties such as Bedfordshire and County Durham, but also the unitary authorities of Luton and Darlington (previously part of these counties). In addition, some counties defined in 1974 have been abolished eg Cleveland and Avon and replaced entirely by unitary authorities. Sadly at least one unitary authority straddles two former county boundaries (eg the single unitary authority of Brighton & Hove covers parts of both East and West Sussex counties) and those formed entirely from old counties have not been allocated any popular title to indicate the collective area eg there is no single name to cover Bath & North East Somerset, Bristol, North Somerset and South Gloucestershire.

To make life more interesting, county councils for metropolitan counties were abolished some years ago and this fact is now being reflected in the administrative maps of the UK. These now show the metropolitan districts rather than counties, all the separate London boroughs, all the unitary authorities in England, Scotland and Wales, and the remaining English shire counties. If the *Yearbook* were to list amateurs by county, it too would have to adopt this cumbersome (and confusing) break down.

Off the mainland, other arrangements for local government apply. Northern Ireland has a local government structure based on 26 separate authorities rather than the more familiar '6' counties. The Channel Islands traditionally divide into Jersey and Guernsey & its Dependencies. The Isle of Man is yet another distinct local government.

A different system has had to be adopted for this edition – that of listing by district authority. As many of these names are unfamiliar to non-residents, a summary of old county to district/unitary authority is given below. This list is not to be used for authoritative purposes and is provided simply to aid the reader in locating an amateur callsign. There are simply too many anomalies and boundary changes for it to be otherwise. Post towns have been dropped as they fragmented the list given the small area covered by districts. Instead a seperate index is included to aid the reader.

It is recognised that this method of sorting is less than ideal such that a simple list by postcode and post town may be more suitable in future. Comments would be appreciated by the Yearbook Editor at RSGB HQ.

Avon
Bath and North East Somerset*
Bristol*
North Somerset*
South Gloucestershire*

Bedfordshire
Bedford
Luton*
Mid Bedfordshire
South Bedfordshire

Berkshire
Bracknell Forest
Newbury
Reading
Slough
Windsor and Maidenhead
Wokingham

Borders Region
Scottish Borders*

Buckinghamsire
Aylesbury Vale
Chiltern
Milton Keynes*
South Bucks
Wycombe

Cambridgeshire
Cambridge
East Cambridgeshire
Fenland
Huntingdon
Peterborough
South Cambridgeshire

Central Region
Clackmannanshire*
Falkirk*
Stirling*

Channel Islands
Guernsey & Dep
Jersey

Cheshire
Chester
Congleton
Crewe and Nantwich
Ellesmere Port and Neston
Halton
Macclesfield
Vale Royal
Warrington

Cleveland
Hartlepool
Middlesborough
Redcar and Cleveland
Stockton-on-Tees

Clwyd
Denbighshire*
Flintshire*
Wrexham*

Co Durham
Chester-le-Street
Darlington*
Derwentside
Durham
Easington
Sedgefield
Teesdale
Wear Valley

Cornwall
Caradon
Carrick
Isles of Scilly
Kerrier
North Cornwall
Penwith
Restormel

Cumbria
Allerdale
Barrow-in-Furness
Carlisle
Copeland
Eden
South Lakeland

Derbyshire
Amber Valley
Bolsover
Chesterfield
Derby*
Derbyshire Dales
Erewash
High Peak
North East Derbyshire
South Derbyshire

Devon
East Devon
Exeter
Mid Devon
North Devon
Plymouth
South Hams
Teignbridge
Torbay
Torridge
West Devon

Dorset
Bournemouth*
Christchurch
East Dorset
North Dorset
Poole*
Purbeck
West Dorset
Weymouth and Portland

Dumfries & Galloway Region
Dumfries & Galloway*

Dyfed
Carmarthenshire*
Ceredigion*
Pembrokeshire*

East Sussex
Brighton and Hove* *(part)*
Eastbourne
Hastings
Lewes
Rother
Wealden

Essex
Basildon
Braintree
Brentwood
Castle Point
Chelmsford
Colchester
Epping Forest
Harlow
Maldon
Rochford
Southend-on-Sea
Tendring
Thurrock
Uttlesford

Fife Region
Fife*

Gloucestershire
Cheltenham
Cotswold
Forest of Dean
Gloucester
Stroud
Tewkesbury

Grampian Region
Aberdeen City*
Aberdeenshire*
Moray*

Greater London
(Former metropolitan county)
Barking and Dagenham
Barnet
Bexley
Brent
Bromley
Camden
City of London
Croydon
Ealing
Enfield
Greenwich
Hackney
Hammersmith and Fulham
Haringey
Harrow
Havering
Hillingdon
Hounslow
Islington
Kensington and Chelsea
Kingston-upon-Thames
Lambeth
Lewisham
Merton
Newham
Redbridge
Richmond-upon-Thames
Southwark
Sutton
Tower Hamlets
Waltham Forest
Wandsworth
Westminster, City of

Greater Manchester
(Former metropolitan county)
Bolton
Bury
Manchester
Oldham
Rochdale
Salford
Stockport
Tameside
Trafford
Wigan

Gwent
Blaenau Gwent*
Monmouthshire*
Newport*
Torfaen*

Gwynedd
Conwy*
Gwynedd*
Isle of Anglesey*

Hampshire
Basingstoke and Deane
East Hampshire
Eastleigh
Fareham

Gosport
Hart
Havant
New Forest
Rushmoor
Test Valley
Winchester
Portsmouth*
Southampton*

Hereford & Worcester
Bromsgrove
Hereford
Leominster
Malvern Hills
Redditch
South Herefordshire
Worcester
Wychavon
Wyre Forest

Hertfordshire
Broxbourne
Dacorum
East Hertfordshire
Hertsmere
North Hertfordshire
St. Albans
Stevenage
Three Rivers
Watford
Welwyn Hatfield

Highland Region
Highland*

Humberside
East Riding of Yorkshire*
Kingston upon Hull*
North East Lincolnshire*
North Lincolnshire*

Isle of Man
Isle of Man

Isle of Wight
Isle of Wight*

Kent
Ashford
Canterbury
Dartford
Dover
Gillingham
Gravesham
Maidstone
Rochester-upon-Medway
Sevenoaks
Shepway
Swale
Thanet
Tonbridge and Malling
Tunbridge Wells

Lancashire
Blackburn
Blackpool
Burnley
Chorley
Fylde
Hyndburn
Lancaster
Pendle
Preston
Ribble Valley
Rossendale
South Ribble
West Lancashire
Wyre

Leicestershire
Blaby
Charnwood
Harborough
Hinckley and Bosworth
Leicester*
Melton
North West Leicestershire
Oadby and Wigston
Rutland*

Lincolnshire
Boston
East Lindsey
Lincoln
North Kesteven
South Holland
South Kesteven
West Lindsey

Lothian Region
City of Edinburgh*
East Lothian*
Midlothian*
West Lothian*

Merseyside
(Former metropolitan county)
Knowsley
Liverpool
Sefton
St. Helens
Wirral

Mid Glamorgan
Bridgend*
Caerphilly*
Merthyr Tydfil*
Rhondda, Cynon, Taff*

Norfolk
Breckland
Broadland
Great Yarmouth
King's Lynn and West Norfolk
North Norfolk
Norwich
South Norfolk

North Yorkshire
Craven
Hambleton
Harrogate
Richmondshire
Ryedale
Scarborough
Selby
York*

Northamptonshire
Corby
Daventry
East Northamptonshire
Kettering
Northampton
South Northamptonshire
Wellingborough

Northern Ireland
NI: Antrim
NI: Ards
NI: Armagh
NI: Ballymena
NI: Ballymoney
NI: Banbridge
NI: Belfast
NI: Carrickfergus
NI: Castlereagh
NI: Coleraine
NI: Cookstown
NI: Craigavon
NI: Down
NI: Dungannon
NI: Fermanagh
NI: Larne
NI: Limavady
NI: Lisburn
NI: Londonderry
NI: Magherafelt
NI: Moyle
NI: Newry and Mourne
NI: Newtownabbey
NI: North Down
NI: Omagh
NI: Strabane

Northumberland
Alnwick
Berwick-upon-Tweed
Blyth Valley
Castle Morpeth
Tynedale
Wansbeck

Nottinghamshire
Ashfield
Bassetlaw
Broxtowe
Gedling
Mansfield
Newark and Sherwood
Nottingham
Rushcliffe

Orkney Islands
Orkney Islands*

Oxfordshire
Cherwell
Oxford
South Oxfordshire
Vale of White Horse
West Oxfordshire

Powys
Powys*

Shetland Islands
Shetland Islands*

Shropshire
Bridgnorth
North Shropshire
Oswestry
Shrewsbury and Atcham
South Shropshire
The Wrekin

Somerset
Mendip
Sedgemoor
South Somerset
Taunton Deane
West Somerset

South Glamorgan
Cardiff*
The Vale of Glamorgan*

South Yorkshire
(Former metropolitan county)
Barnsley
Doncaster
Rotherham
Sheffield

Staffordshire
Cannock Chase
East Staffordshire
Lichfield
Newcastle-under-Lyme
South Staffordshire
Stafford
Staffordshire Moorlands
Stoke-on-Trent*
Tamworth

Strathclyde Region
Argyll & Bute*
City of Glasgow*
East Ayrshire*
East Dunbartonshire*
East Renfrewshire*
Inverclyde*
North Ayrshire*
North Lanarkshire*
Renfrewshire*
South Ayrshire*
South Lanarkshire*
West Dunbartonshire*

Suffolk
Babergh
Forest Heath
Ipswich
Mid Suffolk
St. Edmundsbury
Suffolk Coastal
Waveney

Surrey
Elmbridge
Epsom and Ewell
Guildford
Mole Valley
Reigate and Banstead
Runnymede
Spelthorne
Surrey Heath
Tandridge
Waverley
Woking

Tayside Region
Angus*
Dundee City*
Perth & Kinross*

Tyne & Wear
Gateshead
Newcastle-upon-Tyne
North Tyneside
South Tyneside
Sunderland

Warwickshire
North Warwickshire
Nuneaton and Bedworth
Rugby
Stratford-on-Avon
Warwick

West Glamorgan
Neath Port Talbot*
Swansea*

West Midlands
(Former metropolitan county)
Birmingham
Coventry
Dudley
Sandwell
Solihull
Walsall
Wolverhampton

West Sussex
Adur
Arun
Brighton and Hove* *(part)*
Chichester
Crawley
Horsham
Mid Sussex
Worthing

West Yorkshire
(Former metropolitan county)
Bradford
Calderdale
Kirklees
Leeds
Wakefield

Western Isles
Western Isles*

Wiltshire
Kennet
North Wiltshire
Salisbury
Swindon*
Thamesdown
West Wiltshire

* Indicates a unitary authority created in or before 1997.

Post Town to Local Authority Index

To aid finding the correct local authority, this list of post towns to local authorities has been compiled.

However as Post Towns are determined by the Royal Mail and local authorities by the four Boundaries Commissions, there is no commonality of the two schemes. Consequently a single post town may be served by more than one local authority.

Entries are only given where there is significant overlap between the two geographical schemes. Please refer to a new administrative map, or consult the county look-up on the preceding pages if the address you cannot find an address which is close to a local authority boundary.

Post Town	Local Authority
Aberdare	Rhondda, Cynon, Taff
Aberdeen	Aberdeen City
Aberdeen	Aberdeenshire
Aberdeen City	Aberdeen City
Abergavenny	Monmouthshire
Abergele	Conwy
Abert'y	Blaenau Gwent
Aberystwyth	Ceredigion
Abingdon	Vale of White Horse
Accrington	Hyndburn
Addlestone	Runnymede
Airdrie	North Lanarkshire
Alcester	Stratford-on-Avon
Alderley Edge	Macclesfield
Aldershot	Guildford
Aldershot	Hart
Aldershot	Rushmoor
Alford	East Lindsey
Alfreton	Amber Valley
Alfreton	Bolsover
ALlanrystyd	Ceredigion
Alloa	Clackmannanshire
Alnwick	Alnwick
Alton	East Hampshire
Altrincham	Trafford
Amersham	Chiltern
Amlwch	Isle of Anglesey
Ammanford	Carmarthenshire
Andover	Test Valley
Antrim	NI: Antrim
Appleby in Westm'l'd	Eden
Arbroath	Angus
Armagh	NI: Armagh
Arundel	Arun
Ascot	Windsor and Maidenhead
Ashby-de-la-Zouch	North West Leicestershire
Ashford	Ashford
Ashford	Spelthorne
Ashington	Wansbeck
Ashtead	Mole Valley
Ashton-under-Lyne	Tameside
Atherstone	North Warwickshire
Avoch	Highland
Axminster	East Devon
Aylesbury	Aylesbury Vale
Aylesbury	Wycombe
Aylesford	Tonbridge and Malling
Ayr	East Ayrshire
Ayr	South Ayrshire
Bacup	Rossendale
Bagillt	Flintshire
Bakewell	Derbyshire Dales
Baldock	North Hertfordshire
Balerno	City of Edinburgh
Ballyclare	NI: Antrim
Ballyclare	NI: Newtownabbey
Ballymena	NI: Ballymena
Ballynahinch	NI: Down
Bampton	West Oxfordshire
Banbridge	NI: Banbridge
Banbury	Cherwell
Banchory	Aberdeenshire
Banff	Aberdeenshire
Bangor	Gwynedd
Bangor	NI: North Down
Banstead	Reigate and Banstead
Barking	Barking and Dagenham
Barnet	Barnet
Barnsley	Barnsley
Barnstaple	North Devon
Barrow-in-Furness	Barrow-in-Furness
Barry	The Vale of Glamorgan
Basildon	Basildon
Basingstoke	Basingstoke and Deane
Basingstoke	Hart
Bath	Bath & North East Somerset
Bath	Mendip
Bathgate	West Lothian
Batley	Kirklees
Battle	Rother
Beaworthy	Torridge
Beccles	Waveney
Beckenham	Bromley
Bedale	Richmondshire
Bedford	Bedford
Bedford	Mid Bedfordshire
Bedford	South Bedfordshire
Bedlington	Wansbeck
Belfast	NI: Belfast
Belfast	NI: Castlereagh
Belper	Amber Valley
Belvedere	Bexley
Bembridge	Isle of Wight
Benfleet	Castle Point
Berkeley	Stroud
Berkhamstead	Dacorum
Berwick-upon-Tweed	Berwick-upon-Tweed
Beverley	East Riding of Yorkshire
Bewdley	Wyre Forest
Bexhill	Rother
Bexleyheath	Bexley
Bicester	Cherwell
Bideford	Torridge
Biggleswade	Mid Bedfordshire
Billericay	Basildon
Billingham	Stockton-on-Tees
Bingley	Bradford
Birchington	Thanet
Birkenhead	Wirral
Birmingham	Birmingham
Birmingham	Bromsgrove
Birmingham	North Warwickshire
Birmingham	Sandwell
Birmingham	Solihull
Bishop Auckland	Wear Valley
Bishops Stortford	East Hertfordshire
Bishops Stortford	Uttlesford
Blackburn	Blackburn
Blackburn	Hyndburn
Blackburn	Ribble Valley
Blackpool	Blackpool
Blackwood	Blaenau Gwent
Blackwood	Caerphilly
Blaenavon	Torfaen
Blaina	Blaenau Gwent
Blairgowrie	Perth & Kinross
Blakeney	Forest of Dean
Blanford Forum	North Dorset
Blyth	Blyth Valley
Bodmin	North Cornwall
Bognor Regis	Arun
Bolton	Bolton
Bonnybridge	Falkirk
Bonnyrigg	Midlothian
Bootle/Liverpool	Sefton
Bordon	East Hampshire
Borehamwood	Hertsmere
Boston	Boston
Boston	East Lindsey
Bourne	South Kesteven
Bournemouth	Bournemouth
Brackley	South Northamptonshire
Bracknell	Bracknell Forest
Bradford	Bradford
Bradford	Kirklees
Bradford on Avon	West Wiltshire
Braintree	Braintree
Brandon	Forest Heath
Braunton	North Devon
Brecon	Powys
Brentwood	Brentwood
Bridgend	Bridgend
Bridgend	The Vale of Glamorgan
Bridgnorth	Bridgnorth
Bridgwater	Sedgemoor
Bridlington	East Riding of Yorkshire
Bridport	West Dorset
Brierley Hill	Dudley
Brighouse	Calderdale
Brighton	Brighton and Hove
Bristol	Bath & North East Somerset
Bristol	Bristol
Bristol	North Somerset
Bristol	South Gloucestershire
Brixham	Torbay
Broadstairs	Thanet
Broadstone	Poole
Bromley	Bromley
Bromsgrove	Bromsgrove
Brough	East Riding of Yorkshire
Brynmawr	Blaenau Gwent
Buckingham	Aylesbury Vale
Buckley	Flintshire
Bude	North Cornwall
Burgess Hill	Mid Sussex
Burnham-on-Crouch	Maldon
Burnham-on-Sea	Sedgemoor
Burnley	Burnley
Burntisland	Fife
Burton-on-Trent	East Staffordshire
Burton-on-Trent	South Derbyshire
Bury	Bury
Bury	Rossendale
Bury St Edmunds	Forest Heath
Bury St Edmunds	Mid Suffolk
Bury St Edmunds	St. Edmundsbury
Buxton	High Peak
Caernarfon	Gwynedd
Caerphilly	Caerphilly
Callington	Caradon
Calne	North Wiltshire
Camberley	Hart
Camberley	Surrey Heath
Camborne	Kerrier
Cambridge	Cambridge
Cambridge	East Cambridgeshire
Cambridge	South Cambridgeshire
Cannock	Cannock Chase
Canterbury	Canterbury
Canterbury	Dover
Canvey Island	Castle Point
Cardiff	Cardiff
Cardiff	The Vale of Glamorgan
Cardigan	Ceredigion
Carlisle	Allerdale
Carlisle	Carlisle
Carluke	South Lanarkshire
Carmarthen	Carmarthenshire
Carnforth	Lancaster
Carrickfergus	NI: Carrickfergus
Carshalton	Sutton
Castle Douglas	Dumfries & Galloway
Castleford	Wakefield
Caterham	Tandridge
Chandlers Ford	Eastleigh
Chard	South Somerset
Chatham	Rochester-upon-Medway
Cheadle	Stockport
Chelmsford	Chelmsford
Chelmsford	Maldon
Cheltenham	Cheltenham
Cheltenham	Cotswold
Cheltenham	Tewkesbury
Chepstow	Monmouthshire
Chepstow	Newport
Chertsey	Runnymede
Chesham	Chiltern
Chessington	Kingston-upon-Thames
Chester	Chester
Chester	Flintshire
Chesterfield	Bolsover
Chesterfield	Chesterfield
Chesterfield	North East Derbyshire
Chester-le-Street	Chester-le-Street
Chichester	Chichester
Chigwell	Epping Forest
Chinnor	South Oxfordshire
Chippenham	North Wiltshire
Chipping Norton	West Oxfordshire
Choppington	Wansbeck
Chorley	Chorley
Christchurch	Christchurch
Christchurch	New Forest
Cinderford	Forest of Dean
Cirencester	Cotswold
Clackmannan	Clackmannanshire
Clacton-on-Sea	Tendring
Cleckheaton	Kirklees
Cleethorpes	North East Lincolnshire
Clevedon	North Somerset
Clitheroe	Ribble Valley
Clydebank	West Dunbartonshire
Cobham	Elmbridge
Colchester	Babergh
Colchester	Colchester

Post Town	Area
Colchester	Tendring
Coleford	Forest of Dean
Coleraine	NI: Coleraine
Colne	Pendle
Colwyn Bay	Conwy
Colyton	East Devon
Congleton	Congleton
Consett	Derwentside
Conwy	Conwy
Cookstown	NI: Cookstown
Corby	Corby
Corsham	North Wiltshire
Cottingham	East Riding of Yorkshire
Coulsdon	Croydon
Coventry	Coventry
Coventry	Solihull
Coventry	Warwick
Cowes	Isle of Wight
Craigavon	NI: Craigavon
Cramlington	Blyth Valley
Cranbrook	Tunbridge Wells
Cranleigh	Waverley
Crawley	Crawley
Crawley	Mid Sussex
Crediton	Mid Devon
Crewe	Congleton
Crewe	Crewe and Nantwich
Crewkerne	South Somerset
Cromer	North Norfolk
Crook	Wear Valley
Crowborough	Wealden
Crowthorne	Wokingham
Croydon	Croydon
Cullompton	Mid Devon
Cumnock	East Ayrshire
Cupar	Fife
Currie	City of Edinburgh
Cwmbran	Torfaen
Dagenham	Barking and Dagenham
Dalkeith	Midlothian
Dalton-in-Furness	Barrow-in-Furness
Darlington	Darlington
Dartford	Dartford
Dartford	Sevenoaks
Darwen	Blackburn
Daventry	Daventry
Dawlish	Teignbridge
Deal	Dover
Deeside	Flintshire
Denbigh	Denbighshire
Derby	Amber Valley
Derby	Derby
Derby	Derbyshire Dales
Derby	Erewash
Derby	North West Leicestershire
Derby	South Derbyshire
Derby	Erewash
Dereham	Breckland
Devizes	Kennet
Dewsbury	Kirklees
Didcot	South Oxfordshire
Didcot	Vale of White Horse
Diss	South Norfolk
Doncaster	Bassetlaw
Doncaster	Doncaster
Doncaster	North Lincolnshire
Dorchester	West Dorset
Dorking	Mole Valley
Dover	Dover
Downham Market	King's Lynn and West Norfolk
Driffield	East Riding of Yorkshire
Droitwich	Wychavon
Dudley	Dudley
Dukinfield	Tameside
Dumbarton	West Dunbartonshire
Dumfries	Dumfries & Galloway
Dunblane	Stirling
Dundee	Angus
Dundee	Dundee City
Dunfermline	Fife
Dungannon	NI: Dungannon
Dunoon	Argyll & Bute
Dunstable	South Bedfordshire
Durham	Derwentside
Durham	Durham
Dursley	Stroud
East Cowes	Isle of Wight
East Grinstead	Mid Sussex
East Molesey	Elmbridge
Eastbourne	Eastbourne
Eastbourne	Wealden
Eastleigh	Eastleigh
Edgware	Barnet
Edinburgh	City of Edinburgh
Elgin	Moray
Ellesmere	North Shropshire
Ellon	Aberdeenshire
Ely	East Cambridgeshire
Enfield	Enfield
Enniskillen	NI: Fermanagh
Epping	Epping Forest
Epsom	Epsom and Ewell
Erith	Bexley
Esher	Elmbridge
Evesham	Wychavon
Exeter	East Devon
Exeter	Exeter
Exeter	Teignbridge
Exmouth	East Devon
Eyemouth	Scottish Borders
Fairford	Cotswold
Fakenham	North Norfolk
Falkirk	Falkirk
Falmouth	Carrick
Falmouth	Kerrier
Fareham	Fareham
Faringdon	Vale of White Horse
Farnborough	Rushmoor
Farnham	Waverley
Faversham	Swale
Felixstowe	Suffolk Coastal
Feltham	Hounslow
Ferndown	East Dorset
Ferryhill	Sedgefield
Filey	Scarborough
Fleetwood	Wyre
Flint	Flintshire
Folkestone	Shepway
Fordingbridge	New Forest
Fort William	Highland
Fraserburgh	Aberdeenshire
Frinton-on-Sea	Tendring
Frome	Mendip
Gainsborough	West Lindsey
Galashiels	Scottish Borders
Gateshead	Gateshead
Gerrard's Cross	Chiltern
Gillingham	Gillingham
Gillingham	North Dorset
Girvan	South Ayrshire
Glasgow	City of Glasgow
Glasgow	East Dunbartonshire
Glasgow	North Lanarkshire
Glasgow	South Lanarkshire
Glastonbury	Mendip
Glenrothes	Fife
Glenrothes	Fife
Glossop	High Peak
Gloucester	Forest of Dean
Gloucester	Gloucester
Gloucester	Stroud
Gloucester	Tewkesbury
Godalming	Waverley
Goole	East Riding of Yorkshire
Goole	Selby
Gosport	Gosport
Gosport	Gosport
Gourock	Inverclyde
Grangemouth	Falkirk
Grantham	South Kesteven
Gravesend	Gravesham
Grays	Thurrock
Grays	Thurrock
Great Missenden	Chiltern
Great Yarmouth	Great Yarmouth
Greenford	Ealing
Greenock	Inverclyde
Grimsby	East Lindsey
Grimsby	North East Lincolnshire
Guernsey	Guernsey
Guildford	Guildford
Guildford	Waverley
Guisborough	Redcar and Cleveland
Hailsham	Wealden
Halesowen	Dudley
Halifax	Calderdale
Halstead	Braintree
Hamilton	South Lanarkshire
Hampton	Richmond-upon-Thames
Habrough	North East Lincolnshire
Harlow	Harlow
Harpenden	St. Albans
Harrogate	Harrogate
Harrow	Harrow
Hartlepool	Hartlepool
Harwich	Tendring
Hassocks	Mid Sussex
Hastings	Hastings
Hastings	Rother
Hatfield	Welwyn Hatfield
Haverfordwest	Pembrokeshire
Haverhill	St. Edmundsbury
Hawes	Richmondshire
Hawick	Scottish Borders
Hayes	Hillingdon
Hayle	Penwith
Hayling Island	Havant
Haywards Heath	Mid Sussex
Heanor	Amber Valley
Heanor	Erewash
Heathfield	Wealden
Hebden Bridge	Calderdale
Helensburgh	Argyll & Bute
Helston	Kerrier
Hemel Hempstead	Dacorum
Henley-on-Thames	South Oxfordshire
Henlow	Mid Bedfordshire
Hereford	Hereford
Hereford	Leominster
Hereford	South Herefordshire
Herne Bay	Canterbury
Hertford	East Hertfordshire
Hessle	East Riding of Yorkshire
Hexham	Tynedale
Heywood	Rochdale
High Peak	High Peak
High Wycombe	Wycombe
Highbridge	Sedgemoor
Hillsborough	NI: Lisburn
Hindhead	Waverley
Hinkley	Hinckley and Bosworth
Hitchin	Mid Bedfordshire
Hitchin	North Hertfordshire
Hockley	Rochford
Hoddesdon	Broxbourne
Holsworthy	Torridge
Holt	North Norfolk
Holyhead	Isle of Anglesey
Holywell	Flintshire
Holywood	NI: North Down
Honiton	East Devon
Horley	Reigate and Banstead
Horncastle	East Lindsey
Hornchurch	Havering
Hornsea	East Riding of Yorkshire
Horsham	Horsham
Houghton-le-Spring	Sunderland
Hounslow	Hounslow
Hove	Brighton and Hove
Huddersfield	Kirklees
Hull	East Riding of Yorkshire
Hull	Kingston upon Hull
Hungerford	Newbury
Huntingdon	Huntingdon
Hyde	High Peak
Hyde	Tameside
Ilford	Redbridge
Ilfracombe	North Devon
Ilkeston	Erewash
Ilkley	Bradford
Ilminster	South Somerset
Inverkeithing	Fife
Inverness	Highland
Inverurie	Aberdeenshire
Ipswich	Babergh
Ipswich	Ipswich
Ipswich	Mid Suffolk
Ipswich	Suffolk Coastal
Irvine	North Ayrshire
Isle of Man	Isle of Man
Isleworth	Hounslow
Iver	South Bucks
Ivybridge	South Hams
Jarrow	South Tyneside
Jedburgh	Scottish Borders
Jersey	Jersey
Johnstone	Renfrewshire
Juniper Green	City of Edinburgh
Keighley	Bradford
Keighley	Craven
Keith	Moray
Kelso	Scottish Borders
Kendal	South Lakeland
Kenilworth	Warwick
Kettering	Kettering
Kidderminster	Wyre Forest
Kilmarnock	East Ayrshire
Kilwinning	North Ayrshire
King's Lynn	Breckland
King's Lynn	King's Lynn and West Norfolk
Kingsbridge	South Hams
Kingston-u-Thames	Kingston-upon-Thames
Kingswinford	Dudley
Kinross	Perth & Kinross
Kirkcaldy	Fife
Kirkwall	Orkney Islands
Knaresborough	Harrogate
Knottingley	Wakefield
Knutsford	Macclesfield
Lanark	South Lanarkshire
Lancaster	Lancaster
Lancing	Adur
Langport	South Somerset
Largs	North Ayrshire
Larkhall	South Lanarkshire
Larne	NI: Larne
Launceston	North Cornwall
Leamington Spa	Stratford-on-Avon
Leamington Spa	Warwick
Leatherhead	Guildford
Leatherhead	Mole Valley
Lechlade	Cotswold
Ledbury	Malvern Hills

Post Town	Area
Leeds	Leeds
Leek	Staffordshire Moorlands
Lee-on-Solent	Gosport
Leicester	Blaby
Leicester	Charnwood
Leicester	Harborough
Leicester	Hinckley and Bosworth
Leicester	Leicester
Leicester	Oadby and Wigston
Leigh	Wigan
Leigh-on-Sea	Southend-on-Sea
Leighton Buzzard	Aylesbury Vale
Leighton Buzzard	South Bedfordshire
Leiston	Suffolk Coastal
Leominster	Leominster
Lerwick	Shetland Islands
Letchworth	North Hertfordshire
Leven	Fife
Lewes	Lewes
Leyburn	Richmondshire
Lichfield	Lichfield
Limavady	NI: Limavady
Lincoln	Lincoln
Lincoln	North Kesteven
Lincoln	West Lindsey
Linlithgow	West Lothian
Lisburn	NI: Lisburn
Liskeard	Caradon
Liss	East Hampshire
Littleborough	Rochdale
Littlehampton	Arun
Liverpool	Knowsley
Liverpool	Liverpool
Liverpool	Sefton
Liversedge	Kirklees
Livingston	West Lothian
Llandrindod Wells	Powys
Llandudno	Conwy
Llanelli	Carmarthenshire
Llanfairpwllgwyngyl	Isle of Anglesey
Llangadog	Carmarthenshire
Llangefni	Isle of Anglesey
Llanon	Ceredigion
Lockerbie	Dumfries & Galloway
London E11	Waltham Forest
London E14	Tower Hamlets
London E17	Waltham Forest
London E18	Redbridge
London E4	Waltham Forest
London N1	Islington
London N12	Barnet
London N13	Enfield
London N14	Enfield
London N17	Haringey
London N21	Enfield
London N9	Enfield
London NW1	Camden
London NW11	Barnet
London NW3	Camden
London NW4	Barnet
London NW7	Barnet
London SE19	Croydon
London SE25	Croydon
London SE3	Greenwich
London SE6	Lewisham
London SE9	Greenwich
London SW1	Westminster, City of
London SW14	Richmond-upon-Thames
London SW15	Wandsworth
London SW18	Wandsworth
London SW19	Merton
London SW20	Merton
London SW6	Hammersmith and Fulham
London W13	Ealing
London W3	Ealing
London W4	Hounslow
London W5	Ealing
London W7	Ealing
Londonderry	NI: Londonderry
Longhope	Forest of Dean
Longniddry	East Lothian
Loughborough	Charnwood
Loughborough	Rushcliffe
Loughton	Epping Forest
Louth	East Lindsey
Lowestoft	Waveney
Ludlow	South Shropshire
Luton	Luton
Luton	South Bedfordshire
Lutterworth	Harborough
Lydbrook	Forest of Dean
Lydney	Forest of Dean
Lymington	New Forest
Lymm	Warrington
Lytham St Anne's	Fylde
Mablethorpe	East Lindsey
Macclesfield	Macclesfield
Maesteg	Bridgend
Magherafelt	NI: Magherafelt
Maidenhead	Windsor and Maidenhead
Maidstone	Maidstone
Maldon	Maldon
Malmesbury	North Wiltshire
Malton	Ryedale
Malvern	Malvern Hills
Manchester	Bury
Manchester	Manchester
Manchester	Rochdale
Manchester	Salford
Manchester	Tameside
Manchester	Trafford
Manchester	Wigan
Manningtree	Tendring
Mansfield	Bolsover
Mansfield	Mansfield
Mansfield	Newark and Sherwood
March	Fenland
Margate	Thanet
Market Drayton	North Shropshire
Market Harborough	Harborough
Market Rasen	West Lindsey
Markfield	North West Leicestershire
Marlborough	Kennet
Marlow	Wycombe
Maryport	Allerdale
Matlock	Derbyshire Dales
Melksham	West Wiltshire
Melton Mowbray	Melton
Menai Bridge	Isle of Anglesey
Merthyr Tydfil	Merthyr Tydfil
Mexborough	Doncaster
Mexborough	Rotherham
Micheldean	Forest of Dean
Middlesbrough	Hambleton
Middlesbrough	Middlesborough
Middlesbrough	Redcar and Cleveland
Milford Haven	Pembrokeshire
Millom	Copeland
Milton Keynes	Aylesbury Vale
Milton Keynes	Milton Keynes
Minehead	West Somerset
Mirfield	Kirklees
Mitcham	Merton
Mold	Flintshire
Monmouth	Monmouthshire
Morden	Merton
Morecambe	Lancaster
Moreton-in-Marsh	Cotswold
Morpeth	Alnwick
Morpeth	Castle Morpeth
Motherwell	North Lanarkshire
Mountain Ash	Rhondda, Cynon, Taff
Musselburgh	East Lothian
Nantwich	Crewe and Nantwich
Neath	Neath Port Talbot
Nelson	Pendle
New Malden	Kingston-upon-Thames
New Milton	New Forest
New Quay	Ceredigion
New Romney	Shepway
New Tredegar	Blaenau Gwent
New Tredegar	Caerphilly
Newark	Newark and Sherwood
Newbury	Basingstoke and Deane
Newbury	Newbury
Newcastle-u-Lyme	Newcastle-under-Lyme
Newcastle-upon-Tyne	Castle Morpeth
Newcastle-upon-Tyne	Gateshead
Newcastle-upon-Tyne	Newcastle-upon-Tyne
Newcastle-upon-Tyne	North Tyneside
Newhaven	Lewes
Newmarket	East Cambridgeshire
Newmarket	Forest Heath
Newnham	Forest of Dean
Newport	Caerphilly
Newport	Isle of Wight
Newport	Newport
Newport	The Wrekin
Newport Pagnell	Milton Keynes
Newport-on-Tay	Fife
Newquay	Restormel
Newry	NI: Newry and Mourne
Newton Abbot	Teignbridge
Newton Stewart	Dumfries & Galloway
Newton-le-Willows	St. Helens
Newtown	Powys
Newtownabbey	NI: Newtownabbey
Newtownards	NI: Ards
North Ferriby	East Riding of Yorkshire
North Shields	North Tyneside
North Tawton	West Devon
North Walsham	North Norfolk
Northallerton	Hambleton
Northampton	Daventry
Northampton	Northampton
Northampton	South Northamptonshire
Northampton	Wellingborough
Northolt	Ealing
Northwich	Vale Royal
Northwood	Hillingdon
Norwich	Breckland
Norwich	Broadland
Norwich	North Norfolk
Norwich	Norwich
Norwich	South Norfolk
Nottingham	Ashfield
Nottingham	Ashfield
Nottingham	Broxtowe
Nottingham	Erewash
Nottingham	Gedling
Nottingham	Nottingham
Nottingham	Rushcliffe
Nuneaton	Hinckley and Bosworth
Nuneaton	Nuneaton and Bedworth
Oakham	Rutland
Okehampton	West Devon
Oldham	Oldham
Olney	Milton Keynes
Omagh	NI: Omagh
Orkney	Orkney Islands
Ormskirk	West Lancashire
Orpington	Bromley
Ossett	Wakefield
Oswestry	Oswestry
Other	Angus
Otley	Leeds
Ottery St Mary	East Devon
Oxford	Cherwell
Oxford	Oxford
Oxford	South Oxfordshire
Oxford	Vale of White Horse
Oxford	West Oxfordshire
Paignton	Torbay
Paisley	Renfrewshire
Par	Restormel
Pembroke	Pembrokeshire
Penarth	The Vale of Glamorgan
Penicuik	Midlothian
Penrith	Eden
Pentre	Rhondda, Cynon, Taff
Penzance	Penwith
Pershore	Wychavon
Perth	Perth & Kinross
Peterborough	East Northamptonshire
Peterborough	Fenland
Peterborough	Peterborough
Peterborough	South Kesteven
Peterhead	Aberdeenshire
Peterlee	Easington
Petersfield	East Hampshire
Pevensey	Wealden
Pickering	Ryedale
Pinner	Harrow
Pinner	Hillingdon
Plymouth	Plymouth
Plymouth	South Hams
Polegate	Wealden
Pontefract	Wakefield
Pontypool	Torfaen
Pontypridd	Rhondda, Cynon, Taff
Poole	Poole
Poole	Purbeck
Port Talbot	Neath Port Talbot
Porth	Rhondda, Cynon, Taff
Porthcawl	Bridgend
Portland	Weymouth and Portland
Portsmouth	Portsmouth
Potters Bar	Hertsmere
Poulton-le-Fylde	Wyre
Prescot	Knowsley
Prestatyn	Denbighshire
Preston	Chorley
Preston	Fylde
Preston	Preston
Preston	Ribble Valley
Preston	South Ribble
Preston	West Lancashire
Preston	Wyre
Prestonpans	East Lothian
Prestwick	South Ayrshire
Prudhoe	Tynedale
Pudsey	Leeds
Pulborough	Horsham
Purfleet	Thurrock
Purley	Croydon
Pwlleli	Gwynedd
Rainham	Havering
Ramsgate	Thanet
Rayleigh	Rochford
Reading	Newbury
Reading	Reading
Reading	South Oxfordshire
Reading	Wokingham
Redcar	Redcar and Cleveland
Redditch	Redditch
Redhill	Reigate and Banstead
Redruth	Kerrier
Reigate	Reigate and Banstead
Retford	Bassetlaw
Rhyl	Denbighshire
Richmond	Richmondshire
Richmond	Richmond-upon-Thames
Rickmansworth	Three Rivers

Post Town

Post Town	Area
Ringwood	East Dorset
Ringwood	New Forest
Ripley	Amber Valley
Ripon	Harrogate
Rochdale	Rochdale
Rochester	Rochester-upon-Medway
Rochford	Rochford
Romford	Barking and Dagenham
Romford	Havering
Romney Marsh	Shepway
Romsey	Test Valley
Rossendale	Rossendale
Ross-on-Wye	South Herefordshire
Rotherham	Barnsley
Rotherham	Rotherham
Rowland's Castle	Havant
Royston	North Hertfordshire
Royston	South Cambridgeshire
Ruardean	Forest of Dean
Rugby	Rugby
Rugby	Stratford-on-Avon
Rugeley	Cannock Chase
Rugeley	Lichfield
Ruislip	Hillingdon
Runcorn	Halton
Rushden	East Northamptonshire
Ryde	Isle of Wight
Rye	Rother
Saffron Walden	Uttlesford
Sale	Trafford
Salford	Salford
Salisbury	Salisbury
Saltash	Caradon
Saltburn-by-the-Sea	Redcar and Cleveland
Sandbach	Congleton
Sandown	Isle of Wight
Sandwich	Dover
Sandy	Mid Bedfordshire
Sawbridgeworth	East Hertfordshire
Saxmundham	Suffolk Coastal
Scarborough	Scarborough
Scunthorpe	North Lincolnshire
Seaford	Brighton and Hove
Seaford	Lewes
Seaham	Easington
Seaton	East Devon
Selby	Selby
Sevenoaks	Sevenoaks
Sevenoaks	Tonbridge and Malling
Shaftesbury	North Dorset
Sheerness	Swale
Sheffield	North East Derbyshire
Sheffield	Sheffield
Shefford	Mid Bedfordshire
Shepperton	Spelthorne
Shepton Mallet	Mendip
Sherborne	West Dorset
Sheringham	North Norfolk
Shetland	Shetland Islands
Shipley	Bradford
Shoreham-by-Sea	Adur
Shrewsbury	North Shropshire
Shrewsbury	Shrewsbury and Atcham
Sidcup	Bexley
Sidmouth	East Devon
Sittingbourne	Swale
Skegness	East Lindsey
Skelmersdale	West Lancashire
Skipton	Craven
Sleaford	North Kesteven
Slough	Slough
Slough	South Bucks
Solihull	Solihull
Solihull	Stratford-on-Avon
South Croydon	Croydon
South Shields	South Tyneside
South Wirral	Ellesmere Port and Neston
Southall	Ealing
Southampton	Eastleigh
Southampton	Fareham
Southampton	New Forest
Southampton	Southampton
Southampton	Test Valley
Southampton	Winchester
Southend-on-Sea	Southend-on-Sea
Southminster	Maldon
Southport	Sefton
Southsea	Portsmouth
Sowerby Bridge	Calderdale
Spalding	South Holland
Spennymoor	Sedgefield
Spilsby	East Lindsey
St Albans	St. Albans
St Andrews	Fife
St Austell	Restormel
St Helens	St. Helens
St Ives	Penwith
St Leonards-on-Sea	Hastings
Stafford	South Staffordshire
Stafford	Stafford
Staines	Spelthorne
Stalybridge	Tameside
Stamford	South Kesteven
Stanford-le-Hope	Thurrock
Stanley	Derwentside
Stanmore	Harrow
Stevenage	Stevenage
Steyning	Horsham
Stirling	Stirling
Stockport	Macclesfield
Stockport	Stockport
Stockton-on-Tees	Stockton-on-Tees
Stoke-on-Trent	Congleton
Stoke-on-Trent	Newcastle-under-Lyme
Stoke-on-Trent	Staffordshire Moorlands
Stoke-on-Trent	Stoke-on-Trent
Stone	Stafford
Stonehouse	Stroud
Stourbridge	Dudley
Stourbridge	South Staffordshire
Stourport-on-Severn	Wyre Forest
Stowmarket	Mid Suffolk
Strabane	NI: Strabane
Stranraer	Dumfries & Galloway
Stratford-upon-Avon	Stratford-on-Avon
Street	Mendip
Stroud	Stroud
Sturminster Newton	North Dorset
Sudbury	Babergh
Sunbury-on-Thames	Spelthorne
Sunderland	Sunderland
Surbiton	Kingston-upon-Thames
Sutton	Sutton
Sutton Coldfield	Birmingham
Sutton Coldfield	Walsall
Sutton-in-Ashfield	Ashfield
Swadlincote	South Derbyshire
Swaffham	Breckland
Swanage	Purbeck
Swanlea	Sevenoaks
Swansea	Neath Port Talbot
Swansea	Swansea
Swindon	North Wiltshire
Swindon	Swindon
Tadworth	Reigate and Banstead
Tamworth	Lichfield
Tamworth	North Warwickshire
Tamworth	Tamworth
Taunton	Taunton Deane
Tavistock	West Devon
Tayport	Fife
Teddington	Richmond-upon-Thames
Teignmouth	Teignbridge
Telford	The Wrekin
Tewksbury	Tewkesbury
Tewksbury	Wychavon
Thame	South Oxfordshire
Thatcham	Newbury
Thetford	Breckland
Thirsk	Hambleton
Thornton Heath	Croydon
Thornton-Cleveleys	Blackpool
Thornton-Cleveleys	Wyre
Thurso	Highland
Tipton	Sandwell
Tiverton	Mid Devon
Todmorden	Calderdale
Tonbridge	Maidstone
Tonbridge	Tonbridge and Malling
Tonbridge	Tunbridge Wells
Tonypandy	Rhondda, Cynon, Taff
Torpoint	Caradon
Torquay	Torbay
Totnes	South Hams
Towcester	South Northamptonshire
Tring	Dacorum
Troon	South Ayrshire
Trowbridge	West Wiltshire
Truro	Carrick
Tunbridge Wells	Tunbridge Wells
Turiff	Aberdeenshire
Twickenham	Richmond-upon-Thames
Tywyn	Gwynedd
Uckfield	Wealden
Ulverston	South Lakeland
Upminster	Havering
Uttoxeter	East Staffordshire
Uxbridge	Hillingdon
Ventnor	Isle of Wight
Verwood	East Dorset
Wadebridge	North Cornwall
Wakefield	Leeds
Wakefield	Wakefield
Wallasey	Wirral
Wallingford	South Oxfordshire
Wallington	Sutton
Wallsend	North Tyneside
Walsall	Lichfield
Walsall	South Staffordshire
Walsall	Walsall
Waltham Abbey	Epping Forest
Waltham Cross	Broxbourne
Walton-on-Thames	Elmbridge
Walton-on-the-Naze	Tendring
Wantage	Vale of White Horse
Ware	East Hertfordshire
Wareham	Purbeck
Warley	Sandwell
Warminster	West Wiltshire
Warrington	Vale Royal
Warrington	Warrington
Warrington	Wigan
Warwick	Stratford-on-Avon
Warwick	Warwick
Washington	Sunderland
Waterlooville	East Hampshire
Waterlooville	Havant
Waterlooville	Winchester
Watford	Hertsmere
Watford	Three Rivers
Watford	Watford
Wednesbury	Sandwell
Wednesbury	Walsall
Welling	Bexley
Wellingborough	East Northamptonshire
Wellingborough	Wellingborough
Wellington	Taunton Deane
Wells	Mendip
Wells-next-the-Sea	North Norfolk
Welshpool	Powys
Welwyn	Welwyn Hatfield
Welwyn Garden City	Welwyn Hatfield
Wembley	Brent
West Bromwich	Sandwell
West Drayton	Hillingdon
West Malling	Tonbridge and Malling
West Wickham	Bromley
Westbury	Forest of Dean
Westbury	West Wiltshire
Westcliff-on-Sea	Southend-on-Sea
Westerham	Bromley
Western Isles	Western Isles
Westgate-on-Sea	Thanet
Weston-Super-Mare	North Somerset
Wetherby	Leeds
Weybridge	Elmbridge
Weymouth	West Dorset
Weymouth	Weymouth and Portland
Whitby	Scarborough
Whitchurch	North Shropshire
Whitehaven	Copeland
Whitley Bay	Blyth Valley
Whitley Bay	North Tyneside
Whitstable	Canterbury
Wick	Highland
Wickford	Basildon
Widnes	Halton
Wigan	West Lancashire
Wigan	Wigan
Wigston	Oadby and Wigston
Willenhall	Walsall
Wilmslow	Macclesfield
Wimborne	East Dorset
Wimborne	Poole
Wincanton	South Somerset
Winchester	Winchester
Windsor	Windsor and Maidenhead
Winscombe	North Somerset
Winsford	Vale Royal
Wirral	Wirral
Wisbech	Fenland
Wisbech	King's Lynn and West Norfolk
Wishaw	North Lanarkshire
Witham	Braintree
Witney	West Oxfordshire
Woking	Guildford
Woking	Surrey Heath
Woking	Woking
Wokingham	Wokingham
Wolverhampton	Bridgnorth
Wolverhampton	South Staffordshire
Wolverhampton	Wolverhampton
Woodbridge	Suffolk Coastal
Woolacombe	North Devon
Worcester	Malvern Hills
Worcester	Worcester
Worcester	Wychavon
Worcester Park	Sutton
Workington	Allerdale
Worksop	Bassetlaw
Worthing	Worthing
Wotton-under-Edge	Stroud
Wrexham	Wrexham
Wymondham	South Norfolk
Yarm	Stockton-on-Tees
Yelverton	West Devon
Yeovil	South Somerset
York	East Riding of Yorkshire
York	Hambleton
York	Harrogate
York	Ryedale
York	Selby
York	York

Amateurs Listed by Local Authority District

Aberdeen City

MM0AOF AB10 6QQ
GM4JLZ AB10 6QW
GM2FHH AB10 7AJ
GM3KJE AB10 7JE
G3UPT AB10 7NT
GM6YQA AB11 9AS
GM0CJK AB12 3EG
GM0WRV AB12 3WH
GM1MCN AB12 5BR
GM7GOD AB12 5QR
GM3TLA AB13 0JB
GW4BFX AB15 4WB
GM3FRZ AB15 8LB
GM3LER AB15 8NB
GM0WBW AB16 5JA
MM0BCR AB16 5RP
MM1BOK AB16 5RP
MM1APZ AB16 7TQ
GM0JOV AB22 8GJ
GM0VGI AB22 8HW
GM7MWL AB22 8LJ
GM6MJY AB22 8PH
GM7MMI AB22 8RW
GM0PKX AB22 8SL
GM4JQA AB22 8SY
GM1TDU AB22 8TT
GM1FSU AB22 8XB
GM0MCJ AB22 8XH
MM1BUO AB22 8XR
GM3WRN AB22 8YF
GM4PXB AB22 8ZQ
GM8BNH AB23 8BD
GM0VHC AB23 8EP
GM4JXP AB23 8GD
GM7NQP AB23 8NJ
GM4YWV AB23 8QD
GM1LKD AB23 8QN
GM0DFL AB24 4DR
GM4CAU AB24 4HX
GM6SDV AB24 4NJ
MM1BAH AB25 2RE
M1AWL AB25 3UH
MM1BAE AB25 3XB

Aberdeenshire

GM3NUU AB12 3RL
GM4RAZ AB12 4NY
GM4RGS AB12 4NY
MM0AOY AB12 4QA
GM1HGC AB12 4QF
M1ABA AB12 4QW
GM0WFI AB12 5XN
GM0VFY AB12 5XT
GM0MDB AB21 0XX
G7TUK AB23 8UA
2M1ENI AB23 8UT
2M1ENK AB23 8UT
GM4LYQ AB30 1UT
GM4YRE AB30 1XZ
GM6ENX AB31 4AE
GM6EUC AB31 4EN
GM0TCU AB31 4HG
GM7NXI AB31 4RY
GM0SXQ AB31 4RY
G4BHX AB31 4SD
GM8LYS AB31 5EW
GM8AAN AB31 5QX
GM8SQ AB31 5RE
GM0PKQ AB31 5XA
GM0JKF AB31 5XJ
GM6UHC AB32 6RA
GM8BSQ AB32 6WS
GM1XEA AB32 6WW
GM7DST AB32 6WW
GM0GAT AB32 6XY
GM0GIB AB32 7BE
GM7ITG AB33 8AL
GM0MYV AB33 8NN
GM0KDP AB34 5JZ
GM4RLV AB35 5QH
GM8FVN AB35 5SF
GM3LYA AB35 5YT
G7USH AB39 2EH
GM4SJC AB39 2HG
GM3DNV AB39 3SA
G1KDX AB39 3WX
GM4MBG AB41 0BJ
GM7KRQ AB41 0DU
GM1AIH AB41 0RG
MM1BMK AB41 7JL
GM0GLD AB41 7PU
GM6KDB AB41 7PU
GM4FVS AB41 8BH
GM7RBP AB41 8DH
GM3UAG AB41 8QS
GM4GNR AB41 8QW
GM7LAC AB41 8UJ
GM4YXI AB41 8YH
M1AIO AB41 9EY
GM0JEF AB41 9HF
GM4FUL AB41 9JW
GM4OBD AB41 9LW
GM0DYU AB42 0NW
GM8GXQ AB42 0QL
GM0EUO AB42 1EG
GM1AUZ AB42 1GE
GM1KBZ AB42 1GE
2E1DUL AB42 1GU
GM1OSZ AB42 1HB
GM4UFD AB42 1NX
GM0FHE AB42 1RY
GM7FYB AB42 2UF
GM4WWT AB42 2XL
GM4GLD AB42 2YG
GM4VEE AB42 2YJ
GM1FNY AB42 3HY
GM4EHP AB42 4NL
GM3PIP AB42 5EJ
GM0RSI AB42 5ES
2M0AEL AB42 5FD
GM1KUI AB42 5HR
GM8NMM AB42 5RE
MM0AOQ AB42 5WE
GM7OJJ AB42 8AY
GM7OTT AB42 8AY
GM7NNS AB42 8HX
GM0WIB AB43 6LD
GM0CDV AB43 6LL
MM1AMD AB43 6NN
GM4PMH AB43 7JS
MM1BXZ AB43 7JU
MM1BUE AB43 8QU
GM1BWV AB43 8RW
MM1CAC AB43 9DH
GM3PZG AB43 9NH
GM1INS AB43 9NY
GM8GCY AB43 9PT
GM3ZOT AB43 9PU
GM3UBJ AB44 1RP
2M0ACT AB44 1XY
GM3DZB AB45 1BJ
GM0PYC AB45 1BJ
GM8SVB AB45 1BJ
GM3GG AB45 1GA
GM1VAX AB45 2ES
GM1CCI AB45 2JR
GM0CHM AB45 2LJ
GM0VXB AB45 2NA
GM0VWQ AB45 2NA
GM4KIA AB45 2RD
GM0EJY AB45 2RS
GM4ZJV AB45 2RS
2M1DHI AB45 3YL
GM8XOC AB45 3YT
GM0ULK AB51 0DW
GM4DZM AB51 0ES
GM3ZXH AB51 0HW
GM0TFQ AB51 3WJ
GM0PEO AB51 3XJ
GM0EUK AB51 3YG
GM4DIN AB51 4RQ
GM1RDG AB51 4TB
GM7UPD AB51 5HE
GM1MYF AB51 5NJ
GM3XOQ AB51 5QT
GM7RYT AB51 5QT
GM0FIQ AB51 5QZ
GM7JMO AB51 7SQ
GM0TGE AB51 8SY
G4ERB AB51 8UN
G7OWM AB51 9ND
GM1WWD AB51 9YG
2M1ERO AB52 6PQ
GM7KJL AB52 6TR
MM0ANA AB53 4EU
GM3JGS AB53 4EU
G4FQV AB53 4LQ
GM4TRS AB53 6SL
GM4VHU AB53 6TA
GM0NGJ AB53 6TE
GM7UDI AB53 7GS
G7VXZ AB53 7JB
GM4HWS AB53 7RJ
GM7CPJ AB53 7TQ
GM0BNQ AB53 8HD
G4SIK AB53 8JY
GM4ZEX AB53 8LT
GM8REG AB54 7SY
GM6TAN AB54 7TG
GM4NXT AB54 7XR
GM0OTP AB54 8DX
GM4OMW AB54 8DY
GM4PKJ DD10 0HX
GM8ZEQ DD10 0HX
GM1MLS DD10 0SR
GM3YAO DD10 0SW

Adur

G3KXF BN15 0AE
G3WOR BN15 0AE
G4GPW BN15 0AE
M1AKA BN15 0AE
G0HLY BN15 0BY
G4JEI BN15 0DY
G1SQM BN15 0EJ
G1EOM BN15 0LX
G7FGZ BN15 0ND
G6WOR BN15 0NN
G8XEU BN15 0NN
G8MSQ BN15 0PB
G8MFO BN15 0PH
G0VSI BN15 0RW
G2DPY BN15 8BY
G8FMJ BN15 8DE
G1EBA BN15 8EA
G4MGU BN15 8EW
G0UFP BN15 8HP
G0OAZ BN15 8LN
G1RDU BN15 8LN
G7GMD BN15 8LN
G6BWN BN15 8LZ
G6DIR BN15 8NH
G0WMG BN15 8NY
M0AAN BN15 8RL
G3JSU BN15 8RN
G4EWY BN15 8RX
G4GPX BN15 9BT
G4KIT BN15 9BT
G1WOR BN15 9DE
G8VEH BN15 9DE
G6GAW BN15 9DR
G3LQI BN15 9DZ
G7NQQ BN15 9EL
G3UFS BN15 9PZ
G4FIG BN15 9RQ
G1JPI BN15 9TP
G1PTT BN15 9TS
G3UCW BN15 9UB
G3NDJ BN15 9UP
G6AED BN15 9UQ
G2AUB BN15 9XD
G4YRT BN41 1PL
G1SEM BN42 4AN
G4ZUM BN42 4NT
G0VAJ BN42 4RD
G0NSI BN42 4YA
G1KXY BN42 4YL
G4AWI BN43 5AF
G8IMP BN43 5AG
G4ZHK BN43 5AN
G0SIU BN43 5GB
G4XUZ BN43 5GD
G4OCZ BN43 5HL
G4EUK BN43 5NQ
G3TVO BN43 5NT
G0RPJ BN43 5TD
G0FIG BN43 5TH
G1HJH BN43 5TH
G2HV BN43 5WY
G4JBA BN43 5WY
G4RTX BN43 6AJ
G0FZB BN43 6GP
G0AQH BN43 6GR
G1GGU BN43 6HA
G0TNN BN43 6HF
G0FUI BN43 6HY
G3LLD BN43 6HY
G7CTX BN43 6PJ
G0FIP BN43 6TG
G0HPD BN43 6TG
G0LTD BN43 6TG
G0HLF BN43 6TP
G3XOI BN43 6TP
G0IFQ BN43 6YB

Allerdale

G3ZED CA12 4AP
M0AOT CA12 4AZ
G7NIY CA12 4AZ
G3FPN CA12 4EF
G0UQC CA12 4HS
G4ERL CA12 4NN
G4NEH CA12 5TX
M1ACO CA12 5TX
2E1CPD CA12 5TX
G7OHM CA13 0AB
G0LIU CA13 0BW
G3ZPD CA13 0DG
G4ONI CA13 0PN
G1GWE CA13 0QR
M0AYB CA13 0TJ
G7OZZ CA13 9ED
G0UVC CA13 9JH
G3WCM CA14 1AA
G4BBX CA14 1AA
G1OAE CA14 1EE
G0EKC CA14 1EU
G3WJH CA14 1LP
G3OHK CA14 1PL
G4VVR CA14 1PX
G0EMM CA14 1PY
G1PEN CA14 1PY
G4ZFX CA14 1XJ
G6VCN CA14 3BZ
G4UYI CA14 3EN
G6TAK CA14 3HR
G4PHM CA14 3NL
G1AQI CA14 3RA
G0EBN CA14 3SU
G0OMB CA14 3UR
G1LEH CA14 4HY
G4CJP CA14 4NX
M1BDC CA14 5AW
2E1APW CA14 5DW
G0DPE CA14 5HP
G8RZ CA14 5QG
G3WBZ CA14 5QH
G0MTD CA14 5QR
G4BIU CA15 6BL
G4NDS CA15 6BL
G0NNN CA15 6QW
M0BCV CA15 7BU
G7UPX CA15 7BU
G7WKB CA15 7ED
G0MWE CA15 7EG
2E1EVW CA15 7ER
2E1EWB CA15 7ER
G0GPM CA15 7EW
G0ORM CA15 7QN
G0MTQ CA15 7SS
G7DIS CA15 7SS
G1CJB CA15 8QN
G8DVW CA5 1JF
G1GDB CA5 2EF
G3VIJ CA5 2LU
M0BEE CA5 2QD
G0ORO CA5 2QD
2E1BKM CA5 2QD
G7DXB CA5 2QD
2E1BKP CA5 2QD
G4XBL CA5 3DN
G0BPF CA5 3NW
G0ZMH CA5 3QY
M0AMX CA5 3QY
G0TII CA5 4AA
G0TNL CA5 4JJ
G0IOG CA5 4NR
G7KWQ CA5 4QH
G0TRW CA5 5AN
G0NVA CA5 6NS
G7EKM CA5 6NS
G3KEL CA7 0BZ
G0MVG CA7 9DQ
G7LFB CA7 9JF
G1FBE CA7 9JS
G4XPO CA7 9JS

Alnwick

G8GVN NE65 0EL
G3OGH NE65 0ER
G7OWP NE65 0ER
G1JQQ NE65 0PX
G1XBO NE65 7AE
G8ROO NE65 7BE
G4BCP NE65 7PT
G4WMQ NE65 7QS
M1CCX NE65 7RA
G3SZG NE65 7YG
G3LNM NE65 7YU
G7ANV NE65 8AE
G3VKU NE65 8JT
G7VTB NE65 9BX
G6CBL NE65 9JL
G1RKD NE65 9JN
G7OKB NE65 9JS
G4ZND NE65 9NF
G6YFB NE65 9NG
G7HQB NE66 1AW
G6KXB NE66 1BS
G8JX NE66 1ES
G8PCS NE66 1XR
G3DVF NE66 2QE
G7RKO NE66 2QE
G2FXV NE66 2RS
G3TEP NE66 2SB
G1YAA NE66 2TA
G1AWP NE66 2TW
G3ZUP NE66 2XA
G0DWO NE66 3AQ
G3ZDU NE66 3JE
G1YAE NE66 3JZ
G4EKF NE66 3QN
G7AWC NE66 3TJ

Amber Valley

G4RAR DE21 5BN
G0NPU DE21 5BS
G3GSO DE22 5JS
2E1ECY DE22 5JW
2E1EAA DE22 5JW
2E1DZC DE22 5JY
G4UEE DE4 5BX
G0FVU DE4 5DJ
G1BFK DE4 5DY
G3VLF DE4 5EG
2E1FLO DE4 5ET
G8IQP DE5 0TG
G8ZYC DE5 3AS
G4TLJ DE5 3AZ
2E1FIR DE5 3DA
G4MSG DE5 3EP
2E0AOH DE5 3EX
G4TWW DE5 3GH
G1YPT DE5 3HD
G7EUV DE5 3JX
G0NYM DE5 3LJ
G7SOF DE5 3LX
G4GGL DE5 3PY
G0DGD DE5 3RA
G3NJX DE5 3RE
G3ZYC DE5 3RR
G3XEK DE5 3SS
G3AQX DE5 6BY
G8CNB DE5 8HX
G0NNU DE5 8JG
G7OWZ DE5 8JG
G4DXY DE5 8PQ
2E1AZA DE5 8RF
G8FQV DE5 9QN
G8ZIY DE5 9RB
G0CZS DE5 9RZ
G6PTE DE5 9SH
G3YQL DE5 9SP
G4SKW DE5 9TJ
G8TPK DE5 9TJ
G1IGE DE55 1BL
G0JVX DE55 1BU
G7BGT DE55 1BU
G3MHR DE55 1BW
G4IWW DE55 1DD
G8DXT DE55 1ES
G0GHD DE55 1RU
G4KBI DE55 4AF
G1NWH DE55 4AG
G3RUB DE55 4AG
G7ANO DE55 4AG
G3JWQ DE55 4BP
G3OVZ DE55 4BS
G3OPW DE55 4JF
G4OHV DE55 4LT
G1GHO DE55 4LY
G3MAM DE55 7AH
G4UTN DE55 7DG
G1PWH DE55 7GS
G6YAL DE55 7GS
G4HCQ DE55 7HL
G7SMZ DE55 7HT
G4LSV DE55 7JN
G0VBX DE55 7JY
G7TYP DE55 7JY
M1ATV DE55 7LA
G3OKX DE55 7LB
2E1CCJ DE56 0DH
G7OLE DE56 0EL
G1RYQ DE56 0ER
G0ORC DE56 0HN
G3RKZ DE56 0NE
2E1EIE DE56 0NE
G7BHX DE56 0PF
G7EMZ DE56 0PF
G4ANV DE56 0PG
G0ITL DE56 0PQ
G0RLJ DE56 0PQ
G8IHA DE56 0PY
G4RVL DE56 0QR
G6SKK DE56 0UB
G3IUK DE56 1EA
G3ZYD DE56 1EJ
G6GGD DE56 1FP
G7VEV DE56 1GN
G6ZVR DE56 1NN
G0ROD DE56 1PB
G0FRY DE56 1PD
G3PDD DE56 2AL
G0FDR DE56 2BS
G2DJ DE56 2GG
G3VGW DE56 2GG
G8GTD DE56 2GR
G0UOV DE56 2GR
G0JNK DE56 2GS
G4UWK DE56 2GT
G4UBR DE56 2HP
G0CXD DE56 2LH
G0CXS DE56 2LH
G8ZJM DE56 2SJ
G8SJD DE56 2TZ
G0AEU DE56 2UW
G4DJP DE56 4DP
G3ROD DE56 4DR
2E1DBK DE56 4DR
G7UCB DE56 4FJ
G6GIG DE56 4FS
G4TNZ DE56 4FX
G6TQC DE56 4FX
G4SZT DE56 4GR
G3UBS DE6 4JS
G4LTP DE6 4LP
G7DDR DE6 4NG
G7EUT DE6 4NJ
G3IVF DE6 4NN
G6NLE DE6 4PA
G8EBM DE6 4PF
G4UFX DE7 6AU
G7NPW DE7 6AU
G1GNP DE7 6DD
2E0ADO DE7 6DX
G4OUB DE7 6DX
G3ENZ DE7 6EE
G4NOB DE7 6EG
G7LGY DE75 7AN
G6XTD DE75 7BN
2E1FAT DE75 7BN
G6XMY DE75 7BT
G1WKT DE75 7DB
G1WXO DE75 7DN
2E1CDB DE75 7ER
G7MLT DE75 7FW
G6LYK DE75 7GB
G4DMF DE75 7GQ
G1IQG DE75 7HB
G7JJC DE75 7HF
G3LZC DE75 7HG
G7VUK DE75 7HG
G7EXC DE75 7LY
G8IOW DE75 7LY
G4GBC DE75 7NJ
G4PRF DE75 7PP
G0FEZ DE75 7PQ
G4WAB DE75 7PQ
G4TYY DE75 7PQ
G4ZNI DE75 7PQ
G1GKK DE75 7PZ
G8UCC DE75 7QB
G1SEA DE75 7TL
M1BBF NG16 4AX
G6ZAF NG16 4GP
G1RZR NG16 5NP

Angus

GM4UTK DD10 8EX
GM3KC DD10 8JQ
GM3COQ DD10 8TW
GM4UZY DD10 9BR
GM0ARH DD10 9EJ
GM1XHZ DD10 9RX
GM4YAU DD11 1DR
GM3XFC DD11 1DZ
GM4YWU DD11 2DR
GM4SXJ DD11 2LZ
GM0MIW DD11 2NX
GM1MON DD11 2PD
GM4YWS DD11 4RD
GM3GBZ DD11 4SX
GM0SHD DD11 4SX
GM4JEJ DD11 4TW
GM0RRU DD11 5SP
GM3OBG DD11 5SY
GM3ZXE DD2 5LS
GM4LUD DD2 5PX
GM4FLP DD2 5QN
GM0TAY DD2 5QN
GM3LIW DD2 5RB
GM4API DD3 0PH
GM0NLU DD4 0QU
GM4BNM DD5 3PQ
GM4UGF DD5 3QN
GM4RIW DD5 3QP
GM6BML DD5 4HG
GM7AUX DD5 4HG
GM7CPM DD5 4PG
GM0VOL DD5 4SW
G4FEI DD5 4TH
GM0EOO DD7 6AY
GM6RAK DD7 6HW
GM0EFT DD7 7TB
GM3KEZ DD8 1BR
GM1JWJ DD8 1DQ
GM0DGK DD8 1JR
GM6AAJ DD8 1JY
2M1FGW DD8 2AP
GM1JYV DD8 2BG
GM4XRF DD8 2DE
2M1FGU DD8 2HY
GM2BFV DD8 2JH
GM0OLF DD8 2JN
GM0WNS DD8 2LA
GM6RGU DD8 2PG
GM8RSC DD8 2PQ
GM0TWB DD8 2SP
GM4RWE DD8 2UZ
GM4XKP DD8 3AG
GM0CDC DD8 3AH
GM0DFW DD8 3BH
GM4SEH DD8 3LS
GM0PIY DD8 3PZ
GM1TCN DD8 4DB
2M1CEG DD8 4DE
GM1RPJ DD8 4DJ
GM1KCH DD8 4HR
GM4BVO DD8 4NB
GM8LYO DD8 4PG
GM4TQN DD8 5EE
GM3RKO DD8 5HW
GM0BTK DD8 5JH
GM4HKW DD8 5LB
GM4YHS DD8 5LD
GM4WZY DD9 6BT
GM4PVH DD9 6DW
GM7DAP DD9 6SD
GM0PGG DD9 6SE
GM0JZV DD9 7EH
MM0BDA DD9 7QE
GM1KQK DD9 7QE
GM6RGD DD9 7SZ
GM7IZO DD9 7TE

Argyll & Bute

GM3HSF G84 0ND
GM1FTZ G84 0QR
GM3VNH G84 0QX
MM1ATR G84 0RZ
2M1FKI G84 7JT
GM8VAM G84 7PL
GM6KRD G84 7RF
GM0RTY G84 7TN
GM6JOA G84 8EH
GM7OAF G84 8JP
GM2ACY G84 8PS
GM0IMH G84 9DG
GM4RJX G84 9DN
2M1FKB G84 9DN
GM6KBG G84 9DQ
MM1BXF G84 9DX
GM3OXA G84 9QD
2M1EJK G84 9RX
G4PLM G84 9RX
GM0KCV PA20 9DY
GM0PXR PA21 2DH
GM4EAW PA23 7EW
GM0KBC PA23 7SU
GM4PSW PA23 7SY
GM3LGU PA23 7UB
GM3YLD PA23 8AX
GM1KJF PA23 8JE
GM6HLT PA23 8NA
GM2DWW PA23 8NL
GM0ASY PA23 8QH
GM0COD PA23 8QH
GM4HVR PA23 8SE
GM4HPK PA23 8TB
GM0BUL PA23 9TJ
G7NVG PA24 8AD
GM7BAS PA26 6RH
GM1USN PA27 8DD
GM0GON PA28 6EN
MM0AMW PA28 6PL
GM1PKN PA28 6PL
MM0BED PA28 6QH
GM4HUL PA28 6RZ
GM3LBX PA29 6SX
GM7OSQ PA29 6TF
GM4LNH PA29 6TR
GM3DUS PA29 6XD
GM3BQN PA29 6XZ
GM1PWL PA30 8ER
GM4WMM PA30 8ES
GM4VYQ PA30 8HH
GM4HNK PA31 8QL
GM4VXA PA31 8QL
GM3IJT PA31 8QR
GM3XIJ PA31 8RA
GM3OXU PA31 8UA
GM0DPU PA32 8XU
GM8MLH PA33 1AA
GM4AIE PA34 4EH
MM1BGX PA34 4JN
2M1DHG PA34 4LE
GM1TGS PA34 4NN
GM0EWU PA34 4QT
GM0MNW PA34 4YL
GM4MTI PA34 5AQ
M1ASP PA34 5NA
GM0EWQ PA34 5PG
GM0LRA PA35 1HD
GM3RTJ PA35 1HD
GM6RWW PA35 1HD
GM1RII PA35 1HF
2M1ELU PA35 1HG
GM6UNQ PA35 1HH
GM0EQW PA35 1HY
GM0FHS PA35 1JQ
GM6YPQ PA35 5DU
GM1FPD PA37 1QP
M1AVR PA37 1QQ
GM4AXS PA37 1RA
GM6GFH PA38 4BN
GM3SBE PA48 7TF
GM0PRO PA65 6BG
GM0HCQ PA70 7RE
GM4EHB PA75 6PX
GM0UCB PA75 6QA
GM7NYB PA75 6QN
GM3PGY PA77 6UT
GM1YUH PA78 6TB

Arun

G8PQM BN12 5EG
G0PBV BN12 5NQ
G4KHM BN12 5QA
G6JEY BN12 6PN
G3TNO BN12 6QL
M1BCV BN14 0SF
G4HNX BN16 1AJ
M1BXO BN16 1AJ
G4GRA BN16 1AY
G3OCS BN16 1BL
G1RMW BN16 1DB
G4KYC BN16 1DG
G7KAV BN16 1DT
2E1CAG BN16 1DT
G4GVB BN16 1DU
G8JVE BN16 1DU
G6MIC BN16 1HB
G7IWZ BN16 1HE
G0LOF BN16 1HF
G4TMG BN16 2EF
G0TJX BN16 2EW
G0MHN BN16 2LT
G0SMY BN16 2RE
G3LDO BN16 2TW
G3EHE BN16 2TY
G0VPX BN16 3DY
G6EME BN16 3JD
G8FNH BN16 3LY
G7NVI BN16 3NH
G4UAW BN16 3PW
2E0AFD BN16 3PW
G2FPY BN16 4DA
G4RVE BN16 4HE
G5SD BN17 5DW
G4ALL BN17 5EL
G4JPA BN17 5HP
G4TGZ BN17 5PE
G4IHV BN17 5PF
G4TRP BN17 5PF
G0MOU BN17 5QG
2E1DVB BN17 6BG
G0HKN BN17 6HF
G0BJJ BN17 6HS
G6NQF BN17 6HU
G8KOE BN17 6LU
G6JVT BN17 6NS
G4ZFV BN17 6PA
G4BAQ BN17 6QS
G4RBC BN17 6QX
G6TXY BN17 6QX
G8DOW BN17 6RY
G1MOW BN17 7DF
G3GGN BN17 7JS
G0AGY BN17 7NE
M1BSO BN17 7NE
M1BSP BN17 7NE
G7CCX BN17 7NU
G0CJN BN17 7NY
G1AFK BN17 7PE
G7CCN BN17 7PU
G4JEE BN18 0AP
G3JSW BN18 0AT
G0JZU BN18 0DJ
2E1FNF BN18 0DW
G0RQH BN18 0EG
G7HQM BN18 0ES
G1JLS BN18 0HQ
G2FBU BN18 0HR
G4TSQ BN18 0JE
G4WYL BN18 0JE
2E1AQE BN18 0JE
G0LPF BN18 0JH
G3VZJ BN18 0ND
G7JXB BN18 0PB
G8KFJ BN18 0PJ
G2ARU BN18 0QR
G4ITQ BN18 0SH
G0VCC BN18 0TB
G6FKY BN18 0TD
G4NHU BN18 9AD
G0MFQ PO18 0LG
G0AFN PO20 6QT
2E1GBS PO20 6YR
G8HSH PO21 1NZ
G7TPH PO21 2AH
G0MGS PO21 2DX
G3NCA PO21 2DX
G6NUX PO21 2EN
G0VCJ PO21 2HB
G7PVJ PO21 2NU
G4BDN PO21 2NY
G0IAF PO21 2RB
G4XPT PO21 2RB
G0USE PO21 2UL
G7GJY PO21 2XE
G3SYG PO21 3AH
G4IOV PO21 3BS
G3GUR PO21 3DH
G7SFG PO21 3DL
G8OCM PO21 3EL
G8OCN PO21 3EL
G0CIR PO21 3EQ
G7IMT PO21 3HQ
G3RJS PO21 3LG
G8HVO PO21 3LR
G0KXS PO21 3LT
G7BRV PO21 3LW
G6KUA PO21 3QN
G4RPA PO21 3SL
2E1FFS PO21 4BZ
G3LTM PO21 4ET
G3ORW PO21 4JY
G4XUG PO21 4LB
G7HEV PO21 4LJ
G7EPR PO21 4LW
G8TGH PO21 4NB
G6HTB PO21 4NN
G8REF PO21 4PS
G1ORB PO21 4ST
G4ITY PO21 4TJ
G1XIV PO21 4TN
G6FDU PO21 4UR
G3IJS PO21 4XN
G3SFE PO21 5AN
G2DSP PO21 5LB
G0KJU PO21 5LL
2E1CNG PO21 5LL
2E1FIE PO21 5LL
G0CHI PO21 5LT
G0OSU PO21 5TD
G8ZTD PO21 5TW
G6AII PO22 0AD
G1KRD PO22 0AE
G0SSE PO22 0AE
G0AYV PO22 0BA
2E1EHT PO22 0BA
G4XSD PO22 0EL
G8YMQ PO22 0JD
G0MPL PO22 0JF
G1ITL PO22 0JN
G7BWW PO22 0LH
G1EHA PO22 6DB
G6UTT PO22 6ED
G0NMG PO22 6JL
G4XSB PO22 6JZ
G3DEM PO22 6LX
G0ONW PO22 6QH
G7UEM PO22 6RJ
G0VNT PO22 7EG
G4KLS PO22 7HU
G3AIV PO22 7JS
G4PDY PO22 7NW
G0GYX PO22 7QE
G0DFA PO22 7SL
G3HFX PO22 7SN
G4ZJP PO22 8BU
G8JRW PO22 8HT
G1IOF PO22 8JA
G3YLL PO22 8JJ
G2CIL PO22 8QT
G0IFY PO22 9AY
G0SZK PO22 9EE
G7SQC PO22 9LY

Ashfield

G4XBQ DE55 4PB
G4KJA NG15 6DQ
G4XXS NG15 6DQ
G4WXR NG15 6EP
G6DDP NG15 6FF
G4ITX NG15 6FH
G8YAV NG15 6FN
G7KGN NG15 6FQ
G1HFR NG15 6GA
G3OBJ NG15 6GD
G6PCN NG15 6GG
G4ZII NG15 6GN
G4XDN NG15 6JY
G4EPL NG15 6NE
G6NJF NG15 6NS
G0BXX NG15 6NY
G4SKY NG15 6NY
G4PQG NG15 6SP
G0OBK NG15 6TX
M1ADU NG15 7AG
G0DLQ NG15 7AH
G4JSM NG15 7QA
G5RR NG15 7QA
G3JFW NG15 7QE
G6NHY NG15 7SR
G6ORH NG15 7TF
G3KTX NG15 7TW
G4ODO NG15 7TW
G0RGE NG15 7UB
G0VNW NG15 8BG
G1WJC NG15 8BW
G6UIP NG15 8BZ
G0CEV NG15 8EB
2E1CJB NG15 8EU
G6DJQ NG16 5EH
G0IZL NG16 5FY
G0HMI NG16 5GE
2E1EVH NG16 5HJ
G0WXH NG16 5JH
G7UVL NG16 5JH
G1SGO NG16 5LB
G7JLX NG16 5LQ
G2DWZ NG16 5LZ
G8JGF NG16 6BQ
G0EZW NG16 6BU
G4ROB NG16 6DX
G0CSS NG16 6FN
G4NXB NG16 6GP
G7MMR NG16 6QZ
G4ZUC NG17 1EP
G7PHL NG17 2BQ
G0BTV NG17 2EH
G4DFZ NG17 2FB
G0JDX NG17 2FQ
G3JFD NG17 2HS
G0JDK NG17 2JE
G6HXX NG17 2JP
2E1EOE NG17 2NX
2E1CEX NG17 2NZ
G1XVF NG17 2QD
G0JDG NG17 2QF
G4TSN NG17 2RA
G0UJD NG17 3AB
G0UTN NG17 3AG
M0AJX NG17 3BW
G0DOO NG17 3DL
2E0AGA NG17 3DL
G0IUN NG17 3DP
G0FUD NG17 3FR
G0BCF NG17 3FT
G4HCD NG17 4BE
M0AGA NG17 4BW
G7RXK NG17 4BX
G7JJF NG17 4DR
G0EYN NG17 4EW
G7PBV NG17 4EZ
G0CVA NG17 4HP
G8XXR NG17 4LP
G6NWN NG17 4NL
G7INC NG17 5BD
G3VDF NG17 5GH
G1SIU NG17 5HP
G0GZO NG17 5HU
2E1DZD NG17 5JQ
G0DLJ NG17 5JS
G1VHY NG17 7EH
G1RJD NG17 7FH
G0HMY NG17 7FQ
G6OZH NG17 7HF
G3SQQ NG17 7LY
G1LPQ NG17 7NA
G6MQU NG17 7PA
G4LTU NG17 8BD
G0JPZ NG17 8BT
G0MYZ NG17 8BT
G6SCZ NG17 8DZ
G7LHT NG17 8EJ
G8ZUZ NG17 8GY
G4VOG NG17 8JJ
G0DMN NG17 8JT
G0NRA NG17 8LH
G0NZA NG17 8LH
G1GQC NG17 8LH
G3GQC NG17 8LH
2E1CQM NG17 8LJ
G1EZU NG17 8NQ
G7GME NG17 8NY
G6XNN NG17 9BG
G6XYR NG17 9BN
M0BAE NG17 9BU
G0CVB NG17 9DY
G1GCV NG17 9ED
G0RLK NG17 9FQ
G0EDL NG17 9HD

Ashford

G1UBW CT4 8AS
G3NRU CT4 8BJ
G4BIA TN23 1JP
G7NUG TN23 1LN
G8GZW TN23 3AY
G1EJA TN23 3DY
G6EXU TN23 3EG
G6ZAA TN23 3NJ
G7HKT TN23 5BP
M1BXP TN23 5DB
G0PEK TN23 5DT
G0TJI TN23 5SA
G7TTF TN23 6LS
G0PDP TN23 7UH
G4UKO TN24 0HU
G4HLP TN24 0JD
G0PHO TN24 0JH
G0DLN TN24 0JZ
G4KVE TN24 0RF
G0VGX TN24 0RR
G7OJC TN24 0UZ
G7BYV TN24 8UW
G3ZPI TN24 9AE
G4UUI TN24 9AF
G0KGE TN24 9BD
G7LJL TN24 9BU
G1NNC TN24 9DQ
G0PEM TN24 9DR
G7HJJ TN24 9EU
G8YNF TN24 9EU
G3MMM TN24 9HB
2E1FWM TN24 9JW
G6RAD TN24 9LH
M0AVY TN24 9LU
G7WBQ TN24 9LU
G3ZAJ TN25 4DF

G8MFV TN25 4DW
G0KSG TN25 4HE
G6SRE TN25 4PQ
G6TMK TN25 5BD
G0LZU TN25 5HN
G4FPG TN25 5JB
G3TIS TN25 5LZ
G0JIR TN25 6NE
G7IZY TN25 6NS
G8WLB TN25 6RA
G7EVC TN25 6RW
M1BJN TN25 6RY
G3YBE TN25 6UA
2E1BAE TN25 7EY
G4GLG TN25 7JX
G3FIR TN26 1HW
G3MMN TN26 1HW
G4IDX TN26 1HW
G4RGN TN26 1JA
G0HVP TN26 1JD
G6WID TN26 1JP
G0GOE TN26 1LS
G0MTJ TN26 1LS
G1FFB TN26 2EG
G0CRL TN26 2HL
G3UHY TN26 3DA
G4KUA TN26 3DS
G7FHB TN26 3HA
G0VAI TN26 3HX
G3PSY TN26 3JB
G6HXZ TN26 3LY
G6VLP TN26 3NB
G0CHG TN26 3QS
G3KAP TN27 0LS
G1ZIU TN27 8DW
G7FKZ TN27 8EA
M1ATA TN27 8EA
G1AOQ TN27 8ER
G3UNW TN27 9ER
G0VEC TN30 6RY
G7AYX TN30 6UA
G8VGU TN30 7BA
G8BIS TN30 7EH
G1VZT TN30 7EL

Aylesbury Vale

G3OCP HP17 8AN
G4CYR HP17 8DT
G1MZG HP17 8EU
G3WUN HP17 8EY
G1EOU HP17 8EZ
G3NPL HP17 8HD
G1MGH HP17 8HF
G0DBK HP17 8HG
G1TAL HP17 8HG
G7FEK HP17 8HG
G7NVD HP17 8HG
G0IKP HP17 8HN
G1TRL HP17 8JG
G7AAW HP17 8JS
G3OZF HP17 8RG
G3YSR HP17 8RG
G4IGK HP17 8SH
G6JRM HP18 0BL
G8BCL HP18 0BL
G3GUD HP18 0DD
G6IFE HP18 0DN
G3SGK HP18 0HA
G0MMI HP18 0LH
G0KMC HP18 0LY
G7DXE HP18 0PH
G7BYF HP18 0PS
G3LAC HP18 9AL
G0THH HP18 9BT
G3MAZ HP18 9DG
G6DAP HP18 9HB
G0RAS HP18 9HH
G1SPE HP19 3HU
G0TGM HP19 3LN
G4SPK HP19 3NN
G3ZHX HP19 3NU
G1NMP HP19 3QB
G4CKW HP19 3SB
G4BFV HP19 3WT
G7IMY HP20 1DN
G3KLT HP20 1JA
G4VRS HP20 1JA
G0DFC HP20 1JH
G8ALQ HP20 1JW
G8INN HP20 1RJ
G0BLQ HP20 1XG
G7IMB HP20 2BJ
G1EDD HP20 2HE
G4FXI HP21 7BD
G1FNA HP21 7EP
G1ZBK HP21 7HH
G7FWL HP21 7HH
G8VVZ HP21 7NS
G8MBM HP21 8BN
G0GPW HP21 8BY
G1ATY HP21 8EB
G3YJU HP21 8HZ
G1MZW HP21 8JG
G4DRA HP21 8LL
G0RVJ HP21 8PL
G7VFV HP21 8QG
G0MHZ HP21 8QJ
G1ZJW HP21 8RB
G1FFL HP21 9BS
G1MMR HP21 9BS
G3YAW HP21 9BX
G8AYM HP21 9DS
G7FDL HP21 9HL
G7ITO HP21 9RX
G1JXZ HP21 9UB
G3IGH HP21 9UF
G7ILE HP21 9UW
G7OGJ HP21 9XB
M0AZQ HP21 9YQ
G0TYD HP22 4AY
G4GYO HP22 4AZ
G0GGN HP22 4BX
G8BQH HP22 4EF
G0WZH HP22 4PG
2E1FOC HP22 5EF
G1CDQ HP22 5GB
G6GDI HP22 5HH
G7AAR HP22 5HH
G8HYP HP22 5JS
G8AWY HP22 5LY
G8TJI HP22 5ND
G3SKR HP22 5NQ
G4RQU HP22 5SN
M1BFS HP22 5TX
G1DDR HP22 6AR
G4PFR HP22 6AW
G0BKP HP22 6HS
G8HWJ HP22 6JN
G4UKI HP22 6LG
G4PLS LU6 2EZ
G4CVS LU6 2HL
G4XLG LU6 2HR
G3JZW LU6 2JN
G3XTQ LU7 0EX
G4BPC LU7 0HR
G8GIK LU7 0HZ
G8ZVI LU7 0HZ
G8GZX LU7 0LJ
G3XOP LU7 0LR
G1HMW LU7 0NT
G0MYE LU7 0PD
G6JFN LU7 0QE
G6FZW LU7 0SQ
G6NAG LU7 0SY
G1HZA LU7 0TF
G3RXQ LU7 9DN
G4LRG LU7 9EE
G8VDA MK17 0AH
G0EFN MK17 0BH
G0BMD MK17 0BT
G3ZLX MK17 0BX
G1FXT MK17 0DQ
M0ANS MK17 0EP
G3DBM MK17 0JA
M0ALR MK17 0QG
G7VHA MK17 0QG
G4NUG MK17 0QR
G8FAK MK17 0SB
G7MFZ MK17 0SB
G4UQR MK17 9AJ
G3YLC MK18 1JJ
G3NUB MK18 1PG
G3NPI MK18 1PL
G3NCG MK18 1SN
G8ZUI MK18 1UF
G7PMW MK18 2LZ
G0WBH MK18 2NU
G1UEO MK18 2PF
G0RXH MK18 2QE
G2BSJ MK18 2QR
G1MYO MK18 2QS
G4FYO MK18 3DY
G1GJR MK18 4BA
G8RDB MK18 4BX
G3SHZ MK18 4EL
G0ICC MK18 4LP
G4YTL MK18 5BJ
G7ECO MK18 5JF
G1CIK MK18 7DE
G6TSJ MK18 7HR
G4TGP NN13 5JH
G4BII OX6 0AZ
G8LPN OX6 0BB
G1FXC OX6 0EY
G4JVY OX6 0HA
G4IHX OX6 0HJ
G6PHH OX6 0HP

Babergh

G1NOM CO10 0JH
G0HUG CO10 0JN
G4GGC CO10 0JN
G0SWI CO10 0JN
G1NON CO10 0LA
G6TSQ CO10 0LB
G7NKM CO10 0LD
G1DEU CO10 0NQ
2E1FEZ CO10 0PN
G8AAR CO10 0QP
G6RKG CO10 0RN
G1MYD CO10 0RQ
G6TCQ CO10 0RQ
2E1EGN CO10 0RU
G0PAO CO10 0XY
G8LTY CO10 0YT
G4JAA CO10 0YU
G1YJL CO10 5JA
G8VAF CO10 5JX
G1PMF CO10 6JB
G6DKE CO10 6JB
G6WYQ CO10 6LY
G7HMF CO10 6PJ
G7UTC CO10 6PP
G7MLK CO10 6QX
G1IUW CO10 6TP
G1BWO CO10 7NY
G1FBO CO10 7PH
G4FBQ CO10 7PQ
G3YAI CO10 7RL
G3WRD CO10 7RW
G0PGS CO10 7SJ
G1TWY CO10 9QD
2E1EWI CO10 9QD
G0SNU CO10 9RU
G3PRK CO10 9TS
G1SAQ CO11 1PL
G4XDK CO11 1RN
G0DVJ CO11 1TP
G7SNO CO11 1TP
G0OLM CO6 4LL
G4ZHR CO6 4LZ
2E1BHH CO6 4NY
G7DTT CO6 4PQ
G7BVN CO6 4QF
G4ZVR CO6 4UR
M1ANA CO6 5AD
2E1DXW CO6 5AD
G0IBZ CO6 5AR
G7AIK CO6 5AR
G3LST CO7 6JR
G0WJO CO7 6SE
G8CJD CO7 6SJ
G3KQW CO8 5HZ
G7SMF IP2 0DL
G4FSK IP2 9XQ
G3VTR IP29 4PL
G3ETZ IP30 0HE
G0MGN IP7 5BS
G4XOJ IP7 5EU
G6EYY IP7 5JL
G6XEV IP7 5PQ
G7JVE IP7 5SQ
G1INH IP7 6NH
2E1EGY IP7 6NX
G0NQK IP7 6PN
2E1EAX IP7 6PN
G8DUI IP7 6QH
G0FJB IP7 7AL
G1XVD IP7 7BN
G7NGF IP7 7JH
G0TJY IP8 3AP
G3XAO IP8 3BS
G0RRC IP8 3DN
G3IVH IP8 3HZ
G3KYS IP8 3HZ
G4IJJ IP8 3LF
G4KYU IP8 3NH
G4TVP IP8 3SG
G8PPD IP9 1DX
G0PJO IP9 1HS
G7VPN IP9 1HX
M0AMR IP9 1NY
G0AOR IP9 1PZ
G0DTC IP9 1QH
G4BXZ IP9 1QH
G3GOT IP9 1RW
G3XSA IP9 2BW
G8XRW IP9 2HW
G3WRT IP9 2JB
G0ATN IP9 2JE
G7UOE IP9 2JN
G4DKX IP9 2NZ
2E1CYE IP9 2RJ
M1AOF IP9 2RT
M0AUR IP9 2SQ
G7UAJ IP9 2SQ
2E1FJW IP9 2SQ
G3HMB IP9 2SW
G8BJV IP9 2XD

Barking and Dagenham

G0POY IG11 0NL
G7UXM IG11 7RJ
2E1BGX IG11 8NA
G1NIV IG11 8NW
G0RNU IG11 8QU
G0BIW IG11 9EG
G8ZKO IG11 9JB
G7PRH IG11 9NX
G0OVE IG11 9XB
G6JMJ IG11 9XB
G0RFL RM10 7BT
G7RIA RM10 7QP
G0OVB RM10 8AN
G6OFA RM10 8BG
G0GGH RM10 8BS
G1JQR RM10 8BS
G1YKI RM10 8BT
G0WXE RM10 8DE
G1WUH RM10 8ES
G3UKX RM6 5EX
G7VVK RM6 5EX
G4IEI RM6 5TJ
G0HRZ RM6 6AX
G3SCE RM6 6EB
G0PIS RM6 6RX
G6JHD RM6 6SH
G4SKS RM6 6UB
2E1GAD RM6 6UH
G6ZEL RM6 6UU
G7MFY RM7 0RT
G3SFA RM7 0YT
G7TAX RM8 1DS
G0SZZ RM8 1JS
G0TJE RM8 1JY
G7LIR RM8 1JY
G6XDG RM8 1PP
G1PVS RM8 2AX
G8SLB RM8 2LR
G1DPH RM8 2NA
G4SHN RM8 2TB
G6AIU RM8 3BP
G1PQK RM8 3EJ
G4HOI RM9 4ST
G7SGI RM9 5JL
G0NXL RM9 5LP
G4PQU RM9 5PT
M0BDQ RM9 5PX
2E1CZR RM9 5XH
G7UVW RM9 5XH
G1XZX RM9 6JU
G0TOD RM9 6LD

Barnet

2E1DLS EN4 8EA
G0CEQ EN4 8ER
G4FJW EN4 8HG
G7MHV EN4 8JE
G8DVJ EN4 8PU
G4LXY EN4 8QJ
G6MFT EN4 8RD
G8RPA EN4 8TU
G7MGM EN4 9AS
G0LCS EN4 9DS
G1ELK EN4 9HE
G3GMY EN4 9QT
G1MPU EN4 9RL
G8UDI EN5 2BT
G1ALA EN5 2BW
G0SNO EN5 2JL
G8AIE EN5 2LE
G4BWY EN5 2LE
G4GPR EN5 2LS
G4JAH EN5 2TB
G0TOY EN5 3BB
G4WJT EN5 4AJ
G3HJF EN5 4AL
G3AJD EN5 4EB
G7IJC EN5 4EF
G3WEA EN5 4EY
G0WEU EN5 4QN
G0NAV EN5 5AY
G1HQN EN5 5DP
G7DRF EN5 5HN
G4YSJ EN5 5LU
2E1DXK EN5 5NY
2E1DZI EN5 5NY
G3SUL EN5 5QW
G8TAU EN5 5RH
G7GCB EN5 5SJ
G1BQC HA8 0EP
G1EBU HA8 8HH
G4XEW HA8 8JU
G4GLM HA8 8PS
G4RFI HA8 8RH
G6JLU HA8 8SD
G0STR HA8 8SH
G4SCG HA8 8UA
G1LRJ HA8 9AR
G1GXC HA8 9HE
G6GEG HA8 9QA
G0MOX HA8 9TX
G4WIS HA8 9TX
G1KJH HA8 9UB
G3ASX N10 1LX
G4KPH N10 2HS
G4ZOR N11 1JJ
G3DZW N11 1LA
G3APX N11 3BY
G0AQI N11 3HH
G3GGI N11 3NQ
G8XRP N12 0LS
G8SIK N12 0NY
G3OMR N12 7AH
G4CCT N12 7JG
G6IZK N12 7LT
G7UOU N12 8AR
G8SBA N12 8DZ
G1KGV N12 8EN
G4CCM N12 8EN
G4CTI N12 8EN
G8AKL N12 8EN
G1YSA N12 9AU
G8ORY N12 9PP
G3KPM N14 5LP
G4FQS N14 5LP
G7MNE N14 5RL
G6BLC N14 7NJ
G4EYV N2 0DJ
G0UKA N2 0HP
G6YNT N2 0LP
G7TWC N2 0PX
G4DRO N2 0UR
G8DR N2 8JG
G1IHA N2 9AR
G6TQL N2 9BU
G4IFH N2 9DY
G4XYK N20 0AL
G1AFT N20 0EZ
G4LCB N20 0HT
G0EIQ N20 0QN
G0TVZ N20 8DG
G1FLX N20 8QH
G4BOK N20 9AE
G0RLS N3 1EN
2E1DNM N3 1ND
G4ZLB N3 1UN
G8VMP N3 2SG
G3ZKX N3 3AL
G3OSS N3 3PG
G4SIN NW11 0JX
G3ASV NW11 6DL
G3MCD NW11 6DS
G4IOY NW11 6ET
G8VLN NW11 6LT
G4DBQ NW11 6NT
G8XTF NW11 6SS
G4PLX NW11 6UP
G3GAF NW11 7DL
G2DCU NW11 7QG
G3RDG NW11 8SG
G1INJ NW11 8TE
G6IWZ NW2 1DY
G3JDP NW2 1NL
G1JPD NW2 1QS
G6SNX NW2 2JU
G7WFH NW2 2QP
G4GHZ NW4 1JR
G1GIA NW4 1NP
M0ALQ NW4 1QN
G0VSX NW4 2BX
G3UML NW4 2PN
G3XVH NW4 3NB
G6GML NW4 3TU
G6CKD NW4 4EN
G8JGE NW4 4EN
G3XKC NW4 4JU
G0UQL NW4 4SG
G0ADE NW4 4XB
G3MYG NW7 1JL
G1XSI NW7 1RN
G0JUE NW7 1RS
G1YJH NW7 2AD
G3ZOZ NW7 2LP
G8TRU NW7 2NB
G6XGG NW7 3AJ
G3RRP NW7 3PD
G6HIU NW7 3QB
2E1FST NW7 3RA
2E1FTW NW7 3RA
G8WCX NW7 4HB
G1END NW7 4PY
G1NRM NW7 4QP
G6MUS NW9 5AP
G1IAG NW9 5AZ
G3ERR NW9 5LA
G4IXL NW9 5TN
2E0APD NW9 5TU
G0FAB NW9 6EU

Barnsley

M0AGN S36 4HF
G1YUV S63 0DF
G4FZJ S63 0ED
G4NDP S63 0EF
G3WLV S63 0RZ
G4VNX S63 0TN
G0UEE S63 8AP
G6IIK S63 8BZ
G1EQM S63 8HR
G0UAD S63 8JL
G0GPR S63 9BA
G6WSZ S63 9BY
G4MPV S63 9JA
G4ZTX S63 9PZ
G4MLQ S70 1NT
G3DOI S70 1QB
G4YWJ S70 1QE
G3TIP S70 1TT
G1ANF S70 3AA
G6ZXO S70 3AA
G3TEH S70 3EW
G1WTN S70 3ND
G1YWT S70 3NX
G4DYG S70 3RH
G6ZEW S70 4AE
G3EAE S70 4DW
G3ZQV S70 4HH
G4URI S70 4RN
G4JKW S70 4SB
G4NEW S70 5BH
G4IAY S70 5NR
G8VDP S70 5QY
G7MRZ S70 5QZ
G3JZJ S70 5RJ
G0JEA S70 5RL
G8EGM S70 5SN
G1YZO S70 5UG
G1YZQ S70 5UG
G0OHR S70 5UZ
G6XQT S70 6HR
G4TMZ S70 6JY
G6YOR S70 6JY
G0DUM S70 6LG
G6AUP S70 6PQ
G8BRK S71 1JR
G7WMB S71 1PW
G7TFJ S71 1TA
G0NSC S71 1XB
G1XUH S71 1XB
G7BZE S71 1XQ
G8HTQ S71 2DX
G8ISE S71 2ES
G6DER S71 2EY
G4DSA S71 2HP
G7VYN S71 2LA
G0FKV S71 2NU
G0VMO S71 3AB
G0SZE S71 3BG
2E0APM S71 3BG
G4SQW S71 3EF
G0HZS S71 3PU
G0ANE S71 3SJ
G8VXV S71 3TT
G1ANI S71 4AA
G1BWU S71 4AA
G3OXR S71 4HF
G1BQR S71 4NH
G7NSA S71 4NQ
G0OWB S71 4RY
G7GQD S71 5AJ
G4UFR S71 5BS
G6LWK S71 5BS
G4IHZ S71 5DH
G2AFV S71 5DR
G0JUK S71 5QX
G3AMH S71 5RB
G7NCB S72 0AF
G0PQG S72 0AU
G1JDO S72 0AU
G6CND S72 0AU
G4TZE S72 0AX
M1AQU S72 0BU
G4NAP S72 8BL
G1EFF S72 8DE
G6NXW S72 8DY
G8PLJ S72 8HG
G4LUE S72 8RN
G6AJ S72 8RN
G8LSG S72 8RZ
G0RUX S72 8XG
G1HPG S72 9LL
G0PAZ S72 9LR
M1BOX S73 0AD
G7IDJ S73 0EF
G4JHQ S73 0HJ
G0CUI S73 0LL
G0JEO S73 0RN
G1VGO S73 0YG
G4NSO S73 8AQ
2E0AJO S73 8AS
G8RHL S73 8HQ
G8XFU S73 8HQ
G4YWO S73 8HS
G1VOC S73 8LP
G1OPS S73 8LS
G3YA S73 8TF
G0UQF S73 9AE
G4DUU S73 9DP
G0TKJ S73 9ED
G1YTU S73 9EP
G7JPJ S73 9JA
G4VRT S73 9JL
G8XUH S73 9JL
G8OXA S73 9PL
2E1FNT S74 0BS
G4MZT S74 0HU
G4PIR S74 0NY
G6UQO S74 0QE
G1PEY S74 0QR
G0VGU S74 8DA
2E1FCG S74 9AB
G0PBF S74 9DY
G4YTM S74 9EQ
G1SRN S74 9HE
M1BOW S74 9HU
G4YAP S74 9PW
2E1AQX S75 1DN
G4CUD S75 1HW
G3FLQ S75 1PD
G1BAL S75 1PU
G6YJD S75 1PX
G7SCK S75 1PZ
M0AEE S75 1RA
G0UUC S75 2AY
G1YUU S75 2DT
G7OJM S75 2JW
G7ORU S75 2JW
G4JJ S75 2QH
G8MQX S75 2QJ
G1JAZ S75 3QJ
M1AKG S75 3SE
G5KM S75 4AA
G4DCF S75 4QQ
G4JUR S75 5BE
G7JRC S75 5DF
M1BUH S75 5EL
G7TGE S75 5JQ
G7TGF S75 5JQ
G4KUD S75 5PF
G1NAQ S75 5PW
G6HRF S75 5QA
G0UOS S75 6AY
G4MWF S75 6BJ
G1THO S75 6DY
G0OCV S75 6EP
G0SBO S75 6HH
G1LES S75 6JA
G8GDH S75 6LE
G0TOQ S75 6LY
G3TPX S75 6NS

Barrow-in-Furness

G1BFZ LA12 0LT
G4AGB LA13 0AN
M1BRF LA13 0AN
G4ARF LA13 0AN
G0SKJ LA13 0AU
G4AFR LA13 0BX
G3ZFZ LA13 0EE
G4DAD LA13 0EX
G4ZEG LA13 0HU
G3FOD LA13 0LJ
G6LMW LA13 0RE
G0HIK LA13 0RX
G3VUS LA13 9AR
G6KES LA13 9JA
G3ODL LA13 9LB
G8DGU LA13 9NY
G4ZOI LA13 9PG
G7UKR LA13 9QG
G4WKP LA13 9QH
G1GBF LA13 9RJ
G0MOM LA14 1BQ
G6YXX LA14 1EZ
G0VVH LA14 1NW
M1BHU LA14 2BU
G4RQJ LA14 3AN
G1DNI LA14 3DE
G4NFR LA14 3DE
G0GSJ LA14 3EJ
G6ADR LA14 3JN
G1AGM LA14 3LP
G1CGP LA14 3LP
G6UMN LA14 3QS
G0GJM LA14 4AH
G3IZD LA14 4HE
G0SPG LA14 4HS
G1KRR LA14 4LQ
M0ABU LA14 4ND
G4USW LA14 4NH
G0UXH LA14 4PA
G4XGP LA14 4PF
G4BZR LA14 4PH
G7OXB LA14 4PZ
G7UNS LA14 5ES
G3WTL LA14 5HE
G0PLL LA14 5NB
G4ABU LA14 5NB
G3RFE LA14 5NN
G0LSU LA14 5QQ
M1AJK LA14 5RH
G7UAK LA14 5TX
G7RNX LA15 8BW
M1AVV LA15 8BW
G3APV LA15 8DD
M1BWA LA15 8HR
G4VKE LA15 8LZ
G4DKZ LA15 8NL
G6KDJ LA15 8NP
G7EGQ LA15 8NR
G4SPW LA15 8RZ
G7RMD LA15 8RZ
G3MVD LA15 8SD
G7MCE LA15 8SE
G4BOC LA16 7BY

Basildon

G4EVZ CM11 1AU
G8GZV CM11 1BJ
G0PZG CM11 1EL
G0ORT CM11 1ET
G4KRH CM11 2ER
G8LUP CM11 2LB
G4OJN CM11 2LB
G8GES CM11 2NZ
G8URI CM11 2PD
G3UTC CM11 2QN
G4SIS CM11 2QN
G8HGN CM11 2RQ
G6KVY CM11 2RY
G3MVV CM11 2XA
G4KF CM11 2XA
G4XTS CM11 2XE
G3OCI CM11 2XU
G0ABR CM12 0EU
G7MUG CM12 0EU
G3TTB CM12 0HH
G4DKB CM12 0HL
G7PMI CM12 0JF
G4LPS CM12 0NX
G3AJS CM12 0RB
G4BCV CM12 0RB
G7ORE CM12 0RB
G1UKA CM12 0SU
G8CRX CM12 9EL
G8WXU CM12 9JL
G1DSG CM12 9JN
G8DQN CM12 9LX
G6MON CM12 9XG
G7RRN CM13 3SJ
G6IFH SS11 8AY
G1LYV SS11 8EA
G6DVQ SS11 8ET
G3YPN SS11 8EX
G1OBM SS11 8LL
G8JVD SS11 8QB
G4OAD SS11 8QJ
G0GEC SS11 8QT
G4GUJ SS11 8QT
G8OIV SS11 8QT
G7UVC SS11 8RA
G6SJX SS11 8RF
G6CNQ SS11 8XN
G7JXK SS11 8YA
G4EEZ SS11 8YE
G3IOI SS12 0AR
2E1FZO SS12 0AX
G0HXP SS12 0BQ
G3VQY SS12 0DL
G3YTF SS12 0EJ
G7MBY SS12 0EN
G3LRL SS12 0ER
G3PZZ SS12 0HE
G7AUL SS12 0HG
G4SEV SS12 0HR
G7HGI SS12 0HS
G6OLM SS12 0JX
G4DCB SS12 0LP
G1YHX SS12 0NW
G4PST SS12 9AL
G4YIH SS12 9EQ
G7HCO SS12 9JX
G6ANO SS12 9LL
G0UBL SS12 9RD
G7EID SS13 1HA
G7EIE SS13 1HA
G7SCC SS13 1HA
G6FCL SS13 1JA
G7OTG SS13 1JA
G6AJT SS13 1JU
G8WSS SS13 1NR
G8WUD SS13 1PW
G0SIV SS13 1QU
G6FMW SS13 1RJ
G4XZC SS13 2HU
G0TJW SS13 2HX
G0PKF SS13 3AD
G7JRK SS13 3EW
G8RXG SS13 3JJ
G1KRI SS13 3NN
G1GYQ SS14 1NF
G1YKZ SS14 1PQ
G6TNE SS14 1QU
G7PFG SS14 1TG
G1PXQ SS14 1UA
G7MPH SS14 2AJ
G6HSS SS14 2AZ
G8CUN SS14 2JD
G7KDR SS14 2JQ
G4TER SS14 2NN
G1YKX SS14 2NS
2E1EOZ SS14 2PA
M0ABA SS14 2TN
G8VYK SS14 3AP
G0NPO SS14 3JN
G0KVZ SS14 3LP
G6SNN SS14 3NA
G7OED SS14 3QS
G7KKE SS15 4AH
G8WPE SS15 5DL
G4DES SS15 5HB
G4ZUL SS15 5SJ
M1AYY SS15 5UG
G6LZX SS15 5UY
G7UZX SS15 5UY
G1DMC SS15 5XA
G4HKQ SS15 5XA
G0TRJ SS15 5XJ
G0RTH SS15 5YN
G3UPM SS15 6AL
G0DQF SS15 6HL
G6XXO SS15 6PE
G7WCM SS16 4DS
G1CHV SS16 4NE
G3VYF SS16 4PQ
M1BHC SS16 4PQ
G7ATZ SS16 4SS
G3IFN SS16 4SY
G4EPU SS16 4TL
2E0ACJ SS16 4TT
G4IFD SS16 4TT
G3SCT SS16 4TT
G8PKP SS16 4TT
M1BWK SS16 5AY
G0VDM SS16 5BN
G3WXF SS16 5EN
M1BDR SS16 5HD
G8KIH SS16 5HD
G8OSG SS16 5HL
G3ASH SS16 5HW
G3YCW SS16 5HW
G0ODF SS16 5QF
G1KOT SS16 5QG
G7ALR SS16 5TW
G1MPG SS16 5UR
G4NVT SS16 6AQ
G7PSW SS16 6ED
G6ILD SS16 6LN
G6RAQ SS16 6NA
G4JSC SS16 6RB
G0PAE SS16 6SJ
G8JCV SS16 6SJ

Basingstoke and Deane

G1VXX RG19 8AE
G8XTC RG19 8LA
G8OID RG20 0AT
G0IBB RG20 4TZ
GM4HDE RG20 5ER
G4WPR RG20 5EZ
G3HBI RG20 5QA
G8XBE RG20 5QT
G7GYN RG20 5RH
G0TWJ RG20 5RS
M0AYY RG20 5RY
G8HJF RG20 5SL
G8HJG RG20 5SL
G4PAC RG20 5TD
G4UEF RG20 5TL
G4JAL RG20 9BN
G6GXZ RG20 9BN
G0UMH RG20 9BN
G7VFE RG20 9EY
2E1ESE RG20 9EY
G4DOQ RG20 9TS
2E1ETJ RG20 9TS
G3VLI RG21 3AQ
G8DCA RG21 3HT
G7WCN RG21 3JH
G3VEH RG21 3JU
G8AOO RG21 3JW
G8FMH RG21 3NR
G8JYN RG21 3NR
2E1FSF RG21 4DA
2E1FSG RG21 4DA
G3XNI RG21 4DJ
G7WFU RG21 4DU
G6BBW RG21 5HL
G3TEK RG21 5NR
G7SNB RG21 5NR
G7RPP RG21 5PA
G0GJO RG21 5RS
G8PIY RG21 5SR
G7APY RG21 5UE
G3IDG RG21 7RW
G7SKX RG21 7TG
G7MEQ RG21 8UP
G1YOD RG21 8XS
G3OMU RG21 8YN
2E1FCH RG21 8YY
G3MGZ RG22 4BG
G0EWO RG22 4BQ
G0NWH RG22 4BQ
G4TRI RG22 4DL
G8EOJ RG22 4DL
G1EUJ RG22 4DS
G7MAT RG22 4EA
G8AOK RG22 4EL
2E1CSI RG22 4EW
2E1EHD RG22 4EZ
G4SWF RG22 4HL
G4JIQ RG22 4HN
G8JLB RG22 4HR
M0ACD RG22 4JT
G7UWR RG22 4JT
G6JOR RG22 4NB
G6UZO RG22 4NP
G8NJE RG22 4NZ
G4SZO RG22 4QH
G7LDJ RG22 4SB
G4RHJ RG22 4TA
G8AGJ RG22 4TS
G7PKD RG22 4TY
G1WKK RG22 4UR
G4PND RG22 4UX
G3TUN RG22 4UY
G6MDS RG22 4XD
G1PBU RG22 4XT
G4MQQ RG22 4XT
G8VML RG22 4XT
G7LJB RG22 5BA
G1OQU RG22 5BP
G7DLF RG22 5BT
G0LEP RG22 5DN
G7UJO RG22 5DS
G6OOC RG22 5HP
G7LJR RG22 5JP
G1JHM RG22 5JP
G4PBF RG22 5JP
G8GTZ RG22 5JY
G6KQZ RG22 5LY
G0LQD RG22 5NN
G6ZZR RG22 5QF
G6UTL RG22 5YU
M1AHQ RG22 6AD
G0PTA RG22 6AX
G3PXC RG22 6DR
G0UVQ RG22 6DX
G1PEU RG22 6JA
G0KQA RG22 6NU
2E1EHS RG22 6NZ
2E1FTH RG22 6NZ
G3OAZ RG22 6QP
G7FWM RG22 6QU
G0VVM RG22 6UW
G6HTM RG23 7AA
G3JTQ RG23 7AQ
G4WEN RG23 7BG
G4JGS RG23 7BL
G1EDS RG23 7BQ
2E1CSA RG23 7BY
G3HVA RG23 7DD
2E1DOZ RG23 7DE
G6PJP RG23 7DJ
G4BHE RG23 7EG
G7UOM RG23 7JX
G6CFA RG23 7LB
G8YEQ RG23 7LD
G0XBA RG23 8AD
G7IEA RG23 8BG
G0NIQ RG23 8EX
G4YPK RG23 8EX
2E1DRR RG23 8EX
G4EDH RG23 8JD
G7KAK RG23 8JF
G0ELH RG23 8LJ
G3PFY RG23 8LS
G0JSR RG23 8NG
2E1FCF RG23 8NG
G6ENH RG23 8QW
G8JHH RG24 7EF
G8DMR RG24 7EH
G4OEH RG24 7ER
G6PHQ RG24 7ER
G0BOW RG24 7HU
G4XIP RG24 7JE
G4LXJ RG24 8GR
G4ODM RG24 8LN
G3TCR RG24 8LN
G6JDP RG24 8LN
M0AHC RG24 8RB
G0ILS RG24 8RH
G8UBN RG24 8RL
G4FOL RG24 8RS
G3IVZ RG24 8RW
G8JSF RG24 8SB
G7HOK RG24 8SE
2E1FXJ RG24 8SG
2E1FXK RG24 8SG
G0IAY RG24 8SR
G6ZIO RG24 8TE
G1CTT RG24 8TQ
G0RND RG24 8UF
G4XMP RG24 8XB
G4NCJ RG24 8XR
G7FFB RG24 9DD
G7FFC RG24 9DD
G7GOR RG24 9EH
G4YNH RG24 9HE
G0SEU RG24 9HE
G3IQE RG24 9NR
G8FAD RG25 2BE
G8JMY RG25 2BH
G7NDB RG25 2BZ
G3MJK RG25 2ED
G3CEI RG25 2NH
G4OBN RG25 2PN
G7EVF RG25 2RN
G3UYN RG25 3HH
2E1EHA RG25 3HJ
M1BRW RG25 3HP
2E1EHV RG25 3HP
G4GFM RG25 3LZ
G4CIZ RG26 3DN
G1LIE RG26 3DS
G0CAK RG26 3ED
G8AKA RG26 3EL
G8VGO RG26 3EL
G7DBC RG26 3HP
G3XME RG26 3PJ
G1JRP RG26 3SF
G8DVU RG26 3TL
G8UVU RG26 3TS
G6AOH RG26 3UQ
G4EEE RG26 3UR
G6MNN RG26 3UR
G8GGH RG26 3UR
G4LUA RG26 3YH
G8UUG RG26 3YH
G0TGV RG26 4BN
G1SEI RG26 4EU
G6IOX RG26 4HF
G8MSY RG26 4HH
G0VTP RG26 4JL
G7NZK RG26 4LS
G1CSO RG26 5AN
G7EOX RG26 5BP
G4LVM RG26 5LE
G3AHS RG26 5NT
G4XMO RG26 5NY
G1KLC RG26 5NZ
G1WKE RG26 5PD
G1JWG RG26 5PJ
G7MTW RG26 5UW
G0FEJ RG27 0DQ
G3UQW RG27 0ES
G6AHH RG27 9JS
G7JMZ RG28 7HW
G3NYS RG28 7NF
G4SZR RG28 7NF
G4XXH RG7 2LH
G3SCZ RG7 2PZ
G4MMB RG7 2PZ
G3LTF SP11 6EA

Bassetlaw

G8EGL DN10 4BP
G0OKZ DN10 4BU
G8JET DN10 4BU
G6IJG DN10 4DL
G6KGG DN10 4EN
G0VTL DN10 4JL
G7PBN DN10 4QN
G6GWP DN10 5BS
G3HUD DN10 6AA
G3JOE DN10 6AA
G1PLB DN11 8AZ
G4ZVP DN11 8HP
G4BZG DN11 8LL
G4GBS DN11 8PE
G0MLK DN11 8PF
2E1DBN DN11 8PF
G6NRL DN22 0BY
G4SOI DN22 0JX
G0TGB DN22 0JX
G1EWH DN22 0LN
G4TYS DN22 0NQ

2E1FRW DN22 6LR
G6GKK DN22 6NG
G4XTU DN22 6NW
G4YWX DN22 6SF
G3DXZ DN22 6UF
G1MRT DN22 7AT
G4WEC DN22 7HP
G3NTD DN22 7LJ
G1ORD DN22 7QG
G4TDZ DN22 7TH
G4WBH DN22 7UW
G0IAS DN22 8AJ
G4OCU DN22 8AY
G7EKC DN22 8JU
G4VEO DN22 8NL
G4ZGX DN22 8NL
G4VEP DN22 8NW
G6TRD DN22 8RP
G3HKQ DN22 9JP
G0VUL DN22 9NQ
2E0ANL DN22 9NQ
2E0ANM DN22 9NQ
G4SAA NG20 9EU
G6ZOB NG22 0JB
G7NKE NG22 0JB
G3PRD S80 1AF
G0MZP S80 1PA
2E1FRV S80 2NF
G0DKQ S80 2NQ
G0RSM S80 2NS
G0SZD S80 2NS
G0KVG S80 2SB
G1NXX S80 2SF
G0GLS S80 2SN
G4ZRV S80 2SQ
G7MQW S80 2TT
G7TNZ S80 3DF
G7OBP S80 3HF
G0CEJ S80 3HG
G7SOH S80 3HR
G7TTW S80 3LN
G8TKY S80 3LP
G0VKJ S80 3LP
G7BJD S80 3QJ
G0HMX S81 0AG
G4YJY S81 0AX
G0CEB S81 0BW
G3OZN S81 0EE
G3XXO S81 0LP
G0BDR S81 0QN
G0CJD S81 0QN
G4ZUN S81 0QN
G6EXG S81 0QN
M1BPR S81 0SY
G6TLA S81 0SY
G6TLB S81 0SY
G0UBZ S81 0TB
G4YRZ S81 7BN
G8TBF S81 7DF
G7WNN S81 7DF
G0WTK S81 7DF
G4PYM S81 7DU
G0FMP S81 7ED
G4UFY S81 7JL
G3KIH S81 7JU
G4CRE S81 7LE
G8TWT S81 7PG
G1NSK S81 7QB
G0TKG S81 7QE
G4IBZ S81 8TD
G7JBZ S81 8TU
G1JAX S81 9BB
G3ZMM S81 9BD
G0AGD S81 9BY
G0WQL S81 9DH
G8YPQ S81 9DL
G0MZR S81 9HN
G4MWX S81 9JT
M1BHQ S81 9LE
G3XXN S81 9NW
G6DRG S81 9NX
G0GCY S81 9PB
G6VEP S81 9QW
G0DMA S81 9RH
M1BIS S81 9SA
G0NGW S81 9SL

Bath and North East Somerset

G4YNM BA1 2BL
G7TBR BA1 2PS
G8OTA BA1 2QL
G0EJR BA1 2UU
G6UTK BA1 2XN
G4PVX BA1 3DS
G1JIY BA1 3EW
G3SUR BA1 3JZ
G1BCB BA1 3LP
G7BZB BA1 3LQ
G3LYW BA1 3LU
G4LYW BA1 3LU
M1BCG BA1 3PY
G3TKF BA1 3PY
G4HTV BA1 3PY
G6MZW BA1 3RB
G3UMM BA1 4DZ
G7BFE BA1 4EX
M1BJK BA1 4NQ
G1ZUC BA1 5SY
G6UZG BA1 5TW
G4GCV BA1 6HH
G1MDC BA1 6JR
G7MZY BA1 6NP
G7VSJ BA1 6QN
G7CGV BA1 6QN
G4EBV BA1 7DR
G7MAY BA1 7PX
G6HGK BA1 7RN
G3RVX BA1 7TJ
G0WAW BA1 8AD
G7STQ BA1 8AD
G4YTN BA2 1AE
G4HJT BA2 1DB
G3OYQ BA2 1ED
G8PZE BA2 1HE
G6VIF BA2 1HS
G7GDT BA2 1JZ
G4YCE BA2 1NW
G7ANB BA2 1NW
G3VVO BA2 1RB
G7DZI BA2 2AD
G3JQQ BA2 2AY
G4YDC BA2 2DR
G3LYN BA2 2EU
G4DXL BA2 2EW
G7HOS BA2 2NF
G8CJT BA2 2PG
G8DX BA2 2PP
G4VHI BA2 2QB
G6ZKT BA2 2QB
G6ZKU BA2 2QB
G0LPG BA2 3AL
G0FUW BA2 3BS
G3NAW BA2 3HL
G1AEI BA2 3HZ
G8XZB BA2 3JL
G4EJS BA2 3PP
G1OPW BA2 4DL
G3ZNV BA2 4QG
G7FNN BA2 5AL
G7SDQ BA2 5DE
G4LEX BA2 5LL
G4NDT BA2 5NF
G4NDU BA2 5NF
G4UWZ BA2 5PR
G7CBE BA2 5QP
G4GON BA2 6AL
G3FGH BA2 6BR
G7NOY BA2 6BZ
G3VTO BA2 6DE
2E1CZO BA2 6DE
2E1DAC BA2 6DE
G3VWC BA2 6DF
G3EUK BA2 6RG
G3VIV BA2 6UB
G6MBW BA2 6UU
G6DHU BA2 7AF
G8HKK BA2 7AH
G4FEA BA2 7BA
2E1DAR BA2 7HN
G0SDQ BA2 8AB
G0LIB BA2 8AF
G4KLD BA2 8DG
G1WFU BA2 8EF
M1ALR BA2 8ES
G4GEV BA2 8QQ
G3ISG BA2 9EZ
G8VXZ BA2 9HS
G3ENR BA3 1AP
G0LTE BA3 1DH
G8LJY BA3 1DH
G4BHP BA3 1DZ
G3IGX BA3 1EA
G4OMG BA3 1EA
G6EIY BA3 1ES
G7IRF BA3 1HB
G6ZKC BA3 1LP
G4CBS BA3 1PS
G7IYA BA3 2NB
G1IHI BA3 2PR
G4ZNK BA3 2RH
G4HSA BA3 2RT
G8XWK BA3 3EZ
G8XTO BA3 3EZ
G7VCY BA3 3LB
G0RQT BA3 3NR
G4ZEU BA3 3RL
G0WPB BA3 3UT
G1ZBM BA3 4AE
G1ORL BA3 4AN
G3RHU BA3 4BR
G4PSP BA3 4BR
M0ALZ BA3 4BW
G0CEE BS14 0PF
G0DRX BS14 0PP
G3KMM BS14 0PP
G0CCA BS14 0QS
G3NXU BS18 1BD
G0MPH BS18 1JB
M1AXC BS18 1JH
G0JZH BS18 1QE
G3WJA BS18 1QE
G0DSB BS18 1XD
2E1CCG BS18 1XD
G3XAW BS18 1XH
G1PCA BS18 2DA
G7PHE BS18 2EQ
G7ALN BS18 2JH
G0IUR BS18 2PX
G5FS BS18 3AH
G0KNZ BS18 3AH
G4VPG BS18 3DR
G7TFU BS18 3DU
G8VPG BS18 3EU
G6AYY BS18 3LA
G0KGW BS18 4BG
G0KMV BS18 4DH
G1ZFG BS18 4NT
G1FZV BS18 4PB
G4TVZ BS18 4QQ
G0BLB BS18 4RJ
G4MKX BS18 4XB
G8VDF BS18 5AE
G4EPH BS18 5DZ
G3AHB BS18 5JL
G7SOJ BS18 5JR
G7GNY BS18 5LF
G1DOJ BS18 5LJ
G4LAF BS18 5LJ
G0RWT BS18 5LU
G1OWD BS18 5PN
G7AEQ BS18 5QB
G3HWY BS18 5QT
G1ARZ BS18 5UP
G8OEU BS18 5XA
G0FZP BS18 5YZ
G1LLJ BS18 5YZ
G1KTY BS18 6AP
G0PCQ BS18 6BZ
G0LCX BS18 6EB
G3SDH BS18 6JE
G3KWJ BS18 6XG

Bedford

G6GCO MK40 2BE
G8IRC MK40 2JE
G5BH MK40 2TR
G4FYL MK40 3AN
G1IGX MK40 3BS
G3USD MK40 3DF
G7UNL MK40 3JY
G0PSU MK40 3RJ
G0BMG MK40 3UG
G1JBI MK40 4AT
G4BCS MK40 4AT
G6ECR MK40 4AT
G6ECS MK40 4AT
G6GIH MK40 4HJ
G0ADB MK40 4PZ
G0TCM MK41 0AG
G0JIW MK41 0BQ
G6XZW MK41 6AG
G0GBS MK41 6BS
G6UPA MK41 6DA
G3CMC MK41 6HD
G4GDX MK41 7BS
G4KWH MK41 7BT
G4KYX MK41 7BT
G3UED MK41 7DH
G3USE MK41 7DX
G4JTJ MK41 7HX
G6EDB MK41 7HX
G8CGM MK41 7HY
G4KJU MK41 7JP
G7RVC MK41 7LS
G4FGJ MK41 7PP
G0UBX MK41 7QT
G3UBV MK41 7ST
G4EZQ MK41 7ST
G6CVE MK41 7UH
G0DIL MK41 7YB
G8JKR MK41 7YF
G4UCY MK41 8AP
G4AQS MK41 8AS
G4GIR MK41 8AS
2E1EMI MK41 8AW
M1BVV MK41 8AZ
G3RPL MK41 8BT
G4NZE MK41 8DB
G4XDB MK41 8DR
G8HJH MK41 8DR
G0HER MK41 8EL
G7CBL MK41 8HJ
G6YNW MK41 8NX
M1BED MK41 8QD
G0VJJ MK41 8QD
G3WTP MK41 8QD
G8HGL MK41 8QY
G0BKN MK41 9AL
G0MBR MK41 9AL
G6MBR MK41 9AL
G8SDN MK41 9AL
G7MYO MK41 9AL
G1BYT MK41 9AN
G7LEB MK41 9AS
G4YYC MK41 9DD
M0ASH MK41 9EA
G4PHY MK41 9EL
G1JZK MK41 9EP
G8RTN MK41 9LG
G3WHF MK41 9LL
G6LDH MK41 9LP
G0EYM MK41 9LR
G3WBP MK41 9NE
G1JZT MK41 9QR
2E0ACP MK41 9QR
2E1ASN MK41 9QR
G0VCE MK41 9QR
G7OMG MK41 9QR
G7MDY MK41 9TB
G3FWA MK41 9TQ
G7SNK MK42 0JS
G4IMH MK42 0PX
G0GBI MK42 0SE
G3UQU MK42 0SE
G6HCQ MK42 0SE
G7STG MK42 0UA
G6PAA MK42 0UB
G7RDA MK42 0UB
G3WEP MK42 0UJ
G4OOQ MK42 7BE
G7IJE MK42 7DU
G7MIS MK42 7EH
G4ZPH MK42 7EN
G1NRE MK42 7EU
G7BWH MK42 7EX
G0CPS MK42 7HE
G0DKE MK42 7HF
2E0ACR MK42 7HF
G7SRU MK42 7HX
G8MGP MK42 7PE
G7OZL MK42 7RY
G7WAH MK42 7RY
G8XQN MK42 7TS
G7SPE MK42 8BU
G4JUB MK42 8DR
G0BDK MK42 8DT
G7MMY MK42 8LY
G0VME MK42 8LY
G7SQM MK42 8NL
G7HZQ MK42 8NT
G8WGQ MK42 8RU
G0OXY MK42 9NQ
G4ABQ MK43 7BB
G7EPP MK43 7BG
G3KWA MK43 7DB
G4CEC MK43 7DR
G3IIY MK43 7ED
G3INY MK43 7ED
G0GQZ MK43 7HD
G1XUA MK43 7HD
G3XDU MK43 7HY
G3LDG MK43 7JL
G3SVQ MK43 7JX
G8ADY MK43 7LP
G4PNK MK43 7QF
G0GPV MK43 7SG
2E0ACQ MK43 7SG
G0EKD MK43 7SJ
G4XPJ MK43 7TD
G4XUY MK43 7TD
G7MYM MK43 8DU
G7SWK MK43 8DY
G7MNM MK43 8EA
G4EDJ MK43 8JA
G4MBA MK43 8JA
G0EYG MK43 8JY
G4UMW MK43 8LF
G1KSQ MK43 8NH
G3XTU MK43 8PE
2E0AIE MK43 8PS
G3XYV MK43 8QJ
G3RKP MK43 9BB
G4OGG MK43 9BE
G6HOT MK43 9BT
G0VCW MK43 9DL
G1GAS MK43 9EX
G4YIE MK43 9EX
G0DIM MK43 9EZ
G1WLQ MK43 9HZ
G3YUQ MK43 9JW
G6IEE MK43 9JW
G8XNN MK43 9LB
G1HGE MK43 9LW
M1ATZ MK44 1DG
G1BUV MK44 1DY
G6HMF MK44 1EX
G0RNM MK44 1JE
G4FIJ MK44 1JP
G8IIG MK44 1JP
G8SBV MK44 1JP
G4CEO MK44 1NP
G6EDT MK44 1NX
G8ACQ MK44 1SE
G4FFC MK44 2AU
G0CZY MK44 2BA
G4TGV MK44 2DT
G4MSQ MK44 2EW
G7MZA MK44 2LH
G7MWZ MK44 2RL
G4TXG MK44 3BH
G4PDP MK44 3BN
G4VCF MK44 3DS
G0FQM MK44 3JF
G3RIQ MK45 3AP
G1WDQ MK45 3BU
G7IXM MK45 3DT
G6AJX MK45 3EF
G4NXV NN10 9LG

Berwick-upon-Tweed

M1CCY NE66 4EA
M1CDU NE66 5NG
G4FCC NE67 5AW
G3AQB NE68 7TD
G1VVU NE68 7XR
G1GIT NE70 7NB
GM1XPE NE71 6HU
G3HDT NE71 6HU
G0AVU NE71 6JL
G6VGO TD12 4SJ
G4YQM TD12 4SL
G3YOG TD15 1BY
G2DAU TD15 1LY
G0RFZ TD15 1NS
G3BRA TD15 1PW
G0JCP TD15 2BS
G1IVI TD15 2BS
G1WRP TD15 2BZ
G4OJF TD15 2EB
G0HBY TD15 2FF
G1KRZ TD15 2JY
G0RPZ TD15 2SE
2E1DAT TD15 2SE
G0HBV TD15 2TY

Bexley

G0OIA DA1 4DL
G6CLU DA1 4RY
G8ZHR DA1 4RY
G4APB DA1 4SU
G8MCA DA14 4BE
G6SDO DA14 4JQ
G4OFU DA14 4JR
G7EDA DA14 4LJ
G6INU DA14 4PS
G3LLX DA14 5BZ
G1IGQ DA14 5JF
G7MJH DA14 5LS
G4FAA DA14 5NF
G6GTC DA14 5NG
G0GJW DA14 5NQ
G3LCB DA14 6AU
G3BNE DA14 6JQ
G8XIN DA14 6JT
G0OHQ DA14 6PU
G7GDJ DA14 6PY
M1BXU DA14 6SG
G0IND DA14 6SJ
G4ILH DA15 7NL
G3KGM DA15 8AH
G1TXO DA15 8ER
G0GKI DA15 8HP
G1HYT DA15 8JT
G4BAL DA15 8LY
G4LSU DA15 8PU
G4FVF DA15 8QN
G1PRZ DA15 8TA
G3DCC DA15 9DY
G8OPA DA15 9DZ
G0KYQ DA15 9HZ
G4RPR DA15 9JA
G0JBT DA15 9JJ
G0KPZ DA15 9LF
M1BHL DA15 9LG
G8LLD DA15 9LN
G3WMT DA15 9NJ
G8IWX DA16 1DB
G0ILE DA16 1LG
G3LUH DA16 1PG
G7LTP DA16 1RH
G8SWC DA16 2AW
G6PKS DA16 2BP
G8IUK DA16 2DT
G2DZH DA16 2HJ
M0ARX DA16 2PY
G1IYN DA16 2PY
G7TBE DA16 2RN
G6ODW DA16 2RU
G1NQB DA16 2SQ
G0LZF DA16 3AW
G1FHO DA16 3AW
M1BWE DA16 3HR
G8MIF DA17 5BG
G7UHW DA17 5BT
G8MLP DA17 5DR
G3CUR DA17 5EW
G6FNE DA17 5HS
G4MLY DA17 5NJ
G7MLU DA17 6EU
G4WJH DA17 6HB
G8PPG DA17 6JE
G1IUY DA17 6JT
G6PNJ DA17 6JX
G0VWF DA17 6LP
M1CBK DA18 4BN
G0HVF DA5 1HG
G3TXN DA5 1PJ
G0NDL DA5 2DJ
G0FDZ DA5 3AH
G6YZB DA5 3BT
G1RDI DA5 3HE
G0FHL DA5 3LT
G6HD DA5 3ND
G4MB DA6 7PA
G3AUA DA6 7PY
G0UCH DA6 8EJ
G1EWE DA6 8HA
G4DUM DA6 8JP
G1MAV DA6 8JS
G3CRP DA7 4BW
G4DDP DA7 4EP
G4XKV DA7 4JL
G8PJF DA7 4JX
G4KOW DA7 4JZ
G4DFI DA7 4LF
G4YAG DA7 4LT
G6CUE DA7 4PG
G7GQO DA7 4QT
G4JTZ DA7 4SR
G7MJJ DA7 4UX
G7UUB DA7 5BT
G3IJW DA7 5DG
G6ENO DA7 5DG
G3KHR DA7 5DZ
G4KGM DA7 5HA
G0PXT DA7 5JU
G2AHC DA7 5JW
G8AOL DA7 5NB
G8TNK DA7 5NB
G8JZT DA7 5NP
G8NME DA7 5RD
G8XJE DA7 5SE
2E1CLN DA7 5SG
G7KID DA7 5SL
G4MDG DA7 5SL
G7GPI DA7 5SL
2E1DRV DA7 6AX
G6RLH DA7 6EW
G3VST DA7 6LS
G0UAO DA7 6NL
G4CW DA7 6NL
G0TDJ DA7 6NL
2E1BCW DA7 6QB
2E1BCX DA7 6QB
G4GZN DA7 6QU
G2CCH DA7 6RQ
G8PXG DA8 1DA
2E1BCZ DA8 1JX
G7PHD DA8 1LW
G3BHF DA8 1NL
G4MHJ DA8 1NN
M1AJT DA8 1NN
G4EGU DA8 2JH
G3NRZ DA8 3BL
G3GEV DA8 3EH
G1OLZ DA8 3JN
G6DAW DA8 3QS
M0BCY DA8 3RF
M1BGG DA8 3RF
G7RCE SE2 0DT
G4WNF SE2 0DW
G0EFA SE2 0PP
G8UHS SE2 0QH
G8XTX SE28 8HD
M1BPN SE28 8PG

Birmingham

G3OTY B1 1UJ
G4ACF B1 1UJ
G6HLR B10 9TB
G4UFM B11 3JY
M1BEI B11 3NP
M1BMZ B11 3SA
M1BDD B11 3SH
G0IVD B11 4NJ
G0NHP B12 9ER
G4BJS B12 9NE
G6AWV B12 9NS
G0MZU B12 9RG
G6DSD B13 0BA
G8ENA B13 0BH
G8GKR B13 0BH
G7UGC B13 0BH
G4EBL B13 0DT
G4PUD B13 0EJ
G6DRN B13 0NR
G1NMH B13 0SR
G8LQP B13 8JX
G0EBW B13 8JZ
M0AEJ B13 8NZ
G7OLZ B13 8PG
G6LEL B13 8QJ
G6FKE B13 9EB
M1BML B13 9UD
G7JRL B13 9YW
G1YMC B14 4AD
G4ETZ B14 4DS
G8ACR B14 4JJ
G1POR B14 4LS
G8MYK B14 4TE
G1LZH B14 4TG
G8TUT B14 4TJ
G3OOE B14 4TW
G6KMD B14 5BD
G3CUN B14 5BS
G3PQP B14 5EF
G4MTG B14 5HD
G4MVB B14 5JD
G7PIR B14 5SN
G3NYA B14 5UP
G3EJP B14 6AT
G3YXM B14 6DE
G4ZVS B14 6HH
G4GEU B14 6NZ
G1FDD B14 6TN
G4JGH B14 7AS
G4DFO B14 7AU
G3YHF B14 7EP
G7BRB B14 7PB
G8OPV B14 7RT
G1MIN B15 2DS
G7ARP B15 2HE
G0HVN B15 2TT
G3IUB B15 2TT
G8IUB B15 2TT
G1KUQ B15 2TT
G7AFQ B15 2TT
G8CQH B15 2XA
2E1CSZ B15 3HG
G6NLG B15 3QD
G4NMC B15 3RL
G1JDE B16 0JN
G6AUO B16 0JN
G4WWH B16 9DL
G1XFL B16 9EY
G4BDK B16 9TD
G3SJH B17 0AT
G3KKB B17 0BL
G0KOI B17 0NW
G0LCL B17 0NW
G1STL B17 0QT
G4NZY B17 0QT
G1CUW B17 0TA
G3ZHH B17 0TB
2E1AZI B17 0TN
G1XWK B17 8AU
G3YAP B17 8NJ
G4DBV B17 8PT
G4IAF B17 8PT
G8EFY B17 8PT
G4KXV B17 8PY
G4SKE B17 8PY
G8ZKE B17 8PY
G3FKM B17 8QB
G3ZML B17 8QB
G4XDM B17 8QH
2E1BHB B17 8QH
2E1BHC B17 8QH
M1BYR B17 9EB
G4LQF B17 9JU
2E0ACK B17 9JU
G0EJL B17 9LE
G6PGO B17 9TB
G7END B20 1EL
G1ENP B20 1EX
G8WBL B20 2AA
G6EEB B20 2AQ
G6KHN B20 2HN
G7WKG B20 2NX
G3EVC B20 3ED
G6PLH B21 0EP
G4HDL B21 8AE
G0PUW B23 5DE
G7EKJ B23 5EZ
G8NZO B23 5JE
G7FBY B23 5JX
G8YWU B23 5PW
G7SDP B23 5QJ
G7RYA B23 5RG
G4ZSM B23 5RH
G7MWD B23 5UE
G8LIX B23 5XE
G6VIY B23 5YP
G6DYI B23 6DL
G4TMK B23 6NB
G6IHW B23 6NN
G6BRY B23 6PX
G6DRM B23 6QG
G1KDO B23 7HW
G1JZJ B23 7HX
G0VBQ B23 7JR
G3SBR B23 7LT
G6JAC B23 7LT
G0LNE B23 7PB
G4ORW B23 7RN
G4OXH B24 0BY
G7BWY B24 0ET
G7IXH B24 0PQ
G4JDC B24 0TQ
G8OVA B24 0TQ
G2FLY B24 9AL
G8AKE B24 9AL
G3LIZ B24 9BH
G4CVW B24 9DN
G3TDF B24 9EX
G6GBT B24 9LX
G7PZQ B24 9RQ
G4YQE B24 9RW
G1NNL B24 9SR
G0DZO B25 8BH
G7IZV B25 8JF
G8AFI B25 8JF
G8XXZ B25 8LQ
G8XTQ B25 8NB
G7PSC B25 8NZ
G1GHD B25 8QZ
G4KQU B25 8UL
G6VXC B25 8YT
G3NXC B26 1DY
G7PVA B26 1JF
G6OVO B26 1LA
G8AYY B26 1PL
G3KYE B26 1PR
G4LMF B26 1QP
G6FGW B26 1TT
G0HRX B26 2AF
G3JTT B26 2AF
G1VSD B26 2AW
G1TZF B26 2ES
G6KHM B26 2HT
G1CQF B26 2UL
G3UHQ B26 3HJ
G4NPG B26 3PW
G3GR B26 3RL
G0LBQ B26 3ST
G1OOW B27 6BB
G8EMX B27 6LE
G4EKQ B27 6QD
G3CHU B27 7ND
G1JEP B27 7QA
G6HFZ B27 7RU
G7JOS B27 7TL
G3NVC B27 7UW
G4FPN B27 7XP
G7KHB B28 0AQ
G6ZZS B28 0AT
G4JGV B28 0AU
G3USA B28 0DH
G4FLQ B28 0DS
G0VMI B28 0DU
G3CXT B28 0DZ
G6VDA B28 0EY
G1BLK B28 0HH
G4TUR B28 0HH
G4GIG B28 0HN
G4JTT B28 0HN
G3MHM B28 0PA
G7VNX B28 0QD
G6KVR B28 0QX
G4YZK B28 0RS
G3RGD B28 0TW
G1ATN B28 8AU
G3VPE B28 8AU
M1CAP B28 8HA
G4BQW B28 8JB
G4XVN B28 8LA
G4SUT B28 8QB
G6HLW B28 9EE
G3KLD B28 9EQ
G4FCO B28 9JN
G4OGR B28 9RG
G8GDZ B29 4AG
G3ZKQ B29 4AH
M1BTT B29 4DG
G4KLA B29 4NH
G7EJN B29 4RD
G3DWS B29 4RE
G4YGX B29 5EL
G1FAZ B29 5HA
G0NDI B29 5JS
G0NDJ B29 5JS
G0HDF B29 5NX
G1IRV B29 6PP
G0FPN B29 6PX
G6SL B29 6UB
G4RCZ B29 7QJ
G1XHM B29 7RR
G0TFJ B29 7RR
G1ESL B29 7RZ
G3EMY B29 7SB
G4NCY B29 7SG
G3EGS B30 1LH
G0BWP B30 1NG
G3AAH B30 1QT
G7BZM B30 1RQ
G3MKW B30 1TX
G3CMS B30 1UZ
G4YUI B30 2BA
G4IIP B30 2LX
G6ZJM B30 2QH
G4OMP B30 2SH
G6WMR B30 2SH
G4TKM B30 2TJ
G4KYG B30 2UL
G4ASJ B30 2YT
G1EAX B30 3BZ
G0MQC B30 3QN
G6VGT B31 1AL
G7DNX B31 1NE
G0VJU B31 1TH
G0TEQ B31 1UH
G8XIZ B31 2AX
G8ZXY B31 2BJ
G1JAL B31 2DN
G3OXL B31 2DN
G4IUX B31 2DX
G6IQL B31 2DX
G1FVC B31 2EB
G4YTO B31 2EJ
G6IRJ B31 2HD
M1APP B31 2HD
M0BCN B31 2HD
G8MDM B31 2HZ
G3XVW B31 2JG
G1VLS B31 2LJ
G1JRF B31 2LY
G8VNL B31 2PP
G8KAE B31 2QB
G2AZM B31 2QZ
G0BOT B31 3EH
G3DTG B31 3HJ
M1AYP B31 3HY
G8LKW B31 3PP
G1MLK B31 4AJ
G6WOZ B31 4BE
M0BAZ B31 4BS
G0JGH B31 4ES
G0NVV B31 4HE
G1XXZ B31 4HE
G0GDG B31 4HN
G0DLG B31 4LN
M0ALH B31 4LN
G0NXB B31 4PB
G7FUW B31 4PB
G6TDG B31 4SS
G7VDE B31 4TD
G8MIT B31 4TQ
G6RRZ B31 5DP
2E1FZY B31 5DP
G4ZPI B31 5JP
G0HZT B31 5JU
2E0AQE B31 5QL
G6KAE B31 5RD
G4SWA B32 1DR
G4XUA B32 1EG
G4CGU B32 1EL
G8WMS B32 1EN
G8HQY B32 1EX
2E1DBJ B32 1EY
2E1DGR B32 1EY
G4XEJ B32 1HE
G1MAR B32 1LB
G3MAR B32 1LB
G8BHE B32 1LB
G4LLW B32 1NB
G0PWQ B32 1PG
2E1AIT B32 1PG
G0LGO B32 1QT
G7DCM B32 1QU
G7TPB B32 2AA
G4JCZ B32 2BA
G3KLJ B32 2HA
G0PPX B32 2NW
G6HOV B32 2PE
G7RRP B32 2RQ
G1NHX B32 2SB
G4BGZ B32 2SQ
G4GZZ B32 2SU
G4ZTM B32 2TT
G4CKK B32 2UX
G4EKY B32 3DL
G0UYT B32 3LQ
G0SRA B32 3PR
G1PKG B32 3TA
G4HXI B32 3UG
G4TCC B32 4BP
G1ESS B32 4LN
G0RNL B33 8LE
G3MRP B33 8RT
G7KNP B33 9BU
G8KCB B33 9DL
G6NOW B33 9NG
G6SFW B33 9NZ
G0TNQ B33 9QG
G7VUY B34 6AA
M0BEM B34 7LH
G0EOS B34 7QX
G4LAJ B34 7RS
G0FOC B35 7LD
G8TBW B35 7PD
G6JFV B36 8AL
G3DID B36 8AY
G7LKI B36 8BW
G4XVQ B36 8DD
G0OTF B36 8HA
G1CJH B36 8HD
G3XWX B36 8HD
G3TQE B36 8HJ
G3YIT B36 8JH
G0WKF B36 8JS
G4TLR B36 8LD
G3ZMT B36 8NB
G0OEV B36 8NH
G7HUU B36 8NN
G3NQA B36 8QG
G0IBF B36 8QY
G7HOC B38 0AB
G1KDG B38 8DR
G4AEG B38 8DT
G8AHE B38 8EA
G0FZQ B38 8HZ
G0MKU B38 8LB
G1XPD B38 8LS
M0ADT B38 8ND
G1AOV B38 8PD
G4KRT B38 8PH
G3TGL B38 8PW
G8XUN B38 8PW
G4NPA B38 8RG
G4NPB B38 8RG
G0AHC B38 8TH
G1LWH B38 9HU
G1MZM B38 9QT
G0OKI B38 9RY
G7TVL B42 1EU
G6LGH B42 1EY
GW7VRZ B42 1HE
G0FBQ B42 1HF
G0PPZ B42 1JH
G1HUM B42 1LP
G4WSI B42 1LY
G6HPE B42 1LY
G6UGA B42 1PZ
G7ACM B42 1PZ
G8HEB B42 1QL
G6NHX B42 1RG
G1EDT B42 1RY
M0ABF B42 2AX
G3RJX B42 2BL
G1MOV B42 2BX
G0AGT B42 2BY
G4NTV B42 2DT
G1OKM B42 2DU
G3PJY B42 2EA
G4EUI B42 2HH
G8ASW B42 2HT
G8SFQ B42 2JW
G8SKA B42 2NT
G6NHW B42 2PQ
G4NBW B42 2SQ
G3SWO B43 6DR
G6WOI B44 0AL
G0HHP B44 0JG
G1BJK B44 0LB
G1FQX B44 0TL
G6JQB B44 8BA
G6WIG B44 8LQ
G0WMU B44 8RL
G7OPD B44 8RS
G0GWZ B44 8UG
G4RZD B44 9BY
G7OUF B44 9DA
G4OFN B44 9DR
G7ORV B44 9DR
G0VKA B44 9HL
G8TZW B44 9NY
G7ORT B44 9QH
G6NZW B44 9RP
G3URV B44 9RR
G4BTK B44 9SS
G0HPG B45 0JY
G0HPH B45 0JY
G4LUX B45 0LH
G0UQE B45 8RE
G8TYC B45 8RJ
G1RAY B45 8RN
G0SUS B45 9BL
G7RRJ B45 9NA
G4IVJ B45 9NS
M1BME B45 9PX
G1PHN B45 9RH
2E0AAV B5 7LR
G1ZCY B5 7LR
G6DWS B5 7NE
G2BBC B5 7QQ
G6BBC B5 7QQ
G3MTQ B5 7QY
G1YLQ B5 7XH
G8VBE B5 7XH
G1AAL B72 1AG
G4BEJ B72 1ER
G2FRT B72 1HD
G3MDQ B72 1HD
G3PTM B72 1HD
G0UQJ B72 1JP
G1NFN B72 1JU
G4NQR B72 1JW
G0KTF B72 1NR
G3NCX B72 1ST
G0VXK B72 1YF
G8NFD B72 1YZ
G1XLG B73 5BJ
G8XLG B73 5BJ
G1ZKK B73 5DL
G4LAZ B73 5DX
G4TYR B73 5EA
G6LWZ B73 5EA
G6WJT B73 5EA
G8EZB B73 5EL
G0VZL B73 5LF
G3MZP B73 5LL
G1AKC B73 5LX
G3WEX B73 5PA
2E1AZF B73 5PE
G8NTY B73 5QJ
G4WNB B73 5RJ
G8TRK B73 5SX
G7ICC B73 5UL
G7UHP B73 5XP
G7EPW B73 6BA
G6AGO B73 6BT
G6DGR B73 6BT
G0IBQ B73 6JA
G8GSL B73 6JX
G8YIB B73 6LS
2E1CFF B73 6LZ
G6IX B73 6NS
G4MOJ B73 6PE
M1BON B73 6PE
2E1DOU B73 6PE
G8AMD B73 6PG
G4LBT B73 6QA
G4NBI B73 6QR
G1NPA B73 6UQ
G3YEY B74 2AQ
G4ONO B74 2AW
G3MCB B74 2PS
G3JZF B74 2QA
G7OKE B74 2QA
G4ABW B74 2TB
G8FFD B74 4BW
G1AZE B74 4DQ
G4YZO B74 4JF
G3WWL B74 4JS
G4ORY B74 4LU
G0EWA B74 4PD
G4XNY B74 4PW
G8XMH B74 4SG
G5BR B74 4TF
G1JXX B74 4XR
G1GFA B74 4YD
G7SER B74 4YD
2E1EHI B74 4YD
G6AHA B75 5AQ
M0AZE B75 5AQ
2E1CBE B75 5BD
G7UCJ B75 5BD
G7HRY B75 5EW
2E1CAB B75 5JA
G7RUF B75 5JA
G7RUE B75 5JA
G4CLE B75 5JB
G1NSG B75 5LD
G0KLK B75 5LH
G4ZHH B75 5LN
G3BHT B75 5PX
G6VYM B75 6AQ
G0EVH B75 6AU
G8AVH B75 6AU
G6JOY B75 6BN
G6JOZ B75 6BN
G6HNS B75 6DA
G7NBU B75 6DH
G7UCG B75 6DH
G3AVE B75 6DW
G3GLQ B75 6EN
G3MYC B75 6EN
G8IMN B75 6EN

G0IZH B75 6LP
G4LRZ B75 6RL
G8TBB B75 6SN
G8EFU B75 6SP
G6BBR B75 7BL
G6HUM B75 7BL
G8MAN B75 7DX
G0UUL B75 7DX
G1VVF B75 7ND
G3OMT B75 7RA
G6GPF B76 1AW
G3LNN B76 1EX
G4CPB B76 1HA
G7BGP B76 1HL
G1KFT B76 1JE
G1BUQ B76 1JQ
G1XKN B76 1JR
G8IOS B76 1LZ
G4YUU B76 1NT
G3VHB B76 1PG
G7AVT B76 1PX
G6VEX B76 1QL
G4UDA B76 1QZ
G6UED B76 1XR
G8ERN B76 1XZ
G3VNY B76 2PT
G4AUH B76 2SZ
G0OVS B76 2TF
G1LWP B76 2UB
G1XDG B76 2XB
G1LTE B76 9RP
G7GGT B8 2EA
G0UFX B8 2EH
G0NFR B8 2PD
G6AZL B8 2QE
G0NWX B8 2XQ
G0RSK B8 3BY
G6IMV B8 3QJ
G4SAS B9 4PQ
G7LTG B9 5NG
G7LXV B9 5RY
G1GGK B92 7PW

Blaby

G4ZNL LE18 3TN
G4DQR LE2 9HP
G4GNI LE2 9HR
G6HSI LE2 9NS
G8RFE LE2 9NS
2E1BVO LE2 9NS
G0PBP LE2 9TH
G6VLT LE2 9TT
G0FRV LE3 2JU
G7IEQ LE3 2QD
G7UOS LE3 2SN
G3PMA LE3 2UN
G1AEJ LE3 2UU
G3GXN LE3 2XN
M1APE LE3 2XP
M1BSE LE3 2XQ
G4NUK LE3 3AD
G6SGA LE3 3AD
G1UTY LE3 3BE
G4DJK LE3 3DQ
G4SDZ LE3 3FF
G7GLA LE3 3GN
G3HYH LE3 3HB
G1CKY LE3 3JX
G7AYI LE3 3LA
G7SNA LE3 3PD
G1YEZ LE3 8AF
G1EHR LE3 8AG
G6BBH LE3 8AG
G6BBI LE3 8AG
G6JGM LE3 8AP
G3LYU LE3 8BB
2E1EFS LE3 8DH
G7OBR LE3 8FN
G0LZA LE3 8FQ
G3LMR LE3 8GL
G7RIR LE3 8HA
G8SZX LE3 8JH
G4PDZ LE3 8LT
G8ATE LE3 8PQ
G3BSO LE7 7AG
2E1FAU LE8 4AU
G4WGU LE8 4BE
G6KJH LE8 4DL
G3KYF LE8 4FU
G3PVG LE8 4GD
M0AOU LE8 4GY
G7KOL LE8 4GY
G3SRM LE8 5QB
G8PGI LE8 5QZ
G1IPP LE8 5SU
G6DWW LE8 5SU
G3MXV LE8 5TB
G4YJU LE8 5TG
G4FIE LE8 5TL
G1HEN LE8 5TP
G4YXQ LE8 6ER
G7RWP LE8 6JT
G8ZQB LE8 6NF
G6IM LE9 1RE
G6MFH LE9 1SB
G1IMJ LE9 1SE
G7VLR LE9 1SE
G8VEN LE9 1SE
G1IUT LE9 1SX
G1OOM LE9 1UW
G0VLR LE9 1ZX
G4CSL LE9 2AQ
G4ZIZ LE9 2BP
G0DMB LE9 2DD
G1EBV LE9 2DD
G4FSS LE9 2DE
G1YYL LE9 2DY
G3ONV LE9 2EN
G3WTD LE9 3EB
G8XRG LE9 3GE
G8PEA LE9 3GX
G3RHZ LE9 4BT
G4IQF LE9 4BT
G6ISN LE9 4BT
G4EPN LE9 4BW
G4RCC LE9 4BW
G4SHF LE9 4DN
G0CND LE9 4DZ
G0JSE LE9 4DZ
G3TWY LE9 4FX
G4MYR LE9 4HA
G4RNR LE9 5AB
G7RWQ LE9 5BF
G7NMU LE9 5BW
M1AMC LE9 5EA
G4NQY LE9 5GQ
G1RIP LE9 5HE
G0TPH LE9 5JG
G0AIG LE9 5NW
G2AVF LE9 5PA
G8NTD LE9 5QD
G3ZFQ LE9 5QW
G8KRY LE9 5QW
G1VIN LE9 5QX
G7RXS LE9 5QX
G0TIZ LE9 5QZ
G4SIJ LE9 5RA
G7DPB LE9 5XA
G6PHJ LE9 5YQ

Blackburn

G6HWF BB1 2AZ
G0PXH BB1 2HB
G4XHZ BB1 2HY
2E0AOR BB1 2HY
2E1EYT BB1 2HY
G0DTI BB1 2JQ
G4GXO BB1 2PA
G4EFU BB1 3JD
G1DJQ BB1 3LP
G4BJJ BB1 7EX
G8EMB BB1 7EX
G0HTD BB1 9QY
G0VOF BB1 9SA
G8UYD BB1 9SZ
G1WYB BB2 2NQ
G4UCM BB2 2TH
G0TPE BB2 2TX
G1AVD BB2 3EE
G3SSN BB2 3ET
G1BWI BB2 3NZ
G0WMJ BB2 3RX
G0CBR BB2 4HY
G0BSW BB2 4NJ
G0IYH BB2 4NL
G0VNV BB2 4NQ
G6OWI BB2 4PQ
G4SCQ BB2 4PS
G4UCC BB2 4QJ
G7VPL BB2 4TW
G4UNP BB2 5AD
G0TVN BB2 5AH
G7VIY BB2 5DT
G7DEC BB2 5ER
G6FAZ BB2 5HP
G4JFV BB2 5JL
G0OBU BB2 5LX
G3OZC BB2 5NN
G4JMO BB2 6EB
G4GQO BB2 6EE
G4IAT BB2 6EE
G7VNK BB2 6ET
G3BMV BB2 6JB
G0FWM BB2 6JH
G6GFQ BB2 6JH
G4TMY BB2 6LW
G7BCP BB2 6NE
G6UUS BB2 6PL
G4GVG BB2 6QY
G2HFP BB2 6RN
G8EAW BB2 6SS
G7UAR BB2 6SU
G4LDE BB2 6TA
G0MTO BB2 7BQ
G6MWD BB2 7DT
G0KYO BB3 0AB
G2AKK BB3 0DY
G4JS BB3 0DY
G8WZW BB3 0EH
G0KAO BB3 0HD
G3WYI BB3 0HF
G4JBY BB3 0HZ
G6WGA BB3 0JB
G1EOY BB3 0LB
G8VOT BB3 0LF
G6ETP BB3 0LU
G0SHR BB3 0NF
G8XSU BB3 0NL
G1ZEX BB3 0QT
G1ZFD BB3 0QT
G0EIZ BB3 0RG
G0SQJ BB3 1HH
G1JBE BB3 1LQ
G0HXU BB3 1NS
G3KWO BB3 1NS
G8YLM BB3 1NT
G0IGS BB3 1QJ
G4JGF BB3 2BS
G6CIE BB3 2HP
G6RXP BB3 2LW
G0IPX BB3 2LZ
G3ZQS BB3 2LZ
G7MOB BB3 2LZ
G7VZS BB3 2SA
G0IYG BB3 2SG
G0NPJ BB3 2SQ
G1ECC BB3 2SQ
G4PSE BB3 2SQ
G4IBS BB3 2SS
G8FDG BB3 2ST
G0KXD BB3 3JB
G3YTI BB3 3JH
G0MXH BB3 3NE
G3ZRW BB3 3PT
G0JMR BB3 3PU
G8TJG BB3 3QA
G0FAG BB3 3RA
G2BTO BL7 0BG
G3HCZ BL7 0PA
G0URL BL7 0PA
G4XTG BL7 0PW

Blackpool

G3CTQ FY1 2JS
G1JYJ FY1 2RA
G1TIH FY1 3NN
G1MET FY1 3RB
G0NRE FY1 4DZ
G3IZG FY1 4DZ
G7OEM FY1 4LU
G6GJD FY1 5JZ
G1OLM FY1 5NA
G4PNI FY1 5NJ
G4GOR FY1 5PL
G4SZB FY1 6AP
G0RNA FY1 6BJ
G4EYX FY1 6JP
G0NBW FY1 6LP
G4CKR FY2 0BD
G1BFG FY2 0EN
G0FZL FY2 0PR
G0SLV FY2 0PZ
G6WWV FY2 0RA
G3MCI FY2 0RE
G0PFI FY2 0RP
M0BFU FY2 0SH
2E1ABR FY2 0TR
2E1ABS FY2 0TR
G3WGU FY2 0TR
G8ATG FY2 0TR
G4WYF FY2 0TU
G4TUM FY2 9AQ
G3UIP FY2 9AW
G1UCB FY2 9DG
G4HWK FY2 9DL
G3KXY FY2 9EQ
G6XNU FY2 9EQ
G0KCP FY2 9EW
G0GDA FY2 9HE
2E1FUZ FY2 9JR
G7PDS FY2 9JX
G7WCC FY2 9NA
2E1DZL FY2 9NA
2E1DJD FY2 9NA
G0DCM FY2 9NG
G8YBD FY2 9PL
G1NQN FY2 9QW
G1KCJ FY2 9UE
G8CSY FY3 7JN
G7EPY FY3 7PW
G7OMS FY3 8DP
G1JCW FY3 8HN
G6FCI FY3 8PD
G3SNH FY3 9HU
G4DPI FY3 9PW
G4IGZ FY3 9TN
G4VAL FY3 9TN
G4OSC FY4 1EB
G4MPT FY4 1LB
G0BZO FY4 1PF
G0LLX FY4 1PW
G0HZL FY4 1PY
G1EEA FY4 1QG
G4EZM FY4 1QS
G3TVX FY4 1RR
G4JAJ FY4 2BW
G3GED FY4 2EA
G6YHV FY4 2HF
G1UGG FY4 3AX
G7DIE FY4 3DX
G7POL FY4 3EU
M1CDX FY4 3HH
G6IGV FY4 3JQ
G4AKC FY4 3LF
G6BXO FY4 3LH
G0FYD FY4 3NH
G6BGH FY4 3QA
G3ZMK FY4 3QF
G6VEZ FY4 3QQ
G3YWZ FY4 3RA
G0PES FY4 3RB
G1NQO FY4 4AX
G0UPM FY4 4JP
G0HBQ FY4 4QF
G4YVQ FY4 4QS
G4XQF FY4 4RQ
G0ROU FY4 4TP
G4RXK FY4 4UD
G6XNI FY4 4UZ
M1BUG FY4 4YF
G6VRF FY4 5AE
G6AOS FY4 5BX
G0UHS FY4 5DA
G0BGR FY4 5DS
G6IFQ FY4 5DW
G4XPI FY4 5FD
G4DJY FY4 5HT
G0CEX FY5 1PN
G1RIH FY5 1RA
G6FDS FY5 3AF
G8ZSM FY5 3AQ
G0KMS FY5 3DS
G6EXQ FY5 3DW
G6HMX FY5 3EP
G6SRJ FY5 3EY
G7IGG FY5 3GE
G7MRQ FY5 3QL
G1DEP FY5 3QR
M1AMZ FY5 3QR

Blaenau Gwent

GW4RII NP2 3BJ
GW3YJL NP2 3ET
GW8OUM NP2 3LD
GW4HCV NP2 3RQ
GW4WLT NP2 3SG
GW1OPE NP2 3TA
GW6DEP NP2 4JD
GW6TQH NP2 4JG
GW1NCS NP2 4LE
GW8SPM NP2 4LW
GW3PFV NP3 1BS
GW7NTA NP3 2AD
GW0WVL NP3 2HU
GW8LTV NP3 2NQ
MW0BBM NP3 2RT
GW6ZJG NP3 2RT
GW4YDX NP3 2RY
GW7RSE NP3 3EQ
GW4KFI NP3 4AR
GW7HTU NP3 4LE
MW1AZI NP3 4LZ
GW0JTU NP3 4QY
GW0VPB NP3 4TW
GW0ALM NP3 4UT
GW0JMF NP3 5DY
GW0HCN NP3 5ES
GW0IIW NP3 5ES
GW3TUG NP3 5PR
GW6ZTG NP3 5PR
GW0MNO NP3 5TS
2W1CIP NP3 5TS
GW0VSO NP3 6JR
GW0OQH NP3 6LF
GW7VGB NP3 6LN
GW0JTF NP3 6QY
GW4UTS NP3 6UA

Blyth Valley

G0NHJ NE23 6AS
G0NHK NE23 6AS
G4HIH NE23 6DN
G4XKW NE23 6JN
G0NCZ NE23 6LG
G4HWK NE23 6PF
G0BAU NE23 6SQ
G1SLI NE23 6SX
G4UPA NE23 6TJ
G7VCI NE23 6TL
G7MYT NE23 6TU
G0LCE NE23 6UB
M0ART NE23 7ER
G7OVK NE23 8HG
G1ZJQ NE23 9FD
G7STX NE23 9PQ
G4HJB NE23 9TT
G0TLZ NE23 9UE
G0KYR NE23 9XP
G0BVL NE24 1NR
G0CXG NE24 1NR
G4VKY NE24 2HQ
G4ZTH NE24 2HQ
G1OPJ NE24 2LS
G1ZBN NE24 3DP
M1ARL NE24 3ER
G7PYR NE24 3ER
G1LMC NE24 3HX
G6PNO NE24 3QD
G4YAV NE24 3QH
G0JXC NE24 4AS
G7KYV NE24 4BD
G0HXC NE24 4DT
G1IEK NE24 4DY
G0EVF NE24 4EB
G4ZGH NE24 5JE
M1BMQ NE24 5LX
G4SIE NE24 5NB
G7KDX NE24 5RN
G0GTM NE24 5RR
G0ECQ NE24 5TR
M1BBI NE25 0DR
G0PKR NE25 0NG
G0VRT NE25 0NG
G8ARF NE25 0NP
G3VYZ NE25 0NP
G0PJU NE25 0PA
G0UPJ NE25 0RD
G4TYG NE25 0RH
G4VOU NE25 0SH
G4WVJ NE26 4BL
G0TWV NE26 4DD
G4BMV NE26 4HR
G4VNV NE26 4HU
G3VUD NE26 4RE

Bolsover

G7DGF DE55 2AJ
G1UAZ DE55 2AS
2E1BTR DE55 2AS
2E1DCU DE55 2AY
G1KOK DE55 2BG
G0DGQ DE55 2BJ
G0OKD DE55 2BU
G4CHM DE55 2DN
G0MTS DE55 2DZ
2E1DAE DE55 2HH
G8VSN DE55 2HS
G1INQ DE55 3AE
G4THI DE55 3BQ
G8KNQ DE55 3DB
G4EBX DE55 3HA
G4JDK DE55 3HA
G1BWE DE55 3NW
G4AMF DE55 5HD
G7BHW DE55 5HT
2E0AFI DE55 5HU
2E1CXO DE55 5JN
G7ACQ DE55 5NE
G3PLA DE55 5NR
G7HSP DE55 5PR
G4EKD DE55 5QH
G1ZWX DE55 5TR
2E1FJD DE55 5TR
G6ZDB DE55 5TT
2E1FJJ DE55 5TT
G3RWN DE55 5TU
G6AZE DE55 5TY
G7VVA NG16 6HN
G3RLS NG16 6HR
G7BHU NG16 6LQ
2E1AZG NG16 6LQ
G3ZAT NG16 6PU
G4OIE NG19 8TL
G4PPH NG20 8AZ
G8UST NG20 8AZ
M1BWF NG20 8AZ
G6JCT NG20 8EQ
G6JCV NG20 8EQ
G4VLK NG20 8QH
G1WBM NG20 8QJ
G4ZTL NG20 9AR
G4ODF NG20 9EB
G1KBJ NG20 9RP
G6GOR S43 4AP
G0JXS S43 4AX
G0KPB S43 4AX
G4VDB S43 4BD
G0KUW S43 4BS
G7MOO S43 4DW
G3WEU S43 4EH
G6DAO S43 4HF
G4RTA S43 4NJ
G0WLP S43 4PS
G7LZE S43 4RA
G6YFH S43 4RF
G0TSD S43 4RP
G0RUV S43 4SA
G0EOJ S43 4SX
2E1ERT S43 4TJ
G8KDM S43 4TR
2E1DDV S43 4TR
G1FTU S43 4TT
G4MWU S43 4TT
G4WEG S44 5LU
G0THF S44 5NG
G0LKK S44 5QJ
G0IXZ S44 6BH
2E1FDF S44 6BL
G4AGE S44 6ER
G4RSB S44 6ER
G8YBR S44 6ER
G4WFX S44 6EY
2E1DAP S44 6HN
G1SLE S44 6HT
G7GSX S44 6NX
2E1FIT S44 6NX
G4RDC S44 6PS
G7WAF S44 6RS
G4FEM S44 6SE
G0HMU S44 6SQ
G0OQG S44 6TF
G7UXA S44 6TN
G0RRZ S44 6TU
G4KSY S44 6TX
G6ZVE S44 6XN
G7DJN S44 6XN
2E1FXR S45 8AE
G4UZZ S80 4BN
G0AHD S80 4DE
G3WTY S80 4DW
G6OZT S80 4NP
G1OIZ S80 4PW
G3VRU S80 4SN

Bolton

2E1FZA BL1 2JD
G0WWH BL1 2JF
G0IZJ BL1 3LD
G6VFB BL1 3PE
M0ABO BL1 4JZ
G0VUX BL1 4PA
G3KAC BL1 4RQ
G0WJR BL1 4RQ
G6MRY BL1 4RQ
G6YEA BL1 4RQ
G4NTC BL1 4SA
G0FRL BL1 4UA
G4YZE BL1 5EH
2E1EYV BL1 5JS
2E1BKT BL1 5LU
G4FWP BL1 5NP
G8NVW BL1 5QE
G0UXF BL1 5TQ
G0PVP BL1 6AA
G7IPK BL1 6AZ
G1MIH BL1 6DA
G7VNV BL1 6HZ
G1YNX BL1 6PS
G6FTP BL1 7AF
2E1EAP BL1 7AW
G1NIW BL1 7DQ
G1NXJ BL1 7ER
G4UQK BL1 7JD
G6WIA BL1 7JU
G7EAM BL1 7QF
G0KEV BL1 7RJ
G7ETK BL1 7RW
G3EFI BL1 8BT
G4OZW BL1 8BX
G4RWS BL1 8PA
G4YNK BL1 8SD
M1AWC BL1 8SF
G1LSJ BL1 8TZ
G7ROM BL2 2DL
G0RQU BL2 2PY
G1YQG BL2 2QR
G0JFE BL2 2TA
G1IOO BL2 2TA
G1YQI BL2 3AY
G4XPU BL2 3AY
G1ZWQ BL2 3DU
G0MEP BL2 3FQ
G7BOJ BL2 3NF
G4CMH BL2 3NG
G3XCI BL2 3PU
G6OCA BL2 3PU
G4ZZI BL2 3QJ
G8WVR BL2 3QN
G1RRE BL2 4AQ
G0EWV BL2 4ET
G1ZBW BL2 5BY
G0GRX BL2 5LY
G4MUQ BL2 5LY
G7HVO BL2 5NL
G0MYP BL2 5NP
G7VYZ BL2 6BT
G0TCY BL2 6BX
G4AQB BL2 6LT
G4OQH BL2 6NR
G8YOY BL2 6NX
M0ANQ BL2 6PE
M1AUX BL2 6PE
G6OBE BL2 6PT
G4AOM BL2 6SB
G0FQR BL2 6TH
G6CVY BL2 6UQ
G3ZIK BL2 6US
G7GLS BL3 1DY
G0TBB BL3 1JR
G4YKB BL3 1JU
G1BIF BL3 1NX
G6VPH BL3 1PN
G1EDM BL3 1PW
G1XNR BL3 1QG
G8ROS BL3 1RN
G6CTH BL3 1UD
G6TWX BL3 1UD
G4WFS BL3 1UQ
G1TBV BL3 2DA
G8ZZM BL3 2JP
G0KEN BL3 2PF
G8MKS BL3 3AX
G3ITK BL3 3BB
G7MGC BL3 3DT
G1YYH BL3 3JD
G6KUJ BL3 3JH
G0TYI BL3 3SZ
G0TBW BL3 3TG
M0AVB BL3 4HR
G0DTB BL3 4HR
G4TNA BL3 4LF
G0LHL BL3 4NT
G4GIV BL3 4PB
G4DFP BL3 4PP
G0TEF BL3 4PP
G6TFC BL3 4QE
G6MHC BL3 4QG
G1AEQ BL3 4QH
G7RZW BL3 4QR
2E1FOM BL3 4QR
G0TLW BL3 4RP
G6HFF BL3 4TU
G6MML BL3 4TU
G0KAB BL3 5RN
2E1FGG BL3 6SJ
G7KCB BL3 6SW
G7TSM BL4 0BU
G6KQN BL4 0DS
G1RWX BL4 0HQ
G8XIJ BL4 0HW
G4HYG BL4 0LW
G0WSY BL4 0PA
G0IUA BL4 7LG
G7GLT BL4 8AW
G0JDA BL4 8EB
G4KPB BL4 8LP
G0HEA BL4 8ND
G6QA BL4 8NT
G6HDD BL4 8QU
G0DRL BL4 8QW
G3NDE BL4 8RR
G8XQI BL4 9BJ
G3XUM BL4 9HT
G0KHJ BL4 9LX
G8ZEF BL5 1DW
G1FTT BL5 2BN
G6SSH BL5 2ER
G7BKL BL5 2FD
G2ANC BL5 2HR
G0AIM BL5 2LE
G6GVR BL5 2LY
2E1CKH BL5 3AX
G6GYC BL5 3DN
G4RIE BL5 3HQ
G7IEI BL5 3JT
G7HVR BL5 3SY
G1JZX BL5 3UJ
G8IZR BL5 3UZ
G4EGG BL5 3YB
G4PDD BL6 4AZ
G4IAD BL6 4FA
G4FSN BL6 4JE
G6GMR BL6 4JE
G1AQP BL6 4JF
G8WLL BL6 4LG
G3JNM BL6 4PJ
G1VHW BL6 4RQ
G0MOK BL6 5AG
G0MRL BL6 5BG
G7CIQ BL6 5RH
G1WIR BL6 5RR
G4PAT BL6 5TD
G1VON BL6 5TE
G4CFP BL6 5TW
G8PRH BL6 5UG
2E1CMS BL6 6LH
G6YQI BL6 6NR
G4VPN BL6 6NS
G3SMU BL6 6PZ
G7GFK BL6 6PZ
G3KMS BL6 6QG
G0WBT BL6 7BE
G0KGI BL6 7ED
G6BZP BL6 7TB
G0RGO BL7 0HS
G7PSU BL7 0HS
G0TWD BL7 0HS
G4TQL BL7 9BJ
G7CDU BL7 9BJ
G1HSG BL7 9BN
G1CFC BL7 9BP
G4PMV BL7 9BP
G8UQV BL7 9DX
G1XOL BL7 9LG
G6MAJ BL7 9LL
G0EUZ BL7 9LQ
G0GVG BL7 9NB
G3REM BL7 9QE
G3EGC BL7 9RF
G6EZW BL7 9TG
G4ZEF BL7 9UJ
G0DTO BL7 9UT
G1EUB M26 1HG
G1JUD M26 1HN
G6RXF M26 1HN
G0BOR M46 9GZ

Boston

G0MRA PE20 1JT
G1DZQ PE20 1XG
G7UTE PE20 2BS
G3KRZ PE20 2QE
G3NHF PE20 3AJ
G7NUM PE20 3LL
G3KLC PE20 3QT
G7BPR PE20 3RL
G3SRX PE20 3RQ
G6ADG PE21 0AU
G1DVV PE21 0BZ
G1HDX PE21 0QS
2E1DDN PE21 0RP
G3YLX PE21 0SH
G1SOX PE21 7BQ
G6NUZ PE21 7BZ
G6ZQS PE21 7BZ
G7JMX PE21 7DU
G4OID PE21 7JG
2E1FRQ PE21 7JG
G4RPI PE21 7LS
G4GZL PE21 7PN
G4IDE PE21 7PN
G8MZX PE21 7PN
G7LSP PE21 7QR
G4YKA PE21 7RT
G8ZTR PE21 8BJ
G8OYL PE21 8DA
G7MTE PE21 8DL
G0NWU PE21 8EU
G3WZA PE21 8EU
G0KUC PE21 8HZ
G4FJO PE21 8PA
G0HPV PE21 8UF
G7HNM PE21 8UJ
G0BCE PE21 8UL
G3JRY PE21 9AE
G0RJD PE21 9AH
G7LFU PE21 9AP
G0KQD PE21 9BS
G8AAP PE21 9DE
G0DAF PE21 9HA
G1HJJ PE21 9HR
G6VYS PE21 9LZ
G7IBH PE22 0BG
G2FFD PE22 0BU
G0UFC PE22 0BZ
G7DYD PE22 0JD
G8MWC PE22 0LF
G0KOO PE22 9LS
G7WID PE22 9QQ
G0KZN PE22 9QR
G4DYV PE22 9QX

Bournemouth

G1SDA BH1 1HY
G4UUZ BH1 1HY
G0MPS BH1 1RG
G1IFA BH1 3PJ
G8BLW BH1 3SH
2E1CEQ BH1 4DH
G1JYE BH1 4DX
2E1EZJ BH1 4SQ
G1INB BH10 4EE
G7IWW BH10 4EY
G8NYJ BH10 4EY
G4ZLL BH10 4HR
G1NCG BH10 5AW
G4ZLT BH10 5AW
G3JLH BH10 5BE
G1OQM BH10 5EP
G8CYT BH10 5JT
G8UAD BH10 5LF
G3WZP BH10 5NP
G0AEP BH10 5NX
G4BRQ BH10 6DJ
G7MTT BH10 6DS
2E1DJI BH10 6DS
2E1EXE BH10 6ET
M0ASK BH10 6EY
G1XES BH10 7AA
G7DKR BH11 8ET
G7HLG BH11 8HE
2E1FYQ BH11 8PS
G4RHP BH11 9DY
G0IJE BH11 9DZ
G0IRL BH11 9DZ
G0SKR BH11 9EU
2E1FTA BH11 9EU
2E1ASU BH11 9HJ
G4ONN BH11 9LH
G0UCX BH12 1QE
G4FNI BH2 5NR
G1ABX BH2 5PH
G0NMW BH2 6JG
M1AXG BH2 6NH
G7HEY BH3 7AF
G3ZWD BH3 7JZ
G8WGM BH3 7LZ
G1HDO BH4 8AU
G8CER BH4 8AZ
G3HTB BH4 8BG
G8UFF BH4 8BQ
G7FPR BH4 8BX
G3ZLB BH4 8DH
G7JPP BH4 8HU
G3ZGJ BH4 9LA
G8HHQ BH5 1EJ
G4FUP BH5 1JH
G8PVK BH5 1LY
G4TBI BH5 2DU
G8CQG BH5 2DU
2E1ELO BH5 2DU
G3PKA BH6 3AB
G8ARA BH6 3HJ
2E1DHK BH6 3HR
G1PFY BH6 3LQ
G7SUM BH6 3LU
G4YNO BH6 3NN
G0JLH BH6 3QB
G1VPU BH6 3QB
G3KTU BH6 4DX
G4RGP BH6 4DX
G4YRY BH6 4EW
G3IQX BH6 4HU
G4GVF BH6 4JT
G1FGS BH6 4LU
G6MYT BH6 4NA
G8CRZ BH6 4NB
G4OLE BH6 5BQ
G8GLY BH6 5DT
G6EZM BH6 5HP
G0WXZ BH6 5HW
G7OXG BH6 5JX
G8ASX BH6 5PY
2E1AVT BH6 5PY
G3UBZ BH7 6BQ
G4ERV BH7 6LF
G7NSM BH7 6PN
G4ELC BH7 7AA
G4BNO BH7 7AS
G0OPI BH7 7BD
G4JDW BH7 7BD
G0BXM BH7 7HZ
G4SCB BH7 7JG
2E1FTO BH8 0JP
G6NKJ BH8 0JS
G8IKK BH8 8NS
G0GID BH8 8NX
G7DLC BH8 8RB
M1BUF BH8 8SF
G1BNG BH8 8SR
G4WFZ BH8 8SR
2E1EOL BH8 8UA
G8OWZ BH8 9AE
G0EJO BH8 9HW
G0EGR BH8 9HY
G1GRB BH8 9LT
G7GTP BH9 1BD
G4KLB BH9 1DB
G4LAH BH9 1DB
G4DJG BH9 1DJ
G4ERU BH9 1LH
G1WAW BH9 1LU
G7UNW BH9 1LY
G3IWV BH9 1PH
2E1AQU BH9 1TW
G8RAJ BH9 1TX
G1UTA BH9 2BU
G3JAU BH9 2JE
G6CML BH9 2JQ
G0UPH BH9 2JW
G3AWP BH9 2ND
G0UAP BH9 2QJ
G1BHW BH9 2SU
2E0APU BH9 2TY
G6ZAX BH9 2UD
G6TKH BH9 2UW
G7VIK BH9 3LP
G6FBR BH9 3LU
G1HRH BH9 3LW
G3VEZ BH9 3NT
G3VCE BH9 3NX
G4PTC BH9 3PG
G3BHA BH9 3SB

Bracknell Forest

G8RWG GU15 4PZ
G3NTG GU47 0RH
G4PKE GU47 0TD
M1AEJ GU47 0UP
G3CZM GU47 0XY
G6ACJ GU47 0YY
G4EHJ GU47 8EA
G4SWM GU47 9AU
G4HDK GU47 9RS
G6XUQ RG12 0TH
G0HWT RG12 0TJ
G7ANG RG12 0TX
G7IBU RG12 0UD
G4DDL RG12 0XU
G3RAC RG12 1RG
G3SVD RG12 1WA
G0MZV RG12 2HX
G4MXY RG12 2JR
G8RKO RG12 2LU
G3RXM RG12 2QG
G4YBG RG12 2RA
G7CAH RG12 2RG
G7VTC RG12 2RN
G3WXD RG12 3JX
G0BBB RG12 6HU
G4CVN RG12 7AX
G4BZJ RG12 7DU
G1BBI RG12 7ES
G4RNU RG12 7JA
G4CQH RG12 7LD
2E1FPY RG12 7ND
G6CNR RG12 7NR
G3YEQ RG12 7NU
G8AMK RG12 7QA
G4DDN RG12 7QG
G8LAY RG12 7QG
G0GLW RG12 7QJ
G0IRH RG12 7QQ
2E1AXO RG12 7RD
G8NTZ RG12 7RH
G4AUC RG12 7TF
G3TKS RG12 7WF
G1AOL RG12 7WG
G0CGE RG12 7WH
G8VME RG12 7WH
G3TLH RG12 7WJ
G0SCY RG12 7WL
G0FCT RG12 7YX
G6BRA RG12 7YX
G4KGK RG12 7YY
G6JAZ RG12 7YY
G4KNZ RG12 7ZT
G8ZJP RG12 8DB
G0WND RG12 8FP
G0UIS RG12 8QU
G7SWR RG12 8UF
G0AQU RG12 8UP
2E1AXP RG12 8UQ
G1RXV RG12 8UY
G7JDF RG12 8XE
G1COW RG12 8XH
G4DZA RG12 8XR
G6TNR RG12 8XU
G4PCF RG12 8XY
G6DNL RG12 8ZJ
G6DUB RG12 9BU
G7RUH RG12 9BY
2E1FJU RG12 9BY
G8XKH RG12 9EF
G0VQX RG12 9EQ
G4WYC RG12 9ES
G3YMC RG12 9EX
G7USG RG12 9HT
M1CAX RG12 9JE
2E1FJI RG12 9JE
G3OFW RG12 9JL
G8DUV RG12 9JX
G8HZL RG12 9NP
G8LZS RG12 9PS
G8RQN RG12 9QX
G3PPB RG42 1FL
G6XPB RG42 1RT
G3RMW RG42 1TL
G3MDD RG42 1UE
G4WZJ RG42 2BT
G3UBC RG42 2DT
2E1FLT RG42 2ES
G3NCN RG42 2HL
G7DXC RG42 2JY
G3VG RG42 2LD
G3TAI RG42 2LE
G1EEL RG42 3EF
G4MBV RG42 3RP
G8SBF RG42 3SG
G8FIF RG42 3SZ
G0ATZ RG42 3UN
G4HLF RG42 3XA
G4BUW RG42 4BS
G7FSR RG42 4DG
G4ORB RG42 4DS
G6FXR RG42 4UH
G0GJV RG42 5LG
G4BRA RG42 5LG
G6XSY RG42 5LG
G3IKN RG42 6AS
G6LKS RG42 6PW
G0AQK RG42 7UT
G4NGT RG45 6DU
G4EAH RG45 6EL
G7VWO RG45 7EG
G4DHY RG45 7ER
G0HXN RG45 7HP
G8STR RG45 7JG
G0WCS RG45 7NR
G0UYJ RG45 7QH
G4DGF RG45 7QR
G8LF SL5 8HQ
G1DER SL5 8JS
G1ALW SL5 8JZ

Bradford

G0PCM BD1 4EJ
G8NKN BD10 0NF
G4RSF BD10 0RJ
G4RSG BD10 0RJ
G7PBK BD10 0UL
G3WDW BD10 8BL
G6LJA BD10 8BL
G0LRO BD10 8PD
G1CHQ BD10 8PU
G4TFD BD10 8RS
G4FLJ BD10 8RW
G4GIT BD10 9QR
G6BIU BD12 0JQ
G6CJT BD12 0PL
G0WJC BD12 0UX
G0BVQ BD12 8DN
G1ACJ BD12 8DN
G8NML BD13 1AY
G3JCU BD13 1ET
G4XGN BD13 1LP
G4NTA BD13 1PL
M0AAM BD13 2BH
G1HWM BD13 2EP
G8NWK BD13 2EP
G8AJZ BD13 2HJ
G1NQU BD13 2JH
G0KVM BD13 2JN
G8FJR BD13 3DQ
G4JHS BD13 3PQ
G3LXY BD13 3QX
G4WFC BD13 4LZ
G1HJL BD13 5AD
G7NEC BD13 5EJ
M0AIY BD14 6BL
G0CEF BD14 6BQ
G3HMV BD14 6JZ
G1IEP BD14 6NP
G1UHB BD14 6NP
G3CMN BD14 6RY
G3UCK BD15 0BE
G3USH BD15 0BP
G7HEN BD15 0HB
G4EZX BD15 0HH
G8SX BD15 7AJ
G0ACX BD15 7AQ
G3NN BD15 7AQ
G3NYR BD15 7AU
G4TIV BD15 7AU
G0LGB BD15 7YD
G6BPG BD15 9AT
G0EQF BD15 9BD

G0IFT BD15 9BD
G4ZFT BD15 9BD
G0BBE BD15 9BL
G7KRC BD15 9BL
G0KRG BD15 9BL
G0KRS BD15 9BL
G0RLO BD15 9BL
G6OSJ BD15 9BT
G4THH BD15 9JQ
G6MJR BD15 9JQ
G4JJS BD16 1AT
G0NKN BD16 1AT
G0VPO BD16 1ER
G3UOI BD16 1LW
G4JNN BD16 1QA
G4CTA BD16 1QJ
G4SMK BD16 1RB
G1FOG BD16 1RX
G1CWD BD16 2DY
G7STU BD16 2DZ
G3RXS BD16 3BX
G0SUI BD16 3BX
G6KJT BD16 3DF
G4YCP BD16 3DH
G3EDX BD16 3NJ
G3TXX BD16 4DR
G6MC BD16 4EJ
M0ATN BD16 4EQ
G7UTR BD16 4QG
G2DFP BD16 4QP
G0PVB BD16 4RQ
G4YWR BD16 4RW
G8UPK BD16 4SE
G8ESK BD16 4UB
G6RO BD17 5AJ
G4BNH BD17 5BA
G8FTX BD17 5DD
G0LJM BD17 5ED
G1KAS BD17 5HS
G8GOV BD17 5LZ
G0RJC BD17 5NR
M0AMH BD17 5NR
G3VRX BD17 5NR
G7HKU BD17 5NR
G6XPT BD17 5PL
G6XPU BD17 5PL
G0FVO BD17 6DR
G4GMB BD17 6DR
G4IUP BD17 6UE
G3OTO BD17 7LJ
G7VVX BD17 7LJ
G0UPX BD17 7LQ
G1YWU BD18 1AQ
M0AMA BD18 1DD
G4SON BD18 1LU
G3TJC BD18 1NB
G0OZE BD18 1PD
G7KHE BD18 2JB
G0OEJ BD18 2LT
G0PBA BD18 2LT
G1LZF BD18 3JB
G0GME BD18 3JL
G4ORT BD18 4AG
G8ORM BD18 4DY
G0VJB BD18 4EP
G8CYY BD18 4HR
G7TKH BD18 4JZ
2E1DMY BD2 1QD
G3DKD BD2 2DY
G6XVY BD2 2LL
G6YGJ BD2 2NE
G7DFC BD2 2NJ
G0SNV BD2 2NJ
G1CFZ BD2 3AY
G1ANA BD2 3SR
G1IYA BD2 3SY
G8MVD BD2 4HX
G0DBT BD2 4JD
G1WRG BD2 4RN
G7GLN BD2 4RN
G6PAR BD2 4SA
G7NFG BD2 4SA
G8WXR BD20 0LD
G1VAL BD20 0LG
G4UAS BD20 0LG
G0JFC BD20 0QQ
G8WBK BD20 0QU
G0HOT BD20 5DB
G1OTE BD20 5DD
G7HLV BD20 5LB
G4DAJ BD20 5TE
G0MDO BD20 5UQ
G1HEA BD20 6AY
G0GEQ BD20 6JZ
G1OTA BD20 6SZ
G8WCT BD20 6TP
G4VTY BD20 9JS
G0LVT BD20 9LL
G4YXB BD20 9NR
G3VDK BD21 1HY
G6WHL BD21 2HB
G4JTV BD21 2TR
G8VPX BD21 4NP
G6WAI BD21 4RR
G0AJF BD21 4TF
G0BWY BD21 4TL
G4ZVD BD21 4TT
G0WBI BD21 4UJ
G7JER BD21 4YF
G0MJB BD21 4YG
G6WBR BD21 5BW
M1BVX BD21 5NP
2E1FON BD21 5NP
2E1GAG BD21 5QH
2E1GAH BD22 0EF
M1BUU BD22 0HA
G0WNE BD22 0PD
G4AEE BD22 0QY
G0MEA BD22 6DD
G0EUS BD22 6NB
G7NIC BD22 6QE
M1BZD BD22 7AL
G1AVA BD22 7BP
G0MMC BD22 7DN
G7HJT BD22 7DX
G4TFT BD22 7EX
G0ESA BD22 7JH
G0OSA BD22 7JH
G0GVE BD22 7PD
G7BVV BD22 7QQ
G3UBD BD22 7QS
G0BZH BD22 7RH
G0RLY BD22 8BJ
G0TSJ BD22 8BJ
G3TFF BD22 8HQ
G3BOR BD22 8QE
G8TZT BD22 8RE
G7KRG BD22 9LE
G1SRA BD22 9LE
G1YNF BD3 0DB
G3UDW BD3 0HL
G1BZD BD3 0LG
G4WYD BD3 0RH
G4EIL BD3 8QA
G0GNU BD3 9JT
G0JTP BD3 9NU
G4DCW BD4 0RX
G1BBR BD4 0SW
G0BZV BD4 6BA
G0GMM BD4 7ES
2E1EGV BD4 7JD
G8VSI BD4 7LE
G6LLR BD4 7QU
G8AED BD4 7RT
2E0APP BD4 9AP
G8LCP BD4 9AP
G4KZW BD4 9LX
G6TQS BD5 7HN
G0ITU BD5 9HB
G1WJP BD5 9PL
G3XBE BD6 1BX
G7SKQ BD6 1BY
G3ZMQ BD6 1DE
G8PTK BD6 1DG
G7ONG BD6 1JS
G6ZLY BD6 1JY
G0FUY BD6 1PQ
G4XZE BD6 1PS
G0IQH BD6 1QU
G3VUU BD6 1RN
G4RFO BD6 1RP
M0AMB BD6 1RP
G4YOR BD6 1UU
G3TIX BD6 2LP
G3ZSA BD6 2QS
G7BXN BD6 3BE
G4UNH BD6 3DJ
G0VVK BD6 3JQ
G0FLD BD6 3QF
G7JZM BD6 3SW
G0LGD BD7 2PJ
G1HZH BD7 2QF
G0CDK BD7 4BU
G3YSV BD7 4NU
G0HEO BD7 4PG
G0OWI BD8 0AA
2E1FBY BD8 0AL
M0ARK BD8 0DB
G7TMZ BD8 0DB
G3BDD BD8 0EQ
G3ZNR BD8 0LY
G1OCS BD8 0NW
2E1FBZ BD8 9HU
G8ZMQ BD9 4AX
G4BUH BD9 4DS
G4VDT BD9 4JY
G6FDG BD9 5PA
G1JTD BD9 6AB
G3TBX BD9 6AJ
G8PZF BD9 6BJ
G0TVM BD9 6DQ
G1KWK BD9 6EX
G7PHW BD9 6EZ
G7PHY BD9 6EZ
G0XBN BD9 6QB
G8KSX LS29 0NE
G0HUK LS29 0PU
G8HZJ LS29 0QQ
G4NDQ LS29 0RJ
G7EZT LS29 0RL
G0SNX LS29 6HE
G8SRN LS29 6HH
G3VTY LS29 7NF
G8RGO LS29 7NF
G4BSS LS29 7NR
G3WPZ LS29 7PA
G7VUH LS29 7PA
G8AWN LS29 7QB
G0CMU LS29 7QH
G6NWK LS29 7RG
G0LIW LS29 7RY
G3IBN LS29 8HN
G0DWG LS29 8LN
G3BMX LS29 8NQ
G3WZZ LS29 8QA
G6JWM LS29 9BZ
G4DCY LS29 9JN
G7RDJ LS29 9JN
G4XHG LS29 9QH

Braintree

G4PQY CB9 7ED
G7UHU CB9 7ED
G7LVM CB9 7EE
G7MQC CB9 7EW
G8MRM CB9 7XR
G4VVQ CM3 2AY
G4WVL CM3 2BS
G6GCU CM3 2BS
G8RSX CM3 2DF
G4EPS CM3 2LB
G8LYV CM3 2LB
G4OAU CM3 2LJ
G6OWG CM3 2NQ
G4KTX CM3 2QS
G0HSN CM7 1EG
G8ANR CM7 1EQ
G6EJF CM7 1XW
G6DFZ CM7 2LF
G6OIY CM7 2LF
G3PEN CM7 2PR
G6TAF CM7 2PR
G7DTR CM7 2PR
G6XCU CM7 2RZ
G0IZW CM7 2SL
G8WPO CM7 3NR
G3TGB CM7 3PE
G3KFK CM7 3PP
G6BMV CM7 3QR
G3XG CM7 3YP
G0XBE CM7 3YR
G8PLO CM7 4BY
G4ONH CM7 4JZ
G6XJC CM7 4LE
G8HCZ CM7 5AE
G4MIU CM7 5BU
G4WXT CM7 5BY
G3WVR CM7 5EG
G0EOE CM7 5LH
G7LNU CM7 5LP
G8DJO CM7 5LW
G0EMK CM7 5PY
G4JXG CM7 5PY
G6XCY CM7 5SU
G7JET CM7 5SU
G1XWN CM7 5UQ
G3YTE CM7 8HQ
G6SJY CM7 8LQ
G6IAB CM7 8QH
G6IOW CM7 8UA
G4DML CM7 9EP
G7UDU CM7 9EX
G0DEC CM7 9LL
G8XVO CM7 9LL
G3ITL CM7 9LR
G0DCR CM7 9NF
G1WRH CM7 9NJ
G8NPF CM7 9TG
G0PYW CM7 9TN
G4RPF CM7 9TP
G4VOT CM8 1DR
G0LQP CM8 1EU
G6CSK CM8 1HR
G7CGC CM8 1NQ
G6FBF CM8 1QR
G1TWG CM8 1UB
G8NGM CM8 1XY
G6SPS CM8 1XZ
G0ANZ CM8 2FA
G7LRU CM8 2HH
G1XXW CM8 2JP
G6JUG CM8 2JW
G3KXI CM8 2LE
G8WSC CM8 2LJ
G0DJK CM8 2LL
G3TEC CM8 2LX
G8IFN CM8 2NT
G1SDN CM8 2PE
G1NNA CM8 2SZ
G1NNB CM8 2SZ
G7BIQ CM8 2SZ
G7LEY CM8 2SZ
G0PJZ CM8 2UG
G7BND CM8 2UQ
G4VJE CM8 2XB
G3XVV CM8 3PH
G7TAV CM8 3QP
G4MJW CM8 3QX
G4KQE CM8 3SP
G8WQZ CM8 3SP
G4ICP CM8 3XN
G1HIU CO10 7EQ
G3CQL CO10 7EQ
G0BHW CO10 7EX
G4DHU CO10 7JX
G1EUF CO10 7LS
G8ADZ CO5 9AG
G7BIV CO5 9LX
G0HKG CO6 1AR
G7VGC CO6 1QN
G4YZW CO6 1ST
G8YCG CO6 2ET
G7PLL CO6 2NG
G3SUV CO6 2QB
G0PFV CO8 5EP
G1ZLD CO9 1AS
G0GII CO9 1AY
G6CHJ CO9 1DL
G6LMC CO9 1ED
G7UUD CO9 1EH
G7UUA CO9 1EH
GW6YTY CO9 1JL
G0CQI CO9 1JU
G1GRZ CO9 1JU
G4YAX CO9 1JU
G6NEK CO9 1JU
G6LHG CO9 1JZ
G0GWN CO9 1LB
G0HCF CO9 1LB
G3MMA CO9 1PQ
G1CCW CO9 1TD
G4TEB CO9 2BE
G7JTY CO9 2BG
G6NZY CO9 2DW
G6IAJ CO9 2HN
G6WPJ CO9 2HP
G6KEQ CO9 2LH
G7UVV CO9 2NW
G3PFH CO9 2QS
G0LZY CO9 3AZ
G0KQV CO9 3JH
G8FVC CO9 3NE
G1OFX CO9 3NX
G1WQQ CO9 3RN
G0VIS CO9 3RS
G6WHY CO9 4LN

Breckland

G8MVE IP22 2SB
G6HKF IP22 2SD
2E1FDM IP22 2SD
G3GRQ IP24 1AY
2E1DJR IP24 1DG
2E1DJS IP24 1DG
G4OQK IP24 1NE
G4LXD IP24 1NG
G3SXP IP24 1NH
G4BUV IP24 1PH
G0CLT IP24 1TQ
G2DFY IP24 2DR
G7KMR IP24 2LF
G0EBA IP24 2LW
G0RWQ IP24 2XR
G0DVI IP24 3EX
G0WVG IP24 3HL
2E0ANK IP24 3HL
G4VEL IP24 3HQ
G1LRK IP24 3NF
G8PON IP25 6AS
G3GIB IP25 6BZ
G7JRP IP25 6BZ
G4GRT IP25 6DB
G3LPA IP25 6DP
G4AED IP25 6EA
G3DOV IP25 6EY
G1NAN IP25 6LG
G6YTV IP25 6NL
G0GGR IP25 6QE
G0IMC IP25 6RB
G0NMP IP25 6XB
G7NRT IP25 7AN
G6XTM IP25 7BH
G0KZI IP25 7EH
G4HMQ IP25 7HG
G0UZH IP25 7LG
G0LGF IP25 7LS
G0FMI IP25 7PJ
G1MJE IP25 7QN
G1OMT IP25 7QN
G0JRO IP25 7SJ
G7BRN IP25 7SJ
G1LZQ IP25 7SJ
G6BAF IP26 5EF
G2CNN IP26 5HW
G7BFT IP27 0RH
G6XQP NR16 2DE
2E1FDI NR16 2EQ
G0BOO NR16 2HL
G0ENB NR16 2LF
G0WIJ NR16 2NA
G6HYP NR16 2NA
G3JIE NR16 2PS
G6AVI NR17 1BL
G4NRG NR17 1QU
G5LW NR17 2AN
G0CBO NR17 2DH
G0BEZ NR17 2JP
G3JUU NR17 2NH
G0TYB NR19 1BL
G0CWD NR19 1BP
G0FVE NR19 1HW
G3XPT NR19 1JB
G7HCT NR19 1JY
G8UUV NR19 1LU
G4EYA NR19 1ND
G0TYC NR19 1ND
G7EWK NR19 1PH
G1TLH NR19 2ET
G0TZZ NR19 2HF
G3YHO NR19 2LX
G4UVA NR19 2NS
G7PYL NR19 2QE
G0CHX NR19 2SU
G4URS NR19 2UA
G0LGJ NR19 2UB
G3YLA NR20 3AH
G0AKO NR20 3BH
G4LPW NR20 3DG
G0UYA NR20 3JR
G7NLS NR20 3JR
G3EBJ NR20 3JT
G3IVA NR20 3LJ
G0LGY NR20 3PN
G7RBH NR20 3QZ
G4DYC NR20 3RE
G0UML NR20 3RF
G0VIJ NR20 3SB
G3ZIG NR20 4AA
G4ZIG NR20 4AA
G7HJD NR20 4DW
G4IPN NR20 4LR
G3FKV NR20 4NB
G0TYQ NR20 4NQ
G4BYV NR20 4NU
G4XCR NR20 5TA
G6AER NR21 7JJ
G0JFT NR21 7NN
G3YHP NR21 7NT
G4FKC NR21 7NT
G0DKB NR9 4PE
G3VYG NR9 4QZ
G4VAO NR9 5LN
G3OOK NR9 5PN
G1GYA NR9 5RL
G4CXE PE32 1SY
G1ULE PE32 1TG
M1BYD PE32 1TG
G3DKO PE32 2DU
G7BUN PE32 2NA
G4KLC PE32 2PF
G3UKF PE32 2PS
G1TAG PE32 2QD
G3JKM PE32 2TH
G7GFY PE33 9PS
G1LMQ PE33 9PS
G0TAW PE37 7ET
G7VPS PE37 7EW
G1JBT PE37 7JB
G6SVY PE37 7QZ
G8CDW PE37 7SL
G7MOK PE37 7SL
G7LAX PE37 7TE
G0FVW PE37 8AZ
G0VNN PE37 8EE
G1KIW PE37 8ET
G0SDA PE37 8EX

Brent

G6VIO HA0 1AE
G6PGJ HA0 1JS
G0ECF HA0 3EE
G6MLV HA0 3QG
G0BHA HA0 3TG
G3MNO HA3 0DA
G0TAV HA3 0NB
G3FKI HA3 0PB
G0OWO HA3 0PG
G0RMK HA3 0PG
G7GPT HA3 0RN
G4KAB HA3 0SQ
G4JSG HA3 0TH
G0MWO HA3 0UL
G7TRT HA9 7ET
G6LXJ HA9 7JD
2E1BID HA9 7QH
G4UDT HA9 7QR
G7NMJ HA9 8AS
G3ZZF HA9 8TP
G1APL HA9 9EU
G1GYN HA9 9PU
G7JSV HA9 9SB
G3THQ NW10 1PL
G7DFF NW10 1SU
G6DJE NW10 2BG
G4LRK NW10 3UJ
G7ULM NW10 5UX
M1AHS NW10 9JP
M0AHK NW10 9LN
G4PEF NW2 5EH
G2DOJ NW2 6EU
G4BPN NW2 7SU
G1JKB NW2 7TA
G6DKW NW2 7TA
G7VHK NW6 5BA
G3PHU NW6 7AT
G1AWF NW9 8JL
G1ICO NW9 8YB
G1DMS NW9 9NQ
G6XLL NW9 9NQ

Brentwood

G6BNM CM12 9SL
G3JWI CM13 1BT
G7VZY CM13 1BT
G4ZON CM13 1JX
G0PNO CM13 1SJ
G8NYD CM13 2HZ
G7HRL CM13 2LA
G3VED CM13 2LX
G8CUB CM13 2NF
G7ULT CM13 2RU
G4TTQ CM13 2TJ
G4TTR CM13 2TJ
G4WYI CM13 2UF
G0FMU CM13 3AL
G0LKY CM13 3DZ
G3XQU CM13 3EP
G8TAO CM13 3EX
G8PUY CM13 3PF
G0KSV CM13 3RA
G1JEA CM13 3RG
G7NHK CM13 3RP
G7KME CM14 4JN
G1NGO CM14 4LR
G6VTY CM14 4PT
G0FDS CM14 4YN
2E1FXM CM14 5AQ
G8APZ CM14 5AZ
G8PUB CM14 5AZ
G6PZF CM14 5BN
G7HFL CM14 5BP
G4XFZ CM14 5DG
G6BCL CM14 5HF
G3MCW CM14 5JA
G6WJH CM14 5JB
G4SYR CM15 0BQ
G8HEJ CM15 0DT
G1XUO CM15 0NB
G8DXV CM15 0NB
G8SGM CM15 0NH
G4IGU CM15 0PS
G8SPP CM15 0QX
G4VZF CM15 0TJ
M1BCO CM15 8BA
M0BCZ CM15 8BA
G4HIM CM15 8BE
G3ZRH CM15 8BW
G4NGS CM15 8JL
G1SOT CM15 8LR
G4YVW CM15 8PF
G3MWP CM15 8PP
G0LLA CM15 8PT
G8DWP CM15 8SA
M0ASX CM15 9LN
G6AGN CM15 9NX
G4BLD CM15 9QS
G6VJB CM15 9RB
G3FJO CM4 0AL
G7TPV CM4 0BZ
G8GIG CM4 0RS
G4NLF CM4 0RZ
G8ADX CM4 9AN
G6VF CM4 9HG

Bridgend

GW0DHE CF31 1JY
GW0SZU CF31 1NZ
G0SXZ CF31 1PF
2W1FUD CF31 1PF
GW0PYU CF31 1PQ
GW1TFU CF31 1QF
GW1MGR CF31 1QW
GW6GWK CF31 1QW
GW7MGR CF31 1QW
GW0UMO CF31 1RT
GW0TSX CF31 1TX
GW7AWO CF31 1TX
GW0FJP CF31 2ED
2W1FXX CF31 2HY
MW0AXA CF31 2JR
GW7PRT CF31 2JR
GW3ZTH CF31 2NG
GW6TMW CF31 2NU
GW1ZNC CF31 2PF
GW1UOV CF31 2PR
GW3JUV CF31 4HD
GW4BKG CF31 4JT
GW7NIQ CF31 4NA
GW7NJT CF31 4ND
GW1KQV CF31 4NR
MW0AEL CF31 4PY
GW0VWD CF31 4QJ
GW3SRF CF31 4QS
GW0WIM CF31 4QW
GW1FKY CF31 4QX
GW1ZKM CF31 4RZ
2W0AMY CF31 4SB
GW8ASA CF31 4SY
GW0LXD CF32 0BH
MW1BNY CF32 0DT
GW0TTF CF32 0HL
GW8MXG CF32 0HY
GW1XJJ CF32 0LH
GW0UZK CF32 8AH
GW7VBE CF32 8AH
MW0AML CF32 8BX
G7TNS CF32 8HD
GW0BBQ CF32 8PT
GW0GQC CF32 8SD
GW4IQP CF32 8UU
GW1OUP CF32 8YB
GW1YHA CF32 8YB
GW4VKG CF32 9AF
GW8WZC CF32 9EX
GW0WSU CF32 9JJ
GW3XHJ CF32 9LF
GW4SAE CF32 9LQ
GW4BDV CF32 9LR
GW7NIS CF32 9NH
GW0MNP CF32 9PA
GW0DOX CF32 9SA
GW0KAX CF32 9SB
GW4THK CF32 9UG
GW8CNS CF32 9YH
GW7KYT CF32 9YW
GW4LNP CF32 9YW
MW1AID CF33 4NG
GW4YMJ CF33 4RJ
MW0AGE CF33 6AP
GW0HNK CF33 6LP
GW0CYK CF34 0AB
GW3WRE CF34 0BB
GW4VSE CF34 0BD
GW0VJS CF34 0BD
2W1AID CF34 0LU
GW3SSK CF34 0LU
GW4TGO CF34 0NN
GW0TOM CF34 0NT
GW0VMS CF34 0NT
2W1AKD CF34 0NT
GW0WHU CF34 0PB
MW0AMO CF34 0QA
GW0IRP CF34 0UR
2W1AOD CF34 9DU
GW3XJC CF34 9HP
GW8VQY CF34 9HP
2W1FUL CF34 9HP
GW4JCE CF34 9LT
GW0DIX CF34 9SS
MW0AWO CF34 9ST
MW1BGM CF34 9ST
2W1EYP CF34 9TF
GW2BFD CF35 5EP
GW7LDP CF35 5HD
GW6ZJO CF35 5HX
MW1BUN CF35 5HX
2W1EYN CF35 5HX
GW1YDN CF35 5NF
GW0PND CF35 5PW
GW4HZM CF35 5QA
GW8HWL CF35 6ED
GW3SYL CF35 6HH
GW4DUY CF35 6RP
GW1ERA CF35 6RS
GW3RVG CF35 6SD
GW1YBF CF35 6TY
2W1FUV CF35 6UT
2W1FUW CF35 6UT
MW1BYF CF35 6UW
2W1EYM CF35 6UW
GW0JZN CF35 6YF
MW1AYS CF36 3JA
GW0BAH CF36 3PR
GW4PCJ CF36 3SL
G4STY CF36 3TU
G6MNZ CF36 3TU
GW0BZA CF36 3YS
GW7DTB CF36 5AJ
GW1MNC CF36 5HN
GW7KUK CF36 5RY
GW7NNW CF36 5RY
GW3ARS CF36 5SG

Bridgnorth

G1BDK DY12 3BJ
G4EQK DY14 0UA
G7CJC DY14 8RR
G0LHH DY14 8UJ
G1UWM TF11 8EL
G0VSJ TF11 8HZ
M0AUN TF12 5EF
G0CCK TF12 5NA
G6DBZ TF12 5SH
G6KJY TF13 6BZ
G1DMJ TF13 6EY
G4BZQ TF13 6HQ
G8FUY TF13 6JQ
G3XEI TF13 6JU
G4HFX TF13 6NJ
G0UYE WV15 5DS
G4TCX WV15 5DU
G4JHK WV15 5ED
2E1FNL WV15 5HH
G1WBP WV15 5LF
G6WNG WV15 6BE
G1DKG WV15 6BS
G7VLM WV15 6BW
G0KLS WV15 6HD
G6HPS WV15 6JX
G0IZE WV15 6LT
2E1ETA WV16 4EG
G8UPF WV16 4HU
M1BGN WV16 4HW
G1SVR WV16 4JW
G3SVR WV16 4JW
G3TVR WV16 4JW
G3JZG WV16 4LA
G3ZLR WV16 4NR
G4NKC WV16 4NW
G3LZY WV16 4PU
G4YTX WV16 4RY
G6GUH WV16 4SH
G6SXD WV16 5AH
G6JMG WV16 6EL
G1DIL WV16 6EQ
G4EQR WV16 6JY
G4YGT WV16 6LQ
G8HPR WV16 6LX
G8CBA WV16 6PR
G0RQG WV16 6QN
G1ZJJ WV5 7AS
G0BXD WV5 7BD
G4VYA WV7 3DQ
G0ISI WV7 3EN
G4JLQ WV7 3EX
G7JPN WV7 3HL
G7FEF WV7 3LF
G4EAB WV7 3LS
G4OYJ WV7 3LZ
G0RNX WV7 3ND
2E0ALO WV7 3RB

Brighton and Hove

G0CIM BN1 1NW
G0TLN BN1 2EB
G4GQR BN1 2FB
G0TLU BN1 2FB
G7UPT BN1 3AB
G4CFB BN1 3LU
G6YRJ BN1 3RU
G7MRH BN1 3RU
G4LEV BN1 4AP
G3XHQ BN1 4NT
G4PAQ BN1 4QE
G7SRZ BN1 4SG
G4GYQ BN1 5AD
G8KTG BN1 5AG
G0PAW BN1 5AJ
G1JUP BN1 5AJ
G0SJF BN1 5EB
G0CJZ BN1 5FA
G4FNL BN1 5FN
G3SKI BN1 5GA
G3UWZ BN1 5GB
M0ADY BN1 5GD
G3CUY BN1 5HH
G6CXO BN1 5NH
G3XMW BN1 6DH
G3WBK BN1 6EB
G7VDC BN1 6GR
G8USF BN1 6NG
G6AIK BN1 6NH
M0AGQ BN1 6NL
M1BIP BN1 6NL
G7UTH BN1 6PG
GW7KIL BN1 6WG
G4GDT BN1 6WL
G7VWC BN1 6YG
G0RNN BN1 7ED
G3GZT BN1 7EG
G7UCO BN1 7FB
2E1BSH BN1 7FG
G1KKF BN1 7FR
G6JNW BN1 7GH
G3EWT BN1 7HP
G0EKV BN1 7HT
G4IPR BN1 7HZ
G2DBP BN1 7JD
G0SFV BN1 8DB
G3DVL BN1 8FB
G7PYT BN1 8HA
G6CYZ BN1 8HD
G8GEZ BN1 8HF
G4BVH BN1 8HJ
G0PWH BN1 8JD
G0EXS BN1 8JE
G3TDL BN1 8LR
G3WXG BN1 8NE
G2FSR BN1 8NL
G8VYP BN1 8NL
G8VIC BN1 8NP
G8TTP BN1 8QW
G1GGB BN1 8RE
G1OYG BN1 8RH
2E0ABD BN1 8SP
G7PTZ BN1 8SP
G4ZWD BN1 8TQ
G3XBN BN1 8WE
G3HGQ BN1 8XG
G6ZXF BN1 8XG
G0CKA BN1 8XQ
G3YRF BN2 1AN
G0SIX BN2 1DE
G4HWF BN2 1EH
G3UON BN2 1HU
G1YOI BN2 1JG
G7UDE BN2 1PG
G4WGP BN2 1QD
G6FFH BN2 2GP
M1ATU BN2 2HP
G3XDK BN2 2JL
G4MFR BN2 2YE
G7UZE BN2 2YP
G8EWF BN2 3EH
G3OEM BN2 4DP
G7WBO BN2 4EL
G0RNS BN2 4GF
G7FJC BN2 4HX
G8OMR BN2 4HX
G0VLM BN2 4NX
G8YQB BN2 4RG
G6LMB BN2 5GF
G8JFT BN2 5JS
G3ZSI BN2 5PH
G0JKP BN2 5PJ
G3MMJ BN2 5RE
G4EEJ BN2 5RH
G0NDF BN2 5TD
G6JNV BN2 6BE
G3UNY BN2 6BF
G1GGC BN2 6BJ
G0CIL BN2 6DD
G1SBZ BN2 6DJ
G6IQE BN2 6LB
G3SSE BN2 6LP
G8KHH BN2 6NA
G7UFD BN2 6QH
G0ODI BN2 6RG
G3YY BN2 6RH
G1FZS BN2 6TG
G3ITF BN2 7DP
G7OOS BN2 7GE
G8ZRM BN2 7GG
G4CBZ BN2 7GL
M0ATQ BN2 7HH
G7NKU BN2 8AE
G2CHI BN2 8AF
G1EYS BN2 8AH
G3JFP BN2 8AL
G4MFF BN2 8AN
G0ECW BN2 8FQ
G0ENJ BN2 8HR
G1KAP BN2 8QJ
2E1FGT BN2 8SR
G4KEI BN3 1PD
M1BIL BN3 1RL
G4JZC BN3 1TB
G4HZR BN3 1TD
G2UD BN3 2LA
G3VHT BN3 2NN
G3LRU BN3 3DJ
G1JRC BN3 3WF
G4GUX BN3 3WJ
G0UVK BN3 4GH
G6MPE BN3 5BD
G3VBE BN3 5BE
G1TEY BN3 5BF
G3MOL BN3 5DF
G4YLW BN3 5DP
G1VPC BN3 5FF
G7WHX BN3 5HP
G3ECM BN3 5ND
G6JAY BN3 5SQ
G8BZL BN3 6BJ
G4DXO BN3 6BN
G7SVX BN3 6HG
G3HZT BN3 6NT
G4DGW BN3 6PJ
M1BZU BN3 6PQ
G4KIL BN3 6PT
G6FOL BN3 6QL
G7NCC BN3 6TJ
G3VLC BN3 6UQ
G0IGA BN3 6WQ
G7MGG BN3 7FR
G1GSK BN3 7GX
G3TVK BN3 7JA
G4BWJ BN3 7JS
G1GDJ BN3 7PB
G4ORP BN3 8BA
G4GNX BN3 8JB
G7PMT BN3 8JE
G0KRE BN3 8JS
G0UZD BN3 8LE
G1BIA BN3 8LL
G0SJP BN3 8LN
G6WDR BN3 8PL
G8IQX BN3 8PP
G1JAU BN4 1PW
G3EUE BN4 3WE
G3HVH BN41 1GE
G4XKF BN41 2DN
G4WTL BN41 2FT
G1GID BN41 2HN
G7OBD BN41 2LS
G7OBE BN41 2LS
G3IAV BN41 2PT
G4SLE BN41 2RG
G4LKH BN41 2TA
G1ELE BN41 2YB
G0XAN BN41 2YF
G0HKI BN41 2YG
G0SLW BN41 2YG
G0JUU BN41 2YQ
G7GFX BN41 2YU

Bristol

G7HMQ BS1 4PJ
G0EES BS1 6FD
G8XPO BS10 5EJ
G6GHD BS10 5PN
G0PID BS10 6BP
G7NJX BS10 6BZ
G0JKU BS10 6JN
G0JZA BS10 6JU
G5RQ BS10 7AG
G4EQP BS10 7NU
G0HNO BS10 7PH
G1EIR BS10 7XE
G0MRV BS10 7XL
G4HHL BS11 0EF
G0XAK BS11 0HZ
G8VUM BS11 0QZ
G4DTK BS11 9SP
G1PBX BS11 9SR
G8OTH BS11 9SR
G4BHZ BS11 9SZ
G3YHV BS11 9TX
G4AHG BS11 9TX
G0CYD BS11 9UR
G1RWT BS11 9XB
G1MHP BS13 0AZ
G7VIN BS13 0HS
G1HFJ BS13 0PS
G7VOK BS13 0QA
G3HKA BS13 7BL
G3XED BS13 7DB
G0TDS BS13 7DD
G3WCL BS13 7DF
G6FMN BS13 7LY
G7UWP BS13 7ND
G0AWX BS13 8AT
G0RCU BS13 8AX
M0AKF BS13 8DB
G8JMK BS13 8DP
G1AVB BS13 8HQ
G0BFH BS13 8LF
M1BOB BS13 9HS
G6PJS BS13 9QG
G0CEM BS13 9RB
G4RZY BS14 0EG
G4WAW BS14 0EG
G4YZR BS14 0EG
G4JFX BS14 0HH
M1BHJ BS14 0JB
G7AVD BS14 0RL
G4OJI BS14 0TG
G3ZOX BS14 8DR
G3XFH BS14 8HX
G3GON BS14 8NF
G0KDS BS14 8PZ
G7PVG BS14 8SS
G4UCH BS14 8TN
G4SDR BS14 9DF
G6PRO BS14 9ER
G4PHZ BS14 9EY
G4ZCK BS14 9LJ
G3KUL BS14 9NL
G4TSS BS14 9NN
G4WUB BS14 9PH
G4CQI BS15 1HF
G3EWF BS16 1DQ
M1AFZ BS16 1UP
2E1DPX BS16 1UP
G3FZS BS16 2AQ
G0PDV BS16 2EJ
G4ULH BS16 2EL
G3VWE BS16 2EU
G4TWU BS16 2HP
G3TWT BS16 2PH
G6LOG BS16 2QB
G1IXE BS16 3DR
G1IXF BS16 3DR
G0JLA BS16 3LF
G1FWF BS16 3NG
G4ZBT BS16 3QJ
G4ZCT BS16 3SJ
G6GNG BS16 3XJ
G3UGB BS16 4BU
G1PNF BS16 4HJ
G7HOJ BS16 4HJ
M1BEP BS16 4HW
G7UUC BS16 4JD
G7FBD BS16 4JT
M1ATD BS2 8EP
M1BZC BS2 8JD
G3TVV BS2 9UB
G8RYP BS3 1DT
G4LJW BS3 1LE
G6GQG BS3 1LP
2E1BBG BS3 2AU
G0FXI BS3 2BP
G6NWF BS3 3DN
G4XED BS3 3EA
G8BDZ BS3 3EA
2E1BJE BS3 3EA
G3IUO BS3 3LY
G1HYQ BS3 3PW
G1DBH BS3 4NE
G1ACY BS3 4PL
G8SUV BS3 5BT
G1TNP BS3 5DU
G4EGR BS3 5EG
G4ZXE BS3 5NU
G6HMO BS3 5QT
G4YTH BS4 1BN
G7LPP BS4 1PL
G0UFQ BS4 1ST
G0VJN BS4 1TN
G4YKG BS4 2BG
G6HMV BS4 2DX
G6DEN BS4 2DY
G0IWP BS4 2EG
G6XFC BS4 2HA
G3TTZ BS4 2HE
G6HPC BS4 2LH
G6LPG BS4 2PT
G0EXU BS4 2UP
G0CCU BS4 3EY
G0KOS BS4 3HJ
G4BWB BS4 3NH
G4KKU BS4 3QP
G0IFF BS4 3SY
G1XXE BS4 4AF
G7MHX BS4 4EQ
G3XOD BS4 4JL
G6VPW BS4 4JT
G4LEA BS4 4NA
G8YOF BS4 4QH
G8ZOE BS4 4QJ
G7DUA BS4 4QP
G0JQW BS4 4QS
M1ACQ BS4 4QX
G1SLU BS4 4RA
G4GTD BS4 4RA
G0EWC BS4 4RN
G0SYI BS4 5JA
G1WOS BS5 0JG
G6PDR BS5 0JU
G1JOO BS5 0SE
G6YRQ BS5 6BG
G8YMM BS5 6HW
G4TPV BS5 6RJ

G1SMT BS5 6RQ
G8BIR BS5 6SY
G4GGE BS5 7BQ
G1HPZ BS5 7BT
G0SCK BS5 7EJ
G0GJN BS5 7NG
G0EPJ BS5 7PL
G0HTS BS5 7UB
G1HQO BS5 8AJ
G3TAD BS5 8DX
G3ZKI BS5 8DX
G4JPS BS5 8DX
G8XAA BS5 8DX
G7TQR BS5 8DX
G8CKK BS5 8ST
G0JLE BS5 8SZ
G6XNQ BS5 8SZ
G6YFG BS5 8TA
G0KWF BS5 8TW
G1LUI BS5 9AF
G0EXO BS5 9BL
G0JPS BS5 9BW
G8NNU BS6 5HX
G1AWI BS6 5LU
G1VNK BS6 5LU
G0HMH BS6 5TU
G0TRR BS6 6AL
G3ORV BS6 6BE
G7OQJ BS6 6BZ
G4FVX BS6 6NS
G4TRN BS6 6PD
G8DHZ BS6 6TG
G4DKS BS6 6UD
G4AAJ BS6 7HE
G4LOX BS6 7LG
G0EZE BS6 7QD
G4LYO BS6 7QU
G3XTS BS6 7XS
G8UXB BS6 7XS
G4PQN BS7 0AW
G4DDU BS7 0BU
G6GN BS7 0ED
G1LPN BS7 0EH
G8OQG BS7 0HQ
G0NZT BS7 0HS
G4CSE BS7 0TT
G3FPY BS7 0UH
G3IUV BS7 0UQ
G4CGF BS7 0US
G3MGS BS7 8EX
M1ANL BS7 8EX
G8MHD BS7 8JJ
G0CGJ BS7 8LE
G4YQH BS7 8QG
G4YQR BS7 8QQ
G4NBG BS7 8RN
G4ZBS BS7 8RN
G1HIA BS7 8UP
G0CKQ BS7 9AF
G4DHK BS7 9AZ
G4BYJ BS7 9ET
G0LUT BS7 9EX
G0CRV BS7 9HY
G4ROX BS7 9ST
G0PHZ BS7 9UA
G4VVC BS7 9YF
G3RFX BS8 1BA
G4PII BS8 1SF
G0UHT BS8 1SF
G6ILE BS8 2BN
G1CBG BS8 2DA
G7MML BS8 2JA
G4LQV BS8 2LR
G6GGN BS8 2TX
G7ASK BS8 3AU
G0AQE BS8 3BE
G1YOY BS8 4BH
G8OOQ BS8 4DL
G4BDB BS8 4EB
G0LTG BS8 4TP
G4NYK BS9 1DR
G4WGW BS9 1DW
G4MLR BS9 1JX
2E1DCY BS9 1RE
G0SDW BS9 1SN
M1BDX BS9 2AU
G3TCO BS9 2BW
G4TCO BS9 2BW
G0OBT BS9 2JF
G4WBV BS9 2LU
G3BJJ BS9 2PN
G0KJM BS9 2QP
G8XIM BS9 2QQ
G4ZBQ BS9 2QR
G0GDM BS9 2QT
G8PSO BS9 2QU
G6BGG BS9 2RY
G0GJB BS9 3JX
G8DWW BS9 3LR
G4DPJ BS9 3LU
M1AFH BS9 3NH
M0AWH BS9 3RN
G2HDR BS9 3ST
G8GRD BS9 3SX
G7GTN BS9 3TX
G4HCB BS9 3UU
G8KGH BS9 4AH
G3SJI BS9 4BU
G3ECS BS9 4DB
G4GCT BS9 4DB
G0KKX BS9 4EA
G2BQP BS9 4EL
G0OER BS9 4EU
G7BIL BS9 4HU
G4UIG BS9 4QD
G8GLQ BS9 4QW
G0CJG BS9 4RH
G4GAR BS9 4RH
G4KSR BS9 4RS
2E1DGF BS9 4RX
G4FRO BS9 4TA
G4OEP BS9 4TF

Broadland

G4RSP NR10 3BH
G8BIG NR10 3EA
G4YDB NR10 3EB
G8AWZ NR10 3ED
G7SOQ NR10 3EP
G0TMT NR10 3ES
G8YQH NR10 3HE
G7ETC NR10 3HF
G7PDO NR10 3HS
G1NIG NR10 3LF
G0SGT NR10 3LG
2E0AGV NR10 3NP
G0AUT NR10 3PG
G1ESX NR10 3PP
G4WUT NR10 3PZ
G7JTZ NR10 3QA
G3VPT NR10 3QB
G0VRY NR10 3SE
G0RHZ NR10 4AB
G3TNY NR10 4AH
G0PFJ NR10 4AZ
G1IDJ NR10 4EL
G0GGF NR10 4LP
G0IYK NR10 4LS
G8YJS NR10 4LU
G1TSY NR10 4QU
G0NNS NR10 5HQ
G0ODR NR10 5HQ
G0KYS NR10 5LP
G7UGB NR10 5NT
G3TOZ NR10 5QD
G3KMO NR10 5QW
G1NGE NR10 5QX
G8SMQ NR11 6HD
G7GQB NR11 6JF
G0EVQ NR11 6TS
G8RSQ NR11 6UT
G0TJA NR11 6UU
G1NJG NR12 7AB
G3VZT NR12 7AJ
G7RPK NR12 7AJ
G0VDR NR12 7AJ
G6ESQ NR12 7DT
G6PRI NR12 7DT
G4WUU NR12 7EE
G7RIE NR12 7EG
G1XUW NR12 7EW
G7SNT NR12 7HD
G6PSC NR12 7HG
G8AMJ NR12 7HT
G1EAJ NR12 7HZ
G1HXN NR12 7JU
M1BWQ NR12 7LG
2E1EHZ NR12 7LG
G0DGU NR12 7NB
G3PFS NR12 8TL
G4KAU NR12 8UX
G4JNX NR13 3AB
G4YYL NR13 3AD
G6WIT NR13 3AD
G0OOB NR13 3DH
G7SEH NR13 3DW
G4YGQ NR13 3PL
G7PWD NR13 3PY
2E0AHS NR13 3PY
G4UWY NR13 3RN
2E1FXY NR13 3SH
G0WZR NR13 3TG
G3IWC NR13 3TS
G3OGE NR13 4AB
G4UUB NR13 4AH
G4UAM NR13 4BA
G1MUC NR13 4BB
G0WGY NR13 4BW
G6FLJ NR13 4LG
G8FW NR13 4LP
G1PQJ NR13 4LR
G0KSD NR13 4LZ
G0GQV NR13 4NF
G6WQN NR13 4NT
G7UVY NR13 4QF
2E1FNJ NR13 4QF
M1ADX NR13 5DG
G4MCA NR13 5HR
G0KRU NR13 5NU
G7RBS NR13 5PA
G6DEA NR13 5QR
G7GTQ NR13 6AH
G3WWO NR13 6EL
G4KLM NR13 6LT
G8SXI NR13 6NG
G1BCU NR13 6QH
G6WMG NR13 6RQ
G3PDH NR13 6RR
G7AYQ NR20 5AD
G3JCK NR20 5SE
G0DWV NR6 5AD
G3ASQ NR6 5AU
G3JOC NR6 5DJ
G4ANT NR6 5DJ
G4GVR NR6 5HF
G0DDO NR6 5HQ
G0PFN NR6 5HU
G4EOL NR6 5HU
G6UPI NR6 5NN
G3TPZ NR6 5NW
G4YEJ NR6 5PE
G4ODC NR6 5PG
G0CUL NR6 5PQ
G4YDY NR6 5QJ
G0OPV NR6 5QN
G7SSB NR6 5RW
G4VWE NR6 5SJ
G0BAL NR6 5SQ
G4YGD NR6 5SW
G4XRL NR6 6EF
G0SRV NR6 6EF
G7SVD NR6 6EF
2E1DIO NR6 6EF
G4XNP NR6 6PB
G8SDU NR6 6PL
G7DVO NR6 6UA
G3GPX NR6 6UH
G4RTG NR6 6UP
G7WEW NR6 6UX
M1AEH NR6 6UX
G3IOR NR6 6XD
G4EYT NR6 6XF
G0IXL NR6 6XJ
G3CDY NR6 6XU
G0MJC NR6 7AN
G6TGQ NR6 7BB
G4SOP NR6 7BE
G6EJM NR6 7BG
G0VDO NR6 7BL
G4WEE NR6 7DP
G4JSD NR6 7HE
G4UIY NR6 7HE
G8EEY NR6 7HR
G0UEM NR6 7HZ
G0KRI NR6 7LB
G4AUV NR6 7LG
G4YGA NR6 7LQ
G1NCN NR6 7NL
G7VGQ NR6 7PN
G0CLR NR6 7QQ
G0NFV NR6 7QQ
G0RBG NR6 7RQ
G4TWS NR6 7RQ
G4TWT NR6 7RQ
M1AUW NR7 0DS
G7NFW NR7 0DW
G3HQS NR7 0JJ
G4VYH NR7 0JJ
G0PXC NR7 0LX
G0MBH NR7 0RL
G7UXT NR7 0TE
M1ADR NR7 0TE
G7AJX NR7 8AA
G3HKD NR7 8AD
G4TLK NR7 8BZ
G0GSC NR7 8EG
G0UEB NR7 8EG
G4AJO NR7 8EJ
G7MLR NR7 8EP
G3MFQ NR7 8JN
G3RQY NR7 8JN
G4LGB NR7 8NQ
G7HHL NR7 8QE
G4PCZ NR7 8XH
G0BBV NR7 9LL
G4OLP NR7 9LL
G2AGH NR8 6EJ
G3HEO NR8 6ER
G8AUX NR8 6JZ
G7NIX NR8 6LD
G3JQI NR8 6LL
G4VXR NR8 6LP
G8TAY NR8 6LP
M0AZW NR8 6QE
G7VUU NR8 6QE
G1AMN NR8 6QZ
G7SQW NR8 6XZ
G3YWB NR8 6YW
G7LBO NR9 5SD
G3FEW NR9 5SH
G3KVT NR9 5TF

Bromley

G0MTA BR1 2AF
G2AZP BR1 2HP
G0JBP BR1 2NF
G3JJZ BR1 3AB
G4FXR BR1 3PU
G1OJO BR1 3RG
G4OIM BR1 4DB
G3COF BR1 4HQ
G1GYF BR1 4QY
G3BR BR1 5AX
G4AHT BR1 5BT
G4AXA BR1 5DD
G3IUJ BR2 0EE
G1GDH BR2 0HQ
G7VEI BR2 0NG
G3IZA BR2 0RW
G1HCV BR2 0SH
G3EFS BR2 0UA
G0REE BR2 6BA
G7HCC BR2 6BA
G0SLY BR2 6BY
G6YRY BR2 6DG
G7VYJ BR2 6DX
G4CPQ BR2 7DD
G6TUG BR2 7DL
G8ZYH BR2 7DY
G0BUD BR2 7EX
G4XAT BR2 7HE
G8MVS BR2 7HR
G6SC BR2 7HT
G4TYT BR2 7HU
G8RW BR2 7JA
2E1BLV BR2 7JE
G7USR BR2 7JE
G8FEI BR2 7JR
G4IPF BR2 7JS
G8SXU BR2 7LX
G0BVB BR2 7PE
G4DJN BR2 7PQ
G1BMW BR2 7QA
G8BVQ BR2 7QN
G3OQD BR2 7QN
G3NGK BR2 7QT
G7BHN BR2 7QT
G8BUV BR2 7QT
G3UWY BR2 7QW
G3VFD BR2 8AY
2E1DHL BR2 8AY
G7VFD BR2 8AY
2E1FDP BR2 8AY
G3AEX BR2 8BA
G1DKY BR2 8EL
G4MSK BR2 8EY
G1BYS BR2 8LT
G1ZDR BR2 8PF
G3XDL BR2 9AS
G3UMI BR2 9EL
G0LMQ BR2 9JJ
G3NAT BR2 9QB
G4WGZ BR2 9QB
G4JHW BR3 1JF
G7RUS BR3 1LE
G8BSW BR3 1LY
G4HZX BR3 1NY
G7ATX BR3 1PA
G7KWO BR3 1QS
G7KWP BR3 1QS
G1TMD BR3 3AS
G7TOU BR3 3AY
G3TOK BR3 3BG
G0LSN BR3 3EL
G8NUJ BR3 3JW
G4FAM BR3 3NF
G4CZL BR3 3NN
G4FMA BR3 3NS
G3OKY BR3 3PL
G7LSZ BR3 3QA
G7CVS BR3 3QR
G4KLZ BR3 3QT
G7CXO BR3 3SB
G7HYM BR3 4AE
G4VJT BR3 4JJ
G1EFG BR3 4JU
G1ROK BR3 4LT
G8YNO BR3 4RB
G7ROL BR3 4RJ
G8NKM BR3 4RU
G3BWV BR3 4SD
G3SZR BR3 4SS
G3ZJX BR3 4SZ
G4NPD BR3 4XS
G1SLO BR3 5DB
G6MEH BR3 5DB
G2OS BR3 5HD
G4RTV BR3 6LJ
G4WII BR3 6NJ
G4ZUY BR3 6NJ
G0JOZ BR3 6PZ
G7TLP BR3 6SN
G7UFQ BR4 0BA
G8EKI BR4 0DY
G3MPX BR4 0EA
G8VZS BR4 0HA
G7TQW BR4 0LZ
G1PGS BR4 0QH
G7UZF BR4 0RA
G4IKL BR4 0SQ
G3ZQF BR4 9AH
G0ILW BR4 9AZ
G0JGT BR4 9AZ
G3CTI BR4 9HD
G0CVO BR4 9LF
G6CVP BR4 9NA
G3BWX BR4 9NN
G3CLK BR4 9NP
G8KZJ BR5 1BE
G3FYR BR5 1JE
G0AVY BR5 1JR
G0LWV BR5 1JT
G7URM BR5 1NF
G8KPY BR5 2DL
G1HIG BR5 2HF
G1UPT BR5 2HF
G7RUY BR5 2JR
G6LGR BR5 2LZ
G7RFM BR5 2NT
M1AUC BR5 2PN
G7CEW BR5 2SB
G0OMH BR5 3AT
G3ZCV BR5 3AW
G4MZM BR5 3AW
G4DOE BR5 3LF
G6DEV BR5 3LG
G7FRH BR5 3LY
G6ALK BR5 3WA
G2FQD BR5 4BH
G7BIP BR5 4DJ
G1NQS BR5 4EQ
G6XYD BR5 4EQ
G0AMM BR5 4ES
G3ODB BR5 4HR
G1YRV BR5 4JG
G1VOJ BR5 4LA
G0SZC BR5 4LA
G8SJR BR5 4LQ
G7PMG BR5 4LT
G3BVA BR5 4LU
G8CAM BR5 4NA
G0LXF BR5 4NS
G0JPM BR5 4PF
G4KMF BR5 4PN
G8TNQ BR5 4PP
G0IWH BR5 4PR
G0SIO BR6 0AN
G7PJD BR6 0AQ
G1VII BR6 0BH
G3LXW BR6 0DX
G0TZR BR6 0EJ
G8LSC BR6 0ER
G4JYB BR6 0QE
G6KKM BR6 0RW
G4AND BR6 0SG
G8LSI BR6 0TD
G7DNQ BR6 6AP
2E1FOL BR6 6DE
2E1BCR BR6 6JJ
G7RPS BR6 6JJ
G3SIU BR6 6JU
G8BGR BR6 7BP
G0AMP BR6 7HL
G1OSK BR6 7HL
2E1CFR BR6 7NS
G1UKH BR6 7RS
G4SVS BR6 7TJ
G3SDL BR6 8BN
G6YFF BR6 8DX
G7CEV BR6 8EJ
G3ZOH BR6 8HJ
G3MCA BR6 8HP
G8TBL BR6 8HQ
G6GKL BR6 8JE
G4TQO BR6 8JP
G1RYK BR6 9BN
G0BZX BR6 9BS
G3FAM BR6 9BS
G8HKN BR6 9BZ
G4XVK BR6 9EG
G0AHI BR6 9LN
G1JTX BR6 9LY
G8IKA BR6 9NF
G1EOH BR6 9PN
G6XVQ BR6 9PN
2E1AYE BR6 9PN
G7RPV BR6 9PN
G8AXA BR6 9QA
G4EZR BR6 9RJ
G7AQK BR6 9RS
G1FAA BR6 9UF
G8CQZ BR7 5EQ
G0REN BR7 5QD
G7CXU BR7 6AG
G4MVR BR7 6ED
G7ULL BR7 6JD
G1PIB BR7 6JY
G0GZV BR7 6LA
G8YCJ BR7 6LA
G3WDY SE19 2LQ
G7EWS SE20 7JF
G3VZH SE20 7JG
G4KBY SE20 7UF
G8PMT SE20 7YN
G0UXM SE20 8DH
G3ODU SE26 5LQ
G0AJJ SE9 4JL
G0LYG SE9 4LF
G3KEQ TN16 2JE
G0AUU TN16 3AG
G3ZVN TN16 3BJ
G4ASI TN16 3BN
G3OXS TN16 3BX
M1AJB TN16 3DD
M1BPK TN16 3DD
M0AFN TN16 3HN
G7UDK TN16 3HN
G0VZR TN16 3HN
G3KYV TN16 3LG
G7CJJ TN16 3NX
G4GLN TN16 3PH
G0NSF TN16 3PL
G8JQS TN16 3SG
G3WIA TN16 3TE
G4UOV TN16 3XP

Bromsgrove

G8ADV B14 5SL
G4ZTG B14 5SR
G4RPV B38 0BQ
G3YKO B38 0DN
G3XWV B38 9EY
G7UBA B45 8ET
G4TSB B45 8HP
G2HOS B45 8LY
G7UMY B45 8NZ
G1XKY B45 9DA
G8XWI B45 9DA
G0LEO B45 9EA
G7RLV B45 9HY
G4YTJ B45 9LR
G7LTH B45 9LW
G7MOA B45 9LW
G0TIB B45 9LW
G7FAZ B45 9TA
G8IFT B45 9XB
G0EYO B47 5BS
G3OIC B47 5EN
G4VPD B47 5HY
G0ICJ B47 5NR
G1WAC B47 5NR
G4WAC B47 5NR
G7PGX B47 5PB
G6JYO B47 5PX
G0NES B47 5QE
G3YIU B47 5QG
G4ZWB B47 5RL
G3MZU B47 6BU
G8SH B47 6HE
G0BNZ B47 6HG
G1NJJ B47 6HG
G4SVL B47 6HG
G1PYJ B47 6HP
G0NST B48 7EA
G0WMA B48 7NE
G1JWO B48 7NY
G1DCY B48 7PT
G3HDQ B48 7RX
G8RIN B48 7TL
G4JNV B60 1AL
G1GDR B60 1AW
G4FP B60 1AZ
G1FPY B60 1DF
G8DEC B60 1DG
G3RBL B60 1DR
G4LVK B60 1DS
G4OJS B60 1DY
G4WZA B60 1DY
G6NQB B60 1DZ
G0JPV B60 1HE
G0KHK B60 1HE
G4BBU B60 1HN
G4AHK B60 1HW
G8KNM B60 1HZ
G7RBQ B60 1HZ
G0BLT B60 2EB
G0HAW B60 2EB
G0KIN B60 2HF
G6CHX B60 2LJ
G7LUF B60 2RB
G0EVY B60 3AY
G4NYH B60 3EH
G4UIW B60 3EP
G0BIR B60 3HB
G0EHQ B60 3HB
G0SQI B60 3JE
G8SQH B60 3LD
G7BBY B60 3PL
G4MBW B60 3PN
G6VTA B60 3QH
G6NYG B60 4EB
G4FNQ B60 4NF
G4LXI B61 0EL
G6EET B61 0ER
G7ESV B61 0JG
G4LRL B61 0JT
G0DTG B61 0JY
G0AQF B61 0LQ
G4NZK B61 0LU
G8WJY B61 0NP
G8XUW B61 0PA
G4WFF B61 0PG
G1IEC B61 7BE
G4OAZ B61 7BH
G4ZOZ B61 7BH
2E1DLW B61 7BH
G4CQS B61 7EB
G1DYC B61 7JE
G0BGA B61 7JG
G3ZYT B61 7JY
G7RXX B61 7PJ
G8NUT B61 8LF
G0SUQ B61 8NQ
G4PKW B61 8UA
G6SKU B61 9LH
G7NKW B61 9LH
G0FLJ B62 0LN
G4OHB B62 0LN
G6DZY B62 0PA
G3XPY B62 0PT
G8HQP B90 1EH
G0RCE B97 5SU
G0CLM B97 6QF
G7RLX DY9 0JH
G4IFK DY9 0LJ
G6BEB DY9 0LJ
G6YAK DY9 0LX
G7PVW DY9 0NA
G6DDO DY9 9LX
G3VHA DY9 9QT

Broxbourne

G6XAR EN10 6EE
G4UOI EN10 6JA
G4DMI EN10 6LP
G4XVY EN10 6NY
G8HTA EN11 0JS
G7CNX EN11 0PJ
G0JXR EN11 0QU
G1BWP EN11 8PY
G4OAA EN11 8QX
G8DJU EN11 8QX
G0BQT EN11 9AU
G1MRE EN11 9DG
G4ARY EN11 9DG
G0BUY EN11 9EA
G0VAL EN11 9EA
G3VSJ EN11 9EP
G0BQF EN11 9JR
G1PZP EN11 9JR
G8ZYT EN11 9LD
G4UDZ EN11 9LZ
G1CAY EN11 9NR
G6UBW EN11 9QJ
G7HYG EN11 9QN
G7FCE EN11 9QS
G6FTA EN7 5AU
G4NRT EN7 5DE
G4STV EN7 5EN
G3PID EN7 5EW
G3SSZ EN7 5JT
G3MVM EN7 5NL
G8KKH EN7 6DF
G1AUU EN7 6HU
G4SXF EN7 6LG
G3KSW EN7 6NS
G3YVO EN7 6QU
G3XNP EN7 6RB
G4WIP EN7 6TZ
G3GBN EN8 0DH
G3URA EN8 0HT
G7OZM EN8 0PE
G0KLU EN8 0RF
G1HLO EN8 0SQ
G8RRC EN8 0SX
G0JXN EN8 0TQ
G1EWR EN8 0XE
G7BXB EN8 7BE
2E1DTT EN8 7DW
G4GRN EN8 7JJ
2E1EGI EN8 8DA
G4SGN EN8 8NH
G7SRJ EN8 8NS
G3IMC EN8 8UX
G8VLP EN8 8UX
G1DVX EN8 8XA
G6WEH EN8 9HH
G4SYY EN8 9JA
G1YRK EN8 9JR
G4UUM EN8 9NB
G0NLV EN8 9QY

Broxtowe

G7PNX NG16 1AG
G4XOU NG16 1AR
G4GFA NG16 1AS
G0JWD NG16 1BE
G0NKO NG16 1BE
G4DFU NG16 1BE
G4DVW NG16 1BX
G4TRQ NG16 1BY
G3POJ NG16 1DL
G0EVN NG16 1EG
G0TWA NG16 1EN
G3XRL NG16 1EP
2E1FPZ NG16 1FL
G0FJZ NG16 1HE
G1NBT NG16 1HE
2E1DLZ NG16 1HP
G4NUV NG16 1LE
G6HY NG16 2AH
G6XXJ NG16 2DL
G8XPZ NG16 2DP
2E1DKZ NG16 2DP
G3RTO NG16 2EF
G3VYK NG16 2EN
G0UUX NG16 2JJ
G7UII NG16 2QY
G6MRN NG16 2RA
G0VBS NG16 2RD
G0LIQ NG16 2RG
2E1EIK NG16 2SG
G4CKG NG16 2TH
G4KAQ NG16 3DE
G8SEV NG16 3DE
G4JRJ NG16 3DP
G8ZAU NG16 3DQ
2E1DME NG16 3DQ
G0GYH NG16 3DR
2E1CMC NG16 3DR
2E1CMD NG16 3DR
G6HZX NG16 3DY
G6MAW NG16 3EY
G4XBY NG16 3FP
G1WSD NG16 3FY
G0BXG NG16 3GW
G7HHM NG16 3GW
2E0AMS NG16 3HS
G7BTX NG16 3HS
G1VKY NG16 3JJ
G4VAX NG16 3NL
G4LMT NG16 3PZ
G6KSH NG16 3RF
G7LXQ NG16 5AT
G8BWC NG16 5AU
G4MHB NG9 1AU
G4ZSB NG9 1AX
G3QD NG9 1BD
G3BWN NG9 1EH
G6JRE NG9 1GR
G7TRB NG9 1GX
G6XOD NG9 1HE
G6XOE NG9 1HE
G2SP NG9 1HU
G0TVQ NG9 1HU
G0XBP NG9 1HZ
G7WIQ NG9 1JG
G4VFK NG9 1LA
G8ZK NG9 1LA
G8IGQ NG9 1LA
G0CQT NG9 1LY
G1MDL NG9 1PW
G0OQQ NG9 1PY
2E1EQQ NG9 1PY
G0GRZ NG9 2FJ
G3ZQH NG9 2HR
G7UUQ NG9 2PL
G4JHN NG9 2QR
2E1BKA NG9 2QR
G3TAP NG9 3AD
G7LNV NG9 3BY
M0AHO NG9 3BY
2E1CLL NG9 3EB
G7KSL NG9 3FN
G0PXP NG9 3FP
G3BMM NG9 3FQ
G0RSU NG9 3JG
G4MFV NG9 3JN
G1RZU NG9 3JN
G1PUZ NG9 3JT
G8BFM NG9 3LW
G4GQY NG9 3PX
G4DJA NG9 3PZ
G6XJD NG9 3RA
G6XJE NG9 3RA
M0BCC NG9 3RB
G0UYV NG9 4ES
G1ZJM NG9 4FH
G4KMN NG9 4GU
G8ZSZ NG9 5EB
G7PCZ NG9 5EB
G6AKP NG9 5EU
G4BNX NG9 5EZ
G4AQT NG9 5FU
G7UUP NG9 5GJ
2E1BOC NG9 5GS
G0SPA NG9 5HX
2E1BJG NG9 5HX
2E1CHX NG9 5HY
G7USM NG9 5HY
2E1EQN NG9 5HY
G7HIT NG9 5LA
G6FNI NG9 5LG
2E0AOI NG9 5LN
M0BDS NG9 5LN
G6IPC NG9 5NH
G6VXM NG9 5PA
G3XAZ NG9 5PB
G6RCD NG9 6AB
G8IYZ NG9 6BP
G4KQP NG9 6EW
G7PND NG9 6FZ
G0LAU NG9 6HR
G6UWO NG9 6HS
G4OFL NG9 6JP
G4XEL NG9 6JW
G6MSC NG9 6JW
G1RPP NG9 6NR
G7SHW NG9 6NX
G0NVO NG9 7ET
G1OTQ NG9 7EX
G7WEG NG9 7EZ
G4NBK NG9 7GY
G3VMK NG9 7HB
G6NZG NG9 7HR
G4PMA NG9 8DJ
G3DRP NG9 8DP
G3JTO NG9 8HR
G8PTN NG9 8HU
G8UQP NG9 8JB
G8PBH NG9 8JG
G6BTH NG9 8LN
G1DNZ NG9 8ND

Burnley

G0KMK BB10 1BA
G6EXP BB10 1EQ
G6FTB BB10 1EU
M0AFU BB10 2BX
G8YYX BB10 2HB
G8RXJ BB10 2HE
G0ECG BB10 2JT
G7ELR BB10 2JT
G8BEK BB10 2PZ
M0BEK BB10 2PZ
2E1CTU BB10 2SP
G4YMQ BB10 3AG
G8ZGF BB10 3JG
2E1CTS BB10 3JW
2E1CTT BB10 3JW
G6SYI BB10 3NU
G7STS BB10 3PS
G3XAB BB10 3QH
G0ELR BB10 3QU
G6PPY BB10 4AJ
G3MBU BB10 4JA
G4BYL BB10 4JL
G7CYC BB10 4LH
G0KKK BB10 4TF
G6IKK BB10 4TF
G2FFO BB10 4TT
G8RVY BB10 4UE
G6EVK BB11 2JL
G8LWK BB11 2JS
G0WZL BB11 2NU
G0RKP BB11 2PL
G4RQC BB11 3EJ
G7KWC BB11 3EL
G3IY BB11 3EQ
G1HZL BB11 3LH
M0AIS BB11 3PR
G8XCE BB11 4BP
G0UAA BB11 4EG
G1JXU BB11 4HG
G6AXB BB11 4JQ
G8VFU BB11 4LE
G0RFY BB11 5AE
G0XBB BB11 5DW
G0OCK BB11 5EB
G0OCL BB11 5EB
G1IZU BB11 5HJ
G7BRS BB11 5HP
G6XCD BB11 5LU
G4OPN BB11 5RB
G4UJB BB11 5RE
G7DYO BB12 0BU
2E1CTW BB12 0EA
G0DZC BB12 0ED
G1ZBP BB12 0EF
G8RTO BB12 0EF
G0MUH BB12 0JG
G6YRC BB12 0JJ
G4EGS BB12 0TA
G8UA BB12 6DG
G6PRA BB12 6JT
G6RIM BB12 6NG
G6IFV BB12 6NJ
G3YGC BB12 6NQ
G7DMS BB12 6NZ
G4UMH BB12 6PA
G0LQQ BB12 6RA
G4RQA BB12 7DB
G4GQP BB12 7DN
G7GOQ BB12 7JU
G7KTD BB12 8AF
G7RHX BB12 8AW
2E1CXF BB12 8DR
G8MUY BB12 8TA
G3YGD BB12 8UH

Bury

G4PAS BL0 0DP
G0IEJ BL0 9JX
G7CER BL0 9JX
G4GLH BL0 9LS
M1APV BL0 9NJ
G6SHD BL0 9QG
G4DQP BL0 9QZ
G4APJ BL0 9UT
G1MZC BL0 9YH
G0JOG BL0 9YN
G6ERK BL2 5QU
G0KEX BL2 5RJ
G4YJJ BL2 5RP
G4CIC BL2 6RQ
G4HZB BL2 6RX
G1UTW BL8 1AS
G0HUQ BL8 1BJ
G6DEG BL8 1DW
G6JJF BL8 1ES
G0TGR BL8 1JB
G8UN BL8 1JH
G7CEM BL8 1LR
G7NLP BL8 1NE
G1RAO BL8 1UE
G2DPL BL8 1XW
G3BRS BL8 1XW
G6BRS BL8 1XW
G8NRS BL8 1XW
2E1CDT BL8 1XW
2E1CDU BL8 1XW
G6BIT BL8 1XX
G0BZU BL8 2BQ
G4GSY BL8 2HD
G8LIR BL8 2PP
G0ALQ BL8 2PS
G6DGV BL8 2TS
G0JLL BL8 3BL
G0VYP BL8 3BL
G6PZE BL8 3DB
G8AEN BL8 3DB
G8EZU BL8 3DB
G4UGR BL8 3EN
G0LVN BL8 3EZ
G0WVY BL8 3JA
G3DQQ BL8 3JE
G8PW BL8 3JZ
G4ZSI BL8 3NT
G4IME BL8 4AR
G3BQT BL8 4BG
G0UCV BL8 4BW
G1PKO BL9 5DL
G4KLT BL9 5DL
G0GPH BL9 5DQ
G3FHM BL9 5DY
G3UGC BL9 5JG
2E1FOS BL9 5JT
G3IXC BL9 5LB
G7JXN BL9 6JE
G4UYJ BL9 6NH
G7RBN BL9 6PJ
G4YYD BL9 6PP
G2AYG BL9 6QT
G1YUS BL9 6RN
2E1FPB BL9 6RR
G4KQZ BL9 7BU
G3UQK BL9 7RN
G7JUL BL9 7SG
G4KQI BL9 8BE
G3TNQ BL9 8DN
G6MHO BL9 8HN
G7SPL BL9 8NL
G7DAL BL9 8PD
G3LGN BL9 8PL
G0LVX BL9 9DQ
G6AMX BL9 9HS
G7TYO BL9 9HZ
G8PAL BL9 9PD
M1AGA BL9 9QE
G4KWN BL9 9SZ
G0RAN BL9 9UD
G8BIO M24 4QR
G3AYY M25 0AY
G7PYN M25 0FZ
G3TCA M25 0HR
G8JWC M25 0HR
G3ZVT M25 0HX
G8UPJ M25 0LR
G1PHK M25 0NA
G6YGB M25 1FN
G7NRS M25 1JJ
G4VUK M25 1NB
G1VWP M25 2GP
G6FNA M25 2GP
G7TBM M25 2GP
G2FRZ M25 2RL
G0NVR M25 2RN
G3SSH M25 3HE
G0CSU M25 3HZ
G1JHB M25 3JF
G7VKJ M25 3JJ
G0WEF M25 9LT
G0UAY M25 9NT
G1JZY M25 9TG
G0WJP M25 9TR
G4RIG M26 1BE
G1PKP M26 1JA
G4OLW M26 1JA
G6CLW M26 1JA
G3ZCH M26 1YF
G6OTQ M26 1YF
G8XVX M26 2TD
G0ADL M26 2UX
G3TVT M26 2XE
G4UAU M26 3GL
G7NOQ M26 3GW
G1ESW M26 3UJ
G3ZZA M26 3UN
G3TKI M26 3WU
G7PTE M26 3WU
G7UVE M26 4BQ
G7BYS M26 4FS
G8ZIC M26 4FS
G1EHY M26 4HG
G6UBU M26 4HW
G0VVG M26 4PB
G7PJC M45 6EL
G8AOG M45 6EP
G3GXI M45 6NS
G8KRG M45 6NS
G8DUT M45 6TJ
G0UKM M45 7AZ
G3COA M45 7JN
G1COX M45 7LR
G1ESA M45 7NB
G4ISB M45 7PG
G8NKJ M45 7SS
M1BNH M45 7SS
G8OUT M45 7ST
2E1CEY M45 8ET
G4GOM M45 8GN
G0AHR M45 8WY
G4STL M45 8WY

Caerphilly

GW0PSV CF46 6DU
2W1EKR CF81 8JX
GW0GIQ CF81 9BQ
GW0NKG CF82 7AN
MW1AFW CF82 7FW
GW0PCJ CF82 7RH
GW3YVN CF83 1DA
GW0IRC CF83 1DS
GW0HYU CF83 1SQ
GW0MXG CF83 2NZ
GW4GJT CF83 2NZ
GW0RVC CF83 2NZ
GW6VYE CF83 2RA
2W1EDA CF83 2UE
GW6EFK CF83 2UF
MW0ATK CF83 3NX
GW7VKI CF83 3NX
GW0CKL CF83 4EQ
GW5NF NP1 4BE
GW0NZN NP1 4DF
GW4JJR NP1 4DW
GW1JRM NP1 4JJ
GW4FPX NP1 4JL
GW0NUS NP1 4RE
GW0JWF NP1 4RS
GW3OVD NP1 4TB
GW1NYO NP1 5DD
GW8ZWD NP1 5DH
GW7RZC NP1 5DL
2W1EIN NP1 5DL

MW0ATF NP1 5FZ
MW1AXS NP1 5FZ
2W1EDL NP1 5JY
GW0NPL NP1 5LA
GW3NWC NP1 5LR
GW0LKA NP1 6HA
GW7THP NP1 6PP
2W1GAZ NP1 6QD
2E1EHL NP1 6QT
GW8UXI NP1 7AF
GW1ZKE NP1 7HY
GW0OQN NP1 7LY
GW7MYD NP1 7PB
GW7IAT NP1 7PF
GW0MTI NP1 8FY
GW3EQL NP1 8SS
GW7OCJ NP2 0BD
GW7PBP NP2 0BD
GW0JXG NP2 0DB
GW6GW NP2 0LQ
GW0MAW NP2 0LQ
GW0LKJ NP2 0LX
GW0RQS NP2 0QE
GW3KYA NP2 0UQ
GW8MTJ NP2 1DF
GW1PJP NP2 1DZ
GW6JDJ NP2 1EE
GW0LJW NP2 1EW
GW3YSF NP2 1HB
GW0MOW NP2 1HQ
GW3MMU NP2 1HZ
GW0ARK NP2 1LX
GW3TKZ NP2 1LX
GW0EYH NP2 1NG
GW0ETM NP2 1NY
GW1EHI NP2 1NY
GW3XVQ NP2 1PG
GW8ITI NP2 1QH
GW0IXP NP2 1SJ
GW0JMJ NP2 2AD
GW0LTH NP2 2AF
GW4JKV NP2 2GA
GW4FCV NP2 2HP
GW0SZN NP2 2JB
GW4HBK NP2 2JU
GW0NPM NP2 2ND
GW4KDD NP2 2NR
GW4VRH NP2 2PE
GW4EAI NP2 2PF
GW8TRO NP2 2PF
GW4RGT NP2 2PN
GW8TIX NP2 5BH
GW4CXK NP2 5LF
GW0VCK NP2 6AL

Calderdale

G3SJD HD2 2EB
G4XTE HD2 2EH
G3JWY HD3 3FU
G0DIU HD6 1HH
G7RTA HD6 1QT
G1GTQ HD6 2AT
G1GTR HD6 2AT
G8AUL HD6 2BJ
G3OTE HD6 2DA
M0AJR HD6 2DA
G7PDA HD6 2DA
2E1COL HD6 2DH
M1BIO HD6 2EQ
G7UST HD6 2EQ
G0GRR HD6 2LF
G6KQF HD6 2LL
G7LBD HD6 2NA
G1NYG HD6 2QG
G8MLD HD6 2RS
G4SDX HD6 2RU
G4WAP HD6 3JS
G7KXS HD6 3LD
G3JWN HD6 3LG
2E1FTZ HD6 3NF
G3RSS HD6 3NP
G3WAH HD6 3NP
G1JFQ HD6 3RF
G4KQJ HD6 3RZ
G3HWF HD6 4EB
G4DNB HD6 4EB
G0TJQ HD6 4EQ
G0LDV HD6 4JF
G4IPG HD6 4JQ
G4IPH HD6 4JQ
G4REG HD6 4JZ
G0BXO HX1 2PW
G8AFV HX1 2YJ
G8WFP HX1 2YN
G6FPF HX1 3EA
G0VQQ HX1 3LE
G7SLP HX1 3RB
G7IGJ HX1 3UG
G8EKH HX1 4QD
G6XTT HX1 4QW
G4XYS HX1 4RH
G0CVC HX1 4TA
G7MAJ HX1 4TF
G0KWG HX1 5JL
G4GAK HX2 0BH
G6SZP HX2 0DL
G3ONQ HX2 0EF
G0BFQ HX2 0JB
G1MWT HX2 0LD
G1NDQ HX2 0LU
G0JBI HX2 0NL
G3TAY HX2 0NP
G4MUR HX2 0NW
G6TGD HX2 0NW
G6ACT HX2 0NW
G7RRC HX2 0NW
G7RST HX2 0NW
G0OZP HX2 0PJ
G1RZZ HX2 0PL
G8EXN HX2 0RB
G7OMF HX2 0RH
G4EHD HX2 0RL
G4NOQ HX2 0RP
G3FDC HX2 0UG
G4VOB HX2 0UL
G7FLE HX2 0UT
G8RKX HX2 6EX
G3OTJ HX2 6TT
G6CNL HX2 6UX
G0INK HX2 7HG
G8SDE HX2 7HP
G0TDW HX2 7JX
G6YGV HX2 7PN
G7TDN HX2 7RB
G4NSH HX2 7TX
G3UI HX2 8AA
G1GWX HX2 8DL
G1VVB HX2 8DL
G4YDG HX2 9DA
G0VHU HX2 9DA
G8EYQ HX2 9DS
2E1BBC HX2 9EP
G6GGV HX2 9JD
G0JAQ HX2 9JH
G0JBC HX2 9JJ
G2SU HX2 9JQ
G3TQA HX2 9JQ
G3CXR HX2 9LQ
G7VBT HX2 9SD
G0CBI HX2 9XD
G6COG HX3 0AG
G7SWZ HX3 0BD
G3XSF HX3 0BT
G4OYZ HX3 0DS
G4BZS HX3 0EA
G0DFB HX3 0LT
G1KWX HX3 0NE
G3AAV HX3 0NW
G0SPX HX3 0SR
G8REO HX3 5QF
G1HCU HX3 5SZ
M0AGM HX3 6JP
G1PWI HX3 6NB
G0DKJ HX3 6NX
G0CZL HX3 7AP
G6MDC HX3 7EP
G7PFI HX3 7EX
G3RKH HX3 7HH
G8UTW HX3 7LB
G0PMU HX3 7NA
G0HRJ HX3 7NH
G4XUK HX3 7NR
G3MMK HX3 7PW
G4KEX HX3 7PW
G6YAN HX3 7QS
G8NRY HX3 7RD
G4YRH HX3 7TS
G1BXN HX3 8EA
G3IGW HX3 8HB
G7FZJ HX3 8HB
G1FCR HX3 8NP
G4RAW HX3 8NQ
G2UG HX3 8NQ
G7SZM HX3 8NS
G0DLM HX3 8NU
G1ILO HX3 8UD
G8NQN HX3 9SP
G6FTI HX4 8DY
2E1AGL HX4 8LF
G3TQQ HX4 8NN
G0ECY HX4 8PG
2E1EVN HX4 8QW
G4LIW HX4 9AJ
G4LIX HX4 9AJ
G3YCJ HX4 9HP
G7IGK HX4 9HW
G0IJI HX4 9QL
G4FMM HX5 0BB
G7BLQ HX5 0DN
G6XNJ HX5 0QH
G3YGZ HX5 0QX
G6JSZ HX5 0SP
G6EQN HX5 9HU
G1HQD HX6 1BT
G4AES HX6 1NL
G0DUN HX6 1NS
G0TEH HX6 2NJ
G4UZU HX6 2TY
G7VOL HX6 3EF
G8EUH HX6 3EZ
G0SQA HX6 4AG
G0WFG HX6 4AG
G8EWN HX6 4LQ
G3UBI HX6 4NP
G3NBI HX6 4QJ
G4XNF HX7 5AR
G0AEC HX7 5BH
G0ITI HX7 5BH
G0CMQ HX7 5NF
G3UGF HX7 5PD
G6CQG HX7 5PD
2E1BOO HX7 5PD
2E1BVJ HX7 5PD
G7ISB HX7 5PH
G4HYY HX7 6BQ
G3HZR HX7 7AL
G0FYQ HX7 7AP
G4EFX HX7 7JA
G0WFP HX7 7JX
G6XPZ HX7 7NX
G7CLX HX7 7PB
G3MLS HX7 7PG
G3ZMX OL14 5JB
G0TVO OL14 5JG
G6ZZA OL14 5NW
G3OLP OL14 5NX
G1CBX OL14 5RE
G6MDB OL14 5SQ
G4KDU OL14 6AX
G8RFF OL14 6AX
G4TYW OL14 6ND
G8HHP OL14 6PF
G0TVB OL14 6QX
G6TRG OL14 6QX
M1AIN OL14 6QY
G1JBW OL14 7ET
G4AIT OL14 7HG
G4SGC OL14 7JJ
G4RCJ OL14 7QP
G8IC OL14 8DS
G7THI OL14 8SS

Cambridge

G8ZQO CB1 2HS
G1ERM CB1 2LD
G4EZN CB1 2LL
G1KYV CB1 3AB
G8GML CB1 3AL
G7IAY CB1 3PJ
G4MZN CB1 3PQ
G4ZZJ CB1 3QN
G5DQ CB1 3QU
G3DXB CB1 3RT
G1JKL CB1 3TP
G1OKI CB1 3UB
G0SFA CB1 3UB
G1SFA CB1 3UB
G1KUN CB1 3UF
G8KJP CB1 3UF
2E1DKD CB1 4AD
G7PSF CB1 4DG
G1TOW CB1 4EA
2E0AIZ CB1 4ER
G4GJC CB1 4HD
G3UPJ CB1 4LN
G3UFH CB1 4NA
G3KBR CB1 4RJ
G7BIX CB1 4SA
G4EAG CB1 4SG
G1KTE CB1 4SL
2E1DGZ CB1 4UG
G0GVZ CB1 4UR
G3INR CB1 4UR
G7ORL CB2 1QA
G7RJD CB2 1TA
G4XJU CB2 2HD
G3NHB CB2 2HR
G7SSU CB2 2JS
G4HHJ CB2 2LZ
G0TDT CB2 2RG
2E1DGY CB2 2RG
G4XIL CB3 0HX
G0JEP CB3 0DY
G3MRX CB3 9LQ
G3PYE CB4 1DP
G5PI CB4 1DP
G6AIG CB4 1ET
G6NKM CB4 1FJ
G7ANZ CB4 1FY
G4DNG CB4 1LW
G0UIO CB4 1NL
G0JOE CB4 1NN
G6FKS CB4 1NP
G1VJG CB4 1NZ
G1OXV CB4 1RL
G8IUV CB4 1RT
G3ZAY CB4 1SQ
G6UW CB4 1SQ
G1DVD CB4 1TZ
G8CTX CB4 1TZ
G3URX CB4 1XG
G4DUN CB4 1XJ
G7PLE CB4 1YU
G8DEE CB4 2AE
G3OWB CB4 2AJ
G8JGM CB4 2AN
G1TPV CB4 2AU
G8UDV CB4 2BJ
G3RSE CB4 2DB
G1RTV CB4 2DP
G7RTR CB4 2HJ
G0RDB CB4 2JB
G7KHL CB4 2LD
G8PLZ CB4 2LY
G3XAK CB4 2NL
G6KSV CB4 2NP
G4USM CB4 2PA
G0KFQ CB4 2QR
G6FNB CB4 2SA
G8UDG CB4 2SE
G7NZY CB4 2SH
G6ZUT CB4 2SY
G7HUV CB4 2UD
G1KGC CB4 2UH
G7GSF CB4 2UH
G8XYS CB4 2UH
G1AQX CB4 2UN
G3KNN CB4 3EX
G3HRE CB4 3ND
G4AJW CB4 3PA
G3IDT CB4 3PB
G7JRD CB4 3PL
G6OPO CB4 3QQ
G4MRD CB4 3TA
G0ANV CB4 3XH
G1DAZ CB4 3XH
G8SGF CB5 8LJ
G8OUS CB5 8LZ
G1YPU CB5 8NN
G6WKQ CB5 8UJ

Camden

G7RJO N19 5TR
G3GLV N6 6LH
G7LTQ N6 6NB
G4ZDM NW1 0SY
G7NYW NW1 1DN
G1YPZ NW1 2JL
G1BNV NW1 8TH
M1AOZ NW1 8UD
G4RPK NW1 9BX
G8OIY NW1 9HA
G4MGI NW1 9JH
G7UZN NW1 9QR
G0GNL NW1 9RN
G3AFT NW1 9TL
G3ZKE NW1 9TL
G8DWL NW1 9TL
G1DWC NW2 3TL
G8PDM NW3 1EJ
G0OXH NW3 2QJ
G1VUK NW3 2UA
G0MNS NW3 4BE
G8ACK NW3 4NA
G8UAW NW3 4XR
G0WJF NW3 5SJ
G3UKI NW3 5SU
G0WXV NW3 6PH
G3LSQ NW3 6SJ
G4JDG NW3 6SJ
G4BHY NW3 6YD
G4BTN NW3 7HL
G0ABK NW3 7PG
G3UGX NW3 7RR
G1XZP NW5 1DT
G8KLX NW5 1LR
G7TKW NW5 2PE
G6HEW NW5 3AD
G4WNM NW6 1AN
G7VFY NW6 1EA
G7VHH NW6 1HG
G0WKM NW6 1HG
G0WVJ NW6 1HG
G4VHD NW6 1HT
G3VCK NW6 2DT
GM4SJL NW6 3HP
G1CKK NW6 3LA
G4IOF NW8 6DA
M0AMU NW8 6NN
G7SQH W1P 5AA
GM0MVY WC1E 6BT
G8IKU WC1H 0ND
G2BNY WC1H 8DY
G1AUI WC1N 3HZ
G0GSB WC1N 3XX
G7HMK WC1N 3XX
G7HNF WC1N 3XX

Cannock Chase

G1UUL WS11 1AP
G1BRD WS11 1AR
G3PSU WS11 1BB
G0BHX WS11 1HH
G4XBK WS11 1HH
G4VJQ WS11 1HX
G3URL WS11 1LJ
G8DWA WS11 1NE
G4ICE WS11 1NQ
G7ASY WS11 1PS
G4LQY WS11 1QE
G4HOH WS11 1QT
G0ICW WS11 2DY
G7DWH WS11 2FR
G0CWB WS11 2HD
G0UYW WS11 2JT
G0JJP WS11 2PJ
G0UYH WS11 2SE
G7TCW WS11 3AF
G4YTK WS11 3QZ
G4SQF WS11 3RW
G7IZN WS11 3TN
G4TPC WS11 3TS
G6VXS WS12 4BL
2E1BAI WS12 4DA
G8UUM WS12 4DA
G7AMD WS12 4DD
G0DUQ WS12 4DQ
2E1EXK WS12 4DQ
G4KWR WS12 4LJ
G3SQK WS12 4LR
G4ZBF WS12 4SF
G3ZVK WS12 4SN
G1TOK WS12 5AB
G4RJD WS12 5AW
G6SW WS12 5AW
G3HRR WS12 5DJ
G6JNZ WS12 5DR
G8NTJ WS12 5DS
G0FEC WS12 5EZ
G6PAP WS12 5HZ
G8HRB WS12 5JB
G1SIG WS12 5LB
G0BXN WS12 5ND
G1MSX WS12 5ND
G6ZKL WS12 5NH
G4JVT WS12 5NW
G0VSK WS12 5PR
G1LFR WS12 5QE
G0AXR WS12 5QR
G1ADS WS12 5RN
G7WBH WS12 5RN
G0GPC WS12 5SG
G1FKF WS12 5SW
G0MVV WS12 5SX
2E1CKJ WS12 5YB
G7TEJ WS12 5YJ
G0XAT WS12 5YJ
G4ETA WS15 1AH
G6GOW WS15 1BB
G4NLK WS15 1EP
G7IHW WS15 1EP
M1AYA WS15 1EW
G4IHP WS15 1HN
G4XCW WS15 1HT
G4LAN WS15 1JE
G7ACG WS15 2AR
G7JMB WS15 2AR
G6TUI WS15 2LU
G7VJY WS15 2NS
G0OQJ WS15 2NS
G3LEK WS15 2PS
G0HHJ WS15 2QW
G1PAA WS15 2QW
G1SJJ WS15 2QW
G1WVZ WS15 2SN
G0RLA WS15 2TG
G7TEN WS15 2TL
G1IVY WS15 2XG
G1KQH WS15 2XH
G4BON WS15 2XR
G0JJO WS15 2YG
G1TCY WS15 2YG
G4PWD WS15 2YG
G0SXS WS15 3JZ
G3NSO WS15 4RG
G3TAR WS15 4RG
G6NVS WS15 4TH

Canterbury

2E1EGG CT1 1PX
2E1EHM CT1 1PZ
G7RBB CT1 1QG
2E0AMH CT1 1SJ
G6EGU CT1 1TS
G6YZU CT1 1TS
2E1AKX CT1 1XJ
G4OQD CT1 1YD
G3NFS CT1 1YF
G7IBP CT1 1YF
G8XAJ CT1 3JP
G0VZZ CT1 3NT
G4WDJ CT1 3PY
G4ZME CT1 3QL
G7TDD CT1 3QW
G4KGY CT1 3UL
G6OSW CT1 3XA
G6AOW CT1 3XU
G0GPO CT2 0EA
G7NIN CT2 0JZ
G7ELF CT2 0PD
G3FMU CT2 0QH
G7EOB CT2 0QX
G3KPB CT2 7BL
G7PCY CT2 7NT
G1KSH CT2 7QX
2E1BYC CT2 7RX
G3JES CT2 7TW
G0IEV CT2 8AY
G4LMC CT2 8DP
G7FWF CT2 8ES
G3MDT CT2 8PN
G3SZC CT2 8PP
G4PCN CT2 9BL
G7FCL CT2 9BQ
G8APK CT2 9DH
G8BRD CT2 9DL
G0KDN CT2 9HP
G0EHY CT2 9NB
G7YWF CT2 9NQ
G0BYV CT3 1SY
G6BLB CT3 1UY
G8YMN CT3 3JN
G8UIV CT3 4AH
G7RHV CT3 4DA
G4NEE CT3 4DB
G0TXA CT3 4DS
G2FMW CT4 5AX
G8OGP CT4 5BA
G4ZHN CT4 5EX
G7HYO CT4 5PU
G4AZG CT4 6HX
G6GES CT4 6JB
G7LNB CT4 6JQ
G4ZIF CT4 7AH
G2AJV CT4 7AY
G4BY CT4 7BQ
G0VRW CT4 7ND
G3XAQ CT4 7NN
G0IAA CT5 1EL
G4LQI CT5 1EP
G7EOE CT5 1EP
G7FUM CT5 1NR
G4SIL CT5 1NS
G3CLA CT5 1PA
G4ATX CT5 1PT
2E1DHN CT5 1PT
G4RIS CT5 1QF
G1JAS CT5 2AN
G7JWX CT5 2AT
G8WIR CT5 2DY
G3OXO CT5 2ED
G8NXQ CT5 2LB
G4LTS CT5 2LE
G0KCJ CT5 2LW
G4PYA CT5 2NW
G0FAE CT5 2PH
G4SIA CT5 3AF
G0IQW CT5 3EY
G4SEJ CT5 3JY
G0IFS CT5 3JZ
G6ETA CT5 3LQ
G3KXB CT5 3NG
G1DLU CT5 3NN
G0AUW CT5 3PJ
G1OPK CT5 3QA
G4AKT CT5 3QF
G4BEX CT5 3QL
G4ELP CT5 3QQ
G4ICM CT5 3QQ
G0GCQ CT5 4AH
G3FTE CT5 4BZ
2E1FQX CT5 4DA
G4UPJ CT5 4DD
G4ZKH CT5 4DN
G1BLA CT5 4DP
G4WZZ CT5 4DT
G4XAB CT5 4DT
G0PPY CT5 4EL
G4CZU CT5 4HX
G7UWW CT5 4LY
G0NBB CT5 4NY
G6OJI CT5 4PY
G0ILO CT6 5QT
G0KFA CT6 5SQ
G3WAW CT6 6AF
G1KQE CT6 6BH
G6HYN CT6 6DJ
G0BAB CT6 6HW
2E1AKV CT6 6HY
G4WZQ CT6 6JA
G2AVI CT6 6JP
G4JMP CT6 6JU
G3NNK CT6 6LW
G0EKR CT6 6LZ
G3VJF CT6 6LZ
G4ZZK CT6 6ND
G4AOE CT6 6NN
G6XOB CT6 6PU
G0PQC CT6 6QH
2E0AIT CT6 6RE
G0HVC CT6 6RF
2E1AWS CT6 6SH
G4RKV CT6 6SP
G4SUK CT6 6SR
G7EXQ CT6 6UQ
G4SOT CT6 7BZ
G4EVD CT6 7DW
G6TTL CT6 7EJ
G7EXT CT6 7EJ
G8ELS CT6 7EJ
G8FEZ CT6 7EQ
G1TQH CT6 7EW
G6CVD CT6 7HG
G4DBW CT6 7PY
G0KFO CT6 7QD
G0IHI CT6 7RS
G1UFJ CT6 7RS
G1VNM CT6 7RS
G4MKI CT6 7RS
G1EDK CT6 7SD
G1AGW CT6 7SG
G6YRE CT6 7TA
G7PDH CT6 7TR
2E1DKN CT6 7TR
G4VMZ CT6 7UD
2E1DHQ CT6 7XB
G1DSZ CT6 7XF
G8KOL CT6 8AD
G8URC CT6 8AL
G8ESW CT6 8BN
G4URD CT6 8HG
G1TAN CT6 8HN
G1SDH CT6 8HY
G8PPQ CT6 8JX
G8ELP CT6 8LD
G0LAA CT6 8LS
G0HMK CT6 8LT
G1NEK CT6 8NT
G7GTD CT6 8PU
G8KDU CT6 8QW
G0NFG CT6 8TU

Caradon

G7AWG PL10 1DA
G8UDA PL10 1DA
G8XOX PL10 1JP
G7UGH PL11 2DR
G0EHM PL11 2HT
G3KEC PL11 2JJ
G7BYW PL11 2NA
G4ZZD PL11 2PZ
G7FGD PL11 3DX
G7DVI PL11 3LQ
G6YWC PL11 3LY
G1PRE PL11 3LZ
G4GXK PL12 4BJ
G8SAL PL12 4BJ
G4YXJ PL12 4BN
M1BXB PL12 4BN
G4HZE PL12 4DY
G7WGV PL12 4ER
G3UBY PL12 4JH
G7VBZ PL12 4JJ
M1AYH PL12 4JN
M1BVU PL12 4LL
G3ZYY PL12 4NG
G4KYY PL12 4NG
G0VSM PL12 4NG
M1BFR PL12 4NG
M1BBZ PL12 4NH
G7SSH PL12 4NJ
G1OMZ PL12 4NS
G7WKN PL12 4PP
G0XAW PL12 4PP
G8NSP PL12 4QR
G7NFD PL12 4QR
G4WCB PL12 4SE
G7SOM PL12 5DP
G6SEK PL12 5NH
G6ELG PL12 5NQ
G3EXL PL12 6AZ
G3XLU PL12 6BN
G8HVZ PL12 6DU
G1YHM PL12 6DW
G7FTF PL12 6EF
G1BMP PL12 6EH
G3XCS PL12 6EL
G3SN PL12 6EX
M1AGO PL12 6PS
G0DAV PL12 6RH
G6ION PL12 6SA
G3ZHK PL12 6TD
G4ALY PL12 6TD
G7LZC PL12 6TL
G3ABP PL13 1DE
G0HFB PL13 1ED
G4BRF PL13 1JP
G0LYR PL13 1NL
G6RLS PL13 1NL
G6OIB PL13 1QG
G1YDQ PL13 2JP
G7CLO PL13 2PU
G6YXV PL13 2SD
G4WPP PL14 3BD
G1XDU PL14 3BL
G6FYT PL14 3EX
G3ZHO PL14 3EZ
G0GUT PL14 3HX
G6IJW PL14 3LZ
G4RLZ PL14 3NT
G3UAE PL14 3PT
G4MPQ PL14 3PZ
G7BIK PL14 3PZ
G1NSV PL14 3QG
G3JID PL14 3QJ
G4FRC PL14 3QQ
G3XHX PL14 3SQ
G7PHH PL14 3SQ
M0AVS PL14 3SX
M0BEV PL14 3TD
G1ZHH PL14 4DW
G1ZRQ PL14 4EQ
G4HOL PL14 4EQ
G4YAO PL14 4LL
G3AQF PL14 4LY
G7LPT PL14 4NQ
G0AGF PL14 4QX
G8BCG PL14 4QX
G4MFE PL14 5BW
G0DAU PL14 5DP
G4YOI PL14 5HP
G0EIA PL14 5NP
G1KTZ PL14 5NQ
G0CAY PL14 5PS
G7LJA PL14 5PW
G6NFC PL14 5QU
G3SPB PL14 6DY
G8GLI PL14 6EP
G4RNK PL14 6JB
G1NTV PL14 6JS
G7UXO PL17 7BQ
G1XIC PL17 7BT
G3HHV PL17 7DZ
G1JXS PL17 7EE
G0HHB PL17 7EW
G7NHW PL17 7EZ
G0IVZ PL17 7PA
G1YTV PL17 7PA
G1FMU PL17 7PT
G0AVJ PL17 8EY
G4MTQ PL17 8JE
G0TDN PL17 8JT
G4UMS PL18 9BN
G7CQK PL18 9BX
G4FWN PL18 9NG
G4BVB PL18 9PB
G3HUB PL22 0QH
G0NDB PL23 1NB
G3EHP PL23 1NB
G4AIR PL23 1NB
G7RNB PL23 1NB

Cardiff

GW3YSA CF1 7JT
GW0NXW CF1 7TD
MW0AAP CF1 7TQ
GW3KXX CF1 8BZ
GW7WIX CF1 9NQ
2W0AEQ CF2 1EJ
GW6OSE CF2 1EJ
GW7AGW CF2 1RU
GW7RHV CF2 1RW
GW8HWS CF2 2AG
GW4OWZ CF2 3BW
GW0CDG CF2 3RE
GW3VBP CF2 3SH
GW7ARD CF2 3YF
GW6RVI CF2 4BZ
GW4MOZ CF2 4DF
GW1XZI CF2 4RU
GW8EHQ CF2 5BY
GW4ZAW CF2 5NN
GW3NSP CF2 5QR
GW0ZIG CF2 5QR
2W0AJM CF2 5QT
MW1BOO CF2 5QT
GW0LNZ CF2 6EG
GW6MHV CF2 6EJ
GW3XXB CF2 6EQ
GW3YVC CF2 6ES
GW0WLI CF2 6HA
GW4JJV CF2 6HN
GW4JJW CF2 6HN
GW3UMD CF2 6JN
GW3DDY CF2 6JU
GW6AGS CF2 6LF
GW3UTE CF2 6LQ
GW3YQM CF2 6LR
GW0MLN CF2 6QU
GW3UAY CF2 6QW
GW0WLH CF2 6TA
GW3MKT CF2 7DA
GW0VMD CF2 7HN
GW4KOE CF2 7HP
GW0TPL CF3 0DS
GW7KDU CF3 0JA
GW7JDX CF3 7AZ
MW0AQT CF3 7HX
GW7THM CF3 7HX
GW4OES CF3 7LL
GW4VZJ CF3 8BD
GW7BEY CF3 8BD
GW4IUN CF3 8HQ
GW0KSZ CF3 8JW
GW0POA CF3 8LZ
GW6JNE CF3 8PB
GW7GDH CF3 9AT
GW6YNV CF3 9EY
GW6LZH CF3 9HW
GW3AHN CF3 9RS
GW4NLE CF3 9TB
GW6FLU CF3 9UA
GW3VNO CF3 9XJ
GW4REX CF4 1HJ
GW6UAK CF4 1HY
GW8KNG CF4 1LY
GW3WLN CF4 1NJ
GW4TJQ CF4 1NJ
GW4LFW CF4 1NQ
G7OGP CF4 1SY
GW7TUU CF4 1SY
GW4EEO CF4 1TS
GW4VGB CF4 1UE
GW6GCK CF4 2AH
GW8OKR CF4 2AT
GW1ETH CF4 2BU
GW6DFX CF4 2EH
GW6LSL CF4 2FU
GW7ASL CF4 2LR
GW7LLF CF4 2PL
GW0KRQ CF4 3DU
GW1WZI CF4 3DU
GW3ZHQ CF4 3NQ
GW1LFO CF4 3RB
GW4LFO CF4 3RB
2W1CYC CF4 4AQ
GW3LAD CF4 4AS
GW7EMV CF4 4DJ
GW4UGP CF4 4HF
GW8LKX CF4 4JQ
GW0KWU CF4 4LZ
GW0GDI CF4 4NS
GW1USX CF4 4NS
GW0WMT CF4 4QF
GW8ARC CF4 4SP
GW3NKZ CF4 5EQ
GW3BAZ CF4 5ER
GW5BI CF4 5ER
GW4JOG CF4 5HJ
GW4GZX CF4 5NB
GW0LOI CF4 5NU
GW6JQT CF4 5QG
GW8NNF CF4 5SR
GW0SKO CF4 5TT
GW8GOC CF4 5UE
GW8ITO CF4 5UE
GW4LXO CF4 6AE
GW3EPF CF4 6BS
GW0NIY CF4 6HD
GW1YKT CF4 6HY
GW3RWX CF4 6JZ
GW0BBC CF4 6LA
GW0LAY CF4 6LA
GW8GRY CF4 6LX
GW6ZHM CF4 6RL
GW6ITB CF4 6SS
GW0HZJ CF4 6SW
GW1AQJ CF4 6SW
GW1BCI CF4 6TP
GW7SOK CF4 7AB
2W1CXJ CF4 7AB
GW7WLX CF4 7AB
2W1ELF CF4 7AB
GW3NJW CF4 7AD
GW6LMF CF4 7AN
GW3TKH CF4 7BT
GW4YCU CF4 7DB
GW3UZS CF4 7EH
GW3PPF CF4 7HJ
GW4NHH CF4 7JE
GW3ISJ CF4 7LU
GW4BCB CF4 8AL
GW6MNC CF4 8DD
GW8KTQ CF4 8EA
GW4VHO CF4 8LT
GW3TOB CF4 8PB
GW7SXN CF4 8PN
GW8MHG CF4 8QL
GW7KRY CF4 8QP
GW3VOL CF4 8QR
GW8FNO CF4 8RA
GW0HCB CF4 8RE
2W1FXI CF4 8RE
GW0AJI CF4 8RF
GW3TQI CF4 8RJ
GW6ZDM CF4 8RJ
GW0WHT CF4 8RX
GW4LFV CF4 8SJ
GW6IOA CF4 8SJ
GW8YYF CF4 8TJ
GW3ZXI CF4 9AF
GW4LWL CF4 9AF
GW8WYY CF4 9AF
GW7EYP CF4 9AH
GW6IPR CF4 9AW
GW8XAH CF4 9DF
GW4NHB CF5 1AF
GW1EEZ CF5 1DQ
GW1VRR CF5 1GH
GW1YXR CF5 1GH
GW3GFM CF5 2DL
GW2DHM CF5 2JN
GW7VEU CF5 2JR
GW8RAS CF5 2QP
GW8THM CF5 2QS
GW3WCV CF5 2QT
GW8JLY CF5 2RJ
GW8REV CF5 2RU
GW0LTC CF5 2TD
GW0KIG CF5 3AJ
GW1ESU CF5 3AX
GW1UVM CF5 3DZ
GW6RDV CF5 3EH
GW0PLN CF5 3PT
GW0GFH CF5 3RL
GW6NHB CF5 3SS
GW4PYK CF5 4AJ
GW8VFF CF5 4FG
GW6RCK CF5 4ND
2W1FQG CF5 4ND
2W1FQH CF5 4ND
GW3CKB CF5 4SX
GW4SRO CF5 4SX
GW4FOM CF5 5BQ
GW8MCL CF5 5BQ
GW7ESF CF5 5DD

Carlisle

GM4RQW CA1 1AE
M0AOH CA1 1HQ
G6TSM CA1 1LB
G3XWA CA1 2QJ
G4ARS CA1 2QJ
G8FLE CA1 2QJ
G4KVQ CA1 3HF
G4ISS CA1 3PH
G0WZE CA1 3QH
G0PYQ CA2 4PP
G0NUD CA2 4PZ
G1ELI CA2 4SJ
G3OHL CA2 4UH
M0ADG CA2 6AG
G7WGQ CA2 6AG
2E1CDH CA2 6QU
G8UTQ CA2 7LN
G0JXO CA2 7RD
G3WDV CA3 0HW
G4JYW CA3 0LF
G1FVA CA3 0NU
G0OGG CA3 0NW
G0KWB CA3 0QH
G7NMB CA3 8UR
G7OHE CA3 9LQ
G4DOG CA3 9QA
G6DOX CA4 0RN
G4WOQ CA4 8AT
G0DUT CA4 8AX
G4RL CA4 8BS
G4DBA CA4 8ET
G4BUI CA4 8LD
G7IQA CA4 8QY
G6KYK CA4 8RH
G8IBK CA4 9AA
G7NZR CA4 9JA
G7JKD CA4 9JE
G7LIC CA5 7AF
G3NBB CA5 7BJ
G8KAP CA5 7DP
G4ZYQ CA5 7PA
M0BAF CA6 4NE
G7SEP CA6 4NE
G4LAA CA6 4PG
G3MRV CA6 4PZ
GM3JSX CA6 5AW
G2HKG CA6 5RT
G4KFN CA6 5SS
G4LIL CA6 5SS
G4TUZ CA6 5UR
G8URU CA6 6NT
G0LEJ CA8 1EX
G6KXD CA8 1LE
G7BSG CA8 1SH
G1LZK CA8 2AA
G8TTQ CA8 2LF
G6DPV CA8 2LY
G8YVP CA8 2LY

Carmarthenshire

GW0SEH SA14 6BN
GW0SRF SA14 6DT
GW7RQI SA14 6LY
GW1GXM SA14 6PH
GW4BYA SA14 6SA
GW2ADZ SA14 7DD
GW3MOM SA14 7DR
GW4VDW SA14 7HH
GW4XTY SA14 7NP
MW0AMG SA14 7RU
GW0AKV SA14 7RW
GW4PUC SA14 8AE
GW0IJC SA14 8DL
MW0ATR SA14 8NH
GW4JIY SA14 8TH
GW4KBT SA14 8YT
GW4MZG SA14 9AG
2W1FQL SA14 9AS
G4WYQ SA14 9DT
GW0BVN SA14 9DY
GW0SZW SA14 9DY
GW2DUR SA14 9HF
GW0KJZ SA14 9UP
GW4RUY SA14 9UP
2W1ETV SA15 1BT
2W1ETU SA15 1DF
GW0DZL SA15 1DG
2W1DFW SA15 1EP
2W1DFX SA15 1EP
2W1ETT SA15 1LY
GW1USQ SA15 1NR
GW6HKT SA15 2BG
GW6VFH SA15 2RB
GW0OFH SA15 3RU
GW4TFS SA15 3SN
GW7FXX SA15 3TF
GW0RTR SA15 4BR
GW1NGD SA15 4DB
GW4BDS SA15 4EW
2W1ETW SA15 4LA
GW0EZQ SA15 4NS
GW0RHE SA15 4NS
GW0RNK SA15 4NS
GW0SCN SA15 4NS
GW6TRV SA15 5BE
GW8VUV SA15 5DJ
GW0NLB SA15 5HF
GW0TMU SA15 5RT
GW0SRE SA15 5SF
GW0IVG SA15 5SG
GW6HUD SA15 5UG
GW4LJS SA15 5YY
GW0DWQ SA16 0HF
GW0WGE SA16 0LD
GW0JDW SA16 0NF
GW0VBU SA16 0UF
GW8ZZD SA16 0UR
GW8XMW SA17 4BS
GW1ENG SA17 4EB
GW0VGW SA17 4RE
GW0TDA SA17 5DG
GW3YAF SA17 5HF
GW4ZUS SA17 5JT
GW0HYJ SA17 5LU
GW0WRI SA18 1AU
GW0LDZ SA18 1BD
GW0JDS SA18 1SB
GW4MRM SA18 1SB
GW4IMC SA18 2HF
GW4FKJ SA18 2JD
MW0ANX SA18 2LQ
GW3DJA SA18 2PE
GW0NNB SA18 2SP
GW4TGL SA18 3DQ
GW4OFQ SA18 3JT
GW1ANW SA18 3RA
GW1MLE SA18 3RY
GW1TJK SA18 3RY
GW6YUC SA18 3SP
GW0XLK SA18 3UR
GW4XLK SA18 3UR
GW0VPR SA18 3UR
GW1LFN SA18 3UT
MW0AMI SA19 6EB
GW0IXM SA19 6UL
GW8AWT SA19 7BA

2W1BPS SA19 7BA
G0FJT SA19 7EJ
GW0AIY SA19 7UA
GW7VHW SA19 8NS
GW0AWT SA19 8RF
GW7HDX SA19 9AA
MW0BBJ SA19 9NG
G7PIV SA19 9NG
2E1FAJ SA20 0EH
2E1EDK SA20 0TU
GW0MAP SA20 0UL
G3WMP SA31 1DS
GW0IXJ SA31 1EH
GW4RVA SA31 1JJ
GW0PUM SA31 1LR
GW0CVY SA31 1SY
GW0GMX SA31 2AX
GW8MQ SA31 2DY
GW8HKY SA31 2HH
GW0RAD SA31 2JA
GW1ADY SA31 2JB
GW4WPJ SA31 2JQ
GW0IXK SA31 2LH
GW0TWQ SA31 2LP
GW0UJJ SA31 3BS
GW0UMC SA31 3BZ
GW0BBF SA31 3EJ
GW0GPQ SA31 3ER
GW4YCT SA31 3PQ
GW4ZXL SA31 3PQ
GW4ZZO SA31 3QE
GW4VPK SA32 7AB
GW3KZW SA32 7AH
GW3LJS SA32 7BH
GW0ALR SA32 7ER
GW6RXA SA32 7JT
GW7OIK SA32 7SA
GW0WRW SA32 8AH
GW0VST SA32 8BJ
MW0AMJ SA32 8DQ
GW3NXR SA32 8DX
GW0LNM SA32 8JE
GW0AJU SA32 8LJ
GW0HYL SA32 8LJ
GW1IIZ SA32 8LJ
GW1SGE SA32 8LP
MW0APQ SA32 8NH
2W1DRE SA32 8QH
GW0ADY SA32 8SA
GW0TVX SA33 4ES
GW3XJQ SA33 4PD
GW7FFU SA33 4PD
GW0TSE SA33 5AH
GW0TSF SA33 5AH
GW4KFD SA33 5JZ
GW0COV SA33 5LL
GW3SVY SA33 5NJ
GW0GFN SA33 5NX
GW1NED SA33 5NX
2W1DIK SA33 6XJ
GW0DKG SA34 0QN
GW7IPS SA38 9BA
GW3WCA SA38 9RA
GW0TXP SA39 9AX
GW4VPX SA39 9DH
GW4WWE SA4 1UH
GW0PDB SA44 4AE
MW1BGE SA44 5HB
GW1VMA SA44 5HE
G4MIP SA67 8NR

Carrick

G0NDC TR1 1BW
G1SZD TR1 1EF
M0AUC TR1 1FX
G1UWD TR1 1JD
G1UWE TR1 1JD
G1ODW TR1 1LR
M1ARB TR1 1NH
G0VQN TR1 1QW
M1ASM TR1 1RF
G0NKQ TR1 1RU
G0KSF TR1 1RX
G8PBI TR1 1SX
G0RIZ TR1 1TA
G0WID TR1 1XJ
G0SHY TR1 1YS
G4UOJ TR1 2AQ
G3XFL TR1 2BS
G1GFD TR1 2BY
G1ZRP TR1 2BY
G3LFE TR1 2XX
G4PVV TR1 3DG
2E1CIT TR1 3DG
G4IXM TR1 3ES
2E1BCD TR1 3ES
G8PGU TR1 3JZ
G0OOP TR1 3LX
G1DTS TR1 3LX
G0UWI TR1 3LX
G3CIF TR1 3NQ
G0FHX TR1 3PE
G0FHY TR1 3PE
G7FLI TR1 3PT
G0CUH TR1 3QD
G3VFQ TR1 3RH
2E1BDA TR1 3RL
G7RIO TR1 3RZ
G0OKC TR1 3SE
G1PDH TR1 3TU
G3WKP TR1 3TX
2E1DUF TR1 3UL
G0SDT TR1 3YB
G3UFX TR10 8BH
G7GOP TR10 8BW
G4MYY TR10 8PB
G3RII TR10 8RD
G6CZX TR11 2HE
G0TAJ TR11 2LR
G0OSV TR11 2NB
G4PXN TR11 2NY
G7MSL TR11 2PW
G0ETQ TR11 2QH
G0WMV TR11 2QS
G4FNP TR11 2RW
G4RUA TR11 2SW
G7RKE TR11 3EL
G8END TR11 3JW
G0NIH TR11 3ND
G6CWU TR11 3PU
G6YHK TR11 4HR
G0HQS TR11 4HY
G3SEL TR11 4JY
G4PCW TR11 4PE
G4HVF TR11 4RF
G1IKI TR11 4SW
G0USY TR11 5BE
G7WDS TR11 5NA
2E1ESH TR11 5NA
2E1ESM TR11 5NA
2E1FNK TR11 5NA
G4VXE TR11 5NQ
G3LZN TR11 5TJ
G8JZC TR11 5TR
G4FTO TR11 5TU
G4DHH TR11 5UB
GU3NHL TR11 5UF
G4JGT TR11 5UX
G4WQL TR16 5DT
G3ZZV TR2 4BS
G0JWX TR2 4EP
G0KEM TR2 4EP
G4ADV TR2 4EP
G4STB TR2 4JT
G7TNP TR2 4NE
2E1ADQ TR2 4PQ
G3VWK TR2 4PQ
G0REA TR2 5AR
G6KYR TR2 5HE
G0JWV TR2 5JP
G6LEU TR2 5LE
G0SGF TR2 5NB
G3NJP TR2 5NP
G8MNC TR2 5SQ
G4WCC TR2 5UH
G3PPT TR3 6BB
G3UFJ TR3 6BJ
G3IEZ TR3 6DB
2E1ERJ TR3 6DB
2E1ERK TR3 6DB
G6IBK TR3 6DD
2E1FNR TR3 6DD
G8LIL TR3 6DT
G3MRT TR3 6EB
G3EKX TR3 6HD
G0WFY TR3 6HN
G7REL TR3 6JN
G3VGO TR3 6JY
G4LJY TR3 6LN
G8WQG TR3 6LN
2E0AGJ TR3 6LN
2E1FNY TR3 6LN
G0PGX TR3 6PQ
G0NNR TR3 6PQ
G7TMX TR3 6RJ
G0VZB TR3 6RL
G4TKX TR3 6TX
G7AGZ TR3 6UA
G7NFE TR3 7NW
G6WMT TR4 8AP
G4ZZZ TR4 8QL
G0DHT TR4 8QL
G8RBS TR4 8RJ
M0AWP TR4 8TS
G7NLI TR4 8TS
M0BDH TR4 8TS
G7WER TR4 8TS
G8VCJ TR4 9BA
G0VEJ TR4 9BS
G4BRU TR4 9ED
G0CAM TR4 9EJ
G6WLA TR4 9PF
G8ZYR TR4 9RB
G6ZTP TR4 9RQ
2E1ADT TR5 0TR
G4MQD TR5 0UF
G1ZEI TR6 0DZ
G0GQJ TR6 0EF
G3TYA TR6 0EY
G3YMN TR6 0LL
G0DJL TR8 5PT

Castle Morpeth

G8UZM NE15 0DW
G7VTT NE15 0EA
G0AWA NE20 9AE
G4CHE NE20 9AP
G4DSD NE20 9AW
G4SVE NE20 9EJ
G6STE NE20 9HS
M1ACP NE20 9HS
G0AFU NE20 9QN
G8KPD NE20 9RA
G8RER NE20 9RA
G4ITR NE20 9RR
G1SUM NE20 9RZ
G2AMM NE20 9SZ
G4HUX NE20 9TD
M1ABI NE20 9TD
G4VVO NE20 9TN
G0KNW NE61 1TB
G8ONS NE61 1XF
G0WAD NE61 2AS
G7RXJ NE61 2AS
G3XNG NE61 2AY
G1BWJ NE61 2DW
G0WSX NE61 2JN
G3AGC NE61 2PL
G0EVV NE61 2SG
G7SPP NE61 2SG
G6IIA NE61 2SG
G3YYU NE61 2TA
G4ZTC NE61 2TQ
G7RWC NE61 2UB
M0AUL NE61 2UH
M0AUM NE61 2UT
G3BIK NE61 2XP
G7HHU NE61 2XY
G0KJW NE61 2YA
G7PSL NE61 2YF
G4KBX NE61 3LA
G4FXU NE61 3RB
G7TSR NE61 3RF
G1WSS NE61 3SL
M1AQY NE61 3SX
G6PQW NE61 4AZ
G0JFW NE61 5HS
G4DGQ NE61 5JU
G1FYN NE61 5NW
G4ZTA NE61 5PU
G4GWB NE61 5QZ
G4AAX NE61 5RB
G8PNN NE61 5RB
G0UQD NE61 5RB
G4KHC NE61 6LG
G1ZES NE61 6YJ
G0VHT NE61 6YW
G1ODT NE65 8QH
G1JKX NE65 8UN
G1BBY NE65 8UR
G4ADD NE65 8UW
G6OTW NE65 8UW

Castle Point

G1NPM SS6 7TR
G6YXH SS6 7TR
G1BAR SS7 1DN
G6IFK SS7 1DU
G6FVT SS7 1EB
G4OWQ SS7 1EH
G4EZP SS7 1JL
G4JAR SS7 1JL
G4WPC SS7 1NG
G2FSH SS7 1NL
G4FCX SS7 1QB
G6MVW SS7 1SS
G4WAK SS7 2DL
G3UTA SS7 2JP
G3LUZ SS7 2LN
G3FD SS7 2PP
G7SRC SS7 2PP
G0RAJ SS7 2RN
G6XFB SS7 2SP
M0AOK SS7 2ST
G4ZMN SS7 2TY
G8AVA SS7 2UN
G3OIT SS7 3DN
G0ASN SS7 3EG
G0BBN SS7 3EG
G4RSE SS7 3EG
G6RSE SS7 3EG
G0FPL SS7 3EL
G1VCY SS7 3JP
G1FEH SS7 3JQ
G1FRH SS7 3PW
G1FBZ SS7 3RJ
G8UWH SS7 3RQ
G8YAA SS7 3SH
G8MRS SS7 3TU
G5QK SS7 3TU
G0TTM SS7 3TU
G8UWI SS7 3UU
G4GDK SS7 4AF
G8IXY SS7 4BD
G7TOS SS7 4BS
G0UWQ SS7 4DJ
G4TPK SS7 4EE
G8WUU SS7 4EN
G0BTS SS7 4HU
G1KWH SS7 4HY
G6TKW SS7 4LA
G1ZHN SS7 4LS
G8UYH SS7 4NS
G7EGU SS7 4NT
M1AWG SS7 4NT
G8EOM SS7 5BQ
G4OFK SS7 5DR
G7CDO SS7 5DS
G1GBV SS7 5DT
G0LYX SS7 5DW
M1BTK SS7 5EJ
G4GDS SS7 5ES
G0PKI SS7 5HG
G8VFI SS7 5JQ
G7IIO SS7 5LH
G6ZUE SS7 5NU
G8YPK SS7 5PH
G4PBO SS7 5RD
G1LAW SS7 5SJ
G7VKC SS7 5SN
G0LTO SS8 0BP
G1BGJ SS8 0BU
G7HQF SS8 0DN
G6JBL SS8 0EP
G4MAN SS8 0JG
G6ZQC SS8 0LH
G3JPX SS8 0ND
G4NZD SS8 0QT
G6GZB SS8 0QT
G8TYG SS8 7AT
G1IEO SS8 7EQ
G1ZSG SS8 7HL
G4BQF SS8 7TS
G7EUG SS8 8AW
G0HWI SS8 8HD
G1JJS SS8 8HD
G0JAN SS8 8HX
G1OXO SS8 8HX
G6DOT SS8 8JU
G0WAX SS8 8LX
G0WGA SS8 8LX
G6CNF SS8 9AB
2E1FDZ SS8 9AB
G7PDU SS8 9BL
G4ZQJ SS8 9DJ
G8ZTB SS8 9DS
G4LUM SS8 9EW
G0DEN SS8 9HB
G6LUO SS8 9HB
G0FDJ SS8 9QG
G6SPH SS8 9QL
G1IPX SS8 9QP
G7PRE SS8 9RD
G0KSC SS8 9RS
G0NCT SS8 9SU
G4UVJ SS8 9TD
G1SOB SS8 9YB

Ceredigion

GW4WJU SA38 9EP
GW8GAB SA38 9ET
G0JCT SA38 9JS
GW1CLA SA38 9PH
G1CIY SA38 9QB
GW3ITD SA38 9QL
2W1BIY SA40 9TN
GW4PQE SA40 9YR
GW4OUU SA43 1AF
GW0PNI SA43 1BU
GW7GWO SA43 1PE
GW7WCR SA43 1QW
GW0BPV SA43 1RW
GW6RWJ SA43 2AE
GW0CMI SA43 2DH
GW7LZO SA43 2DJ
GW4HGJ SA43 2DH
GW0PLP SA43 2JE
GW4SZV SA43 2JE
GW0ULC SA43 2JE
GW4SJO SA43 2JH
GW4TVE SA43 2JH
GW3JXN SA43 2JS
GW6ZH SA43 2LG
G7WLM SA43 2LP
GW3EJR SA43 2NJ
GW4JRK SA43 2RJ
GW4XMV SA43 3HZ
GW3UEP SA44 4NA
GW3WXA SA44 5PN
GW0POZ SA44 6JE
GW3KDB SA44 6NP
GW7WEE SA44 6NQ
GW1OSQ SA45 9QB
GW0ADS SA45 9QJ
GW0DDL SA45 9QR
GW7SXU SA45 9RJ
GW3GA SA45 9RL
GW3WVV SA45 9RL
GW4AGV SA45 9RL
GW8WAO SA45 9RL
GW0SYZ SA45 9RL
MW0ALG SA45 9TH
GW7SUC SA46 0BH
GW1GEX SA46 0DA
GW3DRV SA46 0DA
GW3LHK SA46 0ED
GW0HXS SA47 0QN
GW3MZN SA47 0RN
GW0SFO SA47 0RW
GW7SWB SA47 7PG
GW0PNC SA48 8LL
2E1BOJ SA48 8RL
G4DZN SY20 8JH
GW8SIT SY23 1BE
GW4YAW SY23 1HH
GW4TUD SY23 1PU
GW8SFT SY23 1QW
GW7AGG SY23 1SS
GW0ARA SY23 1TE
GW3SON SY23 1TE
GW4LHL SY23 2DX
GW6FOF SY23 2ET
GW1SBO SY23 2HD
GW7HAE SY23 2HY
GW0PDA SY23 3AZ
GW7OZP SY23 3BL
GW3IRK SY23 3EG
GW4CTV SY23 3EZ
GW8LZY SY23 3HF
GW8ONP SY23 3HF
GW6IDK SY23 3JT
GW7LUB SY23 3NB
GW3FRK SY23 4DH
G4XSX SY23 4G
GW7OVV SY23 4NS
GW1XOT SY23 4RD
GW7WLD SY23 5AT
GW7SOA SY23 5BW
GW7ASZ SY23 5DL
G3UYH SY23 5HP
GW0OBB SY23 5NB
GW5AF SY23 5PD
GW0RBZ SY24 5BE
GW3PXY SY24 5BY
GW8GXW SY24 5HU
GW6JWD SY24 5JD
GW4TUC SY24 5NP
MW1BIV SY25 6LW
GW8DVH SY25 6NG
GW7OTU SY25 6PE

Charnwood

G1EEC LE11 1JH
G4PGC LE11 1NS
G1ZSP LE11 1PR
G7IQO LE11 1RD
G4SGY LE11 1SN
G4IAQ LE11 2AA
G4IAR LE11 2AA
G0RIF LE11 2BD
G8LZL LE11 2BD
G3ZUB LE11 2ED
G4TEY LE11 2EF
G4KGG LE11 2HH
G0VAU LE11 2JG
G6OUJ LE11 2JL
G8WLO LE11 2LH
G3LHB LE11 2NU
G0LCU LE11 2PA
G0DMH LE11 2PB
G8IXX LE11 2RU
G3ZKW LE11 2SH
G7NOF LE11 3AG
G7WIK LE11 3AN
G4ZSP LE11 3BU
G3TPI LE11 3DA
G4OAJ LE11 3HX
G7NII LE11 3JB
G1WLW LE11 3JP
G0WTA LE11 3JS
G3OMK LE11 3JT
G6HDJ LE11 3JT
G8SNF LE11 3LT
G7RAL LE11 3LT
G3HPJ LE11 3LT
G3NOB LE11 3LZ
G4COU LE11 3NX
G7HDW LE11 3NX
G6GFG LE11 3PQ
G7LTU LE11 3PS
G4OOL LE11 3RH
G4SBM LE11 3RL
M1ACS LE11 3RY
2E1CGW LE11 3RY
G8UIW LE11 3ST
G7BDR LE11 3TB
G6ACL LE11 4BP
G8DAZ LE11 4BP
G7SCL LE11 4LG
G7WCP LE11 4LL
G4JQV LE11 4PP
G4CCI LE11 4PU
G0LNK LE11 4PY
G0MMJ LE11 4QD
G4OKY LE11 4SN
G7SEK LE11 4TQ
G0PHT LE11 4WA
G1VVT LE11 5DU
G4NTJ LE11 5EZ
G4UAT LE11 5HB
G4AOP LE11 5JN
G8CPK LE11 5JN
G4RVW LE11 5LX
G7IQC LE11 5LX
G7OXA LE11 5UJ
G8HMA LE11 5UU
G8LVL LE11 5UW
G4VCN LE11 5YZ
G6VAW LE11 5YZ
G4IOR LE12 5AP
G3TEG LE12 5HW
G1WUK LE12 5JE
G4BZP LE12 6ST
G8DTM LE12 6ST
G4PZQ LE12 6UA
G3XYC LE12 6UB
G6HVD LE12 7BN
G0LMA LE12 7BP
G3ZEG LE12 7DD
G4SGD LE12 7EN
G8RYE LE12 7ET
G1IFX LE12 7HB
G0MCV LE12 7JJ
G1DHQ LE12 7NX
G6MOI LE12 7PJ
G4VSU LE12 7PR
G4MIV LE12 7PX
G6TZO LE12 7QH
G1HGA LE12 7SB
G0AUV LE12 7SH
G0JHJ LE12 7SX
G4ERT LE12 8BB
G8XYY LE12 8DB
G8IQT LE12 8EJ
G3SQV LE12 8NB
G6FHO LE12 8RD
G7SGG LE12 8SB
G3RDO LE12 8SG
G3XBB LE12 8TN
G3ZNF LE12 9AF
G3KWY LE12 9DT
G8AYG LE12 9DT
2E1FRP LE12 9JB
G4HTH LE12 9LS
G0UGI LE12 9LS
G7BMM LE12 9NL
G1PPH LE12 9NX
G3DXJ LE12 9PP
G3MKU LE12 9PU
G3KQU LE12 9QB
G4JCH LE12 9RW
G8PEO LE12 9SG
G1ETZ LE12 9SQ
G8BUB LE12 9SS
G8OEF LE12 9UA
G4VMM LE4 3AT
G4ZCJ LE4 3HB
G4BJF LE4 3HQ
G4MEF LE4 4DA
G1LGO LE4 4FF
G6TIW LE4 4FU
G1DYT LE4 4GS
G1UZR LE4 4JN
M1BUJ LE4 8BA
G4VWI LE4 8BP
G6SSX LE4 8GY
G4XUH LE4 8JE
G4EUF LE6 0TB
G1VBQ LE7 1GQ
G1CHF LE7 1HJ
G0IPB LE7 1LY
G0PKG LE7 2AB
G0BAI LE7 2BZ
G0IOZ LE7 2EB
G7OKC LE7 2EP
G7DOA LE7 2JE
G1WAR LE7 2JW
2E1FSD LE7 3DW
G0PVE LE7 3FL
G3DAQ LE7 3TU
G0ORL LE7 3ZJ
G8SRK LE7 3ZW
G8HVL LE7 4YE
G6ABP LE7 4YN
G7CZH LE7 7BE
G6WQY LE7 7BW
G3LRS LE7 7DU
G4LRO LE7 7DU
G6XRS LE7 7DU
G7NQU LE7 7FA
G3KH LE7 7HH
G8LDB LE7 7PU
G4NWS LE7 7SQ
G4WRC LE7 7SQ

Chelmsford

G4VYV CM1 1LL
G7KOI CM1 1QB
G0NVM CM1 1RG
G5HF CM1 1RQ
G6ZVV CM1 1TG
G7IRC CM1 1TG
G7RNL CM1 1TG
M0ASJ CM1 1TN
G1JSY CM1 2BS
G8YZU CM1 2EW
G8UOL CM1 2HD
G6PZT CM1 2HL
G3XRC CM1 2JA
G4YTG CM1 2JA
G4TRE CM1 2JX
G7RFT CM1 2NJ
G6EAR CM1 2NQ
G4BNE CM1 2PR
G6FXM CM1 2PR
G3SLT CM1 2PT
G3NAA CM1 2PW
G4FDI CM1 2RS
G4EOR CM1 2RX
G3GUP CM1 2RZ
G3WP CM1 2TT
G1NQY CM1 3EE
G4CCU CM1 3EG
G7RFC CM1 3HQ
G8UDH CM1 3LP
G0PEY CM1 4AF
G4INM CM1 4BN
G1CSR CM1 4DG
G8GFF CM1 4DG
G1IBP CM1 4DU
G4TRF CM1 4DU
G3VCG CM1 4EF
G8PJH CM1 4EF
G8AEU CM1 4EJ
G4GHO CM1 4HB
G8CDB CM1 4HD
G4FKH CM1 4JY
G4TOO CM1 4JY
G6ZWC CM1 4SJ
G4HSK CM1 4UN
G6DJS CM1 4UQ
G3OLK CM1 4UY
G3ZWW CM1 4XH
G7HFC CM1 4XZ
G0LLB CM1 6FD
G4CTS CM1 6FG
2E0ANG CM1 6FG
M0BDY CM1 6FG
G3KPJ CM1 6FJ
G4EOG CM1 6HX
G4JJH CM1 6HY
G7JDA CM1 6JA
G7VBL CM1 6JT
G7LNY CM1 6JX
G4BOQ CM1 6JX
G4DJC CM1 6JX
G7NZH CM1 6JX
G1FOA CM1 6LL
G7RBR CM1 6LL
G3VSI CM1 6QR
G4GUO CM1 6QU
G7EQX CM1 6SS
G7GBY CM1 6SS
G6OQJ CM1 6SX
G7HFP CM1 6SX
G7PXC CM1 6TA
G7RCL CM1 6TX
G3SXK CM1 6UP
G8AAE CM1 6UW
G1EFL CM1 6UX
G3WSV CM1 6XB
G1EXU CM1 6XY
G6AKL CM1 6YU
G3WWC CM1 7BH
M1ADI CM1 7HS
G7NNI CM1 7PY
G1IBO CM1 7QG
G6GOV CM1 7RY
G0NGD CM11 1HB
G6AUD CM11 1JF
G8HSM CM11 1JZ
G0PXA CM11 1LH
G8CCN CM11 1QN
G1JIJ CM2 0ES
G8MYJ CM2 0SA
G4UZT CM2 0TD
G4PVM CM2 0TY
G6BZQ CM2 0UA
G4MDB CM2 6AQ
G7RVY CM2 6BA
G7LIN CM2 6BG
G3VGX CM2 6DQ
G1INA CM2 6EB
G4JDS CM2 6HA
G6WFM CM2 6QE
G4KGL CM2 6QL
G4LAE CM2 6QN
G8UDD CM2 6SQ
G6JYB CM2 7AU
G3CVI CM2 7BU
G3VMJ CM2 7BU
G4ZEW CM2 7BY
G4OLU CM2 7DD
G3BLR CM2 7EL
G8PCA CM2 7HY
G3NOQ CM2 7JG
G1NLZ CM2 7JL
G8BYP CM2 7JL
G3PMW CM2 7JN
G8OQW CM2 7NJ
G4IVU CM2 7NL
G3VNP CM2 7QT
G3JOX CM2 7SF
G3FMO CM2 7TD
G1RPQ CM2 8AL
G7GBZ CM2 8AU
G7RGG CM2 8BH
G8UDS CM2 8BS
G7KLV CM2 8HY
G4ZMU CM2 8HY
G6JPG CM2 8LF
G0RLI CM2 8NF
G7TWE CM2 8NF
G3YDY CM2 8NT
G0BDS CM2 8NY
G8GNZ CM2 8PD
G0KOU CM2 8PT
G3GNQ CM2 8QN
G1MNU CM2 8QP
G2HNF CM2 8QZ
G1OBE CM2 8WA
G3WGE CM2 8YA
G7LVU CM2 8YY
G6FJE CM2 8YY
G6ZNJ CM2 8YY
G4NCS CM2 9BD
G8VU CM2 9BZ
G3PEM CM2 9DD
G1MVI CM2 9DX
G3BGO CM2 9DY
G3UCD CM2 9DZ
G4BMW CM2 9DZ
G0POK CM2 9HA
G6KLK CM2 9HQ
2E1BFF CM2 9JR
G1NNN CM2 9LF
G0SQP CM2 9LW
G6XHI CM2 9NR
G0ODE CM2 9NY
G0SXK CM2 9PW
G3PBT CM2 9RE
G0NEM CM2 9SN
G6FBI CM2 9TZ
G6OOT CM2 9XE
G6ZGI CM2 9XH
G8ZGI CM2 9XH
G3PGN CM3 1AD
G1UZC CM3 1DA
M1BWN CM3 1DA
2E1EWH CM3 1DA
G0MWT CM3 1EL
G3PMX CM3 1EL
G6HKM CM3 1EL
G8LVW CM3 1NP
G6OTE CM3 1RB
G0CZE CM3 1RE
G1GKN CM3 1RT
G6EEE CM3 1RT
G6CMS CM3 3AD
G0SBQ CM3 3BU
G4DBM CM3 3BY
G6KPV CM3 3DD
G0LSY CM3 3DP
G6LUP CM3 3ER
G7SEU CM3 3JP
G2AMQ CM3 4DB
G4LQE CM3 4DX
G8DET CM3 4DX
G7ABL CM3 4EY
G1FOP CM3 4LA
G3ZES CM3 4LD
G4TFI CM3 4LR
G3PJC CM3 4LS
G4EAT CM3 4ND
G3BDV CM3 4NU
G7UVP CM3 4PH
G1EUC CM3 4PN
G0TRM CM3 4PS
G4TNB CM3 4RP
G3SUY CM3 4RQ
G7HCD CM3 4XG
G7KSQ CM3 4XW
2E0ANV CM3 5DS
G4ZPE CM3 5DS
G7HBV CM3 5FS
2E1EQY CM3 5FS
G8OXU CM3 5FZ
G8NAM CM3 5GF
G4GYJ CM3 5GH
G8NRI CM3 5GL
G6OJB CM3 5JB
M0AMZ CM3 5JJ
G6EVI CM3 5JJ
G7WAQ CM3 5JX
2E1BDT CM3 5LR
G0AUR CM3 5NJ
G7VEE CM3 5NY
G7NBR CM3 5PF
G3XMB CM3 5PQ
G3JXV CM3 5PR
G0BCW CM3 5SB
G7PDP CM3 5SN
G0UCU CM3 5TU
G1FHZ CM3 5WU
2E1EDV CM3 5WX
2E1EDW CM3 5WX
G4DUJ CM3 5YG
G8XXL CM3 8AZ
G0LKW CM3 8DS
G4YCH CM3 8RN
G8EGE CM3 8XA
G6ZC CM4 0EJ
G8YQO CM4 0HA
G0AOL CM4 9JW
G4UHM CM4 9JW
G7VVB CM4 9QA
G7WKP CM4 9QA
G3PLB SS11 7BH
G7DAH SS11 7DN
G8BPW SS11 7JE
G0DZQ SS11 7LN
G0BUR SS11 7PT
G1LCS SS11 7PT
G1HHS SS11 7PT

Cheltenham

G3DNS GL50 2AW
G4GVZ GL50 2NG
G4IEY GL50 2SS
G6QM GL50 2SS
G3AUX GL50 2SU
G3JJG GL50 2SU
G4FUJ GL50 2UA
G3XYL GL50 2UG
G8DLH GL50 4AQ
G8GKH GL50 4BY
G4PBY GL50 4HS
G3SMD GL50 4NU
G3RMD GL50 4NW
G8DTA GL50 4NX
G4FZG GL50 4QB
G0NDU GL50 4RG
G0MJL GL50 4RQ
GM6GJW GL50 4RS
G0UTR GL50 4SA
G0WJA GL50 4SA
G4VTA GL50 4SB
G6GJN GL50 4SE
G8XRS GL50 4SP
G3VZP GL50 4SQ
G1SHM GL51 0BP
G4OVB GL51 0LN
G0GZA GL51 0NH
G0PSO GL51 0NY
G0SMM GL51 0NY
G6AFE GL51 0NZ
G0OFS GL51 0PD
G4WLC GL51 0PE
G6MBH GL51 0PP
G6HVE GL51 0TX
G8ZEE GL51 0XE
G6TRY GL51 5BB
G3FPB GL51 5BD
G7HMR GL51 5BD
G1EHT GL51 5EZ
G0TBK GL51 5JE
G4BZU GL51 5JF
G0BHI GL51 5JP
G6GUC GL51 5JR
G0EYP GL51 5LA
G4EDP GL51 5LA
M1AKP GL51 5LG
G4XDL GL51 5LH
G6VQN GL51 5LL
G6XSC GL51 5LL
G6EFC GL51 5LN
G8URP GL51 5QD
G3XMR GL51 5QP
G4BBR GL51 5QP
G3YFT GL51 5RH
G6PYM GL51 5RQ
G4KHB GL51 5RR
G4SGI GL51 5RR
G0TMP GL51 5RY
G0GKD GL51 6AA
G3HCJ GL51 6AA
G4XIG GL51 6BB
G6FPK GL51 6BU
G4KRL GL51 6DG
G1MHM GL51 6DL
G8JAY GL51 6HR
G4BSC GL51 6JE
G4FLS GL51 6JE
G4CJJ GL51 6JP
G0WIG GL51 6JS
G1EDX GL51 6NP
G4WUJ GL51 6NX
G3CNW GL51 6QN
G8DOB GL51 6QZ
G3IFB GL51 6RE
G3LVP GL51 6RL
G8APY GL51 6RL
G4ILI GL51 6RT
G4PBN GL51 6TX
G4KWW GL51 7BJ
G3IER GL51 7DB
G8PZD GL51 7DQ
G7HJZ GL51 7HA
M0AIP GL51 7HS
G3YEU GL51 8AF
G4HQC GL51 8HW
G8XFX GL51 8JF
G4MEP GL51 8LJ
G4NOE GL51 8NE
G8WWC GL51 9BY
G3MZV GL51 9RF
G4MEM GL52 2JA
G4YSF GL52 2QR
G4UTT GL52 3AF
G4MVO GL52 3AR
G8ZFT GL52 3DF
G3XKD GL52 3DG
G4CWM GL52 3DT
G6UJI GL52 3DT
G3UYL GL52 3DU
G4LCM GL52 3DW
G7NCW GL52 3DX
G0VSY GL52 3EH
G0VSZ GL52 3EH
G4PDQ GL52 3ES
G5BK GL52 3ES
G8MZV GL52 3ES
G8LHP GL52 3HF
G3RPQ GL52 3HR
G3KFT GL52 3JE
G8MNT GL52 3JE
G0COI GL52 5AN
G3CON GL52 5DT
G3LME GL52 5EG
G4KFT GL52 5JQ
G3XTP GL52 5LY
G6YCV GL52 6JD
G3BXZ GL52 6LF
M0AUW GL52 6LW
G0SOV GL52 6NN
M0BFJ GL52 6RB
G8UJF GL52 6RF
G4VCL GL52 6SW
G3VKV GL52 6SX
G6DXD GL52 6TX
G4UAZ GL53 0AD
G4RFU GL53 0AZ
G3BNU GL53 0DE
G1GYR GL53 0LU
G4WGK GL53 0LU
G4GAJ GL53 0LX
G3CJ GL53 0NN
G6VVL GL53 0NQ
G8BTV GL53 0NU
G7FCK GL53 0PG
G3ZQJ GL53 7BQ
G8KMR GL53 7DX
G0PTR GL53 7EJ
G3CGD GL53 7JJ
G6BHS GL53 7JT
G8AHB GL53 7QE
M1ASH GL53 8AJ
M0BAM GL53 8AJ
G3IGN GL53 8AP
G4MYW GL53 8BA
G1OEK GL53 8DD
G3BCC GL53 8NH
G4LZY GL53 8NN
G4XXA GL53 8NQ
G4XXB GL53 8NQ
G0LSQ GL53 8NS
G3BZR GL53 8RA
G8CYU GL53 9AJ
G3NKS GL53 9AW
G1NKS GL53 9AW
G4MM GL53 9DP
G0VNJ GL53 9DQ
G3XPR GL53 9ED
G3MOE GL53 9EN
G4HTR GL53 9HH
G7IHN GL53 9HH
G7IPB GL53 9HH
G4INB GL53 9LB
G3AUU GL53 9LL

Cherwell

G6PZN MK18 4AJ
G3JNZ NN13 5RR
G4FCD NN13 5SN
G4YFY OX15 0PD
G8JPW OX15 4BB
G6CIZ OX15 4BP
2E1EZT OX15 4ED
2E1FCX OX15 4EJ
G8UKZ OX15 4SA
G3MYE OX15 5TB
M1BMV OX15 6BJ
2E1FVL OX15 6DY
2E1FJA OX15 6LZ
G6IKQ OX15 6NF
G1XRQ OX16 0LU
G0LDB OX16 0RR
G6TAS OX16 7AZ
G7OIT OX16 7FP
G7UUW OX16 7NF
G4XPY OX16 7PQ
M1AOG OX16 7QW
G7KYH OX16 7YF
G1KKE OX16 8FB
G8KSH OX16 8FD
G6REG OX16 8FS
G4DED OX16 8HY
G0BJI OX16 9AP
G1IIO OX16 9BQ
G7ILJ OX16 9BQ
G8DCX OX16 9DH
G3MXK OX16 9DP
G3HKJ OX16 9EB
G7FKS OX16 9EN
G8THG OX16 9JU
G4RWV OX16 9LB
G8ZRU OX16 9NR
G7NOC OX16 9QA
G0SCS OX16 9QA
M0AXN OX16 9RZ
G1YSY OX16 9SL
G1HXZ OX16 9SR
G7LKC OX16 9TA
G0MEC OX16 9TW
G0RKJ OX16 9UR
G0VHQ OX16 9YA
M1BOL OX16 9YA
G1KIB OX17 1HQ
2E1FWE OX17 2JX
G0HPU OX17 3LB
G7ONO OX17 3LD
G3OZL OX5 1AD
G0THY OX5 1HH
G7UXE OX5 1JT
M1AQN OX5 1JT
G6AON OX5 1PW
G3WTN OX5 1RW
G3UVB OX5 1SF
G6KJM OX5 2AH
G0EDY OX5 2EU
G7KGI OX5 2EU
G0NYT OX5 2HL
G6ZDS OX5 2JQ
G0LUN OX5 2LL
G3MSO OX5 2UB
G8BNB OX5 2XP
G7VRB OX5 2XZ
G0XGM OX5 2XZ
G0LUQ OX6 0DX
G0FCV OX6 0EA
G0LLK OX6 0PE
G6HL OX6 0PT
G4SYV OX6 0XW
G1BYQ OX6 0YP
G8UWD OX6 0YR
G6FPP OX6 0YW
G8FHY OX6 3SH
G3FMT OX6 4JE
G1XTA OX6 4JR
G0REV OX6 4LP
G8NS OX6 7AH
G6EWC OX6 7DZ
G4FTA OX6 7EX

G6EIG OX6 7FL
G1AEB OX6 7HP
G6POQ OX6 7NH
G6BXS OX6 7TU
G6PLF OX6 7UH
G6FJO OX6 7YY
G7ILP OX6 8AR
G8EKN OX6 8BE
G1CEO OX6 8DZ
G6WDF OX6 8FY
G6NB OX6 8GA
G6DB OX6 8GA
G0NUX OX6 8GD
G4BKB OX6 8LR
G0TSK OX6 8YR
G4HTJ OX6 9DF
G4CQJ OX6 9NX
G0THW OX6 9NX
M0BBS OX6 9UJ
G0GLG OX6 9YL

Chester

G4AQP CH1 3DD
G1WTT CH1 3HN
G3SES CH1 4AN
G0OOT CH1 4BX
G6XID CH1 4DA
GW4VFE CH1 4LH
G6LRU CH1 5AF
G0KZS CH1 5AU
G0ULM CH1 5AZ
2E1CRA CH1 5DP
G4OLV CH1 5DT
G1URW CH1 5JQ
G0JIQ CH1 5LQ
G7VPQ CH1 5NW
2E1FKY CH1 5SY
G8UEK CH1 5XH
G6NLQ CH1 6AX
G7SHC CH1 6AX
G8WXL CH1 6DB
G4SMM CH1 6JS
G4WXO CH1 6LU
G0MMT CH1 6PG
G8GWX CH2 1DG
GW0UXU CH2 1HE
G8BMH CH2 1HT
2E1FLU CH2 1LQ
G0NLQ CH2 1LX
G7HGB CH2 1LX
G3ZVH CH2 1NL
G0ONQ CH2 1NQ
G0HEJ CH2 1PF
G0OVP CH2 1QX
G3LPO CH2 1QX
G8RRS CH2 1RD
G0FOR CH2 1SN
G4YCA CH2 1TB
G0OIV CH2 2BX
G3DRB CH2 2DD
G3TOW CH2 2EA
G0DTF CH2 2HS
G0ROY CH2 2LA
G4MOU CH2 3HG
G3HLP CH2 3LS
G6SKP CH2 3RE
G0VRA CH2 4PG
G6LES CH2 4PX
G7IDE CH2 4QJ
G8XPB CH2 4QR
G3TRL CH2 4RA
G1LML CH3 5HA
G4EZO CH3 5HF
G8GIZ CH3 5HF
2E1CKA CH3 5JA
G8KKN CH3 5LA
G1XPW CH3 5LE
G1CZU CH3 5LY
G7BQY CH3 5LY
G4HLM CH3 5PH
G8WWY CH3 5PQ
G8ZRE CH3 5PT
2E1CJW CH3 5PT
G3GIZ CH3 5QX
G6IFA CH3 5QX
G7BSE CH3 5SH
G8SLP CH3 5UL
G3ETH CH3 6AN
G8DOF CH3 6DU
G0VAH CH3 6PE
G7GFC CH3 6RS
G1YNJ CH3 7DD
G3XSR CH3 7DL
G7NCY CH3 7DN
G1DIA CH3 7ES
G3IEG CH3 7JJ
G0HTY CH3 7NP
G0VOA CH3 7NR
G6FDK CH3 7QD
G8KWP CH3 7SY
G6VUX CH3 8EQ
G1DCB CH3 8HA
G3PAK CH3 8HA
G0LHW CH3 8LP
G7NEH CH3 8LP
G6JGL CH4 7DL
G8VAD CH4 7LU
M1BZI CH4 7NF
G3PZH CH4 7PG
G3EON CH4 7RJ
2E1CAR CH4 8AR
G7KLN CH4 8BB
G8RUN CH4 8DY
G4INX CH4 8HS
G4UXD CH4 8LB
G4WXL CH4 8LB
G0UVT CH4 9JR
G0UWA CH4 9JR
G0CTH CH4 9NG
G1SVN CH4 9NG
G6YCW CH4 9NJ
G4VPC CW6 0NL
G3NHP CW6 0PU
G7VOI CW6 0SD
G3TZO SY14 7AX
G4GOO SY14 7HS
G0BYM SY14 7JR
G6XTC SY14 8JB
G4XVR SY14 8JQ
G4XVS SY14 8JQ
G7VOE WA6 0LT
G6IRW WA6 0NW

Chesterfield

G6NZA S40 1HQ
G4LNG S40 1HZ
G4HMW S40 2DR
G3YSY S40 2LT
G1MLO S40 2NB
G1IOR S40 2RH
G7SFF S40 2SL
G6XKZ S40 2UQ
G4TJJ S40 3AG
G3THO S40 3BT
G3ZLF S40 3BT
G3NWP S40 3DA
G4BFR S40 3DF
G4BBI S40 3DR
G3GLX S40 3EY
G7KUM S40 3HT
G6XXC S40 3HX
G0MIL S40 3LS
G1UGL S40 3LW
G3MND S40 3NA
G2HIX S40 3PQ
M0BFW S40 4DG
G7WDL S40 4DG
G6OKU S40 4LG
M1ANW S40 4SD
G0RDF S40 4TF
G1ZWY S40 4UE
G7HRU S40 4UR
G4VHX S40 4XD
G0FMD S40 4XE
G0EWI S40 4XJ
G1KQS S41 0HB
G4HQH S41 0SU
G6AMM S41 7AB
G4RVR S41 8LN
G6FCN S41 8LN
G0WPN S41 8LN
G0FQB S41 8PN
G7NLJ S41 8PN
G1XZJ S41 8QB
G6NDG S41 8QB
G8VQS S41 8RU
G3YNR S41 8RW
G1GNA S41 8SF
G6EYD S41 8SF
G6FTH S41 8SF
G6ZOT S41 9JU
G0LKB S41 9NU
G6NZO S41 9PA
G3RLL S42 7LR
G6OYV S42 7NH
G7ODX S43 1BF
G7ODY S43 1BF
G1OPL S43 1DD
G7SPS S43 1JG
G4PBC S43 1LJ
G6SEJ S43 1LJ
G4PSC S43 1QD
G8YMW S43 2AB
G3ZSZ S43 2AP
G4CVO S43 2BX
G7OQH S43 2BX
G4RIQ S43 2DZ
G3OVK S43 2EA
G1VPE S43 2EZ
M1BCP S43 2HD
M1BVI S43 2LH
G1HFT S43 2NL
G7SPR S43 2NQ
M0APL S43 2NR
G7SMC S43 3AG
G4OFD S43 3AX
G1WRD S43 3DZ
G7VND S43 3HJ
G7LLK S43 3PY
G7SLY S43 3PY
G4OQU S43 3QB
G1IVH S43 3SN
G0MBS S44 5ET
2E1DTS S44 5HE

Chester-le-Street

G3ZIJ DH2 1JJ
2E1DFK DH2 1JU
G7WCD DH2 1JU
G4TQV DH2 1LB
G1NCL DH2 1NR
G0EMF DH2 1NT
G4OCQ DH2 1QH
G0DUG DH2 1RH
G0CUO DH2 1SG
G0MXA DH2 1SX
G1ABM DH2 2LG
G4ODE DH2 2LX
G0DCH DH2 3AL
G8APW DH2 3EU
G6LCL DH2 3HG
G3ZEP DH2 3HG
G7DPZ DH2 3HY
G0LEA DH2 3LU
G7GJU DH2 3LX
G0EBV DH2 3TJ
G8XHW DH3 3ED
G4EUZ DH3 3JZ
G4GBF DH3 3JZ
G7GMY DH3 3NP
G4OSK DH3 3PR
G7MJN DH3 3RS
G4KOT DH3 3UN
G0NHC DH3 4AQ
G1REO DH3 4HU
G1JDP DH3 4JH
2E1EYK DH3 4JH
G4BQM DH3 4LN
G6GLR DH3 4LU
G0FBQ DH3 4LU
G0AAU DH4 6EA
G1HEZ DH9 0QU

Chichester

M1BQK GU27 3JU
G4UWJ GU27 3LA
G4DQZ GU27 3ND
G4YJX GU27 3QN
G0NON GU27 3RG
G6NMQ GU28 0EQ
G8WSX GU28 0EQ
G0HJZ GU28 0NY
G0MWX GU29 9QZ
G3TZL GU31 5AX
G0BUZ GU31 5HJ
G0DDY GU31 5HJ
G4DOK GU31 5NZ
G8HPN GU31 5QB
G7EYS PO10 8AU
G4ALF PO10 8BQ
G0UFY PO10 8BS
G6FVB PO10 8EU
G1MDJ PO10 8HS
G7IKO PO10 8HT
M1BTO PO18 0AY
G3NXO PO18 0HA
G4UCR PO18 0HA
G6AIH PO18 0HT
G0KNU PO18 8AU
GJ0RKM PO18 8AU
G0CHK PO18 8BE
G3AQC PO18 8EU
G3ZQE PO18 8HQ
G3JAX PO18 8JH
G6XZA PO18 8LF
G0SDC PO18 8TZ
G6RGV PO18 8UQ
G3MVZ PO18 9AR
G4ETU PO18 9BL
G7HQU PO18 9EF
G0IOP PO18 9JJ
G0CRG PO19 2AU
G4WIR PO19 2AU
G6UYJ PO19 2AU
G3EWY PO19 2HL
G7KSG PO19 2HX
G6CSX PO19 2NF
G1ZYD PO19 2NW
G0RHK PO19 2PY
G3ZDF PO19 2QR
G8NYK PO19 2TP
G4AKP PO19 2XW
G4SQJ PO19 2YH
G1UFS PO19 3AE
G7BVL PO19 3LY
G0ODJ PO19 3QQ
G0WSD PO19 3QU
G0OMX PO19 4AW
G6DHT PO19 4BG
G0BNT PO19 4DJ
G3YEN PO19 4DZ
2E1FGB PO19 4NE
G4WSX PO19 4NP
G0WBR PO19 4NW
G4EHG PO19 4RL
G4ZTQ PO19 4TW
G3ZEN PO19 4UA
G4VQZ PO19 4UA
2E1DTD PO20 0JB
M1BCY PO20 0JB
2E1ENP PO20 0RF
G4YJD PO20 6DD
G0RJB PO20 6JF
G3KDE PO20 6JY
G0POU PO20 6JZ
M1BNG PO20 6JZ
2E1CDD PO20 6JZ
G8CBO PO20 6JZ
G1NBX PO20 6NY
G0KWD PO20 6NZ
G3NFW PO20 6PA
G4BZB PO20 6PE
G2NM PO20 6PE
M0BDI PO20 7AN
G0SZJ PO20 7EJ
G4GWH PO20 7JG
G0LMJ PO20 7JY
G2CXO PO20 7PE
G3NW PO20 8EX
G3SFB PO20 8RJ
G0PLZ PO20 9EP
M1AIM RH14 0BX
G3PUX RH14 0DY
G3MOZ RH14 0DZ
G4NUX RH14 0TE
G8SLU RH14 0TF
G2HBQ RH20 1ER
G1EAE RH20 1HS

Chiltern

G4CWL HP10 8AR
G0TQX HP10 8NX
G0WTV HP15 6SQ
G3LBM HP15 6TF
G0RBB HP15 6TN
G3XJS HP15 6UG
G0TRE HP15 6XA
G4SNQ HP16 0DF
G6TRP HP16 0HB
G6YOZ HP16 0HB
G4KGT HP16 0LF
G7TQA HP16 0NA
G0GFJ HP16 0NJ
G8XTJ HP16 0NJ
G4HMC HP16 0PE
G0NLJ HP16 0PE
G8FKP HP16 0QD
G7HKZ HP16 0RY
G4PFA HP16 0SP
G4XWP HP16 9DN
G3ZNU HP16 9DS
G6MJA HP16 9JH
M1BWT HP16 9JT
G4UXY HP5 1EZ
G1SHT HP5 1RH
M1BPU HP5 1RL
G4UXA HP5 1RT
G6ZWG HP5 1RT
G1OQK HP5 1RW
G3WPD HP5 1SS
G3XZG HP5 1TA
G6HGM HP5 1XH
G6HPK HP5 2BU
G7PEU HP5 2BU
G7TRA HP5 2DE
G0ABP HP5 2EB
G6LBG HP5 2EB
G3AGZ HP5 2EQ
G0BZN HP5 2EW
G8BLB HP5 2HL
G1JBG HP5 2LH
G7GLZ HP5 2QX
G7IZD HP5 2QX
G6YEC HP5 2RS
G7SLR HP5 2SG
G3VZF HP5 2XL
G1CUZ HP5 2XW
G1MDG HP5 3AD
G3MDG HP5 3AD
G3NCL HP5 3AD
G4HES HP5 3AD
G0WAC HP5 3AD
G3HBW HP5 3AZ
G8XXF HP5 3BE
G6XO HP5 3DA
G3VRY HP5 3DJ
G0NLX HP5 3EA
G6CDV HP5 3JH
G4XRV HP5 3LU
G4CYY HP5 3RJ
G7RGA HP6 5AD
G7TCU HP6 5HS
2E1DGV HP6 5HS
G8MRI HP6 5NW
M1BFY HP6 5NW
G8ETV HP6 5PY
G0KLQ HP6 5RD
G4TBR HP6 5RT
G7TPD HP6 5RU
G7KSG HP6 5RZ
G8RRN HP6 6AQ
G6IQA HP6 6BH
G3HBR HP6 6DR
G8OSF HP6 6DR
G8IXA HP6 6DZ
G4GYN HP6 6HF
G7MMC HP6 6HL
G7EAT HP6 6HP
G1JNG HP6 6JF
G7PTW HP6 6JF
G7WLG HP6 6NL
G7WGE HP6 6QQ
G0PZI HP6 6RF
G4TNU HP6 6RR
G7JAF HP7 0EH
G6YRX HP7 0JL
G2SH HP7 0PY
G1HPU HP7 0RQ
G1EAZ HP7 9AU
G3WLO HP7 9AU
G7EDU HP7 9BG
G3PNV HP7 9BU
G6TQZ HP7 9BU
G3BP HP7 9DT
G0RDI HP7 9DZ
G8SJP HP7 9DZ
G0VRV HP7 9DZ
G6HUH HP7 9HD
G8GDI HP7 9HL
G3PNB HP7 9NQ
G3DTX HP8 4AR
G7MWP HP8 4ER
G1MFR HP8 4EW
G3ZYZ HP8 4HP
G3PBI HP8 4LG
G6DOV HP9 2BA
G3OHX HP9 2QU
G4LHE HP9 2XH
G8AAF HP9 2XN
G3BEX HP9 2XP
G0RTF HP9 2XR
G0YTF HP9 2XR
G4SUG HP9 2YW
G0RAU SL9 0DD
G6XDB SL9 0DD
G6AG SL9 0EZ
G4UNE SL9 0HE
G3ZHY SL9 0JE
G4MDN SL9 0LR
G3NZW SL9 8LJ
G3JIP SL9 8NS
G0UGP SL9 8RL
M0BGB SL9 9AX
G4KEM SL9 9AY
G4AXK SL9 9HS
G8ADM SL9 9LB
G1LIK SL9 9PA
G8VIK SL9 9PW
G6DMV WD3 5PJ

Chorley

G0UCS L40 2QS
G6ZWZ L40 2QS
G3WUC PR5 0DE
G6RMX PR5 0EH
G4OTN PR5 0SA
G0HDL PR5 2PJ
G3XII PR5 2SP
G0BVP PR5 3LP
G0RFJ PR5 3TA
G0PFU PR5 7AP
G0SHS PR5 7HH
G3VYV PR5 7HH
G7FTH PR5 7HH
G0JSJ PR5 7HP
G4WFE PR5 7HP
G6KHP PR5 8DS
G3UCA PR5 8DU
2E1FMB PR5 8EB
G0VMA PR5 8EN
G6YXU PR5 8EY
G0USM PR6 0DG
G7PHM PR6 0DG
G0JMN PR6 0LJ
G7SYY PR6 0PY
M1CEC PR6 0QE
G0CUN PR6 7BJ
G6PMW PR6 7QH
G3ZRE PR6 7TJ
G8VMO PR6 7TT
G4YMZ PR6 7TZ
G6JJI PR6 7TZ
G3WVA PR6 8AZ
G1TSV PR6 8EJ
G7KTQ PR6 8PX
G4BOB PR6 8QQ
G7REB PR6 8QX
G8JZI PR6 8SG
M0AQE PR6 9JS
G0LBT PR6 9JW
G3TXK PR6 9PA
G3GSL PR6 9PL
G4PYC PR6 9PP
G1KVC PR6 9RS
G3DBY PR7 1JX
G4WYZ PR7 1LX
G4TZK PR7 1PH
G7VSI PR7 1RH
G0DAI PR7 1UH
M1ADC PR7 2DL
G0LRE PR7 2JH
G6RMW PR7 2QF
G4WCS PR7 2QH
G7UAY PR7 3HY
G0UZE PR7 3JY
G4IYP PR7 3NH
G3RYY PR7 4AJ
G4DRX PR7 4EY
G3URK PR7 4JU
G7PYW PR7 4NL
G0DVM PR7 4NX
G1JCH PR7 4NX
G4JQE PR7 4NX
G4OAW PR7 4NX
G0NGK PR7 4PJ
G1DAK PR7 4PY
2E1FKH PR7 4PY
G8CFI PR7 4QP
G0UZO PR7 5AE
G4BME PR7 5BD
G4WGT PR7 5EH
G3JMZ PR7 5HH
G8TUB PR7 5JT
G4YSN PR7 5LF
G0OWA PR7 5NY
G1MBE PR7 5NZ
G4JZT PR7 5PF
G1AHM PR7 5PW
G7NPG PR7 5QS
G3AJL PR7 5QW
G8SNQ PR7 5QW
G3YQQ PR7 5RA
G4TZR PR7 5SR
G0WMR PR7 5SR
G4ZYN PR7 5SW
G6OXH PR7 5UA
G1ASX PR7 6BE
G6XVZ PR7 6BT
G3ANG PR7 6BU
G8COR PR7 6DE
G6AHO PR7 6EP
G4IYI PR7 6HN
G7GYR PR7 6LY
G0OFY PR7 6NY
G0WTD PR7 6PD
G0SHU PR7 6PL
G4PQM PR7 6PN
G3WBY PR7 6PP
G0CUB PR7 6PT
G3RPO PR7 6PT
2E1FLP PR7 6PT
G4DHE PR7 6PW
G0WTM PR7 6PW
2E1ENO PR7 6PW
G1ZTG PR7 6QD
G6OJA PR7 6QD

Christchurch

G7UFZ BH23 1JA
2E1CFQ BH23 1JH
G7PNW BH23 1LH
G0RCN BH23 1QL
2E1CDS BH23 1QL
G8ZPW BH23 2AF
G4HEC BH23 2AG
2E1FTJ BH23 2DQ
G1NMR BH23 2DY
G3LUF BH23 2JQ
G8LQV BH23 2JR
G0LOW BH23 2LJ
G6DUN BH23 2LL
2E1CCB BH23 2LL
2E1FTI BH23 2LL
2E1BBV BH23 2LN
M1BEJ BH23 2LN
G3UZD BH23 2NG
M0BAX BH23 2PL
G7NUB BH23 2PL
M1AKX BH23 2RU
G3UXR BH23 2SP
G8PRB BH23 2TE
G4TLQ BH23 3AN
G1GTL BH23 3AP
G4XFB BH23 3BW
2E1DHX BH23 3BY
G6SPB BH23 3EW
G7WSN BH23 3JN
G5SG BH23 3JZ
G7HQR BH23 3LS
G4MUU BH23 3QN
G0FBA BH23 3RZ
G4SNR BH23 3SN
G6FAH BH23 4BP
G3BJR BH23 4DJ
G0TKA BH23 4DR
G7AET BH23 4ED
G1KIM BH23 4HA
G0MUD BH23 4JE
G7MUD BH23 4JE
G3YPT BH23 4JG
2E1EVE BH23 4JG
G6ACD BH23 4JJ
G3SGA BH23 4QR
G4YQT BH23 4SE
G3MPO BH23 5AZ
G5FH BH23 5BL
G7UWZ BH23 5DB
G2DXU BH23 5DN
G4XEV BH23 5DZ
G8BOE BH23 5HJ
G8ZMV BH23 5ND
G8RPI BH23 5RP
G0IXT BH23 6BE
G0DBI BH23 7EY
G4XWT BH23 7HG
G4FND BH23 7JJ
G0IKN BH23 7LD
G6BUF BH23 7LR
G4TKP BH23 7NJ

City of Edinburgh

GM0RWU EH1 1SR
GM7ORX EH1 2HR
GM1BKF EH10 4AN
GM0CQC EH10 4NQ
GM6CMQ EH10 4QH
GM4CAZ EH10 4ST
GM0SCO EH10 4ST
GM3UM EH10 5AW
GM3GBX EH10 5EY
GM4HFM EH10 5JY
GM0VMV EH10 5LY
GM3SRV EH10 5PF
GM4RZW EH10 5PS
GM7RMF EH10 5SN
GM4GJG EH10 5TD
GM8OTI EH10 6ER
2M1FOY EH10 6HF
GM8BJF EH10 6HN
GM4ZOA EH10 6NN
GM3YMX EH10 6PU
2M1DGU EH10 6PU
GM1GEQ EH10 6QR
GM3JDX EH10 6RJ
GM8LRV EH10 6SP
GM8ARV EH10 6UL
GM8IWC EH10 7AE
GM4HRW EH10 7AF
GM4BYF EH10 7AP
GM3HAM EH10 7AP
GM6KAY EH10 7AP
GM0OYV EH10 7BG
GM0WDD EH10 7DD
GM0PGV EH10 7DE
GM4LBN EH10 7DF
2M1ATN EH10 7DG
GM7REG EH10 7DG
GM8GUJ EH10 7HR
GM7WFT EH11 1HA
GM8LYQ EH11 1JZ
GM7WEF EH11 1NX
GM0AXY EH11 1RP
GM4YMM EH11 1RP
GM0SYL EH11 1RP
GM4DIZ EH11 3PF
GM3NTX EH11 3PW
GM6AXZ EH12 5AY
GM8CVN EH12 5BE
GM4DTH EH12 5EA
GM3HNE EH12 5NG
GM4HYR EH12 5RF
GM3HUN EH12 5RP
GM0IMZ EH12 6DF
GM3HOQ EH12 6LE
GM0BPT EH12 6NB
GM0CMO EH12 6NB
GM0CBQ EH12 6UH
GM4DIJ EH12 6XB
GM7UJJ EH12 7EB
GM4AJV EH12 7EW
GM3OYV EH12 7LB
GM1PEL EH12 7PD
GM0VYM EH12 7PS
GM8CSE EH12 7PU
GM1JWC EH12 7PY
GM0ALS EH12 7RZ
GM0SSQ EH12 8DL
GM3VJY EH12 8DL
GM4CUX EH12 8DL
GM4FBR EH12 8EB
GM8DKB EH12 8EE
GM8IOL EH12 8HU
GM3NCO EH12 8JZ
GM8PSV EH12 8LF
GM3RVL EH12 8QW
GM4DHN EH12 8TB
GM1PVD EH12 8UJ
GM0CWR EH12 8XR
GM3XUW EH12 8XW
GM0BCA EH13 0AU
GM0SUH EH13 0ER
GM3BMI EH13 0ET
GM4ZQH EH13 0EW
GM8BQV EH13 0JG
GM4NUN EH13 0RQ
GM7RBW EH13 9DB
GM0AQG EH13 9DR
GM0EWJ EH13 9DR
GM4YLU EH13 9JS
GM0FMW EH13 9LF
GM7GAE EH14 1BJ
GM8PAH EH14 1ET
GM3GIG EH14 1HL
GM4EZJ EH14 1HW
GM6NIA EH14 1JW
GM4YLN EH14 3DZ
GM4JGO EH14 3HH
GM6MUZ EH14 5DX
GM3OWU EH14 5EG
GM4XZN EH14 5EZ
GM4XEP EH14 5JJ
GM8LKL EH14 5JN
GM8NZL EH14 5JN
GM8CHL EH14 5LY
GM4NJD EH14 5NG
GM4REN EH14 5PP
GM8BCB EH14 5PQ
GM3TFY EH14 5SX
GM1FEM EH14 5SY
GM4CAH EH14 5TA
GM1IDP EH14 7BT
GM1GDO EH14 7ED
GM4EAU EH14 7EJ
GM3DJT EH14 7ET
GM4RAH EH14 7HS
GM0CFW EH14 7HT
MM0AWU EH14 7HT
G7VPP EH14 7HT
GM1JKJ EH14 7JE
GM8BHR EH15 1NB
GM7CZC EH15 1NH
GM3JOA EH15 1PA
GM4NSZ EH15 1PA
GM8NSZ EH15 1PA
GM3DIE EH15 1RA
GM0VIV EH15 1SG
GM7RYR EH15 2QR
GM2HDH EH15 3NZ
GM7OLN EH15 3PF
GM4BHU EH15 3QG
GM7IFX EH16 4TG
GM6SAG EH16 5QN
GM3MLW EH16 6AE
GM1FYW EH16 6DN
GM4FH EH16 6NW
GM3LHV EH16 6PD
GM1VTY EH17 7QE
GM4DTJ EH17 8AW
GM4FNV EH17 8BP
2M1ETI EH17 8DX
GM6JAG EH17 8EU
GM4DEK EH17 8HR
GM0CUY EH17 8SP
GM6OUL EH17 8SR
GM4XHQ EH28 8PD
GM4DQK EH29 9AT
2M1EJY EH29 9DP
MM1ASI EH29 9DP
GM7OBI EH29 9DZ
GM7RGC EH3 5LZ
GM8EKF EH3 5NA
GM6KOR EH3 6HB
MM1ALV EH3 6LJ
GM6CZM EH3 7TZ
GM6NIC EH3 7TZ
GM7GNO EH3 9DD
GM0TDO EH30 9NG
GM7RAK EH30 9PL
GM4XWL EH30 9PQ
GM4HCE EH30 9RF
GM0SUY EH30 9XG
GM1YGW EH30 9XN
GM3IDS EH30 9YQ
GM4WZP EH30 9YQ
GM6WTP EH4 1NT
GM5NU EH4 1NY
G8SOK EH4 1PJ
G1BUY EH4 1RA
GM7JGR EH4 2EE
GM4GIO EH4 2EF
GM4IMX EH4 2EF
GM3KIG EH4 2JP
GM4DMQ EH4 2JU
GM4COK EH4 2LX
GM3NHW EH4 2PF
GM1CNH EH4 2PL
GM1CQC EH4 2TT
GM1PHD EH4 2UE
GM8JAV EH4 3BB
GM8MST EH4 3JU
GM6RXQ EH4 4TE
GM0AXX EH4 5AG
GM0RMV EH4 5AW
GM4HWO EH4 5DP
GM3SBC EH4 5JA
MM0AKM EH4 5JE
GM0PXV EH4 5NF
GM6PYD EH4 5PY
GM3KPD EH4 6EL
GM7IQG EH4 6HN
GM8APX EH4 6JY
GM4IQU EH4 6PB
GM3KBP EH4 7AH
GM0SZL EH4 7AL
GM7KTY EH4 7BJ
GM7PZH EH4 7BJ
GM0WFB EH4 7DF
GM4HKH EH4 7EB
GM4YWI EH4 7NG
GM0FWY EH4 8DT
GM4CKM EH5 1AZ
GM0WTP EH5 1NQ
GM0CII EH5 2JJ
GM6JYC EH5 2PU
GM0TEA EH5 3AR
GM3MUM EH5 3PT
MM0AJQ EH6 4RN
GM4RJE EH6 5EP
GM8YIK EH6 6AX
MM1BOH EH6 6RY
GM4IUS EH6 7JH
GM4MCV EH6 8AF
GM8KCK EH6 8ET
GM7AUW EH6 8QP
GM0NVQ EH7 4BY
GM6TYL EH7 4JD
G4PNM EH7 5ET
GM1YME EH7 5JA
GM7OLQ EH7 5RJ
GM7IHZ EH7 5TT
GM0HLK EH7 5UF
GM8LVJ EH7 6HE
GM0JIN EH7 6RH
GM1MFD EH7 6UE
2M1CFS EH7 6UE
MM1BGI EH8 7DW
GM4JEM EH8 7JL
GM0WZO EH8 7LG
MM1ATX EH8 7LG
GM6FCW EH8 8JB
GM7FIE EH8 9XL
GM0EWK EH9 1AQ
GM1MBT EH9 1JB
GM3MYW EH9 1LZ
GM7OCU EH9 1QG
GM6RQU EH9 1UF
GM3IRV EH9 2AJ
GM0KZJ EH9 2EH
GM4XNQ EH9 2EH
GM4GZW EH9 2NZ

City of Glasgow

GM0FRI G1 2BQ
GM3TYQ G11 5AD
GM8TXC G11 5RT
GM7DXT G11 6SB
GM4ACM G11 7LG
2M1DIN G11 7LH
GM7KVU G11 7PP
GM0NOZ G11 7XB
GM1MCA G12 0PU
GM0TET G12 0PZ
GM8YXR G12 0TB
GM0ELL G12 8AF
G6GAF G12 8BT
GM4LSD G12 8QQ
GM3TQH G12 8SP
GM3RUP G12 8TL
MM0AMY G12 9NX
GM8DKG G12 9SD
GM0HSG G13 1PL
GM0EFH G13 1SA
G7UWN G13 1XJ
GM3OAV G13 2JB
GM8CFW G13 2JH
GM0LBM G13 2JX
GM3EXX G13 2XX
GM0GYT G13 3AQ
2M1BAR G13 4EF
GM0FHJ G13 4HL
GM4NWK G13 4QN
GM1MDH G14 9HP
GM7NZI G15 7QE
2M1EKI G15 7QE
GM7PGM G20 0JL
GM0USI G20 6AP
GM4HAO G20 6DA
GM1FML G20 6HD
GM3XLB G20 6NQ
GM0KVD G20 8LL
GM7JDS G21 1RX
GM4PVF G21 1XP
GM0KUJ G21 3HY
GM0VEK G21 3RP
GM4ENN G21 4XY
GM0JHE G23 5BA
GM4AGG G3 6QH
GM3VBT G3 6QN
G3YXH G3 7SD
GM0KTO G31 3NE
GM1FTE G32 0DB
GM8YGI G32 0NX
GM3SPT G32 6TA
GM4DAE G32 8LE
GM3JKC G32 9BW
GM4PCT G33 2DD
GM1POA G33 5HU
2M1EZA G4 0PH
GM7MTK G4 7SA
GM3DIN G41 2AF
GM8PIV G41 3BS
GM4RUP G41 4DE
2M1CHZ G41 5AE
GM0TQB G41 5AF
GM3DSD G42 8RD
GM0GZX G42 9HL
GM0HSC G43 1EQ
GM4JZJ G43 2SU
GM0EPO G43 2TF
GM4ISY G43 2UD
GM0HVD G43 2UP
GM0DQR G44 3DH
GM6KDN G44 4BL
GM0WRR G44 4PA
GM0IMW G44 4QY
MM1BJT G44 4TJ
2M1EDC G44 5AP
GM3YCG G44 5AP
GM0RSE G44 5EN
GM7LJE G44 5PQ
GM0LIM G46 8AB
GM7VUJ G46 8RD
GM1JTJ G51 1SE
GM0AAJ G52 1EY
GM4BGS G52 3HA
GM1HDF G52 3HH
GM1JNC G52 3JY
G4WMN G53 7JP
GM3UTQ G53 7UJ
GM7BOW G69 6JW
GM4RPE G69 6LQ
GM3UQI G69 7AP
GM0OPJ G69 7HW
GM3SAN G69 7HW
GM0LTQ G69 7NB
GM0DVO G76 9BN

City of London

G1ILA EC2Y 8BE
G0PCE EC4V 3EJ

Clackmannanshire

GM0WWX FK10 1DD
GM0UGH FK10 1LU
GM3FRU FK10 1NL
GM6RFG FK10 1SA
2M1FHS FK10 2BT
MM1BYX FK10 2JN
2M1FHR FK10 2JN
GM4WQH FK10 2JU
GM0BFS FK10 2LY
GM4ARJ FK10 2RX
GM0UUB FK10 3AG
GM8GAX FK10 3AW
G8WGU FK10 3BG
GM4WGC FK10 3DZ
GM0TKM FK10 3LD
GM0PEI FK10 3PD
GM4SEW FK10 4ST
GM0BWR FK11 7AR
GM4FXL FK11 7DG
GM4ZGB FK11 7DH
GM0KAE FK12 5EA
GM0LWD FK12 5HG
GM1PST FK12 5HU
GM1BVT FK12 5JU
GM4XHV FK12 5NN
2M1FHT FK13 6EU
GM0TTY FK13 6HF
GM8CIF FK13 6HT
GM0ODW FK13 6JG
GM0UEL FK13 6LN
GM4VCW FK13 6NL
2M1DST FK14 7AS
2M1DRD FK14 7BA
GM1BXG FK14 7BP
GM0SNG FK14 7DU
GM4WBU FK14 7HP
GM4YJL FK14 7JU
GM0LOD FK14 7JW
GM4FSA FK14 7NF

Colchester

G6LRA CO1 1AP
G7OBX CO1 2ES
G7LYH CO1 2HU
G0DZB CO1 2NA
G0VMB CO1 2QZ
G6LTL CO1 2ST
G7IFB CO1 2TL
G0VNY CO1 2YR
G4YJN CO2 0JE
G4LKD CO2 0JT
G7PQX CO2 0LP
G3OMB CO2 0NR
G7WLC CO2 0PR
G6DNA CO2 7EN
G8UUN CO2 7EX
G4ICH CO2 7JH
G3ISK CO2 7JR
G4NNN CO2 7LQ
G4YIR CO2 7PP
G6FBB CO2 7UX
2E1FRM CO2 7UX
G6KNM CO2 8AE
G4URL CO2 8BH
G7PST CO2 8BP
G6RUY CO2 8NS
G6RVP CO2 8NS
G8UBU CO2 8NZ
G7UFT CO2 8PA
G8FYH CO2 8PP
G6XWY CO2 8PX
G7RWF CO2 8QF
G0VAE CO2 8SD
G4DIS CO2 8SJ
G4BHV CO2 8TE
G6FLW CO2 8UD
2E1FRL CO2 9JR
2E1FRO CO2 9JR
M0ALV CO3 3AY
G1TWF CO3 3HF
G3OCQ CO3 3LL
G0LIN CO3 3QN
G0KOY CO3 3RS
G0FQJ CO3 4DT
G7EIF CO3 4DT
G4SOB CO3 4HR
G3PND CO3 4HZ
G7VGF CO3 4HZ
G0WZV CO3 4JJ
M0AJP CO3 4JJ
G4JKC CO3 4JP
G0CKB CO3 4PA
G4AOB CO3 4SP
G8CKW CO3 5AF
G6RDD CO3 5AG
G0HHC CO3 5DP
G0DCL CO3 5DX
G3RBQ CO3 5HL
G4KTI CO3 5HL
G6WYL CO3 5HP
G4PBR CO3 5JQ
G8XDZ CO3 5LB
M1BPY CO3 5RZ
G0WCT CO3 5TR
M1BKH CO3 5XN
G4URA CO3 5XU
M0BCK CO3 5YA
M1BHG CO3 5YA

G0NHB CO3 5YJ
G4AEB CO4 0BN
G8SGB CO4 3AS
G4DAN CO4 3BU
G0TLH CO4 3DY
G6HQI CO4 3EG
G4HKC CO4 3EY
G4HKB CO4 3FD
G0VYC CO4 3FE
G4RKB CO4 3FN
G8DRE CO4 3FP
G4JHP CO4 3HB
G0OSR CO4 3HX
G0PFM CO4 3JP
M1BFQ CO4 3JW
G7EHY CO4 3LX
G1DTE CO4 3NL
M1ALP CO4 3XN
G6CLA CO4 3YA
G3ZOL CO4 3YP
G0OWK CO4 4AJ
G6IIF CO4 4EA
G3MQR CO4 4HX
G1MNA CO4 4JE
G4OYH CO4 4LP
G4TGX CO4 4NH
2E0APS CO4 4NH
G8RFC CO4 4PW
G6AQI CO4 4QJ
G6VCM CO4 4RD
G7RLH CO4 4RE
M1BQZ CO4 4RJ
2E1EWK CO4 4SL
G1GWO CO4 4YU
G1VDC CO4 5AD
G0MBF CO4 5DH
2E1DEL CO4 5DX
G3CO CO4 5JX
G3FIJ CO4 5JX
2E1DXB CO4 5JX
G4UTJ CO4 5LE
2E1EAG CO4 5LQ
G0AHE CO4 5PE
G6XIF CO5 0AA
G6HWM CO5 0AB
G4XQD CO5 0AJ
G4IIK CO5 0AP
G7JMW CO5 0BN
G4KTB CO5 0DT
G0WIZ CO5 0DZ
G4TFP CO5 0EF
G6KVI CO5 0EF
G7NMI CO5 0EF
G7SJF CO5 0EN
G3PQI CO5 0HB
G1AWK CO5 0JA
G4ZZL CO5 0JU
G0IZK CO5 0LR
G0EGX CO5 0PB
G6DBY CO5 0QT
G4PZX CO5 7AS
G7OSL CO5 7HS
G4YK CO5 7LB
G0JPJ CO5 7NH
G7HGV CO5 7PT
G4PAY CO5 8AR
G4PDA CO5 8AR
G8SOI CO5 8DR
G1ODS CO5 8DU
G8WSH CO5 8DX
G0UUD CO5 8JL
G1HPJ CO5 8JU
G0RIS CO5 8LJ
G0JDE CO5 8LN
G0APP CO5 8QP
G6RPF CO5 8QP
2E1DRS CO5 8QP
G2CVO CO5 8RD
M1CAY CO5 8SL
G3NXK CO5 9TD
G4KXF CO5 9TD
G4PYG CO5 9TH
G1LIB CO6 1BP
G6IAW CO6 1BS
G3YEC CO6 1DB
G4JIG CO6 1LZ
G1XUP CO6 1XD
G1IUD CO6 1XG
G7LCD CO6 1XG
G8IUD CO6 1XG
G7OBN CO6 1XG
G4FJC CO6 1XR
2E1FXL CO6 1XR
G7HHN CO6 2ED
G4ZTR CO6 3DB
G8YMZ CO6 3EP
G2ABR CO6 3NA
G6GCE CO6 3NB
G7NZV CO6 3NE
G6XMM CO6 4BJ
G1APT CO6 4TS
G4AFX CO7 6DR
G0VEI CO7 6DR
M1BGA CO7 9DD
G4JIE CO7 9HB
G4MGO CO7 9HB
M1AMI CO7 9HG
G0GGM CO7 9LD
G4SDI CO7 9LG
2E0AIK CO7 9LG
G6CMZ CO7 9LQ
G4FTP CO7 9NH
G4PPX CO7 9NP
M1BQF CO7 9QH
G7TWA CO7 9QZ

Congleton

M1BUT CW10 0DG
G0UMR CW10 0EX
M0AGH CW10 0PH
G4ZKR CW10 0PJ
G0NTG CW10 9BY
G0GKN CW10 9HG
G6EKS CW10 9HP
G3WUG CW10 9HX
G6OKN CW11 1BN
G6ZGB CW11 1BZ
G4YFH CW11 1DR
G6CNB CW11 1FG
G3GMM CW11 1RW
G7OOI CW11 1SG
G4MXE CW11 1SQ
G4GHT CW11 2UB
G6IIU CW11 3BU
G6FDC CW11 3GA
G0JNE CW11 3GF
G0LAZ CW11 3HA
G7HOA CW11 3JF
G7OBW CW11 3JF
2E1CNY CW11 3JF
2E1FIU CW11 3JF
G4VIF CW11 4HL
G7CTE CW11 4PN
G6VPJ CW11 4PQ
GM4BFC CW11 4PX
G4PMY CW11 4RE
G6VMV CW12 1JZ
G4LTZ CW12 1NU
G7PZF CW12 1NY
G1EGE CW12 1RL
G7VTQ CW12 1SD
G4JYK CW12 1SE
G4NVN CW12 1SE
G6DZX CW12 1SH
G8INT CW12 2DQ
G3MCC CW12 2HJ
G7HIO CW12 2HQ
G1DUT CW12 2LL
G3LLJ CW12 3BG
G1PAD CW12 3BH
G7FJK CW12 3BN
G4GMZ CW12 3DE
2E1CPQ CW12 3EP
G4UJO CW12 3HS
G7NCD CW12 3JP
G8JPU CW12 3JY
G1LCE CW12 3PL
G4FSH CW12 3PL
G6ICL CW12 3RA
G8YHH CW12 3RA
G3ZAU CW12 3RB
G4TAG CW12 3RH
G1YBZ CW12 3RL
G8GYQ CW12 3SJ
G8WWF CW12 3TD
2E1CAT CW12 3TD
G6CYX CW12 3TH
G4BER CW12 3TU
G8RXY CW12 3TY
G4WTZ CW12 4DY
G4RPX CW12 4FJ
G6SLD CW12 4FL
G8OTU CW12 4HH
G8UVZ CW12 4HL
G4HRI CW12 4LD
G4JNC CW12 4LP
G4DWW CW12 4LQ
G4RJH CW12 4LY
G0UXO CW12 4NE
G3VAI CW12 4NR
G4SEN CW12 4PG
G0CQY CW12 4PH
G1HSW CW12 4PP
G1SIB CW12 4QH
G6DIS CW12 4QH
G6XYV CW12 4QU
G1VIY CW12 4RF
G6WXI CW12 4RS
G6LQP CW12 4SN
G0PJC CW12 4SX
G4BOH CW12 4TQ
G1OGE CW12 4UF
G6SUN CW4 7AR
G0CTP CW4 7BT
G6WEL CW4 7BX
G1HSC CW4 7EA
G3AKX CW4 7EA
G7VZP CW4 7EE
G1DBL CW4 7LA
G0CSY CW4 8DT
G0EZF CW4 8HH
G4NVA CW4 8JB
2E1CAM CW4 8JE
G7VQT CW4 8JE
G7VTF CW4 8JF
G4ZVA CW4 8JX
G7VQG CW4 8JY
2E1CEK CW4 8NB
2E1CBD CW4 8NJ
G4YTT CW4 8PP
G0IFR ST7 2AP
G4WHT ST7 2BN
G7FZB ST7 2BY
G7APV ST7 2DZ
G6GFC ST7 2EF
G4PUM ST7 2EH
G7VDJ ST7 2HE
G0RME ST7 2JJ
G0OFW ST7 2ND
G4UFU ST7 2NH
G0EIR ST7 2NQ
G0UKK ST7 2NR
G0WLJ ST7 2RL
G8ORG ST7 2RS
G0UVJ ST7 2RS
G8RBR ST7 2RW
G3VBG ST7 2SH
G4ERQ ST7 2SU
G7JGD ST7 2SU
G4TAZ ST7 2XR
G6LMJ ST7 3BL
G8USX ST7 3DL
G3LBZ ST7 3EJ
G8LZO ST7 3HU
G4DLA ST7 3RB
G6NAH ST7 3RY
G6PEJ ST7 3TG
G8UVN ST7 3TL

Conwy

GW0NWR LL18 5AY
GW3TMH LL18 5DL
GW6VEN LL18 5EW
GW0HBZ LL18 5EY
GW4VLU LL18 5HW
GW4DMR LL18 5JE
GW0DSJ LL18 5LN
G0PWI LL21 0NS
GW6JMC LL21 0RB
2W1CGC LL22 7DA
GW3AWC LL22 7DL
GW4TWB LL22 7DS
GW7NTP LL22 7HE
GW0NSZ LL22 7JF
GW4VIB LL22 7LW
GW0PZQ LL22 7TF
2W1BYD LL22 7UG
2W1CUP LL22 8EB
GW6PQT LL22 8ER
GW3WEQ LL22 8JD
GW7TIH LL22 8PJ
GW1CDH LL22 8QA
GW3YIH LL22 9ND
GW4VHS LL22 9NE
GW0OGK LL22 9NY
GW8FOD LL22 9YD
G4DLC LL22 9YH
GW8BIA LL22 9YY
GW7EQC LL24 0EH
GW4PXQ LL25 0NQ
GW4TSG LL26 0RG
GW0HUN LL27 0JE
GW7UYF LL27 0JJ
GW1IHB LL27 0QA
G7OYZ LL27 0QA
GW2FLZ LL28 4AH
GW0MXV LL28 4AW
GW8YAS LL28 4HY
GW4GMI LL28 4LW
GW6HUV LL28 4NS
GW3DMV LL28 4NU
GW2NF LL28 4PJ
GW4JUF LL28 4RY
GW1OIK LL28 4SG
GW6GKP LL28 4TF
GW3HGL LL28 4TT
GW3SMY LL28 4TW
GW7VHD LL28 4YG
GW0UTC LL28 4YY
GW7TED LL28 5AG
GW8RHP LL28 5EW
GW3OYL LL28 5NG
GW0VMW LL28 5NH
GW3DZJ LL28 5NJ
GW3GJQ LL28 5NJ
GW4RJW LL28 5NN
G6IYA LL28 5YS
GW7KGD LL29 6DH
MW0AFD LL29 6DL
GW6NYR LL29 6DL
2E1EED LL29 6DS
GW8BQK LL29 7AX
GW3MDK LL29 7BB
GW6TM LL29 7BB
GW1RHQ LL29 7BU
GW6STK LL29 7HB
GW0EVE LL29 7NB
GW0EVG LL29 7NB
GW0CCR LL29 7TR
GW1RCC LL29 7TR
GW4PUX LL29 7TR
GW4UWI LL29 7YY
GW0PRM LL29 8EX
GW0TVK LL29 8LE
GW3WHU LL29 8PD
GW4NNL LL29 8PW
GW0PNE LL29 8TA
GW4KGR LL29 8UB
GW4ACO LL29 8UN
GW8VUG LL29 8ZA
GW6DNZ LL29 9DF
GW0DYH LL29 9EL
GW7PQG LL29 9HL
GW3UMB LL29 9LJ
GW1GFU LL29 9LJ
GW0WNB LL29 9LJ
GW6VDR LL29 9NP
GW4WBT LL30 1BL
GW0AEZ LL30 1BN
GW1BBH LL30 1ES
G4OSN LL30 1GZ
GW0GST LL30 1HQ
GW4HBS LL30 1JJ
GW6WFW LL30 1LP
MW1AWT LL30 1LT
2W0AMB LL30 1NR
GW3UTL LL30 1PF
GW8WFS LL30 1PY
GW0NEC LL30 1SJ
GW7UUH LL30 1UH
GW7UIZ LL30 1UW
GW1ZTP LL30 1UY
GW4XLP LL30 1YA
GW4RCM LL30 1YY
GW8VVX LL30 2AA
GW7NYR LL30 2AA
GW3QN LL30 2BQ
GW2DNJ LL30 2PQ
GW7DTQ LL30 2QA
GW6HVA LL30 2UU
GW1HAN LL30 3BD
G8FOY LL30 3FD
GW7NYN LL30 3HU
MW0BET LL30 3LL
MW1BJF LL30 3LL
GW4PVU LL30 3LN
GW6NPD LL30 3LT
GW1BDH LL31 9AL
GW3LCQ LL31 9EL
GW7AEL LL31 9HA
G0BZQ LL31 9ND
GW7KAX LL31 9PY
GW4ZXH LL31 9QE
GW3XKB LL31 9UG
GW1KKJ LL31 9UT
G1LED LL32 8GS
GW6MXG LL32 8LF
GW3HUJ LL32 8LU
GW7AIN LL32 8ND
GW3YQP LL32 8NP
GW6SIX LL32 8RL
MW1AHU LL32 8RL
GW0NSR LL32 8RS
2W1DHV LL32 8RS
2W1FQA LL32 8RS
GW3HAI LL32 8RX
GW0MOJ LL32 8SA
GW3MZY LL33 0EG
GW7JLG LL33 0SN
GW0FPY LL33 0SR
GW0LBA LL33 0SS
GW1ACV LL34 6ER
GW0PQI LL34 6HB
GW6VBO LL34 6LH
GW8XAS LL34 6LY
GW3KKC LL34 6PW
GW0AWN LL34 6TN
GW1SGG LL34 6UA
GW6VDY LL34 6YE
GW7RKQ LL34 6YU

Copeland

G6PTF CA14 4PU
G1FYE CA14 5UJ
G0LGQ CA18 1SP
G1KGU CA18 1SW
G0MLR CA20 1BH
G3MSQ CA20 1HU
G0SVK CA20 1NF
G4UJX CA20 1NG
G3OLU CA20 1PB
G0BWF CA20 1PD
G8CBN CA20 1QY
G4JNL CA22 2EL
G0TJZ CA22 2EL
G1GWF CA22 2NA
G6UUY CA22 2NA
G7SIQ CA22 2RL
G3XZQ CA22 2SU
G3BJD CA22 2TJ
G1LMW CA23 3DY
G3YSD CA24 3JL
G7HCI CA24 3LQ
G4RVH CA25 5EU
G0CZU CA25 5LZ
G3XIU CA26 3QQ
G0RZI CA26 3QS
G3JKU CA26 3TD
G6WLP CA26 3XJ
G4DSI CA27 0AN
G0ESI CA27 0AS
G0JSK CA27 0BE
G3LBC CA28 6SG
G8IMM CA28 6SX
G8ORO CA28 6TA
G2BYP CA28 6TJ
G7RKV CA28 7TT
G4VIA CA28 7XG
G0LJB CA28 8EP
M1BEE CA28 8JP
G7MRL CA28 8JP
G0DIG CA28 8PZ
G6TDL CA28 8RB
G1LCH CA28 8YG
G0VYB CA28 9AB
G4XFH CA28 9FP
G6NRA CA28 9PF
G1GTA CA28 9PF
G4FBC CA28 9RH
G4UDO CA28 9RH
G0RVV LA18 4EL
G4VSF LA18 4HA
M0ASU LA18 4HR
G7PKQ LA18 4LS
2E1FNN LA18 4NX
G8RQP LA18 4NZ
G6TCP LA18 4PL
G0KKO LA18 4PQ
G4RBS LA18 5AD
G1NBY LA18 5DB
G7VGU LA18 5HG
M1AOD LA19 5XL

Corby

G4XRD LE16 8TG
G7OLU NN14 1DQ
G7NJZ NN17 1ER
G7PWY NN17 1HS
G6OMA NN17 1LH
G7SCX NN17 1SY
G4VDQ NN17 2AB
G1TQR NN17 2EL
G7VLN NN17 2EL
G1BKI NN17 2LJ
G6RCY NN17 2RP
G0IKR NN17 2RY
G0JEL NN17 3BE
G7CIV NN17 3BU
G6ZYS NN17 3BZ
G4JCJ NN17 3DN
G8JKM NN17 3HT
G6VOE NN17 3JT
G2FYY NN17 3UH
G0NRB NN18 0BN
G0RRO NN18 0LG
G1IIY NN18 0LZ
G4LKF NN18 0LZ
G8JJE NN18 0LZ
G7TQJ NN18 0NR
M1CDP NN18 0PZ
G0CFD NN18 0PZ
G7NBV NN18 0RY
G4OMV NN18 0TA
G7HYH NN18 8BP
G3MTJ NN18 8HX
G4PUZ NN18 8JP
G0GIA NN18 9DG
G4MRA NN18 9EG
G6IBH NN18 9EH
G0IYS NN18 9HH
G6JTC NN18 9HT
G7TIK NN18 9JA
G4XKD NN18 9JL
G4LKU NN18 9JS
G1JNY NN18 9PH

Cotswold

G3SPO GL53 9NN
G4CRB GL54 1DB
G3RWI GL54 1JD
G2DTS GL54 1LJ
M1ASU GL54 2ND
M1BPS GL54 3EJ
G0MXQ GL54 3JJ
G4UWB GL54 3LQ
G4BPD GL54 3NU
G8GME GL54 4AP
G8EZR GL54 4DD
G6HBS GL54 4NQ
G3DWI GL55 6AG
G3HCU GL55 6AL
G6NGF GL55 6RR
G0WEO GL55 6SZ
G0LSK GL55 6TD
G0TPA GL55 6XP
G6XLG GL56 0EF
G2FZO GL56 0EQ
G3NNR GL56 0JH
G4YSW GL56 0JH
G3RYV GL56 0LN
M1CDO GL56 9AG
G3TJA GL56 9BZ
G3LPN GL56 9EQ
G8DCJ GL56 9JZ
G4XRQ GL56 9RJ
G3SUG GL7 1AP
G3VMQ GL7 1BH
M1BHZ GL7 1BL
G0AZD GL7 1BX
G0UIX GL7 1DR
G0FMH GL7 1ER
G4OCJ GL7 1HF
G0FGN GL7 1HG
G0FFE GL7 1PE
G4UYQ GL7 1SX
G0RYR GL7 1TG
G8BAS GL7 1UG
M0AYX GL7 1WN
G4PLT GL7 1XS
G6DAK GL7 2EN
G4EVE GL7 2LS
G1FRD GL7 2NG
G3UJZ GL7 3AR
M1AYG GL7 3EP
G4JTO GL7 3SD
G7AYE GL7 3SD
G7IYI GL7 5BG
G3TSO GL7 5BY
M1BEO GL7 5HT
G0WIY GL7 5JH
G1MYQ GL7 5QS
G3FJY GL7 5UR
G0IZP GL7 5XE
G6EMB GL7 6AN
M0ANO GL7 7BA
G7HJP GL7 7BB
G4KFR GL7 7BJ
M1AUO GL7 7EE
G3TA GL7 7JU
G4KIM GL8 8DD
G1TJW GL8 8JE
G8SFM GL8 8UP
G1RYB GL8 8YG
G3MAV GL8 8YU
G3KAR WR12 7PF
G4PQB WR12 7PJ
G1SNT WR12 7PN

Coventry

G6AJC CV1 2AA
G1HTT CV1 2JQ
G3YGB CV1 4DJ
G8GGR CV1 4DJ
M1CBA CV1 4EA
G6XMT CV1 5DT
G0BJA CV2 1AF
G1NAT CV2 1EF
G7UOI CV2 1HZ
G1AVF CV2 1SA
G1HOP CV2 2DL
G4HHT CV2 2EL
G4FZP CV2 2GE
G8MWR CV2 2GT
G1KUG CV2 2HR
G6FXE CV2 2JD
G4HRY CV2 2JG
G4WKB CV2 2JS
G7DBO CV2 2JS
G1NIO CV2 2LZ
G6FYL CV2 2NE
G7JYY CV2 2QF
G7PEN CV2 2QF
G7LSD CV2 2RA
G1BNX CV2 3BH
G6ULX CV2 3BZ
G1SBI CV2 3DX
G6ENK CV2 3LA
G0VKB CV2 3LA
G8WW CV2 3LW
G1PWJ CV2 3LY
G8JSC CV2 3NH
G6RYF CV2 4AR
G0MTP CV2 4EU
G0UGX CV2 4GL
M1CEB CV2 4JD
G8MIE CV2 4JW
G8SEQ CV2 5BH
G4GEE CV2 5EH
G0DNK CV2 5FL
G7BJL CV2 5FX
G7HGR CV2 5FX
G6WVM CV2 5GJ
G4IQR CV2 5GL
G4ZSN CV2 5GL
G8TGD CV2 5GL
G4HKI CV2 5HA
G4HEN CV2 5HJ
G2JR CV2 5HN
G4ZHY CV2 5JL
G0BJT CV2 5JX
G6EWZ CV2 5LG
G8ZJO CV2 5LL
G1MOF CV2 5NL
G1AMS CV2 5NN
G1OPA CV2 5NQ
G4ZXN CV2 5NU
G7JSQ CV2 5PF
G0DKR CV3 1AU
G0RDU CV3 1EQ
G3CNG CV3 1FQ
G8MJX CV3 1GW
G0KLZ CV3 2AA
G0LHY CV3 2DX
G8PEE CV3 2FD
G3HLI CV3 2HA
G8OHG CV3 2HF
G0GUL CV3 2HG
G1COV CV3 2NW
G8HPV CV3 2NW
G0LXI CV3 2QG
G4DFN CV3 2RS
G7MNT CV3 2RS
2E1DPV CV3 3EQ
G7VKF CV3 3GR
G1HKP CV3 3GS
2E1FRT CV3 4BQ
G0JKJ CV3 5AG
G4ROA CV3 5AU
G3LHA CV3 5BX
G6CTC CV3 5DS
G2ASF CV3 5GP
G2FTK CV3 5GP
M0BBQ CV3 5HH
G1NJC CV3 5HH
G3LMQ CV3 5JT
G0OGL CV3 5LF
G3TZG CV3 5LF
G1CRB CV3 5NW
G8CWU CV3 5PA
G0EJD CV3 5PN
G1EOK CV3 6BT
G3RIF CV3 6BT
G7SVP CV3 6DB
G8RRA CV3 6DF
G4BRN CV3 6DL
G7RWY CV3 6EB
G3DNH CV3 6EH
G7FRW CV3 6EP
G3LYV CV3 6ET
G7TOL CV3 6EW
G4HFF CV3 6GE
G8ZPU CV3 6HW
M0AZY CV3 6JB
G7SXK CV3 6JB
G6RZS CV3 6JH
G6BRB CV3 6LW
G7UEL CV3 6PF
G4CXU CV4 7AS
G3UDD CV4 7AY
G0REO CV4 7BT
G4ZMC CV4 8ED
G4APS CV4 8FD
G0KNX CV4 8HF
G6UCT CV4 9AU
G4VCX CV4 9AX
G8BZT CV4 9AZ
G6PSO CV4 9BW
G0DCZ CV4 9FG
G1YFI CV4 9JH
G6WLM CV4 9LA
G7KPL CV4 9QS
G1UUS CV4 9UJ
G8GBQ CV5 6AU
G3VWD CV5 6AY
G4RUN CV5 6BB
G3PUR CV5 6DJ
G3WCQ CV5 6DR
G0EMX CV5 6FF
G6LFT CV5 6GU
G3XLM CV5 6JQ
G4TPP CV5 6QD
G8BHY CV5 7AD
G3MAU CV5 7EH
G1KFQ CV5 7FR
G4FSR CV5 7GN
G1PPB CV5 7GP
G7DTK CV5 7HP
G7DTL CV5 7HP
G4JYA CV5 7JE
G3NAP CV5 7JU
G2PHH CV5 7JW
G3UKD CV5 7LJ
G8AHR CV5 7NE
G7MZT CV5 7NY
G4MWP CV5 7QE
G4UVW CV5 7QE
G8GEE CV5 8EY
G4DRU CV5 8FL
G3HDO CV5 8FQ
G4YGS CV5 8LD
G8KVU CV5 8LE
G4VCY CV5 8LS
G6VCY CV5 8LS
G6TYU CV5 9EA
G8ALS CV5 9EN
G3JYS CV5 9HL
G4BCG CV5 9JT
G3XQE CV5 9JU
G3NAY CV5 9JY
G4STI CV5 9LN
G4RCS CV5 9LQ
G8RHM CV5 9LQ
G4JDO CV5 9NA
G4KUE CV5 9NS
G7LAW CV6 1BE
G0EWZ CV6 1BY
G6FBC CV6 1GQ
G8KTX CV6 1LF
G6KHA CV6 1NW
G3DKH CV6 1PP
G4UNR CV6 1PT
G6NIS CV6 1PT
G6DKZ CV6 2AA
G1IJT CV6 2EL
G4AOH CV6 2FD
G6JDO CV6 2GU
G8FFN CV6 2HB
G8LGT CV6 2HB
G1ORT CV6 2JH
G4KZU CV6 2NN
G7JWH CV6 3FH
G1ZNK CV6 3GF
G0ITF CV6 4EG
G4LTK CV6 4EY
G3CWH CV6 4FH
G0JZX CV6 5HF
G0LUE CV6 5HH
G1CWO CV6 6EL
G4OQV CV6 6GY
G6ZTF CV6 7GW
G1MSA CV6 7HZ
G6GYN CV7 8HY
G7JYU CV7 8JJ
G0VBM CV7 8JU

Craven

G7KDG BD20 7AH
G7KDH BD20 7AH
G6HBZ BD20 7AS
G4CPA BD20 7DN
M0BCQ BD20 7PR
G0VJL BD20 7PR
G1MCZ BD20 8RU
G0BDN BD20 9ES
G6RIY BD22 0BN
G8BMI BD22 0DQ
G0TYH BD22 0LA
G0XDL BD23 1BD
G0RAX BD23 1BP
G3ODH BD23 1BU
G3RXH BD23 1ND
G6TYZ BD23 1RL
G7WFG BD23 2DA
G0UCD BD23 2RR
G7UTV BD23 2SJ
GM8YSN BD23 2TE
G1LNQ BD23 3TP
G4TGJ BD23 5EN
G6LHA BD23 5EN
G2BOI BD23 5NJ
G6GFI BD23 6BJ
G8DEA BD23 6JS
G7CWZ BD23 6LL
G4YQA BD23 6PN
G2VO BD23 6QN
G3ATH BD23 6QT
G4GLC BD24 0AG
G4HOK BD24 0HE
G4LDD BD24 9DZ
G8DMU HG3 5JL
G0OIE LA2 7AG
G0PAJ LA2 7BA
G6OOK LA2 7DD
G2XK LA2 7DH
G4LIY LA2 7EP
G7VFF LA2 7HN
G1WNY LA2 7HX
G8CSQ LA2 7HX
G4VPS LA6 3AN
G7UQH LA6 3HB
G8RLH LA6 3JF

Crawley

G4MKD RH10 1LP
G4FPD RH10 2AT
G0FPI RH10 2BZ
G3YNP RH10 2DP
G4TVC RH10 2EH
G6GBI RH10 2HS
G3SBJ RH10 2HX
G1OFF RH10 2LD
G6DOW RH10 2NB
G7SFA RH10 3BG
G4LSK RH10 3BJ
G0BYJ RH10 3BW
G3YVR RH10 3DN
G3CTP RH10 3DW
G4XHF RH10 3JP
G0DSU RH10 3NG
G8NVB RH10 3TW
G3ZON RH10 5DQ
G1UMJ RH10 5DW
G6RC RH10 5DW
G4UMJ RH10 5DW
2E1BMF RH10 5DW
G8ITY RH10 5HR
G3MGL RH10 5LD
G6FPH RH10 5LD
G8KCY RH10 5LD
G4FYY RH10 6BJ
G7KUJ RH10 6BU
G0BXV RH10 6DN
G3YAB RH10 6DU
G4BRE RH10 6DU
G3WKI RH10 6HQ
G8KDD RH10 6HS
G8MGZ RH10 6HW
G8MQF RH10 6JE
G7OMI RH10 6JS
G3GDU RH10 6JU
G3JKF RH10 6JU
2E1BLS RH10 6JU
G3NRC RH10 6LU
G0IOE RH10 6LX
G6XVO RH10 6LX
G3MER RH10 6NB
G8HSK RH10 6NZ
G6XAV RH10 6PB
G4SFP RH10 6PN
G3SYD RH10 6SE
G4UBT RH10 7BS
G6LMU RH10 7BS
2E1BML RH10 7ED
G4YCL RH10 7EQ
G4KZG RH10 7EQ
G0ECA RH10 7HU
G0LBL RH10 7HU
G8AWE RH10 7RD
G0VYN RH10 7RL
2E1BLZ RH10 7RQ
G7RWK RH10 7RQ
G0VIQ RH10 7RQ
G0HWO RH10 7SD
G3SFP RH10 7SQ
G6MGZ RH10 7WL
G4MPP RH10 7XG
G5XG RH10 7XH
G4IQM RH10 7YT
G7RUN RH10 7YW
G6FBJ RH10 7YZ
G4ONJ RH11 0BS
G1KSN RH11 0DT
G3GBI RH11 0HW
G4ICS RH11 0LU
G1GHY RH11 0NA
G3KAU RH11 0NA
G0RGM RH11 0NB
G3UNS RH11 0NS
G4KPN RH11 6AD
G6SVX RH11 6AL
G1IPC RH11 7AZ
G3IPP RH11 7BD
G4ZPP RH11 7DL
G4TXI RH11 7DR
G3GRO RH11 7HW
G3WSC RH11 7HW
G6LEB RH11 7HW
G4LEG RH11 7PE
G4TZF RH11 7RF
G4DQC RH11 7RN
2E1FVV RH11 7SR
G4LSG RH11 8DY
G1LTI RH11 8EH
G7LYB RH11 8EU
G0RPL RH11 8HE
G1XFX RH11 8HW
G6YPM RH11 8LZ
G7TCK RH11 8NP
G8AVZ RH11 8SQ
G4XHE RH11 8UF
G6CKE RH11 9EG
G6EDR RH11 9EG
G8XHK RH11 9EG
G6ZHO RH11 9ES
G0EGT RH11 9HB
G1PLP RH11 9HH
G0PVQ RH11 9JH
G1PLV RH11 9NS
G3ZNW RH11 9NY
M1AZB RH11 9QB
G1BZB RH11 9QJ
G8PUH RH11 9QT
G1JNX RH11 9QU
G4KBC RH11 9RB
G0IXF RH11 9SA
GI4RGV RH6 0FW
G4CAA RH6 0LG
G3OKS RH6 9SY

Crewe and Nantwich

G1CIR CW1 2JG
G8DTT CW1 3AX
2E1EHP CW1 3JN
G7UJE CW1 3PJ
2E1FIV CW1 3PJ
G7PEH CW1 3QD
G0HIZ CW1 3RF
G0MMH CW1 3XB
G4OUK CW1 3XJ
G4UCT CW1 4AP
G0ECA CW1 4BD
G7AZK CW1 4BS
G7BMT CW1 4BS
G0TWH CW1 4DY
G1DBI CW1 4ES
G1MIL CW1 4JH
G4DBX CW1 4RA
G0FPS CW1 4TJ
G0CZD CW1 5FU
G3DAC CW1 5JJ
G7UUG CW1 5LF
2E0AMO CW1 5NT
G0CGQ CW1 5NU
G6OBT CW1 5QE
G3ZFC CW1 5SY
G7WHT CW1 5YF
G6CHD CW1 6DY
G4DQG CW1 6HD
G6OGD CW11 3NY
G3YXN CW11 3QN
G1GRN CW11 4RZ
G8UHT CW11 4RZ
G6UKN CW11 4TN
G7GPB CW2 5AP
G0NWF CW2 5AW
G4PJW CW2 5AW
G8SIG CW2 5AZ
G0TMK CW2 5BJ
G4VEF CW2 5BZ
G6CDK CW2 5DA
G2ALZ CW2 5EU
G8ZPK CW2 5HE
GW0BND CW2 5HX
G3XHP CW2 5JN
G4XUM CW2 5JU
G0VFL CW2 5LJ
G0GRO CW2 5LX
G3RIT CW2 5LX
M1BLN CW2 6ET
G8DHQ CW2 6JB
G7DYL CW2 6NQ
G0DRM CW2 6RD
G1LSN CW2 6TG
M0AAQ CW2 7HD
G1XTX CW2 7JR
G3RKX CW2 7JY
G7EUR CW2 7PR
G7GRR CW2 7SW
G4YFO CW2 8BN
G7IFI CW2 8BT
G6WVC CW2 8DW
G3ZBM CW2 8DY
G6IGW CW2 8ED
G4DVS CW2 8EJ
G7GZZ CW2 8EX
G3JJA CW2 8HD
G4MAG CW2 8HG
G4LVR CW2 8HJ
G4RRM CW2 8HJ
G6RRM CW2 8HJ
G0SIE CW2 8HP
G0ZPV CW2 8JR
G8ZSK CW2 8NA
G7SBG CW2 8RE
G6CKM CW2 8RT
G6LVT CW2 8SH
G4BQD CW2 8SZ
G3YMU CW3 0EU
G6LUZ CW3 0HS
G0NBH CW3 0JB
G1XHR CW3 0JB
G3ADV CW3 0JG
M1BSD CW5 5QJ
GM1FLO CW5 5YP
G3KMV CW5 6AL
G7GJM CW5 6BU
G0EEI CW5 6HA
G4PCR CW5 6HJ
G6SNI CW5 6JJ
G6LBV CW5 6NA
G0GHL CW5 6PN
G0VJR CW5 6QF
G3PFE CW5 6SA
G4NIE CW5 7AW
G4ZMR CW5 7DW
G0OEW CW5 7ER
G6MXT CW5 7EY
G7PWU CW5 7LR
G4MHX CW5 7NX
G6VBK CW5 8HG
G3TLV CW5 8HH

Croydon

G0DDT CR0 0AA
G7VTH CR0 0BL
G0VQM CR0 0JE
G7GCF CR0 0JL
G0HSH CR0 0NU
G0RRB CR0 0QR
G1ALR CR0 1XL
G1OIS CR0 2HX
G1ZDY CR0 2PZ
G4WPB CR0 2RS
G4MZJ CR0 3AD
G4MZK CR0 3AD
M0BEH CR0 3AG
G8XDM CR0 3AG
G0GFY CR0 3JF
G1IEY CR0 3NF
G4WAY CR0 3NZ
G8GYM CR0 3QP
G6FGY CR0 3SW
G8SSI CR0 4AD
G4DAF CR0 4EG
G3UAC CR0 4JL
G4DPN CR0 4QA
G0CTE CR0 4QH
G3WBN CR0 5BA
G1ZDG CR0 5BL
G0MWH CR0 5EJ
G4LZE CR0 5HA
G4STD CR0 5HX
G6DAY CR0 5PS
M1CCF CR0 5QA
G0CUU CR0 5SA
G4KQO CR0 5ST
G1MPW CR0 5UN
G1XRY CR0 6BJ
G0ITR CR0 6NE
G8ATP CR0 6PQ
G4ZMT CR0 7EA
G4AVV CR0 7EB
G3NLW CR0 7HY
G4RWW CR0 7HY
G6LX CR0 7HY
G3ZIO CR0 7PR
2E1BXC CR0 8BW
G7VRM CR0 8BW
G7ODG CR0 8LG
G0DXI CR0 8LP
G3RMN CR0 8PN
G3ALG CR0 8SB
G4VTD CR0 8TP
M1BQU CR0 8TP
G0GJF CR0 8TZ
G7TER CR0 9AG
G3WZK CR0 9BE
G7MSF CR0 9DG
G4UIO CR0 9HE
G6KQS CR0 9HL
G7MBX CR0 9HR
G6FPQ CR0 9HU
G0GNQ CR0 9HX
G4FUU CR0 9LB
G0UCT CR0 9LB
G0HWY CR2 0DZ
G0DJT CR2 0JT
G1YRP CR2 0LL
G3OJE CR2 6DJ

G0VVX CR2 6EE
G1WFG CR2 6NE
G3DPW CR2 7DB
G8GOJ CR2 7DG
G8MNY CR2 7ER
G3MCX CR2 7HH
G4DAC CR2 7JJ
G0IZB CR2 7LJ
G6ODE CR2 7LT
G4GFC CR2 7RE
G4ANQ CR2 7SD
G2KU CR2 8BN
G2TV CR2 8BN
G8RMC CR2 8BP
G0THK CR2 8DG
G4CCW CR2 8EP
2E0AHW CR2 8QQ
G7VKT CR2 8QQ
G7VKS CR2 8QQ
G7CEH CR2 9AT
G4ZMX CR2 9BA
M1ACF CR2 9EH
G8JAC CR2 9EH
G4VYC CR2 9EW
G0TCE CR2 9JR
G0VAQ CR2 9JS
G8IYS CR2 9LN
G4WEO CR3 0DU
G3MZA CR5 1BB
G3BFP CR5 1BE
G7NGB CR5 1BE
G4JIU CR5 1BJ
G8EIN CR5 1DF
G4GEW CR5 1DH
G1BQW CR5 1PQ
G0KZT CR5 1QJ
G3CQU CR5 1QP
G4YBQ CR5 1QR
G1TLW CR5 1RF
G0DLP CR5 1RH
G6VYT CR5 1SL
G1PGI CR5 2AD
G1ROO CR5 2AD
G4CJR CR5 2BL
G8KDQ CR5 2DJ
G4UKQ CR5 2DR
G6AHR CR5 2EE
G4FVL CR5 2EJ
G8GAR CR5 2JF
G4GTO CR5 2JS
G7HBJ CR5 2JS
G3ZPB CR5 2LF
G3HAB CR5 3DA
G0FUH CR5 3DD
G3OOU CR5 3DE
G3VCP CR5 3DE
G0VGT CR5 3HQ
G1DWL CR7 6AD
G3YRB CR7 6AD
G6MFM CR7 6BR
G1POM CR7 6BX
G1BYN CR7 6DJ
G3UFY CR7 7AF
G4ELE CR7 7BW
G0FJJ CR7 7DY
G0ELG CR7 7EN
G4KRD CR7 7HQ
G3JAL CR7 7JE
G1HER CR7 7NP
G6KEN CR7 7NP
G7HLW CR7 7NQ
G0TEL CR7 7RX
G0IOO CR7 8DF
G0WBF CR7 8LR
G4REK CR7 8NY
G4LBM CR7 8PD
G0SOC CR7 8PD
G1OTN CR7 8RP
G3FTQ CR8 1ED
G1MLR CR8 1ER
G7PMO CR8 1EU
G3FPK CR8 1EZ
G3ZMN CR8 1HU
G7JAQ CR8 1JA
G7PWV CR8 1JQ
G3GKF CR8 2AZ
G8FOT CR8 2HQ
G0XTT CR8 2JB
G7OKY CR8 2LJ
G6GFJ CR8 2LR
G4HUC CR8 3BF
G4AOJ CR8 3EJ
G7NXQ CR8 3EJ
G4MTM CR8 3JJ
G6AJY CR8 4DF
G3TWJ CR8 4DH
G8TB CR8 4DN
G3DVQ CR8 4EN
G4CDY CR8 4NG
G3ZXV CR8 5JJ
G4JGE CR8 5LB
G3GHI CR8 5LR
G3UGY CR8 5LU
G0GZM CR8 5NT
G1RCE CR8 5NT
G1PKS SE19 2DA
G4SXY SE19 2DS
G4EBI SE19 2SA
G3TAO SE19 3AD
G1XVW SE19 3DL
G4TPB SE19 3HH
G8IVB SE19 3NW
G0NTJ SE19 3PZ
G7HUC SE19 3QT
G4WVK SE19 3TY
G8RNH SE25 4JE
G3BCM SE25 4JJ
G2HBA SE25 4TQ
G1ZDX SE25 4XU
G6FKA SE25 5BU
G8LU SE25 5DP
G7ONI SE25 5HB
G7PWJ SE25 5HB
G7TQE SE25 6HY
G3WRR SE25 6SY
G4ALE SE25 6SY
G7OYN SW16 3LS
G4FDA SW16 3RT
G0WLF SW16 4HT
G3BGY SW16 4JE
G3HFY SW16 4RA
G3JQN SW16 4RA
G3BEQ SW16 4RP
G3DJK SW16 5RE

Dacorum

G1PRM AL3 8EE
G8KLC AL3 8EE
G0OIK AL3 8HW
G0TBJ AL3 8PB
G1AKX HP1 1HU
G7HCU HP1 1SA
G3GRV HP1 1SN
G3EKI HP1 1ST
G0TIW HP1 2BQ
G0WIH HP1 2BQ
G6HIA HP1 2HW
2E1EOI HP1 2HY
M1ABC HP1 2LN
2E1EOB HP1 2RR
G4WGA HP1 3BN
G7HFU HP1 3BZ
G0WIO HP1 3EN
G7SGM HP1 3EW
G6IYJ HP1 3JS
G7IPA HP1 3JU
G0XRO HP1 3LW
G1MVT HP2 4AN
G0IAL HP2 4BA
G0GJA HP2 4BZ
G6ZAO HP2 4BZ
G1IMG HP2 4DB
G7KCT HP2 4DH
G1URZ HP2 4HE
G0BQY HP2 4HQ
G8ASI HP2 4LZ
G1EIZ HP2 4PX
G7HCQ HP2 4RP
G0BLU HP2 5AT
G8VVB HP2 5EL
G4MSC HP2 5ES
G4DWZ HP2 5HU
G8CBE HP2 5JG
G0VBB HP2 5LE
G1JOA HP2 5LL
G6AID HP2 5UY
G8SYD HP2 5UY
G7VMY HP2 6AL
G4BJN HP2 6DS
G8ZGQ HP2 6DX
G2BAP HP2 6HL
G1RFQ HP2 6HU
G1PWF HP2 6LA
G8BBZ HP2 7JP
G7WDO HP2 7JP
G2KG HP2 7LN
G0SUO HP2 7PR
G0TCD HP2 7QW
G7TTZ HP2 7QW
G7TKM HP2 7RG
G6DZJ HP23 4DS
G7UYW HP23 4DS
G8FMC HP23 4HG
G0SEY HP23 4HZ
G1NUO HP23 5DG
G3LNP HP23 5NG
G3WXM HP23 5PB
G3RWV HP23 5PJ
G4SSE HP23 6DL
G3MEH HP23 6EN
G0WSF HP23 6EW
G6XUJ HP23 6HF
G4PMG HP23 6HH
G8MKW HP23 6HH
G1ODZ HP23 6JB
G3NUG HP3 0BN
G3VSQ HP3 0BS
G0VFW HP3 0EA
G1OEB HP3 0HG
G8FAE HP3 0QJ
G8MFH HP3 0QR
G0NNI HP3 8BB
G8OYQ HP3 8EQ
G4SNY HP3 8JH
2E1EXC HP3 8JQ
G6SMZ HP3 8NG
G8LPC HP3 8QS
G4MSW HP3 9JY
2E1EOA HP3 9LE
G6RXD HP3 9NG
G0GVN HP3 9PD
G0NEZ HP3 9PF
G8ERV HP3 9SQ
2E1EOK HP3 9SQ
G1KOG HP3 9UE
2E0ANC HP3 9UL
G0EVS HP3 9UQ
G1VYM HP3 9XA
G8AUM HP4 1DT
G0TPK HP4 1JN
G3JDG HP4 1NY
G4CSI HP4 1QQ
G4KYF HP4 2DR
G3RPA HP4 2PD
G4PKH HP4 2SD
G3PV HP4 3DU
G1WVK HP4 3JH
G1WMN HP4 3PW
G8ZQK HP4 3QT
G1ABW HP4 3UR
G3SGC WD4 8AR
G2BBN WD4 8AW
G0VYA WD4 8HH
G7FQP WD4 9DX
G3XYJ WD4 9HF
G1PAK WD4 9HX

Darlington

G0PMJ DL1 1BW
G0MHE DL1 1EQ
2E0AES DL1 1EQ
G8ANA DL1 1EZ
G8BWH DL1 1HG
G8YSQ DL1 2BB
G3BYX DL1 2DJ
G4YFS DL1 2ES
G0PWZ DL1 2RZ
G0LUY DL1 2SL
G6VPV DL1 2TA
G8XQS DL1 2TA
G4NYJ DL1 2TU
G4LIA DL1 3HT
G6BCG DL1 3SG
G4YCQ DL1 4ES
G8OWS DL1 4LG
G0NWY DL1 4NS
G8YBO DL1 4XH
G7SFQ DL1 5DE
G0TJC DL1 5DF
G4AUN DL1 5DX
G4ADT DL1 5EF
G7TKB DL1 5NP
G8UMM DL2 1HA
G3SGY DL2 1HJ
M1AUA DL2 1QY
G0PNJ DL2 2JH
G3UHJ DL3 0AH
G8HZS DL3 0HX
G2HMK DL3 0JY
G0SBP DL3 0NU
G7TCD DL3 6ES
G4JIR DL3 6HN
G0MSZ DL3 7AR
G1DMN DL3 7HP
G4IFX DL3 7RD
G0GEK DL3 7UX
G6HOW DL3 8BD
G4OXU DL3 8BH
G3JBV DL3 8DX
G4CWA DL3 8HN
G4FVP DL3 8HY
G3OGW DL3 8LX
G0TZU DL3 8RU
G4FEH DL3 9DU
G3CDM DL3 9PB
G4ZVH DL3 9PB
M0AMM DL3 9UQ
G0VQG DL3 9UW
G0LUW DL3 9XL
G8TJR DL3 9XS
G4MEU DL5 6PQ
G3OAL DL5 6RF
G3JKD DL5 6RU

Dartford

G4ZHX DA1 1ND
G6UML DA1 1TR
G6UYY DA1 1YE
G4HWY DA1 2JN
G4BXT DA1 2LW
G7VCM DA1 2QL
G3UBP DA1 3AA
G7IYB DA1 3AG
G7KAO DA1 3BA
G7GRB DA1 3BH
G4WIF DA1 3NF
M1BWM DA1 5DJ
G6TSC DA1 5JW
G8XJB DA1 5JW
G4TDE DA1 5LE
G6SYN DA10 0HP
G0NZR DA10 0LT
2E0ANS DA13 9NF
G1JNQ DA2 6DN
G0IPC DA2 6HD
G8JTG DA2 6HG
G6ZAY DA2 6JX
G7FIA DA2 6LB
G0CGB DA2 6PA
G4VOQ DA2 7BT
G0WAR DA2 7BY
G4IQQ DA2 7ES
G0HRD DA2 7LP
G1LPF DA2 7NE
G0DVL DA2 7NW
G0KUU DA2 7PB
G4LGU DA2 7QQ
G7WLL DA2 7RN
G8CWF DA2 8DX
G3FGP DA3 7HD
G7IYQ DA3 7JL
G0LJC DA3 7JR
G3FIB DA3 7JR
G6XND DA3 7NS
G6YLD DA4 9EX
G3XVC DA4 9HG
G4KCZ DA5 2DT
G3XEW DA5 2ER
G2FNT DA5 2HT
G4CDM DA5 2JB
G0OBH DA9 9DD
G0FDU DA9 9PE
G4NHP DA9 9RA

Daventry

G1OIW CV23 8TN
G6IQM CV23 8UF
G8AHA CV23 8UQ
G0KJF LE16 9TY
M1APX LE16 9TY
G1WVS NN11 3AD
G4TBN NN11 3BN
M1BBK NN11 3BU
G0AUO NN11 3ES
G3YHI NN11 3ES
G3AKF NN11 3NY
G3MRQ NN11 3YT
G4DDH NN11 4DQ
G0IOH NN11 4EE
G7KRB NN11 4EY
G1DLH NN11 4HG
G0DRE NN11 4HJ
G0FPM NN11 4HJ
G7AKV NN11 4HJ
G8PQB NN11 4HJ
G6LZV NN11 4HN
2E1FWW NN11 4JH
G0DXT NN11 4JL
2E1FSV NN11 4JQ
2E1FWQ NN11 4JQ
G8GVV NN11 4NW
2E1AXE NN11 4NW
2E1AXD NN11 4NW
M1ADP NN11 4PT
G4WQJ NN11 4PX
G0CUD NN11 4RE
G7KJX NN11 4RN
G0NNG NN11 4SU
G0IOD NN11 4TT
G6ZZZ NN11 4TT
G0FFI NN11 4TT
G6CMD NN11 4TY
G6EGO NN11 5BX
2E1CZF NN11 5BX
G8SWL NN11 5EX
G6XLB NN11 5HT
G8VQQ NN11 5HY
G0DPA NN11 5JW
G3OXV NN11 5QB
G4XAY NN11 5RW
G1ZJK NN11 5ST
G0FQW NN11 5TY
M1AQX NN11 5XG
G4SFB NN11 6BU
G0XXX NN11 6BU
G8KTA NN11 7HW
G6HSG NN11 7JT
G6MSN NN3 7QU
G6CQR NN3 7RE
G2FIF NN3 7RZ
G3WJY NN3 7ST
G7KJO NN3 7TL
G3SIV NN6 0AH
G4CLA NN6 6EW
G4LRT NN6 6EW
G6EGY NN6 6EZ
G4UUH NN6 6LP
G0LAP NN6 6LU
G4ECO NN6 7AD
G6YPE NN6 7EZ
G4UYM NN6 7HG
G7BQM NN6 7PP
G6SJA NN6 7PU
G0OFC NN6 7QH
G6NW NN6 7QQ
G7HQH NN6 7RW
G0LHE NN6 7SR
G4EPA NN6 7ST
G8HYU NN6 7ST
G6YTB NN6 7TT
G4JSX NN6 7TX
G3IAI NN6 8AB
G3HGW NN6 8AE
G1TUS NN6 8QP
G8GJW NN6 9BS
G0GLY NN6 9HY
G7NRR NN6 9JS
G6TJE NN6 9QB
G4KFV NN7 4JJ
G4PGY NN7 4LL
G0TML NN7 4LS
G7JWN NN7 4PE
G4NVW NN7 4PF
G8FZW NN7 4QU
G8XCD NN7 4RQ
G3ZAW NN7 4SH

Denbighshire

GW4GDM CH7 4PY
G1PKW CH7 4QD
GW4WXM CH7 4QU
GW0GLI LL15 1AQ
GW4WSU LL15 1LZ
GW3YTL LL15 1RR
GW6XJM LL15 1RT
GW6TGR LL15 1UN
GW6JWL LL15 2EY
GW0SYN LL16 3BW
GW3VZO LL16 3DP
GW3RUE LL16 3EU
GW4CQZ LL16 3HE
GW1AKT LL16 3NR
GW4XSP LL16 3TD
GW4HBZ LL16 4AN
GW8SIE LL16 4NN
MW0BFY LL16 4PG
GW4RWR LL16 4RL
GW7NIW LL17 0BH
G6ONZ LL17 0BP
GW7EVG LL17 0DF
GW4AMX LL17 0DP
GW4UYU LL17 0RR
GW1TFB LL17 0SZ
2W1EPL LL18 1DG
GW3FPF LL18 1LE
GW4PFC LL18 1LT
GW7EYB LL18 1LT
GW0PYN LL18 2AN
GW7HEE LL18 2HN
GW7THE LL18 2HU
2W1FHA LL18 2JP
2W1FKX LL18 2LA
GW3RVT LL18 2LU
GW4UJF LL18 2LW
GW0IQT LL18 2NE
GW4XEF LL18 2NU
GW0DFY LL18 2RY
GW4WOJ LL18 2ST
GW6IYP LL18 2TP
GW8XLL LL18 2YA
GW3OIN LL18 3BG
GW3UTG LL18 3EE
GW0DSW LL18 3PB
2W1AKZ LL18 3PE
2W1ALD LL18 3PE
GW4ARC LL18 3PE
GW4HDR LL18 3PE
GW0VFQ LL18 3PE
GW7CEP LL18 3RG
MW0AQG LL18 3RG
GW8WTJ LL18 3RP
GW4NLD LL18 3RR
GW0AYP LL18 3US
GW1NGN LL18 4AD
GW0MOF LL18 4AF
2W1BPC LL18 4AF
GW3FJI LL18 4HH
GW0LAL LL18 4HS
2W1FHF LL18 4HU
2W1FHL LL18 4HU
2W1FPK LL18 4HU
GW8OYT LL18 4JF
GW4TYH LL18 4JR
GW4XWN LL18 4LF
GW1AXG LL18 4LX
GW0VEN LL18 4SD
GW3WPE LL18 4ST
GW8NZN LL18 4ST
GW7EXH LL18 5AG
GW7NFY LL18 5AG
GW0JSX LL18 5RW
2W1BYK LL18 5TT
GW4SLZ LL18 5UE
GW1CJJ LL18 6BB
GW3EMI LL18 6ET
GW4UIR LL18 6HT
2W1BIV LL19 7DA
2W1CEE LL19 7DF
GW7RZN LL19 7LP
GW4EIR LL19 7NL
GW0UDR LL19 7NL
2W1DHZ LL19 7SR
2W1DIG LL19 7SR
GW4DTQ LL19 7TS
2E1EZR LL19 7YR
MW1AJP LL19 7YR
GW4XXP LL19 8LU
2W0ALZ LL19 8RU
GW0NSQ LL19 8RY
GW4ITO LL19 8TL
GW6PMC LL19 8TS
GW3CF LL19 9DT
GW2HIY LL19 9HF
GW3JGA LL19 9HL
GW0DSL LL19 9LS
GW0VUP LL19 9NB
GW3MEO LL19 9NF
GW0CNK LL19 9NN
GW1ARC LL19 9NN
GW0KZW LL19 9PB
GW3HNC LL19 9RD
GW0UDM LL19 9RD
GW4VHP LL19 9SH
GW3XZU LL20 7RL
GW6WAG LL21 0DL

Derby

G7TNX DE1 1GT
G4YNN DE1 2GP
2E1AKN DE1 2HH
G7NFP DE1 2HJ
2E1FYD DE1 3EH
G3KLO DE1 3HH
G0IDL DE2 7LR
G0CPO DE21 2BE
G0NHR DE21 2DG
G4TAY DE21 2DG
G6THB DE21 2DX
G6HSC DE21 2HE
M1CED DE21 2JW
G4FPY DE21 2LL
G8OVV DE21 2RY
G6IQQ DE21 2TD
G1GTG DE21 2XN
G7MUY DE21 4DG
M1AFB DE21 4DX
G7HST DE21 4GY
2E1EDE DE21 4HQ
G8VQO DE21 4JX
G0VOE DE21 4LZ
G6LWY DE21 4ND
G7SZD DE21 4PR
G1RHH DE21 4PW
G0TJF DE21 4QL
G7GXF DE21 4QL
G6VFA DE21 4RF
G4TDY DE21 4SF
G3EMJ DE21 4TA
G4TWV DE21 6BU
G7JCX DE21 6DG
G6RFV DE21 6ER
G8UWJ DE21 6ER
G7RZB DE21 6ER
G0OAF DE21 6EW
G0TLF DE21 6EW
G4TBK DE21 6FX
2E1ENE DE21 6FZ
G0UFT DE21 6GB
G3BYN DE21 6HW
G7AFW DE21 6NX
G4OET DE21 6PE
G1DHY DE21 6PH
G1UJX DE21 6PL
2E1DJC DE21 6QQ
G4LTE DE21 6QU
G4EYN DE21 6SH
G4NHZ DE21 6SY
G6LQY DE21 6WL
G8BFA DE21 6WX
G3ESB DE21 7AB
G7JRI DE21 7AD
G6BEI DE21 7DS
G3ZBI DE21 7EN
G7HHZ DE21 7FZ
G1UYZ DE21 7GL
G3FJN DE21 7JG
2E1FYL DE21 7JT
G6WYD DE21 7JW
G3OXN DE21 7JZ
2E1EBK DE21 7NT
2E1EBS DE21 7NT
G6FVJ DE21 7QA
G1ZLQ DE21 7QE
G8OWM DE22 1BZ
2E0AAL DE22 1DH
G8RBK DE22 1EF
G3SZJ DE22 2BJ
G4EYM DE22 2BJ
G0JVZ DE22 2BW
G4PIM DE22 2BW
G0LOS DE22 2FS
2E1DIA DE22 2GN
M0ADL DE22 2GR
G6DNH DE22 2GW
2E1FBS DE22 2HA
G4EQQ DE22 2HF
G1BEJ DE22 2HF
G0RII DE22 2HG
G4HDP DE22 2HH
G3IFA DE22 2JA
G0MWQ DE22 2JE
G1HNX DE22 2JH
G0GDL DE22 2JN
G0OCG DE22 2JN
G4CWD DE22 2NA
G7ABV DE22 2QH
2E1CXS DE22 2RT
G4VWB DE22 2UN
G0VVA DE22 3AS
G3XOF DE22 3BW
G6EBL DE22 3EF
G6GVF DE22 3ER
G7MHT DE22 3GL
2E1FYX DE22 3GL
G4VSI DE22 3GQ
G8NLP DE22 3LJ
G8PXA DE22 3RG
G3SMV DE22 4BW
G3PWS DE22 4HN
2E1ADJ DE22 4JU
2E1EDD DE22 4JX
G7SJS DE23 6BL
2E1DBT DE23 6BW
G7GXE DE23 6DA
G3URU DE23 6DL
2E1GAR DE23 6DN
2E1CXD DE23 6ES
G3ERD DE23 6GT
G3KQF DE23 6GT
2E1ATT DE23 6GT
G0WQY DE23 6HB
G0IDR DE23 6HL
G6XMG DE23 6JD
G8TNE DE23 6JD
G7OGT DE23 6JD
G8KWW DE23 6LD
G6CWW DE23 6NQ
G1VAB DE23 6PF
G8DBY DE23 6PF
G8BAV DE23 6PS
2E1CZJ DE23 6PS
G1DDI DE23 6TG
G6XGF DE23 6TG
G3FUJ DE23 6XL
G8XOT DE23 7AD
G4BKO DE23 7AN
G4CMZ DE23 7BZ
2E1GAX DE23 7EJ
G0DCV DE23 7ET
G4PZY DE23 7FY
G6OEI DE23 7GA
G2CVV DE23 7GE
G0RUR DE23 7RB
G4COR DE23 7RJ
G4PZH DE23 7RR
G4CAF DE23 7SY
2E0AAS DE23 8DU
G4YNE DE23 8DU
G6LSD DE23 8JA
G7SWW DE23 8PS
G6KKD DE23 8RF
G8HIP DE23 8UT
G7CYF DE24 0AD
G8YCK DE24 0AQ
G1CUH DE24 0AU
G6TZJ DE24 0BG
G0TQF DE24 0BR
G0PRY DE24 0DJ
G3HYR DE24 0FF
G3EEO DE24 0FR
G3OZ DE24 0FR
G0DKX DE24 0FT
G1HEU DE24 0GH
G4RVU DE24 0GX
2E1FYT DE24 0HN
G6KUI DE24 0HZ
G4YVV DE24 0JU
G8BFC DE24 0LG
G7SZG DE24 0LP
G4AKE DE24 0PP
G4RBZ DE24 0QS
2E1CYA DE24 0QX
G4LNQ DE24 0TA
M1ADQ DE24 3BH
G1VAC DE24 3BQ
G0IWF DE24 3DZ
G7BJB DE24 8AZ
G4PDE DE24 8BD
G0TJT DE24 8FT
G6UGT DE24 8NP
G3GQR DE24 8RG
G4SNN DE24 8RN
G7PJJ DE24 8SX
G7VNO DE24 8WD
G7EHN DE24 8XH
G1IWT DE24 9BT
G6FFX DE24 9DA
G7RGX DE24 9EQ
M1AJQ DE24 9LZ
G6CXI DE24 9NG
G4YYI DE24 9PD
G6XEN DE24 9PT
G4CZH DE24 9QA
G7BJC DE3 4LQ
G8GBU DE3 5AH
G8KGC DE3 5AH
G6MWS DE3 5AQ
G4YIZ DE3 5BL
G0CFE DE3 5BW
G4MLL DE3 5BW
G4HJL DE3 5DF
G8YNK DE3 5FN
G6WZE DE3 5FP
G1UZS DE3 5FY
G7GJI DE3 5GH
G0DMK DE3 5GQ
G0CRN DE3 5GT
G0RRE DE3 5GW
G4KOJ DE3 5HB
G4WRM DE3 5HJ
G7GEX DE3 5HZ
G1KKD DE3 5JT
G2DAN DE3 5JW
G7BHY DE3 5PA
G0MGX DE3 5PY
G8CQW DE3 5QP
G4ZHD DE3 5QQ
G7SVQ DE3 5RD
G7PNB DE3 5RF
G3TOV DE3 5RR
G8OWA DE3 5RR
G4XYE DE3 5RT
G0MSS DE3 5SA
G7VEJ DE3 5SD
2E1EZY DE3 5SD
G4KRW DE3 5TX
G0DFS DE3 6EH
G7DFS DE3 6EH
G8MZA DE7 1BD
G3NYZ DE7 1BW
G4NID DE7 3PN
G7DZR DE73 1PF
G4LPZ DE73 1RD
2E1FYS DE73 1RJ
G4XYD DE73 1RP
G1SPT DE73 1SJ
G6FSW DE73 1XA

Derbyshire Dales

G1BGF DE4 2AA
G7ODR DE4 2AA
G1IEX DE4 2AH
G1CJC DE4 2FX
G7VWN DE4 2GG
G0UFM DE4 2GZ
G0OTJ DE4 2HP
G3UZJ DE4 2HR
G4XTK DE4 2JW
G6APZ DE4 2LL
G8ROU DE4 2LL
G6HWR DE4 2PW
G1XAW DE4 2QA
G1WZR DE4 3AG
G4ZEY DE4 3BT
G1GMX DE4 3EP
G8LXN DE4 3EP
G7NBJ DE4 3EU
G1IFY DE4 3GB
G7PGC DE4 3GS
G8ROR DE4 3HG
G0ACZ DE4 3HJ
G4WBK DE4 3HJ
G7JOG DE4 3HJ
G8JTI DE4 3LB
G0ORK DE4 3LL
G1XAX DE4 3NW
G8RBW DE4 3PS
G0FSB DE4 3QG
G0BJC DE4 3QL
G4KLX DE4 3RL
G8TXQ DE4 3RL
G0TKD DE4 3SQ
G0HMZ DE4 3TB
G0KPT DE4 3TB
G1OGC DE4 3TB
G0BJD DE4 3TE
G4UIQ DE4 4AD
G1EKH DE4 4AP
G1SGP DE4 4AR
G0HUI DE4 4BL
G0MMO DE4 4ED
G4GHC DE4 4EL
2E1GAS DE4 4EX
G4RQR DE4 4GL
GM3GKJ DE4 4HN
G8BGW DE4 5FE
G8RNU DE4 5FF
G8LVV DE4 5FP
G4LOW DE4 5LE
G8LOW DE4 5LE
G0CPY DE45 1AJ
G4JIZ DE45 1BA
G6EAZ DE45 1DD
G7ABR DE45 1DX
G8AKQ DE45 1JJ
G3VOT DE45 1NJ
G3SPV DE45 1QQ
M0APH DE45 1RF
G6GMH DE45 1SG
G6ATG DE45 1TB
G4HIJ DE6 1AT
G6YWV DE6 1BB
G1NDL DE6 1BJ
G4PHE DE6 1BJ
G3TVU DE6 1BR
G0DLS DE6 1EX
G4ZJW DE6 1FZ
G4UUQ DE6 1QB
G0BPZ DE6 1SR
G0CQP DE6 2AU
G1MUT DE6 2AW
G3KAG DE6 2EH
G3PTT DE6 3BE
G1MTG DE6 3DT
G1GIF DE6 3EH
G4DKV DE6 3EN
G6RBM DE6 3FE
G0BDW DE6 5HH
G3YRP DE6 5PD
G3LKV DE6 5PU
G3ANI S32 3YF
G4HGN SK17 8JT
G0CWS SK17 8LP
G1EVA SK17 8LP
G1TBI SK17 8SW
G4FZH SK17 9TS
G4KNQ SK17 9TS
G4JIK SK17 9TU

Derwentside

G0BYS DH7 0AG
G6CQC DH7 0BQ
G0HUE DH7 0EA
G0OGB DH7 0HL
G4YMU DH7 0HL
G4IPB DH7 0QD
GW4IWU DH7 0QD
G8AGR DH7 0QD
G3SNT DH7 6TW
G3MEA DH7 9DF
G8MPS DH7 9EL
G7TLR DH7 9NA
G0OGD DH8 0AW
G3UIR DH8 0BJ
G3SOI DH8 0DL
G0VLF DH8 0DY
G0OHA DH8 0DZ
G3KBX DH8 0SD
G0JRT DH8 5LS
G0XVC DH8 5SU
2E1EPA DH8 6HZ
G0GKK DH8 6TL
G0SBU DH8 7AE
G4ZAK DH8 7DG
G3UTS DH8 7DY
G6TSS DH8 7JU
G1NVN DH8 7QT
G8IAJ DH8 7SB
G3EGF DH8 7SJ
G7UXQ DH8 8AT
G7WEU DH8 8AT
2E1DOW DH8 8HX
G4LGA DH8 8LF
G4PFQ DH8 8LF
G1HPI DH8 8PD
G3KMG DH8 9AP
G7JSE DH8 9RG
G1HYG DH9 0QP
G7LQK DH9 0QR
G1HEX DH9 6BE
M1AKO DH9 6DW
G1KOM DH9 6HT
G7TKN DH9 6NP
G3VRP DH9 6UF
G3GDY DH9 6UY
M1BVM DH9 7AD
G7OWN DH9 7BP
G3KRG DH9 7DB
G1JQT DH9 7NU
M1BVL DH9 7PH
G1AIF DH9 7TR
M0AYI DH9 7TZ
G1KRX DH9 8DG
G6UEV DH9 8DG
G0FGG DH9 8JT
G1NTI DH9 8QB
G1OGH DH9 8QY
G0VHB DH9 9EH
G0KYL DH9 9HB
G1GEY DH9 9LW
G0IVX DH9 9NP
G1HAO DH9 9NQ
G6XCO DH9 9NQ
G0VII DH9 9PS
M1AYV NE16 6PQ

Doncaster

G4YUJ DN1 2AH
G0MCC DN1 2NP
G0UQU DN1 2NP
G7BNT DN1 2NP
G4OEM DN1 2PZ
G1PBY DN1 2QU
G3TTU DN1 2TW
G8WZN DN1 3AT
G4AWU DN10 6NQ
G3SFO DN10 6QD
G0PXD DN11 0AA
G7JDH DN11 0JX
G1MSH DN11 0LS
G0FVD DN11 0LT
G7GHI DN11 0LT
G0VHI DN11 0LT
G3SHU DN11 0NH
G4ZLP DN11 0PW
G0PXX DN11 0RP
M1BPO DN11 0RW
G4NQM DN11 0UP
G6DIE DN11 0UP
G8HIQ DN11 0UR
G6OFU DN11 0XX
G6USL DN11 0XX
G3ZVD DN11 0XX
G8JEI DN11 0YB
G1XTN DN11 9EJ
G6HMS DN11 9ET
G7IUT DN11 9PW
G8JJR DN11 9TE
G7EDF DN11 9UF
G4CUC DN12 1HN
M1BPI DN12 1NW
G0DNX DN12 2EW
M1BSG DN12 2HH
G3NXZ DN12 2JW
G3WBG DN12 3DF
G0SSC DN12 3JT
G0TGN DN12 3JZ
G0SDK DN12 4LE
G8LGC DN12 4SB
M1BUX DN2 4BB
G0SQD DN2 4HQ
G0IDZ DN2 4QA
G3HAV DN2 5AJ
G6OSK DN2 5EU
G4PJL DN2 5HN
G3ETQ DN2 5HW
G1IAB DN2 5NG
G0UQV DN2 5NR
G3TJT DN2 5QW
G6YDQ DN2 5SJ
G4LKX DN2 6AN
G8JVJ DN2 6DD
G4ZGM DN2 6DT
G1DUL DN2 6EL
G6DLT DN2 6EP
G0VID DN2 6HF
G7GJO DN2 6HF
M0BDL DN2 6HF
G1MCT DN3 1AN
G3UWT DN3 1DS
G0FXY DN3 1HY
G4KKJ DN3 1JU
G7MRV DN3 1NX
M1BBV DN3 1NX
G4DDS DN3 1NY
G4NZX DN3 1PZ
G0REK DN3 2ES
G4PNT DN3 2HN
G7JTF DN3 2JG
G0PAQ DN3 2LW
G4MDE DN3 2PE
G7GEI DN3 2PQ
G3PTV DN3 3HD
G4YRD DN3 3HD
G7GZK DN3 3HE
G0GOH DN3 3HS
M0AIX DN3 3JD
G4YSM DN3 3LY
G0FNT DN3 3NS
G0HOU DN3 3NS
G0WKA DN3 3PB
G7NWM DN3 3QB
G6NSZ DN4 0BH
M1AAE DN4 0EU
G4WZI DN4 0JW
G3WHL DN4 0PJ
G1NAF DN4 0QE
G3YUF DN4 0TD
G6OCP DN4 0XD
G8XTU DN4 5ED
G6RQJ DN4 5EG
G6ERB DN4 6AA
G6YCE DN4 6AA
G0RFV DN4 6AU
G3NYB DN4 6DQ
G7PUP DN4 6DT
G7PUL DN4 6DX
G4FYE DN4 6EZ
G0GHK DN4 6HE
G1HZR DN4 6HE
G4AWT DN4 6HE
G7HAH DN4 6HE
G6ZKS DN4 6HX
G1IND DN4 6LG
G7LSA DN4 6NH
G0CAS DN4 6NR
G3CFQ DN4 6PF
G0GTI DN4 6QB
G0YZC DN4 6QB
G6YZC DN4 6QB
G7OIC DN4 6QL
2E1BGN DN4 6QL
G0IIF DN4 6RU
G0RON DN4 6SH
G8GHB DN4 6SN
G4OEK DN4 6SQ
G4ZBI DN4 6TE
G0DAM DN4 6TU
G7KEK DN4 6UT
G3ZNQ DN4 6UZ
G7DBD DN4 7EG
G3VHZ DN4 7HU
G6YUD DN4 7HW
G7KMO DN4 7JH
G4NWY DN4 7JN
G7TMO DN4 7JN
G0NMJ DN4 7JY
G3YMO DN4 7PP
2E1CWL DN4 7RT
G3PAF DN4 7RY
G0ILM DN4 8AX
G4OGJ DN4 8LB
G3OFP DN4 9AJ
M0AGJ DN4 9AY
G3SWP DN4 9DA
G1WXM DN4 9JJ
G0LJV DN4 9LA
G0LHM DN4 9QR
G3YS DN5 0DR
G0POE DN5 0ED
G0RLH DN5 0ED
M0BDU DN5 0EG
G7VYY DN5 0LH
G4OVS DN5 7EN
G0MDT DN5 7JJ
G3VLL DN5 7LH
G8LHT DN5 7QF
G7ONV DN5 7RE
G0UQQ DN5 7RS
G1ZSU DN5 7RS
G1SCO DN5 7RY
G0CML DN5 7SG
G0GNP DN5 7TE
G1AGT DN5 7XY
G6NBF DN5 8AU
G1YKO DN5 8BW
G4UXE DN5 8BY
G4UXF DN5 8BY
2E1FCM DN5 8BY
G0KVN DN5 8DH
G4AOO DN5 8DH

Area

G0DQB DN5 8NQ
G4ANP DN5 8QN
G3IGU DN5 8QP
G6GNO DN5 8RR
G4YUK DN5 8SR
G6SVF DN5 8SX
G6CGN DN5 9EW
G4KBS DN5 9HD
G4ZKC DN5 9JY
G7BGZ DN5 9PN
M1ARO DN5 9QU
G0PXE DN5 9RB
G0IXA DN5 9RL
G4PCD DN5 9ST
G0TSH DN6 0EZ
G7AYO DN6 0EZ
G7CTG DN6 7AX
G0UZW DN6 7DB
G4JII DN6 7EA
G0GBQ DN6 7EE
G0WXP DN6 7JX
G6HWI DN6 7JZ
G2CXR DN6 7SE
G3UWR DN6 8EB
G1JTZ DN6 8EH
G1ILF DN6 8HJ
G7VRH DN6 8HT
M1ALK DN6 8JL
G0AHV DN6 8PD
G4KRB DN6 8QT
G3WZW DN6 9BY
G7VMO DN6 9EF
G7UBO DN6 9HJ
G7WFR DN6 9HN
G0DQM DN6 9JX
G3VZE DN7 4BS
G0TJR DN7 4HB
G0LRN DN7 4HP
G0GPA DN7 4JG
G4VVE DN7 5PE
G1IUQ DN7 5PJ
G1NCF DN7 5QF
G3OZD DN7 6AB
G0LCT DN7 6AH
G6HFD DN7 6AN
G7SKW DN7 6PF
G3KPU DN7 6PQ
G7BPF DN7 6QQ
G7SWA DN7 6QY
G4GLF DN7 6SW
G3WDL DN8 4QR
2E1ATS DN8 4SW
G4HZN DN8 5BS
G7MGQ DN8 5EA
2E1FIP DN8 5QS
G4SZX DN8 5TL
G1CAX DN9 3BA
G3JYA DN9 3DL
G6EHV DN9 3DN
G4RHY DN9 3JJ
G4RHZ DN9 3JR
G0TUU DN9 3PQ
G8CXA S64 0AD
G6HPJ S64 0DG
G7BJR S64 0DJ
G4SDE S64 0ER
G7TAO S64 0HT
G3MDL S64 0HY
G4KWM S64 0JD
2E1ERM S64 0JW
G3VJR S64 0NL
G0WXU S64 0NS
2E1FTV S64 9EZ
G7PAG S64 9NJ
G4RPL S64 9PF
G7SZI S64 9RL
G3GOM S66 7BB

Dover

2E1ABW CT13 0AQ
2E1ATH CT13 0AQ
G0DSK CT13 0BH
G8SCZ CT13 0LB
2E1FYB CT13 0LE
G0GBT CT13 0NU
G0WUD CT13 0PA
G0WUE CT13 0PA
G4KPF CT13 9AS
G7MSC CT13 9EL
G3NCB CT13 9JE
G4JYU CT13 9JE
G8MLB CT13 9JE
G0MAZ CT13 9NY
G7BVH CT13 9NY
G2ASX CT14 0AQ
G0LGK CT14 0BT
G1CGH CT14 0DF
G3LCW CT14 0HF
G4VGT CT14 0HN
G4WOS CT14 0JH
G4SUS CT14 0LA
G7DSN CT14 6DG
G7MSS CT14 6HX
G4IDB CT14 6QA
G4DCV CT14 7AL
G0HCX CT14 7BB
G4DDB CT14 7DY
G0RDN CT14 7EZ
2E1APY CT14 7EZ
G4YZX CT14 7JJ
2E0AAK CT14 7QB
G0OXX CT14 7QB
G0UJR CT14 7QB
G7JVP CT14 8BQ
G0BDJ CT14 8DE
G0DQI CT14 8EB
G3BEU CT14 8ES
2E1ACK CT14 9AT
G7DPK CT14 9DD
G3MZI CT14 9DQ
G8SOU CT14 9EF
G3FBU CT14 9EW
G7DXG CT14 9HH
G3ZOR CT14 9LL
G7HIX CT14 9LS
G0DUK CT14 9NB
G0SRR CT14 9NJ
G0WIU CT14 9PD
G1LLU CT14 9QZ
G4GAN CT14 9RG
G3GSY CT14 9TU
G0FAK CT14 9TW
G1URG CT14 9TW
G4GBB CT14 9UY
2E1ABP CT14 9XJ
G4HHX CT15 4BS
G6IOM CT15 4BT
2E1DFB CT15 5EH
G8PZA CT15 5ET
G8SMZ CT15 5JD
G4WIY CT15 5JW
G3IMN CT15 5JY
M1BKI CT15 6AH
G6APV CT15 6BS
G2ACG CT15 6BZ
G4OPR CT15 6EJ
G4UHT CT15 6HL
G8YMD CT15 6HL
G8ZYZ CT15 6HL
2E1BTG CT15 6HL
G6WPK CT15 6JL
G8UWS CT15 7BH
G4OJG CT15 7ES
G4MIX CT15 7EY
G0OJU CT15 7HJ
G4AWW CT15 7JS
G0AXD CT15 7LJ
G3GAZ CT15 7LS
G7IOO CT15 7LW
G0DOV CT16 1EX
G8YXQ CT16 1EX
G1PQX CT16 2AR
2E1CTQ CT16 2AU
G0ICT CT16 2JW
G4BBH CT16 2PQ
G3YQR CT16 3BP
G0VBO CT16 3EB
2E1BUE CT16 3EW
G3ZAC CT16 3HA
G4RLX CT16 3HA
2E1CMJ CT16 3HH
G3ROO CT16 3HZ
2E1DFH CT16 3HZ
G4HL CT16 3JH
G4NPM CT16 3JW
G6RCJ CT16 3LQ
G7JOW CT16 3LQ
2E1BTK CT16 3LU
G4XDW CT16 3NP
G8IYN CT17 0BS
G1IQH CT17 0DA
G0PPI CT17 0LL
2E1ABY CT17 0NP
G4HXE CT17 0NQ
G8YNG CT17 0NT
G0ADK CT17 0NX
G6MQR CT17 0NY
G0SET CT17 0PL
G7BRM CT17 0PR
G3YMD CT17 0QW
G4SAU CT17 0QW
G4FJF CT17 0SF
G0SXU CT17 0TP
G0VLP CT17 9BT
G1PJJ CT17 9BZ
G4VCA CT17 9LD
G7IEX CT17 9NJ
G0WWQ CT17 9NL
G0CRM CT17 9PN
G4SMX CT17 9PN
G0CCG CT17 9RQ
G6WQI CT17 9TB
G4IMP CT18 7AP
G0LGX CT18 7JN
G3OJZ CT18 7LB
G3JEB CT18 7LR
G3RWF CT3 1AU
G3JIX CT3 1JX
G8KUC CT3 1JX
G3UKC CT3 1JX
G3SRE CT3 1JX
G3TAJ CT3 1JX
G7EPS CT3 1JX
G4FLR CT3 1LY
G8DBU CT3 2JG
G1GEI CT3 2LP
G1MXM CT3 2LP
G6EPL CT3 3BX
G7LFQ CT3 3HQ
G6WSF CT3 3HZ
G4BQS CT4 6RT

Dudley

M1BMF B62 0DW
G1WTH B62 0HU
G7DIZ B62 8EU
G7IIS B62 8EU
G1SXF B62 8EZ
M1BOF B62 8HH
G2FXZ B62 8JS
G6GQJ B62 8JS
G0SAA B62 8LU
G3VDM B62 8PZ
G3HAZ B62 8QB
G0MAL B62 8QW
G6IJQ B62 8SH
G3VBV B62 9AW
G8EIO B62 9ED
G4DKP B62 9EL
G4SFG B62 9HP
G4ULB B62 9HP
G8FQC B62 9LP
G4VPZ B62 9NQ
G4DPZ B62 9NR
G0NGZ B62 9PP
G1EBB B62 9PP
G4JSV B62 9QW
G4PVP B62 9RF
G4PTX B62 9TJ
G7CMX B63 1AT
G1FET B63 1BB
G4OQX B63 1BZ
G8TXA B63 1DU
G4LWF B63 1JQ
G4MEB B63 1JQ
G6MEB B63 1JQ
G4AYK B63 1JY
G4NQW B63 1JY
G7VIE B63 2BD
G7VIJ B63 2BD
G0ONH B63 2BE
G1WEX B63 2DP
G0WRD B63 2DP
G6XYO B63 2DW
G7FIJ B63 2JA
G0GUG B63 2JJ
G8UBH B63 2PN
G4EHR B63 2PP
M1BMG B63 2SR
G6DNX B63 2SY
G8TIA B63 2SY
G6LDA B63 2TB
G1KII B63 2UY
G3CQK B63 2XF
G6VVS B63 3DD
G7VIX B63 3DJ
G8XSA B63 3DN
2E1EGR B63 3DP
G4IP B63 3EF
G4UMY B63 3JX
G3NAU B63 3JZ
G8DUM B63 3JZ
G7FLQ B63 3QG
G0OUJ B63 3QR
G7VJT B63 3QT
G7VJT B63 3RA
G6ALW B63 4HG
G1VOB B63 4HQ
G7TXZ B63 4JN
G1PPZ B63 4LX
2E1CRZ B63 4NR
G7VJV B63 4NR
G0NLA B63 4PB
G0TJS B63 4PB
G6VAT B63 4PB
G1IAL B63 4PE
G1TRV B63 4TH
G8ZZT DY1 1SL
G4JCP DY1 2DX
M1BVS DY1 2EF
G0OIY DY1 2GG
G0SOI DY1 2JX
G0SKG DY1 2JZ
G1BTV DY1 2QZ
G6VJC DY1 2SL
G4XME DY1 2SN
G6PPA DY1 2SN
G0NXD DY1 2TA
G6ZSU DY1 2UN
G6IRG DY1 2UZ
G0OWU DY1 3LE
G4MMD DY1 4BU
G7KYZ DY2 0BN
G7WLV DY2 0DG
G0BFZ DY2 0EE
G4WBP DY2 0HZ
G0IZF DY2 0NW
G1HRU DY2 0NW
M0AKZ DY2 0NW
G7DXL DY2 7ES
G0ILH DY2 7EY
G7AXW DY2 7TQ
G8ADG DY2 7TQ
G6NTQ DY2 7TS
G7LEX DY2 8BT
2E1AMT DY2 8DJ
G6DFB DY2 8DY
G0GUC DY2 8ER
G6OXL DY2 8RE
G4RRP DY2 8XL
G7IZM DY2 9EU
G8TEK DY2 9HE
G4OLQ DY2 9HN
G6DKM DY3 1AL
G1JLZ DY3 1HW
G0BZT DY3 1LB
G7WHB DY3 1LB
G4XMT DY3 1LF
G1GST DY3 1PD
G4EIB DY3 1QA
G0DPJ DY3 1RF
G4ICU DY3 1RL
G3XEV DY3 1TG
G0CSW DY3 1XW
G7GEU DY3 2JF
G4OXV DY3 2JL
G0CUZ DY3 2JU
M1BVT DY3 2LR
G7PBX DY3 2NJ
G0RIW DY3 2NL
G4JTL DY3 2PL
G7TAS DY3 2PS
G4FBH DY3 2RA
G4NME DY3 2RH
G4DVE DY3 2TJ
G7FBX DY3 2TJ
G0LQF DY3 2UN
G7GEL DY3 3AS
G0GIE DY3 3DR
G4HSO DY3 3DR
G8BOP DY3 3EF
G7ACJ DY3 3EN
G6DAQ DY3 3LH
G7JIN DY3 3LN
G8OTV DY3 3ND
G7TYS DY3 3NJ
G1ASE DY3 3PG
G3GTW DY3 3SJ
G0OGS DY3 3UF
G0OGS DY3 3XD
G3IIV DY3 3YQ
G0AMW DY5 1BE
G1YMH DY5 1DB
G4THE DY5 1DH
G1DIG DY5 2BU
G4NUS DY5 2DA
G4FAV DY5 2EP
G4PQI DY5 2EP
G1XWR DY5 2ER
G4YVA DY5 2HT
G7FUR DY5 2JT
G4FAH DY5 2QG
G0AUM DY5 2UT
G6HZK DY5 2XY
G4TGM DY5 3DP
G6IWD DY5 3DP
G6LUM DY5 3JE
G6XKE DY5 3JE
G0EEB DY5 3JR
G4TDB DY5 3NY
G4DYU DY5 3PE
G4NLW DY5 3YY
G0JKY DY5 4EF
G0OWJ DY5 4EX
G8WSF DY5 4QL
G7WGD DY5 4QL
G3IMG DY5 4RP
G3RLY DY5 4TE
G4DAR DY5 4TE
G1WGW DY6 0HB
G3SIO DY6 0HL
G4CGB DY6 0JG
G4CVU DY6 0LL
G0SRY DY6 0LN
G1MTU DY6 7AA
2E1FPW DY6 7HG
G6ZUV DY6 7QE
M1CBI DY6 7RD
G4NRA DY6 7RP
G6OES DY6 7RP
G4LVA DY6 7RQ
G1ELJ DY6 7RR
G4OND DY6 7RY
G0BHV DY6 8AA
G1WFJ DY6 8BT
G1BLJ DY6 8DJ
G4TBY DY6 8HL
G3KFD DY6 8JR
G7PLP DY6 8NE
G7TFM DY6 8PB
G8YZF DY6 8PD
G8KPG DY6 8RZ
G4YBT DY6 8SP
G6GTB DY6 8SP
G4VZO DY6 9BL
G3ITH DY6 9DX
G7MON DY6 9EG
G6TXL DY6 9EH
G3ZOM DY6 9ET
G4KEB DY6 9ET
G3APL DY6 9RE
G0ALI DY6 9RG
G4TCI DY6 9RG
G8WYS DY6 9RG
G3HGI DY6 9RG
G6PSZ DY6 9RJ
G4VNE DY6 9RP
G4IDU DY6 9SA
G4NTP DY6 9SH
G4JTE DY6 9SS
G1WJK DY6 9TQ
2E1AJC DY8 1BE
G4CVK DY8 1BE
G4IEB DY8 1BE
G7HEZ DY8 1BJ
G3IVQ DY8 1EX
G8OKB DY8 1JD
G3ZUL DY8 1NA
2E1CLB DY8 1QX
G4HWH DY8 1RQ
G0MWS DY8 1XN
G3LGL DY8 2AN
G8CK DY8 2AN
G8XNL DY8 2DE
G4UOY DY8 2DH
G3ORI DY8 2HL
G3KVH DY8 2JZ
G1BCC DY8 2JZ
G1DWN DY8 2LR
G4VPE DY8 2PF
G4DYK DY8 3AX
G6BDH DY8 3EF
G3NDN DY8 3JE
G7DTV DY8 3JH
G3DUH DY8 3JJ
G7JWJ DY8 3JR
G0UUZ DY8 3NG
G0ASD DY8 3NL
G0ESR DY8 3PJ
G4FYQ DY8 3PJ
G6IRH DY8 3RX
G7FMQ DY8 3UZ
G4WAO DY8 3XT
G6AJU DY8 4DG
G1DEA DY8 4QF
G6DEK DY8 4QF
G8RLN DY8 4QF
G8TMQ DY8 4QF
2E1EAW DY8 4QF
2E1BQS DY8 4QJ
G4XNW DY8 4QN
G0FLW DY8 4QS
G6YAC DY8 4QY
G4OLS DY8 4RN
G8PIP DY8 4SF
G7BFG DY8 4SX
G0ASK DY8 4XS
G0ASL DY8 4XS
G4TFB DY8 4XS
G4TFC DY8 4XS
G0NLT DY8 4XU
G1IHJ DY8 4XX
G6IAK DY8 4YU
G0WBA DY8 5AU
G3SNY DY8 5DB
G4WIG DY8 5JJ
G0KVK DY8 5PH
G4UCI DY8 5QE
G1PKZ DY8 5YE
G4KZJ DY8 5YF
G0OVV DY8 5YJ
G3UXQ DY9 0QZ
G7VFL DY9 0SD
G4ONE DY9 0SP
G6MGX DY9 0TF
G4LTX DY9 0TY
G0TZV DY9 0UP
G4IIC DY9 0YB
G3PWJ DY9 0YE
G1NZZ DY9 0YH
G0ICM DY9 7JA
G4YUD DY9 7JL
M1BQM DY9 7LA
G3XOV DY9 7PS
G1DCU DY9 8BE
G0RXO DY9 8YD
G0HTJ DY9 9AE
G4KZB DY9 9EH
G0EWH DY9 9EL
G0KZM DY9 9EL
G4XOM DY9 9EL
G4YUE DY9 9EL
G7JYZ DY9 9HH
G6RSI DY9 9HT
G7UWV WV14 0SF
G4LUB WV14 8HX
G7HUW WV14 8JG
G1KEB WV14 8SE
G4YUB WV14 8SQ
G0WEB WV14 8XD
G1RBY WV14 8YH
G4YVX WV14 9EU

Dumfries & Galloway

GM1OXQ DG1 1BT
GM3MSG DG1 1PL
GM4JKB DG1 1PP
GM4HAA DG1 1PP
GM1JVU DG1 1QN
GM2AJW DG1 1RZ
GM2BMJ DG1 1RZ
GM1SVQ DG1 1SN
GM6HWZ DG1 2PP
GM1HTI DG1 3BL
GM7WJP DG1 3HE
GM3OXK DG1 3RJ
GM4TNJ DG1 4BU
GM7LWC DG1 4DT
GM4RJF DG1 4DW
GM0BQQ DG1 4EW
GM0NTW DG1 4HR
GM8VBX DG1 4HU
GM0EAS DG1 4NS
GM3BXD DG1 4SJ
GM6GQT DG1 4SY
GM3OFT DG10 9JU
GM3JWV DG10 9LS
GM0RIP DG10 9PG
GM3KAM DG10 9PS
2M0AJW DG11 1HG
MM0AXL DG11 1HG
GM7POK DG11 1HG
GM4RNX DG11 2LL
GM4SZJ DG11 2LL
GM3UHT DG11 3AE
GM1VLA DG11 3BA
GM0CBC DG11 3EQ
GM6KAM DG11 3EQ
GM0TCC DG11 3LR
MM1AWA DG12 5LY
GM0PMW DG12 6BD
GM3SOM DG12 6ER
GM8FHV DG12 6JU
GM0FQV DG12 6QX
GM4BDJ DG13 0AX
GM8ETJ DG13 0AX
MM1AAC DG13 0AX
GM1GES DG13 0JW
GM1ZJI DG13 0QH
GM1CUC DG14 0RR
GM6LJE DG14 0RZ
GM3KAK DG16 5LB
GM4RPO DG2 0AJ
GM7LQR DG2 0HZ
GM0AVB DG2 0LU
GM4NNC DG2 0NL
GM0RJG DG2 0QX
GM1EOA DG2 0QX
GM6PFJ DG2 7LW
GM6BKE DG2 8ES
GM4XUJ DG2 9BP
GM0HZM DG2 9SJ
GM8YFA DG3 5NB
GM6AWA DG4 6JU
GM4WFN DG4 6LQ
GM7AWK DG4 6SN
GM7HBT DG5 4AR
GM2BLC DG5 4LL
GM3HKF DG5 4PZ
GM6KKL DG6 4AN
GM4OKG DG6 4BP
GM4DSO DG6 4HL
GM0NYB DG6 4JW
GM4KOO DG6 4PE
MM1BHO DG6 4UE
GM0BWU DG7 1HH
GM1BZR DG7 1HX
GM4VIS DG7 1HX
GM4TPR DG7 1LW
GM6WNX DG7 1LW
GM4BYC DG7 1QU
GM2DTB DG7 1TH
GM4VIR DG7 2HE
GM4KVY DG7 3AU
MM1AVD DG7 3NX
GM3ZEA DG7 3TQ
GM6FXZ DG8 0PN
GM3MZX DG8 6BH
GM4XTI DG8 6JT
GM0URZ DG8 6LN
GM4DUX DG8 8BA
GM0TFF DG8 8LD
GM3AUE DG8 8LD
2M1FIB DG8 8LD
GM0GCF DG8 8NE
GM6FSG DG8 8PP
GM4GDF DG8 9DT
MM0ABQ DG8 9ED
MM1APO DG8 9ET
2M1DIT DG8 9ET
GM4NKF DG8 9HG
GM4XAW DG8 9JG
GM0HPL DG9 0DX
GM4XUC DG9 0EJ
GM6FPD DG9 0EJ
GM4LQS DG9 0HQ
GM3YAV DG9 0HY
GM4BAE DG9 0HY
GM5UI DG9 0PA
GM4YYF DG9 0RG
GM7AWY DG9 7AF
GM1FMX DG9 7BX
MM0AIW DG9 7NA
2M0ALG DG9 7NA
GM0AGN DG9 7TA
GM4LPT DG9 8BQ
GM3TGG DG9 8DB
GM6KXP DG9 8EY
GM7NSG DG9 8QU
GM0TFE DG9 8QU
GM4BES DG9 8TE
GM0HPK DG9 9BA
GM4RIV DG9 9BA
GM1ZIV DG9 9EX
GM0VUJ DG9 9JR

Dundee City

GM0TOF DD1 4LY
G0TTA DD1 4LY
GM8JVZ DD2 1PJ
GM4ZPC DD2 1TJ
GM1CMF DD2 1UU
GM3WPA DD2 3JR
GM4UZP DD2 4BS
GM4ZQQ DD2 4SG
GM4VXM DD2 4TP
GM8LON DD2 4TW
GM4JCM DD2 4UA
GM4ZFS DD3 0BN
GM8BOW DD3 6AW
GM3ZXB DD3 6DD
GM7IKA DD3 6RG
GM0LDF DD3 6SJ
GM0JWK DD3 8EX
GM0MFE DD3 8HR
GM1BOT DD3 8JW
GM0FSW DD3 9RG
GM0PTP DD3 9RJ
GM3WJE DD4 0TA
GM0VUY DD4 0UJ
GM7OWU DD4 0UJ
GM4MUZ DD4 7DB
GM0PIV DD4 7EL
GM7DPI DD4 7JN
GM8YGB DD4 7UN
GM8YGA DD4 8NY
GM3LTD DD4 8TN
G3ZTP DD4 9DB
GM1VWA DD4 9EX
GM0DNH DD4 9HQ
GM0ISA DD4 9LP
GM4AAF DD4 9LP
GM4JPZ DD5 1BL
GM1SZN DD5 1LB
GM3LCP DD5 1LE
GM0PJP DD5 1PN
GM3NHQ DD5 1QT
GM1JTK DD5 1QW
GM4CUZ DD5 1RH
GM4CAB DD5 2EU
GM1MUY DD5 2RE
GM2AOL DD5 2RT
GM0RKU DD5 3BN
GM0GYQ DD5 3DL
GM0CQL DD5 3EF
GM3JIH DD5 3EL
GM0FXJ DD5 3HT
GM0PSQ DD5 3JH
GM3YVX DD5 3LE
GM7ONJ DD5 3RP
GM0MFD DD5 3TA

Durham

G1FBY DH1 1DF
G4CDB DH1 1HG
G1RPY DH1 2AN
G4LEI DH1 2BD
G3SQR DH1 2BN
G7IXQ DH1 2EA
G7BYE DH1 2JU
G6CQH DH1 2SQ
G4GJH DH1 2TB
M1ARG DH1 3DX
G7REJ DH1 3QE
G6RMJ DH1 4EN
G8AVV DH1 4QX
G4EBN DH1 5DQ
G8YWK DH1 5DQ
G4FOQ DH1 5DW
G8PSS DH1 5FN
G0DGB DH1 5JN
G0DRJ DH1 5JN
G0DCC DH1 5JR
G3JQL DH1 5NL
G8PBV DH1 5NU
G4PFE DH1 5PU
G4RHL DH5 9RG
G0KGQ DH6 1DA
G4XQR DH6 1HA
G0OFO DH6 1HS
G6WML DH6 1LS
G4SYD DH6 1PA
G8SVR DH6 1PA
G0WPM DH6 1RJ
G7DIG DH6 4DG
G1TWQ DH6 4HA
G7GQK DH6 4HJ
G4WUI DH6 4NN
G0PMX DH6 4SE
G0IKX DH6 5DT
G4YGO DH6 5EF
G0MEI DH6 5EL
G6LLD DH6 5ES
G0WYW DH7 6RT
G0WYX DH7 6RT
G8FWY DH7 6SG
2E1CJH DH7 6SG
G1WSZ DH7 6SQ
G0SMF DH7 7DL
G7MBH DH7 7NN
G7WBC DH7 7NT
2E1FBV DH7 7NT
G7UFO DH7 7RL
M0BAR DH7 7RL
G7SSF DH7 7RL
G0WZQ DH7 8AY
G7BLB DH7 8BB
G7UKA DH7 8PQ
G6YAR DH7 8SQ
G0MNH DH7 9AU

Ealing

G6ESM UB1 2LT
G3ING UB1 2NE
G4FKX UB1 2QL
G1LLO UB1 2SB
G6HUI UB1 2SB
G4KFB UB1 2UJ
G7JDE UB1 2UN
G1YES UB1 2UP
G4PPD UB1 2XD
G0IQG UB1 3JF
G3KLK UB1 3JP
G1YRY UB1 3LY
G0MMQ UB2 4HL
2E1ALO UB2 4LP
G3XGM UB2 5QG
G1NLK UB5 4EJ
G1XLT UB5 4NL
G1MRI UB5 4PG
G4ZKT UB5 4QS
G8TPX UB5 4TP
G1DLA UB5 5BX
G6CHO UB5 5EP
G8NPK UB5 5ER
G6VXE UB5 5HN
M1AZQ UB5 5JH
G1SJR UB5 5PJ
G0SLH UB5 6EU
G7GDC UB5 6HP
G1SAJ UB5 6PX
G1VSM UB6 0BH
M0ABI UB6 0BH
G7FBU UB6 0DF
G3PSC UB6 0HQ
M1CDB UB6 0JT
G8BVT UB6 0JX
G0MXY UB6 0JZ
G4IRZ UB6 0QE
G4AWM UB6 0RF
G1APA UB6 0RL
G7KYF UB6 0SE
G4FJX UB6 7AD
G4VCO UB6 7AX
G3MHQ UB6 7DD
G3DKB UB6 8JQ
G0DGG UB6 8NY
G6MRU UB6 8RQ
G4GED UB6 8SA
G4IWD UB6 8SP
G0KIE UB6 8XE
G3MEK UB6 8XE
G0PAV UB6 9EW
G3PZK UB6 9LR
G3MMQ UB6 9TL
G6GIU W13 0AE
G1ZOV W13 0DH
G0PIN W13 0ED
G8KPN W13 0EN
G1ARF W13 0HP
G4CEK W13 0HP
G4ENJ W13 0JG
G3SFZ W13 0JW
G7JHZ W13 0NT
G4SCR W13 8BQ
G3AYC W13 8JR
M1BBC W13 8JR
G3NOH W13 8JR
M0BBC W13 8JR
G3RGP W13 8JZ
M1BMB W13 9DE
G0AIS W13 9DT
M0AZL W13 9DU
G8LVF W13 9EN
M1BVC W13 9HS
G4ERC W13 9PS
G8VDQ W13 9QU
G6JEU W13 9RA
G1HGT W13 9TY
G3YMM W3 0DE
G1CNN W3 0HH
G3HHU W3 0HR
G8LWS W3 0RP
G0SJR W3 6JD
G7TVC W3 6TE
M0AXH W3 7AQ
G0ZHP W3 7AQ
G0UCC W3 7NP
G0WIE W3 8EJ
G7LSW W3 8HB
G4NTB W3 8HB
G6HXB W3 9EJ
G0NGC W3 9RD
G0UQW W3 9RQ
G0ORU W4 1AT
G4TZA W4 5DN
G0OYJ W4 5EN
G4FGW W4 5NN
G0AOS W5 1AZ
G8ZLL W5 1BU
G0VDN W5 1HL
G3TYH W5 1LG
G3UDV W5 1RS
G4IQW W5 2HN
G6VPN W5 2JE
G0WOS W5 2SH
G1DYP W5 3DB
G1MME W5 3LJ
G6DTT W5 3SR
G3UZZ W5 4DN
G0SCG W5 4JW
G6VZM W5 4XA
G8BOJ W5 4XA
2E1DTE W5 5QT
2E1DTF W5 5QT
G1NLJ W7 1AW
G6TJY W7 1BT
G0WQW W7 1LR
G1XAF W7 1LU
2E1DSK W7 2BW
G7IVN W7 2HB
G8RBX W7 3AG
G6TNK W7 3DJ
G4GBV W7 3DR
G3TIG W7 3EJ
G0SRJ W7 3QA
G8MEM W7 3QX
G7KWT W7 3QZ

Easington

G0TBZ DH6 2BW
G8ODV DH6 2JG
G0WIC DH6 2NF
G4CNK DH6 2PD
G0PXQ DH6 2TR
G1TLI DH6 3AP
G6UYU DH6 3EY
G8TDP SR7 0AN
G6AGZ SR7 0BD
G8AGY SR7 0BD
G0MJV SR7 0JT
G1ERU SR7 0LE
G1DLC SR7 0PJ
G6LNL SR7 7BA
G4GSO SR7 7BQ
G0SRT SR7 7BS
G0MLW SR7 7LQ
G0OBN SR7 7TB
G4LOM SR7 7TF
G4NMF SR7 8BG
G0DVP SR7 8DG
G4RVY SR7 8HE
G0MKC SR7 8HZ
G0UFN SR7 8JZ
G4NMK SR7 9PQ
G4NYU SR8 1BZ
G0FBW SR8 1DD
G7GOX SR8 1ER
G0TOK SR8 1LN
G6APN SR8 1LN
2E1DMI SR8 1LN
G4PLU SR8 1LP
G6NXP SR8 2AN
G4RXR SR8 2DT
G0RDD SR8 2EX
G6VHG SR8 2HB
G4GXI SR8 2NH
G0NDD SR8 2NN
G4KGQ SR8 3BA
G4WKT SR8 3NG
G7DBV SR8 3SU
G0EQJ SR8 3UA
G0NXC SR8 4BW
G1WUS SR8 4PQ
G0TOH SR8 5JJ
G0KVJ SR8 5QZ
G3NSI SR8 5QZ
G7LNW TS27 4AY
G7FNV TS27 4NU
G4HDS TS27 4PE
G0GKO TS27 4RT
G0GWD TS28 5JE
G0HJX TS28 5JE

East Ayrshire

GM3CTG KA1 1RU
GM4SMI KA1 1TT
GM4SMH KA1 1UH
GM0AAX KA1 2HP
GM4UDX KA1 2JR
GM3UWO KA1 3DP
GM3BXX KA1 3ED
GM8KXF KA1 3ED
GM3OZB KA1 3LQ
GM4OFZ KA1 3NF
GM0DYF KA1 3RS
GM0FKP KA1 3RY
GM1TXE KA1 4LJ
GM4UYP KA1 4LQ
GM0EUY KA1 4PB
GM0FQQ KA1 4PB
GM0KVI KA1 4RG
GM8FCK KA1 4UD
GM6KPL KA16 9BJ
GM3AKK KA16 9EJ
GM4ZNS KA16 9ER
GM3MAY KA17 0AP
GM4FCP KA17 0BN
MM0ALC KA17 0BT
GM4TPQ KA17 0DQ
GM0VBE KA17 0LP
GM4JRG KA18 1HL
GM1TCP KA18 1NG
GM1YKE KA18 1NG
GM0BKX KA18 1PU
GM0NET KA18 1PU
GM1MDO KA18 1PU
GM7OKX KA18 2RE
GM0SZQ KA18 2RR
GM4XMD KA18 3HS
GM0KXJ KA18 3PY
GM7ANE KA18 4HH
GM1SXZ KA18 4JP
GM7BTL KA18 4JP
GM8XVU KA2 0AR
MM1BRM KA2 0BZ
GM0ECU KA2 0JR
GM0IDJ KA2 0JR
GM0DEQ KA3 1UL
GM6WTH KA3 2EZ
GM0ONX KA3 2JG
GM1FNX KA3 2QB
GM4XGW KA3 2QT
GM4JRP KA3 2TA
GM3AXX KA3 3DQ
GM4VAY KA3 3HG
GM4OSS KA3 3HQ
GM0GOV KA3 4BP
GM0PDQ KA3 5JP
GM4HCO KA3 5JP
GM0OXS KA3 6DB
GM4DLG KA3 6HP
GM0DVH KA3 6JT
GM3OYH KA3 6JT
GM7RPT KA3 6LE
GM4SQM KA3 7DT
GM4UWO KA3 7EA
GM8ZEJ KA3 7JB
GM1MJK KA3 7PL
GM4XLZ KA4 8DB
GM0KAZ KA4 8EJ
GM3YEH KA4 8JY
GM0ADX KA4 8LL
GM3ZRT KA4 8LL
GM4WZL KA4 8NA
GM0GRW KA5 5JE
GM2BUD KA5 6DX
GM1VFQ KA6 6EL
GM4SFA KA6 6ND
2M1FQI KA6 7AU
2M1FQJ KA6 7DG
GM6MD KA6 7LD
GM3PPJ KA6 7LQ
GM0GRD KA6 7PB
GM0JBE KA6 7PS
GM4SDQ KA6 7QL
GM1VDZ KA6 7TY

East Cambridgeshire

G7RUJ CB5 0AN
GW6AHY CB5 0BE
G6XAC CB5 0BF
2E1GBA CB5 0DQ
G1YAS CB5 0DU
G8DQF CB5 0EL
G3XBM CB5 0HF
G8BXA CB5 0HT
G0OKL CB5 0JF
G0SBK CB5 0JX
G6IOB CB5 0LU
G0CLH CB5 0RG
G7CMQ CB5 9BL
G3PTQ CB5 9BQ
G7BNM CB5 9HB
G7VGH CB6 1DD
G0OKK CB6 1DJ
G8GLB CB6 1DP
G3KKC CB6 1HU
G6YQJ CB6 1JS
G0MGI CB6 1JW
G3YZB CB6 1NT
G0MDC CB6 2AW
G1DUI CB6 2HH
G1PEK CB6 2HS
G3PCC CB6 2HX
G7BAD CB6 2JG
G0PPL CB6 2PD
2E1AOB CB6 2PD
G6JPI CB6 2PX
G3AJP CB6 2QJ
G3JYK CB6 2QW
G8DIY CB6 2RB
G8NQR CB6 2RE
G3PMH CB6 2TS
G3PWK CB6 2TS
G0ICE CB6 3AL
G1CQR CB6 3DN
G1YCJ CB6 3LF
G4GND CB6 3PP
G8JHE CB6 3PQ
G4KNS CB6 3PW
G7TKY CB6 3RP
G3SZU CB6 3RT
G4BEL CB6 3ST
G8BBB CB6 3ST
G3EBG CB6 3TW
G4NPH CB6 3UE
G1VLD CB6 3XE
G8BDQ CB7 4AS
G0SOF CB7 4AS
G0DMF CB7 4QY
M1CCH CB7 4SS
GW1KQN CB7 4SY
G4KPX CB7 4TH
G1QLV CB7 4UN
G3AMF CB7 5BJ
G8LWQ CB7 5BL
G3YIJ CB7 5DB
G6BUU CB7 5DH
G0INO CB7 5EG
G8PAG CB7 5EL
G4UJV CB7 5NR
G4FVA CB7 5PB
G8WAV CB7 5QX
G0PIK CB7 5RU
G7AYD CB7 5SD
G0TLQ CB7 5SH
G4ERO CB7 5TN
G3FCM CB7 5TW

G0SPO CB7 5UP
G0EIM CB7 5UU
G7FMV CB7 5XE
G8IDL CB8 0SE
G1ACB CB8 0SQ
G1UAF CB8 0SQ
G7BNL CB8 0SQ
G4SPN CB8 0TG
G0KOJ CB8 8AH
G4ZZM CB8 8DA
G0AHL CB8 8DD
G7VGN CB8 8HU
G3NUL CB8 9DG
G3SZY CB8 9DG
G0ADZ CB8 9DW
G0FRU CB8 9NB
M1BJE CB8 9PD
G7VNP CB8 9RR

East Devon

G3FFH DT7 3RJ
G4WBX DT7 3SF
G0AYY DT7 3SX
G3PGT DT7 3SX
G1ENA DT7 3UR
G3VW DT7 3XA
G0WHN DT7 3XF
G7EQP EX1 3TQ
G6IDG EX10 0EY
G3MSV EX10 0LE
G3DPX EX10 0QZ
G6SNY EX10 8AG
G0AFC EX10 8XW
G3AQM EX10 9EZ
G0AXC EX10 9HU
G0NOC EX10 9HU
G3DCE EX10 9JA
G4TIG EX10 9LR
G1SED EX10 9LS
G6RUM EX10 9LS
G2BIM EX10 9NT
G6BHY EX10 9NY
G7UYK EX10 9QA
G6YWX EX10 9SU
G6XUV EX10 9TJ
G7AGO EX10 9TJ
G6CZT EX10 9XN
G4BNP EX11 1AR
G0WGH EX11 1DT
G7DYB EX11 1EN
G7NBZ EX11 1EP
G6BJL EX11 1LE
G3KPV EX11 1QB
G8NVT EX11 1SY
G3PLP EX11 1UZ
G4WLP EX11 1XH
G4BNV EX11 1XW
G6WZP EX12 2AD
M1BQN EX12 2BN
M1AXM EX12 2BW
G6JJA EX12 2DB
G4XXK EX12 2DJ
G4JFY EX12 2HE
G7VCB EX12 2NJ
G8WCQ EX12 2NT
G1NFQ EX12 2PD
G4WCQ EX12 2PF
G8VXU EX12 2PX
G0CTF EX12 2SB
G3RCO EX12 3EQ
M1AVB EX12 4AG
G0RLX EX13 5BH
G4VHG EX13 5BS
G4RHI EX13 5EF
G6PUA EX13 5HN
G0AOG EX13 5LJ
G7AKJ EX13 5NG
G4CWR EX13 5RE
G0HYX EX13 5RP
G6WWY EX13 5SQ
G1APH EX13 5SY
G7CMP EX13 5SZ
G0GHH EX13 5TT
G7AXE EX13 5TT
G8CA EX13 5TT
G3CYX EX13 6AP
G4FVU EX13 6AQ
G6LUJ EX13 6JU
G3PXM EX13 6QX
G0WDH EX13 7HH
G4OYY EX13 7HL
G1HDQ EX13 7LG
G3GOS EX13 7LU
G0HRH EX13 7PB
G3CFR EX13 7ST
G0ESZ EX14 0DT
G8KRU EX14 0PD
G4HBA EX14 0PD
G8GTV EX14 0QB
G4SLL EX14 0RH
G3RHM EX14 0UN
G7SQI EX14 0UP
G1IAR EX14 0UR
G0DEZ EX14 0XW
G8HOA EX14 0XW
G4PQS EX14 8AX
G0UIL EX14 8BR
G6WZK EX14 8EA
M1BHX EX14 8JB
G8BEA EX14 8JE
G3KJP EX14 8JG
G7VFH EX14 8NS
G0BCO EX14 8QZ
G7RMX EX14 8RJ
G0WWL EX14 8YP
G4DOW EX14 8YQ
G8SHF EX14 8YW
G0TEB EX14 9AL
G4AGT EX14 9DL
G8BVR EX14 9JS
G1KVD EX14 9LX
G0CWK EX14 9SA
G3GFC EX14 9SA
G0BVC EX14 9TA
G4RKE EX14 9TA
G4FDG EX15 2EQ
G1DKE EX5 1AS
G8ZSO EX5 1DR
G4TDV EX5 1EA
G3VRV EX5 1JD
G1OEF EX5 2BY
G4VWK EX5 2EZ
G8JNE EX5 2NX
M1ARP EX5 2SG
G0HHK EX5 2TE
G7VNQ EX5 2TJ
G3ORN EX5 4BB
G7ANX EX5 4EP
G4UUW EX5 5JA
G0RKC EX8 1EH
G3NBR EX8 1JS
G4EBO EX8 1QN
G4RTF EX8 2AD
G7TRF EX8 2JR
G3XDR EX8 2JT
G6SCP EX8 2JU
G3JEP EX8 2LN
G0LHQ EX8 2PE
G4VNI EX8 3AH
G3NC EX8 3AL
GW4VMF EX8 3AN
G3XZX EX8 3BU
G7BWO EX8 3DT
G1GZG EX8 3EJ
G4YRM EX8 3HY
G3UZL EX8 3LD
G3UZM EX8 3NE
G0ONT EX8 3NS
G1YOF EX8 4DY
G8EKW EX8 4EU
G8SBU EX8 4EW
M0AGI EX8 4HG
G7RRW EX8 4HG
G0BOL EX8 4JT
G1XRC EX8 4LF
G8GON EX8 4LF
G8UWE EX8 4LF
G0RIY EX8 4LQ
G3RRK EX8 4NU
G0CGY EX8 4NY
G7RGN EX8 4QQ
G0NRR EX8 4RF
G0XRC EX8 4RF
G0ETZ EX8 5EE
G1ILY EX8 5NZ
G0GHO EX8 5PN
G4GXT EX8 5QN
G3HWL EX9 6BY
G0VEA EX9 6EE
G3EPO EX9 6JF
G6UEI EX9 6JN
G4LJA EX9 6QD
G4TEX EX9 6QG
G8UXL EX9 6SE
G8XRR EX9 6SE
G4WJZ EX9 7AS

East Dorset

G6KSQ BH16 6AF
G3WJJ BH18 9JP
G0DVE BH21 1AE
G1BRS BH21 1DE
G2BDV BH21 1DE
G2BRS BH21 1DE
G4GOY BH21 1HD
G3VMO BH21 1HJ
G3VQR BH21 1JJ
G6EYT BH21 1LH
2E1AES BH21 1LS
G7DKS BH21 1LS
G6XSA BH21 1NY
G4RLM BH21 1PL
2E1FYR BH21 1RA
G8GYZ BH21 1RQ
G0EEF BH21 2AU
G0UPS BH21 2DH
G3UAZ BH21 2EU
G4NHE BH21 2HR
G7MQU BH21 2JB
G0MAX BH21 2JG
M1BAI BH21 2LA
G3BHM BH21 2LD
G0MDK BH21 2LE
G3JBT BH21 2LG
G8SBB BH21 2LG
G0TUL BH21 2NH
G0OFE BH21 2NL
G0WAL BH21 2NZ
G3ZXW BH21 2PQ
M1BSB BH21 2PU
G8SEK BH21 2PY
G0FRR BH21 2UW
G0VBV BH21 2UW
2E0ABS BH21 2UW
G4RFR BH21 2UW
G4YTA BH21 2UW
G0UGS BH21 2UW
G1VIP BH21 3DS
G8UCY BH21 3DS
G0API BH21 3EZ
G3ZTY BH21 3EZ
G7MHO BH21 3EZ
G6ZVJ BH21 3HY
G0FKY BH21 3LG
G1YHV BH21 3LZ
G7AUF BH21 3NH
G7JDW BH21 3PN
G7INA BH21 3PT
G8NVC BH21 3PU
2E0ADM BH21 3QB
G1LHW BH21 3QB
G3RAN BH21 3QL
G8KNS BH21 3SL
G1MXD BH21 3TJ
G8DLT BH21 3TW
G8CRV BH21 3UA
G4GKX BH21 3UB
G4WUK BH21 4BH
G8ZEK BH21 4HD
G3ISQ BH21 4JG
G0RTJ BH21 5LA
G3XCY BH21 5NX
G7MYI BH21 5RD
G7IIC BH21 5RX
G6ELW BH21 6RG
G4MHF BH21 6SR
G3OAF BH21 7AN
G3ODO BH21 7AW
G0GHX BH21 7AZ
2E1BKX BH21 7BQ
G4FRP BH21 7NN
G8IFU BH21 7PB
M0BDF BH21 8LN
2E1AYB BH22 0AL
G7PFT BH22 0AL
G3XFD BH22 0AY
G3BSI BH22 0EN
G3FK BH22 0EU
2E1FSZ BH22 0EY
G0NDH BH22 0JE
G1ULZ BH22 0JS
G4VLP BH22 0LG
G3NRH BH22 0LQ
M0BDE BH22 0ND
M1BKV BH22 0ND
G8RWT BH22 0PD
G7VLV BH22 0PE
G0KQJ BH22 8BJ
G1XXJ BH22 8BJ
G3CPN BH22 8BY
G3XVS BH22 8BZ
G0SRX BH22 8HG
G6PLB BH22 8JY
M0AXI BH22 8LT
2E1DIB BH22 8LW
2E1FSY BH22 8PA
G3XAS BH22 8PQ
G3RGE BH22 8PW
G4RS BH22 8PW
G1NNV BH22 8PX
G3VQN BH22 8RU
G3ZCL BH22 8SB
G3ZGI BH22 8TT
2E1FTS BH22 8XA
2E1FTT BH22 8XA
G0MER BH22 8XE
G3AAO BH22 9AY
G3EXP BH22 9EF
G8YHF BH22 9EQ
G8DJL BH22 9ES
G4ULQ BH22 9HP
G4LJN BH22 9HS
G4CFW BH22 9HW
G3WPB BH22 9JE
G0CDY BH22 9JQ
2E0ADN BH22 9JQ
G4DHA BH22 9JT
G4PAI BH22 9LA
G1WKV BH22 9QP
G6UTN BH22 9RZ
G4ACJ BH22 9SD
G1DNK BH22 9SG
G0NLM BH22 9TL
G0MUK BH24 2AJ
G4DFQ BH24 2BQ
G3BKN BH24 2EP
G6EDU BH24 2HH
G6IMH BH24 2HH
G4HHB BH24 2HX
G3EBK BH24 2HZ
G0SQH BH24 2JA
G1GTM BH24 2JQ
G3ICH BH24 2NY
G3JAA BH24 2NY
G8SXJ BH24 2PF
G3YNJ BH24 2PH
G0DDI BH24 2PZ
G3NWL BH24 2QJ
G4NXR BH24 2QL
G0PRH BH24 2QT
G4ZLI BH24 2QT
G4LIT BH24 2RQ
G4FSU BH24 2SW
G4GTH BH31 6BS
G4YQO BH31 6DN
G0SWF BH31 6EY
G6VRU BH31 6HJ
G0NUM BH31 6HP
G6BJK BH31 6HW
G8NAG BH31 6HX
G4WHO BH31 6JJ
G4WHY BH31 6JJ
G3MRJ BH31 6JT
G4TMF BH31 6LQ
G2CYT BH31 6NW
G8YZL BH31 6NZ
G0LPZ BH31 6QB
G4WPT BH31 6QB
G6SDQ BH31 6QB
G0VPJ BH31 6QB
G4PIJ BH31 6TQ
G6SFR BH31 6TQ
G8GTC BH31 6TQ
G0URR BH31 6TS
G7ESE BH31 6TU
G4YWY BH31 6UG
G6SRV BH31 6XA
G3EVK BH31 7AN
G3XYE BH31 7LE
G4HNJ BH31 7PD
G4MQT BH31 7PD
G0DEJ BH31 7PG
G0DUY SP5 5QZ
G0AFG SP6 3AX
G4KAE SP6 3BN
G3YLR SP6 3BW
G3PUZ SP6 3DG
G0TQZ SP6 3DJ
2E1EVY SP6 3RB

East Dunbartonshire

GM0HCW G61 1AL
GM0GCO G61 1EJ
GM3ZWG G61 1EN
GM6KIW G61 1HS
GM4APK G61 1JY
GM3EZA G61 1PE
GM3RQQ G61 1RE
GM7JFR G61 2AF
GM0TXG G61 2AU
GM7SBB G61 2EP
GM7OHB G61 2LT
GM7SCJ G61 2LT
GM3VQJ G61 3AW
GM7ODA G61 3BH
2M1DIL G61 3BH
GM0HBT G61 3DN
GM3VTB G61 3HD
GM4VTB G61 3HD
GM0ATL G61 3HQ
GM3GTQ G61 3JX
GM0UPE G61 3LG
GM1SXW G61 3LL
GM1SRP G61 4BG
GM1XEB G61 4BG
GM3NMA G61 4DH
GM0RRK G61 4JP
GM4CXM G61 4JU
G0WUK G61 4JU
GB7SDX G61 4JU
GM4LBV G61 4RQ
2M1DZM G62 6LW
GM0HLP G62 7JD
GM3JPF G62 7JD
GM8SAP G62 7JD
GM3KBZ G62 7LH
GM4OTH G62 7NB
GM4KAV G62 7RA
GM4JRF G62 7RL
GM3MAS G62 7RR
GM8ZKF G62 8AT
GM4CNF G62 8HA
2M1DIQ G62 8HE
GM8EJS G62 8NL
GM6ARB G64 1AY
GM0GMO G64 2HP
GM3HOM G64 2NS
GM7FLG G65 7EP
GM0PEX G65 8AY
GM0HNV G65 8AY
GM0MOC G65 8AY
GM0RED G65 8AY
GM3JMM G65 8AY
GM0HJU G65 8DS
GM7BOZ G65 8DS
GM7GBD G65 8ER
GM0HIM G65 8ET
GM4CKP G65 8ET
GM7HIR G65 8ET
GM0ATA G65 8HG
GM0OLD G65 8JA
GM3GRG G66 1AX
GM4JYZ G66 2JQ
GM4LYV G66 2JY
GM1JIU G66 2SB
GM7GTK G66 2SB
GM0EKY G66 3AS
GM3WYL G66 3AS
2M1CDP G66 3HU
GM1BTL G66 3JL
GM1FSZ G66 4BE
GM7SPA G66 4EG
GM6TIB G66 4EL
GM0DAN G66 4HX
GM0XAV G66 4QQ
GM4TYQ G66 4RA
GM3SER G66 5AX
GM0BEL G66 5NG
GM4ZRX G66 5NG

East Hampshire

G8VHP GU10 4DF
G4BGM GU10 4EN
G3FPQ GU10 4LP
G4ZKJ GU10 4LU
G0NFA GU10 5HZ
G0VZS GU10 5HZ
G4OJO GU10 5LD
G7JEJ GU10 5LZ
G7RFV GU30 7BX
G4VKC GU30 7HR
G0DEW GU30 7NY
G8ISI GU30 7SH
G8UYW GU30 7XA
G4AER GU30 7XD
G0BEM GU31 4BU
G4VWO GU31 4EU
G4VWP GU31 4EU
G1GLF GU31 4LB
G6KPR GU31 4LL
G8BKY GU31 4ND
G1LQH GU31 4NR
G1OXF GU31 4NR
G0BTU GU31 4NX
G4ANW GU31 4NX
G8LHZ GU31 4NX
G4ZPV GU32 2AA
G3ELH GU32 2EF
G1RTY GU32 2HN
G4VRC GU32 3AZ
G0MBQ GU32 3BW
2E1DOH GU32 3JZ
G4ACW GU32 3PJ
G8XJO GU33 6HG
G8GBW GU33 6NS
G0RQN GU33 7BW
G4ELM GU33 7DL
M0AIR GU33 7ED
G7WEA GU33 7HL
G1UTC GU33 7LP
M1CCA GU33 7LP
G4WKY GU33 7LR
G6SLZ GU33 7LR
M0AVJ GU34 1HS
G4ASL GU34 1JA
G0WYF GU34 1LA
G4FOY GU34 1NU
G6OSG GU34 1PS
G0JMI GU34 1PW
G1LAO GU34 1QY
G4JNI GU34 2BJ
G0DBS GU34 2EE
G8YFH GU34 2HH
G4JOU GU34 2LD
M0AFO GU34 2NX
G6YBN GU34 2RS
G3DPS GU34 2TA
G8IRL GU34 2TL
G8UML GU34 2TP
G6BXV GU34 3EJ
G4XWV GU34 3EU
M0AFH GU34 3EU
G0DWR GU34 3NP
G7DNF GU34 3QL
G8GOS GU34 4AE
G4ISK GU34 4AJ
G8EAN GU34 4AN
G1KAW GU34 4ET
2E0AJG GU34 4LU
G8MMF GU34 4PD
2E1DMR GU34 4PP
G3YMK GU34 5AL
G3JHM GU34 5AW
G8LES GU34 5AX
G3XVR GU34 5BJ
G7CDP GU34 5BN
2E1FEJ GU34 5BN
G4ZRZ GU34 5BX
G4IPI GU34 5BY
G8WSZ GU34 5BY
G4INI GU34 5DF
G4FVG GU34 5DH
G6UEQ GU34 5LG
G4RVC GU34 5LN
G0NUJ GU34 5LZ
G3YVI GU34 5NP
G6BHH GU34 5PB
G6BHI GU34 5PB
G8CKN GU34 5PF
G3SJX GU34 5PR
2E1FMD GU34 5PT
G4OWS GU34 5RB
G3OTK GU34 5RY
G3ZEO GU34 5SD
G0TNP GU35 0DG
G3ZRM GU35 0HB
G4BCZ GU35 0NR
G3WPV GU35 0RN
G2DDS GU35 0TA
G8NLF GU35 0TH
G3BCY GU35 0UY
G0PVJ GU35 8AJ
G3OFJ GU35 8EZ
G6SNA GU35 8HQ
G0HOY GU35 8HU
G0OOJ GU35 8HU
G4ZEL GU35 8HU
G8OXS GU35 8JR
G4BCY GU35 8JX
G4XSV GU35 8LN
G4LJB GU35 8NA
G4RJK GU35 8NH
G7ECK GU35 8PE
G0NIN GU35 8PS
G8YDU GU35 9AF
G7AKQ GU35 9EH
G8GRQ GU35 9EN
M0ACQ GU35 9EN
G0PVI GU35 9EZ
G1CUB GU35 9EZ
G6SPR GU35 9HH
M0AFA GU35 9ND
G8SXB GU35 9PJ
M0AVC GU35 9PQ
G6OJO GU35 9PZ
G6YIP GU35 9QY
G1TQY GU35 9TA
G0ATG PO8 0HF
G4BQV PO8 0JX
G0VOJ PO8 0LJ
G0JGU PO8 0LT
G1PPD PO8 0NF
G7MFL PO8 0NX
G8KYT PO8 0PA
G0EDM PO8 0PD
G4JCX PO8 0PJ
G8TZE PO8 0PJ
G0RYA PO8 0QD
G3GVC PO8 0TR
G4CHO PO8 9ED
G0BKW PO8 9EW
G0MGH PO8 9HE
G4VJS PO8 9QX
G8AFZ PO8 9QZ
G0UHM PO8 9TJ
G3VPO PO8 9TT
G4PAZ PO8 9TU
G3TSM PO9 6AG
G7NVN PO9 6AZ
G3PXI PO9 6BS
G7DRC PO9 6BU
G6XCV PO9 6DA
G3VYW PO9 6DN
G4JNZ SO24 0DB
G4YDE SO24 0DU
G6VXR SO24 0DU
G3PVS SO24 0EL

East Hertfordshire

G0KFJ AL6 0LT
G8RZD CM21 0AE
G4PGB CM21 0AU
G8WUG CM21 0BT
G1XRT CM21 0DZ
2E1FLV CM21 0EH
G4ERS CM21 0HX
G8PZR CM21 9BZ
G6ENC CM21 9NF
G0PYV CM21 9NN
G3NWX CM21 9NN
G3WOO CM21 9NT
G3WYD CM23 2BL
G3XHZ CM23 2HE
G4CWH CM23 2HX
G0NQN CM23 2PD
G0COE CM23 2PP
G4EVR CM23 2QP
G7CQX CM23 3EU
G3KCG CM23 3EX
M0AQO CM23 3EX
G8PPN CM23 3LS
G6YMU CM23 3NG
G6YMV CM23 3NG
M0AZR CM23 3NG
G8JDT CM23 3NG
G4SUA CM23 3NP
G1NRK CM23 3PX
G1OSH CM23 3PX
G0TVJ CM23 3TP
G0PQF CM23 4AD
G5ZG CM23 4AD
G7VRX CM23 4BW
G7NEK CM23 4DW
G8MDC CM23 4DY
G4HFR CM23 4EF
G0GZU CM23 4EY
G0UPL CM23 4FE
G0XAD CM23 4HU
G1NOL CM23 4JD
G3SUS CM23 4JL
G4OBV CM23 4JP
G8FUL CM23 4JT
G6LIB CM23 4LL
G8THH CM23 4LL
G4SNJ CM23 4PU
G1IBG CM23 5AL
G3ZXF CM23 5AP
G0PZJ CM23 5DS
G4HXH CM23 5DS
G6NGA CM23 5EX
G4GIS CM23 5HY
G7WJV CM23 5JJ
G6DIO CM23 5JR
G4LSE CM23 5LS
G8SEY CM23 5RY
G3OMD SG10 6JG
G0KGR SG11 1HT
G4VMR SG11 1JH
G4VSL SG11 1JH
G0JDV SG11 1NJ
G4TEU SG11 1NJ
G8NSG SG11 1RN
G0LUC SG11 1RX
G0DCW SG11 1TG
G3LAS SG11 2AJ
G3OTR SG11 2DN
G3JBP SG11 2DQ
G1AJA SG11 2RS
G6HVL SG12 0BB
2E1CZU SG12 0BB
G3SZT SG12 0BL
G4ZCX SG12 0HA
G7VRY SG12 0HP
G3ZZQ SG12 0HW
G3OJG SG12 0JQ
G4YFC SG12 0NL
G4YOS SG12 0QW
G8FBB SG12 0RJ
G4PHV SG12 0SY
2E1EMR SG12 0TX
2E1EOS SG12 0TY
2E1FWZ SG12 7EZ
G7VCC SG12 7LE
G0OJY SG12 7ND
G3WMB SG12 7RE
G4VXD SG12 8AZ
G4XUW SG12 8DZ
G0LLR SG12 8NW
G6AJW SG12 9HG
G3TIK SG12 9JN
G6CRC SG12 9JN
G4MGC SG12 9JP
G7OCI SG12 9NE
G6DVO SG13 7JF
G0AUJ SG13 7QU
G3SZF SG13 7QU
2E1FZN SG13 7RG
G0AHB SG13 7RR
G2CKB SG13 7SU
2E1EPZ SG13 7UY
G8OPC SG13 8AD
G0HCC SG13 8DE
G8WXB SG13 8JP
G4TQG SG13 8NA
G8YQJ SG13 8NU
G0IKL SG13 8PF
G4CKA SG14 1NQ
G7UAH SG14 2DY
G6HMJ SG14 2EF
2E1FNA SG14 2EF
G8JWI SG14 2HH
G3IOJ SG14 2JE
G3EPK SG14 2LN
G1BLQ SG14 3DH
G4WXY SG14 3JA
G0OSK SG14 3JU
G7NEA SG14 3JZ
G6WYF SG14 3SF
G7VLD SG14 3SN
G0JGV SG14 3SX
G0NEA SG14 3SX
G4TFH SG2 7EF
G0WYC SG2 7HX
G7IJW SG2 7NZ
G1WFM SG2 7PL
G0RZO SG9 9DG
G6BWE SG9 9DZ
G4MHR SG9 9JF
G3YJE SG9 9LA

East Lindsey

G1XXA DN36 5AQ
G3VIP DN36 5AQ
G0NQW DN36 5BG
2E1CAW DN36 5BG
G4PYD DN36 5BQ
G4LVN DN36 5DR
G4GAB DN36 5DS
GM4BXG DN36 5DT
G7HJR DN36 5ED
G4OII DN36 5JE
G8YOJ DN36 5LP
G7RGY DN36 5LP
G4SEP DN36 5NJ
G4GOJ DN36 5PW
G4WHQ DN36 5QP
M0AEP DN36 5RT
G0NUE DN36 5RT
G1IZB DN36 5TP
G1VMX LN10 6RW
G7PPF LN10 6SB
G4DRF LN10 6SP
2E1FRA LN10 6TD
G3DII LN10 6TL
G7RHG LN10 6TP
G4IJA LN10 6UU
G3OUT LN10 6UY
G3MGX LN10 6YH
M1BUR LN11 0AZ
G3CFG LN11 0DF
G4TON LN11 0DY
G0KKR LN11 0PB
G7VLI LN11 0PF
G0BSY LN11 0PL
G4MXI LN11 0PN
G4CTE LN11 0SR
G1VUG LN11 0XF
G4YVJ LN11 7BH
G3ZPM LN11 7JR
G0LKM LN11 7LN
G1MFW LN11 7LN
G6GZS LN11 7LN
G3PLN LN11 7NH
G0JXY LN11 7NJ
G4SBZ LN11 7NP
G7VIG LN11 7QH
G7KDP LN11 7SS
G4WUY LN11 7TL
G7NLA LN11 8AW
G4AEW LN11 8DN
G7TII LN11 8DX
G6ERQ LN11 8ES
G8OOS LN11 8EW
G3UP LN11 8EY
G4HYU LN11 8NB
G0FUG LN11 8PR
G0DBX LN11 8QH
G4HUD LN11 8RY
G1PRS LN11 8SG
M1BTR LN11 8SP
G6ATL LN11 8TG
G0NQA LN11 9DG
G1UFV LN11 9EJ
G4LRC LN11 9HG
G0GNW LN11 9HT
G7UWU LN11 9HT
G0WIW LN11 9RL
G4YZC LN11 9TF
G3SCD LN11 9XG
G4VUU LN11 9XH
G0VRU LN11 9XQ
G0FMG LN12 1DA
2E1GBO LN12 1HA
G5LL LN12 1HF
G3WNQ LN12 1NX
G0KED LN12 1QB
G1XWD LN12 1QE
2E1FEF LN12 1QE
2E1FEG LN12 1QE
G3VDV LN12 2AJ
G7BUK LN12 2AJ
G3MNS LN12 2BY
2E1GAA LN12 2JD
G3JVR LN12 2LJ
G3XXJ LN12 2LU
G1BVV LN12 2NP
G2ACZ LN12 2NU
G0CQG LN12 2PG
G3LSA LN12 2QS
G4NJW LN12 2QY
G0CBM LN12 2RE
2E1FYZ LN12 2RE
G1HEY LN12 2TX
G3UWA LN12 2UB
G3SYX LN13 0JJ
G0DVY LN13 0JP
G0KUR LN13 0JP
G2FFN LN13 0JP
G3SYB LN13 0JP
G3ZUI LN13 0JW
G4AIQ LN13 0NW
G0KBZ LN13 0PN
G0GIZ LN13 0PW
G0KFB LN13 9EQ
G6XDP LN13 9LQ
G4AK LN13 9NW
G7RAG LN13 9PH
2E1FFJ LN13 9PH
G0RJM LN13 9RL
G0DSH LN13 9SU
G7KDF LN4 4LN
G0MPQ LN4 4PB
G7TGL LN4 4RW
G0TDB LN4 4SY
G0LXE LN4 4YE
2E1FZB LN8 5RB
G3PRR LN8 5RY
G4KIZ LN8 6AD
G3YFU LN8 6AJ
G0TWE LN8 6BJ
G4LAY LN8 6DH
G7VYA LN8 6DP
G1VYC LN8 6LB
G7UGY LN9 5AJ
G0NDY LN9 5HF
G2CR LN9 5JS
G1BPI LN9 5ND
G1JZG LN9 5PT
G7JWB LN9 5RJ
G3SIP LN9 5RJ
G7UHS LN9 5RJ
G7RKH LN9 5RU
G4FOT LN9 5SR
G0VZM LN9 6AW
2E1FRB LN9 6JP
M0AJD LN9 6NN
G4ZWA LN9 6PQ
G0KBX LN9 6QB
G7HPZ PE22 0ST
G1EBZ PE22 7AB
G8LDW PE22 7BS
G8BEH PE22 7EP
M1BQX PE22 7NH
2E1FRE PE22 7NH
G7NYV PE22 7NQ
G8AIJ PE22 7QU
G4MWJ PE22 7TB
G7HZS PE22 8DG
G7GZV PE22 8EX
G4DDI PE22 8EY
G8IEV PE22 8EY
G6IDL PE22 8RW
G4DYJ PE23 4AG
G4GBE PE23 4BE
G6HGN PE23 4BU
G0OHV PE23 4LL
G4MLK PE23 4PF
G4UNW PE23 5DB
G6HYF PE23 5HH
G4NLP PE23 5JB
G3IRW PE23 5LZ
G3CMI PE23 5SE
G4EXX PE24 4AD
G1ZVZ PE24 4AE
G8UNU PE24 4HP
M0ADU PE24 5BB
G0LII PE24 5BB
G6HM PE24 5EN
G0OHF PE24 5HQ
G4LOP PE24 5HT
G7BNN PE24 5HT
G7BNO PE24 5HT
2E1CRX PE24 5HT
G4HAI PE24 5HZ
G0OHN PE24 5JD
2E1FSE PE24 5LZ
G1VWM PE24 5NY
G8SFU PE24 5PU
G1DZF PE24 5SL
G1LLZ PE24 5SQ
G1PLA PE25 1EL
2E1FRR PE25 1HQ
G3NPY PE25 1SE
G4ZQT PE25 1TQ
G1LPC PE25 2AX
G0EPE PE25 2DD
G4XQT PE25 2DN
G6MAE PE25 2ED
G0CHB PE25 2ET
G4XTX PE25 2EW
G6TNH PE25 2JH
G7UVD PE25 2JR
G0SGS PE25 2QS
G4SOR PE25 2QZ
G4MVZ PE25 3ER
G0OTH PE25 3EZ
G0LAG PE25 3QT
G3THX PE25 3RE
G3XZY PE25 3RX
G3OTD PE25 3RZ

East Lothian

GM7VLC EH21 6AR
GM7PPN EH21 6BA
2M1DZW EH21 6BA
GM3DUM EH21 6DF
GM0VYL EH21 6EX
GM6KFO EH21 6SB
G1KFO EH21 7QH
GM0EFQ EH21 7RD
MM1BRT EH21 8AH
GM1FAF EH21 8JT
GM4GVJ EH21 8JT
GM0FZM EH21 8NS
GM0EHL EH31 2DG
GM0CBX EH32 0AN
GM4LRU EH32 0AN
GM4INE EH32 0BB
GM4UYZ EH32 0EE
GM4IKU EH32 0EY
MM0AMV EH32 0HF
GM0WRY EH32 0JT
GM8ZJS EH32 0LQ
GM4KGZ EH32 0NX
2M1DZS EH32 0PG
2M1EUV EH32 0PG
GM0PPE EH32 0SN
GM4TAL EH32 0TA
GM4IKT EH32 0TN
GM4JPG EH32 0TU
GM0CLN EH32 0TY
GM7PSH EH32 9BN
GM3NEX EH32 9JX
GM1RCP EH33 1PF
GM0NTL EH33 2AL
2M1FMX EH33 2AL
GM7TKA EH33 2AR
GM8ZLI EH33 2DN
MM1AVA EH33 2EB
2M1ETM EH33 2EE
MM0ANT EH33 2EN
GM8CFS EH35 5NG
GM4ITH EH39 4HH
GM3AWF EH39 4PW
GM0WJY EH39 4RF
GM3BQA EH39 5NR
GM4HJQ EH40 3BH
GM8KOF EH41 3PJ
GM1ZTA EH41 3QN
MM1BVW EH41 3QP
GM3VEI EH41 3SB
MM1ATY EH41 4RH
GM4FQG EH42 1AY
GM0JPG EH42 1BA
GM0WPW EH42 1RH
GM1SYC EH42 1RT
GM6KEV EH42 1YA

East Northamptonshire

G6VFC NN10 0DJ
G7EQU NN10 0DS
G4LJG NN10 0DT
G7PYQ NN10 0DX
G7CGW NN10 0LY
G7NBG NN10 0NF
G1YWM NN10 0QZ
G8LII NN10 0SN
G4FEV NN10 0SW
G3WDG NN10 0SY
G4KGC NN10 0SY
2E1ESL NN10 0TN
G7TZD NN10 6BA
G3OMS NN10 6BG
G7ESK NN10 6RY
G4GGK NN10 6UT
G0MOR NN10 6XX
G4OMH NN10 6YZ
G4NCA NN10 8EJ
G4ZRM NN10 8HU
G6NHO NN10 8HU
G4WFT NN10 8JP
G4GCQ NN10 9ER
G6SDI NN10 9EZ
G7LNP NN10 9HF
G4ESC NN10 9HH
G1OOS NN10 9HL
G7OCC NN10 9NS
G0CPW NN10 9SQ
G6PUE NN10 9YJ
M1CAG NN14 3JT
G3ICK NN14 3JY
G4HLE NN14 4DD
G4HLW NN14 4DD
G6LJH NN14 4EA
G6LSB NN14 4EG
G6MBG NN14 4PR
G3UQR NN9 5RB
G4PPW NN9 5SY
G6ODT NN9 5TA
G7EWL NN9 5TN
G3NXS NN9 5TY
G6DYU NN9 5TY
G6NFZ NN9 5TY
G0TES NN9 5UB
G3NJF NN9 5US
G7OLO NN9 5UT
G1TLC NN9 6AQ
M0ABE NN9 6AW
G0WTO NN9 6AW
G7CMI NN9 6AW
G1HGY NN9 6BQ
G1JPC NN9 6DQ
G0JXQ NN9 6HE
G0MEO NN9 6HE
G4LAM NN9 6LX
G1KID NN9 6QR
G7HGY NN9 6SX
G4JPB PE8 4DQ
G2CH PE8 4JQ
G3VZL PE8 4JQ
G8YDI PE8 4LN
G6UAM PE8 4QG
G7NDQ PE8 5EN
G3XFA PE8 5PS
G0UIH PE8 5PS
G4CEY PE8 6TT
G8MJT PE8 6YJ
G7DGP PE9 3LZ
G4OZM PE9 3NA
G0IET PE9 3NB
G3MFG PE9 3PW

East Renfrewshire

GM0GDH G44 3QU
GM4JTA G46 6BZ
GM3NGW G46 6DB
GM6FIK G46 6LA
GM0MAC G46 6LW
GM4RIN G46 6TQ
GM0OPS G46 7DP
GM0WRH G46 7LU
GM3BXW G46 7QE
GM4XOI G76 0DA
GM0VIY G76 0EU
GM0PHG G76 0HA
GM4VBE G76 7DU
GM7NOA G76 7HG
GM0SEP G76 7HG
GM4SRL G76 7HG
GM0UIG G76 7HG
GM7GXI G76 7HG
GM8MRW G76 7LL
GM6FT G76 7QW
GM3NEC G76 7TG
GM7AXQ G76 7XA
GM0IVQ G76 7XT
GM6FOT G76 7XT
GM0PHM G76 7XX

GM0DJN G76 8NU
2M1EDM G77 5AX
GM3AWW G77 5DP
GM0RAY G77 5DS
GM0IBM G77 5DS
GM1NET G77 5ND
GM4JMU G77 5NF
GM1NEW G77 5PP
GM3LKY G77 5QD
GM7NNH G77 5QJ
GM7SKB G77 5QJ
GM7OKA G77 5QN
GM4KLO G77 5TQ
GM3GNE G77 6EA
GM3NEQ G77 6HP
GM4YMA G77 6JE
GM0NBA G77 6JU
GM0TEX G77 6LP
GM7UTD G77 6LP
GM6OQN G77 6LQ
GM4ZWJ G77 6PB
GM3NIG G77 6RT
GM0UKZ G77 6UJ
GM7VET G77 6XE
GM3KXQ G77 6XX
MM1ANP G77 6YU
2M1BCA G78 1TT
GM7WGM G78 1TT
GM0GMI G78 2DH
GM1TFZ G78 2DH
GM3CIX G78 2HT
GM0JNB G78 3DE
GM3AEI G78 3JX
GM7NEW G78 3JX
GM4LCP G78 3LP
GM4OSV G78 3QP

East Riding of Yorkshire

G4GNP DN14 5NY
G0WRF DN14 5RZ
G8YYW DN14 5TX
G4CQN DN14 5XR
G0FBR DN14 6AS
G0OLE DN14 6AS
G0PQX DN14 6DH
G0GLZ DN14 6DU
G7TCY DN14 6LX
G8VHL DN14 6QW
G8ZCS DN14 6RF
G0WJQ DN14 6XD
G4NLG DN14 7BH
G4JFH DN14 7BU
G4DBN DN14 7HE
G0FRQ DN14 7HH
G4OUS DN14 7JA
G1RFC DN14 7JL
G3ZGT DN14 7NA
G4GLL DN14 7NY
G3VDE DN14 7QQ
G0FRX DN14 8ET
G8HSG DN14 8ET
G4BDX DN14 8EX
G0IOV DN14 8LA
G0MUX DN14 8PB
G6PXJ DN14 8RP
2E1BGV DN14 9HB
2E1DBZ DN14 9HB
G0NVD DN17 4RZ
G3HJC HU10 6AE
G8DCD HU10 6AJ
G1RVH HU10 6AR
G3TEU HU10 6AW
G0KLG HU10 6BJ
G4LNR HU10 6HS
G7BAC HU10 6HS
G4RXI HU10 6LP
G4NJB HU10 6NG
G7GNM HU10 6PA
G1OSD HU10 6QD
G0IRU HU10 6QR
G1ALC HU10 6QZ
G4AUQ HU10 6QZ
G0UGA HU10 6SF
M1BNA HU10 6SG
G4BYG HU10 6ST
G0VQO HU10 6ST
G0DBN HU10 6UG
G4CGG HU10 7DT
M1ACR HU10 7HT
G0AWP HU10 7HX
G4AQA HU10 7JE
G8JIU HU10 7JJ
G8NIE HU10 7JU
G2CGL HU10 7NE
G3YCC HU10 7PJ
G3ZAR HU10 7PJ
G7PER HU10 7UX
G4FEJ HU11 4DS
G5PQ HU11 4DS
M1BLR HU11 4EN
G4CMT HU11 4EN
G0VRM HU11 4EN
G7CVA HU11 4ER
G0AJX HU11 4HB
G4GZV HU11 4LA
G0HKD HU11 4NE
G1SRJ HU11 4NR
G7JZY HU11 4QE
G6JFU HU11 4QH
M1BLH HU11 4RA
G4LZJ HU11 4RN
G1UCZ HU11 4RY
G0NAA HU11 4SD
G4SKH HU11 4XE
G4YTV HU11 5BH
G6EKT HU11 5BH
G4ZJC HU12 0DG
G0THS HU12 0DY
G4PMJ HU12 0HH
G3DXQ HU12 0LE
G3ULD HU12 0NE
G7DKX HU12 0QN
G3OHT HU12 0RG
G3UIF HU12 0RY
G6AWP HU12 0TE
G7WDM HU12 8LB
G0VAY HU12 8LH
G0NYK HU12 8LX
G7SCT HU12 8NH
G6AXC HU12 8PU
G0VGH HU12 8QB
G7MFO HU12 8UB
G4LHF HU12 9HP
G6BBE HU12 9JS
G3YBU HU12 9NG
G4NXE HU12 9PT
G6FNG HU12 9PT
G1EQU HU12 9QG
G7USB HU13 0AS
G2FS HU13 0BG
G0PZH HU13 0DU
G8FEK HU13 0JN
G0REI HU13 0NN
G4JIO HU13 0RL
G8KKZ HU13 0RN
G0KWE HU13 0RT
G6MPK HU13 9AU
M0AXU HU13 9AX
G6ZTI HU13 9AX
G0UMA HU13 9BN
G0UME HU13 9BN
G0CNV HU13 9BP
G7SAT HU13 9DA
G7MAO HU13 9DB
G1EIB HU13 9DD
G0OYX HU13 9HU
G4VTP HU14 3AU
G3KOG HU14 3BY
G7VRL HU14 3DG
G3YNO HU14 3DW
G3YQA HU14 3DW
G1REL HU14 3DX
G3RSB HU14 3JS
G3SQH HU14 3JT
G7UJY HU14 3PR
G4XWA HU14 3RH
G0SWL HU15 1BE
G6KCE HU15 1BU
G7CLY HU15 1DA
G4NBX HU15 1EH
G6KIA HU15 1EN
G4MJT HU15 1HR
G4FGO HU15 1HY
G0KHZ HU15 1JT
G4MTS HU15 1JU
G8FDR HU15 1NL
G0VXF HU15 1NU
G0GDS HU15 1QL
G4SGU HU15 2AL
G4UVF HU15 2AL
G1SXJ HU15 2HA
M0ARC HU15 2HA
G3LZQ HU15 2HA
G0CZV HU15 2HD
G3VLR HU15 2JG
G0WJK HU15 2LT
G4AON HU15 2QJ
G0LHV HU15 2QN
G4YEB HU15 2TY
G0JNQ HU15 2XU
G0UFZ HU16 4AS
G0MXI HU16 4BH
G1EEQ HU16 4EY
G0PAS HU16 4HN
G4KHT HU16 4HN
G8PSE HU16 4JB
G0KBP HU16 4NF
G4VKK HU16 4RU
G3WWD HU16 4SA
G4EBT HU16 4SD
G3NOP HU16 5JG
G7OWF HU16 5JQ
G3JXG HU16 5LA
G8BQF HU16 5LQ
G1KJY HU16 5ND
G1EFO HU16 5QJ
G4GIY HU16 5RL
G0SXW HU16 5TF
G8EAH HU16 5TR
G1RVS HU16 5UE
2E1BDC HU17 0AJ
G0CFA HU17 0AN
G6XPS HU17 0AN
G7TYT HU17 0BG
G4UOZ HU17 0DW
G7JZD HU17 0EP
G4HBL HU17 0RU
G4DJL HU17 0TH
G4EIM HU17 0TH
G4VNC HU17 5LD
G3YHR HU17 5LG
G7VIZ HU17 5LG
G4SMB HU17 5NE
G8TVT HU17 5NE
G6AXW HU17 5NG
G7EFV HU17 5NR
G3RMX HU17 5NU
G6UGS HU17 7HD
G3YAA HU17 7HE
G4GPY HU17 7HP
G3ZRS HU17 7LU
G0RUF HU17 7NX
G0TAS HU17 7NX
G4EJP HU17 7NY
G4MXS HU17 7RJ
G7SBC HU17 7RJ
G8SGP HU17 8HA
G4ESY HU17 8JL
G7UXF HU17 8JL
G0DMP HU17 8PF
G4CMK HU17 8PJ
G4DEA HU17 8QA
G3TTW HU17 8QL
G3MJM HU17 9GG
G0KVR HU17 9LH
G1NBJ HU17 9PA
G6EUF HU17 9PG
G0LUP HU17 9RH
G1AJD HU17 9RH
G1AJE HU17 9RH
G8UAI HU17 9RH
G0ENO HU17 9RW
G0GBC HU17 9RW
G4DFH HU17 9RW
G0PBK HU17 9UT
G6GHE HU18 1AL
G6VZB HU18 1BX
G4IGY HU18 1DZ
G4FCT HU18 1EF
G8PZX HU18 1EF
M1AKQ HU18 1EF
G6DKF HU18 1ES
G0VET HU18 1ES
G4ADE HU18 1HE
G3XEE HU18 1HN
G0PFD HU18 1HZ
G0IVP HU18 1JG
2E0AHT HU18 1JQ
G7KHV HU18 1JU
G0OZT HU18 1LZ
G0JWY HU18 1RE
G8KFK HU18 1TP
G0GXX HU18 1TT
G8LJQ HU18 1UW
G3TLI HU18 1XX
G4CHH HU18 1XX
G4EKT HU18 1XX
G7CVX HU19 2AY
G6ZBT HU19 2DB
G3HRP HU19 2EP
G1OSG HU19 2EW
G0LVO HU19 2HH
G5ZN HU19 2LZ
G3RJM HU19 2QD
G4KFA HU19 2QD
G7VAR HU19 2QD
G4CRU HU20 3UU
G8HUG HU20 3UX
G0BEC HU20 3XA
G3LIQ HU4 7NR
G0UBY HU4 7NR
G7JLE HU4 7QB
G6VVZ HU4 7QH
G0IFJ HU4 7RG
G8EQZ HU4 7SW
G7NRX HU4 7SW
G0SCQ HU5 4QD
G4COT HU5 5YY
G0PST HU6 8TG
G4USH HU7 5XD
G4UWE HU7 5XD
G0UDD HU7 5XD
G3AGX HU7 5XU
G4DSX HU7 5XW
2E1EWZ HU7 5XW
G3UEN YO15 1AE
G0IAD YO15 1AP
G1NUP YO15 1BE
G4DUP YO15 1EH
G5VO YO15 1HP
G3MYZ YO15 1JJ
G3TVW YO15 1LJ
G4XBU YO15 1LJ
G0VHD YO15 1LJ
G8PWU YO15 1LW
M0ABP YO15 1LY
G0TZI YO15 1NN
G0ETG YO15 1PW
G0VEX YO15 2DS
G7VAY YO15 2NA
G7LYX YO15 2TQ
2E1DQV YO15 3DU
G1SEG YO15 3NJ
G3EZZ YO15 3NJ
G0RMP YO15 3NP
M1BKZ YO16 4AZ
G4LOB YO16 4HN
G6JEF YO16 4JD
G0DEB YO16 4NL
M1AEK YO16 4TP
G3MCF YO16 5DB
G4KHR YO16 5DB
G7TGN YO16 5DZ
G8NDK YO16 5FB
G8ESL YO16 5FW
G2BXZ YO16 5GH
G3ZTR YO16 5GW
G4KNR YO16 5HL
M1BLW YO16 5NH
G1BTG YO16 5NL
G6WNB YO16 5PZ
G3SWW YO16 5SA
G3POC YO16 5SE
M0AHV YO16 5TB
G2FXL YO16 5TQ
G4TKY YO16 5TW
G3PWN YO16 5UA
G8CTA YO16 5UG
M0AQU YO16 5UJ
G7OHD YO16 5YP
G3XYF YO25 0DE
2E0ADL YO25 0DE
G1ILJ YO25 0ED
G4ZNZ YO25 0JW
G4YTD YO25 0LY
G0AWC YO25 0QD
G8MVJ YO25 0TD
G7ONF YO25 0TN
2E0AQF YO25 0YN
2E1FBR YO25 0YN
G7FSH YO25 7AT
G1PJZ YO25 7AY
G3GAW YO25 7BE
G6FMS YO25 7BG
M1BKW YO25 7DJ
G6TNZ YO25 7EN
G1IBF YO25 7EX
G6CPF YO25 7HX
G7TXF YO25 7JW
G0LYZ YO25 7NB
G4BRB YO25 7QW
G4DZT YO25 7RZ
G0RUK YO25 7SG
G1IKT YO25 8DX
2E1DMS YO25 8DX
G4PAA YO25 8HU
G1IMI YO25 8JJ
G6BUV YO25 8LS
G4ZKB YO25 9LP
G7BQX YO25 9LP
G4SBI YO25 9QY
G6ENR YO25 9SQ
G3ZOU YO25 9UL
G1VWZ YO25 9XL
G0WGO YO25 9XT
G3TLF YO25 9YZ
G4XHX YO4 1DP
G4XHY YO4 1DP
G4OMD YO4 1DW
2E1BRT YO4 1EZ
G6GAV YO4 1JJ
M0AYL YO4 1PR
G0FYU YO4 2BX
G4KCF YO4 2BX
G4BBZ YO4 2HB
G1SLF YO4 2QG
G1JMY YO4 2YJ
G3YRU YO4 2YQ
G7FKP YO4 3ET
G3ASJ YO4 3NT
G8XFK YO4 3SF
M1AEQ YO4 4BG
G0EAT YO4 4EX
G0WHS YO4 5HX
G4XBJ YO4 5PL
G2HOX YO4 5PS

East Staffordshire

G6NSQ DE13 0AL
G0SUZ DE13 0AY
G8OZP DE13 0AY
G6ZJJ DE13 0BG
G4YFZ DE13 0DQ
G4GLY DE13 0DU
G4SRB DE13 0DU
G6DRO DE13 0FX
G4RAC DE13 0JD
G1ATA DE13 0JP
G1OXD DE13 0LD
G1OZD DE13 0LD
G1OYZ DE13 0LD
M1CBO DE13 0LR
2E1ELL DE13 0LR
2E1ELM DE13 0LR
M1CDT DE13 0ND
2E1EJX DE13 0NP
G4XKL DE13 0NU
G4IPV DE13 0NY
G0IYC DE13 0PN
G6ZWF DE13 0QX
G4IMJ DE13 0QZ
G4HBY DE13 0RF
G3NFC DE13 0RF
G4VPF DE13 0UU
G4FHO DE13 8AB
G7URX DE13 8HD
G7VSM DE13 8HD
G7UTL DE13 8HU
G4CHI DE13 8NT
G4ICZ DE13 8PT
G7OBY DE13 8SR
G6EIH DE13 9AA
G8VBA DE13 9AB
G0FSF DE13 9AF
G4BPW DE13 9AJ
G0JLU DE13 9AR
G0GZL DE13 9AT
G0FNH DE13 9BJ
G1OYU DE13 9BJ
G8WJQ DE13 9BY
G4SXE DE13 9DE
G0SHO DE13 9EG
G1JKN DE13 9HZ
G0AOD DE13 9JR
G3BZB DE13 9LE
2E1FZK DE13 9LE
2E1FZL DE13 9LE
G3TOY DE13 9RT
G4RWD DE13 9SH
G7PUC DE13 9SX
G6TJK DE14 1BS
G3JWC DE14 2ED
G0SUT DE14 2EE
G6TOC DE14 2JS
G8WSY DE14 2NB
G0NHA DE14 2QR
G1YAN DE14 2QR
2E1FLC DE14 2TA
G0JEE DE14 3DR
G4DIH DE15 0AB
G6OVA DE15 0AR
G4PGX DE15 0AX
G0HIO DE15 0DP
2E0ACU DE15 0DX
2E1AJO DE15 0DX
G0RZA DE15 0EY
G6ZNF DE15 0LA
G7EIK DE15 0LA
G1PRW DE15 0LU
G1IXJ DE15 0QJ
G4PGJ DE15 0QJ
G7TGP DE15 0TR
G6JPQ DE15 0TT
G3MBJ DE15 0TZ
G6BMZ DE15 9AA
G3BSL DE15 9AB
G7IZC DE15 9BJ
M0AGS DE15 9BJ
G1TEW DE15 9BU
G6GNH DE15 9DQ
G3CWT DE15 9EY
G2HNU DE15 9EZ
G6FLY DE15 9LG
G3ACR DE15 9LX
G0IDD DE15 9PT
G4IAW DE15 9QP
G4RJA DE15 9RB
G7KRM DE6 2LP
G4NXN DE6 5HA
G0UUA ST10 4PE
G7MGA ST10 4PT
G3VSB ST14 5DH
G0LJH ST14 5LT
G6VGC ST14 7DN
2E1CKK ST14 7DN
G4EVW ST14 7DZ
2E1BMV ST14 7EN
2E1CLM ST14 7ET
G6SVJ ST14 7LB
G7STC ST14 7NB
2E1ESR ST14 7NB
G7UCN ST14 7NF
G7EJK ST14 7NU
G4TWN ST14 7PN
G6CUY ST14 7QG
G6UOX ST14 7QY
G0CAP ST14 8BT
G3VKH ST14 8JP
G3XJY ST14 8PU
G7LGS ST14 8QH
G8FGQ ST14 8QS
G8YAT ST14 8SE
G4FMO WS15 3EF

Eastbourne

G3TJR BN20 7DN
G4RHS BN20 7EN
G6TIQ BN20 7EU
2E1FCK BN20 7JT
G4ABI BN20 7JW
G1YTR BN20 7QQ
G8GF BN20 7QQ
G4SAN BN20 7TZ
M1BAO BN20 8AE
G6JME BN20 8DA
G4MHK BN20 8HY
G6LPC BN20 8HZ
G0EDS BN20 8LP
G4YJW BN20 8LP
G7ULN BN20 8LP
G4MVN BN20 8PB
G1EFY BN20 8PH
G0DAX BN20 8RL
2E1DDS BN20 8SN
G3CPS BN20 8ST
G1AKV BN20 8UG
2E1DDW BN20 9AN
G1WKH BN20 9BS
G7CJO BN20 9BS
2E1FYA BN20 9BY
2E1BKD BN20 9DY
2E1EFX BN20 9DY
G8BEB BN21 1QE
G8NFZ BN21 1QZ
G0MFS BN21 1SH
2E1DOM BN21 1SN
G6GVM BN21 1UJ
G6ZNO BN21 1UJ
G3LON BN21 1XF
G0EQM BN21 2ED
G4WLV BN21 2HR
G3MHF BN21 2LE
G1SSS BN21 2LQ
G8LQZ BN21 2LQ
G4XNL BN21 2PG
2E1FYC BN21 2QA
2E1FXZ BN21 2UJ
G8BUJ BN21 3AU
G3PUC BN21 3BL
G7PMY BN21 3TJ
G1KPU BN21 3XF
G8NQF BN21 4PA
G0JPY BN22 0AJ
G0TYJ BN22 0ES
G0KUQ BN22 0TL
G0NUG BN22 0TL
G3OSP BN22 0UH
M0ABX BN22 0UQ
G1MNX BN22 0UZ
G7UWG BN22 0XH
G3WXB BN22 7AQ
G0MGA BN22 7HA
G6GVU BN22 7JB
G8NFO BN22 7JS
G8DXU BN22 7NZ
G4KYJ BN22 8AP
G0EBF BN22 8EH
G7PGH BN22 8EH
G3SJV BN22 8NA
G0EEL BN22 8PJ
G7MJI BN22 8RS
G8HGM BN22 8UH
G4SHM BN22 8UQ
M0AWJ BN22 8UU
G7SUV BN22 8UW
G4ZWE BN22 9EX
G7INJ BN22 9LB
G7SZA BN22 9LH
G4YGJ BN22 9NA
G4RUL BN22 9QG
G7JVO BN22 9RH
2E1FHD BN22 9RN
2E1FZE BN22 9RN
G0UAI BN22 9SF
G0OZG BN23 6AB
G4EKS BN23 6AF
G6MAR BN23 6BS
G1SGR BN23 6HY
G6TFJ BN23 6JR
G0EYE BN23 6LN
G1UCD BN23 7AY
G6GVL BN23 7BE
G1LEO BN23 7BH
G6RGA BN23 7EN
G4XXM BN23 7LW
G4PRJ BN23 7PF
G8XCY BN23 7PF
G4ZQS BN23 7QT
G7HFS BN23 7SA
G0KAR BN23 7SD
M1APT BN23 8AL
G4TZH BN23 8BU
G7BSO BN23 8BU
G7AFZ BN23 8DE
G1FBH BN23 8EQ

Eastleigh

G7OOF SO18 3BY
G8ZAX SO18 3PW
G7RAB SO19 6HB
G6WYS SO30 0BP
G8DXK SO30 0JX
G6IKW SO30 0LH
G0TZE SO30 0PE
G8IMR SO30 0PH
G4JNT SO30 0PH
G7UBD SO30 0PH
M0AZD SO30 0QA
G6FRT SO30 2DZ
G4CBJ SO30 2FR
G1FLY SO30 2LB
G3YOM SO30 2NY
G4NHN SO30 2RE
G0UVB SO30 2RF
G0WXS SO30 2SR
G0JFD SO30 2UW
G4NFB SO30 2XJ
G1ETX SO30 3BB
G3VBW SO30 3BB
G3IBY SO30 3EQ
G1APD SO30 3ES
2E1BPB SO30 3FL
G6HNJ SO30 4BA
G7SEZ SO30 4BA
G8RKG SO30 4BH
G3ONG SO30 4EB
G8RWE SO30 4RP
G4GTW SO30 4TP
G3PDP SO31 1AP
G0ERI SO31 4HH
G3KXW SO31 4LX
G0THB SO31 4NZ
G0MLJ SO31 4QU
G6IXE SO31 4RB
G4YWZ SO31 4RQ
G8TEC SO31 5FB
G0WSB SO31 5FE
G4REB SO31 5FQ
G0HAE SO31 5GP
G4HSI SO31 5PT
G4VNK SO31 8AF
G6XQR SO31 8EN
G0LAW SO31 8FN
G3EMF SO31 8FN
G8UAY SO32 2DE
G8YFK SO50 4NH
G0TBC SO50 4PP
G0OFX SO50 4PQ
G7UYZ SO50 4QX
G3VHW SO50 4QY
G4KUL SO50 4RH
G3NML SO50 4RW
G4BGT SO50 5AA
G6JGR SO50 5AA
G1GVM SO50 5BE
G0JFM SO50 5GT
G3NJB SO50 5GT
G4JHC SO50 5HA
G0NJO SO50 5JS
G1TJH SO50 6BD
G0MNK SO50 6BG
G3VRB SO50 6DG
G0WKP SO50 6GT
G8VKS SO50 6NU
G0DXJ SO50 7DH
G3GII SO50 7DN
G3HSS SO50 7DT
G0URA SO50 7GZ
G0VBL SO50 7GZ
GJ3UQM SO50 8DA
G1DYB SO50 8EN
G3XIV SO50 8FL
G8VQA SO50 8PH
G1ETW SO50 8QS
G0TSM SO50 9LD
G4DGY SO53 1EW
G4DZS SO53 1GD
G3BRR SO53 1LE
G7UDM SO53 1LE
G0PRZ SO53 1LN
G0UKB SO53 1LN
M0ACL SO53 1LN
2E1EMS SO53 1LN
G1UDR SO53 1LW
G2HAJ SO53 1PZ
G3WEG SO53 1TW
G2HKW SO53 2BS
G3RWG SO53 2EU
G8ZBN SO53 2FE
M0AYU SO53 2FY
G3IKA SO53 2GG
G7AJE SO53 2GX
G3JRH SO53 2JH
G1THA SO53 2JS
G7BFX SO53 2JZ
G3UYD SO53 2LJ
G6XJZ SO53 2PU
G0BXI SO53 3BN
G4VXI SO53 3BN
G6AMW SO53 3GJ
G7VQV SO53 3GZ
G0TAA SO53 3HW
G6DLJ SO53 4BY
G8CEC SO53 4EA
G4HXT SO53 4TP
G8EJP SO53 5AA
2E1ELI SO53 5AJ
G4HHZ SO53 5AX
G3WSL SO53 5BN
2E1ECV SO53 5BX
2E1EVV SO53 5BX
2E1EEK SO53 5DA
G4ASM SO53 5DT
G4XBZ SO53 5NZ
2E1EVX SO53 5QH
G4MZU SO53 5RJ
G1NPI SO53 5RP
G4NZC SO53 5RP
G8IPQ SO53 5RP
2E1EUI SO53 5RT

Eden

G0MDV CA10 1AJ
G7UNZ CA10 1AJ
G4WKG CA10 1DT
G0IYQ CA10 1PD
G7TES CA10 1PJ
G8VUK CA10 1QT
G3ULY CA10 1QW
G4XET CA10 1TA
G4UWG CA10 2AP
G0TNF CA10 2BW
G0VRZ CA10 2LL
2E0AOK CA10 2RF
G4XTA CA10 2RF
G0VMP CA10 3PF
G6NFU CA10 3SG
G4HYJ CA11 0AY
G4LFU CA11 0BQ
G7BGO CA11 0JH
G7JKC CA11 0TU
G0AUG CA11 0XB
G0AUH CA11 0XB
G4SGH CA11 7DT
G4GQZ CA11 7TG
G0ANT CA11 7UW
G0TDM CA11 7UW
G7GQM CA11 7UW
G3GRX CA11 8AW
G0OFR CA11 8AX
G4GLQ CA11 8EH
G6BSK CA11 8LY
G7WLI CA11 8SQ
G0OYI CA11 8TS
G4WHA CA11 8TU
G3TUK CA11 8TX
G4FUI CA11 8UF
G4WSO CA11 9HA
G4WSP CA11 9HA
G6XWD CA11 9SS
G3ZSK CA11 9SY
G4YLI CA11 9TR
G0KDB CA16 6BD
G7FEO CA16 6BD
G3INI CA16 6DA
G7ITT CA16 6DA
G0TFC CA16 6DT
G3JYP CA16 6HW
G0NYQ CA16 6RD
G7NRG CA16 6TU
G7PLQ CA16 6XT
G0VGJ CA16 6XT
G3FM CA17 4BZ
G3VAL CA17 4HE
G4SPR CA17 4HR
G3GXX CA17 4NG
G4TUA CA17 4NQ
G6ZPJ CA4 0LB
G6EGM CA4 0RZ
G8NEU CA4 9DH
G3WTO CA7 8JA
G0NXS CA9 3LZ

Ellesmere Port and Neston

G6SLN L64 0SG
G1OVY L64 0TB
G3XJZ L64 0TG
G0ICH L64 0TX
G7WGC L64 0UH
G6AFB L64 0UY
G1DEV L64 0XP
G1NUH L64 0XP
G0IZT L64 1SE
G4JZR L64 1TN
G4WLI L64 2XL
G8OKZ L64 2XQ
G1PJL L64 3SQ
G7CEU L64 4AN
G8OJQ L64 4AR
G4HGL L64 4AT
G3NTI L64 4BJ
G3PXX L64 4DZ
G0IHF L64 4EH
G4DHC L64 6QW
G0HAB L64 6SE
G4ELK L64 6SF
G8VOH L64 6TS
G7FVR L64 6UA
G6ADO L64 7TR
G8MMM L64 7TR
2E1CES L64 7TR
G0EXE L64 9QJ
G3GMT L64 9RX
G0EXD L64 9SF
G7DEY L64 9SW
G1AOF L65 2BE
G3TXH L65 2DT
G4RPM L65 3BD
G0HFE L65 5DX
G7JKH L65 6PB
G7VCX L65 6PS
G4JMF L65 6RA
G3YCO L65 6SB
G6ALN L65 6TD
G0ISE L65 7AH
G4JTK L65 7AQ
G0OIP L65 7AY
G1URQ L65 7AZ
G8AWS L65 7EF
G3ABM L65 7EG
G0WVW L65 8DQ
M0AMS L65 8HZ
G7VHR L65 8HZ
G0CTR L65 9EN
G1PRL L66 1JF
G3VZM L66 1JJ
M1BUQ L66 1JJ
2E1EEF L66 1JJ
G0VHJ L66 1JT
G0JZJ L66 1JW
G7CGO L66 1PU
G0CGD L66 1QS
M1AOX L66 1QS
G6ISX L66 1QT
G0AKU L66 1RW
G4DLW L66 2GA
G0JCG L66 2GY
M0AOO L66 2JG
G0HTP L66 2JJ
G0GCM L66 2LF
G0RCW L66 2LH
G0DUB L66 2LH
G1GWS L66 2LH
G8RWS L66 2NY
G4YTY L66 2PA
G3EFZ L66 2PD
G4DBG L66 2RP
G4WUF L66 2RY
M0BAU L66 2SE
G0IEQ L66 2SJ
G4ZKG L66 2SX
G4STZ L66 2TD
G0ULP L66 2TN
G3CSA L66 2YJ
G4WSE L66 2YJ
G0COJ L66 3LL
G4MRJ L66 3NQ
G7IFL L66 3PD
G7GQW L66 3PF
G4UXL L66 3PH
G4KUV L66 3PW
G3CPD L66 3PZ
G0PJY L66 3QH
G7NHV L66 4NA
G0RCY L66 4PD
G3KDN L66 4QJ
G4HNB L66 4RZ
G8IPT L66 4SG
G2AUI L66 4TS
G0HZZ L66 5PD

Elmbridge

G4IYB KT10 0AL
G0DAS KT10 0AQ
G1LZM KT10 0AX
G0EYT KT10 0AZ
G1YWF KT10 0BD
G4IUA KT10 0DT
G3ZUJ KT10 0EB
G1SEU KT10 0EL
G1SCN KT10 0ET
G3RIM KT10 0HS
G1PQT KT10 0JZ
G7WHD KT10 0LU
G4GPB KT10 0QE
G8YXZ KT10 0RZ
G0HHG KT10 8DN
G4ZYY KT10 8DN
G0IDB KT10 9HF
G6LHQ KT11 1AU
G1PCD KT11 1DF
G0UVH KT11 1EL
G0CAN KT11 2AF
G1LKH KT11 2AQ
G0OAS KT11 2BB
G3LNF KT11 2EG
G4SVD KT11 2EZ
G8ETP KT11 2JQ
G3ZNI KT11 2QR
G4FRN KT11 2RL
G8ORX KT11 2RU
G1DDY KT11 2SR
G4SET KT11 2SR
G4YKH KT11 2SX
G1UNB KT11 3HR
2E1COQ KT12 1EW
G3ZTM KT12 1JL
G3LFX KT12 1LE
G8GTR KT12 1NF
G4ARO KT12 2HY
G1JKV KT12 2LD
G3EAO KT12 2LY
G3XHK KT12 2NA
G4HGG KT12 2PP
G1OCY KT12 3BA
G0UHQ KT12 3BA
G1ASW KT12 3BZ
G4AWZ KT12 3DA
G8DXP KT12 3EQ
G8CGT KT12 3HE
G6INX KT12 3SQ
G0FIF KT12 4ED
G6TDR KT12 4EL
G0KEY KT12 4HE
G7KHC KT12 4NS
G4GGI KT12 5BB
G7MFE KT12 5HD
G1KEV KT12 5HF
G1JSS KT12 5HJ
G7DUK KT12 5JT
G6TCK KT12 5QT
G1HTN KT13 0AB
M0AOP KT13 0AJ
G3TXF KT13 0BS
G4VWN KT13 0BS
G0CKV KT13 0JW
G1DLO KT13 0NR
G8HCL KT13 0NR
G4CPM KT13 8AS
G8WHN KT13 8NY
G4RCD KT13 8PN
G6CUV KT13 8PU
G4LQD KT13 8XB
G7OFB KT13 9BD
GW0OJK KT13 9DX
G8VL KT13 9LR
G3ATX KT13 9LT
GW1ZXN KT13 9LW
G0JRH KT13 9RU
G1BHS KT22 0HR
G8UYL KT22 0HR
2E1CVA KT22 0LW
G3BXC KT22 0LZ
G8MOB KT6 5HB
G3JVC KT6 5JD
G8BIQ KT6 5JE
G1SKR KT6 5JN
G7AXM KT6 5QD
G7VSC KT7 0BG
G8MSR KT7 0HF
G7WEP KT7 0YH
G4CPV KT7 0YN
G0NIB KT8 0BL
G3EJC KT8 0JY
G1OEQ KT8 0NN
G6REV KT8 1QE
G3OJX KT8 1RR
G3JGH KT8 1TF
G4AZK KT8 1TN
G1DZB KT8 2ET
G4YKK KT8 2ET
G8CLW KT8 2EX
G0KDI KT8 2HY
G4IRR KT8 2JA
G1KDH KT8 2JB
G6HVQ KT8 2NB
G6BBM KT8 2NW
G4YKF KT8 2PW
G1YVV KT8 2PY
G4GD KT8 9EW
G8WDX KT8 9HE

Enfield

G8NST EN1 1BL
G3WIB EN1 1RL
G8XTR EN1 1XT
G7FIM EN1 1YE
G4OBE EN1 2BZ
2E1AQS EN1 2BZ
G3YJZ EN1 2JU
G7RIM EN1 2PH
G0VAS EN1 2QY
G0NEU EN1 3AE
G7LBT EN1 3DB
G1GTS EN1 3DE
G7UID EN1 3DE
G4RTC EN1 3DW
G6DRF EN1 3HP
G0VPK EN1 3HQ
G4CPT EN1 3JB
G3RWL EN1 3NQ
G8PSF EN1 3NT
G8TWR EN1 3NU
G0DDS EN1 3SF
2E1BRJ EN1 3UA
G8SZR EN1 3UH
G3CPH EN1 3UR
G0NLG EN1 3UT
G2HR EN1 3UT
G3SRA EN1 3UT
G8CSA EN1 3UT
G0NQV EN1 3UU
G7PPK EN1 4BA
G8BCQ EN1 4BD
G4BUB EN1 4HR
G4NLO EN1 4QT
G6AHC EN1 4TS
G3KGX EN2 0EL
G6FJI EN2 0LL
G0NSB EN2 0QP
G0HEN EN2 7BY
G4XIN EN2 7BY
2E1EWL EN2 7BY
G8MOL EN2 7EN
G6RRP EN2 7NB
2E0AHI EN2 8DP
G8PPZ EN2 8EE
G8AQO EN2 8EG
G4KZD EN2 8HS
G3PKZ EN2 9BT
G1HBV EN3 4BD
G7ARF EN3 4BW
2E1FIH EN3 4NY
G3RKJ EN3 4QE
G4TUT EN3 4QF
G4UNL EN3 4UD
G0TSN EN3 4UD
G0WWZ EN3 4UD
G0SHB EN3 5AH
G7OBS EN3 5AN
G3KTZ EN3 5JD
G8YUR EN3 5NT
G4GYP EN3 5QE
G7VZK EN3 6AP
2E1BEL EN3 6DQ
G7WGW EN3 6EN
2E1FQE EN3 6HE
G0PXY EN3 6NF
G7VVF EN3 6NT
2E0AHF EN3 6PE
G3JNJ EN3 7HA
G4GJO EN3 7JF
G7OYP EN3 7LH
G3UJV EN4 0EX
G3VER EN4 0EX
G8VER EN4 0EX
G3YFW EN4 0HT
G0NSO N11 2JY
G3ZVW N13 4BH
G0NMC N13 4DX
G3TXA N13 4EP
G0IGO N13 4LJ
G0WLX N13 4UA
G6SBU N13 5AJ
G6NTW N13 5BS
G3YTY N13 5BT
M1BRB N13 5EL
M1BRC N13 5EL
2E1EUK N13 5EL
M1BRD N13 5EL

2E1EUL N13 5EL
G6LBJ N13 5SS
G0MML N13 5TD
G0KUX N14 4EA
G1YJI N14 4EA
G4IEH N14 4RP
G8FSL N14 4XD
G4AEZ N14 4XN
G6VAL N14 4XN
2E1FMW N14 4XN
G0ASC N14 6DE
G4JYH N14 6DE
G4BAN N14 6QU
G0NYY N18 1RT
G3TIE N21 1EH
G7WAZ N21 1EJ
G3DKR N21 1EL
G3SFG N21 1NP
G4DFB N21 1NP
G8DIS N21 1PA
G0AJO N21 1QU
G6ZZN N21 1QU
G0CUP N21 1RH
G3BWQ N21 2AB
G0RPM N21 2BE
G0OIF N21 2JL
G3TDM N21 2QT
G1EUV N21 3AU
G6RHO N21 3DB
G3MWF N21 3HL
G0VEB N21 3JP
G1BPU N21 3QJ
2E1FNB N9 0HJ
G1ZBA N9 0HR
G0ASA N9 0LL
G4RSM N9 7HL
G1ALD N9 7QG
G4YGH N9 8HD
G4XRT N9 8HE
G6XRT N9 8HE
G4ZIS N9 8LJ
G1RFS N9 8QL
G7COQ N9 9EQ
G4UTR N9 9HP
G3TZZ N9 9HU
G8SZZ N9 9HU
M1BRJ N9 9SU
2E1FFB N9 9SU

Epping Forest

G0PHU CM16 4JB
G0VRB CM16 4JR
G4LGY CM16 5DL
G6MBD CM16 5DP
G4NEX CM16 5DQ
G7IKL CM16 5DQ
G3FDS CM16 5EL
G8WRA CM16 6EH
G6OIX CM16 6HA
G1NGZ CM16 6JD
G4WWY CM16 6LF
G3GWY CM16 6ST
G0LWP CM16 6TD
G1WXS CM16 7BB
G3TUC CM16 7BT
G8LAB CM16 7BU
G8NPM CM16 7BU
G6UFT CM16 7DY
G4ACL CM16 7JX
G6NAX CM16 7PU
G0PTG CM17 0NE
G0SLB CM17 0NE
G3ACB CM17 0PS
2E1FTX CM17 9PT
G6TCJ CM19 5EQ
G6TPO CM19 5HG
G0MGU CM19 5LJ
G7AJW CM21 9PS
G4PJD CM22 7LT
G6ART CM22 7NB
G4IJE CM22 7NH
G7RCW CM4 0LJ
G3PUP CM5 9AS
G4KVR CM5 9BG
G6AXO CM5 9BG
G4GZG CM5 9QL
G4TOX CM5 9QW
G6SKX E4 7SB
G0WDO EN10 6QY
G0IQI EN9 1TJ
G3ZYN EN9 1UH
G4JIN EN9 1UH
G4XVM EN9 2AP
G2FUU EN9 2DD
G6PXV EN9 2ED
G3HYG EN9 2LB
G3TEI EN9 2LB
G3WAG EN9 3AX
G7OKV EN9 3DJ
G3GXG EN9 3DP
G4RWP EN9 3LB
G4MOI EN9 3LD
G1DJI IG10 1BU
G4HSN IG10 1HP
G3AAE IG10 1PZ
G0LWM IG10 1SB
G7EEE IG10 1SB
G7NGM IG10 1SB
G0SGX IG10 1TS
G4WTQ IG10 2AD
G0ABS IG10 2LE
G4ULK IG10 2LE
G4DIR IG10 2LG
G4AUG IG10 2NS
G0TOC IG10 2RE
G3NQT IG10 2SA
G3JAM IG10 3AW
G0PFY IG10 3EP
G7GBK IG10 3JN
G1IZA IG10 3PL
G1IVP IG10 3PR
G7VKB IG10 3QY
G8DPY IG10 4EG
G4IMU IG10 4PY
G3TCL IG10 4QD
G4ONP IG10 4RA
G8DZH IG10 4RA
G7RGF IG7 5AY
G7VGJ IG7 5BG
G4GMN IG7 5ED
G0LWF IG7 5EP
G8AJF IG7 5ET
G4NZB IG7 5HZ
G6BGA IG7 5HZ
G1JGD IG7 5JQ
G6ZQH IG7 5NJ
G7OSB IG7 5RF
G8LGU IG7 6AD
G4YWN IG7 6AH
G8XTD IG7 6EU
G4BMU IG9 5QE
G8MCR IG9 5RU
G4YOA IG9 5TZ
G3VGR IG9 6AQ
G0LWI IG9 6BY
G6SKF IG9 6DY
G3ZWN IG9 6EF
G3WPR IG9 6JT
G4XVH RM4 1AU
G6SVV RM4 1AX
G0NFJ RM4 1DL
G0IXW RM4 1DS
G6DIM RM4 1NH

Epsom and Ewell

G4VHJ KT17 2EF
G0DGF KT17 2HD
G2BCI KT17 2JS
G7PBC KT17 2NG
G4XVG KT17 2NQ
G0ECK KT17 2NW
G3NJM KT17 2NW
G3AOK KT17 4QJ
G6XKF KT17 4QJ
G8ZLG KT18 5JG
G3TVL KT18 5JL
G4XVF KT18 5LZ
G4WNU KT18 6AF
G3OLM KT18 6JB
G3ZDB KT18 7BH
M0AAF KT18 7HZ
G0CYE KT18 7JB
G0TNA KT18 7JT
G4ZSW KT19 0AU
G8KWV KT19 0BJ
G8MTI KT19 0HF
G4ROI KT19 0LB
G2FTB KT19 0NY
G8EBQ KT19 0PQ
G8HPF KT19 0QE
G6YIJ KT19 0RJ
G0PFA KT19 0SZ
G3YGG KT19 8JX
G0WZM KT19 8LU
GW3ZBB KT19 8LY
G1VVY KT19 8RG
G1LQC KT19 8RP
G4SYT KT19 8RP
G7ARI KT19 8TE
G1NYS KT19 9DN
G4WFL KT19 9DP
G6LFW KT19 9DP
G0KAS KT19 9EE
G1IKG KT19 9HH
G6REH KT19 9PA
G1SWF KT19 9RA
G6YAF KT19 9SR
G8YVF KT19 9TQ
G0LRS KT19 9XB
G4CI KT4 7BB
G3IQY KT4 7BZ
G4ADM KT4 7DX
G7LCS KT4 7JQ
G3EWJ SM2 7LJ

Erewash

G0JWA DE21 5AF
G4SQV DE21 5AH
G4AOA DE21 5AP
G4FQZ DE21 5AX
G2HIO DE21 5LF
G8DKV DE21 5LL
G3SNR DE56 4BH
G8YSJ DE56 4BH
G1ZHZ DE7 4DA
2E0AEG DE7 4DF
M0AIT DE7 4DF
G4AIB DE7 4DF
G6CDT DE7 4DL
G3XRD DE7 4JY
G3UQT DE7 4NE
M1AQD DE7 4NE
M0BCJ DE7 4NE
G7NWC DE7 4NY
G0LUI DE7 5AT
G6GJX DE7 5DD
G3IFX DE7 5EF
G4XCK DE7 5EX
2E1BGH DE7 5EX
2E0AOW DE7 5EX
M0BGM DE7 5JG
G7UAD DE7 5JG
G0WGQ DE7 5PL
G6RAJ DE7 5RJ
G4ZJQ DE7 6FL
G4DAM DE7 6GB
G4EDD DE7 6GB
G8KSW DE7 6GU
G0PCX DE7 6HA
G3KTP DE7 6HG
G1EBL DE7 6HW
G8RJF DE7 6HW
G7IRT DE7 8AE
2E1BSJ DE7 8AE
G4PUN DE7 8AJ
G6LHB DE7 8AJ
G8WRY DE7 8AW
G7VPV DE7 8NG
G7VPW DE7 8NG
G6KBC DE7 8PX
2E1BFP DE7 8PX
G0KBN DE7 8PZ
G0WXQ DE7 8PZ
G0RTQ DE7 8QP
G4UFC DE7 8QZ
G4JGA DE7 8TA
G1YEV DE7 9HE
G0RTX DE7 9JZ
G4OIF DE72 3AU
G0UNM DE72 3DE
G8KEA DE72 3GN
G7TLL DE72 3HR
G6ZUW DE72 3JB
G8YZC DE72 3LN
G0HWH DE72 3LP
G0DPQ DE72 3NP
G7UGA DE72 3NR
G4OQL DE72 3PN
G6CHI DE72 3PN
G1DGY DE72 3RP
G4CBQ DE72 3RQ
G6SNJ DE72 3RQ
G1SGZ DE72 3TE
G0JWE DE72 3TL
2E1CFE DE72 3UF
G4EHX DE72 3UG
G3RHG NG10 1AG
G3FWN NG10 1BQ
G4WHN NG10 1DR
G6DRH NG10 1DX
G4TYN NG10 1EA
M1CDK NG10 1LS
G8XAN NG10 1LS
G3REU NG10 1NL
G0CQR NG10 2BY
G0FFQ NG10 2DZ
G7RXE NG10 2EA
G4EAM NG10 2FR
2E1DPF NG10 2FS
G3PRF NG10 2FY
G3WYH NG10 3BB
G4EAX NG10 3BS
G6AGR NG10 3BT
G4YOY NG10 3DD
M0BGJ NG10 3DG
G7TSY NG10 3DG
G1DMH NG10 3EF
G1OPG NG10 3EW
G6TKM NG10 3FP
G0PSI NG10 3GF
G6MDR NG10 3GF
G4OSR NG10 3GH
G0FVZ NG10 3GJ
2E1AKK NG10 3GP
2E1FNM NG10 3LZ
M1AIU NG10 3PG
M1AON NG10 3PG
G8QZ NG10 4BN
G4IVO NG10 4DA
G6PRK NG10 4GD
M1BXL NG10 4GD
2E1FMM NG10 4GD
G4OEE NG10 4GE
M1BPA NG10 4GG
2E1FOG NG10 4GG
G4JAZ NG10 4GR
2E1DWZ NG10 4GW
G4ZWZ NG10 4HW
G7TZW NG10 4JA
G8OCO NG10 4JN
G5OW NG10 4JS
G7BSK NG10 4LD
G7GCX NG10 4LR
G4TYP NG10 4NZ
G0IVI NG10 5BS
2E1CKM NG10 5BS
G4HIC NG10 5EF
G4PXS NG10 5HB
M0ANC NG10 5HW
M0AND NG10 5HW
G7SMJ NG10 5HW
G3HOO NG10 5NF
G4LOV NG10 5PD

Exeter

G3EQM EX1 1RX
G0SZX EX1 1SL
G4SDV EX1 1SL
G8NEI EX1 2BG
G0UTA EX1 2ER
G4KXR EX1 3XP
G3EWH EX2 4LX
G3SXH EX2 4ND
G6HUO EX2 4PL
G3TDW EX2 4SQ
G1EMU EX2 5HB
G8RCZ EX2 5QN
G7HCE EX2 5UH
G0BNW EX2 6AN
G7UYU EX2 6DR
G7UIU EX2 6EA
M1APU EX2 6EA
G4BQH EX2 6JJ
G0JQS EX2 8XN
G0DIE EX3 0DG
G0EUE EX3 0EJ
G4NKP EX3 0HB
2E0AHD EX3 0LJ
G0WUB EX3 0LJ
G4ETO EX3 0NA
G6WDJ EX4 1ES
G8ORU EX4 1EZ
G3EFY EX4 1LP
G0FGE EX4 1NJ
G3VQD EX4 1NT
G4EDG EX4 1NX
G4TRU EX4 1RH
G7UHZ EX4 1SW
G3YBG EX4 1TA
G7BAE EX4 2AP
G3YRX EX4 2DF
G4GVY EX4 2DS
G4HAZ EX4 2DS
M1BAS EX4 2EQ
G4PCB EX4 2EQ
G4TKV EX4 2ER
G1YPM EX4 2EU
G4FYJ EX4 2LR
G0CEL EX4 2NN
G0VVY EX4 2PQ
G0AKH EX4 3RG
G3ZLS EX4 4HR
G4BPV EX4 4SJ
G0TQS EX4 4SJ
G4PXE EX4 5AJ
G6BJS EX4 5BP
G3ZVI EX4 5DE
G4SJU EX4 5DN
2E0AHJ EX4 5DU
G4EBC EX4 6EZ
G6IKC EX4 6HG
G7NUC EX4 6LW
G8EKP EX4 6NA
M0BBV EX4 6ND
G4UPW EX4 6TB
G6YEK EX4 7BG
G0OYH EX4 7DL
G3YBK EX4 7EA
G6FXH EX4 8BB
G3RUV EX4 8ES
G8CKC EX4 8PG
G0BTQ EX4 8QD
G1FPF EX4 9BJ
G4KEE EX4 9DY
G3RUX EX4 9HL
M1AZY EX4 9HP

Falkirk

GM7VPT EH49 6LW
MM0ASB EH49 7ND
2M0APX EH49 7ND
GM0CQQ EH51 0DD
GM4MMM EH51 0EH
G7RLB EH51 7AZ
GM0FSH EH51 9ED
GM0FUA EH51 9JD
GM4HRL EH51 9QD
GM1BVA FK1 1PL
GM6PKP FK1 2AP
MM0AOL FK1 2BX
GM0RMW FK1 2DG
GM6WQH FK1 2LD
GM4LDU FK1 3HN
GM0NQP FK1 4AH
GM3BKC FK1 4AX
GM0KWL FK1 4JF
GM4SNZ FK1 4QB
GM0KUP FK1 5HE
GM3GJB FK1 5HU
GM1LUZ FK1 5JX
GM6VCV FK1 5QJ
GM3MWX FK1 5QN
GM0ONB FK2 0DU
GM3SVE FK2 0DU
GM4MIG FK2 0DU
GM4OGI FK2 0DU
G0TXJ FK2 0SB
GM0KLO FK2 0TF
GM0TKE FK2 0TF
GM6KDD FK2 0TJ
GM4MF FK2 0TL
GM4OPJ FK2 0TS
GM7DZK FK2 0UP
GM3JJQ FK2 0UX
GM4JHG FK2 7DF
GM0MOP FK2 7FL
GM0ZAM FK2 8EE
GM1IEL FK2 8JP
GM0KMD FK2 8NP
GM6RGY FK2 9HA
GM4XQJ FK2 9QQ
GM0PGD FK2 9UF
GM0NJL FK2 9UT
GM3FDN FK3 0BU
GM3YKA FK3 0DA
GM3IWX FK3 0DH
GM0VYI FK3 0DN
GM1KMH FK3 8AB
GM0MZB FK3 8HD
GM6JUA FK3 8JE
GM0FTG FK3 8RF
GM0FRC FK3 8TG
2M1ANY FK3 8TG
GM0KBU FK3 8TG
GM4CAC FK3 8YH
GM8RVC FK3 9DR
GM4OMT FK3 9JD
GM6JWH FK3 9JD
GM0HJS FK3 9JN
GM3NVU FK4 1DE
GM0AEY FK4 1PE
GM1TKI FK4 1JP
GM7RLA FK4 1JX
GM1PGL FK4 1LA
MM1BFE FK4 1LQ
MM0AKX FK4 1PL
GM3OQI FK4 1PL
GM0IYA FK4 1RY
MM1BAB FK4 2BL
GM8JJN FK4 2HG
GM3UCH FK5 3JU
GM4EJX FK5 3LH
GM3HWN FK5 4HQ
GM1RIG FK5 4HQ
GM3AVA FK5 4SG
GM0DKK FK5 4ST
GM0OKJ FK5 4UF
GM1PDL FK6 5AA
GM4XRP FK6 5EG
GM0AZC FK6 5HX
2M1DVQ FK6 6BW
GM4NTX FK6 6RB

Fareham

G8OUH PO13 9NJ
G0BZK PO14 1AB
G1BZU PO14 1AS
G3CRS PO14 1AS
G0RPK PO14 1EG
G0VIX PO14 1EG
G3UWI PO14 1EZ
G3DEQ PO14 1HF
G0CEZ PO14 1RX
G6AVT PO14 2BQ
G0LKO PO14 2ER
G3GVM PO14 2JF
G0CEP PO14 2LF
2E1CZZ PO14 2LF
G4PZV PO14 2PX
G4VVI PO14 2QQ
G0RHU PO14 2QS
G0RHV PO14 2QS
G3IBI PO14 2QS
G7ARJ PO14 2SQ
G1KNI PO14 3AD
G7NYF PO14 3AF
G8PGF PO14 3DR
G0OPD PO14 3EG
G4USA PO14 3HP
G0XGL PO14 3JD
G6HGX PO14 3LB
G3YOS PO14 3LP
G8PO PO14 3LU
G3MJP PO14 3NE
G7HEP PO14 3NN
G0JDY PO14 3RF
G3WAO PO14 3RH
G4NAO PO14 3SQ
G3HKT PO14 4BH
G0RUD PO14 4LF
G0VOT PO14 4LF
2E1DQG PO14 4LF
2E1DQH PO14 4LF
G6CUA PO14 4LG
2E1BRZ PO14 4NB
G7VAP PO14 4NB
G0GFD PO14 4NP
G1SWX PO14 4QN
G8IQG PO14 4QP
G0MAT PO14 4SL
G8EAI PO14 4SL
G0TPB PO14 4SL
G4TLO PO14 4SU
G7LQV PO15 5AJ
G1NCK PO15 5AJ
G7PTH PO15 5AU
G6HHE PO15 5BL
G3RDA PO15 5HP
G3UNB PO15 5HP
M1AXJ PO15 5HQ
G0VAO PO15 5LY
G4IJP PO15 5NA
G3YMS PO15 5NE
G3XPH PO15 5NT
G1MPP PO15 5PF
G2DYF PO15 5PW
G8ACM PO15 5QA
G3UWE PO15 5QB
G6IPB PO15 5QH
G7ECY PO15 6AQ
G0PPH PO15 6BQ
G4VNM PO15 6BQ
G6BAT PO15 6EG
G0AMS PO15 6HD
G3KLF PO15 6JU
G7BQT PO15 6SW
G0SZI PO15 6TA
G0JVE PO16 0QB
2E0ABL PO16 0RX
G4JXK PO16 0SW
G4ZMP PO16 0TR
G1GOZ PO16 7EY
G4UOR PO16 7NL
G3JMK PO16 7PD
G7ILI PO16 7PD
G0OBM PO16 7PY
G4ITG PO16 7QL
G3NPZ PO16 7QP
G3YIW PO16 7QW
G6KTX PO16 7RR
G0JYZ PO16 7XA
G1WKZ PO16 7XR
G1WLD PO16 7XR
2E1FPF PO16 8AF
2E1FPG PO16 8AF
G4XZL PO16 8BX
G8XEZ PO16 8DS
G0FAD PO16 8DY
G4PWG PO16 8JW
G4CYC PO16 8LB
G0LYI PO16 8LF
G3XUF PO16 8LF
G0SFN PO16 8NS
G8IOJ PO16 8PB
G3TZS PO16 8PL
G0RKK PO16 8QD
G6JGT PO16 8QF
G7AUR PO16 8QF
G3YZY PO16 8QU
G8SJA PO16 8QX
G3LMM PO16 8RE
G0VQV PO16 8RU
G6JSI PO16 9BQ
G4YPA PO16 9DN
G3AFF PO16 9HH
G3RDR PO16 9JB
G4MRW PO16 9JU
G7WLR PO16 9NY
G0KCG PO16 9PA
G8BIX PO16 9PY
2E1ETH PO16 9TP
G6FLE PO17 5EP
G6HHF SO31 1BD
G3SED SO31 1BJ
2E1FBW SO31 1BJ
G1WXT SO31 6EP
G4UDY SO31 6EP
2E1BQF SO31 6LF
G7SOR SO31 6LF
G3JTG SO31 6LR
G4SAP SO31 6RN
G8HER SO31 6RR
G4BEQ SO31 6SD
G1KSE SO31 6ST
G4KWY SO31 6SY
G0JYQ SO31 6TH
G3VD SO31 6UX
G3VOR SO31 7BR
G0FBC SO31 7BZ
2E1FKT SO31 7EY
G0PAF SO31 7FB
G6LQI SO31 7FP
G8ACL SO31 7HQ
G4WDH SO31 9AT
G0DDA SO31 9AY
G3HQT SO31 9FG
G3NNW SO31 9FU
G4PXH SO31 9HG
G1HQI SO31 9RW
G7TKD SO31 9SB

Fenland

G4NRU PE13 1JR
G8NIL PE13 1LF
G0LFE PE13 1SA
G0MYA PE13 1SA
G7RAZ PE13 1SD
G4UQN PE13 1SW
G4NFY PE13 2DD
G0IKO PE13 2DX
G0DWI PE13 2ED
G1WRC PE13 2ED
G4ODH PE13 2ED
G4PQL PE13 2ED
G8NED PE13 2ED
G8HQO PE13 2EG
G6NNK PE13 2JD
M1BFM PE13 2JF
M0BFS PE13 2JF
G7TFL PE13 2JR
G3JFQ PE13 2PH
G6YHD PE13 2QL
G6DUQ PE13 2RJ
G0SFQ PE13 3ED
M1BFN PE13 3ED
M1BIG PE13 3HF
G6XMU PE13 3NQ
G8HTH PE13 3QL
G3IPC PE13 3TF
G0FHM PE13 3UN
M1AMJ PE13 3UT
G7WJT PE13 4HS
G8RZN PE13 4JU
G1HJP PE13 4NL
G7DJX PE13 4PQ
G7RUK PE13 4QQ
G7HHQ PE13 4RG
G3JFU PE13 4SE
G8YXJ PE13 5AE
G3XZF PE13 5AF
G0AGS PE13 5AN
G3MFU PE13 5AQ
G1ULP PE13 5AS
G1HQW PE13 5DN
G0XTA PE13 5DR
G0MFT PE13 5LG
M1AKI PE13 5LJ
M0BFT PE13 5LJ
G3YPZ PE13 5NA
G8HDK PE13 5PQ
M1AJN PE13 5PQ
G6RYD PE14 0AP
G0JXJ PE14 0AY
G4FCI PE14 0DF
G0OPC PE14 0HA
G6NHK PE15 0AJ
G1YFE PE15 0HH
G1OQX PE15 0QF
G6SXB PE15 0QF
G8OXE PE15 0QS
G3YLY PE15 0RA
G4AJE PE15 0RW
G1IZP PE15 0SV
G7TBX PE15 0YF
2E1AXZ PE15 8AX
G0VOH PE15 8AX
G1HMT PE15 8BT
G7PJW PE15 8DD
G8PHS PE15 8EL
G7EVP PE15 8HF
G0CRK PE15 8PE
G1LJQ PE15 8SD
2E1DBV PE15 8SL
G0FLP PE15 8SN
G6JVX PE15 9AH
G3VQG PE15 9AP
G1VTO PE15 9DL
M1AIP PE15 9DT
G0BXJ PE15 9EA
G6ESN PE15 9EG
G3VMI PE15 9EQ
G6MMB PE15 9HP
G1HQK PE15 9HT
2E1EAQ PE15 9NU
G0TGF PE15 9QA
G3ESK PE16 6DZ
G7NVT PE16 6HQ
G6RBR PE16 6LQ
G8WUR PE16 6PL
G3SPK PE16 6SS
G6OHM PE16 6TP
G0MSN PE16 6UT
G0HKE PE16 6UX
G0RCH PE7 1BP
G4KIY PE7 1BX
G8MTA PE7 1DL
G1WBL PE7 1DR
2E1EPS PE7 1DW
G6GYM PE7 1HR
G0OYZ PE7 1LP
2E1EOU PE7 1LS
G0IHK PE7 1RL
G4IVH PE7 1RN
G0PGY PE7 1RR
G8BEN PE7 1SU
G4FFN PE7 1UE
G6DCV PE7 1UE
G7VPA PE7 1XX
G3HHT PE7 1YY
G8RAU PE7 1YZ
G7ELG PE7 2AY
G4YIT PE7 2HN
G1JIH PE7 2PD
G3XRK PE7 2RG

Fife

GM0TGG DD6 8AP
GM7FMJ DD6 8AW
GM4FSB DD6 8DT
GM3MOR DD6 8HL
GM4RXW DD6 8HL
GM4AGS DD6 8HP
GM4USY DD6 8JD
MM1BWC DD6 8JH
GM0CNW DD6 8LF
GM8RPE DD6 8ND
GM1BNA DD6 8SL
GM6HGW DD6 8SL
GM1DTJ DD6 8SQ
GM7IIL DD6 9AS
GM3EFH DD6 9DL
GM4RDI DD6 9LG
GM4WPU DD6 9PN
GM0GUJ FK10 4PN
GM4GIL FK10 4QG
GM3ZVF KY1 1DN
GM8SZS KY1 1TN
GM4KNH KY1 2JA
GM4EHO KY1 2LH
GM1AFD KY1 2PW
GM0BUH KY1 2XA
GM7CNW KY1 2XP
GM0LDE KY1 3HQ
GM0CFK KY1 4AA
GM4MTV KY1 4AG
GM3SJY KY1 7BW
GM0OFM KY1 7JX
GM0OFL KY1 7SP
GM6OKJ KY10 2AP
GM4EOU KY10 3HU
GM4JDU KY10 3RB
GM4FQE KY10 3UJ
GM8XZY KY10 3UR
GM3RXU KY10 3XL
GM3CIG KY11 1AH
G4CTP KY11 1DR
GM3KJZ KY11 1ET
GM4ZNA KY11 1HE
GM4UKG KY11 1LD
GM0FWJ KY11 2DF
GM8YAQ KY11 2JD
GM6XQX KY11 2JU
GM0PUN KY11 2LX
GM0RLZ KY11 2QW
GM8BFE KY11 2QW
GM3DPL KY11 2RJ
GM0DYD KY11 2SS
GM0FTL KY11 2SS
GM0GBH KY11 2SS
GM7PXJ KY11 3JQ
GM3TAL KY11 3LG
GM4JKT KY11 3LJ
GM0OVD KY11 4NA
GM3HYX KY11 4PE
GM0XAL KY11 4PY
GM0GUM KY11 4TP
GM0TKC KY11 5BH
GM3MGT KY11 5FG
MM0BAB KY11 5HA
GM3OZJ KY11 5LF
GM1ZBD KY11 5LH
GM4UTP KY11 5LR
GM8IXZ KY11 5LR
GM4ANB KY11 5ND
GM4HVM KY11 5RA
GM4LHM KY11 5RP
GM7VOZ KY11 5SX
GM3SHR KY11 5UA
GM1VKI KY11 5UF
GM3FIZ KY11 5UQ
GM4UJZ KY11 5XF
GM6SYC KY11 5XQ
GM0LLJ KY11 5XR
GM7GLJ KY11 5XU
G4TME KY11 5YE
GM8VKN KY12 0AH
GM3ENJ KY12 0DA
GM1ZCD KY12 0DD
GM3EGU KY12 0DQ
MM1BIM KY12 0DX
GM0IXO KY12 0HN
GM3FAH KY12 0HQ
GM0AJK KY12 0QT
GM4FSF KY12 7HZ
GM7SUW KY12 7XH
GM4OOU KY12 7XU
GM0FTH KY12 8DF
GM4FAU KY12 8NZ
GM8SNE KY12 8QD
GM4JJJ KY12 8QL
GM6FND KY12 8RB
GM0VOU KY12 8SS
GM0SRQ KY12 8YD
GM4ZNX KY12 8YL
GM0SRD KY12 8YW
GM4MPC KY12 9BX
GM3IQL KY12 9DR
GM3NVQ KY12 9EA
GM0RHP KY12 9JQ
GM4NUU KY12 9LP
GM8LRI KY12 9NX
GM8PEV KY12 9TL
GM4IHJ KY12 9UJ
GM6JHH KY12 9YA
GM0URD KY12 9YA
GM6WBV KY12 9YD
GM0CEA KY14 6DY
GM4GFR KY14 7QP
GM0DGL KY15 4AJ
GM7RFN KY15 4BE
GM6GPH KY15 4EF
GM7STI KY15 4NA
GM7VLZ KY15 4NA
GM4HKV KY15 4UG
GM1NDE KY15 5BB
GM0RFK KY15 5BS
GM4IEF KY15 5JS
MM0BGO KY15 5LY
GM0UZV KY15 5QB
GM7CPL KY15 5SE
GM0OGC KY15 5UU
GM3RNZ KY15 7NX
GM0AIR KY15 7RZ
GM3VPN KY15 7UH
GM3LGM KY16 8JL
GM4RGU KY16 8NR
G3YOI KY16 8NS
GM4FVO KY16 8NT
GM4DPC KY16 8PP
GM3FJA KY16 8QP
GM4FFP KY16 8QR
GM3VWY KY16 8QS
GM4XYF KY16 8SR
GM4SQS KY16 8XT
GM4SQT KY16 8XT
GM0HUO KY16 9AF
G4FPZ KY16 9DA
GM1XVI KY16 9NE
GM4ENP KY16 9NE
GM0GKR KY16 9UA
GM4XND KY2 5AD
2M1DVS KY2 5NA
GM0EZG KY2 5NL
GM1BLX KY2 5PH
MM1BGZ KY2 5PZ
2M1ERB KY2 5PZ
GM0MMN KY2 5SE
GM3YBQ KY2 5SQ
2M1BEU KY2 5SQ
GM7UQM KY2 5SQ
GM0HYW KY2 5SR
GM4DOF KY2 5TU
GM0LKT KY2 5YJ
GM4RKM KY2 5YT
GM7IGS KY2 5YU
GM0NXO KY2 6DR
GM4LCJ KY2 6EL
GM1BID KY2 6EP
GM3AEY KY2 6HP
GM0SNT KY2 6HR
GM8YUM KY2 6HW
GM7RDR KY2 6LD
GM7VOP KY2 6NZ
GM0UZM KY2 6SF
GM1ARK KY2 6TS
GM0LOO KY2 6XP
GM7PVV KY2 6XR
GM8ZTV KY2 6YB
GM0IJR KY3 0JW
GM4TRZ KY3 0RZ
GM4GUL KY3 0XJ
GM0RDA KY3 9DL
GM0BUI KY3 9EU
GM4WWJ KY3 9EU
GM3ULG KY3 9HP
GM4AQO KY3 9HP
GM4GRC KY3 9HP
GM0VFD KY3 9JQ
GM0AYW KY3 9RX
GM1SZM KY4 0BG
GM8KSJ KY4 0EQ
GM3TLN KY4 0ET
GM1ATW KY4 8EU
GM6OOA KY4 8EU
GM0EMY KY4 9AH
GM7KVB KY4 9EP
GM0KLP KY5 0AX
GM8JGB KY5 0DG
GM3ZGH KY5 0ND
GM4NNH KY5 0ND
GM4ZNG KY5 8BH
GM1IUA KY5 8NP
GM4HQZ KY5 9BE
GM1EHK KY5 9LD
GM0LDC KY6 1DS
GM7CMR KY6 1JJ
GM3LWS KY6 1JL
GM0MRJ KY6 1JP
GM6SXK KY6 2HH
GM0IOA KY6 2HS
GM6EOP KY6 2JS
GM0REW KY6 2JT
GM0IJD KY6 2JY
GM1EAH KY6 2PX
GM1JZM KY6 2QA
GM7FLM KY6 3DB
GM4DEX KY6 3JD
GM7CQQ KY6 3LH
GM1CEJ KY6 3PJ
GM8ZQY KY6 3QR
GM3NMN KY7 4ES
GM4TNP KY7 4HS
GM7FRC KY7 4HS
GM0MNV KY7 4JG
GM4ZJI KY7 4QN
GM6AOJ KY7 4RU
GM0GTL KY7 4RW
GM3VVF KY7 4SJ
GM3XMY KY7 4SW
GM1ADI KY7 5AX
GM4XAV KY7 5AX
GM0MFU KY7 5BB
GM7GKT KY7 5TD
GM7CTV KY7 6BN
GM3IXW KY7 6EN
GM3OBC KY7 6LA
GM4ALK KY7 6PQ
GM4SBP KY7 6SQ
GM0EZR KY7 6XU
GM8KSQ KY7 7DQ
GM0CFC KY8 1DW
GM4YLY KY8 1HS
GM3HVT KY8 2AY
GM6KRO KY8 2DD
GM4PWQ KY8 2EQ
GM6NYT KY8 2HA
GM4HBQ KY8 3DH
GM0LWW KY8 3PG
G7VQB KY8 4DF
GM7KPE KY8 4RQ
GM3UNJ KY8 4SQ
GM3AHR KY8 5SW
GM4EJI KY8 5TR
GM0ALU KY9 1EB
GM6UJO KY9 1JD
GM6BGQ KY9 1LB

Flintshire

GW6FKB CH4 0AQ
GW3TTR CH4 0HN
G1MMO CH4 0JE
G6XKJ CH4 0JF
GW4EYO CH4 0JU
GW0MDQ CH4 0NB
GW6DBP CH4 0NB
GW0EDC CH4 0NH
GW0PJA CH4 0NJ
GW8VNN CH4 0PT
GW4FGC CH4 0QJ
GW6OKC CH4 0QT
GW4MOK CH4 0RS
G1SXO CH4 0SB
G7TZX CH4 0SP
GW8ZBC CH4 0SQ
G6BZH CH4 0SY
GW0PBJ CH4 9DA
GW4RBA CH5 1AU
GW1ATZ CH5 1BE
GW4YJT CH5 1DU
GW7GKX CH5 1EZ
GW1FJC CH5 1HT
GW6POO CH5 1HT
GW1CEV CH5 1LS
GW4TQU CH5 2EL
GW0WQE CH5 2EL
GW6WQF CH5 2EL
MW0AVQ CH5 2JE
GW6KWU CH5 2JF
GW8ITZ CH5 2LS
GW7KDZ CH5 2PJ
MW0BAJ CH5 2TX
GW0DRS CH5 3DA
G8YHJ CH5 3HS
GW0EHB CH5 3HW
GW0GZR CH5 3HW
GW4SII CH5 3JG
GW3TKD CH5 3LJ
GW4FLZ CH5 3LR
GW0EGQ CH5 3LZ
GW3KFA CH5 3RD
G4SMS CH5 4AT
GW6AEO CH5 4ED
GW7HEC CH5 4HL
GW0TSW CH5 4HL
GW1SXP CH5 4HN
GW7KIU CH5 4HU
GW4ZAG CH5 4HW
MW0AMT CH5 4JH
GW0DSP CH5 4JP
GW7IBZ CH5 4JS
G3KNZ CH5 4LG
G6NCP CH5 4LN
GW4RQS CH5 4RE
GW4VBM CH5 4RE
G6NVJ CH5 4SH
GW7AAU CH5 4SN
GW7AAV CH5 4SN
GW0UEO CH5 4TN
GW7MQE CH5 4XN
GW4RYJ CH6 5AW
GW8JRL CH6 5HL
GW0UIP CH6 5HU
GW6LMI CH6 5JP
GW4ZAR CH6 5PW
GW3PRA CH6 5RQ
MW1AJX CH6 5TG
GW7CMM CH6 5YE
GW6WGY CH6 6DP
GW0AGZ CH6 6EY
GW6XYE CH6 6LS
GW7PYH CH7 1HR
GW6WPL CH7 1HT
GW0TCV CH7 1LD
GW7PUX CH7 1NT
GW8HYI CH7 1QY
GW0PKA CH7 1RA
GW7MVG CH7 1SU
GW3UOO CH7 2AG
GW7AIY CH7 2AG
MW1BLC CH7 2AP
GW8FNX CH7 2BS
GW0UXX CH7 2HQ
GW7AIY CH7 2HQ
GW0GXQ CH7 2JA
GW0GDK CH7 2JN
GW8ICT CH7 2JT
GW4TWE CH7 2NQ
GW3HDF CH7 2PA
GW7MHB CH7 2PB
GW0SXE CH7 2PN
GW3UVA CH7 3BH
GW7BOY CH7 3BR
GW4IEQ CH7 3DY
GW1OTI CH7 3EL
GW4GTE CH7 3ET
GW4TEE CH7 3LB
GW0OHJ CH7 3QA
GW3NQP CH7 4AB
GW1MFY CH7 4AD
GW3CCF CH7 4AP
GW0BWX CH7 4EX

(Flint / Clwyd area, continued)

GW4SDO CH7 4TU
GW1FWE CH7 5AE
GW2FVZ CH7 5BH
GW6RQG CH7 5BN
GW8EQI CH7 5DZ
GW0PJX CH7 5EX
GW0TQM CH7 5HX
GW8RLV CH7 5NH
GW3LAI CH7 5NJ
GW6KAV CH7 5UB
GW0HUS CH7 6DD
GW1XDY CH7 6DD
GW6RNV CH7 6DU
GW3ITT CH7 6DY
GW7NFM CH7 6ED
GW4EVX CH7 6EF
GW0RTA CH7 6HA
GW3FPH CH7 6JF
GW3VBC CH7 6LP
GW1XHG CH7 6PP
GW1HFW CH7 6PY
GW3GSJ CH7 6PY
GW3IVK CH7 6SL
G4BTW CH7 6SW
GW4NEI CH7 6TD
GW0PZU CH7 6TF
GW4RAF CH7 6TF
GW7MGW CH7 6TP
GW7PVD CH7 6TP
GW6FKP CH7 6US
GW3LWU CH7 6UZ
GW2SB CH7 6XY
GW7VKX CH7 6YF
GW3KJN CH7 6YP
G4NQE CH7 6YQ
GW4GSG CH7 6YU
GW0HRG CH8 7BH
GW4VUH CH8 7BH
GW0PFZ CH8 7DF
MW1BIE CH8 7HA
GW7GPD CH8 7PZ
GW4TIZ CH8 7SJ
GW1DRQ CH8 7SW
GW4XNT CH8 7SW
GW6WQJ CH8 7UG
GW7DZA CH8 7UN
GW7DZC CH8 7UN
GW0MMY CH8 8AX
GW4ZXD CH8 8AX
GW0UDJ CH8 8AX
G4ZWS CH8 8BS
GW4JUN CH8 8DL
GW7HGJ CH8 8DT
GW0FEU CH8 8ES
GW1PMQ CH8 8ES
GW0IZR CH8 8HZ
GW0WPT CH8 8LA
GW0HKQ CH8 8NU
GW4KZC CH8 8RP
GW4XYL CH8 8SU
GW4SLK CH8 9DX
MW0ATT CH8 9HY
MW1ARM CH8 9JB
GW7KBI CH8 9JQ
2W1FQF CH8 9JQ
GW4ZWO CH8 9LN
2W1FGK CH8 9LT
2E1FGL CH8 9LT
GW0ADC CH8 9UU
GW6NLP LL11 5YP
2W1FPU LL12 9HN
GW0NOO LL12 9PE
GW0NOP LL12 9PE
GW7COB LL12 9PE
GW7OIV LL12 9PH
GW6IGY LL12 9PJ
GW6PVK LL12 9PJ
GW0UGQ LL19 9UG

Forest Heath

G8DXJ CB8 0AF
G6GQI CB8 0DJ
G1YFF CB8 0EN
G7RVS CB8 0EN
G3GJZ CB8 0QF
G3TWX CB8 7AB
G7SCO CB8 7AR
M1BIB CB8 7AR
G4LVJ CB8 7BD
G0SPK CB8 7BQ
G4ASZ CB8 7DR
G6JJK CB8 7DU
G7BRP CB8 7HL
G6BBK CB8 7SD
G8BBK CB8 7SD
G4DSN CB8 8DT
G8XWR CB8 8HY
G1BJA CB8 8RN
G6HPY CB8 8RN
G4LKI CB8 8SN
G6FTE CB8 8SZ
G0WON IP27 0BS
G6IMM IP27 0DF
G1XHW IP27 0HE
G0FKE IP27 0HZ
G7IZS IP27 0JW
G6MGN IP27 9DU
G6LOJ IP27 9ES
G6EUO IP27 9EZ
G1RFH IP27 9EZ
G6XYX IP27 9EZ
G6DFR IP27 9HF
G8ENY IP28 7AA
G8ATS IP28 7BZ
G0UBU IP28 7HP
G8BCA IP28 7HX
G4XTW IP28 7JW
M0APP IP28 7JY
G6VAZ IP28 7LH
G4HVV IP28 8JQ
G8HVV IP28 8JQ
G4BWP IP28 8LQ
G4MBC IP28 8LQ
G0BRM IP28 8PB
G1SXY IP28 8PB
G7OYF IP28 8QE

Forest of Dean

GI0CUR GL14 1AP
G7HJH GL14 1LN
G7UDN GL14 1QX
G0XAE GL14 1QX
G0DZA GL14 2DW
GW0WBS GL14 2DW
G7KXN GL14 2EB
G0PBB GL14 2EB
G0SNB GL14 2EB
G0ODN GL14 2EF
G7TLG GL14 2PD
G4THC GL14 2QU
G0HHX GL14 2SU
G0IBV GL14 2TA
G7EQY GL14 3AD
G0DAB GL14 3DZ
G8GLV GL14 3NX
G7BAU GL15 4DS
G7VQI GL15 4HR
G3WHW GL15 4JN
G4NIF GL15 4JX
G3NSH GL15 4NS
G4YQF GL15 4NS
G4UHJ GL15 4QU
G4JDE GL15 4RF
G4NNJ GL15 5AT
G7VHJ GL15 5AZ
G4LSB GL15 5DJ
G0ENF GL15 5JD
G4ZFN GL15 5JD
G0FDD GL15 5LP
G6GOY GL15 5SH
G4EPW GL15 5SL
G4IKX GL15 6DN
G8OLL GL15 6JZ
G3CZL GL15 6LQ
G8BXD GL15 6PE
G4CHL GL15 6PT
G4ZYC GL15 6TN
GW6UXY GL15 6TN
G4JYX GL15 6UA
G3NOC GL16 7AG
G0PDD GL16 7AU
G4JIA GL16 7LA
G0SDD GL16 7LG
M1ART GL16 7QD
G8YYR GL16 7QE
G2CIW GL16 7RD
M1AYN GL16 8AY
M1BKE GL16 8AY
G3KTI GL16 8AZ
G1EDP GL16 8BG
M1AWR GL16 8BY
M1BDV GL16 8DG
G8WGD GL16 8PQ
G3TLD GL16 8PT
G4YDA GL16 8PY
G3KKN GL16 8QE
G7MWW GL17 0SB
G0GXT GL17 9AR
G4IVD GL17 9JX
G3SLG GL17 9JZ
G4HVD GL17 9QG
M1AVU GL17 9QS
G7TNH GL17 9QY
G3UOF GL17 9ST
G8LOU GL17 9UB
G7TUS GL17 9XR
G7IRW GL17 9XT
G5BM GL18 1DN
G3ZMD GL18 1EH
G6PMJ GL18 1HN
M1AHN GL18 1LU
G7IHA GL18 1NN
G7JUN GL18 1NN
G3WVQ GL18 1PZ
G4RSQ GL19 3AN
G0IIN GL19 3EP
G1PTZ GL19 3EZ
G0NIZ GL19 3RF
G1HAF GL19 3RF
G1NRX GL19 3RF
G4IGN GL19 3RF
G3PJQ GL2 8DY
G1YZL GL2 8EE
G4PHF GL2 8JT
G7HTS GL2 8LA
G3VZR HR8 1PF
G7EUK HR8 2LH
GW8WCA NP5 4LY
GW4ZTW NP6 7AR
G8PTI NP6 7DN
GW7ERI NP6 7DR
GW4YNR NP6 7EF
G3HCH NP6 7EN
GW8OQV NP6 7NT

Fylde

G3WPT FY3 0BU
G4BVW FY3 0DR
G4VYL FY3 0DR
G4RDY FY3 0DW
G0TFI FY3 0DZ
G0DZF FY6 7SX
G6MQG FY6 7SX
G6UDA FY8 1BS
G7CWI FY8 1HZ
G6KTO FY8 1JL
G0HIJ FY8 1JX
G1TJR FY8 1PQ
GM1MQA FY8 1TW
G2HCP FY8 1XQ
G8LCZ FY8 2DA
G7HJQ FY8 2JF
G3HUI FY8 2NW
G3WBI FY8 2PF
G4WFO FY8 2PH
G7PTM FY8 2RB
G3MYC FY8 3BG
G1AUH FY8 3DQ
G1YUD FY8 3EQ
G8GG FY8 3HD
G7MKT FY8 3HL
G8AMY FY8 3JY
G6EKA FY8 3LF
G0JBS FY8 3QF
2E0ANN FY8 3QG
G7RVF FY8 3QG
G4IHF FY8 3RL
G3FFR FY8 3RN
G6HDI FY8 3RQ
G6MXW FY8 3RQ
G3UUX FY8 3RW
G4AFB FY8 3RW
G8HCQ FY8 3SL
G4MFF FY8 4AR
G0CGV FY8 4LN
G4WLE FY8 4QR
G4DRH FY8 5EW
G2ADM FY8 5HL
G0HLB FY8 5JG
G4AHZ PR4 1AJ
G0OPM PR4 1BP
G1MQF PR4 1BY
G1TBN PR4 1BY
G3MQK PR4 1DN
G4RNF PR4 1HY
G4VAJ PR4 1HY
G1WCQ PR4 1JB
G4RCF PR4 1RL
G1NNJ PR4 1RN
G3UMZ PR4 1RN
G0CPJ PR4 1RX
G7CUL PR4 1SD
G4MEE PR4 1SS
G0GSP PR4 1TD
G6NUQ PR4 1XB
G4AGJ PR4 1YQ
G1GQY PR4 2AY
G1GQZ PR4 2AY
G7BYJ PR4 2DS
G0DQW PR4 2JS
G3GFT PR4 2PL
G7JGW PR4 2XA
G1TUI PR4 2ZA
G6NJL PR4 2ZN
M1ACJ PR4 3DN
G4IHE PR4 3HZ
G0HIU PR4 3ND
M1BNU PR4 3PA
G3TKK PR4 3PN
M0ABK PR4 3SX
2E1ECL PR4 3SX
2E1FJK PR4 3SX
2E1FJL PR4 3SX
G0SGI PR4 3TB
G4MRX PR4 3TU
G8MED PR4 3TX
G8TTD PR4 3YB

Gateshead

G4RXQ DH3 1AR
G1JGE DH3 1HP
G7RAW DH3 1NA
G7VDK DH3 1QJ
G1KYN DH3 2DX
G7RML DH3 2HY
G0CGW DH3 2JG
2E1EJE DH3 2LZ
G0GJQ NE10 0AE
G1NOZ NE10 0NA
G7HEF NE10 0UE
G6WUX NE10 8BE
G7VDD NE10 8EN
G4MSF NE10 8EX
G3YYI NE10 8HG
G3EBL NE10 8LU
G1DZY NE10 8QN
G4YIK NE10 8TE
G7JVG NE10 8UE
G7HDT NE10 9AD
2E1DOO NE10 9BQ
2E1DRX NE10 9BQ
G1OMX NE10 9NB
G3TYU NE10 9RT
G7LNO NE10 9SA
G4VTN NE10 9TZ
G0ESW NE11 0BS
2E0AOP NE11 0EU
M1BZB NE11 0EU
G0NPP NE11 0NR
G0APF NE11 0YE
2E1EDB NE11 9LA
G1AHW NE11 9QH
G0CXX NE11 9XH
G0PUK NE16 3JN
2E1APX NE16 3JN
G3RAR NE16 4PU
G4FBO NE16 4QP
G3VYQ NE16 5JW
G8QM NE16 5LY
G4GFV NE16 5QZ
G0WUI NE16 5RX
2E1EPX NE16 5RX
G6OSH NE16 5SZ
G3OPE NE17 7JR
G4NHQ NE21 4LB
G8CYW NE21 6BW
G7LOG NE21 6JU
2E1EYQ NE21 6LZ
2E1ACU NE21 6QY
G2HMB NE39 1EG
G4NOX NE39 1EQ
G3SEQ NE39 1LT
G3TKB NE39 2AD
G3OPM NE39 2DN
G1VRC NE39 2HD
2E0AJK NE39 2PN
G6PPD NE40 3ED
G7KJP NE40 3RU
G0DHI NE40 4BB
G6BHC NE40 4DW
G3ZVM NE40 4EU
G3OPK NE40 4PG
G4NSC NE40 4TD
2E1DNZ NE8 1RR
G7OOE NE8 1TL
G0NPQ NE8 4AJ
2E1EYU NE8 4EE
G1DUB NE8 4PS
G0CLL NE8 4UH
G6JUP NE8 4UH
G0DSI NE9 5DP
G4WAX NE9 5DP
G6KQJ NE9 5DP
G0SQE NE9 5HN
G7CIT NE9 6AH
G8XLZ NE9 6BS
G0MRO NE9 6DA
G3IPD NE9 6DA
G4PDO NE9 6DA
G0JHU NE9 6DH
2E0AOS NE9 6JJ
M1BYZ NE9 6JJ
2E1EYF NE9 6JJ
G4FLL NE9 6PH
G7KLS NE9 6QS
G0HBO NE9 6UX
G6EBC NE9 6XU
G0EHV NE9 7BW
G7MKB NE9 7JD

Gedling

G3AFR NG14 5AZ
G3WQW NG14 5BL
G8FGZ NG14 5BL
2E1CWQ NG14 5FG
G1PXE NG14 6EB
G8DVN NG14 6FF
G1XYM NG14 6FX
G4IJU NG14 6HL
G3AIN NG14 6LR
G1RHB NG14 6ND
G4WZK NG14 6PH
G6SCG NG14 6QA
G6UUZ NG15 8GD
G8MMP NG15 9AD
G4MFO NG15 9BA
G8EXS NG15 9DG
G0VAT NG15 9DG
G0LUD NG15 9EA
G7ITQ NG15 9GB
G2ATM NG3 5LA
G0JAC NG3 5QB
G8FWH NG3 5QB
G6IZG NG3 5SE
2E1BHT NG3 6AE
G6MHK NG3 6AL
G4IRX NG3 6BL
G1DWR NG3 6DA
G0JOX NG3 6DH
G4PJZ NG3 6DS
G4NRZ NG3 6EF
G6ABU NG3 6EQ
G8IUT NG3 6EQ
G6WHS NG3 6EU
G3MP NG3 6FL
G4ZTY NG3 6FL
G8UQR NG3 6FP
G3TWB NG3 6FT
G4ERY NG3 6HW
G8MIW NG3 6JA
G0HUB NG3 6LP
G6NET NG3 7AJ
G6ZBO NG4 1BU
G6JCI NG4 1DP
G4ANS NG4 1DZ
G1LUU NG4 1FY
G6OBW NG4 1FY
G7NDJ NG4 1JD
G4XWO NG4 1JY
G0IBY NG4 1LE
2E1BHU NG4 1LY
G0FPG NG4 1PP
G3ZUV NG4 1SB
G6UZY NG4 2EL
M1AMM NG4 2GW
M0BCI NG4 2LW
G6ZRV NG4 2QP
G8FNU NG4 3BH
G0FVI NG4 3BH
G0WWW NG4 3BH
G7TMF NG4 3DA
2E1EFK NG4 3DX
G6MBI NG4 3ET
M0ANJ NG4 3JH
G6IOE NG4 3LF
G3UXX NG4 3NB
G8XGW NG4 3NN
G1UMF NG4 3NN
G0AIU NG4 3NT
G6NLU NG4 3QL
G7THF NG4 3RR
G1VKN NG4 3SD
G4EUJ NG4 4AD
G8LGY NG4 4AD
G1EAB NG4 4BL
2E1FOO NG4 4BS
G6YCL NG4 4GE
G7SFS NG4 4GE
G0AGI NG4 4GP
G4CPG NG4 4GS
G4OCS NG4 4JA
G0IOL NG4 4JJ
G0GMT NG4 4PJ
G4EDX NG5 4HU
G0GDQ NG5 4JB
G4WDP NG5 4JX
G4WDQ NG5 4JX
G0KXZ NG5 4LB
G3SQA NG5 4LH
G4SGG NG5 4LT
G6BRP NG5 4LX
G1RGJ NG5 4ND
G4PNX NG5 6FN
G1FDO NG5 6FT
2E1COF NG5 6HJ
G4EAN NG5 6LQ
G7WKH NG5 6QA
G8SSL NG5 6QX
G1HTL NG5 6QZ
G8VBW NG5 6TF
G1YQO NG5 7AS
G7OHK NG5 7AT
G6MVF NG5 7DF
G7PDZ NG5 7GT
G4OOY NG5 7HQ
G7DPV NG5 7LA
G8UGS NG5 7LW
G4EGY NG5 8AG
G8RBU NG5 8FQ
G3XEM NG5 8FT
2E1BSI NG5 8GS
G3OCF NG5 8HQ
G1RGK NG5 8HX
G4JFR NG5 8JJ
G7RDG NG5 8LR
G7IWK NG5 8NZ
G1GQQ NG5 8QF
G0VTK NG5 8QG
G6IBN NG5 8QX
G0RVA NG5 9QB
G1HMZ NG5 9QN
G4PMM NG6 8XG

Gillingham

2E1BNN ME5 8LN
G6OLV ME7 1ND
G6RMA ME7 1QP
G0PJQ ME7 1RJ
GW0HPQ ME7 2DA
G3TCI ME7 2LN
G8STO ME7 2LN
G4IYC ME7 2PX
G1SGS ME7 2QX
G0JLP ME7 2SW
G0VQB ME7 2SX
G4WOP ME7 2TA
G3JRD ME7 2UW
G0VMJ ME7 2YH
G3FCV ME7 3AY
2E1EVI ME7 3EX
G6IVP ME7 3PT
G7IEW ME7 3PT
G0KSY ME7 3PY
G3VTT ME7 3PZ
2E1EVK ME7 3QE
G0TAR ME7 3QQ
G6ITM ME7 3RW
G7TUL ME7 3SJ
G7PEE ME7 3SQ
G0POS ME7 3TJ
G7NJG ME7 3TJ
G3VFC ME7 3TS
2E1EVJ ME7 3TS
G3ZYV ME7 4EP
G2BJW ME7 4EQ
G6TBJ ME7 4JB
G4LHU ME7 4LT
G1XIO ME7 4NN
G8STY ME7 5HY
M1ALM ME7 5QH
G4VRI ME7 5QJ
G7KAT ME8 0DJ
G7MNG ME8 0JX
G6OCO ME8 0LA
G8VVM ME8 0LZ
2E1EVM ME8 0LZ
G4RXH ME8 0ND
G6YLW ME8 0NR
G6AZN ME8 6DU
G0TCW ME8 6JE
G6HST ME8 6SZ
G6KSU ME8 6XT
M1BDS ME8 6XW
G6DVR ME8 7BP
G1SGT ME8 7DR
G7NCV ME8 7JP
2E1DQA ME8 7JP
G8MNL ME8 7QB
G6RXZ ME8 7SW
G1CBK ME8 7TA
G3PQD ME8 8HE
G4ZRY ME8 8JU
G0LNP ME8 8JY
G7AZM ME8 8JY
G1BUB ME8 8RD
G0AMZ ME8 8RP
G0FQD ME8 8SJ
G0DWF ME8 8TA
G1DXU ME8 8TL
G1JAC ME8 8TR
G0BXF ME8 9AQ
G4JJX ME8 9DX
G0APB ME8 9ES
G6PBS ME8 9HQ
G1ZBH ME8 9HX
G7MIM ME8 9JP
G3CRF ME8 9LF
G6YLS ME8 9LG
G0VGR ME8 9LT
G6YNA ME8 9NZ
M0AAK ME8 9PP
G7BRC ME8 9PP
G0BRC ME8 9PP
2E1DPD ME8 9PP
G0LVS ME8 9RG
2E1ESZ ME8 9SR

Gloucester

G7AEA GL1 2TG
G7AEB GL1 2TG
G7AEC GL1 2TG
G7AED GL1 2TG
G7AEE GL1 2TG
G7AEF GL1 2TG
G7AEG GL1 2TG
G7AEH GL1 2TG
G4UPG GL1 4JQ
G3MA GL1 5DY
G0MPZ GL1 5EL
G0EGS GL1 5ER
G0ECJ GL1 5SD
G8PGH GL14 X
G4RRK GL2 0EA
G1JMF GL2 0HA
M1BGO GL2 0HL
G3XMM GL2 0NQ
G0UEF GL2 0PJ
G0UWU GL2 0PX
G0WUW GL2 0PX
G3DXY GL2 0QY
G0EEA GL2 0RX
G1FYC GL2 0SJ
G0TYA GL2 0SJ
G0WNL GL2 0SJ
M0BEZ GL2 0UG
G1AXW GL2 4GD
G6ZES GL2 4GY
GW0OGE GL2 4LF
G6FGO GL2 4LZ
G4OIN GL2 4QP
G8MMG GL2 4SY
G6EFY GL2 4TB
G4LJP GL2 4UH
G0JQX GL2 4UY
G0KVO GL2 4WJ
G1ZND GL2 4WJ
G1YDT GL2 4YJ
G7NGN GL2 5NZ
G3GMN GL2 9DE

Great Yarmouth

G0MYC NR29 3HJ
G8VPE NR29 3HU
G7DXQ NR29 3LT
G1JYD NR29 3NS
G7BMY NR29 4AF
G0GGB NR29 4BH
G3ZEB NR29 4EA
G1EBP NR29 4ES
G4GLI NR29 4ES
G0IVE NR29 4EW
G8PEN NR29 4JD
G8SYV NR29 4JL
G3XAU NR29 4LT
G4MFX NR29 4LY
G1RPV NR29 4PT
G7ANJ NR29 4QF
G5WW NR29 4RQ
G4EMH NR29 4RY
G8UCZ NR29 4SH
2E1FQB NR29 4TW
G4BSH NR29 5EQ
G3XKK NR29 5ER
G3VUE NR29 5HB
G1PVJ NR30 1BP
G3YYQ NR30 1QU
G0OJJ NR30 2EE
G5NB NR30 2EF
2E1EGL NR30 2HS
G7RPJ NR30 2RU
M0ADZ NR30 3DG
2E1BDX NR30 4HU
G4ALC NR30 4LR
G6BXR NR30 4LS
G1VGH NR30 4LU
G4IVN NR30 5HD
G3NHU NR30 5NZ
G3YRC NR30 5NZ
G3WGL NR30 5QW
M1BUC NR30 5RH
G0POH NR30 5SB
M1CDV NR31 0BU
G1XDW NR31 0EJ
G4AIF NR31 0LB
G3OEP NR31 0PA
G4UGV NR31 6EY
G4ZJU NR31 6HT
G0SHJ NR31 6JB
G4OWV NR31 6TB
2E1AGQ NR31 7AS
G4CQC NR31 7NJ
G3AMK NR31 8DF
G3WPG NR31 8DN
G4GBT NR31 8DN
G7LNM NR31 8DR
G1HNU NR31 8DR
G8TAE NR31 8EJ
2E0ABQ NR31 8HP
2E1BNX NR31 8HP
G3HNP NR31 8JF
G3VSV NR31 8JG
G1XKW NR31 8JH
G1GER NR31 8ND
2E1FXS NR31 8ND
G7ATW NR31 8PB
G8VVG NR31 8QD
G0JIZ NR31 8SD
G7UGR NR31 8SY
G3BEY NR31 9AA
G0NFC NR31 9HP
G0FEI NR31 9JT
G1EFM NR31 9JT
2E1AFK NR31 9JU
G6KZI NR31 9LZ
G0VMK NR31 9NF
G0GGE NR31 9NX
G0BAW NR31 9PP
G0CLG NR31 9PP
G7WHM NR31 9PQ
2E1DQE NR31 9RW

Greenwich

G0MLV SE10 8LB
G0EPR SE10 9HG
G8YVQ SE10 9LY
G7GLW SE12 8JN
G3GDB SE12 9BG
G6WCX SE12 9DP
G3HPZ SE18 1PS
G1NPJ SE18 2DB
G7LSY SE18 2EX
G1CJK SE18 2RT
G8GJO SE18 3HF
G7PQO SE18 3HF
G8OOF SE18 6PF
G7OLI SE18 6SU
G4VFH SE18 7LN
G0EFB SE2 0JP
G4JGQ SE2 0PD
G3SPJ SE2 0QX
G4BLG SE2 0SR
G7PKU SE2 9AY
G7PKV SE2 9AY
G0ATB SE2 9HL
G0NZQ SE2 9NP
G6IDD SE28 0ER
G0RTZ SE28 8DU
G0CUA SE3 0EZ
G3RDK SE3 0NF
G3ZYE SE3 0NH
G1UKG SE3 7AH
G4FUG SE3 7BG
G8WBU SE3 7NN
G0NGP SE3 7PE
G1PYR SE3 7QB
G4RFC SE3 7QS
G0ILD SE3 7RQ
G4ZXV SE3 8HF
G6IRE SE3 8LR
G3TDU SE3 8NN
G1NTP SE3 8TT
G1KLW SE3 9PQ
G0BSH SE3 9QN
G1HBR SE7 7DW
G7SIR SE7 7QH
G1FAD SE7 8SH
G0OXE SE7 8TR
G0NRD SE9 1HQ
G7HSQ SE9 1NL
G4ZZN SE9 1PJ
G8XDR SE9 1QH
G0DCI SE9 1SE
G3PQK SE9 1UH
G8MFI SE9 1XQ
G8WNJ SE9 2AE
G4LEH SE9 2AH
G8UMB SE9 2HD
G0FDP SE9 2NZ
G3XCJ SE9 3BW
G0OWV SE9 3EB
G3IGZ SE9 3NX
G8DYN SE9 3NX
G1PDA SE9 3PB
G3TAA SE9 3SF
G3MIQ SE9 3SJ
G0ULL SE9 3XD
G3XMD SE9 3XE
G7GPJ SE9 5AW
G3LSX SE9 6RP

Guernsey

GU3MAT GY1 1EP
GU7CMH GY1 1FP
GU4RUK GY1 1FQ
GU0RAG GY1 1JB
GI7PVI GY1 1PQ
GU7CQN GY1 1QQ
GU4WQP GY1 1QW
GU3YVV GY1 1RP
GU7DSB GY1 1SA
GU4WRP GY1 1SF
GU4EON GY1 1WU
GU0ELF GY1 1XE
GU0JTG GY1 1XE
GU7DHI GY1 2AH
GU3HFN GY1 2BA
GU4SXM GY1 2BA
GU0FZS GY1 2LU
GU6JQF GY1 2PL
G3UZI GY1 2TA
GU0BFE GY1 2UD
GU3NDX GY1 2UD
GU6JSC GY1 2UD
2U1ELN GY2 4DN
GU3HKV GY2 4HJ
GU4HUY GY2 4HJ
GU4XGB GY2 4HJ
GU6AJE GY2 4HW
GU3WIY GY2 4JD
GU4ASO GY2 4UX
GU6TKE GY3 5AF
GU4GNS GY3 5AZ
GU4XEA GY3 5EN
GU4JHH GY3 5JD
GU4YBW GY3 5JG
GU4XGG GY3 5JP
GU0MNQ GY3 5UG
GU7TSI GY4 6DP
GU4WMG GY4 6JW
GU7PMC GY4 6LJ
GU7PMD GY4 6LJ
GU0VNF GY4 6LJ
GU6RWD GY4 6NB
GU1HYN GY4 6NH
GU4NYT GY4 6QJ
GU4SYQ GY4 6QJ
2U1FNQ GY4 6QJ
GU6NCZ GY4 6QS
GU6AYV GY4 6QS
GU7OYU GY5 7BN
GU3UOQ GY5 7DL
GU3UMX GY5 7DT
GU4YOX GY5 7DZ
GU0SUP GY5 7FQ
GU0JCI GY5 7HD
GU7NCZ GY5 7HH
GU0UJC GY5 7XE
GU4CHY GY5 7XZ
GU7NHX GY5 7YB
GU3OPC GY5 7YG
2U1EKH GY5 7YQ
GU0GUX GY5 7YS
GU0NHD GY5 7YS
2U1EJF GY5 7YS
GU3YIZ GY6 8BB
GU7IOJ GY6 8DG
2U1FPL GY6 8EF
GU1JFA GY6 8EX
GU3MBS GY6 8HB
GU7OMJ GY6 8HX
GU4WRO GY6 8LH
GU3WHN GY6 8LN
2U1FOA GY6 8NR
GU3LYC GY6 8RB
GU6EFB GY6 8RT
GU1HTY GY6 8RY
2U1EKE GY6 8RY
GU0BDI GY6 8SJ
GU0BGP GY6 8TT
GU0PYO GY6 8TT
GU8ITE GY6 8UF
GU6XCM GY6 8UG
GU0DXX GY6 8XE
GU7CNI GY6 8YJ
GU8IRF GY7 1HZ
GU1MUP GY7 9HE
GU0HRY GY7 9JD
GU0PSP GY7 9LD
GU6GDO GY7 9UH
GU8FBO GY7 9YJ
GU4WTN GY7 9YX
GU4LJC GY8 0AB
GU1EYP GY8 0ED
GU0TEW GY8 0HQ
GU1DWO GY8 0JB
GU3EJL GY9 3AF
GU3LPV GY9 3TT
G0RBX GY9 3UG
GU4TFM GY9 3XP
GU2RS GY9 9IV
GU3WOW GY9 9LA
GU8FSU GY9 9LE
GU4YZV GY9 9LE
GU3MLR GY9 9SE
GU2FRO GY9 0SA
GU1WJA GY99 1AA
GU4XIT GY99 1AA

Guildford

2E1DAS GU1 1DN
G0SWC GU1 1EP
G0SWE GU1 1EP
G0ADA GU1 1HJ
G7TVS GU1 1HP
G1MDS GU1 1HX
G8MAV GU1 1HX
2E1CML GU1 1JJ
G7MZS GU1 1NB
G6CMF GU1 1NX
G0KTV GU1 1QL
G5RS GU1 1TU
G4JZD GU1 1XT
G3WHM GU1 1XZ
G1ZOD GU1 1YF
G7UXW GU1 2EH
G6TWW GU1 2JB
G0EFO GU1 2JQ
G1RNV GU1 2QF
G8SKK GU1 2QJ
G8MCU GU1 2RR
G8ACJ GU1 2TF
G7DQG GU1 2UH
G7BAI GU1 3BB
G6AFK GU1 3NP
G3YXX GU1 3PA
G7VNL GU1 3PS
G4RSO GU1 3SJ
G8TZN GU1 4NP
G6ZAC GU1 4TL
G0VYQ GU10 1BY
G0CPE GU10 1EU
G7CGT GU10 1YZ
G4GKB GU12 5DT
G8PDP GU12 5HP
G4WEL GU12 5HS
G4WEM GU12 5HS
G8ZAJ GU12 5JG
G8BCO GU12 5JT
G3OST GU12 5NS
G0GMY GU12 5NS
G0NRO GU12 5QS
G8NEF GU12 5QW
G0IZY GU12 6LJ
G6TSX GU12 6PF
G1IYE GU12 6QB
G4WGX GU12 6QG
G0MGM GU12 6RB
G1PXM GU12 6SG
G0MIA GU12 6SP
G7HYA GU12 6ST
G7UBW GU2 5HB
G8UHK GU2 5JF
G6FKN GU2 5LA
G7OML GU2 5NU
G0SUL GU2 5QN
G0ONJ GU2 5RG
G7JVQ GU2 5RT
G8DXW GU2 5SR
2E1AOG GU2 5UT
G1NBR GU2 5XH
G3IGQ GU2 5XH
G7DQE GU2 5XH
G8AHK GU2 5XH
M1BBH GU2 5XH
M0ATD GU2 5XH
M1BCS GU2 5XH
M1BII GU2 5XH
M1BJY GU2 5XH
G7HNW GU2 6AP
G8LOZ GU2 6AP
G8FUH GU2 6AR
G3TYR GU2 6AW
G1MGF GU2 6BL
G0GFP GU2 6BU
G1IPA GU2 6BU
G4IPA GU2 6BU
M1BKP GU2 6GG
G4PMZ GU2 6JU
G8NOB GU2 6PJ
M1AZN GU2 6PL
G4ECF GU2 6PW
G6ROQ GU2 6PW
G4HQQ GU2 6QN
2E1CVO GU2 6QY
G8IMS GU2 6TG
2E1EKU GU2 6TG
G4NBC GU2 6TU
2E1ELJ GU2 6TU
G0JOP GU23 6DQ
2E1BNU GU23 6LP
G4EML GU23 7AP
G8RVV GU23 7AT
G5OD GU23 7BX
G6KXK GU23 7EL
G8AFU GU23 7ET
G4HZA GU23 7HP
G3IUW GU23 7HR
G3XRP GU23 7HS
G7TPM GU24 0JQ
G0SJH GU3 1AZ
G4GIX GU3 1DR
G8YEF GU3 1HQ
G4HZV GU3 2EN
G0PET GU3 2EU
G3VCY GU3 2JW
G4SXL GU3 2JW
G4PNB GU3 3AU
G3TQC GU3 3BX
G4MPH GU3 3EG
G7AWL GU3 3HG
G3UKN GU3 3HR
G6GEK GU3 3NF
G4MPW GU3 3PR
G8CLK GU3 3RD
G3PIZ GU4 7DQ
G7UCR GU4 7FE
G7UPN GU4 7HB
G1LKJ GU4 7HG
G1OYT GU4 7HG
G1IYY GU4 7JD
G4XFW GU4 7JF
G3PMD GU4 7JH
2E1CUY GU4 7JJ
G7VNJ GU4 7JJ
G3CDE GU4 7JQ
G3GFR GU4 7NU
G0KUF GU4 7PD
G7TKF GU4 7TF
M1ACK GU4 7TJ
G1PXG GU4 8AY
G3XON GU4 8DD
G4BHQ GU4 8DE
2E1AEF GU4 8ER
G0JRE GU4 8JL
G1IKL GU4 8JX
G1GQK GU5 9AF
G4ZEM GU5 9LZ
G1SWZ GU7 2RW
G3ENI KT24 5BU
G3TVS KT24 5BU
G4XGL KT24 5HJ
G8GWM KT24 5NR
G0RAL KT24 5SD
2E1GBR KT24 5SN
G7NJI KT24 6ED
M1BFG KT24 6ED
G8VH KT24 6HZ
G3NIW KT24 6QJ
G4PED KT24 6QN

Gwynedd

GW0MYK LL23 7BN
GW0NDA LL23 7BN
GW4WPH LL23 7NL
GW4UFQ LL23 7SF
GW0WUM LL33 0LH
GW4MSI LL35 0HT
GW4EJT LL35 0LP
GW0TWO LL35 0PT
GW1NGX LL36 0DW
GW7TAA LL36 0DW
GW6INF LL36 0TA
GW4IUY LL36 0TF
GW1JPF LL36 9AT
GW4XXF LL36 9DE
GW7RJC LL36 9DE
GW6DHC LL36 9EA
GW0VSC LL36 9EP
GW4UWD LL36 9EP
GW6JDF LL36 9HG
G3LTX LL36 9ND
GW4JPP LL36 9PS
GW4KYK LL36 9PS
GW4XZJ LL36 9UY
GW4LZP LL37 2JQ
GW3SB LL37 2JQ
GW1URD LL37 2JZ
GW6JJV LL37 2JZ
GW6NXL LL37 2UZ
GW4LNK LL38 2BQ
GW3RJR LL38 2RQ
GW4KEV LL39 1LJ
GW0NDP LL40 1HU
GW7VCH LL40 1LR
GW4URJ LL40 1YE
GW3GKZ LL40 1YF
GW7TTX LL40 2BN
GW1ZBE LL40 2EH
G0UCA LL40 2EW
GW0RQM LL40 2EW
GW0MMW LL41 3AQ
GW4KYZ LL41 4EP
MW1BVF LL41 4NE
GW4UJT LL41 4SE
GW0HHD LL42 1DX

GW4KDP LL42 1HT
GW6DDF LL42 1YP
GW8WNB LL43 2AG
GW0AYQ LL43 2BB
GW4ILF LL44 2DN
GW3FDZ LL44 2RQ
GW6JBN LL45 2HH
GW0AGL LL45 2HT
GW6ZAN LL45 2HT
GW0OPY LL45 2LU
GW0UKF LL45 2LU
GW4YBV LL46 2SS
G1TPS LL46 2UA
GW6JLH LL46 2UB
GW0GSW LL46 2UJ
MW1BQO LL48 6EF
GW6IMS LL48 6EG
GW4PHB LL48 6LS
GW1EMZ LL48 6NH
GW7RLZ LL48 6PS
GW7TSO LL48 6RE
GW0MVI LL48 6RU
GW2HCJ LL48 6RU
GW4HLO LL49 9BU
GW0SAU LL49 9BU
GW6AZX LL49 9NB
MW1BTV LL49 9PF
GW4OWX LL49 9SR
MW1AMS LL51 9SX
GW8FVI LL52 0EF
GW7UOH LL52 0PE
GW4IDC LL52 0SR
GW8PKB LL52 0SR
GW0PTX LL53 5AU
GW0TXM LL53 5AU
GW2HFR LL53 5BL
GW0HGC LL53 5HN
GW0SZB LL53 6AD
GW6KLC LL53 6DW
GW3NNB LL53 6EG
GW3LFC LL53 6EH
GW4YSV LL53 6EH
G0DFK LL53 6LS
GW1HPP LL53 6PS
GW7KIO LL53 6RD
GW0RMB LL53 6UB
GW3NLN LL53 6UH
GW0IWD LL53 6UY
GW4KGD LL53 7BT
GW3COI LL53 7BU
GW4TWJ LL53 7EE
GW4UKU LL53 7NW
GW4WKQ LL53 7NW
GW0SEO LL53 7PA
MW0AEV LL53 7PF
GW4GFS LL53 7RF
GW0PZT LL53 8AE
GW3KJW LL53 8AE
G8BZN LL53 8EH
GW4UIL LL53 8NT
GW1IEB LL53 8NW
GW1DKK LL54 5ET
GW4XRW LL54 5HG
GW4XZP LL54 5HG
GW6ZZP LL54 5HG
GW1LEL LL54 5HR
GW6UWV LL54 5NH
GW3IEQ LL54 5TW
GW1OXJ LL54 5UB
MW1BLG LL54 6DH
GW1RQF LL54 6PL
GW4CAT LL54 6RT
GW0CSR LL54 6SG
GW1CAV LL54 7BB
GW1KEU LL54 7BB
GW0PEB LL54 7DH
GW0GLX LL54 7ED
GW8HQM LL54 7LB
GW0DYT LL54 7PA
GW3RRI LL54 7PD
GW0ENT LL54 7YE
GW6IXA LL55 1EY
GW6MUP LL55 1EY
GW4MEI LL55 1HF
GW3HCL LL55 1LN
GW7FAE LL55 1PS
GW1SXN LL55 1SW
GW1PIH LL55 1YB
GW6REQ LL55 1YB
GW4KAZ LL55 1YL
GW7CEQ LL55 2AG
GW0AQR LL55 2UB
GW4RLP LL55 2UR
GW1JIE LL55 3HD
GW6CWZ LL55 3LU
GW6NHL LL55 3LU
GW4JMN LL55 3PP
GW6KFH LL55 3PW
GW8XUM LL55 3PW
GW3SDK LL55 4AH
GW6OSS LL55 4AN
GW4BYY LL55 4EN
GW1XPP LL55 4PS
MW0ANV LL55 4TE
GW7VFP LL55 4YG
GW4GTC LL56 4NZ
GW0MHO LL56 4NZ
GW0DXO LL57 1NG
2W1AZU LL57 1NH
GW0IAU LL57 1TH
GW6PXF LL57 2AX
GW0HGN LL57 2UD
GW6ORE LL57 3BU
GW0EGF LL57 3EG
GW8GOO LL57 3TD
GW3YCD LL57 4AT
GW7EWD LL57 4AX
GW3VFZ LL57 4DP
GW0ETF LL57 4DR
GW6WMB LL57 4DX
GW4FQU LL57 4LD
GW6MYY LL57 4NN
GW4HMR LL57 4NS
GW0FMQ LL57 4PG
GW4TTA LL57 4PG
2W0AMX LL57 4PN
GW0KOD LL57 4RR
GW3REY LL57 4TE
GW3YNM LL57 4TN
GW0ALF LL57 4UR
GW4HHD SY20 9LA
GW0DHQ SY20 9LB
GW4NMQ SY20 9LB
GW7GAH SY20 9RD

Hackney

G6XHM E5 0JR
G3WCK E5 0LW
G4SHO E5 0RG
G3WIP E5 0RR
G1TXQ E5 9AN
2E1FQN E5 9BB
G4CCJ E5 9EQ
G3NBZ E8 1AP
G0PRE E9 5NY
G3RBW E9 5PT
G0TPO E9 6DA
G3SJW E9 7EH
G6WLZ E9 7HG
G7OHI E9 7HN
G3WUB E9 7HU
G7MWH N1 5EH
G4LOU N16 6JU
G6ADK N16 6NT
G1PJI N16 7DA
G0LWL N16 7EA
G1TXW N16 7TB
G0PLC N4 1HN
G0IFD N4 2ES

Halton

2E1BYO L24 5RR
G4OEX WA4 6GY
G3MUX WA4 6UA
G6AGT WA4 6XH
G8YAZ WA7 1DH
G0MQH WA7 1QW
G4VRG WA7 1UB
G1LCC WA7 1UH
G1GJM WA7 2AN
2E1CAE WA7 2DT
G7WFS WA7 2DT
2E1CCZ WA7 2EF
G6FPX WA7 2JW
G0NWE WA7 2LG
2E1DTR WA7 2NH
G1PIX WA7 2QS
G4ZKF WA7 2RF
G0MGD WA7 2RL
G0HES WA7 2TJ
G8HLQ WA7 3AQ
2E1CQI WA7 3AR
G0VQL WA7 3AR
G4LUQ WA7 3JH
G8SIM WA7 4BG
G4WPG WA7 4EH
G0ICD WA7 4ER
G6MPN WA7 4HZ
G7VCP WA7 4JG
G6KLR WA7 4NZ
G7NCU WA7 4SQ
G0MPP WA7 4SX
G4HOD WA7 4XL
G4YZP WA7 4YU
2E1DSM WA7 5FA
G0RLF WA7 5JX
2E1BZH WA7 5JX
2E1BZI WA7 5JX
G4OAB WA7 5LR
G0MPY WA7 5PT
G0HEF WA7 5RF
G0NSL WA7 5RJ
G1MSY WA7 5RW
G0FWR WA7 5YU
G1GFQ WA7 5YU
G1SES WA7 5YU
G0SGU WA7 5YU
G8VNF WA7 6AA
G0RBM WA7 6BN
G0MIX WA7 6DD
G0MSF WA7 6DQ
G6KSO WA7 6DR
G1JMP WA7 6DT
G7IAE WA7 6DW
G3FBH WA7 6HT
2E1FUA WA7 6LE
G0BSD WA7 6LJ
G0RHW WA7 6NN
2E1CCC WA7 6NN
2E1CNX WA7 6NN
G0RVW WA7 6QB
2E1CEP WA7 6QD
G0SBI WA7 6QG
2E1DSF WA7 6QQ
G1LYI WA8 0BX
2E1DTK WA8 0BX
2E1EBL WA8 3HT
G0KDJ WA8 4NP
G8BJQ WA8 4TB
G0ELM WA8 4TR
G0CJW WA8 4XU
2E0AIL WA8 5AS
G7UXU WA8 6HR
G6VTQ WA8 7DP
G1CIV WA8 7EU
G0PHB WA8 7JT
G0TYK WA8 7QR
G1HKQ WA8 7XB
G0FQP WA8 8BG
G1BGH WA8 8JX
G8RIB WA8 8SS
G4YJS WA8 8UY
2E1FKE WA8 8YQ
G8CMP WA8 9BN
G7KHW WA8 9FU

Hambleton

G0GBG DL6 1DZ
G0MJH DL6 1DZ
G0JQA DL6 1ED
G3KJX DL6 1ED
G0LIY DL6 1EE
G8FLV DL6 1EE
G1JST DL6 1EP
G0WGZ DL6 1HD
G1XRE DL6 1HP
G7VIL DL6 1JD
G7UXH DL6 1JS
G7VIM DL6 1QN
G0LEL DL6 1QQ
G7NKJ DL6 1SJ
G7HHK DL6 2AA
G7ITF DL6 2AF
G1JER DL6 2BD
G3MAE DL6 2BE
G0GCK DL6 2QD
G4CAL DL6 2SE
G1HLV DL6 2SN
G4IJO DL6 3AQ
M0BBK DL6 3AQ
G0LLM DL6 3EL
G0AWE DL7 8JJ
G1DZR DL7 8JJ
G3FTU DL7 8JL
G7COC DL7 8NG
2E1CQQ DL7 8NG
G3VIS DL7 8NH
G1EDE DL7 8QY
G0RRQ DL7 8UR
G0EVL DL7 8XN
G7PLR DL7 9QA
G7VIC DL8 1UG
G1WWH DL8 2HT
G3HSG DL8 2JB
G4AFU DL8 2JE
G3LEO DL8 2LP
G4NUY DL8 2QF
G6RJ DL8 2QQ
2E1DZZ DL8 2TQ
G4VUN HG4 4AU
G4JLJ TS15 0HX
G4OSA TS15 0JQ
G4OIW TS9 5AE
2E1DHO TS9 5BN
G0TYM TS9 5BU
G0UYG TS9 5HF
G4RHA TS9 5HX
G7RNQ TS9 5HX
G7UZG TS9 6BH
G7TFG TS9 6JF
G6ESR TS9 6PJ
G1PWD TS9 7AN
G0KTW TS9 7AN
G6BKQ YO6 1ET
G0VYO YO6 1HA
G1AVZ YO6 1HW
G1TLE YO6 1JR
G0GBU YO6 1LF
G6GUW YO6 1LF
G6NIZ YO6 2DF
G3TMN YO6 2ER
G0UTF YO6 2ES
G4NYI YO6 3HH
G0HVG YO6 3JD
G3ZNB YO6 3RE
G6JCM YO6 3SF
G4KNV YO6 3SY
G6XZF YO6 4SW
G8RJZ YO7 1FW
G1FGI YO7 1JN
G4SPC YO7 1JN
G3BQL YO7 1NS
G6JTI YO7 1RT
G7SKH YO7 2DJ
G3HHD YO7 2LL
2E1DOC YO7 3HS
G0LAN YO7 3NL
G4ETM YO7 3NL
G0ASG YO7 4DR
G0VXH YO7 4QP
G7VBN YO7 4RT

Hammersmith and Fulham

G0HFK SW6 1AT
G6JMB SW6 2RR
G1REH SW6 2UU
G3SFX SW6 3TA
G6ZY SW6 5AG
G0ADG SW6 5BG
G0ERC SW6 7LN
G1GXP SW6 7RB
G6LGO SW6 7RB
G7TTM SW6 7TQ
G8EZT W11 4TG
G7AAS W12 0AP
G0JRY W12 0BP
G3SEF W12 0LP
G4HXX W12 0NN
G3IJL W12 7BL
G8MVC W12 8BP
G1HST W12 8NU
G0RKA W12 9TD
G6MUQ W14 0BJ
G4DLK W14 0BL
G7HMU W6 0UA
G7VLF W6 8HN
M0AXP W6 9AY

Harborough

G1WVR LE16 7DD
G4WVR LE16 7DD
G8GMB LE16 7DD
G0SFJ LE16 7DE
G6OFZ LE16 7LQ
G8PAN LE16 7LQ
G3YKS LE16 7PQ
G6FXL LE16 7UN
G4DIA LE16 8BQ
2E0AIH LE16 8BQ
M1AMF LE16 8BS
G3XQZ LE16 8DR
G1NQH LE16 8SJ
G4LNO LE16 8TY
G4JYT LE16 8UJ
G0PGM LE16 9AG
M0BFP LE16 9BJ
G4EOF LE16 9DW
G7PZB LE16 9DX
G1UUX LE16 9EL
G1WIY LE16 9JS
2E0AHN LE16 9JS
G1PHV LE16 9JW
G3SFV LE16 9LW
G1IVF LE16 9NW
G4WTW LE16 9QL
G1WJG LE16 9TW
G4WES LE17 4AP
G4WER LE17 4DR
G0JEW LE17 4DY
G4APD LE17 4DY
G6LGM LE17 4EN
G0AMD LE17 4HZ
G1LVY LE17 4HZ
G4DDW LE17 4LA
G0BDF LE17 4PG
G3ORY LE17 4PS
G1KCR LE17 4QB
G0TTW LE17 4SQ
G0OLS LE17 4TR
G7FFZ LE17 4TR
G0CZN LE17 4TT
G0KAK LE17 4TZ
G1PPU LE17 4TZ
G7DMH LE17 4UR
G3RIR LE17 4XB
G8TKQ LE17 4XB
G3RSF LE17 4YR
G4VSJ LE17 5AS
G6GO LE17 5BZ
G4VOZ LE17 5DE
G8LM LE17 5DE
G0IHO LE17 5DG
G1JOY LE17 5DL
G6TZT LE17 5DL
G8ZUF LE17 5DL
G4IYS LE17 5HW
G4NGF LE17 5HX
G4NVJ LE17 5HX
G8AYN LE17 5PP
G1DPT LE17 5RP
G6FIQ LE17 6BE
2E1AFD LE17 6DU
2E1BGO LE17 6EE
2E1BHQ LE17 6EG
2E1AGF LE17 6EQ
2E1AGG LE17 6EQ
G7LCK LE17 6EQ
2E1BKK LE17 6EQ
G1NWU LE17 6JE
G4VTM LE17 6JW
G6YQU LE17 6NT
2E1BAC LE17 6PE
G8XDQ LE2 2FH
G4ZJR LE7 9DA
G6ISM LE7 9GW
G4JPO LE7 9HA
G7OEY LE7 9HD
2E1CJZ LE7 9HD
G3LVB LE7 9HE
G3URQ LE7 9JS
G0FNO LE7 9JT
G6HQE LE7 9PD
G0TEC LE7 9PF
G3VOV LE7 9PP
G4DR LE7 9PQ
G4NLS LE7 9PU
G1LPJ LE7 9PU
G8DGH LE8 0AP
G8BTU LE8 5WJ
G8CJA LE8 5WJ
G4EQL LE8 8AP
G8FWA LE8 8BD
G7DWC LE8 8BH
G6HAJ LE8 8BN
G3ZPG LE8 8TW
G4IHR LE8 8TX
G8RBI LE8 9FH
G3KQI LE8 9FH
G8HMJ LE8 9FH
G6ZYQ LE8 9FR
2E1BAD LE9 6EN
G0PBY LE9 6NQ
G0LDK LE9 6PL
G0LDO LE9 6PT
G0CRT LE9 6QE
G6NOZ LE9 6RB
G1EDU LE9 6SA

Haringey

G6CNO N10 2EG
G0IPT N10 3EH
G3PNC N10 3TR
G3KOD N10 3UG
G0JUZ N11 2AB
G8YNC N11 2DE
G6FDX N11 2PR
G8MAF N11 2QS
G8OJR N15 4JN
G7KGV N15 6AB
G3YBS N17 0JD
G7MNK N17 0HH
G1GPS N17 0PH
G8PHJ N17 0PH
G0CEU N17 0TE
G8ASC N17 6BP
G0JBB N17 6PH
G3AFY N17 6PH
G1DSY N17 7AH
G3OVX N17 7HT
G3FVL N22 4RU
2E1FJQ N22 4SR
2E1FVJ N22 6LH
G8EWG N22 6NG
2E1FVK N22 6NT
G6HPO N4 3SN
G1JJR N4 4NH
G0FWF N4 4NU
G8EEI N4 4QD
G7UHK N6 4AB
G4HDD N6 4AR
G4NBS N6 4BA
G8MIC N6 4BA
G3XKY N6 4EU
G0RIC N6 5AU
GW4OXG N6 5TH
G4NNQ N8 0NS
G7LHO N8 7LP
G4VCV N8 7RN
G0BYH N8 9HL

Harlow

G8BUI CM17 0AE
G8MHA CM17 0AR
G6ECK CM17 0BJ
G3UXI CM17 0BN
G3UUY CM17 0HQ
G8MHO CM17 0HQ
G8ICC CM17 0HX
G0BXL CM17 0JY
G3UEG CM17 0LH
G7BNF CM17 0LL
G7REF CM17 0LL
G0CPU CM17 0LQ
G7ULS CM17 0SB
G4MQN CM17 9BZ
G7UFP CM17 9DD
G8WTB CM17 9ER
G6TXW CM17 9PA
G0AWY CM17 9PL
G0ULI CM17 9QB
G0IJO CM18 6BE
G7IZW CM18 6EB
G0HDZ CM18 6ES
G3RYK CM18 6EY
G3JWW CM18 6HE
G4BYR CM18 6HR
G3LCX CM18 6JP
G4XUQ CM18 6QJ
G8VQE CM18 6QN
G7PFM CM18 6RH
G1WUA CM18 6RZ
G8NCZ CM18 6SD
G8POE CM18 6ST
G6TED CM18 6XL
G4ITL CM18 6XW
G3WRO CM18 7BY
G0APH CM18 7BY
2E1ERN CM18 7BY
G7RDE CM18 7DB
G3PRN CM18 7DD
G3YEB CM18 7DD
G6OPL CM18 7DF
G3TOF CM18 7EH
G8GDC CM18 7HQ
G3JMA CM18 7JJ
G0HRR CM18 7QD
G1FRL CM18 7QD
M0BEO CM18 7SS
G7VID CM18 7SS
G0TIH CM18 7SU
G1FIM CM18 7SY
G8FEQ CM19 4AA
G8RLF CM19 4JW
M1BTB CM19 4JW
G4ZCF CM19 4NB
G1SFU CM19 4PJ
G4UJN CM19 4PW
G0WDJ CM19 4QY
G7RFZ CM19 4RD
2E1CRH CM19 4RD
G8CUA CM19 5NB
G0SRB CM19 5NB
G6GXO CM19 5NU
G1ILG CM19 5PA
G0REF CM19 5RN
G0CZZ CM20 1TN
G1OSE CM20 1TW
G6BUT CM20 2LE
G3RSP CM20 2PQ
G3YMJ CM20 2PQ
G4GGX CM20 2PZ
G0WUZ CM20 3BJ
G8CHO CM20 3DP
G4MQG CM20 3DY
G3RKN CM20 3EF
G4CDO CM20 3EF
G6UXX CM20 3EF
G7UFF CM20 3EY
G4YBN CM20 3HB
G6UT CM20 3HB
G3VAS CM20 3JU
G4IND CM20 3LD
G8XGO CM20 3LE
G8ICD CM20 3PX
G7AZA CM20 3QN

Harrogate

G0MKK HG1 1TY
G0NHG HG1 2BW
G4CWB HG1 2HA
G7NSP HG1 2JY
G4LQG HG1 2LS
G7FTS HG1 2NE
G4PUK HG1 2PL
G0HCA HG1 2QG
G7WCW HG1 2QG
G7WCZ HG1 2QG
G7WDA HG1 2QG
M1ATQ HG1 2QG
M1BBO HG1 2QG
G7WCX HG1 2QN
G6WKO HG1 2SS
G0LSC HG1 3AL
G0UIW HG1 3BQ
G4JRE HG1 3DB
G1VQK HG1 3EA
G0HUC HG1 3JF
G4FIP HG1 3JW
G6HAT HG1 3LH
G1BQZ HG1 4BG
G7EFG HG1 4BY
G4RWJ HG1 4ER
G4ODS HG1 4ET
M0AVF HG1 4HW
G0KIY HG1 4QR
G1CCX HG1 4QZ
G0IEF HG1 4RU
G1NSL HG1 4RU
G1JLB HG1 4SG
G0WIQ HG1 4SJ
G3GVY HG1 4TF
G4PBA HG1 4TF
G7VGV HG1 5EF
G4LGX HG1 5HS
G1WUC HG1 5JU
M0ACP HG1 5NP
2E1BDU HG2 0AQ
G4AYP HG2 0BG
G3NVX HG2 0BJ
G4GTU HG2 0EB
G3OGZ HG2 0LL
G1TUX HG2 0LS
G4OLB HG2 0PJ
G8XZC HG2 7AJ
G4YOP HG2 7AQ
G7FTD HG2 7BB
G4VZS HG2 7BP
G4XSL HG2 7DT
G4TJI HG2 7ER
2E1EOX HG2 7HB
G1UIB HG2 7HQ
G1ARH HG2 7LJ
G1ROD HG2 7LJ
M1BDZ HG2 7PL
G0IJZ HG2 7RH
G7RZV HG2 8AF
G4DTE HG2 8BB
G7XWH HG2 8DD
G3XWH HG2 8DD
G0MAO HG2 8DL
G7BBJ HG2 8DZ
G3JSB HG2 8EH
G1IAY HG2 8EJ
G1OFY HG2 8EQ
G4MNE HG2 8HW
G0UTW HG2 8HZ
G7CFP HG2 8JG
G4EKJ HG2 8LB
G4ITB HG2 8LE
G4KEN HG2 8LX
G0JTA HG2 8PZ
G0JWT HG2 8QA
G1BPE HG2 9BS
G4ABF HG2 9BY
G0IXC HG2 9HF
G3YHC HG2 9HP
G4TJH HG2 9JU
G3GJV HG3 1AG
G4IUF HG3 1JR
G4STT HG3 1LU
G1BHO HG3 1LY
G3NXL HG3 1NF
G0LUJ HG3 1QH
G0JNG HG3 1QX
G0CVL HG3 2AR
G7JHM HG3 2BU
2E0AGI HG3 2DG
G1ZBO HG3 2DG
2E0AGL HG3 2DG
G7RWZ HG3 2DG
2E1BMI HG3 2DG
2E1BDV HG3 2DG
G3FMW HG3 2DS
G0VSR HG3 2DS
2E1CIX HG3 2DS
G0MVK HG3 2DX
2E1CJC HG3 2DZ
G4LOH HG3 2LH
2E1BLO HG3 2PP
G8MJF HG3 2PX
G0TWY HG3 2RF
G0UDT HG3 2RF
G0VYH HG3 2RF
M0AKC HG3 2RF
M0APA HG3 2RF
M0AYQ HG3 2RF
G8VPE HG3 2SH
G6RII HG3 2UG
G1NSP HG3 2XH
G0KHQ HG3 2XH
G4CWG HG3 2XU
G3RHR HG3 3BX
G4PAH HG3 3DS
G4VTK HG3 3DS
G4ZCZ HG3 3HJ
G4EEV HG3 3JR
G4RVA HG3 3LA
G3JDC HG3 3QL
G0WIF HG3 4AL
G6VIQ HG3 4BH
G4KIC HG3 4EB
G3RYP HG3 4EW
G0KGT HG3 4HX
G4YGN HG3 4HX
G6EQT HG3 4LA
G4JKY HG3 5DR
G4XGD HG3 5DR
G6FXO HG3 5EB
G6FBT HG3 5LR
G7FTW HG4 1HA
2E1DYC HG4 1LF
G7VDP HG4 1LF
G4DSC HG4 1RZ
G0MRQ HG4 1SN
G0MTV HG4 1TD
G4ZVJ HG4 1UA
G7VJR HG4 2DL
G4AFT HG4 2JJ
G4VMY HG4 2LJ
G6MGH HG4 2LJ
G7HIW HG4 2LN
G1VKC HG4 2ND
G3UVY HG4 2PG
G7NSQ HG4 3JZ
G1PWM HG4 3LL
G7VBA HG4 4JH
G7LYN HG4 4LH
G0WQB HG4 5BL
G8TBX HG4 5BY
G4CNG HG4 5HW
G3RGV HG5 0AA
G3HRU HG5 0AJ
G7DMG HG5 0AW
M0AQH HG5 0LW
G7COU HG5 0NY
M0AHU HG5 0TH
G8FVK HG5 8EF
G6GUU HG5 8EL
G3XQJ HG5 8EY
G4KIV HG5 8HQ
G7IPX HG5 8HQ
G6HEB HG5 8JN
G3DUW HG5 8JT
G3XZV HG5 9DE
G3UNA HG5 9EL
G3LHQ HG5 9HN
G4UBL HG5 9HQ
G0GYU HG5 9LD
G3XPM HG5 9LS
G1FOD LS17 0AW
G1RRG LS21 2RS
G8JTD LS21 2RS
G7OAV LS22 4EH
G1PUY LS22 5JW
G4DBP YO5 8AL
G4LCX YO5 8BL
G3BJN YO5 8BP
G0STK YO5 8DE
2E1CUZ YO5 8DE
G0TCR YO5 8DL
G3OXQ YO5 8JA
G1ANV YO5 8LW
G6XNR YO5 8PR
G0LPX YO5 8PY
G0GYP YO5 8QD
G4OVX YO5 8QN
G4KWZ YO5 9DP
G0RHI YO5 9DY
G0MVX YO5 9LP
G1WCY YO5 9LP
G4SJM YO5 9LP
G7AKD YO5 9NF
G8MBO YO5 9QY
G4EPG YO5 9RU
G6NJO YO7 3PF
G8TJH YO7 3PH
G8ZNL YO7 3PX
G0UFI YO7 4PS

Harrow

G0NRI HA1 1PB
G8SUG HA1 1QF
G0AKI HA1 1QP
G7CTP HA1 1QP
G3ZCR HA1 1SB
G3WUX HA1 1TA
G0UHK HA1 3HT
G3VFX HA1 4AJ
G4IQD HA1 4AL
G4MSE HA1 4AL
G6RTY HA1 4AQ
G8BBD HA1 4BS
G1SVL HA1 4PL
G4FVJ HA1 4TY
G4HAB HA2 0PR
G7OZA HA2 0PU
G3UVM HA2 0RX
G4TPM HA2 0SX
G4JMG HA2 6AP
G4IPX HA2 6AX
G3OND HA2 6DG
G6SSV HA2 6DG
G0CAG HA2 6HE
G3LGQ HA2 6HF
G3NQR HA2 6LF
G8LSZ HA2 6LQ
G6MFR HA2 7AX
G4PKV HA2 7EJ
G4AFV HA2 7HW
G3FNJ HA2 7JJ
G4KNT HA2 7QF
G4IRP HA2 7RB
G4UYB HA2 8AT
G6TJJ HA2 8SN
G4KZK HA2 9AZ
G4KEW HA2 9EA
G4WVW HA2 9JL
G4MMA HA2 9LD
G7IEF HA2 9LJ
G4UKD HA2 9QF
G7VVT HA2 9UH
G2FHK HA2 9UJ
G6DOF HA3 5PL
G5HY HA3 6AG
G0TKU HA3 6AG
G0IPK HA3 6AJ
G3VHK HA3 6BX
G7DYU HA3 6SW
G0BSP HA3 6TN
G3PIH HA3 7JF
G0GXM HA3 7JX
G0IOK HA3 7JX
G3LSY HA3 7LZ
G1RNL HA3 7NP
G7JLH HA3 8ED
G7JLI HA3 8ED
G6DOE HA3 8HA
G3BQE HA3 8JB
G0WKJ HA3 8LE
G7DNP HA3 8LZ
G4KEP HA3 8NF
G4ZPR HA3 8NQ
G3BPI HA3 8PA
G6XML HA3 8RY
G6FVF HA3 8RZ
G8FAT HA3 9EJ
G8WKI HA3 9JY
G7HIH HA3 9LB
G1TMW HA3 9NB
G0VWL HA3 9NZ
G0IGP HA3 9PA
G3ASR HA3 9PE
G3SJE HA3 9PE
G7JSX HA3 9PR
G0SQL HA3 9PR
G6ARU HA5 1AQ
G0IRM HA5 1JH
G6MRT HA5 1LE
G4UBB HA5 1LN
G5IJ HA5 1LN
G4PXY HA5 1NB
G1RGT HA5 1NJ
G0KSL HA5 1TL
G4AZL HA5 2BQ
G6XUD HA5 2BQ
G1FBS HA5 3LP
G4KDO HA5 3QN
G3ENV HA5 4EP
G0WEI HA5 4HS
G3UAS HA5 4LN
G0WHV HA5 4NJ
G0NTA HA5 4RA
G0NTB HA5 4RA
G7PCH HA5 4RL
G7RVD HA5 4RL
G1VIO HA5 4TX
G3NAN HA5 5BZ
G7VHO HA5 5JQ
G7VHN HA5 5JQ
G4VFF HA5 5LL
G0HAT HA5 5PJ
G3HB HA5 5ST
G3EYC HA5 5SW
G7LVE HA7 1HY
G7TUI HA7 1JE
G3PPI HA7 1PG
G8CYG HA7 2HL
G8KLH HA7 2LD
G1HLK HA7 2NH
G1ZDU HA7 2PA
G0DDM HA7 2PT
G0BBD HA7 3AG
M0AIU HA7 3HR
G7VRS HA7 3JR
G1DQQ HA7 3PL
G0MTG HA7 3SS
G8XGK HA7 4AG
G1MYM HA7 4EP
G3NDC HA7 4LD
G2MT HA7 4LY
G6NVI HA7 4QN
G8RNM HA8 6DL
G3TLU HA8 6TQ
G4EMZ UB5 4BP

Hart

G3RSI GU10 5BD
G8IXL GU10 5PA
G4JEP GU10 5RG
G3HUK GU13 0AS
G8IKG GU13 0BB
G4AFI GU13 0BN
G8HOP GU13 0EG
G8TSZ GU13 0NH
G7KIT GU13 0NR
G0HKR GU13 0PN
G3RCD GU13 0PW
G3CYL GU13 0QJ
G3VFB GU13 0SH
G1JFR GU13 0SU
G6DPW GU13 0TE
G1GSY GU13 0YF
G3MNB GU13 8AT
G3UZK GU13 8AY
G1CNV GU13 8DP
G3OGM GU13 8EP
G4EFY GU13 8HA
G4FTK GU13 8UR
G7IIH GU13 9AH
G3JJU GU13 9BP
G3NVM GU13 9DE
G4IIY GU13 9DG
M1AEE GU13 9DG
G3MSL GU13 9DY
G4OQZ GU13 9EB
G0LEX GU13 9EG
G7MCS GU13 9EL
G1KAJ GU13 9JD
G3LGT GU13 9LE
G1JPP GU13 9LN
G7AFE GU13 9NB
G3UCT GU13 9QS
G6YWT GU13 9TB
G6FOH GU13 9TG
G3BRQ GU13 9UZ
G4NYY GU13 9XE
G6KRY GU17 0DJ
G7GRN GU17 0DZ
G1FOE GU17 0EN
G4VDF GU17 0JE
G0RVI GU17 0JU
G4AZF GU17 0LA
G0AZY GU17 0LZ
G8CIG GU17 0NE
G8WMJ GU17 0PJ
G8HIO GU17 9BS
G0ADT GU17 9DE
G4CGW GU17 9DL
G3ZYX GU17 9DX
G3REL GU17 9DZ
G3SFU GU17 9DZ
G3TMU GU17 9DZ
G4EJA GU17 9JH
G3XRE GU17 9LP
G8CHI GU46 6EA
G8FWE GU46 6EH
G7CFW GU46 6PE
G7AOA GU46 6YP
G4BWD RG27 0LD
G1RPG RG27 8AA
G4JKK RG27 8DS
G3YGE RG27 8JS
G8DBD RG27 8NZ
G3YFL RG27 8QX
M0AET RG27 8SW
G4NEJ RG27 8TZ
G1OPT RG27 9EJ
G1CSS RG27 9EQ
G7SLV RG27 9EQ
G4SFH RG27 9EY
G4XKQ RG27 9HJ
G6CEK RG27 9NT
G3RXI RG27 9QZ
G4BEZ RG27 9RA
G8VIV RG27 9RN
G6ETX RG27 9RP
G7SOX RG29 1AE
G4HRE RG29 1AL
G1ZBL RG29 1BH
M0AEU RG29 1BH
G7VSP RG29 1EJ
G7FBF RG29 1NR
G7SCE RG29 1NT

Hartlepool

G0PXN TS22 5NH
M1BNK TS24 0DS
G7ZRT TS24 0PZ
G1GZJ TS24 8ET
G7GRQ TS24 8EU
G7VGM TS24 8NN
G4ZCN TS24 8NX
G3IJV TS24 8RB
G4VCJ TS24 9BQ
G4APN TS24 9RP
G0MHC TS24 9RP
G3HSL TS24 9SA
G4XJV TS25 1DU
G8TCH TS25 2HG
M1AIE TS25 2JJ
2E1DOE TS25 2JJ
G1DFZ TS25 2LA
G7VGO TS25 2RD
G1GTP TS25 3DD
G3GIL TS25 3EB
G7TUF TS25 4ES
G0GFF TS25 4HR
G3NWY TS25 5HX
G0NJK TS25 5NA
G0FZZ TS25 5RQ
G1YHP TS26 0PJ
G3NUA TS26 0PJ
G1HXO TS26 0PQ
G7VUN TS26 9ES
G3NWU TS26 9JE
G4PDM TS26 9PD
G4SHJ TS26 9PR
G7SUQ TS27 3EN
G0EGJ TS27 3PP
G7JXY TS27 3QR

Hastings

G6VHL TN34 1TS
G0OOU TN34 1UU
G1HHH TN34 2BD
G4FET TN34 2BD
G6HH TN34 2BD
G2LL TN34 2BD
G4XRN TN34 2ET
G7PIP TN34 2JA
G0GKL TN34 2JE
G0GRK TN34 2JE
2E1ASF TN34 2JE
G6JIF TN34 2NF
G4UWF TN34 2NQ
G6TFE TN34 2PJ
G7WWW TN34 2PS
G3HAO TN34 2QE
G1VNY TN34 2QJ
G4BCO TN34 2QJ
G4KMJ TN34 2RL
G3KMP TN34 2RN
G7VQO TN34 2RT
G3UFI TN34 2SF
G4JWL TN34 2UD
G7KMA TN34 3LT
G4WCP TN34 3LW
G4ITM TN34 3LZ
G0MXE TN34 3NL
G6ZRF TN34 3NS
G4WBZ TN34 3TA
G6HVX TN35 5AX
G0FUU TN35 5EP
G4YMP TN35 5HS
G7AMP TN35 5HS
G4NVQ TN35 5HZ
G4PWE TN35 5JD
G0CJO TN35 5JN
G0RIR TN35 5LE
G4RIT TN35 5LE
2E1AJW TN35 5LW
G4FEF TN35 5NQ
G7MNY TN37 6HB
G7OJY TN37 6JY
G1ORK TN37 6PN
M1BFH TN37 6PU
G8BQX TN37 6QA
G4GKW TN37 6QR
G1GFF TN37 6RS
2E1BDW TN37 6RS
G1IFW TN37 6SD
G1LZS TN37 7AR
G4YZF TN37 7AZ
G6YZF TN37 7AZ
G3ZFX TN37 7BS
G4EOA TN37 7HJ

G6DKB	TN37 7HN
G2DJA	TN37 7HQ
G0OOX	TN37 7JB
G0UZI	TN37 7LS
G1DVU	TN37 7PS
G1EBX	TN37 7QU
G7TCH	TN38 0HX
G0VPC	TN38 0NF
G8CMK	TN38 0QA
G8SBJ	TN38 0QA
G6WKL	TN38 0SN
G0AQT	TN38 0TR
G8XXM	TN38 0XP
G4UAL	TN38 0YP
G6VLE	TN38 8AU
2E1EWX	TN38 8AU
M1BWP	TN38 8BH
G7STL	TN38 8DN
G1EBW	TN38 8DW
G7UMU	TN38 8EQ
G1EHS	TN38 9BU
G1EHU	TN38 9BU
G4VBK	TN38 9DP
G7DME	TN38 9EE
G8AU	TN38 9HP
G7WGA	TN38 9HS
G7OPM	TN38 9LA
G1FTX	TN38 9QW
G7JVN	TN38 9RS

Havant

G2AKY	PO10 7LR
2E1EHJ	PO10 7LR
G1MNP	PO10 7NA
G8RUX	PO10 7ST
G4OZX	PO10 7TR
G4FRZ	PO10 7UL
G0RXQ	PO11 0AE
G3VNU	PO11 0AZ
G4KNL	PO11 0BT
G3JVL	PO11 0DT
G3MCV	PO11 0ER
G0AYI	PO11 0JW
G7KQT	PO11 0JW
G3JMG	PO11 0NJ
G4ZMY	PO11 0PD
G3XUQ	PO11 0PE
G4NQ	PO11 0PQ
G0KCF	PO11 0QE
G8DLL	PO11 0QE
G8VMY	PO11 0QR
G3NDO	PO11 0RL
G6VXZ	PO11 0RL
G8ZJK	PO11 9AX
G7HBS	PO11 9NS
G4JCC	PO11 9PJ
G0RIB	PO11 9PL
G0WEH	PO11 9RP
2E1DPK	PO11 9RP
G0URO	PO11 9SN
G3JOF	PO11 9SU
G3SLD	PO7 5DH
G0ABB	PO7 5ED
G0JSC	PO7 5EX
G0JST	PO7 5EX
G7JST	PO7 5EX
G6HGE	PO7 5HB
G7KZI	PO7 5HH
G6RST	PO7 5HH
G1TDQ	PO7 5HH
G0IEY	PO7 5HW
G0IUY	PO7 5HW
G7HOG	PO7 5LD
G7HOI	PO7 5LD
G4JIH	PO7 5NZ
G0IVW	PO7 5QP
G1WXW	PO7 5QR
G6BHB	PO7 5QT
G8HAT	PO7 5QT
G6ISY	PO7 5QW
G7IUE	PO7 5TB
2E1FPN	PO7 5XB
G1UFA	PO7 6AA
M0AHM	PO7 6AZ
G0JWL	PO7 6BT
G1XPY	PO7 6BW
G0PSF	PO7 6BX
G0RPV	PO7 6DD
G0ASZ	PO7 6DP
G4CRM	PO7 6XA
G4SNL	PO7 7BJ
G8VOI	PO7 7BJ
G7GNA	PO7 7EW
G1OGR	PO7 7JE
G8UXW	PO7 7LA
G3TVI	PO7 7PB
G3KOJ	PO7 7PX
G3WYT	PO7 7QQ
G3YBF	PO7 7QR
G1FMT	PO7 7RG
G4FOW	PO7 7SP
G0III	PO7 7TE
G8PIQ	PO7 7UB
G4VIQ	PO7 7XP
G7HQW	PO7 8BA
G3FJ	PO7 8BX
G3UUS	PO7 8HF
G4PRG	PO7 8HT
G8KOS	PO7 8ND
G1LTZ	PO7 8NE
G8OKE	PO7 8QD
G4TTJ	PO7 8QY
G4SAC	PO7 8RX
G6IOV	PO8 8AX
G0DNF	PO8 8BB
G3RBY	PO8 8BG
G7OAH	PO8 8DY
G3WLY	PO8 8HX
G0NHZ	PO8 8JE
G0KXB	PO8 8JH
G8LNC	PO8 8JJ
G7WGI	PO8 8JL
M0BFK	PO8 8LE
G3VCR	PO8 8PN
G3TKN	PO8 8QH
G3NKO	PO8 8QL
2E1COC	PO8 8RP
G8YTR	PO8 8RU
2E1BOM	PO8 8RX
G0JRN	PO8 8SE
G1YDW	PO8 8SQ
G3LIK	PO8 8SQ
G4JXO	PO8 8SQ
G6EHL	PO8 9BE
G7GWZ	PO8 9EJ
G0KUA	PO8 9QH
G6WXN	PO8 9RW
G0WVM	PO8 9SP
G4YJF	PO9 1DT
G0BSJ	PO9 1NS
G3EIW	PO9 1RT
M1CAD	PO9 1RT
G0FYX	PO9 2AE
G3COO	PO9 2AE
G4FBS	PO9 2AE
G1FFU	PO9 2HR
G0NUL	PO9 2PP
G0VKX	PO9 2QW
G8LWC	PO9 2RW
G4XQZ	PO9 2RZ
G0JEZ	PO9 2UW
G8VEZ	PO9 3AA
G6XRH	PO9 3BW
M0AFG	PO9 3DE
G4HUM	PO9 3DX
M1BSX	PO9 3JZ
2E1DSA	PO9 3JZ
2E1DRW	PO9 3JZ
G7VDN	PO9 3JZ
G4UXJ	PO9 3NJ
G7FWW	PO9 3NX
G4NKX	PO9 3QY
G4RGO	PO9 3RA
G4AYW	PO9 3RG
G6QY	PO9 3TF
G0RGT	PO9 3TG
G0DHZ	PO9 4DG
G3IFF	PO9 4DG
G0RHD	PO9 4DS
G1HED	PO9 4NL
G3KTN	PO9 4NQ
G0DOK	PO9 4PZ
G7EYV	PO9 5BJ
G8XNO	PO9 5DZ
G0ERS	PO9 5PW
G8VIU	PO9 5PW
M1BIY	PO9 5TQ

Havering

G4EPM	RM1 2EU
G6BTX	RM1 4BZ
G4YVH	RM1 4DY
G4FVC	RM1 4EF
G4FQF	RM1 4HD
G6MVS	RM1 4JP
G3XLR	RM1 4LA
G4DXJ	RM1 4PL
G6TGM	RM1 4SP
G8TQZ	RM1 4XR
G8NNJ	RM11 1AG
2E1FVT	RM11 1AG
G4ZST	RM11 1AR
G8HRC	RM11 1AX
G3KXP	RM11 1BD
G7RYM	RM11 1BN
G0PGO	RM11 1EU
G8SOG	RM11 1NA
G4HRC	RM11 1XA
G6WWZ	RM11 2AG
G7IYN	RM11 2AJ
G0BSF	RM11 2BA
G0CBU	RM11 2BS
G1SUW	RM11 2HX
G3LXJ	RM11 2PR
G0TUM	RM11 3AJ
G4DEL	RM11 3BS
G8RXH	RM11 3DJ
G8XBF	RM11 3DJ
G0HTE	RM11 3EA
G0RIQ	RM11 3EN
G0PIU	RM11 3LD
G0PIA	RM11 3PD
G3JHI	RM11 3PY
G1VSH	RM11 3QD
G8WUO	RM11 3QH
G6WKN	RM11 3QT
G4VIX	RM11 3SG
G0TJN	RM11 3SG
G0VHF	RM11 3SG
G3VOF	RM11 3TY
G3KCN	RM12 4BJ
G3SFK	RM12 4EY
G7VVL	RM12 4JS
G0IAP	RM12 4LL
G3JWH	RM12 4NR
G1SQB	RM12 4QT
G4FYW	RM12 4RA
G1GXX	RM12 4SQ
G4NHR	RM12 4YF
G8OCV	RM12 4YF
G0AJH	RM12 5BL
G6BEL	RM12 5EU
G8PMR	RM12 5LS
G3KPZ	RM12 5NX
G6EBI	RM12 5PR
G8UZW	RM12 5QS
G3XYA	RM12 5RD
G4MXF	RM12 6AT
G3ZWF	RM12 6BD
G4BLU	RM12 6HD
G4INT	RM12 6HD
G4DBC	RM12 6JJ
G4EEF	RM12 6TF
G8ZDT	RM13 7BS
G4RZZ	RM13 7EJ
G0PWG	RM13 7QS
G4JRB	RM13 8AJ
G3SFM	RM13 8AL
G0IEO	RM13 8BB
G4YPN	RM13 8JD
G8NWI	RM13 8ND
G1HGQ	RM13 9JP
G1TXR	RM13 9JR
G0SDE	RM13 9LA
G0PBN	RM13 9LW
G1BTF	RM13 9LW
G6TRX	RM14 1AL
G0FDI	RM14 1AS
G0KSJ	RM14 1AT
G8DQJ	RM14 1DH
G4RJC	RM14 1ER
G0IQN	RM14 1HN
G4AZX	RM14 1HP
G0DOM	RM14 1NB
G0PUL	RM14 1NG
G3RJI	RM14 1QU
G3USZ	RM14 1RL
G8XQZ	RM14 1XR
G1EYJ	RM14 2BL
G7OCG	RM14 2ER
G4LMQ	RM14 2HP
G4UNS	RM14 2JH
GW1CLZ	RM14 2LX
G6EVX	RM14 2PS
G4VIW	RM14 3AA
G7GES	RM14 3AH
G3JWF	RM14 3AJ
G7DXX	RM14 3AU
G4AKF	RM14 3AY
G3FFY	RM14 3XD
G4LQX	RM14 3YT
G8FBV	RM14 3YU
M1ASV	RM2 5PL
G3ZQY	RM2 5PP
G4XXT	RM2 5XJ
2E1FVR	RM2 5XL
G3TPJ	RM2 6AP
G4GBW	RM2 6LD
2E1DOA	RM2 6RU
G3FKJ	RM3 0BL
G8VXR	RM3 0BQ
G4TTN	RM3 0DL
G0BCZ	RM3 0EB
G3SKV	RM3 0EJ
G3YUR	RM3 0ND
G8ZKZ	RM3 0QU
G8PP	RM3 0RS
G3NYI	RM3 0TT
G4WIE	RM3 0WL
G4EZL	RM3 0XY
G3SVK	RM3 0YT
G4FNJ	RM3 7AD
G0CBT	RM3 7DX
G3XBF	RM3 7DX
G1WZK	RM3 7UB
G4UXT	RM3 7UB
G6MDN	RM3 8HJ
G8TPP	RM3 8HN
G6WKI	RM3 8HQ
G3RCQ	RM3 9DE
M1BFX	RM3 9LA
G1CSN	RM3 9NA
G4YOT	RM3 9NL
G3YNC	RM5 2DS
G4IKI	RM5 2HB
G6EBO	RM5 2HF
G4MWG	RM5 2QX
G4ZMS	RM5 2SD
G4NOO	RM5 2TJ
G4PBD	RM5 3BE
G0BLO	RM5 3RX
G6HTL	RM5 3YL
G8YIN	RM7 0LA
G0BOF	RM7 0NJ
G7UKH	RM7 0NP
G7HHT	RM7 7LB
G7KLQ	RM7 8AP
G8RQQ	RM7 8AX
G7WJE	RM7 8BD
G0RZT	RM7 8ET
G0GFZ	RM7 8HD
G1LHR	RM7 8HD

Hereford

G0JWJ	HR1 1DG
G3LZM	HR1 1EQ
G7KVT	HR1 1HQ
G3XTD	HR1 1JD
G4MET	HR1 1LX
G1YBG	HR1 1LY
G4SJZ	HR1 1NB
G1HWP	HR1 1NH
G6KWZ	HR1 1QL
G4LZW	HR1 1RB
G4DTB	HR1 1RQ
M1BQQ	HR1 1SD
2E1FZI	HR1 1SD
M1CAR	HR1 1SS
G0DJF	HR1 1TT
G4OJP	HR1 1XB
G8IVO	HR1 1XD
G1CKR	HR1 1XP
M1CEI	HR1 2EY
G0IRD	HR1 2RL
G0RQF	HR1 2ST
2E1CAX	HR1 2ST
2E1GBH	HR1 2ST
G0OGX	HR1 2TS
G3ESY	HR1 2TY
G3RJB	HR2 6DJ
M1BZE	HR2 7AE
G2DFL	HR2 7AN
G1YBB	HR2 7DF
G7PXV	HR2 7DS
G1KGE	HR2 7JQ
G1VZC	HR2 7JQ
G4MBJ	HR2 7LU
G7IAK	HR2 7NZ
G4ZYP	HR2 7QG
G7BLM	HR4 0ET
G4IUO	HR4 0EU
G1INY	HR4 0JJ
G4LNZ	HR4 0LP
G7KZW	HR4 0NT
G0UDF	HR4 0PP
G1NSR	HR4 0QP
2E1GAN	HR4 0QR
G4JSN	HR4 0RP
2E1FOI	HR4 0RR
G4GOG	HR4 7SW
G4HZT	HR4 9NE
G0WJM	HR4 9QJ
G1YFC	HR4 9TD
G6XPY	HR4 9TJ
G8FFA	HR4 9TJ
G4NVX	HR4 9TQ
G3WRA	HR4 9TY

Hertsmere

G4XVV	EN6 1HA
G4GEL	EN6 1QR
G8LFY	EN6 1QR
G3BSF	EN6 2AP
G3SHS	EN6 2AT
G0MDR	EN6 2BB
G7HGS	EN6 2LL
G8NCW	EN6 2QD
G6LKA	EN6 3AB
G4WSL	EN6 3AF
G3RYQ	EN6 3AG
G0VSA	EN6 3AJ
M0ATL	EN6 3AJ
G3RTE	EN6 3DY
G4HTE	EN6 3DZ
G7BHR	EN6 3DZ
G0ANN	EN6 3ES
G0NGH	EN6 3HB
G4HCY	EN6 3HG
G1HZD	EN6 5EG
G0WDL	EN6 5HF
G3WFM	EN6 5HU
G4ECT	EN6 5HU
G3CCM	EN6 5JT
G3ROK	EN6 5LT
G4GPU	EN6 5LX
2E1DBP	EN6 5LZ
G1XPB	EN6 5NB
G7RHI	EN6 5NB
G3UNP	WD2 1AU
G4SWY	WD2 1LU
G3UBL	WD2 1LW
G7KRP	WD2 1NJ
G3KNJ	WD2 2BU
G7DGK	WD2 2EJ
G0LZV	WD2 3AR
G2TA	WD2 3JL
G8XHN	WD2 3NU
G8XXV	WD2 3QT
G4RNW	WD2 3QU
G8RFZ	WD2 3SH
G4GYS	WD2 3SS
G0UUT	WD6 1HH
G4DEY	WD6 1SP
G4TEP	WD6 2DA
G0DDJ	WD6 2ET
G0PQB	WD6 2HE
G6APH	WD6 2HG
G7DFW	WD6 2PB
M1BCB	WD6 2PJ
G8YYC	WD6 2QZ
G8XJN	WD6 2RB
G4GPL	WD6 3LH
G4TIH	WD6 3NJ
G1INI	WD6 3PT
G1SLQ	WD6 4JD
G1JAG	WD6 4ST
G6RHN	WD6 5PE
G8RCO	WD6 5QE
G6OKA	WD7 8JD
G6CXY	WD7 8JF
2E0AID	WD7 8NQ

High Peak

G0BAY	SK13 8DZ
G4YXA	SK13 8HF
G1YKK	SK13 8NW
G6LLL	SK13 8NW
G6XWF	SK13 8QQ
G4IFJ	SK13 8RG
G0WJS	SK13 8RJ
G0MAA	SK13 8SG
G0GLO	SK13 9AJ
G4GNQ	SK13 9AJ
G4LMR	SK13 9AJ
G8BEQ	SK13 9BG
G1GEL	SK13 9DY
G6WTS	SK13 9EB
G6MAO	SK13 9NJ
G8WAH	SK13 9NJ
G0EQH	SK13 9UJ
G8OWB	SK13 9UX
G4DQF	SK14 6EU
G4HJX	SK14 6HJ
G6BIX	SK14 6HP
G1XCB	SK14 7AR
G8FUV	SK14 7HE
G7BRR	SK14 7NB
2E0AGB	SK14 7PN
G6YRK	SK14 8BA
G4KWF	SK14 8BQ
G4YLQ	SK14 8DT
G0UPO	SK14 8DT
G4NJL	SK14 8DX
G4BJT	SK17 6HH
G1OSA	SK17 6HN
G1OHD	SK17 6LH
M1AGN	SK17 6LN
G7TMM	SK17 6NX
G0WHQ	SK17 6XL
G0JNI	SK17 7BW
G8YTX	SK17 7DX
G0KLR	SK17 7EQ
G0GHV	SK17 7JJ
G0NIA	SK17 7NP
G7NFK	SK17 7PE
G0MUR	SK17 7PU
G0RKT	SK17 7PW
G4IKK	SK17 7QN
G0WSP	SK17 7TJ
G1INK	SK17 7TW
G7OGS	SK17 8AU
G4MRQ	SK17 8DW
G4SPA	SK17 8DW
G1WTX	SK17 9AD
G0TGS	SK17 9AG
G7KGZ	SK17 9BE
G0SMP	SK17 9BE
G0JND	SK17 9BG
G8YHX	SK17 9BG
M0AVK	SK17 9HG
G6XMQ	SK17 9HG
G4IHO	SK17 9JS
G7NXA	SK17 9JS
G0XKK	SK17 9JS
G7TZN	SK17 9JU
G6REN	SK17 9JX
G6ZHS	SK17 9LJ
G6SIR	SK17 9LQ
G7NHL	SK17 9PL
G7UBE	SK17 9PL
G4RHB	SK22 1BX
G7OJA	SK22 2JG
G0VOR	SK22 2JL
G7KUB	SK22 3AY
G4KVG	SK22 3DB
G3HQH	SK22 3EA
G0NLL	SK22 3JT
G4YAB	SK22 4DP
G6LJB	SK22 4EJ
G7AUB	SK23 0PR
G0CLP	SK23 6BQ
G4CDZ	SK23 7LH
G6SVH	SK23 9TT

Highland

GM7RJG	IV1 1NS
GM0JFK	IV1 1XD
G3MUA	IV1 1XG
GM7MWX	IV1 1XG
GM4LXM	IV1 1XU
2M0AOF	IV1 1YG
GM3EXS	IV1 2DB
GM4FKD	IV1 2ER
GM4UOD	IV1 2ES
GM4UPL	IV1 2ES
MM1AEL	IV1 2HB
2M1CJO	IV1 2HB
2M1EAU	IV1 2HB
GM2CWL	IV1 2JT
GM0IQD	IV1 2NN
GM3WKM	IV1 2PX
MM1AAY	IV1 2RB
GM0PZV	IV1 2RF
GM1BQP	IV1 2RG
2M1CKB	IV1 2ST
GM4XJX	IV1 2SX
GM0TWK	IV1 2SY
GM0NTI	IV1 2UE
GM0OTI	IV1 2UE
GM8RTI	IV1 2UE
MM1BYY	IV1 2XQ
2M1ECF	IV1 2XR
GM0UDL	IV10 8RA
2M0ALS	IV10 8RA
GM7VZF	IV10 8SN
2M1AHZ	IV10 8TJ
GM4KLN	IV10 8XA
GM4OHY	IV11 8XF
GM3JFG	IV11 8XY
GM4IEO	IV11 8YW
G7CSY	IV12 4TF
G3YZU	IV12 4TU
GM3PIL	IV12 5AF
GM7BCC	IV12 5EW
GM0RML	IV12 5PJ
GM1VAD	IV12 5RA
GM4VIK	IV13 7XY
GM3LVA	IV13 7YR
GM4RRP	IV14 9DJ
GM4ZIT	IV14 9EF
GM7NTJ	IV15 9HY
GM8DRA	IV15 9PF
GM4EWL	IV15 9PF
GM8LEA	IV15 9PG
GM7GIL	IV15 9XE
GM4TTD	IV17 0TU
GM4KKW	IV17 0UJ
GM1FOS	IV17 0YT
GM3SAE	IV18 0BG
GM0KJX	IV18 0ER
GM2MP	IV18 0PE
GM3WOJ	IV18 0PE
GM4DMZ	IV18 0PE
GM4MFL	IV18 0PL
GM4FDT	IV18 0PN
GM3ZSH	IV18 0PY
GM0HMR	IV18 0QA
GM4SUF	IV19 1LB
GM0OMC	IV2 3HT
GM1MYR	IV2 3HT
GM8FMR	IV2 3HW
2M1EBJ	IV2 3SQ
GM4VST	IV2 3SX
GM4JAE	IV2 4AZ
MM0BDK	IV2 4BB
GM0GEE	IV2 4EN
MM0BAG	IV2 4EX
GM8DFC	IV2 4LD
GM0SXP	IV2 4LD
GM4XGA	IV2 4LX
GM4UND	IV2 4LZ
GM7RVR	IV2 4XJ
GM0CSZ	IV20 1RX
GM0WXX	IV22 2HB
GM0UVO	IV22 2NS
GM3VIO	IV25 3HN
GM0HNJ	IV25 3HU
GM4VDL	IV26 2TY
GM7ASN	IV27 4ED
GM4BYT	IV27 4JD
GM4NGY	IV27 4PZ
GM7OQE	IV27 4XL
GM4UZR	IV3 5ES
GM4TPF	IV3 5PH
GM4XKG	IV3 5PH
GM3TMK	IV3 5PZ
GM0HAQ	IV3 5QH
GM0AKJ	IV3 6LH
GM7GVD	IV3 6PF
GM0VPG	IV3 6PN
2M1ECJ	IV3 6PN
GM0RIV	IV3 6SD
GM4JNB	IV3 6SD
GM4VAC	IV3 6TA
GM7HLI	IV3 6TF
GM0TNK	IV3 6TF
G0BZS	IV3 6TN
GM4WBO	IV3 6UR
GM6JJN	IV3 6YJ
GM4GXR	IV4 7AB
GM4LNU	IV4 7BE
GM4TJD	IV4 7EY
GM4UMA	IV4 7EY
GM8PHU	IV4 7HH
2M1AIR	IV4 7HZ
G7ORJ	IV4 7HZ
GM4GZD	IV4 7JQ
GM3YND	IV4 7LP
GM0OZZ	IV40 8EQ
GM6KUL	IV41 8PJ
GM4PUS	IV42 8PY
GM4ATB	IV43 8QS
GM0HBK	IV44 8RD
GM0EZX	IV47 8SL
GM0PNS	IV49 9BP
G4RXD	IV5 5WL
GM3SWK	IV51 9DN
GM0BTL	IV51 9DT
GM0EWX	IV51 9PE
GM0DXE	IV51 9PW
GM4NTL	IV53 8UP
GM4TRH	IV55 8GA
GM3WZV	IV6 7RS
GM0OGZ	IV6 7SB
GM4SFW	IV7 8AJ
2M1DZX	IV7 8AJ
GM0JFL	IV7 8EY
GM0HBM	IV7 8HJ
MM1ATO	IV7 8HL
GM0CJT	IV7 8HY
GM3WED	IV7 8JH
GM4FZT	IV7 8JY
GM1OWV	IV7 8LG
GM2ASU	IV7 8LH
GM7JYW	IV8 8PA
GM4OIJ	IV8 8PG
GM7IBM	IV8 8PG
GM6UYC	IV9 8PX
GM1OVJ	IV9 8QL
GM1OIN	IV9 8QT
GM4WHD	KW1 4HA
GM4JUE	KW1 4NP
GM7WLE	KW1 4NP
GM3SYO	KW1 4NY
GM4MPR	KW1 4RG
GM3JDR	KW1 4XP
GM6KON	KW1 4XY
GM3TCM	KW1 5AD
GM4XLN	KW1 5AS
GM3VCZ	KW1 5SB
GM7JGH	KW1 5SS
GM2DRB	KW1 5TT
GM0HBI	KW10 6RH
GM0IYP	KW10 6RH
MM1AUD	KW10 6SH
GM4NHX	KW12 6XJ
GM6FYY	KW12 6YN
GM1WMS	KW13 6YT
GM6TMH	KW14 7ES
GM4RZJ	KW14 7HS
G1HYO	KW14 7LL
GM6TJD	KW14 7ND
GM3SFH	KW14 7QN
GM3POT	KW14 7QP
GM8YRE	KW14 7XB
GM8YRX	KW14 7XB
GM7UOJ	KW14 8AQ
GM7NUQ	KW14 8BQ
GM1AHC	KW14 8QB
GM3CFS	KW14 8SN
GM7OUR	KW14 8SR
GM1VGZ	KW14 8TD
GM6JRX	KW14 8UB
GM4FUH	KW14 8UB
GM4EFR	KW14 8UT
GM0TKB	KW14 8XN
GM0KHP	KW14 8YD
GM1WIB	KW14 8YD
GM0EXN	KW14 8YE
GM1BAN	KW14 8YT
GM8LFB	KW3 6BA
GM4UGG	KW7 6HA
GM0HLV	KW8 6HP
GM0OAV	KW9 6NG
GM7PXL	PA39 4LA
GM3KTD	PA40 4RQ
2M1CLY	PH20 1BT
GM1NZD	PH20 1DT
GM4XWS	PH21 1JU
GM8ZJL	PH21 1LD
GM4TGC	PH21 1NN
GM4ZRQ	PH22 1RR
GM0JKS	PH23 3AS
GM4LPG	PH26 3PA
GM3FUT	PH31 4AQ
GM0PKB	PH32 4BW
GM3BCX	PH32 4DP
GM0ANG	PH33 6HB
GM4YAT	PH33 6HT
GM3RFA	PH33 6JS
GM7SPB	PH33 6JS
GM1KWA	PH33 6LJ
GM1YGV	PH33 6NX
GM1SMF	PH33 6QA
GM7REY	PH33 6QD
GM4TFJ	PH33 6RQ
G7PKT	PH33 6SE
GM0UMJ	PH33 6SZ
GM4OFC	PH33 6UA
GM8AOB	PH33 6UG
GM4NFI	PH33 6UH
GM4HFD	PH33 7AL
GM6AES	PH33 7AL
GM4UGN	PH33 7AN
GM0ERB	PH33 7ER
GM1YPJ	PH33 7HN
GM3WML	PH33 7HU
GM4OPU	PH33 7LE
GM1IKQ	PH33 7PQ
GM4TJL	PH34 4EX
GM4XJY	PH41 4QF
GM6VYZ	PH41 4QF
GM6VYY	PH41 4QY

Hillingdon

G3KRT	HA4 0AU
G4YGM	HA4 0BT
G8DZK	HA4 0DB
G0ACK	HA4 0DS
G0DVC	HA4 0DW
G4HFU	HA4 0SL
G3SGT	HA4 0TH
G3UUP	HA4 0TH
G3SPZ	HA4 6AA
G0RNF	HA4 6AJ
G7GQS	HA4 6AJ
G3EIE	HA4 6BZ
G0NXE	HA4 6EA
G0DZH	HA4 6ED
G8FKV	HA4 6PN
G0LCY	HA4 6QS
2E1FHY	HA4 6QZ
G0NRN	HA4 6SX
G0ODT	HA4 7LN
G3IKG	HA4 7LS
G0JIM	HA4 7LZ
G6JIM	HA4 7LZ
G0VTC	HA4 7LZ
G8UOZ	HA4 7QD
G6TDX	HA4 7RD
G3MSS	HA4 7SU
G5NZ	HA4 7TG
G4UZE	HA4 7UN
G0AOT	HA4 7UT
G4FFA	HA4 8AJ
G4CLB	HA4 8ED
G8UMY	HA4 8PH
G3EFX	HA4 8QF
G4AUF	HA4 8QF
G3OEC	HA4 8RY
G1DES	HA4 8SB
G0JCF	HA4 8SB
G7KUG	HA4 8SD
G3BPG	HA4 8TA
G0LHC	HA4 8TP
G6JSN	HA4 8TW
G6SYB	HA4 8UA
G3JVM	HA4 9AG
G8KXW	HA4 9AN
G4JUJ	HA4 9AP
G0HOG	HA4 9BL
G3WCY	HA4 9DA
G6PLR	HA4 9HD
G3UPW	HA4 9JF
G1LQV	HA4 9JT
G6SPI	HA4 9LF
G0KFI	HA4 9PR
G4KMH	HA4 9SJ
G1JMG	HA5 1PE
G6JP	HA5 1PE
G7HSO	HA5 1PH
G7VSS	HA5 2AU
G8CAB	HA5 2DA
G3EFP	HA5 2DF
G3LFD	HA5 2EH
G8NTR	HA5 2NB
G3HCY	HA5 2NF
G7MXG	HA5 2SH
G3SWT	HA5 2SR
G3RZS	HA5 2TF
G4EAK	HA5 3TU
G6LFD	HA5 3UP
G4MEX	HA6 1DW
G4JZS	HA6 1EZ
G0BST	HA6 1JE
G4FBK	HA6 1JJ
G3SIK	HA6 1JT
G1BHT	HA6 1NB
G2PT	HA6 1RF
G7VUE	HA6 1SX
G8MAA	HA6 2UJ
G4IUM	HA6 3NH
G6KNK	UB10 0BS
G1DYR	UB10 0HH
G6IRY	UB10 0HP
G4DIG	UB10 0QH
M1ALX	UB10 0SB
G6RJZ	UB10 8BY
G4BVF	UB10 8ED
G0OQK	UB10 8JY
G4IPJ	UB10 8LN
2E1FHX	UB10 8LP
G8VNO	UB10 8NB
G4XAN	UB10 8PR
G4PGF	UB10 8QA
G6DSB	UB10 8QA
G6SND	UB10 8QA
G8LIU	UB10 8QA
G8UBF	UB10 8QA
G6JPJ	UB10 8QY
G6LEY	UB10 8TA
G7KLR	UB10 8UE
G0NPF	UB10 9DP
G8OLI	UB10 9JN
G4PRI	UB10 9JN
G8ABU	UB10 9LZ
G0LQW	UB3 1PZ
G3SUN	UB3 1ST
G0GES	UB3 1TE
G8HOU	UB3 2JE
G0TCK	UB3 2NG
G6WWT	UB3 2NT
G6XDI	UB3 2QX
G4YLC	UB3 2TN
G8GYP	UB3 2TP
G6STI	UB3 2TQ
G1VTS	UB3 3PA
G7JSW	UB3 3PP
G4JQU	UB3 3PX
G0EFS	UB3 4AD
G1MDQ	UB3 4AD
G7BBC	UB3 4JX
G4HIP	UB3 4JX
M1BVD	UB3 4QJ
G3PKR	UB3 5ET
G4XEX	UB3 5HH
G3AHE	UB3 5HZ
G4WYM	UB3 5LR
G0CHR	UB4 0AE
G7DQL	UB4 0AE
G0RET	UB4 0AY
G8UKH	UB4 0BQ
G0ISY	UB4 0DY
G0RGL	UB4 0EF
G0VGY	UB4 0EF
G1LTH	UB4 0HX
G1LLQ	UB4 0JH
G1LQT	UB4 0PD
G0CHQ	UB4 8AT
G1FCT	UB4 8ET
G4OHQ	UB4 8ET
G3CFW	UB4 8EX
2E1EQB	UB4 8NH
G4ZIX	UB4 9DS
G1IQN	UB4 9PB
G1LVR	UB4 9RB
G4KGA	UB4 9SX
G4TGN	UB7 0HY
G4HVE	UB7 0JB
M0AAC	UB7 0LE
G1XGL	UB7 7AB
G4PJN	UB7 7AJ
G0FFK	UB7 7AN
G8WRC	UB7 7NU
G0INZ	UB7 7NX
G0LZP	UB7 8HN
G0BEA	UB7 8LB
G0KYN	UB7 8LB
G7RPW	UB7 8PE
G6URJ	UB7 9AN
G4GTT	UB7 9AX
G0FFN	UB7 9DR
G8MYV	UB7 9NQ
G0VUZ	UB7 9PD
G4ZRC	UB7 9PE
G8FKH	UB8 1BL
2E1CAD	UB8 1BL
G4NWT	UB8 1PH
G3JVP	UB8 1QX
G8EWL	UB8 1UL
G1HIE	UB8 2AE
G4UQH	UB8 2DR
G7CIK	UB8 2EU
G7NFR	UB8 2PQ
G0DUZ	UB8 2UA
G4CVF	UB8 2UL
G6XCZ	UB8 3AR
G1BMC	UB8 3HS
G0KBQ	UB8 3LD
G0ULF	UB8 3PB
G6JHG	UB8 3QU
G4DGX	UB8 3SB
G7IYG	UB8 3TE
G7IYH	UB8 3TE
G4XLM	UB8 3TH
G1ZSK	UB8 3TT
G0PAX	UB9 6BE
G1PAF	UB9 6EL
G7MJF	UB9 6JQ
G4UBM	UB9 6LA
G3VXA	UB9 6QB
G6SSM	UB9 6QB

Hinckley and Bosworth

G0EKG	CV13 0DR
G8XZX	CV13 0JE
G6YZZ	CV13 0JX
G3JLQ	CV13 0LQ
G1ACG	CV13 0LX
G8RJM	CV13 0NP
G4DQB	CV13 0NX
G4MX	CV13 6AD
G1SWI	CV13 6BB
M1BKQ	CV13 6NL
G4KMX	CV9 3LT
G3OUQ	CV9 3QU
G2FXQ	CV9 3RE
G1VNR	LE10 0EH
G8SUN	LE10 0GB
M1BTL	LE10 0HP
G4IWG	LE10 0HT
G0GSH	LE10 0LA
M1AXU	LE10 0LW
G1ETU	LE10 0NE
G4CNZ	LE10 0NH
G3XWD	LE10 0TW
G1WJO	LE10 1HE
G3ZOP	LE10 1JF
G3MFK	LE10 1LT
G6UX	LE10 1RT
G4NNI	LE10 1SJ
G4YFI	LE10 1SL
G8JSE	LE10 1TJ
G7JIQ	LE10 2AS
G8KFD	LE10 2DY
G4OLZ	LE10 2HL
G4PJP	LE10 2JH
G7APU	LE10 2JJ
G4ALB	LE10 2LY
G3XPU	LE10 2ND
G8CTJ	LE10 2NX
G3SVT	LE10 2PN
G3VLG	LE10 2PN
G4OOW	LE10 2RN
G6KZA	LE10 2SS
G3UAA	LE6 0BA
G8LMW	LE6 0BE
G3TQF	LE6 0EZ
G4TKQ	LE6 0JP
G6PGP	LE6 0LH
G4UKW	LE6 0NG
G4IHY	LE6 0YD
G8POS	LE6 0YL
G3SBF	LE67 9RJ
G4RKD	LE67 9RJ
G1HUL	LE67 9SN
G8AWL	LE67 9SN
G4ARI	LE67 9TX
G4CQQ	LE67 9WA
G4DLZ	LE67 9WA
G4BWF	LE67 9WH
2E1CLV	LE67 9WN
G0BSN	LE9 7AF
G8SUM	LE9 7DA
G1LTK	LE9 7DU
G3DHE	LE9 7GH
G6MKL	LE9 7HQ
2E1AZP	LE9 7HU
G3TFV	LE9 7HW
G4DYQ	LE9 7HY
G7HIA	LE9 7QF
G4AFJ	LE9 7QG
G1HVA	LE9 8AA
G6RAU	LE9 8BY
G6SGJ	LE9 8BY
G8CGW	LE9 8DN
G6XXN	LE9 8DR
G1MPD	LE9 8DT
G4CAJ	LE9 8EH
G1WIW	LE9 8EX
G7LZF	LE9 8GB
G6FNQ	LE9 8GP
G0GGA	LE9 8LE
G4IDZ	LE9 9FQ
G0AIZ	LE9 9HR
G1BYK	LE9 9HX
G8CAC	LE9 9HX
G0USV	LE9 9JU
G0EDO	LE9 9LG
G0HZG	LE9 9LQ
G1WSW	LE9 9LQ
G4BCW	LE9 9QB

Horsham

G3REP	BN44 3FP
G0VLC	BN44 3GL
G4YPG	BN44 3HP
G7LGG	BN44 3HT
G0KVX	BN44 3JP
G0BAF	BN44 3LR
G7LJP	BN44 3LR
G4ILY	BN44 3PE
G8OYM	BN44 3PL
G7VBR	BN44 3QG
G1HWY	BN44 3RQ
G0ANS	BN44 3RZ
G8UNP	BN44 3SE
G4RDH	BN44 3TB
G3YZT	BN44 3TQ
G0NRX	BN44 3WH
G4UDU	BN44 3WH
G0UQY	BN44 3WJ
G3ERA	BN44 3YE
G4NMD	BN5 9JA
G6APJ	BN5 9JA
G8XNB	BN5 9SB
G3RAM	BN5 9UT
G3WMY	BN5 9UX
G1LIT	BN5 9YB
G6ZRU	BN5 9YB
G4YIF	RH11 0LQ
G7RQB	RH12 1DQ
G0VPZ	RH12 1DQ
G8GHL	RH12 1NA
G7SYC	RH12 1NB
G1PER	RH12 1SA
G1ZMY	RH12 1SD
G3XEB	RH12 1UB
G7VZQ	RH12 1XR
G3PYC	RH12 2AX
G4CCA	RH12 2DA
G3ZTF	RH12 2HH
2E1ECA	RH12 2HW
G6SBI	RH12 2JP
G4HMO	RH12 2PP
G4JHI	RH12 2PY
G4LLI	RH12 2QL
G3TQY	RH12 3DU
G3ZTU	RH12 3DU
G6EWJ	RH12 3ET
G3SWC	RH12 3HD
G3NPF	RH12 3LD
G0WUU	RH12 3LS
G4PEY	RH12 3ND
G4TLV	RH12 3NT
G7EYL	RH12 3NT
G4CGY	RH12 4AA
G4TMC	RH12 4AR
G0ALH	RH12 4BT
G4TLS	RH12 4BT
2E0AFO	RH12 4BX
G7UEC	RH12 4BX
G0OID	RH12 4DA
G0INX	RH12 4DQ
G3JYL	RH12 4DS
M0BAH	RH12 4DT
G6DMM	RH12 4GR
G8FQS	RH12 4GX
2E1EAC	RH12 4HJ
G6XTJ	RH12 4HR
G8LHS	RH12 4JB

G4EFO RH12 4JE
G6XDH RH12 4JE
G3UQJ RH12 4NW
G4YJA RH12 4RE
2E1DCX RH12 4UF
G1XFM RH12 5EU
G4MWD RH12 5EU
G8RSK RH12 5EW
G7UEH RH12 5FL
G0GMS RH12 5FP
G8MTR RH12 5HX
G0RBQ RH12 5JW
G1ALU RH12 5PJ
G6SRW RH12 5QW
G0GZB RH12 5TL
G4KDR RH12 5WA
G1VJT RH12 5XE
2E1DAQ RH12 5XG
G3VQO RH12 5XW
G7IAS RH12 5YG
G4OBT RH13 5BG
G0DJS RH13 5HB
G4EUG RH13 5HH
G4CII RH13 5JP
G4UYA RH13 5LR
G1RRR RH13 5NF
G4TPO RH13 5NL
G4MSH RH13 5PP
2E1EDP RH13 5PR
G4JHM RH13 5PT
G3OVC RH13 5RN
G8KZZ RH13 5RX
2E1DQN RH13 5UW
G4OIT RH13 6AB
G4MKW RH13 6AE
G3ZBU RH13 6AX
G6MQJ RH13 6DG
G8BAD RH13 6DT
G4FQR RH13 6ED
G4PEO RH13 6EJ
2E1EDY RH13 6EJ
G8YLA RH13 6JH
G6PGQ RH13 6RE
G6FYU RH13 6SB
G0HWS RH13 7AF
G4BVP RH13 7BG
G4VFC RH13 7DQ
G4AZQ RH13 7HP
G3PVH RH13 7HZ
G4HMM RH13 7HZ
2E0AED RH13 7HZ
G7AVF RH13 7PJ
G3VGI RH13 7QT
G6JTK RH13 7SQ
G8TQJ RH13 7TF
G0DLI RH13 7XA
G0IXB RH13 7XF
G0TAO RH13 7XR
G6DID RH13 7XX
G4LRP RH13 7XY
G8KZO RH13 8DP
G3ZQW RH13 8EQ
G4TPW RH13 8ER
G7DFV RH13 8HZ
G8TEF RH13 8JH
G3LEW RH13 8LA
G4BFW RH13 8LT
G3WZT RH13 8LT
G4HRS RH13 8LT
G7KZT RH13 8LU
G0OOC RH13 8QR
G0WVB RH13 8QR
G3WJU RH13 8RT
G3OGP RH14 9BH
G4ZXF RH14 9BX
G0UEY RH14 9HB
G3RDN RH14 9LJ
G0ERF RH14 9NF
2E1EBZ RH14 9NF
G4DDJ RH14 9RP
G4CSD RH14 9TU
G6OSO RH20 1AH
G7PTX RH20 1AW
2E1BSC RH20 1AW
G0INQ RH20 1BZ
2E1DZP RH20 1DP
M1AHW RH20 1DP
G1PMA RH20 1LA
G0LFH RH20 2AT
G4BUE RH20 2HJ
G1BKV RH20 2JJ
G4TNQ RH20 2JR
G3TUJ RH20 2LH
G4LAP RH20 2RB
G7MXT RH20 3AG
G8IPF RH20 3BD
G2ALO RH20 3HJ
G7DQF RH20 3HJ
G0BOP RH20 3HS
G3XJN RH20 3HU
G8OIM RH20 3PA
G7JRV RH20 4AQ
G4OAK RH20 4AU
2E1DZH RH20 4AU
G4LCU RH20 4BS
G0AGB RH20 4EH
G4MXZ RH20 4HH
G3POQ RH20 4LU
G3YWD RH20 4NF
G3FEX RH20 4NS
G3RNM RH20 4NU
G6LBE RH20 4QL
G6YCZ RH20 4QL
G7TMU RH20 4QX

Hounslow

2E1AXK TW13 4AW
G0ONA TW13 4EF
G8DXI TW13 4EQ
G1KCU TW13 4HX
G8ZZR TW13 5HZ
G8MDA TW13 6LD
G0EEZ TW13 6LZ
G7HFX TW13 6NP
G0KXG TW13 6PE
G7VKN TW13 6QF
G0GPT TW13 6RA
G0LSE TW13 6UH
2E1EXT TW13 7JE
G4XOR TW13 7LD
G6RHV TW13 7QQ
G7GAP TW13 7QQ
G3WTU TW14 0JS
G0UJI TW14 8JT
G0VDX TW14 8JT
G7VJA TW14 8LF
G1YVY TW14 9HL
G7UML TW14 9HY
G0DEO TW14 9JE
G7UWS TW14 9LW
G3IUL TW14 9RF
G8RZJ TW14 9RY
G4ZIP TW14 9TQ
G8LWY TW14 9XG
G7OXK TW14 9XG
G6YJJ TW3 1PU
G4CWE TW3 1YH
G3WPK TW3 2NW
G6TME TW3 3LQ
G0MRF TW3 3QX
G1EIH TW4 5BB
G0PER TW4 6LS
G3YAS TW4 7BH
G0WMD TW4 7JB
G0USK TW4 7RA
G0WOX TW4 7RA
G4OHO TW5 0HD
G4VBH TW5 0HD
G6YTX TW5 0ND
G4IXB TW5 0NF
G3AOJ TW5 9AA
G1JRW TW5 9AW
G3ISZ TW5 9AW
G0VLQ TW5 9EX
G3PEC TW5 9HQ
G6CLK TW5 9HQ
G4MZE TW5 9LS
M1AGM TW5 9PG
G0IIK TW5 9PT
G1AWU TW7 4NP
G4DUO TW7 4PQ
G8SJO TW7 5HB
G4JMJ TW7 5PF
G0CPF TW7 6AD
G1UYL TW7 6BX
G4LTC TW7 6HW
G8ZKN TW7 6HW
G4PWS TW7 6HX
G4NJN TW7 6LF
G4DFX TW7 6NP
G4UUP TW7 6NP
G6GAN TW7 7DP
G0EAV TW7 7HU
G1TFM TW7 7QP
G0RUE TW8 0NF
G3XKV TW8 0PL
G4KLY TW8 8HY
G6SUK TW8 9PT
G8MWD TW8 9RD
G4ZMJ UB2 5RQ
G8SPE W3 8HP
G4GPD W3 8LB
G3IGM W3 8LW
M0BFN W4 2DR
G0RNW W4 2EG
G4ELS W4 2JE
G4LIC W4 2JH
G0JYX W4 2LU
G6TBT W4 2RS
G3VKT W4 3LR
G4ZJD W4 3NH
G8ORZ W4 3NP
G6NQO W4 4AR
G8FBK W4 4EU
G4OHV W4 4ND
G3GIQ W5 4NH

Huntingdon

G1BWW PE17 1AS
G7APA PE17 1AS
G8TVC PE17 1HY
G8JWE PE17 1JP
G1ZBB PE17 1LX
2W1AOK PE17 1NB
G3CXP PE17 1NB
G3MQI PE17 1NB
G3NKJ PE17 1NB
G3ROQ PE17 1NB
G4LQJ PE17 1NH
G4KPZ PE17 1QE
G4ZFE PE17 1SB
G4SSO PE17 1SN
G0BNR PE17 1TF
G0BDD PE17 1TJ
G3PXV PE17 1UA
G0BXE PE17 1YA
G3NID PE17 2DQ
G4LKB PE17 2QF
G4LKC PE17 2QF
G3ZYR PE17 2RA
G0MOH PE17 2RT
2E1FQS PE17 2RY
G8FZT PE17 2SD
G1HJD PE17 2TB
G4HIY PE17 2TH
G3WWH PE17 2UQ
G4WMZ PE17 3DJ
G4DLT PE17 3DL
G8LEB PE17 3DL
G4GVN PE17 3EG
G4YND PE17 3EL
2E1BTS PE17 3EY
2E1ECN PE17 3EY
G7PZS PE17 3PN
G1OAU PE17 3PX
G4BIK PE17 3QH
G6GJY PE17 3QH
G7HMI PE17 3QX
G0GXS PE17 3SN
G7JAE PE17 3SW
G4XGT PE17 3TB
G4FCF PE17 3XA
G5RL PE17 4AL
G7JXJ PE17 4FX
G4KJJ PE17 4FX
G7UXS PE17 4FX
G4VBD PE17 4HT
G4ZUH PE17 4HX
G3ASE PE17 4JN
G8JKV PE17 4NE
2E1FGO PE17 4QP
G6VBQ PE17 4SS
G4YDD PE17 4TZ
G0IUM PE17 4UB
G3PYV PE17 4WA
G4XBS PE17 4WX
G6FSU PE17 5BQ
G3ECP PE17 5DZ
M1ANJ PE17 5JH
G3XAV PE17 5SH
G3KMC PE17 5SJ
G7EHR PE17 5SY
G7GGV PE17 5TG
G7KJW PE17 5UA
G6NZC PE17 5UA
G7UOL PE17 5UU
G7ELC PE17 5YE
G6UEC PE17 5YQ
G3YMV PE17 6DQ
G4NKW PE17 6DQ
G1MBA PE17 6EP
G4ECL PE17 6FN
2E1EJU PE17 6HU
G7CSL PE17 6JB
G0RSA PE17 6NL
G6TGW PE17 6XQ
G6THP PE17 6XZ
G1OSJ PE17 6YB
G8XIY PE17 6YJ
G0KHY PE17 6YZ
G8GRT PE18 0AJ
G4HSG PE18 0AU
2E1FUM PE18 0AU
G0IWB PE18 0BB
G3NKG PE18 0BS
G7GJS PE18 0BY
G4JZV PE18 0DZ
G0PYS PE18 0HT
G0EVK PE18 0PA
G4NZG PE18 0RT
G1NOO PE18 0SH
2E1AQH PE18 0SH
G0BTX PE18 6BW
G6PRL PE18 6HJ
G7JAO PE18 6RT
G1XLX PE18 6SN
G7DIU PE18 6UF
G7JPS PE18 6UF
G1MRQ PE18 7BP
G6YHF PE18 7BS
G4UXV PE18 7JE
G4NVS PE18 7JU
G4JKS PE18 7LP
G3MKV PE18 7QH
G7HRM PE18 7QW
G4LHI PE18 7RP
G4WXH PE18 7RZ
G0HSR PE18 7SR
G3RPV PE18 7SR
2E1BOV PE18 7SZ
G7OJX PE18 7TA
2E1ARG PE18 7TA
G1JGT PE18 7TL
G0UFB PE18 7XF
G7MJD PE18 7XP
G4FCR PE18 7XY
G7MXD PE18 7XY
G3AAZ PE18 7YD
M0ALK PE18 7YL
G0JLQ PE18 7YN
G3ZGG PE18 8AL
G4ZGG PE18 8AL
G3TEJ PE18 8DU
G4GWU PE18 8FH
G6GUN PE18 8PR
G8UJV PE18 8PX
2W0AJD PE18 8PY
GW6TAM PE18 8PY
G8LRS PE18 8SG
G4AUJ PE18 8TH
G8RSA PE18 8TQ
G7UAW PE18 8UN
G4YUL PE18 9AN
G4WIA PE18 9BH
G4GHK PE18 9BS
G4GQV PE18 9DS
G3HSU PE18 9EH
G4ZVT PE18 9EH
G4KLE PE18 9JL
G3SIT PE18 9JR
G3XIP PE18 9JX
G6XLQ PE18 9LA
G1KHM PE18 9LP
G4XOH PE18 9RD
G6JAK PE18 9RD
2E1ASP PE18 9RQ
G3RVP PE18 9SB
G3POX PE18 9TG
G0IUG PE18 9TN
G3PHJ PE18 9TT
G8BVJ PE18 9TU
M0AXZ PE18 9TZ
G0SKV PE18 9UR
2E1EAM PE18 9UR
G8OAQ PE18 9UX
G4DPF PE18 9XR
G0CJQ PE18 9YA
G8WRV PE19 1JU
G7HMZ PE19 1LD
G8FDE PE19 1UF
G8FFC PE19 1UF
G1NTB PE19 2DU
G4NVU PE19 2DW
G1OSM PE19 2EE
G3BBW PE19 2NN
G8BBW PE19 2NP
G7JUC PE19 2QF
G6OXI PE19 2QY
G7HRI PE19 2SD
G7ODT PE19 2SZ
G4WDZ PE19 2TL
G6PNM PE19 2UP
G3XNB PE19 3AE
G6TNW PE19 3AN
G4ECU PE19 3AX
G8XGQ PE19 3BU
G1NTH PE19 3HG
G4HTP PE19 3HJ
G3LWJ PE19 3HY
G3AVO PE19 3LF
G4MZV PE19 3LF
G0BYT PE19 3PD
G0EVU PE19 3PJ
G1SVI PE19 3PJ
G7LFZ PE19 4DH
G4PGT PE19 4EZ
M0BGD PE19 4JW
G4ULM PE19 4NE
G8WUZ PE19 4QB
G8TQI PE19 4QW
G3WCD PE19 4RX
G4LIQ PE19 4SD
G0ABW PE19 4UE
G8SQY PE19 4YL
G6JAS PE7 3BU
G0XAH PE7 3ET
G1PLE PE7 3JW
G0HOF PE7 3SS
G4EHW PE7 3SY
G0HPJ PE7 3SY
G4XER PE7 3UR
G7LCW PE7 3XY
G7GWA PE7 3YE
G6CTP SG19 3AP
G3PMR SG19 3AU
G4PFF SG19 3AU
G0WIN SG19 3AU

Hyndburn

G3YWH BB1 2DR
G0PTB BB1 2NN
G4MPI BB1 2PE
G7WJC BB1 4BH
G6SUV BB1 4ES
G7IJL BB1 4ND
G6SZS BB1 4NP
M0ANL BB5 0EX
G0GFT BB5 0EZ
G4RND BB5 0EZ
G0SVP BB5 0HQ
G7SLQ BB5 0JJ
G0SVH BB5 0NQ
G7MYJ BB5 0NT
G0SVV BB5 0QJ
G7WKZ BB5 0SA
G3ILE BB5 0SQ
G0JSA BB5 1BW
G4ZMB BB5 1SL
G6MKQ BB5 2AS
G2FMU BB5 2JT
G6IVE BB5 2LE
G4WJG BB5 2LF
G1OPV BB5 2NF
G1ZED BB5 2NY
G1BQQ BB5 2PA
G4DPU BB5 2PF
G0JWB BB5 2PQ
G6EYS BB5 2PR
G0PFC BB5 2PU
G6HHD BB5 2PZ
G6YJI BB5 2QD
G6VJM BB5 2QP
G7WEC BB5 2QS
G7PQL BB5 2TQ
G4NYL BB5 2TU
G6IKE BB5 2XA
G3VGB BB5 2XD
G4COQ BB5 3AR
G0VGN BB5 3AT
G4XUL BB5 3AX
G1EZW BB5 3RY
G7VCA BB5 3SB
G7PUA BB5 4AF
G7DHD BB5 4AL
G0UNK BB5 4AR
G0HDX BB5 4BG
G7DSP BB5 4BL
G4PHH BB5 4DE
G4GHK BB5 4DX
G4GQV BB5 4JQ
G4FSD BB5 4NW
G3RDV BB5 4NZ
G7WJK BB5 4PH
G7VPE BB5 4QT
G8RBV BB5 4QZ
G2CJK BB5 5AU
G2PB BB5 5DF
G0BMH BB5 5DT
G8AEE BB5 5EF
G3RFN BB5 5GH
G4XEI BB5 5LA
G4FQW BB5 5NQ
G3KEG BB5 5QD
G6ZFO BB5 5QH
G0LQO BB5 5RB
G3PUO BB5 5RX
G1HRF BB5 5WA
M0AFC BB5 5XF
G6LVM BB5 5XG
G7IKU BB5 6AY
G1BQI BB5 6BJ
G0BYO BB5 6BS
G4GLW BB5 6BS
G4PKD BB5 6DA
G1OLF BB5 6DH
G7NGO BB5 6EA
G0CYR BB5 6JQ
G1HIH BB5 6LP
GW3LOD BB5 6PL
G0TZY BB5 6PL
G0AMV BB5 6QG
G1OLD BB5 6QU
G6EVV BB5 6QU
G0PXI BB6 7AS
G3LZO BB6 7JB
G6PXX BB6 7JE
G7VTK BB6 7JH
G8MEC BB6 7JS
G0TPP BB6 7JS
G8VJO BB6 7JU
G7WBZ BB6 7NG
G6LXU BB6 7NL
G3SXC BB6 7PH
G6PWY BB6 7RB
G6KSE BB6 7TF

Inverclyde

GM0ILQ PA13 4DJ
GM0JVC PA13 4NA
GM3PFX PA13 4PL
GM7ADU PA13 4PY
GM7ADY PA13 4PY
GM8TXK PA13 4RE
GM3SUZ PA14 5QG
GM7RYK PA14 5SP
GM0WDF PA14 5XF
2M1CEN PA14 6EW
2M1CEO PA14 6EW
GM0CZM PA14 6LF
GM4CBV PA15 1PH
GM4EGD PA15 2EE
GM7PVT PA15 3AB
GM4FBU PA15 4HH
GM0KYU PA15 4QJ
GM0HJV PA16 0BB
GM8EVF PA16 0DS
GM3XGX PA16 0TE
GM7TFN PA16 7DA
GM0ATQ PA16 7NA
GM3KJI PA16 7QE
GM7TRH PA16 7RN
GM0ADF PA16 7SP
GM0WSR PA16 7SP
GM0EUM PA16 7SP
GM3XTR PA16 7UA
GM6LEZ PA16 8AS
2M1EWA PA16 8EH
GM3ZXG PA16 8ET
GM3LRG PA16 8QG
GM3ZRC PA16 8QG
GM3BGB PA16 8SX
2M1EUO PA16 9DN
GM4JMZ PA16 9HE
GM4YZU PA19 1DY
GM4SVW PA19 1EW
2M1EJI PA19 1HF
MM1AWV PA19 1HF
GM0GMN PA19 1HN
GM8YUI PA19 1HY
MM1AUF PA19 1HY
MM1AUG PA19 1HY
GM4CEA PA19 1JS
GM4JJG PA19 1NP
GM3YOR PA19 1TS
GM4JNF PA19 1UB
GM4ZZW PA19 1XU
GM0IOY PA19 1YA
GM3XNJ PA19 1YN

Ipswich

G7SMN IP1 2PH
G8NKQ IP1 3NP
G3XCO IP1 3QF
G7LKY IP1 3SA
G3JWP IP1 4AG
G1PTH IP1 4AN
G1YAB IP1 4DS
G0JMY IP1 4HA
G7TZM IP1 4LA
G4LRB IP1 4LD
G8MUF IP1 4NZ
G1SDJ IP1 4PQ
G0JVT IP1 5DR
G7OCH IP1 5EA
G1WQY IP1 5HD
G8FTW IP1 5HS
G3TXE IP1 5JJ
G3YWM IP1 5JX
G7PLV IP1 5LR
G0OZS IP1 5PQ
G1HGF IP1 6BD
G1IRC IP1 6BH
G4BAV IP1 6BH
G4IRC IP1 6BH
G3ZLN IP1 6DR
G8XOR IP1 6DU
G7UQV IP1 6EW
G0HJK IP1 6JB
M0AKK IP1 6PQ
G8VNP IP1 6PQ
G7FNN IP1 6QN
G6BJC IP1 6QP
G8BLS IP1 6QP
M0AWS IP1 6QZ
G1SYT IP1 6QZ
G4YGV IP1 6RG
G8ZQA IP1 6RL
G6HTT IP2 0AD
G7TFY IP2 0JS
G4BVI IP2 8AZ
G3XVL IP2 8BQ
G8LBS IP2 9HR
G3WJS IP2 9PD
G8EDL IP2 9PD
G8NYC IP2 9PD
G3SXV IP2 9SX
G8KBB IP2 9SX
G4BQR IP2 9TA
G4LVD IP3 0LA
2E1FNP IP3 0LE
G0INH IP3 0PT
G0IVV IP3 0PT
G6RHK IP3 0QE
G1HSL IP3 0QH
G6XPP IP3 0RF
M1ACB IP3 0RS
G6MMT IP3 8AT
M1BSL IP3 8EA
G0BPU IP3 8JW
G8EHS IP3 8QW
G1EGU IP3 8SL
G3ZZB IP3 8TS
G6PDE IP3 9BD
G4POU IP3 9DE
G4WSR IP3 9EP
G4LVG IP3 9NE
G8TIR IP3 9PD
G4LBU IP3 9RE
M1BOP IP4 1PJ
2E1ERF IP4 1PJ
G8JIF IP4 1PL
G7UWB IP4 1PN
G7UWC IP4 1PN
G4UCX IP4 1PP
2E1FIA IP4 1QE
G6DDK IP4 2ET
G4TVT IP4 2HG
G3SGJ IP4 2HZ
G3DCS IP4 2QS
G8GCO IP4 2TL
G4KGO IP4 3AS
G3YUJ IP4 3BT
G3ZIN IP4 3LJ
G8NGK IP4 3ND
2E1BBX IP4 3NG
G0XJK IP4 3NG
2E1CWN IP4 3NG
G4IUV IP4 3PJ
G6HNN IP4 3PP
G1BEK IP4 4AD
G6MKO IP4 4BG
G8XKT IP4 4BS
G4ILN IP4 4JJ
G8TPC IP4 4JN
G4JNK IP4 4NT
G8WVO IP4 4QP
G6FIL IP4 4SF
G0BHT IP4 5AX
G4WMF IP4 5AX
G7OWR IP4 5AX
M1BXN IP4 5BP
G6BEH IP4 5DW
G4GKE IP4 5EP
G7VLJ IP4 5EX
G4DSH IP4 5JX
G4FHV IP4 5LB
G6MKZ IP4 5LR
G6OGF IP4 5NN
G4YVK IP4 5PP
G0CWW IP4 5RZ

Isle of Anglesey

GW4CNM LL58 8AT
GW1HBU LL58 8ET
GW4MBL LL58 8PG
GW6TOX LL58 8RW
GW8TOX LL58 8RW
GW8FEY LL58 8SP
GW4JUI LL58 8SU
GW3JBJ LL59 5LR
GW4RQQ LL59 5LR
GW7VWY LL59 5LT
GW4CFC LL59 5NB
GW4PNV LL59 5PB
GW7SBO LL59 5QL
G4ZOH LL59 5QY
GW3XRM LL59 5RD
GW3KKG LL59 5TH
GW4IGT LL59 5TH
GW4TAU LL59 5UF
GW3PIO LL60 6BA
GW3TLP LL60 6HD
GW6DOK LL60 6HN
GW8PBX LL61 5AQ
GW4JXN LL61 5AX
GW8UTK LL61 5JB
GW0CWG LL61 5JE
GW0IQZ LL61 5JR
GW0MKP LL61 5JR
GW0ABL LL61 5JY
GW3VVC LL61 5JY
GW8UZL LL61 5QF
GW6IUK LL61 5SZ
GW2DLK LL61 5YX
MW1CAN LL61 6JA
GW4FSY LL61 6LQ
G2OG LL61 6LZ
GW0PZS LL61 6SY
MW1CBM LL61 6TA
GW3PRL LL61 6TZ
GW4NBM LL61 6TZ
GW7FNQ LL62 5AS
GW3EIZ LL62 5AW
GW8FKB LL62 5ED
GW4HKX LL62 5LH
GW0GHG LL63 5PQ
GW8OZO LL63 5RJ
GW0OGI LL63 5SH
GW7JRT LL63 5SR
GW0EAW LL63 5TW
GW6WJM LL64 5JW
GW4WLZ LL65 1DS
GW1HHM LL65 1DT
GW4XAU LL65 1ES
GW0LIS LL65 1EU
GW4IEU LL65 1NW
2W1BNI LL65 1PL
GW1PQE LL65 1SP
GW0KZF LL65 1TN
GW4WJO LL65 2DN
GW7UQN LL65 2DU
GW0KPV LL65 2EF
GW0AVD LL65 2HE
GW4XYI LL65 2PN
GW4VDP LL65 2PP
GW0IJY LL65 2UP
GW4OGC LL65 2US
GW1VYT LL65 3AR
GW0ENU LL65 3ER
MW1BWI LL65 3HN
GW0ONY LL65 3PL
GW0KLC LL65 3RB
GW4VEQ LL65 3RD
GW0VYG LL65 3RD
G6TVD LL65 3RR
GW0MHK LL65 3SY
GW4CWU LL65 4BU
GW0BZE LL65 4UF
G7ACC LL65 4YH
GW4DRR LL65 4YL
GW0SLM LL66 0AX
GW1BAV LL67 0HY
MW1BLD LL67 0LN
GW0FEM LL68 0RE
MW0AER LL68 1AA
GW4ZWN LL68 1AA
GW3RXD LL68 9DG
GW4SXN LL68 9DU
GW1AUO LL68 9HW
GW0OYD LL68 9LS
MW0AUX LL68 9RL
GW0ESK LL68 9RX
2W1DTX LL68 9ST
GW3JRJ LL69 9AQ
GW1GZB LL71 8AL
GW0TPR LL71 8EA
GW3GAH LL72 8HE
2W1AYZ LL72 8HE
GW4EXE LL72 8HN
GW6OAW LL72 8LG
GW8YUJ LL73 8PE
GW3KXC LL74 8RD
GW8ZEI LL74 8RD
GW0MOI LL74 8RG
GW0NUV LL74 8RG
GW0VOG LL74 8RH
GW0BTB LL74 8SR
GW0CTK LL74 8TS
GW6VYD LL74 8UH
GW0WEY LL75 8NQ
GW4URY LL75 8UN
GW0WMY LL75 8UR
GW4XPN LL75 8YG
GW0SSD LL75 8YH
GW1JBF LL75 8YU
GW4TOD LL76 8TZ
GW0GEI LL77 7AJ
GW0PJF LL77 7EZ
GW0UHA LL77 7EZ
GW4GSL LL77 7HR
GW7CMF LL77 7NT
GW6BMP LL77 7PY
GW2DXQ LL77 7RE
2W0APO LL77 7SJ
2W1BCP LL77 7SJ
MW0BER LL77 7SJ
MW1BJV LL77 7SJ
GW6YMS LL77 7SL
GW0ESU LL77 7SU
GW6BUW LL77 7SX
GW7GWM LL77 7TW
GW4CZK LL78 7JL

Isle of Man

GD3YUM IM1 2QD
GD3ESV IM1 4AW
GD7IEH IM1 4BB
MD0ADD IM1 4HH
GD3GBG IM1 4HQ
GD0EOT IM2 1NX
GD0EXM IM2 1PA
MD1BYG IM2 2HA
GD1OOY IM2 2LE
GD6NDE IM2 2LE
GD3YEO IM2 2NP
GD1CRZ IM2 2NR
GD3YDB IM2 3BZ
GD4PFL IM2 3LQ
GD1AQY IM2 3NH
GD0JWR IM2 3RQ
GD7ESM IM2 4AU
GD0BFN IM2 4AZ
GD3FLH IM2 4BP
GD3TNS IM2 4BP
GD1ASB IM2 4HN
GD4HOX IM2 5AE
GD8KWM IM2 5AZ
GD7MAN IM2 5BH
GD0TEP IM2 5BH
GD7HMG IM2 5BQ
GD4IOM IM2 5BQ
GD0PLR IM2 5BQ
GD7LWE IM2 5LG
GD0TFO IM2 5NG
GD3CKO IM2 5PS
GD4PNY IM2 6AX
GD7BMG IM2 6HE
GD1MAN IM2 6JE
GD3LSF IM2 7AT
GD7ESU IM2 7AT
GD4YMY IM2 7BD
GD4IHA IM2 7EB
GD3HQR IM3 1BP
GD0PNK IM3 1LR
GD0BCM IM3 1LS
GD3ZCN IM3 1LS
GD1ETE IM3 1LT
GD0PLT IM3 2AZ
GD4RVQ IM3 3BU
GD6HCB IM3 3BU
GD4DPK IM3 4ET
GD1MFF IM3 4HS
GD0LQA IM3 4LA
GD0IFU IM3 4NU
GD7LAV IM4 1BJ
GD4YON IM4 1HT
GD6TKX IM4 2HJ
GD2HCX IM4 2HP
GD8GRE IM4 2HP
GD7ESR IM4 3AT
GD1XMW IM4 3AU
GD4EBA IM4 3EW
GD4EIP IM4 3HE
GD3MBC IM4 3JB
GD1RHT IM4 3JH
GD4HOZ IM4 3JJ
GD3XPA IM4 3NF
GD3FXN IM4 4AS
GD4FWQ IM4 4ES
GD4GWQ IM4 4ES
GD1GJB IM4 4NJ
GD4XOD IM4 5EA
GD3ZEX IM4 7AP
GD0LQE IM4 7BY
GD3HFC IM4 7JB
GD8COH IM4 7NY
GD3RVQ IM4 7PL
GD4XTT IM4 7PW
GD0AVF IM5 1HJ
GD4MCR IM5 1HP
GD4NTR IM5 1JJ
GD6TWF IM5 1JY
GD4UQV IM5 1PJ
GD4MDY IM5 1PJ
GD6ICR IM5 1PN
GD3HDL IM5 1PS
GD4UQO IM5 1PX
GD4UHB IM5 2AE
GD3EUI IM5 2AF
GD1XMA IM5 3BJ
GD4IHB IM5 3BJ
GD1GHK IM6 1AF
GD8MPF IM6 1HG
GD4BEG IM6 1MP
GD3EFD IM7 1BJ
GD4ELI IM7 2AD
GD4WOW IM7 2AQ
GD0KEO IM7 2AT
GD4XXE IM7 2AU
GD4WBY IM7 2EB
GD4RGR IM7 2EY
GD4XWF IM7 3BL
GD0EEM IM7 3HN
GD4HIT IM7 3HN
GD3RFK IM7 3HP
GD4RFK IM7 3HP
GD4PTV IM7 4AW
GD6ZNL IM7 4HG
GD0KWM IM8 1DZ
GD3VEM IM8 1EJ
GD4AM IM8 1ER
GD3GCE IM8 1NA
GD0HWA IM8 1NF
GD7HSX IM8 2AT
GD3SKH IM8 2BH
GD7RVP IM8 2JD
GD4WXX IM8 2PQ
GD7KHG IM8 2TA
GD4IHC IM8 3EB
GD4XJR IM9 1BU
GD3FOC IM9 1ED
GD4RFW IM9 1EE
GD4HPN IM9 1HJ
GD0JBL IM9 2DW
GD0TFG IM9 2ES
GD0ADV IM9 2HF
GD0BCJ IM9 2HP
GD0IOM IM9 3BA
GD3AHV IM9 3BA
GD4GNH IM9 3BA
GD8KQF IM9 3BA
GD4TVG IM9 4AS
GD0GBA IM9 4BR
GD6OXG IM9 4EF
GD4VBA IM9 4EP
GD7VNS IM9 4ES
GD3SKZ IM9 4HA
GD0MWL IM9 5ED
GD4RAG IM9 5LX
GD4FJI IM9 6EP
GD4HOO IM9 6EY
GD0MAN IM9 6LW
GD4TOW IM9 6LW
GD4XWB IM9 6LW
GD3JIU IM9 6NB
GD8EXI IM9 6TE
GD7ARS IM9 6TJ
GD7DPG IM9 6TJ
GD3XJR IM99 1BN
GD7DUZ IM99 5TH

Isle of Wight

G6CUT PO30 1BU
G4MBD PO30 1DG
G7IRH PO30 1DG
G0NTH PO30 1DP
G8YZS PO30 1DP
G6OOH PO30 1DP
2E1CKN PO30 1DR
G3XOC PO30 1DT
G7ILQ PO30 1HB
G6ORJ PO30 1HG
2E1CNA PO30 1LU
G7UYX PO30 1LU
G7MID PO30 1NL
G0KQO PO30 1PZ
G7NAO PO30 1QT
G1FMC PO30 1RJ
G1JGS PO30 2AT
G8AZM PO30 2DP
G7TQC PO30 2DR
G8RWM PO30 2LL
G1OXL PO30 3AL
G1KAK PO30 3DD
2E1CAI PO30 3DF
G7TAE PO30 3DF
G4MXX PO30 3HQ
G8NHM PO30 4BA
G0FZX PO30 4BX
G0GEZ PO30 4BX
G0PXM PO30 4DJ
G0TZT PO30 4DJ
G3KSH PO30 4HH
G1YXG PO30 4JS
G0EHR PO30 4JT
G7FWG PO30 4NG
G7OPY PO30 4PD
G4FYI PO30 5FH
G4RTW PO30 5HJ
G0OSF PO30 5QP
G0RUT PO30 5QZ
G7ISG PO30 5RH
G7SAP PO30 5SU
G3IMX PO30 5TP
G3XLP PO31 7DR
G6ZUZ PO31 7JN
2E1BND PO31 7LF
G4GRK PO31 7NF
G1GGF PO31 7NQ
2E1FCO PO31 7PT
G6XMI PO31 7QB
G0MIU PO31 7QS
G8AJL PO31 7SJ
M0BDJ PO31 7SQ
G4ZFQ PO31 8AD
G2FJT PO31 8AL
G3WXC PO31 8AL
G4RUC PO31 8AN
G0RSY PO31 8BW
G3XYB PO31 8DP
G3YZK PO31 8DT
G4FJD PO31 8EF
G8RTA PO31 8EF
G3XZB PO31 8HQ
G4ZBH PO31 8JP
G0HDH PO31 8NR
G0CWX PO31 8PE
G3PZB PO31 8PN
G3SKY PO31 8PN
G8MBU PO31 8PN
G3GPZ PO31 8PT
G8RKH PO31 8PT
G4SIQ PO31 8PZ
G4BQO PO32 6BB
G8MLC PO32 6BD
G3PQJ PO32 6BJ
G1EXZ PO32 6BS
G1KIG PO32 6DU
G1RFB PO32 6EE
G4UCZ PO32 6EJ
G6XVN PO32 6HA
G0PQW PO32 6HR
G0PWF PO32 6HR
G8AHF PO32 6HR
G3PRI PO32 6JA
G1JHD PO32 6JZ
G7RAU PO32 6LS
G7RTD PO32 6LT
G0KQR PO32 6PS
G0UES PO32 6QN
G7SVF PO32 6QN
G1EUW PO32 6QQ
G7RCC PO32 6RZ
2E1FEV PO32 6SA
G2AGG PO33 1AB
G8MYF PO33 1AB
2E1FYF PO33 1EB
2E1BQE PO33 1ED
G0MWU PO33 1EL
G1HRQ PO33 1EU
G0MIZ PO33 1JD
G7UHT PO33 1LG
G3KPO PO33 1LN
G1VGM PO33 1LU
2E1EBN PO33 1LU
G4TAT PO33 1NT
G4RGE PO33 1XE
G0SEB PO33 1YF
G1WQC PO33 2BG
2E1CDK PO33 2BH
G7DFD PO33 2QH
G0DRW PO33 2QQ
G7ITM PO33 2TH
G7WBY PO33 2TN
G3XHM PO33 2UH
G7MAR PO33 2UP
G4ZXU PO33 2UZ
G6XYH PO33 3BS
G4SBN PO33 3DL
G6PMS PO33 3DS
G3OWA PO33 3EY
G8FWF PO33 3HP
G3OVI PO33 3HS
2E1DPH PO33 3HS
G4KZI PO33 3HX
G0PRJ PO33 3JA
G0PEC PO33 3JA
G0IMM PO33 3JG
G8YWT PO33 3LH
G4OQR PO33 3PL
G6LVS PO33 3QN
G2CFH PO33 3SD
G4TFW PO33 3SF
G0MFJ PO33 3SU
G4OAG PO33 3TE
G6MQY PO33 3TE
G0UEK PO33 3TL
G4SHH PO33 3UD
G6HZG PO33 3UX
G0WQN PO33 4AY
G3UHX PO33 4ED
G4KKB PO33 4JJ
G7LIO PO33 4LU
G4HQI PO33 4NP
G3DEP PO33 4NQ
G4YNC PO33 4NQ
G7GGH PO33 4PT
G4HAW PO34 5BY
G7MAZ PO34 5EE
G0GNI PO34 5JE
M0AXD PO34 5LH
G7VHL PO34 5LH
G3JLN PO35 5QW
G4AVY PO35 5SG

G4BCH PO35 5UA
G4CQO PO35 5UA
G0IFH PO35 5UE
G6MSW PO35 5UF
G0IWR PO35 5UG
G4LUY PO35 5UG
G7TDJ PO35 5YB
G3LWI PO35 5YP
G3RIO PO36 0AG
G3KFG PO36 0BJ
G3ARL PO36 0BU
G1RIR PO36 0DX
G3GUN PO36 0HE
G4IKW PO36 0JA
G3GEG PO36 0JT
G8TAQ PO36 0JY
G0KXH PO36 0LD
G0IAJ PO36 8DU
G4PRQ PO36 8DX
G4RTY PO36 8NA
G4RUB PO36 8PR
G0WVD PO36 8QE
G6ETC PO36 9BX
G4UNM PO36 9DS
G6WUR PO36 9HQ
G4ULT PO36 9JA
G8DDY PO37 6JJ
G3YZX PO37 6QW
G1SMY PO37 7AF
G4SVY PO37 7BH
G4NOU PO37 7EJ
G1AAQ PO37 7EX
G0HFO PO37 7HH
G4UUJ PO37 7PA
G1JYZ PO38 1AA
G0LGZ PO38 1AN
G0DFU PO38 1AP
G1CPO PO38 1BH
G7VOC PO38 1DB
G4LAL PO38 1DH
G3GLK PO38 1LJ
G0RMJ PO38 1NQ
G0UHN PO38 1QN
G7ONE PO38 1TH
M1BOD PO38 2DZ
2E1BYB PO38 2JQ
G0BAR PO38 2QU
G7BAR PO38 2QU
G0ISF PO38 2QU
G4XIU PO38 2QW
G7PRM PO38 2RD
G1PRX PO38 3AR
G2BY PO38 3DD
G4AHI PO38 3EL
M1BWR PO38 3EQ
G0PEF PO38 3HR
G0ISB PO38 3HX
G3LYD PO38 3HZ
G4RZQ PO38 3NH
G4RQP PO38 3NT
G4NCD PO39 0AH
G0GBD PO39 0AL
G4CSM PO39 0AL
G7WJD PO39 0DL
G7CZW PO39 0EF
G3XZR PO39 0HE
G4GUA PO39 0HP
G3RJK PO40 9BH
G4NAK PO40 9HB
G7RES PO40 9PN
G7DZY PO40 9QS
G7RER PO40 9ST
G3VVZ PO40 9SY
G3IIN PO40 9UA
G6YWW PO41 0PT
G0DKS PO41 0RX
G0GPK PO41 0SL
G4AZC PO41 0TL
G0PTT PO41 0UX
G0UTI PO41 0UX

Isles of Scilly

G3RPC TR21 0NS
G7PZU TR23 5ET
G4VFX TR25 0QL

Islington

G0IIA EC1M 6EU
G0BPB EC1R 0DY
G6BFW EC1V 0AP
G6VHO EC1V 7PD
G0UCZ N1 0YW
G1ETD N1 1BJ
G1GBX N1 1QN
G0PTZ N1 1TN
G0TJD N1 1TN
G0GZZ N1 2JZ
G0DCP N1 2LW
G4JZK N1 2NP
G8UDB N1 2PJ
G0UUO N1 2PJ
G8DEL N1 3BA
M1ACG N1 3QU
G0OOS N1 4NU
G6OUG N1 4RF
G0BQI N1 7AR
G4FIH N1 8NG
G6ODA N19 3JX
G0DFZ N19 3NR
G0WCZ N19 3QX
G7UYN N19 3QX
G4SNK N19 5JY
G0AKM N19 5PZ
G0FVC N19 5PZ
G1IYX N4 3BY
G0JYU N4 3DW
G0JKM N4 3JP
G0WZK N5 1NU
G4AWD N5 2DE
G8ZQJ N5 2DE
G1ONK N5 2DN
M1CDN N5 2DT
G0DBL N6 5LX
G4WQX N7 0BH
M0AOW N7 0EE
G1NWG N7 0EL
G0VBI N7 7ND
G4IZU N7 8HY

Jersey

GJ0KKB JE2 3FT
2J1DGI JE2 3FT
GJ7DGJ JE2 3GJ
MJ0AQJ JE2 3JU
GJ7DNI JE2 3WD
GJ0NAC JE2 3XJ
GJ8GFI JE2 4NJ
GJ1EXC JE2 4PA
GJ0SVZ JE2 4QD
GJ4HXJ JE2 4SD
GJ3YUL JE2 4UG
GJ2CNC JE2 4ZA
GJ3XZE JE2 6FW
GJ4GG JE2 6LQ
GJ6BUK JE2 6NY
GJ4YMX JE2 6QH
GJ1YOT JE2 6QQ
GJ3XOJ JE2 6SD
GJ0BFF JE2 7GN
GJ3SND JE2 7HP
GJ6ENP JE2 7HX
GJ7DTA JE2 7HX
GJ4YCR JE2 7JP
GJ0JSY JE2 7LT
GJ6WDN JE2 7LT
GJ4JVI JE2 7PN
GJ3YLI JE2 7RG
GJ6SNQ JE2 7RL
GJ0FTZ JE2 7RR
GJ6RFE JE2 7RR
GJ7HTV JE2 7RT
GJ6RND JE2 7SA
GJ7DPH JE2 7TJ
GJ6HUL JE2 7TS
2J1CWG JE2 7TW
2J1CWH JE2 7TW
GJ7UIT JE2 7TW
GJ7LJJ JE2 7TX
GJ8RRP JE2 7TZ
GJ4HSW JE2 7UD
GJ2FMV JE3 1EN
GJ4YBM JE3 1EZ
GJ3YHU JE3 1FS
GJ6WDK JE3 1GL
GJ0PDJ JE3 1GT
GJ3LFJ JE3 1LA
GJ5NO JE3 2AA
GJ7AOG JE3 2BJ
2J1EDR JE3 2HA
GJ0FYB JE3 3BQ
M1BFD JE3 4BH
GJ3AME JE3 6AT
GJ1TJP JE3 6ED
2J1CVX JE3 7AD
GJ7SLU JE3 7AD
2J1ECH JE3 7AD
GJ0KYZ JE3 7BB
GJ7VQZ JE3 7BE
GJ4YAD JE3 8BR
GJ0VJP JE3 8DJ
GJ3DVC JE3 8DY
GJ4JVP JE3 8GB
GJ3JGY JE3 8GD
GJ4ODX JE3 8GP
GJ7RWT JE3 8GQ
GJ3IT JE3 8GR
GJ3NCJ JE3 8GR
GJ3ECC JE3 8HP
GJ3EML JE3 8JY
GJ7DNJ JE3 9BA
GJ0LRV JE3 9BB
GJ8CEY JE3 9EF
GJ6FTU JE3 9EJ
GJ8PCY JE3 9EP
GJ0NSG JE4 0PH
GJ4WRR JE4 9VI
GJ8RVT JE4 9YG
GJ4ZFM JE9 9CH
GJ4YLP JE9 9LA
GJ4KBM JE9 9LE
GJ3YLN JE9 9LI
GJ3FXW JE9 9OX
GJ1JTF JE99 1AA

Kennet

G0JYG SN10 1RW
G4RQL SN10 1RW
G0JVF SN10 1SY
G8CBL SN10 1TB
G3WAE SN10 2LD
G7SPM SN10 2NS
G0LCQ SN10 3AN
G3MQD SN10 3AN
G3PZV SN10 3AN
G4GBX SN10 3AN
G4NWR SN10 3BJ
G6EVY SN10 3BJ
G4TIX SN10 3BJ
G4WIK SN10 3BJ
2E1EIB SN10 3BJ
G4XWY SN10 3DJ
G3KSU SN10 3DS
G7LPE SN10 3EG
G6NTE SN10 3EJ
G4SHA SN10 3SL
G1EQF SN10 4BT
G0VNB SN10 4JZ
G3ZJJ SN10 4LP
G0HBE SN10 4RT
G1JMK SN10 5AJ
G6ZFA SN10 5AJ
G8XYA SN10 5BA
G3WZR SN10 5DG
G4VKJ SN10 5HD
G5YN SN10 5HR
G3XKL SN10 5SR
G4EES SN10 5TD
G4TKF SN15 2DJ
G8FLL SN15 2EE
G1VTV SN15 2JB
G1YNP SN15 2JB
G3GRU SN8 1AZ
G4XVW SN8 2AS
G8AQT SN8 2AS
G6EPN SN8 2AZ
G4XYA SN8 2BS
G1FKJ SN8 2DH
G6GPL SN8 2DT
G0MVD SN8 2EG
G6EDJ SN8 2JQ
G0WEP SN8 2PG
G7VMS SN8 2PQ
G4HYI SN8 2QA
G3MFL SN8 3AS
G4DHT SN8 3DZ
G8SXA SN8 3HN
G8SXD SN8 3HN
G3NDS SN8 3NP
G4LMA SN8 3TD
G0SDG SN8 4AJ
G4XXU SN8 4NQ
2E1FTR SN9 5ES
M0AKT SN9 5EX
2E1EMK SN9 5NW
G4TOY SN9 6AE
2E1EMW SN9 6AF
G0JLZ SN9 6DB
G1XSV SP11 9PG
G0WYD SP9 7RE

Kensington and Chelsea

G7GBJ SW10 9BJ
M1BDQ SW10 9RH
G1XGZ SW3 3LB
G3GXW SW3 4SR
G1TQT SW3 4UJ
G8DOH SW3 6BU
G1VVX SW7 2HL
G0RFX SW7 3DQ
G4UAF SW7 4ET
G0OOG SW7 5PF
G3HBN SW7 5PH
G6DBL W10 5QT
G1DPW W11 1NZ
G8EBJ W11 3QT
G0VCF W11 3SJ
G7RMR W14 8BD
G4WQO W8 5QD
G4EHN W8 7SL
G0HTX W8 7SX

Kerrier

G4UKL TR10 9BN
G4UWL TR10 9DR
G0EOH TR10 9DT
G3AYS TR10 9HB
G0BSK TR10 9HL
G7VFA TR10 9HQ
G3NVJ TR10 9HS
G2BLL TR11 5AR
G0IVF TR11 5DJ
G0NQX TR11 5DT
G3FWG TR11 5EH
G4MMZ TR11 5HB
G4VMW TR11 5HF
G4VTR TR11 5HF
G0GRY TR11 5HQ
G4LHY TR11 5HS
G5TU TR11 5HT
G7NWG TR11 5LL
G6ICS TR11 5PA
M0ALL TR11 5RT
M0ACJ TR12 6BU
G4BI TR12 6QF
G0AXK TR12 6SW
G8JPD TR12 6UB
G0XPD TR12 6UB
2E1FGN TR12 6UB
G3GZJ TR12 7AB
G8GYU TR12 7AB
G6QI TR12 7AZ
G3PLE TR12 7BW
G0UPP TR12 7DF
G0JVR TR12 7DN
G0PZE TR12 7DN
G4YQP TR12 7DN
M0ADA TR12 7DN
2E1FDK TR12 7DN
2E1FDL TR12 7DN
G0XAO TR12 7DY
G0RIT TR12 7ET
G8DFA TR12 7EW
G6ZLG TR12 7HW
G7BZP TR12 7HX
G0GUO TR12 7JL
G6BZE TR12 7JZ
G3NJV TR12 7LU
G4NBF TR12 7LW
G7JTO TR12 7LW
G4JYI TR12 7NA
G3KJK TR12 7NN
G0CQJ TR12 7NR
G6MOT TR12 7NR
G3WGZ TR12 7NU
G4WSH TR12 7NZ
G6FQP TR13 0DT
G1AJB TR13 0DY
G1SVP TR13 0DY
G8ZVM TR13 0HE
G4NZF TR13 0JR
G0FKX TR13 0LD
G0GUE TR13 0PL
G4GFY TR13 0SN
G3UWU TR13 8AR
G0HFA TR13 8BP
G3UQE TR13 8DQ
G3KGP TR13 8JZ
G0RJH TR13 8NY
G0WYS TR13 8PD
G1OZV TR13 8PQ
G1ZPJ TR13 8QJ
G7FPG TR13 8UB
G0UDB TR13 9AU
G4WZH TR13 9AW
G0BQH TR13 9HZ
G3AHX TR13 9LP
G4VEZ TR13 9LP
G3KDD TR13 9NB
G4ATR TR13 9NB
G8PPF TR13 9NB
G3ZPW TR13 9NE
G1XJT TR13 9RL
G4RCV TR13 9SB
G4ZYO TR13 9SD
G4YRL TR13 9SY
G8SEE TR14 0EY
G4KES TR14 0JF
G4POT TR14 0JX
G8XCR TR14 0JX
G0CIG TR14 0LD
G8WWW TR14 0QE
G1WHV TR14 0QG
G0CAT TR14 0QZ
G4ROG TR14 0QZ
G4GFB TR14 7HR
G1KXF TR14 7NA
G7GIG TR14 7UT
G4BHC TR14 7XN
G4BHD TR14 8DD
G4LOG TR14 8HG
GW6JFX TR14 8JH
G1HFE TR14 8QF
G1LNA TR14 8QF
G3GQF TR14 8RP
G4XMA TR14 8RQ
G3WJP TR14 8RW
G8VST TR14 8TY
G8ZDS TR14 8UG
G4ZUI TR14 8UP
G4YYM TR14 9AJ
G4DEO TR14 9DE
G7DMU TR14 9EL
G3YKR TR14 9ER
2E1CUX TR15 1BU
G3JYF TR15 1JU
G1ZWH TR15 1NN
2E1BOA TR15 1NW
G0UWO TR15 1PA
G6XKO TR15 1QR
G7URH TR15 1RS
G0FKI TR15 1SE
2E1EVS TR15 2LY
G0FHT TR15 2NY
G4CRC TR15 2NY
G7VIR TR15 2TP
G6ZSQ TR15 2TQ
G6HGR TR15 3AR
G3PUQ TR15 3AW
G4DBZ TR15 3AW
G0IDH TR15 3BP
G4GDU TR15 3BZ
G6KTB TR15 3DF
G4VTS TR15 3EF
G3JQK TR15 3HJ
G3YMA TR15 3LG
G0WYU TR15 3LH
G1OLL TR15 3SE
G8ETD TR15 3TA
G0WGV TR15 3TA
G4JKZ TR15 3TS
G7DUC TR15 3YJ
G4IWC TR16 4BJ
G4KPO TR16 4BJ
G0FHW TR16 4EB
G1FVF TR16 4PE
G7PPQ TR16 4RN
G7ADP TR16 4SG
G7PQY TR16 4SR
M0BEU TR16 4SR
G3UGO TR16 5AG
G3VJB TR16 5AG
G7VOH TR16 5NL
G0FKF TR16 5PN
G0FIC TR16 5QZ
G0UVX TR16 5RL
G4VTL TR16 5SA
G4ODV TR16 5SE
G0MYR TR16 5TQ
2E1FNV TR16 6HN
G7WJJ TR16 6HN
G4ZKH TR16 6HP
G6IMN TR16 6JF
G4USB TR16 6LP
M1BUI TR16 6LQ
G8NXD TR16 6LS
G3RVF TR20 9AR
G3OCB TR3 7AA
G7GUA TR3 7AR
G0AWR TR3 7AT
M1BBR TR3 7BA
G8SLC TR3 7HJ
G0PCV TR3 7HT

Kettering

G4RUS LE16 8QD
G3VUH LE16 8QL
G3NGA LE16 8QW
G3FOZ NN14 1AL
G4KXG NN14 1AY
G8ZIH NN14 1EE
G0GRS NN14 1EL
G4MJN NN14 1LZ
G4RPG NN14 1NE
G4RPT NN14 1NG
G1NEN NN14 1RQ
G0FFB NN14 2LB
G4PDK NN14 2LP
G1LMN NN14 2QB
2E1EFT NN14 2QB
G2BOX NN14 2QL
G4PNL NN14 2QY
G1OET NN14 2RE
G0SOE NN14 2RR
G8ZPE NN14 2SQ
G8EZG NN14 2TH
G1WPR NN14 2UD
G1IHY NN14 2XA
G6DHW NN14 2XG
G1SLG NN14 3AA
G0AYX NN14 4AA
G4AIV NN14 6HD
G4DCD NN14 6HT
G8HBZ NN14 6HT
G6BKD NN14 6HY
G8BKD NN14 6HY
G7JAV NN14 6LR
G1IQA NN14 6TN
G4LTF NN14 6YD
G3WKR NN14 6YG
G7UIR NN15 5BH
G7AJJ NN15 5BN
G8WSV NN15 5BY
G4VID NN15 5DD
G7PQM NN15 5DE
G0NSA NN15 5EG
G4VKX NN15 5EQ
G7GCW NN15 5HB
G8YKE NN15 5HB
G4NZZ NN15 5HD
G7WKV NN15 5HT
G7GMQ NN15 5JB
G4ZVE NN15 5NP
G6OLU NN15 5NT
G7DMK NN15 5SQ
G4HND NN15 5TA
G0ATV NN15 5UF
G0RDV NN15 6BT
G7KRS NN15 6BT
G6DYY NN15 6DY
G0DEF NN15 6HF
G1YLM NN15 6HH
G1XJN NN15 6PF
G7APQ NN15 6PF
G4TTX NN15 6PS
G5KN NN15 6PS
G7SKA NN15 6QH
G4TXM NN15 6QU
G0PAD NN15 6QX
G0PSB NN15 6RZ
G7IVG NN15 6SF
G3VCQ NN15 6UJ
G4OYN NN15 6UW
G3ONT NN15 7BN
G1AZD NN15 7EF
G7KDS NN15 7EF
G7CJG NN15 7EW
G4UIM NN15 7LE
G3SMR NN15 7LG
G0PSG NN15 7LL
G0WWT NN15 7LL
G0RVD NN16 0AU
M0AQP NN16 0LQ
G7TZZ NN16 0PS
G7API NN16 0QB
G7MGP NN16 0QE
G4MPS NN16 0RN
G8FSJ NN16 8EB
G0XAJ NN16 8EE
G4NWH NN16 8EP
G6AKS NN16 8LW
G4BDU NN16 8LZ
G7HIF NN16 8NP
G6EBZ NN16 8PN
G7VIH NN16 8PN
G6OIH NN16 8RW
M1AWN NN16 8TU
G3VJG NN16 8UF
G0ECD NN16 8UP
G4NAC NN16 8UP
G7BAX NN16 9BT
G7AJS NN16 9BU
G4OLI NN16 9ES
G4XRA NN16 9ES
G0GBZ NN16 9EW
G0LDR NN16 9JL
G8WSU NN16 9JR
G4XJK NN16 9JZ
G6PEG NN16 9JZ
G2CIN NN16 9LJ
G4YNG NN16 9NR
G0OOV NN16 9PU
G7APH NN16 9PU
G6OXN NN16 9SB
M1BZT NN16 9ST

Kingston upon Hull

G3MRD HU1 3DG
G4IVB HU11 4AP
G0GBW HU11 4AX
G6ZEN HU11 4BN
M0AYO HU3 1NS
G7UHI HU3 1NS
M0AAR HU3 2LT
G0IRJ HU3 2TZ
G0VVP HU3 3PS
G6CCV HU3 3QH
G0UOJ HU3 5DP
G1FPC HU3 5PP
G0SDI HU3 5PY
G4WCD HU3 5QR
G1HIR HU3 6LG
G3SLU HU3 6QU
G1AZM HU3 6QY
G0SWO HU3 6QY
G3JDY HU3 6QY
G0GBX HU3 6TL
G4IBL HU3 6XQ
G3MLX HU4 6AP
M1ADN HU4 6AP
G4CXZ HU4 6EX
G0OYQ HU4 6LB
G1CFB HU4 6QE
G0ULN HU4 6QJ
G3OQV HU4 6TB
G0EXY HU4 6TE
G3TKA HU4 6TL
G1DQD HU4 6UG
G0POQ HU4 7AH
G3DDH HU4 7BU
G4ERG HU4 7BX
G1FQI HU4 7DU
G4ZPQ HU4 7HB
G0OVT HU5 1ND
G8UVQ HU5 2AE
M0AOA HU5 2BY
G1ZUG HU5 2DH
G0UPH HU5 2RD
G3KDU HU5 3DW
G0UNC HU5 3JY
G7MUV HU5 3LF
G8GBY HU5 3ND
G7DEE HU5 3NF
G0WQZ HU5 3PR
G7OIW HU5 4AZ
G1PLU HU5 4ED
G4HJD HU5 4ED
G1MCI HU5 4HN
G1SQW HU5 4HN
2E0ABJ HU5 4HQ
G3RDP HU5 4LP
G3HOY HU5 4LX
M1BBD HU5 4NX
G6VWF HU5 5DJ
G4CBL HU5 5HS
G4CZV HU5 5HU
G0TGX HU5 5HW
G0GXZ HU5 5HZ
G4CBA HU5 5HZ
G0EAA HU5 5JD
G7JGY HU5 5JW
G0VNH HU5 5LE
M0API HU5 5LE
G6ABG HU5 5NX
G3SSA HU5 5QE
G4DXH HU5 5RN
G3JRX HU5 5SN
G0CHP HU6 7AS
G6IIO HU6 7AW
G4ERZ HU6 7BA
G3NEZ HU6 7BU
G1IUF HU6 7BW
G1LSZ HU6 7DD
G3ZZP HU6 7DZ
G4HXM HU6 7EA
G4PIO HU6 7EE
G4ESJ HU6 7HE
G0ECR HU6 7JT
G0GXF HU6 7JT
M1BRG HU6 7JT
G6JFJ HU6 7JT
G7ECR HU6 7JT
G3UYV HU6 7LX
G4MPL HU6 7XF
G3PQY HU6 8SA
G6LNV HU6 8SB
G0LTZ HU6 8SE
G0LPT HU6 9EB
G0RLB HU6 9ER
G8VHG HU6 9EU
G1TDO HU6 9EZ
G0PHN HU6 9JP
G6JEN HU6 9RA
G8ZFD HU6 9RA
G6LNW HU7 0BX
G3AMW HU7 0DU
G0UKS HU7 0DU
G0VTD HU7 0DU
G1XGE HU7 3EL
G0PHQ HU7 4AY
G7GLM HU7 4HJ
G7CHJ HU7 4LA
G4OBR HU7 4PG
G0BAP HU7 4PQ
G0EPA HU7 4PY
G1ARD HU7 4RF
G7POW HU7 4RU
G8WVB HU7 4UY
G0UGU HU7 4YA
G1CXK HU7 5AB
G7NNT HU7 5AS
G7UCZ HU7 5AS
G7WGU HU7 5AT
M1BLO HU7 5AX
G0PTL HU7 5BU
G0MKZ HU7 6DR
G0WDX HU7 6EE
G4VSP HU8 0HN
G0REC HU8 0JA
G6EZG HU8 0JG
G4VHM HU8 0JP
G6HWH HU8 0NP
G0HMS HU8 0QH
G1LSL HU8 0RE
G0WFF HU8 0ST
G1WQL HU8 8JX
G6GKT HU8 8JX
G1ADE HU8 9BN
G0UTX HU8 9DG
G4RHQ HU8 9LN
G1GGJ HU8 9LZ
G1LGW HU9 1EJ
G0TUJ HU9 2TE
G4MMV HU9 3DX
G6MGR HU9 3JG
G0PSL HU9 3JQ
G7UGW HU9 3JU
G0NAU HU9 3LB
G6EBG HU9 3LB
M1CAV HU9 3LP
G4KXU HU9 4AH
G6DHB HU9 4AH
G4ASA HU9 4JU
G0HMN HU9 4RA
G1MWF HU9 5DT
G4ZLW HU9 5EZ

Kingston-upon-Thames

G7KCR KT1 2AQ
G6NMR KT1 2LL
G4GTP KT1 2QA
G4DA KT1 2SF
G3CJH KT1 3AE
G4XZT KT1 3AU
2E1EHQ KT1 3HE
M0BEJ KT1 3QB
G4GDP KT1 3SB
G4BXM KT2 5BP
G4UBV KT2 5DR
G8WHX KT2 5GD
G4PPN KT2 5GP
G6OBA KT2 5HH
2E1ALV KT2 5JU
M1BRP KT2 5RZ
2E1EWC KT2 5RZ
G7DYA KT2 5TB
G6RFU KT2 5TU
G1MCJ KT2 6JD
G8YJQ KT2 6RA
G1DJD KT2 6RY
G0NZX KT2 7PZ
G4OFO KT3 3BA
G4JUZ KT3 3DY
G3TON KT3 3HP
G4FJP KT3 3HP
G6REA KT3 3HX
G8BPY KT3 3HZ
G0BGY KT3 3JX
G0BQA KT3 3LD
M1AZP KT3 4AY
G1YVZ KT3 4DU
G8GGI KT3 4HR
G4GPZ KT3 5AQ
G1HCM KT3 5BP
G7UVF KT3 5BZ
G8CLF KT3 5NQ
G1DYW KT3 5QT
2E1EWY KT3 6BY
M1CEA KT3 6BY
2E1EXD KT3 6BY
G4ZXO KT3 6DJ
G8BBC KT3 6PQ
G6BQQ KT4 7PE
G0KRT KT4 7SJ
G3OGL KT4 7SJ
G6ZLE KT4 7SN
G4PGA KT4 8BZ
G0IPE KT5 8DD
G6DKS KT5 8EX
G3HFO KT5 8HT
G4LJI KT5 8JY
G8KEJ KT5 8JZ
G1CWI KT5 8PA
G5DS KT5 8PW
G0IRK KT5 8TS
G4ULD KT5 9AA
G6SMJ KT5 9BS
G0JOS KT5 9DX
G4ERW KT5 9DX
G3KQR KT5 9HA
G8PYE KT5 9HD
G4PPU KT5 9JH
G7MLJ KT5 9LW
G1KAG KT5 9NR
2E0APA KT5 9RJ
G0BQV KT5 9RJ
G3KIN KT5 9RJ
G3YIC KT6 4BN
G7DGW KT6 4SW
G0PCY KT6 5AF
G4WGE KT6 6LJ
G0SAC KT6 6LJ
G4JG KT6 6PU
G3ZQT KT6 6QN
G4LWT KT6 6QN
G8DPS KT6 6RF
G1OEP KT6 7NA
G0KEB KT6 7NR
G8KFS KT6 7PP
G3JXA KT6 7TX
2E1ALP KT9 1DP
M0AWN KT9 1EF
G1JRR KT9 1JH
G1SXA KT9 1JJ
G0NCH KT9 1JY
G4HJY KT9 1NL
G0WIT KT9 1PN
G6AVP KT9 1RP
G0CFH KT9 2AB
G1EOO KT9 2AB
G1ADW KT9 2AP
G1SIP KT9 2BU
G7GLR KT9 2DE
G1RMC KT9 2DE
G8RWH KT9 2DE
G6DFM KT9 2DT
G8XCQ KT9 2HA
G0DOE KT9 2HF
G0DOG KT9 2HF
G6KIE KT9 2HG
G6HMG KT9 2NN
G4FKA KT9 2NR
G0SDF KT9 2PR
G3IMK KT9 2PU
2E1CRB KT9 2PU
GM0CIA KT9 2QD
G3SXW KT9 2QD
G0AAA KT9 2QD
G7BHH SW15 3RU

King's Lynn and West Norfolk

G6HGD IP26 4AR
G0CZR IP26 4BD
G7ACN IP26 4DB
G6EZA IP26 4NT
G4AXP IP26 4RB
G6AKC IP26 4RG
M1AZA PE13 3RH
G8GIN PE13 3RP
G7STZ PE14 7AZ
G6OEJ PE14 7DN
G6HKS PE14 7EJ
G4JOA PE14 7ET
G4XSZ PE14 7LH
G0WDK PE14 7LZ
G0FMA PE14 8AN
G1GYI PE14 9AL
G7NCG PE14 9SP
G8BHG PE30 2AQ
G0OFD PE30 2EE
G0BMS PE30 2NA
G4MOC PE30 2PB
G1HYU PE30 2PE
G0VHH PE30 3BJ
G3VET PE30 3BS
G8NSK PE30 3DP
G0IJU PE30 3DT
G3ZCA PE30 3EZ
G3THK PE30 3HA
G8GKL PE30 3HB
G4CDK PE30 3HR
G7VHE PE30 3LH
G8KOC PE30 3LY
G6REC PE30 3NJ
G4CBO PE30 3PX
G8CBB PE30 3QQ
G6BYT PE30 3UD
G1XYZ PE30 4AB
G3HRX PE30 4AB
G3XYZ PE30 4AB
G4OZG PE30 4AB
G4UPQ PE30 4AD
G4PLD PE30 4AE
G7UIA PE30 4AE
G6IIP PE30 4AT
G0TNB PE30 4BA
G1FOO PE30 4HD
G0TNJ PE30 4HD
G8ZLT PE30 4HE
G6RDZ PE30 4NU
G3SZ PE30 4PG
G0WWU PE30 4QA
G8STE PE30 4QE
G1KBH PE30 4QZ
G4TPS PE30 4TQ
M1CCQ PE30 4YH
G7OUZ PE30 5BQ
G1LOK PE30 5LD
G1ONF PE30 5NB
G1DJM PE30 5NY
G7NFQ PE30 5RD
G0BSQ PE31 6AT
G4JNQ PE31 6BT
G3WMN PE31 6JT
G0DIO PE31 6LP
G6FHJ PE31 6LX
G3YLW PE31 6LY
G0MTB PE31 6NH
G3YII PE31 6NH
G6BOS PE31 6NH
G6TVI PE31 6PR
G7WGL PE31 6QN
G1SCQ PE31 7AR
G7RNA PE31 7AR
G6WDC PE31 7BS
G8OSC PE31 7JB
M1CCN PE31 7LA
G4DDT PE31 7LQ
G4EOJ PE31 7QF
G6XTY PE31 7QR
G8DLM PE31 8AE
G7SJW PE31 8AQ
G0IZD PE31 8BP
G3VOD PE31 8LJ
G0CQB PE31 8LN
G4DCJ PE31 8RL
G4LSF PE31 8RL
G8FTP PE31 8RL
G0LRU PE31 8RS
G4ZQC PE31 8RS
G7DNW PE31 8RW
G3OVL PE31 8SF
G4JUV PE32 1AF
G0OJP PE32 1DQ
G7NJP PE32 1DQ
G6FQL PE32 1LW
G7OVE PE32 1NY
G4JQO PE32 1RJ
G1BFQ PE32 1RL
G6CPS PE32 1RL
G6RQA PE32 1RY
G8TUH PE32 1SF
G1ICX PE32 2HR
M0ASQ PE33 0AD
G0JWG PE33 0HE
G0BPL PE33 0HS
G8RES PE33 0HS
G4OKH PE33 0HX
G8MBJ PE33 0HX
G8CWE PE33 0JG
G4RKN PE33 0LB
G0UMM PE33 0PG
G8ULL PE33 0PR
G0MQL PE33 0PW
G3SVS PE33 0QG
G1EMW PE33 9BD
G4IXD PE33 9HP
G4ILV PE33 9HX
G3THS PE33 9JJ
G4EJE PE33 9PJ
G7BIM PE33 9QR
G0FQO PE33 9RP
G1RHE PE33 9RP
G8YQY PE34 3AW
G8OZT PE34 3AY
G6KDP PE34 3BJ
G0UWW PE34 3BX
G7MOD PE34 3HL
G4NJJ PE34 4EA
G6TZG PE34 4PU
G7PTB PE34 4QJ
G0KHF PE36 5BD
G4VUF PE36 5BD
G0XBO PE36 5BZ
2E1EBX PE36 6BX
G3MEZ PE36 6HL
G4HBP PE36 6NE
G7JDQ PE36 6NF
G1OQY PE38 0AG
G3NHR PE38 0BT
G1FXU PE38 0BY
G6TRA PE38 0BY
G4RCP PE38 0DH
G1VIS PE38 0DP
G6NMK PE38 0DP
G3ZOF PE38 0JG
G3UPN PE38 9AX
G4LJ PE38 9ND
G8JAN PE38 9PG
G7TBV PE38 9PG
G4TUO PE38 9QU
G3NEU PE38 9RP
G3ASG PE38 9RQ
G7MUN PE38 9RT

Kirklees

G4HAG BD11 2AD
G8CHN BD11 2EE
G4EZS BD11 2ET
G3SVC BD11 2ET
M0BCE BD11 2JE
G7WKL BD11 2JE
G0FOI BD11 2NN
G4JJC BD12 9DS
G7HSS BD12 9LS
G0JKW BD12 9LU
G0DPX BD19 3BY
G4SKO BD19 3EL
G5PW BD19 4DS
G6GNE BD19 4ET
G3LEJ BD19 4LD
G4OTL BD19 4RP
G0RZP BD19 4SB
G3WYP BD19 4SB
G8WYP BD19 4SB
G1WQI BD19 5AX
G4VKA BD19 5AX
G0IQQ BD19 5JH
G0PDT BD19 5JJ
G7HRK BD19 6HU
G4OSO BD19 6JH
G4OSP BD19 6JH
G3RGN BD19 6LJ
G4IOD BD19 6LW
G4KFP BD19 6LW
G0CPX BD4 6PB
G3BGA HD1 2NU
M1BDT HD1 3HU
G0BFJ HD1 3SL
2E1CQJ HD1 3TP
2E1FVI HD1 4NW
G3VOI HD1 4PE
G8AZF HD1 4PT
G7BNS HD1 5DY
G6STW HD1 6NX
G1CYY HD2 1DA
G3WUI HD2 1DH
G0ODY HD2 1LB
G1EEY HD2 1LG
G0FHA HD2 1PR
G4YDI HD2 1QH
G1AOR HD2 1QP
G8YOC HD2 1QY
G3XXR HD2 1RN
G1GPT HD2 1UP
G3WLW HD2 1UY
G1HMP HD2 2FL
G7RAI HD2 2JP
G4RIA HD2 2LR
G0BWQ HD2 2NE
G1MBM HD2 2NE
G7UBQ HD2 2NF
M0AGR HD2 2NF
G4BYW HD2 2NH
G4MEK HD2 2NL
G4XJO HD2 2NU
G6DLA HD2 2XN
G0SBR HD3 3AD
2E0AEX HD3 3AD
G0IRY HD3 3AP
G3BKJ HD3 3DB
G4KGS HD3 3DD
G0JTX HD3 3DF
G4EIE HD3 3NL
G3TFO HD3 3QX
G6ZDV HD3 3QX
G4TEL HD3 3QZ
G0DLX HD3 3RF
G8MAR HD3 3RJ
G3ZXZ HD3 3SG
G0PRF HD3 3SN
G0DDB HD3 3UE
G8KMK HD3 3UE
G7UNK HD3 3UF
G3MWK HD3 3WW
G6CNZ HD3 3WW
G4LRD HD3 4GH
G1AGA HD3 4LD
G1AGB HD3 4LD
G0PNY HD3 4RF
G0LPV HD3 4RF
G0LVV HD3 4RF
G3LUK HD3 4RF
G4VKM HD3 4RJ

G7SXB HD3 4XP
G4GMT HD3 4YW
G0LUU HD4 5HN
G0TXW HD4 5LG
G4GNA HD4 5NS
G7JMU HD4 5NS
2E1CPR HD4 5NZ
G0DXK HD4 5TD
G7KTC HD4 5UB
G6FMQ HD4 5UD
G8YRN HD4 5UD
G0VVN HD4 6DL
G0EVA HD4 6DY
G0DBU HD4 6QX
G6TGE HD4 6QX
M0AFS HD4 6SS
G4TML HD4 6TE
2E1FUG HD4 7AF
G0BWB HD4 7DA
G0LOQ HD4 7DA
G8XSF HD4 7JS
G1GTH HD4 7LD
G6KNI HD4 7LN
G0FQL HD4 7RF
G0TPM HD4 7RJ
G4JLO HD4 7SR
G1GOG HD4 7WB
G4RLA HD5 0AT
G4BYD HD5 0DW
G0DNP HD5 0HT
G7MHQ HD5 0JB
G0ISX HD5 0JD
G6GSV HD5 0JD
G4BLL HD5 0JY
G6UNC HD5 0LW
2E1EAT HD5 8AG
G3ANE HD5 8DZ
G8TXM HD5 8DZ
G7PAO HD5 8EU
G4SSQ HD5 8LA
G4BQC HD5 8LS
G4UGA HD5 8PH
G0GOJ HD5 8RA
G7KNT HD5 8RB
G3PCS HD5 8RD
M0AOB HD5 8RP
G7TTD HD5 8RP
G0NMH HD5 8SD
G4CEV HD5 8TS
G4KGN HD5 8UE
G4ITV HD5 8UJ
G4RAJ HD5 8UP
G3LLN HD5 8UW
2E1CPM HD5 8XN
G7OOH HD5 8XT
G3YWI HD5 8XU
G4OTE HD5 8XU
G1FYS HD5 9HG
G0BWO HD5 9HX
G4ONR. HD5 9NX
G1NBG HD5 9SF
2E0APW HD5 9UA
2E1CRW HD5 9UA
G6ZVU HD5 9UW
G0HNW HD7 1EH
G1CFA HD7 1EN
G1LFD HD7 1EN
G8PUT HD7 1EN
2E1CGP HD7 1EN
G3RJT HD7 1ES
G0TOJ HD7 1LQ
G3BBD HD7 1LQ
G8SNJ HD7 1NH
G0BAD HD7 1NP
2E1FXO HD7 1PP
G0KHM HD7 1PQ
G3YPE HD7 1PQ
G3LHH HD7 1PY
G8LIK HD7 1PY
2E1FFM HD7 1UL
G7PIG HD7 1XH
G3SPL HD7 1YW
G4LLZ HD7 2DD
G0JTL HD7 2DS
G4CDD HD7 2DS
G3OYI HD7 2ED
2E1FWA HD7 2EG
G6KZR HD7 2EL
G0PLG HD7 2ER
G4UFZ HD7 2FA
G7VDH HD7 2HL
G7UZO HD7 2NT
G8GFY HD7 2RL
G7KPH HD7 2SQ
G4LED HD7 2SQ
2E1EAK HD7 2SQ
G3ZHP HD7 2TJ
G8ZML HD7 2TJ
G4KSS HD7 2UD
G1TKN HD7 2UZ
G6WSX HD7 2YR
G3NAK HD7 3AJ
G7GKQ HD7 3AJ
G3LDJ HD7 3DR
G0BTA HD7 3JH
G1BEB HD7 4AU
G4FSQ HD7 4BT
G4OTC HD7 4JF
G1MQQ HD7 4JR
G4LYY HD7 4JR
G0HRF HD7 4JY
G4MH HD7 4LZ
G4IDR HD7 4NS
G0BBS HD7 4NU
M0AIA HD7 4NU
2E1CRI HD7 4NU
G4UFL HD7 4QP
G4UFJ HD7 4QQ
G3XWN HD7 4RA
G8VMF HD7 4RE
G0PLD HD7 5BW
G3CJD HD7 5DS
G2DBW HD7 5EB
G3SFY HD7 5HU
G0JTM HD7 5LS
G8MQK HD7 5PJ
G1SOY HD7 5QS
G6UJU HD7 5QX
G3CWW HD7 5SY
G4KIE HD7 5UD
G1EVG HD7 6AJ
G3HWQ HD7 6AU
G1XNX HD7 6BN
G0HRB HD7 6LR
G8CW HD7 7LF
G2AND HD8 0AB
G0CVJ HD8 0AL
G6LD HD8 0AN
G3HEQ HD8 0BN
G3TSA HD8 0DE
G4AHJ HD8 0JB
G4XJG HD8 0JG
G7NDC HD8 0JG
G6HBK HD8 0PT
G6APB HD8 0QS
G4RAL HD8 0QZ
G1HCE HD8 0SA
M0ALT HD8 8AG
G4OPY HD8 8AP
2E1FVS HD8 8EY
G1DEN HD8 8HP
G3SDY HD8 8HP
G7WGP HD8 8HP
G7UKK HD8 8LT
G7VBU HD8 8LT
G0PHI HD8 8LX
G0VIK HD8 8NQ
G3ABS HD8 8NY
G7EXD HD8 8QW
G3VQH HD8 8QZ
G4KDM HD8 8RU
2E1GBD HD8 8SG
G4HMU HD8 8SP
G0WNX HD8 8SP
G8RWN HD8 8UX
G0WFE HD8 8UX
G0APS HD8 8XT
G6DHN HD8 9AH
G3ZIF HD8 9AN
G8XEJ HD8 9AR
G4HKY HD8 9BT
G7WGY HD8 9BZ
G0MVF HD8 9DB
G4ATA HD8 9DB
2E1EBV HD8 9DB
G4TKO HD8 9DS
G7UYE HD8 9DW
G4TBZ HD8 9DX
G4JCL HD8 9EH
G8PRN HD8 9ET
2E1BGU HD8 9JJ
G1HEP HD8 9QG
G4EMQ HD8 9QP
G8TIS HD8 9RA
M1BAV WF12 0BD
2E1FLF WF12 0EB
G7LBM WF12 0JZ
G3YZA WF12 0LN
G4GCW WF12 0LW
G0CBW WF12 0NL
2E1FMC WF12 0PS
G0PMP WF12 0QU
2E1FFL WF12 7AW
G8ODA WF12 7JA
G7NPL WF12 7LA
M1AJG WF12 7LS
G8POK WF12 7SQ
G6FTJ WF12 8AJ
G6ZJV WF12 8AQ
G7PCC WF12 8RH
G4GHQ WF12 9DN
G0CBH WF13 2DP
G6ZJI WF13 3LX
G4YAJ WF13 3LZ
M1BHE WF13 3PW
2E1FJO WF13 3PW
G0WXG WF13 3PZ
G6ZUO WF13 3SR
G7UMA WF13 3TG
G1IKF WF13 4BU
G8LLJ WF13 4DU
G6WGE WF13 4EN
G0SZV WF13 4RB
G1KNZ WF13 4RY
G6ITV WF14 0AJ
G6XTZ WF14 0AQ
G6OZR WF14 0HE
G4SUI WF14 0JE
G3UFC WF14 0JN
G1DEX WF14 0LP
G7VNE WF14 0NH
G3HPD WF14 0QN
G3SYK WF14 8HW
G7IFK WF14 8PQ
G2FCP WF14 9AW
G8NZR WF14 9BJ
G7CSV WF14 9PB
G4PHR WF14 9PB
G4WMS WF14 9PB
G0HSE WF14 9PE
G4MLW WF14 9PE
G0HGL WF14 9PL
G1BGQ WF14 9PW
G8SVX WF14 9PY
G8DZW WF15 6QE
G6UZZ WF15 7AL
G6XMH WF15 7AX
G0GMJ WF15 7DN
G4GOX WF15 7DP
G1KJQ WF15 7EE
G4SJR WF15 7EE
G4GCL WF15 7PX
G6ZWY WF15 8BA
G8NYM WF15 8DG
G7MHD WF15 8JU
G3JQC WF16 0BE
2E1CEC WF16 0BN
G0TAL WF16 9JL
2E1GAE WF16 9NP
2E1GAF WF16 9NP
G8GNJ WF16 9PF
G4MLV WF17 0DX
2E1CDQ WF17 0DX
G0SJB WF17 0PF
2E1DCS WF17 0PJ
2E1CUG WF17 0QL
2E1DCR WF17 0QT
G7LTO WF17 0RG
2E1ENR WF17 5HJ
G4LMS WF17 6DG
G8LMS WF17 6DG
G4XZM WF17 6DZ
G4SEQ WF17 6EH
G1NEG WF17 6HG
G2HHV WF17 6HQ
G1UVD WF17 7NT
G0RLL WF17 7SN
G1RIV WF17 7SN
G4ONZ WF17 8BE
2E0AME WF17 8DY
G1HZQ WF17 8EE
G0MZL WF17 8EX
G7RTY WF17 8PX
2E1GAQ WF17 9AT
G0MZZ WF17 9BQ
G0LBY WF17 9DJ
G6XHK WF17 9DL
G6VHW WF17 9HX
G6JLI WF17 9QF
G4XKC WF17 9QU
G0PPQ WF17 9QZ
G0COA WF4 4TE

Knowsley

G4VKV L10 4XL
G0LZX L10 4YT
G0WMO L14 6UF
2E0ANI L14 6UF
G0SXA L14 9PA
G0AJL L25 0QD
G1FAL L26 0TN
G0RFI L26 0UY
G0BVJ L26 6LD
G0DXF L26 7WG
G6UYI L26 7YS
G7NHF L26 7YW
G7UWK L26 9XP
M1BWH L32 1TP
G4PWK L32 4UD
G0XBL L32 7QH
G0JIB L33 1UW
G0LKR L33 1UW
G1YWY L33 1WY
G3YCX L34 1LW
G1KCY L34 1LY
G3REV L34 5SN
G8MKO L35 1QG
G6MMG L35 3JL
G0IEW L35 3PP
G0ILC L35 3RW
G1EVV L35 3SB
G0RPD L35 3SF
G1CKV L35 3SG
G8FHD L35 3TE
G6VBB L35 5BJ
G3BWR L35 5BY
G4EMT L35 5HL
G0FYS L36 0XA
G7RYW L36 1TE
G4EGM L36 1TH
G0GIF L36 1TY
G0NYR L36 1UD
G6XUL L36 1XB
G4IYT L36 2NN
G4PHQ L36 2PL
G4DIT L36 3TY
G1ATG L36 5SF
G3TVN L36 5TN
G3AVJ L36 5TR
G1LUY L36 5YQ
GM8SIQ L36 5YR
G4VCP L36 6AT
G3YHB L36 6DJ
M0ANN L36 6ER
G7AWW L36 8EH
G6URF WA8 5BT
G0GVD WA8 5BU
G7ERH WA8 5BU

Lambeth

G4UVX SE1 7NA
G4KWC SE11 4AL
G0JSV SE11 4HP
G3RGM SE11 5UL
G1FHY SE21 8DA
G0PAR SE21 8JY
G7OOK SE24 9AE
G4CPL SW12 0JJ
M1BJC SW16 1AA
G4CGZ SW16 2LR
G3ARO SW16 5EJ
G6BPN SW16 6JX
G4ILW SW2 1NQ
G7IHV SW2 1PA
G8UYI SW2 3HU
G4MIB SW2 3JX
G7GPZ SW2 4AE
G3KHQ SW2 5LU
G2FAB SW4 0QP
G4PIA SW4 6HB
G7GBF SW4 6NS
G4JZA SW4 8EQ
G8LHQ SW4 9HD
G0PKE SW8 3BG
G0VPH SW9 7UE
G3PPX SW9 8DW
G6NAD SW9 9RH

Lancaster

G8FRO LA1 1SH
G1PEE LA1 2JA
G3VWJ LA1 2TZ
G1OHH LA1 2UQ
G6ZHL LA1 3AW
G7BBN LA1 3DS
G8EXJ LA1 3DU
G8UHO LA1 3DY
G0VGP LA1 3FA
M1BHB LA1 3HJ
M1BOU LA1 3HJ
G6IXK LA1 3LL
G0AZJ LA1 3NB
G1HFA LA1 3ND
G0IIU LA1 3NS
G3ZWZ LA1 3PP
G1YYD LA1 4BD
G0FZA LA1 4BW
G6FKR LA1 4DJ
G0AWM LA1 4ER
G1KMC LA1 4EY
G8TZJ LA1 4LQ
G8ACC LA1 4QY
G4EKZ LA1 4QZ
G1HIB LA1 4RG
G0EIP LA1 4RZ
G3SAQ LA1 4SJ
G0UJE LA1 4YR
G4DLP LA1 5BD
G3XEN LA1 5EB
G8KED LA1 5EB
G3BAP LA1 5HE
G6FGV LA1 5HQ
G0GGJ LA1 5HW
G4UPM LA1 5JA
G6ZGH LA1 5LY
G7VLU LA1 5LZ
G0JHG LA1 5QA
G0LUR LA1 5QA
G7TBQ LA1 5UU
G4DCG LA2 0AB
G1KLO LA2 0BY
G4FM LA2 0EF
G4UJI LA2 0EG
G4TOT LA2 6AG
G3HHR LA2 6EE
G3UCE LA2 6EE
G0VPU LA2 6ER
G0NTT LA2 6HJ
G7VRV LA2 6LB
G7TOZ LA2 6NL
G0UWX LA2 6QB
G6YXG LA2 8NU
G4ZEZ LA2 8QD
G8BLV LA2 9AH
G4KWJ LA2 9AN
G4KWK LA2 9AN
G0AXU LA2 9LF
G4NEQ LA2 9LF
G1CFJ LA2 9NY
G1ZWJ LA2 9PB
G3UVQ LA2 9PB
G4ZJE LA2 9PJ
M1AQA LA2 9PJ
G3CVG LA3 1AY
G6CRV LA3 1NZ
G4ZJO LA3 1QE
G1RLK LA3 1UR
G6BKY LA3 2DJ
G1PWY LA3 2EB
G3LJO LA3 2JN
G3KOQ LA3 2LR
G3RCZ LA3 2LX
G0LIO LA3 2PA
G1MTK LA3 2PG
G6TVW LA3 2QX
G4ZJL LA3 2TB
G4UME LA3 3AA
G0VXS LA3 3DF
G1BMI LA3 3EF
G7IHD LA3 3JZ
G6OMX LA3 3LA
G0LWU LA3 3RA
M1BKB LA4 4HE
G1ROM LA4 4NA
G1ROL LA4 4NA
G6OUT LA4 4PF
G7NLR LA4 4PF
G3CSY LA4 4PZ
G0EBP LA4 4QD
G1PKA LA4 4QS
G1GRP LA4 4RE
G1ONH LA4 4RE
G0LAQ LA4 4RL
G4VAP LA4 4RL
G4VCT LA4 4SP
G7CAF LA4 4TU
G6ZEH LA4 4UN
G6PHF LA4 4UT
M0AZM LA4 5AA
M1BEH LA4 5AA
G3VSE LA4 5BN
G3KSP LA4 5QD
G0PTW LA4 5SB
G1YNU LA4 5SE
G0RDH LA4 5UJ
G4YBS LA4 5UJ
G4XLC LA4 5XP
G0FYH LA4 6BE
G0CHV LA4 6EJ
G4EDZ LA4 6EQ
G0GEU LA4 6HD
G4VOJ LA4 6HY
G0LQX LA4 6JU
G0WJW LA4 6JU
G4IFT LA4 6LT
G7DPC LA4 6LT
G4BWX LA4 6NU
G6BXK LA4 6PA
G4UUX LA4 6PN
G0RSL LA4 6PS
G0AUF LA4 6QG
G8FJA LA4 6QL
G0ASQ LA4 6QR
G4ZBM LA4 6SE
G4PZD LA4 6SR
G0FYW LA4 6TQ
G0ROR LA4 6UG
G3XLS LA5 8AP
G0CWP LA5 8EN
G0NEL LA5 8HH
G0VLV LA5 8HJ
G6ZET LA5 8HQ
G7ECG LA5 9DU
G0NYD LA5 9EJ
G3JAH LA5 9HR
G4ABE LA5 9HS
G4DXX LA5 9LU
G8XXQ LA5 9QP
G7CTH LA5 9TT

Leeds

G7VJH BD11 1DN
G4TCT BD11 1HE
G4YJM BD11 1HR
G6PXA BD11 1HZ
G8LVQ LS1 7WN
G4RCH LS10 3BL
G6YHW LS10 3PZ
G4JPK LS10 3QP
G4LYA LS10 4RT
G1ILT LS10 4SS
G0VRS LS11 0JD
G0WYZ LS11 0JD
G0GVT LS11 5HD
G0GVU LS11 5HD
G4MPO LS11 5RX
G4IDT LS11 5SG
G4PGK LS11 5UR
G7HUJ LS11 6NN
G1MJT LS11 7HY
G7VJF LS11 9LA
G7OCX LS11 9RE
G6COE LS12 2BB
G0RFA LS12 2BD
G8LFN LS12 2QY
G4MOT LS12 2RW
G7RME LS12 3LB
G8SWK LS12 3QJ
G3TDZ LS12 3QL
G8PKJ LS12 3SN
G3PTJ LS12 3TB
G3NSU LS12 3TZ
G6KML LS12 4HF
G6YIN LS12 4HH
G4ICF LS12 4HQ
G8KZY LS12 4HQ
G6YMH LS12 4LA
G4SAX LS12 4PB
G4IDD LS12 4RR
G6SFH LS12 4UW
G4LXW LS12 5EA
M1AVK LS12 5HE
G0OMU LS12 5HU
G4YQW LS12 5LP
G4RYE LS12 5QS
G1OUX LS12 6AY
G4SQG LS13 1BL
G7MZW LS13 1BN
G7LBH LS13 1EA
G7AQA LS13 1NJ
G7ELX LS13 2AN
G1DWT LS13 2BX
G6IFC LS13 2JZ
G4HLI LS13 2LE
G4OFA LS13 2LH
G6XGK LS13 2NJ
G3YEE LS13 3AN
G4YLD LS13 3AN
G7BXA LS13 3DH
G7UEG LS13 3DH
G0URF LS13 3JX
G4NWF LS13 3NN
G4LOS LS13 4AN
G7HRP LS13 4JH
G1OBU LS13 4JL
G1PAW LS13 4JR
G4DMH LS13 4PH
G4OOJ LS13 4QY
M0AVZ LS13 4RN
G7KIA LS13 4SQ
G0JUT LS14 1AP
G3DNF LS14 1EW
G1VHX LS14 1LJ
G0TGH LS14 1NR
G0GAB LS14 1NT
G6CLL LS14 1NT
G1YNH LS14 2EA
G8NXH LS14 2HY
G1ZUB LS14 3AU
G4PTA LS14 3DF
G1EGK LS14 3HA
G7IVC LS14 3HZ
G4VRW LS14 5LE
G7VEB LS14 5PD
M1AUR LS14 6EL
G7PMX LS14 6JZ
M1ATB LS14 6XE
G7KJT LS15 0EZ
G7ELS LS15 0EZ
G3HJP LS15 0LL
G4XZI LS15 0LW
G7NXV LS15 0NQ
G7OZQ LS15 0PZ
G0WNP LS15 4AN
G0BPJ LS15 4AU
G1JQG LS15 4BS
G3XVP LS15 4EZ
G7JQJ LS15 4HB
G4ZRF LS15 4HJ
G0BOX LS15 4HS
G3GXQ LS15 4JD
G3ADQ LS15 4JT
G4OOB LS15 4NY
G4IPP LS15 7BJ
M1CAE LS15 7DN
G4NXY LS15 7EW
G6FFQ LS15 7HA
G0HOB LS15 7QW
G7DCT LS15 7SQ
M1ASR LS15 8EA
G4ATS LS15 8HX
G8ZXT LS15 8JJ
G7ELK LS15 8LN
G0CKY LS15 8LS
G4HSZ LS15 8LW
G7HSR LS15 8RX
G7MTI LS15 8SD
G6INO LS15 8SW
G6OLR LS15 8UA
G1BYB LS15 8XE
G0NPG LS15 9EP
G0ORS LS16 5EB
G8ENO LS16 5JE
G4BDG LS16 6AE
G0SGJ LS16 6BS
G0FWP LS16 6BX
G1LUN LS16 6DA
M1AYK LS16 6DA
G6SYV LS16 6DR
2E1ANQ LS16 6DU
G4XZD LS16 6DX
G1SGM LS16 6EF
G4IEZ LS16 6EH
2E1DOL LS16 6RX
G0PAN LS16 7AB
G0BAK LS16 7BZ
G0IBU LS16 7HB
G3GMV LS16 7HD
G4AEC LS16 7HD
G7RGK LS16 7HX
G4PHP LS16 7JF
G4TIK LS16 7LN
G4BNL LS16 7NU
G3NQZ LS16 7PE
G4TSF LS16 7PG
G4EIQ LS16 7PH
G0WRT LS16 7PN
G0RQZ LS16 7PP
G0SCL LS16 7PP
2E1BFA LS16 7PP
2E1BJL LS16 7PP
2E1CNP LS16 7PP
M1CAA LS16 7QS
G0BPW LS16 7QW
G4HAJ LS16 7QX
G8TEL LS16 7RB
G7BSW LS16 7RH
G4KKQ LS16 7SG
G3NRM LS16 8BL
G1SBK LS16 8DE
G8PQY LS16 8EA
G4CSZ LS16 8EJ
M1AJA LS16 8HL
G1CXQ LS16 8JQ
G0HPT LS16 8LG
G4MVF LS16 9BP
G4HGT LS16 9BU
G7NUO LS16 9DL
G4GOP LS16 9DP
G3MFJ LS16 9DQ
M0BGF LS16 9DQ
G4FKS LS17 5AW
G4RDL LS17 5AW
M1APF LS17 5AX
2E1EMN LS17 5AX
G8UHW LS17 5BH
G3WSZ LS17 5DL
M1AZM LS17 5HW
M1AHR LS17 5JP
G0IEI LS17 5LG
G4NUM LS17 5NS
G4THX LS17 5NS
G3WNR LS17 5PQ
G0AEX LS17 6DD
G2HLL LS17 6HL
G4UZN LS17 6HQ
2E1EMF LS17 6RF
M1APL LS17 6RF
G3AAS LS17 6RY
G8XVI LS17 7BH
G0SBZ LS17 7DU
G0CYL LS17 7DW
G3ZBA LS17 7DW
G7DHM LS17 7EP
2E1COM LS17 7EP
G4EZI LS17 7HY
M1ATT LS17 7JQ
G3WLP LS17 7LB
G3KVJ LS17 7NH
G4FLF LS17 7NS
G4LYM LS17 7PD
G4MSN LS17 7QB
G7RBA LS17 7RB
G7HRH LS17 7TE
G1HHT LS17 7UZ
M1AFV LS17 8BH
G4XZG LS17 8BS
G4NCK LS17 8BY
G4OPM LS17 8EB
G7TFR LS17 8JB
G8BNE LS17 8JT
G0PFT LS17 8LX
G8ASG LS17 8LX
G1OLH LS17 8NS
G4PAJ LS17 8TQ
G0DJM LS17 8XW
G0UMX LS17 9AE
GW0VEM LS17 9AE
G3YFI LS17 9AU
G4ZIU LS17 9DY
G1MPT LS17 9EG
G4ASQ LS18 4EU
G1SBJ LS18 4HY
G7RKJ LS18 4LL
G0KVC LS18 4PJ
G1KFI LS18 4PX
G0LBH LS18 4RL
G4OVT LS18 4RL
G4YDW LS18 4RL
G7BMW LS18 4RL
G3LBO LS18 4SL
G4TJO LS18 5AR
G7VHU LS18 5AW
G4LAD LS18 5HB
G8WYR LS18 5HB
2E1CNO LS18 5JU
2E1COD LS18 5JU
G3TAF LS18 5LG
G4MLN LS18 5NE
G4OAT LS18 5NE
G4HKU LS18 5PG
G4RVO LS18 5PP
G3WZF LS18 5PT
G4OWF LS18 5QU
G4YLM LS18 5QU
G0LHU LS18 5RN
G4YBA LS18 5RQ
G1DPB LS18 5RW
G6OTS LS18 5SJ
G3FCW LS18 5SQ
G4GYL LS18 5SQ
G4FKY LS18 5UN
G1BQV LS19 6AD
G3ZNK LS19 6AR
G3RMQ LS19 6DP
G7BHG LS19 6EH
G7HSV LS19 6EH
G1PFZ LS19 6LB
G0HQU LS19 6NH
G6JXP LS19 6PD
G3XNO LS19 6PJ
G4LZT LS19 6PJ
G4XQV LS19 6RG
G4OWG LS19 6RT
G4FBB LS19 7AS
G4REC LS19 7AU
G0VXV LS19 7NL
G8RTK LS19 7NU
G0JUL LS19 7QA
G4XYR LS19 7QL
G6YRG LS19 7SN
G1ZSO LS19 7TA
G7KHT LS19 7TQ
G0UCY LS2 9JT
G4BZL LS20 8AX
G4GGR LS20 8AY
G4OWD LS20 8EJ
G1NEV LS20 8HA
G3LGS LS20 8JF
G3KKP LS20 8NX
G3MBW LS20 9DH
G7PFY LS20 9EF
M1CAI LS20 9HN
2E1DMQ LS20 9HN
G7VTW LS21 1AH
G4IEN LS21 1BZ
G4LYR LS21 1DB
G4MWQ LS21 1DD
G0CLD LS21 1DH
G8HZO LS21 1DH
G7UDB LS21 1EG
G4RYY LS21 1HE
G0KQG LS21 1JS
G6XOG LS21 1JZ
G8HDN LS21 2AA
G7OXH LS21 2AL
G0ERN LS21 2BL
G0KDQ LS21 2BY
G0NIG LS21 2BY
G4EFV LS21 2DU
G7RCN LS21 2ED
G0CLF LS21 2EJ
G0SNW LS21 3LE
G7LNT LS21 3LW
G7ULH LS21 3LW
G4JVZ LS21 3NR
G4GDL LS21 3NZ
G4IEC LS22 4QY
G4YBD LS22 5AT
G4YZN LS22 5BW
G4CAV LS22 5JR
G1UDB LS22 6NP
G1SBN LS22 6RN
G7RDL LS22 6RN
G4WWL LS22 7RA
G6RDO LS22 7UE
G4JVC LS23 6AY
G3THA LS23 6DN
G1VXT LS23 6HF
2E1FNC LS23 6PD
G8XMU LS23 6PU
G1VAG LS23 6PX
G4OWU LS23 6RJ
G4XYY LS23 6RN
G3VMW LS23 6RP
G8JVS LS23 7AL
G4HKR LS25 1BZ
M0ASO LS25 1DA
G4CRP LS25 1DQ
G4DTT LS25 1EN
G0BPN LS25 1HH
G8MWX LS25 1HN
G3MZB LS25 1JY
G7OYD LS25 1LN
G3JNY LS25 1LQ
M1BYT LS25 1PW
G1PHS LS25 2BP
G1VGP LS25 2EG
G7CCL LS25 2EN
G7ADW LS25 2EP
G1XYT LS25 2JF
G8EEM LS25 2JF
G1FDL LS25 2JR
G7RQU LS25 2LR
G4JKH LS25 2NW
G4ZSS LS25 4AU
G4LLJ LS25 4AX
G4TGU LS25 7DY
G6JSR LS25 7DY
G1UVI LS25 7ET
G4TXK LS25 7ET
2E1AYS LS25 7EX
G1RON LS25 7HQ
G4AAU LS25 7HU
G4XSH LS25 7HY
G8HTV LS25 7JF
G6EHN LS25 7JG
G4DMP LS25 7PB
G1PZA LS25 7PD
G1PZB LS25 7PD
G0KLB LS26 0BJ
G3JQJ LS26 0BJ
G1AMM LS26 0GD
G4OJY LS26 0HP
G3XWI LS26 0HU
G0TUT LS26 0LQ
G3KDO LS26 0LY
G0TPN LS26 0NT
G8TMD LS26 0PP
M1CCW LS26 0SN
G0JZZ LS26 0SR
G4EFZ LS26 8HT
G0DYS LS26 8LH
G4OVR LS26 8LJ
G0RDR LS26 8PD
G3WRS LS26 8SQ
G3WWF LS26 8SQ
G4JMT LS26 8SQ
M0BDC LS26 8TA
G7TCV LS26 8TA
G4MAK LS26 9AF
G6TKC LS26 9HY
G0BVT LS27 0DP
G4VQX LS27 0PU
G7AJN LS27 7BT
G0JVI LS27 7DJ
2E1CPW LS27 7DJ
2E0APZ LS27 7DN
M1BOV LS27 7DN
G4VPO LS27 7DY
G7NHP LS27 8NN
G4HJJ LS27 8NT
G4UNJ LS27 8NY
G0PTI LS27 8TD
G4MBE LS27 9PL
G0MVH LS27 9PW
G6FMF LS28 5AA
2E0AOG LS28 5BF
G7LNL LS28 5BY
G0OPZ LS28 5BY
G4ZHW LS28 5BY
G6LHI LS28 5BY
G1IIX LS28 5DX
G0DPS LS28 5DZ
G7KHY LS28 5EJ
G4BXQ LS28 5LW
G4LVQ LS28 5PL
2E1CQA LS28 5PZ
2E1CQO LS28 5PZ
G7CWR LS28 5QB
G4NZN LS28 5RE
G1NSD LS28 7EP
G4WJF LS28 7HF
G4TIW LS28 7NA
G6HRM LS28 7NA
G7JHS LS28 7NA
M0AYP LS28 7QJ
G0SSZ LS28 8BH
G1NCC LS28 8BX
G4TZV LS28 8DG
G3VXE LS28 8EF
G1EBS LS28 8EN
G1PMG LS28 8JA
G0PQE LS28 8JX
G7EVR LS28 8LH
G3FCS LS28 8NR
G6HBY LS28 9BP
M0APC LS28 9EU
G7VED LS28 9EU
G6OWO LS28 9JW
G4FCQ LS28 9JZ
G3DQG LS28 9ND
G4TFV LS3 1BX
G3YTV LS3 1JS
G4OOH LS4 2LD
G4OOI LS4 2LD
M0ALD LS4 2NQ
G0ETL LS4 2TT
G8HPY LS5 3JP
G6SYX LS5 3RF
G3GGP LS6 1AB
G1JTC LS6 2AP
G3UBM LS6 2AQ
G4XTM LS6 2BB
G0FXK LS6 2RD
G7DCK LS6 2TR
G4YES LS6 3DB
G4MXM LS6 3HB
G3FLV LS6 3NX
G3TRE LS6 3QB
G0MSU LS6 3QE
G4OKW LS6 4BZ
G4IJI LS6 4HD
G0JUI LS6 4JT
2E1CRF LS6 4LA
G4YEW LS6 4PJ
G7VCK LS7 1SL
G6HGT LS7 2LT
G4UPD LS7 2NN
G4PUE LS7 2PT
G1JUL LS7 3DQ
G0WKG LS7 3LE
G4ONX LS7 3LJ
G4ONY LS7 3LJ
G4TCB LS7 3LY
G0VSE LS7 3PH
G0SCV LS7 3QB
G7LUS LS7 3QJ
M0APY LS7 3QJ
G8TGY LS7 3QJ
G3UMH LS7 4NZ
G4YBL LS8 1AD
G1IBX LS8 1BB
G4ZOB LS8 1EW
G7GUK LS8 1NE
G3PKC LS8 1NS
G4HRT LS8 1NY
2E1DSN LS8 1QH
2E1ESC LS8 1QH
G4TIC LS8 1RW
M1AXT LS8 1RW
2E1ENB LS8 1RW
G3ZGA LS8 2AD
G0BTZ LS8 2AW
2E1DNB LS8 2AW
G4KAX LS8 2BS
G4FTI LS8 2EE
G3XEP LS8 2JP
G3PTN LS8 2LA
G3VFH LS8 2LA
G0MYM LS8 2NA
G1MMX LS8 2PD
G6EDE LS8 2PE
G0KQP LS8 2RH
G8HBQ LS8 2SX
G6MTF LS8 2TA
G1UVE LS8 5BY
2E1DLX LS8 5DB
G4HJS LS9 0AE
M1ADM LS9 6BP
G3ZRL LS9 6DP
G7TOJ LS9 6LD
G7LRG LS9 6QH
G8XIG LS9 6QH
G4FVV LS9 7SY
G4FEQ WF10 2AA
G0GYA WF10 2DN
G0CEW WF2 0SE
G6ERZ WF3 1AJ
G7JWY WF3 1AL
G8VJS WF3 1EE
G0DJA WF3 1EF
G8VAT WF3 1ET
G4DZU WF3 1HZ
G0PXF WF3 1JB
G0FLX WF3 1JS
M0AOI WF3 1PF
G8JHA WF3 1RE
2E1EVO WF3 1RX
G4GCP WF3 1TT
G6NPW WF3 1UD
G8KRR WF3 1UE
G7KKJ WF3 1UT
G0IKD WF3 1UX
G1RHW WF3 2AX
G0VTI WF3 2AX
G1FOM WF3 2JG
G0SJZ WF3 3SL
G1HYA WF3 3SS

Leicester

G4CPY LE1 7HB
G4ION LE1 7RH
G3SDC LE1 9BH
G4OAI LE18 1HU
G4WAR LE18 1HU
G1JSP LE2 1PL
G1KOH LE2 1PR
G3OVH LE2 2AQ
G4IHT LE2 2LE
G4ZGL LE2 3AA
G8VA LE2 3AF
G3OOS LE2 3BB
G0HNI LE2 3FD
G8SXS LE2 3FT
G6LTR LE2 3GA
G4BB LE2 3GB
G3TDX LE2 3PJ
G0KPY LE2 3PN
G1UFU LE2 6AG
G0BDE LE2 6DR
G3MCP LE2 6FN
G3LQW LE2 6HQ
G4TFN LE2 6TX
G4MQS LE2 7PG
G0MTF LE2 8DA
G0LMO LE2 8DB
G8FCQ LE2 8DH
G6WMU LE2 8DJ
G1YFT LE2 8DL
G4NBH LE2 8HW
G1PPQ LE2 8NJ
G4TQR LE2 8PH
G4UQS LE2 8QA
G6ITW LE2 8QA
G7CEC LE2 8QE
G5VH LE2 8UP
G1ZYN LE2 9DD
G4WMX LE2 9FL
G6WZM LE2 9GA
G4YXT LE2 9QE
G0JQZ LE2 9QJ
2E1ATL LE2 9QJ
G4GAC LE2 9RU
G8TYF LE3 0ED
G7PPL LE3 0GT
G4AQV LE3 0JF
G6TMQ LE3 0JF
G0ORY LE3 0QR
G3ZCT LE3 0SA
G0EVI LE3 0TN
G0WYT LE3 0UE
G6EJU LE3 1FG
G0ULY LE3 1GJ
G4OOS LE3 1RG
2E1AQF LE3 2AQ
G3WQL LE3 2BW
G0FZC LE3 2EJ
G4AUS LE3 5FA
G7CKV LE3 5FA
G6WSN LE3 5RA
G6JOL LE3 6BD
G7OCY LE3 6FG
G0UHI LE3 6FG
G8ZQG LE3 6FH
G4FZL LE3 6HE
G4KGX LE3 6NA
G4RLC LE3 6NF
G3LTT LE3 6NQ
M1BWY LE3 6PW
2E1FLJ LE3 6PW
G7STN LE3 6UT
2E1FME LE3 9NA
G3OCH LE4 0BJ
G4OZD LE4 0DE
G3KQV LE4 0GS
G3TRA LE4 0GS
G4AGN LE4 0LL
G3UST LE4 0LN
G4ZNT LE4 0QQ
G0AZM LE4 0SU
G7UYJ LE4 1BT
G4HMX LE4 1DL
G4ITP LE4 2BJ

G0MBC	LE4 2FH
G4MIQ	LE4 2FH
2E1CFJ	LE4 6AW
G1OHI	LE4 6ES
G4RMS	LE4 7PA
G8FZZ	LE4 7TJ
G4PAK	LE4 7YD
M1BUK	LE4 9DS
G4UUG	LE4 9GF
G4ITB	LE5 0DL
G5MY	LE5 0LF
2E1CXE	LE5 1AB
G0WBC	LE5 1AB
G6WAY	LE5 1ED
G3ROM	LE5 1FW
G6CVH	LE5 1LR
G4JNM	LE5 1NH
G3FJL	LE5 1PA
G1IWC	LE5 1SR
G0FJI	LE5 1TW
G7VDU	LE5 1UB
G8CCS	LE5 2FF
G4XXZ	LE5 2GG
2E1FJB	LE5 2GH
G3ZZW	LE5 2GP
G7VIP	LE5 2RL
G0ZIP	LE5 2RL
G7DHJ	LE5 3LR
G2DSF	LE5 3QW
G6IFI	LE5 4RA
G6GKO	LE5 4WG
G1UGO	LE5 5RF
G8ANK	LE5 6AB
G1GKA	LE5 6AH
G3RYN	LE5 6DG
G2DLO	LE5 6EA
G4ZDQ	LE5 6JB
G4MGG	LE5 6PT
G1WZQ	LE5 6SA
G7MCD	LE5 6SA
G8GUL	LE5 6SW
G4IGL	LE5 6SY
G4YTF	LE5 6XT

Leominster

G0EXB	DY14 9HP
G0IMK	DY14 9HX
G3WBL	DY14 9LJ
G3WPQ	DY14 9NR
G0TFM	HR1 3DF
G3IXZ	HR1 3LP
G4GVM	HR1 3LP
G0CGO	HR1 3LR
G1JWD	HR1 3LR
G0PMS	HR1 3LT
G7TXE	HR3 6EB
G4EYR	HR3 6HY
G0GXA	HR4 7HQ
G3HVX	HR4 8HS
G4FFD	HR4 8NT
G7HII	HR4 8SZ
G3NPA	HR4 8TA
G7PEG	HR5 3ER
G7IFR	HR5 3JG
G2AHU	HR6 0BQ
G7UUN	HR6 0EF
G3MDN	HR6 0NS
G8BPN	HR6 0PF
G1MUQ	HR6 0RE
G4YPF	HR6 8HE
G6OZU	HR6 8HT
G4ZXQ	HR6 8PR
G3YYC	HR6 8RZ
G0BRH	HR6 8SD
G4OHT	HR6 8SD
G3MWH	HR6 8SF
G7TXL	HR6 8SH
G6CYR	HR6 8SL
G6RAE	HR6 8TL
G2CNO	HR6 9AZ
G0ARF	HR6 9BA
G4YIV	HR6 9DT
G4VOY	HR6 9JQ
G1JOD	HR6 9NT
G0OEA	HR6 9QN
G4BOF	HR6 9QR
G4VVF	HR6 9SR
G4RGB	HR6 9SW
G6KXW	HR6 9UF
G3DFH	SY8 4HU
G4LWU	SY8 4LN
G0EXA	SY8 4NE
M1AGE	SY8 4NW
G4DFE	WR15 8BY
G4NMG	WR15 8BY
G6KSF	WR15 8ET
G1KIV	WR15 8JX
G7RVW	WR15 8LX
2E1CDR	WR15 8LX
G0EXC	WR15 8QX
G3BPF	WR15 8SP

Lewes

G8DWX	BN10 7EA
G0MIB	BN10 7EF
G4KJO	BN10 7JU
M1CBE	BN10 7LS
G8MSM	BN10 7QD
G6FOO	BN10 7QY
G7UBB	BN10 7RP
G4KVV	BN10 7SA
G0OIB	BN10 8DD
G1JYH	BN10 8DW
G4XCA	BN10 8ED
G3MTX	BN10 8HH
G7RUQ	BN10 8NS
G8ZTG	BN10 8PJ
G3VQM	BN10 8QD
G4PNP	BN10 8RP
M0AZT	BN10 8SA
M1BHI	BN10 8SA
G7CQI	BN10 8SH
G0AWG	BN10 8TJ
G0OBI	BN10 8TS
G0BBH	BN10 8XL
G0KVF	BN10 8XL
G8LGS	BN2 8DH
G6CIT	BN2 8PH
G0IHH	BN25 1BB
G3DQT	BN25 1SW
G0CPR	BN25 2DB
G8ERC	BN25 2DB
G1OIO	BN25 2EB
G0GJH	BN25 2HA
G4ATG	BN25 2JZ
G4BMK	BN25 2JZ
G1BAB	BN25 2LX
G4MGN	BN25 2NE
G3KLY	BN25 2NU
G6OHR	BN25 2NZ
2E1BJU	BN25 2QR
M0AHA	BN25 2QY
M0AJA	BN25 2QY
G0KDD	BN25 2RU
G3IPJ	BN25 2RU
G1VAY	BN25 2TX
G3AGF	BN25 2UL
G8KUV	BN25 3DW
G8XXJ	BN25 3EZ
G4FB	BN25 3JT
G7PGZ	BN25 3SR
G0IOF	BN25 3TN
G3HN	BN25 4HL
G6FLH	BN25 4JB
G4ZNM	BN25 4NW
G3BRD	BN25 4PB
G1KVR	BN25 4QR
G4CYZ	BN6 8SB
G1RRW	BN6 8SG
G4PWA	BN6 8SY
G0APZ	BN6 8TH
G4BLX	BN6 8XD
2E1DSD	BN7 1EE
G8EAS	BN7 1LB
G4XBG	BN7 1LB
2E1CCI	BN7 1LT
G0USA	BN7 1NE
G0UFO	BN7 1NE
G3IIO	BN7 1NP
G7TGY	BN7 1NP
G4OWT	BN7 1QG
G7SVV	BN7 1QU
G3CGB	BN7 1UR
G1WSE	BN7 2DL
M0AFR	BN7 2DL
G8WHD	BN7 2DS
G4TBM	BN7 2EJ
G7MOW	BN7 2ET
G4PZU	BN7 2HY
G1MDF	BN7 2RN
G1MXC	BN7 2TX
G4YGE	BN7 2UB
G4UUF	BN7 3EB
G0TZX	BN7 3JU
M1AXQ	BN7 3LF
G4XRU	BN7 3LG
G7POF	BN8 4AU
G1VSQ	BN8 4DD
G7JKW	BN8 4HP
G6DGK	BN8 4NA
G6WIE	BN8 4NY
G7RBF	BN8 4PS
G0XAM	BN8 4QT
G7TOI	BN8 4QX
G0THD	BN8 5HW
G3YXO	BN8 5HX
G7SMT	BN8 5JD
G0JDR	BN8 5QD
G4NVK	BN8 5RL
G4MJC	BN8 6LS
G8ZVX	BN9 0JU
G8BYC	BN9 0NA
G4KZX	BN9 0ND
G4FZS	BN9 0NQ
G4YRA	BN9 0RD
G8RHU	BN9 0RR
G1AAR	BN9 9AH
G7PUW	BN9 9ER
2E1BRW	BN9 9EX
G0SYB	BN9 9NG
G7WHU	BN9 9QB
G3XUS	BN9 9SA
G4RBQ	RH17 7QL

Lewisham

G8OTG	BR1 4LP
G7CRK	BR1 5EG
G0UBA	BR1 5EW
G0UNX	BR1 5EW
G0ONR	BR1 5LP
G0WYG	BR1 5NA
G1WYG	BR1 5NA
G8GHN	BR1 5NJ
G4BOY	BR1 5QU
G6VNZ	SE10 8AJ
G0OVR	SE12 0PU
G6FXN	SE12 0RA
G4XUI	SE12 9LX
G1XFR	SE12 9RS
G3GHN	SE13 5ES
G3JKY	SE13 5ES
G4ZIH	SE13 5EZ
G4JOI	SE13 5QW
G4SWZ	SE13 6EL
G6EAL	SE13 6TR
G0ZZZ	SE13 7XD
G6BSO	SE14 5PP
G4ZFC	SE14 5SU
G3SXA	SE23 1BW
G6BJR	SE23 1RH
G2LW	SE23 2AG
G3TPO	SE23 2RU
G3UNF	SE23 2UQ
G3FZL	SE23 3BN
G3SGN	SE23 3BN
G1ISA	SE26 4BS
G8NHF	SE26 4LS
G6UGO	SE26 5LB
G7IHF	SE3 0BP
G8EXT	SE3 9HL
G8GP	SE4 1PD
G7PSX	SE4 1RG
G0PQZ	SE4 1RH
G0WXL	SE4 1SE
G0DCG	SE4 1XZ
G0HBW	SE4 2JQ
G0VXY	SE4 2LG
G6DJH	SE4 2NL
G1PKU	SE6 1HS
G4NQC	SE6 1LS
G8HCR	SE6 1NJ
G0ASF	SE6 1SA
G7WFV	SE6 2AZ
G0TMI	SE6 2BP
G6FYW	SE6 2LQ
G4YUC	SE6 3HE
G1AJN	SE6 3SJ
G6ZWM	SE6 3TW
G4DKC	SE6 4DH
G0LRM	SE6 4EH
G0TCA	SE6 4UW
G6OWX	SE8 3QA

Lichfield

G3RDW	B74 3AA
G1ZLC	B74 3EL
G3LGW	B78 3DE
G2BZR	B78 3DF
G8OSX	B78 3HW
2E1AWQ	B78 3SW
G7UHG	B78 3SW
M1APQ	B78 3SZ
G4NRX	B78 3TB
G4NRY	B78 3TB
G0ECI	B78 3TG
G0ICA	B78 3TS
G7KIF	B79 9HN
G1RRX	B79 9HT
G8RVZ	B79 9JX
G6DYA	DE13 7AG
G1DQU	DE13 7EZ
G4UJW	DE13 7EZ
G8ZKK	DE13 7EZ
G0RRM	WS13 6AU
G4EHT	WS13 6BH
G0HQX	WS13 6DQ
G3PFT	WS13 6ST
G4ICI	WS13 6TT
G8HYK	WS13 6TY
G0BIH	WS13 7AG
G3NKC	WS13 7BH
G0RVL	WS13 7BU
G0OYM	WS13 7ED
G8XXA	WS13 7LW
G0DRA	WS13 7LZ
G3FZW	WS13 7NA
G0PHE	WS13 7NQ
G3WET	WS13 7PP
G0EVJ	WS13 7PW
G6MQF	WS13 7SJ
G8OTC	WS13 7SL
M0BFL	WS13 8AA
G4CAR	WS13 8AJ
G4ESK	WS13 8AJ
G4GBH	WS13 8AJ
G3MHN	WS13 8AL
G3VTC	WS13 8JB
G8EQC	WS13 8JE
G4GYA	WS13 8NX
G7IXG	WS13 8PW
G1HRY	WS13 8QQ
G4VVY	WS13 8QQ
G0ERH	WS13 8RX
G7ETS	WS14 0JY
G1PGB	WS14 9AH
G0DYQ	WS14 9AT
G3WFD	WS14 9BA
G0PCK	WS14 9BE
G6AQZ	WS14 9BY
G7GMV	WS14 9DB
G4OUM	WS14 9HH
G0DKZ	WS14 9LN
G6DXU	WS14 9LX
G4HCJ	WS14 9NA
G7BMC	WS14 9NA
G3PBK	WS14 9PA
G6NEZ	WS14 9RJ
G6HZH	WS14 9SF
G0FDW	WS14 9SQ
G3YPD	WS14 9SZ
G8ZID	WS14 9UD
G7KHX	WS14 9UG
G0OES	WS14 9UG
G6NVP	WS14 9UJ
G1OBC	WS14 9US
G7CRS	WS14 9US
G8KHV	WS14 9US
G3WBB	WS14 9XB
G2HKS	WS15 1QA
G7HZU	WS15 3LD
G3FIK	WS15 3QL
G4KWP	WS15 3QL
G8YNP	WS15 4AF
G4YVG	WS15 4BG
G0MQQ	WS15 4BX
G7DBR	WS15 4EF
G8JNR	WS15 4EF
2E1FKJ	WS15 4EF
M0ACG	WS15 4YR
G4POR	WS7 0ED
G6JJB	WS7 0EE
G6PXN	WS7 0EQ
G6UCO	WS7 0EQ
G0CXE	WS7 0EY
G1GKH	WS7 0HA
G3HVY	WS7 0HW
G3CRH	WS7 0JR
G1DKW	WS7 8AG
G0AXP	WS7 8AS
G4ZBE	WS7 8AS
G4ZYL	WS7 8AS
G8RDN	WS7 8AS
G4PXV	WS7 8AW
G4PWX	WS7 8BP
G3LXB	WS7 8BT
G7HJX	WS7 8DL
G0EYQ	WS7 8DX
G0VSL	WS7 8EP
G4PPP	WS7 8ER
G0GFC	WS7 8FA
G6DFC	WS7 8FA
G1FCQ	WS7 8LE
G0VTT	WS7 8NB
G7DOS	WS7 8NQ
G0AGU	WS7 8PW
G0BYI	WS7 8PY
G4JAV	WS7 8PZ
G6KDY	WS7 8QA
G8PTL	WS7 8QL
G1AXK	WS7 8QS
G1XYO	WS7 8QS
G3PET	WS7 8QU
G6UQW	WS7 8TB
G4YNY	WS7 8TG
G0PJR	WS7 8TW
G7ATP	WS7 8TW
G0EOG	WS7 8TX
G4NPY	WS7 8UJ
G0MCM	WS7 8YD
G3RTY	WS7 8YD
G4TSD	WS7 9AB
G1WWZ	WS7 9AD
G6NPE	WS7 9AF
G0DAY	WS7 9AQ
G3NLY	WS7 9EA
G3WAS	WS7 9EA
G8BMP	WS7 9EJ
G6OTZ	WS7 9EJ
G3BIF	WS7 9JF
G4EQC	WS7 9JS
G6VPF	WS7 9LL
G8KSZ	WS9 9HU

Lincoln

G4SSV	LN1 3JS
G0VWW	LN1 3LB
G4OEF	LN1 3LU
M1BEV	LN1 3QZ
G0RZR	LN1 3SU
G7DTX	LN1 3TJ
G7MAP	LN1 3TJ
G7KYJ	LN1 3UT
G7ENA	LN1 3XA
G8OPP	LN2 1JD
G0WQA	LN2 2AE
G0VHX	LN2 2DR
G0MZO	LN2 2HU
G0CEG	LN2 2JL
G4RUE	LN2 2JL
G0PKN	LN2 2JR
M1BRZ	LN2 4AY
G0OSO	LN2 4BN
G0TTD	LN2 4DQ
G4CTQ	LN2 4LT
G6EWP	LN2 4LX
G6SWZ	LN2 4LX
G1LSK	LN2 4NA
G4KSA	LN2 4PY
G2BUV	LN2 4QJ
G1EUX	LN2 4SZ
2E1FBA	LN2 4SZ
G7JXX	LN2 5NL
G8VGF	LN2 5NN
G4BU	LN2 7SW
G0MRB	LN5 7UT
G7VFI	LN5 8BA
G0UOZ	LN5 8DA
G0NKA	LN5 8DB
G7KVH	LN5 8DR
G0LEN	LN5 8EL
G7LEN	LN5 8EL
G3GPQ	LN5 8LY
G3VRD	LN5 8SH
G8TSV	LN5 9TN
G4XMQ	LN5 9TY
G7FDD	LN6 0ED
G7FFG	LN6 0HG
G4ZKX	LN6 0JL
G3WUP	LN6 0LU
G1XZG	LN6 0LZ
G3KLQ	LN6 0PS
GW1AHU	LN6 0SJ
G8HMZ	LN6 0SY
G7JQZ	LN6 3NR
G0EUN	LN6 3NU
G8SGI	LN6 3RD
G4WDU	LN6 7AY
G0EDI	LN6 7HX
G7UAV	LN6 7NH
G0VRE	LN6 7NL
G6HUP	LN6 7PA
G3PVU	LN6 7PN
M1CCD	LN6 7RD
G0PMT	LN6 7UQ
G8HWA	LN6 8AD
G0LZS	LN6 8BX
G0IMP	LN6 8SN

Liverpool

G4PKP	L1 4JA
G0MUP	L1 4JN
G0HKZ	L10 7NL
G7SZF	L11 1AN
G6NRK	L11 1AY
G0WRE	L11 1DF
G4VLM	L11 2SQ
G0MXW	L11 2UE
G3MIH	L11 5AB
G3WIC	L11 5AB
G0JSZ	L11 8LR
G1TIQ	L11 8LR
G4XAR	L11 9AG
G7JNX	L12 0PG
G4VYR	L12 4YR
G0TDK	L12 6RL
G0UMB	L12 6RL
G6WIO	L12 7HE
G4SYW	L12 8HU
G8YGM	L12 9NE
G4KVL	L13 3DT
G1AVC	L13 3DU
G6BFM	L13 3DY
GW7KVW	L13 4AY
2E1CYS	L13 4BS
G1TBB	L13 7BN
G3OSI	L13 7DJ
G0SMR	L13 7EH
GW7KIS	L13 8DF
G3RBD	L14 0LG
G7NHQ	L14 3LT
G4RBE	L14 7PE
2E1CPA	L15 0HN
G6RMY	L15 0HP
G0OIT	L15 3HJ
G0ELZ	L15 4LB
G0SJW	L15 4LB
G0OSC	L15 5AX
G3UVP	L15 5BP
G6KXJ	L15 8HX
G0STF	L15 8HY
G3ZNP	L16 1JW
G6IMJ	L16 3NE
2E1AJU	L16 3NJ
2E1AKO	L16 3NJ
G4WWX	L16 6AJ
G0VMT	L16 7PA
G0KBS	L16 7PQ
G7GEE	L16 7QR
G3VYB	L16 7QT
G3RWW	L16 8NZ
G6EQL	L17 0DP
G8GDN	L17 4AD
G0SVM	L17 6AS
G3XCP	L17 6BY
G4RYS	L17 6SG
G1VGC	L17 7AD
G8VQC	L17 7AD
G1JHG	L17 7AL
M1BEN	L18 0HB
G3PDC	L18 1EY
G0MIT	L18 1JZ
G0IFK	L18 3HE
G4RVN	L18 3HX
G0VBD	L18 3LL
G8GZI	L18 4PZ
G1YEP	L18 4QE
G8SDD	L18 4QG
G6IZO	L18 4RF
G1YUX	L18 7HU
G0IIJ	L18 7HW
G7VCN	L18 7JG
G4INV	L18 7JW
G0JUA	L18 9SJ
G8ATC	L18 9TB
G0MSO	L18 9TN
G1RUZ	L19 0LS
G7TUZ	L19 1RF
G1BES	L19 2QZ
G1KOP	L19 2QZ
G4DKQ	L19 3PN
G4YNJ	L19 7NB
G0ATJ	L19 7ND
M0BAI	L19 7ND
GW7KWB	L19 7NW
G1ZBU	L19 9AN
G2BGG	L19 9DG
G0GQX	L19 9EA
G4FYD	L19 9EB
G4CMP	L24 3TH
G4PHL	L24 3XE
G7UEU	L25 0PP
G2BTJ	L25 3PJ
G4GHS	L25 3QE
G0ARV	L25 3QS
G0MFH	L25 4SH
G3XSN	L25 4SQ
G3AHD	L25 4ST
G7SXG	L25 4ST
G8WCL	L25 4ST
G6BOK	L25 4TJ
G3TYE	L25 4UG
G6VNW	L25 4UQ
G3PPN	L25 5JU
G0GQW	L25 5JY
G7UWJ	L25 6DS
G3KRX	L25 6DW
G4GSR	L25 6EJ
G8UCM	L25 6HU
G0NDW	L25 7TF
G0LBG	L25 7UJ
G4MHY	L25 7UJ
2E1DDG	L25 8QJ
G0IHS	L25 8RY
G7KOT	L27 6WQ
2E1CNJ	L27 8YA
G0OIU	L4 2UN
M0AJJ	L4 5PR
G7FMW	L4 5PR
G4WWB	L4 6TX
G0XBI	L4 7XJ
G7UZA	L4 9UG
G3UPX	L5 6PU
G0BOY	L5 6RP
G0XSN	L6 0AJ
G1KON	L6 0BZ
G3LNG	L6 4AG
G6RGO	L6 4AN
G0WWC	L6 6BB
G0EDF	L6 6BN
G0WRE	L7 1QS
M0BFM	L7 6HH
G4KRN	L8 0QU
G8DEY	L8 0UF
G1JUH	L8 4RD
G1MCW	L8 4XR
G4AJZ	L8 6XH
G7LUW	L9 0HE
G4XCY	L9 0NA
G0LPQ	L9 1AN
G7OLG	L9 1AN
G7JDY	L9 1EG
G7VGK	L9 2AH
G7CVC	L9 2BB
GW7CVF	L9 2DE
G4YPD	L9 8DE
G6PBJ	L9 8EG
G0DHJ	L9 9AY
G7JZK	L9 9HH

Luton

G7VZB	LU1 1TX
G7UEV	LU1 3NS
2E1BHF	LU1 3NZ
2E1AUB	LU1 3RW
G4NMJ	LU1 5EY
G8XSD	LU1 5LJ
G7BNW	LU1 5PB
G3NNY	LU1 5PN
G6OUO	LU2 0DN
G0BWK	LU2 0JF
G0LYN	LU2 0JG
G1SVD	LU2 0JG
G7KGR	LU2 0LG
G1OUA	LU2 0LG
G8GWK	LU2 0PR
G6UCQ	LU2 0RE
G6PHU	LU2 0RJ
G6IAT	LU2 0RP
G0ALB	LU2 0RX
G3PNP	LU2 0TP
G3YUI	LU2 0TR
G8ATD	LU2 7BG
G4UVD	LU2 7BY
G8NWP	LU2 7DJ
G3IWH	LU2 7EU
G8DDK	LU2 7EU
G4IET	LU2 7JL
G0SQK	LU2 7NJ
G8NXS	LU2 7SA
G0IGN	LU2 7SL
G4ENB	LU2 7TE
G8DDC	LU2 7TE
G1PZD	LU2 7TX
G3NVL	LU2 7UH
G3TAZ	LU2 7UU
G0KXV	LU2 7UY
G6UHQ	LU2 7YU
G4LBH	LU2 8AF
G1XXP	LU2 8AW
G7CWK	LU2 8AW
G4JCA	LU2 8AZ
G1BMJ	LU2 8BU
G4DSG	LU2 8DP
G7KZH	LU2 8DT
G6PHV	LU2 8EH
G4NMY	LU2 8EJ
G1BDI	LU2 8EP
G4BMM	LU2 8HN
G8LXY	LU2 8HQ
G1JCC	LU2 8QP
G8IZW	LU2 8RU
G1IKD	LU2 8SX
G3AHO	LU2 8TL
G0LIP	LU2 8TR
G4PCS	LU2 8TR
G7EDX	LU2 8TS
G0HWK	LU2 9AE
G3KCJ	LU2 9AE
M0AIL	LU2 9AJ
G0VZH	LU2 9AN
G1UHJ	LU2 9AN
G6EQS	LU2 9AR
G0TNO	LU2 9BL
G7SNW	LU2 9DP
G1NWZ	LU2 9ES
G4EXG	LU2 9HL
G4OIS	LU2 9HL
G7VWA	LU2 9JZ
G7VYB	LU2 9RA
G7URR	LU2 9RB
G4FOR	LU2 9SP
G8JHM	LU3 1HB
G3CGQ	LU3 1HJ
G8KYK	LU3 1HQ
G0OJB	LU3 1JX
G6BCZ	LU3 1PA
G4HPY	LU3 1PJ
G8UOJ	LU3 1PY
G0IYJ	LU3 1TJ
G1AXR	LU3 1UB
G4PIW	LU3 2AT
G3VMD	LU3 2BA
G1ATQ	LU3 2ET
G3PUT	LU3 2EU
G6PPV	LU3 2HJ
G1SCA	LU3 2JJ
G1ILN	LU3 2LF
G8CBU	LU3 2LZ
G6DWM	LU3 2NY
G3NMZ	LU3 2RJ
G0DCS	LU3 2UA
G8CPQ	LU3 2UR
G4CPE	LU3 3AR
G7TMD	LU3 3DG
G6XFR	LU3 3EJ
G4EYL	LU3 3HE
G7HBN	LU3 3JZ
G0SZY	LU3 3QJ
M1BTQ	LU3 3SU
M1ACI	LU3 3UL
G8YNB	LU3 3XJ
G7PKG	LU3 3XL
G3YLV	LU3 3XU
G1BJW	LU4 0AT
G0COQ	LU4 0JA
G3RKI	LU4 0PU
G3BOI	LU4 0QX
G3WBC	LU4 0TS
G6OUM	LU4 0YD
G7EUM	LU4 0YJ
G0HGS	LU4 8EH
G6RPD	LU4 8NU
G8EZV	LU4 8PY
G8UIQ	LU4 9AJ
G3KAA	LU4 9AL
G6KVE	LU4 9AY
G7DBT	LU4 9EN
G3WLM	LU4 9HG
G4DDC	LU4 9HG
G1JIG	LU4 9LT
G1XZW	LU4 9SE
G6MUJ	LU4 9SQ
G4LZF	LU4 9TL

Macclesfield

G3XAE	SK10 1PQ
G3HUR	SK10 1QP
G1KXP	SK10 1QR
G0DMR	SK10 2EL
G4XRG	SK10 2EL
G1AKA	SK10 2HJ
G7NNU	SK10 2JB
G8BHX	SK10 2NS
G6CGO	SK10 2NS
G4JNE	SK10 2PF
G1NUS	SK10 2PN
G7WAB	SK10 2PN
G0DMV	SK10 2PS
G0JNJ	SK10 2PS
G0IUI	SK10 2PW
G4ZNE	SK10 2PW
G7RFK	SK10 2PW
G0DNB	SK10 2QX
G4WGH	SK10 2RF
G8JQG	SK10 2RN
G4MUO	SK10 2SS
G6AKK	SK10 2TT
G8DEX	SK10 2TY
M0ALP	SK10 2UG
G1SHI	SK10 3BS
G7SSK	SK10 3HE
G4WOL	SK10 3JL
G3LIO	SK10 3LN
G8RDJ	SK10 3LN
G7PZG	SK10 3LP
G0BIE	SK10 3NN
G1IZD	SK10 3PE
G0AHJ	SK10 4HY
G1GTY	SK10 4HY
G3UJA	SK10 4QW
G3LDT	SK10 5AH
G6VSQ	SK10 5AH
G3RVN	SK10 5AQ
G8BPU	SK10 5HN
G4XTO	SK10 5HT
G6PEZ	SK10 5HT
G0NYS	SK10 5LJ
G0RUN	SK10 5LN
G3CWI	SK10 5LX
G3YYR	SK10 5LY
G4MCI	SK10 5NX
2E1FUE	SK10 5PY
2E1FUS	SK10 5PY
G0WGJ	SK11 0AG
2E1FLW	SK11 0BU
G3AOS	SK11 0DQ
G1VJH	SK11 0LD
G0AMU	SK11 6RY
G6XKK	SK11 6RY
G4MWS	SK11 6RY
2E1FAD	SK11 6RY
G6TZI	SK11 6UB
M1BYH	SK11 6UN
G0CSX	SK11 7AT
G3AUB	SK11 7EN
G3VKF	SK11 7EN
G7POV	SK11 7PL
G4EEQ	SK11 7RL
G0DMQ	SK11 7UZ
G0KIQ	SK11 7XN
G1VTW	SK11 7YG
G0DJQ	SK11 7YJ
G3JQ	SK11 7YJ
G0DCO	SK11 7YW
G4TJC	SK11 8BE
G0EAZ	SK11 8BW
G6GNN	SK11 8DD
G6LFG	SK11 8EX
G3RWE	SK11 8LL
G3CZO	SK11 8NT
G3GKG	SK11 8QA
G1DOE	SK11 8QB
G0DRO	SK11 8QH
G8XQH	SK11 8QL
G0DMU	SK11 8QP
G0PIL	SK11 8RS
G7BPV	SK11 8RT
G7MSH	SK11 8SD
G1NXV	SK11 9AS
G3HFM	SK11 9BF
G8BVF	SK11 9BX
G4TQF	SK11 9ED
G4IWV	SK11 9LG
G3RNT	SK11 9SS
G3VDB	SK11 9SW
G4UHO	SK12 1AD
G0IMF	SK12 1AT
G6XRL	SK12 1BG
G3GVH	SK12 1BY
G1AFI	SK12 1EF
G6VKL	SK12 1FA
G3SHF	SK12 1HA
G8SRS	SK12 1HA
G6UQ	SK12 1HA
M0AGW	SK12 1HG
G0ONG	SK12 1HQ
G4DBD	SK12 1HU
G1ORR	SK12 1HY
G4PDX	SK12 1JW
G3MNJ	SK12 1LE
G0SNY	SK12 1LQ
G0EVP	SK12 1NN
G4IUJ	SK12 1PU
G6RJH	SK12 1RW
G7UJC	SK12 1RW
G0SKX	SK12 1SL
G1MWS	SK12 1SN
G4FIK	SK12 1SN
G6ECN	SK12 1SN
G3THF	SK12 1SP
G4MEH	SK12 1TG
G4GVI	SK12 1UP
G6HJU	SK12 1XB
G0CQU	SK12 2AE
G8CET	SK12 2AY
G2AAY	SK12 2BU
G0JYD	SK12 2BY
M0AQM	SK12 2DH
G0KTZ	SK12 2LD
G4IRU	SK9 1LU
G7INK	SK9 1PW
G4CQV	SK9 1QE
G1JDQ	SK9 2AD
G0BSU	SK9 2BP
G0BLR	SK9 2HR
G4DMX	SK9 2JW
G8JHL	SK9 2NU
G0OPG	SK9 3AR
G8BHX	SK9 3AY
G0OKR	SK9 3BL
G4IRG	SK9 3JT
G3ZUG	SK9 3JU
G4APA	SK9 3NH
G4GXQ	SK9 3NN
G6EEA	SK9 4AH
G4GMK	SK9 4AL
G7NMZ	SK9 4BB
M0AEA	SK9 4BH
G7RLG	SK9 4DL
G0JAZ	SK9 4DU
G3NLR	SK9 4EN
G8IXP	SK9 4EP
G3WPF	SK9 4HE
G7CQB	SK9 4JW
G1HFH	SK9 5JA
G3TSZ	SK9 5NQ
G1GYH	SK9 5PX
G8PBE	SK9 6BZ
G0AXE	SK9 6EG
G7CMN	SK9 6HD
G1ORS	SK9 6HD
G7RMZ	SK9 6HD
G3ONI	SK9 6HE
G1LDP	SK9 6JQ
G3WTQ	SK9 6LP
G4WKD	SK9 6LS
G3VDQ	SK9 7AB
G7EWH	SK9 7DX
G3FOE	SK9 7HU
G7TNT	SK9 7HW
G3BNW	SK9 7PN
G0TEK	SK9 7QF
G0NHX	SK9 7XB
G0MXR	WA13 0UA
G8BJL	WA13 0UB
G1ZGF	WA16 0DT
G8XMZ	WA16 0HY
G3IJE	WA16 0JX
G6PPT	WA16 0JX
G0AKF	WA16 0NE
G4SAR	WA16 0SF
G3PLS	WA16 0TE
G8RCG	WA16 0TU
G6EXE	WA16 6PS
G3UJE	WA16 7AX
G0AUB	WA16 7HE
G8CJQ	WA16 7RD
G1LEQ	WA16 8BB
G1RAM	WA16 8BB
G3LEQ	WA16 8BB
G4LEQ	WA16 8BB
G8WUY	WA16 8BH
G3ESA	WA16 8BU
G4HZW	WA16 8EP
G7LYL	WA16 8ES
G6FWK	WA16 8JR
G8FIZ	WA16 8SZ
G4MGK	WA16 9DE
M0AEM	WA16 9DZ
G1OQO	WA16 9RS

Maidstone

G7JSA	ME14 2DY
G3YJS	ME14 2ER
G3UNT	ME14 2EX
G3PCX	ME14 2JN
G0EPK	ME14 2NH
G0AES	ME14 2QJ
G4UAQ	ME14 2QR
G3OIP	ME14 3HB
G3YCN	ME14 3HP
2E1BBO	ME14 3HP
G3ZAL	ME14 3JN
GM4DVG	ME14 3JN
G4GAX	ME14 4AR
2E1FSC	ME14 4EA
G4VFQ	ME14 4EF
2E1AUS	ME14 4EY
G7MOX	ME14 4HG
2E1DMT	ME14 4HG
G0BUW	ME14 4JB
G0VIM	ME14 4NE
G0WFN	ME14 4NE
G0SPQ	ME14 4NJ
G4DUT	ME14 5BA
G8HLE	ME14 5HJ
G7HBC	ME14 5JN
G3WXU	ME14 5PG
G4HZI	ME14 5SR
G4ZQB	ME14 5SZ
G1AEY	ME14 5TA
G7DJT	ME15 0AJ
G1LMM	ME15 0AS
G1YPL	ME15 0EG
G4XMS	ME15 0QH
G8TRF	ME15 6BD
G6XAK	ME15 6FF
G8IIM	ME15 6FH
G0OMM	ME15 6RJ
G7VOM	ME15 7AU
G7VON	ME15 7AU
G3LNT	ME15 7EB
G6ZDY	ME15 7EL
G0SFC	ME15 7JB
G3WWI	ME15 7RT
G3WZL	ME15 7RY
G4IWN	ME15 8EG
G0VAR	ME15 8JP
G8LGA	ME15 8TN
2E1FHQ	ME15 8TN
G0JUS	ME15 9AA
G6ALJ	ME15 9AE
G8SQC	ME15 9BJ
G6PEA	ME15 9BU
G8VPD	ME15 9JB
G0NCW	ME15 9PD
G7HON	ME15 9QR
G3RJF	ME15 9RA
G0LCH	ME15 9RP
G4WBA	ME15 9RS
G8RUR	ME15 9UY
G3LUL	ME16 0DD
G3ORP	ME16 0DL
G3TRF	ME16 0DL
G0MIH	ME16 0EA
M1AJM	ME16 0HN
G6RVS	ME16 0JS
G7HOL	ME16 0NE
G4AAW	ME16 0QB
G8PWT	ME16 8EN
G1XJK	ME16 8NP
G8ZOP	ME16 8NR
G0FKK	ME16 8PA
G4AXD	ME16 8UE
G6OAU	ME16 8UQ
G4LMG	ME16 9AJ
G4LMH	ME16 9AJ
G7TUP	ME16 9AS
2E1FAC	ME16 9BA
2E1FAE	ME16 9BA
2E1FLS	ME16 9BA
G8TLP	ME16 9HF
M1AEX	ME16 9HF
G6HTH	ME16 9HH
G3YIF	ME17 1AD
G0UGF	ME17 1EX
G0AEA	ME17 1PL
G1BLB	ME17 1SX
G3VCW	ME17 1SY
G3UOJ	ME17 1TJ
G0IJK	ME17 2AX
G3LRX	ME17 2EJ
G8WHB	ME17 2EU
G7UEO	ME17 2JX
G1HQJ	ME17 2NS
G8ISM	ME17 2QD
G7JDK	ME17 3AG
G3BDU	ME17 3AS
G3YUU	ME17 3BA
M1BWJ	ME17 3DJ
G6NFE	ME17 3JJ
G3ORH	ME17 3LJ
G0WBV	ME17 4AU
G1ZMG	ME17 4DN
G4OPP	ME17 4LG
G7AUS	ME17 4QB
M1ALD	ME5 8XN
2E1DMU	ME5 8XU
G1YQN	ME5 9JD
G6BLK	ME5 9JG
G6NAL	ME5 9JG
G8CVS	ME5 9JN
G1XMP	ME5 9ND
G4ZTF	ME7 3LS
G1ROX	ME9 7QR
G7GMU	TN12 0BT
G4VEG	TN12 0BX
G0EIH	TN12 0LD
G8GYB	TN12 0LD
G0IWJ	TN12 0LQ
G0UXG	TN12 0LR
2E1CTC	TN12 0NU
G3LRA	TN12 0PQ
G8THE	TN12 0PQ
G1NDK	TN12 0RH
G7HRD	TN12 0SG
G8YBH	TN12 9BS
G0RHO	TN12 9JD
G0PUX	TN12 9JQ
G3VLX	TN12 9LG
G7FKX	TN12 9LS
G3XBQ	TN12 9NB
G3UUG	TN12 9TQ
G4ZXI	TN27 9QQ

Maldon

G0UKV	CM0 7BB
G7SDH	CM0 7BB
G0ULB	CM0 7LU
G4MNT	CM0 7PB
G6VQV	CM0 7QQ
G7JDR	CM0 7QT
G4THV	CM0 7RT
G0ATK	CM0 7TH
G7PIX	CM0 7TH
G3HRF	CM0 8DB
G8KWN	CM0 8DN
G7RGR	CM0 8DP
M1CBB	CM0 8EH
G4LDS	CM0 8JH
G6EUW	CM0 8PZ
G7AJU	CM0 8QJ
G1LUX	CM0 8RB
G4KSC	CM0 8RH
G7EHS	CM0 8SW
G7MHK	CM0 8TN
2E0AGW	CM3 6AE
G4XHU	CM3 6AF
G6TXV	CM3 6AJ
G4TWL	CM3 6AQ
G7UIQ	CM3 6AZ
G6IWB	CM3 6BA
G0ULU	CM3 6BB

G1EVD CM3 6BD
G1LQZ CM3 6BD
G0SVI CM3 6BD
G3MAH CM3 6BX
G0IJN CM3 6BY
G0UTT CM3 6BY
G1SNU CM3 6EU
G7LIK CM3 6EU
G7MDR CM3 6EU
G7PEC CM3 6EU
G7JYD CM3 6HU
G3HMQ CM3 6JE
G3HTF CM3 6JE
G8UUL CM3 6JG
G6AHD CM3 6LF
G6IKS CM3 6LF
G4BUP CM3 6NF
G4XBE CM3 6RE
G3VPK CM8 3EJ
G1UZD CM8 3JR
G7NAI CM8 3LT
G3VYD CM8 3LZ
G0EAG CM8 3NN
G3WHR CM8 3NR
G3XAX CM8 3NS
G6RKF CM9 4BL
G4WQI CM9 4NA
G0AAN CM9 4SB
G8GZN CM9 4SE
G8UVG CM9 4SP
G0CGU CM9 4YS
G1NKP CM9 4YU
G8JLM CM9 4YX
G6LJF CM9 5EA
G3UHU CM9 5HF
G1FWR CM9 5HH
G3HDM CM9 5HL
G1AYC CM9 5HZ
G8GLP CM9 5JJ
G3XJL CM9 6AZ
G6HMU CM9 6AZ
G3ZKM CM9 6BD
G4HRB CM9 6DJ
G1NMN CM9 6EW
G4MOV CM9 6EW
G8PYD CM9 6JF
G6LJR CM9 6JH
G6LTD CM9 6PZ
G3UYC CM9 6QP
G2CZS CM9 6SP
G8WTM CM9 6UQ
G0UHD CM9 8BW
G8EEK CM9 8DS
G3KTF CM9 8JS
G6HZW CM9 8JS
G0VDV CM9 8LL
G4NFZ CM9 8NE
G4EMB CM9 8PB
G6OXQ CM9 8PB
G0VDP CM9 8PJ
G0YAE CM9 8PX
G0BFI CM9 8QS
G6CYE CM9 8RJ
G3GLL CM9 8SB
G4GKH CM9 8SF
G4IIH CM9 8SZ
G0LKX CM9 8UD
G1FWY CM9 8UD
2E1DET CM9 8UN
G0IBN CM9 8XB
G0EQN CM9 8XG
G6YWU CO5 0RP

Malvern Hills

G1PDY DY13 0RE
G8SPD DY13 0RH
G8DYG GL19 4NX
G4CIB GL19 4NY
G4RHK GL19 4NY
G4CRN GL20 6BB
G8WJS GL20 6ER
M1BZQ HR1 3QW
G6PXQ HR1 3RF
G6YDD HR1 3RF
G5FD HR7 4BA
G4ZWY HR7 4DZ
G4XNK HR7 4EA
G3OGK HR7 4LY
G7CQD HR7 4TF
G0DNO HR7 4TP
G4XOS HR7 4UJ
G4FTX HR8 1BE
G3WPN HR8 1HG
G1JHY HR8 1SU
GM8MUE HR8 2BG
G3MKB HR8 2DN
G0OBD HR8 2DX
G6SUP HR8 2LN
G1LYO HR8 2PE
G1GCJ HR8 2XD
G4FCL HR8 2XX
G0HLG WR13 5AZ
G1KWF WR13 5DU
G7JNN WR13 5DU
G4FCA WR13 5JA
M1BZN WR13 5JD
2E1EGD WR13 5JD
M1ANY WR13 5PE
G0RWY WR13 6ET
G0MMA WR13 6JW
G3WHJ WR13 6PD
G8TJD WR13 6PG
G4YXU WR13 6RA
G6EOA WR13 6RA
G2AOY WR13 6RE
G4VHV WR13 6SE
G8OTD WR13 6SH
G6VJN WR14 1AA
G4BVY WR14 1AD
G3RNP WR14 1DT
G7PJH WR14 1DZ
G7PRW WR14 1EB
G3FQC WR14 1EQ
G4ZAI WR14 1HA
G3CLE WR14 1HR
G0OWS WR14 1JA
G7TTH WR14 1JF
G3ITP WR14 1LA
G3MXR WR14 1LE
G4EYJ WR14 1LF
G4FNZ WR14 1LP
G0GFI WR14 1NB
G6TWJ WR14 1NN
G8ZKG WR14 1PD
G1PHZ WR14 1PJ
G4FAT WR14 1PJ
G4IKJ WR14 1PU
G7JLC WR14 1PU
G6CMV WR14 1PX
G7EME WR14 1PX
G8AXV WR14 1QH
G6BBG WR14 1QZ
G8IDK WR14 1RQ
G3OOW WR14 1RS
G6MPB WR14 1RS
G7RVM WR14 1RS
2E1DRO WR14 1RS
G1CBL WR14 1SB
G7WHG WR14 1SE
2E1EGJ WR14 1SE
2E1FDB WR14 1SE
G3NKY WR14 1TR
G4BZM WR14 1TY
G6FYE WR14 2AS
G4FPV WR14 2BQ
G6FVH WR14 2DP
G8HB WR14 2DS
G3UIK WR14 2HU
G0WBJ WR14 2HX
G0BVD WR14 2JN
G8WPF WR14 2LF
G0WLG WR14 2LU
G3CPG WR14 2ND
M1CAK WR14 2ND
2E1EFY WR14 2ND
G0FXD WR14 2NG
G8YTU WR14 2NN
G1JHZ WR14 2QF
G1AIB WR14 2QT
2E1EGW WR14 2QT
G3CVK WR14 2SD
G8VUS WR14 2TU
G4VZA WR14 3DN
G4PHD WR14 3DX
G4GHL WR14 3EA
G3IPY WR14 3LG
G3ONW WR14 3NH
G0TWT WR14 3PL
G4GFX WR14 3PT
G3WGY WR14 3QP
G8CMD WR14 3QP
G4IPY WR14 3QW
G0TLE WR14 3RH
G8RHO WR14 4BQ
G0ROB WR14 4BX
G3OAH WR14 4DT
G8LJU WR14 4ES
G1RFX WR14 4ET
G7GAR WR14 4ET
G0RMX WR14 4HL
G3RHH WR14 4HS
G1NFB WR14 4HT
G3FHL WR14 4HT
G6JBY WR14 4JR
G6TEB WR14 4LJ
G3BJB WR14 4LS
G3YPM WR14 4NL
G1FXS WR14 4NL
G6BUY WR14 4PL
G7WIG WR14 4XB
G3TGD WR14 4XB
G8URZ WR2 4SR
G3OCW WR2 4TE
G3TQZ WR2 5SW
G0AZX WR2 5TA
G4RBD WR2 6NQ
G4BAI WR2 6PA
G1JMN WR2 6PQ
G4TCQ WR2 6PW
G0AQB WR2 6RW
G8CKD WR5 3NJ
G8ASJ WR5 3NX
G1XBL WR5 3QB
G8XXY WR6 5HH
G4KTW WR6 5LZ
G3GNA WR6 5NE
G3RYH WR6 5SF
G1MAC WR6 6AY
G4XKX WR6 6BX
G6JJP WR6 6EB
G3WLG WR6 6HQ
G4OPD WR6 6ND
G6NRZ WR6 6ND
G3VGG WR6 6QA
G4AAL WR6 6QA
G4OWK WR6 6QX
G0PJM WR6 6TQ
G3YQC WR6 6XR
G0CAZ WR6 6YU
G6UZT WR8 0HN
G6UGH WR8 0NQ
G0GEP WR8 0PQ
G8ADH WR8 0PQ
G0WNW WR8 0RN
G8XQL WR8 0SJ

Manchester

2E1AZM M11 1EE
2E1DGA M11 1FY
G4YMW M11 3JE
G1NVL M11 4DL
G0UXQ M11 4DQ
G7TZV M11 4GF
G1GMP M11 4PR
G1EJH M12 4AL
G6UQC M12 4WZ
G1HHU M12 5LT
G3DFA M13 0UH
G7VWL M14 5WP
G0BML M14 6PE
2E1FZQ M14 6RS
G4FFW M14 6RX
G0IFX M14 6UD
G8UBP M14 6XH
G1YEK M14 7AP
G1ZST M14 7WT
2E1DPT M16 8LX
G0OVY M16 8PW
G0PKD M16 8PW
G6DHI M16 8PW
G6YES M16 8PW
G0AKA M18 7RE
G6YFL M18 8BP
G0UKG M18 8RU
G1EMR M18 8RU
G7NAL M18 8WW
G0ITP M19 1EU
G8GMK M19 1LG
2E1GAB M19 1NE
G0PXG M19 1QX
G4CSV M19 2FB
G3BZC M19 2FY
G3JRK M19 2LQ
G0PXZ M19 2NL
2E1FLR M19 2NL
2E1FMH M19 2NL
G4BVQ M19 3EH
G8MXB M19 3LA
G3KRS M20 1HZ
G7EAH M20 2QH
G7OQS M20 4AG
G4OQC M20 4AU
G7AZW M20 4GH
G4NYA M20 4PT
M1BEL M20 4PX
G3TEX M20 4QA
G3CCL M20 4WH
G3OJA M20 5GR
G6PFU M20 5NH
G0OVQ M20 5PL
G7SGA M20 6BL
G3YHQ M20 6JA
G1LVN M20 6SZ
G0GDN M20 6UW
G7RLN M21 0SS
G0TWM M21 0TT
G6XYA M21 0UP
G4PWT M21 0US
2E1BYY M21 7LZ
G4IJG M21 7NT
2E1BAM M21 8EG
G7VVQ M21 8TH
G6BHA M21 8XT
G0TMN M21 9JB
2E1FNU M21 9JT
G0CRF M21 9JU
G6CRF M21 9JU
G0WHD M22 0LH
G0IXU M22 0LL
G0JYR M22 1NJ
G1UTM M22 1NN
G1JDU M22 1PX
G1OJB M22 4DP
G0PWA M22 4DZ
G1TGC M22 4PG
G4RWN M22 5JY
G1SUK M22 5QN
2E1FOZ M22 5QN
G7TEA M22 5QW
G1CHA M23 0FE
G0PCN M23 1EP
G0SAY M23 1JX
G6BMY M23 1JY
G8OZD M23 1LT
G1PNC M23 2TZ
G0LNA M23 2UL
G0TNX M23 2YN
G0NID M23 2YQ
G0LNB M23 2YX
G8LPX M23 9AW
G8UQC M23 9BR
G8LUL M23 9BT
G3PJL M23 9BU
G7PXS M23 9NZ
G7FQY M33 3PR
G0IUK M40 0GB
G4UYN M40 1JY
G0VXW M40 1PR
G0JFP M40 2UP
G0VXI M40 3LB
G0UNJ M40 3NB
G8ZHC M40 3PB
M1BJA M40 3PB
G6IRF M40 3SE
G7PMZ M40 3TE
G7ORW M40 3WP
2E1DQS M40 3WP
G7VCG M40 5HL
G3IOA M40 9LD
G7USQ M40 9LF
2E1ELZ M40 9LF
G3OVR M40 9PE
G3ETK M60 1SJ
G6NBE M8 4PY
G0RRX M8 4PY
G4YNI M8 4WQ
G3YYP M8 5SN
G6OGT M9 0RQ
G3KIQ M9 6HR
G0CUJ M9 6JN
M0BFQ M9 6LJ
G1EDO M9 6QZ
G3NSW M9 7AR
G6MDG M9 7EQ
G2ALN M9 8AT
G0SNZ M9 8JD

Mansfield

2E0AIW NG18 2AL
G7TTY NG18 2AU
M1BVP NG18 2LQ
2E1FTE NG18 2LQ
G0EYC NG18 2NJ
G0CYB NG18 2PG
G1IUE NG18 2PG
G0KIU NG18 2RN
G6UDB NG18 2SF
G0SKW NG18 3AQ
G4VQU NG18 3BS
G1KGQ NG18 3DZ
G0DQE NG18 3EN
G4NOR NG18 3EN
2E1FGP NG18 3JG
G1TYU NG18 3JJ
G3EQF NG18 3JJ
G4VWA NG18 3JJ
G4LBY NG18 3JL
M1BPP NG18 3JQ
2E1EEW NG18 3JQ
G0JRC NG18 3NP
G0KJC NG18 3NP
G0FVV NG18 3QS
G0TWP NG18 4BW
G0PTC NG18 4EQ
G4DFV NG18 4ER
G1EIV NG18 4FA
G4GNC NG18 4FB
G7ROI NG18 4FB
G4AAH NG18 4HF
G4SVU NG18 4HF
G0VYT NG18 4HG
G8UMG NG18 4HY
G6RYW NG18 4PG
G1GQI NG18 4PN
G8GNN NG18 4RH
G4GZU NG18 4RT
G4SHU NG18 4RU
G3JUY NG18 4SH
2E0AJC NG18 4SY
G7PMF NG18 4SZ
M1ARF NG18 5EE
G0DQD NG18 5HS
M1BRN NG18 5JH
2E1FND NG18 5JH
2E1EIX NG18 5JT
G4REU NG18 5LA
2E1EDU NG18 5LL
G0RDP NG18 5NB
G7PLM NG18 5NQ
G4TGB NG18 5QS
G4VUM NG18 5SQ
G4SFC NG19 0AR
G7GGG NG19 0BX
G4ZZG NG19 0DX
2E1CPH NG19 0EF
G3VIY NG19 0EJ
G0DAL NG19 0EL
G1HLT NG19 0EY
G6DSQ NG19 0HN
2E1EEX NG19 0HP
G0ELB NG19 0HZ
G3ONB NG19 0JP
G4OJM NG19 0LE
G4LQL NG19 0LS
G0TUP NG19 0PP
G1DZH NG19 0QR
G6CUK NG19 0QR
G0MDS NG19 6DD
G4GYU NG19 6DD
G3XWZ NG19 6EL
G7AVZ NG19 6HZ
G1HGG NG19 6JP
G7POR NG19 6JU
G3AMY NG19 6NB
G4ANU NG19 6NB
G4ZSD NG19 6QQ
G8OEN NG19 6QR
G7VKG NG19 7LX
2E1DZG NG19 7RA
G3SQE NG19 8AN
G0GAG NG19 8HT
G1JGY NG19 8LA
G7ACA NG19 8PY
G1IMY NG19 8QT
2E1FEY NG19 8QZ
G0OYP NG19 9AZ
2E1FER NG19 9DH
G0BYG NG19 9DR
G1WTB NG19 9DW
2E1CXZ NG19 9DW
G7CKL NG19 9HA
G4WPW NG19 9HH
G7UVU NG19 9HS
G3VVE NG19 9NA
2E1EYC NG19 9QW
G0PGA NG20 0DG
G7MNQ NG20 0DN
G0OMG NG20 0PG
G0CAL NG20 0PT
G1OXT NG20 0PT
G3XTL NG20 0PT
G3AWI NG20 0RX
G1KNK NG20 0RZ
M0AJS NG20 0TA
G0ENV NG20 9PY
G0TLT NG20 9QZ
G0HYV NG21 0DF
G8EHX NG21 0ED
G0UYQ NG21 0ED
G4DJM NG21 0EH
G0BZG NG21 9AJ

Melton

G0FEF LE13 0AP
G1YEY LE13 0BA
G0WYM LE13 0DS
G1DPN LE13 0EU
G3XJW LE13 0EW
G7TVH LE13 0HF
G0OAE LE13 0LL
G8AKE LE13 0NN
G2FUM LE13 0NQ
G7PCI LE13 0PF
G1HMF LE13 0QF
G7PCT LE13 0RA
G0LBV LE13 0SH
G4ZVF LE13 0SL
G4PKU LE13 0TB
G6XYP LE13 1EJ
G7ARB LE13 1EN
G7FOX LE13 1EN
G7MRG LE13 1EN
G4MWN LE13 1EW
G6HNY LE13 1HD
G4NNZ LE13 1HG
G4YSP LE13 1HY
G4PTK LE13 1JZ
G8LBM LE13 1LJ
G1MZH LE13 1RT
G6CJK LE13 1RT
G8RBY LE13 1SE
G6BWW LE13 1TB
G3NVK LE13 1UH
G6FDD LE13 1UH
G8NYH LE13 1UR
G4XYB LE14 2AB
G8YEJ LE14 2AP
G2AA LE14 2UH
G4JDI LE14 2UH
G4AMN LE14 3DT
G7UDJ LE14 3EW
G6PQI LE14 3LA
G4HEE LE14 3RY
G6FFM LE14 3UB
G4NRC LE14 3UF
G3STG LE14 3UF
G4FOX LE14 3UF
G4PJV LE14 4AB
G4EVK LE14 4DD
2E1FAH LE14 4NF
2E1FAI LE14 4NF
G7DMM LE15 8LL
G1NPP LE17 4WN
G4WFK NG13 0AR
G8BDO NG13 0EH
G8GUH NG13 0ER
G4LWB NG32 1QB

Mendip

G7VQX BA11 1HZ
G0WOZ BA11 1HZ
G6JNB BA11 1NR
G1JPK BA11 2BD
G1IMF BA11 2DF
G6JKX BA11 2DF
G4UMN BA11 2LB
2E0ANE BA11 2SE
2E1EGM BA11 2SE
G0WUR BA11 2SE
G0WUP BA11 2SE
G0WUQ BA11 2SE
G7BEB BA11 2SH
M1BGB BA11 2UB
M1BGC BA11 2UB
G3XBW BA11 3BX
G8VGI BA11 3DP
G3ZWL BA11 3PF
M0ALW BA11 4AR
G1WFF BA11 4EL
G4XAG BA11 4JA
G4XAH BA11 4JA
G4SCD BA16 0BT
M0AAV BA16 0BY
G4PLY BA16 0BY
G4YTW BA16 0DJ
G1OOB BA16 0HB
G4JBW BA16 0HY
G1ZTM BA16 0NS
G1PVU BA16 0RY
G1TRN BA16 0SE
G1FGK BA16 0TE
G4XLY BA16 0TE
G7AZV BA16 0UB
G4FUS BA16 9QP
G6BJQ BA16 9QQ
G4DZW BA16 9RP
G0HKB BA3 4EX
G7SSA BA3 4HG
G4MKE BA3 4JL
G4WVS BA3 4JL
G4VVP BA3 4JW
G1ORN BA3 4QH
G1SOM BA3 4QH
G8WKK BA3 4QH
G8WKL BA3 4QH
G6KPD BA3 4RE
G0TJP BA3 4RE
G1WQU BA3 4SJ
2E1FPV BA3 5AL
G4YJH BA3 5BB
G7IBL BA3 5DG
G1DOS BA3 5JB
G7JXZ BA3 5PN
G7KEP BA3 5PP
G7MYG BA3 5PR
G0HIC BA3 5PT
G4OWH BA3 5SW
G1OPZ BA3 6LA
G7MRY BA3 6LZ
G1DUU BA3 6NR
G2BUP BA3 6NR
G4GUG BA4 4PU
G0VFI BA4 4PY
G3UJG BA4 4QE
G0BKU BA4 5BH
2E1FLX BA4 5PG
G8CQR BA4 5PP
G7RTU BA4 5RW
G7URZ BA4 5RW
G8KKA BA4 5XR
G6RRY BA4 5YG
G3BOB BA4 6JG
G4LWQ BA4 6LY
G4YLO BA4 6NG
G4STH BA4 6NG
G0KFN BA4 6RY
G7LMX BA4 6TS
G6ENM BA5 1AH
G3OJL BA5 1DG
G0UIT BA5 1DU
G8OEK BA5 1HD
G4UIU BA5 1LB
G4ACK BA5 1LF
G6SIM BA5 1LG
G6YBO BA5 1RJ
G0MZF BA5 2BJ
G7EFM BA5 2BJ
G8BFV BA5 2DG
G4KQQ BA5 2EN
G6KLA BA5 2ES
G3IJU BA5 2FN
G4ZDR BA5 2HZ
G7CON BA5 2NH
G4SFS BA5 3ED
G1ZGC BA5 3HU
G4XWE BA5 3HY
2E1FLZ BA5 3LG
G0MBX BA5 3NH
G7FPW BA5 3NH
G8HCJ BA6 8DA
G0MYH BA6 8EJ
G7UAF BA6 8TJ
G7UAG BA6 8TJ
G0GJX BA6 9AN
G0JLY BA6 9AU
G7LVN BA6 9JE
G0ECZ BA6 9JR
G0MSL BA6 9NH
G3OSJ BA6 9PG
G2BJK BS27 3SF
G0VVR BS27 3UB
G3GTJ TA11 7HA
G6HIQ TA11 7HD

Merthyr Tydfil

GW0JTJ CF46 5HG
GW1IQS CF46 5HQ
GW0OPP CF46 5LB
GW4LKE CF46 5NJ
GW0LBI CF46 6DB
GW0OUV CF46 6DH
GW0JTK CF47 0LA
GW3RNC CF47 0LL
GW4ATY CF47 0RG
GW0UZX CF47 0TB
GW4PWZ CF47 0UX
GW3RYR CF47 8HJ
GW6BMR CF47 8HJ
GW3RTA CF47 9TP
GW3PYD CF47 9YP
GW0UKE CF48 1HG
GW6AYR CF48 1HW
GW0SFI CF48 1SE
GW8VWD CF48 1TH
GW0VMZ CF48 1TP
GW4OZH CF48 1YT
GW3LNR CF48 2BB
GW6VRN CF48 2PU
GW0BYZ CF48 3NU
GW3YUC CF48 3NY
GW0JTE CF48 3PU
GW4WHP CF48 3RP
GW1WRV CF48 3SW
GW3RDB CF48 4EE
GW4ZRW CF48 4EE
GW4BIS CF48 4HD
GW4EJG CF48 4JY
GW0WUL CF48 4RN

Merton

G7UHL CR4 1AX
G0WCR CR4 1LF
2E1FIO CR4 1LF
G7TNV CR4 1QR
G1IZN CR4 1XN
G6XSF CR4 2BZ
G4JWU CR4 2EG
G0LUL CR4 2LF
G6CQF CR4 2ND
G4VWG CR4 3DG
G0VTM CR4 3DZ
G4RBH CR4 3JS
G1KGO CR4 3LW
G1KNH CR4 3QN
G8UG CR4 4LW
G6QN KT3 6JW
G1UFD KT3 6NA
G7HOV KT3 6QD
G4PGS SM3 9HY
G0GOC SM4 4BQ
G0VXX SM4 4DL
G1MJN SM4 4EG
G4KXD SM4 4JF
G8WIM SM4 4JF
G3OML SM4 4JN
2E1FII SM4 4JQ
2E1FIJ SM4 4JQ
G0BXC SM4 4LF
G3WMR SM4 4PU
G0PNT SM4 4QG
G3WHK SM4 4QG
G4ZHA SM4 4QN
G8XMG SM4 4QP
G3DCZ SM4 4RJ
G6EKW SM4 4RP
G4CQR SM4 4TD
G0DGH SM4 5LR
G3PVA SM4 5QB
G3NQE SM4 6QE
G0WCF SW16 5QF
G1WXD SW19 1BX
G8VMS SW19 1EW
G7RGO SW19 1JW
GW1FOF SW19 1TW
G7LXN SW19 3DJ
G6CCE SW19 3DJ
G4ZHT SW19 3JJ
G6GKG SW19 4AW
G4CTA SW19 4PR
G6LKV SW19 5AZ
G6LKW SW19 5AZ
G6ZLS SW19 5AZ
G4CZT SW19 5HL
G4GQA SW19 7EZ
G3NXN SW19 7LW
G4ILP SW19 8DQ
G6ZLT SW19 8EQ
G4KWX SW19 8HD
G8WUF SW19 8NZ
G8TGR SW19 8SW
G2DQU SW20 0EP
G8JEE SW20 0PU
G7UPW SW20 0PU
G3DRN SW20 0UD
G3CNA SW20 0UH
G1LGB SW20 8EL
G3PHM SW20 8LA
G4VCB SW20 8TP
G0CIS SW20 9DL
G8JUG SW20 9EW
G7VDG SW20 9HF
G8IAK SW20 9HG
G8BUZ SW20 9HX
G0VQT SW20 9JT
G0BME SW20 9NB
G1KNB SW20 9NJ

Mid Bedfordshire

G1PII LU5 6LE
G8AZR LU5 6LG
G3NRW LU5 6NF
G1ZNX LU5 6NT
G8UGN LU5 6PH
G0LAX MK17 9ED
G6VNI MK17 9ED
G3VZV MK17 9HF
G6JJT MK43 0DP
G4OPA MK43 0EU
G7BBD MK43 0HA
2E1BTJ MK43 0HA
G4JXZ MK43 0JA
G3UGL MK43 0JN
M0BDT MK43 0JX
G8XER MK43 0LN
G4PMS MK43 0LW
G0GOF MK43 0NH
G7HZP MK43 0PN
G7UYB MK43 0PN
G7IFD MK43 0QZ
G0RAV MK43 0RW
G0COG MK45 1BU
M1BVY MK45 1HS
G8CGX MK45 1HX
G7GZS MK45 1JH
G6JZW MK45 1JY
G4VSB MK45 1LN
M1AFP MK45 1LN
G4VMB MK45 1NY
G4RIK MK45 1PH
G8CTB MK45 1PJ
G4NEC MK45 1PQ
G4OWN MK45 1QB
M1BWD MK45 1QB
G6PHZ MK45 1QD
2E1FRK MK45 1RG
G0WZA MK45 1TB
G7PCF MK45 2AD
G0BVW MK45 2AE
2E1FDJ MK45 2AQ
G4CDT MK45 2AX
G4AHM MK45 2BT
G0XIT MK45 2DH
G0WAS MK45 2DJ
G3VES MK45 2EY
G7SHQ MK45 2LR
G7NBI MK45 2RS
G8FFM MK45 2RS
G0RNI MK45 2SP
G4FKI MK45 2SP
G0JJK MK45 2TP
G7NBL MK45 2TU
G4VXT MK45 2UA
G8PYV MK45 2UB
G4VLA MK45 2UE
G4KSN MK45 2XF
G6KHW MK45 3JP
G7SVT MK45 3JP
G3ARZ MK45 3LG
G0PBZ MK45 3LS
G8HRW MK45 3PJ
G0EUQ MK45 3PL
G4MGX MK45 3WE
G4LOO MK45 4BG
2E1EIC MK45 4BG
G1EVC MK45 4DU
G8XEF MK45 4DU
G4OVZ MK45 4HB
G6BJO MK45 4TB
G7JSB MK45 5BW
G8LQM MK45 5DU
G7WAS MK45 5LB
G6NMI MK45 5LD
G8IJP MK45 5LP
G7CXK SG15 6RG
G0TUX SG15 6RP
G4XBD SG15 6TS
G6UMX SG15 6UF
G7CUT SG15 6UL
G0NGB SG15 6XT
G1IFL SG15 6XT
G8OCA SG15 6XT
G0KOA SG15 6XZ
G8ALR SG16 6AB
G4KQD SG16 6AB
2E0ANY SG16 6AE
M0BDW SG16 6AE
G3JKK SG16 6BJ
G3XTG SG16 6BL
G3ORG SG16 6BN
G8ZIR SG16 6BT
G7TLS SG16 6BT
G7TFZ SG16 6HR
G0PWR SG16 6HW
G1SJO SG16 6JS
G0GWB SG16 6LH
G3NII SG17 5AA
G4LWA SG17 5AD
G6RHL SG17 5AD
2E1EFR SG17 5AE
G4VUI SG17 5AP
G8UFX SG17 5AU
G8XZQ SG17 5AU
G3IXG SG17 5AZ
G4JLP SG17 5BX
G4MCK SG17 5BX
G0GGQ SG17 5EA
2E1BFD SG17 5EJ
G3PNK SG17 5EU
G3YPP SG17 5HA
G6MIS SG17 5HB
G6VBE SG17 5HB
G1CIT SG17 5HW
G7VLP SG17 5LJ
2E1EGF SG17 5LP
G0PQO SG17 5NH
2E1DFC SG17 5QZ
G8BEG SG17 5RZ
G4ZXC SG17 5SG
M1AAG SG18 0DF
G7ABF SG18 0DG
G8HHO SG18 0ED
G7OZU SG18 0HG
G7OME SG18 0HN
G6PVS SG18 0HR
G1RYF SG18 0LS
G8MCY SG18 0NX
G4NNB SG18 0PS
G3ELV SG18 8BG
G8NJO SG18 8BG
G0PHY SG18 8DX
G1GSF SG18 8HD
G1ARJ SG18 8JF
G1FYF SG18 8LY
G3ZZG SG18 8NS
G3XIZ SG18 8PB
G0MND SG18 8QG
G3XVN SG18 8RN
G4KUY SG18 9BJ
G6IYE SG18 9BU
G4YRF SG18 9DF
G2DPQ SG18 9DF
G3MLB SG18 9DL
G4TEH SG18 9ED
G1JVY SG18 9JL
G1GSN SG18 9QP
G4XIM SG18 9QP
G8EKU SG18 9QR
G3XCK SG18 9RS
M0AJF SG18 9RU
G0WII SG18 9RU
2E1FDX SG19 1AQ
G8ZIW SG19 1AT
G0TMJ SG19 1AU
G3FJE SG19 1AU
G4MEO SG19 1AU
G8GHR SG19 1AU
G8TYN SG19 1AU
2E1BEO SG19 1AU
2E1CWJ SG19 1AU
2E1EJM SG19 1AX
G8FRG SG19 1BU
G4ASH SG19 1DA
G1CDZ SG19 1DP
G4CEP SG19 1DT
G7SVE SG19 1HF
M0AJV SG19 1HP
2E1DEP SG19 1HP
2E1FDU SG19 1HP
2E1FEP SG19 1HP
2E1FDG SG19 1NG
G3NEF SG19 1NP
G0LGV SG19 1PL
G3ROZ SG19 1PY
G4FXM SG19 2AB
G8BXM SG19 2AX
2E1EFQ SG19 2JU
G0IKZ SG19 2LB
G1ZPU SG19 2NE
G3XIQ SG19 2NE
G6SWJ SG19 2NS
G3KZE SG19 2PJ
G4MKR SG19 2QB
G1YRM SG19 2QQ
G3OXG SG19 2QQ
G3WSD SG19 2RE
G4PSO SG5 3LS
G7PQP SG5 3LS
2E1EIW SG5 3LS
G8IER SG5 3NH
G4OCX SG5 4DB
G2DUS SG5 4EA
G4BHO SG5 4EQ
G4UKE SG5 4HP
G4THF SG5 4HT
G6DXN SG5 4LA
G7LUR SG5 4LF
G7MJX SG5 4LY

Mid Devon

G4TLL EX15 1BQ
M0AKG EX15 1NJ
M0AKJ EX15 1QD
G1TKQ EX15 1RN
G1LCB EX15 1SY
G0KOF EX15 2HQ
G8HPS EX15 2RN
G4JND EX15 2SD
G4PXU EX15 2SD
G0IFA EX15 2SH
G4CPN EX15 3AX
G3WNV EX15 3BA
G0OED EX15 3DS
G4SCV EX15 3LH
G3ZQL EX15 3QG
G8PXO EX15 3QG
G3WNI EX15 3QY
G4UPS EX15 3RY
G4OOE EX15 3SB
G0UXS EX15 3SL
G4HMG EX15 3TJ
G4MVV EX15 3TJ
G0TAZ EX15 3XA
G0UOL EX16 4AE
G4HGU EX16 4AU
G6XWM EX16 4BE
G6SWD EX16 4BN
G7JGZ EX16 4BW
G4XXD EX16 4ER
G7LJN EX16 4HR
G0JCH EX16 4JF
G4ZAL EX16 4JF
G0MAS EX16 4PN
G7VZD EX16 5BH
G0IFC EX16 6EY
G6SMG EX16 6NS
G0IMJ EX16 6RA
G4YCV EX16 6RJ
G4TSW EX16 6RS
G7TSW EX16 6RS
G6BMQ EX16 7DL
G4FJK EX16 7ED
G4ZTS EX16 7EP
G4PYU EX16 7JH
G6EQI EX16 8LF
G6HV EX16 8NZ
G4DPD EX16 9BT
G8NGJ EX16 9BT
2E1DUI EX16 9RY
G4YAQ EX17 2DH
G6TWD EX17 2DH
G0GKB EX17 2EA
G3HTA EX17 2EP
G0NWQ EX17 3BW
G8GPA EX17 4JB
M1BZL EX17 4QJ
G0JUM EX17 5NA
G3TJW EX17 5NX
G4KXP EX17 5PW
M1AEI EX17 6DH
2E0APR EX17 6DH
G8YOA EX17 6DH
G0GVJ EX17 6NA
G4BOP EX17 6RE
G8BSP EX17 6SQ
G3POD EX18 7LD
G0ABI EX18 7QX
G3XVO EX5 5BJ
G1PHU EX6 6ET

Mid Suffolk

2E1FHH IP13 7EX
G8VOC IP13 7JD
2E1FEI IP13 7JQ
G7JCF IP13 8DT
G3KCF IP14 1BN
G0KCB IP14 1BQ
G4RET IP14 1DA
G3TAQ IP14 1LP
G3ALI IP14 1QX
G1HNH IP14 1RJ
G3ATJ IP14 1RN
G8MYE IP14 1SQ
G3TZE IP14 1TD
G0NFE IP14 1UB
G0JJG IP14 1UF
G8FNJ IP14 2AB
G1XHL IP14 2DN
M1ADV IP14 2DR
G8MJK IP14 2LS
G4AWF IP14 2LZ
G8FAW IP14 2LZ
G0SCM IP14 2NZ
G1UBH IP14 2RE
G3VAJ IP14 3AA
G8KDL IP14 3DU
G0NRU IP14 4AB
G4HFJ IP14 4AF
G6IJN IP14 4DA
G6XGT IP14 4DA
G3TAS IP14 4DJ
G1SUE IP14 4DR
M0BGG IP14 4DR
G6JZV IP14 4ED
G0NMS IP14 4LP
G8SED IP14 4NR
G1EFA IP14 4SE
G4MEY IP14 4SL
G6SUR IP14 5AG
2E1BRC IP14 5AG
G4BJO IP14 5AN
G4THN IP14 5ET
G3ZEQ IP14 5HB
G8FUJ IP14 5HJ
G3ZQU IP14 5JL
2E1FEO IP14 5LR
G0ERL IP14 5LW
2E1FWO IP14 5NL
2E1FVO IP14 5PX
2E1FWY IP14 5SD
G6SYW IP14 5SN
G4GBA IP14 6AJ
2E1FBP IP14 6EH
2E1FBN IP14 6EL
G3ONL IP14 6LB
G3VNT IP14 6LX
G4EVN IP14 6LX
2E1FVM IP14 6PY
G8FMI IP14 6RN
2E1DDH IP21 4AD
G1YDA IP21 5BA
G7UNU IP21 5QQ
G4WGL IP21 5TE
G3XDC IP22 1DG

2E1FZT IP22 1HX
G1GTU IP22 1PU
G8JR IP23 7AW
2E1FXC IP23 7LX
G7OCQ IP23 8BX
2E1FYU IP23 8BX
G1TIJ IP23 8DZ
G3GBB IP23 8JR
G2FXD IP23 8LY
G7MPF IP30 0SZ
G1FTD IP30 9AF
G0KRL IP30 9AJ
G8DQZ IP30 9GE
G6AWO IP30 9HJ
G3MWO IP30 9PX
2E1BLP IP30 9QS
2E1FZW IP30 9QS
G3XLG IP30 9TP
G4JPQ IP30 9UF
G8JFX IP30 9UH
G6GGW IP31 3EL
G7UXD IP31 3EL
G3LPT IP31 3EN
G3FOQ IP31 3LG
G4MID IP31 3PD
G3TQN IP31 3PJ
G6DMY IP31 3PS
G4UZF IP31 3SP
G0SAR IP6 0BR
G1SAR IP6 0BR
G1WSR IP6 0BR
G7IBR IP6 0BR
G8CPH IP6 0BR
G7RSF IP6 0BR
G7MSR IP6 0BR
G7RMP IP6 0BR
G7SCR IP6 0BR
G7RWV IP6 0BR
GE7RLF IP6 0BR
G7RSP IP6 0BR
G7RSS IP6 0BR
G8ZYM IP6 0PY
2E1FRI IP6 0RH
G4TJY IP6 8DA
G7VXS IP6 8DP
G4HRP IP6 8ES
G8VHN IP6 8RZ
G4LSP IP6 8SA
M1ALH IP6 8SA
G6PGM IP6 8UA
2E1FRJ IP6 9RF
G4YQL IP7 7DW
G3MYY IP8 4AU
G0IJS IP8 4EY
G7VGE IP8 4EY
G0JWQ IP8 4EZ
G4FDF IP8 4HD
G8FFU IP8 4PE
G3XOK IP8 4PQ
G3WJI IP8 4SP
G4DMT IP8 4SR

Mid Sussex

G4YCN BN6 8BN
G3XTH BN6 8DD
G0SWS BN6 8DY
G4JCY BN6 8EH
2E1AFA BN6 8HR
G6MJW BN6 8HR
G6YPY BN6 8HR
G8UBJ BN6 8JJ
G8ZKH BN6 8JZ
G8YKV BN6 8LH
G7VNG BN6 8NS
G8XYR BN6 8NU
G0GZE BN6 8PD
G4LYX BN6 8PD
G0GMC BN6 8PR
G1ZMS BN6 8PR
G3ZMS BN6 8PR
G4GKO BN6 8QB
G3SGF BN6 9BZ
G0PCF BN6 9DU
M1ASZ BN6 9LU
G4CIG BN6 9RR
G7KPN BN6 9RW
G3NYX BN6 9RZ
G3VDC BN6 9SD
G4MMI BN6 9UG
G7LFT BN6 9UY
G4TWP RH10 3LS
G3VJM RH10 3PS
2E1DZF RH10 3QA
G1HOG RH10 3QB
G0KKS RH10 3QT
G0ORX RH10 3QZ
G0DKY RH10 3RH
G0PMM RH10 3RL
G4ANN RH10 3XG
G4YMH RH10 4EX
G3RMK RH10 4NG
G7RIY RH10 4NL
G4MVS RH10 4PG
G0GML RH10 4UB
G3YTR RH10 4UJ
G4TTY RH10 4UL
G3VLH RH10 4XA
G3WOX RH10 4XU
G0PYJ RH10 4XZ
G3NZP RH10 4YY
G3LET RH11 9AF
G7MIW RH11 9AW
G3YEZ RH15 0AX
2E1FGX RH15 0BB
G0GNO RH15 0DH
G5RV RH15 0DX
G3YBM RH15 0EU
G3IZW RH15 0LA
2E1EAS RH15 0LA
G3RMY RH15 0NB
G0WGP RH15 0ND
G0GNV RH15 0NW
2E1AFF RH15 0NW
G3XUD RH15 0NZ
G6ZQA RH15 0PH
G8SNH RH15 0PL
2E1FKF RH15 0PS
2E1ELQ RH15 0PS
G0DDE RH15 0PT
G4ULZ RH15 0PT
G6DBX RH15 0PT
G6IHG RH15 0QA
G1SKI RH15 0QB
G0UUP RH15 0QN
2E1DCP RH15 0QQ
G1AZA RH15 0RP
G3XQM RH15 0RW
G7TPW RH15 0RW
G7BLD RH15 0UB
G1DZD RH15 8BD
G0IDP RH15 8BL
G3XUP RH15 8BW
G0LOH RH15 8DE
G0LFF RH15 8DL
G4VTQ RH15 8DZ
G4AKG RH15 8EU
G1XDF RH15 8HG
G7DSZ RH15 8HN
G4ZCS RH15 8QD
G6KBJ RH15 8QW
G4MIZ RH15 9ED
G6CYU RH15 9EZ
G1PJM RH15 9HR
G1POD RH15 9HR
G8KMP RH15 9HZ
G4WEH RH15 9JA
G7CSP RH15 9JD
2E1BYU RH15 9PB
G8ZNU RH15 9PJ
G7OIA RH15 9PL
2E1AOU RH15 9PL
G7ZMS RH15 9PL
2E1BRD RH15 9PL
2E1APQ RH15 9PX
2E1APR RH15 9PX
G0KYX RH15 9QS
G0ONM RH15 9QS
G7TMR RH15 9SJ
G3VAK RH15 9SP
G0SQF RH15 9ST
G8EZE RH15 9US
G4TDN RH15 9XN
G7AII RH16 1ER
G7HHC RH16 1ER
G4JPY RH16 1HD
G4KKO RH16 1HJ
G0OIO RH16 1LB
G4ZWV RH16 1PR
G3ZEK RH16 1TF
G1PSS RH16 1UQ
G7MAG RH16 1UZ
G8ZOV RH16 1UZ
G3FSX RH16 2AB
G3HUA RH16 2LQ
G8MVP RH16 3BH
G7POU RH16 3HZ
G7TNU RH16 3JG
G8MFM RH16 3JS
G0KMR RH16 3JY
G7PWH RH16 3LJ
G0LFJ RH16 3NA
G0LTV RH16 3NB
G4PFV RH16 3NB
G0UJY RH16 3NJ
G0AUI RH16 3QW
G8WYT RH16 3RF
G8ZOU RH16 3SS
2E1FET RH16 4DH
G7FHV RH16 4HL
G1SHN RH16 4JD
G6SOY RH16 4LP
G3VKW RH17 5AL
G0XAA RH17 5AL
G3WYN RH17 5DZ
G3JMB RH17 5ER
G7WBR RH17 6DQ
G2BUY RH17 6HU
G0HXF RH17 6NJ
G0VYK RH17 6TR
G4ZKQ RH17 7PG
G8CSC RH17 7PQ
G0ENA RH17 7PY
G1CDW RH17 7QD
G7DSW RH19 1HS
G3VPS RH19 1JG
G0HIY RH19 1JR
G3KOA RH19 1JR
G7AIF RH19 1SX
G7CEK RH19 1SX
G8VCH RH19 1SX
G8WZO RH19 1SX
G8OYY RH19 1TQ
G3YQW RH19 2DD
G0EID RH19 2ER
G7LJU RH19 2JH
G7NMR RH19 3HP
G3DSK RH19 3JA
G7KBR RH19 3UR
G0RCF RH19 3XF
G1YXT RH19 4BZ
G4TRG RH19 4EA
G6GSF RH19 4JT
G4PFU RH19 4TF

Middlesborough

2E0ADR TS1 4ED
G7VOT TS1 4PR
G6KMG TS3 0LZ
G0APQ TS3 0QH
G4IIN TS3 8LX
G4OOK TS3 8NX
G4YIQ TS3 8QJ
G3RGB TS4 3HX
G0UTP TS4 3NJ
M0AHD TS4 3PP
G4UTV TS5 4AJ
G0NAG TS5 4HA
G1RPX TS5 5JA
G7SMB TS5 5LD
G6THC TS5 5LJ
G4HSB TS5 5NQ
G4VZC TS5 6DP
G4KGV TS5 6PW
G3MXZ TS5 6QE
G7SZC TS5 6RQ
G7NRN TS5 7LA
G1NTL TS5 8BT
G3KXV TS5 8DR
G8WKZ TS5 8EB
G4IJM TS5 8NT
G8DIN TS7 0AA
G1IGW TS7 0LT
G6AJF TS7 0QL
G8YDC TS7 0QL
G4PPS TS7 8AG
G4LVY TS7 8EH
G4NAA TS7 8EQ
G3SXL TS7 8LF
G7HQA TS7 8NF
G3ZZT TS7 8QW
G4JJB TS7 8RA
G7IME TS7 8SL
G0AOO TS8 0RU
G7RXB TS8 0SU
G7PTC TS8 0XB
G4POD TS8 9AB
G4ZML TS8 9BU
G0BQP TS8 9HH
G7MPB TS8 9HW
G4ZQO TS8 9PT
G8LIE TS8 9RF

Midlothian

GM1CHT EH10 7DX
GM4DCL EH18 1AB
GM7PWI EH18 1HT
GM3RFQ EH18 1LW
GM4FIB EH19 2DT
GM4XBR EH19 2DT
GM0OTU EH19 2EH
GM8HSY EH19 2EU
GM6DCU EH19 2JA
GM4SUR EH19 2LD
GM0FQS EH19 2PQ
GM6SEV EH19 3EX
MM1BNS EH19 3EX
GM4GGF EH19 3LQ
GM6RVL EH19 3LU
GM4FGD EH20 9HE
MM1BJO EH20 9LF
GM7LTX EH20 9LU
GM0PQV EH20 9RR
GM1YFO EH20 9SL
MM1BKT EH22 2HT
GM0WHM EH22 2LT
GM1CCN EH22 2NS
GM8DLU EH22 2PG
GM0SUF EH22 3DB
GM4WJL EH22 3HJ
GM0LPK EH22 3LU
GM0TQK EH22 4PU
MM0BAC EH22 4RJ
GM7TAN EH22 4SW
2M1BYW EH22 5ER
GM0PKP EH22 5HX
GM7DRY EH22 5TP
GM0KDC EH23 4BT
2M1FOQ EH23 4HD
GM4HQU EH23 4LX
GM1VUH EH23 4RF
GM4LHW EH25 9NP
GM4VZI EH25 9QJ
GM0MJR EH26 0EL
GM0IDY EH26 0LX
GM1ENI EH26 0QN
GM4HHY EH26 0QN
GM4WLL EH26 0RH
G3YXJ EH26 8DG
GM0LUF EH26 8EF
2M1BBY EH26 8NJ
GM1OVW EH26 9AN
GM6GFL EH26 9EE
GM0OTX EH26 9HE
GM1PSZ EH26 9HS
GM8MNG EH37 5RN
GM8VYZ EH37 5RN
GM3OIB EH37 5SQ

Milton Keynes

G0EYZ MK1 1BA
G0SMK MK1 1BA
2E1BJM MK1 1BA
2E1BJN MK1 1BA
G8ABB MK1 1BL
G8THF MK1 1NF
G0WFH MK1 1NG
G4VVH MK10 9AH
G8XJL MK11 1ET
G3MFE MK11 1EY
G4BXR MK11 1EZ
G8EOW MK11 1HB
G3ZNY MK11 1HE
G6AZR MK11 1HX
G4ZNY MK11 1LB
G0BOQ MK11 1LD
G4NNX MK11 1RB
2E1CKQ MK11 2AA
G0GMB MK11 2AF
G0KKT MK11 2AG
G4WDN MK12 5AY
G6WXM MK12 5AY
G8AQB MK12 5BE
G0MSV MK12 5BH
G6WZL MK12 5BN
G7WKD MK12 5HJ
G6EWX MK12 5JY
G0TGU MK12 6AL
G3NEH MK12 6AN
G3IYX MK13 0BB
G4FYR MK13 0EH
G8OSH MK13 0EP
M0AEQ MK13 0EY
G8KFN MK13 7AJ
G4LIJ MK13 7AY
G8CCV MK13 7AY
G1NRY MK13 7DA
G6INI MK13 7HN
G3SHD MK13 7HS
G4DNI MK13 7LZ
G6ZSR MK13 7QR
G8KPM MK13 7UG
G8EPQ MK13 8AT
G6OBP MK13 8BW
G0ERE MK13 9HZ
G3YYN MK14 5AX
M0ALO MK14 5AX
G7VNW MK14 5AX
G0BUX MK14 5DT
G3OWC MK14 5DT
G6YIG MK14 5DX
G3ZIE MK14 5EX
G0AXF MK14 5HS
G4WBG MK14 5JQ
G0CGC MK14 5JY
G0MKR MK14 5JY
G1CBY MK14 5JY
G7AFL MK14 5JY
G7VYH MK14 6AX
G6ALU MK14 6AY
G8FIK MK14 6BH
M1AGG MK14 6DH
G7UTY MK14 6EJ
G4TXT MK14 6JX
G4TOH MK14 7DY
G1DMR MK14 7LL
G1LRU MK14 7LU
G1RNZ MK14 7PP
G6OTV MK14 7QQ
G3XJO MK15 0BD
G8HYM MK15 8QD
M0AWI MK15 8QD
G7AQN MK15 9AE
G4KSJ MK15 9BJ
G0RKV MK15 9HZ
2E1FOH MK15 9HZ
G1WMV MK15 9JA
G4CGK MK16 0BG
G1BAQ MK16 0EE
G3LFR MK16 0EE
G7IKB MK16 0EF
G0BLL MK16 0HD
G3XLN MK16 0JE
G4XWM MK16 0JP
G0FWD MK16 0LH
M0ATY MK16 0LL
G4MDR MK16 0LT
G3IED MK16 0PA
G7LKG MK16 8DE
G0GWX MK16 8EB
G2FCA MK16 8EB
G8WV MK16 8HT
G4VZB MK16 8JA
G3MRB MK16 8JQ
G8JWT MK16 8NR
G4FIA MK16 8PL
G4VMX MK16 8PP
G1CYQ MK16 8RF
G2XD MK16 9AS
G6IZF MK16 9AT
G4MTF MK16 9DN
G4GJR MK16 9EF
G1XUG MK16 9LR
2E0AON MK16 9LW
2E1EOT MK16 9LW
G3ZCJ MK16 9LZ
G3TGE MK16 9NL
G6ZQU MK16 9NQ
G3ACT MK17 8PA
G1GIE MK17 8PY
G0GQH MK17 8QG
G1MZD MK19 7AF
G3OBY MK19 7BY
G0COH MK19 7BZ
G8ATY MK19 7DA
G0JJH MK19 7ER
G4KST MK19 7LD
G4MGV MK19 7LQ
G0FQU MK19 7LX
G8UEZ MK19 7NF
G1WWP MK19 7NH
G1AAD MK19 7NQ
G3FNK MK2 2HN
G6WZD MK2 2HN
G0WFC MK2 2HT
G0UUF MK2 2HT
G3PUW MK2 2JN
G4NJU MK2 2JW
G0GQP MK2 2LA
G0TBS MK2 2PY
G0OQD MK2 2RN
G0OQE MK2 2RN
G3CAG MK2 2RQ
2E1FOD MK2 2XP
G3CPT MK2 3AD
G4JGG MK2 3LS
G3YJD MK2 3NB
G0UCK MK2 3PP
G3KQS MK3 5AA
G6JDY MK3 5AJ
G0GYW MK3 5BD
G3IMV MK3 5DG
G3PPO MK3 5DQ
2E0APN MK3 5DQ
G6VWZ MK3 5HL
G1BLT MK3 5LL
G6UGX MK3 5NB
G4MXO MK3 6BN
G7EFD MK3 6PX
G6RFH MK3 7PG
G3UDC MK3 7RP
G4GCJ MK3 7RQ
G6ZJD MK3 7TL
G3JXR MK3 7TU
G6OIA MK4 1AR
G7NDO MK4 1BQ
G4WNO MK4 1EG
G4LNC MK4 1LB
G0OGM MK4 2BA
G1PCN MK4 2BT
G6HLL MK4 3AL
2E1BEV MK4 3AL
G7LXT MK46 4BJ
2E0AIQ MK46 4BL
G1AWJ MK46 4BL
G2CYN MK46 4EB
G3PRU MK46 4HE
G8GFA MK46 4HU
G4IBE MK46 5AA
G4NEO MK46 5BJ
G6JFK MK46 5EQ
G0HMF MK46 5ES
G8LKQ MK46 5HE
G4NPE MK46 5JA
G8JIK MK46 5LY
G4MQL MK5 6EE
G0SWT MK5 6HJ
G0UVG MK5 6HT
G3ZPA MK5 7AF
G4CLG MK5 7AJ
G0RXB MK5 7BB
G3MDI MK5 7DN
G6DOI MK5 7HA
G4SSL MK5 8DX
G1YSH MK5 8EU
G7AUE MK5 8EU
M0AIJ MK6 2BE
G1KPC MK6 2DQ
G3XVA MK6 2ES
G0FTI MK6 2HT
G1MKR MK6 2HT
2E1CJN MK6 2HT
G4AML MK6 2JQ
G0FMN MK6 2JR
G0RXJ MK6 2NU
2E1DAK MK6 2NU
G1LSF MK6 2QH
M0ARM MK6 2QS
M1AZH MK6 2QS
G0NSX MK6 2RZ
G1OWZ MK6 3AS
G3LMX MK6 3AY
G7SIV MK6 3AY
G6TTX MK6 3ES
G3HIU MK6 3LE
G8MKC MK6 3LE
G0RDG MK6 3LU
2E1FOB MK6 3LU
2E1FOT MK6 4HL
G0PCD MK6 4HQ
G4WKJ MK6 4JF
2E1CLH MK6 5BN
G7ALW MK6 5EP
G6OCF MK6 5ER
G0OUR MK7 6AA
G4YWC MK7 7SJ
G1JKF MK7 8DP
G1OEN MK7 8DW
G1BXY MK8 0BB
G4NQK MK8 0HA
G4BRH MK8 0HD
G3LIT MK8 8HD
G4BLM MK8 9AF

Mole Valley

G8GGS KT20 7LS
G7DOR KT20 7NB
G4JUM KT21 1BE
G8VNU KT21 1LR
G1HWR KT21 1NE
G4KGE KT21 1NE
G7LSB KT21 1PD
G7EWA KT21 1PZ
G2FSA KT21 1QD
G7RSK KT21 1SG
G6DTW KT21 2EG
G3OVS KT21 2HS
G4CLY KT21 2HU
G4XQE KT21 2HY
G6KVG KT21 2JA
G6YPF KT21 2JA
G1GUB KT21 2JQ
G1HDP KT21 2LY
G3SSM KT21 2NT
G4NHO KT21 2PG
G4MPK KT22 7EF
G1NKN KT22 7HN
G7OHS KT22 7SH
G6PLV KT22 8NL
G8MHN KT22 8NU
G7OFJ KT22 8QJ
G4EYB KT22 8RD
G0IXH KT22 8TL
G4XYC KT22 8UJ
G3NFV KT22 9AZ
G0IPG KT22 9HL
G3UZX KT22 9JR
G4HXN KT22 9LE
G6BTC KT22 9NB
G3VVW KT22 9ND
G7LWF KT22 9NE
G4CMR KT22 9NN
G8HZN KT22 9TZ
G6UQN KT23 3EA
G3UZW KT23 3LS
G8OHP KT23 3NX
G7DND KT23 3PW
G3LHN KT23 3PY
G3GN KT23 4AU
G1RJN KT23 4BA
G3TCT KT23 4BL
G8IWT KT23 4BS
G3SIA KT23 4JX
G0VDK RH2 8PE
G0JMB RH3 7ET
G1THW RH3 7HF
M0AFB RH3 7LT
G3HZJ RH4 1DH
G1XQG RH4 1JG
G0GNA RH4 1LP
G1JLM RH4 1LZ
2E1DGX RH4 1TA
G3JKV RH4 2AN
2E1DRJ RH4 2BY
G0GQY RH4 2DG
G4YFK RH4 2HT
G4VUW RH4 2NR
G6DTH RH4 2RA
G1VNU RH4 3DE
G4VTC RH4 3JH
G3RDH RH5 4AD
G0RWX RH5 4AE
G1OGM RH5 4DP
G3PJX RH5 4DS
G8TBU RH5 4DS
G3CZU RH5 4ES
G1EIL RH5 4HZ
G4YHB RH5 4LU
G4FTQ RH5 4QY
G0LPN RH5 4RQ
G1UCJ RH5 5AH
G4SIB RH5 5AT
G4JEH RH5 5BL
G1ABQ RH5 5BS
G8RYR RH5 5HS
G6KNE RH5 5HT
G6OBJ RH5 5JT
G8KTB RH5 5NW
G3XKG RH5 6AR
G8SBO RH5 6NB
G3RFJ RH6 0AR
G3XPB RH6 0AR
G3HYV RH6 0DE

Monmouthshire

2W1DHM NP4 0UB
GW7TLU NP4 0UB
2W1EBC NP4 0UB
2W1EBD NP4 0UB
GW3RTZ NP5 1AP
GW8AWM NP5 1RB
GW3LDC NP5 1SP
GW8OIG NP5 2EP
GW8PVN NP5 2EQ
GW4JCO NP5 3HR
GW0MSW NP5 3HT
GW3NJG NP5 3JX
GW4RYW NP5 3PL
GW3VFL NP5 3SD
G1OKP NP5 3TF
GW0HUT NP5 4DA
GW8PTS NP5 4DE
GW6RSP NP5 4FH
MW0ARE NP5 4TY
GW7SWN NP5 4TY
GW4DZD NP6 1LR
GW3VXC NP6 1LU
GW0GEV NP6 3AB
GW3NWS NP6 3BZ
GW8GT NP6 3BZ
GW8MZR NP6 3JU
GW7LYD NP6 3LQ
GW3MPP NP6 3PB
GW0DJX NP6 3QD
GW0FBT NP6 3SP
GW0EGH NP6 4BW
GW0FWZ NP6 4BZ
GW1EPR NP6 4GB
GW4FNO NP6 4PP
GW6NQU NP6 4RE
GW1VDT NP6 4SQ
GW4LWO NP6 4UT
GW3JBH NP6 4UW
GW4ALG NP6 5BW
GW4ULG NP6 5BW
GW1SXU NP6 5BX
G4RZE NP6 5DX
GW8XWW NP6 5EA
GW0HBD NP6 5JP
GW4KPD NP6 5LU
GW7UQL NP6 5NA
GW6MWN NP6 5NB
GW4TQD NP6 5NG
GW6ZUQ NP6 5RL
GW0HDY NP6 5TG
GW3HGJ NP6 6DD
G7IRD NP6 6EF
GW8HUS NP6 6HN
GW1FJI NP6 6JW
GW6MKV NP7 0BB
GW6JVB NP7 0BP
GW6XII NP7 0EA
GW1UVN NP7 0ED
GW6VLA NP7 0LZ
GW0JJF NP7 0RD
GW7HLZ NP7 5HN
GW4IHM NP7 5HW
GW4IHN NP7 5HW
GW7WFO NP7 5LG
GW6OIO NP7 6AF
GW6ZZF NP7 6AF
GW7RFP NP7 6BP
GW4KJW NP7 6DF
GW4UHK NP7 6HL
GW0UKG NP7 6HN
GW3ONN NP7 6LJ
GW4GFL NP7 6PF
GW4XQH NP7 6PF
GW8KJK NP7 7AW
GW8NZE NP7 7AW
GW8SCR NP7 7AW
GW8VRS NP7 7DS
GW4RPG NP7 8DN
GW0FXA NP7 8EP
GW7TVM NP7 8HN
GW4KQ NP7 8RS
GW8KCH NP7 9HP
GW0DKF NP7 9NX
GW4EIN NP7 9SA

Moray

GM4TOE AB37 9AR
GM0EIT AB55 5AG
GM0LVK AB55 5AP
GM4TRC AB55 5AP
GM0MGV AB55 5AP
GM4YVR AB55 5EA
GM7TWM AB55 5JD
GM0MGO AB55 5PE
GM1HNZ AB55 6LP
GM4WJA AB55 6QU
GM7SJC AB55 6RJ
GM0ARY AB56 1BE
GM3UKG AB56 1DD
GM0WHF AB56 1DH
GM4MOA AB56 1JX
GM4UWX AB56 4NR
GM4OEZ AB56 4QA
GM4PMT AB56 4QW
GM3KHH AB56 5EP
GM4IZY IV30 1QN
GM6WLJ IV30 1SF
GM4YZI IV30 2BP
GM7LSI IV30 2EG
GM0DQV IV30 2JH
GM8YKT IV30 2NB
GM0LPB IV30 2NS
GM0OTS IV30 2SB
GM7UGV IV30 2SB
GM7LUP IV30 2SY
GM3NCS IV30 2XY
GM0LOK IV30 2YR
GM7CPY IV30 2YR
GM4WKO IV30 3BZ
GM4EWM IV30 3EJ
GM1ASY IV30 3ER
GM7SBR IV30 3JW
GM7FPN IV30 3PQ
GM0LDX IV30 3QE
GM0EMC IV30 3SL
GM6TUE IV30 3SL
GM0GHN IV30 3TU
GM3VBY IV30 3UR
GM4CAI IV30 3YB
GM0GNY IV31 6AU
GM4PGM IV31 6BA
GM0AEG IV31 6JJ
GM4HMN IV31 6NA
GM4MKU IV31 6QH
GM0ONN IV31 6TE
GM4YWQ IV32 7DE
GM0UYZ IV32 7HS
GM0NEG IV32 7NJ
GM8DPV IV36 0JB
GM0KNT IV36 0QR
GM4CFS IV36 0UJ
GM6PLG IV36 0UP

NI: Antrim

MI1BNE BT29 4GJ
GI4WTT BT29 4JL
GI4AID BT29 4QE
GI3ECQ BT29 4RN
GI0WAA BT29 4SR
GI4TRX BT29 4TG
GI3PDN BT39 0BW
GI0SRL BT39 0BW
GI3GSB BT39 0EA
GI8DMX BT39 0HS
GI3ZVZ BT39 0HZ
GI4KQA BT39 0JP
GI4BWM BT39 0SB
GI4KKK BT39 0SB
GI7JKA BT39 0SB
GI3XDD BT39 0TN
GI1XIB BT41 1BD
GI3RNO BT41 1HH
MI1BIW BT41 1HP
GI4KIS BT41 1JB
GI6DNI BT41 1RF
GI4HDJ BT41 2AR
GI6KCX BT41 2AT
GI0MQN BT41 2DR
GI6OQL BT41 2EU
GI1EOS BT41 2EY
GI3YDM BT41 2HJ
GI4PID BT41 2QT
GI7JOJ BT41 2QT
GI7IPO BT41 2TS
GI4VJZ BT41 3BJ
GI3ZSC BT41 3EJ
GI0CXR BT41 3HF
GI0NYO BT41 3JY
GI3YDO BT41 4HD
GI4KUM BT41 4HD
GI8YWR BT41 4HN
GI7IRJ BT41 4HR
GI4NKJ BT41 4LA
GI0VLE BT41 4NP
GI4FUM BT41 4SB
GI4SIW BT41 4SB
GI7EXN BT41 4SB
GI8MIV BT41 4SB

NI: Ards

GI5SJ BT16 0UZ
GI3WFP BT19 7QB
GI1SYM BT21 0BN
GI7FNP BT21 0BQ
GI0SSA BT21 0DY
GI4MKJ BT21 0LL
GI3TLT BT22 1AS
GI7IKP BT22 1EJ
GI4XSF BT22 1HP
GI0UAG BT22 1JX
GI6IES BT22 1JY
GI4YPR BT22 1LT
GI7NMK BT22 2DP
GI0AQD BT22 2HS
GI4PBS BT22 2HY
GI4PBT BT22 2HY
GI1WGK BT22 2JG
GI4EQN BT22 2JP
GI4HCX BT22 2LA
GI7PIZ BT22 2NN
MI1BRO BT22 2NY
GI4NKK BT22 2PU
GI4PGN BT22 2QF
GI1TVH BT22 2RS
GI4BXB BT22 2RS
GI7IMU BT22 2ST
GI0PFL BT23 4BW
MI1BSK BT23 4PD
GI4MCW BT23 4SQ
GI1WLJ BT23 4TB
GI4AAM BT23 4TL
GI4PGH BT23 4TQ
GI4JTS BT23 4UW
GI7FGQ BT23 5HA
GI0SMU BT23 5HE
GI4TSK BT23 5JJ
GI4MBM BT23 5LP
GI6VLY BT23 5LT
GI3NYJ BT23 5QP
GI4OSG BT23 5RJ
GI0EEO BT23 5RN
GI3RKE BT23 5RN
GI1ZYY BT23 5SY
GI0TIE BT23 5TS
GI1IYO BT23 5YX
GI3XZM BT23 6EN
GI4MAC BT23 6HB
GI0VGV BT23 6SD
GI7UGP BT23 7AD
GI7HVC BT23 7AF
GI0LTT BT23 7BE
MI0ABD BT23 7BW
GI0OHT BT23 7BZ
GI7WJY BT23 7ED
GI4OYI BT23 7QP
GI3HNM BT23 8LZ
GI0HHZ BT23 8QT
GI0DUP BT23 8RT
GI7VCR BT23 8SE
MI1BKO BT23 8UF
GI3HXV BT23 8XA
GI0WPV BT23 8YE

NI: Armagh

GI7BET BT60 1AU
GI4FFL BT60 1BL
G0ZER BT60 1BL
GI0MSJ BT60 1DL
GI4XFE BT60 1DS
GI4RYP BT60 1HP
GI0RPS BT60 1LP
GI4SRQ BT60 1LP
GI7IIU BT60 1LP
MI0AJG BT60 1NT
GI8OCR BT60 1NT
GI6TBC BT60 1QR
GI0DWN BT60 2BN
GI0NYI BT60 2DZ
GI0MSI BT60 2JF
GI0KOW BT60 2NA
GI0MSH BT60 3AA
GI0NWG BT60 3JS
GI4ULE BT60 3NS
GI0VHG BT60 3TS
GI6NNP BT60 4AS
GI0GPG BT60 4BL
GI1VMF BT60 4BU
GI8RLG BT61 7DF
GI7DWF BT61 7EN
GI7IEZ BT61 7HZ
GI0LTF BT61 7QU
MI1CEG BT61 7RG
GI0MSG BT61 8DQ
GI3NSV BT61 8EZ
GI4NSV BT61 8EZ
GI0MTE BT61 8NP
GI0OND BT61 8NX
GI0VSG BT61 8QA
GI4SYM BT61 8RW
GI0MSK BT61 9BB
GI6UQU BT61 9BB
GI8RLE BT61 9LD
GI0ADD BT61 9LD
GI7GRY BT61 9LT
GI8RNG BT61 9QB
GI0EBT BT61 9QY
GI3CKF BT62 1SE
GI6DEO BT62 2BE
GI3WWY BT62 2EF
GI0DQJ BT62 2HJ
GI4BQI BT62 2LW
GI4IVI BT62 2NJ
GI3ONF BT62 2NJ
GI7LAA BT62 4HG
GI4MDO BT62 4HP
GI0VIB BT71 7HT

NI: Ballymena

GI1FWK BT39 9TJ
GI8TWB BT41 3DX
GI4VJC BT41 3RR
GI4DCC BT41 3RT
GI3FFF BT42 1DE
GI4HCN BT42 1DE
GI7PBE BT42 1EN
GI0JEV BT42 1JA
GI7NBB BT42 1JW
GI4GNZ BT42 1NR
GI6KBX BT42 1PU
GI7HYU BT42 1QP
GI0AYG BT42 2AU
GI0BGE BT42 2BT
GI4TOR BT42 2DG
GI4OGQ BT42 2DQ
GI4CRL BT42 2NU
GI4KUZ BT42 2RP
GI4LXL BT42 3BE
GI0OHG BT42 3LL
GI7HEW BT42 3LL
GI1XPV BT42 4AR
GI0SFX BT42 4DY
GI4SFZ BT43 5BY
GI4POV BT43 5BZ
GI4ESI BT43 5HE
GI0RBO BT43 5NP
GI3XDX BT43 5PR
GI0ITJ BT43 5PY
GI3TEN BT43 5PY
GI4OCV BT43 6EU
GI4VBZ BT43 6HE
GI6OJC BT43 6NF
GI4NNM BT43 6PB
GI0LMR BT43 6TA
GI4XFX BT44 8AD
GI6KKG BT44 8BX
GI1BSJ BT44 8EF
GI6VCL BT44 8EF

NI: Ballymoney

GI4SZU BT44 9DL
GI7AAH BT44 9HZ
MI0BBF BT44 9QA
GI6GIE BT53 6PZ
GI3OYG BT53 6QF
GI6GAQ BT53 7AH
GI0WYB BT53 7EU
GI8DHW BT53 7QL
GI4PMF BT53 7QN
GI4JRA BT53 8JT
GI8NBW BT53 8LD

NI: Banbridge

GI7VXC BT25 1BF
MI0BAT BT25 1DD
2I1AXG BT25 1PY
GI0UXD BT25 2EQ
GI7PWQ BT25 2PW
MI1BYJ BT25 2PW
GI0VVJ BT31 9BH
GI0USS BT32 3EJ
GI4GUH BT32 3EX
GI7WLS BT32 3HL
GI0WZW BT32 3XJ
GI0HXH BT32 4HF
GI7CBD BT32 4HN
GI3WEM BT32 4LF
GI7THC BT32 4QP
GI8FQB BT32 4RA
GI8RQI BT32 4RA
GI4GPC BT32 5NN
GI1JXE BT32 5RD
MI1AYL BT63 5AT
GI6MTL BT63 6JL

NI: Belfast

GI0JRI BT10 0AN
GI6JOP BT10 0AS
GI3TZF BT10 0AT
MI1BKR BT10 0AW
GI3DXU BT10 0DP
GI7MDP BT10 0HX
GI0JRD BT10 0JQ
GI1LGM BT11 8AL
GI0DPV BT11 8BX
GI1CET BT12 5RN
GI0VWU BT12 6GF
GI0APH BT12 6RD
GI0XBH BT12 7NE
GI0CDM BT13 3DZ
GI1HEK BT13 3HW
GI0ZAK BT13 3HW
GI4CFQ BT13 3LG
GI4PXM BT13 3LR
GI7TSA BT14 6AJ
GI4XFR BT14 6NX
GI0URN BT14 6NX
GI4OZI BT14 6RZ
GI3KOT BT14 6SD
GI4IKF BT14 6TE
GI4WXA BT14 7HW
GI7LQI BT14 8ET
MI0BDZ BT14 8FP
MI1BLA BT14 8FP
GI4EIZ BT14 8HD
GI7UBY BT14 8JN
GI4CUV BT14 8JX
GI7MBP BT14 8JY
GI4UIV BT14 8LD
GI0MXT BT15 3QP
GI7UPQ BT15 3RB
GI4OUC BT15 4AB
GI4CXH BT15 4AS
GI0XPE BT15 4BJ
GI1KDS BT15 4EP
GI7IFW BT15 4JU
GI7KMJ BT15 5AG
GI6EBY BT15 5FS
GI3RQU BT15 5GL
GI4VSC BT15 5HJ
GI4OTZ BT15 5HY
GI7GKC BT28 X
GI0USW BT36 7DZ
GI8SJZ BT4 1LU
GI0RBC BT4 1ND
MI0BDO BT4 1QT
GI7RAM BT4 1QT
GI4IOO BT4 1QU
GI0SRU BT4 2AQ
GI4JLF BT4 2BL
GI3TNK BT4 2BY
GI4ALM BT4 2EE

GI4GOS BT4 2EH
GI8NCX BT4 2EH
GI3SSR BT4 2GQ
GI4SFV BT4 2GT
GI7RAH BT4 2HH
GI4FZD BT4 2HS
MI1BJW BT4 2JR
GI4JDX BT4 2JT
GI3OJO BT4 2JX
GI4ILZ BT4 2JZ
GI4WRJ BT4 2PA
GI4FVM BT4 2WB
GI0VAB BT4 3DE
GI4RXS BT4 3EB
GI3MUS BT4 3ET
GI4NKB BT4 3NL
GI8RPP BT47 GHX
GI7BON BT5 4HT
GI0UTV BT5 4NN
GI4BJK BT5 5FR
GI6BDI BT5 5HS
GI4HQP BT5 5JS
MI1BOE BT5 5JY
GI3KYP BT5 5NT
GI8YTH BT5 5NT
GI4JNS BT5 6AB
GI4UKH BT5 6AE
GI3VHM BT5 6BE
GI4FNU BT5 6BT
GI7UQW BT5 6DD
GI7URC BT5 6FN
GI4MAY BT5 6FW
GI0PFB BT5 6JE
GI1HQU BT5 6JJ
GI8FYP BT5 6PL
GI3XLK BT5 6PU
GI0WCE BT5 6PU
GI8YYM BT5 7HL
GI7VPF BT5 7HR
GI0UQK BT5 7RH
MI1BRA BT6 0AE
GI7CMC BT6 0AE
2E1EGH BT6 0AE
GI4OWB BT6 0AS
GI4IBV BT6 0DG
GI4RCK BT6 8BH
M1BQL BT6 8ER
2I1CMM BT6 8ER
GI3LEG BT6 8NL
GI7UBV BT6 8QL
GI7LCC BT6 9AX
GI3XHL BT6 9DW
GI8HUD BT7 2GF
MI1AWU BT7 2GJ
GI1RWD BT7 2GZ
MI0AEC BT7 3JG
GI1KGZ BT9 5AQ
GI8VTK BT9 5AT
MI1CEF BT9 5BU
GI1MJJ BT9 5PG
GI4LQU BT9 5PQ
GI1HZX BT9 5QP
GI4WVN BT9 5QP
GI4XFS BT9 5QP
GI3ZCK BT9 6AP
GI8SKR BT9 6LJ
GI4ZIQ BT9 6NQ
GI4JIC BT9 6PP
GI3EVU BT9 6RA
GI3PSQ BT9 6TJ
GI4DAH BT9 6UG
GI3VCI BT9 6UH
GI0PEZ BT9 7GU
GI4DOM BT9 7JD

NI: Carrickfergus

GI0USC BT38 7ED
2I1ALE BT38 7JT
GI0UTE BT38 7JT
GI8LCJ BT38 7JT
GI7VIQ BT38 7JY
2I1EXU BT38 7LZ
GI3JRW BT38 7NF
GI3YRL BT38 7NG
GI4SBA BT38 7NG
GI7JEM BT38 7NG
GI1NAV BT38 7PJ
GI4DAV BT38 7QD
GI0JPR BT38 7SL
GI0LIX BT38 7SL
2I1DDF BT38 7SN
M1BQJ BT38 7UE
GI4GVS BT38 8BX
GI0IBC BT38 8BX
GI7VGR BT38 8DH
GI6ECV BT38 8EQ
GI0SSW BT38 8EU
GI6ANC BT38 8FN
GI8KYI BT38 8NA
2I0AHO BT38 8PW
GI0IJB BT38 8RD
GI0RUC BT38 8TX
GI0DFD BT38 9AP
MI0AWL BT38 9AP
GI8WBZ BT38 9AP
MI1AZS BT38 9BA
MI0AFT BT38 9DJ
GI6TFF BT38 9EA
GI0JPW BT38 9EB
GI4FUE BT38 9EP
GI4TEA BT38 9LJ
GI0TWX BT38 9LP
GI7PJF BT38 9LT
GI6IXD BT38 9LT
GI7WCS BT38 9LY
GI6GRV BT38 9ND
GI7POP BT38 9NQ
GI4XHO BT38 9NZ
GI1TFC BT38 9RL
GI0FCC BT38 9SB
GI3CWY BT38 9SP
MI1AYF BT38 9SU

NI: Castlereagh

GI3PKP BT16 0AW
GI4OCL BT16 0BB
GI4TUV BT16 0BE
GI0UZC BT16 0DB
MI1BBE BT16 0ET
GI0BMR BT16 0GW
GI7KHR BT16 0NT
GI4ORG BT16 0PH
GI3WHA BT16 0QP
GI6GDM BT16 0UT
GI4SAM BT16 0UU
GI8SJS BT16 0XU
GI7ULF BT16 0XW
GI6PJD BT16 0YS
GI0BRO BT5 6QA
GI0UEG BT5 6QA
GI4CBG BT5 7DH
GI3MMF BT5 7EQ
GI0OUM BT5 7EY
GI4ZOS BT5 7HW
GI7GXZ BT5 7LU
GI4VRF BT5 7LX
GI7GHM BT5 7LX
MI0ALS BT5 7LY
GI6NIY BT5 7LY
GI0BFO BT5 7LZ
GI3NQH BT5 7NG
GI0BDZ BT5 7NR
GI7KMC BT5 7NX
GI4MYT BT5 7PS
GI5TK BT5 7UB
GI7KEC BT6 0DZ
GI0DHW BT6 0ER
MI0AAW BT6 0FN
GI3XEQ BT6 0NA
GI0BFA BT6 0ND
GI1SMZ BT6 0ND
GI4FGH BT6 0NH
GI3YDH BT6 0NR
GI0KAN BT6 0NT
GI7TEU BT6 0NT
GI4LZR BT6 9JF
GI0YES BT6 9PJ
GI3NEB BT6 9RG
GI4GOV BT6 9RX
GI1BBG BT6 9RY
GI4MDD BT6 9SF
GI7JAM BT6 9SF
GI4WME BT8 4DB
GI7ILU BT8 4DG
GI4CRQ BT8 4DN
GI4XAP BT8 4DN
GI6IRL BT8 4EF
GI1WZA BT8 4HT
GI7IUH BT8 4HT
GI7JLD BT8 4LW
GI8PFB BT8 4LZ
GI3VGL BT8 4NH
GI1ZJF BT8 4NY
GI4JUA BT8 4PQ
GI3YMT BT8 4RG
GI1VAZ BT8 4RJ
GI4KIX BT8 4SB
GI4OMK BT8 4UJ
GI4VIP BT8 4UJ
GI4TAP BT8 8BQ
GI4TAV BT8 8HQ
GI1ZXM BT8 8JF
MI0BDX BT8 8NL
GI7LAI BT8 8QH
GI4JOR BT8 8QS
MI1AZL BT8 8RP
GI4SIP BT8 8TG

NI: Coleraine

GI4GPA BT51 3QD
GI4WWF BT51 3QN
MI1BNO BT51 4AR
GI7LCQ BT51 4BD
2I1FXD BT51 4HN
GI7SOB BT51 4NB
GI3JOZ BT51 4RA
GI4ZAH BT51 4RA
GI8OLH BT51 4TH
GI4HVI BT51 4TS
GI4GOW BT51 4TW
GI4MFM BT51 5AA
GI0MRI BT51 5AA
GI4VIZ BT51 5HB
GI6EIR BT51 5HP
MI1BGH BT51 5JP
GI0STC BT51 5SA
GI6JXG BT51 5TA
GI4JFP BT52 1EW
GI7TEB BT52 1JR
GI6IBL BT52 1JW
GI7ULG BT52 1LA
MI1BJX BT52 1NH
GI4NRQ BT52 2DR
GI4ORI BT52 2DR
MI0AAZ BT52 2HS
GI6WFI BT52 2LJ
GI7TTO BT52 2NY
GI0CMJ BT55 7HY
GI3TDY BT55 7JJ
GI4ORK BT55 7NG
GI4OYM BT56 8HB
GI3ONZ BT56 8HE
GI7TMQ BT56 8HN
GI7SLN BT56 8NJ
GI0AYB BT56 8PJ
GI0OTC BT56 8SP
GI3HHN BT57 8TT
GI0ISQ BT57 8UX
GI8AIR BT57 8YX

NI: Cookstown

GI3TIJ BT45 7NU
GI4NFW BT45 7QF
GI7JKM BT45 7TW
GI0WLW BT70 2QX
GI7TDA BT70 3DY
GI4XAA BT70 3JY
GI4STJ BT71 5AX
GI6NPF BT80 0HP
GI7THY BT80 8BN
GI1IIL BT80 8BQ
GI6UFS BT80 8BQ
GI6JCL BT80 8BW
GI8XSY BT80 8BW
MI1AOO BT80 8DX
GI7EBM BT80 8JA
GI3SOO BT80 8PL
GI6PME BT80 8PL
GI8YGG BT80 8PL
MI0AHH BT80 8PW
MI0AHI BT80 8PW
MI1CBX BT80 8PW
GI6JRY BT80 8QE
GI0JHR BT80 8QJ
MI1AVH BT80 8RL
GI6NDM BT80 8RS
GI8JRE BT80 8XW
MI0AIH BT80 9DD
GI6WHZ BT80 9RJ

NI: Craigavon

GI0RYU BT62 1EY
GI0NNM BT62 1JN
GI7FCP BT62 1JN
GI0HCJ BT62 1JW
GI0AIQ BT62 1JX
MI0AAH BT62 1RN
GI0UJG BT62 1RN
GI0GQA BT62 1TX
GI1PUM BT62 1UG
G7FTO BT62 3BQ
GI7UDV BT62 3ED
GI7IIJ BT62 3HA
2I1BIJ BT62 3HJ
GI4SJQ BT62 3QH
GI1RSR BT62 3QN
GI8XSB BT63 5AR
GI7VBS BT63 5AR
GI6BXY BT63 5DG
GI1ACN BT63 5DQ
MI1BJM BT63 5HD
GI0EEC BT63 5JW
GI0SRN BT63 5JW
GI1PXX BT63 5LP
GI6FTM BT63 5LT
GI7UGJ BT63 5RS
GI6NAQ BT63 5SB
GI4YCZ BT63 5SW
GI6EEH BT63 5TL
GI0PVG BT63 5YD
GI6FHD BT63 5YD
GI4MXW BT63 5YH
GI8JOA BT63 5YJ
GI0NOX BT64 3AF
GI0OXG BT65 4AB
GI6GAG BT65 4AB
GI3VFW BT65 4AH
GI4NKD BT65 4AH
GI0STS BT65 4AL
GI0SZH BT65 5DH
GI0PUZ BT65 5EB
GI0MHB BT66 6LD
GI6FQT BT66 6QW
GI7IUR BT66 6QY
GI4DRY BT66 7BA
GI4ISR BT66 7BH
GI3POS BT66 7EB
GI7MCY BT66 7GD
G7NUX BT66 7LQ
GI4YRP BT66 7PQ
2I1BJB BT66 7QL
GI4CYJ BT66 7RP
GI3WEL BT66 7RU
GI0SQV BT66 7RZ
GI4BDL BT66 7SU
2I1BHV BT66 7SU
GI3SCM BT66 8LE
GI0HDK BT66 8LF
GI7OOM BT66 8LW
GI6EWM BT66 8PF
GI4NMB BT67 0BH
GI8RKC BT67 0GF
GI8HXY BT67 0QD
GI5MV BT67 9AH
GI4ELQ BT67 9JN
GI0UTS BT67 9JU

NI: Down

GI3LFH BT24 7BB
GM8TSI BT24 7BX
GI4KCE BT24 7EW
GI1RXM BT24 7NP
G4SOY BT24 7NT
GI3LQY BT24 8
GI7CNS BT24 8NG
GI7CNT BT24 8NG
GI0WJI BT24 8PT
GI4MHD BT24 8QD
GI4AXV BT24 8QQ
GI6DCX BT24 8UN
GI0LCZ BT24 8YD
GI4SZP BT24 8YS
GI7TPX BT30 6JD
GI1XTK BT30 6NS
GI4NJQ BT30 6PZ
GI8UIU BT30 7DA
GI4RMA BT30 7LY
GI3BUP BT30 7QA
GI4JJD BT30 8SU
GI6OZI BT30 9JD
GI7FNJ BT31 9NH
GI4FCW BT33 0DG
GI3FJX BT33 0NQ

NI: Dungannon

GI8YPG BT70 1LA
GI4XKI BT70 1TH
GI7JPW BT70 1TJ
GI0RDM BT70 1UH
GI3OZW BT70 1UL
GI7FCW BT70 2NR
GI7EZF BT70 3BH
GI4LDN BT70 3LU
GI4LDO BT70 3LU
GI0STM BT70 3PU
GI0KPF BT71 4AP
GI0BDT BT71 4DU
GI6OHA BT71 4NG
GI0IMY BT71 4QS
GI0URI BT71 6DD
GI3NFM BT71 6DR
GI4TBV BT71 6HH
GI4CYK BT71 6LG
GI4CSP BT71 6PS
GI0OKQ BT71 6PZ
GI3NPP BT71 6QG
GI3OQR BT71 6QN
GI7NEB BT71 7DA
GI7KPJ BT71 7EN
GI8ITD BT71 7JN
GI4TED BT71 7QF
GI4XJD BT71 7QF
GI4SLQ BT71 7SJ
GI4CYU BT71 7SY
GI0EWE BT76 0HJ
MI1AWE BT76 0HJ
GI4CQL BT76 0XE

NI: Fermanagh

GI0BFD BT74 5LT
GI4CZW BT74 5NQ
GI1BEW BT74 6BN
GI0BQX BT74 6EP
GI1RBI BT74 6JH
GI8LDM BT74 6JL
GI4PCY BT74 6NG
GI1DTH BT74 6NG
GI6PYP BT74 7FA
GI0LEC BT74 7JN
GI4UHA BT74 7JN
GI4UHP BT75 0SL
GI7UJI BT92 0DW
GI0CAH BT92 3BU
GI7UIM BT92 7DD
GI7JGT BT92 7QY
GI7UIP BT92 8AX
GI1SLZ BT92 9GE
GI0EZS BT93 1TF
GI4SZQ BT93 1UH
GI3GRD BT93 1UN
GI0PTQ BT93 3FT
GI3WBR BT93 8DF
GI6RMO BT94 5BX

NI: Larne

GI8KFG BT38 9HE
G0OZF BT38 9HH
GI4MAJ BT38 9JL
GI4IZF BT38 9LF
GI4JYJ BT39 9QW
GI4UUC BT39 9QW
GI6UUC BT39 9QW
GI6ISQ BT39 9ST
2I1EZB BT40 1EE
GI7USA BT40 1HE
GI4CPP BT40 1NJ
GI6DFU BT40 1PS
GI6EWO BT40 1QL
GI6VCG BT40 1UB
GI7DZE BT40 1UL
GI8WHP BT40 1UZ
GI4MVQ BT40 2EG
GI4RYD BT40 2EN
GI3OAU BT40 2HR
GI7GJX BT40 2HX
G0ACO BT40 2JX
GI6BPF BT40 2QP
GI7MDK BT40 3HL
GI4UPC BT40 3JG
GI4OYE BT40 3RX
GI7NOW BT40 3SF
GI4RVT BT40 3SQ
GI0OTL BT40 3TR
GI4MTZ BT40 3TT
GI7NIL BT40 3UG
GI0XAC BT40 3UG

NI: Limavady

GI0EFW BT47 4PJ
GI0AZA BT47 4PN
GI0AZB BT47 4PN
GI3UYZ BT47 4TW
GI1JJC BT49 0AR
GI0AHZ BT49 0AR
GI1XBI BT49 0AR
GI4TPN BT49 0DN
GI4DQO BT49 0NT
GI4PMP BT49 0NT
GI0RJO BT49 0PP
GI0BZM BT49 0QF
GI3KVD BT49 0QF
GI1WFA BT49 0QJ
GI4SNC BT49 0RD
GI0RBS BT49 0RP
GI0EWP BT49 0SA
GI4EBS BT49 0SD
GI3VAW BT49 0SF
GI4SIZ BT49 9BA
GI4UXW BT49 9BQ
GI6VWS BT49 9BS
GI4GNT BT49 9HQ

NI: Lisburn

GI4CSO BT17 0JG
GI1JQP BT17 9BB
GI4RNP BT17 9EU
GI3VQ BT17 9LT
GI4TUJ BT24 8LB
GI6ATZ BT24 8LF
MI0BEI BT24 8TS
GI7BDJ BT25 2AQ
G8WTR BT26 6AG
GI7DBN BT26 6AQ
GI4KSO BT26 6BH
GI4OZJ BT26 6BX
GI4AHP BT26 6EF
GI3VPV BT26 6ES
GI1CDZ BT26 6HB
GI6UFU BT26 6HB
MI1BRS BT26 6HQ
GI7NFB BT26 6HU
GI4TCR BT26 6LJ
GI4TCS BT26 6LJ
GI6RTB BT26 6LJ
GI1DAO BT26 6NE
GI3KDR BT26 6NS
GI6KJC BT26 6NS
GI4XLB BT26 6NU
GI4RXM BT26 6PW
GI7TFK BT27 4BH
GI4NSS BT27 4BL
GI4NFH BT27 4BS
GI0WYJ BT27 4BT
GI6ETQ BT27 4EF
GI0TSS BT27 4HE
GI6WFX BT27 4JA
GI6KMA BT27 4LD
GI4NTO BT27 4LU
GI4ERM BT27 4PS
GI4NLQ BT27 4QA
GI8PGJ BT27 4QX
GI3JCD BT27 4UB
GI6FOR BT27 5BT
GI4XIR BT27 5DB
2I1CXY BT27 5DB
GI4NRB BT27 5DH
GI7PBQ BT27 5LQ
GI4JER BT27 5PQ
2I1AXH BT27 5RF
GI0NQC BT28 1AU
GI0WOW BT28 1AU
GI0RDJ BT28 1EX
GI0TJV BT28 1EY
GI0RYK BT28 1LD
GI0PGC BT28 1YJ
GI4GID BT28 2DN
GI7ALP BT28 2EY
GI7ALQ BT28 2EY
MI1BYL BT28 2UZ
GI4PES BT28 2XU
GI0TJU BT28 3AH
GI0TIU BT28 3BQ
MI1CAS BT28 3DN
GI4XTC BT28 3DS
GI4LKG BT28 3HG
GI4MEQ BT28 3JH
GI0TJJ BT28 3JP
GI8AYZ BT28 3JS
MI0AYZ BT28 3JS
GI6MZL BT28 3LA
GI6GNA BT28 3LP
GI4RKC BT28 3QB
MI0ASV BT28 3QX
MI0ABN BT28 3QY
GI4SNA BT28 3RE
GI0DVU BT28 3RR
GI4GTY BT28 3RR
GI1YSG BT28 3TQ
GI4MWA BT29 4DS
GI0AIJ BT29 4JQ
GI4TTL BT67 0DS
GI7VCS BT67 0HA
GI0OKU BT67 0JB
GI4MNF BT67 0JB
GI0OEH BT67 0LB
GI3NUM BT67 0LW
GI1FSJ BT67 0NL
GI6CMA BT67 0PY
GI4TBP BT8 8JZ
GI4TGR BT8 8JZ
GI8PDK BT8 8LX

NI: Londonderry

GI0TSA BT47 1FE
GI0AWK BT47 1JP
MI0AQV BT47 1NF
GI0IOT BT47 1QY
GI4XCO BT47 1QY
GI0WGS BT47 1QY
GI0IUP BT47 1UG
GI4OWA BT47 1XS
GI4ZFA BT47 1YA
GI4FHB BT47 1YD
GI4SSF BT47 1YN
GI4YWT BT47 2HW
GI3FTT BT47 2NL
MI0AKU BT47 2QL
GI0WYO BT47 2QL
GI1XGA BT47 2RD
GI4XGQ BT47 3JN
GI0NCA BT47 3JW
GI6BVQ BT47 3PR
GI4AHD BT47 3SW
GI0LWO BT47 3YJ
GI4EQW BT48 0AA
GI4MJD BT48 0AD
GI3TJJ BT48 0AU
MI1CCI BT48 0BD
GI0NWN BT48 0BY
GI6MYQ BT48 0DH
GI4ZLD BT48 0QA
GI0KOV BT48 0QN
GI7MAC BT48 0RU
GI0PXS BT48 7ER
GI4DGI BT48 7JN
GI3TZC BT48 7PJ
GI6DSH BT48 7PJ
GI3KSY BT48 7PN
GI3HXH BT48 7RS
GI0BSO BT48 7SY
GI0DSG BT48 8BA
GI0LRB BT48 8EF
GI0LEK BT48 8HA
GI7KVR BT48 8JD
G0SFT BT48 8JD
GI0IIZ BT48 8JD
2I1DNO BT48 8JD
GI3GGY BT48 8JL
GI1BEU BT48 8JW
GI0TQD BT48 8JW
GI7TOX BT48 8PB
GI4EPK BT48 8PF
GI3TME BT48 8PR
GI0OZQ BT48 9DU
G7UPU BT48 9LA

NI: Magherafelt

GI4LVC BT45 5DR
GI4OMO BT45 5JD
GI0EUG BT45 5JF
GI7CTW BT45 5LY
GI0SRP BT45 5QA
GI4NGP BT45 5QG
GI4FME BT45 5RH
GI0CNI BT45 6BG
GI7MWA BT45 6EX
GI4WNH BT45 6NH
GI0PJH BT45 6PU
GI4EQA BT45 6PY
GI4BGB BT45 7LY
GI3RXV BT45 8AA
GI4BDR BT45 8LW
GI6OCC BT45 8NQ
GI3ZTL BT46 5JR

NI: Moyle

GI4PCQ BT44 0NS
GI8WIU BT54 6DS
MI1BLZ BT54 6RT

NI: Newry and Mourne

GI4SPT BT34 1HL
GI0LRZ BT34 1NZ
MI0AVI BT34 1QN
GI0KVQ BT34 1QN
MI0AQX BT34 2BQ
GI6JJR BT34 2BQ
MI0BES BT34 2BQ
GI0THR BT34 2ND
GI0WAH BT34 2NY
GI0UYY BT34 2PG
GI1YEA BT34 2PG
GI8YJF BT34 3DX
GI4MBQ BT34 3HL
GI0PED BT34 3PG
GI0VVC BT34 3PG
GI0VVB BT34 3PG
GI4BBE BT34 4AQ
GI4UWT BT34 4AR
GI7IVX BT34 4DA
MI1BYV BT34 4JF
GI4XFY BT34 4LP
GI6EBX BT34 4LX
GI3HJH BT34 4NU
GI4WAH BT34 5TJ
GI0VWK BT35 0AL
GI0VGL BT35 0PJ
GI6FXY BT35 6EH
GI4OVE BT35 6NA
GI4NEZ BT35 6PF
GI0JGQ BT35 7EH
GI4SZW BT35 7HD
GI3UOY BT35 8EW
GI6VSK BT35 8PD
GI0CTI BT35 8PN

NI: Newtownabbey

GI0PGU BT36 4PA
GI4OXO BT36 4QT
GI4BUJ BT36 4TP
GI4FHD BT36 4TP
GI4BTG BT36 4WL
GI4SQL BT36 4WQ
GI4VCZ BT36 4ZN
GI7WET BT36 5AZ
GI7MSJ BT36 5BQ
GI7FCM BT36 5DE
2I1AJA BT36 5DJ
MI1AUI BT36 5FG
GI7AQO BT36 5GD
2I1FVW BT36 5JJ
GI7MSY BT36 5JP
GI7JYK BT36 5JU
GI1WYZ BT36 5JZ
GI4RXX BT36 5LP
GI0NMV BT36 5NQ
GI4XJJ BT36 5SJ
GI0ABH BT36 5ZZ
GI0RJU BT36 6AW
GI4OTG BT36 6BA
GI8GJX BT36 6ET
GI1CKU BT36 6HW
GI4RFH BT36 6LA
GI6OIR BT36 6LA
GI4KSH BT36 6LS
GI0ACE BT36 6NL
GI2BZV BT36 6PB
GI0BEB BT36 6QQ
GI7DQI BT36 6RL
GI8RPT BT36 6SP
GI1PHF BT36 6TN
GI4SZZ BT36 6TS
GI8LUR BT36 6TZ
GI7UFX BT36 7UJ
GI4OYL BT36 7UT
GI6VJI BT36 7XA
GI4MCH BT37 0AZ
GI6KVS BT37 0EL
2I1BXH BT37 0FA
GI0XBK BT37 0FA
GI7TDF BT37 0FA
2I1EXO BT37 0FA
GI0BEY BT37 0JN
GI1XLK BT37 0LH
GI0IQA BT37 0RR
GI0BEG BT37 0TS
GI4KBW BT37 0UL
GI4SZY BT37 0XE
GI4SFE BT37 0XH
GI4VWC BT37 0XY
GI4OHI BT37 9EQ
GI0BDU BT37 9HN
G1WNL BT37 9NQ
GI4RYL BT37 9PD
GI0VGQ BT37 9SE
GI0WOO BT37 9SE
GI8UCS BT37 9TJ
GI0GCR BT39 0PT
GI4PRH BT39 0QB
GI6PAZ BT39 0QB
GI8MOV BT39 0QB
GI4JXM BT39 0RJ
GI7JHA BT39 0RY
GI4NKE BT39 0SX
GI7GUT BT39 9HT
GI4TAJ BT39 9HZ
GI4XGO BT39 9HZ
GI3VYY BT39 9JH
GI7DBZ BT39 9JS
GI7GSH BT39 9PS
GI7GVI BT39 9PS
GI7REP BT39 9RH
GI8VKA BT39 9TS

NI: North Down

GI4OVN BT18 0BZ
GI3GTR BT18 0DX
GI4SWS BT18 0DX
MI0AEX BT18 0EY
GI7UQY BT18 0EY
GI0TDP BT18 0HG
GI4JTF BT18 0HH
GI6ETD BT18 0LL
GI6XOV BT18 0NQ
GI4ZTU BT18 0PJ
GI3USK BT18 0PL
GI4NAT BT18 9DY
GI3WUO BT18 9EL
GI0USQ BT18 9EU
GI4NKZ BT18 9LY
GI8YJV BT18 9NB
GI3WSS BT18 9PN
GI7TTJ BT18 9QB
GI4MUE BT19 1AA
GI4MRZ BT19 1AD
GI6IHX BT19 1AE
GI3USS BT19 1BJ
GI3XRQ BT19 1BJ
GI6YM BT19 1BJ
GI6YMC BT19 1BJ
GI6ATD BT19 1DQ
GI3ZKT BT19 1DX
GI4NBO BT19 1ED
GI0EZD BT19 1EG
GI0USB BT19 1EQ
GI7OKJ BT19 1FG
MI1BVO BT19 1HQ
GI8DGB BT19 1LN
GI8YJD BT19 1LP
GI6BNI BT19 1LU
GI4OCK BT19 1NT
GI0VFT BT19 1QJ
GI4BKR BT19 1SN
GI6IHM BT19 1YG
GI3TZB BT19 6AE
GI4TPY BT19 6AF
GI7DIT BT19 6AF
GI3OTU BT19 6AR
GI1VPA BT19 6AY
GI3MBB BT19 6BA
GI0GTJ BT19 6DG
GI6BDN BT19 6DG
GI0VTS BT19 6DJ
GI0PFK BT19 6DT
GI7VIW BT19 6DT
GI3YMY BT19 6DU
GI4SVO BT19 6DY
GI3MMG BT19 6FN
GI1SZC BT19 6NX
GI3VAF BT19 6SD
GI7ONC BT19 6SG
MI1ARY BT19 6XA
GI6JMD BT19 6ZB
GI0FZT BT19 6ZB
GI6PLO BT19 6ZW
GI4VIV BT19 7HF
GI7FFF BT19 7HQ
GI4WYE BT19 7RB
GI4EEB BT19 7XY
GI4NAE BT20 3DA
GI8IJF BT20 3DU
GI4LZS BT20 3EP
GI4JJF BT20 3ER
GI0HSB BT20 3HA
GI0POB BT20 3JD
GI3PQW BT20 3LU
GI3TZX BT20 3PP
GI4TMB BT20 3QD
GI4POC BT20 3RS
GI4NKQ BT20 3SB
GI8VLB BT20 3SD
GI3TJM BT20 3SL
GI7JEB BT20 3TP
GM7RHP BT20 3TW
GI6IVJ BT20 4HS
GI3SHI BT20 4NQ
GI0HHV BT20 4PE
GI7ISX BT20 4PX
GI2AFW BT20 4QS
GI6JGB BT20 4RD
GI2FHN BT20 4RJ
GI3SUM BT20 4RQ
GI3HCP BT20 4TL
GI4MRN BT20 4TX
GI7SAH BT20 4XF
GI0JQH BT20 5ER
GI8ZHW BT20 5NU
GI3UBA BT20 5PN
GI0WVN BT20 5QP
GI4OPH BT20 5QZ
GI4TPI BT20 5RQ
GI8FLQ BT23 7PF

NI: Omagh

GI6PBV BT78 1LD
GI4OHW BT78 1QZ
GI4SXV BT78 1UG
GI0HGP BT78 2AU
GI0LXN BT78 2DL
GI7JHV BT78 4QY
GI4RSI BT79 0EB
GI3NVW BT79 0HF
GI0GQG BT79 0LF
GI4UDI BT79 0PE
GI0DES BT79 0PY
GI3XCZ BT79 0XS
GI6ECD BT79 7DB
GI4LIF BT79 7JJ
GI7RTB BT79 7LA
GI0EJT BT79 7PQ
GI7AUY BT79 7SJ
GI7FHZ BT79 7SJ
GI8BNC BT79 7SS
GI3YBZ BT79 7SS
GI7GAQ BT79 7TB
GI0EJU BT79 7TJ
GI8TST BT79 7XD
GI3HJA BT79 8NH
GI3ZTG BT79 9NL
GI4IWP BT79 9QA

NI: Strabane

GI1CAI BT78 4HX
GI4KJC BT78 4JT
GI0HVJ BT78 4JZ
GI0TMS BT78 4LG
GI4NMZ BT82 0AZ
GI4DXK BT82 0BD
GI4OUP BT82 0DB
GI3CFH BT82 0DP
GI4OUN BT82 0DP
MI1BSJ BT82 0DZ
GI7THH BT82 8EL
GI7VAF BT82 8EL
GI3CSK BT82 8HQ
GI4VQK BT82 8LD
GI4VKS BT82 8NW
GI4ZOJ BT82 8PW
GI0VJE BT82 9DF
GI7FJY BT82 9DJ
GI6NDZ BT82 9HZ
M1AWM BT82 9LA
MI1BMO BT82 9PG

Neath Port Talbot

GW0BBO SA10 6DG
GW0KZK SA10 6DG
GW0CBL SA10 6DS
GW0HPC SA10 6DS
GW0VSW SA10 6DZ
GW4UVN SA10 6EA
GW3CDP SA10 6EG
GW3NDR SA10 6LD
GW6KRK SA10 6SD
GW6KRQ SA10 6UP
GW8DOA SA10 7DD
GW7IBT SA10 7RX
GW0CYG SA10 7UG
GW4FOI SA10 8BG
GW4WWN SA10 8BN
GW1SGH SA10 8DW
GW8VCA SA10 8HD
GW4FXF SA10 8HT
GW4IUK SA11 1DJ
GW0VFF SA11 1JL
2W1DRP SA11 1PP
MW0AES SA11 1YT
GW6NXH SA11 2JU
GW4PCO SA11 2TE
GW1EWW SA11 3AL
GW3WWN SA11 3HX
GW0NKJ SA11 3JW
GW2ABJ SA11 3NJ
GW0HNT SA11 3PJ
GW3EOP SA11 3TR
GW4NZ SA11 3TR
GW3VWT SA11 3TU
GW0TWR SA11 3YQ
GW7JHK SA11 4AA
GW7UAE SA11 4AH
GW0PEA SA11 4DH
GW7TZG SA11 4HS
G4SRE SA11 5BG
GW4UYT SA11 5BG
GW4WYX SA11 5NT
GW3RHC SA11 5TE
GW6EOL SA12 6DF
GW4RKI SA12 6PH
GW4BVN SA12 6SU
GW3XHV SA12 7DE
GW4RML SA12 7DE
GW3VMX SA12 7HS
GW3OPU SA12 7LT
GW3KUY SA12 7RP
GW8JZV SA12 8AL
GW0RLQ SA12 8AP
GW4XWC SA12 8AS
GW8EHK SA12 8ER
2W0ACD SA12 8EU
MW0AHG SA12 8EU
GW4TVQ SA12 8LE
GW4YBE SA12 8LS
GW0RCG SA12 8PP
GW8RBJ SA12 8RF
GW8VHI SA12 8TR
GW3TKG SA12 8YE
GW8JQW SA12 8YE
GW0DJU SA12 9BS
GW6JUL SA12 9NH
GW7GCD SA12 9PD
GW7OTI SA12 9PU
GW6UHY SA12 9TP
GW3SCX SA12 9YE
GW4WAN SA13 1DQ
MW1BAJ SA13 1DQ
GW0KTE SA13 1HA
GW4DOO SA13 1HA
GW0TWF SA13 1SG
GW3HDH SA13 1TA
GW2FRB SA13 1TG
GW4PRP SA13 1YD
MW1BSH SA13 1YG
GW0OLP SA13 2AP
MW0BFD SA13 2EL
GW7LHI SA13 2HL
GW3NKM SA13 2LB
GW0KPD SA13 2LF
GW0NKF SA13 2LF
GW3JUN SA13 2LP
GW3XHD SA13 2US
GW3TMJ SA13 2YE
GW4RDW SA13 3AN
GW0RTP SA13 3EF
GW4TVU SA13 3SD
GW0MHY SA18 1HB
GW0HNE SA18 1PE
GW0BNH SA18 1PT
GW0EJP SA8 3BA
GW4JPN SA8 3EP
GW8SBN SA8 3EP
GW8TBG SA8 3EP
GW6VBN SA8 4AL
GW6UGC SA8 4DB
GW4KYT SA8 4DL
GW7TDK SA8 4LP
GW3UCJ SA8 4QL
GW1IRL SA8 4QU
GW0JZQ SA9 2AE
GW3RYE SA9 2NJ

New Forest

G0JIL BH23 8BS
G0RER BH23 8BS
G8RXA BH23 8BS
2E1AWF BH23 8BS
G8YBZ BH23 8BS
G7RDT BH23 8BS
G4OXK BH23 8BT
G1BLO BH23 8BU
G6WQU BH23 8BW
G0ROF BH23 8DQ
G3PIY BH23 8DU
G3SGL BH23 8DU
G1IOP BH23 8HG
G8JEM BH23 8HG
G0HJL BH23 8NA
M0AUY BH23 8NH
G6DTX BH24 1NY
G6JAR BH24 1PF
G4ZPW BH24 1PQ
G8BDF BH24 1PX
G3OMY BH24 1QU
G3OUI BH24 1RP
G3GPB BH24 1RZ
G3BDY BH24 1SX
G1MLH BH24 1TF
G6PLU BH24 1XD
G1LOU BH24 1XL
G3FYS BH24 1XW
G3ZCI BH24 3DT
G6PBO BH24 3DT
G3MDH BH24 3DX
G1OCH BH24 3ER
G0MYL BH24 3HT
G6NCU BH24 3LJ
G1WSN BH24 3NQ
G3SZA BH24 3RG
G4GQJ BH24 4DA
G4GPI BH24 4HN
G4LUN BH25 5BA
G0DIJ BH25 5BY
G7UCU BH25 5DA
G8REZ BH25 5EX
M1BSC BH25 5GA

Callsign	Postcode
G1ZEC	BH25 5JP
G1ZVC	BH25 5JP
M1BAW	BH25 5JP
2E1DID	BH25 5JP
G8CON	BH25 5JP
G0RIX	BH25 5NA
G4MVP	BH25 5NQ
G1HHO	BH25 5PE
G4VJJ	BH25 5RX
G4DUW	BH25 5SD
2E1DUW	BH25 5SD
G0HGG	BH25 5TB
G0AHK	BH25 5UF
G1JAF	BH25 5UG
G0RNT	BH25 5UT
G1IAV	BH25 5XP
G4HFQ	BH25 6EY
G6LVC	BH25 6NZ
G8WQW	BH25 6NZ
G1UWV	BH25 6QE
G0PZK	BH25 7DF
G8BBN	BH25 7DF
G6IZQ	BH25 7DQ
G7TEW	BH25 7DT
G3BFC	BH25 7DU
G2HCG	BH25 7DX
G7ILD	BH25 7ET
G8BKE	BH25 7HR
G3DPR	BH25 7NA
G0KNF	BH25 7PY
G7IUI	BH25 7PZ
G3ERO	BH25 7TH
G0COC	SO40 2LJ
G8IEM	SO40 2NL
G6ANI	SO40 2NW
G7EOH	SO40 2SD
G6XYF	SO40 3BN
G4KCM	SO40 3GD
G0BOC	SO40 3HP
G0OSD	SO40 3HP
G3PSM	SO40 3HY
G4JYL	SO40 3HY
G3NSB	SO40 3LG
G6FYZ	SO40 3LL
G4VII	SO40 3LP
G0LDJ	SO40 3LZ
G1MVV	SO40 3NE
G0DRG	SO40 3PB
G6DIA	SO40 3QB
G4IXE	SO40 4UT
G6WIM	SO40 4WT
G8YWH	SO40 4XH
G7LBP	SO40 4YG
G6VET	SO40 4YR
G3VSL	SO40 7AL
G6GFA	SO40 7AL
G4ZUQ	SO40 7BW
2E1CZA	SO40 7GE
G8IPG	SO40 7GF
G8JCB	SO40 7GH
G0VNI	SO40 7LA
G0WIL	SO40 7LA
G1HGR	SO40 7PT
G7ESG	SO40 8DX
G0WSL	SO40 8ED
G6WTM	SO40 8GR
G6MCX	SO40 8SR
G1LGY	SO40 8US
M1BGW	SO40 9EN
G1SCX	SO40 9FU
G4BMC	SO41 0FP
G4TOG	SO41 0GG
G4GBZ	SO41 0GP
G1RAX	SO41 0HQ
G0RZD	SO41 0JS
G0TMZ	SO41 0LG
G4RFP	SO41 0LQ
G7JWV	SO41 0PB
G3ZJY	SO41 0TF
G4HYW	SO41 3AR
G0OWG	SO41 3BA
G3VTP	SO41 3NW
M0AYC	SO41 3PE
G7TSN	SO41 3PE
G3AAK	SO41 3PN
G4ZAX	SO41 5RX
G0ORW	SO41 5SA
G1ZYS	SO41 6BB
G0SNK	SO41 6BQ
G4XNA	SO41 6DY
G0AXI	SO41 8AA
G1KFH	SO41 8BD
G0BCS	SO41 8DY
G3DEB	SO41 8DZ
G0TXN	SO41 8ER
G0VLW	SO41 8EW
G8KTV	SO41 8FY
G2AIV	SO41 8GQ
G7EYT	SO41 8HF
G0RWV	SO41 8JN
G7LMY	SO41 8LL
G4PGW	SO41 9BB
G6FRL	SO41 9GA
G0JZW	SO41 9LB
GW4SBB	SO42 7RX
G4FDX	SO42 7TT
G7TQX	SO42 7UH
G3GHY	SO42 7WL
G4BIZ	SO43 7BR
G4DTI	SO43 7EF
2E1FRX	SO43 7HJ
G3NOA	SO43 7JB
G7JTR	SO43 7JB
G0HDI	SO45 1BJ
G0JYL	SO45 1BL
G0EUJ	SO45 1DA
G0UUW	SO45 1DJ
G6LVJ	SO45 1EG
G4CTW	SO45 1FR
G7AYA	SO45 1FZ
G6NXV	SO45 1WF
G0OSG	SO45 1WX
G1UEV	SO45 1WY
G1ITR	SO45 1XD
G6XWZ	SO45 1XD
G3YJJ	SO45 1XU
G6MEE	SO45 1YL
G3WJM	SO45 1ZA
G1RXB	SO45 1ZL
G0SBV	SO45 2HP
G6MNL	SO45 2JT
G6XMA	SO45 2JT
G6GWI	SO45 2LE
G6LNP	SO45 2LN
G3TPV	SO45 2NN
G4JYN	SO45 2NQ
G4YVY	SO45 2NQ
G0BPA	SO45 2NY
G4SBF	SO45 2PP
G0OFP	SO45 2QD
G0GMK	SO45 2QH
G0CVS	SO45 3JW
G4SGJ	SO45 3NB
G0DFN	SO45 4HG
G3OZT	SO45 4HS
G0WCB	SO45 4LE
G3UJI	SO45 4NN
G7LWJ	SO45 4PB
G3AMO	SO45 4PJ
G3KCD	SO45 4RP
G3JLS	SO45 5AQ
G3OUP	SO45 5AQ
G0OSP	SO45 5AU
G8APM	SO45 5BP
G4KNN	SO45 5BS
G1JRU	SO45 5DL
G3OZV	SO45 5FB
G8XRE	SO45 5QY
G7POA	SO45 5ST
G0LKG	SO45 5TG
G1RCW	SO45 5UD
G4JAX	SO45 6AY
G6MCO	SO45 6AY
GW4MOL	SO45 6DL
G8BMQ	SO45 6DW
2E1FRY	SP6 1BW
G0VUU	SP6 1DJ
G6CEZ	SP6 1EQ
G0KXN	SP6 1LH
G8OFO	SP6 1LR
G4POF	SP6 1LW
G8ZME	SP6 1LZ
G7UHE	SP6 1PN
G0PLK	SP6 2AL
G4LTR	SP6 2HT
G6NZB	SP6 2LG
G1MVG	SP6 2LJ
G4YXC	SP6 2LR
2E1DQD	SP6 2NJ
G3SKF	SP6 2QB
G4IBC	SP6 2QB
G0VJD	SP6 2QJ
G4FPP	SP6 2QU
G1HQG	SP6 3EE
G7VOX	SP6 3EE
G1VOQ	SP6 3HA

Newark and Sherwood

Callsign	Postcode
G3OBZ	NG14 5GS
2E1FQO	NG14 7AP
2E1FQP	NG14 7AP
G8GVG	NG14 7AR
G3YKP	NG14 7DE
G8SER	NG14 7EA
G0GDU	NG14 7FW
G4LAV	NG14 7FW
2E1FOR	NG14 7FW
M1APK	NG21 0AH
G4FHK	NG21 0AR
G7KWA	NG21 0AU
G1MMD	NG21 0EU
G1WRO	NG21 0FF
G7BXG	NG21 0JZ
G8KPV	NG21 0NW
G7SVU	NG21 0RN
G7ODW	NG21 0SJ
G1DSC	NG21 0TR
G7JJZ	NG21 0TR
2E1FAL	NG21 0TR
2E1FCN	NG21 0TR
G0PBW	NG21 0TU
G0JVK	NG21 9AL
G0CCV	NG21 9EA
G4WHL	NG21 9LQ
G4WHM	NG21 9LQ
G4XTL	NG21 9NH
G0TYP	NG21 9NR
2E1FFP	NG21 9PW
G1XAV	NG21 9QG
M1ALU	NG21 9QG
M1AJZ	NG21 9QT
M1ANR	NG21 9QT
G0BYX	NG21 9QZ
G0XBU	NG22 0DE
G4ODD	NG22 0HF
G4SCL	NG22 8AP
G4HTO	NG22 8BX
G7LAK	NG22 8DN
G7OPW	NG22 8DN
G8ITU	NG22 8EA
G6FLQ	NG22 8PB
G1YIC	NG22 8PY
G3ZVV	NG22 8QF
G7NDS	NG22 8SJ
2E1FEU	NG22 9AS
M0AXX	NG22 9DG
G4GFT	NG22 9DG
G7JFS	NG22 9HE
G7PRB	NG22 9HJ
G0CSM	NG22 9LR
G4EKL	NG22 9LW
G6EMH	NG22 9RG
2E1DQF	NG22 9RW
M0ADB	NG22 9RZ
2E0AMU	NG22 9SX
G3SBW	NG22 9TH
G0BYQ	NG22 9TJ
G4BZV	NG22 9TJ
G6VGN	NG22 9TJ
M0BDD	NG22 9UU
G4CCB	NG22 9UZ
2E1FRZ	NG22 9UZ
G7DEX	NG22 9XQ
G3MUL	NG23 5NF
G4PCP	NG23 5RP
G1LVH	NG23 5SG
G4CPJ	NG23 6HB
G8DYK	NG23 6JN
G8TNB	NG23 6ST
G0DWB	NG23 7LD
G4KOA	NG23 7NL
G4LGI	NG23 7PQ
G3HLG	NG23 7RA
G7LPK	NG24 1NG
G0HJR	NG24 1NL
G7PQT	NG24 1SH
G4CUO	NG24 2BA
G7UXR	NG24 2BU
G7MUB	NG24 2HT
G1FJH	NG24 2LH
G3YWS	NG24 2NL
G0STT	NG24 3AF
G4YZG	NG24 3DR
G0BVU	NG24 3LY
G8EGU	NG24 3NS
G0LBB	NG24 3NZ
G0DNL	NG24 3PH
G7HNO	NG24 3RE
G7DEH	NG24 4BZ
G4ZHG	NG24 4DY
2E0ANO	NG24 4DZ
G0MCP	NG24 4LU
G7HOT	NG24 4LU
G4APM	NG24 4NE
G3FCB	NG24 4QN
G4YSL	NG24 4SL
G0PUY	NG24 4UA
G4TYI	NG25 0EN
G4CVA	NG25 0HP
G2FZU	NG25 0QP
G4PLA	NG25 0RQ
G4ZUU	NG25 0XL

Newbury

Callsign	Postcode
G4YME	OX12 9NJ
G4CLF	RG14 1BD
G0CKU	RG14 1LL
G7CXV	RG14 1LL
G4KDB	RG14 1NN
G5XV	RG14 1QL
G0HFU	RG14 1TP
G6OEW	RG14 2HA
G6XZS	RG14 2HA
G6PNG	RG14 2LL
G8NHG	RG14 2LL
G1APC	RG14 2LZ
G3OUC	RG14 2ND
G3FCK	RG14 2NF
G6RBP	RG14 2PN
G7RJW	RG14 2PN
2E1CUS	RG14 3BQ
G1FBU	RG14 5JJ
G4RUW	RG14 5JP
G7SLL	RG14 5NR
G8LZU	RG14 6AJ
G4MKF	RG14 6DN
G8NTQ	RG14 6DW
G3MWB	RG14 6ET
G0HBJ	RG14 6HP
G6ZSF	RG14 6HP
G2CXT	RG14 6HT
G0VFE	RG14 6JG
G4WEV	RG14 6JX
G6IDA	RG14 6JX
G8JUS	RG14 6NY
G3RQP	RG14 6PX
G8KHU	RG14 6PY
G3ZGC	RG14 6RU
G1WTS	RG14 6RY
G8AKM	RG14 6RY
G8AYC	RG14 6SX
G0LUK	RG14 6TB
G4ZDP	RG14 7AL
G8MWU	RG14 7DJ
G0WZT	RG14 7EP
G0HPX	RG14 7ET
G6LAE	RG14 7RR
G1DFI	RG14 7TT
G0IUB	RG17 0BY
G3ZPK	RG17 0BZ
G8JRN	RG17 0EU
G1OQV	RG17 0JE
G0LJJ	RG17 0JR
G1FVP	RG17 0JR
G3LWT	RG17 0LA
G8JWD	RG17 0QH
G6IBI	RG17 7BX
G0DTQ	RG17 7DG
G6WLE	RG17 7ED
G0GWK	RG17 7LY
G8HNK	RG17 7TS
G7BWI	RG17 8QA
G6EES	RG17 8YQ
G4WIH	RG17 8YZ
M1ABX	RG17 9QE
G0LTX	RG17 9QE
G7VWG	RG17 9QR
G6RXK	RG17 9SN
G6GCJ	RG17 9UN
G3ZYW	RG17 9UP
G6WEI	RG17 9UY
G4BOO	RG18 3BA
G7TOO	RG18 3BF
G6AMF	RG18 3BN
G3RVM	RG18 3BP
G3WOI	RG18 3BP
G3VMT	RG18 3DT
G7HBF	RG18 3DY
G6LBL	RG18 3FG
G7UPV	RG18 3FQ
G4ORX	RG18 4DL
G1JKP	RG18 4LQ
G0ORH	RG18 4NP
G4MLG	RG18 9EZ
G3VOW	RG18 9HQ
G8LTN	RG18 9HT
G3NVO	RG18 9HX
G4RKO	RG18 9PQ
G3KJC	RG18 9QP
G6SOZ	RG18 9RJ
G7DRR	RG18 9RJ
G1GUC	RG18 9RS
G8MXZ	RG18 9SZ
G0NJF	RG18 9TF
G0NJG	RG18 9TF
G0RTV	RG18 9TG
G7UUF	RG19 3LE
G4FWR	RG19 3RL
G3ICB	RG19 3RS
G0KQT	RG19 3SF
G8NRC	RG19 3TY
G8VYJ	RG19 3UR
G4GNK	RG19 3XA
G7UXK	RG19 3XW
G4BWE	RG19 3XX
G8XEC	RG19 4DY
G3WYW	RG19 4FD
G4XBA	RG19 4FN
G8JIP	RG19 4FQ
G4RTQ	RG19 4GJ
2E0AJZ	RG19 4GW
2E1BZE	RG19 4GW
G7VGL	RG19 4LX
G1PCG	RG19 4QX
G2CPM	RG19 4QX
G4WLG	RG19 4QX
G6IZA	RG19 4WA
G4GZQ	RG19 4YJ
G3IVM	RG19 8DB
G3LLK	RG19 8DD
G0LJQ	RG19 8DX
G0HPN	RG20 0LX
G0LTY	RG20 0NA
G3NAQ	RG20 7BE
G7POB	RG20 7EZ
G4TPH	RG20 8AQ
G6FUT	RG20 8DH
G1YOS	RG20 8SD
M1AUY	RG31 5HW
G0IZI	RG31 5JY
G6YFJ	RG31 6FZ
G7NRM	RG31 6QZ
G8PQZ	RG31 6RH
G4VNU	RG31 7RU
G7PJL	RG7 3DZ
G4XMR	RG7 3EN
G3YGR	RG7 3ES
G4FXJ	RG7 3ES
G4IWS	RG7 3EZ
G8XND	RG7 3HU
G4PUB	RG7 3LZ
G4SVG	RG7 3LZ
G8BGT	RG7 3NA
G0LZR	RG7 3QY
G4MJS	RG7 3SY
G1AWD	RG7 3TL
G3ZOI	RG7 3TU
G8GSQ	RG7 4FB
G0RQY	RG7 4SP
G0HPM	RG7 4TD
G8DGR	RG7 4TH
G1NAB	RG7 5DG
G3LXQ	RG7 5DN
G6HUN	RG7 5HR
G7TKP	RG7 5NS
G8DHL	RG7 5RY
G7AKF	RG7 5SD
G3BGL	RG7 5TH
G3WND	RG7 5TR
G3WUW	RG7 6HN.
G3KIW	RG7 6NU
G4ZSG	RG7 6RA
M1AIB	RG7 6TB
G0KPE	RG7 6TN
G8ZJU	RG8 7AT
G1XMV	RG8 7DZ
G4NSU	RG8 7JL
G4UBH	RG8 8JY
G4CEI	RG8 8LN
G0IRS	RG8 9JB
G3WKZ	RG8 9PS

Newcastle-under-Lyme

Callsign	Postcode
G3GBU	ST4 6DB
G8VSR	ST5 0AB
G8FNG	ST5 0ER
G1UDP	ST5 0JY
G1WVW	ST5 0JY
G8KUZ	ST5 0LB
G4VFJ	ST5 0NG
G6IFS	ST5 0NL
G7NSR	ST5 0NL
G0OQP	ST5 0PB
G3VXS	ST5 0PS
G0NBE	ST5 0QF
G6FSF	ST5 0RS
G1STQ	ST5 1AU
G7BJN	ST5 1DE
G8JBW	ST5 1DS
2E1DOD	ST5 1DS
G4IMV	ST5 1NH
G7UMB	ST5 2DX
G7JXU	ST5 2EF
G4YPQ	ST5 2ER
G0MXD	ST5 2EX
G0WDT	ST5 2JL
G6IXS	ST5 2QT
G6PIR	ST5 2QX
G6HCI	ST5 3BS
G4TQB	ST5 3EF
G4UAY	ST5 3EW
M1AZC	ST5 3JF
G0TVT	ST5 3LD
G4MDM	ST5 3NY
G4UMG	ST5 3NY
G4HRA	ST5 3PH
G3LNK	ST5 3PJ
G7MMK	ST5 3PQ
G6IUH	ST5 3PS
G0OBC	ST5 3QR
G8YEG	ST5 3RH
2E1AMX	ST5 4AZ
G1MRZ	ST5 4AZ
G6MRZ	ST5 4AZ
G0LDS	ST5 4BN
G1LDS	ST5 4BN
G4TVA	ST5 4BN
G8MTB	ST5 4DA
G1TMF	ST5 4LP
G7TDW	ST5 5AX
G3LBS	ST5 5AZ
G3UOK	ST5 5BG
G4NFL	ST5 5EU
G4RNS	ST5 5EU
G3USF	ST5 5HP
G4SRH	ST5 5HP
G0GOI	ST5 6AR
G0NLI	ST5 6AR
G7PNF	ST5 6BB
G7BPG	ST5 6BY
G0UYU	ST5 6DS
G7AOQ	ST5 6HU
G4BQZ	ST5 6PD
G4CBW	ST5 7AH
G7SDF	ST5 7AZ
G7RBL	ST5 7BN
G0AHQ	ST5 7EP
G0HQH	ST5 7PB
G0LSX	ST5 7QB
G7KJE	ST5 7RQ
G4UDG	ST5 7ST
G8YDQ	ST5 7SZ
G4UDH	ST5 7TB
G1PDD	ST5 8BW
G4GNW	ST5 8EL
G7MTV	ST5 8EL
G4UQX	ST5 8RX
G0GZT	ST5 9AJ
G8ZES	ST5 9DD
G7OKO	ST5 9ET
G4DPW	ST5 9JZ
G1OMV	ST5 9LN
G1JDF	ST5 9NA
G1JFL	ST5 9NL
G1XYN	ST5 9NY
G7MRF	ST5 9PS
G6LLF	ST5 9PU
G4MDS	ST7 1AG
G6YQN	ST7 1AT
G0UZP	ST7 1HB
G4WBW	ST7 1NP
G6NGN	ST7 1NQ
G4VEW	ST7 1PD
G4EQZ	ST7 1PF
G6UOZ	ST7 1PF
G7VCT	ST7 1RB
G4VJB	ST7 1RH
G0MVU	ST7 1SG
G8BMJ	ST7 1SZ
M1BTN	ST7 1SZ
G0RDS.	ST7 1TA
G4CWJ	ST7 1TA
G7USI	ST7 1TA
G1TBX	ST7 1UB
G3OHH	ST7 3PH
G3TEY	ST7 3PH
G1CKJ	ST7 4DQ
G0KBI	ST7 4DY
G0KBJ	ST7 4DY
G6XMZ	ST7 4HF
G0UWK	ST7 4HJ
G1SHJ	ST7 4HL
G0HWV	ST7 4JF
G7TXX	ST7 4LW
G4UKP	ST7 4NB
G6AYG	ST7 4PU
G6USG	ST7 4QY
G1VVJ	ST7 4RB
G1ELQ	ST7 4RS
G6BQM	ST7 4RS
G0FZU	ST7 4TA
G8NYZ	ST7 4TA
G1PUV	ST7 4TH
G7MQF	ST7 4XJ
G4DEN	ST7 8BT
G1JDW	ST7 8EP
G7OSO	ST7 8HU
G0MVT	ST7 8LA
G8BAA	ST7 8LB
G0PTD	ST7 8NG
G6UKQ	ST7 8PE
G6YCO	ST7 8PL
G6OXP	TF9 4HW
G7RSM	TF9 4PA
G7UQG	TF9 4PA
G6GTJ	TF9 4QE
G3WEI	TF9 4QR

Newcastle-upon-Tyne

Callsign	Postcode
G7AWZ	NE1 7RU
G7VRU	NE1 8SU
G7RJX	NE13 7ND
G4FFU	NE13 8AR
G4FFV	NE13 8AR
G0UMU	NE13 8AU
G6URX	NE15 6QP
G7UTF	NE15 6QP
G1YDD	NE15 7DQ
G4YMC	NE15 7LP
G0EGK	NE15 8AL
2E1DSP	NE15 8BZ
G0WMX	NE15 8NB
G1ESE	NE15 8PT
G8XGS	NE15 9EB
G1GSR	NE15 9EY
G0OTQ	NE15 9HL
G7PGG	NE15 9JL
G3RXN	NE15 9JU
G0MKG	NE15 9XF
G7EPU	NE2 1AQ
G6JY	NE2 1NQ
G4ETI	NE2 2HD
G0BEV	NE2 2JN
G3ZQM	NE2 2JN
G0TGK	NE2 2JU
G4MRT	NE2 2NP
G3WZH	NE2 2QU
G4WMA	NE2 3DL
2E1AKL	NE2 4BA
2E1ANZ	NE3 1HY
M0ANZ	NE3 1HY
G6OQO	NE3 2DU
G1YHS	NE3 2DX
2E1ACH	NE3 2FJ
G4FLU	NE3 2QY
M1BCN	NE3 3HB
M1BND	NE3 3HR
G0LNS	NE3 3NH
G6UEX	NE3 3NN
G0MZE	NE3 3PQ
G1XYS	NE3 3TN
G6JQE	NE3 3YE
G4ILL	NE3 4AN
G3TFT	NE3 4AQ
G4LX	NE3 4LR
G4LGH	NE3 4NR
G0CQK	NE3 4RN
G1GAD	NE3 4TT
G4MHW	NE3 5AQ
G3LIV	NE3 5BH
G8UEE	NE3 5BH
G1VOP	NE3 5DX
2E1AII	NE4 8DR
G3UKH	NE4 9EP
G1SKQ	NE4 9NH
G3AWR	NE4 9QH
G1YUN	NE4 9QX
G8POG	NE4 9QY
G3FHW	NE4 9TA
G8PHB	NE5 1BT
G7AYF	NE5 1ER
G7SCV	NE5 1SQ
G6VEG	NE5 1TP
G1DNA	NE5 1YL
G1HAB	NE5 1YL
G0DTR	NE5 2DL
G4VMU	NE5 2JS
G7UVN	NE5 2LY
G3NIJ	NE5 2PJ
G4WVH	NE5 3RF
G7HDZ	NE5 4LB
G7LER	NE5 4PP
G0RMO	NE5 4PQ
G0RLE	NE5 4TQ
G0EOY	NE5 5AL
G0DFT	NE5 5BU
G3IOE	NE5 5BU
G6KYW	NE5 5DD
G0WYY	NE5 5EA
M1AQE	NE6 2JY
G7MEI	NE6 2PX
G1IWA	NE6 2RL
G0GXR	NE6 2SX
G1EYD	NE6 2SX
G1EPD	NE6 3JL
G7TPG	NE6 4HA
G7PJT	NE6 4HQ
G3TDV	NE6 4LJ
G0UAC	NE6 4LQ
2E1ENU	NE6 4NP
G6JLL	NE6 4SR
G8GFW	NE6 4SR
G3ZWR	NE6 4SY
G7DAR	NE6 4TN
G0DMW	NE6 5AD
G0IOQ	NE6 5AD
G4UOW	NE6 5AD
G2BCY	NE6 5DU
G7FEE	NE6 5JY
G0VSF	NE6 5LE
G0DZG	NE6 5QR
G8VLS	NE6 5RE
G3YVZ	NE6 5YA
G4KVK	NE7 7AP
G7VLK	NE7 7LP
G7VLL	NE7 7LP
G1XSA	NE7 7PL
G0BZB	NE7 7QE
G8GTE	NE7 7RX

Newham

Callsign	Postcode
G7UHX	E13 8NB
G3PGI	E13 9DS
G4ADK	E13 9JF
G0RIU	E15 1QX
G7IIQ	E15 1QX
G7CXT	E15 3HQ
G7MXL	E15 3LJ
G7UYT	E15 4QD
M0ANG	E16 1SB
2E1CJR	E16 2LH
G6XDN	E16 3RR
G6XCX	E16 3SR
G7NGQ	E6 1DP
G4RFM	E6 1LB
G3IDI	E6 1LL
G1SJU	E6 2BS
G4MVX	E6 2DQ
G6RVZ	E6 2JL
G8TWD	E6 2NA
G0JYJ	E6 3RY
G7KWS	E7 0HN
2E1FZF	E7 0JS
G0LLE	E7 0JS
G4IZR	E7 0LQ

Newport

Callsign	Postcode
GW0WQP	CF3 8UB
GW0ION	NP1 0BP
GW8IQC	NP1 0BT
GW7UMS	NP1 9AF
GW3OCD	NP1 9AH
GW7AVB	NP1 9BF
GW7RIB	NP1 9FB
GW8AGI	NP1 9FS
GW6TUD	NP1 9FT
GW1JFT	NP1 9FU
G4RYQ	NP1 9HN
GW1MGI	NP1 9LL
GW4BVT	NP1 9NW
GW4COJ	NP1 9RB
GW4LCF	NP6 1BZ
GW4UGH	NP6 1DY
GW1TDV	NP6 1EZ
GW4ESL	NP6 1EZ
GW8BNL	NP6 1JL
GW7NFT	NP6 1LP
GW8ZCV	NP6 1SQ
GW8COJ	NP6 1TN
GW7SNF	NP6 2BY
GW7RNC	NP6 2JL
G7PIB	NP6 2JP
GW3ORL	NP6 3AG
GW3GRY	NP6 3BA
GW7MVQ	NP9 0DA
GW1XVM	NP9 0EQ
GW1LXD	NP9 0GU
GW0JBN	NP9 3BN
GW0MSY	NP9 3DE
GW4EZW	NP9 3DJ
GW1NRS	NP9 3DJ
GW4SUE	NP9 3DJ
GW0FXC	NP9 3EE
GW7STV	NP9 3NP
GW3KZT	NP9 3NS
GW8VGG	NP9 3SF
GW4LOD	NP9 3SP
GW8SVN	NP9 4FH
GW6MKI	NP9 4LP
MW1AZR	NP9 4LP
GW7BSC	NP9 4LP
GW4BLE	NP9 5EF
GW7UHM	NP9 5EF
G4JCK	NP9 5HP
GW8KZN	NP9 5RA
GW0KTL	NP9 6BZ
GW4YNP	NP9 6FR
GW3NIN	NP9 6HE
GW3WTZ	NP9 6JN
GW7RTK	NP9 6JT
GW4SDT	NP9 6LL
GW4ZCM	NP9 6LS
GW3YKZ	NP9 6QF
GW8CNF	NP9 6QF
GW4GWS	NP9 6QH
GW4YKW	NP9 6QN
2W1EID	NP9 6UB
GW4JBQ	NP9 6WG
GW8SBK	NP9 6WG
GW0FUR	NP9 6WL
GW1VBA	NP9 6WL
GW0JBH	NP9 7AT
2W1AYO	NP9 7HW
2W1AYP	NP9 7HW
GW3UNH	NP9 7LY
GW4OGO	NP9 7NH
GW8XBY	NP9 7PU
GW4VNS	NP9 7QB
GW8MOZ	NP9 7QP
GW0DHG	NP9 7QW
GW1DLP	NP9 7SA
GW0DXQ	NP9 7SJ
GW3JEZ	NP9 7SP
GW4VES	NP9 7TJ
GW3KWB	NP9 8AG
GW4IED	NP9 8AX
GW3NTR	NP9 8DS
GW6EWQ	NP9 8LQ
MW1BOR	NP9 8LQ
GW4NML	NP9 8NA
GW6KBD	NP9 8NU
GW6NNR	NP9 9DA
GW8WZR	NP9 9EQ
GW4SOC	NP9 9QS
GW8OIJ	NP9 9RY

North Ayrshire

Callsign	Postcode
GM0OAD	KA11 1AQ
GM3USL	KA11 1BD
GM0SEI	KA11 1BD
GM7KSA	KA11 1BD
G7UAC	KA11 1BW
MM1BIF	KA11 1EQ
GM6JOD	KA11 1LR
GM3JOB	KA11 1NJ
GM0JQE	KA11 1PN
GM0DNG	KA11 2EL
GM7KHA	KA11 3AF
GM0WUX	KA11 4ER
GM0KXF	KA12 0JP
GM7VFR	KA12 0QE
GM7CZU	KA12 0QG
GM3NYG	KA12 0SE
GM4PGV	KA12 0SE
GM0FCI	KA12 8DR
GM1ASA	KA12 9DN
GM0DJG	KA12 9PG
GM1MSN	KA13 6EB
GM1MSO	KA13 6EB
MM1BRH	KA13 6JB
GM4KQS	KA13 6JP
GM3YQK	KA13 6PL
MM1BIQ	KA13 6QN
GM0LYH	KA13 6RE
GM7DUG	KA13 6SU
GM0TAE	KA13 7AR
GM0UGG	KA13 7DT
GM8BOM	KA13 7NH
GM0DZE	KA13 7QP
GM0GYL	KA13 7QP
GM8MMW	KA14 3AJ
GM3JVX	KA15 1EY
GM7DTC	KA15 1JE
MM1BDB	KA15 1JJ
GM4VHZ	KA15 2AJ
GM4LIS	KA15 2HG
GM7VCV	KA15 2LF
GM6UHE	KA15 2LN
GM8NVG	KA15 2LN
GM3YAU	KA20 3LF
GM4WNQ	KA20 3LW
GM4LLY	KA20 4ET
GM0SYU	KA21 5LP
GM3YKE	KA21 5NX
GM3DJS	KA21 5RL
GM0OTB	KA21 6AQ
GM7DHA	KA21 6AQ
GM4PSF	KA21 6DA
GM0SYW	KA21 6LX
MM0ABM	KA22 8PL
MM0ARG	KA22 8PL
GM0EFC	KA23 9BX
GM0EFD	KA23 9BX
M0ABL	KA23 9HN
GM1MQE	KA23 9JT
GM0AYT	KA23 9LB
GM3LLP	KA23 9LE
GM4KAT	KA23 9LE
GM3JIG	KA23 9LX
GM3GBY	KA24 4HT
GM6NHF	KA25 7BH
G4BVZ	KA27 8PR
GM3HEN	KA27 8QH
GM3UA	KA27 8RL
GM8BJJ	KA28 0DP
GM0MDD	KA29 0AD
GM0VKG	KA29 0AD
GM3PKV	KA29 0AL
GM3DOD	KA29 0DG
GM3LYY	KA29 0DZ
GM1WMO	KA30 8DN
GM8EUG	KA30 8EF
GM1RJS	KA30 8HU
G0BHG	KA30 8JT
GM0ERT	KA30 8PX
GM0BVK	KA30 8PZ
GM0NUI	KA30 8RH
GM1BKR	KA30 9BZ
GM6OA	KA30 9DN
GM0JSW	KA30 9EL
GM0DWY	KA30 9EQ
MM1AAL	KA30 9ER
GM0RYD	KA30 9EX
GM1JLU	KA30 9NE
GM1XIA	PA17 5BB

North Cornwall

Callsign	Postcode
G4OIA	EX22 6LD
G0OKN	EX23 0BN
2E1EON	EX23 0DT
G0SXM	EX23 0HR
G0KDY	EX23 8DJ
G4KHY	EX23 8EN
G3KMQ	EX23 8HZ
G1SHG	EX23 8JA
G6MAA	EX23 8LR
G8DHA	EX23 8LR
G0DBD	EX23 8NA
G8ULJ	EX23 8PD
G0TDE	EX23 8QP
G3XNE	EX23 8SB
G7TRM	EX23 8SB
G1IJQ	EX23 9AG
G4NLH	EX23 9BB
G6REW	EX23 9DB
M1BTS	EX23 9EW
2E1EOR	EX23 9LE
G1OKT	EX23 9NN
G8HDW	PL15 7DE
G4VRL	PL15 7EL
G6UJB	PL15 7LF
G3LNW	PL15 7PW
G4FIV	PL15 8LR
G3SYM	PL15 8UF
G0UFV	PL15 8XA
G1LJJ	PL15 9AG
G4KKZ	PL15 9BB
G0AAM	PL15 9RR
M0AVP	PL17 8PH
M1APB	PL17 8PH
G4LDA	PL27 6AE
G4YWB	PL27 6NW
G3NKN	PL27 6PL
G6ELZ	PL27 6SX
G1PGG	PL27 6UG
G3YJX	PL27 7HB
G4UAN	PL27 7HD
G7DUB	PL27 7PA
G3EHT	PL27 7RY
G7TLC	PL27 7TB
G7KZG	PL27 7TP
G7VPX	PL27 7XD
G6IVU	PL28 8AX
G3HWD	PL28 8DF
G8XSI	PL28 8EF
G0MWW	PL28 8ES
G3TGC	PL28 8EX
G7NME	PL28 8QH
G4UZO	PL29 3RW
G3IWE	PL30 3AU
G1SHC	PL30 3BJ
G8OOV	PL30 3ED
2E1CEH	PL30 3JE
G4AXC	PL30 3PH
G3XKS	PL30 3PN
G1OFG	PL30 5AF
G8TWA	PL30 5HD
G0TRP	PL30 5NJ
G3IGV	PL31 1BH
G0PNM	PL31 1HY
G4WVD	PL31 1PY
G3FVD	PL31 1QP
G1TWN	PL31 1QX
G0NIT	PL31 2BP
G0EEJ	PL31 2NU
G4YVB	PL32 9PJ
G1VWU	PL32 9RZ
G7BOD	PL32 9UX
G4TPL	PL33 9AR
G6GWX	PL33 9BJ
G8HPD	PL33 9DL
M1CCE	PL34 0AG
G0EDH	PL34 0BH
G3LOV	PL34 0BT
G4MLI	PL34 0DT
G0ATS	PL34 0HH
G1NAK	PL34 0HH
G7RWH	PL35 0AB
G0NFD	PL35 0AE

North Devon

Callsign	Postcode
G3ZEJ	EX16 8EA
G8ZQC	EX18 7BW
G2AJS	EX18 7DA
G3XYG	EX31 1QD
G3AKJ	EX31 1QF
G4RVJ	EX31 1QX
G4SXO	EX31 1RD
G7MLQ	EX31 1RE
G0OQZ	EX31 1RZ
G6RNT	EX31 2BH
G0TXK	EX31 2HR
G0SIK	EX31 2HX
G4OUY	EX31 2LA
G7FIU	EX31 2LH
G8NBO	EX31 3AP
G4JAK	EX31 3AQ
G3PGA	EX31 3BS
G4SJP	EX31 3DS
G4JVW	EX31 3HZ
G4FRV	EX31 3LE
G1DNQ	EX31 3QP
G4ILK	EX31 3RJ
G6TAP	EX31 4HY
G0EEW	EX31 4JH
G0BUJ	EX31 4JR
G3YBP	EX31 4NB
G4RDA	EX31 4NH
G4YCW	EX31 4PR
G0NMD	EX31 4RT
G4JZU	EX32 0AU
G4SGW	EX32 0DF
G0AYM	EX32 0EG
G0WUA	EX32 0NJ
G0HIW	EX32 0SF
G0FTW	EX32 7AH
G3ZFV	EX32 7AS
G0EYF	EX32 7HW
G7HBI	EX32 7NR
G4TNE	EX32 8DN
G0DNV	EX32 8EJ
M1AEO	EX32 8EX
G0NPV	EX32 8PX
G8VCQ	EX32 9BG
G7WCA	EX32 9BU
G2CVY	EX32 9BW
G4XWQ	EX32 9BW
G0GKE	EX32 9DY
G4JKN	EX32 9JJ
G0ERX	EX32 9JQ
G6VRA	EX32 9JU
G3TFG	EX33 1BT
G0DUH	EX33 1DH
G0UPK	EX33 1EB
G3MWG	EX33 1JU
G1LUC	EX33 1JY
G0SHP	EX33 2EH
G3MTD	EX33 2EL
G8EZZ	EX33 2HL
G3CVM	EX33 2LE
G0LKI	EX33 2LT
G0IYB	EX34 0JH
G3ZJU	EX34 0NA
G0HGM	EX34 0PB
G1FEJ	EX34 7BT
2E1GBN	EX34 8DZ
G0PRR	EX34 8JS
G4NGB	EX34 8JS
G7GZW	EX34 8JX
G4RWK	EX34 8NH
G3ATM	EX34 8PP
G1MJI	EX34 9HP
G3JUW	EX34 9HP
G1GBC	EX34 9LG
G0DIZ	EX34 9LN
G7FAI	EX34 9LN
2E1FZR	EX34 9PH
M0AME	EX34 9PQ
G3NGI	EX34 9SS
G4XZF	EX34 9TB
G8VEQ	EX36 3HJ
G0EOP	EX36 3HL
G4JBR	EX36 4BH
G2LV	EX36 4EP
G0UBF	EX36 4EU
G4RVG	EX36 4EW
G8BXO	EX36 4HJ
G4YUO	EX36 4PX
G6ASK	EX36 4RA
G1EEO	EX37 9JG
G3KQP	EX39 4LB
G3UJB	EX39 4LY
G4PTT	EX39 4LY

North Dorset

Callsign	Postcode
G4JBL	DT10 1DF
G3IDC	DT10 1DR
G3DSS	DT10 1HB
G3VXX	DT10 1HB
G2AXU	DT10 1LH

G8MNO DT10 1LQ
G3HEA DT10 1LR
G3DZS DT10 1NT
G0SEN DT10 2AQ
G0HGB DT10 2EE
G1LUM DT10 2EH
G4JOC DT10 2JW
G2ADA DT10 2ND
G4UEY DT10 2NH
G3ZDQ DT11 0AA
G6XAG DT11 0DQ
G8RGU DT11 0EF
G7RMG DT11 0HF
G0NXQ DT11 0HU
G3VOO DT11 0JG
2E1CQH DT11 0LE
2E1CQE DT11 0LJ
G0IJP DT11 0PZ
G6DAI DT11 0SS
G3KWN DT11 7DE
G4XGC DT11 7EL
G0TND DT11 7EL
G3TPH DT11 7LW
2E1EJC DT11 7LX
G7PCE DT11 7LZ
G0ULZ DT11 7LZ
G4TVL DT11 7NG
G4VJI DT11 7NG
G3FXV DT11 7UF
G3FGW DT11 7XE
G7JSU DT11 7XE
G3PCW DT11 7XG
G0IFW DT11 8BL
G0GFL DT11 8EP
G4RBV DT11 8EP
G2KV DT11 8NG
G4VQP DT11 8PB
G7NUN DT11 8TA
G7UPS DT11 8TA
G3MYN DT11 8UB
G4PTU DT11 9BA
G1YHG DT11 9DW
G3TLZ DT11 9ND
G8BXQ DT11 9NN
G6GTU DT11 9NZ
G4ZLX DT11 9PH
2E1DUS SP7 0DF
G3NBC SP7 0QF
G4VRM SP7 8EB
G4OVJ SP7 8EG
2E1DWD SP7 8EH
G4AQ SP7 8JR
G4OVI SP7 8LE
G3BVB SP7 8NF
G0UXZ SP7 8PR
2E1DNT SP7 8PR
G0GWC SP7 8RT
G7UON SP7 9HB
G8LBG SP7 9HD
G0UOD SP7 9HX
G1RJA SP7 9NX
2E1EGU SP7 9PA
G1NWT SP7 9PF
G3XNK SP7 9QH
G4CK SP7 9QQ
G0VNA SP8 4HH
G4ILM SP8 4HY
G4YZZ SP8 4NP
G8PQJ SP8 4NR
G7SUS SP8 4PE
G8WZT SP8 4RU
M1AAB SP8 4UP
G6TER SP8 4UP
G0YLO SP8 4UP
G0ZEE SP8 4UP
G7OMO SP8 4UQ
G0EXI SP8 5DH
M0AKI SP8 5EL
G3FWU SP8 5JR
G1THG SP8 5NB
G7SBV SP8 5NB
G3ZLQ SP8 5NH
G7JIF SP8 5RL
G8DJN SP8 5RZ

North East Derbyshire

M0BAY DE55 6BT
2E1ERW DE55 6BT
G3LGK DE55 6EH
G4JXF DE55 6GQ
2E1FRU DE55 6GU
G1AII DE55 6JG
G4DTP DE55 6JW
G6XZC DE55 6JX
G0XVS S18 5NA
G1EYL S18 5PZ
G0TRN S18 5TE
G8UHM S18 5WA
G6SVZ S18 5YW
G0WYN S18 5YW
2E1ERP S18 5YW
G1VKL S18 6AZ
G1KPZ S18 6BQ
G8PVM S18 6BU
G7BUF S18 6EH
G1EQL S18 6EJ
G0MVC S18 6ER
2E1EEU S18 6ER
G4KXW S18 6FF
G7HRW S18 6FH
G0ABU S18 6GR
G1DFW S18 6LE
G4PXF S18 6LY
G0LUM S18 6ND
G6CTA S18 6RF
G7UDD S18 6RJ
G0XVL S18 6RJ
2E1EIJ S18 6RJ
G0OUT S18 6RP
G0UGJ S18 6RP
2E1BII S18 6RY
M0AKB S18 6ST
G7MOR S18 6ST
G7UAM S18 6UN
G0WWO S18 6UN
G4JUH S18 6UZ
G3YON S18 6WH
G4PMN S18 6YA
G4KCU S18 6YL
G4TQK S18 8XR
G4NKI S40 3HF
G3RCW S41 0NJ
G0UEH S41 0NJ
2E1BFT S42 5EL
G6SZB S42 5HH
G8NSD S42 5HH
2E1FPQ S42 5LL
G1EUT S42 5NG
G6URK S42 5NS
G6PRP S42 5NT
G0GYB S42 5RP
G8CGZ S42 6HH
G0NAO S42 6HS
G4XFF S42 6HS
M1CBH S42 6HU
G3CTZ S42 6JD
G4ROP S42 6JF
G3KMY S42 6LB
G4IJS S42 6ND
G7GGY S42 6PX
2E1BIH S42 6PZ
G4VSR S42 6RX
G3WAM S42 6SP
G8SXV S42 6TG
G1GQB S42 6TP
G3VKK S42 6TY
G8SMC S42 6TY
G3YBO S42 6TY
2E1BIF S42 6TY
2E1BIG S42 6TY
G1XRA S42 6UD
G0JOM S42 6UP
G0NHF S42 6YE
G3VSG S42 7AA
G3MOK S42 7AD
G0BHB S42 7AE
G1WAB S42 7AN
G6PBW S42 7AN
G4BFK S42 7AW
G3PJN S42 7EP
G3TAV S42 7HL
G3VDI S42 7JE
G3UBO S42 7LD
G0TJG S44 5AD
G3GGU S44 5AE
G6INA S44 5AX
G8TFU S44 5BZ
G7UHQ S44 5JE
G1RVT S44 5SY
G3MME S45 0EA
G4LPF S45 8BE
M1ALN S45 8BL
M0BGI S45 8DE
G7SMD S45 8DE
G1RWI S45 8DY
G6POJ S45 8ET
2E1FKM S45 8HN
G0IVB S45 9DF
G1SQA S45 9HB
G7ENM S45 9HB
G0LEM S45 9LD
G4VWU S45 9LD
G3UAF S45 9LY
G0FWH S45 9RU

North East Lincolnshire

G0HOI DN31 1RE
G3NRQ DN31 2BW
G1NYB DN31 2EG
G8XHX DN31 2EL
G1WSC DN31 2HB
G3RPY DN31 2QH
G4VUP DN31 2QQ
G0MUA DN31 2QW
G1HGW DN32 0AJ
M1BXC DN32 0AJ
G4PJX DN32 0AY
G3ELZ DN32 0BP
G0VHE DN32 0DU
G4NPR DN32 0HP
G8YAY DN32 0HP
G4HOF DN32 0JR
G8BMZ DN32 0NG
G0MUY DN32 0NP
G4KVB DN32 7EG
G0HMQ DN32 7HG
G0MNI DN32 7JD
G0CJV DN32 7PL
G6FCX DN32 7QU
G4JHO DN32 8ES
G8VSX DN32 8NS
G4UNC DN32 8QR
G7LIW DN32 8QS
G7PXX DN32 9LX
G7VHM DN32 9NP
G4YSO DN32 9PQ
G0MAF DN32 9PT
G1KBL DN32 9PY
G1RSK DN32 9QG
G3SNE DN32 9QJ
G1CPF DN33 1AB
G4PCH DN33 1AR
G4HZF DN33 1BG
G4IGV DN33 1EG
M0AJT DN33 1JB
G1MIR DN33 1LD
G7WIZ DN33 2AX
G0KUD DN33 2BB
G4NPS DN33 2EA
G3DSZ DN33 2HF
G8RBG DN33 2HW
G0RQV DN33 2HW
G2DWB DN33 2JB
G6ODD DN33 2JB
G6EXZ DN33 2LG
G3AXS DN33 2LW
G6YAT DN33 2LX
G1FVE DN33 2NW
G0CGZ DN33 2PL
G0IOR DN33 2PL
G4TBU DN33 3AU
G0CSV DN33 3BL
G4KKL DN33 3BL
G1BKJ DN33 3BS
G6GZJ DN33 3DB
G4VIM DN33 3EX
G7FCY DN33 3HG
G0JNT DN33 3JP
G3SWU DN33 3JT
2E1FHZ DN34 4AD
M1AKU DN34 4NN
G4DXB DN34 4NU
G0IGC DN34 4PN
G0IGH DN34 4PN
G7TSZ DN34 4SG
G0RUS DN34 5DB
G6RIQ DN34 5PE
G4NVD DN34 5QW
G4PDW DN34 5QY
G7EOG DN34 5RE
G4JMY DN34 5UQ
G4GZ DN35 0AF
G8ULH DN35 0JQ
G0TAB DN35 0LF
M1BYI DN35 0NN
G3RSD DN35 0QW
G0GSY DN35 0RA
G7KPM DN35 0RA
G7RLK DN35 7AY
G0LFA DN35 7DX
G1GAI DN35 7NE
G7VVP DN35 7NQ
M1CDQ DN35 7NQ
2E1FIM DN35 7NQ
G7VKY DN35 7QQ
G4ZVX DN35 7SD
M1BYQ DN35 7TG
G4LOY DN35 7UP
G1LXU DN35 8LX
G4CFO DN35 8NP
G1WYE DN35 8NR
G6VJA DN35 8PL
G7FFW DN35 8PR
G0PBR DN35 8QN
G7UPF DN35 8UN
G4SRF DN35 9DE
G2HOJ DN35 9HU
G7CHB DN35 9NJ
G7PSV DN35 9PP
G4CHZ DN35 9PZ
GM0JPH DN36 4AB
G4EXM DN36 4AY
G8ZLF DN36 4DF
G0BBT DN36 4DS
G7DEU DN36 4EA
G4PPL DN36 4HH
G1JZL DN36 4LJ
G6IAE DN36 4NJ
M0ANU DN36 4RB
G3IYT DN36 4SR
G7IMZ DN36 4TT
G3ZSF DN36 4UF
G7IRE DN36 4XB
G4TNX DN37 0AE
G7MES DN37 0DG
G0SWW DN37 0DG
G3TVC DN37 0JR
G3DOT DN37 0QD
G4UJK DN37 0RF
G4WOH DN37 0UA
G4ZNN DN37 7AU
G8XNY DN37 7BE
G0ATW DN37 7DB
G8XXI DN37 7DX
G6RML DN37 7EZ
G4MSY DN37 7JA
G4YHP DN37 7RD
G1LWL DN37 7RE
G0IVS DN37 7RP
G0WIR DN37 7RR
G4VSD DN37 8AF
G8FKF DN37 8AN
G6PEH DN37 8ER
G0JAP DN37 8HF
G6SXN DN37 8HG
G4UJY DN37 9EA
G3KAW DN37 9HH
G0ITK DN37 9LA
G0CGG DN37 9PL
G4KAL DN37 9PL
G3CNX DN37 9QJ
G4EBK DN37 9QJ
G3RGC DN37 9QL
G0IRR DN37 9RN
G8RIW DN37 9RN
G1BRB DN37 9RZ
G7BTP DN40 1AR
G0HDS DN40 1BH
G4YAM DN40 1DW
G0NBC DN40 1HH
G0BWG DN40 1JU
G4SEE DN40 1LH
G2HLB DN40 1NR
G1UZT DN40 1PD
G0KTX DN40 1PT
G7PLS DN40 1RB
G4SEF DN40 2AY
G6KKO DN40 2BZ
G7BRZ DN40 2EQ
G0TOZ DN40 2HE
G6ENY DN40 3AW

North Hertfordshire

G3XDM AL6 9NW
G8WEV AL6 9ST
G8LKA AL6 9UE
G0NZJ LU2 8LW
G0LQZ LU2 8PP
2E1ARU SG3 6AH
G0FJR SG3 6BQ
M0AZZ SG3 6DS
G7JAW SG3 6DS
G8OKI SG3 6DT
G8AUB SG3 6NL
2E1FXW SG3 6NN
G3BGG SG3 6NU
G0VTE SG3 6PH
G3LQO SG4 0AN
G4OXD SG4 0DP
G4ZJT SG4 0DP
G4KDW SG4 0LG
G7JVD SG4 0QR
M1BVG SG4 0QW
2E1ELT SG4 0QW
G4ARL SG4 0QY
G6JBQ SG4 0RD
G8FCA SG4 0RD
G6DYK SG4 0RQ
G4GXM SG4 7RA
G3BIX SG4 7SP
G3KJT SG4 8AL
G3FNY SG4 8JH
G4PLW SG4 8JH
G3PZE SG4 8NP
2E1EYE SG4 8RX
G8TRR SG4 8SD
G8KDF SG4 8XS
G4SUP SG4 8XX
G6BTP SG4 8XX
G7HMV SG4 8YN
G1AJG SG4 9DG
G4CTM SG4 9EP
G3UYM SG4 9HD
2E1CQY SG4 9HD
G4CEM SG4 9HP
G3EUS SG4 9PX
G8THN SG4 9PX
G8TSM SG4 9PX
G3OVT SG4 9SX
G6CFC SG5 1PA
G8BFH SG5 1PN
G6ERG SG5 1PS
G6LSC SG5 1PT
G7IEG SG5 1QH
G8ITW SG5 1TE
G4RBI SG5 1XB
G8FWI SG5 2AZ
G6XCK SG5 2HP
G4JQB SG5 2JE
G4VXU SG5 2JH
G0BIF SG5 2UF
G3ELI SG5 3QB
G3WRJ SG5 3RD
G6BRW SG5 3RG
G3RZP SG5 3RX
G3CWV SG5 3TP
G1WEG SG5 3UB
G4ACP SG5 3UB
G3UWM SG5 3UP
G7VRE SG6 1AN
G7HRJ SG6 1PP
G6WJS SG6 1RN
G3OHW SG6 1SX
G4SWH SG6 2BL
2E1CRK SG6 2DE
2E1ENL SG6 2DE
G4HUN SG6 2DN
G2DXK SG6 2EQ
G6SXE SG6 2HQ
G8DKK SG6 2JA
G1RQM SG6 2SR
G4MRR SG6 2SX
G6KVK SG6 2SX
G3RZY SG6 2SZ
G8BIT SG6 2TE
G3PYB SG6 3JR
G0UEW SG6 3JS
G3YUZ SG6 3NG
M0ARQ SG6 3PY
G0BIQ SG6 3QT
G8HDD SG6 3RS
G6ADX SG6 3SJ
G0WVO SG6 4BE
G8WET SG6 4DT
G3OLY SG6 4HJ
G3XEY SG6 4HY
G0AMX SG6 4JZ
G3GAV SG6 4PL
G1OKV SG6 4YH
G3WAB SG7 5BG
G6PYH SG7 5LZ
G4NFD SG7 5NQ
G4ISO SG7 6LN
G4HIE SG7 6LT
G6KBS SG7 6LT
G8ABX SG7 6PE
G1XXR SG7 6RZ
G8NHO SG7 6RZ
G6MAY SG7 6TW
G0IYU SG8 5AT
G4HPE SG8 5BW
G6XQY SG8 5BW
G8RQB SG8 5BY
G4NIY SG8 5DR
G8HZP SG8 5PX
G0FUZ SG8 7DJ
G4NTX SG8 7DJ
G3TXO SG8 7DY
G1GIJ SG8 7ED
G6XEF SG8 7ES
G0TIL SG8 7EU
G1AIT SG8 7JS
G0FCU SG8 8AD
G4ZSJ SG8 8EB
G3IUY SG8 8ER
G4MES SG8 8EZ
G0KFT SG8 8LT
G4OAS SG8 9AU
G0SII SG8 9BN
G4BDY SG8 9DS
G4IOX SG8 9DS
G6EDD SG8 9JF
G1HQE SG8 9QL
G1YZI SG8 9SE
G1YZJ SG8 9SE
GW0IEU SG8 9UB
G4TRT SG8 9UG
G1OKF SG8 9UQ
G7UQF SG8 9UQ

North Kesteven

G7GJT LN4 1AE
G4URX LN4 1DZ
G7PAC LN4 1EB
G4DTL LN4 1EG
G1TSL LN4 1HN
G0POJ LN4 1LJ
G1OQF LN4 1NN
G6EFR LN4 1NP
G7HBR LN4 1NP
G4YYE LN4 1NW
G0FFU LN4 1NY
G4RYB LN4 1PG
G4JDD LN4 1PR
G7EJH LN4 1PU
G4DKG LN4 1QP
G7VZA LN4 1RL
G4HCF LN4 1SB
G0GTV LN4 1TU
G8IJC LN4 2EX
G6ELE LN4 2JP
G3WOK LN4 2LE
G6YEX LN4 2NA
G6HZV LN4 2NP
G4JRQ LN4 2PW
G7LJZ LN4 2QS
G8RHN LN4 2RG
G3NXT LN4 3DT
G3AJW LN4 3HN
G4LVF LN4 3NG
G0PXB LN4 4AQ
G7VSN LN4 4EL
G4RPD LN4 4JW
G1EFK LN4 4QA
G7KPS LN5 0AH
G7FAR LN5 0AH
G0RAF LN5 0AH
G3VCA LN5 0AH
G3XYP LN5 0EE
G6HTY LN5 0EE
G4PWM LN5 0ER
G4UNF LN5 0ER
G4WRS LN5 0PA
G7JGI LN5 0RN
G7OVS LN5 0RN
G3SGB LN5 0SF
G4LQH LN5 9DT
G4OOA LN5 9JN
G0LMC LN5 9LZ
G4FQP LN5 9NH
G0GTD LN5 9SL
G4ZOX LN5 9SW
G0MEY LN5 9UT
2E1EKQ LN6 5SE
G0EJV LN6 5TF
G1SPO LN6 5TS
G4HNQ LN6 5TW
G7FGY LN6 5UU
G6KSR LN6 5UW
G4GDQ LN6 8HB
G4OSB LN6 8LY
G7LUV LN6 8PU
G8HYZ LN6 8SX
G4LUU LN6 8UH
G8BGL LN6 9BG
G4AEQ LN6 9DJ
G3IYF LN6 9EX
G0TZJ LN6 9EX
G3WZJ LN6 9HZ
G7IRS LN6 9LW
G7VTN LN6 9PJ
G0UND LN6 9RE
G4JES LN6 9RF
G3RDD LN6 9RR
G0PJS NG32 3NR
G0NXG NG34 0AG
G1UCP NG34 0AG
G0CRE NG34 7ET
2E1FAN NG34 7GJ
G0BTT NG34 7HG
G0WXK NG34 7JF
G6IJX NG34 7QH
G3LSJ NG34 7TD
G4ODG NG34 7TF
G6AWY NG34 7TF
G4GUC NG34 7TL
2E1BPQ NG34 8BJ
2E1BPR NG34 8BJ
G0WMS NG34 8BW
G3EMG NG34 8JQ
G6RPV NG34 8JT
G4KQT NG34 8LR
G3RGO NG34 8LS
G4DV NG34 8LS
G3JFC NG34 8NJ
2E1DPG NG34 8NJ
G3AFK NG34 9AP
G0SGP NG34 9AT
G0LME NG34 9BD
G1PGN NG34 9BT
G1XRP NG34 9FJ
G7JMV NG34 9GZ
G0PBO NG34 9HF
G0MRE NG34 9HS
G0NVY NG34 9HS
G7KAG NG34 9HT
G7UMC NG34 9HW
G3NNQ NG34 9JD
G0POM NG34 9JE
G0OZO NG34 9JE
G1JMC NG34 9JE
G0RHL NG34 9JP
G0RHM NG34 9JP
M0AGF NG34 9JQ
G0DMJ NG34 9JZ
G1PTY NG34 9LE
G6JTW NG34 9LY
G3VHI NG34 9QT
G4UYF NG34 9QY
G0FLV NG34 9RP
G4PQJ NG34 9RQ
G0NHM NG34 9SX

North Lanarkshire

GM3TDS G33 6DG
GM1AYT G65 0PR
GM8IIH G65 0QZ
GM0BRJ G65 0RS
GM0BUE G65 9UL
M1AVT G66 5NS
GM1RRJ G67 2BL
GM7FIS G67 2DA
MM1BJZ G67 2DR
GM3KMN G67 2NN
GM6FLL G67 2QW
GM6FLM G67 2QW
GM1XOG G67 3AA
GM4XLU G67 3LU
GM7WKQ G67 3NL
GM1XOI G67 3NL
GM7KIM G67 3PB
GM7RQK G67 4AD
GM7RWM G67 4DE
GM0GFV G68 0EX
GM0NBM G68 0HX
MM1AOE G68 0JR
GM4EIW G68 9BJ
GM7JUX G68 9DZ
GM7MZZ G68 9EG
GM3PXK G68 9NW
GM4VWV G68 9NW
GM1MRY G69 0JW
GM4LGR G69 0PB
GM4WQQ G69 9EP
GM0LBR G69 9ND
GM7PGT G69 9PA
GM7FGH G71 6ED
GM4GIH G71 6SD
GM4UBJ ML1 2JJ
GM0ARD ML1 2LB
GM0TKQ ML1 3AD
GM1MXW ML1 3AN
GM3ULP ML1 3AS
GM3PAB ML1 3AZ
GM0WNR ML1 3NA
GM0HWB ML1 3QU
GM0EEY ML1 3RA
GM8FYJ ML1 4ST
GM0LIR ML1 5DX
GM1OQT ML1 5PB
GM0VXA ML1 5TD
GM0KMJ ML2 0EH
GM4LQR ML2 0QB
GM4WTJ ML2 0RP
G1WSH ML2 0RQ
GM4OBJ ML2 7RG
GM4JLD ML2 8DT
GM0VXZ ML2 8SE
GM4OYK ML2 8TR
GM3COB ML2 8XS
GM3TCW ML2 9BL
GM0WRU ML2 9DY
GM4NDO ML4 1NQ
GM0SYV ML4 2PR
GM8BBA ML4 2SR
MM1BIT ML4 3DZ
GM3XMS ML5 3QL
GM4SZG ML5 4JQ
GM0NPS ML5 5DF
GM3MTH ML5 5RE
GM0NZM ML6 0HB
GM4UEH ML6 0NW
GM0DEX ML6 0QG
GM4AUP ML6 0QQ
GM0PIQ ML6 6LG
GM0VXQ ML6 6PA
GM0HBV ML6 7DH
GM4AHB ML6 7JF
GM7CPR ML6 7JF
GM0ART ML6 7SZ
GM4WTS ML6 8BL
GM0VWZ ML6 8QG
GM4PRO ML6 8SF
GM7SWX ML6 8XL
GM3ZNC ML6 9DF
GM1OXC ML6 9RT
GM3CFK ML6 9RT
GM3EQZ ML7 4NS
GM3RIJ ML7 4NZ
GM0LEG ML7 5HR
GM6RXJ ML7 5QW

North Lincolnshire

G7MTJ DN15 0AD
G0RMD DN15 7AT
G7FQB DN15 7DZ
G1HKM DN15 7EH
G7FJN DN15 7EH
G7IDP DN15 7EH
G7PAA DN15 7HQ
G0DLL DN15 7PJ
G6MYI DN15 7PJ
G0GWY DN15 8AT
G0RKQ DN15 8BP
G4ZGE DN15 8BT
G7NJF DN15 8BW
G6MNK DN15 8DD
G3CCH DN15 8NU
G4FUH DN15 8NU
G7LUU DN15 8TB
G6JVO DN15 8UG
G4LFY DN15 9AD
G8TIY DN15 9AX
G4CDC DN15 9EW
G0JRR DN15 9HH
G4YZH DN15 9NR
G0EFY DN15 9PW
G6DBC DN15 9QF
G0TTR DN15 9QF
G8TLL DN15 9QY
G3TII DN15 9RG
G8VEL DN15 9RN
G6CMX DN15 9RT
G4WZV DN15 9TP
G4ZGJ DN15 9TP
G0UZJ DN15 9UY
G0UOW DN15 9YE
G0OKF DN16 1NA
G4JRY DN16 2BE
G4HFZ DN16 2BU
2E1EYD DN16 2HL
G1YNQ DN16 2LQ
2E1FIK DN16 3DL
2E1FIN DN16 3DL
G7LHU DN16 3DS
G8GIH DN16 3EN
G0TNS DN16 3LD
G0PAB DN16 3PA
G4EQD DN16 3PB
G0PQY DN16 3PH
G8LIP DN16 3SA
G4NJA DN16 3SW
G4CBY DN17 1AF
G0RUH DN17 1AS
G4JJY DN17 1BD
G1CDY DN17 1DJ
G1ZRT DN17 1DU
G4NFX DN17 1ER
G7JQT DN17 1HQ
G0VHW DN17 1LN
G0TDG DN17 1ND
G0UTM DN17 1SA
G1YUI DN17 1YD
G8YVC DN17 2BD
G0NYL DN17 2EQ
G3KNU DN17 2EQ
G1ZGO DN17 2HA
G7LAL DN17 2NQ
G0ECS DN17 2TP
G4STW DN17 2UH
G7HAJ DN17 3DB
G0AEJ DN17 3DR
G7URL DN17 3HZ
M0ACB DN17 3PB
G6PAJ DN17 3RA
G0JRB DN17 3ST
G0OKY DN17 4AZ
G4FZR DN17 4DG
G0BCK DN17 4EG
G8MOF DN17 4ET
G3RRL DN17 4JG
G3RHQ DN18 5DZ
G0HMD DN18 5HH
G0PHP DN18 5QA
G4TGE DN18 6AE
G4IWR DN18 6AJ
G0KAT DN19 7AU
G8EPH DN19 7QH
G1JFZ DN19 7RB
G3KOC DN19 7RF
G8XFY DN20 0EY
G3HLR DN20 0HP
G0HDV DN20 0HW
G8YAU DN20 0NN
G0AOJ DN20 0PP
M0AUS DN20 9BD
G3MPW DN20 9BE
G0LUG DN20 9EW
G3VBI DN21 4BA
G4SXZ DN21 4BB
G6PHC DN21 4BB
G0MFW DN21 4EL
G4CCH DN21 4NE
G0MPK DN21 4NF
G7MAQ DN21 4PW
G4OKZ DN39 6TG
G0GFX DN39 6TJ
G1OGJ DN39 6TJ
G0FEQ DN39 6UQ
G0DOB DN40 2RG
2E0AQC DN40 3JN
G1OPD DN40 3PR
G4XRO DN9 1DG
G4WJE DN9 1DP
G4GZB DN9 1JS
G4NSM DN9 1JS
G4GZC DN9 1LR
G1FEF DN9 1NR
G4ZWQ DN9 1RG
G6ONE DN9 1RG
G4OGB DN9 2AD
G7IMD DN9 2BT
G7VMH DN9 2BT
G6YVN DN9 2EZ
G6BWA DN9 2LH
G4KFE DN9 2ND
2E1FYH DN9 2PD
G4ZMH LN7 6LA
G3XKI LN7 6LF

North Norfolk

G4RZN NR11 7AA
G0WPS NR11 7AA
G0AZR NR11 7AQ
G1GGN NR11 7AW
G7VRK NR11 7DY
G0ZAA NR11 7DY
M1BOA NR11 7PD
G8VTX NR11 7QD
G0UHR NR11 8BJ
G1SKY NR11 8DD
G3KHK NR11 8DD
G4TNM NR11 8DF
G0TAM NR11 8ED
G0ATY NR11 8HX
G3BGF NR11 8JB
GW1XBF NR11 8JH
G0JLV NR11 8LP
G7HBW NR11 8LW
G4RRN NR11 8PD
G7RNN NR11 8PD
G4LAG NR11 8PF
G3VIG NR11 8SR
G6WME NR12 0DE
G3IPV NR12 0EN
G0AOV NR12 0EW
G4MQK NR12 0NG
G3WQY NR12 0PD
G4OQI NR12 0QT
G6PYR NR12 0RP
G1ANZ NR12 0RU
G4CTT NR12 0RU
G7TUM NR12 0TA
G3ZYF NR12 0TN
G0OIZ NR12 0YU
2E1BMP NR12 8BA
2E1BNA NR12 8BA
G7ONM NR12 8BA
2E1EIV NR12 8BA
G4VUR NR12 8DG
G4ILR NR12 8DJ
G6QQ NR12 8DU
G1YLV NR12 8EP
G4YDZ NR12 8JH
G0AKL NR12 8LZ
G0AIT NR12 8QN
G0WJV NR12 8QR
G0GHB NR12 8UJ
G7PRC NR12 8YL
G7VPD NR12 8YL
G8GGZ NR12 9BQ
G8VSF NR12 9DA
G6JCY NR12 9EH
G1ONC NR12 9EQ
G1KZD NR12 9PJ
G4HXK NR12 9PJ
G1LQM NR12 9PZ
G4SEB NR12 9RB
G4PSH NR12 9RF
M1ABO NR12 9RF
G7VAH NR12 9RF
G4CKB NR12 9SA
G8BQS NR12 9TA
G8WKO NR21 0AD
G4UKZ NR21 0BU
G0WCH NR21 0BX
G7PPW NR21 0DB
G1YMA NR21 0HP
G0PPR NR21 0JS
G7OSJ NR21 0QX
G7OSK NR21 0QX
G6WPE NR21 7EL
G2CVA NR21 7JY
G7IJD NR21 7LG
G1IJC NR21 7LW
G0FVF NR21 7NA
G0MQU NR21 7NA
G6MDM NR21 7NA
G6YLR NR21 7QE
G0HAV NR21 8ER
G0NCC NR21 8ER
G0KYG NR21 8LR
G6YXB NR21 8PH
G0KRD NR21 9ES
G4RRD NR21 9HW
G4VAA NR21 9RG
G8BJN NR21 9RG
G7DVP NR21 9RH
G3FMR NR23 1AZ
G4NRW NR23 1BZ
G8KZP NR23 1EH
G8LGB NR23 1JA
G4KOV NR23 1JY
G4DIC NR23 1LF
G4RAY NR23 1LL
G6VFN NR23 1LR
G0SRZ NR23 1NL
G0EQB NR23 1NP
G1HYC NR23 1NP
G0DSN NR23 1QE
G0NWT NR23 1QE
G7BUS NR23 1QL
G7EVQ NR23 1QL
G7SDL NR23 1QL
G6PBL NR24 2AW
G4IWI NR24 2DE
G4IRY NR24 2LJ
G0FVS NR24 2LQ
G3KHF NR24 2LQ
G8AKP NR24 2PH
G0FVG NR24 2QA
G0ISV NR24 2QA
G3BUF NR24 2QY
G7FMU NR25 6BH
G3PYN NR25 6EG
G8FF NR25 6EH
G3UZN NR25 6JH
G6XQG NR25 6LA
G3RZ NR25 6NW
G3FNT NR25 6QF
G8ZBZ NR25 6RD
G3PXO NR25 6RT
G0SIQ NR25 6RT
G3LQJ NR25 6RU
G0SGQ NR25 6SN
G4CVE NR25 7AE
G0GFQ NR25 7HH
G1ZJR NR25 7HJ
G4EBE NR25 7LL
G1YLX NR25 7XG
G4OLF NR25 7XJ
G4SDF NR25 7XJ
G1IWS NR26 8AB
G7FSI NR26 8BH
G6ZB NR26 8HA
G3VIC NR26 8HR
G4FAI NR26 8JE
G3TWZ NR26 8NA
G7PHT NR26 8PT
G7MOJ NR26 8TB
G3HRK NR26 8UN
G6DFA NR26 8UN
G1IPO NR26 8YD
2E1DFL NR27 0DB
G7AOX NR27 0EE
G0UIQ NR27 0EE
G2BHG NR27 0HQ
G6JXW NR27 0JJ
G3ZQZ NR27 0NT
G3RIS NR27 9AW
G0ULG NR27 9DJ
G4PQP NR27 9HR
G4MIE NR27 9JH
G3HSP NR27 9JT
G8NEO NR27 9LJ
G4DMB NR27 9NA
G4PQQ NR27 9PH
G0KNQ NR27 9QY
G6JCX NR28 0AU
G6RVQ NR28 0DB
G4SFY NR28 0HX
G7TIN NR28 0JP
G6TPB NR28 0LJ
G4ZWM NR28 0LR
G3YOA NR28 0QG
G0GRB NR28 0QU
G1NST NR28 0RZ
G0IYD NR28 0SY
G1XYD NR28 0SY
G4HSL NR28 9AF
G8GCT NR28 9AR
G3DRL NR28 9BA
G0AJJ NR28 9BB
G6VOV NR28 9BB
G4RTH NR28 9DF
G4ZAY NR28 9EY
G6LZM NR28 9HA
G7IIN NR28 9LH
G4TDQ NR28 9SH
G6MYH NR29 5BZ
G0APV NR29 5LG
G7VSG NR29 5NF
G4XWD NR29 5NU
G0VSB NR29 5PZ

North Shropshire

G2WQ SY11 4PB
G3YYM SY12 0BU
G3PYM SY12 0EX
2E1FMV SY12 0EX
G3PLW SY12 0JH
G4EME SY12 0JR
G3XGD SY12 0QJ
GW4RRR SY12 9AA
G6BFE SY12 9DL
G7UAQ SY12 9ER
G6ZMD SY12 9NA
G4GXD SY13 1EP
G4FJQ SY13 1EX
G4HRH SY13 1EX
G3LHP SY13 1SU
G0KEQ SY13 2EF
G6NUL SY13 2EF
G6MEJ SY13 2HD
G8CTD SY13 2HD
G4VQH SY13 2NJ
G3KZY SY13 2NS
G4UJS SY13 2PT
G1GSJ SY13 2PX
G6SLE SY13 2PX
G1BAW SY13 2QP
G4ANY SY13 2RA
G8DSG SY4 2BB
G6ICX SY4 2DP
G4SDU SY4 2EH
G0VKW SY4 2EL
G6DQY SY4 2HY
G4VVD SY4 3JL
G7SLH SY4 4HA
G0RVE SY4 4JX
G0JLS SY4 4LA
2E1FYI SY4 4NU
G8CKM SY4 4NW
G1WVV SY4 5BX
M0BAV SY4 5DE
G1SCR SY4 5DE
G1YJB SY4 5DE
G7IFE SY4 5DE
2E1FYK SY4 5DE
G7IKD SY4 5HX
G7BSP SY4 5NB
G8EWD TF9 1AR
G1CFK TF9 1DQ
G0JNA TF9 1EA
G8PHG TF9 1HP
G1DOR TF9 1ND
G3ZGU TF9 1NG
G1DVH TF9 1NP
G3OKD TF9 2BA
G0PLB TF9 2BD
G1OAP TF9 2DE
G0DHM TF9 2DU
2E1FEM TF9 2RP
G7ELD TF9 2RS
G4MZQ TF9 3BD
G0MBJ TF9 3HX
G0CGA TF9 3QF
G8PSZ TF9 3UH
G0CDS TF9 3UJ
G3JPB TF9 4BE
G7JPB TF9 4BE

North Somerset

G3ZMH BS18 6TL
2E1DWF BS18 7HE
2E1DTN BS18 7LD
G4AYD BS18 7LS
G4YQG BS18 7LS

G4FBE BS18 7LT
G8AGT BS18 8JA
G3KOS BS18 9AQ
G8TYP BS18 9HA
G3YRN BS18 9HY
G4YSR BS18 9LD
G0HOD BS19 1EX
G4DAU BS19 1HZ
G0CCB BS19 1JD
G4OFF BS19 1JU
G7NCR BS19 1NS
G3NBN BS19 1QA
G1VBB BS19 1QF
G4LSX BS19 1QZ
G1ODB BS19 2BG
G0GWF BS19 2DG
G3VLW BS19 2LG
M1BQA BS19 2LW
G8RUP BS19 2NB
G8EIH BS19 2NT
G0GHM BS19 2RA
G0CEN BS19 2RT
G0FGZ BS19 2SN
G4PRR BS19 2TJ
G7PVL BS19 2TN
G4NAQ BS19 2UF
G0LHD BS19 2XD
G1YDK BS19 2XG
G8NQO BS19 2YD
G8HVT BS19 2YH
G6WLX BS19 2YU
G3XSV BS19 3BT
G3XIB BS19 3JR
G1AXS BS19 3ND
G3GAD BS19 3ND
G7UTT BS19 3NR
G3PRH BS19 3PB
G4DEQ BS19 3RX
G1DCG BS19 4BD
G0EZU BS19 4BD
G4DPH BS19 4DA
G3MWL BS19 4DR
G6GVH BS19 4EB
G4FZV BS19 4EB
G4CMC BS19 4HP
G6ANJ BS19 4LN
G0WKW BS19 4LN
G4IXP BS19 5BW
G0BID BS19 5DP
G0HKJ BS19 5DP
G4CYI BS19 5DX
G1OBR BS19 5ER
G4DYM BS19 5ES
G0DDF BS19 5HE
2E1DRN BS19 5HE
G4REH BS19 5HL
G3SWH BS19 5HQ
G0SNP BS20 0EQ
G4WRQ BS20 0HA
G4VYK BS20 0JR
G4YSZ BS20 0JX
G3KUF BS20 0NA
G4NXI BS20 8AF
G8YRC BS20 8BG
G3GZI BS20 8BT
G0FKJ BS20 8DD
G7WKK BS20 8ET
G0JZY BS20 8JD
G3YEP BS20 8JH
G6ZPV BS20 8JQ
G3YLJ BS20 8LE
G4NVV BS20 8LG
G6ZOE BS20 8LG
G7HVN BS20 8LN
G7UTS BS20 8NG
G4FRG BS20 8NQ
G0RWI BS20 8QR
G4JTM BS20 8QR
G3RYC BS20 8RD
M1BZO BS20 8RF
G4VEH BS20 8RQ
G4EJH BS20 9DY
G7AQF BS20 9JR
G4DRZ BS20 9JX
G7GHB BS20 9LQ
G4YNV BS20 9LU
G4UGT BS20 9PF
G4USQ BS20 9QY
G0GRG BS20 9XB
G1HZG BS20 9XB
G6EQZ BS20 9YT
G0ELO BS21 5AQ
G7PXR BS21 5AR
G4BWR BS21 5BU
G6IQF BS21 5EG
G4ONS BS21 5HB
G1CZW BS21 6JJ
G1DAX BS21 6JU
G4IWQ BS21 6LE
G0CJC BS21 6LH
G0IMA BS21 6LQ
G1YDJ BS21 6LQ
G4JCI BS21 6LR
G0AZE BS21 6RD
G0FJM BS21 6RX
G4ULP BS21 7AJ
G0AKS BS21 7DY
G1EFS BS21 7DY
G6GRG BS21 7EX
G8SPC BS21 7EX
G3OQL BS21 7HA
G3WBA BS21 7LU
G7KAC BS21 7NQ
G8LAN BS21 7QL
G6BGY BS21 7SQ
G4WPZ BS21 7UA
G6AEC BS21 7UP
G7RKT BS21 7US
G7JAH BS21 7YJ
G3XUL BS22 0AA
G4OJH BS22 0BJ
G6ATS BS22 0BW
G0JGM BS22 0BZ
G0MFI BS22 0DB
G6IZE BS22 0DS
G6OPD BS22 0FW
G7FEP BS22 0HL
G8VZL BS22 0JP
G7UWI BS22 0LG
G1KHX BS22 0PE
G0ATD BS22 0RL
G4IBO BS22 0RT
G3LPQ BS22 0SS
G1PEI BS22 0TJ
G8PRP BS22 0XP
G0IHE BS22 0YB
G4ABG BS22 0YN
G1VJE BS22 8AD
G4RES BS22 8AS
G7JCC BS22 8JY
G3HOI BS22 8NE
G8VZI BS22 8NW
G3ZUQ BS22 8PG
G0SVA BS22 8QN
G7JXV BS22 8UU
G3XGY BS22 8XH
G7KPB BS22 8XR
G4WAZ BS22 9AL
G7OPJ BS22 9AQ
G1XRO BS22 9AY
G6TJI BS22 9AY
G8OLK BS22 9DE
G1KIC BS22 9DY
G4ZUX BS22 9HT
M1BQC BS22 9HT
M1BQD BS22 9HT
M1BQE BS22 9HT
G4DUQ BS22 9LW
G0KGM BS22 9NG
G0UIJ BS22 9PF
G1FWZ BS22 9PW
G0ADW BS22 9QS
G4TBD BS22 9QW
G3RFO BS22 9SE
G3UTX BS22 9SL
G0PXJ BS22 9SP
G3GMC BS22 9SP
G8SRH BS22 9SZ
G6GXS BS22 9YH
2E1ADV BS23 1EH
G3YNI BS23 1QL
G3PLJ BS23 1RQ
G3LJD BS23 2BH
G3RNX BS23 2HE
G3LQG BS23 2PB
G1DQR BS23 2SL
G4UPR BS23 2UA
G4GLX BS23 2XD
G0IUC BS23 3AS
G6TAL BS23 3BE
G4CXQ BS23 3DF
G3PQE BS23 3DU
G0BRE BS23 3LL
G4RNZ BS23 3RR
G0TCH BS23 3SQ
G4USO BS23 3SX
G8RJO BS23 3TU
G4CDI BS23 3XQ
G0FGF BS23 4BN
G6PZ BS23 4HR
G1CMZ BS23 4HU
G0VJM BS23 4HW
G3SAZ BS23 4JL
G8ZVV BS23 4JP
G0VAZ BS23 4JZ
G4NMV BS23 4QS
G3YTP BS23 4TZ
G0IYE BS23 4YH
G0TAT BS24 0AB
G3PLT BS24 0PA
G3KVR BS24 6AZ
G0KXO BS24 6BE
2E1ADP BS24 6JD
G4TBO BS24 6NY
G7BRX BS24 7AS
G7USJ BS24 8BP
2E1EPF BS24 8BP
G7AQI BS24 8BS
G4HJF BS24 8EH
G3YRM BS24 8EL
G7UUK BS24 9DY
G0BMT BS24 9EH
G4PWP BS24 9JW
G3WXH BS24 9LH
G3MAK BS24 9LL
G3RUD BS24 9LW
2E0APK BS24 9QA
G7EYA BS24 9QL
G0PJI BS24 9RT
G0DKM BS24 9TJ
G3HZW BS24 9TZ
G7PRI BS25 1AR
2E1EMH BS25 1AR
G7SMH BS25 1AT
G4LJO BS25 1AX
2E0ABB BS25 1BJ
G0CHJ BS25 1HL
G8TTX BS25 1HQ
G0FIE BS25 1HT
G3GTA BS25 1LJ
G1CGJ BS25 1NB
G3YOL BS25 1NH
G4KMB BS26 2XZ
G3RUQ BS8 3QP
G3CLL BS8 3RD
G0UQT BS8 3UG
G3NHG BS8 3UU

North Tyneside

G3YRH NE12 0BT
G1USF NE12 0EU
G4WUL NE12 0EW
G1UDL NE12 0GA
G0OMI NE12 0JN
G3LRI NE12 0JP
M0AVU NE12 0JR
G1ERO NE12 0JR
G0AMT NE12 0QE
G8TWL NE12 0WA
G0NBK NE12 8TD
G0WUV NE12 8XQ
G3LDR NE12 9AH
G3LYG NE12 9AW
G1XXF NE12 9RL
G1RHO NE12 9SU
G4OEU NE12 9TY
G4XQN NE13 6AJ
G6YTR NE13 6EY
G1LNW NE13 6QB
G1GNY NE13 7BT
G4TQP NE13 7HG
G0AOK NE13 7HS
G1GDA NE13 7HT
G0GYO NE13 7HW
G0PVT NE23 7BH
G7PAM NE23 7BW
G0JWO NE23 7DE
G0JRW NE23 7LZ
G0JRX NE23 7LZ
G7PWL NE25 8AU
G6FJH NE25 8BD
G0KWS NE25 8BG
G6CEM NE25 8BJ
G4IZQ NE25 8EP
G4VPU NE25 8RU
G8JSW NE25 9AF
G0JGD NE25 9EB
G4KJN NE25 9HB
G3RB NE25 9NW
G0EIJ NE25 9PZ
G3WDS NE25 9QR
G1ODJ NE26 1AF
G3VVU NE26 1HS
G7PLA NE26 1NX
G7VTJ NE26 1SH
G0OBQ NE26 2JR
G6PKM NE26 2NR
G6AVL NE26 2NT
G0GLK NE26 2PD
M1AQZ NE26 2PD
G8FVT NE26 3AD
G3AQS NE26 3AR
G1MHA NE26 3AZ
G3GMS NE26 3BA
G8RNE NE26 3BD
G3YSL NE26 3ED
G5BW NE26 3EF
G7ELV NE26 3PR
G0MMG NE27 0AZ
G6KGB NE27 0DB
G1AHZ NE27 0EF
G1AIA NE27 0EF
G1RUY NE27 0PQ
G7HER NE27 0PQ
G7VKA NE27 0RZ
G4RUI NE27 0UF
G3KTT NE28 7AB
G3ZTJ NE28 7PH
G4PSL NE28 7RZ
G7HYK NE28 8SN
M0ADR NE28 8TL
G6UQX NE28 9AE
G1YHT NE28 9AE
G0UQS NE28 9JQ
G3ZGR NE28 9TB
G6NOY NE28 9TG
G7PNS NE28 9YT
G6ZJW NE29 0JF
G4GWV NE29 0QN
G4FRD NE29 0QX
G0AIO NE29 6XE
G1FNU NE29 7HP
G4PUI NE29 8DH
G0EDK NE29 8HS
G2FXS NE29 8LQ
G1LNR NE29 8LU
G0NWM NE29 8QA
G3PGC NE29 8QA
M1BTG NE29 8SS
G1GYM NE29 9BS
G0MEF NE29 9EN
G4YPT NE29 9EN
G6BRM NE29 9ES
G6FIU NE29 9HG
G0OBO NE29 9HT
G4AXL NE29 9NS
G4UOA NE29 9NS
G8YCP NE29 9NU
G1ZKT NE30 1HD
G4SWT NE30 2DA
G6OZJ NE30 2DA
G7EVW NE30 2NR
G4PEC NE30 2PN
G8PWX NE30 2PN
G7HHO NE30 2RR
G8EZL NE30 3AN
G8LLC NE30 3AQ
G4TLN NE30 3AX
G4MJB NE30 3EW
G1YUP NE30 3NE

North Warwickshire

G4OPE B46 1EP
G4XIQ B46 1EP
G6NTY B46 1HL
G3PVT B46 1NH
G0KBF B46 1QL
G0JXZ B46 1RT
G0CKE B46 1SN
G3ZUM B46 1SN
G1KAT B46 1UF
G0KBF B46 2AN
G4FMC B46 3EE
G0BUV B46 3EH
G1TKH B46 3EL
G4YQD B46 3LZ
G4FTY B46 3NE
G6IDJ B46 3NL
G0FFA B78 1JS
G0FXL B78 1JS
G8NAP B78 1NJ
G0AID B78 1NU
G0GUD B78 1QA
G8XWU B78 1QL
G6MUW B78 1RN
G8ACA B78 2AW
G1DAN B78 2DX
G4WQW B78 2JP
G0FEO B78 2JU
G3RWO B78 2LX
G4IZK B78 2NH
G4UUK B78 2NN
G6VBD B79 0DJ
G4WND B79 0DT
G4YNL B79 0DT
G3SLK CV10 0NF
G7ANQ CV10 0QY
G4MVL CV10 0QY
G0HGZ CV10 0RB
G1IJZ CV10 0RH
G0LLP CV10 0SL
G4KQL CV10 0SN
G4IAG CV7 8AU
G3SCJ CV7 8DX
G3PGZ CV7 8FE
G1UUK CV7 8PR
G6KXN CV7 8PR
G8LQF CV9 1HP
G4IWA CV9 1NZ
G6NWR CV9 1NZ
G3PH CV9 1PS
G4XDE CV9 2DA
G0GEF CV9 2DS
G4YNX CV9 2ET
G4BKA CV9 2HB
G7FFS CV9 2QT
G6SGY CV9 2RP
G4ZUR CV9 3AU
G3NXV CV9 3DP
G0VZK CV9 3EJ
G0HUW CV9 3EX

North West Leicestershire

G4RJO DE11 8AA
G0WHC DE11 8BG
G8VBC DE11 8BG
G1YBM DE12 6HQ
G0HWE DE12 7AB
G2FHF DE12 7BQ
G6JTV DE12 7EG
G4JCG DE12 7JG
G7KVZ DE12 7NA
M1BFV DE12 7NT
G6MYO DE12 7PW
G4FQI DE73 1AG
G8DBO DE73 1AG
G1ZGZ DE74 2DJ
G4IRH DE74 2DJ
G0SOU DE74 2DJ
G0WUG DE74 2DS
M0ACO DE74 2DS
M0ACN DE74 2DS
G0RDY DE74 2DT
G4AL DE74 2DT
G6NZZ DE74 2DX
G6XDS DE74 2DX
G4ISN DE74 2FA
G1ITY DE74 2GG
G3VJK DE74 2JF
G4GEI DE74 2JG
G0LXH DE74 2JW
G3WHV DE74 2LX
G4DTZ DE74 2NJ
G1ZEK DE74 2PJ
G8KZS DE74 2QL
G8PTW DE74 2QQ
G4USI DE74 2QT
G1LCR DE74 2QY
G4UVG DE74 2RX
G7PNG DE74 2SR
G8GBH DE74 2TX
G0VNQ DE74 2XA
G0DBR LE12 9TT
G8TSQ LE12 9TT
G8PFP LE12 9TY
G3AKU LE65 1ED
G0VVF LE65 1EG
G0COT LE65 1ER
G4WYN LE65 1ES
G4LAI LE65 1ES
G6EQB LE65 1EW
G6UZJ LE65 1RA
G4JNH LE65 1TR
G3UBB LE65 1WA
G3JUB LE65 1XD
G3KTC LE65 2FN
G0NXT LE65 2HL
G7NBE LE65 2JR
G1MVQ LE65 2JZ
G6CLP LE65 2JZ
M1AAX LE65 2NN
G0EKK LE65 2NU
G1CWW LE65 2QQ
G1ZLS LE65 2SR
G1UCN LE67 1EJ
G1ODQ LE67 2GL
G6SAL LE67 2HD
G4PLK LE67 2HE
G1YQU LE67 2HU
G7IVE LE67 2NY
G0HME LE67 2QL
G4AEO LE67 2TT
G8YGT LE67 3BE
G3VDW LE67 3BT
G8VFM LE67 3PL
G7WDG LE67 3PW
G8ZZY LE67 4BF
G7LIL LE67 4BH
G1NNF LE67 4DD
G4SOL LE67 4RE
G4JDP LE67 4TG
G8OZQ LE67 4TG
G0VUA LE67 4TH
G8JMG LE67 5AY
G1IWE LE67 5AZ
G4DCE LE67 5BP
G1KBN LE67 5BR
G8TMP LE67 5BT
G0FLG LE67 5DF
G0PXK LE67 5FD
G4UXU LE67 5FY
G4CPI LE67 5GY
M1AYC LE67 5PA
M1BAC LE67 5PA
G1KSC LE67 5PT
G7SEG LE67 5PT
G8YOG LE67 6HP
G3ZJG LE67 6JT
G0FZE LE67 6JW
G1FZE LE67 6JW
G4JKQ LE67 6LF

North Wiltshire

G6HTZ SN11 0EP
G7VEY SN11 0LQ
G4IFB SN11 8LF
M1BQI SN11 8RG
2E1FWG SN11 8RG
G0OIM SN11 9AB
G6TPT SN11 9AQ
M1AZF SN11 9BG
G8LRD SN11 9DU
G3WIW SN11 9EA
G6UHD SN11 9EN
G4NKU SN11 9EW
G1JRZ SN11 9LD
G4UJL SN11 9LR
G0EPM SN11 9LS
G4SUX SN11 9PA
G4GUK SN13 0JD
G3ORX SN13 0JR
G4DIE SN13 0JR
G8NEY SN13 0JZ
G7SNX SN13 0LB
G3JSK SN13 0QR
G3KRW SN13 8DF
G0FGR SN13 9AP
G7JLK SN13 9DU
G0UUV SN13 9DW
G8KLG SN13 9DW
G1LJT SN13 9NE
G0KLJ SN13 9NH
G4JQX SN13 9PZ
G1LRN SN13 9RR
G3MBN SN13 9SH
G4SKN SN13 9TX
G4YNT SN13 9UP
G4EMU SN13 9UT
G4SEZ SN13 9XB
G0WOV SN14 0AR
G1LMU SN14 0AW
G8MPM SN14 0DN
G3MEY SN14 0EB
G4GWR SN14 0FA
G4TCP SN14 0FE
G3LWF SN14 0HT
G4DVY SN14 0HZ
G0RBD SN14 0JB
G4SHI SN14 0LH
G4THG SN14 0PS
G4LQM SN14 0QU
G6DOD SN14 0SY
G1SOG SN14 0UJ
G1OKD SN14 0XZ
2E1EWU SN14 6AN
2E1FUB SN14 6AN
G4HMK SN14 6DY
G0BEH SN14 6LE
G8OPO SN14 6QG
G4IMT SN14 7AB
G6EYA SN14 7JZ
G0BGG SN14 7LT
G0RCP SN14 8BZ
G3PIJ SN14 8DN
G1JYA SN14 8EF
G8WMC SN14 8SA
G4ZUV SN15 1AE
2E1FBK SN15 1AP
G0FGW SN15 1AR
G6JBG SN15 1HD
G3PNH SN15 1PG
G8KIL SN15 1PG
G3UTO SN15 1PJ
G6FNY SN15 1PZ
2E1EGT SN15 1PZ
G7PUR SN15 1RN
G7OAS SN15 1TE
G3UUV SN15 2AF
G8HAM SN15 2BD
M1BPD SN15 2BD
G0MVM SN15 2NF
G8FKZ SN15 2RN
G4SDW SN15 3AA
G4OQG SN15 3EL
G0GJE SN15 3LX
G4SFD SN15 3RX
G6PYF SN15 3SU
G6KZU SN15 3UR
G6IBW SN15 3UZ
G4XPH SN15 3XF
G6VHE SN15 4EZ
G0IAX SN15 4NZ
G7TIV SN15 4QP
G8TTI SN15 4RQ
2E1EII SN15 4RQ
G4PRU SN15 5AQ
G4KJV SN15 5HG
G1WLE SN15 5JD
G0AEV SN15 5NN
G4CDW SN15 5QF
G6UYM SN16 0BL
G4TLY SN16 0DR
G4ORS SN16 0NT
G4SWN SN16 0PZ
G4JKM SN16 0QH
G8DKI SN16 9DL
G8IPS SN16 9DR
G1ISW SN16 9EQ
G8WLV SN16 9HE
G8YMB SN16 9HE
G2BRR SN16 9JE
G4BJC SN16 9JE
G4GPV SN16 9LR
G8KKD SN16 9NR
G8BAZ SN16 9PR
G3IYU SN16 9PT
G8DAB SN16 9RA
G4SWO SN16 9UA
G7CNP SN16 9UA
G4NCF SN4 7AU
G0OWR SN4 7DN
G7KRH SN4 7DX
G8KRD SN4 7ES
G4GKD SN4 7HF
G7TIW SN4 7SH
G0LTP SN4 8AS
G3EYU SN4 8AX
G4RZF SN4 8BB
G4MDH SN4 8DJ
GM6JLM SN4 8LF
G4FTZ SN4 8LL
G4OIQ SN4 8LR
G8HCW SN4 8LW
G4YAL SN4 8NQ
G3RZP SN5 0AD
G4FNC SN5 0AD
G3FEC SN5 9AD
G3YBY SN5 9AD
G4WSB SN5 9AT
G4YEF SN5 9BA
G3TPQ SN5 9BS
G4MQP SN5 9DE
M1AGH SN5 9EJ
M1AGI SN5 9EJ
GW4MGH SN5 9EX
G8IYH SN5 9HQ
G8OFR SN6 6BA
G6HVZ SN6 6BU
G1BZE SN6 6JF
G0TDY SN6 6JG
G8VGQ SN6 6QT

Northampton

G0PYI NN1 3BL
G0GKS NN1 3QR
G3YOV NN1 4HU
G3MPZ NN1 4JB
G1ICH NN1 4LU
M0ARZ NN1 4NA
G7REV NN1 4NA
G4IOJ NN1 5JL
G7KGP NN1 5LT
2E1CQL NN1 5ND
G4LOT NN1 5SD
G3VMU NN1 5ST
2E1BCC NN1 5ST
G1RTX NN1 8HF
G4JFC NN2 6AQ
G6NYH NN2 6EP
G4BZA NN2 6JF
G7LLP NN2 6JN
G7IRI NN2 7AA
G4RKQ NN2 7JQ
G4YJP NN2 7LW
G0GAQ NN2 7PT
2E0AAZ NN2 7QQ
G1IPD NN2 7QX
G6OHY NN2 7RR
G0WYQ NN2 7RY
G0NDS NN2 7TR
G1WCB NN2 7TR
G8WPU NN2 7TR
G0JPD NN2 8BJ
G7OGN NN2 8BJ
2E1BUA NN2 8BJ
G3ZJO NN2 8DZ
G0ONS NN2 8HD
G2HGA NN2 8HP
G1ERY NN2 8HT
2E1ABI NN2 8NR
2E1CCF NN2 8PE
G8PTH NN2 8PH
G7JCE NN2 8PQ
G1NRG NN2 8QU
G6TVB NN2 8QU
G8OZY NN2 8QU
G0JPF NN2 8QX
G4ZER NN2 8TA
G7TEG NN2 8TX
2E1ECM NN2 8TX
G0VFM NN3 2AF
G0MJK NN3 2HE
G8KDO NN3 2JH
G7AAY NN3 2JW
G0OQI NN3 2LJ
G0OKV NN3 2PZ
G0RTC NN3 2SY
G4IVR NN3 2SY
2E1BVC NN3 3AF
G1MMG NN3 3AT
G7ISE NN3 3BX
G1ALX NN3 3DP
G3HDJ NN3 3DY
G0LVP NN3 3JJ
G4MJF NN3 3JL
G1KMS NN3 3JX
G1FOU NN3 3LU
G7SYT NN3 3RX
G7BFH NN3 5AT
G1IRG NN3 5BH
G4WKN NN3 5DA
G0UTB NN3 5HL
G1TRZ NN3 5HL
G1HHW NN3 5JS
G4OIG NN3 5JS
G7THK NN3 5LJ
M1BPZ NN3 6AF
2E1CCE NN3 6AF
G4CCR NN3 6BE
G7KVY NN3 6HP
G1BHL NN3 6JL
G4SCJ NN3 6JQ
M0AKN NN3 6PF
G7ALK NN3 6PF
G3AXK NN3 7DE
G1KOW NN3 7HS
G6IVW NN3 7HU
G1DPJ NN3 7TX
G4USG NN3 7TZ
G6VNO NN3 8AP
M1BQR NN3 8AX
2E1EAY NN3 8AX
G7THZ NN3 8HL
G7RFX NN3 8HY
G0UOB NN3 8JE
G0MHA NN3 8LD
M1BQS NN3 8LX
G4HGB NN3 8QJ
G4IIO NN3 8SS
G4EUR NN3 8TP
G0PTM NN3 8UR
G4IRD NN3 8UX
2E1CAH NN3 8UX
2E1CBU NN3 8UX
2E1CAZ NN3 8YL
2E1EET NN3 9AX
G1VKJ NN3 9DN
2E1AAR NN3 9EE
G8XPQ NN3 9HP
G4MLO NN3 9UF
G8KNU NN3 9UQ
G4EZZ NN3 9UT
G4AHQ NN4 0HP
G4LUW NN4 0NJ
G4OFP NN4 0QG
G0FQI NN4 0RB
G6KIZ NN4 0RT
M1AQC NN4 0SD
G0FTP NN4 0SN
G7LBS NN4 0SN
G7MJC NN4 0SP
G7TEI NN4 0SP
G0LRD NN4 0TR
G6NPZ NN4 6AH
G4RJT NN4 6AX
G6SJQ NN4 6BA
G4DHV NN4 6ER
G7NEG NN4 6HE
G0TMW NN4 6LA
G0GIN NN4 6LX
G3YNF NN4 7AF
G8ZEW NN4 7AP
G8NAL NN4 7BT
G8HTO NN4 8AU
G3KAN NN4 8AY
G6FAL NN4 8AZ
G3TMQ NN4 8LJ
G6ONM NN4 8NS
G1TPN NN4 8PE
G6OXY NN4 8TR
G6ZLP NN4 8UN
G7VQE NN4 9UA
G3EVX NN4 9UE
G1BHR NN4 9UL
G0OGW NN4 9YW
G3NOK NN5 5DA
G6TXZ NN5 5DT
2E1FUQ NN5 5EQ
G0KOG NN5 5PF
G3ZBC NN5 6AP
G7IXK NN5 6BW
G0RDT NN5 6JW
G3JJW NN5 6LF
G8TEB NN5 6NG
G7ENY NN5 6NL
G7HLP NN5 6NL
G0HWC NN5 6NX
G4ICC NN5 6PP
G8KST NN5 6PU
G4NVP NN5 6PY
G1OMI NN5 6RA
G3RLD NN5 6RA
G0VJK NN5 6RF
G0NSN NN5 6TG
G0DLF NN5 6YW
G4PXJ NN5 7BL

Norwich

G7VZL NR1 2AL
G4RRX NR1 2JR
G7UVB NR1 2JX
G0OOR NR1 2NJ
G0WJU NR1 2NL
G4VLS NR1 2NX
2E1FMO NR1 2NX
G7VUD NR1 2PD
G3SEM NR1 2PH
G0REG NR1 2PL
G4HST NR1 2QR
G3PNR NR1 2RG
G7MLC NR1 3HU
G0TXV NR1 3JG
G0USR NR1 4AR
M1BLI NR1 4DB
G0BFM NR1 4HB
G4XPB NR1 4LX
G0KBY NR1 4NJ
G3HYJ NR1 4NJ
G4FHT NR1 4NU
G0WCO NR1 4PD
G4EXN NR2 2AX
G7TTP NR2 2PY
G3VWQ NR2 3EA
G4FZZ NR2 3EE
G1MBX NR2 3EF
G4STX NR2 3EF
G4TKZ NR2 3EF
G6ECJ NR2 3JG
G7FUQ NR2 3NJ
G8GWJ NR2 3NJ
G3HZE NR2 3QR
G3WCE NR2 3RH
M1BQW NR2 3RH
G6TSZ NR2 3TL
M1ANZ NR2 3UT
G4LDG NR2 4BL
G0OPU NR2 4LF
G0DUV NR3 1EY
G0JAG NR3 1NY
G7HIM NR3 2EL
G8AUN NR3 2HW
G6DMN NR3 2JY
G4RCI NR3 2LP
G7VUM NR3 2LW
G0WJZ NR3 2QP
2E1FGA NR3 2RR
G0OOD NR3 2SJ
G7DFP NR3 3HP
G1OAY NR3 3LB
G0SSX NR3 3NN
G0MQI NR3 3NU
G4YDQ NR3 3TB
G4FMK NR3 4BY
G0SOQ NR3 4DE
G4WAQ NR3 4DX
G3ZTV NR3 4HH
G0PEQ NR3 4QW
M1AFQ NR3 4TS
G8ZGS NR4 6HA
G3GLA NR4 6JD
G4NZQ NR4 6LT
G7MKP NR4 6LT
G7IBB NR4 6NF
G0OTZ NR4 6NL
G4KPJ NR4 7EY
M1ACT NR4 7EZ
G4FJY NR4 7HP
G8VLL NR4 7HP
M1BKF NR4 7HP
G3SCV NR4 7JA
G4ULR NR4 7LW
G8UJO NR4 7NJ
G4NCO NR4 7NS
2E1FDA NR4 7NX
G8QR NR4 7QA
G0SMH NR5 8AS
G3BNV NR5 8ND
G4RMN NR5 8RA
G3NJQ NR5 8TE
2E1FOF NR5 8TE
G0MPN NR5 8YF
G7UFV NR5 9AE
G0GSZ NR5 9AR
G4WTD NR5 9AX
G3THV NR5 9BL
G1FLT NR5 9BL
G1DXQ NR5 9LB
G4LFQ NR5 9PE
G4PFZ NR6 5EP
G7WAK NR6 5EP
G6VZU NR6 6DQ
G0IRQ NR6 6DY
G0FKG NR6 6RB

Nottingham

G4DIQ NG1 1DJ
G1YEU NG1 3PW
G4LPD NG11 7AU
G4NSP NG11 7AU
G7HUK NG11 7EQ
G7MMQ NG11 7EB
G8LNG NG11 7FD
G7BGY NG11 8AS
G1PNX NG11 8FD
M1AOK NG11 8FW
M0BCF NG11 8FW
M1AOT NG11 8FW
G0WXF NG11 8GF
G7SWE NG11 8GF
G8SOL NG11 8HF
G7SMQ NG11 8JG
G0PLW NG11 8JS
G4ASN NG11 8PH
G6CIF NG11 8PH
G7RJI NG11 8RE
G0HNZ NG11 8SH
G0JVW NG11 8SL
M0BCW NG11 9AL
G7TYQ NG11 9ED
2E1DEI NG11 9ED
2E1BVY NG11 9FW
G0SOM NG11 9HD
G7IQV NG11 9HD
M1ACV NG11 9HD
G0LXX NG11 9JN
G7IZF NG11 9JZ
G7MKG NG11 9JZ
2E1FED NG2 2EB
M1BHP NG2 2FD
G7HGF NG2 2JH
2E1FRH NG2 3GD
G0HMO NG2 4BB
G1WRQ NG2 4JP
G4NMI NG3 1HJ
2E1FVQ NG3 1HJ
G1SBP NG3 1NR
G4TUX NG3 2BH
M0BCB NG3 2GF
G1WQE NG3 2GF
G6KJO NG3 2LP
G0MLM NG3 2LS
2E1BSW NG3 2LS
G0IEG NG3 3AA
G3EKW NG3 3AA
2E1EFL NG3 3BY
2E1APY NG3 3BY
G0WYI NG3 3EQ
2E1CWA NG3 4JL
2E1CWI NG3 4JL
G1MQB NG3 4PZ
G4UWP NG3 4QG
2E1EKD NG3 4QT
G8CXV NG3 5EN
G4ZDX NG3 5FD
G2FUB NG3 5GD
G8BPQ NG3 5QL
G7FDF NG3 6GR
G7HFE NG3 6GR
G0VQC NG3 7AL
G3SEN NG3 7AP
2E1FRF NG3 7FY
G4ZCW NG3 7GA
G4SJG NG3 7HD
G4VZU NG3 7HD
M0AEB NG3 7HJ
G0VEF NG3 7HJ
G3JSP NG5 1BU
G6SRZ NG5 1EP
G6ZEA NG5 1FW
G6ONW NG5 1GP
G1LOC NG5 1HX
G6CW NG5 1JR
G4EKW NG5 1JR
G6SFF NG5 1NL
G0OAI NG5 2LW
G0DME NG5 3DA
G1HRM NG5 3GS
G7LKV NG5 4BD
G0SCZ NG5 4EE
G3DUL NG5 4HE
G8VCI NG5 5BL
G0EJH NG5 5BU
G6UZA NG5 5DT
G6URR NG5 5DU
G7ULJ NG5 5NP
G7SOP NG5 5RX
2E1FCY NG5 5RX
G0WOL NG5 6EL
G0FKQ NG5 9DW
G1JNM NG5 9HX
M0BCH NG5 9LN
G0KZA NG6 0BH
G1XHO NG6 0BS
G0INA NG6 0LS
2E1EMP NG6 7BH
2E1EMQ NG6 7BH
2E1ERE NG6 7BH
G1ZLA NG6 7FJ
2E1FGD NG6 8DG
G0JOU NG6 8GL
G1SKV NG6 8GZ
G1YBT NG6 8JZ
G0NVS NG6 8NG
2E1DLT NG6 8NL
G8LMX NG6 8NY
G6VWV NG6 8PU
G0VFU NG6 8QY
G6NLD NG6 8SL
G6IQP NG6 8WD
G1GTF NG6 9DT
G8ZZV NG6 9FB
G1YWN NG6 9FX
G0OZA NG6 9GJ
G7BTK NG6 9GQ
G1UDX NG6 9GS
G8OHC NG6 9HN
2E1DOY NG6 9HN
G6GWY NG6 9JE
G0LBW NG6 9JY
G1OXB NG7 2HR
2E1EVA NG7 2JJ
G6UFM NG7 2JN
G4ZDB NG7 2QG
2E1EAZ NG7 5EB
2E1ECG NG7 5EB
G3VVU NG7 5NN
G6ISB NG7 6HU
G6XMP NG7 7EW
G1RAS NG8 1DU
G4OMJ NG8 1HL
G6FHE NG8 1HY
G6UAP NG8 1JU
G7WFM NG8 1QA
G4BWN NG8 2BZ
G6XBH NG8 2ER
G8UUS NG8 2ER
G7KXR NG8 2NA
G8BNG NG8 2NA
M1BYP NG8 2NN
2E1DPW NG8 2NN
G3KDQ NG8 2PE
G4KLJ NG8 2QR
G8NWU NG8 2QR
G0HDM NG8 2QS
G4NXU NG8 2RA
G0FOE NG8 2RE
G1XUQ NG8 2SB
G1XUR NG8 2SB
G8POL NG8 2SB
M1BQH NG8 2TE
2E1DSL NG8 2TE
G0IMI NG8 3DA
2E1DAN NG8 3DD
G4TYO NG8 3NE
G1SMX NG8 4AG
G7NVH NG8 4AN
G0BHO NG8 4AP
G7GVZ NG8 4AU
2E0AQB NG8 4AZ
2E1EPY NG8 4AZ
G0MLI NG8 4BB
2E1FDT NG8 4BE
G8CIK NG8 4DA
G0TSG NG8 4DB
M0BCD NG8 4DN
G1PWW NG8 4DN
G4CNJ NG8 4FH
M0AUK NG8 4GY
G2FRY NG8 4NH
G1RCN NG8 5FJ
G1XUM NG8 5JR
M1BEY NG8 5RA
G7SBK NG8 6LU
M1BSI NG8 6NE
2E1FMQ NG8 6NE
G1YBA NG8 6QE

Nuneaton and Bedworth

G1JWY CV10 0BA
G1WRN CV10 0BA
G4UQU CV10 0BX
G0MDN CV10 0DF
G7HNC CV10 0DH
G0OVG CV10 0DP
G3VDU CV10 0DW
G0VZO CV10 0DW
G4SHY CV10 0HN
G7CEN CV10 0JN
G6LUK CV10 0LB
G4SEU CV10 0LB
G2HAO CV10 0LD
G6XEE CV10 7BJ
G6MTH CV10 7BJ
M1BXT CV10 7BU
G3ZSQ CV10 7DE
G4ZUE CV10 7EE
G1FTH CV10 7EP
G3YTW CV10 7PW
G1UWR CV10 8AP
G7TYJ CV10 8EG
G4YOZ CV10 8EP
G1XGG CV10 8HY
G6DDA CV10 8JF
G0AFZ CV10 8NH
G4NHF CV10 8NH
G4BBQ CV10 8NU
G0LDP CV10 8QN
G1OWR CV10 9DY
G0SZG CV10 9DY
G7JXF CV10 9JH
G1VTU CV10 9NG
G6OQV CV10 9PB
G8OMB CV10 9QJ
G7SSD CV10 9QP
G8WQH CV10 9RR
G0FBG CV11 4NU
G1EXV CV11 5PA
G6MTG CV11 5PB
G1JVL CV11 5UB
G4LMK CV11 6BQ
G4NQZ CV11 6DZ
G0SBC CV11 6EE
G7DDF CV11 6EF
G1WZV CV11 6EZ
G8GXN CV11 6EZ
G4AEH CV11 6FG
G6SFY CV11 6FN
G1FIU CV11 6FW
G8LMN CV11 6GN
G4HXC CV11 6JA
G8UGM CV11 6PL
G0NCQ CV11 6UU
G1AGV CV11 6WB
G0OBE CV12 0AZ
G0JUN CV12 0DE
G0IYW CV12 8BE
G1LBH CV12 8DA
G1OJD CV12 8DG
G0FXO CV12 8SA
G1ZSR CV12 8SG
G1IXU CV12 9AQ
G0BVS CV12 9BX
G1SRD CV12 9BX
G0BIN CV12 9JB
G3YQZ CV12 9ND
G1ORG CV12 9PS
G3YAL CV12 9QY
G1FLW CV12 9SD
G3UVW CV6 6AY
G3ZFR CV6 6AY
G3SPY CV6 6AY
M1AQF CV7 9AD
G7AMQ CV7 9AU
G4LML CV7 9BJ
G4VVM CV7 9GE
G7ASF CV7 9GE
G4WXK CV7 9NJ

Oadby and Wigston

G4LXA LE18 1BA
G1PBD LE18 1BP
G6IFN LE18 1BR
2E1BSY LE18 1DB
G6IHU LE18 1DH
G4UJC LE18 1DS
G0UVP LE18 1DX
G8SSX LE18 1GD
G0HLL LE18 1HY
G4GMD LE18 1ND
G3YOO LE18 2GY
G0BYR LE18 2HN
G1IHN LE18 2HQ
G1VNS LE18 2JB
G1RJE LE18 2JH
G0TCJ LE18 2QX
G1UFT LE18 3QF
G7DPU LE18 3QS
G4XDX LE18 3RJ
G8EVD LE18 3RL
G3DYH LE18 3SA
G7KMK LE18 3SB
G4EMW LE18 3TY
G4OKD LE18 3WD
G1YQP LE18 3XW
G0CFQ LE18 4QA
G6YII LE18 4WH
G2CFC LE2 2RF
G1IGP LE2 4NT
2E1FEC LE2 4NY
G4TNG LE2 4US
G6PFN LE2 5HF
G3ZJS LE2 5JB
G5UM LE2 5PF
G3XKX LE2 5PF
M1BMC LE2 5RB
G8TYX LE2 5TL
G8HHZ LE2 5TN
G1HWO LE2 5TR
G6IJL LE2 5TS
G4CWY LE2 5UB
G3HAN LE2 5UE

Oldham

G7TYX M35 0AW
G7BAB M35 0PX
G4WYA M35 0SA
G3PYH M35 9EJ
G0ICY M35 9PD
G4YPS M35 9PY
G0FZF OL1 2JS
G1EVS OL1 2PU
G7EBF OL1 4DN
G1UVR OL1 4NH
G6HJD OL1 4NW
G3WXN OL2 5AS
G0KHI OL2 5BP
G5XJ OL2 5DD
G8YKO OL2 5HL
G6TPQ OL2 5NF
G6PBI OL2 5SD
G6GLD OL2 5UT
G1UTP OL2 6BH
G1GKR OL2 6LR
G7MXQ OL2 6NJ
G4TQW OL2 6RR
G0BJR OL2 6RW
G1ORC OL2 6RW
G4ORC OL2 6RW
G1NFE OL2 6SS
G0KUY OL2 6TW
G1UKW OL2 6TW
G4GNG OL2 7AY
G3TSJ OL2 7DR
M0AUG OL2 7EG
G0UJF OL2 7EQ
G1POG OL2 7QY
G6TCD OL2 8DA
G0MMX OL2 8DR
G7MCT OL2 8EA
G6TVE OL2 8HL
G6NUK OL2 8HU
G0PVO OL2 8JF
G0GTC OL3 5DS
G2HJT OL3 5LZ
G0EOX OL3 5NU
G1GZK OL3 5PL
G1KFG OL3 5PL
2E1BHR OL3 5PL
2E1BRY OL3 5PL
G7ENC OL3 5PW
G3SNA OL3 5PX
G3KLL OL3 5RJ
G4UQL OL3 5RJ
G4AIW OL3 6EB
G0HFC OL3 7AW
G4HFG OL4 1AS
G4WHF OL4 1AS
G3TJX OL4 1LX
G1JZU OL4 2DJ
G4UFG OL4 2PW
G8UZU OL4 2QD
G1AXE OL4 3EL
G0RFM OL4 3PR
G7VHZ OL4 3PZ
G1WYA OL4 3QW
G0EYU OL4 4AZ
G4EBQ OL4 4AZ
G8YAG OL4 4DW
G4SSC OL4 4ED
2E1BOB OL4 4EP
G7UJL OL4 4EP
G7WFD OL4 4ES
G4NIR OL4 4JX
G4VYB OL4 4JX
G6DAD OL4 4RY
G6USN OL4 4RZ
G6TAI OL4 4SJ
G4PLV OL4 5LF
G0ULA OL4 5LG
G7GQJ OL4 5PH
G0RKE OL4 5SH
G4GLV OL4 5SH
G0KTQ OL8 1BE
G1XGW OL8 1LT
G0IMB OL8 1LW
G4URG OL8 2EW
G1HAC OL8 2LP
M1BGF OL8 2PD
G4OTM OL8 2PX
G1UTR OL8 2QG
G3LGV OL8 3BA
G6ALM OL8 3BA
G0DKL OL8 3DG
G7WAE OL8 3NP
G7OAL OL8 3PR
G2JT OL8 4AL
G0WNY OL8 4AL
M1BEZ OL8 4HG
G0MGL OL8 4HT
G0COX OL8 4HX
G1LUR OL8 4NX
G0KTR OL8 4SF
G0SQU OL8 4SF
G8JRZ OL9 0EJ
G6EJI OL9 0HJ
G0SIY OL9 0NS
G6HNF OL9 0RA
G1SLA OL9 0RF
G6MMS OL9 7JW
G7NPJ OL9 7LS
G8YEV OL9 7RX
G4TLW OL9 8AP
G3OMC OL9 8DB
G3SMZ OL9 8HH
G1SGW OL9 8PU
G4BKH OL9 8QD
M1BGR OL9 9LX
G3IGC OL9 9QR
G4AVF OL9 9RA
G0HSM OL9 9TA
G1VSJ OL9 9TB

Orkney Islands

GM1RQD KW15 1EE
GM6VQZ KW15 1ER
2M1CYF KW15 1JT
GM4ZZH KW15 1NA
GM6WOF KW15 1NA
GM7TMT KW15 1NA
GM3MTS KW15 1PE
GM0JAV KW15 1PL
GM7UGZ KW15 1QF
GM0WFA KW15 1QJ
GM0MHS KW15 1RL
GM8NFG KW15 1SL
GM0WED KW15 1SX
GM0OWM KW15 1SZ
GM3IBU KW15 1SZ
GM4TYU KW15 1SZ
GM0TLX KW15 1SZ
MM0AUP KW15 1TB
GM1MWK KW15 1TE
GM1VZG KW15 1XJ
GM3PLO KW16 3DJ
GM4IOB KW16 3EP
GM7GMC KW16 3HR
GM4EWZ KW17 2AT
GM0LDT KW17 2BE
GM4PSX KW17 2BJ
GM4JEF KW17 2BL
GM4IDV KW17 2DW
GM7RDY KW17 2EB
GM4DZX KW17 2EZ
GM0HTT KW17 2HR
GM0HTH KW17 2JU
GM7RDH KW17 2LQ
GI0PMO KW17 2ND
GM0IDV KW17 2NN
GM0HTG KW17 2PB
GM4YBJ KW17 2PB
GM3POI KW17 2QL
GM0IJV KW17 2QQ
GM1MXE KW17 2RD
GM0HQG KW17 2RF
GM8WJK KW17 2RN
GM6OPK KW17 2RT

Oswestry

GW0SFP LL14 5DP
GW7LXI SY10 7AE
G6YGI SY10 7AS
G1ORA SY10 7BB
G4TTO SY10 7BB
G6MIJ SY10 7BG
G0MQA SY10 7BQ
G4TQE SY10 7BX
G0NFI SY10 7HF
GW1PLJ SY10 7PQ
G4UDE SY10 7RQ
G0JJU SY10 8EN
G4LU SY10 8LD
G7PHF SY10 8LG
G1MAB SY10 8LL
G3BOC SY10 8LL
GW4JUW SY10 8QH
G3DGR SY10 9AP
G8AJA SY10 9AU
G1VIR SY10 9DF
G4IOQ SY10 9DQ
G0VEU SY10 9ES
G1JVH SY10 9HN
G7JHC SY11 1TB
G6BSX SY11 2JZ
G2FDF SY11 2LS
G3JPU SY11 2NB
G6AUS SY11 2XA
G6HUR SY11 3DG
G0CWZ SY11 3DH
G1ENS SY11 3JS
G7SFI SY11 3NP
G1WZX SY11 3PJ
G7OTQ SY11 4PN
G4CLZ SY11 4SG
GW4UBQ SY22 6NF
G3NSS SY4 1JD
G0DNI SY4 1JH

Oxford

G0KUB OX1 3PG
G8HXU OX1 3TN
G3BLS OX2 0BE
G0RNZ OX2 0HH
G4FON OX2 0HS
G1OWM OX2 6AY
G4ZSV OX2 6HZ
G4XVJ OX2 6JQ
G7TTL OX2 6LE
G4ZHE OX2 6TX
G3NSM OX2 6UU
G0FGP OX2 7EN
G8PX OX2 8LP
G0BPX OX3 0EX
G6DVE OX3 0JG
G1OYF OX3 0JJ
G4OBB OX3 0LS
G3BTV OX3 0NY
G7UZQ OX3 0QA
G0AGJ OX3 0RD
G3UGJ OX3 0RJ
G1CSZ OX3 0SJ
G0LBZ OX3 0US
G6ZLM OX3 7NA
G6FGZ OX3 7RR
G4KSQ OX3 8ED
G7EXP OX3 8HB
G4AYR OX3 8JA
G4XZU OX3 8JB
G4RKF OX3 8NL
G2HMY OX3 8PB
G3UJO OX3 8PD
G4MAQ OX3 8TS
G3ZTC OX3 9EW
G1KEO OX3 9PJ
G0AFY OX3 9QF
G0CPP OX4 1BY
G3LOF OX4 1LT
G3PMI OX4 1NY
G6LFJ OX4 1QS
G6MSQ OX4 1SS
G4AZT OX4 2NU
G7IVM OX4 3AR
G6AQC OX4 3BX
G0AZP OX4 3HN
G1PCU OX4 3HT
M0AEO OX4 3HU
G0KNV OX4 3LW
G7IMH OX4 3NU
M0ACU OX4 3QU
G3HJS OX4 3SW
G1YDI OX4 3YG
G0XBG OX4 4AR
G6BTB OX4 4NE
G8OMW OX4 4PW
G7ARG OX4 4UD
G7TLD OX4 5DL
G4KON OX4 5EL
G6ZPL OX4 5ER
G0LUO OX4 5PJ

Pembrokeshire

GW6NMO SA37 0JS
GW0TMV SA41 3RR
GW3FPC SA41 3UY
GW3PKO SA42 0QG
GW6CJJ SA42 0QS
GW6OZA SA42 0QY
GW4UEP SA42 0RQ
GW4OQB SA42 0XS
GW0GZQ SA43 3BY
GW0GUY SA61 1BW
GW4TIE SA61 1LD
GW1HCW SA61 1RX
GW0XAQ SA61 1ST
GW8ZMU SA61 1UA
GW6KHH SA61 2NU
GW6ZSK SA61 2PS
GW8YJN SA61 2SQ
GW0UQH SA62 3BE
GW4CMW SA62 3EU
GW4PCX SA62 3HX
MW1BHY SA62 3PT
GW7VSO SA62 3PW
GW0VYF SA62 3RN
G7OXN SA62 3RS
GW7NGU SA62 3TR
GW1RJU SA62 4DN
G3WGB SA62 4ET
GW3PDW SA62 4JD
MW0AWM SA62 4LL
GW1MVZ SA62 4LR
GW3IGG SA62 4LR
GW8UKW SA62 5AT
GW0EJE SA62 5DP
GW4UZL SA62 5DP
GW3KRD SA62 5SQ
GW6VKD SA62 5XX
G7RRZ SA62 6AG
GW1JFV SA62 6AU
GW4RZU SA62 6AU
GW8NAC SA62 6EN
GW0GTK SA62 6JE
GW0DDK SA62 6JZ
GW7FLA SA62 6TQ
GW4HXO SA62 6TX
MW0ARD SA62 6XT
GW3OSV SA63 4SA
GW3OGC SA63 4TE
GW4UEJ SA64 0ES
GW1JWN SA64 0HB
GW0BOD SA65 9DS
GW0CXK SA65 9LH
GW6NKG SA65 9LN
GW0AQC SA65 9RL
GW0HMC SA65 9RL
GW8YLK SA65 9SL
GW4LLC SA66 7HS
GW3RKV SA66 7QS
GW8SPH SA66 7QY
GW3VEW SA67 8JE
GW3VEP SA67 8JF
MW0BEY SA67 8JL
GW7UWM SA67 8JL
GW0WGG SA67 8SX
GW0AJY SA67 8UG
GW4WMD SA68 0SX
GW3TUD SA68 9EJ
GW3WBH SA69 9PE
GW0JRF SA70 7DP
GW0VJO SA70 7DP
GW0WBP SA70 7PF
GW0WBQ SA70 7QD
GW3VEN SA70 8DH
GW4CBR SA70 8SG
GW4AKO SA71 4BB
GW4TEJ SA71 4ER
2W1DGL SA71 4EW
GW4WMK SA71 4NB
GW0BTW SA71 4NU
GW4RGI SA71 5AX
GW3WWB SA71 5DJ
MW0AIE SA71 5ET
GW6TEO SA71 5HJ
GW4OZU SA71 5HY
GW4VTG SA71 5JH
GW4OXL SA71 5LG
GW3RUL SA71 5NP
GW4QJX SA72 6BL
GW3LXI SA72 6ED
GW4TGT SA72 6JP
GW4VRO SA72 6LH
2W1DGK SA72 6NZ
GW4REI SA72 6QN
2W1DGM SA72 6RJ
M1BXX SA72 6TL
GW4YVN SA72 6TP
G1ZJO SA73 1BP
GW3VGY SA73 1EH
GW0LIK SA73 1HD
GW4ZYV SA73 1HD
2W1CRR SA73 1NR
GW1WQM SA73 1PA
GW3MOP SA73 1PA
GW0VND SA73 1SA
GW4MTU SA73 1SF
GW4XQK SA73 1TB
GW7HGU SA73 1TB
GW0SYG SA73 1TB
GW4HGS SA73 1TR
MW1BCQ SA73 1TR
GW4OZJ SA73 2DS
GW4OKF SA73 2ED
GW1PND SA73 2HR
GW7VMT SA73 2JR
GW0TLJ SA73 2LH
GW0GUB SA73 2LU
MW0BBU SA73 2NU
MW1BGP SA73 2NU
GW4ODN SA73 3HN
GW3DEX SA73 3HT
GW6MWG SA73 3SP

Pendle

G3UEU BB12 9BA
G6FEP BB12 9ED
G8YVU BB12 9QA
G0RTU BB12 9QA
G6KGW BB8 0DH
G0RFQ BB8 0HN
G6PDM BB8 0ND
G4GMU BB8 0QG
G0UGM BB8 0RQ
G0UIR BB8 0RQ
G4YMO BB8 5BQ
G4GOZ BB8 5JB
G3KJY BB8 5JS
G7WAW BB8 5NH
G4LWG BB8 5NW
G7SNQ BB8 5PD
G7WEK BB8 5PD
GW6XZI BB8 5QF
G1KSV BB8 5SH
G3NSG BB8 6AY
G1MDE BB8 6AZ
G0HFG BB8 6EB
G4ILG BB8 6HB
G6RRB BB8 6HB
G4GIZ BB8 6PB
G8RMX BB8 6PE
G4VMA BB8 6PX
G4XSK BB8 6QX
G0VON BB8 6RD
G4KAK BB8 7AA
G4CPS BB8 7DP
G1WAP BB8 7HW
G4HCH BB8 7JR
G6SMK BB8 7JT
G3ZCX BB8 7JY
G6HMN BB8 8BW
M0ANP BB8 8JG
G4DUI BB8 8JX
G0IQM BB8 8QS
G0WQM BB8 8QS
G0HBN BB8 8SJ
G4WUZ BB8 9DD
G6DFV BB8 9DF
G4HCC BB8 9QA
G4RTS BB8 9QJ
G0VQJ BB8 9QR
G7VZM BB8 9RS
G7WGB BB8 9SD
M1BPH BB8 9SD
2E1CUL BB8 9SN
G0TFK BB9 0AP
G4KTR BB9 0DA
G4ZLU BB9 0HZ
M0BFZ BB9 0JB
G4UEA BB9 0PP
G0BPR BB9 0QD
G4MJX BB9 0RP
G4MLB BB9 0ST
G0ECB BB9 0SU
G4KFF BB9 0YH
2E1CUA BB9 0YH
G4ZFD BB9 5BY
G8PZU BB9 5DS
G4OFE BB9 5DX
G6PEI BB9 5EU
G4BLH BB9 5EW
G8CME BB9 5EW
G3OGU BB9 5EX
G4MYU BB9 5RS
G0LLL BB9 6BQ
G4UCU BB9 6EA
G4WZM BB9 6LD
G8TUL BB9 6LU
G4SEG BB9 6LZ
G1ZWK BB9 6NN
G4TSV BB9 7BD
G4UUA BB9 7BD
G0AQZ BB9 7BP
G6PBQ BB9 7DB
G0ABV BB9 7RA
2E1BOZ BB9 7RW
G0TZF BB9 7XU
G4XGV BB9 8AB
2E1CTZ BB9 8AB
G3UUM BB9 8DF
G0WOJ BB9 8DF
M1BSV BB9 8JW
G0WXY BB9 8QU
G0JAF BB9 8SD
2E1AZV BB9 9HR
G6PVU BB9 9JA
G6YHS BB9 9JL
G3TRK BB9 9LN
M1BZV BB9 9LW
G0GGT BB9 9NQ
G7SYQ BB9 9PA

Penwith

G0ARL TR17 0BU
G8VDU TR17 0DR
G7CCS TR18 2BA
G0BQU TR18 2DG
G4PEM TR18 2HH
G7SBL TR18 2HQ
G0ANF TR18 3AB
G4YWP TR18 3AF
G3NRD TR18 3BB
G1IVO TR18 3LD
G3RID TR18 3NA
G0PZR TR18 3PD
G0WZC TR18 3PD
G7SBP TR18 3QR
G1JAH TR18 3RH
G0PGB TR18 4BH
G0EFV TR18 4LD
G2JL TR18 4SJ
G4HVK TR18 4SN
G8EJC TR18 4SP
G0AJG TR18 5AY
G4BPJ TR18 5DQ
G8GFS TR18 5EA
G0MWZ TR18 5QW
G1ZME TR19 6DZ
G3XRJ TR19 6LX
G4RRQ TR19 6TT
M0AWT TR19 7AR
G0THQ TR19 7BS
G4AMT TR19 7BT
G3HRD TR19 7DS
G3OYB TR19 7DZ
G4YAF TR19 7ET
G7SBN TR19 7HB
G4XAM TR19 7HH
M1BKS TR19 7SS
G3UUZ TR19 7ST
G4XCV TR19 7TA
G3PZN TR19 7UT
G8ARM TR20 8AJ
G7FKG TR20 8BG
G3YSN TR20 8DJ
G3UUC TR20 8RG
G7TBJ TR20 8RU
G7TJD TR20 8UL
G1GKF TR20 8XA
G3GYE TR20 8XD
G0AIX TR20 8XW
G3ZPJ TR20 9BW
G1RND TR20 9HB
G4RXZ TR20 9HJ
G4SOK TR20 9HN
G3AGA TR20 9JP
G3MPD TR20 9JP
G6VDU TR20 9LQ
G3IUE TR20 9PA
G8JAB TR26 1AX
G7HVD TR26 1BL
G4EIS TR26 1BQ
G6AKE TR26 1BQ
G0KYI TR26 1HA
G8LUB TR26 1HA
G4EXJ TR26 1QJ
G3DDX TR26 1QW
G6LKY TR26 1RJ
G0KVP TR26 1SE
G4HTY TR26 1SE
G1UAY TR26 2BH
G0ALA TR26 2BP
G0PGJ TR26 2DN
G0BBA TR26 2DU
G0CIO TR26 2DU
G3DLH TR26 2PT
G4BKI TR26 2PT
G0RFC TR26 2TT
G3NPB TR26 2TT
G8PMB TR26 2TT
G4YQS TR26 3AL
G0PMN TR26 3EB
G7MJM TR26 3JF
G4TXD TR27 4AQ
G1STA TR27 4DH
G0JFJ TR27 4EF
G4KYH TR27 4EN
G1CHM TR27 4LY
G4KYO TR27 4NJ
G1LCY TR27 4PJ
G6YWK TR27 4QD
G4HUG TR27 4QT
G4GKY TR27 5AB
G0PUA TR27 5AF
G3IQI TR27 5DA
G3HFS TR27 5DP
G4ZQD TR27 5DZ
G3UHZ TR27 5EQ
G3YNK TR27 5HA
G4MSV TR27 5HA
G4RVP TR27 5HA
G6STD TR27 5JA
G6YVD TR27 5JF
G4TXE TR27 6HJ
G3ZMP TR27 6HU
G3EGQ TR27 6HZ
G1ZOB TR27 6JB

Perth & Kinross

GM2CPC DD2 5AA
GM0KKE DD2 5DL
GM0MXP FK15 9PX
GM4HJO KY13 7DN
GM0KVE KY13 7ES
GM7FGF KY13 7HF
GM0GNT KY13 7JX
GM0KDO KY13 7PQ
GM0BKS KY13 7QT
GM1PKY KY13 7RP
GM4UIJ KY13 7RR
GM4WLN KY13 7UH
GM0VCN KY13 7XG
GM4MSL PH1 1BX
GM4HJK PH1 1DT
GM8JCR PH1 1EL
GM4BVD PH1 1JD
GM8VBP PH1 1JD
GM8UGO PH1 1LE
GM6JKU PH1 1QB
GM3TRI PH1 2BG
GM0BBR PH1 2LN
GM4YZT PH1 2NA
GM6UJG PH1 2NF
GM4YXK PH1 3BL
GM1AHF PH1 3DD
GM1AHG PH1 3DD
GM4DYZ PH1 3HN
GM3DAP PH1 3NL
GM1DSK PH1 4NA
GM4EAF PH1 4QS
GM4ZRH PH1 4QS
GM1FIX PH1 5DD
GM4XLI PH10 6JS
2M1EPV PH10 6PF
GM4OWR PH10 6XY
GM3BHY PH10 7EP
GM0VIT PH10 7JL
GM3WFJ PH10 7NY
GM1ZQF PH11 8AS
GM0LYT PH11 8BU
2M0AHB PH11 8BU
GM3WAP PH11 8DW
GM0GSG PH11 8JN
GM0HIG PH11 8JN
GM7ILO PH13 9BP
GM6YRN PH15 2EJ
GM4FOZ PH15 2QY
GM3KSD PH16 5EP
GM7JGE PH16 5HH
GM8KPH PH16 5HH
GM4ODW PH16 5JS
GM0OYE PH16 5SR
GM1PFU PH2 0BP
GM0FTX PH2 0HH
GM6UWF PH2 0LD
GM1EAF PH2 0NF
GM8VPT PH2 0NF
GM4EVS PH2 6BL
GM4AWA PH2 6DX
GM3JUD PH2 6HT
GM4ULS PH2 6PW
GM3EOB PH2 6RR
GM4DQJ PH2 6SD
GM1GHZ PH2 6TE
GM3UYR PH2 7BY
GM1FBM PH2 7LL
GM0GAV PH2 7ND
GM8YRT PH2 7NS
GM0IST PH2 7QA
GM0NYP PH2 7QA
GM1RGM PH2 7QA
GM0LVI PH2 7QQ
GM4WSY PH2 7QU
GM4LFK PH2 7SY
GM0FVJ PH2 8HR
GM3THI PH2 8PT
GM0PSS PH2 9EG
GM7NSS PH2 9EG
GM0MXZ PH2 9HS
GM3YEW PH2 9LW
GM4ZET PH2 9PF
GM1BXI PH3 1DD
GM4JZB PH3 1EW
GM4VNQ PH3 1HB
GM4SKB PH3 1JS
GM0SGH PH4 1QE
GM1MVK PH5 2AH
GM6JDZ PH7 3AY
GM8LJ PH7 3EY
GM4YRO PH7 3ND
GM6OFO PH7 3NY
GM0AOF PH7 3RP
GM1LTM PH7 3RP
GM4YDC PH7 4LE
GM3GNM PH8 0HY
GM4NGJ PH9 0LG

Peterborough

G8HXR PE1 2AU
G1PJO PE1 2NE
G3MLP PE1 2PW
G1YAW PE1 2QW
G4FMG PE1 2TH
G6VVB PE1 3EH
G0KHL PE1 3HH
M1CDL PE1 3JJ
G0DSQ PE1 3PF
G0REM PE1 3QJ
G0SEA PE1 3QJ
G3EEL PE1 4DX
G4FKQ PE1 4HA
G1KLI PE1 4JJ
G8CKV PE1 4LT
G7AFO PE1 4NH
G3RDZ PE1 4PG
G1YAX PE1 4SX
G7JJP PE1 4SZ
G1ZHL PE1 4UR
2E1AXF PE1 4YQ
G4FKG PE1 5HF
G1XVT PE1 5HG
G6VRM PE1 5HH
G0WEZ PE2 5ER
G1FWU PE2 5ER
G0ZEC PE2 5EX
G3EHQ PE2 5LG
M1AKH PE2 5NL
M0AMC PE2 5NN
G4VNG PE2 5NS
G4UGK PE2 5NT
G0MHQ PE2 5PD
G7IBX PE2 5PR
G7AMW PE2 5PS
2E1CQX PE2 5PS
G4TVX PE2 5QU
G0XBJ PE2 5RY
G8VJF PE2 5RZ
G0PUV PE2 5SB
M1AAF PE2 5SB
G3CCA PE2 5SB
G7JWD PE2 5SL
G4XGR PE2 5SL
2E1DPS PE2 5SL
G0HQQ PE2 5SN
2E1DQW PE2 5SW
G0WLA PE2 5UB
G6KUN PE2 5UZ
G8ILG PE2 6YP
G8BYB PE2 6YR
G1NIT PE2 6YY
G6AKF PE2 7DA
G4HCG PE2 7DN
G3HJY PE2 7HT
G6TAV PE2 8BH
G0HZE PE2 8LH
G4WJB PE2 8LQ
G4CNW PE2 8LS
G4FVK PE2 8LS
G8AQP PE2 8LY
G8XLH PE2 8UP
2E1ATR PE2 8UP
M0AOG PE2 9JG
G8VWU PE2 9JN
G4DXW PE2 9PD
G4LBQ PE3 4LS
G2NJ PE3 6BD
G2CBC PE3 6BY
G3ZJW PE3 6LB
G7CLQ PE3 7JG
G8NGZ PE3 8ES
G4UQF PE3 8PA
G4IQC PE3 8QB
G6MGA PE3 9AU
G4VAM PE3 9ED
2E1EJR PE3 9NP
G4LMY PE3 9RH
G7CSZ PE3 9SH
G4FFS PE3 9UH
G6GNC PE3 9UH
G4DCS PE3 9YT
G3GMW PE4 5AU
G1DRI PE4 5BJ
G7SQY PE4 5DB
G3WRL PE4 5DD
G7DWO PE4 5DF
2E1EJO PE4 5DQ
G1MLC PE4 5ED
G6AYU PE4 6AQ
G0NTR PE4 6GB
G0CNL PE4 6JS
G3XMP PE4 6JU
G4EOD PE4 6LY
G3LOC PE4 6QP
G4WIN PE4 6QZ
G4ZKM PE4 6RL
G3DQW PE4 6SJ
G3TGO PE4 6SJ
G4DJZ PE4 6ZH
M0AWB PE4 7BL
G3RED PE4 7TW
G7PPX PE4 7UQ
G3VYU PE4 7ZD
2E1CSY PE5 7AX
G4ULI PE5 7BH
G4EOO PE6 0PF
G0TZO PE6 0PF
G0NUU PE6 0QG
G8VQK PE6 0SP
G0MBM PE6 7LT
G0GXP PE6 7NF
G3YOY PE6 7NW
G4FXY PE6 7RG
G6CZO PE6 7RG
G4SQA PE6 7SF
G0BLV PE6 7XG
G4HLB PE6 7XT
G0CVZ PE6 7YF
G8KQZ PE6 7YF
G7NKI PE6 7YG
G4UQA PE6 9BN
G6CKR PE6 9DE
G6IDW PE6 9EP
G4IXI PE8 6AH
G0TQJ PE8 6AN

Plymouth

M1BCA PL1 3AT
G7USE PL1 3HJ
G0UGZ PL1 3JW
G0CFZ PL1 3RJ
G6BJJ PL1 4HL
G3TSE PL1 4QP
G7FKY PL1 5DN
G7OGM PL1 5HE
G3VVP PL2 1RR
G0GAJ PL2 2AQ
G7LUL PL2 2BU
G0WGB PL2 2ER
G6MCT PL2 2LB
G1KCW PL2 2RG
G3XLZ PL2 3AZ
G8EHD PL2 3ET
G4MED PL2 3HF
G4IML PL2 3LA
G3XCQ PL2 3NP
G7DQA PL2 3NT
G0EFR PL2 3PB
G7ANY PL2 3QZ
G7LLD PL2 3RS
G1SQI PL2 3RS
G7GYY PL2 3RS
G1RAB PL2 3RY
G0WTQ PL2 3SX
G0UMN PL2 3SY
G4DZP PL3 4AE
G7UDX PL3 4BD
G7URV PL3 4BD
G1LTC PL3 4HH
G3SGV PL3 4JB
G8SIC PL3 4LP
G4KXH PL3 4PR
G6GEX PL3 4RB
G3GOV PL3 4RW
G4OXM PL3 4SB
G4WIQ PL3 5AN
G4BZF PL3 5AY
G0FCG PL3 5BP
G8LER PL3 5DA
G8RMP PL3 5DA
G4BCX PL3 5DU
G3SQN PL3 5NP
G4ADP PL3 5RF
G4GWK PL3 5RY
G8BOI PL3 6BY
G0GVX PL3 6DH
G3HPC PL3 6HA
G0LSJ PL3 6JZ
G4XZS PL3 6LX
G6ZGK PL3 6LX
G7DIR PL3 6NQ
G6ALR PL3 6PP
G7UXL PL3 6PS
M1AJW PL3 6QY
G0BDB PL3 6SU
G0NAP PL4 0JW
G7BAV PL4 6BG
G4YGZ PL4 6PS
G4EKV PL4 6QA
G0INP PL4 6RB
G1RXR PL4 7ER
G7JEQ PL4 8AA
G0UOP PL4 8AA
G7TTK PL4 8JP
G7GOK PL4 8PS
G7ART PL4 9BT
G0MKO PL4 9DQ
G0TQT PL4 9ET
G7KII PL4 9EZ
G7VDW PL4 9HH
M1AHF PL4 9NE
M1BZK PL4 9NE
G7DQC PL5 1DS
G7NVS PL5 1EZ
G0NCS PL5 1PD
G3YJQ PL5 1SR
G6JEM PL5 1UL
G0WOU PL5 1UX
G4DMM PL5 2EA
G4JHY PL5 2EA
G7SOY PL5 2EG
G3XSG PL5 2EX
G8IDE PL5 2NW
G4EJQ PL5 2PJ
G1JFN PL5 3LQ
G7NIA PL5 4AZ
G4GKQ PL5 4JQ
G3KDP PL5 4NX
G3RYZ PL5 4QE
G7RTO PL5 4TR
G7ESZ PL6 5DH
G7CNC PL6 5DP
G0KZQ PL6 5ET
G0RLM PL6 5EX
G3NBX PL6 5JW
G0JNZ PL6 5NR
G1PMD PL6 5PG
G3ENX PL6 5PG
G0KYE PL6 5SN
M1BUW PL6 5TD
G0WZN PL6 5TQ
G7BXS PL6 5TR
G0IIC PL6 5TZ
G0KIK PL6 5YB
G3JTJ PL6 6AH
M0BFR PL6 6AR
G7UUS PL6 6BA
G3ZGF PL6 6BT
M1BVA PL6 6DT
G4CGM PL6 6JP
M1AWK PL6 6LS
G6IPQ PL6 6RE
G3UVS PL6 7DJ
G4GTG PL6 7DT
G7OOV PL6 7DT
G3GWH PL6 7DX
G3YDU PL6 7HU
G3ZZS PL6 7HX
G4YDR PL6 7JA
G0VKN PL6 7JG
G7GVH PL6 7PH
G8WZJ PL6 8RQ
G0LCP PL6 8SB
G7TLK PL6 8SY
G7AUQ PL6 8TD
G7ESO PL6 8TD
G6RSB PL6 8UB
G0OBP PL6 8XF
G3ARE PL7 1JR
G3LCY PL7 1JY
G7UCQ PL7 1PE
G7IZT PL7 1PN
G0VUG PL7 1PN
G4PBK PL7 1PY
G3VCN PL7 1QZ
G0INU PL7 1RB
G0TXQ PL7 1SA
G0MQD PL7 1SJ
G3TZQ PL7 1TX
G4LOI PL7 2DY
G4NLU PL7 2EY
G7HHW PL7 2EZ
G7NMA PL7 2QP
G7NHB PL7 2RU
G1SYZ PL7 2ZA
G7UTI PL7 4HR
G0UMI PL7 4HY
G0SFD PL7 4JE
G0MSM PL7 4JG
G0LSF PL7 4NT
G8CMG PL7 4QY

G6CYH PL9 7DP
G6TDB PL9 7DP
G3PRC PL9 7LA
G8PRC PL9 7LA
G3VNG PL9 7LA
G4NDD PL9 7LD
G3KHU PL9 7LF
G7UWO PL9 7LU
G7LLE PL9 7NW
G4BLI PL9 8DR
G0AHU PL9 8EU
G0SHZ PL9 8EX
G0VZX PL9 8HU
G3ULN PL9 8PJ
G3RMZ PL9 8RB
G3SPI PL9 8UR
G0SJC PL9 9AQ
G4OPZ PL9 9BZ
G1TPK PL9 9DD
G6XZK PL9 9HL
G0DHU PL9 9JB
G0RMC PL9 9JD
G7NGH PL9 9JZ
G8VRE PL9 9LU
G3TGR PL9 9NG
G0LRJ PL9 9NH
G0OJW PL9 9NN
G8VLZ PL9 9NN
M0ASI PL9 9NT
G4SJD PL9 9RR
G7IZU PL9 9UY

Poole

G4GJX BH11 9PE
G4WDW BH11 9RW
G0TUC BH11 9SG
G7PEZ BH11 9SG
G7JNS BH11 9SJ
G0VOD BH11 9ST
G6JKK BH11 9TZ
2E1FKG BH12 1EF
G4UDN BH12 1PP
G3KCE BH12 2DW
G3XBZ BH12 2HN
G4FDS BH12 2JF
G6XYT BH12 2ND
2E1CCT BH12 2ND
G0CRY BH12 3HB
G4AMW BH12 3LF
G0GZN BH12 3LP
G0ISO BH12 3LP
G1UIO BH12 4BD
G0AEL BH12 4DJ
G4SMZ BH12 4EA
G1LLA BH12 4HU
G4WAI BH12 4JF
G4JQW BH12 4JQ
G0ISC BH12 4LT
G3OBD BH12 5ED
G8CKA BH13 6HJ
G0KKL BH13 7AJ
G4PVY BH13 7BE
G0ROZ BH14 0BD
G6EJH BH14 0DB
G6CGQ BH14 0LL
M0ALB BH14 0LQ
G8NKY BH14 0PF
G8YYA BH14 0PF
G0GFR BH14 0PP
G3BLN BH14 0QF
G6NLC BH14 0QS
G4HJN BH14 0RW
G6ABA BH14 8AD
G8TNU BH14 8AU
G3RGJ BH14 8EG
G3KWE BH14 8HQ
G8GAT BH14 8JQ
G8DEU BH14 8LN
G4CTR BH14 8LT
G1GOR BH14 8SQ
G4HBD BH14 8SW
G4WUN BH14 8TJ
G1TEX BH14 9EL
G4UWS BH14 9LW
G4DWA BH14 9NP
G7GPY BH14 9NX
G0KCC BH14 9QP
G0KQH BH15 1QT
G4UTG BH15 1TU
G3ELQ BH15 2BA
G0CBP BH15 2DB
G0PRS BH15 2DW
G4WCK BH15 2DW
G6OAI BH15 2DW
G1XDO BH15 2ED
G0KZD BH15 2EX
G6NIO BH15 2HQ
G7TZC BH15 2JA
G7VMC BH15 2JJ
G0WTG BH15 2JJ
G3WCU BH15 2JQ
G0SOH BH15 2LA
G4WCJ BH15 2LP
G3NIL BH15 3AQ
G1SMD BH15 3AR
G7TZB BH15 3BG
G4SHV BH15 3DA
G0SJT BH15 3DD
G0HKT BH15 3DN
G4WRT BH15 3DU
G3WLK BH15 3DW
G3HUO BH15 3ET
G3UEE BH15 3ET
G8HEL BH15 3HA
G3FTK BH15 3NE
G3MXF BH15 3NL
G3JTK BH15 3NR
G4XGM BH15 3QH
G0IBJ BH15 3QU
G3RZV BH15 3QZ
G4GHP BH15 3RS
G6MXL BH15 4DH
G1ZPZ BH15 4DP
2E1FIS BH15 4EA
G1UEQ BH15 4HP
G4CVX BH15 4HP
G0JJI BH15 4JS
G0BAO BH15 4PN
2E1BVG BH16 5BA
G0UPG BH16 5BZ
2E1DQL BH16 5BZ
G3ZPR BH17 7AH
G3MEC BH17 7EU
G4YKN BH17 7HJ
G0HHI BH17 7HZ
G4RFV BH17 7LF
G6WEX BH17 7NX
2E1DFZ BH17 7QZ
G1YHE BH17 7UW
2E1FYP BH17 8BG
G7SVI BH17 8BP
G3XYD BH17 8BU
G1HBF BH17 8DB
G4NAG BH17 8QR
G0EQV BH17 8SB
G8DT BH17 8SD
M1BKN BH17 8SF
M0AKY BH17 8SR
G3ZCG BH17 9EF
G4YKC BH17 9HS
G7GNU BH17 9WE
M0AHB BH18 8AP
G7VQR BH18 8DE
G6XIK BH18 8HS
2E0AMW BH18 8JS
G3SWM BH18 8PW
G4BYO BH18 9AE
G4IBM BH18 9AE
G1YHJ BH18 9AJ
G0JCJ BH18 9BU
G8HNA BH18 9DF
G8HNB BH18 9DF
G8JMB BH18 9ED
G6CZQ BH18 9EX
G0CZG BH18 9HQ
G4BKE BH18 9HQ
G6PIM BH18 9HY
G0WKH BH18 9ND
G0BRQ BH18 9NR
G4VCQ BH18 9NU
G3JSJ BH18 9NX
G6NZN BH21 1QT
G1JHR BH21 1RJ
G0SKN BH21 1RR
G6BER BH21 1SJ
G8CEZ BH21 1SL
G6SHS BH21 1TB
G0GOX BH21 1TP
G4WEY BH21 1TU
G8MXW BH21 1UQ
G0HOA BH21 1UT
G8PWA BH21 1XR
2E1FWH BH21 1XW
2E1DBW BH21 3AX
2E1DTM BH21 3AX
G3RJH BH21 3BG
G4VZT BH21 3BJ
G3BCI BH4 9LF
G4PRS BH4 9LF

Portsmouth

M1AQG PO1 3JB
G0LFI PO1 5AR
G0LQB PO1 5DR
G6XDY PO1 5DS
G4PSJ PO1 5QE
G4YEO PO1 5RZ
G1MGU PO2 0HB
2E1BPA PO2 0NJ
G0HCI PO2 0QP
M1BZY PO2 0SB
G4DIU PO2 0TN
G4MKQ PO2 0UF
G1LFG PO2 7AT
G0DAE PO2 7JN
G8UYY PO2 7LB
G4EFB PO2 7PG
G6ATK PO2 8NF
G0OYN PO2 8NH
G6ERI PO2 8PS
G4AVX PO2 9BD
G7OKI PO2 9BS
G7MET PO2 9JR
G4JMR PO3 5PU
G6SGW PO3 5TG
G8KQV PO3 5TR
G8NVZ PO3 5TR
M0ASE PO3 6BG
M1API PO3 6BX
G3RCE PO3 6DZ
G7DWV PO3 6HB
2E1FPE PO3 6HT
2E1DBH PO3 6JB
2E1EFI PO3 6LL
G3ZBP PO3 6LR
G7MJB PO3 6LZ
G3VTN PO3 6NA
G6ESJ PO3 6NB
G6XJB PO4 0AU
G0AXS PO4 0BA
G3CNO PO4 0LJ
G6ZKM PO4 0QZ
G6NZ PO4 8AS
G4YCG PO4 8AS
G8OEJ PO4 8AU
G3XWM PO4 8DJ
G0JMG PO4 8HN
2E1FPH PO4 8JG
G1ERF PO4 8JS
G0IWN PO4 8JU
G3VXM PO4 8NP
G6RSV PO4 8QJ
G7EYG PO4 8QJ
G1LTQ PO4 8YS
G3CAJ PO4 9BG
G6BAA PO4 9EJ
G3ADR PO4 9EZ
G4MSS PO4 9LN
G6GEY PO4 9TS
G0HIP PO4 9YA
G8BKX PO5 1JH
M1BFC PO5 1ND
G4WNJ PO5 1RD
G4WDD PO5 2AE
G7VWH PO5 2DS
G6IMG PO5 2LU
G0LFN PO5 2NL
G8ZON PO5 4DN
G0EQS PO6 1BE
G0WMM PO6 1EL
G7NYY PO6 1LW
G0VFH PO6 1LW
G4GZO PO6 1LZ
G3SSC PO6 1NB
G6HJV PO6 1NB
G0KAH PO6 1ND
G0SSF PO6 1NF
G4OVM PO6 1NG
G4ZPA PO6 1NR
G0WQK PO6 1PH
G8AFG PO6 1PJ
G4UZS PO6 2BQ
G7OWQ PO6 2ES
G0OYB PO6 2JT
G7DRT PO6 2JU
G6UXW PO6 2PU
G7TYW PO6 2RL
G7TXD PO6 2SA
G0RPX PO6 2TX
G3OQC PO6 3DG
G7HQP PO6 3DG
G0OBA PO6 3EU
G1JAB PO6 3HD
G4EFA PO6 3JJ
G1OVG PO6 3QT
M0ALF PO6 3RD
G1XZQ PO6 4AE
G4PYS PO6 4BB
M1AFN PO6 4BS
G7AAQ PO6 4EJ
G8CXF PO6 4ET
G8LVB PO6 4EX
G3ZJB PO6 4NT
G3VEF PO6 4PA
G8KGI PO6 4PA
G4LIO PO6 4QL
G6GYF PO6 4QL
M1BST PO6 4SA
G1MCO PO6 4SL
G7PMA PO6 4SS
G1TDP PO7 5BT

Powys

GW6OJK HR3 5AN
GW0RIL LD1 5EB
GW0RBH LD1 5NL
GW1FBI LD1 5NL
GW4XXJ LD1 5NL
GW7UNV LD1 5UR
2W1CUN LD1 6AP
2W1CVH LD1 6AP
GW4KOS LD1 6RN
GW4VVL LD1 6TA
GW3RMJ LD1 6UE
GW0RKH LD2 3AR
GW7JYJ LD2 3EP
GW4EKE LD2 6EW
GW3LJP LD2 6EW
GW6DQH LD3 0AT
GW8YPR LD3 0AT
GW0DIB LD3 0BP
GW0BKJ LD3 0DU
GW8EUF LD3 0EU
GW1SMJ LD3 0HN
GW7RFA LD3 0PW
GW4NKR LD3 0UR
GW3OKM LD3 7BY
GW0IVT LD3 7JF
GW0ABT LD3 7UW
GW3ECH LD3 7UW
GW3SPC LD3 7UW
GW4ISF LD3 7YP
GW6RUE LD3 8DH
GW0XAP LD3 8PD
GW0GIH LD3 9HT
MW1CBF LD3 9HT
GW7LIY LD3 9LF
GW6JJX LD3 9LP
GW6NSK LD3 9SY
GW7SDB LD3 9TT
MW1BAP LD4 4DP
GW7KIV LD4 4DR
GW8VLD LD6 5DB
GW0KQX LD6 5LW
GW4IIA LD7 1BL
GW7UCA LD7 1PA
GW8GQE LD7 1RW
GW8ARR LD7 1YL
GW4GAF LD7 1YT
GW0CTG LD8 2BU
GW8CAK LD8 2EP
GW4AHO LD8 2UB
GW4FKW NP8 1AN
GW0BKK NP8 1DD
GW0XYL NP8 1DJ
GW4ZUW NP8 1DJ
GW4FRH NP8 1DQ
G7RLO NP8 1DX
GW4LEU NP8 1HD
GW6KGR NP8 1LN
GW4UZC NP8 1SY
GW4JLK SA11 5UN
GW4YWM SA9 1AT
GW4JCD SA9 1EQ
GW4YUX SA9 1ES
GW0LML SA9 1SE
GW3AGB SA9 2TT
GW6AMK SA9 2XD
GW0BXZ SY10 0JX
G0AJA SY10 0PF
GW1APU SY10 9JQ
GW4RYK SY15 6JH
GW0RJV SY15 6RS
GW4NQJ SY15 6SB
2W1CEZ SY15 6SB
GW4AZW SY16 1EZ
GW0EHS SY16 1HY
GW0IQP SY16 1HY
GW1JNI SY16 1HY
GW1HKY SY16 1JW
GW7FYG SY16 1PT
GW7JPC SY16 1QL
GW7DJL SY16 1QT
GW1NGL SY16 1RA
GW4NFF SY16 2BP
GW7TIX SY16 2ER
GW1HAX SY16 2LJ
2W1DRM SY16 3DB
GW0JAI SY16 3EH
2W1DRL SY16 3ER
G7VZU SY16 3ER
GW0MMB SY16 3ER
G1EKX SY16 3NT
GW7DRX SY16 4PZ
GW3FXI SY17 5JG
GW2BHS SY17 5JT
GW7DVB SY17 5LJ
GW3KAJ SY18 6DQ
GW0CYI SY18 6HG
GW4DTU SY18 6NP
GW3XPK SY18 6SN
GW0NEJ SY20 8RZ
GW4LPU SY20 9EZ
GW1JNR SY21 0AE
2W1DQI SY21 0AE
GW4DEP SY21 0JT
GW3KGV SY21 0PW
GW4DWX SY21 7NQ
GW3JPT SY21 7RD
MW1ABT SY21 7TP
GW4MZB SY21 8HL
GW1EWQ SY21 8HZ
GW4DYY SY21 8NJ
GW8OXZ SY21 8NJ
GW3TCV SY21 8SG
GW6EUT SY21 8TH
GW1KTW SY21 9PN
GW1EAV SY21 9PX
GW4GNY SY21 9PX
GW4HVN SY21 9PX
GW7RCR SY21 9PX
G3XDH SY22 5ND
GW3KVX SY22 6BE
GW4RIX SY22 6DE
GW3XET SY22 6LA
GW0DLW SY22 6LT
G4CRH SY22 6SL
G4YLX SY5 9AT
GW4FRX SY5 9BN

Preston

G1PHB PR1 1TS
G1VTQ PR1 2YL
G0DAG PR1 3JJ
G0OBR PR1 3JJ
G6ADY PR1 4JJ
G7PZE PR1 4NH
G3NQX PR1 4UD
G6NKI PR1 4UD
G4WYH PR1 5HJ
G0RRY PR1 5SN
G7ING PR1 5TA
G4WXI PR1 5TR
G6WXS PR1 5UY
G4THA PR1 6PY
G4MFJ PR1 7XP
G6MCC PR1 8PP
G4PZT PR2 1AD
G6EUG PR2 1JD
2E0ACV PR2 1PB
G4WQT PR2 1PB
G4ZKA PR2 1PB
G4LWM PR2 1YP
G0SBY PR2 2AP
G4ZCG PR2 2EJ
G7OFU PR2 2HB
G1IWG PR2 2HR
G6WBS PR2 2LR
G7ILX PR2 3AP
G4GVD PR2 3DB
G3URP PR2 3DE
2E1AFH PR2 3EX
G4UQI PR2 3EX
G7NNC PR2 3JQ
G7AQD PR2 3LP
G3ZAZ PR2 3ND
2E1ENQ PR2 3QQ
2E1AFI PR2 3RY
G0LCR PR2 3RY
G4RPW PR2 3RY
G7CLR PR2 3RY
G0KSN PR2 3SX
G7WEM PR2 3TT
G7WEL PR2 3TT
G7LPF PR2 3UU
G7IFM PR2 3UU
G0BGK PR2 3XE
G7BBM PR2 3XR
G4LHR PR2 3YR
M0AKQ PR2 3YS
G7FQL PR2 3ZP
GM4UTC PR2 3ZT
G7PSZ PR2 4XN
2E1FLI PR2 5LN
G3SYA PR2 6DH
G4HQA PR2 6EY
2E1EEG PR2 6LA
G6GLV PR2 6QF
2E1FJP PR2 6TH
G8ZAC PR2 6TU
G4UUS PR2 8AX
G7KRU PR2 8DE
2E0AFZ PR2 8DH
G8DTN PR2 8GT
G3ZXC PR2 8JN
M1BEW PR2 8XN
G1VZW PR2 9QX
G7OET PR2 9QX
G4DBU PR2 9RY
G3RSM PR2 9SQ
G8LQL PR2 9SR
G4PNH PR2 9SS
G4SHX PR2 9TD
G7LTL PR2 9TG
G3OWO PR2 9XH
G8CQV PR3 2BQ
G0KMP PR3 5HB
G4PWJ PR3 5LE
G0HYN PR4 0AG
G3SSU PR4 0AQ

Purbeck

G8MCW BH16 5ED
M0ATB BH16 5EJ
G7OGB BH16 5EJ
G2HKQ BH16 5LA
G0RPA BH16 5LS
G4NFT BH16 5LS
G0JBZ BH16 5PB
G6AKG BH16 5QT
G0ICG BH16 6AP
G7WBU BH16 6BW
G3EBO BH16 6DQ
G4XMZ BH16 6DR
G3PFM BH16 6EQ
G3PKL BH16 6ET
G1JUI BH16 6HJ
G3TJY BH16 6JD
G1YHI BH16 6JH
G0CGL BH19 1AN
G1HEJ BH19 1HY
G4ZPB BH19 1HY
G7FXO BH19 1LG
G3HFZ BH19 1LL
G6CAC BH19 1NF
G3WAR BH19 1QF
G0TGP BH19 1QF
G7OYX BH19 2PZ
G3YPQ BH19 2SB
G0TOT BH19 3DA
G3WSN BH19 3DS
G2HLU BH19 3HG
G1YOH BH19 3HH
G6NA BH19 3HN
G7EYD BH19 3LA
G7OVM BH19 3LQ
M1BKL BH20 4AG
G7MUT BH20 4DN
G4GHA BH20 4EL
G7HVK BH20 4RE
G1IER BH20 4SB
G1HXR BH20 4SG
G1WPG BH20 5BX
M0AAL BH20 5NG
GW4JWV BH20 5PP
G6HCH BH20 5SF
G7VXQ BH20 6AG
G7MND BH20 6EQ
G4RAM BH20 6HF
G6NGK BH20 6HF
M0AUI BH20 6NG
G4FUB BH20 6NL
G3KXT BH20 6NP
G4FWF BH20 7AA
G1KKG BH20 7BE
G4MFB BH20 7BS
G1AAH BH20 7BX
G7EXG BH20 7DF
G0JVD DT2 8JZ
G3EFK DT2 8JZ

Reading

G4FYM RG1 2PY
G6ZTZ RG1 3LP
G7IYF RG1 4RF
G6FBA RG1 5LR
G6BRU RG1 5QL
G3WNP RG1 5RD
G4GWJ RG1 5SN
G4HKZ RG1 5SN
G8CZJ RG1 5SN
2E1BDO RG1 6DG
2E1BZJ RG1 6DG
2E0AGM RG1 6EY
G1MWI RG1 6HE
2E1BDR RG1 6JE
G3DMQ RG1 6NG
G6SBN RG1 7HT
G4OKA RG1 7PD
G8VWJ RG1 7TT
G4JPE RG1 7TY
G4FLY RG1 7UG
G4JQT RG1 8HA
G4OPK RG1 8HA
GM6AMP RG2 0ES
G7KWD RG2 0LN
G0BMU RG2 7EJ
G3PWU RG2 7HE
G4OMN RG2 7JR
G4JXH RG2 7JU
G0MLD RG2 7ND
G4REZ RG2 7RQ
G7REH RG2 7RT
G6BPB RG2 7TT
G1ASD RG2 8HJ
G1NYZ RG2 8LJ
G4SWE RG2 8LL
G4DBF RG2 8QP
G0NAH RG2 8SD
G6ZLQ RG2 8SG
G6GBL RG30 2HA
G2HAX RG30 6XS
G4EFE RG31 5DZ
G3ZIB RG31 5NS
G1HNW RG31 5PN
G7UET RG31 6JJ
G4AWY RG4 5AL
G4BXY RG4 5AY
G7ENS RG4 5DG
G4PKM RG4 6DD
G1HSM RG4 6DH
G8NQC RG4 6QB
G8MZG RG4 6SG
G0NCO RG4 6UA
G0CCC RG4 7HH
G4CCC RG4 7HH
G4DWP RG4 7HL
G4JTR RG4 7HR
M0BFE RG4 7HR
G5HN RG4 7HT
G4JNU RG4 7NE
G3SLI RG4 7NT
G4CNN RG4 7QA
G4EJK RG4 7QP
G4VSQ RG4 8BZ
G8MIA RG4 8EH
G3TBJ RG4 8EN
G0OIW RG4 8HH
G4ZSZ RG4 8HJ
G7JJG RG4 8HJ
G7KSS RG4 8HJ
G3OYN RG4 8JH
G4TXA RG4 8LE
G3YZZ RG4 8PA
G3CMY RG4 8SX
G5JR RG4 8TA
G0MQW RG4 8TT
G8VWV RG4 8UA
G4VSO RG4 8UY
G3YEG RG4 8XH
M0AAA RG6 1NN
G0VKT RG6 1NN
G0PUB RG6 1PL
G4YFB RG6 1PL

Redbridge

G4USL E11 1PZ
G4GTZ E11 2JD
G0KVU E11 2JN
G4WRU E11 2NJ
G8ZHS E11 2NJ
G4GGT E11 2PP
G1SEF E12 5DZ
G3ARU E12 5DZ
G0AUK E12 5EQ
G3AAJ E12 5EQ
G4XMY E12 5EQ
G0VPV E12 5EQ
2E1FZC E12 5HD
G0FMB E18 1AP
G8PUJ E18 1DJ
G3XXC E18 2AB
G8NWL E18 2BU
G4HUE E18 2HE
G0GSL E18 2LL
G6SPN E18 2NJ
G0PGF E18 2NT
G0KRX E18 2PL
G4LKT E18 2QA
G3JTS IG1 1ET
G0KCD IG1 1HE
G7IUL IG1 1JE
M0ANB IG1 1TT
G6ZQJ IG1 2ER
G3PCA IG1 2SX
G3XRT IG1 2SX
G0IQK IG1 2UF
G0CUQ IG1 2UH
G0CQE IG1 2XT
G8JAP IG1 3AW
G3BDH IG1 3JF
G6RJS IG1 3JW
G0WHW IG1 3LQ
G4MLD IG1 3NF
G1UOL IG1 3NZ
G3LVN IG1 3PG
G6XCR IG1 3SF
G4BCJ IG1 3TQ
G3THY IG1 4QX
G8FAR IG1 4UX
G3OHS IG11 9DD
G3CAZ IG2 6AS
G4PQW IG2 6EQ
G1OYM IG2 6YG
G3ETX IG2 7AN
G0NAX IG2 7LF
G4JRD IG2 7NQ
G3LJQ IG2 7RZ
G7JUZ IG2 7SF
G8TND IG2 7TN
G3MDP IG2 7TQ
G6GJV IG3 8DU
G7HBO IG3 8JR
G0VKH IG3 8NN
M1BDY IG3 8QW
G7PRJ IG3 9AH
G4HQK IG3 9DG
G8ORQ IG3 9DG
G6TNJ IG3 9EJ
G4EDL IG3 9EN
G4LRM IG3 9LB
G3IJA IG3 9SA
G4UOX IG3 9SB
G7VBF IG3 9SW
G0MEW IG3 9TB
G3KHC IG3 9XW
G0JMM IG4 5AA
G3ALK IG4 5AE
G4GMG IG4 5AE
G4KHN IG4 5JR
G7VBW IG4 5LZ
G3TUA IG5 0AQ
G4FXQ IG5 0AQ
G1ICK IG5 0HP
G8JFL IG5 0NP
G4PSR IG5 0RR
G6XSB IG5 0RZ
G8VBK IG5 0TZ
G8YAE IG5 0XA
G0ATP IG6 1BJ
G3ZCS IG6 1BS
G4JLW IG6 1LA
G1NUG IG6 1LG
G1BMK IG6 1PB
G4YUF IG6 1PJ
2E1FOE IG6 2EB
G8PPA IG6 2LH
G8HST IG6 2QU
G8PRJ IG6 2QU
G2AGO IG6 3AE
G1HEQ IG6 3TF
G7VMJ IG6 3XP
G4RHM IG8 0EG
G1NTE IG8 7NR
G4AJG IG8 7QU
G8ZMI IG8 8LN
G0VXR IG8 8NL
G6SWT IG8 9AA
G8FRH IG8 9AA
G6BUS IG8 9HH
G4BNB IG8 9HY
G0FEK RM6 4EB
G1HKS RM6 4EB
G3JSR RM6 4RL
G4YGL RM6 4SL
G0WXN RM6 6JU

Redcar and Cleveland

G0MBV TS10 1BB
G6ZGC TS10 1BW
G3FMZ TS10 1RS
G0ETD TS10 2BP
G0GKY TS10 2BP
G4XXG TS10 2BP
2E1DMZ TS10 2DG
G4XFO TS10 2HN
G6UJR TS10 2HR
G3UZB TS10 2JZ
G4MBT TS10 2LL
G0AJZ TS10 2LT
G3INP TS10 3NU
G3NVP TS10 4AG
G0ASV TS10 4PG
G0VLK TS10 5EB
G3JMO TS10 5EL
G0BIA TS11 6BE
G3PSG TS11 6BE
G7AXN TS11 6DB
G4WNA TS11 6DF
2E1CME TS11 6DF
G4HWC TS11 6JU
G4NUO TS11 6JX
G6LCH TS11 6NH
G0VZT TS11 6NW
G1ZPG TS11 7BP
G3BRU TS11 7ER
G0NUT TS11 7HT
G4DGO TS11 7LH
G4RHX TS11 8AU
G0HJJ TS11 8AW
G1GMF TS11 8BP
G4GCU TS11 8BT
2E1CIO TS11 8DB
G4CRS TS11 8DU
G4OLK TS11 8DU
G8JLA TS12 1AL
G4NOP TS12 1LR
G4ULW TS12 2AQ
G3XAG TS12 2DQ
G0PWN TS12 2RB
G3KBI TS12 2TJ
G6ICZ TS12 2TJ
G0EWT TS12 2XG
G4AQN TS12 2XW
2E1FFZ TS12 3AW
G7SPV TS12 3JH
G6MVQ TS12 3JU
G0FSG TS12 3LE
G0PZM TS12 3LS
G7EBX TS13 4DT
G4WUS TS13 4DW
G4OXZ TS13 4LW
G3OLZ TS13 4NT
G4JCS TS13 4UG
G4KIR TS13 4XJ
G4HEB TS14 6DJ
M0AWK TS14 6QL
G7VUO TS14 6QL
G4AIZ TS14 7LN
G4KUU TS14 7LP
G1AAG TS14 7LZ
M1AMP TS14 7LZ
G4HRU TS14 7NB
G0IBW TS14 7PF
G7TXV TS14 7PR
G4YKT TS14 8AP
G3NOD TS14 8BP
G6GNY TS14 8BX
M0AGO TS14 8JG
G7ION TS14 8LD
G4YMB TS14 8LN
G0VAP TS14 8LW
2E1EXZ TS14 8PB
G0NYZ TS6 0BN
G7TWU TS6 0BQ
G4MCF TS6 0BS
2E1CIQ TS6 0BS
M1BZA TS6 0PP
G0FXR TS6 0QQ
G0LWE TS6 0QQ
G1DAT TS6 0QY
G1CLT TS6 7NA
G1CCM TS6 8AF
G7ECE TS6 8DP
G7FVH TS6 8EH
G7PZP TS6 9DF
G4ORQ TS6 9LY
G4XMM TS6 9LY
G0PDK TS6 9QW
G7PAE TS6 9SZ
G7NQX TS6 9SZ
G7WEN TS7 0EZ
G0LYD TS7 9AX

Redditch

G6DZH B96 6AG
G6NJH B96 6AG
G3TBW B96 6AY
G3RZI B96 6ED
G0MYD B97 4JJ
G0TOX B97 4JL
G3WF B97 4LX
G4NTG B97 4NA
G4NYZ B97 4NH
G3KWK B97 4NP
G3SAH B97 4NQ
G8MGK B97 4PN
G8DSM B97 4RL
G8RCE B97 4RP
G2FXJ B97 5AA
G3TNI B97 5AY
G4ZWR B97 5DF
G2FTY B97 5EB
G4TUI B97 5EB
G4SGV B97 5EP
G1JJA B97 5JA
G6EAY B97 5JA
G0TPG B97 5LX
G0ORE B97 5NG
G0RMG B97 5NW
G0LYK B97 5NY
G1XVY B97 5PZ
G1TQU B97 5QT
G8NJQ B97 5RN
G6HCW B97 5RT
G4SRV B97 5UN
G6LPS B97 5XD
G4WKZ B97 5YR
G1NEB B97 6EL
G4WXG B97 6EN
G8IYK B97 6EP
G2HHH B97 6NJ
G0TNH B97 6NS
G1AJV B97 6PH
G4MUV B98 0AS
G1JLQ B98 0BD
G4SWR B98 0BH
G7AWV B98 0BL
G1AVW B98 0BX
G1VXU B98 0BX
G7GFM B98 0JE
G6MTB B98 0JP
G0PLA B98 0JQ
G1HOL B98 0JR
G1DOA B98 0NA
G1RAG B98 0NF
G1ZUU B98 0NF
G6OTX B98 0NJ
G4KNX B98 0PD
G0CAX B98 0PJ
G1CGQ B98 0PL
G4DFC B98 0SB
G0ESJ B98 7EJ
G0UWH B98 7EJ
G6VSC B98 7JU
G0AOI B98 7LU
G1DHM B98 7NH
G6FVL B98 7PD
G7UMK B98 7PL
G1XHJ B98 7PW
G3WJN B98 7QA
M1ANN B98 7RH
G4UDK B98 7UT
G1XVR B98 7UU
G4WXS B98 7XB
G7AAF B98 7XP
G1MZT B98 7XT
2E1DLQ B98 7YD
G7WHV B98 7YD
G4LDB B98 7YL
G1AJW B98 7YR
G3GIY B98 7YS
G8WWD B98 7YS
G3TQD B98 7YU
G1INC B98 7YY
M0AMF B98 7YZ
G0NSW B98 8DJ
G6UIC B98 8HJ
G0KWQ B98 8PL
G0LFY B98 8QL
G0LFZ B98 8QL
G1DUO B98 8RD
G6VUN B98 8RW
G3HHM B98 9AD
G6ERJ B98 9JH
G7IVR B98 9JU

Reigate and Banstead

G1PDO CR5 3RG
G3GWC CR5 3SX
G1ZVE KT17 3LN
G7JSS KT17 3NL
G4IQV KT18 5PY
G4GFI KT18 5QT
G8GD KT18 5RE
G0BOE KT18 5RP
G8AAI KT18 5TE
G0CIW KT18 5TL
G0XTM KT20 5AL
G6TZE KT20 5HZ
G4WNV KT20 5JF
G7PHK KT20 5JQ
G4NLB KT20 5JY
G3WIM KT20 5NE
G0OLX KT20 5PR
G3SUX KT20 5RZ
G3UCR KT20 5SE
G3DNJ KT20 5TF
G4XXI KT20 5UA
G6ALG KT20 5UT
G3NPC KT20 6BS
G6XTK KT20 6DP
G2ACN KT20 6LL
G8KYX KT20 6NR
G4HVO KT20 6XE
G6GGY KT20 6XE
G7KWF KT20 7AD
G4BYZ KT20 7BA
G4WNR KT20 7EF
G4FIT KT20 7EW
G8CCQ KT20 7UD
G3BBR RH1 1JS
M1BCM RH1 2BE
G6YAH RH1 2BN
G3VGE RH1 2BZ
G8ZFQ RH1 2DY
G0RHG RH1 2EZ
G0LXV RH1 2HA
G3MPB RH1 2HH
G7LZM RH1 2JB
G1WIS RH1 2JB
G8XOV RH1 2LA
G1YRR RH1 3BN
G8UNO RH1 3EY
G3MO RH1 3PS
G0KCH RH1 3PX
G3JRC RH1 3QA
G3NIQ RH1 4AF
G3YSX RH1 4AS
G0HMG RH1 4DE
G5LK RH1 4DE
G3TRC RH1 4DE
G7RAT RH1 4DE
G4VFL RH1 4DQ
G7KYY RH1 5AS
G3YQO RH1 5BD
G4LJU RH1 5BJ
G4FHE RH1 5HS
G7LUO RH1 5JB
G6ZDP RH1 5LD
G4ASX RH1 6AW
G8HMG RH1 6AW
G8OFJ RH1 6NF
G1YXJ RH1 6PB
G4PFX RH1 6PB
G8LAM RH1 6PQ
G7OOU RH1 6QR
G0VQD RH1 6QR
G2DW RH2 0EF
G4VUY RH2 0EH
G0AFB RH2 0JA
G8CTO RH2 0PA
G0LLD RH2 0PZ
G6YRV RH2 0QA
G8PZI RH2 0RE
G7CVM RH2 7BS
G0ROH RH2 7DA
G6AUE RH2 7DT
G4AVE RH2 7HJ
G7OBF RH2 7RJ
G4UFF RH2 8DF
G8ZRV RH2 8DS
G8YEP RH2 8EL
G4YXF RH2 8EY
G3WOQ RH2 8HX
G3CNC RH2 8JB
G6VMI RH2 8NE
G3KAX RH2 8PN
G1AML RH6 7AP
G8JXV RH6 7BX
G1HLY RH6 7JF
G4WLS RH6 8AX
G8AMU RH6 8BS
G3JHP RH6 8JR
G0SOG RH6 8LB
G7KYW RH6 8LB
G6DCH RH6 8LF
G8FBM RH6 8LF
G6VKX RH6 8RE
G4EIV RH6 8RX
G4NHA RH6 9LS
G4FBI RH6 9TX
G6YWI SM7 1AE
G3OLX SM7 1AJ
G0ARG SM7 1JJ
G4DFA SM7 1JJ
G8OTS SM7 1JJ
G6XZM SM7 1LB
G4UFO SM7 1LN
G8CBH SM7 2AZ
G3MES SM7 2BA
G0WRX SM7 2BA
G1DPX SM7 2DG
G0IUH SM7 2QJ
G3KTA SM7 3JR
G4RYG SM7 3PG
G8GNX SM7 3PN
G7DNY SM7 3RF

Renfrewshire

GM6PCW PA1 2JE
GM0PYM PA1 2JR
GM0TQA PA1 2JR
GM0HUU PA1 2SL
GM3EDZ PA1 3AY
GM4VLX PA1 3NG
GM4MPY PA1 3RT
GM0SCW PA1 3TX
GM0HRT PA10 2DB
GM1BLM PA11 3AY
MM0APF PA11 3EL
GM0NAI PA11 3EL
GM3ZFU PA12 4BT
GM4FLX PA12 4EG
GM4NSL PA12 4HN
GM4BZI PA12 4NB
GM4HRJ PA14 6XP
GM0KTJ PA2 0AR
GM1TDT PA2 0BD
GM0TUR PA2 0NQ
GM7SAR PA2 0NQ
GM0BLX PA2 6AD
GM0UOU PA2 6AN
GM8AVM PA2 6AQ
GM4EQY PA2 6RA
GM4YAA PA2 7NU
2M1AVY PA2 7NU
GM6KEC PA2 7NU

2M1AVZ PA2 7NU
GM0BCY PA2 7QY
GM4BRM PA2 7RP
GM3DDL PA2 8AW
2M0AAY PA2 8BY
GM4LHQ PA2 8BY
GM3RPM PA2 8EB
GM0NAE PA2 8QW
GM4BNZ PA2 9BD
GM3DEE PA2 9NU
GM7KSC PA2 9QJ
GM8TVV PA2 9RD
GM3VVM PA3 3LF
GM3WNB PA3 3ST
GM7UFN PA3 3SU
GM7UVS PA3 3SU
GM0GLV PA3 3TB
GM0UKD PA3 3TF
GM4KHI PA3 4LL
GM1SXX PA3 4SQ
GM1FLQ PA4 0PN
GM8IID PA4 0SL
GM7MBB PA4 0XA
GM7HHB PA4 8TD
GM7VHQ PA4 9DD
GM1AXI PA5 8AR
GM1RHA PA5 8JT
GM1VBG PA5 8NF
GM0EDJ PA5 8RJ
GM7DFI PA5 8UB
2M0AAW PA5 9AD
2M1OGZ PA5 9AJ
GM1JHU PA5 9BH
GM3VAR PA6 7JX
GM4LFA PA6 7LG
GM0ODB PA6 7NH
GM7OAW PA7 5DT
G7WOS PA7 5DT
GM8OAH PA7 5ES
GM3UWX PA7 5HE
GM3NLB PA7 5HW
GM1TGA PA7 5JT
GM1XBK PA7 5JT
GM3JUF PA7 5LJ
MM1BTJ PA8 6AY
GM0GDD PA8 6HJ
GM7GDE PA8 7AN
GM0BBU PA8 7AS
GM4AGL PA8 7HX
GM4UGB PA8 7HY
2M1AZJ PA8 7JE
GM1VYG PA9 1BP

Restormel

G1WUP PL22 0EP
G3GKK PL23 1DT
G3SXR PL23 1JH
G0AEW PL24 2AT
M1AZJ PL24 2BB
G8XNA PL24 2DG
G6VOG PL24 2EE
G0RJI PL24 2EF
G4OKS PL24 2HA
G7MBD PL24 2HJ
G4ZEB PL24 2LH
G0GMA PL24 2LN
G3SRZ PL24 2NJ
G4UBK PL24 2SR
G0KTD PL25 3AU
G4HYV PL25 3DH
G0NNO PL25 3DR
G0VDU PL25 3DW
G4LTY PL25 3EB
G8WBG PL25 3HB
G1OFI PL25 3HJ
G1SDX PL25 3HS
G3LQU PL25 3HX
G4JXS PL25 3JA
G1DDK PL25 3NJ
G4XPK PL25 3PH
G4UON PL25 3QE
G4JOD PL25 3QG
G3LDY PL25 3SJ
G4ZZY PL25 3TB
G4RLN PL25 3TR
G1ZSX PL25 4BJ
G4IAP PL25 4BZ
G0WAM PL25 4DH
G1XMI PL25 4HR
G8NYR PL25 4HR
G4XOP PL25 4HT
G0BRR PL25 4JA
G4XBC PL25 4JA
G4TKT PL25 4PN
M1APJ PL25 4QD
G0RCO PL25 4QW
G4MFQ PL25 4RJ
G6GAB PL25 4UW
G0ECC PL25 5EA
G4TRV PL25 5EA
G4YNZ PL25 5HD
G3MFW PL26 6AU
G4POP PL26 6HT
G3GHS PL26 6JA
G0ENZ PL26 6NA
G3SLZ PL26 6PS
G6HLH PL26 6QS
G3YIY PL26 6QZ
G1JBB PL26 6TF
G7FYZ PL26 6UR
G6LGN PL26 7BN
G4HFO PL26 7EH
G4TJX PL26 7EH
G8ZQM PL26 7ER
G3JSY PL26 7HA
G8TNA PL26 7HG
G4IKD PL26 7JN
G8VRV PL26 7NN
G7PPY PL26 7NZ
G4KNI PL26 7PF
G7FLX PL26 7PN
M1BXH PL26 7PN
2E1FFE PL26 7PN
G3KYM PL26 7UX
G8YKZ PL26 7XL
M1AEG PL26 7XN
G0FCB PL26 7XQ
G1TTK PL26 8JH
G6HHS PL26 8QJ
G4JYF PL26 8QL
G0IGD PL26 8RJ
G4MXB PL26 8TJ
2E1CWE PL26 8UA
G0GAW PL26 8UH
G3WKF PL26 8UH
G6ELR PL26 8UR
G0HZD PL26 8XX
G0UUQ PL26 8YT
G3XKW PL30 5AR
G1TAZ PL30 5AW
G0AXH PL30 5EN
G1FXD PL30 5PS
G3DDN TR7 1DX
G3ZZY TR7 1ND
G4XGF TR7 1PJ
M0BFB TR7 1TY
G7JOE TR7 1TY
G0LCJ TR7 2DG
G4WOI TR7 2EY
G8SCY TR7 2JN
G3JVN TR7 2LE
G0UPZ TR7 2QD
G4VPJ TR7 2RB
M1AWX TR7 2RH
G0FLU TR7 2RW
G2KF TR7 2SP
G6ZWI TR7 2SU
G3WBW TR7 2SW
G7BVX TR7 3AE
G8TMC TR7 3AF
G0HEW TR7 3AW
G6CEP TR7 3BN
2E1CXC TR7 3BN
G0ANY TR7 3BX
G7CRN TR7 3EB
G8GOR TR7 3EJ
G0OAJ TR7 3HW
G8XNH TR7 3JL
G6AFA TR7 3JT
G7VES TR8 4EZ
G7VER TR8 4EZ
M1ASQ TR8 4NP
G0AZI TR8 5BX
G1HSE TR8 5RU
M0AGY TR9 6AT
G4DND TR9 6BG
G6HER TR9 6EB
G8FMD TR9 6EB
G7OQL TR9 6PD
G3WJO TR9 6PP

Rhondda, Cynon, Taff

GW0ONU CF37 1HT
GW0JUJ CF37 1HY
GW1VVK CF37 1NH
GW1YSM CF37 1RN
GW1FKL CF37 1SS
GW4ZVV CF37 2DA
GW3YBN CF37 2HF
GW8PBM CF37 2LD
GW8LRO CF37 3AD
GW0SLC CF37 3DF
GW8ICS CF37 3EF
GW0JHH CF37 3EU
GW0BNN CF37 4DD
G7VOO CF37 4DD
MW0ATG CF37 4DP
GW0SQY CF37 4LR
GW0WLN CF37 4ND
GW4KHQ CF38 1AA
GW3MOV CF38 1AN
GW4KCV CF38 1AP
GW8SJN CF38 1AU
GW0MOQ CF38 1AU
GW1KQY CF38 1BW
GW4CNL CF38 1BY
GW7HOM CF38 1HL
GW0NNY CF38 1LS
GW7OLL CF38 1TD
GW4NPC CF38 1UB
GW7VTI CF38 2AU
GW4PAF CF38 2NH
GW4LZL CF38 2NP
GW7KIP CF38 2PD
GW6TTK CF38 2PD
GW1AYA CF38 2PQ
GW1AYB CF38 2PQ
GW4RKZ CF39 8AL
GW6MIH CF39 8AL
GW7CSK CF39 8DU
GW0LHO CF39 8NG
GW0UHJ CF39 8PL
GW0WLQ CF39 8PP
2W1FWI CF39 8TW
2W1FWJ CF39 8TW
GW3JVW CF39 9DB
GW4JAD CF39 9NH
GW3DRK CF39 9SD
GW6HJO CF39 9UP
GW0KLY CF39 9YD
2W1EDJ CF39 9YU
GW4WQC CF4 7PD
GW4HWR CF4 7RS
GW7MPW CF4 7RS
GW1BDF CF40 1DQ
GW1BDG CF40 1DQ
GW4KBG CF40 1HP
GW4BUZ CF40 1HR
GW0KQU CF40 1JT
GW1TDW CF40 1LE
GW6ZYI CF40 1LX
GW4YCO CF40 1QY
MW1AJI CF40 1SA
GW1XXL CF40 2DA
GW4UVC CF40 2DH
GW4SYO CF40 2HN
GW0LYF CF40 2JD
GW0OET CF40 2LY
MW0AQD CF40 2NX
2W1FVH CF40 2NX
GW0KLN CF40 2RY
GW0BEW CF41 7AS
GW3XHG CF41 7HF
GW3PWA CF41 7JR
GW0KKN CF41 7NG
GW0DLA CF41 7NJ
GW1SQT CF41 7SJ
GW7LJG CF41 7SJ
GW0FJQ CF41 7TW
GW1NZF CF41 7UD
GW4NOS CF41 7UG
GW6ZRX CF42 5BH
GW8PNE CF42 5PL
GW0NDZ CF42 6DF
GW1TDX CF42 6DY
GW4VWY CF42 6DY
GW1ABP CF42 6EL
GW1VAW CF42 6RL
GW4LPC CF43 3ES
MW1BTM CF43 3LJ
GW7VPJ CF43 3RN
GW3SUH CF43 4EU
GW4YLF CF43 4HT
GW0UKC CF44 0HR
GW6XZU CF44 0LD
GW3PEX CF44 0LL
GW3SFC CF44 0PB
GW6MKR CF44 0PX
GW4SUN CF44 0PY
GW4SCK CF44 0RG
GW6MXB CF44 0RG
GW4ZUA CF44 0TH
GW0OUH CF44 6DD
G1HNG CF44 6DE
GW7FSF CF44 6DT
GW0WWY CF44 6EW
GW4RUX CF44 6LD
MW1AND CF44 6YG
GW0HXB CF44 8EG
GW4OPW CF44 8RP
GW0KZE CF44 8UB
GW7SQP CF44 8UB
GW4KCY CF44 8YB
GW1XRV CF44 9BQ
GW7UFH CF44 9ND
GW4VEI CF44 9RN
GW0UHO CF45 3BS
GW4ZNU CF45 3HR
GW4SXA CF45 3NL
GW0LKH CF45 3RD
GW4XMI CF45 3RH
GW0VQZ CF45 3YL
GW4ZUJ CF45 3YW
GW4SXB CF45 4BG
GW8ZTP CF45 4DU
GW0SGL CF45 4EJ
GW8PMJ CF45 4PQ
GW0TZG CF45 4YP
GW4AZE CF45 4YU
GW7VSF CF72 9ND
GW4YDF CF72 9RU
2W1FBD CF72 9SQ
G4DVB CF72 9TX

Ribble Valley

G3SQO BB1 9EE
G1YJW BB1 9HH
G1MXO BB1 9HX
G8IEZ BB1 9JD
G0GSN BB1 9PW
G3ROS BB12 7HT
G0FNJ BB12 7NX
G3XAC BB12 7QG
G6RZG BB12 7QR
G3RNR BB12 7QT
G0DLT BB12 7ST
G3LPS BB2 7EP
G4FOJ BB2 7ES
G0DKT BB2 7HB
G6JPN BB2 7JP
G6ILH BB2 7NY
G4HUQ BB2 7PA
G0YNM BB6 8AP
G8CZG BB6 8AT
G0BPQ BB6 8EF
G6VCR BB6 8EP
G4VEY BB6 8ET
G0WNU BB6 8ET
G3DMO BB6 8HB
G6HMA BB6 8HB
G4YXE BB7 1EY
G0TOU BB7 1JL
G0AZH BB7 1ND
G3EJV BB7 1PD
G3INL BB7 1PE
G1NEJ BB7 2AS
G0KXX BB7 2HP
G4YWK BB7 2PG
M1BOZ BB7 2QH
G4LKZ BB7 2QW
G0NLP BB7 3BB
G0AZG BB7 3DA
G0NLO BB7 3DA
G4IJD BB7 3DH
G1JBV BB7 3JL
G3YFG BB7 9BJ
G4XSG BB7 9JN
2E1DKE BB7 9RL
G3SHJ BB7 9SA
G0SDJ PR3 2XB
G6JWU PR3 2YB
G3WGK PR3 3AA
G7NOI PR3 3FH
G0MUB PR3 3HL
G2MJ PR3 3SB
G3NNA PR3 3SY
G1CFG PR3 3TE
G6MAC PR3 3TQ
G3NKL PR3 3TY
G0HJB PR3 3UA
G6AJA PR3 3UD
G6RTD PR3 3WN
G1MDA PR3 3XA

Richmondshire

G7OAP DL10 4BE
2E1BAP DL10 4DB
2E1CFG DL10 4DB
G8PYX DL10 4PS
G0SKZ DL10 4PS
G8YPN DL10 4SE
2E1EZH DL10 4TR
G3RLV DL10 5DB
2E1AOY DL10 5DB
G0MQV DL10 5DJ
G8ZNB DL10 5ED
G7CIU DL10 5ND
G7HBB DL10 5PE
2E1DPA DL10 6JP
2E1DTP DL10 6LZ
2E1DVD DL10 6PP
G3EYO DL10 6QP
G4FZN DL10 6SB
G8HQW DL10 6SB
2E1DZK DL10 6SB
G4VGY DL10 7AG
2E1CKV DL10 7BG
G0RYS DL10 7BQ
G4DBY DL10 7DL
2E1DRI DL10 7EX
G4RPB DL10 7HN
2E1EAH DL10 7HT
G7LLY DL10 7JP
2E1DUG DL11 6LF
G3DJE DL11 6QG
G8TOQ DL11 6QX
G7PTD DL11 6TR
G7SLO DL11 6TR
2E1DBR DL11 7AR
G7RNT DL11 7BJ
2E1DPO DL11 7RA
2E1DPL DL11 7RD
G8BAG DL11 7SU
2E1CPI DL11 7SU
G4NLL DL2 2PX
G8NKA DL2 2QQ
2E1CQR DL2 2SN
G8MTV DL2 2SQ
G3EJF DL8 1PX
G3JZP DL8 1PX
G4HVA DL8 1SX
G7HSN DL8 1SX
G8LNQ DL8 1SX
2E0AGR DL8 3LA
M0BDR DL8 3LA
G0OLR DL8 4QN
2E1AZW DL8 4QN
G0FBU DL8 4QU
G1EPZ DL8 5DY
G0BAQ DL8 5HU
M1BPM DL8 5LS
G4LSL DL9 3NB
G1YQY DL9 3NJ
G3EKL DL9 3NJ
2E1COG DL9 3NJ
M1ABS DL9 4PD
G4FRQ DL9 4QZ

Richmond-upon-Thames

G1RQZ KT8 9AX
G1SGU SW13 9DX
G0WLD SW13 9JS
G2FRI SW14 7AR
G3MPY SW14 7JW
G0TRD SW14 7QF
G3RJU SW14 7QG
G1XPF SW14 7RP
G4AVK SW14 8JH
G6JMX SW14 8JJ
G8ANN SW14 8JU
G6URT SW14 8NG
G4XVE SW14 8NY
G1HBO SW14 8PB
G8IUA TW1 1DD
G4RBR TW1 1PY
G3CPC TW1 1QL
G7CSS TW1 1RB
G4KKM TW1 2AZ
G3URE TW1 2LB
M0ARR TW1 3DH
G1LRR TW1 3EW
G1TOB TW1 4BQ
G8DSU TW1 4QZ
G6PJW TW1 4RT
2E1DQM TW10 5DU
G4FZY TW10 5EF
G6YDS TW10 5HQ
G6RQY TW10 6EW
G1CRT TW10 6PD
G3YIQ TW10 6QD
G4SIG TW10 7DZ
G1PHA TW10 7ED
G1IRW TW10 7LG
G0ROE TW10 7LR
G0SLI TW10 7NQ
G8SNM TW10 7TY
G8CYE TW10 7YG
G4RWA TW11 0AW
G4LGO TW11 0BG
G7PMV TW11 0DH
G3PPC TW11 8AS
GW1IKN TW11 8QL
G8ZWA TW11 8SH
G8FMZ TW11 8SN
G3PZC TW11 8UD
G8WGE TW11 9DW
G0DEH TW11 9HA
G8AUU TW11 9LN
G3ZLD TW11 9NH
G0JME TW11 9QS
G0WGD TW11 9QS
G6OHT TW12 1DW
G4PPE TW12 1HX
G0CMN TW12 1LE
G0NFT TW12 1SL
G0SPF TW12 1SL
G0DHN TW12 2DE
G8SNV TW12 2JH
G8GOP TW12 2JT
G3RCB TW12 2LU
G7DWM TW12 2PZ
G7BYD TW12 2RA
G0AHM TW12 2RE
G4UXB TW12 3BW
G4ZOU TW12 3DH
G6RSQ TW12 3DS
G6NIX TW12 3YN
G1NSQ TW2 5BY
G6AWM TW2 5BY
G7JSH TW2 5HA
G7APM TW2 5HB
G7TOF TW2 5JJ
G1FKS TW2 5JP
G8YYB TW2 5PD
G2AIW TW2 6AA
G8WRL TW2 6BL
G3ZSB TW2 6EJ
G4LRI TW2 6HA
G1LVZ TW2 6HW
G0DNU TW2 7DF
G0IIP TW2 7DY
G1OVO TW2 7EA
G0OFN TW2 7JE
G4OQN TW2 7LD
G0OAY TW2 7NH
G6LKH TW2 7SN
G7GZC TW3 2LZ
G7GOX TW4 5KH
G4TSH TW4 5LZ
G0RHB TW4 5PF
G3UIJ TW4 5QH
G6XRK TW7 7HG
G6AGA TW7 7NL
G1THY TW9 1GU
G3ZUF TW9 1QF
G2YK TW9 1UH
G1TDK TW9 1UL
G4JT TW9 2DN
G0ACD TW9 2TJ
G0LYJ TW9 3AY
G3YXB TW9 3HB
G0OOI TW9 4AS
G3CIK TW9 4AS
G3XSC TW9 4ED
G0OHW TW9 4EE
G4GRM TW9 4JD

Rochdale

G2FWZ BL9 7TX
G0ELK M24 1DZ
G6CVW M24 1HE
2E1FKP M24 1JG
G3PJK M24 2BW
G0TRK M24 2EQ
G1EWB M24 2JP
G1YMY M24 2NJ
G6GCY M24 2NN
G3SLN M24 2PL
G4SHC M24 2QL
G4VVT M24 2TQ
G0KBV M24 2WF
G4ZQL M24 4FW
G4TZB M24 5JA
G0FCZ M24 5RL
G0OGN M24 5TZ
G4YVU M24 6HY
G4MNZ OL10 2LS
G1NXR OL10 2QP
G7KXP OL10 3EE
G0IQV OL10 3HY
G8JCL OL10 3JB
G7NIO OL10 3LS
G0RFE OL10 3RS
G0UAF OL10 4DB
G6NID OL10 4DB
G7NJE OL10 4QD
G7NIH OL11 1DT
G1PSL OL11 1PN
G1TES OL11 1QD
G8EEA OL11 2BN
G0JSF OL11 2EB
G4TDW OL11 2LN
G7WFF OL11 2NG
G1ITS OL11 2QB
G0SPY OL11 2TU
G4BIH OL11 2UW
G1JEM OL11 3BZ
G0OWH OL11 3HE
G6BVO OL11 3HE
G4TDU OL11 3JG
2E1EZN OL11 3JX
G0EUP OL11 3JZ
G4HHS OL11 3LF
G7BRJ OL11 3LG
G3DTP OL11 3LP
G7BPM OL11 4DD
G8HDS OL11 4DD
G0XPI OL11 4PP
G4CKH OL11 4PP
2E1DFQ OL11 5AN
G0WHL OL11 5JD
2E1DPC OL11 5JD
G0SVL OL11 5JG
G3VC OL11 5JN
G3RLE OL11 5NB
G7CBY OL11 5PL
G7CBZ OL11 5PL
G3ETY OL11 5PU
G0LSP OL11 5QA
G6GXK OL11 5QS
G0TFL OL11 5QW
G7TOB OL11 5RT
G7TOD OL11 5RT
G0GQT OL11 5TG
G6YCU OL11 5TG
G7UCT OL11 5TG
G6EFO OL11 5XB
G0WTB OL11 5XW
G6HAE OL11 5YG
2E1DEN OL11 5YH
G3WYV OL11 5YS
M1AIV OL12 0BG
G7EVN OL12 0NX
G1ZUH OL12 0SH
G4HRN OL12 0UY
G6BWK OL12 6HB
G1SVF OL12 6JH
G1CIA OL12 6LZ
G7UCP OL12 7AR
G4OAC OL12 7HF
G0RBW OL12 7HZ
G4GAI OL12 7JN
G0PUD OL12 7JQ
2E1FIX OL12 7JQ
2E1DFD OL12 7NA
G6CSN OL12 7ND
G0RZC OL12 7PF
2E1CBQ OL12 7RG
G7LKR OL12 7RT
G0WWE OL12 7SR
G3SHL OL12 7UX
G0VRX OL12 9AJ
G7RQL OL12 9AJ
G0UQA OL12 9AJ
2E1FKK OL12 9AJ
2E1FKL OL12 9AJ
G7OAI OL12 9QP
G3RIK OL12 9TY
2E1DKU OL15 0BP
G8NQI OL15 0NG
G3DNN OL15 8EB
G0BLM OL15 8EN
G6NCL OL15 8EN
G6NCM OL15 8EN
G1NDU OL15 8QR
G1ISB OL15 8QS
G7OEW OL15 9BL
G4OUJ OL15 9EF
M1BBY OL15 9HL
G8TTU OL15 9JE
G0IPN OL16 2DD
G0WPO OL16 2PB
G0GSQ OL16 2XL
G7AAP OL16 3EX
G7FYM OL16 3EX
G1FHI OL16 3QZ
G7LHV OL16 3TD
G4PJE OL16 3UJ
G0HXQ OL16 4BA
G4YLJ OL16 4EA
G7EQK OL16 4EZ
G3FLR OL16 4RY
G7LFM OL16 4SH
G0FNM OL16 5BB
G0GVS OL16 5BB
G0KZH OL16 5BB
G4UTA OL16 5BB
G8ZEN OL16 5HP

Rochester-upon-Medway

G7FZN ME1 1NB
G8WKX ME1 1QH
2E1FPT ME1 1RY
G3TSI ME1 2BE
G7VFU ME1 2BL
G0WFM ME1 2DL
G0WYV ME1 2HZ
G4SAW ME1 2NN
G6OTG ME1 2PE
G3YSC ME1 2TZ
G4YTU ME1 2TZ
G3ZJP ME1 2UH
G7KCK ME1 2XP
G0DKO ME1 3JU
G1ERZ ME1 3NJ
G0ANK ME1 3PD
G8RXV ME1 3PJ
G3FTH ME2 1HB
G5MW ME2 1HB
G8MWA ME2 1HB
G7JJD ME2 1HW
G6ABW ME2 2HE
G2DOH ME2 2HX
G4PVZ ME2 2LD
M1BOS ME2 2ND
2E1ENX ME2 2ND
G7LEG ME2 2PU
G0MIF ME2 2QY
M0AZS ME2 2QY
G1AEV ME2 2RD
2E1DEC ME2 2TZ
G7UTB ME2 2TZ
G7MIF ME2 2XW
G4VSN ME2 2YR
G4TVW ME2 2YW
G8ZJX ME2 3AP
G0LAD ME2 3BA
G4ETD ME2 3BY
G6OPP ME2 3EP
G0LJD ME2 3HW
G0HHQ ME2 3JH
G4EVY ME2 3JW
G6YTC ME2 3ND
M1BNV ME2 3ND
G0VMQ ME2 3PN
G7XYL ME2 3QT
G4PPV ME2 3SN
G8DBV ME2 3SN
G3FNZ ME2 3SP
G3VUN ME2 3TR
G8ZRQ ME2 3TW
G4BDD ME2 3XA
G0VKL ME2 4EB
G0GQT ME2 4NE
G4RFN ME2 4NL
G1JYT ME2 4NU
G4CWV ME2 4NU
G7SUU ME2 4PB
G1DOT ME2 4PG
G3NZR ME2 4PN
G8WWN ME3 0EE
G7RVX ME3 7QN
G6ZEQ ME3 7RY
G3OHP ME3 7SH
G1HRV ME3 7TN
G6KSK ME3 8BE
G4UXS ME3 8DH
G7VQJ ME3 8HT
G4INO ME3 8JW
G0GUI ME3 8LF
G3BSN ME3 8LR
G4IRV ME3 8SG
G7JYL ME3 8UL
G1UXZ ME3 8UL
G0NCE ME3 9BZ
G4TAM ME3 9EU
G3UXH ME3 9HX
G6GEC ME3 9HX
G3PBY ME3 9JD
G0PCA ME3 9PR
G0PCB ME3 9PR
G4VMD ME3 9SH
G6UPH ME3 9SH
G1LHL ME3 9SP
G1ALK ME3 9ST
G4EHQ ME3 9TH
G1YXH ME4 6HL
G8ZLK ME4 6LY
G1FZY ME4 6QG
G7DSU ME4 6SE
G8CJM ME4 6UU
G4HJE ME5 0BH
G7MFW ME5 0DY
G8VJU ME5 0JS
G4AKQ ME5 0JY
G3RE ME5 0JY
G6TXB ME5 0PA
G7AEY ME5 0RY
G8TGB ME5 0SB
G1JQA ME5 0TB
G1JUO ME5 0TB
G4LJT ME5 0TB
G0PZX ME5 7BU
G1IBJ ME5 7HJ
G7LAN ME5 7JY
G0FWO ME5 7LW
G6UVL ME5 7PD
G0ABN ME5 7PT
G6ULD ME5 7RF
G6FHK ME5 8JD
G6POF ME5 8JU
G1LHD ME5 8NY
G7IBN ME5 8NY
G8WMK ME5 8PG
G3PSR ME5 8PL
G6XXQ ME5 8QA
G6TOT ME5 8RD
G7ICV ME5 8RW
G7KNM ME5 8TN
G0VWV ME5 8TN
G4HIZ ME5 8UL
G7MPZ ME5 9AA
M1BGS ME5 9EH
G6DJV ME5 9LF
G6SQS ME5 9LF
G3JHU ME5 9TD
G7VKO ME7 3DJ

Rochford

G6OLY SS11 8TB
G1HPV SS3 0AR
G7ENI SS3 0AR
G0NXN SS3 0DR
G7PMK SS3 0DR
G4ZFJ SS3 0ER
G1EXR SS3 0EX
G3EQX SS3 0LZ
G1BMQ SS4 1DH
M1AHT SS4 1DP
G1HJW SS4 1HY
G6HNI SS4 1JE
G2OR SS4 1QJ
G8SGH SS4 1QJ
G0KTC SS4 1RS
G4SKT SS4 1RS
G8GHK SS4 1SH
M1BHW SS4 1SH
G3UVH SS4 1TG
G1RAP SS4 3AH
G7FFI SS4 3BX
G6KFD SS4 3DQ
M0AGB SS4 3HZ
G6XYU SS4 3LY
G1GCY SS4 3PY
G7PUZ SS4 3PY
G1JRD SS4 3QQ
G0GBY SS4 3TB
G1KHS SS5 4DN
G0PEP SS5 4EY
G3OJV SS5 4EY
G0DFE SS5 4JN
G0DTP SS5 4JY
G6WFS SS5 4JY
G7WHZ SS5 4NT
G4TWH SS5 4PJ
G3FHG SS5 4QE
G4MCU SS5 4SW
G1FXM SS5 5AS
G1ZYB SS5 5AT
G7GBN SS5 5AT
G7MEY SS5 5AT
G1KVO SS5 5BG
G0PEJ SS5 5BP
G4FQT SS5 5EB
G7ENT SS5 5EE
G3ZJZ SS5 5EL
G0JWS SS5 5HE
G7WHP SS5 5HN
G4KDH SS5 6AA
G0EBG SS5 6BG
G7TBW SS5 6BN
G4RDS SS5 6DD
G3LUI SS5 6HG
G1FSM SS5 6JB
G7JPU SS5 6JF
G0UPW SS5 6LF
G7FPY SS5 6LP
G4MUS SS5 6LR
G8RAN SS5 6LT
G0MGT SS5 6LU
G3HWM SS5 6LZ
G0CIJ SS5 6PY
G3TRH SS6 7DT
G1TWS SS6 7JR
G1ZHM SS6 7NP
G4FSE SS6 7NS
G8FAX SS6 7QH
G1WER SS6 7QU
G0DRH SS6 7RG
G8LHW SS6 7TD
G0MLO SS6 8AB
G0HSK SS6 8AR
G6XAT SS6 8AR
G4JQJ SS6 8AW
G3OGX SS6 8BN
G0FKS SS6 8BP
G7PPS SS6 8LQ
G6PCP SS6 8PL
G7MFH SS6 8SG
G4FQA SS6 8SL
G0DJB SS6 8SP
G0DRV SS6 8TQ
G0WIX SS6 8UA
G4KIH SS6 8YF
G8UVB SS6 8YF
G6LXP SS6 9AD
G4PWB SS6 9AL
G8RDL SS6 9BD
G3UOG SS6 9DH
G4KRX SS6 9HS
G4XEO SS6 9JZ
G3ZHA SS6 9LY
G6AVK SS6 9PD
G3MHC SS6 9PZ
G4ENW SS9 5DN

Rossendale

G0OCW BB4 4AN
G8HBR BB4 4BG
G6FGE BB4 4BT
G6AUX BB4 4DQ
G7JXD BB4 4DZ
G0LRR BB4 4EF
G4TWG BB4 4EF
G6UOU BB4 4JB
G0KIM BB4 4PB
G7IAW BB4 5BU
G0PGW BB4 5NG
G7OXF BB4 5NR
G8FGB BB4 5TE
G0FCA BB4 6AZ
G4ZLJ BB4 6BE
G7KGL BB4 6BW
G8JCN BB4 6DS
G3XWB BB4 6EE
G4DNE BB4 6EU
G4LPO BB4 6JA
2E1DGS BB4 6QN
G7WGO BB4 6QN
G0MBE BB4 6RR
G4UUE BB4 6RX
G4KLV BB4 6TQ
G0MEX BB4 7AY
G3FEV BB4 7QX
G7CZL BB4 7TH
G6VZZ BB4 7TS
G8PWG BB4 7UH
G0WAO BB4 8PG
G4VVK BB4 8PY
G1RKP BB4 8QE
G6TKB BB4 9HG
G4UNB BB4 9NZ
G0FQF BB4 9PX
G0DEI BB4 9TQ
G3IVG BB4 9TY
M0AKS BB4 9UQ
G7SML BB5 2RY
G0EAD BB5 2SR
G3GYU BL0 0HE
G6GLT BL0 0JD
G4YVO BL0 0LD
G6ZVO BL0 0LD
M1ADS BL0 0LX
G3VSH BL0 0QA
2E1EPD BL0 0QA
2E1EPE BL0 0QA
G4JBK BL0 0RY
G4PVS BL0 0RY
G0KWX OL12 8NP
G4TMV OL12 8PX
G0GND OL12 8PZ
G8TKD OL12 8SU
M1ATL OL12 8TH
G0UMQ OL12 8TP
G7DNM OL13 0EQ
G3BNF OL13 0PA
M1BIR OL13 8BT
G7MOF OL13 8LB
G8TYY OL13 8RN
G4LNE OL13 9LN
G7KLT OL13 9LS
G1XAR OL13 9PY
G0AEB OL13 9RH
G7JQF OL13 9RL

Rother

G3VFO TN19 7BE
G0OZY TN19 7DP
G6VYV TN19 7PH
G4NVM TN19 7PJ
G7MQL TN19 7QG
G7IGF TN19 7QY
G4RNO TN21 9JJ
G0THA TN31 6AN
G1IVN TN31 6DY
G6IPH TN31 6EN
G0OZM TN31 6EP
G1FTK TN31 6HU
G0FLA TN31 6NJ
G3ROC TN31 6NJ
G1KNF TN31 6QG
G1KNG TN31 6QG
G0THE TN31 6TR
G4TLE TN31 6UL
G0TFV TN31 6UR
G4VRA TN31 6YJ
G1OZR TN31 7BW
G1RNP TN31 7DJ
G0NNP TN31 7PT
G1RYE TN31 7PU
G0ISM TN31 7RU
G6HAI TN31 7UL
G6NLM TN31 7UU
2E1FAY TN31 7XR
G0WNK TN32 5DR
G0WFV TN32 5EF
G6MXO TN32 5EJ
G4YNW TN32 5JN
G6KGU TN32 5QS
G0URX TN32 5QS
G4HLK TN32 5SA
G8SVO TN32 5SA
G7CKS TN33 0EU
G7ERC TN33 0EU
G8UQJ TN33 0HG
G0JHK TN33 0JE
G3HZI TN33 0JF
G0BFG TN33 0LT
2E1BSF TN33 0NA
G3IVC TN33 0QE
G7MYQ TN33 0RZ
G4CUS TN33 0TP
G3YYF TN33 9AH
G8OGE TN33 9AU
G8MYG TN33 9AY
G0AHX TN35 4AQ
G4ERA TN35 4AQ
G3GYZ TN35 4BG
G3FHN TN35 4BN
G8ZVC TN35 4BQ
G3SYZ TN35 4DN
G3OZZ TN35 4DP
M1AKL TN35 4DP
G3BDQ TN35 4EP
G7SJV TN35 4ET
2E1DCC TN35 4ET
G3JSF TN35 4HE
G1OZO TN35 4NB
G0VTB TN35 4NG
G4XFG TN35 4QT
G4NER TN35 4QU
G1IKV TN35 4QZ
G1GFZ TN35 4SL
G6VQC TN36 4BN
G7UCM TN36 4BU
G8FJF TN36 4DA
G8GQG TN39 3AX
G3MRS TN39 3BG
G0MWF TN39 3EE
2E1CYT TN39 3LB
G4MMG TN39 3LU
2E1BEI TN39 3NE
G7MQQ TN39 3PA
G3WUH TN39 3PB
G3WGF TN39 3ST
G7OOP TN39 3SW
G3SVL TN39 3TD
G7TAJ TN39 3TD
G3JOR TN39 3UA
G7OGK TN39 3UQ
G6MUV TN39 3YA
2E1BUR TN39 4AG
G7LEL TN39 4AS
G8HJV TN39 4BG
G7UCL TN39 4BG
G1PDS TN39 4BN
G4HGK TN39 4DY
G0THV TN39 4HZ
G1COT TN39 4JB
G1WLO TN39 4JY
G5LY TN39 4JY
G7JCU TN39 4LP
M0AEK TN39 4NR
G0JEX TN39 4QG
G1RQH TN39 4SS
G4KUN TN39 4SZ
G4CWT TN39 4TA
2E1BEB TN39 4XD
G1TBO TN39 5DA
G8KXJ TN39 5EG
G1MTB TN39 5EQ
M1BWV TN39 5HL
G0JBM TN39 5HN
G7GHP TN39 5HU
G7EZD TN40 1LS
G8KPE TN40 1QH
G8FEJ TN40 1QH
G3NMJ TN40 1TU
G3NZO TN40 1TW
G3GGH TN40 2AZ
G4SIF TN40 2AZ
2E1ATF TN40 2AZ
GW4LPA TN40 2EL
G4ELB TN40 2ET
G0AKY TN40 2RF
G1VNZ TN40 2RH
G3MVX TN40 2SD
G0ILN TN40 2SH
G3MUF TN40 2SY

G8TUU TN40 2TD
G4XST TN5 7EG
G7IRN TN5 7JA

Rotherham

G4TZN S25 2SF
G4JVX S25 2TA
G4ZDD S26 2EP
G1HQC S26 5QT
M1ALG S60 1JU
G7ORI S60 2BA
2E1FCJ S60 2UR
G4UPT S60 3AW
G3VSK S60 3AZ
G0RTT S60 3HF
G1IDF S60 3JN
G7PFD S60 3PX
G3FCD S60 4BN
G7BXO S60 4DB
G0FZD S60 4DS
G7TLH S60 4DS
G6BKL S60 4DZ
G3IXA S60 4JH
G4NLC S60 4NH
G0ISK S60 5ES
G0LFS S60 5ES
G1DEZ S60 5HG
G0CGS S60 5JU
G0UVM S60 5JY
G6PFD S60 5NS
G0OEU S60 5QF
G6CPY S60 5RE
G6NKD S60 5RR
G8AHN S60 5TH
G0UOM S61 1BJ
G6TUX S61 1NT
G6LXW S61 2BL
G0SHG S61 2DD
G0OWD S61 2EQ
G1OHU S61 2EX
G8EQD S61 2HB
G7CBG S61 2JU
G7VPM S61 2JU
G6MLS S61 2JW
G0TLA S61 2JZ
G0RQE S61 2LE
G4SVN S61 2NT
G4NXS S61 2SS
G6XGJ S61 2SS
G4EBG S61 2TT
G7OFV S61 3HN
G4GES S61 3NP
G4GET S61 3NP
G4NMP S61 3PQ
G0MTT S61 4DL
G7ULZ S61 4DL
G4HVW S61 4LP
M0BEQ S61 4PA
G7OGR S61 4PD
G1GBR S62 5JR
G7BQS S62 5NH
G0SGB S62 5QQ
G0VPS S62 5TY
G6VIN S62 5UG
G7MST S62 6LN
G6JDC S62 7BX
2E1FTK S62 7BZ
G1IFH S62 7DQ
G1DWI S62 7ED
G3USW S62 7EN
G0DZX S62 7JG
G6HGU S62 7JZ
G3DVK S62 7LP
G6NLW S62 7PR
G0TKF S62 7UF
G1KZA S63 6NU
G4PJT S63 6QJ
G7COD S63 6RQ
G6VBR S63 7AP
G3MWN S63 7JB
G4BTS S63 7JB
G4TZL S63 7JB
2E1FCA S63 7JB
G7VIU S63 7JE
G7EVT S63 7JU
G3YOC S63 7LR
G6CUI S63 7SR
G1YPR S63 7ST
G4PEN S63 7TG
G7MUE S64 5UQ
G0VXC S64 5UU
2E1FCE S64 8AH
G6MFU S64 8DS
G8BIW S64 8DU
G7PBL S64 8HG
G4EHM S64 8HY
G0VJY S64 8NU
G0CSK S64 8NX
G6GYK S64 8PL
G1KIK S64 8PX
G0KSK S64 8TE
G7MEX S64 8TE
2E1FTF S64 8TE
G6OET S64 8UH
G4IDL S64 8UL
G4BQX S64 8UW
G1BOP S65 1RP
G0DSC S65 2AA
G1RWR S65 2RR
G7BDW S65 2RR
G4NJI S65 2UA
G0RHA S65 3AS
G3JDK S65 3BY
G4RHF S65 3JF
2E1FCS S65 3LJ
G0JPQ S65 3NX
G1KKZ S65 3PD
G4OQM S65 3QJ
G8NVX S65 4HH
G0JIY S65 4HZ
G6YSK S65 4HZ
G4APO S65 4NZ
G4RGH S65 4QR
G8DRQ S65 4QR
G4UHQ S66 3YW
G6ODO S66 7DB
G0HDG S66 7EG
G1BHQ S66 7EU
G0GOO S66 7HA
G0EIB S66 7HA
G6PCX S66 7HB
G0DWW S66 7HG
G7OFI S66 7HG
G4BVV S66 7HJ
G4SKM S66 7HJ
G1POC S66 7HX
G1CAQ S66 7LP
G1ILC S66 7LU
G0DXB S66 7PD
G6PMP S66 8AR
G0DCY S66 8BG
G6TJC S66 8BL
G4ZJJ S66 8DX
G3ZHI S66 8DY
G1PQW S66 8EJ
G7SKM S66 8EJ
G0LBK S66 8HB
G1YLH S66 8HG
G1OJT S66 8JU
G7HAR S66 8JU
G7KQW S66 8QU
G4JMC S66 9DL
G6THH S73 0TN
G7OEI S73 0TN
G3WLS S73 0TT
G0GCS S73 0XR
G8ZTO S81 8JW
G4ZYI S81 8RW

Rugby

G8HQC CV21 1EW
G3OQO CV21 1HW
G6GND CV21 1JB
G1BIM CV21 1JP
G0REQ CV21 1NH
G1VAJ CV21 1NJ
G0ZMC CV21 1PG
G4TDA CV21 1PP
G1DQB CV21 1RT
G1GUL CV21 1RT
G3TYP CV21 2ES
G1FLI CV21 2QU
G1ICA CV21 2TE
M0BBH CV21 3AB
G8UTN CV21 3AB
G7BNI CV21 3BA
G0NZI CV21 3HW
G7TOY CV21 3LG
G3IKL CV21 3QH
G4ZCP CV21 3QU
M1BYC CV21 3RR
G0COY CV21 3SZ
G7EQZ CV21 3TG
G0FIN CV21 3TS
G4IZI CV21 4AJ
G4HVB CV21 4AL
G8EYY CV21 4BP
G7KRE CV21 4HG
2E0ALQ CV21 4HN
G7NFO CV21 4JY
M0ASD CV21 4LT
G3RDC CV21 4PE
G0UFW CV22 5BG
G7APD CV22 5ET
G8LYB CV22 5ET
G7KGS CV22 5EY
G0DLB CV22 5HN
G0RLV CV22 5HN
G7KOG CV22 5JN
G4OIH CV22 5JU
2E1FSI CV22 5LF
G7KXU CV22 5QJ
G8AQN CV22 5QJ
G8FFZ CV22 5RW
G1MKT CV22 5SD
G3OBV CV22 6HB
G0HWX CV22 6HQ
G4GZS CV22 6HU
G3BXF CV22 6JR
G3CYH CV22 6JR
G4ACY CV22 6LG
G6DRX CV22 6LR
M0ANH CV22 6NS
G3NVV CV22 6PZ
G0HZF CV22 6QH
G8MFP CV22 6RG
G4RCU CV22 6SA
G6CYT CV22 7AP
G7FXU CV22 7AQ
G0BMF CV22 7BZ
G8DLX CV22 7BZ
G4SSW CV22 7EW
G8JCD CV22 7HY
G4IZM CV22 7LA
G1XAS CV22 7PF
M1AXE CV22 7PG
G7LIH CV22 7PT
G4AUB CV22 7PU
2E1FBU CV22 7RA
G6IQT CV22 7RW
G7JBW CV22 7TT
G1BIN CV23 0DB
G3THU CV23 0DQ
G4CFG CV23 0NN
G0TFN CV23 0NS
G0GSA CV23 0PE
G4AZH CV23 0PH
G0TZH CV23 0PH
G0UXP CV23 0PH
G4DYH CV23 0SD
G4IXT CV23 0UJ
G8HLF CV23 0UR
G4SFO CV23 8DE
G3ZJK CV23 8DN
G1ERG CV23 8DQ
G3VOQ CV23 8EN
G0TNV CV23 9AZ
G7GAB CV23 9DL
G4PEQ CV23 9JS
G3UUR CV23 9NX
G8IHF CV23 9RB
G4GDY CV3 2AJ
G8YKY CV3 2AX
G0PTN CV7 9JA
G4ZDG CV7 9JA
G8ESH CV8 3EL
G6WTD CV8 3ET
G0HOV CV8 3FR
G3JZL CV8 3JF
G4ZXA LE10 3LF
G1VCE LE10 3LP
G3YLP LE10 3LQ

Runnymede

G3FBN GU25 4AL
G0TTG GU25 4EZ
G6PMT GU25 4HA
G8MLK GU25 4LD
G0OSE KT15 1AS
G7DNE KT15 1DF
G4WMP KT15 1DH
G4HUR KT15 1EL
G0OVW KT15 1EQ
G8OTP KT15 1LW
G4ZRB KT15 1PJ
G4EVA KT15 1PN
G6EVA KT15 1PN
G7HIN KT15 1SR
G8HFW KT15 2DE
G7HFQ KT15 2PL
G1BDZ KT15 2SR
G7HXN KT15 2UQ
G3NQK KT15 3BS
2E0ABT KT15 3DJ
G4CCZ KT15 3DJ
G1SMB KT15 3DL
G4YSI KT15 3DN
G8JSN KT15 3ET
G0PVF KT15 3HS
G8FSZ KT15 3RJ
G8CUG KT15 3SE
G8KQH KT15 3TU
G0RMU KT15 3UA
G3ORB KT15 3UD
G4SSB KT15 3UD
G6IHY KT16 0AJ
G1UMY KT16 0ER
G8BWA KT16 8BU
G0PCZ KT16 8EP
G0BZF KT16 8LX
G1AEA KT16 8QB
G7VOS KT16 8RH
G6YRI KT16 9ED
G7VJP KT16 9EZ
G7NEN KT16 9JH
G8NYB KT16 9JJ
G0RNJ KT16 9LP
G8CYL KT16 9PF
G3OLH KT16 9PH
G3HPO KT16 9PR
G0OEK TW18 3HH
G4GSC TW18 3NQ
G8NIU TW20 0BJ
G6KBI TW20 0EE
G6JWV TW20 0HT
G4JDR TW20 0QU
G6EUP TW20 0SX
G4OJB TW20 0TJ

Rushcliffe

G1KBC LE12 5EE
G3TTH LE12 5HB
G4ZMA LE12 6LD
G8HVF LE12 6LN
G3URG LE12 6NG
G4RDF LE12 6NJ
G3UXF LE12 6NU
G3ZJV LE12 6PP
G7KNQ LE12 6QR
G3JHS LE12 6RH
G8AXN LE12 6SZ
G7TIZ LE12 6SZ
M0BCO LE12 6SZ
G0INJ LE12 6UZ
G0WFO LE14 3AF
2E1BJO LE14 3AF
2E1BDN LE14 3AF
G0TYG LE14 3AF
G2UT LE14 3AF
G4CWC LE14 3QA
G3JRS NG11 0DB
G4BKQ NG11 0HP
G4NPT NG11 0JW
G0IXR NG11 6DH
G4XSW NG11 6DL
G6GEA NG11 6DS
G7JSK NG11 6GZ
G3LCR NG11 6JE
G4NNY NG11 6QD
G7HZZ NG11 6QE
2E1CFC NG11 6QE
M1BAL NG12 1AX
G7ITW NG12 1AY
G4SUZ NG12 1DA
G4ZDF NG12 1DG
G1EGO NG12 2AP
G3YCE NG12 2BX
M1BKA NG12 2FF
2E1FTC NG12 2GA
G1VBP NG12 2HT
G4BPK NG12 2LA
G8AXH NG12 3AB
G0KDG NG12 3DE
G6OAN NG12 3DJ
G1GSG NG12 3DP
G3EDT NG12 3DT
G3ZUD NG12 3ES
G0MXX NG12 3FD
G0MAY NG12 3JS
G3TNX NG12 3JY
G6PKY NG12 3LA
G3FRE NG12 3NE
G1ZDC NG12 3TH
G8IQM NG12 4BX
G0NPC NG12 4DN
2E1CCN NG12 4DN
G4XRB NG12 4EA
G4UEK NG12 4ES
G3LXN NG12 4FQ
2E1CEU NG12 4FW
M1AZV NG12 4FY
G8ENB NG12 5BN
G7ODB NG12 5DN
G4NXM NG12 5EA
G0GWP NG12 5ET
G4MSR NG12 5HA
G3XVZ NG12 5HG
G7UQJ NG12 5HU
G0CRZ NG12 5JN
G3KZX NG12 5LQ
G3SJJ NG12 5NX
G1ZLB NG12 5PY
G4XZA NG12 5RA
G7OLX NG13 8SZ
G8MZW NG13 8TG
G4TTS NG13 8TL
M0BBE NG13 8TN
G7NZJ NG13 8TN
G4FAB NG13 8TY
G0USJ NG13 9HZ
G6XSS NG13 9JF
G4ZSO NG13 9PW
G8KBD NG2 5DN
G1TTB NG2 5DS
G1HXP NG2 5FU
G4VOW NG2 5GB
G3YUT NG2 5GG
G8GWP NG2 5HZ
M0AVH NG2 5LA
G7THU NG2 5LA
G0RMN NG2 6DF
G1HOD NG2 6DH
G1OTJ NG2 6DJ
G7HCJ NG2 6DJ
G3WNC NG2 6EB
G3VLN NG2 6FQ
G0FOG NG2 6HQ
G8IMC NG2 6JE
G8ZSD NG2 6PE
G8STJ NG2 6QG
G4DCI NG2 6RB
G6ZOS NG2 7AD
G8KJJ NG2 7FL
G8FRB NG2 7GA
G4NZU NG2 7GG
G3RLO NG2 7GQ
G4ABT NG2 7HJ
G6RNF NG2 7HW
G0PIB NG2 7QS
G4LOF NG2 7TE
G8TVM NG2 7UQ

Rushmoor

G6RQZ GU11 1YY
M0AHJ GU11 3DB
G4GVV GU11 3EL
G8YGD GU11 3HQ
G4GGZ GU11 3JX
G3FOR GU11 3LX
G0BCH GU11 3PX
G4UEL GU11 3QB
G8UEM GU11 3QB
G7EXO GU11 3SL
G0HWL GU12 4AT
G0LOX GU12 4BB
G0ORN GU12 4EA
G4YFU GU12 4HZ
G8OLY GU12 4LD
G1OQG GU12 4PR
G4SBU GU12 4PS
G3KND GU12 4RD
G4SPD GU12 4SF
G7TGM GU14 0AD
G6MRW GU14 0DW
G8MTK GU14 0JZ
G1AAP GU14 0PB
2E1EIG GU14 0PL
2E1EIH GU14 0PL
G8IBC GU14 0RF
G7VTS GU14 6DA
G0TBV GU14 6HX
G1NTX GU14 6LX
G4JFN GU14 6NU
G4VAH GU14 6NU
G4KCC GU14 6QA
G2BPF GU14 6RE
G3WOS GU14 6RF
G6VUG GU14 6RR
G4PUG GU14 6RT
G3PQC GU14 7AR
G4APU GU14 7AR
G5GC GU14 7AZ
G0DWM GU14 7DA
G0HNA GU14 7DA
G3IZJ GU14 7DD
G3UHW GU14 7EU
G8MY GU14 7EX
G1OER GU14 7EY
G3DJY GU14 7EZ
G0LVR GU14 8AF
G7FIF GU14 8DH
G4PDR GU14 8NN
G1FVH GU14 8PH
G3PQF GU14 8QJ
G3AID GU14 8RG
G8ELX GU14 8RS
G0GCA GU14 8SD
G0HNL GU14 8TA
G6KPX GU14 8UG
G4MIF GU14 9BS
G1CBS GU14 9DJ
G1KXQ GU14 9EA
G8HUF GU14 9EJ
G8OEY GU14 9EJ
G6OHD GU14 9HG
M0AJZ GU14 9HJ
G1XIE GU14 9HJ
G0MPI GU14 9JQ
GW4YEG GU14 9LN
G7JXT GU14 9NR
G0MCQ GU14 9PJ
G0EMR GU14 9PT
G4SHQ GU14 9RJ
G4BNK GU14 9RL
G1PET GU14 9TG
G1VLB GU14 9TR
G6VCH GU14 9XA
G4EMV GU17 9ET
G3HVE GU17 9HB
G0EFL GU17 9HH
G6AAB GU17 9HH

Rutland

G0IVY LE15 6BH
G4GWW LE15 6BH
G8TNS LE15 6DA
G6ZCX LE15 6JW
G6ZCY LE15 6JW
G6LEI LE15 6LD
G0CWU LE15 6LL
G3ALC LE15 6LT
G4FDP LE15 6LT
G0WKY LE15 6LX
G4LPL LE15 6PR
G8VGY LE15 6PR
G7BCW LE15 6SE
G3WMA LE15 6SJ
G4PJY LE15 6SL
G3PLL LE15 7BZ
G3PKD LE15 7DB
G4IVF LE15 7DY
G6RAF LE15 7DZ
G3TCQ LE15 7DZ
G3ZDW LE15 7DZ
G0VIF LE15 7EY
G4PKZ LE15 7JL
G4LFE LE15 7LJ
G7FGR LE15 7NG
G3OKB LE15 8JS
G6UYT LE15 8JX
G2BPW LE15 8JZ
G1CPM LE15 8NP
G4EWB LE15 8QZ
G4OSJ LE15 8RH
G8TTE LE15 8SU
G4HDM LE15 9RY
G4FIS LE15 9TS
G0GRU PE9 3RP
G8VDJ PE9 3UQ
G0FPZ PE9 4AJ
G0INV PE9 4AJ
G1CUG PE9 4LR

Ryedale

G0UVR YO13 9PA
G3UWP YO17 0BE
G0WZJ YO17 0HJ
G4GQW YO17 0HL
G1VDO YO17 0LB
G0FNS YO17 0LE
G8HCK YO17 0NJ
G3ISU YO17 0PL
G3VVR YO17 0QX
G7KMH YO17 0SR
G3TPW YO17 0SY
G7CGS YO17 0SY
G4HNW YO17 0YN
G8YVT YO17 8HA
G4CVG YO17 8LZ
G7WLJ YO17 8TB
G7WLN YO17 8TQ
G8JYS YO17 9AE
M1AVM YO17 9AE
G6OJV YO17 9AR
G0KOE YO17 9SJ
G0PWB YO18 7HF
G4OBK YO18 7HJ
G3EEH YO18 7HN
G3GJY YO18 7HQ
G0KGA YO18 7HX
G4BNS YO18 7HZ
G1VDS YO18 7JT
G0FNP YO18 7JY
G8ZMJ YO18 7LE
G1MRD YO18 7QU
G6WOB YO18 7SG
G4LUH YO18 7SN
G6AQG YO18 7TE
G6AQH YO18 7TE
G3FYP YO18 8AF
G4GWC YO18 8DS
M1AVW YO4 1LU
G1YHO YO6 1PQ
G4CSN YO6 1PQ
G0HKH YO6 4LA
G4FWI YO6 4NH
G1NBK YO6 4QB
G7HAF YO6 5EZ
G3LOJ YO6 6BS
G3OMJ YO6 6DH
G0GXI YO6 6EF
G1SZT YO6 6EL
M0AGL YO6 7HH
G3ZEM YO6 7JJ

Salford

G3VOM M27 0DR
G6EYH M27 0FD
G6YGW M27 0FH
2E1ENW M27 0GF
G1RNU M27 0HA
G1WGF M27 0HA
G0TVA M27 0HZ
G0LVH M27 0YA
G0MKA M27 0YH
M1BAQ M27 4ED
G8RAX M27 4FA
G0HQP M27 5FP
G4IRB M27 5GZ
G8SMR M27 5GZ
G0KLF M27 5NA
G0NPI M27 5TD
G0VPN M27 5UX
G1OAN M27 6PY
G0SFR M27 6QB
G4FLA M27 6WF
G4FXA M27 6WH
G0CVH M27 8FS
G4VDJ M27 8QS
G0MZY M27 8RE
G1SYV M27 8RU
G0LVG M27 8XP
G1PAL M27 9QE
G4EFD M27 9RD
G3YAE M27 9RW
M0APZ M28 0AY
G0HQN M28 0HQ
G8KOQ M28 0SL
G3PMJ M28 0SX
G7NCF M28 0TT
G3GAG M28 1DD
G7ELA M28 1DE
G8XOD M28 1FX
G0UNP M28 1FX
G7VEM M28 1JP
G7DPW M28 1LP
G3MXT M28 1LQ
G3XLX M28 1YS
G6MX M28 2PA
G1MPZ M28 2PF
G8USV M28 2PS
G0TEE M28 2RJ
2E0AHH M28 2RJ
G8UEA M28 2RT
G0RGQ M28 2RW
G1VJQ M28 2RW
G8DTF M28 2SH
G0RHY M28 2SL
G6YJA M28 3BL
G0OHY M28 3DT
G6MEI M28 3HG
G0PAG M28 3LU
G3OAX M28 3PW
G4WIV M28 3QR
G4NTY M28 3RF
G4UZG M28 3RP
G7VPY M28 7EN
G0FXX M28 7EW
G4UPU M28 7EX
G7FFK M28 7JD
G4TJU M28 7JW
G4JLG M28 7QF
G7WCF M28 7UQ
G7FPS M28 7UW
G7NKB M30 0DW
G8IUQ M30 0GA
G6YAQ M30 0HW
G1ASN M30 0LT
G6SPG M30 7HQ
G4GPJ M30 7SA
G4VPW M30 8AG
G6EOS M30 8AG
M1ABV M30 8AJ
G0BVM M30 8BP
G0FPO M30 8DD
G7PWP M30 8JB
M1ANX M30 8LH
G4UOT M30 8LT
G0IZO M30 8QE
G8MOK M30 8WA
G8VF M30 9AU
G7AOW M30 9DL
G6ZGF M30 9FA
G0KFP M30 9FD
G6FUY M30 9GF
G4EYQ M30 9LA
G4ZFF M30 9LY
G0WIA M38 0DU
G0SFS M38 0PU
G6ZTC M38 9GQ
G8MLF M38 9WL
G6PKV M38 9XU
G7YEW M44 5AR
G0GSK M44 5AU
G1OAJ M44 5EA
M1ALZ M44 5UA
G6HCL M44 5YW
G8TXW M44 6AQ
G7REK M44 6DA
G0JYK M44 6EN
G0HQO M44 6JP
G0IFV M44 6JP
G4OMM M44 6ND
G6LBQ M44 6WX
2E1EML M44 6ZD
G8GAJ M5 2UE
2E1CEA M6 5AB
2E1CJF M6 5EN
G1IXN M6 5PZ
G8FYK M6 7AL
G6HKH M6 7DZ
G0PLX M6 7LA
G6FEI M6 7PQ
G8YPH M6 8FZ
G0IYM M6 8QQ
G7JOF M7 2BZ
G0MRX M7 2FX
G7BYG M7 3ST
G7VTA M7 3TJ
G7BQD M7 4AP
G7LDR M7 4EZ
G3ETU M7 4LP
G4SBV M7 4NB

Salisbury

G0AYD BA12 0RN
G0BGI BA12 6HR
G4SSP BA12 6JX
G3BHK BA12 6NG
G0DZU SO51 6FU
G0HLA SO51 6FU
G3ZNH SP1 1LA
G0DJV SP1 1NU
G3VQZ SP1 1NU
G3MXP SP1 1QL
G8OFA SP1 1QZ
G8RHC SP1 1QZ
G2AQJ SP1 1SA
G7GWF SP1 1SH
G8WJB SP1 2JH
G6PGV SP1 3AA
G7DNG SP1 3BH
G6CZB SP1 3DN
G4MOE SP1 3HS
G6JUJ SP1 3HS
G0MZI SP1 3LB
G3MML SP1 3NG
G7WAA SP1 3PR
G4TRD SP1 3PZ
G8ODM SP1 3QQ
G3XTF SP1 3RA
G0ETH SP2 0BQ
G2FIX SP2 0DY
G4RLF SP2 0LW
G4SXQ SP2 0NF
G3VYS SP2 2BE
G0AJE SP2 7AJ
G8CWJ SP2 7EX
G2ZU SP2 7HG
G8WAM SP2 7JT
G0GCN SP2 7LG
G7OAM SP2 7LX
G6JEK SP2 8AT
G6JEL SP2 8AT
G0BDP SP2 8DN
G0VVO SP2 8DR
G4DUA SP2 8DT
G8WBO SP2 8DU
G7AGY SP2 8EF
G6RPW SP2 8EH
G3KOZ SP2 8EP
G6TVK SP2 8JJ
G4SDY SP2 8NB
G7UHN SP2 8NL
2E0AAX SP2 8NR
G3OIL SP2 8NR
G3XXI SP2 8NZ
G0TGW SP2 8PB
G4HQM SP2 8QS
G1ELZ SP2 8QU
G0VHZ SP2 9BH
G6MHU SP2 9BX
G8NDV SP2 9BY
G1UBA SP2 9HR
G1BMT SP2 9LF
G6DRW SP2 9LH
G7GVC SP2 9PS
G4THW SP3 4DN
G8PMA SP3 4JU
G3UJU SP3 6EE
G8PCB SP3 6PP
G4NWJ SP4 0BX
G8ALR SP4 0ER
G8HUV SP4 6AU
G6JGP SP4 6DJ
G7WBW SP4 6HU
G1SSZ SP4 6NB
G4WPI SP4 7AD
G6YUX SP4 7AS
G0YBU SP4 7BP
G3YFE SP4 7EE
G4YQV SP4 7HZ
G4UED SP4 7NN
G4SXR SP4 7PJ
G8WUJ SP4 7PJ
2E1CMZ SP4 7PJ
G6DHY SP4 7PU
G7GKD SP4 7RE
G1WRY SP4 7RG
G4YRV SP4 7RG
G0HKC SP4 8DB
M1BNF SP4 9NN
G4MNP SP5 1AA
G6BVR SP5 1AY
G4OBS SP5 1EU
G4YIS SP5 1PP
G4LDR SP5 1PX
G3CMJ SP5 1QR
G1ITJ SP5 1RS
G3MDM SP5 1SH
G3BKL SP5 1SN
G4VQJ SP5 2AL
G7WBA SP5 2BZ
G0OSW SP5 2ED
G4MBS SP5 2ES
G6IDQ SP5 2ES
G0DKN SP5 2HN
G4GXZ SP5 2JT
G7RYG SP5 2LG
G0RZB SP5 2LN
G0TSY SP5 2LN
G7OYW SP5 2LR
G3SHK SP5 2NQ
G3NZK SP5 2PR
G3WZ SP5 2SE
G3YWT SP5 3AT
G0CDZ SP5 3AZ
G6MB SP5 3AZ
G1EWT SP5 3BD
G6ILC SP5 3JN
G7BOR SP5 3LX
G6ITH SP5 3TH
G6RLG SP5 3TH
G0MLX SP5 4LE
G3KLH SP5 4LR
G4UNG SP5 5ED
G4MGZ SP7 0HS
G0MPM SP7 9NB

Sandwell

G0UAU B43 5AW
G6NUO B43 5BN
G1CYD B43 5HP
G4VJO B43 5JD
G7IKG B43 5JR
G6NPL B43 5LE
G8GAZ B43 5LE
G4KXE B43 5LS
G0NOV B43 5LY
G3APZ B43 5RU
G3KPT B43 5TJ
G3FIA B43 6BB
G6NAN B43 6BB
G4PMW B43 6JE
G1MRP B43 6JR
G8ODT B43 6NQ
G6YSB B43 6QE
G3WPY B43 7AP
G0NNE B62 8TA
G4YAH B62 8TB
G7TUV B64 5EX
G0PPJ B64 6DU
G4JFF B64 6RB
G3VPX B64 7EZ
G7BRA B64 7EZ
G6NNO B64 7HJ
G4ISQ B64 7LE
G1ENR B64 7PJ
G0WYA B65 0NP
G1OMP B65 0RL
G1OOZ B65 9BQ
G0PAI B65 9DZ
G0OYF B65 9HZ
G0KJG B65 9JN
G3SFT B65 9JX
G0MJT B65 9LG
G4FGQ B65 9ND
G3TRG B65 9NT
G8TRG B65 9NT
G6ZQL B65 9QW
G4FJJ B65 9SD
G1JVM B67 5AY
G1XFO B67 5BB
G6IQY B67 5BB
2E1EHF B67 5BL
2E1EHG B67 5BL
2E1EHH B67 5BL
G0CDP B67 5DD
G1UCO B67 5DH
G3DFL B67 5EA
2E1FGC B67 5EJ
G0VCR B67 5JE
G7TBS B67 5JR
G8IWE B67 5RS
G0CCF B67 6AN
G0JAA B67 6AS
M0BBW B67 6HU
G7WIM B67 6HU
G1UUZ B67 6LA
G3URI B67 6NX
G1FFO B67 6PR
G0TVR B67 6QX
G4KVC B67 7BX
G3WJC B68 0BU
G8DEM B68 0NA
G8PTF B68 0NU
G4RWY B68 0PU
G4JBU B68 8AP
G7UJK B68 8AS
M1AED B68 8BE
G4HWN B68 8ED
G0CWC B68 8LT
G1SAN B68 8LT
G4OJJ B68 8LT
G4OYT B68 8PR
G2BXP B68 8PT
G3NAI B68 8QH
G4JSR B68 9DB
G8SBZ B68 9DL
G7SIB B68 9DR
G8MKE B68 9ES
G8YLB B68 9JL
G5KS B68 9LG
G4VRX B68 9LU
G1TBL B68 9NX
G0NNF B68 9PW
G8RAO B68 9PW
G3TVF B68 9QE
G7RTQ B68 9TB
G0WXA B68 9UL
G6VXD B69 1AF
G0NJZ B69 1BA
G7IYX B69 1BA
G7IHC B69 1DQ
G3NZS B69 1NT
G4YFT B69 1PW
G1WHY B69 1QA
G4ORD B69 1QA
G7UIL B69 1RB
G0OYY B69 1RP
G6EOR B69 1SE
G7JOC B69 1SG
G8SPU B69 1TP
G0LYY B69 1TQ
G8TFB B69 1UD
G7UNB B69 1US
G0CBF B69 2DX
G4JVH B69 2EE
G1XWM B69 3LA
G7LOE B69 4QA
G3VFF B70 0DD
G4DVM B70 0SL
G1ZOE B70 6TL
G7POG B70 7AZ
G1LOA B70 7HR
G1TBK B70 7LS
G4WBC B70 7LS
G7BRG B70 7SP
G0TEM B70 7SP
G7RWG B70 8AJ
G0BZP B70 8LD
G7UMW B70 8LD
G6ZYZ B70 8PE
G8WEG B70 9JF
M1AIJ B70 9TJ
G0RCX B70 9TL
G0BRG B71 1EW
G0WCK B71 1NJ
G6UKV B71 2AE
G0DQO B71 2BL
G6UZL B71 2EH
G4BBO B71 2ES
G7RSA B71 2HU
G7IPP B71 2NT
G0WWA B71 2NX
G4LWN B71 2QW
G6IZP B71 2QZ
G1AJS B71 3BD
G8GUN B71 3BD
G6VFO B71 3BT
G1DTB B71 3HH
G6MPT B71 3HX
G3THZ B71 3JS
G0HKK B71 3NF
G6VAA DY4 0AJ
G0PYK DY4 0DF
G6KSN DY4 0UU
G7NUK DY4 7ND
G7JCD DY4 8HZ
M0AXO DY4 8JT
M1BAX DY4 8JT
G7CBW DY4 8UZ
G7TDP DY4 9HW
G4OUH DY4 9PQ
G4OUI DY4 9PQ
G7EKT WS10 0DN
G1DHA WS10 0HX
G0EAN WS10 0JF
G0WJJ WS10 0PN
G7FPO WS10 0QP
G7VUB WS10 0RW
G0CMR WS10 0TB
G1GZM WS10 7HY
G7CNZ WS10 7HY
G0KNM WS10 9AU
G1SHQ WS10 9AU
G1PQY WS10 9BT
G6JAM WS10 9DJ
G1PVW WS10 9NX
G4TDP WS10 9PH
G6CSR WS10 9RQ
G7RCP WS10 9SX
G3AGW WS5 4EE
G4TCM WS5 4HE
G4XEZ WS5 4RY

Scarborough

G0TRH YO11 1DB
G0OII YO11 2BJ
2E1EMX YO11 2DG
G0PHD YO11 2HF
G0IDI YO11 2RN
G0RCS YO11 2TP
G3LCG YO11 2TP
G4OVW YO11 2TU
2E1FFY YO11 2XQ
G4VDH YO11 3AP
G4DEW YO11 3AW
G0FBM YO11 3DR
G1WJQ YO11 3JG
G2AQN YO11 3JP
G0KDA YO11 3JS
G6AIB YO11 3PB
G3DHY YO11 3PZ
G8WVZ YO11 3PZ
G8AZA YO11 3RE
G0AIF YO11 3RW
G3GKI YO11 3SD
G3DIF YO11 3SE
G4KBQ YO11 3SJ
G8MVO YO11 3TF
G4AKR YO11 3TS
G7LFV YO11 3TS
G3HFW YO11 3TW
G6EBQ YO11 3TW
G0NXX YO11 3UU
G0FLZ YO12 4AA
G0FDK YO12 4DF
G4JJQ YO12 4EW
G7OWX YO12 4EY
G0RPU YO12 4JE
G0NNZ YO12 4JR
G4WXJ YO12 4JY
G0FXT YO12 4LA
G8WYB YO12 4LL
G4IAJ YO12 4LZ
G3RIX YO12 4RJ
G0COL YO12 4RL
G0OOO YO12 4RN
G7OOO YO12 4RN
G7ROY YO12 4RN
G4SSH YO12 4RN
G4UUU YO12 4RN
2E0ABI YO12 4RQ
G0UUU YO12 4RQ
G4YSS YO12 4RQ
G7SHM YO12 4RW
G0TTH YO12 4TW
G4PNC YO12 4TX
G4HPV YO12 5AF
G4DPO YO12 5BN
G4YKO YO12 5DB
G0WJB YO12 5DP
G0PRI YO12 5JA
G8SGX YO12 5JA
G7RRY YO12 5JA
2E1DJT YO12 5LD
G0FBV YO12 5NG
G0WHO YO12 5NJ
G0VAF YO12 5PD
G0LEU YO12 5PG
G3HTD YO12 5PH
G0VAG YO12 5PX
G1PPO YO12 5QG
G3PEJ YO12 5QJ
G3XIH YO12 5QJ
G4DWU YO12 5QL
G0DOA YO12 5RG
G0JNR YO12 5RQ
G0KEU YO12 5SB
G0TOS YO12 6DQ
G4JZO YO12 6ED

G0NSE YO12 6HH
G4ZAO YO12 6HL
G3JBR YO12 6JW
G4BP YO12 6JW
G6AFZ YO12 6JW
G0SHM YO12 6LW
G0RCL YO12 6NT
G4EGB YO12 6RA
G3ENB YO12 6RN
G4FCH YO12 6RN
G3HKO YO12 6RQ
G4NSE YO12 6SF
G0CDR YO12 6SG
G0VXE YO12 6SP
G3WOD YO12 6TF
G8ETS YO12 6TG
G4ZGP YO12 6TL
G0KFG YO12 6TW
G3KEV YO12 6UA
G0PFE YO12 7HL
G0LSD YO12 7HT
G1OSP YO12 7JT
G8PIC YO12 7LF
G0DHV YO12 7PT
G1VYB YO12 7QG
G3LSW YO12 7QZ
G2CP YO12 7RR
G7RPM YO12 7TH
G0VYD YO12 7TH
G0OPQ YO13 0DQ
G0VFV YO13 0EZ
G6PWZ YO13 0HE
G7TTV YO13 0HW
G3TMC YO13 0JA
G4SQY YO13 0JD
G4FNG YO13 0LW
G3ISL YO13 0PJ
G4UQP YO13 0QA
G0IEB YO13 0ST
G4MGP YO13 9AR
G4MGQ YO13 9AR
G4MVA YO13 9AY
G4AJJ YO13 9DY
G4LSS YO13 9DY
2E1AZN YO13 9DY
G0PWP YO13 9ER
G0NUP YO13 9ER
G0SIG YO13 9ER
G4FLM YO13 9ET
G4JAQ YO13 9ET
G4JFI YO13 9EU
G0FUE YO13 9EU
G7SGK YO13 9HB
G0NXK YO13 9HS
G3KAE YO13 9HW
G6SDY YO13 9JA
G4CCV YO13 9JE
G4OPI YO13 9JQ
G0PSK YO13 9JW
2E1DLP YO13 9JY
2E1EZM YO13 9JY
G7BAY YO13 9LH
G0NVU YO14 0AL
G8UOW YO14 0DB
G4WCY YO14 0DQ
G8ZFS YO14 0LB
G0VKK YO14 0LX
G4KZZ YO14 0NB
G4OY YO14 0NH
G1OZP YO14 0PU
G1GDU YO14 9BE
G7MZE YO14 9ER
G4KTL YO14 9EW
2E1FJE YO14 9NE
G7LOV YO14 9NL
G4EDR YO14 9NY
G3JNW YO21 1JP
M0AZC YO21 1JS
G3NTA YO21 1NA
G0UYS YO21 1NB
G6HTA YO21 1QX
G0UOF YO21 1SA
G1VOY YO21 1UE
G0WDC YO21 2BL
G8KHT YO21 2EN
G6HPU YO21 3BG
G7RTJ YO21 3JB
G0UBK YO21 3LR
G4IZE YO21 3LS
G3VGN YO21 3NQ
G1LPL YO21 3SD
G0WNF YO22 4AS
G3FFL YO22 4DS
G0EQI YO22 4HL
G0EGE YO22 4JU
G0DSO YO22 4PB
G4HHH YO22 4QG
G4LRH YO22 4QH
G3VEO YO22 4TH
2E1DLK YO22 4TH
G0EBL YO22 5AN
G4DAX YO22 5AN
G0ANH YO22 5EH
G4UIA YO22 5EL
G0AOP YO22 5EP
G4NIQ YO22 5EP
G0ABQ YO22 5HG
G3DAV YO22 5HG
G0SUC YO22 5HG
G0UZQ YO22 5JT
G3VNI YO22 5LY
G0WNJ YO22 5QQ

Scottish Borders

MM0ABJ EH44 6JT
GM0JCS EH44 6PD
GM1OPO EH44 6RD
GM0ICP EH45 8DA
GM1RKI EH45 8HS
GM6JFP EH45 8JX
GM8LLY EH45 8NA
GM0UTD EH45 8NU
GM4WTK EH45 8PP
GM4LDX EH45 8PW
GM0JCX TD1 1EZ
GM7GIS TD1 1HD
GM1IXW TD1 2AT
GM8UUW TD1 2DE
GM1JLP TD1 2EE
GM4VEJ TD1 2EJ
GM7GPG TD1 2HW
GM8CJW TD1 2RB
G7WJL TD1 2RJ
GM6WRY TD1 3JW
GM7LUN TD1 3ND
GM0HQT TD1 3RF
GM0JFB TD1 3TP
GM4FIZ TD10 6XA
GM4CXP TD10 6XE
MM1CDC TD10 6XF
GM8IIO TD11 3EE
GM0TUS TD11 3HG
GM4IWK TD11 3HH
GM0RDZ TD11 3HQ
GM4NQT TD11 3LA
GM1PGO TD11 3TP
GM0HNP TD12 4HE
GM0RXP TD12 4HE
GM4SYF TD12 4JQ
G0REY TD12 4NE
GM8BDX TD12 4NG
G0HNX TD12 4NG
GM3BNX TD12 4NS
GM3PAE TD13 5XB
GM4VJV TD13 5XY
GM4XZZ TD13 5YD
GM4HVU TD13 5YS
GM1REY TD14 5DH
GM1REZ TD14 5DH
GM4ZXJ TD14 5DX
GM1JFF TD14 5EU
GM3KAI TD14 5JN
GM0BPY TD14 5JQ
GM1DVO TD14 5QF
GM3WKB TD14 5SP
GM4WWU TD14 5SP
GM0IGF TD14 5TY
GM0BCQ TD15 1SU
GM0BPO TD15 1XW
GM0BRS TD15 1XW
GM4WDO TD15 1XW
GM0ALW TD5 7AS
GM0ALX TD5 7AS
GM4KHS TD5 7BJ
GM4UIB TD5 7BS
GM0SIA TD5 7EU
GM0EUA TD5 7NB
GM6PTX TD5 7NN
GM4TXN TD5 7NQ
GM0BPH TD5 7RB
GM6TBE TD5 7SJ
GM3VLB TD5 8BB
GM4DNS TD5 8BW
GM4NLJ TD5 8LB
GM6YRZ TD5 8LB
GM0AMB TD6 0SZ
GM4YEQ TD6 0SZ
GM0RDC TD6 9EP
GM6IZU TD6 9QZ
GM1FMV TD6 9RH
GM0MHV TD7 4AR
MM1APS TD7 4BB
GM7GRH TD7 4BG
GM3PSJ TD7 5AL
GM0KCN TD7 5BP
GM0FTJ TD7 5DD
MM1BOT TD7 5DL
GM4UFP TD7 5EY
GM4VYU TD7 5JB
GM1ZTB TD8 6DU
GM0NKX TD8 6HT
GM0VHR TD8 6HT
GM3KJA TD8 6JY
GM0KEL TD8 6NP
GM4UPX TD8 6NP
GM7WLO TD8 6SD
GM4IAO TD8 6SU
GM4DMK TD8 6TS
GM6MCV TD8 6TZ
GM6MYA TD8 6TZ
GM1RHX TD9 0LT
GM1JVI TD9 0QX
GM4LPJ TD9 0SN
GM8JFE TD9 7HZ
GM0IGJ TD9 7PR
GM8GUX TD9 8BA
GM0JCR TD9 8EP
GM6BKK TD9 8HB
GM3WIG TD9 9JJ
GM1AQV TD9 9ST

Sedgefield

G7ESY DL16 6DW
G6HWA DL16 6DY
G7ESX DL16 6HP
G1LPS DL16 6LX
G0KNN DL16 6RN
G0PWK DL16 7BA
G1TDR DL16 7DR
G0OBF DL16 7DT
G0RNY DL16 7DY
G4SZD DL16 7DY
G0JEC DL16 7HN
G7KJR DL16 7HU
G0BVO DL16 7QS
G1IPK DL16 7ST
G0LFD DL17 0HP
G4PIX DL17 0RL
G4CAY DL17 0RW
G4MQV DL17 8DA
G6CII DL17 8DA
G8UYM DL17 8DA
G4XHP DL17 8PW
G1LGQ DL17 8QE
G4SUL DL17 8QH
G0EEK DL17 8RN
G3LUC DL4 1BH
G7TKJ DL4 1EU
M1AYZ DL4 2EE
G0IED DL5 4SZ
G0EHH DL5 5AQ
G3ZWB DL5 5HX
G0EJB DL5 7AY
G0IDS DL5 7DE
G0IXE DL5 7ET
G7OSM DL5 7LA
G1ZEU DL5 7PS
G0WXC TS21 2DS
G0SJU TS21 2EZ

Sedgemoor

G3IXO BS25 1SA
G3RXG BS25 1SA
G4XYH BS25 1TG
G7DRO BS25 1UE
G4MQX BS26 2AN
G4MCE BS26 2AX
G7FAD BS26 2TR
G2CJL BS27 3AG
G6UVQ BS27 3AP
G6TKR BS27 3BS
G3CRR BS27 3DT
G0VSS BS27 3HZ
2E0AJS BS27 3HZ
M0BDG BS27 3HZ
G3RBJ BS28 4BZ
G0HVB BS28 4SW
G6PPU TA3 5AU
G3OJK TA5 2AP
G3IBK TA5 2HH
G1BPM TA5 2NH
G4WMY TA5 2PZ
G3FRN TA5 2QT
G1IHO TA5 2QU
G0NBN TA5 2RB
G4SBX TA6 3EG
G4ETN TA6 3HY
G4BRW TA6 3LP
G0OXB TA6 3NU
G7SYS TA6 3QE
G3RHW TA6 4DW
G7CXZ TA6 4HJ
G6JTT TA6 4JL
G6UNN TA6 4LD
G7GMZ TA6 4NA
G0IFI TA6 4QJ
G1BKL TA6 4RY
G0MIK TA6 4SH
G7KHZ TA6 4UT
G3PNF TA6 4XR
G6HWO TA6 5NP
G0LCV TA6 5NS
G0COM TA6 5PF
G0CNA TA6 5PH
G4WMV TA6 6AZ
G8PVG TA6 6DZ
G1MCY TA6 6JU
G7AHE TA6 6JZ
G0MSP TA6 6LB
G8TVO TA6 6LR
G3PCG TA6 6NB
G0GUK TA6 6QJ
G4DPS TA6 6SL
G7XOZ TA6 6UR
G1XOZ TA6 6UR
G0WRQ TA6 7AJ
G3TWO TA6 7EB
G4QK TA6 7EE
G0MFR TA6 7EJ
G4TFF TA6 7EJ
G0KEA TA6 7HA
G4YJZ TA6 7JH
G1BTI TA6 7LJ
G4EHU TA6 7NT
G7WBL TA6 7NZ
G1HLP TA6 7PD
G7SEQ TA6 7QA
G7URS TA6 7QY
G8YPV TA6 7RF
G3FSA TA7 0DE
G0OYA TA7 8BB
G1IOU TA7 8BB
G1OFO TA7 8BD
G7VWM TA7 8EA
G0KZV TA7 8EG
G4FOB TA7 8ES
G1RKM TA7 8EU
G1PKM TA7 8HA
G6RRV TA7 8QR
G1HSF TA7 9BT
G4WZF TA7 9EW
G4VHQ TA7 9HB
G8SUW TA7 9LR
G4NQQ TA7 9LS
G6WOV TA7 9NJ
G0PQD TA8 1DH
G4EFK TA8 1DH
G3GLO TA8 1DQ
G1FAJ TA8 1EW
G3WNG TA8 1HE
G4CKO TA8 1HG
G0OQT TA8 1HY
G4DVK TA8 1JA
G4EJW TA8 1LT
G0GNJ TA8 1NF
G6YMQ TA8 1QW
G3GIW TA8 2DD
G4ASG TA8 2DE
G6TAH TA8 2EJ
G4HLN TA8 2FA
G1OKU TA8 2HF
G0FYP TA8 2HP
G3NMI TA8 2JJ
G3WUZ TA8 2JT
G1OUR TA8 2LT
G8AYK TA8 2NG
G3GZZ TA8 2NL
G3WLA TA8 2NW
G4CGH TA8 2NW
G3NYD TA8 2PH
G7VEG TA8 2QH
G1ENN TA8 2RN
G1XWZ TA8 2TU
G4FMH TA9 3JR
G4UOS TA9 3LD
G4NFO TA9 3LE
G4TKS TA9 3LG
G0VYV TA9 3QZ
G7GJZ TA9 3RF
G0PAY TA9 3SF
G8HUH TA9 4BU
G1JGM TA9 4DQ
G3TJE TA9 4DT
G7VIV TA9 4DU
G0GOB TA9 4DY
G3FIT TA9 4JQ
G3SXQ TA9 4QT

Sefton

G0PVZ L10 3JN
G3XAN L10 3LD
G6YPV L10 6LP
G0FIV L20 3DY
G3YXS L20 4PW
G4KRF L20 4QG
G1SPK L20 4QR
G8BYH L20 5BB
G6YNX L20 6DF
G0OCC L20 6EJ
G4ITI L20 7BB
G3KVE L20 9ET
G0LBF L20 9LG
G7BGL L21 1DD
2E1FAS L21 1EG
G3JUB L21 4NR
G1PJK L21 5JT
G1SWE L21 5JT
G0DBE L21 7NN
G1DFP L21 7QH
G4VIL L21 7QH
G1LOV L21 9HX
G3KFC L21 9JS
G7RZT L22 1RR
G4OJK L22 4QT
G4XRX L22 8QB
G1MTJ L23 0RF
G1JJG L23 0RG
G6FDI L23 0RG
G8JYV L23 0RQ
G6TGI L23 0SB
M1BTD L23 0SB
G3XMG L23 1US
G0FVY L23 1XL
G4GKR L23 2SG
G8NEM L23 2SQ
G0HNU L23 2UT
G3IZF L23 3BZ
G0JID L23 5TU
G4AYA L23 6XS
G3VSR L23 7UG
G1KLK L23 7UL
G7OEA L23 9UD
M1BXS L23 9UD
G3WBF L23 9UE
G8WMG L23 9UF
G3VXK L23 9XJ
G6GOG L23 9XS
G0URU L23 9XY
G0MJG L23 9XY
G3LNL L23 9YU
G7UMR L30 1SA
G7IKM L30 5QD
G3PQX L30 8RF
G8ZTF L31 0BW
G4OVD L31 0DA
G4WGN L31 1DY
G1IUM L31 2LW
G0PWL L31 2NN
G7VJU L31 2NN
G0HMP L31 3DW
G4KIN L31 3DW
G8YPL L31 3DX
G1MRU L31 4DQ
G0DOR L31 5JJ
G4LBJ L31 5JU
G1DGW L31 5NL
G0FEG L31 5PA
G3SGQ L31 6BS
G4HDU L31 6BS
G3WDD L31 6BY
G8GTI L31 8EG
M1BWZ L31 9DE
G3LTV L37 1PX
G3MIP L37 2DG
G6TEX L37 2EG
G6OEM L37 2JD
G4NWG L37 3HP
G6CIP L37 3JL
G3POG L37 3JU
G8DUF L37 6AD
G4MZZ L37 6BQ
G8XOS L37 6DT
G3VQS L37 6ED
G4SMT L37 7EY
G8CSP L38 0BA
G0OWP L38 0BG
2E1AEM L38 9EN
G0JCD PR8 1HW
G6EZY PR8 1JH
G2AXY PR8 1QY
G4IQK PR8 1RT
G0EKX PR8 2HA
G2ART PR8 2HF
G3WTB PR8 2JR
G4BEU PR8 2LW
G7UQE PR8 2PB
G0ROT PR8 2PG
G3FNQ PR8 2PH
G4JYQ PR8 2PX
G6WJR PR8 2RJ
G6ILX PR8 2RS
2E1CVS PR8 2SF
G1GBE PR8 3AQ
G0RHR PR8 3AW
G7SUR PR8 3BP
G3ZII PR8 3DB
2E1DVY PR8 3DQ
G0JGY PR8 3EF
G7MJS PR8 3HZ
G7OAA PR8 3HZ
G7JRX PR8 3NB
G4RNE PR8 3NY
G1UCI PR8 3RS
G1GSX PR8 3RU
G6WZO PR8 3RY
G1DFT PR8 3SG
G0TSI PR8 3SG
2E1DDQ PR8 3SH
2E1DDR PR8 3SH
G4IMQ PR8 3SU
G4JZM PR8 3TJ
G4YYV PR8 3TT
2E1DVK PR8 4AN
G4NMU PR8 4DF
G8BJW PR8 4EN
G0MPW PR8 4HA
G7CCH PR8 4HA
G3HWS PR8 4HH
G0FOO PR8 4JT
M1AJL PR8 4JU
2E1DBL PR8 4JX
2E1DBM PR8 4JX
G7LZB PR8 4SY
G4TUP PR8 5DA
G3KJM PR8 5DJ
G7NEV PR8 5EQ
G6IVC PR8 6JA
G0FNA PR8 6JW
G7OMN PR8 6JX
G6BZV PR8 6LE
G7TIM PR8 6NA
G4UOF PR8 6NJ
2E1CZV PR8 6QN
G7OUQ PR8 6ST
G6IXQ PR8 6SX
G3ORK PR8 6XH
G8YBW PR9 0PW
G6FVM PR9 0RF
G3LTB PR9 7AA
G3STT PR9 7AA
G0RLT PR9 7AS
2E1DSH PR9 7BP
M1AUZ PR9 7DR
G7RQZ PR9 7EU
M1BUL PR9 7HR
G0JCQ PR9 7PN
G3HAA PR9 7PQ
G4NPW PR9 7QR
G6JUQ PR9 8HY
G4EVC PR9 8JX
G1PJH PR9 8NP
G8EGC PR9 8PW
G2DQX PR9 8QB
G4UPK PR9 8QW
G0IKV PR9 9DL
G7NHD PR9 9FA
G7VHY PR9 9FB
G4SCO PR9 9GA
G0PVU PR9 9GJ
G1ZMJ PR9 9GJ
G7IFA PR9 9HA
G4FQN PR9 9LU
G0FOW PR9 9NU
G7HRQ PR9 9QY
G8RON PR9 9RN
G8TZL PR9 9TQ
G0RBK PR9 9UR
G1EKU PR9 9UY
G4RQX PR9 9XL
G4FMQ PR9 9XZ
G4YPH PR9 9YX

Selby

G4FPO DN14 0JX
G4III DN14 0LN
G3GWR DN14 0NS
G6LUE DN14 0NW
G4REQ DN14 0PD
G6JKP DN14 0PG
G4AFZ DN14 0QY
G8FCT DN14 0RD
G8OCW DN14 0RD
G7NVL DN14 9PR
G0PEV LS24 8AG
G4FUO LS24 8JF
G4FFM LS24 9BR
G4YXZ LS24 9EQ
G7BEA LS24 9LQ
G4TDC LS24 9PQ
G4KVS LS25 5AQ
G4ZSA LS25 5BT
G8INL LS25 5PJ
G4NCB LS25 6AF
G0RXV LS25 6LR
G0DIR WF11 9HT
G8NDF WF11 9NN
M1AXX YO2 1UJ
G0KQM YO2 3XQ
G7KJD YO2 3YD
G3OHC YO4 6LZ
G1YRC YO4 6NU
G4YRC YO4 6NU
G7AUP YO4 6NU
G4XTV YO4 6PH
G0DLK YO4 6PW
G4VXV YO4 6QG
G1HID YO4 6RG
G8DBM YO5 7AD
G3PRO YO5 7DA
G1KNA YO8 0AZ
G3NKH YO8 0LU
G4MWH YO8 0NY
G4CTZ YO8 0TF
G6XIH YO8 0TP
G3WQ YO8 7HF
G4VSV YO8 7HP
G0DWC YO8 7PW
G3ODD YO8 7QL
G6YYN YO8 7QP
2E1DRU YO8 7QP
G6JYX YO8 7RA
G6ZOI YO8 7RZ
G3ZIV YO8 7TG
G0GFA YO8 8ES
G4YPV YO8 8HD
M1CCG YO8 8HZ
G6LLT YO8 8PB
G4NPQ YO8 9BH
G4HBW YO8 9DD
G6AHQ YO8 9DD
G8TLA YO8 9EU
G0BGV YO8 9LL
G8ZMN YO8 9PG
G6JRL YO8 9QD
G4LYE YO8 9XH
G7DRM YO8 9XH

Sevenoaks

G8NDL BR8 7BP
M1BTE BR8 7BS
G0WMC BR8 7BX
M1BQG BR8 7BX
G7FVP BR8 7DA
M1ABU BR8 7EF
G4CVC BR8 7EJ
G7SXJ BR8 7HP
G1PUO BR8 7JA
G8BHD BR8 7JG
G7MIE BR8 7JG
G0HUA BR8 7JH
2E1FOJ BR8 7JH
G4IKQ BR8 7JY
G4GSE BR8 7LS
G8MIN BR8 7QT
G4WYG BR8 7RE
G7LIT BR8 7RJ
G0NQB BR8 7RR
G3IKQ BR8 7SB
G6SYA BR8 7SE
G7AKM BR8 7UA
G6LAX BR8 7YA
G6YYQ BR8 7YR
G3PJB BR8 8AQ
G4ZCV BR8 8DD
G7MJP BR8 8DT
G7PEB BR8 8DT
G1XWV BR8 8ER
2E1DQK BR8 8HP
2E1DQQ BR8 8HP
M1AEY BR8 8JL
G8VJG BR8 8LP
G0KDV BR8 8NU
G4AGC BR8 8NZ
M1AEP BR8 8TN
G6UV DA3 7EY
G3WBS DA3 8AP
G3IZU DA3 8AR
G6RHP DA3 8AR
G6HIG DA3 8AS
G1FJJ DA3 8DD
G0JBQ DA3 8DR
G4NRV DA3 8EU
G0HLC DA3 8JN
2E1FPI DA4 0DF
G7FUV DA4 0EB
G0ARQ DA4 9DQ
G1RCV DA4 9DQ
G3RCV DA4 9DQ
G4BWV DA4 9DQ
G4WTE DA4 9NA
G0MFO TN11 8EX
G4BUO TN11 8EX
G3YJW TN11 8LE
G7GFD TN11 8PZ
G8KVO TN13 1EL
G1YSR TN13 1HD
G7EIC TN13 1HG
G0SEV TN13 1HG
G3THM TN13 2AH
G7TIA TN13 2EB
G0IPH TN13 2BB
G1TGZ TN13 2HE
G6FMU TN13 2UU
G3GPG TN13 3DJ
M0ATS TN13 3JD
G0RLT TN13 3RB
G0CDQ TN13 3XA
G0ITQ TN13 8PZ
G4AKY TN14 5JX
G0SLU TN14 5LD
G8OMQ TN14 5PN
G6BNJ TN14 5PT
G0WQC TN14 5QP
G6KLF TN14 5RR
G7JQY TN14 5RR
G0OBL TN14 5RX
G4RWB TN14 6EB
G3SAR TN14 6HT
G6BYF TN14 6JA
G7IET TN14 7AU
G7MQS TN14 7AU
G1KVW TN14 7EY
G0UKH TN14 7JE
G7JYG TN14 7JU
G7KTR TN14 7NA
G7NKC TN14 7NS
G7IPN TN14 7RY
G7PVF TN14 7TD
G1EHB TN15 0AQ
G8HJD TN15 6AB
G8PUX TN15 6AB
G3IAR TN15 6AH
G1KRY TN15 6AT
G3GJW TN15 6BL
G3ZRX TN15 6BL
G4BRC TN15 6BL
G1MNY TN15 6BP
G4UCO TN15 6DD
G4RVV TN15 6DT
G0KFW TN15 6NU
G6EDM TN15 6PG
G8PWO TN15 6QL
G4IJH TN15 6QR
G3MWS TN15 6RH
G7ERN TN15 6RH
G7MYY TN15 6RU
G7WIR TN15 6UD
G8VZR TN15 6XJ
G8GJQ TN15 6XX
G6SEE TN16 1RU
G6SEF TN16 1RU
G3ZUO TN3 0SA
G0TKZ TN8 5DR
G4ARZ TN8 5EE
G3JMJ TN8 5EL
G8EGF TN8 5HU
G6MXV TN8 6AF
G4GMS TN8 6HG
G0ILL TN8 6JW
G3RQJ TN8 7NN
G3UYB TN8 7NU

Sheffield

G0INF S1 1BY
G3UOS S1 3JD
G2AS S10 1HG
G0IYV S10 1HT
G8YXI S10 1PB
G3YJR S10 1ST
G7ISD S10 1UY
G0JKA S10 1WL
G1YCV S10 1WL
M1BEF S10 2FG
G4SOM S10 3PT
G0BSX S10 3RF
G4BOJ S10 3RH
G8AGN S10 4BN
G4CUI S10 4DN
G6YFY S10 4ED
M0AQR S10 4EF
G3WDM S10 4GF
G4EJL S10 4GS
G4BBL S10 4LN
G7DSD S10 5BY
G1JVN S10 5RE
G1JVO S10 5RE
G7LQO S10 5RR
G4ISW S10 5SN
G3YQJ S10 5TF
G3YBA S10 5UA
G0RPQ S11 7ED
G0TQE S11 7ED
G4CYA S11 7EE
G0CVT S11 7HA
G0KBD S11 7HA
G0MGP S11 7RJ
G4MBN S11 7RJ
G0MGW S11 8AQ
G1CPU S11 8PT
G0UBM S11 8XL
G0HHA S11 9BR
G0LNV S11 9BR
G4ILX S11 9BX
G7RMO S11 9EA
G8BLP S11 9GG
G4IJB S11 9LQ
G4APV S11 9NT
G0OUK S11 9QU
G6NVE S11 9QZ
G0HGI S11 9RA
G4HIA S11 9RP
G4HIB S11 9RP
G3NHE S11 9RR
G3JON S11 9RS
G8RFP S11 9SB
G3RVI S11 9SP
G3UXM S12 2AY
M1ASK S12 2BL
G4VLZ S12 2BU
G1HFK S12 2FN
G1YBU S12 2GG
G7GHH S12 2QG
G8SWF S12 2SF
G8ZMB S12 3AB
G4OBF S12 3FH
G1GPM S12 3HB
G8VQU S12 3HE
G4MJL S12 3HJ
G1BWZ S12 3HR
G4RVS S12 3HW
G4DYB S12 4QE
G0BQW S12 4XE
G7UEX S12 4XH
G0NRM S12 4YF
G3CJJ S13 7RH
G7NSU S13 8SQ
M1BXA S13 8TG
G0HYS S13 8TH
2E1ASC S13 9AN
G0VBN S13 9HX
G1HHP S13 9QL
G0PBH S13 9WT
G7SMO S14 1BT
G7MTU S14 1JG
G8UCR S14 1PD
G3IMF S17 3DY
G0VNE S17 3EZ
G0TVW S17 3GF
G4AYO S17 3GZ
G0LVA S17 3HE
G4WWA S17 3NB
M0BAK S17 3NB
G6ULI S17 4AF
G4HOP S17 4DL
G7KBD S17 4HN
G8QX S17 4LY
G4KPU S17 4PF
G0LMI S17 4QR
G1JNT S2 1EB
G1VWH S2 2AW
G7SWQ S2 3EH
G1YED S2 3LB
G1DBZ S2 3LQ
2E0AJT S2 3QJ
G7SCP S2 3SG
G0EXP S2 3WA
2E1AQR S2 5EZ
G4MWL S2 5QW
G0WPC S20 5BL
G4MRB S3 9JF
G6DIZ S3 9JF
M1AGR S3 9QN
G1OKB S35 0GD
G4DJD S4 7DE
G7WBV S4 8GH
G0SFH S4 8JZ
2E1EXP S5 0DB
2E1FKW S5 0DB
G3CSP S5 0GE
2E1FDD S5 0JL
G4TCE S5 0JP
G7WGX S5 0JY
G7RYZ S5 0TQ
2E1FCI S5 0TQ
G6ADD S5 0TY
G4SLF S5 0UD
G0VUH S5 6AZ
G7DMP S5 6LX
G0UHF S5 6SQ
G0SJY S5 7LL
G0OLL S5 7TL
G3MFX S5 7TX
G0UQZ S5 7WE
G0HXX S5 8LQ
G2FLH S5 8TJ
G6DCT S5 8TR
2E1FXN S5 8UH
G4PCL S5 9AP
G8BLF S5 9DE
G0SEW S5 9GA
G8AOZ S5 9GS
G0UFS S5 9LS
G7PSR S6 1AB
G8CWC S6 2AU
M0AXE S6 2PN
G3PHO S6 3NU
G7BMO S6 3TD
G4VYI S6 4AH
G0BKE S6 4AY
G2CDT S6 4AY
2E1DUM S6 4BA
G0CGT S6 4FG
G0JKE S6 4NB
G0NDV S6 4SX
G1BRE S6 5BS
G5NV S6 5HS
G4TFZ S6 5LN
G0JTT S6 5PA
G1MLV S6 6AB
G4UMQ S6 6BH
G7IAN S6 6BT
G6UJC S6 6BY
G0JKH S6 6DJ
G8FDJ S6 6EW
G1WWI S6 6JL
G4RNA S6 6LG
G8KB S6 6SG
G4LJQ S7 1HE
G0WLR S7 1LB
G7GFH S7 1NE
G4FAL S7 1PB
G0JKG S7 1RW
G4GRY S7 1SD
G0WFL S7 2EB
G3XQB S7 2GS
G4PBI S7 2LY
G7SSC S8 0GA
G0WWK S8 0GA
G7TVY S8 0GA
G0NFY S8 0RA
G0PEW S8 7BW
G1LJL S8 7DP
G6SWW S8 7DZ
G4ZJB S8 7EA
G3RKQ S8 7FF
G1ASG S8 7FZ
G3TNM S8 7RA
G1KFF S8 7RJ
G3HAG S8 7UF
G8LIH S8 8DQ
G8YVW S8 8FT
G7OEV S8 8GD
G3WXJ S8 8GP
G0WLV S8 8GP
G8JLE S8 8GX
G4RVS S8 8HE
G8OCE S8 8JA
G4DYT S8 8JJ
G3XSI S8 8LB
G4PPR S8 8LW
G0NQY S8 8PJ
2E1AOX S8 8PR
G0OPE S8 8PX
G8LMO S8 8QJ
G0NJD S8 8QR
G8OFN S8 8RD
G3RKL S8 8SE
G0UUE S8 8SF
G3LMB S8 9EB
G2FKT S8 9EJ
G4UEQ S8 9HR
G3KVM S8 9HU
G0SND S8 9JB
G3KVG S8 9JL
G1HGB S8 9NG
G7DFX S8 9RL
G6FWT S9 1AF
G0BYP S9 1JX
2E1AGV S9 1LQ
2E1BKF S9 1LQ
G6VUE S9 1LU
G7ACR S9 1NQ
G4RBU S9 4AU
G4VBP S9 4BU
G0MNC S9 4QJ
M0AFY S9 5EH
G8SVT S9 5GJ

Shepway

G8MBV CT15 7HE
G3ZHU CT15 7HR
G1IUB CT18 7AA
G3XPO CT18 7BW
G4MDZ CT18 7DE
M1ABJ CT18 7EG
G0BPS CT18 7EG
G0ROO CT18 7EG
G8KSD CT18 8BY
G4TJZ CT18 8EY
G4EZK CT18 8HU
G6SXM CT18 8JL
G4HSU CT19 4AU
G0NOH CT19 4EQ
G4IOT CT19 4JA
G0PEG CT19 4QF
G0VWB CT19 4QH
G3EEW CT19 5BQ
G0TYV CT19 5BQ
2E1BSZ CT19 5BZ
G6GXE CT19 5JF
G0MCY CT19 5LF
G8SUQ CT19 5NA
G4EQJ CT19 5NB
G1RQI CT19 5PR
M1AZO CT19 5PW
G1DME CT19 5QE
MM0ACR CT19 5QE
MM0ACT CT19 5QE
G6AUK CT19 5QS
G3XVY CT19 6BA
G4HQJ CT19 6HN
G8YNH CT19 6JR
G5KW CT19 6NX
G7IYO CT19 6NY
G4MHS CT19 6PR
G7HIY CT20 1HY
G2FA CT20 2HW
G0SLJ CT20 2HW
G4ZMQ CT20 2NP
G0IXV CT20 2NZ
M1ATG CT20 2RN
G7THG CT20 3BE
G4OG CT20 3DL
G1ABO CT20 3NH
G3BRV CT20 3PF
G4FVI CT20 3PH
G4EGQ CT20 3SA
G3XHW CT20 3TA
G7SZO CT21 4EA
G3LWD CT21 4JP
G3SBV CT21 4JP
G3ZHT CT21 4JQ
G7BKM CT21 5RL
G6FTC CT21 5SP
G4SSZ CT21 5XA
G7WFP CT21 6BT
G4OAN CT21 6DN
G8NKW CT4 6AU
G3MLO CT4 6DN
G7EEG CT4 6XT
G2QT TN25 6JX
G0OIN TN28 8EW
G1NQX TN28 8JE
G4TGK TN28 8JL
G1HJS TN28 8LB
G0IWC TN28 8LR
G4MUD TN28 8NB
G0BGH TN28 8NX
G8IIK TN28 8PB
G6AGI TN28 8QR
G4YAZ TN28 8RX
G0KRH TN28 8SW
G1UTG TN28 8SY
G0IMD TN28 8UL
G1UTF TN29 0BU
G4UDW TN29 0DZ
G1SYR TN29 0JR
G0LFR TN29 0LA
G0LNN TN29 0LA
G7DNB TN29 0LA
G4WVT TN29 0NL
G0IEE TN29 0QX
G6DZT TN29 0RD
G6OJZ TN29 0RD
G4TZX TN29 0RU
G7UBK TN29 0TY
G7PWS TN29 0XB
G1HSX TN29 9BD
G6NLZ TN29 9EJ
G8XCL TN29 9LE
G1ZLL TN29 9PQ
G1HSI TN29 9RG
G1AEX TN29 9RT
G4RLU TN29 9YF
2E1DOR TN29 9YG

Shetland Islands

GM8YEC ZE1 0BB
GM4WXQ ZE1 0PA
GM4LER ZE1 0PJ
GM3ZET ZE1 0PJ
GM4SRU ZE1 0RE
2M1ASQ ZE1 0RE
2M1ANT ZE1 0SE
GM4CAQ ZE1 0SE
GM4LBE ZE1 0SE
GM0MZD ZE1 0SR
GM4PXG ZE1 0SR
GM3ZNM ZE1 0SU
GM0DJI ZE2 9AX
GM1CBQ ZE2 9BS
GM8MMA ZE2 9DA
GM4FNE ZE2 9DD
GM4GQM ZE2 9DL
GM3KLA ZE2 9ED
GM3STU ZE2 9EF
GM4GPP ZE2 9EF
GM4AGX ZE2 9ER
GM3WHT ZE2 9HH
GM0EKM ZE2 9HL
GM4ENK ZE2 9JB
GM4IPK ZE2 9JJ

GM3WCH ZE2 9JX
GM8LNH ZE2 9LG
GM1KKI ZE2 9LL
GM3XPQ ZE2 9LU
G3SKN ZE2 9PE
GM0ILB ZE2 9QN
GM0VFA ZE2 9QN
GM4SSA ZE2 9RS
GM4SWU ZE2 9SF
GM4DQD ZE2 9SQ
GM1ZNR ZE2 9SX
GM4ZHL ZE2 9TH
GM7RKD ZE3 9JW
GM7CHX ZE3 9JZ

Shrewsbury and Atcham

G3JTH SY1 1NL
G4WSN SY1 2TN
G0BUF SY1 2UG
G3UDA SY1 3BY
G0EML SY1 3PY
G4CBT SY1 3PZ
G1DGM SY1 3RH
G4XBI SY1 3RH
G4TPD SY1 3SB
G0JIX SY1 3SX
M1ADA SY1 4AY
M1ADB SY1 4ED
G0IRI SY1 4ER
G6VGY SY1 4EY
2E1FZU SY1 4LD
G7WGK SY1 4LG
G7SBD SY1 4LP
G8SIU SY1 4RB
G8RTB SY1 4RP
G7LGH SY1 4TJ
G1CLD SY1 4TJ
G0IGM SY2 5EF
G3TSV SY2 5EF
G0HCU SY2 5NR
G7LRB SY2 5NW
G3SRT SY2 5PF
G3VZG SY2 5PF
M1AXW SY2 5UQ
G0GTN SY2 5UQ
G0EBD SY2 5YD
G3WOP SY2 6BY
G0JWZ SY2 6EL
G3IDY SY2 6HQ
G3XUE SY2 6JU
G4RLO SY2 6SH
GW1ORP SY21 8ER
G4JSZ SY3 0BE
G8CNI SY3 0BT
G7JTD SY3 0ER
G4HAC SY3 0EX
G4MYZ SY3 0JH
G7NLY SY3 0LE
G7TPK SY3 0PN
G1CWZ SY3 0QF
G0EIY SY3 5AH
G6TGJ SY3 5BJ
G6CJW SY3 5BW
M0AMP SY3 5DF
M0AAU SY3 5DF
G3NRX SY3 6AB
G0RQX SY3 6BB
G3OWJ SY3 7AB
G7DWX SY3 7AP
G8OFZ SY3 7NB
G0ARP SY3 8QJ
G4ZZP SY3 8RU
G1EBT SY3 8RX
G4AZS SY3 8SU
G7VYI SY3 8SW
G3NFQ SY3 8YQ
G3VWH SY3 9DB
G0ALV SY3 9LU
G0SST SY3 9LU
2E1FYV SY3 9QS
G8RIK SY4 1EE
G0WDW SY4 2HT
G4CBM SY4 3AX
G4YKX SY4 3AX
G3TUO SY4 3DD
G8ZWF SY4 4AB
G3XQL SY4 4PU
G3SOA SY4 4SN
G4AZV SY5 0AW
G8DIR SY5 0EH
G6UDG SY5 0PY
G3NSY SY5 0QB
G0MCA SY5 0YE
G6FHM SY5 7AP
G0KRF SY5 7JD
G3JPE SY5 8AN
2E1FYW SY5 8LE
G8DIQ SY5 8NH
G2CQX SY5 8RA
G3VKX SY5 9AG
G3YFK SY5 9DJ
G8JQB SY5 9RT
G7NBP SY6 6LX
G7UMF SY6 6LX
2E1GAK SY6 6LX
G7MZB SY6 6NH

Slough

G6CGD SL1 1YD
G7HID SL1 2HX
G3RZF SL1 2LQ
G7PEX SL1 2XY
G4FKE SL1 3JU
G4ZAC SL1 3RE
G8HGX SL1 3RE
G7HUO SL1 3TR
G7HUP SL1 3TR
G3WIR SL1 3XG
G4XGD SL1 3XG
G6WIR SL1 3XG
G0MTR SL1 4XP
G4KVI SL1 4XP
G4CIJ SL1 5DF
G1YRX SL1 5JQ
G7ROC SL1 5LR
G8VKO SL1 5PQ
G0MWV SL1 5QY
G6DHD SL1 5QY
G7HNX SL1 5QY
G4HIN SL1 6AH
G4RAA SL1 6HD
G4KTY SL1 6LD
G2BZQ SL2 1DA
G4XJN SL2 1DH
M0AGZ SL2 1DP
G6POC SL2 1HG
G0BON SL2 1PF
G4XTZ SL2 1SL
G1EUN SL2 1UQ
G6XJI SL2 2JY
G1OIF SL2 2NF
G8BUE SL2 2QD
G8AZB SL2 5EN
G1SPJ SL2 5PN
G1NMQ SL2 5QL
G0JAM SL3 0LU
G1LMI SL3 7BB
G6TSF SL3 7ET
G6FCJ SL3 7HA
G4HTB SL3 7HN
G3RNN SL3 7NB
2E1DFM SL3 7NB
G0HZK SL3 7QF
G6GEN SL3 7QY
M0AJM SL3 7TN
G0TGQ SL3 8AU
G6RTM SL3 8DZ
G8GGM SL3 8ED
G0WEC SL3 8ED
G0HXM SL3 8JH
G3VHH SL3 8JH
G3VKB SL3 8RJ
G3MJH SL3 8TF
G4CFT SL3 9NU
G4ZWX SL4 6HL

Solihull

G1SWU B28 9NT
G8BFW B36 0AD
G6COK B36 0AR
G7UBX B36 0DY
G4RIO B36 0EL
G6JQH B36 0EL
G6VJR B36 0HR
G0UUG B36 0HS
G6AXR B36 0JP
G4YUW B36 0JR
G3OKH B36 0LG
2E1AWO B36 0LJ
G4OOX B36 0PB
G7RSR B36 0RD
2E1FZM B36 0RD
G0ELJ B36 0RT
G7UEJ B36 0UH
G4YIJ B36 9BH
G0AVI B36 9DR
G0GOD B36 9EY
G4SGA B36 9HX
M0AZK B36 9JB
G4IMB B36 9JD
G8PXU B36 9JD
G3OAB B36 9JE
G4ZPJ B36 9JZ
G4ZCR B36 9LA
G6VXL B36 9LL
G6YCI B36 9LL
G7OZE B36 9LL
G1JWL B36 9NU
G4LRN B36 9SH
G4JQK B36 9SZ
G1GFW B36 9TW
G4XPV B36 9TY
G0FCQ B37 5HX
G4FGF B37 5LD
G8GRC B37 5LD
G4SPY B37 6DL
G8YPT B37 6EA
G1RLB B37 6QR
G6APQ B37 6SB
G4HOC B37 7EE
G4HPT B37 7EE
M0BBN B37 7HT
G7BSA B37 7HT
G2FFK B37 7HW
G0MLH B37 7JJ
G7AWR B37 7PS
G1SKM B37 7RD
G4DZQ B90 1HP
G6ZDQ B90 1JA
G1PZH B90 1JN
G4AQJ B90 1LF
G7RRR B90 1PF
G4EQV B90 1QT
G1CMW B90 2AB
G3XBY B90 2BB
G7GFP B90 2BQ
G0STY B90 2DJ
G1INW B90 2DQ
G1MJO B90 2EJ
G4ZVZ B90 2HB
G7NCT B90 2HL
G4RTI B90 2HY
G2AGR B90 2JZ
G3BSK B90 2LR
G1JUR B90 2PP
G4TVR B90 2PU
G7OKF B90 2PU
G1PSJ B90 2QE
G3FGT B90 2QE
G3JYO B90 2QF
2E1ACG B90 2QP
G4KUR B90 2QP
G3FPI B90 2QR
G3WPM B90 2QW
G6AQW B90 2QW
G0HBC B90 2RY
G0IHU B90 3DF
G0MTN B90 3EX
G7WAC B90 3EX
G0WRC B90 3EX
G4NRR B90 3HL
G6DHD B90 3HL
G4KOR B90 3HX
G6FIO B90 3JZ
2E1ACC B90 3NN
G7BZD B90 3NN
G8AVX B90 3PB
G8SRG B90 3PB
G8XQA B90 3PL
G1RVA B90 3PT
G8ZUW B90 3RS
G3MRZ B90 3SA
G7HRF B90 3SA
G8OMI B90 4AT
G6BWT B90 4BU
G1ZQE B90 4BX
2E1BVK B90 4HL
G8PFT B90 4JR
G6HNR B90 4PN
G7HNR B90 4PN
G4ZVW B90 4PR
G0NFZ B90 4QR
G6UUR B90 4RU
G0PDN B90 4TD
G7JHX B90 4XS
G4EDW B90 4YF
G4OGH B90 4YF
G4NYG B90 4YP
2E1CHR B91 1AF
G1NAM B91 1AG
G6VUJ B91 1BS
G4MPG B91 1DD
G7KZV B91 1DQ
G7MAB B91 1DQ
G4AMI B91 1DX
G6JEB B91 1DY
G7DDD B91 1HJ
G4ABX B91 1HN
G4WMH B91 1LN
G8VXQ B91 1LW
G4JPU B91 1PR
G4LOE B91 1PR
G1FMW B91 1QB
G3LUA B91 1TJ
G8GBM B91 1TQ
G4DLQ B91 1TS
G3AOV B91 1UF
G2VJ B91 2AX
G1ITN B91 2DB
G3DEJ B91 2EG
G3WAD B91 2EH
G1RBX B91 3GA
G1BHB B91 3JY
G6KMQ B91 3LL
G4KSG B91 3ND
G8EYT B91 3UB
G3WZI B91 3UD
G1GVT B91 3XY
G8SAN B92 0AH
G4EOV B92 7AZ
G6RZY B92 7EE
G6BJG B92 7EY
G6DFH B92 7HB
G6PVA B92 7HD
G1STK B92 7HE
G6MCY B92 7HE
G4FJB B92 7HH
G8YMU B92 7JA
G0HLW B92 7JB
G4VMO B92 7JH
G4KQV B92 7NU
2E1FCW B92 8AB
G6EDF B92 8AL
G6HXV B92 8AL
G8CLV B92 8BA
G8LOC B92 8BS
G7LED B92 8DP
G4CEX B92 8DX
G4BBT B92 8EE
G3GEI B92 8EE
G0LJT B92 8EF
G4TDF B92 8HS
G8ULQ B92 8HS
G7TEE B92 8HU
G4LLV B92 8HX
2E1ACP B92 8NH
G4ABV B92 8NN
G4PCE B92 8NN
G1STP B92 8QH
G3VJP B92 8QH
G6SHF B92 8QH
M1BMY B92 8RB
G6PTJ B92 8SJ
G1FUG B92 9DB
G4VQL B92 9DD
G0PHR B92 9DQ
G3ZDP B92 9JE
G4KWO B92 9ND
G6WPO B92 9NE
G7JVF B92 9NW
G7GQX B92 9PT
G4EIG B92 9QH
G0IZQ B93 0HX
G3OIF B93 0PT
G8YVA B93 8AL
G4GIK B93 8DA
G4VMN B93 8EH
G3UFQ B93 8QP
G8AHW B93 8QS
G8IK B93 8RA
G0WKT B93 8ST
G6VKS B93 9AW
G0EFZ B93 9EJ
G3KEK B93 9EQ
G7GWI B93 9EQ
G3MGB B93 9HU
G4MAU B93 9JL
G1JYR B93 9LQ
G1VIW B93 9LA
G4CVM B93 9PP
G4XFL B93 9PP
G3SMK B94 5LP
G1UTJ B94 6PF
G6CXM B94 6PF
G0ENY CV5 9AS
G6NXR CV5 9QJ
G6DKA CV7 7AA
G4LMV CV7 7AB
G3SIB CV7 7DU
M1BHA CV7 7FQ
G4IEV CV7 7FX
G6SCM CV7 7FX
G0TDL CV7 7GA
G0BIP CV7 7LB
G0EPV CV7 7LU
2E1ALB CV7 7ND
G3UKM CV7 7QF
G7TIO CV7 7QL
G4JYE CV7 7QW
2E1BRM CV7 7QW
G8JCT CV7 7UG

South Ayrshire

2M1DLH KA10 6HX
GM4IGS KA10 6LL
GM4XRY KA10 6SD
GM0WOB KA10 6SE
GM7VXR KA10 6TT
GM0JHF KA10 6TU
GM6BAO KA10 6UG
GM4SLY KA10 6UN
GM4ZNC KA10 6XD
GM6JEP KA10 7DY
GM0FRH KA10 7ED
GM4CXF KA10 7EE
GM4BIT KA10 7HA
GM3YDN KA19 7SG
GM3LTW KA19 8AD
GM4YED KA19 8AG
GM7IHJ KA19 8AG
GM3LAW KA19 8BJ
GM4VKI KA2 9EX
GM8JUY KA2 9EZ
GM0FSZ KA26 0BP
GM0NBG KA26 0DB
MM0AVV KA26 0DB
GM0RTS KA26 0DE
MM0BEF KA26 0DE
GM7VVJ KA26 0DE
GM0UDY KA26 0DE
GM7UMG KA26 0EF
GM7LDU KA26 0NG
GM4ZTO KA26 0NQ
GM0CJY KA26 0NW
GM4WEW KA26 0PA
GM4WEX KA26 0PA
GM3JKS KA26 0QA
GM4ZFG KA26 0RT
GM4AWP KA26 0RY
GM0JMO KA26 9AH
GM7SAK KA26 9AH
GM7CDI KA26 9DT
GM6OFB KA26 9DZ
GM3VNW KA26 9EL
GM0KWW KA26 9JH
GM0LYO KA6 5PA
GM0ONV KA6 5AU
GM7SXI KA6 5AX
GM3KJF KA6 5EW
GM7APK KA6 6JE
GM4PPT KA6 6LB
GM4FGS KA7 2LW
GM4WFV KA7 2SU
GM4HSR KA7 3DT
GM0TBH KA7 3JB
GM7OIN KA7 3NF
GM8ZGC KA7 3PA
GM4SQO KA7 3PE
G7VOV KA7 3PW
GM4XEQ KA7 3RJ
GM4CUB KA7 4DN
GM4DWS KA7 4DN
GM3PMB KA7 4EG
GM5VG KA7 4EG
GM7KXJ KA7 4JB
GM4XMF KA7 4JB
2M1FQK KA7 4JE
GM4CAM KA7 4JR
2M1FXF KA7 4JR
MM0AUB KA7 4LU
GM3CSO KA7 4PA
MM1ATS KA7 4PY
GM6BHR KA7 4QD
GM3FAO KA7 4QQ
GM0AYR KA7 4RE
GM4SUC KA7 4RE
GM3TNT KA7 4UA
GM4SNP KA8 0SP
MM1BJJ KA8 8AU
GM0IKY KA8 9SQ
GM3MIE KA9 1JA
GM3WIL KA9 1TT
GM3WIV KA9 2BW
GM4TPX KA9 2DB
GM4LVW KA9 2DL
GM4OSQ KA9 2DW
GM3MQO KA9 2HE
GM4DOZ KA9 2HN
GM4OBG KA9 2JJ
GM4RSJ KA9 2JT
GM3ZAS KA9 2LP
GM6ZDW KA9 2QW

South Bedfordshire

G8ADC LU1 4AN
G0EZY LU1 4AX
G8NVF LU1 4DD
G8OBB LU1 4DJ
G6CQB LU1 4DU
G6NDJ LU1 4DU
G8LCE LU1 4EJ
G3HGM LU1 4HZ
G8LSS LU1 4JA
G4FAX LU1 4JG
2E1FFG LU3 3PN
G8ADS LU5 4AB
G1FXB LU5 4AL
G3ZVO LU5 4BN
G8NZD LU5 4EA
G3ZNT LU5 4LF
G3SGS LU5 4LP
G1DGV LU5 4PF
G4RZR LU5 4PJ
G1ANS LU5 4PR
G1ZWB LU5 4PR
G4UOO LU5 4PW
G6LGW LU5 4QL
G3LNC LU5 4QW
G6BCS LU5 4RB
G3BIO LU5 4SP
G3LAZ LU5 4SP
G4TZI LU5 4SP
G4OER LU5 5AS
G4HGZ LU5 5DN
G7FDN LU5 5EP
G8IGE LU5 5HZ
G1ALL LU5 5PR
M1BCU LU5 5ST
G6YTP LU5 5UQ
G4DUL LU5 6AZ
G8TLU LU5 6BB
G8GUU LU5 6ED
G0WFT LU5 6EL
G3XKN LU5 6EP
M1BFI LU6 1DN
G8NRB LU6 1HB
G3HOH LU6 1LD
G0FGJ LU6 1LZ
G7CHN LU6 1NX
G0ODU LU6 1QJ
M0AAB LU6 2AF
G7VWK LU6 2AG
G0WTL LU6 2AG
G3MNK LU6 2BQ
G4WJH LU6 2DZ
G3PTG LU6 2FH
G2DBI LU6 2LG
G3GZH LU6 2LP
G7NRV LU6 2NE
G3ZFP LU6 2NN
G7TVT LU6 3AR
G1OFA LU6 3BB
G6IAN LU6 3DD
G0FDV LU6 3EE
G1INL LU6 3JF
G1HIJ LU6 3JT
G8ION LU6 3NB
G0AME LU6 3NS
G6IQH LU6 3PD
G4EKB LU6 3PT
G4MTL LU6 3PT
G8MTL LU6 3PT
G4WYO LU6 3RS
G8PJD LU6 3RS
G0LCW LU6 3RW
G6KHD LU7 0AX
G1HUE LU7 0EB
G3TQU LU7 7HX
G4XQQ LU7 7LA
G0KRR LU7 7QQ
G0SMQ LU7 7RJ
G4CAK LU7 7SH
G6POV LU7 7TS
G3YZW LU7 7TY
G8XTW LU7 7XD
G1EOC LU7 7XJ
G3YYG LU7 7YG
G0AYC LU7 8DQ
G3PHZ LU7 8DQ
G0SWN LU7 8DQ
G8KKU LU7 8DU
G0GPN LU7 8EN
G1MMI LU7 8HS
G4ATQ LU7 8LH
GW4BUS LU7 8LS
G3MBO LU7 8LU
G6IXH LU7 8UJ
G1EOJ LU7 8UP
M1BVN LU7 8XP
G6HLP LU7 9JE
M1BYM MK45 4LF
2E1FHP MK45 4LF
2E1FDC MK45 4LN
G6WVS MK45 4LT
G4DAQ MK45 4NE
G4SXH MK45 4NP
G4EIY MK45 4PF
G0KOH MK45 4QJ
2E1FDR MK45 4QJ
2E1FDS MK45 4QJ

South Bucks

G8JZJ HP9 1AJ
G0AFT HP9 1AW
G8FIG HP9 1BD
G8KCC HP9 1BH
G0BKR HP9 1DE
G4VVG HP9 1DP
G3BII HP9 2LA
2E1DRC SL0 0DT
G3CUI SL0 0EW
G1EKP SL0 0JS
G4JFG SL0 0QD
G8VWP SL0 0RS
G4KDN SL0 9DJ
G0GVO SL0 9DR
G4OTQ SL0 9DS
G3SGX SL0 9EU
G3OZK SL0 9LF
G3WCB SL0 9NJ
G8RSL SL0 9RN
G3YFO SL1 7BQ
G4SYE SL1 7EH
G3DOJ SL1 7EL
G6DRP SL1 7EW
G6WZC SL1 7LY
G4ZGC SL1 7NA
G7EUL SL1 7NF
G8TEQ SL1 8AT
G7FWV SL1 8BQ
G0IWU SL1 8BT
G0AXM SL1 8DD
G8MCS SL1 8DY
G4LLN SL2 3HE
G0SKA SL2 3LZ
G7VKM SL2 3SY
G0RKS SL2 3XD
2E1AXI SL2 3XL
2E1AVX SL2 3XL
G0GUN SL2 3XT
G1DSA SL2 3XT
G3PFO SL2 4AX
G4XPR SL2 4DH
G7UFI SL2 4DH
2E1CQS SL2 4JF
G0KLW SL3 6DH
M1AEZ SL3 6HJ
M1BVR SL3 6HJ
G1LCI SL6 0DR
G6ZEZ SL6 0PB
G8HNI SL6 0PQ
G7VGD SL9 7EJ
G4YIU SL9 7HR
G4ARK SL9 7JE
G3UWL SL9 7LR
G8UYK SL9 7NT
G4HQE SL9 8DJ
G4AVW UB9 4AA
G4NSN UB9 5BW
G0VOV UB9 5DJ
G0PBS UB9 5LL

South Cambridgeshire

G1EIX CB1 4XQ
G7VCE CB1 4YL
G6EUI CB1 5AW
G1JLE CB1 5BG
G4YJK CB1 5BT
G8RYL CB1 5DT
G4WSZ CB1 5HQ
G3EDD CB1 5JG
G4HJW CB1 5JY
G3UDI CB1 5LD
G3JQS CB1 5LR
G4FTW CB1 5LY
G3ZUK CB1 5PE
G4SCE CB1 5PE
G6ITY CB1 6AZ
G4YDJ CB1 6BU
G4PDI CB1 6DW
G7RFY CB1 6EJ
G7SQX CB1 6ER
G0DEU CB1 6HS
G3BTL CB1 6LG
G7IGQ CB1 6LS
G4DYW CB1 6LX
G1CQA CB1 6LY
G0CHC CB1 6RY
G7ILR CB1 6SX
G6HYI CB1 6TW
G0LXZ CB1 6UJ
G4ASY CB10 1RF
G1GFC CB2 4BN
G6XQE CB2 4BN
G4WEZ CB2 4DH
G6XCE CB2 4DL
2E1FFU CB2 4DW
G7TKI CB2 4DW
M1AQP CB2 4DW
G6ASH CB2 4HT
G0FQE CB2 4HW
G6XJJ CB2 4HW
G3TOS CB2 4HX
G7SRK CB2 4HY
G3WFF CB2 4HZ
G0PGQ CB2 4NG
G3UUU CB2 4NT
G3MJX CB2 4PS
G3BTM CB2 4QG
G6BOX CB2 4QP
G4EIK CB2 4QU
G6WJX CB2 4RT
G4IVV CB2 4SF
G8PK CB2 4UW
G8MEI CB2 5AE
G3WLD CB2 5BX
G4KNO CB2 5DU
G6BBN CB2 5EA
G3UUT CB2 5EG
G7UUT CB2 5EG
G3VKP CB2 5EH
G4EBW CB2 5EJ
G6TJP CB2 5HJ
G0SHA CB2 5JJ
G3TVM CB2 5JW
G1IMM CB2 5LG
G7VBJ CB2 5PR
G1XAG CB2 5QD
G2PU CB2 5QQ
G1TFK CB2 5QT
G8VCN CB2 5QT
G8SHC CB2 5RZ
G6UFL CB2 6RT
2E1FLN CB3 0NS
G0ECL CB3 0PP
G8MEX CB3 7AE
G4LNY CB3 7AQ
G8UKG CB3 7DF
G8MLA CB3 7DH
G3PJT CB3 7DY
G3KRL CB3 7ES
G2XV CB3 7JT
G3BYW CB3 7JT
G8EVY CB3 7JT
2E1EKJ CB3 7JW
G6KWA CB3 7LS
G4VHK CB3 7NL
G3GGK CB3 7NX
G3RUH CB3 7PS
G3XZP CB3 7PW
G3TAG CB3 7RE
G4VYG CB3 7RE
G7SPZ CB3 7RY
G4IPM CB3 7TH
G0OEL CB3 7UN
G4NBS CB3 7XJ
G4NXO CB3 7XJ
G3YMW CB3 7XL
G6HFS CB3 7XR
G6NIW CB3 7XR
G4YAC CB3 7XS
G8GJV CB3 7XT
G4GNN CB3 7XX
M1AUE CB3 7XX
G4AKD CB3 8BT
G0AWH CB3 8DT
2E1AEJ CB3 8EP
G0NJJ CB3 8EP
G6KIB CB3 8HH
G4UGQ CB3 8JQ
G4VWS CB3 8JQ
G0DAH CB3 8ND
G0GWA CB3 8NJ
G7AZJ CB3 8SD
G8HRF CB3 8SD
2E0ADF CB3 8SF
2E1AEK CB3 8SF
G0TYL CB3 8SF
G0KRB CB3 8SF
G0FYV CB3 8SG
G6VTN CB3 8SU
G6HSW CB3 8TD
G6TUS CB3 8TW
G8PKM CB3 8UN
G4RQH CB3 9LH
G0VFJ CB3 9NL
G3MBM CB4 4DR
G0CCL CB4 4DW
G3NIE CB4 4EA
G0ILI CB4 4ED
G8XMS CB4 4EU
G3XXX CB4 4HY
G1SAA CB4 4JR
G0THI CB4 4JU
M0AAJ CB4 4LQ
G8XLE CB4 4LZ
G3YAC CB4 4NE
G0XAS CB4 4NP
M0AQF CB4 4PP
G1VRA CB4 4QB
G1KIJ CB4 4QG
G0GKP CB4 4QR
G3XJE CB4 4QS
G1DEQ CB4 4TG
G3TUL CB4 4TJ
G4SUM CB4 4UN
G3RFP CB4 4UW
G0CNF CB4 4XS
G7PLX CB4 5AF
G6TQM CB4 5AL
G8NPR CB4 5AY
G0GPX CB4 5BA
G7LBL CB4 5DJ
G0LKZ CB4 5DT
G6IPU CB4 5HB
G1JZN CB4 5HY
G1XAA CB4 5HY
G8OZJ CB4 5JF
G3HZP CB4 5JG
G3RGQ CB4 5JT
G4NEY CB4 5JU
G4XXX CB4 5JX
G6XRY CB4 5LS
G6GZH CB4 5NE
G8GJU CB4 5NF
G1JKR CB4 5PL
G4BHJ CB4 5PZ
G1KJX CB4 5QR
G1DMF CB4 5RG
G7CPQ CB4 5RT
G8JDQ CB4 5RT
G6HWB CB4 6AB
G0OIR CB4 6BS
G1NWM CB4 6DA
G6MCI CB4 6DG
G0WZD CB4 6EB
G4VTU CB4 6ZG
G1DRM CB4 6ZL
G4NGL CB4 6ZL
G0TZQ CB5 8SZ
G3KKD CB5 9AD
G7TSB CB5 9NQ
G8ZRD CB5 9NY
G0BGX CB5 9PS
G1DGS CB5 9PS
G1MKP CB5 9PS
G4BAO CB5 9PX
G7RQE CB5 9QJ
G6DNO CB5 9QJ
G1RTF CB5 9RQ
G0WXI CB5 9RQ
G4RLR PE19 4SX
G3TDT SG19 3HZ
G0FOT SG19 3LG
G1UWH SG19 3LG
G0CMB SG8 0DJ
G1HND SG8 0DJ
G4JZQ SG8 0EU
G6PCE SG8 0PF
G8RZL SG8 0QS
G6VIK SG8 0RB
G4JNA SG8 5JR
G1HDR SG8 5NG
G0LZD SG8 5PT
G8UBX SG8 6DR
G7CDK SG8 6NR
G7SVJ SG8 7QX
G1PUQ SG8 7SD
G3HPM SG8 7TF

South Derbyshire

G1PKR DE11 0DS
G4XHH DE11 0JW
G4OXF DE11 0LN
M0APK DE11 0LY
G6PCJ DE11 0LY
G1PKV DE11 0NB
G8VDX DE11 0NH
G1XOP DE11 0NQ
G4EWK DE11 0RX
G6BSF DE11 0SL
G0OAP DE11 0TB
G6DCS DE11 0TE
G7EJO DE11 0UL
G8UZQ DE11 0UU
G6NJW DE11 7BZ
2E1FLD DE11 7DX
G3BJY DE11 7HN
G7ASM DE11 7JJ
G1ZMW DE11 7LY
G4UWA DE11 7LY
G8UHF DE11 7NT
G6NZL DE11 7QU
G4WPE DE11 8BY
G6SLH DE11 8BY
G4HZO DE11 9DU
G1OWK DE11 9DU
G0KVB DE11 9EW
G7CYZ DE11 9NS
G1SNQ DE11 9SJ
G4ZDE DE11 9SJ
G6JWZ DE12 6BL
G4AYS DE12 6BL
G4CRT DE12 6LU
G0SRC DE12 6LU
G4UWQ DE12 6PP
G3AXI DE12 6QB
M0APD DE12 6RT
G3INB DE12 8DA
G1VQH DE12 8ES
G7EHU DE12 8HA
G4HIW DE12 8NJ
G3HKN DE15 0PS
G0CQS DE15 0PT
G0TZM DE15 0QW
G4ELG DE15 0SL
G3WSM DE24 3AA
G1LKK DE24 3BS
2E1BUM DE65 5FH
G6KAA DE65 5GL
G6YWN DE65 5PS
G1TKY DE65 5QQ
G4WRN DE65 6DQ
G0HLI DE65 6EB
G0BDL DE65 6EF
G2DLJ DE65 6FR
G8WSQ DE65 6GF
G0NYN DE65 6GG
G0MRY DE65 6GP
G0DMS DE65 6GR
G8ODX DE65 6GR
G1NUN DE65 6LA
M1BZX DE65 6LA
2E1AUU DE65 6SA
G0HHL DE65 6WA
G3URR DE72 2AA
G3GWU DE72 2AB
G4XAQ DE72 2AB
G4AJF DE72 2AE
G0PBM DE72 2BA
G1GNL DE72 2BA
G4NYB DE72 2BJ
G7HLT DE72 2BL
G3ZOW DE72 2GS
2E1GBP DE72 3FG
G0MZJ DE73 1BX
G8NFM DE73 1BX
G6OBY DE73 1DA
G6OCE DE73 1DP
G0JPU DE73 1DQ
G4FZK DE73 1FT
G6CRM DE73 1GT
G0JCK DE73 1JJ
G3XER DE73 1JJ
G6RRU DE73 1JZ
2E1FKU DE73 1LY

South Gloucestershire

G4SQQ BS10 7TE
G4GOA BS12 0BA
G0PKV BS12 0BJ
G0NQI BS12 0BJ
G0RVM BS12 0DJ
G4GSA BS12 0DZ
G0LOJ BS12 0EG
G3BRT BS12 0EN
G4ZOG BS12 1AE
G1USV BS12 1AY
G1PXW BS12 1BX
G4UGX BS12 1ER
G0BXA BS12 1EU
G7FCO BS12 1HB
G3XNN BS12 1JF
G4AGH BS12 1JH
G8AZT BS12 1SR
G4UOA BS12 1SW
G4ABC BS12 2BA
G4YZD BS12 2EW
G6RAZ BS12 2EW
M1ACN BS12 2EZ
G7FVN BS12 2HA
G4NDF BS12 2HF
G7UYR BS12 2JG
G0WMB BS12 2JG
G0GZW BS12 2JL
G1FND BS12 2JL
G4UGO BS12 2LQ
G4GMW BS12 2LZ
G3ZWY BS12 2NG
2E1BXB BS12 2NG
G0EDG BS12 2PL
G3HTJ BS12 2PR
2E1DFF BS12 2QT
G7NAR BS12 2QX
G6YNL BS12 2SB
G8DFN BS12 2UA
G6OLJ BS12 2YD
G3XIY BS12 2YE
G4RBW BS12 2YN
G7GZU BS12 2YN
G1KTU BS12 3AE
G0EBZ BS12 3AQ
G4YHG BS12 3DZ
G1PAS BS12 3JW
G0JYO BS12 3LH
G4KXO BS12 3LQ
G0WRN BS12 3PF
G3ZUT BS12 3RE
2E1CZY BS12 4DB
G1AVJ BS12 4DU
G4RLK BS12 4EZ
G7ISR BS12 4HD
2E1CYP BS12 4HH
G4OST BS12 4LQ
G0MGC BS12 4NS
G7IOS BS12 4PL
G7IOT BS12 4PL
G4BNN BS12 4PR
G0MIG BS12 4PT
G8MXD BS12 5EN
G8SYC BS12 5ER
G1XUC BS12 5HD
G0UMP BS12 5HH
G4RAE BS12 5HQ
G4IYE BS12 5HX
M1BFU BS12 5HX
G0NFH BS12 5NP
G3VVI BS12 5NX
G7JUO BS12 5PU
G3ZZU BS12 5PW
G6ZRS BS12 5PY
G7RRA BS12 5RB
G8YJM BS12 5RL
G0ECM BS12 5RN
G6RUP BS12 5SA
G2FXO BS12 6BT
G6PNB BS12 6BT
G8CLS BS12 6BT
G4DVV BS12 6EB
G1YXA BS12 6EF
G6BQH BS12 6NY
G8EKZ BS12 6NZ
G3ZJH BS12 6PH
G8IMB BS12 6QH
G6PTY BS12 6RG
G4LAW BS12 6RZ
G4AEL BS12 6UD
G6TVJ BS12 6XA
G7RHT BS12 6XA
G3CJR BS12 6XN
G0RFD BS12 7HZ
G8RFD BS12 7HZ
G7MNO BS12 7LJ
G0CDH BS12 7PS
G7IYM BS12 7QX
G4YQQ BS12 7RD
G4BOL BS12 8BB
G7AJC BS12 8BP
G1NQM BS12 8DP
G0IFL BS12 8EB
G0LEB BS12 8EB
G0RAT BS12 9AR
G8VKI BS12 9AR
G4HBT BS12 9BF
G6JID BS12 9BJ
G3OYF BS15 1PZ
G0TDV BS15 1XB
G4OPO BS15 2DQ
G6AUR BS15 2EN
G4BVK BS15 2ES
G6RQP BS15 2ET
G0WIP BS15 2NT
G4WOD BS15 2NT
2E1CPC BS15 2NT
G1DFM BS15 2NZ
G1ODD BS15 2NZ
G1XYR BS15 2QJ
G4RAB BS15 2QP
G1IHL BS15 2QP
G7VVM BS15 2SH
G0RFB BS15 2UE
G4HBU BS15 2UT
G0CFM BS15 2ZE
G4ZYF BS15 3AF
G7IRP BS15 3BE
G4CYE BS15 3BL
G0THT BS15 3BL
G3KZC BS15 3BY
G1SDP BS15 3BZ
G4DEM BS15 3EW
G4EIA BS15 3EX
G0WPH BS15 3HH
G7MWJ BS15 3JR
G8DBP BS15 3JX
G4WJY BS15 3PA
G8VFL BS15 3RB
G4DCX BS15 3RN
G8PQA BS15 3RW
G7CWN BS15 3SJ
G0JYX BS15 3TB
G0GOR BS15 4BL
G4ZMW BS15 4BL
G4ULV BS15 4BQ
G4VJL BS15 4DR
G6NQM BS15 4HN
G7BME BS15 4HT
G6XEB BS15 4HU
G0VBK BS15 4JA
G0FMJ BS15 4LR
G7IPH BS15 4LT
G4UBI BS15 4QJ
G1NUA BS15 4QU

G1FUJ BS15 4RT
G2FQP BS15 4XR
G8TEO BS15 5BT
G4NKT BS15 5DD
G6YB BS15 5DD
G4MQF BS15 5EJ
G4VRP BS15 5PP
G7VVO BS15 5RL
G4MCQ BS15 5UT
G4FJH BS15 5YD
G1AIG BS15 5YL
G4YOC BS15 6DU
G8VZB BS15 6EZ
G0VDJ BS15 6JA
G8ZFL BS15 6PX
G0JJS BS15 6RH
G7NBW BS15 6UH
G7JVK BS15 6XB
G4EXZ BS15 6YQ
2E1ETS BS15 7AL
M1BEQ BS15 7AL
G1ABA BS15 7BS
G8YML BS16 2SL
G4CXW BS16 2SW
G1IHT BS16 4QQ
M1BGK BS16 4SQ
G4TAE BS16 5HS
G3IZM BS16 5LE
G8SPI BS16 5RE
G4UGU BS16 5TN
G8GPF BS16 6AQ
G6CWF BS16 6HN
G4OEQ BS16 6JG
G4UHE BS16 6JR
G7VYM BS16 6LF
G8RRR BS16 6PL
G7IHL BS16 6PT
G1TYK BS16 6PZ
G8FGD BS16 6PZ
G3RUJ BS16 6SQ
G3FYX BS17 1EP
G6TJZ BS17 1JW
G4PHK BS17 1LG
G3JMY BS17 1LU
G3GBD BS17 1NA
G4LTL BS17 1NR
G1ISR BS17 1PX
G4YCD BS17 1RH
G0NQG BS17 2HL
G4RKG BS17 2NQ
G7HYS BS17 2RL
G0KGL BS17 3AZ
G0NQJ BS17 3DR
G8XSP BS17 3EZ
G4RXF BS17 3HN
G4ACU BS17 3JG
G7NSZ BS17 3PS
M1AHH BS17 3PS
2E1DSW BS17 3RN
G7ITD BS17 4EB
G0IUD BS17 4EG
G0SYE BS17 4ET
G3YAD BS17 4EY
G4OHR BS17 4HX
G7IBF BS17 4JX
G6IZZ BS17 4NN
G4TVD BS17 4NP
G7ORR BS17 4RD
G7CJD BS17 4SA
G0XJS BS17 4TE
G0JTR BS17 4UA
G1JOR BS17 4UE
G0JZF BS17 4UR
G0JMD BS17 4YE
G0XAF BS17 4YJ
G6HKZ BS17 4YW
GW4WRD BS17 5BW
G7KMZ BS17 5DZ
G1WVM BS17 5EX
G7TZO BS17 5LR
G7AMU BS17 5QQ
G3SZH BS17 5TG
M1AEU BS17 6DP
G1XSO BS17 6JP
G0WOI BS17 6JU
G1ZFF BS17 6LA
G7BYN BS17 6XB
G0LXL BS17 6XD
G4TAH BS17 6XL
G3PTO BS17 6XT
G7DSQ BS18 2DD
G6YCG BS7 0LP
G0TME BS7 0QJ
G0NZU BS7 0RG
G4NWU BS7 0RG
G7GDI BS7 0RG
G6BZG BS7 0RH
G8ZRN BS7 0RH
G8VYT BS7 0RP
G3XPJ BS7 0RT
G7GLQ BS7 0SF
G4EUV BS7 0SL
G7DRU BS7 0SR
G1HXT GL12 8AS
G4BCA GL12 8BU
G3NXI GL12 8DE
G1USW GL12 8NB
G1XAK GL12 8NB
G1VNL GL12 8TN
G4TRB GL12 8TP
G6FFL GL12 8UU
G1KKS GL9 1HH
G0XAY GL9 1HP
G8WKT GL9 1HU
G3MUN SN14 8HG
G4UFV SN14 8LD

South Hams

G4OUZ PL21 0AD
G4RIM PL21 0AD
G0VAC PL21 0AZ
G6GRU PL21 0DD
G4VFG PL21 0ET
G0TQR PL21 0HT
M0AVN PL21 0HZ
G4NDL PL21 0LT
G1CKT PL21 0NH
G6WDZ PL21 0QB
G0TIM PL21 0RX
G4XFM PL21 9BD
G7NJH PL21 9BE
G7LNG PL21 9BH
G4ONC PL21 9BT
G3CJG PL21 9BU
G7HIK PL21 9BZ
G0EOZ PL21 9DH
G4ETP PL21 9JE
G0WFJ PL21 9NB
G0IWY PL21 9PT
G7CMU PL21 9SL
G8XTE PL21 9SN
G0OYO PL21 9TD
G4DCH PL21 9TH
G7VYQ PL6 7AT
G3KFN PL6 7AY
M1BFK PL6 7AY
G0DRC PL6 7BY
G1HHC PL6 7BY
G1HHD PL6 7BY
G1RCD PL6 7BY
G8PSC PL6 7SH
G4GHR PL6 7UB
G8PVR PL7 5EG
G4BVS PL8 1BZ
G0ESL PL8 2EN
G8OHA PL8 2HR
G7JLZ PL8 2NT
G0JFX PL9 0DZ
G0MQK PL9 0EU
G4SCH PL9 0HA
G4LJX PL9 0HT
G2BCB PL9 0JY
G3YQF PL9 0LB
G4XTR TQ10 9EU
G7BEP TQ10 9EU
G0JAH TQ10 9JW
G3FYF TQ13 7RU
G7RKX TQ13 7SS
G4OYC TQ3 1NH
G6XXB TQ3 1NH
G3CJP TQ3 1NQ
G7ULD TQ3 1NZ
G6UPR TQ3 1SQ
G4AQR TQ3 1TA
G4RJ TQ5 0HG
G3FIY TQ6 0HY
G4WGO TQ6 0PJ
G3XHC TQ6 9EG
G4LZD TQ6 9JN
G7XPC TQ6 9NZ
G4RD TQ6 9PL
G6HKK TQ7 1BB
G0MJU TQ7 1DP
G8KWJ TQ7 1LR
G4RYO TQ7 1QU
G3ENO TQ7 2AA
G3LHC TQ7 2AQ
G3VTG TQ7 2BW
G0NOQ TQ7 2HF
G2RO TQ7 2HW
G8PQ TQ7 2HY
G4VMP TQ7 2JL
M1BXD TQ7 2SY
G3NSN TQ7 3AF
G8EUT TQ7 3BX
G6STJ TQ7 3EE
G8EMA TQ7 3JG
G7IMV TQ7 3JZ
G3JII TQ7 4AS
G3XLW TQ7 4QD
G6TOI TQ7 4SA
G7ARV TQ8 8AR
G4NLM TQ8 8HQ
G0MRU TQ8 8JP
G8TQH TQ8 8PJ
G3HST TQ8 8PW
G3PNO TQ9 5EF
G1WZG TQ9 5FQ
G0UBG TQ9 5QY
G0MDU TQ9 5WG
G7RZF TQ9 5YJ
G2DOT TQ9 6ER
G7KCE TQ9 6SR
G7SSS TQ9 6SY
G4VHL TQ9 7BA
G3ZAE TQ9 7BU
G6JPM TQ9 7EH
G0DIH TQ9 7EY
G7DHW TQ9 7NQ
G4LUF TQ9 7NY
G4DNC TQ9 7PD
G3YFZ TQ9 7TA

South Herefordshire

G4BJD HR1 3AY
G1AYP HR1 3BJ
G0PRB HR1 3DA
G0BOK HR1 3HA
G0JMH HR1 3RY
M1BYO HR1 3SJ
2E1FFO HR1 3SJ
G6TID HR1 4DL
G6PFX HR1 4DQ
G7LVG HR1 4LR
G4XCS HR1 4PN
2E1FFN HR2 0AH
G4ZEO HR2 0DB
G4ASR HR2 0HP
G1FVN HR2 0JB
G7BTI HR2 7UG
G1DYQ HR2 7UG
G4WXF HR2 7UT
G1DRW HR2 7XZ
2E1BCJ HR2 8DE
G0SYT HR2 8HJ
G4OGW HR2 8JT
2E1AZO HR2 8JT
2E0AFP HR2 8PU
G4XTF HR2 9BE
G3RMF HR2 9BS
G1LSX HR2 9LT
G0SAT HR2 9NH
G0JLI HR2 9NS
G0WZY HR2 9NX
G0DHB HR2 9QN
G0DYN HR2 9QN
G1RLF HR2 9RH
G0IAH HR3 6BY
G1UMS HR3 6BY
G4BSO HR4 7RB
G3CSS HR4 7RW
2E1FHK HR4 7SA
G4FAD HR4 8AZ
G4XRS HR4 8DG
M1BYN HR4 8LH
2E1FGF HR4 8LH
G3EPV HR4 8NG
G3YDD HR4 8NG
G3WGN HR8 2NW
G8RZS HR9 5BL
G0BUQ HR9 5JP
G3DYO HR9 5JY
G3TWS HR9 5QS
G8TWS HR9 5QS
2E1GBC HR9 5QU
G7NUV HR9 5RZ
G3FED HR9 6AJ
G6GQL HR9 6BH
G3LCK HR9 6BQ
G0TRI HR9 6DH
G3JUC HR9 6EY
G0LJP HR9 6EZ
G8CGK HR9 6LD
G1EHX HR9 6QJ
G0IMV HR9 7BE
G4RRT HR9 7DL
G7RVI HR9 7DN
G0GQI HR9 7JH
G3EFL HR9 7JR
G4FHY HR9 7LN
G7RIU HR9 7NF
G0EHT HR9 7UF

South Holland

G8ELV PE11 1HD
G1LSB PE11 1HQ
G1XDT PE11 1NJ
G6NHT PE11 1PD
G4TWR PE11 2AF
G1OUK PE11 2AX
G4XWW PE11 2BJ
G0JUR PE11 2HE
G0JLF PE11 2LE
G4ODA PE11 2LW
G1ZJP PE11 2PX
G0UOQ PE11 2UN
G6AHV PE11 2UU
G4NQS PE11 3AF
G7FML PE11 3AP
G8UKB PE11 3BX
G8NCT PE11 3DH
G0NVC PE11 3LL
G6UHS PE11 3PN
G3TMA PE11 3TB
G7UIO PE11 3XB
G4NBR PE11 4AX
G8EQI PE11 4BQ
G3UYE PE11 4DA
G0GJR PE11 4DB
G7CWM PE11 4DQ
G3SJR PE11 4EN
G1DSP PE11 4EU
G4DSP PE11 4EU
G4OO PE11 4EU
M1BMW PE11 4HF
G0RTL PE11 4HN
G7RTL PE11 4NE
G7ORB PE11 4PU
M1BET PE11 4PU
G0MVO PE11 4QH
G3UDP PE11 4QQ
G0FLC PE11 4XQ
G1ZLF PE11 4XU
G3LYT PE12 0BE
G0YKC PE12 0EE
G1JBJ PE12 0EU
G3REH PE12 0EZ
G7STM PE12 0LT
G0OME PE12 0PS
G0RNV PE12 0SF
G8FPW PE12 0TY
G3XDA PE12 6BG
G4CEQ PE12 6BX
G3VPR PE12 6DA
G4EXP PE12 6DB
G1PAT PE12 6PN
G4SJV PE12 6PN
G4VGQ PE12 6PN
G0LWC PE12 6PT
G0MHH PE12 6QG
G7OQO PE12 6SF
G4NSF PE12 6TA
G3YIG PE12 6TA
G1MLP PE12 6TG
G7RRD PE12 6TJ
G3WPL PE12 7AX
G7UYP PE12 7BT
G1IRQ PE12 7HL
2E1BOE PE12 7HR
G1WYC PE12 7JE
G6DDU PE12 7LQ
G7CUP PE12 7NB
G7HCR PE12 7PP
G3NSF PE12 8AG
G4SCU PE12 9AZ
G4ZID PE12 9HS
G0SMV PE12 9JN
G4KHF PE12 9LQ
G0WKN PE12 9LQ
M0AHQ PE12 9LY
G4RMJ PE12 9PJ
G0SHC PE12 9XP
G7BZC PE12 9YR
G6LTN PE13 5QR
G0VJX PE13 5RB
G0EUD PE34 4HQ
G0EJC PE6 0AL
M1BXJ PE6 0BA
G8MEE PE6 0JB
G0TFR PE6 0LH
G4ZXZ PE6 0LU
G6MTY PE6 9QB

South Kesteven

G6PII NG23 5AU
G4BPE NG23 5BA
G1UMK NG23 5DU
G4NSW NG23 5ES
G2BNI NG23 5EW
G6JWO NG23 5HN
G0ORP NG24 2SA
G4KTG NG31 7AW
2E1EBR NG31 7BW
G3YOQ NG31 7EA
G0GRC NG31 7HH
G8WWJ NG31 7HH
G6IPW NG31 7JL
G0RCI NG31 7NN
G1UTZ NG31 7PH
G6SSN NG31 7QS
G0TNC NG31 7QY
2E1DCV NG31 7RH
G7SJX NG31 7RH
G0JTO NG31 7SJ
2E1AWI NG31 7XG
G4DZC NG31 8BN
G8OPS NG31 8JY
G3VSX NG31 8JZ
G0IBX NG31 8LA
G4WZU NG31 8LN
G4VUA NG31 8LP
G6JUT NG31 8RB
G7GRC NG31 8RB
G6UOH NG31 8RN
G7IXS NG31 8RT
2E1CSD NG31 9JD
G7BYA NG31 9JS
G6VYK NG31 9PE
G3MAN NG31 9PL
G0EJJ NG31 9PY
2E1EFG NG31 9QA
G3VIP NG31 9QN
2E1ANE NG31 9RG
G4VMI NG31 9RW
G1YWL NG31 9SY
G3PJR NG32 1DJ
G4MQM NG32 1HX
G3SKT NG32 1JG
G1WQA NG32 1TA
G6PIF NG32 2EA
GM3PTI NG32 2NS
G3TBK NG32 3DF
G4HVC NG32 3DF
G4HOJ NG32 3DL
G7TJX NG32 3QL
G1DGN NG33 4DT
G4FGY NG33 4HA
G7REA NG33 4HW
G0KTB NG33 4LL
G0KQK NG33 4RT
G3KHZ NG33 4SB
G3NJK NG33 4SB
G6CTK NG33 4ST
G1GET NG33 5HB
G8VUR NG33 5JD
G7PBA NG33 5QE
G3MHH NG34 0NH
G3KVP NG34 0SE
G4RIP NG34 0SP
G4WJM PE10 0EE
G3PVM PE10 0ES
G4NIZ PE10 0HT
G3GQO PE10 0JB
G3YWO PE10 0JE
G8NWM PE10 0NT
G7BAP PE10 0RB
G1IXV PE10 0RB
G3FYS PE10 0RB
G8VVC PE10 0RB
G7RFD PE10 0RN
G7RFE PE10 0RN
G8YOX PE10 9BN
G3HCQ PE10 9BX
G4FHU PE10 9BX
G6VEY PE10 9HB
G1XGS PE10 9HD
G0OTE PE10 9JX
G4RQK PE10 9LY
G4TEI PE10 9LY
G4DHF PE10 9PA
G6KLO PE10 9QS
G6SKS PE10 9QT
G6SKT PE10 9QT
G0BHU PE10 9SB
G4EMK PE10 9SQ
G8ILU PE10 9SQ
G1JBO PE10 9TT
G8PUE PE10 9XL
G8TXX PE10 9XL
G6HEE PE6 8EA
G1AUM PE6 8HD
G4UAC PE6 8HH
G7HKW PE6 8HN
G7SAU PE6 8HP
G1FJD PE6 8HR
G0GBV PE6 8LA
G6ENZ PE6 8NA
G6CVV PE6 8NT
G6KEZ PE6 8NW
G4FIQ PE6 8NY
G8ETW PE6 8QA
G7LTM PE6 8QX
G1HOJ PE6 8RD
G6NYL PE6 8RD
G1ESC PE6 8SZ
G1RVK PE6 8TF
G6BHL PE6 9HP
G7SOE PE6 9LP
G8OAD PE6 9NP
G0IXD PE6 9QL
G1ERC PE6 9RA
G0AFV PE6 9RB
G6ENN PE6 9RB
G4NHL PE9 1HF
G0CUG PE9 1HG
G6HNP PE9 1SN
G3HES PE9 1UJ
G6WCT PE9 1UT
G3OJQ PE9 2DU
G7VEX PE9 2FF
G1FSW PE9 2JY
G1DCX PE9 2SG
G6VYR PE9 2SQ
G3KWC PE9 2SU
G4PZB PE9 2TA
G1JGR PE9 2TS
G3THC PE9 2UB
G4WJN PE9 2UF
G7IGV PE9 2YX
G4PEL PE9 4JJ
G1IZH PE9 4PE

South Lakeland

G4VGG LA10 5AW
G4LWW LA10 5NT
G7UKP LA10 5NT
G6NQL LA10 5PJ
G1GVJ LA10 5QR
G4MIS LA10 5QR
G6LZZ LA10 5TQ
G6SSQ LA10 5TQ
G3BW LA11 6AT
G3FDW LA11 6AT
G3LZZ LA11 6DP
G3JTI LA11 6EB
G8GEF LA11 6RB
G8ATA LA11 7AB
G3NEP LA11 7JA
G4CEJ LA11 7LJ
G8MGG LA11 7RJ
G6SDG LA12 0HT
G1BMB LA12 0JF
G8ALE LA12 0QH
G3KKJ LA12 0RG
G4DVH LA12 0RT
G6CAT LA12 7EG
G4UCL LA12 7ES
G0AOM LA12 8AW
G6NVF LA12 9EA
G7BZH LA12 9EY
G4EOB LA12 9HY
G1MFX LA12 9NH
G7GUB LA12 9NU
G6LKB LA12 9PD
G4BDE LA12 9PF
G4FHA LA12 9PF
G8VRH LA12 9PU
G4PXR LA12 9QZ
G3DLO LA20 6ER
G8YGO LA21 8EQ
G4BNG LA22 9RL
G0JGX LA23 1DF
M1BKG LA23 1NW
G3JJJ LA23 2DW
G0VEG LA23 2EZ
G7NFF LA23 2JS
G0FLL LA23 2LA
G8IRN LA23 3PW
G1NTG LA5 0BU
G7ODM LA5 0HT
G8KGK LA6 1PJ
G6POI LA6 1QN
G7CFG LA6 2AJ
G8JAI LA6 2LJ
G3GEF LA6 2PG
G0BDA LA7 7EB
G1ILV LA7 7JB
G6MGP LA7 7JB
G1YMV LA7 7JP
G4UXH LA7 7PT
G3UEC LA8 0JL
G3PLX LA8 8BN
G0PVX LA8 8HX
M1BLB LA8 9DH
G1YHF LA8 9JP
G3CRJ LA8 9NS
G6IHF LA9 4HP
G0SEQ LA9 4HZ
G3VVT LA9 4QR
G2BAH LA9 5QP
G0LKU LA9 6AJ
G3VJI LA9 6AJ
G1IGC LA9 6AU
G6YCM LA9 6AU
G3WRI LA9 6EA
G1JMM LA9 6HJ
G4ZBO LA9 6HT
G3HMR LA9 7AQ
G0KSS LA9 7DW
G7NJD LA9 7ED
G6IPN LA9 7HA
G0KHA LA9 7JG
G0RTM LA9 7JG
G0HIF LA9 7NA
G3XJI LA9 7PE
G6UYK LA9 7RE

South Lanarkshire

GM4REF G44 4NA
GM7VYR G71 7ET
GM4VPA G71 7LY
GM1VBE G71 8AR
G7VSR G71 8AR
GM0XFK G72 0RQ
GM7SFE G72 7TP
GM4NDV G72 8NL
G4TTC G72 8RD
GM0MYS G72 9NP
GM0AZU G72 9NX
GM4MKY G73 1DP
GM0BCI G73 2LF
GM4BRD G73 2TA
GM0LBN G73 3EN
GM7AYW G73 3QP
GM8SNB G73 3QP
GM0JVV G73 4AE
GM2CRV G73 4NE
GM4GLE G73 4RD
GM4HYF G73 4RN
GM4TPE G73 4RN
GM1FUD G73 5AZ
MM0AYE G73 5RG
GM8VWC G73 5RG
GM4DLU G74 1BQ
GM0TPI G74 2BN
GM8NET G74 2BS
GM3LNI G74 2JU
GM8HBB G74 3AF
GM0LYM G74 3EA
GM0WLE G74 3UY
GM4XGY G74 4LG
GM4BOA G74 4QS
GM3ZDH G74 4RS
GM4TRY G74 4TH
GM3NZJ G75 0HA
GM4PQV G75 0HJ
GM6IOU G75 0HJ
GM3ITE G75 0LN
GM1SLW G75 8TA
GM1SQZ G75 8TN
GM0UET G75 8YG
GM6AMI G75 9AW
GM0SBJ G75 9JP
GM1YLB ML10 6HA
GM0SXD ML10 6HA
GM4DSS ML10 6HF
GM8JFZ ML10 6HG
GM3GVD ML10 6HW
GM3UDJ ML10 6RP
GM6PER ML11 0AL
GM7UPY ML11 0DF
GM0DQC ML11 0ED
GM7GHL ML11 0ED
GM6YRH ML11 0HY
GM0LTJ ML11 0PJ
GM3GDS ML11 0PT
GM4IIR ML11 7PS
GM7DSC ML11 7RU
GM3YXY ML11 7SE
GM6GWB ML11 7SS
GM6LSG ML11 7SS
GM0CTY ML11 8HP
M1BJG ML11 8RG
GM3NNZ ML11 9AE
GM8BSE ML11 9AY
GM3EZI ML11 9JJ
GM4GM ML11 9QA
GM6BIG ML11 9QA
GM4VGU ML11 9SR
GM3ZTW ML11 9ST
GM0GYN ML11 9XS
GM4KKV ML12 6DD
GM7GIO ML12 6LB
GM0EUU ML3 6PP
GM4UBF ML3 6PP
GM6FBZ ML3 7DE
GM0KDF ML3 7HB
GM0RNR ML3 7HB
GM0ELP ML3 7HG
GM8NGG ML3 7NF
GM4ILE ML3 7PE
GM0POF ML3 7PW
GM0MDX ML3 7PW
GM4UQG ML3 7PY
GM0STB ML3 7PY
GM8LBC ML3 7YJ
GM0CSN ML3 8QH
GM6KLL ML3 9DS
GM4BVU ML3 9UX
GM4ARU ML8 4NR
GM0VYY ML8 4NR
GM4XXO ML8 4QT
GM6GVE ML8 5BH
GM8JYJ ML8 5DE
GM4FGI ML8 5EB
GM3UCI ML8 5HB
GM4COX ML8 5HR
GM4UXX ML8 5HR
GM7OHQ ML8 5JD
GM6VVX ML8 5JZ
GM7NAA ML8 5LF
GM4LAO ML8 5LX
GM0RZY ML8 5NE
GM8FHK ML8 5TS
GM0SYY ML9 1DN
GM0UEQ ML9 1PU
GM3HVK ML9 1QB
GM3HBT ML9 2AU
GM7AOM ML9 2DA
GM3MXN ML9 2HJ
GM6PMG ML9 2HX
GM3NKG ML9 2LG
GM4ISM ML9 2TD
GM6HFH ML9 3EN
GM3LLB ML9 3LU

South Norfolk

G4RAV IP20 9HF
G3JPZ IP20 9HY
G3ICG IP20 9JY
G4TUW IP20 9LW
G4PFG IP20 9NQ
G6ZYM IP20 9PU
G0YAP IP21 4DG
G3IPG IP21 4TQ
G3KRY IP22 2DJ
G0MTM IP22 2DZ
G7VNN IP22 3BL
G0RPY IP22 3DJ
G8RML IP22 3DS
2E1FWX IP22 3HE
G8BOB IP22 3HL
G7JIP IP22 3LP
G0TZL IP22 3LQ
G0WCJ IP22 3NA
G3XLL IP22 3NF
G6ILT IP22 3NP
M0AJI IP22 3TW
G0VDC IP22 3UD
G4YFV IP22 3UY
G3AYZ IP22 3XD
G7NEM NR14 6AA
G1MSK NR14 6BQ
G3ITB NR14 6BY
G6JWX NR14 6HD
G0TTV NR14 6HX
G3LTG NR14 6LW
G0GQQ NR14 6PP
G3VKM NR14 6PP
2E1DRY NR14 6PP
2E1DSE NR14 6PP
G4TVJ NR14 6UT
G4SOZ NR14 7BU
G0JIV NR14 7DN
G0RSS NR14 7DP
G7OZH NR14 7ER
G4XQW NR14 7EW
G4AXN NR14 7HQ
G0MYT NR14 7LN
G0VJH NR14 7LU
G4LOJ NR14 7PB
G4WUO NR14 7PN
G3LQX NR14 7QX
2E1CAQ NR14 7RY
2E1CBK NR14 7RY
2E1CPE NR14 7RY
2E1CPV NR14 7RY
G7JUA NR14 7SP
G4VCE NR14 8AL
G4LEP NR14 8AS
G3LIA NR14 8AU
G6LVN NR14 8BT
G3PMT NR14 8EF
G6VAD NR14 8HX
G3LDI NR14 8LQ
G4MQU NR14 8LQ
G3CWC NR14 8NA
G6WMV NR14 8PF
G4ZOK NR14 8RB
G4UYR NR14 8RG
G4VBX NR14 8RG
G4GRV NR14 8SS
G0IYY NR15 1HR
G4GUS NR15 1NJ
G7SQV NR15 2AL
G7SST NR15 2DR
2E1DSB NR15 2HP
G3VYN NR15 2NT
G3KVA NR15 2TA
G0SNR NR15 2XE
G7TIE NR15 2XU
G4SGS NR16 1AR
G6ARM NR16 1DG
G0WGR NR16 1NA
G8ZFX NR16 1RN
G3SDT NR16 1RS
G3XYO NR16 1RW
G1GSB NR16 1SL
G3JNB NR16 1SY
G7VZI NR18 0AR
2E1DQT NR18 0AR
G3TKQ NR18 0DB
G6JJE NR18 0DB
2E1EMD NR18 0DB
2E1EME NR18 0DB
G1FPK NR18 0DE
G3YIA NR18 0DN
G1EGL NR18 0EA
G4OJV NR18 0EN
G8DYA NR18 0EX
G0VZD NR18 0HR
G6LQM NR18 0HS
G6YYU NR18 0HX
G7AZI NR18 0ND
G3MPN NR18 0NT
G7REE NR18 0NU
G6DAN NR18 0NZ
G6YFZ NR18 0QP
G0UEI NR18 0SE
G4BDW NR18 0SJ
G4KID NR18 0SJ
G7UCC NR18 0SJ
G0HCR NR18 0TY
G8YSX NR18 9BH
G8ZKW NR18 9BH
G7EOA NR18 9DD
G0PSY NR18 9PR
G4ZDT NR18 9PR
G7SNR NR18 9PR
G7TXQ NR18 9PR
G7UBC NR18 9PR
G4RKK NR18 9TB
G4RKL NR18 9TB
G7DIB NR18 9TB
2E1BLJ NR34 0LB
2E1BMJ NR34 0LB
2E1BMK NR34 0LB
G1RSO NR34 0LF
G3RK NR35 2PY
G3CIM NR4 6AA
G0KYA NR4 6UB
G4SVI NR4 6UW
G4XVI NR4 6UW
G0CKH NR4 7RY
G3HUL NR5 0ES
G4ONF NR5 0HB
G6KCG NR5 0HG
G3BNG NR5 0HZ
G4KQY NR5 0JB
G3PZQ NR5 0LR
G6KAI NR5 0NH
G0OPR NR5 0NJ
G0UYC NR5 0SW
G8XBZ NR5 0SY
G3PZX NR8 5AU
G3YIR NR8 5AU
2E1EEZ NR8 5BX
2E1EFA NR8 5BX
G3EZB NR8 5DP
G3WHG NR9 3AB
G7PJI NR9 3EP
G3TZT NR9 3HA
G3WBX NR9 3HQ
G0DEE NR9 3JF
G4NST NR9 3JP
G4YKZ NR9 3JY
G3LZV NR9 3LL
G4TUK NR9 3NR
G1UDA NR9 3QJ
G3PTB NR9 4AB
G0TCO NR9 5BY
2E1FNZ NR9 5EJ
2E1EMG NR9 5HB

South Northamptonshire

G4MDU MK19 6BY
G0VKU MK19 6LW
M0AFJ MK19 7JH
G4XJE MK19 7JQ
G4KQH NN11 3PR
G3EDK NN11 6UH
G3UMT NN12 6AW
G8CHK NN12 6AW
G1YMJ NN12 6DN
G6ZGA NN12 6DN
G6RHJ NN12 6ED
G8EUX NN12 6JB
G4BKC NN12 6LT
G4GWT NN12 6QQ
G8MAG NN12 6RA
G0EPI NN12 7DL
G4MBB NN12 7JR
G3DGW NN12 7LL
G8ZOW NN12 7NH
G4IUH NN12 7PH
G7LUQ NN12 7PX
G1TAI NN12 7RW
G1DJU NN12 7SA
G2AXO NN12 7SB
G7FLR NN12 7SD
G7HMA NN12 7TS
G4KFC NN12 7TX
G4EQX NN12 7TY
G6BOI NN12 7UG
G4DMS NN12 8AL
G0VCB NN12 8AL
G6EPJ NN12 8AP
G0REB NN12 8DB
G8YJZ NN12 8JZ
G4UXO NN12 8LU
G8NMA NN12 8TT
G6BMG NN12 8UH
G4DQQ NN12 8UT
G4AHH NN12 8XW
G0UID NN13 5HU
G3ZUS NN13 5LL
G4EBF NN13 5LX
G3GWB NN13 5QJ
G8LED NN13 5QJ
G0TNM NN13 5QJ
G8JOX NN13 5SJ
G4PPJ NN13 5TW
G7NIU NN13 6AG
G1FER NN13 6BT
G7GOV NN13 6DT
G1WTY NN13 6ES
M0BBG NN13 6JA
G8MQY NN13 6LR
G4DWC NN13 6PB
G7HJU NN13 7BW
G0RJQ NN13 7DA
G8OZH NN13 7ET
G1EPO NN13 7TY
G1WWY NN13 7TY
G7NDT NN7 1AE
G6KPJ NN7 1HB
G6PHT NN7 1LX
G3RLW NN7 1LZ
G4BHT NN7 1NT
G6ZER NN7 1NT
G3XEC NN7 2AL
G3JCZ NN7 2AN
G4HWA NN7 2BJ
G6UVN NN7 2BW
G7SGL NN7 2HU
G0ETP NN7 2JD
G1GOY NN7 2JE
G0NRG NN7 2LF
G4MZX NN7 2LF
G8DHU NN7 2NR
G3TXQ NN7 3AT
G4KHX NN7 3AW
G4YLZ NN7 3BA
G4XKM NN7 3DF
M1CAW NN7 3DF
G7PTA NN7 3EL
G8EDX NN7 3JB
G3RWR NN7 3NA
G1VUY NN7 3QT
G3IPL NN7 4BX
G4RZH NN7 4BX
G4NES NN7 4BY
G4CZB NN7 4DD
G4TGW NN7 4DN
G4IPL NN7 4DP
G8TMM NN7 4DR
G6HRL NN7 4SF
G7KJY OX17 1LU
M1AHY OX17 2AB
G3LTN OX17 2PQ
G4DKD OX17 2QQ
G1AHQ OX17 3AF
G8SXQ OX17 3AG
G7VYT OX17 3DG
G1BJE OX17 3QT
G0BIV OX17 3XB

South Oxfordshire

G1ULQ HP18 9JZ
G6KLE OX10 0DJ
G1BZW OX10 0QG
G3RDF OX10 6AE
G0MUZ OX10 6HP
G4DQW OX10 6LT
G3OEB OX10 6LU
G7RLQ OX10 6LX
G4SOH OX10 6RY
G0CAD OX10 7AR
G6YQT OX10 7DB
G1EAP OX10 7DF
G4ITC OX10 7LB
G1URJ OX10 7NL
G7TYH OX10 8AD
G0ADH OX10 8HP
G4YSQ OX10 8LH
G0LUA OX10 9AD
G0BDG OX10 9BH
G6TPC OX10 9BH
G8FKL OX10 9DR
G4DNX OX10 9JH
G0PRU OX10 9JH
M1AFF OX10 9JH
G6ZHC OX10 9LB
G7TIB OX10 9PA
G6UXM OX10 9PE
M1BRU OX10 9PE
G6IMQ OX10 9PY
G4CUR OX10 9QB
G6DKI OX10 9QH
M1BQT OX10 9QT
G8SMH OX11 0AE
G8FYD OX11 0AU
G6MSH OX11 0BQ
G0KNJ OX11 7AA
G0KOC OX11 7AQ
G4PWI OX11 7AX
G4ZKI OX11 7DB
G1KRB OX11 7DD
G6UGP OX11 7DD
G0DPK OX11 7DF
G1HKT OX11 7JP
G6PKZ OX11 7JP
G3GGG OX11 7JQ
G6MGY OX11 7RU
G7KYL OX11 7SN
G1GQH OX11 7SQ
G4SYL OX11 8BP
G7VIA OX11 8EJ
G7VIB OX11 8EJ
G6ORN OX11 8HA
G0NIJ OX11 8HD
G1OWI OX11 8HD
G1SSL OX11 8HD
G7AKP OX11 8HD
G0BSA OX11 8HP
G0MQM OX11 8HP
G6KFY OX11 8HR
G7PVX OX11 8JT
G4EQM OX11 8LG
G0SWG OX11 8NA
G0JYW OX11 8NT
G7AJB OX11 8NT
G1YHB OX11 8QT
G4XYM OX11 8TE
G8CUL OX11 9JX
G8NVI OX11 9JX
G4ZLQ OX11 9LE
G0SXY OX11 9RN
2E1AIZ OX14 4NL
G3MZC OX14 4NL
G4PMK OX14 4NL
2E1EHR OX14 4NL
G7LUK OX14 4PU
G0FLB OX14 4QH
G4HJI OX14 4RF
G3UMF OX3 8TA
G8KNJ OX33 1EN
G0VLI OX33 1SX
G8MCI OX33 1XJ
G3DQC OX33 1YD
G4LLL OX33 1YL
G7IVF OX4 4YT
G0GXB OX44 7QX
G6IQU OX44 7TA
G7BGM OX44 7TB
G3SBM OX44 7TJ
G4DAP OX44 7TN
G0MCO OX44 7YN
G0IDF OX44 9AW
G6DBU OX44 9BJ
G1YSX OX44 9DL
G7BKZ OX44 9DL
G8ECG OX44 9HP
G0MVW OX9 2NE
G4FEO OX9 3BE
G6GHU OX9 3DG
G8NRR OX9 3JF
G0FOK OX9 3JS
G4HXU OX9 3NH
G3INN OX9 3NQ
G4ADR OX9 3NQ
G0UUR OX9 3NQ

G8JK	OX9 4DR
G8KMH	OX9 4EA
G6SFT	OX9 4EL
G1ZRR	OX9 4EU
G0IKI	OX9 4EW
G0IIB	OX9 4HB
G1ECE	OX9 4JY
G1GCJ	OX9 4JY
G7AGR	OX9 4JY
G8GCJ	OX9 4JY
G4FRL	OX9 4PE
M1BXQ	OX9 4PE
G3AKN	OX9 4PL
G0ODQ	OX9 4PW
G6ZTV	OX9 4QZ
G3ZRY	OX9 4RA
G8AAD	OX9 4SJ
G0DVF	OX9 4TJ
G0FRN	OX9 4TP
G8ZWE	OX9 4TT
G7IKS	OX9 4UD
G8LGM	OX9 5AD
G7GXO	OX9 5AQ
G6CYN	OX9 5EJ
G0DUF	OX9 5RF
G3YRE	OX9 5SX
G0SWK	OX9 7AJ
G0LGG	RG4 9AD
G4DAT	RG4 9BT
G7JOO	RG4 9RS
G6NEA	RG4 9RY
G8PDY	RG4 9SA
G0JMS	RG4 9TF
G4ZHI	RG8 0JH
G3NGX	RG8 0JL
M0AZG	RG8 0PL
G1VRF	RG8 0QJ
G4CLD	RG8 0QY
G3LLG	RG8 0SE
G3UK	RG8 7EW
G7FTJ	RG8 7NY
G4CDJ	RG8 7PP
G1GMH	RG8 7PX
G4CUE	RG8 7QG
G3TRY	RG8 7QG
G4MDF	RG8 7QG
G8NMO	RG8 7QG
G0LHZ	RG8 7QL
G0VLX	RG8 9BS
G8VOE	RG9 1AP
G3BBX	RG9 1LP
G3TFL	RG9 1LT
G1XET	RG9 1RB
G4WBI	RG9 1UU
G8PJC	RG9 1UU
G3DCO	RG9 3JG
G4FDB	RG9 3LP
G3WOE	RG9 4DH
G6CHA	RG9 4DH
G6RKJ	RG9 4ED
G8PJL	RG9 4QW
G3XTT	RG9 5HJ
G3PZL	RG9 5PS

South Ribble

G6LOC	PR1 0BN
G0TYW	PR1 0JS
G1PIC	PR1 0JT
G4LKM	PR1 0LL
G1INU	PR1 0LU
2E1FQQ	PR1 0NE
G3ODK	PR1 0TA
G3VKN	PR1 0TJ
M1AMW	PR1 0UR
G4RFA	PR1 0UX
G7NZF	PR1 0XL
2E1BHZ	PR1 0XN
G7RNG	PR1 0XN
G6EPX	PR1 0XQ
G0EIF	PR1 0YE
G4FSJ	PR1 9AD
G7EFC	PR1 9AJ
G7SDR	PR1 9AJ
G0JJL	PR1 9BH
G7JZJ	PR1 9DD
G1HMY	PR1 9DR
G3BWI	PR1 9EL
G1JCQ	PR1 9LA
2E0AJH	PR1 9LB
G6VUK	PR1 9LT
G4RSK	PR1 9QT
M1AKF	PR4 4AF
G1PJX	PR4 4GE
2E1BSV	PR4 4GE
G3VBL	PR4 4JD
2E0AFL	PR4 4JX
2E1FOK	PR4 4JX
G0UNQ	PR4 4LE
G0LEE	PR4 4RB
G4YSU	PR4 4RJ
G0LQK	PR4 4XJ
G1MBN	PR4 5BH
G0LZQ	PR4 5BL
G1IPY	PR4 5BX
G4XNS	PR4 5FL
G4WAL	PR4 5NP
G6TNA	PR4 5NP
G3PMO	PR4 5QD
G7CUA	PR4 5ZH
2E1BSU	PR4 5ZY
G3PS	PR5 0DT
G8RIP	PR5 0JN
G3PVD	PR5 0JQ
2E1ECR	PR5 0JR
G4BGL	PR5 0UQ
G6LUF	PR5 1HE
G3UPY	PR5 1JD
G3BPJ	PR5 1QU
G4YWG	PR5 1RJ
G3UDZ	PR5 1SF
G0FDX	PR5 1UH
G0GVA	PR5 1UH
G0MAH	PR5 1WD
G3ZOQ	PR5 1YB
G3GGS	PR5 2AA
G4SBQ	PR5 2AF
G7BWD	PR5 2BT
G1BUN	PR5 2GA
G3HKU	PR5 2JP
G1IQF	PR5 2NR
G1RBZ	PR5 2NS
G4YIA	PR5 2QZ
G4XEB	PR5 2UU
G0FQN	PR5 2UX
G0EIG	PR5 2XB
G3KQY	PR5 2XT
G1BMN	PR5 2XY
G0NGE	PR5 2ZN
G0JSL	PR5 3JB
G0JSM	PR5 3JB
2E1BVQ	PR5 3JB
G6WFK	PR5 3JL
G0VHO	PR5 3LB
G0KLT	PR5 3QJ
G7CYM	PR5 3SE
M1CAL	PR5 3SQ
G7VOQ	PR5 3SQ
G7RAY	PR5 3UT
G1JHP	PR5 3WB
M0AAE	PR5 3XT
G0JHC	PR5 3XT
G7ATY	PR5 4EE
G3DWQ	PR5 4HS
G3KUE	PR5 4HS
G6IMY	PR5 4LT
G3STB	PR5 4RA
2E1AGD	PR5 4RJ
G0OLG	PR5 5AJ
G0EHW	PR5 5RA
G4HNR	PR5 5SJ
G4KMC	PR5 5TY
G3AZI	PR5 5UP
G3ZVQ	PR5 5YB
G3EXU	PR5 5YX
G3ZOC	PR5 5YX
G3XUH	PR5 6AH
G0FQC	PR5 6EP
G1KNC	PR5 6RJ
G0ASH	PR5 6UY
G0KSH	PR5 6UY
G8GLS	PR5 8AE

South Shropshire

G4EMD	DY14 0NB
G8IYE	DY14 0QB
G8DFI	DY14 8EB
G3TBU	DY14 8PD
GW1KJE	SY15 6HU
GW4ZWC	SY21 8JL
G3UQH	SY5 0HG
2E1FZS	SY5 0NS
G3URZ	SY6 6AD
G4JKA	SY6 6AD
G4ZUD	SY6 6AF
G3OMX	SY6 6EF
G5IX	SY6 6JE
G4SMA	SY6 6JW
G0KLH	SY6 6TA
G3WMO	SY7 0AY
G7EYM	SY7 9DX
G3ZZX	SY7 9ER
G0LIA	SY7 9LT
G0CGI	SY7 9PG
G3VWX	SY8 1DW
G7HSA	SY8 1HD
G8MYO	SY8 1PB
G1XHA	SY8 1TN
G8WDT	SY8 2AD
G3LFY	SY8 2DA
G4OYX	SY8 2PH
G6LXL	SY8 3AE
G1DFR	SY8 3BJ
G3TFZ	SY8 3LF
G8HMV	SY8 3NJ
G7JKL	SY8 4DG
G4AIJ	SY8 4JT

South Somerset

G3KZR	BA10 0BS
G3MIZ	BA10 0EQ
G7VNC	BA10 0JD
G4WTX	BA10 0RJ
G7SFY	BA20 2AZ
G4JBH	BA20 2DA
G8AFN	BA20 2DB
G1FZL	BA20 2EH
G4ERN	BA20 2EH
G0PWJ	BA20 2HE
G7AHZ	BA20 2JA
G4ZYX	BA20 2PE
G0UMS	BA21 3AH
G7LNJ	BA21 3BT
G1FGT	BA21 3DX
G1PAH	BA21 3EX
G3CFV	BA21 3EY
G3OMH	BA21 3HF
G0LHX	BA21 3HS
G0PCS	BA21 3JD
G3NOF	BA21 3JR
G4FYS	BA21 3PD
G3ABH	BA21 3RJ
G3UXL	BA21 3SL
G4NLI	BA21 3SN
G7MSK	BA21 3TW
G4FAZ	BA21 4AH
G3GC	BA21 4AW
G7GGJ	BA21 4AW
G8WBT	BA21 4BA
G6GLZ	BA21 4BD
G4EVI	BA21 4DD
G3BEC	BA21 4EY
G4PJO	BA21 4JN
G4ZFB	BA21 4LX
G1PZK	BA21 4NA
G3KCV	BA21 4NN
G3XFW	BA21 4PR
G4KKG	BA21 4QF
G0TIJ	BA21 4RS
G0WRB	BA21 4YF
G0ZDX	BA21 4YF
G7ZDX	BA21 4YF
G6NUR	BA21 5AL
G7RUR	BA21 5DF
G3MYM	BA21 5JE
G0LNI	BA21 5JQ
G0NKC	BA21 5LB
G0MVN	BA21 5NN
G7AIC	BA21 5NY
G0AIL	BA21 5RP
G4YEH	BA21 5RP
G3CMH	BA21 5SP
G3ICO	BA21 5SP
G6IUQ	BA21 5SU
G3OBL	BA21 5XA
G8DXO	BA22 7AG
G3UGR	BA22 7NH
M1ARI	BA22 7NN
G3ATK	BA22 8DB
G4PZR	BA22 8EX
G1XNK	BA22 8NS
G0WRK	BA22 8NY
G7WBE	BA22 8RB
G3MLH	BA22 8UR
G3YPL	BA22 8UR
G1GAN	BA22 9AH
G3AST	BA22 9EN
G0EON	BA22 9LF
G4GBN	BA22 9LF
G3PRQ	BA22 9LR
G4BMO	BA22 9LY
G3VLQ	BA22 9TZ
G0ANP	BA7 7HE
G6ZJK	BA7 7JY
G7VCJ	BA7 7LT
G6TKA	BA7 7NF
G0ENW	BA8 0BP
G6NDA	BA8 0DN
2E0APV	BA8 0ED
G4KHU	BA8 0JG
G4WJW	BA8 0JP
G3OOL	BA8 0LF
G0WRA	BA8 0LF
G7WRA	BA8 0LF
G4CJL	BA8 0RJ
G7VEN	BA9 8AN
G1RSF	BA9 8HF
G8GJA	BA9 9BS
G4YXX	BA9 9BZ
2E1EXL	BA9 9DW
G7PBT	BA9 9HB
G4BIN	BA9 9LS
G7SSG	BA9 9LT
G8NPY	BA9 9PT
G6IRX	BA9 9RB
G6RCT	BA9 9SB
G8OPG	BA9 9SG
G4HDY	BA9 9TA
G8TTY	DT9 4PJ
G7SDD	DT9 5AR
G8YEO	DT9 5AR
M0ARO	DT9 5BD
G7UPP	DT9 5ED
G3RSU	TA10 0HH
G0PNF	TA10 0JH
G4WLF	TA10 0JL
G6FHR	TA10 0NF
G7PKY	TA10 0NF
G6AYD	TA10 0NG
G1EME	TA10 0PP
G6JNS	TA10 0PP
G0DGA	TA10 9AE
G0DGE	TA10 9AE
G3THG	TA10 9AJ
G8UIG	TA10 9EY
G7AZU	TA10 9HH
G3PCJ	TA10 9NJ
G3USC	TA10 9NL
G4YTP	TA10 9PS
G0SZT	TA10 9RH
G5GZ	TA10 9SA
G0SLG	TA10 9SE
G7IJQ	TA10 9SE
G0FZH	TA10 9TF
G7LPO	TA10 9TH
G0HDJ	TA11 6BW
G3TFM	TA11 6JW
G7IHP	TA11 6LG
G6VVC	TA11 6LZ
G1DCZ	TA11 6PX
G4CDL	TA11 7AP
G3WMX	TA11 7AU
G7RRL	TA11 7BQ
G8GVW	TA12 6BG
G1AKD	TA12 6DT
G3AIK	TA12 6JU
G4PDG	TA13 5AP
G4VPM	TA13 5AR
M1AIK	TA13 5BN
G0JVG	TA13 5DX
G7RVG	TA14 6PZ
2E1ELV	TA14 6PZ
G6EER	TA14 6QN
G8LKP	TA14 6QN
M0AGT	TA14 6QX
M1BSN	TA14 6SH
G7LBV	TA14 6UG
G3EDW	TA16 5NS
G6AFL	TA18 7AH
G1PVZ	TA18 7AW
G4UTY	TA18 7BE
G3IWT	TA18 7DH
G2KI	TA18 7PU
G3BPM	TA18 7QD
M0BAO	TA18 7RJ
G7WFL	TA18 7RJ
G3ESF	TA18 8BL
G4FVW	TA18 8BL
G8FAS	TA18 8DB
2E1FHI	TA18 8DY
M0BBO	TA18 8JB
M1BGD	TA18 8JB
G3CQE	TA18 8LW
G8LZA	TA18 8NG
G6DSJ	TA19 0AT
G0MEQ	TA19 0HH
G4ECC	TA19 0NS
G4OXR	TA19 0NZ
G6AOV	TA19 9AH
G6LVG	TA19 9AH
G0SCC	TA19 9AN
G3KSK	TA19 9BB
G4HTD	TA19 9HX
G4KJD	TA19 9NH
G0WSC	TA19 9QL
M0AOD	TA19 9QN
2E1DLC	TA19 9QU
G7WAL	TA19 9QU
G4NBN	TA19 9RJ
G3OSH	TA19 9RW
G6NLI	TA20 1AH
G4IRS	TA20 1BA
G0VCD	TA20 1BG
G8VFR	TA20 1BX
G3HAL	TA20 1BZ
G4JBG	TA20 1DR
G6WZA	TA20 1JU
G1TMQ	TA20 1LL
G0FGX	TA20 1QS
G3DIC	TA20 1RD
G4VHB	TA20 2BL
G1ECV	TA20 2BU
G8RCF	TA20 2JH
G0GAX	TA20 2NB
G8GFZ	TA20 2PY
G1ECD	TA20 2QG
G4RRI	TA20 2RR
G4SZS	TA20 3DN
G8GZC	TA20 3PT
G3BON	TA20 4HU
G8XOE	TA3 6PU
G1LVK	TA3 6QN
G8XET	TA3 6QN

South Staffordshire

G1ZDF	DY3 4NQ
G0TMF	DY3 4NQ
G3YZP	DY3 4PQ
G8KBG	DY3 4RA
G4XKS	DY7 5AF
G7KKC	DY7 5HA
G4CYB	DY7 5HU
G4WDS	DY7 6AD
G0PDF	DY7 6AF
G4TOA	DY7 6AL
G3OQU	DY7 6AU
G0TBI	DY7 6BN
G4JY	DY7 6DT
G3TDC	DY7 6DX
G4IES	DY7 6ED
G0TDR	DY7 6EL
G3KZG	DY7 6HF
G0UZB	DY7 6QP
G0DBJ	DY7 6RR
G1DKF	DY7 6RR
G0AOW	DY7 6RW
G0AOX	DY7 6RW
G4WRA	DY7 6RW
G7FXZ	DY7 6SP
G4EZE	ST17 0QY
G4IVK	ST17 0RF
G1EQB	ST17 0RJ
G1ELX	ST17 0RL
2E1BWF	ST17 0RQ
G7SNN	ST17 0RQ
G4RWQ	ST17 0SJ
G6DGX	ST18 9DA
G3JBF	ST19 5AH
G3TYT	ST19 5AH
M1AQS	ST19 5BT
G3PIN	ST19 5BX
G6YZK	ST19 5BY
G0VJC	ST19 5EH
G0KCM	ST19 5EX
G0GJG	ST19 5HD
G8UYZ	ST19 5HF
M0AFM	ST19 5HQ
G7UIX	ST19 5HQ
G8DBS	ST19 5JY
G0BFK	ST19 5QH
M1BFJ	ST19 5QT
2E1ECX	ST19 5QT
G3XFB	ST19 9HZ
G4IDG	ST19 9JN
G0NEN	ST19 9NQ
G3JIK	ST19 9QB
G3LUP	WS12 4QP
G0BEO	WS6 6AX
2E0AGG	WS6 6DF
G7JCR	WS6 6DF
G4CJK	WS6 6DT
G4FYP	WS6 6DT
G3URM	WS6 6HE
G4PFO	WS6 6HE
G4JUK	WS6 6LD
G8MFU	WS6 6PT
2E1AOR	WS6 7BE
G0GPB	WS6 7BX
G4TQC	WS6 7DP
G6RDL	WS6 7EF
G6RIJ	WS6 7EJ
G8TRQ	WS6 7EJ
G7IPI	WS6 7EP
G1KRF	WS6 7EQ
G0EZO	WS6 7ES
G6RIG	WS6 7EU
G4PFM	WS6 7LB
G4ENG	WS6 7NA
G4CHJ	WV10 7BB
G0UIU	WV10 7HS
G7NZM	WV10 7HS
G4URM	WV10 7JS
G2DTQ	WV10 7LD
G6WJC	WV10 7LJ
G1TNR	WV10 7TY
G0DVD	WV11 2AJ
G2AQH	WV11 2AZ
G6ALZ	WV11 2DB
G4WAS	WV11 2DW
G8TA	WV11 2DW
G6LYE	WV11 2JG
G6RRP	WV11 2RQ
G4VZL	WV11 2TE
G0TQP	WV3 8JT
G3KQJ	WV4 4TT
G8BCF	WV4 4XX
G2DXH	WV4 5LF
G3ZGY	WV5 0BD
G6MZN	WV5 0BQ
G0TUN	WV5 0EA
G4WLK	WV5 0HW
G0ESP	WV5 0JG
G1DFF	WV5 0JX
G3SOE	WV5 0LH
M1AUN	WV5 0LX
G8JVW	WV5 0PA
G4RVK	WV5 7HF
G6AKN	WV5 7HF
G0HMJ	WV5 7HT
G3TPP	WV5 8BH
G4EAJ	WV5 8BS
G1ZTK	WV5 8EF
G4FIF	WV5 8HQ
G4WRX	WV5 8HQ
G6CRD	WV5 8LD
G4PNE	WV5 9AD
G4OTS	WV5 9AW
G6MSD	WV5 9BN
G6CQT	WV5 9EH
G6GTA	WV5 9HG
G4MZI	WV5 9HX
G4DFG	WV6 7AG
G1GTK	WV6 7AQ
G8CRR	WV6 7DE
G4CYO	WV6 7DU
G1DKI	WV6 7LX
G6EEF	WV6 7RZ
G6HDF	WV6 7SB
G4ZXS	WV6 7SP
G7RXW	WV6 7SX
G7PYG	WV6 7TD
G4DDZ	WV6 7TP
G3IOB	WV6 7YY
G4XSJ	WV6 7YY
G0BGB	WV7 3AP
G4MPN	WV8 1AG
G6HCV	WV8 1BJ
G4TCF	WV8 1BL
G4SQP	WV8 1BW
G3CAQ	WV8 1DA
G8JIS	WV8 1EH
G4AUD	WV8 1ER
G8IMH	WV8 1ES
G8XUE	WV8 1JP
G3ZPE	WV8 1NT
G1LGJ	WV8 1PG
G3KNG	WV8 1PG
G4EVP	WV8 1PG
G6NBK	WV8 1PG
G8IWQ	WV8 2AB
G6YIU	WV8 2BE
G8JBT	WV8 2DJ
G0LDY	WV8 2DS
G8RF	WV8 2EL
G4ZVA	WV8 2EY
G0AQJ	WV8 2JJ
G4GZK	WV9 5DE

South Tyneside

G3ISP	NE31 1UH
G7NAN	NE31 1UQ
G1SUI	NE31 2AU
G4SFN	NE31 2HT
G0WKQ	NE31 2LJ
G0WZB	NE31 2LJ
G1NZP	NE31 2XS
G7MPJ	NE32 3AS
G6ZEY	NE32 3TX
G0TFT	NE32 3TX
G0DIP	NE32 4AE
G0BFK	NE32 4DR
G3JDO	NE32 4HX
G1NHC	NE32 4LL
G7MWU	NE32 4LP
G7NQN	NE32 4RZ
G4NHC	NE32 4XL
G4WUM	NE32 5QT
2E1FNE	NE32 5YJ
G3SEG	NE33 2AN
G1FCK	NE33 2BY
G3IAB	NE33 2TB
G4LDT	NE33 2TG
G6UWI	NE33 3AQ
G4LFG	NE33 3NN
G0DSX	NE33 4HP
G6UXL	NE33 4JY
G7PZY	NE33 4LB
G3PRE	NE33 4UE
G1NZL	NE33 5HX
G6LIK	NE33 5PH
G3KZZ	NE34 0DR
G3EJO	NE34 0EQ
G0SPW	NE34 0HB
G4XWR	NE34 0NJ
G3HIF	NE34 0RT
G0ZAT	NE34 6AY
G3HRB	NE34 6DH
G0WUH	NE34 6SQ
G6BIA	NE34 7AQ
G0RBT	NE34 7HA
G4KSL	NE34 7JA
G0SLP	NE34 7JF
G4NS	NE34 7JQ
G7HSY	NE34 7PT
G1OHX	NE34 7RF
G4NTW	NE34 8BU
M1BXK	NE34 8NS
G8GIU	NE34 9BU
G3UXG	NE34 9DF
G7TFV	NE34 9DF
G7PHG	NE34 9JJ
G4YXV	NE34 9JR
G7PHI	NE34 9TA
G8PDE	NE35 9HE
G8JUC	NE35 9LG
G3XWU	NE35 9NG
G7WLY	NE36 0NJ
G0RCJ	NE36 0RF
G3IXB	SR6 7DG
G7BTW	SR6 7NE
G0SPP	SR6 7TZ
G3SFL	SR6 7UD

Southampton

G3NHV	SO1 1QL
G0GAU	SO1 4FD
G1ZXF	SO1 6DY
G6NJT	SO14 0LB
G3UVC	SO14 0YN
G8EQA	SO14 3HS
G8WBN	SO14 6FZ
G4YRX	SO14 6RG
G0PCW	SO14 6WD
G4AEU	SO15 2DT
G0DKU	SO15 3EH
G7BQI	SO15 3SE
2E0AQG	SO15 3SE
G7DYV	SO15 3SE
M0AFE	SO15 4HU
2E1ESQ	SO15 4HU
G6LOB	SO15 4JF
2E1EVQ	SO15 4ZA
G4MYS	SO15 5DS
G8UVS	SO15 5DS
G7SOU	SO15 5DS
G0PZF	SO15 5EF
G0TLG	SO15 5FY
G1CRP	SO15 5JH
G6HBN	SO15 5NZ
2E1EOO	SO15 5PJ
G3XUO	SO15 7QF
G3FRX	SO15 7RL
G4LDC	SO15 7RN
G6JVP	SO15 8NY
M1ABR	SO16 2NU
G3IXN	SO16 2NZ
G4HWJ	SO16 2PF
G4NMS	SO16 2PF
G0VNX	SO16 3BA
G0TGO	SO16 3EE
G7SON	SO16 3GA
G4VAS	SO16 3GZ
G4ZAS	SO16 3HY
G0SWY	SO16 3LQ
2E0AJP	SO16 3LQ
G3ZQC	SO16 3QD
G0IQY	SO16 3QQ
G3ION	SO16 3QX
G1XBR	SO16 4BN
G0POT	SO16 4DA
G8HWF	SO16 4EU
2E1EOP	SO16 4LZ
G7BOH	SO16 4NN
G7UAX	SO16 4NR
G4ZQI	SO16 5DJ
G3CGE	SO16 5DW
G4DLD	SO16 5DZ
G4ZFY	SO16 5FD
2E0AJQ	SO16 5FG
M0APN	SO16 5FG
G8KVN	SO16 5FG
G6NUD	SO16 5GY
2E1EPG	SO16 5HF
G3VQF	SO16 5NQ
G6TGB	SO16 5SB
G2BFC	SO16 6US
2E1BUO	SO16 7AE
G7GIJ	SO16 7AW
G6RNW	SO16 7EE
G3KJ	SO16 7HG
G3JTX	SO16 7LW
G3ZIL	SO16 7PE
G3FLJ	SO16 8DD
G6CRG	SO16 8EH
G6HHN	SO16 8HL
M1ANE	SO16 9BJ
G1UWL	SO16 9EH
G0WFQ	SO16 9EH
G3SOU	SO16 9EH
G8FAB	SO16 9EH
2E1EUE	SO16 9EH
G0AQW	SO16 9EW
G3GMK	SO16 9FX
2E1FNI	SO16 9HF
G0XAZ	SO16 9JF
2E1FLQ	SO16 9QE
G0DHR	SO17 1NU
G0MSR	SO17 1NU
G0TEI	SO17 1RH
G7VTY	SO17 1SD
G0OOW	SO17 1SS
M0AIF	SO17 1TU
G2HNI	SO17 2EY
G8BSK	SO17 2LS
G3WIE	SO17 3RQ
2E1BQJ	SO17 3RT
2E1BJS	SO17 3RZ
G7VGY	SO17 3RZ
2E0AEZ	SO17 3TG
G7JTT	SO17 3TG
G0DTK	SO17 3TH
G4SEM	SO18 1GS
G8CYJ	SO18 1HJ
G1VGA	SO18 1JS
G3TUF	SO18 1LS
G7PHV	SO18 1NW
G3OFX	SO18 1QR
G3CFA	SO18 2AP
G0TIA	SO18 2DD
G7UYI	SO18 2FP
G3FLN	SO18 2GQ
M1AFM	SO18 2JQ
G0EYB	SO18 2LZ
G1ISE	SO18 2PQ
G1LWF	SO18 2PQ
G4KMU	SO18 2PX
G3IDB	SO18 2QE
G1KAC	SO18 2QE
G4IWB	SO18 2QE
G4UJR	SO18 4FJ
G7EQB	SO18 4FJ
G8WYX	SO18 5EJ
G3VCL	SO18 5FS
G4YYS	SO18 5PG
G4PTF	SO18 5QJ
G3KXE	SO18 5QL
G6GNS	SO18 6AJ
G0UJP	SO18 6BB
G6NPK	SO18 6BJ
G8IOK	SO18 6NY
G7BZU	SO19 0NQ
G6XYS	SO19 1BX
G8BJB	SO19 1DP
G4UDB	SO19 1FW
G6HHH	SO19 2HF
G4BBD	SO19 2NX
G6VDT	SO19 4DD
G3NZL	SO19 4DE
G8GUS	SO19 4DE
G3MRA	SO19 4DX
G1LDN	SO19 5JP
G3GLW	SO19 5JP
G4GFN	SO19 5JS
G0LCN	SO19 5LF
G0ILA	SO19 5LN
G7BLX	SO19 6PS
G0UFG	SO19 7AF
G8NNQ	SO19 7AS
G8RNT	SO19 7GG
G7MTA	SO19 7JN
G4JHD	SO19 7PW
G4UEN	SO19 8EY
G0IOI	SO19 8FF
G0DXC	SO19 8FT
G0SER	SO19 8GE
G0EZM	SO19 8LN
G1YOM	SO19 8NT
G4OLY	SO19 8NT
G8WWM	SO19 8NT
G8OUD	SO19 8RP
G0XRN	SO19 9FJ
G3YHJ	SO19 9HD
G3GRW	SO19 9NU
G8GBP	SO19 9PE
G6KYE	SO2 3RT
G3KSF	SO4 1HF
G1COB	SO4 2NL
G4WKU	SO4 4UD
G0ORV	SO4 4WQ
G1UBN	SO4 6BQ
G3KMI	SO9 5NH

Southend-on-Sea

G0KFR	SS0 0AD
G0UKP	SS0 0DY
G4SOQ	SS0 0JD
G1JHK	SS0 0JU
G1POV	SS0 0LE
G3RLJ	SS0 0NT
G6CVB	SS0 0NY
G6WCI	SS0 0QL
G0ENN	SS0 0RA
G8JDW	SS0 0SN
G1PVT	SS0 7DR
G6AXE	SS0 7DR
G0TZP	SS0 7HZ
G0MBP	SS0 7LA
G2BBI	SS0 7PU
G7PNK	SS0 7QU
G3MAW	SS0 7SG
G4DVJ	SS0 8BN
G4UOL	SS0 8DB
G3MII	SS0 8EA
G0FNB	SS0 8NL
G0GHP	SS0 9EX
G0UAW	SS0 9EY
G4YPW	SS0 9HS
G8CEX	SS0 9PS
G8IWI	SS0 9RA
2E1DVG	SS0 9SB
2E1FQR	SS0 9SB
G3MJN	SS0 9SB
G1XRM	SS0 9SZ
G5VQ	SS0 9TY
G1JOT	SS0 9XD
G8PST	SS1 1EN
G0VCY	SS1 1QE
G6IWC	SS1 2QL
G4RSZ	SS1 2QU
G1HNN	SS1 2SX
G4UAI	SS1 2SX
G4UTF	SS1 2SX
G0UXN	SS1 2UZ
G1IQU	SS1 3AD
G3KWG	SS1 3AJ
G8ZPO	SS1 3DG
G1JLG	SS1 3HD
G0NNJ	SS1 3LB
G0PDM	SS1 3LN
G0TIG	SS1 3NW
G3RMC	SS1 3PF
G4YAK	SS1 3PX
G3YVK	SS1 3QU
G4OGY	SS1 3QY
G4KDE	SS1 3SG
G6TLN	SS1 3SS
G0SAV	SS1 3UE
G8BXV	SS2 4AA
G0AZF	SS2 4DG
G3UUI	SS2 4ED
G0SKI	SS2 4HR
G0ESO	SS2 4HU
G4DEZ	SS2 4HU
G4ZDA	SS2 4HU
G4EDN	SS2 4JE
G7PUI	SS2 4JN
G7CRY	SS2 4JS
G3TJU	SS2 4LA
G0XOX	SS2 4LU
G0TUI	SS2 4LW
G1YCR	SS2 4NN
G1UAE	SS2 4QA
G8OQR	SS2 4QA
G7SSR	SS2 4QA
G0DPC	SS2 4SP
G3IWB	SS2 4TG
G4ZGO	SS2 5HQ
G1XTD	SS2 6HP
G1DXD	SS2 6HR
G8GYH	SS2 6QP
G1FEV	SS2 6SQ
G6BHE	SS2 6TE
G0TTI	SS2 6TF
G3GYW	SS2 6UA
G0MYJ	SS2 6UJ
G6IDU	SS3 8TR
G7CQA	SS3 9AR
G1KJG	SS3 9BE
G7VZG	SS3 9EJ
G1NNQ	SS3 9EY
G4CEU	SS3 9HF
G1AEW	SS3 9HL
G6XNK	SS3 9LP
G8LYW	SS3 9LP
G8OPY	SS3 9NU
G1HQQ	SS3 9NY
G1URL	SS3 9PB
G7BNE	SS3 9PB
G8XHD	SS3 9RZ
G3SCY	SS3 9SB
M1ANK	SS9 1HQ
M0BDB	SS9 1HQ
G6LYM	SS9 1JY
G1XGP	SS9 1NE
G0URY	SS9 1NJ
G8IAR	SS9 1PX
G0KMF	SS9 1RU
G0JAO	SS9 2HT
G1NMF	SS9 2JR
G3ESL	SS9 2ND
G3MVU	SS9 2NL
G4SNV	SS9 2PB
G3ELW	SS9 2PQ
G8CJG	SS9 2PR
G0JYI	SS9 2PX
G8DQX	SS9 2TB
G4OQ	SS9 2XD
G3OA	SS9 3BB
G0RSW	SS9 3DN
G3YZS	SS9 3EF
G3SOL	SS9 3EH
G1VJJ	SS9 3HF
G4XGE	SS9 3JD
G0JDD	SS9 3JT
G4NGW	SS9 3JZ
G3KGN	SS9 3NT
G1GXF	SS9 3QS
G6OAV	SS9 4AZ
G8WSW	SS9 4DA
G0DCN	SS9 4DE
G0FPV	SS9 4DE
G3XRA	SS9 4DS
G1YAF	SS9 4HA
G1YUG	SS9 4HA
G6GGZ	SS9 4HG
G0DBM	SS9 4LE
G1GXW	SS9 4PL
G6DAH	SS9 4PW
G6NLS	SS9 4PW
G3PHL	SS9 4QY
G0SCT	SS9 4RZ
G8HVS	SS9 4SL
G3RPZ	SS9 5EL
G0EFI	SS9 5HR
G6XBS	SS9 5PW
G3SVI	SS9 5PX
G7DTS	SS9 5RF
G4PRK	SS9 5SB
M1BOY	SS9 5SF
G3RCX	SS9 5SW

Southwark

G0NVJ	SE1 2XR
G7CVY	SE1 4YT
G3KYL	SE1 5PQ
G8LOT	SE1 5QN
G1UFX	SE1 8JJ
G7KUX	SE11 4AY
G3XWK	SE15 3TR
G1BYO	SE15 3UH
G3XBH	SE15 4AN
G7KYD	SE15 4JX
G3HMO	SE15 4PY
G0WSW	SE16 1GD
G7PUJ	SE16 2PW
G3CTR	SE16 2QF
G7AHO	SE17 2DY
G0GBK	SE17 3EA
G3UOB	SE19 1HA
G7VML	SE22 0DG
G0FAH	SE22 0SB
G8TOM	SE22 8ND
G3VA	SE22 8SS
G3EYY	SE24 9ND
G1ZHQ	SE24 9NR
G8UEI	SE5 0PR
G0EDV	SE5 8TW

Spelthorne

G8FVJ	TW15 1AB
G4ESG	TW15 1DR
G0MWP	TW15 1EN
G0OSX	TW15 1PW
G1WWB	TW15 1QE
G0GIY	TW15 1QP
G0JXP	TW15 1RS
G8ZZG	TW15 1SQ
G4RSU	TW15 1UQ
G4ZVU	TW15 2AB
G6UMK	TW15 2AB
G8CAH	TW15 2AP
G4YDN	TW15 2LB
G3XTZ	TW15 2LP
G6JQD	TW15 2LR
G8LPA	TW15 2LU
G1WZB	TW15 2SJ
M1CBR	TW15 2SL
G1ZDT	TW15 2TD
G7AYP	TW15 3AD
G0PUR	TW15 3AY
G1IBS	TW15 3DF
G2HS	TW15 3ET
GW0ECN	TW15 3JH
M0AHS	TW15 3PF
G1PXL	TW15 3PG
G3JUL	TW15 3QA
G1RYO	TW16 5JL
G3HOK	TW16 5LA
G4LDW	TW16 5NB
G3PSW	TW16 5PT
G1SBT	TW16 6HF
G3HTC	TW16 6HF
M1AWF	TW16 6HF
G3HBZ	TW16 6HW
G3UQS	TW16 6NG
G8BQZ	TW16 6PD
G7PAK	TW16 6PG
G3ZXA	TW16 6SG
G0NLW	TW16 7HX
G4YAS	TW16 7NA
2E1AOF	TW16 7TL
G0HYT	TW16 7UA
G0VDZ	TW17 0EN
G6IFR	TW17 0JB
G8BIJ	TW17 0JZ
G4WPD	TW17 0QA
G6KLH	TW17 0RP
G4XGI	TW17 0SH
G0LGA	TW17 8EU
G2FKP	TW17 8QT
G7MII	TW17 8RR
G3YCQ	TW17 8RX
G4CKQ	TW17 8RX
G8YRY	TW18 1AG
G0MIJ	TW18 1DG
G0JSP	TW18 1DJ
G1SXB	TW18 1EE
G7RZQ	TW18 1JW
G0NIF	TW18 1LX
G3TDR	TW18 1NG
G3UES	TW18 1NG
G0FPP	TW18 1NQ
G4MTN	TW18 1PA
G1ARR	TW18 1QS
G1PPW	TW18 1QX
G6PVC	TW18 2DD
G3MCK	TW18 2DF
G0NKY	TW18 2EH
G0TNR	TW18 2HA
G6PVV	TW18 2HA
GM4ZRR	TW18 2NW
G4GJV	TW18 2PF
G4SGL	TW18 2PF
G3YWX	TW18 2PW
M0BBR	TW18 2QD
G3OEG	TW18 4HR
G0SZS	TW18 4LD
G0RWM	TW18 4NN
G3KKQ	TW18 4NR
G3NTM	TW18 4NR
G3WWT	TW18 4YZ
G4AKA	TW19 6AX
G4VLF	TW19 7AW
G0TID	TW19 7JE
G4MTH	TW19 7LF
G8FTI	TW19 7PF

Stafford

G8DKS	ST10 4PG
G1BHF	ST11 9QT
G4CWN	ST12 9BE
G6ITO	ST12 9JA
G0RJT	ST12 9JQ
G8VYF	ST15 0AA
G0KJP	ST15 0DH
2E1BRA	ST15 0DW
G0SKQ	ST15 0DX
G0TED	ST15 0EB
G1LAP	ST15 0EG
G8DQH	ST15 0EJ
G7KEE	ST15 0EP
G1DSF	ST15 0EP
G3JUX	ST15 0JA
G4TTM	ST15 0JB
G1HSJ	ST15 0JN
G6TDM	ST15 0LA
M1AIH	ST15 0LF
M1BDP	ST15 0LG
G8KNX	ST15 0NA
G6YYA	ST15 0RH
GW7GPU	ST15 8AZ
G0VWT	ST15 8LF
G4CRK	ST15 8LW
G6JAF	ST15 8PR
G4TOF	ST15 8SB
G4FKB	ST15 8UU
G4EJD	ST15 8XL
G8MZZ	ST15 8YP
G6EMJ	ST16 1DH
G8MOH	ST16 1DH

2E1DCM ST16 1EB
G3ZHS ST16 1HJ
G4TMD ST16 1LD
G0VNU ST16 1LG
G6YLZ ST16 1NL
G0VGK ST16 1NW
G0ITS ST16 1PG
G6ZLJ ST16 1QP
G0LJN ST16 1SL
G4YFF ST16 1TB
G3KNB ST16 1TW
G4PMR ST16 1UJ
G3XPD ST16 2DZ
G0EYX ST16 3HQ
G0GAP ST16 3NX
G0GOZ ST16 3PL
G1SPU ST16 3QT
G3CST ST16 3TU
M1BCD ST16 3YE
G6OAS ST16 3YE
G1ITP ST16 3YN
G4GKZ ST17 0AD
G4NVH ST17 0AQ
G1FNQ ST17 0AX
G3ESW ST17 0EJ
G4PKF ST17 0HJ
G3WAC ST17 0HN
G0HVZ ST17 0NF
G6KQD ST17 0NU
G3LOE ST17 0PA
G6LXF ST17 0PA
G0TFD ST17 0TW
G8MRZ ST17 0TZ
2E1ESD ST17 0TZ
G4POH ST17 0XY
2E1BLN ST17 0YS
G6RFM ST17 4BP
G2HNA ST17 4BX
2E1BYQ ST17 4DE
G0LCI ST17 4LJ
G3SBL ST17 4LX
G1XWO ST17 4NR
G0GAR ST17 4QP
2E1EXJ ST17 4QQ
G4PWV ST17 4QS
G4VMC ST17 4RW
G0LYC ST17 4SE
G8KUA ST17 9BE
G6KJK ST17 9FT
G1SJG ST17 9HP
G1UDT ST17 9HP
G6JUO ST17 9JB
M1ARR ST17 9JW
G3GYC ST17 9NA
2E1ENY ST17 9NH
G7ANP ST17 9PB
G3VUL ST17 9QJ
G3TBH ST17 9QQ
G4RSW ST17 9RL
G0BYA ST17 9UE
G0FXS ST17 9UL
G1UDS ST17 9UR
G3JDM ST17 9YA
G0LKN ST18 0AT
G8IYG ST18 0BG
G8JVU ST18 0EW
G4ZMM ST18 0HJ
G0DDN ST18 0JN
G7PHP ST18 0NZ
G0PVS ST18 0PQ
G0GUF ST18 0QJ
M1AQR ST18 0RA
G4PET ST18 0RD
G4PEU ST18 0RD
2E1ELE ST18 0SP
G7PHQ ST18 0UB
G4OUT ST18 0UR
G4THY ST18 0UR
G6KGA ST18 0XB
G3IHB ST18 9BJ
G4KQK ST18 9DQ
G7RCU ST18 9EB
G4FJA ST18 9EX
G3RLH ST18 9HS
G4YVD ST18 9JF
2E1AHO ST18 9LR
G7THV ST18 9LR
G0MQT ST18 9ND
G7NBX ST18 9QP
G4UKA ST18 9QR
G3KGW ST20 0DD
G4RKU ST20 0DS
G4IVG ST20 0HA
G4OTB ST20 0JD
G4LSA ST20 0LR
G3WUA ST21 6BW
G6EKM ST21 6DS
G4JKF ST21 6QZ
G0UUM ST21 6RG
G0UUN ST21 6RG
2E0AQA ST21 6RG
G3EHM ST3 7JQ
G3HVI ST3 7JQ
G3MFH ST3 7JY
G0NED ST3 7NS
G6BCV TF10 8DA

Staffordshire Moorlands

G0DOC CW12 3QG
G3ALA SK17 0LU
2E1AZQ SK17 0NF
G7OJD ST10 1AT
M1AMB ST10 1DP
G4SXK ST10 1DT
2E1AGK ST10 1EJ
G0LWA ST10 1EJ
G3EPG ST10 1LZ
G7SVL ST10 1NA
G7IMO ST10 1NL
G4OIB ST10 1PH
G6GA ST10 1PY
G4CHG ST10 1QG
G7PIJ ST10 1QQ
G0TPY ST10 1QS
G3UNM ST10 1RE
2E1AWK ST10 1RE
G7MWS ST10 1RU
G8KMG ST10 1RU
G3JGE ST10 1TA
G7PHB ST10 1XB
G4NHT ST10 1XB
G4OUG ST10 1XB
G4PGG ST10 1XB
G6FFU ST10 2AE
G6KTE ST10 2AG
G6YCN ST10 2PT
G3IDQ ST10 2QF
G4OTX ST10 3EA
M1AEN ST10 3HD
G1RIY ST10 3HP
G4YYO ST10 4BH
G8HVA ST10 4BX
G0VBT ST10 4DZ
G0CBV ST10 4JF
G3JZI ST10 4LS
G6GVZ ST10 4LY
G3IUS ST10 4NW
G0VHY ST11 9AP
G6IXN ST11 9AU
G0VRK ST11 9BH
G4HKG ST11 9BL
G1FXW ST11 9DW
G1MAD ST11 9DW
G7VLA ST11 9HA
G1DOG ST11 9NT
G7UKN ST11 9NZ
G4YRR ST11 9PL
G4HUO ST11 9RN
G3HZL ST11 9SL
G3IBE ST13 5BG
G4MDJ ST13 5BP
G4NLN ST13 5DB
G8YQA ST13 5LU
G4POE ST13 5PJ
M1BIX ST13 5PZ
G7HIJ ST13 5RR
M1AHZ ST13 6LZ
G4XIZ ST13 7AN
G0WOM ST13 7HJ
M1ALE ST13 8DD
G3FKY ST13 8DP
M1BPQ ST13 8DQ
G4TGS ST13 8ES
G8JGL ST13 8JA
M1ABF ST13 8JN
G7CFT ST13 8LL
G4UQM ST13 8LN
G8ZWL ST13 8RX
G8SDX ST13 8RX
G8YMS ST13 8UU
G6CCQ ST2 7NG
G7VUL ST2 9DS
G0FYC ST2 9DZ
G0BRL ST6 8QH
G0UTZ ST6 8RX
G3UHV ST6 8TG
G6OBD ST6 8TQ
G4YKC ST6 8NT
G6PFB ST6 8QN
G7GKH ST8 6QZ
G4WTP ST8 6RA
G6CYV ST8 6RA
G0BYU ST8 6SE
G6ZHU ST8 7AL
G0UHW ST8 7BJ
G8BMG ST8 7BL
G0GWV ST8 7BU
G7IAM ST8 7LS
G4WAM ST8 7LY
G7EHD ST8 7LZ
G4UQW ST8 7NG
G0CHL ST8 7NW
G6PJC ST8 7PF
G7LYS ST8 7SW
G8APB ST8 7SW
G4GVE ST8 7SZ
G4RXB ST8 7SZ
G0DBC ST8 7UF
G3LHG ST9 0BD
G0RKN ST9 0DG
M0ASR ST9 0DG
G4DVN ST9 0JN
G1EWC ST9 0LF
G0UMA ST9 0LS
G1IWX ST9 0NB
G8NSS ST9 0PF
G4KME ST9 9BW
G4LCL ST9 9LW

Stevenage

G3INU SG1 1JS
G3MHX SG1 1LS
G3HNB SG1 1PZ
G6XSZ SG1 1RP
G0OVO SG1 1SS
G3SAD SG1 1SS
G2BKZ SG1 1TE
G7SQU SG1 1TR
G0IBK SG1 1TX
G7DNV SG1 1UJ
G0NMA SG1 2BA
G1JMS SG1 2JJ
2E1CRE SG1 2RS
G1MHV SG1 3AU
G6TEZ SG1 3DS
G7CYQ SG1 3EZ
2E1ARS SG1 3EZ
2E1DJF SG1 3EZ
G7CAV SG1 3RT
G0ETA SG1 3TS
G8WWI SG1 4AY
G4SPV SG1 4EP
G1PAB SG1 4NQ
G7HCB SG1 4NR
G0GTE SG1 4PJ
G6ANR SG1 4RN
G3FAU SG1 5HF
G7DRG SG1 5HF
G0SVU SG1 5HH
G4BYE SG1 5NF
G8KMG SG1 5PD
G7HCL SG1 5QS
G8SAD SG1 5QS
G7WAT SG1 5RX
G4IKY SG1 5SF
G8FMA SG1 5SF
G8NDP SG1 5SQ
G6SEM SG1 5SU
2E1DGW SG1 5TR
G4ZZS SG2 0AA
G3JLZ SG2 0EQ
G0HGA SG2 0ER
G7PNE SG2 0JH
G0HOP SG2 0JJ
G6BYK SG2 0NZ
G0BXH SG2 0QG
G1YFK SG2 0QG
G1FWS SG2 7BH
G3UKB SG2 7BJ
G4CZZ SG2 7QS
G4ETG SG2 8AN
G0RAM SG2 8ND
G0HMT SG2 8PA
G1LIQ SG2 8RW
G3VBD SG2 8RX
G8ZFI SG2 8RY
G4JVG SG2 8RZ
G0WPX SG2 8RZ
2E1FHJ SG2 8RZ
G0UVN SG2 8UG
M0ARH SG2 9HR
G6WHH SG2 9HX
G4NMA SG2 9NJ
G8HGP SG2 9NL
G6YMA SG2 9NP
G7PQI SG2 9NP
G0EVZ SG2 9RN
G8KHI SG2 9RR
G4NEA SG2 9TP
G0MFY SG2 9TY

Stirling

GM0DUX FK15 0DF
GM7SAQ FK15 0ED
GM2BWF FK15 0JL
MM0AXR FK15 0NX
GM0XCW FK15 9ED
GM3YTS FK15 9ED
G0WSO FK15 9ED
GM0BWS FK15 9EQ
GM1PGQ FK15 9EQ
GM4VZY FK15 9JL
GM0GMD FK15 9NA
GM3DGD FK17 8AY
GM4LVV FK17 8ES
G6OYU FK21 8TH
GM0JOL FK21 8TN
GM1KSB FK6 5JJ
GM0VRP FK7 0DT
GM4YMD FK7 0LH
GM4VGR FK7 0LJ
GM2FVV FK7 0LL
GM6UNL FK7 0QE
GM4TMS FK7 0QP
GM4UYE FK7 0QP
GM0GRL FK7 7UA
GM0NWI FK7 7UL
GM0FNE FK7 7XA
GM1BSG FK7 8PN
GM1NSU FK7 9JW
GM1SBD FK7 9LP
GM0BWT FK7 9LU
GM4RDB FK7 9LW
GM8MOI FK7 9PR
GM3RWP FK8 1EL
GM6NX FK8 1XD
GM0LZC FK8 3BY
GM0HZO FK8 3JY
MM0AIK FK8 3PW
GM0EGI FK8 3PW
GM3FFQ FK9 4JR
2M1CEJ FK9 4LZ
GM4PJR FK9 4PS
2M1DWG FK9 4PU
GM0SVS FK9 4SA
GM1XJE FK9 4SA
GM0JML FK9 4SN
GM7VYG FK9 5DN
GM4PYJ FK9 5EU
GM0IXN FK9 5HY
GM4FXX FK9 5LR
GM4RCN FK9 5LR
GM4RTN FK9 5LR
MM1CCR FK9 5PX
GM1RZB FK9 5RD
GM3ODP G63 0EX
GM3PXG G63 0PL
GM6AYW G63 0PZ
GM4CRV G63 0YG
GM8VJV G63 9BX
GM1SRR G63 9SQ

Stockport

G0OGP SK1 2QE
M1ANQ SK1 4AB
G0HVS SK1 4AX
G8WNK SK1 4BL
G4DSR SK2 5BS
G8MFR SK2 5BS
G8JFC SK2 5DR
G0OBW SK2 5HX
G8ILD SK2 5HZ
G3WEC SK2 5JR
G7ILS SK2 5RN
M1ANT SK2 5SF
G0NZL SK2 5SR
G1FFR SK2 5UR
G8RIC SK2 5XP
G8CHY SK2 6BN
G0KHR SK2 6BX
G6IRR SK2 6DP
G3URW SK2 6ER
G4ORV SK2 6HA
G4WWO SK2 6JQ
G4FRM SK2 6JS
G1CXE SK2 7DJ
G6DXH SK2 7DT
G1JVF SK2 7EB
G3UHF SK2 7EB
G4HON SK2 7EB
G0ASI SK2 7HH
G3RZJ SK2 7LQ
G7SRH SK3 0JJ
G1ZPA SK3 0NB
G3JZT SK3 0PX
G0PTO SK3 0QS
G0MLC SK3 0TX
G1HWJ SK3 0UJ
G7TQT SK3 0UN
M1AQJ SK3 8DR
G8ZAD SK3 8HQ
G7VFQ SK3 8PA
G3KAH SK3 8PB
M1BAT SK3 8QU
G3ENL SK3 8SR
G3SHW SK3 8TJ
G7RKU SK3 8UT
G6LIY SK3 9BZ
G4VLB SK3 9EL
G8UWM SK3 9HF
G0NKU SK3 9JN
G6PWQ SK3 9JU
G0NZV SK3 9LA
G7CPN SK3 9LF
G7RCO SK3 9LS
G1NKZ SK3 9NG
G4ZGQ SK3 9QD
G0JNH SK3 9QJ
G0OQR SK3 9QL
G8NRU SK4 1NH
G0OZK SK4 1QL
G6LQE SK4 2AA
G7LHS SK4 2BH
G8YTP SK4 2DB
G1CQT SK4 2DF
G4HNO SK4 2PU
G0KOQ SK4 2QR
G4DVI SK4 2RJ
G3TZV SK4 2RR
G7VTD SK4 3AD
G1TAU SK4 3DJ
G1JFU SK4 3HD
G4RPJ SK4 3HD
G4ZDO SK4 3PP
G0AMY SK4 3QF
G0NKV SK4 3RG
G4NKK SK4 4AE
G3RRG SK4 4AX
GM0EDR SK4 4BU
G3WFW SK4 4BX
G7KSP SK4 4BX
G7RYC SK4 4BX
G7PRD SK4 4DF
G4JBF SK4 4LB
G3YFD SK4 4NE
G4HWW SK4 4NL
G4XQI SK4 4PL
G4JQY SK4 4PU
G0WOP SK4 4RL
G0DTT SK4 5AH
G3ANH SK4 5AR
M0AIC SK4 5AT
G7UEN SK4 5AT
G1FMA SK4 5AW
M1BSM SK4 5DJ
M1BBU SK4 5HS
G1PYU SK4 5JU
G6DCM SK5 6EX
G0IES SK5 6SH
G7NYO SK5 6SL
G0JER SK5 6UH
G8WPL SK5 6UH
G7RPC SK5 6UR
G0LJF SK5 6UU
G4CXA SK5 7AT
G4ILA SK5 7AX
G1JQK SK5 7EU
G3RNV SK5 7EX
G4WPL SK5 7JB
G4SNG SK5 7JW
G0LSM SK5 7LB
G7MVJ SK5 7LB
G8ZLU SK5 7LE
M1BAR SK5 8DE
2E1FXU SK5 8JQ
G0TIP SK5 8JY
G7LBA SK6 1DF
G0GSM SK6 1EF
G8NOS SK6 1EW
G0WMZ SK6 1HR
2E1CXP SK6 1LH
G1HRI SK6 1LQ
G4KUC SK6 1PG
G8VHF SK6 1PG
G6GUT SK6 1PT
G0IRX SK6 1PW
G4OZC SK6 1PY
G3XDO SK6 1QN
G0VIE SK6 1QZ
G6OXW SK6 2AJ
G8NRV SK6 2DY
G1TNK SK6 2ED
G1YZH SK6 3ES
G1LHZ SK6 3JA
G0VOX SK6 4AF
G0URT SK6 4AF
G0EOM SK6 4DH
G7KEI SK6 4DN
G7DZH SK6 4DX
G1GSM SK6 4HD
G3PTX SK6 4JF
M0ASF SK6 4NF
M0ASG SK6 4NF
G8IAN SK6 4NH
G6RIC SK6 4PT
M0ABW SK6 4QE
M1CAQ SK6 4QE
G0ORD SK6 5AE
G7VEP SK6 5AE
G1BTT SK6 5AQ
G3PMV SK6 5BR
G3XGH SK6 5BT
G7GGN SK6 5EB
G4WVF SK6 5ND
G4BUX SK6 5NH
G8BUX SK6 5NH
G3VO SK6 5NQ
G4GQC SK6 5NS
G4NPU SK6 5PG
G7VQA SK6 6AH
G4RLD SK6 6HG
G3MUO SK6 6HJ
G4BEV SK6 6HJ
M1ARV SK6 6JL
M0BEX SK6 6JL
M1ASD SK6 6JL
M0AWV SK6 6JL
G7BYU SK6 6LL
G3SNG SK6 6PJ
G0PDH SK6 7BG
G0LWZ SK6 7JA
G7CNQ SK6 7JA
G8YUP SK6 7JA
G0HHR SK6 7NU
G4FBG SK6 8AB
M1ASB SK6 8AN
G8MUX SK6 8AT
G6WWA SK6 8BJ
G0UOE SK6 8EJ
G3ZDZ SK6 8ET
G3NXX SK6 8HY
G3SMT SK7 1EH
G0IGR SK7 1ET
G4BIC SK7 1HR
G7UNY SK7 1HT
G8OBK SK7 1LE
G3TDH SK7 1LG
G4GRU SK7 1PP
G4EWE SK7 1QD
G0TDZ SK7 1QQ
M1AGC SK7 2AH
G3NUQ SK7 2DT
G1LQF SK7 2EA
G3KAF SK7 2ER
G0VRC SK7 2ER
G0ADI SK7 2JR
G0ADJ SK7 2JR
G7VOA SK7 2LD
G0PJW SK7 2NU
G4PLZ SK7 2PF
G4GEO SK7 3AA
G4SWB SK7 3DY
G3ZOD SK7 3HZ
G8VLR SK7 3JU
G0KZO SK7 3JW
G4SYC SK7 3JW
G7OJQ SK7 4BH
G7GIV SK7 4BP
G7SGD SK7 4EB
G0FLQ SK7 4QE
G4GEY SK7 4QG
G0CJA SK7 4QP
G0HEX SK7 4QW
G0BRZ SK7 5BH
G4OQP SK7 5JN
G7OQG SK7 5JS
G0LDU SK7 5LR
G1GYC SK7 5PW
G4VSW SK7 5QY
M0AFF SK7 6AY
G0JQU SK7 6BG
G4NIV SK7 6BH
G4BUN SK7 6DL
G6WZG SK7 6EP
G6DBQ SK7 6HG
G8WPV SK7 6HU
G4ALP SK7 6JA
G8EVR SK7 6JB
G0JQI SK7 6JG
G4DQL SK8 1DX
G7GOZ SK8 1HP
G7TKT SK8 1HY
G3MTR SK8 1HZ
G3USO SK8 1JQ
G2FZN SK8 2AD
G6TRN SK8 2AQ
M1ASA SK8 2EE
2E1ATV SK8 2HH
G0FAU SK8 2JA
G4SSN SK8 2JW
G6LDW SK8 2JW
G4FAS SK8 3AJ
G0DGN SK8 3AQ
G6AQL SK8 3AQ
G0WOA SK8 3BG
G3YJF SK8 3DW
G0HVT SK8 3DZ
G0KCE SK8 3DZ
G3HJK SK8 3ET
G6XDC SK8 3HF
G0TVE SK8 3HP
G3NZU SK8 3HW
G6AOB SK8 3HW
M0AGV SK8 3JA
G3ZBZ SK8 3LL
G8PEU SK8 3PF
G0DOU SK8 3PJ
G8BCU SK8 3RG
G8ZCJ SK8 3UD
G3UYG SK8 4AL
G8ONF SK8 4BU
G3NYE SK8 4NA
G3MNT SK8 4PE
G1EPJ SK8 4PF
G3IPO SK8 4PH
G0HFT SK8 5DD
2E1BVH SK8 5DD
G0NBQ SK8 5EN
G6AXK SK8 5HP
G0NOM SK8 5HX
G0NKM SK8 5PF
G0PRV SK8 5QJ
G3JHC SK8 5RT
G3VSY SK8 5RT
G3VOU SK8 6AX
G4NFA SK8 6EQ
G0IQO SK8 6ER
G7GDQ SK8 6EX
G8TMJ SK8 6EX
G4KJK SK8 6JG
G4FRW SK8 6JH
M0ARA SK8 6LS
G8BRF SK8 6ND
2E1FZZ SK8 6NU
G4LPX SK8 6PD
G0OZJ SK8 6PW
G8GRP SK8 7AJ
G8LDC SK8 7AL
G4SFU SK8 7DB
G6YDO SK8 7EH
G8GHO SK8 7EP
G4NII SK8 7HP
G6TKV SK8 7HU
G4EUC SK8 7LG
G0HZN SK8 7LP
G0PHF SK8 7LX
G4AHX SK8 7NJ
G6FOI SK8 7PB
G0IGB SK8 7PZ

Stockton-on-Tees

M1AFU TS15 9EZ
G6LIJ TS15 9JA
G4LIH TS15 9JG
G1FIZ TS15 9JQ
G4ZXT TS15 9JQ
G7RAJ TS15 9ND
G8KIK TS15 9NL
G8MBK TS15 9NL
G4JJM TS15 9NQ
G3VGZ TS15 9RZ
G4EEH TS15 9RZ
G3CBW TS16 0ER
G0JGE TS16 0JB
G7NQZ TS16 0JB
G0MLA TS16 0LN
G7HBH TS16 9AS
G6JOS TS16 9AZ
G8ILI TS16 9EH
G1KSS TS16 9EP
G3RUG TS16 9HP
M0AVW TS17 0LT
G1HSH TS17 0PA
G0MSX TS17 0QT
G6CKH TS17 8LX
G0MLB TS18 1PQ
G3EID TS18 3QQ
G4ZFU TS18 4DL
G0DOD TS18 4DR
G4UJE TS18 4HF
G4VEU TS18 4HU
G3WWG TS18 4JA
G4ANL TS18 4PW
G4MXH TS18 5AT
G4UVL TS18 5AT
G3NOG TS18 5DF
G3ASM TS18 5HD
G4JXR TS18 5LH
G4SUY TS18 5PX
G6NRY TS18 5QE
G0ECP TS19 0LU
G3UFV TS19 0RA
G6DJO TS19 0RW
G4SKX TS19 0UE
G4YOV TS19 0XH
G3LCZ TS19 7DJ
G0ELC TS19 7JL
G1ZRS TS19 7JT
G1HCO TS19 7JW
G4WZG TS19 7LA
G4RRW TS19 7PU
G3JZU TS19 7PW
G7VCO TS19 7QX
G1UMA TS19 7SN
G3XDI TS19 8AL
G8KSA TS19 8XF
G4YFD TS2 1SL
G4GGP TS20 1AZ
G3EKT TS20 1AZ
G3YVY TS20 1EW
G4YZQ TS20 1EW
G7CLG TS20 1JQ
G4GQG TS20 1LU
G7FFM TS20 1PZ
G0NAQ TS20 1QL
G0VGB TS20 1SF
G7CLH TS20 1SJ
G0OPH TS20 1SL
G0HAA TS20 1TE
G4ZWU TS20 1TJ
M1BTC TS20 2DN
G3KGF TS20 2EP
G7OSW TS20 2ET
G0BZC TS20 2EX
G8VFE TS20 2JJ
G0IIL TS20 2PP
G4LIM TS20 2SP
G4DXP TS21 1DZ
G8LTD TS21 1HT
G3NBL TS21 1LQ
G6HKE TS21 1PJ
G8ZIA TS21 3LT
G3KQN TS22 5EN
G4PVN TS23 1DL
G0HJM TS23 2AN
G4ABM TS23 2BQ
G6WWW TS23 2EA
G6LDJ TS23 2HX
G7NRO TS23 2PG
G7LOA TS23 2PJ
M0ASN TS23 2QB
G0ENP TS23 2QU
G0IFN TS23 2QU
G0CVD TS23 2RA
G4WZD TS23 2RA
M0BEB TS23 2RZ
G0EJX TS23 3BY
G8OKS TS23 3DJ
G0BKC TS23 3HN
G0AIH TS23 3HQ
G8ILB TS23 3QQ
G4HFH TS23 3RD
G0JON TS23 3SY
G4MYN TS23 3UA
2E1FEW TS23 3XJ
G1HVZ TS23 3XW
G2CZO TS8 0BJ

Stoke-on-Trent

G7LNX ST1 2EE
G1DMM ST1 3HD
G6ING ST1 3NE
G7ELB ST1 6BG
G6KKN ST1 6BY
G8NTG ST1 6DE
G6PAY ST1 6EN
G1MTC ST1 6QN
G4FMJ ST1 6SL
G1YQL ST2 0BX
G0TBF ST2 0QH
G4GRZ ST2 0QY
G1IFS ST2 7AR
G4BEM ST2 7AR
G4DPV ST2 7AR
G8KXM ST2 7AS
G4CKT ST2 7BZ
G6YGQ ST2 7EF
G4OSI ST2 7HJ
G1STO ST2 7JJ
M1BRE ST2 7JZ
G1SQC ST2 7NB
G3ISX ST2 7NF
M0ABZ ST2 7PF
G4DVA ST2 8AQ
G8JXC ST2 8BW
G4PGR ST2 8DE
G7VSW ST2 8DQ
G3ALP ST2 8NG
G4TTP ST2 9BT
G7OAE ST2 9HH
G0ODH ST2 9PU
G3ZBD ST3 1AT
G7SRN ST3 1JH
G0TTO ST3 1SJ
G7KDJ ST3 1SJ
G0TAC ST3 1SJ
G8PNZ ST3 2AU
G6OPU ST3 2DT
2E1ESJ ST3 2EG
G4AYY ST3 2EN
G0CWO ST3 2HA
G4DUC ST3 3HT
G7UAT ST3 3HT
G1VDY ST3 3LG
G7SRM ST3 3NR
G6TPI ST3 4NF
G0RSX ST3 5DD
G7EWX ST3 5DX
G3VTE ST3 5EG
G7NEE ST3 5JA
G1JII ST3 5JH
G6AYH ST3 5JL
G3XLC ST3 5NL
G0VZE ST3 5QW
G7VRR ST3 5RL
G7FSA ST3 5RP
G8NAI ST3 5RQ
G4WJX ST3 5ST
G6INM ST3 5TP
G1FHH ST3 5UD
G0SKM ST3 5UG
G7VKR ST3 6BH
G3TQS ST3 6EF
G0SMN ST3 6HA
G6THM ST3 6NS
G6THN ST3 6NS
G1DSE ST3 6PF
G4RQG ST3 6PQ
G1PAG ST3 6RG
G1UYT ST3 6RG
G0WLC ST3 7EN
G6YIW ST3 7GB
G4ZBW ST3 7HD
G0UPV ST3 7HN
G0NZH ST3 7RD
G6KTF ST3 7RU
G4XEE ST3 7UT
G6LQG ST3 7UT
G1WHP ST3 7YE
G8DZJ ST3 7YE
G6UDI ST4 1ED
G0NMY ST4 2EX
G1YPH ST4 3HH
G7JWO ST4 3HJ
G1EZJ ST4 4DY
G4CJM ST4 4PH
G0VKI ST4 4PL
G3UD ST4 4QP
G0JGB ST4 5AE
G7DHP ST4 5AH
G0JZS ST4 5AN
G1VDI ST4 5AQ
G6WFF ST4 5DW
G1OQW ST4 5EF
M1BFF ST4 5HA
G3RIC ST4 5HL
G7SXP ST4 5RL
G8BFO ST4 6EE
G1JIW ST4 6HY
G0JGF ST4 6LA
G0DTW ST4 6LD
M1BCC ST4 6LF
G4OYA ST4 6LL
G1ZNZ ST4 6PX
G1UCC ST4 7HE
G1XZA ST4 7HF
G1XZB ST4 7HF
G3COY ST4 7JJ
G4ATC ST4 7JJ
G7PFR ST4 8BT
M0ACA ST4 8DN
G7WHA ST4 8DN
G8SAR ST4 8EL
G6HBQ ST4 8HP
G3SLX ST4 8NJ
M1AJU ST4 8NJ
G7EVK ST4 8NW
G6TGY ST4 8QE
G0PXL ST4 8TF
G3XMK ST4 8TN
G4DUB ST4 8UY
G6AZP ST4 8XY
G4EJM ST4 8YD
G6VTX ST4 8YG
G8DYI ST4 8YQ
G1HMO ST6 1RP
G7CKX ST6 1RP
G7LKP ST6 1RX
G7LNZ ST6 1RX
G6HFK ST6 2JX
G0UNT ST6 2PD
G7SSW ST6 3BX
G0JHD ST6 4BE
G6CKG ST6 4BZ
G6DSG ST6 5LG
G1VVA ST6 5PJ
G1XLN ST6 5SJ
G1FLA ST6 6DX
G4WEP ST6 6DZ
G8WBP ST6 6HE
M1AUP ST6 6LP
G7UIE ST6 6NG
G3ZRQ ST6 6RA
G0UDO ST6 6RN
G7KLU ST6 6SW
G6ETI ST6 6TB
G1USH ST6 6TZ
G0MGJ ST6 6UR
G8MVV ST6 6XL
G7MSQ ST6 7AL
G6NWO ST6 7BS
G7LOB ST6 7DR
G1SAM ST6 7EU
G7DFQ ST6 7HR
G7EUF ST6 7JY
G3ICZ ST6 8BZ
G7HMB ST6 8DX
G3PIK ST6 8LZ
G1VYS ST6 8PL
G8OSJ ST7 4US

Stratford-on-Avon

G1YFD B49 5DD
G4GYI B49 5DD
G3LLS B49 5DY
G0KXT B49 5JB
G1VKA B49 5QE
G1MCX B49 6BW
M1AUB B49 6BW
G4STE B49 6JZ
G3EVT B49 6LF
G4ACZ B49 6LF
G1WLU B49 6LX
G8GAG B49 6NS
2E1FRN B49 6PE
G3FVV B49 6QN
M0AIZ B49 6QP
G4OHJ B50 4AN
G0VBZ B50 4AR
G3VRF B50 4AR
G8UCW B50 4DN
G7TEX B50 4EB
G6UHL B80 7BF
G4YJC B80 7HN
G7APL B80 7JJ
G3VXH B80 7NZ
G0UGY B80 7PD
G3BJS B80 7RD
G6FDO B80 7RD
G3GBS B94 5DP
G8NOF B94 5EB
G3LNS B94 5HL
G6HSD B94 5RZ
G0AST B94 5SJ
G0EPL B95 5BA
G4PIP B95 5NN
G3HCT B95 5NW
2E1FRC B95 6AB
G3UOC B95 6BH
G4CGR B95 6BH
G3XTI B95 6JA
G3KFS B98 9EH
G0WAZ B98 9EN
G0RMI CV23 8JA
G3IYB CV23 8LT
G3SNO CV23 8LX
G1KCS CV23 8NE
G3OJI CV23 8NN
G6EVC CV23 8NN
G1GPE CV23 8RB
G6NMA CV23 8RB
G0FKW CV23 8RR
G4NXA CV23 8RR
G8OKN CV31 1XH
G4NQN CV33 0AR
2E1CNQ CV33 0AR
2E1FML CV33 0BG
G3ZST CV33 0BN
G1NVG CV33 0DE
G4TIL CV33 0EQ
G4TJT CV33 0EQ
G6JZN CV33 0GA
G7EQM CV33 0GL
2E1ELH CV33 0HN
G6NMH CV33 0HS
G7GXX CV33 0HS
G4MYH CV33 0HY
G3TFA CV33 0NF
G4MVM CV33 0NS
G0GNF CV33 0QS
G7WKS CV33 0QS
G1WWK CV33 0RH
G4EWW CV33 0TL
G1VRP CV33 0TT
G8PLR CV33 0TY
G1EAN CV33 9HG
G6OMY CV33 9HX
G7PIO CV33 9JL
G4GEP CV33 9JQ
G3YJN CV33 9LN
G7VUX CV33 9LT
G4YMT CV33 9ND
G1FEO CV33 9TH
G6FEO CV33 9TH
M1ATN CV35 0HH
M0AYA CV35 0HH
G7PKZ CV35 0LD
G0MLE CV35 0PJ
G0JUQ CV35 0PQ
G7JUH CV35 0PT
G1YYP CV35 0RE
M0BBT CV35 0UE
G7OCV CV35 0UE
G3NWH CV35 8BN
G3OAY CV35 8BN
M0AWE CV35 8LH
G7VPG CV35 8PZ
G7VQS CV35 9AT
G8GCU CV35 9BG
G0SUX CV35 9ER
G3YKI CV35 9HP
G4YBX CV35 9TQ
G6BPK CV35 9TQ
G4KZS CV35 9TW
G6GZR CV36 4HH
G8AIQ CV36 4HQ
G0CXJ CV36 4PE
G8HGG CV36 5DA
G0CDO CV36 5DB
G0SDX CV36 5DB
G3III CV36 5EF
G1HPB CV37 0AP
G8JXP CV37 0DN
G2HCA CV37 0DU
G4RYM CV37 0PP
G6MMD CV37 0PP
G7EWR CV37 0RE
G0VKV CV37 0RE
G0STG CV37 0SE
G0TVC CV37 6DD
G0OGN CV37 6EP
G6HOB CV37 6GP
G3WDU CV37 6GR
G0REP CV37 6TD
G0JJY CV37 6TF
G0IPF CV37 6TL
2E1FAZ CV37 6UN
2E1BEF CV37 6UW
2E1BEG CV37 6UW
G7SIS CV37 6UW
G7SDT CV37 6UW
G0EDT CV37 6XG
G0MNJ CV37 6XG
G0CCJ CV37 7DD
G3OOQ CV37 7DE
G3YED CV37 7EW
M0BEL CV37 7JA
G8FRS CV37 7JS
G3MXH CV37 7LR
G3XZO CV37 7TL
G0IHM CV37 8HF
M0AFV CV37 8RA
2E1DMH CV37 8RA
G0CHO CV37 8TS
G3MGJ CV37 9AD
G4MMF CV37 9AN
2E1FLB CV37 9AU
G8HJS CV37 9BA
2E1EYG CV37 9FE
G6FIP CV37 9JU
G1CRD CV37 9PF
G7BRF CV37 9PN
G4ABS CV37 9PP
G0BKA CV37 9QL
G0BKB CV37 9QL
G8OCS CV37 9QW
G1TFY CV37 9SA
G4LRV CV37 9SD
M0AWA CV37 9SJ
G8RDT CV37 9ST
2E1FCC CV37 9ST
2E1FCD CV37 9ST
G3OJJ CV37 9XJ
G4AXW CV37 9XT
G3UXJ GL56 9SA
G0UWB OX15 5AP
G0OKO OX15 5HZ
G6SLG OX15 6DS
G1PJE OX17 1DH

Stroud

G7POI GL10 2PT
G4DC GL10 2PX
G0LJE GL10 2PZ
G7MWC GL10 2PZ
G8ND GL10 2QG
G6XKV GL10 2QH
2E1CZK GL10 2RD
G7VEK GL10 2RD
G0PDE GL10 3AU
G3VCV GL10 3HN
G4XWZ GL10 3HS
G4EXF GL10 3HX
G7NWU GL10 3LB
G4EDY GL10 3LD
G3TBF GL10 3LJ
G0RUY GL10 3NB
G4UBC GL10 3NL
G4SHB GL10 3QW

G4CRG	GL10 3RT
G4FRR	GL10 3SN
G4CIO	GL10 3TU
G7GKA	GL10 3WP
G4FGX	GL11 4AG
G6HKL	GL11 4AP
G4JXC	GL11 4EW
G6GLO	GL11 4NT
G1YXF	GL11 4NT
G7GVP	GL11 4PB
G4KYI	GL11 5DA
G4VZR	GL11 5EL
G8ZPF	GL11 5EL
G7AHT	GL11 5EW
G4IZS	GL11 5JF
G8ZTM	GL11 5JQ
G0ATX	GL11 5PQ
G1GDT	GL11 5SW
G0AFP	GL11 5US
G1VUE	GL11 5US
G4YIC	GL11 6DX
G4FQH	GL11 6HB
G0DQS	GL11 6JE
G0BRW	GL11 6LF
G4FOD	GL11 6LT
G8KOV	GL11 6PE
G4SPQ	GL11 6PF
G6DGA	GL11 6PN
G3JCW	GL12 7BE
G7ARQ	GL12 7BL
G0CBK	GL12 7EF
G3YBR	GL12 7EX
G4TRA	GL12 7HQ
G6CSW	GL12 7JF
G8WTA	GL12 7JF
G8GLX	GL12 7LH
G0SYF	GL12 7LQ
G3XSJ	GL12 7NG
G2DTD	GL12 7PZ
G7FEQ	GL12 7RH
G7JWE	GL12 8SG
G0SCB	GL13 9AF
G0NVX	GL13 9BU
G7FDG	GL13 9HS
G0RYV	GL13 9LE
G4NYM	GL13 9LR
M1ACM	GL13 9PL
G0UGR	GL13 9PY
G4TZZ	GL13 9TG
G0IWW	GL13 9TQ
G7FPU	GL13 9TQ
G4RLT	GL13 9US
G0IHC	GL13 9UT
G0OIS	GL2 4QD
G4UQJ	GL2 4QF
G8RMI	GL2 4QQ
G8IEW	GL2 7DF
G1WZO	GL2 7ED
G4LZQ	GL2 7ET
G1XAL	GL2 7LB
G3STZ	GL2 7LH
G0NRZ	GL4 8AL
G2DAD	GL4 8AL
G4MQB	GL4 8AL
G4OYU	GL4 8HU
G0FBX	GL4 8NH
G4UXN	GL5 1BW
G6SQT	GL5 1HS
G8JTQ	GL5 1NU
G7EUW	GL5 1PE
G3GYF	GL5 1PL
G8VLY	GL5 1PL
G0MZK	GL5 1RD
G4GHX	GL5 1RD
G4ITS	GL5 1RH
2E1AFN	GL5 1RU
M0ATX	GL5 1RU
M0ACW	GL5 1RU
G0OOF	GL5 1RU
G4PLF	GL5 1RU
G0NUN	GL5 1ST
G6UTC	GL5 1TT
G8PNM	GL5 1UN
G0FCO	GL5 1US
G3ELF	GL5 2AR
G7KCG	GL5 2BJ
G3IWA	GL5 2DF
M1AJO	GL5 2DG
G1NKF	GL5 2EA
G7URN	GL5 2PS
G0PKJ	GL5 2QF
G3REB	GL5 2RF
G8AER	GL5 2UG
G4YYR	GL5 3PQ
G3OCL	GL5 3QT
G4GXB	GL5 3QZ
G6LED	GL5 3TR
G4GWZ	GL5 3TS
G7JWQ	GL5 3TU
G4PLE	GL5 3UF
G4RJG	GL5 4DF
G0BUS	GL5 4DJ
G7HLD	GL5 4DQ
M1BFA	GL5 4PG
G0GNM	GL5 4PG
G0TNG	GL5 4PX
G0IUW	GL5 4SA
G8SGT	GL5 4ST
G3RRD	GL5 5AP
G3YNY	GL5 5BW
G7HUZ	GL5 5EL
G3ICA	GL5 5JS
G3YIE	GL5 5JY
G4RTR	GL5 5LL
G4ENA	GL5 5LN
G0FDE	GL6 0AP
G3WMQ	GL6 0DR
G0RRD	GL6 0EQ
G0PPM	GL6 0HE
G0DZM	GL6 0LD
G3ZTX	GL6 0NJ
G3MGF	GL6 0PE
G3OYX	GL6 0PU
G7KUU	GL6 6AD
G3YQV	GL6 6AJ
G0EAB	GL6 6BA
G4GXA	GL6 6BA
G0SUB	GL6 6BH
G3LHU	GL6 6BL
G3EKD	GL6 6HE
G8UVM	GL6 6NR
M0BAP	GL6 6PH
G4VLV	GL6 6QJ
G0VPT	GL6 6SA
G3OAD	GL6 6TJ
G4WPN	GL6 7DW
G3CEG	GL6 7EP
G4SJN	GL6 7NY
G3KYZ	GL6 8AA
G0EZA	GL6 8AU
G4UXM	GL6 8BU
G4KDQ	GL6 8BW
G1VDE	GL6 8DG
G6GSC	GL6 8DG
G3VBQ	GL6 8FB
G8VVY	GL6 8JY
G3NQF	GL6 8LZ
G3TEV	GL6 8ND
G4OHA	GL6 8NY
G8VG	GL6 9BY
G4BSM	GL6 9BZ
G7EXX	GL6 9BZ
G6CYO	GL6 9EP
G4IJV	GL6 9HB
G7PSQ	GL6 9LD
G0OVU	GL7 6LQ

St. Albans

2E1BDB	AL1 1DW
2E1BWA	AL1 1EA
G4LBF	AL1 1HY
G8HXT	AL1 1JL
G4ETK	AL1 1LT
G0NVZ	AL1 1UA
G3MFO	AL1 2DX
G6PFP	AL1 2NQ
G6BDW	AL1 2PG
G3SSQ	AL1 2QW
G8ZZS	AL1 3QZ
2E1FWU	AL1 4BG
2E1DUQ	AL1 4EW
G4IIS	AL1 4PN
G3XXF	AL1 4PZ
G3YCY	AL1 4RD
G6WAO	AL1 4UN
G0LQU	AL1 4XZ
G3LMP	AL1 4YA
G4BIX	AL1 5EX
G3UFB	AL1 5LD
G4DDV	AL1 5NJ
G4COL	AL1 5NS
G7TIJ	AL1 5QB
G4FTL	AL2 1AT
G0FVH	AL2 1TD
G1ARL	AL2 2AH
G1IEV	AL2 2AN
G8UCP	AL2 2QL
G7GJN	AL2 3DD
G1WVD	AL2 3PU
G4DPP	AL2 3PY
G8RRU	AL2 3QX
G0HTL	AL2 3SJ
G0KVA	AL2 3SJ
G4LLT	AL2 3SN
G1IHS	AL2 3SR
G6CKW	AL2 3ST
2E1BVN	AL2 3XH
G7SDG	AL2 3XN
G4WMB	AL3 4AS
G3ZJF	AL3 4JL
G4CBE	AL3 4LR
G3YGA	AL3 4LU
G3IDZ	AL3 4NA
G4HVG	AL3 4NJ
G0RWS	AL3 4TA
G0LAE	AL3 5AX
G3YHS	AL3 5AX
G6PWL	AL3 5LX
G8TVL	AL3 6DN
G3GEX	AL3 6JB
G3VJO	AL3 7EL
2E1AVU	AL3 7EW
2E1AWG	AL3 7EW
G3TXP	AL3 7HB
M0AQW	AL3 7HD
G0MVE	AL3 7HD
G1AXF	AL3 7HD
2E1EWN	AL3 7HD
G0EHO	AL3 7JB
G6USO	AL3 7PF
G0LZW	AL4 0DH
G8BNR	AL4 0DH
G8MEA	AL4 0DN
G4PVB	AL4 0DR
G6MGQ	AL4 0HB
G6POD	AL4 0PX
G6DYM	AL4 0QA
G1GJK	AL4 0SA
G8HZQ	AL4 0UP
G3UFF	AL4 0UY
G4XJS	AL4 0XA
G4WNP	AL4 8JD
G6AHE	AL4 8LW
G0HGO	AL4 8PR
G8AZV	AL4 8PR
M0ABY	AL4 8TP
G0XBC	AL4 8TP
G4ZRA	AL4 9AF
G2PA	AL4 9BZ
G5AL	AL4 9DJ
G6PWS	AL4 9EH
G7ODL	AL4 9EX
G7VNZ	AL4 9JF
G7ORK	AL4 9JQ
G4BEO	AL4 9JU
G6RZQ	AL4 9NN
G4CZA	AL4 9NZ
G0GJC	AL4 9PJ
G7PKH	AL4 9PW
G4BPR	AL4 9QG
G3LXP	AL4 9QS
G6BRL	AL4 9RP
G3XYH	AL4 9SJ
G4DJX	AL4 9TG
G3NKA	AL4 9TH
G4ZES	AL4 9XB
G0OMY	AL5 1AD
G4JOV	AL5 1AD
G3PLR	AL5 1BP
G8EPK	AL5 1BP
2E1BZB	AL5 1EF
G7EEU	AL5 1ER
G7HDR	AL5 1EZ
G4DOC	AL5 1HD
2E1CVJ	AL5 1JL
G3UHN	AL5 1JQ
G1XEH	AL5 1LL
G8CLJ	AL5 1RF
G3UXO	AL5 1RJ
G8KPF	AL5 1SD
G7EXM	AL5 1SN
G8CLY	AL5 1SR
2E1EZX	AL5 2AE
G0DFF	AL5 2AL
M0AYS	AL5 2HJ
2E1FWT	AL5 2JP
G0PMK	AL5 2JW
G4GEZ	AL5 2JW
2E1AQI	AL5 2NP
G4HKA	AL5 2PY
2E1EXB	AL5 2QH
2E1FTU	AL5 2QX
2E1CXW	AL5 2RQ
G1BRP	AL5 3AS
G4OGZ	AL5 3NX
2E1FIZ	AL5 3PP
2E1EYW	AL5 4AX
G1ZYJ	AL5 4BT
G0CPN	AL5 4DB
G3SBA	AL5 4HE
G4KCN	AL5 4HN
G6KHC	AL5 4JD
G3YWA	AL5 4LP
G4ZOW	AL5 4LZ
G3NND	AL5 4PG
2E1FXG	AL5 4PP
G3UTW	AL5 4QY
G6BQJ	AL5 4SE
G3JVO	AL5 4SE
2E1EIP	AL5 4XE
G7UTZ	AL5 4XE
G6IXT	AL5 5BL
G4OAV	AL5 5BT
G3MSW	AL5 5DW
G8XOB	AL5 5DW
G7JUB	AL5 5HE
G0FFG	AL5 5HS
G7KFQ	AL5 5HZ
G7RIH	AL5 5HZ
2E1FOX	AL5 5HZ
2E1EBF	AL5 5PT
G7EFL	AL5 5SJ
M1CBV	AL5 5ST
G0RVH	AL5 5SU
G3SDG	AL5 5SU
G6FRA	AL5 5SU
G8KGV	AL5 5UA

St. Edmundsbury

G8SVZ	CB8 8YG
G7VGZ	CB8 9YB
G7MVU	CB9 0AP
G4LYF	CB9 0DL
G8LHD	CB9 0EQ
G7EQG	CB9 0JR
G0CUV	CB9 0LL
G8CRN	CB9 0NN
G3JPM	CB9 7JH
G0DWA	CB9 7JN
G8SGK	CB9 7LS
G6MQN	CB9 7NA
G4KVU	CB9 7NN
G1BBA	CB9 8EG
G7DLI	CB9 8JW
G4HQD	CB9 9AU
G4LYB	CB9 9DF
G7SGB	CB9 9EJ
G4PZW	CB9 9JX
G8XOM	CB9 9NG
G7ADS	CO10 8AA
G1BWX	CO10 8HA
G3EQJ	CO10 8HS
G8CDG	CO10 8NN
G4BKP	IP22 2PN
G4KJF	IP22 2QE
G4AZJ	IP22 2RP
G3GIH	IP24 2QG
G6ULS	IP28 6EL
G3PFJ	IP28 6ES
G1FXG	IP28 6LD
G3TPM	IP28 6TN
G0FIU	IP28 6UA
G4PXC	IP28 6UG
G4XRK	IP28 6UG
G3MBL	IP29 4RF
G1XAM	IP29 5AD
G8KMM	IP29 5AD
2E1GBE	IP29 5AP
2E1GBF	IP29 5AP
G3ZKZ	IP29 5BS
G4CDP	IP29 5DD
G0WVE	IP29 5DD
G4ERF	IP29 5DX
G0DZY	IP29 5EW
G7WKC	IP29 5HE
G4VBS	IP29 5HR
G8CRM	IP29 5QL
G4CBN	IP29 5QP
G2FVX	IP29 5RD
G0JRM	IP29 5SE
G1UGJ	IP30 0ED
G4BSA	IP30 0EL
G7MLO	IP30 0TS
G7SRA	IP30 0TS
G0BXP	IP30 9JB
G7MYB	IP31 1EZ
G4UDD	IP31 1JJ
G6TJH	IP31 1NP
G0WZU	IP31 1NP
2E1ARW	IP31 2AP
G7OOG	IP31 2AP
2E1BPL	IP31 2AP
2E1DMA	IP31 2AP
G1UGK	IP31 2DX
G4ENS	IP31 2DX
G0MEV	IP31 2EE
G0MEZ	IP31 2EE
G0MXK	IP31 2EE
G1OUT	IP31 2EE
G4SBW	IP31 2EF
G4CIM	IP31 2HB
G1RPR	IP31 2HE
2E1FZV	IP31 2JE
G0HEV	IP31 2JL
G2TO	IP31 2JL
G6BSE	IP31 2JL
G8JSL	IP31 2LB
G3PHW	IP31 2PB
G4UCW	IP31 2PB
G0DUA	IP31 2RP
G6FXG	IP32 6AP
G6XSI	IP32 6PS
G6LPD	IP32 6PU
G6VPK	IP32 6QA
G1CFK	IP32 6RE
G7SDC	IP32 6RS
2E1FZH	IP32 6RS
G1FWA	IP32 7DD
G3IJH	IP32 7JJ
G3OWQ	IP33 1JH
G3TMX	IP33 1RF
G6NLR	IP33 2DP
G4CGV	IP33 2HF
G8XJK	IP33 2HZ
G0DVT	IP33 2JA
G6IGK	IP33 2JA
G7HMC	IP33 2JA
G3TQX	IP33 2LT
G6CYL	IP33 2LW
G8RJY	IP33 2LY
G0XEG	IP33 2LY
G8FRL	IP33 2QZ
G1TWX	IP33 2TE
G1UGH	IP33 3DU
G6PCC	IP33 3JP
G0BYY	IP33 3LT
M1AKN	IP33 3NX
G0ARU	IP33 3QF
G3IRM	IP33 3QF
G1VGI	IP33 3QJ
G6CZS	IP33 3QW
G6BJP	IP33 3XB
G4XSM	IP33 3XQ

St. Helens

G4WCO	L34 2RS
G6HIJ	L34 2RS
G1YCN	L35 0QR
G6FIT	L35 0QR
G3RJE	L35 4PA
G1XTJ	L35 4QF
G3WOH	L35 6PJ
G0SLK	L35 8NE
G7KZN	L35 8PE
G8XLI	WA10 1HY
G6OOY	WA10 2AY
G0KQI	WA10 2HA
G7MXP	WA10 3JH
M0AQK	WA10 3SN
M1BWS	WA10 3XT
G0VAM	WA10 4HJ
G1OSL	WA10 4LN
G4MWM	WA10 4NA
M0ACK	WA10 4RH
G1DKC	WA10 5JR
G8MHE	WA10 5PB
G7MMA	WA10 6DR
G6RQL	WA10 6JD
G7FGO	WA10 6LX
G0BCX	WA10 6QA
G4IFU	WA10 6RB
G1DDD	WA10 6SH
G4MWO	WA10 6SH
G7KEJ	WA10 6TH
G1NVE	WA10 6TP
G8CXZ	WA11 0BL
G1JTY	WA11 0DY
G1XAP	WA11 0DY
G0DZI	WA11 0ES
G1VGE	WA11 0HN
G4WGJ	WA11 0NB
G0AFJ	WA11 0PY
G1NZR	WA11 0QD
G4YKV	WA11 0RB
G0SSL	WA11 0RS
G4WCH	WA11 0SW
G4XIE	WA11 0UD
G0HBU	WA11 0YQ
G0END	WA11 7JQ
G4KIP	WA11 8AG
G3JIR	WA11 8AT
G1KIL	WA11 8BB
G4WTH	WA11 8BL
M1BOJ	WA11 8DF
G4FKP	WA11 8DJ
G4EST	WA11 8HU
G4UAE	WA11 8JJ
M0AQQ	WA11 8JW
G0BHH	WA11 8JY
G3CSC	WA11 8NR
G1ISJ	WA11 9AN
G4MYA	WA11 9BZ
G4DIY	WA11 9EL
G6YSN	WA11 9HU
M1AWB	WA11 9LD
2E1FIW	WA11 9LD
G0NEB	WA11 9RE
G0EAM	WA11 9RW
G1IOQ	WA12 0AZ
G0VVQ	WA12 0DR
G4KHG	WA12 0DU
G4XOL	WA12 0JT
G7PUN	WA12 0LN
G7TPL	WA12 0LQ
G1HIO	WA12 0LY
G1HIP	WA12 0LY
G6OBG	WA12 0NN
G3XRI	WA12 8HF
G1HSA	WA12 8JE
M1AJF	WA12 8LY
G0LNY	WA12 8NE
G7EKX	WA12 8PY
G6IKM	WA12 8RA
G6JPT	WA12 8SQ
G6TWL	WA12 9JN
2E1FSH	WA12 9LF
G6DWB	WA12 9LT
G7MFA	WA12 9LT
G1JPT	WA12 9NR
G8RII	WA12 9PW
G1MMA	WA12 9UD
G6III	WA12 9UG
G3YVH	WA12 9YG
G0KFE	WA9 2JT
G7OEB	WA9 2JX
G1WAE	WA9 2PP
2E1FBG	WA9 2PU
G8HLH	WA9 3NF
G1XXI	WA9 3QX
2E1BGY	WA9 3QX
M0AEI	WA9 3TU
G8STF	WA9 3XL
G1WDO	WA9 4AP
G1OMY	WA9 4BS
G3MWM	WA9 4JR
G7IFO	WA9 4JR
G0AWS	WA9 4LZ
G4TMU	WA9 4NZ
G8YAO	WA9 4NZ
G8OUI	WA9 4PW
G1ZNT	WA9 4RY
G4ROU	WA9 4XH
G0STH	WA9 5AR
G4WGB	WA9 5AR
G4KZQ	WA9 5LZ
G7JWU	WA9 5NP
G6NSM	WN4 0RG
G3ZUZ	WN4 0RH
G8CXW	WN4 0SJ
G0HBR	WN4 0SP
G0FZN	WN5 7HF
G4LON	WN5 7PX
G0OQS	WN5 7QL
G8LBJ	WN5 7QL
G4VFR	WN5 7TF

Suffolk Coastal

G4VBQ	IP10 0DU
G4SYG	IP10 0EU
G3BOK	IP10 0PA
2E1ALG	IP10 0PB
2E0ACE	IP10 0PF
G4LYD	IP10 0PF
G4RHR	IP10 0PF
2E1CZS	IP10 0PF
2E1AJE	IP10 0PP
G4DDK	IP10 0PP
G4GVW	IP10 0PP
2E1AVS	IP10 0PP
G0JVU	IP10 0TJ
G7ANH	IP10 0TJ
G8EAX	IP10 0TZ
G3OJ	IP10 0UU
G3XIX	IP10 0UU
G8MXV	IP10 0XL
G8JPA	IP10 0YN
G6DNJ	IP11 0YQ
G0BEU	IP11 7EG
G0VWE	IP11 7HA
2E1GBL	IP11 7JS
2E1EIY	IP11 7PA
G0FDY	IP11 7PG
G3YHK	IP11 7PG
M0BAQ	IP11 7PU
G3XYK	IP11 7QT
G7SMX	IP11 7RN
G3XKU	IP11 7RX
G4RW	IP11 7SA
G3ADZ	IP11 7SP
G6AYX	IP11 8HS
G0CFT	IP11 8NE
G1JRL	IP11 8NY
G8TMB	IP11 8NY
G8BBV	IP11 8PN
G7SOO	IP11 8PW
G7OPS	IP11 8UH
G4YQC	IP11 8UH
G4ZFR	IP11 8UH
G7TUG	IP11 8UL
G7MNS	IP11 8XZ
G3TRD	IP11 9AF
G6OQC	IP11 9AF
G6PXK	IP11 9AS
G1BAI	IP11 9BP
2E1GBK	IP11 9DG
G0KBM	IP11 9DX
G3XFF	IP11 9DX
M1BGL	IP11 9DY
2E1GBM	IP11 9HH
G4IVC	IP11 9HS
2E1FVB	IP11 9JJ
G4PM	IP11 9NQ
G3WDR	IP11 9NS
G4FBV	IP11 9PL
G8XRL	IP11 9PS
G3WYG	IP11 9RQ
G0UPD	IP11 9RZ
G8NXU	IP11 9SA
G0FJL	IP11 9SG
G6MCG	IP11 9SS
G7BNB	IP11 9TE
G1PAJ	IP11 9TF
M1BIU	IP11 9TJ
G8VZZ	IP12 1AH
M0ADM	IP12 1DJ
G7CIY	IP12 1HA
G7DWN	IP12 1HA
M1AGU	IP12 1HU
G8LQB	IP12 1JD
G0OZR	IP12 1JS
G8XYQ	IP12 1JT
M0BCT	IP12 1JY
G7LBJ	IP12 1JY
2E1FSR	IP12 1LB
G1AGK	IP12 1LD
G8HQV	IP12 1LF
G0DDZ	IP12 1LQ
2E1DYS	IP12 1LQ
2E1DYV	IP12 1LQ
2E1FSL	IP12 1LQ
G2HFD	IP12 1LW
M1CAU	IP12 1PE
G7JXR	IP12 1QR
G8CID	IP12 1QX
G8VCU	IP12 1RL
2E1EXY	IP12 1SE
2E1FSS	IP12 1SE
G7NMP	IP12 2EF
G4AVS	IP12 2HZ
G7RMQ	IP12 2JR
M1BKJ	IP12 2PJ
G3PAI	IP12 2QG
G4JWY	IP12 2QW
G0ILZ	IP12 3JY
G0NQU	IP12 3LA
G6DMR	IP12 3LA
M1AGP	IP12 3QL
2E1EWV	IP12 3QL
G1XWS	IP12 3QU
G4DFD	IP12 3QY
2E1FSJ	IP12 4HE
2E1FUI	IP12 4HE
G7OPO	IP12 4HP
G7OOW	IP12 4HR
G3TLY	IP12 4HW
2E1DYU	IP12 4JB
G3UYX	IP12 4JN
G4FSG	IP12 4JP
G4AKW	IP12 4JR
G8BRL	IP12 4JW
G0OZV	IP12 4LG
G8GHQ	IP12 4LL
G4ASB	IP12 4NR
G4FAW	IP12 4NS
G1DIK	IP12 4PT
G3WTS	IP12 4PT
M1BPB	IP12 4PY
2E1FIC	IP12 4PY
M1BPC	IP12 4PY
2E1FID	IP12 4PY
2E1FVD	IP12 4PY
G8ITX	IP12 4QL
2E1EXV	IP12 4QR
2E1EXW	IP12 4QR
2E1EXX	IP12 4QR
G1MOS	IP12 4SL
G4SYA	IP12 4SU
G7EHT	IP12 4TG
2E1DYT	IP12 4TJ
G4PUQ	IP13 0EH
2E1ALQ	IP13 0LN
G4HUP	IP13 0LN
2E0ADZ	IP13 0LN
G3RPB	IP13 0LR
G6KQ	IP13 0LR
2E1EJD	IP13 0ND
G8TNZ	IP13 0SS
G4CCN	IP13 6AX
2E1FJC	IP13 6RF
G0AOY	IP13 6SU
G3UWB	IP13 6TS
M1AGT	IP13 6TS
M0AHL	IP13 6TS
G4RSD	IP13 7PP
G3PYW	IP13 8LZ
2E1DCF	IP13 8NE
G7SQE	IP13 8NE
2E1EFD	IP13 8NE
G8IQF	IP13 9HA
G3LQR	IP13 9PQ
G0VDE	IP13 9PU
G3PHT	IP15 5EL
2E1GAT	IP15 5QH
G1YRF	IP16 4AR
G8NT	IP16 4BZ
G0UEA	IP16 4DT
2E0AKF	IP16 4DT
2E0AKE	IP16 4DT
G3MYA	IP16 4EX
G1MJV	IP16 4HU
G6FS	IP16 4JY
G0TUQ	IP16 4JY
G4INP	IP16 4JZ
G8AXO	IP16 4LE
2E1GAY	IP16 4LS
G8ACO	IP16 4PS
G0CFI	IP16 4QZ
M0BCS	IP17 1DQ
G0CJX	IP17 1JY
G3TWQ	IP17 1NP
G3IIX	IP17 1SA
G6IBZ	IP17 1XX
G0KDR	IP17 2AS
2E1DTH	IP17 2AU
2E1DTJ	IP17 2AU
2E1EKG	IP17 2DN
M1APD	IP17 2DN
G0FSP	IP17 2PH
G0AJC	IP17 3DD
G0BYK	IP18 6UJ
G0CFB	IP19 0PY
M0ARY	IP19 0PY
G7REM	IP19 0PY
G4IHI	IP19 9DY
G8EUE	IP19 9DZ
G0PVV	IP3 8RB
G4ZYZ	IP3 8TR
G0DID	IP3 8TZ
G4ROH	IP4 5RH
G4ETC	IP4 5UE
G8CJL	IP4 5UQ
G3PWB	IP4 5UU
G8VOK	IP4 5XD
2E1BEM	IP4 5XD
M1ADT	IP5 2GB
G1YLE	IP5 7AQ
G3WIU	IP5 7AY
G4AEY	IP5 7EB
G8XPC	IP5 7EB
G3UKW	IP5 7ED
G3XDY	IP5 7ED
G8VVR	IP5 7EL
G0EBQ	IP5 7FB
G7WCJ	IP5 7GJ
G4LSQ	IP5 7GW
G3SYJ	IP5 7PA
M1CAH	IP5 7PB
2E1FIG	IP5 7PB
G3NYK	IP5 7QR
G8HUE	IP5 7QZ
G4MRS	IP5 7RE
G4BAH	IP5 7SA
G4CXT	IP5 7SE
G6XWK	IP5 7SH
G8VQH	IP5 7SN
G3RVC	IP5 7ST
G3ZID	IP5 7SU
G4DWF	IP5 7SU
2E1BGG	IP5 7SU
2E1BIM	IP5 7SU
2E1FSM	IP5 7SU
2E1FSP	IP5 7SU
G4BTX	IP5 7SY
M1ATJ	IP5 7SY
G0HJW	IP5 7TA
G4XCT	IP5 7TD
G7ABP	IP5 7TH
G7ABQ	IP5 7TH
G6BNW	IP5 7TP
2E1FFD	IP5 7TZ
G4EQE	IP5 7UA
2E1FWD	IP5 7UF
2E1FTG	IP5 7UG
G4OPB	IP5 7UQ
2E0AML	IP5 7UQ
2E1EMJ	IP5 7UQ
G8FDC	IP5 7UQ
G4PBE	IP5 7UX
G0WTR	IP5 7XF
G0ORG	IP5 7XF
G4FNR	IP5 7YN
2E1FDW	IP5 7YR
G0BQO	IP5 7YT
G0TCP	IP6 9AR
G0DPP	IP6 9AU
G7MIP	IP6 9BJ
G3OKT	IP6 9DS
G1YRE	IP6 9HT
G3IRQ	IP6 9NG
G3RHP	IP6 9PD

Sunderland

M1ADK	DH4 4EF
G7CZV	DH4 4PF
G7BEE	DH4 4PQ
M0AVX	DH4 4PZ
G1FCY	DH4 4PZ
G8RCU	DH4 4XH
G0ABF	DH4 5DH
G3NMD	DH4 5DH
G8PJX	DH4 5NW
G0FRZ	DH4 6HT
G0FDA	DH4 6JG
G4VOK	DH4 6JG
G0FBL	DH4 7LP
G4OTF	DH4 7QT
G0OLO	DH4 7RD
G6VMF	DH5 0EF
G4RHC	DH5 0HT
G0UYF	DH5 8DD
G4WJV	DH5 8EX
G0CKI	DH5 8HR
G0TCF	DH5 8HR
G0CEK	DH5 9AW
M0AHT	DH5 9EL
G4SAL	DH5 9LX
G7GJV	DH5 9LX
G4HBR	DH5 9RR
G3AOT	NE37 1AQ
G4XYP	NE37 1EE
G4ZXR	NE37 1LP
G3VRO	NE37 1NR
G0KIJ	NE37 1QB
G0DHS	NE37 1QD
G4YEX	NE37 1QQ
G4YDM	NE37 2BT
G0YCA	NE37 2LZ
G4YGF	NE37 2NU
M1BVK	NE37 3AY
G0EPP	NE37 3LA
G3BEJ	NE38 0DE
G6BIM	NE38 0EP
G0UYX	NE38 0ES
M1BVJ	NE38 0NE
G1TWT	NE38 0NR
G0SLN	NE38 0PJ
G6ZIC	NE38 0QE
G4WNI	NE38 7AL
G8VYQ	NE38 7DB
G0NSP	NE38 7JN
G0OHK	NE38 7TA
G4RTJ	NE38 8BP
G0UQP	NE38 8DR
G1HCC	NE38 8DS
G8TFT	NE38 8PF
G7NHE	NE38 8PF
G0ITM	NE38 8RX
G3ZMO	NE38 9HE
G0PZW	NE38 9HS
G0SLQ	NE9 7PG
G1YLA	NE9 7SJ
G4LPK	SR1 3SQ
G4ZSK	SR2 0BH
G7KPC	SR2 0JY
G4GTX	SR2 7TS
G7PQD	SR2 8QB
G6LMR	SR2 8RS
G6INK	SR2 8RX
G6FZC	SR2 8SA
G0KFY	SR2 9BB
G7SWD	SR2 9DU
G0ASM	SR2 9DX
G4MSJ	SR2 9EE
G4TOI	SR2 9EJ
G3YJG	SR2 9HQ
G3XID	SR2 9SF
G4HPS	SR3 1AN
G4OBX	SR3 1AN
G0TAX	SR3 1JF
G3ZMG	SR3 1JQ
G7KOF	SR3 1LG
G4MTW	SR3 1LX
G3RDI	SR3 1NT
G1YHQ	SR3 1SG
G0BWJ	SR3 1UJ
G0VUS	SR3 3ES
2E1EKC	SR3 3JQ
G4MLF	SR3 3PG
G3ZOG	SR3 3PX
G4PIF	SR3 3RF
G8NLS	SR3 3SN
G0AOE	SR3 4AA
G4VLT	SR3 4EZ
G1YIZ	SR4 0AB
G4HIX	SR4 0AE
G7RIN	SR4 0DS
G7MXM	SR4 6XG
G1RLT	SR4 7NB
G0NIC	SR4 7NN
G1YUL	SR4 7SA
G1ZJU	SR4 7TD
G0EHX	SR4 8HE
G8HPW	SR4 8HT
G0CZP	SR4 8JT
G3WOM	SR4 8NP
G4WMW	SR4 9LS
G4EKM	SR4 9NQ
G0SRG	SR4 9NQ
G4DGB	SR4 9NU
G6FDB	SR4 9RQ
M1BTP	SR5 2EJ
G6EHO	SR5 2PA
G0IWG	SR5 3QW
G1CSA	SR5 4BU
G7PTT	SR5 4LX
G0UNE	SR5 5QA
G6BKT	SR5 5SS
G0WKZ	SR6 0AQ
G0BAN	SR6 0JP
G4TMX	SR6 0NT
G3ZTB	SR6 8AN
G4MRK	SR6 8ET
G4KKP	SR6 8EX
G8EQB	SR6 9HJ
G0IID	SR6 9HP
G7DDI	SR6 9HU
G0KNB	SR6 9NS
G0ARZ	SR6 9PF
G3EEQ	SR6 9PN

Surrey Heath

G3YTX	GU15 1BF
G6ZAM	GU15 1BT
G8GQS	GU15 1DE
G3HEJ	GU15 1DG
G4GRR	GU15 1DH
G1EVW	GU15 1DL
G6FYC	GU15 1DL
G1PTO	GU15 1EF
G1TTG	GU15 1EF
G3BEG	GU15 1EF
G8BVB	GU15 1EF
G7BCS	GU15 1EG
G0HIN	GU15 1EJ
G3OKU	GU15 1HP
G8SMG	GU15 1JG
G0RGX	GU15 1PS
G3ZTZ	GU15 1RB
G8EAD	GU15 1RU
G3RKK	GU15 2BH
G3SEP	GU15 2HN
G4REE	GU15 2LD
G4ELJ	GU15 2LT
G6LWA	GU15 2NR
G0WZX	GU15 2SD
G3VKI	GU15 2SP
G6ENU	GU15 2SP
G7SEJ	GU15 3EF
G4ZHZ	GU15 3EQ
M0AYT	GU15 3EU
G4GLP	GU15 3NP
G0VTQ	GU15 4BT
G7GKN	GU15 4HT
G4XYW	GU15 4QW
G6GS	GU15 4QW
G6UAW	GU15 4RU
G8PMQ	GU15 4TP
G0AZQ	GU15 4TR
M1AOL	GU15 4UE
G1VIG	GU15 4UG
G8EII	GU15 4UP
G4JQS	GU15 4XY
G8LMY	GU15 4YQ
G0RSV	GU15 4YY
G7LTW	GU16 5EH
G7WDD	GU16 5LJ
G7KNK	GU16 5LP
G3PPU	GU16 5LR
G8XNC	GU16 5NA
G3PSV	GU16 5QD
G3YAG	GU16 5QU
G3RRA	GU16 5RA
G4MEA	GU16 5RW
G3KFU	GU16 5SA
G6UIL	GU16 5SA
G3KOB	GU16 5ST
G4KJI	GU16 5TE
G4MBZ	GU16 5UZ
G1NAA	GU16 5XA
G4DJB	GU16 5XQ
G3TJI	GU16 5XX
G0VES	GU16 5YJ
G3VIR	GU16 5YJ
G4CDH	GU16 5YJ
G1RUR	GU16 6BS
G1OSO	GU16 6BZ
G6JDH	GU16 6DH
G6ZBY	GU16 6DN
G6BLU	GU16 6ER
G0FRS	GU16 6ET
G4BJQ	GU16 6ET
G7MUF	GU16 6HH
G4SYB	GU16 6JJ
G6MJB	GU16 6PH
G4CFN	GU16 6QH
G3TJS	GU18 5RZ
G0KDL	GU18 5TA
G1TLA	GU18 5TE
G8JMP	GU18 5TE
G3UKE	GU18 5TP
G8HXD	GU18 5TR
G4SUO	GU18 5UJ
G6CBU	GU19 5JR
G8RSB	GU19 5LY
G0HEE	GU19 5NF
G8MZD	GU19 5NU
G8WSP	GU20 6BU
G1ZRJ	GU20 6DB
G7JFU	GU20 6JP
G4DRK	GU20 6LH
G4FML	GU24 8AP
G3KMA	GU24 8AR
G3HTP	GU24 8LA
G7CSI	GU24 8PR
G0SXN	GU24 8PY
G0RVK	GU24 9DG
G6VLV	GU24 9ES
G8GJI	GU24 9HH
G8UDT	GU24 9LL
G0TAN	GU24 9NG
G4PIY	GU24 9NR
G0DER	GU24 9QE

Sutton

G3FP	CR0 4PS
G4FFX	CR0 4PU
G6UKB	CR0 4UB
G0TXL	CR4 4JX
G0UQO	CR8 3PB
G7IQZ	KT4 8AH
G4VLN	KT4 8EZ
G4BQY	KT4 8NB
G0BVV	KT4 8PH
G4PGE	KT4 8PJ
G7VBI	KT4 8PL
G7RBC	KT4 8SN
2E1CSH	KT4 8SR
M1BER	KT4 8UD
G8DIU	KT4 8UJ
G6VAX	KT4 8XG
G7VNM	KT4 8XU
G3OFF	SM1 1RD
G4HSD	SM1 2BL
G8PNQ	SM1 2JY
G0BWV	SM1 2PA
G2XP	SM1 2PA
G7SAC	SM1 2PA
G1POK	SM1 2QW
G7KIN	SM1 2SQ
G2BHQ	SM1 2TQ
G4KFZ	SM1 3BA
G0CPZ	SM1 3DF
G8GIJ	SM1 3JL
G4FFY	SM1 3LD
G4XOC	SM1 3PY
G3NHX	SM1 3QB
G8GLC	SM1 3SL
G1DTC	SM1 4AT
G0SWU	SM1 4BG
G6OWD	SM1 4DR
G3WYB	SM1 4EA
G7MTF	SM1 4JY
G0EXG	SM1 4LR
G3LQP	SM1 4RX
G3TPT	SM1 4RX
G4VET	SM1 4TH
G7CVK	SM1 4TJ
G7NNN	SM2 5BY
G1XKD	SM2 5ER
G1GBI	SM2 5HP
G8NFP	SM2 5HW
G8TQK	SM2 5HW
G1PJC	SM2 5JR
G0KBL	SM2 5LF
G1EAM	SM2 5NN
G0PUQ	SM2 5TA
G7EKA	SM2 6BA
G8NXB	SM2 6EQ
G0GZC	SM2 6RW
G1DQL	SM2 6RW
G3XTC	SM3 8ES
G3ORE	SM3 8HF
G0BLF	SM3 8LB
M1ABY	SM3 8NN
G0AXA	SM3 8QR
G4DSY	SM3 9EG
G4FMU	SM3 9EG
G7TKG	SM3 9EQ
G7NPY	SM3 9JP
G6SGM	SM3 9NE
G7RAD	SM3 9NJ
G8VCL	SM3 9NL
G7UZI	SM3 9QJ
G6XGV	SM3 9RF

G6XJN SM3 9RF
G6YOY SM3 9SD
G1HWK SM3 9SG
G7KEA SM3 9SQ
G4ZKE SM3 9SQ
G3TOM SM3 9SY
G8XAO SM3 9TH
G8CVQ SM3 9TZ
G0EAU SM3 9UB
G7DAC SM3 9UB
G6TPH SM3 9UJ
G1UBV SM3 9UP
G3IRP SM4 6LQ
G0BKD SM5 1AW
2E1CHK SM5 1AW
G6CAF SM5 1BH
G1PYL SM5 1ER
G6ZTM SM5 1QT
G6UCY SM5 2BY
G6CTV SM5 2HD
G1EGZ SM5 2PB
G7BDK SM5 3DZ
G4VAV SM5 3EJ
G7NKH SM5 3HA
G8LSE SM5 3HQ
G4FDN SM5 3NG
G8PAT SM5 3NG
2E1EPT SM5 3QH
G8WXP SM5 3QY
G4FQQ SM5 3RS
G0JNU SM5 4AG
G4PLH SM5 4PD
G4BVG SM5 4QH
G7ITZ SM6 0BW
G7CRQ SM6 0PB
G1KGA SM6 0PZ
G6XZP SM6 0QY
G3BMQ SM6 0TN
G4XSA SM6 7AG
G0LMG SM6 7EN
G7NHY SM6 7JU
G4YZL SM6 7LY
G4ZSR SM6 7LY
G3RJW SM6 8AE
G3SRC SM6 8AE
G4CCY SM6 8AE
G4DDY SM6 8AE
G6SAQ SM6 8EP
G8ZOY SM6 8EP
G7MQZ SM6 8NA
G6ZLD SM6 8PT
G6BME SM6 8RW
G3JCL SM6 9NH
2E1EPH SM6 9QG
G3OFZ SM7 1PW
G3TCZ SM7 1PW

Swale

G2DLX CT2 9JX
G0VTV ME10 1BB
G3ISD ME10 1BY
G8LDJ ME10 1EH
G4GJA ME10 1JA
G8LLU ME10 1JA
G0JUW ME10 1JU
G1REU ME10 1JY
G3EHW ME10 1LB
G0UAS ME10 1QT
G3MXA ME10 1QY
G3VPA ME10 1RD
G4TWC ME10 1RF
G6DUI ME10 1TS
G6SZG ME10 1UD
G3GET ME10 1XF
G1BOX ME10 1YL
G8ONY ME10 1YL
G0UKJ ME10 2DG
G4IBH ME10 2EY
G0JBA ME10 2LQ
G1YTO ME10 2LR
G7RUC ME10 2LS
M0AZP ME10 2LZ
G1CQX ME10 2LZ
M1CAT ME10 2NB
G7MHZ ME10 2NB
G3AFV ME10 2QE
G1LZW ME10 2SR
G7OQA ME10 2SR
G4MAD ME10 3AJ
G3ZAV ME10 3AS
G7MEE ME10 3AT
2E1DHY ME10 3AT
G7EIA ME10 3BH
G0AXQ ME10 3BP
2E1AUN ME10 3BP
G7FXH ME10 3DA
G1VJN ME10 3NR
2E1CLC ME10 3QY
G4DXI ME10 4PS
G4LBS ME10 4PS
G0DRQ ME10 4QD
G1DLJ ME10 4QD
2E1AUA ME10 4QD
G4SRC ME10 4QE
G4VEC ME10 4QE
G6SRC ME10 4QE
G7BWV ME10 4RX
G0RNQ ME10 4SL
G1YTC ME10 4TP
G7ENQ ME10 4TP
G8DPW ME11 5AG
G1JCL ME11 5AW
G6ILN ME11 5BU
G1AFW ME12 2BS
G0PEH ME12 2DJ
G8WQM ME12 2DJ
G0JYH ME12 2HG
G7GAZ ME12 2HG
G0DRT ME12 2NH
G7NJB ME12 2NL
G4YVP ME12 2QE
G4KSU ME12 2QS
G2HKU ME12 2RP
G7LZY ME12 3BB
M1ARU ME12 3BG
G8XGB ME12 3DH
M0AKA ME12 3EL
G7LLC ME12 3HU
G3GDH ME12 3HX
G6NUI ME12 3JX
G1JQH ME12 3LH
G7UMH ME12 3LH
G3AGR ME12 3LN
G3ZYQ ME12 3LR
G7ECQ ME12 3QR
G7IAU ME12 3QW
G3OXH ME12 4JU
2E1FVZ ME12 4RL
G4AVP ME13 0LA
G4ZKY ME13 7BG
G6AQV ME13 7DW
G3FCT ME13 7EJ
G4VXB ME13 7JZ
G1HDK ME13 8DZ
G3FUN ME13 8EH
G1EUQ ME13 8HP
G8LUK ME13 8PS
G4YGP ME13 9RS
G1GCZ ME13 9RS
G4AWG ME13 9SE
G4MKG ME9 0PL
G4TQS ME9 7AG
G6PGT ME9 7AG
G3UIB ME9 7EX
G4IOG ME9 7JU
G6HWT ME9 7NR
G3PSS ME9 7NU
G4LUO ME9 7QA
G6TCW ME9 7QA
G3UKL ME9 7SY
G4IYA ME9 8RY
G6OFT ME9 9AH
G0VVZ ME9 9DU
G0OZH ME9 9HU
G4MPA ME9 9HU
G0FKL ME9 9JE
G6SIQ ME9 9LF
G0ISL ME9 9NL
G6HOS ME9 9SP
G0RBR TN27 0HE

Swansea

GW4OXB SA1 2HF
GW0JFQ SA1 3QZ
GW7RLS SA1 4PH
GW7SKC SA1 4PH
GW1XBG SA1 6HN
GW0SAJ SA1 6HZ
MW1AUV SA1 6NH
GW4PHT SA1 6PD
GW4YPO SA1 6PJ
GW0ULX SA1 6TN
GW4VBV SA1 7AD
GW4WFM SA1 7EF
GW4UVT SA1 7HP
GW4YID SA1 7JY
GW0DST SA1 7LG
GW6MLL SA1 8EJ
GW4HDB SA1 8HQ
GW0ALR SA1 8NE
GW0INN SA2 0BP
GW0OLN SA2 0DN
GW1WGR SA2 0FZ
GW3OMN SA2 0FZ
GW3TYI SA2 0QE
G7NVM SA2 7AU
GW4IFE SA2 7AX
GW6ZWH SA2 7BY
GW1FBL SA2 7DF
2W1ETN SA2 7DJ
GW4BNJ SA2 7EH
GW6UFH SA2 7EQ
GW3RGL SA2 7HH
GW6JFM SA2 7PD
GW8VFQ SA2 7PQ
GW2FYV SA2 7QQ
GW8HDH SA2 7RP
GW3VPL SA2 7TS
GW1SVG SA2 7UH
GW8TVX SA2 7UH
GW3XIS SA2 8BE
GW0SGG SA2 8HE
GW4IOI SA2 8HJ
GW6AAG SA2 8LL
GW3INW SA2 8LR
GW3TSQ SA2 8LT
GW3KGI SA2 8LX
GW3UWS SA2 8LX
GW6MMM SA2 8LX
GW6WDS SA2 8PP
GW0UIZ SA2 8RE
GW4PNZ SA2 9BP
GW4ADL SA2 9BW
GW4JGU SA2 9DT
GW4SMG SA2 9DT
GW4UNV SA2 9DT
G4JUC SA2 9GR
GW3ARP SA2 9HA
GW3TMS SA2 9HY
GW4NFJ SA2 9JW
GW3WEZ SA2 9LY
GW4HAT SA2 9LY
GW4KTT SA3 1AE
GW0FYO SA3 1AZ
GW4AZI SA3 1BA
GW6OTD SA3 1BR
GW8VGB SA3 1LB
GW4SLS SA3 2BR
GW3ZCF SA3 3BA
GW6AYM SA3 3JJ
GW0VRL SA3 3JR
GW1AXU SA3 3JW
G4RKX SA3 3JW
GW0BNO SA3 3LA
GW8RDI SA3 4EF
GW4EPF SA3 4HF
GW1WWE SA3 4PD
GW3SRG SA3 4PD
GW3YGH SA3 4PD
GW0KWA SA3 4QW
GW6KQC SA3 4RE
GW1AUT SA3 4SA
GW3GLY SA3 4ST
GW6WEU SA3 4SW
GW0GJD SA3 4UB
GW4PEX SA3 5AN
GW6VKY SA3 5AZ
GW6JGE SA3 5BU
GW8SRW SA3 5DS
GW4TGA SA3 5HB
MW0AJH SA3 5HT
GW4LDP SA3 5NL
GW4HSH SA3 5NQ
GW3SIY SA3 5PE
GW4CC SA3 5PE
GW4EVL SA3 5PR
GW1PFK SA3 5QJ
GW4TGF SA3 5QJ
GW7JVS SA3 5QL
GW0RHC SA3 5QL
GW3YGM SA3 5TQ
GW1GJS SA4 1BT
GW3DKE SA4 1EW
GW3XYW SA4 1HU
2W0APT SA4 1HX
2W1EST SA4 1HX
GW0VEW SA4 1HX
GW4JDZ SA4 1JF
GW7FGL SA4 1LQ
G7UXY SA4 1TF
GW0NFN SA4 1UG
2W1BPJ SA4 1UG
GW3MMT SA4 3AR
GW4TOU SA4 3DN
GW4XES SA4 3DN
GW4ZBU SA4 3DN
M1AAH SA4 3HB
GW8TYS SA4 3JQ
GW7TGZ SA4 3LJ
GW1PZZ SA4 3PG
GW4UCK SA4 3PU
GW7MMH SA4 3PU
GW3HOJ SA4 3SG
GW4JQQ SA4 3TY
GW8SZC SA4 3TY
GW6WRP SA4 3YX
GW0EHN SA4 4DH
GW0SWD SA4 4DN
GW4TES SA4 4FP
GW3JBZ SA4 4GE
GW0DIQ SA4 4GZ
GW0DRI SA4 4GZ
GW1TIU SA4 4LR
GW7LMW SA4 4UX
GW0BLZ SA4 4WJ
GW0TWS SA4 4XQ
GW4JPC SA4 4XZ
G7PQS SA4 4YG
GW8PNR SA4 6BB
MW1BJD SA4 6QF
GW3EHN SA4 6QN
GW0BBG SA4 6QY
GW0JKB SA4 6SZ
GW3IHN SA4 6UA
GW7SDE SA4 6UH
GW4UNY SA5 4BL
GW4WFP SA5 4BL
GW6VMB SA5 4BL
GW4NAW SA5 4DH
GW0BMI SA5 4TP
GW0NKI SA5 7DJ
GW4MII SA5 7DW
2W1FBX SA5 7HY
GW7ODP SA5 8BR
GW0NLY SA5 8DW
GW0DGO SA5 8EH
GW0BUA SA5 8LN
GW4KUS SA5 8QN
GW4URB SA5 9BS
GW0AZW SA5 9DG
GW4MTD SA5 9HN
GW0NCU SA5 9NH
GW7TZI SA5 9NP
GW8NXK SA5 9PG
GW0COU SA6 5AX
GW7MMG SA6 5DX
GW4SPL SA6 5HF
GW4EVJ SA6 5JP
GW7RQW SA6 5LB
G0VOQ SA6 5PL
GW0DXZ SA6 5RN
GW4ZRK SA6 5RN
GW4TFX SA6 6AS
GW3HDR SA6 6AT
GW0UIZ SA6 6BA
GW1NWF SA6 6BY
GW6ZUS SA6 6BY
2W1CNN SA6 6BY
2W1EYZ SA6 6BY
GW4HZH SA6 6DA
GW4ZQY SA6 6ER
GW6FXB SA6 6HH
GW1DTA SA6 6LU
GW4SKC SA6 6NA
GW0MVS SA6 6QB
GW0LDQ SA6 6QD
GW0KYY SA6 6SW
2E1FGR SA6 6TB
GW4HOQ SA6 7AG
GW1WTZ SA6 7DU
2W1GAC SA6 7LN
GW0TXS SA6 7NU
GW3NUO SA6 7PB
GW1EOI SA6 7PN
GW4MVY SA6 7PQ
GW4WPA SA6 7PS
GW0CKX SA6 8HU
GW6TYJ SA7 9HF
GW4LKS SA7 9HS
G6VPZ SA7 9NR
GW4JGW SA7 9QX
2W1COT SA7 9QY
GW4RXO SA7 9RH
GW0TCN SA7 9RY
GW8DUP SA7 9SX
GW7FRD SA7 9XB

Swindon

G1FTV SN1 1DT
G3SNV SN1 2BL
G1ZGT SN1 2EP
G7WIC SN1 2JU
G3YYK SN1 3JJ
G0TLI SN1 3NL
G3YKC SN1 4DP
G4DWM SN1 4EB
2E1FJV SN1 4ED
G1KMN SN1 4EY
G3WSJ SN1 4JX
G0VTA SN2 1AU
G7UCE SN2 1AU
G0WDQ SN2 1AU
2E1EQI SN2 1AU
G4SQR SN2 1AX
G0HBG SN2 1DG
G7ROQ SN2 1DY
G4ZAZ SN2 1NJ
G2BUJ SN2 1PS
G7FEA SN2 1RX
G7PVM SN2 2AQ
G7NUR SN2 2EN
G7VJG SN2 2HG
G3KXU SN2 2HS
G3IHR SN2 2JG
M1BTU SN2 2LR
G6OMH SN2 2PE
G0HOX SN2 2SF
G4ZAM SN2 3BT
2E1FSO SN2 3DY
M1AOU SN2 3LZ
G3KEU SN2 3PA
G8ELH SN2 3PP
2E1FSN SN2 3PU
G6ARO SN2 3QB
G4CST SN2 4AA
G3GYQ SN2 4AQ
G8NTS SN2 4AR
G1WYP SN2 4AZ
G0UNZ SN2 4BB
G4HGV SN2 4DE
G8LYG SN2 4DE
G4FQX SN2 4DP
G4SHK SN2 4EA
G7DFE SN2 5BU
G0VCV SN2 5DE
M0BFA SN2 5JN
G7GRA SN2 5LL
G7HPI SN2 6BT
G4MNB SN2 6JN
G7PYV SN2 6LJ
G0JTD SN2 6LL
M0ACM SN2 6LP
G8SRC SN2 6LP
G7FLK SN2 6LW
G4MDT SN2 6QA
G0KUS SN2 6TH
G6VDK SN3 1AZ
G7MGY SN3 1BX
G7CRM SN3 1BZ
G0FCH SN3 1DJ
G8VJY SN3 1DR
G3LWX SN3 1DZ
G8IWO SN3 1ER
G7MTX SN3 1HZ
G3FLB SN3 1JA
G4IYW SN3 1LH
G8FWK SN3 1NB
G0EET SN3 1NH
G7KIW SN3 1NJ
G7TAF SN3 1PY
G8PIL SN3 1RY
G6BTR SN3 2AH
M1ACC SN3 2DR
G1EJB SN3 2EU
G3NPM SN3 3NB
G4ZYH SN3 3PJ
G7III SN3 3TD
G3LLZ SN3 4AA
G3SIR SN3 4AY
G4XIB SN3 4BW
G7VMD SN3 4BX
G0JOK SN3 4DX
G4HRV SN3 4DZ
G6ALS SN3 4EF
G0BQK SN3 4HB
M1CBS SN3 4HH
G3HSV SN3 4HU
G3MAI SN3 4LA
G3IDW SN3 4NF
G4BPO SN3 4RD
G7BPO SN3 4RD
M0BCG SN3 4SN
G8ETI SN3 5AG
G3WEF SN3 5AU
G4LDL SN3 5AX
G8KWD SN3 5AX
G0RRG SN3 5BA
G8XNG SN3 5BA
G0CPA SN3 5BN
G0SNM SN3 5DE
G0TLS SN3 5DR
G0RNH SN3 5DY
G4XUT SN3 6DU
G4JVJ SN3 6HG
G6NYC SN3 6JB
G1MOE SN3 6LA
G1YGY SN3 6LA
G8JEG SN3 6LH
G8UKY SN3 6LS
G4ENR SN3 6NF
G7AIL SN3 6NJ
G4AQK SN3 6NL
G0DJR SN4 0AJ
G3BGM SN4 0AR
G0WGI SN4 0ET
G4NHD SN4 0LR
G4WQG SN4 0LU
G3IRA SN4 0LW
G6ZCH SN4 0PH
G7KGM SN4 0PW
M1CDJ SN4 0RL
G3JOT SN4 9AR
G0HOJ SN4 9AT
G8IYD SN4 9AU
G7VHG SN4 9EN
G0DVB SN4 9HY
G3AIU SN4 9LA
G1MUM SN5 5BH
G0BCT SN5 5BJ
G8IYJ SN5 6DB
G6PCA SN5 6NE
G7EAK SN5 7AE
G1VRJ SN5 7AF
G8SGW SN5 7BB
G3KUD SN5 7BG
G1TOL SN5 7BL
G4XDC SN5 7BU
G4GXW SN5 7BX
G1GUI SN5 7EJ
G0IER SN5 8AG
G0BZJ SN5 8AJ
G0VQW SN5 8DG
G6LBR SN5 8PU
G4SOA SN5 9FG
G6KAW SN5 9FG
G8CPA SN5 9PH
G4OED SN5 9QT
G0HCO SN5 9QU
G0RHH SN5 9QU
G3KWH SN5 9RN
G8EFK SN5 9SA
G6UZM SN5 9SH
G8NVQ SN5 9SX
G4DSF SN5 9TL
G3OMA SN5 9TS
G4DOA SN5 9UQ
G4AJA SN6 7BZ
M0ATC SN6 7BZ
G4GDR SN6 7EB
M1ATC SN6 7HN
G4LZZ SN6 7HR
G0NPN SN6 7LA
G0NUH SN6 7NE
G6MVP SN6 7NU
G4BRK SN6 7NZ
G8JHC SN6 7PD
G8JRF SN6 7PG

Tameside

G0ROW M34 2DQ
G3HMF M34 2GF
G7SPH M34 2HG
G8OKD M34 2HZ
G0WYP M34 2NT
G3DXS M34 2PU
G6TXR M34 3GL
G7GXR M34 3GL
G7TYB M34 3NS
G3XVG M34 3NX
G6TCV M34 3RQ
G0BOH M34 3TE
G4WGR M34 3TH
G8AKN M34 5AZ
G8HTN M34 5BD
2E1FMP M34 5GH
2E1FXT M34 5HE
G1JDT M34 5LJ
G7PJN M34 5LL
G6FBV M34 5NS
G0NHO M34 5QX
G6ZDT M34 5RX
G0THL M34 5SZ
G6DVH M34 5TE
G1BBK M34 6AF
G6NVU M34 6EA
G7HRZ M34 6EA
G4KKE M34 6FG
G3HZM M34 6FG
G7DOY M34 6FT
G0YRT M34 6JY
G1EJK M34 6LR
G1UTS M34 7TX
G1WPL M43 6DS
G8CZE M43 6EF
G0RJA M43 6FH
G4HBI M43 6FS
G3HIA M43 6HJ
G6KRS M43 6HL
G0WBK M43 6HU
G1JAA M43 6QB
G3ELY M43 7HQ
G0TKW M43 7HT
G0MRD M43 7HX
G0NHS M43 7JJ
G1YCM M43 7JL
G8OTZ M43 7JU
G8OXX M43 7QY
G4AHF M43 7WL
G3RII M43 7XA
G4ING OL5 0AG
G4HXV OL5 0BS
G1DVT OL5 0PG
G7MVO OL5 0PG
G3RKM OL5 0SL
G3TXU OL5 9EA
G0FCX OL5 9JN
G8GVZ OL5 9LQ
G3WGQ OL6 6SD
G0NMW OL6 7DU
G6IML OL6 8BU
G4EAP OL6 8DS
G0JFV OL6 8DS
G0AQS OL6 8QR
G7MRO OL6 8RJ
G6BRD OL6 8RP
G4IFI OL6 8XW
G4MKT OL6 8XW
G3BPQ OL6 9BZ
2E1FAP OL6 9DJ
G7DGC OL6 9DU
G3FTP OL6 9EW
G0KYB OL6 9PZ
G7SEV OL6 9QY
G0UPA OL6 9QY
G1GWJ OL7 0DU
G7NHC OL7 0DU
G7JTL OL7 0DY
G3DUI OL7 0HD
G3HOX OL7 0HD
G4GHB OL7 0JD
M0AAD OL7 9NL
G8INC SK14 1DT
G4HLA SK14 3BW
G0RGN SK14 4BS
G6JLP SK14 4DT
G0RST SK14 4HU
G6PZS SK14 4RT
G0DOH SK14 4SB
G6CKN SK14 5AN
G8LCS SK14 5BB
G4USX SK14 5DD
G3AEF SK14 5EG
G6XYW SK14 5EG
G0FQZ SK14 5EX
G3CEL SK14 5JA
G4PYQ SK14 5JX
G1KAX SK14 5LB
G0WVV SK14 5NS
G0WWF SK14 5QA
G4LIG SK14 6BJ
G4KQM SK14 6JP
G4EZF SK14 6LE
G7JGF SK14 8HT
G0NPK SK14 8LE
G6SLY SK14 8QH
G0PVR SK15 1BL
G4YMN SK15 1DJ
G7FEG SK15 1HD
G0NIL SK15 1HG
G0WCP SK15 1LU
G8YIG SK15 1US
G4SEO SK15 2QJ
G6LBO SK15 2QQ
2E1EKL SK15 2SG
G0VJZ SK15 2SL
G6YSZ SK15 2TH
G4NHW SK15 2TR
G7TSQ SK15 2UE
G1TYP SK15 3DZ
G0TIX SK15 3HJ
G0ITZ SK15 3HU
G0ONE SK16 4AU
G4NPI SK16 4AY
G7MBI SK16 4EB
G1BSY SK16 4EW
G7BCK SK16 4JJ
GW7JUJ SK16 4LP
G3LTS SK16 4QE
G0HRQ SK16 4XA
G0NAJ SK16 5DN
M0AHF SK16 5DP
2E0AOO SK16 5DY
2E1BUX SK16 5DY
2E1BVS SK16 5DY
2E1BVV SK16 5DY
2E1CWP SK16 5HR
2E1CVM SK16 5HR
2E1CVU SK16 5HR
2E1CWX SK16 5HR
G0XAU SK16 5HR
G4JMW SK16 5HS
G4LTM SK16 5HS
G4SDH SK16 5HT
G1HBE SK16 5JL
G1HTF SK16 5JL
G8DEW SK16 5LB
G0LOE SK16 5NW
G0CKM SK16 5RT

Tamworth

G0WKI B77 1BT
G6EOO B77 1BY
G7KHN B77 1DB
G4MFN B77 1JD
G7UKF B77 1LY
G1AUX B77 1NJ
G6PTZ B77 1NQ
G1MGN B77 1PE
G3YTT B77 2DZ
2E1FFF B77 2HH
G3YUX B77 2HQ
G0PFQ B77 2NA
G4JBX B77 2QD
G8MKT B77 2RH
2E1FFX B77 3BE
G4NWM B77 3BT
G4NWN B77 3BT
G4LZV B77 3BT
G4HZG B77 3BT
G0WRM B77 3BT
2E1FES B77 3ES
G4NWO B77 3JH
G4SRI B77 3JQ
G4ZGS B77 3JU
G7SXW B77 3JW
G8RFL B77 3JZ
G6LPX B77 3LA
G0LHR B77 4BZ
G6NHG B77 4DL
G4VPP B77 4EJ
G6ZCI B77 4EP
2E1END B77 4ES
G4OCH B77 4ET
G0IMN B77 4HA
G0KFF B77 4HP
G0LTR B77 4HP
G0TRB B77 4JL
G7LYQ B77 4LA
G7EKD B77 4NG
G4VWQ B77 4PG
G0VTU B77 5BB
G0HAY B77 5LD
G1GAW B77 5PL
G8VPH B77 5PQ
G6MOD B77 5QF
G6ZLV B77 5QF
G6FBH B77 5TD
G4INA B78 3XE
G4SBS B78 3YA
2E1FKV B79 7JE
2E1FPS B79 7JE
G6IYR B79 7SZ
G6IYS B79 7SZ
G6KWC B79 7UB
G8KRV B79 8DE
G8NCK B79 8EJ
G7SED B79 8EW
G0MWI B79 8HL
G3PVR B79 8LD
G7WFK B79 8QB
G0VUC B79 8SA
G4UPH B79 8TJ
G0HYR B79 8TQ
G4FWC B79 8TQ
G6MER B79 8TQ
G8III B79 8TQ
G8TRS B79 8TQ

Tandridge

G7VFN CR3 0AJ
G7VFX CR3 0AJ
G8LFC CR3 0EH
G1BQJ CR3 5AZ
G6LJ CR3 5DD
G3CJJ CR3 5DE
G4APL CR3 5EL
G7BSF CR3 5EL
G0SCR CR3 5EL
G6LTK CR3 5JN
G4DTC CR3 5JP
G3VN CR3 5LE
G4LXR CR3 5LP
G7DAY CR3 5NY
G7ETY CR3 5PP
G7BWE CR3 5RB
G8DTQ CR3 5RT
G0SYR CR3 5RT
G0IEC CR3 5SG
G4DJR CR3 5TF
G8LMI CR3 5UG
G2RD CR3 6AP
G6CDW CR3 6DQ
G7NBM CR3 6DX
G4BBW CR3 6HE
G4IWE CR3 6HF
G6YUB CR3 6NH
G1RKR CR3 6NP
G5BZ CR3 7BB
G8CPB CR3 7DL
G3ODY CR3 7LJ
G1CKB CR6 9BT
G8HDP CR6 9BT
G3POZ CR6 9HP
G7MDV CR6 9JF
G6YOG CR6 9JQ
G0JGO CR6 9LB
G3DFD CR6 9LL
G6NPJ RH10 3HX
G3DUN RH6 9NQ
G3PWW RH6 9PA
G3PFT RH6 9PP
G0HUF RH6 9QP
G4RTD RH6 9RP
G3OYU RH7 6DG
G7APO RH7 6JR
GI7UHJ RH7 6JY
G0MHD RH7 6JY
2E1BGS RH7 6JY
G3LHZ RH7 6LL
G7TWJ RH7 6NQ
G4TXC RH7 6PP
G3IKB RH8 0JE
G1DQS RH8 0LS
G7TXT RH8 0LS
G6OTB RH8 0LZ
G3BST RH8 0RG
2E0ALW RH8 9AU
G8CNN RH8 9PF
G1NTW RH9 8BP
G3NEI RH9 8BT
G0ALE RH9 8ER
G3RQZ RH9 8ER
G8LSN RH9 8HB
G7KXY RH9 8HD
G0FTU RH9 8JW
G6JIY TN16 2BQ
G4NSD TN16 2BT
G1FKP TN16 2LA

Taunton Deane

G4XUR TA1 1LB
M0AON TA1 2DS
G4NQI TA1 2EY
M0BCL TA1 2HF
G7UYA TA1 2HF
G1NZK TA1 2JQ
G0UOG TA1 2JQ
G7CCV TA1 2NA
G4ASK TA1 2QJ
G6EIO TA1 2QT
G0FNF TA1 2RP
G0PNB TA1 2TA
G0WKU TA1 2XF
G1FED TA1 2XN
GM7SLZ TA1 2YA
G1BCX TA1 3DF
G4VBO TA1 3DT
G0PPD TA1 3EG
G0EYR TA1 3EH
G4RNT TA1 3EJ
G8FUT TA1 3JE
G4BJX TA1 3PA
G4NUJ TA1 3SA
G0UFU TA1 3XN
G4AVJ TA1 3YB
G1YLN TA1 4DY
G8YFF TA1 4HZ
G4MYE TA1 4JF
G3PFF TA1 4JW
G3JKE TA1 4JX
G4WOC TA1 4NE
G1VVE TA1 4NS
G4AUM TA1 4RE
M0AID TA1 4RG
G7SHU TA1 4SD
2E0AKD TA1 4SG
G4MRO TA1 4SY
G3RWC TA1 4YH
G3OSY TA1 4YQ
G4UVV TA1 5AQ
G4WTA TA1 5DS
G0RIE TA1 5EH
G4DII TA1 5HN
G7KIQ TA1 5JH
M1BXE TA1 5JH
2E1BFW TA1 5JH
G3AZ TA1 5JJ
G3BEH TA1 5JN
G1NTK TA1 5LL
G7DVD TA1 5NL
G4VVS TA1 5PQ
G8ZSP TA1 5PQ
G3TJH TA1 5PW
G0CPG TA2 6LP
G7GGA TA2 6SG
G6EED TA2 6TF
G1AJZ TA2 7EN
G0PQH TA2 7JE
G4DKL TA2 7LN
G4XKK TA2 7PB
G0PGL TA2 7PS
G1UDN TA2 7SA
G7LDD TA2 7SP
G8JXK TA2 7SY
G0RWA TA2 8DX
G0PWV TA2 8DX
G7KYG TA2 8DZ
G1AMG TA2 8EB
G4UTM TA2 8NA
G8BTY TA2 8NA
G8TJF TA2 8NA
G7SXL TA2 8NR
G4LYP TA2 8SQ
G8UXY TA21 0AY
G0COZ TA21 0BE
G0DOI TA21 0DY
G3AZY TA21 0JS
G4VFD TA21 0LG
G8FZI TA21 0RE
G0MJF TA21 8BB
G8ZVK TA21 8BD
G3ZVU TA21 8ED
G4ZLF TA21 8EL
G1CQG TA21 8ER
G4RGA TA21 8EX
G6TWA TA21 8HR
G8HNM TA21 8HY
G4OLA TA21 8PY
G3TRU TA21 8QF
G4YXR TA21 8RU
G3LCA TA21 9AP
G4RGF TA21 9AT
G4AIU TA21 9AX
G0UKX TA21 9BD
G8TZK TA21 9HT
G8XHU TA21 9PE
G4KEL TA3 5AG
G4WWQ TA3 5JZ
G3MTG TA3 5LU
G0MZQ TA3 5QP
G4WXD TA3 5QP
G4JZL TA3 5TE
G1DRY TA3 6HA
G1MEG TA3 6HA
G0WYK TA3 6LP
G0MLT TA3 6TN
G7DKZ TA3 6TN
G4NIL TA3 7HB
G4BFM TA3 7HF
G4MFD TA3 7HH
G4KIK TA3 7LG
G3XZW TA3 7PU
G5JJ TA3 7PU
G6TOY TA3 7RE
G4UVZ TA3 7SD
G1XQP TA4 1AQ
G8YCN TA4 1ET
G0GWS TA4 2BU
G4NYE TA4 2LY
G3UCC TA4 2NT
G4KPT TA4 2UD
G3UWH TA4 3AQ
G4LZU TA4 3JP
G0UOU TA4 3UD
G8FDF TA7 0RF

Teesdale

G4FUT DL12 0DH
G0ADO DL12 0QU
G4FVZ DL12 0RG
G3GEA DL12 0RP
G7LIE DL12 0ST
G4AFE DL12 0UU
G0NRK DL12 8EB
G0MLP DL12 9EW
G7STO DL13 5JY
G0LLC DL13 5NF
G4KUX DL13 5RH
G0OCB DL14 0DH
G1XNI DL14 0DH
G0NTZ DL14 9QB
G0ADQ DL2 3AG
G4MIJ DL2 3DN
G6VOC DL2 3HS
G7DVG DL2 3RN
G3GJJ DL2 3RY

Teignbridge

G3IMW EX2 9QH
G8NZB EX4 2JP
G7JAK EX6 6DB
G3DSV EX6 6DE
G6BLA EX6 6EQ
G4WTU EX6 7AD
G4ZAV EX6 7AD
G7RFB EX6 7DE
G4BZE EX6 7SR
G4ARE EX6 7SS
G0KYV EX6 7SS
G3KQG EX6 7TZ
G8TVB EX6 8AF
G3MIR EX6 8QH
G0JIS EX7 0BP
G4ONV EX7 0EA
G1NMY EX7 0HB
G1ZVO EX7 0JZ
G3ID EX7 0LR
G4VXH EX7 0NA
G0KDT EX7 0NN
G4OZO EX7 9AY
G4KPP EX7 9DB
G3APN EX7 9HF
G0RGC EX7 9JG
G0KCI EX7 9NE
G4WLA EX7 9NQ
G3AVN EX7 9ST
G3GZQ TQ11 0AS
G7RYL TQ11 0BP
G1GMG TQ11 0EA
G0KUV TQ11 0EH
M1BKU TQ12 1BJ
G3SXS TQ12 1BX
G1NNU TQ12 1DG
G0EKH TQ12 1DZ
G4LCO TQ12 1EG
G0EOR TQ12 1LD
G0OFA TQ12 1PN
G3YAR TQ12 1RQ
G0NUZ TQ12 1UP
G1FON TQ12 1YJ
G1WQN TQ12 1YJ
M1BJL TQ12 2BY
G0RDO TQ12 2ND
G7RVZ TQ12 2PU
G0WAE TQ12 2PU
G7AQV TQ12 2PX
M1AIY TQ12 3BP
G0WWD TQ12 3JE
G6BAL TQ12 3LQ
G6YJO TQ12 3PE
G7MEA TQ12 3QX
G4DTW TQ12 3TE
M1AKV TQ12 3TJ
G0KEC TQ12 4EA
M1ALY TQ12 4ER
G4VUD TQ12 4HE
G4LAK TQ12 4JG
G4MNA TQ12 4JZ
G3LHJ TQ12 4LF
G3NJA TQ12 4LF
G8NJA TQ12 4LF
G0RTW TQ12 4LW
G0IGK TQ12 4NE
G4VTT TQ12 4NE
G0NXI TQ12 4NH
G4ELZ TQ12 4NJ
G7ULU TQ12 4NJ
G4GZM TQ12 4NQ
G3XXE TQ12 4NS
G4XJL TQ12 4QS
G4XJM TQ12 4QS
G0SDL TQ12 4RE
G3TLK TQ12 4SB
G0OXT TQ12 5BA
G8OXR TQ12 5BA
G6FIN TQ12 5BZ
G0BNJ TQ12 5EW
G4NRH TQ12 5JG
G4VTO TQ12 5LF
G0BWE TQ12 5LH
G6YAY TQ12 5LH
G7NFX TQ12 5PB
G8VSV TQ12 5QJ
G4FCN TQ12 5QS
G8GCS TQ12 5QS
G7MNL TQ12 5QS
G0SBM TQ12 5QS
G6CLD TQ12 5QS
G6GLP TQ12 5QS
G2YD TQ12 5RH
G8YEN TQ12 5TT
G4NTS TQ12 6BZ
G1ZBJ TQ12 6GX
G3PVB TQ12 6HE
G8SLM TQ12 6HE
G8YBY TQ12 6HE
G0MKF TQ12 6JU
G7IIV TQ12 6JU
G6TEQ TQ12 6NW
G4RYH TQ13 0EW
G3XTM TQ13 0JB
G7OZI TQ13 0JN
G7OZJ TQ13 0JN
G1NQQ TQ13 0LG
G7VPC TQ13 0LT
G4YPU TQ13 7AD
G4HMA TQ13 7AZ
G1BGK TQ13 7DR
M0APB TQ13 7JF
G7JDZ TQ13 7JS
G3BVW TQ13 8NW
G7VUP TQ13 8SJ
G6WIC TQ13 8SS
G0WTU TQ13 9EJ
G6ICH TQ13 9EP
G3RPD TQ13 9NB
G4TEM TQ13 9RE
G1XBX TQ13 9RE
G1XEF TQ13 9TA
G3WLT TQ14 0DG
G1USI TQ14 0DJ

G3FML TQ14 0EQ
G8KTC TQ14 8EF
G4BVA TQ14 8FE
G3MLD TQ14 8LB
G1MKE TQ14 8LE
G3TIR TQ14 8NE
G0CWQ TQ14 8PN
G6XD TQ14 8QG
G2AZC TQ14 8QR
G4BG TQ14 8QX
G3YHH TQ14 8UF
G1GIG TQ14 8UW
G7ZZY TQ14 9BB
G0BXT TQ14 9DR
G1AKI TQ14 9EW
G3MOA TQ14 9HG
G7FRL TQ14 9JP
G7MYK TQ14 9JP
G8HNO TQ14 9NA
G0PWU TQ14 9NF
G1IEJ TQ14 9NF
G0PNA TQ14 9RB
G3BLO TQ14 9TS
G8RAA TQ14 9UR
G1EUA TQ14 9UY
G1HFI TQ14 9UY
G7MQN TQ14 9UY
G8LXS TQ9 6DB

Tendring

2E1EBO CO11 1JQ
G4TZM CO11 1NF
G4UWW CO11 1UE
G0NXH CO11 2AL
G4JVM CO11 2AL
M1BEB CO11 2HE
M0BGE CO11 2HE
G4AVL CO11 2HS
G1CHN CO11 2HW
G3YAJ CO11 2HX
G7VWE CO11 2HX
G1ERB CO11 2LZ
G6KUS CO11 2PY
M0AFX CO11 2QY
G1LBM CO11 2RU
2E1AUM CO11 2RU
G4ZFK CO11 2SB
G6UWK CO11 2SP
G0RGH CO11 2TR
G3PED CO11 2TR
G4WHK CO12 3NR
G0OHL CO12 3SB
G6YJK CO12 3UA
G8XAX CO12 4AT
G7OEF CO12 4EF
G1VCZ CO12 4EQ
G8ZTS CO12 4HY
M1BQB CO12 4LA
G0STW CO12 4TS
G0RDX CO12 5BQ
G4YJQ CO12 5EE
G0SHH CO12 5EF
G3YYZ CO12 5EJ
2E1ACF CO12 5JB
G0OEY CO12 5JB
G4EYE CO12 5JF
G0WFR CO12 5JF
G4WFR CO12 5JF
G7VHF CO12 5LS
G8PAI CO12 5NE
G8MUZ CO13 0AW
G6PTI CO13 0BG
M1BGJ CO13 0BQ
2E1FJY CO13 0BQ
G6DCJ CO13 0EQ
M1BRR CO13 0HZ
2E1EWG CO13 0HZ
M1AAZ CO13 0HZ
G4MYG CO13 0JG
G6CYA CO13 0JJ
G6NHU CO13 0LQ
G4HLS CO13 0NE
G0KYP CO13 0NQ
G3LWM CO13 9DX
G1YJK CO13 9HH
G3OJS CO13 9HR
2E1DND CO13 9JA
G4IDH CO13 9JY
G1IKE CO13 9PA
G0GYY CO14 8AY
G8HSI CO14 8DF
G6CDU CO14 8NL
G0MTU CO14 8NN
G3VLD CO14 8PX
G0OFF CO14 8QZ
G6XOU CO14 8RG
G3TUU CO14 8RL
G7BWF CO14 8RL
G8ZJE CO14 8RR
G0NMB CO14 8SJ
G4ZBP CO14 8SJ
G7LIP CO15 1AU
G7PMU CO15 1HB
G7UTN CO15 1HP
G6WZQ CO15 1HU
G8TPT CO15 2ET
G7MKW CO15 2LH
G4OAX CO15 2PH
G8DMT CO15 2QT
G0NIP CO15 2SQ
G0UHU CO15 3AR
2E1FLY CO15 3JA
M1BPE CO15 3JR
2E1EWR CO15 3JR
G7AJG CO15 3NF
G7HMN CO15 3QL
G7DNS CO15 3SZ
G4LVE CO15 4DF
G0KSX CO15 4EB
G7NWS CO15 4EZ
G3YUS CO15 4HX
G3ZEZ CO15 4JE
G0LQT CO15 4ND
G8TOI CO15 4PJ
G0FED CO15 4PX
G8ATL CO15 4RJ
G7FER CO15 4SG
2E1DEX CO15 4TY
G4ZCD CO15 4UX
G3CJI CO15 4UX
G1WXR CO15 5LF
G3CRC CO15 5NA
G0SMJ CO15 5NA
G0WJD CO15 5NA
G3JNI CO15 5PZ
G6UDE CO15 5QX
G1SDO CO15 5TR
G6DLZ CO15 6DA
G6CTY CO15 6DP
G1YJJ CO15 6JU
G3BLG CO15 6NF
G4ZJK CO16 0AR
G0UBJ CO16 0AX
G4CYF CO16 0EP
G1JCP CO16 0HT
G1XAJ CO16 0HU
G8POZ CO16 0LA
G4GFH CO16 7AT
G1WDF CO16 7DA
G0PKT CO16 7HF
G7UES CO16 7LE
G8RRL CO16 8BD
G4RUJ CO16 8BN
G0SCP CO16 8BY
G4WHZ CO16 8DB
G7HJK CO16 8FE
G8LHF CO16 8FP
G0BOI CO16 8JP
G1WJR CO16 8PR
G0MBA CO16 8PT
G2CMW CO16 8TT
G0MJA CO16 8UQ
G0MSE CO16 8UR
G4FJT CO16 8US
G6MGE CO16 8US
G0TKH CO16 8YH
G6TNQ CO16 8YX
G4PIQ CO16 9AA
G3VMP CO16 9AN
G3KIC CO16 9DR
G3PQM CO16 9ES
G1DSQ CO16 9HN
G3JKT CO16 9HN
G3LZR CO16 9HZ
G0EBI CO16 9NE
G8CVP CO16 9NE
G4LFD CO16 9PD
G0FIW CO16 9PS
G1HBW CO16 9PU
G8JPJ CO16 9PU
G7BVZ CO16 9RS
G7TBT CO7 0BE
G7TBU CO7 0BE
G1GCT CO7 0DG
G6AEB CO7 0DU
G3MGW CO7 0JD
G4IZX CO7 0LA
G0SOX CO7 0LB
G4CIA CO7 0LB
G4EUW CO7 0NN
G6PMD CO7 0NZ
G4JAC CO7 0QR
G7KRO CO7 0RU
G1BFF CO7 0SJ
G1SVV CO7 7BT
G7USX CO7 7EG
G4HPU CO7 7PA
G6IGU CO7 7SY
G0GYI CO7 8DH
G1WEQ CO7 8EE
G8EWC CO7 8HL
G4AZD CO7 8JA
G1GSD CO7 8JH
G1TWH CO7 8JH
G3PFP CO7 8JH
G4ZKS CO7 8LJ
G8HOR CO7 8PN
G1SVW CO7 9NS
G4YBI CO7 9RB

Test Valley

G3WPI SO16 8AZ
G4YHQ SO20 6AD
G8IUP SO20 6AH
G3RDQ SO20 6BA
G6MCN SO20 6BA
G8ROZ SO20 6BG
G3ZZL SO20 6JE
G3FWI SO20 6PN
G3LGA SO20 8DB
M1BDJ SO20 8EA
G0VCO SO21 3RL
G8BZR SO51 0GP
G0OZD SO51 0GQ
G4GSK SO51 0LF
G8OQP SO51 0QB
G4YEP SO51 5AU
G1TWW SO51 5PQ
G6AIQ SO51 5QR
G4EOW SO51 5SQ
G6IVR SO51 5SQ
2E1DCW SO51 5SQ
2E1CYZ SO51 5SQ
G0BHK SO51 5SS
2E1DBA SO51 5UU
G8DXF SO51 6AR
G3UFW SO51 6BG
G8KHN SO51 6BN
G3DUZ SO51 6DF
G3YXR SO51 6DF
G0FHC SO51 6EE
G3NAE SO51 6EX
G3JHL SO51 6GA
2E1CCW SO51 6JW
G3XSH SO51 6RH
G3JFY SO51 7HU
G3OSQ SO51 7JY
G6CPE SO51 7JY
G8AUJ SO51 7JZ
GW8NXC SO51 7JZ
G8OFX SO51 7JZ
G4ZGD SO51 7LE
G8OQO SO51 7LG
G4YVE SO51 7LL
G8YWA SO51 7QE
G7LXH SO51 7QT
G6LYA SO51 7RP
G1JOG SO51 7TE
G6GXG SO51 7TE
G1NTN SO51 7TJ
G5CW SO51 7TP
G6VZF SO51 7UQ
G0KHB SO51 8AF
G6GCI SO51 8EQ
G8ZMM SO51 8FB
G2DSY SO51 8FN
G4HWM SO51 8GF
G4CFX SO51 8PA
G3OGJ SO51 8PN
G3TUQ SO51 9AD
G4YEE SO51 9BT
M0AZI SO52 9AT
2E1FNO SO52 9AT
G6TRW SO52 9EU
G7WEZ SO52 9FB
G3WII SO52 9FS
G6IVT SO52 9FT
G1TOT SO52 9FX
G1IPQ SO52 9GR
G0IVR SO52 9GT
G3ABA SO52 9GT
G6AZV SO52 9JP
G7AJR SO52 9JT
G6XJU SO53 3PL
G0DQH SO53 4QJ
G4RXE SO53 5NG
G7GJC SO53 5PB
G7GJD SO53 5PB
G7GJE SO53 5PB
G8LVC SO53 5PB
G3MDR SO53 5PD
G1GCS SP10 2AG
G0OMD SP10 2AH
G1LMA SP10 2AU
G3OZY SP10 2BS
G2DKI SP10 2HB
G4YSB SP10 2HE
M0ALE SP10 2PZ
G7PVE SP10 2QT
G7TCC SP10 2QT
G0RBL SP10 3AB
G6YEY SP10 3DS
G1ZMV SP10 3EN
M1BCE SP10 3EP
G7VRJ SP10 3EP
G4VMJ SP10 3HD
G0KIC SP10 3LJ
G4UVO SP10 3QX
G7SXX SP10 3RL
G6PKT SP10 3TT
G0OGA SP10 3UE
G0RRJ SP10 3UH
G7DLE SP10 3UQ
G1HPS SP10 4AR
G0UJW SP10 4BQ
G0AMO SP10 4EN
G0ARC SP10 4EN
G6GFO SP10 5HD
G8CEP SP10 5HT
G8NDN SP10 5NA
G7TFB SP10 5NA
G7EBI SP11 0JD
G7JKI SP11 0JN
G6SDE SP11 0RE
G8FHI SP11 0RS
G3AEO SP11 6JE
G6SJG SP11 6JY
G4OZL SP11 7ER
G4BLF SP11 7RG
G6JSF SP11 8LG
G3ZDG SP11 8LH
G8GYS SP11 8NE
G4NNS SP11 9AN
G3YGF SP5 1HJ

Tewkesbury

G6VKA GL19 4BT
G3OTX GL19 4HW
G3SNN GL19 4LL
GW0BAZ GL2 8DL
G0FGS GL2 8DW
G1ISY GL2 8EH
G6EEV GL2 8EL
G0MXM GL2 8EW
G4PJJ GL2 8JP
G4ZYR GL2 8LJ
G4BGW GL2 9NW
G7GVJ GL2 9PS
G3PGW GL20 5HF
G0FCM GL20 5HL
G1KNX GL20 5PD
M0ADJ GL20 5RN
M1AXD GL20 5TH
G8AZN GL20 5TZ
G0VFZ GL20 5TZ
G3LZK GL20 6DF
G3PKX GL20 6HE
G6AHX GL20 6JW
M1AMA GL20 8AS
G4EAZ GL20 8AT
G6GWV GL20 8AU
G4YGY GL20 8BA
G8YMR GL20 8DY
G8TQP GL20 8ES
G0BXK GL20 8JD
G0VGA GL20 8PY
G8VSH GL20 8QP
G1NVS GL20 8RB
G0HKO GL20 8RD
G7UXP GL20 8RE
G6UWM GL20 8TN
2E1BWM GL3 1EZ
G0VCZ GL3 1LY
G4CHD GL3 1NE
G4ENZ GL3 2AU
G0VIG GL3 2BT
G3SZS GL3 2DS
G6XQO GL3 2DW
G0UHG GL3 2DW
G4UCN GL3 2DX
G1CMH GL3 2HT
G3SUA GL3 2LD
G3ZKN GL3 2LZ
G4FRI GL3 2PN
G6UGW GL3 2PU
G0EJF GL3 2QS
G4CDQ GL3 2QT
G7TIU GL3 2TR
G6USD GL3 3TZ
G8DHF GL3 3TZ
G6SFB GL3 4AS
G8IWB GL3 4DB
G1NVO GL3 4ES
G0JFU GL3 4LT
G4NVY GL3 4NG
G7DMZ GL3 4PD
G6OTP GL3 4PS
G3KII GL51 5SN
G3JPP GL51 5SZ
G7URT GL51 5SZ
G4ERR GL51 5TG
G4RFL GL51 5TG
G6DPS GL51 5WA
G0UPB GL51 9SN
G8TOD GL51 9TH
G8DV GL52 3PA
G8ENW GL52 3PS
GW6BDM GL52 3QE
GW6BOQ GL52 3QE
G0MTW GL52 4AB
G3XKH GL52 4AG
G3LGF GL52 4AJ
G3RQS GL52 4AQ
G6LTB GL52 4BG
G3CXI GL52 4BN
G3PYI GL52 4BN
G0EVX GL52 4BW
M1CBY GL52 4BY
M1CBZ GL52 4BY
G4LJE GL52 4DD
G0OEZ GL52 4ES
G7TLM GL52 4EW
G4EUN GL52 4EY
G8MNR GL52 4HH
G0SFE GL52 4HU
G3YZ GL52 4HZ
G6SKN GL52 4LA
G0VCA GL52 4LA
G3JFH GL52 4LX
G3AKI GL52 4NA
G7PTV GL52 4NU
G0IYO GL52 4NX
G4KWD GL52 4NX
G1GOO GL52 4NZ
G4GKK GL52 4NZ
G4MEN GL52 4NZ
G8UJG GL52 4NZ
G4BRX GL52 4PB
G0EFP GL52 4PT
GM3KGT GL52 4PY
G4INC GL52 4PZ
G6UYN GL52 4QD
G3NOI GL52 4QP
G3ZTI GL52 4QS
G4IOA GL52 4RH
G4YQZ GL52 4RW
G4ERP GL52 4SA
G6EEU GL52 4SA
G0MBZ GL52 4TG
G4MPJ GL52 4TG
G8MGD GL52 4TZ
G3LRM GL52 4UA
G4BMP GL52 4UD
G4HOW GL52 4WH
G4CKX GL52 4XP
G0UPU GL54 5ER
G4MUW GL54 5JX
G3OHA GL54 5LE
G0TPD GL54 5LQ
G3DPM GL54 5QR
M1BZP GL54 5QU
G7AWP GL54 5QX

Thanet

G3REZ CT10 1BQ
G4GSP CT10 1BQ
G8FME CT10 1DR
G8GCL CT10 1HU
G6EQY CT10 1PG
2E1EKM CT10 1PG
2E1EKN CT10 1PG
M1BAN CT10 1PG
G3RAD CT10 1QT
G1VID CT10 1RB
G0LGW CT10 1RP
G3ACQ CT10 1SH
G8WZX CT10 1SL
G4GAT CT10 1TL
G0CIX CT10 2AF
M0ASC CT10 2DT
G4ADS CT10 2EP
M0AQA CT10 2HA
G4WUP CT10 2JF
G4AQE CT10 2JG
G7FNU CT10 2ND
G3DNR CT10 2PJ
G0RJJ CT10 2SD
G0KCA CT10 2TX
G7OHO CT10 2XN
G7UPG CT10 2XN
G0CLO CT10 2XR
G2IC CT10 3AZ
G0VUT CT10 3AZ
G8VR CT10 3BH
G8LVI CT10 3DP
G1KMJ CT10 3DR
G4GAP CT10 3ES
G0GUW CT10 3EY
G0RXU CT10 3NE
G4VBI CT10 3QP
G4VPI CT10 3SD
G0AHA CT11 0AW
G0CSD CT11 0BG
G3RGR CT11 0DS
M1CCL CT11 0LQ
G3VSU CT11 0NJ
G7FMI CT11 0PB
2E1CPF CT11 0PB
G0DVS CT11 0PQ
G4GUD CT11 0PS
G0MBL CT11 0RN
M1AJV CT11 7AS
G3LHS CT11 7AT
G0TBO CT11 7AT
G8WQT CT11 7DY
G6ENA CT11 7EF
G0TFB CT11 7HJ
G2HBC CT11 7HU
G0DVS CT11 7JU
GM7ORS CT11 7NW
2E1CPL CT11 8AX
G0CPV CT11 8EG
M1AJU CT11 8EP
2E1GAV CT11 8QX
G0WMH CT11 8RD
G3TVD CT11 9DE
G7IUV CT11 9LF
G0HRS CT11 9LU
G0JIF CT11 9LU
G3WYU CT11 9PL
G4BXI CT11 9QE
G7IQM CT11 9QZ
G3BD CT11 9SJ
G6UXG CT12 4BE
G6OKB CT12 4HR
G1KEH CT12 4JS
G7REO CT12 5AF
G1LFM CT12 5AW
G6NVL CT12 5BE
G0LFL CT12 5ED
2E1FYM CT12 5EH
G7FGI CT12 5JA
G0EPQ CT12 5JT
G0WWP CT12 5LD
G1AIO CT12 5LX
G7FBO CT12 6DX
G7ITS CT12 6DX
G3OPL CT12 6DZ
G3ZBF CT12 6EZ
G7SFD CT12 6JQ
G0CBY CT12 6LP
G4BXU CT12 6NW
G7FMT CT12 6RD
G7JSO CT12 6RN
G4HEZ CT12 6SQ
G8UHJ CT12 6SW
G7JIO CT12 6TD
G0AKE CT12 6TG
M1CDA CT7 0DU
G0ANW CT7 0HD
G1HWH CT7 0PY
M1AIS CT7 0PY
2E1EGS CT7 0PY
G1UEA CT7 9AS
G0INT CT7 9AZ
G1JEH CT7 9ED
G3SRU CT7 9HN
G3TPK CT7 9JA
G3CTQ CT7 9QH
G4RNJ CT7 9QH
G0MQB CT7 9QN
G0AFS CT8 8AG
G4SBD CT8 8AN
G4SEK CT8 8AN
G1NLQ CT8 8AP
G7MEZ CT8 8AP
2E1DDZ CT8 8AP
G1UBT CT8 8BT
G4KPV CT8 8BT
G4ACQ CT8 8BY
G6NRU CT8 8QG
M1AGL CT8 8QX
G0CEY CT8 8RJ
2E1CPB CT8 8RJ
G8BJP CT8 8SJ
G0WFU CT9 1NR
G4ZVC CT9 1TR
G3WZG CT9 2AN
M0AJC CT9 2EJ
2E1EHY CT9 2TJ
G4DTA CT9 2UH
M0APV CT9 3AP
G3VID CT9 3AR
G0FTB CT9 3DB
G0LGE CT9 3DY
G1DRX CT9 3EF
G1WWR CT9 3EF
2E1COV CT9 3EF
G3DDA CT9 3EJ
G4PTE CT9 3EJ
G7HVG CT9 3ES
G4HUS CT9 3ES
G1IEM CT9 3HG
G6FSD CT9 3HG
G1KRK CT9 3HR
G0DFI CT9 3JB
G1DUJ CT9 3JB
2E1DKM CT9 3NN
M1BBS CT9 3NN
G0KRM CT9 3PN
G8OBH CT9 3PN
M1CBT CT9 3PT
G1AOC CT9 3QX
G0CHN CT9 3RX
G0RPC CT9 3UJ
M1CDE CT9 3XB
2E1CTG CT9 3XB
G3SHX CT9 3XE
G0NEP CT9 3XJ
G7HFH CT9 3XX
G0HGH CT9 4BT
G3YUH CT9 4DH
G0UZA CT9 4DH
G7CQT CT9 5DT
G7EYE CT9 5DT
G3BHW CT9 5EW
G3TRX CT9 5HT
G4BSW CT9 5JA
G4RQZ CT9 5JY
2E1EKF CT9 5LY
G7MKF CT9 5PA
G4MGF CT9 5QA
G8DHJ CT9 5QA
G3ZZZ CT9 5QB
G8GHH CT9 5QB
2E1BTX CT9 5QW
G4UKS CT9 5RD
G6SRQ CT9 5RD
G0KOK CT9 5RW
G0TBU CT9 5SS
2E1CQP CT9 5SS
G0UAK CT9 5SS
M1AHK CT9 5TR

The Vale of Glamorgan

GW4SUD CF31 3LN
GW8ZIL CF31 3LN
GW4NBY CF31 3LR
GW0SIS CF31 3LR
GW7SSI CF32 0PF
GW4KWV CF32 0PH
GW4ZEA CF32 0PJ
GW4NGU CF32 0SF
GW4SKP CF32 0SL
GW3MFY CF35 5AE
GW4ZXG CF35 5BH
GW7DHX CF5 6AB
GW0PPG CF5 6AF
GW4SGR CF5 6AF
G6SGR CF5 6AF
GW0LST CF5 6AF
GW7BTC CF5 6AF
GW4WVO CF5 6BQ
GW0DV CF5 6DA
GW3RIH CF5 6LQ
GW4CSY CF5 6SS
MW0AMQ CF5 6TS
MW0AMN CF5 6TS
GW7DWR CF61 1GX
GW8MFQ CF61 1GX
GW4VLQ CF61 1UF
GW4LFF CF61 1ZF
GW8YTO CF61 2DQ
GW0CNJ CF61 2LN
GW8ZHN CF61 2XB
GW4ZCL CF62 3DZ
GW6ZMN CF62 3EA
GW8CMU CF62 3EH
GW0JZR CF62 3FY
GW1UIK CF62 3FY
G7CTT CF62 3HU
2W1BPH CF62 4JN
G7WKM CF62 4JR
MW1AML CF62 4JR
GW7DZB CF62 4NR
2W1BPG CF62 4PD
2W1EPN CF62 6FF
GW1JVB CF62 6RA
GW1XFB CF62 6RA
GW8LJJ CF62 6RA
GW1KDE CF62 6RR
GW1TKO CF62 6RR
GW1PKX CF62 7JA
GW7PFK CF62 7JG
GW1VRW CF62 7PW
GW1XUD CF62 8BJ
GW6CNS CF62 8BL
GW8KZA CF62 8BR
GW3VKL CF62 9HJ
GW3WSU CF62 9HJ
GW4BRS CF62 9HJ
GW4GSH CF62 9HJ
GW6BRC CF62 9HJ
GW6DGU CF62 9HJ
GW3CBA CF62 9SW
GW0AGA CF63 1BJ
GW2DDX CF63 1BJ
GW7DWE CF63 1EP
GW0YEY CF63 2NR
GW8YEY CF63 2NR
GW4ANK CF63 3QD
2W1BQW CF63 3RE
GW6ATT CF63 4EF
GW7VST CF63 4EF
GW0TSL CF63 4HW
GW0GRQ CF63 4PQ
GW8HEZ CF64 1BJ
GW0EWY CF64 1HP
GW3ENN CF64 1JU
GW3LQE CF64 1WW
GW4UMX CF64 2LA
GW7LCP CF64 2LA
GW0SQT CF64 2LA
GW3VNZ CF64 2PF
GW0GHF CF64 2PG
GW3WBU CF64 2QZ
GW4BCF CF64 2RD
GW0NQQ CF64 2RR
GW0KZG CF64 2RT
GW0PZZ CF64 2RX
GW7EXW CF64 2RY
GW0OTY CF64 2SA
GW4UGI CF64 2TZ
GW3SPA CF64 3DD
GW1MBV CF64 3HH
GW4IUL CF64 3JJ
GW0ICU CF64 3LZ
GW0APL CF64 3NB
GW4IXC CF64 3NQ
GW4HDW CF64 3QN
GW0ROL CF64 3QX
GW3VLU CF64 3QZ
GW3PYX CF64 3RB
GW3SFQ CF64 4LR
GW4UZW CF64 4TE
G0PPS CF64 4TG
G8PRU CF64 4TG
GW4XKE CF64 4TG
GW4MOG CF64 4UH
GW3IVR CF64 5BR
GW1DQV CF64 5BW
GW4JVE CF64 5RY
GW8WEY CF64 5SR
2W1CHM CF64 5TE
GW0LBJ CF64 5TH
GW0CKK CF64 5UG
GW4BVJ CF71 7NL

The Wrekin

G3VVL SY4 4QN
2E1FZP SY4 4QX
G1VIT TF1 1JH
G8BKF TF1 1JX
2E0ANB TF1 2AR
G8VZT TF1 2AS
G1BHV TF1 2BY
G0AGC TF1 2DT
G3MWQ TF1 2DX
G3BQQ TF1 2EA
G3HXG TF1 2EL
G0ASP TF1 2ER
G0DHL TF1 2ER
G8BKH TF1 2JF
G4AUY TF1 2LE
2E0AOC TF1 3AR
2E1EQT TF1 3AR
G3MHY TF1 3DA
G1JJK TF1 3DB
G6HKP TF1 3ED
G7TPA TF1 3HJ
G6KTK TF1 3HL
G1OMU TF1 3HU
2E0AOV TF1 3NN
G0CYN TF1 3NN
G7VDS TF1 3PF
2E1DEM TF1 4BN
2E0AMV TF1 4BS
M0AOR TF1 4BS
G1ZPQ TF1 4FZ
G1IRX TF1 4PQ
G6NOL TF1 4PQ
G2RSA TF1 4QR
2E1AVM TF1 4QW
G7OQT TF1 4QW
2E1FDE TF1 4XD
G6ZFU TF1 4XU
G7ACD TF10 7DX
G4EYU TF10 7EF
G4IFR TF10 7EF
G7KJA TF10 7HH
G8HDA TF10 7LS
G0UCI TF10 7LS
G0SSU TF10 7LY
G3JXT TF10 7RB
G1KRU TF10 8JL
G4IUT TF10 8NF
G7LPB TF10 8NN
G7VRW TF10 8PF
G4OLO TF10 9AQ
G7CII TF10 9HD
G0WJH TF10 9JQ
G0OEN TF10 9LN
G3HRN TF10 9LN
G3MVK TF10 9LN
2E1EQK TF2 0AJ
2E0APQ TF2 0BD
2E1EQU TF2 0BD
G1OAR TF2 0DX
G7DXM TF2 0EF
G4ZHU TF2 6BJ
2E1EQF TF2 6NU
2E0ANU TF2 6PA
G3DMC TF2 6RA
G0OPL TF2 6RS
G0BMB TF2 6SJ
G0WQH TF2 7DA
2E1ACW TF2 7DA
2E0AOD TF2 7DA
G4AGA TF2 7HT
G0VXJ TF2 7LJ
G1SKW TF2 7LQ
2E1FQV TF2 8BP
2E1FQW TF2 8BP
G0CSJ TF2 8BX
2E1EQJ TF2 8DD
G6ELO TF2 8DL
G0HDC TF2 8HY
2E0AKP TF2 8PE
G7WKX TF2 8PF
G0FMX TF2 8PF
GM0BCP TF2 9DJ
G3TQP TF2 9LF
G6ILZ TF2 9NG
G4ZJY TF2 9NU
G3JKX TF2 9RB
G0PBT TF2 9RY
G6GSI TF2 9SL
G4UMM TF2 9TG
G7SEY TF2 9TJ
G0VWS TF3 1DR
G4UMF TF3 1DY
G8ZIP TF3 1ED
2E1DZJ TF3 1ED
G0UFE TF3 1EG
2E1ASI TF3 1EG
G7DRL TF3 1LJ
G0WCM TF3 1LY
G1THE TF3 1PR
G1JBN TF3 1QW
G8BMD TF3 1RW
G1TKT TF3 1SE
2E1FPM TF3 1SX
2E1BZS TF3 1YG
2E1CBG TF3 1YG
G0GQO TF3 1YN
G6UMF TF3 2AH
G7TYK TF3 2BU
2E1CND TF3 2DJ
G0HQK TF3 2DQ
G1UWN TF3 2DQ
2E1DJM TF3 2EE
2E1FOV TF3 2EJ
2E0AMN TF3 2HR
G4FBZ TF3 2JX
G3VUV TF3 2LE
G6FFR TF3 2NQ
2E1EQW TF3 5EN
G0GAL TF3 5HE
G0MQX TF4 2EW
G6NWT TF4 2HS
G7RBX TF4 2JN
G0VNO TF4 2QE
G8JVM TF4 3DD
G0RUJ TF4 3DF
2E1EQD TF4 3EB
G0VPW TF4 3JF
2E1FNX TF4 3NG
2E1CNU TF4 3RJ
2E1DYM TF4 3RJ
2E1DYN TF4 3RJ
G4TCK TF4 3TP
G8OYB TF5 0LL
G8RXZ TF5 0NY
G7BCO TF5 0PG
G8ZXU TF6 5BL
G6SGD TF6 5ER
G7BWQ TF6 6AR
G0JBO TF6 6DH
G4WRK TF6 6HH
2E1AEC TF6 6HQ
G3UKV TF6 6HQ
G3ZME TF6 6HQ
G8UKV TF6 6HQ
G4AUZ TF6 6NB
2E1DED TF6 6NP
G2FSP TF6 6PZ
M1ACA TF7 4AB
2E1FEB TF7 4AY
G7UIV TF7 4DF
G4BFT TF7 4DX
G0VMY TF7 4HE
2E1CJU TF7 4HQ
G6ZME TF7 4HT
G8UGL TF7 4HT
G0OCP TF7 4JH
G4VZK TF7 4NH
G1TBR TF7 5DJ
G0UCN TF7 5LT
2E0APG TF7 5NA
G4GSB TF7 5PB
G0JAL TF7 5QR
G0EGD TF7 5QU
2E1AIY TF7 5SU
G4EIX TF7 5SU
G0WKE TF7 5SY
G0WOQ TF7 5SY
G0CER TF8 7ET
M0AXV TF8 7EW
G0VXG TF8 7ND
2E1ESW TF8 7ND

Three Rivers

G7NBF HA6 2HN
G4HMD HA6 3AU
G4KEO HA6 3DS
G6ZRL HA6 3LR
G6YLV HP3 8RY
G1BSZ HP3 8SH
G6MNI HP3 8SH
G6MNJ HP3 8SH
G8BXH WD1 4LW
G3MED WD1 4PA
G3NDK WD1 5AA
G1ICP WD1 5AP
G0DDV WD1 5EH
G0OJT WD1 6JH
G0JKZ WD1 6NJ
G1ISX WD1 6NW
G1XEP WD1 6QQ
G0OMJ WD1 8YH
G0WBL WD3 1HL
G8ZTY WD3 1HT
G4DUS WD3 2BT
G3ZER WD3 2EN
G1AOE WD3 2GP
G6IBD WD3 2GW
G0WCU WD3 2LL
G3LD WD3 2NQ
G7HGQ WD3 2QA
G7CFF WD3 2QJ
G4IPT WD3 2XN
G6AJG WD3 3AU
G4VUX WD3 3DE
G7LXP WD3 3DF
G6ZFV WD3 3EE
G3TVH WD3 3FE
G8NFB WD3 3JF
G3EBV WD3 3JQ
G3TZA WD3 3LU
G8MCJ WD3 3PH
G6DFL WD3 3PX
G1MPC WD3 3RW
M1AQV WD3 3RY
G4HTF WD3 3SP
G4GCE WD3 4BG
G6JJ WD3 4DW
G4JHU WD3 4EB
G3OQB WD3 4EE
G3VOS WD3 5BD
G4COV WD3 5BJ
G1WPF WD3 5HN
G0XDI WD3 5HN
G4ORE WD3 5JD
G1DNY WD3 5JJ
G0VJI WD3 5JJ
G0ETU WD3 5LW
G8NIK WD3 5NH
GW7KZS WD3 5PP
G0FSY WD3 5PR
G4WQB WD3 5PR
G6YJR WD3 5PX
G7NDN WD3 5QE
G3WJG WD3 5QH
G4TAO WD3 5QQ
G4AEM WD3 5RG
G3HU WD3 5RJ
G8YYL WD3 5SB
G7GTG WD4 8JE
G7GTH WD4 8JE
G0SGV WD4 8LB
G6IDZ WD4 8NB
G7TMH WD4 8NJ
G6WPR WD4 8PP
G8ELG WD5 0BA
G8VRO WD5 0BL
G4KUF WD5 0DA
G3XLI WD5 0DD
G3JCR WD5 0DH
G3PMF WD5 0DR
G4PFH WD5 0DR
G3DCN WD5 0DS
G4AQI WD5 0EE
G3OYT WD5 0EU

Thurrock

G6TMF CM13 3HE
G8LUN RM14 3RX
G4NLV RM15 4BU
G8UAZ RM15 4NA
G7EGX RM15 4NB
GM1AMC RM15 4PJ
G6AMY RM15 4RT
G0PNN RM15 5BZ
G4KKD RM15 6EJ
G6LCU RM16 1YR
G0MTL RM16 2DA
G4HCK RM16 2GB
G6WNA RM16 2GW
G3KMD RM16 2HX
G1WEF RM16 2QL
G6DXP RM16 2RB
G1ZAY RM16 2RS
G6DLO RM16 2RW
G7FCJ RM16 2TH
G1DXH RM16 2TP
G3VGU RM16 2YB
G6TYF RM16 3AU
G4BAU RM16 3DF
G0DEK RM16 3LT
G8WTN RM16 4JB
G0EKN RM16 4XD
G1FAN RM16 5UD
G8BCJ RM16 5UR
G4YOD RM16 6QH
2E1FJM RM17 5LW
G0BJN RM17 5QN
G0BKL RM17 5QR
G6EFE RM17 5RG
G6OBO RM17 5RG
G4KDZ RM17 5RP
G1BSN RM17 5RW
G4LUT RM17 5SF
G6EHB RM17 5UW
G8VZJ RM17 5UW
G1NOR RM17 5YN
G0VDL RM17 5YW
G6CRX RM17 5YW
G4ROC RM17 6DQ
2E1FGJ RM17 6LD
G6DLX RM18 7DB
G4GXF RM18 7RL
G3PHD RM18 8BA
G0UXK RM18 8BA
G1EHM RM18 8DT
G0UFF RM18 8PT
G6FQN RM18 8RW
G1VWB RM18 8SX
G1SID RM18 8XB
G7RVH RM19 1GP
2E1DLO RM19 1GP
2E1DWM RM19 1GP
G4BFS RM19 1SL
G1DVP RM19 1TW
G1GDQ SS16 5JX
G4HXY SS17 0DF
G1EUM SS17 0DT
G2FGB SS17 0DT
G3EYZ SS17 0DT
G0NGG SS17 0EF
G7JVB SS17 0NH
G0BJO SS17 0PB
G0VVU SS17 0QP
G1OOG SS17 0RS
G4XNI SS17 0RS
M0AKE SS17 7BD
2E1EXA SS17 7BD
2E1FZX SS17 7BD
G4KIQ SS17 7BQ
G8YOE SS17 7BS
G4OVG SS17 7ER
G0KKH SS17 7NG
G1FBW SS17 7NG
G4WDM SS17 7QB
G4LTH SS17 7QL
G4LNT SS17 7QZ
G4PAD SS17 7QZ
G4SLH SS17 7QZ
G1VSE SS17 8BP
M1ANM SS17 8DD
G1ERQ SS17 8DE
G8EXQ SS17 8DH
G1YAH SS17 8DT
G4ZXM SS17 8EB
G7OUT SS17 8HH
G4JHE SS17 8HP
G6WXZ SS17 8LY
G8LWA SS17 8LZ
G0WML SS17 8LZ
G3IUC SS17 8NT
G7USP SS17 8PH
G7DXV SS17 8QU
G7FEL SS17 8RG
G6WOC SS17 9AD

G4ZPS SS17 9AE
G3SMF SS17 9AJ
G1NSW SS17 9HN
M1BZR SS17 9NQ

Tonbridge and Malling

G7KOY ME1 3SZ
G3TCG ME18 5PD
G3DEY ME18 5PR
G4LNM ME18 5PX
G8NVH ME18 5QH
G0OLT ME19 4TT
G4GFU ME19 5AW
G7JSD ME19 5DP
G3PAG ME19 5EH
G0VUN ME19 5JX
G4LXC ME19 5LT
G7ECI ME19 5NX
G4MLS ME19 5PY
G1JJQ ME19 5QE
G6XMR ME19 5QJ
G1AJY ME19 5QY
G7LMT ME19 6ES
G0IRW ME19 6RR
G0IRV ME19 6RW
M1BJS ME20 6AE
G3XCB ME20 6EB
G0PWX ME20 6EF
G3PYO ME20 6ET
2E0ANZ ME20 6EU
G0VGD ME20 6EU
G6HTS ME20 6HF
G4RIU ME20 6LA
G3KKM ME20 6LW
G4EMC ME20 6LY
G7WJQ ME20 6NL
G7LJQ ME20 6QZ
G8GKC ME20 6RF
G1FHR ME20 6SZ
2E1FTY ME20 6TP
G8PFZ ME20 6TS
G4YLT ME20 6TZ
G1XUE ME20 7SF
G7CEL ME5 9BT
G8IXC ME5 9HA
G0BIX ME5 9HE
M1CAO ME5 9HR
G0AEN ME5 9HX
G4DQN ME5 9LA
G0AKR ME5 9RD
G4MXQ ME6 5EE
G0MLF ME6 5EX
G6HXR ME6 5HJ
G3ZWH ME6 5LS
G6ZMG ME6 5PA
G0JAR ME6 5RL
M1BCH TN10 3BG
G3GVV TN10 3DG
G0DJC TN10 3HG
G7TYR TN10 3HG
G8WZK TN10 3HG
G1OBQ TN10 3LS
G1OGV TN10 3LW
G0HAU TN10 3NA
G0EBS TN10 3TF
G1WTU TN10 3TJ
G1FQK TN10 3TS
G7KNS TN10 4HD
G1BXT TN10 4HR
G4FYG TN10 4LG
G0FSX TN10 4QY
G4HGR TN11 0HU
G3XCT TN11 9BL
G6IUS TN11 9DA
G7GYH TN11 9DE
G7THJ TN11 9ED
G4AUL TN11 9HB
G8CAA TN11 9HD
G4TPJ TN11 9HQ
G4JED TN11 9HU
G3ZDT TN11 9PL
G0KDU TN12 5AW
G4YIM TN12 5EQ
G4CTC TN15 7NR
G4BKW TN15 7SY
G8KW TN15 7SY
G3AMG TN15 8HT
G0CSF TN15 8JU
G1HLH TN15 8JU
G8ZYI TN15 8LL
G4DFY TN15 8LQ
G6AUQ TN15 8LU
G6SFC TN15 8NB
G4TYD TN15 8PE
G6OWT TN15 8PE
G1KEY TN15 8PZ
G3POY TN15 8QF
G3EWM TN15 8RS
G7TPZ TN15 8SA
G4AJS TN9 1JP
G0CKP TN9 1NH
M0AFZ TN9 1UX
G4SLD TN9 2BX
G0RJE TN9 2DU
G8SAS TN9 2DU
G7GDV TN9 2EJ
G6TQ TN9 2JX
G3YPY TN9 2NQ
G4OYW TN9 2QD
G0JUY TN9 2UY

Torbay

G4OVO TQ1 1HU
G4FKU TQ1 1JZ
G1GHU TQ1 1TT
G0CRJ TQ1 2ED
G3BIT TQ1 2HD
G0PQA TQ1 2ND
G6IYM TQ1 2NL
G0NOB TQ1 2PD
G4NWI TQ1 2PD
G4IAS TQ1 2QW
G3YBX TQ1 3BL
G7UHH TQ1 3HB
G1PED TQ1 3RF
G6DDJ TQ1 3RQ
G4OBW TQ1 3TY
G0FGI TQ1 4DT
G0EDE TQ1 4HZ
G1YHL TQ1 4JR
G0SUU TQ1 4LD
G0MHU TQ1 4LW
G3GAO TQ1 4NJ
G1XBY TQ1 4NP
G0ODO TQ1 4PF
G4LEM TQ1 4QS
G8XST TQ1 4QY
G3PXL TQ2 5QX
G7UEK TQ2 5SG
G3SCH TQ2 6DS
G4NOA TQ2 6HF
G4NOD TQ2 6HF
G0SIH TQ2 6JU
G3SDW TQ2 6JX
G7AHP TQ2 6NH
G3YDE TQ2 6UT
G7JHE TQ2 7AG
G3ICQ TQ2 7ED
G0HSW TQ2 7NR
G7SXZ TQ2 7PR
G7VFS TQ2 8EF
G0MRZ TQ2 8HB
G8GRB TQ2 8LF
G3MEP TQ2 8QE
G8UKI TQ2 8QE
G7DKB TQ3 1AB
G0LQM TQ3 1AE
G0UWF TQ3 1AX
G4DDA TQ3 1DB
G6YXT TQ3 1DW
G3XJH TQ3 1EL
G7DMX TQ3 1LE
M0AHW TQ3 1QB
G7UYM TQ3 2AH
G4CLN TQ3 2BN
G7UYO TQ3 2DZ
G8BKQ TQ3 2EN
G0BAJ TQ3 2HT
G4VQI TQ3 2JP
G3YQG TQ3 2LT
G4LPY TQ3 2PT
G7DRQ TQ3 2PU
G7JDV TQ3 2PU
G7VMI TQ3 2PU
G7MCK TQ3 2QS
G4APF TQ3 2RT
G5GW TQ3 2TX
G3GCW TQ3 3BT
G0OJX TQ3 3DN
G7LYG TQ3 3JN
G8ZEV TQ3 3JU
G3CLW TQ3 3JX
G6DPA TQ3 3LZ
G0CGM TQ3 3NA
G4FBF TQ3 3NN
G0CDB TQ3 3NU
G4SSD TQ3 3NU
G7FDC TQ3 3NU
G3RMA TQ3 3PT
G1WUU TQ3 3QN
G1XKQ TQ3 3QN
G4MAW TQ3 3SG
G3HTX TQ3 3XG
G7DVJ TQ4 5AR
G4AKB TQ4 5AU
G4GSZ TQ4 5AU
G0KPA TQ4 5ES
G4UUT TQ4 5HU
G0ODD TQ4 5JQ
G0BMP TQ4 5NG
G4ZPN TQ4 5NS
G1KPI TQ4 5PG
G4TSO TQ4 6HU
G4FLW TQ4 6JU
G4HZU TQ4 6JW
G3VND TQ4 6LF
G6UIM TQ4 6LH
G7LMR TQ4 7AA
G0IBH TQ4 7AD
G4NGR TQ4 7BD
M1ANO TQ4 7BZ
G3ZJE TQ4 7ED
G0UQI TQ4 7HG
G0SWM TQ4 7LF
G0HZB TQ4 7LW
G1RJG TQ4 7LZ
G0KTU TQ4 7ND
G0OSH TQ4 7RL
G4DZH TQ4 7RL
G7JGL TQ4 7RL
GE7OSH TQ4 7RL
G7TNY TQ4 7RU
G0BTH TQ4 7RW
G7JFM TQ4 7TB
G7VQL TQ5 0AG
G7VQK TQ5 0AG
G4SXX TQ5 0AT
G0OJS TQ5 0AY
G3CMT TQ5 0JY
G0TQC TQ5 0LF
G3YCH TQ5 0LF
G0UWS TQ5 0LF
2E1CZQ TQ5 0LF
G3WFC TQ5 0NA
G0RGP TQ5 0PB
G3HJD TQ5 0PB
G3UFZ TQ5 0PB
G8ETU TQ5 0PB
G3RVA TQ5 0RJ
G4WZN TQ5 8JN
G4WVM TQ5 8JY
G2AUK TQ5 8JZ
G4YQN TQ5 8LB
G7SVZ TQ5 8LB
G7UVZ TQ5 8LB
M1CBJ TQ5 8LJ
G3MQX TQ5 8PY
G0WCC TQ5 8QB
G4XPA TQ5 8QD
G6ZOJ TQ5 8QG
G3XNX TQ5 8QR
G4UXP TQ5 8RF
G6YTZ TQ5 8RS
G7SVH TQ5 9AA
G1KPV TQ5 9DU
G0KEK TQ5 9EE
G0VQY TQ5 9EL
G2CKQ TQ5 9HA
G1EFT TQ5 9HF
G1GED TQ5 9JW
G0TDQ TQ5 9JW
G1EZI TQ5 9LJ
G3CBF TQ5 9LJ
G3REW TQ5 9LS
G1AEU TQ5 9PJ
G0TSB TQ5 9PJ
G3KZJ TQ5 9RH
G0HSL TQ5 9RH
G0VUD TQ5 9SH
G3KDV TQ5 9SN
G0PTK TQ5 9UD
G4OKC TQ5 9UD
G0NPY TQ5 9UN
G7VSL TQ5 9UR

Torfaen

GW0FJH NP4 0DG
GW7LPM NP4 0NB
GW0DQT NP4 0NJ
GW4NXD NP4 0PT
GW0KPM NP4 0PW
GW8XZJ NP4 5AJ
GW1VJB NP4 5BX
GW1SXT NP4 5BZ
GW0TTN NP4 5DB
GW7CKR NP4 5DH
GW8CKJ NP4 5EZ
GW4HYZ NP4 5LT
GW4OCN NP4 5LU
2W1ERX NP4 5LU
GW0OSQ NP4 5RU
GW4ZQV NP4 5XJ
GW0KNC NP4 6BP
MW0AUH NP4 6HH
GW0PUH NP4 6NR
GW3RNH NP4 6NR
GW8ERA NP4 7QS
GW7VEH NP4 7TH
GW4UZH NP4 8AF
GW8LHO NP4 8DG
GW4AYQ NP4 8DQ
GW0OLZ NP4 8JW
2W1DEA NP4 8JW
GW6VBC NP4 9ED
GW3XJA NP4 9LQ
2W1AJZ NP4 9LT
GW8XRU NP4 9PA
GW6STS NP4 9QL
GW4IZJ NP44 1AT
GW6WSY NP44 1BN
GW0VAW NP44 1EE
MW0ARB NP44 1EE
GW0UKT NP44 1EE
GW7SIT NP44 1EE
GW0FGO NP44 1NB
2W1ERY NP44 1NB
2E1ERZ NP44 1NB
GW0FYF NP44 1QJ
GW4TUL NP44 1QQ
GW7DIL NP44 1RB
GW6VZW NP44 1SA
GW1XVC NP44 2AH
GW0KAM NP44 2AN
GW1YKY NP44 2AN
MW0AZN NP44 2EJ
GW7UKJ NP44 2EJ
GW3MSY NP44 2NB
GW7PRK NP44 2PR
2W1FEH NP44 2PW
GW4OKI NP44 3JY
2W1FMG NP44 3NR
GW8IAD NP44 4JZ
GW1YZF NP44 4LB
GW7PBR NP44 4LF
GW0HIR NP44 4NW
GW0BVY NP44 4PX
GW4UWR NP44 4QQ
GW0BNP NP44 4QX
GW7VEL NP44 4TE
2W1FMZ NP44 4TE
2W1BHX NP44 4TF
GW6RGT NP44 5EJ
GW8JOY NP44 5HN
GW4IGQ NP44 5HZ
GW0DQY NP44 5TQ
GW1ZFX NP44 5TQ
GW7SSQ NP44 5UD
GW0OZB NP44 5UN
GW4RIB NP44 6HZ
GW0DHA NP44 6JH
2W1DRB NP44 6JH
GW8MER NP44 6JW
GW4RGL NP44 7JG
GW7CBU NP44 7LL
G1BPB NP44 7NZ
GW6BHQ NP44 8HN
2W1DFT NP44 8JE
GW4VUC NP44 8TA
GW7RQV NP44 8TQ
GW4ITJ NP6 1GT
GW6BAH NP6 1PE

Torridge

2E1DOB EX19 8LR
G8CEQ EX21 5EJ
G0LSI EX21 5HA
G7FHW EX21 5JG
G0GHT EX21 5RU
G0NZF EX21 5SZ
G7IQU EX21 5SZ
G0IKC EX21 5UD
G0LAR EX21 5UD
G3BYG EX21 5UF
G7NWB EX21 5XX
M1BNR EX22 6DA
G7VTE EX22 6DA
G8MWW EX22 6QN
G1GZI EX22 6RS
G3VDH EX22 6XX
G6LAU EX22 7AA
G1BHM EX22 7BT
G0CLC EX22 7DL
G0RYO EX22 7DL
G4CQM EX22 7EW
G0RQL EX22 7NY
G8FXG EX22 7RZ
G0EBH EX22 7SP
G8MXI EX22 7TQ
G0PNZ EX31 3LW
G4NCU EX31 3LX
M1BBW EX38 7HY
G4KXQ EX38 7JE
G0OKA EX38 8AL
G1ZUS EX38 8AS
G4UFK EX38 8AS
G4COW EX38 8BY
G5HD EX38 8LU
G0DLC EX39 1BD
G2FKO EX39 1BD
G4ELU EX39 1BD
G0XCF EX39 1BS
G6YSL EX39 1DF
G4NCV EX39 1ET
G0JXD EX39 1HQ
G0OXW EX39 1NX
G4XHK EX39 1NX
G0PGK EX39 1PE
G7MWI EX39 1PW
G4SRP EX39 1RD
G0BGL EX39 1SD
G0UNB EX39 1SG
G6WZN EX39 1TP
G7FJZ EX39 1TX
G1ZTJ EX39 2LL
G0PBC EX39 2RR
G3MNV EX39 2RR
G8SFD EX39 3BN
G7VPB EX39 3DF
G3YGJ EX39 3DJ
G4PUA EX39 3EE
G4UJM EX39 3EE
G4SOF EX39 3ET
G0FCL EX39 3PD
G4ETJ EX39 3QX
G0UMT EX39 3QZ
G0JRZ EX39 3RB
G3YIN EX39 3RW
G4PMB EX39 3SB
G4SJT EX39 4AP
G0AOB EX39 4DL
G4PEK EX39 4DY
G1ZRN EX39 4EU
G6EMP EX39 4HA
G0GFK EX39 4HG
G4MXT EX39 4PQ
G4YBB EX39 5DJ
G0UUI EX39 5DY
G3APU EX39 5PG
G6XFN EX39 5RH
G8TYD EX39 6AQ

Tower Hamlets

G6USX E1 0EE
G8XCJ E14 3AJ
G7KUA E14 3DT
G4ZLH E14 6DW
G6DFY E14 6DX
G7KXG E14 7DW
G0PUO E14 7SH
G0NTM E14 8DY
G0KBO E14 8EY
G4LXH E14 8LH
G1TJT E14 8LL
G7DRK E14 8NN
G1RYS E14 8SR
G0NXM E3 2BB
G8VJR E3 2UF
M0AIG E3 4HA

Trafford

G7RMA M16 0HR
G4UXR M16 7WR
G3NJU M31 2SY
G1SCL M31 3FJ
G1NYJ M31 3HP
G8AFC M31 3PF
G8BRU M31 4DL
G3VYA M31 4DY
G4HAK M31 4NR
G0LZL M31 4PT
G7KRI M32 0AN
M1BBP M32 0AW
G7JKK M32 0AW
G1TJL M32 0UQ
G1PKQ M32 8DQ
M0ANR M32 8PB
G0JZM M32 8QN
G7EKU M32 8QN
G7FKJ M32 9BS
G6KLQ M32 9BX
G0MAI M32 9DP
G4IRM M32 9HA
G4HTS M32 9JB
G6YOP M32 9JB
G1JHX M32 9JH
G0MUL M32 9LJ
G0HUR M32 9QD
G0BJK M32 9QF
G3RIV M32 9RB
G1UFM M32 9SJ
G3HCO M33 2AR
G4URV M33 2EA
G6LCS M33 2FJ
G8LQO M33 2LD
G0KKF M33 2PE
G7GUW M33 2TJ
G1NCR M33 2TL
G7RGZ M33 2XD
G2HW M33 3DF
G1YGH M33 3EF
G3YQD M33 3GR
G3FNM M33 3GS
G3SVW M33 3GT
G3SMM M33 3GU
G4ERX M33 3HP
G3HLM M33 3JB
G0RFG M33 3LH
G0TKL M33 3LQ
G1BOO M33 3QP
G0GBF M33 3QR
2E1EBM M33 3QT
G7NND M33 3WL
G4AOK M33 3WS
G4JZZ M33 4AE
G3LJF M33 4AF
G4EDM M33 4FS
G4KGU M33 4HP
G7TUH M33 4HR
G7FWE M33 4JZ
G4YLK M33 4LD
G4AIH M33 4LP
G0IIM M33 4NU
G4AUR M33 4RF
G0NWJ M33 4RG
G4MYB M33 4WG
G8ILW M33 5DL
G3KCB M33 5FB
G0MRK M33 5NY
G0BHP M33 5QJ
G0TRC M33 5QJ
G0AOU M33 5UF
G3FVA M33 5UF
G7NAT M33 5WN
G3ZDM M33 6EZ
2E1FAF M33 6EZ
G6PFE M33 6FY
G4CLP M33 6LL
G3WFT M33 6NF
G3SUI M33 6NW
G2AKR M33 6QW
G4FPA M33 6QW
G0FRB M33 6WU
G6XBV M33 7FN
G7HKQ M41 0GF
G6PSA M41 0PY
G7SHI M41 0QS
G7VNR M41 0RE
G3JIS M41 0RJ
G3ZKL M41 0TB
G3XGE M41 0TY
G0TON M41 0TY
G1GXH M41 0XR
G0ONF M41 0XY
G7LKL M41 5BZ
G7THA M41 5DS
G0KJQ M41 5SJ
G3YHD M41 5TQ
M1BTI M41 6EL
G3XXG M41 6LE
G8YUK M41 6WA
G1TUN M41 7AZ
G4WXB M41 7BA
G4SDL M41 7BT
G4WLJ M41 7WH
G4EXK M41 7WX
G4MFI M41 7WX
G6LDV M41 8BY
G4RCT M41 8GG
G7MPL M41 8PH
G8HXE M41 8RN
G7EMD M41 8RQ
G7RNO M41 9EL
G4XHT M41 9ER
G0TRG M41 9FG
G4ZRU M41 9FG
G4XDR M41 9FW
G7HMW M41 9JZ
G1YEL M41 9LD
M1ALF M41 9LD
G0CTM M41 9LS
G7AGC M41 9NG
G1LWE M41 9NJ
G1TRC M41 9NN
G4NCZ M41 9PT
G1WOJ M41 9PU
G4UGM WA14 1JL
G7WBJ WA14 1NW
G3WFH WA14 2ND
G4YNS WA14 3AH
G7RHD WA14 3LQ
G2CAZ WA14 4AZ
G7TIR WA14 4QA
G6HLU WA14 4RQ
G4PYH WA14 4UE
G4NAV WA14 5AZ
G1GLS WA14 5HA
G7AGA WA14 5JF
G4HUF WA14 5LT
G4HVP WA14 5LT
G7EYR WA14 5LT
G4UGW WA15 0AN
G4OQE WA15 0LF
G4CNI WA15 0LT
G4FMB WA15 0LT
G4ZAB WA15 0LT
G0KTT WA15 0LY
G0WEV WA15 6AB
G3EOO WA15 6BW
G6RGN WA15 6BX
G7MVY WA15 6BX
G4VBJ WA15 6DY
G6VAH WA15 6DY
G4WQD WA15 6HQ
G6MCE WA15 6HR
G0CBJ WA15 6LF
G7FOV WA15 6LQ
G0FTS WA15 6NG
G4TFU WA15 6NG
G7CAQ WA15 6NG
G7JSF WA15 6NG
G6DOQ WA15 6RS
G7IXI WA15 6RS
G5ZK WA15 7AU
G6DOM WA15 7JD
G4KZE WA15 7NA
G4ZJA WA15 7NA
G3YBD WA15 7QJ
G1HPQ WA15 7RL
2E1EAI WA15 7RT
G1AIV WA15 7SP
G8JH WA15 7TE
G0HAL WA15 7YH
G4YWE WA15 8BW
G1HCX WA15 8BX
G1GNX WA15 8DN
G4UKV WA15 8DN
G6GTH WA15 8DN
G4SVR WA15 8ET
G7VCF WA15 8HA
G6MNB WA15 8LY
G0UXR WA15 8PD
G0GWI WA15 8PW
G3WKD WA15 8QD
G2FUD WA15 8SQ
G0NHQ WA15 9DE
G4ZEN WA15 9HH
G4RVX WA15 9JA
G3FBI WA15 9NT
G0KJK WA15 9RL

Tunbridge Wells

G4XCZ TN1 2JJ
G4ZFP TN1 2LH
G3WTR TN1 2SH
G0XKT TN12 6AJ
G6YCT TN12 6AR
G8MTO TN12 6DY
M0ACZ TN12 6DY
G0MID TN12 6HP
G3TLG TN12 6JN
G0IPO TN12 6WB
G3KOM TN12 7HA
G0GCI TN12 7HL
G0HAX TN12 7JJ
G0IKB TN12 8JX
G2BGU TN12 8LA
G4XYO TN12 8NJ
2E1CQT TN17 1DR
G4WCI TN17 1JR
G6VGA TN17 1LE
G0LNX TN17 2AX
G3GUE TN17 2DD
G0INY TN17 2LA
2E1FVY TN17 3DZ
G7UHB TN17 3LU
2E1DUT TN17 3ND
G3KNI TN17 3NH
G4KZO TN17 4AA
G8PBJ TN17 4AN
G0OEQ TN17 4BT
G1XEX TN17 4HH
G3XWL TN18 4JE
M1AOB TN2 3BL
G4OSH TN2 3DA
G1VTN TN2 3NH
G4KIU TN2 3NH
G6CRR TN2 4BT
G1WKS TN2 4BX
G3AIO TN2 4BX
G3GWD TN2 4DE
G8SFR TN2 4EY
G0WHY TN2 4JB
G6AHK TN2 4LF
G8JXG TN2 4NS
G4KGF TN2 4PB
G3WKS TN2 4SU
G4FDC TN2 4SU
G4HOM TN2 4SU
G3STF TN2 4TJ
G4JFD TN2 5DD
G3EDV TN2 5DQ
G0AOA TN2 5HG
G4CTY TN2 5NB
G6ZXV TN2 5NT
G7DLY TN2 5NZ
G4JZP TN2 5QU
G8HGI TN2 5UN
G4BWH TN3 0AA
G3XPX TN4 0BA
G3VHN TN4 0BU
G1EUD TN4 0DX
G4MIK TN4 0EQ
G0VQA TN4 0HD
G0AAK TN4 0LF
G6NZS TN4 0QB
G7MZL TN4 0QE
G1ZSY TN4 0TQ
G3LMS TN4 8EA
G4GTN TN4 8ED
G3HOU TN4 8EP
G0TBM TN4 8ES
G0OPF TN4 8JS
G7RUX TN4 8PY
G0SLD TN4 8QB
G0ABM TN4 8UN
G4MXL TN4 9AR
G0BMN TN4 9DJ
G3KIP TN4 9DR
G4BMD TN4 9HS
G0VPY TN4 9RP
G6MWB TN5 7PQ

Tynedale

G0NGF CA6 7HA
G1PPG CA6 7JS
G6AUC DH8 9JP
G4AOS NE19 1TA
G3HSW NE41 8ES
G0NEE NE42 6AT
2E1BUJ NE42 6AT
2E1DDO NE42 6AU
G1MOB NE42 6EH
G4STK NE42 6EH
G4XSS NE42 6EH
G0EQC NE42 6JH
G8SFA NE42 6PU
G8RQF NE42 6PZ
G0OYL NE42 6QA
G7JTK NE42 6QA
G2DML NE43 7JX
G3UVU NE43 7RQ
G1JYK NE43 7SF
G7HES NE44 6HU
G1SIO NE45 5BA
G8POO NE45 5EX
G3TSS NE45 5JH
G0NCV NE45 5JR
G7UME NE45 5LE
G7SNJ NE46 1LE
G4IIB NE46 1SX
G3DVY NE46 1UG
G5RI NE46 1UL
G1GLZ NE46 2JY
G6DNV NE46 2QB
G8XHH NE46 3AG
G3KBQ NE46 3DP
G3KMH NE46 3LE
G0CNU NE46 4DD
G1SWP NE46 4LG
G1PIF NE46 4RX
G0GXO NE47 6HG
G4UEO NE47 6HW
G0AXZ NE47 9DR
G7JAU NE48 1EG
G6RTN NE48 1EP
G4IZW NE48 2AP
G3TTI NE48 2RZ
G7OPN NE48 3DS
G4IWJ NE48 4AN
G1TIF NE48 4BD
G3VUF NE49 0QS
G4EUY NE49 9DT

Uttlesford

G3EAY CB10 1QG
G8FFF CB10 1XF
G1OKJ CB10 2AH
G1WSF CB10 2AH
G4YHN CB10 2AJ
G4UHR CB10 2AP
G0UUS CB10 2BG
G4AXX CB10 2DP
G0UPY CB10 2ED
G3TQO CB10 2EF
G6GOL CB10 2LF
G8VAE CB10 2LL
G0TCI CB10 2TZ
G0WRZ CB10 2YE
G8XLB CB11 3BJ
G0GWM CB11 3DN
G0ICB CB11 3ED
G3RGA CB11 3EZ
G0IMG CB11 3HD
G4DUV CB11 3LH
G7CXA CB11 3PT
G7DXK CB11 3QW
G3WMS CB11 3RU
G4KEY CB11 3SJ
G0UEU CB11 4AY
G0IUL CB11 4BE
G7RSW CB11 4BE
G4EZH CB11 4DF
G0HEM CB11 4DJ
G1LTL CB11 4DJ
2E1FIF CB11 4DR
G1PXH CB11 4LA
G0HED CB11 4PE
G8BFS CB11 4QL
G4FZM CB11 4QS
G3TXC CB11 4SJ
G8EBX CB11 4TH
G3NOX CB11 4UU
G0FYG CB11 4XH
G6JRH CM22 6BY
G6IWT CM22 6LG
G6YBH CM22 6QJ
G7LQP CM22 6QW
G4PFJ CM22 6SW
G8BPH CM22 7AS
G8ACT CM22 7JR
G8XFO CM22 7RA
G4WWP CM22 7SJ
G7VJE CM22 7TP
G4UFN CM23 1BZ
G3XSE CM23 5QF
G8LNS CM23 5QG
G8VDG CM23 5QP
G3WXW CM24 8HA
G4YWA CM24 8HG
G3PEW CM24 8LR
G3ONE CM3 1QQ
G8OPI CM6 1BX
G7DPR CM6 1BY
G7JXL CM6 1HN
G6PHY CM6 1NB
G3WCO CM6 1PQ
G7TYU CM6 2AP
G8MYI CM6 2HA
G0IAG CM6 2JN
G7SZZ CM6 3HY
G4FOH SG8 8RP

Vale Royal

G4UGD CW6 9BN
G0RBA CW7 1BY
G1ILH CW7 1EU
G3JWK CW7 1EY
G7VFO CW7 1HA
G6COB CW7 1HE
G1GQH CW7 1JR
G8ATB CW7 1JX
G4TEZ CW7 1LS
G7SOZ CW7 1LT
G7CRG CW7 1NA
G0IRA CW7 1NA
G8ZTT CW7 1NA
G8HAV CW7 1NA
G7RYN CW7 1NE
M1ATI CW7 1NE
G0BYF CW7 1PJ
M1ARS CW7 1QL
G0EOL CW7 2HL
G0GZI CW7 2LY
G7DSA CW7 2LY
G0IHB CW7 2PF
G0TEN CW7 2PF
G0SPH CW7 2SA
2E1BZL CW7 2SA
G4XFD CW7 2UE
G1DBK CW7 3DD
G0ADU CW7 3EN
G0OZI CW7 3LA
G1GWG CW7 3LA
G1DBR CW7 3LE
G0VOK CW8 1HJ
G0FQG CW8 1NF
G0PVA CW8 1NG
G6HXU CW8 1PL
G7HXU CW8 1PL
G3ZTT CW8 1PY
G4WPO CW8 1PY
G4XUV CW8 1PY
G7SUF CW8 1RB
G1GYJ CW8 1RD
G4CJU CW8 2AJ
G0JIT CW8 2BQ
G3MCN CW8 2HX
G3YWU CW8 2JB
G3CVW CW8 2JS
2E1FLM CW8 2NW
G1YJR CW8 2PB
G3SIQ CW8 2PB
G6AFG CW8 2PP
2E1FLK CW8 2QL
2E1FLL CW8 2QL
G7JJJ CW8 2TY
G6DQO CW8 2UR
G4JYP CW8 2XJ
G0HHU CW8 2XR
G1UJV CW8 2XR
G1GSW CW8 3BN
G7VDQ CW8 3BS
G4YVI CW8 3HD
G3JAS CW8 3NL
G4NOC CW8 3NL
G4IAB CW8 3PT
G4XDG CW8 4AA
G0HXD CW8 4DF
G0BKH CW8 4ES
G1MCG CW8 4ES
G7MCP CW8 4HH
G6GAK CW8 4HR
G8UYB CW8 4JQ
G4JJN CW8 4JX
G4ROK CW8 4JX
G6LXI CW8 4PE
G8NCS CW8 4XA
G0DND CW9 5LJ
G1LDC CW9 5PZ
G6PGG CW9 6BA
G8BJA CW9 6DA
G1MVE CW9 6DS
G3MAX CW9 6DX
G1URR CW9 6EB
G6RMC CW9 6EF
G6SHZ CW9 6EJ
G3WCS CW9 6JB
G6LDM CW9 6PH
G4RRJ CW9 6PN
G8VNX CW9 6PX
G4SEC CW9 7AY
G0ONO CW9 7DN
G1JHL CW9 7EJ
M1BQP CW9 7ES
G1GOP CW9 7JB
G1GOQ CW9 7JB
G0OGJ CW9 7QH
G4GIQ CW9 7QH
G4PKX CW9 7QH
G4BVE CW9 7SZ
G7LQD CW9 7XN
G6ZTT CW9 7XN
G0THJ CW9 8BB
M1BTZ CW9 8BH
2E1FSU CW9 8BH
G0UOC CW9 8DS
G0LBO CW9 8QA
G4CAX CW9 8QQ
G4YTB CW9 8QQ
M1BXM CW9 8QS
G1HUX WA4 4EN
M0AWD WA4 4EN
G8KTE WA6 0QX
G0LZJ WA6 0QX
G3GKS WA6 6DN
G3MGK WA6 6PY
G4BYP WA6 6SN
G6VGH WA6 6YA
G1OXM WA6 7BS
G2AOZ WA6 7LE
G2DVA WA6 7LT
G8DVF WA6 7QJ
G4VYJ WA6 7RE
G4OTI WA6 7RU
2E1FUY WA6 8DY
G1IVV WA6 8HP
2E1FKD WA6 8HP
G8MMN WA6 8JS
G8SFS WA6 8JS
G3PFR WA6 8LL
G4ZTT WA6 8LL
G6DP WA6 9EW

Vale of White Horse

G7OGO OX1 5HW
G7VCZ OX1 5LD
G4AZA OX1 5LS
G6EHP OX1 5LT
G3KCZ OX1 5LX
G3JFR OX1 5NF
2E0ADA OX1 5NH
G0RKG OX1 5NH
G0REL OX1 5NH
G0RJX OX1 5NH
G0SXO OX1 5NH
G6YTW OX1 5NZ
2E1AAI OX1 5PB
G7KGH OX1 5PE
G0SPS OX1 5PE
2E1ABE OX1 5PE
G4MFP OX11 0DX
G4XYU OX11 0ED
G0FRO OX11 0ES
G1GGT OX11 0HB
G1HDM OX11 0PW
G3RRS OX11 0QX
G8BFK OX11 0SA
G1TRF OX11 0SU
G4SAJ OX11 9PU
G6DPL OX11 9PU
G8PTY OX12 0AT
G1NBP OX12 0BD
G3VHE OX12 0BN
G0UUH OX12 0HL
G1YDG OX12 0HT
G4DSE OX12 0JA
G0LVF OX12 0JN
G7AKO OX12 0JN
G0ANX OX12 0LD
G3YSQ OX12 0LD
G6SRY OX12 0LD
G7THB OX12 0LR
G1VSO OX12 0NJ
G0MRR OX12 0NL
G0LCB OX12 0NY
G1NDV OX12 0PY
G0OAB OX12 7DF
G8PKN OX12 7DG
G4DPA OX12 7DN
G3CU OX12 7DP
G6LNU OX12 7EP
G0HLS OX12 7EW
G7TGB OX12 7EW
G4LRQ OX12 7JA
G0JEK OX12 7JN
G3TEL OX12 7NP
G1IYB OX12 7PN
G7JIJ OX12 7PT
G4XIW OX12 7QF
G0CEI OX12 8QW
G6YGR OX12 9BT
G8DQK OX12 9EE
G0OQX OX12 9HU
G0MCK OX12 9SD
M0ADN OX12 9XQ
G3VPW OX12 9XS
G3OGO OX12 9YA
G8BGM OX12 9YH
G8AGY OX13 5DB
G3LBA OX13 5DZ
G3ZHL OX13 5HL
G0AOZ OX13 5HS
G3RBP OX13 5HT
G0SXG OX13 5JJ
G1FBQ OX13 5LN
G3VHS OX13 5LR
G3AZT OX13 5QQ
G0MCN OX13 6AS
G7WJZ OX13 6BL
G8UDJ OX13 6BX
G6ZKY OX13 6DU
G4IWZ OX13 6JX
M1BWL OX13 6LB
2E1CVV OX13 6LB
G7TDM OX13 6LB
G0KMW OX13 6LP
G4AEV OX13 6NU
G4DXT OX13 6PH
G0LMX OX13 6QJ
M0ABG OX13 6QJ
G0POR OX13 6SU
G8JFF OX14 1AL
G1KEI OX14 1DW
G0NAL OX14 1EL
G6DYX OX14 1JU
G4WXC OX14 1NJ
G3GCI OX14 1NZ
G1RJW OX14 1PR
G3IWW OX14 1PR
M1AOV OX14 1QU
G7KQY OX14 1SN
G7GNE OX14 1SN
G3ZSX OX14 1XA
G4HFV OX14 1XQ
G0LNL OX14 1XX
G4DNA OX14 2AB
G4EXW OX14 2BN
M1BQV OX14 2BN
G0RRT OX14 2BW
G1PVQ OX14 2DQ
G7TRV OX14 2EQ
G7CAG OX14 2ES
G8IB OX14 2HQ
G7LXU OX14 2LU
G1IVK OX14 2LY
M0AEN OX14 2ND
2E1DCN OX14 2ND
2E1DCO OX14 2ND
G0ROS OX14 2NG
G6NPP OX14 2NS
G7ERO OX14 2NS
G1PUC OX14 3AW
G3HKH OX14 3JF
G4AZN OX14 3PL
G5LO OX14 3PL

G6ZHB OX14 3QB
G6DBJ OX14 3RS
G6YLP OX14 3SQ
G0OYS OX14 3SR
G7VGA OX14 3SW
G0NPA OX14 3SX
G6MRP OX14 3TD
G8NRP OX14 3TE
G7PES OX14 3TJ
G4SQI OX14 3TR
G6VLC OX14 3UL
G3XGZ OX14 3UZ
G8KIG OX14 3XB
G1RLZ OX14 3YD
G4JJP OX14 3YD
G0UZS OX14 4BD
G0EDZ OX14 4BH
G1JTN OX14 4BN
G6GQP OX14 4DH
G3SEK OX14 4HP
G5RP OX14 4HP
G1BDP OX14 4JX
G4BGH OX14 4LA
M1BVB OX14 4LD
G7WDY OX14 5DR
G7TNQ OX14 5JA
GM0UGO OX14 5LJ
G7EWF OX14 5LN
G4JEO OX14 5LW
G6PBZ OX14 5NG
G0RWJ OX14 5NW
G0OMZ OX14 5PP
G4MTA OX14 5PX
G3YFM OX14 5RL
G1OQB OX14 5RN
G7DOE OX14 5RN
G7DOF OX14 5RN
G6PEP OX14 5RQ
G8LEO OX2 9HE
G3UOM OX2 9HZ
G6ASA OX2 9JG
G1LHQ OX2 9NL
G1LVW OX2 9PA
G8POC OX2 9PQ
G4MQR OX2 9QA
G0CFN OX2 9SN
G3XEY SN6 8AZ
G4VBC SN6 8EB
G3ONU SN6 8ER
G0GXV SN6 8LZ
M0AMK SN6 8SQ
G0JBJ SN7 7DG
G3NNG SN7 7EY
G3PIA SN7 7EY
2E1BUN SN7 7EY
G8NRL SN7 7HB
G4RUZ SN7 7NG
G8GKQ SN7 7TU
G4VLW SN7 8AT
G6GNW SN7 8BE
G4PFY SN7 8LE
G6HIX SN7 8LE
G3DDS SN7 8LQ
G4HLX SN7 8LX
G8EQX SN7 8LX
G0MYX SN7 8ND
G3ZEF SN7 8ND
G0ITO SN7 8QR
G0VBH SN7 8RP
G0VBG SN7 8RP

Wakefield

G7VDY S72 9BQ
G7VDZ S72 9BQ
M1AHM S72 9BQ
2E1FSK S72 9BQ
G3GRT WF1 2BT
M1CBL WF1 2JB
2E1FXP WF1 2JB
G4VQR WF1 2LX
G6LIC WF1 2PA
M0BCU WF1 2PX
2E1FHC WF1 2PX
G7WHS WF1 2PX
G4DSM WF1 2RA
G3XIE WF1 3DF
G4LXV WF1 3DN
G0CYU WF1 3DY
G7PNM WF1 3HE
G4GUI WF1 3NQ
G4VRJ WF1 3NW
G4TLM WF1 3PR
G8LGE WF1 3TL
G7BYH WF1 4NY
G0UII WF1 4TG
G7PNA WF1 4TG
G4XUX WF1 4TJ
G7EEI WF1 5AT
G8HTB WF1 5NX
G7BSL WF1 5TD
G7POT WF10 1PF
G4TCG WF10 2RB
G4DTO WF10 3EB
G7MTG WF10 3HY
G3OAR WF10 4LF
G0PMB WF10 4PH
G7MYR WF10 4PH
G7MVE WF10 4QL
G4RQI WF10 5AU
G1SKA WF10 5HU
M1AKW WF10 5RH
G0AGM WF11 0EG
G6OJX WF11 0ER
G3HCW WF11 0EX
G0VXD WF11 0HB
G3FYQ WF11 0JH
G0NQE WF11 0JH
G4FBA WF11 8JF
G0TVL WF11 8JL
G4TFA WF11 8LZ
G8NDQ WF11 8PW
G4IBN WF11 8RN
G0FVN WF11 8RP
G0FMO WF11 8SR
G0VXM WF11 8TD
G0BQB WF2 0BS
G4SMV WF2 0JZ
G4RCG WF2 0PP
G4OOC WF2 0PP
G6XXE WF2 0QR
G4LJK WF2 0RS
G4ROS WF2 0RU
G3TRV WF2 0SP
G6LDP WF2 0TJ
G6NYF WF2 6AA
G1VOD WF2 6DT
2E1DMP WF2 6EY
2E1DMO WF2 6EY
M0AKD WF2 6HJ
G4BLT WF2 6JP
G8DVS WF2 6JP
G0KNR WF2 6JP
G7OKZ WF2 6RY
G0WHH WF2 6RY
G0DIS WF2 6SQ
G7JTH WF2 6SR
2E1DGD WF2 6SR
G8YHI WF2 6SU
G8IJI WF2 7EF
G1OOU WF2 7HW
2E1FEX WF2 7HW
G1YUQ WF2 7HZ
M1AAR WF2 7LF
G2GC WF2 7NF
G7SWH WF2 7PR
G7VDT WF2 7SF
G0LLO WF2 8AP
2E0ALT WF2 8AR
G0OMT WF2 8AR
G1BRR WF2 8AR
G6YTH WF2 8BP
2E0APJ WF2 8EH
G0ISJ WF2 8JG
G0JMZ WF2 8LD
G7GLO WF2 8NF
M0AOS WF2 8UF
G3NDV WF2 9EE
G6AJS WF2 9JZ
G6MZJ WF2 9NX
G4VYZ WF2 9PB
G0UPF WF3 4BD
G6YMZ WF3 4EJ
2E1EPM WF3 4HY
G1VOX WF3 4JA
G0EVT WF3 4JJ
2E1EPC WF3 4JJ
G8OJK WF4 1EF
G1XBE WF4 1EH
G8UCH WF4 1EX
G4AYL WF4 1PF
G4MSZ WF4 1PQ
M1CEJ WF4 1QX
2E1EHX WF4 1QX
G7HYZ WF4 1RZ
G3LRP WF4 2BG
G0TYS WF4 2EH
G4GJB WF4 2JJ
G0DXP WF4 2NF
G0HDP WF4 3BL
M1CBU WF4 3EG
G4XDV WF4 3JR
G7IOF WF4 3NL
G2FKZ WF4 3NW
2E1DML WF4 3PB
G4IZH WF4 3PG
G0FEV WF4 3PR
G1FUI WF4 4JQ
G8XVK WF4 4JX
G4IAU WF4 5AN
G4NOK WF4 5AN
G0BWL WF4 5AZ
G4SPI WF4 5BN
G3PXF WF4 5JF
G4PZM WF4 5LN
G4MMY WF4 5LS
G1TJX WF4 5NQ
G1AHS WF4 6AE
G3RZX WF4 6BX
G4DWO WF4 6EQ
G6LYZ WF4 6JW
G4XNR WF5 0DL
2E1ENM WF5 0DL
G4ZQG WF5 0DS
G1WXK WF5 0PP
G0MVA WF5 0QE
G1HVL WF5 0RF
G0MJZ WF5 0RU
G0RJF WF5 8BA
G4TDI WF5 8EJ
G3SEY WF5 8LF
G1ISP WF5 8LH
G0LKP WF5 8LP
2E0AOT WF5 8NA
G7VLQ WF5 8QN
G4MVE WF5 8QP
G1VAO WF5 8RY
G3VQQ WF5 8RY
M0BFO WF5 9EU
G8OHW WF5 9EU
2E1FZD WF5 9EU
G7NTI WF5 9EW
G6AAK WF5 9SJ
G0PHS WF6 1EE
G7DJY WF6 1HB
G4SAV WF6 1SS
G4SBG WF6 1SX
G6MID WF6 2AA
G6XEX WF6 2AA
G0TBY WF6 2NT
M0ADK WF7 5EB
M1ADL WF7 5EX
G4WYP WF7 5LR
G7IJY WF7 5LW
G4NRF WF7 5NJ
G1ZGH WF7 6AH
G0HPA WF7 6BL
M1BVQ WF7 6DT
G0LES WF7 6EZ
G6LVE WF7 6HD
G8GKX WF7 6HD
G3ESP WF7 7BS
G3US WF7 7DF
G4ICB WF7 7DS
G0KNH WF7 7HE
M0AGU WF7 7HU
G0KMN WF7 7LA
G1LFI WF7 7LA
G4KCI WF7 7PE
G6MYG WF7 7PW
G4ISU WF8 1SB
G0RLN WF8 2AJ
G7HQC WF8 2DJ
G1FYQ WF8 2QQ
G4KMW WF8 2QQ
G0FTA WF8 2QT
G3HCX WF8 2RA
G0SEG WF8 2RT
G4FPE WF8 2TF
2E0AOZ WF8 2TW
G6WBT WF8 3NJ
G0TVU WF8 3NT
G0GEA WF8 3NW
G4WKC WF8 3PN
G4AAQ WF8 3QT
G0RAE WF8 3RA
G6VZS WF8 4BX
G0SNS WF8 4LG
G4YNU WF8 4ND
G3SYC WF8 4QH
G3VTD WF8 4SJ
G0BPK WF9 1AZ
G8YCQ WF9 1ED
G1CNY WF9 1NU
G8HRA WF9 2BT
G0CYX WF9 2DD
G0FYM WF9 2DD
G1CMT WF9 2LN
G6OXZ WF9 2TL
G1DKV WF9 2TL
2E1FSX WF9 2TL
G4SPM WF9 3EW
2E1CIY WF9 3NG
G4ZVB WF9 3QE
G0GEH WF9 4AY
G1PPX WF9 4JA
G7PLH WF9 4JD
G1MLJ WF9 4JQ
G4PLL WF9 5AJ
G1FFH WF9 5HE

Walsall

G3NAV B43 7LQ
G4PFK B43 7PG
G1FFA B43 7QR
G4RDD B74 2BG
G4DDE B74 2BQ
G6HOC B74 2BS
G8SYS B74 2DA
2E1EQE B74 2DA
G7JJW B74 2EA
2E1DSU B74 2EG
G8GWR B74 2EP
G8KFF B74 2JE
G4DDD B74 2LA
G7DRI B74 3DL
G4RJM B74 3JD
G8KYM B74 3JF
G6VPL B74 3JU
G6RBO B74 3LR
G0AXV B74 3LS
G8LTW B74 3NF
G1MGZ B74 3NP
G1OBA B74 3NP
G0CMH B74 3PE
G3XFN B74 3PG
G4IWF B74 3PG
G7OBC B74 3PQ
G6UKZ B74 3QH
G1EYT B74 3QP
G3DJJ B74 3SH
G0FSL WS1 2DA
G0GUW WS1 2DA
G3FKU WS1 2LD
G3JMR WS1 2LD
G1LMR WS1 2NQ
G4EJU WS1 2PJ
G7TTE WS1 2PJ
G4KBA WS1 3AL
G0HUD WS1 3HB
G1MMN WS1 3LB
G0WLK WS1 4DX
G4YVF WS1 4DZ
G4FCB WS1 4HF
G1RBA WS10 7RQ
G6RTG WS10 7RR
G4YLS WS10 7SF
G8DLP WS10 7SG
G1WYK WS10 8BN
G6PUH WS10 8NT
G0WCI WS10 8RN
G1MEZ WS10 8TN
G6ZYX WS10 8UB
G1WRU WS10 8XY
G6XVM WS2 0AQ
2E0AAM WS2 0DY
G6ORQ WS2 0EE
G0CLX WS2 0EG
G0RDL WS2 0EU
G0JVN WS2 0HJ
G4VML WS2 0HT
G6YET WS2 0HT
G4EFG WS2 7BB
G7PHN WS2 7EZ
G1YFH WS2 7HN
G3KUG WS2 7HY
G0HWP WS2 8RX
G4RLI WS2 8TT
G8ICJ WS2 9BQ
G8GMT WS2 9BY
G7TGG WS2 9HZ
G8DGX WS2 9QB
M0AVL WS2 9QX
G7NLL WS2 9QX
G0ENM WS2 9TF
G6STL WS2 9TJ
G6EOX WS3 1AW
G8ZHA WS3 1AW
G4HHP WS3 1DR
2E1CLQ WS3 1DT
G1ZDA WS3 1PU
G4PPC WS3 2EZ
G7PPC WS3 2EZ
G0WRG WS3 2EZ
G7WRG WS3 2EZ
G7KMD WS3 2HT
G0MKV WS3 2HT
G0MKW WS3 2HT
G7DLD WS3 2HT
M1AGW WS3 2NG
G7GCU WS3 2RS
G0VFB WS3 2SS
G6YLT WS3 3EQ
G1LUF WS3 3NB
G4VFO WS3 3PF
G6SRN WS3 3PF
G0GBE WS3 3PG
G1SJB WS3 3PG
G0KFS WS3 3QA
G0CXO WS3 3QD
G1YWI WS3 3QD
G8IJE WS3 3RF
G0CNG WS3 3XB
G1WQH WS3 4BD
G6VKI WS3 4BD
G6BYL WS3 4DH
G4CBK WS3 4EA
G4WAF WS3 4EP
G1XLW WS3 4ET
G6ERN WS3 4EZ
G6LRD WS3 4EZ
G0MLY WS3 4HJ
G1DKV WS3 4HJ
2E1ATE WS3 4HJ
G1BCQ WS3 4JL
G4EQG WS3 4LB
G0DAC WS3 4LS
G3GGR WS3 4LW
G8AGR WS3 4PB
G7BNK WS3 4QN
G1SHU WS4 1BB
G0IMX WS4 1DQ
G4PVR WS4 1EB
G0JAB WS4 1EP
G4VHT WS4 1HT
G4YZM WS4 1HT
G6EQD WS4 1HW
G7DOB WS4 1JB
G7FWD WS4 1PP
G0UYP WS4 1QU
G4DXZ WS4 1QU
G1BJJ WS4 1RP
G4BXC WS4 1RP
G3KOP WS4 1RS
G4UDV WS4 1XA
G6DMW WS4 2EN
G4GJE WS4 2HB
G6JJO WS5 3AA
M0AST WS5 3AD
G1WBG WS5 3AY
G3YHN WS5 3DH
G8MLW WS5 3DH
G1RLR WS5 3DT
G1UBC WS5 3DT
G3WVJ WS5 3EF
G6JVA WS5 3EH
G4GKC WS5 3ES
G6HLL WS5 3ES
G3IGI WS5 3LX
G4SEL WS5 3NQ
G8PID WS5 3NS
G0NOL WS5 3PF
G8NQY WS5 3PN
G0EYV WS5 3QB
G4CIF WS5 3QL
G6UEU WS5 3QW
G3ZHC WS5 3QX
G3TIN WS5 4DN
G4SSR WS5 4DN
G8HTM WS5 4DN
G3VMQ WS5 4DN
G0OYK WS5 4LR
G7NQV WS5 4NE
G8XXU WS5 4NH
G8KXO WS8 6DL
G0MPO WS8 6JF
G6UFP WS8 6JG
G6PJM WS8 7AF
G0OEI WS8 7AY
G0FBO WS8 7BZ
G1JMW WS8 7BZ
G0BSM WS8 7DN
G3OHN WS8 7PL
G0NEQ WS9 0BA
G4MVJ WS9 0BB
G3NAL WS9 0BP
G1EMT WS9 0DQ
G7IXP WS9 0DQ
M1AZD WS9 0ES
2E1CBA WS9 0ES
G4NNU WS9 0HW
G4RLJ WS9 0JR
G4GJU WS9 0QA
G4XFT WS9 0QE
G0HYZ WS9 0SY
G4FCE WS9 0TD
G8NZP WS9 0YY
G0UNR WS9 8HD
G3VTI WS9 8HP
G4FGP WS9 8JA
G4EWI WS9 8JE
G4FAP WS9 8JG
G4PJK WS9 8LA
G6LRL WS9 8RX
G0OEB WS9 8XE
G7WCB WS9 9BD
G4SMP WS9 9DS
G0ALC WS9 9JR
G4KBO WS9 9JW
G7APS WS9 9LS
G7ANK WS9 9NX
G4FAJ WS9 9PD
G8LPI WS9 9RD
G6IFT WS9 9RN
G4JZF WV12 4AN
G4CFR WV12 4DY
G0KDK WV12 4EE
G0IJJ WV12 4EY
G0KBK WV12 4JD
G3TQL WV12 4LW
G1XII WV12 4SL
G4CFH WV12 5AU
G6NAJ WV12 5DT
G0KRY WV12 5DZ
G6IHV WV12 5EJ
G7ISV WV12 5PL
2E1AGJ WV12 5QF
2E1ATD WV12 5QN
G0BJZ WV12 5RF
G4VIT WV12 5TD
G1YLJ WV12 5YJ
G3RJD WV13 1BZ
G0PJG WV13 1EB
G6BLS WV13 1HN
G8GMA WV13 1JQ
G8CBC WV13 2EB
G0OVK WV13 2LP
G4OGK WV13 2PS
G6HRX WV13 2RY
G0TMR WV13 3BZ
G0TMG WV13 3BZ
G8GMC WV13 3DD
G0KRK WV13 3DG
G4ETW WV13 3DG

Waltham Forest

G0BJS E10 6DL
G4VXP E10 6EE
G8VZD E10 6QT
G7JRJ E11 1BA
G4ZJX E11 1DT
G8ZNE E11 1EE
G0VEH E11 1JX
G4KGW E11 1JZ
G8TEX E11 3JH
G3AXF E11 3LP
G7LAF E11 4AA
G4FAG E11 4AW
G1JPS E11 4HY
G7FBT E11 4QT
G8NGF E17 3EJ
G6HRK E17 3QJ
G3PKQ E17 3RJ
G1HLS E17 4BD
G4RNB E17 4BH
G8PWK E17 4HH
G1ISS E17 4JN
G4HMS E17 4RE
G4HUK E17 4RE
G7HMS E17 4RE
M1ADZ E17 5AF
G0CKD E17 5AQ
G6CJN E17 5AX
G8KNF E17 5AX
M1AYT E17 5DL
G4YOF E17 5PL
G8WYI E17 7HN
G0FVB E17 8EH
G0OVC E17 9DN
G8BVY E17 9EH
G0VUB E17 9HB
G3WVW E4 6BQ
G0CGH E4 6DQ
G6EXN E4 6DR
G4HNC E4 6EF
G4YOH E4 6EU
G4DZV E4 6LU
G4CGA E4 6NA
G0WMP E4 6PX
G4DAL E4 6QW
G3JPG E4 7BP
G0IPL E4 7DB
G4ZLK E4 7DB
G8IUC E4 7DX
G2XG E4 7HN
G4SWQ E4 7HX
G7ITX E4 7HX
M1BLJ E4 7JG
G8JYX E4 7LG
G4BRG E4 7LL
G8JM E4 7PF
G7TET E4 7PJ
G7IQM E4 8BA
G7AHR E4 8BG
G4NEL E4 8DZ
G8MJH E4 8DZ
G4RSX E4 8HD
G0NQF E4 8YE
G7VAB E4 9AL
G0HXR E4 9DF
G0JPI E4 9DF
G0LWN E4 9DU
G6SKR E4 9HE
G7JEF E4 9HX
G1JOL E4 9JP
G0DPT E4 9LW
G1MBW E4 9LW
G6XRF E4 9PS
G1JOI E4 9RW
M1BXR E4 9SS
G3WKG IG8 9QY

Wandsworth

G4CKS SW11 1RH
G3ZLY SW11 2AY
2E1EXN SW11 3AJ
G6AYS SW11 3NN
G1EGB SW11 4AA
G0VMC SW11 4AA
G2CUJ SW11 5LN
G4LCS SW11 5SJ
G7OLH SW11 6NB
G7IWU SW12 9LU
G6CKY SW12 9SS
G8CES SW15 1LA
G7WCI SW15 1PQ
G4IZO SW15 2EG
M0AUD SW15 2PW
M0AGX SW15 3AD
M1AFK SW15 3AQ
G3TTY SW15 3LP
G1HRD SW15 4EQ
G1SAK SW15 6EX
G7OXV SW15 6NW
G4IKO SW16 1RH
G8ZOJ SW17 0EB
G6MDF SW17 0LN
G1LAR SW17 0QQ
G0KUE SW17 8HP
G0WBZ SW17 8HP
G7NLF SW17 8PR
G3LCH SW17 8QZ
G8TQV SW17 8SF
G8OXI SW18 1PN
G3PAQ SW18 2AJ
G4PZJ SW18 3BB
M1BUA SW18 3JU
G8JIC SW18 4QX
G0TEO SW18 4QX
G4PKK SW18 5AN
G0AQN SW18 5JJ
G3TSW SW18 5LZ
G6XZL SW18 5LZ
G0HSV SW18 5PW
G8HYL SW19 6EF
G4GYY SW19 6HY
G4UNI SW19 6JA
G1HJA SW19 6JW
G4DHX SW19 6JW
G4MRP SW19 6PF
G4CGD SW19 6QU
G8ILZ SW19 6SJ

Wansbeck

G4TMQ NE22 5BU
G1DDS NE22 5DZ
G4UTQ NE22 5ER
G3XAI NE22 5JD
G4SLN NE22 5RE
G0FBH NE22 5YB
G1TIL NE22 5YB
G4WNG NE22 6DF
G3OTH NE22 6DR
G0OWE NE22 6HW
G1DSB NE22 6JJ
G4AXF NE22 6LD
G0GSV NE22 6LE
G0DJO NE22 6NE
G1LMX NE22 7DY
G0ACR NE22 7EE
G7SNP NE22 7EH
G0NDG NE22 7HD
G4WOX NE62 5AW
G1NOS NE62 5DN
G4TQA NE62 5DN
G4ZOY NE62 5LD
G7VBD NE62 5QP
G4OLC NE62 5UE
G4URW NE62 5UE
G6WJD NE62 5UE
G0WZG NE62 5UF
G0KUI NE62 5UL
G3ZKG NE62 5XF
G0NDK NE62 5XG
G4VYX NE63 0BS
G0PWE NE63 0BT
G0KGY NE63 0LE
G4OIV NE63 0LE
G8CTL NE63 0LY
G7FFR NE63 0PL
G7SRE NE63 0UB
G4XXY NE63 8EL
G0CQH NE63 8JF
G4AVO NE63 8JF
G7UZS NE63 8JF
G0AKW NE63 8JT
G8VIB NE63 8JT
G0PJV NE63 8RZ
G7VDA NE63 9BB
G4AVN NE63 9BG
G7VDO NE63 9BG
2E0AOU NE63 9JT
G1BNN NE63 9JT
G3CQO NE63 9LL
G6JXS NE63 9QJ
G0ILR NE63 9RZ
G7VDV NE63 9SD
M0AHZ NE63 9TJ
G3KLP NE63 9TP
G0UDZ NE64 6PU
G0VAD NE64 6XN
G0LEF NE64 6XT

Warrington

G4FMI WA1 3EN
G7ONZ WA1 3ET
G4KAG WA1 3HT
G6DSW WA1 3SE
G1VSK WA1 3UN
G1NWA WA1 4BJ
G7JOA WA1 4BJ
G7JXC WA1 4DP
G0ANL WA1 4DY
G4TZT WA1 4ED
G0PBE WA1 4HA
G0CDA WA1 4LU
G6DMJ WA1 4LU
G6HRZ WA1 4ND
G7PCA WA1 4NF
G0IHY WA1 4NU
G0PZP WA1 4NW
G1NXT WA1 4NY
G3VJV WA13 0JS
G4JME WA13 0PE
G3NKH WA13 0RD
G0STA WA13 0TR
G8VMQ WA13 9BA
2E1FXV WA13 9LU
G8TYH WA13 9NL
G4AUX WA13 9NQ
G4VSX WA13 9QA
G0BAA WA13 9QE
G4WCE WA13 9QE
G4UFW WA13 9QG
G7OJE WA13 9RB
G3YRQ WA2 0AG
G1KSK WA2 0BE
G3NFB WA2 0BL
G4VSS WA2 0EW
G0WRS WA2 0EW
G6OSV WA2 0EZ
G6ORD WA2 0QF
G6MKD WA2 0QW
G1KFR WA2 0QZ
G1OND WA2 0SQ
G4AVD WA2 0UJ
G7BXE WA2 7AX
G7VAG WA2 7QB
G1VPH WA2 7QH
G1FOW WA2 7SE
M1AGJ WA2 8AU
G8JSM WA2 8QN
G4XDH WA2 9AJ
G0RFT WA2 9BQ
G7NEX WA2 9BX
G0NOR WA2 9NE
G0VQK WA2 9PS
2E1FSB WA2 9RW
2E1FUT WA2 9RW
G7HKN WA2 9TW
G0TDX WA3 4DD
G4LWY WA3 4DF
G4ASP WA3 4DL
G0UTU WA3 4DR
G1IQK WA3 4ES
G7IXD WA3 4HY
G0RPG WA3 4LD
G1NXS WA3 4NW
G0WJX WA3 4NW
G7GZB WA3 4NW
G8NDE WA3 4PD
G0ACP WA3 5EB
G7KXA WA3 5HX
G0IZN WA3 5JJ
G3XDP WA3 5QY
G6HKN WA3 5RU
G6DSP WA3 5RW
G4YKR WA3 6AQ
G6RIB WA3 6ER
G1NEE WA3 6JH
G1XOA WA3 6LU
M0AAS WA3 6NG
G6HLL WA3 6PB
G1ZOY WA3 6RU
G4HGF WA3 6TH
G0BCU WA3 6TZ
G0JPC WA3 7LR
G4SSJ WA3 7LY
G4SCI WA3 7NB
G0TDC WA3 7NB
G7KEL WA3 7NS
G7KEM WA3 7NS
G0YSS WA4 1DW
G7CED WA4 1EX
G1KOD WA4 1RJ
G0MWY WA4 1RU
G3SBI WA4 2BL
M0AKR WA4 2DU
G6MQI WA4 2EU
G4XQA WA4 2HA
G0KXW WA4 2HE
G8VLJ WA4 2HH
G4YPI WA4 2HQ
G4ZSL WA4 2HR
G0CKZ WA4 2HS
G1FFV WA4 2JE
G7HRN WA4 2JF
G4PBZ WA4 2RE
G0RPF WA4 2RW
G6GEL WA4 3BJ
2E1EDQ WA4 3HF
2E1ETC WA4 3HF
G3JDT WA4 4BY
G3ZRN WA4 4BY
G4TQH WA4 4NF
G4EAQ WA4 5AW
G3OPT WA4 5BW
G8UAS WA4 5BY
G6MLM WA4 5DB
G6PKA WA4 5EA
G0MYN WA4 5EJ
MW0AWC WA4 5JR
G0GSO WA4 5NP
G4LLG WA4 5QD
G0KXY WA4 6AF
G1DVA WA4 6AP
G3EZX WA4 6AZ
M1AUH WA4 6DZ
G6VDL WA4 6JE
G1PIY WA4 6LE
G1KPY WA4 6LJ
G6NBP WA4 6LT
G4BQJ WA4 6QU
G3OGQ WA4 6SZ
G6EII WA4 6XA
G6MCS WA5 1EW
G0CWA WA5 1HN
G8GTW WA5 1JY
G7FFT WA5 1NB
G8HTP WA5 1XA
G0CVM WA5 1XG
M1ASC WA5 1XH
M0BBA WA5 1XH
G3ZHE WA5 2AQ
G6UCI WA5 2AR
G0TYZ WA5 2AR
G0NVT WA5 2BY
G3ZKD WA5 2DD
G6FJA WA5 2DP
G7UXN WA5 2HS
G4TGQ WA5 2HX
G6JOX WA5 2NN
G7GGM WA5 2PA
G0KKU WA5 2PH
G0SLR WA5 2QE
G6DPH WA5 2QE
G1GNS WA5 2QY
G8WAJ WA5 2RB
G4HSS WA5 2SG
G6WRC WA5 2SR
G8NRF WA5 2SR
G3DTJ WA5 2TA
G3RRM WA5 2TW
G3XUB WA5 3AN
G1DWU WA5 3BQ
G0APY WA5 3DA
G0OON WA5 3EE
G4KCB WA5 3EJ
G1VVN WA5 3EN
G8VVP WA5 3ET
M0ANM WA5 3ET
G0RBI WA5 3FA
G3VZU WA5 3LU
G6XRE WA5 3NW
G7DQX WA5 3NZ
G8TLI WA5 3NZ
G4WNW WA5 4AX
G7VFW WA5 4HN
G4OMZ WA5 4JD
G6YHL WA5 5AG
2E1FTL WA5 5PS
M1CBP WA5 5PS
G8RCL WA5 5QR
G8FWM WA5 5RL
G4LPH WA5 5SJ
G7MEU WA5 5SR
G0CBN WA5 5SX
G0CPD WA5 5SX
G8NHD WA5 5XA

Warwick

G4KRO B94 6LE
G3UAL CV31 1JT
G3XKE CV31 1RJ
G6YMW CV31 1UL
G8TXF CV31 1UL
M1APY CV31 1UQ
G0GIL CV31 2JZ
G3OYS CV31 2NF
G0RTI CV31 2PB
G4XVO CV31 2QB
G7SCM CV31 2QJ
G3YFV CV32 5FA
G4PKT CV32 5JG
G6ISG CV32 5LU
G7ANF CV32 5PG
G8WCZ CV32 5XW
G8EYP CV32 5YZ
G8UIO CV32 6BA
G3EHA CV32 6BN
G8SFB CV32 6ED
G0BNE CV32 6HE
G7GPA CV32 6HE
M1BHS CV32 6HE
2E1EYH CV32 6HE
2E1FRD CV32 6HE
G0GRM CV32 6HF
2E1FQY CV32 6HY
G7DIW CV32 6LR
G3MID CV32 6NE
G6ESK CV32 6NN
G4TYA CV32 6NT
G8FOZ CV32 6PP
G8CDC CV32 6QR
G7JXQ CV32 7AB
G8DF CV32 7BU
G7JOD CV32 7DQ
G0NUF CV32 7DY
M1AUK CV32 7DY
G1YMN CV32 7DZ
G4DXN CV32 7ED
G0FEH CV32 7EL
G8HRI CV32 7EY
G8TFF CV32 7HN
G8CXL CV32 7HP
G8WRG CV32 7HP
G7TEC CV32 7LG
G4RHT CV32 7NB
G0KPH CV32 7QS
G0DWJ CV32 7SN
G3XNU CV32 7TS
G4SWK CV32 7TT
G0LXG CV32 7TT
G6WAR CV32 7TT
G3UDN CV32 7TT
G0DXN CV32 7UR
G6VQW CV33 9AQ
G0HIH CV33 9BW
G4HGM CV33 9RW
G0ASX CV33 9RZ
G0GLQ CV34 4NU
G7TMA CV34 5EX
G0KAQ CV34 5HW
G4WKS CV34 5HW
G8XDL CV34 5NL
G7RMW CV34 5NL
G0CRB CV34 5PA
G1FIP CV34 5PB
G4RTU CV34 5PL
G0OYR CV34 5PU
G8AIM CV34 5RB
G4TIF CV34 5RQ
G1HRL CV34 5TG
G4JFS CV34 5TS
G6YLB CV34 5UJ
G0GLU CV34 5XD
G1WXF CV34 5XD
G0FGC CV34 5XP
G3NAR CV34 6AN
G8OVO CV34 6AR
G0PDO CV34 6AT
G8XJU CV34 6LR
G6RAV CV34 6NE
G3VRW CV34 6PF
2E1EKY CV34 6PP
G3TTC CV34 6QA
G3PXU CV34 6QL
G0LVD CV34 6SL
G2LO CV34 6TN
G4RCR CV35 7DX
G1WUM CV35 7HF
G4NRP CV35 7HN
G3XTN CV35 7QL
G6BPY CV35 8BT
G8NMK CV35 8BT
G7GUG CV35 8QP
G4MQW CV35 8QR
2E1FRG CV35 8RS
G4BUD CV35 8SZ
G6XCC CV35 8TY
G4HAD CV35 8XJ
G4JUY CV35 8XJ
2E1DFJ CV8 1AD
G0AJB CV8 1AH
G7RYO CV8 1AH
G1UUT CV8 1BE
M0AZJ CV8 1DJ
G6UMP CV8 1DJ
G0KPG CV8 1EZ
G0OGV CV8 1EZ
G4XLF CV8 1EZ
G4GJY CV8 1FL
G0IZZ CV8 1FQ
G6GQF CV8 1HX
G3VTL CV8 1PL
G7JAI CV8 1QP
G8CYA CV8 1RB
G6XRI CV8 2BE
G4DF CV8 2BJ
G0INS CV8 2DS
G0DKV CV8 2DU
G3PQ CV8 2DU
G0JMW CV8 2EE
G0PYD CV8 2FE
G0FBY CV8 2HD
G1HGD CV8 2HG
G7IBA CV8 2HG
G4NUH CV8 2JR
G8YDA CV8 2QF
G6VGZ CV8 2TT
G3TFC CV8 3AP

Watford

G6JGF WD1 2RQ
G3EBP WD1 3AU
G4BYS WD1 3BB
G4IVT WD1 3DA
G6ELH WD1 3DB
G4TJA WD1 3DD
G8KXI WD1 3DJ
G8LDY WD1 3DP
G3ZRG WD1 3DX
G0PAU WD1 3PJ
G8FLX WD1 3SR
G3SHY WD1 3SU
G4KUJ WD1 3SW
G3WSB WD1 3SZ
G3WDX WD1 4DH
G0HTM WD1 7JD
G0NUR WD1 7JD
G7RTM WD1 7JD
G6LZB WD1 7PT
G1DZO WD1 7RT
G0FIB WD1 7RY
G4CNH WD1 7SB
G4YRN WD1 7SB
G7VEF WD1 8AY
G8BJK WD1 8DH
G0IDC WD1 8ES
G8IIZ WD1 8HQ
G3YXW WD1 8JY
G0OYC WD1 8NX
G7NLW WD2 4NU
M1ASX WD2 4RH
G3YHY WD2 5JQ
G1IGA WD2 5RP
G6CXQ WD2 5SE
G0MDM WD2 6EH
G6FAF WD2 6HG
2E1EQC WD2 6PS
G4RMC WD2 6QH
G7COT WD2 6QW
G6OBU WD2 7EW
2E1EVG WD2 7LY

Waveney

G7BLK IP18 6RE
G8TBV IP19 8DB
G6AGF IP19 8DS
G0JJE IP19 8EG
G0TLV IP19 8JQ
G3DVA IP19 8LS
G4HMH IP19 8QE
G4JWF IP19 8TG
G0HSI IP19 8TH
G0FJK IP20 0NY
G6IUF IP20 0NY
G7PXY NR32 2AD
G0RRI NR32 2AL
2E1BPN NR32 2BY
G3JRM NR32 2BZ
G4RLS NR32 2BZ
G4XAE NR32 2LB
G7OIL NR32 2NH
G3ZZI NR32 2QP
G7PGQ NR32 2SA
G8KPI NR32 2SU
G0FIY NR32 3DA
G1MPL NR32 3EL
G3DDK NR32 3ET
G0VMF NR32 3HS
G7BZN NR32 3JJ
G1MSG NR32 3JZ
G0FJN NR32 3NS

G2UK NR32 3PJ
G7TJZ NR32 4AU
2E1BNV NR32 4DH
G4CPW NR32 4DR
2E1CTF NR32 4HF
G8JBD NR32 4LW
G6HU NR32 4NW
G3TSF NR32 4PF
G8VOJ NR32 4PH
2E1BMC NR32 4PH
G3AER NR32 4QB
G3RXF NR32 4RB
G0IIY NR32 4RW
G0JSG NR32 4SU
G4RMT NR32 4UE
G4ROJ NR32 5AN
G3PPD NR32 5AT
G4TAD NR32 5BL
G8JUK NR32 5DW
G4VYN NR32 5EZ
G8MRQ NR32 5EZ
M1BUS NR32 5HF
M1BMX NR32 5HF
G4UKX NR32 5HZ
G4RKP NR32 5LB
G4FCZ NR32 5LL
G0VDQ NR32 5LR
G0PRN NR33 0DB
G1IAD NR33 0JW
G8HAU NR33 0LG
G4KDL NR33 0TZ
G8RHQ NR33 7AP
2E1DQY NR33 7DP
G6WJW NR33 7HY
2E1ERI NR33 7LJ
G7POZ NR33 7PS
2E1BMU NR33 7PS
G8BTX NR33 7QZ
G6YBM NR33 7TE
G8FGY NR33 8BQ
G4SAI NR33 8BZ
G1RGV NR33 8HN
G3WDI NR33 8JZ
G3LYX NR33 8LF
G0FDT NR33 8NG
G6MCB NR33 8PJ
G7BYI NR33 8PJ
G1ZCS NR33 8QG
G0WLS NR33 8QN
M1ATH NR33 8RF
G7SDM NR33 8SD
G6NKS NR33 8TL
G0JDL NR33 8TP
G3JMX NR33 9AQ
G7PQB NR33 9DL
G1XTP NR33 9DN
G3XGK NR33 9DT
G3XSK NR33 9EE
G1CYN NR33 9HG
G3GNK NR33 9JD
M1ATP NR33 9JU
G3ECI NR33 9NT
G0ULS NR33 9NX
G3YDZ NR33 9PG
G7MWM NR33 9QL
G7RXI NR34 7DJ
G4RCL NR34 7QW
G3RAE NR34 7RD
G3WDN NR34 7RZ
G0SBH NR34 8AT
G0KGC NR34 8NE
G0MIV NR34 9LF
G0WBG NR34 9NE
2E1BLA NR34 9QT
G4KFS NR34 9RT
2E1BPP NR34 9UB
G3SKK NR34 9UX
G3KIJ NR34 9YW
G1UES NR35 1DF
2E1FDV NR35 1LH
G0LRH NR35 1QZ
G0JKI NR35 1RE

Waverley

G3ZSS GU10 1RH
G3IAF GU10 2EP
G1RUG GU10 2JG
G4MZC GU10 2PA
G4SXC GU10 3HR
G4IVQ GU10 3PB
G8RYJ GU10 4JW
G8UMA GU10 4PA
G1EML GU10 4PQ
G1EMM GU10 4PQ
G4VQT GU10 4QP
G7SYU GU10 4QX
G4AHN GU10 4RL
G4LES GU10 4TF
G6BOF GU10 4TP
G4VCG GU10 4TR
G3ZFT GU26 6NY
G3YER GU26 6PA
G4NOT GU26 6PA
G7UEI GU26 6PZ
G7WBM GU26 6PZ
G8EGG GU26 6QE
G3VXF GU26 6QX
G8FQM GU26 6QX
G4DNN GU26 6QZ
G4CMG GU26 6RP
G6XN GU26 6SJ
G0OAR GU26 6SN
G4ROM GU26 6SN
G3WLH GU26 6SX
G1OTZ GU27 1JD
2E1AEE GU27 2JR
G4VIE GU27 2NB
G1MAL GU27 2NY
G3TUX GU27 2RF
G1IDZ GU5 0BB
G0UNF GU5 0BL
G0DVW GU5 0PU
G4VQF GU5 0SA
G4YXP GU5 0SA
G6APF GU5 0SD
G6ZPR GU5 0SE
G6PXZ GU5 0SF
G2AAN GU5 0SU
G6IVQ GU5 0UN
G4CJT GU6 7ET
G7TRU GU6 7ET
G3AGO GU6 7HR
G4IBI GU6 7HR
G6BCQ GU6 7HR
G0DRR GU6 8DT
G6LWO GU6 8PD
G0FGB GU6 8PQ
G1DSM GU6 8PQ
G7VBY GU6 8PX
G3ZDD GU7 1LW
G4MFW GU7 1NS
G6NWW GU7 1SG
G4CWP GU7 1SY
G4VIY GU7 1SY
G7HHG GU7 1TQ
G0GZF GU7 1YF
G0LPW GU7 2EF
GI8KUO GU7 2HU
G3TCU GU7 2LD
G1OQI GU7 2LH
G1WTW GU7 2NE
G4VRN GU7 2NT
G0PFF GU7 2NX
G8IBL GU7 3AG
G8ZMC GU7 3EU
G4OMI GU7 3HS
G4URO GU7 3JW
G3MBK GU7 3NT
G0MUQ GU7 3NU
G6XKC GU7 3NU
G1FJF GU7 3NZ
G6NAV GU7 3QJ
G0EMT GU7 3RW
G1ERS GU7 3SL
G0WDP GU8 4BA
G0SUA GU8 4ND
2E1AFS GU8 4RG
G7CND GU8 5AB
G1SWY GU8 5BE
G3KZB GU8 5BG
G8KZB GU8 5BG
G3EPP GU8 5HJ
G3GJX GU8 5QU
G4XBF GU8 5RN
G4WWR GU8 5TD
G4ZEJ GU8 5TD
G6WWR GU8 5TD
G4DUF GU8 5TU
G3TGN GU8 5UA
G0GNE GU8 6DE
G3YZN GU8 6DU
2E1AFR GU8 6DZ
G4FRS GU9 0BS
G6FRS GU9 0BS
G8ATK GU9 0BS
G0BRI GU9 0BT
M1BRI GU9 0BT
G8FQN GU9 0HY
G6BEN GU9 0JG
G4RRA GU9 0LY
G7PWA GU9 0NU
G7NMT GU9 0QL
G0YYY GU9 0RZ
G6XNP GU9 7AQ
G4WUV GU9 7BP
G4WUW GU9 7BP
G7HOE GU9 7BT
G3VYI GU9 7BX
G6CTR GU9 7DA
G0KIH GU9 8BX
2E1DIM GU9 8BY
G3TAX GU9 8BZ
2E1CRC GU9 8DY
G8OGR GU9 8JQ
G4RFS GU9 8NT
G7NDL GU9 8RP
G0BIT GU9 8SA
G0CTZ GU9 8SA
G0XBQ GU9 8SA
G0CXU GU9 8TA
G8BCT GU9 8UE
G7MXX GU9 8UW
G3XMQ GU9 9AT
G7JUR GU9 9AY
G7UWL GU9 9BA
G4FSI GU9 9BS
G4INQ GU9 9BS
G3ADS GU9 9DJ
G3LDU GU9 9DZ
G1COP GU9 9LP
G3XQX GU9 9NL
G4VFW GU9 9RQ

Wealden

G4CUG BN20 0AS
G8GTH BN20 0ES
G0WZS BN20 0ET
G8GEA BN20 0EU
G1RMN BN20 9JA
G8CVV BN20 9JA
G2ATZ BN20 9NB
G3ZQB BN20 9NJ
G1UUP BN20 9NS
G3ZHZ BN20 9QD
G4UPY BN20 9RD
G8LEI BN20 9RH
G7GSC BN20 9RJ
G8CRC BN20 9RQ
G8SD BN20 9SB
M1ABM BN24 5DY
G0BMJ BN24 5HL
G1SIV BN24 5HW
G1SIW BN24 5HW
G8IQO BN24 5JF
2E1EGE BN24 5LJ
G1NEZ BN24 6DE
G6GVS BN24 6DE
G6TDW BN24 6HJ
G6YME BN24 6LA
2E1EWM BN24 6SB
G0HVO BN24 6SL
G8OXD BN26 5AG
G8RJH BN26 5DL
G3DQY BN26 5DT
M0AOM BN26 5HS
G3CMU BN26 5NS
G7JUD BN26 5SG
G4BJP BN26 6DA
G4CSG BN26 6HU
G0UNN BN26 6LT
G1ZGM BN26 6LU
G0UOI BN26 6PN
2E1BSQ BN26 6UW
G0VVV BN27 1HG
G0THX BN27 1LJ
G1IFV BN27 1NA
M1BSY BN27 1NA
G4RWM BN27 1SR
2E1BJW BN27 2BE
G8JLD BN27 2BT
G4XUU BN27 3AA
G1GQE BN27 3BJ
G3JYG BN27 3BZ
G4BLS BN27 3DH
G0UZY BN27 3DR
G1RPA BN27 3EL
G4PIT BN27 3EL
G0LKF BN27 3HP
G4CLV BN27 3HP
G0RFF BN27 3HU
G1ATL BN27 3LS
G4DRD BN27 3LW
G4KAR BN27 3NP
G8TNH BN27 3NP
G4LYU BN27 3TL
G3WQK BN27 3TL
M0AWY BN27 3TP
M1BMR BN27 3XB
G3UEH BN27 3XG
G4AQL BN27 4DY
G7DQZ BN27 4HG
G0CRD BN27 4JL
G7JTQ BN27 4PD
G7WHN BN27 4QU
2E1FIL BN27 4RD
G4HEJ BN27 4TX
2E1CIB BN8 6AW
G3TTJ BN8 6DT
G4SVB BN8 6HW
2E1ESP BN8 6PR
G3EKJ BN8 6PS
G8VTB RH17 7LG
G7GBC RH18 5DA
2E1CGR RH18 5JF
G2ACK RH19 3PH
G6BTO RH19 4JL
G3IIW TN2 5LU
G3MUJ TN21 0AD
G3WGG TN21 0DE
G3JQV TN21 0EB
G3WWS TN21 0EF
G7CSX TN21 0HW
G0JDM TN21 0RH
G0FUV TN21 0RL
G3EHV TN21 0XA
G7VAE TN21 0XL
G7VAD TN21 0XL
G4GW TN21 8AG
G1ITE TN21 8BB
G0NTF TN21 8BT
G7JLL TN21 8DD
G7HQY TN21 8EX
G4AWJ TN21 8HE
G6GPM TN21 8LA
G0KTS TN21 8NS
M1BWU TN21 8NS
M0AWQ TN21 8RY
G8KQA TN21 8SA
G1PJT TN21 8TG
G4WRZ TN21 8TT
G3GSI TN21 8UR
G3FBR TN21 8YD
G3BBK TN21 9AL
G1BFS TN21 9DS
2E1CCU TN21 9DT
2E0ACA TN21 9NR
2E1ANK TN21 9NR
G3SGR TN21 9NR
G0UNA TN21 9NR
2E1DLA TN21 9NR
2E1EGA TN21 9NX
G2BHY TN21 9RA
G3TGF TN21 9RA
G4ZOQ TN22 1BZ
G0VZN TN22 1RY
G8BTC TN22 1TU
G3MXJ TN22 1UB
G4FOC TN22 1UB
G3ZUN TN22 2BA
G0LFX TN22 2ED
G0BUK TN22 3AP
G4GEN TN22 3HW
G4MMH TN22 3LR
G1LIF TN22 3NQ
G8FCD TN22 4AG
G4NAJ TN22 4HT
G1GDD TN22 4JW
G6ILM TN22 4JZ
G3YMH TN22 4LB
G1KAR TN22 5AH
G1SHH TN22 5AH
G0MAR TN22 5BU
G3LUW TN22 5DB
G3JZE TN22 5RG
G8SC TN22 5SX
G3BWY TN3 9DL
G3OJW TN3 9DL
G0CHY TN3 9QU
G4PUP TN33 9EW
G4SVK TN33 9EW
G3RZC TN33 9LD
G6MXE TN5 6RY
G4CBC TN6 1BH
G4LZK TN6 1BS
G4DRB TN6 1DU
G8TBR TN6 1EJ
G7TDL TN6 1EZ
G3GTF TN6 1HA
G8RAC TN6 1HU
G3ZUU TN6 1LR
G7EVE TN6 1QQ
G7JAN TN6 1QQ
2E1EHE TN6 1RE
G4SGZ TN6 1TX
G6GPE TN6 2AD
G4XCE TN6 2AD
G3TLB TN6 2AG
G4OTV TN6 2BG
G4UPI TN6 2BG
G0PNV TN6 2BN
G6GWE TN6 2BN
G0CRW TN6 2ED
G3TXZ TN6 2ED
M1ARX TN6 2EN
G0WTC TN6 2ET
G0NOA TN6 2HR
G7HYN TN6 2HR
G6UUO TN6 2JB
G3TLB TN6 2NB
G6ZRC TN6 2NB
G3CYU TN6 2NY
G3RST TN6 2RY
G8LSD TN6 2SB
G3KCR TN6 2SL
G6PS TN6 2SY
G0HWQ TN6 2TJ
G3OHV TN6 2UH
G3FET TN6 2UN
G0OSS TN6 3BD
G4TIS TN6 3BE
G4LDJ TN6 3BQ
G8SCH TN6 3JU
G3JDA TN6 3ND
G4HXQ TN6 3NH
G0NAR TN7 4DH
G0MSA TN7 4DH

Wear Valley

G3WUE DL13 1AJ
M1BPX DL13 2RH
G6OTL DL13 2XY
G4AIK DL13 3LF
2E0AIM DL13 4DS
G6KVX DL14 0RG
2E1EYS DL14 0RP
G4TDG DL14 6AW
G0FBK DL14 6EH
G0RLW DL14 6LQ
G7TZS DL14 6TU
G0EQD DL14 6UT
G0MMV DL14 7SA
G0GWL DL14 8AE
G4ZOF DL14 8PD
G4TYF DL14 8RW
G1FMJ DL14 8TL
M0ACV DL14 9LG
G4TTF DL14 9LG
G4OON DL14 9LL
G1YNO DL14 9LQ
G0GFG DL15 0AA
G6RKS DL15 0DA
G4STP DL15 0DA
G6LCM DL15 0DA
G4XPP DL15 0DX
G6LQR DL15 0HN
G1YUB DL15 0LQ
G0BNY DL15 0NP
G1DEO DL15 0QD
G0OXR DL15 8QU
G7OCK DL15 9DB
G0HCD DL15 9QX
G0PRQ DL15 9SA
G4PSI DL15 9TS
G0OXN DL15 9TW
G7BPN DL15 9UT

Wellingborough

G0NFL NN14 1HG
G8AMG NN29 7AT
G7CTN NN29 7BY
G6MKJ NN29 7DH
M0ASM NN29 7DP
G0VOB NN29 7DP
G4ORJ NN29 7DZ
G7ERB NN29 7DZ
G8TFY NN29 7EA
G4RST NN29 7EF
G8COM NN29 7EF
G4PAV NN29 7EF
G7HBU NN29 7LP
G6NUS NN29 7LT
G3MJW NN29 7LZ
G0PBU NN29 7NF
G0PSZ NN29 7NS
G7OGL NN29 7PW
G6FJF NN29 7PW
2E1BVU NN29 7PW
G1DAV NN29 7QA
G1EUG NN29 7TP
G3UNV NN29 7TT
G3KLV NN6 0AW
G8LOP NN6 0AX
G0ERR NN6 0BB
G8HOE NN6 0HX
2E1CHF NN6 0JG
G7OXY NN6 0LA
G4TNL NN6 0LP
G0DSM NN6 0LZ
G1XXH NN6 0NT
G0RFW NN6 0NX
G4LYC NN6 0NY
G4UHW NN6 0PB
G6OVC NN6 0PF
G7RTC NN6 0PQ
G7UPZ NN8 1EE
G8NWZ NN8 1EL
G0KEJ NN8 1PS
G7VIT NN8 2DA
G4ZIW NN8 2DX
G6UWS NN8 2EN
G1FXE NN8 2JA
G0VRF NN8 2LR
G6TRO NN8 2NB
G4HME NN8 2NQ
G4RLL NN8 2UA
G8JPV NN8 3DW
G0HBA NN8 3JJ
G6CPX NN8 3JZ
G4PZF NN8 3NH
G6CZE NN8 3NL
G0ING NN8 3PJ
G5LP NN8 3PJ
G1ONQ NN8 3TP
G6BBD NN8 3ZA
G0SVW NN8 4AG
G4WZB NN8 4AT
G0JPL NN8 4AU
G4YFX NN8 4EN
G7DTY NN8 4EQ
G8GOM NN8 4ET
G3ZAG NN8 4RU
G1IJJ NN8 4SF
G1RVF NN8 4TQ
G1IQE NN8 5BA
G0FQA NN8 5BZ
G7HLU NN8 5FB
G4RDM NN8 5NX
G0EAE NN8 5PG
G6HGG NN8 5PQ
G3CYI NN8 5UB
G8OSZ NN8 5YJ
G0RUA NN8 5YN
G0RRF NN8 5ZL
G0HNG NN9 5LP
G6MNA NN9 5NY
G6UOJ NN9 5NY
G4XEM NN9 5YQ
G4XEN NN9 5YQ
G6FEJ NN9 5YS

Welwyn Hatfield

G8AWO AL10 0AU
G0EVD AL10 0HF
G2BSI AL10 0HH
G8XOQ AL10 0PE
G6ZGD AL10 0PS
G4IUZ AL10 0PX
G0MEE AL10 0RQ
G3XEG AL10 0SH
G6UVO AL10 0SX
G1ROH AL10 0UB
G4UEV AL10 0UQ
G3IGP AL10 8AJ
G0DCU AL10 8AR
G6LOY AL10 8DD
G4RMD AL10 8DQ
G6INV AL10 8HE
G4LWV AL10 8HH
G8WQK AL10 8PW
G1IHE AL10 8QF
G0MHR AL10 8TG
G3WGC AL10 8TG
G4KGP AL10 8TG
G6BOB AL10 9AB
G0AMG AL10 9ED
G3VIX AL10 9LE
G0SVD AL10 9LJ
G4SNI AL10 9PB
G7CMB AL10 9SA
G4WTB AL10 9SB
G6UBH AL10 9SB
G0PSW AL10 9SD
G1MMZ AL6 0DB
G7OVO AL6 0DB
G0URC AL6 0DH
G3UFA AL6 0DL
G0VKM AL6 0PW
G2BLA AL6 0QQ
G3KOX AL6 0RW
G0TAG AL6 0XG
G4HGD AL6 9BW
2E1EWT AL6 9PB
G0WTF AL6 9PN
G4LPP AL6 9RF
M0ASZ AL6 9RJ
G4SZI AL6 9TF
G4TUH AL7 1BY
G0IIG AL7 1DU
G7BDS AL7 1NL
G0NSH AL7 1QT
G0WAT AL7 1SD
G7JUY AL7 1SR
G6GAD AL7 2AN
G6POE AL7 2HF
G1HBC AL7 2JP
G6YIQ AL7 2JS
M1ACE AL7 2JS
M0ABC AL7 2PW
G0SAH AL7 2PW
G8FXM AL7 2QJ
G4LWO AL7 3HS
G3ZUA AL7 3JG
2E1DNR AL7 3RA
G8TVW AL7 3RJ
G0MKN AL7 3RL
G1PGJ AL7 3TD
G4GYD AL7 3TG
G8TWC AL7 3TG
G0AII AL7 3TN
G7MQB AL7 3TU
2E1DGJ AL7 3TU
G0HVX AL7 3TW
G4LFB AL7 4AJ
G0JJM AL7 4AQ
G7JTN AL7 4HG
G0NLU AL7 4JT
G3RML AL7 4LU
G0NJS AL7 4ND
G3ZQI AL7 4RL
G3CSE AL7 4SB
G8BIZ AL7 4TE
M1AKS AL8 6EL
G0FRM AL8 6HR
G0GWW AL8 6HZ
G8LOK AL8 6QZ
G1LLW AL8 6RH
G7TXP AL8 6SS
G7TXO AL8 6SS
G0CYC AL8 6TA
G6AOI AL8 6YQ
G7JKN AL8 7BD
G4MDC AL8 7BG
G0UBO AL8 7DQ
G0ODS AL8 7DW
G7MGH AL8 7DW
G0VDT AL8 7EH
G4POB AL8 7HZ
G4JVA AL8 7LG
G4DRI AL8 7LH
G8TVZ AL8 7NN
G3WER AL8 7PS
2E1DAV AL8 7QR
2E1DBI AL8 7QR
2E1GBI AL8 7QY
G8AXK AL8 7RX
G4TIM AL8 7SG
G8CYX AL8 7ST
G7BLJ AL9 5DL
G1EKM AL9 5HE
G7UVO AL9 5PH
G3LUO AL9 6JB
G6SBG AL9 7AY
G4HAY AL9 7UQ
G8AIF AL9 7UQ
G0ICK EN6 4DZ
G4JIJ EN6 4DZ
G4ZSC EN6 4JE
G4UAA EN6 4JH
G7MYN EN6 4JS
G8LAX EN6 4LH
G4POI EN6 4NP
G6NEP SG13 8RF

West Devon

G3ZQR EX19 8SL
G3PSZ EX20 1EA
G7VMQ EX20 1EA
G4NJK EX20 1NH
G0FJA EX20 1PR
G0FYL EX20 1PR
G1JXA EX20 2AQ
G3VDL EX20 2JF
G1BOB EX20 2NG
G4NFP EX20 3AJ
G3YET EX20 3QZ
G4AVZ EX20 3QZ
G4XIX EX20 4DQ
G0OAW EX21 5BQ
G3WDK EX6 6QW
M1BBB PL16 0AJ
G3SBP PL16 0HH
G3ZNE PL19 0ER
G4COF PL19 0HZ
G0NIE PL19 0LR
G4WDR PL19 0LR
G1LZY PL19 0NA
G0GSR PL19 0PN
G0LSG PL19 8DP
G1KKA PL19 8DR
G1OJL PL19 8DX
G3MWZ PL19 8HA
G7VXP PL19 8JE
G3XOU PL19 9AL
G8ONR PL19 9BW
G6FOV PL19 9DJ
G3RSK PL19 9DL
G7IIB PL19 9DL
G3TYO PL19 9DN
G7CUB PL19 9JL
G0NUO PL19 9LJ
G8AKC PL19 9QD
G0NJB PL19 9QP
G7LLG PL20 6AT
G7MOH PL20 6AT
G7ILN PL20 6BD
G8BWB PL20 6BU
G4BXS PL20 6DH
G2CQJ PL20 6DN
M0AED PL20 6DU
G7VQF PL20 6DZ
G6IIZ PL20 6HP
G4CYW PL20 6HY
G3TXL PL20 6LJ
G8LKZ PL20 6LX
G4GUY PL20 6NG
G0IAI PL20 6PT
G6URM PL20 6RB
G3SGZ PL20 7AR
G4HPX PL20 7AR
G0WEQ PL20 7AR
G0UOA PL20 7DA
G1SQG PL20 7DD
G0CYF PL20 7JS
G8MWN PL20 7JS
G0PGI PL20 7LH
G3ZXO PL20 7LR
G3VVG PL20 7NA
G8LAK PL20 7TR
G4UTX PL6 7AP
G0OAO TQ13 8HT

West Dorset

G2AMG BA22 9RR
G1DZZ DT1 1LL
G1GMV DT1 1NZ
G4ZPO DT1 1XJ
G7VVC DT1 1YB
G4XSC DT1 2AH
G6KAC DT1 2AH
G2DGB DT1 2BA
G4EDK DT1 2DE
G8DJW DT1 2DP
G4CFY DT1 2EF
G0DOQ DT1 2JB
G1BJN DT1 2LB
G3YUD DT1 2LZ
G4EAS DT1 2NR
G3XHR DT2 0JF
G3LOX DT2 0JJ
G4LOA DT2 7BB
G6AOF DT2 7DT
G0PUU DT2 7TE
G7JIM DT2 7TF
G8SDS DT2 7TF
G1HYX DT2 8BQ
2E1CQG DT2 8EW
G0HVA DT2 8PE
G0FIT DT2 8QJ
G0PZN DT2 8QR
G4SXD DT2 8QR
G7VMP DT2 8RT
G1RPT DT2 8TS
G7TNO DT2 8TX
G4DRS DT2 8UN
G4JQL DT2 8UN
G4GOK DT2 9AF
G4AXU DT2 9DD
M1AFE DT2 9DU
G8JKD DT2 9HP
G0KYH DT2 9JN
G6DOR DT2 9JU
G8GYL DT2 9LN
G3ZGN DT2 9QX
G4GUV DT2 9RG
G2FNK DT2 9TD
G7HNY DT3 4AG
G1NLG DT3 4AN
G3VPF DT3 4AS
G0NRQ DT3 4BA
G3ETA DT3 4BT
G0PGT DT3 4DY
G3TN DT3 4EB
G4SNU DT3 4JA
G7LPN DT3 4JP
G7GOH DT3 6EY
G4SDJ DT3 6HA
G0FAJ DT4 9EP
G4VOC DT6 3AH
G4JZI DT6 3AR
G0LEV DT6 3DE
G8MCC DT6 3DR
G4NIA DT6 3LZ
G1BDS DT6 4AN
G3PXH DT6 4AN
G7NZC DT6 4AN
G4UHU DT6 4AX
G1YIL DT6 4DQ
G0EZJ DT6 4ES
G4MZL DT6 4ET
G8LCV DT6 4HG
G7KLZ DT6 4QU
G8UZY DT6 4RF
G4JWA DT6 4RG
G3MTP DT6 5HX
G6UZR DT6 5LD
G6RMV DT6 5LS
G7RZZ DT6 5QD
G3XTY DT6 5QS
G4HNG DT6 6BB
G0HET DT6 6DF
G1EFX DT6 6DF
G3ZUE DT6 6DU
G3EZE DT6 6RA
G3OFB DT7 3AE
G0WNI DT7 3EL
G1DXG DT7 3HS
G1HTO DT7 3JE
G4HHU DT7 3JE
G0BEP DT7 3PU
G8NIC DT8 3BA
G0CFF DT8 3HU
G8JBQ DT8 3HU
G8LOJ DT8 3JT
G1SQH DT8 3NH
G4ZJH DT8 3PW
GI4BQN DT8 3RF
G7RGI DT8 3RF
G7KBE DT9 3NZ
G0FAA DT9 4BJ
G7IKK DT9 4BL
G8AWB DT9 4BS
G3CQR DT9 4HZ
G0WDG DT9 4PZ
G3BCE DT9 5LS
G3MJS DT9 6AJ
G1EOQ DT9 6AL
G3PPR DT9 6DT
G2UH DT9 6RG
G3OPJ DT9 6RN
G3KAZ DT9 6RR
G0JCY TA18 8PL
G0EGC TA18 8QE
G4TIA TA20 4PJ

West Dunbartonshire

GM0KWR G60 5DP
GM3YCB G60 5LE
GM6AQB G60 5LE
GM7UKS G81 1AW
GM3ACL G81 2EE
GM0GYM G81 2LL
GM4PLI G81 2ST
GM3ZTA G81 5ED
GM4ZMK G81 5EG
2M1EDT G81 5HJ
GM6KTP G81 5PD
GM3KCY G81 6AW
MM1BMS G81 6BJ
GM4KHE G81 6HZ
GM0NUQ G81 6JL
GM3RGU G81 6LW
GM3ITN G81 6NR
GM4TOQ G81 6PU
MM1AHL G82 1EH
GM6OML G82 1EH
MM1APC G82 2LH
GM7NHU G82 2PN
GM7BYB G82 3AU
GM0KZX G82 3ER
GM4HEL G82 3ER
GM8FFH G82 3QW
GM7AON G82 4BG
GM0GAK G82 4LF
GM0POD G83 0DJ
GM7MYF G83 0DU
GM0BFW G83 8EX
GM3VCO G83 8HL
GM4LGM G83 9BQ
GM1MLW G83 9NL
MM1BJP G83 9PQ
2M1FLG G83 9PQ
GM7OBM G83 9QA

West Lancashire

G7RGV L39 0EG
G0OXV L39 1LR
G3SZV L39 1NN
G4UEW L39 1PJ
G4IGX L39 2DX
G1ZUF L39 2EY
G6BKR L39 3AZ
G8UGH L39 3LJ
G4OHP L39 3NN
G4LBX L39 3NT
G0RPI L39 3NY
G7DGH L39 3PJ
G8TUN L39 4RE
G7LFC L39 4TF
G8FUB L39 4UD
G4ATU L39 5AJ
G6ODU L39 5AP
G0BLW L39 5AT
G3VHC L39 5AT
G3FIC L39 5AY
G3NNT L39 5BG
G1YWX L39 5BY
G4MHO L39 5BY
G4UVB L39 5QJ
G8SAX L39 6RF
G1JFE L39 6SH
G7SWV L39 7JL
G1KVI L39 7JP
G7HSB L39 7JU
G1DQZ L39 8RT
G4VYP L39 8RT
G8ALG L39 8SG
G3HWX L39 8SX
G7CVZ L40 0SN
G7JPL L40 0SN
G4VDD L40 1SY
G0CJS L40 1TX
G7EVY L40 1TX
G1ZYM L40 1UP
G1JZZ L40 5SE
G1EUH L40 5TJ
G0DDU L40 5TN
G6YGP L40 5TY
G1XKL L40 6JG
G6NJJ L40 6JQ
G4LTI L40 6JR
G1RVV L40 7RA
G0AFQ L40 7RB
G4CF L40 7RD
G0CBB L40 7RP
G6ITU L40 7TA
G4ZPY L40 7TG
G4IQJ L40 9RS
G6TMN PR4 6DS
G4MGB PR4 6HD
M1BDG PR4 6HH
G0VAV PR4 6JX
G0KMU PR4 6PA
G0NEI PR4 6PA
G4OWS PR4 6RB
G6WLI PR4 6RZ
G0JJD PR4 6TR
G0EHK PR4 6UD
G6PLT PR4 6UD
G3MPF PR4 6UL
G4POY PR4 6YB
G4SBE PR8 5HS
G3OEI PR8 5LP
G6CIO PR9 8ET
G4PJS WN6 9DD
G4ZRI WN6 9DD
G0MRM WN8 0AA
G6TKY WN8 0AA
G4ACI WN8 0AE
G6SGQ WN8 0AX
G0NBJ WN8 0DE
G4FEU WN8 0DE
G3RCA WN8 0EP
G7VJW WN8 0HH
G0ZAP WN8 0JG
G4ENF WN8 0QE
G1AGQ WN8 0QE
G4WWG WN8 6AS
G0RSQ WN8 6BT
G4MFH WN8 6EP
M1AEB WN8 6JQ
G0GBP WN8 6JX
G6OMN WN8 6PA
G6HXL WN8 6TA
G4JTP WN8 6TB
G7DUI WN8 6TB
G3PNQ WN8 7AR
2E1DVR WN8 7HR
G6ORS WN8 7NT
G0RQW WN8 7PE
G1DMW WN8 7RA
G4AVT WN8 7TG
M1BAF WN8 7TU
G6NLH WN8 8AH
G6YRB WN8 8EG
G6WEW WN8 8EN
G6UNX WN8 8PA
G4WJR WN8 8PR
G4ZAF WN8 8RP
G0RPT WN8 8TR
G0LAK WN8 9BP
G4SME WN8 9BP
G8XEI WN8 9BQ
G4XUO WN8 9DS
G1SMI WN8 9DU
G0SSM WN8 9HQ
G1UOR WN8 9JZ
G6UGE WN8 9NB
G3KTJ WN8 9QQ

West Lindsey

G1EWI DN21 1BS
G1HKF DN21 1DD
G7AVU DN21 1DH
G4NNT DN21 1HA
G1MXV DN21 1JF
G8EVI DN21 1LP
M0AYF DN21 1RG
G8WEI DN21 1RG
G4VUE DN21 1RG
G7OWE DN21 1RH
G8VRN DN21 1TT
G8SYT DN21 2HD
G3YPS DN21 2JA
G1HKD DN21 2JD
G3UHS DN21 2JZ
G1NPU DN21 2PU
G6YVS DN21 2TU
G0FMY DN21 3BY
G3OQX DN21 3ED
G4XHC DN21 3JY
G4YET DN21 3QZ
G3TMD DN21 3SR
2E1CZB DN21 3UJ
G4GZA DN21 4AF
G4MBK DN21 4ER
G3RAU DN21 5BZ
G8SFF DN21 5LB
G4LTG DN21 5QT
G4VHH DN21 5QU
G7JVC DN21 5UT
G0SQS DN21 5UT
G8HDJ DN38 6AP
G7TRG DN38 6AU
G4FHF DN38 6BD
G3MHT DN38 6ER
2E1FVU LN1 2AD
G4OZN LN1 2AE
G6COL LN1 2AS
G5FZ LN1 2AS
G4STO LN1 2AS
G1AUQ LN1 2DN
G3FJV LN1 2DN
G3OS LN1 2DN
G4ZKW LN1 2DY
G0KAU LN1 2EL
G7OJU LN1 2EP
G4VSK LN1 2EZ
G0PIO LN1 2HH
G7KIE LN1 2HZ
G0BEN LN1 2LU
G0BUC LN1 2NH
G7OLF LN1 2NJ
G1WVO LN1 2QA
G3KAY LN1 2QL
G3UPI LN1 2QL
G6BPH LN1 2QN
G4CM LN1 2SG
G4GDC LN1 2SG
G3NIC LN2 2NE
G4ADA LN2 2PL
G0BUP LN2 2PR
G0POC LN2 2PU
G3RRN LN2 2QU
G3EBH LN2 2RW
G0TCS LN2 2TL
G1LQB LN2 3JS
G0NXF LN2 3JX
2E1CSC LN2 3JX
G4RAP LN2 3LL
G3WWX LN2 3RT
G4YQK LN2 3SR
2E1DES LN2 3SW
G0JOD LN2 3TU
G3ATC LN2 3TU
G1ATC LN2 3TU
G0LPD LN2 4RB
G6DKK LN3 4AN
G0VKF LN3 4AX
G3PTZ LN3 4BE
M1AYR LN3 4BH
G3AWK LN3 4BJ
G6RFR LN3 4DH
G7NSK LN3 4EB
G4LPV LN3 4ED
G0TQV LN3 4EJ
G4WMO LN3 4HU
G6RCH LN3 4JG
G1FLL LN3 4JQ
G7PMQ LN3 4LS
G4SLG LN3 4NA
G7PBU LN3 5AB
G2FHM LN3 5DF
G7PHR LN3 5XS
G0HXL LN7 6BB
G0CVN LN7 6EL
G3MCS LN7 6HY
G3ZRP LN7 6JP
G0GDJ LN7 6NB
G3KHS LN7 6PH
G0LZM LN7 6PN
G3RXP LN7 6RB
G8XML LN7 6TB
G7TWV LN7 6UG
G4CLL LN8 2HL
G4UHZ LN8 3AG
G3AIZ LN8 3EE

M1BPL	LN8 3UB
2E1FYY	LN8 3UB
G6NGE	LN8 3UE
M1CAJ	LN8 3UR
2E1FDQ	LN8 3UR
G0MCB	LN8 3XL
M1BRY	LN8 3XZ
G1YFQ	LN8 3YS
G7AJP	LN8 6EF
G0GOQ	LN8 6EG
G3PDL	LN8 6HB

West Lothian

GM4OGM	EH27 8AD
GM7KZL	EH27 8AN
GM1OXE	EH47 7JW
GM1SZH	EH47 7LS
GM0GLH	EH47 8BY
GM6KJQ	EH47 9JZ
GM3RXZ	EH48 1AP
GM4LHJ	EH48 1DF
GM0JUB	EH48 1EU
GM6RQW	EH48 1ST
GM4VZW	EH48 3DH
GM0EEH	EH48 3RD
GM3OXX	EH48 3RX
GM3LQH	EH48 4DA
GM3END	EH48 4DQ
GM0EDQ	EH48 4HG
GM3PPE	EH48 4LD
GM4LZO	EH48 4LD
GM1VKG	EH48 4NF
GM4BQD	EH48 4NU
G4XFV	EH49 6AQ
GM3PDX	EH49 6BS
GM1FAI	EH49 6DD
GM4LUS	EH49 6HA
GM4XUS	EH49 6HA
GM0KZU	EH49 6HX
GM4YPL	EH49 6QT
GM0MWJ	EH49 6SH
GM4MIM	EH49 7BH
GM3VTH	EH49 7JU
GM0RUW	EH49 7LD
GM0MWR	EH49 7RU
GM0EWF	EH52 5EB
2M1GBG	EH52 5HQ
GM4VDG	EH52 5PN
GM0UHC	EH52 6LY
GM0DHD	EH52 6PL
GM3OIV	EH52 6QB
GM4DGT	EH52 6UP
GM0NBO	EH53 0AD
GM4EFL	EH53 0BY
GM0FCP	EH53 0DZ
GM7UKL	EH53 0DZ
2M1AOL	EH53 0LE
GM1PSU	EH53 0ST
GM0NAZ	EH54 5AD
GM6OGN	EH54 5JP
GM4SVM	EH54 6BS
GM0CBA	EH54 6JJ
GM4FKO	EH54 6LQ
MM1APG	EH54 6LQ
GM7GTS	EH54 6LT
2M1CJG	EH54 6LY
MM1AXK	EH54 6PE
GM0REZ	EH54 6RJ
GM4HML	EH54 6RJ
GM0RSF	EH54 6UR
GM7AHA	EH54 7BZ
GM7LVV	EH54 8ED
GM0LEW	EH54 8HY
GM1PZT	EH54 8JB
GM6PSK	EH54 8RW
GM3DKW	EH54 8RY
2M1FXH	EH54 9AA
2M1FXQ	EH54 9AA
GM4BBF	EH54 9AX
2M1CGG	EH54 9AX
GM8NVE	EH54 9BN
GM8JME	EH54 9BP
GM0LOT	EH54 9DT
MM0ACS	EH54 9EN
GM0JCN	EH54 9EN
GM4MHE	EH55 8LE

West Oxfordshire

G4YCJ	GL7 3LW
G4VMV	OX18 1BH
G7KDM	OX18 1DP
G7SRB	OX18 1ED
G3GPE	OX18 2DE
G0NZE	OX18 2HX
2E1ESK	OX18 2LH
G4RJY	OX18 2NN
G3LQC	OX18 2NW
G4FZB	OX18 3AQ
G3MTK	OX18 3JF
G3OPH	OX18 3PH
G8KOD	OX18 3QA
G7OPB	OX18 3QE
G3XMU	OX18 3SD
G4LQZ	OX18 3SJ
G0RBN	OX18 3SJ
2E1EUM	OX18 3TA
G0MFV	OX18 3XT
G4POL	OX20 1JY
G0LXW	OX20 1RG
G6HRA	OX20 1RZ
G0UPN	OX20 1YJ
G3TXT	OX5 3BW
G6GEV	OX7 3EP
G7TMS	OX7 3EP
G4LLQ	OX7 3QR
G6BPJ	OX7 5BA
G0NYH	OX7 5DZ
G0JZE	OX7 5JS
G8UFN	OX7 5LE
G7BZI	OX7 5TG
G3DVV	OX7 5UT
G6KCJ	OX7 6AD
G8BII	OX7 6ER
G0MVB	OX7 6JG
G8GJF	OX7 6JQ
G4IER	OX7 6JZ
G0BII	OX8 1EL
G6LDO	OX8 1EW
G8YMO	OX8 1EW
G8RFN	OX8 1EX
G7OQQ	OX8 1LU
G1RCX	OX8 1NP
G0OBJ	OX8 1NS
G6ANV	OX8 1NT
G7AHB	OX8 1NW
G0BQZ	OX8 1QJ
G8JVI	OX8 1QY
G3ZKO	OX8 1SX
G8RDA	OX8 5EG
G7LGV	OX8 5FP
G4IOK	OX8 5HQ
G0VZC	OX8 5JW
G7UKM	OX8 5LS
G8GJG	OX8 5NJ
G3JKZ	OX8 5NQ
G6UPQ	OX8 5RQ
G1FZQ	OX8 5US
G3BVU	OX8 6AE
G4CXJ	OX8 6ED
G4ZHS	OX8 6EY
G4GUN	OX8 6JB
G1VBL	OX8 6JH
G4EGN	OX8 6NS
G7RRM	OX8 6NU
G0PYF	OX8 6PZ
G0MKY	OX8 6PZ
G8TIM	OX8 6RQ
G6OM	OX8 6TZ
G8EWT	OX8 6UW
G0CFR	OX8 7HX
G0CFS	OX8 7HX
M0ATJ	OX8 7JB
G4YUA	OX8 7NL
G3GDA	OX8 7NQ
G7CKP	OX8 7NQ
2E1ETR	OX8 7YP
G0GRT	OX8 8DD
G6ULP	OX8 8DD
G1XFP	OX8 8EQ
G8PKG	OX8 8EU
G1ZBG	OX8 8HH
G1DOL	OX8 8JY

West Somerset

G0JFA	TA22 9PT
G8CCF	TA22 9RU
G3FVC	TA23 0DG
G0NWS	TA23 0HN
M0AOE	TA23 0HW
G8KVP	TA23 0PY
G4WSF	TA23 0UB
G6IIG	TA23 5BW
G1CGU	TA24 5DY
M0AOC	TA24 5HB
G4PIB	TA24 5HR
G6IHB	TA24 5HW
G3NFY	TA24 5JU
G8YZY	TA24 5NZ
M0AOJ	TA24 5NZ
G0JLD	TA24 5QB
G0OWX	TA24 5QB
G4WRL	TA24 5ST
G6HOR	TA24 6AF
G8BGG	TA24 6EQ
G4PHC	TA24 6BZ
G3AXN	TA24 6EQ
G7OJB	TA24 6JX
G0WMN	TA24 8AF
G8EAM	TA24 8AQ
G0ACA	TA24 8AW
G3PWQ	TA24 8BE
G3RDU	TA24 8EG
G4CJG	TA24 8JH
G4FHN	TA24 8NP
G7HQQ	TA24 8TQ
G7LJM	TA24 8TQ
G4BQB	TA4 3TR
G0NKZ	TA4 4DE
G0SZO	TA4 4DE
G3RNB	TA4 4LN
G3HMG	TA4 4PD
G3LZE	TA4 4PG
G4AJU	TA4 4PG
G8AYV	TA5 1QG
G3YDC	TA5 1RB
G4OEC	TA5 1SF

West Wiltshire

G0DQQ	BA12 0AQ
G8TMA	BA12 0AS
G3LCL	BA12 0AW
G4ZUP	BA12 0JW
G7FHA	BA12 0PW
G7HVL	BA12 7AG
G7VHC	BA12 7AG
G3NFJ	BA12 7AP
G1KWS	BA12 7DY
G1BBJ	BA12 7EX
GI4RNC	BA12 7EX
G3UPV	BA12 7NR
G7VKW	BA12 7RN
G0BFU	BA12 8DJ
G0GGG	BA12 8EB
G1NMW	BA12 8HR
G7COA	BA12 8LY
G4YMG	BA12 8TF
G3ZHR	BA12 9JZ
G3MHV	BA12 9PN
G4WHV	BA12 9PN
G3YQL	BA12 9QD
G0BOM	BA12 9QJ
G1TQN	BA12 9QJ
M1BPW	BA13 3AQ
G4IQZ	BA13 3ES
G6APD	BA13 3ET
G0VFS	BA13 3HP
G4YSE	BA13 3HQ
M0ADW	BA13 3LQ
G1WFO	BA13 3SH
G2BQY	BA13 3UF
G3BPE	BA13 3UF
G4YXS	BA13 4AT
G6ZDE	BA13 4HJ
G4JQN	BA13 4LG
G4CLC	BA13 4NY
2E1EYI	BA14 0DN
G0RKB	BA14 0HG
G0TFX	BA14 0HG
G4KHK	BA14 0LH
G6CKL	BA14 0LH
G7FXY	BA14 0LJ
G7NGI	BA14 0RE
G0HAS	BA14 0SS
G4KDK	BA14 0ST
G1UGV	BA14 0TD
G0KCZ	BA14 0TX
G0SYK	BA14 0UG
G6PAE	BA14 0UJ
G6HQX	BA14 0UX
2E1FTN	BA14 6EH
G7PUB	BA14 6JG
G0LJG	BA14 6JQ
G6FNJ	BA14 6LQ
G4UJJ	BA14 6NA
G0GHR	BA14 6QP
G7JQW	BA14 6QP
G4SPE	BA14 6SA
G8CPF	BA14 7BR
G0RMA	BA14 7HG
G7MOM	BA14 7HR
G4SFL	BA14 7JL
G3BJC	BA14 7LD
G0BBL	BA14 7LE
G8GUA	BA14 7PE
G1HFY	BA14 7PG
G7NWR	BA14 7PG
G4GFJ	BA14 7PH
G0VYU	BA14 7PR
G3AZW	BA14 7RS
G7MZN	BA14 8LU
G0GKH	BA14 8RY
G3VRE	BA14 8RY
G4LW	BA14 9AR
G0WPL	BA14 9BQ
G4VUG	BA14 9BU
G6POW	BA14 9DA
G0JYF	BA14 9ES
G3PYF	BA14 9LQ
G7PQW	BA14 9PH
G0HFX	BA14 9PW
G8BYI	BA14 9PW
M1AIW	BA14 9QP
G0TOE	BA14 9RB
G0EUR	BA14 9TB
G0GHW	BA14 9TH
G0IAK	BA15 1AX
G4BMR	BA15 1JE
G4YPE	BA15 1JF
G1OOJ	BA15 1NF
G3SGM	BA15 1NZ
G8SQN	BA15 1PR
G0OFT	BA15 1RJ
G3BKG	BA15 1RJ
G4VVZ	BA15 1SE
G4ZAP	BA15 1SE
G8MGQ	BA15 1SS
G6BIF	BA15 1SW
G3GKC	BA15 1TB
G0BNG	BA15 1UD
G1LCN	BA15 2BH
G1FJS	BA15 2BX
M1BQY	BA15 2DL
G0GRI	BA15 2DL
G3SBU	BA15 2EL
G3SIH	BA15 2EQ
G4LYG	BA15 2HG
M0AXW	BA15 2HL
G7WJX	BA15 2HL
G4OTY	BA15 2JZ
G7TXN	SN10 1SW
G4DMC	SN12 6AG
G3SPU	SN12 6AL
G0HEL	SN12 6BE
G0HAD	SN12 6HN
G7TKO	SN12 6HS
G0YEF	SN12 6QL
G4SRD	SN12 6TQ
G7EPX	SN12 6UH
G1PWU	SN12 6UL
G0XAR	SN12 6XN
G4ZTK	SN12 6XW
G3RVY	SN12 7AB
G8JLV	SN12 7AR
G0KTP	SN12 7DT
G1BHX	SN12 7HW
M0ACF	SN12 7RW
G0KTN	SN12 7RX
G6EJD	SN12 7TE
G7HOP	SN12 8BG
G0IUE	SN12 8BQ
2E1EYL	SN12 8BQ
G3VMZ	SN12 8EJ
G7KNU	SN12 8JS
G3PYP	SN12 8JS
G4NDC	SN12 8LA
G1AJC	SN12 8NB
G4SCZ	SN12 8QZ
G0LJS	SN15 2JW

Western Isles

GM0DRU	HS1 2DS
GM7JED	HS1 2PG
GM4PTQ	HS1 2PT
2M0APB	HS1 2TL
MM0BFF	HS2 0BG
GM0HMM	HS2 0HD
2M1EOV	HS2 0HD
G7OQC	HS2 0PP
G4JNW	HS2 0XA
GM7KUN	HS2 9AU
GM3JIJ	HS6 0QB
GM0HBF	HS6 5EU
MM1BID	HS8 5TU

Westminster, City of

G1TIE	NW1 6NR
GM3LAU	NW1 6UE
G3JXW	NW8 7AJ
G4BHB	NW8 7BA
G3XQP	NW8 7DY
G0UED	NW8 9LL
G0BNU	SW1A 2AH
G4ASF	SW1A 2AH
G4MBH	SW1A 2AH
G4MMT	SW1A 2AH
G4MZY	SW1A 2AH
GM4OLH	SW1A 2AH
G0MJO	SW1P 2NE
G4MRL	SW1V 2DB
G1AIP	SW1V 2SU
G6ZFK	SW1V 4AS
G5OG	SW1W 8JS
G4XXW	SW1X 7LD
G0SPJ	SW1X 8PA
G0PZC	SW7 1BL
G5YC	SW7 2BZ
G7SZL	W11 1AY
G0APW	W1H 3HL
G4ZDS	W1N 5RE
G3KCT	W2 2AF
G4AJ	W2 2PB
M0ACY	W2 3ER
G4CUQ	W2 3JA
G6IKN	W2 3NN
G6FJT	W2 4AN
G7VVR	W2 5DT
G4YAR	W2 5LJ
G0WYL	W9 1DT
G4YEI	W9 1EY
G4VEA	W9 1LG

Weymouth and Portland

G7JCI	DT3 5BD
G3ZGP	DT3 5BG
G0VLJ	DT3 5BQ
2E1EBH	DT3 5DQ
2E1AHC	DT3 5DT
M1BTH	DT3 5DT
G4UUR	DT3 5DU
G4VBY	DT3 5EP
G3OWE	DT3 5HE
G1RSE	DT3 5JU
G3YWW	DT3 5NG
G3SDO	DT3 5PF
G3RZG	DT3 5QA
G4BNY	DT3 5QE
G6ZMU	DT3 5QU
G0VHS	DT3 5RJ
2E1EXG	DT3 6AP
G4TRW	DT3 6AS
G0LQI	DT3 6BG
G0RIK	DT3 6BX
G3CVF	DT3 6BY
2E1AEI	DT3 6DS
G3KDA	DT3 6HT
G8IKP	DT3 6JX
G3AVV	DT3 6JZ
G0NEV	DT3 6LA
G0FBS	DT3 6LE
G0BMQ	DT3 6LF
G3JRL	DT3 6LF
G3OEW	DT3 6LL
G3EGV	DT3 6NE
G0SEC	DT3 6NG
G2CO	DT3 6NL
G3LAG	DT3 6PT
M1BDW	DT3 6PT
G3OPD	DT3 6PW
G7BRU	DT3 6PZ
G6XSK	DT3 6RH
M1BLE	DT3 6RW
G6OYF	DT3 6SG
G8PME	DT4 0BB
G4RWL	DT4 0EB
G4ZXP	DT4 0ET
G1WIK	DT4 0EZ
G0ROX	DT4 0JX
G0SUK	DT4 0JX
M1AUQ	DT4 0LR
G4OWY	DT4 0NA
G0EVW	DT4 0PE
G3SDS	DT4 0PE
G8BCH	DT4 0PE
G0JCA	DT4 0PP
G1CAN	DT4 0QY
G0MFK	DT4 0SN
G0ECX	DT4 7HJ
G6LNF	DT4 7QN
G3DLG	DT4 7TE
G3KWW	DT4 8AL
G3PGK	DT4 8LW
G4FYT	DT4 9AL
G0HOS	DT4 9BH
G4DOL	DT4 9BN
G7FZD	DT4 9BP
G4NHY	DT4 9BW
G4JVQ	DT4 9DB
2E1FWP	DT4 9JN
G1NLU	DT4 9JU
G8ZVS	DT4 9LE
2E1DGT	DT4 9LJ
G2XQ	DT4 9LU
G3EAT	DT4 9NF
G6GYG	DT4 9PA
G8WXC	DT4 9QZ
G3VNB	DT4 9RB
G7DOW	DT4 9RR
G8JQA	DT4 9RT
G1BBT	DT4 9UE
G0RYL	DT5 1AS
G3KES	DT5 1BS
2E1FGE	DT5 1DE
G0CAE	DT5 1EJ
G4ZQA	DT5 1HP
M1CCK	DT5 1JE
M1CCJ	DT5 1JE
G8HJT	DT5 1JQ
G4MFS	DT5 1JW
G0UIK	DT5 1LG
G0VOP	DT5 2AA
G4ZIY	DT5 2AA
2E1DTB	DT5 2AA
G0BZW	DT5 2AB
G0BQD	DT5 2AW
G0ACQ	DT5 2AY
G4RAK	DT5 2EA
G4CFZ	DT5 2EE
M1BCZ	DT5 2EP
G7VQP	DT5 2HJ
2E1APT	DT5 2HJ
2E1DQZ	DT5 2HJ
G1YRS	DT5 2JZ
G7EIS	DT5 2JZ

Wigan

M1AQO	M28 1AL
G4GLJ	M29 7AH
G4JHA	M29 7BH
2E1FFT	M29 7DD
2E1FFV	M29 7DD
G8ZWH	M29 7DG
2E0AOX	M29 7EG
G8NSE	M29 7EJ
G0NAS	M29 7ER
G7DCQ	M29 7ER
G3BSA	M29 7ES
G6BVN	M29 7FP
M0BFX	M29 7FP
2E1FEQ	M29 7FP
G4GKT	M29 7HH
G4DZK	M29 7HS
G8FQZ	M29 7HS
G0TFP	M29 7PH
G7CQW	M29 7RE
G8SYM	M29 7RT
G6MIU	M29 8LU
G6HFW	M29 8LZ
G6EJR	M29 8PQ
2E1EEM	M29 8PW
G4NRM	M29 8QS
G8ZOB	M29 8QS
G3LBL	M29 8WL
G4EQT	M29 8WT
G3KQE	M46 0AW
G1LBQ	M46 0LQ
G6SKM	M46 0LQ
G6YBC	M46 0LX
G0DYB	M46 0RF
G1DGP	M46 9AB
G4JSB	M46 9AP
G4ZVK	M46 9DW
G6TZD	M46 9LL
G6TFV	M46 9LN
G4HZJ	M46 9LQ
G4MWC	M46 9LQ
G4NRO	M46 9LQ
G6FSA	M46 9LQ
2E1DKQ	M46 9LQ
G6HGQ	M46 9LX
G0OMF	M46 9PQ
G0KBA	M46 9QB
G4GBK	M46 9QB
G4UPO	WA3 1EB
G4UPP	WA3 1EB
M1BRX	WA3 1EB
G6LWE	WA3 1EE
G6AHF	WA3 1EU
G4VZX	WA3 2BG
G4SYP	WA3 2DD
G4VDX	WA3 2EE
G1EFU	WA3 2EP
G0LVJ	WA3 2EP
G1JEO	WA3 2HH
G4ONG	WA3 2QL
G4XNV	WA3 2SE
G3CSL	WA3 2TF
G4JPX	WA3 3DF
G1XCL	WA3 3EZ
G0TQG	WA3 3NA
G4FDL	WA3 3QL
G0RPO	WA3 3QX
G0LVY	WA3 3TZ
G6WVL	WA3 3UX
M0ADF	WN1 1YE
G1SUH	WN1 2BA
2E1CXG	WN1 2EY
G2HFC	WN1 2NE
G1VYA	WN1 2QJ
2E1ECW	WN1 2QL
G3YNU	WN1 2RR
G4WGF	WN1 2SS
G6ZIY	WN1 3NF
G3TTL	WN1 3PG
G4SOO	WN1 2DU
G1PBB	WN2 1HW
G8OMC	WN2 1HW
G4MEG	WN2 1LA
M0AWX	WN2 1NA
G1SWH	WN2 1NA
M1AWS	WN2 1RL
G7PMB	WN2 1RR
G0FYE	WN2 2NA
2E1DEK	WN2 2NA
G8JCZ	WN2 3EW
G3RSW	WN2 3HL
G1YMP	WN2 3HR
G1TAR	WN2 3SB
G1OVZ	WN2 4AB
G4WIL	WN2 4ET
G6CLX	WN2 4EY
G0AYF	WN2 4HF
G1ECI	WN2 4HL
G4EHY	WN2 4LD
G1EIO	WN2 4LZ
G7JZI	WN2 4RG
G1BDU	WN2 4RP
G0CDJ	WN2 4RQ
G4PGQ	WN2 4SZ
G0CTN	WN2 4TL
G4TZS	WN2 4TU
G4ZOO	WN2 4TW
G7LVS	WN2 4TY
G0DBH	WN2 5HT
G0IZG	WN2 5HY
G7FMB	WN2 5TF
2E1EHB	WN2 5XA
2E1EHC	WN2 5XA
M1AIX	WN2 5XA
G4VDZ	WN3 4PR
2E1DFE	WN3 5HG
G8DPE	WN3 5PR
G4OVV	WN3 5PS
G7GVB	WN3 5PS
G0DPI	WN3 5QN
G0HRW	WN3 5QN
G7EDK	WN3 5QN
G7IXC	WN3 5QN
2E1FMR	WN3 5QQ
G3BPK	WN3 6AA
G4GWG	WN3 6AA
G0KHH	WN3 6AJ
G3PFZ	WN3 6EJ
G4GSM	WN3 6EN
G4WZS	WN3 6FA
2E1DNJ	WN3 6HG
G6LST	WN3 6JY
G4WDC	WN3 6LD
G6TYB	WN3 6LD
G4HSC	WN3 6TQ
G4WXX	WN4 0AZ
G0SJS	WN4 0NJ
G7IHE	WN4 8AY
G4MKS	WN4 8ED
2E1EQG	WN4 8QS
G0SIW	WN4 8QT
G7LTR	WN4 8RX
G0UNG	WN4 8ST
G0ULO	WN4 8UQ
2E1ERD	WN4 8XS
G4HSM	WN4 9EX
G3HUX	WN4 9LD
G6MRK	WN4 9LX
G4CUF	WN4 9LZ
G6OCB	WN4 9LZ
2E1FPJ	WN4 9NJ
G6NTM	WN4 9PJ
G1DON	WN5 0AJ
M1BZJ	WN5 0DZ
G4SVT	WN5 0JA
G0DSR	WN5 0PU
G0NEO	WN5 7AX
G4HAP	WN5 7BG
G3PJW	WN5 7EB
G7VJX	WN5 7EH
G0FXV	WN5 7EQ
G4EII	WN5 7HT
G6RZJ	WN5 7UA
G3YVP	WN5 8HR
G7SOV	WN5 8LX
G0UMY	WN5 8NP
G0UCQ	WN5 8PQ
G7CWE	WN5 8RH
G6ORO	WN5 9BH
G0TRY	WN5 9HZ
G4KTU	WN5 9PA
G8GVS	WN5 9TX
G0UVL	WN5 9XY
G6WTA	WN6 0BH
G4PPG	WN6 0LW
G8EHF	WN6 0NR
G8EHG	WN6 0NR
G1MSR	WN6 0QR
G7NQB	WN6 0SG
G0EOK	WN6 0SQ
G0AAV	WN6 0TG
G6XKY	WN6 0UA
2E1CXK	WN6 7EQ
M1AET	WN6 7EQ
G4EOT	WN6 7RF
2E1DUB	WN6 8DE
G8JMA	WN6 8DF
G4KCP	WN6 8HR
G0PPK	WN6 8LH
G1EPF	WN6 8NJ
G4PPB	WN6 8NJ
G4SFJ	WN6 8QA
G6PUV	WN6 8QA
G4BCT	WN6 8QX
G4WSK	WN6 8QY
G4EHK	WN6 9JH
G0SQX	WN7 1HN
2E1ECB	WN7 1JA
G0MNY	WN7 1JZ
G0MKB	WN7 1LS
G0BED	WN7 1LT
G6HNQ	WN7 1NB
G0KMB	WN7 1SG
G1HFS	WN7 1TS
G3WIS	WN7 2AE
G7SKR	WN7 2JP
G1PYC	WN7 2LE
G6TET	WN7 2NG
G3XPZ	WN7 2PB
G4SVA	WN7 2PB
G4JXI	WN7 2UU
G4UDF	WN7 3DA
G4COE	WN7 3EB
G1HAW	WN7 3HL
G1WDJ	WN7 3JX
G8PUN	WN7 3NE
G4GFD	WN7 3NU
2E1CKI	WN7 3QG
G6XFQ	WN7 3QJ
G6CHC	WN7 4BY
G0SSK	WN7 4DQ
G0KGX	WN7 4DR
G4GWF	WN7 4DU
G8SNI	WN7 4EE
G7IER	WN7 4HZ
G0LGC	WN7 4SX
G1FLV	WN7 4TA
G4VLY	WN7 4TT
G1VKT	WN7 5BT
G6LDY	WN7 5DG
2E1FFQ	WN7 5ES
M0AVA	WN7 5HF
2E1CON	WN7 5HT
M1ALB	WN7 5JZ
M1AMR	WN7 5NU
G4VXW	WN7 5PN
G1DUS	WN7 5PW
G4NTT	WN7 5PW
G1XPA	WN7 5QA

Winchester

G1EHE	GU32 1LH
G1ALN	PO15 7DJ
G6WBG	PO15 7EJ
G7RUV	PO15 7HN
G7RUT	PO15 7HN
G0NXP	PO15 7JQ
G8LWO	PO17 5BZ
2E1EIR	PO17 6HD
G3XUX	PO17 6HS
G8CKS	PO7 4RU
G0SSY	PO7 4SP
M1AHA	PO7 4SW
G2BFO	PO7 6HF
G6XBG	PO7 6LJ
G0WKL	PO7 6NA
G0WEJ	PO7 6NE
G8LNU	PO7 6PA
G8FXV	PO7 6PE
G4WQZ	PO7 6PR
G4TST	PO7 6PS
G6CZZ	PO7 6SH
G4DCP	PO7 6UA
G3RTP	PO7 6UE
G2DBT	SO21 1LN
G3UPD	SO21 1US
G3HRH	SO21 2DE
G8IDJ	SO21 2EG
G3WYR	SO21 2LH
G6SFX	SO21 2LJ
G6PWT	SO21 2QE
G6UQI	SO21 2QW
G0LMD	SO21 3DU
G0GWG	SO21 3EB
G3ZHV	SO21 3EB
G6KRN	SO21 3EB
G4SCM	SO21 3EH
G4DBL	SO21 3EW
G1NBU	SO21 3EY
G3LMH	SO21 3EY
G3HQX	SO21 3HA
G3SJK	SO21 3HP
G4XRJ	SO21 3HP
G6FVZ	SO21 3SG
G4FPC	SO22 4AF
M0AYV	SO22 4BJ
G3MCL	SO22 4EY
G0FUS	SO22 4HJ
G3VPG	SO22 4HS
G4JXE	SO22 4HS
G4CEW	SO22 4JJ
G0UMD	SO22 4JL
G4NBU	SO22 4JN
G3PTS	SO22 4JY
G4SVC	SO22 4LF
G4MCM	SO22 4LQ
G6AAZ	SO22 4PQ
2E1FLA	SO22 4PQ
2E1FKZ	SO22 4PQ
G4AFA	SO22 4PZ
G3XJM	SO22 4QH
G8NEH	SO22 4QL
M1AVC	SO22 4RU
G3RQR	SO22 5AT
G0DZV	SO22 5AX
G1DAU	SO22 5BX
G1THP	SO22 5DY
G3RUZ	SO22 5JH
G3XSZ	SO22 5JJ
G4CJO	SO22 5JX
G2PS	SO22 5LF
G0JMK	SO22 5ND
G3UYK	SO22 5ND
G4ULU	SO22 5QA
G3AJX	SO22 6BW
G0NPE	SO22 6ED
G0BES	SO22 6EU
G0WNZ	SO22 6EU
G3YSK	SO22 6HE
2E1EDX	SO22 6HS
G7LNI	SO22 6JG
G0EBK	SO22 6JH
G4FKR	SO22 6QH
G4XGO	SO22 6QH
G3ROG	SO23 0HE
G8DWI	SO23 0JE
G6CAA	SO23 0JE
G4HPJ	SO23 0NP
G4AXY	SO23 0NU
G0IHA	SO23 0PP
G3AZU	SO23 0PX
G4LYL	SO23 0QQ
G3VNN	SO23 7HD
G1HRA	SO23 7HT
G1YIQ	SO23 7LL
G0TAH	SO23 7ND
G1MBG	SO23 7NY
G4DKH	SO23 7QQ
G8RKV	SO23 8UY
G3JIZ	SO24 9EQ
G4YFJ	SO24 9ND
G1OOX	SO24 9NP
G6SDC	SO24 9PE
G7FQE	SO24 9PJ
G0JHT	SO24 9PL
G4DMG	SO24 9PP
G7JKY	SO24 9SW
G4EWT	SO32 1EE
G0JLX	SO32 1GE
G7VTZ	SO32 1HY
G1GXB	SO32 1JR
G4NYD	SO32 1JS
G7IWA	SO32 1LW
G8DNL	SO32 1RT
G1MFG	SO32 2AA
G4COM	SO32 2AR
G4IDW	SO32 2AR
G4ZRT	SO32 2TL
G6NXM	SO32 2TR
G1PRP	SO32 3QN

Windsor and Maidenhead

G1ZQR	RG10 0JB
G8KOM	RG10 9YT
G4CRW	SL3 9BA
G3BXS	SL3 9DJ
G4XBV	SL4 1LD
G6PWF	SL4 2LT
G8IUM	SL4 2PD
G7IXT	SL4 2PT
G3ZCD	SL4 3HA
G6CQO	SL4 3LP
G8FUO	SL4 3LP
G1HDG	SL4 3RD
G0HTK	SL4 4AE
G8ODP	SL4 4BG
G3GGO	SL4 4EH
G0AJS	SL4 4EP
G4FVD	SL4 4PA
G3WHB	SL4 4PZ
G4WOB	SL4 4YY
G8DPH	SL4 5DX
G4BSK	SL4 5JR
G6XWU	SL4 5LD
G4UAH	SL4 5PP
G3SID	SL4 5RE
G0WYR	SL4 5RE
2E1EGC	SL4 5RE
G4WQS	SL4 5TD
G6SKZ	SL4 5TT
M1BDH	SL4 6BW
M1BEC	SL4 6DU
G7VME	SL4 6DU
G1NLS	SL4 6JA
G1NWO	SL4 6LE
G7CYD	SL4 6NF
G3KWP	SL4 6PB
M1BUP	SL4 6PL
G6XAU	SL5 0AA
G4JMB	SL5 0EA
G0WLO	SL5 0EA
G3PLY	SL5 0HU
G0CKT	SL5 0QY
G0CTX	SL5 0QY
G4WED	SL5 8AB
G6EIZ	SL5 8QB
G7JTS	SL5 9JL
G4RSN	SL5 9PQ
G3VXZ	SL6 1BZ
G6CJB	SL6 1DD
G4GGV	SL6 1DF
G6GFR	SL6 1XE
G0VKE	SL6 1XL
G0TNY	SL6 2BJ
G4USK	SL6 2DD
G4ZQF	SL6 2DS
G8AJM	SL6 2DW
G7RWI	SL6 2EL
G0OIQ	SL6 2LH
G3ONR	SL6 2YP
G3VMR	SL6 3EL
G4FTG	SL6 3EP
G4TDL	SL6 3ET
G4XOW	SL6 3NR
G4MYQ	SL6 3NY
G8WXQ	SL6 3PZ
M1BAD	SL6 3YS
G6NVY	SL6 4DE
G6VEQ	SL6 4LL
G8FXX	SL6 4LL
G4XYN	SL6 4NG
G0SVN	SL6 4NG
G8ZGO	SL6 4PH
G7RAO	SL6 4PP
G1MRX	SL6 4QT
G3UXY	SL6 5AR
G8EUD	SL6 5AU
G0LZG	SL6 5AW
G0JPE	SL6 5DN
2E1BPV	SL6 5DN
G3HKZ	SL6 5DY
GM4SZA	SL6 5EG
G1VJF	SL6 5JA
G8GHT	SL6 5JD
G6LKQ	SL6 6AD
G4XDU	SL6 6DA
G3UHK	SL6 6DB
M0AIV	SL6 6DY
G7JCP	SL6 6EG
G8NMQ	SL6 6EP
G8RYW	SL6 6JB
G4HIF	SL6 6JL
G8TWE	SL6 6JL
G0GLA	SL6 6LG
G8RFK	SL6 6LT
G6NBI	SL6 6SJ
2E1BBH	SL6 6SP
G7RFS	SL6 6SQ
G4STF	SL6 7DS
2E1ANN	SL6 7EJ
G1IBM	SL6 7EJ
G0UZU	SL6 7EJ
G8EOX	SL6 7EJ
2E1ERR	SL6 7EJ
G4HWI	SL6 7EZ
G1DSI	SL6 7JX
G0PMZ	SL6 7LF
G3OTN	SL6 7LY
G0FFL	SL6 7SH
G6BJY	SL6 7TZ
G3YSG	SL6 7UE
G0AYA	SL6 7UH
G0BQE	SL6 7UH
G7OCL	SL6 7UL
2E1BAW	SL6 7UT
G4YHZ	SL6 7UU
G7HCF	SL6 8DH
G0RKY	SL6 8HD
G6ETZ	SL6 8HN
G8VPO	SL6 8HU
G4BOV	SL6 8JZ
G6GBC	SL6 8LB
G6PNI	SL6 8RZ
G7EBR	SL6 9DH
G3PQA	SL6 9DN
G0CMW	SL6 9JF
G3WKX	SL6 9JT
G3WYK	SL6 9JT
G1LIG	SL6 9NF
G4NIO	SL6 9PJ
G0NJQ	SL6 9SN
G7TWW	SL7 1SA
G4GBM	TW19 5AS
G3BYY	TW19 5BY
G4SVQ	TW19 5HB

Wirral

M1AAS	L41 0AX
M1BTW	L41 4HQ
2E1BBP	L41 7DR
G7CQZ	L41 8BX
G7GIN	L41 8HE
G8UWG	L42 0LB
M1BBA	L42 1RT
G0DYG	L42 2BR
G1XHT	L42 2DP
G6XIO	L42 2DR
G3ZMV	L42 4NY
G3OTW	L42 4PD
M1AVZ	L42 6QF
M1AOS	L42 6QN
M0AQS	L42 6QN
G4EWJ	L42 6QY
G6KJF	L42 6QZ
G8UWL	L42 7HB
G1RSP	L42 7JW
G4PKO	L42 7LA
M0AZF	L42 7LB
G0PZO	L42 7LB
G0KQS	L42 8NB
G6JZE	L42 8QA
G0KXL	L42 9LJ
G6RVH	L42 9NZ
G4OAR	L43 0TT
G4WCG	L43 0TT
G1MHF	L43 0XB
G6NFB	L43 1UA
G7MMM	L43 1UT
G4MUP	L43 2HZ
G0TDD	L43 2JF
G4SYJ	L43 2JG
G7SFM	L43 2JP
G4MWI	L43 2LF
G0LZI	L43 2LZ
M1ARH	L43 2NQ
G6ZFW	L43 5XE
G6XHF	L43 6TE
G0UCP	L43 6UT
G0MHF	L43 6UY
M1AQT	L43 6XS
G8FPU	L43 8SU
G0DBG	L43 8TF
G3NMT	L43 9SW
G1RCI	L44 0AQ
G6CGF	L44 0EB
G4KVP	L44 0EL
G1GUT	L44 1DW
G3YJY	L44 2AZ
G1PMJ	L44 3AF
G1MCR	L44 3DA
G8RXB	L44 3DA
G8NNS	L44 3DZ
G8TFW	L44 3DZ
G0JQK	L44 3ED
G7VQM	L44 4AG
M1AVS	L44 4DF
G0DPO	L44 4DL
G0WMQ	L44 4EU
G7KOS	L44 5RQ
G4ZCA	L44 5RQ
G3EGX	L44 5UL
G0OJN	L44 6QE
G7IIF	L44 7BH
G4AHC	L44 8AE
G8VDS	L44 8AN
G6LWC	L44 9AA
G0DYC	L44 9BP
G1NXB	L44 9DZ
G0JEG	L44 9EN
G0ELI	L44 9ER
G1MYX	L44 9ER
G3GZX	L45 0NQ
G0NBD	L45 1HA

G6ABO L45 1JE
2E1CFH L45 1JL
G4BTO L45 3JL
G0PIT L45 4RS
G1BYI L45 5BX
G4FPB L45 5DB
G1FEP L45 5HN
G0KNG L45 5JA
G7HJV L45 6TA
G3SEJ L45 6TD
G3STD L45 6TD
G8STD L45 6TD
G3XCD L45 6TE
G6IHD L45 6UQ
G3VEB L45 7NZ
G8TRY L45 7QF
G8WDC L45 7QF
G4VQS L45 7QG
G4KOY L45 7QQ
M0BFV L45 8PF
M1BMM L45 8PF
G8UWO L45 8QA
G1JEZ L45 8QL
G4ZEV L45 9JW
G6IIN L45 9LJ
G3PEZ L46 0QT
G0KMM L46 0RQ
G3TBQ L46 0RR
G0TYN L46 0SL
G7OND L46 0SQ
G0OTD L46 0SU
G1XYP L46 0SZ
G7NHR L46 1QZ
G7NLE L46 1RQ
G0WUO L46 2QZ
G3LCI L46 2RE
G0PVY L46 2SB
G4UUV L46 2SB
G0GMZ L46 3RD
M0AJB L46 3SW
G4NXG L46 3SW
G0DNQ L46 6BW
G8JZL L46 6BY
G4WUA L46 6DH
G1YKV L46 6HU
G0CLV L46 7SJ
M0AXJ L46 7UA
G8UIY L46 7UE
G7FND L46 7UP
G0EQE L46 8TX
G0LHN L46 9PF
G6NZT L46 9RP
G3OKA L47 0LB
G4ELA L47 0LQ
G7OFM L47 0NA
G0AVS L47 2AN
G4RVZ L47 2AZ
G3PQT L47 2DS
G3PYU L47 2DS
G6UWW L47 3AQ
G8THZ L47 3DD
G4OVY L47 7BT
G6BDS L47 8XZ
G8UZZ L47 9RN
G0IAW L47 9RT
G6NII L48 0QN
G1NTO L48 0RL
G4YXO L48 2HA
G4KPY L48 4DH
G8NDD L48 5DS
G0RXI L48 5DW
2E0AGE L48 5HE
G8ODK L48 5HQ
G6HQS L48 6DF
G3OVE L48 7EX
G8LEM L48 7HR
G7IND L48 8BH
G4MSA L48 8BJ
G0DVZ L48 8BR
G7NIR L48 9UF
G0SSG L48 9UL
G1CKQ L48 9UP
G0GSF L48 9XA
G3HFA L48 9XB
M1AOI L48 9XS
G4OKP L48 9YA
G4BKF L49 0TD
G0VAX L49 0XD
G8ZJH L49 1RY
G3YSM L49 1SG
G6PRE L49 1SS
G6XTV L49 2NJ
G7HQE L49 2PE
G6USU L49 2PN
G0TXF L49 2PU
G7IUW L49 2PZ
G6KBQ L49 2RQ
G3MAJ L49 3AW
G0MQJ L49 3NG
G8KWU L49 3PP
G4CVZ L49 3PR
G3FLG L49 3PS
G0LPU L49 3QH
G1XYV L49 4GS
G3VUY L49 4GY
G4TZW L49 4NN
G8AIV L49 4NT
G0TPJ L49 4PF
G0DED L49 4PN
G3VVA L49 4PP
G4EYP L49 4RG
G3RLB L49 5LB
G7MIZ L49 6JD
G4MIA L49 6LA
G7HUD L49 6LZ
G4NCI L49 6NF
G0NZS L49 6NP
G3UZU L49 7NA
G0NDM L49 8HY
G7EBA L60 0BB
G4XCM L60 0BD
G2AMV L60 0BP
G3WIO L60 0BP
G6LKG L60 1UH
G0MQR L60 1UL
G6AGP L60 1YD
G4DBE L60 1YH
G3QX L60 2TL
G3NPJ L60 2UA
M0AUA L60 2UA
G0TKN L60 3SS
G4EIC L60 5RY
G4VBW L60 5RY
G0KTY L60 6BB
G3UVR L60 7RA
G4MGR L60 7RA
G4PUH L60 7RA
G6XHG L60 7RD
G7RFH L60 7RU
M1BXV L60 8QQ
G3RLA L60 9JT
G4GKG L61 0HG
G7OKR L61 1BA
G8MGY L61 1BJ
G4UHX L61 2XP
M1ANG L61 4UT
G6HWD L61 4UZ
G0EVB L61 4XB
G4YJV L61 4XH
G8MVX L61 4YG
G4NOY L61 6UZ
G8DTS L61 6XT
G3OKF L61 7XF
G7IOH L61 7XF
G4GGB L61 8RW
G4EFP L61 8SH
G8REQ L61 8SX
G8NZC L61 9NB
G6XPM L61 9QN
G3ODC L62 0BD
G1UAR L62 0BT
M1AOR L62 1AU
G7ELI L62 1DR
2E0APL L62 2AQ
G3MKH L62 2AW
G7IIP L62 2BZ
M0ATZ L62 2DG
G6IIM L62 3LF
G0UYD L62 5EP
G3NNV L62 6AL
G6HKY L62 6AN
G7IGU L62 6AW
G6PYL L62 6BE
G8WHR L62 6BR
G0GBN L62 6BU
G8IWJ L62 6DT
G4YWD L62 7EX
M0AUF L62 7HB
G6AVY L62 8BB
G3YGL L62 8BR
G0VEO L62 8BS
G6HQD L62 8BT
G8CVF L62 8DJ
G4UCE L62 8DL
G4VWL L62 8DY
G0PZD L62 8EX
G7PZS L62 8EZ
G3AVL L62 9AA
G8SME L62 9AF
G0CTU L62 9AZ
G3TWN L63 0EP
G8TSG L63 0JJ
G7RXQ L63 0NL
2E1DUV L63 0NL
G0OWY L63 0NP
G4UAX L63 0NP
G6VS L63 0PW
G6NOI L63 2JU
G1GUW L63 2NL
G0CFG L63 2QT
G1KEP L63 4JS
G0OGY L63 4LB
G7CZA L63 5JN
G4NSA L63 5JR
G3HAC L63 5LN
G4KJX L63 5LZ
G1FOO L63 5NE
G3NWR L63 5NE
G4JGP L63 5NR
G0OXA L63 5PA
G4NNO L63 7LL
G0IAV L63 7LS
G7CHH L63 7NS
G3JXI L63 7QD
2E1AUR L63 7QD
G6ZNR L63 7QE
G8IQY L63 7RU
G7TIY L63 8LB
G0TCL L63 8LR
G3NSZ L63 8NZ
G3XCG L63 9AJ
G1OKY L63 9HD
G3FZR L63 9JR
G0PXO L63 9LP
2E1BAU L63 9LP
G7RFW L63 9LU
M1BZZ L63 9YJ

Woking

G4SXT GU21 1AL
G7CSJ GU21 1PN
G3XKA GU21 1QY
G1WNW GU21 1XF
M1BKY GU21 1YP
2E1CSB GU21 1YP
G0SPZ GU21 2DE
G1LIJ GU21 2JS
G1EGM GU21 2RA
G8GS GU21 2RL
G6TWR GU21 2RN
G1OSN GU21 3DD
G8JMU GU21 3DD
G0VYR GU21 3DU
G8YSV GU21 3HX
G8BTL GU21 3NZ
G7WIY GU21 3PJ
G1HLQ GU21 3QH
G6YLA GU21 3QH
G8GZZ GU21 3QH
G8iLK GU21 3QH
G3USX GU21 4BG
G8SSY GU21 4JG
G4CZP GU21 4JR
G4UMI GU21 4LR
G4KXL GU21 4LX
G3NR GU21 4LZ
G0ADM GU21 4PS
G0PQN GU21 4RH
G1ILS GU21 4RZ
G7RZA GU21 4UR
G0KQY GU21 4UY
G0NIX GU21 5EG
G0NGI GU21 5PY
G0SOO GU21 5QZ
G0VTN GU21 5QZ
G4LXT GU21 5SJ
G0CDF GU22 0DB
G8KYK GU22 0DF
G6VBJ GU22 0JL
G6BZ GU22 0JX
G7HHI GU22 0NQ
G3ZHB GU22 0NT
G6BQC GU22 0NT
2E1BWJ GU22 0NT
G3AEU GU22 7EY
G3JZX GU22 7LZ
G0MIC GU22 7NX
G3GAQ GU22 7UE
G8AZC GU22 8EF
M0BGN GU22 8LF
G3SXI GU22 8LS
G3WVG GU22 8PE
G1XBV GU22 8RY
2E1BNL GU22 8RY
G0NRJ GU22 9BQ
G4YPC GU22 9BQ
G1SKL GU22 9EZ
G0XGK GU24 0AB
G0KEP GU24 0EU
G8MMO GU24 0HG
G6MQZ GU4 7QB
G6JRZ KT14 6DT
G7WDE KT14 6PJ
G7GLY KT14 6QG
G6XAN KT14 7EF
G3GLB KT14 7HY
G6KTG KT14 7JF
G3OKQ KT14 7JQ
G0PBL KT14 7NG
G0OBG KT14 7TG

Wokingham

G8LVU RG10 0AX
G8DZC RG10 0BA
G3PGM RG10 0DP
G3RVD RG10 0EY
G1BSX RG10 0LP
G8KZG RG10 8BJ
G3WPH RG10 8BN
M1CAZ RG10 9AX
G3LRQ RG10 9AY
G4HFN RG10 9BA
G8NXJ RG10 9BN
G3VKQ RG10 9PY
M1BFO RG10 9QD
M1BGT RG10 9QH
2E1EQS RG10 9QH
GM7GEF RG10 9RP
G8EPZ RG10 9TS
G2DSB RG2 9AS
G1HCR RG2 9DG
G8IJG RG2 9HA
G3IHH RG2 9NH
G6IHH RG2 9NH
G0ROV RG2 9NR
G7KWN RG2 9PU
G8KJ RG4 6UR
G0IZV RG4 6XH
G4LFX RG40 1DP
G7LZZ RG40 1QG
G8CIT RG40 1QG
G8IEI RG40 1QG
G3RGG RG40 1RH
G1PQO RG40 1RY
G4LCE RG40 1SA
G4MKM RG40 1SP
G8SZB RG40 1TA
G4YRL RG40 1TS
G4ORU RG40 1TW
G7OJT RG40 1XE
G4XAL RG40 1XF
G1KEM RG40 1XX
G8FBF RG40 1XX
G4BTI RG40 2BT
G8BSF RG40 2EL
G4IRJ RG40 2HX
G4WQU RG40 2LE
G3GNB RG40 2LT
G4LRY RG40 3EB
G3LTP RG40 3EY
G4OKM RG40 3HT
G7JTV RG40 3LB
G3XPC RG40 3LG
G6AXY RG40 3NF
G3OFK RG40 3PR
G4UXG RG40 3PR
G8NZK RG40 3QJ
M0AIM RG40 3QJ
G3WGV RG40 3QJ
G8AKB RG40 3QS
G3WME RG40 3RL
G1KTF RG40 4HB
G8RKZ RG40 4HP
G4FUY RG40 4JA
G8NPZ RG40 4JE
G8LCL RG40 4JE
2E1BLG RG40 4JE
G1ZSF RG40 4RA
G4NIP RG40 4RA
G4DYO RG40 4RD
G0NRF RG40 4UA
G0EER RG40 4UD
G4HOU RG40 4UD
G6AAU RG40 4UG
G3ZBS RG40 5YB
G1BAX RG41 1JN
G6CGC RG41 1NR
G3YHG RG41 1PH
2E1AUQ RG41 1PH
G3XZJ RG41 1QH
G1OUM RG41 2PS
G8OVH RG41 2PZ
G0JHW RG41 2RJ
G0VLZ RG41 2SD
G3VIE RG41 2ST
G3APO RG41 2XT
G8GJM RG41 3AG
G3MGU RG41 3HG
G8DUO RG41 3TF
G8PKV RG41 3UA
G0NMN RG41 3UL
G0RMR RG41 3UL
G3ZKH RG41 3XF
G4KNJ RG41 3YY
G4MAE RG41 3YY
G6TLX RG41 4AR
G6ZIP RG41 4AS
G8IBP RG41 4AW
G4LJF RG41 4AX
G6IJK RG41 4BD
G3TUY RG41 4ED
G4DHZ RG41 4SS
G4AEP RG41 4SY
G1RVP RG41 4UX
G7SEO RG41 4UY
G8MFF RG41 5AU
G6ANW RG41 5EU
G1EEM RG41 5HJ
G0GFU RG41 5JE
G4GBI RG41 5JF
G4YFN RG41 5JJ
G7COP RG41 5JZ
G6IVB RG41 5NW
G1HBD RG41 5PE
G6MHM RG41 5TF
G7UCV RG41 5UR
G6ZLU RG45 6BZ
G1MSS RG45 6DB
G4RZI RG45 6EF
G6LLG RG45 6EF
2E1ANA RG45 6HE
G6LKZ RG45 6HE
2E1DIH RG45 6HE
G4EPX RG45 6HJ
G4NJC RG45 6HJ
G8ZDU RG45 6NG
G1DPY RG45 6PP
G1WGM RG45 6QF
G3KBS RG45 6QF
G1LHE RG45 6QN
G0FPT RG45 6QR
G4LGE RG45 6QU
G3ZWK RG45 6SS
G0AZS RG45 6UX
G4HNF RG5 3AD
G4NIK RG5 3AY
G3OAK RG5 3BH
G3WTV RG5 3DA
G0BXQ RG5 3DG
G0DWD RG5 3HA
G0ISH RG5 3JQ
G7MER RG5 3LJ
G1SEC RG5 3PD
G7VVZ RG5 3QR
G0VQR RG5 4AP
2E0ACY RG5 4AP
G4ELY RG5 4AW
G0OPB RG5 4BL
G4DNH RG5 4EG
G3JSG RG5 4EX
G6LLP RG5 4HL
G7AQL RG5 4HL
G1DNP RG5 4JN
G8GZR RG5 4LA
G4JVD RG5 4LH
G0FIJ RG5 4LR
G8MBQ RG5 4LR
G1ION RG5 4NB
G1SEO RG5 4NL
G0LPI RG5 4NP
G8FRC RG5 4NT
G8JON RG5 4PF
G7UOQ RG5 4QB
G3UUB RG5 4QT
M1BHH RG5 4TT
G4KWT RG5 4UN
G0VPE RG5 4UN
G6SFE RG5 4UW
G8CSK RG5 4XR
G0ILT RG6 1AP
2E1BZG RG6 1BA
G6LGC RG6 1BU
G0KIA RG6 1DX
G0DQG RG6 1HR
G3VMY RG6 1HW
G6LAJ RG6 3AF
G7NZO RG6 3AQ
G4AOL RG6 3BW
G6ENS RG6 3XB
G4AEI RG6 4AA
G7UZY RG6 4AB
G6SXO RG6 4AG
G3TEB RG6 4AZ
G7EIX RG6 4BB
G4XEG RG6 4DA
G7VHV RG6 4HX
G6YLN RG6 4HZ
G6IEI RG6 4XA
G4KTC RG6 5AA
G3UPZ RG6 5PN
G6EXI RG6 5PY
G6PFC RG6 5QX
G4FXT RG6 5QZ
G4KWL RG6 5RA
G4YFE RG6 5SG
G6BFN RG6 5SL
G8VRW RG6 5SN
G0CEC RG6 5TG
G4OAE RG6 5TH
G0MZN RG6 5XG
G3ULT RG6 5XG
G8VWH RG6 5XG
G4LMX RG6 5YH
G6JVK RG6 5YH
G1POJ RG6 6BG
G7LWY RG6 7LJ
G7AOK RG6 7LQ
G4OQJ RG6 7PA
G0JTN RG6 7PD
G4DYI RG6 7RT
G7FBE RG7 1BQ
G7GEB RG7 1BQ
G3SQU RG7 1DS
G8JZO RG7 1HH
G8MVY RG7 1TW

Wolverhampton

G7WCQ WV1 2AD
G4SGE WV1 2AR
G4JUD WV1 4PL
G4PVZ WV1 4PL
G7WEV WV1 4SY
G7MEG WV10 0BB
G0EGP WV10 0NA
G1KIE WV10 0SA
G4ZBC WV10 0SH
G4YUT WV10 0ST
G7RFF WV10 6AF
G0GEL WV10 6BQ
G0GEO WV10 6BQ
G4IGG WV10 6DX
G7AWU WV10 6EL
G4OSU WV10 6EN
G1AEF WV10 6EZ
G4DHL WV10 6EZ
G1YOB WV10 6LX
G1VAN WV10 6NN
G3JJR WV10 6NU
G8ZWU WV10 6NW
G6JTL WV10 6PD
G3ONP WV10 6SB
G0HYH WV10 6TL
G3OCK WV10 6UA
G3FWD WV10 6XH
G1MSB WV10 8AY
G6ZFZ WV10 8AY
G4SUJ WV10 8QH
G3CVX WV10 8RS
G0BET WV10 8TD
G6UVU WV10 8TH
M0AXG WV10 9HN
G0SMO WV10 9QE
G0OJR WV10 9SN
G0OQY WV10 9TL
G6TC WV11 1AA
G6MJM WV11 1AP
G8GKT WV11 1BH
G4EYY WV11 1HF
G8ZIK WV11 1JG
G0WAY WV11 1JH
G8BHH WV11 1YD
G4PBP WV11 1YD
G0OKT WV11 2JW
G1BCE WV11 2PD
G8JBC WV11 3AN
G8PSL WV11 3AN
G0KYK WV11 3EG
M1BZH WV11 3QU
G0EGG WV11 3RT
G7VKZ WV11 3XB
2E1CLT WV13 3EL
2E1EHW WV13 3EL
G0EDD WV13 3PA
G1PBF WV13 3PB
G4BXC WV13 3PB
G0UYI WV14 6QH
G6CKJ WV14 7BD
G8JHO WV14 7BN
G7GJA WV14 7ER
G4ZAD WV14 8BJ
M1AIZ WV14 8DS
G7ACJ WV2 1HR
G7BZQ WV2 2AU
G7SRG WV2 2AU
G4DGM WV2 3HU
G1LEX WV3 0PZ
G3GRB WV3 0QN
G4KUW WV3 0UT
G8DKF WV3 7AF
G3VZZ WV3 7AH
G0EZT WV3 7BL
G1GVP WV3 7DN
G3NOW WV3 7HA
G7GXM WV3 7JP
G7KPD WV3 7LU
G7LTE WV3 7LU
G7JTB WV3 7LU
G4JSQ WV3 7NG
G7CXB WV3 7NJ
G3ZLJ WV3 8HJ
G8DJC WV3 8HN
G3RSX WV3 8LY
G4WRB WV3 9BL
M0AYG WV3 9DP
G4BTE WV3 9DQ
G0CYO WV3 9HZ
G8EBD WV3 9JF
G4TEC WV3 9JY
G3URJ WV3 9LW
G1LZT WV4 4AY
G3RQX WV4 4AY
G4LWC WV4 4JJ
G6SXJ WV4 4JJ
G4IFM WV4 4NL
G0NWV WV4 4QY
G4XDO WV4 4QY
M1AHX WV4 5BZ
G4GMU WV4 5LN
G4LLU WV4 5PZ
G1RLI WV4 5RF
G3UBX WV4 5RF
G7DDQ WV4 5RP
G7JRN WV4 5RT
G0OCR WV4 5UD
G0EEN WV4 6DA
G4WFI WV4 6PS
G7SVM WV4 6QY
2E1EUY WV4 6RJ
G1JXR WV4 6TA
G4POK WV6 0BH
G8UPM WV6 0HY
G4TDO WV6 0SF
G6BRJ WV6 8AY
G3THW WV6 8DF
G4TCA WV6 8DF
G1SGA WV6 8LA
G8NJF WV6 8NL
G4SGQ WV6 8NU
G3SEZ WV6 8PQ
G1HMI WV6 8RF
G7LPY WV6 8RQ
G0HDD WV6 8TR
G4HRG WV6 8TS
G8GXF WV6 9AZ
G3EEZ WV6 9LD
G8UYR WV6 9LF
GW0VBC WV6 9LS
G8SWZ WV6 9NJ
G6JJG WV6 9NQ
G4NTE WV6 9PW
G1RIE WV8 1RS
M0AFQ WV8 1XZ
G7OLC WV9 5LL
G1BWG WV9 5PH
G0MCE WV9 5QF
G4FAQ WV9 5RN
G4UIT WV9 5SD

Worcester

2E1AZS WR1 1JZ
G6TRS WR1 1QN
G1MAK WR1 3EH
G8XMO WR1 3NY
G4RPC WR2 4DJ
G3TKE WR2 4DJ
G0UAB WR2 4DZ
G4RJU WR2 4EB
G6FRB WR2 4HE
G6CBP WR2 4JQ
G1XIY WR2 4XN
G4ROR WR2 4XT
2E1DLD WR2 4XT
G1IVL WR2 5LQ
G6EQF WR2 5ND
G8RGJ WR2 5PT
G1EIP WR2 5PX
G0BAM WR2 5QY
G1AQF WR2 5RW
G7HNL WR2 5SJ
G4VWX WR2 5UH
G3NL WR2 6AA
G4MFU WR2 6BY
G1ICQ WR2 6DA
G0WPK WR2 6EH
G4DEV WR3 7AN
G4VZH WR3 7EH
G0CCQ WR3 7JS
G4BYB WR3 7LR
G8PHV WR3 7ND
G4NUZ WR3 7PP
G6BCE WR3 7PS
G4ETQ WR3 7PT
G0WXJ WR3 8EJ
G1VKB WR3 8EZ
M0AFK WR3 8JS
G6NAP WR3 8NT
G6TYT WR3 8NT
G1GAP WR3 8XA
G8EPC WR4 0DN
G7SAI WR4 0DN
G2PK WR4 0HQ
G6RVB WR4 0JF
G6DMG WR4 0PE
G4XVL WR4 0RS
G8PWD WR4 0RS
G1DOK WR4 9ET
G7EPM WR4 9EX
G3SPP WR4 9HS
G4DLM WR4 9HX
G3NXJ WR4 9JX
G0HUV WR4 9LB
G3NXQ WR4 9SD
G1UVB WR4 9TS
2E1COI WR4 9TW
G3VQW WR4 9UG
G0DEP WR4 9YU
G0PMH WR5 1BT
G7WGH WR5 1HZ
G4IDF WR5 1JJ
G4MHC WR5 1JJ
G3NDQ WR5 1QA
G4KPL WR5 1SD
G6AHI WR5 1SD
G6PMO WR5 2AL
G1JGF WR5 2BD
G8STI WR5 2BD
G7BHB WR5 2DL
G8ASO WR5 2EG
G4URN WR5 2ET
G0WJN WR5 2HE
G4HNZ WR5 2HL
M0AYK WR5 2HP
G7WIL WR5 2HP
G3WPP WR5 2HR
G8RDY WR5 2HU
G4ELL WR5 2JU
G4GIM WR5 2LN
G3GSC WR5 2LR
G0SMZ WR5 2NG
G3BGR WR5 3BD
G1WPH WR5 3HD
G6NFJ WR5 3HD
G0AAT WR5 3HT
G7CRR WR5 3SH
G6UXF WR5 3SJ
G4NMR WR5 3SY
G6FZV WR5 3SZ
G3REA WR5 3TD

Worthing

G4EHZ BN11 1DS
G7GSD BN11 2DN
G1JTM BN11 2DT
G4FFE BN11 2DW
G8YRF BN11 2DW
G4AOC BN11 2LA
G4DYI BN11 2LT
G0ASB BN11 2NA
G1BGC BN11 2PB
G7HJN BN11 2SS
G8ZGU BN11 3HP
G3LCF BN11 4HE
G0VUF BN11 4PD
G4KFU BN11 5AY
G4VBR BN11 5HF
G3PAX BN11 5PB
G8RJB BN11 5QS
2E1CPP BN12 4BH
G7SAL BN12 4BH
G3XJG BN12 4BJ
G4IVM BN12 4EE
G3BSU BN12 4HJ
M0APW BN12 4HR
G3SZM BN12 4LD
G4TWK BN12 4LH
G7RZU BN12 4ND
G1IJY BN12 4NE
M1BLM BN12 4PG
G1EXG BN12 4QB
G4OHC BN12 4RT
G7VMB BN12 4TU
G6IEQ BN12 5BU
G1DKX BN12 5DJ
G8ZNK BN12 6AR
G0DMY BN12 6EF
G2DZF BN12 6EF
G7TLY BN12 6EF
G0IGI BN12 6JR
G1WSM BN12 6JR
G4GOT BN12 6LA
G6HIE BN12 6LD
G3UNK BN12 6LN
G4HNU BN13 1AE
G1SPW BN13 1BT
G1OLE BN13 1DA
G3YSW BN13 1DA
G6BWP BN13 1DR
G4CZX BN13 1DZ
G6HXW BN13 1JS
G0FJY BN13 1NZ
G1ZZC BN13 1PX
M1AHC BN13 1QN
G1HIM BN13 2BS
G0FUM BN13 2BT
G1KIZ BN13 2DH
G4LKW BN13 2DH
G7MZI BN13 2DH
G3UEQ BN13 2NT
G1JDM BN13 2PA
G3SPN BN13 2PH
G3YHM BN13 2QY
G0RCB BN13 2RN
G1AYL BN13 2TD
G0IVM BN13 2TN
G0JXX BN13 3AG
G3HPB BN13 3BB
G3SXE BN13 3BH
G4ALZ BN13 3DG
G7EQR BN13 3DS
G4RRU BN13 3HG
G4KOU BN13 3HN
G3VWL BN13 3HZ
G8YGK BN13 3JD
G2FUF BN13 3JR
G4OAY BN13 3JY
G0AKK BN13 3LG
G4FPM BN13 3LG
G0OMN BN13 3LN
G0CIP BN13 3QF
G1YVI BN13 3QF
M0AIB BN13 3YL
G3ZBG BN14 0AH
G2BMI BN14 0BF
2E1AMB BN14 0EY
G1NRT BN14 0HD
G3KGA BN14 0HR
G7UPI BN14 0HS
G0IQL BN14 7BX
G8YGS BN14 7BY
G0ELN BN14 7EE
G8MIH BN14 7HH
G8ZWC BN14 7HW
G7GOA BN14 7LL
G7JRM BN14 7NE
G1KKH BN14 7PE
G6MBL BN14 7PY
G8DHE BN14 7QW
G0SWH BN14 7SB
G7PIK BN14 8AH
G6MMJ BN14 8AZ
G3UQD BN14 8DG
G6KHG BN14 8EL
G8FNV BN14 8ER
G2ALM BN14 9AU
G8BTK BN14 9DG
G1RFO BN14 9DS
G8RDK BN14 9JE
G0ELV BN14 9LD
G8TMV BN14 9LQ
G8VHK BN14 9LY
G3VXJ BN14 9NJ
G3GAI BN14 9NZ
G6IDF BN14 9PR
G3NLX BN14 9RG
G3UDH BN14 9RJ
G3PNW BN14 9RT
G3PNX BN14 9RT

Wrexham

GW6FBM LL11 2BB
GW6WVD LL11 2BG
GW3ZNN LL11 2HR
GW6NSG LL11 2LU
GW6MPX LL11 2LW
2W1FJT LL11 2YA
GW1MPR LL11 3EN
MW1ANU LL11 3HL
MW0ATI LL11 3HL
2W1FPX LL11 3HL
GW8GKS LL11 3LJ
GW1IAW LL11 3LR
GW7KNN LL11 3PB
MW0ARL LL11 3PZ
GW6OHX LL11 3YB
GW1LHV LL11 3YT
GW8ASD LL11 4AF
GW7DKI LL11 4UF
GW8KSE LL11 5AH
GW8KSF LL11 5AH
GW4TJN LL11 5UF
GW4GSS LL11 5UH
G6NOO LL11 5UR
MW0ATP LL11 5UR
GW4OVH LL11 5YF
GW0VMR LL11 6NS
GW1ZHI LL11 6RH
GW4IGF LL12 0BP
GW0WXW LL12 0BP
G8HPL LL12 0HN
GW0KPU LL12 0NW
GW4AVC LL12 7AS
GW3HHF LL12 7BA
GW0ABE LL12 7PD
GW6SBD LL12 7PP
GW4LMU LL12 7TP
GW6GMF LL12 7UG
2W1DAM LL12 7UH
2W1DAO LL12 7UH
GW0EMB LL12 8AF
GW6FED LL12 8BE
2W1EEN LL12 8BN
GW6HQA LL12 8DE
MW1ASG LL12 8HL
2W1FNS LL12 8HL
GW3LDH LL12 8JY
GW0VSH LL12 8JY
GW8RUA LL12 8LS
GW3JAZ LL12 8NR
GW6JTX LL12 8PU
GW0CES LL12 8ST
GW7WXM LL12 8XT
GW6YDT LL12 8YG
2W0AQD LL12 8YL
2W1FJH LL12 8YL
GW4TEQ LL13 0AU
GW4VAG LL13 0AY
G6ILY LL13 0BH
GW6UFO LL13 0LP
GW6HRG LL13 0TQ
G6FES LL13 0UJ
GW7DRN LL13 7AS
GW3GWA LL13 7DH
GW1ZHX LL13 7PB
GW0KYT LL13 7PD
GW7FNR LL13 7PD
GW0GWE LL13 7QE
MW0BEN LL13 7QH
GW6ZCS LL13 7QW
GW8OBW LL13 8ES
GW8VEE LL13 8QH
2W1GAU LL13 9BA
GW3HEU LL13 9LA
GW3XQO LL13 9NQ
GW4TNF LL13 9QH
GW4ZYM LL13 9YH
GW1MVL LL14 1AN
GW0VML LL14 1AN
MW0ARV LL14 1BB
GW4WUR LL14 1LW
GW0CVP LL14 1LW
GW4WVB LL14 1ST
GW8WYW LL14 1LW
MW1BLT LL14 2BA
GW4XLS LL14 2DH
GW6ZCR LL14 2EA
GW6GTS LL14 2EP
GW8GGW LL14 2EP
GW7PCX LL14 2HT
MW0ASL LL14 2ND
GW8HBP LL14 2RL
GW0WZZ LL14 2RL
GW0WER LL14 3EH
GW4WCM LL14 3ND
GW0TBT LL14 3PN
2W1FJG LL14 4DA
MW1BJB LL14 4DN
GW7MDH LL14 4JB
GW6OWQ LL14 4LG
MW0ALU LL14 5BJ
GW7NTZ LL14 5BJ
GW0EHA LL14 5TH
GW1MKV LL14 6DD
GW1GXQ LL14 6HS
GW1ZJC LL14 6HS
2W1FJZ LL20 7AD
2W1FWB LL20 7AD
2W1EEP LL20 7AR
MW0AYM LL20 7AT
GW7GHE LL20 7AT
2W1FJN LL20 7AT
G3TZU SY13 2JX
G0BSB SY13 2LP
G0KGD SY14 7NE

Wychavon

G4LZA B49 5LD
G3UMV B49 5LJ
G4UMV B49 5LJ
G6LRT B49 5LJ
G0UMV B49 5LJ
G6CUQ B96 6LT
G0VFX B96 6QU
G4DDF B96 6SW
G3IKR B96 6TB
2E1DJQ CV37 8XH
2E1DKO CV37 8XR
2E1FHV CV37 8XS
2E0AMM CV37 8XW
G3XGV CV37 8XY
G6DQZ DY10 4EL
2E1ARF DY10 4JA
G4NPN DY11 7LA
G1JRX DY11 7XU
G0HFR GL20 7AT
G1IME GL20 7AT
G3XGW GL20 7AU
G3WIK GL20 7HB
G7STT GL20 7NL
2E1ESS GL20 7NQ
G7FGK GL20 7QG
2E1ERG GL20 7QG
G0HDB GL20 7QL
2E1DRT GL20 7QL
2E1EQZ GL20 7QQ
G3OLW GL20 8HS
G3HBV WR10 1HF
G1IPE WR10 1JH
G7TGK WR10 1JY
G0IXX WR10 1LW
G1AFJ WR10 1LW
G1PYQ WR10 1LW
G0TSR WR10 1LY
G0ENR WR10 1NW
G3UEY WR10 1PW
G4TXF WR10 1QL
G4OZQ WR10 2AX
G4YIG WR10 2JE
G4RMV WR10 2LA
G1VGN WR10 2LL
G4NIJ WR10 2NY
G6CMP WR10 2NY
2E1EOC WR10 3BB
2E1EOD WR10 3BB
G4TDR WR10 3HJ
G7LGQ WR11 4JQ
G3PGQ WR11 4LX
G3EDS WR11 4NE
G6DRC WR11 4NL
2E1AWR WR11 4NU
2E1DHD WR11 4NU
G4URU WR11 4QL
G3XCW WR11 5BJ
G7AXI WR11 5BP
G0XTL WR11 5BQ
2E1DXV WR11 5DL
G4XQG WR11 5EU
G2BKO WR11 5EY
G0GBL WR11 5HQ
G8GPH WR11 5HQ
G6YMY WR11 5JP
G4GTS WR11 5LG
G4PCM WR11 5LL
G1GQW WR11 5LP
G1JMD WR11 5NA
2E1DFS WR11 5QN
G6HAA WR11 5RB
G0NDE WR11 5TH
2E1BAY WR11 5TJ
G4UXC WR11 5TY
G4WET WR11 5TY
G4MFK WR11 5UF
2E1DGG WR11 5XS
G3KLZ WR11 6BE
G8GSQ WR11 6BE
G4YJB WR11 6BU
2E1DGH WR11 6BU
G3DEF WR11 6BX
G6NBL WR11 6DE
G4EKG WR11 6EQ
G7JSC WR11 6NB
G8EPB WR11 6NP
G6XKX WR11 6QA
G1POS WR11 6QQ
G3CUF WR11 6UA
G2HDU WR11 6UL
G3WY WR11 6XA
G8XDX WR11 6XS
G8OWO WR11 6XY
G0ERA WR11 6YW
G4NRD WR11 6YW
G0VVW WR12 7DG
G8HFL WR12 7EP
G3IXI WR12 7HB
G0TWL WR12 7PB
G3WDQ WR3 7XG
G6VVE WR7 4AP
G0CRX WR7 4BT
G3GHB WR7 4DH
G1RLD WR7 4HD
G6TVC WR7 4NA
G4THT WR7 4NH
G4NXP WR7 4QA
G8NWR WR8 9BL
G8LCM WR8 9LP
G7WEB WR9 0DG
G3VJX WR9 0ND
G1YBI WR9 0RP
G3RLF WR9 7AF
G3ELS WR9 7AH
G3KTH WR9 7AY
G3TRB WR9 7AZ
G8GYO WR9 7BP
G4RCB WR9 7DF
G3LFF WR9 7DP
G8RQO WR9 7DQ
G0WFK WR9 7EB
G0BLS WR9 7QE
G4THU WR9 7QU
G4HDO WR9 7SE
G0BDM WR9 8DD
G3VPH WR9 8HG
G8XGG WR9 8HW
G4PQZ WR9 8HZ
G4PVO WR9 8HZ
G8TFO WR9 8NP
G3BZS WR9 8NR
G4TID WR9 8QR
G7VDX WR9 8SS
2E1COB WR9 8ST
G7SHT WR9 8SZ

G4AEK WR9 8TQ
G7VJM WR9 8TQ
G8HHR WR9 9BZ
G0GGV WR9 9DB
G0AXB WR9 9EG
G0WIS WR9 9HE
G7RVT WR9 9HU
G4LVO WR9 9LA
2E1AQY WR9 9LA
2E1ARA WR9 9LA
2E1CBH WR9 9LA
G4BYM WR9 9LG

Wycombe

G8ZRG HP10 0HJ
G4PGZ HP10 0HW
G0NCL HP10 0JP
G0SYA HP10 0QH
G4BIO HP10 8AE
G8PFR HP10 8BA
G4XVP HP10 8BL
G7AKZ HP10 8BQ
G3GAA HP10 8DU
G3ZIC HP10 8HB
G7PGY HP10 8HX
G7SFC HP10 8HX
G3RAF HP10 8LN
G4DDM HP10 8LN
G8FC HP10 8LN
G8RAF HP10 8LN
G3COJ HP10 9AB
G3PWY HP10 9AX
G4KCX HP10 9DW
G8NMT HP10 9DW
2E1FMF HP10 9ES
G8VRG HP10 9JL
G0RNO HP10 9JY
G3WQG HP10 9LH
G8ECJ HP10 9LH
G3INZ HP10 9PL
G8JQV HP10 9RH
G1PLT HP10 9RW
G7BOB HP11 1JL
G8XQT HP11 1JW
G8DJF HP11 1QT
G6FGJ HP11 1RB
G0OCY HP11 2JL
G4TBG HP11 2JN
G1EUI HP11 2PL
G3UJK HP11 2TJ
G4TJM HP11 2TX
G4CJY HP11 2UA
G3CDJ HP11 2UD
G1URO HP12 3BP
G4UMB HP12 3DD
G7GUO HP12 3DS
G8DTE HP12 3HT
GM0IZC HP12 3LL
G4COS HP12 3LN
G1LQX HP12 3NN
G3JMH HP12 3NS
G4FAK HP12 3NT
G8VJW HP12 3PA
G0CPT HP12 3PF
G3PVJ HP12 3PG
G8WCH HP12 4AA
G7OST HP12 4AU
G6BFP HP12 4BB
G0UDK HP12 4BY
G0ANO HP12 4DD
G4ZOC HP12 4DD
G4AOQ HP12 4LG
G4VXN HP12 4NL
G0UOH HP12 4NR
G1RDX HP12 4NY
G8DLZ HP12 4PG
G0IKE HP12 4QR
G3OOP HP12 4SA
G3ZGQ HP12 4SP
G8KFQ HP13 5FA
G3FNO HP13 5JN
G4TZQ HP13 5JN
G8HDL HP13 5NG
G7RTI HP13 5NW
G2CGF HP13 5PX
G8AAU HP13 5QA
G1EYG HP13 5RH
G6XQW HP13 5RU
G7GQH HP13 5SA
G3FSN HP13 5SN
M1BDL HP13 5SP
G4LMM HP13 5SS
G4MUI HP13 5TA
G6HBJ HP13 5UD
G3LOA HP13 5UN
G8JAW HP13 5UN
G1CPC HP13 5XN
M1ARQ HP13 6JA
G4XBW HP13 6JX
G2DBA HP13 6QP
G0ACL HP13 6TJ
G1GAR HP13 6TJ
G6GTZ HP13 6XW
G7GUS HP13 7AN
G4LQN HP13 7EW
G1ECA HP13 7HR
G7DPE HP13 7LL
G0WEK HP13 7PQ
G8RLJ HP13 7PU
G7SXC HP13 7RD
G1MIY HP13 7UE
G3CAR HP13 7XN
G4UKM HP13 7XN
G4YAN HP13 7XN
G8CAR HP13 7XN
G6HYJ HP13 7YA
G4KBB HP14 3BN
G3UNI HP14 3BP
G7WFZ HP14 3JD
G4PUO HP14 3JN
M1BFL HP14 3JP
G3LYP HP14 3JW
G1MPJ HP14 3NB
G3TZP HP14 3NN
G4HFS HP14 3PH
G3SYW HP14 3QL
G1CUJ HP14 3RP
G4WJS HP14 3RP
G7IZA HP14 3SX
G0ALY HP14 3TB
G0GGL HP14 3TF
G0NOG HP14 3TW
G1EKC HP14 4AP
G8XII HP14 4AX
G3SKD HP14 4BY
G0MDJ HP14 4DJ
2E1FIQ HP14 4DZ
G4YBH HP14 4JG
G2BWW HP14 4JJ
G4YKQ HP14 4JN
G8EBT HP14 4JN
G3MGH HP14 4LN
G3EFE HP14 4LP
G0ZDL HP14 4NN
G6XQB HP14 4QB
G0HND HP14 4SG
G4JFZ HP14 4TR
G3SRJ HP14 4UH
G0IOU HP14 4UY
G6AHN HP15 6EG
G7MLX HP15 6EY
G3VCT HP15 6JR
G3WNS HP15 6LJ
G8AKU HP15 6LJ
G3XZK HP15 7AT
G6AVF HP15 7BY
G6UDF HP15 7ED
G4UIH HP15 7EE
G4UII HP15 7EE
G8DOR HP15 7JY
G8XDS HP15 7QN
G3TYG HP15 7RE
G4MQC HP15 7TD
G7VCU HP15 7TD
G4PPK HP15 7TF
G6GUD HP15 7TF
G7DZD HP15 7TF
G6EYJ HP15 7TP
G0HZQ HP16 9PT
G4GZH HP16 9RF
G3VKO HP17 0LL
G1IJM HP17 0UE
G6MVO HP17 0UE
G3XKF HP17 8SP
G4HFL HP27 0DH
G1DAE HP27 0EB
G0SPB HP27 0EB
G7ADE HP27 0EB
G4XNO HP27 0EE
G0WQR HP27 0EE
G7AYS HP27 0JA
G4ZMO HP27 0JB
G0SNF HP27 0JQ
G4GBR HP27 0JW
G3SNP HP27 0NB
G3LUB HP27 0NX
G0MBB HP27 0PG
G3YQB HP27 0QQ
G7DPF HP27 0QY
G4KZT HP27 9AA
G3XJP HP27 9AS
G1THD HP27 9AY
G8NPO HP27 9DD
G0AGR HP27 9DT
G0WQQ HP27 9HE
G1SJT HP27 9HZ
G4JBE HP27 9JZ
G4ODI HP27 9LY
G6YXY HP27 9NG
G4CGE HP27 9QA
G3NBS HP27 9RP
G7VBV HP27 9SJ
G3XXM HP27 9SL
G7LSF RG9 6HZ
G3SQM RG9 6SA
G3TOP SL7 1BW
G3DXD SL7 1DR
G7JDN SL7 1JW
G8CJH SL7 1QQ
G6GIF SL7 1TX
G3MYU SL7 1TZ
G3SMW SL7 1UJ
G4KCD SL7 1UW
G4FTJ SL7 1XT
G0DCE SL7 2AT
2E1CFT SL7 2DE
G7RWO SL7 2DE
G3SET SL7 2EL
G7OSC SL7 2JL
G0CXV SL7 2NZ
G3OOZ SL7 3BD
G4YSH SL7 3BZ
G0AAP SL7 3DA
G4VLL SL7 3EG
G8KJU SL7 3LF
G1SNO SL7 3LU
G4RNN SL7 3LZ
G3LVW SL7 3PY
G4BEB SL7 3PY
G3RIY SL7 3QJ
G0UYM SL7 3RB
G3IQF SL7 3RH
M0ANF SL8 5JP
G3TWG SL8 5RW

Wyre

G3MLA FY3 7LX
G4HWU FY3 7SW
G0SVQ FY4 5EZ
G4KQC FY5 1JW
G3OSR FY5 2AR
G4GFQ FY5 2ET
G4AV FY5 2HP
G0FYY FY5 2HS
G0LQN FY5 2JA
G4ATH FY5 2JD
G4BFH FY5 2JD
G6GMW FY5 2JD
G1PPK FY5 2JS
G0VQI FY5 2NS
G0NCG FY5 2NW
G6DQD FY5 2RD
G0VMH FY5 2RG
G4ANE FY5 2RT
G3CCC FY5 2SW
G0MWM FY5 2ZA
G0TAK FY5 2ZB
G0GER FY5 3QB
G8YKG FY5 4DE
G3YEI FY5 4DX
G3ZRZ FY5 4NT
G4APP FY5 5AP
G3YNG FY5 5AW
G4UHI FY5 5AY
G2CBH FY5 5BL
G4CSH FY5 5BY
G0NXU FY5 5EH
2E1ADR FY5 5HH
G3ZNG FY5 5HP
M0ATU FY6 0AH
G7IED FY6 0BT
G7PDR FY6 0BT
G0JZT FY6 0DD
G0RRP FY6 0DX
G0MTY FY6 0HA
G4RFJ FY6 0HE
G6PMX FY6 0HE
G8FFW FY6 0HG
G3NBQ FY6 0HJ
G0PNG FY6 0HQ
G6GVK FY6 0HQ
G0BBI FY6 0LA
G7UWE FY6 0PB
G1KBA FY6 0QB
G4KHJ FY6 7DX
G3PLD FY6 7EA
G3UIS FY6 7EG
G3IWP FY6 7EW
2E1EIU FY6 7EW
G0AAY FY6 7JZ
G4TMA FY6 7LP
GW3NJY FY6 7NA
G0WKS FY6 7NT
G6KOE FY6 7PB
G0AJW FY6 7PF
G3BDT FY6 7PN
G4SXI FY6 7PP
G3XCE FY6 7PW
G1OCK FY6 7QE
G1OVB FY6 7QF
G6SOE FY6 7QL
G1TXV FY6 7RB
2E1FHW FY6 7SJ
G8KBH FY6 7UB
G8YOK FY6 7UB
G4LOR FY6 7UX
G3MCE FY6 8AD
G3PHX FY6 8BX
G3RDS FY6 8BX
G4FXG FY6 8BZ
G7FED FY6 8BZ
G3HNY FY6 8DE
G3DZV FY6 8DN
G3VDO FY6 8EB
M1ADJ FY6 8EH
G3OFQ FY6 8ES
G0PMY FY6 9AP
G4XKR FY6 9AP
G3GJU FY6 9BH
G0RHF FY6 9DN
G0ETV FY6 9DR
G7NYD FY6 9DZ
M0BAL FY7 6BG
G8YPY FY7 6QA
G0NCY FY7 6QQ
2E1FUH FY7 6QQ
2E1FUJ FY7 6QQ
G0IEP FY7 7AY
G0GDT FY7 7BW
G0CUX FY7 7EA
G0LRK FY7 7HA
G4TTG FY7 7HH
G4NCX FY7 7HR
G0KMT FY7 7JB
G0IAE FY7 7NH
G4GYF FY7 8BQ
G3YGU FY7 8BZ
2E1FVE FY7 8DY
G8RDP FY7 8EG
G4FWM FY7 8HG
G4KHL FY7 8HY
G6RHQ FY7 8LU
G1ETA FY7 8NL
G1NSB FY7 8PE
G0EPY FY7 8QH
G6LNS PR3 0BB
G0TUB PR3 0TD
G0AJQ PR3 0XD
G8CWQ PR3 1AD
G3VNA PR3 1EE
G0LXP PR3 1FJ
G4AMY PR3 1NL
G4TVN PR3 1PD
G4IAL PR3 1PL
G1HKR PR3 1RD
G4BSD PR3 1RD
G7FNM PR3 1RD
G6AIZ PR3 1RF
G1WVL PR3 1WB
G0VWX PR3 1YL
G3LPL PR3 6AB
G3UOL PR3 6AL
G6HCF PR3 6BN
G6GVJ PR3 6SQ
G0IYT PR3 6SS

Wyre Forest

G1HRE DY10 1LG
G0NFO DY10 1LR
G4ALT DY10 1LY
G0GKU DY10 1NS
G1ALZ DY10 1NT
G4AKK DY10 1NW
G3MRC DY10 1SS
2E1AUG DY10 1SS
G3HWZ DY10 1TN
G3TJB DY10 1XT
G7ABZ DY10 1YH
G0MBG DY10 2BZ
G0MJX DY10 2HB
G1LOP DY10 2QP
G1UXJ DY10 2RA
G4MD DY10 2RA
G6BFN DY10 2RF
G0ISG DY10 2ST
G1XJZ DY10 2TH
G4JXJ DY10 2TH
G4XCX DY10 2TP
G1HWZ DY10 2TS
G6BAM DY10 2UN
G4EFS DY10 2XG
G0MJY DY10 2YB
G0WVT DY10 2YB
G6WI DY10 2YF
G7OJZ DY10 2YG
G0NTC DY10 2YJ
2E1DLR DY10 3AG
G4OBC DY10 3BH
G0TLP DY10 3JA
G7RDQ DY10 3JZ
G1JDR DY10 3LY
G0DKP DY10 3LZ
G8OXG DY10 3QR
G0EYW DY10 3QS
G4AFY DY10 3QS
G4GXP DY10 3QS
G6KRC DY10 3QS
G7CHW DY10 3QS
G8WOX DY10 3QS
G4TXV DY10 3QW
G2FSS DY10 3QY
G1DNX DY10 3RX
G8TPM DY10 3TL
G0UQB DY10 3UB
G7PBO DY10 3UR
G4OIK DY10 3XL
G3YPU DY10 4AE
G4ZLN DY10 4HG
G0HHH DY10 4JQ
G7RBZ DY10 4LW
G0LOM DY10 4NR
G7EWZ DY10 4RZ
G1YAO DY10 4TG
G0GPS DY10 4TS
G1RWS DY11 5AW
G4ACS DY11 5BG
G3ETJ DY11 5DD
G1YQS DY11 5DH
G8AKX DY11 5DL
G0FJD DY11 5DW
G6ORM DY11 5DY
G8UZV DY11 5EB
G7TNK DY11 5EB
G6BDY DY11 5HJ
G1AJU DY11 5JP
G8DKD DY11 5JY
G8TLH DY11 5LU
G0VBP DY11 5LZ
2E1CGZ DY11 5NH
2E1DVM DY11 5NH
G0IHP DY11 5PE
G3NXD DY11 5QR
G3EMK DY11 5RA
G6TQD DY11 5RY
G8UEF DY11 5SX
G4ZIB DY11 5TZ
G0HRL DY11 5UA
G0OWT DY11 5UA
G3TAW DY11 6AA
G1SBG DY11 6BD
G4CTU DY11 6BX
G4PRD DY11 6DQ
G0EOF DY11 6JU
G7GJP DY11 6NP
G0HTF DY11 6PZ
G1BDT DY11 6QS
G3ZQQ DY11 6RL
G8GYI DY11 6TE
G7DCJ DY11 7BY
G0GFE DY11 7EW
G0UDI DY12 1DB
G1OZB DY12 1DD
G4OBA DY12 1LF
G3LZT DY12 1NL
G0JMX DY12 1PB
G4KTS DY12 1QD
G8EPR DY12 2HD
G4OIL DY12 2HT
G0TKT DY12 2HX
G0LOZ DY12 2JX
G0RRV DY12 2JX
G6ODI DY12 2JY
G4USN DY12 2PA
G4SPZ DY12 2PU
G3GGL DY12 2RJ
G7BMZ DY12 2RJ
G4KLQ DY12 2UG
G4SNO DY12 3AA
G4JNG DY12 3AH
G0KRC DY13 0AH
G0RJP DY13 0AH
2E1AJV DY13 0BA
2E1AQG DY13 0DA
2E1CVD DY13 0DW
G0TUW DY13 0EB
G7ESI DY13 0EL
G7JWL DY13 0EQ
G4OPV DY13 0EW
G1CDJ DY13 0HJ
G0IBT DY13 0JT
G1LBK DY13 0LL
G7SCZ DY13 0NU
G7WDC DY13 0NU
G0PMF DY13 0NY
G0SSR DY13 0NY
2E1DLN DY13 0NY
G7WCK DY13 0RB
G8TDG DY13 8JX
G6CBB DY13 8LP
G3VHL DY13 8LR
2E1AJD DY13 8PJ
G0PMG DY13 8QD
G0PWC DY13 8QP
G4SND DY13 8QX
G0FSR DY13 8RA
G0LJK DY13 8RZ
G8DUW DY13 8SR
G0MKL DY13 8TF
G0KJA DY13 8TG
2E1CVE DY13 8TH
G7KPF DY13 8UQ
G6CPO DY13 9BE
G4GXU DY13 9DB
G8BKL DY13 9EU
2E1DHJ DY13 9NA
G7RBT DY13 9ND
G7IQD DY13 9NE
G7IEO DY13 9PB
G0FML DY13 9PD
2E1EPQ DY13 9RX
G4ZKD DY14 9DB

York

G8LZG HU13 0JR
G6VWP HU13 0JW
G3NPT HU13 0LB
2E1DLF YO1 1DB
G0WUY YO1 3BZ
G0PMI YO1 3HE
G7NZU YO1 3JZ
GW4RCE YO1 3LF
2E1BRG YO1 3PD
M1AQI YO1 3PD
G1UVL YO1 3PP
G1UVK YO1 3PZ
G4KCT YO1 3QB
G8SFI YO1 3QB
G8FIS YO1 3QB
2E1DYF YO1 3QB
2E1DYG YO1 3QB
2E1AYX YO1 3QH
G6MCQ YO1 3QJ
G7PME YO1 3TN
2E1CIK YO1 4AY
G1CIH YO1 4BP
G4KDX YO1 4BQ
G0KFV YO1 4DR
G4RNV YO1 4HH
G4RLQ YO1 4LS
G3OZE YO1 4PB
G0IEH YO1 4QQ
G0UKO YO1 5AT
G4PZS YO1 5BU
G0UOY YO1 5DD
G7UOY YO1 5DD
M1APH YO1 5EU
G4IIX YO1 5JE
G8MBB YO1 5PQ
G4TXO YO2 1AN
G0DRF YO2 1AQ
G1JBZ YO2 1AU
G3MGI YO2 1DG
G4YEK YO2 1LE
G0EFU YO2 1LW
G3GRF YO2 1QW
G5KC YO2 2DT
G4NON YO2 2JT
G3ZQA YO2 2RP
G0PWM YO2 3BD
G4CRY YO2 3JB
G0ICQ YO2 3JP
G0PWO YO2 3LF
G7HCV YO2 3LF
G7HZR YO2 3LN
G8JYR YO2 3QB
G8HYF YO2 3RE
G8TZY YO2 3UD
G7WAR YO2 3YR
G0SGA YO2 4DW
G4DRW YO2 4LE
G7SUA YO2 4LE
G1CBB YO2 4LX
G0SCD YO2 4SN
G4NFE YO2 4TU
G8RBQ YO2 4TU
G4TOM YO2 4XP
G0FLH YO2 5AD
G0VYS YO2 5EZ
G6DDH YO2 5FF
G7FGA YO2 5HQ
G0SEF YO2 5HY
G4EYF YO2 5JA
G8IMZ YO2 5LH
G4KTF YO2 5NW
G3TEZ YO2 5PG
G0WBU YO2 5QQ
G3WQM YO2 5QY
G1BYP YO2 5QZ
G4LKP YO2 5RP
G3JME YO2 6AW
G4FDD YO2 6JB
G2ADR YO2 6JE
G4LKV YO2 6NN
G0UDP YO2 6PZ
G3HWW YO3 0AF
G3WVO YO3 0AF
G3DTA YO3 0BP
G7SBZ YO3 0BR
2E1ENN YO3 0DF
G6YXO YO3 0ED
G4FRA YO3 0JD
G7LXK YO3 0LN
G0HVH YO3 0LU
M1AQB YO3 0LW
G6JRI YO3 0NE
G0FDH YO3 0PX
G1DRG YO3 0PZ
G7CRA YO3 0QS
G3HWE YO3 0RX
G7SWU YO3 0TY
G1VIZ YO3 0UR
G0WPF YO3 3DT
2E1DPU YO3 3DT
2E1ENA YO3 3EL
G1GCF YO3 3FE
G4YMS YO3 3LY
G4PBJ YO3 3NL
G0XAB YO3 3NS
M1AVL YO3 3PB
G4JQF YO3 3QE
G3YZR YO3 3QZ
G8BVL YO3 3QZ
G7PHC YO3 3QZ
G8INO YO3 3RP
G0VZI YO3 3RR
G7PJZ YO3 3RR
G3RYX YO3 3ST
G0SED YO3 3YN
G4YCS YO3 4SZ
G0TXY YO3 5TE
G4ELR YO3 5YN
G0FZO YO3 5ZN
G0VWP YO3 6DZ
G0VKZ YO3 6JJ
G0LOP YO3 6NA
G4HEV YO3 6PE
G1MTP YO3 6PR
G4EMA YO3 6PZ
G4NEM YO3 6QD
G4ESU YO3 6QG
G1YQZ YO3 6RT
G3ZKS YO3 6RT
G4IUE YO3 6SA
G0GYJ YO3 6SU
G0NLK YO3 6TX
G8RLW YO3 6YH
M1BMU YO3 6YH
G0EVO YO3 7EJ
G4VRU YO3 7EJ
G8OYF YO3 7NZ
G0ULQ YO3 7SQ
G0TTS YO3 9BD
G1KAO YO3 9DB
G3VJZ YO3 9DY
G1PJV YO3 9EQ
G7BXJ YO3 9LY
G3BMO YO3 9NJ
G3ORD YO3 9QA
G6NRF YO3 9RS
G3VGH YO3 9SH
G3YHA YO3 9TG
G0AWZ YO3 9UA
G3TEE YO3 9UJ
G1GHG YO3 9YG

Area